CHILTON'S™
IMPORT CAR REPAIR
MANUAL 1995–99

President   Dean F. Morgantini, S.A.E.
Vice President–Finance   Barry L. Beck
Vice President–Sales   Glenn D. Potere

Executive Editor   Kevin M. G. Maher, A.S.E.
Production Manager   Ben Greisler, S.A.E.
Production Assistant   Melinda Possinger

Project Managers   George B. Heinrich III, A.S.E., S.A.E., Will Kessler, A.S.E., S.A.E., James R. Marotta, A.S.E., S.T.S., Richard Schwartz, A.S.E., Todd W. Stidham

Schematics Editor   Christopher G. Ritchie

Editors   Christopher Bishop, Leonard Davis, A.S.E., S.T.S., Dawn M. Hoch, S.A.E., Matthew E. Frederick, A.S.E., S.A.E., Frank Keytanjian, A.S.E., S.A.E., Thomas A. Mellon, A.S.E., S.A.E., Eric Michael Mihalyi, A.S.E., S.A.E., S.T.S., Christine L. Nuckowski, S.A.E.

CHILTON™ *Automotive Books*

PUBLISHED BY **W. G. NICHOLS, INC.**

Manufactured in USA
© 1998 W. G. Nichols
1020 Andrew Drive
West Chester, PA 19380
ISBN 0-8019-7923-4
Library of Congress Catalog Card No. 98-73246
1234567890 7654321098

Table of Contents

Table of Contents

Import Car Sections

Model Index

HOW TO USE THIS MANUAL

Model Specific Sections

The model specific sections are grouped by manufacturer and arranged in alphabetical order. The text and illustrations that comprise the service procedures in each model specific section are arranged in the following order of systems and components: Engine Repair (Gasoline, then Diesel if applicable), Fuel System (Gasoline, then Diesel if applicable), Drive Train, Steering and Suspension.

All illustrations are located as close as possible to the applicable procedure. Procedures are for all models in the particular section unless specifically noted otherwise.

Unit Repair Sections

The Unit Repair Sections (URS's) are written to cover all applicable 1995-99 models for the specific URS system or component, unless specifically noted otherwise. The procedures covered in the 10 URS's are not repeated in the model specific sections; therefore, refer to the URS's for the service procedures for the applicable systems or components. Refer to the Table of Contents for URS coverage.

Locating Information

The Table of Contents, located at the front of the book, lists each Unit Repair Section (URS) and model specific section in this manual.

To find where a particular model specific section is located in the book, you need only look in the Table of Contents. Once you have found the proper section, you may wish to find where specific procedures located in that section. Turn to the Index at the front of the model specific section. At the upper left-hand side is a listing of the main topics within that section and the page number on which they may be found. Following the main topics is an alphabetical listing of all of the procedures within the section and their page numbers.

The Model Index, located just after the Table of Contents in the beginning of this manual, may also be used to locate the specific section for any vehicle model covered in this manual.

Safety Notice

Proper service and repair procedures are vital to the safe, reliable operation of all motor vehicles, as well as the personal safety of those performing the repairs. This manual outlines procedures for servicing and repairing vehicles using safe effective methods. The procedures contain many NOTES, WARNINGS and CAUTIONS which should be followed along with standard safety procedures to eliminate the possibility of personal injury or improper service which could damage the vehicle or compromise its safety.

It is important to note that repair procedures and techniques, tools and parts for servicing vehicles, as well as the skill and experience of the individual performing the work vary widely. It is not possible to anticipate all of the conceivable ways or conditions under which vehicles may be serviced, or to provide cautions as to all of the possible hazards that may result. Standard and accepted safety precautions and equipment should be used when handling toxic or flammable fluids, and safety goggles or other protection should be used during cutting, grinding, chiseling, prying, or any other process that can cause material removal or projectiles.

Some procedures require the use of tools specially designed for a specific purpose. Before substituting another tool or procedure, you must be completely satisfied that neither your personal safety, nor the performance of the vehicle will be endangered.

Although information in this manual is based on industry sources and is as complete as possible at the time of publication, the possibility exists that some vehicle manufacturers made later changes which could not be included here. Information on very late models may not be available in some circumstances. While striving for total accuracy, NP/Chilton cannot assume responsibility for any errors, changes, or omissions that may occur in the compilation of this data.

Part Numbers

Part numbers listed in this book are not recommendations by NP/Chilton for any product by brand name. They are references that can be used with interchanges manuals and aftermarket supplier catalogs to locate each brand supplier's discrete part number.

Special Tools

Special tools are recommended by the vehicle manufacturer to perform their specific job. Use has been kept to a minimum, but where absolutely necessary, they are referred to in the text by the part number of the tool manufacturer. These tools may be purchased, under the appropriate part number, from your local dealer or regional distributor, or an equivalent tool can be purchased locally from a tool supplier or parts outlet. Before substituting any tool for the one recommended, read the previous Safety Notice.

Copyright Notice

NP/Chilton would like to thank all manufacturer's involved for their generous assistance.

Get ready for ASE testing with Motor Age Self Study Guides.

Each training unit contains a complete description of the ASE Task Analysis and Test Specifications, and covers the subject areas of the corresponding ASE test question group. Also included are sample ASE test questions. In addition, each book includes a special glossary, sample questions and an expanded answer analysis to increase your knowledge of the subject.

AA Car & Light Truck
A1 Engine Repair
A2 Automatic
Transmission/Transaxle
A3 Manual Drive Train & Axles
A4 Suspension & Steering
A5 Brakes
A6 Electrical/Electronic Systems
A7 Heating & A/C
A8 Engine Performance

Parts Specialist
P1 Medium/Heavy Parts Specialist
P2 Automobile Parts Specialist

Advanced Level
L1 Advanced Engine Performance Specialist
L2 Med/Hvy Vehicle Electronic Diesel Engine Diagnosis Specialist
F1 Light Vehicle Compressed Natural Gas

ALSO AVAILABLE:
TT Medium/Heavy Truck Service: T1 Gasoline Engines, T2 Diesel Engines, T3 Drive Train, T4 Brakes,
    T5 Suspension & Steering, T6 Electrical/Electronic Systems, T7 Heating, Ventilation & A/C, T8
Preventive Maintenance Inspection (PMI), & MM Engine Machinist (M1, M2, M3),
BB Collision Repair/Paint & Refinish: B2 Paint & Refinishing, B3 Non-Structural Analysis & Damage
    Repair, B4 Structural Analysis & Damage Repair, B5 Mechanical & Electrical Components,
    B6 Damage Analysis & Estimating

Name: _____
         first   middle   last

Company _____

Address: _____ Apt. # _____

City: _____ State: _____ Zip: _____

Phone (DAYTIME): _____ Fax _____

For pricing and shipping information,
fax this form to Trudy Kolb, 610-964-4251

SPECIFICATIONS

1

ACURA
2.2CL • 2.3CL • 2.5TL • 3.0CL • 3.2TL • 3.5RL • Integra • Legend

VEHICLE IDENTIFICATION CHART

Engine Code							Model Year	
Code	Liters	Cu. In. (cc)	Cyl.	Fuel Sys.	Eng. Mfg.		Code	Year
B18B1	1.8	112 (1834)	4	PGM-FI	Honda		S	1995
B18C1	1.8	110 (1797)	4	PGM-FI	Honda		T	1996
C30A1	3.0	183 (2977)	6	PGM-FI	Honda		V	1997
C32A1	3.2	196 (3206)	6	PGM-FI	Honda		W	1998
C32A6	3.2	196 (3206)	6	PGM-FI	Honda		X	1999
C32B1	3.2	196 (3206)	6	PGM-FI	Honda			
C35A1	3.5	212 (3494)	6	PGM-FI	Honda			
F22B1	2.2	132 (2156)	4	PGM-FI	Honda			
F23A1	2.3	138 (2254)	4	PGM-FI	Honda			
G25A1	2.5	150 (2451)	5	PGM-FI	Honda			
G25A4	2.5	150 (2451)	5	PGM-FI	Honda			
J30A1	3.0	183 (2997)	6	PGM-FI	Honda			

PGM-FI - Programmed Fuel Injection

79231CG6

ENGINE IDENTIFICATION

Year	Model	Engine Displacement Liters (cc)	Engine Series (ID/VIN)	Fuel System	No. of Cylinders	Engine Type
1995	Integra	1.8 (1834)	B18B1	PGM-FI	4	DOHC
	Integra GSR	1.8 (1797)	B18C1	PGM-FI	4	DOHC
	Legend	3.2 (3206)	C32A1	PGM-FI	6	SOHC
	NSX	3.0 (2977)	C30A1	PGM-FI	6	DOHC
	Vigor	2.5 (2451)	G25A1	PGM-FI	5	SOHC
1996	3.5RL	3.5 (3494)	C35A1	PGM-FI	6	SOHC
	2.5TL	2.5 (2451)	G25A4	PGM-FI	5	SOHC
	3.2TL	3.2 (3206)	C32A6	PGM-FI	6	SOHC
	Integra	1.8 (1834)	B18B1	PGM-FI	4	DOHC
	Integra GSR	1.8 (1797)	B18C1	PGM-FI	4	DOHC
	NSX	3.0 (2977)	C30A1	PGM-FI	6	DOHC
1997	2.2CL	2.2 (2156)	F22B1	PGM-FI	4	SOHC
	3.0CL	3.0 (2997)	J30A1	PGM-FI	6	SOHC
	3.5RL	3.5 (3494)	C35A1	PGM-FI	6	SOHC
	2.5TL	2.5 (2451)	G25A4	PGM-FI	5	SOHC
	3.2TL	3.2 (3206)	C32A6	PGM-FI	6	SOHC
	Integra	1.8 (1834)	B18B1	PGM-FI	4	DOHC
	Integra GSR	1.8 (1797)	B18C1	PGM-FI	4	DOHC
	NSX	3.0 (2977)	C30A1	PGM-FI	6	DOHC
	NSX	3.2 (3206)	C32B1	PGM-FI	6	DOHC
1998-99	2.3CL	2.3 (2254)	F23A1	PGM-FI	4	SOHC
	3.0CL	3.0 (2997)	J30A1	PGM-FI	6	SOHC
	3.5RL	3.5 (3494)	C35A1	PGM-FI	6	SOHC
	2.5TL	2.5 (2451)	G25A4	PGM-FI	5	SOHC
	3.2TL	3.2 (3206)	C32A6	PGM-FI	6	SOHC
	Integra	1.8 (1834)	B18B1	PGM-FI	4	DOHC
	Integra GSR	1.8 (1797)	B18C1	PGM-FI	4	DOHC
	NSX	3.0 (2977)	C30A1	PGM-FI	6	DOHC
	NSX	3.2 (3206)	C32B1	PGM-FI	6	DOHC

PGM-FI - Programmed Fuel Injection
DOHC - Double Overhead Camshafts
SOHC - Single Overhead Camshaft

79231CG7

GENERAL ENGINE SPECIFICATIONS

Year	Engine ID/VIN	Engine Displacement Liters (cc)	Fuel System Type	Net Horsepower @ rpm	Net Torque @ rpm (ft. lbs.)	Bore x Stroke (in.)	Compression Ratio	Oil Pressure @ rpm
1995	B18B1	1.8 (1834)	PGM-FI	142@6300	127@5000	3.19x3.50	9.2:1	50@3000
	B18C1	1.8 (1797)	PGM-FI	170@7600	128@6200	3.19x3.43	10.0:1	50@3000
	C30A1	3.0 (2977)	PGM-FI	252@6600	210@5300	3.54x3.07	10.2:1	50@3000
	C32A1	3.2 (3206)	PGM-FI	200@5500	210@4500	3.54x3.31	9.6:1	50@3000
	C32A1	3.2 (3206)	PGM-FI	230@6200	206@5000	3.54x3.31	9.6:1	50@3000
	G25A1	2.5 (2451)	PGM-FI	176@6300	170@3900	3.35x3.40	9.0:1	50@3000
1996	B18B1	1.8 (1834)	PGM-FI	142@6300	127@5200	3.19x3.50	9.2:1	50@3000
	B18C1	1.8 (1797)	PGM-FI	170@7600	128@6200	3.19x3.43	10.0:1	50@3000
	C30A1	3.0 (2977)	PGM-FI	252@6600	210@5300	3.54x3.07	10.2:1	50@3000
	C32A6	3.2 (3206)	PGM-FI	200@5300	210@4500	3.54x3.31	9.6:1	50@3000
	G25A4	2.5 (2451)	PGM-FI	176@6300	170@3900	3.53x3.40	9.6:1	50@3000
1997	B18B1	1.8 (1834)	PGM-FI	142@6300	127@5200	3.19x3.50	9.2:1	50@3000
	B18C1	1.8 (1797)	PGM-FI	170@7600	128@6200	3.19x3.43	10.0:1	50@3000
	C30A1	3.0 (2977)	PGM-FI	252@6600	210@5300	3.54x3.07	10.2:1	50@3000
	C32A6	3.2 (3206)	PGM-FI	200@5300	210@4500	3.54x3.31	9.6:1	50@3000
	C32B1	3.2 (3206)	PGM-FI	290@7100	224@5500	3.66x3.07	10.2:1	50@3000
	C35A1	3.5 (3494)	PGM-FI	210@5200	224@2800	3.54x3.58	9.6:1	50@3000
	F22B1	2.2 (2156)	PGM-FI	145@5500	147@4500	3.35x3.74	8.8:1	50@3000
	G25A4	2.5 (2451)	PGM-FI	176@6300	170@3900	3.53x3.40	9.6:1	50@3000
	J30A1	3.0 (2997)	PGM-FI	200@5600	195@4800	3.39x3.82	9.4:1	50@3000
1998-99	B18B1	1.8 (1834)	PGM-FI	142@6300	127@5200	3.19x3.50	9.2:1	50@3000
	B18C1	1.8 (1797)	PGM-FI	170@7600	128@6200	3.19x3.43	10.0:1	50@3000
	C30A1	3.0 (2977)	PGM-FI	252@6600	210@5300	3.54x3.07	10.2:1	50@3000
	C32A6	3.2 (3206)	PGM-FI	200@5300	210@4500	3.54x3.31	9.6:1	50@3000
	C32B1	3.2 (3206)	PGM-FI	290@7100	224@5500	3.66x3.07	10.2:1	50@3000
	C35A1	3.5 (3494)	PGM-FI	210@5200	224@2800	3.54x3.58	9.6:1	50@3000
	G25A4	2.5 (2451)	PGM-FI	176@6300	170@3900	3.53x3.40	9.6:1	50@3000
	J30A1	3.0 (2997)	PGM-FI	200@5600	195@4800	3.39x3.82	9.4:1	50@3000
	F23A1	2.3 (2254)	PGM-FI	150@5700	152@4800	3.39x3.82	9.3:1	50@3000

PGM-FI - Programmed Fuel Injection

79231CG8

GASOLINE ENGINE TUNE-UP SPECIFICATIONS

Year	Engine ID/VIN	Engine Displacement Liters (cc)	Spark Plugs Gap (in.)	Ignition Timing (deg.) MT	Ignition Timing (deg.) AT	Fuel Pump (psi)		Idle Speed (rpm) MT	Idle Speed (rpm) AT	Valve Clearance In.	Valve Clearance Ex.
1995	B18B1	1.8 (1834)	0.041	16B	16B	40-47	①	700-800	700-800	0.003-0.005	0.006-0.008
	B18C1	1.8 (1797)	0.049	16B	16B	48-55	①	700-800	700-800	0.006-0.007	0.007-0.008
	C30A1	3.0 (2977)	0.041	15B	15B	46-53	①	750-850	700-800	0.006-0.007	0.007-0.008
	C32A1	3.2 (3206)	0.043	15B	15B	44-51	①	600-700	550-650	HYD	HYD
	C32A6	3.2 (3206)	0.043	15B	15B	44-51	①	630-730	580-680	HYD	HYD
	G25A1	2.5 (2451)	0.043	15B	15B	43-50	①	650-750	650-750	0.009-0.011	0.011-0.013
1996	B18C1	1.8 (1797)	0.051	16B	16B	48-55	①	700-800	—	0.006-0.007	0.007-0.008
	B18B1	1.8 (1834)	0.039-0.043	16B	16B	40-47	①	700-800	700-800	0.003-0.005	0.006-0.008
	C30A1	3.0 (2977)	0.043-0.047	15B	15B	47-53	①	750-850	730-830	0.006-0.007	0.007-0.008
	C32A6	3.2 (3206)	0.039-0.043	—	15B	37-48	①	—	600-700	HYD	HYD
	G25A4	2.5 (2451)	0.039-0.043	—	15B	43-50	①	—	650-750	0.009-0.011	0.011-0.013
1997	B18B1	1.8 (1834)	0.039-0.043	16B	16B	40-47	①	700-800	700-800	0.003-0.005	0.006-0.008
	B18C1	1.8 (1797)	0.051	16B	16B	48-55	①	700-800	—	0.006-0.007	0.007-0.008
	C30A1	3.0 (2977)	0.043-0.047	15B	15B	47-53	①	750-850	730-830	0.006-0.007	0.007-0.008
	C32A6	3.2 (3206)	0.039-0.043	—	15B	37-48	①	—	600-700	HYD	HYD
	C32B1	3.2 (3206)	0.043-0.047	—	15B	47-53	①	—	750-850	0.006-0.007	0.007-0.008
	C35A1	3.5 (3494)	0.039-0.043	—	15B	43-50	①	—	600-700	HYD	HYD
	F22B1	2.2 (2156)	0.039-0.043	15B	15B	30-37	①	650-750	650-750	0.009-0.011	0.011-0.013
	G25A4	2.5 (2451)	0.039-0.043	—	15B	43-50	①	—	650-750	0.009-0.011	0.011-0.013
	J30A1	3.0 (2997)	0.039-0.043	—	15B	41-48	①	—	650-750	0.008-0.009	0.011-0.013
1998-99	B18B1	1.8 (1834)	0.039-0.043	16B	16B	40-47	①	700-800	700-800	0.003-0.005	0.006-0.008
	B18C1	1.8 (1797)	0.051	16B	16B	48-55	①	700-800	—	0.006-0.007	0.007-0.008
	C30A1	3.0 (2977)	0.043-0.047	15B	15B	47-53	①	750-850	730-830	0.006-0.007	0.007-0.008
	C32A6	3.2 (3206)	0.039-0.043	—	15B	37-48	①	—	600-700	HYD	HYD

79231CG9

GASOLINE ENGINE TUNE-UP SPECIFICATIONS

Year	Engine ID/VIN	Engine Displacement Liters (cc)	Spark Plugs Gap (in.)	Ignition Timing (deg.) MT	AT	Fuel Pump (psi)	Idle Speed (rpm) MT	AT	Valve Clearance In.	Ex.
1998-99 (cont.)	C32B1	3.2 (3206)	0.043-0.047	—	15B	47-53 ①	—	750-850	0.006-0.007	0.007-0.008
	C35A1	3.5 (3494)	0.039-0.043	—	15B	43-50 ①	—	600-700	HYD	HYD
	F23A1	2.3 (2254)	0.039-0.043	12B	12B	47-54 ①	700-750	700-750	0.009-0.011	0.011-0.013
	G25A4	2.5 (2451)	0.039-0.043	—	15B	43-50 ①	—	650-750	0.009-0.011	0.011-0.013
	J30A1	3.0 (2997)	0.039-0.043	—	15B	41-48 ①	—	650-750	0.008-0.009	0.011-0.013

NOTE: The Vehicle Emission Control Information label often reflects specification changes made during production. The label figures must be used if they differ from those in this chart.

B - Before Top Dead Center

HYD - Hydraulic

① At idle, pressure regulator vacuum hose disconnected

79231CG0

CAPACITIES

Year	Model	Engine ID/VIN	Engine Displacement Liters (cc)	Engine Oil with Filter	Transmission (pts.) 5-Spd	6-Spd	Auto.	Transfer Case (pts.)	Drive Axle Front (pts.)	Rear (pts.)	Fuel Tank (gal.)	Cooling System (qts.)
1995	Integra	B18B1	1.8 (1834)	4.0	4.6	—	5.8	—	—	—	13.2	①
	Integra	B18C1	1.8 (1797)	4.2	4.6	-	5.8	—	—	—	13.2	5.0
	Legend	C32A1	3.2 (3206)	4.8	4.8	4.8	7.0	—	2.2	—	18.0	8.0
	NSX	C30A1	3.0 (2977)	5.3	5.8	—	6.2	—	—	—	18.5	12.7
	Vigor	G25A1	2.5 (2451)	4.5	3.8	—	5.2	—	1.9	—	17.2	6.3
1996	2.5TL	G25A1	2.5 (2451)	4.5	—	—	5.8	—	—	—	17.2	5.5
	3.2TL	C32A6	3.2 (3206)	5.0	—	—	7.0	—	—	—	17.2	3.4
	Integra	B18B1	1.8 (1834)	4.0	4.6	—	5.8	—	—	—	13.2	①
	Integra GSR	B18C1	1.8 (1797)	4.2	4.6	—	—	—	—	—	13.2	5.0
	NSX	C30A1	3.0 (2977)	4.4	5.8	—	6.2	—	—	—	18.5	12.7
1997	2.2CL	F22B1	2.2 (2156)	4.5	4.2	—	5.8	—	—	—	17.0	②
	2.5TL	G25A1	2.5 (2451)	4.5	—	—	5.8	—	—	—	17.2	5.5
	3.0CL	J30A1	3.0 (2997)	4.5	—	—	5.8	—	—	—	17.2	5.5
	3.2TL	C32A6	3.2 (3206)	5.0	—	—	7.0	—	—	—	17.2	3.4
	3.5RL	C35A1	3.5 (3494)	5.0	—	—	7.0	—	—	—	17.2	3.4
	Integra	B18B1	1.8 (1834)	4.0	4.6	—	5.8	—	—	—	13.2	①
	Integra GSR	B18C1	1.8 (1797)	4.2	4.6	—	—	—	—	—	13.2	5.0
	NSX	C30A1	3.0 (2977)	4.4	—	5.8	—	—	—	—	18.5	12.7
	NSX	C32B1	3.2 (3206)	5.3	—	—	6.2	—	—	—	18.5	12.7
1998-99	2.3CL	F23A1	2.3 (2254)	4.5	4.2	—	5.8	—	—	—	17.0	②
	2.5TL	G25A1	2.5 (2451)	4.5	—	—	5.8	—	—	—	17.2	5.5
	3.0CL	J30A1	3.0 (2997)	4.5	—	—	5.8	—	—	—	17.2	5.5
	3.2TL	C32A6	3.2 (3206)	5.0	—	—	7.0	—	—	—	17.2	3.4
	3.5RL	C35A1	3.5 (3494)	5.0	—	—	7.0	—	—	—	17.2	3.4
	Integra	B18B1	1.8 (1834)	4.0	4.6	—	5.8	—	—	—	13.2	①
	Integra GSR	B18C1	1.8 (1797)	4.2	4.6	—	—	—	—	—	13.2	5.0
	NSX	C30A1	3.0 (2977)	4.4	—	5.8	—	—	—	—	18.5	12.7
	NSX	C32B1	3.2 (3206)	5.3	—	—	6.2	—	—	—	18.5	12.7

NOTE: All capacities are approximate. Add fluid gradually and ensure a proper fluid level is obtained.

NOTE: Capacities given are service, not overhaul capacities

① Automatic transmission: 5.0
   Manual transmission: 4.6

② Automatic transmisson: 5.6
   Manual Transmission: 5.7

79231CH1

VALVE SPECIFICATIONS

Year	Engine ID/VIN	Engine Displacement Liters (cc)	Seat Angle (deg.)	Face Angle (deg.)	Spring Test Pressure (lbs. @ in.)	Spring Installed Height (in.)	Stem-to-Guide Clearance (in.)		Stem Diameter (in.)	
							Intake	Exhaust	Intake	Exhaust
1995	B18B1	1.8 (1834)	45	45	NA	NA	0.0010-0.0020	0.0020-0.0030	0.2591-0.2594	0.2579-0.2583
	B18C1	1.8 (1797)	45	45	NA	NA	0.0010-0.0022	0.0020-0.0031	0.2156-0.2159	0.2146-0.2150
	C30A1	3.0 (2977)	45	45	NA	NA	0.0010-0.0020	0.0020-0.0030	0.2156-0.2159	0.2146-0.2150
	C32A1	3.2 (3206)	45	45	NA	NA	0.0010-0.0020	0.0020-0.0030	0.2157-0.2161	0.2146-0.2150
	G25A1	2.5 (2451)	45	45	NA	NA	0.0008-0.0018	0.0020-0.0030	0.2156-0.2159	0.2146-0.2150
1996	B18B1	1.8 (1834)	45	45	NA	NA	0.0010-0.0020	0.0020-0.0030	0.2591-0.2594	0.2579-0.2583
	B18C1	1.8 (1797)	45	45	NA	NA	0.0010-0.0022	0.0020-0.0031	0.2156-0.2159	0.2146-0.2150
	C30A1	3.0 (2977)	45	45	NA	NA	0.0010-0.0020	0.0020-0.0030	0.2156-0.2159	0.2146-0.2150
	C32A6	3.2 (3206)	45	45	NA	NA	0.0010-0.0020	0.0020-0.0030	0.2157-0.2181	0.2146-0.2150
	G25A4	2.5 (2451)	45	45	NA	NA	0.0008-0.0018	0.0020-0.0030	0.2156-0.2159	0.2146-0.2150
1997	B18B1	1.8 (1834)	45	45	NA	NA	0.0010-0.0020	0.0020-0.0030	0.2591-0.2594	0.2579-0.2583
	B18C1	1.8 (1797)	45	45	NA	NA	0.0010-0.0022	0.0020-0.0031	0.2156-0.2159	0.2146-0.2150
	C30A1	3.0 (2977)	45	45	NA	NA	0.0010-0.0020	0.0020-0.0030	0.2156-0.2159	0.2146-0.2150
	C32A6	3.2 (3206)	45	45	NA	NA	0.0010-0.0020	0.0020-0.0030	0.2157-0.2181	0.2146-0.2150
	C32B1	3.2 (3206)	45	45	NA	NA	0.0010-0.0029	0.0020-0.0030	0.2156-0.2159	0.2146-0.2150
	C35A1	3.5 (3494)	45	45	NA	NA	0.0010-0.0029	0.0020-0.0030	0.2157-0.2161	0.2146-0.2150
	F22B1	2.2 (2156)	45	45	NA	NA	0.0008-0.0030	0.0022-0.0050	0.2148-0.2163	0.2134-0.2150
	G25A4	2.5 (2451)	45	45	NA	NA	0.0008-0.0018	0.0020-0.0030	0.2156-0.2159	0.2146-0.2150
	J30A1	3.0 (2997)	45	45	NA	NA	0.0010-0.0020	0.0020-0.0030	0.2156-0.2159	0.2146-0.2150
1998-99	B18B1	1.8 (1834)	45	45	NA	NA	0.0010-0.0020	0.0020-0.0030	0.2591-0.2594	0.2579-0.2583
	B18C1	1.8 (1797)	45	45	NA	NA	0.0010-0.0022	0.0020-0.0031	0.2156-0.2159	0.2146-0.2150
	C30A1	3.0 (2977)	45	45	NA	NA	0.0010-0.0020	0.0020-0.0030	0.2156-0.2159	0.2146-0.2150
	C32A6	3.2 (3206)	45	45	NA	NA	0.0010-0.0020	0.0020-0.0030	0.2157-0.2181	0.2146-0.2150

79231CH2

VALVE SPECIFICATIONS

Year	Engine ID/VIN	Engine Displacement Liters (cc)	Seat Angle (deg.)	Face Angle (deg.)	Spring Test Pressure (lbs. @ in.)	Spring Installed Height (in.)	Stem-to-Guide Clearance (in.)		Stem Diameter (in.)	
							Intake	Exhaust	Intake	Exhaust
1998-99 (cont.)	C32B1	3.2 (3206)	45	45	NA	NA	0.0010-0.0029	0.0020-0.0030	0.2156-0.2159	0.2146-0.2150
	C35A1	3.5 (3494)	45	45	NA	NA	0.0010-0.0029	0.0020-0.0030	0.2157-0.2161	0.2146-0.2150
	F23A1	2.3 (2254)	45	45	NA	NA	0.0008-0.0018	0.0022-0.0030	0.2159-0.2163	0.2146-0.2150
	G25A4	2.5 (2451)	45	45	NA	NA	0.0008-0.0018	0.0020-0.0030	0.2156-0.2159	0.2146-0.2150
	J30A1	3.0 (2997)	45	45	NA	NA	0.0010-0.0020	0.0020-0.0030	0.2156-0.2159	0.2146-0.2150

NA - Not Available

79231CH3

TORQUE SPECIFICATIONS
All readings in ft. lbs.

Year	Engine ID/VIN	Engine Displacement Liters (cc)	Cylinder Head Bolts	Main Bearing Bolts	Rod Bearing Bolts	Crankshaft Damper Bolts	Flywheel Bolts	Manifold Intake	Manifold Exhaust	Spark Plugs	Lug Nut
1995	B18B1	1.8 (1834)	①	②	③	130	④	17	23	13	80
	B18C1	1.8 (1797)	①	⑤	⑥	130	④	17	23	13	80
	C30A1	3.0 (2977)	56	⑦	⑧	⑨	④	16	25	13	80
	C32A1	3.2 (3206)	⑩	⑪	33	174	④	16	22	13	80
	G25A1	2.5 (2451)	72	⑫	24	181	④	16	23	13	80
1996	B18B1	1.8 (1834)	①	56	③	130	④	17	23	13	80
	B18C1	1.8 (1797)	①	⑤	⑥	130	④	17	23	13	80
	C30A1	3.0 (2977)	56	⑦	⑧	181	④	16	25	13	80
	C32A6	3.2 (3206)	⑩	⑪	33	174	④	16	22	13	80
	G25A4	2.5 (2451)	⑬	⑭	24	181	54	16	23	13	80
1997	B18B1	1.8 (1834)	①	56	③	130	④	17	23	13	80
	B18C1	1.8 (1797)	①	⑤	⑥	130	④	17	23	13	80
	C30A1	3.0 (2977)	56	⑦	⑧	181	④	16	25	13	80
	C32A6	3.2 (3206)	⑩	⑪	33	174	④	16	22	13	80
	C32B1	3.2 (3206)	56	⑮	29	181	④	16	22	13	80
	C35A1	3.5 (3494)	56	⑯	33	181	54	16	22	13	80
	F22B1	2.2 (2156)	⑰	⑱	34	181	④	16	23	13	80
	G25A4	2.5 (2451)	⑬	⑭	24	181	54	16	23	13	80
	J30A1	3.0 (2997)	⑲	⑲	⑲	181	54	16	22	13	80
1998-99	B18B1	1.8 (1834)	①	56	③	130	④	17	23	13	80
	B18C1	1.8 (1797)	①	⑤	⑥	130	④	17	23	13	80
	C30A1	3.0 (2977)	56	⑦	⑧	181	④	16	25	13	80
	C32A6	3.2 (3206)	⑩	⑪	33	174	④	16	22	13	80
	C32B1	3.2 (3206)	56	⑮	29	181	④	16	22	13	80
	C35A1	3.5 (3494)	56	⑯	33	181	54	16	22	13	80
	F23A1	2.3 (2254)	⑳	⑳	⑳	181	⑳	16	22	13	80
	G25A4	2.5 (2451)	⑬	⑭	24	181	54	16	23	13	80
	J30A1	3.0 (2997)	⑲	⑲	⑲	181	54	16	22	13	80

① Step 1: 22 ft. lbs.
Step 2: 61 ft. lbs.

② Step 1: 22 ft. lbs.
Step 2: 54 ft. lbs.

③ Step 1: 14 ft. lbs.
Step 2: 23 ft. lbs.

④ Manual transmission: 76 ft. lbs.
Automatic transmission: 54 ft. lbs.

⑤ Step 1: 22 ft. lbs.
Step 2: Cap Nos. 1, 5: 56 ft. lbs.
      Cap Nos. 2-4: 49 ft. lbs.

⑥ Step 1: 14 ft. lbs.
Step 2: 33 ft. lbs.

⑦ Cap bolts: 29 ft. lbs.
Cap bridge bolts: 48 ft. lbs.

⑧ Tighten to 14 ft. lbs. plus an additional 95 degrees

⑨ Step 1: Tighten to 203 ft. lbs.
Step 2: Loosen completely
Step 3: Tighten to 181 ft. lbs.

⑩ Step 1: 29 ft. lbs.
Step 2: 56 ft. lbs.

⑪ Inner: 57 ft. lbs.
Outer: 29 ft. lbs.
Side: 36 ft. lbs.

⑫ Step 1: 22 ft. lbs.
Step 2: 49 ft. lbs.

⑬ Step 1: 29 ft. lbs.
Step 2: 51 ft. lbs.
Step 3: 72 ft. lbs.

⑭ Step 1: 22 ft. lbs.
Step 2: 54 ft. lbs.

⑮ Step 1: Cap bolts 48 ft. lbs.
Step 2: Side bolts 36 ft. lbs.

⑯ Step 1: Outer (9mm) 29 ft. lbs.
Step 2: Inner (11mm) 56 ft. lbs.
Step 3: Side (10mm) 36 ft. lbs.

⑰ Step 1: 29 ft. lbs.
Step 2: 51 ft. lbs.
Step 3: 72 ft. lbs.

⑱ Step 1: 22 ft. lbs.
Step 2: 53 ft. lbs.

⑲ Cylinder head
    Step 1: 29 ft. lbs.
    Step 2: 51 ft. lbs.
    Step 3: 72.3 ft. lbs
Main bearings:
    Step 1: Cap bolts 56 ft. l;bs.
    Step 2: Side bolts 36 ft. lbs.
Rod bearings:
    Step 1: 14 ft. lbs.
    Step 2: Rotate 90 degrees

⑳ Cylinder head:
    Step 1: 22 ft. lbs.
    Step 2: Rotate 90 degrees
    Step 3: Rotate 90 degrees
    Step 4: If new bolt, rotate additional 90 degrees
Main bearings:
    Step 1: 22 ft. lbs.
    Step 2: 58 ft. lbs.
    Step 3: 6mm bolts 8.7 ft. lbs.
Rod bearings:
    Step 1: 14 ft. lbs.
    Step 2: Rotate 90 degrees
Flywheel: MT 76 ft. lbs.
    AT 54 ft. lbs.

79231CH4

SCHEDULED MAINTENANCE INTERVALS
(ACURA 3.0CL)

TO BE SERVICED	TYPE OF SERVICE	VEHICLE MILEAGE INTERVAL (x1000)															
		7.5	15	22.5	30	37.5	45	52.5	60	67.5	75	82.5	90	97.5	105	112.5	120
Engine oil	R	✓	✓	✓	✓	✓	✓	✓	✓	✓	✓	✓	✓	✓	✓	✓	✓
Engine oil filter	R		✓		✓		✓		✓		✓		✓		✓		✓
Air cleaner element	R				✓				✓				✓				✓
Valve clearance	S/I	Inspect only if noisy, and every 30,000 miles thereafter.															
Spark plugs	R														✓		
Timing Belt	R														✓		
Water pump	S/I														✓		
Accessory drive belts	S/I ①				✓				✓				✓				✓
Idle speed	S/I														✓		
Engine coolant	R						✓						✓				
Transmission fluid	R												✓				
Front and rear brakes	S/I		✓		✓		✓		✓		✓		✓		✓		✓
Brake fluid	R						✓						✓				
Parking brake	A		✓		✓		✓		✓		✓		✓		✓		✓
Rotate tires	S/I	✓	✓	✓	✓	✓	✓	✓	✓	✓	✓	✓	✓	✓	✓	✓	✓
Tie rod ends, steeirng gear box and boots	S/I		✓		✓		✓		✓		✓		✓		✓		✓
Suspension components	S/I		✓		✓		✓		✓		✓		✓		✓		✓
CV-joint boots	S/I		✓		✓		✓		✓		✓		✓		✓		✓
Brake hoses and lines	S/I		✓		✓		✓		✓		✓		✓		✓		✓
Fluid levels and condition	S/I		✓		✓		✓		✓		✓		✓		✓		✓
Cooling system hoses and connections	S/I		✓		✓		✓		✓		✓		✓		✓		✓
Exhaust system	S/I		✓		✓		✓		✓		✓		✓		✓		✓
Fuel lines and connections	S/I		✓		✓		✓		✓		✓		✓		✓		✓
Supplemental Restraint System (SRS) ②	S/I	Ten years after production.															

① Inspect accessory drive belt tension and general condition.

② Note that this should only be performed by a qualified automotive technician.

R - Replace   S/I - Inspect and service, if needed   L - Lubricate   A - Adjust   C - Clean

FREQUENT OPERATION MAINTENANCE (SEVERE SERVICE)

If a vehicle is operated under any of the following conditions it is considered severe service:

- Towing a trailer or using a camper or car-top carrier.

- Repeated short trips of less than 5 miles in temperatures below freezing, or trips of less than 10 miles in any temperature.

- Extensive idling or low-speed driving for long distances as in heavy commercial use, such as delivery, taxi or police cars.

79231CH5

SCHEDULED MAINTENANCE INTERVALS
(ACURA 3.0CL)

- **Operating on rough, muddy or salt-covered roads.**

- **Operating on unpaved or dusty roads.**

- **Driving in extremely hot (over 90°) conditions.**

Engine oil & filter - replace every 3,750 miles or 6 months, whichever occurs first.

Air cleaner element - clean every 30,000 miles, starting with the 15,000 mile interval.

               - replace every 30,000 miles, starting with the 30,000 mile interval.

Timing belt - replace at the 60,000 mile interval, if the vehicle is driven in very high temperatures (over 110°F/43°C) or in very low temperatures (under -20°F/-29°C). Otherwise, refer to the scheduled maintenance chart.

Transmission fluid - replace every 30,000 miles.

Front and rear brakes - inspect every 7,500 miles or 6 months, whichever occurs first.

Parking brake - adjust every 7,500 miles.

Locks and hinges - lubricate every 15,000 miles.

Tie rod ends, steering gear box and boots - inspect every 7,500 miles or 6 months, whichever occurs first.

Suspension components - inspect every 7,500 miles or 6 months, whichever occurs first.

CV-joint boots - inspect every 7,500 miles or 6 months, whichever occurs first.

Lights and controls - inspect every 15,000 miles.

Vehicle underbody - inspect every 15,000 miles.

79231CH6

SCHEDULED MAINTENANCE INTERVALS
(ACURA 2.2CL, 2.5TL, 3.2TL, 3.5RL, INTEGRA, LEGEND, NSX)

TO BE SERVICED	TYPE OF SERVICE	VEHICLE MILEAGE INTERVAL (x1000)												
		7.5	15	22.5	30	37.5	45	52.5	60	67.5	75	82.5	90	97.5
Engine oil & filter	R	✓	✓	✓	✓	✓	✓	✓	✓	✓	✓	✓	✓	✓
Rear brake discs, calipers & pads	S/I		✓		✓		✓		✓		✓		✓	
Rotate tires	S/I	✓	✓	✓	✓	✓	✓	✓	✓	✓	✓	✓	✓	✓
A/C filter (Legend)	R		✓		✓		✓		✓		✓		✓	
A/C filter (3.5RL)	R				✓				✓				✓	
Brake hoses & lines (including ABS)	S/I		✓		✓		✓		✓		✓		✓	
Cooling system hoses & connections (2.2CL, 2.5TL, 3.2TL, 3.5RL, 1995 Integra & 1995-99 NSX)	S/I		✓		✓		✓		✓		✓		✓	
Driveshaft boots (2.2CL, 2.5TL, 3.2TL, 3.5RL & 1995 Integra)	S/I		✓		✓		✓		✓		✓		✓	
Exhaust system (2.2CL, 2.5TL, 3.2TL, 3.5RL, 1995 Integra & 1995-99 NSX)	S/I		✓		✓		✓		✓		✓		✓	
Front brake discs & calipers	S/I		✓		✓		✓		✓		✓		✓	
Fuel pipes, hoses & connections (2.2CL, 2.5TL, 3.2TL, 3.5RL, 1995 Integra & 1995-99 NSX)	S/I		✓		✓		✓		✓		✓		✓	
Fuel pipes, hoses & connections (1993-94)	S/I				✓				✓				✓	
Suspension components (2.2CL, 2.5TL, 3.2TL, 3.5RL & 1995-99 NSX)	S/I		✓		✓		✓		✓		✓		✓	
Suspension mounting bolts	S/I		✓		✓		✓		✓		✓		✓	

79231CH7

SCHEDULED MAINTENANCE INTERVALS
(ACURA 2.2CL, 2.5TL, 3.2TL, 3.5RL, INTEGRA, LEGEND, NSX) (Cont.)

TO BE SERVICED	TYPE OF SERVICE	VEHICLE MILEAGE INTERVAL (x1000)												
		7.5	15	22.5	30	37.5	45	52.5	60	67.5	75	82.5	90	97.5
Tie rods, steering gear box & boots (2.2CL, 2.5TL, 3.2TL, 3.5RL, 1995 Integra & 1995-99 NSX)	S/I		✓		✓		✓		✓		✓		✓	
Steering operation, tie rod ends, steering gearbox & boots	S/I		✓		✓				✓				✓	
Valve clearance (2.5TL)	S/I		✓		✓		✓		✓		✓		✓	
Valve clearance (2.2CL, 1995 Integra & 1995-99 NSX)	S/I				✓				✓				✓	
Parking brake (2.2CL, 2.5TL, 3.2TL, 3.5RL, 1995 Integra & 1995-99 NSX)	S/I		✓		✓		✓		✓		✓		✓	
Air cleaner element	R				✓				✓				✓	
Automatic transmission fluid	R				✓				✓				✓	
Brake fluid (including ABS) (Integra, Legend, Vigor, 2.5TL & 1995 NSX)	R				✓				✓				✓	
Brake fluid (including ABS) (3.5RL)	R								✓				✓	
Brake fluid (including ABS) (2.2CL, 3.2TL & 1996-99 NSX)	R						✓				✓			
Front differential oil (2.5TL, Legend)	R				✓				✓				✓	

79231CH9

SCHEDULED MAINTENANCE INTERVALS
(ACURA 2.2CL, 2.5TL, 3.2TL, 3.5RL, INTEGRA, LEGEND, NSX) (Cont.)

TO BE SERVICED	TYPE OF SERVICE	VEHICLE MILEAGE INTERVAL (x1000)												
		7.5	15	22.5	30	37.5	45	52.5	60	67.5	75	82.5	90	97.5
Front differential fluid (3.2TL & 3.5RL)	R								✓				✓	
Manual transmission oil	R				✓				✓				✓	
ABS operation	S/I				✓				✓				✓	
Drive belt(s)	S/I				✓				✓				✓	
Parking brake drums & linings (Legend)	S/I				✓				✓				✓	
Spark plugs (2.2CL, 2.5TL, Integra except GSR)	R				✓				✓				✓	
Spark plugs (3.2TL, Integra GSR, Legend & NSX)	R								✓					
Spark plugs (3.5RL)①	R													
Engine coolant	R						✓				✓			
ABS high pressure hose (NSX)	R								✓					
Fuel filter	R								✓					
PCV valve	S/I								✓					
Timing belt (except as noted below)	R												✓	
Timing belt & timing balancer belt (2.2CL)	R												✓	
Timing belt & timing balancer belt (3.5RL)①	R													
Transmission fluid (2.2CL, 2.5TL, 3.2TL, 3.5RL & 1996-99 NSX)	R												✓	
Distributor, ignition cap & rotor (2.2CL, 2.5TL, Integra)	S/I								✓					
Idle speed (2.2CL, 2.5TL, 3.2TL, Integra, Legend & NSX)	S/I								✓					

79231CH0

SCHEDULED MAINTENANCE INTERVALS
(ACURA 2.2CL, 2.5TL, 3.2TL, 3.5RL, INTEGRA, LEGEND, NSX) (Cont.)

TO BE SERVICED	TYPE OF SERVICE	VEHICLE MILEAGE INTERVAL (x1000)												
		7.5	15	22.5	30	37.5	45	52.5	60	67.5	75	82.5	90	97.5
Idle speed (3.5RL)②	S/I													
Ignition wires	S/I								✓					
TWC converter heat shield	S/I								✓					
Water pump	S/I												✓	
Water pump (3.5RL)②	S/I													

① Replace at 105,000 miles.
② Service or inspect at 105,000 miles.
R – Replace   S/I – Service or Inspect

FREQUENT OPERATION MAINTENANCE (SEVERE SERVICE)

If a vehicle is operated under any of the following conditions it is considered severe service:
- Extremely dusty areas.
- 50% or more of the vehicle operation is in 32°C (90°F) or higher temperatures, or constant operation in temperatures below 0°C (32°F).
- Prolonged idling (vehicle operation in stop and go traffic).
- Frequent short running periods (engine does not warm to normal operating temperatures).
- Police, taxi, delivery usage or trailer towing usage.

Oil & oil filter change – change every 3750 miles.
Brake hoses & lines (including ABS) (3.5RL) - service or inspect every 7500 miles.
Cooling system hoses & connections (3.5RL) - service or inspect every 7500 miles.
Driveshaft boots (2.2CL, 2.5TL, 3.2TL & 3.5RL) - check every 7500 miles.
Exhaust system (3.5RL) - check every 7500 miles.
Brake discs, calipers & pads - service or inspect every 7500 miles.
Fuel pipes, hoses & connections (3.5RL) - check every 7500 miles.
Power steering system - service or inspect every 7500 miles.
Suspension components - service or inspect every 7500 miles.
Tie rod ends, steering gear box & boots (2.2CL, 2.5TL, 3.2TL & 3.5RL) - service or inspect every 7500 miles.
Air cleaner element (1995-99 NSX) – service or inspect every 7500 miles.
Air cleaner element (except 1995-99 NSX) - service or inspect every 15,000 miles.
Front differential oil (2.5TL, 3.2TL, 3.5RL, Legend) - replace every 15,000 miles.
Transmission oil (Legend, NSX) - replace every 15,000 miles.
Transmission oil (2.2CL, 2.5TL, 3.2TL & 3.5RL) - replace every 30,000 miles.
Timing belt (2.5TL, 3.2TL & 1995 Integra) - replace every 60,000 miles.
Water pump (2.5TL & 3.2TL) - service or inspect every 60,000 miles.

79231CJ1

SCHEDULED MAINTENANCE
ACURA
INTEGRA, INTEGRA GSR, LEGEND, NSX
2.2CL, 2.3CL, 2.5TL, 3.0CL, 3.2TL, 3.5RL

The following should be used as a guide when determining the amount of work required for a particular service if taken to a repair shop. In estimating how long a particular Scheduled Maintenance Service should take, please observe the following:

- **Chilton Time** is time based on field research and data supplied by the vehicle manufacturer.
- Labor time operations are given in hours and tenths of an hour.
- All labor operations, are to be used as a guide.

Mechanic Skill Level Codes:
- **(G) GENERAL:** Normally skilled with certification.
- **(M) MAINTENANCE:** Semi-skilled working on certicication.
- **(P) PRECISION:** Really skilled with multiple certification.

	Chilton Time
(G) 7500 Mile Service	
1995-99	.8
Inspect front brake pads, add	
NSX	.1
(G) 15000 Mile Service	
1997-99 2.2CL, 3.0CL	2.0
1995-99 2.5TL	2.1
1996-99 3.5RL	1.9
1995-99 Integra, GSR	1.9
1995 Legend	1.5
1995-99 NSX	1.9
(G) 22500 Mile Service	
1995-99	.8
Inspect front brake pads, add	
NSX	.1
(G) 30000 Mile Service	
1997-99 2.2CL, 3.0CL	3.0
1995-99 2.5TL	4.2
1996-99 3.2TL	2.8
1996-99 3.5RL	3.0
1995-99 Integra, GSR	3.8
1995 Legend	3.3
1995-99 NSX	3.4
w/AT add	.5
(G) 37500 Mile Service	
1995-99	.8
Inspect front brake pads, add	
NSX	.1

	Chilton Time
(G) 45000 Mile Service	
1997-99 2.2CL, 3.0CL	2.7
1995-99 2.5TL	2.4
1996-99 3.2TL	2.9
1996-99 3.5RL	2.4
1995-99 Integra, GSR	2.3
1995 Legend	1.9
1995-99 NSX	2.8
(G) 52500 Mile Service	
1995-99	.8
Inspect front brake pads, add	
NSX	.1
(G) 60000 Mile Service	
1997-99 2.2CL, 3.0CL	5.0
1995-99 2.5TL	5.5
1996-99 3.2TL	5.3
1996-99 3.5RL	5.0
1995-99 Integra, GSR	5.3
1995 Legend	5.6
1995-96 NSX	5.6
w/AT add	.5

	Chilton Time
(G) 67500 Mile Service	
1995-99	.8
Inspect front brake pads, add	
NSX	.1
(G) 75000 Mile Service	
1997-99 2.2CL, 3.0CL	2.7
1995-99 2.5TL	2.4
1996-99 3.2TL	2.9
1996-99 3.5RL	2.4
1995-99 Integra, GSR	2.3
1995 Legend	1.9
1995-99 NSX	2.8
(G) 82500 Mile Service	
1995-99	.8
Inspect front brake pads, add	
NSX	.1
(G) 90000 Mile Service	
1997-99 2.2CL, 3.0CL	7.1
1995-99 2.5TL	8.4
1996-99 3.2TL	7.3
1996-99 3.5RL	4.5
1995-99 Integra, GSR	7.5
1995 Legend	6.0
1995-99 NSX	7.0
w/AT add	.5
(G) 97500 Mile Service	
1995-99	.8
Inspect front disc brake pads, add	
NSX	.1

79231CJ2

AUDI
90 • A6 • Cabriolet • S6

VEHICLE IDENTIFICATION CHART

Engine Code							Model Year	
Code	Liters	Cu. In. (cc)	Cyl.	Fuel Sys.	Eng. Mfg.		Code	Year
AAH	2.8	169 (2771)	6	MFI	Audi		S	1995
AAN	2.2	136 (2226)	5	MFI	Audi		T	1996
ACK	2.8	169 (2771)	6	MFI	Audi		V	1997
ABZ	4.2	255 (4172)	8	MFI	Audi		W	1998
AEB	1.8	107 (1781)	4	MFI-Turbo	Audi		X	1999
AEW	3.7	226 (3685)	8	MFI	Audi			
AFC	2.8	169 (2771)	6	MFI	Audi			
AHA	2.8	169 (2771)	6	MFI	Audi			

MFI - Multi-point Fuel Injection

79231CM4

ENGINE IDENTIFICATION

Year	Model	Engine Displacement Liters (cc)	Engine Series (ID/VIN)	Fuel System	No. of Cylinders	Engine Type
1995	90 Sedan	2.8 (2771)	ACK	MFI	6	DOHC
	90 Sport Sedan	2.8 (2771)	ACK	MFI	6	DOHC
	90 Quattro	2.8 (2771)	ACK	MFI	6	DOHC
	A6 Sedan	2.8 (2771)	ACK	MFI	6	DOHC
	A6 Wagon	2.8 (2771)	ACK	MFI	6	DOHC
	A6 Quattro	2.8 (2771)	ACK	MFI	6	DOHC
	Cabriolet	2.8 (2771)	ACK	MFI	6	DOHC
	S6 Quattro Sedan	2.2 (2226)	AAN	MFI	5	DOHC
	S6 Quattro Wagon	2.2 (2226)	AAN	MFI	5	DOHC
1996	A4 Sedan	2.8 (2771)	AFC	MFI	6	DOHC
	A4 Quattro	2.8 (2771)	ACK	MFI	6	DOHC
	A6 Sedan	2.8 (2771)	AFC	MFI	6	DOHC
	A6 Wagon	2.8 (2771)	AFC	MFI	6	DOHC
	A6 Quattro	2.8 (2771)	AFC	MFI	6	DOHC
	Cabriolet	2.8 (2771)	AFC	MFI	6	DOHC
1997	A4 Sedan	1.8 (1781)	AEB	MFI	4	DOHC
	A4 Sedan	2.8 (2771)	AFC	MFI	6	DOHC
	A4 Quattro	2.8 (2771)	AHA	MFI	6	DOHC
	A6 Sedan	2.8 (2771)	AHA	MFI	6	DOHC
	A6 Wagon	2.8 (2771)	AFC	MFI	6	DOHC
	A6 Quattro	2.8 (2771)	AHA	MFI	6	DOHC
	A8 Sedan	3.7 (3685)	AEW	MFI	8	DOHC
	A8 Quattro	4.2 (4172)	ABZ	MFI	8	DOHC
	Cabriolet	2.8 (2771)	AFC	MFI	6	DOHC
1998-99	A4 Sedan	1.8 (1781)	AEB	MFI	4	DOHC
	A4 Sedan	2.8 (2771)	AFC	MFI	6	DOHC
	A4 Wagon	2.8 (2771)	AFC	MFI	6	DOHC
	A4 Quattro	2.8 (2771)	AFC	MFI	6	DOHC
	A6 Sedan	2.8 (2771)	AFC	MFI	6	DOHC
	A6 Wagon	2.8 (2771)	AFC	MFI	6	DOHC
	A6 Quattro	2.8 (2771)	AFC	MFI	6	DOHC
	A8 Sedan	3.7 (3685)	AEW	MFI	8	DOHC
	A8 Quattro	4.2 (4172)	ABZ	MFI	8	DOHC
	Cabriolet	2.8 (2771)	AFC	MFI	6	DOHC

MFI - Multi-point Fuel Injection
DOHC - Double Overhead Camshafts

79231CM5

GENERAL ENGINE SPECIFICATIONS

Year	Engine ID/VIN	Engine Displacement Liters (cc)	Fuel System Type	Net Horsepower @ rpm	Net Torque @ rpm (ft. lbs.)	Bore x Stroke (in.)	Compression Ratio	Oil Pressure @ rpm
1995	AAN	2.2 (2226)	MFI	227@5900	258@1950	3.19x3.40	9.3:1	29@2000
	ACK	2.8 (2771)	MFI	190@6000	207@3200	3.25x3.40	10.3:1	41@3000
1996	ACK	2.8 (2771)	MFI	190@6000	207@3200	3.25x3.40	10.0:1	41@3000
	AFC	2.8 (2771)	MFI	174@5500	180@3000	3.25x3.40	10.0:1	41@3000
1997	ABZ	4.2 (4172)	MFI	300@6000	295@3300	3.32x3.66	10.8:1	41@3000
	AEB	1.8 (1781)	MFI	150@5700	155@1750	3.18x3.40	9.5:1	41@3000
	AEW	3.7 (3685)	MFI	230@5800	229@2300	3.32x3.24	10.8:1	41@3000
	AFC	2.8 (2771)	MFI	174@5500	180@3000	3.25x3.40	10.0:1	41@3000
	AHA	2.8 (2771)	MFI	190@6000	207@3200	3.25x3.40	10.0:1	41@3000
1998-99	ABZ	4.2 (4172)	MFI	300@6000	295@3300	3.32x3.66	10.8:1	41@3000
	AEB	1.8 (1781)	MFI	150@5700	155@1750	3.18x3.40	9.5:1	41@3000
	AEW	3.7 (3685)	MFI	230@5800	229@2300	3.32x3.24	10.8:1	41@3000
	AFC	2.8 (2771)	MFI	174@5500	180@3000	3.25x3.40	10.0:1	41@3000
	AHA	2.8 (2771)	MFI	190@6000	207@3200	3.25x3.40	10.0:1	41@3000

MFI - Multiport fuel injection

79231CM6

GASOLINE ENGINE TUNE-UP SPECIFICATIONS

Year	Engine ID/VIN	Engine Displacement Liters (cc)	Spark Plugs Gap (in.)	Ignition Timing (deg.) MT	Ignition Timing (deg.) AT	Fuel Pump (psi)	Idle Speed (rpm) MT	Idle Speed (rpm) AT	Valve Clearance In.	Valve Clearance Ex.
1995	AAN	2.2 (2226)	0.024	①	①	58-61	770-830	770-830	HYD	HYD
	ACK	2.8 (2771)	0.039	①	①	52-61	650-750	650-750	HYD	HYD
1996	ACK	2.8 (2771)	0.039	①	①	46-55	650-750	650-750	HYD	HYD
	AFC	2.8 (2771)	0.039	①	①	52-61	650-750	650-750	HYD	HYD
1997	ABZ	4.2 (4172)	0.039	①	①	52-61	—	720-800	HYD	HYD
	AEB	1.8 (1781)	0.039	①	①	50-58	820-900	820-900	HYD	HYD
	AEW	3.7 (3685)	0.039	①	①	52-61	680-760	—	HYD	HYD
	AFC	2.8 (2771)	0.039	①	①	46-55	650-750	650-750	HYD	HYD
	AHA	2.8 (2771)	0.039	①	①	55-61	700-800	700-800	HYD	HYD
1998-99	ABZ	4.2 (4172)	0.039	①	①	52-61	—	720-800	HYD	HYD
	AEB	1.8 (1781)	0.039	①	①	50-58	820-900	820-900	HYD	HYD
	AEW	3.7 (3685)	0.039	①	①	52-61	680-760	—	HYD	HYD
	AFC	2.8 (2771)	0.039	①	①	46-55	650-750	650-750	HYD	HYD
	AHA	2.8 (2771)	0.039	①	①	55-61	700-800	700-800	HYD	HYD

NOTE: The Vehicle Emission Control Information label often reflects specification changes made during production. The label figures must be used if they differ from those in this chart.

HYD - Hydraulic

① The basic setting is controlled by the ECU and is not adjustable

79231CM7

CAPACITIES

Year	Model	Engine ID/VIN	Engine Displacement Liters (cc)	Engine Oil with Filter	Transmission (pts.)			Transfer Case (pts.)	Drive Axle		Fuel Tank (gal.)	Cooling System (qts.)
					4-Spd	5-Spd	Auto.		Front (pts.)	Rear (pts.)		
1995	90 Quattro	ACK	2.8 (2771)	5.3	—	5.0	①	—	②	—	21.1	8.4
	90 Sedan	ACK	2.8 (2771)	4.7	—	5.6	—	—	—	3.2	21.1	9.5
	90 Sport Sedan	ACK	2.8 (2771)	5.3	—	5.0	①	—	②	—	21.1	8.4
	A6 Quattro	ACK	2.8 (2771)	5.3	—	5.0	①	—	—	—	17.4	11.6
	A6 Sedan	ACK	2.8 (2771)	5.3	—	5.0	③	—	—	④	21.1	8.4
	A6 Wagon	ACK	2.8 (2771)	5.3	—	5.0	③	—	—	④	21.1	8.4
	Cabriolet	ACK	2.8 (2771)	5.3	—	5.0	①	—	—	—	17.4	11.6
	S6 Quattro Sedan	AAN	2.2 (2226)	5.3	—	—	⑤	—	—	2.0	16.9	11.6
	S6 Quattro Wagon	AAN	2.2 (2226)	5.3	—	—	⑤	—	—	—	17.4	11.6
1996	A4 Sedan	AFC	2.8 (2771)	5.3	—	4.8	5.4	—	1.6	—	16.0	6.9
	A4 Quattro	ACK	2.8 (2771)	5.3	—	5.8	5.4	—	1.6	⑥	16.0	6.9
	A6 Sedan	AFC	2.8 (2771)	5.3	—	—	5.5	—	②	—	21.1	12.7
	A6 Wagon	AFC	2.8 (2771)	5.3	—	—	5.5	—	1.5	—	21.1	12.7
	A6 Quattro	AFC	2.8 (2771)	5.3	—	—	5.5	—	1.5	⑦	21.1	12.7
	Cabriolet	AFC	2.8 (2771)	5.3	—	—	5.5	—	1.6	—	17.4	11.6
1997	A4 Sedan	AEB	1.8 (1781)	4.6	—	4.8	5.4	—	1.6	—	16.4	6.9
	A4 Sedan	AFC	2.8 (2771)	5.3	—	5.8	5.4	—	1.6	—	16.4	6.9
	A4 Quattro	AHA	2.8 (2771)	5.3	—	—	5.4	—	1.6	⑥	16.4	6.9
	A6 Sedan	AHA	2.8 (2771)	5.3	—	—	5.5	—	1.6	—	21.1	12.7
	A6 Wagon	AFC	2.8 (2771)	5.3	—	—	5.5	—	1.6	—	21.1	12.7
	A6 Quattro	AHA	2.8 (2771)	5.3	—	—	5.5	—	1.5	⑦	21.1	12.7
	A8 Sedan	AEW	3.7 (3685)	8.0	—	—	5.5	—	1.6	—	23.7	11.6
	A8 Quattro	ABZ	4.2 (4172)	8.0	—	—	8.0	—	1.6	⑧	23.7	11.6
	Cabriolet	AFC	2.8 (2771)	5.3	—	—	5.5	—	1.6	—	17.4	11.6
1998-99	A4 Sedan	AEB	1.8 (1781)	4.6	—	4.8	5.4	—	1.6	—	16.4	6.9
	A4 Sedan	AFC	2.8 (2771)	5.3	—	5.0	5.4	—	1.6	—	16.4	6.9
	A4 Wagon	AFC	2.8 (2771)	5.3	—	—	5.4	—	1.6	—	16.4	6.9
	A4 Quattro	AHA	2.8 (2771)	5.3	—	—	5.4	—	1.6	⑥	16.4	.6.9
	A6 Sedan	AHA	2.8 (2771)	6.9	—	—	5.5	—	1.6	—	21.1	12.7
	A6 Wagon	AFC	2.8 (2771)	6.9	—	—	5.5	—	1.5	—	21.1	12.7
	A6 Quattro	AHA	2.8 (2771)	6.9	—	—	5.5	—	1.5	⑦	21.1	12.7
	A8 Sedan	AEW	3.7 (3685)	8.0	—	—	5.5	—	1.6	—	23.7	11.6
	A8 Quattro	ABZ	4.2 (4172)	8.0	—	—	8.0	—	1.6	⑧	23.7	11.6
	Cabriolet	AFC	2.8 (2771)	5.3	—	—	5.5	—	1.6	—	17.4	11.6

NOTE: All capacities are approximate. Add fluid gradually and ensure a proper fluid level is obtained.

① Initial fill: 11.4
   Change: 6.4
② 097 transmission: 2.0
   01K transmission: 1.4
③ Initial fill: 14.8
   Change: 5.8
④ Center: 2.2
   Rear: 3.0
⑤ Initial Fill: 11.4
   Change: NA
⑥ Center: 1.7
   Rear: 3.8
⑦ Center: 1.7
   Rear: 3.2
⑧ Center: 1.6
   Rear: 3.0

79231CM8

VALVE SPECIFICATIONS

Year	Engine ID/VIN	Engine Displacement Liters (cc)	Seat Angle (deg.)	Face Angle (deg.)	Spring Test Pressure (lbs. @ in.)	Spring Installed Height (In.)	Stem-to-Guide Clearance (in.) Intake		Stem Diameter (in.) Intake	Exhaust
1995	AAN	2.2 (2226)	45	45	NA	NA	0.039 ①	0.051 ①	0.2744	0.2732
	ACK	2.8 (2771)	45	45	NA	NA	0.039 ①	0.051 ①	0.2346	0.2338
1996	ACK	2.8 (2771)	45	45	NA	NA	0.039 ①	0.051 ①	0.2346	0.2338
	AFC	2.8 (2771)	45	45	NA	NA	0.039 ①	0.051 ①	NA	NA
1997	ABZ	4.2 (4172)	45	45	NA	NA	0.039 ①	0.051 ①	NA	NA
	AEB	1.8 (1781)	45	45	NA	NA	0.039 ①	0.051 ①	0.2339	0.2339
	AEW	3.7 (3685)	45	45	NA	NA	0.039 ①	0.051 ①	NA	NA
	AFC	2.8 (2771)	45	45	NA	NA	0.039 ①	0.051 ①	NA	NA
	AHA	2.8 (2771)	45	45	NA	NA	0.039 ①	0.051 ①	0.2346	0.2338
1998-99	ABZ	4.2 (4172)	45	45	NA	NA	0.039 ①	0.051 ①	NA	NA
	AEB	1.8 (1781)	45	45	NA	NA	0.039 ①	0.051 ①	0.2339	0.2339
	AEW	3.7 (3685)	45	45	NA	NA	0.039 ①	0.051 ①	NA	NA
	AFC	2.8 (2771)	45	45	NA	NA	0.039 ①	0.051 ①	NA	NA
	AHA	2.8 (2771)	45	45	NA	NA	0.039 ①	0.051 ①	0.2346	0.2338

NA - Not Available

① To measure Stem-to-Guide clearance, insert new valve into guide until end of valve is flush with end of guide. Use dial indicator to measure valve head movement.
   Specification given is for maximum wear.

79231CM9

TORQUE SPECIFICATIONS
All readings in ft. lbs.

Year	Engine ID/VIN	Engine Displacement Liters (cc)	Cylinder Head Bolts	Main Bearing Bolts	Rod Bearing Bolts	Crankshaft Damper Bolts	Flywheel Bolts	Manifold Intake	Exhaust	Spark Plugs	Lug Nut
1995	AAN	2.2 (2226)	①	48	②	③	②	15	18	28	89
	ACK	2.8 (2771)	④	⑤	⑥	⑦	⑧	15	18	22	89
1996	ACK	2.8 (2771)	④	⑤	⑥	⑦	⑧	15	18	22	89
	AFC	2.8 (2771)	④	⑤	⑥	⑦	⑧	15	18	22	89
1997	ABZ	4.2 (4172)	⑨	NA	NA	NA	NA	NA	NA	22	89
	AEB	1.8 (1781)	④	⑩	⑥	30	⑪	7	18	22	89
	AEW	3.7 (3685)	⑨	NA	NA	NA	NA	NA	NA	22	89
	AFC	2.8 (2771)	④	⑤	⑥	⑦	⑧	15	18	22	89
	AHA	2.8 (2771)	④	⑤	⑥	⑦	⑧	15	18	22	89
1998-99	ABZ	4.2 (4172)	⑨	NA	NA	NA	NA	NA	NA	22	89
	AEB	1.8 (1781)	④	⑩	⑥	30	⑪	7	18	22	89
	AEW	3.7 (3685)	⑨	NA	NA	NA	NA	NA	NA	22	89
	AFC	2.8 (2771)	④	⑤	⑥	⑦	⑧	15	18	22	89
	AHA	2.8 (2771)	④	⑤	⑥	⑦	⑧	15	18	22	89

NA - Not Available

① Step 1: 30 ft. lbs.
   Step 2: 44 ft. lbs.
   Step 3: +180 degrees
② Step 1: 22 ft. lbs.
   Step 2: +90 degrees
③ 258 ft. lbs. with special tool
   332 ft. lbs. without special tool
④ Step 1: 44 ft. lbs.
   Step 2: Rotate 90 degrees
   Step 3: Rotate 90 degrees
⑤ Step 1: 44 ft. lbs.
   Step 2: Rotate 180 degrees

⑥ Step 1: 22 ft. lbs.
   Step 2: Rotate 90 degrees
⑦ Center Bolt:
   Step 1: 148 ft. lbs.
   Step 2: Rotate 180 degrees
   Damper Bolts: 15 ft. lbs.
⑧ Flywheel MT:
   Step 1: 30 ft. lbs.
   Step 2; Rotate 90 degrees
   Step 3: Rotate 90 degrees
   Flexplate AT;
   Step 1: 22 ft. lbs.
   Step 2: Rotate 90 degrees

⑨ Step 1: 30 ft. lbs.
   Step 2: 44 ft. lbs.
   Step 3: Rotate 180 degrees
⑩ Step 1: 48 ft. lbs.
   Step 2: Rotate 90 degrees
⑪ Flywheel MT:
   Step 1: 44 ft. lbs.
   Step 2: Rotate 180 degrees
   Flexplate AT: 22 ft. lbs.

79231CM0

SCHEDULED MAINTENANCE INTERVALS
(AUDI 90, S6, A4, A6, A8, & CABRIOLET)

TO BE SERVICED	TYPE OF SERVICE	VEHICLE MILEAGE INTERVAL (x1000)												
		7.5	15	22.5	30	37.5	45	52.5	60	67.5	75	82.5	90	97.5
Engine oil & filter②	R	✓	✓	✓	✓	✓	✓	✓	✓	✓	✓	✓	✓	✓
Automatic shiftlock operation	S/I	✓	✓	✓	✓	✓	✓	✓	✓	✓	✓	✓	✓	✓
Cooling system	S/I	✓	✓	✓	✓	✓	✓	✓	✓	✓	✓	✓	✓	✓
Passenger compartment air filter	R		✓		✓		✓		✓		✓		✓	
Automatic transmission fluid, filter & final drive	S/I		✓		✓		✓				✓		✓	
Battery electrolyte level	S/I		✓		✓		✓		✓		✓		✓	
Brake system (brake pads & fluid level)	S/I		✓		✓		✓		✓		✓		✓	
Drive axle shaft boots	S/I		✓		✓		✓				✓		✓	
Engine (check for leaks)	S/I		✓		✓		✓				✓		✓	
Exhaust system	S/I		✓		✓		✓		✓		✓		✓	
Idle speed⑤	S/I		✓		✓		✓		✓		✓		✓	
Manual transmission oil	S/I		✓		✓		✓		✓		✓		✓	
ODB system check for codes③	S/I		✓		✓		✓		✓		✓		✓	
V-belts④	S/I			✓		✓				✓		✓		
Air cleaner element	R			✓					✓				✓	
Spark plugs	R			✓					✓				✓	
Power steering fluid level	S/I			✓					✓				✓	
Automatic transmission fluid⑦	R						✓						✓	
Timing belt	R												✓	
Brake fluid①	R													
Front axle dust seals on ball joints & tie rod ends	S/I							✓						

79231CN1

SCHEDULED MAINTENANCE INTERVALS
(AUDI 90, S6, A4, A6, A8, & CABRIOLET) (Cont.)

TO BE SERVICED	TYPE OF SERVICE	VEHICLE MILEAGE INTERVAL (x1000)												
		7.5	15	22.5	30	37.5	45	52.5	60	67.5	75	82.5	90	97.5
Poly-ribbed belt	R												✓	
Rotate tires	S/I	✓												

① Replace every 2 years regardless of mileage.
② Reset service interval display, if equipped.
③ If equipped.
④ Replace at 45,000 & 90,000 miles.
⑤ except California models.
⑥ Replace at mileage interval or every 2 years, whichever comes first.
⑦ A6, Cabriolet.
R – Replace   S/I – Service or Inspect

FREQUENT OPERATION MAINTENANCE (SEVERE SERVICE)

If a vehicle is operated under any of the following conditions it is considered severe service:
- Extremely dusty areas.
- 50% or more of the vehicle operation is in 32°C (90°F) or higher temperatures, or constant operation in temperatures below 0°C (32°F).
- Prolonged idling (vehicle operation in stop and go traffic).
- Frequent short running periods (engine does not warm to normal operating temperatures).
- Police, taxi, delivery usage or trailer towing usage.
Oil & oil filter change – change every 3750 miles.
Air filter element – service or inspect every 15,000 miles.
Automatic transmission fluid - replace every 30,000 miles.

79231CN2

SCHEDULED MAINTENANCE
AUDI
90, A4, A6, A8, S6, CABRIOLET

The following should be used as a guide when determining the amount of work required for a particular service if taken to a repair shop. In estimating how long a particular Scheduled Maintenance Service should take, please observe the following:

- **Chilton Time** is time based on field research and data supplied by the vehicle manufacturer.
- Labor time operations are given in hours and tenths of an hour.
- All labor operations, are to be used as a guide.

Mechanic Skill Level Codes:
- **(G)** GENERAL: Normally skilled with certification.
- **(M)** MAINTENANCE: Semi-skilled working on certicication.
- **(P)** PRECISION: Really skilled with multiple certification.

	Chilton Time
(M) 7500 Mile Service	
1995 90	1.0
1995-99 A6	1.2
1995-99 Cabriolet	1.0
1996-99 A4	1.0
1997-99 A8	1.2
1999 S6	1.2
(G) 15000 Mile Service	
1995 90	2.0
1995-99 A6	2.2
1995-99 Cabriolet	2.0
1996-99 A4	1.8
1997-99 A8	1.9
1999 S6	2.2
(M) 22500 Mile Service	
1995 90	.5
1995-99 A6	.8
1995-99 Cabriolet	.5
1996-99 A4	.8
1997-99 A8	.8
1999 S6	.8
(G) 30000 Mile Service	
1995 90	2.4
1995-99 A6	2.6
1995-99 Cabriolet	2.4
1996-99 A4	2.3
1997-99 A8	2.7
1999 S6	2.6

	Chilton Time
(M) 37500 Mile Service	
1995 90	.5
1995-99 A6	.8
1995-99 Cabriolet	.5
1996-99 A4	.8
1997-99 A8	.8
1999 S6	.8
(G) 45000 Mile Service	
1995 90	1.7
1995-99 A6	1.9
1995-99 Cabriolet	1.7
1996-99 A4	1.5
1997-99 A8	2.0
1999 S6	1.9
(M) 52500 Mile Service	
1995 90	1.0
1995-99 A6	1.2
1995-99 Cabriolet	1.0
1996-99 A4	1.0
1997-99 A8	1.2
1999 S6	1.2
(G) 60000 Mile Service	
1995 90	2.4
1995-99 A6	2.6
1995-99 Cabriolet	2.4
1996-99 A4	2.3
1997-99 A8	2.7
1999 S6	2.6
(M) 67500 Mile Service	
1995 90	.5
1995-99 A6	.8
1995-99 Cabriolet	.5
1996-99 A4	.8
1997-99 A8	.8
1999 S6	.8

	Chilton Time
(G) 75000 Mile Service	
1995 90	2.0
1995-99 A6	2.2
1995-99 Cabriolet	2.0
1996-99 A4	1.8
1997-99 A8	1.9
1999 S6	2.2
(M) 82500 Mile Service	
1995 90	.5
1995-99 A6	.8
1995-99 Cabriolet	.5
1996-99 A4	.8
1997-99 A8	.8
1999 S6	.8
(G) 90000 Mile Service	
1995 90	2.9
1995-99 A6	3.1
1995-99 Cabriolet	2.9
1996-99 A4	2.8
1997-99 A8	3.2
1999 S6	3.1
Replace timing belt (2.8L) add	1.5
(M) 97500 Mile Service	
1995 90	1.0
1995-99 A6	1.2
1995-99 Cabriolet	1.0
1996-99 A4	1.0
1997-99 A8	1.2
1999 S6	1.2

79231CN3

BMW
318i • 318iS • 318iC • 323is • 323iC • 325i • 325iS • 325iC • 328i • 525i • 525ti • 528i • 530i • 530ti • 540i • 740i • 740iL • 840Ci • 750iL • 850Ci • 850CSi • M3 • Z3

VEHICLE IDENTIFICATION CHART

Code	Liters	Cu. In. (cc)	Cyl.	Fuel Sys.	Eng. Mfg.
M42B18	1.8L	116 (1796)	4	M1.7	BMW
M44B19	1.9L	116 (1895)	4	①	BMW
M50B25	2.5L	152 (2494)	6	②	BMW
M50B25	2.5L	152 (2494)	6	M3.1	BMW
M52B28	2.8L	170 (2793)	6	②	BMW
M60B30	3.0L	183 (2997)	8	M3.3	BMW
M60B40	4.0L	243 (3982)	8	M3.3	BMW
M62B44	4.4L	268 (4398)	8	M5.2	BMW
M73B54	5.4L	328 (5379)	12	M5.3	BMW
S50B30	3.0L	182 (2990)	6	M3.3	BMW
S52B32	3.2L	192 (3152)	6	M3.3	BMW
S70B56	5.6L	340 (5576)	12	M3.3	BMW

Model Year	
Code	Year
S	1995
T	1996
V	1997
W	1998
X	1999

① Bosch ML-Motronic w/knock control (2 sensors)

② 1996-97 Siemens MS 41.0 w/knock control (2 sensors)
   1998-99 Siemens MS 41.1 w/knock control (2 sensors)

79231C01

ENGINE IDENTIFICATION

Year	Model	Engine Displacement Liters (cc)	Engine Series (ID/VIN)	Fuel System	No. of Cylinders	Engine Type
1995	318i	1.8 (1796)	M42B18	M1.7	4	DOHC
	318iS	1.8 (1796)	M42B18	M1.7	4	DOHC
	318iC	1.8 (1796)	M42B18	M1.7	4	DOHC
	325i	2.5 (2494)	M50B25	M3.1	6	DOHC
	325iS	2.5 (2494)	M50B25	M3.1	6	DOHC
	325iC	2.5 (2494)	M50B25	M3.1	6	DOHC
	525i	2.5 (2494)	M50B25	M3.1	6	DOHC
	525ti	2.5 (2494)	M50B25	M3.1	6	DOHC
	M3	3.0 (2990)	S50B30	M3.3	6	DOHC
	530i	3.0 (2997)	M60B30	M3.3	8	DOHC
	530ti	3.0 (2997)	M60B30	M3.3	8	DOHC
	540i	4.0 (3982)	M60B40	M3.3	8	DOHC
	740i	4.0 (3982)	M60B40	M3.3	8	DOHC
	740iL	4.0 (3982)	M60B40	M3.3	8	DOHC
	840Ci	4.0 (3982)	M60B40	M3.3	8	DOHC
	750iL	5.4 (5379)	M73B54	M5.2	12	SOHC
	850Ci	5.4 (5379)	M73B54	M5.2	12	SOHC
	850CSi	5.6 (5576)	S70B56	M3.3	12	SOHC
1996	318i	1.8 (1796)	M42B18	M1.7	4	DOHC
	318iS	1.8 (1796)	M42B18	M1.7	4	DOHC
	318iC	1.8 (1796)	M42B18	M1.7	4	DOHC
	325i	2.5 (2494)	M50B25	M3.1	6	DOHC
	325iS	2.5 (2494)	M50B25	M3.1	6	DOHC
	325iC	2.5 (2494)	M50B25	M3.1	6	DOHC
	525i	2.5 (2494)	M50B25	M3.1	6	DOHC
	525ti	2.5 (2494)	M50B25	M3.1	6	DOHC
	M3	3.0 (2990)	S50B30	M3.3	6	DOHC
	530i	3.0 (2997)	M60B30	M3.3	8	DOHC
	530ti	3.0 (2997)	M60B30	M3.3	8	DOHC
	540i	4.0 (3982)	M60B40	M3.3	8	DOHC
	740i	4.0 (3982)	M60B40	M3.3	8	DOHC
	740iL	4.0 (3982)	M60B40	M3.3	8	DOHC
	840Ci	4.0 (3982)	M60B40	M3.3	8	DOHC
	750iL	5.4 (5379)	M73B54	M5.2	12	SOHC
	850Ci	5.4 (5379)	M73B54	M5.2	12	SOHC
	850CSi	5.6 (5576)	S70B56	M3.3	12	SOHC
1997	318i	1.8 (1796)	M42B18	M1.7	4	DOHC
	318iS	1.8 (1796)	M42B18	M1.7	4	DOHC
	318iC	1.8 (1796)	M42B18	M1.7	4	DOHC
	325i	2.5 (2494)	M50B25	M3.1	6	DOHC
	325iS	2.5 (2494)	M50B25	M3.1	6	DOHC
	325iC	2.5 (2494)	M50B25	M3.1	6	DOHC
	525i	2.5 (2494)	M50B25	M3.1	6	DOHC
	525ti	2.5 (2494)	M50B25	M3.1	6	DOHC
	M3	3.0 (2990)	S50B30	M3.3	6	DOHC

79231C02

ENGINE IDENTIFICATION

Year	Model	Engine Displacement Liters (cc)	Engine Series (ID/VIN)	Fuel System	No. of Cylinders	Engine Type
1997 (cont.)	530i	3.0 (2997)	M60B30	M3.3	8	DOHC
	530ti	3.0 (2997)	M60B30	M3.3	8	DOHC
	540i	4.0 (3982)	M60B40	M3.3	8	DOHC
	740i	4.0 (3982)	M60B40	M3.3	8	DOHC
	740iL	4.0 (3982)	M60B40	M3.3	8	DOHC
	840Ci	4.0 (3982)	M60B40	M3.3	8	DOHC
	750iL	5.4 (5379)	M73B54	M5.2	12	SOHC
	850Ci	5.4 (5379)	M73B54	M5.2	12	SOHC
	850CSi	5.6 (5576)	S70B56	M3.3	12	SOHC
1998-99	318ti	1.9 (1895)	M44B19	①	4	DOHC
	318i	1.9 (1895)	M44B19	①	4	DOHC
	323is	2.5 (2494)	M50B25	②	6	DOHC
	323iC	2.5 (2494)	M50B25	②	6	DOHC
	328i	2.8 (2793)	M52B28	②	6	DOHC
	328iS	2.8 (2793)	M52B28	②	6	DOHC
	328iC	2.8 (2793)	M52B28	②	6	DOHC
	M3 coupe	3.2 (3152)	S52B32	M3.3	6	DOHC
	M3 sedan	3.2 (3152)	S52B32	M3.3	6	DOHC
	Z3	1.9 (1895)	M44B19	M5.2	4	DOHC
	Z3	2.8 (2793)	M52B28	②	6	DOHC
	528i	2.8 (2793)	M52B28	②	6	DOHC
	540i	4.4 (4398)	M62B44	M3.3	8	DOHC
	740i	4.4 (4398)	M62B44	M3.3	8	DOHC
	740iL	4.4 (4398)	M62B44	M3.3	8	DOHC
	750iL	5.4 (5379)	M73B54	M5.2	12	SOHC

DOHC - Double Overhead Camshaft
SOHC - Single Overhead Camshaft

① Bosch ML-Motronic w/knock control (2 sensors)
② 1996-97 Siemens MS 41.0 w/knock control (2 sensors)

79231C03

GENERAL ENGINE SPECIFICATIONS

Year	Engine ID/VIN	Engine Displacement Liters (cc)	Fuel System Type	Net Horsepower @ rpm	Net Torque @ rpm (ft. lbs.)	Bore x Stroke (In.)	Com- pression Ratio	Oil Pressure @ rpm
1995	M42B18	1.8 (1796)	M1.7	138@6000	129@4500	3.31x3.19	10.0:1	18@idle
	M50B25	2.5 (2494)	M3.1	189@5900	181@4200	3.31x2.95	10.5:1	28@idle
	S50B30	3.0 (2990)	M3.1	240@6000	225@4200	3.39x3.38	10.5:1	28@idle
	M60B30	3.0 (2997)	M3.3	215@5800	214@4500	2.94x2.66	10.5:1	18@idle
	M60B40	4.0 (3982)	M3.3	282@5800	295@4500	3.50x3.15	10.0:1	18@idle
	M73B54	5.4 (5379)	M5.2	322@5000	361@3900	3.35x3.11	10.0:1	18@idle
	S70B56	5.6 (5576)	M1.7	372@5300	402@4000	3.93x3.15	9.8:1	18@idle
1996	M44B19	1.9(1895)	①	138@6000	133@4300	3.35x3.29	10.0:1	8@idle
	M52B28	2.8 (2793)	②	190@5300	204@3950	3.31x3.31	10.2:1	8@idle
	S52B32	3.2 (3152)	NA	240@6000	236@3800	3.40x3.53	NA	NA
	M60B44	4.4 (4398)	③	282@5700	310@3900	3.62x3.26	10.0:1	8@idle
	M73B54	5.4 (5379)	M5.2	322@5000	361@3900	3.35x3.11	10.0:1	18@idle
1997	M44B19	1.9(1895)	①	138@6000	133@4300	3.35x3.29	10.0:1	8@idle
	M52B28	2.8 (2793)	②	190@5300	204@3950	3.31x3.31	10.2:1	8@idle
	S52B32	3.2 (3152)	NA	240@6000	236@3800	3.40x3.53	NA	NA
	M60B44	4.4 (4398)	③	282@5700	310@3900	3.62x3.26	10.0:1	8@idle
	M73B54	5.4 (5379)	M5.2	322@5000	361@3900	3.35x3.11	10.0:1	18@idle
1998-99	M44B19	1.9(1895)	①	138@6000	133@4300	3.35x3.29	10.0:1	8@idle
	M50B25	2.5 (2494)	②	168@5500	181@3950	3.31x2.95	10.5:1	8@idle
	M52B28	2.8 (2793)	②	190@5300	204@3950	3.31x3.31	10.2:1	8@idle
	S52B32	3.2 (3152)	NA	240@6000	236@3800	3.40x3.53	10.5:1	8@idle
	M62B44	4.4 (4398)	③	282@5700	310@3900	3.62x3.26	10.0:1	8@idle
	M73B54	5.4 (5379)	M5.2	322@5000	361@3900	3.35x3.11	10.0:1	18@idle

NA - Not Available

① Bosch ML-Motronic w/knock control (2 sensors)

② 1996-97 Siemens MS 41.0 w/knock control (2 sensors)
  1998-99 Siemens MS 41.1 w/knock control (2 sensors)

③ Bosch HFM-Motronic 5.2 with adaptive knock control system

79231C04

GASOLINE ENGINE TUNE-UP SPECIFICATIONS

Year	Engine ID/VIN	Engine Displacement Liters (cc)	Spark Plugs Gap (in.)	Ignition Timing (deg.) MT	AT	Fuel Pump (psi)	Idle Speed (rpm) MT	AT	Valve Clearance In.	Ex.
1995	M42B18	1.8 (1796)	0.030	①	①	43	850	850	HYD	HYD
	M50B25	2.5 (2494)	0.030	①	①	51	700	700	HYD	HYD
	S50B30	3.0 (2990)	0.030	①	①	51	700	700	HYD	HYD
	M60B30	3.0 (2997)	0.030	①	①	51	600	600	HYD	HYD
	M60B40	4.0 (3982)	0.030	①	①	51	600	600	HYD	HYD
	M73B54	5.4 (5379)	0.030	①	①	43	700	700	HYD	HYD
	S70B56	5.6 (5576)	0.030	①	①	43	700	800	HYD	HYD
1996	M44B19	1.9(1895)	②	①	①	43	810-890	810-890	HYD	HYD
	M52B28	2.8 (2793)	②	①	①	43	700	700	HYD	HYD
	S52B32	3.2 (3152)	②	①	①	43	660-740	660-740	HYD	HYD
	M60B44	4.4 (4398)	②	①	①	43	660-740	660-740	HYD	HYD
	M73B54	5.4 (5379)	②	①	①	43	700	700	HYD	HYD
1997	M44B19	1.9(1895)	②	①	①	43	810-890	810-890	HYD	HYD
	M52B28	2.8 (2793)	②	①	①	43	660-740	660-740	HYD	HYD
	S52B32	3.2 (3152)	②	①	①	43	660-740	660-740	HYD	HYD
	M60B44	4.4 (4398)	②	①	①	43	750-850	750-850	HYD	HYD
	M73B54	5.4 (5379)	②	①	①	43	700	700	HYD	HYD
1998-99	M44B19	1.9(1895)	②	①	①	43	810-890	810-890	HYD	HYD
	M50B25	2.5 (2494)	②	①	①	43	660-740	660-740	HYD	HYD
	M52B28	2.8 (2793)	②	①	①	43	660-740	660-740	HYD	HYD
	S52B32	3.2 (3152)	②	①	①	43	660-740	660-740	HYD	HYD
	M62B44	4.4 (4398)	②	①	①	43	750-850	750-850	HYD	HYD
	M73B54	5.4 (5379)	②	①	①	43	700	700	HYD	HYD

NOTE: The Vehicle Emission Control Information label often reflects specification changes made during production. The label figures must be used if they differ from those in this chart.

HYD - Hydraulic

① Computer-controlled; No adjustment or verification possible

② Except M engines: 0.028

M engines: 0.024

All models with dual mass electrode: 0.035

All models with three-mass and four-mass electrode cannot be adjusted

79231C05

CAPACITIES

Year	Model	Engine ID/VIN	Engine Displacement Liters (cc)	Engine Oil with Filter	Transmission (pts.)			Transfer Case (pts.)	Drive Axle		Fuel Tank (gal.)	Cooling System (qts.)
					4-Spd	5-Spd	Auto.		Front (pts.)	Rear (pts.)		
1995	318i	M42B18	1.8 (1796)	5.0	—	2.4	6.4	—	—	3.3	17.2	13.5
	318iS	M42B18	1.8 (1796)	5.0	—	2.4	6.4	—	—	3.3	17.2	13.5
	318iC	M42B18	1.8 (1796)	5.0	—	2.4	6.4	—	—	3.3	17.2	13.5
	325i	M50B25	2.5 (2494)	6.9	—	2.6	6.4	—	—	3.3	17.2	11.0
	325iS	M50B25	2.5 (2494)	6.9	—	2.6	6.4	—	—	3.3	17.2	11.0
	325iC	M50B25	2.5 (2494)	6.9	—	2.6	6.4	—	—	3.3	17.2	11.0
	525i	M50B25	2.5 (2494)	6.9	—	2.6	6.4	—	—	3.3	21.1	12.7
	525ti	M50B25	2.5 (2494)	6.9	—	2.6	6.4	—	—	3.3	21.1	12.7
	M3	S50B30	3.0 (2990)	6.9	—	2.6	—	—	—	3.3	17.2	13.5
	530i	M60B30	3.0 (2997)	7.9	—	2.6	6.4	—	—	3.3	21.1	12.7
	530ti	M60B30	3.0 (2997)	7.9	—	2.6	6.4	—	—	3.3	21.1	12.7
	540i	M60B40	4.0 (3982)	7.9	—	2.6 ①	6.4	—	—	3.7	21.1	13.4
	740i	M60B40	4.0 (3982)	7.9	—	—	6.4	—	—	3.7	22.5	13.4
	740iL	M60B40	4.0 (3982)	7.9	—	—	6.4	—	—	3.7	25.1	13.4
	840Ci	M60B40	4.0 (3982)	7.9	—	—	6.4	—	—	3.7	23.8	13.4
	750iL	M73B54	5.4 (5379)	7.9	—	—	6.4	—	—	3.7	25.1	14.8
	850Ci	M73B54	5.4 (5379)	7.9	—	—	6.4	—	—	3.7	23.8	13.7
	850CSi	S70B56	5.6 (5576)	7.9	—	2.6	6.4	—	—	3.7	23.8	13.7
1996	318i	M44B19	1.9 (1895)	5	—	2.6	6	—	—	3	14.5	7
	318iC	M44B19	1.9 (1895)	5	—	2.6	6	—	—	3	16.4	7
	318is	M44B19	1.9 (1895)	5	—	2.6	6	—	—	3	16.4	7
	318ti	M44B19	1.9 (1895)	5	—	2.6	6	—	—	3	16.4	7
	Z3	M44B19	1.9 (1895)	6	—	2.6	6	—	—	3	13.5	7
	328i	M52B28	2.8 (2793)	6	—	2.6	6	—	—	3	16.4	10
	328is	M52B28	2.8 (2793)	6	—	2.6	6	—	—	3	16.4	10
	328iC	M52B28	2.8 (2793)	6	—	2.6	6	—	—	3	16.4	10
	M3	S52B32	3.2 (3152)	6	—	2.6	6	—	—	3	16.4	10
	740iL	M60B44	4.4 (4398)	7.9	—	—	6.4	—	—	3.7	25.1	13.4
	750iL	M73B54	5.4 (5379)	7.9	—	—	6.4	—	—	3.7	25.1	14.8
	840Ci	M60B44	4.4 (4398)	7.9	—	—	6.4	—	—	3.7	23.8	13.4
	850Ci	M73B54	5.4 (5379)	7.9	—	—	6.4	—	—	3.7	23.8	13.7
1997	318i	M44B19	1.9 (1895)	5	—	2.6	6	—	—	3	14.5	7
	318iC	M44B19	1.9 (1895)	5	—	2.6	6	—	—	3	16.4	7
	318is	M44B19	1.9 (1895)	5	—	2.6	6	—	—	3	16.4	7
	318ti	M44B19	1.9 (1895)	5	—	2.6	6	—	—	3	16.4	7
	Z3	M44B19	1.9 (1895)	6	—	2.6	6	—	—	3	13.5	7
	328i	M52B28	2.8 (2793)	6	—	2.6	6	—	—	3	16.4	10
	328is	M52B28	2.8 (2793)	6	—	2.6	6	—	—	3	16.4	10
	328iC	M52B28	2.8 (2793)	6	—	2.6	6	—	—	3	16.4	10
	M3	S52B32	3.2 (3152)	6	—	2.6	6	—	—	3	16.4	10
	740iL	M60B44	4.4 (4398)	7.9	—	—	6.4	—	—	3.7	25.1	13.4
	750iL	M73B54	5.4 (5379)	7.9	—	—	6.4	—	—	3.7	25.1	14.8
	840Ci	M60B44	4.4 (4398)	7.9	—	—	6.4	—	—	3.7	23.8	13.4
	850Ci	M73B54	5.4 (5379)	7.9	—	—	6.4	—	—	3.7	23.8	13.7

79231C06

CAPACITIES

Year	Model	Engine ID/VIN	Engine Displacement Liters (cc)	Engine Oil with Filter	Transmission (pts.)			Transfer Case (pts.)	Drive Axle		Fuel Tank (gal.)	Cooling System (qts.)
					4-Spd	5-Spd	Auto.		Front (pts.)	Rear (pts.)		
1998-99	318ti	M44B19	1.9(1895)	5	—	2.6	6	—	—	3	14.5	7
	318i	M44B19	1.9(1895)	5	—	2.6	6	—	—	3	16.4	7
	323is	M50B25	2.5 (2494)	5.5	—	2.6	6	—	—	3	16.4	11
	323iC	M50B25	2.5 (2494)	5.5	—	2.6	6	—	—	3	16.4	11
	328i	M52B28	2.8 (2793)	6	—	2.6	6	—	—	3	16.4	10
	328iS	M52B28	2.8 (2793)	6	—	2.6	6	—	—	3	16.4	10
	328iC	M52B28	2.8 (2793)	6	—	2.6	6	—	—	3	16.4	10
	M3 coupe	S52B32	3.2 (3152)	6	—	2.6	6	—	—	3	16.4	10
	M3 sedan	S52B32	3.2 (3152)	6	—	2.6	6	—	—	3	16.4	10
	Z3	M44B19	1.9(1895)	6	—	2.6	6	—	—	3	13.5	7
	Z3	M52B28	2.8 (2793)	6	—	2.6	6	—	—	—	13.5	10
	528i	M52B28	2.8 (2793)	6	—	2.6	6	—	—	3	18.5	10.5
	540i	M62B44	4.4 (4398)	6	—	2.6 ①	6	—	—	3	18.5	12.5
	740i	M62B44	4.4 (4398)	7	—	— ①	6	—	—	3	22.5	12.5
	740iL	M62B44	4.4 (4398)	7	—	—	6	—	—	3	22.5	12.5
	750iL	M73B54	5.4 (5379)	7.9	—	—	6.4	—	—	3.7	25.1	14.8

NOTE: All capacities are approximate. Add fluid gradually and check to be sure a proper fluid level is obtained.

① 5 or 6 speed

79231C07

VALVE SPECIFICATIONS

Year	Engine ID/VIN	Engine Displacement Liters (cc)	Seat Angle (deg.)	Face Angle (deg.)	Spring Test Pressure (lbs. @ in.)	Spring Installed Height (in.)	Stem-to-Guide Clearance (in.)		Stem Diameter (in.)	
							Intake	Exhaust	Intake	Exhaust
1995	M42B18	1.8 (1796)	45	NA	NA	NA	0.020 ①	0.020 ①	0.275	0.275
	M50B25	2.5 (2494)	45	NA	NA	NA	0.020 ①	0.020 ①	0.275	0.275
	S50B30	3.0 (2990)	45	NA	NA	NA	0.020 ①	0.020 ①	0.275	0.275
	M60B30	3.0 (2997)	45	NA	NA	NA	0.020 ①	0.020 ①	0.235	0.235
	M60B40	4.0 (3982)	45	NA	NA	NA	0.020 ①	0.020 ①	0.235	0.235
	M73B54	5.4 (5379)	45	NA	NA	NA	0.020 ①	0.020 ①	0.280	0.280
	S70B56	5.6 (5576)	45	NA	NA	NA	0.021 ①	0.021 ①	0.280	0.280
1996	M44B19	1.9 (1895)	45	NA	NA	NA	0.020 ①	0.020 ①	NA	NA
	M52B28	2.8 (2793)	45	NA	NA	NA	0.020 ①	0.020 ①	NA	NA
	S52B32	3.2 (3152)	45	NA	NA	NA	0.020 ①	0.020 ①	NA	NA
	M60B44	4.4 (4398)	45	NA	NA	NA	0.020 ①	0.020 ①	NA	NA
	M73B54	5.4 (5379)	45	NA	NA	NA	0.020 ①	0.020 ①	0.280	0.280
1997	M44B19	1.9 (1895)	45	NA	NA	NA	0.020 ①	0.020 ①	NA	NA
	M52B28	2.8 (2793)	45	NA	NA	NA	0.020 ①	0.020 ①	NA	NA
	S52B32	3.2 (3152)	45	NA	NA	NA	0.020 ①	0.020 ①	NA	NA
	M60B44	4.4 (4398)	45	NA	NA	NA	0.020 ①	0.020 ①	NA	NA
	M73B54	5.4 (5379)	45	NA	NA	NA	0.020 ①	0.020 ①	0.280	0.280
1998-99	M44B19	1.9 (1895)	45	NA	NA	NA	0.020 ①	0.020 ①	NA	NA
	M50B25	2.5 (2494)	45	NA	NA	NA	0.020 ①	0.020 ①	NA	NA
	M52B28	2.8 (2793)	45	NA	NA	NA	0.020 ①	0.020 ①	NA	NA
	S52B32	3.2 (3152)	45	NA	NA	NA	0.020 ①	0.020 ①	NA	NA
	M62B44	4.4 (4398)	45	NA	NA	NA	0.020 ①	0.020 ①	NA	NA
	M73B54	5.4 (5379)	45	NA	NA	NA	0.020 ①	0.020 ①	0.280	0.280

NA - Not Available

① To measure Stem-to-Guide clearance, insert new valve into guide until end of valve is flush with end of guide. Use dial indicator to measure valve head movement. Specification given is for maximum wear.

79231C08

TORQUE SPECIFICATIONS
All readings in ft. lbs.

Year	Engine ID/VIN	Engine Displacement Liters (cc)	Cylinder Head Bolts	Main Bearing Bolts	Rod Bearing Bolts	Crankshaft Damper Bolts	Flywheel Bolts	Manifold Intake	Manifold Exhaust	Spark Plugs	Lug Nut
1995	M42B18	1.8 (1796)	①	②	③	217-231	90	④	④	14-22	65-79
	M50B25	2.5 (2494)	①	②	③	303	78	④	④	14-22	65-79
	S50B30	3.0 (2990)	①	②	③	303	78	④	④	14-22	65-79
	M60B30	3.0 (2997)	⑤	⑥	⑦	⑧	78	④	④	14-22	65-79
	M60B40	4.0 (3982)	⑤	⑥	⑦	⑧	78	④	④	14-22	65-79
	M73B54	5.4 (5379)	⑨	⑥	③	⑧	78	④	④	14-22	65-79
	S70B56	5.6 (5576)	⑨	⑥	③	⑧	78	④	④	14-22	65-79
1996	M44B19	1.9(1895)	NA	NA	NA	NA	77-88	④	④	14-22	65-79
	M52B28	2.8 (2793)	NA	NA	NA	NA	77-88	④	④	14-22	65-79
	S52B32	3.2 (3152)	NA	NA	NA	NA	77-88	④	④	14-22	65-79
	M60B44	4.4 (4398)	NA	NA	NA	NA	77-88	④	④	14-22	65-79
	M73B54	5.4 (5379)	⑨	⑥	③	⑧	78	④	④	14-22	65-79
1997	M44B19	1.9(1895)	NA	NA	NA	NA	77-88	④	④	14-22	65-79
	M52B28	2.8 (2793)	NA	NA	NA	NA	77-88	④	④	14-22	65-79
	S52B32	3.2 (3152)	NA	NA	NA	NA	77-88	④	④	14-22	65-79
	M60B44	4.4 (4398)	NA	NA	NA	NA	77-88	④	④	14-22	65-79
	M73B54	5.4 (5379)	⑨	⑥	③	⑧	78	④	④	14-22	65-79
1998-99	M44B19	1.9(1895)	NA	NA	NA	NA	77-88	④	④	14-22	65-79
	M50B25	2.5 (2494)	NA	NA	NA	NA	77-88	④	④	14-22	65-79
	M52B28	2.8 (2793)	NA	NA	NA	NA	77-88	④	④	14-22	65-79
	S52B32	3.2 (3152)	NA	NA	NA	NA	77-88	④	④	14-22	65-79
	M62B44	4.4 (4398)	NA	NA	NA	NA	77-88	④	④	14-22	65-79
	M73B54	5.4 (5379)	⑨	⑥	③	⑧	78	④	④	14-22	65-79

NA - Not Available

① Step 1: 22 ft. lbs.
Step 2: Turn an additional 90 degrees
Step 3: Repeat Step 2

② Step 1: 14-18 ft. lbs.
Step 2: Turn an additional 50 degrees

③ Step 1: 17 ft. lbs.
Step 2: Turn an additional 70 degrees

④ M8 bolts: 16 ft. lbs.
M7 bolts: 11 ft. lbs.
M6 bolts: 7 ft. lbs.

⑤ Step 1: 22 ft. lbs.
Step 2: Turn an additional 80 degrees
Step 3: Repeat Step 2

⑥ Step 1: 14.5 ft. lbs.
Step 2: Turn an additional 47-53 degrees

⑦ Step 1: 3 ft. lbs.
Step 2: 15 ft. lbs.
Step 3: Turn an additional 80 degrees

⑧ Step 1: 78 ft. lbs.
Step 2: Turn an additional 60 degrees
Step 3: Repeat Step 2
Step 4: Turn an additional 30 degrees

⑨ Hex head bolts:
Step 1: 22 ft. lbs., No set time
Step 2: Turn an additional 120 degrees
Torx bolts:
Step 1: 22 ft. lbs., 15 minute settling time
Step 2: Turn an additional 120 degrees

79231C09

SCHEDULED MAINTENANCE INTERVALS
(BMW 3 SERIES, 5 SERIES, 7 SERIES, 8 SERIES, M3 & Z3)

Note: BMW does not rely solely on vehicle mileage to determine service intervals. An on-board diagnostic center, monitors engine operating conditions, along with mileage, to determine the most effective maintenance intervals. The information is then conveyed to the driver through the service indicator lights, located in the center of the instrument panel.

TO BE SERVICED	TYPE OF SERVICE	SERVICE INTERVALS			
		INITIAL 1200 MILES	OIL SERVICE	INSPECTION I	INSPECTION II
Oil level	S/I	✓			
Engine oil	R	✓①			
Engine oil & filter	R②		✓	✓	✓
Engine air cleaner element	R③				✓
Spark plugs	R				✓
Fuel filter	R④				✓
Fuel, vapor lines & fuel cap	S/I	✓		✓	✓
Cooling system	S/I	✓		✓	✓
Exhaust pipe & muffler	S/I	✓		✓	✓
Catalytic converter & shielding	S/I	✓		✓	✓
Throttle linkage	S/I			✓	✓
Engine (check for leakage)	S/I	✓			
Engine drive belts	S/I				✓
Maintenance Indicators	RE		⑤	✓	✓
Engine coolant	R			⑥	⑥
Oxygen sensor	R⑦				
Intake air dust separators	S/I⑧				✓
Brake & clutch fluids ⑥	S/I			✓	✓
Brake pads & discs	S/I			✓	✓
Parking brake system	S/I			✓	✓
Power steering system	S/I			✓	✓

79231C10

SCHEDULED MAINTENANCE INTERVALS
(BMW 3 SERIES, 5 SERIES, 7 SERIES, 8 SERIES, M3 & Z3) (Cont.)

Note: BMW does not rely solely on vehicle mileage to determine service intervals. An on-board diagnostic center, monitors engine operating conditions, along with mileage, to determine the most effective maintenance intervals. The information is then conveyed to the driver through the service indicator lights, located in the center of the instrument panel.

TO BE SERVICED	TYPE OF SERVICE	SERVICE INTERVALS			
		INITIAL 1200 MILES	OIL SERVICE	INSPECTION I	INSPECTION II
Rear axle oil	S/I			✓	✓
Steering play, suspension track rods, front axle joints, steering linkage & joint disc	S/I			✓	✓
Transmission fluid/oil	S/I			✓	✓ ⑩
Wheel centering hubs	S/I			✓	✓
Rear axle oil ⑨	R		✓		✓
OBD system for codes	S/I	✓		✓	✓

R – Replace   S/I – Service or Inspect   RE – Reset

① Service is not required for 318 & 325 models.
② On vehicle operated less than 6200 miles per year, more frequent service may be required.
③ Replace more frequently if vehicle is operated in dusty conditions.
④ Recommend service for California models, required for all other models.
⑤ Reset the oil service indicator lights only.
⑥ Replace every 2 years with inspection service.
⑦ Replace every 100,000 miles on all models.
⑧ Required service for 850 models only.
⑨ At first oil service, then at each inspection II.
⑩ Change fluid (A/T) or oil (M/T) at inspection II.

79231C11

SCHEDULED MAINTENANCE
BMW
3 SERIES, 5 SERIES, 7 SERIES, 8 SERIES
M3, Z3

The following should be used as a guide when determining the amount of work required for a particular service if taken to a repair shop.
In estimating how long a particular Scheduled Maintenance Service should take, please observe the following:

- **Chilton Time** is time based on field research and data supplied by the vehicle manufacturer.
- Labor time operations are given in hours and tenths of an hour.
- All labor operations, are to be used as a guide.

Mechanic Skill Level Codes:
(G) GENERAL: Normally skilled with certification.
(M) MAINTENANCE: Semi-skilled working on certicication.
(P) PRECISION: Really skilled with multiple certification.

	Chilton Time
(G) 12000 Mile Service	
1995-99	1.3
(G) Engine Oil Service	
1995-99	.6
(G) Inspection I	
1995-99	3.1
(G) Inspection II	
1995-99	4.8

79231C12

CHRYSLER IMPORTS
Colt • Expo • Summit • Summit Wagon

VEHICLE IDENTIFICATION CHART

Engine Code						Model Year	
Code	Liters	Cu. In. (cc)	Cyl.	Fuel Sys.	Eng. Mfg.	Code	Year
A	1.5	89.6 (1468)	4	MFI	Mitsubishi	S	1995
C	1.8	111.9 (1834)	4	MFI	Mitsubishi	T	1996
G	2.3	143.5 (2351)	4	MFI'	Mitsubishi	V	1997
						W	1998
						X	1999

79231C13

ENGINE IDENTIFICATION

Year	Model	Engine Displacement Liters (cc)	Engine Series (ID/VIN)	Fuel System	No. of Cylinders	Engine Type
1995	Summit	1.5 (1468)	A	MFI	4	SOHC
	Summit	1.8 (1834)	C	MFI	4	SOHC
	Summit Wagon	1.8 (1834)	C	MFI	4	SOHC
	Summit Wagon	2.3 (2351)	G	MFI	4	SOHC
1996	Summit	1.5 (1468)	A	MFI	4	SOHC
	Summit	1.8 (1834)	C	MFI	4	SOHC
	Summit Wagon	1.8 (1834)	C	MFI	4	SOHC
	Summit Wagon	2.4 (2351)	G	MFI	4	SOHC

MFI - Multipoint fuel injection
SOHC - Single overhead camshaft

79231C14

GENERAL ENGINE SPECIFICATIONS

Year	Engine ID/VIN	Engine Displacement Liters (cc)	Fuel System Type	Net Horsepower @ rpm	Net Torque @ rpm (ft. lbs.)	Bore x Stroke (in.)	Compression Ratio	Oil Pressure @ rpm
1995	A	1.5 (1468)	MFI	92@6000	93@3000	2.97x3.23	9.2:1	①
	C	1.8 (1834)	MFI	113@6000 ②	116@4500	3.19X3.50	9.5:1	①
	G	2.4 (2351)	MFI	136@5500	145@4250	3.41x3.94	9.5:1	①
1996	A	1.5 (1468)	MFI	92@6000	93@3000	2.97x3.23	9.2:1	①
	C	1.8 (1834)	MFI	113@6000 ②	116@4500	3.19x3.50	9.5:1	①
	G	2.4 (2351)	MFI	136@5500	145@4250	3.41x3.94	9.5:1	①

MFI - Multiport fuel injection
① 11.4 psi or more at curb idle speed
② Manual: 119@6000

79231C15

GASOLINE ENGINE TUNE-UP SPECIFICATIONS

Year	Engine ID/VIN	Engine Displacement Liters (cc)	Spark Plugs Gap (in.)	Ignition Timing (deg.) MT	Ignition Timing (deg.) AT	Fuel Pump (psi)	Idle Speed (rpm) MT	Idle Speed (rpm) AT	Valve Clearance In.	Valve Clearance Ex.
1995	A	1.5 (1468)	0.039-0.043	5B	5B	38 ①	750	750	0.008	0.010
	C	1.8 (1834)	0.039-0.043	5B	5B	38 ①	700	700	0.008	0.012
	G	2.4 (2351)	0.039-0.043	5B	5B	38 ①	750	750	HYD	HYD
1996	A	1.5 (1468)	0.039-0.043	5B	5B	38 ①	750	70	0.008	0.010
	C	1.8 (1934)	0.039-0.043	5B	5B	38 ①	700	700	0.008	0.012
	G	2.4 (2351)	0.039-0.043	5B	5B	38 ①	750	750	HYD	HYD

NOTE: The Vehicle Emission Control Information label often reflects specification changes made during production. The label figures must be used if they differ from those in this chart.

B - Before top dead center

HYD - Hydraulic

① Pressure at idle with vacuum applied to fuel pressure regulator

79231C16

CAPACITIES

Year	Model	Engine ID/VIN	Engine Displacement Liters (cc)	Engine Oil with Filter	Transmission (pts.) 4-Spd	Transmission (pts.) 5-Spd	Transmission (pts.) Auto.	Transfer Case (pts.)	Drive Axle Front (pts.)	Drive Axle Rear (pts.)	Fuel Tank (gal.)	Cooling System (qts.)
1995	Summit	A	1.5 (1468)	3.7	—	3.8	12.6	—	—	—	13.2	5.3
	Summit	C	1.8 (1834)	4.2	—	3.8	12.6	—	—	—	13.2	6.3
	Summit Wagon	C	1.8 (1834)	4.0	—	①	②	1.20	—	1.5	14.5	6.3
	Summit Wagon	G	2.4 (2351)	4.5	—	4.8	②	1.20	—	1.5	14.5	6.8
1996	Summit	A	1.5 (1468)	3.7	—	3.8	12.6	—	—	—	13.2	5.3
	Summit	C	1.8 (1834)	4.2	—	3.8	12.6	—	—	—	13.2	6.3
	Summit Wagon	C	1.8 (1834)	4.0	—	①	②	1.20	—	1.5	14.5	6.3
	Summit Wagon	G	2.4 (2351)	4.5	—	3.8	②	1.20	—	1.5	14.5	6.8

① 2WD models: 3.8 pts.; 4WD models: 4.8 pts.

② 2WD models: 12.8 qts.; 4WD models: 13.8 qts.

79231C17

VALVE SPECIFICATIONS

Year	Engine ID/VIN	Engine Displacement Liters (cc)	Seat Angle (deg.)	Face Angle (deg.)	Spring Test Pressure (lbs. @ in.)	Spring Installed Height (in.)	Stem-to-Guide Clearance (in.) Intake	Stem-to-Guide Clearance (in.) Exhaust	Stem Diameter (in.) Intake	Stem Diameter (in.) Exhaust
1995	A	1.5 (1468)	44-44.5	45-45.5	①	②	0.0008-0.0020	0.0020-0.0035	0.2585-0.2591	0.2571-0.2579
	C	1.8 (1834)	43.5-44	45-45.5	③	⑤	0.0008-0.0016	0.0012-0.0024	0.2350-0.2354	0.2343-0.2350
	G	2.4 (2351)	43.5-44	45-45.5	④	1.740	0.0008-0.0020	0.0012-0.0028	0.2350-0.2354	0.2343-0.2350
1996	A	1.5 (1468)	44	45-45.5	①	②	0.0008-0.0039	0.0020-0.0059	0.2587-0.2591	0.2571-0.2579
	C	1.8 (1834)	43.5-44	45-45.5	132 ⑥	⑤	0.0008-0.0039	0.0012-0.0059	0.2350-0.2354	0.2343-0.2350
	G	2.4 (2351)	43-44	45-45.5	60 ⑥	1.740	0.0008-0.0039	0.0012-0.0059	0.2350-0.2354	0.2343-0.2350

① Intake: 51@ installed height; Exhaust: 64@ installed height

② Intake free length: 1.776-1.815; Exhaust free length: 1.803-1.843

③ 132@ installed height

④ Free length: 1.898-1.937

⑤ 60@ installed height

⑥ Free length: 1.965-2.004

⑦ 73 @ installed height

⑧ At installed height

79231C18

TORQUE SPECIFICATIONS
All readings in ft. lbs.

Year	Engine ID/VIN	Engine Displacement Liters (cc)	Cylinder Head Bolts	Main Bearing Bolts	Rod Bearing Bolts	Crankshaft Damper Bolts	Flywheel Bolts	Manifold Intake	Manifold Exhaust	Spark Plugs	Lug Nut
1995	A	1.5 (1468)	①	38	14.5	62	94-101	13	13	15-21	65-80
	C	1.8 (1834)	④	⑤	⑥	134	72	14	②	15-21	65-80
	G	2.4 (2351)	③	⑤	14.5	—	98	13	13	18	65-80
1996	A	1.5 (1468)	53	38	⑥	62	98	13	13	15-21	65-80
	C	1.8 (1834)	④	⑤	⑥	134	72	14	11-14 ⑦	18	65-80
	G	2.4 (2351)	③	⑤	⑥	—	98	13	18-22	18	65-80

① Cold: 51-54 ft. lbs.;
   Hot: 58-61 ft. lbs.
② Upper bolts: 11-14 ft. lbs.
   Lower bolts: 22 ft. lbs.
③ Step 1: 58 ft. lbs.
   Step 2: Fully loosen
   Step 3: 14 ft. lbs.
   Step 4: +90 degrees
   Step 5: +90 degrees
④ Step 1: 54 ft. lbs.
   Step 2: Fully loosen all bolts
   Step 3: 14.5 ft. lbs.
   Step 4: +90 degrees
   Step 5: +90 degrees
⑤ Step 1: 18 ft. lbs.
   Step 2: +90 degrees
⑥ Step 1: 14.5 ft. lbs.
   Step 2: +90 degrees
⑦ Lower exhaust manifold nuts: 22 ft. lbs.

79231C19

SCHEDULED MAINTENANCE INTERVALS
(CHRYSLER SUMMIT, SUMMIT WAGON & EXPO)

TO BE SERVICED	TYPE OF SERVICE	VEHICLE MILEAGE INTERVAL (x1000)												
		7.5	15	22.5	30	37.5	45	52.5	60	67.5	75	82.5	90	97.5
Engine oil & filter	R	✓	✓	✓	✓	✓	✓	✓	✓	✓	✓	✓	✓	✓
Coolant level, hoses & clamps	S/I	✓	✓	✓	✓	✓		✓	✓	✓	✓	✓	✓	✓
Rotate tires	S/I	✓	✓	✓	✓	✓	✓	✓	✓	✓	✓	✓	✓	✓
Brake hoses	S/I		✓		✓		✓		✓		✓		✓	
Disc brake pads	S/I		✓		✓		✓		✓		✓		✓	
Drive shaft boots	S/I		✓		✓		✓		✓		✓		✓	
Valve clearances (1.5L & 1.8L)	S/I		✓		✓		✓		✓		✓		✓	
Air cleaner element	R				✓				✓				✓	
Engine Coolant	R				✓				✓				✓	
Spark plugs	R				✓				✓				✓	
Automatic transaxle fluid & filter①	R				✓				✓				✓	
Ball joints & steering linkage seals	S/I				✓				✓				✓	
Drive belt(s)	S/I				✓				✓				✓	
Exhaust system	S/I				✓				✓				✓	
Fuel hoses	S/I				✓				✓				✓	
Manual transaxle oil	S/I				✓				✓				✓	
Rear axle oil (Colt Vista & Summit Wagon)	S/I				✓				✓				✓	
Rear drum brake linings & rear wheel cylinders	S/I				✓				✓				✓	
Ignition cables	R								✓					
Timing belt	R								✓					
Distributor cap & rotor	S/I								✓					
EVAP system (except canister)	S/I								✓					
Fuel system	S/I								✓					

① Automatic transaxle fluid & filter - service or inspect every 15,000 miles.

R – Replace S/I – Service or Inspect

79231C20

SCHEDULED MAINTENANCE INTERVALS
(CHRYSLER SUMMIT, SUMMIT WAGON & EXPO) (Cont.)

FREQUENT OPERATION MAINTENANCE (SEVERE SERVICE)
If a vehicle is operated under any of the following conditions it is considered severe service:
- Extremely dusty areas.
- 50% or more of the vehicle operation is in 32°C (90°F) or higher temperatures, or constant operation in temperatures below 0°C (32°F).
- Prolonged idling (vehicle operation in stop and go traffic).
- Frequent short running periods (engine does not warm to normal operating temperatures).
- Police, taxi, delivery usage or trailer towing usage.
Oil & oil filter change – change every 3000 miles.
Disc brake pads - service or inspect every 6000 miles (1996-97).
Air cleaner element – service or inspect every 15,000 miles.
Rear drum brake linings & rear wheel cylinders - service or inspect every 15,000 miles.
Spark plugs - replace every 15,000 miles.

79231C21

SCHEDULED MAINTENANCE
CHRYSLER IMPORTS
COLT, EXPO, SUMMIT, SUMMIT WAGON

The following should be used as a guide when determining the amount of work required for a particular service if taken to a repair shop.
In estimating how long a particular Scheduled Maintenance Service should take, please observe the following:

- **Chilton Time** is time based on field research and data supplied by the vehicle manufacturer.
- Labor time operations are given in hours and tenths of an hour.
- All labor operations, are to be used as a guide.

Mechanic Skill Level Codes:
- **(G) GENERAL:** Normally skilled with certification.
- **(M) MAINTENANCE:** Semi-skilled working on certicication.
- **(P) PRECISION:** Really skilled with multiple certification.

	Chilton Time			Chilton Time			Chilton Time
(M) 7500 Mile Service			**(G) 45000 Mile Service**			**(M) 67500 Mile Service**	
1995-96	.3		1995-96	1.0		1995-96	.3
(G) 15000 Mile Service			**(G) 50000 Mile Service**			**(G) 75000 Mile Service**	
1995-96	1.0		1995-96	4.0		1995-96	1.0
(M) 22500 Mile Service			**(M) 52500 Mile Service**			**(M) 82500 Mile Service**	
1995-96	.3		1995-96	.3		1995-96	.3
(G) 30000 Mile Service			**(G) 60000 Mile Service**			**(G) 90000 Mile Service**	
1995-96	6.0		1995-96	9.2		1995-96	6.0
(M) 37500 Mile Service						**(M) 97500 Mile Service**	
1995-96	.3					1995-96	.3

79231C22

HONDA
Accord • Civic • Del Sol • Prelude

VEHICLE IDENTIFICATION CHART

Code	Liters	Cu. In. (cc)	Cyl.	Fuel Sys.	Eng. Mfg.
B16A2	1.6	97 (1595)	4	PGM-FI	Honda
B16A3	1.6	97 (1595)	4	PGM-FI	Honda
C27A4	2.7	157 (2675)	6	PGM-FI	Honda
D15B7	1.5	95 (1493)	4	PGM-FI	Honda
D15B8	1.5	95 (1493)	4	PGM-FI	Honda
D15Z1	1.5	95 (1493)	4	PGM-FI	Honda
D16Y5	1.6	97 (1590)	4	PGM-FI	Honda
D16Y7	1.6	97 (1590)	4	PGM-FI	Honda
D16Y8	1.6	97 (1590)	4	PGM-FI	Honda
D16Z6	1.6	97 (1590)	4	PGM-FI	Honda
F22A1	2.2	132 (2156)	4	PGM-FI	Honda
F22B1	2.2	132 (2156)	4	PGM-FI	Honda
F22B2	2.2	132 (2156)	4	PGM-FI	Honda
F23A1	2.3	137 (2254)	4	PGM-FI	Honda
F23A4	2.3	137 (2254)	4	PGM-FI	Honda
F23A5	2.3	137 (2254)	4	PGM-FI	Honda
H22A1	2.2	132 (2157)	4	PGM-FI	Honda
H22A4	2.2	132 (2157)	4	PGM-FI	Honda
H23A1	2.3	132 (2259)	4	PGM-FI	Honda
J30A1	3.0	183 (2997)	6	PGM-FI	Honda

Code	Year
S	1995
T	1996
V	1997
W	1998
X	1999

PGM-FI - Programmed Fuel Injection

79231CE0

ENGINE IDENTIFICATION

Year	Model	Engine Displacement Liters (cc)	Engine Series (ID/VIN)	Fuel System	No. of Cylinders	Engine Type
1995	Accord DX/LX	2.2 (2156)	F22B2	PGM-FI	4	SOHC 16V
	Accord EX	2.2 (2156)	F22B1	PGM-FI	4	SOHC 8V
	Accord V-6	2.7 (2675)	C27A4	PGM-FI	4	SOHC 16V
	Civic	1.5 (1493)	D15B7	PGM-FI	4	SOHC 16V
	Civic	1.5 (1493)	D15B8	PGM-FI	4	SOHC 16V
	Civic	1.5 (1493)	D15Z1	PGM-FI	4	SOHC 16V
	Civic	1.6 (1590)	D16Z6	PGM-FI	4	DOHC 16V
	del Sol	1.6 (1595)	B16A3	PGM-FI	4	SOHC 16V
	del Sol	1.5 (1493)	D15B7	PGM-FI	4	SOHC 16V
	del Sol	1.6 (1590)	D16Z6	PGM-FI	6	SOHC 24V
	Prelude Si VTEC	2.2 (2157)	H22A1	PGM-FI	4	SOHC 16V
	Prelude S	2.2 (2156)	F22A1	PGM-FI	4	DOHC 16V
	Prelude Si	2.3 (2259)	H23A1	PGM-FI	4	DOHC 16V
1996	Accord DX/LX	2.2 (2156)	F22B2	PGM-FI	4	SOHC 16V
	Accord EX	2.2 (2156)	F22B1	PGM-FI	4	SOHC 16V
	Accord V-6	2.7 (2675)	C27A4	PGM-FI	6	SOHC 24V
	Civic	1.6 (1590)	D16Y5	PGM-FI	4	SOHC 16V
	Civic	1.6 (1590)	D16Y7	PGM-FI	4	SOHC 16V
	Civic	1.6 (1590)	D16Y8	PGM-FI	4	SOHC 8V
	del Sol	1.6 (1595)	B16A2	PGM-FI	4	DOHC 16V
	del Sol	1.6 (1590)	D16Y7	PGM-FI	4	SOHC 16V
	del Sol	1.6 (1590)	D16Y8	PGM-FI	4	SOHC 16V
	Prelude S	2.2 (2156)	F22A1	PGM-FI	4	SOHC 16V
	Prelude Si	2.3 (2259)	H23A1	PGM-FI	4	DOHC 16V
	Prelude Si VTEC	2.2 (2157)	H22A1	PGM-FI	4	DOHC 16V
1997	Accord Coupe	2.2 (2156)	F22B1	PGM-FI	4	SOHC 16V
	Accord Coupe	2.2 (2156)	F22B2	PGM-FI	4	SOHC 16V
	Accord Sedan	2.7 (2675)	C27A4	PGM-FI	6	SOHC 24V
	Accord Sedan	2.2 (2156)	F22B1	PGM-FI	4	SOHC 16V
	Accord Sedan	2.2 (2156)	F22B2	PGM-FI	4	SOHC 16V
	Accord Wagon	2.2 (2156)	F22B1	PGM-FI	4	SOHC 16V
	Accord Wagon	2.2 (2156)	F22B2	PGM-FI	4	SOHC 16V
	Civic	1.6 (1590)	D16Y5	PGM-FI	4	SOHC 16V
	Civic	1.6 (1590)	D16Y7	PGM-FI	4	SOHC 16V
	Civic	1.6 (1590)	D16Y8	PGM-FI	4	SOHC 8V
	del Sol	1.6 (1595)	B16A2	PGM-FI	4	DOHC 16V
	del Sol	1.6 (1590)	D16Y7	PGM-FI	4	SOHC 16V
	del Sol	1.6 (1590)	D16Y8	PGM-FI	4	SOHC 16V
	Prelude	2.2 (2156)	H22A4	PGM-FI	4	DOHC 16V
	Prelude SH	2.2 (2156)	H22A4	PGM-FI	4	DOHC 16V
1998-99	Accord Coupe (EX, LX)	2.3 (2254)	F23A1	PGM-FI	4	SOHC 16V
	Accord Coupe (EX, LX)	2.3 (2254)	F23A4	PGM-FI	4	SOHC 16V
	Accord Coupe (EX, LX)	3.0 (2997)	J30A1	PGM-FI	6	SOHC 24V
	Accord Sedan (DX)	2.3 (2254)	F23A5	PGM-FI	4	SOHC 16V
	Accord Sedan (EX, LX)	2.3 (2254)	F23A1	PGM-FI	4	SOHC 16V
	Accord Sedan (EX, LX)	2.3 (2254)	F23A4	PGM-FI	4	SOHC 16V
	Accord Sedan(EX, LX)	3.0 (2997)	J30A1	PGM-FI	6	SOHC 24V

79231CF1

ENGINE IDENTIFICATION

Year	Model	Engine Displacement Liters (cc)	Engine Series (ID/VIN)	Fuel System	No. of Cylinders	Engine Type
1998-99 (cont.)	Civic	1.6 (1590)	D16Y5	PGM-FI	4	SOHC 16V
	Civic	1.6 (1590)	D16Y7	PGM-FI	4	SOHC 16V
	Civic	1.6 (1590)	D16Y8	PGM-FI	4	SOHC 8V
	Prelude	2.2 (2156)	H22A4	PGM-FI	4	DOHC 16V
	Prelude SH	2.2 (2156)	H22A4	PGM-FI	4	DOHC 16V

PGM-FI - Programmed Fuel Injection
SOHC - Single Overhead Camshaft
DOHC - Double Overhead Camshafts

79231CF2

GENERAL ENGINE SPECIFICATIONS

Year	Engine ID/VIN	Engine Displacement Liters (cc)	Fuel System Type	Net Horsepower @ rpm	Net Torque @ rpm (ft. lbs.)	Bore x Stroke (in.)	Compression Ratio	Oil Pressure @ rpm
1995	B16A3	1.6 (1595)	PGM-FI	160@7600	111@7000	3.19x3.05	10.2:1	50@3000
	C27A4	2.7 (2675)	PGM-FI	170@5600	165@4500	3.43x2.95	9.0:1	63@3000
	D15B7	1.5 (1493)	PGM-FI	102@5900	98@5000	2.95x3.33	9.2:1	50@3000
	D15B8	1.5 (1493)	PGM-FI	70@5000	91@2000	2.95x3.33	9.1:1	50@3000
	D15Z1	1.5 (1493)	PGM-FI	92@5500	97@4500	2.95x3.33	9.3:1	50@3000
	D16Z6	1.6 (1590)	PGM-FI	125@6600	106@5200	2.95x3.54	9.2:1	50@3000
	F22A1	2.2 (2156)	PGM-FI	135@5200	142@4000	3.35x3.74	8.8:1	50@3000
	F22B1	2.2 (2156)	PGM-FI	145@5500	147@4500	3.35x3.74	8.8:1	50@3000
	F22B2	2.2 (2156)	PGM-FI	130@5300	139@4200	3.35x3.74	8.8:1	50@3000
	H22A1	2.2 (2157)	PGM-FI	190@6800	158@5500	3.43x3.57	10.0:1	50@3000
	H23A1	2.3 (2259)	PGM-FI	160@5800	156@4500	3.43x3.74	9.8:1	50@3000
1996	B16A2	1.6 (1595)	PGM-FI	160@7600	111@7000	3.19x3.05	10.2:1	50@3000
	C27A4	2.7 (2675)	PGM-FI	170@5600	165@4500	3.43x2.95	9.0:1	63@3000
	D16Y5	1.6 (1590)	PGM-FI	115@6300	104@5400	2.95x3.54	9.6:1	50@3000
	D16Y7	1.6 (1590)	PGM-FI	106@6200	103@4600	2.95x3.54	9.4:1	50@3000
	D16Y8	1.6 (1590)	PGM-FI	127@6600	107@5500	2.95x3.54	9.4:1	50@3000
	F22A1	2.2 (2156)	PGM-FI	135@5200	142@4000	3.35x3.74	8.8:1	50@3000
	F22B1	2.2 (2156)	PGM-FI	145@5500	147@4500	3.35x3.74	8.8:1	50@3000
	F22B2	2.2 (2156)	PGM-FI	130@5300	139@4200	3.35x3.74	8.8:1	50@3000
	H22A1	2.2 (2157)	PGM-FI	190@6800	158@5500	3.43x3.57	10.0:1	50@3000
	H23A1	2.3 (2259)	PGM-FI	160@5800	156@4500	3.43x3.74	9.8:1	50@3000
1997	B16A2	1.6 (1595)	PGM-FI	160@7600	111@7000	3.19x3.05	10.2:1	50@3000
	C27A4	2.7 (2675)	PGM-FI	170@5600	165@4500	3.43x2.95	9.0:1	63@3000
	D16Y5	1.6 (1590)	PGM-FI	115@6300	104@5400	2.95x3.54	9.6:1	50@3000
	D16Y7	1.6 (1590)	PGM-FI	106@6200	103@4600	2.95x3.54	9.4:1	50@3000
	D16Y8	1.6 (1590)	PGM-FI	127@6600	107@5500	2.95x3.54	9.4:1	50@3000
	F22B1	2.2 (2156)	PGM-FI	145@5500	147@4500	3.35x3.74	8.8:1	50@3000
	F22B2	2.2 (2156)	PGM-FI	130@5300	139@4200	3.35x3.74	8.8:1	50@3000
	H22A4	2.2 (2157)	PGM-FI	190@6800	158@5500	3.43x3.57	10.0:1	50@3000
1998-99	D16Y5	1.6 (1590)	PGM-FI	115@6300	104@5400	2.95x3.54	9.6:1	50@3000
	D16Y7	1.6 (1590)	PGM-FI	106@6200	103@4600	2.95x3.54	9.4:1	50@3000
	D16Y8	1.6 (1590)	PGM-FI	127@6600	107@5500	2.95x3.54	9.4:1	50@3000
	F23A1	2.3 (2254)	PGM-FI	150@5700	152@4900	3.39x3.82	9.3:1	50@3000
	F23A4	2.3 (2254)	PGM-FI	150@5700	152@4900	3.39x3.82	9.3:1	50@3000
	F23A5	2.3 (2254)	PGM-FI	150@5700	152@4900	3.39x3.82	9.3:1	50@3000
	H22A4	2.2 (2157)	PGM-FI	①	158@5500	3.43x3.57	10.0:1	50@3000
	J30A1	3.0 (2997)	PGM-FI	200@5500	195@4700	3.39x3.39	9.4:1	50@3000

PGM-FI: Programmed Fuel Injection
① Manual ttransaxle: 195@7000
   Automatic transaxle: 190@6600

79231CF3

GASOLINE ENGINE TUNE-UP SPECIFICATIONS

Year	Engine ID/VIN	Engine Displacement Liters (cc)	Spark Plugs Gap (in.)	Ignition Timing (deg.) MT	Ignition Timing (deg.) AT	Fuel Pump (psi)	Idle Speed (rpm) MT	Idle Speed (rpm) AT	Valve Clearance In.	Valve Clearance Ex.
1995	B16A3	1.6 (1595)	0.047-0.051	16B	—	31-38	650-750	—	0.006-0.007	0.007-0.008
	C27A4	2.7 (2675)	0.039-0.043	—	15B	30-37	—	650-750	0.009-0.011	0.011-0.013
	D15B7	1.5 (1493)	0.039-0.043	16B	16B	31-38	620-720	650-750	0.007-0.009	0.009-0.011
	D15B8	1.5 (1493)	0.039-0.043	12B	—	31-38	620-720	650-750	0.007-0.009	0.009-0.011
	D15Z1	1.5 (1493)	0.039-0.043	16B	—	31-38	550-650	—	0.007-0.009	0.009-0.011
	D16Z6	1.6 (1590)	0.039-0.043	16B	16B	31-38	620-720	650-750	0.007-0.009	0.009-0.011
	F22A1	2.2 (2156)	0.039-0.043	15B	15B	28-35	650-750	650-750	0.009-0.011	0.011-0.013
	F22B1	2.2 (2156)	0.039-0.043	15B	15B	30-37	650-750	650-750	0.009-0.011	0.011-0.013
	F22B2	2.2 (2156)	0.039-0.043	15B	15B	30-37	650-750	650-750	0.009-0.011	0.011-0.013
	H22A1	2.2 (2157)	0.039-0.043	15B	—	24-31	650-750	—	0.006-0.007	0.007-0.008
	H23A1	2.3 (2259)	0.039-0.043	15B	15B	28-35	650-750	650-750	0.003-0.004	0.006-0.007
1996	B16A2	1.6 (1595)	0.047-0.051	16B	—	31-38	650-750	—	0.006-0.007	0.007-0.008
	C27A4	2.7 (2675)	0.039-0.043	—	15B	30-37	—	650-750	0.009-0.011	0.011-0.013
	D16Y5	1.6 (1590)	0.039-0.043	12B	12B	28-36	620-720	650-750	0.007-0.009	0.009-0.011
	D16Y7	1.6 (1590)	0.039-0.043	12B	12B	28-36	620-720	650-750	0.007-0.009	0.009-0.011
	D16Y8	1.6 (1590)	0.039-0.043	12B	12B	28-36	620-720	650-750	0.007-0.009	0.009-0.011
	F22A1	2.2 (2156)	0.039-0.043	15B	15B	28-35	650-750	650-750	0.009-0.011	0.011-0.013
	F22B1	2.2 (2156)	0.039-0.043	15B	15B	30-37	650-750	650-750	0.009-0.011	0.011-0.013
	F22B2	2.2 (2156)	0.039-0.043	15B	15B	30-37	650-750	650-750	0.009-0.011	0.011-0.013
	H22A1	2.2 (2157)	0.039-0.043	15B	—	24-31	650-750	—	0.006-0.007	0.007-0.008
	H23A1	2.3 (2259)	0.039-0.043	15B	15B	28-35	650-750	650-750	0.003-0.004	0.006-0.007

79231CF4

GASOLINE ENGINE TUNE-UP SPECIFICATIONS

Year	Engine ID/VIN	Engine Displacement Liters (cc)	Spark Plugs Gap (in.)	Ignition Timing (deg.) MT	AT	Fuel Pump (psi)	Idle Speed (rpm) MT	AT	Valve Clearance In.	Ex.
1997	B16A2	1.6 (1595)	0.047-0.051	16B	—	31-38	650-750	—	0.006-0.007	0.007-0.008
	C27A4	2.7 (2675)	0.039-0.043	—	15B	30-37	—	650-750	0.009-0.011	0.011-0.013
	D16Y5	1.6 (1590)	0.039-0.043	12B	12B	28-36	620-720	650-750	0.007-0.009	0.009-0.011
	D16Y7	1.6 (1590)	0.039-0.043	12B	12B	28-36	620-720	650-750	0.007-0.009	0.009-0.011
	D16Y8	1.6 (1590)	0.039-0.043	12B	12B	28-36	620-720	650-750	0.007-0.009	0.009-0.011
	F22A1	2.2 (2156)	0.039-0.043	15B	15B	28-35	650-750	650-750	0.009-0.011	0.011-0.013
	F22B1	2.2 (2156)	0.039-0.043	15B	15B	30-37	650-750	650-750	0.009-0.011	0.011-0.013
	F22B2	2.2 (2156)	0.039-0.043	15B	15B	30-37	650-750	650-750	0.009-0.011	0.011-0.013
	H22A4	2.2 (2157)	0.039-0.043	15B	15B	47-54	650-750	650-750	0.006-0.007	0.007-0.008
1998-99	D16Y5	1.6 (1590)	0.039-0.043	12B	12B	28-36	620-720	650-750	0.007-0.009	0.009-0.011
	D16Y7	1.6 (1590)	0.039-0.043	12B	12B	28-36	620-720	650-750	0.007-0.009	0.009-0.011
	D16Y8	1.6 (1590)	0.039-0.043	12B	12B	28-36	620-720	650-750	0.007-0.009	0.009-0.011
	F23A1	2.3 (2254)	0.039-0.0043	12B	12B	40-47	650-750	650-750	0.009-0.011	0.011-0.013
	F23A4	2.3 (2254)	0.039-0.0043	12B	12B	40-47	650-750	650-750	0.009-0.011	0.011-0.013
	F23A5	2.3 (2254)	0.039-0.0043	12B	12B	40-47	650-750	650-750	0.009-0.011	0.011-0.013
	H22A4	2.2 (2157)	0.039-0.043	15B	15B	47-54	650-750	650-750	0.006-0.007	0.007-0.008
	J30A1	3.0 (2997)	0.039-0.043	—	10B	41-48	—	630-730	0.008-0.009	0.011-0.013

NOTE: The Vehicle Emission Control Information label often reflects specification changes made during production. The label figures must be used if they differ from those in this chart.

B - Before Top Dead Center

79231CF5

CAPACITIES

Year	Model	Engine ID/VIN	Engine Displacement Liters (cc)	Engine Oil with Filter	Transmission (pts.)			Transfer Case (pts.)	Drive Axle		Fuel Tank (gal.)	Cooling System (qts.)
					4-Spd	5-Spd	Auto.		Front (pts.)	Rear (pts.)		
1995	Accord DX/LX	F22B2	2.2 (2156)	4.0	—	4.0	5.0	—	—	—	17.0	①
	Accord EX	F22B1	2.2 (2156)	4.5	—	4.0	5.0	—	—	—	17.0	①
	Accord V-6	C27A4	2.7 (2675)	4.6	—	—	6.2	—	—	—	17.0	7.2
	Civic	D15B7	1.5 (1493)	3.5	—	3.8	5.8	—	—	—	11.9	②
	Civic	D15B8	1.5 (1493)	3.5	—	3.8	—	—	—	—	11.9	3.8
	Civic	D15Z1	1.5 (1493)	3.5	—	3.8	—	—	—	—	11.9	3.7
	Civic	D16Z6	1.6 (1590)	3.5	—	3.8	5.8	—	—	—	11.9	②
	del Sol	B16A3	1.6 (1595)	4.2	—	4.8	—	—	—	—	11.9	4.1
	del Sol	D15B7	1.5 (1493)	3.5	—	4.0	5.6	—	—	—	11.9	②
	del Sol	D16Z6	1.6 (1590)	3.5	—	4.0	5.6	—	—	—	11.9	②
	Prelude S	F22A1	2.2 (2156)	4.0	—	4.0	5.0	—	—	—	15.9	②
	Prelude Si	H23A1	2.3 (2259)	4.5	—	4.0	5.0	—	—	—	15.9	②
	Prelude Si VTEC	H22A1	2.2 (2157)	5.1	—	4.0	—	—	—	—	15.9	4.6
1996	Accord DX/LX	F22B2	2.2 (2156)	4.0	—	4.0	5.0	—	—	—	17.0	①
	Accord EX	F22B1	2.2 (2156)	4.5	—	4.0	5.0	—	—	—	17.0	①
	Accord V-6	C27A4	2.7 (2675)	4.6	—	—	6.2	—	—	—	17.0	7.2
	Civic	D16Y5	1.6 (1590)	3.5	—	3.8	5.8	—	—	—	11.9	4.5
	Civic	D16Y7	1.6 (1590)	3.5	—	3.8	5.8	—	—	—	11.9	4.4
	Civic	D16Y8	1.6 (1590)	3.5	—	3.8	5.8	—	—	—	11.9	4.3
	del Sol	D16Y7	1.6 (1590)	3.5	—	4.0	5.6	—	—	—	11.9	①
	del Sol	D16Y8	1.6 (1590)	3.5	—	4.0	5.6	—	—	—	11.9	8.2
	del Sol	B16A2	1.6 (1595)	4.2	—	4.8	—	—	—	—	11.9	4.1
	Prelude S	F22A1	2.2 (2156)	4.0	—	4.0	5.0	—	—	—	15.9	②
	Prelude Si	H23A1	2.3 (2259)	4.5	—	4.0	5.0	—	—	—	15.9	②
	Prelude Si VTEC	H22A1	2.2 (2157)	5.1	—	4.0	—	—	—	—	15.9	4.6
1997	Accord Coupe	F22B1	2.2 (2156)	4.5	—	4.0	5.0	—	—	—	17.0	①
	Accord Coupe	F22B2	2.2 (2156)	4.0	—	4.0	5.0	—	—	—	17.0	①
	Accord Sedan	C27A4	2.7 (2675)	4.6	—	—	6.2	—	—	—	17.0	7.2
	Accord Sedan	F22B1	2.2 (2156)	4.5	—	4.0	5.0	—	—	—	17.0	①
	Accord Sedan	F22B2	2.2 (2156)	4.0	—	4.0	5.0	—	—	—	17.0	①
	Accord Wagon	F22B1	2.2 (2156)	4.5	—	4.0	5.0	—	—	—	17.0	①
	Accord Wagon	F22B2	2.2 (2156)	4.0	—	4.0	5.0	—	—	—	17.0	①
	Civic	D16Y5	1.6 (1590)	3.5	—	3.8	5.8	—	—	—	11.9	4.5
	Civic	D16Y7	1.6 (1590)	3.5	—	3.8	5.8	—	—	—	11.9	4.4
	Civic	D16Y8	1.6 (1590)	3.5	—	3.8	5.8	—	—	—	11.9	4.3
	del Sol	B16A2	1.6 (1595)	4.2	—	4.8	—	—	—	—	11.9	4.1
	del Sol	D16Y7	1.6 (1590)	3.5	—	4.0	5.6	—	—	—	11.9	①
	del Sol	D16Y8	1.6 (1590)	3.5	—	4.0	5.6	—	—	—	11.9	8.2
	Prelude	H22A4	2.2 (2156)	5.1	—	4.0	5.0	—	—	—	15.9	4.6
	Prelude SH	H22A4	2.2 (2156)	5.1	—	4.0	—	—	—	—	15.9	4.6

79231CF6

CAPACITIES

Year	Model	Engine ID/VIN	Engine Displacement Liters (cc)	Engine Oil with Filter	Transmission (pts.)			Transfer Case (pts.)	Drive Axle		Fuel Tank (gal.)	Cooling System (qts.)
					4-Spd	5-Spd	Auto.		Front (pts.)	Rear (pts.)		
1998-99	Accord Coupe (EX, LX)	F23A1	2.3 (2254)	4.0	—	4.0	5.0	—	—	—	17.0	③
	Accord Coupe (EX, LX)	F23A4	2.3 (2254)	4.5	—	4.0	5.0	—	—	—	17.0	③
	Accord Coupe (EX, LX)	J30A1	3.0 (2997)	4.6	—	—	6.2	—	—	—	17.1	5.9
	Accord Sedan (DX)	F23A5	2.3 (2254)	4.5	—	4.0	5.2	—	—	—	17.0	③
	Accord Sedan (EX, LX)	F23A1	2.3 (2254)	4.0	—	4.0	5.0	—	—	—	17.0	③
	Accord Sedan (EX, LX)	F23A4	2.3 (2254)	4.5	—	4.0	5.2	—	—	—	17.0	③
	Accord Sedan (EX, LX)	J30A1	3.0 (2997)	4.6	—	—	6.2	—	—	—	17.1	5.9
	Civic	D16Y5	1.6 (1590)	3.5	—	3.8	5.8	—	—	—	11.9	4.5
	Civic	D16Y7	1.6 (1590)	3.5	—	3.8	5.8	—	—	—	11.9	4.4
	Civic	D16Y8	1.6 (1590)	3.5	—	3.8	5.8	—	—	—	11.9	4.3
	Prelude	H22A4	2.2 (2156)	5.1	—	4.0	—	—	—	—	15.9	4.6
	Prelude SH	H22A4	2.2 (2156)	5.1	—	4.0	—	—	—	—	15.9	4.6

NOTE: All capacities are approximate. Add fluid gradually and ensure a proper fluid level is obtained.

NOTE: Capacities given are service, not overhaul capacities

① Automatic transaxle: 5.6
  Manual transaxle: 5.7

② Automatic transaxle: 4.0
  Manual transaxle: 3.8

③ Automatic Transaxle: 5.7
  Manual Transaxle: 5.8

79231CF7

VALVE SPECIFICATIONS

Year	Engine ID/VIN	Engine Displacement Liters (cc)	Seat Angle (deg.)	Face Angle (deg.)	Spring Test Pressure (lbs. @ in.)	Spring Installed Height (in.)	Stem-to-Guide Clearance (in.)		Stem Diameter (in.)	
							Intake	Exhaust	Intake	Exhaust
1995	B16A3	1.6 (1595)	45	45	NA	NA	0.0010-0.0030	0.0020-0.0040	0.2144-0.2459	0.2134-0.2150
	C27A4	2.7 (2675)	45	45	NA	NA	0.0008-0.0030	0.0020-0.0040	0.2580-0.2594	0.2570-0.2583
	D15B7	1.5 (1493)	45	45	NA	NA	0.0010-0.0030	0.0020-0.0040	0.2150-0.2160	0.2130-0.2150
	D15B8	1.5 (1493)	45	45	NA	NA	0.0010-0.0030	0.0020-0.0040	0.2150-0.2160	0.2130-0.2150
	D15Z1	1.5 (1493)	45	45	NA	NA	0.0010-0.0030	0.0020-0.0050	0.2150-0.2160	0.2130-0.2150
	D16Z6	1.6 (1590)	45	45	NA	NA	0.0010-0.0030	0.0020-0.0050	0.2150-0.2160	0.2130-0.2150
	F22A1	2.2 (2156)	45	45	NA	NA	0.0008-0.0030	0.0022-0.0050	0.2148-0.2163	0.2134-0.2150
	F22B1	2.2 (2156)	45	45	NA	NA	0.0008-0.0030	0.0022-0.0050	0.2148-0.2163	0.2134-0.2150
	F22B2	2.2 (2156)	45	45	NA	NA	0.0008-0.0030	0.0022-0.0050	0.2148-0.2163	0.2134-0.2150
	H22A1	2.2 (2157)	45	45	NA	NA	0.0010-0.0030	0.0020-0.0040	0.2144-0.2459	0.2144-0.2459
	H23A1	2.3 (2259)	45	45	NA	NA	0.0010-0.0030	0.0020-0.0040	0.2580-0.2594	0.2570-0.2583
1996	B16A2	1.6 (1595)	45	45	NA	NA	0.0010-0.0030	0.0020-0.0040	0.2144-0.2459	0.2134-0.2150
	C27A4	2.7 (2675)	45	45	NA	NA	0.0008-0.0030	0.0020-0.0040	0.2580-0.2594	0.2570-0.2583
	D15B7	1.5 (1493)	45	45	NA	NA	0.0010-0.0030	0.0020-0.0040	0.2150-0.2160	0.2130-0.2150
	D15B8	1.5 (1493)	45	45	NA	NA	0.0010-0.0030	0.0020-0.0040	0.2150-0.2160	0.2130-0.2150
	D15Z1	1.5 (1493)	45	45	NA	NA	0.0010-0.0030	0.0020-0.0050	0.2150-0.2160	0.2130-0.2150
	D16Y5	1.6 (1590)	45	45	NA	NA	0.0010-0.0020	0.0020-0.0030	0.2157-0.2161	0.2146-0.2150
	D16Y7	1.6 (1590)	45	45	NA	NA	0.0010-0.0020	0.0020-0.0030	0.2157-0.2161	0.2146-0.2150
	D16Y8	1.6 (1590)	45	45	NA	NA	0.0010-0.0020	0.0020-0.0030	0.2157-0.2161	0.2146-0.2150
	D16Z6	1.6 (1590)	45	45	NA	NA	0.0010-0.0030	0.0020-0.0050	0.2150-0.2160	0.2130-0.2150
	F22A1	2.2 (2156)	45	45	NA	NA	0.0008-0.0030	0.0022-0.0050	0.2148-0.2163	0.2134-0.2150
	F22B1	2.2 (2156)	45	45	NA	NA	0.0008-0.0030	0.0022-0.0050	0.2148-0.2163	0.2134-0.2150

79231CF8

VALVE SPECIFICATIONS

Year	Engine ID/VIN	Engine Displacement Liters (cc)	Seat Angle (deg.)	Face Angle (deg.)	Spring Test Pressure (lbs. @ in.)	Spring Installed Height (in.)	Stem-to-Guide Clearance (in.)		Stem Diameter (in.)	
							Intake	Exhaust	Intake	Exhaust
1996 (cont.)	F22B2	2.2 (2156)	45	45	NA	NA	0.0008-0.0030	0.0022-0.0050	0.2148-0.2163	0.2134-0.2150
	H22A1	2.2 (2157)	45	45	NA	NA	0.0010-0.0030	0.0020-0.0040	0.2144-0.2459	0.2144-0.2459
	H23A1	2.3 (2259)	45	45	NA	NA	0.0010-0.0030	0.0020-0.0040	0.2580-0.2594	0.2570-0.2583
1997	B16A2	1.6 (1595)	45	45	NA	NA	0.0010-0.0022	0.0020-0.0031	0.2156-0.2159	0.2146-0.2150
	C27A4	2.7 (2675)	45	45	NA	NA	0.0008-0.0030	0.0020-0.0040	0.2580-0.2594	0.2570-0.2583
	D16Y5	1.6 (1590)	45	45	NA	NA	0.0010-0.0020	0.0020-0.0030	0.2157-0.2161	0.2146-0.2150
	D16Y7	1.6 (1590)	45	45	NA	NA	0.0010-0.0020	0.0020-0.0030	0.2157-0.2161	0.2146-0.2150
	D16Y8	1.6 (1590)	45	45	NA	NA	0.0010-0.0020	0.0020-0.0030	0.2157-0.2161	0.2146-0.2150
	F22B1	2.2 (2156)	45	45	NA	NA	0.0008-0.0030	0.0020-0.0050	0.2148-0.2163	0.2134-0.2150
	F22B2	2.2 (2156)	45	45	NA	NA	0.0008-0.0030	0.0020-0.0050	0.2148-0.2163	0.2134-0.2150
	H22A4	2.2 (2156)	45	45	NA	NA	0.0010-0.0022	0.0020-0.0031	0.2156-0.2159	0.2156-0.2159
1998-99	D16Y5	1.6 (1590)	45	45	NA	NA	0.0010-0.0020	0.0020-0.0030	0.2157-0.2161	0.2146-0.2150
	D16Y7	1.6 (1590)	45	45	NA	NA	0.0010-0.0020	0.0020-0.0030	0.2157-0.2161	0.2146-0.2150
	D16Y8	1.6 (1590)	45	45	NA	NA	0.0010-0.0020	0.0020-0.0030	0.2157-0.2161	0.2146-0.2150
	F23A1	2.3 (2254)	45	45	NA	NA	0.0008-0.0018	0.0022-0.0031	0.2159-0.2163	0.2146-0.2150
	F23A4	2.3 (2254)	45	45	NA	NA	0.0008-0.0018	0.0022-0.0031	0.2159-0.2163	0.2146-0.2150
	F23A5	2.3 (2254)	45	45	NA	NA	0.0008-0.0018	0.0022-0.0031	0.2159-0.2163	0.2146-0.2150
	H22A4	2.2 (2157)	45	45	NA	NA	0.0010-0.0022	0.0020-0.0031	0.2156-0.2159	0.2156-0.2159
	J30A1	3.0 (2997)	45	45	NA	NA	0.0008-0.0018	0.0022-0.0031	0.2159-0.2163	0.2146-0.2150

NA - Not Available

79231CF9

TORQUE SPECIFICATIONS
All readings in ft. lbs.

Year	Engine ID/VIN	Engine Displacement Liters (cc)	Cylinder Head Bolts	Main Bearing Bolts	Rod Bearing Bolts	Crankshaft Damper Bolts	Flywheel Bolts	Manifold Intake	Manifold Exhaust	Spark Plugs	Lug Nut
1995	B16A3	1.6 (1595)	①	②	30	130	87	17	23	13	80
	C27A4	2.7 (2675)	③	④	33	181	54	16	22	13	80
	D15B7	1.5 (1493)	⑤	⑥	23	134	87 ⑦	17	23	13	80
	D15B8	1.5 (1493)	⑤	⑥	23	134	87	17	23	13	80
	D15Z1	1.5 (1493)	⑧	⑥	23	134	87	17	23	13	80
	D16Z6	1.6 (1590)	⑧	⑪	23	134	87 ⑦	17	23	13	80
	F22A1	2.2 (2156)	⑫	⑧	34	181	76 ⑦	16	23	13	80
	F22B1	2.2 (2156)	⑫	⑧	34	181	76 ⑦	16	23	13	80
	F22B2	2.2 (2156)	⑫	⑧	34	181	76 ⑦	16	23	13	80
	H22A1	2.2 (2157)	⑫	⑧	34	181	76	16	23	13	80
	H23A1	2.3 (2259)	⑫	⑧	34	181	76 ⑦	16	23	13	80
1996	B16A2	1.6 (1595)	①	②	30	130	76	17	23	13	80
	C27A4	2.7 (2675)	③	④	33	181	54	16	22	13	80
	D15B7	1.5 (1493)	⑤	⑥	23	134	87 ⑦	17	23	13	80
	D15B8	1.5 (1493)	⑤	⑥	23	134	87	17	23	13	80
	D15Z1	1.5 (1493)	⑧	⑥	23	134	87	17	23	13	80
	D16Z6	1.6 (1590)	⑧	⑨	23	134	87 ⑦	17	23	13	80
	D16Y7	1.6 (1590)	⑩	⑨	23	134	87	17	23	13	80
	D16Y8	1.6 (1590)	⑩	⑨	23	134	87	17	23	13	80
	F22A1	2.2 (2156)	⑧	⑧	34	181	76	16	23	13	80
	F22B1	2.2 (2156)	⑪	⑧	34	181	76 ⑦	16	23	13	80
	F22B2	2.2 (2156)	⑪	⑧	34	181	76 ⑦	16	23	13	80
	H22A1	2.2 (2157)	⑧	⑧	34	181	76	16	23	13	80
	H23A1	2.3 (2259)	⑧	⑧	34	181	76	16	23	13	80
1997	B16A2	1.6 (1595)	①	②	30	130	76	17	23	13	80
	C27A4	2.7 (2675)	③	④	33	181	54	16	22	13	80
	D16Y5	1.6 (1590)	⑩	⑨	23	134	87	17	23	13	80
	D16Y7	1.6 (1590)	⑩	⑨	23	134	87	17	23	13	80
	D16Y8	1.6 (1590)	⑩	⑨	23	134	87	17	23	13	80
	F22A1	2.2 (2156)	⑧	⑧	34	181	76	16	23	13	80
	F22B1	2.2 (2156)	⑪	⑧	34	181	76 ⑦	16	23	13	80
	F22B2	2.2 (2156)	⑪	⑧	34	181	76 ⑦	16	23	13	80
	H22A4	2.2 (2157)	⑧	⑧	34	181	76	16	23	13	80
1998-99	D16Y5	1.6 (1590)	⑩	⑨	23	134	87	17	23	13	80
	D16Y7	1.6 (1590)	⑩	⑨	23	134	87	17	23	13	80
	D16Y8	1.6 (1590)	⑩	⑨	23	134	87	17	23	13	80
	F23A1	2.3 (2254)	⑫	⑬	⑭	181	76 ⑦	16	23	13	80

79231CF0

TORQUE SPECIFICATIONS
All readings in ft. lbs.

Year	Engine ID/VIN	Engine Displacement Liters (cc)	Cylinder Head Bolts	Main Bearing Bolts	Rod Bearing Bolts	Crankshaft Damper Bolts	Flywheel Bolts	Manifold		Spark Plugs	Lug Nut
								Intake	Exhaust		
1998-99 (cont.)	F23A4	2.3 (2254)	⑫	⑬	⑭	181	76 ⑦	16	23	13	80
	F23A5	2.3 (2254)	⑫	⑬	⑭	181	76 ⑦	16	23	13	80
	H22A4	2.2 (2157)	⑮	54	181	181	76 ⑦	16	23	13	80
	J30A1	3.0 (2997)	⑮	⑯	181	181	76 ⑦	16	23	13	80

① Step 1: 22 ft. lbs.
Step 2: 61 ft. lbs.

② Step 1: 18 ft. lbs.
Step 2: 54 ft. lbs.

③ Step 1: 29 ft. lbs.
Step 2: 58 ft. lbs.

④ Inner: 48 ft. lbs.
Outer: 29 ft. lbs.
Side: 36 ft. lbs.

⑤ Step 1: 22 ft. lbs.
Step 2: 47 ft. lbs.

⑥ Step 1: 18 ft. lbs.
Step 2: 33 ft. lbs.

⑦ Automatic transaxle: 54 ft. lbs.

⑧ Step 1: 22 ft. lbs.
Step 2: 53 ft. lbs.

⑨ Step 1: 18 ft. lbs.
Step 2: 38 ft. lbs.

⑩ Cylinder head bolts to be tightened in four steps:
Step 1: 14 ft. lbs.
Step 2: 36 ft. lbs.
Step 3: 49 ft. lbs.
Step 4: Bolts 1-2, tighten an additional 49 ft. lbs.

⑪ Step 1: 29 ft. lbs.
Step 2: 51 ft. lbs.
Step 3: 72 ft. lbs.

⑫ Step 1: 22 Ft. lbs.
Step 2: Rotate 90 degrees
Step 3: Rotate 90 degrees
Step 4: If new bolt rotate additional 90 degrees

⑬ Step 1: 11mm bolts, 29 ft. lbs.
Step 2: 11mm bolts, 58 ft. lbs.
Step 3: 6mm bolts, 8.7 ft. lbs.

⑭ Step 1: 14 ft. lbs.
Step 2: Rotate 90 degrees

⑮ Step 1: 29 ft. lbs.
Step 2: 51 ft. lbs.
Step 3: 72.3 ft. lbs.

⑯ Step 1: Cap bolts, 56 ft. lbs.
Step 2: Side bolts, 36 ft. lbs.

79231CG1

SCHEDULED MAINTENANCE INTERVALS
(HONDA ACCORD, CIVIC, DEL SOL & PRELUDE)

TO BE SERVICED	TYPE OF SERVICE	VEHICLE MILEAGE INTERVAL (x1000)												
		7.5	15	22.5	30	37.5	45	52.5	60	67.5	75	82.5	90	97.5
Engine oil & filter	R	✓	✓	✓	✓	✓	✓	✓	✓	✓	✓	✓	✓	✓
Front brake pads	S/I	✓	✓	✓	✓	✓	✓	✓	✓	✓	✓	✓	✓	✓
Rotate tires	S/I	✓	✓	✓	✓	✓	✓	✓	✓	✓	✓	✓	✓	✓
Brake hoses & lines (including ABS)	S/I		✓		✓		✓		✓		✓		✓	
Cooling system, hoses & connections	S/I		✓		✓		✓		✓		✓		✓	
Driveshaft boots	S/I		✓		✓		✓		✓		✓		✓	
Exhaust system	S/I		✓		✓		✓		✓		✓		✓	
Front brake discs & calipers	S/I		✓		✓		✓		✓		✓		✓	
Front wheel alignment	S/I		✓		✓		✓		✓		✓		✓	
Front & rear wheel alignment (Prelude w/4WS)	S/I		✓		✓		✓		✓		✓		✓	
Fuel pipes, hoses & connections	S/I		✓		✓		✓		✓		✓		✓	
Parking brake adjustment	S/I		✓		✓		✓		✓		✓		✓	
Power steering system	S/I		✓		✓		✓		✓		✓		✓	
Rear brake discs, calipers & pads	S/I		✓		✓		✓		✓		✓		✓	
Suspension components	S/I		✓		✓		✓		✓		✓		✓	
Suspension mounting bolts	S/I		✓		✓		✓		✓		✓		✓	
Tie rods, steering gear box & boots	S/I		✓		✓		✓		✓		✓		✓	
Valve clearance (1995 Civic & 1995-99 Prelude VTEC)	S/I		✓		✓		✓		✓		✓		✓	
Valve clearance (1995-99 Accord L4, 1996-99 Civic, 1995 Del Sol non-VTEC & 1995-99 Prelude non-VTEC)	S/I				✓				✓				✓	

79231CG2

SCHEDULED MAINTENANCE INTERVALS
(HONDA ACCORD, CIVIC, DEL SOL & PRELUDE) (Cont.)

TO BE SERVICED	TYPE OF SERVICE	VEHICLE MILEAGE INTERVAL (x1000)												
		7.5	15	22.5	30	37.5	45	52.5	60	67.5	75	82.5	90	97.5
Valve clearance (1996-99 Del Sol VTEC)	S/I				✓				✓				✓	
Valve clearance (1996-99 Del Sol non-VTEC)	S/I				✓									
Parking brake	S/I		✓		✓				✓				✓	
Air cleaner element	R				✓				✓				✓	
Transmission fluid (1996-99 Civic CVT)	R				✓		✓		✓		✓		✓	
Transmission fluid (A/T or M/T) (except as noted below)	R				✓				✓				✓	
Transmission fluid (1996-99 Prelude L4 & Del Sol A/T or M/T)	R												✓	
Brake fluid (including ABS) (all 1995 & 1996-99 Accord V6)	R				✓				✓				✓	
Brake fluid (including ABS) (1996-99 Accord L4, Civic, Del Sol & Prelude)	R						✓						✓	
Spark plugs (non-VTEC)	R				✓				✓				✓	
Spark plugs (VTEC)	R								✓					
ABS operation	S/I				✓				✓				✓	
Alternator drive belt	S/I				✓				✓				✓	
Power steering pump belt	S/I				✓				✓				✓	
Rear brake drums, wheel cylinders & linings (except Prelude)	S/I				✓				✓				✓	

79231CG3

SCHEDULED MAINTENANCE INTERVALS
(HONDA ACCORD, CIVIC, DEL SOL & PRELUDE) (Cont.)

TO BE SERVICED	TYPE OF SERVICE	VEHICLE MILEAGE INTERVAL (x1000)												
		7.5	15	22.5	30	37.5	45	52.5	60	67.5	75	82.5	90	97.5
Engine coolant	R						✓		.		✓			
ABS high pressure hose	R								✓					
Fuel filter	R								✓					
Timing belt	R												✓	
Timing balancer belt	R												✓	
Distributor, ignition cap & rotor	S/I								✓					
Idle speed	S/I								✓					
Ignition wires	S/I								✓					
PCV valve	S/I								✓					
TWC converter heat shield	S/I								✓					
Water pump	S/I												✓	

R – Replace   S/I – Service or Inspect

FREQUENT OPERATION MAINTENANCE (SEVERE SERVICE)

If a vehicle is operated under any of the following conditions it is considered severe service:
- Extremely dusty areas.
- 50% or more of the vehicle operation is in 32°C (90°F) or higher temperatures, or constant operation in temperatures below 0°C (32°F).
- Prolonged idling (vehicle operation in stop and go traffic).
- Frequent short running periods (engine does not warm to normal operating temperatures).
- Police, taxi, delivery usage or trailer towing usage.

Oil & oil filter change – change every 3750 miles.
Driveshaft boots - service or inspect every 7500 miles.
Front brake discs & calipers, & rear brake discs, calipers & pads - service or inspect every 7500 miles.
Power steering system - service or inspect every 7500 miles.
Suspension components - service or inspect every 7500 miles.
Tie rods, steering gear box & boots - service or inspect every 7500 miles.
Air cleaner element – service or inspect every 15,000 miles.
Transmission fluid (1996-99 Accord V6 & 1996-99 Civic CVT) - replace every 15,000 miles.
Transmission fluid (1996-99 Accord L4, Civic, Del Sol & Prelude) - replace every 30,000 miles.
Timing balancer belt - replace every 60,000 miles.
Timing belt - replace every 60,000 miles.
Water pump - service or inspect every 60,000 miles.

79231CG4

SCHEDULED MAINTENANCE
HONDA
ACCORD, CIVIC, DEL SOL, PRELUDE

The following should be used as a guide when determining the amount of work required for a particular service if taken to a repair shop. In estimating how long a particular Scheduled Maintenance Service should take, please observe the following:

- **Chilton Time** is time based on field research and data supplied by the vehicle manufacturer.
- Labor time operations are given in hours and tenths of an hour.
- All labor operations, are to be used as a guide.

Mechanic Skill Level Codes:
- **(G)** GENERAL: Normally skilled with certification.
- **(M)** MAINTENANCE: Semi-skilled working on certicication.
- **(P)** PRECISION: Really skilled with multiple certification.

	Chilton Time
(G) 7500 Mile Service	
1995-99	1.1
(G) 15000 Mile Service	
1995-99	2.0
adjust valves add	1.1
(G) 22500 Mile Service	
1995-99	1.1
(G) 30000 Mile Service	
1995-99	3.3
adjust valves add	1.1
replace trans fluid add	.5
replace spark plugs add	.5
replace brake fluid add	.6
(G) 37500 Mile Service	
1995-99	1.1

	Chilton Time
(G) 45000 Mile Service	
1995-99	2.0
replace trans fluid add	.5
replace brake fluid add	.6
replace engine coolant add	.7
(G) 52500 Mile Service	
1995-99	1.1
(G) 60000 Mile Service	
1995-99	3.5
adjust valves add	1.1
replace trans fluid add	.5
replace spark plugs add	.5
replace brake fluid add	.6
replace fuel filter add	.3
(G) 67500 Mile Service	
1995-99	1.1

	Chilton Time
(G) 75000 Mile Service	
1995-99	2.6
adjust valves add	1.1
replace trans fluid add	.5
replace engine coolant add	.7
(G) 82500 Mile Service	
1995-99	1.1
(G) 90000 Mile Service	
1995-99	3.5
adjust valves add	1.1
replace trans fluid add	.5
replace spark plugs add	.5
replace brake fluid add	.6
replace timing belt add	2.2
replace timing belt V6 add	3.5
(G) 97500 Mile Service	
1995-99	1.1

79231CG5

HYUNDAI
Accent • Elantra • Scoupe • Sonata • Tiburon

VEHICLE IDENTIFICATION CHART

Engine Code						Model Year	
Code	Liters	Cu. In. (cc)	Cyl.	Fuel Sys.	Eng. Mfg.	Code	Year
E	1.5	91.17 (1495)	4	MFI	Hyundai	S	1995
E	1.5	91.17 (1495)	4	MFI-T	Hyundai	T	1996
K	1.5	91.17 (1495)	4	MFI	Hyundai	V	1997
R	1.6	97.29 (1595)	4	MFI	Hyundai	W	1998
M	1.8	109.54 (1795)	4	MFI	Hyundai	X	1999
M	1.8	112.04 (1836)	4	MFI	Hyundai		
F	2.0	120.52 (1975)	4	MFI	Hyundai		
P	2.0	121.90 (1997)	4	MFI	Hyundai		
T	3.0	181.40 (2972)	6	MFI	Hyundai		

MFI : Multi-Port Fuel Injection

MFI-T: Multi-Port Fuel Injection Turbocharged

79231CD9

ENGINE IDENTIFICATION

Year	Model	Engine Displacement Liters (cc)	Engine Series (ID/VIN)	Fuel System	No. of Cylinders	Engine Type
1995	Scoupe	1.5 (1495)	E	MFI	4	SOHC
	Scoupe	1.5 (1495)	E	MFI-T	4	SOHC
	Accent	1.5 (1495)	K	MFI	4	SOHC
	Elantra	1.6 (1595)	R	MFI	4	DOHC
	Elantra	1.8 (1836)	M	MFI	4	DOHC
	Sonata	2.0 (1997)	P	MFI	4	DOHC
	Sonata	3.0 (2972)	T	MFI	6	SOHC
1996	Accent	1.5 (1495)	K	MFI	4	SOHC
	Accent	1.5 (1495)	K	MFI	4	DOHC
	Elantra	1.8 (1795)	M	MFI	4	DOHC
	Sonata	2.0 (1997)	P	MFI	4	DOHC
	Sonata	3.0 (2972)	T	MFI	6	SOHC
1997	Accent	1.5 (1495)	K	MFI	4	SOHC
	Accent	1.5 (1495)	K	MFI	4	DOHC
	Elantra	1.8 (1795)	M	MFI	4	DOHC
	Tiburon	1.8 (1795)	M	MFI	4	DOHC
	Tiburon	2.0 (1975)	F	MFI	4	DOHC
	Sonata	2.0 (1997)	P	MFI	4	DOHC
	Sonata	3.0 (2972)	T	MFI	6	SOHC
1998-99	Accent	1.5 (1495)	K	MFI	4	SOHC
	Elantra	1.8 (1795)	M	MFI	4	DOHC
	Tiburon	1.8 (1795)	M	MFI	4	DOHC
	Tiburon	2.0 (1975)	F	MFI	4	DOHC
	Sonata	2.0 (1997)	P	MFI	4	DOHC
	Sonata	3.0 (2972)	T	MFI	6	SOHC

MFI : Multi-Port Fuel Injection
MFI-T: Multi-Port Fuel Injection Turbocharged
SOHC: Single Overhead Camshaft
DOHC: Double Overhead Camshafts

79231CD0

GENERAL ENGINE SPECIFICATIONS

Year	Engine ID/VIN		Engine Displacement Liters (cc)	Fuel System Type	Net Horsepower @ rpm	Net Torque @ rpm (ft. lbs.)	Bore x Stroke (in.)	Compression Ratio	Oil Pressure @ rpm
1995	E		1.5 (1495)	MFI	92@5500	97@4500	2.97 x 3.29	10.0:1	21@Idle
	E		1.5 (1495)	MFI-T	115@5500	123@4500	2.97 x 3.29	7.5:1	21@Idle
	K		1.5 (1495)	MFI	92@5500	96@3000	2.97 x 3.29	10.0:1	21@Idle
	R		1.6 (1595)	MFI	113@6000	102@5000	3.24 x 2.95	9.2:1	12@Idle
	M		1.8 (1836)	MFI	124@6000	116@5000	3.21 x 3.46	9.2:1	12@Idle
	P		2.0 (1997)	MFI	137@6000	129@4000	3.35 x 3.46	9.0:1	12@Idle
	T		3.0 (2972)	MFI	142@5000	168@2500	3.59 x 2.99	8.9:1	12@Idle
1996	K	①	1.5 (1495)	MFI	92@5500	97@4000	2.97 x 3.29	10.0:1	21@Idle
	K	②	1.5 (1495)	MFI	105@6000	101@4500	2.97 x 3.29	9.5:1	21@Idle
	M		1.8 (1795)	MFI	124@6000	116@5000	3.23 x 3.35	10.0:1	24@Idle
	P		2.0 (1997)	MFI	137@6000	129@4000	3.35 x 3.46	9.0:1	12@Idle
	T		3.0 (2972)	MFI	142@5000	168@2500	3.59 x 2.99	8.9:1	12@Idle
1997	K	①	1.5 (1495)	MFI	92@5500	97@4000	2.97 x 3.29	10.0:1	21@Idle
	K	②	1.5 (1495)	MFI	105@6000	101@4500	2.97 x 3.29	9.5:1	21@Idle
	M		1.8 (1795)	MFI	124@6000	116@5000	3.23 x 3.35	10.0:1	24@Idle
	F		2.0 (1975)	MFI	140@6000	133@4800	3.23 x 3.68	10.3:1	24@Idle
	P		2.0 (1997)	MFI	137@6000	129@4000	3.35 x 3.46	9.0:1	12@Idle
	T		3.0 (2972)	MFI	142@5000	168@2500	3.59 x 2.99	8.9:1	12@Idle
1998-99	K		1.5 (1495)	MFI	92@5500	97@4000	2.97 x 3.29	10.0:1	21@Idle
	M		1.8 (1795)	MFI	124@6000	116@5000	3.23 x 3.35	10.0:1	24@Idle
	F		2.0 (1975)	MFI	140@6000	133@4800	3.23 x 3.68	10.3:1	24@Idle
	P		2.0 (1997)	MFI	137@6000	129@4000	3.35 x 3.46	9.0:1	12@Idle
	T		3.0 (2972)	MFI	142@5000	168@2500	3.59 x 2.99	8.9:1	12@Idle

MFI : Multi-Port Fuel Injection

MFI-T: Multi-Port Fuel Injection Turbocharged

① Single Overhead Camshaft (SOHC)

② Double Overhead Camshafts (DOHC)

79231CE1

GASOLINE ENGINE TUNE-UP SPECIFICATIONS

Year	Engine ID/VIN		Engine Displacement Liters (cc)	Spark Plugs Gap (in.)	Ignition Timing (deg.)		Fuel Pump (psi)	Idle Speed (rpm)		Valve Clearance	
					MT	AT		MT	AT	In. ③	Ex. ③
1995	E		1.5 (1495)	0.039-0.043	4-14B	4-14B	43	700-900	700-900	0.010	0.012
	K		1.5 (1495)	0.039-0.043	6-16B	6-16B	43	700-900	700-900	HYD	HYD
	R		1.6 (1595)	0.039-0.043	3-7B	3-7B	48	650-850	650-850	HYD	HYD
	M		1.8 (1836)	0.039-0.043	3-7B	3-7B	48	600-800	600-800	HYD	HYD
	P		2.0 (1997)	0.039-0.043	3-7B	3-7B	48	650-850	650-850	HYD	HYD
	T		3.0 (2972)	0.039-0.043	3-7B	3-7B	48	600-800	600-800	HYD	HYD
1996	K	①	1.5 (1495)	0.039-0.043	6-16B	6-16B	43	700-900	700-900	HYD	HYD
	K	②	1.5 (1495)	0.039-0.043	4-14B	4-14B	43	700-900	700-900	HYD	HYD
	M		1.8 (1795)	0.039-0.043	5-15B	5-15B	43	700-900	700-900	HYD	HYD
	P		2.0 (1997)	0.039-0.043	3-7B	3-7B	48	650-850	650-850	HYD	HYD
	T		3.0 (2972)	0.039-0.043	3-7B	3-7B	48	600-800	600-800	HYD	HYD
1997	K	①	1.5 (1495)	0.039-0.043	6-16B	6-16B	43	700-900	700-900	HYD	HYD
	K	②	1.5 (1495)	0.039-0.043	4-14B	4-14B	43	700-900	700-900	HYD	HYD
	M		1.8 (1795)	0.039-0.043	5-15B	5-15B	43	700-900	700-900	HYD	HYD
	F		2.0 (1975)	0.039-0.043	5-15B	5-15B	43	700-900	700-900	HYD	HYD
	P		2.0 (1997)	0.039-0.043	3-7B	3-7B	48	650-850	650-850	HYD	HYD
	T		3.0 (2972)	0.039-0.043	3-7B	3-7B	48	600-800	600-800	HYD	HYD
1998-99	K		1.5 (1495)	0.039-0.043	6-16B	6-16B	43	700-900	700-900	HYD	HYD
	M		1.8 (1795)	0.039-0.043	5-15B	5-15B	43	700-900	700-900	HYD	HYD
	F		2.0 (1975)	0.039-0.043	5-15B	5-15B	43	700-900	700-900	HYD	HYD
	P		2.0 (1997)	0.039-0.043	3-7B	3-7B	48	650-850	650-850	HYD	HYD
	T		3.0 (2972)	0.039-0.043	3-7B	3-7B	48	600-800	600-800	HYD	HYD

HYD: Hydraulic Valve Lifters

B: Before Top Dead Center

① Single Overhead Camshaft (SOHC)

② Double Overhead Camshafts (DOHC)

③ Valve clearance is checked with engine hot

79231CE2

CAPACITIES

| Year | Model | Engine ID/VIN | Engine Displacement Liters (cc) | Engine Oil with Filter | Transmission (pts.) | | | Transfer Case (pts.) | Drive Axle | | Fuel Tank (gal.) | Cooling System (qts.) |
					4–Spd	5–Spd	Auto.		Front (pts.)	Rear (pts.)		
1995	Accent	K	1.5 (1495)	3.5	—	4.6	13.6	—	—	—	11.9	5.8
	Elantra	R	1.6 (1595)	3.5	—	3.8	12.8	—	—	—	13.8	5.4
	Elantra	M	1.8 (1836)	4.6	—	3.8	12.8	—	—	—	13.8	6.3
	Scoupe	E	1.5 (1495)	3.4	—	4.5	12.8	—	—	—	11.9	5.6
	Sonata	P	2.0 (1997)	3.9	—	3.8	12.8	—	—	—	17.2	7.7
	Sonata	T	3.0 (2972)	4.2	—	—	15.8	—	—	—	17.2	9
1996	Accent	K	1.5 (1495)	3.5	—	4.6	13.6	—	—	—	11.9	5.8
	Elantra	M	1.8 (1795)	4.2	—	4.5	12.8	—	—	—	14.5	6.3
	Sonata	P	2.0 (1997)	3.9	—	3.8	12.8	—	—	—	17.2	7.7
	Sonata	T	3.0 (2972)	4.2	—	—	15.8	—	—	—	17.2	9
1997	Accent	K	1.5 (1495)	3.5	—	4.6	13.6	—	—	—	11.9	6.3
	Elantra	M	1.8 (1795)	4.2	—	4.5	12.8	—	—	—	14.5	6.3
	Sonata	P	2.0 (1997)	3.9	—	5.2	12.8	—	—	—	17.2	7.7
	Sonata	T	3.0 (2972)	4.2	—	—	15.8	—	—	—	17.2	9
	Tiburon	M	1.8 (1795)	4.2	—	4.5	13.8	—	—	—	14.5	6.3
	Tiburon	F	2.0 (1975)	4.2	—	4.5	13.8	—	—	—	14.5	6.3
1998-99	Accent	K	1.5 (1495)	3.5	—	4.6	13.6	—	—	—	11.9	6.3
	Elantra	M	1.8 (1795)	4.2	—	4.5	12.8	—	—	—	14.5	6.3
	Sonata	P	2.0 (1997)	3.9	—	5.2	12.8	—	—	—	17.2	7.7
	Sonata	T	3.0 (2972)	4.2	—	—	15.8	—	—	—	17.2	9
	Tiburon	M	1.8 (1795)	4.2	—	4.5	13.8	—	—	—	14.5	6.3
	Tiburon	F	2.0 (1975)	4.2	—	4.5	13.8	—	—	—	14.5	6.3

NOTE: All capacities are approximate. Add fluid gradually and check to be sure a proper fluid level is obtained.

79231CE3

VALVE SPECIFICATIONS

Year	Engine ID/VIN		Engine Displacement Liters (cc)	Seat Angle (deg.)	Face Angle (deg.)	Spring Test Pressure (lbs. @ in.)	Spring Installed Height (in.)	Stem-to-Guide Clearance (in.)		Stem Diameter (in.)	
								Intake	Exhaust	Intake	Exhaust
1995	E	①	1.5 (1495)	45	45	44@1.261	1.261	0.0012-0.0024	0.0020-0.0035	0.2364	0.2364
	E	②	1.5 (1495)	45	45	44@1.261	1.261	0.0012-0.0024	0.0020-0.0035	0.2364	0.2364
	K		1.5 (1495)	45	45	54@1.358	1.358	0.0012-0.0024	0.0014-0.0026	0.3920	0.3960
	R		1.6 (1595)	44-44.5	45-45.5	66@1.575	⑤	0.0008-0.0019	0.0020-0.0033	0.2585-0.2591	0.2571-0.2579
	M		1.8 (1836)	45	45	56@1.458	1.358	0.0008-0.0019	0.0019-0.0033	0.2348-0.2354	0.2334-0.2342
	P		2.0 (1997)	45-45.5	45-45.5	66@1.575	⑤	0.0008-0.0019	0.0020-0.0033	0.2585-0.2891	0.2571-0.2579
	T		3.0 (2972)	44-44.5	45	74@1.591	1.590	0.0012-0.0024	0.0020-0.0035	0.3150	0.3134
1996	K	③	1.5 (1495)	45	45	54@1.358	1.358	0.0012-0.0024	0.0014-0.0026	0.3920	0.3960
	K	④	1.5 (1495)	45	45	48@1.378	1.378	0.0012-0.0024	0.0020-0.0031	0.2344-0.2350	0.2337-0.2343
	M		1.8 (1795)	45	45	56@1.458	1.358	0.0008-0.0019	0.0019-0.0033	0.2348-0.2354	0.2334-0.2342
	P		2.0 (1997)	45-45.5	45-45.5	66@1.575	⑤	0.0008-0.0019	0.0020-0.0033	0.2585-0.2891	0.2571-0.2579
	T		3.0 (2972)	44-44.5	45	74@1.591	1.590	0.0012-0.0024	0.0020-0.0035	0.3150	0.3134
1997	K	③	1.5 (1495)	45	45	54@1.358	1.358	0.0012-0.0024	0.0014-0.0026	0.3920	0.3960
	K	④	1.5 (1495)	45	45	48@1.378	1.378	0.0012-0.0024	0.0020-0.0031	0.2344-0.2350	0.2337-0.2343
	M		1.8 (1795)	45	45	56@1.458	1.358	0.0008-0.0019	0.0019-0.0033	0.2348-0.2354	0.2334-0.2342
	F		2.0 (1975)	45	45	56@1.457	1.358	0.0008-0.0019	0.0019-0.0033	0.2348-0.2354	0.2334-0.2342
	P		2.0 (1997)	45-45.5	45-45.5	66@1.575	⑤	0.0008-0.0019	0.0033	0.2585-0.2891	0.2571-0.2579
	T		3.0 (2972)	44-44.5	45	74@1.591	1.590	0.0012-0.0024	0.0020-0.0035	0.3150	0.3134

79231CE4

VALVE SPECIFICATIONS

Year	Engine ID/VIN	Engine Displacement Liters (cc)	Seat Angle (deg.)	Face Angle (deg.)	Spring Test Pressure (lbs. @ in.)	Spring Installed Height (in.)	Stem-to-Guide Clearance (in.)		Stem Diameter (in.)	
							Intake	Exhaust	Intake	Exhaust
1998-99	K	1.5 (1495)	45	45	54@1.358	1.358	0.0012-0.0024	0.0014-0.0026	0.3920	0.3960
	M	1.8 (1795)	45	45	56@1.458	1.358	0.0008-0.0019	0.0019-0.0033	0.2348-0.2354	0.2334-0.2342
	F	2.0 (1975)	45	45	56@1.457	1.358	0.0008-0.0019	0.0019-0.0033	0.2348-0.2354	0.2334-0.2342
	P	2.0 (1997)	45-45.5	45-45.5	66@1.575	⑤	0.0008-0.0019	0.0020-0.0033	0.2585-0.2891	0.2571-0.2579
	T	3.0 (2972)	44-44.5	45	74@1.591	1.590	0.0012-0.0024	0.0020-0.0035	0.3150	0.3134

① Non-Turbocharged
② Turbocharged
③ Single Overhead Camshaft (SOHC)
④ Double Overhead Camshafts (DOHC)
⑤ Free Length - 1.902 in.

79231CE5

TORQUE SPECIFICATIONS
All readings in ft. lbs.

Year	Engine ID/VIN		Engine Displacement Liters (cc)	Cylinder Head Bolts	Main Bearing Bolts	Rod Bearing Bolts	Crankshaft Damper Bolts	Flywheel Bolts	Manifold		Spark Plugs	Lug Nut
									Intake	Exhaust		
1995	E	①	1.5 (1495)	⑦	40-43	25-28	140-148	94-101	11-14	11-14	18	65-80
	E	②	1.5 (1495)	⑦	40-43	25-28	140-148	94-101	11-14	18-20	18	65-80
	K		1.5 (1495)	⑦	40-44	25-28	110-118	94-101	11-14	11-14	18	65-80
	R		1.6 (1595)	76-83 ⑧	47-51	36-38	80-94	94-101	18-22	18-22	18	65-80
	M		1.8 (1836)	76-83 ⑧	47-51	36-38	80-94	94-101	18-22	18-22	18	65-80
	P		2.0 (1997)	76-83 ⑧	47-51	36-38	80-94	94-101	18-22	18-22	18	65-80
	T		3.0 (2972)	⑨	55-61	36-38	109-115	65-70	11-14	11-16	18	65-80
1996	K	③	1.5 (1495)	⑦	40-44	25-28	110-118	94-101	11-14	11-14	18	65-80
	K	④	1.5 (1495)	⑦	40-43	23-26	103-110	88-96	13-18	22-30	18	65-80
	M		1.8 (1795)	②	⑪	34-39	125-133	88-95	13-18	22-30	18	65-80
	P		2.0 (1997)	76-83 ⑧	47-51	36-38	80-94	94-101	18-22	18-22	18	65-80
	T		3.0 (2972)	⑨	55-61	36-38	109-115	65-70	11-14	11-16	18	65-80
1997	K	③	1.5 (1495)	⑦	40-44	25-28	110-118	94-101	11-14	11-14	18	65-80
	K	④	1.5 (1495)	⑦	40-43	23-26	103-110	88-96	13-18	22-30	18	65-80
	M	⑤	1.8 (1795)	⑩	⑪	34-39	125-133	88-95	13-18	22-30	18	65-80
	M	⑥	1.8 (1795)	⑩	⑪	34-39	125-133	88-95	11-14	17-22	18	65-80
	F		2.0 (1975)	⑩	⑪	34-39	125-133	88-95	11-14	17-22	18	65-80
	P		2.0 (1997)	76-83 ⑧	47-51	36-38	80-94	94-101	18-22	18-22	18	65-80
	T		3.0 (2972)	⑨	55-61	36-38	109-115	65-70	11-14	11-16	18	65-80
1998-99	K		1.5 (1495)	⑦	40-44	25-28	110-118	94-101	11-14	11-14	18	65-80
	M	⑤	1.8 (1795)	⑩	⑪	34-39	125-133	88-95	13-18	22-30	18	65-80
	M	⑥	1.8 (1795)	⑩	⑪	34-39	125-133	88-95	11-14	17-22	18	65-80
	F		2.0 (1975)	⑩	⑪	34-39	125-133	88-95	11-14	17-22	18	65-80
	P		2.0 (1997)	76-83 ⑧	47-51	36-38	80-94	94-101	18-22	18-22	18	65-80
	T		3.0 (2972)	⑨	55-61	36-38	109-115	65-70	11-14	11-16	18	65-80

① Non-Turbocharged
② Turbocharged
③ Single Overhead Camshaft (SOHC)
④ Double Overhead Camshafts (DOHC)
⑤ Elantra
⑥ Tiburon
⑦ Cold: 51-54 ft. lbs.; Warm: 58-61 ft. lbs.
⑧ Cold
⑨ Cold: 65-72 ft. lbs.; Warm: 72-80 ft. lbs.
⑩ M10 Bolts - Step 1: 22 ft. lbs.; Step 2: Plus 60-65 degrees; Step 3: Plus 60-65 degrees
   M12 Bolts - Step 1: 26 ft. lbs.; Step 2: Plus 60-65 degrees; Step 3: Plus 60-65 degrees
⑪ Step 1: 20-24 ft. lbs.; Step 2: Plus 60-65 degrees

79231CE6

SCHEDULED MAINTENANCE INTERVALS
(HYUNDAI ACCENT, ELANTRA, SCOUPE, SONATA & TIBURON)

TO BE SERVICED	TYPE OF SERVICE	VEHICLE MILEAGE INTERVAL (x1000)												
		7.5	15	22.5	30	37.5	45	52.5	60	67.5	75	82.5	90	97.5
Engine oil & filter	R	✓	✓	✓	✓	✓	✓	✓	✓	✓	✓	✓	✓	✓
Automatic transaxle fluid	S/I		✓		✓		✓		✓		✓		✓	
Brake pads, calipers & rotors	S/I		✓		✓		✓		✓		✓		✓	
Brake hoses & lines	S/I		✓		✓		✓		✓		✓		✓	
Driveshafts & boots	S/I		✓		✓		✓		✓		✓		✓	
Valve clearance (Scoupe)	S/I		✓		✓		✓		✓		✓		✓	
Wheel bearing grease	S/I				✓				✓				✓	
Air cleaner filter	R				✓				✓				✓	
Automatic transaxle fluid & filter	R				✓				✓				✓	
Brake fluid	R				✓				✓				✓	
Engine Coolant	R				✓				✓				✓	
Fuel hose, vapor hose & fuel filler cap	S/I							✓						
Spark plugs	R				✓				✓				✓	
Spark plugs (Sonata 3.0L V6)	R								✓					
Bolts & nuts on chassis & body (Accent)	S/I				✓				✓				✓	
Drive belts	S/I				✓				✓				✓	
Exhaust pipe connections, muffler & suspension bolts	S/I				✓				✓				✓	
Manual transaxle oil	S/I				✓				✓				✓	
Rear brake drums, linings & parking brake	S/I				✓				✓				✓	
Steering gear rack, linkage & boots	S/I				✓				✓				✓	
Suspension ball joints & dust covers (Accent)	S/I				✓				✓				✓	

79231CE7

SCHEDULED MAINTENANCE INTERVALS
(HYUNDAI ACCENT, ELANTRA, SCOUPE, SONATA & TIBURON) (Cont.)

TO BE SERVICED	TYPE OF SERVICE	VEHICLE MILEAGE INTERVAL (x1000)												
		7.5	15	22.5	30	37.5	45	52.5	60	67.5	75	82.5	90	97.5
Timing belt (1996-99 Accent & Elantra)	S/I				✓				✓				✓	
Timing belt (except 1996-99 Accent & Elantra)	R								✓					
Fuel filter	R							✓						
Crankcase emission control system (carburetor)	S/I								✓					
Fuel lines & connections	S/I								✓					
Vacuum & crankcase ventilation hoses	S/I								✓					

R – Replace   S/I – Service or Inspect

FREQUENT OPERATION MAINTENANCE (SEVERE SERVICE)

If a vehicle is operated under any of the following conditions it is considered severe service:
- Extremely dusty areas.
- 50% or more of the vehicle operation is in 32°C (90°F) or higher temperatures, or constant operation in temperatures below 0°C (32°F).
- Prolonged idling (vehicle operation in stop and go traffic).
- Frequent short running periods (engine does not warm to normal operating temperatures).
- Police, taxi, delivery usage or trailer towing usage.

Oil & oil filter change – change every 3000 miles.
Brake pads, calipers & rotors - service or inspect every 7500 miles.
Driveshaft boots - service or inspect every 7500 miles.
Steering gear rack, linkage & boots - service or inspect every 7500 miles.
Air cleaner filter – service or inspect every 15,000 miles.
Automatic transaxle fluid & filter - replace every 15,000 miles.
Rear brake drums & linings - service or inspect every 15,000 miles.
Spark plugs - replace every 24,000 miles.
Crankcase emission control system (carburetor) - service or inspect every 30,000 miles.

79231CE8

SCHEDULED MAINTENANCE
HYUNDAI
ACCENT, ELANTRA, SCOUPE, SONATA, TIBURON

The following should be used as a guide when determining the amount of work required for a particular service if taken to a repair shop. In estimating how long a particular Scheduled Maintenance Service should take, please observe the following:

- **Chilton Time** is time based on field research and data supplied by the vehicle manufacturer.
- Labor time operations are given in hours and tenths of an hour.
- All labor operations, are to be used as a guide.

Mechanic Skill Level Codes:
(G) GENERAL: Normally skilled with certification.
(M) MAINTENANCE: Semi-skilled working on certication.
(P) PRECISION: Really skilled with multiple certification.

	Chilton Time		Chilton Time		Chilton Time
(M) 7500 Mile Service		**(M) 45000 Mile Service**		**(M) 67500 Mile Service**	
1995-99	.3	1995-99 Accent, Elantra	.7	1995-99	.3
(M) 15000 Mile Service		1995 Scoupe	.9	**(M) 75000 Mile Service**	
1995-99 Accent, Elantra	.6	1995-99 Sonata	.7	1995-99 Accent, Elantra	.7
1995 Scoupe	.8	1997-99 Tiburon.............	.7	1995 Scoupe	.9
1995-99 Sonata	.7	w/AT add	.1	1995-99 Sonata	.7
1997-99 Tiburon.............	.6	**(M) 52500 Mile Service**		1997-99 Tiburon.............	.7
w/AT add	.1	1995-99	.7	w/AT add	.1
(M) 22500 Mile Service		**(G) 60000 Mile Service**		**(M) 82500 Mile Service**	
1995-99	.3	1995-99 Accent, Elantra	4.3	1995-99	.3
(G) 30000 Mile Service		1995 Scoupe	5.7	**(G) 90000 Mile Service**	
1995-99 Accent, Elantra	3.4	1995-99 Sonata	5.5	1995-99 Accent, Elantra	3.6
1995 Scoupe	3.4	w/3.0L add	.5	1995 Scoupe	3.4
1995-99 Sonata	3.2	1997-99 Tiburon.............	5.1	1995-99 Sonata	3.2
1997-99 Tiburon.............	3.4	w/AT add	.6	1997-99 Tiburon.............	3.6
w/AT add	.6			w/AT add	.6
(M) 37500 Mile Service				**(M) 97500 Mile Service**	
1995-99	.3			1995-99	.3

79231CE9

INFINITI
G20 • I30 • J30 • Q45

VEHICLE IDENTIFICATION CHART

Engine Code						Model Year	
Code	Liters	Cu. In. (cc)	Cyl.	Fuel Sys.	Eng. Mfg.	Code	Year
SR20DE	2.0	122 (1998)	4	MFI	Nissan	S	1995
VG30DE	3.0	181 (2960)	6	MFI	Nissan	T	1996
VH41DE	4.1	252 (4130)	8	MFI	Nissan	V	1997
VH45DE	4.5	274 (4494)	8	MFI	Nissan	W	1998
VQ30DE	3.0	182 (2988)	6	MFI	Nissan	X	1999

MFI - Multi-port Fuel Injection

79231CB9

ENGINE IDENTIFICATION

Year	Model	Engine Displacement Liters (cc)	Engine Series (ID/VIN)	Fuel System	No. of Cylinders	Engine Type
1995	G20	2.0 (1998)	SR20DE (C)	MFI	4	DOHC
	J30	3.0 (2960)	VG30DE (A)	MFI	6	DOHC
	Q45	4.5 (4494)	VH45DE (N)	MFI	8	DOHC
1996	G20	2.0 (1998)	SR20DE (C)	MFI	4	DOHC
	J30	3.0 (2960)	VG30DE (A)	MFI	6	DOHC
	I30	3.0 (2988)	VQ30DE (C)	MFI	6	DOHC
	Q45	4.5 (4494)	VH45DE (N)	MFI	8	DOHC
1997	J30	3.0 (2960)	VG30DE (A)	MFI	6	DOHC
	I30	3.0 (2988)	VQ30DE (C)	MFI	6	DOHC
	Q45	4.1 (4130)	VH41DE (B)	MFI	8	DOHC
1998-99	I30	3.0 (2988)	VQ30DE (C)	MFI	6	DOHC
	Q45	4.1 (4130)	VH41DE (B)	MFI	8	DOHC

MFI - Multi-port Fuel Injection
DOHC - Double Overhead Camshaft

79231CB0

GENERAL ENGINE SPECIFICATIONS

Year	Engine ID/VIN	Engine Displacement Liters (cc)	Fuel System Type	Net Horsepower @ rpm	Net Torque @ rpm (ft. lbs.)	Bore x Stroke (in.)	Compression Ratio	Oil Pressure @ rpm
1995	C	2.0 (1998)	MFI	140@6400	132@4800	3.39x3.39	9.5:1	46-57@3200
	A	3.0 (2960)	MFI	210@6400	193@4800	3.43x3.27	10.5:1	51-65@3000
	N	4.5 (4494)	MFI	278@6000	292@4000	3.66x3.26	10.2:1	67-81@3000
1996	C	2.0 (1998)	MFI	140@6400	132@4800	3.39x3.39	9.5:1	46-57@3200
	A	3.0 (2960)	MFI	210@6400	193@4800	3.43x3.27	10.5:1	51-65@3000
	C	3.0 (2988)	MFI	190@5600	205@4000	3.66x2.89	10.1:1	63-80@3000
	N	4.5 (4494)	MFI	278@6000	292@4000	3.66x3.26	10.2:1	67-81@3000
1997	A	3.0 (2960)	MFI	210@6400	193@4800	3.43x3.27	10.5:1	51-65@3000
	C	3.0 (2988)	MFI	190@5600	205@4000	3.66x2.89	10.1:1	63-80@3000
	B	4.1 (4130)	MFI	266@5600	278@4000	3.66x2.99	10.2:1	67-81@3000
1998-99	C	3.0 (2988)	MFI	190@5600	205@4000	3.66x2.89	10.1:1	63-80@3000
	B	4.1 (4130)	MFI	266@5600	278@4000	3.66x2.99	10.2:1	67-81@3000

79231CC1

GASOLINE ENGINE TUNE-UP SPECIFICATIONS

Year	Engine ID/VIN	Engine Displacement Liters (cc)	Spark Plugs Gap (in.)	Ignition Timing (deg.) MT	Ignition Timing (deg.) AT	Fuel Pump (psi)	Idle Speed (rpm) MT	Idle Speed (rpm) AT	Valve Clearance In.	Valve Clearance Ex.
1995	C	2.0 (1998)	0.039-0.041	15B	15B	34 ①	800	800	HYD	HYD
	A	3.0 (2960)	0.039-0.041	—	15B	34 ①	—	720	HYD	HYD
	N	4.5 (4494)	0.039-0.041	—	15B	34 ①	—	650	HYD	HYD
1996	C	2.0 (1998)	0.039-0.041	15B	15B	34 ①	800	800	HYD	HYD
	A	3.0 (2960)	0.039-0.041	—	15B	34 ①	—	720	HYD	HYD
	C	3.0 (2988)	0.039-0.043	15B	15B	34 ①	625	700	HYD	HYD
	N	4.5 (4494)	0.039-0.041	—	15B	34 ①	—	650	HYD	HYD
1997	A	3.0 (2960)	0.039-0.041	—	15B	34 ①	—	720	HYD	HYD
	C	3.0 (2988)	0.039-0.043	15B	15B	34 ①	625	700	HYD	HYD
	B	4.1 (4130)	0.039-0.041	—	15B	34 ①	—	650	HYD	HYD
1998-99	C	3.0 (2988)	0.039-0.043	15B	15B	34 ①	625	700	HYD	HYD
	B	4.1 (4130)	0.039-0.041	—	15B	34 ①	—	650	HYD	HYD

NOTE: The Vehicle Emission Control Information label often reflects specification changes made during production. The label figures must be used if they differ from those in this chart.

B - Before top dead center

① 43 psi with regulator vacuum hose disconnected

79231CC2

CAPACITIES

Year	Model	Engine ID/VIN	Engine Displacement Liters (cc)	Engine Oil with Filter	Transmission (pts.) 4-Spd	5-Spd	Auto.	Transfer Case (pts.)	Drive Axle Front (pts.)	Rear (pts.)	Fuel Tank (gal.)	Cooling System (qts.)
1995	G20	C	2.0 (1998)	3.60	—	7.60	7.40	—	—	—	15.9	6.25
	J30	A	3.0 (2960)	4.50	—	—	8.75	—	—	3.20	19.0	9.75
	Q45	N	4.5 (4494)	6.35	—	—	10.8	—	—	3.20	22.5	10.9
1996	G20	C	2.0 (1998)	3.60	—	7.6	14.8	—	—	—	15.9	6.25
	J30	A	3.0 (2960)	4.50	—	—	17.5	—	—	3.20	19.0	9.8
	I30	C	3.0 (2988)	4.25	—	①	19.8	—	—	—	18.5	9.0
	Q45	N	4.5 (4494)	6.35	—	—	22.2	—	—	2.80	22.5	10.9
1997	J30	A	3.0 (2960)	4.50	—	—	17.5	—	—	3.20	19.0	9.8
	I30	C	3.0 (2988)	4.25	—	①	19.8	—	—	—	18.5	9.0
	Q45	B	4.1 (4130)	6.35	—	—	22.2	—	—	2.80	22.5	10.9
1998-99	I30	C	3.0 (2988)	4.25	—	①	19.8	—	—	—	18.5	9.0
	Q45	B	4.1 (4130)	6.35	—	—	22.2	—	—	2.80	22.5	10.9

NOTE: All capacities are approximate. Add fluid gradually and check to be sure a proper fluid level is obtained.

① RSF50V: 9.13-9.50

RSF50A: 9.50-10.13

79231CC3

VALVE SPECIFICATIONS

Year	Engine ID/VIN	Engine Displacement Liters (cc)	Seat Angle (deg.)	Face Angle (deg.)	Spring Test Pressure (lbs. @ in.)	Spring Free Height (in.)	Stem-to-Guide Clearance (in.)		Stem Diameter (in.)	
							Intake	Exhaust	Intake	Exhaust
1995	C	2.0 (1998)	45.25-45.75	44.85-45.10	127.9-144.3@1.181	1.943	0.0008-0.0021	0.0016-0.0029	0.2348-0.2354	0.2341-0.2346
	A	3.0 (2960)	45.25-45.75	45	120.6@1.043	1.697	0.0008-0.0021	0.0016-0.0029	0.2348-0.2354	0.2341-0.2346
	N	4.5 (4494)	45.25-45.75	44.85-45.10	120.4@1.055	1.862	0.0011-0.0020	0.0014-0.0020	0.2743-0.2744	0.3134-0.3136
1996	C	2.0 (1998)	45.25-45.75	44.85-45.10	127.9-144.3@1.181	1.943	0.0008-0.0021	0.0016-0.0029	0.2348-0.2354	0.2341-0.2346
	A	3.0 (2960)	45.25-45.75	45	120.6@1.043	1.697	0.0008-0.0021	0.0016-0.0029	0.2348-0.2354	0.2341-0.2346
	C	3.0 (2988)	45.25-45.75	NA	120.1@1.085	1.845	0.0008-0.0021	0.0016-0.0029	0.2348-0.2354	0.2341-0.2346
	N	4.5 (4494)	45.25-45.75	44.85-45.10	120.4@1.055	1.946	0.0011-0.0020	0.0014-0.0020	0.2743-0.2744	0.3134-0.3136
1997	A	3.0 (2960)	45.25-45.75	45	120.6@1.043	1.697	0.0008-0.0021	0.0016-0.0029	0.2348-0.2354	0.2341-0.2346
	C	3.0 (2988)	45.25-45.75	NA	120.1@1.085	1.845	0.0008-0.0021	0.0016-0.0029	0.2348-0.2354	0.2341-0.2346
	B	4.1 (4130)	45.25-45.75	44.85-45.10	120.4@1.055	1.946	0.0011-0.0020	0.0014-0.0020	0.2743-0.2744	0.3134-0.3136
1998-99	C	3.0 (2988)	45.25-45.75	NA	120.1@1.085	1.845	0.0008-0.0021	0.0016-0.0029	0.2348-0.2354	0.2341-0.2346
	B	4.1 (4130)	45.25-45.75	44.85-45.10	120.4@1.055	1.946	0.0011-0.0020	0.0014-0.0020	0.2743-0.2744	0.3134-0.3136

NA: Not Available

① Inner: 57.3@0.984

Outer: 117.7@1.181

79231CC4

TORQUE SPECIFICATIONS
All readings in ft. lbs.

Year	Engine ID/VIN	Engine Displacement Liters (cc)	Cylinder Head Bolts	Main Bearing Bolts	Rod Bearing Bolts	Crankshaft Damper Bolts	Flywheel Bolts	Manifold		Spark Plugs	Lug Nut
								Intake	Exhaust		
1995	C	2.0 (1998)	①	②	③	105-112	61-69	13-15	27-35	14-22	72-87
	A	3.0 (2960)	⑥	67-74	⑦	159-174	61-69	12-15	17-20	14-22	72-87
	N	4.5 (4494)	④	⑤	③	260-275	61-69	12-15	20-23	14-22	72-87
1996	C	2.0 (1998)	①	②	③	105-112	61-69	13-15	27-35	14-22	72-87
	A	3.0 (2960)	⑥	67-74	⑦	159-174	61-69	12-15	17-20	14-22	72-87
	C	3.0 (2988)	⑧	⑨	⑩	⑪	61-69	20-23 ⑫	21-24	14-22	72-87
	N	4.5 (4494)	④	⑬	⑦	260-275	61-69	12-15	20-23	14-22	72-87
1997	A	3.0 (2960)	⑥	67-74	⑦	159-174	61-69	12-15	17-20	14-22	72-87
	C	3.0 (2988)	⑧	⑨	⑩	⑪	61-69	20-23 ⑫	21-24	14-22	72-87
	B	4.1 (4130)	④	⑬	⑦	260-275	61-69	12-15	20-23	14-22	72-87
1998-99	C	3.0 (2988)	⑧	⑨	⑩	⑪	61-69	20-23 ⑫	21-24	14-22	72-87
	B	4.1 (4130)	④	⑬	⑦	260-275	61-69	12-15	20-23	14-22	72-87

① Step 1: 29 ft. lbs.
Step 2: 58 ft. lbs.
Step 3: Loosen bolts completely
Step 4: 25-33 ft. lbs.
Step 5: Tighten an additional 90-100 degrees
Step 6: Repeat Step 5.

② Step 1: 24-28 ft. lbs.
Step 2: Tighten an additional 45-50
degrees or 54-61 ft. lbs.

③ Step 1: 10-12 ft. lbs.
Step 2: Tighten an additi
degrees or 28-33 ft. lbs.

④ Step 1: 22 ft. lbs.
Step 2: 69 ft. lbs.
Step 3: Loosen bolts completely
Step 4: 18-25 ft. lbs.
Step 5: Tighten an additional 90-95
degrees or 69-72 ft. lbs.

⑤ See text

⑥ Step 1: 29 ft. lbs.
Step 2: 90 ft. lbs.
Step 3: Loosen bolts completely
Step 4: 25-33 ft. lbs.
Step 5: Tighten an additional 70-75
degrees or 90 ft. lbs.

⑦ Step 1: 10-12 ft. lbs.
Step 2: Tighten an additional 60-65
degrees or 43-48 ft. lbs.

⑧ Step 1: 72 ft. lbs.
Step 2: Loosen bolts completely
Step 3: 25-33 ft. lbs.
Step 4: Tighten an additional 90-95 degrees
Step 5: Repeat Step 4

⑨ Step 1: Shift crankshaft to align the bearing beam
Step 2: Tighten all bolts to 24-28 ft. lbs.
Step 3: Tighten an additional 90-95 degrees

⑩ Step 1: Tighten all nuts to 15 ft. lbs.
Step 2: Tighten an additional 90-95 degrees

⑪ Step 1: 29-36 ft. lbs.
Step 2: Tighten an additional 60-66 degrees

⑫ Intake collector: 13-16 ft. lbs.

⑬ Step 1: Shift crankshaft back and forth to seat bearing caps
Step 2: Tighten inner cap bolts to 27-31 ft. lbs.
Step 3: Tighten outer cap bolts to 20-24 ft. lbs.
Step 4: Tighten No. 1-3, 5 inner cap bolts an additional 60 degrees
Step 5: Tighten No. 4 inner cap bolt and additional 35 degrees
Step 6: Tighten outer cap bolts an additional 35 degrees
Step 7: Tighten bearing cap side bolts to 34-38 ft. lbs.

79231CC5

SCHEDULED MAINTENANCE INTERVALS
(INFINITI G20, I30, J30 & Q45)

TO BE SERVICED	TYPE OF SERVICE	7.5	15	22.5	30	37.5	45	52.5	60	67.5	75	82.5	90	97.5
Engine oil & filter	R	✓	✓	✓	✓	✓	✓	✓	✓	✓	✓	✓	✓	✓
Automatic transaxle fluid	S/I		✓		✓		✓		✓		✓		✓	
Brake lines & cables	S/I		✓		✓		✓		✓		✓		✓	
Brake pads & discs	S/I		✓		✓		✓		✓		✓		✓	
Differential gear oil (J30 & Q45)	S/I		✓		✓		✓		✓		✓		✓	
Driveshaft boots (I30 & G20)	S/I		✓		✓		✓		✓		✓		✓	
Full-active suspension fluid (Q45)①	S/I		✓		✓		✓		✓		✓		✓	
Manual transaxle oil (G20 & I30)	S/I		✓		✓		✓		✓		✓		✓	
Air filter element	R				✓				✓				✓	
Exhaust system	S/I				✓				✓				✓	
Fuel lines	S/I				✓				✓				✓	
Steering gear linkage axle & suspension parts	S/I				✓				✓				✓	
SUPER HICAS linkage (J30 & Q45)	S/I				✓				✓				✓	
Vapor lines	S/I				✓				✓				✓	
Engine Coolant	R								✓				✓	
Spark plugs	R								✓					
Timing belt	R								✓					
Drive belts	S/I								✓					

① Replace at 60,000 miles (if not previously replaced).
R – Replace   S/I – Service or Inspect

FREQUENT OPERATION MAINTENANCE (SEVERE SERVICE)
If a vehicle is operated under any of the following conditions it is considered severe service:
- Extremely dusty areas.
- 50% or more of the vehicle operation is in 32°C (90°F) or higher temperatures, or constant operation in temperatures below 0°C (32°F).
- Prolonged idling (vehicle operation in stop and go traffic).
- Frequent short running periods (engine does not warm to normal operating temperatures).
- Police, taxi, delivery usage or trailer towing usage.
Oil & oil filter change – change every 3750 miles.
Brake pads & discs - service or inspect every 7500 miles.
Driveshaft boots (G20 & I30) - service or inspect every 7500 miles.
Exhaust system - service or inspect every 7500 miles.
Steering gear, linkage, axle & suspension parts - service or inspect every 7500 miles.
Steering linkage ball joints & front suspension ball joints - service or inspect every 7500 miles.
SUPER HICAS linkage (J30 & Q45) - service or inspect every 7500 miles.

79231CC6

SCHEDULED MAINTENANCE
INFINITI
G20, I30, J30, Q45,

The following should be used as a guide when determining the amount of work required for a particular service if taken to a repair shop. In estimating how long a particular Scheduled Maintenance Service should take, please observe the following:

- **Chilton Time** is time based on field research and data supplied by the vehicle manufacturer.
- Labor time operations are given in hours and tenths of an hour.
- All labor operations, are to be used as a guide.

Mechanic Skill Level Codes:
- **(G)** GENERAL: Normally skilled with certification.
- **(M)** MAINTENANCE: Semi-skilled working on certication.
- **(P)** PRECISION: Really skilled with multiple certification.

	Chilton Time
(M) 7500 Mile Service	
1995-99	.5
(G) 15000 Mile Service	
1995-99 G20, Q45	.9
1995-99 J30, I30	.8
(G) 22500 Mile Service	
1995-99	.3
(G) 30000 Mile Service	
1995-99 G20, I30, J30	1.6
1995-99 Q45	1.5

	Chilton Time
(M) 37500 Mile Service	
1995-99	.5
(G) 45000 Mile Service	
1995-99 G20, Q45	.9
1995-99 J30, I30	.8
(M) 52500 Mile Service	
1995-99	.5
(G) 60000 Mile Service	
1995-99 G20, I30, J30	7.2
1995-99 Q45	7.8
(M) 67500 Mile Service	
1995-99	.5

	Chilton Time
(G) 75000 Mile Service	
1995-99 G20, Q45	.9
1995-99 J30, I30	.8
(M) 82500 Mile Service	
1995-99	.5
(G) 90000 Mile Service	
1995-99 G20, I30, J30	2.1
1995-99 Q45	2.2
(M) 97500 Mile Service	
1995-99	.5

79231CC8

KIA
Sephia

VEHICLE IDENTIFICATION CHART

Engine Code							Model Year	
Code	Liters	Cu. In. (cc)	Cyl.	Fuel Sys.	Eng. Mfg.		Code	Year
1	1.6	97.4 (1597)	4	MFI	Mazda		S	1995
1	1.8	109 (1793)	4	EGI	KIA		T	1996
3	1.6	97.4 (1597)	4	MFI	Mazda		V	1997
4	1.6	97.4 (1597)	4	MFI	Mazda		W	1998
5	1.8	112.2 (1839)	4	MFI	Mazda		X	1999

MFI - Multi-port Fuel Injection
EGI - Electronic Gasoline Injection

79231C23

ENGINE IDENTIFICATION

Year	Model	Engine Displacement Liters (cc)	Engine Series (ID/VIN)	Fuel System	No. of Cylinders	Engine Type
1995	Sephia	1.6 (1597)	B6 / 1	MFI	4	SOHC
		1.6 (1597)	B6 / 3	MFI	4	DOHC
		1.8 (1839)	BP / 5	MFI	4	DOHC
1996	Sephia	1.6 (1597)	B6 / 4	MFI	4	DOHC
		1.8 (1839)	BP / 5	MFI	4	DOHC
1997	Sephia	1.6 (1597)	B6 / 4	MFI	4	DOHC
		1.8 (1839)	BP / 5	MFI	4	DOHC
1998-99	Sephia	1.8 (1793)	TE / 1	EGI	4	DOHC

EGI - Electronic Gasoline Injection
MFI - Multi-port Fuel Injection
SOHC - Single Overhead Camshaft
DOHC - Double Overhead Camshafts

79231C24

GENERAL ENGINE SPECIFICATIONS

Year	Engine ID/VIN	Engine Displacement Liters (cc)	Fuel System Type	Net Horsepower @ rpm	Net Torque @ rpm (ft. lbs.)	Bore x Stroke (in.)	Compression Ratio	Oil Pressure @ rpm
1995	B6 / 1	1.6 (1597)	MFI	82@5000 ①	92@2500 ②	3.07x3.29	9.3:1	43-57@3000
	B6 / 3	1.6 (1597)	MFI	105@6200	100@3600	3.07x3.29	9.0:1	43-57@3000
	BP / 5	1.8 (1839)	MFI	126@6500	114@4500	3.27x3.35	9.0:1	43-57@3000
1996	B6 / 4	1.6 (1597)	MFI	105@6200	100@3600	3.07x3.29	9.0:1	43-57@3000
	BP / 5	1.8 (1839)	MFI	126@6500	114@4500	3.27x3.35	9.0:1	43-57@3000
1997	B6 / 4	1.6 (1597)	MFI	105@6200	100@3600	3.07x3.29	9.0:1	43-57@3000
	BP / 5	1.8 (1839)	MFI	126@6500	114@4500	3.27x3.35	9.0:1	43-57@3000
1998-99	TE / 1	1.8 (1793)	EGI	125@6000	108@4500	3.19x3.43	9.4:1	64-78@3000

EGI - Electronic Gasoline Injection
MFI - Multi-port Fuel Injection

① California: 88@5000
② California: 98@4000
③ California: 114@5500
④ California: 129@4500
⑤ California: 128@6000
⑥ California: 160@5500
⑦ California and New York: 156@5000
    Except California and New York: 160@4800
⑧ California, New York and Massachusetts: 90@5500
    Except California, New York and Massachusetts: 92@5500
⑨ Protege: 122@6000, Miata: 133@6500
⑩ Protege: 117@4000, Miata: 114@5000

79241C25

GASOLINE ENGINE TUNE-UP SPECIFICATIONS

Year	Engine ID/VIN	Engine Displacement Liters (cc)	Spark Plug Gap (in.)	Ignition Timing (deg.) MT	Ignition Timing (deg.) AT	Fuel Pump (psi)	Idle Speed (rpm) MT	Idle Speed (rpm) AT	Valve Clearance In.	Valve Clearance Ex.
1995	B6 / 1	1.6 (1597)	0.041	6-8B	6-8B	30-38 ①	750	750	HYD	HYD
	B6 / 3	1.6 (1597)	0.041	9-11B	9-11B	30-38 ①	700	750	HYD	HYD
	BP / 5	1.8 (1839)	0.041	9-11B	9-11B	30-38 ①	750	750	HYD	HYD
1996	B6 / 4	1.6 (1597)	0.041	9-11B	9-11B	30-38 ①	700	750	HYD	HYD
	BP / 5	1.8 (1839)	0.041	9-11B	9-11B	39-45 ①	750	750	HYD	HYD
1997	B6 / 4	1.6 (1597)	0.041	9-11B	9-11B	30-38 ①	700	750	HYD	HYD
	BP / 5	1.8 (1839)	0.041	9-11B	9-11B	39-45 ①	750	750	HYD	HYD
1998-99	TE / 1	1.8 (1793)	0.028-0.032	3-13B	3-13B	64	750-850	750-850	HYD	HYD

NOTE: The Vehicle Emission Control Information label often reflects specification changes made during production. The label figures must be used if they differ from those in this chart.

B - Before top dead center

HYD - Hydraulic

① Pressure indicated is with gauge in-line, regulator vacuum hose connected and engine idling

79231C26

CAPACITIES

Year	Model	Engine ID/VIN	Engine Displacement Liters (cc)	Engine Oil with Filter	Transmission (pts.) 4-Spd	Transmission (pts.) 5-Spd	Transmission (pts.) Auto.	Transfer Case (pts.)	Drive Axle Front (pts.)	Drive Axle Rear (pts.)	Fuel Tank (gal.)	Cooling System (qts.)
1995	Sephia	B6 / 1	1.6 (1597)	3.6	—	5.6	11.4	—	—	—	13.2	6.3
	Sephia	B6 / 3	1.6 (1597)	3.6	—	5.6	11.4	—	—	—	13.2	6.3
	Sephia	BP / 5	1.8 (1839)	4.0	—	5.6	11.4	—	—	—	12.7	6.3
1996	Sephia	B6 / 4	1.6 (1597)	3.6	—	5.6	11.4	—	—	—	13.2	6.3
	Sephia	BP / 5	1.8 (1839)	4.0	—	5.6	11.4	—	—	—	12.7	6.3
1997	Sephia	B6 / 4	1.6 (1597)	3.6	—	5.6	11.4	—	—	—	13.2	6.3
	Sephia	BP / 5	1.8 (1839)	4.0	—	5.6	11.4	—	—	—	①	6.3
1998-99	Sephia	TE / 1	1.8 (1793)	4.0	—	5.6	11.4	—	—	—	13.2	6.3

NOTE: All capacities are approximate. Add fluid gradually and ensure a proper level is obtained.

① Federal models: 13.2 gal.

Ca lifornia models: 12.7 gal.

79231C27

VALVE SPECIFICATIONS

Year	Engine ID/VIN	Engine Displacement Liters (cc)	Seat Angle (deg.)	Face Angle (deg.)	Maximum out of Square (in.)	Spring Free Length (in.)	Stem-to-Guide Clearance (in.)		Stem Diameter (in.)	
							Intake	Exhaust	Intake	Exhaust
1995	B6 / 1	1.6 (1597)	45	45	0.0670	1.570	0.0010-0.0023	0.0012-0.0025	0.2351-0.2356	0.2349-0.2354
	B6 / 3	1.6 (1597)	45	45	0.0670	1.570	0.0010-0.0023	0.0012-0.0025	0.2351-0.2356	0.2349-0.2354
	BP / 5	1.8 (1839)	45	45	0.0640	1.560	0.0010-0.0023	0.0012-0.0025	0.2351-0.2356	0.2349-0.2354
1996	BP / 5	1.8 (1839)	45	45	0.059-0.066	①	0.0010-0.0023	0.0012-0.0025	0.2351-0.2356	0.2349-0.2354
	B6 / 4	1.6 (1597)	45	45	0.0670	1.570	0.0010-0.0023	0.0012-0.0025	0.2351-0.2356	0.2349-0.2354
1997	BP / 5	1.8 (1839)	45	45	0.059-0.066	①	0.0010-0.0023	0.0012-0.0025	0.2351-0.2356	0.2349-0.2354
	B6 / 4	1.6 (1597)	45	45	0.0670	1.570	0.0010-0.0023	0.0012-0.0025	0.2351-0.2356	0.2349-0.2354
1998-99	TE / 1	1.8 (1793)	45	45	0.0638	1.840	B	C	0.2350-0.2356	0.2348-0.2354

NA - Not Available
① Intake: 1.80 in.
   Exhaust: 1.903 in.
② Standard range: 0.0010-0.0023 in.
   Maximum value: 0.0080 in.
③ Standard range: 0.0012-0.0025 in.
   Maximum value: 0.0080 in.

79231C28

TORQUE SPECIFICATIONS
All readings in ft. lbs.

Year	Engine ID/VIN	Engine Displacement Liters (cc)	Cylinder Head Bolts	Main Bearing Bolts	Rod Bearing Bolts	Crankshaft Damper Bolts	Flywheel Bolts	Manifold		Spark Plugs	Lug Nut
								Intake	Exhaust		
1995	B6 / 1	1.6 (1597)	56-60	40-43	35-37	116-123	71-76	14-19	12-17	11-17	65-87
	B6 / 3	1.6 (1597)	56-60	40-43	37-40	116-123	71-76	14-19	28-34	11-17	65-87
	BP / 5	1.8 (1839)	56-60	40-43	35-37	116-123	71-76	14-19	28-34	11-17	65-87
1996	BP / 5	1.8 (1839)	56-60	40-43	35-36	116-122	71-76	14-18	29-34	11-16	65-87
	B6 / 4	1.6 (1597)	56-60	40-43	37-40	116-123	71-76	14-19	28-34	11-17	65-87
1997	BP / 5	1.8 (1839)	56-60	40-43	35-36	116-122	71-76	14-18	29-34	11-16	65-87
	B6 / 4	1.6 (1597)	56-60	40-43	37-40	116-123	71-76	14-19	28-34	11-17	65-87
1998-99	TE / 1	1.8 (1793)	①	②	35-37	9-13 ③	71-76	14-19	28-34	17-Nov	65-87

① Step 1: 36 ft. lbs.
   Step 2: Loosen fully
   Step 3: 29 ft. lbs.
   Step 4: Tighten 90 degrees
   Step 5: Additional 90 degrees.

② Step 1: 29 ft. lbs.
   Step 2: Loosen fully
   Step 3: 14.5 ft. lbs.
   Step 4: Tighten 90 degrees
   Step 5: Tighten 60 degrees

③ Crankshaft pulley

79231C29

SCHEDULED MAINTENANCE INTERVALS
(KIA SEPHIA)

TO BE SERVICED	TYPE OF SERVICE	VEHICLE MILEAGE INTERVAL (x1000)															
		7.5	15	22.5	30	37.5	45	52.5	60	67.5	75	82.5	90	97.5	105	112.5	120
Accessory drive belts	S/I				✓				✓				✓				✓
Air cleaner element	R				✓				✓				✓				✓
Air conditioner system ①	S/I	Inspect the system operation and refrigerant amount annually.															
Brake lines, hoses and connections	S/I				✓				✓				✓				✓
Chassis and body fasteners	T				✓				✓				✓				✓
Clutch pedal height, free-play and operation	S/I				✓				✓				✓				✓
Cooling system hoses and coolant level	S/I				✓				✓				✓				✓
CV-joint boots	S/I				✓				✓				✓				✓
Engine coolant	R				✓				✓				✓				✓
Engine oil and filter	R	✓	✓	✓	✓	✓	✓	✓	✓	✓	✓	✓	✓	✓	✓	✓	✓
Exhaust system heat shields	S/I				✓				✓				✓				✓
Front and rear brakes	S/I				✓				✓				✓				✓
Front ball joints	S/I				✓				✓				✓				✓
Fuel filter	R								✓								✓
Fuel lines and hoses	S/I				✓				✓				✓				✓
Idle speed	A				✓				✓				✓				✓
Locks and hinges	L	✓	✓	✓	✓	✓	✓	✓	✓	✓	✓	✓	✓	✓	✓	✓	✓
Spark plugs	R				✓				✓				✓				✓
Steering operation and linkage	S/I				✓				✓				✓				✓
Timing belt (California models)	R														✓		
Timing belt (California models)	S/I								✓				✓				
Timing belt (except California models)	R								✓								✓

① This must only be done by a qualified automotive technician trained on MVAC systems.

R - Replace   S/I - Inspect and service, if needed   L - Lubricate   A - Adjust   T - Tighten

FREQUENT OPERATION MAINTENANCE (SEVERE SERVICE) ADDITIONS

If a vehicle is operated under any of the following conditions it is considered severe service:

- Towing a trailer or using a camper or car-top carrier.
- Repeated short trips of less than 5 miles in temperatures below freezing, or trips of less than 10 miles in any temperature.
- Extensive idling or low-speed driving for long distances as in heavy commercial use, such as delivery, taxi or police cars.
- Operating on rough, muddy or salt-covered roads.

79231C30

SCHEDULED MAINTENANCE INTERVALS
(KIA SEPHIA)

- **Operating on unpaved or dusty roads.**
- **Driving in extremely hot (over 90°) conditions.**

Engine oil and filter - replace every 5,000 miles or 5 months, whichever occurs first.

Air cleaner element - inspect every 15,000 miles or 15 months and replace every 30,000 miles or 30 months, whichever occurs first.

Fuel system hoses (California models only) - replace every 105,000 miles.

Emission system hoses (non-California models) - inspect every 55,000 miles or 55 months, whichever occurs first.

Emission system hoses (California models) - inspect every 60,000 miles or 60 months, whichever occurs first.

Front and rear disc brakes - inspect every 15,000 miles or 15 months, whichever occurs first.

Chassis and body fasteners - tighten every 15,000 miles or 15 months, whichever occurs first.

Locks and hinges - lubricate every 5,000 miles or 5 months, whichever occurs first.

79231C31

SCHEDULED MAINTENANCE
KIA
SEPHIA

The following should be used as a guide when determining the amount of work required for a particular service if taken to a repair shop. In estimating how long a particular Scheduled Maintenance Service should take, please observe the following:

- **Chilton Time** is time based on field research and data supplied by the vehicle manufacturer.
- Labor time operations are given in hours and tenths of an hour.
- All labor operations, are to be used as a guide.

Mechanic Skill Level Codes:
(G) GENERAL: Normally skilled with certification.
(M) MAINTENANCE: Semi-skilled working on certication.
(P) PRECISION: Really skilled with multiple certification.

	Chilton Time		Chilton Time		Chilton Time
(M) 7500 Mile Service		**(M) 52500 Mile Service**		**(G) 90000 Mile Service**	
1995-99	.5	1995-99	.5	1995-99	2.3
(M) 15000 Mile Service		**(G) 60000 Mile Service**		**(M) 97500 Mile Service**	
1995-99	.5	1995-99	4.0	1995-99	.5
(M) 22500 Mile Service		Calif.	2.2	**(G) 105000 Mile Service**	
1995-99	.5	**(M) 67500 Mile Service**		1995-99	.5
(G) 30000 Mile Service		1995-99	.5	Calif. models replace timing	
1995-99	2.0	**(M) 75000 Mile Service**		belt add	2.0
(M) 37500 Mile Service		1995-99	.5	**(M) 112500 Mile Service**	
1995-99	.5	**(M) 82500 Mile Service**		1995-99	.5
(M) 45000 Mile Service		1995-99	.5	**(G) 120000 Mile Service**	
1995-99	.5			1995-99	4.0
				Calif.	2.2

79231C32

LEXUS
ES300 · GS300 · GS400 · LS400 · SC300 · SC400

7920 Porsche - Ch1

VEHICLE IDENTIFICATION CHART

Engine Code						Model Year	
Code	Liters	Cu. In. (cc)	Cyl.	Fuel Sys.	Eng. Mfg.	Code	Year
1MZ-FE	3.0	183 (2997)	6	SFI	Toyota	S	1995
1UZ-FE	4.0	242 (3969)	8	MFI	Toyota	T	1996
1UZ-FE	4.0	242 (3969)	8	SFI	Toyota	V	1997
2JZ-GE	3.0	183 (2995)	6	SFI	Toyota	W	1998
						X	1999

MFI - Multi-port Fuel Injection

SFI - Sequential Multi-port Fuel Injection

79231C33

ENGINE IDENTIFICATION

Year	Model	Engine Series (ID/VIN)	Engine Displacement Liters (cc)	Fuel System	No. of Cylinders	Engine Type
1995	ES300	1MZ-FE	3.0 (2959)	SFI	6	DOHC
	GS300	2JZ-GE	3.0 (2997)	SFI	6	DOHC
	LS400	1UZ-FE	4.0 (3969)	SFI	8	DOHC
	SC300	2JZ-GE	3.0 (2997)	SFI	6	DOHC
	SC400	1UZ-FE	4.0 (3969)	MFI	8	DOHC
1996	ES300	1MZ-FE	3.0 (2995)	SFI	6	DOHC
	GS300	2JZ-GE	3.0 (2997)	SFI	6	DOHC
	LS400	1UZ-FE	4.0 (3969)	SFI	8	DOHC
	SC300	2JZ-GE	3.0 (2997)	SFI	6	DOHC
	SC400	1UZ-FE	4.0 (3969)	SFI	8	DOHC
1997	ES300	1MZ-FE	3.0 (2995)	SFI	6	DOHC
	GS300	2JZ-GE	3.0 (2997)	SFI	6	DOHC
	LS400	1UZ-FE	4.0 (3969)	SFI	8	DOHC
	SC300	2JZ-GE	3.0 (2997)	SFI	6	DOHC
	SC400	1UZ-FE	4.0 (3969)	SFI	8	DOHC
1998-99	ES300	1MZ-FE	3.0 (2995)	SFI	6	DOHC
	GS300	2JZ-GE	3.0 (2997)	SFI	6	DOHC
	GS400	1UZ-FE	4.0 (3969)	SFI	8	DOHC
	LS400	1UZ-FE	4.0 (3969)	SFI	8	DOHC
	SC300	2JZ-GE	3.0 (2997)	SFI	6	DOHC
	SC400	1UZ-FE	4.0 (3969)	SFI	8	DOHC

DOHC - Double Overhead Camshaft

MFI - Multi-port Fuel Injection

SFI - Sequential Multi-port Fuel Injection

79231C34

GENERAL ENGINE SPECIFICATIONS

Year	Engine ID/VIN	Engine Displacement Liters (cc)	Fuel System Type	Net Horsepower @ rpm	Net Torque @ rpm (ft. lbs.)	Bore x Stroke (in.)	Compression Ratio	Oil Pressure @ rpm
1995	1MZ-FE	3.0 (2995)	SFI	188@5200	203@4400	3.44x3.27	10.5:1	43-78@3000
	1UZ-FE ①	4.0 (3969)	MFI	250@5600	260@4400	3.44x3.25	10.0:1	43-85@3000
	1UZ-FE ②	4.0 (3969)	SFI	260@5300	270@4500	3.45x3.25	10.4:1	43-85@3000
	2JZ-GE	3.0 (2997)	SFI	220@5800	210@4800	3.39x3.39	10.0:1	47-84@3000
1996	1MZ-FE	3.0 (2995)	SFI	188@5200	203@4400	3.44x3.27	10.5:1	43-78@3000
	1UZ-FE	4.0 (3969)	SFI	260@5300	270@4500	3.45x3.25	10.4:1	43-85@3000
	2JZ-GE	3.0 (2997)	SFI	220@5800	210@4800	3.39x3.39	10.0:1	47-84@3000
1997	1MZ-FE	3.0 (2995)	SFI	188@5200	203@4400	3.44x3.27	10.5:1	43-78@3000
	1UZ-FE	4.0 (3969)	SFI	260@5300	270@4500	3.45x3.25	10.4:1	43-85@3000
	2JZ-GE	3.0 (2997)	SFI	220@5800	210@4800	3.39x3.39	10.0:1	47-84@3000
1998-99	1MZ-FE	3.0 (2995)	SFI	200@5200	214@4400	3.44X3.27	10.5:1	43-78@3000
	1UZ-FE ③	4.0 (3969)	SFI	290@6000	300@4000	3.44x3.25	10.5:1	43-85@3000
	1UZ-FE ④	4.0 (3969)	SFI	300@6000	310@4000	3.44x3.25	10.5:1	43-85@3000
	2JZ-GE	3.0 (2997)	SFI	225@6000	220@4000	3.39x3.39	10.5:1	47-84@3000

MFI - Multi-port Fuel Injection

SFI - Sequential Multi-port Fuel Injection

① SC400

② LS400

③ LS400 and SC400

④ GS400

79231C35

GASOLINE ENGINE TUNE-UP SPECIFICATIONS

Year	Engine ID/VIN	Engine Displacement Liters (cc)	Spark Plugs Gap (in.)	Ignition Timing (deg.) MT	Ignition Timing (deg.) AT	Fuel Pump (psi)	Idle Speed (rpm) MT	Idle Speed (rpm) AT	Valve Clearance In.	Valve Clearance Ex.
1995	1MZ-FE	3.0 (2995)	0.043	—	10B ①	38-44	—	650-750	0.006-0.010	0.010-0.014
	2JZ-GE	3.0 (2997)	0.043	10B ①	10B ①	38-44	650-750	650-750	0.006-0.010	0.010-0.014
	1UZ-FE	4.0 (3969)	0.043	—	10B ①	38-44	—	600-700	0.006-0.010	0.010-0.014
1996	1MZ-FE	3.0 (2995)	0.043	—	10B ①	38-44	—	650-750	0.006-0.010	0.010-0.014
	2JZ-GE	3.0 (2997)	0.043	10B ①	10B ①	38-44	650-750	650-750	0.006-0.010	0.010-0.014
	1UZ-FE	4.0 (3969)	0.043	—	10B ①	38-44	—	600-700	0.006-0.010	0.010-0.014
1997	1MZ-FE	3.0 (2995)	0.043	—	10B ①	38-44	—	650-750	0.006-0.010	0.010-0.014
	2JZ-GE	3.0 (2997)	0.043	10B ①	10B ①	38-44	650-750	650-750	0.006-0.010	0.010-0.014
	1UZ-FE	4.0 (3969)	0.043	—	10B ①	38-44	—	600-700	0.006-0.010	0.010-0.014
1998-99	1MZ-FE	3.0 (2995)	0.043	—	8-12B ①	44-50	—	650-750	0.006-0.010	0.010-0.014
	2JZ-GE	3.0 (2997)	0.043	8-12B ②	8-12B ②	44-50	650-750	650-750	0.006-0.010	0.010-0.014
	1UZ-FE	4.0 (3969)	0.043	—	8-12B ②	44-50	—	700-800	0.006-0.010	0.010-0.014

NOTE: The Vehicle Emission Control Information label often reflects specification changes made during production. The label figures must be used if they differ from those in this chart.

B - Before top dead center

79231C36

① Terminals TE1 and E1 of check connector must be connected

② Terminals TC and E1 of check connector must be connected

CAPACITIES

Year	Model	Engine ID/VIN	Engine Displacement Liters (cc)	Engine Oil with Filter	Transmission (pts.)			Drive Axle		Fuel Tank (gal.)	Cooling System (qts.)
					4-Spd	5-Spd	Auto. ①	Front (pts.)	Rear (pts.)		
1995	ES300	1MZ-FE	3.0 (2995)	5.1	—	—	7.4	1.8	—	18.5	9.2
	GS300	2JZ-GE	3.0 (2997)	5.7	—	—	4.0	—	2.8	21.1	7.9
	LS400	1UZ-FE	4.0 (3969)	5.6	—	—	4.0	—	2.8	22.5	11.4
	SC300	2JZ-GE	3.0 (2997)	5.1	—	5.4	3.4	—	2.8	20.6	8.9
	SC400	1UZ-FE	4.0 (3969)	5.6	—	—	4.0	—	2.8	20.6	12.0
1996	ES300	1MZ-FE	3.0 (2995)	5.1	—	—	7.4	—	—	18.5	9.2
	GS300	2JZ-GE	3.0 (2997)	5.7	—	—	4.0	—	2.8	21.1	7.9
	LS400	1UZ-FE	4.0 (3969)	5.6	—	—	4.0	—	2.8	22.5	11.4
	SC300	2JZ-GE	3.0 (2997)	5.1	—	5.4	3.4	—	2.8	20.6	8.9
	SC400	1UZ-FE	4.0 (3969)	5.6	—	—	4.0	—	2.8	20.6	12.0
1997	ES300	1MZ-FE	3.0 (2995)	5.1	—	—	7.4	—	—	18.5	9.2
	GS300	2JZ-GE	3.0 (2997)	5.7	—	—	4.0	—	2.8	21.1	7.9
	LS400	1UZ-FE	4.0 (3969)	5.6	—	—	4.0	—	2.8	22.5	11.4
	SC300	2JZ-GE	3.0 (2997)	5.1	—	5.4	3.4	—	2.8	20.6	8.9
	SC400	1UZ-FE	4.0 (3969)	5.6	—	—	4.0	—	2.8	20.6	12.0
1998-99	ES300	1MZ-FE	3.0 (2995)	5.0	—	—	8.2	—	—	18.5	9.7
	GS300	2JZ-GE	3.0 (2997)	5.7	—	—	4.0	—	2.8	19.8	8.1
	GS400	1UZ-FE	4.0 (3969)	5.5	—	—	4.0	—	2.8	19.8	9.8
	LS400	1UZ-FE	4.0 (3969)	5.9	—	—	4.0	—	2.8	22.5	11.4
	SC300	2JZ-GE	3.0 (2997)	5.5	—	—	3.4	—	2.8	20.6	8.9
	SC400	1UZ-FE	4.0 (3969)	5.1	—	—	4.0	—	2.8	20.6	11.5

NOTE: All capacities are approximate. Add fluid gradually and check to be sure a proper fluid level is obtained.

① Specification is for transmission drain and refill, not overhaul.

79231C37

VALVE SPECIFICATIONS

Year	Engine ID/VIN	Engine Displacement Liters (cc)	Seat Angle (deg.)	Face Angle (deg.)	Spring Test Pressure (lbs. @ in.)	Spring Free Length (in.)	Stem-to-Guide Clearance (in.)		Stem Diameter (in.)	
							Intake	Exhaust	Intake	Exhaust
1995	1MZ-FE	3.0 (2995)	NA	44.5	41.9-46.3@ 1.331	1.791	0.0010- 0.0024	0.0012- 0.0026	0.2154- 0.2159	0.2152- 0.2157
	2JZ-GE	3.0 (2997)	NA	44.5	41.9-46.3@ 1.358	①	0.0010- 0.0024	0.0012- 0.0026	0.2350- 0.2356	0.2348- 0.2354
	1UZ-FE ②	4.0 (3969)	NA	44.5	41.9-46.3@ 1.295	1.717	0.0010- 0.0024	0.0012- 0.0026	0.2350- 0.2356	0.2348- 0.2354
	1UZ-FE ③	4.0 (3969)	NA	44.5	41.9-46.3@ 1.295	2.039	0.0010- 0.0024	0.0012- 0.0026	0.2350- 0.2356	0.2348- 0.2354
1996	1MZ-FE	3.0 (2995)	NA	44.5	41.9-46.3@ 1.331	1.791	0.0010- 0.0024	0.0012- 0.0026	0.2154- 0.2159	0.2152- 0.2157
	2JZ-GE	3.0 (2997)	NA	44.5	41.9-46.3@ 1.358	1.642	0.0010- 0.0024	0.0012- 0.0026	0.2350- 0.2356	0.2348- 0.2354
	1UZ-FE	4.0 (3969)	NA	44.5	41.9-46.3@ 1.295	2.039	0.0010- 0.0024	0.0012- 0.0026	0.2350- 0.2356	0.2348- 0.2354
1997	1MZ-FE	3.0 (2995)	NA	44.5	41.9-46.3@ 1.331	1.791	0.0010- 0.0024	0.0012- 0.0026	0.2154- 0.2159	0.2152- 0.2157
	2JZ-GE	3.0 (2997)	NA	44.5	41.9-46.3@ 1.358	1.642	0.0010- 0.0024	0.0012- 0.0026	0.2350- 0.2356	0.2348- 0.2354
	1UZ-FE	4.0 (3969)	NA	44.5	41.9-46.3@ 1.295	2.039	0.0010- 0.0024	0.0012- 0.0026	0.2350- 0.2356	0.2348- 0.2354
1998-99	1MZ-FE	3.0 (2995)	NA	44.5	41.9-46.3@ 1.331	1.791	0.0010- 0.0024	0.0012- 0.0026	0.2154- 0.2159	0.2152- 0.2157
	2JZ-GE	3.0 (2997)	NA	44.5	41.9-46.3@ 1.358	④	0.0010- 0.0024	0.0012- 0.0026	0.2350- 0.2356	0.2348- 0.2354
	1UZ-FE	4.0 (3969)	NA	44.5	⑤	2.130	0.0010- 0.0024	0.0012- 0.0026	0.2154- 0.2159	0.2152- 0.2157

NA - Not Available

① Blue - 1.6433
   Yellow - 1.6417
② SC400
③ LS400
④ Pink - 1.7209
   Yellow - 1.7362
⑤ GS400 - 45.9-50.7@1.3795
   SC400 & LS400 - 45.9-50.7@1.378

79231C38

TORQUE SPECIFICATIONS
All readings in ft. lbs.

Year	Engine ID/VIN	Engine Displacement Liters (cc)	Cylinder Head Bolts	Main Bearing Bolts	Rod Bearing Bolts	Crankshaft Damper Bolts	Flywheel Bolts	Manifold		Spark Plugs	Lug Nut
								Intake	Exhaust		
1995	1MZ-FE	3.0 (2995)	①	②	③	159	61	11	36	13	76
	1UZ-FE ④	4.0 (3969)	⑤	⑥	③	181	72	13	29	13	76
	1UZ-FE ⑦	4.0 (3969)	⑤	⑥	③	181	61	13	32	13	76
	2JZ-GE	3.0 (2997)	⑧	⑨	⑩	239	⑪	15	29	13	76
1996	1MZ-FE	3.0 (2995)	①	②	⑤	159	61	11	36	13	76
	1UZ-FE	4.0 (3969)	⑤	⑥	③	181	61	13	32	13	76
	2JZ-GE	3.0 (2997)	⑧	⑨	⑩	239	⑪	20	29	13	76
1997	1MZ-FE	3.0 (2995)	①	②	⑤	159	61	11	36	13	76
	1UZ-FE	4.0 (3969)	⑤	⑥	③	181	61	13	32	13	76
	2JZ-GE	3.0 (2997)	⑧	⑨	⑩	239	⑪	20	29	13	76
1998-99	1MZ-FE	3.0 (2995)	①	②	③	159	61	11	36	13	76
	1UZ-FE	4.0 (3969)	⑤	⑥	⑤	181	61	13	32	13	76
	2JZ-GE	3.0 (2997)	⑧	⑨	⑩	243	61	21	30	13	76

① Head bolt:
Step 1: 40 ft. lbs.
Step 2: Plus 90 degrees
Recessed head bolt: 13 ft. lbs.

② 6-point bolts - 20 ft. lbs.
12-point bolts
Step 1: 16 ft. lbs.
Step 2: Plus an additional 90 degrees

③ Step 1: 18 ft. lbs.
Step 2: Plus 90 degrees

④ SC400

⑤ Step 1: Tighten to 29 ft. lbs.
Step 2: Plus 90 degrees

⑥ Nuts:
Step 1: 20 ft. lbs.
Step 2: Plus 90 degrees
Bolts: 36 ft. lbs.

⑦ LS400

⑧ Step 1: 25 ft. lbs.
Step 2: Tighten an additional 90 degrees
Step 3: Tighten an additional 90 degrees

⑨ Step 1: 33 ft. lbs.
Step 2: Plus 90 degrees

⑩ Step 1: 22 ft. lbs.
Step 2: Plus 90 degrees

⑪ Driveplate: 61 ft. lbs.
Flywheel:
Step 1: 36 ft. lbs.
Step 2: Plus 90 degrees

⑫ Step 1: 26 ft. lbs.
Step 2: Plus 90 degrees
Step 3: Plus 90 degrees

79231C39

SCHEDULED MAINTENANCE INTERVALS
(LEXUS ES300, SC300, LS400, SC400, GS300 & GS400)

TO BE SERVICED	TYPE OF SERVICE	VEHICLE MILEAGE INTERVAL (x1000)												
		7.5	15	22.5	30	37.5	45	52.5	60	67.5	75	82.5	90	97.5
Engine oil & filter	R	✓	✓	✓	✓	✓	✓	✓	✓	✓	✓	✓	✓	✓
Air conditioning filter (LS400)②	S/I	✓	✓	✓	✓	✓	✓	✓	✓	✓	✓	✓	✓	✓
Automatic transmission fluid & filter	S/I		✓		✓		✓		✓		✓		✓	
Ball joints & dust covers	S/I		✓		✓		✓		✓		✓		✓	
Bolts & nuts on chassis & body	S/I		✓		✓		✓		✓		✓		✓	
Brake fluid①	S/I		✓		✓		✓		✓		✓		✓	
Brake line pipes & hoses	S/I		✓		✓		✓		✓		✓		✓	
Brake linings & drums	S/I		✓		✓		✓		✓		✓		✓	
Brake pads & discs (front & rear)	S/I		✓		✓		✓		✓		✓		✓	
Differential oil	S/I		✓		✓		✓		✓		✓		✓	
Driveshaft boots (ES300)	S/I		✓		✓		✓		✓		✓		✓	
Manual transmission oil (SC300 & SC400)	S/I		✓		✓		✓		✓		✓		✓	
Steering gear housing oil	S/I		✓		✓		✓		✓		✓		✓	
Steering linkage	S/I		✓		✓		✓		✓		✓		✓	
Air filter	R				✓				✓				✓	
Exhaust pipes & mountings	S/I				✓				✓				✓	
Fuel lines & connections	S/I				✓				✓				✓	
Engine Coolant	R						✓					✓		
Fuel tank cap gasket	R								✓					
Spark plugs	R								✓					
Charcoal canister	S/I								✓					
Drive belts	S/I								✓					

79231C40

SCHEDULED MAINTENANCE INTERVALS
(LEXUS ES300, SC300, LS400, SC400, GS300 & GS400) (Cont.)

TO BE SERVICED	TYPE OF SERVICE	VEHICLE MILEAGE INTERVAL (x1000)												
		7.5	15	22.5	30	37.5	45	52.5	60	67.5	75	82.5	90	97.5
Valve clearance	S/I								✓					

① Replace every 30,000 miles (unless previously replaced).
② Replace every 15,000 miles.
R – Replace   S/I – Service or Inspect

FREQUENT OPERATION MAINTENANCE (SEVERE SERVICE)

If a vehicle is operated under any of the following conditions it is considered severe service:
- Extremely dusty areas.
- 50% or more of the vehicle operation is in 32°C (90°F) or higher temperatures, or constant operation in temperatures below 0°C (32°F).
- Prolonged idling (vehicle operation in stop and go traffic).
- Frequent short running periods (engine does not warm to normal operating temperatures).
- Police, taxi, delivery usage or trailer towing usage.
Oil & oil filter change – change every 3750 miles.
Ball joints & dust covers - service or inspect every 7500 miles.
Bolts & nuts on chassis & body - service or inspect every 7500 miles.
Brake linings & drums - service or inspect every 7500 miles.
Brake pads & discs (front & rear) - service or inspect every 7500 miles.
Driveshaft boots (ES300) - service or inspect every 7500 miles.
Steering linkage - service or inspect every 7500 miles.
Air filter - service or inspect every 15,000 miles.
Automatic transmission fluid & filter - replace every 15,000 miles.
Differential oil - replace every 15,000 miles.
Exhaust pipes & mountings - service or inspect every 15,000 miles.
Manual transmission oil - replace every 15,000 miles.
Drive belts - service or inspect at 60,000 miles & every 7500 miles thereafter.
Timing belt - replace every 60,000 miles.

79231C41

SCHEDULED MAINTENANCE
LEXUS
ES300, GS300, GS400, LS400, SC300, SC400,

The following should be used as a guide when determining the amount of work required for a particular service if taken to a repair shop. In estimating how long a particular Scheduled Maintenance Service should take, please observe the following:

- **Chilton Time** is time based on field research and data supplied by the vehicle manufacturer.
- Labor time operations are given in hours and tenths of an hour.
- All labor operations, are to be used as a guide.

Mechanic Skill Level Codes:
- **(G) GENERAL:** Normally skilled with certification.
- **(M) MAINTENANCE:** Semi-skilled working on certicication.
- **(P) PRECISION:** Really skilled with multiple certification.

	Chilton Time
(M) 7500 Mile Service	
1995-99	.3
LS400 add	.2
(G) 15000 Mile Service	
1995-99 ES300, SC300	1.5
1995-99 LS400	1.6
1995-99 SC400	1.5
1995-99 GS300/400	1.4
(M) 22500 Mile Service	
1995-99	.3
LS400 add	.2
(G) 30000 Mile Service	
1995-99 ES300, SC300	2.3
1995-99 LS400	2.4
1995-99 SC400	2.3
1995-99 GS300/400	2.2

	Chilton Time
(M) 37500 Mile Service	
1995-99	.3
LS400 add	.2
(G) 45000 Mile Service	
1995-99 ES300, SC300	2.0
1995-99 LS400	2.1
1995-99 SC400	2.0
1995-99 GS300/400	1.9
(M) 52500 Mile Service	
1995-99	.3
LS400 add	.2
(G) 60000 Mile Service	
1995-99 ES300, SC300	3.8
1995-99 LS400	4.1
1995-99 SC400	3.8
1995-99 GS300/400	3.7
(M) 67500 Mile Service	
1995-99	.3
LS400 add	.2

	Chilton Time
(G) 75000 Mile Service	
1995-99 ES300, SC300	1.5
1995-99 LS400	1.6
1995-99 SC400	1.5
1995-99 GS300/400	1.4
(M) 82500 Mile Service	
1995-99	.3
LS400 add	.2
(G) 90000 Mile Service	
1995-99 ES300, SC300	2.3
1995-99 LS400	2.4
1995-99 SC400	2.3
1995-99 GS300/400	2.2
(M) 97500 Mile Service	
1995-99	.3
LS400 add	.2

79231C42

MAZDA
323 • 626 • Miata • Millenia • MX6 • Protege

VEHICLE IDENTIFICATION CHART

Engine Code							Model Year	
Code	Liters	Cu. In. (cc)	Cyl.	Fuel Sys.	Eng. Mfg.		Code	Year
13B	1.3	80.0 (1308)	4	EFI	Mazda		S	1995
B6E	1.6	97.4 (1597)	4	MPFI	Mazda		T	1996
B6ZE	1.6	97.4 (1597)	4	MPFI	Mazda		V	1997
BPD	1.8	112.2 (1839)	4	MPFI	Mazda		W	1998
BPE	1.8	112.2 (1839)	4	MPFI	Mazda		X	1999
F2	2.2	133.2 (2184)	4	MPFI	Mazda			
FSD	2.0	121.5 (1991)	4	MPFI	Mazda			
JE-ZE	3.0	180.2 (2954)	6	EFI	Mazda			
K8D	1.8	112.4 (1844)	6	MPFI	Mazda			
KJS	2.3	137.2 (2254)	6	MPFI	Mazda			
KLD	2.5	152.3 (2496)	6	MPFI	Mazda			
Z5D	1.5	90.8 (1489)	4	MPFI	Mazda			

MPFI: Multi-Point Fuel Injection

EFI: Electronic Fuel Injection

79231C43

ENGINE IDENTIFICATION

Year	Model	Engine Displacement Liters (cc)	Engine Series (ID/VIN)	Fuel System	No. of Cylinders	Engine Type
1995	323	1.6 (1597)	B6E	EFI	4	SOHC
	626	2.0 (1991)	FS	EFI	4	DOHC
	626	2.5 (2497)	KL	EFI	6	DOHC
	929	3.0 (2954)	JE-ZE	EFI	6	DOHC
	Miata	1.8 (1839)	BPD	EFI	4	DOHC
	Millenia	2.5 (2497)	KLD	EFI	6	DOHC
	Millenia S	2.3 (2255)	KJS	EFI	6	DOHC
	MX3	1.6 (1597)	B6ZE	EFI	4	DOHC
	MX3	1.8 (1844)	K8D	EFI	6	DOHC
	MX6	2.0 (1991)	FS	EFI	4	DOHC
	MX6	2.5 (2497)	KL	EFI	6	DOHC
	Protege	1.8 (1839)	Z5D	EFI	4	SOHC
	Protege LX	1.8 (1839)	BPD	EFI	4	DOHC
	RX7	1.3 (1308)	13B	EFI	-	Rotary Turbo
1996	626 DX	2.0 (1991)	FS	EFI	4	DOHC
	626 ES	2.5 (2497)	KL	EFI	6	DOHC
	626 LX	2.0 (1991)	FS	EFI	4	DOHC
	626 LX-V6	2.5 (2497)	KL	EFI	6	DOHC
	Miata	1.8 (1839)	BPD	EFI	4	DOHC
	Millenia	2.5 (2497)	KLD	EFI	6	DOHC
	Millenia S	2.3 (2255)	KJS	EFI	6	DOHC
	MX6	2.0 (1991)	FS	EFI	4	DOHC
	MX6 LS	2.5 (2497)	KL	EFI	6	DOHC
	Protege DX	1.5 (1489)	Z5D	EFI	4	DOHC
	Protege ES	1.8 (1839)	BPD	EFI	4	DOHC
	Protege LX	1.5 (1489)	Z5D	EFI	4	DOHC
1997	626 DX	2.0 (1991)	FS	EFI	4	DOHC
	626 ES	2.5 (2497)	KL	EFI	6	DOHC
	626 LX	2.0 (1991)	FS	EFI	4	DOHC
	626 LX-V6	2.5 (2497)	KL	EFI	6	DOHC
	Miata	1.8 (1839)	BPD	EFI	4	DOHC
	Millenia	2.5 (2497)	KLD	EFI	6	DOHC
	Millenia S	2.3 (2255)	KJS	EFI	6	DOHC
	MX6	2.0 (1991)	FS	EFI	4	DOHC
	MX6 LS	2.5 (2497)	KL	EFI	6	DOHC
	Protege DX	1.5 (1489)	Z5D	EFI	4	DOHC
	Protege ES	1.8 (1839)	BPD	EFI	4	DOHC
	Protege LX	1.5 (1489)	Z5D	EFI	4	DOHC

79231C44

ENGINE IDENTIFICATION

Year	Model	Engine Displacement Liters (cc)	Engine Series (ID/VIN)	Fuel System	No. of Cylinders	Engine Type
1998-99	626 DX	2.0 (1991)	FS	EFI	4	DOHC
	626 ES	2.5 (2497)	KL	EFI	6	DOHC
	626 LX	2.0 (1991)	FS	EFI	4	DOHC
	626 LX-V6	2.5 (2497)	KL	EFI	6	DOHC
	Miata	1.8 (1839)	BPD	EFI	4	DOHC
	Millenia	2.5 (2497)	KLD	EFI	6	DOHC
	Millenia S	2.3 (2255)	KJS	EFI	6	DOHC
	Protege DX	1.5 (1489)	Z5D	EFI	4	DOHC
	Protege ES	1.8 (1839)	BPD	EFI	4	DOHC
	Protege LX	1.5 (1489)	Z5D	EFI	4	DOHC

EFI - Electronic Fuel Injection   SOHC - Single Overhead Camshaft   DOHC - Double Overhead Camshafts

79231C45

GENERAL ENGINE SPECIFICATIONS

Year	Engine ID/VIN	Engine Displacement Liters (cc)	Fuel System Type	Net Horsepower @ rpm	Net Torque @ rpm (ft. lbs.)	Bore x Stroke (in.)	Com-pression Ratio	Oil Pressure @ rpm
1995	13B	1.3 (1308)	EFI Turbo	255 @ 6500	217 @ 5000	3.07x3.29	⑥	43-57@3000
	B6E	1.6 (1597)	EFI	82@5000 ①	92@2500 ②	3.07x3.29	9.3:1	43-57@3000
	B6ZE	1.6 (1597)	EFI	105@6200	100@3600	3.07x3.29	9.0:1	43-57@3000
	BPD	1.8 (1839)	EFI	126@6500	114@4500	3.27x3.35	9.0:1	43-57@3000
	BPE	1.8 (1839)	EFI	103@5500	111@4000	3.27x3.35	8.9:1	43-57@3000
	FS	2.0 (1991)	EFI	118@5500 ③	127@4500 ④	3.27x3.62	9.0:1	57-71@3000
	JE-ZE	3.0 (2954)	EFI	195@5750	200@3500	3.54x3.05	9.2:1	53-75@3000
	K8D	1.8 (1844)	EFI	130@6500 ⑤	115@4500	2.95x2.74	9.2:1	48-71@3000
	KJS	2.3 (2255)	EFI	210@5300	210@3500	3.16x2.92	10.0:1	44-66@3000
	KL	2.5 (2497)	EFI	164@5600 ⑥	160@4800 ⑦	3.33x2.92	9.2:1	49-71@3000
	KLD	2.5 (2497)	EFI	170@5800	160@4800	3.33x2.92	9.2:1	49-71@3000
	Z5D	1.5 (1489)	EFI	⑧	96@4000	2.96x3.29	9.4:1	43-57@3000
1996	BPD	1.8 (1839)	EFI	⑨	⑩	3.27x3.35	9.0:1	43-57@3000
	FS	2.0 (1991)	EFI	118@5500	127@4500	3.27x3.62	9.0:1	57-71@3000
	KJS	2.3 (2255)	EFI	210@5300	210@3500	3.16x2.92	10.0:1	44-66@3000
	KL	2.5 (2497)	EFI	164@5500	⑦	3.33x2.92	9.2:1	49-71@3000
	KLD	2.5 (2497)	EFI	170@5800	160@4800	3.33x2.92	9.2:1	49-71@3000
	Z5D	1.5 (1489)	EFI	⑧	96@4000	2.96x3.29	9.4:1	43-57@3000
1997	BPD	1.8 (1839)	EFI	⑨	⑩	3.27x3.35	9.0:1	43-57@3000
	FS	2.0 (1991)	EFI	118@5500	127@4500	3.27x3.62	9.0:1	57-71@3000
	KJS	2.3 (2255)	EFI	210@5300	210@3500	3.16x2.92	10.0:1	44-66@3000
	KL	2.5 (2497)	EFI	164@5500	⑦	3.33x2.92	9.2:1	49-71@3000
	KLD	2.5 (2497)	EFI	170@5800	160@4800	3.33x2.92	9.2:1	49-71@3000
	Z5D	1.5 (1489)	EFI	⑧	96@4000	2.96x3.29	9.4:1	43-57@3000
1998-99	BPD	1.8 (1839)	EFI	⑨	⑩	3.27x3.35	9.0:1	43-57@3000
	FS	2.0 (1991)	EFI	125@5000	127@5000	3.27x3.62	9.0:1	57-71@3000
	KJS	2.3 (2255)	EFI	210@5300	210@3500	3.16x2.92	10.0:1	44-66@3000
	KL	2.5 (2497)	EFI	170@5000	163 @ 5000	3.33x2.92	9.5:1	49-71@3000
	KLD	2.5 (2497)	EFI	170@5800	160@4800	3.33x2.92	9.2:1	49-71@3000
	Z5D	1.5 (1489)	EFI	⑧	96@4000	2.96x3.29	9.4:1	43-57@3000

EFI - Electronic Fuel Injection

① California: 88@5000

② California: 98@4000

③ California: 114@5500

④ California: 129@4500

⑤ California: 128@6000

⑥ California: 160@5500

⑦ California and New York: 156@5000

　Except California and New York: 160@4800

⑧ California, New York and Massachusetts: 90@5500

　Except California, New York and Massachusetts: 92@5500

⑨ Protege: 122@6000, Miata: 133@6500

⑩ Protege: 117@4000, Miata: 114@5000

79231C46

GASOLINE ENGINE TUNE-UP SPECIFICATIONS

Year	Engine ID/VIN	Engine Displacement Liters (cc)	Spark Plug Gap (in.)	Ignition Timing (deg.) MT		Ignition Timing (deg.) AT		Fuel Pump (psi)		Idle Speed (rpm) MT		Idle Speed (rpm) AT		Valve Clearance In.	Valve Clearance Ex.
1995	13B	1.3 (1308)	0.056	①		①		NA		750		750		NA	NA
	B6E	1.6 (1597)	0.041	6-8B	②	6-8B	②	30-38	③	750	④	750	④	HYD	HYD
	B6ZE	1.6 (1597)	0.041	9-11B	②	9-11B	②	30-38	③	700	⑤	750	⑤	HYD	HYD
	BPD	1.8 (1839)	0.041	9-11B	⑥	9-11B	⑥	30-38	③	750	④	750	④	HYD	HYD
	BPE	1.8 (1839)	0.041	4-6B	②	4-6B	②	30-38	③	750	④	750	④	HYD	HYD
	FS	2.0 (1991)	0.041	11-13B	②	11-13B	②	30-38	③	700	②	700	②	HYD	HYD
	JE-ZE	3.0 (2954)	0.041	11-13B	②	11-13B	②	30-38	③	700	⑦	700	⑦	HYD	HYD
	K8D	1.8 (1844)	0.041	9-11B	②	9-11B	②	30-38	③	700	⑤	750	⑤	HYD	HYD
	KJS	2.3 (2255)	0.030	6B	②	6B	②	39-48	②	670		650		0.011	0.011
	KL	2.5 (2497)	0.041	9-11B	②	9-11B	②	30-38	③	670	②	650	②	HYD	HYD
	KLD	2.5 (2497)	0.041	9-11B	②	9-11B	②	39-45	③	670		650		HYD	HYD
	Z5D	1.5 (1489)	0.041	6-18B		6-18B		29-34	③	650-750		700-800		0.011	0.011
1996	BPD	1.8 (1839)	0.041	⑧		⑧		39-45	③	⑨		⑨		HYD	HYD
	FS	2.0 (1991)	0.041	11-13B	②	6-18B	②	37-46	③	650-750	②	650-750	②	HYD	HYD
	KJS	2.3 (2255)	0.030	6B	②	6B	②	39-48	③	600-700		600-700		0.011	0.011
	KL	2.5 (2497)	0.041	9-11B	②	9-11B	②	39-45	③	600-700	②	600-700	②	HYD	HYD
	KLD	2.5 (2497)	0.041	9-11B	②	9-11B	②	39-45	③	600-700		600-700		HYD	HYD
	Z5D	1.5 (1489)	0.041	6-18B		6-18B		29-34	③	650-750		700-800		0.011	0.011
1997	BPD	1.8 (1839)	0.041	⑧		⑧		39-45	③	⑨		⑨		HYD	HYD
	FS	2.0 (1991)	0.041	11-13B	②	6-18B	②	37-46	③	650-750	②	650-750	②	HYD	HYD
	KJS	2.3 (2255)	0.030	6B	②	6B	②	39-48	③	600-700		600-700		0.011	0.011
	KL	2.5 (2497)	0.041	9-11B	②	9-11B	②	39-45	③	600-700	②	600-700	②	HYD	HYD
	KLD	2.5 (2497)	0.041	9-11B	②	9-11B	②	39-45	③	600-700		600-700		HYD	HYD
	Z5D	1.5 (1489)	0.041	6-18B		6-18B		29-34	③	650-750		700-800		0.011	0.011
1998-99	BPD	1.8 (1839)	0.041	⑧		⑧		39-45	③	⑨		⑨		HYD	HYD
	FS	2.0 (1991)	0.041	11-13B	②	6-18B	②	37-46	③	650-750	②	650-750	②	HYD	HYD
	KJS	2.3 (2255)	0.030	6B	②	6B	②	39-48	③	600-700		600-700		0.011	0.011
	KL	2.5 (2497)	0.041	9-11B	②	9-11B	②	39-45	③	600-700	②	600-700	②	HYD	HYD
	KLD	2.5 (2497)	0.041	9-11B	②	9-11B	②	39-45	③	600-700		600-700		HYD	HYD
	Z5D	1.5 (1489)	0.041	6-18B		6-18B		29-34	③	650-750		700-800		0.011	0.011

NOTE: The Vehicle Emission Control Information label often reflects specification changes made during production. The label figures must be used if they differ from those in this chart

A - After top dead center   B - Before top dead center   HYD - Hydraulic

① Electronic Distributor: Leading side 5°A, Trailing side 20° A with data link connector terminal 10 grounded

② Data link connector terminal 10 grounded

③ Pressure indicated is with gauge in-line, regulator vacuum hose connected and engine idling

④ Canadian vehicles: With parking brake applied and the data link connector terminal 10 grounded

⑤ With system selector test switch on self test

⑥ Miata/MX5: 10 before top dead center with data link connector terminal 10 grounded

⑦ Vehicle in park with data link connector terminal 10 grounded
Protege: 0

⑧ Miata: 9-11B with data link connector terminal 10 grounded
Protege: 700-800
Miata: 800-900 with data link connector terminal 10 grounded

79231C47

CAPACITIES

| Year | Model | Engine ID/VIN | Engine Displacement Liters (cc) | Engine Oil with Filter | Transmission (pts.) | | | Transfer Case (pts.) | Drive Axle | | Fuel Tank (gal.) | Cooling System (qts.) |
					4-Spd	5-Spd	Auto.		Front (pts.)	Rear (pts.)		
1995	323	B6E	1.6 (1597)	3.6	-	5.6	13.4	-	①	-	13.2	5.3 ②
	626	FS	2.0 (1991)	3.7	-	5.8	17.6	-	①		15.5	7.4
	626	KL	2.5 (2497)	4.2	-	5.8	18.6	-	①		15.5	7.9
	929	JE	3.0 (2954)	5.3	-	-	18.2	-	-	2.8	18.5	③
	Miata	BP	1.8 (1839)	4.0	-	4.2	15.4	-	-	2.1	12.7	6.3
	Millenia	KL	2.5 (2497)	4.2	-	-	18.6	-	①		18.0	7.4
	Millenia S	KJ	2.3 (2255)	4.3	-	-	18.6	-	①		18.0	7.4
	MX3	B6	1.6 (1598)	3.5	-	5.6	12.2	-	①	-	13.2	6.3
	MX3	K8	1.8 (1844)	5.2	-	5.6	12.2	-	①	-	13.2	7.9
	MX6	FS	2.0 (1991)	3.7	-	5.8	17.6	-	①		15.5	7.4
	MX6	KL	2.5 (2497)	4.2	-	5.8	18.6	-	①		15.5	7.9
	Protege	Z5	1.5 (1498)	3.7	-	5.6	10.4	-	①	-	13.2	6.3
	Protege LX	BP	1.8 (1839)	4.0	-	5.6	10.4	-	①	-	13.2	6.3
	RX7	13B	1.3 (1308)	4.0	-	5.2	18.2	-	-	2.7	④	9.2 ②
1996	626 DX	FS	2.0 (1991)	3.7	-	5.8	18.4	-	①	-	15.5	7.4
	626 ES	KL	2.5 (2497)	4.2	-	5.8	18.6	-	①	-	15.5	7.9
	626 LX	FS	2.0 (1991)	3.7	-	5.8	18.4	-	①	-	15.5	7.4
	626 LX-V6	KL	2.5 (2497)	4.2	-	5.8	18.6	-	①	-	15.5	7.9
	Miata	BP	1.8 (1839)	4.0	-	4.2	13.5	-	①	2.1	12.7	6.3
	Millenia	KL	2.5 (2497)	4.2	-	-	16.9	-	①	-	18.0	7.4
	Millenia S	KJ	2.3 (2255)	4.3	-	-	16.9	-	①	-	18.0	7.4
	MX6	FS	2.0 (1991)	3.7	-	5.8	18.4	-	①	-	15.5	7.4
	MX6	KL	2.5 (2497)	4.2	-	5.8	18.6	-	①	-	15.5	7.9
	Protege DX	Z5	1.5 (1498)	3.7	-	5.6	11.3	-	①	-	13.2	6.3
	Protege ES	BP	1.8 (1839)	4.0	-	5.6	11.3	-	①	-	13.2	6.3
	Protege LX	Z5	1.5 (1498)	3.7	-	5.6	11.3	-	①	-	13.2	6.3
1997	626 DX	FS	2.0 (1991)	3.7	-	5.8	18.4	-	①	-	15.5	7.4
	626 ES	KL	2.5 (2497)	4.2	-	5.8	18.6	-	①	-	15.5	7.9
	626 LX	FS	2.0 (1991)	3.7	-	5.8	18.4	-	①	-	15.5	7.4
	626 LX-V6	KL	2.5 (2497)	4.2	-	5.8	18.6	-	①	-	15.5	7.9
	Miata	BP	1.8 (1839)	4.0	-	4.2	13.5	-	①	2.1	12.7	6.3
	Millenia	KL	2.5 (2497)	4.2	-	-	16.9	-	①	-	18.0	7.4
	Millenia S	KJ	2.3 (2255)	4.3	-	-	16.9	-	①	-	18.0	7.4
	MX6	FS	2.0 (1991)	3.7	-	5.8	18.4	-	①	-	15.5	7.4
	MX6	KL	2.5 (2497)	4.2	-	5.8	18.6	-	①	-	15.5	7.9
	Protege DX	Z5	1.5 (1498)	3.7	-	5.6	11.3	-	①	-	13.2	6.3
	Protege ES	BP	1.8 (1839)	4.0	-	5.6	11.3	-	①	-	13.2	6.3
	Protege LX	Z5	1.5 (1498)	3.7	-	5.6	11.3	-	①	-	13.2	6.3

79231C48

CAPACITIES

Year	Model	Engine ID/VIN	Engine Displacement Liters (cc)	Engine Oil with Filter	Transmission (pts.)			Transfer Case (pts.)	Drive Axle		Fuel Tank (gal.)	Cooling System (qts.)
					4-Spd	5-Spd	Auto.		Front (pts.)	Rear (pts.)		
1998-99	626 DX	FS	2.0 (1991)	3.7	-	5.8	18.4	-	①	-	15.5	7.4
	626 ES	KL	2.5 (2497)	4.2	-	5.8	18.6	-	①	-	15.5	7.9
	626 LX	FS	2.0 (1991)	3.7	-	5.8	18.4	-	①	-	15.5	7.4
	626 LX-V6	KL	2.5 (2497)	4.2	-	5.8	18.6	-	①	-	15.5	7.9
	Miata	BP	1.8 (1839)	4.0	-	4.2	13.5	-	①	2.1	12.7	6.3
	Millenia	KL	2.5 (2497)	4.2	-	-	16.9	-	①	-	18.0	7.4
	Millenia S	KJ	2.3 (2255)	4.3	-	-	16.9	-	①	-	18.0	7.4
	Protege DX	Z5	1.5 (1498)	3.7	-	5.6	11.3	-	①	-	13.2	6.3
	Protege ES	BP	1.8 (1839)	4.0	-	5.6	11.3	-	①	-	13.2	6.3
	Protege LX	Z5	1.5 (1498)	3.7	-	5.6	11.3	-	①	-	13.2	6.3

NOTE: All capacities are approximate. Add fluid gradually and check to be sure a proper fluid level is obtained.

① Included in transaxle

② Automatic transmission: 6.3 qts.

③ With heater: 9.9 qts.
   Without heater: 9.3 qts.

④ Manual transmission: 20.1 gals.
   Automatic transmission: 18.5 gals.

79231C49

VALVE SPECIFICATIONS

Year	Engine ID/VIN	Engine Displacement Liters (cc)	Seat Angle (deg.)	Face Angle (deg.)	Maximum out of Square (in.)	Spring Free Length (in.)	Stem-to-Guide Clearance (in.)		Stem Diameter (in.)	
							Intake	Exhaust	Intake	Exhaust
1995	B6	1.6 (1597)	45	45	0.0670	1.570	0.0010-0.0023	0.0012-0.0025	0.2351-0.2356	0.2349-0.2354
	BPD	1.8 (1839)	45	45	0.0640	1.560	0.0010-0.0023	0.0012-0.0025	0.2351-0.2356	0.2349-0.2354
	FS	2.0 (1991)	45	45	0.0610	1.437	0.0010-0.0023	0.0012-0.0025	0.2351-0.2356	0.2349-0.2354
	JE-ZE	3.0 (2954)	45	45	0.0600	1.723	0.0010-0.0023	0.0012-0.0025	0.2351-0.2356	0.2349-0.2354
	K8D	1.8 (1844)	45	45	0.064	1.847	0.0010-0.0024	0.0012-0.0026	0.2350-0.2356	0.2348-0.2354
	KJS	2.3 (2255)	NA	45	0.062	1.413	0.0010-0.0023	0.0012-0.0025	0.2351-0.2356	0.2349-0.2354
	KL	2.5 (2497)	45	45	0.0642	1.437	0.0010-0.0023	0.0012-0.0025	0.2351-0.2356	0.2349-0.2354
	Z5D	1.5 (1489)	45	45	0.0520	1.240	0.0010-0.0023	0.0012-0.0025	0.2154-0.2159	0.2152-0.2157
1996	BPD	1.8 (1839)	45	45	0.059-0.066	①	0.0010-0.0023	0.0012-0.0025	0.2351-0.2356	0.2349-0.2354
	FS	2.0 (1991)	45	45	0.061	1.732	0.0010-0.0023	0.0012-0.0025	0.2351-0.2356	0.2349-0.2354
	KJS	2.3 (2255)	NA	45	0.062	1.413	0.0010-0.0023	0.0012-0.0025	0.2351-0.2356	0.2349-0.2354
	KL	2.5 (2497)	45	45	0.064	1.847	0.0010-0.0023	0.0012-0.0025	0.2351-0.2356	0.2349-0.2354
	KLD	2.5 (2497)	45	45	0.064	1.847	0.0010-0.0023	0.0012-0.0025	0.2351-0.2356	0.2349-0.2354
	Z5D	1.5 (1489)	45	45	0.0520	1.240	0.0010-0.0023	0.0012-0.0025	0.2154-0.2159	0.2152-0.2157
1997	BPD	1.8 (1839)	45	45	0.059-0.066	①	0.0010-0.0023	0.0012-0.0025	0.2351-0.2356	0.2349-0.2354
	FS	2.0 (1991)	45	45	0.061	1.732	0.0010-0.0023	0.0012-0.0025	0.2351-0.2356	0.2349-0.2354
	KJS	2.3 (2255)	NA	45	0.062	1.413	0.0010-0.0023	0.0012-0.0025	0.2351-0.2356	0.2349-0.2354
	KL	2.5 (2497)	45	45	0.064	1.847	0.0010-0.0023	0.0012-0.0025	0.2351-0.2356	0.2349-0.2354
	KLD	2.5 (2497)	45	45	0.064	1.847	0.0010-0.0023	0.0012-0.0025	0.2351-0.2356	0.2349-0.2354
	Z5D	1.5 (1489)	45	45	0.0520	1.240	0.0010-0.0023	0.0012-0.0025	0.2154-0.2159	0.2152-0.2157

79231C50

VALVE SPECIFICATIONS

Year	Engine ID/VIN	Engine Displacement Liters (cc)	Seat Angle (deg.)	Face Angle (deg.)	Maximum out of Square (in.)	Spring Free Length (in.)	Stem-to-Guide Clearance (in.)		Stem Diameter (in.)	
							Intake	Exhaust	Intake	Exhaust
1998-99	BPD	1.8 (1839)	45	45	0.059-0.066	①	0.0010-0.0023	0.0012-0.0025	0.2351-0.2356	0.2349-0.2354
	FS	2.0 (1991)	45	45	0.061	1.732	0.0010-0.0023	0.0012-0.0025	0.2351-0.2356	0.2349-0.2354
	KJS	2.3 (2255)	NA	45	0.062	1.413	0.0010-0.0023	0.0012-0.0025	0.2351-0.2356	0.2349-0.2354
	KL	2.5 (2497)	45	45	0.064	1.847	0.0010-0.0023	0.0012-0.0025	0.2351-0.2356	0.2349-0.2354
	KLD	2.5 (2497)	45	45	0.064	1.847	0.0010-0.0023	0.0012-0.0025	0.2351-0.2356	0.2349-0.2354
	Z5D	1.5 (1489)	45	45	0.0520	1.240	0.0010-0.0023	0.0012-0.0025	0.2154-0.2159	0.2152-0.2157

NA - Not Available

① Intake: 1.80 in.
　 Exhaust: 1.903 in.

79231C51

TORQUE SPECIFICATIONS
All readings in ft. lbs.

Year	Engine ID/VIN	Engine Displacement Liters (cc)	Cylinder Head Bolts	Main Bearing Bolts	Rod Bearing Bolts	Crankshaft Damper Bolts	Flywheel Bolts	Manifold Intake	Manifold Exhaust	Spark Plugs	Lug Nut
1995	13B	1.3 (1308)	NA	NA	NA	NA	NA	12-16	48-57	NA	NA
	B6E	1.6 (1597)	56-60	40-43	①	116-123	71-76	14-19	12-17	11-17	65-87
	B6ZE	1.6 (1597)	56-60	40-43	37-40	116-123	71-76	14-19	28-34	11-17	65-87
	BPD	1.8 (1839)	56-60	40-43	35-37	116-123	71-76	14-19	28-34	11-17	65-87
	BPE	1.8 (1839)	56-60	40-43	36-38	116-123	71-76	14-19	12-17	11-17	65-87
	FS	2.0 (1991)	②	③	④	116-123	71-76	14-18	⑤	11-16	65-87
	JE-ZE	3.0 (2954)	⑥	⑦	⑧	116-123	76-81	14-19	16-21	10-13	65-87
	K8D	1.8 (1844)	⑨	⑩	⑪	116-123	45-50	14-19	14-19	11-17	65-87
	KJS	2.3 (2255)	⑨	NA	NA	116-122	45-49	14-18	12-17	11-16	65-87
	KL	2.5 (2496)	⑨	⑩	④	116-123	45-49	14-18	14-18	11-16	65-87
	KLD	2.5 (2496)	⑨	⑩	④	116-122	45-49	14-18	14-18	11-16	65-87
	Z5D	1.5 (1489)	②	40-43	22-25	116-122	71-76	14-18	14-16	11-16	65-87
1996	BPD	1.8 (1839)	56-60	40-43	35-36	116-122	71-76	14-18	29-34	11-16	65-87
	FS	2.0 (1991)	②	③	④	116-122	71-76	14-18	⑤	11-16	65-87
	KJS	2.3 (2255)	⑨	⑫	④	116-122	45-49	14-18	14-18	11-16	65-87
	KL	2.5 (2496)	⑨	⑫	④	116-122	45-49	14-18	14-18	11-16	65-87
	KLD	2.5 (2496)	⑨	⑩	④	116-122	45-49	14-18	14-18	11-16	65-87
	Z5D	1.5 (1489)	②	40-43	22-25	116-122	71-76	14-18	14-16	11-16	65-87
1997	BPD	1.8 (1839)	56-60	40-43	35-36	116-122	71-76	14-18	29-34	11-16	65-87
	FS	2.0 (1991)	②	③	④	116-122	71-76	14-18	⑤	11-16	65-87
	KJS	2.3 (2255)	⑨	⑫	④	116-122	45-49	14-18	14-18	11-16	65-87
	KL	2.5 (2496)	⑨	⑫	④	116-122	45-49	14-18	14-18	11-16	65-87
	KLD	2.5 (2496)	⑨	⑩	④	116-122	45-49	14-18	14-18	11-16	65-87
	Z5D	1.5 (1489)	②	40-43	22-25	116-122	71-76	14-18	14-16	11-16	65-87
1998-99	BPD	1.8 (1839)	56-60	40-43	35-36	116-122	71-76	14-18	29-34	11-16	65-87
	FS	2.0 (1991)	②	③	④	116-122	71-76	14-18	15-20	11-16	65-87
	KJS	2.3 (2255)	⑨	⑫	④	116-122	45-49	14-18	14-18	11-16	65-87
	KL	2.5 (2496)	⑨	⑫	④	116-122	45-49	14-18	14-18	11-16	65-87
	KLD	2.5 (2496)	⑨	⑩	④	116-122	45-49	14-18	14-18	11-16	65-87
	Z5D	1.5 (1489)	②	40-43	22-25	116-122	71-76	14-18	12-17	11-16	65-87

① 323: 35-38 ft. lbs.
MX3: 35-37 ft. lbs.
② Step 1: 16 ft. lbs.
Step 2: Turn each bolt 90 degrees
Step 3: Repeat Step 2
③ Step 1: 16 ft. lbs.
Step 2: Turn each bolt 90 degrees
④ Step 1: 19 ft. lbs.
Step 2: Turn each bolt 90 degrees
⑤ Nuts: 15-20 ft. lbs
Bolts: 12-16 ft. lbs.

⑥ Step 1: 14 ft. lbs.
Step 2: Turn each bolt 90 degrees
Step 3: Repeat Step 2
⑦ Step 1: 14 ft. lbs.
Step 2: Turn each bolt 90 degrees
Step 3: Turn each bolt 45 degrees
⑧ Step 1: 22 ft. lbs.
Step 2: Turn each nut 90 degrees
⑨ Step 1: 17-19 ft. lbs.
Step 2: Turn each bolt 90 degrees
Step 3: Repeat Step 2

⑩ Step 1: Inner bolts: 17-18 ft. lbs.; Outer bolts: 13-15 ft. lbs.
Step 2: Inner bolt Nos. 1-3: Turn each bolt 70 degrees
Step 3: Inner bolt No. 4: Turn each bolt 80 degrees
Step 4: Outer bolts: Turn each bolt 60 degrees
Step 5: Repeat Step 2
⑪ Step 1: 16-19 ft. lbs.
Step 2: Turn each bolt 90 degrees
Step 3: Repeat Step 2
⑫ Step 1: Inner bolts: 17-19 ft. lbs.;
Step 2: Outer bolts: 13.5-15.5 ft. lbs.
Step 3: Inner bolt Nos. 1-3: Turn each bolt 70 degrees
Step 4: Inner bolt No. 4: Turn each bolt 80 degrees
Step 5: Outer bolts: Turn each bolt 60 degrees
Step 5: Repeat Step 3-5

79231C52

SCHEDULED MAINTENANCE INTERVALS
(MAZDA 323, 626, 929, MX3, MX6, RX-7, MIATA, MILLENIA & PROTEGE)

TO BE SERVICED	TYPE OF SERVICE	VEHICLE MILEAGE INTERVAL (x1000)												
		7.5	15	22.5	30	37.5	45	52.5	60	67.5	75	82.5	90	97.5
Engine oil & filter④	R	✓	✓	✓	✓	✓	✓	✓	✓	✓	✓	✓	✓	✓
Air cleaner element	R				✓				✓				✓	
Engine coolant③	R				✓				✓				✓	
Spark plugs (except Millenia KJ engine)	R				✓				✓				✓	
Spark plugs (Millenia KJ engine)	R								✓					
Automatic transaxle fluid	S/I				✓				✓				✓	
Bolts & nuts on chassis & body	S/I				✓				✓				✓	
Brake lines, hoses & connections	S/I				✓				✓				✓	
Cooling system	S/I				✓				✓				✓	
Disc brakes	S/I				✓				✓				✓	
Drive belts (Millenia①)	S/I				✓				✓				✓	
Drive shaft dust boots	S/I				✓				✓				✓	
Exhaust system heat shield	S/I				✓				✓				✓	
Front & rear suspension ball joints	S/I				✓				✓				✓	
Fuel lines & hoses	S/I				✓				✓				✓	
Idle speed⑤	S/I				✓				✓				✓	
Steering operation & linkages	S/I				✓				✓				✓	
Engine timing belt②	R								✓					
Fuel filter	R								✓					
Manual transmission oil (Miata)	R								✓					
Hose & tube for emission	S/I								✓					

① (Millenia KJ engine) - replace every 105,000 miles.
② (Calif.) - inspect every 30,000 miles & replace at 105,000 miles (if not replaced previously).
③ (Millenia) - replace initially at 45,000 miles, & every 30,000 miles thereafter.
④ (RX-7) - change every 5000 miles.
⑤ (RX-7 & 929) - check every 15,000 miles.
R – Replace   S/I – Service or Inspect

79231C53

SCHEDULED MAINTENANCE INTERVALS
(MAZDA 323, 626, 929, MX3, MX6, RX-7, MIATA, MILLENIA & PROTEGE) (Cont.)

FREQUENT OPERATION MAINTENANCE (SEVERE SERVICE)

If a vehicle is operated under any of the following conditions it is considered severe service:

- Extremely dusty areas.
- 50% or more of the vehicle operation is in 32°C (90°F) or higher temperatures, or constant operation in temperatures below 0°C (32°F).
- Prolonged idling (vehicle operation in stop and go traffic).
- Frequent short running periods (engine does not warm to normal operating temperatures).
- Police, taxi, delivery usage or trailer towing usage.

Oil & oil filter change – change every 5000 miles.
Oil & oil filter change (Puerto Rico) - change every 3000 miles.
Air cleaner element – service or inspect every 15,000 miles.
Automatic transaxle fluid - service or inspect every 15,000 miles.
Bolts & nuts on chassis & body - tighten every 15,000 miles.
Disc brakes - service or inspect every 15,000 miles.

79231C54

SCHEDULED MAINTENANCE
MAZDA
323, 626, MX3, MX6
MIATA, MILLENIA, PROTEGE, RX7

The following should be used as a guide when determining the amount of work required for a particular service if taken to a repair shop.
In estimating how long a particular Scheduled Maintenance Service should take, please observe the following:

- **Chilton Time** is time based on field research and data supplied by the vehicle manufacturer.
- Labor time operations are given in hours and tenths of an hour.
- All labor operations, are to be used as a guide.

Mechanic Skill Level Codes:
    (G) GENERAL: Normally skilled with certification.
    (M) MAINTENANCE: Semi-skilled working on certicication.
    (P) PRECISION: Really skilled with multiple certification.

	Chilton Time		Chilton Time		Chilton Time
(M) 7500 Mile Service		**(M) 37500 Mile Service**		**(M) 75000 Mile Service**	
1995-96 MX-3	.3	1995-96 MX-3	.3	1995-96 MX-3	.3
1995-99 Miata	.3	1995-99 Miata	.3	1995-99 Miata	.3
1995 323	.3	1995 323	.3	1995 323	.3
1995-99 Protege	.3	1995-99 Protege	.3	1995-99 Protege	.3
1995-99 626, MX-6	.3	1995-99 626, MX-6	.3	1995-99 626, MX-6	.3
1995-99 Millenia	.3	1995-99 Millenia	.3	1995-99 Millenia	.3
1995 RX-7	.3	1995 RX-7	.3	1995 RX-7	.3
(M) 15000 Mile Service		**(M) 45000 Mile Service**		**(M) 82500 Mile Service**	
1995-96 MX-3	.3	1995-96 MX-3	.3	1995-96 MX-3	.3
1995-99 Miata	.3	1995-99 Miata	.3	1995-99 Miata	.3
1995 323	.3	1995 323	.3	1995 323	.3
1995-99 Protege	.3	1995-99 Protege	.3	1995-99 Protege	.3
1995-99 626, MX-6	.3	1995-99 626, MX-6	.3	1995-99 626, MX-6	.3
1995-99 Millenia	.3	1995-99 Millenia	.3	1995-99 Millenia	.3
1995 RX-7	.3	1995 RX-7	.3	1995 RX-7	.3
(M) 22500 Mile Service		**(M) 52500 Mile Service**		**(G) 90000 Mile Service**	
1995-96 MX-3	.3	1995-96 MX-3	.3	1995-96 MX-3	3.1
1995-99 Miata	.3	1995-99 Miata	.3	1995-99 Miata	3.5
1995 323	.3	1995 323	.3	1995 323	3.1
1995-99 Protege	.3	1995-99 Protege	.3	1995-99 Protege	3.1
1995-99 626, MX-6	.3	1995-99 626, MX-6	.3	1995-99 626, MX-6	3.7
1995-99 Millenia	.3	1995-99 Millenia	.3	1995-99 Millenia	3.7
1995 RX-7	.3	1995 RX-7	.3	1995 RX-7	3.5
(G) 30000 Mile Service		**(G) 60000 Mile Service**		**(M) 97500 Mile Service**	
1995-96 MX-3	3.1	1995-96 MX-3	6.2	1995-96 MX-3	.3
1995-99 Miata	3.5	1995-99 Miata	7.0	1995-99 Miata	.3
1995 323	3.1	1995 323	6.7	1995 323	.3
1995-99 Protege	3.1	1995-99 Protege	6.7	1995-99 Protege	.3
1995-99 626, MX-6	3.7	1995-99 626, MX-6	6.5	1995-99 626, MX-6	.3
1995-99 Millenia	3.7	1995-99 Millenia	6.5	1995-99 Millenia	.3
1995 RX-7	3.5	1995 RX-7	4.0	1995 RX-7	.3
		(M) 67500 Mile Service			
		1995-96 MX-3	.3		
		1995-99 Miata	.3		
		1995 323	.3		
		1995-99 Protege	.3		
		1995-99 626, MX-6	.3		
		1995-99 Millenia	.3		
		1995 RX-7	.3		

79231C55

MERCEDES-BENZ
C220 • C280 • CL500 • CLK320 • E320 • S320 • SL320 • S350 • E420 • S420 • E500 • S500 • S600 • SL500 • SL600 • SLK230

VEHICLE IDENTIFICATION CHART

Engine Code						Model Year	
Code	Liters	Cu. In. (cc)	Cyl.	Fuel Sys.	Eng. Mfg.	Code	Year
104.941	2.8L	171 (2799)	6	HFM	MB	S	1995
104.941	3.6L	220 (3606)	6	HFM	MB	T	1996
104.991	3.2L	195.1 (3199)	6	HFM	MB	V	1997
104.992	3.2L	195.1 (3199)	6	HFM	MB	W	1998
104.994	3.2L	195.1 (3199)	6	HFM - Motronic	MB	X	1999
104.995	3.2L	195.1 (3199)	6	ME2.1	MB		
111.961	2.2L	134 (2199)	4	HFM	MB		
111.973	2.3L	140 (2295)	4	ME2.1	MB		
111.974	2.3L	140 (2295)	4	ME2.1	MB		
112.920	2.8L	171 (2799)	6	ME 2.0	MB		
112.940	3.2L	195 (3199)	6	ME 2.0	MB		
112.941	3.2L	195 (3199)	6	ME 2.0	MB		
113.940	4.3L	260 (4265)	8	ME 2.0	MB		
119.970	5.0L	303 (4973)	8	LH	MB		
119.971	4.2L	256 (4196)	8	LH	MB		
119.972	5.0L	303 (4973)	8	ME-1	MB		
119.975	4.2L	256 (4196)	8	LH	MB		
119.980	5.0L	303 (4973)	8	ME-1	MB		
119.981	4.2L	256 (4196)	8	ME-1	MB		
119.982	5.0L	303 (4973)	8	ME-1	MB		
119.985	4.2L	256 (4196)	8	ME-1	MB		
120.980	6.0L	365 (5987)	12	LH	MB		
120.981	6.0L	365 (5987)	12	ME-1	MB		
120.982	6.0L	365 (5987)	12	ME-1	MB		
120.983	6.0L	365 (5987)	12	ME-1	MB		
603.971	3.5L	210 (3449)	6	EDC	MB		
606.910	3.0L	182.7 (2996)	6	EDC	MB		
606.962	3.0L	182.7 (2996)	6	EDC	MB		

HFM - Hot Film engine Management with sequential fuel injection

HFM - Motronic - Hot Film engine Management with Motronic controls

ME - Motronic Engine management

LH - Bosch Hot Wire fuel system

EDC - Electronic Diesel Control

79231CK3

ENGINE IDENTIFICATION

Year	Model	Engine Displacement Liters (cc)	Engine Series (ID/VIN)	Fuel System	No. of Cylinders	Engine Type
1995	C220	2.2 (2199)	111.961	HFM	4	DOHC
	C280	2.8 (2799)	104.941	HFM	6	DOHC
	E320	3.2 (3199)	104.992	HFM	6	DOHC
	S320	3.2 (3199)	104.994	HFM	6	DOHC
	SL320	3.2 (3199)	104.991	HFM	6	DOHC
	S350	3.5 (3449)	603.971	EDS	6	SOHC
	E420	4.2 (4196)	119.975	LH	8	DOHC
	S420	4.2 (5196)	119.971	LH	8	DOHC
	E500	5.0 (4973)	119.974	LH	8	DOHC
	S500	5.0 (4973)	119.970	LH	8	DOHC
	SL500	5.0 (4973)	119.972	LH	8	DOHC
	S600	6.0 (5987)	120.980	LH	12	DOHC
	SL600	6.0 (5987)	120.981	LH	12	DOHC
1996	C220	2.2 (2199)	111.961	HFM	4	DOHC
	C280	2.8 (2799)	104.941	HFM	6	DOHC
	C36	2.8 (2799)	104.941	HFM	6	DOHC
	E300	3.0 (2996)	606.912	EDS	6	DOHC
	E320	3.2 (3199)	104.995	HFM	6	DOHC
	S320	3.2 (3199)	104.994	HFM	6	DOHC
	SL320	3.2 (2199)	104.991	HFM	6	DOHC
	S420	4.2 (5196)	119.981	ME	8	DOHC
	S500	5.0 (4973)	119.980	ME	8	DOHC
	SL500	5.0 (4973)	119.982	ME	8	DOHC
	S600	6.0 (5987)	120.982	ME	12	DOHC
	SL600	6.0 (5987)	120.983	ME	12	DOHC
1997	C220	2.2 (2199)	111.961	HFM	4	DOHC
	C280	2.8 (2799)	104.941	HFM	6	DOHC
	C36	2.8 (2799)	104.941	HFM	6	DOHC
	E300	3.0 (2996)	606.912	EDS	6	DOHC
	E320	3.2 (3199)	104.995	HFM	6	DOHC
	S320	3.2 (3199)	104.994	HFM	6	DOHC
	SL320	3.2 (2199)	104.991	HFM	6	DOHC
	S420	4.2 (5196)	119.981	ME	8	DOHC
	S500	5.0 (4973)	119.980	ME	8	DOHC
	SL500	5.0 (4973)	119.982	ME	8	DOHC
	S600	6.0 (5987)	120.982	ME	12	DOHC
	SL600	6.0 (5987)	120.983	ME	12	DOHC
1998-99	C230	2.3 (2295)	111.974	ME2.1	4	DOHC
	C280	2.8 (2799)	112.920	ME 2.0	6	DOHC
	E300	3.0 (2996)	606.962	EDS	6	DOHC
	E320	3.2 (3199)	112.941	ME 2.0	6	DOHC
	E430	4.3 (4265)	113.940	ME 2.0	8	DOHC

79231CK4

ENGINE IDENTIFICATION

Year	Model	Engine Displacement Liters (cc)	Engine Series (ID/VIN)	Fuel System	No. of Cylinders	Engine Type
1998-99 (cont.)	S320	3.2 (3199)	104.994	HFM	6	DOHC
	S420	4.2 (5196)	119.981	ME	8	DOHC
	S500	5.0 (4973)	119.980	ME-1	8	DOHC
	S600	6.0 (5987)	120.982	ME	12	DOHC
	CL500	5.0 (4973)	119.980	ME-1	8	DOHC
	CL600	6.0 (5987)	120.982	ME	12	DOHC
	SL500	5.0 (4973)	119.982	ME	8	DOHC
	SL600	6.0 (5987)	120.983	ME	12	DOHC
	CLK320	3.2 (3199)	112.940	ME 2.0	6	DOHC
	SLK 230	2.3 (2295)	111.973	ME2.1	4	DOHC

CIS-E - Continuous Injection System with electronic controls
EDS - Electronic Diesel System
HFM - Multiport Fuel Injection

SOHC - Single overhead camshaft
DOHC - Double overhead camshaft
ME - Multiport Fuel Injection

79231CK5

GENERAL ENGINE SPECIFICATIONS

Year	Engine ID/VIN	Engine Displacement Liters (cc)	Fuel System Type	Net Horsepower @ rpm	Net Torque@rpm (ft. lbs.)	Bore x Stroke (in.)	Compression Ratio	Oil Pressure @ rpm
1995	111.961	2.2 (2199)	HFM	148@5500	155@4000	3.54x3.41	9.8:1	43.5-58@2000
	104.941	2.8 (2799)	HFM	194@5500	199@3750	3.54x2.89	10.0:1	69.6@2000
	104.991	3.2 (3199)	HFM	229@5600	232@3750	3.54x3.30	10.0:1	69.6@2000
	104.992	3.2 (3199)	HFM	217@5500	229@3750	3.54x3.30	10.0:1	69.6@2000
	104.994	3.2 (3199)	HFM	228@5600	229@3750	3.54x3.30	10.0:1	69.6@2000
	603.971	3.5 (3449)	EDS	148@4000	232@2000	3.50x3.60	22.0:1	①
	119.971	3.5 (3449)	EDS	275@5700	302@3900	3.62x3.11	11.0:1	23.2-72.5@2000
	119.975	4.2 (5196)	LH	275@5700	295@3900	3.62x3.11	11.0:1	23.2-72.5@2000
	119.970	5.0 (4973)	LH	315@5600	345@3900	3.80x3.35	10.0:1	23.2-72.5@2000
	119.972	5.0 (4973)	LH	315@5600	345@3900	3.80x3.35	10.0:1	23.2-72.5@2000
	119.974	5.0 (4973)	LH	315@5600	347@3900	3.80x3.35	10.0:1	23.2-72.5@2000
	120.980	6.0 (5987)	LH	389@5200	421@3800	3.50x3.16	10.0:1	①
	120.981	6.0 (5987)	LH	389@5200	420@3800	3.50x3.16	10.0:1	①
1996	111.961	2.2 (2199)	HFM	148@5500	155@4000	3.54x3.41	9.8:1	43.5-58@2000
	104.941	2.8 (2799)	HFM	194@5500	199@3750	3.54x2.89	10.0:1	69.6@2000
	606.912	3.0 (2966)	EDS	134@5000	155@2600	3.43x3.31	22.0:1	①
	104.991	3.2 (3199)	HFM	229@5600	232@3750	3.54x3.30	10.0:1	69.6@2000
	104.994	3.2 (3199)	HFM	228@5600	232@3750	3.54x3.30	10.0:1	69.6@2000
	104.995	3.2 (3199)	HFM	217@5500	229@5750	3.54x3.30	10.0:1	NA
	104.941 ②	3.6 (3606)	HFM	268@5750	280@4000	3.64x3.58	10.5:1	NA
	119.981	4.2 (5196)	ME	275@5700	295@3900	3.62x3.11	11.0:1	23.2-72.5@2000
	119.980	5.0 (4973)	ME	315@5600	347@3900	3.80x3.35	10.0:1	23.2-72.5@2000
	119.982	5.0 (4973)	ME	315@5600	345@3900	3.80x3.35	10.0:1	23.2-72.5@2000
	120.982	6.0 (5987)	ME	389@5200	420@3800	3.50x3.16	10.0:1	①
	120.983	6.0 (5987)	ME	389@5200	420@3800	3.50x3.16	10.0:1	①
1997	111.961	2.2 (2199)	HFM	148@5500	155@4000	3.54x3.41	9.8:1	43.5-58@2000
	104.941	2.8 (2799)	HFM	194@5500	199@3750	3.54x2.89	10.0:1	69.6@2000
	606.912	3.0 (2966)	EDS	134@5000	155@2600	3.43x3.31	22.0:1	①
	104.991	3.2 (3199)	HFM	229@5600	232@3750	3.54x3.30	10.0:1	69.6@2000
	104.994	3.2 (3199)	HFM	228@5600	232@3750	3.54x3.30	10.0:1	69.6@2000
	104.995	3.2 (3199)	HFM	217@5500	229@5750	3.54x3.30	10.0:1	NA
	104.941 ②	3.6 (3606)	HFM	268@5750	280@4000	3.64x3.58	10.5:1	NA
	119.981	4.2 (5196)	ME	275@5700	295@3900	3.62x3.11	11.0:1	23.2-72.5@2000
	119.980	5.0 (4973)	ME	315@5600	347@3900	3.80x3.35	10.0:1	23.2-72.5@2000
	119.982	5.0 (4973)	ME	315@5600	345@3900	3.80x3.35	10.0:1	23.2-72.5@2000
	120.982	6.0 (5987)	ME	389@5200	420@3800	3.50x3.16	10.0:1	①
	120.983	6.0 (5987)	ME	389@5200	420@3800	3.50x3.16	10.0:1	①
1998-99	111.974	2.3 (2295)	ME2.1	148@5500	162@4000	3.58x3.48	10.4:1	23.2-72.5@2000
	112.920	2.8 (2799)	ME 2.0	194@5800	195@30-4600	3.54x2.89	10.0:1	23.2-72.5@2000
	606.962	3.0 (2996)	EDS	174@5000	244@16-3000	3.43x3.31	22.0:1	①
	112.941	3.2 (3199)	ME 2.0	221@5500	232@30-4800	3.54x3.30	10.0:1	③
	113.940	4.3 (4265)	ME 2.0	275@5750	295@30-4400	3.54x3.31	10.0:1	③
	104.994	3.2 (3199)	HFM	228@5600	232@3750	3.54x3.30	10.0:1	69.6@2000

79231CK6

GENERAL ENGINE SPECIFICATIONS

Year	Engine ID/VIN	Engine Displacement Liters (cc)	Fuel System Type	Net Horsepower @ rpm	Net Torque@rpm (ft. lbs.)	Bore x Stroke (in.)	Compression Ratio	Oil Pressure @ rpm
1998-99 (cont.)	119.981	4.2 (5196)	ME	275@5700	295@3900	3.62x3.11	11.0:1	23.2-72.5@2000
	119.980	5.0 (4973)	ME	315@5600	347@3900	3.80x3.35	10.0:1	23.2-72.5@2000
	119.982	5.0 (4973)	ME	315@5600	345@3900	3.80x3.35	10.0:1	23.2-72.5@2000
	120.983	6.0 (5987)	ME	389@5200	420@3800	3.50x3.16	10.0:1	①
	112.940	3.2 (3199)	ME 2.0	215@5500	229@30-4600	3.54x3.30	10.0:1	③
	111.973	2.3 (2295)	ME2.1	185@5300	200@25-4800	3.58x3.48	8.8:1	③

CIS-E - Continuous Injection System with electronic controls

EDS - Electronic Diesel System

HFM - Multiport Fuel Injection and Ignition System

ME - Multiport Fuel Injection and Ignition System

NA - Not Available

① With engine at operating temperature, oil pressure should be a minimum of 43.5 psi @ idle. When engine is accelerated, oil pressure should increase immediately and reach a minimum pressure of 43.5 psi @ 3000 rpm.

② C36

③ 43.5@3000

   10@700

79231CK7

GASOLINE ENGINE TUNE-UP SPECIFICATIONS

Year	Engine ID/VIN	Engine Displacement Liters (cc)	Spark Plug Gap (in.)	Ignition Timing (deg.) MT	Ignition Timing (deg.) AT	Fuel Pump (psi)	Idle Speed (rpm) MT	Idle Speed (rpm) AT	Valve Clearance In.	Valve Clearance Ex.
1995	111.961	2.2 (2199)	0.031	—	①	③	—	700-800	HYD	HYD
	104.941	2.8 (2799)	0.031	—	7-11B	③	—	650-750	HYD	HYD
	104.991	3.2 (3199)	0.031	—	①	③	—	650-750	HYD	HYD
	104.992	3.2 (3199)	0.031	—	①	③	—	650-750	HYD	HYD
	104.994	3.2 (3199)	0.031	—	①	③	—	650-750	HYD	HYD
	119.975	4.2 (4196)	0.031	—	①	③	—	600-750	HYD	HYD
	119.971	4.2 (5196)	0.031	—	①	③	—	600-750	HYD	HYD
	119.970	5.0 (4973)	0.031	—	①	③	—	600-750	HYD	HYD
	119.972	5.0 (4973)	0.031	—	①	③	—	600-750	HYD	HYD
	119.974	5.0 (4973)	0.031	—	①	③	—	600-750	HYD	HYD
	120.980	6.0 (5987)	0.031	—	①	③	—	600-750	HYD	HYD
	120.981	6.0 (5987)	0.031	—	①	③	—	600-750	HYD	HYD
1996	111.961	2.2 (2199)	0.031	—	①	③	—	700-800	HYD	HYD
	104.941	2.8 (2799)	0.031	—	7-11B	③	—	650-750	HYD	HYD
	104.991	3.2 (3199)	0.031	—	①	③	—	650-750	HYD	HYD
	104.994	3.2 (3199)	0.031	—	①	③	—	650-750	HYD	HYD
	104.995	3.2 (3199)	0.031	—	①	③	—	650-750	HYD	HYD
	104.941 ②	3.6 (3606)	0.031	—	①	③	—	650-750	HYD	HYD
	119.981	4.2 (4196)	0.031	—	①	③	—	600-750	HYD	HYD
	119.980	5.0 (4973)	0.031	—	①	③	—	600-750	HYD	HYD
	119.982	5.0 (4973)	0.031	—	①	③	—	600-750	HYD	HYD
	120.982	6.0 (5987)	0.031	—	①	③	—	600-750	HYD	HYD
	120.983	6.0 (5987)	0.031	—	①	③	—	600-750	HYD	HYD
1997	111.961	2.2 (2199)	0.031	—	①	③	—	700-800	HYD	HYD
	104.941	2.8 (2799)	0.031	—	7-11B	③	—	650-750	HYD	HYD
	104.991	3.2 (3199)	0.031	—	①	③	—	650-750	HYD	HYD
	104.994	3.2 (3199)	0.031	—	①	③	—	650-750	HYD	HYD
	104.995	3.2 (3199)	0.031	—	①	③	—	650-750	HYD	HYD
	104.941 ②	3.6 (3606)	0.031	—	①	③	—	650-750	HYD	HYD
	119.981	4.2 (4196)	0.031	—	①	③	—	600-750	HYD	HYD
	119.980	5.0 (4973)	0.031	—	①	③	—	600-750	HYD	HYD
	119.982	5.0 (4973)	0.031	—	①	③	—	600-750	HYD	HYD
	120.982	6.0 (5987)	0.031	—	①	③	—	600-750	HYD	HYD
	120.983	6.0 (5987)	0.031	—	①	③	—	600-750	HYD	HYD
1998-99	111.974	2.3 (2295)	0.031	—	①	③	—	700-800	HYD	HYD
	112.920	2.8 (2799)	0.031	—	7-11B	③	—	650-750	HYD	HYD
	112.941	3.2 (3199)	0.031	—	①	③	—	650-750	HYD	HYD
	113.940	4.3 (4265)	0.031	—	①	③	—	650-750	HYD	HYD
	104.994	3.2 (3199)	0.031	—	①	③	—	650-750	HYD	HYD
	119.981	4.2 (5196)	0.031	—	①	③	—	650-750	HYD	HYD

79231CK8

GASOLINE ENGINE TUNE-UP SPECIFICATIONS

Year	Engine ID/VIN	Engine Displacement Liters (cc)	Spark Plug Gap (in.)	Ignition Timing (deg.)		Fuel Pump (psi)	Idle Speed (rpm)		Valve Clearance	
				MT	AT		MT	AT	In.	Ex.
1998-99 (cont.)	119.980	5.0 (4973)	0.031	—	①	③	—	650-750	HYD	HYD
	119.982	5.0 (4973)	0.031	—	①	③	—	600-750	HYD	HYD
	120.983	6.0 (5987)	0.031	—	①	③	—	600-750	HYD	HYD
	112.940	3.2 (3199)	0.031	—	①	③	—	600-750	HYD	HYD
	111.973	2.3 (2295)	0.031	—	①	③	—	600-750	HYD	HYD

NOTE: The Vehicle Emission Control Information label often reflects specification changes made during production. The label figures must be used if they differ from those in this chart.

B - Before top dead center

HYD - Hydraulic

① Timing controlled by engine control module. Adjustment is not possible.

② C36

③ 46-52 psi without vacuum applied

53-61 psi with vacuum applied

36 psi retention after 30 minutes

79231CK9

CAPACITIES

Year	Model	Engine ID/VIN	Engine Displacement Liters (cc)	Engine Oil with Filter	Transmission (pts.) 5-Spd	Transmission (pts.) Auto.	Drive Axle Front (pts.)	Drive Axle Rear (pts.)	Fuel Tank (gal.)	Cooling System (qts.)
1995	C220	111.961	2.2 (2199)	6.1	—	11.7	—	2.3	16.4	8.8
	C280	104.941	2.8 (2799)	7.9	—	11.7	—	2.3	16.4	10.5
	E320	104.992	3.2 (3199)	7.9	—	13.1	—	2.7	18.5	9.5
	S320	104.994	3.2 (3199)	7.9	—	13.1	—	2.7	26.4	15.3
	SL320	104.991	3.2 (3199)	7.9	—	13.1	—	2.7	21.1	11.6
	S350	603.971	3.5 (3449)	8.5	—	13.1	—	2.7	26.4	10.6
	E420	119.975	4.2 (4196)	8.5	—	16.3	—	2.7	23.7	13.2
	S420	119.971	4.2 (5196)	8.5	—	16.3	—	2.7	26.4	17.4
	E500	119.974	5.0 (4973)	8.5	—	16.3	—	2.7	23.7	13.2
	S500	119.970	5.0 (4973)	8.5	—	16.3	—	2.9	26.4	17.4
	SL500	119.972	5.0 (4973)	8.9	—	16.3	—	2.9	21.1	15.9
	S600	120.980	6.0 (5987)	10.0	—	16.3	—	2.9	26.4	19.6
	SL600	120.981	6.0 (5987)	10.6	—	16.3	—	2.9	21.1	21.1
1996	C220	111.961	2.2 (2199)	6.2	—	11.7	—	2.3	16.4	8.8
	C280	104.941	2.8 (2799)	7.9	—	11.7	—	2.3	16.4	10.5
	E300	606.912	3.0 (2996)	NA	—	NA	—	NA	17.2	NA
	E320	104.995	3.2 (3199)	NA	—	NA	—	NA	21.1	NA
	S320	104.994	3.2 (3199)	7.9	—	13.1	—	2.7	26.4	15.3
	SL320	104.991	3.2 (3199)	7.9	—	13.1	—	2.7	21.1	11.6
	C36	104.941	3.6 (3606)	7.9	—	NA	—	NA	16.4	7.9
	S420	119.981	4.2 (5196)	8.5	—	16.3	—	2.7	26.4	17.4
	S500	119.980	5.0 (4973)	8.5	—	16.3	—	2.9	26.4	17.4
	SL500	119.982	5.0 (4973)	8.9	—	16.3	—	2.9	21.1	15.9
	S600	120.982	6.0 (5987)	10.0	—	16.3	—	2.9	26.4	19.6
	SL600	120.983	6.0 (5987)	10.6	—	16.3	—	2.9	21.1	21.1
1997	C220	111.961	2.2 (2199)	6.2	—	11.7	—	2.3	16.4	8.8
	C280	104.941	2.8 (2799)	7.9	—	11.7	—	2.3	16.4	10.5
	E300	606.912	3.0 (2996)	NA	—	NA	—	NA	17.2	NA
	E320	104.995	3.2 (3199)	NA	—	NA	—	NA	21.1	NA
	S320	104.994	3.2 (3199)	7.9	—	13.1	—	2.7	26.4	15.3
	SL320	104.991	3.2 (3199)	7.9	—	13.1	—	2.7	21.1	11.6
	C36	104.941	3.6 (3606)	7.9	—	NA	—	NA	16.4	7.9
	S420	119.981	4.2 (5196)	8.5	—	16.3	—	2.7	26.4	17.4
	S500	119.980	5.0 (4973)	8.5	—	16.3	—	2.9	26.4	17.4
	SL500	119.982	5.0 (4973)	8.9	—	16.3	—	2.9	21.1	15.9
	S600	120.982	6.0 (5987)	10.0	—	16.3	—	2.9	26.4	19.6
	SL600	120.983	6.0 (5987)	10.6	—	16.3	—	2.9	21.1	21.1
1998-99	C230	111.974	2.3 (2295)	6	—	19.7	—	2.9	16.4	8.8
	C280	112.920	2.8 (2799)	8.5	—	19.7	—	2.9	16.4	10
	E300	606.962	3.0 (2996)	7	—	19.7	—	2.9	21.1	9.5
	E320	112.941	3.2 (3199)	6.5	—	19.7	—	2.9	①	9

79231CK0

CAPACITIES

Year	Model	Engine ID/VIN	Engine Displacement Liters (cc)	Engine Oil with Filter	Transmission (pts.) 5-Spd	Transmission (pts.) Auto.	Drive Axle Front (pts.)	Drive Axle Rear (pts.)	Fuel Tank (gal.)	Cooling System (qts.)
1998-99 (cont.)	E430	113.940	4.3 (4265)	8	—	19.7	—	2.9	23.0	11.6
	S320	104.994	3.2 (3199)	7.5	—	17	—	2.9	26.4	15.3
	S420	119.981	4.2 (5196)	8.5	—	16.3	—	2.7	26.4	17.4
	S500	119.980	5.0 (4973)	8.5	—	16.3	—	2.9	26.4	17.4
	S600	120.982	6.0 (5987)	10.0	—	16.3	—	2.9	26.4	19.6
	CL500	119.980	5.0 (4973)	8.5	—	19.7	—	2.9	26.4	21.1
	CL600	120.982	6.0 (5987)	10	—	19.7	—	2.9	26.4	21.1
	SL500	119.982	5.0 (4973)	8.9	—	16.3	—	2.9	21.1	15.9
	SL600	120.983	6.0 (5987)	10.6	—	16.3	—	2.9	21.1	21.1
	CLK320	112.940	3.2 (3199)	6.5	—	19.7	—	2.9	16.4	17.5
	SLK 230	111.973	2.3 (2295)	6	—	19.7	—	2.9	14.0	9

NOTE: All capacities are approximate. Add fluid gradually and check to be sure a proper fluid level is obtained.

① Sedan 21.1
　 Wagon 18.5

79231CL1

TORQUE SPECIFICATIONS
All readings in ft. lbs.

Year	Engine ID/VIN	Engine Displacement Liters (cc)	Cylinder Head Bolts	Main Bearing Bolts	Rod Bearing Bolts	Crankshaft Damper Bolts	Flywheel Bolts	Manifold Intake	Manifold Exhaust	Spark Plugs	Lug Nut
1995	111.961	2.2 (2199)	①	②	③	221	③	15	29.5	④	81
	104.941	2.8 (2799)	①	②	③	⑤	⑥	18.4	29.5	④	81
	104.991	3.2 (3199)	①	②	③	⑤	⑥	18.4	29.5	④	81
	104.992	3.2 (3199)	①	②	③	⑤	⑥	18.4	29.5	④	81
	104.994	3.2 (3199)	①	②	③	⑤	⑥	18.4	29.5	④	95
	603.971	3.5 (3449)	⑦	⑧	NA	⑤	⑨	18.4	18.4	④	95
	119.975	4.2 (4196)	①	⑩	⑪	295	⑫	18.4	22.1	④	81
	119.971	4.2 (5196)	①	⑩	⑪	295	⑫	18.4	22.1	④	95
	119.970	5.0 (4973)	①	⑩	⑪	295	⑫	18.4	22.1	④	95
	119.972	5.0 (4973)	①	⑩	⑪	295	⑫	18.4	22.1	④	81
	119.974	5.0 (4973)	①	⑩	⑪	295	⑫	18.4	22.1	④	81
	120.980	6.0 (5987)	⑬	⑩	⑪	295	⑫	18.4	29.5	④	95
	120.981	6.0 (5987)	⑬	⑩	⑪	295	⑫	18.4	29.5	④	81
1996	111.961	2.2 (2199)	⑭	②	③	300	③	14.7	29.5	④	81
	104.941	2.8 (2799)	①	②	③	⑤	⑥	18.4	29.5	④	81
	606.912	3.0 (2996)	NA	NA	NA	NA	NA	NA	NA	NA	81
	104.991	3.2 (3199)	①	②	③	⑤	⑥	18.4	29.5	④	81
	104.994	3.2 (3199)	①	②	③	⑤	⑥	18.4	29.5	④	110
	104.995	3.2 (3199)	NA	NA	NA	NA	NA	NA	NA	NA	81
	104.941 ⑮	3.6 (3606)	NA	NA	NA	NA	NA	NA	NA	NA	81
	119.981	4.2 (4196)	①	⑩	⑪	400	⑫	18.4	22.1	④	81
	119.980	5.0 (4973)	①	⑩	⑪	400	⑫	18.4	22.1	④	110
	119.982	5.0 (4973)	①	⑩	⑪	400	⑫	18.4	22.1	④	81
	120.982	6.0 (5987)	⑬	⑩	⑪	400	⑫	18.4	29.5	④	110
	120.983	6.0 (5987)	⑬	⑩	⑪	400	⑫	18.4	29.5	④	81
1997	111.961	2.2 (2199)	⑭	②	③	300	③	14.7	29.5	④	81
	104.941	2.8 (2799)	①	②	③	⑤	⑥	18.4	29.5	④	81
	606.912	3.0 (2996)	NA	NA	NA	NA	NA	NA	NA	NA	81
	104.991	3.2 (3199)	①	②	③	⑤	⑥	18.4	29.5	④	81
	104.994	3.2 (3199)	①	②	③	⑤	⑥	18.4	29.5	④	110
	104.995	3.2 (3199)	NA	NA	NA	NA	NA	NA	NA	NA	81
	104.941 ⑮	3.6 (3606)	NA	NA	NA	NA	NA	NA	NA	NA	81
	119.981	4.2 (4196)	①	⑩	⑪	400	⑯	18.4	22.1	④	81
	119.980	5.0 (4973)	①	⑩	⑪	400	⑯	18.4	22.1	④	110
	119.982	5.0 (4973)	①	⑩	⑪	400	⑯	18.4	22.1	④	81
	120.982	6.0 (5987)	⑬	⑩	⑪	400	⑯	18.4	29.5	④	110
	120.983	6.0 (5987)	⑬	⑩	⑪	400	⑯	18.4	29.5	④	81

79231CL2

TORQUE SPECIFICATIONS
All readings in ft. lbs.

Year	Engine ID/VIN	Engine Displacement Liters (cc)	Cylinder Head Bolts	Main Bearing Bolts	Rod Bearing Bolts	Crankshaft Damper Bolts	Flywheel Bolts	Manifold		Spark Plugs	Lug Nut
								Intake	Exhaust		
1998-99	104.994	3.2 (3199)	①	②	③	⑤	⑬	18.4	29.5	④	110
	111.973	2.3 (2295)	⑭	②	③	300	⑩	14.7	29.5	④	81
	111.974	2.3 (2295)	⑭	②	③	300	⑩	14.7	29.5	④	81
	112.920	2.8 (2799)	⑯	⑰	⑱	⑲	⑳	15	12	21	100
	112.940	3.2 (3199)	⑯	⑰	⑱	⑲	⑳	15	12	21	100
	112.941	3.2 (3199)	⑯	⑰	⑱	⑲	⑳	15	12	21	100
	113.940	4.3 (4265)	⑯	⑰	⑱	⑲	⑳	15	12	21	100
	119.980	5.0 (4973)	①	⑩	⑪	400	⑮	18.4	22.1	④	110
	119.981	4.2 (5196)	①	⑩	⑪	400	⑮	18.4	22.1	④	81
	119.982	5.0 (4973)	①	⑩	⑪	400	⑮	18.4	22.1	④	81
	120.983	6.0 (5987)	⑬	⑩	⑪	400	⑮	18.4	29.5	④	81

① Step 1: 40.5 ft. lbs.
Step 2: + 90 degrees
Step 3: + 90 degrees
Step 4: M8 bolts: 18.4 ft. lbs.

② Step 1: 40.5 ft. lbs.
Step 2: 90-100 degrees

③ Step 1: 22.1 ft. lbs.
Step 2: + 90-100 degrees

④ Spark plug with conical seat: 7.3-14.7 ft. lbs.
Spark plug with flat seat: 14.7-22.1 ft. lbs.

⑤ Step 1: 147.5 ft. lbs.
Step 2: + 90 degrees

⑥ Without dual mass flywheel:
Step 1: 22.1 ft. lbs.
Step 2: + 90-100 degrees
With dual mass flywheel:
Step 1: 29.5 ft. lbs.
Step 2: + 90-100 degrees

⑦ Step 1: 11 ft. lbs.
Step 2: 25.8 ft. lbs.
Step 3: + 90 degrees
Step 4: Wait ten minutes
Step 5: + 90 degrees
Step 6: M8 bolts: 18.4 ft. lbs.

⑧ M11 bolts:
Step 1: 40.5 ft. lbs.
Step 2: + 90-100 degrees

⑨ Without dual mass flywheel:
Step 1: 25.8 ft. lbs.
Step 2: + 90-100 degrees
With dual mass flywheel:
Step 1: 29.5 ft. lbs.
Step 2: + 90-100 degrees

⑩ M8 bolts: 22.1 ft. lbs.
M10 bolts: 36.8 ft. lbs.

⑪ Step 1: 33.1 ft. lbs.
Step 2: + 90-100 degrees

⑫ Step 1: 25.8 ft. lbs.
Step 2: + 90-100 degrees

⑬ Step 1: 40.5 ft. lbs.
Step 2: + 90 degrees
Step 3: + 90 degrees

⑭ Step 1: 51.6 ft. lbs.
Step 2: + 90 degrees
Step 3: + 90 degrees

⑮ C36

⑯ Step 1: 15 ft. lbs.
Step 2: 37 ft. lbs.
Step 3: 65 degrees
Step 4: 65 degrees

⑰ M8x40: 18 ft. lbs.
M8x75
Step 1: 10 ft. lbs.
Step 2: 90-100 degrees
M10x90
Step 1: 15 ft. lbs.
Step 2: 90-100 degrees

⑱ Step 1: 44 inch lbs.
Step 2: 18 ft. lbs
Step 3. 90 degrees

⑲ Step 1: 148 ft. lbs.
Step 2: 95 degrees

⑳ Step 1: 33 ft. lbs.
Step 2: 90 degrees

79231CL3

MITSUBISHI
3000GT • Diamante • Eclipse • Galant • Mirage

VEHICLE IDENTIFICATION CHART

Engine Code						Model Year	
Code	Liters	Cu. In. (cc)	Cyl.	Fuel Sys.	Eng. Mfg.	Code	Year
6G74	3.5	213 (3497)	6	MFI	Mitsubishi	S	1995
6G72	3.0	181 (2972)	6	MFI	Mitsubishi	T	1996
6G72	3.0	181 (2972)	6	MFI-Turbo	Mitsubishi	V	1997
4G64	2.4	143 (2351)	4	MFI	Mitsubishi	W	1998
4G15	1.5	87 (1468)	4	MFI	Mitsubishi	X	1999
4G93	1.8	112 (1834)	4	MFI	Mitsubishi		
420A	2.0	122 (1996)	4	MFI	Mitsubishi		
4G63	2.0	122 (1997)	4	MFI-Turbo	Mitsubishi		

79231C56

ENGINE IDENTIFICATION

Year	Model	Engine Displacement Liters (cc)	Engine Series (ID/VIN)	Fuel System	No. of Cylinders	Engine Type
1995	3000GT	3.0 (2972)	6G72	MFI	6	DOHC
	Diamante	3.0 (2972)	6G72	MFI	6	SOHC
	Diamante	3.0 (2972)	6G72	MFI	6	DOHC
	Eclipse	2.0 (1996)	420A	MFI	4	DOHC
	Eclipse	2.0 (1997)	4G63	MFI	4	DOHC
	Galant	2.4 (2350)	4G64	MFI	4	SOHC
	Mirage	1.5 (1468)	4G15	MFI	4	SOHC
	Mirage	1.8 (1834)	4G93	MFI	4	SOHC
1996	3000GT	3.0 (2972)	6G72	MFI	6	DOHC
	Diamante	3.0 (2972)	6G72	MFI	6	SOHC
	Diamante	3.0 (2972)	6G72	MFI	6	DOHC
	Eclipse	2.0 (1996)	420A	MFI	4	DOHC
	Eclipse	2.0 (1997)	4G63	MFI	4	DOHC
	Eclipse	2.4 (2350)	4G64	MFI	4	SOHC
	Galant	2.4 (2350)	4G64	MFI	4	SOHC
	Mirage	1.5 (1468)	4G15	MFI	4	SOHC
	Mirage	1.8 (1834)	4G93	MFI	4	SOHC
1997	3000GT	3.0 (2972)	6G72	MFI	6	DOHC
	Diamante	3.0 (2972)	6G72	MFI	6	SOHC
	Diamante	3.0 (2972)	6G72	MFI	6	DOHC
	Eclipse	2.0 (1996)	420A	MFI	4	DOHC
	Eclipse	2.0 (1997)	4G63	MFI	4	DOHC
	Eclipse	2.4 (2350)	4G64	MFI	4	SOHC
	Galant	2.4 (2350)	4G64	MFI	4	SOHC
	Mirage	1.5 (1468)	4G15	MFI	4	SOHC
	Mirage	1.8 (1834)	4G93	MFI	4	SOHC
1998-99	3000GT	3.0 (2972)	6G72	MFI	6	SOHC
	3000GT	3.0 (2972)	6G72	MFI	6	DOHC
	3000GT	3.0 (2972)	6G72	MFI-Turbo	6	DOHC
	Diamante	3.5 (3497)	6G74	MFI	6	SOHC
	Eclipse	2.0 (1996)	420A	MFI	4	DOHC
	Eclipse	2.0 (1997)	4G63	MFI-Turbo	4	DOHC
	Eclipse Spyder	2.0 (1997)	4G63	MFI-Turbo	4	DOHC
	Eclipse Spyder	2.4 (2351)	4G64	MFI	4	SOHC
	Galant	2.4 (2351)	4G64	MFI	4	SOHC
	Mirage	1.5 (1468)	4G15	MFI	4	SOHC
	Mirage	1.8 (1834)	4G93	MFI	4	SOHC

MFI - Multiport Fuel Injection
Turbo - Turbocharged
SOHC - Single Overhead Camshaft
DOHC - Double Overhead Camshaft

79231C57

GENERAL ENGINE SPECIFICATIONS

Year	Engine ID/VIN		Engine Displacement Liters (cc)	Fuel System Type	Net Horsepower @ rpm	Net Torque @ rpm (ft. lbs.)	Bore x Stroke (In.)	Compression Ratio	Oil Pressure @ rpm
1995	420A		2.0 (1996)	MFI	140@6000	130@4800	3.44x3.27	9.6:1	11@idle
	4G15		1.5 (1468)	MFI	92@6000	93@3000	2.97x3.23	9.2:1	54@2000
	4G63		2.0 (1997)	MFI	①	②	3.35x3.46	8.5:1	11@idle
	4G64		2.4 (2350)	MFI	③	148@3000	3.41x3.94	9.5:1	41@2000
	4G93		1.8 (1834)	MFI	④	116@4500	3.19x3.50	9.5:1	41@2000
	6G72	⑤	3.0 (2972)	MFI	175@5500	185@3000	3.59x2.99	10.0:1	30-80@2000
	6G72	⑥	3.0 (2972)	MFI	202@6000	201@3500	3.59x2.99	10.0:1	30-80@2000
	6G72	⑦	3.0 (2972)	MFI	⑧	⑨	3.59x2.99	9.0:1	30-80@2000
	6G72	⑩	3.0 (2972)	MFI	⑪	205@4500	3.59x2.99	10.0:1	30-80@2000
	6G72	⑫	3.0 (2972)	MFI	320@6000	315@2500	3.59x2.99	8.0:1	30-80@2000
	6G74		3.5 (3497)	MFI	214@5000	228@3000	3.66x3.38	9.5:1	30-80@2000
1996	420A		2.0 (1996)	MFI	140@6000	130@4800	3.44x3.27	9.6:1	11@idle
	4G15		1.5 (1468)	MFI	92@6000	93@3000	2.97x3.23	9.2:1	54@2000
	4G63		2.0 (1997)	MFI	①	②	3.35x3.46	8.5:1	11@idle
	4G64		2.4 (2350)	MFI	③	148@3000	3.41x3.94	9.5:1	41@2000
	4G93		1.8 (1834)	MFI	④	116@4500	3.19x3.50	9.5:1	41@2000
	6G72	⑤	3.0 (2972)	MFI	175@5500	185@3000	3.59x2.99	10.0:1	30-80@2000
	6G72	⑥	3.0 (2972)	MFI	202@6000	201@3500	3.59x2.99	10.0:1	30-80@2000
	6G72	⑦	3.0 (2972)	MFI	⑧	⑨	3.59x2.99	9.0:1	30-80@2000
	6G72	⑩	3.0 (2972)	MFI	⑪	205@4500	3.59x2.99	10.0:1	30-80@2000
	6G72	⑫	3.0 (2972)	MFI	320@6000	315@2500	3.59x2.99	8.0:1	30-80@2000
	6G74		3.5 (3497)	MFI	214@5000	228@3000	3.66x3.38	9.5:1	30-80@2000
1997	420A		2.0 (1996)	MFI	140@6000	130@4800	3.44x3.27	9.6:1	11@idle
	4G15		1.5 (1468)	MFI	92@6000	93@3000	2.97x3.23	9.2:1	54@2000
	4G63		2.0 (1997)	MFI	①	②	3.35x3.46	8.5:1	11@idle
	4G64		2.4 (2350)	MFI	③	148@3000	3.41x3.94	9.5:1	41@2000
	4G93		1.8 (1834)	MFI	④	116@4500	3.19x3.50	9.5:1	41@2000
	6G72	⑤	3.0 (2972)	MFI	175@5500	185@3000	3.59x2.99	10.0:1	30-80@2000
	6G72	⑥	3.0 (2972)	MFI	202@6000	201@3500	3.59x2.99	10.0:1	30-80@2000
	6G72	⑦	3.0 (2972)	MFI	⑧	⑨	3.59x2.99	9.0:1	30-80@2000
	6G72	⑩	3.0 (2972)	MFI	⑪	205@4500	3.59x2.99	10.0:1	30-80@2000
	6G72	⑫	3.0 (2972)	MFI	320@6000	315@2500	3.59x2.99	8.0:1	30-80@2000
	6G74		3.5 (3497)	MFI	214@5000	228@3000	3.66x3.38	9.5:1	30-80@2000
1998-99	420A		2.0 (1996)	MFI	140@6000	130@4800	3.45x3.27	9.6:1	11@idle
	4G15		1.5 (1468)	MFI	92@5500	93@3000	2.97x3.23	9.0:1	54@2000
	4G63		2.0 (1997)	MFI-Turbo	210@6000	214@3000	3.35x3.46	8.5:1	11@idle
	4G64		2.4 (2351)	MFI	141@5500	148@3000	3.41x3.94	9.5:1	41@2000
	4G93		1.8 (1834)	MFI	113@5500	116@4500	3.19x3.50	9.5:1	41@2000
	6G72	⑬	3.0 (2972)	MFI	161@5500	185@4000	3.59x2.99	8.9:1	30-80@2000
	6G72	⑩	3.0 (2972)	MFI	218@6000	205@4500	3.59x2.99	10.0:1	30-80@2000
	6G72	⑫	3.0 (2972)	MFI	320@6000	315@2500	3.59x2.99	8.0:1	30-80@2000
	6G74		3.5 (3497)	MFI	214@5000	228@3000	3.65x3.37	9.0:1	30-80@2000

MFI - Multiport fuel injection

① Manual transaxle: 210@6000
  Automatic transaxle: 205@6000
② Manual transaxle: 214@3000
  Automatic transaxle: 220@3000
③ California: 138@5500
  Except California: 141@5500
④ California: 111@6000
  Except California: 113@6000
⑤ Diamante SOHC, 2 valves per cylinder
⑥ Diamante DOHC, 4 valves per cylinder
⑦ Montero SOHC, 4 valves per cylinder

⑧ California: 168@5500
  Except California: 177@5500
⑨ California: 183@4500
  Except California: 188@4500
⑩ 3000GT DOHC Engine
⑪ California: 218@6000
  Except California: 222@6000
⑫ 3000GT Spyder DOHC, 4 valves
  per cylinder, Twin Turbochargers
⑬ 3000GT SOHC Engine

79231C58

GASOLINE ENGINE TUNE-UP SPECIFICATIONS

Year	Engine ID/VIN	Engine Displacement Liters (cc)	Spark Plugs Gap (in.)	Ignition Timing (deg.) MT	AT	Fuel Pump (psi)	Idle Speed (rpm) MT	AT	Valve Clearance In.	Ex.
1995	4G15	1.5 (1468)	0.039-0.043	5B	5B	38	750	750	0.008	0.010
	4G63	2.0 (1997)	0.039-0.043	5B	5B	38	750	750	HYD	HYD
	4G63 ③	2.0 (1997)	0.028-0.031	5B	5B	①	750	750	HYD	HYD
	4G64 ⑦	2.4 (2350)	0.039-0.043	5B	5B	38	750	750	HYD	HYD
	4G64 ⑧	2.4 (2350)	0.039-0.043	5B	5B	38	800	800	HYD	HYD
	4G93	1.8 (1834)	0.039-0.043	5B	5B	38	750	750	0.008	0.012
	6G72 ⑦	3.0 (2972)	0.039-0.043	5B	5B	38	700	700	HYD	HYD
	6G72 ⑧	3.0 (2972)	0.039-0.043	5B	5B	38	700	700	HYD	HYD
	6G72 ③	3.0 (2972)	0.039-0.043	5B	5B	34	700	700	HYD	HYD
1996	420A	2.0 (1996)	0.033-0.038	12B	12B	38 ④	700-900	700-900	HYD	HYD
	4G15	1.5 (1468)	0.039-0.043	5B	5B	38 ④	650-850	650-850	0.008 ⑤	0.010 ⑤
	4G63 ③	2.0 (1997)	0.028-0.031	5B	5B	38 ④	650-850	650-850	HYD	HYD
	4G64	2.4 (2350)	0.039-0.043	5B	5B	38 ④	650-850	650-850	HYD	HYD
	4G93	1.8 (1834)	0.039-0.043	5B	5B	38 ④	600-800	650-850	0.008 ⑤	0.012 ⑤
	6G72	3.0 (2972)	0.039-0.043	5B	5B	38 ④	600-800	600-800	HYD	HYD
	6G72	3.0 (2972)	0.039-0.043	5B	5B	38 ④	600-800	600-800	HYD	HYD
	6G72 ③	3.0 (2972)	0.039-0.043	5B	5B	34 ④	600-800	600-800	HYD	HYD
	6G74	3.5 (3497)	0.039-0.043	5B	5B	50 ⑥	600-800	600-800	HYD	HYD
1997	420A	2.0 (1996)	0.033-0.038	12B	12B	38 ④	700-900	700-900	HYD	HYD
	4G15	1.5 (1468)	0.039-0.043	5B	5B	38 ④	650-850	650-850	0.008 ⑤	0.010 ⑤
	4G63 ③	2.0 (1997)	0.028-0.031	5B	5B	38 ④	650-850	650-850	HYD	HYD
	4G64	2.4 (2350)	0.039-0.043	5B	5B	38 ④	650-850	650-850	HYD	HYD
	4G93	1.8 (1834)	0.039-0.043	5B	5B	38 ④	600-800	650-850	0.008 ⑤	0.012 ⑤
	6G72	3.0 (2972)	0.039-0.043	5B	5B	38 ④	600-800	600-800	HYD	HYD
	6G72	3.0 (2972)	0.039-0.043	5B	5B	38 ④	600-800	600-800	HYD	HYD
	6G72 ③	3.0 (2972)	0.039-0.043	5B	5B	34 ④	600-800	600-800	HYD	HYD
	6G74	3.5 (3497)	0.039-0.043	5B	5B	50 ⑥	600-800	600-800	HYD	HYD
1998-99	420A	2.0 (1996)	0.048-0.053	⑨	⑨	47-50	700-900	700-900	HYD	HYD
	4G15	1.5 (1468)	0.039-0.043	2-8B	2-8B	38	600-800	600-800	HYD	HYD
	4G63	2.0 (1997)	0.028-0.031	2-8B	2-8B	33	650-850	650-850	HYD	HYD
	4G64	2.4 (2351)	0.039-0.043	2-8B	2-8B	38	650-850	650-850	HYD	HYD
	4G93	1.8 (1834)	0.039-0.043	2-8B	2-8B	38	600-800	600-800	HYD	HYD
	6G72	3.0 (2972)	0.039-0.043	2-8B	2-8B	38 ⑨	600-800	600-800	HYD	HYD
	6G72 ③	3.0 (2972)	0.039-0.043	2-8B	2-8B	34 ⑩	600-800	600-800	HYD	HYD
	6G74	3.5 (3497)	0.039-0.043	2-8B	2-8B	38 ⑨	600-800	600-800	HYD	HYD

NOTE: The Vehicle Emission Control Information label often reflects specification changes made during production. The label figures must be used if they differ from those in this chart.

B - Before top dead center

HYD - Hydraulic

① Manual transmission: 36
   Automatic transmission: 43
② California: 700
③ Turbocharged
④ With vacuum hose connected
⑤ Hot engine
⑥ With vacuum hose disconnected
⑦ Single Overhead Camshaft
⑧ Double Overhead Camshaft
⑨ Automatically controlled by the PCM and not manually adjustable

79231C59

CAPACITIES

Year	Model	Engine ID/VIN	Engine Displacement Liters (cc)	Engine Oil with Filter	Transmission (pts.) 4-Spd	5 or 6-Spd	Auto.	Transfer Case (pts.)	Drive Axle Front (pts.)	Rear (pts.)	Fuel Tank (gal.)	Cooling System (qts.)
1995	3000GT	6G72	3.0 (1972)	①	—	②	15.8	③	NA	2.3	19.8	8.5
	Diamante	6G72	3.0 (1972)	4.5	—	—	15.8	NA	NA	NA	19.0	8.5
	Eclipse	4G37	1.8 (1755)	4.1	—	3.8	12.8	NA	NA	NA	15.9	6.6
	Eclipse	4G63	2.0 (1997)	④	—	⑤	⑥	1.2	NA	1.48	15.9	7.6
	Galant	4G64	2.4 (2350)	4.7	—	4.4	⑦	NA	NA	NA	16.9	7.4
	Mirage	4G15	1.5 (1468)	3.5	—	3.8	12.6	NA	NA	NA	13.2	5.3
	Mirage	4G93	1.8 (1834)	4.0	—	3.8	12.6	NA	NA	NA	13.2	5.3
	Precis	G4DJ	1.5 (1468)	3.6	3.6	3.8	12.8	NA	NA	NA	11.9	5.6
1996	3000GT	6G72	3.0 (2972)	4.5	—	②	15.8	⑧	NA	2.3	19.8	8.5
	Diamante	6G72	3.0 (2972)	4.5	—	—	15.8	NA	NA	NA	19.0	8.5
	Eclipse	420A	2.0 (1996)	4.5	—	4.2	18.2	NA	NA	NA	17.0	7.4
	Eclipse	4G63	2.0 (1997)	4.5	—	⑨	14.2	1.2	NA	1.48	17.0	7.4
	Eclipse	4G64	2.4 (2350)	4.5	—	4.2	12.8	NA	NA	NA	17.0	14.8
	Galant	4G64	2.4 (2350)	4.5	—	4.6	12.6	NA	NA	NA	16.9	7.4
	Mirage	4G15	1.5 (1468)	3.7	—	3.8	12.6	NA	NA	NA	13.2	5.3
	Mirage	4G93	1.8 (1834)	4.2	—	3.8	12.6	NA	NA	NA	13.2	6.3
1997	3000GT	6G72	3.0 (2972)	4.5	—	②	15.8	⑧	NA	2.3	19.8	8.5
	Diamante	6G72	3.0 (2972)	4.5	—	—	15.8	NA	NA	NA	19.0	8.5
	Eclipse	420A	2.0 (1996)	4.5	—	4.2	18.2	NA	NA	NA	17.0	7.4
	Eclipse	4G63	2.0 (1997)	4.5	—	⑨	14.2	1.2	NA	1.48	17.0	7.4
	Eclipse	4G64	2.4 (2350)	4.5	—	4.2	12.8	NA	NA	NA	17.0	14.8
	Galant	4G64	2.4 (2350)	4.5	—	4.6	12.6	NA	NA	NA	16.9	7.4
	Mirage	4G15	1.5 (1468)	3.7	—	3.8	12.6	NA	NA	NA	13.2	5.3
	Mirage	4G93	1.8 (1834)	4.2	—	3.8	12.6	NA	NA	NA	13.2	6.3
1998-99	3000GT	6G72	3.0 (2972)	①	—	⑩	15.8	1.26 ⑪	NA	2.4 ⑪	19.8	8.5
	Diamante	6G74	3.5 (3497)	4.7	—	—	18.0	—	—	—	18.7	10.0
	Eclipse	420A	2.0 (1996)	4.5	—	4.2	18.2	—	—	—	17.0	7.4
	Eclipse	4G63	2.0 (1997)	4.6	—	⑫	14.2	1.06	—	1.8	17.0	7.4
	Eclipse Spyder	4G63	2.0 (1997)	4.6	—	⑫	14.2	1.06	—	1.8	17.0	7.4
	Eclipse Spyder	4G64	2.4 (2351)	4.5	—	4.8	12.8	1.06	—	1.8	17.0	7.4
	Galant	4G64	2.4 (2351)	4.5	—	4.6	12.6	—	—	—	16.9	7.4
	Mirage	4G15	1.5 (1468)	3.5	—	4.4	16.4	—	—	—	13.2	5.3
	Mirage	4G93	1.8 (1834)	4	—	4.6	16.4	—	—	—	13.2	6.3

NOTE: All capacities are approximate. Add fluid gradually and ensure a proper fluid level is obtained.

NA - Not Available

① Without Turbocharger: 4.5 qts.
   With Turbocharger: 4.9 qts.
② FWD: 4.8 pts.
   AWD: 5.0 pts.
③ 5 speed: 0.58 pts.
   6 speed: 1.26 pts.
④ Without Turbocharger: 4.6 qts.
   With Turbocharger: 5.1 qts.
⑤ Without Turbocharger: 3.8 pts.
   With Turbocharger: 4.9 pts.
⑥ Without Turbocharger: 12.8 pts.
   With Turbocharger: 14.8 pts.

⑦ Single overhead camshaft: 12.6 pts.
   Double overhead camshaft: 15.8 pts.
⑧ M/T: 0.58 pts.
   A/T: 0.64 pts.
⑨ FWD: 4.2 pts.
   AWD: 4.6 pts.
⑩ FWD: 4.2 pts.
   AWD: 5.0 pts.
⑪ AWD models only.
⑫ FWD: 4.2 pts.
   AWD: 4.8 pts.

79231C60

VALVE SPECIFICATIONS

Year	Engine ID/VIN	Engine Displacement Liters (cc)	Seat Angle (deg.)	Face Angle (deg.)	Spring Test Pressure (lbs. @ in.)	Spring Installed Height (in.)	Stem-to-Guide Clearance (in.)		Stem Diameter (in.)	
							Intake	Exhaust	Intake	Exhaust
1995	4G15	1.5 (1468)	44-44.5	45-45.5	①	1.57	0.001-0.002	0.002-0.004	0.260	0.256
	4G37	1.8 (1755)	44-44.5	45-45.5	68@1.47	1.47	0.001-0.002	0.002-0.004	0.315	0.315
	4G63	2.0 (1997)	44-44.5	45-45.5	66@1.57	1.57	0.001-0.002	0.002-0.004	0.260	0.256
	4G64 ②	2.4 (2350)	44-44.5	45-45.5	60@1.74	1.74	0.001-0.002	0.002-0.003	0.236	0.232
	4G64 ③	2.4 (2350)	44-44.5	45-45.5	54@1.57	1.57	0.001-0.002	0.002-0.004	0.260	0.256
	4G93	1.8 (1834)	44-44.5	45-45.5	49@1.74	1.74	0.001-0.002	0.001-0.002	0.236	0.236
	6G72	3.0 (2972)	44-44.5	45-45.5	72.5@1.59	1.59	0.001-0.002	0.002-0.004	0.315	0.311
	6G72 ③	3.0 (2972)	44-44.5	45-45.5	52.9@1.49	1.49	0.001-0.002	0.002-0.004	0.260	0.256
	G4DJ	1.5 (1468)	44-44.5	45-45.5	①	1.57	0.001-0.002	0.002-0.004	0.260	0.256
1996	420A	2.0 (1996)	45	45-45.5	38@1.50	1.50	0.0019-0.0026	0.0029-0.0037	0.234	0.233
	4G15	1.5 (1468)	44-44.5	45-45.5	①	1.57	0.0008-0.0020	0.0020-0.0035	0.260	0.256
	4G63	2.0 (1997)	44-44.5	45-45.5	54@1.57	1.57	0.0008-0.0020	0.0020-0.0035	0.260	0.256
	4G64 ④	2.4 (2350)	44-44.5	45-45.5	73@1.59	1.59	0.0008-0.0020	0.0020-0.0035	0.315	0.311
	4G64 ⑤	2.4 (2350)	44-44.5	45-45.5	60@1.74	1.74	0.0008-0.0020	0.0008-0.0028	0.236	0.232
	4G93	1.8 (1834)	44-44.5	45-45.5	49@1.74	1.74	0.0008-0.0016	0.0012-0.0024	0.236	0.236
	6G72 ⑥	3.0 (2972)	44-44.5	45-45.5	72.5@1.59	1.59	0.0012-0.0024	0.0020-0.0035	0.315	0.311
	6G72 ⑦	3.0 (2972)	44-44.5	45-45.5	60@1.74	1.74	0.0008-0.0020	0.0016-0.0028	0.236	0.236
	6G72 ⑧	3.0 (2972)	44-44.5	45-45.5	52.9@1.49	1.49	0.0008-0.0020	0.0020-0.0035	0.260	0.256
	6G74	3.5 (3497)	44-44.5	45-45.5	52.9@1.49	1.49	0.0008-0.0020	0.0020-0.0035	0.260	0.256
1997	420A	2.0 (1996)	45	45-45.5	38@1.50	1.50	0.0019-0.0026	0.0029-0.0037	0.234	0.233
	4G15	1.5 (1468)	44-44.5	45-45.5	①	1.57	0.0008-0.0020	0.0020-0.0035	0.260	0.256
	4G63	2.0 (1997)	44-44.5	45-45.5	54@1.57	1.57	0.0008-0.0020	0.0020-0.0035	0.260	0.256
	4G64 ④	2.4 (2350)	44-44.5	45-45.5	73@1.59	1.59	0.0008-0.0020	0.0020-0.0035	0.315	0.311
	4G64 ⑤	2.4 (2350)	44-44.5	45-45.5	60@1.74	1.74	0.0008-0.0020	0.0008-0.0028	0.236	0.232
	4G93	1.8 (1834)	44-44.5	45-45.5	49@1.74	1.74	0.0008-0.0016	0.0012-0.0024	0.236	0.236

79231C61

VALVE SPECIFICATIONS

Year	Engine ID/VIN	Engine Displacement Liters (cc)	Seat Angle (deg.)	Face Angle (deg.)	Spring Test Pressure (lbs. @ in.)	Spring Installed Height (in.)	Stem-to-Guide Clearance (in.)		Stem Diameter (in.)	
							Intake	Exhaust	Intake	Exhaust
1997 (cont.)	6G72 ⑥	3.0 (2972)	44-44.5	45-45.5	72.5@1.59	1.59	0.0012-0.0024	0.0020-0.0035	0.315	0.311
	6G72 ⑦	3.0 (2972)	44-44.5	45-45.5	60@1.74	1.74	0.0008-0.0020	0.0016-0.0028	0.236	0.236
	6G72 ⑧	3.0 (2972)	44-44.5	45-45.5	52.9@1.49	1.49	0.0008-0.0020	0.0020-0.0035	0.260	0.256
	6G74	3.5 (3497)	44-44.5	45-45.5	52.9@1.49	1.49	0.0008-0.0020	0.0020-0.0035	0.260	0.256
1998-99	420A	2.0 (1996)	44-44.5	45-45.5	60@1.74	1.740	0.0008-0.0039	0.0012-0.0059	0.233-0.234	0.232
	4G15	1.5 (1468)	44-44.5	45-45.5	⑧	1.570	0.0008-0.0039	0.0020-0.0059	0.260	0.260
	4G63	2.0 (1997)	44-44.5	45-45.5	54@1.57	1.570	0.0008-0.0040	0.0020-0.0060	NA	NA
	4G64 ⑨	2.4 (2351)	45-45.5	44.5-45	123-137@1.153	1.496	0.0009-0.0100	0.0020-0.0100	0.233-0.234	0.233
	4G64 ⑩	2.4 (2351)	44-44.5	45-45.5	60@1.74	1.740	0.0008-0.0039	0.0012-0.0059	0.236	0.232
	4G93	1.8 (1834)	44-44.5	45-45.5	59@1.74	1.740	0.0008-0.0039	0.0020-0.0059	0.234	0.234
	6G72 ①	3.0 (2972)	NA	45-45.5	72.5@1.591	1.591	0.0012-0.0040	0.0020-0.0060	0.3150	0.3110
	6G72 ②	3.0 (2972)	NA	45-45.5	52.9@1.492	1.492	0.0008-0.0040	0.0020-0.0060	0.2600	0.2560
	6G74	3.5 (3497)	44-44.5	45-45.5	60@1.740	1.740	0.0008-0.0040	0.0016-0.0060	0.236	0.236

① Intake: 51 @ 1.57
  Exhaust: 64 @ 1.57
② Single overhead camshaft
③ Double overhead camshaft
④ 8 valve SOHC
⑤ 16 valve SOHC
⑥ 12 valve SOHC
⑦ 24 valve SOHC
⑧ Intake: 51@1.57
  Exhaust: 64@1.57
⑨ Eclipse Spyder models
⑩ Galant models

79231C62

TORQUE SPECIFICATIONS
All readings in ft. lbs.

Year	Engine ID/VIN	Engine Displacement Liters (cc)	Cylinder Head Bolts	Main Bearing Bolts	Rod Bearing Bolts	Crankshaft Damper Bolts	Flywheel Bolts	Manifold Intake	Manifold Exhaust	Spark Plugs	Lug Nut
1995	G4DJ	1.5 (1468)	51-54	36-39	23-25	51-72	94-101	11-14	11-14	18	65-80
	4G15	1.5 (1468)	53	38	14.5 (1)	61	98	13	13	18	65-80
	4G37	1.8 (1755)	53	38	25	87	98	13	13	18	87-101
	4G93	1.8 (1834)	(2)	18 (1)	14.5 (1)	134	72	14	(3)	18	65-80
	4G63	2.0 (1997)	(2)	18 (1)	14.5 (1)	87	98	26	20	18	87-101
	4G93	1.8 (1834)	(2)	18 (1)	14.5 (1)	134	72	14	(3)	18	65-80
	4G93	1.8 (1834)	(2)	18 (1)	14.5 (1)	134	72	14	(3)	18	65-80
	4G93	1.8 (1834)	(2)	18 (1)	14.5 (1)	134	72	14	(3)	18	65-80
	4G93	1.8 (1834)	(2)	18 (1)	14.5 (1)	134	72	14	(3)	18	65-80
	4G93	1.8 (1834)	(2)	18 (1)	14.5 (1)	134	72	14	(3)	18	65-80
1996	4G15	1.5 (1468)	53	38	14.5 (1)	83	98	13	13	18	65-80
	4G93	1.8 (1834)	(2)	18 (1)	14.5 (1)	134	72	14	(3)	18	65-80
	420A	2.0 (1996)	(4)	55	20 (1)	45	94-101	17	17	18	65-80
	4G63	2.0 (1997)	(2)	18 (1)	14.5 (1)	87	98	26	(5)	18	65-80
	4G64	2.4 (2350)	(2)	14.5 (1)	14.5 (1)	87	98	13	(6)	18	(7)
	6G72 (8)	3.0 (2972)	80	57	38	136	54	10	14	18	(7)
	6G72 (9)	3.0 (2972)	80	67	38	134	54	16	22	18	(7)
	6G72 (10)	3.0 (2972)	80	67	38	134	54	10	33	18	(7)
1997	4G15	1.5 (1468)	53	38	14.5 (1)	83	98	13	13	18	65-80
	4G93	1.8 (1834)	(2)	18 (1)	14.5 (1)	134	72	14	(3)	18	65-80
	420A	2.0 (1996)	(4)	55	20 (1)	45	94-101	17	17	18	65-80
	4G63	2.0 (1997)	(2)	18 (1)	14.5 (1)	87	98	26	(5)	18	65-80
	4G64	2.4 (2350)	(2)	14.5 (1)	14.5 (1)	87	98	13	(6)	18	(7)
	6G72 (8)	3.0 (2972)	80	57	38	136	54	10	14	18	(7)
	6G72 (9)	3.0 (2972)	80	67	38	134	54	16	22	18	(7)
	6G72 (10)	3.0 (2972)	80	67	38	134	54	10	33	18	(7)
1998-99	6G72 (11)	3.0 (2972)	80	57	38	136	54	13	14	18	87-101
	6G72 (10)	3.0 (2972)	80	67	38	136	54	16	33	18	87-101
	6G72 (12)	3.0 (2972)	(13)	54	38	136	54	16	16	18	87-101
	4G64	2.4 (2351)	(14)	(15)	(15)	87	98	13	(16)	18	(17)
	4G15	1.5 (1468)	(18)	37	(19)	76	95	12	12	18	65-80
	4G93	1.8 (1834)	(18)	(20)	(21)	131	71	15	(3)	18	65-80
	420A	2.0 (1996)	(22)	(23)	(24)	45	NA	17	17	20	87-101
	4G63	2.0 (1997)	(14)	(20)	(15)	87	98	(25)	20-21	18	87-101
	6G74	3.5 (3497)	80	67	38	134	54	16	22	18	65-80

(1) Torque to specification plus an additional 1/4 turn

(2) Step 1: 54 ft. lbs.
Step 2: Loosen fully
Step 3: Retighten to 14.5 ft. lbs.
Step 4: additional 90°
Step 5: additional 90°

(3) M10 bolts: 22-30 ft. lbs.
M8 bolts: 11-14 ft. lbs.

(4) Step 1:
Bolts 1-6: 24 ft. lbs.
Bolts 7-10: 20 ft. lbs.
Step 2:
Bolts 1-6: 48 ft. lbs.
Bolts 7-10: 20 ft. lbs.
Step 3: Repeat Step 2
Step 4: additional 90°

(5) Bolts: 14 ft. lbs.
Nuts: 26 ft. lbs.

(6) M8 bolts: 20 ft. lbs.
M10 bolts: 22 ft. lbs.

(7) Diamante/Galant: 65-80 ft. lbs.
Eclipse/3000GT: 85-100 ft. lbs.

(8) 12 valve, SOHC

(9) 24 valve, SOHC

(10) Double overhead camshaft

(11) Single overhead camshaft

(12) Turbocharged

(13) Step 1: 91 ft. lbs.
Step 2: loosen fully
Step 3: 91 ft. lbs.

(14) Step 1: 14.5 ft. lbs.
Step 2: additional 90°
Step 3: additional 90°

(15) Step 1: 14.5 ft. lbs.
Step 2: additional 90-100°

(16) M8 bolts: 20 ft. lbs.
M10 bolts: 22 ft. lbs.

(17) Galant: 65-80 ft. lbs.
Eclipse Spyder: 87-101 ft. lbs.
Step 3: additional 90°

(18) Step 1: 15 ft. lbs.
Step 2: additional 90°

(19) Step 1: 12 ft. lbs.
Step 2: additional 90°

(20) Step 1: 18 ft. lbs.
Step 2: additional 90-100°

(21) Step 1: 15 ft. lbs.
Step 2: additional 90-100°

(22) Long bolts -
Step 1: 48 ft. lbs.
Step 2: additional 90 degrees
Short bolts -
Step 1: 20 ft. lbs.
Step 2: additional 90 degrees

(23) Bolts 1-10: 55 ft. lbs.
Bolts A-K: 20 ft. lbs.

(24) Step 1: 20 ft. lbs.
Step 2: additional 90 degrees

(25) Bolts: 14 ft. lbs.
Nuts: 26 ft. lbs.

79231C63

SCHEDULED MAINTENANCE INTERVALS
(MITSUBISHI DIAMANTE, ECLIPSE, GALANT, MIRAGE, PRECIS & 3000GT)

TO BE SERVICED	TYPE OF SERVICE	VEHICLE MILEAGE INTERVAL (x1000)												
		7.5	15	22.5	30	37.5	45	52.5	60	67.5	75	82.5	90	97.5
Engine oil & filter③	R	✓	✓	✓	✓	✓	✓	✓	✓	✓	✓	✓	✓	✓
Automatic transaxle fluid & filter	S/I		✓		✓		✓		✓		✓		✓	
Brake hoses	S/I		✓		✓		✓		✓		✓		✓	
Disc brake pads	S/I		✓		✓		✓		✓		✓		✓	
Driveshaft boots	S/I		✓		✓		✓		✓		✓		✓	
Valve clearance (Mirage)	S/I		✓		✓		✓		✓		✓		✓	
Air cleaner element	R				✓				✓				✓	
Engine coolant	R				✓				✓				✓	
Spark plugs (except Diamante & 3000GT w/platinum tip)	R				✓				✓				✓	
Spark plugs (Diamante & 3000GT w/platinum tip)	R								✓					
Ball joints & steering linkage seals	S/I				✓				✓				✓	
Drive belt(s)	S/I				✓				✓				✓	
Exhaust system	S/I				✓				✓				✓	
Fuel hoses	S/I				✓				✓				✓	
Manual transaxle oil (Mirage)	S/I				✓				✓				✓	
Manual transaxle oil (including transfer) (Eclipse, 3000GT)	S/I				✓				✓				✓	
Manual transaxle oil (Galant)	S/I				✓				✓				✓	
Rear axle oil (Eclipse, & 3000GT AWD)②	S/I				✓				✓				✓	
Rear drum brake linings & rear wheel cylinders (Eclipse, Galant & Mirage)	S/I				✓				✓				✓	
Ignition cables	R								✓					

79231C64

SCHEDULED MAINTENANCE INTERVALS
(MITSUBISHI DIAMANTE, ECLIPSE, GALANT, MIRAGE, PRECIS & 3000GT) (Cont.)

TO BE SERVICED	TYPE OF SERVICE	VEHICLE MILEAGE INTERVAL (x1000)												
		7.5	15	22.5	30	37.5	45	52.5	60	67.5	75	82.5	90	97.5
Timing belt(s)	R								✓					
Distributor cap & rotor (except 3000GT)	S/I								✓					
EVAP system (except canister)	S/I								✓					
Fuel system (tank, pipe line, connection & fuel tank filler tube cap)	S/I								✓					

① Replace every 30,000 miles.
② With LSD - replace every 30,000 miles.
③ 3000GT turbo - replace every 5000 miles.
R – Replace   S/I – Service or Inspect

FREQUENT OPERATION MAINTENANCE (SEVERE SERVICE)
If a vehicle is operated under any of the following conditions it is considered severe service:
- Extremely dusty areas.
- 50% or more of the vehicle operation is in 32°C (90°F) or higher temperatures, or constant operation in temperatures below 0°C (32°F).
- Prolonged idling (vehicle operation in stop and go traffic).
- Frequent short running periods (engine does not warm to normal operating temperatures).
- Police, taxi, delivery usage or trailer towing usage.
Oil & oil filter change – change every 3000 miles.
Disc brake pads - service or inspect every 6000 miles.
Air filter element – service or inspect every 15,000 miles.
Automatic transaxle fluid & filter - replace every 15,000 miles.
Rear drum brake linings & rear wheel cylinders (Eclipse, Galant & Mirage).
Spark plugs (except Diamante & 3000GT w/platinum tip) - replace every 15,000 miles.
Manual transaxle oil (including transfer (Galant, Mirage & 3000GT) - replace every 30,000 miles.

79231C65

SCHEDULED MAINTENANCE
MITSUBISHI
DIAMANTE, ECLIPSE, GALANT, MIRAGE, 3000GT

The following should be used as a guide when determining the amount of work required for a particular service if taken to a repair shop. In estimating how long a particular Scheduled Maintenance Service should take, please observe the following:

● **Chilton Time** is time based on field research and data supplied by the vehicle manufacturer.

● Labor time operations are given in hours and tenths of an hour.

● All labor operations, are to be used as a guide.

Mechanic Skill Level Codes:
- **(G) GENERAL:** Normally skilled with certification.
- **(M) MAINTENANCE:** Semi-skilled working on certicication.
- **(P) PRECISION:** Really skilled with multiple certification.

	Chilton Time
(M) 7500 Mile Service	
1995-99	.3
(M) 15000 Mile Service	
1995-99	.8
w/AT add	.1
(M) 22500 Mile Service	
1995-99	.3
(G) 30000 Mile Service	
1995-99	3.0
Inspect rear brake lining & wheel cyls., Eclipse, Galant & Mirage add	.2
w/AT add	.1
w/AWD add	.1
(M) 37500 Mile Service	
1995-99	.3

	Chilton Time
(M) 45000 Mile Service	
1995-99	.8
w/AT add	.1
(M) 52500 Mile Service	
1995-99	.3
(G) 60000 Mile Service	
1995-99	6.9
Inspect rear brake lining & wheel cyls., Eclipse, Galant & Mirage add	.2
w/AT add	.1
w/AWD add	.1
Inspect dist. cap & rotor, add	.2

	Chilton Time
(M) 67500 Mile Service	
1995-99	.3
(M) 75000 Mile Service	
1995-99	.8
w/AT add	.1
(M) 82500 Mile Service	
1995-99	.3
(G) 90000 Mile Service	
1995-99	3.0
Inspect rear brake lining & wheel cyls., Eclipse, Galant & Mirage add	.2
w/AT add	.1
w/AWD add	.1
(M) 97500 Mile Service	
1995-99	.3

79231C66

NISSAN
200SX • 240SX • 300ZX • Altima • Maxima • Sentra

VEHICLE IDENTIFICATION CHART

Engine Code						Model Year	
Code	Liters	Cu. In. (cc)	Cyl.	Fuel Sys.	Eng. Mfg.	Code	Year
GA16DE	1.6	97 (1597)	4	MFI	Nissan	S	1995
SR20DE	2.0	122 (1998)	4	MFI	Nissan	T	1996
KA24DE	2.4	146 (2389)	4	MFI	Nissan	V	1997
VG30DE	3.0	181 (2960)	6	MFI	Nissan	W	1998
VG30DETT	3.0	181 (2960)	6	MFI	Nissan	X	1999
VQ30DE	3.0	182 (2988)	6	MFI	Nissan		

79231CC9

ENGINE IDENTIFICATION

Year	Model	Engine Displacement Liters (cc)	Engine Series (ID/VIN)	Fuel System	No. of Cylinders	Engine Type
1995	240SX	2.4 (2389)	KA24DE	MFI	4	DOHC
	300ZX	3.0 (2960)	VG30DE	MFI	6	DOHC
	300ZX	3.0 (2960)	VG30DETT	MFI	6	DOHC
	Altima	2.4 (2389)	KA24DE	MFI	4	DOHC
	Maxima	3.0 (2988)	VQ30DE	MFI	6	DOHC
	Sentra	1.6 (1597)	GA16DE	MFI	4	DOHC
	Sentra	2.0 (1998)	SR20DE	MFI	4	DOHC
1996	240SX	2.4 (2389)	KA24DE	MFI	4	DOHC
	300ZX	3.0 (2960)	VG30DE	MFI	6	DOHC
	300ZX	3.0 (2960)	VG30DETT	MFI	6	DOHC
	Altima	2.4 (2389)	KA24DE	MFI	4	DOHC
	Maxima	3.0 (2988)	VQ30DE	MFI	6	DOHC
	Sentra	1.6 (1597)	GA16DE	MFI	4	DOHC
	Sentra	2.0 (1998)	SR20DE	MFI	4	DOHC
1997	200SX	1.6 (1597)	GA16DE	MFI	4	DOHC
	200SX	2.0 (1998)	SR20DE	MFI	4	DOHC
	240SX	2.4 (2389)	KA24DE	MFI	4	DOHC
	Altima	2.4 (2389)	KA24DE	MFI	4	DOHC
	Maxima	3.0 (2988)	VQ30DE	MFI	6	DOHC
	Sentra	1.6 (1597)	GA16DE	MFI	4	DOHC
	Sentra	2.0 (1998)	SR20DE	MFI	4	DOHC
1998-99	200SX	1.6 (1597)	GA16DE	MFI	4	DOHC
	200SX	2.0 (1998)	SR20DE	MFI	4	DOHC
	240SX	2.4 (2389)	KA24DE	MFI	4	DOHC
	Altima	2.4 (2389)	KA24DE	MFI	4	DOHC
	Maxima	3.0 (2988)	VQ30DE	MFI	6	DOHC
	Sentra	1.6 (1597)	GA16DE	MFI	4	DOHC

MFI - Multi-Port Fuel Injection
DOHC - Double Overhead Camshaft
SOHC - Single Overhead Camshaft

79231CC0

GENERAL ENGINE SPECIFICATIONS

Year	Engine ID/VIN	Engine Displacement Liters (cc)	Fuel System Type	Net Horsepower @ rpm	Net Torque @ rpm (ft. lbs.)	Bore x Stroke (in.)	Compression Ratio	Oil Pressure @ rpm
1995	GA16DE	1.6 (1597)	MFI	110@6000	108@4000	2.99x3.46	9.5:1	50@3000
	SR20DE	2.0 (1998)	MFI	140@6400	132@4800	3.39x3.39	9.5:1	46@3200
	KA24DE	2.4 (2389)	MFI	①	②	3.50x3.78	10.5:1	60@3000
	VG30DE	3.0 (2960)	MFI	222@6400	198@4800	3.43x3.27	10.5:1	51@3000
	VG30DETT	3.0 (2960)	MFI	②	283@3600	3.43x3.27	8.5:1	51@3000
	VQ30DE	3.0 (2988)	MFI	190@5600	205@4000	3.66x2.89	10.0:1	63@3000
1996	GA16DE	1.6 (1597)	MFI	110@6000	108@4000	2.99x3.46	9.5:1	50@3000
	SR20DE	2.0 (1998)	MFI	140@6400	132@4800	3.39x3.39	9.5:1	46@3200
	KA24DE	2.4 (2389)	MFI	①	②	3.50x3.78	10.5	60@3000
	VG30DE	3.0 (2960)	MFI	222@6400	198@4800	3.43x3.27	10.5:1	51@3000
	VG30DETT	3.0 (2960)	MFI	③	283@3600	3.43x3.27	8.5:1	51@3000
	VQ30DE	3.0 (2988)	MFI	190@5600	205@4000	3.66x2.89	10.0:1	63@3000
1997	GA16DE	1.6 (1597)	MFI	110@6000	108@4000	2.99x3.46	9.5:1	50@3000
	SR20DE	2.0 (1998)	MFI	140@6400	132@4800	3.39x3.39	9.5:1	46@3200
	KA24DE	2.4 (2389)	MFI	③	②	3.50x3.78	④	60@3000
	VQ30DE	3.0 (2988)	MFI	190@5600	205@4000	3.66x2.89	10.0:1	63@3000
1998-99	GA16DE	1.6 (1597)	MFI	115@6000	108@4000	2.99x3.46	9.9:1	50@3000
	SR20DE	2.0 (1998)	MFI	140@6400	132@4800	3.39x3.39	9.5:1	46@3200
	KA24DE	2.4 (2389)	MFI	①	②	3.50x3.78	④	60@3000
	VQ30DE	3.0 (2988)	MFI	190@5600	205@4000	3.66x2.89	10.0:1	63@3000

MFI - Mutli-port Fuel Injection

① 240SX: 155@5600
   Altima: 150@5600
② 240SX: 160@4400
   Altima: 154@4400
③ Manual transmission: 300@6400
   Automatic transmission: 280@6400
④ 240SX: 9.5:1
   Altima: 9.2:1

79231CD1

GASOLINE ENGINE TUNE-UP SPECIFICATIONS

Year	Engine ID/VIN	Engine Displacement Liters (cc)	Spark Plugs Gap (in.)	Ignition Timing (deg.) MT	Ignition Timing (deg.) AT	Fuel Pump (psi)	Idle Speed (rpm) MT	Idle Speed (rpm) AT	Valve Clearance In.	Valve Clearance Ex.
1995	GA16DE	1.6 (1597)	0.041	10B	10B	36 ①	700	800 ②	0.015 ③	0.016 ③
	SR20DE	2.0 (1998)	0.033 ④	15B	15B	36 ①	800	800 ②	HYD	HYD
	KA24DE	2.4 (2389)	0.041	20B	20B	33 ①	700	700 ②	0.014 ③	0.015 ③
	VE30DE	3.0 (2960)	0.041	15B	15B	36 ①	750	750 ②	HYD	HYD
	VG30DE	3.0 (2960)	0.041	15B	15B	36 ①	700	770 ②	HYD	HYD
	VG30DETT	3.0 (2960)	0.041	15B	15B	36 ①	700	750 ②	HYD	HYD
1996	GA16DE	1.6 (1597)	0.041	10B	10B	36 ①	700	800 ②	0.015 ③	0.016 ③
	SR20DE	2.0 (1998)	0.033 ④	15B	15B	36 ①	800	800 ②	HYD	HYD
	KA24DE	2.4 (2389)	0.041	20B	20B	33 ①	700	700 ②	0.014 ③	0.015 ③
	VG30DE	3.0 (2960)	0.041	15B	15B	36 ①	700	770 ②	HYD	HYD
	VG30DETT	3.0 (2960)	0.041	15B	15B	36 ①	700	750 ②	HYD	HYD
	VQ30DE	3.0 (2988)	0.041 ⑥	15B	15B	34 ①	650	700 ②	0.014 ③	0.015 ③
1997	GA16DE	1.6 (1597)	0.041	10B	10B	36 ①	700	800 ②	0.015 ③	0.016 ③
	SR20DE	2.0 (1998)	0.033 ④	15B	15B	36 ①	800	800 ②	HYD	HYD
	KA24DE	2.4 (2389)	0.041	20B	20B	33 ①	700	700 ②	0.014 ③	0.015 ③
	VQ30DE	3.0 (2988)	0.041 ⑥	15B	15B	34 ①	650	700 ②	0.014 ③	0.015 ③
1998-99	GA16DE	1.6 (1597)	0.041	8B	8B	36 ①	625 ⑤	725 ②	0.015 ③	0.016 ③
	SR20DE	2.0 (1998)	0.033	15B	15B	36 ①	800	800 ②	HYD	HYD
	KA24DE	2.4 (2389)	0.041 ⑥	20B	20B	33 ①	650	650 ②	0.015 ③	0.015 ③
	VQ30DE	3.0 (2988)	0.041 ⑥	15B	15B	34 ①	650	700 ②	0.014 ③	0.015 ③

NOTE: The Vehicle Emission Control Information label often reflects specification changes made during production. The label figures must be used if they differ from those in this chart.

B - Before top dead center

HYD - Hydraulic

① 1 System pressure at idle with vacuum hose connected; should increase to 43 psi when disconnected

② 2 Automatic transmission in neutral

③ Engine warm

④ Conventional - 0.033
Platinum - 0.041

⑤ Canada: 750

⑥ Do not check or adjust gap on platinum-tipped spark plugs

79231CD2

79231CC9

CAPACITIES

Year	Model	Engine ID/VIN	Engine Displacement Liters (cc)	Engine Oil with Filter (qts.)	Transmission (pts.) 4-Spd	Transmission (pts.) 5-Spd	Transmission (pts.) Auto.	Drive Axle Front (pts.)	Drive Axle Rear (pts.)	Fuel Tank (gal.)	Cooling System (qts.)
1995	240SX	KA24DE	2.4 (2389)	3.8	—	5.1	17.5	—	①	16.0	7.1
	300ZX	VG30DE	3.0 (2960)	3.6	—	5.9	17.5	—	3.1	19.0	9.5
	300ZX	VG30DETT	3.0 (2960)	3.6	—	5.9	17.2	—	3.9	19.0	9.5
	Altima	KA24DE	2.4 (2389)	4.1	—	10.0	20.0	—	—	16.0	8.2
	Maxima	VG30E	3.0 (2960)	4.1	—	10.0	15.5	—	—	18.0	9.3
	Maxima	VE30DE	3.0 (2960)	4.0	—	10.0	18.0	—	—	18.0	11.2
	Sentra	GA16DE	1.6 (1597)	3.4	5.9	6.1	14.7	—	—	13.0	5.6
	Sentra	SR20DE	2.0 (1998)	3.4	—	7.5	14.7	—	—	13.0	6.0
1996	240SX	KA24DE	2.4 (2389)	3.8	—	5.1	17.5	—	①	16.0	7.1
	300ZX	VG30DE	3.0 (2960)	3.6	—	5.9	17.5	—	3.1	19.0	9.5
	300ZX	VG30DETT	3.0 (2960)	3.6	—	5.9	17.3	—	3.9	19.0	9.5
	Altima	KA24DE	2.4 (2389)	4.1	—	10.0	20.0	—	—	16.0	8.2
	Maxima	VQ30DE	3.0 (2988)	4.3	—	9.5	20.0	—	—	18.5	9.4
	Sentra	GA16DE	1.6 (1597)	3.4	—	6.1	14.8	—	—	13.0	5.6
	Sentra	SR20DE	2.0 (1998)	3.4	—	7.5	14.8	—	—	13.0	6.0
1997	200SX	GA16DE	1.6 (1597)	3.4	—	8.2	14.8	—	—	13.2	②
	200SX	SR20DE	2.0 (1998)	3.6	—	8.2	14.8	—	—	13.2	6.5
	240SX	KA24DE	2.4 (2389)	4.0	—	5.1	17.5	—	①	17.2	7.3
	Altima	KA24DE	2.4 (2389)	4.1	—	10.0	20.0	—	—	15.9	8.3
	Maxima	VQ30DE	3.0 (2988)	4.3	—	9.5	20.0	—	—	18.5	9.4
	Sentra	GA16DE	1.6 (1597)	3.4	—	④	14.8	—	—	13.2	②
	Sentra	SR20DE	2.0 (1998)	3.4	—	7.5	14.8	—	—	13.0	6.0
1998-99	200SX	GA16DE	1.6 (1597)	3.5	—	④	14.8	—	—	13.2	⑤
	200SX	SR20DE	2.0 (1998)	3.6	—	④	14.8	—	—	13.2	6.5
	240SX	KA24DE	2.4 (2389)	4.0	—	5.3	17.5	—	①	17.2	7.3
	Altima	KA24DE	2.4 (2389)	4.1	—	10.0	20.0	—	—	15.9	8.3
	Maxima	VQ30DE	3.0 (2988)	4.3	—	9.5	20.0	—	—	18.5	9.4
	Sentra	GA16DE	1.6 (1597)	3.5	—	④	14.8	—	—	13.2	⑤
	Sentra	SR20DE	2.0 (1998)	3.4	—	7.5	14.8	—	—	13.0	6.0

NOTE: All capacities are approximate. Add fluid gradually and check to be sure a proper fluid level is obtained.

① With limited slip: 3.1 pts.

Standard: 2.8 pts.

② GA16DE with MT: 5.4 qts.

GA16DE with AT: 5.6 qts.

③ 2 seaters and 2+2: 18.7 gals.

Convertible: 18.2 gals.

④ RS5F31A: 6.5 pts.

RS5F32V: 8.0 pts.

⑤ GA16DE with MT: 5.5 qts.
GA16DE with AT: 6.0 qts.

79231CD3

VALVE SPECIFICATIONS

Year	Engine ID/VIN	Engine Displacement Liters (cc)	Seat Angle (deg.)	Face Angle (deg.)	Spring Test Pressure (lbs. @ in.)	Spring Installed Height (in.)	Stem-to-Guide Clearance (in.)		Stem Diameter (in.)	
							Intake	Exhaust	Intake	Exhaust
1995	GA16DE	1.6 (1597)	45	45.25-45.75	77@0.9945	NA	0.0008-0.0020	0.0016-0.0028	0.2152-0.2157	0.2144-0.2150
	SR20DE	2.0 (1998)	45	45.25-45.75	144@1.181	NA	0.0008-0.0021	0.0016-0.0029	0.2348-0.2354	0.2341-0.2346
	KA24DE	2.4 (2389)	45	45.25-45.75	123@1.024	NA	0.0008-0.0021	0.0016-0.0029	0.2742-0.2748	0.2734-0.2740
	VG30DE	3.0 (2960)	45	45.25-45.75	120@1.043		0.0008-0.0021	0.0016-0.0029	0.2348-0.2354	0.2341-0.2346
	VG30DETT	3.0 (2960)	45	45.25-45.75	120@1.043		0.0008-0.0021	0.0016-0.0029	0.2348-0.2354	0.2341-0.2346
	VQ30DE	3.0 (2988)	45	45.25-45.75	102@1.085	NA	0.0008-0.0021	0.0016-0.0029	0.2348-0.2354	0.2341-0.2346
1996	GA16DE	1.6 (1597)	45	45.25-45.75	77@0.9945	NA	0.0008-0.0020	0.0016-0.0028	0.2152-0.2157	0.2144-0.2150
	SR20DE	2.0 (1998)	45	45.25-45.75	144@1.181	NA	0.0008-0.0021	0.0016-0.0029	0.2348-0.2354	0.2341-0.2346
	KA24DE	2.4 (2389)	45	45.25-45.75	106@1.026	NA	0.0008-0.0021	0.0016-0.0029	0.2742-0.2748	0.2734-0.2740
	VG30DE	3.0 (2960)	45	45.25-45.75	120@1.043	NA	0.0008-0.0021	0.0016-0.0029	0.2348-0.2354	0.2341-0.2346
	VG30DETT	3.0 (2960)	45	45.25-45.75	120@1.043	NA	0.0008-0.0021	0.0016-0.0029	0.2348-0.2354	0.2341-0.2346
	VQ30DE	3.0 (2988)	45	45.25-45.75	102@1.085	NA	0.0008-0.0021	0.0016-0.0029	0.2348-0.2354	0.2341-0.2346
1997	GA16DE	1.6 (1597)	45	45.25-45.75	77@0.9945	NA	0.0008-0.0020	0.0016-0.0028	0.2152-0.2157	0.2144-0.2150
	SR20DE	2.0 (1998)	45	45.25-45.75	144@1.181	NA	0.0008-0.0021	0.0016-0.0029	0.2348-0.2354	0.2341-0.2346
	KA24DE	2.4 (2389)	45	45.25-45.75	106@1.026	NA	0.0008-0.0021	0.0016-0.0029	0.2742-0.2748	0.2734-0.2740
	VQ30DE	3.0 (2988)	45	45.25-45.75	102@1.085	NA	0.0008-0.0021	0.0016-0.0029	0.2348-0.2354	0.2341-0.2346
1998-99	GA16DE	1.6 (1597)	45	45.25-45.75	77@0.995	NA	0.0008-0.0020	0.0016-0.0028	0.2152-0.2157	0.2144-0.2150
	SR20DE	2.0 (1998)	45	45.25-45.75	137@1.181	NA	0.0008-0.0021	0.0016-0.0029	0.2348-0.2354	0.2341-0.2346
	KA24DE	2.4 (2389)	45	45.25-45.75	123@1.024	NA	0.0008-0.0021	0.0016-0.0029	0.2742-0.2748	0.2734-0.2740
	VQ30DE	3.0 (2988)	45	45.25-45.75	102@1.085	NA	0.0008-0.0021	0.0016-0.0029	0.2348-0.2354	0.2341-0.2346

NA - Not Available

79231CD4

TORQUE SPECIFICATIONS
All readings in ft. lbs.

Year	Engine ID/VIN	Engine Displacement Liters (cc)	Cylinder Head Bolts	Main Bearing Bolts	Rod Bearing Bolts	Crankshaft Damper Bolts	Flywheel Bolts	Manifold Intake	Manifold Exhaust	Spark Plugs	Lug Nut
1995	GA16DE	1.6 (1597)	①	34-38	②	98-112	③	14	14	18	75
	SR20DE	2.0 (1998)	④	⑤	⑥	105-112	61-69	14	30	18	75
	KA24DE	2.4 (2389)	⑦	34-41	⑥	105-112	105-112	14	32	18	75
	VG30DE	3.0 (2960)	⑧	67-74	⑥	123-130	61-69	⑨	18	18	75
	VG30DETT	3.0 (2960)	⑧	67-74	⑩	159-174	61-69	⑨	22	18	75
	VQ30DE	3.0 (2988)	⑪	⑫	⑬	⑭	61-69	⑮	23	18	80
1996	GA16DE	1.6 (1597)	①	34-38	②	98-112	③	14	14	18	75
	SR20DE	2.0 (1998)	④	⑤	⑥	105-112	61-69	14	30	18	75
	KA24DE	2.4 (2389)	⑦	34-41	⑥	105-112	105-112	14	32	18	75
	VG30DE	3.0 (2960)	⑧	67-74	⑥	123-130	61-69	⑨	18	18	75
	VG30DETT	3.0 (2960)	⑧	67-74	⑩	159-174	61-69	⑨	22	18	75
	VQ30DE	3.0 (2988)	⑪	⑫	⑬	⑭	61-69	⑮	23	18	80
1997	GA16DE	1.6 (1597)	①	34-38	②	98-112	③	14	19	18	79
	SR20DE	2.0 (1998)	④	⑤	⑥	105-112	61-69	14	30	18	79
	KA24DE	2.4 (2389)	⑦	34-41	⑥	105-112	105-112	14	32	18	80
	VQ30DE	3.0 (2988)	⑪	⑫	⑬	⑭	61-69	⑮	23	18	80
1998-99	GA16DE	1.6 (1597)	①	34-38	②	98-112	③	14	19	18	79
	SR20DE	2.0 (1998)	④	⑨	⑥	105-112	61-69	14	30	18	79
	KA24DE	2.4 (2389)	⑦	34-41	⑥	105-112	105-112	14	32	18	80
	VQ30DE	3.0 (2988)	⑪	⑫	⑬	⑭	61-69	⑮	23	18	80

① Bolt Nos. 1-10:
   Step 1: 22 ft. lbs.
   Step 2: 43 ft. lbs.
   Step 3: Loosen completely then retorque to 22 ft. lbs.
   Step 4: 43 ft. lbs. or an additional 50-55 degrees
   Bolt Nos. 11-15: Torque last, to 72 inch lbs.

② Step 1: 12 ft. lbs.
   Step 2: 19 ft. lbs. or an additional 35-40 degrees

③ Manual transmission: 61-69 ft. lbs.
   Automatic transmission: 69-76 ft. lbs.

④ Step 1: 29 ft. lbs.
   Step 2: 58 ft. lbs.
   Step 3: Loosen completely then retorque to 30 ft. lbs.
   Step 4: Turn each bolt, in sequence,
       an additional 90-100 degrees
   Step 5: Repeat Step 4

⑤ Step 1: 28 ft. lbs.
   Step 2: 54-61 ft. lbs. or an additional 45-50 degrees

⑥ 12 ft. lbs. plus an additional 60-65 degrees

⑦ Step 1: 22 ft. lbs.
   Step 2: 58 ft. lbs.
   Step 3: Loosen completely then retorque to 22 ft. lbs.
   Step 4: 58 ft. lbs. or an additional 80-85 degrees

⑧ Step 1: 29 ft. lbs.
   Step 2: 90 ft. lbs.
   Step 3: Loosen completely then retorque to 29 ft. lbs.
   Step 4: 90 ft. lbs. or an additional 70 degrees

⑨ Step 1: 20-24 ft. lbs.
   Step 2: 75-80 degrees
   Step 3: Loosen completely and retorque to 24-28 ft. lbs.
   Step 4: 45-50 degree turn

⑩ Step 1: 12 ft. lbs.
   Step 2: 43-48 ft. lbs. or an additional 60-65 degrees

⑪ Step 1: 29-36 ft. lbs.
   Step 2: Plus 60-65 degrees

⑫ Step 1: 3.6-7.2 ft. lbs.
   Step 2: 20-23 ft. lbs.

⑬ Step 1: 29 ft. lbs.
   Step 2: 90 ft. lbs.
   Step 3: Loosen completely and retorque to 25-33 ft. lbs.
   Step 4: Plus 90 ft. lbs. or 70 degrees
   Step 5: Tighten two bolts marked with an "X" to 7-9 ft. lbs.

⑭ Step 1: 29-36 ft. lbs.
   Step 2: 60-66 degrees

⑮ Step 1: 10-12 ft. lbs.
   Step 2: 43-48 ft. lbs. or an additional 60-65 degrees

79231CD5

SCHEDULED MAINTENANCE INTERVALS
(NISSAN 200SX, 240SX, 300ZX, ALTIMA, MAXIMA & SENTRA/NX)

TO BE SERVICED	TYPE OF SERVICE	VEHICLE MILEAGE INTERVAL (x1000)												
		7.5	15	22.5	30	37.5	45	52.5	60	67.5	75	82.5	90	97.5
Engine oil & filter①	R	✓	✓	✓	✓	✓	✓	✓	✓	✓	✓	✓	✓	✓
Brake lines & cables	S/I		✓		✓		✓		✓		✓		✓	
Brake pads, discs, drums & linings	S/I		✓		✓		✓		✓		✓		✓	
Differential gear oil (240SX & 300ZX)	S/I		✓		✓		✓		✓		✓		✓	
Driveshaft boots (Altima, Maxima, Sentra/NX & 200SX)	S/I		✓		✓		✓		✓		✓		✓	
Exhaust system (300ZX)	S/I		✓		✓		✓		✓		✓		✓	
Exhaust system (except 300ZX)	S/I				✓				✓				✓	
Transmission or transaxle oil	S/I		✓		✓		✓		✓		✓		✓	
Air cleaner filter	R				✓				✓				✓	
Spark plugs (except below)	R				✓				✓				✓	
Spark plugs (platinum tip) (Sentra/NX, 200SX SR20DE, Maxima VE30DE, 300ZX & 240SX)	R								✓					
Idle RPM (Sentra/NX & 200SX GA16DE)	S/I				✓				✓				✓	
Steering gear & linkage, axle & suspension parts	S/I				✓				✓				✓	
SUPER HICAS linkage (300ZX turbo)	S/I				✓				✓				✓	
Engine Coolant	R								✓					
Timing belt (1993-94 Maxima VE30DE & 1993-95 300ZX)	R								✓					
Drive belts	S/I								✓					
Fuel lines	S/I								✓					
Vapor lines	S/I								✓					

① 300ZX turbo - replace every 5000 miles.
R – Replace   S/I – Service or Inspect

79231CD6

FREQUENT OPERATION MAINTENANCE (SEVERE SERVICE)

If a vehicle is operated under any of the following conditions it is considered severe service:

- Extremely dusty areas.
- 50% or more of the vehicle operation is in 32°C (90°F) or higher temperatures, or constant operation in temperatures below 0°C (32°F).
- Prolonged idling (vehicle operation in stop and go traffic).
- Frequent short running periods (engine does not warm to normal operating temperatures).
- Police, taxi, delivery usage or trailer towing usage.

Oil & oil filter change (300ZX turbo) – change every 3000 miles.
Oil & oil filter change (except 300ZX turbo) – change every 3750 miles.
Brake pads & discs - service or inspect every 7500 miles.
Driveshaft boots (Altima, Maxima, Sentra/NX & 200SX) - service or inspect every 7500 miles.
Exhaust system - service or inspect every 7500 miles.
Steering gear & linkage, axle & suspension parts - service or inspect every 7500 miles.
Steering linkage ball joints & front suspension ball joints - service or inspect every 7500 miles.
SUPER HICAS linkage (300ZX turbo) - service or inspect every 7500 miles.
Air cleaner filter - service or inspect every 15,000 miles.

79231CD7

SCHEDULED MAINTENANCE
NISSAN
200SX, 240SX, 300ZX, ALTIMA, MAXIMA, SENTRA, NX

**The following should be used as a guide when determining the amount of work required for a particular service if taken to a repair shop.
In estimating how long a particular Scheduled Maintenance Service should take, please observe the following:**

- **Chilton Time** is time based on field research and data supplied by the vehicle manufacturer.
- Labor time operations are given in hours and tenths of an hour.
- All labor operations, are to be used as a guide.

Mechanic Skill Level Codes:
         (G) GENERAL: Normally skilled with certification.
         (M) MAINTENANCE: Semi-skilled working on certicication.
         (P) PRECISION: Really skilled with multiple certification.

	Chilton Time
(M) 7500 Mile Service	
1995-99	.3
(M) 15000 Mile Service	
1995-99 200SX, 240SX	.9
1995-96 300ZX	1.0
1995-99 Altima, Maxima	.8
1995-99 Sentra, NX	.8
(M) 22500 Mile Service	
1995-99	.3
(G) 30000 Mile Service	
1995-99 200SX	2.0
w/GA16DE engine add	.2
1995-97 240SX	1.8
1995-96 300ZX	2.3
1995-99 Altima, Maxima	1.6
1995-99 Sentra, NX	1.9

	Chilton Time
(M) 37500 Mile Service	
1995-99	.3
(M) 45000 Mile Service	
1995-99 200SX, 240SX	.9
1995-96 300ZX	1.0
1995-99 Altima, Maxima	.8
1995-99 Sentra, NX	.8
(M) 52500 Mile Service	
1995-99	.3
(G) 60000 Mile Service	
1995-99 200SX	2.6
w/GA16DE engine add	.2
1995-96 300ZX	3.0
1995-99 Altima	2.8
1995-99 Maxima	2.8
1995-99 Sentra, NX	3.0
(M) 67500 Mile Service	
1995-99	.3

	Chilton Time
(M) 75000 Mile Service	
1995-99 200SX, 240SX	.9
1995-96 300ZX	1.0
1995-99 Altima, Maxima	.8
1995-99 Sentra, NX	.8
(M) 82500 Mile Service	
1995-99	.3
(G) 90000 Mile Service	
1995-99 200SX	2.0
w/GA16DE engine add	.2
1995-99 240SX	1.8
1995-96 300ZX	2.3
1995-99 Altima, Maxima	1.6
1995-99 Sentra, NX	1.9
(M) 97500 Mile Service	
1995-99	.3

79231CD8

PORSCHE
911 • 928 • 968 • Boxster

VEHICLE IDENTIFICATION CHART

Code	Liters	Cu. In. (cc)	Cyl.	Fuel Sys.	Eng. Mfg.
M28/49, 50	5.4	329 (5397)	8	LH–Jetronic	Porsche
M44/43, 44	3.0	182 (2990)	4	DME	Porsche
M64/07, 08	3.6	220 (3602)	6	DME	Porsche
M64/23, 24	3.6	220 (3600)	6	DME	Porsche
M64/50	3.6	220 (3602)	6	KE	Porsche
M64/60	3.6	220 (3600)	6	DME	Porsche
M96/20	2.5	151 (2480)	6	DME	Porsche

Model Year	
Code	Year
S	1995
T	1996
V	1997
W	1998
X	1999

DME - Digital Motor Electronic
LH - Bosch air flow-controlled
KE - Bosch electronic CIS

79231C67

ENGINE IDENTIFICATION

Year	Model	Engine Displacement Liters (cc)	Engine Series (ID/VIN)	Fuel System	No. of Cylinders	Engine Type
1995	968	3.0 (2990)	M44/43, 44	DME	4	DOHC
	911 Carrera	3.6 (3600)	M64/07, 08	DME	6	SOHC
	911 Carrera 4	3.6 (3600)	M64/07	DME	6	SOHC
	911 Turbo	3.6 (3600)	M64/50	KE	6	SOHC
	928 GTS	5.4 (5397)	M28/49, 50	LH-Jetronic	8	DOHC
1996	911 Carrera	3.6 (3600)	M64/23, 24	DME	6	SOHC
	911 Carrera 4	3.6 (3600)	M64/23	DME	6	SOHC
	911 Turbo	3.6 (3600)	M64/60	DME	6	SOHC
1997	911 Carrera	3.6 (3600)	M64/23, 24	DME	6	SOHC
	911 Carrera 4	3.6 (3600)	M64/23	DME	6	SOHC
	911 Turbo	3.6 (3600)	M64/60	DME	6	SOHC
1998-99	911 Carrera	3.6 (3600)	M64/23, 24	DME	6	SOHC
	911 Carrera 4	3.6 (3600)	M64/23	DME	6	SOHC
	Boxster	2.5 (2480)	M96/20	DME	6	DOHC

SOHC - Single Overhead Camshaft
DOHC - Double Overhead Camshaft
DME - Digital Motor Electronic
LH - Bosch air flow-controlled
KE - Bosch electronic CIS

79231C68

GENERAL ENGINE SPECIFICATIONS

Year	Engine ID/VIN	Engine Displacement Liters (cc)	Fuel System Type	Net Horsepower @ rpm	Net Torque @ rpm (ft. lbs.)	Bore x Stroke (in.)	Compression Ratio	Oil Pressure @ rpm
1995	M44/43, 44	3.0 (2990)	DME	236@6200	225@4100	4.09x3.46	11.0:1	44@3000
	M64/07, 08	3.6 (3600)	DME	270@6100	243@4800	3.94x3.01	11.3:1	73@5000
	M64/50	3.6 (3600)	KE	355@5500	383@4500	3.94x3.01	7.5:1	73@5000
	M28/49, 50	5.4 (5397)	LH-Jetronic	345@5700	369@4250	3.94x3.38	10.4:1	73@5000
1996	M64/23, 24	3.6 (3600)	DME	282@6300	250@5250	3.94x3.01	11.3:1	95@5000
	M64/60	3.6 (3600)	DME	400@5700	400@4500	3.94x3.01	8.0:1	95@5000
1997	M64/23, 24	3.6 (3600)	DME	282@6300	250@5250	3.94x3.01	11.3:1	95@5000
	M64/60	3.6 (3600)	DME	400@5700	400@4500	3.94x3.01	8.0:1	95@5000
1998-99	M64/23, 24	3.6 (3600)	DME	282@6300	250@5250	3.94x3.01	11.3:1	95@5000
	M96/20	2.5 (2480)	DME	201@6000	181@4500	3.40x2.80	11.0:1	73@5000

DME - Digital motor electronic
LH - Bosch air flow-controlled
KE - Bosch electronic CIS
NA - Not Available

79231C69

GASOLINE ENGINE TUNE-UP SPECIFICATIONS

Year	Engine ID/VIN	Engine Displacement Liters (cc)	Spark Plugs Gap (In.)	Ignition Timing (deg.)		Fuel Pump (psi)	Idle Speed (rpm)		Valve Clearance	
				MT	AT		MT	AT	In.	Ex.
1995	M44/43, 44	3.0 (2990)	0.028	10B	10B	53-59	840	880	HYD	HYD
	M64/50	3.6 (3600)	0.031	①	①	29-58	950	—	0.004	0.004
	M64/07, 08	3.6 (3600)	0.032	①	①	45-52	880	880	0.004	0.004
	M28/49, 50	5.4 (5397)	0.028	10B	10B	75-80	700	700	HYD	HYD
1996	M64/23, 24	3.6 (3600)	0.026	①	①	53-59	800	750	HYD	HYD
	M64/60	3.6 (3600)	0.026	①	①	53-59	800	—	HYD	HYD
1997	M64/23, 24	3.6 (3600)	0.026	①	①	53-59	800	750	HYD	HYD
	M64/60	3.6 (3600)	0.026	①	①	53-59	800	—	HYD	HYD
1998-99	M64/23, 24	3.6 (3600)	0.026	①	①	53-59	800	750	HYD	HYD
	M96/20	2.5 (2480)	0.026	①	①	53-59	790	750	HYD	HYD

NOTE: The Vehicle Emission Control Information label often reflects specification changes made during production. The label figures must be used if they differ from those in this chart.

B - Before top dead center
HYD - Hydraulic
TDC - Top Dead Center
① Not adjustable

79231C70

CAPACITIES

Year	Model	Engine ID/VIN	Engine Displacement Liters (cc)	Engine Oil with Filter	Transmission (pts.) 4-Spd	5-Spd	Auto.	Transfer Case (pts.)	Drive Axle Front (pts.)	Rear (pts.)	Fuel Tank (gal.)	Cooling System (qts.)
1995	968	M44/43, 44	3.0 (2990)	7.4	—	5.8	14.8	—	—	1.4	19.6	8.4
	911 Turbo	M64/50	3.6 (3600)	13.8	—	—	—	—	—	①	19.4	—
	911 Carrera 2	M64/07, 08	3.6 (3600)	12.1	—	7.5	7.4	—	—	—	19.4	—
	911 Carrera 4	M64/07, 08	3.6 (3600)	12.1	—	8.0	—	—	2.5	①	19.4	—
	928 GTS	M28/49, 50	5.4 (5397)	8.0	—	8.5	19.6	—	—	2.0	22.7	16.9
1996	911 Carrera	M64/23	3.6 (3600)	12.1	—	3.75 ②	—	—	—	—	19.4	—
	911 Carrera	M64/24	3.6 (3600)	12.1	—	—	9.5	—	—	1.0	19.4	—
	911 Carrera 4	M64/23	3.6 (3600)	12.1	—	3.75 ②	—	—	1.2	—	19.4	—
	911 Turbo	M64/60	3.6 (3600)	12.1	—	3.75 ②	—	—	1.2	—	19.4	—
1997	911 Carrera	M64/23	3.6 (3600)	12.1	—	3.75 ②	—	—	—	—	19.4	—
	911 Carrera	M64/24	3.6 (3600)	12.1	—	—	9.5	—	—	1.0	19.4	—
	911 Carrera 4	M64/23	3.6 (3600)	12.1	—	3.75 ②	—	—	1.2	—	19.4	—
	911 Turbo	M64/60	3.6 (3600)	12.1	—	3.75 ②	—	—	1.2	—	19.4	—
1998-99	911 Carrera	M64/23	3.6 (3600)	12.1	—	3.75 ②	—	—	—	—	19.4	—
	911 Carrera	M64/24	3.6 (3600)	12.1	—	—	9.5	—	—	1.0	19.4	—
	911 Carrera 4	M64/23	3.6 (3600)	12.1	—	3.75 ②	—	—	1.2	—	19.4	—
	Boxster	M96/20	2.5 (2480)	8.2	—	2.25	9.5	—	—	0.8	15.3	NA

NOTE: All capacities are approximate. Add fluid gradually and check to be sure a proper fluid level is obtained.

NA - Not Available

① Manual transaxle is included in transmission

② 6 speed manual transaxle

79231C71

VALVE SPECIFICATIONS

Year	Engine ID/VIN	Engine Displacement Liters (cc)	Seat Angle (deg.)	Face Angle (deg.)	Spring Test Pressure (lbs. @ in.)	Spring Installed Height (in.)	Stem-to-Guide Clearance (in.)		Stem Diameter (in.)	
							Intake	Exhaust	Intake	Exhaust
1995	M44/43, 44	3.0 (2990)	45	45	-	①	0.032 ⑥	0.032 ⑥	0.275	0.274
	M64/07, 08	3.6 (3602)	45	45	-	④	0.032 ⑥	0.032 ⑥	0.353	0.352
	M64/50	3.6 (3600)	45	45	②	③	0.030-0.060	0.050-0.080	0.353	0.352
	M28/49, 50	5.4 (5397)	45	45	-	⑤	0.032 ⑥	0.032 ⑥	0.274	0.273
1996	M64/23, 24	3.6 (3600)	45	45	-	④	0.032 ⑥	0.032 ⑥	0.353	0.352
	M64/60	3.6 (3600)	45	45	-	④	0.032 ⑥	0.032 ⑥	0.353	0.352
1997	M64/23, 24	3.6 (3600)	45	45	-	④	0.032 ⑥	0.032 ⑥	0.353	0.352
	M64/60	3.6 (3600)	45	45	-	④	0.032 ⑥	0.032 ⑥	0.353	0.352
1998-99	M64/23, 24	3.6 (3600)	45	45	-	④	0.032 ⑥	0.032 ⑥	0.353	0.352
	M96/20	2.5 (2480)	45	45	-	NA	NA	NA	NA	NA

① Intake: 1.496
   Exhaust: 1.467
② Intake: 176@1.21
   Exhaust: 165@1.25
③ Intake: 1.378
   Exhaust: 1.398
④ Intake: 1.358
   Exhaust: 1.319
⑤ Intake: 1.398
   Exhaust: 1.358
⑥ With valve 0.39 in. off seat, rock back and forth with dial indicator set perpendicular to valve stem on valve head. Maximum dial indicator runout: 0.032 in.

79231C72

TORQUE SPECIFICATIONS
All readings in ft. lbs.

Year	Engine ID/VIN	Engine Displacement Liters (cc)	Cylinder Head Bolts	Main Bearing Bolts	Rod Bearing Bolts	Crankshaft Damper Bolts	Flywheel Bolts	Manifold		Spark Plugs	Lug Nut
								Intake	Exhaust		
1995	M44/43, 44	3.0 (2990)	①	②	③	210	④	15	15	18-22	96
	M64/07, 08	3.6 (3602)	⑦	⑨	⑤	173	④	15	15	18-22	96
	M28/49, 50	5.4 (5397)	⑤	⑥	③	218	④	15	15	18-22	96
1996	M64/23, 24	3.6 (3600)	⑦	⑨	①	173	④	15	15	18-22	96
	M64/60	3.6 (3600)	⑦	⑨	①	173	④	15	15	18-22	96
1997	M64/23, 24	3.6 (3600)	⑦	⑨	①	173	④	15	15	18-22	96
	M64/60	3.6 (3600)	⑦	⑨	①	173	④	15	15	18-22	96
1998-99	M64/23, 24	3.6 (3600)	⑦	⑨	①	173	④	15	15	18-22	96
	M96/20	2.5 (2480)	⑩	⑪	⑪	NA	③	7	NA	18-22	96

① Step 1: 15 ft. lbs.
　Step 2: Turn an additional 60 degrees
　Step 3: Turn an additional 90 degrees
② M12 bolts:
　Step 1: 22 ft. lbs.
　Step 2: Turn an additional 60 degrees
　M10 bolts:
　Step 1: 15 ft. lbs.
　Step 2: 37 ft. lbs.
　M8 bolts: 15 ft. lbs.
　M6 bolts: 7 ft. lbs.

③ Step 1: 18 ft. lbs.
　Step 2: Turn an additional 90 degrees
④ Step 1: 29 ft. lbs.
　Step 2: 66 ft. lbs.
⑤ Step 1: 15 ft. lbs.
　Step 2: Turn an additional 90 degrees
　Step 3: Turn an additional 90 degrees
⑥ M12 bolts:
　Step 1: 22 ft. lbs.
　Step 2: 41 ft. lbs.
　Step 3: 59 ft. lbs.

M10 bolts:
　Step 1: 15 ft. lbs.
　Steo 2: 41 ft. lbs.
M8 bolts:
　Step 1: 11 ft. lbs.
　Step 2: 15 ft. lbs.
⑦ Step 1: 11 ft. lbs.
　Step 2: Turn an additional 90 degrees
⑧ Step 1: 14 ft. lbs.
　Step 2: Turn an additional 90 degrees
⑨ M10 bolts: 29 ft. lbs.
　M8 bolts: 17 ft. lbs.

⑩ Step 1: 18 ft. lbs.
　Step 2: Loosen Completely
　Step 3: 15 ft. lbs.
　Step 4: Turn an additional 90 degrees
　Step 5: Turn an additional 90 degrees
⑪ Step 1: 15 ft. lbs.
　Step 2: Turn an additional 90 degrees

79231C73

SCHEDULED MAINTENANCE INTERVALS
(PORSCHE BOXSTER, 911, 928, & 968)

TO BE SERVICED	TYPE OF SERVICE	VEHICLE MILEAGE INTERVAL (x1000)												
		7.5	15	22.5	30	37.5	45	52.5	60	67.5	75	82.5	90	97.5
Engine oil & filter	R	✓	✓	✓	✓	✓	✓	✓	✓	✓	✓	✓	✓	✓
Battery	S/I		✓		✓		✓		✓		✓		✓	
Brake system	S/I		✓		✓		✓		✓		✓		✓	
Chassis lubrication	S/I		✓		✓		✓		✓		✓		✓	
Clutch adjustment	S/I		✓		✓		✓		✓		✓		✓	
Clutch hydraulic system	S/I		✓		✓		✓		✓		✓		✓	
Coolant level, hoses & connections	S/I		✓		✓		✓		✓		✓		✓	
Crankcase ventilation system	S/I		✓		✓		✓		✓		✓		✓	
Drive belts	S/I		✓		✓		✓		✓		✓		✓	
Fuel system	S/I		✓		✓		✓		✓		✓		✓	
Intake air system	S/I		✓		✓		✓		✓		✓		✓	
Parking brake system	S/I		✓		✓		✓		✓		✓		✓	
Steering system	S/I		✓		✓		✓		✓		✓		✓	
Suspension components & wheel bearings	S/I		✓		✓		✓		✓		✓		✓	
Throttle linkage	S/I		✓		✓		✓		✓		✓		✓	
Transmission fluid level	S/I		✓		✓		✓		✓		✓		✓	
Valve clearances	S/I		✓		✓		✓		✓		✓		✓	
Air filter element	R				✓				✓				✓	
Automatic transmission fluid & filter	R				✓				✓				✓	
Auxiliary air pump filter element	R				✓				✓				✓	
Brake fluid	R				✓				✓				✓	
Engine Coolant	R				✓				✓				✓	
Fuel filter	R				✓				✓				✓	
Limited slip differential fluid	R				✓				✓				✓	
Spark plugs	R				✓				✓				✓	

79231C74

SCHEDULED MAINTENANCE INTERVALS
(PORSCHE BOXSTER, 911, 928, & 968) (Cont.)

TO BE SERVICED	TYPE OF SERVICE	VEHICLE MILEAGE INTERVAL (x1000)												
		7.5	15	22.5	30	37.5	45	52.5	60	67.5	75	82.5	90	97.5
Automatic transmission differential oil	R								✓					
Axle drive fluid	R								✓					
Camshaft timing belt	R								✓					

R – Replace   S/I – Service or Inspect

FREQUENT OPERATION MAINTENANCE (SEVERE SERVICE)

If a vehicle is operated under any of the following conditions it is considered severe service:
- Extremely dusty areas.
- 50% or more of the vehicle operation is in 32°C (90°F) or higher temperatures, or constant operation in temperatures below 0°C (32°F).
- Prolonged idling (vehicle operation in stop and go traffic).
- Frequent short running periods (engine does not warm to normal operating temperatures).
- Police, taxi, delivery usage or trailer towing usage.

Oil & oil filter change – change every 3750 miles.
Air filter element – service or inspect every 15,000 miles.

79231C75

SCHEDULED MAINTENANCE
PORSCHE
911, 928, 968, BOXSTER

The following should be used as a guide when determining the amount of work required for a particular service if taken to a repair shop.
In estimating how long a particular Scheduled Maintenance Service should take, please observe the following:

- **Chilton Time** is time based on field research and data supplied by the vehicle manufacturer.
- Labor time operations are given in hours and tenths of an hour.
- All labor operations, are to be used as a guide.

Mechanic Skill Level Codes:
        (G) GENERAL: Normally skilled with certification.
        (M) MAINTENANCE: Semi-skilled working on certicication.
        (P) PRECISION: Really skilled with multiple certification.

	Chilton Time		Chilton Time		Chilton Time
(M) 7500 Mile Service		**(M) 37500 Mile Service**		**(G) 75000 Mile Service**	
1995-99 911, Boxster	.3	1995-99 911, Boxster	.3	1995-99 911, Boxster	2.4
1995 928, 968	.3	1995 928, 968	.3	1995 928, 968	2.5
(G) 15000 Mile Service		**(G) 45000 Mile Service**		**(M) 82500 Mile Service**	
1995-99 911, Boxster	2.4	1995-99 911, Boxster	2.4	1995-99 911, Boxster	.3
1995 928, 968	2.5	1995 928, 968	2.5	1995 928, 968	.3
(M) 22500 Mile Service		**(M) 52500 Mile Service**		**(G) 90000 Mile Service**	
1995-99 911, Boxster	.3	1995-99 911, Boxster	.3	1995-99 911, Boxster	6.1
1995 928, 968	.3	1995 928, 968	.3	1995 928, 968	6.7
(G) 30000 Mile Service		**(G) 60000 Mile Service**		w/AT add	.5
1995-99 911, Boxster	6.1	1995-99 911, Boxster	12.8	**(M) 97500 Mile Service**	
1995 928, 968	6.7	1995 928, 968	13.4	1995-99 911, Boxster	.3
w/AT add	.5	w/AT add	.8	1995 928, 968	.3
		(M) 67500 Mile Service			
		1995-99 911, Boxster	.3		
		1995 928, 968	.3		

79231C76

SAAB
9-3 • 900 • 9000

VEHICLE IDENTIFICATION CHART

VIN Code	Eng. Series	Liters	Cu. In. (cc)	Cyl.	Fuel Sys.	Eng. Mfg.		Code	Year
			Engine Code					**Model Year**	
B	B234I/V	2.3	140 (2290)	I4	MFI	Saab		S	1995
M	B234L/M	2.3	140 (2290)	I4	MFI-Turbo	Saab		T	1996
N	B204L/N	2.0	121 (1985)	I4	MFI-Turbo	Saab		V	1997
R	B234R/R	2.3	140 (2290)	I4	MFI-Turbo	Saab		W	1998
U	B234E/U	2.3	140 (2290)	I4	MFI-Turbo	Saab		X	1999
V	B258I/V	2.5	152 (2498)	V6	MFI	GM			
W	B308I/W	3.0	180 (2961)	V6	MFI	GM			

MFI - Multiport Fuel Injection

79231C77

ENGINE IDENTIFICATION

Year	Model	Engine Displacement Liters (cc)	Engine Series (ID/VIN)	Fuel System	No. of Cylinders	Engine Type
1995	900	2.0 (1985)	B204L/N	MFI-Turbo	4	DOHC
	900	2.3 (2290)	B234I/B	MFI	4	DOHC
	900	2.5 (2498)	B258I/V	MFI	6	DOHC
	9000	2.3 (2290)	B234I/B	MFI	4	DOHC
	9000	2.3 (2290)	B234L/M	MFI-Turbo	4	DOHC
	9000	2.3 (2290)	B234R/R	MFI-Turbo	4	DOHC
	9000	2.3 (2290)	B234E/U	MFI-Turbo	4	DOHC
1996	900	2.0 (1985)	B204L/N	MFI-Turbo	4	DOHC
	900	2.3 (2290)	B234I/B	MFI	4	DOHC
	900	2.5 (2498)	B258I/V	MFI	6	DOHC
	9000	2.3 (2290)	B234L/M	MFI-Turbo	4	DOHC
	9000	2.3 (2290)	B234R/R	MFI-Turbo	4	DOHC
	9000	2.3 (2290)	B234E/U	MFI-Turbo	4	DOHC
	9000	3.0 (2961)	B308I/W	MFI	6	DOHC
1997	900	2.0 (1985)	B204L/N	MFI-Turbo	4	DOHC
	900	2.3 (2290)	B234I/B	MFI	4	DOHC
	900	2.5 (2498)	B258I/V	MFI	6	DOHC
	9000	2.3 (2290)	B234L/M	MFI-Turbo	4	DOHC
	9000	2.3 (2290)	B234R/R	MFI-Turbo	4	DOHC
	9000	2.3 (2290)	B234E/U	MFI-Turbo	4	DOHC
	9000	3.0 (2961)	B308I/W	MFI	6	DOHC
1998-99	900	2.0 (1985)	B204L/N	MFI-Turbo	4	DOHC
	900	2.3 (2290)	B234I/B	MFI	4	DOHC
	9000	2.3 (2290)	B234R/R	MFI-Turbo	4	DOHC
	9-3	2.0 (1985)	B204L/N	MFI-Turbo	4	DOHC

MFI - Multiport Fuel Injection
DOHC - Double Overhead Camshaft

79231C78

GENERAL ENGINE SPECIFICATIONS

Year	Engine ID/VIN	Engine Displacement Liters (cc)	Fuel System Type	Net Horsepower @ rpm	Net Torque @ rpm (ft. lbs.)	Bore x Stroke (in.)	Compression Ratio	Oil Pressure @ rpm
1995	B204L/N	2.0 (1985)	MFI-Turbo	185@5500	194@2100	3.54x3.07	9.2:1	39@2000
	B234I/B	2.3 (2290)	MFI	150@5700	155@4300	3.54x3.54	10.5:1	39@2000
	B234L/M	2.3 (2290)	MFI-Turbo	200@5500 ①	240@1800 ②	3.54x3.54	9.25:1	39@2000
	B234R/R	2.3 (2290)	MFI-Turbo	225@5500	253@1800	3.54x3.54	9.25:1	39@2000
	B234E/U	2.3 (2290)	MFI-Turbo	170@5700	192@3200	3.54x3.54	9.25:1	39@2000
	B258I/V	2.5 (2498)	MFI	170@5900	167@4200	3.21x3.13	10.8:1	NA
1996	B204L/N	2.0 (1985)	MFI-Turbo	185@5500	194@2100	3.54x3.07	9.2:1	39@2000
	B234I/B	2.3 (2290)	MFI	150@5700	155@4300	3.54x3.54	10.5:1	39@2000
	B234E/U	2.3 (2290)	MFI-Turbo	170@5700	192@3200	3.54x3.54	9.25:1	39@2000
	B234R/R	2.3 (2290)	MFI-Turbo	225@5500	153@1800	3.54x3.54	9.25:1	39@2000
	B234L/M	2.3 (2290)	MFI-Turbo	200@5500	240@1800	3.54x3.54	9.25:1	39@2000
	B258I/V	2.5 (2498)	MFI	170@5900	167@4200	3.21x3.13	10.8:1	NA
	B308I/W	3.0 (2961)	MFI	210@6200	200@3300	3.38x3.34	10.8:1	NA
1997	B204L/N	2.0 (1985)	MFI-Turbo	185@5500	194@2100	3.54x3.07	9.2:1	39@2000
	B234I/B	2.3 (2290)	MFI	150@5700	155@4300	3.54x3.54	10.5:1	39@2000
	B234E/U	2.3 (2290)	MFI-Turbo	170@5700	192@3200	3.54x3.54	9.25:1	39@2000
	B234R/R	2.3 (2290)	MFI-Turbo	225@5500	153@1800	3.54x3.54	9.25:1	39@2000
	B234L/M	2.3 (2290)	MFI-Turbo	200@5500	240@1800	3.54x3.54	9.25:1	39@2000
	B258I/V	2.5 (2498)	MFI	170@5900	167@4200	3.21x3.13	10.8:1	NA
	B308I/W	3.0 (2961)	MFI	210@6200	200@3300	3.38x3.34	10.8:1	NA
1998-99	B204L/N	2.0 (1985)	MFI-Turbo	185@5500	194@2100	3.54x3.07	9.2:1	39@2000
	B234I/B	2.3 (2290)	MFI	150@5700	155@4300	3.54x3.54	10.5:1	39@2000
	B234R/R	2.3 (2290)	MFI-Turbo	225@5500	153@1800	3.54x3.54	9.25:1	39@2000

MFI - Multiport Fuel Injection

NA - Not Available

① 9000 Aero, manual: 225@5500

② 9000 Aero, manual: 258@1950

79231C79

GASOLINE ENGINE TUNE-UP SPECIFICATIONS

Year	Engine ID/VIN	Engine Displacement Liters (cc)	Spark Plugs Gap (in.)	Ignition Timing (deg.)		Fuel Pump (psi)	Idle Speed (rpm)		Valve Clearance	
				MT	AT		MT	AT	In.	Ex.
1995	B204L/N	2.0 (1985)	0.039	①	①	43 ②	850	850	HYD	HYD
	B234I/B	2.3 (2290)	0.024	①	①	43 ②	850	850	HYD	HYD
	B234L/M	2.3 (2290)	0.024	①	①	43 ②	850	850	HYD	HYD
	B234R/R	2.3 (2290)	0.024	①	①	43 ②	850	850	HYD	HYD
	B234E/U	2.3 (2290)	0.024	①	①	43 ②	850	850	HYD	HYD
	B258I/V	2.5 (2498)	0.031	①	①	43 ②	800	800	HYD	HYD
1996	B204L/N	2.0 (1985)	0.039	①	①	43 ②	850	850	HYD	HYD
	B234I/B	2.3 (2290)	0.024	①	①	43 ②	850	850	HYD	HYD
	B234L/M	2.3 (2290)	0.040	①	①	43 ②	850	850	HYD	HYD
	B234R/R	2.3 (2290)	0.040	①	①	43 ②	850	850	HYD	HYD
	B234E/U	2.3 (2290)	0.040	①	①	43 ②	850	850	HYD	HYD
	B258I/V	2.5 (2498)	0.031	①	①	43 ②	800	800	HYD	HYD
	B308I/W	3.0 (2961)	0.031	①	①	44 ②	800	800	HYD	HYD
1997	B204L/N	2.0 (1985)	0.039	①	①	43 ②	850	850	HYD	HYD
	B234I/B	2.3 (2290)	0.024	①	①	43 ②	850	850	HYD	HYD
	B234L/M	2.3 (2290)	0.040	①	①	43 ②	850	850	HYD	HYD
	B234R/R	2.3 (2290)	0.040	①	①	43 ②	850	850	HYD	HYD
	B234E/U	2.3 (2290)	0.040	①	①	43 ②	850	850	HYD	HYD
	B258I/V	2.5 (2498)	0.031	①	①	43 ②	800	800	HYD	HYD
	B308I/W	3.0 (2961)	0.031	①	①	44 ②	800	800	HYD	HYD
1998-99	B204L/N	2.0 (1985)	0.039	①	①	43 ②	850	850	HYD	HYD
	B234I/B	2.3 (2290)	0.024	①	①	43 ②	850	850	HYD	HYD
	B234R/R	2.3 (2290)	0.040	①	①	43 ②	850	850	HYD	HYD

NOTE: The Vehicle Emission Control Information label often reflects specification changes made during production. The label figures must be used if they differ from those in this chart.

B - Before top dead center

HYD - Hydraulic

① Pre-programmed in ECU and cannot be adjusted

② Fuel line pressure regulator before the control pressure regulator

79231C80

CAPACITIES

Year	Model	Engine ID/VIN	Engine Displacement Liters (cc)	Engine Oil with Filter (qts.)	Transmission (pts.)			Transfer Case (pts.)	Drive Axle		Fuel Tank (gal.)	Cooling System (qts.)
					4-Spd	5-Spd	Auto.		Front (pts.)	Rear (pts.)		
1995	900	B204L/N	2.0 (1985)	5.5	—	3.6	14.8	—	—	—	18.0	8.7
	900	B234I/B	2.3 (2290)	5.1	—	3.6	14.8	—	—	—	18.0	8.7
	900	B258I/V	2.5 (2498)	4.6	—	3.6	14.8	—	—	—	18.0	8.2
	9000	B234I/B	2.3 (2290)	4.5	—	5.3	16.5	—	—	—	17.4	9.5
	9000	B234L/M	2.3 (2290)	4.5	—	5.3	16.5	—	—	—	17.4	9.5
	9000	B234R/R	2.3 (2290)	4.5	—	5.3	16.5	—	—	—	17.4	9.5
	9000	B234E/U	2.3 (2290)	5.1	—	7.2	14.0	—	—	—	17.4	9.5
1996	900	B204L/N	2.0 (1985)	4.1	—	7.2	13.6	—	—	—	18.0	8.7
	900	B234I/B	2.3 (2290)	4.1	—	7.2	13.6	—	—	—	18.0	8.7
	900	B258I/V	2.5 (2498)	4.6	—	—	13.6	—	—	—	18.0	8.2
	9000	B234L/M	2.3 (2290)	5.1	—	7.2	14.0	—	—	—	17.4	9.5
	9000	B234R/R	2.3 (2290)	5.1	—	7.2	—	—	—	—	17.4	9.5
	9000	B234E/U	2.3 (2290)	5.1	—	7.2	14.0	—	—	—	17.4	9.5
	9000	B308I/W	3.0 (2961)	4.9	—	—	14.0	—	—	—	17.4	9.0
1997	900	B204L/N	2.0 (1985)	4.1	—	7.2	13.6	—	—	—	18.0	8.7
	900	B234I/B	2.3 (2290)	4.1	—	7.2	13.6	—	—	—	18.0	8.7
	900	B258I/V	2.5 (2498)	4.6	—	—	13.6	—	—	—	18.0	8.2
	9000	B234L/M	2.3 (2290)	5.1	—	7.2	14.0	—	—	—	17.4	9.5
	9000	B234R/R	2.3 (2290)	5.1	—	7.2	—	—	—	—	17.4	9.5
	9000	B234E/U	2.3 (2290)	5.1	—	7.2	14.0	—	—	—	17.4	9.5
	9000	B308I/W	3.0 (2961)	4.9	—	—	14.0	—	—	—	17.4	9.0
1998-99	900	B204L/N	2.0 (1985)	4.1	—	7.2	13.6	—	—	—	18.0	8.7
	900	B234I/B	2.3 (2290)	4.1	—	7.2	13.6	—	—	—	18.0	8.7
	9000	B234R/R	2.3 (2290)	5.1	—	7.2	—	—	—	—	17.4	9.5
	9-3	B204L/N	2.0 (1985)	4.1	—	7.2	13.6	—	—	—	18.0	8.7

NOTE: All capacities are approximate. Add fluid gradually and check to be sure a proper fluid level is obtained.

79231C81

VALVE SPECIFICATIONS

Year	Engine ID/VIN	Engine Displacement Liters (cc)	Seat Angle (deg.)	Face Angle (deg.)	Spring Test Pressure (lbs. @ in.)	Spring Installed Height (in.)	Stem-to-Guide Clearance (in.)		Stem Diameter (in.)	
							Intake	Exhaust	Intake	Exhaust
1995	B204L/N	2.0 (1985)	45	44.5	138-150@ 1.12	1.46	0.020	0.020	0.2740- 0.2746	0.2738- 0.2748
	B234I/B	2.3 (2290)	45	44.5	138-150@ 1.12	1.46	0.020	0.020	0.2740- 0.2746	0.2738- 0.2748
	B234L/M	2.3 (2290)	45	44.5	138-141@ 1.12	1.46	0.020	0.020	0.2740- 0.2746	0.2738- 0.2748
	B234R/R	2.3 (2290)	45	44.5	138-141@ 1.12	1.46	0.020	0.020	0.2740- 0.2746	0.2738- 0.2748
	B234E/U	2.3 (2290)	45	44.5	138-141@ 1.12	1.46	0.020	0.020	0.2740- 0.2746	0.2738- 0.2748
	B258I/V	2.5 (2498)	45	45.3	142-153@ 0.94	NA	NA	NA	0.2344- 0.2350	0.2340- 0.2346
1996	B204L/N	2.0 (1985)	45	44.5	138-150@ 1.12	1.46	0.020	0.020	0.2740- 0.2746	0.2738- 0.2748
	B234I/B	2.3 (2290)	45	44.5	138-150@ 1.12	1.46	0.020	0.020	0.2740- 0.2746	0.2738- 0.2748
	B234L/M	2.3 (2290)	45	44.5	138-141@ 1.12	1.46	0.020	0.020	0.2740- 0.2746	0.2738- 0.2748
	B234R/R	2.3 (2290)	45	44.5	138-141@ 1.12	1.46	0.020	0.020	0.2740- 0.2746	0.2738- 0.2748
	B234E/U	2.3 (2290)	45	44.5	138-141@ 1.12	1.46	0.020	0.020	0.2740- 0.2746	0.2738- 0.2748
	B258I/V	2.5 (2498)	45	45.3	142-153@ 0.94	NA	NA	NA	0.2344- 0.2350	0.2340- 0.2346
	B308I/W	3.0 (2961)	45	45.3	142-153@ 0.94	NA	NA	NA	0.2344- 0.2350	0.2340- 0.2346
1997	B204L/N	2.0 (1985)	45	44.5	138-150@ 1.12	1.46	0.020	0.020	0.2740- 0.2746	0.2738- 0.2748
	B234I/B	2.3 (2290)	45	44.5	138-150@ 1.12	1.46	0.020	0.020	0.2740- 0.2746	0.2738- 0.2748
	B234L/M	2.3 (2290)	45	44.5	138-141@ 1.12	1.46	0.020	0.020	0.2740- 0.2746	0.2738- 0.2748
	B234R/R	2.3 (2290)	45	44.5	138-141@ 1.12	1.46	0.020	0.020	0.2740- 0.2746	0.2738- 0.2748
	B234E/U	2.3 (2290)	45	44.5	138-141@ 1.12	1.46	0.020	0.020	0.2740- 0.2746	0.2738- 0.2748
	B258I/V	2.5 (2498)	45	45.3	142-153@ 0.94	NA	NA	NA	0.2344- 0.2350	0.2340- 0.2346
	B308I/W	3.0 (2961)	45	45.3	142-153@ 0.94	NA	NA	NA	0.2344- 0.2350	0.2340- 0.2346
1998-99	B204L/N	2.0 (1985)	45	44.5	138-150@ 1.12	1.46	0.020	0.020	0.2740- 0.2746	0.2738- 0.2748
	B234I/B	2.3 (2290)	45	44.5	138-150@ 1.12	1.46	0.020	0.020	0.2740- 0.2746	0.2738- 0.2748
	B234R/R	2.3 (2290)	45	44.5	138-141@ 1.12	1.46	0.020	0.020	0.2740- 0.2746	0.2738- 0.2748

NA - Not Available

79231C82

TORQUE SPECIFICATIONS
All readings in ft. lbs.

Year	Engine ID/VIN	Engine Displacement Liters (cc)	Cylinder Head Bolts	Main Bearing Bolts	Rod Bearing Bolts	Crankshaft Damper Bolts	Flywheel Bolts	Manifold Intake	Manifold Exhaust	Spark Plugs	Lug Nut
1995	B204L/N	2.0 (1985)	③	81	35	130	59	16	19	20	80-90
	B234I/B	2.3 (2290)	③	81	35	130	59	16	13	20	80-90
	B234L/M	2.3 (2290)	③	81	35	130	59	16	19	20	80-90
	B234R/R	2.3 (2290)	③	81	35	130	59	16	19	20	80-90
	B234E/U	2.3 (2290)	①	81	35	130	59	16	19	20	80-90
	B258I/V	2.5 (2498)	④	⑤	⑥	⑦	⑧	15	15	19	80-90
1996	B204L/N	2.0 (1985)	①	81	35	130	59	16	19	20	80-90
	B234I/B	2.3 (2290)	①	81	35	130	59	16	13	20	80-90
	B234L/M	2.3 (2290)	①	81	35	130	59	16	19	20	80-90
	B234R/R	2.3 (2290)	①	81	35	130	59	16	19	20	80-90
	B234E/U	2.3 (2290)	①	81	35	130	59	16	19	20	80-90
	B258I/V	2.5 (2498)	④	⑤	⑥	⑦	⑧	15	15	19	80-90
	B308I/W	3.0 (2961)	④	⑤	⑥	⑦	⑧	15	15	19	80-90
1997	B204L/N	2.0 (1985)	①	81	35	130	59	16	19	20	80-90
	B234I/B	2.3 (2290)	①	81	35	130	59	16	13	20	80-90
	B234L/M	2.3 (2290)	①	81	35	130	59	16	19	20	80-90
	B234R/R	2.3 (2290)	①	81	35	130	59	16	19	20	80-90
	B234E/U	2.3 (2290)	①	81	35	130	59	16	19	20	80-90
	B258I/V	2.5 (2498)	④	⑤	⑥	⑦	⑧	15	15	19	80-90
	B308I/W	3.0 (2961)	④	⑤	⑥	⑦	⑧	15	15	19	80-90
1998-99	B204L/N	2.0 (1985)	①	81	35	130	59	16	19	20	80-90
	B234I/B	2.3 (2290)	①	81	35	130	59	16	13	20	80-90
	B234R/R	2.3 (2290)	①	81	35	130	59	16	19	20	80-90

① Step 1: 44 ft. lbs.
Step 2: 59 ft. lbs. If 17mm head bolts, torque to 70 ft. lbs.
Step 3: Tighten each bolt an additional 90 degrees

② Step 1: 44 ft. lbs.
Step 2: Tighten 19mm head bolts to 63 ft. lbs.

③ Step 1: 44 ft. lbs.
Step 2: 59 ft. lbs.
Step 3: Tighten each bolt an additional 90 degrees

④ Step 1: 19 ft. lbs.
Step 2: Tighten each bolt an additional 90 degrees
Step 3: Repeat Step 2
Step 4: Repeat Step 2

⑤ Step 1: 37 ft. lbs.
Step 2: Tighten each bolt an additional 60 degrees
Step 3: Tighten each bolt an additional 15 degrees

⑥ Step 1: 26 ft. lbs.
Step 2: Tighten each bolt an additional 45 degrees
Step 3: Tighten each bolt an additional 15 degrees

⑦ Step 1: 185 ft. lbs.
Step 2: Turn bolt an additional 45 degrees

⑧ Step 1: 48 ft. lbs.
Step 2: Turn bolt an additional 30 degrees

79231C83

SCHEDULED MAINTENANCE INTERVALS
(SAAB 900, 9000, 9-3, & 9-5)

TO BE SERVICED	TYPE OF SERVICE	VEHICLE MILEAGE INTERVAL (x1000)												
		5	10	15	20	25	30	35	40	45	50	55	60	65
Engine oil & filter	R	✓	✓	✓	✓	✓	✓	✓	✓	✓	✓	✓	✓	✓
Battery electrolyte level	S/I	✓		✓		✓		✓		✓		✓		✓
Brake fluid①	S/I	✓		✓		✓		✓		✓		✓		✓
Brake lines & hoses	S/I	✓		✓		✓		✓		✓		✓		✓
Brake pads & discs	S/I	✓		✓		✓		✓		✓		✓		✓
Drive belts	S/I	✓		✓		✓		✓		✓		✓		
Engine coolant strength	S/I	✓		✓		✓		✓		✓		✓		✓
Exhaust system	S/I	✓		✓		✓		✓		✓		✓		✓
Final drive oil level (900 A/T)	S/I	✓		✓		✓		✓		✓		✓		✓
Gearbox oil	S/I	✓		✓		✓		✓		✓		✓		✓
Outer & inner drive joint boots	S/I	✓		✓		✓		✓		✓		✓		✓
Rotate tires (front to rear)	S/I	✓		✓		✓		✓		✓		✓		
Automatic transmission fluid & filter	R	✓						✓						✓
Air cleaner element	R							✓						✓
Engine coolant	R							✓						✓
Spark plugs	R							✓						✓
Ventilation air filter	R							✓						✓
Ball joint clearance	S/I							✓						✓
Engine cooling system, hoses & cap	S/I	✓						✓						
Front wheel alignment	S/I							✓						✓
Fuel lines	S/I							✓						✓
Shock absorbers & bushings	S/I							✓						✓
Fuel filter	R													✓
Power steering fluid	R							✓						
Crankcase ventilation & vacuum lines	S/I													✓

79231C84

SCHEDULED MAINTENANCE INTERVALS
(SAAB 900, 9000, 9-3 & 9-5) (Cont.)

TO BE SERVICED	TYPE OF SERVICE	VEHICLE MILEAGE INTERVAL (x1000)												
		5	10	15	20	25	30	35	40	45	50	55	60	65
Distributor cap & rotor	S/I													✓
EVAP system	S/I													✓
Front suspension, rear axle mountings	S/I	✓												
Parking brake adjustment	S/I	✓												
Spark plug wires	S/I													✓

① replace every 35,000 miles.
R – Replace   S/I – Service or Inspect

FREQUENT OPERATION MAINTENANCE (SEVERE SERVICE)
If a vehicle is operated under any of the following conditions it is considered severe service:
- Extremely dusty areas.
- 50% or more of the vehicle operation is in 32°C (90°F) or higher temperatures, or constant operation in temperatures below 0°C (32°F).
- Prolonged idling (vehicle operation in stop and go traffic).
- Frequent short running periods (engine does not warm to normal operating temperatures).
- Police, taxi, delivery usage or trailer towing usage.
Oil & oil filter change – change every 2500 miles.
Air filter element – service or inspect every 15,000 miles.

79231C85

SCHEDULED MAINTENANCE
SAAB
900, 9000, 9-3, 9-5

The following should be used as a guide when determining the amount of work required for a particular service if taken to a repair shop.
In estimating how long a particular Scheduled Maintenance Service should take, please observe the following:

- **Chilton Time** is time based on field research and data supplied by the vehicle manufacturer.
- Labor time operations are given in hours and tenths of an hour.
- All labor operations, are to be used as a guide.

Mechanic Skill Level Codes:
  (G) GENERAL: Normally skilled with certification.
  (M) MAINTENANCE: Semi-skilled working on certicication.
  (P) PRECISION: Really skilled with multiple certification.

	Chilton Time		Chilton Time		Chilton Time
(G) 5000 Mile Service		**(G) 25000 Mile Service**		**(M) 50000 Mile Service**	
1995-98 900	2.1	1995-98 900	1.8	1995-99	.3
w/AT add	.6	w/AT add	.1	**(G) 55000 Mile Service**	
1995-98 9000	2.1	1995-98 9000	1.8	1995-98 900	1.8
w/AT add	.5	1999 9-3, 9-5	1.8	w/AT add	.1
1999 9-3, 9-5	2.1	**(M) 30000 Mile Service**		1995-98 9000	1.8
(M) 10000 Mile Service		1995-99	.3	1999 9-3, 9-5	1.8
1995-99	.3	**(G) 35000 Mile Service**		**(M) 60000 Mile Service**	
(G) 15000 Mile Service		1995-98 900	5.4	1995-99	.3
1995-98 900	1.8	w/AT add	.6	**(G) 65000 Mile Service**	
w/AT add	.1	1995-98 9000	5.4	1995-98 900	5.1
1995-98 9000	1.8	w/AT add	.5	w/AT add	.6
1999 9-3, 9-5	1.8	1999 9-3, 9-5	5.4	1995-98 9000	5.1
(M) 20000 Mile Service		**(G) 40000 Mile Service**		w/AT add	.5
1995-99	.3	1995-99	.3	1999 9-3, 9-5	5.1
		(G) 45000 Mile Service			
		1995-98 900	1.8		
		w/AT add	.1		
		1995-98 9000	1.8		
		1999 9-3, 9-5	1.8		

79231C86

SUBARU
Impreza • Legacy • SVX

VEHICLE IDENTIFICATION CHART

Engine Code							Model Year	
Code	Liters	Cu. In. (cc)	Cyl.	Fuel Sys.	Eng. Mfg.		Code	Year
1 ①	1.8	111 (1829)	4	MFI	Subaru		S	1995
2 ②	1.8	111 (1829)	4	MFI	Subaru		T	1996
3 ①	2.2	135 (2212)	4	MFI	Subaru		V	1997
3	3.3	202 (3318)	6	MFI	Subaru		W	1998
4	2.2	135 (2212)	4	MFI	Subaru		X	1999
6	2.2	135 (2212)	4	MFI	Subaru			
6	2.5	150 (2457)	4	MFI	Subaru			

MFI - Multiport fuel injection

① 2WD

② 4WD

79231CL4

ENGINE IDENTIFICATION

Year	Model		Engine Displacement Liters (cc)	Engine Series (ID/VIN)	Fuel System	No. of Cylinders	Engine Type
1995	Impreza - 2wd		1.8 (1820)	1	MFI	4	SOHC
	Impreza - 4wd		1.8 (1820)	2	MFI	4	SOHC
	Impreza	①	2.2 (2212)	4	MFI	4	SOHC
	Legacy	①	2.2 (2212)	6	MFI	4	SOHC
	SVX		3.3 (3318)	3	MFI	6	DOHC
1996	Impreza - 2wd		1.8 (1820)	1	MFI	4	SOHC
	Impreza - 4wd		1.8 (1820)	2	MFI	4	SOHC
	Impreza	①	2.2 (2212)	4	MFI	4	SOHC
	Legacy 2wd		2.2 (2212)	3	MFI	4	SOHC
	Legacy	①	2.2 (2212)	4	MFI	4	SOHC
	Legacy	①	2.5 (2457)	6	MFI	4	DOHC
	SVX		3.3 (3318)	3	MFI	6	DOHC
1997	Impreza		1.8 (1820)	2	MFI	4	SOHC
	Impreza	①	2.2 (2212)	4	MFI	4	SOHC
	Legacy		2.2 (2212)	4	MFI	4	SOHC
	Legacy	①	2.5 (2457)	6	MFI	4	DOHC
	SVX		3.3 (3318)	3	MFI	6	DOHC
1998-99	Impreza	①	2.2 (2212)	4	MFI	4	SOHC
	Impreza RS		2.5 (2457)	6	MFI	4	DOHC
	Legacy		2.2 (2212)	4	MFI	4	SOHC
	Legacy	①	2.5 (2457)	6	MFI	4	DOHC

DOHC - Double overhead camshaft
SOHC - Single overhead camshaft
MFI - Multiport fuel injection
① Includes Outback trim levels

79231CL5

GENERAL ENGINE SPECIFICATIONS

Year	Engine ID/VIN	Engine Displacement Liters (cc)	Fuel System Type	Net Horsepower @ rpm	Net Torque @ rpm (ft. lbs.)	Bore x Stroke (in.)	Compression Ratio	Oil Pressure @ rpm	
1995	1	1.8 (1820)	MFI	110@5600	110@4400	3.46x2.95	9.5:1	14@600	①
	2	1.8 (1820)	MFI	110@5600	110@4400	3.46x2.95	9.5:1	14@600	①
	3	3.3 (3318)	MFI	230@5400	228@4400	3.82x2.95	10.1:1	14@600	①
	4	2.2 (2212)	MFI	135@5400	140@4400	3.82x2.95	9.5:1	14@600	①
	6	2.2 (2212)	MFI	135@5400	140@4400	3.82x2.95	9.5:1	14@600	①
1996	1	1.8 (1820)	MFI	110@5600	110@4400	3.46x2.95	9.5:1	14@600	①
	2	1.8 (1820)	MFI	110@5600	110@4400	3.46x2.95	9.5:1	14@600	①
	3	2.2 (2212)	MFI	135@5400	140@4400	3.82x2.95	9.5:1	14@600	①
	3	3.3 (3318)	MFI	230@5400	228@4400	3.82x2.95	10.1:1	14@600	①
	4	2.2 (2212)	MFI	135@5400	140@4400	3.82x2.95	9.5:1	14@600	①
	6	2.5 (2457)	MFI	155@5600	155@2800	3.92x3.11	9.5:1	14@600	①
1997	2	1.8 (1820)	MFI	115@5600	120@4000	3.46x2.95	9.7:1	14@600	①
	3	3.3 (3318)	MFI	230@5400	228@4400	3.82x2.95	10.1:1	14@600	①
	4	2.2 (2212)	MFI	137@5400	145@4000	3.82x2.95	9.7:1	14@600	①
	6	2.5 (2457)	MFI	165@5600	162@4000	3.92x3.11	9.7:1	14@600	①
1998-99	4	2.2 (2212)	MFI	137@5400	145@4000	3.82x2.95	9.7:1	14@600	①
	6	2.5 (2457)	MFI	165@5600	162@4000	3.92x3.11	9.7:1	14@600	①

MFI - Multiport fuel injection

① Test with engine at normal operating temperature

79231CL6

GASOLINE ENGINE TUNE-UP SPECIFICATIONS

Year	Engine ID/VIN	Engine Displacement Liters (cc)	Spark Plugs Gap (in.)	Ignition Timing (deg.) ① MT	AT	Fuel Pump (psi)	Idle Speed (rpm) ② MT	AT	Valve Clearance ③ In.	Ex.
1995	1	1.8 (1820)	0.039-0.043	20B	20B	36	700	700	HYD	HYD
	2	1.8 (1820)	0.039-0.043	20B	20B	36	700	700	HYD	HYD
	3	3.3 (3318)	0.039-0.043	—	20B	43	—	610	HYD	HYD
	4	2.2 (2212)	0.039-0.043	20B	20B	36	600-800	600-800	HYD	HYD
	6	2.2 (2212)	0.039-0.043	20B	20B	36	600-800	600-800	HYD	HYD
1996	1	1.8 (1820)	0.039-0.043	12-28B	12-28B	34-38	600-800	600-800	HYD	HYD
	2	1.8 (1820)	0.039-0.043	12-28B	12-28B	34-38	600-800	600-800	HYD	HYD
	3	2.2 (2212)	0.039-0.043	6-22B	12-28B	34-38	600-800	600-800	HYD	HYD
	3	3.3 (3318)	0.039-0.043	—	20B	43	—	600-800	HYD	HYD
	4	2.2 (2212)	0.039-0.043	6-22B	12-28B	34-38	600-800	600-800	HYD	HYD
	6	2.5 (2457)	0.039-0.043	7-23B	7-23B	34-38	600-800	600-800	HYD	HYD
1997	2	1.8 (1820)	0.039-0.043	8-24 BTDC	8-24 BTDC	34-38	600-800	600-800	0.0071-0.0087	0.0090-0.0106
	3	3.3 (3318)	0.039-0.043	20B	20B	43	—	600-800	0.0071-0.0087	0.0090-0.0106
	4	2.2 (2212)	0.039-0.043	6-22 BTDC	12-28 BTDC	34-38	600-800	600-800	0.0071-0.0087	0.0090-0.0106
	6	2.5 (2457)	0.039-0.043	7-23 BTDC	7-23 BTDC	34-38	600-800	600-800	0.0071-0.0087	0.0090-0.0106
1998-99	4	2.2 (2212)	0.039-0.043	6-22 BTDC	12-28 BTDC	34-38	600-800	600-800	0.0071-0.0087	0.0090-0.0106
	6	2.5 (2457)	0.039-0.043	7-23 BTDC	7-23 BTDC	34-38	600-800	600-800	0.0071-0.0087	0.0090-0.0106

NOTE: The Vehicle Emission Control Information lable often reflects specification changes made during production. The lable fugures must be used if they differ from those in this chart.

B - Before top dead center

HYD - Hudraulic

① At idle speed.

② With engine under no load.

③ With engine cold.

79231CL7

CAPACITIES

Year	Model		Engine ID/VIN	Engine Displacement Liters (cc)	Engine Oil with Filter	Transmission (pts.)			Transfer Case (pts.)	Drive Axle		Fuel Tank (gal.)	Cooling System (qts.)
						4-Spd	5-Spd	Auto.		Front ③ (pts.)	Rear (pts.)		
1995	Impreza		1	1.8 (1820)	4.2	—	8.4	16.8	—	2.6	-	13.2	6.6
	Impreza		2	1.8 (1820)	4.2	—	8.4	16.8	—	2.6	1.6	13.2	6.6
	Impreza	①	4	2.2 (2212)	4.4	—	8.4	16.8	—	2.6	1.6	13.2	6.1
	Legacy	①	6	2.2 (2212)	4.4	—	②	16.8	—	2.6	1.6	15.9	6.4
	SVX		3	3.3 (3318)	6.3	—	—	20.0	—	2.8	1.8	18.5	7.4
1996	Impreza		1	1.8 (1820)	4.2	—	7.4	16.8	—	2.6	-	13.2	6.6
	Impreza		2	1.8 (1820)	4.2	—	7.4	16.8	—	2.6	1.6	13.2	6.6
	Impreza	①	4	2.2 (2212)	4.4	—	7.4	16.8	—	2.6	1.6	13.2	6.1
	Legacy		3	2.2 (2212)	4.4	—	7.0	16.8	—	2.6	-	15.9	6.1
	Legacy	①	4	2.2 (2212)	4.4	—	7.4	16.8	—	2.6	1.6	15.9	6.1
	Legacy	①	6	2.5 (2457)	4.7	—	—	20.0	—	2.6	1.6	15.9	6.3
	SVX		3	3.3 (3318)	6.3	—	—	20.0	—	2.8	1.8	18.5	7.4
1997	Impreza		2	1.8 (1820)	4.2	—	7.4	16.8	—	2.6	1.6	13.2	6.6
	Impreza	①	4	2.2 (2212)	4.4	—	7.4	16.8	—	2.6	1.6	13.2	6.1
	Legacy		4	2.2 (2212)	4.4	—	7.4	16.8	—	2.6	1.6	15.9	6.1
	Legacy	①	6	2.5 (2457)	4.7	—	7.4	20.0	—	2.6	1.6	15.9	6.3
	SVX		3	3.3 (3318)	6.3	—	—	20.0	—	2.8	1.8	18.5	7.4
1998-99	Impreza	①	4	2.2 (2212)	4.2	—	7.4	16.8	—	2.6	1.6	13.2	6.1
	Impreza RS		6	2.5 (2457)	4.7	—	7.4	20.0	—	2.6	1.6	13.2	6.3
	Legacy		4	2.2 (2212)	4.4	—	7.4	15.0	—	2.6	1.6	15.9	6.1
	Legacy	①	6	2.5 (2457)	4.7	—	7.4	20.0	—	2.6	1.6	15.9	6.3

Note: All capacities are approximate. Add fluid gradually and check to be sure a proper fluid level is obtained.

① Includes Outback trim levels

② 2WD: 7.0
4WD: 7.4

③ A/T differential only

79231CL8

VALVE SPECIFICATIONS

Year	Engine ID/VIN	Engine Displacement Liters (cc)	Seat Angle (deg.)	Face Angle (deg.)	Spring Test Pressure (lbs. @ in.)	Spring Installed Height (in.)	Stem-to-Guide Clearance (in.)		Stem Diameter (in.)	
							Intake	Exhaust	Intake	Exhaust
1995	1	1.8 (1820)	45	45	90-103@ 1.150	1.150	0.0014- 0.0059	0.0016- 0.0059	0.2343- 0.2348	0.2341- 0.2346
	2	1.8 (1820)	45	45	90-103@ 1.150	1.150	0.0014- 0.0059	0.0016- 0.0059	0.2343- 0.2348	0.2341- 0.2346
	3	3.3 (3318)	45	45	④	⑤	0.0012- 0.0039	0.0016- 0.0059	0.2344- 0.2350	0.2341- 0.2346
	4	2.2 (2212)	45	45	90-103@ 1.150	1.150	0.0014- 0.0059	0.0016- 0.0059	0.2343- 0.2348	0.2341- 0.2346
	6	2.2 (2212)	45	45	90-103@ 1.150	1.150	0.0014- 0.0059	0.0016- 0.0059	0.2343- 0.2348	0.2341- 0.2346
1996	1	1.8 (1820)	45	45	90-103@ 1.150	1.150	0.0014- 0.0059	0.0016- 0.0059	0.2343- 0.2348	0.2341- 0.2346
	2	1.8 (1820)	45	45	90-103@ 1.150	1.150	0.0014- 0.0059	0.0016- 0.0059	0.2343- 0.2348	0.2341- 0.2346
	3	2.2 (2212)	45	45	90-103@ 1.150	1.150	0.0014- 0.0059	0.0016- 0.0059	0.2343- 0.2348	0.2341- 0.2346
	3	3.3 (3318)	45	45	④	⑤	0.0012- 0.0039	0.0016- 0.0059	0.2344- 0.2350	0.2341- 0.2346
	4	2.2 (2212)	45	45	90-103@ 1.150	1.150	0.0014- 0.0059	0.0016- 0.0059	0.2343- 0.2348	0.2341- 0.2346
	6	2.5 (2457)	45	45	104-120@ 1.315	①	0.0014- 0.0059	0.0016- 0.0059	0.2343- 0.2348	0.2341- 0.2346
1997	2	1.8 (1820)	45	45	91-103@ 1.110	②	0.0014- 0.0059	0.0016- 0.0059	0.2343- 0.2348	0.2341- 0.2346
	3	3.3 (3318)	45	45	④	⑤	0.0012- 0.0039	0.0016- 0.0059	0.2344- 0.2350	0.2341- 0.2346
	4	2.2 (2212)	45	45	91-103@ 1.110	②	0.0014- 0.0059	0.0016- 0.0059	0.2343- 0.2348	0.2341- 0.2346
	6	2.5 (2457)	45	45	102-118@ 7	③	0.0014- 0.0059	0.0016- 0.0059	0.2343- 0.2348	0.2341- 0.2346
1998-99	4	2.2 (2212)	45	45	91-103@ 1.110	②	0.0014- 0.0059	0.0016- 0.0059	0.2343- 0.2348	0.2341- 0.2346
	6	2.5 (2457)	45	45	102-118@ 1.315	③	0.0014- 0.0059	0.0016- 0.0059	0.2343- 0.2348	0.2341- 0.2346

① Free length: 1.567
② Free length: 1.7342
③ Free length: 1.8913
④ Inner spring: 33-37@0.772
   Outer spring: 70-80@0.831
⑤ Inner spring: 0.772

79231CL9

TORQUE SPECIFICATIONS
All readings in ft. lbs.

Year	Engine ID/VIN	Engine Displacement Liters (cc)	Cylinder Head Bolts	Main ④ Bearing Bolts	Rod Bearing Bolts	Crankshaft Damper Bolts	Flywheel Bolts	Manifold		Spark Plugs	Lug Nut
								Intake	Exhaust ⑤		
1995	1	1.8 (1820)	①	③	31-34	87-104	51-55	17-20	19-26	13-17	58-72
	2	1.8 (1820)	①	③	31-34	87-104	51-55	17-20	19-26	13-17	58-72
	3	3.3 (3318)	②	22	31-34	108-123	51-55	17-20	19-26	13-17	58-72
	4	2.2 (2212)	①	③	31-34	87-104	51-55	17-20	19-26	13-17	58-72
	6	2.2 (2212)	①	③	31-34	87-104	51-55	17-20	19-26	13-17	58-72
1996	1	1.8 (1820)	①	③	31-34	87-104	51-55	17-20	19-26	13-17	58-72
	2	1.8 (1820)	①	③	31-34	87-104	51-55	17-20	19-26	13-17	58-72
	3	2.2 (2212)	①	③	31-34	87-104	51-55	17-20	19-26	13-17	58-72
	3	3.3 (3318)	②	22	31-34	108-123	51-55	17-20	19-26	13-17	58-72
	4	2.2 (2212)	①	③	31-34	87-104	51-55	17-20	19-26	13-17	58-72
	6	2.5 (2547)	①	③	31-34	123-137	51-55	17-20	19-26	13-17	58-72
1997	2	1.8 (1820)	①	③	31-34	87-104	51-55	17-20	19-26	13-17	58-72
	3	3.3 (3318)	②	22	31-34	108-123	51-55	17-20	19-26	13-17	58-72
	4	2.2 (2212)	①	③	31-34	87-104	51-55	17-20	19-26	13-17	58-72
	6	2.5 (2547)	①	③	31-34	123-137	51-55	17-20	19-26	13-17	58-72
1998-99	4	2.2 (2212)	①	③	31-34	87-101	51-55	17-20	19-26	13-17	58-72
	6	2.5 (2547)	①	③	31-34	123-137	51-55	17-20	19-26	13-17	58-72

① Oil All Bolts
  Step 1: 22 ft. lbs.
  Step 2: 51 ft. lbs.
  Step 3: Loosen all bolts 180 degrees
  Step 4: Repeat Step 3
  Step 5: Bolts 1-2: 25 ft. lbs.
  Step 6: Bolts 3-6: 11 ft. lbs.
  Step 7: Tighten all bolts 80-90 degrees
  Step 8: Tighten all bolts an additional 80-90 degrees
  Note: Do NOT exceed 180 degrees total tightening
② Step 1: 22 ft. lbs.
  Step 2: 51 ft. lbs.
  Step 3: Loosen all bolts 180 degrees
  Step 4: Repeat Step 3
  Step 5: 20 ft. lbs.
  Step 6: Bolts 1-4: 80-90 degrees
  Step 7: Bolts 5-8: 33 ft. lbs.
③ Split engine case connecting bolts:
  Short bolts: 17-20 ft. lbs.
  Long bolts: 33-37 ft. lbs.
  Smaller short bolts (if used): 5 ft. lbs.
④ Engine block connecting bolts

79231CL0

SCHEDULED MAINTENANCE INTERVALS
(SUBARU IMPREZA, IMPREZA OUTBACK, IMPREZA OUTBACK SPORT, LEGACY, LEGACY OUTBACK & SVX)

TO BE SERVICED	TYPE OF SERVICE	VEHICLE MILEAGE INTERVAL (x1000)												
		7.5	15	22.5	30	37.5	45	52.5	60	67.5	75	82.5	90	97.5
Engine oil & filter	R	✓	✓	✓	✓	✓	✓	✓	✓	✓	✓	✓	✓	✓
Brake lines	S/I		✓		✓		✓		✓		✓		✓	
Clutch & hill holder system	S/I		✓		✓		✓		✓		✓		✓	
Disc brake pads & discs, front & rear axle boots & axle shaft joint portions	S/I		✓		✓		✓		✓		✓		✓	
Parking brake	S/I		✓		✓		✓		✓		✓		✓	
Steering & suspension	S/I		✓		✓		✓		✓		✓		✓	
Air filter element	R				✓				✓				✓	
Engine Coolant	R				✓				✓				✓	
Fuel filter	R				✓				✓				✓	
Spark plugs (except below)	R				✓				✓				✓	
Spark plugs (SVX & 1996-99 Legacy)	R								✓					
Automatic transmission fluid & filter	S/I				✓				✓				✓	
Brake fluid	S/I				✓				✓				✓	
Brake linings & drums	S/I				✓				✓				✓	
Camshaft drive belt①	S/I				✓				✓				✓	
Coolant level, hoses & clamps	S/I				✓				✓				✓	
Drive belts	S/I				✓				✓				✓	
Fuel system, hoses & connections	S/I				✓				✓				✓	
Differential gear oil (front & rear) (SVX)	S/I				✓								✓	
Transmission and/or differential gear oil (except SVX)	S/I				✓								✓	
Front & rear wheel bearing repack	S/I								✓					

① Non-California vehicles - replace every 60,000 miles.

R – Replace   S/I – Service or Inspect

79231CM1

SCHEDULED MAINTENANCE INTERVALS
(SUBARU IMPREZA, IMPREZA OUTBACK, IMPREZA OUTBACK SPORT, LEGACY, LEGACY OUTBACK & SVX) (Cont.)

FREQUENT OPERATION MAINTENANCE (SEVERE SERVICE)

If a vehicle is operated under any of the following conditions it is considered severe service:

- Extremely dusty areas.
- 50% or more of the vehicle operation is in 32°C (90°F) or higher temperatures, or constant operation in temperatures below 0°C (32°F).
- Prolonged idling (vehicle operation in stop and go traffic).
- Frequent short running periods (engine does not warm to normal operating temperatures).
- Police, taxi, delivery usage or trailer towing usage.

Oil & oil filter change – change every 3750 miles.
Clutch & hill holder system - service or inspect every 7500 miles.
Disc brake pads & discs, front & rear axle boots & axle shaft joint portions - service or inspect every 7500 miles.
Steering & suspension - service or inspect every 7500 miles.
Air filter element – service or inspect every 15,000 miles.
Automatic transmission fluid – service or inspect every 15,000 miles.
Brake linings & drums – service or inspect every 15,000 miles.
Coolant level, hoses & clamps - service or inspect every 15,000 miles.
Differential gear oil (front & rear) (SVX) - service or inspect every 15,000 miles.
Drive belts - service or inspect every 15,000 miles.
Transmission/differential gear oil (except SVX) - service or inspect every 15,000 miles.
Front & rear wheel bearing repack - service or inspect every 30,000 miles.

79231CM2

SCHEDULED MAINTENANCE
SUBARU
IMPREZA, IMPREZA OUTBACK, IMPREZA OUTBACK SPORT
LEGACY, LEGACY OUTBACK, SVX

The following should be used as a guide when determining the amount of work required for a particular service if taken to a repair shop. In estimating how long a particular Scheduled Maintenance Service should take, please observe the following:

- **Chilton Time** is time based on field research and data supplied by the vehicle manufacturer.
- Labor time operations are given in hours and tenths of an hour.
- All labor operations, are to be used as a guide.

Mechanic Skill Level Codes:
      (G) GENERAL: Normally skilled with certification.
      (M) MAINTENANCE: Semi-skilled working on certication.
      (P) PRECISION: Really skilled with multiple certification.

	Chilton Time
(M) 7500 Mile Service	
1995-99	.3
(M) 15000 Mile Service	
1995-99	.9
(M) 22500 Mile Service	
1995-99	.3
(G) 30000 Mile Service	
1995-99	6.0
SVX add	.2

	Chilton Time
(M) 37500 Mile Service	
1995-99	.3
(M) 45000 Mile Service	
1995-99	.9
(M) 52500 Mile Service	
1995-99	.3
(G) 60000 Mile Service	
1995-99	6.6
(M) 67500 Mile Service	
1995-99	.3

	Chilton Time
(M) 75000 Mile Service	
1995-99	.9
(M) 82500 Mile Service	
1995-99	.3
(G) 90000 Mile Service	
1995-99	6.0
SVX add	.2
(M) 97500 Mile Service	
1995-99	.3

79231CM3

SUZUKI
Esteem • Swift

VEHICLE IDENTIFICATION

Engine Code						Model Year	
Code	Liters	Cu. in. (cc)	Cyl.	Fuel Sys.	Eng. Mfg.	Code	Year
2	1.3	79.3 (1298)	4	TFI	Suzuki	S	1995
3	1.6	97.7 (1590)	4	MFI	Suzuki	T	1996
						V	1997
						W	1998
						X	1999

MFI - Multi-port Fuel Injection

TFI - Throttle body Fuel Injection

79231C87

ENGINE IDENTIFICATION

Year	Model	Engine Displacement Liters (cc)	Engine Series (ID/VIN)	Fuel System	No. of Cylinders	Engine Type
1995	Swift	1.3 (1298)	2	TFI	4	SOHC
	Esteem	1.6 (1590)	3	MFI	4	SOHC
1996	Swift	1.3 (1298)	2	TFI	4	SOHC
	Esteem	1.6 (1590)	3	MFI	4	SOHC
1997	Swift	1.3 (1298)	2	TFI	4	SOHC
	Esteem	1.6 (1590)	3	MFI	4	SOHC
1998-99	Swift	1.3 (1298)	2	TFI	4	SOHC
	Esteem	1.6 (1590)	3	MFI	4	SOHC

MFI - Multiport Fuel Injection
TFI - Throttlebody Fuel Injection
SOHC - Single Overhead Camshaft
DOHC - Double Overhead Camshaft

79231C88

GENERAL ENGINE SPECIFICATIONS

Year	Engine ID/VIN	Engine Displacement Liters (cc)	Fuel System Type	Net Horsepower @ rpm	Net Torque @ rpm (ft. lbs.)	Bore x Stroke (in.)	Compression Ratio	Oil Pressure @ rpm
1995	2	1.3 (1298)	TFI	70@5500	74@3000	2.91x2.97	9.5:1	47-61@3000
	3	1.6 (1590)	MFI	98@6000	94@3200	2.95x3.54	9.5:1	47-61@4000
1996	2	1.3 (1298)	TFI	70@5500	74@3000	2.91x2.97	9.5:1	47-61@3000
	3	1.6 (1590)	MFI	98@6000	94@3200	2.95x3.54	9.5:1	47-61@4000
1997	2	1.3 (1298)	TFI	70@5500	74@3000	2.91x2.97	9.5:1	47-61@3000
	3	1.6 (1590)	MFI	98@6000	94@3200	2.95x3.54	9.5:1	47-61@4000
1998-99	2	1.3 (1298)	TFI	70@5500	74@3000	2.91x2.97	9.5:1	47-61@3000
	3	1.6 (1590)	MFI	98@6000	94@3200	2.95x3.54	9.5:1	47-61@4000

MFI - Multiport Fuel Injection
TFI - Throttle body Fuel Injection

79231C89

GASOLINE ENGINE TUNE-UP SPECIFICATIONS

Year	Engine ID/VIN	Engine Displacement Liters (cc)	Spark Plugs Gap (in.)	Ignition Timing (deg.) MT	Ignition Timing (deg.) AT	Fuel Pump (psi)	Idle Speed (rpm) MT	Idle Speed (rpm) AT	Valve Clearance In.	Valve Clearance Ex.
1995	2	1.3 (1298)	0.029	5B	5B	13-20 ①	750	850	HYD	HYD
	3	1.6 (1590)	0.029	5B	5B	28-34 ①	750-800	750-800	②	②
1996	2	1.3 (1298)	0.029	5B	5B	13-20 ①	750	850	HYD	HYD
	3	1.6 (1590)	0.029	5B	5B	28-34 ①	750-800	750-800	②	②
1997	2	1.3 (1298)	0.029	5B	5B	13-20 ①	750	850	HYD	HYD
	3	1.6 (1590)	0.029	5B	5B	28-34 ①	750-800	750-800	②	②
1998-99	2	1.3 (1298)	0.029	5B	5B	13-20 ①	750	850	HYD	HYD
	3	1.6 (1590)	0.029	5B	5B	28-34 ①	750-800	750-800	②	②

NOTE: The Vehicle Emission Control Information label often reflects specification changes made during production. The label figures must be used if they differ from those in this chart.

HYD - Hydraulic

B - Before top dead center

① At idle

② When cold: 0.005-0.007
   When hot: 0.007-0.008

79231C90

CAPACITIES

Year	Model	Engine ID/VIN	Engine Displacement Liters (cc)	Engine Oil with Filter (qts)	Transmission (pts.) 4-Spd	Transmission (pts.) 5-Spd	Transmission (pts.) Auto.	Transfer Case (pts.)	Drive Axle Front (pts.)	Drive Axle Rear (pts.)	Fuel Tank (gal.)	Cooling System (qts.)
1995	Esteem	3	1.6 (1590)	3.3	—	5.0	10.4 ①	—	—	—	13.5	③
	Swift	2	1.3 (1298)	3.3	—	5.0	10.4 ①	—	—	—	10.6	②
1996	Esteem	3	1.6 (1590)	3.3	—	5.0	10.4 ①	—	—	—	13.5	③
	Swift	2	1.3 (1298)	3.3	—	5.0	10.4 ①	—	—	—	10.6	②
1997	Esteem	3	1.6 (1590)	3.3	—	5.0	10.4 ①	—	—	—	13.5	③
	Swift	2	1.3 (1298)	3.3	—	5.0	10.4 ①	—	—	—	10.6	②
1998-99	Esteem	3	1.6 (1590)	3.3	—	5.0	10.4 ①	—	—	—	13.5	③
	Swift	2	1.3 (1298)	3.3	—	5.0	10.4 ①	—	—	—	10.6	②

NOTE: All capacities are approximate. Add fluid gradualy and check to be sure a proper fluid level is obtained.

① Specification for automatic transaxle is after complete overhaul. Drain and fill will be less

② Manual transmission: 4.8 qts.
   Automatic transmission: 4.9 qts.

③ Manual transmission: 4.8 qts.
   Automatic transmission: 4.7 qts.

79231C91

VALVE SPECIFICATIONS

Year	Engine ID/VIN	Engine Displacement Liters (cc)	Seat Angle (deg.)	Face Angle (deg.)	Spring Test Pressure (lbs. @ in.)	Spring Installed Height (in.)	Stem-to-Guide Clearance (in.)		Stem Diameter (in.)	
							Intake	Exhaust	Intake	Exhaust
1995	2	1.3 (1298)	45	45	55-64@1.63	1.941	0.0008-0.0019	0.0014-0.0025	0.2742-0.2748	0.2737-0.2742
	3	1.6 (1590)	45	45	24-28@1.24	1.450	0.0008-0.0018	0.0018-0.0028	0.2152-0.2157	0.2142-0.2148
1996	2	1.3 (1298)	45	45	55-64@1.63	1.941	0.0008-0.0019	0.0014-0.0025	0.2742-0.2748	0.2737-0.2742
	3	1.6 (1590)	45	45	24-28@1.24	1.450	0.0008-0.0018	0.0018-0.0028	0.2152-0.2157	0.2142-0.2148
1997	2	1.3 (1298)	45	45	55-64@1.63	1.941	0.0008-0.0019	0.0014-0.0025	0.2742-0.2748	0.2737-0.2742
	3	1.6 (1590)	45	45	24-28@1.24	1.450	0.0008-0.0018	0.0018-0.0028	0.2152-0.2157	0.2142-0.2148
1998-99	2	1.3 (1298)	45	45	55-64@1.63	1.941	0.0008-0.0019	0.0014-0.0025	0.2742-0.2748	0.2737-0.2742
	3	1.6 (1590)	45	45	24-28@1.24	1.450	0.0008-0.0018	0.0018-0.0028	0.2152-0.2157	0.2142-0.2148

79231C92

TORQUE SPECIFICATIONS
All readings in ft. lbs.

Year	Engine ID/VIN	Engine Displacement Liters (cc)	Cylinder Head Bolts	Main Bearing Bolts	Rod Bearing Bolts	Crankshaft Damper Bolts	Flywheel Bolts	Manifold Intake	Manifold Exhaust	Spark Plugs	Lug Nut
1995	2	1.3 (1298)	51-54	36-41	24-26	76-83 ①	41-47	13-20	13-20	14-21	36-57
	3	1.6 (1590)	48-51	36-41	24-26	76-83 ①	57	13-20	13-20	14-21	58-80
1996	2	1.3 (1298)	51-54	36-41	24-26	76-83 ①	41-47	13-20	13-20	14-21	36-57
	3	1.6 (1590)	48-51	36-41	24-26	76-83 ①	57	13-20	13-20	14-21	58-80
1997	2	1.3 (1298)	51-54	36-41	24-26	76-83 ①	41-47	13-20	13-20	14-21	36-57
	3	1.6 (1590)	48-51	36-41	24-26	76-83 ①	57	13-20	13-20	14-21	58-80
1998-99	2	1.3 (1298)	51-54	36-41	24-26	76-83 ①	41-47	13-20	13-20	14-21	36-57
	3	1.6 (1590)	48-51	36-41	24-26	76-83 ①	57	13-20	13-20	14-21	58-80

① Specification shown is for crankshaft timing sprocket nut

79231C93

SCHEDULED MAINTENANCE INTERVALS
(SUZUKI SWIFT & ESTEEM)

TO BE SERVICED	TYPE OF SERVICE	VEHICLE MILEAGE INTERVAL (x1000)												
		7.5	15	22.5	30	37.5	45	52.5	60	67.5	75	82.5	90	97.5
Engine oil & filter	R	✓	✓	✓	✓	✓	✓	✓	✓	✓	✓	✓	✓	✓
Automatic transmission fluid & filter③	S/I	✓	✓	✓	✓	✓	✓	✓	✓	✓	✓	✓	✓	✓
Clutch pedal free travel	S/I	✓	✓	✓	✓	✓	✓	✓	✓	✓	✓	✓	✓	✓
Drive axle boots	S/I	✓	✓	✓	✓	✓	✓	✓	✓	✓	✓	✓	✓	✓
Gear shift control lever/shift operation	S/I	✓	✓	✓	✓	✓	✓	✓	✓	✓	✓	✓	✓	✓
Inspect & rotate tires	S/I	✓	✓	✓	✓	✓	✓	✓	✓	✓	✓	✓	✓	✓
Manual transmission oil②	S/I	✓	✓	✓	✓	✓	✓	✓	✓	✓	✓	✓	✓	✓
Power steering system	S/I	✓	✓	✓	✓	✓	✓	✓	✓	✓	✓	✓	✓	✓
Suspension system	S/I	✓	✓	✓	✓	✓	✓	✓	✓	✓	✓	✓	✓	✓
Brake discs, pads, drums & shoes	S/I	✓		✓		✓		✓		✓		✓		✓
Brake hoses, pipes, brake lever & cable	S/I	✓		✓		✓		✓		✓		✓		✓
Brake fluid①	S/I		✓		✓		✓		✓		✓		✓	
Brake pedal	S/I		✓		✓		✓		✓		✓		✓	
Cooling system, hoses & connections	S/I		✓		✓		✓		✓		✓		✓	
Fuel tank, cap & lines	S/I		✓		✓		✓		✓		✓		✓	
Valve lash (clearance)	S/I		✓		✓		✓		✓		✓		✓	
Air cleaner filter element	R				✓				✓				✓	
Engine Coolant	R				✓				✓				✓	

79231C94

SCHEDULED MAINTENANCE INTERVALS
(SUZUKI SWIFT & ESTEEM) (Cont.)

TO BE SERVICED	TYPE OF SERVICE	VEHICLE MILEAGE INTERVAL (x1000)											
		7.5	15	22.5	30	37.5	45	52.5	60	67.5	75	82.5	90
Spark plugs	R				✓				✓				✓
Drive belts	S/I				✓				✓				✓
Exhaust system	S/I				✓				✓				✓
Automatic transmission fluid hose	R						✓						✓
Camshaft timing belt	R								✓				
Ignition wiring	S/I								✓				

① Replace every 60,000 miles.   ② Replace every 15,000 miles.   ③ Replace every 100,000 miles.

R – Replace   S/I – Service or Inspect

FREQUENT OPERATION MAINTENANCE (SEVERE SERVICE)

If a vehicle is operated under any of the following conditions it is considered severe service:
- **Extremely dusty areas.**
- **50% or more of the vehicle operation is in 32°C (90°F) or higher temperatures, or constant operation in temperatures below 0°C**
- **Prolonged idling (vehicle operation in stop and go traffic).**
- **Frequent short running periods (engine does not warm to normal operating temperatures).**
- **Police, taxi, delivery usage or trailer towing usage.**

Oil & oil filter change – change every 3000 miles.
Brake discs, pads, drums & shoes - service or inspect initially at 3000 miles, 6000 miles, & every 12,000 miles thereafter.
Brake hoses & pipes - service or inspect initially at 3000 miles, 6000 miles, & every 12,000 miles thereafter.
Air cleaner filter element - service or inspect every 3000 miles & replace every 30,000 miles (if not replaced previously).
Automatic transmission fluid & filter - service or inspect every 6000 miles & replace every 15,000 miles (if not replaced previously).
Clutch pedal free travel - service or inspect every 6000 miles.
Gear shift control lever/shift operation - service or inspect every 6000 miles.
Inspect & rotate tires - service or inspect every 6000 miles.
Manual transmission oil - service or inspect every 6000 miles & replace every 12,000 miles (if not replaced previously).
Power steering system - service or inspect every 6000 miles.
Steering system - service or inspect every 6000 miles.
Suspension system - service or inspect every 6000 miles.
Drive belts - service or inspect every 15,000 miles.
Exhaust system - service or inspect every 15,000 miles.

79231C95

SCHEDULED MAINTENANCE
SUZUKI
ESTEEM, SWIFT

The following should be used as a guide when determining the amount of work required for a particular service if taken to a repair shop. In estimating how long a particular Scheduled Maintenance Service should take, please observe the following:

- **Chilton Time** is time based on field research and data supplied by the vehicle manufacturer.
- Labor time operations are given in hours and tenths of an hour.
- All labor operations, are to be used as a guide.

Mechanic Skill Level Codes:
    (G) GENERAL: Normally skilled with certification.
    (M) MAINTENANCE: Semi-skilled working on certicication.
    (P) PRECISION: Really skilled with multiple certification.

	Chilton Time
(G) 7500 Mile Service	
1995-99	1.7
w/AT add	.1
(G) 15000 Mile Service	
1995-99	1.9
w/AT add	.1
(G) 22500 Mile Service	
1995-99	1.7
w/AT add	.1
(G) 30000 Mile Service	
1995-99	3.6
w/AT add	.1

	Chilton Time
(G) 37500 Mile Service	
1995-99	1.7
w/AT add	.1
(G) 45000 Mile Service	
1995-99	2.4
w/AT add	.1
(G) 52500 Mile Service	
1995-99	1.7
w/AT add	.1
(G) 60000 Mile Service	
1995-99	4.0
w/AT add	.1
(G) 67500 Mile Service	
1995-99	1.7
w/AT add	.1

	Chilton Time
(G) 75000 Mile Service	
1995-99	2.1
w/AT add	.1
(G) 82500 Mile Service	
1995-99	1.7
w/AT add	.1
(G) 90000 Mile Service	
1995-99	
1995-99	3.9
w/AT add	.1
(G) 97500 Mile Service	
1995-99	1.7
w/AT add	.1

79231C96

TOYOTA
Avalon • Camry • Celica • Corolla • MR2 • Paseo • Supra • Tercel

VEHICLE IDENTIFICATION CHART

Engine Code						Model Year	
Code	Liters	Cu. In. (cc)	Cyl.	Fuel Sys.	Eng. Mfg.	Code	Year
1MZ-FE	3.0	180 (2952)	6	EFI	Toyota	S	1995
2JZ-GE	3.0	183 (2997)	6	EFI	Toyota	T	1996
5S-FE	2.2	138 (2264)	4	EFI	Toyota	V	1997
7A-FE	1.8	107 (1762)	4	EFI	Toyota	W	1998
2JZ-GTE	3.0	183 (2997)	6	EFI	Toyota	X	1999
3E-E	1.5	89 (1456)	4	EFI	Toyota		
3S-GTE	2.0	122 (1998)	4	EFI	Toyota		
4A-FE	1.6	96 (1587)	4	EFI	Toyota		
5E-FE	1.5	91 (1495)	4	EFI	Toyota		
1ZZ-FE	1.8	109 (1794)	4	EFI	Toyota		
1MZ-FE	3.0	183 (2995)	6	EFI	Toyota		

79231C97

ENGINE IDENTIFICATION

Year	Model	Engine Displacement Liters (cc)		Engine Series (ID/VIN)	Fuel System	No. of Cylinders	Engine Type
1995	Avalon	3.0 (2995)		1MZ-FE	EFI	6	DOHC
	Camry	2.2 (2164)		5S-FE	EFI	4	DOHC
	Camry	3.0 (2952)		1MZ-FE	EFI	6	DOHC
	Celica	1.8 (1762)		7A-FE	EFI	4	DOHC
	Celica	2.2 (2164)		5S-FE	EFI	4	DOHC
	Corolla	1.6 (1587)		4A-FE	EFI	4	DOHC
	Corolla	1.8 (1762)		7A-FE	EFI	4	DOHC
	MR2	2.0 (1998)	①	3S-GTE	EFI	4	DOHC
	MR2	2.2 (2164)		5S-FE	EFI	4	DOHC
	Paseo	1.5 (1495)		5E-FE	EFI	4	DOHC
	Supra	3.0 (2997)		2JZ-GE	EFI	6	DOHC
	Supra	3.0 (2997)	②	2JZ-GTE	EFI	6	DOHC
	Tercel	1.5 (1457)		3E-E	EFI	4	SOHC
1996	Avalon	3.0 (2995)		1MZ-FE	EFI	6	DOHC
	Camry	2.2 (2164)		5S-FE	EFI	4	DOHC
	Camry	3.0 (2995)		1MZ-FE	EFI	6	DOHC
	Celica	1.8 (1762)		7A-FE	EFI	4	DOHC
	Celica	2.2 (2164)		5S-FE	EFI	4	DOHC
	Corolla	1.6 (1587)		4A-FE	EFI	4	DOHC
	Corolla	1.8 (1762)		7A-FE	EFI	4	DOHC
	Paseo	1.5 (1497)		5E-FE	EFI	4	DOHC
	Supra	3.0 (2997)		2JZ-GE	EFI	6	DOHC
	Supra	3.0 (2997)	②	2JZ-GTE	EFI	6	DOHC
	Tercel	1.5 (1497)		5E-FE	EFI	4	DOHC
1997	Avalon	3.0 (2995)		1MZ-FE	EFI	6	DOHC
	Camry	2.2 (2164)		5S-FE	EFI	4	DOHC
	Camry	3.0 (2995)		1MZ-FE	EFI	6	DOHC
	Celica	1.8 (1762)		7A-FE	EFI	4	DOHC
	Celica	2.2 (2164)		5S-FE	EFI	4	DOHC
	Corolla	1.6 (1587)		4A-FE	EFI	4	DOHC
	Corolla	1.8 (1762)		7A-FE	EFI	4	DOHC
	Paseo	1.5 (1497)		5E-FE	EFI	4	DOHC
	Supra	3.0 (2997)		2JZ-GE	EFI	6	DOHC
	Supra	3.0 (2997)	②	2JZ-GTE	EFI	6	DOHC
	Tercel	1.5 (1497)		5E-FE	EFI	4	DOHC
1998-99	Avalon	3.0 (2995)		1MZ-FE	EFI	6	DOHC
	Camry	2.2 (2164)		5S-FE	EFI	4	DOHC
	Camry	3.0 (2995)		1MZ-FE	EFI	6	DOHC
	Celica	2.2 (2164)		5S-FE	EFI	4	DOHC
	Corolla	1.8 (1794)		1ZZ-FE	EFI	4	DOHC
	Supra	3.0 (2997)		2JZ-GE	EFI	6	DOHC
	Supra	3.0 (2997)	②	2JZ-GTE	EFI	6	DOHC
	Tercel	1.5 (1497)		5E-FE	EFI	4	DOHC

EFI - Electronic Fuel Injection
SOHC - Single Overhead Camshaft
DOHC - Double Overhead Camshaft

① . Turbocharged
② Twin Turbocharged

79231C98

GENERAL ENGINE SPECIFICATIONS

Year	Engine ID/VIN	Engine Displacement Liters (cc)	Fuel System Type	Net Horsepower @ rpm	Net Torque @ rpm (ft. lbs.)	Bore x Stroke (in.)	Compression Ratio	Oil Pressure @ idle
1995	1MZ-FE	3.0 (2952)	EFI	185@5200	195@4400	3.45x3.45	9.6:1	4.3
	2JZ-GE	3.0 (2997)	EFI	220@5800	210@4800	3.39x3.39	10.0:1	7
	2JZ-GTE	3.0 (2997)	EFI	320@5600	315@4000	3.39x3.39	8.5:1	7
	3E-E	1.5 (1456)	EFI	82@5200	89@4400	2.87x3.43	9.3:1	4.3
	3S-GTE	2.0 (1998)	EFI	200@6000	200@3200	3.39x3.39	8.8:1	4.3
	4A-FE	1.6 (1587)	EFI	102@5800	101@4800	3.19x3.03	9.5:1	4.3
	5E-FE	1.5 (1495)	EFI	100@6400	91@3200	2.91x3.43	9.4:1	4.3
	5S-FE	2.2 (2164)	EFI	135@5400	145@4400	3.43x3.58	9.5:1	4.3
	7A-FE	1.8 (1762)	EFI	115@5600	115@2800	3.19x3.37	9.5:1	4.3
1996	1MZ-FE	3.0 (2995)	EFI	192@5200	210@4400	3.44x3.27	10.5:1	4.3
	2JZ-GE	3.0 (2997)	EFI	220@5800	210@4800	3.39x3.39	10.0:1	7
	2JZ-GTE	3.0 (2997)	EFI	320@5600	315@4000	3.39x3.39	8.5:1	7
	4A-FE	1.6 (1587)	EFI	100@5600	105@4400	3.19x3.03	9.5:1	4.3
	5E-FE	1.5 (1497)	EFI	93@5400	100@4400	2.91x3.43	9.4:1	4.3
	5S-FE	2.2 (2164)	EFI	130@5400	145@4400	3.43x3.58	9.5:1	4.3
	7A-FE	1.8 (1762)	EFI	105@5600	117@2800	3.19x3.37	9.5:1	4.3
1997	1MZ-FE	3.0 (2995)	EFI	192@5200	210@4400	3.44x3.27	10.5:1	4.3
	2JZ-GE	3.0 (2997)	EFI	220@5800	210@4800	3.39x3.39	10.0:1	7
	2JZ-GTE	3.0 (2997)	EFI	320@5600	315@4000	3.39x3.39	8.5:1	7
	4A-FE	1.6 (1587)	EFI	100@5600	105@4400	3.19x3.03	9.5:1	4.3
	5E-FE	1.5 (1497)	EFI	93@5400	100@4400	2.91x3.43	9.4:1	4.3
	5S-FE	2.2 (2164)	EFI	130@5400	145@4400	3.43x3.58	9.5:1	4.3
	7A-FE	1.8 (1762)	EFI	105@5600	117@2800	3.19x3.37	9.5:1	4.3
1998-99	1MZ-FE	3.0 (2995)	EFI	192@5200	210@4400	3.44x3.27	10.5:1	4.3
	1ZZ-FE	1.8 (1794)	EFI	120@5600	122@4400	3.11x3.60	10.0:1	4.3
	2JZ-GE	3.0 (2997)	EFI	220@5800	210@4800	3.39x3.39	10.0:1	7
	2JZ-GTE	3.0 (2997)	EFI	320@5600	315@4000	3.39x3.39	8.5:1	7
	5E-FE	1.5 (1497)	EFI	93@5400	100@4400	2.91x3.43	9.4:1	4.3
	5S-FE	2.2 (2164)	EFI	130@5400	145@4400	3.43x3.58	9.5:1	4.3

EFI - Electronic fuel injection

79231C99

GASOLINE ENGINE TUNE-UP SPECIFICATIONS

Year	Engine ID/VIN	Engine Displacement Liters (cc)	Spark Plugs Gap (in.)	Ignition Timing (deg.) MT	Ignition Timing (deg.) AT	Fuel Pump (psi)	Idle Speed (rpm) MT	Idle Speed (rpm) AT	Valve Clearance In.	Valve Clearance Ex.
1995	3E-E	1.5 (1456)	0.043	10B	10B	41-45	750	800	0.008	0.008
	5E-FE	1.5 (1495)	0.043	10B	10B	41-42	750	750	0.012-0.010	0.012-0.016
	4A-FE	1.6 (1587)	0.031	10B	10B	38-44	①	①	0.006-0.010	0.008-0.012
	7A-FE	1.8 (1762)	0.031	10B	10B	38-44	800	800	0.006-0.010	0.008-0.012
	3S-GTE	2.0 (1998)	0.031	10B	10B	33-38	750-850	750-850	0.006-0.010	④
	5S-FE	2.2 (2164)	0.043	10B	10B	38-44	②	③	0.007-0.011	0.011-0.015
	2JZ-GE	3.0 (2997)	0.043	⑤	⑤	38-44	650-750	650-750	0.006-0.010	0.010-0.014
	2JZ-GTE	3.0 (2997)	0.043	⑤	⑤	33-40	600-700	600-700	0.006-0.010	0.010-0.014
	1MZ-FE	3.0 (2952)	0.043	⑤	⑤	38-44	650-750	650-750	0.006-0.010	0.010-0.014
1996	5E-FE	1.5 (1497)	0.043	⑤	⑤	41-42	700-800	700-800	0.006-0.010	0.012-0.016
	4A-FE	1.6 (1587)	0.031	⑤	⑤	38-44	650-750	650-750	0.006-0.010	0.010-0.014
	7A-FE	1.8 (1762)	0.031	⑤	⑤	38-44	650-750	650-750	0.006-0.010	0.010-0.014
	5S-FE	2.2 (2164)	0.043	⑤	⑤	38-44	700-800	700-800	0.007-0.011	0.011-0.015
	1MZ-FE	3.0 (2952)	0.043	⑤	⑤	38-44	650-750	650-750	0.006-0.010	0.010-0.014
	2JZ-GE	3.0 (2997)	0.043	⑤	⑤	38-44	650-750	650-750	0.006-0.010	0.010-0.014
	2JZ-GTE	3.0 (2997)	0.043	⑤	⑤	33-40	600-700	600-700	0.006-0.010	0.010-0.014
1997	5E-FE	1.5 (1497)	0.043	⑤	⑤	41-42	700-800	700-800	0.006-0.010	0.012-0.016
	4A-FE	1.6 (1587)	0.031	⑤	⑤	38-44	650-750	650-750	0.006-0.010	0.010-0.014
	7A-FE	1.8 (1762)	0.031	⑤	⑤	38-44	650-750	650-750	0.006-0.010	0.010-0.014
	5S-FE	2.2 (2164)	0.043	⑤	⑤	38-44	700-800	700-800	0.007-0.011	0.011-0.015
	1MZ-FE	3.0 (2952)	0.043	⑤	⑤	38-44	650-750	650-750	0.006-0.010	0.010-0.014
	2JZ-GE	3.0 (2997)	0.043	⑤	⑤	38-44	650-750	650-750	0.006-0.010	0.010-0.014
	2JZ-GTE	3.0 (2997)	0.043	⑤	⑤	33-40	600-700	600-700	0.006-0.010	0.010-0.014
1998-99	5E-FE	1.5 (1497)	0.043	⑤	⑤	41-42	700-800	700-800	0.006-0.010	0.012-0.016
	1ZZ-FE	1.8 (1794)	0.043	⑤	⑤	44-50	650-750	650-750	0.006-0.010	0.010-0.014

79231C00

GASOLINE ENGINE TUNE-UP SPECIFICATIONS

Year	Engine ID/VIN	Engine Displacement Liters (cc)	Spark Plugs Gap (in.)	Ignition Timing (deg.)		Fuel Pump (psi)	Idle Speed (rpm)		Valve Clearance	
				MT	AT		MT	AT	In.	Ex.
1998-99	5S-FE	2.2 (2164)	0.043	⑤	⑤	44-50	700-800	700-800	0.007-0.011	0.011-0.015
	1MZ-FE	3.0 (2952)	0.043	⑤	⑤	44-50	650-750	650-750	0.006-0.010	0.010-0.014
	2JZ-GE	3.0 (2997)	0.043	⑤	⑤	44-50	650-750	650-750	0.006-0.010	0.010-0.014
	2JZ-GTE	3.0 (2997)	0.043	⑤	⑤	33-40	600-700	600-700	0.006-0.010	0.010-0.014

NOTE: The Vehicle Emission Control Information label often reflects specification changes made during production. The label figures must be used if they differ from those in this chart.

B - Before top dead center

① 2WD Federal and Canada: 700
   2WD California and 4WD: 800

② USA: 750
   Canada: 850

③ USA: 700
   Canada: 750

④ MR2: 0.008-0.012
   Celica: 0.011-0.015

⑤ 10B at idle, with terminal TE1 and E1 connected of DLC1

79231CA1

CAPACITIES

Year	Model	Engine ID/VIN	Engine Displacement Liters (cc)	Engine Oil with Filter	Transmission (pts.) 4-Spd	5-Spd	Auto.	Transfer Case (pts.)	Drive Axle Front (pts.)	Rear (pts.)	Fuel Tank (gal.)	Cooling System (qts.)
1995	Avalon	1MZ-FE	3.0 (2952)	5.0	-	-	7.4	-	1.8	-	18.5	9.8
	Camry	1MZ-FE	3.0 (2952)	4.5	-	8.8	6.2	-	3.4	-	18.5	9.0
	Camry	5S-FE	2.2 (2164)	4.3	-	5.4	5.2	-	3.4	-	18.5	6.7
	Celica	4AFE	1.6 (1587)	3.4	-	5.4	5.2	-	3.4	-	⑥	③
	Celica	5S-FE	2.2 (2164)	①	-	11.0	-	-	3.4	-	15.9	6.9
	Corolla	4A-FE	1.6 (1587)	3.4	-	5.4	5.2	-	-	-	13.2	6.5
	Corolla	7A-FE	1.8 (1762)	3.4	-	5.4	5.2	-	-	-	13.2	6.6
	MR2	3S-GTE	2.0 (1998)	4.1	-	8.8	7.0	-	-	3.4	14.3	14.4
	MR2	5S-FE	2.2 (2164)	4.5	-	5.4	7.0	-	-	3.4	14.3	13.7
	Paseo	5E-FE	1.5 (1495)	3.4	-	5.0	6.6	-	3.0	-	11.9	⑤
	Supra	2JZ-GE	3.0 (2997)	5.5	-	5.4	15.2	-	-	2.9	18.5	⑦
	Supra	2JZ-GTE	3.0 (2997)	5.3	-	5.4	17.4	-	-	2.9	18.5	⑧
	Tercel	3E-E	1.5 (1457)	3.4	5.0	5.0	5.2	-	3.0	-	11.9	②
1996	Avalon	1MZ-FE	3.0 (2995)	5.0	-	-	7.4	-	1.8	-	18.5	9.8
	Camry	1MZ-FE	3.0 (2995)	5.0	-	-	7.4	-	1.8	-	18.5	9.8
	Camry	5S-FE	2.2 (2164)	3.8	-	5.4	5.2	-	3.4	-	18.5	6.7
	Celica	5S-FE	2.2 (2164)	4.1	-	5.4	5.2	-	3.4	-	15.9	⑫
	Celica	7A-FE	1.8 (1762)	3.9	-	4.0	6.6	-	-	-	15.9	⑬
	Corolla	4A-FE	1.6 (1587)	3.2	-	4.0	5.2	-	3.0	-	13.2	⑪
	Corolla	7A-FE	1.8 (1762)	3.9	-	4.0	6.6	-	-	-	13.2	⑨
	Paseo	5E-FE	1.5 (1497)	3.0	-	4.0	6.6	-	-	-	11.9	⑤
	Supra	2JZ-GE	3.0 (2997)	5.5	-	5.4	3.4	-	-	2.9	18.5	⑦
	Supra	2JZ-GTE	3.0 (2997)	5.3	-	3.8	4.0	-	-	2.9	18.5	⑧
	Tercel	5E-FE	1.5 (1497)	3.0	5.0	5.0	⑨	-	3.0 ⑩	-	11.9	②
1997	Avalon	1MZ-FE	3.0 (2995)	5.0	-	-	7.4	-	1.8	-	18.5	9.8
	Camry	1MZ-FE	3.0 (2995)	5.0	-	-	7.4	-	1.8	-	18.5	9.8
	Camry	5S-FE	2.2 (2164)	3.8	-	5.4	5.2	-	3.4	-	18.5	6.7
	Celica	5S-FE	2.2 (2164)	4.1	-	5.4	5.2	-	3.4	-	15.9	⑫
	Celica	7A-FE	1.8 (1762)	3.9	-	4.0	6.6	-	-	-	15.9	⑬
	Corolla	4A-FE	1.6 (1587)	3.2	-	4.0	5.2	-	3.0	-	13.2	⑪
	Corolla	7A-FE	1.8 (1762)	3.9	-	4.0	6.6	-	-	-	13.2	⑭
	Paseo	5E-FE	1.5 (1497)	3.0	-	4.0	6.6	-	-	-	11.9	⑤
	Supra	2JZ-GE	3.0 (2997)	5.5	-	5.4	3.4	-	-	2.9	18.5	⑦
	Supra	2JZ-GTE	3.0 (2997)	5.3	-	3.8	4.0	-	-	2.9	18.5	⑧
	Tercel	5E-FE	1.5 (1497)	3.0	5.0	5.0	⑨	-	3.0 ⑩	-	11.9	②
1998-99	Avalon	1MZ-FE	3.0 (2995)	5.0	-	-	7.4	-	1.8	-	18.5	9.8
	Camry	1MZ-FE	3.0 (2995)	5.0	-	9.8	7.4	-	1.8	-	18.5	9.6
	Camry	5S-FE	2.2 (2164)	3.8	-	4.6	5.2	-	3.4	-	18.5	7.3
	Celica	5S-FE	2.2 (2164)	4.1	-	5.4	5.2	-	3.4	-	15.9	⑬
	Corolla	1ZZ-FE	1.8 (1794)	3.2	-	4.0	5.2	-	3.0	-	13.2	⑪
	Supra	2JZ-GE	3.0 (2997)	5.5	-	-	3.4	-	-	2.9	18.5	8.5
	Supra	2JZ-GTE	3.0 (2997)	5.3	-	3.8	4.0	-	-	2.9	18.5	④
	Tercel	5E-FE	1.5 (1497)	3.0	5.0	5.0	⑨	-	3.0 ⑩	-	11.9	②

Note: All capacities are approximate. Add fluid gradually and check to be sure a proper fluid level is obtained.

① With oil cooler: 4.4
Without oil cooler: 4.3
② Manual transmission: 5.2
Automatic transmission: 5.7
③ Manual transmission: 5.5
Automatic transmission: 5.8
④ Manual transmission: 9.4
Automatic transmission: 9.3

⑤ Manual transmission: 5.3
Automatic transmission: 5.7
⑥ 2WD: 15.9; 4WD: 18.0
⑦ Manual transmission: 7.7
Automatic transmission: 8.8
⑧ Manual transmission: 10.0
Automatic transmission: 9.9

⑨ A132L transmission: 5.2
A242L transmission: 6.6
⑩ A132L transmission
⑪ M/T with Nippodenso radiator: 5.6
A/T with Nippodenso radiator: 6.2
M/T with Harrison radiator: 6.3
A/T with Harrison radiator: 6.2

⑫ Manual trans.: 7.1
Automatic trans.: 7.5
⑬ Manual trans.: 6.4
Automatic trans.: 7.0
⑭ M/T with Nippodenso radiator: 5.8
A/T with Nippodenso radiator: 6.2
M/T with Harrison radiator: 6.6
A/T with Harrison radiator: 6.4

79231CA2

VALVE SPECIFICATIONS

Year	Engine ID/VIN	Engine Displacement Liters (cc)	Seat Angle (deg.)	Face Angle (deg.)	Spring Test Pressure (lbs. @ in.)	Spring Installed Height (in.)	Stem-to-Guide Clearance (in.)		Stem Diameter (in.)	
							Intake	Exhaust	Intake	Exhaust
1995	3E-E	1.5 (1456)	45	44.5	35.1	1.384	0.0010-0.0024	0.0012-0.0026	0.2350-0.2356	0.2348-0.2354
	5E-FE	1.5 (1495)	45	44.5	37-37	1.252	0.0010-0.0024	0.0012-0.0026	0.2350-0.2356	0.2348-0.2354
	4A-FE	1.6 (1587)	45	44.5	37.3	1.248	0.0010-0.0024	0.0012-0.0026	0.2350-0.2356	0.2348-0.2354
	7A-FE	1.8 (1762)	45	44.5	37.3	1.248	0.0010-0.0023	0.0012-0.0025	0.2346-0.2352	0.2344-0.2350
	3S-GTE	2.0 (1998)	45	45.5	53.1	1.354	0.0010-0.0023	0.0012-0.0025	0.2346-0.2352	0.2344-0.2350
	5S-FE	2.2 (2164)	45	45.5	42.5	1.366	0.0010-0.0024	0.0012-0.0026	0.2350-0.2356	0.2348-0.2354
	1MZ-FE	3.0 (2952)	45	44.5	42-46	1.331	0.0010-0.0024	0.0012-0.0026	0.2154-0.2159	0.2152-0.2157
	2JZ-GE	3.0 (2997)	45	44.5	42-46@1.358	1.358	0.0010-0.0024	0.0012-0.0026	0.2350-0.2356	0.2348-0.2358
	2JZ-GTE	3.0 (2997)	45	44.5	42-46@1.358	1.358	0.0010-0.0024	0.0012-0.0026	0.2350-0.2356	0.2348-0.2358
1996	5E-FE	1.5 (1497)	45	44.5	33.3-36.8@1.252	1.252	0.0010-0.0024	0.0012-0.0026	0.2350-0.2356	0.2348-0.2354
	4A-FE	1.6 (1587)	45	44.5	35.5-39.0@1.248	1.248	0.0010-0.0024	0.0012-0.0026	0.2350-0.2356	0.2348-0.2354
	7A-FE	1.8 (1762)	45	44.5	35.5-39.0@1.248	1.248	0.0010-0.0024	0.0012-0.0025	0.2346-0.2352	0.2344-0.2350
	5S-FE	2.2 (2164)	45	44.5	36.8-42.5@1.366	1.366	0.0010-0.0024	0.0012-0.0026	0.2350-0.2356	0.2348-0.2354
	1MZ-FE	3.0 (2995)	45	44.5	41.9-46.3@1.331	1.331	0.0010-0.0024	0.0012-0.0026	0.2154-0.2159	0.2152-0.2157
	2JZ-GE	3.0 (2997)	45	44.5	42-46@1.358	1.358	0.0010-0.0024	0.0012-0.0026	0.2350-0.2356	0.2348-0.2358
	2JZ-GTE	3.0 (2997)	45	44.5	42-46@1.358	1.358	0.0010-0.0024	0.0012-0.0026	0.2350-0.2356	0.2348-0.2358
1997	5E-FE	1.5 (1497)	45	44.5	33.3-36.8@1.252	1.252	0.0010-0.0024	0.0012-0.0026	0.2350-0.2356	0.2348-0.2354
	4A-FE	1.6 (1587)	45	44.5	35.5-39.0@1.248	1.248	0.0010-0.0024	0.0012-0.0026	0.2350-0.2356	0.2348-0.2354
	7A-FE	1.8 (1762)	45	44.5	35.5-39.0@1.248	1.248	0.0010-0.0024	0.0012-0.0025	0.2346-0.2352	0.2344-0.2350
	5S-FE	2.2 (2164)	45	44.5	36.8-42.5@1.366	1.366	0.0010-0.0024	0.0012-0.0026	0.2350-0.2356	0.2348-0.2354
	1MZ-FE	3.0 (2995)	45	44.5	41.9-46.3@1.331	1.331	0.0010-0.0024	0.0012-0.0026	0.2154-0.2159	0.2152-0.2157
	2JZ-GE	3.0 (2997)	45	44.5	42-46@1.358	1.358	0.0010-0.0024	0.0012-0.0026	0.2350-0.2356	0.2348-0.2358
	2JZ-GTE	3.0 (2997)	45	44.5	42-46@1.358	1.358	0.0010-0.0024	0.0012-0.0026	0.2350-0.2356	0.2348-0.2358
1998-99	5E-FE	1.5 (1497)	45	44.5	33.3-36.8@1.252	1.252	0.0010-0.0024	0.0012-0.0026	0.2350-0.2356	0.2348-0.2354
	1ZZ-FE	1.8 (1794)	45	44.5	31.3-34.8@1.323	1.323	0.0010-0.0024	0.0012-0.0026	0.2154-0.2159	0.2152-0.2157

79231CA3

VALVE SPECIFICATIONS

Year	Engine ID/VIN	Engine Displacement Liters (cc)	Seat Angle (deg.)	Face Angle (deg.)	Spring Test Pressure (lbs. @ In.)	Spring Installed Height (in.)	Stem-to-Guide Clearance (in.)		Stem Diameter (in.)	
							Intake	Exhaust	Intake	Exhaust
1998-99	5S-FE	2.2 (2164)	45	44.5	36.8-42.5@ 1.366	1.366	0.0010- 0.0024	0.0012- 0.0026	0.2350- 0.2356	0.2348- 0.2354
	1MZ-FE	3.0 (2995)	45	44.5	41.9-46.3@ 1.331	1.331	0.0010- 0.0024	0.0012- 0.0026	0.2154- 0.2159	0.2152- 0.2157
	2JZ-GE	3.0 (2997)	45	44.5	42-46@ 1.358	1.358	0.0010- 0.0024	0.0012- 0.0026	0.2350- 0.2356	0.2348- 0.2358
	2JZ-GTE	3.0 (2997)	45	44.5	42-46@ 1.358	1.358	0.0010- 0.0024	0.0012- 0.0026	0.2350- 0.2356	0.2348- 0.2358

79231CA4

TORQUE SPECIFICATIONS
All readings in ft. lbs.

Year	Engine ID/VIN	Engine Displacement Liters (cc)	Cylinder Head Bolts	Main Bearing Bolts	Rod Bearing Bolts	Crankshaft Damper Bolts	Flywheel Bolts	Manifold Intake	Manifold Exhaust	Spark Plugs	Lug Nut
1995	1MZ-FE	3.0 (2952)	①	⑤	⑧	159	30	14	36	13	76
	2JZ-GE	3.0 (2997)	②	⑥	⑨	239	⑪	20	29	13	76
	2JZ-GTE	3.0 (2997)	②	⑥	⑨	239	⑪	20	29	13	76
	3E-E	1.5 (1456)	③	⑥	29	112	88	14	38	13	76
	3S-GTE	2.0 (1998)	④	43	49	80	80	14	38	13	76
	4A-FE	1.6 (1587)	44	44	36	87	58	14	18	13	76
	5E-FE	1.5 (1495)	④	42	29	112	65	14	35	13	76
	5S-FE	2.2 (2164)	④	43	⑧	80	⑬	14	36	13	76
	7A-FE	1.8 (1762)	⑦	44	⑨	87	⑭	14	29	13	76
1996	1MZ-FE	3.0 (2995)	①	⑩	⑧	159	61	11	36	13	76
	2JZ-GE	3.0 (2997)	②	⑥	⑨	239	⑪	20	29	13	76
	2JZ-GTE	3.0 (2997)	②	⑥	⑨	239	⑪	20	29	13	76
	4A-FE	1.6 (1587)	⑦	44	⑨	87	⑭	14	25	14	76
	5E-FE	1.5 (1497)	⑥	42	29	112	65	14	35	13	76
	5S-FE	2.2 (2164)	④	43	⑧	80	⑬	14	36	13	76
	7A-FE	1.8 (1762)	⑦	44	⑧	87	⑭	14	25	14	76
1997	1MZ-FE	3.0 (2995)	①	⑩	⑧	159	61	11	36	13	76
	2JZ-GE	3.0 (2997)	②	⑥	⑨	239	④	20	29	13	76
	2JZ-GTE	3.0 (2997)	②	⑥	⑨	239	④	20	29	13	76
	4A-FE	1.6 (1587)	⑦	44	⑨	87	⑭	14	25	14	76
	5E-FE	1.5 (1497)	⑥	42	29	112	65	14	35	13	76
	5S-FE	2.2 (2164)	④	43	⑧	80	⑬	14	36	13	76
	7A-FE	1.8 (1762)	⑦	44	⑧	87	⑭	14	25	14	76
1998-99	1MZ-FE	3.0 (2995)	①	⑩	⑧	159	61	11	36	13	76
	1ZZ-FE	1.8 (1794)	④	⑫	⑪	102	④	14	27	13	76
	2JZ-GE	3.0 (2997)	②	⑥	⑨	243	–	21	30	13	76
	2JZ-GTE	3.0 (2997)	②	⑥	⑨	243	④	21	30	13	76
	5E-FE	1.5 (1497)	⑥	42	29	112	65	14	35	13	76
	5S-FE	2.2 (2164)	④	43	⑧	80	⑬	14	36	13	76

① Step 1: 40 ft. lbs.
Step 2: 90 degree turn
Recessed head bolt: 13 ft. lbs.

② Step 1: 25 ft. lbs.
Step 2: 90 degree turn
Step 3: 90 degree turn
Recessed head bolt: 13 ft. lbs.

③ Step 1: 22 ft. lbs.
Step 2: 36 ft. lbs.
Step 3: 90 degree turn

④ Step 1: 36 ft. lbs.
Step 2: 90 degree turn

⑤ Step 1: 16 ft. lbs.
Step 2: 90 degree turn
Recessed head bolt: 20 ft. lbs.

⑥ Step 1: 33 ft. lbs.
Step 2: 90 degree turn

⑦ Step 1: 22 ft. lbs.
Step 2: 90 degree turn
Step 3: 90 degree turn

⑧ Step 1: 18 ft. lbs.
Step 2: 90 degree turn

⑨ Step 1: 22 ft. lbs.
Step 2: 90 degree turn

⑩ 12 pointed head: 16 ft. lbs. + 90 degrees
6 pointed head: 20 ft. lbs.

⑪ Step 1: 15 ft. lbs.
Step 2: 90 degree turn

⑫ Step 1: 16 ft. lbs.
Step 2: 32 ft. lbs.
Steps 3 and 4: 45 degree turn

⑬ Manual transmission: 65 ft. lbs.
Automatic transmission: 61 ft. lbs.

⑭ Manual transmission: 58 ft. lbs.
Automatic transmission: 47 ft. lbs.

79231CA5

SCHEDULED MAINTENANCE INTERVALS
(TOYOTA AVALON, CAMRY, CELICA, COROLLA, MR2, PASEO, SUPRA & TERCEL)

TO BE SERVICED	TYPE OF SERVICE	VEHICLE MILEAGE INTERVAL (x1000)												
		7.5	15	22.5	30	37.5	45	52.5	60	67.5	75	82.5	90	97.5
Engine oil & filter②	R	✓	✓	✓	✓	✓	✓	✓	✓	✓	✓	✓	✓	✓
Idle speed (Paseo)	S/I	✓		✓		✓		✓		✓		✓		✓
Drive belts	S/I								✓	✓	✓	✓	✓	✓
Automatic transaxle fluid & filter	S/I		✓		✓		✓		✓		✓		✓	
Ball joints & dust covers	S/I		✓		✓		✓		✓		✓		✓	
Bolts & nuts on body & chassis	S/I		✓		✓		✓		✓		✓		✓	
Brake line pipes & hoses	S/I		✓		✓		✓		✓		✓		✓	
Brake linings & drums (except MR2)	S/I		✓		✓		✓		✓		✓		✓	
Brake pads & discs (front & rear if equipped)	S/I		✓		✓		✓		✓		✓		✓	
Differential oil (Camry, Celica, Corolla & Supra①)	S/I		✓		✓		✓		✓		✓		✓	
Drive shaft boots (except Supra)	S/I		✓		✓		✓		✓		✓		✓	
Manual transaxle oil	S/I		✓		✓		✓		✓		✓		✓	
Steering gear housing oil	S/I		✓		✓		✓		✓		✓		✓	
Steering linkage	S/I		✓		✓		✓		✓		✓		✓	

79231CA6

SCHEDULED MAINTENANCE INTERVALS
(TOYOTA AVALON, CAMRY, CELICA, COROLLA, MR2, PASEO, SUPRA & TERCEL) (Cont.)

TO BE SERVICED	TYPE OF SERVICE	VEHICLE MILEAGE INTERVAL (x1000)												
		7.5	15	22.5	30	37.5	45	52.5	60	67.5	75	82.5	90	97.5
Air filter	R				✓				✓				✓	
Rear wheel bearings (Paseo & Tercel)	R				✓				✓				✓	
Spark plugs (Corolla, Paseo & Tercel)	R				✓				✓				✓	
Spark plugs (platinum tip) (Avalon, Camry, Celica, MR2 & Supra)	R								✓					
Exhaust system	S/I				✓				✓				✓	
Fuel lines & connections	S/I				✓				✓				✓	
Valve clearance	S/I				✓				✓				✓	
Engine Coolant	R						✓				✓			
Fuel tank cap gasket	R								✓					
Charcoal canister	S/I								✓					

① Supra w/LSD - replace every 30,000 miles.
② MR2 3S-GTE & Supra 3JZ-GTE - change every 5000 miles.
R – Replace   S/I – Service or Inspect

FREQUENT OPERATION MAINTENANCE (SEVERE SERVICE)

If a vehicle is operated under any of the following conditions it is considered severe service:
- Extremely dusty areas.
- 50% or more of the vehicle operation is in 32°C (90°F) or higher temperatures, or constant operation in temperatures below 0°C (32°F).
- Prolonged idling (vehicle operation in stop and go traffic).
- Frequent short running periods (engine does not warm to normal operating temperatures).
- Police, taxi, delivery usage or trailer towing usage.
Oil & oil filter change – change every 6000 miles.
Oil & oil filter change (MR2 3S-GTE & Supra 3JZ-GTE) – change every 2500 miles.
Bolts & nuts on chassis & body - tighten every 7500 miles.
Ball joints & dust covers - service or inspect every 12,000 miles.
Brake linings & drums (except MR2) - service or inspect every 12,000 miles.
Brake pads & discs (front & rear if equipped) - service or inspect every 12,000 miles.
Drive shaft boots (except Supra) - service or inspect every 12,000 miles.
Steering linkage - service or inspect every 12,000 miles.
Air filter - service or inspect every 15,000 miles.
Exhaust system - service or inspect every 15,000 miles.
Timing belt - replace every 60,000 miles.

79231CA7

SCHEDULED MAINTENANCE
TOYOTA
AVALON, CAMRY, CELICA, COROLLA, PASEO
SUPRA, TERCEL

The following should be used as a guide when determining the amount of work required for a particular service if taken to a repair shop.
In estimating how long a particular Scheduled Maintenance Service should take, please observe the following:

- **Chilton Time** is time based on field research and data supplied by the vehicle manufacturer.
- Labor time operations are given in hours and tenths of an hour.
- All labor operations, are to be used as a guide.

Mechanic Skill Level Codes:
>
> **(G) GENERAL:** Normally skilled with certification.
>
> **(M) MAINTENANCE:** Semi-skilled working on certication.
>
> **(P) PRECISION:** Really skilled with multiple certification.

	Chilton Time
(M) 7500 Mile Service	
1995-99	.3
Paseo add	.2
(M) 15000 Mile Service	
1995-99	1.1
Inspect differential oil Camry, Celica, Corolla & Supra, add	.1
Inspect drive shaft boots, all models except Supra add	.1
w/AT add	.1
(M) 22500 Mile Service	
1995-99	.3
Paseo add	.2
(G) 30000 Mile Service	
1995-99 Avalon	2.1
1995-99 Camry	2.1
1995-99 Celica	2.0
1995-99 Corolla	2.6
1995-98 Paseo	3.1
1995-98 Tercel	3.1
1995-99 Supra	2.0
Supra w/LSD add	.2
w/AT add	.1

	Chilton Time
(M) 37500 Mile Service	
1995-99	.3
Paseo add	.2
(M) 45000 Mile Service	
1995-99	1.1
Inspect differential oil Camry, Celica, Corolla & Supra, add	.1
Inspect drive shaft boots, all models except Supra add	.1
w/AT add	.1
(M) 52500 Mile Service	
1995-99	.3
Paseo add	.2
(G) 60000 Mile Service	
1995-99 Avalon	3.3
1995-99 Camry, Celica	3.3
1995-99 Corolla	3.1
1995-98 Paseo	3.0
1995-99 Supra	3.2
Supra w/LSD add	.2
1995-98 Tercel	3.0
w/AT add	.1

	Chilton Time
(M) 67500 Mile Service	
1995-99	.4
Paseo add	.2
(M) 75000 Mile Service	
1995-99	1.2
Inspect differential oil Camry, Celica, Corolla & Supra, add	.1
Inspect drive shaft boots, all models except Supra add	.1
(M) 82500 Mile Service	
1995-99	.4
Paseo add	.2
(G) 90000 Mile Service	
1995-99 Avalon	2.2
1995-99 Camry, Celica	2.2
1995-99 Corolla	2.7
1995-98 Paseo, Tercel	2.7
1995-99 Supra	2.1
Supra w/LSD add	.2
w/AT add	.1
(M) 97500 Mile Service	
1995-99	.4
Paseo add	.2

79231CA8

VOLKSWAGEN
Beetle • Cabrio • Golf • GTI • Jetta • Passat

VEHICLE IDENTIFICATION CHART

Engine Code							Model Year	
Code	Liters	Cu. In. (cc)	Cyl.	Fuel Sys.	Eng. Mfg.		Code	Year
AAA	2.8	170 (2792)	6	Motronic	Volkswagen		S	1995
AAZ	1.9	116 (1896)	4	Diesel	Volkswagen		T	1996
ABA	2.0	121 (1984)	4	Motronic	Volkswagen		V	1997
ACC	1.8	109 (1780)	4	Mono Motronic	Volkswagen		W	1998
AEB	1.8	109 (1781)	4	Motronic	Volkswagen		X	1999
AEG	2.0	121 (1984)	4	Motronic	Volkswagen			
AHA	2.8	170 (2792)	6	Motronic	Volkswagen			
AHH	1.9	116 (1896)	4	Diesel	Volkswagen			
ALH	1.9	116 (1986)	4	Diesel	Volkswagen			

79231CJ3

ENGINE IDENTIFICATION

Year	Model	Engine Displacement Liters (cc)	Engine Series (ID/VIN)	Fuel System	No. of Cylinders	Engine Type
1995	Cabrio	2.0 (1984)	ABA	Motronic	4	SOHC
	Golf III	1.8 (1780)	ACC	Mono Motronic	4	SOHC
	Golf III	2.0 (1984)	ABA	Motronic	4	SOHC
	GTI	2.0 (1984)	ABA	Motronic	4	SOHC
	GTI	2.8 (2792)	AAA	Motronic	6	DOHC
	Jetta III	1.8 (1780)	ACC	Mono Motronic	4	SOHC
	Jetta III	2.8 (2792)	AAA	Motronic	6	DOHC
	Jetta III	2.0 (1984)	ABA	Motronic	4	SOHC
	Passat	2.0 (1984)	ABA	Motronic	4	SOHC
	Passat	2.8 (2792)	AAA	Motronic	6	DOHC
1996	Cabrio	2.0 (1984)	ABA	Motronic	4	SOHC
	Golf	1.8 (1780)	ACC	Mono Motronic	4	SOHC
	Golf	2.0 (1984)	ABA	Motronic	4	SOHC
	GTI	2.0 (1984)	ABA	Motronic	4	SOHC
	GTI	2.8 (2792)	AAA	Motronic	6	DOHC
	Jetta	2.0 (1984)	ABA	Motronic	4	SOHC
	Jetta	2.8 (2792)	AAA	Motronic	6	DOHC
	Passat	2.0 (1984)	ABA	Motronic	4	SOHC
	Passat	2.8 (2792)	AAA	Motronic	6	DOHC
1997	Cabrio	2.0 (1984)	ABA	Motronic	4	SOHC
	Golf	1.8 (1780)	ACC	Mono Motronic	4	SOHC
	Golf	2.0 (1984)	ABA	Motronic	4	SOHC
	GTI	2.0 (1984)	ABA	Motronic	4	SOHC
	GTI	2.8 (2792)	AAA	Motronic	6	DOHC
	Jetta	2.0 (1984)	ABA	Motronic	4	SOHC
	Jetta	2.8 (2792)	AAA	Motronic	6	DOHC
	Passat	1.9 (1896)	AAZ	DSL	4	SOHC
	Passat	2.0 (1984)	ABA	Motronic	4	SOHC
	Passat	2.8 (2792)	AAA	Motronic	6	DOHC
1998-99	Beetle	1.9 (1896)	ALH	DSL	4	SOHC
	Beetle	2.0 (1984)	AEG	Motronic	4	SOHC
	Cabrio	2.0 (1984)	ABA	Motronic	4	SOHC
	Golf	1.8 (1780)	ACC	Mono Motronic	4	SOHC
	Golf	2.0 (1984)	ABA	Motronic	4	SOHC
	GTI	2.0 (1984)	ABA	Motronic	4	SOHC
	GTI	2.8 (2792)	AAA	Motronic	6	DOHC
	Jetta	1.9 (1896)	AAZ	DSL	4	SOHC
	Jetta	2.0 (1984)	ABA	Motronic	4	SOHC
	Jetta	2.8 (2792)	AAA	Motronic	6	DOHC
	Passat	1.8 (1781)	AEB	Motronic	4	DOHC
	Passat	1.9 (1896)	AHH	DSL	4	SOHC
	Passat	2.8 (2792)	AHA	Motronic	6	DOHC

DSL - Diesel
CIS-E - Continuous Injection System with Electronic controls

SOHC - Single Overhead Camshaft
DOHC - Double Overhead Camshaft

79231CJ4

GENERAL ENGINE SPECIFICATIONS

Year	Engine ID/VIN	Engine Displacement Liters (cc)	Fuel System Type	Net Horsepower @ rpm	Net Torque@rpm (ft. lbs.)	Bore x Stroke (in.)	Compression Ratio	Oil Pressure @ rpm
1995	AAA	2.8 (2792)	Motronic	178@5800	177@4200	3.19x3.56	10.0:1	29@2000
	ABA	2.0 (1984)	Motronic	115@5400	122@3200	3.25x3.65	9.4:1	29@2000
	ACC	1.8 (1780)	Mono Motronic	90@5000	106@2500	3.19x3.40	9.0:1	29@2000
1996	AAA	2.8 (2792)	Motronic	178@5800	177@4200	3.19x3.56	10.0:1	29@2000
	ABA	2.0 (1984)	Motronic	115@5400	122@3200	3.25x3.65	9.4:1	29@2000
	ACC	1.8 (1780)	Mono Motronic	90@5000	106@2500	3.19x3.40	9.0:1	29@2000
1997	AAA	2.8 (2792)	Motronic	178@5800	177@4200	3.19x3.56	10.0:1	29@2000
	AAZ	1.9 (1896)	DSL	75@4200	111@3400	3.13x3.76	22.5:1	29@2000
	ABA	2.0 (1984)	Motronic	115@5400	122@3200	3.25x3.65	9.4:1	29@2000
	ACC	1.8 (1780)	Mono Motronic	90@5000	106@2500	3.19x3.40	9.0:1	29@2000
1998-99	AAA	2.8 (2792)	Motronic	172@5700	173@4200	3.19x3.56	10.0:1	29@2000
	AAZ	1.9 (1896)	DSL	75@4200	111@3400	3.13x3.76	22.5:1	29@2000
	ABA	2.0 (1984)	Motronic	115@5400	122@3200	3.25x3.65	10.0:1	29@2000
	ACC	1.8 (1780)	Mono Motronic	90@5000	106@2500	3.19x3.40	9.0:1	29@2000
	AEB	1.8 (1781)	Motronic	150@5700	155@4600	3.19x3.40	9.5:1	29@2000
	AEG	2.0 (1984)	Motronic	115@5400	125@2400	3.25x3.65	10.0:1	29@2000
	AHA	2.8 (1984)	Motronic	NA	NA	NA	NA	NA
	AHH	1.9 (1896)	DSL	90@3750	154@1900	3.13x3.76	19.5:1	29@2000
	ALH	1.9 (1896)	DSL	90@3750	154@1900	3.13x3.76	19.5:1	29@2000

DSL - Diesel
CIS-E - Continuous Injection System with Electronic controls
NA - Information Not Available

79231CJ5

GASOLINE ENGINE TUNE-UP SPECIFICATIONS

Year	Engine ID/VIN	Engine Displacement Liters (cc)	Spark Plugs Gap (in.)	Ignition Timing (deg.) MT	AT	Fuel Pump (psi)	Idle Speed (rpm) MT	AT	Valve Clearance In.	Ex.
1995	AAA	2.8 (2782)	0.028	5-7B	5-7B	58	650-750	650-750	HYD	HYD
	ABA	2.0 (1984)	0.024	5-7B	5-7B	43.5	800-880	800-880	HYD	HYD
	ACC	1.8 (1780)	0.030	5-7B	5-7B	17.4	800-1000	800-1000	HYD	HYD
1996	AAA	2.8 (2792)	0.028	5-7B	5-7B	58	650-750	650-750	HYD	HYD
	ABA	2.0 (1984)	0.024	5-7B	5-7B	43.5	800-880	800-880	HYD	HYD
	ACC	1.8 (1780)	0.030	5-7B	5-7B	17.4	800-1000	800-1000	HYD	HYD
1997	AAA	2.8 (2792)	0.028	5-7B	5-7B	58	650-750	650-750	HYD	HYD
	ABA	2.0 (1984)	0.024	5-7B	5-7B	43.5	800-880	800-880	HYD	HYD
	ACC	1.8 (1780)	0.030	5-7B	5-7B	17.4	800-1000	800-1000	HYD	HYD
1998-99	AAA	2.8 (2792)	0.028	5-7B ①	5-7B ①	58 ②	650-750	650-750	HYD	HYD
	ABA	2.0 (1984)	0.024	5-7B ①	5-7B ①	43.5 ②	800-880	800-880	HYD	HYD
	ACC	1.8 (1780)	0.030	5-7B ①	5-7B ①	17.4 ②	800-1000	800-1000	HYD	HYD
	AEB	1.8 (1781)	0.035-0.043	5-7B ①	5-7B ①	51.5 ②	820-920	820-920	HYD	HYD
	AEG	2.0 (1984)	0.035-0.043	5-7B ①	5-7B ①	51.5 ②	760-880	760-880	HYD	HYD
	AHA	2.8 (2792)	0.063	5-7B ①	5-7B ①	51.5 ②	620-740	620-740	HYD	HYD

NOTE: The Vehicle Emission Control Information label often reflects specification changes made during production. The label figures must be used if they differ from those in this chart.
B - Before top dead center.
HYD - Hydraulic
① Specifications for reference only. The ignition timing is controlled bt the ECM and is not adjustable.
② System pressure at idle.

79231CJ6

CAPACITIES

Year	Model	Engine ID/VIN	Engine Displacement Liters (cc)	Engine Oil with Filter	Transaxle (pts.) 4-Spd	Transaxle (pts.) 5-Spd	Transaxle (pts.) Auto.	Drive Axle Front (pts.)	Drive Axle Rear (pts.)	Fuel Tank (gal.)	Cooling System (qts.)
1995	Cabrio	ABA	2.0 (1984)	4.3	—	4.2	6.4 ③	—	—	14.5	6.5
	Golf III	ACC	1.8 (1780)	4.2	—	4.2	6.4 ③	—	—	14.5	6.5
	Golf III	ABA	2.0 (1984)	4.3	—	4.2	6.4 ③	—	—	14.5	6.5
	Golf III	AAA	2.8 (2792)	5.8	—	4.2	6.4 ③	—	—	14.5	9.5
	GTI	AAA	2.8 (2792)	5.8	—	4.2	6.4 ③	—	—	14.5	9.5
	Jetta III	ACC	1.8 (1780)	4.2	—	4.2	6.4 ③	—	—	14.5	6.5
	Jetta III	ABA	2.0 (1984)	4.3	—	4.2	6.4 ③	—	—	14.5	6.5
	Jetta III	AAA	2.8 (2792)	5.8	—	4.2	6.4 ③	—	—	14.5	9.5
	Passat	AAA	2.0 (2792)	4.3	—	4.2	6.4 ③	—	—	18.5	7.1
	Passat	AAA	2.8 (2792)	5.8	—	4.2	6.4 ③	—	—	18.5	9.5
1996	Cabrio	ABA	2.0 (1984)	4.8	—	4.2	6.4 ③	—	—	14.5	6.7
	Golf	ACC	1.8 (1780)	4.2	—	4.2	6.4 ③	—	—	14.5	6.5
	Golf	ABA	2.0 (1984)	4.3	—	4.2	6.4 ③	—	—	14.5	6.5
	Golf	AAA	2.8 (2792)	5.8	—	4.2	6.4 ③	—	—	14.5	8.5
	GTI	ABA	2.0 (1984)	4.3	—	4.2	6.4 ③	—	—	14.5	6.5
	GTI	AAA	2.8 (2792)	5.8	—	4.2	6.4 ③	—	—	14.5	8.5
	Jetta	ABA	2.0 (1984)	4.8	—	4.2	6.4 ③	—	—	14.5	6.7
	Jetta	AAA	2.8 (2792)	5.8	—	4.2	6.4 ③	—	—	14.5	8.5
	Passat	ABA	2.0 (1984)	4.2	—	4.2	6.4 ③	—	—	18.5	7.1
	Passat	AAA	2.8 (2792)	5.8	—	4.2	6.4 ③	—	—	18.5	9.7
1997	Cabrio	ABA	2.0 (1984)	4.8	—	4.2	6.4 ③	—	—	14.5	6.7
	Golf	ABA	2.0 (1984)	4.8	—	4.2	6.4 ③	—	—	14.5	6.7
	Golf	AAA	2.8 (2792)	5.8	—	4.2	6.4 ③	—	—	14.5	8.5
	GTI	ABA	2.0 (1984)	4.8	—	4.2	6.4 ③	—	—	14.5	6.7
	GTI	AAA	2.8 (2792)	5.8	—	4.2	6.4 ③	—	—	14.5	8.5
	Jetta	ABA	2.0 (1984)	4.8	—	4.2	6.4 ③	—	—	14.5	6.7
	Jetta	AAA	2.8 (2792)	5.8	—	4.2	6.4 ③	—	—	14.5	8.5
	Passat	ABA	2.0 (1984)	4.8	—	4.2	6.4 ③	—	—	18.5	8.5
	Passat	AAA	2.8 (2792)	5.8	—	4.2	6.4 ③	—	—	18.5	9.7
	Passat	AAZ	1.9 (1896)	4.2 ①	—	4.2	6.4 ③	—	—	18.5	4.4 ②
1998-99	Beetle	ALH	1.9 (1896)	4.8	—	4.2	6.8 ⑤	—	—	14.5	6.7
	Beetle	AEG	2.0 (1984)	4.2	—	4.2	6.8 ⑤	—	—	14.5	6.7
	Cabrio	ABA	2.0 (1984)	4.8	—	4.2	6.4 ③	—	—	14.5	6.7
	Golf	ABA	2.0 (1984)	4.8	—	4.2	6.4 ③	—	—	14.5	6.7
	Golf	AAA	2.8 (2792)	5.8	—	4.2	6.4 ③	—	—	14.5	8.5
	GTI	ABA	2.0 (1984)	4.8	—	4.2	6.4 ③	—	—	14.5	6.7
	GTI	AAA	2.8 (2792)	5.8	—	4.2	6.4 ③	—	—	14.5	8.5
	Jetta	AAZ	1.9 (1896)	4.4	—	4.2	6.4 ③	—	—	14.5	6.4
	Jetta	ABA	2.0 (1984)	4.8	—	4.2	6.4 ③	—	—	14.5	6.7
	Jetta	AAA	2.8 (2792)	5.8	—	4.2	6.4 ③	—	—	14.5	8.5
	Passat	AEB	1.8 (1781)	4.2	—	4.8	7.4 ④	—	—	18.5	6.9
	Passat	AHA	2.8 (2792)	5.8	—	4.8	6.4 ④	—	—	N/A	N/A
	Passat	AHH	1.9 (1896)	3.7	—	4.8	7.4 ④	—	—	N/A	N/A

NOTE: All capacities are approximate. Add fluid gradually and check often to avoid overfilling.
① Volkswagen specifies a capacity of 4.4 quarts for the Passat TDI wagon.
② Volkswagen specifies a capacity of 6.4 quarts for the Passat TDI wagon.
③ Drain and refill only. A refill after a transaxle overhaul may require up to 11.8 pints.
④ The 01V transaxle requires approximately 5.5 pints for a refill. Refill after transaxle overhaul may require up to 19 pints.
⑤ Drain and refill only. A refill after a transaxle overhaul may require up to 11.2 pints.

79231CJ7

VALVE SPECIFICATIONS

Year	Engine ID/VIN	Engine Displacement Liters (cc)	Seat Angle (deg.)	Face Angle (deg.)	Spring Test Pressure (lbs. @ in.)	Spring Installed Height (in.)	Stem-to-Guide Clearance (in.)		Stem Diameter (in.)	
							Intake	Exhaust	Intake	Exhaust
1993	2H	1.8 (1780)	45	45	NA	NA	0.039	0.051	0.3138	0.3130
	ABG	1.8 (1780)	45	45	NA	NA	0.039	0.059	0.3138	0.3130
	9A	2.0 (1984)	45	45	NA	NA	0.039	0.051	0.2744	0.2732
	ABA	2.0 (1984)	45	45	NA	NA	0.039	0.051	0.2744	0.2736
	AAA	2.8 (2792)	45	45	NA	NA	0.039	0.051	0.2744	0.2736
1994	9A	2.0 (1984)	45	45	NA	NA	0.039	0.051	0.2744	0.2732
	ABA	2.0 (1984)	45	45	NA	NA	0.039	0.051	0.2744	0.2736
	AAA	2.8 (2792)	45	45	NA	NA	0.039	0.051	0.2744	0.2736
1995	ACC	1.8 (1780)	45	45	NA	NA	0.039	0.051	0.3130	0.3130
	ABA	2.0 (1984)	45	45	NA	NA	0.039	0.051	0.2744	0.2736
	AAA	2.8 (2782)	45	45	NA	NA	0.039	0.051	0.2744	0.2736
1996-97	ACC	1.8 (1780)	45	45	NA	NA	0.039	0.051	0.3138	0.3130
	ABA	2.0 (1984)	45	45	NA	NA	0.039	0.051	0.2744	0.2736
	AAA	2.8 (2782)	45	45	NA	NA	0.039	0.051	0.2744	0.2736

NA - Not Available

79231CJ8

TORQUE SPECIFICATIONS
All readings in ft. lbs.

Year	Engine ID/VIN	Engine Displacement Liters (cc)	Cylinder Head Bolts	Main Bearing Bolts	Rod Bearing Bolts	Crankshaft Damper Bolt	Flywheel Bolts	Manifold		Spark Plugs	Lug Nut
								Intake	Exhaust		
1995	AAA	2.8 (2782)	(1)	(2)	(3)	(4)	(5)	18	18	22	81
	ABA	2.0 (1984)	(1)	48	(6)	(7)	(5)	15	15	22	81
	ACC	1.8 (1780)	(1)	48	(3)	(7)	(5)	22	22	18	81
1996	AAA	2.8 (2782)	(1)	(2)	(3)	(4)	(5)	18	18	22	81
	ABA	2.0 (1984)	(1)	48	(6)	(7)	(5)	15	15	22	81
	ACC	1.8 (1780)	(1)	48	(3)	(7)	(5)	22	22	18	81
1997	AAA	2.8 (2782)	(1)	(2)	(3)	(7)	(5)	18	18	22	81
	AAZ	1.9 (1896)	(1)	48 (8)	(3)	(4)	(5)	18	18	22	81
	ABA	2.0 (1984)	(1)	48	(6)	(7)	(5)	15	15	22	81
	ACC	1.8 (1780)	(1)	48	(3)	(7)	(5)	22	22	18	81
1998-99	AAA	2.8 (2792)	(1)	(2)	(6)	(7)	(5)	15	15	22	89
	AAZ	1.9 (1896)	(1)	48 (8)	(3)	(7)	(5)	18	18	NA	89
	ABA	2.0 (1984)	(1)	48	(6)	(7)	(5)	15	15	22	89
	AEB	1.8 (1781)	(1)	48 (8)	(3)	(7)	(5)	18	18	22	89
	AEG	2.0 (1984)	(9)	48 (8)	(3)	(7)	(10)	18	18	22	89
	AHA	2.8 (2792)	NA	NA	NA	NA	NA	18	18	22	89
	AHH	1.9 (1896)	(1)	48 (8)	(3)	(7)	(5)	18	18	NA	89
	ALH	1.9 (1896)	(1)	48 (8)	(3)	(11)	(5)	18	18	NA	89

(1) Torque in four steps: (use new bolts on all engines).
    Step 1: 30 ft. lbs.
    Step 2: 44 ft. lbs.
    Step 3: + 90 degrees
    Step 4: + 90 degrees
(2) 22 ft. lbs. plus an additional 1/2 turn Use new bolts.
(3) Torque to 22 ft. lbs. plus an additional 90 degrees
(4) 74 ft. lbs. plus an additional 1/4 turn. Use new bolts.
(5) 44 ft. lbs. plus an additional 90 degrees;
    except Passat GLX: 52 ft. lbs. plus 90 degrees.

(6) 22 ft. lbs. plus an additional 1/4 turn. Use new bolts.
(7) Step 1: 66 ft. lbs.
    Step 2: plus 90 degrees
(8) Plus an additional 90 degrees. Always replace bolt.
(9) Torque in four steps. Use new bolts.
    Step 1: 30 ft. lbs.
    Step 2: +90 degrees
    Step 3: +90 degrees
(10) 30 ft. lbs. + 90 degrees. Always replace bolt.
(11) 88 ft. lbs. +90 degrees. Always replace bolt.

79231CJ9

SCHEDULED MAINTENANCE INTERVALS
(VOLKSWAGEN BEETLE, CABRIO, GOLF, JETTA, PASSAT & GTI)

TO BE SERVICED	TYPE OF SERVICE	7.5	15	22.5	30	37.5	45	52.5	60	67.5	75	82.5	90	97.5
Engine oil & filter	R	✓	✓	✓	✓	✓	✓	✓	✓	✓	✓	✓	✓	✓
Brake pad thickness①	R	✓	✓	✓	✓	✓	✓	✓	✓	✓	✓	✓	✓	✓
A/T final drive fluid level	S/I		✓		✓		✓		✓		✓		✓	
Battery	S/I		✓		✓		✓		✓		✓		✓	
Brake system	S/I		✓		✓		✓		✓		✓		✓	
Cooling system	S/I		✓		✓		✓		✓		✓		✓	
Driveshaft boots	S/I		✓		✓		✓		✓		✓		✓	
Engine (check for leaks)	S/I		✓		✓		✓		✓		✓		✓	
Engine coolant level	S/I		✓		✓		✓		✓		✓		✓	
Exhaust system	S/I		✓		✓		✓		✓		✓		✓	
Fuel system	S/I		✓		✓		✓		✓		✓		✓	
Idle speed (gasoline)	S/I		✓		✓		✓		✓		✓		✓	
Idle speed (diesel)	S/I				✓				✓				✓	
Intake air system	S/I		✓		✓		✓		✓		✓		✓	
OBD system - check for codes	S/I		✓		✓		✓		✓		✓		✓	
Power steering fluid level	S/I		✓		✓		✓		✓		✓		✓	
Steering system	S/I		✓		✓		✓		✓		✓		✓	
Timing belt (diesel)	S/I				✓				✓				✓	
Transaxle fluid level	S/I		✓		✓		✓		✓		✓		✓	
Water separator (diesel)	S/I		✓		✓		✓		✓		✓		✓	
Air filter element	R				✓				✓				✓	
Engine coolant	R				✓				✓				✓	
Fuel filter (diesel)	R				✓				✓				✓	
Spark plugs (w/o supercharger)	R				✓				✓				✓	
Spark plugs (w/supercharger)	R								✓					

79231CJ0

SCHEDULED MAINTENANCE INTERVALS
(VOLKSWAGEN BEETLE, CABRIO, GOLF, JETTA, PASSAT & GTI) (Cont.)

TO BE SERVICED	TYPE OF SERVICE	VEHICLE MILEAGE INTERVAL (x1000)												
		7.5	15	22.5	30	37.5	45	52.5	60	67.5	75	82.5	90	97.5
Passenger compartment air filter	R				✓				✓				✓	
Drive belts	S/I				✓				✓				✓	
Dust seals on ball joints, tie rod ends & tie rods	S/I				✓				✓				✓	
Brake fluid③	R													

① Diesel
② Replace every 60,000 miles.
③ Replace every two years regardless of mileage.
R – Replace   S/I – Service or Inspect

FREQUENT OPERATION MAINTENANCE (SEVERE SERVICE)

If a vehicle is operated under any of the following conditions it is considered severe service:
- Extremely dusty areas.
- 50% or more of the vehicle operation is in 32°C (90°F) or higher temperatures, or constant operation in temperatures below 0°C (32°F).
- Prolonged idling (vehicle operation in stop and go traffic).
- Frequent short running periods (engine does not warm to normal operating temperatures).
- Police, taxi, delivery usage or trailer towing usage.

Oil & oil filter change – change every 3750 miles.
Air filter element – service or inspect every 15,000 miles.
Automatic transaxle fluid & filter - replace every 30,000 miles.

79231CK1

SCHEDULED MAINTENANCE
VOLKSWAGEN
BEETLE, CABRIO, GOLF, JETTA, PASSAT, GTI

The following should be used as a guide when determining the amount of work required for a particular service if taken to a repair shop. In estimating how long a particular Scheduled Maintenance Service should take, please observe the following:

- **Chilton Time** is time based on field research and data supplied by the vehicle manufacturer.
- Labor time operations are given in hours and tenths of an hour.
- All labor operations, are to be used as a guide.

Mechanic Skill Level Codes:
- **(G) GENERAL:** Normally skilled with certification.
- **(M) MAINTENANCE:** Semi-skilled working on certication.
- **(P) PRECISION:** Really skilled with multiple certification.

	Chilton Time
(M) 7500 Mile Service	
1995-99	.3
w/Diesel add	.2
(G) 15000 Mile Service	
1995-99	2.0
w/AT add	.1
w/Diesel add	.2
w/16V engine add	.2
(M) 22500 Mile Service	
1995-99	.3
w/Diesel add	.2
(G) 30000 Mile Service	
1995-99	3.1
w/AT add	.1
w/Diesel add	.7
w/16V engine add	.2

	Chilton Time
(M) 37500 Mile Service	
1995-99	.3
w/Diesel add	.2
(G) 45000 Mile Service	
1995-99	2.0
w/AT add	.1
w/Diesel add	.2
w/16V engine add	.2
(M) 52500 Mile Service	
1995-99	.3
w/Diesel add	.2
(G) 60000 Mile Service	
1995-99	3.1
w/AT add	.1
w/Diesel add	.7
w/Cabriolet add	.1
w/Supercharger add	.3
(M) 67500 Mile Service	
1995-99	.3
w/Diesel add	.2

	Chilton Time
(G) 75000 Mile Service	
1995-99	2.0
w/AT add	.1
w/Diesel add	.2
w/16V engine add	.2
(M) 82500 Mile Service	
1995-99	.3
w/Diesel add	.2
(G) 90000 Mile Service	
1995-99	3.1
w/AT add	.1
w/Diesel add	.7
w/16V engine add	.2
(M) 97500 Mile Service	
1995-99	.3
w/Diesel add	.2

79231CK2

VOLVO
240 • 850 • 940 • 960

VEHICLE IDENTIFICATION CHART

Engine Code							Model Year	
Code	Liters	Cu. in. (cc)	Cyl.	Fuel Sys.	Eng. Mfg.		Code	Year
B5234T3/53	2.3	144 (2319)	5	EFI	Volvo		S	1995
B5254S/55	2.4	151 (2435)	5	EFI	Volvo		T	1996
B5254T/56	2.4	151 (2435)	5	EFI	Volvo		V	1997
B5234T/57	2.3	144 (2319)	5	EFI	Volvo		W	1998
B5254FT/58	2.3	144 (2319)	5	EFI	Volvo		X	1999
B230FT/87	2.3	144 (2319)	4	EFI	Volvo			
B230F/88	2.3	144 (2319)	4	EFI	Volvo			
B6304F/95	2.9	181 (2922)	6	EFI	Volvo			
B6304S/96	2.9	181 (2922)	6	EFI	Volvo			

EFI - Electronic Fuel Injection

79231CA9

ENGINE IDENTIFICATION

Year	Model	Engine Displacement Liters (cc)	Engine Series (ID/VIN)	Fuel System	No. of Cylinders	Engine Type
1993	240	2.3 (2316)	B230F/88	EFI	4	SOHC
	940	2.3 (2316)	B230F/88	EFI	4	SOHC
	940 ②	2.3 (2316)	B230FT/87	EFI	4	SOHC
	850	2.3 (2435)	B5254FS/55	EFI	5	DOHC ③
	960	2.9 (2922)	B6304F/95	EFI	6	DOHC ①
1994	940	2.3 (2316)	B230F/88	EFI	4	SOHC
	940 ②	2.3 (2316)	B230FT/87	EFI	4	SOHC
	850 ②	2.3 (2319)	B5234T/57	EFI	5	DOHC ③
	850	2.4 (2435)	B5254S/55	EFI	5	DOHC ③
	960	2.9 (2922)	B6304F/95	EFI	6	DOHC ①
1995	940 Turbo	2.3 (2316)	B230FT/86 or 87	EFI	4	SOHC
	940 Sedan	2.3 (2316)	B230F/88	EFI	4	SOHC
	940 Wagon	2.3 (2316)	B20F/88	EFI	4	SOHC
	T-5R	2.3 (2319)	B5234T5/58	EFI	5	DOHC ③
	850 Turbo	2.3 (2319)	B5234T/57	EFI	5	DOHC ③
	850	2.4 (2435)	B5254S/55	EFI	5	DOHC ③
	850 GLT	2.4 (2435)	B5254S/55	EFI	5	DOHC ③
	960 Sedan	2.9 (2922)	B6304F/95 ④	EFI	6	DOHC ①
	960 Wagon	2.9 (2922)	B6304F/95 ④	EFI	6	DOHC ①
1996-97	850 Turbo	2.3 (2319)	B5234T/57	EFI	5	DOHC ③
	850 R	2.3 (2319)	B5234FT/58	EFI	5	DOHC ③
	850	2.4 (2435)	B5254S/55	EFI	5	DOHC ③
	960	2.9 (2922)	B6304F/96	EFI	6	DOHC ①

NOTE: Since all 850 engines are now in principle F-engines, this symbol is no longer needed. Engine are now denoted with S or T.

B - Petrol (gasoline)

F - Fuel-injected with catalytic converter   EFI - Electronic fuel injection   ① 24 valve engine   ④ 95 with air pump

S - Normally-aspirated   SOHC - Single overhead camshaft   ② Turbocharged   96 without air pump

T - Turbocharged engine   DOHC - Double overhead camshaft   ③ 20 valve engine

79231CA0

GENERAL ENGINE SPECIFICATIONS

Year	Engine ID/VIN	Engine Displacement Liters (cc)	Fuel System Type	Net Horsepower @ rpm	Net Torque @ rpm (ft. lbs.)	Bore x Stroke (In.)	Com-pression Ratio	Oil Pressure @ rpm
1995	B230F/88	2.3 (2316)	EFI	114@5400	136@2150	3.78x3.15	9.8:1	35-85@2000
	B230FT/86 or 87	2.3 (2316)	EFI	162@4800	195@3450	3.78x3.15	8.7:1	35-85@2000
	B5234T/57	2.3 (2319)	EFI	222@5200	221@2100	3.19x3.54	8.5:1	49.8@4000
	B5234T5/58	2.3 (2319)	EFI	240@5600	221@2100	3.19x3.54	8.5:1	49.8@4000
	B5254S/55	2.4 (2435)	EFI	168@6200	162@4700	3.27x3.54	10.5:1	49.8@4000
	B6304F/95	2.9 (2922)	EFI	181@5200	199@4100	3.27x3.54	10.7:1	36@2000
1996	B5254S/55	2.4(2435)	EFI	168@6100	162@4700	3.27 x 3.54	10.5:1	49.8@4000
	B5254T/56	2.4(2435)	EFI	190@5200	191@1800	3.27 x 3.54	10.5:1	49.8@4000
	B5234T/57	2.3(2319)	EFI	222@5200	221@2100	3.19 x 3.54	8.5:1	49.8@4000
	B5254FT/58	2.3(2319)	EFI	240@5600	221@2100	3.19 x 3.54	8.5:1	49.8@4000
	B6304S/96	2.9(2922)	EFI	181@5200	199@4100	3.27 x 3.54	10.7:1	36@2000
1997	B5254S/55	2.4(2435)	EFI	168@6100	162@4700	3.27 x 3.54	10.5:1	49.8@4000
	B5254T/56	2.4(2435)	EFI	190@5200	191@1800	3.27 x 3.54	10.5:1	49.8@4000
	B5234T/57	2.3(2319)	EFI	222@5200	221@2100	3.19 x 3.54	8.5:1	49.8@4000
	B5254FT/58	2.3(2319)	EFI	240@5600	221@2100	3.19 x 3.54	8.5:1	49.8@4000
	B6304S/96	2.9(2922)	EFI	181@5200	199@4100	3.27 x 3.54	10.7:1	36@2000
1998-99	B5254S/55	2.4(2435)	EFI	168@6100	162@4700	3.27 x 3.54	10.5:1	49.8@4000
	B5254T/56	2.4(2435)	EFI	190@5200	199@1800	3.27 x 3.54	10.5:1	49.8@4000
	B5234T/57	2.3(2319)	EFI	236@5100	243@2100	3.19 x 3.54	8.5:1	49.8@4000
	B6304S/96	2.9(2922)	EFI	181@5200	199@4100	3.27 x 3.54	10.7:1	36@2000
	B5234T3/53	2.3(2319)	EFI	236@5100	243@2700	3.19 x 3.54	8.5:1	49.8@4000

EFI - Electronic Fuel Injection

79231CB1

GASOLINE ENGINE TUNE-UP SPECIFICATIONS

	Engine ID/VIN	Engine Displacement Liters (cc)	Spark Plugs Gap (in.)	Ignition Timing (deg.)		Fuel Pump (psi)	Idle Speed (rpm)		Valve Clearance	
				MT	AT		MT	AT	In.	Ex.
1995	B-230F/88	2.3 (2316)	0.028	—	12B	43	775	775	0.014-0.018	0.014-0.018
	B-230FT/87	2.3 (2316)	0.028-0.032	—	12B	43	—	750	0.014-0.018	0.014-0.018
	B-6304F/95	2.9 (2922)	0.024-0.028	—	16B	43	—	700-800	HYD	HYD
	B-5254S/55	2.4 (2435)	0.028	—	10B	43	—	750-850	HYD	HYD
	B-5234T/57	2.3 (2319)	0.028	—	6B	58	—	800-900	HYD	HYD
1996	B-5254S/55	2.4 (2435)	0.028	3-7B	10B	43	750-850	750-850	HYD	HYD
	B-5254T/56	2.4 (2435)	0.028	3-7B	6B	58	800-900	800-900	HYD	HYD
	B-5234T/57	2.3 (2319)	0.028	3-7B	6B	58	800-900	800-900	HYD	HYD
	B-5254FT/58	2.3 (2319)	0.028	3-7B	6B	58	800-900	800-900	HYD	HYD
	B-6304F/96	2.9 (2922)	0.024-0.028	—	16B	43	—	700-800	HYD	HYD
1997	B-5254S/55	2.4 (2435)	0.028	3-7B	10B	43	750-850	750-850	HYD	HYD
	B-5254T/56	2.4 (2435)	0.028	3-7B	6B	58	800-900	800-900	HYD	HYD
	B-5234T/57	2.3 (2319)	0.028	3-7B	6B	58	800-900	800-900	HYD	HYD
	B-5254FT/58	2.3 (2319)	0.028	3-7B	6B	58	800-900	800-900	HYD	HYD
	B-6304F/96	2.9 (2922)	0.024-0.028	—	16B	43	—	700-800	HYD	HYD
1998-99	B-5234T3/53	2.3 (2319)	0.028	3-7B	6B	58	800-900	800-900	HYD	HYD
	B-5254S/55	2.4 (2435)	0.028	3-7B	10B	43	750-850	750-850	HYD	HYD
	B-5254T/56	2.4 (2435)	0.028	3-7B	6B	58	800-900	800-900	HYD	HYD
	B-5234T/57	2.3 (2319)	0.028	3-7B	6B	58	800-900	800-900	HYD	HYD
	B-6304F/96	2.9 (2922)	0.024-0.028	—	16B	43	—	700-800	HYD	HYD

HYD- Hydraulic lash adjusters
B- Before top dead center

79231CB2

CAPACITIES

Year	Model	Engine ID/VIN	Engine Displacement Liters (cc)	Engine Oil with Filter (qts.) ①	Transmission (pts.) 4:Spd	5:Spd	Auto.	Transfer Case (pts.)	Drive Axle Front (pts.)	Rear (pts.)	Fuel Tank (gal.)	Cooling System (qts.)
1995	940	B:230F/88	2.3 (2316)	4	—	—	15.6 ②	—	—	③	19.8	10
	940Turbo	B:230FT/87	2.3 (2316)	4	—	—	15.6 ②	—	—	③	19.8	10
	850	B:5254S/55	2.4 (2435)	5.6	—	4.4	8.4	—	—	—	19.3	7.6
	850Turbo	B:5234T/57	2.3 (2319)	5.6	—	4.4	8.4	—	—	—	19.3	7.6
	960	B6304F/95	2.9 (2922)	6	—	—	16.4	—	—	③	21.8	10
1996	850	B:5254S/55	2.4 (2435)	5.6	—	4.4	8.4	—	—	—	19.3	7.6
	850GLT	B:5254T/56	2.4 (2435)	5.6	—	4.4	8.4	—	—	—	19.3	7.6
	850T:5	B:5234T/57	2.3 (2319)	5.6	—	4.4	8.4	—	—	③	19.3	7.6
	850R	B:5254FT/58	2.3 (2319)	5.6	—	4.4	8.4	—	—	—	19.3	7.6
	960	B:6304S/96	2.9 (2922)	6	—	—	16.4	—	—	③	21.8	10
1997	850	B:5254S/55	2.4 (2435)	5.6	—	4.4	8.4	—	—	—	19.3	7.6
	850GLT	B:5254T/56	2.4 (2435)	5.6	—	4.4	8.4	—	—	—	19.3	7.6
	850T:5	B:5234T/57	2.3 (2319)	5.6	—	4.4	8.4	—	—	—	19.3	7.6
	850R	B:5254FT/58	2.3 (2319)	5.6	—	4.4	8.4	—	—	—	19.3	7.6
	960	B:6304S/96	2.9 (2922)	6	—	—	16.4	—	—	③	21.8	10
1998:99	S70	B5254S/55	2.4 (2435)	6.1	—	4.4	8.4	—	—	—	18.5	7.6
	S70GLT	B:5254T/56	2.4 (2435)	6.1	—	4.4	8.4	—	—	—	18.5	7.6
	S70T:5	B:5234T/57	2.3 (2319)	6.1	—	4.4	8.4	—	—	—	18.5	7.6
	S90	B:6304S/95	2.9 (2922)	6	—	—	16.4	—	—	③	21.8	10
	C70	B:5234T3/53	2.3 (2319)	6.1	—	4.4	8.4	—	—	—	18.5	7.6
	V70	B:5254S/55	2.4 (2435)	6.1	—	4.4	8.4	—	—	—	18.5	7.6
	V70GLT	B:5254T/56	2.4 (2435)	6.1	—	4.4	8.4	—	—	—	18.5	7.6
	V70T:5	B:5234T/57	2.3 (2319)	6.1	—	4.4	8.4	—	—	—	18.5	7.6
	V70 AWD	B:5254T/56	2.4 (2435)	6.1	—	—	8.4	1.7	—	2.9	18.5	7.6
	V70R AWD	B:5234T/57	2.3 (2319)	6.1	—	—	8.4	1.7	—	2.9	18.5	7.6
	V70XC AWD	B:5254T/56	2.4 (2435)	6.1	—	—	8.4	1.7	—	2.9	18.5	7.6
	V90	B:6304S/96	2.9 (2922)	6.1	—	—	16.4	—	—	③	21.8	10

NOTE: All capacities are approximate. Add fluid gradualy and check to be sure a proper fluid level is obtained.

① On turbocharged
engines, add 0.7
US qts. if the
cooler is drained

② Total fluid capacity
cannot be drained.
3.6 qts (3.4 liters)
remains in the torque
converter and control systems

③ 1030 axle:2.8
1031 axle:3.4
1035 axle:2.9
1041 axle:3.1
1045 axle:2.8
1055 axle:3.2
1065 axle:2.9

79231CB3

VALVE SPECIFICATIONS

Year	Engine ID/VIN	Engine Displacement Liters (cc)	Seat Angle (deg.)	Face Angle (deg.)	Spring Test Pressure (lbs. @ in.)	Spring Installed Height (in.)	Stem-to-Guide Clearance (in.)		Stem Diameter (in.)	
							Intake	Exhaust	Intake	Exhaust
1995	B230F	2.3 (2316)	45	44.5	158@1.08	1.79	0.0012-0.0024	0.0024-0.0036	0.3132-0.3138	0.3128-0.3134
	B230FT	2.3 (2316)	45	44.5	158@1.08	1.79	0.0012-0.0024	0.0024-0.0036	0.3132-0.3138	0.3128-0.3134
	B5234T	2.3 (2319)	45	NA	①	NA	0.0012-0.0024	0.0012-0.0024	0.2738-0.2750	0.2734-0.2746
	B5234T5	2.3 (2319)	45	NA	①	NA	0.0012-0.0024	0.0012-0.0024	0.2738-0.2750	0.2734-0.2746
	B5254S	2.3 (2319)	45	NA	①	NA	0.0012-0.0024	0.0012-0.0024	0.2738-0.2750	0.2734-0.2746
	B6304F	2.9 (2922)	45.25	45.5	①	NA	0.0012-0.0024	0.0012-0.0024	0.2738-0.2744	0.2738-0.2744
1996	B5234T	2.3 (2319)	45	44.5	①	NA	0.0012-0.0024	0.0012-0.0024	0.2738-0.2750	0.2734-0.2746
	B5234FT	2.3 (2319)	45	44.5	①	NA	0.0012-0.0024	0.0012-0.0024	0.2738-0.2750	0.2734-0.2746
	B5254S	2.4 (2435)	45	44.5	①	NA	0.0012-0.0024	0.0012-0.0024	0.2738-0.2750	0.2734-0.2746
	B6304F	2.9 (2922)	45	44.5	①	NA	0.0012-0.0024	0.0012-0.0024	0.2738-0.2744	0.2738-0.2744
1997	B5234T	2.3 (2319)	45	44.5	①	NA	0.0012-0.0024	0.0012-0.0024	0.2738-0.2750	0.2734-0.2746
	B5234FT	2.3 (2319)	45	44.5	①	NA	0.0012-0.0024	0.0012-0.0024	0.2738-0.2750	0.2734-0.2746
	B5254S	2.4 (2435)	45	44.5	①	NA	0.0012-0.0024	0.0012-0.0024	0.2738-0.2750	0.2734-0.2746
	B6304F	2.9 (2922)	45	44.5	①	NA	0.0012-0.0024	0.0012-0.0024	0.2738-0.2744	0.2738-0.2744
1998-99	B5234T	2.3 (2319)	45	44.5	①	NA	0.0012-0.0024	0.0012-0.0024	0.2738-0.2750	0.2734-0.2746
	B5234FT	2.3 (2319)	45	44.5	①	NA	0.0012-0.0024	0.0012-0.0024	0.2738-0.2750	0.2734-0.2746
	B5254S	2.4 (2435)	45	44.5	①	NA	0.0012-0.0024	0.0012-0.0024	0.2738-0.2750	0.2734-0.2746
	B6304F	2.9 (2922)	45	44.5	①	NA	0.0012-0.0024	0.0012-0.0024	0.2738-0.2744	0.2738-0.2744

NOTE: Exhaust valves for turbocharged engines are stellite-coated and must not be machined. They may be ground against the valve seat.

NA - Not Available

① Intake valve: 150@1.00
Exhaust valve: 61@1.34

79231CB4

TORQUE SPECIFICATIONS
All readings in ft. lbs.

Year	Engine ID/VIN	Engine Displacement Liters (cc)	Cylinder Head Bolts	Main Bearing Bolts	Rod Bearing Bolts	Crankshaft Damper Bolts	Flywheel Bolts	Manifold Intake	Manifold Exhaust	Spark Plugs	Lug Nut
1995	B230FT	2.3 (2316)	①	80	②	③	51	12	12	18	④
	B230F	2.3 (2316)	①	80	②	③	51	12	12	18	④
	B5234T	2.3 (2319)	⑤	⑥	⑦	133	⑧	15	18	18	81
	B5234T5	2.3 (2319)	⑤	⑥	⑦	133	⑧	15	18	18	81
	B5254S	2.4 (2435)	⑤	⑥	⑦	133	⑧	15	18	18	81
	B6304F	2.9 (2922)	⑤	⑥	⑦	221	⑧	15	18	18	④
1996	B5254T	2.4 (2435)	①	⑥	⑦	133	⑨	15	18	18	81
	B5234T	2.3 (2319)	①	⑥	⑦	133	⑨	15	18	18	81
	B5254S	2.4 (2435)	①	⑥	⑦	133	⑨	15	18	18	81
	B6304S	2.9 (2922)	①	⑥	⑦	221	⑨	15	18	18	④
1997	B5254T	2.4 (2435)	①	⑥	⑦	133	⑨	15	18	18	81
	B5234T	2.3 (2319)	①	⑥	⑦	133	⑨	15	18	18	81
	B5254S	2.4 (2435)	①	⑥	⑦	133	⑨	15	18	18	81
	B6304S	2.9 (2922)	①	⑥	⑦	221	⑨	15	18	18	④
1998-99	B5254T	2.4 (2435)	①	⑥	⑦	133	⑨	15	18	18	81
	B5234T	2.3 (2319)	①	⑥	⑦	133	⑨	15	18	18	81
	B5254S	2.4 (2435)	①	⑥	⑦	133	⑨	15	18	18	81
	B6304S	2.9 (2922)	①	⑥	⑦	221	⑨	15	18	18	④

① Step 1: 14 ft. lbs.
  Step 2: 43 ft. lbs.
  Step 3: + 90 degrees
② Step 1: 14 ft. lbs.
  Step 2: + 90 degrees
③ Step 1: 43 ft. lbs.
  Step 2: + 90 degrees
④ Torque lugs in a diagonal pattern
  P20: 85 ft. lbs. (115nm)
  P70/90: 63 ft. lbs. (85nm)
⑤ Step 1: 15 ft. lbs.
  Step 2: 44 ft. lbs.
  Step 3: + 130 degrees
  Bolts should be tightened in sequence from center towards the ends
⑥ Tighten cylinder block, intermediate section, in stages:
  Step 1: M10 bolts: 15 ft. lbs. (20mm)
  Step 2: M10 bolts: 33 ft. lbs. (45mm)
  Step 3: M8 bolts: 18 ft. lbs. (25mm)
  Step 4: M7 bolts: 13 ft. lbs. (17mm)
  Step 5: M10 bolts: + 90 degrees
⑦ Step 1: 15 ft. lbs.
  Step 2: + 90 degrees
⑧ Step 1: 33 ft. lbs.
  Step 2: + 65 degrees
⑨ Step 1: 33 ft. lbs.
  Step 2: + 50 degrees

79231CB5

SCHEDULED MAINTENANCE INTERVALS
(VOLVO 800 SERIES, 900 SERIES, C70, S70, S90, V70 & V90)

TO BE SERVICED	TYPE OF SERVICE	VEHICLE MILEAGE INTERVAL (x1000)												
		5	10	20	30	40	50	60	70	80	90	100	110	120
Engine oil & filter①	R	✓	✓	✓	✓	✓	✓	✓	✓	✓	✓	✓	✓	✓
Automatic transmission fluid	S/I	✓	✓	✓	✓	✓	✓	✓	✓	✓	✓	✓	✓	✓
Fluid levels (all)	S/I	✓	✓	✓	✓	✓	✓	✓	✓	✓	✓	✓	✓	✓
Rotate tires	S/I		✓	✓	✓	✓	✓	✓	✓	✓	✓	✓	✓	✓
Automatic transmission shift control	S/I		✓	✓	✓	✓	✓	✓	✓	✓	✓	✓	✓	✓
Brake pads & parking brake	S/I		✓	✓	✓	✓	✓	✓	✓	✓	✓	✓	✓	✓
Driveshaft boots (850)	S/I		✓	✓	✓	✓	✓	✓	✓	✓	✓	✓	✓	✓
Engine & transmission (check for leaks)	S/I		✓	✓	✓	✓	✓	✓	✓	✓	✓	✓	✓	✓
Exhaust system	S/I		✓	✓	✓	✓	✓	✓	✓	✓	✓	✓	✓	✓
Grease link arm stops (1995-97 850)	S/I		✓	✓	✓	✓	✓	✓	✓	✓	✓	✓	✓	✓
Reset service reminder①	S/I		✓	✓	✓	✓	✓	✓	✓	✓	✓	✓	✓	✓
Driveshaft, U-joints	S/I				✓	✓	✓	✓	✓	✓	✓	✓	✓	✓
Driveshaft joints (850)	S/I							✓	✓	✓	✓	✓	✓	✓
Clutch	S/I			✓		✓		✓		✓		✓		✓
Kickdown cable (940)	S/I			✓		✓		✓		✓		✓		✓
Brake & fuel lines & hoses	S/I				✓		✓		✓		✓			✓
Steering & suspension	S/I				✓		✓		✓		✓			✓
Air cleaner filter	R				✓			✓			✓			✓
Spark plugs	R				✓			✓			✓			✓
Timing belt (1995 960)	R					✓							✓	
Timing belt (850 & 1996-97 960)	R								✓					

79231CB6

SCHEDULED MAINTENANCE INTERVALS
(VOLVO 800 SERIES, 900 SERIES, C70, S70, S90, V70 & V90) (Cont.)

TO BE SERVICED	TYPE OF SERVICE	VEHICLE MILEAGE INTERVAL (x1000)												
		5	10	20	30	40	50	60	70	80	90	100	110	120
Timing belt (B230F, FT)	R						✓					✓		
Timing belt (B230F, FT	S/I							✓					✓	
Timing belt (B230FD)	S/I		✓										✓	
Timing belt (B230FD)	R											✓		
Valve clearance (740)	S/I				✓			✓			✓			✓
Drive belt tensioner (850)	S/I				✓			✓			✓			✓
Drive belts⑤	S/I				✓			✓			✓			✓
Fuel line filter④	R							✓						✓
EGR system	S/I							✓				✓		
PCV nipple (orifice)/hoses	S/I							✓				✓		
Check suspension torques③	S/I		✓											
Brake fluid②	R													

① Perform operation every 5000 miles on turbocharged models.
② Replace every 2 years or 30,000 miles, whichever comes first under normal conditions, more frequently in mountainous areas or moist climates.
③ Except 850 shown. 850 (1993) - perform at 1500 miles.
④ 1995 shown; replace every 100,000 miles (1996-97).
⑤ 850, 960: replace at 60,000 miles.

R – Replace   S/I – Service or Inspect

FREQUENT OPERATION MAINTENANCE (SEVERE SERVICE)
If a vehicle is operated under any of the following conditions it is considered severe service:
- Extremely dusty areas.
- 50% or more of the vehicle operation is in 32°C (90°F) or higher temperatures, or constant operation in temperatures below 0°C (32°F).
- Prolonged idling (vehicle operation in stop and go traffic).
- Frequent short running periods (engine does not warm to normal operating temperatures).
- Police, taxi, delivery usage or trailer towing usage.

Oil & oil filter change – (all models) change every 5000 miles.
Air filter element – service or inspect every 15,000 miles.

79231CB7

SCHEDULED MAINTENANCE
VOLVO
850, 940, 960, C70
S70, S90, V70, V90

The following should be used as a guide when determining the amount of work required for a particular service if taken to a repair shop. In estimating how long a particular Scheduled Maintenance Service should take, please observe the following:

- **Chilton Time** is time based on field research and data supplied by the vehicle manufacturer.
- Labor time operations are given in hours and tenths of an hour.
- All labor operations, are to be used as a guide.

Mechanic Skill Level Codes:
    (G) GENERAL: Normally skilled with certification.
    (M) MAINTENANCE: Semi-skilled working on certicication.
    (P) PRECISION: Really skilled with multiple certification.

	Chilton Time		Chilton Time		Chilton Time
(M) 5000 Mile Service		**(M) 50000 Mile Service**		**(G) 90000 Mile Service**	
1995-99	.5	1995-97 850	1.9	1995-97 850	3.8
		1995-99 940	1.7	1995-99 940	3.0
(M) 10000 Mile Service		1995-97 960	3.7	1995-97 960	3.1
1995-99	1.6	1998-99 C70, S70, V70	1.9	1998-99 C70, S70, V70	3.8
		1998-99 S90, V90	3.7	1998-99 S90, V90	3.1
(M) 20000 Mile Service		Renew timing belt add	2.0		
1995-97 850	1.9			**(G) 100000 Mile Service**	
1995-99 940	1.8	**(G) 60000 Mile Service**		1995-97 850	2.5
1995-97 960	1.7	1995-97 850	4.9	1995-99 940	2.3
1998-99 C70, S70, V70	1.9	1995-99 940	3.7	1996-97 960	2.2
1998-99 S90, V90	1.7	1995-97 960	4.1	1998-99 C70, S70, V70	2.5
		1998-99 C70, S70, V70	4.9	1998-99 S90, V90	2.2
(G) 30000 Mile Service		1998-99 S90, V90	4.1	Renew timing belt, add	2.0
1995-97 850	3.4	Inspect timing belt add	.2		
1995-99 940	2.8			**(M) 110000 Mile Service**	
1995-97 960	2.7	**(G) 70000 Mile Service**		1995-97 850	2.0
1998-99 C70, S70, V70	3.4	1995-97 850	4.0	1995-99 940	1.7
1998-99 S90, V90	2.7	1995-99 940	1.9	1995-97 960	1.7
		1996-97 960	3.7	1998-99 C70, S70, V70	2.0
(G) 40000 Mile Service		1998-99 C70, S70, V70	4.0	1998-99 S90, V90	1.7
1995-97 850	2.2	1998-99 S90, V90	3.7	Inspect timing belt add	.2
1995-99 940	2.1				
w/AT add	.4	**(G) 80000 Mile Service**		**(G) 120000 Mile Service**	
1995-97 960	2.0	1995-97 850	2.3	1995-97 850	4.3
1998-99 C70, S70, V70	2.2	1995-99 940	2.1	1995-99 940	3.6
1998-99 S90, V90	2.0	1995-97 960	2.0	1995-97 960	3.7
		1998-99 C70, S70, V70	2.3	1998-99 C70, S70, V70	4.3
		1998-99 S90, V90	2.0	1998-99 S90, V90	3.7

79231CB8

MAINTENANCE LIGHT RESETTING AND DTC RETRIEVAL

2

MAINTENANCE INDICATOR LIGHT RESETTING

This section describes reset procedures for maintenance lights. Maintenance lights are used to indicate to the operator that some type of routine maintenance should be performed. Unlike a Check Engine light that will be displayed when there is a fault with the engine management system, the maintenance light will be displayed when an engine or transmission oil change is recommended according to driving conditions. Also, the light will be displayed to indicate when the emission control system is in need of servicing.

Acura

RESETTING

1995 Legend and Vigor, and 1995–96 2.5TL and 3.2TL

The Maintenance Reminder Indicator informs you when it is time for scheduled maintenance. When it is near 7,500 miles (12,000 km) since the last maintenance, the indicator will turn yellow. If you exceed 7,500 miles (12,000 km), the indicator will turn red. The indicator can be reset by inserting the ignition key or other similar object into the slot below the indicator. This will extinguish the indicator for the next 7500 miles (12,000 km).

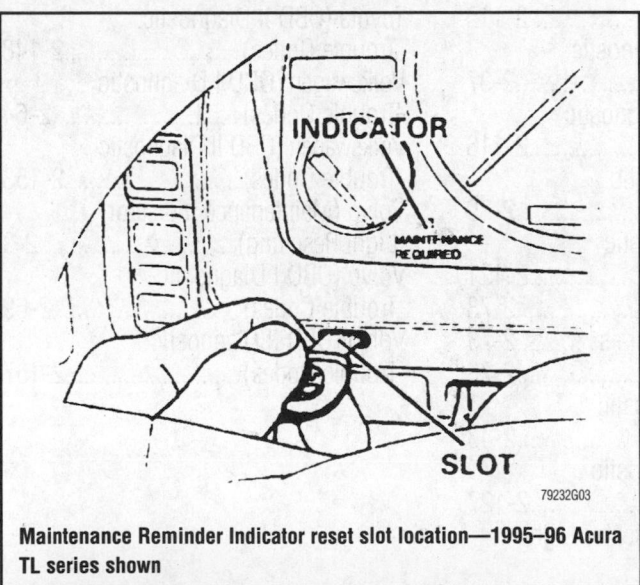

Maintenance Reminder Indicator reset slot location—1995–96 Acura TL series shown

1995–97 Integra and 1996–97 3.5 RL

The Maintenance Reminder Indicator reminds you that it is time for scheduled maintenance. For the first 6,000 miles (9,600 km) after the Maintenance Required Indicator is reset, it will come on for 2 seconds when you turn the ignition **ON**. Between 6,000 miles (9,600 km) and 7,500 miles (12,000 km) this indicator will light for two seconds when you first turn the ignition **ON**, then flash for ten seconds. If you exceed 7,500 miles (12,000 km) without having the scheduled maintenance performed, this indicator will remain on as a constant reminder. Reset the indicator by pressing the reset button. This button is located on the bottom of the dashboard to the right of the steering column.

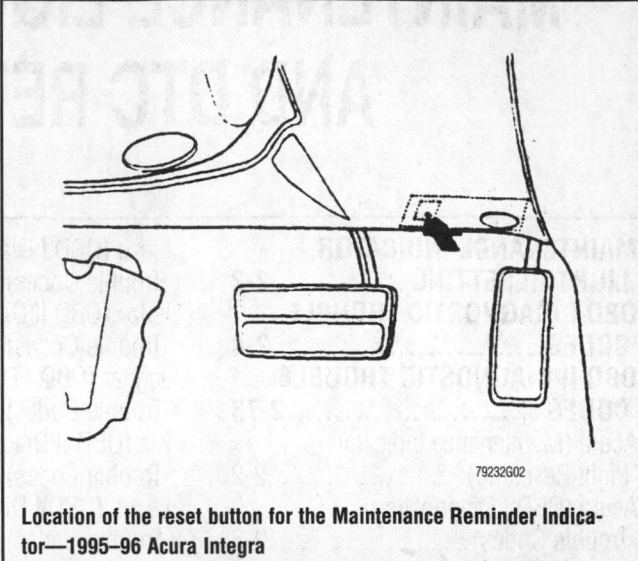

Location of the reset button for the Maintenance Reminder Indicator—1995–96 Acura Integra

Audi

RESETTING

1995–96 100, A4, A6, S4 and S6

Some models are equipped with a Service Reminder Indicator located in the trip recorder. Whenever a routine scheduled service is due, the particular type of service flashes for a few seconds in place of the trip recorder. The Service Reminder Indicator will alert the driver approximately 620 miles (1000 km) or 10 days prior to the actual service time is required. The display will continue to flash for about 60 seconds after the engine has been started. If desired, it can be switched to the trip recorder before this time has elapsed by pressing the reset button next to the speedometer. The following service displays will be shown when required. When the required service has be performed, the indicator must be reset by the dealer. A special VAG 1551/1 scan tool is required to perform the reset procedure.

OEL -Oil change service required
In 1 -Inspection service required
In 2 -Additional service work required

BMW

RESETTING

All 1995–96 Models

The on-board computer is used to evaluate mileage, average engine speed, engine and coolant temperatures, as well as other computer input factors that determine maintenance intervals. There are 5 green, 1 yellow and 1 red LED's used to remind the driver of oil changes and other maintenance services. The green LED's will be illuminated when the ignition is in the **ON** position and the engine is not running. There will not be as many green LED's illuminated when maintenance time gets closer. A yellow LED that is illuminated when the engine is running, will indicate maintenance is now due. The red LED will be illuminated when the service interval has been exceeded by approximately 1000 miles (1600 km). There is a service interval reset tool manufactured by the Assenmacher Tool Company, tool number 62–1–100 and with the aid of an additional adapter, the tool can be used on the 1995–96 models.

Honda

RESETTING

Vehicles equipped with a Maintenance Reminder Indicator will illuminate when it is time for scheduled maintenance. When it is near 7,500 miles (12,000 km) since the last maintenance, the indicator will turn yellow. If you exceed 7,500 miles (12,000 km), the indicator will turn red. The indicator can be reset by inserting the ignition key, or other similar object, into the slot below the indicator. This will extinguish the indicator for the next 7,500 miles (12,000 km).

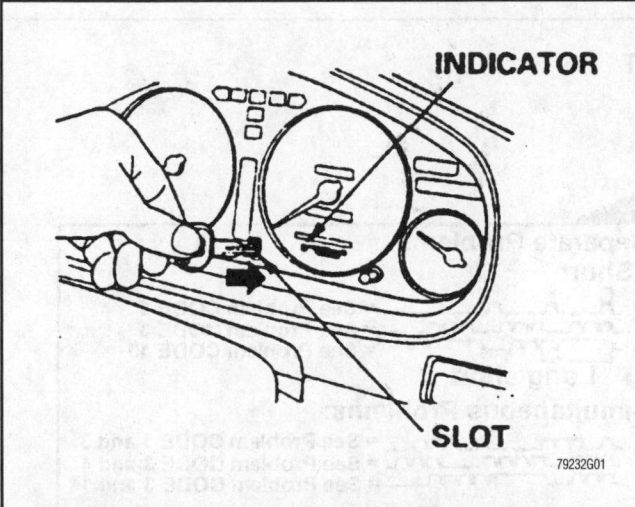

Maintenance Reminder Indicator reset slot location—1996 Honda Accord shown

OBD I DIAGNOSTIC TROUBLE CODE

Acura

GENERAL INFORMATION

Programmed Fuel Injection (PGM-FI) System is a fully electronic microprocessor based engine management system. The Electronic Control Unit (ECU) is given responsibility for control of injector timing and duration, intake air control, ignition timing, cold start enrichment, fuel pump control, fuel cut-off, NC compressor operation, alternator control as well as EGR function and canister purge cycles.

The ECU receives electric signals from many sensors and sources on and around the engine. The signals are processed against pre-programmed values; correct output signals from the ECU are determined by these calculations. The ECU contains additional memories, back-up and fail-safe functions as well as self diagnostic capabilities.

Saab

RESETTING

1995–96 900 Series

The 900 series uses an information center called the Saab Information Display (SID). The TIME FOR SERVICE light will illuminate in the SID when it is time for scheduled service. Reset the light after service is completed as follows: press and hold the CLEAR button for at least 8 seconds until TIME FOR SERVICE appears on the display and an audible signal is heard. After 21 starts of the engine the TIME FOR SERVICE reminder message will be canceled.

Volvo

RESETTING

On some 1995–96 Volvo models, an oil service interval reminder light is located on the instrument cluster and will illuminate at 5000 mile (8000 km) intervals. The light will continue to illuminate for 2 minutes after each engine start or until the counter is reset. After completing the necessary service, reset the mileage counter as follows:

1995 740, 940 and 960

1. To reset the mileage counter, remove the rubber plug located between the speedometer and the clock.
2. Remove the rubber plug and depress the reset button, using a small, suitable rod.
3. Verify the service indicator light is out and replace the rubber plug.

1996 850 and 960

The service reminder light can only be reset using a special scanner tool. The scanner tool is connected to the Diagnostic Link Connector (DLC). Consult the manufacturer's instructions for resetting the service light.

SELF-DIAGNOSTICS

Service Precautions

- Do not operate the fuel pump when the fuel lines are empty.
- Do not operate the fuel pump when removed from the fuel tank.
- Do not reuse fuel hose clamps.
- The washer(s) below any fuel system bolt (banjo fittings, service bolt, fuel filter, etc.) must be replaced whenever the bolt is loosened. Do not reuse the washers; a high-pressure fuel leak may result.
- Make sure all ECU harness connectors are fastened securely. A poor connection can cause an extremely high voltage surge and result in damage to integrated circuits.
- Keep all ECU parts and harnesses dry during service. Protect the ECU and all solid-state components from rough handling or extremes of temperature.

• Use extreme care when working around the ECU or other components; the airbag or SRS wiring may be in the vicinity. On these vehicles, the SRS wiring and connectors are yellow; do not cut or test these circuits.

• Before attempting to remove any parts, turn the ignition switch **OFF** and disconnect the battery ground cable.

• Always use a 12 volt battery as a power source for the engine, never a booster or high-voltage charging unit.

• Do not disconnect the battery cables with the engine running.

• Do not disconnect any wiring connector with the engine running or the ignition **ON** unless specifically instructed to do so.

• Do not apply battery power directly to injectors.

• Whenever possible, use a flashlight instead of a drop light.

• Keep all open flame and smoking material out of the area.

• Use a shop cloth or similar to catch fuel when opening a fuel system. Consider the fuel-soaked rag to be a flammable solid and dispose of it in the proper manner.

• Relieve fuel system pressure before servicing any fuel system component.

• Always use eye or full-face protection when working around fuel lines, fittings or components.

• Always keep a dry chemical (class B-C) fire extinguisher near the area.

Reading Codes

1986–90 LEGEND, 1986–91 INTEGRA

When a fault is noted, the ECU stores an identifying code and illuminates the CHECK ENGINE light. The code will remain in memory until cleared; the dashboard warning lamp may not illuminate during the next ignition cycle if the fault is no longer present. Not all faults noted by the ECU will trigger the dashboard warning lamp although the fault code will be set in memory. For this reason, troubleshooting should be based on the presence of stored codes, not the illumination of the warning lamp while the car is operating.

Stored codes are displayed by a flashing LED on the ECU. When the CHECK ENGINE warning lamp has been on or reported on, lift or remove the carpet from the right front passenger footwell. The ECU is below a protective cover; the LED may be viewed through a

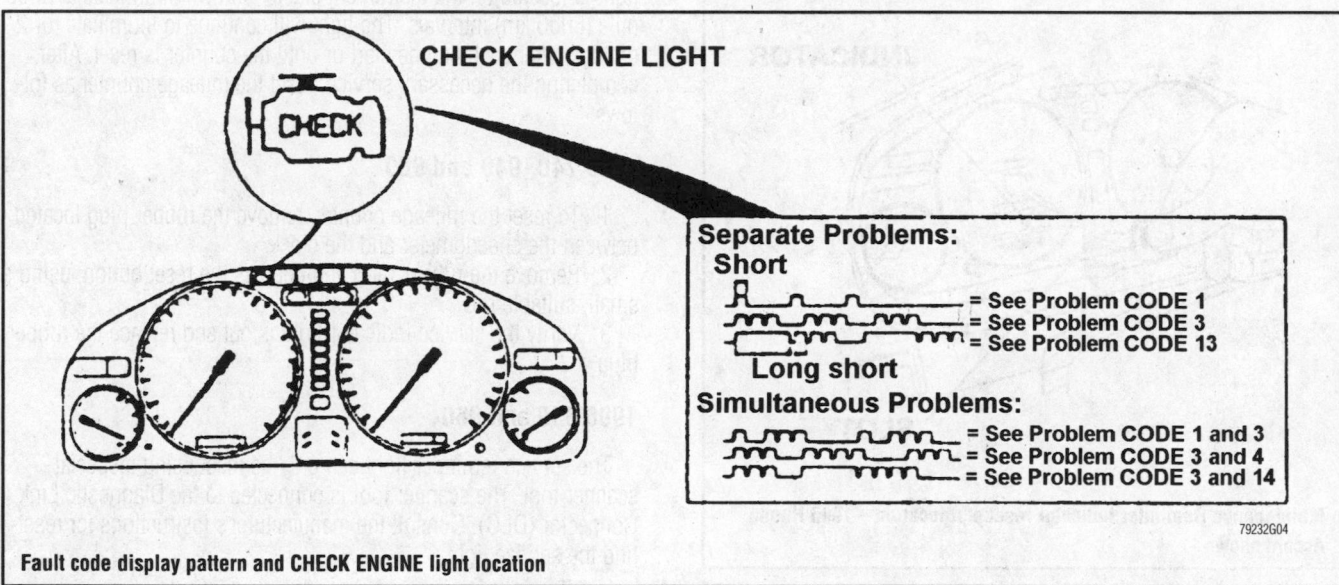

Fault code display pattern and CHECK ENGINE light location

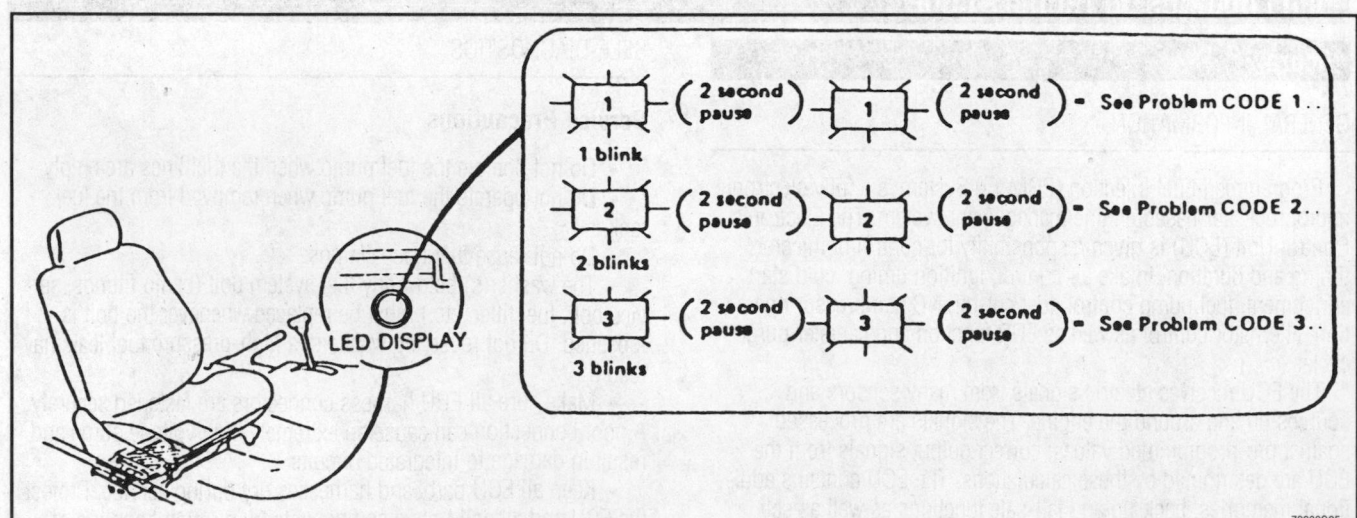

ECU location and self-diagnostic LED display—1986–91 Integra

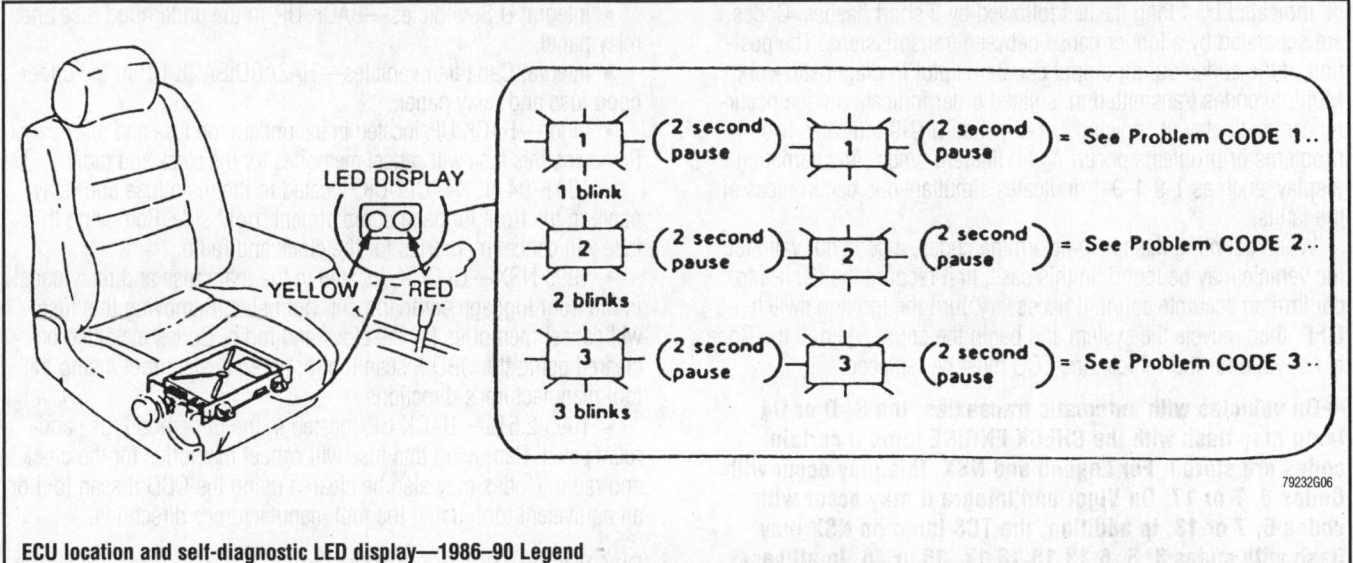

ECU location and self-diagnostic LED display—1986–90 Legend

small window without removing the ECU cover. Turn the ignition switch **ON**; the LED will display any stored codes by rhythmic flashing. Note that 1986–90 Legends have two LEDs on the controller; one is red and one is amber. The red one will flash the fault codes; the amber one is used during idle adjustment and is not related to this procedure. 1986–91 Integra use a single LED which is used to display codes only.

Codes 1–9 are indicated by a series of short flashes; two-digit codes use a number of long flashes for the first digit followed by the appropriate number of short flashes. For example, Code 43 would be indicated by 4 long flashes followed by 3 short flashes. Codes are separated by a longer pause between transmissions. The position of the codes during output can be helpful in diagnostic work. Multiple codes transmitted in isolated order indicate unique occurrences; a display of showing 1-1-1-pause-9-9-9 indicates two problems or problems occurring at different times. An alternating display, such as 1-9-1-9-1, indicates simultaneous occurrences of the faults.

When counting flashes to determine codes, a code not valid for the vehicle may be found. In this case, first recount the flashes to confirm an accurate count. If necessary, turn the ignition switch **OFF**, then recycle the system and begin the count again. If the Code is not valid for the vehicle, the ECU must be replaced.

1992–95 INTEGRA, 1991–95 LEGEND, NSX, VIGOR AND 1995 2.5TL

When a fault is noted, the ECU stores an identifying code and illuminates the CHECK ENGINE light. The code will remain in memory until cleared; the dashboard warning lamp may not illuminate during the next ignition cycle if the fault is no longer present. Not all faults noted by the ECU will trigger the dashboard warning lamp although the fault code will be set in memory. For this reason, troubleshooting should be based on the presence of stored codes, not the illumination of the warning lamp while the car is operating.

Beginning in 1991 on the Legend and in 1992 on Integra, codes are read thorough the use of the CHECK ENGINE light or more commonly know today as the Malfunction Indicator Lamp (MIL). NSX and Vigor are read in the same manner. The 1995 Legend equipped with the 2.7L V6 engine and the 2.5TL utilize OBD II trouble codes. These codes may be read either through the CHECK ENGINE light or by using a special OBD II scan tool.

Additionally, all models are equipped with a service connector in side the cabin of the vehicle. If the service connector is jumped, the CHECK ENGINE lamp will display the stored codes in the same fashion.

The 2-pin service connector is located under the extreme right dashboard on Integra, Legend, 2.5TL and NSX; on Vigor models, it is found behind the right side of the center console well under the dashboard.

Codes 1–9 are indicated by a series of short flashes; two-digit codes use a number of long flashes for the first digit followed by the appropriate number of short flashes. For example, Code 43 would

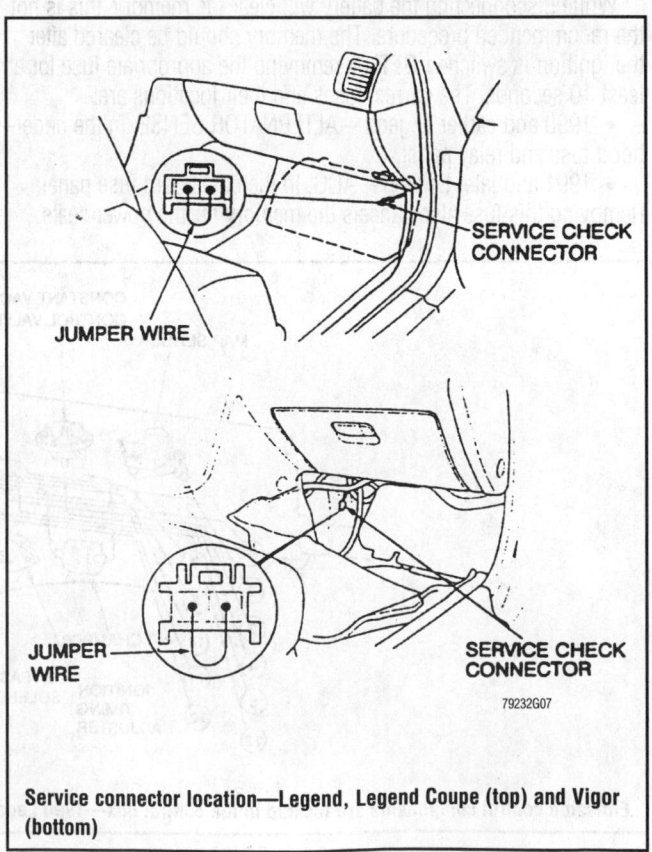

Service connector location—Legend, Legend Coupe (top) and Vigor (bottom)

be indicated by 4 long flashes followed by 3 short flashes. Codes are separated by a longer pause between transmissions. The position of the codes during output can be helpful in diagnostic work. Multiple codes transmitted in isolated order indicate unique occurrences; a display of showing 1-1-1-pause-9-9-9 indicates two problems or problems occurring at different times. An alternating display, such as 1-9-1-9-1, indicates simultaneous occurrences of the faults.

When counting flashes to determine codes, a code not valid for the vehicle may be found. In this case, first recount the flashes to confirm an accurate count. If necessary, turn the ignition switch **OFF**, then recycle the system and begin the count again. If the Code is not valid for the vehicle, the ECU must be replaced.

➡**On vehicles with automatic transaxles, the S, D or D4 lamp may flash with the CHECK ENGINE lamp if certain codes are stored. For Legend and NSX, this may occur with Codes 6, 7 or 17. On Vigor and Integra it may occur with codes 6, 7 or 13. In addition, the TCS lamp on NSX may flash with codes 3, 5, 6,13,15,16,17, 35 or 36. In all cases, proceed with the diagnosis based on the engine code shown. After repairs, recheck the lamp. If the additional warning lamp is still lit, proceed with diagnosis for that system.**

Clearing Codes

1986–95 VEHICLES

Stored codes are removed from memory by removing power to the ECU. Disconnecting the power may also clear the memories used for other solid-state equipment such as the clock and radio. For this reason, always make note of the radio presets before clearing the system. Additionally, some radios contain anti-theft programming; obtain the owner's code number before clearing the codes.

While disconnecting the battery will clear the memory, this is not the recommended procedure. The memory should be cleared after the ignition is switched **OFF** by removing the appropriate fuse for at least 10 seconds. The correct fuses and their locations are:
- 1990 and earlier Legend—ALTERNATOR SENSE, in the underhood fuse and relay panel.
- 1991 and later Legend—ACG, in the dashboard fuse panel. Removing this fuse also cancels the memory for the power seats.

- Integra, U.S. vehicles—BACK UP, in the underhood fuse and relay panel.
- Integra, Canadian vehicles—HAZARDIBACK UP in the underhood fuse and relay panel.
- Vigor—BACK UP, located in the underhood fuse and relay panel. Removing this fuse will cancel memories for the clock and radio.
- 1991–94 NSX—CLOCK, located in the main fuse and relay panel in the front luggage compartment, right side. Removing this fuse will cancel memories for the clock and radio.
- 1995 NSX—CLOCK, located in the main fuse and relay panel in the front luggage compartment, right side. Removing this fuse will cancel memories for the clock and radio. Codes may also be cleared using the OBD II scan tool or an equivalent tool, using the tool manufacturer's directions.
- 1995 2.5TL—BACK UP, located in the underhood fuse and relay panel. Removing this fuse will cancel memories for the clock and radio. Codes may also be cleared using the OBD II scan tool or an equivalent tool, using the tool manufacturer's directions.

DIAGNOSTIC TROUBLE CODES

1986–93 Integra 1.6L (1590cc)

Code 0 Electronic Control Unit
Code 1 Oxygen Content
Code 2 Replace Engine Control Unit with known-good unit
Code 3 Manifold Absolute Pressure
Code 4 Replace Engine Control Unit with known-good unit
Code 5 Manifold Absolute Pressure
Code 6 Coolant Temperature
Code 7 Throttle Angle
Code 8 Crank Angle (TDC)
Code 9 Crank Angle (Cyl)
Code 10 Intake Air Temperature
Code 12 EGR system (if equipped)
Code 13 Atmospheric Pressure
Code 14 Electronic Air Control
Code 15 Ignition output signal
Code 16 Fuel injector
Code 17 Vehicle speed sensor

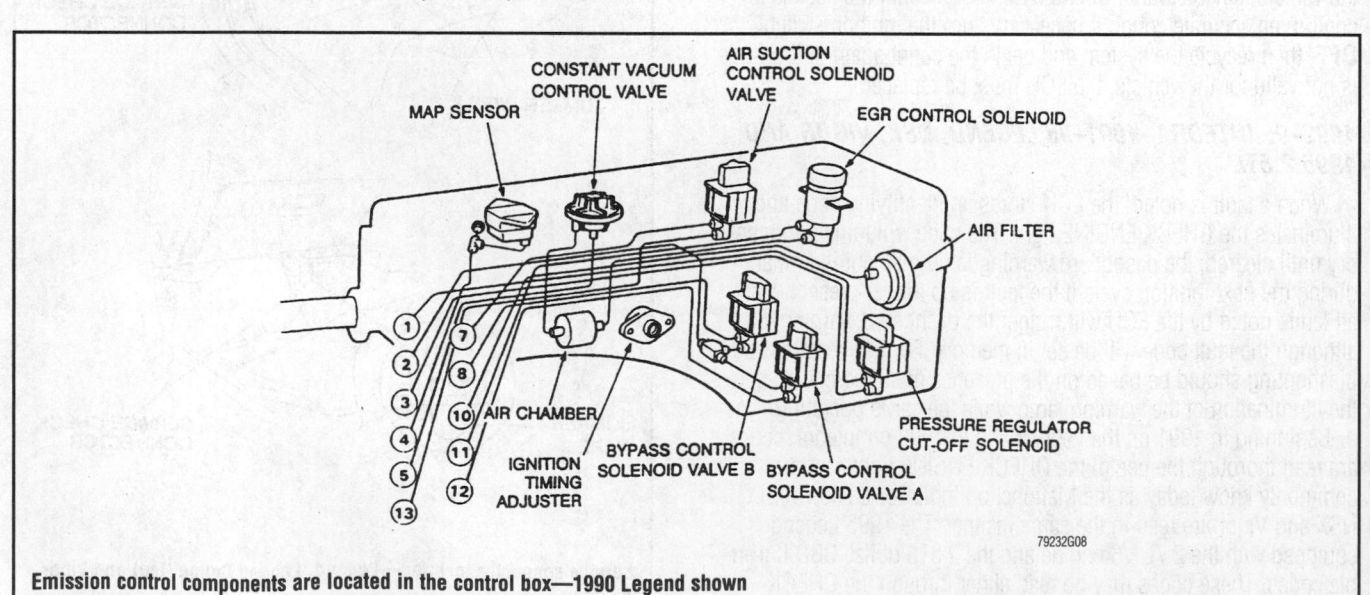

Emission control components are located in the control box—1990 Legend shown

Code 19 Lock up solenoid valve
Code 20 Electric load (to 1989)
Code 43 Fuel supply system

1986–87 Legend 2.5L (2494cc) C25A1

Code 0 Electronic Control Unit
Code 1 Oxygen Content
Code 2 Replace Engine Control Unit with known-good unit
Code 3 Manifold Absolute Pressure
Code 4 Replace Engine Control Unit with known-good unit
Code 5 Manifold Absolute Pressure
Code 6 Coolant Temperature
Code 7 Throttle Angle
Code 8 Crank Angle (TDC)
Code 9 Crank Angle (Cyl)
Code 10 Intake Air Temperature
Code 12 EGR system (if equipped)
Code 13 Atmospheric Pressure
Code 14 Electronic Air Control
Code 15 Ignition output signal
Code 16 Fuel injector
Code 17 Vehicle speed sensor
Code 20 Electric load (to 1989)
Code 43 Fuel supply system

1987–90 Legend 2.7L (2675cc) C27A1

Code 0 Electronic Control Unit (ECU)
Code 1 Front Oxygen Content
Code 2 Rear Oxygen Content
Code 3 Manifold Absolute Pressure (MAP)
Code 4 Crank angle
Code 5 Manifold Absolute Pressure (MAP)
Code 6 Coolant Temperature
Code 7 Throttle angle
Code 8 Crank Angle-Top Dead Center
Code 9 Crank Angle-Number 1 Cylinder
Code 10 Intake Air temperature
Code 12 Exhaust Gas Recirculation system
Code 13 Atmospheric pressure
Code 14 Electronic Idle Control
Code 15 Ignition output signal
Code 17 Vehicle speed pulser
Code 18 Ignition timing adjustment
Code 30 A/T FI Signal A (if equipped)
Code 31 A/T FI Signal B (if equipped)

1991–93 Integra 1.8L (1834cc) B18A1, 1992–93 Integra 1.7L (1678cc) B17A1, 1994–95 Integra 1.8L (1834cc) B18B1 and 1994–95 Integra 1.8L (1797cc) B18C1

Code 0 Electronic Control Unit
Code 1 Oxygen Content
Code 3 Manifold Absolute Pressure
Code 4 Crank Angle sensor
Code 5 Manifold Absolute Pressure
Code 6 Coolant Temperature
Code 7 Throttle Angle
Code 8 TDC Position
Code 9 No. 1 Cylinder Position
Code 10 Intake Air Temperature
Code 12 EGR system

Code 13 Atmospheric Pressure
Code 14 Electronic Air Control
Code 15 Ignition output signal
Code 16 Fuel injector
Code 17 Vehicle speed sensor
Code 20 Electric Load Detector
Code 21 VTEC Solenoid Valve (1.8L GS-R)
Code 22 VTEC Oil Pressure Switch (1.8L GS-R)
Code 30 TCM Signal A
Code 31 TCM Signal B
Code 41 H025 Heater
Code 43 Fuel supply system

1991–95 Legend 3.2L (3206cc) C32A1

Code 0 Electronic Control Unit (ECU)
Code 1 Left Oxygen Sensor
Code 2 Right Oxygen Sensor
Code 3 Manifold Absolute Pressure (MAP)
Code 4 Crank angle 1
Code 5 Manifold Absolute Pressure (MAP)
Code 6 Coolant Temperature
Code 7 Throttle angle
Code 9 Crank Angle-Number 1 Cylinder
Code 10 Intake Air temperature
Code 12 Exhaust Gas Recirculation (EGR) system
Code 13 Atmospheric pressure
Code 14 Electronic Air Control (EACV)
Code 15 Ignition output signal
Code 17 Vehicle speed pulser
Code 18 Ignition timing adjustment
Code 23 Left Knock Sensor
Code 30 A/T FI Signal A
Code 35 Traction Control System Circuit
Code 36 Traction Control System Circuit
Code 41 Left Oxygen Sensor Heater
Code 42 Right Oxygen Sensor Heater
Code 43 Left Fuel Supply System
Code 44 Right Fuel Supply System
Code 45 Left Fuel Supply Metering
Code 46 Right Fuel Supply Metering
Code 53 Right Knock Sensor
Code 54 Crank angle 2
Code 59 No. 1 Cylinder Position 2 (Cylinder Sensor)

1991–95 NSX 3.0L (2977cc) C30A1

Code 0 CU
Code 1 Front Oxygen Sensor
Code 2 Rear Oxygen Sensor
Code 3 Manifold Absolute Pressure (MAP)
Code 4 Crank angle A
Code 5 Manifold Absolute Pressure (MAP)
Code 6 Coolant Temperature
Code 7 Throttle angle
Code 9 Crank Angle-Number 1 Cylinder/Position A
Code 10 Intake Air temperature
Code 12 Exhaust Gas Recirculation (EGR) system
Code 13 Atmospheric pressure
Code 14 Electronic Air Control (EACV)
Code 15 Ignition output signal
Code 16 Fuel Injector

Code 17 Vehicle speed pulser
Code 18 Ignition timing adjustment
Code 22 VTEC System; front, bank 2
Code 23 Front Knock Sensor
Code 30 A/T FI Signal A
Code 31 NT FI Signal B
Code 35 TC STB signal
Code 36 TCFC signal
Code 37 Accelerator Position; Sensors 1, 2 or 1 and 2 circuits
Code 40 Throttle Position or Throttle Valve Control Motor Circuits 1 or 2
Code 41 Front Oxygen Sensor Heater; (circuit malfunction; bank 2 sensor 1)
Code 42 Rear Primary Heated Oxygen Sensor Heater (circuit malfunction)
Code 43 Front Fuel Supply System
Code 44 Rear Fuel Supply System
Code 45 Front Fuel Supply Metering; front bank 2
Code 46 Rear Fuel Supply Metering; rear bank 1
Code 47 Fuel Pump
Code 51 Rear Spool Solenoid Valve
Code 52 VTEC System; rear, bank 1
Code 53 Rear Knock Sensor
Code 54 Crank Angle B
Code 59 No. 1 Cylinder Position B (Cylinder Sensor)
Code 61 Front Heated Oxygen Sensor (slow response; bank 2 sensor 1)
Code 62 Rear Primary Heated Oxygen Sensor (slow response; bank 1 sensor 1)
Code 63 Front Secondary Oxygen Sensor (slow response or circuit voltage high or low)
Code 65 Front Secondary Heated Oxygen Sensor (circuit malfunction; bank 2 sensor 2)
Code 64 Rear Secondary Oxygen Sensor (slow response or circuit voltage high or low)
Code 66 Rear Secondary Heated Oxygen Sensor (circuit malfunction; bank 1 sensor 2)
Code 67 Front Catalytic Converter System
Code 68 Rear Catalytic Converter System
Code 80 Exhaust Gas Recirculation (EGR) system
Code 86 Coolant temperature
Code 70 Automatic Transaxle; the D indicator light and MIL may come on simultaneously.
Code 71 Misfire detected; cylinder No. 1 or random misfire
Code 72 Misfire detected; cylinder No. 2 or random misfire
Code 73 Misfire detected; cylinder No. 3 or random misfire
Code 74 Misfire detected; cylinder No. 4 or random misfire
Code 75 Misfire detected; cylinder No. 5 or random misfire
Code 76 Misfire detected; cylinder No. 6 or random misfire
Code 79 Spark Plug Voltage Detection; circuit malfunction; (Front Bank (Bank 2) or (Rear Bank (Bank 1)
Code 79 Spark Plug Voltage Detection; circuit malfunction; (Front Bank (Bank 2) or (Rear Bank (Bank 1)
Code 79 Spark Plug Voltage Detection Module; reset circuit malfunction; (Front Bank (Bank 2)) or (Rear Bank (Bank 1)
Code 92 Evaporative Emission Control System

1992–94 Vigor 2.5L (G25A1)

Code 0 Electronic Control Unit
Code 1 HO25 circuit
Code 3 Manifold Absolute Pressure

Code 4 Crank Angle Sensor
Code 5 Manifold Absolute Pressure
Code 6 Coolant Temperature
Code 7 Throttle Angle
Code 8 TDC and or Crankshaft Position sensors
Code 9 No. 1 Cylinder Position
Code 10 Intake Air Temperature
Code 12 EGR system
Code 13 Atmospheric Pressure
Code 14 Electronic Air Control
Code 15 Ignition output signal
Code 16 Fuel injector
Code 17 Vehicle speed sensor
Code 18 Ignition Timing Adjuster
Code 20 Electric Load Detector
Code 30 NT FI Signal
Code 31 NT FI Signal
Code 41 HO25 Heater
Code 43 Fuel supply system
Code 45 Fuel Supply Metering
Code 50 Mass Air Flow (MAF) circuit—2.5TL
Code 53 Rear Knock Sensor
Code 54 Crankshaft Speed Fluctuation sensor—2.5TL
Code 61 HO25 sensor heater—2.5TL
Code 65 Secondary HO25 sensor—2.5TL
Code 67 Catalytic Converter System—2.5TL
Code 70 Automatic transaxle or NT FI Data line—2.5TL
Code 71 Misfire detected; cylinder No. 1 or random misfire
Code 72 Misfire detected; cylinder No. 2 or random misfire
Code 73 Misfire detected; cylinder No. 3 or random misfire
Code 74 Misfire detected; cylinder No. 4 or random misfire
Code 75 Misfire detected; cylinder No. 5 or random misfire
Code 76 Random misfire detected—2.5TL
Code 80 EGR system—2.5TL
Code 86 Coolant Temperature circuit—2.5TL
Code 92 Evaporative Emission Control System—2.5TL

Audi

GENERAL INFORMATION

Motronic and Multi-Point Injection (MPI) Systems

The Motronic and MPI fuel injection systems are similar and share most components and modes of operation. Audi uses Motronic to describe the fuel injection system on the V8 Quattro, S4, 200 Quattro and the 200 Quattro Wagon. Audi uses MPI to describe the systems used on the 90 Quattro, Coupe Quattro and the 2.8L V6 equipped 100 series vehicles.

The Motronic and MPI fuel injection systems are self-learning adaptive systems. They continuously learn using a sophisticated feedback system that readjusts various control settings. These new values are then stored in the ECU memory. The adaptive capability allows the systems to compensate for changes in the engine's operating conditions, such as intake leaks, altitude changes or any other system malfunction. If the battery or ECU is disconnected, the vehicle must be driven so ECU can 're-learn' its operating conditions.

Operation of the fuel injection system is based on the information received by the various sensors. This keeps the system constantly updated on engine speed, coolant temperature, throttle position and the intake air volume.

On the V8, the power supply to ECU is at terminal 18, through a 5 amp fuse (S27) in the main fuse/relay panel. Power from fuse S27 energizes the power supply relay in the ECU when engine speed reaches 25 rpm. The main fuse/relay panel is located behind the side kick panel cover on the passenger's side.

A Hall effect signal from the right distributor helps the ECU establish a reference point to start the fuel injection process. After the engine is running, the reference sender and speed sensor provide the necessary information to the ECU for ignition and fuel injection.

The ECU has a self-diagnostic feature. Any faults detected by the sensors are sent to the ECU and are recorded in the ECU memory. Fault codes can be displayed using LED tester US 1115 and a jumper wire.

Continuous Injection System (CIS-E)

The CIS-E system incorporates 2 control units. An Ignition Control Unit (ICU) or Knock Sensor Control Unit (KSCU, on SOQOS only) and a Fuel Injection Control Unit (FICU).

The CIS-E system also has self-diagnosis and troubleshooting capabilities. Input and output signals from various sensors, switches and signaling devices are constantly monitored for faults. These faults are stored in the control unit memory. Faults can be displayed by a flashing 4 digit code sequence from an LED light located on the instrument panel.

CIS-Motronic Fuel Injection System

The CIS Motronic system used on Audi 80 and 90 models use a single Electronic Control Unit (ECU), located behind the NC evaporator assembly. The ECU controls the fuel delivery, ignition system and operation of the emission control components. The CIS Motronic system also incorporates self-diagnostic capabilities. The CIS-Motronic system consists of the following components:

- Ignition coil with power stage
- Differential pressure regulator
- Cold start valve
- Idle stabilizer valve
- Ignition distributor with Hall sender
- Knock sensor
- Coolant temperature sensor
- Idle/Full throttle switches
- Air sensor potentiometer
- Oxygen sensor
- Carbon canister frequency valve
- Carbon canister ON/OFF valve
- CIS Motronic control unit

The ECU receives signals from various sensors, switches and signaling components which are constantly monitored for faults. These faults are stored in the ECU memory. Faults can be displayed by using a suitable test light connected between the battery positive terminal and the test lead, located next to the fuel distributor in the engine compartment. Characteristics of the CIS-Motronic system are as follows:

- Fuel injection control
- Oxygen sensor regulation with adaptive learning capability
- MAP type ignition control with individual cylinder knock regulation
- Idle speed control
- Fuel tank ventilation control
- Permanent fault memory for self-diagnosis

SELF-DIAGNOSTICS

Service Precautions

- Do not disconnect the battery or power to the control module before reading the fault codes. On the Motronic SMPI and Audi SMPI systems, fault code memory is erased when power is interrupted.
- Make sure the ignition switch is **OFF** before disconnecting any wiring.
- Before removing or installing a control module, disconnect the negative battery cable. The unit receives power through the main connector at all times and will be permanently damaged if improperly powered up or down.
- Keep all parts and harnesses dry during service. Protect the control module and all solid-state components from rough handling or extremes of temperature.
- Do not apply voltage to engine control module to simulate output signals.
- When coil wire, terminal 4, is disconnected from distributor, always ground using a jumper wire.
- Do not try to start the engine with the fuel injectors removed.
- In emergency starting situations, use a fast charge for cranking up to 15 seconds only and not more than 16.5 volts; allow at least 1 minute between attempts.

Generating Codes

Before attempting to read trouble codes, the vehicle must be driven for at least 5 minutes to set codes in the computer's memory. This procedure is referred to as 'generating codes'.

CIS SYSTEMS

1. The engine must be running to generate fault codes. Read all the Steps in this procedure before starting.

2. On 1985–88 models, turn the ignition switch **ON** without starting the engine to make sure the engine warning light works, if equipped. If it does not light with the ignition **ON**, engine not running, but does light when attempting to retrieve fault codes, either the wiring between the control units is faulty or the ignition control unit is faulty.

3. On 1989–92 California models only, turn the ignition switch **ON** without starting the engine to make sure the engine warning light On Board Diagnostics (OBD) works. If it does not light with the ignition **ON**, engine not running, then turn ignition **OFF** and bridge terminals of diagnostic connectors using adapter (cable 357 971 51 4E) or an equivalent tool. Now turn ignition **ON**, but do not start the engine. The engine warning light OBD will light up, if it does not light the wiring between the control modules is faulty or the ignition control module is faulty. Consult a wiring diagram and complete the necessary repairs before continuing.

4. Fuel pump relay and fuses 13, 19, 24 and 28 on 1985–88 models or 13, 19, 21, 24, 27 and 28 on 1989–92 models, must be good and all ground connections in the engine compartment must be good. Also, make sure the air conditioning is OFF.

5. To generate fault codes, the vehicle must be driven with air conditioner OFF, for at least 5 minutes at normal operating temperature. The engine must be kept above 3000 RPM for the majority of the test drive with at least one full throttle application. On turbocharged engines, full boost should be reached during the full throttle acceleration. After the test drive, allow the engine to idle for at least 2 minutes before retrieving codes.

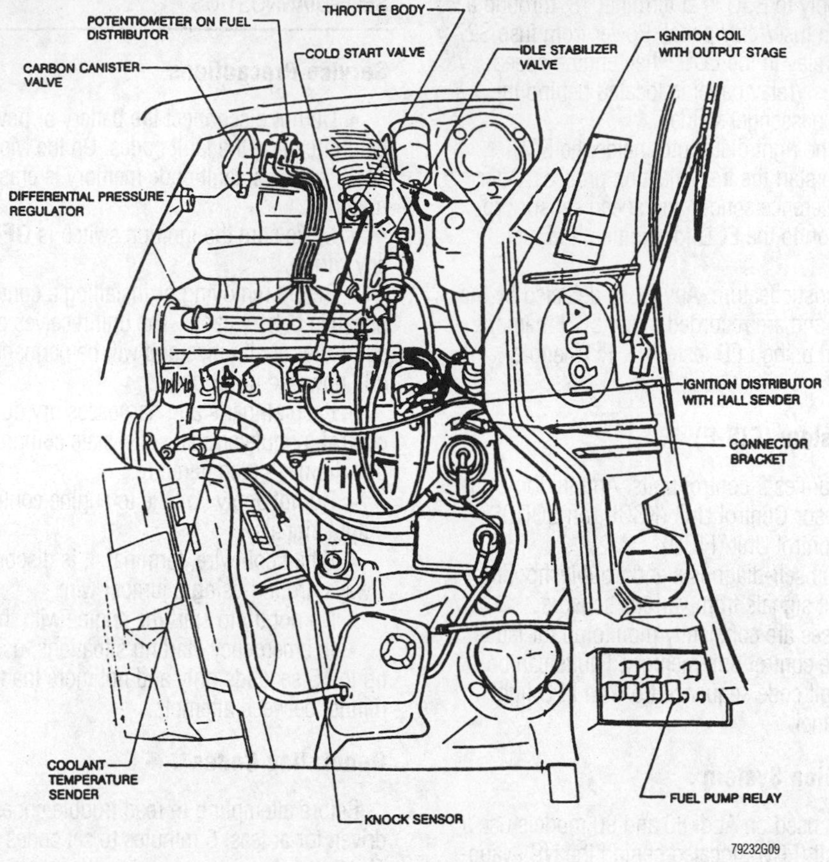

POTENTIOMETER ON FUEL DISTRIBUTOR — THROTTLE BODY

CARBON CANISTER VALVE — COLD START VALVE — IDLE STABILIZER VALVE — IGNITION COIL WITH OUTPUT STAGE

DIFFERENTIAL PRESSURE REGULATOR

IGNITION DISTRIBUTOR WITH HALL SENDER

CONNECTOR BRACKET

COOLANT TEMPERATURE SENDER

KNOCK SENSOR — FUEL PUMP RELAY

79232G09

CIS-E components location —5-cylinder engine shown, 4-cylinder engine is similar

6. Do not turn the ignition switch **OFF** or the temporary memory will be erased. If the engine stalls, do not restart it. The codes will still be in memory.

7. If the engine will not run, operate the starter for at least 6 seconds and leave the ignition switch **ON**.

Reading Codes

➡Vehicles that do not have a CHECK ENGINE light or Malfunction Indicator Lamp (MIL) codes can only be accessed by the use of special equipment. US 1115 LED tester, VAG 1551 diagnostic tester or equivalent special testers can only be used to retrieve diagnostic codes from these vehicles. When using special diagnostic equipment, always observe the tool manufacturer's instructions.

WITH FLASH TESTER

1985–94 Vehicles

1. On California models equipped with a engine warning light, an LED tester is not required. On models not equipped with a engine warning light, connect the US 1115 LED tester or equivalent to the test connectors under the left-hand side of dash and above the pedals.

2. On 1985–88 models with a engine warning light, locate the fuel pump relay on the main fuse/relay panel. Insert a spare fuse into the terminals on top of the relay for at least 4 seconds, then remove the fuse to activate the diagnostic program. The engine warning light on the instrument panel or the LED tester will begin to flash the first code. It will continue to flash this code until the fuse is installed.

3. On all models, codes should be retrieved with the engine running at idle. If engine will not start, operate the starter for approximately 6 seconds and leave the ignition switch **ON**. On CIS systems the engine should be left running after the generating codes procedure.

4. On 1989–94 models, so equipped, locate the test connectors above the pedals and connect the tester. To connect the LED tester, connect the positive terminal of the LED tester to the positive terminal in connector A. Connect the negative terminal of the LED tester

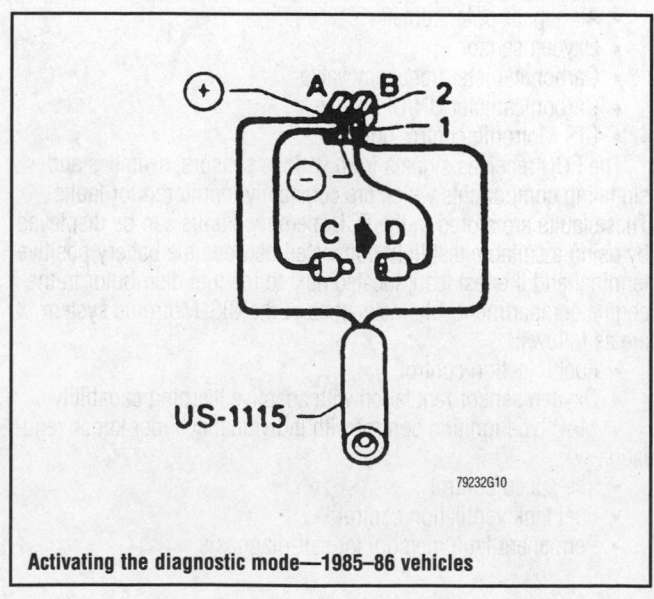

79232G10

Activating the diagnostic mode—1985–86 vehicles

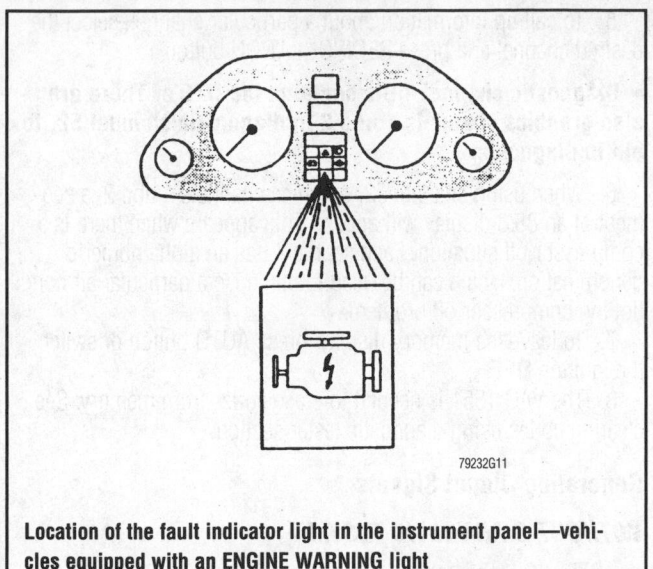

Location of the fault indicator light in the instrument panel—vehicles equipped with an ENGINE WARNING light

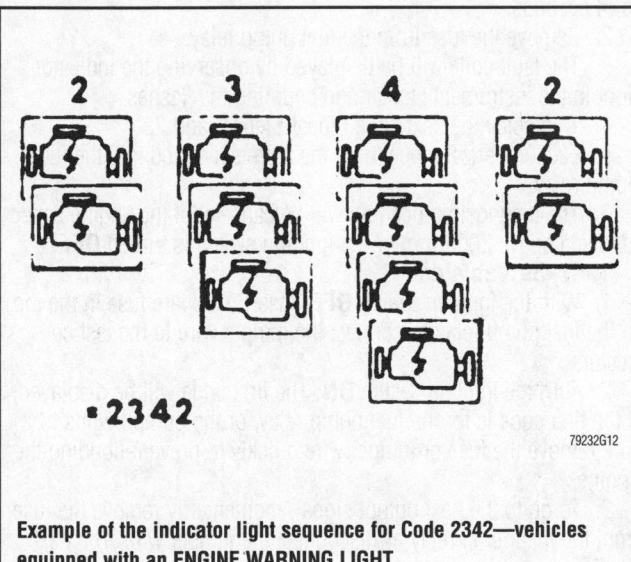

Example of the indicator light sequence for Code 2342—vehicles equipped with an ENGINE WARNING LIGHT

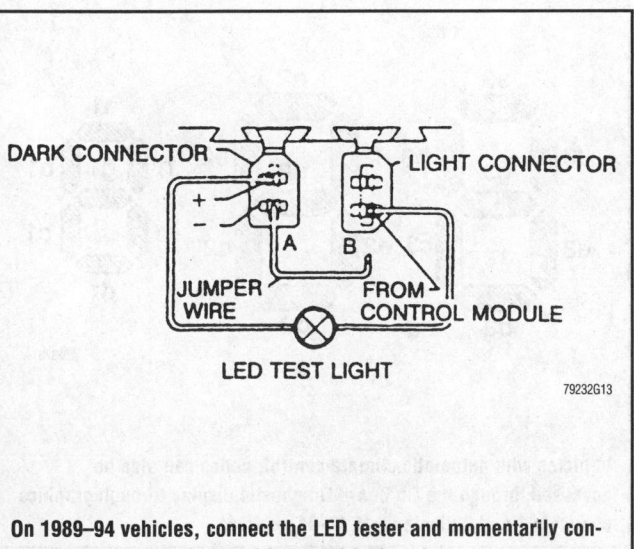

On 1989–94 vehicles, connect the LED tester and momentarily connect jumper terminals A and B to read fault codes

to the only terminal in connector B. Connect one end of a jumper wire to the negative terminal in connector A, touch the other end of the jumper wire to the terminal in connector B for at least 4 seconds.

5. Fault codes will now be displayed as flashing by the tester or by the engine warning light on California models. Touch the jumper wire to the terminal in connector B for another 4 seconds to advance to the next code. Do not leave jumper wire connected for ten seconds or memory will be erased. Engine idle speed may increase slightly when reading injection control module codes.

6. All flash codes are 4 digits, with about 2.5 seconds between digits. Codes are displayed in order of importance, usually beginning with ignition system codes. Count the flashes and write down the code, then proceed onto the next code. Read all codes, before starting any repairs. If the first code is 4444 or 0000 there are no faults present. 0000 is represented by the light ON for 2.5 seconds with 2.5 second intervals between. When all fault codes have been reported (0000 or 4444 displayed), turn the ignition **OFF**.

➡**On CIS systems codes are erased when the ignition is turned OFF.**

WITH DIAGNOSTIC TESTER

1987–89 Vehicles

1. On all 1987–88 models, the VAG 1551 or equivalent diagnostic tester can be connected to the terminals on top of the fuel pump relay. For power to the tester, a separate power supply wire must be connected to the positive battery terminal. On 1989 models, the tester can be connected to the diagnostic terminals above the pedals. The terminals are shaped so they cannot be connected incorrectly. Power for the tester is supplied through fuse 21.

2. With the tester connected, turn the ignition switch **ON**, but do not start the engine or the codes will be erased.

3. Select menu option 2, Blink Code Output. Press and release, then press and hold the run (arrow) key until the program starts, then release the key. An asterisk (*) will appear and flash the codes, which the tester will count and report on the screen as numbers. If Code 4444 is displayed, no faults are found in memory.

4. Press the run key to advance to the next code. Read all the fault codes before starting repairs.

5. All engine system fault codes will be displayed first. If there are other control units on the vehicle, press the run key again to access those codes.

6. When the End of the Report Code 0000 is displayed and there are no other control units on the vehicle, pressing the run key again may return to the main menu or it may erase the codes. To stop the program without erasing the codes, turn the ignition switch **OFF** and press the clear button once.

1990–91 Vehicles

1. Connect the VAG 1551 or equivalent diagnostic tester to the diagnostic terminals above the pedals or in the passenger side foot well. The terminals are shaped so they cannot be connected incorrectly. Power for the tester is supplied through fuse 21 or 27.

2. On all models, codes should be retrieved with the engine running at idle. If the engine will not start, operate the starter for at least 5 seconds and leave the ignition switch **ON**. On CIS systems, the engine should be left running after the procedure for generating codes.

3. Operate the tester to select menu option 2, Blink Code Output or Fault Memory Recall. An asterisk (*) will appear and flash the

codes, which the tester will count and report on the screen as numbers. If Code 4444 is displayed, no faults are found in memory.

4. If the engine is not running, some codes will be displayed. These codes can be ignored if the engine has been intentionally stalled but should be investigated if the engine will not start.

5. Press the run key to advance to the next code. Read all the fault codes before starting repairs.

6. When the End of the Report Code 0000 is displayed and there are no other control units on the vehicle, pressing the run key again may return to the main menu or it may erase the codes. To stop the program without erasing the codes, turn the ignition switch **OFF** and press the clear button once.

1992–94 Vehicles with 2.2L and 2.8L Engines

Diagnostic trouble codes (DTCs) may be accessed using the VAG 1551 scan tool.

1. Turn the ignition switch to the **OFF** position.

2. Connect the VAG1551/1 diagnostic lead to the data link connectors (DLCs) in the underhood relay box.

➡**Observe the connector shape when connecting diagnostic leads.**

3. Connect the black lead of the VAG1551/1 diagnostic lead to the DLC 1, and the white lead to the DLC 4.

4. Connect the VAG15S1/1 diagnostic lead to the VAG 1551 scan tool. The scan tool should read—VAG self diagnosis—1 Rapid Data Transmission or 2 Flash Code Output.

5. Additional operating instructions may be accessed by pressing the help key on the VAG 1551 scan tool. press the arrow key to continue fault tracing.

1992–94 Vehicles with 4.2L Engine

Diagnostic trouble codes (DTCs) may be accessed using the VAG 1551 scan tool.

1. Turn the ignition switch to the **OFF** position.

2. Connect the VAG1551/1 diagnostic lead to the data link connectors (DLCs) under the passenger side footwell carpet.

➡**Observe the connector shape when connecting diagnostic leads.**

3. Connect the black lead of the VAG1551/1 diagnostic lead to the DLC 1, and the white lead to the DLC 2 and the blue lead to the DLC 4.

4. Connect the VAG1551/1 diagnostic lead to the VAG 1551 scan tool. The scan tool should read—VAG self diagnosis—Rapid Data Transmission or 2 Flash Code Output.

5. Additional operating instructions may be accessed by pressing the help key on the VAG 1551 scan tool. Press the arrow key to continue fault tracing.

WITH ON BOARD DIAGNOSTIC (OBD) DISPLAY

1992–94 Vehicles

The air conditioning system On-Board Diagnostic (OBD) can be accessed without the need of a scan tool. The air conditioning control head contains a 61 channel OBD display.

1. To start the display, turn the ignition **ON** or start the engine.

2. Press and hold down RECIRCULATION button 1 and press and hold down upper AIR DISTRIBUTION button 2.

3. Release both buttons and 'Qic' will be displayed, Olc indicates channel 1, 02c indicates channel 2, etc.

4. To change to a different channel, press the temperature + button to go to the next higher channel or the temperature -button to go to the next lower channel.

5. To call up information about a particular channel, select the desired channel and press RECIRCULATION button 1.

➡**Diagnostic channel 1 Olc contains the DTC's. There are also graphics channels 1 and 2 in diagnostic channel 52, to aid in diagnosis.**

6. When using channel 52, graphics channels 1 and 2, a segment of an 88.8 display will appear. This appears when there is a compressor off situation. Each segment has an alpha numeric denomination, which can be used to diagnose a particular air conditioning compressor off problem.

7. To leave the memory display, press AUTO button or switch the ignition **OFF**.

8. The VAG 1551 is needed to erase codes from memory. See clearing codes using diagnostic tester section.

Generating Output Signals

WITHOUT DIAGNOSTIC TESTER

1985–86 Vehicles

1. Insert the fuse in the opening on top of the fuel pump relay for 4 seconds.

2. Remove the fuse from the fuel pump relay.

3. The fault code will be displayed by observing the indicator light in the instrument cluster and counting the flashes.

4. To display the next code repeat Steps 1 and 2.

5. Each code will repeat until the fuse is inserted into the fuel pump relay.

6. The diagnosis procedure will be canceled if the engine speed is raised above 2000 rpm or the ignition switch is turned **OFF**.

1987–89 Vehicles

1. With the ignition switch **OFF**, insert the spare fuse in the top of the fuel pump relay or connect the jumper wire to the test connectors.

2. Turn the ignition switch **ON**. The first code will be displayed. If the first code is for the fuel pump relay, or the pump begins to run, remove the fuse or jumper wire quickly to prevent flooding the engine.

3. To go to the next output signal, momentarily remove the fuse from the fuel pump relay or disconnect the jumper wire. The next

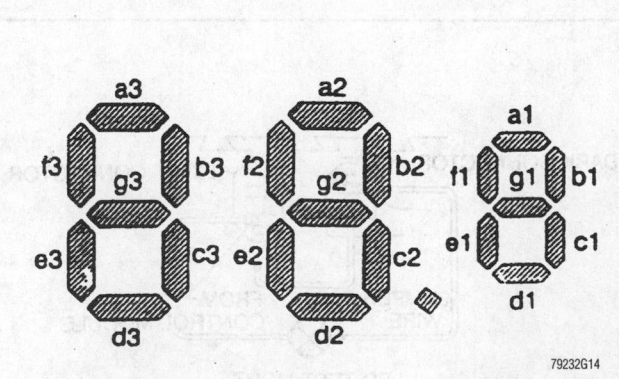

79232G14

Vehicles with automatic climate control, codes can also be accessed through the On Board Diagnostic display through graphics channels 1 and 2 display—1992–94 vehicles

item on the output code list will be activated when the full throttle switch is closed. Be sure to use the correct output code list, the sequence is not the same on all engines.

4. When the full throttle switch is closed, each solenoid or frequency valve can be checked by listening or touching the valve to detect operation. Cold start valves are operated for only 10 seconds.

5. When the last output test has been completed, Code 0000 will be displayed. At this time, the fault code memory can be erased or the test can be repeated by turning the ignition switch **OFF** and **ON** again.

6. If the starter is operated at any point in the output test, the control unit will switch to reporting input signal codes.

1990–94 Vehicles

1. With the ignition switch **OFF**, connect the LED test light and the jumper wire to the test connectors.

2. Turn the ignition switch **ON**. The first code will be displayed. If the first code is for the fuel pump relay, or the pump begins to run, remove the jumper wire quickly to prevent flooding the engine.

3. To go to the next output signal, momentarily disconnect the jumper wire. The next item on the output code list will be activated when the full throttle switch is closed. Be sure to use the correct output code list, the sequence is not the same on all engines.

4. When the full throttle switch is closed, each solenoid or frequency valve can be checked by listening or touching the valve to detect operation. Cold start valves on CIS systems are operated for only 10 seconds.

5. When the last output test has been completed, Code 0000 will be displayed. At this time, the fault code memory can be erased.

6. If the starter is operated at any point in the output test, the control module will switch to reporting input signal codes.

WITH DIAGNOSTIC TESTER

1987–91 Vehicles

1. Follow the procedure for retrieving fault codes. After all codes have been reported, turn the ignition switch **OFF** and press the clear button.

2. Select the Blink Code Output program on the tester. Press and hold the run key until the Continuous Short Circuit message appears on the screen, then turn the ignition switch **ON** without starting the engine.

3. Press and release the run key. The first output code should appear on the display. If the first output signal is for the fuel pump relay, remove the fuse for the fuel pump quickly after that test to avoid flooding the engine.

4. The code for each item in the output signal test will appear in the order listed. Press the run key to change to the next item on the list.

5. Except for the cold start valve, the output signal will be activated, as long as the code appears on the screen. As each item is activated, touch each valve to physically check that it is vibrating or humming.

6. When the test is completed, turn the ignition switch **OFF** to stop the test. The test can be repeated by turning the ignition switch **ON**.

1992–94 Vehicles

1. Go to Step 3, if not using Rapid Data Transfer. Turn the ignition **ON**, but do not start the car.

➡ **The engine must not be running when output checks are being performed. The output check mode will not work if the car is running.**

2. After selecting mode 1, Rapid Data Transfer, and the Select Function display appears, enter 01 for engine electronics. Now the control module coding and engine identification numbers will appear. After the coding is deciphered and the information matches your engine, press the Run key to continue. If the Fault In Communication display appears, one of the four displays indicating an open/short wire will appear or possibly the control module may be defective. Pushing the Help button will give you a list of possible causes for this problem. This problem must be corrected before continuing.

3. When the Select Function message appears, select 03 for Output Check Diagnosis. Each output check being tested will be displayed on the VAG 1551. Press the Run key to advance the next output check. The output check displayed on the screen will be performed until the next output check is selected.

➡ **When testing the fuel pump, do not run the test too long or the engine could become flooded. When the output checks are finished the Select Function Menu will appear. The output tests can be run again by selecting function 03. Turn the ignition OFF for approximately 20 seconds, before selecting Output Check Diagnosis again.**

4. Follow the procedure for retrieving fault codes. After all codes have been reported, turn the ignition switch **OFF** and press the clear C button.

5. Select the Blink Code Output program on the tester, then turn the ignition **ON**, but do not start the engine. Output checks can only be performed with the engine NOT RUNNING. The tests will be stopped if the engine is started or a speed impulse is recognized.

6. Press and release the Run key. The first output code should appear on the display. If the first output signal is for the fuel pump relay, remove the fuse for the fuel pump quickly after that test to avoid flooding the engine.

➡ **During output checks diagnosis, the carbon canister solenoid valve, idle stabilizer valve and cold start valve are checked audibly or by touch. Avoid background noise while audibly checking these components.**

7. The code for each item in the output signal test will appear in the order listed for your particular engine. Press the Run key to change to the next item on the list.

8. Except for the cold start valve, the output signal will be activated, as long as the code appears on the screen. As each item is activated, touch each valve to physically check that it is vibrating or humming.

9. When the test is completed, turn the ignition switch **OFF** to stop the test. The test can be repeated by turning the ignition switch **ON**.

Clearing Codes

WITHOUT DIAGNOSTIC TESTER

1985–88 Vehicles

1. On California vehicles, the output signals must be tested before codes can be erased. Leave the ignition switch **ON**.

2. After the last output signal code is displayed, install the fuse for at least 4 seconds and remove it again. The engine warning light should come **ON** for 2.5 seconds, then go OFF for 2.5 seconds, displaying Code 0000.

3. Install the fuse again for at least 10 seconds, then remove it. If the engine warning light stays ON, all codes have been erased.

4. On Federal vehicles, after activating the fault code memory, the codes will automatically be erased when the ignition switch is turned **OFF** or when the engine is started.

1989–94 Vehicles

1 The output signals must be tested before codes can be erased. Leave the ignition switch **ON**.

2. After the last output signal code is displayed, connect the test connector jumper wire for at least 4 seconds and remove it again. The engine warning light or flash tester light should come ON for 2.5 seconds, then go OFF for 2.5 seconds, displaying Code 0000.

3. Connect the jumper wire again for at least 10 seconds, then remove it. If the engine warning light or tester light stays ON, all codes have been erased.

WITH DIAGNOSTIC TESTER

1985–88 Vehicles

When all control unit memory codes have been retrieved and Code 0000 is displayed, press and hold the run key with the ignition switch **ON** to clear all codes.

1989–91 Vehicles—Except CIS-E III

When all control module memory codes have been retrieved and code 0000 or End of output is displayed, press the run key, now press 05 and the codes will be erased using mode 1 (Rapid Data Transfer). Mode 1 will display a message saying: Fault memory is erased! after erasing the codes. In mode 2 (Blink Code Output), press and hold the run key with the ignition **ON** and all codes will be cleared.

➡**This procedure erases all control module codes, make sure you have checked all control modules for DTC's before performing the erasing codes procedure.**

1992–94 Vehicles—Except CIS-E III

Diagnostic trouble codes (DTCs) may be erased after they are retrieved using the VAG 1551 scan tool.

1. Turn the ignition switch to the **OFF** position.

2. Connect the VAG 1551 scan tool as outlined in Generating Output Signals.

3. Press 01 on the VAG scan tool to select VAG address 'Engine Electronics'.

4. Press the arrow button until the display reads 'Select Function XX'.

5. Press the Q on the VAG scan tool and view all DTCs.

6. Press the 05 on the VAG scan tool to select 02—Cancel Fault Code Memory. Press Q to erase all DTCs.

7. Road test the vehicle and reactivate DTC memory to ensure all faults have been eliminated.

1989–92 Vehicles—CIS-E III

After activating the fault code memory, the codes will automatically be erased when the ignition switch is turned **OFF** or when the engine is started.

➡**CIS-E III California models have permanent memory and will retain the fault codes after the ignition is turned off. This memory can be erased by following the erasure method using the VAG 1551 diagnostic tester. Control module part numbers can be used to identify the control modules and determine whether the CIS-E III system has permanent or temporary memory.**

DIAGNOSTIC TROUBLE CODES

1985–89 Vehicles

2.1L (MC), 2.2L (MC), 2.0L (3A) AND 2.3L (NG, NF) ENGINES

Code 1111 Ignition control unit or fuel injection control unit
Code 1231 Transmission speed sensor
Code 2111 Engine speed sensor
Code 2112 Ignition reference sensor
Code 2113 Hall sensor
Code 2121 Idle switch
Code 2122 Engine speed/Hall sensor
Code 2123 Full throttle switch
Code 2132 No data being transmitted from fuel injection control unit to ignition control unit
Code 2141 Knock control 1, knock sensor 1 for cylinder 2; knock control 2, knock sensor 2 for cylinder 4
Code 2142 Knock sensor 1 on cylinder 2; knock sensor 2 on cylinder 4
Code 2143 Knock control 1, knock sensor 1 for cylinder 2; knock control 2, knock sensor 2 for cylinder 4
Code 2144 Knock sensor 1 on cylinder 2; knock sensor 2 on cylinder 4
Code 2212 Throttle valve potentiometer (position sensor)
Code 2221 Vacuum hose to pressure sensor in control unit
Code 2222 Pressure sensor in control unit
Code 2223 Altitude sensor
Code 2232 Air sensor potentiometer (position sensor)
Code 2233 Reference (supply) voltage
Code 2234 MPI control unit supply voltage
Code 2242 CO potentiometer
Code 2312 Engine coolant temperature (ECT) sensor
Code 2322 Intake air temperature (IAT) sensor
Code 2341 Oxygen sensor control unit is at its limit
Code 2342 Oxygen sensor (does not control)
Code 4431 Idle stabilizer valve
Code 4444 No faults stored in memory
Code 0000 End of diagnosis

1990–94 Vehicles

2.0L (3A), 2.3L (NG, NF, 7A), 2.2L (MC, 3B, AAN), 2.8L (AAH), 3.6L (PT) AND 4.2L (ABH) ENGINES

Code 00000 or 0000 No faults in memory (1992–93 all engines except 2.8L and 1992 2.3L)
Code 00000 or 0000 End of diagnosis (1990–91 all engines; 1992–94 2.8L and 1992 2.3L)
Code 00000 or 4444 No faults in memory (1990–91 all engines; 1992–94 2.8L and 1992 2.3L)
Code 00281 or 1231 Vehicle speed sender signal is missing
Code 00513 or 2111 Engine speed (RPM) sensor has no change in signal
Code 00513 or 2231 Air mass sensor has open/short in circuit
Code 00514 or 2112 Crankshaft position (CKP) sensor has no change in signal
Code 00515 or 2113 Hall sender has fault in basic setting or open/short in circuit

Code 2114 Hall sender is not on reference point or out of adjustment

Code 00516 or 2121 Idle switch (closed throttle position switch 4.2L ABH) has open/short in circuit

Code 00517 or 2123 Full throttle switch

Code 00518 or 2212 Throttle position (TP) sensor has open/short in circuit

Code 00519 or 2222 Manifold vacuum sensor signal is out of range

Code 00520 or 2232 Air mass sensor signal is missing/signal out of limit

Code 00521 or 2242 CO potentiometer position sensor (2.3L 7A 1990–91)

Code 00522 or 2312 Engine coolant temperature (ECT) sensor signal is out of range

Code 00523 or 2322 Intake air temperature (IAT) sensor has open/short in circuit

Code 00524 or 2142 Knock sensor (KS) 1 has no change in signal, possible open/short between KS and ECM

Code 00525 or 2342 Oxygen sensor signal is out of range

Code 00528 or 2223 Pressure sensor (altitude sensor 1990–91) has open/short in circuit

Code 00529 or 2122 Engine RPM signal missing (2.3L NG,NF)

Code 00531 or 2233 Air mass sensor reference voltage signal missing; voltage high (2.3L 7A 1990–91)

Code 00532 or 2234 Supply voltage signal is too high or low

Code 00533 or 2231 Idle speed regulation, the idle speed is too low or too high

Code 00535 or 2141 Knock sensor regulation has exceeded its maximum control limit (1992 2.3L NG)

Code 00536 or 2141 First & second knock regulation, the maximum control limits have been exceeded (4.2L ABH engine)

Code 00536 or 2143 Second knock control has exceeded its control limits (1992–93 2.8L AAH)

Code 00537 or 2341 Oxygen sensor signal is out of range

Code 2343 Air/fuel mixture rich (2.0L engine)

Code 2344 Air/fuel mixture lean (2.0L engine)

Code 00538 or 2241 Second knock control (1991 2.2L 3B)

Code 00540 or 2144 Knock sensor 2 has no change in signal, possible open/short between KS and ECM

Code 00543 or 2214 Engine speed signal is too high, the RPM exceeds maximum limit

Code 00544 or 2224 Wastegate frequency valve has exceeded maximum boost pressure (Manifold dump valve 1990–91 2.2L MC)

Code 00545 or 2314 Engine/Transmission electrical connection has ground between ECM and TCM

Code 00546 or 2132 Fuel injection/ignition control data link (2.3L NG, NF)

Code 00553 or 2324 Mass air flow (MAF) sensor signal

Code 00554 or 2331 Oxygen control for cylinders (4-6) exceeded control limits (1992–93 2.8L AAH)

Code 00555 or 2332 Oxygen sensor 2 (G108) signal is missing (1992–93 2.8L AAH)

Code 00560 or 2411 EGR system not working properly

Code 00560 or 2441 EGR system has false readings (1992–93 2.8L AAH, California)

Code 00561 or 2413 Fuel mixture too rich

Code 00575 or 222 Manifold pressure signal missing (1990–91 2.2L MC)

Code 00577 or 2141 Knock regulation cylinder 1 has exceeded control limit

Code 00578 or 2141 Knock exceeded control limit

Code 00579 or 2141 Knock exceeded control limit

Code 00580 or 2143 Knock exceeded control limit

Code 00581 or 2143 Knock regulation cylinder 5 has exceeded control limit

Code 00824 or 3424 Engine warning light is defective

Code 4312 EGR frequency valve (2.3L 7A 1990–91)

Code 4331 Carbon canister solenoid valve 2

Code 01242 or 4332 Ignition final control circuit problem (1992–93 2.8L AAH)

Code 01247 or 4343 Carbon canister solenoid has short/open in circuit

Code 01249 or 4411 Fuel Injector cylinder 1 (& 5 on 3.6/4.2L) open/short injector circuit, fuse 23 open (fuse 13 on 2.8L)

Code 01250 or 4412 Fuel Injector cylinder 2 (& 7 on 3.6/4.2L) open/short injector circuit, fuse 23 open (fuse 13 on 2.8L)

Code 01251 or 4413 Fuel Injector cylinder 3 (& 6 on 3.6/4.2L) open/short injector circuit, fuse 23 open (fuse 13 on 2.8L)

Code 01252 or 4414 Fuel Injector cylinder 4 (& 8 on 3.6/4.2L) open/short injector circuit, fuse 23 open (fuse 13 on 2.8L)

Code 01253 or 4415 Fuel Injector cylinder 5 (code applies to 2.8L) has open/short in injector circuit, fuse 13 is open, ECM

Code 01253 or 4416 Fuel Injector cylinder 6 (code applies to 2.8L) has open/short in injector circuit, fuse 13 is open, ECM

Code 01253 or 4421 Fuel Injector cylinder 5 (code does not apply to 6 & 8 cylinder engines) has open/short in injector circuit, fuse 23 is open

Code 01254 or 4422 Fuel Injector cylinder 6 has open/short in circuit (1992–93 2.8L AAH)

Code 01257 or 4431 Idle air control (IAC) has open/short in circuit, fuse 2 is open

Code 01262 or 4442 Boost pressure limiting valve has open/short in circuit, thermofuse S75 for EVAP frequency valve is blown

Code 01265 or 4312 EGR valve has open/short in circuit (1992–93 2.8L AAH)

Code 65535 or 1111 Control module is defective

Code 65535 or 2324 Engine control module (ECM) is defective (1992–93 4.2L ABH) Ignore this code if displayed as an intermittent

Honda

GENERAL INFORMATION

Honda utilizes 2 types of fuel systems. The first is the feedback carburetor system of which there are 2 types; a 2 barrel down draft-fixed venturi type, and 2 side draft carburetors variable venturi type. The feedback carburetor was in use up to 1991 in Honda vehicles.

The second type fuel system is Programmed Fuel Injection (PGM-FI) system. This system began in 1985 and was available in the Accord and Civic. As of 1992 all Hondas are fuel injected.

SELF-DIAGNOSTICS

Service Precautions

• Make sure all ECM harness connectors are fastened securely. A poor connection can cause an extremely high voltage surge and result in damage to integrated circuits.

• Keep all ECM parts and harnesses dry during service. Protect the ECM and all solid-state components from rough handling or extremes of temperature.

• Use extreme care when working around the ECM or other components. The airbag or SRS wiring may be in the vicinity. On these vehicles, the SRS wiring and connectors are yellow. Do not cut or test these circuits.

• Before attempting to remove any parts, turn the ignition switch **OFF** and disconnect the battery ground cable.

• Always use a 12 volt battery as a power source for the engine, never a booster or high-voltage charging unit.

• Do not disconnect the battery cables with the engine running.

• Do not disconnect any wiring connector with the engine running or the ignition **ON** unless specifically instructed.

• Do not apply battery power directly to injectors.

• Whenever possible, use a flashlight instead of a drop light.

• Relieve fuel system pressure before servicing any fuel system component.

• Always use eye or full-face protection when working around fuel lines, fittings or components.

Reading Codes

1985–89 VEHICLES

When a fault is noted, the ECU stores an identifying code and illuminates the CHECK ENGINE light. The code will remain in memory until cleared; the dashboard warning lamp may not illuminate during the next ignition cycle if the fault is no longer present. Not all faults noted by the ECU will trigger the dashboard warning lamp although the fault code will be set in memory. For this reason, troubleshooting should be based on the presence of stored codes, not the illumination of the warning lamp while the car is operating.

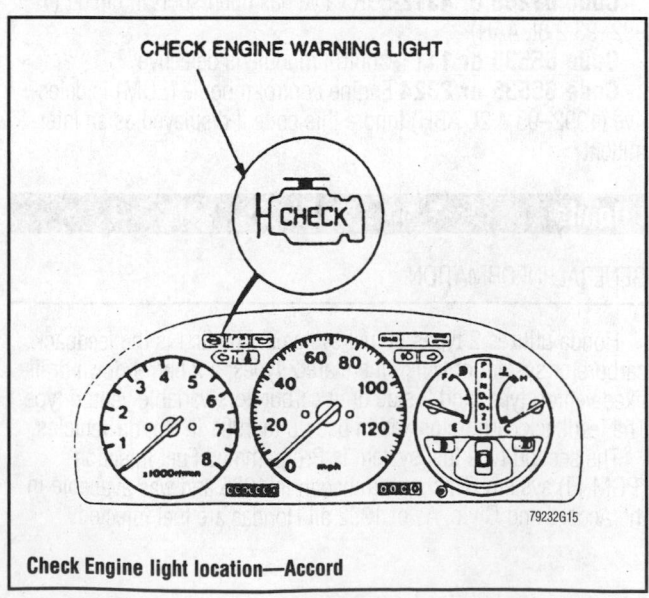

CHECK ENGINE WARNING LIGHT

CHECK

79232G15

Check Engine light location—Accord

Stored codes are displayed by either a single flashing LED (Light Emitting Diode) light, or an illuminated light pattern of 4 LED lights on the ECU. When the CHECK ENGINE warning lamp has been on or reported on, check the ECU LED for presence of codes.

The location of the malfunction is determined by observing the LED display. Earlier Hondas used 2 types of LED displays: a single LED and a 4 LED display. After 1987 all models use the single LED display.

Systems with a single LED indicate the malfunction with a series of flashes. The number of flashes indicates a code which identifies the location of the component or system malfunction. The code will flash, followed by a 2 second pause, repeat, followed by another 2 second pause, then move to the next code.

On systems with 4 LED's a display pattern identifies the malfunction. The LED's are numbered 1, 2, 4 and 8 as counted from right-to-left. The code is determined by observing which LED's are lit on the display. Each code is displayed once, followed by a 2 second pause, then the next code is displayed.

The LED's are part of the Electronic Control Module (ECM). Depending on the vehicles, the ECU is located in the following places:

• 1985–89 Accord—Under the driver side front seat
• 1985–87 Civic and CRX—Under the passenger side seat
• 1988–89 Civic and CRX—Under the passenger side foot-well, below the dashboard
• 1987 Prelude—Behind driver side rear seat trim panel
• 1988–89 Prelude—Under the passenger side footwell, below the dash. (The LED may be viewed through a small window without removing the ECU cover).

Turn the ignition switch **ON**; the LED will display any stored codes.

On the 1985 Accord and 1985–87 Civic/CRX having the 4 LED display, codes are indicated by a specific pattern of LED lights illuminated on the ECU.

On 1986–89 Accord, 1988–89 Civic, and Prelude having the single LED display, codes 1–9 are indicated by a series of short flashes; two-digit codes use a number of long flashes for the first digit followed by the appropriate number of short flashes. For example, Code 43 would be indicated by 4 long flashes followed by 3 short flashes. Codes are separated by a longer pause between transmissions. The position of the codes during output can be helpful in diagnostic work. Multiple codes transmitted in isolated order indicate unique occurrences; a display of showing 1-1-1 pause 9-9-9 indicates two problems or problems occurring at different times. An alternating display, such as 1-9-1-9-1, indicates simultaneous occurrences of the faults.

When counting flashes to determine codes, a code not valid for the vehicle may be found. In this case, first recount the flashes to confirm an accurate count. If necessary, turn the ignition switch **OFF**, then recycle the system and begin the count again. If the Code is not valid for the vehicle, the ECU must be replaced.

➡**On vehicles with electronically controlled automatic transaxles, the 5, D or D4 lamp may flash with the CHECK ENGINE lamp if certain codes are stored. If this does occur, proceed with the diagnosis based on the engine code shown. After repairs, recheck the lamp. If the additional warning lamp is still lit, proceed with diagnosis for that system.**

1990–95 VEHICLES

When a fault is noted, the ECM stores an identifying code and illuminates the CHECK ENGINE light. The code will remain in memory until cleared. The dashboard warning lamp may not illuminate during the next ignition cycle if the fault is no longer present. Not all faults noted by the ECM will trigger the dashboard warning lamp although the fault code will be set in memory. For this reason, troubleshooting should be based on the presence of stored codes, not the illumination of the warning lamp while the car is operating.

In 1990, the Accord and Prelude were equipped with a 2-pin service connector in addition to the LED. if the service connector is jumpered, with the ignition key in the **ON** position, the CHECK ENGINE lamp will display the stored codes in a series of flashes. The 2-pin service connector is located under the passenger side of dash on the Accord and behind the center console on the Prelude.

As of 1992, the LED on the ECU was eliminated and all vehicles obtain codes by jumping the 2-pin connector when the ignition switch is **ON**. The CHECK ENGINE light will then flash codes present in the ECU memory.

Diagnostic Codes 1–9 are indicated by a series of short flashes; two-digit codes use a number of long flashes for the first digit followed by the appropriate number of short flashes. For example, Code 43 would be indicated by 4 long flashes followed by 3 short flashes. Codes are separated by a longer pause between transmissions. The position of the codes during output can be helpful in diagnostic work. Multiple codes transmitted in isolated order indicate unique occurrences; a display of showing 1-1-1 pause 9-9-9 indicates two problems or problems occurring at different times. An alternating display, such as 1-9-1-9-1, indicates simultaneous occurrences of the faults.

When counting flashes to determine codes, a code not valid for the vehicle may be found. In this case, first recount the flashes to confirm an accurate count. If necessary, turn the ignition switch **OFF**, then recycle the system and begin the count again. If the code is not valid for the vehicle, the ECM must be replaced.

➡**On vehicles with electronically controlled automatic transaxles, the D4 lamp may flash with the CHECK ENGINE lamp if certain codes are stored. If this does occur, proceed with the diagnosis based on the engine code shown. After repairs, recheck the lamp. If the additional warning lamp is still lit, proceed with diagnosis for that system.**

Clearing Codes

1985–87 VEHICLES

The memory for the PGM-FI CHECK ENGINE lamp on the dashboard will be erased when the ignition switch is turned OFF; however, the memory for the LED display will not be canceled. Thus, the CHECK ENGINE lamp will not come on when the ignition switch is again turned **ON** unless the trouble is once more detected. Troubleshooting should be done according to the LED display even if the CHECK ENGINE lamp is off.

After making repairs, disconnect the battery negative cable from the battery negative terminal for at least 10 seconds and reset the ECU memory. After reconnecting the cable, check that the LED display is turned off.

Turn the ignition switch **ON**. The PGM-FI CHECK ENGINE lamp should come on for about 2 seconds. If the CHECK ENGINE lamp

won't come on, check for:—Blown CHECK ENGINE lamp bulb—Blown fuse (causing faulty back up light, seat belt alarm, clock, memory function of the car radio) —Open circuit in Yellow wire—Open circuit in wiring and control unit.

After the PGM-FI CHECK ENGINE lamp and self-diagnosis indicators have been turned on, turn the ignition switch **OFF**. If the LED display fails to come on when the ignition switch is turned **ON** again, check for:—Blown fuses, especially No. 10 fuse—Open circuit in wire between ECU fuse.

Replace the ECU only after making sure that all couplers and connectors are connected securely.

1988–90 VEHICLES

The memory for the PGM-CARB and PGM-FI CHECK ENGINE lamp on the dashboard will be erased when the ignition switch is turned **OFF**; however, the memory for the LED display will not be canceled. Thus, the CHECK ENGINE lamp will not come on when the ignition switch is again turned **ON** unless the trouble is once

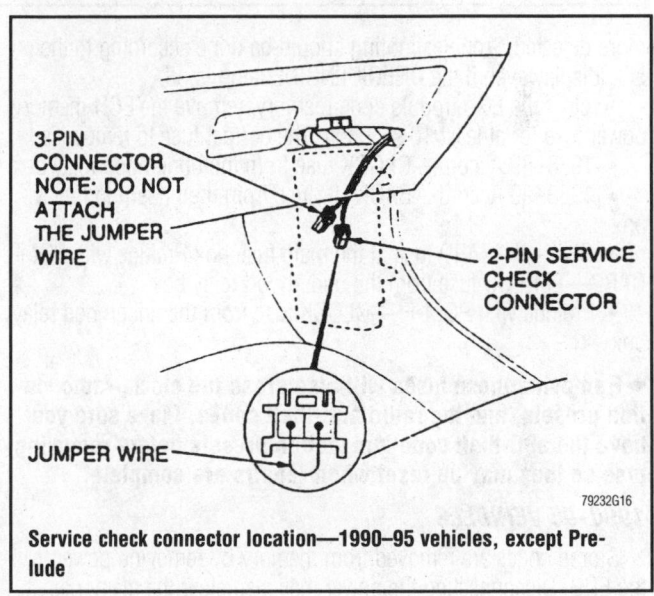

Service check connector location—1990–95 vehicles, except Prelude

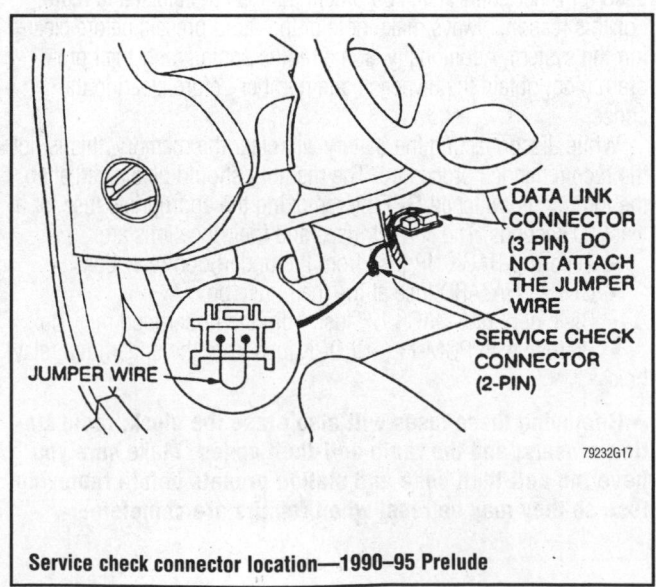

Service check connector location—1990–95 Prelude

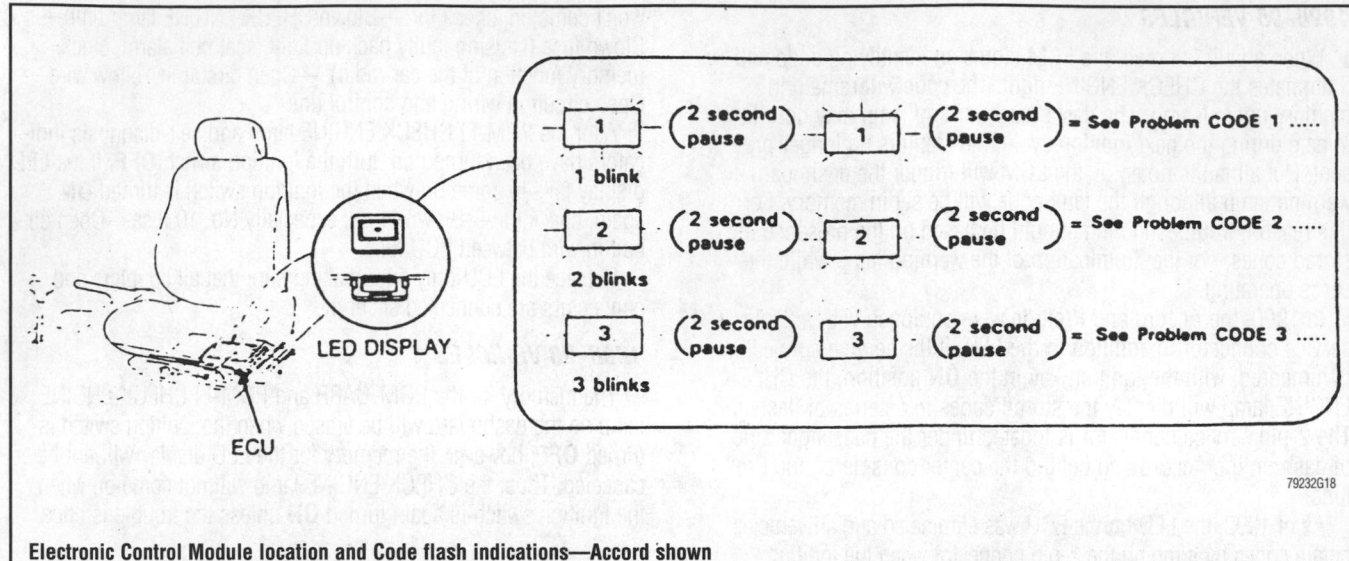

Electronic Control Module location and Code flash indications—Accord shown

more detected. Troubleshooting should be done according to the LED display even if the CHECK ENGINE lamp is off.

To clear the ECU trouble code memory, remove the ECU memory power fuse for at least 10 seconds. The correct fuse to remove is:

- 1988–89 Accord—CLOCK fuse from the underhood relay box
- 1988–90 Accord—BACK UP fuse from the underhood relay box
- Civic—HAZARD fuse at the main fuse box Prelude with PGM-CARB—EFI-ECU fuse from the underhood relay box
- Prelude with PGM-FI—CLOCK fuse from the underhood relay box

➡**Removing these fuses will also erase the clock, radio station presets, and the radio anti-theft codes. Make sure you have the anti-theft code and station presets before removing fuse so they may be reset when repairs are complete.**

1990–95 VEHICLES

Stored codes are removed from memory by removing power to the ECU. Disconnecting the power may also clear the memories used for other solid-state equipment such as the clock and radio. For this reason, always make note of the radio presets before clearing the system. Additionally, some radios contain anti-theft programming; obtain the owner's code number before clearing the codes.

While disconnecting the battery will clear the memory, this is not the recommended procedure. The memory should be cleared after the ignition is switched **OFF** by removing the appropriate fuse for at least 10 seconds. The correct fuses and their locations are:

- Accord—BACK UP fuse from the underhood relay box
- Civic—HAZARD fuse at the main fuse box
- Civic del Sol—BACK UP fuse from the underhood relay box
- Prelude with PGM-FI—CLOCK fuse from the underhood relay box

➡**Removing these fuses will also erase the clock, radio station presets, and the radio anti-theft codes. Make sure you have the anti-theft code and station presets before removing fuse so they may be reset when repairs are complete.**

DIAGNOSTIC TROUBLE CODES

1985–87 4-LED System

ACCORD

➡**This list includes vehicles with fuel injection engines: 1 .5L (EW3), 1 .5L (DI 5A3), 1 .8L (E53) Engine. Code definition as shown in illustration for the 4-LED type ECM.**

CIVIC AND CRX

This list includes the following carbureted engines: 1 .5L (EW1 and D15A2) engines. Code definition as shown in illustration for the 4-LED type ECM.

1986–93 Single LED System

1988–91 PRELUDE

This list includes all of the following carbureted engines: 2.0L (B20A5, B20A3) Engines

Code 1 Oxygen count
Code 2 Vehicle speed pulser
Code 3 Manifold Absolute Pressure (MAP)
Code 4 Vacuum switch signal
Code 5 Manifold solute Pressure (MAP)
Code 6 Coolant temperature
Code 7 Coolant switch signal (MT) or Shift position switch signal (AT) (1990–91 engines only)
Code 8 Ignition coil signal
Code 9 No. 1 cylinder position sensor
Code 10 Intake air temperature sensor (IAT sensor)
Code 12 Exhaust Gas Recirculation (EGR) System (except Del Sol and Civic & CRX 1.6L D16A6)
Code 13 Barometric pressure sensor (BARO sensor)
Code 14 Idle air control (IAC valve) except 1987- A20A3 engine or—1986 BS, BT and 1987 A20A3 Engines, Code 14 or high is possible faulty Electronic Control Module (ECM)
Code 14 Electronic Air Control
Code 15 Ignition output signal

Code 16 Fuel Injector
Code 17 Vehicle Speed sensor (VSS)
Code 19 NT lock-up control solenoid valve NB (DI 5B1, D15B2, D15B6, D15B7, D15B8, D15Z1, D16A6, D16Z6)

Code 20 Electric load detector (ELD)
Code 21 V-TEC control solenoid (D15Z1, D16Z6, H22A1)
Code 22 V-TEC pressure switch (D15Z1, D16Z6, H22A1)
Code 23 Knock sensor (H22A1-DOHC—VTEC)

ECM TROUBLE CODES

Code	Explanation
○ ○ ○ ○ (Dash Warning Light ON only)	Loose or poorly connected power line to Electronic Control Unit (ECU). Short circuit in combination meter or warning light wire. Faulty ECU
○ ○ ○ ● (1)	Disconnected oxygen sensor coupler. Spark plug misfire. Short or open circuit in oxygen sensor circuit. Faulty oxygen sensor
○ ○ ● ○ (2)	Faulty Electronic Control Unit (ECU)
○ ○ ● ● (2 1)	Disconnected Manifold Absolute Pressure (MAP) sensor coupler. Short or open circuit in MAP sensor wire. Faulty MAP sensor
○ ● ○ ○ (4)	Faulty Electronic Control Unit (ECU)
○ ● ○ ● (4 1)	Disconnected Manifold Absolute Pressure (MAP) sensor piping
○ ● ● ○ (4 2)	Disconnected coolant temperature sensor coupler. Open circuit in coolant temperature sensor wire. Faulty coolant temperature sensor (thermostat housing)
○ ● ● ● (4 2 1)	Disconnected throttle angle sensor coupler. Open or short circuit in throttle angle sensor wire. Faulty throttle angle sensor
● ○ ○ ○ (8)	Short or open circuit in crank angle sensor wire. Crank angle sensor wire interfering with high tension wire. Crank angle sensor at fault
● ○ ○ ● (8)	Short or open circuit in crank angle sensor wire. Crank angle sensor wire interfering with high tension wire. Crank angle sensor at fault
● ○ ● ○ (8 2)	Disconnected intake air temperature sensor. Open circuit in intake air temperature sensor wire. Faulty intake air temperature sensor
● ○ ● ● (8 2 1)	Disconnected idle mixture adjuster sensor coupler. Shorted or disconnected idle mixture adjuster sensor wire. Faulty idle mixture adjuster sensor
● ● ○ ○ (8 4)	Disconnected Exhaust Gas Recirculation (EGR) control system coupler. Shorted or disconnected EGR control wire. Faulty EGR control system
● ● ○ ● (8 4 1)	Disconnected atmospheric pressure sensor coupler. Shorted or disconnected atmospheric pressure sensor wire. Faulty atmospheric pressure sensor
● ● ● ○ (8 4 2)	Faulty Electronic Control Unit (ECU)
● ● ● ● (8 4 2 1)	Faulty Electronic Control Unit (ECU)

79232G19

4-LED code definition—1985 Accord 18L (E53), 1985–86 Civic Si 1.5L (EW3), 1987 Civic CRX Si 1.5L (D15A3) fuel injected engines

ENGINE CODES

Code (8 4 2 1)	Explanation
○ ○ ○ ●	Short circuit in frequency solenoid valve B (brown/black) wire
○ ○ ● ○	Short circuit in frequency solenoid valve A (green/white) wire
○ ○ ● ●	Disconnected Manifold Absolute Pressure (MAP) sensor connector. Short or open circuit in MAP sensor (black/white, green/white, yellow/white) wires. Faulty MAP sensor
○ ● ○ ○	Short circuit in ignition timing control unit (red/white, blue/white) wires. Faulty ignition timing control unit
○ ● ● ○	Disconnected Coolant Temperature Sensor (CTS) A connector. Short or open circuit in coolant temperature sensor A (light blue) wire. Faulty coolant temperature sensor A
● ○ ○ ○	Short or open circuit in ignition coil (blue) wire
● ● ○ ○	Disconnected Exhaust Gas Recirculation (EGR) lift sensor connector. Short or open circuit in EGR lift sensor (yellow, green/white, yellow/white) wires. Faulty EGR lift sensor
● ● ○ ●	Disconnected atmospheric pressure sensor connector. Short or open circuit in atmospheric pressure sensor (green/black, green/white, yellow/white) wires. Faulty atmospheric pressure sensor
15+	If the LED display pattern differs from those listed above, the Electronic Control Unit (ECU) is faulty

79232G20

4-LED code definition—1985–87 Civic/CRX with feedback carburetor—1.5 (EW1 and D15A2) engines

Code 30 NT FI Signal A (F22A1, F22A4, F22A6)
Code 31 A/T FI Signal B (F22A1, F22A4, F22A6)
Code 41 Heated Oxygen Sensor Heater (F22A1, F22A4)
Code 43 Fuel supply system (except DiSBi, D15B2, D15B6, B20A5, B21A, D16A6)
Code 48 Heated oxygen sensor (D15Z1 engine only, except Calif. emission)

1985–95 Vehicles

ACCORD, CIVIC, DEL SOL, ODYSSEY AND PRELUDE

This list includes all of the following fuel injected engines:
1.5L (DiSBi, D15B2, D15B6, D15B7, D15B8, D15Z1), 1.6L (B16A3, D16A6, D16Z6), 2.0L (A20A3, BS, BT, B20A5) 2.1L (B21A) 2.2L (F22A1, F22A4, F22A6, F22B1, F22B2, H22A1), 2.3L (H23A1) and 2.7L (C27A4) Engines
Code 0 Electronic Control Module (ECM)
Code 1 Heated oxygen sensor (or Oxygen content) or—Oxygen content A (A20A3, B20A5)
Code 2 Oxygen content B (A20A3, B20A5) or—Electronic Control Module (ECM) (BS, BT—1986 only) and (A20A3—1987 only)
Code 3 Manifold Absolute Pressure (MAP)
Code 4 Crankshaft position sensor or—Faulty ECU (BS, BT—1986 only) and (A20A3 -1987 only, B20A5, B21A1)
Code 5 Manifold Absolute Pressure (MAP)
Code 6 Engine coolant temperature (ECT)
Code 7 Throttle position sensor (TP sensor)
Code 8 Top dead center sensor (TDC sensor)

Code 9 No. 1 cylinder position sensor
Code 10 Intake air temperature sensor (IAT sensor)
Code 11 Electronic Control Module (ECM) (BS, BT —1986 only) and (A20A3—1987 only)
Code 12 Exhaust Gas Recirculation (EGR) System (except Del Sol and Civic & CRX 1 .6L DI 6A6)
Code 13 Barometric pressure sensor (BARO sensor)
Code 14 Idle air control (IAC valve) except 1987 —A20A3 engine. or—1986 BS, BT and 1987 A20A3 Engines, Code 14 or high is possible faulty Electronic Control Module (ECM)
Code 15 Ignition output signal
Code 16 Fuel Injector
Code 17 Vehicle Speed sensor (VSS)
Code 19 NT lock-up control solenoid valve NB (DiSBi, D15B2, D15B6, D15B7, D15B8, D15Z1, D16A6, D16Z6)
Code 20 Electric load detector (ELD)
Code 21 V-TEC control solenoid (D15Z1, D16Z6, H22A1)
Code 22 V-TEC pressure switch (D15Z1, D16Z6, H22A1)
Code 23 Knock sensor (H22A1-DOHC—VTEC)
Code 30 NT FI Signal A (F22A1, F22A4, F22A6)
Code 31 NT FI Signal B (F22A1, F22A4, F22A6)
Code 41 Heated Oxygen Sensor Heater (F22A1, F22A4)
Code 43 Fuel supply system (except D1SB1, D15B2, D15B6, B20A5, B21A, D16A6)
Code 45 Fuel supply metering
Code 48 Heated oxygen sensor (D15Z1 engine only, except Calif. emission)
Code 61 Front Heated Oxygen Sensor
Code 63 Rear Heated Oxygen Sensor

Code 65 Rear Heated Oxygen Sensor Heater
Code 67 Catalytic Converter System
Code 70 Automatic Transaxle or A/F FI Data line
Code 71 Misfire detected; cylinder No. 1 or random misfire
Code 72 Misfire detected; cylinder No. 2 or random misfire
Code 73 Misfire detected; cylinder No. 3 or random misfire
Code 74 Misfire detected; cylinder No. 4 or random misfire
Code 75 Misfire detected; cylinder No. 5 or random misfire
Code 76 Misfire detected; cylinder No. 6 or random misfire
Code 80 Exhaust Gas Recirculation (EGR) system
Code 86 Coolant temperature
Code 92 Evaporative Emission Control System

Hyundai

GENERAL INFORMATION

Hyundai utilizes 2 fuel system types. The first is a feedback carburetor and the second is fuel injection. The feedback carburetor is used on 1986–93 Excel.

Multi-point Fuel Injection (MFI) was introduced in 1989 on the Sonata. In 1990 Excel was available with fuel injection also. Scoupe, Elantra and Accent came only with MFI.

SELF-DIAGNOSTICS

Service Precautions

• Keep the ECU parts and harnesses dry during service. Protect the ECU and all solid-state components from rough handling or temperature extremes.
• Use extreme care when working around the ECU or other components.
• Disconnect the negative battery cable before attempting to disconnect or remove any electronic parts.
• Disconnect the negative battery cable and ECU connector before performing arc welding on the vehicle.
• Disconnect and remove the ECU from the vehicle before subjecting the vehicle to the temperatures experienced in a heated paint booth.

Reading Codes

➡**Hyundai utilized a Feedback carburetor system in the Excel. However, the self-diagnostic system is not used. Self-diagnosis pertains to fuel injected vehicles only. 1993–94 Scoupe and 1995 Accent have the ability to read diagnostic codes through the Malfunction Indicator Lamp (MIL) and therefore special equipment is not required to retrieve codes. All other vehicles however, do not process this ability, therefore either a Multi-Use Tester or and Analog voltmeter must be used to retrieve codes. When using special diagnostic equipment, always observe the tool manufacturer's instructions.**

1989–95 VEHICLES

Using Multi-Use Tester

1. Turn the ignition switch to the **OFF** position.
2. Connect the multi-use tester to the diagnosis connector in the fuse box.
3. Connect the power-source terminal of the multi-use tester to the cigar lighter.
4. Turn the ignition switch to the **ON** position.

5. Follow the manufacturer's instructions to retrieve the trouble codes. The codes will be displayed in numerical order.

Using Analog Voltmeter

1. Connect the voltmeter to the self-diagnosis connector.
2. Turn the ignition switch to the **ON** position.
3. Observe the voltmeter to read the trouble codes. The code is determined by noting the duration of the voltmeter sweeps. A sweep of long duration indicates the multiple of ten digit, while a sweep of short duration indicates the single digit. For example, the code number 12 is indicated by 1 sweep of long duration followed by 2 sweeps of short duration and so on. The trouble codes will be displayed in numerical order.

Using Engine (MIL) Lamp

The 1993–94 Scoupe and 1995 Accent has the ability to read codes by using the Maintenance Indicator Lamp (MIL).

1. Turn the ignition switch **ON** but do not start the vehicle.
2. Ground the L-wire (PIN 10) in the diagnostic terminal for 2½ seconds.
3. The first code to flash should be (4444) which will flash until the L-wire is disconnected.

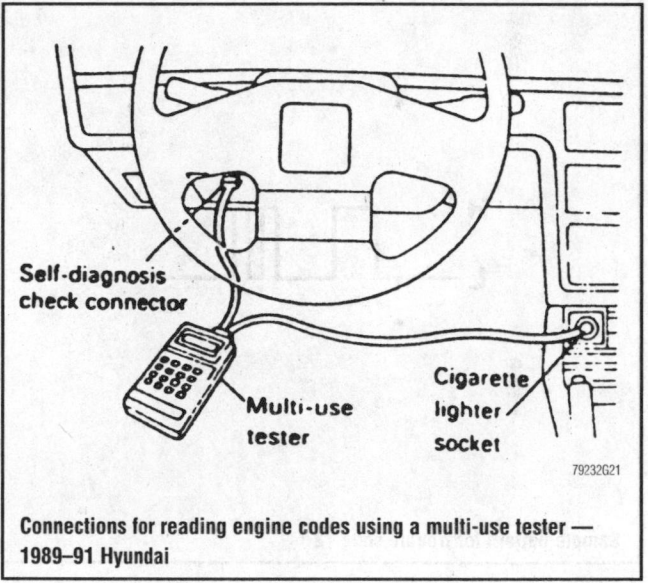

Connections for reading engine codes using a multi-use tester — 1989–91 Hyundai

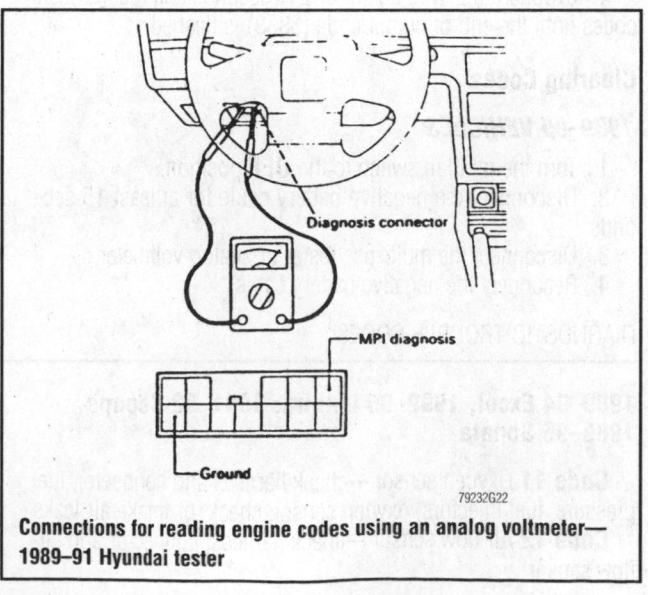

Connections for reading engine codes using an analog voltmeter— 1989–91 Hyundai tester

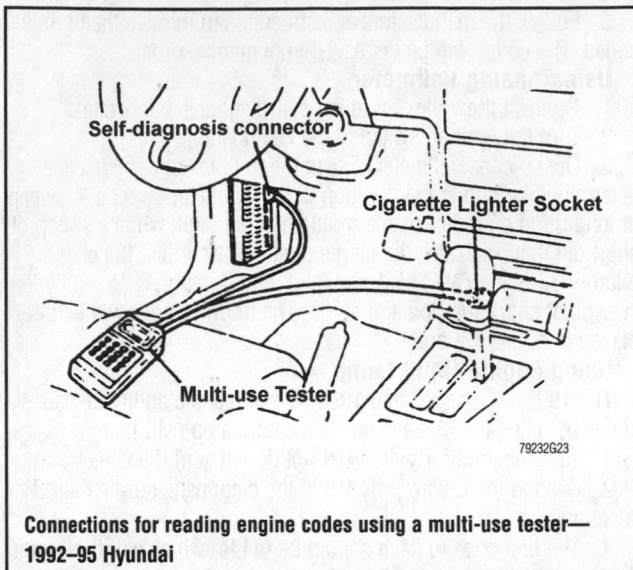

Connections for reading engine codes using a multi-use tester—1992–95 Hyundai

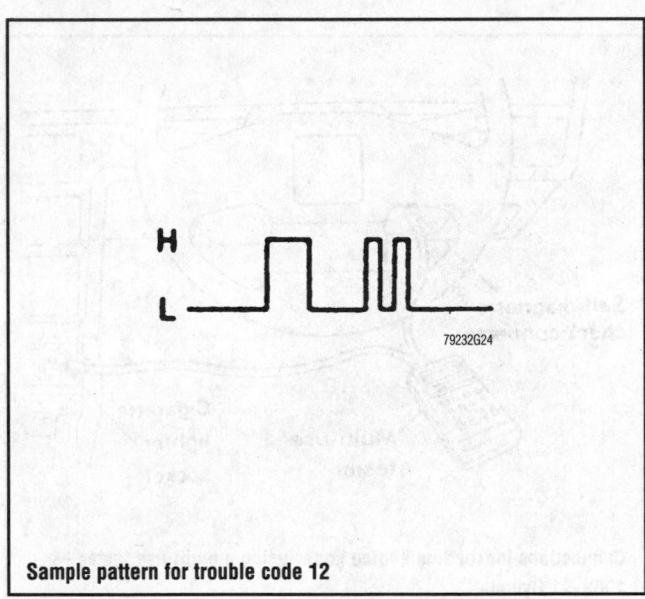

Sample pattern for trouble code 12

4. Ground the L-wire again for 2½ seconds and record flash codes until the end of output code (3333) is flashed.

Clearing Codes

1989–95 VEHICLES

1. Turn the ignition switch to the **OFF** position.
2. Disconnect the negative battery cable for at least 15 seconds.
3. Disconnect the multi-use tester or analog voltmeter.
4. Reconnect the negative battery cable.

DIAGNOSTIC TROUBLE CODES

1989–94 Excel, 1992–95 Elantra, 1991–92 Scoupe, 1989–95 Sonata

Code 11 Oxygen sensor—check harness and connector, fuel pressure, fuel injectors, oxygen sensor, check for intake air leaks

Code 12 Air flow sensor—check harness, connector and air flow sensor

Code 13 Air temperature sensor—check harness, connector and air temperature sensor

Code 14 Throttle position sensor—check harness and connector, throttle position sensor, idle position switch

Code 15 Motor position sensor—check harness and connector, motor position sensor

Code 21 Engine coolant temperature sensor—check harness and connector, engine coolant temperature sensor

Code 22 Crank angle sensor—check harness, connector and distributor assembly

Code 23 No. 1 cylinder top dead center sensor—check harness, connector and distributor assembly

Code 24 Vehicle speed sensor. (reed switch)—check harness and connector, and vehicle speed sensor

Code 25 Barometric pressure sensor—check harness, connector and barometric pressure sensor

Code 41 Injector—check harness and connector, injector coil resistance

Code 42 Fuel pump—check harness and connector, control relay

Code 43 EGR—check harness and connector, EGR temperature sensor, EGR valve, EGR control solenoid valve, EGR valve control vacuum (California)

Code 44 Ignition coil fault, faulty power transistor

Code 59 Oxygen (H025) sensor fault

1993–95 Scoupe, 1995 Accent

Code 1122 Electronic Control Unit failure-RAM/ROM
Code 1169 Electronic Control Unit failure
Code 1233 Electronic Control Unit failure-ROM
Code 1234 Electronic Control Unit failure-RAM
Code 2121 Boost sensor control valve
Code 3112 Injector No.1
Code 3114 AC opening failure
Code 3116 Injector No. 3
Code 3117 Airflow sensor
Code 3121 Boost pressure sensor failure
Code 3122 AC closing failure
Code 3128 H025 sensor
Code 3135 Evaporative purge control solenoid valve
Code 3137 Alternator output/low battery
Code 3145 Coolant temperature sensor
Code 3149 NC compressor
Code 3152 Turbo boost to high
Code 3153 Throttle position sensor
Code 3159 Vehicle speed sensor
Code 3211 Knock sensor
Code 3222 Phase sensor
Code 3224 ECM-knock evolution sensor
Code 3232 Crankshaft position sensor
Code 3233 ECM-knock evolution sensor
Code 3234 Injector No. 2
Code 3235 Injector No. 4
Code 3241 ECM-injector or purge control valve
Code 3242 ECM-IAC motor or NC relay
Code 3243 Electronic Control Unit failure
Code 4133 Electronic Control Unit failure
Code 4151 Air/fuel control
Code 4152 Air/fuel adaptive failure
Code 4153 Air/fuel adaptive (multiple) failure
Code 4154 Air/fuel adaptive (additive) failure

Code 4155 ECM-a/c relay, IAC motor, injector or PCV
Code 4156 Boost sensor control deviation failure

Infiniti

GENERAL INFORMATION

The Infiniti Electronic Concentrated Control System (ECCS) is an air flow controlled, sequential port fuel injection and engine control system. It is used on all models equipped with 2.0L, 3.0L and 4.5L engines. The ECCS electronic control unit consists of a microcomputer, an inspection lamp, a diagnostic mode selector and connectors for signal input and output, powers and grounds.

The safety relay prevents electrical damage to the electronic control unit, or ECU, and the injectors in case the battery terminals are accidentally connected in reverse. The safety relay is built into the fuel pump control circuit.

Ignition timing is controlled in response to engine operating conditions. The optimum ignition timing in each driving condition is pre-programmed in the computer. The signal from the control unit is transmitted to the power transistor and this signal controls when the transistor turns the ignition coil primary circuit on and off (hence, the ignition timing). The idle speed is also controlled according to engine operating conditions, temperature and gear position. On manual transmission models, if battery voltage is less than 12 volts for a few seconds, a higher idle speed will be maintained by the control unit to improve charging function.

There is a fail-safe system built into the ECCS control unit. This system makes engine starting possible if a portion of the ECU's central processing unit circuit fails. Also, if a major component such as the crank angle sensor or the air flow meter were to malfunction, the ECU substitutes or borrows data to compensate for the fault. For example, if the output voltage of the air flow meter is extremely low, the ECU will substitute a pre-programmed value for the air flow meter signal and allows the vehicle to be driven as long as the engine speed is kept below 2000 rpm. Or, if the cylinder head temperature sensor circuit is open, the control unit clamps the warm-up enrichment at a certain amount. This amount is almost the same as that when the cylinder head temperature is between 68–176°F (20–80°C).

If the fuel pump circuit malfunctions, the fuel pump relay comes on until the engine stops. This allows the fuel pump to receive power from the relay.

The electronic control unit controls the following functions:
- Injector pulse width
- Ignition timing
- Intake valve timing control (045)
- Air regulator control (G20)
- Exhaust gas recirculation (EGR) solenoid valve operation
- Exhaust gas sensor heater operation
- Idle speed
- FICD solenoid valve operation (G20 and M30)
- Fuel pump relay operation
- Fuel pump voltage (M30 and 045)
- Fuel pressure regulator control (M30)
- AIV control (G20)
- Carbon canister control solenoid valve operation
- Air conditioner relay operation (During early wide-open throttle)
- Radiator fan operation (G20)
- Traction control system (TCS) operation (045, if equipped)

- Self-diagnosis
- Fail-safe mode operation

SELF-DIAGNOSTICS

Service Precautions

- Do not disconnect the injector harness connectors with the engine running.
- Do not apply battery power directly to the injectors.
- Do not disconnect the ECU harness connectors before the battery ground cable has been disconnected.
- Make sure all ECU connectors are fastened securely. A poor connection can cause an extremely high surge voltage in the coil and condenser and result in damage to integrated circuits.
- When testing the ECU with a DVOM make sure that the probes of the tester never touch each other as this will result in damage to a transistor in the ECU.
- Keep the ECCS harness at least 4 in. away from adjacent harnesses to prevent an ECCS system malfunction due to external electronic noise.
- Keep all parts and harnesses dry during service.
- Before attempting to remove any parts, turn **OFF** the ignition switch and disconnect the battery ground cable.
- Always use a 12 volt battery as a power source.
- Do not attempt to disconnect the battery cables with the engine running or the ignition key **ON**.
- Do not clean the air flow meter with any type of detergent.
- Do not attempt to disassemble the ECCS control unit under any circumstances.
- Avoid static electricity build-up by properly grounding yourself prior to handling any ECU or related parts.

Reading Codes

➡**Diagnostic codes may be retrieved by observing code flashes through the LED lights located on the Electronic Control Module (ECM). A special Nissan Consult monitor tool can be used, but is not required. When using special diagnostic equipment, always observe the tool manufacturer's instructions.**

1990–95 VEHICLES

2-Mode Diagnostic System

Infiniti vehicles use a 2-mode diagnostic system incorporated in the ECU which uses inputs from various sensors to determine the correct air/fuel ratio. If any of the sensors malfunction the ECU will store the code in memory.

An Infiniti/Nissan Consult monitor may be used to retrieve these codes by simply connecting the monitor to the diagnosis connector located on the driver's side near the hood release.

Turn the ignition switch **ON** and press START, ENGINE and then SELF-DIAG RESULTS, the results will then be output to the monitor.

The conventional CHECK ENGINE or red LED ECU light may be used for self-diagnostics. The conventional 2-mode diagnostic system is broken into 2 separate modes each capable of 2 tests, an ignition switch **ON** or engine running test as outlined below:

Mode I—Bulb Check

In this mode the RED indicator light on the ECU and the CHECK ENGINE light should be ON. To enter this mode simply turn the ignition switch **ON** and observe the light.

Mode 1—Malfunction Warning

In this mode the ECU is acknowledging if there is a malfunction

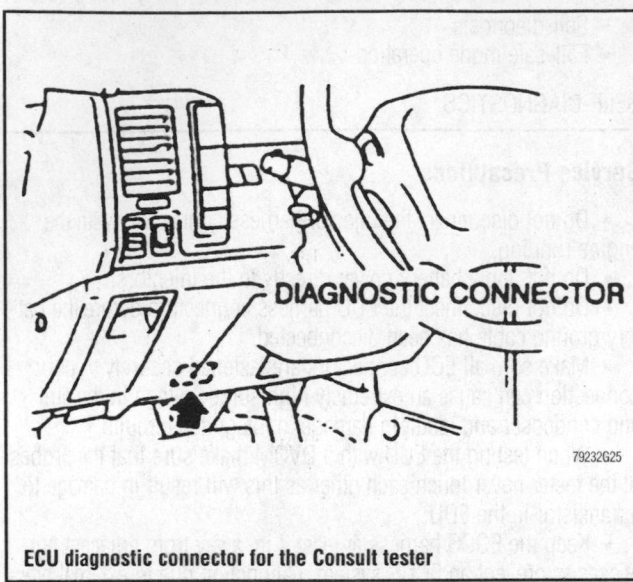

ECU diagnostic connector for the Consult tester

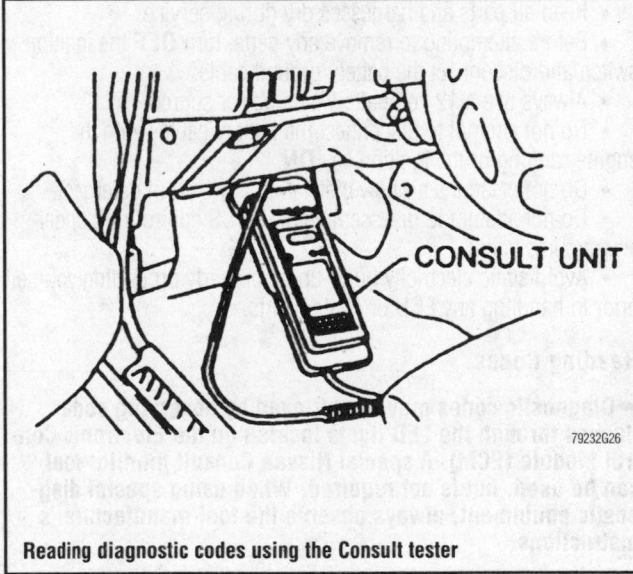

Reading diagnostic codes using the Consult tester

by illuminating the RED indicator light on the ECU and the CHECK ENGINE light. If the light turns OFF, the system is normal. To enter this mode, simply start the engine and observe the light.

Mode 2—Self-Diagnostic Codes

In this mode the ECU will output all malfunctions via the CHECK ENGINE light or the red LED on the ECU. The code may be retrieved by counting the number of flashes. The longer flashes indicate the first digit and the shorter flashes indicate the second digit. To enter this mode proceed as follows:

1. Turn the ignition switch **ON**, but do not start the vehicle.
2. Turn the ECU diagnostic mode selector fully clockwise for 2 seconds, then turn it back fully counterclockwise.
3. Observe the red LED on the ECU or CHECK ENGINE light for stored codes.

Mode 2—Exhaust Gas Sensor Monitor

In this mode the red LED on the ECU or CHECK ENGINE light will display the condition of the fuel mixture and whether the system is in closed loop or open loop. When the light flashes ON, the exhaust gas sensor is indicating a lean mixture. When the light stays OFF, the sensor is indicating a rich mixture. If the light

remains ON or OFF, it is indicating an open loop system. If the system is equipped with 2 exhaust gas sensors, the left side will operate first. If already in Mode 2, proceed to Step 3 for exhaust gas sensor monitor.

1. Turn the ignition switch **ON**.
2. Turn the diagnostic switch ON, by turning the switch fully clockwise for 2 seconds and then fully counterclockwise.
3. Start the engine and run until thoroughly warm. Raise the idle to 2,000 rpm and hold for approximately 2 minutes. Ensure the red LED or CHECK ENGINE light flash ON and OFF more than 5 times every 10 seconds with the engine speed at 2,000 rpm.

➡**If equipped with 2 exhaust gas sensors, switch to the right sensor by turning the ECU mode selector fully clockwise for 2 seconds and then fully counterclockwise with the engine running.**

Clearing Codes

1990–95 VEHICLES

All control unit diagnostic codes may be cleared by disconnecting the negative battery for a period of 15 seconds. The codes will be cleared when mode 1 is re-entered from mode 2. The Nissan Consult Monitor or equivalent can also be used to clear codes.

DIAGNOSTIC TROUBLE CODES

1990–95 Vehicles

Code 16 TCS Signal
Code 21 Ignition signal missing in primary coil
Code 31 ECM (engine ECCS control unit)
Code 32 EGR circuit
Code 33 Heated oxygen sensor circuit
Code 34 Knock Sensor (KS) circuit
Code 35 EGR temperature sensor circuit
Code 42 Fuel temperature sensor circuit
Code 43 Throttle sensor circuit
Code 45 Injector leak
Code 46 Secondary throttle sensor circuit
Code 51 Injector circuit
Code 53 Heated oxygen sensor circuit (right bank)
Code 54 NT controller circuit
Code 55 No malfunctioning in the above circuit
Code 11 Crankshaft position sensor
Code 12 Mass Air flow sensor
Code 13 Engine coolant temperature sensor circuit
Code 14 Vehicle speed sensor

Isuzu

GENERAL INFORMATION

Isuzu vehicles may be fitted with either a Feedback Carburetor (FBC), a Throttle Body Fuel Injection (TBI) system or a Multi-port Fuel Injection System (MFI).

The Feedback Carburetor System (FBC) is primarily used on 1985 and earlier normally aspirated engines, although some engines may use this system up to 1993. The system Electronic Control Module (ECM) constantly monitors and controls engine operation by reading data from various sensors and outputting signals to the carburetor. This helps lower emissions while maintaining the fuel economy, driveability and performance of the vehicle.

The Throttle Body (TBI) fuel injection system was put into production in 1989. It is used on 2.8L and 3.1L engines. The system functions much the same as a multi-port fuel injection system but with one exception—fuel is injected into the intake manifold rather than into each individual cylinder. This system may also control the ignition system.

The 1-TEC Multi-port Fuel Injection (MFI) system was first used in 1985 and continues to be used today. The system constantly monitors and controls engine operation through the use of data sensors, an Electronic Control Module (ECM) and other components. Individual fuel injectors are mounted at each cylinder and provide a metered amount of fuel as required by current operating conditions. This system may also control the ignition system and, as equipped, the turbocharger system.

SELF-DIAGNOSTICS

All vehicles covered in this section have self-diagnostic capabilities. The ECM diagnostics are in the form of trouble codes stored in the system's memory. When a trouble code is detected by the control module, it will turn the malfunction indicator lamp ON until the code is cleared. An intermittent problem will set a code. The lamp will turn OFF if the problem goes away, but the trouble code will stay in memory until ECM power is interrupted.

Service Precautions

• Keep all ECM parts and harnesses dry during service. Protect the ECM and all solid-state components from rough handling or extremes of temperature.

• Use extreme care when working around the ECM or other solid-state components. Do not allow any open circuit to short or ground in the ECM circuit. Voltage spikes may cause damage to solid-state components.

• Before attempting to remove any parts, turn the ignition switch **OFF** or disconnect the negative battery cable.

• Remove the ECM before any arc welding is performed to the vehicle.

• Electronic components are very susceptible to damage caused by electrostatic discharge (static electricity). To prevent electronic

component damage, do not touch the control module connector pins or soldered components on the control module circuit board.

Reading Codes

➡**Diagnostic codes may be retrieved through the use of the CHECK ENGINE light or Malfunction Indicator Lamp (MIL). A special Scan tool can be used, but is not required. When using special diagnostic equipment, always observe the tool manufacturer's instructions.**

1982–86 VEHICLES

The trouble code system is actuated by connecting a diagnostic lead to ground. The location of the diagnostic lead differs from model-to-model and, in some cases from year-to-year within the same model.

I-Mark RWD models for 1982–86; the trouble code test leads are usually taped to the wiring harness under the instrument panel, at the right side of the steering column and just above the accelerator pedal.

I-Mark FWD models for 1985–86; a 3 terminal connector, also known as the Assembly Line Diagnostic Link (ALDL) or Assembly Line Communications Link (ALCL) is located near the ECM connector. Connect a jumper wire between A and C.

Impulse for 1983–86; connect diagnostic lead terminals together (1 male and 1 female). Terminals are located under dash near the top of the driver's side kick panel.

Trooper II for 1985; connect the diagnostic lead terminals together (1 male and 1 female). The terminals are located under dash, on the passenger's side, behind the radio. The terminal leads for 1986 models are located near the ALDL connector, under dash, on the driver's side behind the cigarette lighter.

Pick-up truck for 1982; connect the diagnostic lead terminals together (1 male and 1 female). The terminals are branched from the harness near the ECM, under the dash on the driver's side behind the hood release.

The trouble code is determined by counting the flashes of the 'Check Engine lamp. Trouble Code 12 will flash first, indicating that the self-diagnostic system is working. Code 12 consists of 1 flash, short pause, then 2 flashes. There will be a longer pause and Code 12 will repeat 2 more times. Each code flashes 3 times. The cycle will then repeat itself until the engine is started or the ignition switch is turned **OFF**. In most cases, the codes will be checked with the

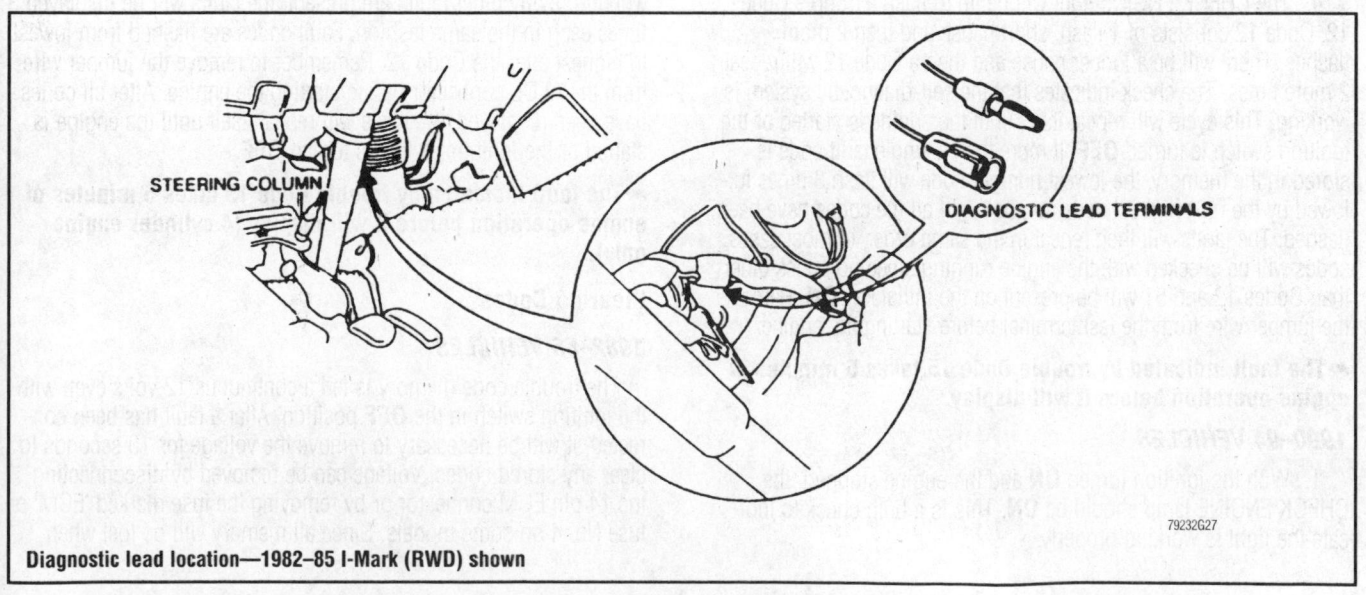

STEERING COLUMN

DIAGNOSTIC LEAD TERMINALS

79232G27

Diagnostic lead location—1982–85 I-Mark (RWD) shown

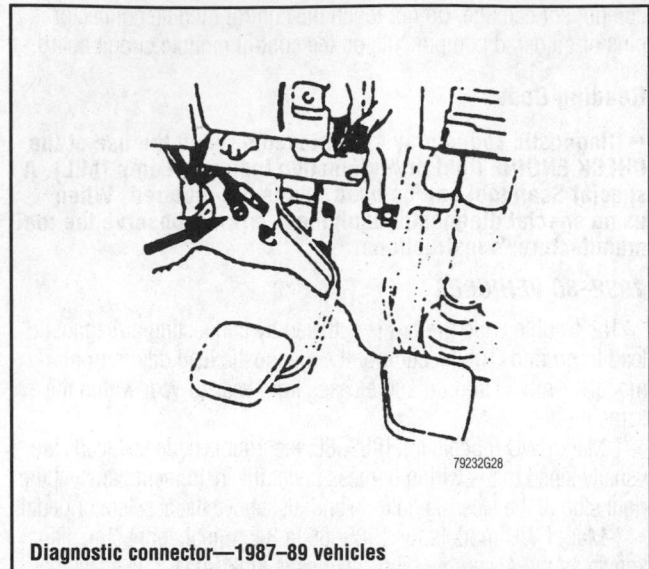

Diagnostic connector—1987–89 vehicles

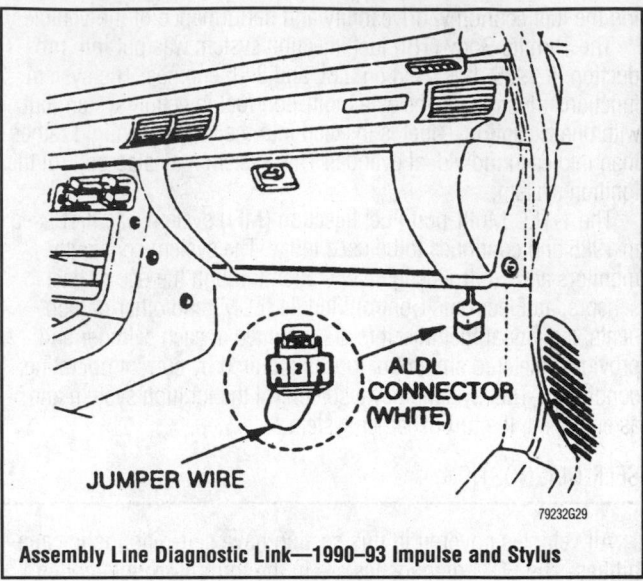

Assembly Line Diagnostic Link—1990–93 Impulse and Stylus

engine running since no codes other than 12 or 51 will be present on initial key **ON**.

1987–94 VEHICLES

With Scan Tool

1. Turn the ignition key to the **OFF** position.
2. Connect the scan tool to the Assembly Line Diagnostic Link (ALDL).
3. Turn **ON** the ignition for scan tool to access engine computer.

Without Scan Tool

1987–89 VEHICLES

1. With the ignition turned **ON**, and the engine stopped, the CHECK ENGINE lamp should be ON. This is a bulb check to indicate the light is working properly.
2. For the I-Mark, connect a jumper wire between the A and C terminals of the Assembly Line Diagnostic Link (ALDL). The connector is located next to the ECM near the heater blower motor.
3. For the Impulse, Trooper and Pickup; connect the trouble code TEST lead (white cable) to the ground lead (black cable). It is located 8 in. from the ECM connector next to the clutch pedal or center console.
4. The CHECK ENGINE light will begin to flash a trouble Code 12. Code 12 consists of 1 flash, short pause and then 2 more flashes. There will be a longer pause and then a Code 12 will repeat 2 more times. The check indicates that the self-diagnostic system is working. This cycle will repeat itself until the engine is started or the ignition switch is turned **OFF**. If more than a single fault code is stored in the memory, the lowest number code will flash 3 times followed by the next highest code number until all the codes have been flashed. The faults will then repeat in the same order. In most cases, codes will be checked with the engine running since no codes other than Codes 12 and 51 will be present on the initial key **ON**. Remove the jumper wire from the test terminal before starting the engine.

➡ **The fault indicated by trouble Code 15 takes 5 minutes of engine operation before it will display.**

1990–94 VEHICLES

1. With the ignition turned **ON** and the engine stopped, the CHECK ENGINE lamp should be **ON**. This is a bulb check to indicate the light is working properly.

2. Enter the diagnostic modes as follows:

   a. For the Impulse, Stylus, and Trooper; jumper the 1 and 3 terminals (outer terminals) of the white Assembly Line Diagnostic Link (ALDL). The connector for Impulse and Stylus is located behind the kick panel on the passenger side of the vehicle. On Trooper, the ALDL connector is located behind the left side of the center console.

   b. For the Amigo, Pickup, and Rodeo with 4 cylinder engine; connect the trouble code TEST lead (white cable) and a ground lead (black cable) together. It is located 8 in. from the ECM connector (next to the clutch pedal or brake pedal).

   c. For the Amigo, Pickup, and Rodeo with 6 cylinder engine; jumper wire the A and B terminals together of the Assembly Line Diagnostic Link (ALDL). The ALDL is located in the center console and is sometimes covered by a plastic cover labeled DIAGNOSTIC CONNECTOR. Read the trouble codes with the ignition switch **ON** and the engine OFF.

3. The CHECK ENGINE light will begin to flash a trouble Code 12. Code 12 consists of 1 flash, a short pause and then 2 more flashes. There will be a longer pause and a Code 12 will repeat 2 more times. Code 12 indicates that the self-diagnostic system is working. If any other faults are present, the faults will be displayed 3 times each in the same fashion. Fault codes are flashed from lowest to highest after the Code 12. Remember to remove the jumper wire from the ALDL connector before starting the engine. After all codes have been displayed, the cycle will repeat itself until the engine is started or the ignition switch is turned **OFF**.

➡ **The fault indicated by trouble Code 15 takes 5 minutes of engine operation before it will display (4 cylinder engine only).**

Clearing Codes

1982–86 VEHICLES

The trouble code memory is fed a continuous 12 volts even with the ignition switch in the **OFF** position. After a fault has been corrected, it will be necessary to remove the voltage for 10 seconds to clear any stored codes. Voltage can be removed by disconnecting the 14 pin ECM connector or by removing the fuse marked 'ECM' or fuse No. 4 on some models. Since all memory will be lost when removing the fuse, it will be necessary to reset the clock and other electrical equipment.

1987–94 VEHICLES

The trouble code memory is fed a continuous 12 volts even with the ignition switch in the **OFF** position. After a fault has been corrected, it will be necessary to remove the voltage for 30 seconds to clear any stored codes. The quickest way to remove the voltage is to remove the ECM fuse from the fuse block or the MAIN 60A fuse for 10 seconds. The voltage can also be removed by disconnecting the negative battery cable. This will mean electronic instrumentation, such as a clock and radio, would have to be reset.

1987-89 I-Mark; to clear the trouble codes; turn the ignition switch **OFF** and remove the ECM (30 amp) slow blow fuse located in the engine compartment fuse block for 30 seconds.

1987–89 Impulse; turn the ignition switch **OFF** and disconnect the ECM 13-pin connector or remove the No. 4 fuse from the fuse block for 30 seconds. The electronic functions with memory have to be reset after removing the No. 4 fuse.

1987–89 Pickup and Trooper; turn the ignition switch **OFF** and disconnect the ECM 13-pin connector or remove the No. 4 fuse from the fuse block for 30 seconds.

The 60 amp slow blow fuse may be removed from the fuse block in the engine compartment. However, the electronic functions with memory have to be reset after removing the No. 4 fuse for 30 seconds.

1990–93 I-Mark, Impulse and Stylus; to clear the trouble codes, turn the ignition switch **OFF** and remove the ECM (30 amp) fusible link (FL 3) located in the engine compartment relay and fuse block for 30 seconds.

1992–94 Amigo, Pickup, Rodeo and Trooper; To clear the trouble codes; turn the ignition switch **OFF** and remove the ECM fuse from the under-dash fuse block for 30 seconds. Removing the number 3 fuse from the under dash fuse panel will result in having to reset all the electronic functions with memory in the vehicle. This applies to trucks with 4-cylinder engines.

Removing the 60 amp slow blow fuse from the fuse block in the engine compartment will also erase codes.

DIAGNOSTIC TROUBLE CODES

1982–94 Vehicles

1982–84 1.8L FBC Engine
Code 12 Idle switch is not turned ON
Code 13 Idle switch is not turned OFF
Code 14 Wide Open Throttle (WOT) switch is not turned ON
Code 15 Wide Open Throttle (WOT) switch is not turned ON
Code 21 Output transistor is not turned ON
Code 22 Output transistor is not turned OFF
Code 23 Abnormal oxygen sensor
Code 24 Abnormal Water Temperature Sensor (WTS) switch
Code 25 Abnormal Random Access Memory (RAM)
Code 12, 13, 14 and 15 'Check Engine' lamp not ON
Code 21, 22, 23, 24 and 25 'Check Engine' lamp ON

1985–89 1.SL FBC (Vin 7), 1983–86 2.0L FBC (Vin A) and 1986–94 2.3L FBC (Vin L) Engines
Code 12 Normal
Code 13 Oxygen sensor circuit
Code 14 Coolant Temperature Sensor (CTS)—circuit shorted
Code 15 Coolant Temperature Sensor (CTS)—circuit open
Code 21 Idle switch—circuit open or Wide Open Throttle (WOT) switch—circuit shorted

Code 22 Fuel Cut Solenoid (FCS)—circuit open or grounded
Code 23 Mixture Control (M/C) solenoid—circuit open or grounded, or Vacuum Control Solenoid (VCS)—circuit open or grounded (1983 1.8L Truck, 1983–86 2.0L Truck, 1986–88 2.3L Truck)
Code 24 Vehicle Speed Sensor (VSS) circuit
Code 25 Air Switching Solenoid (ASS)—circuit open or grounded
Code 25 Vacuum Switching Valve (VSV)—circuit open or grounded (1987–89 1 .5L Car)
Code 26 Vacuum Switching Valve (VSV) system for canister purge -circuit open or grounded
Code 27 Vacuum Switching Valve (VSV)-constant high voltage to ECM
Code 31 No ignition reference pulses to ECM
Code 32 EGR temperature sensor—system malfunction
Code 34 EGR temperature sensor—circuit failure electronic idle control
Code 42 Fuel Cut Relay and/or circuit shorted
Code 44 Oxygen Sensor circuit—lean indication
Code 45 Oxygen Sensor circuit—rich indication
Code 51 Shorted Fuel Cut Solenoid (FCS) circuit and/or faulty Electronic Control Module (ECM), or faulty calibration unit (PROM) or installation on 1985–89 1.5L I-Mark
Code 52 Faulty Electronic Control Module (ECM)—Random Access Memory (RAM) problem in ECM
Code 53 Shorted Air Switching Solenoid (ASS) or Air Injection System and/or faulty Electronic Control Module (ECM)
Code 54 Shorted Vacuum Control Solenoid (VCS) and/or faulty Electronic Control Module (ECM)
Code 54 Shorted Mixture Control Solenoid circuit and/or faulty Electronic Control Module (ECM) (1987–89 1.5L I-Mark)
Code 55 Faulty Electronic Control Module (ECM)

1985–87 2.0L Turbo EFI (Vin F), 1983–89 2.0L EFI (Vin A), 1988–89 2.3L EFI (Vin L), and 1988–94 2.6L EFI (Vin E) Engines
Code 12 Normal
Code 13 Oxygen sensor circuit
Code 14 Engine Coolant Temperature (ECT) sensor -grounded
Code 15 Engine Coolant Temperature (ECT) sensor—incorrect signal (open circuit on 1988–94 2.6L)
Code 16 Engine Coolant Temperature (ECT) sensor -open circuit
Code 21 Throttle Valve Switch (TVS) system—idle contact and full contact made simultaneously
Code 22 Starter—no signal input
Code 23 Ignition power transistor—output terminal grounded
Code 25 Vacuum Switching Valve (VSV)—output terminal grounded or open
Code 26 Canister purge Vacuum Switching Valve (VSV)—open or grounded
Code 27 Canister purge Vacuum Switching Valve (VSV)—faulty transistor or bad ground circuit
Code 32 EGR temperature sensor—faulty sensor or harness
Code 33 Fuel injector system—output terminal grounded or open
Code 34 EGR Vacuum switching valve—output terminal grounded or open
Code 35 Ignition power transistor—open circuit
Code 41 Crank Angle sensor (CAS)—no signal or faulty signal
Code 43 Throttle Valve Switch—idle contact closed continuously

Code 44 Fuel metering system—lean signal (Oxygen sensor-low voltage)

Code 45 Fuel metering system—rich signal (Oxygen sensor-high voltage)

Code 51 Faulty ECM

Code 52 Faulty ECM

Code 53 Vacuum Switching Valve (VSV)—grounded or faulty power transistor

Code 54 Ignition power transistor—grounded or faulty power transistor

Code 55 Faulty ECM

Code 61 Air Flow Sensor (AFS)—grounded, shorted, open or broken HOT wire

Code 62 Air Flow Sensor (AFS)—broken COLD wire

Code 63 Vehicle Speed Sensor (VSS)—no signal input

Code 64 Fuel injector system—grounded or faulty transistor

Code 65 Throttle Valve Switch (TVS)—full contact closed continuously

Code 66 Knock sensor—grounded or open circuit

Code 71—Throttle Position Sensor (TPS)—turbo control system—abnormal signal

Code 72 EGR Vacuum switching valve—output terminal grounded or open

Code 73 EGR Vacuum switching valve—faulty transistor or grounded system

1987–89 1.5L Turbo EFI (Vin 9), 1989 1.6L EFI (Vin 5), 1991–92 1.6L Turbo EFI (Vin 4), 1989–91 2.8L TBI (Vin R) and 1991–94 3.1L TBI (Vin Z) Engines

Code 12 Normal

Code 13 Oxygen sensor circuit

Code 14 Engine Coolant Temperature (ECT) sensor —high temperature indicated

Code 15 Engine Coolant Temperature (ECT) sensor —low temperature indicated

Code 21 Throttle Position Sensor (TPS)—voltage high

Code 22 Throttle Position Sensor (TPS)—voltage low

Code 23 Intake Air Temperature (IAT)—low temperature indicated

Code 24 Vehicle Speed Sensor (VSS)—no input signal

Code 25 Intake Air Temperature (IAT)—high temperature indicated

Code 31 Turbocharger wastegate control

Code 32 EGR system fault

Code 33 Manifold Absolute Pressure (MAP) sensor—voltage high

Code 34 Manifold Absolute Pressure (MAP) sensor—voltage low

Code 42 Electronic Spark Timing (EST) circuit fault

Code 43 Electronic Spark Control (ESC)—knock failure circuit

Code 44 Oxygen sensor circuit—lean exhaust

Code 45 Oxygen sensor circuit—rich exhaust

Code 51- PROM error—faulty or incorrect PROM

Code 52- CALPAK error—faulty or incorrect CALPAK

Code 54 Fuel Pump Circuit—low voltage

Code 55 ECM error

1990–91 1.6L EFI (Vin 5), 1992–94 1.8L EFI (Vin 8), 1991–94 2.3L EFI (Vin 5/6), 1992–94 3.2L EFI (Vin V/W) Engines

Code 23 Intake Air Temperature (IAT)—out of range

Code 24 Vehicle Speed Sensor (VSS)—no input signal

Code 32 EGR system fault

Code 33 Manifold Absolute Pressure (MAP) sensor—out of range

Code 44 Oxygen sensor circuit—lean exhaust

Code 45 Oxygen sensor circuit—rich exhaust

Code 51 ECM failure

Code 13 Oxygen sensor circuit

Code 14 Engine Coolant Temperature (ECT) sensor -out of range

Code 21 Throttle Position Sensor (TPS)—out of range

Jaguar

GENERAL INFORMATION

Fuel metering is controlled by regulating the time that the injectors are open during the engine operating cycle. Constant fuel pressure is maintained within the fuel rail; injector duration or operating time controls the volume of fuel admitted to the cylinders.

The injection system is managed by a digital Engine Control Unit (ECU). This micro-processor based unit controls the electrical signals to the injectors, triggering them for the correct time period. The ECU relies primarily on the manifold pressure and rpm signals from the engine. Once these signals, indicating engine speed and load, are received, the controller uses them to choose proper injector operating periods. This basic pulse length will be slightly modified by the signals from other engine sensors. These secondary control factors include engine coolant temperature, inlet air temperature, throttle position and battery condition.

The injectors are triggered 6 times per engine cycle; on V-12 engines, the injectors of both banks are operated 6 times per cycle.

SELF-DIAGNOSTICS

Service Precautions

• Keep all PCME parts and harnesses dry during service. Protect the PCME and all solid-state components from rough handling or extremes of temperature.

• Before attempting to remove any parts, turn the ignition switch OFF and disconnect the battery ground cable.

• Make sure all harness connectors are fastened securely. A poor connection can cause an extremely high voltage surge, resulting in damage to integrated circuits.

• Always use a 12 volt battery as a power source.

• Do not disconnect the battery cables with the engine running; never run the engine with battery cables loose or disconnected.

• Do not attempt to disassemble the PCME unit under any circumstances.

• When performing PCME input/output signal diagnosis, remove the water-proofing rubber plug, if equipped, from the connectors to make it easier to insert tester probes into the connector. Always reinstall it after testing.

• When connecting or disconnecting pin connectors from the PCME, take care not to bend or break any pin terminals. Check that there are no bends or breaks on PCME pin terminals before attempting any connections.

• When measuring supply voltage of PCME-controlled components, keep the tester probes separated from each other and from accidental grounding. If the tester probes accidentally make contact with each other during measurement, a short circuit will damage the PCME.

• Use great care when working on or around air bag systems. Due to back-up circuitry, the system may stay armed for a period of time without battery power.

- When working on or around air bag wiring or components, always disarm the system using the correct procedures. Once disarmed, attach a flag or note to the steering wheel. Re-arm the system when repairs are completed and remove the wheel marker.
- Never attempt to measure the resistance of the air bag squib; detonation may occur.

Reading Codes

1987–94 VEHICLES

For 1987–91 vehicles, only the 6 cylinder engines provide fault codes. For 1992–94 vehicles, all engines will provide fault codes. The codes may be read on the trip computer display on the dashboard. Generally, codes will be held in memory and the dash warning lamp will be lit when a fault is sensed by the PCME.

➡Not every stored code will cause the warning lamp to light after the first occurrence. Always check for codes during diagnosis, including cases where the dash warning lamp was not reported lit.

To read the stored codes on the XJ6 models, bring the vehicle to a complete stop. Switch the ignition **OFF** and wait at least 10 seconds. Turn the ignition to the II or **ON** position but do not start or crank the engine. Press the trip computer button (VCM). After a short period of time, the stored codes will be displayed graphically on the trip computer panel. The display may include the designation FF, an abbreviation for FUEL system FAILURE. The codes on XJS models will be displayed 5 seconds after the ignition key is turned to the II or **ON** position but do not start or crank the engine.

On both XJ6 and XJS models, the fault codes are displayed in order of priority, one at a time. The next is only displayed when the preceding one is cleared. To retrieve next code, turn the ignition key to the **OFF** position; now repeat the read and clear procedure again for as many times as necessary.

Clearing

1987–91 VEHICLES

Codes will not be cleared from memory until all stored codes have been displayed. Interrupting the display cycle will allow some codes to be retained and the memory will not clear.

To clear stored codes, drive the vehicle in excess of 19 mph (20 km/h); the memories will be cleared electronically. Each system will perform its self-check function and any new codes will be set. If no faults are present in a particular system, the failure memory will remain cleared.

1992–94 VEHICLES

To clear engine codes on XJ6 & XJS models find the Diagnostic Trouble Code (DTC) reset connector. On XJ6 models the DTC connector is a red round econo seal connector with a pink/red wire. The connector is located behind the passenger side under dash panel next to the PCME. On XJS models the DTC connector is a purple male PM5 connector with a yellow/green wire. The connector is located behind the passenger side center console footwell panel. To clear the codes use a jumper wire to connect the DTC connector to ground for 3 seconds. This will clear one code at a time so you can proceed to the next trouble code. Repeat this procedure until no more trouble codes are present.

DIAGNOSTIC TROUBLE CODES

1987–89 3.6L Engine

Code 1 Cranking signal failure—crankshaft signal missing after cranking for 6 seconds or cranking signal line from Li 2-8 is active above 2000 rpm

Code 2 Airflow meter circuit—open or shorted to ground

Code 3 Coolant temperature sensor failure

Code 4 Feedback failure (where applicable)

Code 5 Airflow meter/throttle potentiometer failure—low throttle potentiometer voltage with high airflow meter voltage

Code 6 Airflow meter/throttle potentiometer failure—high throttle potentiometer voltage with low airflow meter voltage

Code 7 Idle fuel adjustment potentiometer failure

Code 8 Not allocated—If this code appears, a 6.8 kilo-ohm resistor installed in place of a hot start sensor is faulty

1989–94 4.0L, 1992–94 5.3L Engine

Code 11 Idle potentiometer TPS out of range

Code 12 Airflow meter circuit signal out range

Code 13 PCME pressure sensor loss of vacuum signal, incorrect fuel pressure or faulty PCME

Code 14 Engine Coolant Temperature (ECT)—sensor signal out of range or static during engine warm-up

Code 16 Intake Air Temperature (IAT)—sensor resistance out of range or faulty thermistor

Code 17 Throttle Position Sensor (TPS), open or short

Code 18 Throttle Position Sensor (TPS)/Mass Air Flow Meter (MAFS) calibration TPS voltage signal is low with high air flow

Code 19 Throttle Position Sensor (TPS)/Mass Air Flow Meter (MAFS) calibration TPS voltage signal is high with low air flow

Code 22 PCME output to fuel pump relay

Code 23 Poor feedback control in rich direction

Code 23 Fuel supply (rich or lean)—open or short in fuel supply circuit, faulty or restricted fuel line or injectors (5.3L)

Code 24 Ignition drive—PCME output to ignition amplifier module

Code 26 Air leak—poor feedback control in lean direction

Code 29 PCME self check problem

Code 33 Injector drive fault—PCME output to injectors

Code 34 Injector drive circuit—injector leakage

Code 34 Bank A (Right) injectors—open or short circuit / faulty or restricted fuel injectors (5.3L)

Code 36 Bank B (Left) injectors—open or short circuit / faulty or restricted fuel injectors (5.3L)

Code 37 EGR drive—PCME output to EGR switch valve

Code 39 EGR check sensor—checks EGR operation

Code 44 Oxygen (Lambda) sensor—feedback out of control, fuel mixture rich or lean

Code 44 Right Oxygen (Lambda) sensor—resistance out of range (5.3L)

Code 45 Left Oxygen (Lambda) sensor—resistance out of range (5.3L)

Code 46 Idle speed control coil 1 drive—PCME output to idle speed control stepper motor

Code 47 Idle speed control coil 2 drive—PCME output to idle speed control stepper motor

Code 48 Idle speed control motor/valve—stepper motor out of position, temperature less than 860F (300C)

Code 49 Fuel injection ballast resistor—open circuit or faulty resistor (5.3L)

Code 66 Secondary Air Injection Relay—voltage out of operating range

Code 68 Road speed sensor—PCME senses vehicle travel at greater than 5 km/h with high engine air flow

Code 69 Park/Neutral/Park Switch

Code 89 Purge valve drive

Lexus

GENERAL INFORMATION

This system is broken down into 3 major systems: the Fuel System, Air Induction System and the Electronic Control System.

The air induction system provides sufficient clean air for the engine operation. This system includes the throttle body, air intake ducting and cleaner and idle control components. Lexus equips the E5300, 5C300 and 5C400 with a system of induction tuning that changes the induction path length. The intake Air Control Valve (IACV) changes the length of the induction path to broaden the power curve by matching the resonance characteristics of the intake charge with the engine speed. When the IACV opens, the effective length of the intake tract is shortened, boosting top end power. With the IACV closed, the intake tract is long, boosting the low end torque.

Lexus engines are equipped with a computer which centrally controls the electronic fuel injection, electronic spark advance and the exhaust gas recirculation valve. The systems can be diagnosed by means of an Electronic Control Unit (ECU).

The ECU receives signals from the various sensors indicating changing engine operations conditions such as:

- Intake air flow
- Intake air temperature
- Coolant temperature sensor
- Engine rpm
- Acceleration/deceleration
- Exhaust oxygen content

These signals are utilized by the ECU to determine the injection duration necessary for an optimum air/fuel ratio.

SELF-DIAGNOSTICS

Service Precautions

- Keep all ECU parts and harnesses dry during service. Protect the ECU and all solid-state components from rough handling or extremes of temperature.
- Before attempting to remove any parts, turn the ignition switch **OFF** and disconnect the battery ground cable.
- Make sure all harness connectors are fastened securely. A poor connection can cause an extremely high voltage surge, resulting in damage to integrated circuits.
- Always use a 12 volt battery as a power source.
- Do not attempt to disconnect the battery cables with the engine running.
- Do not attempt to disassemble the ECU unit under any circumstances.
- If installing a 2-way or CB radio, mobile phone or other radio equipment, keep the antenna as far as possible away from the electronic control unit. Keep the antenna feeder line at least 8 in. away

from the EEI harness and do not run the lines parallel for a long distance. Be sure to ground the radio to the vehicle body.

- When performing ECU input/output signal diagnosis, remove the water-proofing rubber plug, if equipped, from the connectors to make it easier to insert tester probes into the connector. Always reinstall it after testing.
- Always insert test probes into a connector from the wiring side when checking continuity, amperage or voltage.
- When connecting or disconnecting pin connectors from the ECU, take care not to bend or break any pin terminals. Check that there are no bends or breaks on ECU pin terminals before attempting any connections.
- When measuring supply voltage of ECU-controlled components, keep the tester probes separated from each other and from accidental grounding. If the tester probes accidentally make contact with each other during measurement, a short circuit will damage the ECU.
- Use great care when working on or around air bag systems. Wait at least 20 seconds after turning the ignition switch to LOCK and disconnecting the negative battery cable before performing any other work. The air bag system is equipped with a back-up power system which will keep the system functional for 20 seconds without battery power.
- All air bag connectors are a standard yellow color. The related wiring is encased in standard yellow sheathing. Testing and diagnostic procedures must be followed exactly when performing diagnosis on this system. Improper procedures may cause accidental deployment or disable the system when needed.
- Never attempt to measure the resistance of the air bag squib. Detonation may occur.

Reading Codes

1990–95

All models contain a self-dignostic system. Stored fault codes are transmitted through the blinking of the CHECK ENGINE warning lamp. This occurs when the system is placed in normal diagnostic mode or in test mode. Normal diagnostic mode is used to read stored codes while the vehicle is stopped. The test mode is used after the vehicle is driven under certain conditions. In test mode,

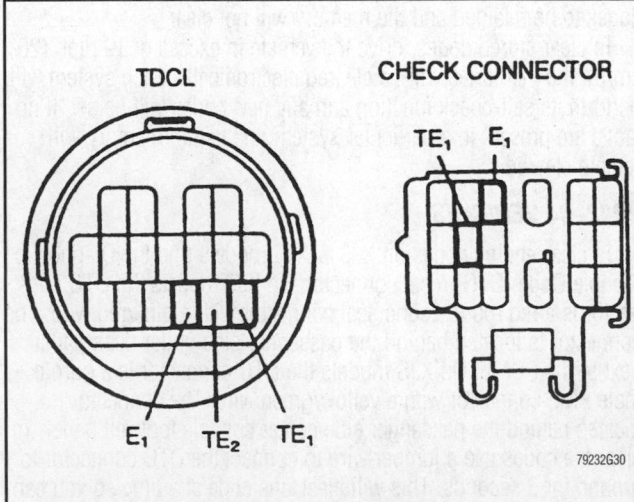

Jump terminals TE_1 and E_1 to enter normal diagnostic mode, jump terminals TE_2 and E_1 to enter test mode

while the ECU monitors, the technician will simulate conditions of the suspected fault in an attempt to cause the malfunction. When a malfunction is found, the CHECK ENGINE lamp will illuminate to alert the technician that the fault is presently occurring.

When troubleshooting 1995 E5300 and L5400 models, an OBD II scan tool or LEXUS hand held tester is necessary to access codes and read data output from the ECM. This is special tool, some after-market tools may be available from your autoparts store for purchase or rent.

WITHOUT OBD II SYSTEM

To read the fault codes, the following initial conditions must be met:

1. Battery voltage at or above 11 volts.
2. Throttle fully closed.
3. Transmission in N.
4. All electrical systems and accessories OFF.

Normal Diagnostic Mode

5. Turn the ignition **ON** but do not start the engine.

6. Use a jumper wire to connect terminals TE1 and E1 of the check connector in the engine compartment or of the TDCL connector below the left side of the dash, if so equipped.

7. Fault codes will be transmitted through the controlled flashing of the CHECK ENGINE warning lamp.

8. If no malfunction was found or no code was stored, the lamp will flash 2 times per second with no other pauses or patterns. This confirms that the diagnostic system is working but has nothing to report. This light pattern may be referred to as the system normal signal. It should be present when no other codes are stored.

9. If faults are present, the CHECK lamp will blink the number of the code(s). All codes are 2 digits; the pulsing of the light represents the digits, not the count. For example, Code 25 is displayed as 2 flashes, a pause and 5 flashes.

10. If more than 1 code is stored, the next will be transmitted after a 2½ second pause.

➡ **If multiple codes are stored, they will be transmitted in numerical order from lowest to highest. This does not indicate the order of fault occurrence.**

11. When all codes have been transmitted, the entire pattern will repeat after a 4½ second pause. The cycle will continue as long as the diagnostic terminals are connected.

12. After recording the codes, disconnect the jumper at the diagnostic connector and turn the ignition **OFF**.

Test Mode

1. Turn the ignition switch **OFF**.

2. Use a jumper wire to connect the TE2 and E1 terminals of the check connector or TDCL. The test mode cannot be initiated if the connection between TE2 and E1 is made with the key in the **ON** position.

3. Turn the ignition switch **ON**, but do not start the engine. The CHECK ENGINE light should flash. If the light does not flash check the TE1 terminal circuit.

4. Start the engine and simulate the conditions of the problem or malfunction.

5. When the road test is complete, connect the TE1 and E1 terminals of the TDCL or check connector and read trouble codes.

6. After the codes are read and noted, disconnect the jumpers from the connector. When the engine is not cranked, a code for starter signal and cam position sensors will be set, but this is not abnormal. If any of the sensed switches are used, the transmission shift lever, throttle or air conditioner, the switch condition code will be set, but this is not abnormal either.

Clearing Codes

1990–95 VEHICLES

Although the CHECK ENGINE lamp will reset itself after a repair is made, the original fault code will still be stored in memory. It is therefore necessary to clear the code after repairs are completed.

1. Turn the ignition **OFF**
2. Remove the 20 amp EFI fuse from junction fuse box No. 2
3. Wait at least 10 seconds before reinstalling the EFI fuse.
4. Road test vehicle and check to see that no fault codes are present.

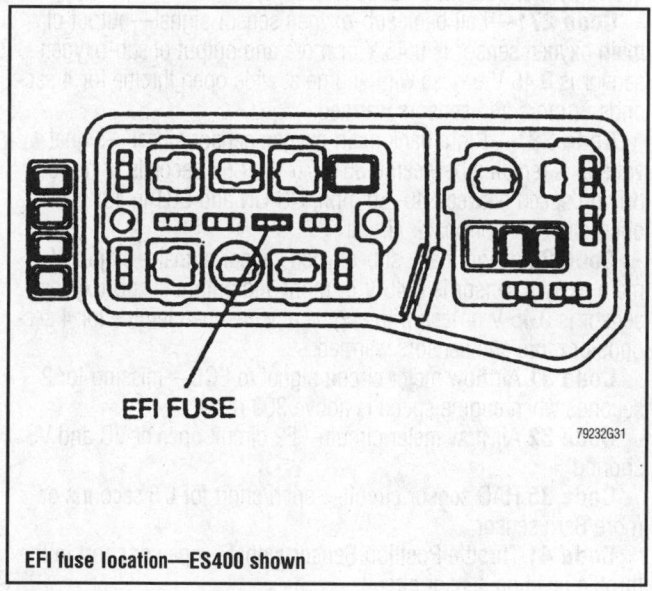

EFI fuse location—ES400 shown

79232G31

DIAGNOSTIC TROUBLE CODES

1990–95 Vehicles Without OBD II System

Constant blinking of CHECK ENGINE light: Normal system operation

Code 12 Rpm NE, Gi or G2 signal to ECU—missing for 2 seconds or more after STA turns ON

Code 13 Rpm NE signal to ECU—missing for 50 msec. or more with engine speed above 1000 rpm, between 2 pulses of the G signal, NE signal of other than 12 pulses to ECU, or deviance of Gi, G2 and NE signal continues for 1 second with engine warm and idling.

Code 14 Igniter IGF1 signal to ECU—missing for 8 successive ignitions

Code 15 Igniter IGF2 signal to ECU—missing for 8 successive ignitions

Code 16 ECT control signal—normal signal missing from ECT CPU (1990–94)

Code 16 NT control system—normal signal missing from between the engine CPU and NT CPU in the ECM (1995)

Code 17 No. 1 cam position sensor—Gi signal to ECU missing

Code 18 No. 2 cam position sensor—G2 signal to ECU—missing

Code 21* Left bank main oxygen sensor signal—signal voltage is remains between 0.35–0.70 V for 60 seconds or more at driving speed between 40–50 mph, NC ON and ECT in 4th gear or open/short sensor heater circuit

Code 22 Engine Coolant Temperature sensor circuit -open/short for 0.5 seconds or more

Code 24 Intake air temperature sensor circuit -open/short for 0.5 seconds or more

Code 25*—Air/fuel ratio LEAN malfunction—voltage output from oxygen sensor is less than 0.45 V for 90 seconds with engine racing at 2000 rpm, feedback frequency 5 Hz or more with main oxygen sensor signal centered at 0.45 V and idle switch ON, or feedback value of right and left banks differs by more than a certain percentage.

Code 26*—Air/fuel ratio RICH malfunction—feedback frequency 5 Hz or more with main oxygen sensor signal centered at 0.45 V and idle switch ON, or feedback value of right and left banks differs by more than a certain percentage.

Code 27*—Left bank sub-oxygen sensor signal—output of main oxygen sensor is 0.45 V or more and output of sub-oxygen sensor is 0.45 V or less with engine at wide open throttle for 4 seconds or more and sensors warmed.

Code 28*—Right bank main oxygen sensor signal—signal voltage is remains between 0.35–0.70 V for 60 seconds or more at driving speed between 40–50 mph, NC ON and ECT in 4th gear or open/short sensor heater circuit

Code 29* Right bank sub-oxygen sensor signal—output of main oxygen sensor is 0.45 V or more and output of sub-oxygen sensor is 0.45 V or less with engine at wide open throttle for 4 seconds or more and sensors warmed.

Code 31 Air flow meter circuit signal to ECU—missing for 2 seconds when engine speed is above 300 rpm

Code 32 Air flow meter circuit—E2 circuit open or VC and VS shorted

Code 35 HAC sensor circuit—open/short for 0.5 seconds or more/Baro sensor

Code 41 Throttle Position Sensor signal—open or short in the throttle position sensor circuit.

Code 42 Vehicle speed sensor circuit—engine RPM over 2350, and VSS shows zero miles per hour.

Code 43 Starter signal to ECU—missing

Code 47 Sub-throttle position sensor signal (VTA2) -open/short for at least 0.5 seconds or signal outputs exceed 1.45 V with idle contacts ON.

Code 51 NC signal ON, IDL contacts off, or shift in R, D, 2 or 1 range—during check mode

Code 52 No. 1 knock sensor signal—missing from ECU for 3 revolutions when engine speed is between 1600–5200 rpm

Code 53 Knock control signal—ECU knock control malfunction detected with engine speed between 650–5200 rpm

Code 55 No. 2 knock sensor signal—missing from ECU for 3 revolutions when engine speed is between 1600–5200 rpm

Code 71* EGR gas temperature below 149°F (65°C) for 90 seconds or more during EGR control

Code 78* Fuel pump control signal—open or short in the fuel pump control circuit

➡ ***2 trip detection logic code: A single occurrence of this fault will be temporarily stored in memory. CHECK ENGINE light will NOT illuminate until fault is detected a second time (during a separate ignition cycle).**

Mazda

GENERAL INFORMATION

Mazda utilizes 2 types of fuel systems between the years 1984–94. The Feedback Carburetor (FBC) system and Electronic Gas Injection (EGI) system, or fuel injection system. The feedback carburetor system was used in the GLC, 323, 626 and RX-7 between years 1984–87.

It was also used in the B2000, B2200 and B2600 pickup trucks between years 1984–92.

Electronic Gas Injection (EGI) was first available in the 1984 RX-7. 626 picked it up in 1986, 323 in 1987. In 1988 all models except B2200 and B2600 pickup trucks came equipped with fuel injection. Mazda uses various variations of EGI. Navajo uses the Ford EEC-IV system. However, the EEC-IV system will not be covered in this section.

SELF-DIAGNOSTICS

Service Precautions

• Before connecting or disconnecting the ECU harness connectors, make sure the ignition switch is OFF and the negative battery cable is disconnected to avoid the possibility of damage to the control unit.

• When performing ECU input/output signal diagnosis, remove the pin terminal retainer from the connectors to make it easier to insert tester probes into the connector.

• When connecting or disconnecting pin connectors from the ECU, take care not to bend or break any pin terminals. Check that there are no bends or breaks on ECU pin terminals before attempting any connections.

• Before replacing any ECU, perform the ECU input/output signal diagnosis to make sure the ECU is functioning properly or not.

• After checking through EGI troubleshooting, perform the EFI self-diagnosis and driving test.

• When measuring supply voltage of ECU controlled components with a circuit tester, separate 1 tester probe from another. If the 2 tester probes accidentally make contact with each other during measurement, a short circuit will result and may damage the ECU.

Reading Codes

➡ **Diagnostic codes may be retrieved through the use of the CHECK ENGINE light or Malfunction Indicator Lamp (MIL). Special System Checker No. 83, Digital Code Checker and a Self-diagnosis Checker are all special diagnostic equipment used to retrieve codes, however these tools are not required. When using special diagnostic equipment, always observe the tool manufacturer's instructions.**

1984–86 VEHICLES WITH SYSTEM CHECKER TOOL 83

On 1984–85 GLC, 626 and RX-7, 1986 323 and 1986 B2000 Pick-up, the System Checker No. 83 (tool No. 49-G030–920), is used to detect and indicate any problems of each sensor, damaged wiring, poor contact or a short circuit between each of the sensor control units. Trouble is indicated by a red lamp and a buzzer. If there are more than 2 problems at a time, the indicator lamp turns ON in the numerical order of the code number. Even if the problem is corrected during indication, 1 cycle will be indicated, If after a malfunction has occurred and the ignition key is switched **OFF**, the malfunction indicator for the feedback system will not be displayed on the checker.

Read engine trouble codes using the following procedures:
1984–85 GLC

1. Operate the engine until normal operating temperatures are reached. Allow the engine to run at idle.

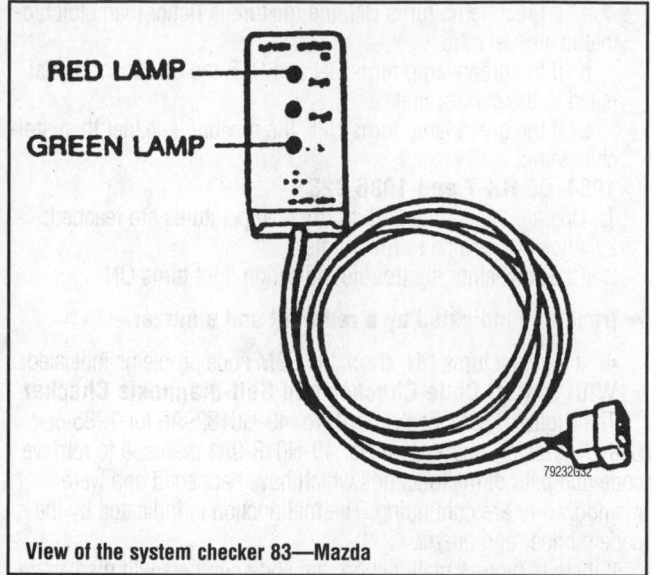

View of the system checker 83—Mazda

2. Connect System Checker tool No. 83 (49-G030–920) to the check connector, located near the ECU.

3. Check whether the trouble indication light turns ON.

➡Trouble is indicated by a red light and a buzzer.

4. If the light turns ON, check for cause of problems.

➡If the trouble code is code number 3 (feedback system), proceed as follows:

5. Start the engine, letting it run until it reaches normal operating temperature. Connect a tachometer to the engine.

6. Connect a dwell meter (90 degrees, 4 cylinder) to the yellow wire in the service (check) connector of the air/fuel solenoid valve.

7. Run the engine at idle and note the reading on the dwell meter.

8. If the dwell meter reading is 00 degrees, the probable causes are as follows:

   a. The wiring hamess from the IG to the check connector BrY terminal lis open.

   b. The wiring harness from the check connector Y terminal to the control unit (F) terminal is grounded.

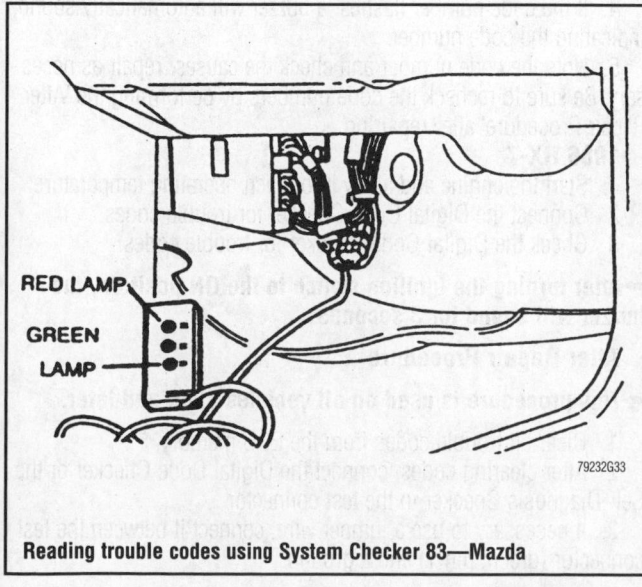

Reading trouble codes using System Checker 83—Mazda

   c. The transistor in the control unit for the air/fuel solenoid is short circuit.

9. If the dwell meter reading is 2711, check whether the green lamp (feedback signal) illuminates or does not illuminate.

10. If the oxygen sensor signal lamp does not illuminate, proceed as follows:

   a. If the green lamp does not illuminate, the air is sucked from the intake system or the air is sucked from the exhaust manifold.

   b. Carburetor jets are clogged.

   c. The valve of the air/fuel solenoid is stuck to the lower position, giving a lean air/fuel mixture condition.

11. If the oxygen sensor signal lamp illuminates, proceed as follows:

   a. If green lamp turns ON, the mixture is richer than stoichiometric air/fuel ratio.

   b. If the green lamp turns ON and OFF, the 02 sensor signal is fed to the control unit.

   c. If the green lamp turns OFF, the mixture is leaner than stoichiometric.

1984–85 626

1. Operate the engine until normal operating temperatures are reached. Allow the engine to run at idle.

2. Connect System Checker tool No. 83 (49-G030–920) to the check connector, located near the ECU.

3. Check whether the trouble indication light turns ON.

➡If there is more than 2 problems at the same time, the indicator lamp lights on in the numerical order of the code number. Even if the problem is corrected during indication, 1 cycle will be Indicated. If after a malfunction has occurred the Ignition key is switched off, the malfunction indicator for the feedback system will not be displayed on the checker. The control unit has a built In fall-safe mechanism. If a malfunction occurs during driving, the control unit will on its own initiative, send out a command and driving performance will be affected. The commands are as follows:

   a. Water Thermo-Sensor—the control unit outputs a constant 1760F (800C) command.

   b. Feed-Back Sensor—the control unit holds air/fuel solenoid to dwell meter reading 180 (duty 0%) for 626 or 270 (duty 30%) for B2200.

   c. Vacuum Sensor—the control unit prevents operation of the EGR valve, and holds the air/fuel solenoid to a duty of 0%.

   d. EGR Position Sensor—the control unit prevents operation of the EGR valve.

➡If the trouble code is code number 3 (feedback system), proceed as follows:

4. Start the engine, letting it run until it reaches normal operating temperature. Connect a tachometer to the engine.

5. Connect a dwell meter (90 degrees, 4 cylinder) to the yellow wire in the service (check) connector of the air/fuel solenoid valve.

6. Run the engine at idle and note the reading on the dwell meter.

7. If the dwell meter reading is 0~ degrees, the probable causes are as follows:

   a. The wiring harness from the IG to the check connector BrY terminal is open.

   b. The wiring harness from the check connector Y terminal to the control unit (F) terminal is grounded.

   c. The transistor in the control unit for the air/fuel solenoid is open.

8. If the dwell meter reading is 90°, the probable causes are as follows:

    a. The wiring harness from the IG to the check connector BrY terminal is open.

    b. The wiring harness from the check connector BrY ter-. minal to the control unit (F) terminal is grounded.

    c. The transistor in the control unit for the air/fuel solenoid is short circuited.

9. If the dwell meter reading is 18~, check whether the green lamp (feedback signal) illuminates or does not illuminate.

10. If the oxygen sensor signal lamp does not illuminate, proceed as follows:

    a. If the green lamp does not illuminate, the air is sucked from the intake system or the air is sucked from the exhaust manifold.

    b. Carburetor jets are clogged.

    c. The valve of the air/fuel solenoid is stuck to the lower position, giving a lean air/fuel mixture condition.

11. If the oxygen sensor signal lamp illuminates, proceed as follows:

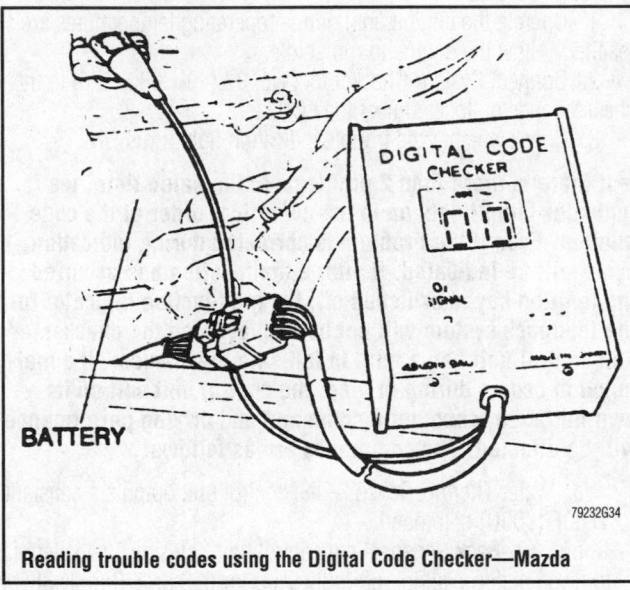

Reading trouble codes using the Digital Code Checker—Mazda

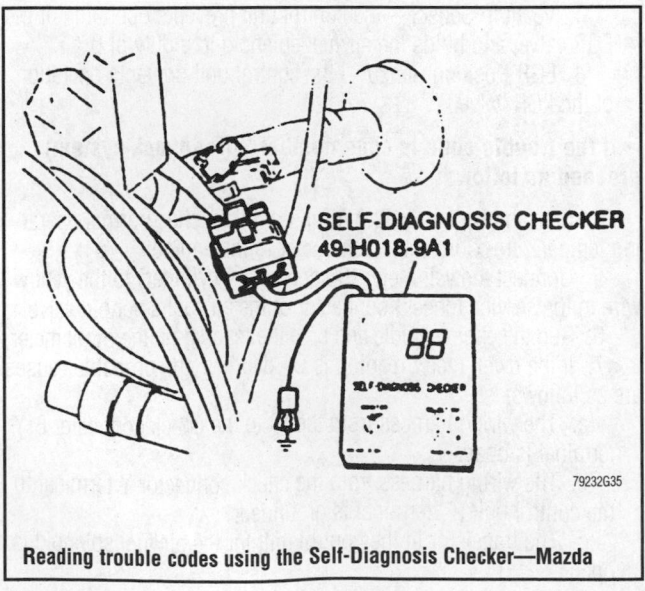

Reading trouble codes using the Self-Diagnosis Checker—Mazda

    a. If green lamp turns ON, the mixture is richer than stoichiometric air/fuel ratio.

    b. If the green lamp turns ON and OFF, the 02 sensor signal is fed to the control unit.

    c. If the green lamp turns OFF, the mixture is leaner than stoichiometric.

1984–85 RX-7 and 1986 323

1. Operate the engine until normal temperatures are reached.
2. Allow the engine to run at idle.
3. Check whether the trouble indication light turns ON.

➡**Trouble is indicated by a red light and a buzzer.**

4. If the light turns ON, check the ECM code problems indicated.
With Digital Code Checker and Self-diagnosis Checker
The Digital Code Checker tool No. 49-G01829A0 for 1986 or Self-Diagnosis Checker tool No. 49-H018-9A1 are used to retrieve code numbers of malfunctions which have happened and were memorized or are continuing. The malfunction is indicated by the code number and buzzer.

If there is more 1 malfunction, the code numbers will display on the self diagnosis checker 1 by 1 in numerical order. In the case of malfunctions, 09, 13 and 01, the code numbers are displayed in order of 01, 09 and then 13.

The ECU has a built in fail-safe mechanism for the main input sensors. If a malfunction occurs, the emission control unit will substitute values; this will slightly effect the driving performance but the vehicle may still be driven.

The ECU continuously checks for malfunctions of the input devices within 2 seconds after turning the ignition switch to the **ON** position and the test connector is grounded.

The malfunction indicator light indicates a pattern the same as the buzzer of the self-diagnosis checker when the self-diagnosis check connector is grounded. When the self-diagnosis check connector is not grounded, the lamp illuminates steady while the malfunction recovers. However, the malfunction code is memorized in the emission control unit.

Read engine trouble codes using the following procedures:
1986 Vehicles Except RX-7

1. Warm the engine to normal operating temperatures, by keeping the engine speed below 400 rpm.
2. Connect the Digital Code Checker No.
3. Wait for 3 minutes for the code(s) to register.
4. If the code number flashes, a buzzer will automatically sound, indicating the code number.
5. Note the code number and check the causes, repair as necessary. Be sure to recheck the code numbers by performing the 'After Repair Procedure' after repairing.
1986 RX-7

1. Start the engine and allow it to reach operating temperature.
2. Connect the Digital Code Checker for trouble codes.
3. Check the Digital Code Checker for trouble codes.

➡**After turning the ignition switch to the ON position, the buzzer will sound for 3 seconds.**

After Repair Procedure

➡**This procedure is used on all vehicles 1986 and later.**

1. Clear all trouble codes from the ECU memory.
2. After clearing codes, connect the Digital Code Checker or the Self-Diagnosis Checker to the test connector.
3. If necessary to use a jumper wire, connect it between the test connector (green: pin 1) and a ground.

4. Turn the ignition switch **ON**, but do not start the engine for 6 seconds.

5. Operate the engine until normal operating temperatures are reached, then, run it at 2000 rpm for 2 minutes.

6. Verify that no code numbers are displayed.

1987–94 VEHICLES WITH SELF-DIAGNOSIS CHECKER

The self-diagnosis checker (49-HOI 8-9A1) and System Selector (49-BOI 9—9A0), are used to retrieve code numbers of malfunctions which have happened and were memorized or are continuing. The malfunction is indicated by a code number.

If there is more than 1 malfunction, the code numbers will display on the self-diagnosis checker in numerical order. The ECU has a built in fail-safe mechanism for the main input sensors. If a malfunction occurs, the emission control unit will substitute values. This will affect driving performance, but the vehicle may still be driven.

The ECU continuously checks for malfunctions of the input devices. But the ECU checks for malfunctions of the output devices within 3 seconds after the green (1 pin) test connector or TEN terminal of the diagnosis connector is grounded and the ignition switch is turned to the **ON** position.

Read engine trouble codes using the following procedures:

1987–91 323, Miata and Protege

1. Connect the tester to the check connector at the rear of the left side wheel housing and to the negative battery cable.

2. Set the tester select switch to the A setting.

3. With a jumper wire, ground the 1-pin test connector.

4. Turn the ignition switch **ON**.

5. Make sure that 88 flashes on the monitor and that the audible buzzer sounds for 3 seconds after turning the ignition switch **ON**.

6. If 88 does not flash, check the main relay, power supply circuit and the check connector wiring.

7. If 88 flashes and the buzzer sounds for more than 20 seconds, replace the engine control unit and repeat Steps 3 and 4.

8. Note any other code numbers that are present and refer to the code chart. Repair if necessary.

1992 626 and MX-6

The check connector is located at the rear of the left side wheel house on 626/MX-6.

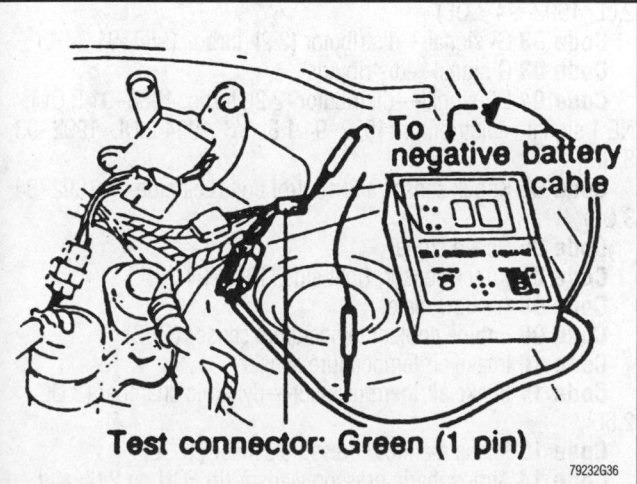

Test connector: Green (1 pin)

79232G36

Reading trouble codes using the Self-diagnosis Checker—1987–94 Mazda

1. Connect the tester to the check connector and to ground.

2. Set the tester select switch to the A setting.

3. With a jumper wire, ground the 1 pin test connector.

4. Turn the ignition switch **ON**.

5. Make sure that 88 flashes on the monitor and that the audible buzzer sounds for 3 seconds after turning the ignition switch **ON**.

6. If 88 does not flash, check the main relay, power supply circuit and the check connector wiring.

7. If 88 flashes and the buzzer sounds for more than 20 seconds, replace the engine control unit and perform steps number 1 through 6 again.

➡**Before replacing the ECU on the MPV or B2600i, check for a short circuit between ECU terminal IB for JE engine and 1F for G6 engine and the 6 pin check connector.**

8. Note and record any other code numbers that are present.

1992–94 323, Protege, 929, Miata (MX-5), MX-3 and 1993–94 RX-7

1. Connect the system selector to the diagnosis connector at the rear of the left side wheel housing.

2. Set the SYSTEM SELECT switch to the 1 setting.

3. Set the TEST switch to the SELF—TEST position.

4. Connect the self-diagnosis checker, the system selector and ground.

5. Set the self-diagnosis checker SELECT switch to the A position.

6. Turn the ignition switch **ON**.

7. Make sure that 88 flashes on the monitor and that the audible buzzer sounds for 3 seconds after turning the ignition switch **ON**.

8. If 88 does not flash, check the main relay, power supply circuit and the diagnosis connector wiring.

9. If 88 flashes and the buzzer sounds for more than 20 seconds, check for a short between ECU terminal 1F and the FEN terminal of the diagnosis connector. Replace the engine control unit if necessary and perform Steps 1 through 7 again.

10. Note and record any other code numbers that are present.

1993–94 626 and MX-6

1. Connect the system selector to the diagnosis connector at the rear of the left side wheel housing.

2. Set the SYSTEM SELECT switch to the 1 setting.

3. Set the TEST switch to the SELF—TEST position.

4. Connect the self-diagnosis checker the system selector and ground.

5. Set the self-diagnosis checker SELECT switch to the A position.

6. Turn the ignition switch **ON**.

7. Make sure that 88 flashes on the monitor and that the audible buzzer sounds for 3 seconds after turning the ignition switch **ON**.

8. If 88 does not flash, check the main relay, power supply circuit and the diagnosis connector wiring.

9. If 88 flashes and the buzzer sounds for more than 20 seconds, check for a short between PCM terminal iF (manual trans.), (iG auto trans.) and the FEN terminal of the diagnosis connector. Replace the engine control unit if necessary and perform Steps 1 through 7 again.

10. Note and record any other code numbers that are present.

1984–94 VEHICLES WITHOUT SELF-DIAGNOSIS CHECKER

The malfunction indicator light indicates a pattern the same as the buzzer of the self-diagnosis checker when the green (1 pin) test connector or FEN terminal of the diagnosis connector is grounded.

When the green (1 pin) test connector or FEN terminal of the diagnosis connector is not grounded, the lamp illuminates steady while malfunction of the main input sensor occurs and goes out if the malfunction recovers. However, the malfunction code is memorized in the control unit.

Clearing Codes

1984–86 VEHICLES

All Vehicles
1. Turn the ignition switch OFF.
2. Disconnect the negative battery cable.
3. Depress the brake pedal for at least 5 seconds.
4. Reconnect the negative battery cable.

1987–91 VEHICLES

All Vehicles
1. Cancel the memory of the malfunction by disconnecting the negative battery cable and depressing the brake pedal for at least 20 seconds, then reconnect the negative battery cable.
2. Except Miata, MX-3, 323 and Protege, connect the Self-Diagnosis Checker 49-H018–9A1 to the check connector. Ground the test connector (green: 1 pin) using a jumper wire.
3. On Miata, MX-3, 323 and Protege, connect Self-Diagnosis Checker (49-BOI 9–9A0) to the diagnosis connector.
4. Turn the ignition switch ON, but do not start the engine for approximately 6 seconds.
5. Start the engine and allow it to reach normal operating temperature. Then run the engine at 2000 rpm for 2 minutes. Check that no code numbers are displayed.

1992–94 VEHICLES

323, MX-3 (B6 engine), MX-5/Miata
1. Disconnect the negative battery cable.
2. Press the brake pedal for at least 20 seconds.
3. Connect the negative battery cable.
4. Connect the self-diagnosis tester to the diagnosis connector.
5. Turn the ignition switch ON.
6. Start and warm-up the engine.
7. Run engine at 2,000 rpm for 3 minutes.
8. Verify that no more codes are stored.

1992 626 and MX-6
1. Disconnect the negative battery cable.
2. Press the brake pedal for at least 5 seconds.
3. Connect the negative battery cable.
4. Connect the self-diagnosis tester and ground the test connector.
5. Turn the ignition switch to the ON position for 6 seconds.
6. Start and warm-up the engine.
7. Run engine at 2,000 rpm for 2 minutes (3 minutes on truck).
8. Verify that no more codes are stored.

1993–94 626 and MX-6 (FS Engine)
1. Disconnect the negative battery cable.
2. Press the brake pedal for at least 20 seconds.
3. Connect the negative battery cable.
4. Connect the self-diagnosis tester to the diagnosis connector.
5. Turn the ignition switch ON.
6. Start and warm-up the engine.
7. Run engine at 2,000 rpm for 2 minutes.
8. Verify that no more codes are stored.

1993–94 626 and MX-6 (KL Engine), MX-3 (K8 engine) and 1993–94 RX-7
1. Disconnect the negative battery cable.
2. Press the brake pedal for at least 20 seconds.
3. Connect the negative battery cable.
4. Connect the self-diagnosis tester to the diagnosis connector.
5. Turn the ignition switch ON.
6. Verify that no more codes are stored.

929
1. Turn the ignition switch OFF.
2. Disconnect the negative battery cable for 20 seconds.

DIAGNOSTIC TROUBLE CODES

1984–94 Except RX-7

1984–85 VEHICLES

2.0L ENGINE (CODE FE)
Code 01 Engine speed
Code 02 Water thermosensor
Code 03 Oxygen sensor
Code 04 Vacuum sensor
Code 05 EGR position sensor

1986–87 VEHICLES

1.6L, 2.0L AND 2.2L ENGINES
Code 01 Ignition pulse
Code 02 Air flow meter
Code 03 Water thermosensor
Code 04 Intake air thermo or Temperature sensor
Code 05 Feedback system
Code 06 Atmospheric pressure sensor (1986 1 .6L)
Code 08 EGR position sensor
Code 09 Atmospheric pressure sensor
Code 22 No. 1 Cylinder sensor (2.2L turbocharged)

1988–94 VEHICLES

16L, 1.8L, 2.0L, 2.2L, 25L, 26L AND 30L ENGINES
Code 01 Ignition pulse
Code 02- Ne signal—distributor
Code 02 NE 2 signal—crankshaft (1992–93 1.8L V6, 1994 2.0L, 1992–94 3.0L)
Code 03 Gi signal—distributor (2.2L turbo, 1988–91 3.0L)
Code 03 G signal—distributor
Code 04 G2 signal—distributor (2.2L turbo, 1988–91 3.0L); NE 1 signal—distributor (1992–94 1.8L V6, 1994 2.0L, 1992–93 3.0L)
Code 05 Knock sensor and control unit (Left side on 1992–94 3.0L)
Code 06 Speed signal
Code 07 Knock sensor; right side (1992–94 3.0L)
Code 08 Air flow meter
Code 09 Engine coolant temperature sensor (C IS)
Code 10 Intake air temperature sensor
Code 11 Intake air thermosensor—dynamic chamber (3.0L, 2.6L)
Code 13 Intake manifold pressure sensor (1 .3L)
Code 14 Atmospheric pressure sensor (in ECU on 2.6L and 1994 2.5L)
Code 15 Oxygen sensor

Code 15 Oxygen sensor; left side on 1992–94 1.8L V6, 1994 2.5L, 1990–94 3.0L

Code 16 EGR position sensor

Code 17 Closed loop system

Code 17 Closed loop system; left side on 1992–94 1.8L V6, 1993–94 2.5L 1990–94 3.0L

Code 23 Heated oxygen sensor; right side on 1992–94 1.8L V6, 1994 2.5L 1990–91 3.0L

Code 24 Closed loop system; right side on 1992–94 1.8L V6, 1993 2.5L 1990–91 3.0L

Code 25 Solenoid valve—pressure regulator

Code 26 Solenoid valve—purge control

Code 26 Solenoid valve—purge control No. 2 (1988–89 3.0L)

Code 27 Solenoid valve—purge control No. 1 (1988–89 3.0L)

Code 27 Solenoid valve—No. 2 purge control (1989 1.6L)

Code 28 Solenoid valve—EGR vacuum

Code 29 Solenoid valve—EGR vent

Code 30 Relay (cold start injector 3.0L)

Code 34 ISC valve

Code 34 Idle air control valve (1993–94 2.0L and 2.5L, 1.6L, 1.8L, 2.6L, 3.1L)

Code 36 Oxygen sensor heater relay (1990 3.0L)

Code 36 Right side oxygen sensor heater (1992–94 3.0L)

Code 37 Left side oxygen sensor heater (1992–94 3.0L)

Code 37 Coolant fan relay

Code 40 Oxygen sensor heater relay (1991 3.0L)

Code 40 Solenoid (triple induction system) and oxygen sensor relay (1988–89 3.0L)

Code 41 Solenoid valve—VRIS (1989–94 MPV 3.0L)

Code 41 Solenoid valve—VRIS 1 (1992–94 1.8L V6, 1993 2.5L)

Code 41 Solenoid valve—VICS (3.0L)

Code 42 Solenoid valve—Waste gate (turbocharged)

Code 46 Solenoid valve—VRIS 2 (1992–94 1.8L V6, 1993 2.5L)

Code 65 NC signal—PCMT (1992–94 3.0L)

Code 67 Coolant fan relay No. 1 (1993 2.5L)

Code 67 Coolant fan relay No. 2 (1992–94 1.8L V6)

Code 68 Coolant fan relay No. 2, No.3 with ATX (1993 2.5L)

Code 69 Engine coolant temperature sensor—fan (1992–94 1.8L V6, 1993 2.0L and 2.5L)

1984–94 RX-7

1.3L ROTARY ENGINE

Code 01 Crank angle sensor (1984–87)

Code 01 Ignition coil—trailing (1988–91)

Code 02 Air flow meter (1984–87)

Code 02 Ne signal—crank angle sensor (1988–91)

Code 03 Water thermosensor (1984–87)

Code 03 G signal—crank angle sensor (1988–91)

Code 04 Intake air temperature sensor—in the air flow meter (1984–87)

Code 05 Oxygen sensor (1984–87)

Code 05 Knock sensor (1993)

Code 06 Throttle sensor (1984–87)

Code 06 Speedometer sensor (1993)

Code 07 Boost sensor / Pressure sensor (1984–87 turbo)

Code 08 Air flow meter (1988–91)

Code 09 Atmospheric pressure sensor (1984–87)

Code 09 Water thermosensor (1988–94)

Code 10 Intake air thermosensor—in air flow meter (1988–91)

Code 11 Intake air thermosensor (1988–93)

Code 12 Coil with igniter—trailing (1984–87)

Code 12 Throttle sensor—wide open throttle (1988–94)

Code 13 Intake manifold pressure sensor (1988–94)

Code 14 Atmospheric pressure sensor (1988–94, in ECU on 1993)

Code 15 Intake air temperature sensor—in dynamic chamber (1984–87)

Code 15 Oxygen sensor (1988–94)

Code 16 EGR switch (1993 California)

Code 17 Closed loop system (1988–93)

Code 18 Throttle sensor—closed or narrow throttle (1988–94)

Code 20 Metering oil pump position sensor (1988–94)

Code 23 Fuel thermosensor (1993)

Code 25 Solenoid valve—pressure regulator control (1993)

Code 26 Metering oil pump stepping motor (1993)

Code 27 Step motor—metering oil pump (1988–91)

Code 27 Metering oil pump (1993)

Code 28 Solenoid valve—EGR (1993)

Code 30 Solenoid valve—split air bypass (1988–94)

Code 31 Solenoid valve—relief No. 1 (1988–94)

Code 32 Solenoid valve—switching (1988–94)

Code 33 Solenoid valve—port air bypass (1988–94)

Code 34 Solenoid valve—bypass air control (1988–91)

Code 34 Solenoid valve—idle speed control (1993)

Code 37 Metering oil pump (1988–93)

Code 38 Solenoid valve—accelerated warm-up system (1988–94)

Code 39 Solenoid valve—relief No. 2 (1993)

Code 40 Auxiliary port valve (1988–91)

Code 40 Solenoid valve—purge control (1993)

Code 41 Solenoid valve—variable dynamic effect Intake control (1988–91)

Code 42 Solenoid valve—turbo boost pressure regulator (1988–91)

Code 42 Solenoid valve—turbo pre-control (1993)

Code 43 Solenoid valve—wastegate control (1993)

Code 44 Solenoid valve—turbo control (1993)

Code 45 Solenoid valve—charge control (1993)

Code 46 Solenoid valve—charge relief (1993)

Code 50 Solenoid valve—double throttle control (1993)

Code 51 Fuel pump relay (1988–94)

Code 54 Air pump relay (1993)

Code 71 Injector—front secondary (1988–94)

Code 73 Injector—rear secondary (1988–94)

Code 76 Slip lockup off signal—EC-AT CU (1993)

Code 77 Torque reduced—EC-AT CU (1993)

Mitsubishi

GENERAL INFORMATION

Mitsubishi uses 2 types of fuel systems. Feedback carburetor system and fuel injection. The type of fuel injection system is known as Electronic Controlled Injection (ECI).

Mitsubishi uses a conventional downdraft two-barrel compound type carburetor which incorporates an automatic choke, accelerator pump, and enrichment system. In addition, a deceleration device is provided.

The Electronic Fuel Injection (EFI) system, used on Mitsubishi vehicles, is classified as a Multi-Point Injection (MPI) system. The MPI system controls the fuel flow, idle speed, and ignition timing. The basic function of the MPI system is to control the air/fuel ratio in accordance with all engine operating conditions. An Electronic Control Unit (ECU) is the heart of the MPI system. Based on data from various sensors, the ECU computes the desired air/fuel ratio.

SELF-DIAGNOSTICS

Service Precautions

- Before connecting or disconnecting the ECU harness connectors, make sure the ignition switch is OFF and the negative battery cable is disconnected to avoid the possibility of damage to the control unit.
- When performing ECU input/output signal diagnosis, remove the pin terminal retainer from the connectors to make it easier to insert tester probes into the connector.
- When connecting or disconnecting pin connectors from the ECU, take care not to bend or break any pin terminals. Check that there are no bends or breaks on ECU pin terminals before attempting any connections.
- Before replacing any ECU, perform the ECU input/output signal diagnosis to make sure the ECU is functioning properly.
- When measuring supply voltage of ECU-controlled components with a circuit tester, separate 1 tester probe from another. If the 2 tester probes accidentally make contact with each other during measurement, a short circuit will result and damage the ECU.

Reading Codes

➡**All though the CHECK ENGINE light or Malfunction Indicator Lamp (MIL) will illuminate when there is trouble detected, diagnostic codes can only be retrieved with the use of either a analog voltmeter or a Multi-use Tester. When using diagnostic equipment, always observe the tool manufacturer's instructions.**

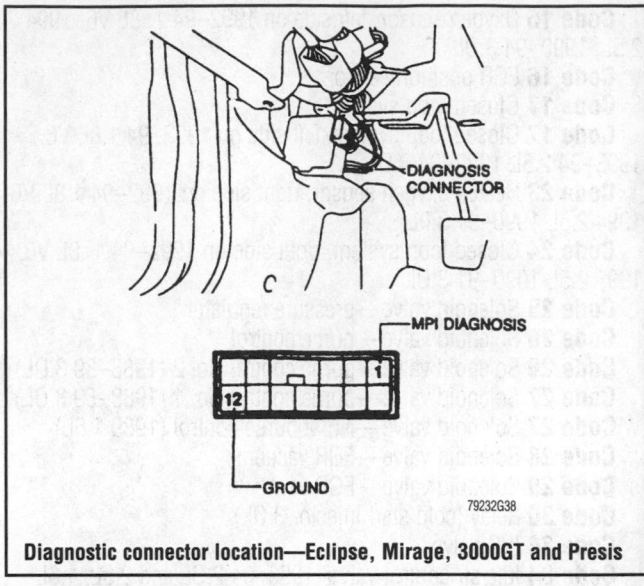

Diagnostic connector location—Eclipse, Mirage, 3000GT and Presis

1984–86 VEHICLES

With ECI/MPI Tester

Refer to manufacturer's tester manual regarding diagnosis with this tester.

1985–94 VEHICLES

With Analog Voltmeter

The voltmeter can be used to retrieve code numbers of malfunctions which have happened and were memorized or are continuing to happen. On the voltmeter, the malfunction is indicated by a sweep of the needle. The voltmeter should be connected to the data link connector located under the driver side dashboard. Connect the voltmeter between the Multi-Point Injection (MPI) terminal and the ground terminal. Turn the ignition switch **ON** if the normal condition exists, the voltmeter pointer will indicate a normal pattern. A normal pattern is indicated by constant needle sweeps. If a problem exists in the system the voltmeter pointer will indicate it in a series of pointer sweeps. For example, a Code 3 would be 3 consecutive short sweeps of the voltmeter needle.

If there is more than 1 malfunction, the low code numbers will

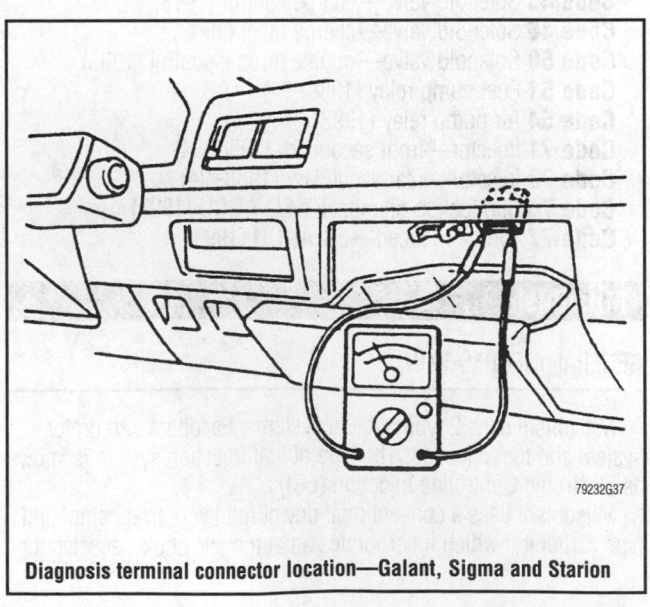

Diagnosis terminal connector location—Galant, Sigma and Starion

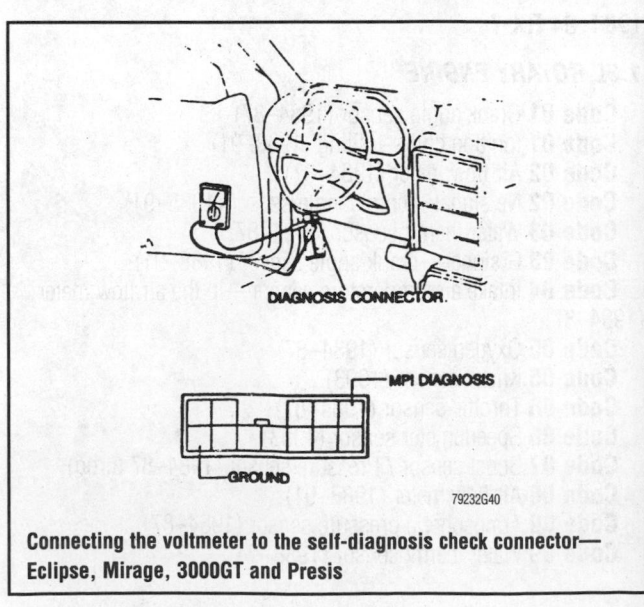

Connecting the voltmeter to the self-diagnosis check connector— Eclipse, Mirage, 3000GT and Presis

first be indicated and after a 2 second pause (no code indication) the higher code will be indicated.

With Multi-Use Tester

To read the trouble codes using the Multi-Use Tester (MB991341 or equivalent) follow the steps below:

1. Turn the ignition switch **OFF**.
2. Insert the power supply terminal to the cigarette lighter socket.
3. Connect the tester connector to the diagnosis connector in the glove compartment, under the 'hood or under the driver side dashboard.
4. Turn the ignition switch **ON** and push the DIAG key.
5. Observe the trouble code and make the necessary repairs.

On most models the CHECK ENGINE malfunction indicator light will light up and remain illuminated to indicate that there is a problem in the system. After this light has been reported to be ON, the system should be checked for malfunction codes.

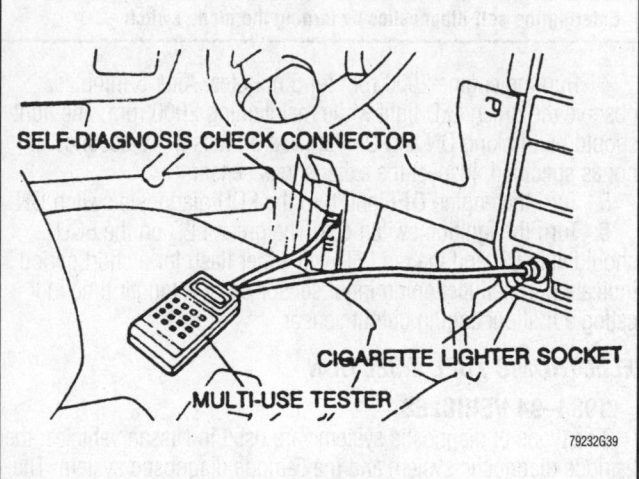

SELF-DIAGNOSIS CHECK CONNECTOR

CIGARETTE LIGHTER SOCKET

MULTI-USE TESTER

79232G39

Connecting the multi-use tester to the self-diagnosis check connector location—Eclipse, Mirage, 3000GT and Presis

Clearing Codes

Without Multi-Use Tester

1984–86 vehicles—engine codes can be cleared by disconnecting the negative battery terminal or by disconnecting ECU connector for 15 seconds or longer.

1987–94 vehicles—engine codes can be cleared by disconnecting the negative battery terminal for 10 seconds or longer.

With Multi-Use Tester

Engine codes may also be cleared by setting the ignition switch to the **ON** position and using the malfunction code ERASE signal.

DIAGNOSTIC TROUBLE CODES

1989–94 Vehicles

DIAMANTE, ECLIPSE, 3000GT, GALANT, MIRAGE, MONTERO, PRECIS, SIGMA, STARION, EXPO:

Code 11 Oxygen sensor
Code 12 Air flow sensor
Code 13 Intake Air Temperature Sensor
Code 14 Throttle Position Sensor (TPS)
Code 15 SC Motor Position Sensor (MPS)
Code 21 Engine Coolant Temperature Sensor

Code 22 Crank angle sensor
Code 23 No. 1 cylinder TDC (Camshaft position) Sensor
Code 24 Vehicle speed sensor
Code 25 Barometric pressure sensor
Code 31 Knock (KS) sensor
Code 32 Manifold pressure sensor
Code 36 Ignition timing adjustment signal
Code 39 Oxygen sensor (rear—turbocharged)
Code 41 Injector
Code 42 Fuel pump
Code 43 EGR-California
Code 44 Ignition Coil; power transistor unit (No. 1 and No. 4 cylinders) on 3.0L
Code 52 Ignition Coil; power transistor unit (No. 2 and No. 5 cylinders) on 3.0L
Code 53 Ignition Coil; power transistor unit (No. 3 and No. 6 cylinders)
Code 55 AC valve position sensor
Code 59 Heated oxygen sensor
Code 61 Transaxle control unit cable (automatic transmission)
Code 62 Warm-up control valve position sensor (non-turbo)

Nissan

GENERAL INFORMATION

Nissan uses 2 types of fuel systems. Electronic Control Carburetor (ECC) system and Electronic Concentrated Control System (ECCS). The ECC system is a Feedback carburetor system. The ECCS is a fuel injected system which may be either throttle body injection or Multi-port injection. Both ECC and ECCS systems were available as of 1984.

SELF-DIAGNOSTICS

Service Precautions

- Do not disconnect the injector harness connectors with the engine running.
- Do not apply battery power directly to the injectors.
- Do not disconnect the ECU harness connectors before the battery ground cable has been disconnected.
- Make sure all ECU connectors are fastened securely. A poor connection can cause an extremely high surge voltage in the coil and condenser and result in damage to integrated circuits.
- When testing the ECU with a DVOM make sure that the probes of the tester never touch each other as this will result in damage to a transistor in the ECU.
- Keep the ECCS harness at least 4 in. away from adjacent harnesses to prevent an ECCS system malfunction due to external electronic noise.
- Keep all parts and harnesses dry during service.
- Before attempting to remove any parts, turn **OFF** the ignition switch and disconnect the battery ground cable.
- Always use a 12 volt battery as a power source.
- Do not attempt to disconnect the battery cables with the engine running or the ignition key **ON**.
- Do not clean the air flow meter with any type of detergent.
- Do not attempt to disassemble the ECCS control unit under any circumstances.
- Avoid static electricity build-up by properly grounding yourself prior to handling any ECU or related parts.

Reading Codes

➡Diagnostic codes may be retrieved by observing the code flashes through the LED lights located on the Electronic Control Module (ECM). A special Nissan Consult monitor tool can be used, but is not required. When using special diagnostic equipment, always observe the tool manufacturer's instructions.

ELECTRONIC CONTROLLED CARBURETOR

1987–88 VEHICLES

The 1 .6L (EI 6S) carbureted engine utilizes a duty-controlled solenoid valve for fuel enrichment and an Idle Speed Control (ISC) actuator for basic controls instead of the conventional choke valve plate and fast idle cam. There are several other inputs which further affect the air/fuel ratio. The system is controlled in 2 ways: open or closed loop. To inspect the system for malfunctions, proceed as follows:

1. Position the ECU so the red and green LED's are visible.
2. Run the engine until it is at normal operating temperature.
3. Verify the diagnosis switch on the ECU is OFF.

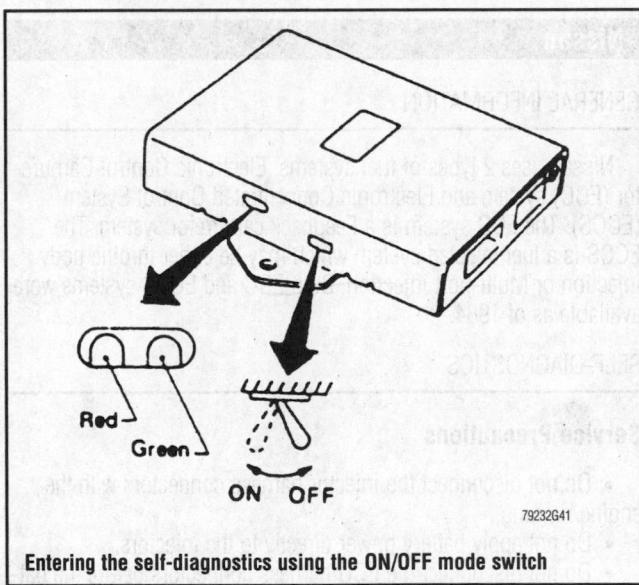

Entering the self-diagnostics using the ON/OFF mode switch

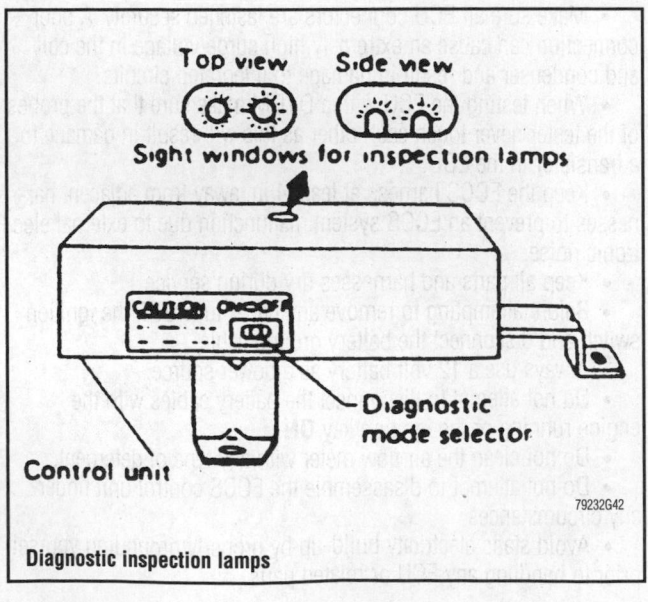

Diagnostic inspection lamps

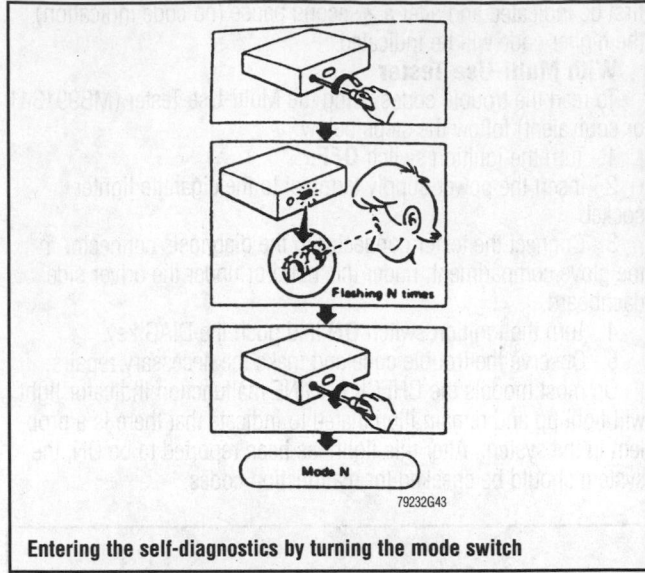

Entering the self-diagnostics by turning the mode switch

4. Run the engine 2000 rpm for 5 minutes. After 5 minutes, observe the green LED light while maintaining 2000 rpm. The light should be blinking ON and OFF at least 5 times in 10 seconds. If not as specified, inspect the exhaust gas sensor.
5. Turn the engine **OFF** and turn the ECU diagnosis switch **ON**.
6. Turn the ignition switch **ON**. The green LED on the ECU should stay ON and the red LED will either flash for a short period indicating a malfunctioning input sensor or for a longer time indicating a malfunctioning output sensor.

ELECTRONIC FUEL INJECTION

1984–94 VEHICLES

Two types of diagnostic systems are used in Nissan vehicles: the 2-mode diagnostic system and the 5-mode diagnostic system. The 2 mode system is used in some vehicles starting in 1990, ultimately, all vehicles used the 2-mode system after 1991 with the exception of 1991–94 Maxima (VG30E engine), Pathfinder and Truck. These vehicles continued to use the 5-mode system. The 5-mode system began in 1984.

The 5-mode diagnostic system is incorporated in the ECU which uses inputs from various sensors to determine the correct air/fuel ratio. If any of the sensors malfunction the ECU will store the code in memory. The 5-mode diagnostic system is capable of various tests as outlined below. When using these modes, the ECM may have to be removed from its mounting bracket to better access the mode selector switch.

➡Vehicles are equipped with a CHECK ENGINE light on the instrument panel. If any systems are malfunctioning, the light will illuminate the same time as the red lamp while the engine is running and the system is in Mode 1.

Mode 1—Heated Oxygen Sensor

During closed loop operation the green lamp turns ON when a lean condition is detected and turns OFF under a rich condition. During open loop the green lamp remains ON or OFF. This mode is used to check Heated Oxygen sensor functions for correct operation. To enter Mode 1, proceed as follows:

1. Turn the ignition switch **ON**.
2. Turn the diagnostic switch located on the side of the ECU ON by either flipping the switch to the ON position or turning the screw switch fully clockwise.

3. Turn the diagnostic switch OFF or fully counterclockwise as soon as the inspection lamps flash 1 time.

4. The self-diagnostic system is now in Mode 1.

Mode 2—Mixture Ratio Feedback Control Monitor

The green inspection lamp is operating in the same manner as in Mode 1. During closed loop operation the red inspection lamp turns ON and OFF simultaneously with the green lamp when the mixture ratio is controlled within the specified value. During open loop the red lamp remains ON or OFF. Mode 2 is used for checking that optimum control of the fuel mixture is obtained. To enter Mode 2, proceed as follows:

1. Turn the ignition switch **ON**.

2. Turn the diagnostic switch ON, by either flipping the switch to the ON position or use a screwdriver and turn the switch fully clockwise.

3. Turn the diagnostic switch OFF or fully counterclockwise as soon as the inspection lamps flash 2 times.

4. The self-diagnostic system is now in Mode 2.

Mode 3—Self-Diagnosis System

This mode of the self-diagnostics is for stored code retrieval. To enter Mode 3, proceed as follows:

1. Thoroughly warm the engine before proceeding. With the engine OFF, turn the ignition switch **ON**.

2. Turn the diagnostic switch located on the side of the ECU ON by either flipping the switch to the ON position or using a screwdriver, turn the switch fully clockwise.

3. Turn the diagnostic switch OFF or fully counterclockwise as soon as the inspection lamps flash 3 times.

4. The self-diagnostic system is now in Mode 3.

➡**When the battery is disconnected or self-diagnostic Mode 4 is selected after using Mode 3, all stored codes will be cleared. However if the ignition key is turned OFF and then the procedure is followed to enter Mode 4 directly, the stored codes will not be cleared.**

5. The codes will now be displayed by the red and green inspection lamps flashing. The red lamp will flash first and the green lamp will follow. The red lamp is the tens and the green lamp is the units, that is, the red lamp flashes 1 time and the green lamp flashes 2 times, this would indicate a Code 12.

Mode 4—On/Off Switches

This mode checks the operation of the Vehicle Speed Sensor (VSS), Closed Throttle Position (CTP) and starter switches. Entering this mode will also clear all stored codes in the ECU. To enter Mode 4, proceed as follows:

1. Turn the ignition switch **ON**.

2. Turn the diagnostic switch located on the side of the ECU ON by either flipping the switch to the ON position or turning the mode switch fully clockwise.

3. Turn the diagnostic switch OFF or fully counterclockwise as soon as the inspection lamps flash 4 times.

4. The self-diagnostic system is now in Mode 4.

5. Turn the ignition switch to the START position and verify the red inspection lamp illuminates. This verifies that the starter switch is working.

6. Depress the accelerator and verify the red inspection lamp goes OFF. This verifies that the CTP switch is working.

7. Raise and properly support the vehicle and verify the lamp goes ON when the vehicle speed is above 12 mph (20 km/h). This verifies that the VSS is working

8. Turn the ignition switch **OFF**.

Mode 5—Real Time Diagnostics

In this mode the ECU is capable of detecting and alerting the technician the instant a malfunction in the crank angle sensor, air flow meter, ignition signal or the fuel pump occurs while operating/driving the vehicle. Items which are noted to be malfunctioning are not stored in the ECU's memory. To enter Mode 5, proceed as follows:

1. Turn the ignition switch **ON**.

2. Turn the diagnostic switch located on the side of the ECU ON by either flipping the switch to the ON position or by turning the switch fully clockwise.

3. Turn the diagnostic switch OFF or fully counterclockwise as soon as the inspection lamps flash 5 times.

4. The self-diagnostic system is now in Mode 5.

5. Ensure the inspection lamps are not flashing. If they are, count the number of flashes within a 3.2 second period:

- 1 Flash = Crank angle sensor
- 2 Flashes = Air flow meter
- 3 Flashes = Fuel pump
- 4 Flashes = Ignition signal

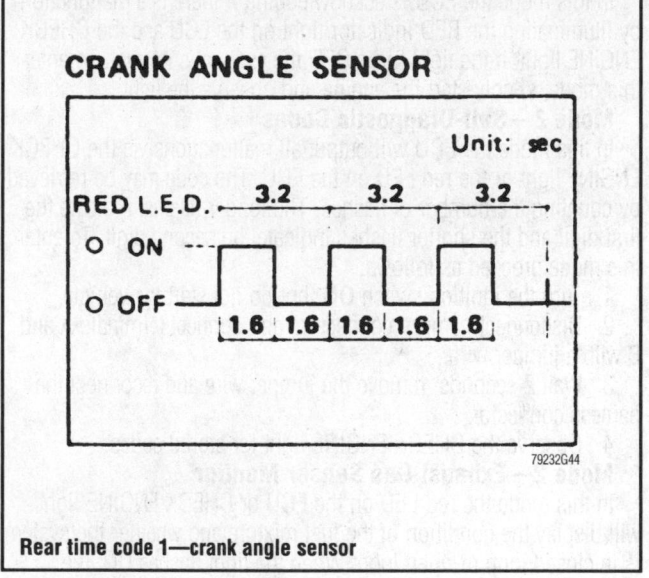

Rear time code 1—crank angle sensor

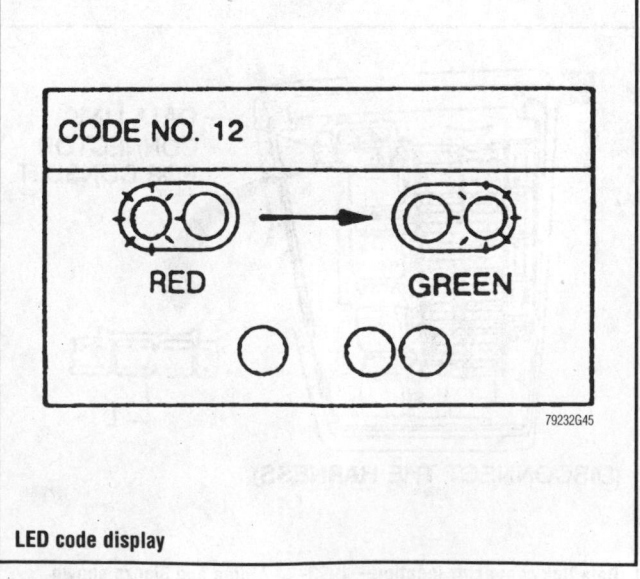

LED code display

2-MODE DIAGNOSTIC SYSTEM

The 1992–94 300ZX, Stanza, 240ZX, Sentra/NX Coupe, Maxima (VE30DE engine), and the 1993–94 Altima uses a 2-mode diagnostic system incorporated in the ECU which uses inputs from various sensors to determine the correct air/fuel ratio. If any of the sensors malfunction the ECU will store the code in memory.

A Nissan Consult monitor, or equivalent may be used to retrieve these codes by simply connecting the monitor to the diagnostic connector located on the driver's side near the hood release. Turn the ignition switch to **ON** and press START, ENGINE and then SELF-DIAG RESULTS, the results will then be output to the monitor.

The conventional CHECK ENGINE or red LED ECU light may also be used for self-diagnostics. The conventional 2-Mode diagnostic system is broken into 2 separate modes each capable of 2 tests, an ignition switch **ON** or engine running test as outlined below:

Mode 1—Bulb Check

In this mode the RED indicator light on the ECU and the CHECK ENGINE light should be ON. To enter this mode simply turn the ignition switch **ON** and observe the light.

Mode I—Malfunction Warning

In this mode the ECU is acknowledging if there is a malfunction by illuminating the RED indicator light on the ECU and the CHECK ENGINE light. If the light turns OFF, the system is normal. To enter this mode, simply start the engine and observe the light.

Mode 2—Self-Diagnostic Codes

In this mode the ECU will output all malfunctions via the CHECK ENGINE light or the red LED on the ECU. The code may be retrieved by counting the number of flashes. The longer flashes indicate the first digit and the shorter flashes indicate the second digit. To enter this mode proceed as follows:

1. Turn the ignition switch **ON**, but do not start the vehicle.
2. Disconnect harness connectors and connect terminals A and B with a jumper wire.
3. Wait 2 seconds, remove the jumper wire and reconnect the harness connector.
4. Observe the CHECK ENGINE light for stored codes.

Mode 2—Exhaust Gas Sensor Monitor

In this mode the red LED on the ECU or CHECK ENGINE light will display the condition of the fuel mixture and whether the system is in closed loop or open loop. When the light flashes ON, the exhaust gas sensor is indicating a lean mixture. When the light stays OFF, the sensor is indicating a rich mixture. If the light remains ON or OFF, it is indicating an open loop system. If the system is equipped with 2 exhaust gas sensors, the left side will operate first. If already in Mode 2, proceed to Step C to enter the exhaust gas sensor monitor.

1. On all models perform the following steps:
   a. Turn the ignition switch **ON**.
   b. Turn the diagnostic switch ON, by turning the switch fully clockwise for 2 seconds and then fully counterclockwise.
   c. Start the engine and run until thoroughly warm. Raise the idle to 2,000 rpm and hold for approximately 2 minutes. Ensure the red LED or CHECK ENGINE light flashes ON and OFF more than 5 times every 10 seconds with the engine speed at 2,000 rpm.

➤ **If equipped with 2 exhaust gas sensors, switch to the right sensor by turning the ECU mode selector fully clockwise for 2 seconds and then fully counterclockwise with the engine running.**

2. On Quest models, perform the following steps:
   a. Turn the ignition switch **ON**.
   b. Disconnect harness connectors and connect terminals A and B with a jumper wire.
   c. Wait 2 seconds, remove the jumper wire and reconnect the harness connectors.
   d. Start the engine and run until thoroughly warm. Raise the idle to 2,000 rpm and hold for approximately 2 minutes. Ensure the red LED or CHECK ENGINE light flashes ON and OFF more than 5 times every 10 seconds with the engine speed at 2,000 rpm.

Clearing Codes

ENGINE CODES

Except Mode 5, 3 and 2 Systems

All control unit diagnostic codes may be cleared by disconnecting the negative battery cable for a period of 15 seconds. Entering Mode 4 of the Electronic Fuel Injection system diagnostics will also clear stored ECM engine codes.

Mode 5, 3 and 2 Systems

On 5-mode systems, enter mode 4 immediately after using mode 3 and the codes will be cleared. On 2-mode systems, the codes will be cleared when mode 1 is re-entered from mode 2. The Nissan Consult Monitor or equivalent can also be used to clear codes on 2-mode systems.

DIAGNOSTIC TROUBLE CODES

1984–87 Vehicles

Code 11 Crankshaft position sensor circuit
Code 12 Mass Air flow sensor circuit
Code 13 Engine coolant temperature sensor circuit
Code 21 Ignition signal circuit
Code 22 Fuel pump circuit
Code 23 Idle switch circuit
Code 24 Transmission switch
Code 31 AC switch, fast idle control of load signal
Code 32 Starter signal
Code 33 EGR gas sensor
Code 34 Detonation (Knock) sensor
Code 41 Air or Fuel temperature sensor

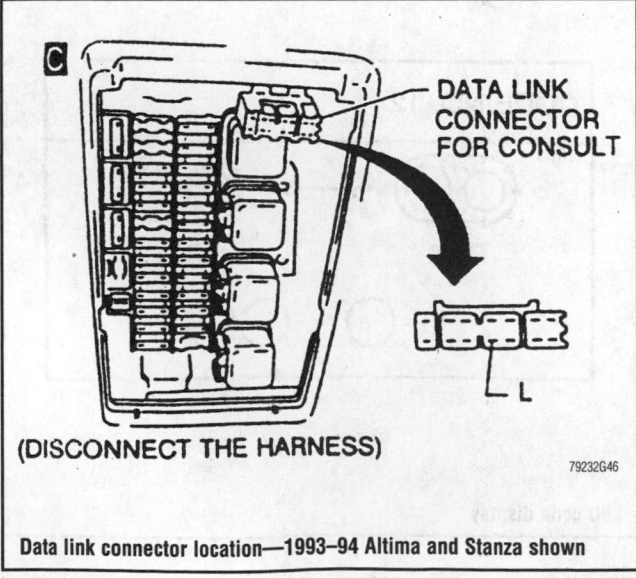

C DATA LINK CONNECTOR FOR CONSULT

L

(DISCONNECT THE HARNESS)

79232G46

Data link connector location—1993–94 Altima and Stanza shown

Code 42 Throttle sensor (or BP sensor in Canada)
Code 43 Mixture feedback control slips out (or low battery in Canada)
Code 44 No Malfunctioning circuits

1988–94 Vehicles

Code 11 Crankshaft position sensor circuit
Code 12 Mass Air flow sensor circuit
Code 13 Engine coolant temperature sensor circuit
Code 14 Vehicle speed sensor circuit
Code 15 Mixture ratio feedback control slips out (1988)
Code 21 Ignition signal circuit
Code 22 Fuel pump circuit (to 1991)
Code 23 Idle switch circuit (to 1991)
Code 24 Fuel Switch circuit or OD. switch circuit (to 1990)
Code 25 AAC valve circuit (to 1991)
Code 31 Electronic Control Module (ECM) or A/C circuit
Code 32 Exhaust Gas Recirculation (EGR) function
Code 33 Oxygen sensor circuit (left side, if two)
Code 34 Knock sensor circuit
Code 35 Exhaust gas temperature sensor circuit
Code 41 Air temperature sensor circuit
Code 42 Fuel temperature sensor circuit
Code 43 Throttle position sensor circuit
Code 44 No malfunctioning circuits
Code 45 Injector leak
Code 51 Injector circuit
Code 53 Heated oxygen sensor circuit (right side)
Code 54 Signal circuit from NT control unit to ECM
Code 55 No malfunctioning in the above circuits

Porsche

GENERAL INFORMATION

Except 911 Turbo

Porsche vehicle use 2 forms of electronic fuel injection. The 928 uses LH-Jetronic fuel injection. The 911, 944 and the 968 use Digital Motor Electronics (DME) fuel injection. Both systems are advanced versions of their fuel injection systems and provide excellent control of emissions, fuel economy and performance.

911 Turbo

The Porsche 911 Turbo has always been a specialized vehicle. The latest version is a hybrid of the 911 Carrera 2 body and the proven 3.3L turbocharged engine. While the 3.6L engine of the Carrera 2 uses Digital Motor Electronics as the fuel injection system, the 911 Turbo has used since its introduction the traditional Continuous Injection System (CIS).

SELF-DIAGNOSTICS

Service Precautions

• Do not disconnect the battery or power to the control unit before reading the fault codes. Fault code memory is erased when power is interrupted.
• Make sure the ignition switch is **OFF** before disconnecting any wiring.
• Before removing or installing a control unit, disconnect the

negative battery cable. The unit receives power through the main connector at all times and will be permanently damaged if improperly powered up or down.
• Keep all parts and harnesses dry during service. Protect the control unit and all solid-state components from rough handling or extremes of temperature.
• All air bag system wiring is in a yellow harness. Use extreme care when working around this wiring. Do not test these circuits without first disconnecting the air bag units.

Reading Codes

Vehicles prior to 1991 with self diagnostic capability require the use of special tool 9288 Tester or 9268 flash tester to retrieve diagnostics codes. On 1991 and later models, the flash codes can be read on the CHECK ENGINE light on the instrument panel. When using special diagnostic equipment, always observe the tool manufacturer's instructions.

➡**1987 vehicles with LH and EZK systems did not have OBD capability.**

1987–93 VEHICLES

With 9288 Diagnostic Tester

When the tester is attached to the diagnostic connector, the control units will present country and application codes to the tester for display on the screen. The tester will then provide a menu with instructions for retrieving fault codes from the control unit memory. If the tester display shows fault not present, this indicates the fault is intermittent or the conditions under which the fault occurs do not exist at this time. The necessary conditions will be displayed on the screen. If the display shows signal not plausible, the input or output signal does exist but is out of the correct operating range.

Without 9288 Diagnostic Tester

The control unit is equipped with a self diagnostic program that will detect emissions related malfunctions and turn the CHECK ENGINE light ON while the engine is running. Emissions related fault codes stored in the control unit can be read with the 9268 flash tester. On 1991 and later models, the flash codes can be read on the CHECK ENGINE light on the instrument panel. Only faults that may effect exhaust emissions are reported as flash codes. All other codes are only accessible with the 9288 Diagnostic Tester.

1. If required, connect the flash tester to the diagnostic connector using the adapter connector.
2. Turn the ignition switch **ON** without starting the engine.
3. Fully press the accelerator pedal to close the full load switch. After about 3 seconds the CHECK ENGINE light or tester will flash.
4. When the pedal is released, flash codes will be reported. All codes are 4 digits. Each digit will be flashed with about 2.5 seconds between digits. When the whole code has been displayed, the light will stay ON or OFF. Count the flashes and write the numbers down. If Code 1500 or 2500 is displayed, no codes are stored in memory.
5. Repeat Steps 3 and 4 until Code 1000 appears, indicating all codes have been reported. On 928 models, the EZK unit will display Code 2000 when all codes have been reported.

➡**On all 928 models, if the first digit is 2, the fault is in the EZK ignition system control unit.**

If the second digit is 1, the detected fault is current. If the second digit is 2, the detected fault has not occurred during the last running of the vehicle but did occur within the last 50 engine starts. The remaining 2 digits indicate which component or circuit is at fault.

Clearing Codes

1987–93 VEHICLES

The fault code memory should be cleared before returning the vehicle to service. All Codes in memory and the idle control adaptation are lost when the control unit or the battery is disconnected. To avoid the loss of the learned idle program, use the instructions on the 9288 tester to clear the memory. If this tester is not available, disconnect the battery or control unit to clear the memory. It will be necessary to drive the vehicle for at least 6 minutes and run the engine at idle for about 10 minutes so the control unit can learn idle speed, timing and mixture parameters. Make sure the engine is at operating temperature and that all accessories are OFF. The throttle-at-idle position switch must be closed and functioning or system adaptation will not take place.

DIAGNOSTIC TROUBLE CODES

1223 Coolant temperature sensor
1224 Air temperature sensor
1231 Battery voltage
1232 Throttle idle switch
1233 Throttle full load switch
1251 Fuel injector group 1, even numbers
1252 Fuel injector group 2, odd numbers
1261 Fuel pump relay
1262 Idle speed control actuator
1263 Carbon canister purge valve
1264 Oxygen sensor heater relay
1221 Control unit self test

1215 Airflow sensor
1221 Oxygen sensor
1222 Oxygen regulation

Saab

GENERAL INFORMATION

The LH-Jetronic fuel injection system was introduced in the Saab 900 in 1985. The 9000 picked it up in 1986. The 1990 models and in some markets 1991–94 models are equipped with an LH 2.4 fuel system Electronic Control Unit (ECU). Most 1991–94 models, are equipped with an LH 2.4.2 fuel system ECU, except the 1991 9000 with the B234 engine, which has an LH 2.4.1 fuel system ECU. Turbocharged 900 models are also equipped with an Automatic Performance Control (APC) ECU and 9000 models are equipped with an integrated Direct Ignition-Automatic Performance Control (DI/APC) control unit, which control ignition functions and turbocharger-related functions.

The LH-system ECU has the particular fuel system identification marked on it for identification. Visually, the LH 2.4 fuel system can be differentiated from the LH 2.4.2 fuel system by the pins on the Automatic Idle Control (AIC) valve. The LH 2.4 fuel system AIC valve has 2 pins and the LH 2.4.2 fuel system AIC valve has 3 pins.

The LH 2.4.1 fuel system differs from the LH 2.4 fuel system in that the cold start injector has been discontinued. The direct ignition system has taken over this function. Also, the vehicle speed sensor is used in the control of the AIC function to tell the fuel system ECU whether the car is moving or at a standstill.

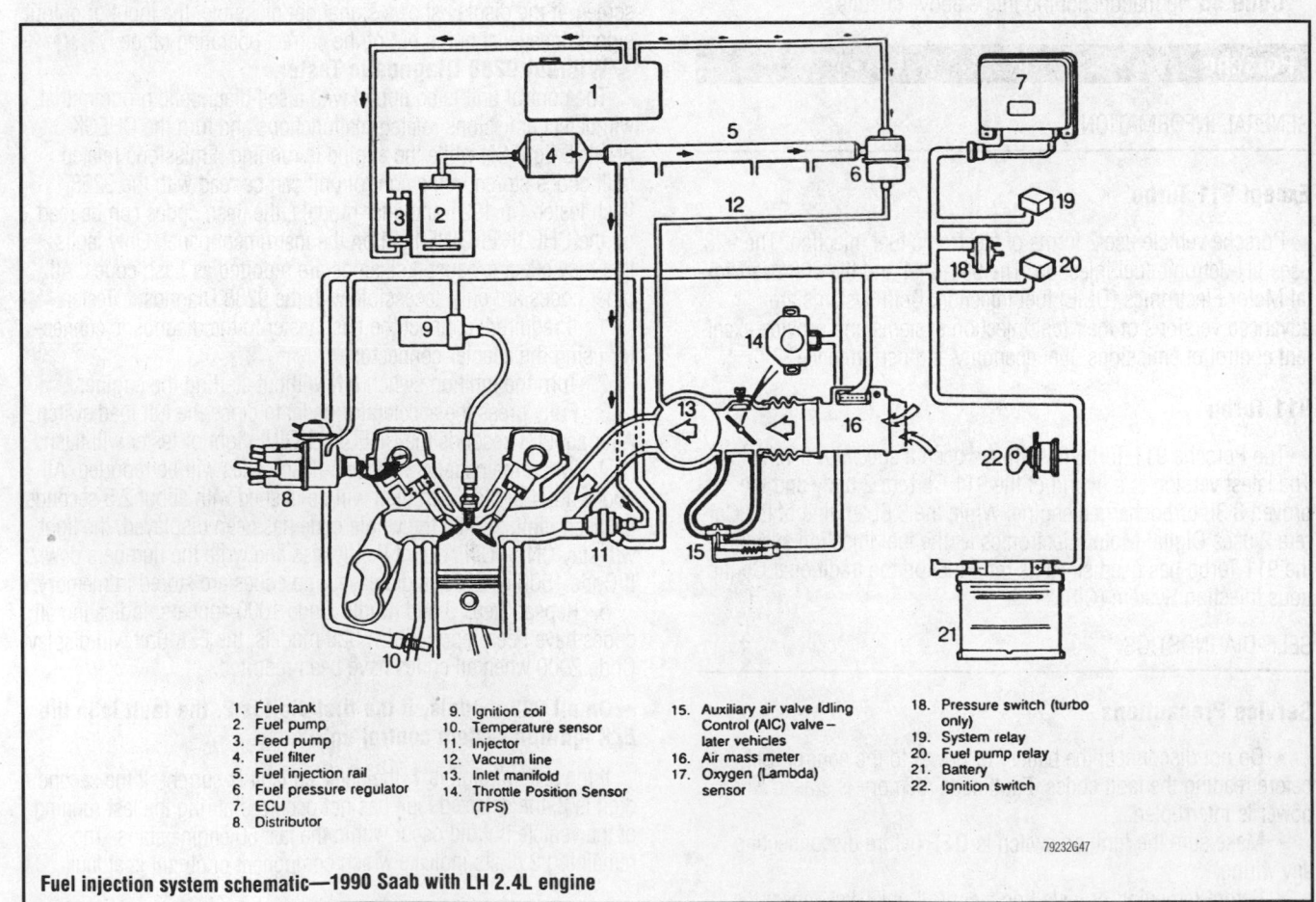

1. Fuel tank
2. Fuel pump
3. Feed pump
4. Fuel filter
5. Fuel injection rail
6. Fuel pressure regulator
7. ECU
8. Distributor
9. Ignition coil
10. Temperature sensor
11. Injector
12. Vacuum line
13. Inlet manifold
14. Throttle Position Sensor (TPS)
15. Auxiliary air valve Idling Control (AIC) valve – later vehicles
16. Air mass meter
17. Oxygen (Lambda) sensor
18. Pressure switch (turbo only)
19. System relay
20. Fuel pump relay
21. Battery
22. Ignition switch

Fuel injection system schematic—1990 Saab with LH 2.4L engine

79232G47

The LH 2.4 (some markets) and LH 2.4.2 fuel systems with the Electronic Throttle System (ETS), used on the 1992–94 9000 with traction control, differ from the systems without ETS in the following ways: The electronically controlled throttle eliminates the need for automatic idling and load control throughout the load range, the throttle angle transmitter is located in the actuator motor, which is integrated with the throttle housing and the electronically controlled throttle system carries out compensation for the air conditioning. Vehicles with ETS also have an Automatic Slip Reduction (ASR) control unit that carries out other traction control system functions.

Visually, the throttle housing on engines with ETS is larger, is vacuum operated and has an emergency cable, but in other respects this system operates in the same way as the LH 2.4 and 2.4.2 systems without ETS.

The central component of the LH-Jetronic fuel injection system is the air mass meter that measures the mass of air flow instead of the volume. The microprocessor measures how much electrical energy is used when air flow passes an electrically heated platinum wire in the air mass meter. The higher the rate of air flow, the higher the energy necessary to keep the temperature of the wire constant. At the same time, the microprocessor monitors the engine speed and temperature, calculating the exact amount of fuel needed for optimum performance. The microprocessor also incorporates an rpm limiter that ensures that no opening signals will be transmitted to the injectors at engine speeds above 6000 rpm.

The LH-Jetronic fuel injection system provides the air mass meter with a self-cleaning function. During burn-off the platinum wire in the air mass meter is quickly heated to about 1800°F (1000°C) for a 1 second duration, 4 seconds after the ignition is switched **OFF**. This burns away any deposits on the wire that would be detrimental to efficient operation.

The APC system on turbocharged vehicles enables the engine to achieve optimum performance and good fuel economy, regardless of the grade of fuel being used. A knock sensor, in conjunction with the pressure transducer and ignition system information, detects knocking in the engine and sends an electrical signal to the microprocessor inside the ECU. The ECU processes these signals and sends electrical pulses to a solenoid valve that controls the charging pressure in the intake manifold. The turbocharger is designed to come into operation at fairly low engine speeds, thereby providing a high torque within the speed range of normal driving. It is water cooled and the coolant for the bearing housing is supplied by a pipe connected to the cooling system.

Charging pressure is regulated by a pressure regulator valve (known as a wastegate). The charging pressure regulator is fitted to the exhaust side of the engine and regulates the flow of exhaust gas to the compressor. The valve remains closed when the engine load is low. As the demand on the engine is increased, the wastegate opens.

The DI/APC system was updated in 1991 to include an air temperature sensor, located upstream of the throttle housing. The boost pressure is governed by the position of the throttle valve, but it is subject to temperature compensation based on information supplied by the new air temperature sensor. The separate pressure-switch function is discontinued, with the pressure-sensing function now being regulated by the DI/APCsystem ECU. The load signal provided by the air mass meter is sent to the LH-system ECU, which will keep the DI/APC system ECU informed of boost status.

Starting in 1991 there is also a spark plug burn-off function that occurs when the engine is stopped. The burn-off function, which operates in all cylinders simultaneously, lasts for 5 seconds at a frequency corresponding to 6000 rpm.

Turbocharged California vehicles, except the 1991–94 9000 Turbo with the B234 engine, are equipped with an electronically controlled EGR system. A modulating valve functions as a 3-way valve as it controls the vacuum to the EGR valve. A vacuum regulator is incorporated in the modulating valve to maintain a constant vacuum to the EGR valve. A vacuum storage tank is connected via a vacuum check valve to the intake manifold. The check valve prevents the loss of vacuum in the tank during acceleration. An EGR temperature sensor provides information to the LH-system ECU. If the temperature deviates from a normal range, a problem is indicated due to improper exhaust gas flow in the EGR pipe and a fault code will be set.

The LH-Jetronic system also incorporates an emergency system known as a limp home function. If a malfunction is detected, the limp home feature of the ECU is actuated, enabling the vehicle to continue its journey, but with somewhat diminished performance. If the vehicle is operated in this mode, the Check Engine Light (CEL) on the display panel will be illuminated. An integrated fault-storing capability enables diagnosis to be carried out efficiently.

SELF DIAGNOSTICS

Service Precautions

• Before connecting or disconnecting ECU harness connectors, make sure the ignition switch is **OFF** and the negative battery cable is disconnected to avoid the possibility of damage to the control unit.

• When connecting or disconnecting pin connectors from the ECU, take care not to bend or break any pin terminals. Check that there are no bends or breaks on ECU pin terminals before attempting any connections.

• Before replacing any ECU, perform ECU input/output signal diagnosis to determine if the ECU is functioning properly.

• When measuring supply voltage of ECU controlled components with a circuit tester, separate 1 tester probe from another. If the 2 tester probes accidentally make contact with each other during measurement, a short circuit may result and damage the ECU.

• Always disarm the airbag (SRS) system when working on the airbag or ABS system.

• Always verify the ignition is switched **OFF** before connecting or disconnecting any electrical connections, especially connections to a control unit.

Reading Codes

➡**Diagnostic codes may be retrieved by observing the code flashes through the CHECK ENGINE light or Malfunction Indicator Lamp (MIL) only on vehicles listed under 'Without Diagnostic Tester' procedure. With this procedure a basic jumper switch (momentary type) is required to activate the computer. Other vehicles would require the use of special diagnostic tools: LH System tester or a ISAT Tester to retrieve codes. Read all procedures before attempting to perform checks. When using special diagnostic tools always observe the tool manufacturer's instructions.**

79232G48

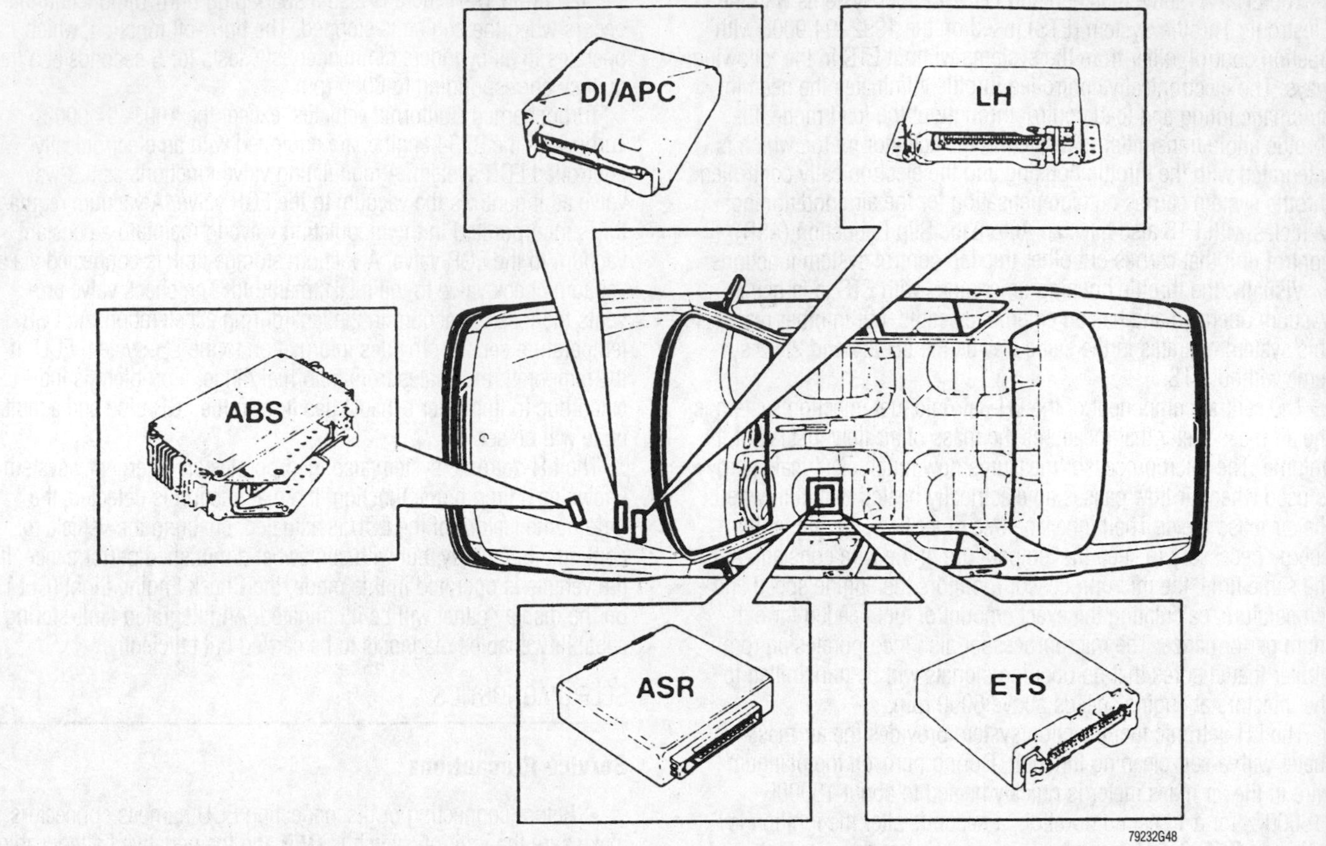

Locations of the electronic control units—1992 9000 shown

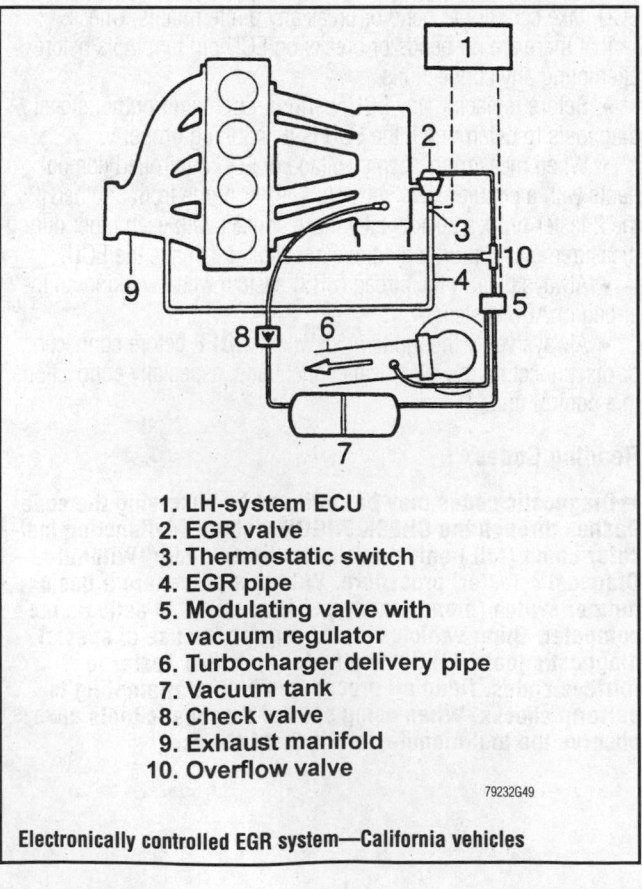

1. LH-system ECU
2. EGR valve
3. Thermostatic switch
4. EGR pipe
5. Modulating valve with vacuum regulator
6. Turbocharger delivery pipe
7. Vacuum tank
8. Check valve
9. Exhaust manifold
10. Overflow valve

79232G49

Electronically controlled EGR system—California vehicles

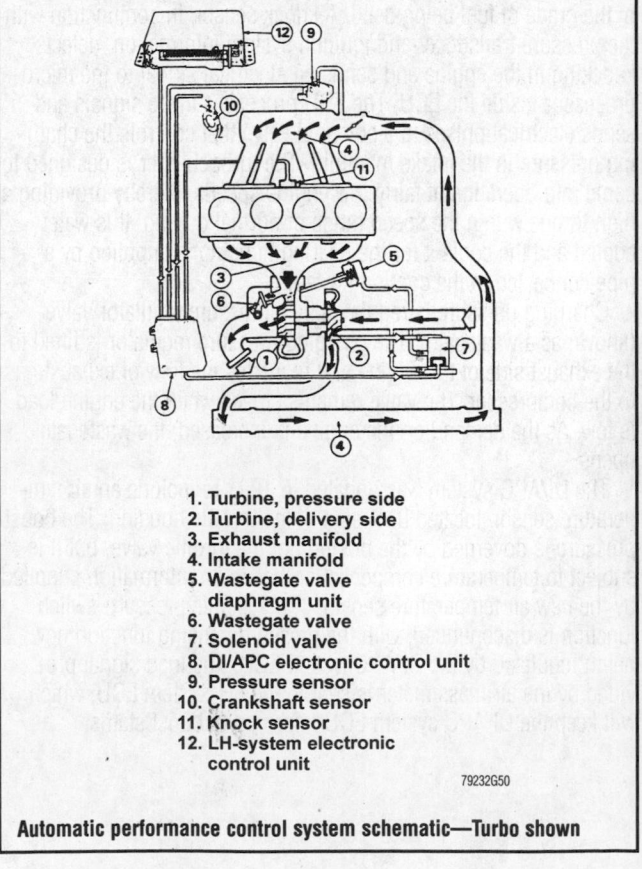

1. Turbine, pressure side
2. Turbine, delivery side
3. Exhaust manifold
4. Intake manifold
5. Wastegate valve diaphragm unit
6. Wastegate valve
7. Solenoid valve
8. DI/APC electronic control unit
9. Pressure sensor
10. Crankshaft sensor
11. Knock sensor
12. LH-system electronic control unit

79232G50

Automatic performance control system schematic—Turbo shown

1985–94 VEHICLES

With LH System Tester

The Saab LH system tester 8394223 has been developed to simplify service and fault diagnosis work on the LH fuel injection system. The tester consists of a test unit, power supply lead, test lead incorporating a 2-way 35-pin connector and a pressure sensor with magnetic base.

The tester is equipped with an automatic program for diagnosing faults, both permanent and intermittent, while the vehicle is operating. Faults detected are then stored in memo for recall after the vehicle is road-tested.

Connect the diagnostic tester, or equivalent, as follows:

1. Insert the power supply lead between the door and body where there is a break in the seal, then run it under the back of the hood on the left hand side.

2. Clean the battery terminals to ensure proper contact with lead clips.

3. Connect the power supply lead to the tester first, then connect the lead clips to the battery.

4. Remove the cover on the left side over the space behind the false bulkhead panel.

5. Remove the ABS system ECU and bracket, if equipped.

6. Remove the LH system ECU connector. Connect the test lead between the LH-system ECU and the vehicle's wiring loom. Fit a couple of ties around the connector and ECU to hold them tightly together.

The tester, designed to perform 3 basic functions, monitor mode, test mode and fuel mode, is now in start mode. If at any time the

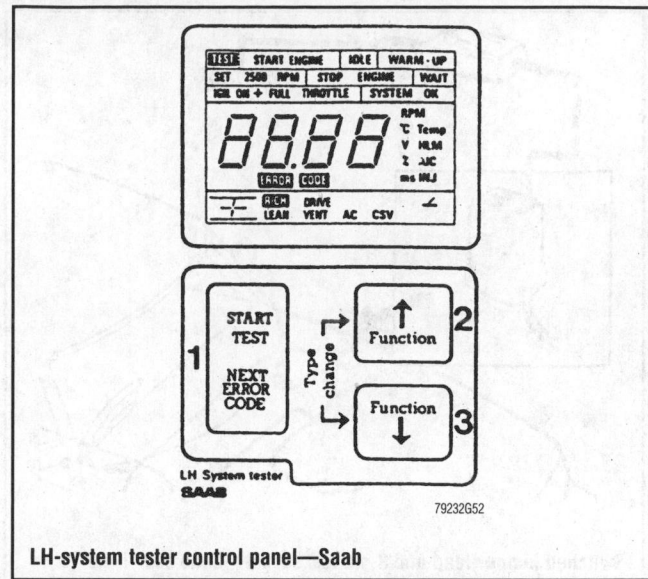

LH-system tester control panel—Saab

operation in progress must be interrupted, simultaneously press all 3 control buttons and the tester will revert to the starting point.

Monitor Mode

The monitor mode can be selected either by switching the ignition **ON**, or if the ignition is **OFF**, by pressing the START TEST button when MON appears on the tester display.

When in monitor mode, the tester is used to manually control parameter and functional checks.

Test Mode

Test mode can only be selected from the monitor mode. Once the LH system version has been selected, the test mode can be activated. Press the START TEST button to select the test mode. TEST will now appear on the display.

In the test mode, the program instructs by way of prompts in the upper part of the display.

Fuel Mode

To select the fuel mode, the ignition must be **OFF**. Initially, MON will be displayed on the tester for approximately 5 seconds. During this time, if required, the monitor mode can be selected. If none of the tester buttons are activated, FUEL will appear on the display for approximately 2 seconds. To activate the fuel mode, press the START button when FUEL is displayed.

When the tester is in fuel mode the following checks can be performed:

- Fuel pump delivery flow
- Fuel pressure and fuel-pressure regulator
- Residual pressure
- Fuel pump delivery pressure Delivery flow from injectors

Without Diagnostic Tester

This method can be used to retrieve fault codes from 1988–94 Saab models equipped with LH 2.4, LH 2.4.1 and LH 2.4.2 fuel injection systems. These systems are capable of an internal self-diagnostic checks and have the ability to store up to 3 intermittent faults at a time. Serious malfunctions are always given priority and must be rectified before the memory can store information on minor faults. The built in diagnosis function also has the capability to manually test the components and signals of the LH system.

The Saab switched jumper lead 8393886, or equivalent, is necessary to conduct these tests.

Connect the switched jumper lead for as follows:

For the Saab 900, Use the switched jumper lead to connect the

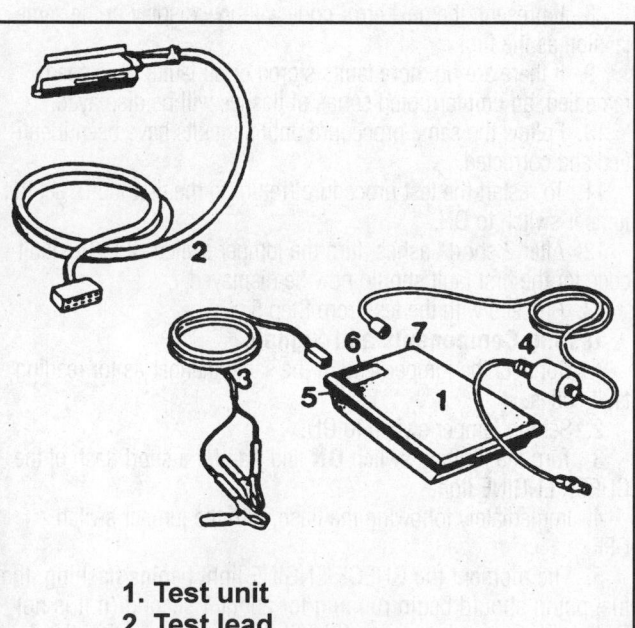

1. Test unit
2. Test lead
3. Power supply lead
4. Pressure sensor
5. Port for 12V power supply
6. Port for test lead
7. Port for pressure sensor

LH-Jetronic system tester and connectors—Saab

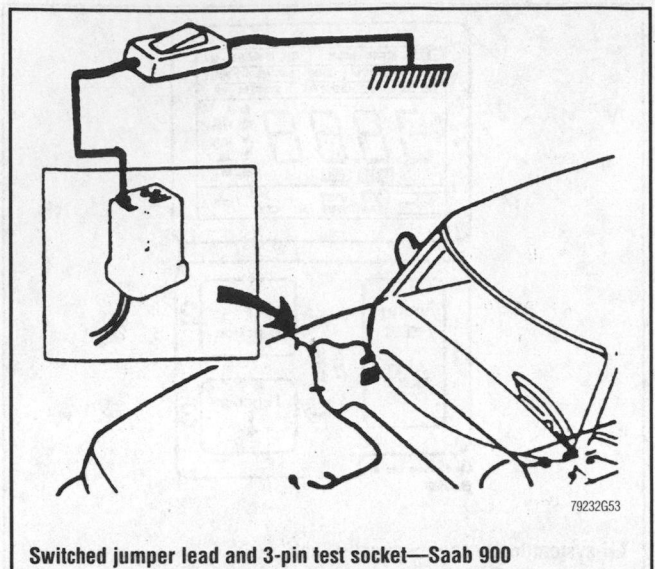

Switched jumper lead and 3-pin test socket—Saab 900

79232G53

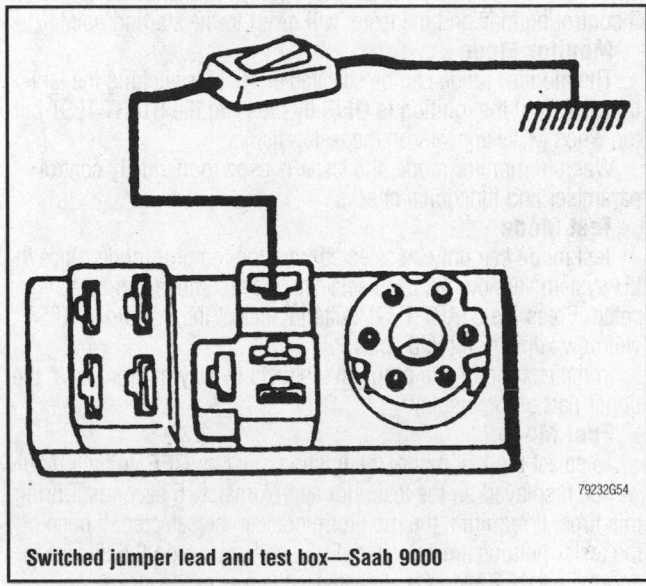

Switched jumper lead and test box—Saab 9000

79232G54

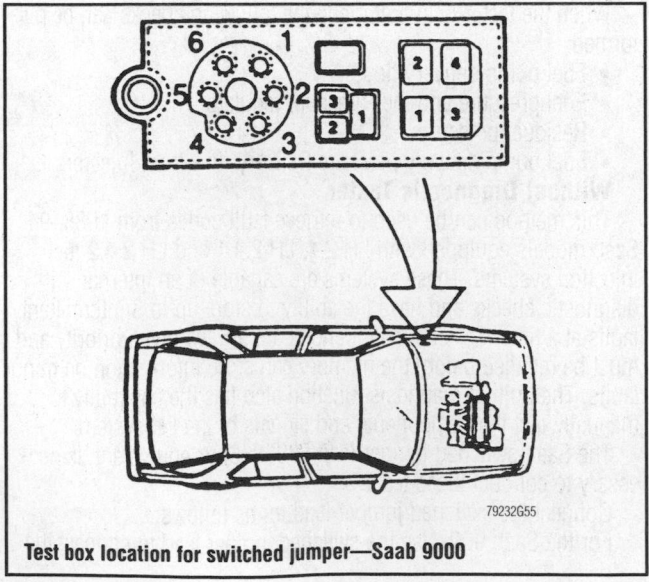

Test box location for switched jumper—Saab 9000

79232G55

No. 3 pin in the 3-pin test socket, on the right-hand side in the engine compartment, to the battery ground (negative terminal).

For the Saab 9000, use the switched jumper lead to connect the 3-pin socket, in the test box on the left-hand side of the engine compartment, to the battery ground (negative terminal).

1. Switch the ignition **ON**. The CHECK ENGINE light should now illuminate.

2. Set the jumper switch to ON (grounding ECU pin 16). The CHECK ENGINE light should now be extinguished.

3. Watch the CHECK ENGINE light carefully. After about 2.5 seconds, it will flash briefly, signifying that the first error code is about to display.

4. As soon as the light has flashed, turn the jumper switch **OFF**.

5. The first of a possible 3 error codes will now be displayed by a series of short flashes. The number 1 is represented by a single flash followed by a long pause. The number 2 is represented by 2 flashes separated from each other by a short pause, but separated from the next number by a long pause. The number 3 would consist of 3 flashes separated by short pauses and followed by a long pause, and so on. For example Code 12112 would consist of: flash-long pause, flash-short pause, flash-long pause, flash-long pause, flash-long pause, flash-short pause, flash-long pause. The code will be displayed repeatedly until the next test step is taken.

6. To check for any additional error codes, turn the jumper switch ON.

7. Watch the CHECK ENGINE light carefully. After a short flash, turn the jumper switch OFF.

8. If present, the next error code will now display in the same fashion as the first.

9. If there are no more faults stored or all faults have been remedied, an uninterrupted series of flashes will be displayed.

10. Follow the same procedure until all faults have been identified and corrected.

11. To restart the test procedure (return to the first fault), set the jumper switch to ON.

12. After 2 short flashes, turn the jumper switch OFF. The fault code for the first fault should now be displayed.

13. Proceed with the test from Step 5.

Testing Components and Signals

1. Connect the jumper lead in the same manner as for reading fault codes.

2. Set the jumper switch to ON.

3. Turn the ignition switch **ON** and wait for a short flash of the CHECK ENGINE light.

4. Immediately following the flash, turn the jumper switch OFF.

5. The moment the CHECK ENGINE light begins flashing, the fuel pump should begin running for about 1 second (if it is not faulty). There will be no identification codes sent during this test.

6. To move on to the next test, set the jumper switch to ON.

7. After a short flash, set the jumper switch to OFF. A test code (NOT FAULT CODE) will be displayed and the corresponding component will activate.

8. Continue through the remaining items in the test sequence in the same method—set switch to ON, wait for a short flash, set the switch to OFF. Components and signals are checked in the following order:

Fuel pump (no code displayed).

Injection valves (1.5 ms-10 Hz).

AIC valve (switches between open and closed positions).

EVAP Canister Purge valve (switches between closed and open)—CHECK ENGINE flashing stops.

EGR valve operates—CHECK ENGINE flashing stops. Drive signal (changes when shifting from D to N)—CHECK

ENGINE flashing stops. Air conditioning operates—CHECK ENGINE flashing stops.

Throttle position switch position (changes as accelerator is depressed) -CHECK ENGINE flashing stops. Throttle position switch WOT. position (changes as accelerator is pressed down to the floor)—CHECK ENGINE flashing stops.

Fuel pump operates—CHECK ENGINE flashing stops.

1990–94 VEHICLES

With SAAB ISAT Tester

The Saab SAT tester is also available to extract fault codes, both constant and intermittent, or to issue command codes.

Read codes with the ISAT tester, or equivalent, as follows:

1. Never unplug connector from the ECU or disconnect a battery lead before the fault data stored in the ECU has been transferred to the tester.

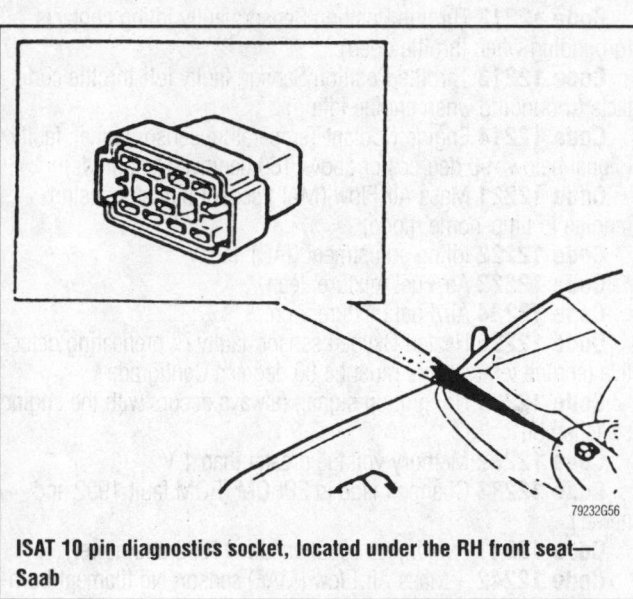

ISAT 10 pin diagnostics socket, located under the RH front seat—Saab

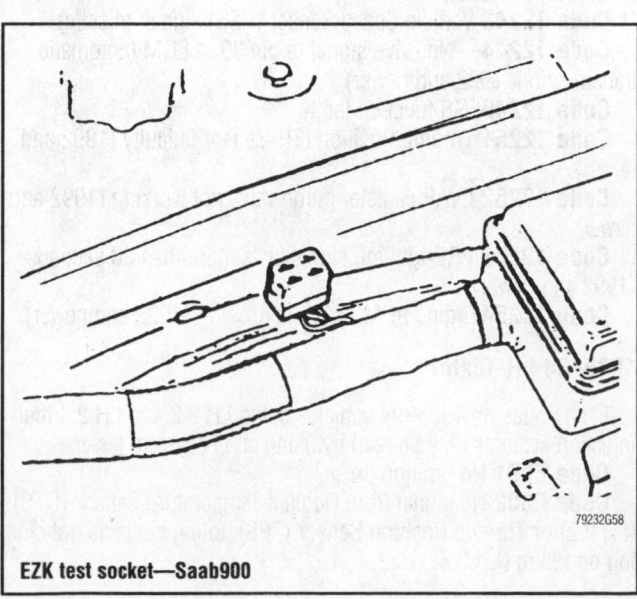

EZK test socket—Saab900

ISAT tester—Saab

2. Connect the diagnostic tester to the diagnostic socket. The diagnosis socket is a black 10-pin connector located under the RH front seat.

3. Turn the ignition to the **ON** position.

4. Press the No. 1 on the SAT to identify that you are checking the LH system.

The Trionic system engine fault codes can be read in the same manner, but the Saab adapter # 8611188 must be used with SAT and current EPROM update.

EZK Ignition Codes

The EZK ignition system is capable of self diagnostics only with the aid of the system tester 8394058, or equivalent. The test should be performed only in the event that a malfunction is suspected or when adjusting ignition timing.

Read fault codes as follows:

5. With the ignition switch **OFF**, connect the tester to the test box on the left-hand side of the engine compartment on the Saab 9000 or to the test socket located forward of the electrical distribution box on the Saab 900.

6. Turn the ignition switch **ON** and start engine. The fault indication LED (green) on the tester should illuminate for about 2 seconds while the starter motor is running.

7. Warm the engine to normal operating temperature, making sure that the engine is briefly run above 2300 rpm at some point during warm-up.

8. Run engine at idling speed and check tester LEDs for flashing. (CHECK ENGINE light will flash at a corresponding rate to any green LED fault indication.) The red LED light indicates spark knocking.

9. The fault code is determined by counting the number of LED flashes.

➡ **The EZK ignition fault codes can also be read with the Saab ISAT tester. To pull the ignition codes, follow the procedure for reading LH fuel system codes with ISAT. The EZK codes will be displayed with the fuel system fault codes. The check engine light will illuminate steady if the EZK or fuel system has a fault in memory.**

DL/APC Ignition Codes

The Saab combined Direct Ignition and Automatic Performance Control (DI/APC) system is controlled by the LH fuel injection ECU.

The DI/APC system is an adaptive system which compensates for engine wear and other conditions which would adversely affect engine performance. With the aid of the Saab SAT tester, or equivalent, it is possible to extract fault codes, both constant and intermittent or to issue command codes.

Read codes with the SAT tester as follows:

1. Never unplug connector from the ECU or disconnect a battery lead before the fault data stored in the ECU has been transferred to the ISAT.

2. The diagnosis socket is a black 10-pin connector located under the RH front seat.

3. Turn the ignition switch to the **ON** position.

4. Press No. 2 on the SAT tester to identify the DI/APC system and follow the instructions in the ISAT manual that accompany the tester to read fault codes or issue command codes.

Electronic Throttle Control (ETS) Codes

A Saab ISAT tester must be used to access the fault codes. Make sure to use the correct code chart when diagnosing fault codes, the codes differ for automatic and manual transmissions. Read codes with SAT tester as follows:

1. Never unplug connector from the ECU or disconnect a battery lead before the fault data stored in the ECU has been transferred to the tester.

2. Connect the diagnostic tester to the diagnostic socket. The diagnosis socket is a black 10-pin diagnostic connector located under the RH front seat.

3. Turn the ignition to the **ON** position.

4. Press the No. 3 on the ISAT to identify that you are checking the ETS system.

Clearing Codes

1985–94 VEHICLES

With LH System Tester

After completing repairs, reset the tester to the start mode by pressing all 3 control buttons at the same time. If fault codes are still found, disconnect the battery for at least 10 minutes or road test vehicle for 10 minutes or until CHECK ENGINE light extinguishes. The fault code should erase itself after extended operation with the repaired component.

With SAAB ISAT Tester

After repairs are completed, the SAT command code 900 is used to clear diagnostic memory.

Without System Tester

1. Set the jumper switch to ON.
2. After 3 short flashes, turn the jumper switch OFF.
3. The CHECK ENGINE light will now either flash in a continuous series of long flashes (this represents Code 00000) or display the Code 12444, indicating that the contents of the memory have been erased.

EZK IGNITION CODES

The EZK ignition does not store intermittent fault codes. All codes should clear when repair procedures are carried out.

DI/APC IGNITION CODES

After repairs are completed, the ISAT command Code 900 is used to clear diagnostic memory.

ELECTRONIC THROTTLE CONTROL (ETS) SYSTEM CODES

After repairs are completed, the SAT command Code 900 is used to clear diagnostic memory. A confirmation code '11111' will be displayed after the codes have been successfully erased. If this display is not received, repeat the code erasure procedure.

DIAGNOSTIC TROUBLE CODES

1988–94 Flash Codes

Fault codes on 1988–94 vehicles using LH 2.4, 2.4.1 and LH 2.4.2 fuel injected systems may be read without the use of a diagnostic tester.

Code 00000 No more faults or faults not detected

Code 12111 Oxygen sensor adaptation fault; air/fuel mixture during idling

Code 12112 Oxygen sensor adaptation fault; air/fuel mixture with engine running

Code 12113 Idling control (IAC) adaptation fault; pulse ratio too low

Code 12114 Idling control (IAC) adaptation fault; pulse ratio too high

Code 12211 Incorrect battery voltage with engine running (below 10V or over 16V)

Code 12212 Throttle Position Sensor; faulty idling contacts (grounding when throttle open)

Code 12213 Throttle Position Sensor; faulty full-throttle contacts (grounding when engine idling)

Code 12214 Engine Coolant Temperature sensor signal; faulty (signal below -90 degrees or above 160 degrees Centigrade)

Code 12221 Mass Air Flow (MAF) sensor signal; missing (engine in limp-home mode)

Code 12222 Idling adjustment (IAC); faulty

Code 12223 Air/Fuel mixture; lean

Code 12224 Air/Fuel mixture; rich

Code 12225 Heated Oxygen sensor; faulty or preheating defective (engine temperature must be 80 degrees Centigrade

Code 12231 No ignition signal; (always occurs with the engine switched off)

Code 12232 Memory voltage greater than 1 V

Code 12233 Change made in EPROM (ROM fault 1992 and newer)

Code 12241 Fuel injector malfunction (1992 and newer)

Code 12242—Mass Air Flow (MAF) sensor; No filament burn-off (1992 and newer)

Code 12243 Vehicle Speed Sensor (VSS) signal; missing

Code 12244—No drive signal to pin 30 in ECM (automatic transmission, 1992 and newer)

Code 12245 EGR function faulty

Code 12251 Throttle Position (TP) sensor is faulty (1992 and newer)

Code 12252 EVAP canister purge valve not working (1992 and newer)

Code 12253 PRE-Ignition signal lasts more than 20 seconds (1992 and newer)

Code 12254 Engine RPM signal is missing (1992 and newer)

1985–94 LH-Tester

Fault codes on 1985–94 vehicles using LH 2.2 and LH 2.4 fuel injection systems may be read by using an LH system tester.

Code E001 No ignition pulse

Code E002 No signal from Coolant Temperature Sensor (CTS) (LH 2.2) or Throttle Position Sensor (TPS); idling contacts not closing on idling (LH 2.4)

Code E003 Throttle Position Sensor (TPS); idling contacts not closing on idling (LH 2.2) or Throttle Position Sensor (TPS); full load contacts constantly open (LH 2.4)

Code E004 Battery voltage to Electronic Control Unit (ECU) memory; missing

Code E005 Electronic Control Unit (ECU) pin 5 not grounding

Code E006 Air Mass Meter (AMM) not grounding

Code E007 No signal from Air Mass Meter (AMM)

Code E008 Air Mass Meter (AMM); no filament burn-off function

Code E009 No power to system relay

Code E010 No signal from Electronic Control Unit (ECU) pin 10 to Automatic Idle Control (AIC) valve

Code E011 Electronic Control Unit (ECU) pin 11 not grounding

Code E012 Throttle Position sensor (TPS)—full throttle contacts constantly open

Code E013 No injection pulse (LH 2.2) or No signal from temperature sensor (LH 2.4)

Code E014 Air Mass Meter (AMM)—break in CO—adjusting circuit

Code E017 Fuel pump relay—control circuit faulty (LH 2.2) or break in ground circuit continuity (LH 2.4)

Code E018 No power at + 15 supply terminal

Code E020 Faulty signal from Oxygen sensor (LH 2.2) or Fuel pump relay; faulty control circuit (LH 2.4)

Code E021 System relay; faulty control circuit

Code E023 No signal from Automatic Idle Control (AIC) valve

Code E024 No load signal (LH 2.2) or Lambda sensor; faulty signal (LH 2.4)

Code E025 Electronic Control Unit (ECU) pin 25 not grounding

Code E033ignal to Automatic Idle Control (AIC) valve from Electronic Control Unit (ECU); missing

Code E035 No power at +15 supply terminal

Code EI0I Starter motor revolutions too low

Code E102 Short in Coolant Temperature Sensor (CTS) circuit (LH 2.2) or Throttle Position Sensor (TPS); idling contacts not opening on increase from idling to 2500 rpm (LH 2.4)

Code E103 Throttle Position Sensor (TPS); idling contacts not opening on increase from idling to 2500 rpm (LH 2.2) or Throttle Position Sensor (TPS); full load contacts constantly closed (LH 2.4)

Code E107 Low signal from Air Mass Meter (AMM)

Code E108—Air Mass Meter (AMM); filament burn-off function constantly actuated

Code E109 Low voltage from system relay

Code E112 Throttle Position Sensor (TPS)—full load contacts constantly closed

Code E113 Erratic or No Injection Pulse

Code E120 Lambda sensor—signal too low

Code E207 High signal from Air Mass Meter (AMM)

Code E213 Continuous pulses to injectors

Code E218 Continuous pulses from injectors

Code E220 Lambda sensor—signal too high

Code E320 DI/APC system Electronic Control Unit (ECU)—pre-ignition signal constantly actuated

Code E328 Pre-ignition signal constantly grounded Code AICO—Automatic Idle Control (AIC) valve pulse ratio—faulty

Code GLOU (Glow) Air Mass Meter (AMM) filament burn-off function operating

Code CI Turbo

Code C2 Turbo with AIC

Code C3 Turbo with AIC and catalytic converter

Code C4 Turbo with AIC and Saab DI

Code CS Non-Turbo with AIC

Code FPU Fuel pump relay and system relay operating

Code FUEL Fuel mode

Code OFF Starting point for injection valve test

Code FIn Injection valve open

Code MON Monitor mode

Subaru

GENERAL INFORMATION

Feedback Carburetor

The DFC328 feedback carburetor is a 2 barrel, downdraft type which consists of the following systems:

Float system—Provided with the fuel return system. Primary side—Which consists of a slow system, main system, accelerating pump system and a choke system.

Secondary side—Which consists of step system and main system.

The primary and secondary side use the same float system. Fuel in the fuel tank is routed through the fuel pump and the needle valve, into the float chamber. Fuel level in the float chamber is maintained constant by the function of the needle and float. Fuel level height is adjusted by adjusting the float seat.

The float system consists of a Float Chamber Ventilation (FCV) system. When the engine is started and the coolant temperature is above 680F (200C), the FCV solenoid valve turns on. This allows air from the air filter to flow through the float chamber to the air vent. This ventilates the float chamber.

The choke system consists of a choke valve linked to a bimetal through a choke lever, so that the choke valve is kept opened at a suitable angle relative to ambient temperature by means of the bi-metal force. When the engine is started, the main vacuum diaphragm is operated by the vacuum sensed in the downstream portion of the secondary throttle valve, so that the choke valve is opened through a vacuum piston and a connecting rod. This allows an appropriate amount of air to be inducted, and the over-choke is prevented.

The auxiliary vacuum diaphragm is also operated by the vacuum, and allows the setting angle of the bi-metal force through a setting piston and a connecting rod. This operation allows the choke valve to be kept open at a moderate position to prevent an over-rich mixture. After the engine is started, the heater warms the bi-metal which adjusts the opening of the choke valve automatically.

A Coasting Fuel Cut (CFC) system is used to activate the anti-dieseling switch on the carburetor during deceleration. This closes the slow system passage for improved fuel economy. The control unit detects deceleration when the following conditions are met:

1. When intake manifold pressure is below -21.65 in. Hg, and the NC system is OFF; or -17.72 in. Hg and the air conditioning system is ON

2. When the vehicle is operated at 25 mph

3. Clutch pedal is released

4. When engine reaches 2500 rpm

When the control unit determines that the vehicle is decelerating, current to the anti-dieseling switch is interrupted. This closes the slow system passage so that the fuel flow is shut off. However, when coolant temperature is below 176°F (80°C), the fuel flow will not be shut off.

Single Point Fuel Injection

The SPFI is used on the Loyale 1.8L engine only. The system electronically controls the amount of injection from the fuel injector, and supplies the optimum air/fuel mixture under all operating conditions of the engine. Features of the SPFI system are as follows:

1. Precise control of of the air/fuel mixture is accomplished by an increased number of input signals transmitting engine operating conditions to the control unit.

2. The use of hot wire type air flow meter not only eliminates the need for high altitude compensation, but improves driving performance at high altitudes.

3. The air control valve automatically regulates the idle speed to the set value under all engine operating conditions.

4. Ignition timing is electrically controlled, thereby allowing the use of complicated spark advances characteristics.

5. Wear of the air flow meter and fuel injector is automatically corrected so that they maintain their original performance.

6. Troubleshooting can easily be accomplished by the built-in self-diagnosis function.

Multi-Point Fuel Injection

The MPFI system supplies the optimum air/fuel mixture to the engine under all various operating conditions.

System fuel, which is pressurized at a constant pressure, is injected into the intake air passage of the cylinder head. The amount of fuel injected is controlled by the intermittent injection system where the electro-magnetic injection valve (fuel injector) opens only for a short period of time, depending on the amount of fuel required for 1 cycle of operation. During system operation, the amount of injection is determined by the duration of an electric pulse sent to the fuel injector, which permits precise metering of the fuel.

Each of the operating conditions of the engine are converted into electric signals, resulting in additional features of the system, such as improved adaptability and easier addition of compensating element. The MPFI system also incorporates the following features:

- Reduced emission of exhaust gases
- Reduction in fuel consumption
- Increased engine output
- Superior acceleration and deceleration
- Superior starting and warm-up performance in cold weather since compensation is made for coolant and intake air temperature
- Good performance with turbocharger, if equipped.

SELF-DIAGNOSTICS

Service Precautions

- Before connecting or disconnecting ECU harness connectors, make sure the ignition switch is **OFF** and the negative battery cable is disconnected to avoid the possibility of damage to the control unit.
- When connecting or disconnecting pin connectors from the ECU, take care not to bend or break any pin terminals. Check that there are no bends or breaks on ECU pin terminals before attempting any connections.
- Before replacing any ECU, perform ECU input/output signal diagnosis to determine if the ECU is functioning properly.

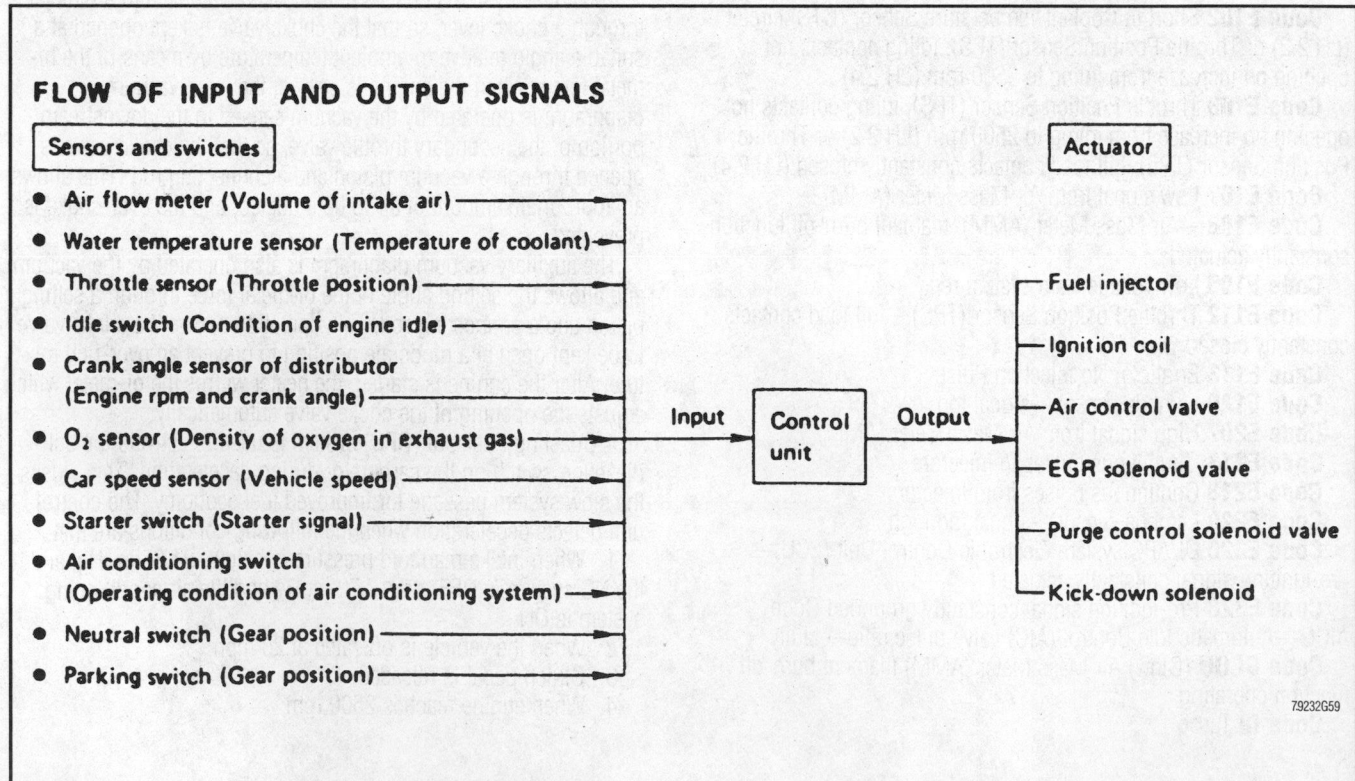

FLOW OF INPUT AND OUTPUT SIGNALS

Sensors and switches

- Air flow meter (Volume of intake air)
- Water temperature sensor (Temperature of coolant)
- Throttle sensor (Throttle position)
- Idle switch (Condition of engine idle)
- Crank angle sensor of distributor (Engine rpm and crank angle)
- O₂ sensor (Density of oxygen in exhaust gas)
- Car speed sensor (Vehicle speed)
- Starter switch (Starter signal)
- Air conditioning switch (Operating condition of air conditioning system)
- Neutral switch (Gear position)
- Parking switch (Gear position)

Input → Control unit → Output

Actuator
- Fuel injector
- Ignition coil
- Air control valve
- EGR solenoid valve
- Purge control solenoid valve
- Kick-down solenoid

79232G59

Inputs and outputs from the control unit—Subaru (SPFI)

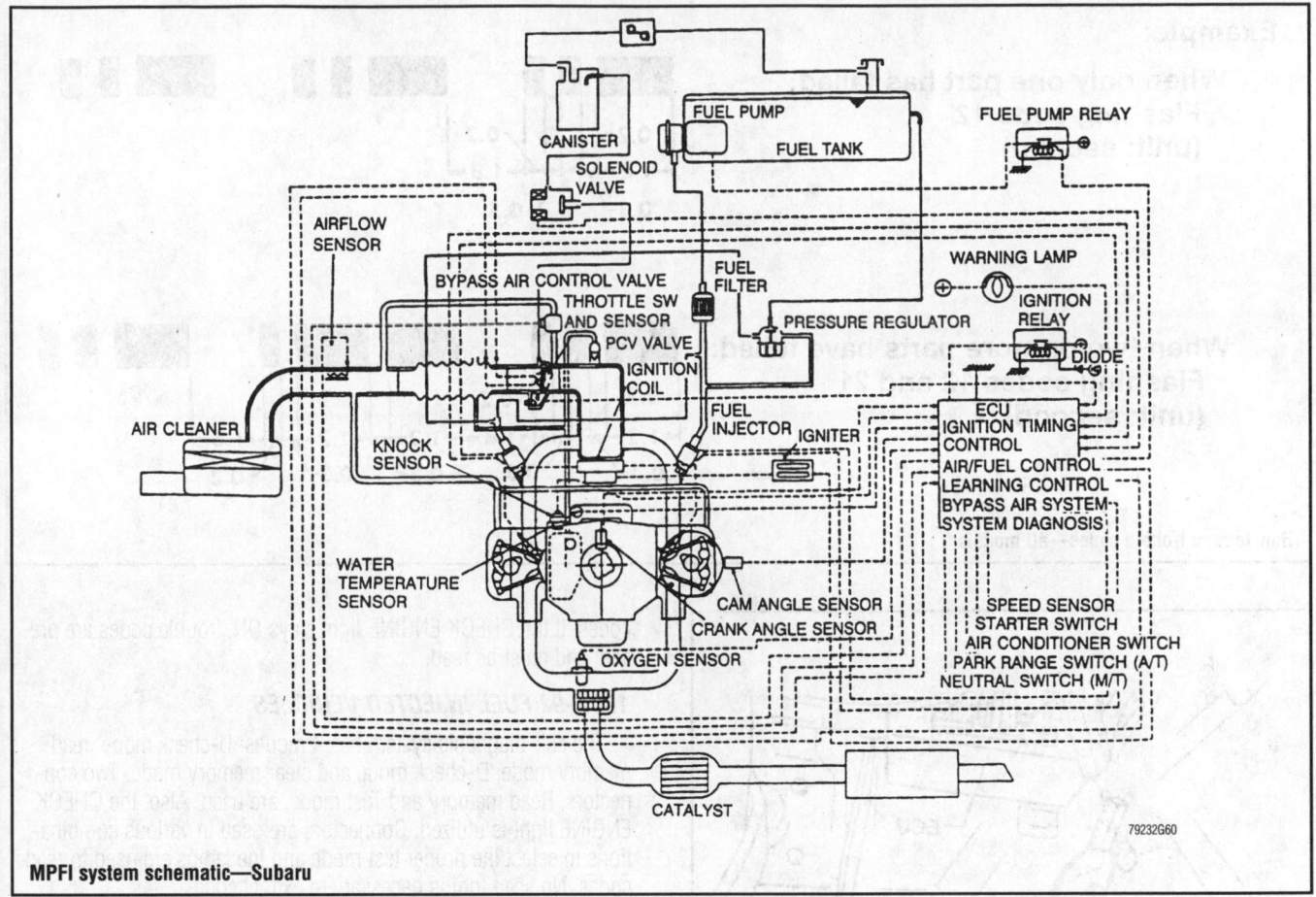

MPFI system schematic—Subaru

• When measuring supply voltage of ECU-controlled components with a circuit tester, separate 1 tester probe from another. If the 2 tester probes accidentally make contact with each other during measurement, a short circuit may result and damage the EC U.

Reading Codes

1987–92 FEEDBACK CARBURETOR VEHICLES

The self-diagnosis system has 4 modes: U-check mode, Read memory mode, D-check mode and Clear memory mode. Two connectors, Read memory and Test mode, are used. Also, the CHECK ENGINE light is utilized. Connectors are used in various combinations to select the proper test mode and the lamps are used to read codes. No scan tool is necessary to extract codes.

➡The engine should be running when in the D-check or clear memory modes.

U-Check Mode

The U-check is a user-oriented mode in which only the components necessary for start-up and drive are diagnosed. On occurrence of a fault the CHECK ENGINE light is turned ON to indicate that system inspection is necessary. The diagnosis of less significant components which do not adversely effect start-up and driving are excluded from this mode.

Read Memory Mode

The Read memory mode is used to detect faults which recently occurred but are not currently present.

1. Turn the ignition switch OFF.
2. Connect the Read memory connector.

3. Turn the ignition switch ON with the engine OFF.
4. If the CHECK ENGINE light turns ON, trouble code(s) are present.
5. If the oxygen monitor lamp turns ON, trouble code(s) are being produced; confirm the trouble code(s).
6. Disconnect the read memory connector.
7. Perform the D-check mode.

Mode	Read memory connector	Test mode connector
U-check	DISCONNECT	DISCONNECT
Read memory	CONNECT	DISCONNECT
D-check	DISCONNECT	CONNECT
(Clear memory)	CONNECT	CONNECT

Mode change connectors for different modes

Example:

    **When only one part has failed:
Flashing code 12
(unit: second)**

    **When two or more parts have failed:
Flashing codes 12 and 21
(unit: second)**

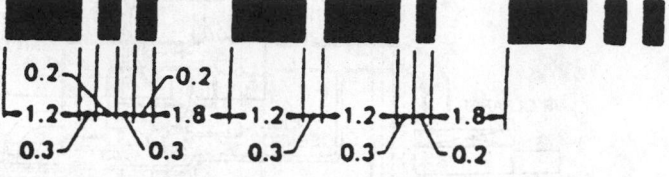

79232G62

How to read trouble codes—all models

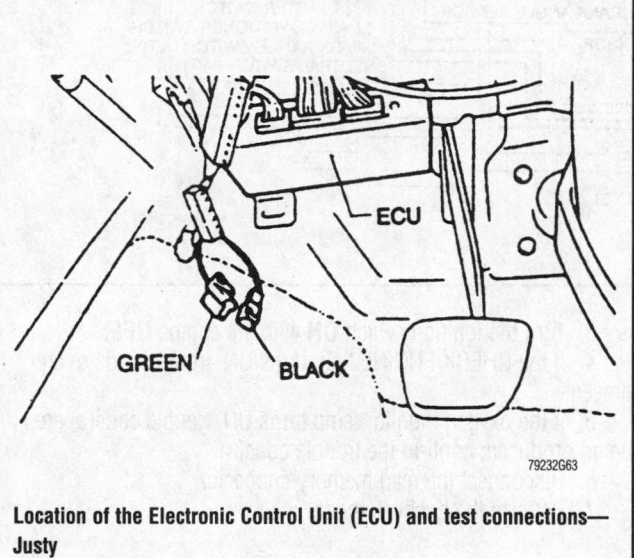

Location of the Electronic Control Unit (ECU) and test connections—Justy

D-Check Mode

The D-check mode is used to check the current status of the entire system.

1. Start the engine and warm it to normal operating temperatures.
2. Turn the ignition switch **OFF**.
3. Connect the test mode connector.
4. Turn the ignition switch **ON** with the engine **OFF**.
5. Make sure the CHECK ENGINE light turns ON; there should also be noise from the operation of the fuel pump.
6. Depress the accelerator pedal completely. Return it to ½ throttle position and hold it there for 2 seconds, then release the pedal completely.
7. Start the engine: If the CHECK ENGINE light indicates a trouble code, confirm code. If the CHECK ENGINE light turns OFF, continue test.
8. Race the engine briefly with the throttle fully opened.
9. Drive the vehicle above 5 mph, at engine speeds above 1500 rpm, for at least 1 minute.
10. If the CHECK ENGINE light blinks, there are no trouble

codes. If the CHECK ENGINE light stays ON, trouble codes are present and must be read.

1987–92 FUEL INJECTED VEHICLES

The self-diagnosis system has 4 modes: U-check mode, read memory mode, D-check mode and clear memory mode. Two connectors, Read memory and Test mode, are used. Also, the CHECK ENGINE light is utilized. Connectors are used in various combinations to select the proper test mode and the lamps are used to read codes. No scan tool is necessary to extract codes.

➡ **The engine should be running when in the D-check or clear memory modes.**

U-Check Mode

The U-check is a user-oriented mode in which only the components necessary for start-up and drive are diagnosed. On occurrence of a fault, the CHECK ENGINE light is turned ON to indicate that system inspection is necessary. The diagnosis of less significant components which do not adversely effect start-up and driving are excluded from this mode.

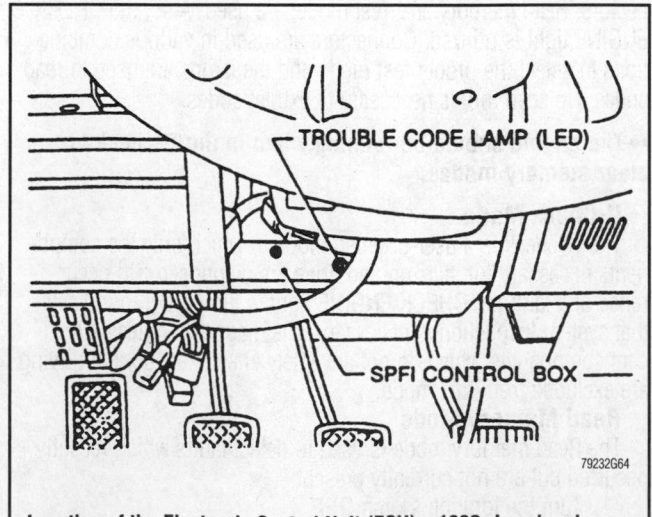

Location of the Electronic Control Unit (ECU)—1800, Loyal and Legacy

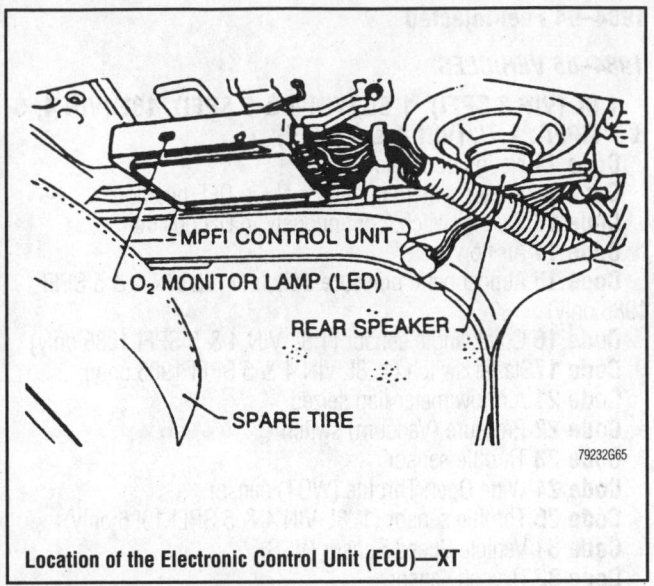

Location of the Electronic Control Unit (ECU)—XT

Read Memory Mode

The Read memory mode is used to detect faults which recently occurred but are not currently present.

1. Turn the ignition switch **OFF**.
2. Connect the Read memory connector.
3. Turn the ignition switch **ON** with the engine **OFF**.
4. If the CHECK ENGINE light turns ON, trouble code(s) are present.
5. If the oxygen monitor lamp turns ON, trouble code(s) are being produced; confirm the trouble code(s).
6. Disconnect the read memory connector.
7. Perform the D-check mode.

D-Check Mode

The D-check mode is used to check the current status of the entire system.

1. Start the engine and warm it to normal operating temperatures.
2. Turn the ignition switch **OFF**.
3. Connect the test mode connector.
4. Turn the ignition switch **ON** with the engine **OFF**.
5. Make sure the CHECK ENGINE light turns ON; there should also be noise from the operation of the fuel pump.
6. Depress the accelerator pedal completely. Return it to ½throttle position and hold it there for 2 seconds, then release the pedal completely.
7. Start the engine: If the CHECK ENGINE light indicates a trouble code, confirm code. If the CHECK ENGINE light turns OFF, continue test.
8. Race the engine briefly with the throttle fully opened.
9. Drive the vehicle above 5 mph, at engine speeds above 1500 rpm, for at least 1 minute.
10. If the CHECK ENGINE light blinks, there are no trouble codes. If the CHECK ENGINE light stays ON, trouble codes are present and must be read.

Clearing Codes

1987-92 FEEDBACK CARBURETOR VEHICLES

1. Start the engine and warm it to normal operating temperatures.
2. Turn the ignition switch **OFF**.

3. Connect the test mode connector and the read memory connector.
4. Turn the ignition switch **ON** with the engine **OFF**.
5. Make sure the CHECK ENGINE light turns ON.
6. Depress the accelerator pedal completely. Return it to ½ throttle position and hold it there for 2 seconds, then release the pedal completely.
7. Start the engine; the CHECK ENGINE light should turn OFF.
8. Race the engine with the throttle fully opened for a second or two.
9. Drive the vehicle above 5 mph, at engine speeds above 1500 rpm, for at least 1 minute.
10. The CHECK ENGINE light should blink showing that there are no trouble codes. If the CHECK ENGINE light stays ON, read the trouble codes and re-perform the D-check mode.

1984-94 FUEL INJECTED VEHICLES

1. Start the engine and warm it to normal operating temperatures.
2. Turn the ignition switch **OFF**.

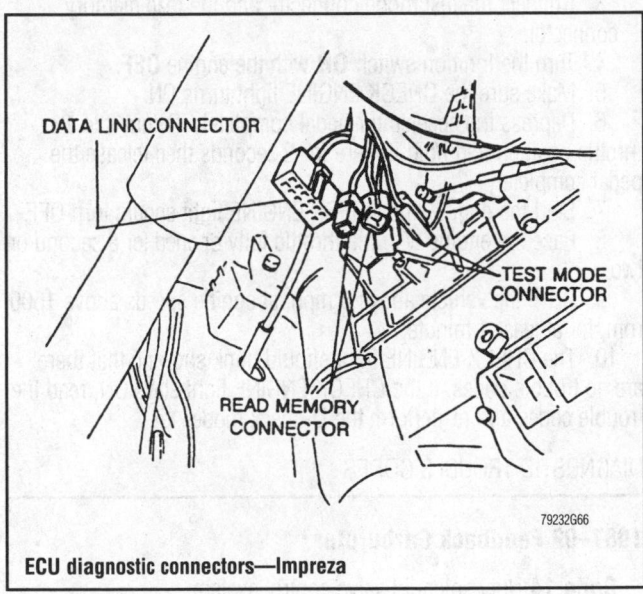

ECU diagnostic connectors—Impreza

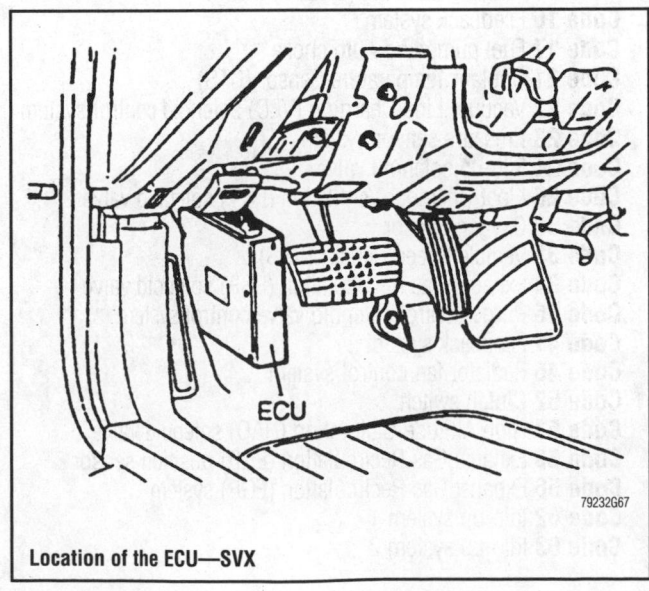

Location of the ECU—SVX

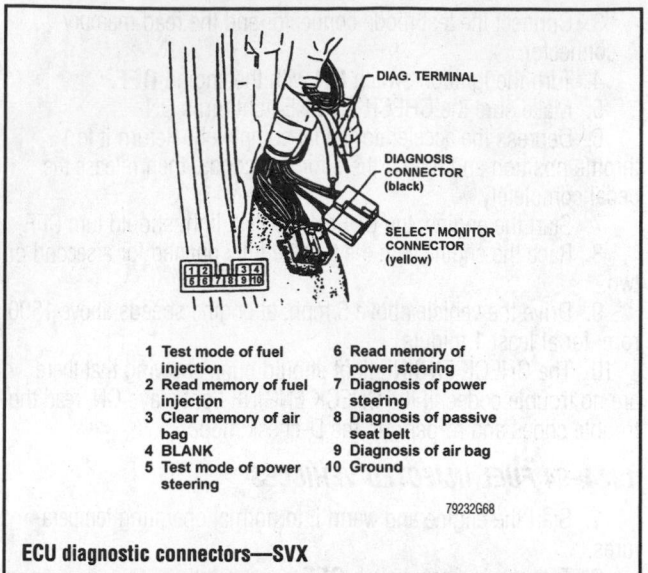

1 Test mode of fuel injection
2 Read memory of fuel injection
3 Clear memory of air bag
4 BLANK
5 Test mode of power steering
6 Read memory of power steering
7 Diagnosis of power steering
8 Diagnosis of passive seat belt
9 Diagnosis of air bag
10 Ground

79232G68

ECU diagnostic connectors—SVX

3. Connect the test mode connector and the read memory connector.

4. Turn the ignition switch **ON** with the engine **OFF**.

5. Make sure the CHECK ENGINE light turns ON.

6. Depress the accelerator pedal completely. Return it to ½ throttle position and hold it there for 2 seconds then release the pedal completely.

7. Start the engine; the CHECK ENGINE light should turn OFF.

8. Race the engine with the throttle fully opened for a second or two.

9. Drive the vehicle above 5 mph, at engine speeds above 1500 rpm, for at least 1 minute.

10. The CHECK ENGINE light should blink showing that there are no trouble codes. If the CHECK ENGINE light stays ON, read the trouble codes and re-perform the D-check mode.

DIAGNOSTIC TROUBLE CODES

1987–92 Feedback Carburetor

Code 14 Duty solenoid valve control system
Code 15 Coasting Fuel Cut (CFC) system
Code 16 Feedback system
Code 17 Fuel pump and Auto choke
Code 21 Coolant Temperature Sensor (CTS)
Code 22 Vacuum Line Charging (VLC) solenoid control system
Code 23 Pressure sensor system
Code 24 Idle-up solenoid valve
Code 25 Float Chamber Ventilation (FCV) solenoid valve
Code 32 Oxygen Sensor
Code 33 Vehicle Speed Sensor (VSS)
Code 34 Exhaust Gas Recirculation (EGR) solenoid valve
Code 35 Purge control solenoid valve control system
Code 41 Feedback system
Code 46 Radiator fan control system
Code 52 Clutch switch
Code 53 High Altitude Calibration (HAC) solenoid valve
Code 55 Exhaust Gas Recirculation (EGR) position sensor
Code 56 Exhaust Gas Recirculation (EGR) system
Code 62 Idle-up system 1
Code 63 Idle-up system 2

1984–94 Fuel Injected

1984–86 VEHICLES

1.6L (VIN 2 SPFI), 1.8L (VIN 4 & 5 SPFI), 18L (VIN 4, 5 & 7 MPFI), 2.7L (VIN 8 & 9 MPFI)
Code 11 No ignition pulse
Code 12 Starter switch; continuously in OFF position
Code 13 Starter switch; continuously in ON position
Code 14 Air flow meter
Code 15 Atmospheric pressure switch (1 .8L VIN 4 & 5 SPFI 1986 only)
Code 16 Crank angle sensor (1.8L VIN 4 & 5 SPFI 1986 only)
Code 17 Starter switch (1 .8L VIN 4 & 5 SPFI 1986 only)
Code 21 Air flow meter flap seized
Code 22 Pressure (Vacuum) switch
Code 23 Throttle sensor
Code 24 Wide Open Throttle (WOT) sensor
Code 25 Throttle sensor (1 .8L VIN 4 & 5 SPEI 1986 only)
Code 31 Vehicle Speed Sensor (VSS)
Code 32 Oxygen sensor
Code 33 Coolant Temperature Sensor (CTS)
Code 34 Intake Air Thermosensor (IAT)
Code 35 EGR solenoid or -Air flow meter (1.8L VIN 4 & 5 SPFI 1986 only)
Code 41 Open or ground in sensor
Code 42 Fuel injector
Code 43 Kickdown Low Hold (KDLH) relay
Code 46 Neutral safety switch (1 .8L VIN 4 & 5 SPFI 1986 only)
Code 53 Fuel pump (1 .8L VIN 4 & 5 SPFI 1986 only)
Code 55 Kickdown Low Hold (KDLH) relay (1 .8L VIN 4 & 5 SPFI 1986 only)
Code 57 Canister purge control (1 .8L VIN 4 & 5 SPFI 1986 only)
Code 58 Air control valve (1 .8L VIN 4 & 5 SPFI 1986 only)
Code 62 EGR control system (1.8L VIN 4 & 5 SPFI 1986 only)
Code 88 Faulty ECU (1 .8L VIN 4 & 5 SPFI 1986 only)

1987–94 VEHICLES

1.2L (VIN 7 & 8 MPFI), 1.8L (VIN 4 & 5 SPFI), 1.8L (VIN 2, 4, 5 & 7 MPFI), 2.2L (VIN 6 MPFI), 2.7L (VIN 8 & 9 MPFI) and 3.3L (VIN 3)
Code 11 Crank angle sensor
Code 12 Starter switch
Code 13 Crank angle (Cam or Cylinder distinction) sensor
Code 14 Fuel injector (1 .8L VIN 4 & 5 SPFI) or—Fuel injector No. 1 (1 .2L VIN 7 & 8 MPFI), (1 .8L VIN 2 MPFI), (2.2L VIN 6 MPFI) and (3.3L VIN 3 MPFI) or—Fuel injectors No. 1 and No. 2 (1.8L VIN 4 & 5 MPFI) or—Fuel injectors No. 5 and No. 6 (2.7L VIN 8 & 9 MPFI)
Code 15 Fuel injector No. 2 (1.2L VIN 7 & 8 MPFI), (1.8L VIN 2 MPFI), (2.2L VIN 6 MPFI) and (3.3L VIN 3 MPFI) or—Fuel injectors No. 3 and No. 4 (1 .8L VIN 4 & 5 MPFI) or—Fuel injectors No. 1 and No. 2 (2.7L VIN 8 & 9 MPFI)
Code 16 Fuel injector No. 3
Code 17 Fuel injector No. 4
Code 18 Fuel injector No. 5
Code 19 Fuel injector No. 6
Code 21 Coolant Temperature Sensor (CTS)
Code 22 Knock sensor (1.8L VIN 2 MPFI), (2.2L VIN 6 MPFI), (2.7L VIN 8 & 9 MPFI)
Code 22 Knock sensor 1; right (3.3L VIN 3 MPFI)

Code 23 Air flow meter

Code 24 Air control valve

Code 25 Fuel injectors No. 3 and No. 4; abnormal injector output (2.7L VIN 8 & 9 MPFI)

Code 26 Air temperature sensor; abnormal signal (1 .2L VIN 7 & 8 MPFI)

Code 28 Knock sensor 2; left (3.3L VIN 3 MPFI)

Code 29 Crank angle sensor 2 (3.3L VIN 3 MPFI)

Code 31 Throttle sensor

Code 32 Oxygen sensor 1; right (3.3L VIN 3 MPFI)

Code 33 Vehicle speed sensor 2 (1 .8L VIN 2 MPFI), (3.3L VIN 3 MPFI)

Code 34 EGR solenoid valve or (California) clogged EGR line

Code 35 Purge control solenoid valve

Code 36 Air suction valve; faulty valve function or—Igniter; abnormal signal

Code 37 Oxygen sensor 2; left (3.3L VIN 3 MPFI)

Code 38 Engine torque control (3.3L VIN 3 MPFI)

Code 41 AF (Air/Fuel) learning control or—System too lean (1.8 VIN 4 & 5 MPFI), (2.7L VIN 8 & 9 MPFI)

Code 42 Idle switch

Code 43 Power switch

Code 44 Duty solenoid valve (wastegate control); valve inoperative

Code 45 Kickdown control relay (1 .8L VIN 4 & 5 SPFI)

Code 45 A: Atmospheric pressure sensor; faulty sensor (2.2L VIN 6 MPFI 1991 only)

Code 45 B: Pressure exchange solenoid valve; valve inoperative (2.2L VIN 6 MPFI 1991 only)

Code 45 Atmospheric pressure sensor (1 .2L VIN 7 & 8 MPFI), (2.2L VIN 6 MPFI 1992 1993), (3.3L VIN 3 MPFI)

Code 49 Air flow sensor

Code 51 Neutral switch

Code 52 Clutch switch; signal remains ON or OFF (Front Wheel Drive/Manual Transaxle only) (1 .2L VIN 7 & 8 MPFI)

Code 52 Parking switch (2.2L VIN 6 MPFI), (3.3L VIN 3 MPFI)

Code 55 EGR gas temperature sensor

Code 56 EGR System; faulty EGR function (1 .8L VIN 2 MPFI)

Code 56 EGR system (California) (3.3L VIN 3 MPFI)

Code 61 Parking switch; continuously in ON position

Code 62 Electric load signal; headlight HI/LO signal or rear defogger signal remains ON or OFF

Code 63 Blower fan switch; signal remains ON or OFF

Code 65 Vacuum pressure sensor; abnormal signal

➡**If more than one definition is listed for a code or the code is not listed here, consult your 'Chilton Total Car Care' manual to obtain the specific meaning for your vehicle. This list is for reference and does not mean that a component is defective. The code identifies the circuit and component that require further testing.**

Suzuki

GENERAL INFORMATION

Suzuki used both feedback carburetor and fuel injection systems. Suzuki uses the Hitachi 2-barrel, downdraft type carburetor, which has both a primary and secondary system. A feedback system

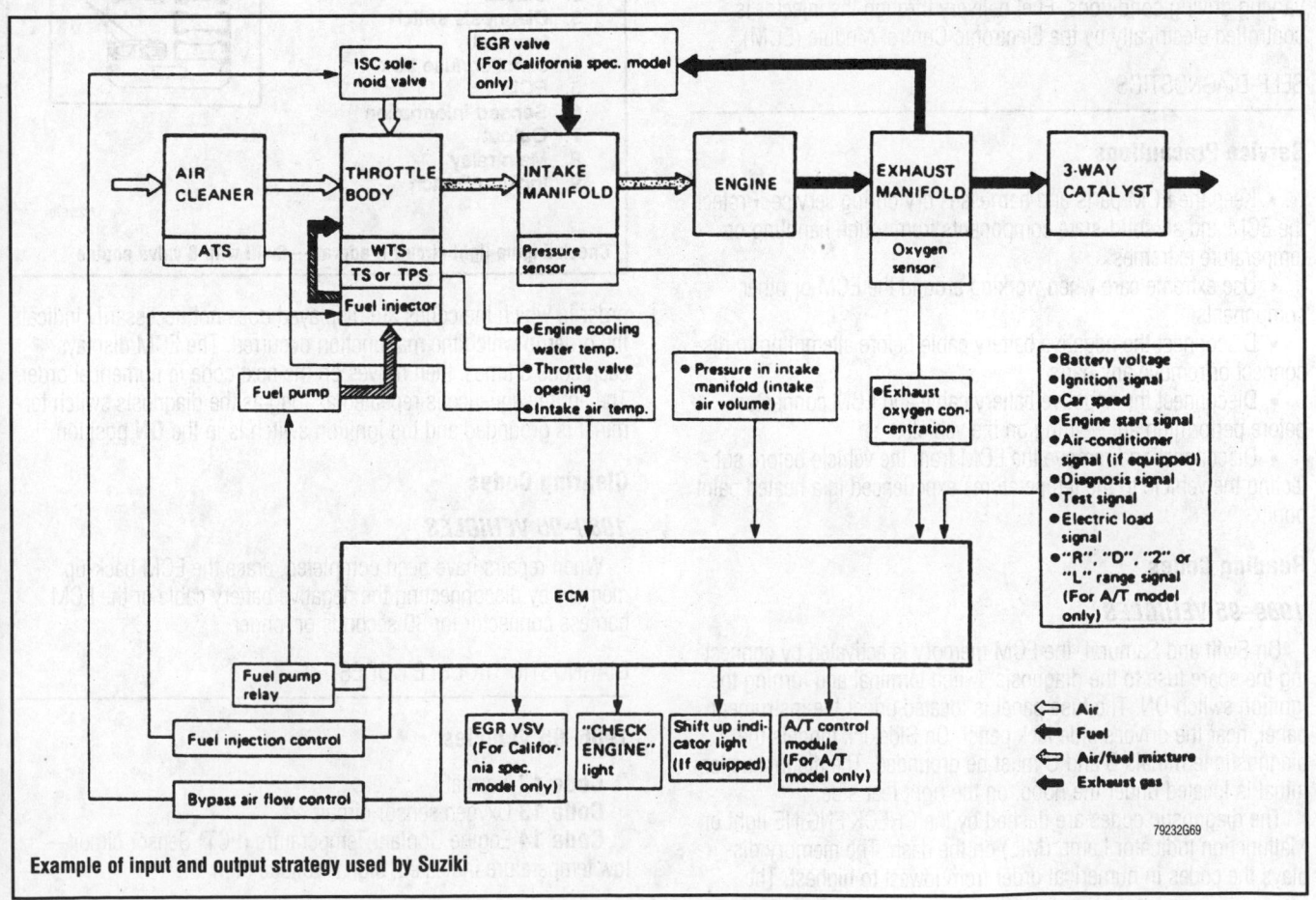

Example of input and output strategy used by Suziki

79232G69

is provided to maintain the air/fuel ratio, to reduce emission levels and to improve fuel economy simultaneously.

The primary system operates under normal driving conditions and the secondary system operates under high speed-high load driving conditions. A choke valve is provided in the primary system.

The primary system is equipped with a choke system. The choke system is a fully automatic type using a thermo-wax. A mixture control solenoid valve is also incorporated which is operated by an electrical signal from the Electronic Control Module (ECM). The acceleration pump system and a fuel cutoff solenoid are also part of the primary system.

The secondary system is equipped with a secondary diaphragm through which vacuum is supplied from the primary side, via a Vacuum Switching Valve (VSV) and a Vacuum Transmitting Valve (VTV), to operate the secondary throttle vavle. The VSV and VTV are used only on the Suzuki Samurai and have been eliminated on the Suzuki Sidekick 1300cc.

A Switch vent solenoid valve is provided on the top of the float chamber. Its purpose is to reduce the evaporative emissions.The 2-barrel, downdraft type carburetor is also equipped with a idle-up system. This system operates at idle and compensates the idle speed when any of the following conditions exist:

- When any electrical load (lights, rear defogger, heater fan, etc.) is operating.
- When the vehicle is at a high altitude.
- When the engine temperature is below 44°F (7°C).
- When the engine speed is lower than 1500 rpm after the engine is started.

The Electronic Fuel Injection (EFI) system supplies the vehicle's combustion chambers with air/fuel mixture of optimized ratio under varying driving conditions. Fuel delivery through the injector is controlled electrically by the Electronic Control Module (ECM).

SELF-DIAGNOSTICS

Service Precautions

- Keep the ECM parts and harnesses dry during service. Protect the ECM and all solid-state components from rough handling or temperature extremes.
- Use extreme care when working around the ECM or other components.
- Disconnect the negative battery cable before attempting to disconnect or remove any parts.
- Disconnect the negative battery cable and ECM connector before performing arc welding on the vehicle.
- Disconnect and remove the ECM from the vehicle before subjecting the vehicle to the temperatures experienced in a heated paint booth.

Reading Codes

1989–95 VEHICLES

On Swift and Samurai, the ECM memory is activated by connecting the spare fuse to the diagnosis switch terminal and turning the ignition switch ON. The fuse panel is located under the instrument panel, near the driver's side kick panel. On Sidekick models the diagnostic terminals B and C must be grounded. The diagnostic terminal is located under the hood, on the right rear side.

The diagnostic codes are flashed by the CHECK ENGINE light or Malfunction Indicator Lamp (MIL) on the dash. The memory displays the codes in numerical order from lowest to highest. The

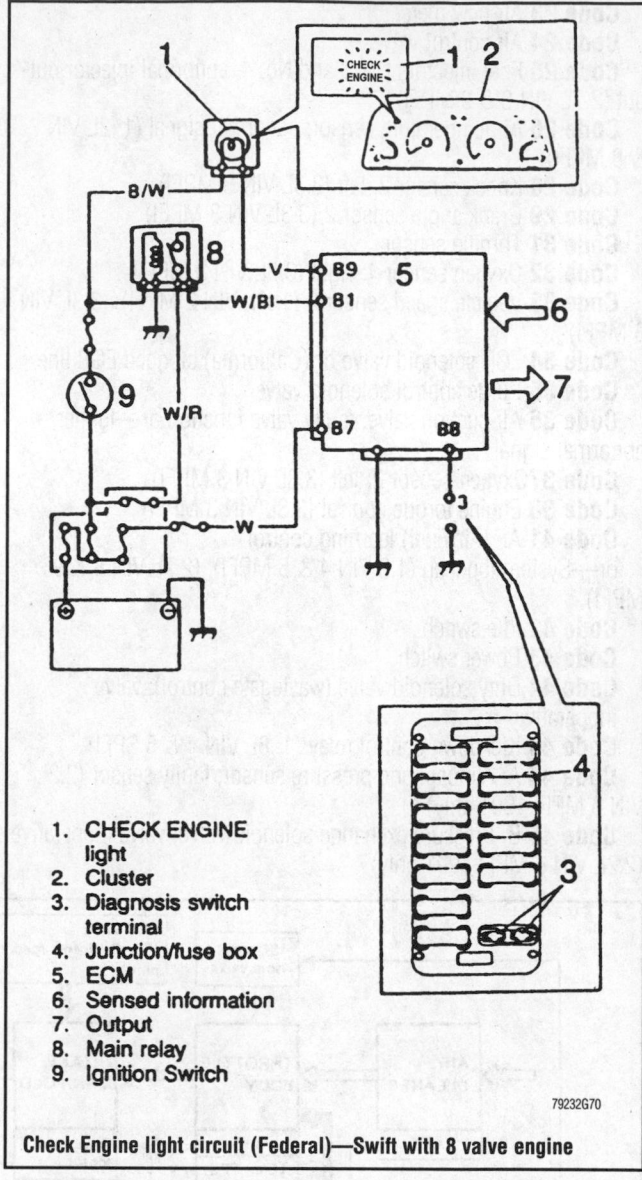

1. **CHECK ENGINE light**
2. **Cluster**
3. **Diagnosis switch terminal**
4. **Junction/fuse box**
5. **ECM**
6. **Sensed information**
7. **Output**
8. **Main relay**
9. **Ignition Switch**

79232G70

Check Engine light circuit (Federal)—Swift with 8 valve engine

order in which the codes are displayed does not necessarily indicate the order in which the malfunction occurred. The ECM displays each code 3 times, then moves on the next code in numerical order. The entire sequence is repeated as long as the diagnosis switch terminal is grounded and the ignition switch is in the **ON** position.

Clearing Codes

1989–95 VEHICLES

When repairs have been completed, erase the ECM back-up memory by disconnecting the negative battery cable or the ECM harness connector for 30 seconds or longer.

DIAGNOSTIC TROUBLE CODES

1986–95 Vehicles

Code 12 Normal
Code 13 Oxygen sensor circuit
Code 14 Engine Coolant Temperature (ECT) Sensor circuit—low temperature indicated, signal voltage high

Code 15 Engine Coolant Temperature (ECT) Sensor circuit—high temperature indicated, signal voltage low

Code 21 Throttle Position Sensor (TPS) circuit—signal voltage high

Code 22 Throttle Position Sensor (TPS) circuit—signal voltage low

Code 23 Air Temperature Sensor (ATS) circuit—low temperature indicated, signal voltage high

Code 24 Vehicle Speed Sensor (VSS) circuit

Code 25 Air Temperature Sensor (ATS) circuit—high temperature indicated, signal voltage low

Code 31 Pressure Sensor (PS) circuit—high pressure indicated, signal voltage high

Code 32 Pressure Sensor (PS) circuit—low pressure indicated, signal voltage low

Code 33 Mass Air Flow Sensor (MAS) circuit—signal voltage high

Code 34 Mass Air Flow Sensor (MAS) circuit—signal voltage low

Code 41 Ignition signal

Code 42 Crank Angle Sensor (CAS) circuit (except 1989–90 Sidekick) or Fifth switch circuit, Lock-up signal circuit (1989–90 Sidekick)

Code 44 Idle switch of Throttle Position Sensor (TPS)—open circuit

Code 45 Idle switch of Throttle Position Sensor (TPS)—shorted circuit

Code 51 Exhaust Gas Recirculation (EGR) system and/or Recirculated Exhaust Gas Temperature Sensor (REGTS) system—California vehicle

Code 52 Fuel Injector—California vehicle

Code 53 Ground circuit—California vehicle

Code 54 Fifth gear switch circuit

Code 71 Test switch circuit

Toyota

GENERAL INFORMATION

Toyota vehicles may be fitted with either a Feedback Carburetor (FBC) or Multi-port Fuel Injection (MEI) system. The Toyota Feedback Carburetor (FBC) system was used on selected engines from 1983 1990. Two types of carburetors were used. The 3E engine used a variable venturi carburetor, while all other engines used a more typical down draft style carburetor. The Multi-port Fuel Injection system was first used in 1980 and continues in use today.

SELF-DIAGNOSTICS

As the engine control computers became capable of more functions, self-diagnostic and memory circuits were added.

These systems allow the ECU to note a fault, assign an identity code and store the code in memory for later retrieval.

All fuel injected control engine units possess the ability to provide fault codes during diagnosis. The number, type and meaning of engine codes vary by year and model.

While most fault codes are held in an electronic memory and are retained even after the ignition is switched **OFF**, certain codes are

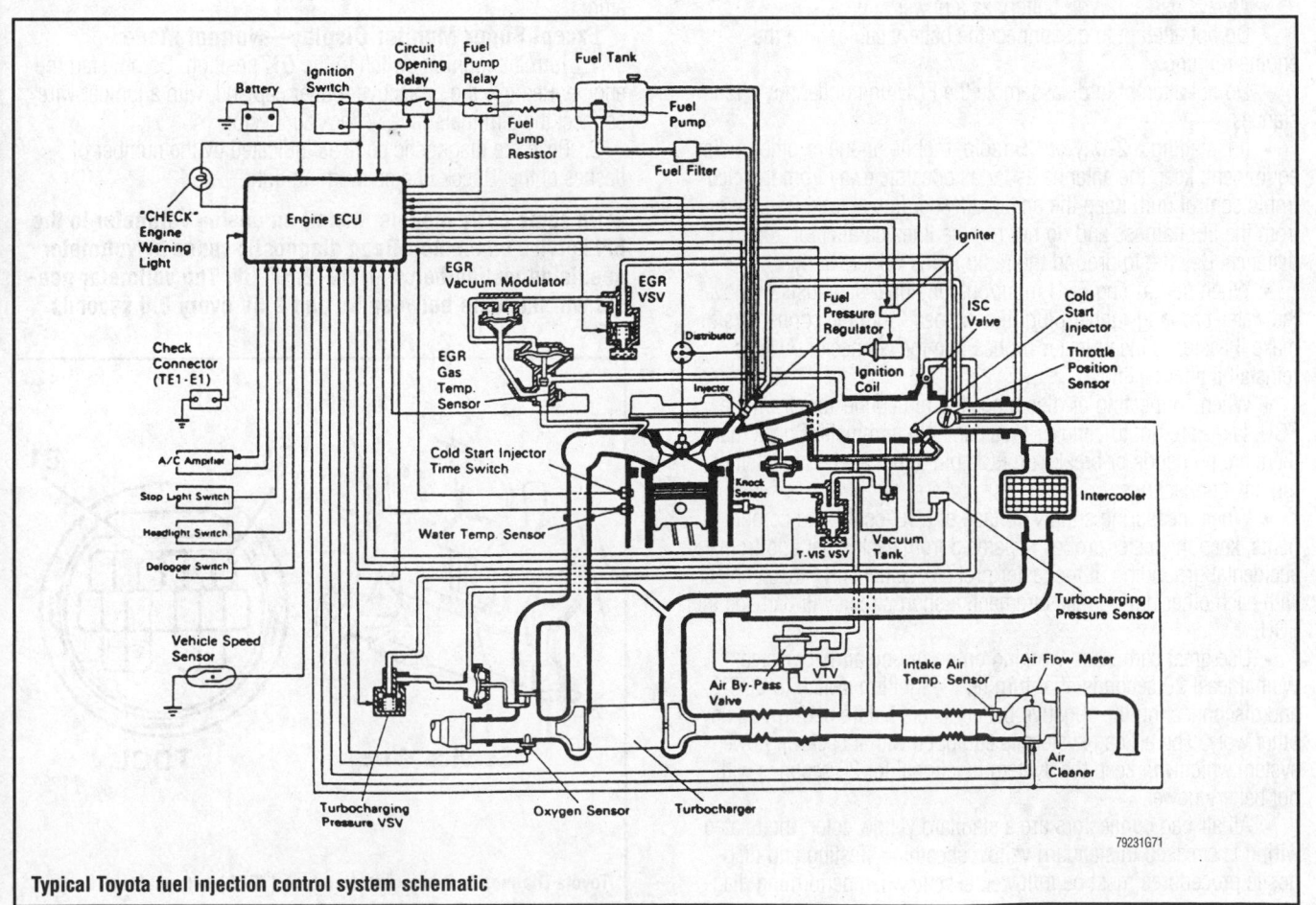

Typical Toyota fuel injection control system schematic

only held or displayed as long as the ignition is **ON**. If the fault is present at the next restart, the code will reset.

When a controller or ECU notes a fault, the dash warning lamp for the appropriate system will be lit to advise the operator. If the dash lamp is normally lit during system operation, as in the case of cruise control, the lamp will flash when a fault is found. The illumination or flashing of the dash lamp indicates that the controller has detected a fault and placed itself into the back-up or default mode.

Beginning in 1995 some models were equipped with an on board diagnostic system known as OBD II. To diagnose this system an OBD II scan tool, complying with SAE J1978 or TOYOTA hand held tester is necessary to access codes and read data output from the ECM. This is a rather expensive tool and not cost effective for the general public. The following model and engine applications are equipped with the OBD II system:

- 1995 Tercel
- 1995 Camry with a 1 MZ-FE engine
- 1995 Avalon

Service Precautions

- Keep all ECU parts and harnesses dry during service. Protect the ECU and all solid-state components from rough handling or extremes of temperature.
- Before attempting to remove any parts, turn the ignition switch **OFF** and disconnect the battery ground cable.
- Make sure all harness connectors are fastened securely. A poor connection can cause an extremely high voltage surge, resulting in damage to integrated circuits.
- Always use a 12 volt battery as a power source.
- Do not attempt to disconnect the battery cables with the engine running.
- Do not attempt to disassemble the ECU unit under any circumstances.
- If installing a 2-way or CB radio, mobile phone or other radio equipment, keep the antenna as far as possible away from the electronic control unit. Keep the antenna feeder line at least 8 in. away from the EEI harness and do not run the lines parallel for a long distance. Be sure to ground the radio to the vehicle body.
- When performing ECU input/output signal diagnosis, remove the water-proofing rubber plug, if equipped, from the connectors to make it easier to insert tester probes into the connector. Always reinstall it after testing.
- When connecting or disconnecting pin connectors from the ECU, take care not to bend or break any pin terminals. Check that there are no bends or breaks on ECU pin terminals before attempting any connections.
- When measuring supply voltage of ECU-controlled components, keep the tester probes separated from each other and from accidental grounding. If the tester probes accidentally make contact with each other during measurement, a short circuit will damage the ECU.
- Use great care when working on or around air bag systems. Wait at least 20 seconds after turning the ignition switch to LOCK and disconnecting the negative battery cable before performing any other work. The air bag system is equipped with a back-up power system which will keep the system functional for 20 seconds without battery power.
- All air bag connectors are a standard yellow color; the related wiring is encased in standard yellow sheathing. Testing and diagnostic procedures must be followed exactly when performing diag-

nosis on this system. Improper procedures may cause accidental deployment or disable the system when needed.
- Never attempt to measure the resistance of the air bag squib; detonation may occur.

Reading Codes

The following procedures are for all vehicles except those equipped with the OBD II system. Accessing OBD II system codes can only be accomplished with the use of a OBD II scan tool, complying with SAE J1978 or TOYOTA hand held tester. This is a rather expensive tool and not cost effective for the general public. The following models are equipped with the OBD II system:

- 1995 Tercel
- 1995 Camry with a 1 MZ-FE engine
- 1995 Avalon

1983–86 VEHICLES

The diagnostic codes can be read by the number of blinks of the 'Check Engine' warning light when the proper terminals of the check connector are short-circuited. If the vehicle is equipped with a super monitor display, the diagnostic code is indicated on the display screen. The initial conditions for entering the self-diagnostics are as follows:

1. The battery voltage of the vehicle should be above 11 volts. The throttle valve must be in a fully closed position (throttle position sensor IDL points closed).
2. If equipped with an automatic transmission, place it in P or N.
3. Turn the air conditioning switch OFF.
4. Start the engine and allow it reach normal operating temperature.

Except Super Monitor Display—Normal Mode

1. Turn the ignition switch to the **ON** position. Do not start the engine. Remove the protective rubber cap and, with a jumper wire connect the terminals of the check connector.
2. Read the diagnostic code as indicated by the number of flashes of the 'Check Engine' warning light.

➡**On some early models, install an analog voltmeter to the EFI service connector. Read diagnostic codes by voltmeter needle deflection between 0V–2.5V–5V. The voltmeter needle will fluctuate between 5V and 2.5V every 0.6 seconds.**

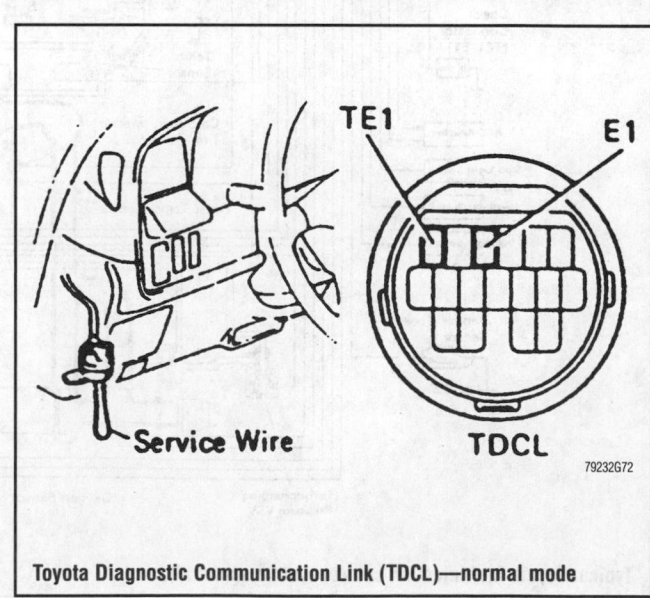

Toyota Diagnostic Communication Link (TDCL)—normal mode

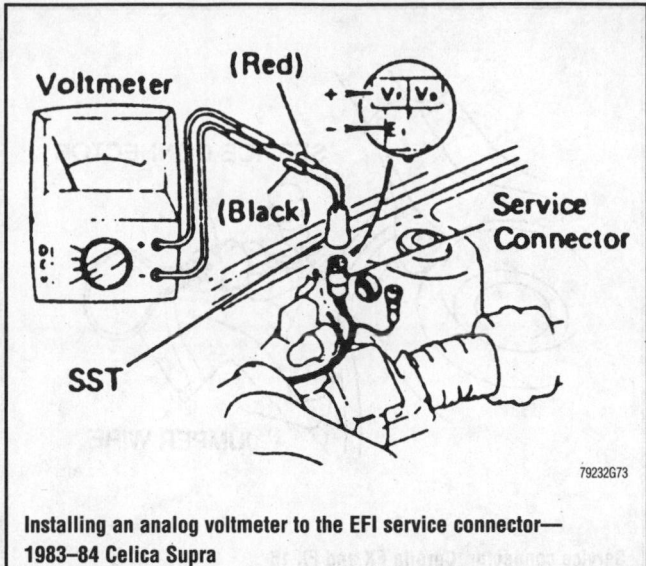

Installing an analog voltmeter to the EFI service connector—1983–84 Celica Supra

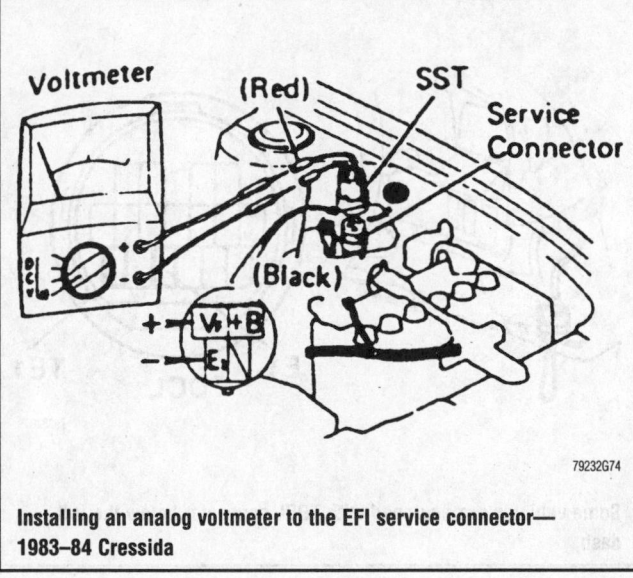

Installing an analog voltmeter to the EFI service connector—1983–84 Cressida

3. If the system is operating normally (no malfunction), the light will blink once every ¼ second. On single digit code number systems, the light will blink once every 3 or 4.5 seconds.

4. In the event of a malfunction, the light will blink once every ½ second (on some models it may be 1, 2 or 3 seconds). The 1st number of blinks will equal the 1st digit of a 2-digit diagnostic code. After a 1.5 second pause, the 2nd number of blinks will equal the 2nd number of a 2-digit diagnostic code. If there are 2 or more codes, there will be a 2.5 second pause between each. On single digit code number systems the light will blink a number of times equal to the malfunction code indication every 2 or 4.5 seconds.

5. After all the codes have been output, there will be a 4.5 second pause and they will be repeated as long as the terminals of the check connector are shorted.

➡ **In event of multiple trouble codes, indication will begin from the smaller value and continue to the larger in order.**

6. After the diagnosis check, remove the jumper wire from the check connector and install the protective rubber cap.

Test Mode

1. Using a jumper wire, connect the TE2 and E1 terminals of the Toyota Diagnostic Communication Link (TDCL), then turn the ignition switch **ON** to begin the diagnostic test mode.

2. Start the engine and drive the vehicle at a speed of 10 mph or more. Simulate the conditions where the malfunction has been reported to happen.

3. Using a jumper wire, connect the TE2 and E1 terminals of the TDCL connector.

4. Read the diagnosis code as indicated by the number of 'Check Engine' light flashes.

5. After diagnosis check remove the jumper wires.

Super Monitor Display

The super monitor display system was offered as an option on some late model Toyota vehicles.

1. Turn the ignition switch **ON** but do not start the engine.

2. Simultaneously push and hold in the SELECT and INPUT M keys for at least 3 seconds. The letters DIAG will appear on the screen.

3. After a short pause, hold the SET key in for at least 3 seconds. If the system is normal (no malfunctions), ENG-OK will appear on the screen.

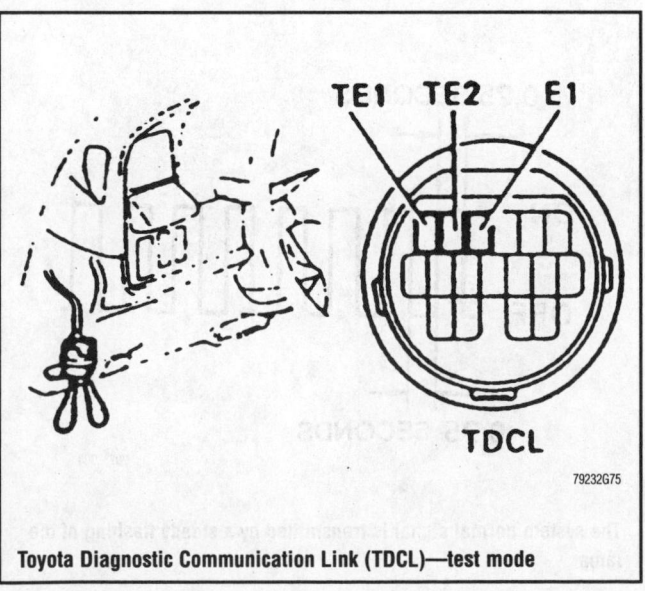

Toyota Diagnostic Communication Link (TDCL)—test mode

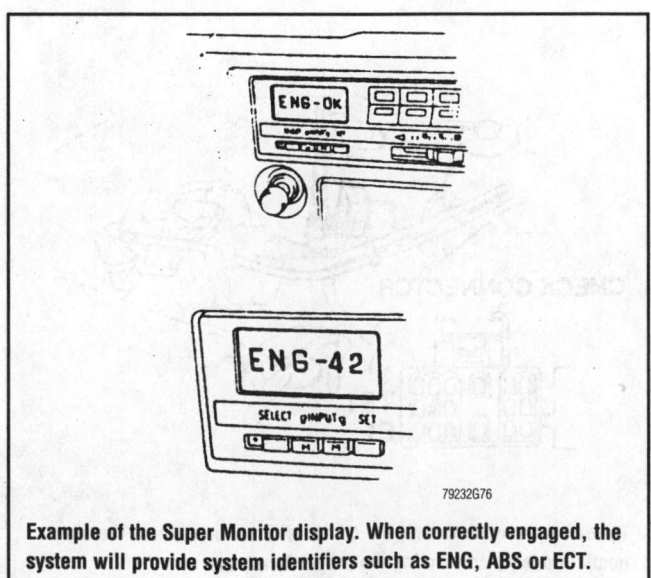

Example of the Super Monitor display. When correctly engaged, the system will provide system identifiers such as ENG, ABS or ECT.

4. If there is a malfunction, the code number for it will appear on the screen. In event of 2 or more numbers, there will be a 3 second pause between each (example:EN-42).

1987–89 VEHICLES

Stored fault codes are transmitted through the blinking of the CHECK engine warning lamp. This occurs only when the system is placed into the diagnostic mode; it does not occur while the vehicle is being driven.

To read the fault codes:

1. The following initial conditions must be met:
   a. Battery voltage at or above 11 volts.
   b. Throttle fully closed.
   c. Transmission in N or P.
   d. All electrical systems and accessories OFF.
2. Turn the ignition **ON** but do not start the engine.
3. Use a jumper wire to connect terminals T and E1 at the diagnostic connector. On all 1989 vehicles except Corolla, MR2 and Tercel, connect terminals TE1 and Et. For 1988–89 Vans, jumper the 2 pins of the service connector. On 1989 Corolla, MR2 and Tercel, connect terminals T and E1.
4. The fault codes will be transmitted through the controlled flashing of the CHECK engine warning lamp.
5. If no malfunction was found or no code was stored, the lamp will flash 2 times per second with no other pauses or patterns. This confirms that the diagnostic system is working but has nothing to report. This light pattern may be referred to as the System Normal signal; it should be present when no other codes are stored.
6. The CHECK lamp will blink the number of the code(s). All codes are 2 digits; the pulsing of the light represents the digits, not the count. For example, Code 25 is displayed as 2 flashes a pause and 5 flashes.
7. If more than 2 codes are stored, the next will be transmitted after a 21/2 second pause.

➡️**If multiple codes are stored, they will be transmitted in numerical order from lowest to highest. This does not indicate the order of fault occurrence.**

8. When all codes have been transmitted, the entire pattern will repeat after a 4½ second pause. The repeats continue as long as the diagnostic terminals are connected.

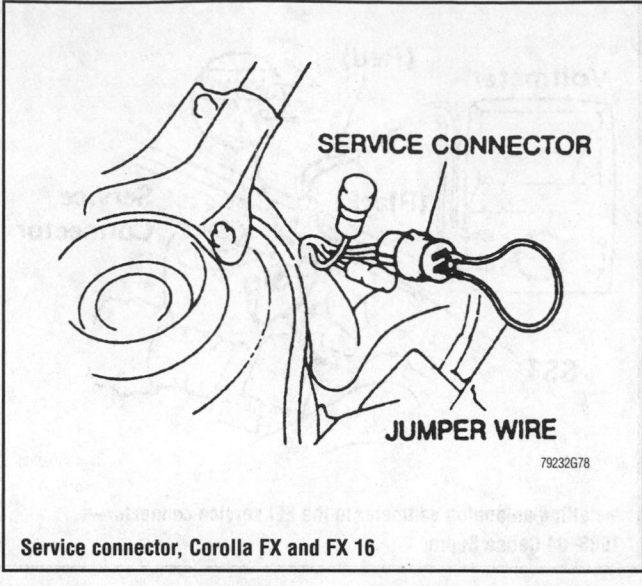

Service connector, Corolla FX and FX 16

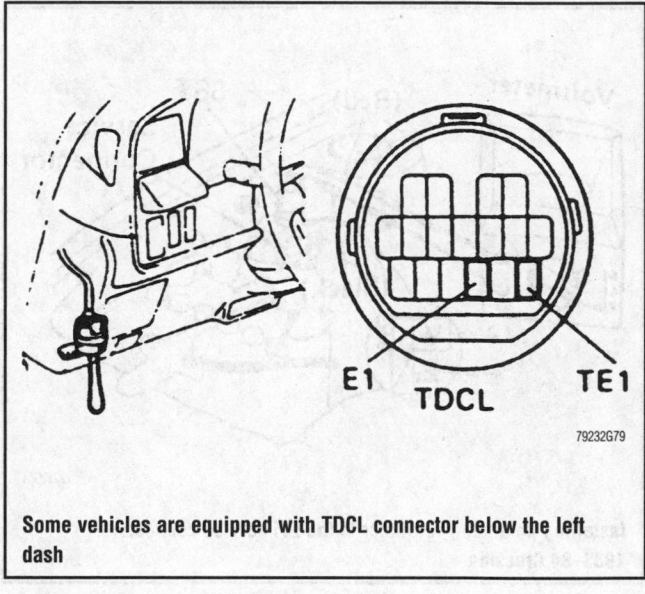

Some vehicles are equipped with TDCL connector below the left dash

To read the engine codes, the computer must be put into the diagnostic mode by connecting the proper terminals

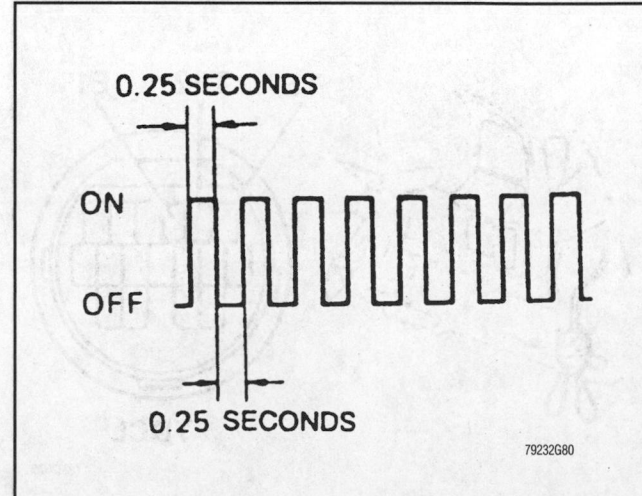

The system normal signal is transmitted by a steady flashing of the lamp

9. After recording the codes, disconnect the jumper at the diagnostic connector and turn the ignition **OFF**.

Super Monitor System

This procedure is used on Cressida and Supra equipped with Super Monitor.

1. The following initial conditions must be met:
   a. Battery voltage at or above 11 volts.
   b. Throttle fully closed.
   c. Transmission in N or P.
   d. All electrical systems and accessories OFF.
2. Turn the ignition **ON** but do not start the engine.
3. Simultaneously press and hold the SELECT and INPUT M keys for at least 2 seconds. The letters DIAG will appear on the screen, showing that the system is in the diagnostic mode.
4. After a short pause, hold in the SET key for at least 2 seconds.
5. If the system is normal, with no faults stored, the message ENG OK will appear on the screen. If faults are stored, the code number will appear on the screen with a system designator; for example, ENG-42. If 2 or more codes are stored, each will appear after a 3 second pause.

1990–95 VEHICLES

Stored fault codes are transmitted through the blinking of the CHECK engine warning lamp. This occurs only when the system is placed into the diagnostic mode; it does not occur while the vehicle is being driven.

To read the fault codes:

1. The following initial conditions must be met:
   a. Battery voltage at or above 11 volts.
   b. Throttle fully closed.
   c. Transmission in N or P.
   d. All electrical systems and accessories OFF.
2. Turn the ignition **ON** but do not start the engine.
3. Use a jumper wire to connect terminals TE1 and E1 at the diagnostic connector in the engine compartment or at the TDCL connector below the left dashboard if so equipped.
4. The fault codes will be transmitted through the controlled flashing of the CHECK ENGINE warning lamp.
5. If no malfunction was found or no code was stored, the lamp will flash 2 times per second with no other pauses or patterns. This confirms that the diagnostic system is working but has nothing to report. This light pattern may be referred to as the System Normal signal; it should be present when no other codes are stored.
6. The CHECK lamp will blink the number of the code(s). All codes are 2-digit; the pulsing of the light represents the digits, not the count. For example, Code 25 is displayed as 2 flashes, a pause and 5 flashes.
7. If more than 1 code is stored, the next will be transmitted after a 2½ second pause.

➡ **If multiple codes are stored, they will be transmitted in numerical order from lowest to highest. This does not indicate the order of fault occurrence.**

8. When all codes have been transmitted, the entire pattern will repeat after a 4½ second pause. The repeats continue as long as the diagnostic terminals are connected.
9. After recording the codes, disconnect the jumper at the diagnostic connector and turn the ignition **OFF**.

Clearing Codes

1986–95

Stored codes will remain in memory until cleared. The correct method of clearing codes is to turn the ignition switch **OFF**, then remove the proper fuse. On all vehicles except as noted below, remove the EEI fuse. Each fuse must be removed for at least 10 seconds. The time required may be longer in cold weather.

Disconnecting the negative battery cable will also clear the memory but is not recommended due to other on-board memories being cleared as well. Once the system power is restored, re-check for stored codes. Only the System Normal indication should be present. If any other code is stored, the clearing procedure must be repeated or additional repairs performed; the old code will remain stored along with any new ones.

After repairs, it is recommended to clear the memory before test driving the vehicle. Upon returning from the drive, interrogate the memory; if the original code is again present, the repair was unsuccessful.

DIAGNOSTIC TROUBLE CODES

1983–1984 Engine

Code 1 Normal operation
Code 2 Open or shorted air flow meter circuit—defective air flow meter or Electronic Control Unit (ECU)
Code 3 Open or shorted air flow meter circuit—defective air flow meter or Electronic Control Unit (ECU)
Code 4 Open Water Thermosnsor (THW) circuit—defective Water Thermosensor (THW) or Electronic Control Unit (ECU)
Code 5 Open or shorted oxygen sensor circuit—lean or rich indication—defective oxygen sensor or Electronic Control Unit (ECU)
Code 6 No ignition signal—defective ignition system circuit, Integrated Ignition Assembly (IIA) or Electronic Control Unit (ECU)
Code 7 Defective Throttle Position Sensor (TPS) circuit, Throttle Position Sensor (TPS) or Electronic Control Unit (ECU)

1985–1987 Engines

➡ **The 1985 2.0L (25-E and 3Y-EC) engines use 1984 Codes**

Code 1 Normal operation
Code 2 Open or shorted air flow meter circuit—defective air flow meter or Electronic Control Unit (ECU)
Code 3 No signal from igniter 4 times in succession—defective igniter or main relay circuit, igniter or Electronic Control Unit (ECU)
Code 4 Open Water Thermosensor (THW) circuit—defective Water Thermosensor (THW) or Electronic Control Unit (ECU)
Code 5 Open or shorted oxygen sensor circuit—lean or rich indication—defective oxygen sensor or Electronic Control Unit (ECU)
Code 6 No engine revolution sensor (Ne) signal to Electronic Control Unit (ECU) or Ne value being over 1000 rpm in spite of no Ne signal to ECU—defective igniter circuit, igniter, distributor or Electronic Cont rol Unit (ECU)
Code 7 Open or shorted Throttle Position Sensor (TPS) circuit, Throttle Position Sensor (TPS) or Electronic Control Unit (ECU)
Code 8 Open or shorted intake air thermosensor circuit—

defective intake air thermosensor circuit or Electronic Control Unit (ECU)

Code 10 No starter switch signal to Electronic Control Unit (ECU) with vehicle speed at 0 and engine speed over 800 rpm—defective speed sensor circuit, main relay circuit, igniter switch to starter circuit, igniter switch or Electronic Control Unit (ECU)

Code 11 Short circuit in check connector terminal T with the air conditioning switch ON or throttle switch (IDL) contact point OFF—defective air conditioner switch, Throttle Position Sensor (TPS) circuit, Throttle Position Sensor (TPS) or Electronic Control Unit (ECU)

Code 12 Knock control sensor signal has not reached judgement level in succession—defective knock control sensor circuit, knock control sensor or Electronic Control Unit (ECU)

Code 13 Knock CPU faulty

1988–95 Engines except 1995 OBD II

Constant blinking of indicator light: No faults detected

Code 11 Momentary interruption in power supply to ECU; up to 1991

Code 12 Engine revolution (NE or G) signal to ECU; missing within several seconds after engine is cranked

Code 13 Rpm NE signal to ECU; missing when engine speed is above 1000 rpm

Code 14 Igniter (IGE) signal to ECU; missing 4–11 times in succession

Code 16 ECT control signal—normal signal missing from ECT CPU (1990–94)

Code 16 A/T control system—normal signal missing from between the engine CPU and A/T CPU in the ECM (1995)

Code 21 Main oxygen sensor signal; voltage output does not exceed a set value on the lean and rich sides continuously for a certain period of time or open/short sensor heater circuit

Code 22 Water temperature sensor circuit (THW); open/short for 500 msec. or more

Code 23 Intake air temperature signal (THA)

Code 24 Intake air temperature sensor circuit (THA); open/short for 500 msec. or more

Code 25 Air/fuel ratio LEAN malfunction; Oxygen sensor output is less than 0.45 V for at least 90 seconds when oxygen sensor is warmed up (engine racing at 2000 rpm). California only: air/fuel ratio feedback compensat ion/adaptive control: feedback value continues at upper (LEAN) limit, or is not renewed, for a certain period of time.

Code 26 Air/fuel ratio RICH malfunction; California only: Air/fuel ratio feedback compensation/adaptive control: feedback value continues at lower (RICH) limit, or is not renewed, for a certain period of time.

Code 27 Sub-oxygen sensor signal; detection of sensor/signal deterioration or open/short sensor heater circuit (California only)

Code 28 No. 2 oxygen sensor signal/heater signal

Code 31 Air flow meter circuit; open or shorted when idle contacts are closed

Code 31 Vacuum (Manifold absolute pressure) sensor signal; open/short circuit

Code 32 Air flow meter circuit; circuit open or shorted when idling

Code 34 Turbocharging pressure signal; excessive pressure

Code 35 Altitude compensation (HAC) sensor signal; open/short

Code 35 Turbocharging pressure sensor signal; open/short

Code 36 Turbocharging pressure sensor signal; open or short detected for 0.5 sec or more in the turbocharging pressure sensor signal circuit; 1992–94

Code 41 Throttle position sensor circuit (VTA); open/short

Code 42 Vehicle speed sensor circuit

Code 43 No starter switch (STA) signal to ECU until engine speed reaches 800 rpm when cranking

Code 51 NC signal ON, DL contact OFF, or shift position in R, D, 2 or 1 range; with check terminals T and El connected

Code 52 Knock sensor signal (KNK); open/short

Code 53 Knock control signal in ECU; ECU knock control faulty

Code 55 Knock sensor (rear side) signal in ECU; ECU knock control faulty

Code 71 EGR system malfunction; EGR gas temperature signal (THG) is below water temperature sensor signal or below intake air temperature sensor signal plus 86°F (30°C), after driving for 240 seconds in EGR operation range (California only)

Code 72 Fuel cut solenoid signal circuit (FCS) open; up to 1991

Code 78 Fuel pump control signal input circuit to pump (FPC) open

Code 81 TCM communication; open detected in ECT1 circuit for 2 or more seconds

Code 83 TCM communication; open detected in ESA1 circuit 0.5 sec after idle

Code 84 TCM communication; open in ESA2 circuit for 0.5 seconds after idle

Code 85 TCM communication; open in ESA3 circuit for more than 0.5 seconds after idle

---

Volkswagen

GENERAL INFORMATION

CIS-E Fuel Injection

The CIS-E Motronic system is the latest development of the electronically controlled mechanical Continuous Injection System (CIS). This system uses injectors, fuel pump and air flow sensor that are similar to those on earlier systems. The fuel distributor is equipped with an electronically controlled differential pressure regulator. This is operated by the ECU to control the fuel pressure in the lower chamber of the fuel distributor, which controls air/fuel mixture.

The ECU is now equipped with an adaptive learning program which allows it to learn and remember the normal operating range of the mixture control output signal. This gives the system the capability to compensate for changes in altitude, slight vacuum leaks or other changes due to things such as engine wear. Cold engine driveability and emissions are improved. The new ECU also is capable of cold start enrichment without the use of a thermo-time switch. The Fox still uses the thermo-time switch on CIS equipped vehicles.

The fuel injector pressure has been increased for better fuel atomization and residual pressure. The threads on the new injectors are different so they cannot be interchanged with older units. Some of the other components used on the CIS-E system are similar to those used on the fully electronic engine management systems. Some of the testing procedures are the same but the parts are not necessarily interchangeable.

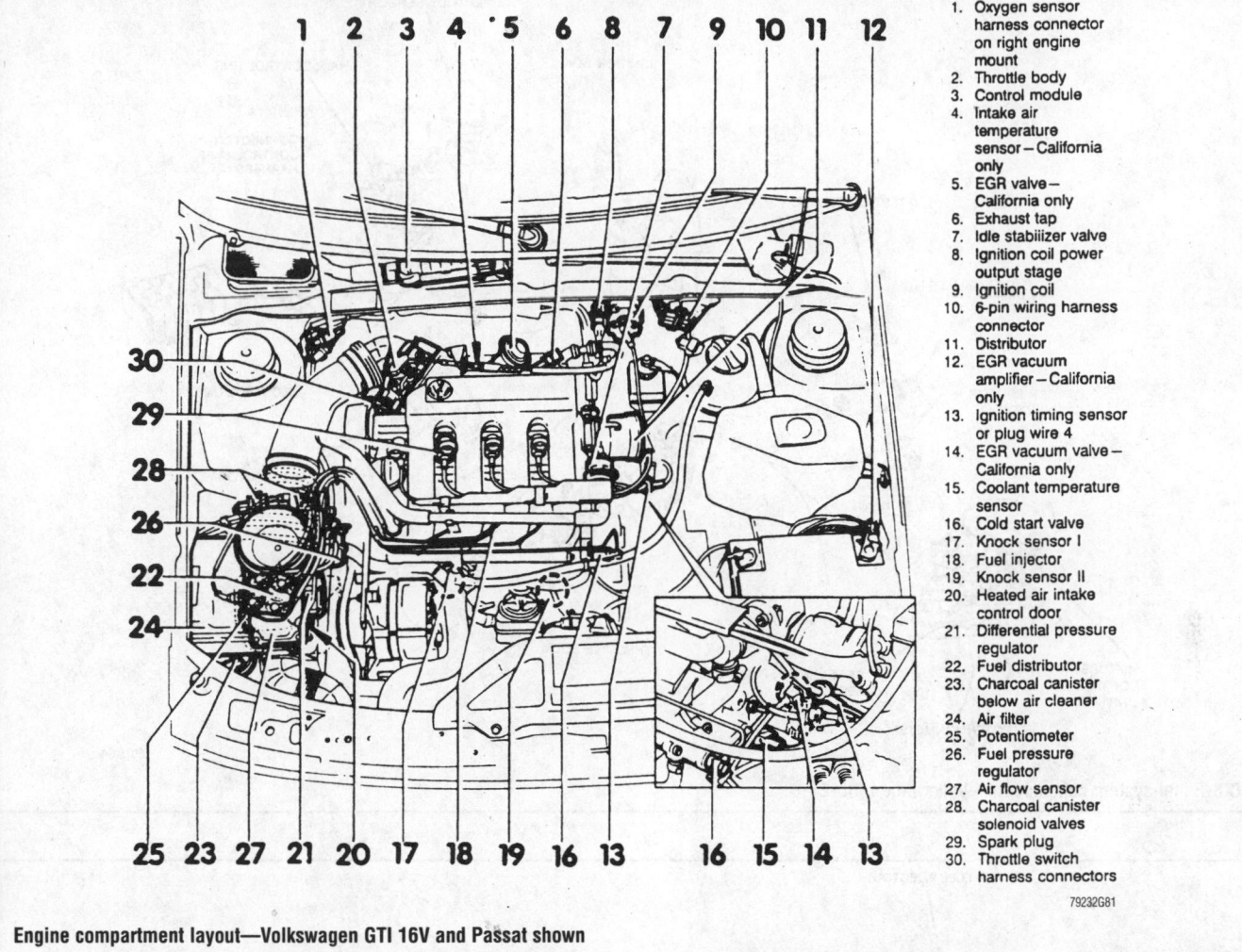

1. Oxygen sensor harness connector on right engine mount
2. Throttle body
3. Control module
4. Intake air temperature sensor—California only
5. EGR valve—California only
6. Exhaust tap
7. Idle stabilizer valve
8. Ignition coil power output stage
9. Ignition coil
10. 6-pin wiring harness connector
11. Distributor
12. EGR vacuum amplifier—California only
13. Ignition timing sensor or plug wire 4
14. EGR vacuum valve—California only
15. Coolant temperature sensor
16. Cold start valve
17. Knock sensor I
18. Fuel injector
19. Knock sensor II
20. Heated air intake control door
21. Differential pressure regulator
22. Fuel distributor
23. Charcoal canister below air cleaner
24. Air filter
25. Potentiometer
26. Fuel pressure regulator
27. Air flow sensor
28. Charcoal canister solenoid valves
29. Spark plug
30. Throttle switch harness connectors

79232G81

Engine compartment layout—Volkswagen GTI 16V and Passat shown

Motronic and Motronic 2.9 Multiport Fuel Injection (MFI) Systems

Motronic and Motronic 2.9 systems are developments of the electronically controlled Multiport Fuel Injection (MEI) system. The two systems are almost identical. The engine control module (ECM) monitors engine intake air quantity using the Mass Air Flow (MAF) sensor. This is a true mass air measurement system. Using input from the MAE and other sensors, the ECM can calculate the length of time the injectors should be opened, and also controls the ignition system timing.

The ECM is equipped with an adaptive learning program which allows it to learn and remember the normal operating range of the mixture control output signal. This allows the system to compensate for changes in altitude, slight vacuum leaks or other changes due to other things such as engine wear. Cold engine driveability and emissions are improved.

The ECM is also equipped with a fault memory. If the sensor signal or output solenoid feedback signal is outside preprogrammed parameters, the ECM will store a fault code representing the fault and sensor involved. The ECM will also illuminate the Malfunction Indicator Lamp (MIL) to inform the vehicle operator that the vehicle requires service.

Mono-Motronic Throttle Body Fuel Injection (TBI) System

The Mono-Motronic system is development of the electronically controlled Throttle Body Fuel Injection (TBI) system. Based on out-

puts from the Throttle Position (TP), Engine Coolant Temperature (ECT) and Intake Air Temperature/Fuel Injector Temperature (IAT/FIT) sensors, the Engine Control Module (ECM) can infer intake air flow by air temperature and throttle position. This is a speed-density type control system. The ECM calculates the length of time the injector(s) should be opened, and with input from other sensors, also controls ignhion timing.

The ECM is equipped with an adaptive learning program which allows it to learn and remember the normal operating range of the mixture control output signal. This allows the system to compensate for changes in altitude, slight vacuum leaks or other changes due to other things such as engine wear. Cold engine driveability and emissions are improved. The ECM is also equipped with a fault memory. If the sensor signal or output solenoid feedback signal is outside preprogrammed parameters, the ECM will store a fault code representing the fault and sensor involved. The ECM will also illuminate the Malfunction Indicator Lamp (MIL) to inform the vehicle operator that the vehicle requires service.

Digifant Multiport Fuel Injection (MFI) System

The Digifant Motronic system is development of the electronically controlled Multiport Fuel Injection (MEI) system. This system is quite similar on all models, but there is one significant difference among the three engines, intake air flow is measured using one of two systems. The 1.8L NA engine is equipped with a Vane Air Flow

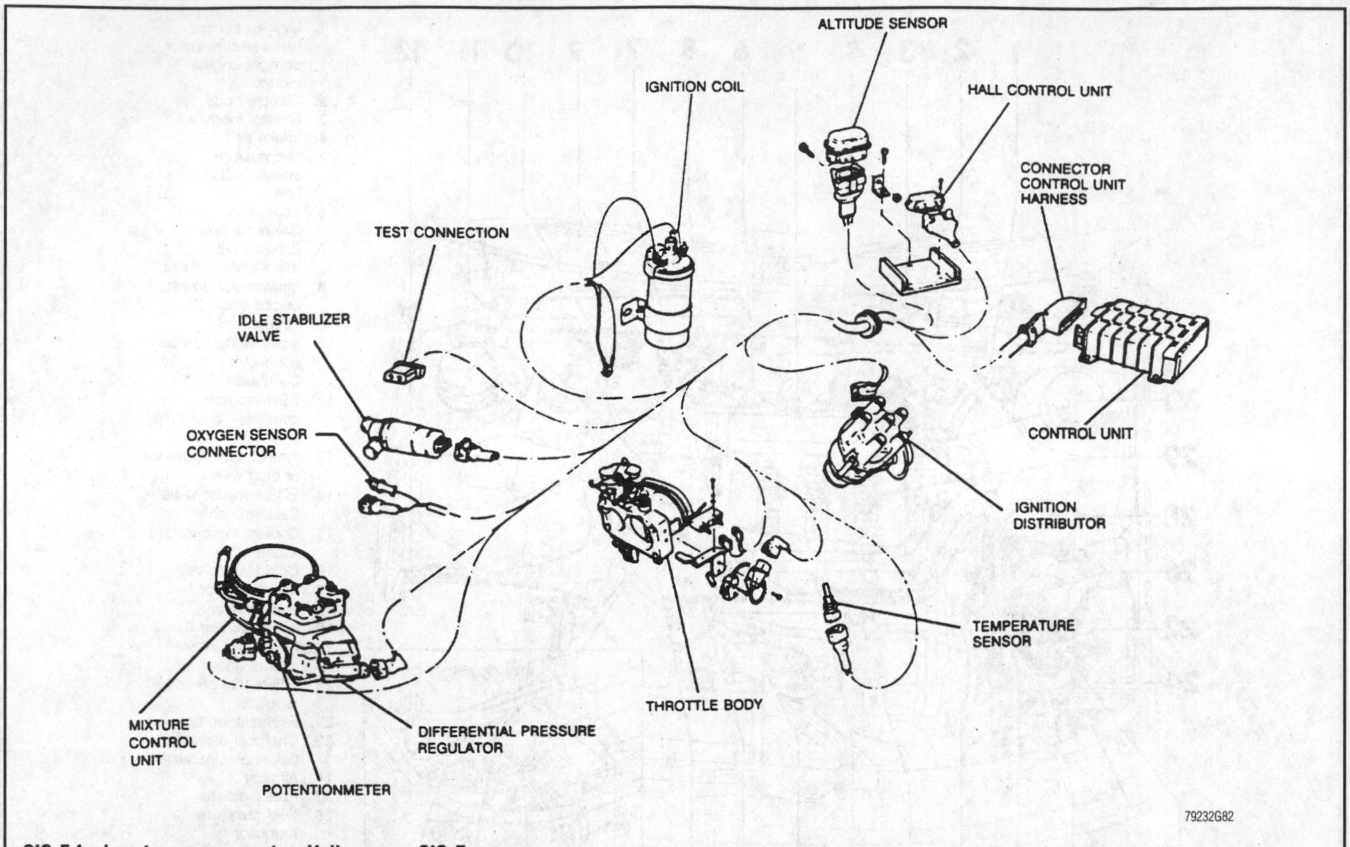

CIS-E fuel system components—Volkswagen CIS-E

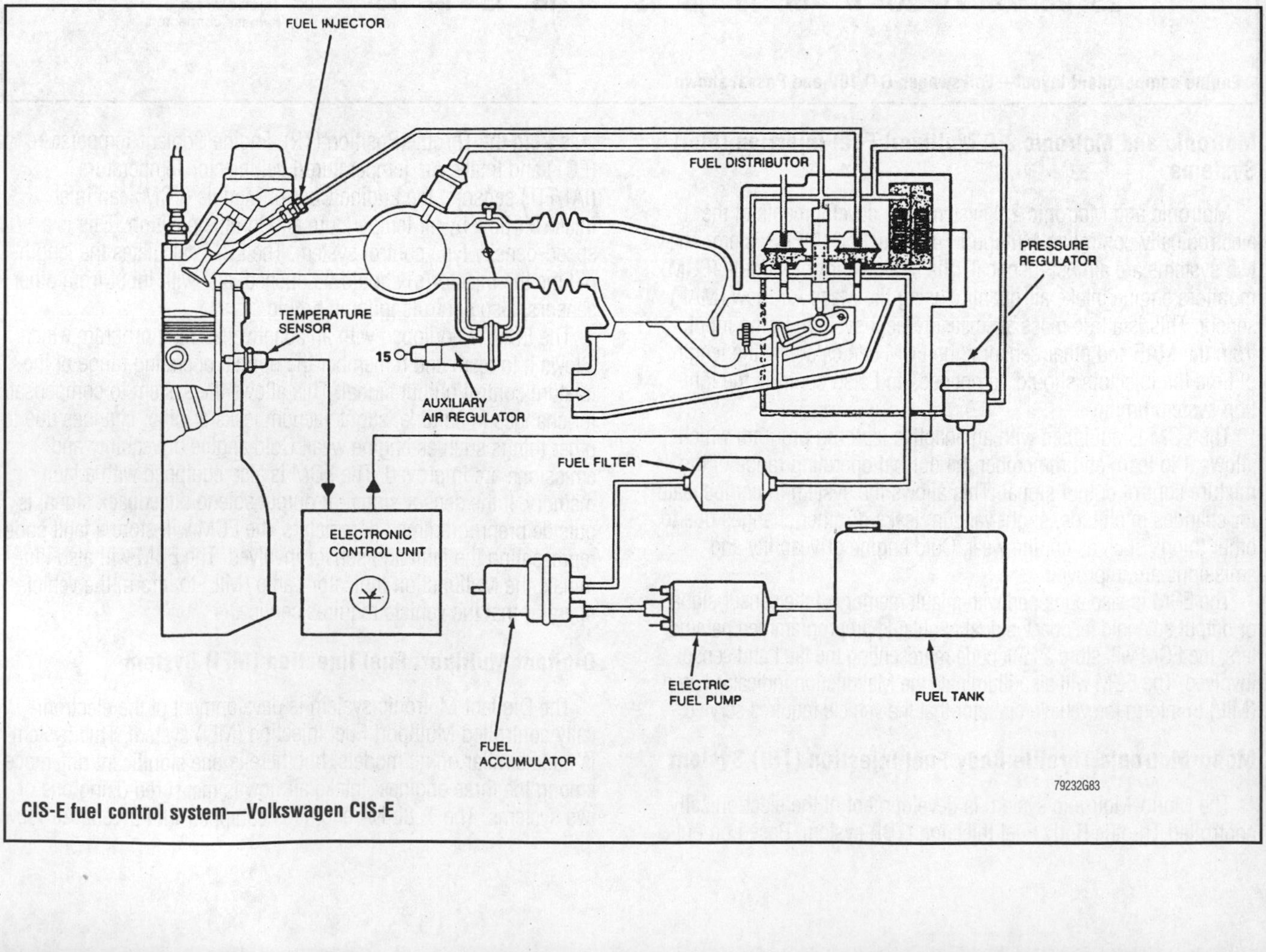

ALTITUDE SENSOR

IGNITION COIL

HALL CONTROL UNIT

CONNECTOR CONTROL UNIT HARNESS

TEST CONNECTION

IDLE STABILIZER VALVE

OXYGEN SENSOR CONNECTOR

CONTROL UNIT

IGNITION DISTRIBUTOR

TEMPERATURE SENSOR

THROTTLE BODY

MIXTURE CONTROL UNIT

DIFFERENTIAL PRESSURE REGULATOR

POTENTIONMETER

79232G82

CIS-E fuel control system—Volkswagen CIS-E

FUEL INJECTOR

FUEL DISTRIBUTOR

PRESSURE REGULATOR

TEMPERATURE SENSOR

15

AUXILIARY AIR REGULATOR

FUEL FILTER

ELECTRONIC CONTROL UNIT

ELECTRIC FUEL PUMP

FUEL TANK

FUEL ACCUMULATOR

79232G83

(VAF) sensor. This is a true mass air measurement system. The 1.8L SC and 2.5L NA engines use a Manifold Absolute Pressure (MAP) sensor.

Along with output from the Intake Air Temperature (IAT) sensor, the ECM can infer intake air flow by air temperature and pressure. This is a speed-density type control system. Using either measurement system, the ECM calculates the length of time the injectors should be opened, and with input from other sensors, also controls ignition timing.

The ECM is equipped with an adaptive learning program which allows it to learn and remember the normal operating range of the mixture control output signal. This allows the system to compensate for changes in altitude, slight vacuum leaks or other changes due to other things such as engine wear. Cold engine driveability and emissions are improved.

The ECM is also equipped with a fault memory. If the sensor signal or output solenoid feedback signal is outside preprogrammed parameters, the ECM will store a fault code representing the fault and sensor involved. The ECM will also illuminate the Malfunction Indicator Lamp (MIL) to inform the vehicle operator that the vehicle requires service. Some vehicles equipped with both the Digifant Motronic system and California specification emissions equipment also have the capability of flashing diagnostic codes.

SELF-DIAGNOSTICS

Service Precautions

• Do not disconnect the battery or the control unit before reading the fault codes. On the Motronic system, fault code memory is erased when power is interrupted.

• Make sure the ignition switch is **OFF** before disconnecting any wiring.

• Before removing or installing a control unit, disconnect the negative battery cable. The unit receives power through the main connector at all times and will be permanently damaged if improperly powered up or down.

• Keep all parts and harnesses dry during service. Protect the control unit and all solid-state components from rough handling or extremes of temperature.

Reading Codes

Only California vehicles with the Digifant II and Digifant I systems were equipped with the capability of flashing diagnostic codes.

On the Digifant II system codes were viewed through a combination rocker switch/indicator light. The following California vehicles were equipped with the Digifant II system:

1988–90 Golf, Jetta and GTI with 2.0L 16 valve engine
1990 Cabriolet with engine code 2H

On the Digifant I system a jumper cable would have to be connected and then the codes would flash from the CHECK engine light on the dash. The following California vehicles were equipped with the Digifant I system:

1990–93 Fox—Digifant I
1991–93 Cabriolet—Digifant I
1991–92 Corado—Digifant I

On all other systems, codes can only be retrieved with the use of a special diagnostic tester, the VAG 1551. This tester is available at car dealerships. The VAG 1551 tester can be used on all vehicles that have code capability.

➡**Some diagnostic codes may be retrieved by connecting special jumper cable 357 971 514E or an equivalent to the check connectors. Others for the most part are going to require the use of a special VAG 1551 tester and adapter to retrieve any remaining codes.**

DIGIFANT II SYSTEM

ROCKER SWITCH METHOD

California models are the only vehicles equipped with the On-Board Diagnostic (OBD) lamp. On these models codes may be accessed by 2 methods. The first is the use of a combination rocker switch/lamp located on the instrument panel. The second is used by the dealers, they use a special tool called the VAG 1551.

An indicator light labeled CHECK is located in a rocker switch on the instrument panel. Each time the engine is started, the indicator light will flash once to inform the operator the bulb is working. The light will come on and stay on if a fault develops in the engine management system. It will also display diagnostic codes to assist in trouble diagnosis. A diagnostic code consists of 4 groups of flashes. There is a 2.5 second pause (light OFF), between each group of flashes.

The indicator light will come on for two and half seconds prior to displaying a fault code when the diagnostic procedure has been activated. The fault code will continue repeating while the ignition is **ON**.

If the fault is not repaired the indicator light will come on and stay on when the ignition is turned **ON** to signify the fault still exists.

The following California vehicles were equipped with the Digifant II system:

1988–90 Golf, Jetta and GTI with 2.0L 16 valve engine
1990 Cabriolet with engine code 2H

Prior to checking for codes, drive the vehicle for 10 minutes or more.

1. Turn the ignition to the **ON** position, do not start the engine.

2. Press and hold down the rocker switch for 4–6 seconds then release the swtich. The CHECK indicator lamp will begin flashing a diagnostic code.

3. Press and hold down the rocker switch again for 4–6 seconds then release it. The indicator lamp will flash the next diagnostic code and will continue until all codes have been displayed. When all diagnostic codes have been displayed, the indicator lamp will flash a series of 2.5 second flashes ON and 2.5 seconds OFF. This is an 'End Of Fault Sequence' code. If there are no faults stored in the control unit memory, the indicator lamp will flash Code 4444.

➡**Occasionally the control unit will sense various deviations or changes in the air/fuel mixture. Because of the sensitivity of this system a fault code may set without any apparent problem showing up. This is a normal function with systems of this type.**

DIGIFANT I SYSTEM

JUMPER CABLE METHOD

California models are the only vehicles equipped with the On-Board Diagnostic (OBD) lamp. On these models codes may be access by 2 methods. The first is the use of a special jumper cable connected to the diagnostic connector. The second is used by the dealers, they use a special tool called the VAG 1551.

The following California vehicles were equipped with the Digifant I system:

1990–93 Fox—Digifant I
1991–93 Cabriolet—Digifant I
1991–92 Corado—Digifant I

1. Verify that all the fuses and grounds in the engine compartment are good.

2. Turn the ignition key **ON**.

3. Connect jumper cable 357 971 514E or equivalent to the connectors located in the center console. The black end of jumper wire connects to the black diagnostic connector in the console. The white end of the jumper wire connects to white diagnostic connector in the console.

4. Connect the jumper wire for about S seconds. When the OBD light begins flashing, remove the jumper wire.

5. Count the flashes of the light to get codes, each separate flash, in one code, will be have a short interval in between. Each interval between codes will be about 2.5 seconds. Count flashes until either Code 4444 or 0000 appears. To end this procedure turn the ignition switch **OFF**.

6. Output checks can not be performed without a diagnostic tester.

With Diagnostic Tester

This method is used by the Dealers and requires the use of the special tester VAG 1551. The following vehicles can only be accessed using this tester:

1991–92 Golf, Jetta—Digifant I system
1993–94 Corado—with VR6 Engine—Motronic system
All CIS-E Motronic and Motronic systems

1. Make sure ignition switch is **OFF** and that all fuses are good. Make sure all grounds in engine compartment are good, especially those for the battery and control module.

2. Make sure the air conditioning system is OFF.

3. Connect the VAG 1551 diagnostic tester using VAG 1551/1 Adapter cable or equivalent. The diagnostic connectors are located in the center console, under the shifter. The shifter knob and console cover must be removed to access diagnostic connectors. Connect diagnostic connector 1 (black) to the black connector on the scan tool. Connect diagnostic connector 2 (white) to the white connector on the scan tool. The blue connector is not required.

4. Turn the ignition switch **ON**; now start the vehicle and let it idle. If vehicle will not start, crank engine for 6 seconds and leave ignition **ON**.

5. Turn the tester ON and make sure it is receiving power. The screen will display 2 menu options; Rapid Data Transfer and Blink Code Output:

6. If you choose to use mode 02, Blink Code Output, skip to Step 10.

7. Select mode 01, Rapid Data Transfer, and address word 01. Now press the 0 button to enter your selection. The tester will display a control unit part number, the system it controls and an application (country) code.

   a. If the information is displayed and is correct, press the (run) key to continue. The display 'Select function XX' will appear.

   b. If 'Control unit does not answer' is displayed, use the Help key to display a list of possible causes. When the problem is repaired, return to step 1 and start over again.

8. When function 02 is selected, the control module will report fault codes to the diagnostic tester.

9. When all codes have been reported, proceed to the Output Check diagnosis or select function 06 to exit the fault code memory

without erasing the codes. Repair and erase the faults, then check and see if all faults have been corrected.

10. The following steps will retrieve engine codes by using Blink Code Output.

11. To operate the VAG 1551 tester in Blink Code Output, select menu option # 2. An asterisk will appear and flash the codes, which the tester will count and report on the screen as numbers. If Code 4444 or 0000 is displayed, no faults are found in memory.

➡ **On vehicles that use 4444 for no codes present, the 0000 will stand for output ended.**

12. If the engine is not running, some codes may be displayed. These can be ignored if the engine was intentionally stalled, but should be investigated if the engine will not start.

13. Press the (run) key to advance to the next code. Read through entire code list before starting repairs.

14. When the 0000 (output ended) code is displayed, pressing the (run) key again will proceed to another control module. If no other control modules are to be tested, the following display will appear: Blink Code Output is ended. To stop the program without erasing the codes, turn the ignition key **OFF** and press the clear C button once.

15. Repair and erase the faults, then check to see if all faults have been corrected.

Output Check Diagnosis

Only the Motronic system is equipped with this program. It allows testing most of the engine output devices without running the engine. The program cannot be run without the VAG 1551 Diagnostic Tester or equivalent. During the test, four output devices are activated in the following order:

- Differential pressure regulator
- Carbon canister frequency valve
- Idle stabilizer valve
- Cold start valve

Testing the differential pressure regulator requires a multi-meter that will read milliamps. The other items can be checked with a voltmeter, test light or by listening and feeling for valve activation. The cold start valve is activated for a limited time to avoid flooding the engine.

1. Connect the diagnostic tester, turn the ignition switch **ON** and confirm that the tester will communicate with the control unit. See the procedure for retrieving fault codes.

2. Select Rapid Data Transfer and Function 03. When the test is started by pressing the Q button (enter), the first output signal is generated.

3. Each time the Run button is pressed, the tester will send an output signal to the next device on the list.

4. When the last item has been tested, select Function 06 to exit the program. To repeat the test, turn the ignition switch **OFF** and **ON** again.

➡ **Leave the ignition OFF for approximately 20 seconds, before selecting Output Check diagnosis again.**

Clearing Codes

1988–95 VEHICLES

Without Diagnostic Tester

1. To erase codes, wait until Code 4444 or 0000 is displayed.

2. Turn ignition switch **OFF** and connect jumper wire to diagnostic connectors again.

3. Turn the ignition switch **ON** and leave connectors jumpered for about 5 seconds, When Code 4444 or 0000 appears the codes will be erased.

4. Turn the ignition switch **OFF** and remove jumper wire.

With Diagnostic Tester

For both engine and automatic transaxle, after all fault codes have been retrieved, select Function 05 and press the Q button to enter the selection. The memory will be erased only if all fault codes have been retrieved. Test drive the vehicle for at least 10 minutes, including at least 1 full throttle application above 3000 rpm. Check the fault code memory again to make sure all faults have been repaired.

DIAGNOSTIC TROUBLE CODES

1988–95 Vehicles

➡ **The 5 digit code groups are used with a diagnostic tester. The 4 digit code groups are the flashing codes.**

00000 or 4444 No faults in memory
00281 or 1231 Vehicle Speed Sensor (VSS) signal is missing
00282 or 1232 Throttle actuator solenoid or wiring harness
00513 or 2111 Engine RPM sensor signal is missing
00514 or 2112 Ignition reference sensor signal is missing
00515 or 2113 Hall sender signal is missing
00516 or 2121 Idle switch has open short in circuit
00517 or 2123 Full throttle switch
00518 or 2212 Throttle position sensor
00519 or 2222 Manifold absolute pressure (MAP) sensor
00520 or 2232 Air flow sensor signal is missing
00521 or 2242 CO potentiometer
00522 or 2312 Engine coolant temperature (ECT) sensor
00523 or 2322 Intake air temperature (IAT) sensor
00524 or 2142 Knock sensor 1 signal is missing
00525 or 2342 Oxygen sensor signal missing
00527 or 2412 Intake air temperature (IAT) sensor has open/short in circuit
00532 or 2234 Supply voltage is too high
00533 or 2231 Idle speed regulation out of limit
00535 or Both 2141/2142 Knock sensor or control program
00537 or 2341 Oxygen sensor signal out of limit
00540 or 2144 Knock sensor 2 signal is missing
00543 or 2214 RPM exceeds maximum limit
00545 or 2314 Engine/Transmission electrical connection
00549 or 2314 Fuel consumption signal
00552 or 2323 Air flow sensor signal missing
00553 or 2324 Mass Air Flow (MAE) sensor signal is out of range
00558 or NA Adaptive mixture control lean (Fuel injector leak, EVAP purge system)
00559 or NA Adaptive mixture control rich (vacuum leak)
00560 or 2411 EGR temperature sensor circuit
00561 or 2413 Mixture adaptation limits are out of range
00585 or 2411 EGR temperature sensor circuit (2.8L AAA engine only)
00586 EGR controlling system, EGR valve is sticking or false signals
00587 Adjustment limit mixture regulator is lean
00609 Ignition output 1 circuit
00624 A/C compressor engagement circuit has mechanical or electrical malfunction

00640 or 3434 Heated Oxygen sensor relay has open/short circuit
01025—Malfunction indicator lamp (MIL) circuit
01242 or 4332 Output stages in engine control module (ECM)
01247 or 4343 EVAP frequency valve 1 has open/short in circuit
01249 or 4411 Fuel injector #1 circuit has open/short
01250 or 4412 Fuel injector #2 circuit has open/short
01251 or 4413 Fuel injector #3 circuit has open/short
01252 or 4414 Fuel injector #4 circuit has open/short
01253 or 4421 Fuel injector #5 circuit has open/short
01254 or 4422 Fuel injector #6 circuit has open/short
01257 or 4431 Idle Air Control (IAC) valve has open/short in circuit or a mechanical malfunction
01259 or 4433 Fuel pump relay is faulty or short circuited
01265 or 4312 EGR frequency valve has open/short in circuit
65535 or 1111 Engine Control Module (ECM) is defective
0000 End of output
NA Not Available

Volvo

GENERAL INFORMATION

The LH-Jetronic 2.2 was used on models through 1990. The LH-Jetronic 2.4 and 3.1 fuel injection systems are used on 1990 and newer 240, 700 and 900 series vehicles. The LHJetronic 3.2 fuel injection system is used on the 1993 and later 850. The Motronic 1.8 was used on the 1992 and later 960. The Motronic 4.3 is used on 1994 and later 850 Turbo. On all fuel systems except the LH 2.2 system are monitored by a self-diagnostic system that lights up a warning lamp on the instrument panel. The LH-Jetronic 2.2 does not have self-diagnostic ability. Many different fault codes can be set, however only three can be stored at any one time. Fault tracing can be carried out by utilizing the diagnostic unit.

SELF-DIAGNOSTICS

Service Precautions

- Do not operate the fuel pump when the fuel lines are empty.
- Do not operate the fuel pump when removed from the fuel tank.
- Do not reuse fuel hose clamps.
- The washer(s) below any fuel system bolt (banjo fittings, service bolt, fuel filter, etc.) must be replaced whenever the bolt is loosened. Do not reuse the washers; a high-pressure fuel leak may result.
- Make sure all ECU harness connectors are fastened securely. A poor connection can cause an extremely high voltage surge and result in damage to integrated circuits.
- Keep all ECU parts and harnesses dry during service. Protect the ECU and all solid-state components from rough handling or extremes of temperature.
- Use extreme care when working around the ECU or other components; the airbag or SRS wiring may be in the vicinity. On these vehicles, the SRS wiring and connectors are yellow; do not cut or test these circuits.
- Before attempting to remove any parts, turn the ignition switch **OFF** and disconnect the battery ground cable.
- Always use a 12 volt battery as a power source for the engine, never a booster or high-voltage charging unit.

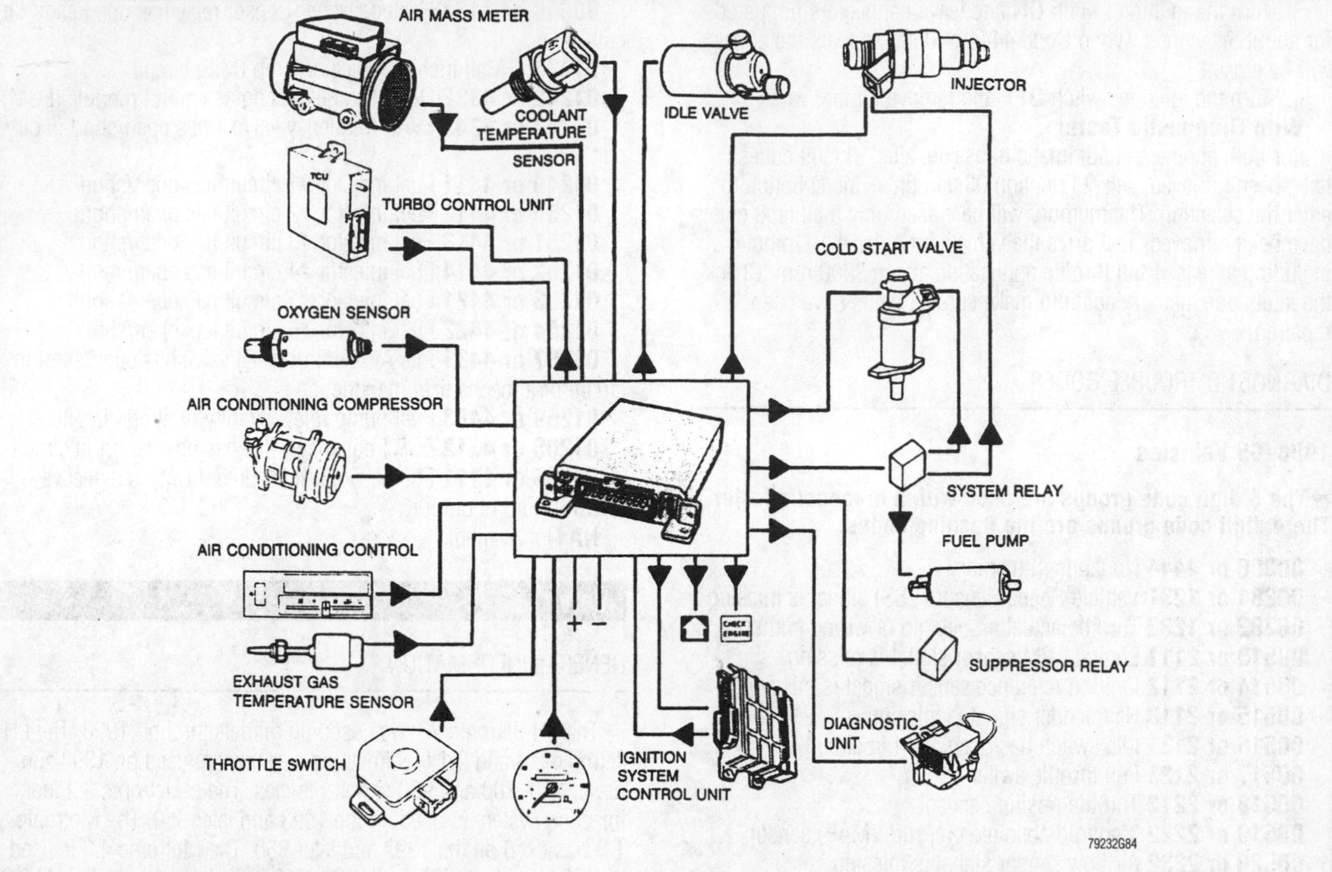

AIR MASS METER

COOLANT TEMPERATURE SENSOR

IDLE VALVE

INJECTOR

TCU

TURBO CONTROL UNIT

COLD START VALVE

OXYGEN SENSOR

AIR CONDITIONING COMPRESSOR

SYSTEM RELAY

FUEL PUMP

AIR CONDITIONING CONTROL

SUPPRESSOR RELAY

CHECK ENGINE

EXHAUST GAS TEMPERATURE SENSOR

DIAGNOSTIC UNIT

THROTTLE SWITCH

IGNITION SYSTEM CONTROL UNIT

79232G84

Fuel system components—LH–Jetronic

• Do not disconnect the battery cables with the engine running.
• Do not disconnect any wiring connector with the engine running or the ignition **ON** unless specifically instructed to do so.
• Do not apply battery power directly to injectors.
• Whenever possible, use a flashlight instead of a drop light.
• Keep all open flame and smoking material out of the area. a Use a shop cloth or similar to catch fuel when opening a fuel system. Consider the fuel-soaked rag to be a flammable solid and dispose of it in the proper manner.
• Relieve fuel system pressure before servicing any fuel system component.
• Always use eye or full-face protection when working around fuel lines, fittings or components.
• Always keep a dry chemical (class B-C) fire extinguisher near the area.

Reading Codes

➡**On-board engine diagnostics were not available on vehicles prior to 1988.**

1. Open diagnostic socket cover and install selector cable into socket No. 2 for fuel injection codes or socket No. 6 (except Motronic systems) for ignition codes.
2. Turn ignition to the **ON** position.
3. Enter control system 1 by pressing the button once. Hold the button for at least 1 second, but not more than 3.
4. Watch the diode light and count the number of flashes in the 3 flash series indicating a fault code. The flash series are separated by 3 second intervals. Note fault codes.

➡**If there are no fault codes in the diagnostic unit, the diode will flash 1-1-1 and the fuel system is operating correctly.**

5. If diode light does not flash when button is pressed, or no code is flashed there is a problem with the soft-diagnostic system, proceed as follows:

a. Check ground connections on the intake manifold, and the ground connection for the Lambda-sond at the right front mudguard.

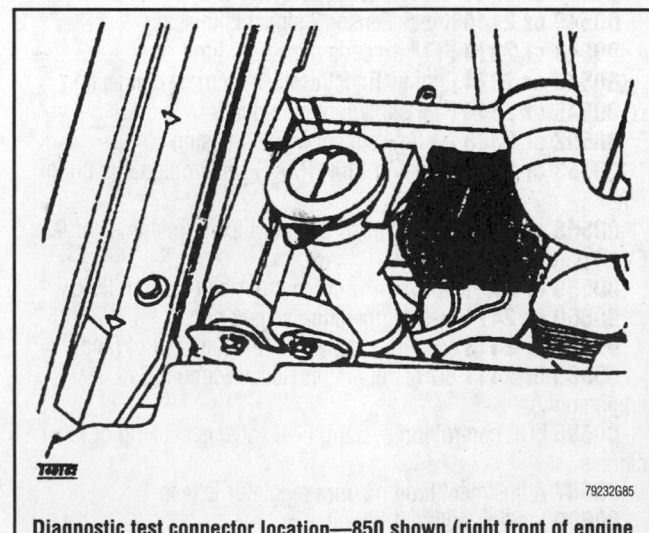

79232G85

Diagnostic test connector location—850 shown (right front of engine compartment)

b. Check the fuses for the pump relay and the primary pump. On 240 models, fuses are located inside the engine compartment on the left side wheel well housing. On 760/780 models, fuses are located in the center console, just below the radio. On 740/940 models, fuses are located behind the ashtray. Access can be gained by removing the ashtray, and pressing upward on the tab marked 'electrical fuses press'. On 850 models, the fuses are located on the left side of the engine compartment behind the strut mount plate. The fuses on 960 models are located on the far left side of the dashboard. The driver's door must be open to gain access to the fuses.

c. Remove glove compartment, and check control unit ground connections.

d. Turn ignition switch to the **OFF** position. Remove control unit connector and connector protective sleeve.

e. Check diagnostic socket, (Steps 5e–5j), by connecting a voltmeter between ground and No. 4 connection on the control unit connector. Reading should be 12 volts. If no voltage is present, check lead between control unit connector and fuse No. 1 in the fuse/relay box.

f. Turn the ignition to the **ON** position, and install selector cable into the No. 2 socket on the diagnostic socket. Connect a voltmeter between ground and No. 12 connection on the control unit connector. Reading should be 12 volts. Press the button on diagnostic socket and note reading. Reading on the voltmeter should be 0 volts. If no voltage at the control unit is present, take reading at the diagnostic socket connector. If reading remains at 12 volts when button is pressed, check diagnostic socket.

g. Connect a voltmeter between ground and the red/black lead on the diagnostic socket connector. Reading should be 12 volts.

h. Connect a suitable ohmmeter between ground and the brown/black lead in the diagnostic socket connector. Reading should be 0 ohms.

i. Turn ignition to the **OFF** position. Connect ohmmeter between diagnostic socket selector cable and the pin under selector button. Ohmmeter should read infinity. Press button, and note reading. Reading should be 0 ohms.

j. Connect a suitable diode/multimeter tester, or equivalent, between the diagnostic socket diode light and the selector cable.

Connect red test pin from the tester to pin under diode light and black test pin from tester to selector cable. A reading on the tester indicates correct diode light function. With no reading on tester, replace diagnostic socket.

k. Check the system relay/primary relay by connecting a voltmeter between ground and the No. 9 connection on the control unit connector, then connect a jumper wire between ground and No. 21 connection on the control unit connector. The relay should activate and the reading should be 12 volts.

6. Press the diagnostic socket button. Note any additional fault codes.

➡ **The diagnostic system memory is full when it contains 3 fault codes. Until those codes are corrected and the memory erased, the system cannot give information on any other problems.**

7. Press the diagnostic socket button for the third time to see if a third fault code is stored in the memory. If the diode light flashes the same Code 1-1-1, there are no other codes in the memory.

Clearing Codes

1989–94 VEHICLES

1. Turn the ignition switch to the **ON** position.
2. Read fault codes.
3. Press diagnostic socket button 1 time and hold for approximately 5 seconds. Release button. After 3 seconds the diode light should light up. While the light is still lit, press the button again and hold for approximately 5 seconds. After releasing the button, the diode light should go off.
4. To ensure that the memory is erased, press the button 1 time, for 1 second but not more than 3 seconds. The diode light should flash Code 1-1-1.
5. Start and run engine. If engine will not start, correct the problem before proceeding and start over with step 1.
6. Check to see if new fault codes have been stored in the memory by pressing the diagnostic socket button 1 time, for 1 second but not more than 3 seconds.
7. If fault Code 1-1-1 flashes, it indicates that there are no additional fault codes stored in its memory.

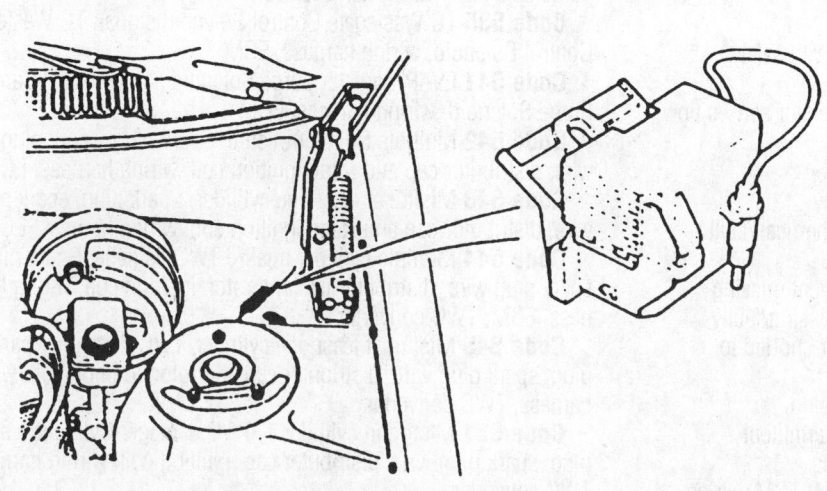
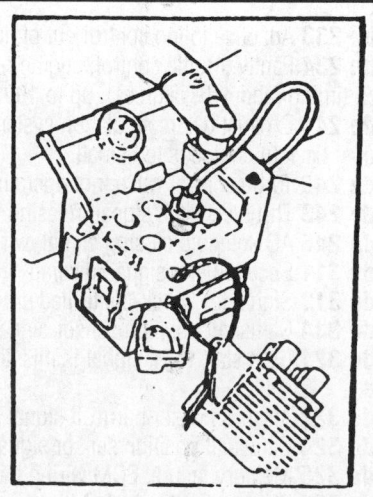

Diagnostic socket box location—240, 740, 940 and 960 shown

79232G86

DIAGNOSTIC TROUBLE CODES

1989–94 Vehicles

Code 111 No fault

Code 112 Fault in control module

Code 113 Heated oxygen sensor at maximum enrichment limit; injector clogged, break in lead, etc.

Code 115 Fuel injector 1; wiring harness, ECM

Code 121 Air Mass Meter (MAF) or air pressure sensor signal; missing/faulty

Code 122 Air temperature sensor signal; missing/faulty

Code 123 Engine temperature sensor signal faulty or missing

Code 125 Fuel injector 2; wiring harness, ECM

Code 131 Engine Speed Signal

Code 132 Battery voltage; too low or too high

Code 133 Throttle (Shutter) switch; idle setting faulty

Code 135 Fuel injector 3; wiring harness, ECM

Code 142 Fault in control module

Code 143 Front knock sensor signal missing/faulty

Code 144 Load signal missing from fuel system control module

Code 145 Fuel injector 4; wiring harness, ECM

Code 153 Rear H025 signal; wiring harness, ECM

Code 154 EGR system leakage; EGR valve, EGR transfer pipes

Code 155 Fuel injector 5; wiring harness, ECM

Code 212 Oxygen sensor (Lambda-sond) signal; missing/faulty

Code 213 Throttle switch; full load setting faulty

Code 214 Timing pick-up signal; missing intermittently

Code 214 Vehicle speed sensor signal intermittent

Code 221 Adaptive Heated Oxygen sensor running rich at part load

Code 222 System relay; signal is missing/faulty

Code 223 Idling valve signal; missing/faulty

Code 224 Missing/faulty temperature sensor circuit; signal missing/faulty

Code 225 NC pressure sensor signal; sensor/wiring harness faulty

Code 231 Adaptive Heated Oxygen sensor running lean at part load

Code 232 Adaptive Oxygen sensor (Lambda-sond) control; lean or rich, idle

Code 233 Adaptive idling control out of limits

Code 234 Faulty throttle control; engine runs with safety retarded timing (about 10 degrees); up to 1991

Code 241 Exhaust gas recirculation system; sensor senses flow of exhaust back to engine is too small

Code 242 Turbo control valve; not operating

Code 243 Throttle switch signal; missing/faulty

Code 245 AC solenoid closing signal; wiring harness/faulty

Code 311 Speedometer signal missing

Code 312 Signal for knock-controlled enrichment missing

Code 314 Camshaft position sensor signal; missing/faulty

Code 321 Cold start valve signal is missing or shorted to ground

Code 322 Air mass meter burn-off signal missing

Code 324 Camshaft position sensor signal intermittent

Code 325 Memory failure; ECM wiring harness

Code 335 Request to illuminate MIL from TCM; TCM, wiring harness

Code 342 NC blocking relay; current too high

Code 411 Throttle switch signal; missing/faulty

Code 416 Boost pressure reduction request from TCM; TCM wiring harness/faulty

Code 413 EGR temperature sensor signal; missing or incorrect

Code 421 Boost pressure sensor in control module

Code 423 Throttle position sensor signal; missing/faulty

Code 424 Load signal from fuel system; RPM too low for boost

Code 431 Coolant temperature sensor signal; missing

Code 432 High temperature warning in control box (temp. above 185°F (85°C))

Code 433 Rear knock sensor signal; missing/faulty

Code 435 Front H025 slow response; front H025

Code 436 Rear H025 compensation; rear H025

Code 443 TWC efficiency; TWC converter

Code 444 Acceleration sensor signal; acceleration sensor wiring harness

Code 451 Misfire, cylinder 1; Spark plug, spark plug wire, distributor, ignition coil, wiring harness

Code 452 Misfire, cylinder 2; Spark plug, spark plug wire, distributor, ignition coil, wiring harness

Code 453 Misfire, cylinder 3; Spark plug, spark plug wire, distributor, ignition coil, wiring harness

Code 454 Misfire, cylinder 4; Spark plug, spark plug wire, distributor, ignition coil, wiring harness

Code 455 Misfire, cylinder 5; Spark plug, spark plug wire, distributor, ignition coil, wiring harness

Code 512 Heated oxygen sensor at maximum lean running limit

Code 513 High temperature warning in control box (temp. above 203°F (95°C))

Code 514 Engine cooling fan, Low speed signal; engine cooling fan relay wiring harness

Code 521 Front H025 preheating; Front H025, wiring harness

Code 522 Rear H025 preheating; Rear H025, wiring harness

Code 531 Power stage group A; fuel injectors, EVAP canister purge solenoid, wiring harness, ECM

Code 532 Power stage group B; fuel injectors, EVAP canister purge solenoid, wiring harness, ECM

Code 533 Power stage group C; fuel injectors, EVAP canister purge solenoid, wiring harness, ECM

Code 534 Power stage group D; fuel injectors, EVAP canister purge solenoid, wiring harness, ECM

Code 535 TC Wastegate Control Solenoid signal; TC Wastegate Control Solenoid, wiring harness, ECM

Code 541 EVAP Canister purge Solenoid signal; EVAP Canister Purge Solenoid, wiring harness, ECM

Code 542 Multiple cylinder misfire; Spark plug, spark plug wire, distributor cap and rotor, ignition coil, wiring harness, ECM

Code 543 Misfire at least one cylinder; Spark plug, spark plug wire, distributor cap and rotor, ignition coil, wiring harness, ECM

Code 544 Multiple cylinder misfire TWC damage; Spark plug, spark plug wire, distributor cap and rotor, ignition coil, wiring harness, ECM, TWC converter

Code 545 Misfire at least one cylinder, TWC damage; Spark plug, spark plug wire, distributor cap and rotor, ignition coil, wiring harness, TWC converter

Code 551 Misfire in cylinder 1, TWC damage; Cylinder 1 spark plug, spark plug wire, distributor cap, ignition coil, wiring harness, TWC converter

Code 552 Misfire in cylinder 2, TWC damage; Cylinder 1 spark plug, spark plug wire, distributor cap, ignition coil, wiring harness, TWC converter

Code 553 Misfire in cylinder 3, TWC damage; Cylinder 1 spark plug, spark plug wire, distributor cap, ignition coil, wiring harness, TWC converter

➡ **Code combination explanations are as follows:**

Code 113, 221 & 232 In the part load range; Air/Fuel mixture is lean and on idle

Code 113 & 221 Air/Fuel mixture is lean in the part load range

Code 113 & 231 Air/Fuel mixture is probably rich in the part load range

OBD II DIAGNOSTIC TROUBLE CODES

Introduction

To comply with OBD II Regulations, the Control Module is equipped with software designed to allow it to monitor vehicle emission control systems and components. Once the ignition is turned on or the engine is started, and certain test conditions are met, the PCM runs a series of monitors to test the emission control systems and components. Test conditions include different inputs such as time since startup, run-time, engine speed and temperature, transaxle gear position, and the engine open or closed loop status. Once the monitor is started, the control module attempts to run it to completion. If a particular monitor fails a test, a code is set and operating conditions at that time are recorded in memory. If the same component or system fails twice in succession, the Malfunction Indicator Lamp (MIL) is activated.

Monitors are divided into two types: Main Monitors and the Comprehensive Component Monitors.

- Catalyst Monitor
- EGR Monitor
- EVAP Monitor
- Fuel System Monitor
- Misfire Monitor
- Oxygen Sensor Monitor
- Oxygen Sensor Heater Monitor

Certain monitors, in particular the fuel system and misfire monitors, have limitations that are different from any of the other monitors. The first time either of these monitors fail, the MIL is activated, and engine conditions at the time of the fault are recorded. In order for the control module to turn off an MIL related to these two monitors, it must determine that no faults are present with engine operating conditions similar to when it detected the fault. To qualify, the engine must be operated within a specified speed range, engine load range and temperature range.

A warm-up cycle is considered to be vehicle operation after the engine has been turned off for a period of time, with the ECT input rising a specified amount and reaching normal operating temperature. When the MIL is turned off because a fault is no longer present, most OBD II codes will be erased after a minimum of 40 warm-up cycles. Misfire and fuel system codes require a minimum of 80 warm-up cycles before they clear.

OBD II Systems use a standardized test connector, called the Data Link Connector (DLC). It is located beneath the left side of the instrument panel. The DLC is located out of the line of sight of vehicle passengers, but is easily viewable from a kneeling position outside the vehicle. The connector is rectangular in design and contains up to 16 terminals. It has keying features to allow for easy connection. Both the DLC and Scan Tool connectors have latching features that ensure the scan tool will remain properly connected.

Some common uses of the Scan Tool are to identify and clear Diagnostic Trouble Codes (DTCs) and to read control module freeze frame.

The Malfunction Indicator Lamp (MIL) looks similar to the "Check Engine" lamp. However, on OBD II Systems, it is controlled under a strict set of guidelines that dictate when the MIL is illuminated. If any of the control module monitors detects a fault that could impact vehicle emissions, a fault code is set. A One-Trip Monitor requires that a test fail once, a Two-Trip Monitor requires a test fail twice in succession, and a Three-Trip Monitor requires that a test fail three times in succession to activate the MIL.

The MIL is mounted in the instrument panel and has two functions: To act as a bulb check at key On and to inform the driver that an emissions fault has occurred.

Once the engine is started, if no faults are detected, the control module should extinguish the MIL after a few seconds. If the MIL remains On or flashes with the engine running a driveability symptom is present.

Non-OBD II Trouble Codes

For years, vehicles have been capable of storing diagnostic trouble codes. Codes prior to the 1996 OBD II legislation have been proprietary to the vehicle manufacturer. In some cases, the codes are specific to the individual make and model.

Furthermore, some manufacturers have developed specialized devices to read their codes. This complicates code reading and clearing.

For further information on Non-OBD II trouble codes, please refer to the Chilton's "Total Car Care" manual for your particular make and model of vehicle.

OBD II Trouble Codes

Federal law required all vehicle manufacturers to meet On Board Diagnostics, Second Generation or OBD II standards by 1996. In order to meet this standard, the automobile's on-board computer must monitor and perform diagnostic tests on vehicle emissions to ensure that the vehicle is operating at an acceptable (legal) emission level. The maximum allowable emission level is set by the Federal Test Procedure (FTP).

Some 1995 and all 1996–99 vehicles are OBD II compliant. All OBD II vehicles have the same 16 pin diagnostic connector or DLC. This eliminates the need to have a manufacturer specific connector to plug a scan tool into your vehicle.

➡ **Many 1995 vehicles have a 16 pin OBD II connector, however, this does not mean that the vehicle is OBD II compliant.**

TROUBLE CODE DESCRIPTION

In the past, trouble code numbers varied between manufacturers, years, makes and models. OBD II requires that all vehicle manufacturers use a common Diagnostic Trouble Code (DTC) numbering system. Since the generic listing was not specific enough, most manufacturers came up with their own DTC listings which are called manufacturer specific codes. Both generic and manufacturer specific codes are 5 digits. The numbers can be decoded as follows:

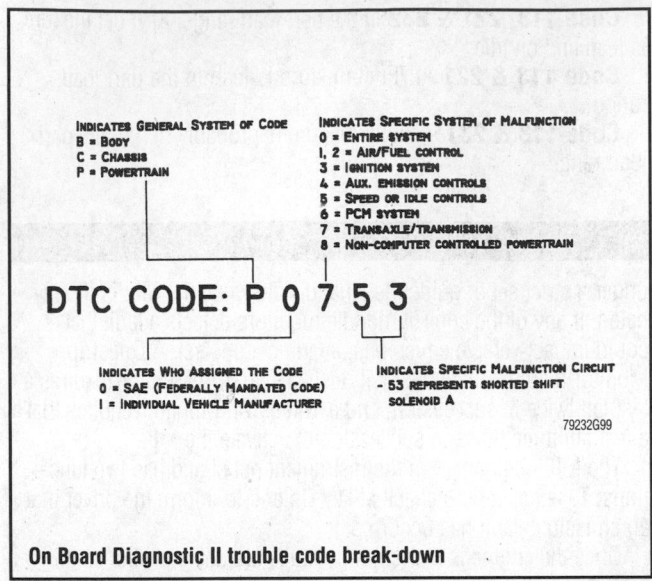

On Board Diagnostic II trouble code break-down

The first digit is a letter which identifies the function of the device or circuit which has the fault. This digit can be either:

- P—Powertrain
- B—Body
- C—Chassis
- U—Network or data link code

The second digit is either a 0 or 1 and indicates whether the code is generic or manufacturer specific.

- 0—Generic
- 1—Manufacturer Specific

The third digit represents the specific vehicle circuit or system that has the fault. Listed below are the number identifiers for the powertrain system.

- 1—Fuel and Air Metering
- 2—Fuel and Air Metering (Injector Circuit Malfunctions Only)
- 3—Ignition System or Misfire
- 4—Auxiliary Emission Control
- 5—Vehicle Speed Control and Idle Control System
- 6—Computer and Auxiliary Outputs
- 7—Transmission
- 8—Transmission

The last two digits indicate the specific trouble code.

On OBD II vehicles there are two different types of DTCs: Stored and Pending. For a DTC to become Stored, certain malfunction conditions must occur. The condition(s) required to Store codes are different for every DTC and vary by vehicle manufacturer.

In order for some DTCs to become Stored, a malfunction condition has to happen more than once. If the malfunction conditions are required to occur more than once, the potential malfunction is called a Pending DTC. The DTC remains pending until the malfunction condition occurs the required number of times to make the code stored. If the malfunction condition does not occur again after a set time the pending DTC will be cleared.

Acura

READING CODES

Reading the control module memory is on of the first steps in OBD II system diagnostics. This step should be initially performed

to determine the general nature of the fault. Subsequent readings will determine if the fault has been cleared.

Reading codes can be performed by any of the methods below:
- Read the control module memory with the Generic Scan Tool (GST)
- Read the control module memory with the vehicle manufacturer's specific tester

To read the fault codes, connect the scan tool or tester according to the manufacturer's instructions. Follow the manufacturer's specified procedure for reading the codes.

CLEARING CODES

Control module reset procedures are a very important part of OBD II system diagnostics. This step should be done at the end of any fault code repair and at the end of any driveability repair.

Clearing codes can be performed by any of the methods below:
- Clear the control module memory with the Generic Scan Tool (GST)
- Clear the control module memory with the vehicle manufacturer's specific tester
- Turn the ignition **OFF** and remove the negative battery cable for at least 1 minute.

Removing the negative battery cable may cause other systems in the vehicle to loose their memory. Prior to removing the cable, ensure you have the proper reset codes for radios and alarms.

➡ **The MIL may also be de-activated for some codes if the vehicle completes three consecutive trips without a fault detected with vehicle conditions similar to those present during the fault.**

OBD II TROUBLE CODES

P0100 Mass or Volume Air Flow Circuit Malfunction
P0101 Mass or Volume Air Flow Circuit Range/Performance Problem
P0102 Mass or Volume Air Flow Circuit Low Input
P0103 Mass or Volume Air Flow Circuit High Input
P0104 Mass or Volume Air Flow Circuit Intermittent
P0105 Manifold Absolute Pressure/Barometric Pressure Circuit Malfunction
P0106 Manifold Absolute Pressure/Barometric Pressure Circuit Range/Performance Problem
P0107 Manifold Absolute Pressure/Barometric Pressure Circuit Low Input
P0108 Manifold Absolute Pressure/Barometric Pressure Circuit High Input
P0109 Manifold Absolute Pressure/Barometric Pressure Circuit Intermittent
P0110 Intake Air Temperature Circuit Malfunction
P0111 Intake Air Temperature Circuit Range/Performance Problem
P0112 Intake Air Temperature Circuit Low Input
P0113 Intake Air Temperature Circuit High Input
P0114 Intake Air Temperature Circuit Intermittent
P0115 Engine Coolant Temperature Circuit Malfunction
P0116 Engine Coolant Temperature Circuit Range/Performance Problem
P0117 Engine Coolant Temperature Circuit Low Input
P0118 Engine Coolant Temperature Circuit High Input

P0119 Engine Coolant Temperature Circuit Intermittent

P0120 Throttle/Pedal Position Sensor/Switch "A" Circuit Malfunction

P0121 Throttle/Pedal Position Sensor/Switch "A" Circuit Range/Performance Problem

P0122 Throttle/Pedal Position Sensor/Switch "A" Circuit Low Input

P0123 Throttle/Pedal Position Sensor/Switch "A" Circuit High Input

P0124 Throttle/Pedal Position Sensor/Switch "A" Circuit Intermittent

P0125 Insufficient Coolant Temperature For Closed Loop Fuel Control

P0126 Insufficient Coolant Temperature For Stable Operation

P0130 O2 Circuit Malfunction (Bank no. 1 Sensor no. 1)

P0131 O2 Sensor Circuit Low Voltage (Bank no. 1 Sensor no. 1)

P0132 O2 Sensor Circuit High Voltage (Bank no. 1 Sensor no. 1)

P0133 O2 Sensor Circuit Slow Response (Bank no. 1 Sensor no. 1)

P0134 O2 Sensor Circuit No Activity Detected (Bank no. 1 Sensor no. 1)

P0135 O2 Sensor Heater Circuit Malfunction (Bank no. 1 Sensor no. 1)

P0136 O2 Sensor Circuit Malfunction (Bank no. 1 Sensor no. 2)

P0137 O2 Sensor Circuit Low Voltage (Bank no. 1 Sensor no. 2)

P0138 O2 Sensor Circuit High Voltage (Bank no. 1 Sensor no. 2)

P0139 O2 Sensor Circuit Slow Response (Bank no. 1 Sensor no. 2)

P0140 O2 Sensor Circuit No Activity Detected (Bank no. 1 Sensor no. 2)

P0141 O2 Sensor Heater Circuit Malfunction (Bank no. 1 Sensor no. 2)

P0142 O2 Sensor Circuit Malfunction (Bank no. 1 Sensor no. 3)

P0143 O2 Sensor Circuit Low Voltage (Bank no. 1 Sensor no. 3)

P0144 O2 Sensor Circuit High Voltage (Bank no. 1 Sensor no. 3)

P0145 O2 Sensor Circuit Slow Response (Bank no. 1 Sensor no. 3)

P0146 O2 Sensor Circuit No Activity Detected (Bank no. 1 Sensor no. 3)

P0147 O2 Sensor Heater Circuit Malfunction (Bank no. 1 Sensor no. 3)

P0150 O2 Sensor Circuit Malfunction (Bank no. 2 Sensor no. 1)

P0151 O2 Sensor Circuit Low Voltage (Bank no. 2 Sensor no. 1)

P0152 O2 Sensor Circuit High Voltage (Bank no. 2 Sensor no. 1)

P0153 O2 Sensor Circuit Slow Response (Bank no. 2 Sensor no. 1)

P0154 O2 Sensor Circuit No Activity Detected (Bank no. 2 Sensor no. 1)

P0155 O2 Sensor Heater Circuit Malfunction (Bank no. 2 Sensor no. 1)

P0156 O2 Sensor Circuit Malfunction (Bank no. 2 Sensor no. 2)

P0157 O2 Sensor Circuit Low Voltage (Bank no. 2 Sensor no. 2)

P0158 O2 Sensor Circuit High Voltage (Bank no. 2 Sensor no. 2)

P0159 O2 Sensor Circuit Slow Response (Bank no. 2 Sensor no. 2)

P0160 O2 Sensor Circuit No Activity Detected (Bank no. 2 Sensor no. 2)

P0161 O2 Sensor Heater Circuit Malfunction (Bank no. 2 Sensor no. 2)

P0162 O2 Sensor Circuit Malfunction (Bank no. 2 Sensor no. 3)

P0163 O2 Sensor Circuit Low Voltage (Bank no. 2 Sensor no. 3)

P0164 O2 Sensor Circuit High Voltage (Bank no. 2 Sensor no. 3)

P0165 O2 Sensor Circuit Slow Response (Bank no. 2 Sensor no. 3)

P0166 O2 Sensor Circuit No Activity Detected (Bank no. 2 Sensor no. 3)

P0167 O2 Sensor Heater Circuit Malfunction (Bank no. 2 Sensor no. 3)

P0170 Fuel Trim Malfunction (Bank no. 1)

P0171 System Too Lean (Bank no. 1)

P0172 System Too Rich (Bank no. 1)

P0173 Fuel Trim Malfunction (Bank no. 2)

P0174 System Too Lean (Bank no. 2)

P0175 System Too Rich (Bank no. 2)

P0176 Fuel Composition Sensor Circuit Malfunction

P0177 Fuel Composition Sensor Circuit Range/Performance

P0178 Fuel Composition Sensor Circuit Low Input

P0179 Fuel Composition Sensor Circuit High Input

P0180 Fuel Temperature Sensor "A" Circuit Malfunction

P0181 Fuel Temperature Sensor "A" Circuit Range/Performance

P0182 Fuel Temperature Sensor "A" Circuit Low Input

P0183 Fuel Temperature Sensor "A" Circuit High Input

P0184 Fuel Temperature Sensor "A" Circuit Intermittent

P0185 Fuel Temperature Sensor "B" Circuit Malfunction

P0186 Fuel Temperature Sensor "B" Circuit Range/Performance

P0187 Fuel Temperature Sensor "B" Circuit Low Input

P0188 Fuel Temperature Sensor "B" Circuit High Input

P0189 Fuel Temperature Sensor "B" Circuit Intermittent

P0190 Fuel Rail Pressure Sensor Circuit Malfunction

P0191 Fuel Rail Pressure Sensor Circuit Range/Performance

P0192 Fuel Rail Pressure Sensor Circuit Low Input

P0193 Fuel Rail Pressure Sensor Circuit High Input

P0194 Fuel Rail Pressure Sensor Circuit Intermittent

P0195 Engine Oil Temperature Sensor Malfunction

P0196 Engine Oil Temperature Sensor Range/Performance

P0197 Engine Oil Temperature Sensor Low

P0198 Engine Oil Temperature Sensor High

P0199 Engine Oil Temperature Sensor Intermittent

P0200 Injector Circuit Malfunction

P0201 Injector Circuit Malfunction—Cylinder no. 1

P0202 Injector Circuit Malfunction—Cylinder no. 2

P0203 Injector Circuit Malfunction—Cylinder no. 3

P0204 Injector Circuit Malfunction—Cylinder no. 4

P0205 Injector Circuit Malfunction—Cylinder no. 5

P0206 Injector Circuit Malfunction—Cylinder no. 6

P0207 Injector Circuit Malfunction—Cylinder no. 7

P0208 Injector Circuit Malfunction—Cylinder no. 8

P0209 Injector Circuit Malfunction—Cylinder no. 9

P0210 Injector Circuit Malfunction—Cylinder no. 10

P0211 Injector Circuit Malfunction—Cylinder no. 11

P0212 Injector Circuit Malfunction—Cylinder no. 12

P0213 Cold Start Injector no. 1 Malfunction

P0214 Cold Start Injector no. 2 Malfunction

P0215 Engine Shutoff Solenoid Malfunction

P0216 Injection Timing Control Circuit Malfunction

P0217 Engine Over Temperature Condition

P0218 Transmission Over Temperature Condition

P0219 Engine Over Speed Condition

P0220 Throttle/Pedal Position Sensor/Switch "B" Circuit Malfunction

P0221 Throttle/Pedal Position Sensor/Switch "B" Circuit Range/Performance Problem

P0222 Throttle/Pedal Position Sensor/Switch "B" Circuit Low Input

P0223 Throttle/Pedal Position Sensor/Switch "B" Circuit High Input

P0224 Throttle/Pedal Position Sensor/Switch "B" Circuit Intermittent

P0225 Throttle/Pedal Position Sensor/Switch "C" Circuit Malfunction

P0226 Throttle/Pedal Position Sensor/Switch "C" Circuit Range/Performance Problem

P0227 Throttle/Pedal Position Sensor/Switch "C" Circuit Low Input

P0228 Throttle/Pedal Position Sensor/Switch "C" Circuit High Input

P0229 Throttle/Pedal Position Sensor/Switch "C" Circuit Intermittent

P0230 Fuel Pump Primary Circuit Malfunction
P0231 Fuel Pump Secondary Circuit Low
P0232 Fuel Pump Secondary Circuit High
P0233 Fuel Pump Secondary Circuit Intermittent
P0234 Engine Over Boost Condition
P0261 Cylinder no. 1 Injector Circuit Low
P0262 Cylinder no. 1 Injector Circuit High
P0263 Cylinder no. 1 Contribution/Balance Fault
P0264 Cylinder no. 2 Injector Circuit Low
P0265 Cylinder no. 2 Injector Circuit High
P0266 Cylinder no. 2 Contribution/Balance Fault
P0267 Cylinder no. 3 Injector Circuit Low
P0268 Cylinder no. 3 Injector Circuit High
P0269 Cylinder no. 3 Contribution/Balance Fault
P0270 Cylinder no. 4 Injector Circuit Low
P0271 Cylinder no. 4 Injector Circuit High
P0272 Cylinder no. 4 Contribution/Balance Fault
P0273 Cylinder no. 5 Injector Circuit Low
P0274 Cylinder no. 5 Injector Circuit High
P0275 Cylinder no. 5 Contribution/Balance Fault
P0276 Cylinder no. 6 Injector Circuit Low
P0277 Cylinder no. 6 Injector Circuit High
P0278 Cylinder no. 6 Contribution/Balance Fault
P0279 Cylinder no. 7 Injector Circuit Low
P0280 Cylinder no. 7 Injector Circuit High
P0281 Cylinder no. 7 Contribution/Balance Fault
P0282 Cylinder no. 8 Injector Circuit Low
P0283 Cylinder no. 8 Injector Circuit High
P0284 Cylinder no. 8 Contribution/Balance Fault
P0285 Cylinder no. 9 Injector Circuit Low
P0286 Cylinder no. 9 Injector Circuit High
P0287 Cylinder no. 9 Contribution/Balance Fault
P0288 Cylinder no. 10 Injector Circuit Low
P0289 Cylinder no. 10 Injector Circuit High
P0290 Cylinder no. 10 Contribution/Balance Fault
P0291 Cylinder no. 11 Injector Circuit Low
P0292 Cylinder no. 11 Injector Circuit High
P0293 Cylinder no. 11 Contribution/Balance Fault
P0294 Cylinder no. 12 Injector Circuit Low
P0295 Cylinder no. 12 Injector Circuit High
P0296 Cylinder no. 12 Contribution/Balance Fault

P0300 Random/Multiple Cylinder Misfire Detected
P0301 Cylinder no. 1—Misfire Detected
P0302 Cylinder no. 2—Misfire Detected
P0303 Cylinder no. 3—Misfire Detected
P0304 Cylinder no. 4—Misfire Detected
P0305 Cylinder no. 5—Misfire Detected
P0306 Cylinder no. 6—Misfire Detected
P0307 Cylinder no. 7—Misfire Detected
P0308 Cylinder no. 8—Misfire Detected
P0309 Cylinder no. 9—Misfire Detected
P0310 Cylinder no. 10—Misfire Detected
P0311 Cylinder no. 11—Misfire Detected
P0312 Cylinder no. 12—Misfire Detected
P0320 Ignition/Distributor Engine Speed Input Circuit Malfunction

P0321 Ignition/Distributor Engine Speed Input Circuit Range/Performance

P0322 Ignition/Distributor Engine Speed Input Circuit No Signal

P0323 Ignition/Distributor Engine Speed Input Circuit Intermittent

P0325 Knock Sensor no. 1—Circuit Malfunction (Bank no. 1 or Single Sensor)

P0326 Knock Sensor no. 1—Circuit Range/Performance (Bank no. 1 or Single Sensor)

P0327 Knock Sensor no. 1—Circuit Low Input (Bank no. 1 or Single Sensor)

P0328 Knock Sensor no. 1—Circuit High Input (Bank no. 1 or Single Sensor)

P0329 Knock Sensor no. 1—Circuit Input Intermittent (Bank no. 1 or Single Sensor)

P0330 Knock Sensor no. 2—Circuit Malfunction (Bank no. 2)

P0331 Knock Sensor no. 2—Circuit Range/Performance (Bank no. 2)

P0332 Knock Sensor no. 2—Circuit Low Input (Bank no. 2)
P0333 Knock Sensor no. 2—Circuit High Input (Bank no. 2)
P0334 Knock Sensor no. 2—Circuit Input Intermittent (Bank no. 2)

P0335 Crankshaft Position Sensor "A" Circuit Malfunction
P0336 Crankshaft Position Sensor "A" Circuit Range/Performance

P0337 Crankshaft Position Sensor "A" Circuit Low Input
P0338 Crankshaft Position Sensor "A" Circuit High Input
P0339 Crankshaft Position Sensor "A" Circuit Intermittent
P0340 Camshaft Position Sensor Circuit Malfunction
P0341 Camshaft Position Sensor Circuit Range/Performance
P0342 Camshaft Position Sensor Circuit Low Input
P0343 Camshaft Position Sensor Circuit High Input
P0344 Camshaft Position Sensor Circuit Intermittent
P0350 Ignition Coil Primary/Secondary Circuit Malfunction
P0351 Ignition Coil "A" Primary/Secondary Circuit Malfunction
P0352 Ignition Coil "B" Primary/Secondary Circuit Malfunction
P0353 Ignition Coil "C" Primary/Secondary Circuit Malfunction
P0354 Ignition Coil "D" Primary/Secondary Circuit Malfunction
P0355 Ignition Coil "E" Primary/Secondary Circuit Malfunction
P0356 Ignition Coil "F" Primary/Secondary Circuit Malfunction
P0357 Ignition Coil "G" Primary/Secondary Circuit Malfunction
P0358 Ignition Coil "H" Primary/Secondary Circuit Malfunction
P0359 Ignition Coil "I" Primary/Secondary Circuit Malfunction
P0360 Ignition Coil "J" Primary/Secondary Circuit Malfunction
P0361 Ignition Coil "K" Primary/Secondary Circuit Malfunction
P0362 Ignition Coil "L" Primary/Secondary Circuit Malfunction

P0370 Timing Reference High Resolution Signal "A" Malfunction

P0371 Timing Reference High Resolution Signal "A" Too Many Pulses

P0372 Timing Reference High Resolution Signal "A" Too Few Pulses

P0373 Timing Reference High Resolution Signal "A" Intermittent/Erratic Pulses

P0374 Timing Reference High Resolution Signal "A" No Pulses

P0375 Timing Reference High Resolution Signal "B" Malfunction

P0376 Timing Reference High Resolution Signal "B" Too Many Pulses

P0377 Timing Reference High Resolution Signal "B" Too Few Pulses

P0378 Timing Reference High Resolution Signal "B" Intermittent/Erratic Pulses

P0379 Timing Reference High Resolution Signal "B" No Pulses

P0380 Glow Plug/Heater Circuit "A" Malfunction

P0381 Glow Plug/Heater Indicator Circuit Malfunction

P0382 Glow Plug/Heater Circuit "B" Malfunction

P0385 Crankshaft Position Sensor "B" Circuit Malfunction

P0386 Crankshaft Position Sensor "B" Circuit Range/Performance

P0387 Crankshaft Position Sensor "B" Circuit Low Input

P0388 Crankshaft Position Sensor "B" Circuit High Input

P0389 Crankshaft Position Sensor "B" Circuit Intermittent

P0400 Exhaust Gas Recirculation Flow Malfunction

P0401 Exhaust Gas Recirculation Flow Insufficient Detected

P0402 Exhaust Gas Recirculation Flow Excessive Detected

P0403 Exhaust Gas Recirculation Circuit Malfunction

P0404 Exhaust Gas Recirculation Circuit Range/Performance

P0405 Exhaust Gas Recirculation Sensor "A" Circuit Low

P0406 Exhaust Gas Recirculation Sensor "A" Circuit High

P0407 Exhaust Gas Recirculation Sensor "B" Circuit Low

P0408 Exhaust Gas Recirculation Sensor "B" Circuit High

P0410 Secondary Air Injection System Malfunction

P0411 Secondary Air Injection System Incorrect Flow Detected

P0412 Secondary Air Injection System Switching Valve "A" Circuit Malfunction

P0413 Secondary Air Injection System Switching Valve "A" Circuit Open

P0414 Secondary Air Injection System Switching Valve "A" Circuit Shorted

P0415 Secondary Air Injection System Switching Valve "B" Circuit Malfunction

P0416 Secondary Air Injection System Switching Valve "B" Circuit Open

P0417 Secondary Air Injection System Switching Valve "B" Circuit Shorted

P0418 Secondary Air Injection System Relay "A" Circuit Malfunction

P0419 Secondary Air Injection System Relay "B" Circuit Malfunction

P0420 Catalyst System Efficiency Below Threshold (Bank no. 1)

P0421 Warm Up Catalyst Efficiency Below Threshold (Bank no. 1)

P0422 Main Catalyst Efficiency Below Threshold (Bank no. 1)

P0423 Heated Catalyst Efficiency Below Threshold (Bank no. 1)

P0424 Heated Catalyst Temperature Below Threshold (Bank no. 1)

P0430 Catalyst System Efficiency Below Threshold (Bank no. 2)

P0431 Warm Up Catalyst Efficiency Below Threshold (Bank no. 2)

P0432 Main Catalyst Efficiency Below Threshold (Bank no. 2)

P0433 Heated Catalyst Efficiency Below Threshold (Bank no. 2)

P0434 Heated Catalyst Temperature Below Threshold (Bank no. 2)

P0440 Evaporative Emission Control System Malfunction

P0441 Evaporative Emission Control System Incorrect Purge Flow

P0442 Evaporative Emission Control System Leak Detected (Small Leak)

P0443 Evaporative Emission Control System Purge Control Valve Circuit Malfunction

P0444 Evaporative Emission Control System Purge Control Valve Circuit Open

P0445 Evaporative Emission Control System Purge Control Valve Circuit Shorted

P0446 Evaporative Emission Control System Vent Control Circuit Malfunction

P0447 Evaporative Emission Control System Vent Control Circuit Open

P0448 Evaporative Emission Control System Vent Control Circuit Shorted

P0449 Evaporative Emission Control System Vent Valve/Solenoid Circuit Malfunction

P0450 Evaporative Emission Control System Pressure Sensor Malfunction

P0451 Evaporative Emission Control System Pressure Sensor Range/Performance

P0452 Evaporative Emission Control System Pressure Sensor Low Input

P0453 Evaporative Emission Control System Pressure Sensor High Input

P0454 Evaporative Emission Control System Pressure Sensor Intermittent

P0455 Evaporative Emission Control System Leak Detected (Gross Leak)

P0460 Fuel Level Sensor Circuit Malfunction

P0461 Fuel Level Sensor Circuit Range/Performance

P0462 Fuel Level Sensor Circuit Low Input

P0463 Fuel Level Sensor Circuit High Input

P0464 Fuel Level Sensor Circuit Intermittent

P0465 Purge Flow Sensor Circuit Malfunction

P0466 Purge Flow Sensor Circuit Range/Performance

P0467 Purge Flow Sensor Circuit Low Input

P0468 Purge Flow Sensor Circuit High Input

P0469 Purge Flow Sensor Circuit Intermittent

P0470 Exhaust Pressure Sensor Malfunction

P0471 Exhaust Pressure Sensor Range/Performance

P0472 Exhaust Pressure Sensor Low

P0473 Exhaust Pressure Sensor High

P0474 Exhaust Pressure Sensor Intermittent

P0475 Exhaust Pressure Control Valve Malfunction

P0476 Exhaust Pressure Control Valve Range/Performance

P0477 Exhaust Pressure Control Valve Low

P0478 Exhaust Pressure Control Valve High

P0479 Exhaust Pressure Control Valve Intermittent

P0480 Cooling Fan no. 1 Control Circuit Malfunction

P0481 Cooling Fan no. 2 Control Circuit Malfunction

P0482 Cooling Fan no. 3 Control Circuit Malfunction

P0483 Cooling Fan Rationality Check Malfunction
P0484 Cooling Fan Circuit Over Current
P0485 Cooling Fan Power/Ground Circuit Malfunction
P0500 Vehicle Speed Sensor Malfunction
P0501 Vehicle Speed Sensor Range/Performance
P0502 Vehicle Speed Sensor Circuit Low Input
P0503 Vehicle Speed Sensor Intermittent/Erratic/High
P0505 Idle Control System Malfunction
P0506 Idle Control System RPM Lower Than Expected
P0507 Idle Control System RPM Higher Than Expected
P0510 Closed Throttle Position Switch Malfunction
P0520 Engine Oil Pressure Sensor/Switch Circuit Malfunction
P0521 Engine Oil Pressure Sensor/Switch Range/Performance
P0522 Engine Oil Pressure Sensor/Switch Low Voltage
P0523 Engine Oil Pressure Sensor/Switch High Voltage
P0530 A/C Refrigerant Pressure Sensor Circuit Malfunction
P0531 A/C Refrigerant Pressure Sensor Circuit Range/Performance
P0532 A/C Refrigerant Pressure Sensor Circuit Low Input
P0533 A/C Refrigerant Pressure Sensor Circuit High Input
P0534 A/C Refrigerant Charge Loss
P0550 Power Steering Pressure Sensor Circuit Malfunction
P0551 Power Steering Pressure Sensor Circuit Range/Performance
P0552 Power Steering Pressure Sensor Circuit Low Input
P0553 Power Steering Pressure Sensor Circuit High Input
P0554 Power Steering Pressure Sensor Circuit Intermittent
P0560 System Voltage Malfunction
P0561 System Voltage Unstable
P0562 System Voltage Low
P0563 System Voltage High
P0565 Cruise Control On Signal Malfunction
P0566 Cruise Control Off Signal Malfunction
P0567 Cruise Control Resume Signal Malfunction
P0568 Cruise Control Set Signal Malfunction
P0569 Cruise Control Coast Signal Malfunction
P0570 Cruise Control Accel Signal Malfunction
P0571 Cruise Control/Brake Switch "A" Circuit Malfunction
P0572 Cruise Control/Brake Switch "A" Circuit Low
P0573 Cruise Control/Brake Switch "A" Circuit High
P0574 Through P0580 Reserved for Cruise Codes
P0600 Serial Communication Link Malfunction
P0601 Internal Control Module Memory Check Sum Error
P0602 Control Module Programming Error
P0603 Internal Control Module Keep Alive Memory (KAM) Error
P0604 Internal Control Module Random Access Memory (RAM) Error
P0605 Internal Control Module Read Only Memory (ROM) Error
P0606 PCM Processor Fault
P0608 Control Module VSS Output "A" Malfunction
P0609 Control Module VSS Output "B" Malfunction
P0620 Generator Control Circuit Malfunction
P0621 Generator Lamp "L" Control Circuit Malfunction
P0622 Generator Field "F" Control Circuit Malfunction
P0650 Malfunction Indicator Lamp (MIL) Control Circuit Malfunction
P0654 Engine RPM Output Circuit Malfunction
P0655 Engine Hot Lamp Output Control Circuit Malfunction
P0656 Fuel Level Output Circuit Malfunction
P0700 Transmission Control System Malfunction
P0701 Transmission Control System Range/Performance

P0702 Transmission Control System Electrical
P0703 Torque Converter/Brake Switch "B" Circuit Malfunction
P0704 Clutch Switch Input Circuit Malfunction
P0705 Transmission Range Sensor Circuit Malfunction (PRNDL Input)
P0706 Transmission Range Sensor Circuit Range/Performance
P0707 Transmission Range Sensor Circuit Low Input
P0708 Transmission Range Sensor Circuit High Input
P0709 Transmission Range Sensor Circuit Intermittent
P0710 Transmission Fluid Temperature Sensor Circuit Malfunction
P0711 Transmission Fluid Temperature Sensor Circuit Range/Performance
P0712 Transmission Fluid Temperature Sensor Circuit Low Input
P0713 Transmission Fluid Temperature Sensor Circuit High Input
P0714 Transmission Fluid Temperature Sensor Circuit Intermittent
P0715 Input/Turbine Speed Sensor Circuit Malfunction
P0716 Input/Turbine Speed Sensor Circuit Range/Performance
P0717 Input/Turbine Speed Sensor Circuit No Signal
P0718 Input/Turbine Speed Sensor Circuit Intermittent
P0719 Torque Converter/Brake Switch "B" Circuit Low
P0720 Output Speed Sensor Circuit Malfunction
P0721 Output Speed Sensor Circuit Range/Performance
P0722 Output Speed Sensor Circuit No Signal
P0723 Output Speed Sensor Circuit Intermittent
P0724 Torque Converter/Brake Switch "B" Circuit High
P0725 Engine Speed Input Circuit Malfunction
P0726 Engine Speed Input Circuit Range/Performance
P0727 Engine Speed Input Circuit No Signal
P0728 Engine Speed Input Circuit Intermittent
P0730 Incorrect Gear Ratio
P0731 Gear no. 1 Incorrect Ratio
P0732 Gear no. 2 Incorrect Ratio
P0733 Gear no. 3 Incorrect Ratio
P0734 Gear no. 4 Incorrect Ratio
P0735 Gear no. 5 Incorrect Ratio
P0736 Reverse Incorrect Ratio
P0740 Torque Converter Clutch Circuit Malfunction
P0741 Torque Converter Clutch Circuit Performance or Stuck Off
P0742 Torque Converter Clutch Circuit Stuck On
P0743 Torque Converter Clutch Circuit Electrical
P0744 Torque Converter Clutch Circuit Intermittent
P0745 Pressure Control Solenoid Malfunction
P0746 Pressure Control Solenoid Performance or Stuck Off
P0747 Pressure Control Solenoid Stuck On
P0748 Pressure Control Solenoid Electrical
P0749 Pressure Control Solenoid Intermittent
P0750 Shift Solenoid "A" Malfunction
P0751 Shift Solenoid "A" Performance or Stuck Off
P0752 Shift Solenoid "A" Stuck On
P0753 Shift Solenoid "A" Electrical
P0754 Shift Solenoid "A" Intermittent
P0755 Shift Solenoid "B" Malfunction
P0756 Shift Solenoid "B" Performance or Stuck Off
P0757 Shift Solenoid "B" Stuck On
P0758 Shift Solenoid "B" Electrical
P0759 Shift Solenoid "B" Intermittent

P0760 Shift Solenoid "C" Malfunction
P0761 Shift Solenoid "C" Performance Or Stuck Off
P0762 Shift Solenoid "C" Stuck On
P0763 Shift Solenoid "C" Electrical
P0764 Shift Solenoid "C" Intermittent
P0765 Shift Solenoid "D" Malfunction
P0766 Shift Solenoid "D" Performance Or Stuck Off
P0767 Shift Solenoid "D" Stuck On
P0768 Shift Solenoid "D" Electrical
P0769 Shift Solenoid "D" Intermittent
P0770 Shift Solenoid "E" Malfunction
P0771 Shift Solenoid "E" Performance Or Stuck Off
P0772 Shift Solenoid "E" Stuck On
P0773 Shift Solenoid "E" Electrical
P0774 Shift Solenoid "E" Intermittent
P0780 Shift Malfunction
P0781 1–2 Shift Malfunction
P0782 2–3 Shift Malfunction
P0783 3–4 Shift Malfunction
P0784 4–5 Shift Malfunction
P0785 Shift/Timing Solenoid Malfunction
P0786 Shift/Timing Solenoid Range/Performance
P0787 Shift/Timing Solenoid Low
P0788 Shift/Timing Solenoid High
P0789 Shift/Timing Solenoid Intermittent
P0790 Normal/Performance Switch Circuit Malfunction
P0801 Reverse Inhibit Control Circuit Malfunction
P0803 1–4 Upshift (Skip Shift) Solenoid Control Circuit Malfunction
P0804 1–4 Upshift (Skip Shift) Lamp Control Circuit Malfunction
P1106 Map Sensor Circuit Intermittent High Voltage
P1107 MAP Sensor Circuit Intermittent Low Voltage
P1111 IAT Sensor Circuit Intermittent High Voltage
P1112 IAT Sensor Circuit Intermittent Low Voltage
P1114 ECT Sensor Circuit Intermittent Low Voltage
P1115 ECT Sensor Circuit Intermittent High Voltage
P1121 TP Sensor Circuit Intermittent High Voltage
P1122 TP Sensor Circuit Intermittent Low Voltage
P1133 HO2S-11 Insufficient Switching (Bank 1 Sensor 1)
P1134 HO2S-11 Transition Time Ratio (Bank 1 Sensor 1)
P1153 HO2S-21 Insufficient Switching (Bank 2 Sensor I)
P1154 HO2S-21 Transition Time Ratio (Bank 2 Sensor 1)
P1171 Fuel System Lean During Acceleration
P1391 G-Acceleration Sensor Intermittent Low Voltage
P1390 G-Acceleration (Low G) Sensor Performance
P1392 Rough Road G-Sensor Circuit Low Voltage
P1393 Rough Road G-Sensor Circuit High Voltage **P1394** G-Acceleration Sensor Intermittent High Voltage
P1406 EGR Valve Pintle Position Sensor Circuit Fault
P1441 EVAP System Flow During Non-Purge
P1442 EVAP System Flow During Non-Purge **P1508** Idle Speed Control System-Low
P1509 Idle Speed Control System-High
P1618 Serial Peripheral Interface Communication Error
P1640 Output Driver Module 'A' Fault
P1790 PCM ROM (Transmission Side) Check Sum Error
P1792 PCM EEPROM (Transmission Side) Check Sum Error
P1835 Kick Down Switch Always On
P1850 Brake Band Apply Solenoid Electrical Fault
P1860 TCC PWM Solenoid Electrical Fault
P1870 Transmission Component Slipping

Audi

READING CODES

Reading the control module memory is one of the first steps in OBD II system diagnostics. This step should be initially performed to determine the general nature of the fault. Subsequent readings will determine if the fault has been cleared.

Reading codes can be performed by any of the methods below:
- Read the control module memory with the Generic Scan Tool (GST)
- Read the control module memory with the VAG 1550 tester

To read the fault codes, connect the scan tool or tester according to the manufacturer's instructions. Follow the manufacturer's specified procedure for reading the codes.

CLEARING CODES

Control module reset procedures are a very important part of OBD II System diagnostics. This step should be done at the end of any fault code repair and at the end of any driveability repair.

Clearing codes can be performed by any of the methods below:
- Clear the control module memory with the Generic Scan Tool (GST)
- Clear the control module memory with the vehicle manufacturer's specific tester
- Turn the ignition **OFF** and remove the negative battery cable for at least 1 minute.

Removing the negative battery cable may cause other systems in the vehicle to loose their memory. Prior to removing the cable, ensure you have the proper reset codes for radios and alarms.

➡**The MIL will may also be de-activated for some codes if the vehicle completes three consecutive trips without a fault detected with vehicle conditions similar to those present during the fault.**

OBD II TROUBLE CODES

P0102 Mass or Volume Air Flow Circuit Low Input
P0103 Mass or Volume Air Flow Circuit High Input
P0107 Manifold Absolute Pressure or Barometric Pressure Low Input
P0108 Manifold Absolute Pressure or Barometric Pressure High Input
P0112 Intake Air Temperature Circuit Low Input
P0113 Intake Air Temperature Circuit High Input
P0116 Engine Coolant Temperature Circuit Range/Performance
P0117 Engine Coolant Temperature Circuit Low Input
P0118 Engine Coolant Temperature Circuit High Input
P0120 Throttle Position Sensor A Circuit Malfunction
P0121 Throttle/Pedal Position Sensor A Circuit Range/Performance
P0122 Throttle/Pedal Position Sensor A Circuit Low Input **P0123** Throttle/Pedal Position Sensor A Circuit High Input
P0125 Insufficient Coolant Temperature For Closed Loop Fuel Control
P0130 Oxygen Sensor Circuit, Bank 1-Sensor 1 Malfunction
P0131 Oxygen Sensor Circuit, Bank 1-Sensor 1 Low Voltage
P0132 Oxygen Sensor Circuit, Bank 1-Sensor 1 High Voltage

P0133 Oxygen Sensor Circuit, Bank 1-Sensor 1 Slow Response
P0134 Oxygen Sensor Circuit, Bank 1-Sensor 1 No Activity Detected
P0135 Oxygen Sensor Heater Circuit, Bank 1-Sensor 1 Malfunction
P0136 Oxygen Sensor Circuit, Bank 1-Sensor 2 Malfunction
P0137 Oxygen Sensor Circuit, Bank 1-Sensor 2 Low Voltage
P0138 Oxygen Sensor Circuit, Bank 1-Sensor 2 High Voltage
P0140 Oxygen Sensor Circuit, Bank 1-Sensor 2 No Activity Detected
P0141 Oxygen Sensor Heater Circuit, Bank 1-Sensor 2 Malfunction
P0150 Oxygen Sensor Circuit, Bank 2-Sensor 1 Malfunction
P0151 Oxygen Sensor Circuit, Bank 2-Sensor 1 Low Voltage
P0152 Oxygen Sensor Circuit, Bank 2-Sensor 1 High Voltage
P0153 Oxygen Sensor Circuit, Bank 2-Sensor 1 Slow Response
P0154 Oxygen Sensor Circuit, Bank 2-Sensor 1 No Activity Detected
P0156 Oxygen Sensor Circuit, Bank 2-Sensor 2 Malfunction
P0157 Oxygen Sensor Circuit, Bank 2-Sensor 2 Low Voltage
P0158 Oxygen Sensor Circuit, Bank 2-Sensor 2 High Voltage
P0160 Oxygen Sensor Circuit, Bank 2-Sensor 2 No Activity Detected
P0171 System Too Lean, Bank 1
P0172 System Too Rich, Bank 1
P0300 Random Multiple Misfire Detected
P0301 Cylinder 1 Misfire Detected
P0302 Cylinder 2 Misfire Detected
P0303 Cylinder 3 Misfire Detected
P0304 Cylinder 4 Misfire Detected
P0305 Cylinder 5 Misfire Detected
P0306 Cylinder 6 Misfire Detected
P0321 Ignition Distributor Engine Speed Input Circuit Range/Performance
P0322 Ignition /Distributor Engine Speed Input Circuit No Signal
P0327 Knock sensor 1 Circuit Low Input
P0328 Knock Sensor Circuit, High Input
P0332 Knock sensor 2 Circuit Low Input
P0332 Knock Sensor Circuit, Low Input
P0333 Knock Sensor 2 Circuit, High Input
P0341 Camshaft Position Sensor Circuit Range/Performance
P0411 Secondary Air Injection System Incorrect Flow Detected
P0422 Main Catalyst Efficiency Below Threshold (Bank 1)
P0422 Main Catalyst, Bank I Efficiency Below Threshold
P0440 Evaporative Emission Control System Malfunction
P0441 EVAP Emission Contr. Sys. Incorrect Purge Flow
P0442 EVAP Emission Contr. Sys. (Small Leak) Leak Detected
P0455 EVAP Emission Contr. Sys. (Gross Leak) Leak Detected
P0501 Vehicle Speed Sensor Range/Performance
P0506 Idle Control System RPM Lower Than Expected
P0507 Idle Control System RPM Higher Than Expected
P0510 Closed Throttle Position Switch Malfunction
P0560 System Voltage Malfunction
P0562 System Voltage Low Voltage
P0563 System Voltage High Voltage
P0601 Internal Contr. Module Memory Check Sum Error
P0604 Internal Contr. Module Random Access Memory (RAM) Error
P0605 Internal Control Module Read Only Memory (ROM) Error
P0707 Transmission Range Sensor Circuit Low Input

P0708 Transmission Range Sensor Circuit High Input
P0715 Input Turbine Speed Sensor Circuit Malfunction
P0722 Output Speed Sensor Circuit No Signal
P0725 Engine Speed Input Circuit Malfunction
P0748 Pressure Control Solenoid Electrical
P0753 Shift Solenoid A Electrical
P0758 Shift Solenoid B Electrical
P0763 Shift Solenoid C Electrical
P0768 Shift Solenoid D Electrical
P0773 Shift Solenoid E Electrical
P0300 Random/Multiple Cylinder Misfire Detected
P0301 Cylinder I Misfire Detected
P0302 Cylinder 2 Misfire Detected
P1102 Oxygen Sensor Heating Circuit, Bank 1-Sensor 1 Short to B+
P1105 Oxygen Sensor Heating Circuit, Bank 1-Sensor 2 Short to B+
P1107 Oxygen Sensor Heating Circuit, Bank 2-Sensor I Short to B+
P1110 Oxygen Sensor Heating Circuit, Bank 2-Sensor 2 Short to B+
P1127 Long Term Fuel Trim Multiplicative, Bank 1 System Too Rich
P1128 Long Term Fuel Trim Multiplicative, Bank 1 System Too Lean
P1129 Long Term Fuel Trim Multiplicative, Bank2 System too Rich
P1130 Long Term Fuel Trim Multiplicative, Bank2 System too Lean
P1136 Long Term Fuel Trim Additive, Bank 1 System Too Lean
P1137 Long Term Fuel Trim Additive, Bank 1 System Too Rich
P1138 Long Term Fuel Trim Additive Fuel, Bank I System too Lean
P1139 Long Term Fuel Trim Additive Fuel, Bank 1 System too Rich
P1141 Load Calculation Cross Check Range/Performance
P1176 Oxygen Correction Behind Catalyst, B1 Limit Attained
P1177 Oxygen Correction Behind Catalyst. 82 Limit Attained
P1196 Oxygen Sensor Heater Circuit, Bank 1-Sensor 1 Electrical Malfunction
P1197 Oxygen Sensor Heater Circuit, Bank 2-Sensor I Electrical Malfunction
P1198 Oxygen Sensor Heater Circuit, Bankl-Sensor2 Electrical Malfunction
P1198 Oxygen Sensor Heater Circuit, Bank 1-Sensor 2 Electrical Malfunction
P1199 Oxygen Sensor Heater Circuit, Bank 2-Sensor 2 Electrical Malfunction
P1201 Cylinder 1, Fuel Injection Circuit Electrical Malfunction
P1202 Cylinder 2, Fuel Injection Circuit Electrical Malfunction
P1203 Cylinder 3, Fuel Injection Circuit Electrical Malfunction
P1204 Cylinder 4, Fuel Injection Circuit Electrical Malfunction
P1205 Cylinder 5, Fuel Injection Circuit Electrical Malfunction
P1206 Cylinder 6, Fuel Injection Circuit Electrical Malfunction
P1207 Cylinder 7, Fuel Injection Circuit Electrical Malfunction
P1208 Cylinder 8, Fuel Injection Circuit Electrical Malfunction
P1213 Cylinder I-Fuel Injection Circuit Short to B+
P1214 Cylinder 2-Fuel Injection Circuit Short to B+
P1215 Cylinder 3 Fuel Injection Circuit Short to B+
P1216 Cylinder 4 Fuel Injection Circuit Short to B+
P1217 Cylinder 5 Fuel Injection Circuit Short to B+

P1218 Cylinder 6 Fuel Injection Circuit Short to B+
P1219 Cylinder 7, Fuel Injection Circuit Short to B+
P1219 Cylinder 8, Fuel Injection Circuit Short to B+
P1225 Cylinder I Fuel Injection Circuit Short to Ground
P1226 Cylinder 2 Fuel Injection Circuit Short to Ground
P1227 Cylinder 3 Fuel Injection Circuit Short to Ground
P1228 Cylinder 4 Fuel Injection Circuit Short to Ground
P1229 Cylinder 5 Fuel Injection Circuit Short to Ground
P1230 Cylinder 6 Fuel Injection Circuit Short to Ground
P1237 Cylinder I Fuel Injection Circuit Open Circuit
P1238 Cylinder 2 Fuel Injection Circuit Open Circuit
P1239 Cylinder 3 Fuel Injection Circuit Open Circuit
P1240 Cylinder 4 Fuel Injection Circuit Open Circuit
P1241 Cylinder 5 Fuel Injection Circuit Open Circuit
P1242 Cylinder 6 Fuel Injection Circuit Open Circuit
P1250 Fuel Level Too Low
P1280 Fuel Injection Air Control Valve Circuit Flow too Low
P1283 Fuel Injection Air Control Valve Circuit Electrical Malfunction
P1325 Cylinder I Knock Control Limit Attained
P1326 Cylinder 2 Knock Control Limit Attained
P1327 Cylinder 3 Knock Control Limit Attained
P1328 Cylinder 4 Knock Control Limit Attained
P1329 Cylinder 5 Knock Control Limit Attained
P1330 Cylinder 6 Knock Control Limit Attained
P1331 Cylinder 7, Knock Control Limit Attained
P1332 Cylinder 8, Knock Control Limit Attained
P1337 Camshaft Position Sensor, Bank 1 Short to Ground
P1338 Camshaft Position Sensor, Bank 1 Open Circuit/Short to B+
P1340 Boost Pressure Control Valve Short to B+
P1386 Internal Control Module Knock Control Circuit Error
P1391 Camshaft Position Sensor, Bank 2 Short to Ground
P1392 Camshaft Position Sensor, Bank 2 Open Circuit/Short to B+
P1410 Tank Ventilation Valve Circuit Short to B+
P1420 Secondary Air Injection Module Short To B+
P1421 Secondary Air Injection Module Short To Ground
P1421 Secondary Air Injection Valve Circuit Short to Ground
P1422 Secondary Air Injection System Control Valve Circuit Short To B+
P1422 Secondary Air Injection Valve Circuit Short to B+
P1425 Tank Vent Valve Short To Ground
P1425 Tank Vent Valve Short to Ground
P1426 Tank Vent Valve Open
P1426 Tank Vent Valve Open
P1432 Secondary Air Injection Valve Open
P1433 Secondary Air Injection System Pump Relay Circuit Open
P1434 Secondary Air Injection System Pump Relay Circuit Short to B+
P1435 Secondary Air Injection System Pump Relay Circuit Short to Ground
P1436 Secondary Air Injection System Pump Relay Circuit Electrical Malfunction
P1450 Secondary Air Injection System Circuit Short To B+
P1451 Secondary Air Injection System Circuit Short To Ground
P1452 Secondary Air Injection System Open Circuit
P1471 EVAP Emission Control LDP Circuit Short to B+
P1472 EVAP Emission Control LDP Circuit Short to Ground
P1473 EVAP Emission Control LDP Circuit Open Circuit
P1475 EVAP Emission Control LDP Circuit Malfunction/Signal Circuit Open

P1476 EVAP Emission Control LDP Circuit Malfunction/Insufficient Vacuum
P1477 EVAP Emission Control LDP Circuit Malfunction
P1500 Fuel Pump Relay Circuit Electrical Malfunction
P1501 Fuel Pump Relay Circuit Short to Ground
P1502 Fuel Pump Relay Circuit Short to B+
P1505 Closed Throttle Position Switch Does Not Close/Open Circuit
P1506 Closed Throttle Position Switch Does Not Open/Short to Ground
P1507 Idle System Learned Value Lower Limit Attained
P1508 Idle System Learned Value Upper Limit Attained
P1512 Intake Manifold Changeover Valve Circuit, Short to B+
P1515 Intake Manifold Changeover Valve Circuit, Short to Ground
P1516 Intake Manifold Changeover Valve Circuit, Open
P1519 Intake Camshaft Control, Bank I Malfunction
P1522 Intake Camshaft Control, Bank 2 Malfunction
P1543 Throttle Actuation Potentiometer Signal Too Low
P1544 Throttle Actuation Potentiometer Signal Too High
P1545 Throttle Position Control Malfunction
P1547 Boost Pressure Control Valve Short to Ground
P1548 Boost Pressure Control Valve Open
P1555 Charge Pressure Upper Limit Exceeded
P1556 Charge Pressure Negative Deviation
P1557 Charge Pressure Positive Deviation
P1558 Throttle Actuator Electrical Malfunction
P1559 Idle Speed Control Throttle Position Adaptation Malfunction
P1560 Maximum Engine Speed Exceeded
P1564 Idle Speed Control, Throttle Position Low Voltage During Adaptation
P1580 Throttle Actuator (B1) Malfunction
P1582 Idle Adaptation At Limit
P1602 Power Supply (B+) Terminal 30 Low Voltage
P1606 Rough Road Spec Engine Torque ABS-ECU Electrical Malfunction
P1611 MIL Call-up Circuit/Transmission Control Module Short to Ground
P1612 Electronic Control Module Incorrect Coding
P1613 MIL Call-up Circuit Open/Short to B+
P1624 MIL Request Signal Active
P1625 CAN-Bus Implausible Message from Transmission Control
P1626 CAN-Bus Missing Message from Transmission Control
P1640 Internal Control Module (EEPROM) Error
P1640 Internal Control Module (EEPROM) Error
P1681 Control Unit Programming not Finished
P1690 Malfunction Indicator Light Malfunction
P1693 Malfunction Indicator Light Short to B+
P1778 Solenoid EV7 Electrical Malfunction
P1780 Engine Intervention Readable

OBD II TROUBLE CODE EQUIVALENTS

16486 Mass or Volume Air Flow Circuit Low Input
16487 Mass or Volume Air Flow Circuit High Input
16491 Manifold Absolute Pressure or Barometric Pressure Low Input
16492 Manifold Absolute Pressure or Barometric Pressure High Input
16496 Intake Air Temperature Circuit Low Input

16497 Intake Air Temperature Circuit High Input
16500 Engine Coolant Temperature Circuit Range/Performance
16501 Engine Coolant Temperature Circuit Low Input
16502 Oxygen Engine Coolant Temperature Circuit High Input
16504 Throttle Position Sensor A Circuit Malfunction
16505 Throttle/Pedal Position Sensor A Circuit Range/Performance
16506 Throttle/Pedal Position Sensor A Circuit Low Input
16507 Throttle/Pedal Position Sensor A Circuit High Input
16509 Insufficient Coolant Temperature For Closed Loop Fuel Control
16514 Oxygen Sensor Circuit, Bank 1-Sensor 1 Malfunction
16515 Oxygen Sensor Circuit, Bank 1-Sensor 1 Low Voltage
16516 Oxygen Sensor Circuit, Bank 1-Sensor 1 High Voltage
16517 Oxygen Sensor Circuit, Bank 1-Sensor 1 Slow Response
16518 Oxygen Sensor Circuit, Bank 1-Sensor 1 No Activity Detected
16519 Oxygen Sensor Heater Circuit, Bank 1-Sensor 1 Malfunction
16520 Oxygen Sensor Circuit, Bank 1-Sensor 2 Malfunction
16521 Oxygen Sensor Circuit, Bank 1-Sensor 2 Low Voltage
16522 Oxygen Sensor Circuit, Bank 1-Sensor 2 High Voltage
16524 Oxygen Sensor Circuit, Bank 1-Sensor 2 No Activity Detected
16525 Oxygen Sensor Heater Circuit, Bank 1-Sensor 2 Malfunction
16534 Oxygen Sensor Circuit, Bank 2-Sensor 1 Malfunction
16535 Oxygen Sensor Circuit, Bank 2-Sensor 1 Low Voltage
16536 Oxygen Sensor Circuit, Bank 2-Sensor 1 High Voltage
16537 Oxygen Sensor Circuit, Bank 2-Sensor 1 Slow Response
16538 Oxygen Sensor Circuit, Bank 2-Sensor 1 No Activity Detected
16540 Oxygen Sensor Circuit, Bank 2-Sensor 2 Malfunction
16541 Oxygen Sensor Circuit, Bank 2-Sensor 2 Low Voltage
16542 Oxygen Sensor Circuit, Bank 2-Sensor 2 High Voltage
16544 Oxygen Sensor Circuit, Bank 2-Sensor 2 No Activity Detected
16555 Oxygen System Too Lean, Bank 1
16556 Oxygen System Too Rich, Bank 1
16684 Random Multiple Misfire Detected
16685 Cylinder 1 Misfire Detected
16686 Cylinder 2 Misfire Detected
16687 Cylinder 3 Misfire Detected
16688 Cylinder 4 Misfire Detected
16689 Cylinder 5 Misfire Detected
16690 Cylinder 6 Misfire Detected
16705 Ignition Distributor Engine Speed Input Circuit Range/Performance
16706 Ignition /Distributor Engine Speed Input Circuit No Signal
16711 Knock sensor 1 Circuit Low Input
16712 Knock Sensor Circuit, High Input
16716 Knock sensor 2 Circuit Low Input
16716 Knock Sensor Circuit, Low Input
16717 Knock Sensor 2 Circuit, High Input
16725 Camshaft Position Sensor Circuit Range/Performance
16795 Secondary Air Injection System Incorrect Flow Detected
16806 Main Catalyst Efficiency Below Threshold (Bank 1)
16806 Main Catalyst, Bank I Efficiency Below Threshold
16824 Evaporative Emission Control System Malfunction
16825 EVAP Emission Contr. Sys. Incorrect Purge Flow
16826 EVAP Emission Contr. Sys. (Small Leak) Leak Detected

16839 EVAP Emission Contr. Sys. (Gross Leak) Leak Detected
16885 Vehicle Speed Sensor Range/Performance
16890 Idle Control System RPM Lower Than Expected
16891 Idle Control System RPM Higher Than Expected
16894 Closed Throttle Position Switch Malfunction
16944 System Voltage Malfunction
16946 System Voltage Low Voltage
16947 System Voltage High Voltage
16985 Internal Contr. Module Memory Check Sum Error
16988 Internal Contr. Module Random Access Memory (RAM) Error
16989 Internal Control Module Read Only Memory (ROM) Error
17091 Transmission Range Sensor Circuit Low Input
17092 Transmission Range Sensor Circuit High Input
17099 Input Turbine Speed Sensor Circuit Malfunction
17106 Output Speed Sensor Circuit No Signal
17109 Engine Speed Input Circuit Malfunction
17132 Pressure Control Solenoid Electrical
17137 Shift Solenoid A Electrical
17142 Shift Solenoid B Electrical
17147 Shift Solenoid C Electrical
17152 Shift Solenoid D Electrical
17157 Shift Solenoid E Electrical
16684 Random/Multiple Cylinder Misfire Detected
16685 Cylinder I Misfire Detected
16686 Cylinder 2 Misfire Detected
17510 Oxygen Sensor Heating Circuit, Bank 1-Sensor 1 Short to B+
17513 Oxygen Sensor Heating Circuit, Bank 1-Sensor 2 Short to B+
17515 Oxygen Sensor Heating Circuit, Bank 2-Sensor I Short to B+
17518 Oxygen Sensor Heating Circuit, Bank 2-Sensor 2 Short to B+
17535 Long Term Fuel Trim Multiplicative, Bank 1 System Too Rich
17536 Long Term Fuel Trim Multiplicative, Bank 1 System Too Lean
17537 Long Term Fuel Trim Multiplicative, Bank2 System too Rich
17538 Long Term Fuel Trim Multiplicative, Bank2 System too Lean
17544 Long Term Fuel Trim Additive, Bank 1 System Too Lean
17545 Long Term Fuel Trim Additive, Bank 1 System Too Rich
17546 Long Term Fuel Trim Additive Fuel, Bank I System too Lean
17547 Long Term Fuel Trim Additive Fuel, Bank 1 System too Rich
17549 Load Calculation Cross Check Range/Performance
17584 Oxygen Correction Behind Catalyst, B1 Limit Attained
17585 Oxygen Correction Behind Catalyst. 82 Limit Attained
17604 Oxygen Sensor Heater Circuit, Bank 1-Sensor 1 Electrical Malfunction
17605 Oxygen Sensor Heater Circuit, Bank 2-Sensor I Electrical Malfunction
17606 Oxygen Sensor Heater Circuit, BankI-Sensor2 Electrical Malfunction
17606 Oxygen Sensor Heater Circuit, Bank 1-Sensor 2 Electrical Malfunction
17607 Oxygen Sensor Heater Circuit, Bank 2-Sensor 2 Electrical Malfunction
17609 Cylinder 1, Fuel Injection Circuit Electrical Malfunction

17610 Cylinder 2, Fuel Injection Circuit Electrical Malfunction
17611 Cylinder 3, Fuel Injection Circuit Electrical Malfunction
17612 Cylinder 4, Fuel Injection Circuit Electrical Malfunction
17613 Cylinder 5, Fuel Injection Circuit Electrical Malfunction
17614 Cylinder 6, Fuel Injection Circuit Electrical Malfunction
17615 Cylinder 7, Fuel Injection Circuit Electrical Malfunction
17616 Cylinder 8, Fuel Injection Circuit Electrical Malfunction
17621 Cylinder I-Fuel Injection Circuit Short to B+
17622 Cylinder 2-Fuel Injection Circuit Short to B+
17623 Cylinder 3 Fuel Injection Circuit Short to B+
17624 Cylinder 4 Fuel Injection Circuit Short to B+
17625 Cylinder 5 Fuel Injection Circuit Short to B+
17626 Cylinder 6 Fuel Injection Circuit Short to B+
17627 Cylinder 7, Fuel Injection Circuit Short to B+
17628 Cylinder 8, Fuel Injection Circuit Short to B+
17633 Cylinder I Fuel Injection Circuit Short to Ground
17634 Cylinder 2 Fuel Injection Circuit Short to Ground
17635 Cylinder 3 Fuel Injection Circuit Short to Ground
17636 Cylinder 4 Fuel Injection Circuit Short to Ground
17637 Cylinder 5 Fuel Injection Circuit Short to Ground
17638 Cylinder 6 Fuel Injection Circuit Short to Ground
17645 Cylinder I Fuel Injection Circuit Open Circuit
17646 Cylinder 2 Fuel Injection Circuit Open Circuit
17647 Cylinder 3 Fuel Injection Circuit Open Circuit
17648 Cylinder 4 Fuel Injection Circuit Open Circuit
17649 Cylinder 5 Fuel Injection Circuit Open Circuit
17650 Cylinder 6 Fuel Injection Circuit Open Circuit
17658 Fuel Level Too Low
17688 Fuel Injection Air Control Valve Circuit Flow too Low
17691 Fuel Injection Air Control Valve Circuit Electrical Malfunction
17733 Cylinder I Knock Control Limit Attained
17734 Cylinder 2 Knock Control Limit Attained
17735 Cylinder 3 Knock Control Limit Attained
17736 Cylinder 4 Knock Control Limit Attained
17737 Cylinder 5 Knock Control Limit Attained
17738 Cylinder 6 Knock Control Limit Attained
17739 Cylinder 7, Knock Control Limit Attained
17740 Cylinder 8, Knock Control Limit Attained
17745 Camshaft Position Sensor, Bank 1 Short to Ground
17746 Camshaft Position Sensor, Bank 1 Open Circuit/Short to B+
17954 Boost Pressure Control Valve Short to B+
17794 Internal Control Module Knock Control Circuit Error
17799 Camshaft Position Sensor, Bank 2 Short to Ground
17800 Camshaft Position Sensor, Bank 2 Open Circuit/Short to B+
17818 Tank Ventilation Valve Circuit Short to B+
17818 Tank Ventilation Valve Circuit Short to B+
17828 Secondary Air Injection Module Short To B+
17829 Secondary Air Injection Module Short To Ground
17829 Secondary Air Injection Valve Circuit Short to Ground
17830 Secondary Air Injection System Control Valve Circuit Short To B+
17830 Secondary Air Injection Valve Circuit Short to B+
17833 Tank Vent Valve Short To Ground
17833 Tank Vent Valve Short to Ground
17834 Tank Vent Valve Open
17834 Tank Vent Valve Open
17840 Secondary Air Injection Valve Open
17841 Secondary Air Injection System Pump Relay Circuit Open

17842 Secondary Air Injection System Pump Relay Circuit Short to B+
17843 Secondary Air Injection System Pump Relay Circuit Short to Ground
17844 Secondary Air Injection System Pump Relay Circuit Electrical Malfunction
17858 Secondary Air Injection System Circuit Short To B+
17859 Secondary Air Injection System Circuit Short To Ground
17860 Secondary Air Injection System Open Circuit
17879 EVAP Emission Control LDP Circuit Short to B+
17880 EVAP Emission Control LDP Circuit Short to Ground
17881 EVAP Emission Control LDP Circuit Open Circuit
17883 EVAP Emission Control LDP Circuit Malfunction/Signal Circuit Open
17884 EVAP Emission Control LDP Circuit Malfunction/Insufficient Vacuum
17885 EVAP Emission Control LDP Circuit Malfunction
17908 Fuel Pump Relay Circuit Electrical Malfunction
17909 Fuel Pump Relay Circuit Short to Ground
17910 Fuel Pump Relay Circuit Short to B+
17913 Closed Throttle Position Switch Does Not Close/Open Circuit
17914 Closed Throttle Position Switch Does Not Open/Short to Ground
17915 Idle System Learned Value Lower Limit Attained
17916 Idle System Learned Value Upper Limit Attained
17920 Intake Manifold Changeover Valve Circuit, Short to B+
17923 Intake Manifold Changeover Valve Circuit, Short to Ground
17924 Intake Manifold Changeover Valve Circuit, Open
17927 Intake Camshaft Control, Bank 2 Malfunction
17951 Throttle Actuation Potentiometer Signal Too Low
17952 Throttle Actuation Potentiometer Signal Too High
17953 Throttle Position Control Malfunction
17955 Boost Pressure Control Valve Short to Ground
17956 Boost Pressure Control Valve Open
17963 Charge Pressure Upper Limit Exceeded
17964 Charge Pressure Negative Deviation
17965 Charge Pressure Positive Deviation
17966 Throttle Actuator Electrical Malfunction
17967 Idle Speed Control Throttle Position Adaptation Malfunction
17968 Maximum Engine Speed Exceeded
17972 Idle Speed Control, Throttle Position Low Voltage During Adaptation
17988 Throttle Actuator (B1) Malfunction
17990 Idle Adaptation At Limit
18010 Power Supply (B+) Terminal 30 Low Voltage
18014 Rough Road Spec Engine Torque ABS-ECU Electrical Malfunction
18019 MIL Call-up Circuit/Transmission Control Module Short to Ground
18020 Electronic Control Module Incorrect Coding
18021 MIL Call-up Circuit Open/Short to B+
18032 MIL Request Signal Active
18033 CAN-Bus Implausible Message from Transmission Control
18034 CAN-Bus Missing Message from Transmission Control
18048 Internal Control Module (EEPROM) Error
18048 Internal Control Module (EEPROM) Error
18089 Control Unit Programming not Finished

18098 Malfunction Indicator Light Malfunction
18101 Malfunction Indicator Light Short to B+
18186 Solenoid EV7 Electrical Malfunction
18188 Engine Intervention Readable

Honda

READING CODES

With Scan Tool

Reading the control module memory is on of the first steps in OBD II system diagnostics. This step should be initially performed to determine the general nature of the fault. Subsequent readings will determine if the fault has been cleared.

Reading codes can be performed by any of the methods below:

• Read the control module memory with the Generic Scan Tool (GST)

• Read the control module memory with the vehicle manufacturer's specific tester

To read the fault codes, connect the scan tool or tester according to the manufacturer's instructions. Follow the manufacturer's specified procedure for reading the codes.

Without Scan Tool

Honda also provides a way of reading OBD II trouble code equivalents using a service connector and viewing the MIL. This method is similar to the flash codes from non-OBD II vehicles.

To read codes, plug the service connector into the service check connector and turn the ignition on. The MIL will flash any stored trouble codes.

CLEARING CODES

Control module reset procedures are a very important part of OBD II System diagnostics. This step should be done at the end of any fault code repair and at the end of any driveability repair.

Clearing codes can be performed by any of the methods below:

• Clear the control module memory with the Generic Scan Tool (GST)

• Clear the control module memory with the vehicle manufacturer's specific tester

• Turn the ignition **OFF** and remove the negative battery cable for at least 1 minute.

Removing the negative battery cable may cause other systems in the vehicle to loose their memory. Prior to removing the cable, ensure you have the proper reset codes for radios and alarms.

➡ **The MIL will may also be de-activated for some codes if the vehicle completes three consecutive trips without a fault detected with vehicle conditions similar to those present during the fault.**

OBD II TROUBLE CODES

P0100 Mass or Volume Air Flow Circuit Malfunction
P0101 Mass or Volume Air Flow Circuit Range/Performance Problem
P0102 Mass or Volume Air Flow Circuit Low Input
P0103 Mass or Volume Air Flow Circuit High Input

P0104 Mass or Volume Air Flow Circuit Intermittent
P0105 Manifold Absolute Pressure/Barometric Pressure Circuit Malfunction
P0106 Manifold Absolute Pressure/Barometric Pressure Circuit Range/Performance Problem
P0107 Manifold Absolute Pressure/Barometric Pressure Circuit Low Input
P0108 Manifold Absolute Pressure/Barometric Pressure Circuit High Input
P0109 Manifold Absolute Pressure/Barometric Pressure Circuit Intermittent
P0110 Intake Air Temperature Circuit Malfunction
P0111 Intake Air Temperature Circuit Range/Performance Problem
P0112 Intake Air Temperature Circuit Low Input
P0113 Intake Air Temperature Circuit High Input
P0114 Intake Air Temperature Circuit Intermittent
P0115 Engine Coolant Temperature Circuit Malfunction
P0116 Engine Coolant Temperature Circuit Range/Performance Problem
P0117 Engine Coolant Temperature Circuit Low Input
P0118 Engine Coolant Temperature Circuit High Input
P0119 Engine Coolant Temperature Circuit Intermittent
P0120 Throttle/Pedal Position Sensor/Switch "A" Circuit Malfunction
P0121 Throttle/Pedal Position Sensor/Switch "A" Circuit Range/Performance Problem
P0122 Throttle/Pedal Position Sensor/Switch "A" Circuit Low Input
P0123 Throttle/Pedal Position Sensor/Switch "A" Circuit High Input
P0124 Throttle/Pedal Position Sensor/Switch "A" Circuit Intermittent
P0125 Insufficient Coolant Temperature For Closed Loop Fuel Control
P0126 Insufficient Coolant Temperature For Stable Operation
P0130 O2 Circuit Malfunction (Bank no. 1 Sensor no. 1)
P0131 O2 Sensor Circuit Low Voltage (Bank no. 1 Sensor no. 1)
P0132 O2 Sensor Circuit High Voltage (Bank no. 1 Sensor no. 1)
P0133 O2 Sensor Circuit Slow Response (Bank no. 1 Sensor no. 1)
P0134 O2 Sensor Circuit No Activity Detected (Bank no. 1 Sensor no. 1)
P0135 O2 Sensor Heater Circuit Malfunction (Bank no. 1 Sensor no. 1)
P0136 O2 Sensor Circuit Malfunction (Bank no. 1 Sensor no. 2)
P0137 O2 Sensor Circuit Low Voltage (Bank no. 1 Sensor no. 2)
P0138 O2 Sensor Circuit High Voltage (Bank no. 1 Sensor no. 2)
P0139 O2 Sensor Circuit Slow Response (Bank no. 1 Sensor no. 2)
P0140 O2 Sensor Circuit No Activity Detected (Bank no. 1 Sensor no. 2)
P0141 O2 Sensor Heater Circuit Malfunction (Bank no. 1 Sensor no. 2)
P0142 O2 Sensor Circuit Malfunction (Bank no. 1 Sensor no. 3)
P0143 O2 Sensor Circuit Low Voltage (Bank no. 1 Sensor no. 3)
P0144 O2 Sensor Circuit High Voltage (Bank no. 1 Sensor no. 3)
P0145 O2 Sensor Circuit Slow Response (Bank no. 1 Sensor no. 3)

P0146 02 Sensor Circuit No Activity Detected (Bank no. 1 Sensor no. 3)

P0147 02 Sensor Heater Circuit Malfunction (Bank no. 1 Sensor no. 3)

P0150 02 Sensor Circuit Malfunction (Bank no. 2 Sensor no. 1)

P0151 02 Sensor Circuit Low Voltage (Bank no. 2 Sensor no. 1)

P0152 02 Sensor Circuit High Voltage (Bank no. 2 Sensor no. 1)

P0153 02 Sensor Circuit Slow Response (Bank no. 2 Sensor no. 1)

P0154 02 Sensor Circuit No Activity Detected (Bank no. 2 Sensor no. 1)

P0155 02 Sensor Heater Circuit Malfunction (Bank no. 2 Sensor no. 1)

P0156 02 Sensor Circuit Malfunction (Bank no. 2 Sensor no. 2)

P0157 02 Sensor Circuit Low Voltage (Bank no. 2 Sensor no. 2)

P0158 02 Sensor Circuit High Voltage (Bank no. 2 Sensor no. 2)

P0159 02 Sensor Circuit Slow Response (Bank no. 2 Sensor no. 2)

P0160 02 Sensor Circuit No Activity Detected (Bank no. 2 Sensor no. 2)

P0161 02 Sensor Heater Circuit Malfunction (Bank no. 2 Sensor no. 2)

P0162 02 Sensor Circuit Malfunction (Bank no. 2 Sensor no. 3)

P0163 02 Sensor Circuit Low Voltage (Bank no. 2 Sensor no. 3)

P0164 02 Sensor Circuit High Voltage (Bank no. 2 Sensor no. 3)

P0165 02 Sensor Circuit Slow Response (Bank no. 2 Sensor no. 3)

P0166 02 Sensor Circuit No Activity Detected (Bank no. 2 Sensor no. 3)

P0167 02 Sensor Heater Circuit Malfunction (Bank no. 2 Sensor no. 3)

P0170 Fuel Trim Malfunction (Bank no. 1)

P0171 System Too Lean (Bank no. 1)

P0172 System Too Rich (Bank no. 1)

P0173 Fuel Trim Malfunction (Bank no. 2)

P0174 System Too Lean (Bank no. 2)

P0175 System Too Rich (Bank no. 2)

P0176 Fuel Composition Sensor Circuit Malfunction

P0177 Fuel Composition Sensor Circuit Range/Performance

P0178 Fuel Composition Sensor Circuit Low Input

P0179 Fuel Composition Sensor Circuit High Input

P0180 Fuel Temperature Sensor "A" Circuit Malfunction

P0181 Fuel Temperature Sensor "A" Circuit Range/Performance

P0182 Fuel Temperature Sensor "A" Circuit Low Input

P0183 Fuel Temperature Sensor "A" Circuit High Input

P0184 Fuel Temperature Sensor "A" Circuit Intermittent

P0185 Fuel Temperature Sensor "B" Circuit Malfunction

P0186 Fuel Temperature Sensor "B" Circuit Range/Performance

P0187 Fuel Temperature Sensor "B" Circuit Low Input

P0188 Fuel Temperature Sensor "B" Circuit High Input

P0189 Fuel Temperature Sensor "B" Circuit Intermittent

P0190 Fuel Rail Pressure Sensor Circuit Malfunction

P0191 Fuel Rail Pressure Sensor Circuit Range/Performance

P0192 Fuel Rail Pressure Sensor Circuit Low Input

P0193 Fuel Rail Pressure Sensor Circuit High Input

P0194 Fuel Rail Pressure Sensor Circuit Intermittent

P0195 Engine Oil Temperature Sensor Malfunction

P0196 Engine Oil Temperature Sensor Range/Performance

P0197 Engine Oil Temperature Sensor Low

P0198 Engine Oil Temperature Sensor High

P0199 Engine Oil Temperature Sensor Intermittent

P0200 Injector Circuit Malfunction

P0201 Injector Circuit Malfunction—Cylinder no. 1

P0202 Injector Circuit Malfunction—Cylinder no. 2

P0203 Injector Circuit Malfunction—Cylinder no. 3

P0204 Injector Circuit Malfunction—Cylinder no. 4

P0205 Injector Circuit Malfunction—Cylinder no. 5

P0206 Injector Circuit Malfunction—Cylinder no. 6

P0207 Injector Circuit Malfunction—Cylinder no. 7

P0208 Injector Circuit Malfunction—Cylinder no. 8

P0209 Injector Circuit Malfunction—Cylinder no. 9

P0210 Injector Circuit Malfunction—Cylinder no. 10

P0211 Injector Circuit Malfunction—Cylinder no. 11

P0212 Injector Circuit Malfunction—Cylinder no. 12

P0213 Cold Start Injector no. 1 Malfunction

P0214 Cold Start Injector no. 2 Malfunction

P0215 Engine Shutoff Solenoid Malfunction

P0216 Injection Timing Control Circuit Malfunction

P0217 Engine Over Temperature Condition

P0218 Transmission Over Temperature Condition

P0219 Engine Over Speed Condition

P0220 Throttle/Pedal Position Sensor/Switch "B" Circuit Malfunction

P0221 Throttle/Pedal Position Sensor/Switch "B" Circuit Range/Performance Problem

P0222 Throttle/Pedal Position Sensor/Switch "B" Circuit Low Input

P0223 Throttle/Pedal Position Sensor/Switch "B" Circuit High Input

P0224 Throttle/Pedal Position Sensor/Switch "B" Circuit Intermittent

P0225 Throttle/Pedal Position Sensor/Switch "C" Circuit Malfunction

P0226 Throttle/Pedal Position Sensor/Switch "C" Circuit Range/Performance Problem

P0227 Throttle/Pedal Position Sensor/Switch "C" Circuit Low Input

P0228 Throttle/Pedal Position Sensor/Switch "C" Circuit High Input

P0229 Throttle/Pedal Position Sensor/Switch "C" Circuit Intermittent

P0230 Fuel Pump Primary Circuit Malfunction

P0231 Fuel Pump Secondary Circuit Low

P0232 Fuel Pump Secondary Circuit High

P0233 Fuel Pump Secondary Circuit Intermittent

P0234 Engine Over Boost Condition

P0261 Cylinder no. 1 Injector Circuit Low

P0262 Cylinder no. 1 Injector Circuit High

P0263 Cylinder no. 1 Contribution/Balance Fault

P0264 Cylinder no. 2 Injector Circuit Low

P0265 Cylinder no. 2 Injector Circuit High

P0266 Cylinder no. 2 Contribution/Balance Fault

P0267 Cylinder no. 3 Injector Circuit Low

P0268 Cylinder no. 3 Injector Circuit High

P0269 Cylinder no. 3 Contribution/Balance Fault

P0270 Cylinder no. 4 Injector Circuit Low

P0271 Cylinder no. 4 Injector Circuit High

P0272 Cylinder no. 4 Contribution/Balance Fault

P0273 Cylinder no. 5 Injector Circuit Low

P0274 Cylinder no. 5 Injector Circuit High
P0275 Cylinder no. 5 Contribution/Balance Fault
P0276 Cylinder no. 6 Injector Circuit Low
P0277 Cylinder no. 6 Injector Circuit High
P0278 Cylinder no. 6 Contribution/Balance Fault
P0279 Cylinder no. 7 Injector Circuit Low
P0280 Cylinder no. 7 Injector Circuit High
P0281 Cylinder no. 7 Contribution/Balance Fault
P0282 Cylinder no. 8 Injector Circuit Low
P0283 Cylinder no. 8 Injector Circuit High
P0284 Cylinder no. 8 Contribution/Balance Fault
P0285 Cylinder no. 9 Injector Circuit Low
P0286 Cylinder no. 9 Injector Circuit High
P0287 Cylinder no. 9 Contribution/Balance Fault
P0288 Cylinder no. 10 Injector Circuit Low
P0289 Cylinder no. 10 Injector Circuit High
P0290 Cylinder no. 10 Contribution/Balance Fault
P0291 Cylinder no. 11 Injector Circuit Low
P0292 Cylinder no. 11 Injector Circuit High
P0293 Cylinder no. 11 Contribution/Balance Fault
P0294 Cylinder no. 12 Injector Circuit Low
P0295 Cylinder no. 12 Injector Circuit High
P0296 Cylinder no. 12 Contribution/Balance Fault
P0300 Random/Multiple Cylinder Misfire Detected
P0301 Cylinder no. 1—Misfire Detected
P0302 Cylinder no. 2—Misfire Detected
P0303 Cylinder no. 3—Misfire Detected
P0304 Cylinder no. 4—Misfire Detected
P0305 Cylinder no. 5—Misfire Detected
P0306 Cylinder no. 6—Misfire Detected
P0307 Cylinder no. 7—Misfire Detected
P0308 Cylinder no. 8—Misfire Detected
P0309 Cylinder no. 9—Misfire Detected
P0310 Cylinder no. 10—Misfire Detected
P0311 Cylinder no. 11—Misfire Detected
P0312 Cylinder no. 12—Misfire Detected
P0320 Ignition/Distributor Engine Speed Input Circuit Malfunction
P0321 Ignition/Distributor Engine Speed Input Circuit Range/Performance
P0322 Ignition/Distributor Engine Speed Input Circuit No Signal
P0323 Ignition/Distributor Engine Speed Input Circuit Intermittent
P0325 Knock Sensor no. 1—Circuit Malfunction (Bank no. 1 or Single Sensor)
P0326 Knock Sensor no. 1—Circuit Range/Performance (Bank no. 1 or Single Sensor)
P0327 Knock Sensor no. 1—Circuit Low Input (Bank no. 1 or Single Sensor)
P0328 Knock Sensor no. 1—Circuit High Input (Bank no. 1 or Single Sensor)
P0329 Knock Sensor no. 1—Circuit Input Intermittent (Bank no. 1 or Single Sensor)
P0330 Knock Sensor no. 2—Circuit Malfunction (Bank no. 2)
P0331 Knock Sensor no. 2—Circuit Range/Performance (Bank no. 2)
P0332 Knock Sensor no. 2—Circuit Low Input (Bank no. 2)
P0333 Knock Sensor no. 2—Circuit High Input (Bank no. 2)
P0334 Knock Sensor no. 2—Circuit Input Intermittent (Bank no. 2)
P0335 Crankshaft Position Sensor "A" Circuit Malfunction

P0336 Crankshaft Position Sensor "A" Circuit Range/Performance
P0337 Crankshaft Position Sensor "A" Circuit Low Input
P0338 Crankshaft Position Sensor "A" Circuit High Input
P0339 Crankshaft Position Sensor "A" Circuit Intermittent
P0340 Camshaft Position Sensor Circuit Malfunction
P0341 Camshaft Position Sensor Circuit Range/Performance
P0342 Camshaft Position Sensor Circuit Low Input
P0343 Camshaft Position Sensor Circuit High Input
P0344 Camshaft Position Sensor Circuit Intermittent
P0350 Ignition Coil Primary/Secondary Circuit Malfunction
P0351 Ignition Coil "A" Primary/Secondary Circuit Malfunction
P0352 Ignition Coil "B" Primary/Secondary Circuit Malfunction
P0353 Ignition Coil "C" Primary/Secondary Circuit Malfunction
P0354 Ignition Coil "D" Primary/Secondary Circuit Malfunction
P0355 Ignition Coil "E" Primary/Secondary Circuit Malfunction
P0356 Ignition Coil "F" Primary/Secondary Circuit Malfunction
P0357 Ignition Coil "G" Primary/Secondary Circuit Malfunction
P0358 Ignition Coil "H" Primary/Secondary Circuit Malfunction
P0359 Ignition Coil "I" Primary/Secondary Circuit Malfunction
P0360 Ignition Coil "J" Primary/Secondary Circuit Malfunction
P0361 Ignition Coil "K" Primary/Secondary Circuit Malfunction
P0362 Ignition Coil "L" Primary/Secondary Circuit Malfunction
P0370 Timing Reference High Resolution Signal "A" Malfunction
P0371 Timing Reference High Resolution Signal "A" Too Many Pulses
P0372 Timing Reference High Resolution Signal "A" Too Few Pulses
P0373 Timing Reference High Resolution Signal "A" Intermittent/Erratic Pulses
P0374 Timing Reference High Resolution Signal "A" No Pulses
P0375 Timing Reference High Resolution Signal "B" Malfunction
P0376 Timing Reference High Resolution Signal "B" Too Many Pulses
P0377 Timing Reference High Resolution Signal "B" Too Few Pulses
P0378 Timing Reference High Resolution Signal "B" Intermittent/Erratic Pulses
P0379 Timing Reference High Resolution Signal "B" No Pulses
P0380 Glow Plug/Heater Circuit "A" Malfunction
P0381 Glow Plug/Heater Indicator Circuit Malfunction
P0382 Glow Plug/Heater Circuit "B" Malfunction
P0385 Crankshaft Position Sensor "B" Circuit Malfunction
P0386 Crankshaft Position Sensor "B" Circuit Range/Performance
P0387 Crankshaft Position Sensor "B" Circuit Low Input
P0388 Crankshaft Position Sensor "B" Circuit High Input
P0389 Crankshaft Position Sensor "B" Circuit Intermittent
P0400 Exhaust Gas Recirculation Flow Malfunction
P0401 Exhaust Gas Recirculation Flow Insufficient Detected
P0402 Exhaust Gas Recirculation Flow Excessive Detected
P0403 Exhaust Gas Recirculation Circuit Malfunction
P0404 Exhaust Gas Recirculation Circuit Range/Performance
P0405 Exhaust Gas Recirculation Sensor "A" Circuit Low
P0406 Exhaust Gas Recirculation Sensor "A" Circuit High
P0407 Exhaust Gas Recirculation Sensor "B" Circuit Low
P0408 Exhaust Gas Recirculation Sensor "B" Circuit High
P0410 Secondary Air Injection System Malfunction
P0411 Secondary Air Injection System Incorrect Flow Detected

P0412 Secondary Air Injection System Switching Valve "A" Circuit Malfunction

P0413 Secondary Air Injection System Switching Valve "A" Circuit Open

P0414 Secondary Air Injection System Switching Valve "A" Circuit Shorted

P0415 Secondary Air Injection System Switching Valve "B" Circuit Malfunction

P0416 Secondary Air Injection System Switching Valve "B" Circuit Open

P0417 Secondary Air Injection System Switching Valve "B" Circuit Shorted

P0418 Secondary Air Injection System Relay "A" Circuit Malfunction

P0419 Secondary Air Injection System Relay "B" Circuit Malfunction

P0420 Catalyst System Efficiency Below Threshold (Bank no. 1)

P0421 Warm Up Catalyst Efficiency Below Threshold (Bank no. 1)

P0422 Main Catalyst Efficiency Below Threshold (Bank no. 1)

P0423 Heated Catalyst Efficiency Below Threshold (Bank no. 1)

P0424 Heated Catalyst Temperature Below Threshold (Bank no. 1)

P0430 Catalyst System Efficiency Below Threshold (Bank no. 2)

P0431 Warm Up Catalyst Efficiency Below Threshold (Bank no. 2)

P0432 Main Catalyst Efficiency Below Threshold (Bank no. 2)

P0433 Heated Catalyst Efficiency Below Threshold (Bank no. 2)

P0434 Heated Catalyst Temperature Below Threshold (Bank no. 2)

P0440 Evaporative Emission Control System Malfunction

P0441 Evaporative Emission Control System Incorrect Purge Flow

P0442 Evaporative Emission Control System Leak Detected (Small Leak)

P0443 Evaporative Emission Control System Purge Control Valve Circuit Malfunction

P0444 Evaporative Emission Control System Purge Control Valve Circuit Open

P0445 Evaporative Emission Control System Purge Control Valve Circuit Shorted

P0446 Evaporative Emission Control System Vent Control Circuit Malfunction

P0447 Evaporative Emission Control System Vent Control Circuit Open

P0448 Evaporative Emission Control System Vent Control Circuit Shorted

P0449 Evaporative Emission Control System Vent Valve/Solenoid Circuit Malfunction

P0450 Evaporative Emission Control System Pressure Sensor Malfunction

P0451 Evaporative Emission Control System Pressure Sensor Range/Performance

P0452 Evaporative Emission Control System Pressure Sensor Low Input

P0453 Evaporative Emission Control System Pressure Sensor High Input

P0454 Evaporative Emission Control System Pressure Sensor Intermittent

P0455 Evaporative Emission Control System Leak Detected (Gross Leak)

P0460 Fuel Level Sensor Circuit Malfunction

P0461 Fuel Level Sensor Circuit Range/Performance

P0462 Fuel Level Sensor Circuit Low Input

P0463 Fuel Level Sensor Circuit High Input

P0464 Fuel Level Sensor Circuit Intermittent

P0465 Purge Flow Sensor Circuit Malfunction

P0466 Purge Flow Sensor Circuit Range/Performance

P0467 Purge Flow Sensor Circuit Low Input

P0468 Purge Flow Sensor Circuit High Input

P0469 Purge Flow Sensor Circuit Intermittent

P0470 Exhaust Pressure Sensor Malfunction

P0471 Exhaust Pressure Sensor Range/Performance

P0472 Exhaust Pressure Sensor Low

P0473 Exhaust Pressure Sensor High

P0474 Exhaust Pressure Sensor Intermittent

P0475 Exhaust Pressure Control Valve Malfunction

P0476 Exhaust Pressure Control Valve Range/Performance

P0477 Exhaust Pressure Control Valve Low

P0478 Exhaust Pressure Control Valve High

P0479 Exhaust Pressure Control Valve Intermittent

P0480 Cooling Fan no. 1 Control Circuit Malfunction

P0481 Cooling Fan no. 2 Control Circuit Malfunction

P0482 Cooling Fan no. 3 Control Circuit Malfunction

P0483 Cooling Fan Rationality Check Malfunction

P0484 Cooling Fan Circuit Over Current

P0485 Cooling Fan Power/Ground Circuit Malfunction

P0500 Vehicle Speed Sensor Malfunction

P0501 Vehicle Speed Sensor Range/Performance

P0502 Vehicle Speed Sensor Circuit Low Input

P0503 Vehicle Speed Sensor Intermittent/Erratic/High

P0505 Idle Control System Malfunction

P0506 Idle Control System RPM Lower Than Expected

P0507 Idle Control System RPM Higher Than Expected

P0510 Closed Throttle Position Switch Malfunction

P0520 Engine Oil Pressure Sensor/Switch Circuit Malfunction

P0521 Engine Oil Pressure Sensor/Switch Range/Performance

P0522 Engine Oil Pressure Sensor/Switch Low Voltage

P0523 Engine Oil Pressure Sensor/Switch High Voltage

P0530 A/C Refrigerant Pressure Sensor Circuit Malfunction

P0531 A/C Refrigerant Pressure Sensor Circuit Range/Performance

P0532 A/C Refrigerant Pressure Sensor Circuit Low Input

P0533 A/C Refrigerant Pressure Sensor Circuit High Input

P0534 A/C Refrigerant Charge Loss

P0550 Power Steering Pressure Sensor Circuit Malfunction

P0551 Power Steering Pressure Sensor Circuit Range/Performance

P0552 Power Steering Pressure Sensor Circuit Low Input

P0553 Power Steering Pressure Sensor Circuit High Input

P0554 Power Steering Pressure Sensor Circuit Intermittent

P0560 System Voltage Malfunction

P0561 System Voltage Unstable

P0562 System Voltage Low

P0563 System Voltage High

P0565 Cruise Control On Signal Malfunction

P0566 Cruise Control Off Signal Malfunction

P0567 Cruise Control Resume Signal Malfunction

P0568 Cruise Control Set Signal Malfunction

P0569 Cruise Control Coast Signal Malfunction

P0570 Cruise Control Accel Signal Malfunction

P0571 Cruise Control/Brake Switch "A" Circuit Malfunction

P0572 Cruise Control/Brake Switch "A" Circuit Low
P0573 Cruise Control/Brake Switch "A" Circuit High
P0574 Through P0580 Reserved for Cruise Codes
P0600 Serial Communication Link Malfunction
P0601 Internal Control Module Memory Check Sum Error
P0602 Control Module Programming Error
P0603 Internal Control Module Keep Alive Memory (KAM) Error
P0604 Internal Control Module Random Access Memory (RAM) Error
P0605 Internal Control Module Read Only Memory (ROM) Error
P0606 PCM Processor Fault
P0608 Control Module VSS Output "A" Malfunction
P0609 Control Module VSS Output "B" Malfunction
P0620 Generator Control Circuit Malfunction
P0621 Generator Lamp "L" Control Circuit Malfunction
P0622 Generator Field "F" Control Circuit Malfunction
P0650 Malfunction Indicator Lamp (MIL) Control Circuit Malfunction
P0654 Engine RPM Output Circuit Malfunction
P0655 Engine Hot Lamp Output Control Circuit Malfunction
P0656 Fuel Level Output Circuit Malfunction
P0700 Transmission Control System Malfunction
P0701 Transmission Control System Range/Performance
P0702 Transmission Control System Electrical
P0703 Torque Converter/Brake Switch "B" Circuit Malfunction
P0704 Clutch Switch Input Circuit Malfunction
P0705 Transmission Range Sensor Circuit Malfunction (PRNDL Input)
P0706 Transmission Range Sensor Circuit Range/Performance
P0707 Transmission Range Sensor Circuit Low Input
P0708 Transmission Range Sensor Circuit High Input
P0709 Transmission Range Sensor Circuit Intermittent
P0710 Transmission Fluid Temperature Sensor Circuit Malfunction
P0711 Transmission Fluid Temperature Sensor Circuit Range/Performance
P0712 Transmission Fluid Temperature Sensor Circuit Low Input
P0713 Transmission Fluid Temperature Sensor Circuit High Input
P0714 Transmission Fluid Temperature Sensor Circuit Intermittent
P0715 Input/Turbine Speed Sensor Circuit Malfunction
P0716 Input/Turbine Speed Sensor Circuit Range/Performance
P0717 Input/Turbine Speed Sensor Circuit No Signal
P0718 Input/Turbine Speed Sensor Circuit Intermittent
P0719 Torque Converter/Brake Switch "B" Circuit Low
P0720 Output Speed Sensor Circuit Malfunction
P0721 Output Speed Sensor Circuit Range/Performance
P0722 Output Speed Sensor Circuit No Signal
P0723 Output Speed Sensor Circuit Intermittent
P0724 Torque Converter/Brake Switch "B" Circuit High
P0725 Engine Speed Input Circuit Malfunction
P0726 Engine Speed Input Circuit Range/Performance
P0727 Engine Speed Input Circuit No Signal
P0728 Engine Speed Input Circuit Intermittent
P0730 Incorrect Gear Ratio
P0731 Gear no. 1 Incorrect Ratio
P0732 Gear no. 2 Incorrect Ratio
P0733 Gear no. 3 Incorrect Ratio
P0734 Gear no. 4 Incorrect Ratio

P0735 Gear no. 5 Incorrect Ratio
P0736 Reverse Incorrect Ratio
P0740 Torque Converter Clutch Circuit Malfunction
P0741 Torque Converter Clutch Circuit Performance or Stuck Off
P0742 Torque Converter Clutch Circuit Stuck On
P0743 Torque Converter Clutch Circuit Electrical
P0744 Torque Converter Clutch Circuit Intermittent
P0745 Pressure Control Solenoid Malfunction
P0746 Pressure Control Solenoid Performance or Stuck Off
P0747 Pressure Control Solenoid Stuck On
P0748 Pressure Control Solenoid Electrical
P0749 Pressure Control Solenoid Intermittent
P0750 Shift Solenoid "A" Malfunction
P0751 Shift Solenoid "A" Performance or Stuck Off
P0752 Shift Solenoid "A" Stuck On
P0753 Shift Solenoid "A" Electrical
P0754 Shift Solenoid "A" Intermittent
P0755 Shift Solenoid "B" Malfunction
P0756 Shift Solenoid "B" Performance or Stuck Off
P0757 Shift Solenoid "B" Stuck On
P0758 Shift Solenoid "B" Electrical
P0759 Shift Solenoid "B" Intermittent
P0760 Shift Solenoid "C" Malfunction
P0761 Shift Solenoid "C" Performance Or Stuck Off
P0762 Shift Solenoid "C" Stuck On
P0763 Shift Solenoid "C" Electrical
P0764 Shift Solenoid "C" Intermittent
P0765 Shift Solenoid "D" Malfunction
P0766 Shift Solenoid "D" Performance Or Stuck Off
P0767 Shift Solenoid "D" Stuck On
P0768 Shift Solenoid "D" Electrical
P0769 Shift Solenoid "D" Intermittent
P0770 Shift Solenoid "E" Malfunction
P0771 Shift Solenoid "E" Performance Or Stuck Off
P0772 Shift Solenoid "E" Stuck On
P0773 Shift Solenoid "E" Electrical
P0774 Shift Solenoid "E" Intermittent
P0780 Shift Malfunction
P0781 1–2 Shift Malfunction
P0782 2–3 Shift Malfunction
P0783 3–4 Shift Malfunction
P0784 4–5 Shift Malfunction
P0785 Shift/Timing Solenoid Malfunction
P0786 Shift/Timing Solenoid Range/Performance
P0787 Shift/Timing Solenoid Low
P0788 Shift/Timing Solenoid High
P0789 Shift/Timing Solenoid Intermittent
P0790 Normal/Performance Switch Circuit Malfunction
P0801 Reverse Inhibit Control Circuit Malfunction
P0803 1–4 Upshift (Skip Shift) Solenoid Control Circuit Malfunction
P0804 1–4 Upshift (Skip Shift) Lamp Control Circuit Malfunction
P0505 Idle Control System Malfunction
P0700 Automatic Transaxle
P0715 Automatic Transaxle
P0720 Automatic Transaxle
P0730 Automatic Transaxle
P0740 Automatic Transaxle
P0753 Automatic Transaxle

P0758 Automatic Transaxle
P1106 Barometric Pressure Circuit Range/Performance Problem
P1107 Barometric Pressure Circuit Low Input
P1108 Barometric Pressure Circuit High Input
P1121 Throttle Position Lower Than Expected
P1122 Throttle Position Higher Than Expected
P1128 Manifold Absolute Pressure Lower Than Expected
P1129 Manifold Absolute Pressure Higher Than Expected
P1259 VTEC System Malfunction
P1297 Electrical Load Detector Circuit Low Input
P1298 Electrical Load Detector Circuit High Input
P1297 Electrical Load Detector Circuit Low Input
P1298 Electrical Load Detector Circuit High Input
P1336 Crankshaft Speed Fluctuation Sensor Intermittent Interruption
P1337 Crankshaft Speed Fluctuation Sensor No Signal
P1359 Crankshaft Position Top Dead Center Sensor/Cylinder Position Connector Disconnection
P1361 Top Dead Center Sensor Intermittent Interruption
P1362 Top Dead Center Sensor No Signal
P1381 Cylinder Position Sensor Intermittent Interruption
P1382 Cylinder Position Sensor No Signal
P1456 Evaporative Emission Control System Leak Detected (Fuel Tank System)
P1457 Evaporative Emission Control System Leak Detected (EVAP Control Canister Leak)
P1491 EGR Valve Lift Insufficient Detected
P1498 EGR Valve Lift Sensor High Voltage
P1519 Idle Air Control Valve Circuit Failure
P1508 Idle Air Control Valve Circuit Failure
P1607 Powertrain Control Module Internal Circuit Failure A
P1705 Automatic Transaxle
P1706 Automatic Transaxle
P1753 Automatic Transaxle
P1768 Automatic Transaxle
P1790 Automatic Transaxle
P1791 Automatic Transaxle

OBD II TROUBLE CODE EQUIVALENTS

1 O2 Sensor Circuit High Voltage (Bank no. 1 Sensor no. 1)
1 O2 Sensor Circuit Low Voltage (Bank no. 1 Sensor no. 1)
3 Manifold Absolute Pressure/Barometric Pressure Circuit Low Input
3 Manifold Absolute Pressure/Barometric Pressure Circuit High Input
4 Crankshaft Position Sensor "A" Circuit Malfunction
4 Crankshaft Position Sensor "A" Circuit Range/Performance
5 Manifold Absolute Pressure Higher Than Expected
5 Manifold Absolute Pressure Lower Than Expected
6 Engine Coolant Temperature Circuit High Input
6 Engine Coolant Temperature Circuit Low Input
7 Throttle Position Higher Than Expected
7 Throttle Position Lower Than Expected
7 Throttle/Pedal Position Sensor/Switch "A" Circuit High Input
7 Throttle/Pedal Position Sensor/Switch "A" Circuit Low Input
8 Crankshaft Position Top Dead Center Sensor/Cylinder Position Connector Disconnection
8 Top Dead Center Sensor Intermittent Interruption
8 Top Dead Center Sensor No Signal
9 Cylinder Position Sensor Intermittent Interruption
9 Cylinder Position Sensor No Signal

10 Intake Air Temperature Circuit High Input
10 Intake Air Temperature Circuit Low Input
12 EGR Valve Lift Insufficient Detected
12 EGR Valve Lift Sensor High Voltage
13 Barometric Pressure Circuit High Input
13 Barometric Pressure Circuit Low Input
13 Barometric Pressure Circuit Range/Performance Problem
14 Idle Air Control Valve Circuit Failure
14 Idle Air Control Valve Circuit Failure
14 Idle Control System Malfunction
20 Electrical Load Detector Circuit High Input
20 Electrical Load Detector Circuit High Input
20 Electrical Load Detector Circuit Low Input
20 Electrical Load Detector Circuit Low Input
22 VTEC System Malfunction
23 Knock Sensor no. 1—Circuit Malfunction (Bank no. 1 or Single Sensor)
80 Exhaust Gas Recirculation Flow Insufficient Detected
41 O2 Sensor Heater Circuit Malfunction (Bank no. 1 Sensor no. 1)
45 System Too Lean (Bank no. 1)
45 System Too Rich (Bank no. 1)
54 Crankshaft Speed Fluctuation Sensor Intermittent Interruption
54 Crankshaft Speed Fluctuation Sensor No Signal
61 O2 Sensor Circuit Slow Response (Bank no. 1 Sensor no. 1)
63 O2 Sensor Circuit High Voltage (Bank no. 1 Sensor no. 2)
63 O2 Sensor Circuit Low Voltage (Bank no. 1 Sensor no. 2)
63 O2 Sensor Circuit Slow Response (Bank no. 1 Sensor no. 2)
65 O2 Sensor Heater Circuit Malfunction (Bank no. 1 Sensor no. 2)
67 Catalyst System Efficiency Below Threshold (Bank no. 1)
70 Automatic Transaxle
70 Transmission Control System Malfunction
70 Input/Turbine Speed Sensor Circuit Malfunction
70 Output Speed Sensor Circuit Malfunction
70 Incorrect Gear Ratio
70 Torque Converter Clutch Circuit Malfunction
70 Shift Solenoid "A" Electrical
70 Shift Solenoid "B" Electrical
71 Cylinder no. 1—Misfire Detected
72 Cylinder no. 2—Misfire Detected
73 Cylinder no. 3—Misfire Detected
74 Cylinder no. 4—Misfire Detected
86 Engine Coolant Temperature Circuit Range/Performance Problem
90 Evaporative Emission Control System Leak Detected (EVAP Control Canister Leak)
90 Evaporative Emission Control System Leak Detected (Fuel Tank System)
91 Evaporative Emission Control System Pressure Sensor Low Input
91 Evaporative Emission Control System Pressure Sensor High Input

Infiniti

READING CODES

Reading the control module memory is on of the first steps in OBD II system diagnostics. This step should be initially performed to determine the general nature of the fault. Subsequent readings will determine if the fault has been cleared.

Reading codes can be performed by any of the methods below:
- Read the control module memory with the Generic Scan Tool (GST)
- Read the control module memory with the vehicle manufacturer's specific tester

To read the fault codes, connect the scan tool or tester according to the manufacturer's instructions. Follow the manufacturer's specified procedure for reading the codes.

CLEARING CODES

Control module reset procedures are a very important part of OBD II System diagnostics. This step should be done at the end of any fault code repair and at the end of any driveability repair.

Clearing codes can be performed by any of the methods below:
- Clear the control module memory with the Generic Scan Tool (GST)
- Clear the control module memory with the vehicle manufacturer's specific tester

➡**The MIL will may also be de-activated for some codes if the vehicle completes three consecutive trips without a fault detected with vehicle conditions similar to those present during the fault.**

OBD II TROUBLE CODES

P0000 No Self Diagnostic Failure Indicated
P0100 Mass or Volume Air Flow Circuit Malfunction
P0101 Mass or Volume Air Flow Circuit Range/Performance Problem
P0102 Mass or Volume Air Flow Circuit Low Input
P0103 Mass or Volume Air Flow Circuit High Input
P0104 Mass or Volume Air Flow Circuit Intermittent
P0105 Manifold Absolute Pressure/Barometric Pressure Circuit Malfunction
P0106 Manifold Absolute Pressure/Barometric Pressure Circuit Range/Performance Problem
P0107 Manifold Absolute Pressure/Barometric Pressure Circuit Low Input
P0108 Manifold Absolute Pressure/Barometric Pressure Circuit High Input
P0109 Manifold Absolute Pressure/Barometric Pressure Circuit Intermittent
P0110 Intake Air Temperature Circuit Malfunction
P0111 Intake Air Temperature Circuit Range/Performance Problem
P0112 Intake Air Temperature Circuit Low Input
P0113 Intake Air Temperature Circuit High Input
P0114 Intake Air Temperature Circuit Intermittent
P0115 Engine Coolant Temperature Circuit Malfunction
P0116 Engine Coolant Temperature Circuit Range/Performance Problem
P0117 Engine Coolant Temperature Circuit Low Input
P0118 Engine Coolant Temperature Circuit High Input
P0119 Engine Coolant Temperature Circuit Intermittent
P0120 Throttle/Pedal Position Sensor/Switch "A" Circuit Malfunction
P0121 Throttle/Pedal Position Sensor/Switch "A" Circuit Range/Performance Problem
P0122 Throttle/Pedal Position Sensor/Switch "A" Circuit Low Input
P0123 Throttle/Pedal Position Sensor/Switch "A" Circuit High Input
P0124 Throttle/Pedal Position Sensor/Switch "A" Circuit Intermittent
P0125 Insufficient Coolant Temperature For Closed Loop Fuel Control
P0126 Insufficient Coolant Temperature For Stable Operation
P0130 O2 Circuit Malfunction (Bank no. 1 Sensor no. 1)
P0131 O2 Sensor Circuit Low Voltage (Bank no. 1 Sensor no. 1)
P0132 O2 Sensor Circuit High Voltage (Bank no. 1 Sensor no. 1)
P0133 O2 Sensor Circuit Slow Response (Bank no. 1 Sensor no. 1)
P0134 O2 Sensor Circuit No Activity Detected (Bank no. 1 Sensor no. 1)
P0135 O2 Sensor Heater Circuit Malfunction (Bank no. 1 Sensor no. 1)
P0136 O2 Sensor Circuit Malfunction (Bank no. 1 Sensor no. 2)
P0137 O2 Sensor Circuit Low Voltage (Bank no. 1 Sensor no. 2)
P0138 O2 Sensor Circuit High Voltage (Bank no. 1 Sensor no. 2)
P0139 O2 Sensor Circuit Slow Response (Bank no. 1 Sensor no. 2)
P0140 O2 Sensor Circuit No Activity Detected (Bank no. 1 Sensor no. 2)
P0141 O2 Sensor Heater Circuit Malfunction (Bank no. 1 Sensor no. 2)
P0142 O2 Sensor Circuit Malfunction (Bank no. 1 Sensor no. 3)
P0143 O2 Sensor Circuit Low Voltage (Bank no. 1 Sensor no. 3)
P0144 O2 Sensor Circuit High Voltage (Bank no. 1 Sensor no. 3)
P0145 O2 Sensor Circuit Slow Response (Bank no. 1 Sensor no. 3)
P0146 O2 Sensor Circuit No Activity Detected (Bank no. 1 Sensor no. 3)
P0147 O2 Sensor Heater Circuit Malfunction (Bank no. 1 Sensor no. 3)
P0150 O2 Sensor Circuit Malfunction (Bank no. 2 Sensor no. 1)
P0151 O2 Sensor Circuit Low Voltage (Bank no. 2 Sensor no. 1)
P0152 O2 Sensor Circuit High Voltage (Bank no. 2 Sensor no. 1)
P0153 O2 Sensor Circuit Slow Response (Bank no. 2 Sensor no. 1)
P0154 O2 Sensor Circuit No Activity Detected (Bank no. 2 Sensor no. 1)
P0155 O2 Sensor Heater Circuit Malfunction (Bank no. 2 Sensor no. 1)
P0156 O2 Sensor Circuit Malfunction (Bank no. 2 Sensor no. 2)
P0157 O2 Sensor Circuit Low Voltage (Bank no. 2 Sensor no. 2)
P0158 O2 Sensor Circuit High Voltage (Bank no. 2 Sensor no. 2)
P0159 O2 Sensor Circuit Slow Response (Bank no. 2 Sensor no. 2)
P0160 O2 Sensor Circuit No Activity Detected (Bank no. 2 Sensor no. 2)
P0161 O2 Sensor Heater Circuit Malfunction (Bank no. 2 Sensor no. 2)
P0162 O2 Sensor Circuit Malfunction (Bank no. 2 Sensor no. 3)
P0163 O2 Sensor Circuit Low Voltage (Bank no. 2 Sensor no. 3)

P0164 O2 Sensor Circuit High Voltage (Bank no. 2 Sensor no. 3)

P0165 O2 Sensor Circuit Slow Response (Bank no. 2 Sensor no. 3)

P0166 O2 Sensor Circuit No Activity Detected (Bank no. 2 Sensor no. 3)

P0167 O2 Sensor Heater Circuit Malfunction (Bank no. 2 Sensor no. 3)

P0170 Fuel Trim Malfunction (Bank no. 1)

P0171 System Too Lean (Bank no. 1)

P0172 System Too Rich (Bank no. 1)

P0173 Fuel Trim Malfunction (Bank no. 2)

P0174 System Too Lean (Bank no. 2)

P0175 System Too Rich (Bank no. 2)

P0176 Fuel Composition Sensor Circuit Malfunction

P0177 Fuel Composition Sensor Circuit Range/Performance

P0178 Fuel Composition Sensor Circuit Low Input

P0179 Fuel Composition Sensor Circuit High Input

P0180 Fuel Temperature Sensor "A" Circuit Malfunction

P0181 Fuel Temperature Sensor "A" Circuit Range/Performance

P0182 Fuel Temperature Sensor "A" Circuit Low Input

P0183 Fuel Temperature Sensor "A" Circuit High Input

P0184 Fuel Temperature Sensor "A" Circuit Intermittent

P0185 Fuel Temperature Sensor "B" Circuit Malfunction

P0186 Fuel Temperature Sensor "B" Circuit Range/Performance

P0187 Fuel Temperature Sensor "B" Circuit Low Input

P0188 Fuel Temperature Sensor "B" Circuit High Input

P0189 Fuel Temperature Sensor "B" Circuit Intermittent

P0190 Fuel Rail Pressure Sensor Circuit Malfunction

P0191 Fuel Rail Pressure Sensor Circuit Range/Performance

P0192 Fuel Rail Pressure Sensor Circuit Low Input

P0193 Fuel Rail Pressure Sensor Circuit High Input

P0194 Fuel Rail Pressure Sensor Circuit Intermittent

P0195 Engine Oil Temperature Sensor Malfunction

P0196 Engine Oil Temperature Sensor Range/Performance

P0197 Engine Oil Temperature Sensor Low

P0198 Engine Oil Temperature Sensor High

P0199 Engine Oil Temperature Sensor Intermittent

P0200 Injector Circuit Malfunction

P0201 Injector Circuit Malfunction—Cylinder no. 1

P0202 Injector Circuit Malfunction—Cylinder no. 2

P0203 Injector Circuit Malfunction—Cylinder no. 3

P0204 Injector Circuit Malfunction—Cylinder no. 4

P0205 Injector Circuit Malfunction—Cylinder no. 5

P0206 Injector Circuit Malfunction—Cylinder no. 6

P0207 Injector Circuit Malfunction—Cylinder no. 7

P0208 Injector Circuit Malfunction—Cylinder no. 8

P0209 Injector Circuit Malfunction—Cylinder no. 9

P0210 Injector Circuit Malfunction—Cylinder no. 10

P0211 Injector Circuit Malfunction—Cylinder no. 11

P0212 Injector Circuit Malfunction—Cylinder no. 12

P0213 Cold Start Injector no. 1 Malfunction

P0214 Cold Start Injector no. 2 Malfunction

P0215 Engine Shutoff Solenoid Malfunction

P0216 Injection Timing Control Circuit Malfunction

P0217 Engine Over Temperature Condition

P0218 Transmission Over Temperature Condition

P0219 Engine Over Speed Condition

P0220 Throttle/Pedal Position Sensor/Switch "B" Circuit Malfunction

P0221 Throttle/Pedal Position Sensor/Switch "B" Circuit Range/Performance Problem

P0222 Throttle/Pedal Position Sensor/Switch "B" Circuit Low Input

P0223 Throttle/Pedal Position Sensor/Switch "B" Circuit High Input

P0224 Throttle/Pedal Position Sensor/Switch "B" Circuit Intermittent

P0225 Throttle/Pedal Position Sensor/Switch "C" Circuit Malfunction

P0226 Throttle/Pedal Position Sensor/Switch "C" Circuit Range/Performance Problem

P0227 Throttle/Pedal Position Sensor/Switch "C" Circuit Low Input

P0228 Throttle/Pedal Position Sensor/Switch "C" Circuit High Input

P0229 Throttle/Pedal Position Sensor/Switch "C" Circuit Intermittent

P0230 Fuel Pump Primary Circuit Malfunction

P0231 Fuel Pump Secondary Circuit Low

P0232 Fuel Pump Secondary Circuit High

P0233 Fuel Pump Secondary Circuit Intermittent

P0234 Engine Over Boost Condition

P0261 Cylinder no. 1 Injector Circuit Low

P0262 Cylinder no. 1 Injector Circuit High

P0263 Cylinder no. 1 Contribution/Balance Fault

P0264 Cylinder no. 2 Injector Circuit Low

P0265 Cylinder no. 2 Injector Circuit High

P0266 Cylinder no. 2 Contribution/Balance Fault

P0267 Cylinder no. 3 Injector Circuit Low

P0268 Cylinder no. 3 Injector Circuit High

P0269 Cylinder no. 3 Contribution/Balance Fault

P0270 Cylinder no. 4 Injector Circuit Low

P0271 Cylinder no. 4 Injector Circuit High

P0272 Cylinder no. 4 Contribution/Balance Fault

P0273 Cylinder no. 5 Injector Circuit Low

P0274 Cylinder no. 5 Injector Circuit High

P0275 Cylinder no. 5 Contribution/Balance Fault

P0276 Cylinder no. 6 Injector Circuit Low

P0277 Cylinder no. 6 Injector Circuit High

P0278 Cylinder no. 6 Contribution/Balance Fault

P0279 Cylinder no. 7 Injector Circuit Low

P0280 Cylinder no. 7 Injector Circuit High

P0281 Cylinder no. 7 Contribution/Balance Fault

P0282 Cylinder no. 8 Injector Circuit Low

P0283 Cylinder no. 8 Injector Circuit High

P0284 Cylinder no. 8 Contribution/Balance Fault

P0285 Cylinder no. 9 Injector Circuit Low

P0286 Cylinder no. 9 Injector Circuit High

P0287 Cylinder no. 9 Contribution/Balance Fault

P0288 Cylinder no. 10 Injector Circuit Low

P0289 Cylinder no. 10 Injector Circuit High

P0290 Cylinder no. 10 Contribution/Balance Fault

P0291 Cylinder no. 11 Injector Circuit Low

P0292 Cylinder no. 11 Injector Circuit High

P0293 Cylinder no. 11 Contribution/Balance Fault

P0294 Cylinder no. 12 Injector Circuit Low

P0295 Cylinder no. 12 Injector Circuit High

P0296 Cylinder no. 12 Contribution/Balance Fault

P0300 Random/Multiple Cylinder Misfire Detected

P0301 Cylinder no. 1—Misfire Detected

P0302 Cylinder no. 2—Misfire Detected

P0303 Cylinder no. 3—Misfire Detected

P0304 Cylinder no. 4—Misfire Detected

P0305 Cylinder no. 5—Misfire Detected
P0306 Cylinder no. 6—Misfire Detected
P0307 Cylinder no. 7—Misfire Detected
P0308 Cylinder no. 8—Misfire Detected
P0309 Cylinder no. 9—Misfire Detected
P0310 Cylinder no. 10—Misfire Detected
P0311 Cylinder no. 11—Misfire Detected
P0312 Cylinder no. 12—Misfire Detected
P0320 Ignition/Distributor Engine Speed Input Circuit Malfunction
P0321 Ignition/Distributor Engine Speed Input Circuit Range/Performance
P0322 Ignition/Distributor Engine Speed Input Circuit No Signal
P0323 Ignition/Distributor Engine Speed Input Circuit Intermittent
P0325 Knock Sensor no. 1—Circuit Malfunction (Bank no. 1 or Single Sensor)
P0326 Knock Sensor no. 1—Circuit Range/Performance (Bank no. 1 or Single Sensor)
P0327 Knock Sensor no. 1—Circuit Low Input (Bank no. 1 or Single Sensor)
P0328 Knock Sensor no. 1—Circuit High Input (Bank no. 1 or Single Sensor)
P0329 Knock Sensor no. 1—Circuit Input Intermittent (Bank no. 1 or Single Sensor)
P0330 Knock Sensor no. 2—Circuit Malfunction (Bank no. 2)
P0331 Knock Sensor no. 2—Circuit Range/Performance (Bank no. 2)
P0332 Knock Sensor no. 2—Circuit Low Input (Bank no. 2)
P0333 Knock Sensor no. 2—Circuit High Input (Bank no. 2)
P0334 Knock Sensor no. 2—Circuit Input Intermittent (Bank no. 2)
P0335 Crankshaft Position Sensor "A" Circuit Malfunction
P0336 Crankshaft Position Sensor "A" Circuit Range/Performance
P0337 Crankshaft Position Sensor "A" Circuit Low Input
P0338 Crankshaft Position Sensor "A" Circuit High Input
P0339 Crankshaft Position Sensor "A" Circuit Intermittent
P0340 Camshaft Position Sensor Circuit Malfunction
P0341 Camshaft Position Sensor Circuit Range/Performance
P0342 Camshaft Position Sensor Circuit Low Input
P0343 Camshaft Position Sensor Circuit High Input
P0344 Camshaft Position Sensor Circuit Intermittent
P0350 Ignition Coil Primary/Secondary Circuit Malfunction
P0351 Ignition Coil "A" Primary/Secondary Circuit Malfunction
P0352 Ignition Coil "B" Primary/Secondary Circuit Malfunction
P0353 Ignition Coil "C" Primary/Secondary Circuit Malfunction
P0354 Ignition Coil "D" Primary/Secondary Circuit Malfunction
P0355 Ignition Coil "E" Primary/Secondary Circuit Malfunction
P0356 Ignition Coil "F" Primary/Secondary Circuit Malfunction
P0357 Ignition Coil "G" Primary/Secondary Circuit Malfunction
P0358 Ignition Coil "H" Primary/Secondary Circuit Malfunction
P0359 Ignition Coil "I" Primary/Secondary Circuit Malfunction
P0360 Ignition Coil "J" Primary/Secondary Circuit Malfunction
P0361 Ignition Coil "K" Primary/Secondary Circuit Malfunction
P0362 Ignition Coil "L" Primary/Secondary Circuit Malfunction
P0370 Timing Reference High Resolution Signal "A" Malfunction
P0371 Timing Reference High Resolution Signal "A" Too Many Pulses
P0372 Timing Reference High Resolution Signal "A" Too Few Pulses

P0373 Timing Reference High Resolution Signal "A" Intermittent/Erratic Pulses
P0374 Timing Reference High Resolution Signal "A" No Pulses
P0375 Timing Reference High Resolution Signal "B" Malfunction
P0376 Timing Reference High Resolution Signal "B" Too Many Pulses
P0377 Timing Reference High Resolution Signal "B" Too Few Pulses
P0378 Timing Reference High Resolution Signal "B" Intermittent/Erratic Pulses
P0379 Timing Reference High Resolution Signal "B" No Pulses
P0380 Glow Plug/Heater Circuit "A" Malfunction
P0381 Glow Plug/Heater Indicator Circuit Malfunction
P0382 Glow Plug/Heater Circuit "B" Malfunction
P0385 Crankshaft Position Sensor "B" Circuit Malfunction
P0386 Crankshaft Position Sensor "B" Circuit Range/Performance
P0387 Crankshaft Position Sensor "B" Circuit Low Input
P0388 Crankshaft Position Sensor "B" Circuit High Input
P0389 Crankshaft Position Sensor "B" Circuit Intermittent
P0400 Exhaust Gas Recirculation Flow Malfunction
P0401 Exhaust Gas Recirculation Flow Insufficient Detected
P0402 Exhaust Gas Recirculation Flow Excessive Detected
P0403 Exhaust Gas Recirculation Circuit Malfunction
P0404 Exhaust Gas Recirculation Circuit Range/Performance
P0405 Exhaust Gas Recirculation Sensor "A" Circuit Low
P0406 Exhaust Gas Recirculation Sensor "A" Circuit High
P0407 Exhaust Gas Recirculation Sensor "B" Circuit Low
P0408 Exhaust Gas Recirculation Sensor "B" Circuit High
P0410 Secondary Air Injection System Malfunction
P0411 Secondary Air Injection System Incorrect Flow Detected
P0412 Secondary Air Injection System Switching Valve "A" Circuit Malfunction
P0413 Secondary Air Injection System Switching Valve "A" Circuit Open
P0414 Secondary Air Injection System Switching Valve "A" Circuit Shorted
P0415 Secondary Air Injection System Switching Valve "B" Circuit Malfunction
P0416 Secondary Air Injection System Switching Valve "B" Circuit Open
P0417 Secondary Air Injection System Switching Valve "B" Circuit Shorted
P0418 Secondary Air Injection System Relay "A" Circuit Malfunction
P0419 Secondary Air Injection System Relay "B" Circuit Malfunction
P0420 Catalyst System Efficiency Below Threshold (Bank no. 1)
P0421 Warm Up Catalyst Efficiency Below Threshold (Bank no. 1)
P0422 Main Catalyst Efficiency Below Threshold (Bank no. 1)
P0423 Heated Catalyst Efficiency Below Threshold (Bank no. 1)
P0424 Heated Catalyst Temperature Below Threshold (Bank no. 1)
P0430 Catalyst System Efficiency Below Threshold (Bank no. 2)
P0431 Warm Up Catalyst Efficiency Below Threshold (Bank no. 2)
P0432 Main Catalyst Efficiency Below Threshold (Bank no. 2)
P0433 Heated Catalyst Efficiency Below Threshold (Bank no. 2)
P0434 Heated Catalyst Temperature Below Threshold (Bank no. 2)

P0440 Evaporative Emission Control System Malfunction

P0441 Evaporative Emission Control System Incorrect Purge Flow

P0442 Evaporative Emission Control System Leak Detected (Small Leak)

P0443 Evaporative Emission Control System Purge Control Valve Circuit Malfunction

P0444 Evaporative Emission Control System Purge Control Valve Circuit Open

P0445 Evaporative Emission Control System Purge Control Valve Circuit Shorted

P0446 Evaporative Emission Control System Vent Control Circuit Malfunction

P0447 Evaporative Emission Control System Vent Control Circuit Open

P0448 Evaporative Emission Control System Vent Control Circuit Shorted

P0449 Evaporative Emission Control System Vent Valve/Solenoid Circuit Malfunction

P0450 Evaporative Emission Control System Pressure Sensor Malfunction

P0451 Evaporative Emission Control System Pressure Sensor Range/Performance

P0452 Evaporative Emission Control System Pressure Sensor Low Input

P0453 Evaporative Emission Control System Pressure Sensor High Input

P0454 Evaporative Emission Control System Pressure Sensor Intermittent

P0455 Evaporative Emission Control System Leak Detected (Gross Leak)

P0460 Fuel Level Sensor Circuit Malfunction

P0461 Fuel Level Sensor Circuit Range/Performance

P0462 Fuel Level Sensor Circuit Low Input

P0463 Fuel Level Sensor Circuit High Input

P0464 Fuel Level Sensor Circuit Intermittent

P0465 Purge Flow Sensor Circuit Malfunction

P0466 Purge Flow Sensor Circuit Range/Performance

P0467 Purge Flow Sensor Circuit Low Input

P0468 Purge Flow Sensor Circuit High Input

P0469 Purge Flow Sensor Circuit Intermittent

P0470 Exhaust Pressure Sensor Malfunction

P0471 Exhaust Pressure Sensor Range/Performance

P0472 Exhaust Pressure Sensor Low

P0473 Exhaust Pressure Sensor High

P0474 Exhaust Pressure Sensor Intermittent

P0475 Exhaust Pressure Control Valve Malfunction

P0476 Exhaust Pressure Control Valve Range/Performance

P0477 Exhaust Pressure Control Valve Low

P0478 Exhaust Pressure Control Valve High

P0479 Exhaust Pressure Control Valve Intermittent

P0480 Cooling Fan no. 1 Control Circuit Malfunction

P0481 Cooling Fan no. 2 Control Circuit Malfunction

P0482 Cooling Fan no. 3 Control Circuit Malfunction

P0483 Cooling Fan Rationality Check Malfunction

P0484 Cooling Fan Circuit Over Current

P0485 Cooling Fan Power/Ground Circuit Malfunction

P0500 Vehicle Speed Sensor Malfunction

P0501 Vehicle Speed Sensor Range/Performance

P0502 Vehicle Speed Sensor Circuit Low Input

P0503 Vehicle Speed Sensor Intermittent/Erratic/High

P0505 Idle Control System Malfunction

P0506 Idle Control System RPM Lower Than Expected

P0507 Idle Control System RPM Higher Than Expected

P0510 Closed Throttle Position Switch Malfunction

P0520 Engine Oil Pressure Sensor/Switch Circuit Malfunction

P0521 Engine Oil Pressure Sensor/Switch Range/Performance

P0522 Engine Oil Pressure Sensor/Switch Low Voltage

P0523 Engine Oil Pressure Sensor/Switch High Voltage

P0530 A/C Refrigerant Pressure Sensor Circuit Malfunction

P0531 A/C Refrigerant Pressure Sensor Circuit Range/Performance

P0532 A/C Refrigerant Pressure Sensor Circuit Low Input

P0533 A/C Refrigerant Pressure Sensor Circuit High Input

P0534 A/C Refrigerant Charge Loss

P0550 Power Steering Pressure Sensor Circuit Malfunction

P0551 Power Steering Pressure Sensor Circuit Range/Performance

P0552 Power Steering Pressure Sensor Circuit Low Input

P0553 Power Steering Pressure Sensor Circuit High Input

P0554 Power Steering Pressure Sensor Circuit Intermittent

P0560 System Voltage Malfunction

P0561 System Voltage Unstable

P0562 System Voltage Low

P0563 System Voltage High

P0565 Cruise Control On Signal Malfunction

P0566 Cruise Control Off Signal Malfunction

P0567 Cruise Control Resume Signal Malfunction

P0568 Cruise Control Set Signal Malfunction

P0569 Cruise Control Coast Signal Malfunction

P0570 Cruise Control Accel Signal Malfunction

P0571 Cruise Control/Brake Switch "A" Circuit Malfunction

P0572 Cruise Control/Brake Switch "A" Circuit Low

P0573 Cruise Control/Brake Switch "A" Circuit High

P0574 **Through P0580** Reserved for Cruise Codes

P0600 Serial Communication Link Malfunction

P0601 Internal Control Module Memory Check Sum Error

P0602 Control Module Programming Error

P0603 Internal Control Module Keep Alive Memory (KAM) Error

P0604 Internal Control Module Random Access Memory (RAM) Error

P0605 Internal Control Module Read Only Memory (ROM) Error

P0606 PCM Processor Fault

P0608 Control Module VSS Output "A" Malfunction

P0609 Control Module VSS Output "B" Malfunction

P0620 Generator Control Circuit Malfunction

P0621 Generator Lamp "L" Control Circuit Malfunction

P0622 Generator Field "F" Control Circuit Malfunction

P0650 Malfunction Indicator Lamp (MIL) Control Circuit Malfunction

P0654 Engine RPM Output Circuit Malfunction

P0655 Engine Hot Lamp Output Control Circuit Malfunction

P0656 Fuel Level Output Circuit Malfunction

P0700 Transmission Control System Malfunction

P0701 Transmission Control System Range/Performance

P0702 Transmission Control System Electrical

P0703 Torque Converter/Brake Switch "B" Circuit Malfunction

P0704 Clutch Switch Input Circuit Malfunction

P0705 Transmission Range Sensor Circuit Malfunction (PRNDL Input)

P0706 Transmission Range Sensor Circuit Range/Performance

P0707 Transmission Range Sensor Circuit Low Input

P0708 Transmission Range Sensor Circuit High Input
P0709 Transmission Range Sensor Circuit Intermittent
P0710 Transmission Fluid Temperature Sensor Circuit Malfunction
P0711 Transmission Fluid Temperature Sensor Circuit Range/Performance
P0712 Transmission Fluid Temperature Sensor Circuit Low Input
P0713 Transmission Fluid Temperature Sensor Circuit High Input
P0714 Transmission Fluid Temperature Sensor Circuit Intermittent
P0715 Input/Turbine Speed Sensor Circuit Malfunction
P0716 Input/Turbine Speed Sensor Circuit Range/Performance
P0717 Input/Turbine Speed Sensor Circuit No Signal
P0718 Input/Turbine Speed Sensor Circuit Intermittent
P0719 Torque Converter/Brake Switch "B" Circuit Low
P0720 Output Speed Sensor Circuit Malfunction
P0721 Output Speed Sensor Circuit Range/Performance
P0722 Output Speed Sensor Circuit No Signal
P0723 Output Speed Sensor Circuit Intermittent
P0724 Torque Converter/Brake Switch "B" Circuit High
P0725 Engine Speed Input Circuit Malfunction
P0726 Engine Speed Input Circuit Range/Performance
P0727 Engine Speed Input Circuit No Signal
P0728 Engine Speed Input Circuit Intermittent
P0730 Incorrect Gear Ratio
P0731 Gear no. 1 Incorrect Ratio
P0732 Gear no. 2 Incorrect Ratio
P0733 Gear no. 3 Incorrect Ratio
P0734 Gear no. 4 Incorrect Ratio
P0735 Gear no. 5 Incorrect Ratio
P0736 Reverse Incorrect Ratio
P0740 Torque Converter Clutch Circuit Malfunction
P0741 Torque Converter Clutch Circuit Performance or Stuck Off
P0742 Torque Converter Clutch Circuit Stuck On
P0743 Torque Converter Clutch Circuit Electrical
P0744 Torque Converter Clutch Circuit Intermittent
P0745 Pressure Control Solenoid Malfunction
P0746 Pressure Control Solenoid Performance or Stuck Off
P0747 Pressure Control Solenoid Stuck On
P0748 Pressure Control Solenoid Electrical
P0749 Pressure Control Solenoid Intermittent
P0750 Shift Solenoid "A" Malfunction
P0751 Shift Solenoid "A" Performance or Stuck Off
P0752 Shift Solenoid "A" Stuck On
P0753 Shift Solenoid "A" Electrical
P0754 Shift Solenoid "A" Intermittent
P0755 Shift Solenoid "B" Malfunction
P0756 Shift Solenoid "B" Performance or Stuck Off
P0757 Shift Solenoid "B" Stuck On
P0758 Shift Solenoid "B" Electrical
P0759 Shift Solenoid "B" Intermittent
P0760 Shift Solenoid "C" Malfunction
P0761 Shift Solenoid "C" Performance Or Stuck Off
P0762 Shift Solenoid "C" Stuck On
P0763 Shift Solenoid "C" Electrical
P0764 Shift Solenoid "C" Intermittent
P0765 Shift Solenoid "D" Malfunction
P0766 Shift Solenoid "D" Performance Or Stuck Off

P0767 Shift Solenoid "D" Stuck On
P0768 Shift Solenoid "D" Electrical
P0769 Shift Solenoid "D" Intermittent
P0770 Shift Solenoid "E" Malfunction
P0771 Shift Solenoid "E" Performance Or Stuck Off
P0772 Shift Solenoid "E" Stuck On
P0773 Shift Solenoid "E" Electrical
P0774 Shift Solenoid "E" Intermittent
P0780 Shift Malfunction
P0781 1–2 Shift Malfunction
P0782 2–3 Shift Malfunction
P0783 3–4 Shift Malfunction
P0784 4–5 Shift Malfunction
P0785 Shift/Timing Solenoid Malfunction
P0786 Shift/Timing Solenoid Range/Performance
P0787 Shift/Timing Solenoid Low
P0788 Shift/Timing Solenoid High
P0789 Shift/Timing Solenoid Intermittent
P0790 Normal/Performance Switch Circuit Malfunction
P0801 Reverse Inhibit Control Circuit Malfunction
P0803 1–4 Upshift (Skip Shift) Solenoid Control Circuit Malfunction
P0804 1–4 Upshift (Skip Shift) Lamp Control Circuit Malfunction
P1120 Secondary Throttle Position Sensor Circuit Fault
P1125 Tandem Throttle Position Sensor Circuit Fault
P1210 Traction Control System Signal Fault
P1220 Fuel Pump Control Module Fault
P1320 Ignition Control Signal Fault
P1336 Crankshaft Position Sensor Circuit Fault
P1400 EGR/EVAP Control Solenoid Circuit Fault
P1401 EGR Temperature Sensor Circuit Fault
P1443 EVAP Canister Control Vacuum Switch Circuit Fault
P1445 EVAP Purge Volume Control Valve Circuit Fault
P1605 TCM A/T Diagnosis Communication Line Fault
P1705 Throttle Position Sensor (Switch) Circuit Fault
P1760 Overrun Clutch Solenoid Valve Circuit Fault
P1900 Cooling Fan Control Circuit Fault

Isuzu

READING CODES

Reading the control module memory is on of the first steps in OBD II system diagnostics. This step should be initially performed to determine the general nature of the fault. Subsequent readings will determine if the fault has been cleared.

Reading codes can be performed by any of the methods below:

• Read the control module memory with the Generic Scan Tool (GST)

• Read the control module memory with the vehicle manufacturer's specific tester

To read the fault codes, connect the scan tool or tester according to the manufacturer's instructions. Follow the manufacturer's specified procedure for reading the codes.

CLEARING CODES

Control module reset procedures are a very important part of OBD II System diagnostics. This step should be done at the end of any fault code repair and at the end of any driveability repair.

Clearing codes can be performed by any of the methods below:
• Clear the control module memory with the Generic Scan Tool (GST)
• Clear the control module memory with the vehicle manufacturer's specific tester
• Turn the ignition **OFF** and remove the negative battery cable for at least 1 minute.

Removing the negative battery cable may cause other systems in the vehicle to loose their memory. Prior to removing the cable, ensure you have the proper reset codes for radios and alarms.

➡**The MIL will may also be de-activated for some codes if the vehicle completes three consecutive trips without a fault detected with vehicle conditions similar to those present during the fault.**

OBD II TROUBLE CODES

P0100 Mass or Volume Air Flow Circuit Malfunction
P0101 Mass or Volume Air Flow Circuit Range/Performance Problem
P0102 Mass or Volume Air Flow Circuit Low Input
P0103 Mass or Volume Air Flow Circuit High Input
P0104 Mass or Volume Air Flow Circuit Intermittent
P0105 Manifold Absolute Pressure/Barometric Pressure Circuit Malfunction
P0106 Manifold Absolute Pressure/Barometric Pressure Circuit Range/Performance Problem
P0107 Manifold Absolute Pressure/Barometric Pressure Circuit Low Input
P0108 Manifold Absolute Pressure/Barometric Pressure Circuit High Input
P0109 Manifold Absolute Pressure/Barometric Pressure Circuit Intermittent
P0110 Intake Air Temperature Circuit Malfunction
P0111 Intake Air Temperature Circuit Range/Performance Problem
P0112 Intake Air Temperature Circuit Low Input
P0113 Intake Air Temperature Circuit High Input
P0114 Intake Air Temperature Circuit Intermittent
P0115 Engine Coolant Temperature Circuit Malfunction
P0116 Engine Coolant Temperature Circuit Range/Performance Problem
P0117 Engine Coolant Temperature Circuit Low Input
P0118 Engine Coolant Temperature Circuit High Input
P0119 Engine Coolant Temperature Circuit Intermittent
P0120 Throttle/Pedal Position Sensor/Switch "A" Circuit Malfunction
P0121 Throttle/Pedal Position Sensor/Switch "A" Circuit Range/Performance Problem
P0122 Throttle/Pedal Position Sensor/Switch "A" Circuit Low Input
P0123 Throttle/Pedal Position Sensor/Switch "A" Circuit High Input
P0124 Throttle/Pedal Position Sensor/Switch "A" Circuit Intermittent
P0125 Insufficient Coolant Temperature For Closed Loop Fuel Control
P0126 Insufficient Coolant Temperature For Stable Operation
P0130 O2 Circuit Malfunction (Bank no. 1 Sensor no. 1)
P0131 O2 Sensor Circuit Low Voltage (Bank no. 1 Sensor no. 1)

P0132 O2 Sensor Circuit High Voltage (Bank no. 1 Sensor no. 1)
P0133 O2 Sensor Circuit Slow Response (Bank no. 1 Sensor no. 1)
P0134 O2 Sensor Circuit No Activity Detected (Bank no. 1 Sensor no. 1)
P0135 O2 Sensor Heater Circuit Malfunction (Bank no. 1 Sensor no. 1)
P0136 O2 Sensor Circuit Malfunction (Bank no. 1 Sensor no. 2)
P0137 O2 Sensor Circuit Low Voltage (Bank no. 1 Sensor no. 2)
P0138 O2 Sensor Circuit High Voltage (Bank no. 1 Sensor no. 2)
P0139 O2 Sensor Circuit Slow Response (Bank no. 1 Sensor no. 2)
P0140 O2 Sensor Circuit No Activity Detected (Bank no. 1 Sensor no. 2)
P0141 O2 Sensor Heater Circuit Malfunction (Bank no. 1 Sensor no. 2)
P0142 O2 Sensor Circuit Malfunction (Bank no. 1 Sensor no. 3)
P0143 O2 Sensor Circuit Low Voltage (Bank no. 1 Sensor no. 3)
P0144 O2 Sensor Circuit High Voltage (Bank no. 1 Sensor no. 3)
P0145 O2 Sensor Circuit Slow Response (Bank no. 1 Sensor no. 3)
P0146 O2 Sensor Circuit No Activity Detected (Bank no. 1 Sensor no. 3)
P0147 O2 Sensor Heater Circuit Malfunction (Bank no. 1 Sensor no. 3)
P0150 O2 Sensor Circuit Malfunction (Bank no. 2 Sensor no. 1)
P0151 O2 Sensor Circuit Low Voltage (Bank no. 2 Sensor no. 1)
P0152 O2 Sensor Circuit High Voltage (Bank no. 2 Sensor no. 1)
P0153 O2 Sensor Circuit Slow Response (Bank no. 2 Sensor no. 1)
P0154 O2 Sensor Circuit No Activity Detected (Bank no. 2 Sensor no. 1)
P0155 O2 Sensor Heater Circuit Malfunction (Bank no. 2 Sensor no. 1)
P0156 O2 Sensor Circuit Malfunction (Bank no. 2 Sensor no. 2)
P0157 O2 Sensor Circuit Low Voltage (Bank no. 2 Sensor no. 2)
P0158 O2 Sensor Circuit High Voltage (Bank no. 2 Sensor no. 2)
P0159 O2 Sensor Circuit Slow Response (Bank no. 2 Sensor no. 2)
P0160 O2 Sensor Circuit No Activity Detected (Bank no. 2 Sensor no. 2)
P0161 O2 Sensor Heater Circuit Malfunction (Bank no. 2 Sensor no. 2)
P0162 O2 Sensor Circuit Malfunction (Bank no. 2 Sensor no. 3)
P0163 O2 Sensor Circuit Low Voltage (Bank no. 2 Sensor no. 3)
P0164 O2 Sensor Circuit High Voltage (Bank no. 2 Sensor no. 3)
P0165 O2 Sensor Circuit Slow Response (Bank no. 2 Sensor no. 3)
P0166 O2 Sensor Circuit No Activity Detected (Bank no. 2 Sensor no. 3)
P0167 O2 Sensor Heater Circuit Malfunction (Bank no. 2 Sensor no. 3)
P0170 Fuel Trim Malfunction (Bank no. 1)
P0171 System Too Lean (Bank no. 1)
P0172 System Too Rich (Bank no. 1)

P0173 Fuel Trim Malfunction (Bank no. 2)
P0174 System Too Lean (Bank no. 2)
P0175 System Too Rich (Bank no. 2)
P0176 Fuel Composition Sensor Circuit Malfunction
P0177 Fuel Composition Sensor Circuit Range/Performance
P0178 Fuel Composition Sensor Circuit Low Input
P0179 Fuel Composition Sensor Circuit High Input
P0180 Fuel Temperature Sensor "A" Circuit Malfunction
P0181 Fuel Temperature Sensor "A" Circuit Range/Performance
P0182 Fuel Temperature Sensor "A" Circuit Low Input
P0183 Fuel Temperature Sensor "A" Circuit High Input
P0184 Fuel Temperature Sensor "A" Circuit Intermittent
P0185 Fuel Temperature Sensor "B" Circuit Malfunction
P0186 Fuel Temperature Sensor "B" Circuit Range/Performance
P0187 Fuel Temperature Sensor "B" Circuit Low Input
P0188 Fuel Temperature Sensor "B" Circuit High Input
P0189 Fuel Temperature Sensor "B" Circuit Intermittent
P0190 Fuel Rail Pressure Sensor Circuit Malfunction
P0191 Fuel Rail Pressure Sensor Circuit Range/Performance
P0192 Fuel Rail Pressure Sensor Circuit Low Input
P0193 Fuel Rail Pressure Sensor Circuit High Input
P0194 Fuel Rail Pressure Sensor Circuit Intermittent
P0195 Engine Oil Temperature Sensor Malfunction
P0196 Engine Oil Temperature Sensor Range/Performance
P0197 Engine Oil Temperature Sensor Low
P0198 Engine Oil Temperature Sensor High
P0199 Engine Oil Temperature Sensor Intermittent
P0200 Injector Circuit Malfunction
P0201 Injector Circuit Malfunction—Cylinder no. 1
P0202 Injector Circuit Malfunction—Cylinder no. 2
P0203 Injector Circuit Malfunction—Cylinder no. 3
P0204 Injector Circuit Malfunction—Cylinder no. 4
P0205 Injector Circuit Malfunction—Cylinder no. 5
P0206 Injector Circuit Malfunction—Cylinder no. 6
P0207 Injector Circuit Malfunction—Cylinder no. 7
P0208 Injector Circuit Malfunction—Cylinder no. 8
P0209 Injector Circuit Malfunction—Cylinder no. 9
P0210 Injector Circuit Malfunction—Cylinder no. 10
P0211 Injector Circuit Malfunction—Cylinder no. 11
P0212 Injector Circuit Malfunction—Cylinder no. 12
P0213 Cold Start Injector no. 1 Malfunction
P0214 Cold Start Injector no. 2 Malfunction
P0215 Engine Shutoff Solenoid Malfunction
P0216 Injection Timing Control Circuit Malfunction
P0217 Engine Over Temperature Condition
P0218 Transmission Over Temperature Condition
P0219 Engine Over Speed Condition
P0220 Throttle/Pedal Position Sensor/Switch "B" Circuit Malfunction
P0221 Throttle/Pedal Position Sensor/Switch "B" Circuit Range/Performance Problem
P0222 Throttle/Pedal Position Sensor/Switch "B" Circuit Low Input
P0223 Throttle/Pedal Position Sensor/Switch "B" Circuit High Input
P0224 Throttle/Pedal Position Sensor/Switch "B" Circuit Intermittent
P0225 Throttle/Pedal Position Sensor/Switch "C" Circuit Malfunction
P0226 Throttle/Pedal Position Sensor/Switch "C" Circuit Range/Performance Problem

P0227 Throttle/Pedal Position Sensor/Switch "C" Circuit Low Input
P0228 Throttle/Pedal Position Sensor/Switch "C" Circuit High Input
P0229 Throttle/Pedal Position Sensor/Switch "C" Circuit Intermittent
P0230 Fuel Pump Primary Circuit Malfunction
P0231 Fuel Pump Secondary Circuit Low
P0232 Fuel Pump Secondary Circuit High
P0233 Fuel Pump Secondary Circuit Intermittent
P0234 Engine Over Boost Condition/Injector)
P0261 Cylinder no. 1 Injector Circuit Low
P0262 Cylinder no. 1 Injector Circuit High
P0263 Cylinder no. 1 Contribution/Balance Fault
P0264 Cylinder no. 2 Injector Circuit Low
P0265 Cylinder no. 2 Injector Circuit High
P0266 Cylinder no. 2 Contribution/Balance Fault
P0267 Cylinder no. 3 Injector Circuit Low
P0268 Cylinder no. 3 Injector Circuit High
P0269 Cylinder no. 3 Contribution/Balance Fault
P0270 Cylinder no. 4 Injector Circuit Low
P0271 Cylinder no. 4 Injector Circuit High
P0272 Cylinder no. 4 Contribution/Balance Fault
P0273 Cylinder no. 5 Injector Circuit Low
P0274 Cylinder no. 5 Injector Circuit High
P0275 Cylinder no. 5 Contribution/Balance Fault
P0276 Cylinder no. 6 Injector Circuit Low
P0277 Cylinder no. 6 Injector Circuit High
P0278 Cylinder no. 6 Contribution/Balance Fault
P0279 Cylinder no. 7 Injector Circuit Low
P0280 Cylinder no. 7 Injector Circuit High
P0281 Cylinder no. 7 Contribution/Balance Fault
P0282 Cylinder no. 8 Injector Circuit Low
P0283 Cylinder no. 8 Injector Circuit High
P0284 Cylinder no. 8 Contribution/Balance Fault
P0285 Cylinder no. 9 Injector Circuit Low
P0286 Cylinder no. 9 Injector Circuit High
P0287 Cylinder no. 9 Contribution/Balance Fault
P0288 Cylinder no. 10 Injector Circuit Low
P0289 Cylinder no. 10 Injector Circuit High
P0290 Cylinder no. 10 Contribution/Balance Fault
P0291 Cylinder no. 11 Injector Circuit Low
P0292 Cylinder no. 11 Injector Circuit High
P0293 Cylinder no. 11 Contribution/Balance Fault
P0294 Cylinder no. 12 Injector Circuit Low
P0295 Cylinder no. 12 Injector Circuit High
P0296 Cylinder no. 12 Contribution/Balance Fault
P0300 Random/Multiple Cylinder Misfire Detected
P0301 Cylinder no. 1—Misfire Detected
P0302 Cylinder no. 2—Misfire Detected
P0303 Cylinder no. 3—Misfire Detected
P0304 Cylinder no. 4—Misfire Detected
P0305 Cylinder no. 5—Misfire Detected
P0306 Cylinder no. 6—Misfire Detected
P0307 Cylinder no. 7—Misfire Detected
P0308 Cylinder no. 8—Misfire Detected
P0309 Cylinder no. 9—Misfire Detected
P0310 Cylinder no. 10—Misfire Detected
P0311 Cylinder no. 11—Misfire Detected
P0312 Cylinder no. 12—Misfire Detected
P0320 Ignition/Distributor Engine Speed Input Circuit Malfunction

P0321 Ignition/Distributor Engine Speed Input Circuit Range/Performance

P0322 Ignition/Distributor Engine Speed Input Circuit No Signal

P0323 Ignition/Distributor Engine Speed Input Circuit Intermittent

P0325 Knock Sensor no. 1—Circuit Malfunction (Bank no. 1 or Single Sensor)

P0326 Knock Sensor no. 1—Circuit Range/Performance (Bank no. 1 or Single Sensor)

P0327 Knock Sensor no. 1—Circuit Low Input (Bank no. 1 or Single Sensor)

P0328 Knock Sensor no. 1—Circuit High Input (Bank no. 1 or Single Sensor)

P0329 Knock Sensor no. 1—Circuit Input Intermittent (Bank no. 1 or Single Sensor)

P0330 Knock Sensor no. 2—Circuit Malfunction (Bank no. 2)

P0331 Knock Sensor no. 2—Circuit Range/Performance (Bank no. 2)

P0332 Knock Sensor no. 2—Circuit Low Input (Bank no. 2)

P0333 Knock Sensor no. 2—Circuit High Input (Bank no. 2)

P0334 Knock Sensor no. 2—Circuit Input Intermittent (Bank no. 2)

P0335 Crankshaft Position Sensor "A" Circuit Malfunction

P0336 Crankshaft Position Sensor "A" Circuit Range/Performance

P0337 Crankshaft Position Sensor "A" Circuit Low Input

P0338 Crankshaft Position Sensor "A" Circuit High Input

P0339 Crankshaft Position Sensor "A" Circuit Intermittent

P0340 Camshaft Position Sensor Circuit Malfunction

P0341 Camshaft Position Sensor Circuit Range/Performance

P0342 Camshaft Position Sensor Circuit Low Input

P0343 Camshaft Position Sensor Circuit High Input

P0344 Camshaft Position Sensor Circuit Intermittent

P0350 Ignition Coil Primary/Secondary Circuit Malfunction

P0351 Ignition Coil "A" Primary/Secondary Circuit Malfunction

P0352 Ignition Coil "B" Primary/Secondary Circuit Malfunction

P0353 Ignition Coil "C" Primary/Secondary Circuit Malfunction

P0354 Ignition Coil "D" Primary/Secondary Circuit Malfunction

P0355 Ignition Coil "E" Primary/Secondary Circuit Malfunction

P0356 Ignition Coil "F" Primary/Secondary Circuit Malfunction

P0357 Ignition Coil "G" Primary/Secondary Circuit Malfunction

P0358 Ignition Coil "H" Primary/Secondary Circuit Malfunction

P0359 Ignition Coil "I" Primary/Secondary Circuit Malfunction

P0360 Ignition Coil "J" Primary/Secondary Circuit Malfunction

P0361 Ignition Coil "K" Primary/Secondary Circuit Malfunction

P0362 Ignition Coil "L" Primary/Secondary Circuit Malfunction

P0370 Timing Reference High Resolution Signal "A" Malfunction

P0371 Timing Reference High Resolution Signal "A" Too Many Pulses

P0372 Timing Reference High Resolution Signal "A" Too Few Pulses

P0373 Timing Reference High Resolution Signal "A" Intermittent/Erratic Pulses

P0374 Timing Reference High Resolution Signal "A" No Pulses

P0375 Timing Reference High Resolution Signal "B" Malfunction

P0376 Timing Reference High Resolution Signal "B" Too Many Pulses

P0377 Timing Reference High Resolution Signal "B" Too Few Pulses

P0378 Timing Reference High Resolution Signal "B" Intermittent/Erratic Pulses

P0379 Timing Reference High Resolution Signal "B" No Pulses

P0380 Glow Plug/Heater Circuit "A" Malfunction

P0381 Glow Plug/Heater Indicator Circuit Malfunction

P0382 Glow Plug/Heater Circuit "B" Malfunction

P0385 Crankshaft Position Sensor "B" Circuit Malfunction

P0386 Crankshaft Position Sensor "B" Circuit Range/Performance

P0387 Crankshaft Position Sensor "B" Circuit Low Input

P0388 Crankshaft Position Sensor "B" Circuit High Input

P0389 Crankshaft Position Sensor "B" Circuit Intermittent

P0400 Exhaust Gas Recirculation Flow Malfunction

P0401 Exhaust Gas Recirculation Flow Insufficient Detected

P0402 Exhaust Gas Recirculation Flow Excessive Detected

P0403 Exhaust Gas Recirculation Circuit Malfunction

P0404 Exhaust Gas Recirculation Circuit Range/Performance

P0405 Exhaust Gas Recirculation Sensor "A" Circuit Low

P0406 Exhaust Gas Recirculation Sensor "A" Circuit High

P0407 Exhaust Gas Recirculation Sensor "B" Circuit Low

P0408 Exhaust Gas Recirculation Sensor "B" Circuit High

P0410 Secondary Air Injection System Malfunction

P0411 Secondary Air Injection System Incorrect Flow Detected

P0412 Secondary Air Injection System Switching Valve "A" Circuit Malfunction

P0413 Secondary Air Injection System Switching Valve "A" Circuit Open

P0414 Secondary Air Injection System Switching Valve "A" Circuit Shorted

P0415 Secondary Air Injection System Switching Valve "B" Circuit Malfunction

P0416 Secondary Air Injection System Switching Valve "B" Circuit Open

P0417 Secondary Air Injection System Switching Valve "B" Circuit Shorted

P0418 Secondary Air Injection System Relay "A" Circuit Malfunction

P0419 Secondary Air Injection System Relay "B" Circuit Malfunction

P0420 Catalyst System Efficiency Below Threshold (Bank no. 1)

P0421 Warm Up Catalyst Efficiency Below Threshold (Bank no. 1)

P0422 Main Catalyst Efficiency Below Threshold (Bank no. 1)

P0423 Heated Catalyst Efficiency Below Threshold (Bank no. 1)

P0424 Heated Catalyst Temperature Below Threshold (Bank no. 1)

P0430 Catalyst System Efficiency Below Threshold (Bank no. 2)

P0431 Warm Up Catalyst Efficiency Below Threshold (Bank no. 2)

P0432 Main Catalyst Efficiency Below Threshold (Bank no. 2)

P0433 Heated Catalyst Efficiency Below Threshold (Bank no. 2)

P0434 Heated Catalyst Temperature Below Threshold (Bank no. 2)

P0440 Evaporative Emission Control System Malfunction

P0441 Evaporative Emission Control System Incorrect Purge Flow

P0442 Evaporative Emission Control System Leak Detected (Small Leak)

P0443 Evaporative Emission Control System Purge Control Valve Circuit Malfunction

P0444 Evaporative Emission Control System Purge Control Valve Circuit Open

P0445 Evaporative Emission Control System Purge Control Valve Circuit Shorted

P0446 Evaporative Emission Control System Vent Control Circuit Malfunction

P0447 Evaporative Emission Control System Vent Control Circuit Open

P0448 Evaporative Emission Control System Vent Control Circuit Shorted

P0449 Evaporative Emission Control System Vent Valve/Solenoid Circuit Malfunction

P0450 Evaporative Emission Control System Pressure Sensor Malfunction

P0451 Evaporative Emission Control System Pressure Sensor Range/Performance

P0452 Evaporative Emission Control System Pressure Sensor Low Input

P0453 Evaporative Emission Control System Pressure Sensor High Input

P0454 Evaporative Emission Control System Pressure Sensor Intermittent

P0455 Evaporative Emission Control System Leak Detected (Gross Leak)

P0460 Fuel Level Sensor Circuit Malfunction

P0461 Fuel Level Sensor Circuit Range/Performance

P0462 Fuel Level Sensor Circuit Low Input

P0463 Fuel Level Sensor Circuit High Input

P0464 Fuel Level Sensor Circuit Intermittent

P0465 Purge Flow Sensor Circuit Malfunction

P0466 Purge Flow Sensor Circuit Range/Performance

P0467 Purge Flow Sensor Circuit Low Input

P0468 Purge Flow Sensor Circuit High Input

P0469 Purge Flow Sensor Circuit Intermittent

P0470 Exhaust Pressure Sensor Malfunction

P0471 Exhaust Pressure Sensor Range/Performance

P0472 Exhaust Pressure Sensor Low

P0473 Exhaust Pressure Sensor High

P0474 Exhaust Pressure Sensor Intermittent

P0475 Exhaust Pressure Control Valve Malfunction

P0476 Exhaust Pressure Control Valve Range/Performance

P0477 Exhaust Pressure Control Valve Low

P0478 Exhaust Pressure Control Valve High

P0479 Exhaust Pressure Control Valve Intermittent

P0480 Cooling Fan no. 1 Control Circuit Malfunction

P0481 Cooling Fan no. 2 Control Circuit Malfunction

P0482 Cooling Fan no. 3 Control Circuit Malfunction

P0483 Cooling Fan Rationality Check Malfunction

P0484 Cooling Fan Circuit Over Current

P0485 Cooling Fan Power/Ground Circuit Malfunction

P0500 Vehicle Speed Sensor Malfunction

P0501 Vehicle Speed Sensor Range/Performance

P0502 Vehicle Speed Sensor Circuit Low Input

P0503 Vehicle Speed Sensor Intermittent/Erratic/High

P0505 Idle Control System Malfunction

P0506 Idle Control System RPM Lower Than Expected

P0507 Idle Control System RPM Higher Than Expected

P0510 Closed Throttle Position Switch Malfunction

P0520 Engine Oil Pressure Sensor/Switch Circuit Malfunction

P0521 Engine Oil Pressure Sensor/Switch Range/Performance

P0522 Engine Oil Pressure Sensor/Switch Low Voltage

P0523 Engine Oil Pressure Sensor/Switch High Voltage

P0530 A/C Refrigerant Pressure Sensor Circuit Malfunction

P0531 A/C Refrigerant Pressure Sensor Circuit Range/Performance

P0532 A/C Refrigerant Pressure Sensor Circuit Low Input

P0533 A/C Refrigerant Pressure Sensor Circuit High Input

P0534 A/C Refrigerant Charge Loss

P0550 Power Steering Pressure Sensor Circuit Malfunction

P0551 Power Steering Pressure Sensor Circuit Range/Performance

P0552 Power Steering Pressure Sensor Circuit Low Input

P0553 Power Steering Pressure Sensor Circuit High Input

P0554 Power Steering Pressure Sensor Circuit Intermittent

P0560 System Voltage Malfunction

P0561 System Voltage Unstable

P0562 System Voltage Low

P0563 System Voltage High

P0565 Cruise Control On Signal Malfunction

P0566 Cruise Control Off Signal Malfunction

P0567 Cruise Control Resume Signal Malfunction

P0568 Cruise Control Set Signal Malfunction

P0569 Cruise Control Coast Signal Malfunction

P0570 Cruise Control Accel Signal Malfunction

P0571 Cruise Control/Brake Switch "A" Circuit Malfunction

P0572 Cruise Control/Brake Switch "A" Circuit Low

P0573 Cruise Control/Brake Switch "A" Circuit High

P0574 Through P0580 Reserved for Cruise Codes

P0600 Serial Communication Link Malfunction

P0601 Internal Control Module Memory Check Sum Error

P0602 Control Module Programming Error

P0603 Internal Control Module Keep Alive Memory (KAM) Error

P0604 Internal Control Module Random Access Memory (RAM) Error

P0605 Internal Control Module Read Only Memory (ROM) Error

P0606 PCM Processor Fault

P0608 Control Module VSS Output "A" Malfunction

P0609 Control Module VSS Output "B" Malfunction

P0620 Generator Control Circuit Malfunction

P0621 Generator Lamp "L" Control Circuit Malfunction

P0622 Generator Field "F" Control Circuit Malfunction

P0650 Malfunction Indicator Lamp (MIL) Control Circuit Malfunction

P0654 Engine RPM Output Circuit Malfunction

P0655 Engine Hot Lamp Output Control Circuit Malfunction

P0656 Fuel Level Output Circuit Malfunction

P0700 Transmission Control System Malfunction

P0701 Transmission Control System Range/Performance

P0702 Transmission Control System Electrical

P0703 Torque Converter/Brake Switch "B" Circuit Malfunction

P0704 Clutch Switch Input Circuit Malfunction

P0705 Transmission Range Sensor Circuit Malfunction (PRNDL Input)

P0706 Transmission Range Sensor Circuit Range/Performance

P0707 Transmission Range Sensor Circuit Low Input

P0708 Transmission Range Sensor Circuit High Input

P0709 Transmission Range Sensor Circuit Intermittent

P0710 Transmission Fluid Temperature Sensor Circuit Malfunction

P0711 Transmission Fluid Temperature Sensor Circuit Range/Performance

P0712 Transmission Fluid Temperature Sensor Circuit Low Input

P0713 Transmission Fluid Temperature Sensor Circuit High Input

P0714 Transmission Fluid Temperature Sensor Circuit Intermittent

P0715 Input/Turbine Speed Sensor Circuit Malfunction

P0716 Input/Turbine Speed Sensor Circuit Range/Performance

P0717 Input/Turbine Speed Sensor Circuit No Signal

P0718 Input/Turbine Speed Sensor Circuit Intermittent

P0719 Torque Converter/Brake Switch "B" Circuit Low

P0720 Output Speed Sensor Circuit Malfunction

P0721 Output Speed Sensor Circuit Range/Performance

P0722 Output Speed Sensor Circuit No Signal

P0723 Output Speed Sensor Circuit Intermittent

P0724 Torque Converter/Brake Switch "B" Circuit High

P0725 Engine Speed Input Circuit Malfunction

P0726 Engine Speed Input Circuit Range/Performance

P0727 Engine Speed Input Circuit No Signal

P0728 Engine Speed Input Circuit Intermittent

P0730 Incorrect Gear Ratio

P0731 Gear no. 1 Incorrect Ratio

P0732 Gear no. 2 Incorrect Ratio

P0733 Gear no. 3 Incorrect Ratio

P0734 Gear no. 4 Incorrect Ratio

P0735 Gear no. 5 Incorrect Ratio

P0736 Reverse Incorrect Ratio

P0740 Torque Converter Clutch Circuit Malfunction

P0741 Torque Converter Clutch Circuit Performance or Stuck Off

P0742 Torque Converter Clutch Circuit Stuck On

P0743 Torque Converter Clutch Circuit Electrical

P0744 Torque Converter Clutch Circuit Intermittent

P0745 Pressure Control Solenoid Malfunction

P0746 Pressure Control Solenoid Performance or Stuck Off

P0747 Pressure Control Solenoid Stuck On

P0748 Pressure Control Solenoid Electrical

P0749 Pressure Control Solenoid Intermittent

P0750 Shift Solenoid "A" Malfunction

P0751 Shift Solenoid "A" Performance or Stuck Off

P0752 Shift Solenoid "A" Stuck On

P0753 Shift Solenoid "A" Electrical

P0754 Shift Solenoid "A" Intermittent

P0755 Shift Solenoid "B" Malfunction

P0756 Shift Solenoid "B" Performance or Stuck Off

P0757 Shift Solenoid "B" Stuck On

P0758 Shift Solenoid "B" Electrical

P0759 Shift Solenoid "B" Intermittent

P0760 Shift Solenoid "C" Malfunction

P0761 Shift Solenoid "C" Performance Or Stuck Off

P0762 Shift Solenoid "C" Stuck On

P0763 Shift Solenoid "C" Electrical

P0764 Shift Solenoid "C" Intermittent

P0765 Shift Solenoid "D" Malfunction

P0766 Shift Solenoid "D" Performance Or Stuck Off

P0767 Shift Solenoid "D" Stuck On

P0768 Shift Solenoid "D" Electrical

P0769 Shift Solenoid "D" Intermittent

P0770 Shift Solenoid "E" Malfunction

P0771 Shift Solenoid "E" Performance Or Stuck Off

P0772 Shift Solenoid "E" Stuck On

P0773 Shift Solenoid "E" Electrical

P0774 Shift Solenoid "E" Intermittent

P0780 Shift Malfunction

P0781 1–2 Shift Malfunction

P0782 2–3 Shift Malfunction

P0783 3–4 Shift Malfunction

P0784 4–5 Shift Malfunction

P0785 Shift/Timing Solenoid Malfunction

P0786 Shift/Timing Solenoid Range/Performance

P0787 Shift/Timing Solenoid Low

P0788 Shift/Timing Solenoid High

P0789 Shift/Timing Solenoid Intermittent

P0790 Normal/Performance Switch Circuit Malfunction

P0801 Reverse Inhibit Control Circuit Malfunction

P0803 1–4 Upshift (Skip Shift) Solenoid Control Circuit Malfunction

P0804 1–4 Upshift (Skip Shift) Lamp Control Circuit Malfunction

P1106 Map Sensor Circuit Intermittent High Voltage

P1107 MAP Sensor Circuit Intermittent Low Voltage

P1111 IAT Sensor Circuit Intermittent High Voltage

P1112 IAT Sensor Circuit Intermittent Low Voltage

P1114 ECT Sensor Circuit Intermittent Low Voltage

P1115 ECT Sensor Circuit Intermittent High Voltage

P1121 TP Sensor Circuit Intermittent High Voltage

P1122 TP Sensor Circuit Intermittent Low Voltage

P1133 HO2S-11 Insufficient Switching (Bank 1 Sensor 1)

P1134 HO2S-11 Transition Time Ratio (Bank 1 Sensor 1)

P1153 HO2S-21 Insufficient Switching (Bank 2 Sensor I)

P1154 HO2S-21 Transition Time Ratio (Bank 2 Sensor 1)

P1171 Fuel System Lean During Acceleration

P1391 G-Acceleration Sensor Intermittent Low Voltage

P1390 G-Acceleration (Low G) Sensor Performance

P1392 Rough Road G-Sensor Circuit Low Voltage

P1393 Rough Road G-Sensor Circuit High Voltage

P1394 G-Acceleration Sensor Intermittent High Voltage

P1406 EGR Valve Pintle Position Sensor Circuit Fault

P1441 EVAP System Flow During Non-Purge

P1442 EVAP System Flow During Non-Purge

P1508 Idle Speed Control System-Low

P1509 Idle Speed Control System-High

P1618 Serial Peripheral Interface Communication Error

P1640 Output Driver Module 'A' Fault

P1790 PCM ROM (Transmission Side) Check Sum Error

P1792 PCM EEPROM (Transmission Side) Check Sum Error

P1835 Kick Down Switch Always On

P1850 Brake Band Apply Solenoid Electrical Fault

P1860 TCC PWM Solenoid Electrical Fault

P1870 Transmission Component Slipping

Kia

READING CODES

Reading the control module memory is on of the first steps in OBD II system diagnostics. This step should be initially performed to determine the general nature of the fault. Subsequent readings will determine if the fault has been cleared.

Reading codes can be performed by any of the methods below:

- Read the control module memory with the Generic Scan Tool (GST)
- Read the control module memory with the vehicle manufacturer's specific tester

To read the fault codes, connect the scan tool or tester according to the manufacturer's instructions. Follow the manufacturer's specified procedure for reading the codes.

CLEARING CODES

Control module reset procedures are a very important part of OBD II System diagnostics. This step should be done at the end of any fault code repair and at the end of any driveability repair.

Clearing codes can be performed by any of the methods below:

• Clear the control module memory with the Generic Scan Tool (GST)

• Clear the control module memory with the vehicle manufacturer's specific tester

• Turn the ignition **OFF** and remove the negative battery cable for at least 1 minute.

Removing the negative battery cable may cause other systems in the vehicle to loose their memory. Prior to removing the cable, ensure you have the proper reset codes for radios and alarms.

➡**The MIL will may also be de-activated for some codes if the vehicle completes three consecutive trips without a fault detected with vehicle conditions similar to those present during the fault.**

OBD II TROUBLE CODES

P0100 Mass or Volume Air Flow Circuit Malfunction
P0101 Mass or Volume Air Flow Circuit Range/Performance Problem
P0102 Mass or Volume Air Flow Circuit Low Input
P0103 Mass or Volume Air Flow Circuit High Input
P0104 Mass or Volume Air Flow Circuit Intermittent
P0105 Manifold Absolute Pressure/Barometric Pressure Circuit Malfunction
P0106 Manifold Absolute Pressure/Barometric Pressure Circuit Range/Performance Problem
P0107 Manifold Absolute Pressure/Barometric Pressure Circuit Low Input
P0108 Manifold Absolute Pressure/Barometric Pressure Circuit High Input
P0109 Manifold Absolute Pressure/Barometric Pressure Circuit Intermittent
P0110 Intake Air Temperature Circuit Malfunction
P0111 Intake Air Temperature Circuit Range/Performance Problem
P0112 Intake Air Temperature Circuit Low Input
P0113 Intake Air Temperature Circuit High Input
P0114 Intake Air Temperature Circuit Intermittent
P0115 Engine Coolant Temperature Circuit Malfunction
P0116 Engine Coolant Temperature Circuit Range/Performance Problem
P0117 Engine Coolant Temperature Circuit Low Input
P0118 Engine Coolant Temperature Circuit High Input
P0119 Engine Coolant Temperature Circuit Intermittent
P0120 Throttle/Pedal Position Sensor/Switch "A" Circuit Malfunction
P0121 Throttle/Pedal Position Sensor/Switch "A" Circuit Range/Performance Problem
P0122 Throttle/Pedal Position Sensor/Switch "A" Circuit Low Input

P0123 Throttle/Pedal Position Sensor/Switch "A" Circuit High Input
P0124 Throttle/Pedal Position Sensor/Switch "A" Circuit Intermittent
P0125 Insufficient Coolant Temperature For Closed Loop Fuel Control
P0126 Insufficient Coolant Temperature For Stable Operation
P0130 O2 Circuit Malfunction (Bank no. 1 Sensor no. 1)
P0131 O2 Sensor Circuit Low Voltage (Bank no. 1 Sensor no. 1)
P0132 O2 Sensor Circuit High Voltage (Bank no. 1 Sensor no. 1)
P0133 O2 Sensor Circuit Slow Response (Bank no. 1 Sensor no. 1)
P0134 O2 Sensor Circuit No Activity Detected (Bank no. 1 Sensor no. 1)
P0135 O2 Sensor Heater Circuit Malfunction (Bank no. 1 Sensor no. 1)
P0136 O2 Sensor Circuit Malfunction (Bank no. 1 Sensor no. 2)
P0137 O2 Sensor Circuit Low Voltage (Bank no. 1 Sensor no. 2)
P0138 O2 Sensor Circuit High Voltage (Bank no. 1 Sensor no. 2)
P0139 O2 Sensor Circuit Slow Response (Bank no. 1 Sensor no. 2)
P0140 O2 Sensor Circuit No Activity Detected (Bank no. 1 Sensor no. 2)
P0141 O2 Sensor Heater Circuit Malfunction (Bank no. 1 Sensor no. 2)
P0142 O2 Sensor Circuit Malfunction (Bank no. 1 Sensor no. 3)
P0143 O2 Sensor Circuit Low Voltage (Bank no. 1 Sensor no. 3)
P0144 O2 Sensor Circuit High Voltage (Bank no. 1 Sensor no. 3)
P0145 O2 Sensor Circuit Slow Response (Bank no. 1 Sensor no. 3)
P0146 O2 Sensor Circuit No Activity Detected (Bank no. 1 Sensor no. 3)
P0147 O2 Sensor Heater Circuit Malfunction (Bank no. 1 Sensor no. 3)
P0150 O2 Sensor Circuit Malfunction (Bank no. 2 Sensor no. 1)
P0151 O2 Sensor Circuit Low Voltage (Bank no. 2 Sensor no. 1)
P0152 O2 Sensor Circuit High Voltage (Bank no. 2 Sensor no. 1)
P0153 O2 Sensor Circuit Slow Response (Bank no. 2 Sensor no. 1)
P0154 O2 Sensor Circuit No Activity Detected (Bank no. 2 Sensor no. 1)
P0155 O2 Sensor Heater Circuit Malfunction (Bank no. 2 Sensor no. 1)
P0156 O2 Sensor Circuit Malfunction (Bank no. 2 Sensor no. 2)
P0157 O2 Sensor Circuit Low Voltage (Bank no. 2 Sensor no. 2)
P0158 O2 Sensor Circuit High Voltage (Bank no. 2 Sensor no. 2)
P0159 O2 Sensor Circuit Slow Response (Bank no. 2 Sensor no. 2)
P0160 O2 Sensor Circuit No Activity Detected (Bank no. 2 Sensor no. 2)
P0161 O2 Sensor Heater Circuit Malfunction (Bank no. 2 Sensor no. 2)
P0162 O2 Sensor Circuit Malfunction (Bank no. 2 Sensor no. 3)
P0163 O2 Sensor Circuit Low Voltage (Bank no. 2 Sensor no. 3)
P0164 O2 Sensor Circuit High Voltage (Bank no. 2 Sensor no. 3)

P0165 O2 Sensor Circuit Slow Response (Bank no. 2 Sensor no. 3)
P0166 O2 Sensor Circuit No Activity Detected (Bank no. 2 Sensor no. 3)
P0167 O2 Sensor Heater Circuit Malfunction (Bank no. 2 Sensor no. 3)
P0170 Fuel Trim Malfunction (Bank no. 1)
P0171 System Too Lean (Bank no. 1)
P0172 System Too Rich (Bank no. 1)
P0173 Fuel Trim Malfunction (Bank no. 2)
P0174 System Too Lean (Bank no. 2)
P0175 System Too Rich (Bank no. 2)
P0176 Fuel Composition Sensor Circuit Malfunction
P0177 Fuel Composition Sensor Circuit Range/Performance
P0178 Fuel Composition Sensor Circuit Low Input
P0179 Fuel Composition Sensor Circuit High Input
P0180 Fuel Temperature Sensor "A" Circuit Malfunction
P0181 Fuel Temperature Sensor "A" Circuit Range/Performance
P0182 Fuel Temperature Sensor "A" Circuit Low Input
P0183 Fuel Temperature Sensor "A" Circuit High Input
P0184 Fuel Temperature Sensor "A" Circuit Intermittent
P0185 Fuel Temperature Sensor "B" Circuit Malfunction
P0186 Fuel Temperature Sensor "B" Circuit Range/Performance
P0187 Fuel Temperature Sensor "B" Circuit Low Input
P0188 Fuel Temperature Sensor "B" Circuit High Input
P0189 Fuel Temperature Sensor "B" Circuit Intermittent
P0190 Fuel Rail Pressure Sensor Circuit Malfunction
P0191 Fuel Rail Pressure Sensor Circuit Range/Performance
P0192 Fuel Rail Pressure Sensor Circuit Low Input
P0193 Fuel Rail Pressure Sensor Circuit High Input
P0194 Fuel Rail Pressure Sensor Circuit Intermittent
P0195 Engine Oil Temperature Sensor Malfunction
P0196 Engine Oil Temperature Sensor Range/Performance
P0197 Engine Oil Temperature Sensor Low
P0198 Engine Oil Temperature Sensor High
P0199 Engine Oil Temperature Sensor Intermittent
P0200 Injector Circuit Malfunction
P0201 Injector Circuit Malfunction—Cylinder no. 1
P0202 Injector Circuit Malfunction—Cylinder no. 2
P0203 Injector Circuit Malfunction—Cylinder no. 3
P0204 Injector Circuit Malfunction—Cylinder no. 4
P0205 Injector Circuit Malfunction—Cylinder no. 5
P0206 Injector Circuit Malfunction—Cylinder no. 6
P0207 Injector Circuit Malfunction—Cylinder no. 7
P0208 Injector Circuit Malfunction—Cylinder no. 8
P0209 Injector Circuit Malfunction—Cylinder no. 9
P0210 Injector Circuit Malfunction—Cylinder no. 10
P0211 Injector Circuit Malfunction—Cylinder no. 11
P0212 Injector Circuit Malfunction—Cylinder no. 12
P0213 Cold Start Injector no. 1 Malfunction
P0214 Cold Start Injector no. 2 Malfunction
P0215 Engine Shutoff Solenoid Malfunction
P0216 Injection Timing Control Circuit Malfunction
P0217 Engine Over Temperature Condition
P0218 Transmission Over Temperature Condition
P0219 Engine Over Speed Condition
P0220 Throttle/Pedal Position Sensor/Switch "B" Circuit Malfunction
P0221 Throttle/Pedal Position Sensor/Switch "B" Circuit Range/Performance Problem
P0222 Throttle/Pedal Position Sensor/Switch "B" Circuit Low Input

P0223 Throttle/Pedal Position Sensor/Switch "B" Circuit High Input
P0224 Throttle/Pedal Position Sensor/Switch "B" Circuit Intermittent
P0225 Throttle/Pedal Position Sensor/Switch "C" Circuit Malfunction
P0226 Throttle/Pedal Position Sensor/Switch "C" Circuit Range/Performance Problem
P0227 Throttle/Pedal Position Sensor/Switch "C" Circuit Low Input
P0228 Throttle/Pedal Position Sensor/Switch "C" Circuit High Input
P0229 Throttle/Pedal Position Sensor/Switch "C" Circuit Intermittent
P0230 Fuel Pump Primary Circuit Malfunction
P0231 Fuel Pump Secondary Circuit Low
P0232 Fuel Pump Secondary Circuit High
P0233 Fuel Pump Secondary Circuit Intermittent
P0234 Engine Over Boost Condition
P0261 Cylinder no. 1 Injector Circuit Low
P0262 Cylinder no. 1 Injector Circuit High
P0263 Cylinder no. 1 Contribution/Balance Fault
P0264 Cylinder no. 2 Injector Circuit Low
P0265 Cylinder no. 2 Injector Circuit High
P0266 Cylinder no. 2 Contribution/Balance Fault
P0267 Cylinder no. 3 Injector Circuit Low
P0268 Cylinder no. 3 Injector Circuit High
P0269 Cylinder no. 3 Contribution/Balance Fault
P0270 Cylinder no. 4 Injector Circuit Low
P0271 Cylinder no. 4 Injector Circuit High
P0272 Cylinder no. 4 Contribution/Balance Fault
P0273 Cylinder no. 5 Injector Circuit Low
P0274 Cylinder no. 5 Injector Circuit High
P0275 Cylinder no. 5 Contribution/Balance Fault
P0276 Cylinder no. 6 Injector Circuit Low
P0277 Cylinder no. 6 Injector Circuit High
P0278 Cylinder no. 6 Contribution/Balance Fault
P0279 Cylinder no. 7 Injector Circuit Low
P0280 Cylinder no. 7 Injector Circuit High
P0281 Cylinder no. 7 Contribution/Balance Fault
P0282 Cylinder no. 8 Injector Circuit Low
P0283 Cylinder no. 8 Injector Circuit High
P0284 Cylinder no. 8 Contribution/Balance Fault
P0285 Cylinder no. 9 Injector Circuit Low
P0286 Cylinder no. 9 Injector Circuit High
P0287 Cylinder no. 9 Contribution/Balance Fault
P0288 Cylinder no. 10 Injector Circuit Low
P0289 Cylinder no. 10 Injector Circuit High
P0290 Cylinder no. 10 Contribution/Balance Fault
P0291 Cylinder no. 11 Injector Circuit Low
P0292 Cylinder no. 11 Injector Circuit High
P0293 Cylinder no. 11 Contribution/Balance Fault
P0294 Cylinder no. 12 Injector Circuit Low
P0295 Cylinder no. 12 Injector Circuit High
P0296 Cylinder no. 12 Contribution/Balance Fault
P0300 Random/Multiple Cylinder Misfire Detected
P0301 Cylinder no. 1—Misfire Detected
P0302 Cylinder no. 2—Misfire Detected
P0303 Cylinder no. 3—Misfire Detected
P0304 Cylinder no. 4—Misfire Detected
P0305 Cylinder no. 5—Misfire Detected
P0306 Cylinder no. 6—Misfire Detected

P0307 Cylinder no. 7—Misfire Detected

P0308 Cylinder no. 8—Misfire Detected

P0309 Cylinder no. 9—Misfire Detected

P0310 Cylinder no. 10—Misfire Detected

P0311 Cylinder no. 11—Misfire Detected

P0312 Cylinder no. 12—Misfire Detected

P0320 Ignition/Distributor Engine Speed Input Circuit Malfunction

P0321 Ignition/Distributor Engine Speed Input Circuit Range/Performance

P0322 Ignition/Distributor Engine Speed Input Circuit No Signal

P0323 Ignition/Distributor Engine Speed Input Circuit Intermittent

P0325 Knock Sensor no. 1—Circuit Malfunction (Bank no. 1 or Single Sensor)

P0326 Knock Sensor no. 1—Circuit Range/Performance (Bank no. 1 or Single Sensor)

P0327 Knock Sensor no. 1—Circuit Low Input (Bank no. 1 or Single Sensor)

P0328 Knock Sensor no. 1—Circuit High Input (Bank no. 1 or Single Sensor)

P0329 Knock Sensor no. 1—Circuit Input Intermittent (Bank no. 1 or Single Sensor)

P0330 Knock Sensor no. 2—Circuit Malfunction (Bank no. 2)

P0331 Knock Sensor no. 2—Circuit Range/Performance (Bank no. 2)

P0332 Knock Sensor no. 2—Circuit Low Input (Bank no. 2)

P0333 Knock Sensor no. 2—Circuit High Input (Bank no. 2)

P0334 Knock Sensor no. 2—Circuit Input Intermittent (Bank no. 2)

P0335 Crankshaft Position Sensor "A" Circuit Malfunction

P0336 Crankshaft Position Sensor "A" Circuit Range/Performance

P0337 Crankshaft Position Sensor "A" Circuit Low Input

P0338 Crankshaft Position Sensor "A" Circuit High Input

P0339 Crankshaft Position Sensor "A" Circuit Intermittent

P0340 Camshaft Position Sensor Circuit Malfunction

P0341 Camshaft Position Sensor Circuit Range/Performance

P0342 Camshaft Position Sensor Circuit Low Input

P0343 Camshaft Position Sensor Circuit High Input

P0344 Camshaft Position Sensor Circuit Intermittent

P0350 Ignition Coil Primary/Secondary Circuit Malfunction

P0351 Ignition Coil "A" Primary/Secondary Circuit Malfunction

P0352 Ignition Coil "B" Primary/Secondary Circuit Malfunction

P0353 Ignition Coil "C" Primary/Secondary Circuit Malfunction

P0354 Ignition Coil "D" Primary/Secondary Circuit Malfunction

P0355 Ignition Coil "E" Primary/Secondary Circuit Malfunction

P0356 Ignition Coil "F" Primary/Secondary Circuit Malfunction

P0357 Ignition Coil "G" Primary/Secondary Circuit Malfunction

P0358 Ignition Coil "H" Primary/Secondary Circuit Malfunction

P0359 Ignition Coil "I" Primary/Secondary Circuit Malfunction

P0360 Ignition Coil "J" Primary/Secondary Circuit Malfunction

P0361 Ignition Coil "K" Primary/Secondary Circuit Malfunction

P0362 Ignition Coil "L" Primary/Secondary Circuit Malfunction

P0370 Timing Reference High Resolution Signal "A" Malfunction

P0371 Timing Reference High Resolution Signal "A" Too Many Pulses

P0372 Timing Reference High Resolution Signal "A" Too Few Pulses

P0373 Timing Reference High Resolution Signal "A" Intermittent/Erratic Pulses

P0374 Timing Reference High Resolution Signal "A" No Pulses

P0375 Timing Reference High Resolution Signal "B" Malfunction

P0376 Timing Reference High Resolution Signal "B" Too Many Pulses

P0377 Timing Reference High Resolution Signal "B" Too Few Pulses

P0378 Timing Reference High Resolution Signal "B" Intermittent/Erratic Pulses

P0379 Timing Reference High Resolution Signal "B" No Pulses

P0380 Glow Plug/Heater Circuit "A" Malfunction

P0381 Glow Plug/Heater Indicator Circuit Malfunction

P0382 Glow Plug/Heater Circuit "B" Malfunction

P0385 Crankshaft Position Sensor "B" Circuit Malfunction

P0386 Crankshaft Position Sensor "B" Circuit Range/Performance

P0387 Crankshaft Position Sensor "B" Circuit Low Input

P0388 Crankshaft Position Sensor "B" Circuit High Input

P0389 Crankshaft Position Sensor "B" Circuit Intermittent

P0400 Exhaust Gas Recirculation Flow Malfunction

P0401 Exhaust Gas Recirculation Flow Insufficient Detected

P0402 Exhaust Gas Recirculation Flow Excessive Detected

P0403 Exhaust Gas Recirculation Circuit Malfunction

P0404 Exhaust Gas Recirculation Circuit Range/Performance

P0405 Exhaust Gas Recirculation Sensor "A" Circuit Low

P0406 Exhaust Gas Recirculation Sensor "A" Circuit High

P0407 Exhaust Gas Recirculation Sensor "B" Circuit Low

P0408 Exhaust Gas Recirculation Sensor "B" Circuit High

P0410 Secondary Air Injection System Malfunction

P0411 Secondary Air Injection System Incorrect Flow Detected

P0412 Secondary Air Injection System Switching Valve "A" Circuit Malfunction

P0413 Secondary Air Injection System Switching Valve "A" Circuit Open

P0414 Secondary Air Injection System Switching Valve "A" Circuit Shorted

P0415 Secondary Air Injection System Switching Valve "B" Circuit Malfunction

P0416 Secondary Air Injection System Switching Valve "B" Circuit Open

P0417 Secondary Air Injection System Switching Valve "B" Circuit Shorted

P0418 Secondary Air Injection System Relay "A" Circuit Malfunction

P0419 Secondary Air Injection System Relay "B" Circuit Malfunction

P0420 Catalyst System Efficiency Below Threshold (Bank no. 1)

P0421 Warm Up Catalyst Efficiency Below Threshold (Bank no. 1)

P0422 Main Catalyst Efficiency Below Threshold (Bank no. 1)

P0423 Heated Catalyst Efficiency Below Threshold (Bank no. 1)

P0424 Heated Catalyst Temperature Below Threshold (Bank no. 1)

P0430 Catalyst System Efficiency Below Threshold (Bank no. 2)

P0431 Warm Up Catalyst Efficiency Below Threshold (Bank no. 2)

P0432 Main Catalyst Efficiency Below Threshold (Bank no. 2)

P0433 Heated Catalyst Efficiency Below Threshold (Bank no. 2)

P0434 Heated Catalyst Temperature Below Threshold (Bank no. 2)

P0440 Evaporative Emission Control System Malfunction

P0441 Evaporative Emission Control System Incorrect Purge Flow

P0442 Evaporative Emission Control System Leak Detected (Small Leak)

P0443 Evaporative Emission Control System Purge Control Valve Circuit Malfunction

P0444 Evaporative Emission Control System Purge Control Valve Circuit Open

P0445 Evaporative Emission Control System Purge Control Valve Circuit Shorted

P0446 Evaporative Emission Control System Vent Control Circuit Malfunction

P0447 Evaporative Emission Control System Vent Control Circuit Open

P0448 Evaporative Emission Control System Vent Control Circuit Shorted

P0449 Evaporative Emission Control System Vent Valve/Solenoid Circuit Malfunction

P0450 Evaporative Emission Control System Pressure Sensor Malfunction

P0451 Evaporative Emission Control System Pressure Sensor Range/Performance

P0452 Evaporative Emission Control System Pressure Sensor Low Input

P0453 Evaporative Emission Control System Pressure Sensor High Input

P0454 Evaporative Emission Control System Pressure Sensor Intermittent

P0455 Evaporative Emission Control System Leak Detected (Gross Leak)

P0460 Fuel Level Sensor Circuit Malfunction

P0461 Fuel Level Sensor Circuit Range/Performance

P0462 Fuel Level Sensor Circuit Low Input

P0463 Fuel Level Sensor Circuit High Input

P0464 Fuel Level Sensor Circuit Intermittent

P0465 Purge Flow Sensor Circuit Malfunction

P0466 Purge Flow Sensor Circuit Range/Performance

P0467 Purge Flow Sensor Circuit Low Input

P0468 Purge Flow Sensor Circuit High Input

P0469 Purge Flow Sensor Circuit Intermittent

P0470 Exhaust Pressure Sensor Malfunction

P0471 Exhaust Pressure Sensor Range/Performance

P0472 Exhaust Pressure Sensor Low

P0473 Exhaust Pressure Sensor High

P0474 Exhaust Pressure Sensor Intermittent

P0475 Exhaust Pressure Control Valve Malfunction

P0476 Exhaust Pressure Control Valve Range/Performance

P0477 Exhaust Pressure Control Valve Low

P0478 Exhaust Pressure Control Valve High

P0479 Exhaust Pressure Control Valve Intermittent

P0480 Cooling Fan no. 1 Control Circuit Malfunction

P0481 Cooling Fan no. 2 Control Circuit Malfunction

P0482 Cooling Fan no. 3 Control Circuit Malfunction

P0483 Cooling Fan Rationality Check Malfunction

P0484 Cooling Fan Circuit Over Current

P0485 Cooling Fan Power/Ground Circuit Malfunction

P0500 Vehicle Speed Sensor Malfunction

P0501 Vehicle Speed Sensor Range/Performance

P0502 Vehicle Speed Sensor Circuit Low Input

P0503 Vehicle Speed Sensor Intermittent/Erratic/High

P0505 Idle Control System Malfunction

P0506 Idle Control System RPM Lower Than Expected

P0507 Idle Control System RPM Higher Than Expected

P0510 Closed Throttle Position Switch Malfunction

P0520 Engine Oil Pressure Sensor/Switch Circuit Malfunction

P0521 Engine Oil Pressure Sensor/Switch Range/Performance

P0522 Engine Oil Pressure Sensor/Switch Low Voltage

P0523 Engine Oil Pressure Sensor/Switch High Voltage

P0530 A/C Refrigerant Pressure Sensor Circuit Malfunction

P0531 A/C Refrigerant Pressure Sensor Circuit Range/Performance

P0532 A/C Refrigerant Pressure Sensor Circuit Low Input

P0533 A/C Refrigerant Pressure Sensor Circuit High Input

P0534 A/C Refrigerant Charge Loss

P0550 Power Steering Pressure Sensor Circuit Malfunction

P0551 Power Steering Pressure Sensor Circuit Range/Performance

P0552 Power Steering Pressure Sensor Circuit Low Input

P0553 Power Steering Pressure Sensor Circuit High Input

P0554 Power Steering Pressure Sensor Circuit Intermittent

P0560 System Voltage Malfunction

P0561 System Voltage Unstable

P0562 System Voltage Low

P0563 System Voltage High

P0565 Cruise Control On Signal Malfunction

P0566 Cruise Control Off Signal Malfunction

P0567 Cruise Control Resume Signal Malfunction

P0568 Cruise Control Set Signal Malfunction

P0569 Cruise Control Coast Signal Malfunction

P0570 Cruise Control Accel Signal Malfunction

P0571 Cruise Control/Brake Switch "A" Circuit Malfunction

P0572 Cruise Control/Brake Switch "A" Circuit Low

P0573 Cruise Control/Brake Switch "A" Circuit High

P0574 Through P0580 Reserved for Cruise Codes

P0600 Serial Communication Link Malfunction

P0601 Internal Control Module Memory Check Sum Error

P0602 Control Module Programming Error

P0603 Internal Control Module Keep Alive Memory (KAM) Error

P0604 Internal Control Module Random Access Memory (RAM) Error

P0605 Internal Control Module Read Only Memory (ROM) Error

P0606 PCM Processor Fault

P0608 Control Module VSS Output "A" Malfunction

P0609 Control Module VSS Output "B" Malfunction

P0620 Generator Control Circuit Malfunction

P0621 Generator Lamp "L" Control Circuit Malfunction

P0622 Generator Field "F" Control Circuit Malfunction

P0650 Malfunction Indicator Lamp (MIL) Control Circuit Malfunction

P0654 Engine RPM Output Circuit Malfunction

P0655 Engine Hot Lamp Output Control Circuit Malfunction

P0656 Fuel Level Output Circuit Malfunction

P0700 Transmission Control System Malfunction

P0701 Transmission Control System Range/Performance

P0702 Transmission Control System Electrical

P0703 Torque Converter/Brake Switch "B" Circuit Malfunction

P0704 Clutch Switch Input Circuit Malfunction

P0705 Transmission Range Sensor Circuit Malfunction (PRNDL Input)

P0706 Transmission Range Sensor Circuit Range/Performance

P0707 Transmission Range Sensor Circuit Low Input

P0708 Transmission Range Sensor Circuit High Input

P0709 Transmission Range Sensor Circuit Intermittent

P0710 Transmission Fluid Temperature Sensor Circuit Malfunction

P0711 Transmission Fluid Temperature Sensor Circuit Range/Performance

P0712 Transmission Fluid Temperature Sensor Circuit Low Input

P0713 Transmission Fluid Temperature Sensor Circuit High Input

P0714 Transmission Fluid Temperature Sensor Circuit Intermittent

P0715 Input/Turbine Speed Sensor Circuit Malfunction

P0716 Input/Turbine Speed Sensor Circuit Range/Performance

P0717 Input/Turbine Speed Sensor Circuit No Signal

P0718 Input/Turbine Speed Sensor Circuit Intermittent

P0719 Torque Converter/Brake Switch "B" Circuit Low

P0720 Output Speed Sensor Circuit Malfunction

P0721 Output Speed Sensor Circuit Range/Performance

P0722 Output Speed Sensor Circuit No Signal

P0723 Output Speed Sensor Circuit Intermittent

P0724 Torque Converter/Brake Switch "B" Circuit High

P0725 Engine Speed Input Circuit Malfunction

P0726 Engine Speed Input Circuit Range/Performance

P0727 Engine Speed Input Circuit No Signal

P0728 Engine Speed Input Circuit Intermittent

P0730 Incorrect Gear Ratio

P0731 Gear no. 1 Incorrect Ratio

P0732 Gear no. 2 Incorrect Ratio

P0733 Gear no. 3 Incorrect Ratio

P0734 Gear no. 4 Incorrect Ratio

P0735 Gear no. 5 Incorrect Ratio

P0736 Reverse Incorrect Ratio

P0740 Torque Converter Clutch Circuit Malfunction

P0741 Torque Converter Clutch Circuit Performance or Stuck Off

P0742 Torque Converter Clutch Circuit Stuck On

P0743 Torque Converter Clutch Circuit Electrical

P0744 Torque Converter Clutch Circuit Intermittent

P0745 Pressure Control Solenoid Malfunction

P0746 Pressure Control Solenoid Performance or Stuck Off

P0747 Pressure Control Solenoid Stuck On

P0748 Pressure Control Solenoid Electrical

P0749 Pressure Control Solenoid Intermittent

P0750 Shift Solenoid "A" Malfunction

P0751 Shift Solenoid "A" Performance or Stuck Off

P0752 Shift Solenoid "A" Stuck On

P0753 Shift Solenoid "A" Electrical

P0754 Shift Solenoid "A" Intermittent

P0755 Shift Solenoid "B" Malfunction

P0756 Shift Solenoid "B" Performance or Stuck Off

P0757 Shift Solenoid "B" Stuck On

P0758 Shift Solenoid "B" Electrical

P0759 Shift Solenoid "B" Intermittent

P0760 Shift Solenoid "C" Malfunction

P0761 Shift Solenoid "C" Performance Or Stuck Off

P0762 Shift Solenoid "C" Stuck On

P0763 Shift Solenoid "C" Electrical

P0764 Shift Solenoid "C" Intermittent

P0765 Shift Solenoid "D" Malfunction

P0766 Shift Solenoid "D" Performance Or Stuck Off

P0767 Shift Solenoid "D" Stuck On

P0768 Shift Solenoid "D" Electrical

P0769 Shift Solenoid "D" Intermittent

P0770 Shift Solenoid "E" Malfunction

P0771 Shift Solenoid "E" Performance Or Stuck Off

P0772 Shift Solenoid "E" Stuck On

P0773 Shift Solenoid "E" Electrical

P0774 Shift Solenoid "E" Intermittent

P0780 Shift Malfunction

P0781 1–2 Shift Malfunction

P0782 2–3 Shift Malfunction

P0783 3–4 Shift Malfunction

P0784 4–5 Shift Malfunction

P0785 Shift/Timing Solenoid Malfunction

P0786 Shift/Timing Solenoid Range/Performance

P0787 Shift/Timing Solenoid Low

P0788 Shift/Timing Solenoid High

P0789 Shift/Timing Solenoid Intermittent

P0790 Normal/Performance Switch Circuit Malfunction

P0801 Reverse Inhibit Control Circuit Malfunction

P0803 1–4 Upshift (Skip Shift) Solenoid Control Circuit Malfunction

P0804 1–4 Upshift (Skip Shift) Lamp Control Circuit Malfunction

P0740 Torque Converter Clutch System Fault

P0750 TCM Shift Solenoid 'A' Electrical Fault

P0755 TCM Shift Solenoid 'B' Electrical Fault

P0760 TCM Shift Solenoid 'C' Electrical Fault

P1102 H02S-11 Heater Circuit High Voltage

P1105 H02S-12 Heater Circuit High Voltage

P1115 H02S-11 Heater Circuit Low Voltage

P1117 H02S-12 Heater Circuit Low Voltage

P1123 Long Term Fuel Trim Adaptive Air System Low

P1124 Long Term Fuel Trim Adaptive Air System High

P1127 Long Term Fuel Trim Multiplicative Air System Low

P1128 Long Term Fuel Trim Multiplicative Air System High

P1140 Load Calculation Cross Check

P1170 H02S-11 Circuit Voltage Stuck At Mid-Range

P1195 EGR Boost Or Pressure Sensor Circuit Fault

P1196 Ignition Switch Start Circuit Fault

P1213 Fuel Injector 1, 2, 3 Or 4 Circuit High Voltage

P1214 Fuel Injector 1, 2, 3 Or 4 Circuit High Voltage

P1215 Fuel Injector 1, 2, 3 Or 4 Circuit High Voltage

P1216 Fuel Injector 1, 2, 3 Or 4 Circuit High Voltage

P1225 Fuel Injector 1, 2, 5 Or 4 Circuit Low Voltage

P1226 Fuel Injector 1, 2, 5 Or 4 Circuit Low Voltage

P1227 Fuel Injector 1, 2, 5 Or 4 Circuit Low Voltage

P1228 Fuel Injector 1, 2, 5 Or 4 Circuit Low Voltage

P1250 Pressure Regulator Control Solenoid Circuit Fault

P1345 No SGC (CMP) Signal To PCM

P1386 Knock Sensor Control Zero Test

P1401 EGR Control Solenoid Circuit Signal Low

P1402 EGR Control Solenoid Circuit Signal High

P1402 EGR Valve Position Sensor Circuit Fault

P1410 EVAP Purge Control Solenoid Circuit High Voltage

P1412 EGR Differential Pressure Sensor Signal Low

P1413 EGR Differential Pressure Sensor Signal High

P1425 EVAP Purge Control Solenoid Circuit Low Voltage

P1449 Canister Drain Cut Valve Solenoid Circuit Fault

P1455 Fuel Tank Sending Unit Circuit Fault

P1458 Air Conditioning Compressor Clutch Signal Fault

P1485 EGR Vent Control Solenoid Circuit Fault

P1486 EGR Vacuum Control Solenoid Circuit Fault

P1487 EGR Boost Sensor Solenoid Circuit Fault

P1510 Idle Air Control Valve Closing Coil High Voltage

P1513 Idle Air Control Valve Closing Coil Low Voltage
P1515 A/T To M/T Codification
P1523 VICS Solenoid Valve Circuit Fault
P1552 Idle Air Control Valve Opening Coil Low Voltage
P1553 Idle Air Control Valve Opening Coil High Voltage
P1606 Chassis Accelerator Sensor Signal Circuit Fault
P1608 PCM Internal Fault
P1611 MIL Request Circuit Low Voltage
P1614 MIL Request Circuit High Voltage
P1616 Chassis Accelerator Sensor Signal Low Voltage
P1617 Chassis Accelerator Sensor Signal High Voltage
P1624 TCM to PCM MIL Request Circuit Fault
P1655 Unused Power Stage "B"
P1660 Unused Power Stage 'A'
P1660 Unused Power Stage 'B'
P1665 Power Stage Group 'A'
P1743 Torque Converter Clutch Solenoid Circuit Fault
P1794 Battery Or Circuit Fault
P1797 Clutch Pedal Switch (MT) Or PIN Switch Circuit Fault

Lexus

READING CODES

Reading the control module memory is on of the first steps in OBD II system diagnostics. This step should be initially performed to determine the general nature of the fault. Subsequent readings will determine if the fault has been cleared.

Reading codes can be performed by any of the methods below:

• Read the control module memory with the Generic Scan Tool (GST)

• Read the control module memory with the vehicle manufacturer's specific tester

To read the fault codes, connect the scan tool or tester according to the manufacturer's instructions. Follow the manufacturer's specified procedure for reading the codes.

CLEARING CODES

Control module reset procedures are a very important part of OBD II System diagnostics. This step should be done at the end of any fault code repair and at the end of any driveability repair.

Clearing codes can be performed by any of the methods below:

• Clear the control module memory with the Generic Scan Tool (GST)

• Clear the control module memory with the vehicle manufacturer's specific tester

• Turn the ignition **OFF** and remove the negative battery cable for at least 1 minute.

Removing the negative battery cable may cause other systems in the vehicle to loose their memory. Prior to removing the cable, ensure you have the proper reset codes for radios and alarms.

➡ The MIL will may also be de-activated for some codes if the vehicle completes three consecutive trips without a fault detected with vehicle conditions similar to those present during the fault.

OBD II TROUBLE CODES

P0100 Mass or Volume Air Flow Circuit Malfunction
P0101 Mass or Volume Air Flow Circuit Range/Performance Problem
P0102 Mass or Volume Air Flow Circuit Low Input
P0103 Mass or Volume Air Flow Circuit High Input
P0104 Mass or Volume Air Flow Circuit Intermittent
P0105 Manifold Absolute Pressure/Barometric Pressure Circuit Malfunction
P0106 Manifold Absolute Pressure/Barometric Pressure Circuit Range/Performance Problem
P0107 Manifold Absolute Pressure/Barometric Pressure Circuit Low Input
P0108 Manifold Absolute Pressure/Barometric Pressure Circuit High Input
P0109 Manifold Absolute Pressure/Barometric Pressure Circuit Intermittent
P0110 Intake Air Temperature Circuit Malfunction
P0111 Intake Air Temperature Circuit Range/Performance Problem
P0112 Intake Air Temperature Circuit Low Input
P0113 Intake Air Temperature Circuit High Input
P0114 Intake Air Temperature Circuit Intermittent
P0115 Engine Coolant Temperature Circuit Malfunction
P0116 Engine Coolant Temperature Circuit Range/Performance Problem
P0117 Engine Coolant Temperature Circuit Low Input
P0118 Engine Coolant Temperature Circuit High Input
P0119 Engine Coolant Temperature Circuit Intermittent
P0120 Throttle/Pedal Position Sensor/Switch "A" Circuit Malfunction
P0121 Throttle/Pedal Position Sensor/Switch "A" Circuit Range/Performance Problem
P0122 Throttle/Pedal Position Sensor/Switch "A" Circuit Low Input
P0123 Throttle/Pedal Position Sensor/Switch "A" Circuit High Input
P0124 Throttle/Pedal Position Sensor/Switch "A" Circuit Intermittent
P0125 Insufficient Coolant Temperature For Closed Loop Fuel Control
P0126 Insufficient Coolant Temperature For Stable Operation
P0130 O2 Circuit Malfunction (Bank no. 1 Sensor no. 1)
P0131 O2 Sensor Circuit Low Voltage (Bank no. 1 Sensor no. 1)
P0132 O2 Sensor Circuit High Voltage (Bank no. 1 Sensor no. 1)
P0133 O2 Sensor Circuit Slow Response (Bank no. 1 Sensor no. 1)
P0134 O2 Sensor Circuit No Activity Detected (Bank no. 1 Sensor no. 1)
P0135 O2 Sensor Heater Circuit Malfunction (Bank no. 1 Sensor no. 1)
P0136 O2 Sensor Circuit Malfunction (Bank no. 1 Sensor no. 2)
P0137 O2 Sensor Circuit Low Voltage (Bank no. 1 Sensor no. 2)
P0138 O2 Sensor Circuit High Voltage (Bank no. 1 Sensor no. 2)
P0139 O2 Sensor Circuit Slow Response (Bank no. 1 Sensor no. 2)

P0140 O2 Sensor Circuit No Activity Detected (Bank no. 1 Sensor no. 2)

P0141 O2 Sensor Heater Circuit Malfunction (Bank no. 1 Sensor no. 2)

P0142 O2 Sensor Circuit Malfunction (Bank no. 1 Sensor no. 3)

P0143 O2 Sensor Circuit Low Voltage (Bank no. 1 Sensor no. 3)

P0144 O2 Sensor Circuit High Voltage (Bank no. 1 Sensor no. 3)

P0145 O2 Sensor Circuit Slow Response (Bank no. 1 Sensor no. 3)

P0146 O2 Sensor Circuit No Activity Detected (Bank no. 1 Sensor no. 3)

P0147 O2 Sensor Heater Circuit Malfunction (Bank no. 1 Sensor no. 3)

P0150 O2 Sensor Circuit Malfunction (Bank no. 2 Sensor no. 1)

P0151 O2 Sensor Circuit Low Voltage (Bank no. 2 Sensor no. 1)

P0152 O2 Sensor Circuit High Voltage (Bank no. 2 Sensor no. 1)

P0153 O2 Sensor Circuit Slow Response (Bank no. 2 Sensor no. 1)

P0154 O2 Sensor Circuit No Activity Detected (Bank no. 2 Sensor no. 1)

P0155 O2 Sensor Heater Circuit Malfunction (Bank no. 2 Sensor no. 1)

P0156 O2 Sensor Circuit Malfunction (Bank no. 2 Sensor no. 2)

P0157 O2 Sensor Circuit Low Voltage (Bank no. 2 Sensor no. 2)

P0158 O2 Sensor Circuit High Voltage (Bank no. 2 Sensor no. 2)

P0159 O2 Sensor Circuit Slow Response (Bank no. 2 Sensor no. 2)

P0160 O2 Sensor Circuit No Activity Detected (Bank no. 2 Sensor no. 2)

P0161 O2 Sensor Heater Circuit Malfunction (Bank no. 2 Sensor no. 2)

P0162 O2 Sensor Circuit Malfunction (Bank no. 2 Sensor no. 3)

P0163 O2 Sensor Circuit Low Voltage (Bank no. 2 Sensor no. 3)

P0164 O2 Sensor Circuit High Voltage (Bank no. 2 Sensor no. 3)

P0165 O2 Sensor Circuit Slow Response (Bank no. 2 Sensor no. 3)

P0166 O2 Sensor Circuit No Activity Detected (Bank no. 2 Sensor no. 3)

P0167 O2 Sensor Heater Circuit Malfunction (Bank no. 2 Sensor no. 3)

P0170 Fuel Trim Malfunction (Bank no. 1)

P0171 System Too Lean (Bank no. 1)

P0172 System Too Rich (Bank no. 1)

P0173 Fuel Trim Malfunction (Bank no. 2)

P0174 System Too Lean (Bank no. 2)

P0175 System Too Rich (Bank no. 2)

P0176 Fuel Composition Sensor Circuit Malfunction

P0177 Fuel Composition Sensor Circuit Range/Performance

P0178 Fuel Composition Sensor Circuit Low Input

P0179 Fuel Composition Sensor Circuit High Input

P0180 Fuel Temperature Sensor "A" Circuit Malfunction

P0181 Fuel Temperature Sensor "A" Circuit Range/Performance

P0182 Fuel Temperature Sensor "A" Circuit Low Input

P0183 Fuel Temperature Sensor "A" Circuit High Input

P0184 Fuel Temperature Sensor "A" Circuit Intermittent

P0185 Fuel Temperature Sensor "B" Circuit Malfunction

P0186 Fuel Temperature Sensor "B" Circuit Range/Performance

P0187 Fuel Temperature Sensor "B" Circuit Low Input

P0188 Fuel Temperature Sensor "B" Circuit High Input

P0189 Fuel Temperature Sensor "B" Circuit Intermittent

P0190 Fuel Rail Pressure Sensor Circuit Malfunction

P0191 Fuel Rail Pressure Sensor Circuit Range/Performance

P0192 Fuel Rail Pressure Sensor Circuit Low Input

P0193 Fuel Rail Pressure Sensor Circuit High Input

P0194 Fuel Rail Pressure Sensor Circuit Intermittent

P0195 Engine Oil Temperature Sensor Malfunction

P0196 Engine Oil Temperature Sensor Range/Performance

P0197 Engine Oil Temperature Sensor Low

P0198 Engine Oil Temperature Sensor High

P0199 Engine Oil Temperature Sensor Intermittent

P0200 Injector Circuit Malfunction

P0201 Injector Circuit Malfunction—Cylinder no. 1

P0202 Injector Circuit Malfunction—Cylinder no. 2

P0203 Injector Circuit Malfunction—Cylinder no. 3

P0204 Injector Circuit Malfunction—Cylinder no. 4

P0205 Injector Circuit Malfunction—Cylinder no. 5

P0206 Injector Circuit Malfunction—Cylinder no. 6

P0207 Injector Circuit Malfunction—Cylinder no. 7

P0208 Injector Circuit Malfunction—Cylinder no. 8

P0209 Injector Circuit Malfunction—Cylinder no. 9

P0210 Injector Circuit Malfunction—Cylinder no. 10

P0211 Injector Circuit Malfunction—Cylinder no. 11

P0212 Injector Circuit Malfunction—Cylinder no. 12

P0213 Cold Start Injector no. 1 Malfunction

P0214 Cold Start Injector no. 2 Malfunction

P0215 Engine Shutoff Solenoid Malfunction

P0216 Injection Timing Control Circuit Malfunction

P0217 Engine Over Temperature Condition

P0218 Transmission Over Temperature Condition

P0219 Engine Over Speed Condition

P0220 Throttle/Pedal Position Sensor/Switch "B" Circuit Malfunction

P0221 Throttle/Pedal Position Sensor/Switch "B" Circuit Range/Performance Problem

P0222 Throttle/Pedal Position Sensor/Switch "B" Circuit Low Input

P0223 Throttle/Pedal Position Sensor/Switch "B" Circuit High Input

P0224 Throttle/Pedal Position Sensor/Switch "B" Circuit Intermittent

P0225 Throttle/Pedal Position Sensor/Switch "C" Circuit Malfunction

P0226 Throttle/Pedal Position Sensor/Switch "C" Circuit Range/Performance Problem

P0227 Throttle/Pedal Position Sensor/Switch "C" Circuit Low Input

P0228 Throttle/Pedal Position Sensor/Switch "C" Circuit High Input

P0229 Throttle/Pedal Position Sensor/Switch "C" Circuit Intermittent

P0230 Fuel Pump Primary Circuit Malfunction

P0231 Fuel Pump Secondary Circuit Low

P0232 Fuel Pump Secondary Circuit High

P0233 Fuel Pump Secondary Circuit Intermittent

P0234 Engine Over Boost Condition

P0261 Cylinder no. 1 Injector Circuit Low

P0262 Cylinder no. 1 Injector Circuit High

P0263 Cylinder no. 1 Contribution/Balance Fault

P0264 Cylinder no. 2 Injector Circuit Low
P0265 Cylinder no. 2 Injector Circuit High
P0266 Cylinder no. 2 Contribution/Balance Fault
P0267 Cylinder no. 3 Injector Circuit Low
P0268 Cylinder no. 3 Injector Circuit High
P0269 Cylinder no. 3 Contribution/Balance Fault
P0270 Cylinder no. 4 Injector Circuit Low
P0271 Cylinder no. 4 Injector Circuit High
P0272 Cylinder no. 4 Contribution/Balance Fault
P0273 Cylinder no. 5 Injector Circuit Low
P0274 Cylinder no. 5 Injector Circuit High
P0275 Cylinder no. 5 Contribution/Balance Fault
P0276 Cylinder no. 6 Injector Circuit Low
P0277 Cylinder no. 6 Injector Circuit High
P0278 Cylinder no. 6 Contribution/Balance Fault
P0279 Cylinder no. 7 Injector Circuit Low
P0280 Cylinder no. 7 Injector Circuit High
P0281 Cylinder no. 7 Contribution/Balance Fault
P0282 Cylinder no. 8 Injector Circuit Low
P0283 Cylinder no. 8 Injector Circuit High
P0284 Cylinder no. 8 Contribution/Balance Fault
P0285 Cylinder no. 9 Injector Circuit Low
P0286 Cylinder no. 9 Injector Circuit High
P0287 Cylinder no. 9 Contribution/Balance Fault
P0288 Cylinder no. 10 Injector Circuit Low
P0289 Cylinder no. 10 Injector Circuit High
P0290 Cylinder no. 10 Contribution/Balance Fault
P0291 Cylinder no. 11 Injector Circuit Low
P0292 Cylinder no. 11 Injector Circuit High
P0293 Cylinder no. 11 Contribution/Balance Fault
P0294 Cylinder no. 12 Injector Circuit Low
P0295 Cylinder no. 12 Injector Circuit High
P0296 Cylinder no. 12 Contribution/Balance Fault
P0300 Random/Multiple Cylinder Misfire Detected
P0301 Cylinder no. 1—Misfire Detected
P0302 Cylinder no. 2—Misfire Detected
P0303 Cylinder no. 3—Misfire Detected
P0304 Cylinder no. 4—Misfire Detected
P0305 Cylinder no. 5—Misfire Detected
P0306 Cylinder no. 6—Misfire Detected
P0307 Cylinder no. 7—Misfire Detected
P0308 Cylinder no. 8—Misfire Detected
P0320 Ignition/Distributor Engine Speed Input Circuit Malfunction
P0321 Ignition/Distributor Engine Speed Input Circuit Range/Performance
P0322 Ignition/Distributor Engine Speed Input Circuit No Signal
P0323 Ignition/Distributor Engine Speed Input Circuit Intermittent
P0325 Knock Sensor no. 1—Circuit Malfunction (Bank no. 1 or Single Sensor)
P0326 Knock Sensor no. 1—Circuit Range/Performance (Bank no. 1 or Single Sensor)
P0327 Knock Sensor no. 1—Circuit Low Input (Bank no. 1 or Single Sensor)
P0328 Knock Sensor no. 1—Circuit High Input (Bank no. 1 or Single Sensor)
P0329 Knock Sensor no. 1—Circuit Input Intermittent (Bank no. 1 or Single Sensor)
P0330 Knock Sensor no. 2—Circuit Malfunction (Bank no. 2)
P0331 Knock Sensor no. 2—Circuit Range/Performance (Bank no. 2)

P0332 Knock Sensor no. 2—Circuit Low Input (Bank no. 2)
P0333 Knock Sensor no. 2—Circuit High Input (Bank no. 2)
P0334 Knock Sensor no. 2—Circuit Input Intermittent (Bank no. 2)
P0335 Crankshaft Position Sensor "A" Circuit Malfunction
P0336 Crankshaft Position Sensor "A" Circuit Range/Performance
P0337 Crankshaft Position Sensor "A" Circuit Low Input
P0338 Crankshaft Position Sensor "A" Circuit High Input
P0339 Crankshaft Position Sensor "A" Circuit Intermittent
P0340 Camshaft Position Sensor Circuit Malfunction
P0341 Camshaft Position Sensor Circuit Range/Performance
P0342 Camshaft Position Sensor Circuit Low Input
P0343 Camshaft Position Sensor Circuit High Input
P0344 Camshaft Position Sensor Circuit Intermittent
P0350 Ignition Coil Primary/Secondary Circuit Malfunction
P0351 Ignition Coil "A" Primary/Secondary Circuit Malfunction
P0352 Ignition Coil "B" Primary/Secondary Circuit Malfunction
P0353 Ignition Coil "C" Primary/Secondary Circuit Malfunction
P0354 Ignition Coil "D" Primary/Secondary Circuit Malfunction
P0355 Ignition Coil "E" Primary/Secondary Circuit Malfunction
P0356 Ignition Coil "F" Primary/Secondary Circuit Malfunction
P0357 Ignition Coil "G" Primary/Secondary Circuit Malfunction
P0358 Ignition Coil "H" Primary/Secondary Circuit Malfunction
P0359 Ignition Coil "I" Primary/Secondary Circuit Malfunction
P0360 Ignition Coil "J" Primary/Secondary Circuit Malfunction
P0361 Ignition Coil "K" Primary/Secondary Circuit Malfunction
P0362 Ignition Coil "L" Primary/Secondary Circuit Malfunction
P0370 Timing Reference High Resolution Signal "A" Malfunction
P0371 Timing Reference High Resolution Signal "A" Too Many Pulses
P0372 Timing Reference High Resolution Signal "A" Too Few Pulses
P0373 Timing Reference High Resolution Signal "A" Intermittent/Erratic Pulses
P0374 Timing Reference High Resolution Signal "A" No Pulses
P0375 Timing Reference High Resolution Signal "B" Malfunction
P0376 Timing Reference High Resolution Signal "B" Too Many Pulses
P0377 Timing Reference High Resolution Signal "B" Too Few Pulses
P0378 Timing Reference High Resolution Signal "B" Intermittent/Erratic Pulses
P0379 Timing Reference High Resolution Signal "B" No Pulses
P0380 Glow Plug/Heater Circuit "A" Malfunction
P0381 Glow Plug/Heater Indicator Circuit Malfunction
P0382 Glow Plug/Heater Circuit "B" Malfunction
P0385 Crankshaft Position Sensor "B" Circuit Malfunction
P0386 Crankshaft Position Sensor "B" Circuit Range/Performance
P0387 Crankshaft Position Sensor "B" Circuit Low Input
P0388 Crankshaft Position Sensor "B" Circuit High Input
P0389 Crankshaft Position Sensor "B" Circuit Intermittent
P0400 Exhaust Gas Recirculation Flow Malfunction
P0401 Exhaust Gas Recirculation Flow Insufficient Detected
P0402 Exhaust Gas Recirculation Flow Excessive Detected
P0403 Exhaust Gas Recirculation Circuit Malfunction
P0404 Exhaust Gas Recirculation Circuit Range/Performance
P0405 Exhaust Gas Recirculation Sensor "A" Circuit Low
P0406 Exhaust Gas Recirculation Sensor "A" Circuit High

P0407 Exhaust Gas Recirculation Sensor "B" Circuit Low
P0408 Exhaust Gas Recirculation Sensor "B" Circuit High
P0410 Secondary Air Injection System Malfunction
P0411 Secondary Air Injection System Incorrect Flow Detected
P0412 Secondary Air Injection System Switching Valve "A" Circuit Malfunction
P0413 Secondary Air Injection System Switching Valve "A" Circuit Open
P0414 Secondary Air Injection System Switching Valve "A" Circuit Shorted
P0415 Secondary Air Injection System Switching Valve "B" Circuit Malfunction
P0416 Secondary Air Injection System Switching Valve "B" Circuit Open
P0417 Secondary Air Injection System Switching Valve "B" Circuit Shorted
P0418 Secondary Air Injection System Relay "A" Circuit Malfunction
P0419 Secondary Air Injection System Relay "B" Circuit Malfunction
P0420 Catalyst System Efficiency Below Threshold (Bank no. 1)
P0421 Warm Up Catalyst Efficiency Below Threshold (Bank no. 1)
P0422 Main Catalyst Efficiency Below Threshold (Bank no. 1)
P0423 Heated Catalyst Efficiency Below Threshold (Bank no. 1)
P0424 Heated Catalyst Temperature Below Threshold (Bank no. 1)
P0430 Catalyst System Efficiency Below Threshold (Bank no. 2)
P0431 Warm Up Catalyst Efficiency Below Threshold (Bank no. 2)
P0432 Main Catalyst Efficiency Below Threshold (Bank no. 2)
P0433 Heated Catalyst Efficiency Below Threshold (Bank no. 2)
P0434 Heated Catalyst Temperature Below Threshold (Bank no. 2)
P0440 Evaporative Emission Control System Malfunction
P0441 Evaporative Emission Control System Incorrect Purge Flow
P0442 Evaporative Emission Control System Leak Detected (Small Leak)
P0443 Evaporative Emission Control System Purge Control Valve Circuit Malfunction
P0444 Evaporative Emission Control System Purge Control Valve Circuit Open
P0445 Evaporative Emission Control System Purge Control Valve Circuit Shorted
P0446 Evaporative Emission Control System Vent Control Circuit Malfunction
P0447 Evaporative Emission Control System Vent Control Circuit Open
P0448 Evaporative Emission Control System Vent Control Circuit Shorted
P0449 Evaporative Emission Control System Vent Valve/Solenoid Circuit Malfunction
P0450 Evaporative Emission Control System Pressure Sensor Malfunction
P0451 Evaporative Emission Control System Pressure Sensor Range/Performance
P0452 Evaporative Emission Control System Pressure Sensor Low Input
P0453 Evaporative Emission Control System Pressure Sensor High Input

P0454 Evaporative Emission Control System Pressure Sensor Intermittent
P0455 Evaporative Emission Control System Leak Detected (Gross Leak)
P0460 Fuel Level Sensor Circuit Malfunction
P0461 Fuel Level Sensor Circuit Range/Performance
P0462 Fuel Level Sensor Circuit Low Input
P0463 Fuel Level Sensor Circuit High Input
P0464 Fuel Level Sensor Circuit Intermittent
P0465 Purge Flow Sensor Circuit Malfunction
P0466 Purge Flow Sensor Circuit Range/Performance
P0467 Purge Flow Sensor Circuit Low Input
P0468 Purge Flow Sensor Circuit High Input
P0469 Purge Flow Sensor Circuit Intermittent
P0470 Exhaust Pressure Sensor Malfunction
P0471 Exhaust Pressure Sensor Range/Performance
P0472 Exhaust Pressure Sensor Low
P0473 Exhaust Pressure Sensor High
P0474 Exhaust Pressure Sensor Intermittent
P0475 Exhaust Pressure Control Valve Malfunction
P0476 Exhaust Pressure Control Valve Range/Performance
P0477 Exhaust Pressure Control Valve Low
P0478 Exhaust Pressure Control Valve High
P0479 Exhaust Pressure Control Valve Intermittent
P0480 Cooling Fan no. 1 Control Circuit Malfunction
P0481 Cooling Fan no. 2 Control Circuit Malfunction
P0482 Cooling Fan no. 3 Control Circuit Malfunction
P0483 Cooling Fan Rationality Check Malfunction
P0484 Cooling Fan Circuit Over Current
P0485 Cooling Fan Power/Ground Circuit Malfunction
P0500 Vehicle Speed Sensor Malfunction
P0501 Vehicle Speed Sensor Range/Performance
P0502 Vehicle Speed Sensor Circuit Low Input
P0503 Vehicle Speed Sensor Intermittent/Erratic/High
P0505 Idle Control System Malfunction
P0506 Idle Control System RPM Lower Than Expected
P0507 Idle Control System RPM Higher Than Expected
P0510 Closed Throttle Position Switch Malfunction
P0520 Engine Oil Pressure Sensor/Switch Circuit Malfunction
P0521 Engine Oil Pressure Sensor/Switch Range/Performance
P0522 Engine Oil Pressure Sensor/Switch Low Voltage
P0523 Engine Oil Pressure Sensor/Switch High Voltage
P0530 A/C Refrigerant Pressure Sensor Circuit Malfunction
P0531 A/C Refrigerant Pressure Sensor Circuit Range/Performance
P0532 A/C Refrigerant Pressure Sensor Circuit Low Input
P0533 A/C Refrigerant Pressure Sensor Circuit High Input
P0534 A/C Refrigerant Charge Loss
P0550 Power Steering Pressure Sensor Circuit Malfunction
P0551 Power Steering Pressure Sensor Circuit Range/Performance
P0552 Power Steering Pressure Sensor Circuit Low Input
P0553 Power Steering Pressure Sensor Circuit High Input
P0554 Power Steering Pressure Sensor Circuit Intermittent
P0560 System Voltage Malfunction
P0561 System Voltage Unstable
P0562 System Voltage Low
P0563 System Voltage High
P0565 Cruise Control On Signal Malfunction
P0566 Cruise Control Off Signal Malfunction
P0567 Cruise Control Resume Signal Malfunction

P0568 Cruise Control Set Signal Malfunction
P0569 Cruise Control Coast Signal Malfunction
P0570 Cruise Control Accel Signal Malfunction
P0571 Cruise Control/Brake Switch "A" Circuit Malfunction
P0572 Cruise Control/Brake Switch "A" Circuit Low
P0573 Cruise Control/Brake Switch "A" Circuit High
P0574 **Through P0580** Reserved for Cruise Codes
P0600 Serial Communication Link Malfunction
P0601 Internal Control Module Memory Check Sum Error
P0602 Control Module Programming Error
P0603 Internal Control Module Keep Alive Memory (KAM) Error
P0604 Internal Control Module Random Access Memory (RAM) Error
P0605 Internal Control Module Read Only Memory (ROM) Error
P0606 PCM Processor Fault
P0608 Control Module VSS Output "A" Malfunction
P0609 Control Module VSS Output "B" Malfunction
P0620 Generator Control Circuit Malfunction
P0621 Generator Lamp "L" Control Circuit Malfunction
P0622 Generator Field "F" Control Circuit Malfunction
P0650 Malfunction Indicator Lamp (MIL) Control Circuit Malfunction
P0654 Engine RPM Output Circuit Malfunction
P0655 Engine Hot Lamp Output Control Circuit Malfunction
P0656 Fuel Level Output Circuit Malfunction
P0700 Transmission Control System Malfunction
P0701 Transmission Control System Range/Performance
P0702 Transmission Control System Electrical
P0703 Torque Converter/Brake Switch "B" Circuit Malfunction
P0704 Clutch Switch Input Circuit Malfunction
P0705 Transmission Range Sensor Circuit Malfunction (PRNDL Input)
P0706 Transmission Range Sensor Circuit Range/Performance
P0707 Transmission Range Sensor Circuit Low Input
P0708 Transmission Range Sensor Circuit High Input
P0709 Transmission Range Sensor Circuit Intermittent
P0710 Transmission Fluid Temperature Sensor Circuit Malfunction
P0711 Transmission Fluid Temperature Sensor Circuit Range/Performance
P0712 Transmission Fluid Temperature Sensor Circuit Low Input
P0713 Transmission Fluid Temperature Sensor Circuit High Input
P0714 Transmission Fluid Temperature Sensor Circuit Intermittent
P0715 Input/Turbine Speed Sensor Circuit Malfunction
P0716 Input/Turbine Speed Sensor Circuit Range/Performance
P0717 Input/Turbine Speed Sensor Circuit No Signal
P0718 Input/Turbine Speed Sensor Circuit Intermittent
P0719 Torque Converter/Brake Switch "B" Circuit Low
P0720 Output Speed Sensor Circuit Malfunction
P0721 Output Speed Sensor Circuit Range/Performance
P0722 Output Speed Sensor Circuit No Signal
P0723 Output Speed Sensor Circuit Intermittent
P0724 Torque Converter/Brake Switch "B" Circuit High
P0725 Engine Speed Input Circuit Malfunction
P0726 Engine Speed Input Circuit Range/Performance
P0727 Engine Speed Input Circuit No Signal
P0728 Engine Speed Input Circuit Intermittent
P0730 Incorrect Gear Ratio

P0731 Gear no. 1 Incorrect Ratio
P0732 Gear no. 2 Incorrect Ratio
P0733 Gear no. 3 Incorrect Ratio
P0734 Gear no. 4 Incorrect Ratio
P0735 Gear no. 5 Incorrect Ratio
P0736 Reverse Incorrect Ratio
P0740 Torque Converter Clutch Circuit Malfunction
P0741 Torque Converter Clutch Circuit Performance or Stuck Off
P0742 Torque Converter Clutch Circuit Stuck On
P0743 Torque Converter Clutch Circuit Electrical
P0744 Torque Converter Clutch Circuit Intermittent
P0745 Pressure Control Solenoid Malfunction
P0746 Pressure Control Solenoid Performance or Stuck Off
P0747 Pressure Control Solenoid Stuck On
P0748 Pressure Control Solenoid Electrical
P0749 Pressure Control Solenoid Intermittent
P0750 Shift Solenoid "A" Malfunction
P0751 Shift Solenoid "A" Performance or Stuck Off
P0752 Shift Solenoid "A" Stuck On
P0753 Shift Solenoid "A" Electrical
P0754 Shift Solenoid "A" Intermittent
P0755 Shift Solenoid "B" Malfunction
P0756 Shift Solenoid "B" Performance or Stuck Off
P0757 Shift Solenoid "B" Stuck On
P0758 Shift Solenoid "B" Electrical
P0759 Shift Solenoid "B" Intermittent
P0760 Shift Solenoid "C" Malfunction
P0761 Shift Solenoid "C" Performance Or Stuck Off
P0762 Shift Solenoid "C" Stuck On
P0763 Shift Solenoid "C" Electrical
P0764 Shift Solenoid "C" Intermittent
P0765 Shift Solenoid "D" Malfunction
P0766 Shift Solenoid "D" Performance Or Stuck Off
P0767 Shift Solenoid "D" Stuck On
P0768 Shift Solenoid "D" Electrical
P0769 Shift Solenoid "D" Intermittent
P0770 Shift Solenoid "E" Malfunction
P0771 Shift Solenoid "E" Performance Or Stuck Off
P0772 Shift Solenoid "E" Stuck On
P0773 Shift Solenoid "E" Electrical
P0774 Shift Solenoid "E" Intermittent
P0780 Shift Malfunction
P0781 1–2 Shift Malfunction
P0782 2–3 Shift Malfunction
P0783 3–4 Shift Malfunction
P0784 4–5 Shift Malfunction
P0785 Shift/Timing Solenoid Malfunction
P0786 Shift/Timing Solenoid Range/Performance
P0787 Shift/Timing Solenoid Low
P0788 Shift/Timing Solenoid High
P0789 Shift/Timing Solenoid Intermittent
P0790 Normal/Performance Switch Circuit Malfunction
P0801 Reverse Inhibit Control Circuit Malfunction
P0803 1–4 Upshift (Skip Shift) Solenoid Control Circuit Malfunction
P0804 1–4 Upshift (Skip Shift) Lamp Control Circuit Malfunction
P1100 Barometric Pressure Sensor Circuit Fault
P1200 Fuel Pump Relay Circuit Fault
P1300 Igniter Circuit Fault (Bank 1)

P1305 Igniter Circuit Fault (Bank 2)
P1335 Crankshaft Position Sensor Circuit Fault
P1400 Sub-Throttle Position Sensor Circuit Fault
P1401 Sub-Throttle Position Sensor Performance
P1500 Starter Signal Circuit Fault
P1510 Air Volume Too Low With Supercharger On
P1600 PCM Battery Back-up Circuit Fault
P1605 Knock Control CPU Fault
P1700 Vehicle Speed Sensor Circuit Fault
P1705 Direct Clutch Speed Sensor Circuit Fault
P1765 Linear Shift Solenoid Circuit Fault
P1780 Park Neutral Position Switch Fault

Mazda

READING CODES

Reading the control module memory is on of the first steps in OBD II system diagnostics. This step should be initially performed to determine the general nature of the fault. Subsequent readings will determine if the fault has been cleared.

Reading codes can be performed by any of the methods below:

• Read the control module memory with the Generic Scan Tool (GST)

• Read the control module memory with the vehicle manufacturer's specific tester

To read the fault codes, connect the scan tool or tester according to the manufacturer's instructions. Follow the manufacturer's specified procedure for reading the codes.

CLEARING CODES

Control module reset procedures are a very important part of OBD II System diagnostics. This step should be done at the end of any fault code repair and at the end of any driveability repair.

Clearing codes can be performed by any of the methods below:

• Clear the control module memory with the Generic Scan Tool (GST)

• Clear the control module memory with the vehicle manufacturer's specific tester

• Turn the ignition **OFF** and remove the negative battery cable for at least 1 minute.

Removing the negative battery cable may cause other systems in the vehicle to loose their memory. Prior to removing the cable, ensure you have the proper reset codes for radios and alarms.

➡**The MIL will may also be de-activated for some codes if the vehicle completes three consecutive trips without a fault detected with vehicle conditions similar to those present during the fault.**

OBD II TROUBLE CODES

P0100 Mass or Volume Air Flow Circuit Malfunction
P0101 Mass or Volume Air Flow Circuit Range/Performance Problem
P0102 Mass or Volume Air Flow Circuit Low Input
P0103 Mass or Volume Air Flow Circuit High Input
P0104 Mass or Volume Air Flow Circuit Intermittent

P0105 Manifold Absolute Pressure/Barometric Pressure Circuit Malfunction
P0106 Manifold Absolute Pressure/Barometric Pressure Circuit Range/Performance Problem
P0107 Manifold Absolute Pressure/Barometric Pressure Circuit Low Input
P0108 Manifold Absolute Pressure/Barometric Pressure Circuit High Input
P0109 Manifold Absolute Pressure/Barometric Pressure Circuit Intermittent
P0110 Intake Air Temperature Circuit Malfunction
P0111 Intake Air Temperature Circuit Range/Performance Problem
P0112 Intake Air Temperature Circuit Low Input
P0113 Intake Air Temperature Circuit High Input
P0114 Intake Air Temperature Circuit Intermittent
P0115 Engine Coolant Temperature Circuit Malfunction
P0116 Engine Coolant Temperature Circuit Range/Performance Problem
P0117 Engine Coolant Temperature Circuit Low Input
P0118 Engine Coolant Temperature Circuit High Input
P0119 Engine Coolant Temperature Circuit Intermittent
P0120 Throttle/Pedal Position Sensor/Switch "A" Circuit Malfunction
P0121 Throttle/Pedal Position Sensor/Switch "A" Circuit Range/Performance Problem
P0122 Throttle/Pedal Position Sensor/Switch "A" Circuit Low Input
P0123 Throttle/Pedal Position Sensor/Switch "A" Circuit High Input
P0124 Throttle/Pedal Position Sensor/Switch "A" Circuit Intermittent
P0125 Insufficient Coolant Temperature For Closed Loop Fuel Control
P0126 Insufficient Coolant Temperature For Stable Operation
P0130 O2 Circuit Malfunction (Bank no. 1 Sensor no. 1)
P0131 O2 Sensor Circuit Low Voltage (Bank no. 1 Sensor no. 1)
P0132 O2 Sensor Circuit High Voltage (Bank no. 1 Sensor no. 1)
P0133 O2 Sensor Circuit Slow Response (Bank no. 1 Sensor no. 1)
P0134 O2 Sensor Circuit No Activity Detected (Bank no. 1 Sensor no. 1)
P0135 O2 Sensor Heater Circuit Malfunction (Bank no. 1 Sensor no. 1)
P0136 O2 Sensor Circuit Malfunction (Bank no. 1 Sensor no. 2)
P0137 O2 Sensor Circuit Low Voltage (Bank no. 1 Sensor no. 2)
P0138 O2 Sensor Circuit High Voltage (Bank no. 1 Sensor no. 2)
P0139 O2 Sensor Circuit Slow Response (Bank no. 1 Sensor no. 2)
P0140 O2 Sensor Circuit No Activity Detected (Bank no. 1 Sensor no. 2)
P0141 O2 Sensor Heater Circuit Malfunction (Bank no. 1 Sensor no. 2)
P0142 O2 Sensor Circuit Malfunction (Bank no. 1 Sensor no. 3)
P0143 O2 Sensor Circuit Low Voltage (Bank no. 1 Sensor no. 3)
P0144 O2 Sensor Circuit High Voltage (Bank no. 1 Sensor no. 3)
P0145 O2 Sensor Circuit Slow Response (Bank no. 1 Sensor no. 3)

P0146 O2 Sensor Circuit No Activity Detected (Bank no. 1 Sensor no. 3)

P0147 O2 Sensor Heater Circuit Malfunction (Bank no. 1 Sensor no. 3)

P0150 O2 Sensor Circuit Malfunction (Bank no. 2 Sensor no. 1)

P0151 O2 Sensor Circuit Low Voltage (Bank no. 2 Sensor no. 1)

P0152 O2 Sensor Circuit High Voltage (Bank no. 2 Sensor no. 1)

P0153 O2 Sensor Circuit Slow Response (Bank no. 2 Sensor no. 1)

P0154 O2 Sensor Circuit No Activity Detected (Bank no. 2 Sensor no. 1)

P0155 O2 Sensor Heater Circuit Malfunction (Bank no. 2 Sensor no. 1)

P0156 O2 Sensor Circuit Malfunction (Bank no. 2 Sensor no. 2)

P0157 O2 Sensor Circuit Low Voltage (Bank no. 2 Sensor no. 2)

P0158 O2 Sensor Circuit High Voltage (Bank no. 2 Sensor no. 2)

P0159 O2 Sensor Circuit Slow Response (Bank no. 2 Sensor no. 2)

P0160 O2 Sensor Circuit No Activity Detected (Bank no. 2 Sensor no. 2)

P0161 O2 Sensor Heater Circuit Malfunction (Bank no. 2 Sensor no. 2)

P0162 O2 Sensor Circuit Malfunction (Bank no. 2 Sensor no. 3)

P0163 O2 Sensor Circuit Low Voltage (Bank no. 2 Sensor no. 3)

P0164 O2 Sensor Circuit High Voltage (Bank no. 2 Sensor no. 3)

P0165 O2 Sensor Circuit Slow Response (Bank no. 2 Sensor no. 3)

P0166 O2 Sensor Circuit No Activity Detected (Bank no. 2 Sensor no. 3)

P0167 O2 Sensor Heater Circuit Malfunction (Bank no. 2 Sensor no. 3)

P0170 Fuel Trim Malfunction (Bank no. 1)

P0171 System Too Lean (Bank no. 1)

P0172 System Too Rich (Bank no. 1)

P0173 Fuel Trim Malfunction (Bank no. 2)

P0174 System Too Lean (Bank no. 2)

P0175 System Too Rich (Bank no. 2)

P0176 Fuel Composition Sensor Circuit Malfunction

P0177 Fuel Composition Sensor Circuit Range/Performance

P0178 Fuel Composition Sensor Circuit Low Input

P0179 Fuel Composition Sensor Circuit High Input

P0180 Fuel Temperature Sensor "A" Circuit Malfunction

P0181 Fuel Temperature Sensor "A" Circuit Range/Performance

P0182 Fuel Temperature Sensor "A" Circuit Low Input

P0183 Fuel Temperature Sensor "A" Circuit High Input

P0184 Fuel Temperature Sensor "A" Circuit Intermittent

P0185 Fuel Temperature Sensor "B" Circuit Malfunction

P0186 Fuel Temperature Sensor "B" Circuit Range/Performance

P0187 Fuel Temperature Sensor "B" Circuit Low Input

P0188 Fuel Temperature Sensor "B" Circuit High Input

P0189 Fuel Temperature Sensor "B" Circuit Intermittent

P0190 Fuel Rail Pressure Sensor Circuit Malfunction

P0191 Fuel Rail Pressure Sensor Circuit Range/Performance

P0192 Fuel Rail Pressure Sensor Circuit Low Input

P0193 Fuel Rail Pressure Sensor Circuit High Input

P0194 Fuel Rail Pressure Sensor Circuit Intermittent

P0195 Engine Oil Temperature Sensor Malfunction

P0196 Engine Oil Temperature Sensor Range/Performance

P0197 Engine Oil Temperature Sensor Low

P0198 Engine Oil Temperature Sensor High

P0199 Engine Oil Temperature Sensor Intermittent

P0200 Injector Circuit Malfunction

P0201 Injector Circuit Malfunction—Cylinder no. 1

P0202 Injector Circuit Malfunction—Cylinder no. 2

P0203 Injector Circuit Malfunction—Cylinder no. 3

P0204 Injector Circuit Malfunction—Cylinder no. 4

P0205 Injector Circuit Malfunction—Cylinder no. 5

P0206 Injector Circuit Malfunction—Cylinder no. 6

P0207 Injector Circuit Malfunction—Cylinder no. 7

P0208 Injector Circuit Malfunction—Cylinder no. 8

P0209 Injector Circuit Malfunction—Cylinder no. 9

P0210 Injector Circuit Malfunction—Cylinder no. 10

P0211 Injector Circuit Malfunction—Cylinder no. 11

P0212 Injector Circuit Malfunction—Cylinder no. 12

P0213 Cold Start Injector no. 1 Malfunction

P0214 Cold Start Injector no. 2 Malfunction

P0215 Engine Shutoff Solenoid Malfunction

P0216 Injection Timing Control Circuit Malfunction

P0217 Engine Over Temperature Condition

P0218 Transmission Over Temperature Condition

P0219 Engine Over Speed Condition

P0220 Throttle/Pedal Position Sensor/Switch "B" Circuit Malfunction

P0221 Throttle/Pedal Position Sensor/Switch "B" Circuit Range/Performance Problem

P0222 Throttle/Pedal Position Sensor/Switch "B" Circuit Low Input

P0223 Throttle/Pedal Position Sensor/Switch "B" Circuit High Input

P0224 Throttle/Pedal Position Sensor/Switch "B" Circuit Intermittent

P0225 Throttle/Pedal Position Sensor/Switch "C" Circuit Malfunction

P0226 Throttle/Pedal Position Sensor/Switch "C" Circuit Range/Performance Problem

P0227 Throttle/Pedal Position Sensor/Switch "C" Circuit Low Input

P0228 Throttle/Pedal Position Sensor/Switch "C" Circuit High Input

P0229 Throttle/Pedal Position Sensor/Switch "C" Circuit Intermittent

P0230 Fuel Pump Primary Circuit Malfunction

P0231 Fuel Pump Secondary Circuit Low

P0232 Fuel Pump Secondary Circuit High

P0233 Fuel Pump Secondary Circuit Intermittent

P0234 Engine Over Boost Condition

P0261 Cylinder no. 1 Injector Circuit Low

P0262 Cylinder no. 1 Injector Circuit High

P0263 Cylinder no. 1 Contribution/Balance Fault

P0264 Cylinder no. 2 Injector Circuit Low

P0265 Cylinder no. 2 Injector Circuit High

P0266 Cylinder no. 2 Contribution/Balance Fault

P0267 Cylinder no. 3 Injector Circuit Low

P0268 Cylinder no. 3 Injector Circuit High

P0269 Cylinder no. 3 Contribution/Balance Fault

P0270 Cylinder no. 4 Injector Circuit Low

P0271 Cylinder no. 4 Injector Circuit High

P0272 Cylinder no. 4 Contribution/Balance Fault

P0273 Cylinder no. 5 Injector Circuit Low

P0274 Cylinder no. 5 Injector Circuit High
P0275 Cylinder no. 5 Contribution/Balance Fault
P0276 Cylinder no. 6 Injector Circuit Low
P0277 Cylinder no. 6 Injector Circuit High
P0278 Cylinder no. 6 Contribution/Balance Fault
P0279 Cylinder no. 7 Injector Circuit Low
P0280 Cylinder no. 7 Injector Circuit High
P0281 Cylinder no. 7 Contribution/Balance Fault
P0282 Cylinder no. 8 Injector Circuit Low
P0283 Cylinder no. 8 Injector Circuit High
P0284 Cylinder no. 8 Contribution/Balance Fault
P0285 Cylinder no. 9 Injector Circuit Low
P0286 Cylinder no. 9 Injector Circuit High
P0287 Cylinder no. 9 Contribution/Balance Fault
P0288 Cylinder no. 10 Injector Circuit Low
P0289 Cylinder no. 10 Injector Circuit High
P0290 Cylinder no. 10 Contribution/Balance Fault
P0291 Cylinder no. 11 Injector Circuit Low
P0292 Cylinder no. 11 Injector Circuit High
P0293 Cylinder no. 11 Contribution/Balance Fault
P0294 Cylinder no. 12 Injector Circuit Low
P0295 Cylinder no. 12 Injector Circuit High
P0296 Cylinder no. 12 Contribution/Balance Fault
P0300 Random/Multiple Cylinder Misfire Detected
P0301 Cylinder no. 1—Misfire Detected
P0302 Cylinder no. 2—Misfire Detected
P0303 Cylinder no. 3—Misfire Detected
P0304 Cylinder no. 4—Misfire Detected
P0305 Cylinder no. 5—Misfire Detected
P0306 Cylinder no. 6—Misfire Detected
P0307 Cylinder no. 7—Misfire Detected
P0308 Cylinder no. 8—Misfire Detected
P0309 Cylinder no. 9—Misfire Detected
P0310 Cylinder no. 10—Misfire Detected
P0311 Cylinder no. 11—Misfire Detected
P0312 Cylinder no. 12—Misfire Detected
P0320 Ignition/Distributor Engine Speed Input Circuit Malfunction
P0321 Ignition/Distributor Engine Speed Input Circuit Range/Performance
P0322 Ignition/Distributor Engine Speed Input Circuit No Signal
P0323 Ignition/Distributor Engine Speed Input Circuit Intermittent
P0325 Knock Sensor no. 1—Circuit Malfunction (Bank no. 1 or Single Sensor)
P0326 Knock Sensor no. 1—Circuit Range/Performance (Bank no. 1 or Single Sensor)
P0327 Knock Sensor no. 1—Circuit Low Input (Bank no. 1 or Single Sensor)
P0328 Knock Sensor no. 1—Circuit High Input (Bank no. 1 or Single Sensor)
P0329 Knock Sensor no. 1—Circuit Input Intermittent (Bank no. 1 or Single Sensor)
P0330 Knock Sensor no. 2—Circuit Malfunction (Bank no. 2)
P0331 Knock Sensor no. 2—Circuit Range/Performance (Bank no. 2)
P0332 Knock Sensor no. 2—Circuit Low Input (Bank no. 2)
P0333 Knock Sensor no. 2—Circuit High Input (Bank no. 2)
P0334 Knock Sensor no. 2—Circuit Input Intermittent (Bank no. 2)
P0335 Crankshaft Position Sensor "A" Circuit Malfunction

P0336 Crankshaft Position Sensor "A" Circuit Range/Performance
P0337 Crankshaft Position Sensor "A" Circuit Low Input
P0338 Crankshaft Position Sensor "A" Circuit High Input
P0339 Crankshaft Position Sensor "A" Circuit Intermittent
P0340 Camshaft Position Sensor Circuit Malfunction
P0341 Camshaft Position Sensor Circuit Range/Performance
P0342 Camshaft Position Sensor Circuit Low Input
P0343 Camshaft Position Sensor Circuit High Input
P0344 Camshaft Position Sensor Circuit Intermittent
P0350 Ignition Coil Primary/Secondary Circuit Malfunction
P0351 Ignition Coil "A" Primary/Secondary Circuit Malfunction
P0352 Ignition Coil "B" Primary/Secondary Circuit Malfunction
P0353 Ignition Coil "C" Primary/Secondary Circuit Malfunction
P0354 Ignition Coil "D" Primary/Secondary Circuit Malfunction
P0355 Ignition Coil "E" Primary/Secondary Circuit Malfunction
P0356 Ignition Coil "F" Primary/Secondary Circuit Malfunction
P0357 Ignition Coil "G" Primary/Secondary Circuit Malfunction
P0358 Ignition Coil "H" Primary/Secondary Circuit Malfunction
P0359 Ignition Coil "I" Primary/Secondary Circuit Malfunction
P0360 Ignition Coil "J" Primary/Secondary Circuit Malfunction
P0361 Ignition Coil "K" Primary/Secondary Circuit Malfunction
P0362 Ignition Coil "L" Primary/Secondary Circuit Malfunction
P0370 Timing Reference High Resolution Signal "A" Malfunction
P0371 Timing Reference High Resolution Signal "A" Too Many Pulses
P0372 Timing Reference High Resolution Signal "A" Too Few Pulses
P0373 Timing Reference High Resolution Signal "A" Intermittent/Erratic Pulses
P0374 Timing Reference High Resolution Signal "A" No Pulses
P0375 Timing Reference High Resolution Signal "B" Malfunction
P0376 Timing Reference High Resolution Signal "B" Too Many Pulses
P0377 Timing Reference High Resolution Signal "B" Too Few Pulses
P0378 Timing Reference High Resolution Signal "B" Intermittent/Erratic Pulses
P0379 Timing Reference High Resolution Signal "B" No Pulses
P0380 Glow Plug/Heater Circuit "A" Malfunction
P0381 Glow Plug/Heater Indicator Circuit Malfunction
P0382 Glow Plug/Heater Circuit "B" Malfunction
P0385 Crankshaft Position Sensor "B" Circuit Malfunction
P0386 Crankshaft Position Sensor "B" Circuit Range/Performance
P0387 Crankshaft Position Sensor "B" Circuit Low Input
P0388 Crankshaft Position Sensor "B" Circuit High Input
P0389 Crankshaft Position Sensor "B" Circuit Intermittent
P0400 Exhaust Gas Recirculation Flow Malfunction
P0401 Exhaust Gas Recirculation Flow Insufficient Detected
P0402 Exhaust Gas Recirculation Flow Excessive Detected
P0403 Exhaust Gas Recirculation Circuit Malfunction
P0404 Exhaust Gas Recirculation Circuit Range/Performance
P0405 Exhaust Gas Recirculation Sensor "A" Circuit Low
P0406 Exhaust Gas Recirculation Sensor "A" Circuit High
P0407 Exhaust Gas Recirculation Sensor "B" Circuit Low
P0408 Exhaust Gas Recirculation Sensor "B" Circuit High
P0410 Secondary Air Injection System Malfunction
P0411 Secondary Air Injection System Incorrect Flow Detected

P0412 Secondary Air Injection System Switching Valve "A" Circuit Malfunction

P0413 Secondary Air Injection System Switching Valve "A" Circuit Open

P0414 Secondary Air Injection System Switching Valve "A" Circuit Shorted

P0415 Secondary Air Injection System Switching Valve "B" Circuit Malfunction

P0416 Secondary Air Injection System Switching Valve "B" Circuit Open

P0417 Secondary Air Injection System Switching Valve "B" Circuit Shorted

P0418 Secondary Air Injection System Relay "A" Circuit Malfunction

P0419 Secondary Air Injection System Relay "B" Circuit Malfunction

P0420 Catalyst System Efficiency Below Threshold (Bank no. 1)

P0421 Warm Up Catalyst Efficiency Below Threshold (Bank no. 1)

P0422 Main Catalyst Efficiency Below Threshold (Bank no. 1)

P0423 Heated Catalyst Efficiency Below Threshold (Bank no. 1)

P0424 Heated Catalyst Temperature Below Threshold (Bank no. 1)

P0430 Catalyst System Efficiency Below Threshold (Bank no. 2)

P0431 Warm Up Catalyst Efficiency Below Threshold (Bank no. 2)

P0432 Main Catalyst Efficiency Below Threshold (Bank no. 2)

P0433 Heated Catalyst Efficiency Below Threshold (Bank no. 2)

P0434 Heated Catalyst Temperature Below Threshold (Bank no. 2)

P0440 Evaporative Emission Control System Malfunction

P0441 Evaporative Emission Control System Incorrect Purge Flow

P0442 Evaporative Emission Control System Leak Detected (Small Leak)

P0443 Evaporative Emission Control System Purge Control Valve Circuit Malfunction

P0444 Evaporative Emission Control System Purge Control Valve Circuit Open

P0445 Evaporative Emission Control System Purge Control Valve Circuit Shorted

P0446 Evaporative Emission Control System Vent Control Circuit Malfunction

P0447 Evaporative Emission Control System Vent Control Circuit Open

P0448 Evaporative Emission Control System Vent Control Circuit Shorted

P0449 Evaporative Emission Control System Vent Valve/Solenoid Circuit Malfunction

P0450 Evaporative Emission Control System Pressure Sensor Malfunction

P0451 Evaporative Emission Control System Pressure Sensor Range/Performance

P0452 Evaporative Emission Control System Pressure Sensor Low Input

P0453 Evaporative Emission Control System Pressure Sensor High Input

P0454 Evaporative Emission Control System Pressure Sensor Intermittent

P0455 Evaporative Emission Control System Leak Detected (Gross Leak)

P0460 Fuel Level Sensor Circuit Malfunction

P0461 Fuel Level Sensor Circuit Range/Performance

P0462 Fuel Level Sensor Circuit Low Input

P0463 Fuel Level Sensor Circuit High Input

P0464 Fuel Level Sensor Circuit Intermittent

P0465 Purge Flow Sensor Circuit Malfunction

P0466 Purge Flow Sensor Circuit Range/Performance

P0467 Purge Flow Sensor Circuit Low Input

P0468 Purge Flow Sensor Circuit High Input

P0469 Purge Flow Sensor Circuit Intermittent

P0470 Exhaust Pressure Sensor Malfunction

P0471 Exhaust Pressure Sensor Range/Performance

P0472 Exhaust Pressure Sensor Low

P0473 Exhaust Pressure Sensor High

P0474 Exhaust Pressure Sensor Intermittent

P0475 Exhaust Pressure Control Valve Malfunction

P0476 Exhaust Pressure Control Valve Range/Performance

P0477 Exhaust Pressure Control Valve Low

P0478 Exhaust Pressure Control Valve High

P0479 Exhaust Pressure Control Valve Intermittent

P0480 Cooling Fan no. 1 Control Circuit Malfunction

P0481 Cooling Fan no. 2 Control Circuit Malfunction

P0482 Cooling Fan no. 3 Control Circuit Malfunction

P0483 Cooling Fan Rationality Check Malfunction

P0484 Cooling Fan Circuit Over Current

P0485 Cooling Fan Power/Ground Circuit Malfunction

P0500 Vehicle Speed Sensor Malfunction

P0501 Vehicle Speed Sensor Range/Performance

P0502 Vehicle Speed Sensor Circuit Low Input

P0503 Vehicle Speed Sensor Intermittent/Erratic/High

P0505 Idle Control System Malfunction

P0506 Idle Control System RPM Lower Than Expected

P0507 Idle Control System RPM Higher Than Expected

P0510 Closed Throttle Position Switch Malfunction

P0520 Engine Oil Pressure Sensor/Switch Circuit Malfunction

P0521 Engine Oil Pressure Sensor/Switch Range/Performance

P0522 Engine Oil Pressure Sensor/Switch Low Voltage

P0523 Engine Oil Pressure Sensor/Switch High Voltage

P0530 A/C Refrigerant Pressure Sensor Circuit Malfunction

P0531 A/C Refrigerant Pressure Sensor Circuit Range/Performance

P0532 A/C Refrigerant Pressure Sensor Circuit Low Input

P0533 A/C Refrigerant Pressure Sensor Circuit High Input

P0534 A/C Refrigerant Charge Loss

P0550 Power Steering Pressure Sensor Circuit Malfunction

P0551 Power Steering Pressure Sensor Circuit Range/Performance

P0552 Power Steering Pressure Sensor Circuit Low Input

P0553 Power Steering Pressure Sensor Circuit High Input

P0554 Power Steering Pressure Sensor Circuit Intermittent

P0560 System Voltage Malfunction

P0561 System Voltage Unstable

P0562 System Voltage Low

P0563 System Voltage High

P0565 Cruise Control On Signal Malfunction

P0566 Cruise Control Off Signal Malfunction

P0567 Cruise Control Resume Signal Malfunction

P0568 Cruise Control Set Signal Malfunction

P0569 Cruise Control Coast Signal Malfunction

P0570 Cruise Control Accel Signal Malfunction

P0571 Cruise Control/Brake Switch "A" Circuit Malfunction

P0572 Cruise Control/Brake Switch "A" Circuit Low
P0573 Cruise Control/Brake Switch "A" Circuit High
P0574 Through P0580 Reserved for Cruise Codes
P0600 Serial Communication Link Malfunction
P0601 Internal Control Module Memory Check Sum Error
P0602 Control Module Programming Error
P0603 Internal Control Module Keep Alive Memory (KAM) Error
P0604 Internal Control Module Random Access Memory (RAM) Error
P0605 Internal Control Module Read Only Memory (ROM) Error
P0606 PCM Processor Fault
P0608 Control Module VSS Output "A" Malfunction
P0609 Control Module VSS Output "B" Malfunction
P0620 Generator Control Circuit Malfunction
P0621 Generator Lamp "L" Control Circuit Malfunction
P0622 Generator Field "F" Control Circuit Malfunction
P0650 Malfunction Indicator Lamp (MIL) Control Circuit Malfunction
P0654 Engine RPM Output Circuit Malfunction
P0655 Engine Hot Lamp Output Control Circuit Malfunction
P0656 Fuel Level Output Circuit Malfunction
P0700 Transmission Control System Malfunction
P0701 Transmission Control System Range/Performance
P0702 Transmission Control System Electrical
P0703 Torque Converter/Brake Switch "B" Circuit Malfunction
P0704 Clutch Switch Input Circuit Malfunction
P0705 Transmission Range Sensor Circuit Malfunction (PRNDL Input)
P0706 Transmission Range Sensor Circuit Range/Performance
P0707 Transmission Range Sensor Circuit Low Input
P0708 Transmission Range Sensor Circuit High Input
P0709 Transmission Range Sensor Circuit Intermittent
P0710 Transmission Fluid Temperature Sensor Circuit Malfunction
P0711 Transmission Fluid Temperature Sensor Circuit Range/Performance
P0712 Transmission Fluid Temperature Sensor Circuit Low Input
P0713 Transmission Fluid Temperature Sensor Circuit High Input
P0714 Transmission Fluid Temperature Sensor Circuit Intermittent
P0715 Input/Turbine Speed Sensor Circuit Malfunction
P0716 Input/Turbine Speed Sensor Circuit Range/Performance
P0717 Input/Turbine Speed Sensor Circuit No Signal
P0718 Input/Turbine Speed Sensor Circuit Intermittent
P0719 Torque Converter/Brake Switch "B" Circuit Low
P0720 Output Speed Sensor Circuit Malfunction
P0721 Output Speed Sensor Circuit Range/Performance
P0722 Output Speed Sensor Circuit No Signal
P0723 Output Speed Sensor Circuit Intermittent
P0724 Torque Converter/Brake Switch "B" Circuit High
P0725 Engine Speed Input Circuit Malfunction
P0726 Engine Speed Input Circuit Range/Performance
P0727 Engine Speed Input Circuit No Signal
P0728 Engine Speed Input Circuit Intermittent
P0730 Incorrect Gear Ratio
P0731 Gear no. 1 Incorrect Ratio
P0732 Gear no. 2 Incorrect Ratio
P0733 Gear no. 3 Incorrect Ratio
P0734 Gear no. 4 Incorrect Ratio

P0735 Gear no. 5 Incorrect Ratio
P0736 Reverse Incorrect Ratio
P0740 Torque Converter Clutch Circuit Malfunction
P0741 Torque Converter Clutch Circuit Performance or Stuck Off
P0742 Torque Converter Clutch Circuit Stuck On
P0743 Torque Converter Clutch Circuit Electrical
P0744 Torque Converter Clutch Circuit Intermittent
P0745 Pressure Control Solenoid Malfunction
P0746 Pressure Control Solenoid Performance or Stuck Off
P0747 Pressure Control Solenoid Stuck On
P0748 Pressure Control Solenoid Electrical
P0749 Pressure Control Solenoid Intermittent
P0750 Shift Solenoid "A" Malfunction
P0751 Shift Solenoid "A" Performance or Stuck Off
P0752 Shift Solenoid "A" Stuck On
P0753 Shift Solenoid "A" Electrical
P0754 Shift Solenoid "A" Intermittent
P0755 Shift Solenoid "B" Malfunction
P0756 Shift Solenoid "B" Performance or Stuck Off
P0757 Shift Solenoid "B" Stuck On
P0758 Shift Solenoid "B" Electrical
P0759 Shift Solenoid "B" Intermittent
P0760 Shift Solenoid "C" Malfunction
P0761 Shift Solenoid "C" Performance Or Stuck Off
P0762 Shift Solenoid "C" Stuck On
P0763 Shift Solenoid "C" Electrical
P0764 Shift Solenoid "C" Intermittent
P0765 Shift Solenoid "D" Malfunction
P0766 Shift Solenoid "D" Performance Or Stuck Off
P0767 Shift Solenoid "D" Stuck On
P0768 Shift Solenoid "D" Electrical
P0769 Shift Solenoid "D" Intermittent
P0770 Shift Solenoid "E" Malfunction
P0771 Shift Solenoid "E" Performance Or Stuck Off
P0772 Shift Solenoid "E" Stuck On
P0773 Shift Solenoid "E" Electrical
P0774 Shift Solenoid "E" Intermittent
P0780 Shift Malfunction
P0781 1–2 Shift Malfunction
P0782 2–3 Shift Malfunction
P0783 3–4 Shift Malfunction
P0784 4–5 Shift Malfunction
P0785 Shift/Timing Solenoid Malfunction
P0786 Shift/Timing Solenoid Range/Performance
P0787 Shift/Timing Solenoid Low
P0788 Shift/Timing Solenoid High
P0789 Shift/Timing Solenoid Intermittent
P0790 Normal/Performance Switch Circuit Malfunction
P0801 Reverse Inhibit Control Circuit Malfunction
P0803 1–4 Upshift (Skip Shift) Solenoid Control Circuit Malfunction
P0804 1–4 Upshift (Skip Shift) Lamp Control Circuit Malfunction
P1000 OBD II Monitor Testing Not Complete More Driving Required
P1001 Key On Engine Running (KOER) Self-Test Not Able To Complete, KOER Aborted
P1100 Mass Air Flow (MAF) Sensor Intermittent
P1101 Mass Air Flow (MAF) Sensor Out Of Self-Test Range
P1110 Intake Air Temperature (IAT) Sensor Signal Circuit Fault

P1112 Intake Air Temperature (IAT) Sensor Intermittent

P1113 Intake Air Temperature (IAT) Sensor Intermittent

P1116 Engine Coolant Temperature (ECT) Sensor Out Of Self-Test Range

P1117 Engine Coolant Temperature (ECT) Sensor Intermittent

P1120 Throttle Position (TP) Sensor Out Of Range (Low)

P1121 Throttle Position (TP) Sensor Inconsistent With MAF Sensor

P1124 Throttle Position (TP) Sensor Out Of Self-Test Range

P1125 Throttle Position (TP) Sensor Circuit Intermittent

P1127 Exhaust Not Warm Enough, Downstream Heated Oxygen Sensors (HO2S) Not Tested

P1128 Upstream Heated Oxygen Sensors (HO2S) Swapped From Bank To Bank

P1129 Downstream Heated Oxygen Sensors (HO2S) Swapped From Bank To Bank

P1130 Lack Of Upstream Heated Oxygen Sensor (HO2S 11) Switch, Adaptive Fuel At Limit (Bank #1)

P1131 Lack Of Upstream Heated Oxygen Sensor (HO2S 11) Switch, Sensor Indicates Lean (Bank #1)

P1132 Lack Of Upstream Heated Oxygen Sensor (HO2S 11) Switch, Sensor Indicates Rich (Bank#1)

P1137 Lack Of Downstream Heated Oxygen Sensor (HO2S 12) Switch, Sensor Indicates Lean (Bank#1)

P1138 Lack Of Downstream Heated Oxygen Sensor (HO2S 12) Switch, Sensor Indicates Rich (Bank#1)

P1150 Lack Of Upstream Heated Oxygen Sensor (HO2S 21) Switch, Adaptive Fuel At Limit (Bank #2)

P1151 Lack Of Upstream Heated Oxygen Sensor (HO2S 21) Switch, Sensor Indicates Lean (Bank#2)

P1152 Lack Of Upstream Heated Oxygen Sensor (HO2S 21) Switch, Sensor Indicates Rich (Bank #2)

P1170 (HO2S 11) Signal Remained Unchanged For More Than 20 Seconds After Closed Loop

P1173 Feedback A/F Mixture Control (HO2S 21) Signal Remained Unchanged For More Than 20 Seconds After Closed Loop

P1195 Barometric (BARO) Pressure Sensor Circuit Malfunction (Signal Is From EGR Boost Sensor)

P1196 Starter Switch Circuit Malfunction

P1235 Fuel Pump Control Out Of Range (MIL DTC)

P1236 Fuel Pump Control Out Of Range (No MIL)

P1250 Fuel Pressure Regulator Control (FPRC) Solenoid Malfunction

P1252 Fuel Pressure Regulator Control (FPRC) Solenoid Malfunction

P1260 THEFT Detected—Engine Disabled

P1270 Engine RPM Or Vehicle Speed Limiter Reached

P1345 No Camshaft Position Sensor Signal

P1351 Ignition Diagnostic Monitor (IDM) Circuit Input Malfunction

P1351 Indicates Ignition System Malfunction

P1352 Indicates Ignition System Malfunction

P1353 Indicates Ignition System Malfunction

P1354 Indicates Ignition System Malfunction

P1358 Ignition Diagnostic Monitor (IDM) Signal Out Of Self-Test Range

P1359 Spark Output Circuit Malfunction

P1360 Ignition Coil "A" Secondary Circuit Fault

P1361 Ignition Coil "A" Secondary Circuit Fault

P1362 Ignition Coil "A" Secondary Circuit Fault

P1364 Spark Output Circuit Malfunction

P1365 Ignition Coil Secondary Circuit Fault

P1390 Octane Adjust (OCT ADJ) Out Of Self-Test Range

P1400 Differential Pressure Feedback EGR (DPFE) Sensor Circuit Low Voltage Detected

P1401 Differential Pressure Feedback EGR (DPFE) Sensor Circuit High Voltage Detected/EGR Temperature Sensor

P1402 EGR Valve Position Sensor Open Or Short

P1405 Differential Pressure Feedback EGR (DPFE) Sensor Upstream Hose Off Or Plugged

P1406 Differential Pressure Feedback EGR (DPFE) Sensor Downstream Hose Off Or Plugged

P1407 Exhaust Gas Recirculation (EGR) No Flow Detected (Valve Stuck Closed Or Inoperative)

P1408 Exhaust Gas Recirculation (EGR) Flow Out Of Self-Test Range

P1409 Electronic Vacuum Regulator (EVR) Control Circuit Malfunction

P1443 Evaporative Emission Control System—Vacuum System, Purge Control Solenoid Or Purge Control Valve Malfunction

P1444 Purge Flow Sensor (PFS) Circuit Low Input

P1445 Purge Flow Sensor (PFS) Circuit High Input

P1449 Evaporative Emission Control System Unable To Hold Vacuum

P1455 Evaporative Emission Control System Control Leak Detected (Gross Leak)

P1460 Wide Open Throttle Air Conditioning Cut-Off Circuit Malfunction

P1464 Air Conditioning (A/C) Demand Out Of Self-Test Range/A/C On During KOER Or CCT Test

P1474 Low Fan Control Primary Circuit Malfunction

P1485 EGR Control Solenoid Open Or Short

P1486 EGR Vent Solenoid Open Or Short

P1487 EGR Boost Check Solenoid Open Or Short

P1500 Vehicle Speed Sensor (VSS) Circuit Intermittent

P1501 Vehicle Speed Sensor (VSS) Out Of Self-Test Range/Vehicle Moved During Test

P1502 Invalid Self Test—Auxiliary Powertrain Control Module (APCM) Functioning

P1504 Idle Air Control (IAC) Circuit Malfunction

P1505 Idle Air Control (IAC) System At Adaptive Clip

P1506 Idle Air Control (IAC) Overspeed Error

P1507 Idle Air Control (IAC) Underspeed Error

P1508 Bypass Air Solenoid "1" Circuit Fault

P1509 Bypass Air Solenoid "2" Circuit Fault

P1521 Variable Resonance Induction System (VRIS) Solenoid #1 Open Or Short

P1522 Variable Resonance Induction System (VRIS) Solenoid #2 Open Or Short

P1523 High Speed Inlet Air (HSIA) Solenoid Open Or Short

P1524 Charge Air Cooler Bypass Solenoid Circuit Fault

P1525 ABV Vacuum Solenoid Circuit Fault

P1526 ABV Vent Solenoid Circuit Fault

P1529 Atmospheric balance Air Control Valve Fault

P1540 ABV System Fault

P1601 Serial Communication Error

P1602 Serial Communication Error

P1605 Powertrain Control Module (PCM)—Keep Alive Memory (KAM) Test Error

P1608 PCM Internal Circuit Malfunction

P1609 PCM Internal Circuit Malfunction

P1627 Serial Communication Error

P1628 Serial Communication Error

P1650 Power Steering Pressure (PSP) Switch Out Of Self-Test Range

P1651 Power Steering Pressure (PSP) Switch Input Malfunction

P1701 Reverse Engagement Error

P1703 Brake On/Off (BOO) Switch Out Of Self-Test Range

P1705 Transmission Range (TR) Sensor Out Of Self-Test Range

P1706 High Vehicle Speed In Park

P1709 Park Or Neutral Position (PNP) Or Clutch Pedal Position (CPP) Switch Out Of Self-Test Range

P1711 Transmission Fluid Temperature (TFT) Sensor Out Of Self-Test Range

P1720 Vehicle Speed Sensor (VSS) Circuit Malfunction

P1729 4x4 Low Switch Error

P1741 Torque Converter Clutch (TCC) Control Error

P1742 Torque Converter Clutch (TCC) Solenoid Failed On (Turns On MIL)

P1743 Torque Converter Clutch (TCC) Solenoid Failed On (Turns On TCIL)

P1746 Electronic Pressure Control (EPC) Solenoid Open Circuit (Low Input)

P1747 Electronic Pressure Control (EPC) Solenoid Short Circuit (High Input)

P1749 Electronic Pressure Control (EPC) Solenoid Failed Low

P1751 Shift Solenoid#1 (SS1) Performance

P1754 Coast Clutch Solenoid (CCS) Circuit Malfunction

P1756 Shift Solenoid#2 (SS2) Performance

P1761 Shift Solenoid #(SS2) Performance

P1780 Transmission Control Switch (TCS) Circuit Out Of Self-Test Range

P1781 4x4 Low Switch, Out Of Self-Test Range

P1783 Transmission Over Temperature Condition

P1794 PCM Battery Direct Power Circuit Fault

P1797 P/N Switch Open or Short Circuit Fault

Mitsubishi

READING CODES

Reading the control module memory is on of the first steps in OBD II system diagnostics. This step should be initially performed to determine the general nature of the fault. Subsequent readings will determine if the fault has been cleared.

Reading codes can be performed by any of the methods below:

• Read the control module memory with the Generic Scan Tool (GST)

• Read the control module memory with the vehicle manufacturer's specific tester

To read the fault codes, connect the scan tool or tester according to the manufacturer's instructions. Follow the manufacturer's specified procedure for reading the codes.

CLEARING CODES

Control module reset procedures are a very important part of OBD II System diagnostics. This step should be done at the end of any fault code repair and at the end of any driveability repair.

Clearing codes can be performed by any of the methods below:

• Clear the control module memory with the Generic Scan Tool (GST)

• Clear the control module memory with the vehicle manufacturer's specific tester

• Turn the ignition **OFF** and remove the negative battery cable for at least 1 minute.

Removing the negative battery cable may cause other systems in the vehicle to loose their memory. Prior to removing the cable, ensure you have the proper reset codes for radios and alarms.

➡**The MIL will may also be de-activated for some codes if the vehicle completes three consecutive trips without a fault detected with vehicle conditions similar to those present during the fault.**

OBD II TROUBLE CODES

P0100 Mass or Volume Air Flow Circuit Malfunction

P0101 Mass or Volume Air Flow Circuit Range/Performance Problem

P0102 Mass or Volume Air Flow Circuit Low Input

P0103 Mass or Volume Air Flow Circuit High Input

P0104 Mass or Volume Air Flow Circuit Intermittent

P0105 Manifold Absolute Pressure/Barometric Pressure Circuit Malfunction

P0106 Manifold Absolute Pressure/Barometric Pressure Circuit Range/Performance Problem

P0107 Manifold Absolute Pressure/Barometric Pressure Circuit Low Input

P0108 Manifold Absolute Pressure/Barometric Pressure Circuit High Input

P0109 Manifold Absolute Pressure/Barometric Pressure Circuit Intermittent

P0110 Intake Air Temperature Circuit Malfunction

P0111 Intake Air Temperature Circuit Range/Performance Problem

P0112 Intake Air Temperature Circuit Low Input

P0113 Intake Air Temperature Circuit High Input

P0114 Intake Air Temperature Circuit Intermittent

P0115 Engine Coolant Temperature Circuit Malfunction

P0116 Engine Coolant Temperature Circuit Range/Performance Problem

P0117 Engine Coolant Temperature Circuit Low Input

P0118 Engine Coolant Temperature Circuit High Input

P0119 Engine Coolant Temperature Circuit Intermittent

P0120 Throttle/Pedal Position Sensor/Switch "A" Circuit Malfunction

P0121 Throttle/Pedal Position Sensor/Switch "A" Circuit Range/Performance Problem

P0122 Throttle/Pedal Position Sensor/Switch "A" Circuit Low Input

P0123 Throttle/Pedal Position Sensor/Switch "A" Circuit High Input

P0124 Throttle/Pedal Position Sensor/Switch "A" Circuit Intermittent

P0125 Insufficient Coolant Temperature For Closed Loop Fuel Control

P0126 Insufficient Coolant Temperature For Stable Operation

P0130 O2 Circuit Malfunction (Bank no. 1 Sensor no. 1)

P0131 O2 Sensor Circuit Low Voltage (Bank no. 1 Sensor no. 1)

P0132 O2 Sensor Circuit High Voltage (Bank no. 1 Sensor no. 1)

P0133 O2 Sensor Circuit Slow Response (Bank no. 1 Sensor no. 1)

P0134 O2 Sensor Circuit No Activity Detected (Bank no. 1 Sensor no. 1)

P0135 O2 Sensor Heater Circuit Malfunction (Bank no. 1 Sensor no. 1)

P0136 O2 Sensor Circuit Malfunction (Bank no. 1 Sensor no. 2)

P0137 O2 Sensor Circuit Low Voltage (Bank no. 1 Sensor no. 2)

P0138 O2 Sensor Circuit High Voltage (Bank no. 1 Sensor no. 2)

P0139 O2 Sensor Circuit Slow Response (Bank no. 1 Sensor no. 2)

P0140 O2 Sensor Circuit No Activity Detected (Bank no. 1 Sensor no. 2)

P0141 O2 Sensor Heater Circuit Malfunction (Bank no. 1 Sensor no. 2)

P0142 O2 Sensor Circuit Malfunction (Bank no. 1 Sensor no. 3)

P0143 O2 Sensor Circuit Low Voltage (Bank no. 1 Sensor no. 3)

P0144 O2 Sensor Circuit High Voltage (Bank no. 1 Sensor no. 3)

P0145 O2 Sensor Circuit Slow Response (Bank no. 1 Sensor no. 3)

P0146 O2 Sensor Circuit No Activity Detected (Bank no. 1 Sensor no. 3)

P0147 O2 Sensor Heater Circuit Malfunction (Bank no. 1 Sensor no. 3)

P0150 O2 Sensor Circuit Malfunction (Bank no. 2 Sensor no. 1)

P0151 O2 Sensor Circuit Low Voltage (Bank no. 2 Sensor no. 1)

P0152 O2 Sensor Circuit High Voltage (Bank no. 2 Sensor no. 1)

P0153 O2 Sensor Circuit Slow Response (Bank no. 2 Sensor no. 1)

P0154 O2 Sensor Circuit No Activity Detected (Bank no. 2 Sensor no. 1)

P0155 O2 Sensor Heater Circuit Malfunction (Bank no. 2 Sensor no. 1)

P0156 O2 Sensor Circuit Malfunction (Bank no. 2 Sensor no. 2)

P0157 O2 Sensor Circuit Low Voltage (Bank no. 2 Sensor no. 2)

P0158 O2 Sensor Circuit High Voltage (Bank no. 2 Sensor no. 2)

P0159 O2 Sensor Circuit Slow Response (Bank no. 2 Sensor no. 2)

P0160 O2 Sensor Circuit No Activity Detected (Bank no. 2 Sensor no. 2)

P0161 O2 Sensor Heater Circuit Malfunction (Bank no. 2 Sensor no. 2)

P0162 O2 Sensor Circuit Malfunction (Bank no. 2 Sensor no. 3)

P0163 O2 Sensor Circuit Low Voltage (Bank no. 2 Sensor no. 3)

P0164 O2 Sensor Circuit High Voltage (Bank no. 2 Sensor no. 3)

P0165 O2 Sensor Circuit Slow Response (Bank no. 2 Sensor no. 3)

P0166 O2 Sensor Circuit No Activity Detected (Bank no. 2 Sensor no. 3)

P0167 O2 Sensor Heater Circuit Malfunction (Bank no. 2 Sensor no. 3)

P0170 Fuel Trim Malfunction (Bank no. 1)

P0171 System Too Lean (Bank no. 1)

P0172 System Too Rich (Bank no. 1)

P0173 Fuel Trim Malfunction (Bank no. 2)

P0174 System Too Lean (Bank no. 2)

P0175 System Too Rich (Bank no. 2)

P0176 Fuel Composition Sensor Circuit Malfunction

P0177 Fuel Composition Sensor Circuit Range/Performance

P0178 Fuel Composition Sensor Circuit Low Input

P0179 Fuel Composition Sensor Circuit High Input

P0180 Fuel Temperature Sensor "A" Circuit Malfunction

P0181 Fuel Temperature Sensor "A" Circuit Range/Performance

P0182 Fuel Temperature Sensor "A" Circuit Low Input

P0183 Fuel Temperature Sensor "A" Circuit High Input

P0184 Fuel Temperature Sensor "A" Circuit Intermittent

P0185 Fuel Temperature Sensor "B" Circuit Malfunction

P0186 Fuel Temperature Sensor "B" Circuit Range/Performance

P0187 Fuel Temperature Sensor "B" Circuit Low Input

P0188 Fuel Temperature Sensor "B" Circuit High Input

P0189 Fuel Temperature Sensor "B" Circuit Intermittent

P0190 Fuel Rail Pressure Sensor Circuit Malfunction

P0191 Fuel Rail Pressure Sensor Circuit Range/Performance

P0192 Fuel Rail Pressure Sensor Circuit Low Input

P0193 Fuel Rail Pressure Sensor Circuit High Input

P0194 Fuel Rail Pressure Sensor Circuit Intermittent

P0195 Engine Oil Temperature Sensor Malfunction

P0196 Engine Oil Temperature Sensor Range/Performance

P0197 Engine Oil Temperature Sensor Low

P0198 Engine Oil Temperature Sensor High

P0199 Engine Oil Temperature Sensor Intermittent

P0200 Injector Circuit Malfunction

P0201 Injector Circuit Malfunction—Cylinder no. 1

P0202 Injector Circuit Malfunction—Cylinder no. 2

P0203 Injector Circuit Malfunction—Cylinder no. 3

P0204 Injector Circuit Malfunction—Cylinder no. 4

P0205 Injector Circuit Malfunction—Cylinder no. 5

P0206 Injector Circuit Malfunction—Cylinder no. 6

P0207 Injector Circuit Malfunction—Cylinder no. 7

P0208 Injector Circuit Malfunction—Cylinder no. 8

P0209 Injector Circuit Malfunction—Cylinder no. 9

P0210 Injector Circuit Malfunction—Cylinder no. 10

P0211 Injector Circuit Malfunction—Cylinder no. 11

P0212 Injector Circuit Malfunction—Cylinder no. 12

P0213 Cold Start Injector no. 1 Malfunction

P0214 Cold Start Injector no. 2 Malfunction

P0215 Engine Shutoff Solenoid Malfunction

P0216 Injection Timing Control Circuit Malfunction

P0217 Engine Over Temperature Condition

P0218 Transmission Over Temperature Condition

P0219 Engine Over Speed Condition

P0220 Throttle/Pedal Position Sensor/Switch "B" Circuit Malfunction

P0221 Throttle/Pedal Position Sensor/Switch "B" Circuit Range/Performance Problem

P0222 Throttle/Pedal Position Sensor/Switch "B" Circuit Low Input

P0223 Throttle/Pedal Position Sensor/Switch "B" Circuit High Input

P0224 Throttle/Pedal Position Sensor/Switch "B" Circuit Intermittent

P0225 Throttle/Pedal Position Sensor/Switch "C" Circuit Malfunction

P0226 Throttle/Pedal Position Sensor/Switch "C" Circuit Range/Performance Problem

P0227 Throttle/Pedal Position Sensor/Switch "C" Circuit Low Input

P0228 Throttle/Pedal Position Sensor/Switch "C" Circuit High Input

P0229 Throttle/Pedal Position Sensor/Switch "C" Circuit Intermittent

P0230 Fuel Pump Primary Circuit Malfunction

P0231 Fuel Pump Secondary Circuit Low

P0232 Fuel Pump Secondary Circuit High

P0233 Fuel Pump Secondary Circuit Intermittent

P0234 Engine Over Boost Condition

P0261 Cylinder no. 1 Injector Circuit Low

P0262 Cylinder no. 1 Injector Circuit High

P0263 Cylinder no. 1 Contribution/Balance Fault

P0264 Cylinder no. 2 Injector Circuit Low

P0265 Cylinder no. 2 Injector Circuit High

P0266 Cylinder no. 2 Contribution/Balance Fault

P0267 Cylinder no. 3 Injector Circuit Low

P0268 Cylinder no. 3 Injector Circuit High

P0269 Cylinder no. 3 Contribution/Balance Fault

P0270 Cylinder no. 4 Injector Circuit Low

P0271 Cylinder no. 4 Injector Circuit High

P0272 Cylinder no. 4 Contribution/Balance Fault

P0273 Cylinder no. 5 Injector Circuit Low

P0274 Cylinder no. 5 Injector Circuit High

P0275 Cylinder no. 5 Contribution/Balance Fault

P0276 Cylinder no. 6 Injector Circuit Low

P0277 Cylinder no. 6 Injector Circuit High

P0278 Cylinder no. 6 Contribution/Balance Fault

P0279 Cylinder no. 7 Injector Circuit Low

P0280 Cylinder no. 7 Injector Circuit High

P0281 Cylinder no. 7 Contribution/Balance Fault

P0282 Cylinder no. 8 Injector Circuit Low

P0283 Cylinder no. 8 Injector Circuit High

P0284 Cylinder no. 8 Contribution/Balance Fault

P0285 Cylinder no. 9 Injector Circuit Low

P0286 Cylinder no. 9 Injector Circuit High

P0287 Cylinder no. 9 Contribution/Balance Fault

P0288 Cylinder no. 10 Injector Circuit Low

P0289 Cylinder no. 10 Injector Circuit High

P0290 Cylinder no. 10 Contribution/Balance Fault

P0291 Cylinder no. 11 Injector Circuit Low

P0292 Cylinder no. 11 Injector Circuit High

P0293 Cylinder no. 11 Contribution/Balance Fault

P0294 Cylinder no. 12 Injector Circuit Low

P0295 Cylinder no. 12 Injector Circuit High

P0296 Cylinder no. 12 Contribution/Balance Fault

P0300 Random/Multiple Cylinder Misfire Detected

P0301 Cylinder no. 1—Misfire Detected

P0302 Cylinder no. 2—Misfire Detected

P0303 Cylinder no. 3—Misfire Detected

P0304 Cylinder no. 4—Misfire Detected

P0305 Cylinder no. 5—Misfire Detected

P0306 Cylinder no. 6—Misfire Detected

P0307 Cylinder no. 7—Misfire Detected

P0308 Cylinder no. 8—Misfire Detected

P0309 Cylinder no. 9—Misfire Detected

P0310 Cylinder no. 10—Misfire Detected

P0311 Cylinder no. 11—Misfire Detected

P0312 Cylinder no. 12—Misfire Detected

P0320 Ignition/Distributor Engine Speed Input Circuit Malfunction

P0321 Ignition/Distributor Engine Speed Input Circuit Range/Performance

P0322 Ignition/Distributor Engine Speed Input Circuit No Signal

P0323 Ignition/Distributor Engine Speed Input Circuit Intermittent

P0325 Knock Sensor no. 1—Circuit Malfunction (Bank no. 1 or Single Sensor)

P0326 Knock Sensor no. 1—Circuit Range/Performance (Bank no. 1 or Single Sensor)

P0327 Knock Sensor no. 1—Circuit Low Input (Bank no. 1 or Single Sensor)

P0328 Knock Sensor no. 1—Circuit High Input (Bank no. 1 or Single Sensor)

P0329 Knock Sensor no. 1—Circuit Input Intermittent (Bank no. 1 or Single Sensor)

P0330 Knock Sensor no. 2—Circuit Malfunction (Bank no. 2)

P0331 Knock Sensor no. 2—Circuit Range/Performance (Bank no. 2)

P0332 Knock Sensor no. 2—Circuit Low Input (Bank no. 2)

P0333 Knock Sensor no. 2—Circuit High Input (Bank no. 2)

P0334 Knock Sensor no. 2—Circuit Input Intermittent (Bank no. 2)

P0335 Crankshaft Position Sensor "A" Circuit Malfunction

P0336 Crankshaft Position Sensor "A" Circuit Range/Performance

P0337 Crankshaft Position Sensor "A" Circuit Low Input

P0338 Crankshaft Position Sensor "A" Circuit High Input

P0339 Crankshaft Position Sensor "A" Circuit Intermittent

P0340 Camshaft Position Sensor Circuit Malfunction

P0341 Camshaft Position Sensor Circuit Range/Performance

P0342 Camshaft Position Sensor Circuit Low Input

P0343 Camshaft Position Sensor Circuit High Input

P0344 Camshaft Position Sensor Circuit Intermittent

P0350 Ignition Coil Primary/Secondary Circuit Malfunction

P0351 Ignition Coil "A" Primary/Secondary Circuit Malfunction

P0352 Ignition Coil "B" Primary/Secondary Circuit Malfunction

P0353 Ignition Coil "C" Primary/Secondary Circuit Malfunction

P0354 Ignition Coil "D" Primary/Secondary Circuit Malfunction

P0355 Ignition Coil "E" Primary/Secondary Circuit Malfunction

P0356 Ignition Coil "F" Primary/Secondary Circuit Malfunction

P0357 Ignition Coil "G" Primary/Secondary Circuit Malfunction

P0358 Ignition Coil "H" Primary/Secondary Circuit Malfunction

P0359 Ignition Coil "I" Primary/Secondary Circuit Malfunction

P0360 Ignition Coil "J" Primary/Secondary Circuit Malfunction

P0361 Ignition Coil "K" Primary/Secondary Circuit Malfunction

P0362 Ignition Coil "L" Primary/Secondary Circuit Malfunction

P0370 Timing Reference High Resolution Signal "A" Malfunction

P0371 Timing Reference High Resolution Signal "A" Too Many Pulses

P0372 Timing Reference High Resolution Signal "A" Too Few Pulses

P0373 Timing Reference High Resolution Signal "A" Intermittent/Erratic Pulses

P0374 Timing Reference High Resolution Signal "A" No Pulses

P0375 Timing Reference High Resolution Signal "B" Malfunction

P0376 Timing Reference High Resolution Signal "B" Too Many Pulses

P0377 Timing Reference High Resolution Signal "B" Too Few Pulses

P0378 Timing Reference High Resolution Signal "B" Intermittent/Erratic Pulses

P0379 Timing Reference High Resolution Signal "B" No Pulses

P0380 Glow Plug/Heater Circuit "A" Malfunction
P0381 Glow Plug/Heater Indicator Circuit Malfunction
P0382 Glow Plug/Heater Circuit "B" Malfunction
P0385 Crankshaft Position Sensor "B" Circuit Malfunction
P0386 Crankshaft Position Sensor "B" Circuit Range/Performance
P0387 Crankshaft Position Sensor "B" Circuit Low Input
P0388 Crankshaft Position Sensor "B" Circuit High Input
P0389 Crankshaft Position Sensor "B" Circuit Intermittent
P0400 Exhaust Gas Recirculation Flow Malfunction
P0401 Exhaust Gas Recirculation Flow Insufficient Detected
P0402 Exhaust Gas Recirculation Flow Excessive Detected
P0403 Exhaust Gas Recirculation Circuit Malfunction
P0404 Exhaust Gas Recirculation Circuit Range/Performance
P0405 Exhaust Gas Recirculation Sensor "A" Circuit Low
P0406 Exhaust Gas Recirculation Sensor "A" Circuit High
P0407 Exhaust Gas Recirculation Sensor "B" Circuit Low
P0408 Exhaust Gas Recirculation Sensor "B" Circuit High
P0410 Secondary Air Injection System Malfunction
P0411 Secondary Air Injection System Incorrect Flow Detected
P0412 Secondary Air Injection System Switching Valve "A" Circuit Malfunction
P0413 Secondary Air Injection System Switching Valve "A" Circuit Open
P0414 Secondary Air Injection System Switching Valve "A" Circuit Shorted
P0415 Secondary Air Injection System Switching Valve "B" Circuit Malfunction
P0416 Secondary Air Injection System Switching Valve "B" Circuit Open
P0417 Secondary Air Injection System Switching Valve "B" Circuit Shorted
P0418 Secondary Air Injection System Relay "A" Circuit Malfunction
P0419 Secondary Air Injection System Relay "B" Circuit Malfunction
P0420 Catalyst System Efficiency Below Threshold (Bank no. 1)
P0421 Warm Up Catalyst Efficiency Below Threshold (Bank no. 1)
P0422 Main Catalyst Efficiency Below Threshold (Bank no. 1)
P0423 Heated Catalyst Efficiency Below Threshold (Bank no. 1)
P0424 Heated Catalyst Temperature Below Threshold (Bank no. 1)
P0430 Catalyst System Efficiency Below Threshold (Bank no. 2)
P0431 Warm Up Catalyst Efficiency Below Threshold (Bank no. 2)
P0432 Main Catalyst Efficiency Below Threshold (Bank no. 2)
P0433 Heated Catalyst Efficiency Below Threshold (Bank no. 2)
P0434 Heated Catalyst Temperature Below Threshold (Bank no. 2)
P0440 Evaporative Emission Control System Malfunction
P0441 Evaporative Emission Control System Incorrect Purge Flow
P0442 Evaporative Emission Control System Leak Detected (Small Leak)
P0443 Evaporative Emission Control System Purge Control Valve Circuit Malfunction
P0444 Evaporative Emission Control System Purge Control Valve Circuit Open
P0445 Evaporative Emission Control System Purge Control Valve Circuit Shorted

P0446 Evaporative Emission Control System Vent Control Circuit Malfunction
P0447 Evaporative Emission Control System Vent Control Circuit Open
P0448 Evaporative Emission Control System Vent Control Circuit Shorted
P0449 Evaporative Emission Control System Vent Valve/Solenoid Circuit Malfunction
P0450 Evaporative Emission Control System Pressure Sensor Malfunction
P0451 Evaporative Emission Control System Pressure Sensor Range/Performance
P0452 Evaporative Emission Control System Pressure Sensor Low Input
P0453 Evaporative Emission Control System Pressure Sensor High Input
P0454 Evaporative Emission Control System Pressure Sensor Intermittent
P0455 Evaporative Emission Control System Leak Detected (Gross Leak)
P0460 Fuel Level Sensor Circuit Malfunction
P0461 Fuel Level Sensor Circuit Range/Performance
P0462 Fuel Level Sensor Circuit Low Input
P0463 Fuel Level Sensor Circuit High Input
P0464 Fuel Level Sensor Circuit Intermittent
P0465 Purge Flow Sensor Circuit Malfunction
P0466 Purge Flow Sensor Circuit Range/Performance
P0467 Purge Flow Sensor Circuit Low Input
P0468 Purge Flow Sensor Circuit High Input
P0469 Purge Flow Sensor Circuit Intermittent
P0470 Exhaust Pressure Sensor Malfunction
P0471 Exhaust Pressure Sensor Range/Performance
P0472 Exhaust Pressure Sensor Low
P0473 Exhaust Pressure Sensor High
P0474 Exhaust Pressure Sensor Intermittent
P0475 Exhaust Pressure Control Valve Malfunction
P0476 Exhaust Pressure Control Valve Range/Performance
P0477 Exhaust Pressure Control Valve Low
P0478 Exhaust Pressure Control Valve High
P0479 Exhaust Pressure Control Valve Intermittent
P0480 Cooling Fan no. 1 Control Circuit Malfunction
P0481 Cooling Fan no. 2 Control Circuit Malfunction
P0482 Cooling Fan no. 3 Control Circuit Malfunction
P0483 Cooling Fan Rationality Check Malfunction
P0484 Cooling Fan Circuit Over Current
P0485 Cooling Fan Power/Ground Circuit Malfunction
P0500 Vehicle Speed Sensor Malfunction
P0501 Vehicle Speed Sensor Range/Performance
P0502 Vehicle Speed Sensor Circuit Low Input
P0503 Vehicle Speed Sensor Intermittent/Erratic/High
P0505 Idle Control System Malfunction
P0506 Idle Control System RPM Lower Than Expected
P0507 Idle Control System RPM Higher Than Expected
P0510 Closed Throttle Position Switch Malfunction
P0520 Engine Oil Pressure Sensor/Switch Circuit Malfunction
P0521 Engine Oil Pressure Sensor/Switch Range/Performance
P0522 Engine Oil Pressure Sensor/Switch Low Voltage
P0523 Engine Oil Pressure Sensor/Switch High Voltage
P0530 A/C Refrigerant Pressure Sensor Circuit Malfunction
P0531 A/C Refrigerant Pressure Sensor Circuit Range/Performance

P0532 A/C Refrigerant Pressure Sensor Circuit Low Input
P0533 A/C Refrigerant Pressure Sensor Circuit High Input
P0534 A/C Refrigerant Charge Loss
P0550 Power Steering Pressure Sensor Circuit Malfunction
P0551 Power Steering Pressure Sensor Circuit Range/Performance
P0552 Power Steering Pressure Sensor Circuit Low Input
P0553 Power Steering Pressure Sensor Circuit High Input
P0554 Power Steering Pressure Sensor Circuit Intermittent
P0560 System Voltage Malfunction
P0561 System Voltage Unstable
P0562 System Voltage Low
P0563 System Voltage High
P0565 Cruise Control On Signal Malfunction
P0566 Cruise Control Off Signal Malfunction
P0567 Cruise Control Resume Signal Malfunction
P0568 Cruise Control Set Signal Malfunction
P0569 Cruise Control Coast Signal Malfunction
P0570 Cruise Control Accel Signal Malfunction
P0571 Cruise Control/Brake Switch "A" Circuit Malfunction
P0572 Cruise Control/Brake Switch "A" Circuit Low
P0573 Cruise Control/Brake Switch "A" Circuit High
P0574 Through P0580 Reserved for Cruise Codes
P0600 Serial Communication Link Malfunction
P0601 Internal Control Module Memory Check Sum Error
P0602 Control Module Programming Error
P0603 Internal Control Module Keep Alive Memory (KAM) Error
P0604 Internal Control Module Random Access Memory (RAM) Error
P0605 Internal Control Module Read Only Memory (ROM) Error
P0606 PCM Processor Fault
P0608 Control Module VSS Output "A" Malfunction
P0609 Control Module VSS Output "B" Malfunction
P0620 Generator Control Circuit Malfunction
P0621 Generator Lamp "L" Control Circuit Malfunction
P0622 Generator Field "F" Control Circuit Malfunction
P0650 Malfunction Indicator Lamp (MIL) Control Circuit Malfunction
P0654 Engine RPM Output Circuit Malfunction
P0655 Engine Hot Lamp Output Control Circuit Malfunction
P0656 Fuel Level Output Circuit Malfunction
P0700 Transmission Control System Malfunction
P0701 Transmission Control System Range/Performance
P0702 Transmission Control System Electrical
P0703 Torque Converter/Brake Switch "B" Circuit Malfunction
P0704 Clutch Switch Input Circuit Malfunction
P0705 Transmission Range Sensor Circuit Malfunction (PRNDL Input)
P0706 Transmission Range Sensor Circuit Range/Performance
P0707 Transmission Range Sensor Circuit Low Input
P0708 Transmission Range Sensor Circuit High Input
P0709 Transmission Range Sensor Circuit Intermittent
P0710 Transmission Fluid Temperature Sensor Circuit Malfunction
P0711 Transmission Fluid Temperature Sensor Circuit Range/Performance
P0712 Transmission Fluid Temperature Sensor Circuit Low Input
P0713 Transmission Fluid Temperature Sensor Circuit High Input
P0714 Transmission Fluid Temperature Sensor Circuit Intermittent

P0715 Input/Turbine Speed Sensor Circuit Malfunction
P0716 Input/Turbine Speed Sensor Circuit Range/Performance
P0717 Input/Turbine Speed Sensor Circuit No Signal
P0718 Input/Turbine Speed Sensor Circuit Intermittent
P0719 Torque Converter/Brake Switch "B" Circuit Low
P0720 Output Speed Sensor Circuit Malfunction
P0721 Output Speed Sensor Circuit Range/Performance
P0722 Output Speed Sensor Circuit No Signal
P0723 Output Speed Sensor Circuit Intermittent
P0724 Torque Converter/Brake Switch "B" Circuit High
P0725 Engine Speed Input Circuit Malfunction
P0726 Engine Speed Input Circuit Range/Performance
P0727 Engine Speed Input Circuit No Signal
P0728 Engine Speed Input Circuit Intermittent
P0730 Incorrect Gear Ratio
P0731 Gear no. 1 Incorrect Ratio
P0732 Gear no. 2 Incorrect Ratio
P0733 Gear no. 3 Incorrect Ratio
P0734 Gear no. 4 Incorrect Ratio
P0735 Gear no. 5 Incorrect Ratio
P0736 Reverse Incorrect Ratio
P0740 Torque Converter Clutch Circuit Malfunction
P0741 Torque Converter Clutch Circuit Performance or Stuck Off
P0742 Torque Converter Clutch Circuit Stuck On
P0743 Torque Converter Clutch Circuit Electrical
P0744 Torque Converter Clutch Circuit Intermittent
P0745 Pressure Control Solenoid Malfunction
P0746 Pressure Control Solenoid Performance or Stuck Off
P0747 Pressure Control Solenoid Stuck On
P0748 Pressure Control Solenoid Electrical
P0749 Pressure Control Solenoid Intermittent
P0750 Shift Solenoid "A" Malfunction
P0751 Shift Solenoid "A" Performance or Stuck Off
P0752 Shift Solenoid "A" Stuck On
P0753 Shift Solenoid "A" Electrical
P0754 Shift Solenoid "A" Intermittent
P0755 Shift Solenoid "B" Malfunction
P0756 Shift Solenoid "B" Performance or Stuck Off
P0757 Shift Solenoid "B" Stuck On
P0758 Shift Solenoid "B" Electrical
P0759 Shift Solenoid "B" Intermittent
P0760 Shift Solenoid "C" Malfunction
P0761 Shift Solenoid "C" Performance Or Stuck Off
P0762 Shift Solenoid "C" Stuck On
P0763 Shift Solenoid "C" Electrical
P0764 Shift Solenoid "C" Intermittent
P0765 Shift Solenoid "D" Malfunction
P0766 Shift Solenoid "D" Performance Or Stuck Off
P0767 Shift Solenoid "D" Stuck On
P0768 Shift Solenoid "D" Electrical
P0769 Shift Solenoid "D" Intermittent
P0770 Shift Solenoid "E" Malfunction
P0771 Shift Solenoid "E" Performance Or Stuck Off
P0772 Shift Solenoid "E" Stuck On
P0773 Shift Solenoid "E" Electrical
P0774 Shift Solenoid "E" Intermittent
P0780 Shift Malfunction
P0781 1–2 Shift Malfunction
P0782 2–3 Shift Malfunction
P0783 3–4 Shift Malfunction
P0784 4–5 Shift Malfunction

P0785 Shift/Timing Solenoid Malfunction
P0786 Shift/Timing Solenoid Range/Performance
P0787 Shift/Timing Solenoid Low
P0788 Shift/Timing Solenoid High
P0789 Shift/Timing Solenoid Intermittent
P0790 Normal/Performance Switch Circuit Malfunction
P0801 Reverse Inhibit Control Circuit Malfunction
P0803 1–4 Upshift (Skip Shift) Solenoid Control Circuit Malfunction
P0804 1–4 Upshift (Skip Shift) Lamp Control Circuit Malfunction
P1100 Induction Control Motor Position Sensor Fault
P1101 Traction Control Vacuum Solenoid Circuit Fault
P1102 Traction Control Ventilation Solenoid Circuit Fault
P1103 Turbocharger Waste Gate Actuator Circuit Fault
P1104 Turbocharger Waste Gate Solenoid Circuit Fault
P1105 Fuel Pressure Solenoid Circuit Fault
P1294 Target Idle Speed Not Reached
P1295 No 5-Volt Supply To TP Sensor
P1296 No 5-Volt Supply To MAP Sensor
P1297 No Change In MAP From Start To Run
P1300 Ignition Timing Adjustment Circuit
P1390 Timing Belt Skipped One Tooth Or More
P1391 Intermittent Loss Of CMP Or CKP Sensor Signals
P1400 Manifold Differential Pressure Sensor Fault
P1443 EVAP Purge Control Solenoid "2" Circuit Fault
P1486 EVAP Leak Monitor Pinched Hose Detected
P1487 High Speed Radiator Fan Control Relay Circuit Fault
P1989 High Speed Condenser Fan Control Relay Fault
P1490 Low Speed Fan Control Relay Fault
P1492 Battery Temperature Sensor High Voltage
P1494 EVAP Ventilation Switch Or Mechanical Fault
P1495 EVAP Ventilation Solenoid Circuit Fault
P1496 5-Volt Supply Output Too Low
P1500 Generator FR Terminal Circuit Fault
P1600 PCM-TCM Serial Communication Link Circuit Fault
P1696 PCM Failure- EEPROM Write Denied
P1715 No CCD Messages From TCM
P1750 TCM Pulse Generator Circuit Fault
P1791 Pressure Control, Shift Control, TCC Solenoid Fault
P1899 PCM ECT Level Signal to TCM Circuit Fault

Nissan

➡**This section also provides coverage for the Mercury Villager since it shares a platform with the Nissan Quest**

READING CODES

With Scan Tool

Reading the control module memory is on of the first steps in OBD II system diagnostics. This step should be initially performed to determine the general nature of the fault. Subsequent readings will determine if the fault has been cleared.

Reading codes can be performed by any of the methods below:

• Read the control module memory with the Generic Scan Tool (GST)

• Read the control module memory with the vehicle manufacturer's specific tester

To read the fault codes, connect the scan tool or tester according to the manufacturer's instructions. Follow the manufacturer's specified procedure for reading the codes.

Without Scan Tool

The ECM is capable of outputting data in four different modes, depending on the position of the mode switch and the ignition key. Modes are switched by turning the mode screw on the side of the ECM, near the red LED. Additional modes are accessed by turning the ignition key on or off.

The ECM is located forward of the center console, behind an access panel on the Altima, and in the passenger's side kick panel on the 240SX.

With the ECM set in Mode 1 and the ignition in the **ON** position, a malfunction indicator lamp bulb check may be performed. When the engine is started, the ECM will illuminate the indicator lamps as a warning of a fault in the system.

Mode 2 is set by turning the mode selector screw fully clockwise, waiting 2 seconds, then turning the screw fully counterclockwise. With the ignition in the **ON** position, self-diagnostic results will be output as a series of lamp flashes. When the engine is started, the oxygen sensor monitor function is enabled and the red LED on the ECM is used to determine proper oxygen sensor function.

1. Remove the access cover and locate the mode adjusting screw and LED on the ECM.

2. Turn the ignition switch **ON**, but do not start the engine. Both the LED and the malfunction indicator lamp on the instrument panel should be illuminated. This is a bulb check.

3. Start the engine.

➡**Switching modes is not possible while the engine is running.**

4. If the LED or malfunction indicator lamp illuminates, there is a fault in the system.

5. Turn the mode selector screw fully clockwise. Wait 2 seconds, then turn the screw fully counterclockwise.

6. The diagnostic trouble codes will now be read from the ECM memory. They will appear as flashes of the malfunction indicator lamp, or the ECM's LED.

7. After all codes have been read, turn the mode selector screw fully clockwise to erase the codes.

➡**Turn the mode adjusting screw to the fully counterclockwise position whenever the vehicle is in use.**

8. Turn the ignition **OFF**.

➡**When the ignition switch is turned OFF during diagnosis, power to the ECM will drop after approximately 5 seconds. The diagnosis will automatically return to Mode 1 at this time.**

CLEARING CODES

With Scan Tool

Control module reset procedures are a very important part of OBD II System diagnostics. This step should be done at the end of any fault code repair and at the end of any driveability repair.

Clearing codes can be performed by any of the methods below:

• Clear the control module memory with the Generic Scan Tool (GST)

• Clear the control module memory with the vehicle manufacturer's specific tester

➡**The MIL will may also be de-activated for some codes if the vehicle completes three consecutive trips without a fault detected with vehicle conditions similar to those present during the fault.**

Without Scan Tool

The easiest way to clear trouble codes without a scan tool is to turn the mode selector screw fully clockwise after all codes have been read.

➡**Turn the mode adjusting screw to the fully counterclockwise position whenever the vehicle is in use.**

Codes may also be erased by turning the ignition **OFF** and remove the negative battery cable for at least 1 minute. However, removing the negative battery cable may cause other systems in the vehicle to loose their memory. Prior to removing the cable, ensure you have the proper reset codes for radios and alarms.

OBD II TROUBLE CODES

P0000 No Self Diagnostic Failure Indicated
P0100 Mass or Volume Air Flow Circuit Malfunction
P0101 Mass or Volume Air Flow Circuit Range/Performance Problem
P0102 Mass or Volume Air Flow Circuit Low Input
P0103 Mass or Volume Air Flow Circuit High Input
P0104 Mass or Volume Air Flow Circuit Intermittent
P0105 Manifold Absolute Pressure/Barometric Pressure Circuit Malfunction
P0106 Manifold Absolute Pressure/Barometric Pressure Circuit Range/Performance Problem
P0107 Manifold Absolute Pressure/Barometric Pressure Circuit Low Input
P0108 Manifold Absolute Pressure/Barometric Pressure Circuit High Input
P0109 Manifold Absolute Pressure/Barometric Pressure Circuit Intermittent
P0110 Intake Air Temperature Circuit Malfunction
P0111 Intake Air Temperature Circuit Range/Performance Problem
P0112 Intake Air Temperature Circuit Low Input
P0113 Intake Air Temperature Circuit High Input
P0114 Intake Air Temperature Circuit Intermittent
P0115 Engine Coolant Temperature Circuit Malfunction
P0116 Engine Coolant Temperature Circuit Range/Performance Problem
P0117 Engine Coolant Temperature Circuit Low Input
P0118 Engine Coolant Temperature Circuit High Input
P0119 Engine Coolant Temperature Circuit Intermittent
P0120 Throttle/Pedal Position Sensor/Switch "A" Circuit Malfunction
P0121 Throttle/Pedal Position Sensor/Switch "A" Circuit Range/Performance Problem
P0122 Throttle/Pedal Position Sensor/Switch "A" Circuit Low Input
P0123 Throttle/Pedal Position Sensor/Switch "A" Circuit High Input

P0124 Throttle/Pedal Position Sensor/Switch "A" Circuit Intermittent
P0125 Insufficient Coolant Temperature For Closed Loop Fuel Control
P0126 Insufficient Coolant Temperature For Stable Operation
P0130 O2 Circuit Malfunction (Bank no. 1 Sensor no. 1)
P0131 O2 Sensor Circuit Low Voltage (Bank no. 1 Sensor no. 1)
P0132 O2 Sensor Circuit High Voltage (Bank no. 1 Sensor no. 1)
P0133 O2 Sensor Circuit Slow Response (Bank no. 1 Sensor no. 1)
P0134 O2 Sensor Circuit No Activity Detected (Bank no. 1 Sensor no. 1)
P0135 O2 Sensor Heater Circuit Malfunction (Bank no. 1 Sensor no. 1)
P0136 O2 Sensor Circuit Malfunction (Bank no. 1 Sensor no. 2)
P0137 O2 Sensor Circuit Low Voltage (Bank no. 1 Sensor no. 2)
P0138 O2 Sensor Circuit High Voltage (Bank no. 1 Sensor no. 2)
P0139 O2 Sensor Circuit Slow Response (Bank no. 1 Sensor no. 2)
P0140 O2 Sensor Circuit No Activity Detected (Bank no. 1 Sensor no. 2)
P0141 O2 Sensor Heater Circuit Malfunction (Bank no. 1 Sensor no. 2)
P0142 O2 Sensor Circuit Malfunction (Bank no. 1 Sensor no. 3)
P0143 O2 Sensor Circuit Low Voltage (Bank no. 1 Sensor no. 3)
P0144 O2 Sensor Circuit High Voltage (Bank no. 1 Sensor no. 3)
P0145 O2 Sensor Circuit Slow Response (Bank no. 1 Sensor no. 3)
P0146 O2 Sensor Circuit No Activity Detected (Bank no. 1 Sensor no. 3)
P0147 O2 Sensor Heater Circuit Malfunction (Bank no. 1 Sensor no. 3)
P0150 O2 Sensor Circuit Malfunction (Bank no. 2 Sensor no. 1)
P0151 O2 Sensor Circuit Low Voltage (Bank no. 2 Sensor no. 1)
P0152 O2 Sensor Circuit High Voltage (Bank no. 2 Sensor no. 1)
P0153 O2 Sensor Circuit Slow Response (Bank no. 2 Sensor no. 1)
P0154 O2 Sensor Circuit No Activity Detected (Bank no. 2 Sensor no. 1)
P0155 O2 Sensor Heater Circuit Malfunction (Bank no. 2 Sensor no. 1)
P0156 O2 Sensor Circuit Malfunction (Bank no. 2 Sensor no. 2)
P0157 O2 Sensor Circuit Low Voltage (Bank no. 2 Sensor no. 2)
P0158 O2 Sensor Circuit High Voltage (Bank no. 2 Sensor no. 2)
P0159 O2 Sensor Circuit Slow Response (Bank no. 2 Sensor no. 2)
P0160 O2 Sensor Circuit No Activity Detected (Bank no. 2 Sensor no. 2)
P0161 O2 Sensor Heater Circuit Malfunction (Bank no. 2 Sensor no. 2)
P0162 O2 Sensor Circuit Malfunction (Bank no. 2 Sensor no. 3)
P0163 O2 Sensor Circuit Low Voltage (Bank no. 2 Sensor no. 3)
P0164 O2 Sensor Circuit High Voltage (Bank no. 2 Sensor no. 3)
P0165 O2 Sensor Circuit Slow Response (Bank no. 2 Sensor no. 3)

P0166 O2 Sensor Circuit No Activity Detected (Bank no. 2 Sensor no. 3)

P0167 O2 Sensor Heater Circuit Malfunction (Bank no. 2 Sensor no. 3)

P0170 Fuel Trim Malfunction (Bank no. 1)

P0171 System Too Lean (Bank no. 1)

P0172 System Too Rich (Bank no. 1)

P0173 Fuel Trim Malfunction (Bank no. 2)

P0174 System Too Lean (Bank no. 2)

P0175 System Too Rich (Bank no. 2)

P0176 Fuel Composition Sensor Circuit Malfunction

P0177 Fuel Composition Sensor Circuit Range/Performance

P0178 Fuel Composition Sensor Circuit Low Input

P0179 Fuel Composition Sensor Circuit High Input

P0180 Fuel Temperature Sensor "A" Circuit Malfunction

P0181 Fuel Temperature Sensor "A" Circuit Range/Performance

P0182 Fuel Temperature Sensor "A" Circuit Low Input

P0183 Fuel Temperature Sensor "A" Circuit High Input

P0184 Fuel Temperature Sensor "A" Circuit Intermittent

P0185 Fuel Temperature Sensor "B" Circuit Malfunction

P0186 Fuel Temperature Sensor "B" Circuit Range/Performance

P0187 Fuel Temperature Sensor "B" Circuit Low Input

P0188 Fuel Temperature Sensor "B" Circuit High Input

P0189 Fuel Temperature Sensor "B" Circuit Intermittent

P0190 Fuel Rail Pressure Sensor Circuit Malfunction

P0191 Fuel Rail Pressure Sensor Circuit Range/Performance

P0192 Fuel Rail Pressure Sensor Circuit Low Input

P0193 Fuel Rail Pressure Sensor Circuit High Input

P0194 Fuel Rail Pressure Sensor Circuit Intermittent

P0195 Engine Oil Temperature Sensor Malfunction

P0196 Engine Oil Temperature Sensor Range/Performance

P0197 Engine Oil Temperature Sensor Low

P0198 Engine Oil Temperature Sensor High

P0199 Engine Oil Temperature Sensor Intermittent

P0200 Injector Circuit Malfunction

P0201 Injector Circuit Malfunction—Cylinder no. 1

P0202 Injector Circuit Malfunction—Cylinder no. 2

P0203 Injector Circuit Malfunction—Cylinder no. 3

P0204 Injector Circuit Malfunction—Cylinder no. 4

P0205 Injector Circuit Malfunction—Cylinder no. 5

P0206 Injector Circuit Malfunction—Cylinder no. 6

P0207 Injector Circuit Malfunction—Cylinder no. 7

P0208 Injector Circuit Malfunction—Cylinder no. 8

P0209 Injector Circuit Malfunction—Cylinder no. 9

P0210 Injector Circuit Malfunction—Cylinder no. 10

P0211 Injector Circuit Malfunction—Cylinder no. 11

P0212 Injector Circuit Malfunction—Cylinder no. 12

P0213 Cold Start Injector no. 1 Malfunction

P0214 Cold Start Injector no. 2 Malfunction

P0215 Engine Shutoff Solenoid Malfunction

P0216 Injection Timing Control Circuit Malfunction

P0217 Engine Over Temperature Condition

P0218 Transmission Over Temperature Condition

P0219 Engine Over Speed Condition

P0220 Throttle/Pedal Position Sensor/Switch "B" Circuit Malfunction

P0221 Throttle/Pedal Position Sensor/Switch "B" Circuit Range/Performance Problem

P0222 Throttle/Pedal Position Sensor/Switch "B" Circuit Low Input

P0223 Throttle/Pedal Position Sensor/Switch "B" Circuit High Input

P0224 Throttle/Pedal Position Sensor/Switch "B" Circuit Intermittent

P0225 Throttle/Pedal Position Sensor/Switch "C" Circuit Malfunction

P0226 Throttle/Pedal Position Sensor/Switch "C" Circuit Range/Performance Problem

P0227 Throttle/Pedal Position Sensor/Switch "C" Circuit Low Input

P0228 Throttle/Pedal Position Sensor/Switch "C" Circuit High Input

P0229 Throttle/Pedal Position Sensor/Switch "C" Circuit Intermittent

P0230 Fuel Pump Primary Circuit Malfunction

P0231 Fuel Pump Secondary Circuit Low

P0232 Fuel Pump Secondary Circuit High

P0233 Fuel Pump Secondary Circuit Intermittent

P0234 Engine Over Boost Condition

P0261 Cylinder no. 1 Injector Circuit Low

P0262 Cylinder no. 1 Injector Circuit High

P0263 Cylinder no. 1 Contribution/Balance Fault

P0264 Cylinder no. 2 Injector Circuit Low

P0265 Cylinder no. 2 Injector Circuit High

P0266 Cylinder no. 2 Contribution/Balance Fault

P0267 Cylinder no. 3 Injector Circuit Low

P0268 Cylinder no. 3 Injector Circuit High

P0269 Cylinder no. 3 Contribution/Balance Fault

P0270 Cylinder no. 4 Injector Circuit Low

P0271 Cylinder no. 4 Injector Circuit High

P0272 Cylinder no. 4 Contribution/Balance Fault

P0273 Cylinder no. 5 Injector Circuit Low

P0274 Cylinder no. 5 Injector Circuit High

P0275 Cylinder no. 5 Contribution/Balance Fault

P0276 Cylinder no. 6 Injector Circuit Low

P0277 Cylinder no. 6 Injector Circuit High

P0278 Cylinder no. 6 Contribution/Balance Fault

P0279 Cylinder no. 7 Injector Circuit Low

P0280 Cylinder no. 7 Injector Circuit High

P0281 Cylinder no. 7 Contribution/Balance Fault

P0282 Cylinder no. 8 Injector Circuit Low

P0283 Cylinder no. 8 Injector Circuit High

P0284 Cylinder no. 8 Contribution/Balance Fault

P0285 Cylinder no. 9 Injector Circuit Low

P0286 Cylinder no. 9 Injector Circuit High

P0287 Cylinder no. 9 Contribution/Balance Fault

P0288 Cylinder no. 10 Injector Circuit Low

P0289 Cylinder no. 10 Injector Circuit High

P0290 Cylinder no. 10 Contribution/Balance Fault

P0291 Cylinder no. 11 Injector Circuit Low

P0292 Cylinder no. 11 Injector Circuit High

P0293 Cylinder no. 11 Contribution/Balance Fault

P0294 Cylinder no. 12 Injector Circuit Low

P0295 Cylinder no. 12 Injector Circuit High

P0296 Cylinder no. 12 Contribution/Balance Fault

P0300 Random/Multiple Cylinder Misfire Detected

P0301 Cylinder no. 1—Misfire Detected

P0302 Cylinder no. 2—Misfire Detected

P0303 Cylinder no. 3—Misfire Detected

P0304 Cylinder no. 4—Misfire Detected

P0305 Cylinder no. 5—Misfire Detected

P0306 Cylinder no. 6—Misfire Detected

P0307 Cylinder no. 7—Misfire Detected

P0308 Cylinder no. 8—Misfire Detected

P0309 Cylinder no. 9—Misfire Detected
P0310 Cylinder no. 10—Misfire Detected
P0311 Cylinder no. 11—Misfire Detected
P0312 Cylinder no. 12—Misfire Detected
P0320 Ignition/Distributor Engine Speed Input Circuit Malfunction
P0321 Ignition/Distributor Engine Speed Input Circuit Range/Performance
P0322 Ignition/Distributor Engine Speed Input Circuit No Signal
P0323 Ignition/Distributor Engine Speed Input Circuit Intermittent
P0325 Knock Sensor no. 1—Circuit Malfunction (Bank no. 1 or Single Sensor)
P0326 Knock Sensor no. 1—Circuit Range/Performance (Bank no. 1 or Single Sensor)
P0327 Knock Sensor no. 1—Circuit Low Input (Bank no. 1 or Single Sensor)
P0328 Knock Sensor no. 1—Circuit High Input (Bank no. 1 or Single Sensor)
P0329 Knock Sensor no. 1—Circuit Input Intermittent (Bank no. 1 or Single Sensor)
P0330 Knock Sensor no. 2—Circuit Malfunction (Bank no. 2)
P0331 Knock Sensor no. 2—Circuit Range/Performance (Bank no. 2)
P0332 Knock Sensor no. 2—Circuit Low Input (Bank no. 2)
P0333 Knock Sensor no. 2—Circuit High Input (Bank no. 2)
P0334 Knock Sensor no. 2—Circuit Input Intermittent (Bank no. 2)
P0335 Crankshaft Position Sensor "A" Circuit Malfunction
P0336 Crankshaft Position Sensor "A" Circuit Range/Performance
P0337 Crankshaft Position Sensor "A" Circuit Low Input
P0338 Crankshaft Position Sensor "A" Circuit High Input
P0339 Crankshaft Position Sensor "A" Circuit Intermittent
P0340 Camshaft Position Sensor Circuit Malfunction
P0341 Camshaft Position Sensor Circuit Range/Performance
P0342 Camshaft Position Sensor Circuit Low Input
P0343 Camshaft Position Sensor Circuit High Input
P0344 Camshaft Position Sensor Circuit Intermittent
P0350 Ignition Coil Primary/Secondary Circuit Malfunction
P0351 Ignition Coil "A" Primary/Secondary Circuit Malfunction
P0352 Ignition Coil "B" Primary/Secondary Circuit Malfunction
P0353 Ignition Coil "C" Primary/Secondary Circuit Malfunction
P0354 Ignition Coil "D" Primary/Secondary Circuit Malfunction
P0355 Ignition Coil "E" Primary/Secondary Circuit Malfunction
P0356 Ignition Coil "F" Primary/Secondary Circuit Malfunction
P0357 Ignition Coil "G" Primary/Secondary Circuit Malfunction
P0358 Ignition Coil "H" Primary/Secondary Circuit Malfunction
P0359 Ignition Coil "I" Primary/Secondary Circuit Malfunction
P0360 Ignition Coil "J" Primary/Secondary Circuit Malfunction
P0361 Ignition Coil "K" Primary/Secondary Circuit Malfunction
P0362 Ignition Coil "L" Primary/Secondary Circuit Malfunction
P0370 Timing Reference High Resolution Signal "A" Malfunction
P0371 Timing Reference High Resolution Signal "A" Too Many Pulses
P0372 Timing Reference High Resolution Signal "A" Too Few Pulses
P0373 Timing Reference High Resolution Signal "A" Intermittent/Erratic Pulses
P0374 Timing Reference High Resolution Signal "A" No Pulses

P0375 Timing Reference High Resolution Signal "B" Malfunction
P0376 Timing Reference High Resolution Signal "B" Too Many Pulses
P0377 Timing Reference High Resolution Signal "B" Too Few Pulses
P0378 Timing Reference High Resolution Signal "B" Intermittent/Erratic Pulses
P0379 Timing Reference High Resolution Signal "B" No Pulses
P0380 Glow Plug/Heater Circuit "A" Malfunction
P0381 Glow Plug/Heater Indicator Circuit Malfunction
P0382 Glow Plug/Heater Circuit "B" Malfunction
P0385 Crankshaft Position Sensor "B" Circuit Malfunction
P0386 Crankshaft Position Sensor "B" Circuit Range/Performance
P0387 Crankshaft Position Sensor "B" Circuit Low Input
P0388 Crankshaft Position Sensor "B" Circuit High Input
P0389 Crankshaft Position Sensor "B" Circuit Intermittent
P0400 Exhaust Gas Recirculation Flow Malfunction
P0401 Exhaust Gas Recirculation Flow Insufficient Detected
P0402 Exhaust Gas Recirculation Flow Excessive Detected
P0403 Exhaust Gas Recirculation Circuit Malfunction
P0404 Exhaust Gas Recirculation Circuit Range/Performance
P0405 Exhaust Gas Recirculation Sensor "A" Circuit Low
P0406 Exhaust Gas Recirculation Sensor "A" Circuit High
P0407 Exhaust Gas Recirculation Sensor "B" Circuit Low
P0408 Exhaust Gas Recirculation Sensor "B" Circuit High
P0410 Secondary Air Injection System Malfunction
P0411 Secondary Air Injection System Incorrect Flow Detected
P0412 Secondary Air Injection System Switching Valve "A" Circuit Malfunction
P0413 Secondary Air Injection System Switching Valve "A" Circuit Open
P0414 Secondary Air Injection System Switching Valve "A" Circuit Shorted
P0415 Secondary Air Injection System Switching Valve "B" Circuit Malfunction
P0416 Secondary Air Injection System Switching Valve "B" Circuit Open
P0417 Secondary Air Injection System Switching Valve "B" Circuit Shorted
P0418 Secondary Air Injection System Relay "A" Circuit Malfunction
P0419 Secondary Air Injection System Relay "B" Circuit Malfunction
P0420 Catalyst System Efficiency Below Threshold (Bank no. 1)
P0421 Warm Up Catalyst Efficiency Below Threshold (Bank no. 1)
P0422 Main Catalyst Efficiency Below Threshold (Bank no. 1)
P0423 Heated Catalyst Efficiency Below Threshold (Bank no. 1)
P0424 Heated Catalyst Temperature Below Threshold (Bank no. 1)
P0430 Catalyst System Efficiency Below Threshold (Bank no. 2)
P0431 Warm Up Catalyst Efficiency Below Threshold (Bank no. 2)
P0432 Main Catalyst Efficiency Below Threshold (Bank no. 2)
P0433 Heated Catalyst Efficiency Below Threshold (Bank no. 2)
P0434 Heated Catalyst Temperature Below Threshold (Bank no. 2)
P0440 Evaporative Emission Control System Malfunction
P0441 Evaporative Emission Control System Incorrect Purge Flow

P0442 Evaporative Emission Control System Leak Detected (Small Leak)

P0443 Evaporative Emission Control System Purge Control Valve Circuit Malfunction

P0444 Evaporative Emission Control System Purge Control Valve Circuit Open

P0445 Evaporative Emission Control System Purge Control Valve Circuit Shorted

P0446 Evaporative Emission Control System Vent Control Circuit Malfunction

P0447 Evaporative Emission Control System Vent Control Circuit Open

P0448 Evaporative Emission Control System Vent Control Circuit Shorted

P0449 Evaporative Emission Control System Vent Valve/Solenoid Circuit Malfunction

P0450 Evaporative Emission Control System Pressure Sensor Malfunction

P0451 Evaporative Emission Control System Pressure Sensor Range/Performance

P0452 Evaporative Emission Control System Pressure Sensor Low Input

P0453 Evaporative Emission Control System Pressure Sensor High Input

P0454 Evaporative Emission Control System Pressure Sensor Intermittent

P0455 Evaporative Emission Control System Leak Detected (Gross Leak)

P0460 Fuel Level Sensor Circuit Malfunction

P0461 Fuel Level Sensor Circuit Range/Performance

P0462 Fuel Level Sensor Circuit Low Input

P0463 Fuel Level Sensor Circuit High Input

P0464 Fuel Level Sensor Circuit Intermittent

P0465 Purge Flow Sensor Circuit Malfunction

P0466 Purge Flow Sensor Circuit Range/Performance

P0467 Purge Flow Sensor Circuit Low Input

P0468 Purge Flow Sensor Circuit High Input

P0469 Purge Flow Sensor Circuit Intermittent

P0470 Exhaust Pressure Sensor Malfunction

P0471 Exhaust Pressure Sensor Range/Performance

P0472 Exhaust Pressure Sensor Low

P0473 Exhaust Pressure Sensor High

P0474 Exhaust Pressure Sensor Intermittent

P0475 Exhaust Pressure Control Valve Malfunction

P0476 Exhaust Pressure Control Valve Range/Performance

P0477 Exhaust Pressure Control Valve Low

P0478 Exhaust Pressure Control Valve High

P0479 Exhaust Pressure Control Valve Intermittent

P0480 Cooling Fan no. 1 Control Circuit Malfunction

P0481 Cooling Fan no. 2 Control Circuit Malfunction

P0482 Cooling Fan no. 3 Control Circuit Malfunction

P0483 Cooling Fan Rationality Check Malfunction

P0484 Cooling Fan Circuit Over Current

P0485 Cooling Fan Power/Ground Circuit Malfunction

P0500 Vehicle Speed Sensor Malfunction

P0501 Vehicle Speed Sensor Range/Performance

P0502 Vehicle Speed Sensor Circuit Low Input

P0503 Vehicle Speed Sensor Intermittent/Erratic/High

P0505 Idle Control System Malfunction

P0506 Idle Control System RPM Lower Than Expected

P0507 Idle Control System RPM Higher Than Expected

P0510 Closed Throttle Position Switch Malfunction

P0520 Engine Oil Pressure Sensor/Switch Circuit Malfunction

P0521 Engine Oil Pressure Sensor/Switch Range/Performance

P0522 Engine Oil Pressure Sensor/Switch Low Voltage

P0523 Engine Oil Pressure Sensor/Switch High Voltage

P0530 A/C Refrigerant Pressure Sensor Circuit Malfunction

P0531 A/C Refrigerant Pressure Sensor Circuit Range/Performance

P0532 A/C Refrigerant Pressure Sensor Circuit Low Input

P0533 A/C Refrigerant Pressure Sensor Circuit High Input

P0534 A/C Refrigerant Charge Loss

P0550 Power Steering Pressure Sensor Circuit Malfunction

P0551 Power Steering Pressure Sensor Circuit Range/Performance

P0552 Power Steering Pressure Sensor Circuit Low Input

P0553 Power Steering Pressure Sensor Circuit High Input

P0554 Power Steering Pressure Sensor Circuit Intermittent

P0560 System Voltage Malfunction

P0561 System Voltage Unstable

P0562 System Voltage Low

P0563 System Voltage High

P0565 Cruise Control On Signal Malfunction

P0566 Cruise Control Off Signal Malfunction

P0567 Cruise Control Resume Signal Malfunction

P0568 Cruise Control Set Signal Malfunction

P0569 Cruise Control Coast Signal Malfunction

P0570 Cruise Control Accel Signal Malfunction

P0571 Cruise Control/Brake Switch "A" Circuit Malfunction

P0572 Cruise Control/Brake Switch "A" Circuit Low

P0573 Cruise Control/Brake Switch "A" Circuit High

P0574 Through P0580 Reserved for Cruise Codes

P0600 Serial Communication Link Malfunction

P0601 Internal Control Module Memory Check Sum Error

P0602 Control Module Programming Error

P0603 Internal Control Module Keep Alive Memory (KAM) Error

P0604 Internal Control Module Random Access Memory (RAM) Error

P0605 Internal Control Module Read Only Memory (ROM) Error

P0606 PCM Processor Fault

P0608 Control Module VSS Output "A" Malfunction

P0609 Control Module VSS Output "B" Malfunction

P0620 Generator Control Circuit Malfunction

P0621 Generator Lamp "L" Control Circuit Malfunction

P0622 Generator Field "F" Control Circuit Malfunction

P0650 Malfunction Indicator Lamp (MIL) Control Circuit Malfunction

P0654 Engine RPM Output Circuit Malfunction

P0655 Engine Hot Lamp Output Control Circuit Malfunction

P0656 Fuel Level Output Circuit Malfunction

P0700 Transmission Control System Malfunction

P0701 Transmission Control System Range/Performance

P0702 Transmission Control System Electrical

P0703 Torque Converter/Brake Switch "B" Circuit Malfunction

P0704 Clutch Switch Input Circuit Malfunction

P0705 Transmission Range Sensor Circuit Malfunction (PRNDL Input)

P0706 Transmission Range Sensor Circuit Range/Performance

P0707 Transmission Range Sensor Circuit Low Input

P0708 Transmission Range Sensor Circuit High Input

P0709 Transmission Range Sensor Circuit Intermittent

P0710 Transmission Fluid Temperature Sensor Circuit Malfunction

P0711 Transmission Fluid Temperature Sensor Circuit Range/Performance

P0712 Transmission Fluid Temperature Sensor Circuit Low Input

P0713 Transmission Fluid Temperature Sensor Circuit High Input

P0714 Transmission Fluid Temperature Sensor Circuit Intermittent

P0715 Input/Turbine Speed Sensor Circuit Malfunction

P0716 Input/Turbine Speed Sensor Circuit Range/Performance

P0717 Input/Turbine Speed Sensor Circuit No Signal

P0718 Input/Turbine Speed Sensor Circuit Intermittent

P0719 Torque Converter/Brake Switch "B" Circuit Low

P0720 Output Speed Sensor Circuit Malfunction

P0721 Output Speed Sensor Circuit Range/Performance

P0722 Output Speed Sensor Circuit No Signal

P0723 Output Speed Sensor Circuit Intermittent

P0724 Torque Converter/Brake Switch "B" Circuit High

P0725 Engine Speed Input Circuit Malfunction

P0726 Engine Speed Input Circuit Range/Performance

P0727 Engine Speed Input Circuit No Signal

P0728 Engine Speed Input Circuit Intermittent

P0730 Incorrect Gear Ratio

P0731 Gear no. 1 Incorrect Ratio

P0732 Gear no. 2 Incorrect Ratio

P0733 Gear no. 3 Incorrect Ratio

P0734 Gear no. 4 Incorrect Ratio

P0735 Gear no. 5 Incorrect Ratio

P0736 Reverse Incorrect Ratio

P0740 Torque Converter Clutch Circuit Malfunction

P0741 Torque Converter Clutch Circuit Performance or Stuck Off

P0742 Torque Converter Clutch Circuit Stuck On

P0743 Torque Converter Clutch Circuit Electrical

P0744 Torque Converter Clutch Circuit Intermittent

P0745 Pressure Control Solenoid Malfunction

P0746 Pressure Control Solenoid Performance or Stuck Off

P0747 Pressure Control Solenoid Stuck On

P0748 Pressure Control Solenoid Electrical

P0749 Pressure Control Solenoid Intermittent

P0750 Shift Solenoid "A" Malfunction

P0751 Shift Solenoid "A" Performance or Stuck Off

P0752 Shift Solenoid "A" Stuck On

P0753 Shift Solenoid "A" Electrical

P0754 Shift Solenoid "A" Intermittent

P0755 Shift Solenoid "B" Malfunction

P0756 Shift Solenoid "B" Performance or Stuck Off

P0757 Shift Solenoid "B" Stuck On

P0758 Shift Solenoid "B" Electrical

P0759 Shift Solenoid "B" Intermittent

P0760 Shift Solenoid "C" Malfunction

P0761 Shift Solenoid "C" Performance Or Stuck Off

P0762 Shift Solenoid "C" Stuck On

P0763 Shift Solenoid "C" Electrical

P0764 Shift Solenoid "C" Intermittent

P0765 Shift Solenoid "D" Malfunction

P0766 Shift Solenoid "D" Performance Or Stuck Off

P0767 Shift Solenoid "D" Stuck On

P0768 Shift Solenoid "D" Electrical

P0769 Shift Solenoid "D" Intermittent

P0770 Shift Solenoid "E" Malfunction

P0771 Shift Solenoid "E" Performance Or Stuck Off

P0772 Shift Solenoid "E" Stuck On

P0773 Shift Solenoid "E" Electrical

P0774 Shift Solenoid "E" Intermittent

P0780 Shift Malfunction

P0781 1–2 Shift Malfunction

P0782 2–3 Shift Malfunction

P0783 3–4 Shift Malfunction

P0784 4–5 Shift Malfunction

P0785 Shift/Timing Solenoid Malfunction

P0786 Shift/Timing Solenoid Range/Performance

P0787 Shift/Timing Solenoid Low

P0788 Shift/Timing Solenoid High

P0789 Shift/Timing Solenoid Intermittent

P0790 Normal/Performance Switch Circuit Malfunction

P0801 Reverse Inhibit Control Circuit Malfunction

P0803 1–4 Upshift (Skip Shift) Solenoid Control Circuit Malfunction

P0804 1–4 Upshift (Skip Shift) Lamp Control Circuit Malfunction

P1120 Secondary Throttle Position Sensor Circuit Fault

P1125 Tandem Throttle Position Sensor Circuit Fault

P1210 Traction Control System Signal Fault

P1220 Fuel Pump Control Module Fault

P1320 Ignition Control Signal Fault

P1336 Crankshaft Position Sensor Circuit Fault

P1400 EGR/EVAP Control Solenoid Circuit Fault

P1401 EGR Temperature Sensor Circuit Fault

P1443 EVAP Canister Control Vacuum Switch Circuit Fault

P1445 EVAP Purge Volume Control Valve Circuit Fault

P1605 TCM A~T Diagnosis Communication Line Fault

P1705 Throttle Position Sensor (Switch) Circuit Fault

P1760 Overrun Clutch Solenoid Valve Circuit Fault

P1900 Cooling Fan Control Circuit Fault

OBD II TROUBLE CODE EQUIVALENTS

0505 No Self Diagnostic Failure Indicated

0102 Mass or Volume Air Flow Circuit Malfunction

0401 Intake Air Temperature Circuit Malfunction

0103 Engine Coolant Temperature Circuit Malfunction

0403 Throttle/Pedal Position Sensor/Switch "A" Circuit Malfunction

0908 Insufficient Coolant Temperature For Closed Loop Fuel Control

0303 O2 Circuit Malfunction

0307 Closed Loop Control

0901 O2 Sensor Heater Circuit Malfunction (Bank no. 1 Sensor no. 1)

0707 O2 Sensor Circuit Malfunction (Bank no. 1 Sensor no. 2)

0902 O2 Sensor Heater Circuit Malfunction (Bank no. 1 Sensor no. 2)

0115 System Too Lean (Bank no. 1)

0114 System Too Rich (Bank no. 1)

0701 Random/Multiple Cylinder Misfire Detected

0608 Cylinder no. 1—Misfire Detected

0607 Cylinder no. 2—Misfire Detected

0606 Cylinder no. 3—Misfire Detected

0605 Cylinder no. 4—Misfire Detected

0304 Knock Sensor no. 1—Circuit Malfunction (Bank no. 1 or Single Sensor)

0802 Crankshaft Position Sensor "A" Circuit Malfunction

0101 Camshaft Position Sensor Circuit Malfunction

0302 Exhaust Gas Recirculation Flow Malfunction

0306 Exhaust Gas Recirculation Flow Excessive Detected

0702 Catalyst System Efficiency Below Threshold (Bank no. 1)

0104 Vehicle Speed Sensor Malfunction

0205 Idle Control System Malfunction

0301 Internal Control Module Read Only Memory (ROM) Error

1003 Transmission Range Sensor Circuit Malfunction (PRNDL Input)

1101 Inhibitor Switch Circuit

1208 Transmission Fluid Temperature Sensor Circuit Malfunction

1102 Output Speed Sensor Circuit Malfunction

1207 Engine Speed Input Circuit Malfunction

1103 Gear no. 1 Incorrect Ratio

1104 Gear no. 2 Incorrect Ratio

1105 Gear no. 3 Incorrect Ratio

1106 Gear no. 4 Incorrect Ratio

1204 Torque Converter Clutch Circuit Malfunction

1205 Pressure Control Solenoid Malfunction

1108 Shift Solenoid "A" Malfunction

1201 Shift Solenoid "B" Malfunction

0201 Ignition Control Signal Fault

0905 Crankshaft Position Sensor Circuit Fault

1005 EGR/EVAP Control Solenoid Circuit Fault

0305 EGR Temperature Sensor Circuit Fault

0804 TCM A~T Diagnosis Communication Line Fault

1206 Throttle Position Sensor (Switch) Circuit Fault

1203 Overrun Clutch Solenoid Valve Circuit Fault

1308 Cooling Fan Control Circuit Fault

Porsche

READING CODES

Reading the control module memory is on of the first steps in OBD II system diagnostics. This step should be initially performed to determine the general nature of the fault. Subsequent readings will determine if the fault has been cleared.

Reading codes can be performed by any of the methods below:

• Read the control module memory with the Generic Scan Tool (GST)

• Read the control module memory with the vehicle manufacturer's specific tester

To read the fault codes, connect the scan tool or tester according to the manufacturer's instructions. Follow the manufacturer's specified procedure for reading the codes.

CLEARING CODES

Control module reset procedures are a very important part of OBD II System diagnostics. This step should be done at the end of any fault code repair and at the end of any driveability repair.

Clearing codes can be performed by any of the methods below:

• Clear the control module memory with the Generic Scan Tool (GST)

• Clear the control module memory with the vehicle manufacturer's specific tester

• Turn the ignition **OFF** and remove the negative battery cable for at least 1 minute.

Removing the negative battery cable may cause other systems in the vehicle to loose their memory. Prior to removing the cable, ensure you have the proper reset codes for radios and alarms.

➡**The MIL will may also be de-activated for some codes if the vehicle completes three consecutive trips without a fault detected with vehicle conditions similar to those present during the fault.**

OBD II TROUBLE CODES

P0100 Mass or Volume Air Flow Circuit Malfunction

P0101 Mass or Volume Air Flow Circuit Range/Performance Problem

P0102 Mass or Volume Air Flow Circuit Low Input

P0103 Mass or Volume Air Flow Circuit High Input

P0104 Mass or Volume Air Flow Circuit Intermittent

P0105 Manifold Absolute Pressure/Barometric Pressure Circuit Malfunction

P0106 Manifold Absolute Pressure/Barometric Pressure Circuit Range/Performance Problem

P0107 Manifold Absolute Pressure/Barometric Pressure Circuit Low Input

P0108 Manifold Absolute Pressure/Barometric Pressure Circuit High Input

P0109 Manifold Absolute Pressure/Barometric Pressure Circuit Intermittent

P0110 Intake Air Temperature Circuit Malfunction

P0111 Intake Air Temperature Circuit Range/Performance Problem

P0112 Intake Air Temperature Circuit Low Input

P0113 Intake Air Temperature Circuit High Input

P0114 Intake Air Temperature Circuit Intermittent

P0115 Engine Coolant Temperature Circuit Malfunction

P0116 Engine Coolant Temperature Circuit Range/Performance Problem

P0117 Engine Coolant Temperature Circuit Low Input

P0118 Engine Coolant Temperature Circuit High Input

P0119 Engine Coolant Temperature Circuit Intermittent

P0120 Throttle/Pedal Position Sensor/Switch "A" Circuit Malfunction

P0121 Throttle/Pedal Position Sensor/Switch "A" Circuit Range/Performance Problem

P0122 Throttle/Pedal Position Sensor/Switch "A" Circuit Low Input

P0123 Throttle/Pedal Position Sensor/Switch "A" Circuit High Input

P0124 Throttle/Pedal Position Sensor/Switch "A" Circuit Intermittent

P0125 Insufficient Coolant Temperature For Closed Loop Fuel Control

P0126 Insufficient Coolant Temperature For Stable Operation

P0130 O2 Circuit Malfunction (Bank no. 1 Sensor no. 1)

P0131 O2 Sensor Circuit Low Voltage (Bank no. 1 Sensor no. 1)

P0132 O2 Sensor Circuit High Voltage (Bank no. 1 Sensor no. 1)

P0133 O2 Sensor Circuit Slow Response (Bank no. 1 Sensor no. 1)

P0134 O2 Sensor Circuit No Activity Detected (Bank no. 1 Sensor no. 1)

P0135 O2 Sensor Heater Circuit Malfunction (Bank no. 1 Sensor no. 1)

P0136 O2 Sensor Circuit Malfunction (Bank no. 1 Sensor no. 2)

P0137 O2 Sensor Circuit Low Voltage (Bank no. 1 Sensor no. 2)

P0138 O2 Sensor Circuit High Voltage (Bank no. 1 Sensor no. 2)

P0139 O2 Sensor Circuit Slow Response (Bank no. 1 Sensor no. 2)

P0140 O2 Sensor Circuit No Activity Detected (Bank no. 1 Sensor no. 2)

P0141 O2 Sensor Heater Circuit Malfunction (Bank no. 1 Sensor no. 2)

P0142 O2 Sensor Circuit Malfunction (Bank no. 1 Sensor no. 3)

P0143 O2 Sensor Circuit Low Voltage (Bank no. 1 Sensor no. 3)

P0144 O2 Sensor Circuit High Voltage (Bank no. 1 Sensor no. 3)

P0145 O2 Sensor Circuit Slow Response (Bank no. 1 Sensor no. 3)

P0146 O2 Sensor Circuit No Activity Detected (Bank no. 1 Sensor no. 3)

P0147 O2 Sensor Heater Circuit Malfunction (Bank no. 1 Sensor no. 3)

P0150 O2 Sensor Circuit Malfunction (Bank no. 2 Sensor no. 1)

P0151 O2 Sensor Circuit Low Voltage (Bank no. 2 Sensor no. 1)

P0152 O2 Sensor Circuit High Voltage (Bank no. 2 Sensor no. 1)

P0153 O2 Sensor Circuit Slow Response (Bank no. 2 Sensor no. 1)

P0154 O2 Sensor Circuit No Activity Detected (Bank no. 2 Sensor no. 1)

P0155 O2 Sensor Heater Circuit Malfunction (Bank no. 2 Sensor no. 1)

P0156 O2 Sensor Circuit Malfunction (Bank no. 2 Sensor no. 2)

P0157 O2 Sensor Circuit Low Voltage (Bank no. 2 Sensor no. 2)

P0158 O2 Sensor Circuit High Voltage (Bank no. 2 Sensor no. 2)

P0159 O2 Sensor Circuit Slow Response (Bank no. 2 Sensor no. 2)

P0160 O2 Sensor Circuit No Activity Detected (Bank no. 2 Sensor no. 2)

P0161 O2 Sensor Heater Circuit Malfunction (Bank no. 2 Sensor no. 2)

P0162 O2 Sensor Circuit Malfunction (Bank no. 2 Sensor no. 3)

P0163 O2 Sensor Circuit Low Voltage (Bank no. 2 Sensor no. 3)

P0164 O2 Sensor Circuit High Voltage (Bank no. 2 Sensor no. 3)

P0165 O2 Sensor Circuit Slow Response (Bank no. 2 Sensor no. 3)

P0166 O2 Sensor Circuit No Activity Detected (Bank no. 2 Sensor no. 3)

P0167 O2 Sensor Heater Circuit Malfunction (Bank no. 2 Sensor no. 3)

P0170 Fuel Trim Malfunction (Bank no. 1)

P0171 System Too Lean (Bank no. 1)

P0172 System Too Rich (Bank no. 1)

P0173 Fuel Trim Malfunction (Bank no. 2)

P0174 System Too Lean (Bank no. 2)

P0175 System Too Rich (Bank no. 2)

P0176 Fuel Composition Sensor Circuit Malfunction

P0177 Fuel Composition Sensor Circuit Range/Performance

P0178 Fuel Composition Sensor Circuit Low Input

P0179 Fuel Composition Sensor Circuit High Input

P0180 Fuel Temperature Sensor "A" Circuit Malfunction

P0181 Fuel Temperature Sensor "A" Circuit Range/Performance

P0182 Fuel Temperature Sensor "A" Circuit Low Input

P0183 Fuel Temperature Sensor "A" Circuit High Input

P0184 Fuel Temperature Sensor "A" Circuit Intermittent

P0185 Fuel Temperature Sensor "B" Circuit Malfunction

P0186 Fuel Temperature Sensor "B" Circuit Range/Performance

P0187 Fuel Temperature Sensor "B" Circuit Low Input

P0188 Fuel Temperature Sensor "B" Circuit High Input

P0189 Fuel Temperature Sensor "B" Circuit Intermittent

P0190 Fuel Rail Pressure Sensor Circuit Malfunction

P0191 Fuel Rail Pressure Sensor Circuit Range/Performance

P0192 Fuel Rail Pressure Sensor Circuit Low Input

P0193 Fuel Rail Pressure Sensor Circuit High Input

P0194 Fuel Rail Pressure Sensor Circuit Intermittent

P0195 Engine Oil Temperature Sensor Malfunction

P0196 Engine Oil Temperature Sensor Range/Performance

P0197 Engine Oil Temperature Sensor Low

P0198 Engine Oil Temperature Sensor High

P0199 Engine Oil Temperature Sensor Intermittent

P0200 Injector Circuit Malfunction

P0201 Injector Circuit Malfunction—Cylinder no. 1

P0202 Injector Circuit Malfunction—Cylinder no. 2

P0203 Injector Circuit Malfunction—Cylinder no. 3

P0204 Injector Circuit Malfunction—Cylinder no. 4

P0205 Injector Circuit Malfunction—Cylinder no. 5

P0206 Injector Circuit Malfunction—Cylinder no. 6

P0207 Injector Circuit Malfunction—Cylinder no. 7

P0208 Injector Circuit Malfunction—Cylinder no. 8

P0209 Injector Circuit Malfunction—Cylinder no. 9

P0210 Injector Circuit Malfunction—Cylinder no. 10

P0211 Injector Circuit Malfunction—Cylinder no. 11

P0212 Injector Circuit Malfunction—Cylinder no. 12

P0213 Cold Start Injector no. 1 Malfunction

P0214 Cold Start Injector no. 2 Malfunction

P0215 Engine Shutoff Solenoid Malfunction

P0216 Injection Timing Control Circuit Malfunction

P0217 Engine Over Temperature Condition

P0218 Transmission Over Temperature Condition

P0219 Engine Over Speed Condition

P0220 Throttle/Pedal Position Sensor/Switch "B" Circuit Malfunction

P0221 Throttle/Pedal Position Sensor/Switch "B" Circuit Range/Performance Problem

P0222 Throttle/Pedal Position Sensor/Switch "B" Circuit Low Input

P0223 Throttle/Pedal Position Sensor/Switch "B" Circuit High Input

P0224 Throttle/Pedal Position Sensor/Switch "B" Circuit Intermittent

P0225 Throttle/Pedal Position Sensor/Switch "C" Circuit Malfunction

P0226 Throttle/Pedal Position Sensor/Switch "C" Circuit Range/Performance Problem

P0227 Throttle/Pedal Position Sensor/Switch "C" Circuit Low Input

P0228 Throttle/Pedal Position Sensor/Switch "C" Circuit High Input

P0229 Throttle/Pedal Position Sensor/Switch "C" Circuit Intermittent

P0230 Fuel Pump Primary Circuit Malfunction
P0231 Fuel Pump Secondary Circuit Low
P0232 Fuel Pump Secondary Circuit High
P0233 Fuel Pump Secondary Circuit Intermittent
P0261 Cylinder no. 1 Injector Circuit Low
P0262 Cylinder no. 1 Injector Circuit High
P0263 Cylinder no. 1 Contribution/Balance Fault
P0264 Cylinder no. 2 Injector Circuit Low
P0265 Cylinder no. 2 Injector Circuit High
P0266 Cylinder no. 2 Contribution/Balance Fault
P0267 Cylinder no. 3 Injector Circuit Low
P0268 Cylinder no. 3 Injector Circuit High
P0269 Cylinder no. 3 Contribution/Balance Fault
P0270 Cylinder no. 4 Injector Circuit Low
P0271 Cylinder no. 4 Injector Circuit High
P0272 Cylinder no. 4 Contribution/Balance Fault
P0273 Cylinder no. 5 Injector Circuit Low
P0274 Cylinder no. 5 Injector Circuit High
P0275 Cylinder no. 5 Contribution/Balance Fault
P0276 Cylinder no. 6 Injector Circuit Low
P0277 Cylinder no. 6 Injector Circuit High
P0278 Cylinder no. 6 Contribution/Balance Fault
P0279 Cylinder no. 7 Injector Circuit Low
P0280 Cylinder no. 7 Injector Circuit High
P0281 Cylinder no. 7 Contribution/Balance Fault
P0282 Cylinder no. 8 Injector Circuit Low
P0283 Cylinder no. 8 Injector Circuit High
P0284 Cylinder no. 8 Contribution/Balance Fault
P0285 Cylinder no. 9 Injector Circuit Low
P0286 Cylinder no. 9 Injector Circuit High
P0287 Cylinder no. 9 Contribution/Balance Fault
P0288 Cylinder no. 10 Injector Circuit Low
P0289 Cylinder no. 10 Injector Circuit High
P0290 Cylinder no. 10 Contribution/Balance Fault
P0291 Cylinder no. 11 Injector Circuit Low
P0292 Cylinder no. 11 Injector Circuit High
P0293 Cylinder no. 11 Contribution/Balance Fault
P0294 Cylinder no. 12 Injector Circuit Low
P0295 Cylinder no. 12 Injector Circuit High
P0296 Cylinder no. 12 Contribution/Balance Fault
P0300 Random/Multiple Cylinder Misfire Detected
P0301 Cylinder no. 1—Misfire Detected
P0302 Cylinder no. 2—Misfire Detected
P0303 Cylinder no. 3—Misfire Detected
P0304 Cylinder no. 4—Misfire Detected
P0305 Cylinder no. 5—Misfire Detected
P0306 Cylinder no. 6—Misfire Detected
P0307 Cylinder no. 7—Misfire Detected
P0308 Cylinder no. 8—Misfire Detected
P0309 Cylinder no. 9—Misfire Detected
P0310 Cylinder no. 10—Misfire Detected
P0311 Cylinder no. 11—Misfire Detected
P0312 Cylinder no. 12—Misfire Detected
P0320 Ignition/Distributor Engine Speed Input Circuit Malfunction
P0321 Ignition/Distributor Engine Speed Input Circuit Range/Performance
P0322 Ignition/Distributor Engine Speed Input Circuit No Signal
P0323 Ignition/Distributor Engine Speed Input Circuit Intermittent
P0325 Knock Sensor no. 1—Circuit Malfunction (Bank no. 1 or Single Sensor)

P0326 Knock Sensor no. 1—Circuit Range/Performance (Bank no. 1 or Single Sensor)
P0327 Knock Sensor no. 1—Circuit Low Input (Bank no. 1 or Single Sensor)
P0328 Knock Sensor no. 1—Circuit High Input (Bank no. 1 or Single Sensor)
P0329 Knock Sensor no. 1—Circuit Input Intermittent (Bank no. 1 or Single Sensor)
P0330 Knock Sensor no. 2—Circuit Malfunction (Bank no. 2)
P0331 Knock Sensor no. 2—Circuit Range/Performance (Bank no. 2)
P0332 Knock Sensor no. 2—Circuit Low Input (Bank no. 2)
P0333 Knock Sensor no. 2—Circuit High Input (Bank no. 2)
P0334 Knock Sensor no. 2—Circuit Input Intermittent (Bank no. 2)
P0335 Crankshaft Position Sensor "A" Circuit Malfunction
P0336 Crankshaft Position Sensor "A" Circuit Range/Performance
P0337 Crankshaft Position Sensor "A" Circuit Low Input
P0338 Crankshaft Position Sensor "A" Circuit High Input
P0339 Crankshaft Position Sensor "A" Circuit Intermittent
P0340 Camshaft Position Sensor Circuit Malfunction
P0341 Camshaft Position Sensor Circuit Range/Performance
P0342 Camshaft Position Sensor Circuit Low Input
P0343 Camshaft Position Sensor Circuit High Input
P0344 Camshaft Position Sensor Circuit Intermittent
P0350 Ignition Coil Primary/Secondary Circuit Malfunction
P0351 Ignition Coil "A" Primary/Secondary Circuit Malfunction
P0352 Ignition Coil "B" Primary/Secondary Circuit Malfunction
P0353 Ignition Coil "C" Primary/Secondary Circuit Malfunction
P0354 Ignition Coil "D" Primary/Secondary Circuit Malfunction
P0355 Ignition Coil "E" Primary/Secondary Circuit Malfunction
P0356 Ignition Coil "F" Primary/Secondary Circuit Malfunction
P0357 Ignition Coil "G" Primary/Secondary Circuit Malfunction
P0358 Ignition Coil "H" Primary/Secondary Circuit Malfunction
P0359 Ignition Coil "I" Primary/Secondary Circuit Malfunction
P0360 Ignition Coil "J" Primary/Secondary Circuit Malfunction
P0361 Ignition Coil "K" Primary/Secondary Circuit Malfunction
P0362 Ignition Coil "L" Primary/Secondary Circuit Malfunction
P0370 Timing Reference High Resolution Signal "A" Malfunction
P0371 Timing Reference High Resolution Signal "A" Too Many Pulses
P0372 Timing Reference High Resolution Signal "A" Too Few Pulses
P0373 Timing Reference High Resolution Signal "A" Intermittent/Erratic Pulses
P0374 Timing Reference High Resolution Signal "A" No Pulses
P0375 Timing Reference High Resolution Signal "B" Malfunction
P0376 Timing Reference High Resolution Signal "B" Too Many Pulses
P0377 Timing Reference High Resolution Signal "B" Too Few Pulses
P0378 Timing Reference High Resolution Signal "B" Intermittent/Erratic Pulses
P0379 Timing Reference High Resolution Signal "B" No Pulses
P0380 Glow Plug/Heater Circuit "A" Malfunction
P0381 Glow Plug/Heater Indicator Circuit Malfunction
P0382 Glow Plug/Heater Circuit "B" Malfunction
P0385 Crankshaft Position Sensor "B" Circuit Malfunction

P0386 Crankshaft Position Sensor "B" Circuit Range/Performance

P0387 Crankshaft Position Sensor "B" Circuit Low Input

P0388 Crankshaft Position Sensor "B" Circuit High Input

P0389 Crankshaft Position Sensor "B" Circuit Intermittent

P0400 Exhaust Gas Recirculation Flow Malfunction

P0401 Exhaust Gas Recirculation Flow Insufficient Detected

P0402 Exhaust Gas Recirculation Flow Excessive Detected

P0403 Exhaust Gas Recirculation Circuit Malfunction

P0404 Exhaust Gas Recirculation Circuit Range/Performance

P0405 Exhaust Gas Recirculation Sensor "A" Circuit Low

P0406 Exhaust Gas Recirculation Sensor "A" Circuit High

P0407 Exhaust Gas Recirculation Sensor "B" Circuit Low

P0408 Exhaust Gas Recirculation Sensor "B" Circuit High

P0410 Secondary Air Injection System Malfunction

P0411 Secondary Air Injection System Incorrect Flow Detected

P0412 Secondary Air Injection System Switching Valve "A" Circuit Malfunction

P0413 Secondary Air Injection System Switching Valve "A" Circuit Open

P0414 Secondary Air Injection System Switching Valve "A" Circuit Shorted

P0415 Secondary Air Injection System Switching Valve "B" Circuit Malfunction

P0416 Secondary Air Injection System Switching Valve "B" Circuit Open

P0417 Secondary Air Injection System Switching Valve "B" Circuit Shorted

P0418 Secondary Air Injection System Relay "A" Circuit Malfunction

P0419 Secondary Air Injection System Relay "B" Circuit Malfunction

P0420 Catalyst System Efficiency Below Threshold (Bank no. 1)

P0421 Warm Up Catalyst Efficiency Below Threshold (Bank no. 1)

P0422 Main Catalyst Efficiency Below Threshold (Bank no. 1)

P0423 Heated Catalyst Efficiency Below Threshold (Bank no. 1)

P0424 Heated Catalyst Temperature Below Threshold (Bank no. 1)

P0430 Catalyst System Efficiency Below Threshold (Bank no. 2)

P0431 Warm Up Catalyst Efficiency Below Threshold (Bank no. 2)

P0432 Main Catalyst Efficiency Below Threshold (Bank no. 2)

P0433 Heated Catalyst Efficiency Below Threshold (Bank no. 2)

P0434 Heated Catalyst Temperature Below Threshold (Bank no. 2)

P0440 Evaporative Emission Control System Malfunction

P0441 Evaporative Emission Control System Incorrect Purge Flow

P0442 Evaporative Emission Control System Leak Detected (Small Leak)

P0443 Evaporative Emission Control System Purge Control Valve Circuit Malfunction

P0444 Evaporative Emission Control System Purge Control Valve Circuit Open

P0445 Evaporative Emission Control System Purge Control Valve Circuit Shorted

P0446 Evaporative Emission Control System Vent Control Circuit Malfunction

P0447 Evaporative Emission Control System Vent Control Circuit Open

P0448 Evaporative Emission Control System Vent Control Circuit Shorted

P0449 Evaporative Emission Control System Vent Valve/Solenoid Circuit Malfunction

P0450 Evaporative Emission Control System Pressure Sensor Malfunction

P0451 Evaporative Emission Control System Pressure Sensor Range/Performance

P0452 Evaporative Emission Control System Pressure Sensor Low Input

P0453 Evaporative Emission Control System Pressure Sensor High Input

P0454 Evaporative Emission Control System Pressure Sensor Intermittent

P0455 Evaporative Emission Control System Leak Detected (Gross Leak)

P0460 Fuel Level Sensor Circuit Malfunction

P0461 Fuel Level Sensor Circuit Range/Performance

P0462 Fuel Level Sensor Circuit Low Input

P0463 Fuel Level Sensor Circuit High Input

P0464 Fuel Level Sensor Circuit Intermittent

P0465 Purge Flow Sensor Circuit Malfunction

P0466 Purge Flow Sensor Circuit Range/Performance

P0467 Purge Flow Sensor Circuit Low Input

P0468 Purge Flow Sensor Circuit High Input

P0469 Purge Flow Sensor Circuit Intermittent

P0470 Exhaust Pressure Sensor Malfunction

P0471 Exhaust Pressure Sensor Range/Performance

P0472 Exhaust Pressure Sensor Low

P0473 Exhaust Pressure Sensor High

P0474 Exhaust Pressure Sensor Intermittent

P0475 Exhaust Pressure Control Valve Malfunction

P0476 Exhaust Pressure Control Valve Range/Performance

P0477 Exhaust Pressure Control Valve Low

P0478 Exhaust Pressure Control Valve High

P0479 Exhaust Pressure Control Valve Intermittent

P0480 Cooling Fan no. 1 Control Circuit Malfunction

P0481 Cooling Fan no. 2 Control Circuit Malfunction

P0482 Cooling Fan no. 3 Control Circuit Malfunction

P0483 Cooling Fan Rationality Check Malfunction

P0484 Cooling Fan Circuit Over Current

P0485 Cooling Fan Power/Ground Circuit Malfunction

P0500 Vehicle Speed Sensor Malfunction

P0501 Vehicle Speed Sensor Range/Performance

P0502 Vehicle Speed Sensor Circuit Low Input

P0503 Vehicle Speed Sensor Intermittent/Erratic/High

P0505 Idle Control System Malfunction

P0506 Idle Control System RPM Lower Than Expected

P0507 Idle Control System RPM Higher Than Expected

P0510 Closed Throttle Position Switch Malfunction

P0520 Engine Oil Pressure Sensor/Switch Circuit Malfunction

P0521 Engine Oil Pressure Sensor/Switch Range/Performance

P0522 Engine Oil Pressure Sensor/Switch Low Voltage

P0523 Engine Oil Pressure Sensor/Switch High Voltage

P0530 A/C Refrigerant Pressure Sensor Circuit Malfunction

P0531 A/C Refrigerant Pressure Sensor Circuit Range/Performance

P0532 A/C Refrigerant Pressure Sensor Circuit Low Input

P0533 A/C Refrigerant Pressure Sensor Circuit High Input

P0534 A/C Refrigerant Charge Loss

P0550 Power Steering Pressure Sensor Circuit Malfunction

P0551 Power Steering Pressure Sensor Circuit Range/Performance

P0552 Power Steering Pressure Sensor Circuit Low Input

P0553 Power Steering Pressure Sensor Circuit High Input

P0554 Power Steering Pressure Sensor Circuit Intermittent

P0560 System Voltage Malfunction

P0561 System Voltage Unstable

P0562 System Voltage Low

P0563 System Voltage High

P0565 Cruise Control On Signal Malfunction

P0566 Cruise Control Off Signal Malfunction

P0567 Cruise Control Resume Signal Malfunction

P0568 Cruise Control Set Signal Malfunction

P0569 Cruise Control Coast Signal Malfunction

P0570 Cruise Control Accel Signal Malfunction

P0571 Cruise Control/Brake Switch "A" Circuit Malfunction

P0572 Cruise Control/Brake Switch "A" Circuit Low

P0573 Cruise Control/Brake Switch "A" Circuit High

P0574 Through P0580 Reserved for Cruise Codes

P0600 Serial Communication Link Malfunction

P0601 Internal Control Module Memory Check Sum Error

P0602 Control Module Programming Error

P0603 Internal Control Module Keep Alive Memory (KAM) Error

P0604 Internal Control Module Random Access Memory (RAM) Error

P0605 Internal Control Module Read Only Memory (ROM) Error

P0606 PCM Processor Fault

P0608 Control Module VSS Output "A" Malfunction

P0609 Control Module VSS Output "B" Malfunction

P0620 Generator Control Circuit Malfunction

P0621 Generator Lamp "L" Control Circuit Malfunction

P0622 Generator Field "F" Control Circuit Malfunction

P0650 Malfunction Indicator Lamp (MIL) Control Circuit Malfunction

P0654 Engine RPM Output Circuit Malfunction

P0655 Engine Hot Lamp Output Control Circuit Malfunction

P0656 Fuel Level Output Circuit Malfunction

P0700 Transmission Control System Malfunction

P0701 Transmission Control System Range/Performance

P0702 Transmission Control System Electrical

P0703 Torque Converter/Brake Switch "B" Circuit Malfunction

P0704 Clutch Switch Input Circuit Malfunction

P0705 Transmission Range Sensor Circuit Malfunction (PRNDL Input)

P0706 Transmission Range Sensor Circuit Range/Performance

P0707 Transmission Range Sensor Circuit Low Input

P0708 Transmission Range Sensor Circuit High Input

P0709 Transmission Range Sensor Circuit Intermittent

P0710 Transmission Fluid Temperature Sensor Circuit Malfunction

P0711 Transmission Fluid Temperature Sensor Circuit Range/Performance

P0712 Transmission Fluid Temperature Sensor Circuit Low Input

P0713 Transmission Fluid Temperature Sensor Circuit High Input

P0714 Transmission Fluid Temperature Sensor Circuit Intermittent

P0715 Input/Turbine Speed Sensor Circuit Malfunction

P0716 Input/Turbine Speed Sensor Circuit Range/Performance

P0717 Input/Turbine Speed Sensor Circuit No Signal

P0718 Input/Turbine Speed Sensor Circuit Intermittent

P0719 Torque Converter/Brake Switch "B" Circuit Low

P0720 Output Speed Sensor Circuit Malfunction

P0721 Output Speed Sensor Circuit Range/Performance

P0722 Output Speed Sensor Circuit No Signal

P0723 Output Speed Sensor Circuit Intermittent

P0724 Torque Converter/Brake Switch "B" Circuit High

P0725 Engine Speed Input Circuit Malfunction

P0726 Engine Speed Input Circuit Range/Performance

P0727 Engine Speed Input Circuit No Signal

P0728 Engine Speed Input Circuit Intermittent

P0730 Incorrect Gear Ratio

P0731 Gear no. 1 Incorrect Ratio

P0732 Gear no. 2 Incorrect Ratio

P0733 Gear no. 3 Incorrect Ratio

P0734 Gear no. 4 Incorrect Ratio

P0735 Gear no. 5 Incorrect Ratio

P0736 Reverse Incorrect Ratio

P0740 Torque Converter Clutch Circuit Malfunction

P0741 Torque Converter Clutch Circuit Performance or Stuck Off

P0742 Torque Converter Clutch Circuit Stuck On

P0743 Torque Converter Clutch Circuit Electrical

P0744 Torque Converter Clutch Circuit Intermittent

P0745 Pressure Control Solenoid Malfunction

P0746 Pressure Control Solenoid Performance or Stuck Off

P0747 Pressure Control Solenoid Stuck On

P0748 Pressure Control Solenoid Electrical

P0749 Pressure Control Solenoid Intermittent

P0750 Shift Solenoid "A" Malfunction

P0751 Shift Solenoid "A" Performance or Stuck Off

P0752 Shift Solenoid "A" Stuck On

P0753 Shift Solenoid "A" Electrical

P0754 Shift Solenoid "A" Intermittent

P0755 Shift Solenoid "B" Malfunction

P0756 Shift Solenoid "B" Performance or Stuck Off

P0757 Shift Solenoid "B" Stuck On

P0758 Shift Solenoid "B" Electrical

P0759 Shift Solenoid "B" Intermittent

P0760 Shift Solenoid "C" Malfunction

P0761 Shift Solenoid "C" Performance Or Stuck Off

P0762 Shift Solenoid "C" Stuck On

P0763 Shift Solenoid "C" Electrical

P0764 Shift Solenoid "C" Intermittent

P0765 Shift Solenoid "D" Malfunction

P0766 Shift Solenoid "D" Performance Or Stuck Off

P0767 Shift Solenoid "D" Stuck On

P0768 Shift Solenoid "D" Electrical

P0769 Shift Solenoid "D" Intermittent

P0770 Shift Solenoid "E" Malfunction

P0771 Shift Solenoid "E" Performance Or Stuck Off

P0772 Shift Solenoid "E" Stuck On

P0773 Shift Solenoid "E" Electrical

P0774 Shift Solenoid "E" Intermittent

P0780 Shift Malfunction

P0781 1–2 Shift Malfunction

P0782 2–3 Shift Malfunction

P0783 3–4 Shift Malfunction

P0784 4–5 Shift Malfunction

P0785 Shift/Timing Solenoid Malfunction

P0786 Shift/Timing Solenoid Range/Performance

P0787 Shift/Timing Solenoid Low
P0788 Shift/Timing Solenoid High
P0789 Shift/Timing Solenoid Intermittent
P0790 Normal/Performance Switch Circuit Malfunction
P0801 Reverse Inhibit Control Circuit Malfunction
P0803 1–4 Upshift (Skip Shift) Solenoid Control Circuit Malfunction
P0804 1–4 Upshift (Skip Shift) Lamp Control Circuit Malfunction
P1102 Oxygen Sensor Heating
P1105 Oxygen Sensor Heating
P1107 Oxygen Sensor Heating
P1110 Oxygen Sensor Heating
P1115 Oxygen Sensor Heating
P1117 Oxygen Sensor Heating
P1119 Oxygen Sensor Heating
P1121 Oxygen Sensor Heating
P1123 Oxygen Sensing Heating
P1124 Oxygen Sensing
P1125 Oxygen Sensing
P1126 Oxygen Sensing
P1127 Oxygen Sensing
P1128 Oxygen Sensing
P1129 Oxygen Sensing
P1130 Oxygen Sensing
P1136 Oxygen Sensing
P1137 Oxygen Sensing
P1138 Oxygen Sensing
P1139 Oxygen Sensing
P1140 Load Signal
P1157 Engine Compartment Temperature
P1158 Engine Compartment Temperature
P1213 Fuel Injector, Cylinder 1
P1214 Fuel Injector, Cylinder 2
P1215 Fuel Injector, Cylinder 3
P1216 Fuel Injector, Cylinder 4
P1217 Fuel Injector, Cylinder 5
P1218 Fuel Injector, Cylinder 6
P1225 Fuel Injector, Cylinder 1
P1226 Fuel Injector, Cylinder 2
P1227 Fuel Injector, Cylinder 3
P1228 Fuel Injector, Cylinder 4
P1229 Fuel Injector, Cylinder 5
P1230 Fuel Injector, Cylinder 6
P1237 Fuel Injector, Cylinder 1
P1238 Fuel Injector, Cylinder 2
P1239 Fuel Injector, Cylinder 3
P1240 Fuel injector, Cylinder 4
P1241 Fuel Injector, Cylinder 5
P1242 Fuel Injector, Cylinder 6
P1265 Airbag Signal
P1275 Oxygen Sensor Aging Ahead of Three Way Catalytic Converter
P1276 Oxygen Sensor Aging Ahead of Three Way Catalytic Converter
P1313 Misfire Cylinder 1, Emission Related
P1314 Misfire Cylinder 2, Emission Related
P1315 Misfire Cylinder 3, Emission Related
P1316 Misfire, Cylinder 4, Emission Related
P1317 Misfire, Cylinder 5, Emission Related
P1318 Misfire, Cylinder 6, Emission Related

P1319 Misfire Emission Related
P1324 Timing Chain out of Position, Bank 2
P1340 Timing Chain out of Position, Bank 1
P1384 Knock Sensor 1
P1385 Knock Sensor 2
P1386 Knock Sensor Test Pulse
P1386 Knock Control Test Pulse
P1397 Camshaft Position Sensor 2
P1411 Secondary Air Injection System
P1455 A/C Compressor Control
P1456 A/C Compressor Control
P1457 A/C Compressor Control
P1458 A/C Compressor Signal
P1501 Fuel Pump Relay End-Stage
P1502 Fuel Pump Relay End-Stage
P1510 Idle Air Control Valve
P1513 Idle Air Control Valve
P1514 Idle Control Valve
P1515 Intake Manifold Resonance Flap
P1516 Intake Manifold Resonance Flap
P1524 Camshaft Adjustment, Bank 2
P1530 Camshaft Adjustment, Bank 1
P1531 Camshaft Adjustment, Bank 1
P1539 Camshaft Adjustment, Bank 2
P1541 Fuel Pump Relay End-Stage
P1551 Idle Air Control Valve
P1552 Idle Air Control Valve
P1553 Idle Air Control Valve
P1555 Charge Pressure Characteristics
P1556 Charge Deviations
P1557 Charge Deviations
P1570 Immobilizer
P1571 Immobilizer
P1585 Misfire With Empty Fuel Tank
P1593 Intake Manifold Length Tuning 2
P1594 Intake Manifold Length Tuning 2
P1595 Intake Manifold Length Tuning 2
P1600 Voltage Supply
P1601 Voltage Supply
P1602 Voltage Supply
P1610 MIL Activated Externally
P1611 MIL Activated Externally
P1614 MIL Activated Externally
P1640 Engine Control Module
P1656 Coolant Shutoff Valve
P1671 Engine Compartment Purge Fan End-Stage
P1673 Fan End-Stage
P1689 Engine Control Module
P1691 Malfunction Indicator Lamp
P1692 Malfunction Indicator Lamp
P1693 Malfunction Indicator Lamp
P1704 Kickdown Switch
P1710 Speed Signal, Right Front
P1715 Speed Signal, Left Front
P1744 Manual Program Switch
P1746 Control Unit Defective (Relay)
P1748 Control unit defective (relay sticks)
P1749 Version coding
P1750 Voltage supply, solenoid valve pressure regulators 1
P1761 Shiftlock P/N
P1762 Shiftlock P/N

P1764 Instrument cluster triggering
P1765 Throttle-valve information error
P1770 Load signal from ECM
P1782 Engine Engagement
P1813 Pressure regulator 1
P1818 Pressure regulator 2
P1823 Pressure regulator 3
P1828 Pressure regulator 4

Saab

READING CODES

Reading the control module memory is on of the first steps in OBD II system diagnostics. This step should be initially performed to determine the general nature of the fault. Subsequent readings will determine if the fault has been cleared.

Reading codes can be performed by any of the methods below:
- Read the control module memory with the Generic Scan Tool (GST)
- Read the control module memory with the vehicle manufacturer's specific tester

To read the fault codes, connect the scan tool or tester according to the manufacturer's instructions. Follow the manufacturer's specified procedure for reading the codes.

CLEARING CODES

Control module reset procedures are a very important part of OBD II System diagnostics. This step should be done at the end of any fault code repair and at the end of any driveability repair.

Clearing codes can be performed by any of the methods below:
- Clear the control module memory with the Generic Scan Tool (GST)
- Clear the control module memory with the vehicle manufacturer's specific tester
- Turn the ignition **OFF** and remove the negative battery cable for at least 1 minute.

Removing the negative battery cable may cause other systems in the vehicle to loose their memory. Prior to removing the cable, ensure you have the proper reset codes for radios and alarms.

➡**The MIL will may also be de-activated for some codes if the vehicle completes three consecutive trips without a fault detected with vehicle conditions similar to those present during the fault.**

OBD II TROUBLE CODES

P0100 Mass or Volume Air Flow Circuit Malfunction
P0101 Mass or Volume Air Flow Circuit Range/Performance Problem
P0102 Mass or Volume Air Flow Circuit Low Input
P0103 Mass or Volume Air Flow Circuit High Input
P0104 Mass or Volume Air Flow Circuit Intermittent
P0105 Manifold Absolute Pressure/Barometric Pressure Circuit Malfunction
P0106 Manifold Absolute Pressure/Barometric Pressure Circuit Range/Performance Problem
P0107 Manifold Absolute Pressure/Barometric Pressure Circuit Low Input

P0108 Manifold Absolute Pressure/Barometric Pressure Circuit High Input
P0109 Manifold Absolute Pressure/Barometric Pressure Circuit Intermittent
P0110 Intake Air Temperature Circuit Malfunction
P0111 Intake Air Temperature Circuit Range/Performance Problem
P0112 Intake Air Temperature Circuit Low Input
P0113 Intake Air Temperature Circuit High Input
P0114 Intake Air Temperature Circuit Intermittent
P0115 Engine Coolant Temperature Circuit Malfunction
P0116 Engine Coolant Temperature Circuit Range/Performance Problem
P0117 Engine Coolant Temperature Circuit Low Input
P0118 Engine Coolant Temperature Circuit High Input
P0119 Engine Coolant Temperature Circuit Intermittent
P0120 Throttle/Pedal Position Sensor/Switch "A" Circuit Malfunction
P0121 Throttle/Pedal Position Sensor/Switch "A" Circuit Range/Performance Problem
P0122 Throttle/Pedal Position Sensor/Switch "A" Circuit Low Input
P0123 Throttle/Pedal Position Sensor/Switch "A" Circuit High Input
P0124 Throttle/Pedal Position Sensor/Switch "A" Circuit Intermittent
P0125 Insufficient Coolant Temperature For Closed Loop Fuel Control
P0126 Insufficient Coolant Temperature For Stable Operation
P0130 O2 Circuit Malfunction (Bank no. 1 Sensor no. 1)
P0131 O2 Sensor Circuit Low Voltage (Bank no. 1 Sensor no. 1)
P0132 O2 Sensor Circuit High Voltage (Bank no. 1 Sensor no. 1)
P0133 O2 Sensor Circuit Slow Response (Bank no. 1 Sensor no. 1)
P0134 O2 Sensor Circuit No Activity Detected (Bank no. 1 Sensor no. 1)
P0135 O2 Sensor Heater Circuit Malfunction (Bank no. 1 Sensor no. 1)
P0136 O2 Sensor Circuit Malfunction (Bank no. 1 Sensor no. 2)
P0137 O2 Sensor Circuit Low Voltage (Bank no. 1 Sensor no. 2)
P0138 O2 Sensor Circuit High Voltage (Bank no. 1 Sensor no. 2)
P0139 O2 Sensor Circuit Slow Response (Bank no. 1 Sensor no. 2)
P0140 O2 Sensor Circuit No Activity Detected (Bank no. 1 Sensor no. 2)
P0141 O2 Sensor Heater Circuit Malfunction (Bank no. 1 Sensor no. 2)
P0142 O2 Sensor Circuit Malfunction (Bank no. 1 Sensor no. 3)
P0143 O2 Sensor Circuit Low Voltage (Bank no. 1 Sensor no. 3)
P0144 O2 Sensor Circuit High Voltage (Bank no. 1 Sensor no. 3)
P0145 O2 Sensor Circuit Slow Response (Bank no. 1 Sensor no. 3)
P0146 O2 Sensor Circuit No Activity Detected (Bank no. 1 Sensor no. 3)
P0147 O2 Sensor Heater Circuit Malfunction (Bank no. 1 Sensor no. 3)
P0150 O2 Sensor Circuit Malfunction (Bank no. 2 Sensor no. 1)
P0151 O2 Sensor Circuit Low Voltage (Bank no. 2 Sensor no. 1)

P0152 O2 Sensor Circuit High Voltage (Bank no. 2 Sensor no. 1)

P0153 O2 Sensor Circuit Slow Response (Bank no. 2 Sensor no. 1)

P0154 O2 Sensor Circuit No Activity Detected (Bank no. 2 Sensor no. 1)

P0155 O2 Sensor Heater Circuit Malfunction (Bank no. 2 Sensor no. 1)

P0156 O2 Sensor Circuit Malfunction (Bank no. 2 Sensor no. 2)

P0157 O2 Sensor Circuit Low Voltage (Bank no. 2 Sensor no. 2)

P0158 O2 Sensor Circuit High Voltage (Bank no. 2 Sensor no. 2)

P0159 O2 Sensor Circuit Slow Response (Bank no. 2 Sensor no. 2)

P0160 O2 Sensor Circuit No Activity Detected (Bank no. 2 Sensor no. 2)

P0161 O2 Sensor Heater Circuit Malfunction (Bank no. 2 Sensor no. 2)

P0162 O2 Sensor Circuit Malfunction (Bank no. 2 Sensor no. 3)

P0163 O2 Sensor Circuit Low Voltage (Bank no. 2 Sensor no. 3)

P0164 O2 Sensor Circuit High Voltage (Bank no. 2 Sensor no. 3)

P0165 O2 Sensor Circuit Slow Response (Bank no. 2 Sensor no. 3)

P0166 O2 Sensor Circuit No Activity Detected (Bank no. 2 Sensor no. 3)

P0167 O2 Sensor Heater Circuit Malfunction (Bank no. 2 Sensor no. 3)

P0170 Fuel Trim Malfunction (Bank no. 1)

P0171 System Too Lean (Bank no. 1)

P0172 System Too Rich (Bank no. 1)

P0173 Fuel Trim Malfunction (Bank no. 2)

P0174 System Too Lean (Bank no. 2)

P0175 System Too Rich (Bank no. 2)

P0176 Fuel Composition Sensor Circuit Malfunction

P0177 Fuel Composition Sensor Circuit Range/Performance

P0178 Fuel Composition Sensor Circuit Low Input

P0179 Fuel Composition Sensor Circuit High Input

P0180 Fuel Temperature Sensor "A" Circuit Malfunction

P0181 Fuel Temperature Sensor "A" Circuit Range/Performance

P0182 Fuel Temperature Sensor "A" Circuit Low Input

P0183 Fuel Temperature Sensor "A" Circuit High Input

P0184 Fuel Temperature Sensor "A" Circuit Intermittent

P0185 Fuel Temperature Sensor "B" Circuit Malfunction

P0186 Fuel Temperature Sensor "B" Circuit Range/Performance

P0187 Fuel Temperature Sensor "B" Circuit Low Input

P0188 Fuel Temperature Sensor "B" Circuit High Input

P0189 Fuel Temperature Sensor "B" Circuit Intermittent

P0190 Fuel Rail Pressure Sensor Circuit Malfunction

P0191 Fuel Rail Pressure Sensor Circuit Range/Performance

P0192 Fuel Rail Pressure Sensor Circuit Low Input

P0193 Fuel Rail Pressure Sensor Circuit High Input

P0194 Fuel Rail Pressure Sensor Circuit Intermittent

P0195 Engine Oil Temperature Sensor Malfunction

P0196 Engine Oil Temperature Sensor Range/Performance

P0197 Engine Oil Temperature Sensor Low

P0198 Engine Oil Temperature Sensor High

P0199 Engine Oil Temperature Sensor Intermittent

P0200 Injector Circuit Malfunction

P0201 Injector Circuit Malfunction—Cylinder no. 1

P0202 Injector Circuit Malfunction—Cylinder no. 2

P0203 Injector Circuit Malfunction—Cylinder no. 3

P0204 Injector Circuit Malfunction—Cylinder no. 4

P0205 Injector Circuit Malfunction—Cylinder no. 5

P0206 Injector Circuit Malfunction—Cylinder no. 6

P0207 Injector Circuit Malfunction—Cylinder no. 7

P0208 Injector Circuit Malfunction—Cylinder no. 8

P0209 Injector Circuit Malfunction—Cylinder no. 9

P0210 Injector Circuit Malfunction—Cylinder no. 10

P0211 Injector Circuit Malfunction—Cylinder no. 11

P0212 Injector Circuit Malfunction—Cylinder no. 12

P0213 Cold Start Injector no. 1 Malfunction

P0214 Cold Start Injector no. 2 Malfunction

P0215 Engine Shutoff Solenoid Malfunction

P0216 Injection Timing Control Circuit Malfunction

P0217 Engine Over Temperature Condition

P0218 Transmission Over Temperature Condition

P0219 Engine Over Speed Condition

P0220 Throttle/Pedal Position Sensor/Switch "B" Circuit Malfunction

P0221 Throttle/Pedal Position Sensor/Switch "B" Circuit Range/Performance Problem

P0222 Throttle/Pedal Position Sensor/Switch "B" Circuit Low Input

P0223 Throttle/Pedal Position Sensor/Switch "B" Circuit High Input

P0224 Throttle/Pedal Position Sensor/Switch "B" Circuit Intermittent

P0225 Throttle/Pedal Position Sensor/Switch "C" Circuit Malfunction

P0226 Throttle/Pedal Position Sensor/Switch "C" Circuit Range/Performance Problem

P0227 Throttle/Pedal Position Sensor/Switch "C" Circuit Low Input

P0228 Throttle/Pedal Position Sensor/Switch "C" Circuit High Input

P0229 Throttle/Pedal Position Sensor/Switch "C" Circuit Intermittent

P0230 Fuel Pump Primary Circuit Malfunction

P0231 Fuel Pump Secondary Circuit Low

P0232 Fuel Pump Secondary Circuit High

P0233 Fuel Pump Secondary Circuit Intermittent

P0261 Cylinder no. 1 Injector Circuit Low

P0262 Cylinder no. 1 Injector Circuit High

P0263 Cylinder no. 1 Contribution/Balance Fault

P0264 Cylinder no. 2 Injector Circuit Low

P0265 Cylinder no. 2 Injector Circuit High

P0266 Cylinder no. 2 Contribution/Balance Fault

P0267 Cylinder no. 3 Injector Circuit Low

P0268 Cylinder no. 3 Injector Circuit High

P0269 Cylinder no. 3 Contribution/Balance Fault

P0270 Cylinder no. 4 Injector Circuit Low

P0271 Cylinder no. 4 Injector Circuit High

P0272 Cylinder no. 4 Contribution/Balance Fault

P0273 Cylinder no. 5 Injector Circuit Low

P0274 Cylinder no. 5 Injector Circuit High

P0275 Cylinder no. 5 Contribution/Balance Fault

P0276 Cylinder no. 6 Injector Circuit Low

P0277 Cylinder no. 6 Injector Circuit High

P0278 Cylinder no. 6 Contribution/Balance Fault

P0279 Cylinder no. 7 Injector Circuit Low

P0280 Cylinder no. 7 Injector Circuit High

P0281 Cylinder no. 7 Contribution/Balance Fault
P0282 Cylinder no. 8 Injector Circuit Low
P0283 Cylinder no. 8 Injector Circuit High
P0284 Cylinder no. 8 Contribution/Balance Fault
P0285 Cylinder no. 9 Injector Circuit Low
P0286 Cylinder no. 9 Injector Circuit High
P0287 Cylinder no. 9 Contribution/Balance Fault
P0288 Cylinder no. 10 Injector Circuit Low
P0289 Cylinder no. 10 Injector Circuit High
P0290 Cylinder no. 10 Contribution/Balance Fault
P0291 Cylinder no. 11 Injector Circuit Low
P0292 Cylinder no. 11 Injector Circuit High
P0293 Cylinder no. 11 Contribution/Balance Fault
P0294 Cylinder no. 12 Injector Circuit Low
P0295 Cylinder no. 12 Injector Circuit High
P0296 Cylinder no. 12 Contribution/Balance Fault
P0300 Random/Multiple Cylinder Misfire Detected
P0301 Cylinder no. 1—Misfire Detected
P0302 Cylinder no. 2—Misfire Detected
P0303 Cylinder no. 3—Misfire Detected
P0304 Cylinder no. 4—Misfire Detected
P0305 Cylinder no. 5—Misfire Detected
P0306 Cylinder no. 6—Misfire Detected
P0307 Cylinder no. 7—Misfire Detected
P0308 Cylinder no. 8—Misfire Detected
P0309 Cylinder no. 9—Misfire Detected
P0310 Cylinder no. 10—Misfire Detected
P0311 Cylinder no. 11—Misfire Detected
P0312 Cylinder no. 12—Misfire Detected
P0320 Ignition/Distributor Engine Speed Input Circuit Malfunction
P0321 Ignition/Distributor Engine Speed Input Circuit Range/Performance
P0322 Ignition/Distributor Engine Speed Input Circuit No Signal
P0323 Ignition/Distributor Engine Speed Input Circuit Intermittent
P0325 Knock Sensor no. 1—Circuit Malfunction (Bank no. 1 or Single Sensor)
P0326 Knock Sensor no. 1—Circuit Range/Performance (Bank no. 1 or Single Sensor)
P0327 Knock Sensor no. 1—Circuit Low Input (Bank no. 1 or Single Sensor)
P0328 Knock Sensor no. 1—Circuit High Input (Bank no. 1 or Single Sensor)
P0329 Knock Sensor no. 1—Circuit Input Intermittent (Bank no. 1 or Single Sensor)
P0330 Knock Sensor no. 2—Circuit Malfunction (Bank no. 2)
P0331 Knock Sensor no. 2—Circuit Range/Performance (Bank no. 2)
P0332 Knock Sensor no. 2—Circuit Low Input (Bank no. 2)
P0333 Knock Sensor no. 2—Circuit High Input (Bank no. 2)
P0334 Knock Sensor no. 2—Circuit Input Intermittent (Bank no. 2)
P0335 Crankshaft Position Sensor "A" Circuit Malfunction
P0336 Crankshaft Position Sensor "A" Circuit Range/Performance
P0337 Crankshaft Position Sensor "A" Circuit Low Input
P0338 Crankshaft Position Sensor "A" Circuit High Input
P0339 Crankshaft Position Sensor "A" Circuit Intermittent
P0340 Camshaft Position Sensor Circuit Malfunction
P0341 Camshaft Position Sensor Circuit Range/Performance

P0342 Camshaft Position Sensor Circuit Low Input
P0343 Camshaft Position Sensor Circuit High Input
P0344 Camshaft Position Sensor Circuit Intermittent
P0350 Ignition Coil Primary/Secondary Circuit Malfunction
P0351 Ignition Coil "A" Primary/Secondary Circuit Malfunction
P0352 Ignition Coil "B" Primary/Secondary Circuit Malfunction
P0353 Ignition Coil "C" Primary/Secondary Circuit Malfunction
P0354 Ignition Coil "D" Primary/Secondary Circuit Malfunction
P0355 Ignition Coil "E" Primary/Secondary Circuit Malfunction
P0356 Ignition Coil "F" Primary/Secondary Circuit Malfunction
P0357 Ignition Coil "G" Primary/Secondary Circuit Malfunction
P0358 Ignition Coil "H" Primary/Secondary Circuit Malfunction
P0359 Ignition Coil "I" Primary/Secondary Circuit Malfunction
P0360 Ignition Coil "J" Primary/Secondary Circuit Malfunction
P0361 Ignition Coil "K" Primary/Secondary Circuit Malfunction
P0362 Ignition Coil "L" Primary/Secondary Circuit Malfunction
P0370 Timing Reference High Resolution Signal "A" Malfunction
P0371 Timing Reference High Resolution Signal "A" Too Many Pulses
P0372 Timing Reference High Resolution Signal "A" Too Few Pulses
P0373 Timing Reference High Resolution Signal "A" Intermittent/Erratic Pulses
P0374 Timing Reference High Resolution Signal "A" No Pulses
P0375 Timing Reference High Resolution Signal "B" Malfunction
P0376 Timing Reference High Resolution Signal "B" Too Many Pulses
P0377 Timing Reference High Resolution Signal "B" Too Few Pulses
P0378 Timing Reference High Resolution Signal "B" Intermittent/Erratic Pulses
P0379 Timing Reference High Resolution Signal "B" No Pulses
P0380 Glow Plug/Heater Circuit "A" Malfunction
P0381 Glow Plug/Heater Indicator Circuit Malfunction
P0382 Glow Plug/Heater Circuit "B" Malfunction
P0385 Crankshaft Position Sensor "B" Circuit Malfunction
P0386 Crankshaft Position Sensor "B" Circuit Range/Performance
P0387 Crankshaft Position Sensor "B" Circuit Low Input
P0388 Crankshaft Position Sensor "B" Circuit High Input
P0389 Crankshaft Position Sensor "B" Circuit Intermittent
P0400 Exhaust Gas Recirculation Flow Malfunction
P0401 Exhaust Gas Recirculation Flow Insufficient Detected
P0402 Exhaust Gas Recirculation Flow Excessive Detected
P0403 Exhaust Gas Recirculation Circuit Malfunction
P0404 Exhaust Gas Recirculation Circuit Range/Performance
P0405 Exhaust Gas Recirculation Sensor "A" Circuit Low
P0406 Exhaust Gas Recirculation Sensor "A" Circuit High
P0407 Exhaust Gas Recirculation Sensor "B" Circuit Low
P0408 Exhaust Gas Recirculation Sensor "B" Circuit High
P0410 Secondary Air Injection System Malfunction
P0411 Secondary Air Injection System Incorrect Flow Detected
P0412 Secondary Air Injection System Switching Valve "A" Circuit Malfunction
P0413 Secondary Air Injection System Switching Valve "A" Circuit Open
P0414 Secondary Air Injection System Switching Valve "A" Circuit Shorted
P0415 Secondary Air Injection System Switching Valve "B" Circuit Malfunction

P0416 Secondary Air Injection System Switching Valve "B" Circuit Open

P0417 Secondary Air Injection System Switching Valve "B" Circuit Shorted

P0418 Secondary Air Injection System Relay "A" Circuit Malfunction

P0419 Secondary Air Injection System Relay "B" Circuit Malfunction

P0420 Catalyst System Efficiency Below Threshold (Bank no. 1)

P0421 Warm Up Catalyst Efficiency Below Threshold (Bank no. 1)

P0422 Main Catalyst Efficiency Below Threshold (Bank no. 1)

P0423 Heated Catalyst Efficiency Below Threshold (Bank no. 1)

P0424 Heated Catalyst Temperature Below Threshold (Bank no. 1)

P0430 Catalyst System Efficiency Below Threshold (Bank no. 2)

P0431 Warm Up Catalyst Efficiency Below Threshold (Bank no. 2)

P0432 Main Catalyst Efficiency Below Threshold (Bank no. 2)

P0433 Heated Catalyst Efficiency Below Threshold (Bank no. 2)

P0434 Heated Catalyst Temperature Below Threshold (Bank no. 2)

P0440 Evaporative Emission Control System Malfunction

P0441 Evaporative Emission Control System Incorrect Purge Flow

P0442 Evaporative Emission Control System Leak Detected (Small Leak)

P0443 Evaporative Emission Control System Purge Control Valve Circuit Malfunction

P0444 Evaporative Emission Control System Purge Control Valve Circuit Open

P0445 Evaporative Emission Control System Purge Control Valve Circuit Shorted

P0446 Evaporative Emission Control System Vent Control Circuit Malfunction

P0447 Evaporative Emission Control System Vent Control Circuit Open

P0448 Evaporative Emission Control System Vent Control Circuit Shorted

P0449 Evaporative Emission Control System Vent Valve/Solenoid Circuit Malfunction

P0450 Evaporative Emission Control System Pressure Sensor Malfunction

P0451 Evaporative Emission Control System Pressure Sensor Range/Performance

P0452 Evaporative Emission Control System Pressure Sensor Low Input

P0453 Evaporative Emission Control System Pressure Sensor High Input

P0454 Evaporative Emission Control System Pressure Sensor Intermittent

P0455 Evaporative Emission Control System Leak Detected (Gross Leak)

P0460 Fuel Level Sensor Circuit Malfunction

P0461 Fuel Level Sensor Circuit Range/Performance

P0462 Fuel Level Sensor Circuit Low Input

P0463 Fuel Level Sensor Circuit High Input

P0464 Fuel Level Sensor Circuit Intermittent

P0465 Purge Flow Sensor Circuit Malfunction

P0466 Purge Flow Sensor Circuit Range/Performance

P0467 Purge Flow Sensor Circuit Low Input

P0468 Purge Flow Sensor Circuit High Input

P0469 Purge Flow Sensor Circuit Intermittent

P0470 Exhaust Pressure Sensor Malfunction

P0471 Exhaust Pressure Sensor Range/Performance

P0472 Exhaust Pressure Sensor Low

P0473 Exhaust Pressure Sensor High

P0474 Exhaust Pressure Sensor Intermittent

P0475 Exhaust Pressure Control Valve Malfunction

P0476 Exhaust Pressure Control Valve Range/Performance

P0477 Exhaust Pressure Control Valve Low

P0478 Exhaust Pressure Control Valve High

P0479 Exhaust Pressure Control Valve Intermittent

P0480 Cooling Fan no. 1 Control Circuit Malfunction

P0481 Cooling Fan no. 2 Control Circuit Malfunction

P0482 Cooling Fan no. 3 Control Circuit Malfunction

P0483 Cooling Fan Rationality Check Malfunction

P0484 Cooling Fan Circuit Over Current

P0485 Cooling Fan Power/Ground Circuit Malfunction

P0500 Vehicle Speed Sensor Malfunction

P0501 Vehicle Speed Sensor Range/Performance

P0502 Vehicle Speed Sensor Circuit Low Input

P0503 Vehicle Speed Sensor Intermittent/Erratic/High

P0505 Idle Control System Malfunction

P0506 Idle Control System RPM Lower Than Expected

P0507 Idle Control System RPM Higher Than Expected

P0510 Closed Throttle Position Switch Malfunction

P0520 Engine Oil Pressure Sensor/Switch Circuit Malfunction

P0521 Engine Oil Pressure Sensor/Switch Range/Performance

P0522 Engine Oil Pressure Sensor/Switch Low Voltage

P0523 Engine Oil Pressure Sensor/Switch High Voltage

P0530 A/C Refrigerant Pressure Sensor Circuit Malfunction

P0531 A/C Refrigerant Pressure Sensor Circuit Range/Performance

P0532 A/C Refrigerant Pressure Sensor Circuit Low Input

P0533 A/C Refrigerant Pressure Sensor Circuit High Input

P0534 A/C Refrigerant Charge Loss

P0550 Power Steering Pressure Sensor Circuit Malfunction

P0551 Power Steering Pressure Sensor Circuit Range/Performance

P0552 Power Steering Pressure Sensor Circuit Low Input

P0553 Power Steering Pressure Sensor Circuit High Input

P0554 Power Steering Pressure Sensor Circuit Intermittent

P0560 System Voltage Malfunction

P0561 System Voltage Unstable

P0562 System Voltage Low

P0563 System Voltage High

P0565 Cruise Control On Signal Malfunction

P0566 Cruise Control Off Signal Malfunction

P0567 Cruise Control Resume Signal Malfunction

P0568 Cruise Control Set Signal Malfunction

P0569 Cruise Control Coast Signal Malfunction

P0570 Cruise Control Accel Signal Malfunction

P0571 Cruise Control/Brake Switch "A" Circuit Malfunction

P0572 Cruise Control/Brake Switch "A" Circuit Low

P0573 Cruise Control/Brake Switch "A" Circuit High

P0574 **Through P0580** Reserved for Cruise Codes

P0600 Serial Communication Link Malfunction

P0601 Internal Control Module Memory Check Sum Error

P0602 Control Module Programming Error

P0603 Internal Control Module Keep Alive Memory (KAM) Error

P0604 Internal Control Module Random Access Memory (RAM) Error

P0605 Internal Control Module Read Only Memory (ROM) Error

P0606 PCM Processor Fault

P0608 Control Module VSS Output "A" Malfunction

P0609 Control Module VSS Output "B" Malfunction

P0620 Generator Control Circuit Malfunction

P0621 Generator Lamp "L" Control Circuit Malfunction

P0622 Generator Field "F" Control Circuit Malfunction

P0650 Malfunction Indicator Lamp (MIL) Control Circuit Malfunction

P0654 Engine RPM Output Circuit Malfunction

P0655 Engine Hot Lamp Output Control Circuit Malfunction

P0656 Fuel Level Output Circuit Malfunction

P0700 Transmission Control System Malfunction

P0701 Transmission Control System Range/Performance

P0702 Transmission Control System Electrical

P0703 Torque Converter/Brake Switch "B" Circuit Malfunction

P0704 Clutch Switch Input Circuit Malfunction

P0705 Transmission Range Sensor Circuit Malfunction (PRNDL Input)

P0706 Transmission Range Sensor Circuit Range/Performance

P0707 Transmission Range Sensor Circuit Low Input

P0708 Transmission Range Sensor Circuit High Input

P0709 Transmission Range Sensor Circuit Intermittent

P0710 Transmission Fluid Temperature Sensor Circuit Malfunction

P0711 Transmission Fluid Temperature Sensor Circuit Range/Performance

P0712 Transmission Fluid Temperature Sensor Circuit Low Input

P0713 Transmission Fluid Temperature Sensor Circuit High Input

P0714 Transmission Fluid Temperature Sensor Circuit Intermittent

P0715 Input/Turbine Speed Sensor Circuit Malfunction

P0716 Input/Turbine Speed Sensor Circuit Range/Performance

P0717 Input/Turbine Speed Sensor Circuit No Signal

P0718 Input/Turbine Speed Sensor Circuit Intermittent

P0719 Torque Converter/Brake Switch "B" Circuit Low

P0720 Output Speed Sensor Circuit Malfunction

P0721 Output Speed Sensor Circuit Range/Performance

P0722 Output Speed Sensor Circuit No Signal

P0723 Output Speed Sensor Circuit Intermittent

P0724 Torque Converter/Brake Switch "B" Circuit High

P0725 Engine Speed Input Circuit Malfunction

P0726 Engine Speed Input Circuit Range/Performance

P0727 Engine Speed Input Circuit No Signal

P0728 Engine Speed Input Circuit Intermittent

P0730 Incorrect Gear Ratio

P0731 Gear no. 1 Incorrect Ratio

P0732 Gear no. 2 Incorrect Ratio

P0733 Gear no. 3 Incorrect Ratio

P0734 Gear no. 4 Incorrect Ratio

P0735 Gear no. 5 Incorrect Ratio

P0736 Reverse Incorrect Ratio

P0740 Torque Converter Clutch Circuit Malfunction

P0741 Torque Converter Clutch Circuit Performance or Stuck Off

P0742 Torque Converter Clutch Circuit Stuck On

P0743 Torque Converter Clutch Circuit Electrical

P0744 Torque Converter Clutch Circuit Intermittent

P0745 Pressure Control Solenoid Malfunction

P0746 Pressure Control Solenoid Performance or Stuck Off

P0747 Pressure Control Solenoid Stuck On

P0748 Pressure Control Solenoid Electrical

P0749 Pressure Control Solenoid Intermittent

P0750 Shift Solenoid "A" Malfunction

P0751 Shift Solenoid "A" Performance or Stuck Off

P0752 Shift Solenoid "A" Stuck On

P0753 Shift Solenoid "A" Electrical

P0754 Shift Solenoid "A" Intermittent

P0755 Shift Solenoid "B" Malfunction

P0756 Shift Solenoid "B" Performance or Stuck Off

P0757 Shift Solenoid "B" Stuck On

P0758 Shift Solenoid "B" Electrical

P0759 Shift Solenoid "B" Intermittent

P0760 Shift Solenoid "C" Malfunction

P0761 Shift Solenoid "C" Performance Or Stuck Off

P0762 Shift Solenoid "C" Stuck On

P0763 Shift Solenoid "C" Electrical

P0764 Shift Solenoid "C" Intermittent

P0765 Shift Solenoid "D" Malfunction

P0766 Shift Solenoid "D" Performance Or Stuck Off

P0767 Shift Solenoid "D" Stuck On

P0768 Shift Solenoid "D" Electrical

P0769 Shift Solenoid "D" Intermittent

P0770 Shift Solenoid "E" Malfunction

P0771 Shift Solenoid "E" Performance Or Stuck Off

P0772 Shift Solenoid "E" Stuck On

P0773 Shift Solenoid "E" Electrical

P0774 Shift Solenoid "E" Intermittent

P0780 Shift Malfunction

P0781 1–2 Shift Malfunction

P0782 2–3 Shift Malfunction

P0783 3–4 Shift Malfunction

P0784 4–5 Shift Malfunction

P0785 Shift/Timing Solenoid Malfunction

P0786 Shift/Timing Solenoid Range/Performance

P0787 Shift/Timing Solenoid Low

P0788 Shift/Timing Solenoid High

P0789 Shift/Timing Solenoid Intermittent

P0790 Normal/Performance Switch Circuit Malfunction

P0801 Reverse Inhibit Control Circuit Malfunction

P0803 1–4 Upshift (Skip Shift) Solenoid Control Circuit Malfunction

P0804 1–4 Upshift (Skip Shift) Lamp Control Circuit Malfunction

P1102 Front heated oxygen sensor, bank 1, control module input. Current in preheating circuit much too high.

P1105 Rear heated oxygen sensor, bank 1, control module input. Current in preheating circuit much too high.

P1115 Front heated oxygen sensor, bank 1, control module input. Current in preheating circuit much too low.

P1117 Rear heated oxygen sensor, bank 1, control module input. Current in preheating circuit much too low.

P1123 Additive adaptation, bank 1. Min value.

P1124 Additive adaptation, bank 1. Max value.

P1125 Additive adaptation, bank 2. Min value.

P1126 Additive adaptation, bank 2. Max value.

P1127 Multiplicative adaptation, bank 1. Min value.

P1128 Multiplicative adaptation, bank 1. Max value.

P1129 Multiplicative adaptation, bank 2. Min value.

P1130 Multiplicative adaptation, bank 2. Max value.

P1170 Closed loop. Malfunction.

P1171 Closed loop. Lean mixture.

P1172 Closed loop. Rich mixture.

P1213 Injector, cylinder 1, control module output. Shorting to battery positive (B+).

P1214 Injector, cylinder 2, control module output. Shorting to battery positive (B+).

P1215 Injector, cylinder 3, control module output. Shorting to battery positive (B+).

P1216 Injector, cylinder 4, control module output. Shorting to battery positive (B+).

P1217 Injector, cylinder 5, control module output. Shorting to battery positive (B+).

P1218 Injector, cylinder 6, control module output. Shorting to battery positive (B+).

P1225 Injector, cylinder 1, control module output. Open circuit or shorting to ground.

P1226 Injector, cylinder 2, control module output. Open circuit or shorting to ground.

P1227 Injector, cylinder 3, control module output. Open circuit or shorting to ground.

P1228 Injector, cylinder 4, control module output. Open circuit or shorting to ground.

P1229 Injector, cylinder 5, control module output. Open circuit or shorting to ground.

P1230 Injector, cylinder 6, control module output. Open circuit or shorting to ground.

P1386 Control module, electronic circuitry for processing knock sensor signals Internal fault.

P1396 Crankshaft position sensor, control module input. Malfunctioning, slotted ring has too many ribs.

P1410 EVAP canister purge valve, control module output. Shorting to battery positive (B+).

P1416 Tank level. Low level in conjunction with misfiring or fault in fuel system.

P1425 EVAP canister purge module output. Shorting to ground.

P1426 EVAP canister purge module output. Open circuit.

P1500 Battery voltage outside limits.

P1501 Fuel pump relay, control module output. Shorting to ground.

P1502 Fuel pump relay, control module output. Shorting to battery positive (B+)

P1510 Idle air control valve, open control module output. Shorting to battery positive (B+).

P1513 Idle air control valve, open control module output. Shorting to ground.

P1514 Idle air control valve, open function, control module output. Open circuit.

P1541 Fuel pump relay, control module out put.

P1549 Boost pressure control. Malfunction.

P1551 Idle air control valve, close function, control module output. Open circuit.

P1552 Idle air control valve, close function, control module output. Shorting to ground.

P1553 Idle air control valve, close function, control module output. Shorting to battery positive (B+).

P1576 Brake light switch. Shorting to battery positive (B+).

P1577 Brake light switch. Open circuit.

P1585 Fuel less than 10 liters.

P1611 CHECK ENGINE request, input signal to control module. Shorting to ground.

P1616 Rough road sensor. Control module input low, shorting to ground.

P1617 Rough road sensor. Control module input high; open circuit or shorting to battery positive (B+).

P1624 The automatic transmission has a stored emission-related fault.

P1664 Shift up, output signal from control module. Malfunction.

P1665 Intermittent fault which cannot identify any other specific diagnostic trouble code.

P1669 TCS active, input signal to control module.

P1670 Intermittent fault which cannot identify any other specific diagnostic trouble code.

P1675 Intermittent fault which cannot identify any other specific diagnostic trouble code.

P1680 Relay, secondary air injection. Control module output, open circuit or short circuit.

P1691 CHECK ENGINE, output signal from On control module. Open circuit.

P1692 CHECK ENGINE, output signal from On control module. Shorting to ground.

P1693 CHECK ENGINE, output signal from On control module. Open circuit or shorting to ground or battery positive (B+).

Subaru

READING CODES

Reading the control module memory is on of the first steps in OBD II system diagnostics. This step should be initially performed to determine the general nature of the fault. Subsequent readings will determine if the fault has been cleared.

Reading codes can be performed by any of the methods below:

• Read the control module memory with the Generic Scan Tool (GST)

• Read the control module memory with the vehicle manufacturer's specific tester

To read the fault codes, connect the scan tool or tester according to the manufacturer's instructions. Follow the manufacturer's specified procedure for reading the codes.

CLEARING CODES

Control module reset procedures are a very important part of OBD II System diagnostics. This step should be done at the end of any fault code repair and at the end of any driveability repair.

Clearing codes can be performed by any of the methods below:

• Clear the control module memory with the Generic Scan Tool (GST)

• Clear the control module memory with the vehicle manufacturer's specific tester

• Turn the ignition **OFF** and disconnect the negative battery cable for at least 1 minute.

Removing the negative battery cable may cause other systems in the vehicle to loose their memory. Prior to removing the cable, ensure you have the proper reset codes for radios and alarms.

➡The MIL will may also be de-activated for some codes if the vehicle completes three consecutive trips without a fault detected with vehicle conditions similar to those present during the fault.

OBD II TROUBLE CODES

P0100 Mass or Volume Air Flow Circuit Malfunction
P0101 Mass or Volume Air Flow Circuit Range/Performance Problem
P0102 Mass or Volume Air Flow Circuit Low Input
P0103 Mass or Volume Air Flow Circuit High Input
P0104 Mass or Volume Air Flow Circuit Intermittent
P0105 Manifold Absolute Pressure/Barometric Pressure Circuit Malfunction
P0106 Manifold Absolute Pressure/Barometric Pressure Circuit Range/Performance Problem
P0107 Manifold Absolute Pressure/Barometric Pressure Circuit Low Input
P0108 Manifold Absolute Pressure/Barometric Pressure Circuit High Input
P0109 Manifold Absolute Pressure/Barometric Pressure Circuit Intermittent
P0110 Intake Air Temperature Circuit Malfunction
P0111 Intake Air Temperature Circuit Range/Performance Problem
P0112 Intake Air Temperature Circuit Low Input
P0113 Intake Air Temperature Circuit High Input
P0114 Intake Air Temperature Circuit Intermittent
P0115 Engine Coolant Temperature Circuit Malfunction
P0116 Engine Coolant Temperature Circuit Range/Performance Problem
P0117 Engine Coolant Temperature Circuit Low Input
P0118 Engine Coolant Temperature Circuit High Input
P0119 Engine Coolant Temperature Circuit Intermittent
P0120 Throttle/Pedal Position Sensor/Switch "A" Circuit Malfunction
P0121 Throttle/Pedal Position Sensor/Switch "A" Circuit Range/Performance Problem
P0122 Throttle/Pedal Position Sensor/Switch "A" Circuit Low Input
P0123 Throttle/Pedal Position Sensor/Switch "A" Circuit High Input
P0124 Throttle/Pedal Position Sensor/Switch "A" Circuit Intermittent
P0125 Insufficient Coolant Temperature For Closed Loop Fuel Control
P0126 Insufficient Coolant Temperature For Stable Operation
P0130 O2 Circuit Malfunction (Bank no. 1 Sensor no. 1)
P0131 O2 Sensor Circuit Low Voltage (Bank no. 1 Sensor no. 1)
P0132 O2 Sensor Circuit High Voltage (Bank no. 1 Sensor no. 1)
P0133 O2 Sensor Circuit Slow Response (Bank no. 1 Sensor no. 1)
P0134 O2 Sensor Circuit No Activity Detected (Bank no. 1 Sensor no. 1)
P0135 O2 Sensor Heater Circuit Malfunction (Bank no. 1 Sensor no. 1)
P0136 O2 Sensor Circuit Malfunction (Bank no. 1 Sensor no. 2)
P0137 O2 Sensor Circuit Low Voltage (Bank no. 1 Sensor no. 2)
P0138 O2 Sensor Circuit High Voltage (Bank no. 1 Sensor no. 2)

P0139 O2 Sensor Circuit Slow Response (Bank no. 1 Sensor no. 2)
P0140 O2 Sensor Circuit No Activity Detected (Bank no. 1 Sensor no. 2)
P0141 O2 Sensor Heater Circuit Malfunction (Bank no. 1 Sensor no. 2)
P0142 O2 Sensor Circuit Malfunction (Bank no. 1 Sensor no. 3)
P0143 O2 Sensor Circuit Low Voltage (Bank no. 1 Sensor no. 3)
P0144 O2 Sensor Circuit High Voltage (Bank no. 1 Sensor no. 3)
P0145 O2 Sensor Circuit Slow Response (Bank no. 1 Sensor no. 3)
P0146 O2 Sensor Circuit No Activity Detected (Bank no. 1 Sensor no. 3)
P0147 O2 Sensor Heater Circuit Malfunction (Bank no. 1 Sensor no. 3)
P0150 O2 Sensor Circuit Malfunction (Bank no. 2 Sensor no. 1)
P0151 O2 Sensor Circuit Low Voltage (Bank no. 2 Sensor no. 1)
P0152 O2 Sensor Circuit High Voltage (Bank no. 2 Sensor no. 1)
P0153 O2 Sensor Circuit Slow Response (Bank no. 2 Sensor no. 1)
P0154 O2 Sensor Circuit No Activity Detected (Bank no. 2 Sensor no. 1)
P0155 O2 Sensor Heater Circuit Malfunction (Bank no. 2 Sensor no. 1)
P0156 O2 Sensor Circuit Malfunction (Bank no. 2 Sensor no. 2)
P0157 O2 Sensor Circuit Low Voltage (Bank no. 2 Sensor no. 2)
P0158 O2 Sensor Circuit High Voltage (Bank no. 2 Sensor no. 2)
P0159 O2 Sensor Circuit Slow Response (Bank no. 2 Sensor no. 2)
P0160 O2 Sensor Circuit No Activity Detected (Bank no. 2 Sensor no. 2)
P0161 O2 Sensor Heater Circuit Malfunction (Bank no. 2 Sensor no. 2)
P0162 O2 Sensor Circuit Malfunction (Bank no. 2 Sensor no. 3)
P0163 O2 Sensor Circuit Low Voltage (Bank no. 2 Sensor no. 3)
P0164 O2 Sensor Circuit High Voltage (Bank no. 2 Sensor no. 3)
P0165 O2 Sensor Circuit Slow Response (Bank no. 2 Sensor no. 3)
P0166 O2 Sensor Circuit No Activity Detected (Bank no. 2 Sensor no. 3)
P0167 O2 Sensor Heater Circuit Malfunction (Bank no. 2 Sensor no. 3)
P0170 Fuel Trim Malfunction (Bank no. 1)
P0171 System Too Lean (Bank no. 1)
P0172 System Too Rich (Bank no. 1)
P0173 Fuel Trim Malfunction (Bank no. 2)
P0174 System Too Lean (Bank no. 2)
P0175 System Too Rich (Bank no. 2)
P0176 Fuel Composition Sensor Circuit Malfunction
P0177 Fuel Composition Sensor Circuit Range/Performance
P0178 Fuel Composition Sensor Circuit Low Input
P0179 Fuel Composition Sensor Circuit High Input
P0180 Fuel Temperature Sensor "A" Circuit Malfunction
P0181 Fuel Temperature Sensor "A" Circuit Range/Performance
P0182 Fuel Temperature Sensor "A" Circuit Low Input
P0183 Fuel Temperature Sensor "A" Circuit High Input
P0184 Fuel Temperature Sensor "A" Circuit Intermittent
P0185 Fuel Temperature Sensor "B" Circuit Malfunction

P0186 Fuel Temperature Sensor "B" Circuit Range/Performance
P0187 Fuel Temperature Sensor "B" Circuit Low Input
P0188 Fuel Temperature Sensor "B" Circuit High Input
P0189 Fuel Temperature Sensor "B" Circuit Intermittent
P0190 Fuel Rail Pressure Sensor Circuit Malfunction
P0191 Fuel Rail Pressure Sensor Circuit Range/Performance
P0192 Fuel Rail Pressure Sensor Circuit Low Input
P0193 Fuel Rail Pressure Sensor Circuit High Input
P0194 Fuel Rail Pressure Sensor Circuit Intermittent
P0195 Engine Oil Temperature Sensor Malfunction
P0196 Engine Oil Temperature Sensor Range/Performance
P0197 Engine Oil Temperature Sensor Low
P0198 Engine Oil Temperature Sensor High
P0199 Engine Oil Temperature Sensor Intermittent
P0200 Injector Circuit Malfunction
P0201 Injector Circuit Malfunction—Cylinder no. 1
P0202 Injector Circuit Malfunction—Cylinder no. 2
P0203 Injector Circuit Malfunction—Cylinder no. 3
P0204 Injector Circuit Malfunction—Cylinder no. 4
P0205 Injector Circuit Malfunction—Cylinder no. 5
P0206 Injector Circuit Malfunction—Cylinder no. 6
P0207 Injector Circuit Malfunction—Cylinder no. 7
P0208 Injector Circuit Malfunction—Cylinder no. 8
P0209 Injector Circuit Malfunction—Cylinder no. 9
P0210 Injector Circuit Malfunction—Cylinder no. 10
P0211 Injector Circuit Malfunction—Cylinder no. 11
P0212 Injector Circuit Malfunction—Cylinder no. 12
P0213 Cold Start Injector no. 1 Malfunction
P0214 Cold Start Injector no. 2 Malfunction
P0215 Engine Shutoff Solenoid Malfunction
P0216 Injection Timing Control Circuit Malfunction
P0217 Engine Over Temperature Condition
P0218 Transmission Over Temperature Condition
P0219 Engine Over Speed Condition
P0220 Throttle/Pedal Position Sensor/Switch "B" Circuit Malfunction
P0221 Throttle/Pedal Position Sensor/Switch "B" Circuit Range/Performance Problem
P0222 Throttle/Pedal Position Sensor/Switch "B" Circuit Low Input
P0223 Throttle/Pedal Position Sensor/Switch "B" Circuit High Input
P0224 Throttle/Pedal Position Sensor/Switch "B" Circuit Intermittent
P0225 Throttle/Pedal Position Sensor/Switch "C" Circuit Malfunction
P0226 Throttle/Pedal Position Sensor/Switch "C" Circuit Range/Performance Problem
P0227 Throttle/Pedal Position Sensor/Switch "C" Circuit Low Input
P0228 Throttle/Pedal Position Sensor/Switch "C" Circuit High Input
P0229 Throttle/Pedal Position Sensor/Switch "C" Circuit Intermittent
P0230 Fuel Pump Primary Circuit Malfunction
P0231 Fuel Pump Secondary Circuit Low
P0232 Fuel Pump Secondary Circuit High
P0233 Fuel Pump Secondary Circuit Intermittent
P0261 Cylinder no. 1 Injector Circuit Low
P0262 Cylinder no. 1 Injector Circuit High
P0263 Cylinder no. 1 Contribution/Balance Fault

P0264 Cylinder no. 2 Injector Circuit Low
P0265 Cylinder no. 2 Injector Circuit High
P0266 Cylinder no. 2 Contribution/Balance Fault
P0267 Cylinder no. 3 Injector Circuit Low
P0268 Cylinder no. 3 Injector Circuit High
P0269 Cylinder no. 3 Contribution/Balance Fault
P0270 Cylinder no. 4 Injector Circuit Low
P0271 Cylinder no. 4 Injector Circuit High
P0272 Cylinder no. 4 Contribution/Balance Fault
P0273 Cylinder no. 5 Injector Circuit Low
P0274 Cylinder no. 5 Injector Circuit High
P0275 Cylinder no. 5 Contribution/Balance Fault
P0276 Cylinder no. 6 Injector Circuit Low
P0277 Cylinder no. 6 Injector Circuit High
P0278 Cylinder no. 6 Contribution/Balance Fault
P0279 Cylinder no. 7 Injector Circuit Low
P0280 Cylinder no. 7 Injector Circuit High
P0281 Cylinder no. 7 Contribution/Balance Fault
P0282 Cylinder no. 8 Injector Circuit Low
P0283 Cylinder no. 8 Injector Circuit High
P0284 Cylinder no. 8 Contribution/Balance Fault
P0285 Cylinder no. 9 Injector Circuit Low
P0286 Cylinder no. 9 Injector Circuit High
P0287 Cylinder no. 9 Contribution/Balance Fault
P0288 Cylinder no. 10 Injector Circuit Low
P0289 Cylinder no. 10 Injector Circuit High
P0290 Cylinder no. 10 Contribution/Balance Fault
P0291 Cylinder no. 11 Injector Circuit Low
P0292 Cylinder no. 11 Injector Circuit High
P0293 Cylinder no. 11 Contribution/Balance Fault
P0294 Cylinder no. 12 Injector Circuit Low
P0295 Cylinder no. 12 Injector Circuit High
P0296 Cylinder no. 12 Contribution/Balance Fault
P0300 Random/Multiple Cylinder Misfire Detected
P0301 Cylinder no. 1—Misfire Detected
P0302 Cylinder no. 2—Misfire Detected
P0303 Cylinder no. 3—Misfire Detected
P0304 Cylinder no. 4—Misfire Detected
P0305 Cylinder no. 5—Misfire Detected
P0306 Cylinder no. 6—Misfire Detected
P0307 Cylinder no. 7—Misfire Detected
P0308 Cylinder no. 8—Misfire Detected
P0309 Cylinder no. 9—Misfire Detected
P0310 Cylinder no. 10—Misfire Detected
P0311 Cylinder no. 11—Misfire Detected
P0312 Cylinder no. 12—Misfire Detected
P0320 Ignition/Distributor Engine Speed Input Circuit Malfunction
P0321 Ignition/Distributor Engine Speed Input Circuit Range/Performance
P0322 Ignition/Distributor Engine Speed Input Circuit No Signal
P0323 Ignition/Distributor Engine Speed Input Circuit Intermittent
P0325 Knock Sensor no. 1—Circuit Malfunction (Bank no. 1 or Single Sensor)
P0326 Knock Sensor no. 1—Circuit Range/Performance (Bank no. 1 or Single Sensor)
P0327 Knock Sensor no. 1—Circuit Low Input (Bank no. 1 or Single Sensor)
P0328 Knock Sensor no. 1—Circuit High Input (Bank no. 1 or Single Sensor)

P0329 Knock Sensor no. 1—Circuit Input Intermittent (Bank no. 1 or Single Sensor)

P0330 Knock Sensor no. 2—Circuit Malfunction (Bank no. 2)

P0331 Knock Sensor no. 2—Circuit Range/Performance (Bank no. 2)

P0332 Knock Sensor no. 2—Circuit Low Input (Bank no. 2)

P0333 Knock Sensor no. 2—Circuit High Input (Bank no. 2)

P0334 Knock Sensor no. 2—Circuit Input Intermittent (Bank no. 2)

P0335 Crankshaft Position Sensor "A" Circuit Malfunction

P0336 Crankshaft Position Sensor "A" Circuit Range/Performance

P0337 Crankshaft Position Sensor "A" Circuit Low Input

P0338 Crankshaft Position Sensor "A" Circuit High Input

P0339 Crankshaft Position Sensor "A" Circuit Intermittent

P0340 Camshaft Position Sensor Circuit Malfunction

P0341 Camshaft Position Sensor Circuit Range/Performance

P0342 Camshaft Position Sensor Circuit Low Input

P0343 Camshaft Position Sensor Circuit High Input

P0344 Camshaft Position Sensor Circuit Intermittent

P0350 Ignition Coil Primary/Secondary Circuit Malfunction

P0351 Ignition Coil "A" Primary/Secondary Circuit Malfunction

P0352 Ignition Coil "B" Primary/Secondary Circuit Malfunction

P0353 Ignition Coil "C" Primary/Secondary Circuit Malfunction

P0354 Ignition Coil "D" Primary/Secondary Circuit Malfunction

P0355 Ignition Coil "E" Primary/Secondary Circuit Malfunction

P0356 Ignition Coil "F" Primary/Secondary Circuit Malfunction

P0357 Ignition Coil "G" Primary/Secondary Circuit Malfunction

P0358 Ignition Coil "H" Primary/Secondary Circuit Malfunction

P0359 Ignition Coil "I" Primary/Secondary Circuit Malfunction

P0360 Ignition Coil "J" Primary/Secondary Circuit Malfunction

P0361 Ignition Coil "K" Primary/Secondary Circuit Malfunction

P0362 Ignition Coil "L" Primary/Secondary Circuit Malfunction

P0370 Timing Reference High Resolution Signal "A" Malfunction

P0371 Timing Reference High Resolution Signal "A" Too Many Pulses

P0372 Timing Reference High Resolution Signal "A" Too Few Pulses

P0373 Timing Reference High Resolution Signal "A" Intermittent/Erratic Pulses

P0374 Timing Reference High Resolution Signal "A" No Pulses

P0375 Timing Reference High Resolution Signal "B" Malfunction

P0376 Timing Reference High Resolution Signal "B" Too Many Pulses

P0377 Timing Reference High Resolution Signal "B" Too Few Pulses

P0378 Timing Reference High Resolution Signal "B" Intermittent/Erratic Pulses

P0379 Timing Reference High Resolution Signal "B" No Pulses

P0380 Glow Plug/Heater Circuit "A" Malfunction

P0381 Glow Plug/Heater Indicator Circuit Malfunction

P0382 Glow Plug/Heater Circuit "B" Malfunction

P0385 Crankshaft Position Sensor "B" Circuit Malfunction

P0386 Crankshaft Position Sensor "B" Circuit Range/Performance

P0387 Crankshaft Position Sensor "B" Circuit Low Input

P0388 Crankshaft Position Sensor "B" Circuit High Input

P0389 Crankshaft Position Sensor "B" Circuit Intermittent

P0400 Exhaust Gas Recirculation Flow Malfunction

P0401 Exhaust Gas Recirculation Flow Insufficient Detected

P0402 Exhaust Gas Recirculation Flow Excessive Detected

P0403 Exhaust Gas Recirculation Circuit Malfunction

P0404 Exhaust Gas Recirculation Circuit Range/Performance

P0405 Exhaust Gas Recirculation Sensor "A" Circuit Low

P0406 Exhaust Gas Recirculation Sensor "A" Circuit High

P0407 Exhaust Gas Recirculation Sensor "B" Circuit Low

P0408 Exhaust Gas Recirculation Sensor "B" Circuit High

P0410 Secondary Air Injection System Malfunction

P0411 Secondary Air Injection System Incorrect Flow Detected

P0412 Secondary Air Injection System Switching Valve "A" Circuit Malfunction

P0413 Secondary Air Injection System Switching Valve "A" Circuit Open

P0414 Secondary Air Injection System Switching Valve "A" Circuit Shorted

P0415 Secondary Air Injection System Switching Valve "B" Circuit Malfunction

P0416 Secondary Air Injection System Switching Valve "B" Circuit Open

P0417 Secondary Air Injection System Switching Valve "B" Circuit Shorted

P0418 Secondary Air Injection System Relay "A" Circuit Malfunction

P0419 Secondary Air Injection System Relay "B" Circuit Malfunction

P0420 Catalyst System Efficiency Below Threshold (Bank no. 1)

P0421 Warm Up Catalyst Efficiency Below Threshold (Bank no. 1)

P0422 Main Catalyst Efficiency Below Threshold (Bank no. 1)

P0423 Heated Catalyst Efficiency Below Threshold (Bank no. 1)

P0424 Heated Catalyst Temperature Below Threshold (Bank no. 1)

P0430 Catalyst System Efficiency Below Threshold (Bank no. 2)

P0431 Warm Up Catalyst Efficiency Below Threshold (Bank no. 2)

P0432 Main Catalyst Efficiency Below Threshold (Bank no. 2)

P0433 Heated Catalyst Efficiency Below Threshold (Bank no. 2)

P0434 Heated Catalyst Temperature Below Threshold (Bank no. 2)

P0440 Evaporative Emission Control System Malfunction

P0441 Evaporative Emission Control System Incorrect Purge Flow

P0442 Evaporative Emission Control System Leak Detected (Small Leak)

P0443 Evaporative Emission Control System Purge Control Valve Circuit Malfunction

P0444 Evaporative Emission Control System Purge Control Valve Circuit Open

P0445 Evaporative Emission Control System Purge Control Valve Circuit Shorted

P0446 Evaporative Emission Control System Vent Control Circuit Malfunction

P0447 Evaporative Emission Control System Vent Control Circuit Open

P0448 Evaporative Emission Control System Vent Control Circuit Shorted

P0449 Evaporative Emission Control System Vent Valve/Solenoid Circuit Malfunction

P0450 Evaporative Emission Control System Pressure Sensor Malfunction

P0451 Evaporative Emission Control System Pressure Sensor Range/Performance

P0452 Evaporative Emission Control System Pressure Sensor Low Input

P0453 Evaporative Emission Control System Pressure Sensor High Input

P0454 Evaporative Emission Control System Pressure Sensor Intermittent

P0455 Evaporative Emission Control System Leak Detected (Gross Leak)

P0460 Fuel Level Sensor Circuit Malfunction

P0461 Fuel Level Sensor Circuit Range/Performance

P0462 Fuel Level Sensor Circuit Low Input

P0463 Fuel Level Sensor Circuit High Input

P0464 Fuel Level Sensor Circuit Intermittent

P0465 Purge Flow Sensor Circuit Malfunction

P0466 Purge Flow Sensor Circuit Range/Performance

P0467 Purge Flow Sensor Circuit Low Input

P0468 Purge Flow Sensor Circuit High Input

P0469 Purge Flow Sensor Circuit Intermittent

P0470 Exhaust Pressure Sensor Malfunction

P0471 Exhaust Pressure Sensor Range/Performance

P0472 Exhaust Pressure Sensor Low

P0473 Exhaust Pressure Sensor High

P0474 Exhaust Pressure Sensor Intermittent

P0475 Exhaust Pressure Control Valve Malfunction

P0476 Exhaust Pressure Control Valve Range/Performance

P0477 Exhaust Pressure Control Valve Low

P0478 Exhaust Pressure Control Valve High

P0479 Exhaust Pressure Control Valve Intermittent

P0480 Cooling Fan no. 1 Control Circuit Malfunction

P0481 Cooling Fan no. 2 Control Circuit Malfunction

P0482 Cooling Fan no. 3 Control Circuit Malfunction

P0483 Cooling Fan Rationality Check Malfunction

P0484 Cooling Fan Circuit Over Current

P0485 Cooling Fan Power/Ground Circuit Malfunction

P0500 Vehicle Speed Sensor Malfunction

P0501 Vehicle Speed Sensor Range/Performance

P0502 Vehicle Speed Sensor Circuit Low Input

P0503 Vehicle Speed Sensor Intermittent/Erratic/High

P0505 Idle Control System Malfunction

P0506 Idle Control System RPM Lower Than Expected

P0507 Idle Control System RPM Higher Than Expected

P0510 Closed Throttle Position Switch Malfunction

P0520 Engine Oil Pressure Sensor/Switch Circuit Malfunction

P0521 Engine Oil Pressure Sensor/Switch Range/Performance

P0522 Engine Oil Pressure Sensor/Switch Low Voltage

P0523 Engine Oil Pressure Sensor/Switch High Voltage

P0530 A/C Refrigerant Pressure Sensor Circuit Malfunction

P0531 A/C Refrigerant Pressure Sensor Circuit Range/Performance

P0532 A/C Refrigerant Pressure Sensor Circuit Low Input

P0533 A/C Refrigerant Pressure Sensor Circuit High Input

P0534 A/C Refrigerant Charge Loss

P0550 Power Steering Pressure Sensor Circuit Malfunction

P0551 Power Steering Pressure Sensor Circuit Range/Performance

P0552 Power Steering Pressure Sensor Circuit Low Input

P0553 Power Steering Pressure Sensor Circuit High Input

P0554 Power Steering Pressure Sensor Circuit Intermittent

P0560 System Voltage Malfunction

P0561 System Voltage Unstable

P0562 System Voltage Low

P0563 System Voltage High

P0565 Cruise Control On Signal Malfunction

P0566 Cruise Control Off Signal Malfunction

P0567 Cruise Control Resume Signal Malfunction

P0568 Cruise Control Set Signal Malfunction

P0569 Cruise Control Coast Signal Malfunction

P0570 Cruise Control Accel Signal Malfunction

P0571 Cruise Control/Brake Switch "A" Circuit Malfunction

P0572 Cruise Control/Brake Switch "A" Circuit Low

P0573 Cruise Control/Brake Switch "A" Circuit High

P0574 Through P0580 Reserved for Cruise Codes

P0600 Serial Communication Link Malfunction

P0601 Internal Control Module Memory Check Sum Error

P0602 Control Module Programming Error

P0603 Internal Control Module Keep Alive Memory (KAM) Error

P0604 Internal Control Module Random Access Memory (RAM) Error

P0605 Internal Control Module Read Only Memory (ROM) Error

P0606 PCM Processor Fault

P0608 Control Module VSS Output "A" Malfunction

P0609 Control Module VSS Output "B" Malfunction

P0620 Generator Control Circuit Malfunction

P0621 Generator Lamp "L" Control Circuit Malfunction

P0622 Generator Field "F" Control Circuit Malfunction

P0650 Malfunction Indicator Lamp (MIL) Control Circuit Malfunction

P0654 Engine RPM Output Circuit Malfunction

P0655 Engine Hot Lamp Output Control Circuit Malfunction

P0656 Fuel Level Output Circuit Malfunction

P0700 Transmission Control System Malfunction

P0701 Transmission Control System Range/Performance

P0702 Transmission Control System Electrical

P0703 Torque Converter/Brake Switch "B" Circuit Malfunction

P0704 Clutch Switch Input Circuit Malfunction

P0705 Transmission Range Sensor Circuit Malfunction (PRNDL Input)

P0706 Transmission Range Sensor Circuit Range/Performance

P0707 Transmission Range Sensor Circuit Low Input

P0708 Transmission Range Sensor Circuit High Input

P0709 Transmission Range Sensor Circuit Intermittent

P0710 Transmission Fluid Temperature Sensor Circuit Malfunction

P0711 Transmission Fluid Temperature Sensor Circuit Range/Performance

P0712 Transmission Fluid Temperature Sensor Circuit Low Input

P0713 Transmission Fluid Temperature Sensor Circuit High Input

P0714 Transmission Fluid Temperature Sensor Circuit Intermittent

P0715 Input/Turbine Speed Sensor Circuit Malfunction

P0716 Input/Turbine Speed Sensor Circuit Range/Performance

P0717 Input/Turbine Speed Sensor Circuit No Signal

P0718 Input/Turbine Speed Sensor Circuit Intermittent

P0719 Torque Converter/Brake Switch "B" Circuit Low

P0720 Output Speed Sensor Circuit Malfunction

P0721 Output Speed Sensor Circuit Range/Performance

P0722 Output Speed Sensor Circuit No Signal

P0723 Output Speed Sensor Circuit Intermittent

P0724 Torque Converter/Brake Switch "B" Circuit High
P0725 Engine Speed Input Circuit Malfunction
P0726 Engine Speed Input Circuit Range/Performance
P0727 Engine Speed Input Circuit No Signal
P0728 Engine Speed Input Circuit Intermittent
P0730 Incorrect Gear Ratio
P0731 Gear no. 1 Incorrect Ratio
P0732 Gear no. 2 Incorrect Ratio
P0733 Gear no. 3 Incorrect Ratio
P0734 Gear no. 4 Incorrect Ratio
P0735 Gear no. 5 Incorrect Ratio
P0736 Reverse Incorrect Ratio
P0740 Torque Converter Clutch Circuit Malfunction
P0741 Torque Converter Clutch Circuit Performance or Stuck Off
P0742 Torque Converter Clutch Circuit Stuck On
P0743 Torque Converter Clutch Circuit Electrical
P0744 Torque Converter Clutch Circuit Intermittent
P0745 Pressure Control Solenoid Malfunction
P0746 Pressure Control Solenoid Performance or Stuck Off
P0747 Pressure Control Solenoid Stuck On
P0748 Pressure Control Solenoid Electrical
P0749 Pressure Control Solenoid Intermittent
P0750 Shift Solenoid "A" Malfunction
P0751 Shift Solenoid "A" Performance or Stuck Off
P0752 Shift Solenoid "A" Stuck On
P0753 Shift Solenoid "A" Electrical
P0754 Shift Solenoid "A" Intermittent
P0755 Shift Solenoid "B" Malfunction
P0756 Shift Solenoid "B" Performance or Stuck Off
P0757 Shift Solenoid "B" Stuck On
P0758 Shift Solenoid "B" Electrical
P0759 Shift Solenoid "B" Intermittent
P0760 Shift Solenoid "C" Malfunction
P0761 Shift Solenoid "C" Performance Or Stuck Off
P0762 Shift Solenoid "C" Stuck On
P0763 Shift Solenoid "C" Electrical
P0764 Shift Solenoid "C" Intermittent
P0765 Shift Solenoid "D" Malfunction
P0766 Shift Solenoid "D" Performance Or Stuck Off
P0767 Shift Solenoid "D" Stuck On
P0768 Shift Solenoid "D" Electrical
P0769 Shift Solenoid "D" Intermittent
P0770 Shift Solenoid "E" Malfunction
P0771 Shift Solenoid "E" Performance Or Stuck Off
P0772 Shift Solenoid "E" Stuck On
P0773 Shift Solenoid "E" Electrical
P0774 Shift Solenoid "E" Intermittent
P0780 Shift Malfunction
P0781 1–2 Shift Malfunction
P0782 2–3 Shift Malfunction
P0783 3–4 Shift Malfunction
P0784 4–5 Shift Malfunction
P0785 Shift/Timing Solenoid Malfunction
P0786 Shift/Timing Solenoid Range/Performance
P0787 Shift/Timing Solenoid Low
P0788 Shift/Timing Solenoid High
P0789 Shift/Timing Solenoid Intermittent
P0790 Normal/Performance Switch Circuit Malfunction
P0801 Reverse Inhibit Control Circuit Malfunction
P0803 1–4 Upshift (Skip Shift) Solenoid Control Circuit Malfunction

P0804 1–4 Upshift (Skip Shift) Lamp Control Circuit Malfunction
P1100 Starter Switch Circuit Fault
P1101 Neutral Position Switch Circuit Fault (MT)

Suzuki

READING CODES

Reading the control module memory is on of the first steps in OBD II system diagnostics. This step should be initially performed to determine the general nature of the fault. Subsequent readings will determine if the fault has been cleared.

Reading codes can be performed by any of the methods below:
• Read the control module memory with the Generic Scan Tool (GST)
• Read the control module memory with the vehicle manufacturer's specific tester

To read the fault codes, connect the scan tool or tester according to the manufacturer's instructions. Follow the manufacturer's specified procedure for reading the codes.

CLEARING CODES

Control module reset procedures are a very important part of OBD II System diagnostics. This step should be done at the end of any fault code repair and at the end of any driveability repair.

Clearing codes can be performed by any of the methods below:
• Clear the control module memory with the Generic Scan Tool (GST)
• Clear the control module memory with the vehicle manufacturer's specific tester
• Turn the ignition **OFF** and remove the negative battery cable for at least 1 minute.

Removing the negative battery cable may cause other systems in the vehicle to loose their memory. Prior to removing the cable, ensure you have the proper reset codes for radios and alarms.

➡**The MIL will may also be de-activated for some codes if the vehicle completes three consecutive trips without a fault detected with vehicle conditions similar to those present during the fault.**

OBD II TROUBLE CODES

P0100 Mass or Volume Air Flow Circuit Malfunction
P0101 Mass or Volume Air Flow Circuit Range/Performance Problem
P0102 Mass or Volume Air Flow Circuit Low Input
P0103 Mass or Volume Air Flow Circuit High Input
P0104 Mass or Volume Air Flow Circuit Intermittent
P0105 Manifold Absolute Pressure/Barometric Pressure Circuit Malfunction
P0106 Manifold Absolute Pressure/Barometric Pressure Circuit Range/Performance Problem
P0107 Manifold Absolute Pressure/Barometric Pressure Circuit Low Input
P0108 Manifold Absolute Pressure/Barometric Pressure Circuit High Input
P0109 Manifold Absolute Pressure/Barometric Pressure Circuit Intermittent

P0110 Intake Air Temperature Circuit Malfunction
P0111 Intake Air Temperature Circuit Range/Performance Problem
P0112 Intake Air Temperature Circuit Low Input
P0113 Intake Air Temperature Circuit High Input
P0114 Intake Air Temperature Circuit Intermittent
P0115 Engine Coolant Temperature Circuit Malfunction
P0116 Engine Coolant Temperature Circuit Range/Performance Problem
P0117 Engine Coolant Temperature Circuit Low Input
P0118 Engine Coolant Temperature Circuit High Input
P0119 Engine Coolant Temperature Circuit Intermittent
P0120 Throttle/Pedal Position Sensor/Switch "A" Circuit Malfunction
P0121 Throttle/Pedal Position Sensor/Switch "A" Circuit Range/Performance Problem
P0122 Throttle/Pedal Position Sensor/Switch "A" Circuit Low Input
P0123 Throttle/Pedal Position Sensor/Switch "A" Circuit High Input
P0124 Throttle/Pedal Position Sensor/Switch "A" Circuit Intermittent
P0125 Insufficient Coolant Temperature For Closed Loop Fuel Control
P0126 Insufficient Coolant Temperature For Stable Operation
P0130 O2 Circuit Malfunction (Bank no. 1 Sensor no. 1)
P0131 O2 Sensor Circuit Low Voltage (Bank no. 1 Sensor no. 1)
P0132 O2 Sensor Circuit High Voltage (Bank no. 1 Sensor no. 1)
P0133 O2 Sensor Circuit Slow Response (Bank no. 1 Sensor no. 1)
P0134 O2 Sensor Circuit No Activity Detected (Bank no. 1 Sensor no. 1)
P0135 O2 Sensor Heater Circuit Malfunction (Bank no. 1 Sensor no. 1)
P0136 O2 Sensor Circuit Malfunction (Bank no. 1 Sensor no. 2)
P0137 O2 Sensor Circuit Low Voltage (Bank no. 1 Sensor no. 2)
P0138 O2 Sensor Circuit High Voltage (Bank no. 1 Sensor no. 2)
P0139 O2 Sensor Circuit Slow Response (Bank no. 1 Sensor no. 2)
P0140 O2 Sensor Circuit No Activity Detected (Bank no. 1 Sensor no. 2)
P0141 O2 Sensor Heater Circuit Malfunction (Bank no. 1 Sensor no. 2)
P0142 O2 Sensor Circuit Malfunction (Bank no. 1 Sensor no. 3)
P0143 O2 Sensor Circuit Low Voltage (Bank no. 1 Sensor no. 3)
P0144 O2 Sensor Circuit High Voltage (Bank no. 1 Sensor no. 3)
P0145 O2 Sensor Circuit Slow Response (Bank no. 1 Sensor no. 3)
P0146 O2 Sensor Circuit No Activity Detected (Bank no. 1 Sensor no. 3)
P0147 O2 Sensor Heater Circuit Malfunction (Bank no. 1 Sensor no. 3)
P0150 O2 Sensor Circuit Malfunction (Bank no. 2 Sensor no. 1)
P0151 O2 Sensor Circuit Low Voltage (Bank no. 2 Sensor no. 1)
P0152 O2 Sensor Circuit High Voltage (Bank no. 2 Sensor no. 1)
P0153 O2 Sensor Circuit Slow Response (Bank no. 2 Sensor no. 1)

P0154 O2 Sensor Circuit No Activity Detected (Bank no. 2 Sensor no. 1)
P0155 O2 Sensor Heater Circuit Malfunction (Bank no. 2 Sensor no. 1)
P0156 O2 Sensor Circuit Malfunction (Bank no. 2 Sensor no. 2)
P0157 O2 Sensor Circuit Low Voltage (Bank no. 2 Sensor no. 2)
P0158 O2 Sensor Circuit High Voltage (Bank no. 2 Sensor no. 2)
P0159 O2 Sensor Circuit Slow Response (Bank no. 2 Sensor no. 2)
P0160 O2 Sensor Circuit No Activity Detected (Bank no. 2 Sensor no. 2)
P0161 O2 Sensor Heater Circuit Malfunction (Bank no. 2 Sensor no. 2)
P0162 O2 Sensor Circuit Malfunction (Bank no. 2 Sensor no. 3)
P0163 O2 Sensor Circuit Low Voltage (Bank no. 2 Sensor no. 3)
P0164 O2 Sensor Circuit High Voltage (Bank no. 2 Sensor no. 3)
P0165 O2 Sensor Circuit Slow Response (Bank no. 2 Sensor no. 3)
P0166 O2 Sensor Circuit No Activity Detected (Bank no. 2 Sensor no. 3)
P0167 O2 Sensor Heater Circuit Malfunction (Bank no. 2 Sensor no. 3)
P0170 Fuel Trim Malfunction (Bank no. 1)
P0171 System Too Lean (Bank no. 1)
P0172 System Too Rich (Bank no. 1)
P0173 Fuel Trim Malfunction (Bank no. 2)
P0174 System Too Lean (Bank no. 2)
P0175 System Too Rich (Bank no. 2)
P0176 Fuel Composition Sensor Circuit Malfunction
P0177 Fuel Composition Sensor Circuit Range/Performance
P0178 Fuel Composition Sensor Circuit Low Input
P0179 Fuel Composition Sensor Circuit High Input
P0180 Fuel Temperature Sensor "A" Circuit Malfunction
P0181 Fuel Temperature Sensor "A" Circuit Range/Performance
P0182 Fuel Temperature Sensor "A" Circuit Low Input
P0183 Fuel Temperature Sensor "A" Circuit High Input
P0184 Fuel Temperature Sensor "A" Circuit Intermittent
P0185 Fuel Temperature Sensor "B" Circuit Malfunction
P0186 Fuel Temperature Sensor "B" Circuit Range/Performance
P0187 Fuel Temperature Sensor "B" Circuit Low Input
P0188 Fuel Temperature Sensor "B" Circuit High Input
P0189 Fuel Temperature Sensor "B" Circuit Intermittent
P0190 Fuel Rail Pressure Sensor Circuit Malfunction
P0191 Fuel Rail Pressure Sensor Circuit Range/Performance
P0192 Fuel Rail Pressure Sensor Circuit Low Input
P0193 Fuel Rail Pressure Sensor Circuit High Input
P0194 Fuel Rail Pressure Sensor Circuit Intermittent
P0195 Engine Oil Temperature Sensor Malfunction
P0196 Engine Oil Temperature Sensor Range/Performance
P0197 Engine Oil Temperature Sensor Low
P0198 Engine Oil Temperature Sensor High
P0199 Engine Oil Temperature Sensor Intermittent
P0200 Injector Circuit Malfunction
P0201 Injector Circuit Malfunction—Cylinder no. 1
P0202 Injector Circuit Malfunction—Cylinder no. 2
P0203 Injector Circuit Malfunction—Cylinder no. 3
P0204 Injector Circuit Malfunction—Cylinder no. 4
P0205 Injector Circuit Malfunction—Cylinder no. 5
P0206 Injector Circuit Malfunction—Cylinder no. 6

P0207 Injector Circuit Malfunction—Cylinder no. 7
P0208 Injector Circuit Malfunction—Cylinder no. 8
P0209 Injector Circuit Malfunction—Cylinder no. 9
P0210 Injector Circuit Malfunction—Cylinder no. 10
P0211 Injector Circuit Malfunction—Cylinder no. 11
P0212 Injector Circuit Malfunction—Cylinder no. 12
P0213 Cold Start Injector no. 1 Malfunction
P0214 Cold Start Injector no. 2 Malfunction
P0215 Engine Shutoff Solenoid Malfunction
P0216 Injection Timing Control Circuit Malfunction
P0217 Engine Over Temperature Condition
P0218 Transmission Over Temperature Condition
P0219 Engine Over Speed Condition
P0220 Throttle/Pedal Position Sensor/Switch "B" Circuit Malfunction
P0221 Throttle/Pedal Position Sensor/Switch "B" Circuit Range/Performance Problem
P0222 Throttle/Pedal Position Sensor/Switch "B" Circuit Low Input
P0223 Throttle/Pedal Position Sensor/Switch "B" Circuit High Input
P0224 Throttle/Pedal Position Sensor/Switch "B" Circuit Intermittent
P0225 Throttle/Pedal Position Sensor/Switch "C" Circuit Malfunction
P0226 Throttle/Pedal Position Sensor/Switch "C" Circuit Range/Performance Problem
P0227 Throttle/Pedal Position Sensor/Switch "C" Circuit Low Input
P0228 Throttle/Pedal Position Sensor/Switch "C" Circuit High Input
P0229 Throttle/Pedal Position Sensor/Switch "C" Circuit Intermittent
P0230 Fuel Pump Primary Circuit Malfunction
P0231 Fuel Pump Secondary Circuit Low
P0232 Fuel Pump Secondary Circuit High
P0233 Fuel Pump Secondary Circuit Intermittent
P0234 Engine Over Boost Condition
P0261 Cylinder no. 1 Injector Circuit Low
P0262 Cylinder no. 1 Injector Circuit High
P0263 Cylinder no. 1 Contribution/Balance Fault
P0264 Cylinder no. 2 Injector Circuit Low
P0265 Cylinder no. 2 Injector Circuit High
P0266 Cylinder no. 2 Contribution/Balance Fault
P0267 Cylinder no. 3 Injector Circuit Low
P0268 Cylinder no. 3 Injector Circuit High
P0269 Cylinder no. 3 Contribution/Balance Fault
P0270 Cylinder no. 4 Injector Circuit Low
P0271 Cylinder no. 4 Injector Circuit High
P0272 Cylinder no. 4 Contribution/Balance Fault
P0273 Cylinder no. 5 Injector Circuit Low
P0274 Cylinder no. 5 Injector Circuit High
P0275 Cylinder no. 5 Contribution/Balance Fault
P0276 Cylinder no. 6 Injector Circuit Low
P0277 Cylinder no. 6 Injector Circuit High
P0278 Cylinder no. 6 Contribution/Balance Fault
P0279 Cylinder no. 7 Injector Circuit Low
P0280 Cylinder no. 7 Injector Circuit High
P0281 Cylinder no. 7 Contribution/Balance Fault
P0282 Cylinder no. 8 Injector Circuit Low
P0283 Cylinder no. 8 Injector Circuit High

P0284 Cylinder no. 8 Contribution/Balance Fault
P0285 Cylinder no. 9 Injector Circuit Low
P0286 Cylinder no. 9 Injector Circuit High
P0287 Cylinder no. 9 Contribution/Balance Fault
P0288 Cylinder no. 10 Injector Circuit Low
P0289 Cylinder no. 10 Injector Circuit High
P0290 Cylinder no. 10 Contribution/Balance Fault
P0291 Cylinder no. 11 Injector Circuit Low
P0292 Cylinder no. 11 Injector Circuit High
P0293 Cylinder no. 11 Contribution/Balance Fault
P0294 Cylinder no. 12 Injector Circuit Low
P0295 Cylinder no. 12 Injector Circuit High
P0296 Cylinder no. 12 Contribution/Balance Fault
P0300 Random/Multiple Cylinder Misfire Detected
P0301 Cylinder no. 1—Misfire Detected
P0302 Cylinder no. 2—Misfire Detected
P0303 Cylinder no. 3—Misfire Detected
P0304 Cylinder no. 4—Misfire Detected
P0305 Cylinder no. 5—Misfire Detected
P0306 Cylinder no. 6—Misfire Detected
P0307 Cylinder no. 7—Misfire Detected
P0308 Cylinder no. 8—Misfire Detected
P0309 Cylinder no. 9—Misfire Detected
P0310 Cylinder no. 10—Misfire Detected
P0311 Cylinder no. 11—Misfire Detected
P0312 Cylinder no. 12—Misfire Detected
P0320 Ignition/Distributor Engine Speed Input Circuit Malfunction
P0321 Ignition/Distributor Engine Speed Input Circuit Range/Performance
P0322 Ignition/Distributor Engine Speed Input Circuit No Signal
P0323 Ignition/Distributor Engine Speed Input Circuit Intermittent
P0325 Knock Sensor no. 1—Circuit Malfunction (Bank no. 1 or Single Sensor)
P0326 Knock Sensor no. 1—Circuit Range/Performance (Bank no. 1 or Single Sensor)
P0327 Knock Sensor no. 1—Circuit Low Input (Bank no. 1 or Single Sensor)
P0328 Knock Sensor no. 1—Circuit High Input (Bank no. 1 or Single Sensor)
P0329 Knock Sensor no. 1—Circuit Input Intermittent (Bank no. 1 or Single Sensor)
P0330 Knock Sensor no. 2—Circuit Malfunction (Bank no. 2)
P0331 Knock Sensor no. 2—Circuit Range/Performance (Bank no. 2)
P0332 Knock Sensor no. 2—Circuit Low Input (Bank no. 2)
P0333 Knock Sensor no. 2—Circuit High Input (Bank no. 2)
P0334 Knock Sensor no. 2—Circuit Input Intermittent (Bank no. 2)
P0335 Crankshaft Position Sensor "A" Circuit Malfunction
P0336 Crankshaft Position Sensor "A" Circuit Range/Performance
P0337 Crankshaft Position Sensor "A" Circuit Low Input
P0338 Crankshaft Position Sensor "A" Circuit High Input
P0339 Crankshaft Position Sensor "A" Circuit Intermittent
P0340 Camshaft Position Sensor Circuit Malfunction
P0341 Camshaft Position Sensor Circuit Range/Performance
P0342 Camshaft Position Sensor Circuit Low Input
P0343 Camshaft Position Sensor Circuit High Input
P0344 Camshaft Position Sensor Circuit Intermittent

P0350 Ignition Coil Primary/Secondary Circuit Malfunction
P0351 Ignition Coil "A" Primary/Secondary Circuit Malfunction
P0352 Ignition Coil "B" Primary/Secondary Circuit Malfunction
P0353 Ignition Coil "C" Primary/Secondary Circuit Malfunction
P0354 Ignition Coil "D" Primary/Secondary Circuit Malfunction
P0355 Ignition Coil "E" Primary/Secondary Circuit Malfunction
P0356 Ignition Coil "F" Primary/Secondary Circuit Malfunction
P0357 Ignition Coil "G" Primary/Secondary Circuit Malfunction
P0358 Ignition Coil "H" Primary/Secondary Circuit Malfunction
P0359 Ignition Coil "I" Primary/Secondary Circuit Malfunction
P0360 Ignition Coil "J" Primary/Secondary Circuit Malfunction
P0361 Ignition Coil "K" Primary/Secondary Circuit Malfunction
P0362 Ignition Coil "L" Primary/Secondary Circuit Malfunction
P0370 Timing Reference High Resolution Signal "A" Malfunction
P0371 Timing Reference High Resolution Signal "A" Too Many Pulses
P0372 Timing Reference High Resolution Signal "A" Too Few Pulses
P0373 Timing Reference High Resolution Signal "A" Intermittent/Erratic Pulses
P0374 Timing Reference High Resolution Signal "A" No Pulses
P0375 Timing Reference High Resolution Signal "B" Malfunction
P0376 Timing Reference High Resolution Signal "B" Too Many Pulses
P0377 Timing Reference High Resolution Signal "B" Too Few Pulses
P0378 Timing Reference High Resolution Signal "B" Intermittent/Erratic Pulses
P0379 Timing Reference High Resolution Signal "B" No Pulses
P0380 Glow Plug/Heater Circuit "A" Malfunction
P0381 Glow Plug/Heater Indicator Circuit Malfunction
P0382 Glow Plug/Heater Circuit "B" Malfunction
P0385 Crankshaft Position Sensor "B" Circuit Malfunction
P0386 Crankshaft Position Sensor "B" Circuit Range/Performance
P0387 Crankshaft Position Sensor "B" Circuit Low Input
P0388 Crankshaft Position Sensor "B" Circuit High Input
P0389 Crankshaft Position Sensor "B" Circuit Intermittent
P0400 Exhaust Gas Recirculation Flow Malfunction
P0401 Exhaust Gas Recirculation Flow Insufficient Detected
P0402 Exhaust Gas Recirculation Flow Excessive Detected
P0403 Exhaust Gas Recirculation Circuit Malfunction
P0404 Exhaust Gas Recirculation Circuit Range/Performance
P0405 Exhaust Gas Recirculation Sensor "A" Circuit Low
P0406 Exhaust Gas Recirculation Sensor "A" Circuit High
P0407 Exhaust Gas Recirculation Sensor "B" Circuit Low
P0408 Exhaust Gas Recirculation Sensor "B" Circuit High
P0410 Secondary Air Injection System Malfunction
P0411 Secondary Air Injection System Incorrect Flow Detected
P0412 Secondary Air Injection System Switching Valve "A" Circuit Malfunction
P0413 Secondary Air Injection System Switching Valve "A" Circuit Open
P0414 Secondary Air Injection System Switching Valve "A" Circuit Shorted
P0415 Secondary Air Injection System Switching Valve "B" Circuit Malfunction
P0416 Secondary Air Injection System Switching Valve "B" Circuit Open

P0417 Secondary Air Injection System Switching Valve "B" Circuit Shorted
P0418 Secondary Air Injection System Relay "A" Circuit Malfunction
P0419 Secondary Air Injection System Relay "B" Circuit Malfunction
P0420 Catalyst System Efficiency Below Threshold (Bank no. 1)
P0421 Warm Up Catalyst Efficiency Below Threshold (Bank no. 1)
P0422 Main Catalyst Efficiency Below Threshold (Bank no. 1)
P0423 Heated Catalyst Efficiency Below Threshold (Bank no. 1)
P0424 Heated Catalyst Temperature Below Threshold (Bank no. 1)
P0430 Catalyst System Efficiency Below Threshold (Bank no. 2)
P0431 Warm Up Catalyst Efficiency Below Threshold (Bank no. 2)
P0432 Main Catalyst Efficiency Below Threshold (Bank no. 2)
P0433 Heated Catalyst Efficiency Below Threshold (Bank no. 2)
P0434 Heated Catalyst Temperature Below Threshold (Bank no. 2)
P0440 Evaporative Emission Control System Malfunction
P0441 Evaporative Emission Control System Incorrect Purge Flow
P0442 Evaporative Emission Control System Leak Detected (Small Leak)
P0443 Evaporative Emission Control System Purge Control Valve Circuit Malfunction
P0444 Evaporative Emission Control System Purge Control Valve Circuit Open
P0445 Evaporative Emission Control System Purge Control Valve Circuit Shorted
P0446 Evaporative Emission Control System Vent Control Circuit Malfunction
P0447 Evaporative Emission Control System Vent Control Circuit Open
P0448 Evaporative Emission Control System Vent Control Circuit Shorted
P0449 Evaporative Emission Control System Vent Valve/Solenoid Circuit Malfunction
P0450 Evaporative Emission Control System Pressure Sensor Malfunction
P0451 Evaporative Emission Control System Pressure Sensor Range/Performance
P0452 Evaporative Emission Control System Pressure Sensor Low Input
P0453 Evaporative Emission Control System Pressure Sensor High Input
P0454 Evaporative Emission Control System Pressure Sensor Intermittent
P0455 Evaporative Emission Control System Leak Detected (Gross Leak)
P0460 Fuel Level Sensor Circuit Malfunction
P0461 Fuel Level Sensor Circuit Range/Performance
P0462 Fuel Level Sensor Circuit Low Input
P0463 Fuel Level Sensor Circuit High Input
P0464 Fuel Level Sensor Circuit Intermittent
P0465 Purge Flow Sensor Circuit Malfunction
P0466 Purge Flow Sensor Circuit Range/Performance
P0467 Purge Flow Sensor Circuit Low Input
P0468 Purge Flow Sensor Circuit High Input
P0469 Purge Flow Sensor Circuit Intermittent

P0470 Exhaust Pressure Sensor Malfunction
P0471 Exhaust Pressure Sensor Range/Performance
P0472 Exhaust Pressure Sensor Low
P0473 Exhaust Pressure Sensor High
P0474 Exhaust Pressure Sensor Intermittent
P0475 Exhaust Pressure Control Valve Malfunction
P0476 Exhaust Pressure Control Valve Range/Performance
P0477 Exhaust Pressure Control Valve Low
P0478 Exhaust Pressure Control Valve High
P0479 Exhaust Pressure Control Valve Intermittent
P0480 Cooling Fan no. 1 Control Circuit Malfunction
P0481 Cooling Fan no. 2 Control Circuit Malfunction
P0482 Cooling Fan no. 3 Control Circuit Malfunction
P0483 Cooling Fan Rationality Check Malfunction
P0484 Cooling Fan Circuit Over Current
P0485 Cooling Fan Power/Ground Circuit Malfunction
P0500 Vehicle Speed Sensor Malfunction
P0501 Vehicle Speed Sensor Range/Performance
P0502 Vehicle Speed Sensor Circuit Low Input
P0503 Vehicle Speed Sensor Intermittent/Erratic/High
P0505 Idle Control System Malfunction
P0506 Idle Control System RPM Lower Than Expected
P0507 Idle Control System RPM Higher Than Expected
P0510 Closed Throttle Position Switch Malfunction
P0520 Engine Oil Pressure Sensor/Switch Circuit Malfunction
P0521 Engine Oil Pressure Sensor/Switch Range/Performance
P0522 Engine Oil Pressure Sensor/Switch Low Voltage
P0523 Engine Oil Pressure Sensor/Switch High Voltage
P0530 A/C Refrigerant Pressure Sensor Circuit Malfunction
P0531 A/C Refrigerant Pressure Sensor Circuit Range/Performance
P0532 A/C Refrigerant Pressure Sensor Circuit Low Input
P0533 A/C Refrigerant Pressure Sensor Circuit High Input
P0534 A/C Refrigerant Charge Loss
P0550 Power Steering Pressure Sensor Circuit Malfunction
P0551 Power Steering Pressure Sensor Circuit Range/Performance
P0552 Power Steering Pressure Sensor Circuit Low Input
P0553 Power Steering Pressure Sensor Circuit High Input
P0554 Power Steering Pressure Sensor Circuit Intermittent
P0560 System Voltage Malfunction
P0561 System Voltage Unstable
P0562 System Voltage Low
P0563 System Voltage High
P0565 Cruise Control On Signal Malfunction
P0566 Cruise Control Off Signal Malfunction
P0567 Cruise Control Resume Signal Malfunction
P0568 Cruise Control Set Signal Malfunction
P0569 Cruise Control Coast Signal Malfunction
P0570 Cruise Control Accel Signal Malfunction
P0571 Cruise Control/Brake Switch "A" Circuit Malfunction
P0572 Cruise Control/Brake Switch "A" Circuit Low
P0573 Cruise Control/Brake Switch "A" Circuit High
P0574 Through P0580 Reserved for Cruise Codes
P0600 Serial Communication Link Malfunction
P0601 Internal Control Module Memory Check Sum Error
P0602 Control Module Programming Error
P0603 Internal Control Module Keep Alive Memory (KAM) Error
P0604 Internal Control Module Random Access Memory (RAM) Error
P0605 Internal Control Module Read Only Memory (ROM) Error

P0606 PCM Processor Fault
P0608 Control Module VSS Output "A" Malfunction
P0609 Control Module VSS Output "B" Malfunction
P0620 Generator Control Circuit Malfunction
P0621 Generator Lamp "L" Control Circuit Malfunction
P0622 Generator Field "F" Control Circuit Malfunction
P0650 Malfunction Indicator Lamp (MIL) Control Circuit Malfunction
P0654 Engine RPM Output Circuit Malfunction
P0655 Engine Hot Lamp Output Control Circuit Malfunction
P0656 Fuel Level Output Circuit Malfunction
P0700 Transmission Control System Malfunction
P0701 Transmission Control System Range/Performance
P0702 Transmission Control System Electrical
P0703 Torque Converter/Brake Switch "B" Circuit Malfunction
P0704 Clutch Switch Input Circuit Malfunction
P0705 Transmission Range Sensor Circuit Malfunction (PRNDL Input)
P0706 Transmission Range Sensor Circuit Range/Performance
P0707 Transmission Range Sensor Circuit Low Input
P0708 Transmission Range Sensor Circuit High Input
P0709 Transmission Range Sensor Circuit Intermittent
P0710 Transmission Fluid Temperature Sensor Circuit Malfunction
P0711 Transmission Fluid Temperature Sensor Circuit Range/Performance
P0712 Transmission Fluid Temperature Sensor Circuit Low Input
P0713 Transmission Fluid Temperature Sensor Circuit High Input
P0714 Transmission Fluid Temperature Sensor Circuit Intermittent
P0715 Input/Turbine Speed Sensor Circuit Malfunction
P0716 Input/Turbine Speed Sensor Circuit Range/Performance
P0717 Input/Turbine Speed Sensor Circuit No Signal
P0718 Input/Turbine Speed Sensor Circuit Intermittent
P0719 Torque Converter/Brake Switch "B" Circuit Low
P0720 Output Speed Sensor Circuit Malfunction
P0721 Output Speed Sensor Circuit Range/Performance
P0722 Output Speed Sensor Circuit No Signal
P0723 Output Speed Sensor Circuit Intermittent
P0724 Torque Converter/Brake Switch "B" Circuit High
P0725 Engine Speed Input Circuit Malfunction
P0726 Engine Speed Input Circuit Range/Performance
P0727 Engine Speed Input Circuit No Signal
P0728 Engine Speed Input Circuit Intermittent
P0730 Incorrect Gear Ratio
P0731 Gear no. 1 Incorrect Ratio
P0732 Gear no. 2 Incorrect Ratio
P0733 Gear no. 3 Incorrect Ratio
P0734 Gear no. 4 Incorrect Ratio
P0735 Gear no. 5 Incorrect Ratio
P0736 Reverse Incorrect Ratio
P0740 Torque Converter Clutch Circuit Malfunction
P0741 Torque Converter Clutch Circuit Performance or Stuck Off
P0742 Torque Converter Clutch Circuit Stuck On
P0743 Torque Converter Clutch Circuit Electrical
P0744 Torque Converter Clutch Circuit Intermittent
P0745 Pressure Control Solenoid Malfunction
P0746 Pressure Control Solenoid Performance or Stuck Off

P0747 Pressure Control Solenoid Stuck On
P0748 Pressure Control Solenoid Electrical
P0749 Pressure Control Solenoid Intermittent
P0750 Shift Solenoid "A" Malfunction
P0751 Shift Solenoid "A" Performance or Stuck Off
P0752 Shift Solenoid "A" Stuck On
P0753 Shift Solenoid "A" Electrical
P0754 Shift Solenoid "A" Intermittent
P0755 Shift Solenoid "B" Malfunction
P0756 Shift Solenoid "B" Performance or Stuck Off
P0757 Shift Solenoid "B" Stuck On
P0758 Shift Solenoid "B" Electrical
P0759 Shift Solenoid "B" Intermittent
P0760 Shift Solenoid "C" Malfunction
P0761 Shift Solenoid "C" Performance Or Stuck Off
P0762 Shift Solenoid "C" Stuck On
P0763 Shift Solenoid "C" Electrical
P0764 Shift Solenoid "C" Intermittent
P0765 Shift Solenoid "D" Malfunction
P0766 Shift Solenoid "D" Performance Or Stuck Off
P0767 Shift Solenoid "D" Stuck On
P0768 Shift Solenoid "D" Electrical
P0769 Shift Solenoid "D" Intermittent
P0770 Shift Solenoid "E" Malfunction
P0771 Shift Solenoid "E" Performance Or Stuck Off
P0772 Shift Solenoid "E" Stuck On
P0773 Shift Solenoid "E" Electrical
P0774 Shift Solenoid "E" Intermittent
P0780 Shift Malfunction
P0781 1–2 Shift Malfunction
P0782 2–3 Shift Malfunction
P0783 3–4 Shift Malfunction
P0784 4–5 Shift Malfunction
P0785 Shift/Timing Solenoid Malfunction
P0786 Shift/Timing Solenoid Range/Performance
P0787 Shift/Timing Solenoid Low
P0788 Shift/Timing Solenoid High
P0789 Shift/Timing Solenoid Intermittent
P0790 Normal/Performance Switch Circuit Malfunction
P0801 Reverse Inhibit Control Circuit Malfunction
P0803 1–4 Upshift (Skip Shift) Solenoid Control Circuit Malfunction
P0804 1–4 Upshift (Skip Shift) Lamp Control Circuit Malfunction
P1250 EFI Heater Circuit Fault
P1408 Manifold Differential Pressure Sensor Circuit Fault
P1410 Fuel Tank Pressure Control Solenoid Circuit Fault
P1450 Barometric Pressure Sensor Circuit Fault
P1451 Barometric Pressure Sensor Performance
P1460 Cooling Fan Control System Fault
P1500 Starter Signal Circuit Fault
P1510 Back-up Power Supply Fault
P1530 Ignition Timing Adjustment Switch Circuit
P1600 PCM Battery Circuit Fault
P1700 TCM Throttle Position Sensor Circuit Fault
P1705 TCM ECT Circuit Fault
P1715 PNP Switch Circuit Fault
P1717 AT Drive Range Signal Circuit Fault

Toyota

READING CODES

Reading the control module memory is on of the first steps in OBD II system diagnostics. This step should be initially performed to determine the general nature of the fault. Subsequent readings will determine if the fault has been cleared.

Reading codes can be performed by any of the methods below:
- Read the control module memory with the Generic Scan Tool (GST)
- Read the control module memory with the vehicle manufacturer's specific tester

To read the fault codes, connect the scan tool or tester according to the manufacturer's instructions. Follow the manufacturer's specified procedure for reading the codes.

CLEARING CODES

Control module reset procedures are a very important part of OBD II System diagnostics. This step should be done at the end of any fault code repair and at the end of any driveability repair.

Clearing codes can be performed by any of the methods below:
- Clear the control module memory with the Generic Scan Tool (GST)
- Clear the control module memory with the vehicle manufacturer's specific tester
- Turn the ignition **OFF** and remove the negative battery cable for at least 1 minute.

Removing the negative battery cable may cause other systems in the vehicle to loose their memory. Prior to removing the cable, ensure you have the proper reset codes for radios and alarms.

➡**The MIL will may also be de-activated for some codes if the vehicle completes three consecutive trips without a fault detected with vehicle conditions similar to those present during the fault.**

OBD II TROUBLE CODES

P0100 Mass or Volume Air Flow Circuit Malfunction
P0101 Mass or Volume Air Flow Circuit Range/Performance Problem
P0102 Mass or Volume Air Flow Circuit Low Input
P0103 Mass or Volume Air Flow Circuit High Input
P0104 Mass or Volume Air Flow Circuit Intermittent
P0105 Manifold Absolute Pressure/Barometric Pressure Circuit Malfunction
P0106 Manifold Absolute Pressure/Barometric Pressure Circuit Range/Performance Problem
P0107 Manifold Absolute Pressure/Barometric Pressure Circuit Low Input
P0108 Manifold Absolute Pressure/Barometric Pressure Circuit High Input
P0109 Manifold Absolute Pressure/Barometric Pressure Circuit Intermittent
P0110 Intake Air Temperature Circuit Malfunction

P0111 Intake Air Temperature Circuit Range/Performance Problem

P0112 Intake Air Temperature Circuit Low Input

P0113 Intake Air Temperature Circuit High Input

P0114 Intake Air Temperature Circuit Intermittent

P0115 Engine Coolant Temperature Circuit Malfunction

P0116 Engine Coolant Temperature Circuit Range/Performance Problem

P0117 Engine Coolant Temperature Circuit Low Input

P0118 Engine Coolant Temperature Circuit High Input

P0119 Engine Coolant Temperature Circuit Intermittent

P0120 Throttle/Pedal Position Sensor/Switch "A" Circuit Malfunction

P0121 Throttle/Pedal Position Sensor/Switch "A" Circuit Range/Performance Problem

P0122 Throttle/Pedal Position Sensor/Switch "A" Circuit Low Input

P0123 Throttle/Pedal Position Sensor/Switch "A" Circuit High Input

P0124 Throttle/Pedal Position Sensor/Switch "A" Circuit Intermittent

P0125 Insufficient Coolant Temperature For Closed Loop Fuel Control

P0126 Insufficient Coolant Temperature For Stable Operation

P0130 O2 Circuit Malfunction (Bank no. 1 Sensor no. 1)

P0131 O2 Sensor Circuit Low Voltage (Bank no. 1 Sensor no. 1)

P0132 O2 Sensor Circuit High Voltage (Bank no. 1 Sensor no. 1)

P0133 O2 Sensor Circuit Slow Response (Bank no. 1 Sensor no. 1)

P0134 O2 Sensor Circuit No Activity Detected (Bank no. 1 Sensor no. 1)

P0135 O2 Sensor Heater Circuit Malfunction (Bank no. 1 Sensor no. 1)

P0136 O2 Sensor Circuit Malfunction (Bank no. 1 Sensor no. 2)

P0137 O2 Sensor Circuit Low Voltage (Bank no. 1 Sensor no. 2)

P0138 O2 Sensor Circuit High Voltage (Bank no. 1 Sensor no. 2)

P0139 O2 Sensor Circuit Slow Response (Bank no. 1 Sensor no. 2)

P0140 O2 Sensor Circuit No Activity Detected (Bank no. 1 Sensor no. 2)

P0141 O2 Sensor Heater Circuit Malfunction (Bank no. 1 Sensor no. 2)

P0142 O2 Sensor Circuit Malfunction (Bank no. 1 Sensor no. 3)

P0143 O2 Sensor Circuit Low Voltage (Bank no. 1 Sensor no. 3)

P0144 O2 Sensor Circuit High Voltage (Bank no. 1 Sensor no. 3)

P0145 O2 Sensor Circuit Slow Response (Bank no. 1 Sensor no. 3)

P0146 O2 Sensor Circuit No Activity Detected (Bank no. 1 Sensor no. 3)

P0147 O2 Sensor Heater Circuit Malfunction (Bank no. 1 Sensor no. 3)

P0150 O2 Sensor Circuit Malfunction (Bank no. 2 Sensor no. 1)

P0151 O2 Sensor Circuit Low Voltage (Bank no. 2 Sensor no. 1)

P0152 O2 Sensor Circuit High Voltage (Bank no. 2 Sensor no. 1)

P0153 O2 Sensor Circuit Slow Response (Bank no. 2 Sensor no. 1)

P0154 O2 Sensor Circuit No Activity Detected (Bank no. 2 Sensor no. 1)

P0155 O2 Sensor Heater Circuit Malfunction (Bank no. 2 Sensor no. 1)

P0156 O2 Sensor Circuit Malfunction (Bank no. 2 Sensor no. 2)

P0157 O2 Sensor Circuit Low Voltage (Bank no. 2 Sensor no. 2)

P0158 O2 Sensor Circuit High Voltage (Bank no. 2 Sensor no. 2)

P0159 O2 Sensor Circuit Slow Response (Bank no. 2 Sensor no. 2)

P0160 O2 Sensor Circuit No Activity Detected (Bank no. 2 Sensor no. 2)

P0161 O2 Sensor Heater Circuit Malfunction (Bank no. 2 Sensor no. 2)

P0162 O2 Sensor Circuit Malfunction (Bank no. 2 Sensor no. 3)

P0163 O2 Sensor Circuit Low Voltage (Bank no. 2 Sensor no. 3)

P0164 O2 Sensor Circuit High Voltage (Bank no. 2 Sensor no. 3)

P0165 O2 Sensor Circuit Slow Response (Bank no. 2 Sensor no. 3)

P0166 O2 Sensor Circuit No Activity Detected (Bank no. 2 Sensor no. 3)

P0167 O2 Sensor Heater Circuit Malfunction (Bank no. 2 Sensor no. 3)

P0170 Fuel Trim Malfunction (Bank no. 1)

P0171 System Too Lean (Bank no. 1)

P0172 System Too Rich (Bank no. 1)

P0173 Fuel Trim Malfunction (Bank no. 2)

P0174 System Too Lean (Bank no. 2)

P0175 System Too Rich (Bank no. 2)

P0176 Fuel Composition Sensor Circuit Malfunction

P0177 Fuel Composition Sensor Circuit Range/Performance

P0178 Fuel Composition Sensor Circuit Low Input

P0179 Fuel Composition Sensor Circuit High Input

P0180 Fuel Temperature Sensor "A" Circuit Malfunction

P0181 Fuel Temperature Sensor "A" Circuit Range/Performance

P0182 Fuel Temperature Sensor "A" Circuit Low Input

P0183 Fuel Temperature Sensor "A" Circuit High Input

P0184 Fuel Temperature Sensor "A" Circuit Intermittent

P0185 Fuel Temperature Sensor "B" Circuit Malfunction

P0186 Fuel Temperature Sensor "B" Circuit Range/Performance

P0187 Fuel Temperature Sensor "B" Circuit Low Input

P0188 Fuel Temperature Sensor "B" Circuit High Input

P0189 Fuel Temperature Sensor "B" Circuit Intermittent

P0190 Fuel Rail Pressure Sensor Circuit Malfunction

P0191 Fuel Rail Pressure Sensor Circuit Range/Performance

P0192 Fuel Rail Pressure Sensor Circuit Low Input

P0193 Fuel Rail Pressure Sensor Circuit High Input

P0194 Fuel Rail Pressure Sensor Circuit Intermittent

P0195 Engine Oil Temperature Sensor Malfunction

P0196 Engine Oil Temperature Sensor Range/Performance

P0197 Engine Oil Temperature Sensor Low

P0198 Engine Oil Temperature Sensor High

P0199 Engine Oil Temperature Sensor Intermittent

P0200 Injector Circuit Malfunction

P0201 Injector Circuit Malfunction—Cylinder no. 1

P0202 Injector Circuit Malfunction—Cylinder no. 2

P0203 Injector Circuit Malfunction—Cylinder no. 3

P0204 Injector Circuit Malfunction—Cylinder no. 4

P0205 Injector Circuit Malfunction—Cylinder no. 5

P0206 Injector Circuit Malfunction—Cylinder no. 6

P0207 Injector Circuit Malfunction—Cylinder no. 7

P0208 Injector Circuit Malfunction—Cylinder no. 8

P0209 Injector Circuit Malfunction—Cylinder no. 9
P0210 Injector Circuit Malfunction—Cylinder no. 10
P0211 Injector Circuit Malfunction—Cylinder no. 11
P0212 Injector Circuit Malfunction—Cylinder no. 12
P0213 Cold Start Injector no. 1 Malfunction
P0214 Cold Start Injector no. 2 Malfunction
P0215 Engine Shutoff Solenoid Malfunction
P0216 Injection Timing Control Circuit Malfunction
P0217 Engine Over Temperature Condition
P0218 Transmission Over Temperature Condition
P0219 Engine Over Speed Condition
P0220 Throttle/Pedal Position Sensor/Switch "B" Circuit Malfunction
P0221 Throttle/Pedal Position Sensor/Switch "B" Circuit Range/Performance Problem
P0222 Throttle/Pedal Position Sensor/Switch "B" Circuit Low Input
P0223 Throttle/Pedal Position Sensor/Switch "B" Circuit High Input
P0224 Throttle/Pedal Position Sensor/Switch "B" Circuit Intermittent
P0225 Throttle/Pedal Position Sensor/Switch "C" Circuit Malfunction
P0226 Throttle/Pedal Position Sensor/Switch "C" Circuit Range/Performance Problem
P0227 Throttle/Pedal Position Sensor/Switch "C" Circuit Low Input
P0228 Throttle/Pedal Position Sensor/Switch "C" Circuit High Input
P0229 Throttle/Pedal Position Sensor/Switch "C" Circuit Intermittent
P0230 Fuel Pump Primary Circuit Malfunction
P0231 Fuel Pump Secondary Circuit Low
P0232 Fuel Pump Secondary Circuit High
P0233 Fuel Pump Secondary Circuit Intermittent
P0234 Engine Over Boost Condition
P0261 Cylinder no. 1 Injector Circuit Low
P0262 Cylinder no. 1 Injector Circuit High
P0263 Cylinder no. 1 Contribution/Balance Fault
P0264 Cylinder no. 2 Injector Circuit Low
P0265 Cylinder no. 2 Injector Circuit High
P0266 Cylinder no. 2 Contribution/Balance Fault
P0267 Cylinder no. 3 Injector Circuit Low
P0268 Cylinder no. 3 Injector Circuit High
P0269 Cylinder no. 3 Contribution/Balance Fault
P0270 Cylinder no. 4 Injector Circuit Low
P0271 Cylinder no. 4 Injector Circuit High
P0272 Cylinder no. 4 Contribution/Balance Fault
P0273 Cylinder no. 5 Injector Circuit Low
P0274 Cylinder no. 5 Injector Circuit High
P0275 Cylinder no. 5 Contribution/Balance Fault
P0276 Cylinder no. 6 Injector Circuit Low
P0277 Cylinder no. 6 Injector Circuit High
P0278 Cylinder no. 6 Contribution/Balance Fault
P0279 Cylinder no. 7 Injector Circuit Low
P0280 Cylinder no. 7 Injector Circuit High
P0281 Cylinder no. 7 Contribution/Balance Fault
P0282 Cylinder no. 8 Injector Circuit Low
P0283 Cylinder no. 8 Injector Circuit High
P0284 Cylinder no. 8 Contribution/Balance Fault
P0285 Cylinder no. 9 Injector Circuit Low
P0286 Cylinder no. 9 Injector Circuit High

P0287 Cylinder no. 9 Contribution/Balance Fault
P0288 Cylinder no. 10 Injector Circuit Low
P0289 Cylinder no. 10 Injector Circuit High
P0290 Cylinder no. 10 Contribution/Balance Fault
P0291 Cylinder no. 11 Injector Circuit Low
P0292 Cylinder no. 11 Injector Circuit High
P0293 Cylinder no. 11 Contribution/Balance Fault
P0294 Cylinder no. 12 Injector Circuit Low
P0295 Cylinder no. 12 Injector Circuit High
P0296 Cylinder no. 12 Contribution/Balance Fault
P0300 Random/Multiple Cylinder Misfire Detected
P0301 Cylinder no. 1—Misfire Detected
P0302 Cylinder no. 2—Misfire Detected
P0303 Cylinder no. 3—Misfire Detected
P0304 Cylinder no. 4—Misfire Detected
P0305 Cylinder no. 5—Misfire Detected
P0306 Cylinder no. 6—Misfire Detected
P0307 Cylinder no. 7—Misfire Detected
P0308 Cylinder no. 8—Misfire Detected
P0320 Ignition/Distributor Engine Speed Input Circuit Malfunction
P0321 Ignition/Distributor Engine Speed Input Circuit Range/Performance
P0322 Ignition/Distributor Engine Speed Input Circuit No Signal
P0323 Ignition/Distributor Engine Speed Input Circuit Intermittent
P0325 Knock Sensor no. 1—Circuit Malfunction (Bank no. 1 or Single Sensor)
P0326 Knock Sensor no. 1—Circuit Range/Performance (Bank no. 1 or Single Sensor)
P0327 Knock Sensor no. 1—Circuit Low Input (Bank no. 1 or Single Sensor)
P0328 Knock Sensor no. 1—Circuit High Input (Bank no. 1 or Single Sensor)
P0329 Knock Sensor no. 1—Circuit Input Intermittent (Bank no. 1 or Single Sensor)
P0330 Knock Sensor no. 2—Circuit Malfunction (Bank no. 2)
P0331 Knock Sensor no. 2—Circuit Range/Performance (Bank no. 2)
P0332 Knock Sensor no. 2—Circuit Low Input (Bank no. 2)
P0333 Knock Sensor no. 2—Circuit High Input (Bank no. 2)
P0334 Knock Sensor no. 2—Circuit Input Intermittent (Bank no. 2)
P0335 Crankshaft Position Sensor "A" Circuit Malfunction
P0336 Crankshaft Position Sensor "A" Circuit Range/Performance
P0337 Crankshaft Position Sensor "A" Circuit Low Input
P0338 Crankshaft Position Sensor "A" Circuit High Input
P0339 Crankshaft Position Sensor "A" Circuit Intermittent
P0340 Camshaft Position Sensor Circuit Malfunction
P0341 Camshaft Position Sensor Circuit Range/Performance
P0342 Camshaft Position Sensor Circuit Low Input
P0343 Camshaft Position Sensor Circuit High Input
P0344 Camshaft Position Sensor Circuit Intermittent
P0350 Ignition Coil Primary/Secondary Circuit Malfunction
P0351 Ignition Coil "A" Primary/Secondary Circuit Malfunction
P0352 Ignition Coil "B" Primary/Secondary Circuit Malfunction
P0353 Ignition Coil "C" Primary/Secondary Circuit Malfunction
P0354 Ignition Coil "D" Primary/Secondary Circuit Malfunction
P0355 Ignition Coil "E" Primary/Secondary Circuit Malfunction
P0356 Ignition Coil "F" Primary/Secondary Circuit Malfunction
P0357 Ignition Coil "G" Primary/Secondary Circuit Malfunction

P0358 Ignition Coil "H" Primary/Secondary Circuit Malfunction
P0359 Ignition Coil "I" Primary/Secondary Circuit Malfunction
P0360 Ignition Coil "J" Primary/Secondary Circuit Malfunction
P0361 Ignition Coil "K" Primary/Secondary Circuit Malfunction
P0362 Ignition Coil "L" Primary/Secondary Circuit Malfunction
P0370 Timing Reference High Resolution Signal "A" Malfunction
P0371 Timing Reference High Resolution Signal "A" Too Many Pulses
P0372 Timing Reference High Resolution Signal "A" Too Few Pulses
P0373 Timing Reference High Resolution Signal "A" Intermittent/Erratic Pulses
P0374 Timing Reference High Resolution Signal "A" No Pulses
P0375 Timing Reference High Resolution Signal "B" Malfunction
P0376 Timing Reference High Resolution Signal "B" Too Many Pulses
P0377 Timing Reference High Resolution Signal "B" Too Few Pulses
P0378 Timing Reference High Resolution Signal "B" Intermittent/Erratic Pulses
P0379 Timing Reference High Resolution Signal "B" No Pulses
P0380 Glow Plug/Heater Circuit "A" Malfunction
P0381 Glow Plug/Heater Indicator Circuit Malfunction
P0382 Glow Plug/Heater Circuit "B" Malfunction
P0385 Crankshaft Position Sensor "B" Circuit Malfunction
P0386 Crankshaft Position Sensor "B" Circuit Range/Performance
P0387 Crankshaft Position Sensor "B" Circuit Low Input
P0388 Crankshaft Position Sensor "B" Circuit High Input
P0389 Crankshaft Position Sensor "B" Circuit Intermittent
P0400 Exhaust Gas Recirculation Flow Malfunction
P0401 Exhaust Gas Recirculation Flow Insufficient Detected
P0402 Exhaust Gas Recirculation Flow Excessive Detected
P0403 Exhaust Gas Recirculation Circuit Malfunction
P0404 Exhaust Gas Recirculation Circuit Range/Performance
P0405 Exhaust Gas Recirculation Sensor "A" Circuit Low
P0406 Exhaust Gas Recirculation Sensor "A" Circuit High
P0407 Exhaust Gas Recirculation Sensor "B" Circuit Low
P0408 Exhaust Gas Recirculation Sensor "B" Circuit High
P0410 Secondary Air Injection System Malfunction
P0411 Secondary Air Injection System Incorrect Flow Detected
P0412 Secondary Air Injection System Switching Valve "A" Circuit Malfunction
P0413 Secondary Air Injection System Switching Valve "A" Circuit Open
P0414 Secondary Air Injection System Switching Valve "A" Circuit Shorted
P0415 Secondary Air Injection System Switching Valve "B" Circuit Malfunction
P0416 Secondary Air Injection System Switching Valve "B" Circuit Open
P0417 Secondary Air Injection System Switching Valve "B" Circuit Shorted
P0418 Secondary Air Injection System Relay "A" Circuit Malfunction
P0419 Secondary Air Injection System Relay "B" Circuit Malfunction
P0420 Catalyst System Efficiency Below Threshold (Bank no. 1)
P0421 Warm Up Catalyst Efficiency Below Threshold (Bank no. 1)

P0422 Main Catalyst Efficiency Below Threshold (Bank no. 1)
P0423 Heated Catalyst Efficiency Below Threshold (Bank no. 1)
P0424 Heated Catalyst Temperature Below Threshold (Bank no. 1)
P0430 Catalyst System Efficiency Below Threshold (Bank no. 2)
P0431 Warm Up Catalyst Efficiency Below Threshold (Bank no. 2)
P0432 Main Catalyst Efficiency Below Threshold (Bank no. 2)
P0433 Heated Catalyst Efficiency Below Threshold (Bank no. 2)
P0434 Heated Catalyst Temperature Below Threshold (Bank no. 2)
P0440 Evaporative Emission Control System Malfunction
P0441 Evaporative Emission Control System Incorrect Purge Flow
P0442 Evaporative Emission Control System Leak Detected (Small Leak)
P0443 Evaporative Emission Control System Purge Control Valve Circuit Malfunction
P0444 Evaporative Emission Control System Purge Control Valve Circuit Open
P0445 Evaporative Emission Control System Purge Control Valve Circuit Shorted
P0446 Evaporative Emission Control System Vent Control Circuit Malfunction
P0447 Evaporative Emission Control System Vent Control Circuit Open
P0448 Evaporative Emission Control System Vent Control Circuit Shorted
P0449 Evaporative Emission Control System Vent Valve/Solenoid Circuit Malfunction
P0450 Evaporative Emission Control System Pressure Sensor Malfunction
P0451 Evaporative Emission Control System Pressure Sensor Range/Performance
P0452 Evaporative Emission Control System Pressure Sensor Low Input
P0453 Evaporative Emission Control System Pressure Sensor High Input
P0454 Evaporative Emission Control System Pressure Sensor Intermittent
P0455 Evaporative Emission Control System Leak Detected (Gross Leak)
P0460 Fuel Level Sensor Circuit Malfunction
P0461 Fuel Level Sensor Circuit Range/Performance
P0462 Fuel Level Sensor Circuit Low Input
P0463 Fuel Level Sensor Circuit High Input
P0464 Fuel Level Sensor Circuit Intermittent
P0465 Purge Flow Sensor Circuit Malfunction
P0466 Purge Flow Sensor Circuit Range/Performance
P0467 Purge Flow Sensor Circuit Low Input
P0468 Purge Flow Sensor Circuit High Input
P0469 Purge Flow Sensor Circuit Intermittent
P0470 Exhaust Pressure Sensor Malfunction
P0471 Exhaust Pressure Sensor Range/Performance
P0472 Exhaust Pressure Sensor Low
P0473 Exhaust Pressure Sensor High
P0474 Exhaust Pressure Sensor Intermittent
P0475 Exhaust Pressure Control Valve Malfunction
P0476 Exhaust Pressure Control Valve Range/Performance
P0477 Exhaust Pressure Control Valve Low
P0478 Exhaust Pressure Control Valve High
P0479 Exhaust Pressure Control Valve Intermittent

P0480 Cooling Fan no. 1 Control Circuit Malfunction
P0481 Cooling Fan no. 2 Control Circuit Malfunction
P0482 Cooling Fan no. 3 Control Circuit Malfunction
P0483 Cooling Fan Rationality Check Malfunction
P0484 Cooling Fan Circuit Over Current
P0485 Cooling Fan Power/Ground Circuit Malfunction
P0500 Vehicle Speed Sensor Malfunction
P0501 Vehicle Speed Sensor Range/Performance
P0502 Vehicle Speed Sensor Circuit Low Input
P0503 Vehicle Speed Sensor Intermittent/Erratic/High
P0505 Idle Control System Malfunction
P0506 Idle Control System RPM Lower Than Expected
P0507 Idle Control System RPM Higher Than Expected
P0510 Closed Throttle Position Switch Malfunction
P0520 Engine Oil Pressure Sensor/Switch Circuit Malfunction
P0521 Engine Oil Pressure Sensor/Switch Range/Performance
P0522 Engine Oil Pressure Sensor/Switch Low Voltage
P0523 Engine Oil Pressure Sensor/Switch High Voltage
P0530 A/C Refrigerant Pressure Sensor Circuit Malfunction
P0531 A/C Refrigerant Pressure Sensor Circuit Range/Performance
P0532 A/C Refrigerant Pressure Sensor Circuit Low Input
P0533 A/C Refrigerant Pressure Sensor Circuit High Input
P0534 A/C Refrigerant Charge Loss
P0550 Power Steering Pressure Sensor Circuit Malfunction
P0551 Power Steering Pressure Sensor Circuit Range/Performance
P0552 Power Steering Pressure Sensor Circuit Low Input
P0553 Power Steering Pressure Sensor Circuit High Input
P0554 Power Steering Pressure Sensor Circuit Intermittent
P0560 System Voltage Malfunction
P0561 System Voltage Unstable
P0562 System Voltage Low
P0563 System Voltage High
P0565 Cruise Control On Signal Malfunction
P0566 Cruise Control Off Signal Malfunction
P0567 Cruise Control Resume Signal Malfunction
P0568 Cruise Control Set Signal Malfunction
P0569 Cruise Control Coast Signal Malfunction
P0570 Cruise Control Accel Signal Malfunction
P0571 Cruise Control/Brake Switch "A" Circuit Malfunction
P0572 Cruise Control/Brake Switch "A" Circuit Low
P0573 Cruise Control/Brake Switch "A" Circuit High
P0574 Through P0580 Reserved for Cruise Codes
P0600 Serial Communication Link Malfunction
P0601 Internal Control Module Memory Check Sum Error
P0602 Control Module Programming Error
P0603 Internal Control Module Keep Alive Memory (KAM) Error
P0604 Internal Control Module Random Access Memory (RAM) Error
P0605 Internal Control Module Read Only Memory (ROM) Error
P0606 PCM Processor Fault
P0608 Control Module VSS Output "A" Malfunction
P0609 Control Module VSS Output "B" Malfunction
P0620 Generator Control Circuit Malfunction
P0621 Generator Lamp "L" Control Circuit Malfunction
P0622 Generator Field "F" Control Circuit Malfunction
P0650 Malfunction Indicator Lamp (MIL) Control Circuit Malfunction
P0654 Engine RPM Output Circuit Malfunction
P0655 Engine Hot Lamp Output Control Circuit Malfunction
P0656 Fuel Level Output Circuit Malfunction

P0700 Transmission Control System Malfunction
P0701 Transmission Control System Range/Performance
P0702 Transmission Control System Electrical
P0703 Torque Converter/Brake Switch "B" Circuit Malfunction
P0704 Clutch Switch Input Circuit Malfunction
P0705 Transmission Range Sensor Circuit Malfunction (PRNDL Input)
P0706 Transmission Range Sensor Circuit Range/Performance
P0707 Transmission Range Sensor Circuit Low Input
P0708 Transmission Range Sensor Circuit High Input
P0709 Transmission Range Sensor Circuit Intermittent
P0710 Transmission Fluid Temperature Sensor Circuit Malfunction
P0711 Transmission Fluid Temperature Sensor Circuit Range/Performance
P0712 Transmission Fluid Temperature Sensor Circuit Low Input
P0713 Transmission Fluid Temperature Sensor Circuit High Input
P0714 Transmission Fluid Temperature Sensor Circuit Intermittent
P0715 Input/Turbine Speed Sensor Circuit Malfunction
P0716 Input/Turbine Speed Sensor Circuit Range/Performance
P0717 Input/Turbine Speed Sensor Circuit No Signal
P0718 Input/Turbine Speed Sensor Circuit Intermittent
P0719 Torque Converter/Brake Switch "B" Circuit Low
P0720 Output Speed Sensor Circuit Malfunction
P0721 Output Speed Sensor Circuit Range/Performance
P0722 Output Speed Sensor Circuit No Signal
P0723 Output Speed Sensor Circuit Intermittent
P0724 Torque Converter/Brake Switch "B" Circuit High
P0725 Engine Speed Input Circuit Malfunction
P0726 Engine Speed Input Circuit Range/Performance
P0727 Engine Speed Input Circuit No Signal
P0728 Engine Speed Input Circuit Intermittent
P0730 Incorrect Gear Ratio
P0731 Gear no. 1 Incorrect Ratio
P0732 Gear no. 2 Incorrect Ratio
P0733 Gear no. 3 Incorrect Ratio
P0734 Gear no. 4 Incorrect Ratio
P0735 Gear no. 5 Incorrect Ratio
P0736 Reverse Incorrect Ratio
P0740 Torque Converter Clutch Circuit Malfunction
P0741 Torque Converter Clutch Circuit Performance or Stuck Off
P0742 Torque Converter Clutch Circuit Stuck On
P0743 Torque Converter Clutch Circuit Electrical
P0744 Torque Converter Clutch Circuit Intermittent
P0745 Pressure Control Solenoid Malfunction
P0746 Pressure Control Solenoid Performance or Stuck Off
P0747 Pressure Control Solenoid Stuck On
P0748 Pressure Control Solenoid Electrical
P0749 Pressure Control Solenoid Intermittent
P0750 Shift Solenoid "A" Malfunction
P0751 Shift Solenoid "A" Performance or Stuck Off
P0752 Shift Solenoid "A" Stuck On
P0753 Shift Solenoid "A" Electrical
P0754 Shift Solenoid "A" Intermittent
P0755 Shift Solenoid "B" Malfunction
P0756 Shift Solenoid "B" Performance or Stuck Off
P0757 Shift Solenoid "B" Stuck On
P0758 Shift Solenoid "B" Electrical

P0759 Shift Solenoid "B" Intermittent
P0760 Shift Solenoid "C" Malfunction
P0761 Shift Solenoid "C" Performance Or Stuck Off
P0762 Shift Solenoid "C" Stuck On
P0763 Shift Solenoid "C" Electrical
P0764 Shift Solenoid "C" Intermittent
P0765 Shift Solenoid "D" Malfunction
P0766 Shift Solenoid "D" Performance Or Stuck Off
P0767 Shift Solenoid "D" Stuck On
P0768 Shift Solenoid "D" Electrical
P0769 Shift Solenoid "D" Intermittent
P0770 Shift Solenoid "E" Malfunction
P0771 Shift Solenoid "E" Performance Or Stuck Off
P0772 Shift Solenoid "E" Stuck On
P0773 Shift Solenoid "E" Electrical
P0774 Shift Solenoid "E" Intermittent
P0780 Shift Malfunction
P0781 1–2 Shift Malfunction
P0782 2–3 Shift Malfunction
P0783 3–4 Shift Malfunction
P0784 4–5 Shift Malfunction
P0785 Shift/Timing Solenoid Malfunction
P0786 Shift/Timing Solenoid Range/Performance
P0787 Shift/Timing Solenoid Low
P0788 Shift/Timing Solenoid High
P0789 Shift/Timing Solenoid Intermittent
P0790 Normal/Performance Switch Circuit Malfunction
P0801 Reverse Inhibit Control Circuit Malfunction
P0803 1–4 Upshift (Skip Shift) Solenoid Control Circuit Malfunction
P0804 1–4 Upshift (Skip Shift) Lamp Control Circuit Malfunction
P1100 Barometric Pressure Sensor Circuit Fault
P1200 Fuel Pump Relay Circuit Fault
P1300 Igniter Circuit Fault (Bank 1)
P1305 Igniter Circuit Fault (Bank 2)
P1335 Crankshaft Position Sensor Circuit Fault
P1400 Sub-Throttle Position Sensor Circuit Fault
P1401 Sub-Throttle Position Sensor Performance
P1500 Starter Signal Circuit Fault
P1510 Air Volume Too Low With Supercharger On
P1600 PCM Battery Back-up Circuit Fault
P1605 Knock Control CPU Fault
P1700 Vehicle Speed Sensor Circuit Fault
P1705 Direct Clutch Speed Sensor Circuit Fault
P1765 Linear Shift Solenoid Circuit Fault
P1780 Park Neutral Position Switch Fault

Volkswagen

READING CODES

Reading the control module memory is one of the first steps in OBD II system diagnostics. This step should be initially performed to determine the general nature of the fault. Subsequent readings will determine if the fault has been cleared.

Reading codes can be performed by any of the methods below:
• Read the control module memory with the Generic Scan Tool (GST)
• Read the control module memory with the Vag 1550 tester

To read the fault codes, connect the scan tool or tester according to the manufacturer's instructions. Follow the manufacturer's specified procedure for reading the codes.

CLEARING CODES

Control module reset procedures are a very important part of OBD II System diagnostics. This step should be done at the end of any fault code repair and at the end of any driveability repair.

Clearing codes can be performed by any of the methods below:
• Clear the control module memory with the Generic Scan Tool (GST)
• Clear the control module memory with the vehicle manufacturer's specific tester
• Turn the ignition **OFF** and remove the negative battery cable for at least 1 minute.

Removing the negative battery cable may cause other systems in the vehicle to loose their memory. Prior to removing the cable, ensure you have the proper reset codes for radios and alarms.

➡**The MIL will may also be de-activated for some codes if the vehicle completes three consecutive trips without a fault detected with vehicle conditions similar to those present during the fault.**

OBD II TROUBLE CODES

P0102 Mass or Volume Air Flow Circuit Low Input
P0103 Mass or Volume Air Flow Circuit High Input
P0107 Manifold Absolute Pressure or Barometric Pressure Low Input
P0108 Manifold Absolute Pressure or Barometric Pressure High Input
P0112 Intake Air Temperature Circuit Low Input
P0113 Intake Air Temperature Circuit High Input
P0116 Engine Coolant Temperature Circuit Range/Performance
P0117 Engine Coolant Temperature Circuit Low Input
P0118 Engine Coolant Temperature Circuit High Input
P0120 Throttle Position Sensor A Circuit Malfunction
P0121 Throttle/Pedal Position Sensor A Circuit Range/Performance
P0122 Throttle/Pedal Position Sensor A Circuit Low Input**P0123** Throttle/Pedal Position Sensor A Circuit High Input
P0125 Insufficient Coolant Temperature For Closed Loop Fuel Control
P0130 Oxygen Sensor Circuit, Bank 1-Sensor 1 Malfunction**P0131** Oxygen Sensor Circuit, Bank 1-Sensor 1 Low Voltage**P0132** Oxygen Sensor Circuit, Bank 1-Sensor 1 High Voltage**P0133** Oxygen Sensor Circuit, Bank 1-Sensor 1 Slow Response
P0134 Oxygen Sensor Circuit, Bank 1-Sensor 1 No Activity Detected
P0135 Oxygen Sensor Heater Circuit, Bank 1-Sensor 1 Malfunction
P0136 Oxygen Sensor Circuit, Bank 1-Sensor 2 Malfunction
P0137 Oxygen Sensor Circuit, Bank 1-Sensor 2 Low Voltage
P0138 Oxygen Sensor Circuit, Bank 1-Sensor 2 High Voltage
P0140 Oxygen Sensor Circuit, Bank 1-Sensor 2 No Activity Detected
P0141 Oxygen Sensor Heater Circuit, Bank 1-Sensor 2 Malfunction

P0150 Oxygen Sensor Circuit, Bank 2-Sensor 1 Malfunction
P0151 Oxygen Sensor Circuit, Bank 2-Sensor 1 Low Voltage
P0152 Oxygen Sensor Circuit, Bank 2-Sensor 1 High Voltage
P0153 Oxygen Sensor Circuit, Bank 2-Sensor 1 Slow Response
P0154 Oxygen Sensor Circuit, Bank 2-Sensor 1 No Activity Detected
P0156 Oxygen Sensor Circuit, Bank 2-Sensor 2 Malfunction
P0157 Oxygen Sensor Circuit, Bank 2-Sensor 2 Low Voltage
P0158 Oxygen Sensor Circuit, Bank 2-Sensor 2 High Voltage
P0160 Oxygen Sensor Circuit, Bank 2-Sensor 2 No Activity Detected
P0171 System Too Lean, Bank 1
P0172 System Too Rich, Bank 1
P0300 Random Multiple Misfire Detected
P0301 Cylinder 1 Misfire Detected
P0302 Cylinder 2 Misfire Detected
P0303 Cylinder 3 Misfire Detected
P0304 Cylinder 4 Misfire Detected
P0305 Cylinder 5 Misfire Detected
P0306 Cylinder 6 Misfire Detected
P0321 Ignition Distributor Engine Speed Input Circuit Range/Performance
P0322 Ignition /Distributor Engine Speed Input Circuit No Signal
P0327 Knock sensor 1 Circuit Low Input
P0328 Knock Sensor Circuit, High Input
P0332 Knock sensor 2 Circuit Low Input
P0332 Knock Sensor Circuit, Low Input
P0333 Knock Sensor 2 Circuit, High Input
P0341 Camshaft Position Sensor Circuit Range/Performance
P0411 Secondary Air Injection System Incorrect Flow Detected
P0422 Main Catalyst Efficiency Below Threshold (Bank 1)
P0422 Main Catalyst, Bank I Efficiency Below Threshold
P0440 Evaporative Emission Control System Malfunction
P0441 EVAP Emission Contr. Sys. Incorrect Purge Flow
P0442 EVAP Emission Contr. Sys. (Small Leak) Leak Detected
P0455 EVAP Emission Contr. Sys. (Gross Leak) Leak Detected
P0501 Vehicle Speed Sensor Range/Performance
P0506 Idle Control System RPM Lower Than Expected
P0507 Idle Control System RPM Higher Than Expected
P0510 Closed Throttle Position Switch Malfunction
P0560 System Voltage Malfunction
P0562 System Voltage Low Voltage
P0563 System Voltage High Voltage
P0601 Internal Contr. Module Memory Check Sum Error
P0604 Internal Contr. Module Random Access Memory (RAM) Error
P0605 Internal Control Module Read Only Memory (ROM) Error
P0707 Transmission Range Sensor Circuit Low Input
P0708 Transmission Range Sensor Circuit High Input
P0715 Input Turbine Speed Sensor Circuit Malfunction
P0722 Output Speed Sensor Circuit No Signal
P0725 Engine Speed Input Circuit Malfunction
P0748 Pressure Control Solenoid Electrical
P0753 Shift Solenoid A Electrical
P0758 Shift Solenoid B Electrical
P0763 Shift Solenoid C Electrical
P0768 Shift Solenoid D Electrical
P0773 Shift Solenoid E Electrical
P0300 Random/Multiple Cylinder Misfire Detected
P0301 Cylinder I Misfire Detected
P0302 Cylinder 2 Misfire Detected

P1102 Oxygen Sensor Heating Circuit, Bank 1-Sensor 1 Short to B+
P1105 Oxygen Sensor Heating Circuit, Bank 1-Sensor 2 Short to B+
P1107 Oxygen Sensor Heating Circuit, Bank 2-Sensor I Short to B+
P1110 Oxygen Sensor Heating Circuit, Bank 2-Sensor 2 Short to B+
P1127 Long Term Fuel Trim Multiplicative, Bank 1 System Too Rich
P1128 Long Term Fuel Trim Multiplicative, Bank 1 System Too Lean
P1129 Long Term Fuel Trim Multiplicative, Bank2 System too Rich
P1130 Long Term Fuel Trim Multiplicative, Bank2 System too Lean
P1136 Long Term Fuel Trim Additive, Bank 1 System Too Lean
P1137 Long Term Fuel Trim Additive, Bank 1 System Too Rich
P1138 Long Term Fuel Trim Additive Fuel, Bank I System too Lean
P1139 Long Term Fuel Trim Additive Fuel, Bank 1 System too Rich
P1141 Load Calculation Cross Check Range/Performance
P1176 Oxygen Correction Behind Catalyst, B1 Limit Attained
P1177 Oxygen Correction Behind Catalyst. 82 Limit Attained
P1196 Oxygen Sensor Heater Circuit, Bank 1-Sensor 1 Electrical Malfunction
P1197 Oxygen Sensor Heater Circuit, Bank 2-Sensor I Electrical Malfunction
P1198 Oxygen Sensor Heater Circuit, BankI-Sensor2 Electrical Malfunction
P1198 Oxygen Sensor Heater Circuit, Bank 1-Sensor 2 Electrical Malfunction
P1199 Oxygen Sensor Heater Circuit, Bank 2-Sensor 2 Electrical Malfunction
P1201 Cylinder 1, Fuel Injection Circuit Electrical Malfunction
P1202 Cylinder 2, Fuel Injection Circuit Electrical Malfunction
P1203 Cylinder 3, Fuel Injection Circuit Electrical Malfunction
P1204 Cylinder 4, Fuel Injection Circuit Electrical Malfunction
P1205 Cylinder 5, Fuel Injection Circuit Electrical Malfunction
P1206 Cylinder 6, Fuel Injection Circuit Electrical Malfunction
P1207 Cylinder 7, Fuel Injection Circuit Electrical Malfunction
P1208 Cylinder 8, Fuel Injection Circuit Electrical Malfunction
P1213 Cylinder I-Fuel Injection Circuit Short to B+
P1214 Cylinder 2-Fuel Injection Circuit Short to B+
P1215 Cylinder 3 Fuel Injection Circuit Short to B+
P1216 Cylinder 4 Fuel Injection Circuit Short to B+
P1217 Cylinder 5 Fuel Injection Circuit Short to B+
P1218 Cylinder 6 Fuel Injection Circuit Short to B+
P1219 Cylinder 7, Fuel Injection Circuit Short to B+
P1219 Cylinder 8, Fuel Injection Circuit Short to B+
P1225 Cylinder I Fuel Injection Circuit Short to Ground
P1226 Cylinder 2 Fuel Injection Circuit Short to Ground
P1227 Cylinder 3 Fuel Injection Circuit Short to Ground
P1228 Cylinder 4 Fuel Injection Circuit Short to Ground
P1229 Cylinder 5 Fuel Injection Circuit Short to Ground
P1230 Cylinder 6 Fuel Injection Circuit Short to Ground
P1237 Cylinder I Fuel Injection Circuit Open Circuit
P1238 Cylinder 2 Fuel Injection Circuit Open Circuit
P1239 Cylinder 3 Fuel Injection Circuit Open Circuit
P1240 Cylinder 4 Fuel Injection Circuit Open Circuit
P1241 Cylinder 5 Fuel Injection Circuit Open Circuit

P1242 Cylinder 6 Fuel Injection Circuit Open Circuit
P1250 Fuel Level Too Low
P1280 Fuel Injection Air Control Valve Circuit Flow too Low
P1283 Fuel Injection Air Control Valve Circuit Electrical Malfunction
P1325 Cylinder I Knock Control Limit Attained
P1326 Cylinder 2 Knock Control Limit Attained
P1327 Cylinder 3 Knock Control Limit Attained
P1328 Cylinder 4 Knock Control Limit Attained
P1329 Cylinder 5 Knock Control Limit Attained
P1330 Cylinder 6 Knock Control Limit Attained
P1331 Cylinder 7, Knock Control Limit Attained
P1332 Cylinder 8, Knock Control Limit Attained
P1337 Camshaft Position Sensor, Bank 1 Short to Ground
P1338 Camshaft Position Sensor, Bank 1 Open Circuit/Short to B+
P1340 Boost Pressure Control Valve Short to B+
P1386 Internal Control Module Knock Control Circuit Error
P1391 Camshaft Position Sensor, Bank 2 Short to Ground
P1392 Camshaft Position Sensor, Bank 2 Open Circuit/Short to B+
P1410 Tank Ventilation Valve Circuit Short to B+
P1420 Secondary Air Injection Module Short To B+
P1421 Secondary Air Injection Module Short To Ground
P1421 Secondary Air Injection Valve Circuit Short to Ground
P1422 Secondary Air Injection System Control Valve Circuit Short To B+
P1422 Secondary Air Injection Valve Circuit Short to B+
P1425 Tank Vent Valve Short To Ground
P1425 Tank Vent Valve Short to Ground
P1426 Tank Vent Valve Open
P1426 Tank Vent Valve Open
P1432 Secondary Air Injection Valve Open
P1433 Secondary Air Injection System Pump Relay Circuit Open
P1434 Secondary Air Injection System Pump Relay Circuit Short to B+
P1435 Secondary Air Injection System Pump Relay Circuit Short to Ground
P1436 Secondary Air Injection System Pump Relay Circuit Electrical Malfunction
P1450 Secondary Air Injection System Circuit Short To B+
P1451 Secondary Air Injection System Circuit Short To Ground
P1452 Secondary Air Injection System Open Circuit
P1471 EVAP Emission Control LDP Circuit Short to B+
P1472 EVAP Emission Control LDP Circuit Short to Ground
P1473 EVAP Emission Control LDP Circuit Open Circuit
P1475 EVAP Emission Control LDP Circuit Malfunction/Signal Circuit Open
P1476 EVAP Emission Control LDP Circuit Malfunction/Insufficient Vacuum
P1477 EVAP Emission Control LDP Circuit Malfunction
P1500 Fuel Pump Relay Circuit Electrical Malfunction
P1501 Fuel Pump Relay Circuit Short to Ground
P1502 Fuel Pump Relay Circuit Short to B+
P1505 Closed Throttle Position Switch Does Not Close/Open Circuit
P1506 Closed Throttle Position Switch Does Not Open/Short to Ground
P1507 Idle System Learned Value Lower Limit Attained
P1508 Idle System Learned Value Upper Limit Attained
P1512 Intake Manifold Changeover Valve Circuit, Short to B+

P1515 Intake Manifold Changeover Valve Circuit, Short to Ground
P1516 Intake Manifold Changeover Valve Circuit, Open
P1519 Intake Camshaft Control, Bank I Malfunction
P1522 Intake Camshaft Control, Bank 2 Malfunction
P1543 Throttle Actuation Potentiometer Signal Too Low
P1544 Throttle Actuation Potentiometer Signal Too High
P1545 Throttle Position Control Malfunction
P1547 Boost Pressure Control Valve Short to Ground
P1548 Boost Pressure Control Valve Open
P1555 Charge Pressure Upper Limit Exceeded
P1556 Charge Pressure Negative Deviation
P1557 Charge Pressure Positive Deviation
P1558 Throttle Actuator Electrical Malfunction
P1559 Idle Speed Control Throttle Position Adaptation Malfunction
P1560 Maximum Engine Speed Exceeded
P1564 Idle Speed Control, Throttle Position Low Voltage During Adaptation
P1580 Throttle Actuator (B1) Malfunction
P1582 Idle Adaptation At Limit
P1602 Power Supply (B+) Terminal 30 Low Voltage
P1606 Rough Road Spec Engine Torque ABS-ECU Electrical Malfunction
P1611 MIL Call-up Circuit/Transmission Control Module Short to Ground
P1612 Electronic Control Module Incorrect Coding
P1613 MIL Call-up Circuit Open/Short to B+
P1624 MIL Request Signal Active
P1625 CAN-Bus Implausible Message from Transmission Control
P1626 CAN-Bus Missing Message from Transmission Control
P1640 Internal Control Module (EEPROM) Error
P1640 Internal Control Module (EEPROM) Error
P1681 Control Unit Programming not Finished
P1690 Malfunction Indicator Light Malfunction
P1693 Malfunction Indicator Light Short to B+
P1778 Solenoid EV7 Electrical Malfunction
P1780 Engine Intervention Readable

OBD II TROUBLE CODE EQUIVALENTS

16486 Mass or Volume Air Flow Circuit Low Input
16487 Mass or Volume Air Flow Circuit High Input
16491 Manifold Absolute Pressure or Barometric Pressure Low Input
16492 Manifold Absolute Pressure or Barometric Pressure High Input
16496 Intake Air Temperature Circuit Low Input
16497 Intake Air Temperature Circuit High Input
16500 Engine Coolant Temperature Circuit Range/Performance
16501 Engine Coolant Temperature Circuit Low Input
16502 Oxygen Engine Coolant Temperature Circuit High Input
16504 Throttle Position Sensor A Circuit Malfunction
16505 Throttle/Pedal Position Sensor A Circuit Range/Performance
16506 Throttle/Pedal Position Sensor A Circuit Low Input
16507 Throttle/Pedal Position Sensor A Circuit High Input
16509 Insufficient Coolant Temperature For Closed Loop Fuel Control
16514 Oxygen Sensor Circuit, Bank 1-Sensor 1 Malfunction

16515 Oxygen Sensor Circuit, Bank 1-Sensor 1 Low Voltage

16516 Oxygen Sensor Circuit, Bank 1-Sensor 1 High Voltage

16517 Oxygen Sensor Circuit, Bank 1-Sensor 1 Slow Response

16518 Oxygen Sensor Circuit, Bank 1-Sensor 1 No Activity Detected

16519 Oxygen Sensor Heater Circuit, Bank 1-Sensor 1 Malfunction

16520 Oxygen Sensor Circuit, Bank 1-Sensor 2 Malfunction

16521 Oxygen Sensor Circuit, Bank 1-Sensor 2 Low Voltage

16522 Oxygen Sensor Circuit, Bank 1-Sensor 2 High Voltage

16524 Oxygen Sensor Circuit, Bank 1-Sensor 2 No Activity Detected

16525 Oxygen Sensor Heater Circuit, Bank 1-Sensor 2 Malfunction

16534 Oxygen Sensor Circuit, Bank 2-Sensor 1 Malfunction

16535 Oxygen Sensor Circuit, Bank 2-Sensor 1 Low Voltage

16536 Oxygen Sensor Circuit, Bank 2-Sensor 1 High Voltage

16537 Oxygen Sensor Circuit, Bank 2-Sensor 1 Slow Response

16538 Oxygen Sensor Circuit, Bank 2-Sensor 1 No Activity Detected

16540 Oxygen Sensor Circuit, Bank 2-Sensor 2 Malfunction

16541 Oxygen Sensor Circuit, Bank 2-Sensor 2 Low Voltage

16542 Oxygen Sensor Circuit, Bank 2-Sensor 2 High Voltage

16544 Oxygen Sensor Circuit, Bank 2-Sensor 2 No Activity Detected

16555 Oxygen System Too Lean, Bank 1

16556 Oxygen System Too Rich, Bank 1

16684 Random Multiple Misfire Detected

16685 Cylinder 1 Misfire Detected

16686 Cylinder 2 Misfire Detected

16687 Cylinder 3 Misfire Detected

16688 Cylinder 4 Misfire Detected

16689 Cylinder 5 Misfire Detected

16690 Cylinder 6 Misfire Detected

16705 Ignition Distributor Engine Speed Input Circuit Range/Performance

16706 Ignition /Distributor Engine Speed Input Circuit No Signal

16711 Knock sensor 1 Circuit Low Input

16712 Knock Sensor Circuit, High Input

16716 Knock sensor 2 Circuit Low Input

16716 Knock Sensor Circuit, Low Input

16717 Knock Sensor 2 Circuit, High Input

16725 Camshaft Position Sensor Circuit Range/Performance

16795 Secondary Air Injection System Incorrect Flow Detected

16806 Main Catalyst Efficiency Below Threshold (Bank 1)

16806 Main Catalyst, Bank I Efficiency Below Threshold

16824 Evaporative Emission Control System Malfunction

16825 EVAP Emission Contr. Sys. Incorrect Purge Flow

16826 EVAP Emission Contr. Sys. (Small Leak) Leak Detected

16839 EVAP Emission Contr. Sys. (Gross Leak) Leak Detected

16885 Vehicle Speed Sensor Range/Performance

16890 Idle Control System RPM Lower Than Expected

16891 Idle Control System RPM Higher Than Expected

16894 Closed Throttle Position Switch Malfunction

16944 System Voltage Malfunction

16946 System Voltage Low Voltage

16947 System Voltage High Voltage

16985 Internal Contr. Module Memory Check Sum Error

16988 Internal Contr. Module Random Access Memory (RAM) Error

16989 Internal Control Module Read Only Memory (ROM) Error

17091 Transmission Range Sensor Circuit Low Input

17092 Transmission Range Sensor Circuit High Input

17099 Input Turbine Speed Sensor Circuit Malfunction

17106 Output Speed Sensor Circuit No Signal

17109 Engine Speed Input Circuit Malfunction

17132 Pressure Control Solenoid Electrical

17137 Shift Solenoid A Electrical

17142 Shift Solenoid B Electrical

17147 Shift Solenoid C Electrical

17152 Shift Solenoid D Electrical

17157 Shift Solenoid E Electrical

16684 Random/Multiple Cylinder Misfire Detected

16685 Cylinder I Misfire Detected

16686 Cylinder 2 Misfire Detected

17510 Oxygen Sensor Heating Circuit, Bank 1-Sensor 1 Short to B+

17513 Oxygen Sensor Heating Circuit, Bank 1-Sensor 2 Short to B+

17515 Oxygen Sensor Heating Circuit, Bank 2-Sensor I Short to B+

17518 Oxygen Sensor Heating Circuit, Bank 2-Sensor 2 Short to B+

17535 Long Term Fuel Trim Multiplicative, Bank 1 System Too Rich

17536 Long Term Fuel Trim Multiplicative, Bank 1 System Too Lean

17537 Long Term Fuel Trim Multiplicative, Bank2 System too Rich

17538 Long Term Fuel Trim Multiplicative, Bank2 System too Lean

17544 Long Term Fuel Trim Additive, Bank 1 System Too Lean

17545 Long Term Fuel Trim Additive, Bank 1 System Too Rich

17546 Long Term Fuel Trim Additive Fuel, Bank I System too Lean

17547 Long Term Fuel Trim Additive Fuel, Bank 1 System too Rich

17549 Load Calculation Cross Check Range/Performance

17584 Oxygen Correction Behind Catalyst, B1 Limit Attained

17585 Oxygen Correction Behind Catalyst. 82 Limit Attained

17604 Oxygen Sensor Heater Circuit, Bank 1-Sensor 1 Electrical Malfunction

17605 Oxygen Sensor Heater Circuit, Bank 2-Sensor I Electrical Malfunction

17606 Oxygen Sensor Heater Circuit, Bankl-Sensor2 Electrical Malfunction

17606 Oxygen Sensor Heater Circuit, Bank 1-Sensor 2 Electrical Malfunction

17607 Oxygen Sensor Heater Circuit, Bank 2-Sensor 2 Electrical Malfunction

17609 Cylinder 1, Fuel Injection Circuit Electrical Malfunction

17610 Cylinder 2, Fuel Injection Circuit Electrical Malfunction

17611 Cylinder 3, Fuel Injection Circuit Electrical Malfunction

17612 Cylinder 4, Fuel Injection Circuit Electrical Malfunction

17613 Cylinder 5, Fuel Injection Circuit Electrical Malfunction

17614 Cylinder 6, Fuel Injection Circuit Electrical Malfunction

17615 Cylinder 7, Fuel Injection Circuit Electrical Malfunction

17616 Cylinder 8, Fuel Injection Circuit Electrical Malfunction

17621 Cylinder I-Fuel Injection Circuit Short to B+

17622 Cylinder 2-Fuel Injection Circuit Short to B+

17623 Cylinder 3 Fuel Injection Circuit Short to B+

17624 Cylinder 4 Fuel Injection Circuit Short to B+

17625 Cylinder 5 Fuel Injection Circuit Short to B+

17626 Cylinder 6 Fuel Injection Circuit Short to B+

17627 Cylinder 7, Fuel Injection Circuit Short to B+
17628 Cylinder 8, Fuel Injection Circuit Short to B+
17633 Cylinder I Fuel Injection Circuit Short to Ground
17634 Cylinder 2 Fuel Injection Circuit Short to Ground
17635 Cylinder 3 Fuel Injection Circuit Short to Ground
17636 Cylinder 4 Fuel Injection Circuit Short to Ground
17637 Cylinder 5 Fuel Injection Circuit Short to Ground
17638 Cylinder 6 Fuel Injection Circuit Short to Ground
17645 Cylinder I Fuel Injection Circuit Open Circuit
17646 Cylinder 2 Fuel Injection Circuit Open Circuit
17647 Cylinder 3 Fuel Injection Circuit Open Circuit
17648 Cylinder 4 Fuel Injection Circuit Open Circuit
17649 Cylinder 5 Fuel Injection Circuit Open Circuit
17650 Cylinder 6 Fuel Injection Circuit Open Circuit
17658 Fuel Level Too Low
17688 Fuel Injection Air Control Valve Circuit Flow too Low
17691 Fuel Injection Air Control Valve Circuit Electrical Malfunction
17733 Cylinder I Knock Control Limit Attained
17734 Cylinder 2 Knock Control Limit Attained
17735 Cylinder 3 Knock Control Limit Attained
17736 Cylinder 4 Knock Control Limit Attained
17737 Cylinder 5 Knock Control Limit Attained
17738 Cylinder 6 Knock Control Limit Attained
17739 Cylinder 7 Knock Control Limit Attained
17740 Cylinder 8 Knock Control Limit Attained
17745 Camshaft Position Sensor, Bank 1 Short to Ground
17746 Camshaft Position Sensor, Bank 1 Open Circuit/Short to B+
17954 Boost Pressure Control Valve Short to B+
17794 Internal Control Module Knock Control Circuit Error
17799 Camshaft Position Sensor, Bank 2 Short to Ground
17800 Camshaft Position Sensor, Bank 2 Open Circuit/Short to B+
17818 Tank Ventilation Valve Circuit Short to B+
17818 Tank Ventilation Valve Circuit Short to B+
17828 Secondary Air Injection Module Short To B+
17829 Secondary Air Injection Module Short To Ground
17829 Secondary Air Injection Valve Circuit Short to Ground
17830 Secondary Air Injection System Control Valve Circuit Short To B+
17830 Secondary Air Injection Valve Circuit Short to B+
17833 Tank Vent Valve Short To Ground
17833 Tank Vent Valve Short to Ground
17834 Tank Vent Valve Open
17834 Tank Vent Valve Open
17840 Secondary Air Injection Valve Open
17841 Secondary Air Injection System Pump Relay Circuit Open
17842 Secondary Air Injection System Pump Relay Circuit Short to B+
17843 Secondary Air Injection System Pump Relay Circuit Short to Ground
17844 Secondary Air Injection System Pump Relay Circuit Electrical Malfunction
17858 Secondary Air Injection System Circuit Short To B+
17859 Secondary Air Injection System Circuit Short To Ground
17860 Secondary Air Injection System Open Circuit
17879 EVAP Emission Control LDP Circuit Short to B+
17880 EVAP Emission Control LDP Circuit Short to Ground
17881 EVAP Emission Control LDP Circuit Open Circuit
17883 EVAP Emission Control LDP Circuit Malfunction/Signal Circuit Open

17884 EVAP Emission Control LDP Circuit Malfunction/Insufficient Vacuum
17885 EVAP Emission Control LDP Circuit Malfunction
17908 Fuel Pump Relay Circuit Electrical Malfunction
17909 Fuel Pump Relay Circuit Short to Ground
17910 Fuel Pump Relay Circuit Short to B+
17913 Closed Throttle Position Switch Does Not Close/Open Circuit
17914 Closed Throttle Position Switch Does Not Open/Short to Ground
17915 Idle System Learned Value Lower Limit Attained
17916 Idle System Learned Value Upper Limit Attained
17920 Intake Manifold Changeover Valve Circuit, Short to B+
17923 Intake Manifold Changeover Valve Circuit, Short to Ground
17924 Intake Manifold Changeover Valve Circuit, Open
17927 Intake Camshaft Control, Bank 2 Malfunction
17951 Throttle Actuation Potentiometer Signal Too Low
17952 Throttle Actuation Potentiometer Signal Too High
17953 Throttle Position Control Malfunction
17955 Boost Pressure Control Valve Short to Ground
17956 Boost Pressure Control Valve Open
17963 Charge Pressure Upper Limit Exceeded
717964 Charge Pressure Negative Deviation
17965 Charge Pressure Positive Deviation
17966 Throttle Actuator Electrical Malfunction
17967 Idle Speed Control Throttle Position Adaptation Malfunction
17968 Maximum Engine Speed Exceeded
17972 Idle Speed Control, Throttle Position Low Voltage During Adaptation
17988 Throttle Actuator (B1) Malfunction
17990 Idle Adaptation At Limit
18010 Power Supply (B+) Terminal 30 Low Voltage
18014 Rough Road Spec Engine Torque ABS-ECU Electrical Malfunction
18019 MIL Call-up Circuit/Transmission Control Module Short to Ground
18020 Electronic Control Module Incorrect Coding
18021 MIL Call-up Circuit Open/Short to B+
18032 MIL Request Signal Active
18033 CAN-Bus Implausible Message from Transmission Control
18034 CAN-Bus Missing Message from Transmission Control
18048 Internal Control Module (EEPROM) Error
18048 Internal Control Module (EEPROM) Error
18089 Control Unit Programming not Finished
18098 Malfunction Indicator Light Malfunction
18101 Malfunction Indicator Light Short to B+
18186 Solenoid EV7 Electrical Malfunction
18188 Engine Intervention Readable

Volvo

READING CODES

Reading the control module memory is on of the first steps in OBD II system diagnostics. This step should be initially performed to determine the general nature of the fault. Subsequent readings will determine if the fault has been cleared.

Reading codes can be performed by any of the methods below:

- Read the control module memory with the Generic Scan Tool (GST)
- Read the control module memory with the vehicle manufacturer's specific tester

To read the fault codes, connect the scan tool or tester according to the manufacturer's instructions. Follow the manufacturer's specified procedure for reading the codes.

CLEARING CODES

Control module reset procedures are a very important part of OBD II System diagnostics. This step should be done at the end of any fault code repair and at the end of any driveability repair.

Clearing codes can be performed by any of the methods below:

- Clear the control module memory with the Generic Scan Tool (GST)
- Clear the control module memory with the vehicle manufacturer's specific tester
- Turn the ignition **OFF** and remove the negative battery cable for at least 1 minute.

Removing the negative battery cable may cause other systems in the vehicle to loose their memory. Prior to removing the cable, ensure you have the proper reset codes for radios and alarms.

➡**The MIL will may also be de-activated for some codes if the vehicle completes three consecutive trips without a fault detected with vehicle conditions similar to those present during the fault.**

OBD II TROUBLE CODES

P0100 Mass or Volume Air Flow Circuit Malfunction
P0101 Mass or Volume Air Flow Circuit Range/Performance Problem
P0102 Mass or Volume Air Flow Circuit Low Input
P0103 Mass or Volume Air Flow Circuit High Input
P0104 Mass or Volume Air Flow Circuit Intermittent
P0105 Manifold Absolute Pressure/Barometric Pressure Circuit Malfunction
P0106 Manifold Absolute Pressure/Barometric Pressure Circuit Range/Performance Problem
P0107 Manifold Absolute Pressure/Barometric Pressure Circuit Low Input
P0108 Manifold Absolute Pressure/Barometric Pressure Circuit High Input
P0109 Manifold Absolute Pressure/Barometric Pressure Circuit Intermittent
P0110 Intake Air Temperature Circuit Malfunction
P0111 Intake Air Temperature Circuit Range/Performance Problem
P0112 Intake Air Temperature Circuit Low Input
P0113 Intake Air Temperature Circuit High Input
P0114 Intake Air Temperature Circuit Intermittent
P0115 Engine Coolant Temperature Circuit Malfunction
P0116 Engine Coolant Temperature Circuit Range/Performance Problem
P0117 Engine Coolant Temperature Circuit Low Input
P0118 Engine Coolant Temperature Circuit High Input
P0119 Engine Coolant Temperature Circuit Intermittent
P0120 Throttle/Pedal Position Sensor/Switch "A" Circuit Malfunction

P0121 Throttle/Pedal Position Sensor/Switch "A" Circuit Range/Performance Problem
P0122 Throttle/Pedal Position Sensor/Switch "A" Circuit Low Input
P0123 Throttle/Pedal Position Sensor/Switch "A" Circuit High Input
P0124 Throttle/Pedal Position Sensor/Switch "A" Circuit Intermittent
P0125 Insufficient Coolant Temperature For Closed Loop Fuel Control
P0126 Insufficient Coolant Temperature For Stable Operation
P0130 O2 Circuit Malfunction (Bank no. 1 Sensor no. 1)
P0131 O2 Sensor Circuit Low Voltage (Bank no. 1 Sensor no. 1)
P0132 O2 Sensor Circuit High Voltage (Bank no. 1 Sensor no. 1)
P0133 O2 Sensor Circuit Slow Response (Bank no. 1 Sensor no. 1)
P0134 O2 Sensor Circuit No Activity Detected (Bank no. 1 Sensor no. 1)
P0135 O2 Sensor Heater Circuit Malfunction (Bank no. 1 Sensor no. 1)
P0136 O2 Sensor Circuit Malfunction (Bank no. 1 Sensor no. 2)
P0137 O2 Sensor Circuit Low Voltage (Bank no. 1 Sensor no. 2)
P0138 O2 Sensor Circuit High Voltage (Bank no. 1 Sensor no. 2)
P0139 O2 Sensor Circuit Slow Response (Bank no. 1 Sensor no. 2)
P0140 O2 Sensor Circuit No Activity Detected (Bank no. 1 Sensor no. 2)
P0141 O2 Sensor Heater Circuit Malfunction (Bank no. 1 Sensor no. 2)
P0142 O2 Sensor Circuit Malfunction (Bank no. 1 Sensor no. 3)
P0143 O2 Sensor Circuit Low Voltage (Bank no. 1 Sensor no. 3)
P0144 O2 Sensor Circuit High Voltage (Bank no. 1 Sensor no. 3)
P0145 O2 Sensor Circuit Slow Response (Bank no. 1 Sensor no. 3)
P0146 O2 Sensor Circuit No Activity Detected (Bank no. 1 Sensor no. 3)
P0147 O2 Sensor Heater Circuit Malfunction (Bank no. 1 Sensor no. 3)
P0150 O2 Sensor Circuit Malfunction (Bank no. 2 Sensor no. 1)
P0151 O2 Sensor Circuit Low Voltage (Bank no. 2 Sensor no. 1)
P0152 O2 Sensor Circuit High Voltage (Bank no. 2 Sensor no. 1)
P0153 O2 Sensor Circuit Slow Response (Bank no. 2 Sensor no. 1)
P0154 O2 Sensor Circuit No Activity Detected (Bank no. 2 Sensor no. 1)
P0155 O2 Sensor Heater Circuit Malfunction (Bank no. 2 Sensor no. 1)
P0156 O2 Sensor Circuit Malfunction (Bank no. 2 Sensor no. 2)
P0157 O2 Sensor Circuit Low Voltage (Bank no. 2 Sensor no. 2)
P0158 O2 Sensor Circuit High Voltage (Bank no. 2 Sensor no. 2)
P0159 O2 Sensor Circuit Slow Response (Bank no. 2 Sensor no. 2)
P0160 O2 Sensor Circuit No Activity Detected (Bank no. 2 Sensor no. 2)
P0161 O2 Sensor Heater Circuit Malfunction (Bank no. 2 Sensor no. 2)
P0162 O2 Sensor Circuit Malfunction (Bank no. 2 Sensor no. 3)
P0163 O2 Sensor Circuit Low Voltage (Bank no. 2 Sensor no. 3)

P0164 O2 Sensor Circuit High Voltage (Bank no. 2 Sensor no. 3)

P0165 O2 Sensor Circuit Slow Response (Bank no. 2 Sensor no. 3)

P0166 O2 Sensor Circuit No Activity Detected (Bank no. 2 Sensor no. 3)

P0167 O2 Sensor Heater Circuit Malfunction (Bank no. 2 Sensor no. 3)

P0170 Fuel Trim Malfunction (Bank no. 1)
P0171 System Too Lean (Bank no. 1)
P0172 System Too Rich (Bank no. 1)
P0173 Fuel Trim Malfunction (Bank no. 2)
P0174 System Too Lean (Bank no. 2)
P0175 System Too Rich (Bank no. 2)
P0176 Fuel Composition Sensor Circuit Malfunction
P0177 Fuel Composition Sensor Circuit Range/Performance
P0178 Fuel Composition Sensor Circuit Low Input
P0179 Fuel Composition Sensor Circuit High Input
P0180 Fuel Temperature Sensor "A" Circuit Malfunction
P0181 Fuel Temperature Sensor "A" Circuit Range/Performance
P0182 Fuel Temperature Sensor "A" Circuit Low Input
P0183 Fuel Temperature Sensor "A" Circuit High Input
P0184 Fuel Temperature Sensor "A" Circuit Intermittent
P0185 Fuel Temperature Sensor "B" Circuit Malfunction
P0186 Fuel Temperature Sensor "B" Circuit Range/Performance
P0187 Fuel Temperature Sensor "B" Circuit Low Input
P0188 Fuel Temperature Sensor "B" Circuit High Input
P0189 Fuel Temperature Sensor "B" Circuit Intermittent
P0190 Fuel Rail Pressure Sensor Circuit Malfunction
P0191 Fuel Rail Pressure Sensor Circuit Range/Performance
P0192 Fuel Rail Pressure Sensor Circuit Low Input
P0193 Fuel Rail Pressure Sensor Circuit High Input
P0194 Fuel Rail Pressure Sensor Circuit Intermittent
P0195 Engine Oil Temperature Sensor Malfunction
P0196 Engine Oil Temperature Sensor Range/Performance
P0197 Engine Oil Temperature Sensor Low
P0198 Engine Oil Temperature Sensor High
P0199 Engine Oil Temperature Sensor Intermittent
P0200 Injector Circuit Malfunction
P0201 Injector Circuit Malfunction—Cylinder no. 1
P0202 Injector Circuit Malfunction—Cylinder no. 2
P0203 Injector Circuit Malfunction—Cylinder no. 3
P0204 Injector Circuit Malfunction—Cylinder no. 4
P0205 Injector Circuit Malfunction—Cylinder no. 5
P0206 Injector Circuit Malfunction—Cylinder no. 6
P0207 Injector Circuit Malfunction—Cylinder no. 7
P0208 Injector Circuit Malfunction—Cylinder no. 8
P0209 Injector Circuit Malfunction—Cylinder no. 9
P0210 Injector Circuit Malfunction—Cylinder no. 10
P0211 Injector Circuit Malfunction—Cylinder no. 11
P0212 Injector Circuit Malfunction—Cylinder no. 12
P0213 Cold Start Injector no. 1 Malfunction
P0214 Cold Start Injector no. 2 Malfunction
P0215 Engine Shutoff Solenoid Malfunction
P0216 Injection Timing Control Circuit Malfunction
P0217 Engine Over Temperature Condition
P0218 Transmission Over Temperature Condition
P0219 Engine Over Speed Condition
P0220 Throttle/Pedal Position Sensor/Switch "B" Circuit Malfunction
P0221 Throttle/Pedal Position Sensor/Switch "B" Circuit Range/Performance Problem

P0222 Throttle/Pedal Position Sensor/Switch "B" Circuit Low Input
P0223 Throttle/Pedal Position Sensor/Switch "B" Circuit High Input
P0224 Throttle/Pedal Position Sensor/Switch "B" Circuit Intermittent
P0225 Throttle/Pedal Position Sensor/Switch "C" Circuit Malfunction
P0226 Throttle/Pedal Position Sensor/Switch "C" Circuit Range/Performance Problem
P0227 Throttle/Pedal Position Sensor/Switch "C" Circuit Low Input
P0228 Throttle/Pedal Position Sensor/Switch "C" Circuit High Input
P0229 Throttle/Pedal Position Sensor/Switch "C" Circuit Intermittent
P0230 Fuel Pump Primary Circuit Malfunction
P0231 Fuel Pump Secondary Circuit Low
P0232 Fuel Pump Secondary Circuit High
P0233 Fuel Pump Secondary Circuit Intermittent
P0234 Engine Over Boost Condition
P0261 Cylinder no. 1 Injector Circuit Low
P0262 Cylinder no. 1 Injector Circuit High
P0263 Cylinder no. 1 Contribution/Balance Fault
P0264 Cylinder no. 2 Injector Circuit Low
P0265 Cylinder no. 2 Injector Circuit High
P0266 Cylinder no. 2 Contribution/Balance Fault
P0267 Cylinder no. 3 Injector Circuit Low
P0268 Cylinder no. 3 Injector Circuit High
P0269 Cylinder no. 3 Contribution/Balance Fault
P0270 Cylinder no. 4 Injector Circuit Low
P0271 Cylinder no. 4 Injector Circuit High
P0272 Cylinder no. 4 Contribution/Balance Fault
P0273 Cylinder no. 5 Injector Circuit Low
P0274 Cylinder no. 5 Injector Circuit High
P0275 Cylinder no. 5 Contribution/Balance Fault
P0276 Cylinder no. 6 Injector Circuit Low
P0277 Cylinder no. 6 Injector Circuit High
P0278 Cylinder no. 6 Contribution/Balance Fault
P0279 Cylinder no. 7 Injector Circuit Low
P0280 Cylinder no. 7 Injector Circuit High
P0281 Cylinder no. 7 Contribution/Balance Fault
P0282 Cylinder no. 8 Injector Circuit Low
P0283 Cylinder no. 8 Injector Circuit High
P0284 Cylinder no. 8 Contribution/Balance Fault
P0285 Cylinder no. 9 Injector Circuit Low
P0286 Cylinder no. 9 Injector Circuit High
P0287 Cylinder no. 9 Contribution/Balance Fault
P0288 Cylinder no. 10 Injector Circuit Low
P0289 Cylinder no. 10 Injector Circuit High
P0290 Cylinder no. 10 Contribution/Balance Fault
P0291 Cylinder no. 11 Injector Circuit Low
P0292 Cylinder no. 11 Injector Circuit High
P0293 Cylinder no. 11 Contribution/Balance Fault
P0294 Cylinder no. 12 Injector Circuit Low
P0295 Cylinder no. 12 Injector Circuit High
P0296 Cylinder no. 12 Contribution/Balance Fault
P0300 Random/Multiple Cylinder Misfire Detected
P0301 Cylinder no. 1—Misfire Detected
P0302 Cylinder no. 2—Misfire Detected
P0303 Cylinder no. 3—Misfire Detected
P0304 Cylinder no. 4—Misfire Detected

P0305 Cylinder no. 5—Misfire Detected
P0306 Cylinder no. 6—Misfire Detected
P0307 Cylinder no. 7—Misfire Detected
P0308 Cylinder no. 8—Misfire Detected
P0309 Cylinder no. 9—Misfire Detected
P0310 Cylinder no. 10—Misfire Detected
P0311 Cylinder no. 11—Misfire Detected
P0312 Cylinder no. 12—Misfire Detected
P0320 Ignition/Distributor Engine Speed Input Circuit Malfunction
P0321 Ignition/Distributor Engine Speed Input Circuit Range/Performance
P0322 Ignition/Distributor Engine Speed Input Circuit No Signal
P0323 Ignition/Distributor Engine Speed Input Circuit Intermittent
P0325 Knock Sensor no. 1—Circuit Malfunction (Bank no. 1 or Single Sensor)
P0326 Knock Sensor no. 1—Circuit Range/Performance (Bank no. 1 or Single Sensor)
P0327 Knock Sensor no. 1—Circuit Low Input (Bank no. 1 or Single Sensor)
P0328 Knock Sensor no. 1—Circuit High Input (Bank no. 1 or Single Sensor)
P0329 Knock Sensor no. 1—Circuit Input Intermittent (Bank no. 1 or Single Sensor)
P0330 Knock Sensor no. 2—Circuit Malfunction (Bank no. 2)
P0331 Knock Sensor no. 2—Circuit Range/Performance (Bank no. 2)
P0332 Knock Sensor no. 2—Circuit Low Input (Bank no. 2)
P0333 Knock Sensor no. 2—Circuit High Input (Bank no. 2)
P0334 Knock Sensor no. 2—Circuit Input Intermittent (Bank no. 2)
P0335 Crankshaft Position Sensor "A" Circuit Malfunction
P0336 Crankshaft Position Sensor "A" Circuit Range/Performance
P0337 Crankshaft Position Sensor "A" Circuit Low Input
P0338 Crankshaft Position Sensor "A" Circuit High Input
P0339 Crankshaft Position Sensor "A" Circuit Intermittent
P0340 Camshaft Position Sensor Circuit Malfunction
P0341 Camshaft Position Sensor Circuit Range/Performance
P0342 Camshaft Position Sensor Circuit Low Input
P0343 Camshaft Position Sensor Circuit High Input
P0344 Camshaft Position Sensor Circuit Intermittent
P0350 Ignition Coil Primary/Secondary Circuit Malfunction
P0351 Ignition Coil "A" Primary/Secondary Circuit Malfunction
P0352 Ignition Coil "B" Primary/Secondary Circuit Malfunction
P0353 Ignition Coil "C" Primary/Secondary Circuit Malfunction
P0354 Ignition Coil "D" Primary/Secondary Circuit Malfunction
P0355 Ignition Coil "E" Primary/Secondary Circuit Malfunction
P0356 Ignition Coil "F" Primary/Secondary Circuit Malfunction
P0357 Ignition Coil "G" Primary/Secondary Circuit Malfunction
P0358 Ignition Coil "H" Primary/Secondary Circuit Malfunction
P0359 Ignition Coil "I" Primary/Secondary Circuit Malfunction
P0360 Ignition Coil "J" Primary/Secondary Circuit Malfunction
P0361 Ignition Coil "K" Primary/Secondary Circuit Malfunction
P0362 Ignition Coil "L" Primary/Secondary Circuit Malfunction
P0370 Timing Reference High Resolution Signal "A" Malfunction
P0371 Timing Reference High Resolution Signal "A" Too Many Pulses
P0372 Timing Reference High Resolution Signal "A" Too Few Pulses

P0373 Timing Reference High Resolution Signal "A" Intermittent/Erratic Pulses
P0374 Timing Reference High Resolution Signal "A" No Pulses
P0375 Timing Reference High Resolution Signal "B" Malfunction
P0376 Timing Reference High Resolution Signal "B" Too Many Pulses
P0377 Timing Reference High Resolution Signal "B" Too Few Pulses
P0378 Timing Reference High Resolution Signal "B" Intermittent/Erratic Pulses
P0379 Timing Reference High Resolution Signal "B" No Pulses
P0380 Glow Plug/Heater Circuit "A" Malfunction
P0381 Glow Plug/Heater Indicator Circuit Malfunction
P0382 Glow Plug/Heater Circuit "B" Malfunction
P0385 Crankshaft Position Sensor "B" Circuit Malfunction
P0386 Crankshaft Position Sensor "B" Circuit Range/Performance
P0387 Crankshaft Position Sensor "B" Circuit Low Input
P0388 Crankshaft Position Sensor "B" Circuit High Input
P0389 Crankshaft Position Sensor "B" Circuit Intermittent
P0400 Exhaust Gas Recirculation Flow Malfunction
P0401 Exhaust Gas Recirculation Flow Insufficient Detected
P0402 Exhaust Gas Recirculation Flow Excessive Detected
P0403 Exhaust Gas Recirculation Circuit Malfunction
P0404 Exhaust Gas Recirculation Circuit Range/Performance
P0405 Exhaust Gas Recirculation Sensor "A" Circuit Low
P0406 Exhaust Gas Recirculation Sensor "A" Circuit High
P0407 Exhaust Gas Recirculation Sensor "B" Circuit Low
P0408 Exhaust Gas Recirculation Sensor "B" Circuit High
P0410 Secondary Air Injection System Malfunction
P0411 Secondary Air Injection System Incorrect Flow Detected
P0412 Secondary Air Injection System Switching Valve "A" Circuit Malfunction
P0413 Secondary Air Injection System Switching Valve "A" Circuit Open
P0414 Secondary Air Injection System Switching Valve "A" Circuit Shorted
P0415 Secondary Air Injection System Switching Valve "B" Circuit Malfunction
P0416 Secondary Air Injection System Switching Valve "B" Circuit Open
P0417 Secondary Air Injection System Switching Valve "B" Circuit Shorted
P0418 Secondary Air Injection System Relay "A" Circuit Malfunction
P0419 Secondary Air Injection System Relay "B" Circuit Malfunction
P0420 Catalyst System Efficiency Below Threshold (Bank no. 1)
P0421 Warm Up Catalyst Efficiency Below Threshold (Bank no. 1)
P0422 Main Catalyst Efficiency Below Threshold (Bank no. 1)
P0423 Heated Catalyst Efficiency Below Threshold (Bank no. 1)
P0424 Heated Catalyst Temperature Below Threshold (Bank no. 1)
P0430 Catalyst System Efficiency Below Threshold (Bank no. 2)
P0431 Warm Up Catalyst Efficiency Below Threshold (Bank no. 2)
P0432 Main Catalyst Efficiency Below Threshold (Bank no. 2)
P0433 Heated Catalyst Efficiency Below Threshold (Bank no. 2)
P0434 Heated Catalyst Temperature Below Threshold (Bank no. 2)

P0440 Evaporative Emission Control System Malfunction

P0441 Evaporative Emission Control System Incorrect Purge Flow

P0442 Evaporative Emission Control System Leak Detected (Small Leak)

P0443 Evaporative Emission Control System Purge Control Valve Circuit Malfunction

P0444 Evaporative Emission Control System Purge Control Valve Circuit Open

P0445 Evaporative Emission Control System Purge Control Valve Circuit Shorted

P0446 Evaporative Emission Control System Vent Control Circuit Malfunction

P0447 Evaporative Emission Control System Vent Control Circuit Open

P0448 Evaporative Emission Control System Vent Control Circuit Shorted

P0449 Evaporative Emission Control System Vent Valve/Solenoid Circuit Malfunction

P0450 Evaporative Emission Control System Pressure Sensor Malfunction

P0451 Evaporative Emission Control System Pressure Sensor Range/Performance

P0452 Evaporative Emission Control System Pressure Sensor Low Input

P0453 Evaporative Emission Control System Pressure Sensor High Input

P0454 Evaporative Emission Control System Pressure Sensor Intermittent

P0455 Evaporative Emission Control System Leak Detected (Gross Leak)

P0460 Fuel Level Sensor Circuit Malfunction

P0461 Fuel Level Sensor Circuit Range/Performance

P0462 Fuel Level Sensor Circuit Low Input

P0463 Fuel Level Sensor Circuit High Input

P0464 Fuel Level Sensor Circuit Intermittent

P0465 Purge Flow Sensor Circuit Malfunction

P0466 Purge Flow Sensor Circuit Range/Performance

P0467 Purge Flow Sensor Circuit Low Input

P0468 Purge Flow Sensor Circuit High Input

P0469 Purge Flow Sensor Circuit Intermittent

P0470 Exhaust Pressure Sensor Malfunction

P0471 Exhaust Pressure Sensor Range/Performance

P0472 Exhaust Pressure Sensor Low

P0473 Exhaust Pressure Sensor High

P0474 Exhaust Pressure Sensor Intermittent

P0475 Exhaust Pressure Control Valve Malfunction

P0476 Exhaust Pressure Control Valve Range/Performance

P0477 Exhaust Pressure Control Valve Low

P0478 Exhaust Pressure Control Valve High

P0479 Exhaust Pressure Control Valve Intermittent

P0480 Cooling Fan no. 1 Control Circuit Malfunction

P0481 Cooling Fan no. 2 Control Circuit Malfunction

P0482 Cooling Fan no. 3 Control Circuit Malfunction

P0483 Cooling Fan Rationality Check Malfunction

P0484 Cooling Fan Circuit Over Current

P0485 Cooling Fan Power/Ground Circuit Malfunction

P0500 Vehicle Speed Sensor Malfunction

P0501 Vehicle Speed Sensor Range/Performance

P0502 Vehicle Speed Sensor Circuit Low Input

P0503 Vehicle Speed Sensor Intermittent/Erratic/High

P0505 Idle Control System Malfunction

P0506 Idle Control System RPM Lower Than Expected

P0507 Idle Control System RPM Higher Than Expected

P0510 Closed Throttle Position Switch Malfunction

P0520 Engine Oil Pressure Sensor/Switch Circuit Malfunction

P0521 Engine Oil Pressure Sensor/Switch Range/Performance

P0522 Engine Oil Pressure Sensor/Switch Low Voltage

P0523 Engine Oil Pressure Sensor/Switch High Voltage

P0530 A/C Refrigerant Pressure Sensor Circuit Malfunction

P0531 A/C Refrigerant Pressure Sensor Circuit Range/Performance

P0532 A/C Refrigerant Pressure Sensor Circuit Low Input

P0533 A/C Refrigerant Pressure Sensor Circuit High Input

P0534 A/C Refrigerant Charge Loss

P0550 Power Steering Pressure Sensor Circuit Malfunction

P0551 Power Steering Pressure Sensor Circuit Range/Performance

P0552 Power Steering Pressure Sensor Circuit Low Input

P0553 Power Steering Pressure Sensor Circuit High Input

P0554 Power Steering Pressure Sensor Circuit Intermittent

P0560 System Voltage Malfunction

P0561 System Voltage Unstable

P0562 System Voltage Low

P0563 System Voltage High

P0565 Cruise Control On Signal Malfunction

P0566 Cruise Control Off Signal Malfunction

P0567 Cruise Control Resume Signal Malfunction

P0568 Cruise Control Set Signal Malfunction

P0569 Cruise Control Coast Signal Malfunction

P0570 Cruise Control Accel Signal Malfunction

P0571 Cruise Control/Brake Switch "A" Circuit Malfunction

P0572 Cruise Control/Brake Switch "A" Circuit Low

P0573 Cruise Control/Brake Switch "A" Circuit High

P0574 Through P0580 Reserved for Cruise Codes

P0600 Serial Communication Link Malfunction

P0601 Internal Control Module Memory Check Sum Error

P0602 Control Module Programming Error

P0603 Internal Control Module Keep Alive Memory (KAM) Error

P0604 Internal Control Module Random Access Memory (RAM) Error

P0605 Internal Control Module Read Only Memory (ROM) Error

P0606 PCM Processor Fault

P0608 Control Module VSS Output "A" Malfunction

P0609 Control Module VSS Output "B" Malfunction

P0620 Generator Control Circuit Malfunction

P0621 Generator Lamp "L" Control Circuit Malfunction

P0622 Generator Field "F" Control Circuit Malfunction

P0650 Malfunction Indicator Lamp (MIL) Control Circuit Malfunction

P0654 Engine RPM Output Circuit Malfunction

P0655 Engine Hot Lamp Output Control Circuit Malfunction

P0656 Fuel Level Output Circuit Malfunction

P0700 Transmission Control System Malfunction

P0701 Transmission Control System Range/Performance

P0702 Transmission Control System Electrical

P0703 Torque Converter/Brake Switch "B" Circuit Malfunction

P0704 Clutch Switch Input Circuit Malfunction

P0705 Transmission Range Sensor Circuit Malfunction (PRNDL Input)

P0706 Transmission Range Sensor Circuit Range/Performance

P0707 Transmission Range Sensor Circuit Low Input

P0708 Transmission Range Sensor Circuit High Input

P0709 Transmission Range Sensor Circuit Intermittent

P0710 Transmission Fluid Temperature Sensor Circuit Malfunction

P0711 Transmission Fluid Temperature Sensor Circuit Range/Performance

P0712 Transmission Fluid Temperature Sensor Circuit Low Input

P0713 Transmission Fluid Temperature Sensor Circuit High Input

P0714 Transmission Fluid Temperature Sensor Circuit Intermittent

P0715 Input/Turbine Speed Sensor Circuit Malfunction

P0716 Input/Turbine Speed Sensor Circuit Range/Performance

P0717 Input/Turbine Speed Sensor Circuit No Signal

P0718 Input/Turbine Speed Sensor Circuit Intermittent

P0719 Torque Converter/Brake Switch "B" Circuit Low

P0720 Output Speed Sensor Circuit Malfunction

P0721 Output Speed Sensor Circuit Range/Performance

P0722 Output Speed Sensor Circuit No Signal

P0723 Output Speed Sensor Circuit Intermittent

P0724 Torque Converter/Brake Switch "B" Circuit High

P0725 Engine Speed Input Circuit Malfunction

P0726 Engine Speed Input Circuit Range/Performance

P0727 Engine Speed Input Circuit No Signal

P0728 Engine Speed Input Circuit Intermittent

P0730 Incorrect Gear Ratio

P0731 Gear no. 1 Incorrect Ratio

P0732 Gear no. 2 Incorrect Ratio

P0733 Gear no. 3 Incorrect Ratio

P0734 Gear no. 4 Incorrect Ratio

P0735 Gear no. 5 Incorrect Ratio

P0736 Reverse Incorrect Ratio

P0740 Torque Converter Clutch Circuit Malfunction

P0741 Torque Converter Clutch Circuit Performance or Stuck Off

P0742 Torque Converter Clutch Circuit Stuck On

P0743 Torque Converter Clutch Circuit Electrical

P0744 Torque Converter Clutch Circuit Intermittent

P0745 Pressure Control Solenoid Malfunction

P0746 Pressure Control Solenoid Performance or Stuck Off

P0747 Pressure Control Solenoid Stuck On

P0748 Pressure Control Solenoid Electrical

P0749 Pressure Control Solenoid Intermittent

P0750 Shift Solenoid "A" Malfunction

P0751 Shift Solenoid "A" Performance or Stuck Off

P0752 Shift Solenoid "A" Stuck On

P0753 Shift Solenoid "A" Electrical

P0754 Shift Solenoid "A" Intermittent

P0755 Shift Solenoid "B" Malfunction

P0756 Shift Solenoid "B" Performance or Stuck Off

P0757 Shift Solenoid "B" Stuck On

P0758 Shift Solenoid "B" Electrical

P0759 Shift Solenoid "B" Intermittent

P0760 Shift Solenoid "C" Malfunction

P0761 Shift Solenoid "C" Performance Or Stuck Off

P0762 Shift Solenoid "C" Stuck On

P0763 Shift Solenoid "C" Electrical

P0764 Shift Solenoid "C" Intermittent

P0765 Shift Solenoid "D" Malfunction

P0766 Shift Solenoid "D" Performance Or Stuck Off

P0767 Shift Solenoid "D" Stuck On

P0768 Shift Solenoid "D" Electrical

P0769 Shift Solenoid "D" Intermittent

P0770 Shift Solenoid "E" Malfunction

P0771 Shift Solenoid "E" Performance Or Stuck Off

P0772 Shift Solenoid "E" Stuck On

P0773 Shift Solenoid "E" Electrical

P0774 Shift Solenoid "E" Intermittent

P0780 Shift Malfunction

P0781 1–2 Shift Malfunction

P0782 2–3 Shift Malfunction

P0783 3–4 Shift Malfunction

P0784 4–5 Shift Malfunction

P0785 Shift/Timing Solenoid Malfunction

P0786 Shift/Timing Solenoid Range/Performance

P0787 Shift/Timing Solenoid Low

P0788 Shift/Timing Solenoid High

P0789 Shift/Timing Solenoid Intermittent

P0790 Normal/Performance Switch Circuit Malfunction

P0801 Reverse Inhibit Control Circuit Malfunction

P0803 1–4 Upshift (Skip Shift) Solenoid Control Circuit Malfunction

P0804 1–4 Upshift (Skip Shift) Lamp Control Circuit Malfunction

P1307 Accelerometer signal

P1308 Accelerometer signal

P1326 Fault in engine control module (ECM) knock control circuit.

P1327 Fault in engine control module (ECM) knock control circuit.

P1328 Fault in engine control module (ECM) knock control circuit.

P1329 Fault in engine control module (ECM) knock control circuit.

P1401 Fault in engine control module (ECM) engine coolant temperature (ECT) sensor circuit NTC switching

P1403 Fault in engine control module (ECM) control module box temperature sensor

P1404 Fault in engine control module (ECM) control module box temperature sensor

P1405 Temperature warning greater than 230 degrees F

P1406 Temperature warning greater than 212 degrees F

P1505 Idle air control (IAC) valve opening signal

P1506 Idle air control (IAC) valve opening signal

P1507 Idle air control (IAC) valve closing signal

P1508 Idle air control (IAC) valve closing signal

P1604 Ignition discharge module (IDM) group D

P1605 Ignition discharge module (IDM) group E

P1617 Cable fault between AW 50–42 transmission control module (TCM) and Motronic 4.4 engine control module (ECM) (lamp lights)

PI618 Cable fault between AW 50-Q2 transmission control module (TCM) and Motronic 4.4 engine control module (ECM) (lamp lights)

P1619 Engine cooling fan (FC) low-speed, signal

P1620 Engine cooling fan (FC) low-speed, signal

P1621 Diagnostic trouble code (DTC) in automatic transmission control module(TCM)

FIRING ORDERS

3

FIRING ORDERS

On every vehicle manufactured between 1995 and 1999, there are essentially only two basic methods for distributing the ignition system spark to the spark plugs: distributor system and Distributorless Ignition System (DIS). The distributor system uses a rotating rotor within the distributor cap to dispense the system's spark to the applicable spark plug. DIS systems use one of three general set-ups for spark distribution: remote coil pack(s), waste spark system, and direct ignition system (also often referred to as DIS). All DIS systems are controlled by the engine control computer, which computes the proper ignition timing based upon incoming reference signals from engine sensors.

The remote coil pack set-up uses one or more coil packs connected to the spark plugs via plug wires. The waste spark system is actually a sub-type of the remote coil pack system. The only difference being that two spark plugs are fired simultaneously because they share one coil. Many waste spark systems are designed as a hybrid of a direct ignition system and a remote coil pack system, because a coil is mounted directly on top of one spark plug and attached to another spark plug via a plug wire. Direct ignition does away with the spark plug wires completely and uses a single coil pack mounted directly on top of the spark plug for each cylinder.

Firing orders are most important for vehicles equipped with distributor ignition systems because the distributor can be rotated (which can lead to confusion as to which plug tower is what). If the distributor is rotated and the spark plug wires are installed on the original cap towers, the ignition timing will be adversely affected. DIS systems are not adjustable in the same manner as distributor systems. Therefore, if you connect the wires (when applicable) to the proper coil pack towers, the ignition timing will always be correct. Thus, if your vehicle is equipped with a DIS system and the firing order illustration does not contain a specific firing order, simply attach the wires to the proper coil pack towers and the ignition system will function properly.

➡ **To avoid confusion, remove and tag the spark plug wires one at a time, for replacement.**

If your vehicle is equipped with a distributor which is not keyed for installation with only one orientation, it could have been removed previously and rewired. The resultant wiring would hold the correct firing order, but could change the relative placement of the plug towers in relation to the engine. For this reason it is imperative that you label all wires before disconnecting any of them. Also, before removal, compare the current wiring with the accompanying illustrations. If the current wiring does not match, make notes in your book to reflect how your engine is wired.

FIRING ORDER INDEX

79233C01

FIRING ORDER INDEX

79233C02

FIRING ORDER INDEX

FIRING ORDER INDEX

MANUFACTURER

79233C04

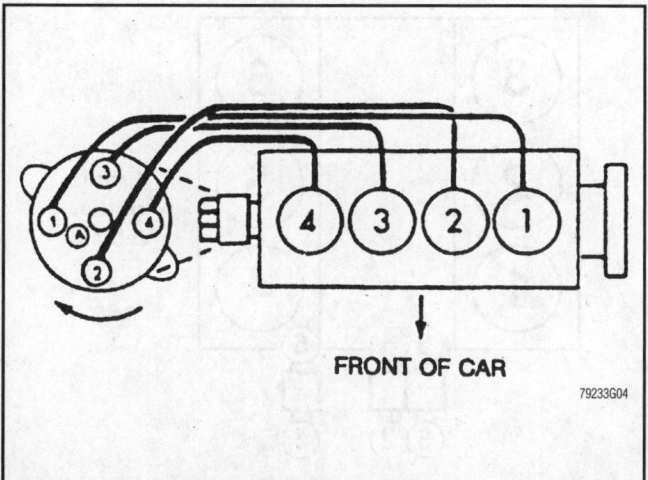

Fig. 1 Acura 1.8L (B18B1 and B18C1) Engines
Firing order: 1–3–4–2
Distributor rotation: Clockwise

79233G04

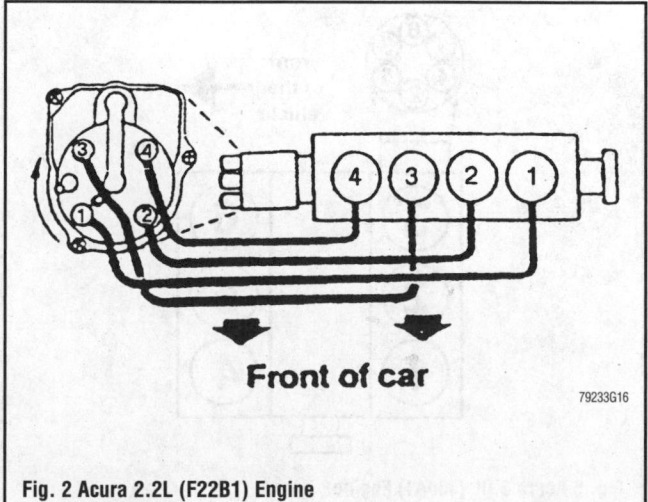

Fig. 2 Acura 2.2L (F22B1) Engine
Firing order: 1–3–4–2
Distributor rotation: Clockwise

79233G16

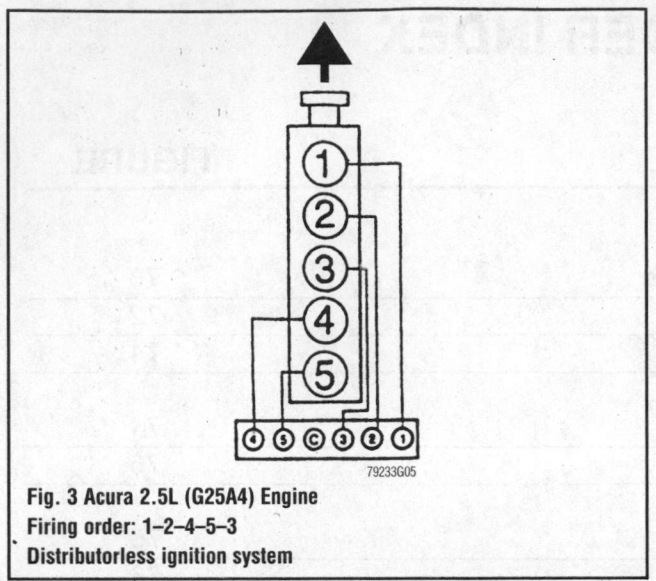

Fig. 3 Acura 2.5L (G25A4) Engine
Firing order: 1–2–4–5–3
Distributorless ignition system

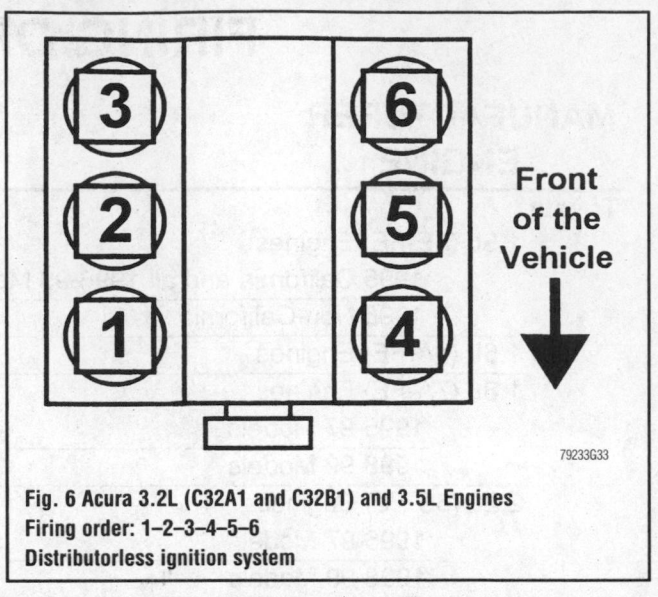

Fig. 6 Acura 3.2L (C32A1 and C32B1) and 3.5L Engines
Firing order: 1–2–3–4–5–6
Distributorless ignition system

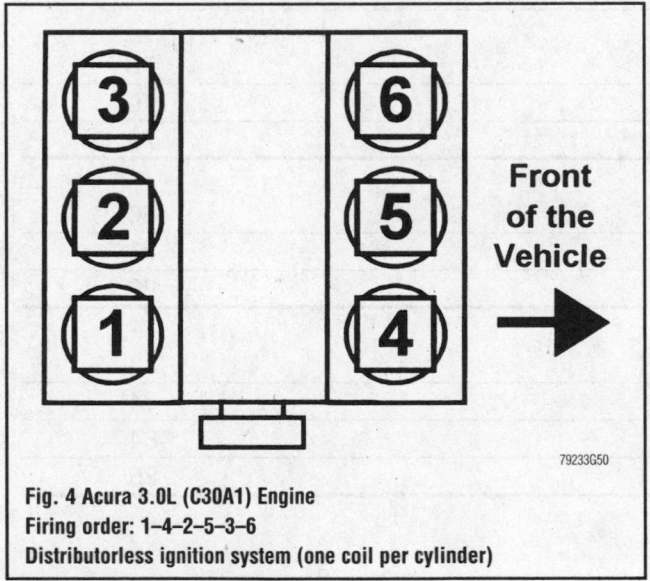

Fig. 4 Acura 3.0L (C30A1) Engine
Firing order: 1–4–2–5–3–6
Distributorless ignition system (one coil per cylinder)

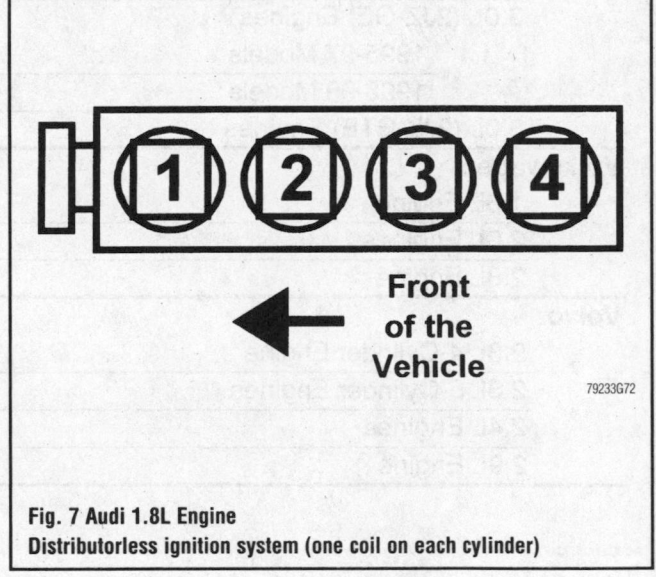

Fig. 7 Audi 1.8L Engine
Distributorless ignition system (one coil on each cylinder)

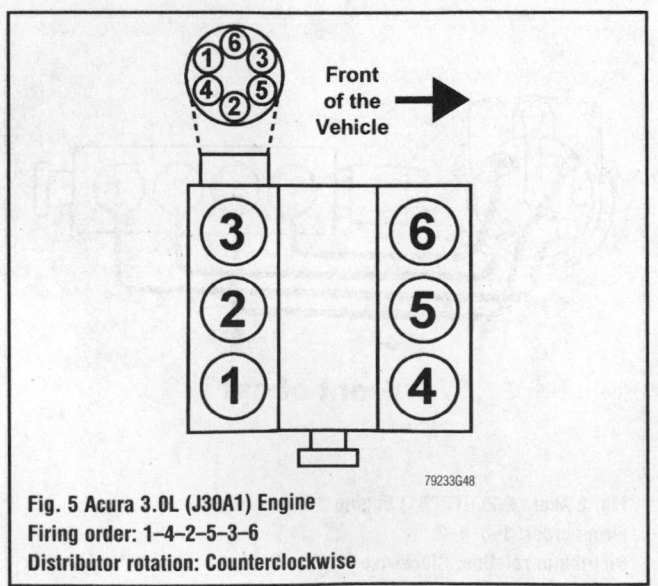

Fig. 5 Acura 3.0L (J30A1) Engine
Firing order: 1–4–2–5–3–6
Distributor rotation: Counterclockwise

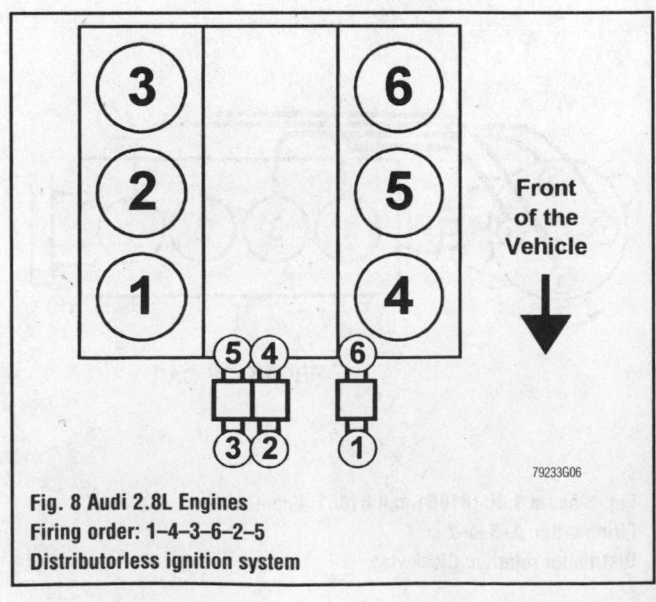

Fig. 8 Audi 2.8L Engines
Firing order: 1–4–3–6–2–5
Distributorless ignition system

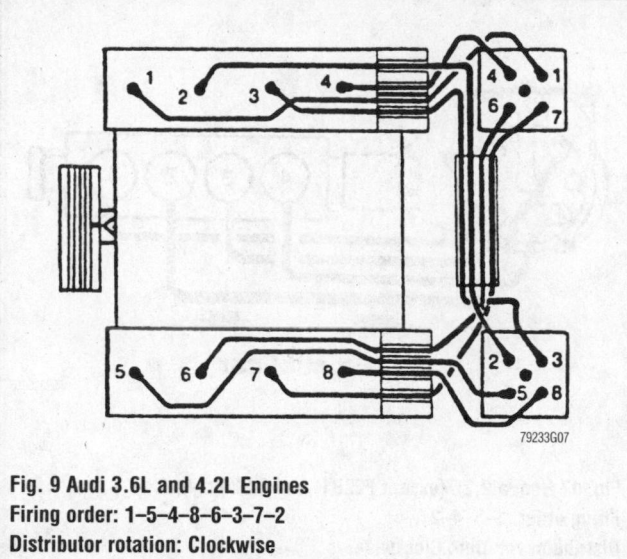

Fig. 9 Audi 3.6L and 4.2L Engines
Firing order: 1–5–4–8–6–3–7–2
Distributor rotation: Clockwise

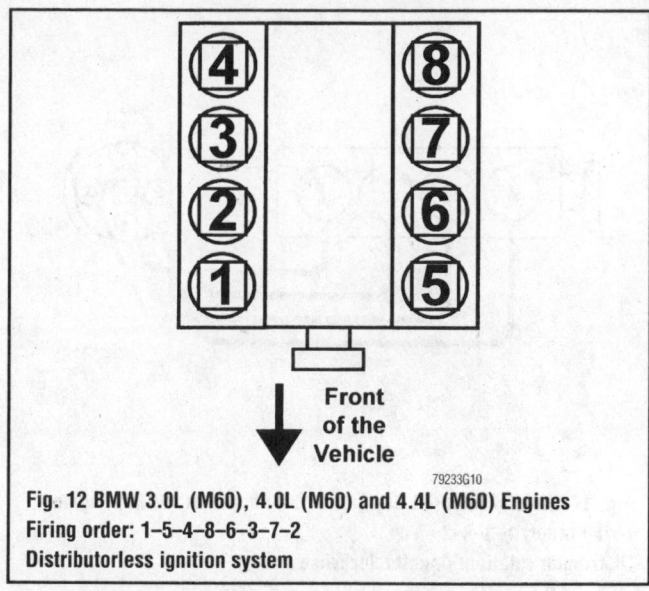

Fig. 12 BMW 3.0L (M60), 4.0L (M60) and 4.4L (M60) Engines
Firing order: 1–5–4–8–6–3–7–2
Distributorless ignition system

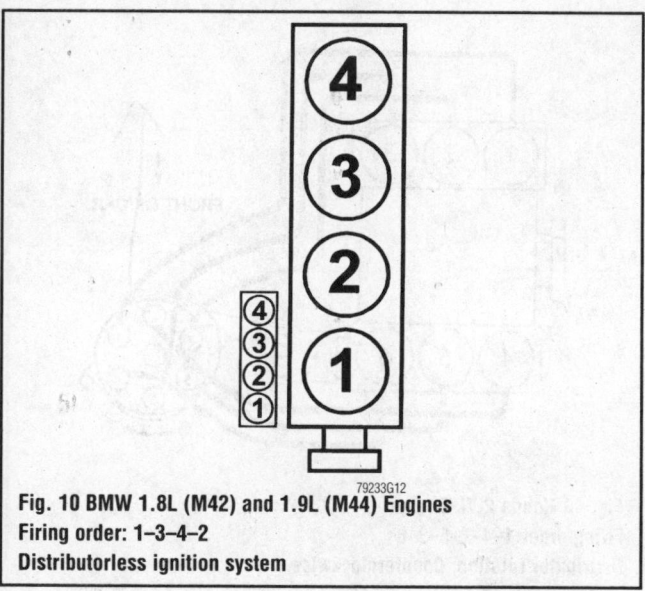

Fig. 10 BMW 1.8L (M42) and 1.9L (M44) Engines
Firing order: 1–3–4–2
Distributorless ignition system

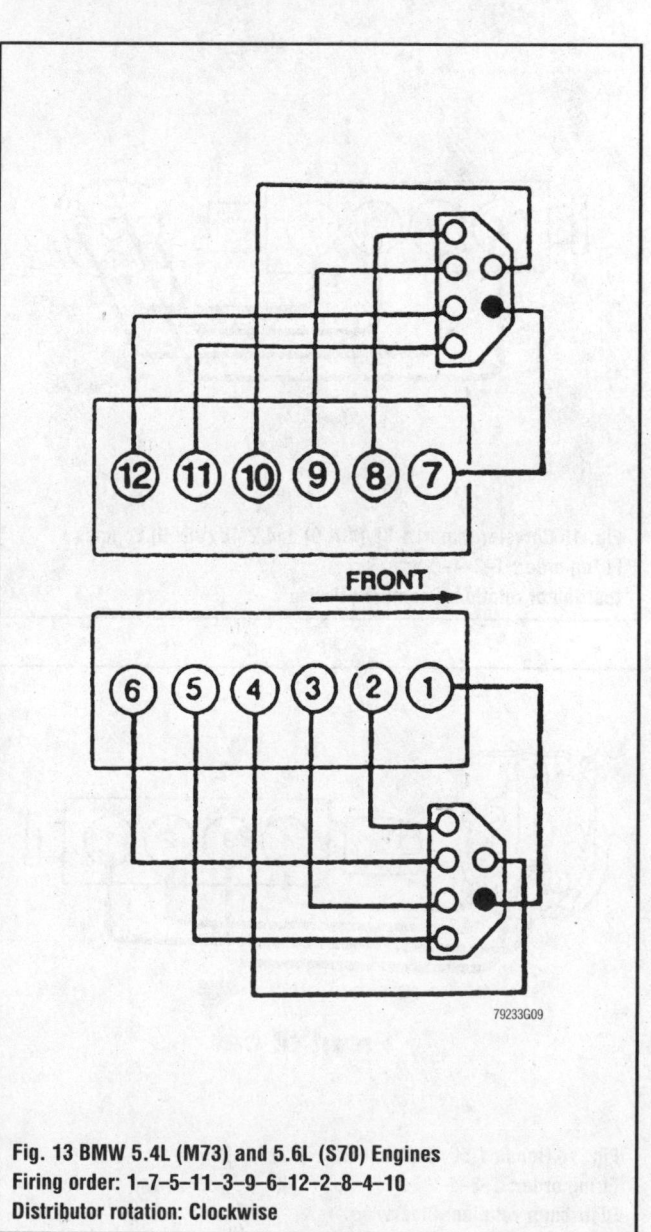

Fig. 13 BMW 5.4L (M73) and 5.6L (S70) Engines
Firing order: 1–7–5–11–3–9–6–12–2–8–4–10
Distributor rotation: Clockwise

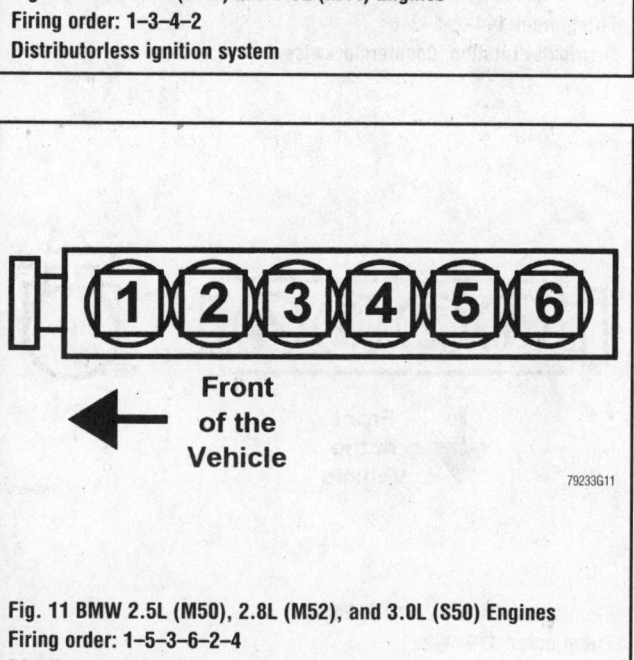

Fig. 11 BMW 2.5L (M50), 2.8L (M52), and 3.0L (S50) Engines
Firing order: 1–5–3–6–2–4
Distributorless ignition system

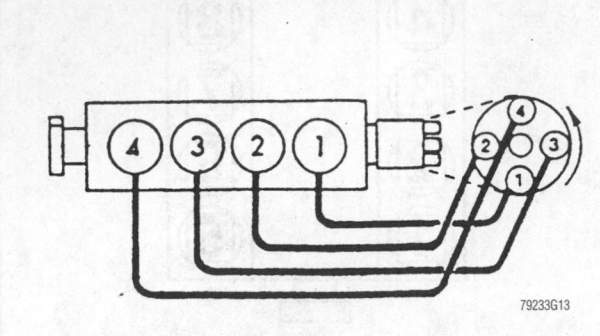

Fig. 14 Chrysler Import 1.5L, 1.8L (4G93) and 2.4L (4G64) Engines
Firing order: 1–3–4–2
Distributor rotation: Counterclockwise

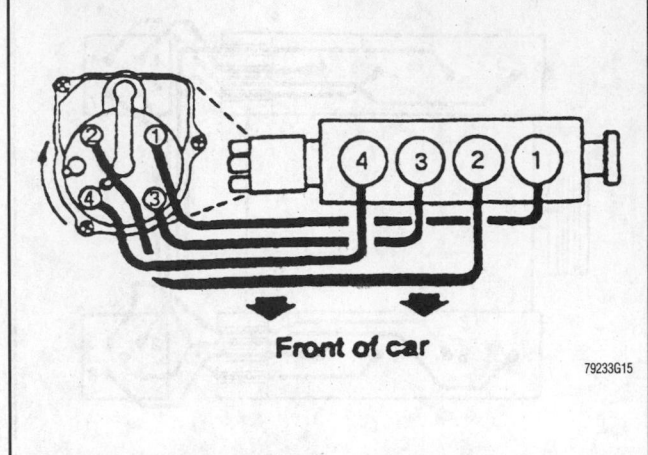

Fig. 17 Honda 2.2L (except F22B1) and 2.3L Engines
Firing order: 1–3–4–2
Distributor rotation: Clockwise

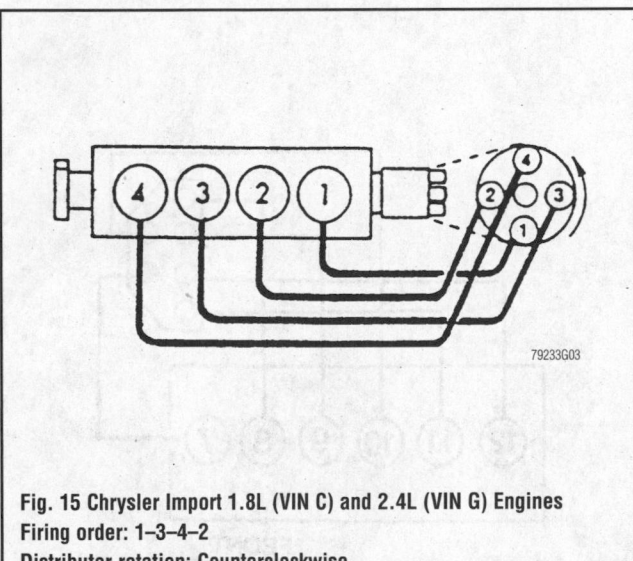

Fig. 15 Chrysler Import 1.8L (VIN C) and 2.4L (VIN G) Engines
Firing order: 1–3–4–2
Distributor rotation: Counterclockwise

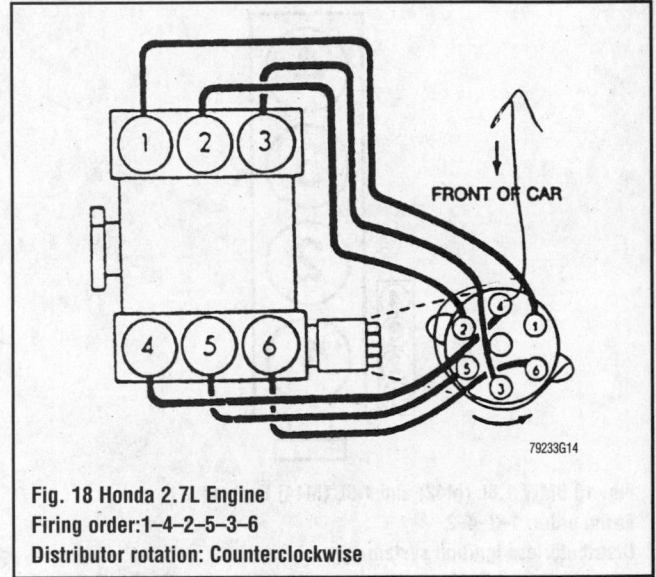

Fig. 18 Honda 2.7L Engine
Firing order: 1–4–2–5–3–6
Distributor rotation: Counterclockwise

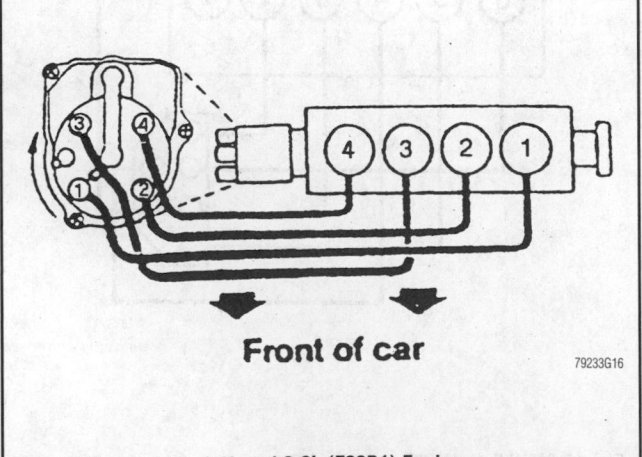

Fig. 16 Honda 1.5L, 1.6L and 2.2L (F22B1) Engines
Firing order: 1–3–4–2
Distributor rotation: Clockwise

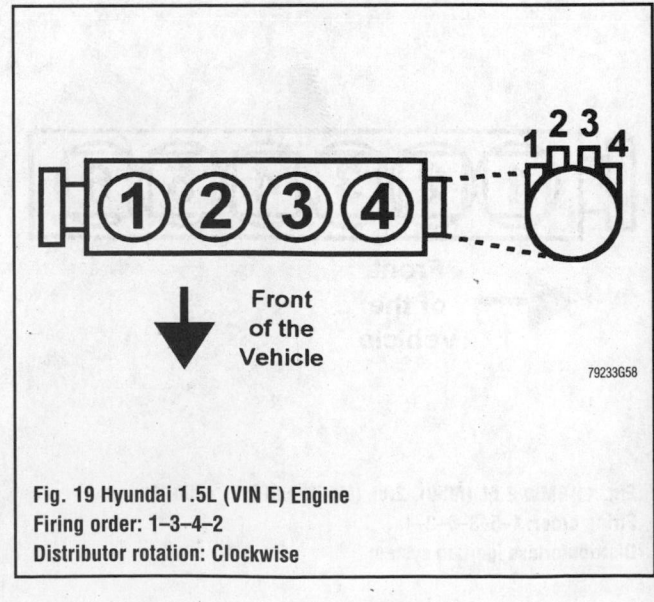

Fig. 19 Hyundai 1.5L (VIN E) Engine
Firing order: 1–3–4–2
Distributor rotation: Clockwise

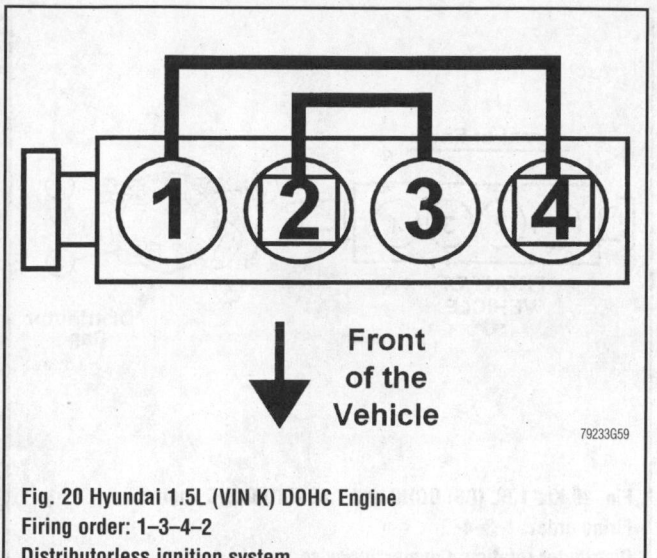

Fig. 20 Hyundai 1.5L (VIN K) DOHC Engine
Firing order: 1–3–4–2
Distributorless ignition system

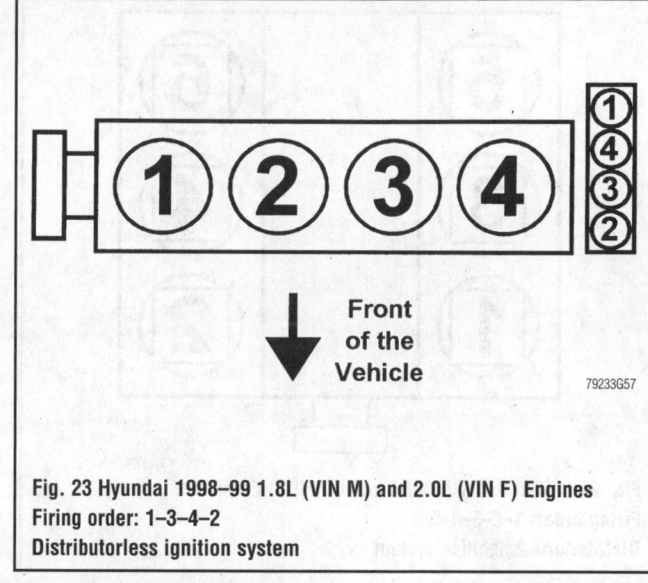

Fig. 23 Hyundai 1998–99 1.8L (VIN M) and 2.0L (VIN F) Engines
Firing order: 1–3–4–2
Distributorless ignition system

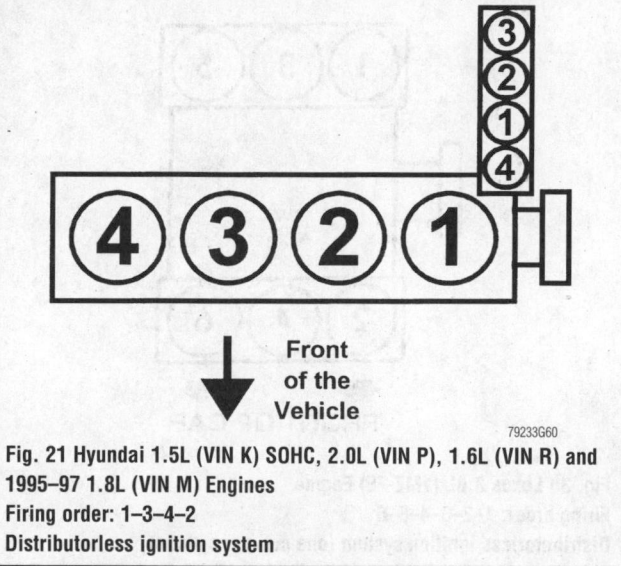

Fig. 21 Hyundai 1.5L (VIN K) SOHC, 2.0L (VIN P), 1.6L (VIN R) and
1995–97 1.8L (VIN M) Engines
Firing order: 1–3–4–2
Distributorless ignition system

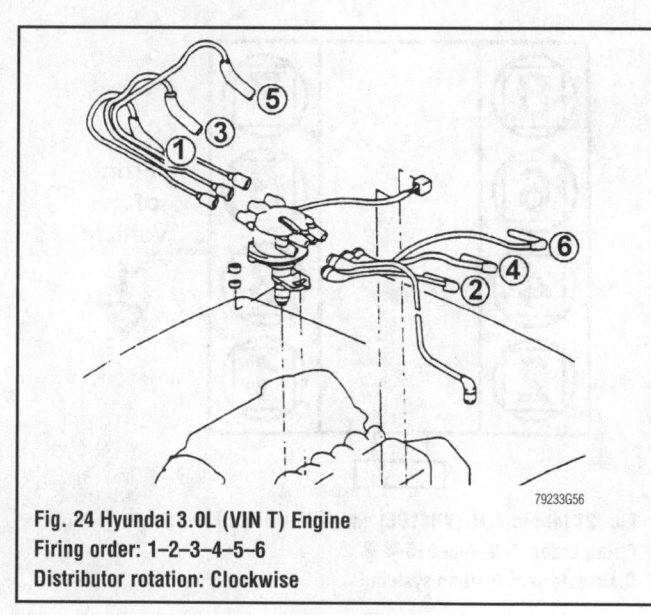

Fig. 24 Hyundai 3.0L (VIN T) Engine
Firing order: 1–2–3–4–5–6
Distributor rotation: Clockwise

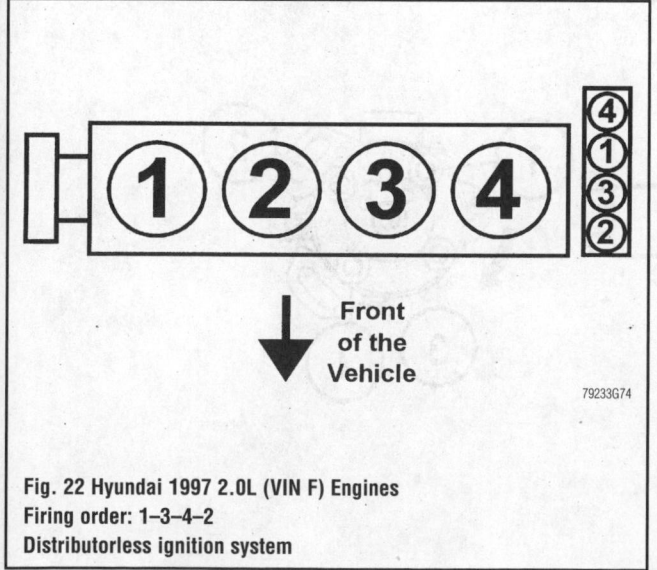

Fig. 22 Hyundai 1997 2.0L (VIN F) Engines
Firing order: 1–3–4–2
Distributorless ignition system

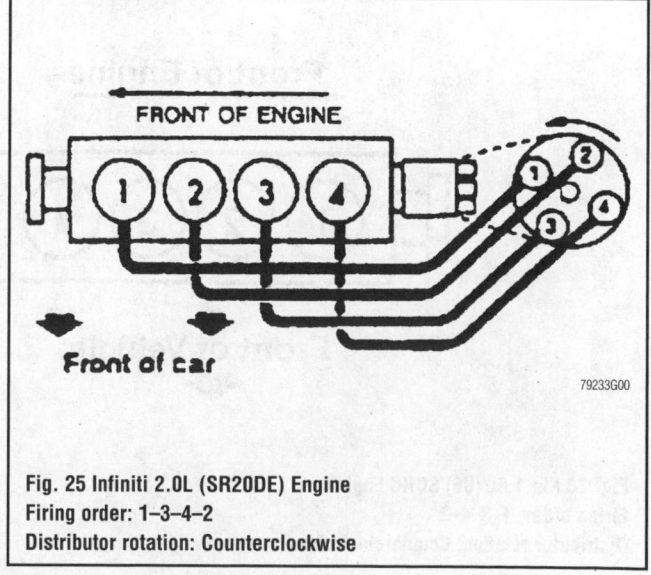

Fig. 25 Infiniti 2.0L (SR20DE) Engine
Firing order: 1–3–4–2
Distributor rotation: Counterclockwise

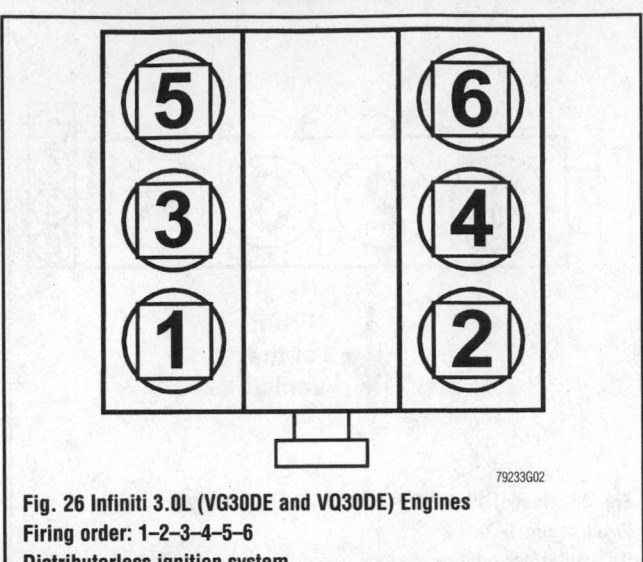

Fig. 26 Infiniti 3.0L (VG30DE and VQ30DE) Engines
Firing order: 1–2–3–4–5–6
Distributorless ignition system

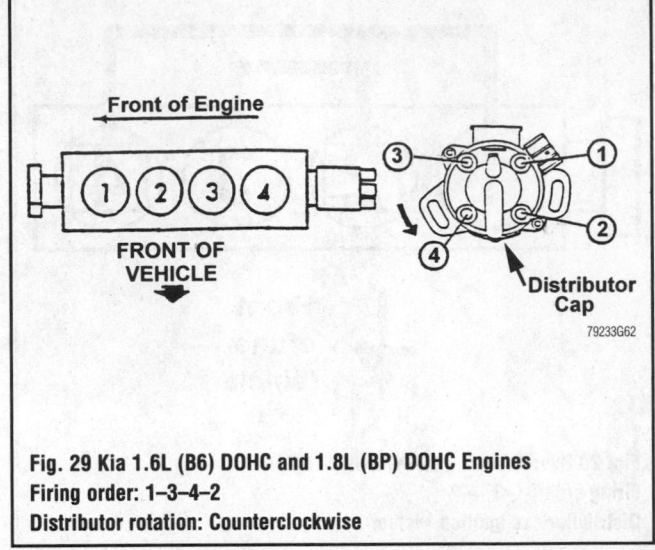

Fig. 29 Kia 1.6L (B6) DOHC and 1.8L (BP) DOHC Engines
Firing order: 1–3–4–2
Distributor rotation: Counterclockwise

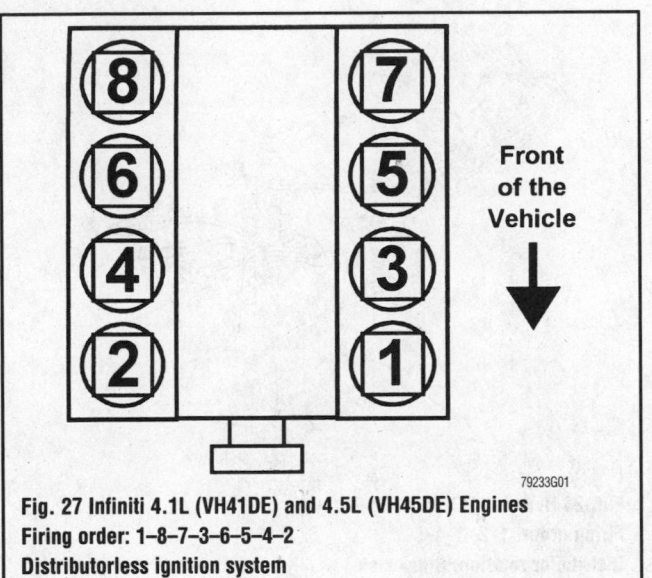

Fig. 27 Infiniti 4.1L (VH41DE) and 4.5L (VH45DE) Engines
Firing order: 1–8–7–3–6–5–4–2
Distributorless ignition system

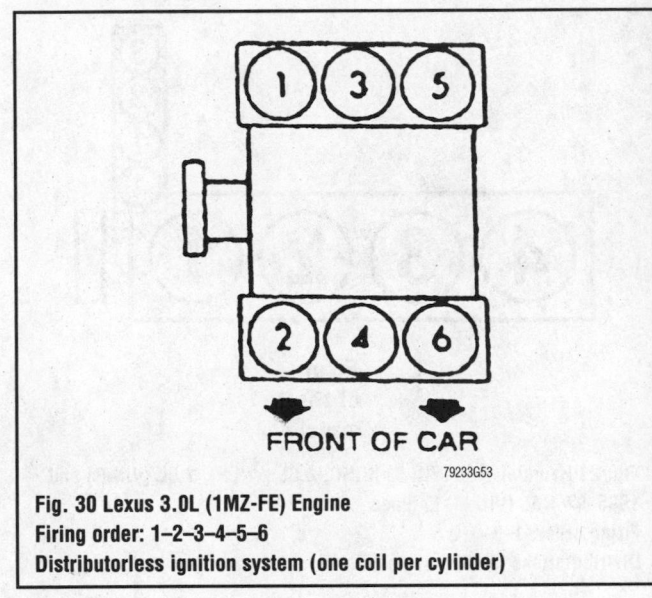

Fig. 30 Lexus 3.0L (1MZ-FE) Engine
Firing order: 1–2–3–4–5–6
Distributorless ignition system (one coil per cylinder)

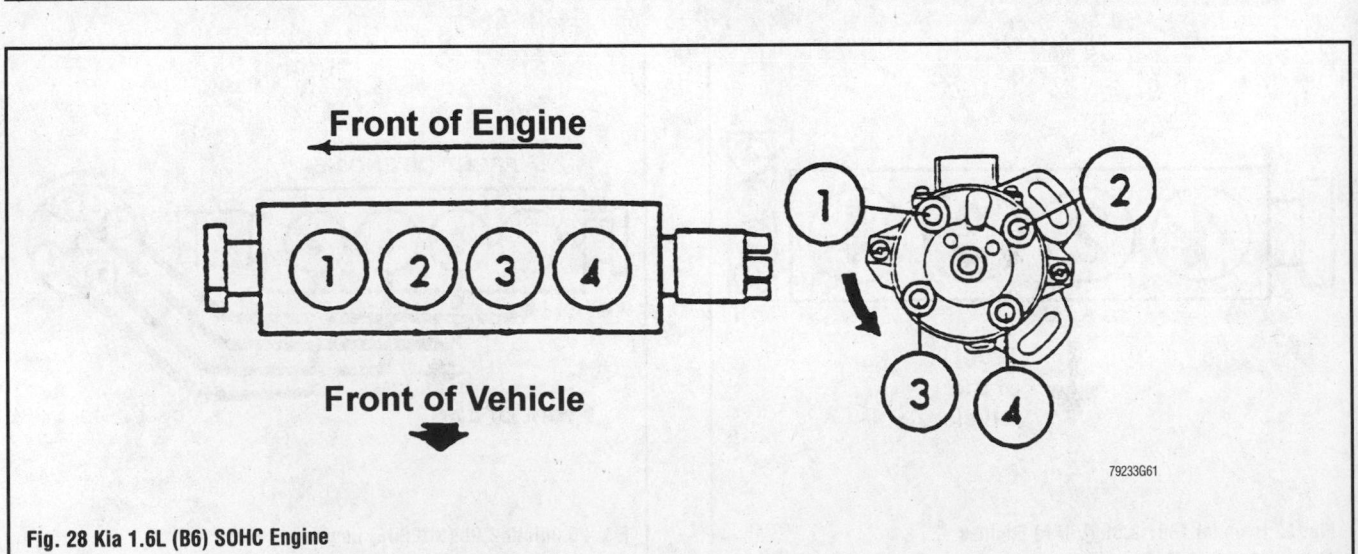

Fig. 28 Kia 1.6L (B6) SOHC Engine
Firing order: 1–3–4–2
Distributor rotation: Counterclockwise

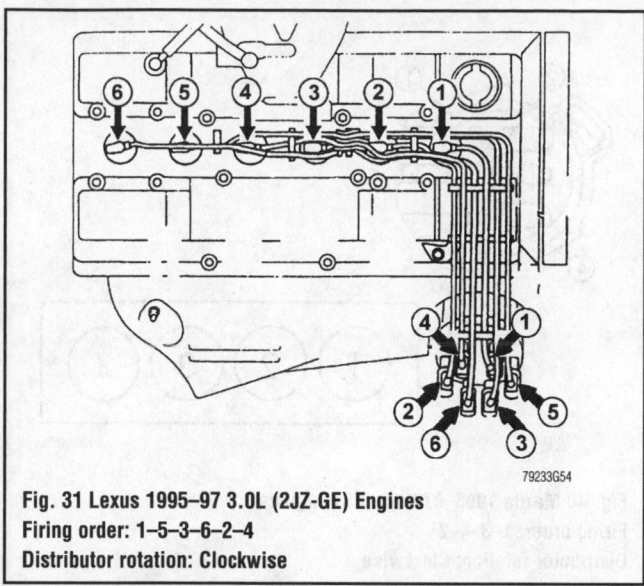

Fig. 31 Lexus 1995–97 3.0L (2JZ-GE) Engines
Firing order: 1–5–3–6–2–4
Distributor rotation: Clockwise

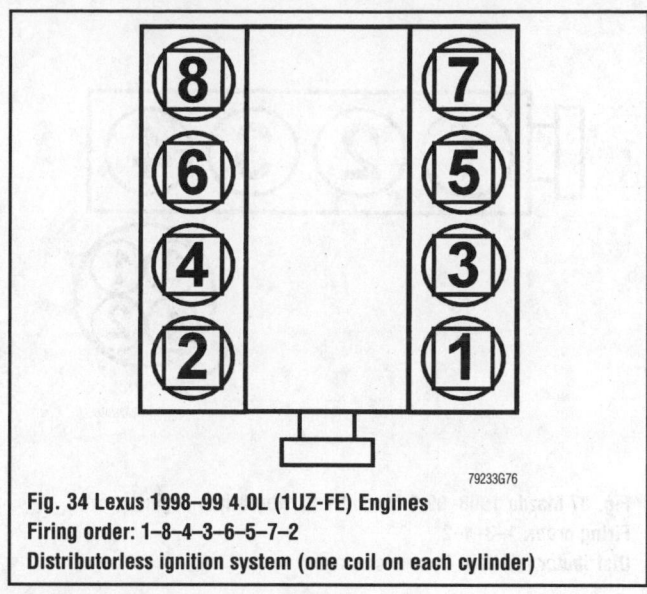

Fig. 34 Lexus 1998–99 4.0L (1UZ-FE) Engines
Firing order: 1–8–4–3–6–5–7–2
Distributorless ignition system (one coil on each cylinder)

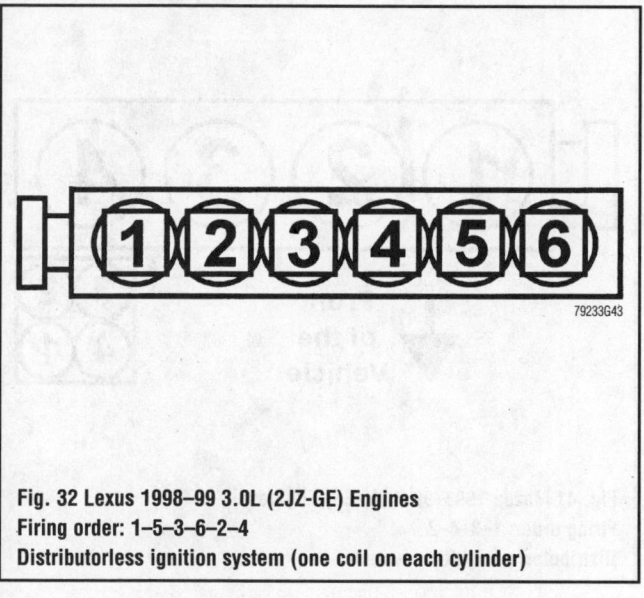

Fig. 32 Lexus 1998–99 3.0L (2JZ-GE) Engines
Firing order: 1–5–3–6–2–4
Distributorless ignition system (one coil on each cylinder)

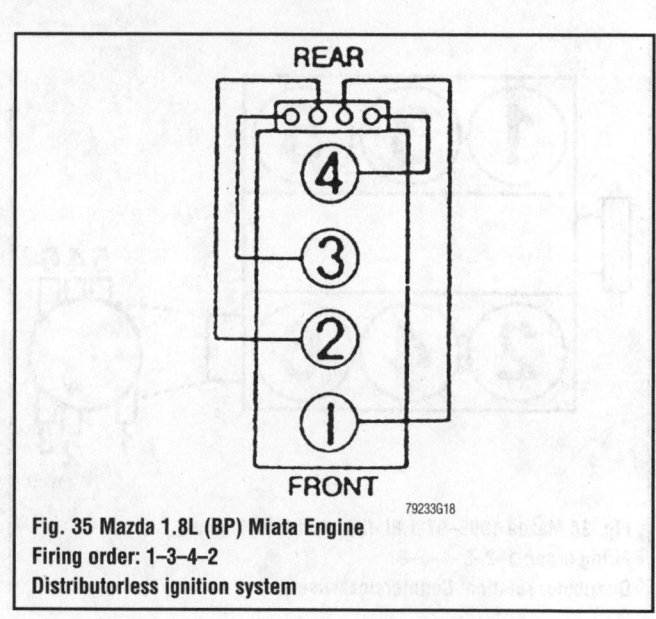

Fig. 35 Mazda 1.8L (BP) Miata Engine
Firing order: 1–3–4–2
Distributorless ignition system

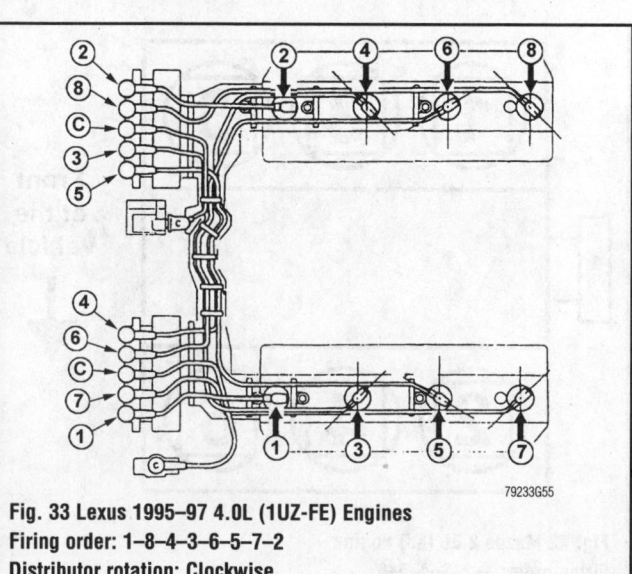

Fig. 33 Lexus 1995–97 4.0L (1UZ-FE) Engines
Firing order: 1–8–4–3–6–5–7–2
Distributor rotation: Clockwise

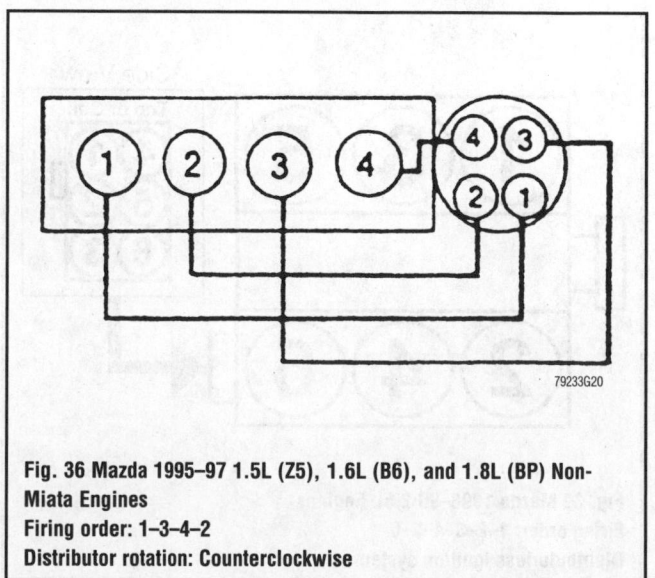

Fig. 36 Mazda 1995–97 1.5L (Z5), 1.6L (B6), and 1.8L (BP) Non-Miata Engines
Firing order: 1–3–4–2
Distributor rotation: Counterclockwise

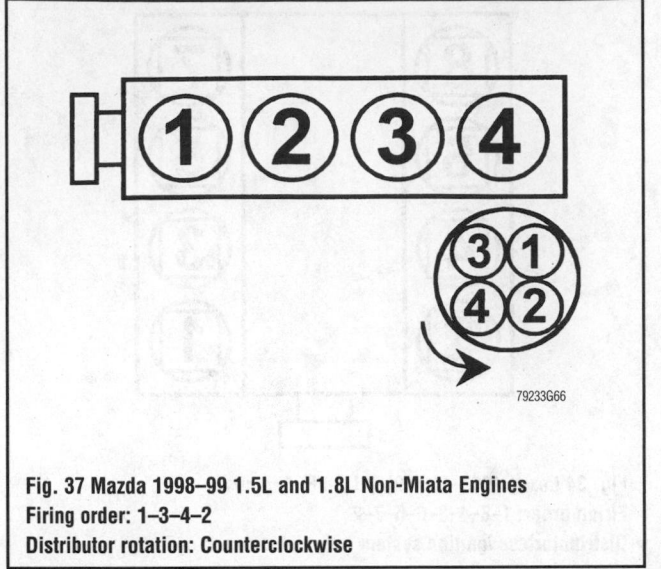

Fig. 37 Mazda 1998–99 1.5L and 1.8L Non-Miata Engines
Firing order: 1–3–4–2
Distributor rotation: Counterclockwise

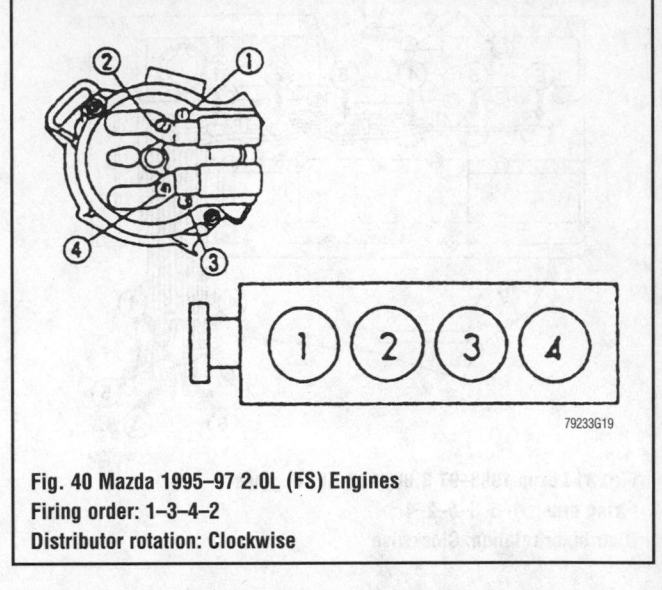

Fig. 40 Mazda 1995–97 2.0L (FS) Engines
Firing order: 1–3–4–2
Distributor rotation: Clockwise

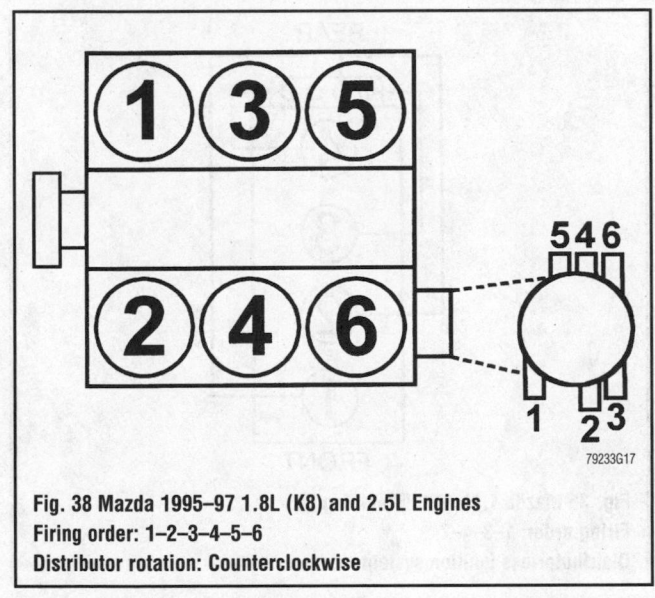

Fig. 38 Mazda 1995–97 1.8L (K8) and 2.5L Engines
Firing order: 1–2–3–4–5–6
Distributor rotation: Counterclockwise

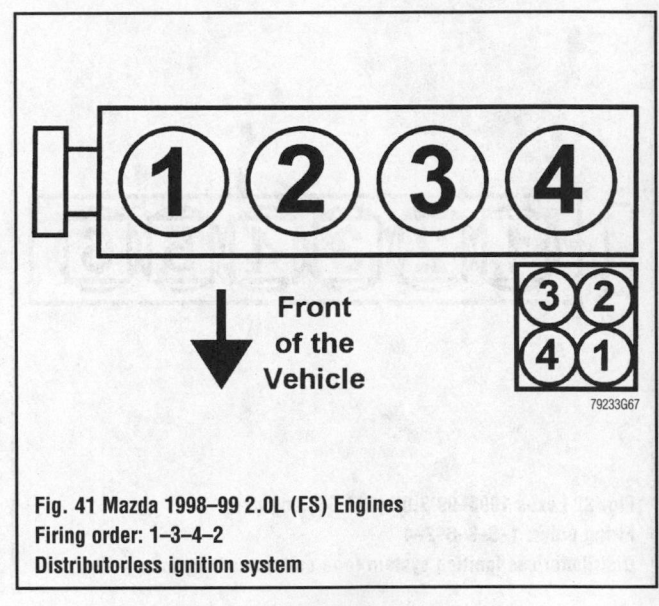

Fig. 41 Mazda 1998–99 2.0L (FS) Engines
Firing order: 1–3–4–2
Distributorless ignition system

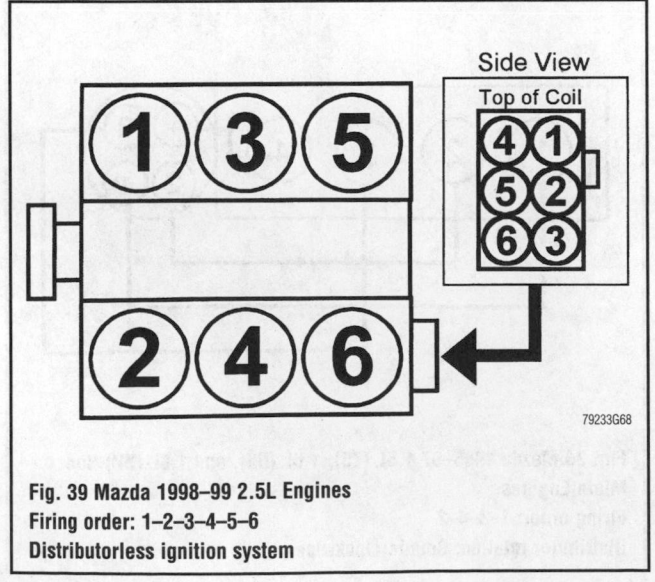

Fig. 39 Mazda 1998–99 2.5L Engines
Firing order: 1–2–3–4–5–6
Distributorless ignition system

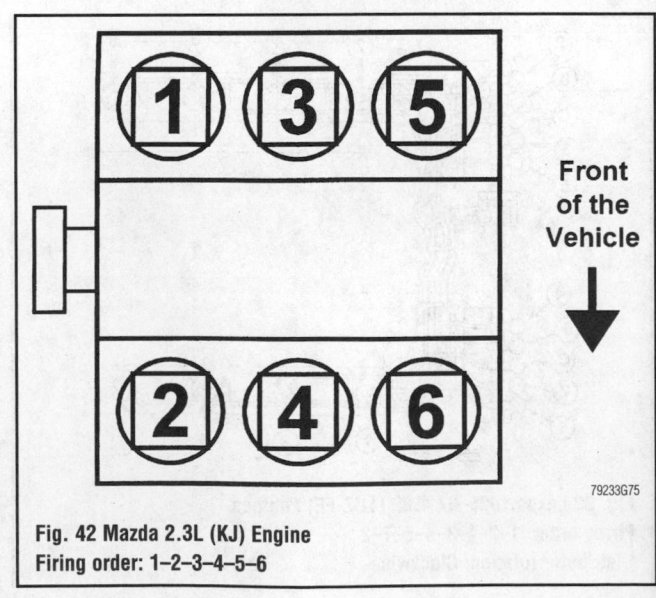

Fig. 42 Mazda 2.3L (KJ) Engine
Firing order: 1–2–3–4–5–6

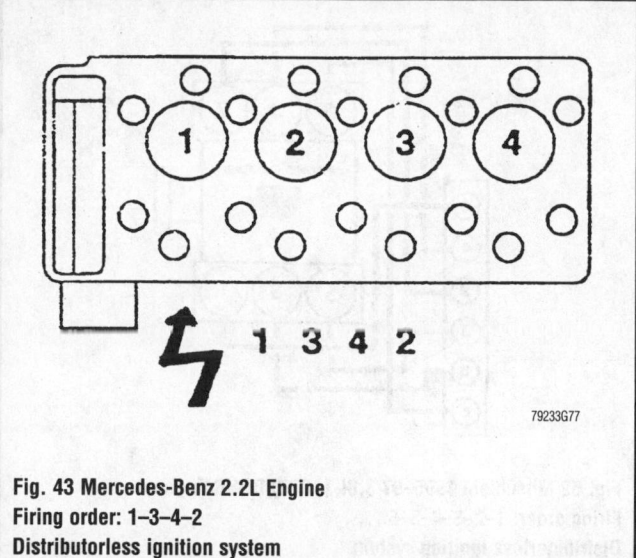

Fig. 43 Mercedes-Benz 2.2L Engine
Firing order: 1–3–4–2
Distributorless ignition system

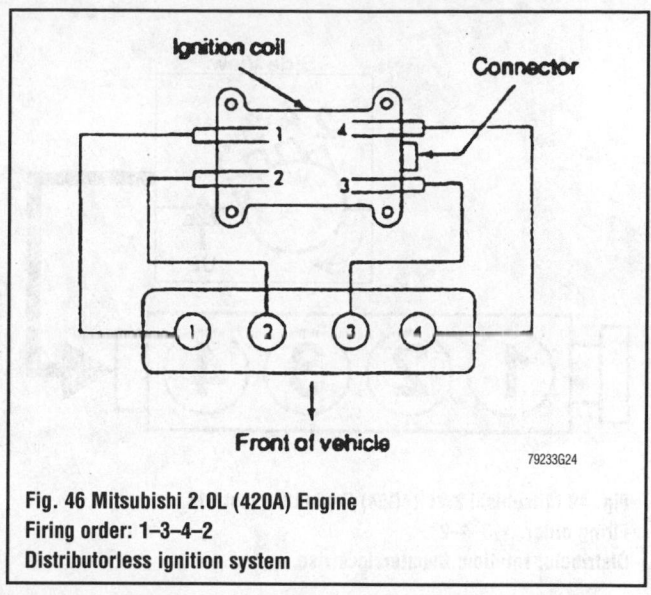

Fig. 46 Mitsubishi 2.0L (420A) Engine
Firing order: 1–3–4–2
Distributorless ignition system

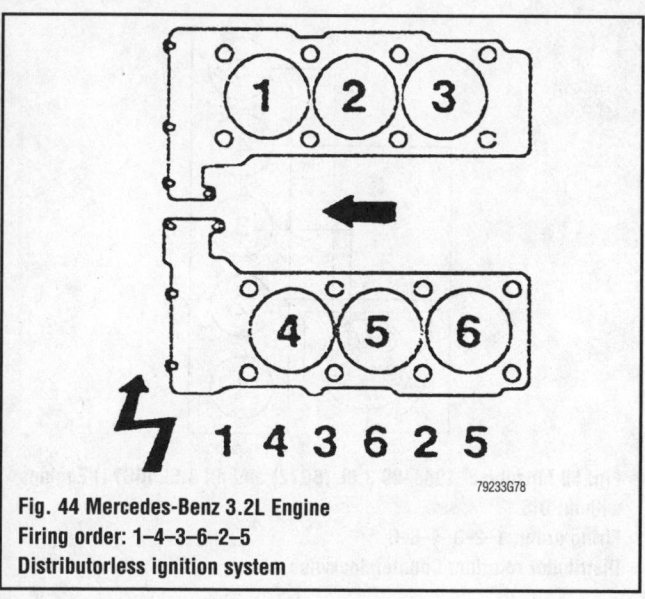

Fig. 44 Mercedes-Benz 3.2L Engine
Firing order: 1–4–3–6–2–5
Distributorless ignition system

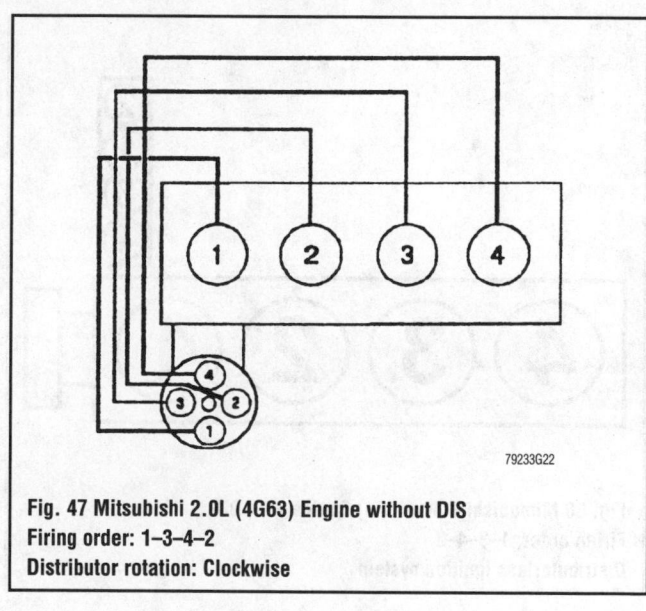

Fig. 47 Mitsubishi 2.0L (4G63) Engine without DIS
Firing order: 1–3–4–2
Distributor rotation: Clockwise

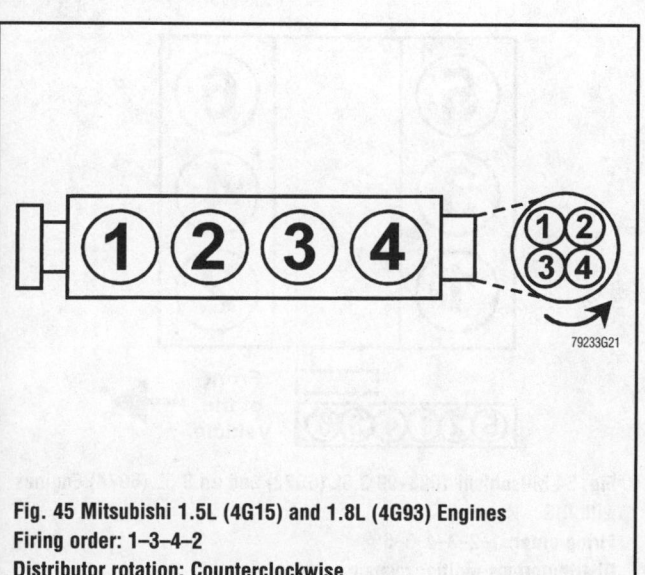

Fig. 45 Mitsubishi 1.5L (4G15) and 1.8L (4G93) Engines
Firing order: 1–3–4–2
Distributor rotation: Counterclockwise

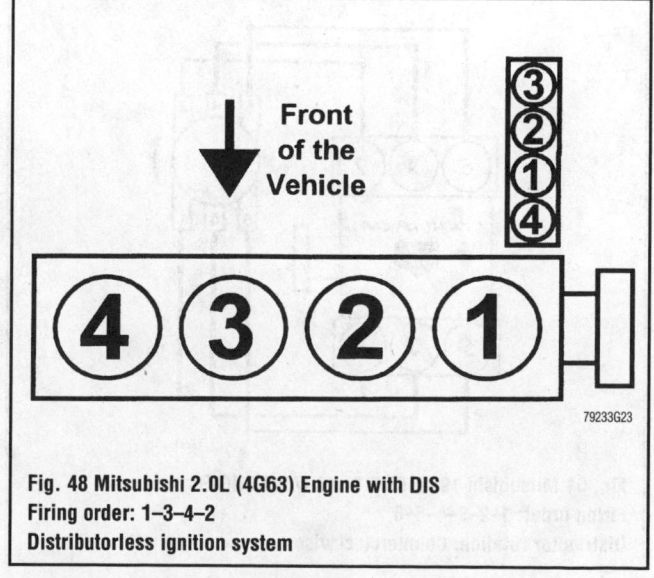

Fig. 48 Mitsubishi 2.0L (4G63) Engine with DIS
Firing order: 1–3–4–2
Distributorless ignition system

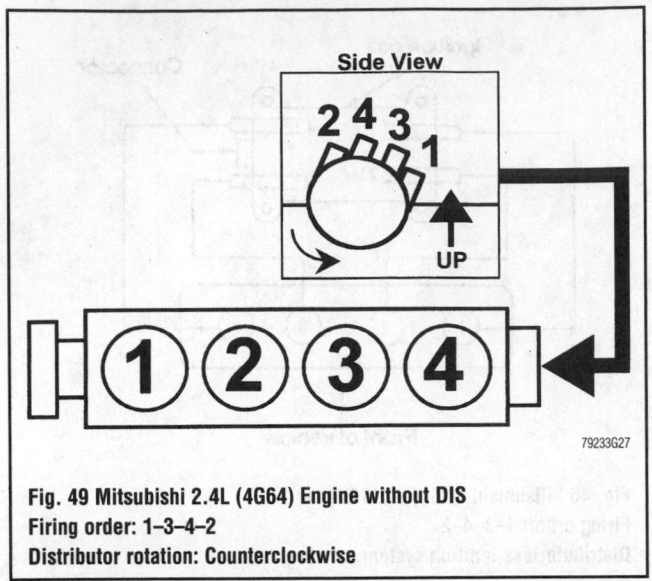

Fig. 49 Mitsubishi 2.4L (4G64) Engine without DIS
Firing order: 1–3–4–2
Distributor rotation: Counterclockwise

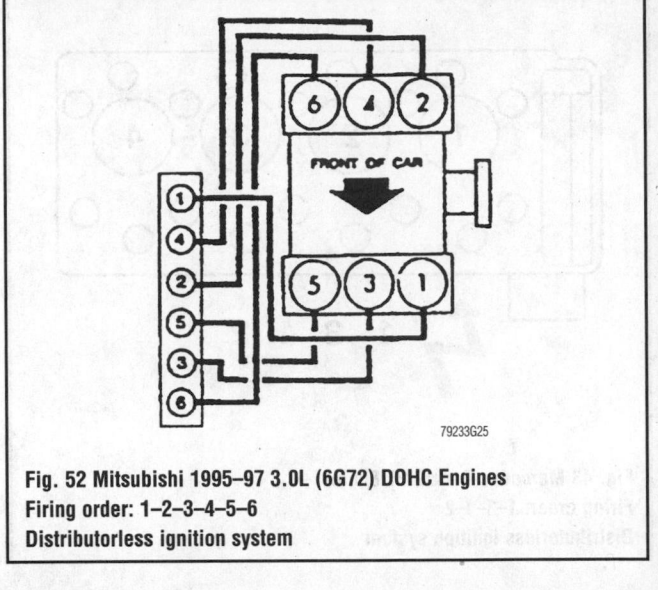

Fig. 52 Mitsubishi 1995–97 3.0L (6G72) DOHC Engines
Firing order: 1–2–3–4–5–6
Distributorless ignition system

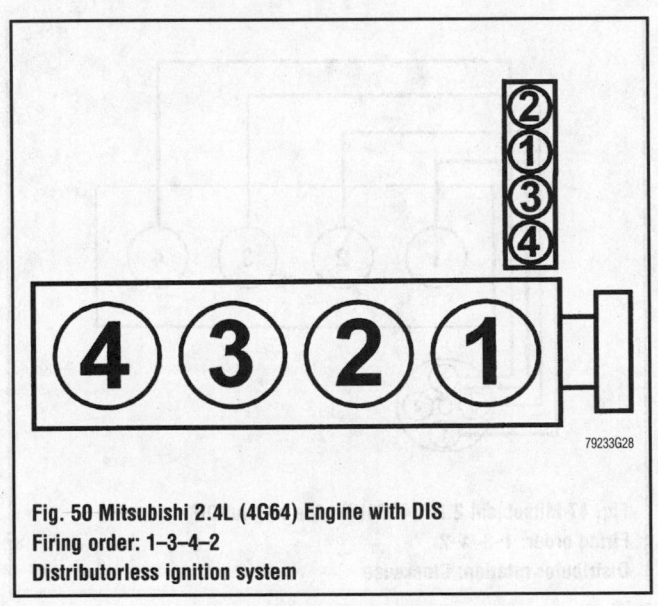

Fig. 50 Mitsubishi 2.4L (4G64) Engine with DIS
Firing order: 1–3–4–2
Distributorless ignition system

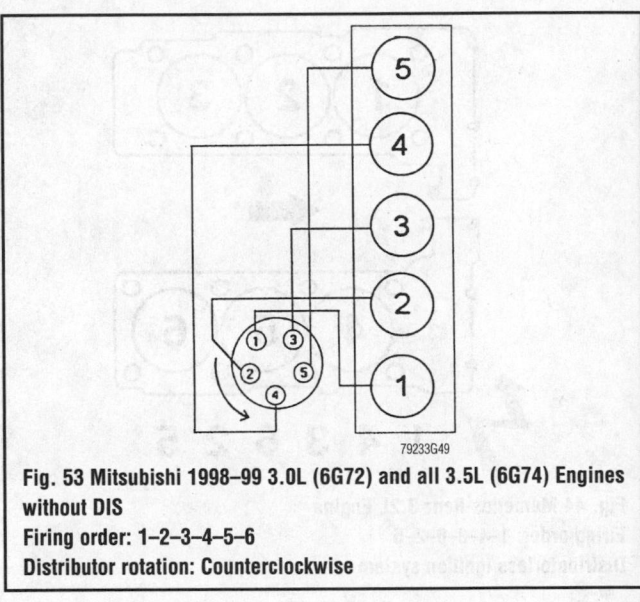

Fig. 53 Mitsubishi 1998–99 3.0L (6G72) and all 3.5L (6G74) Engines without DIS
Firing order: 1–2–3–4–5–6
Distributor rotation: Counterclockwise

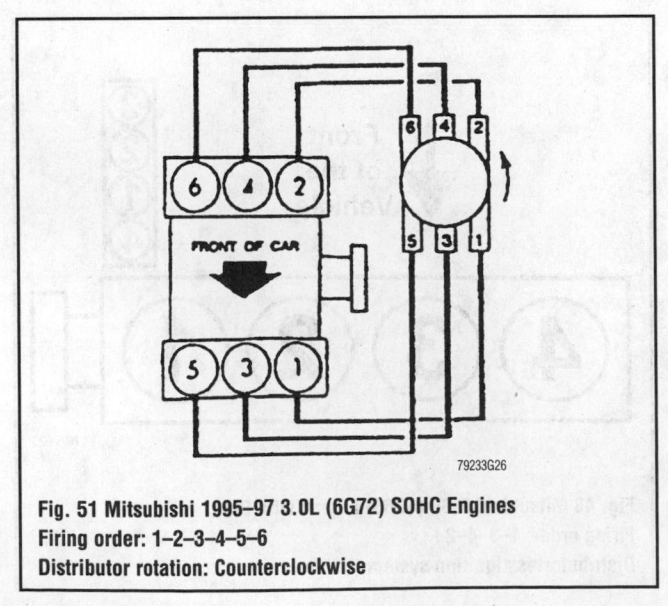

Fig. 51 Mitsubishi 1995–97 3.0L (6G72) SOHC Engines
Firing order: 1–2–3–4–5–6
Distributor rotation: Counterclockwise

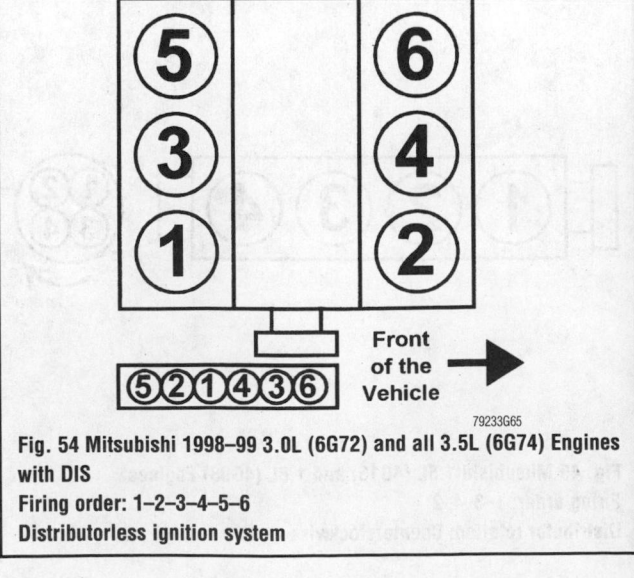

Fig. 54 Mitsubishi 1998–99 3.0L (6G72) and all 3.5L (6G74) Engines with DIS
Firing order: 1–2–3–4–5–6
Distributorless ignition system

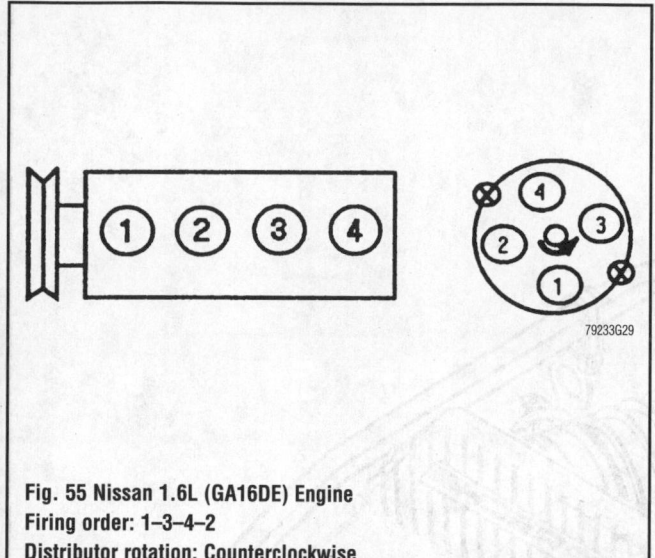

Fig. 55 Nissan 1.6L (GA16DE) Engine
Firing order: 1–3–4–2
Distributor rotation: Counterclockwise

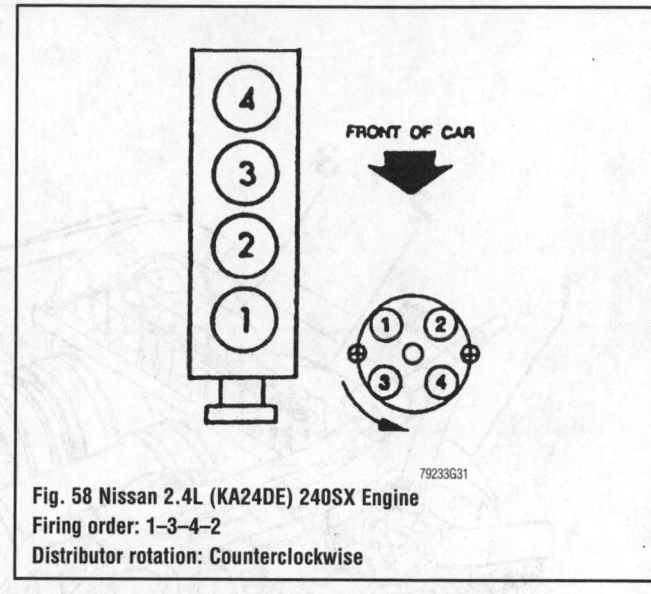

Fig. 58 Nissan 2.4L (KA24DE) 240SX Engine
Firing order: 1–3–4–2
Distributor rotation: Counterclockwise

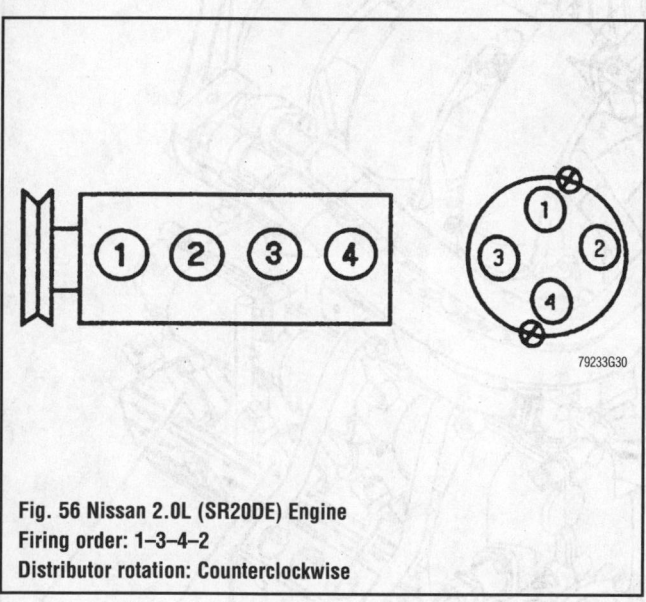

Fig. 56 Nissan 2.0L (SR20DE) Engine
Firing order: 1–3–4–2
Distributor rotation: Counterclockwise

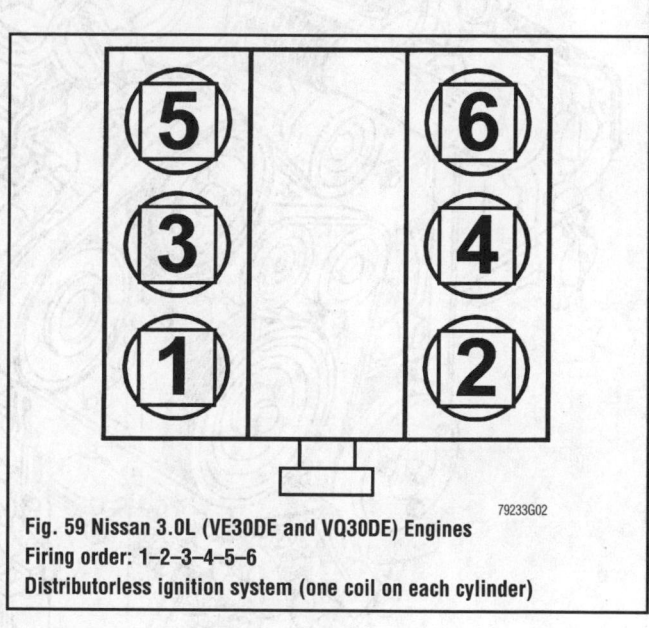

Fig. 59 Nissan 3.0L (VE30DE and VQ30DE) Engines
Firing order: 1–2–3–4–5–6
Distributorless ignition system (one coil on each cylinder)

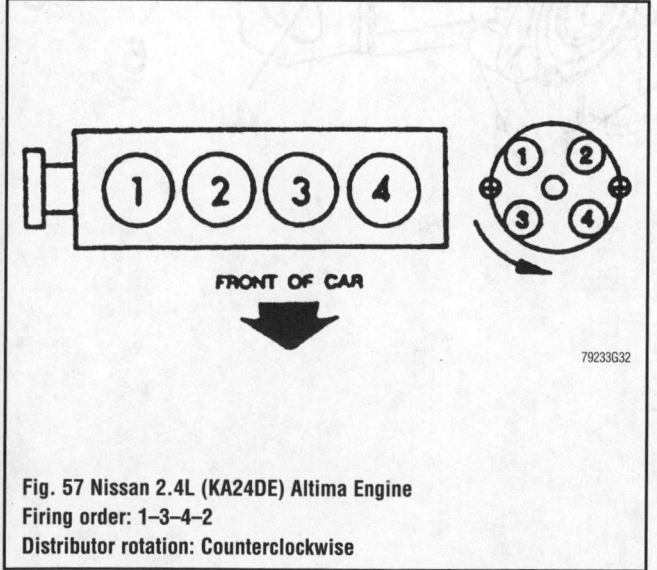

Fig. 57 Nissan 2.4L (KA24DE) Altima Engine
Firing order: 1–3–4–2
Distributor rotation: Counterclockwise

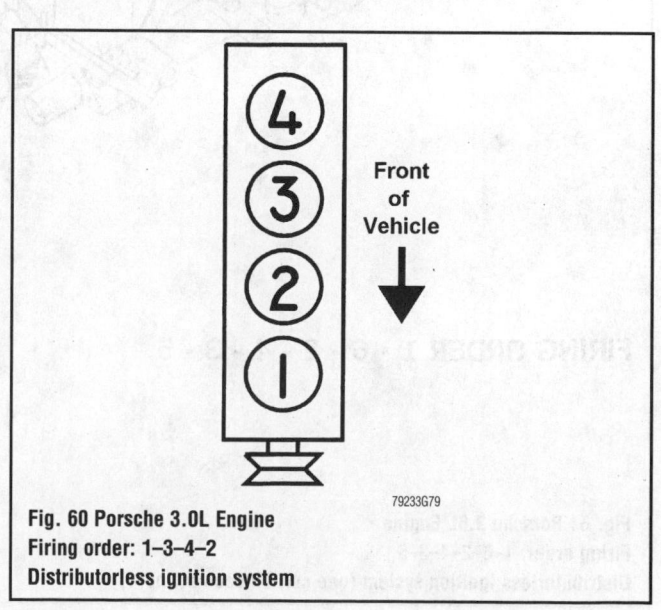

Fig. 60 Porsche 3.0L Engine
Firing order: 1–3–4–2
Distributorless ignition system

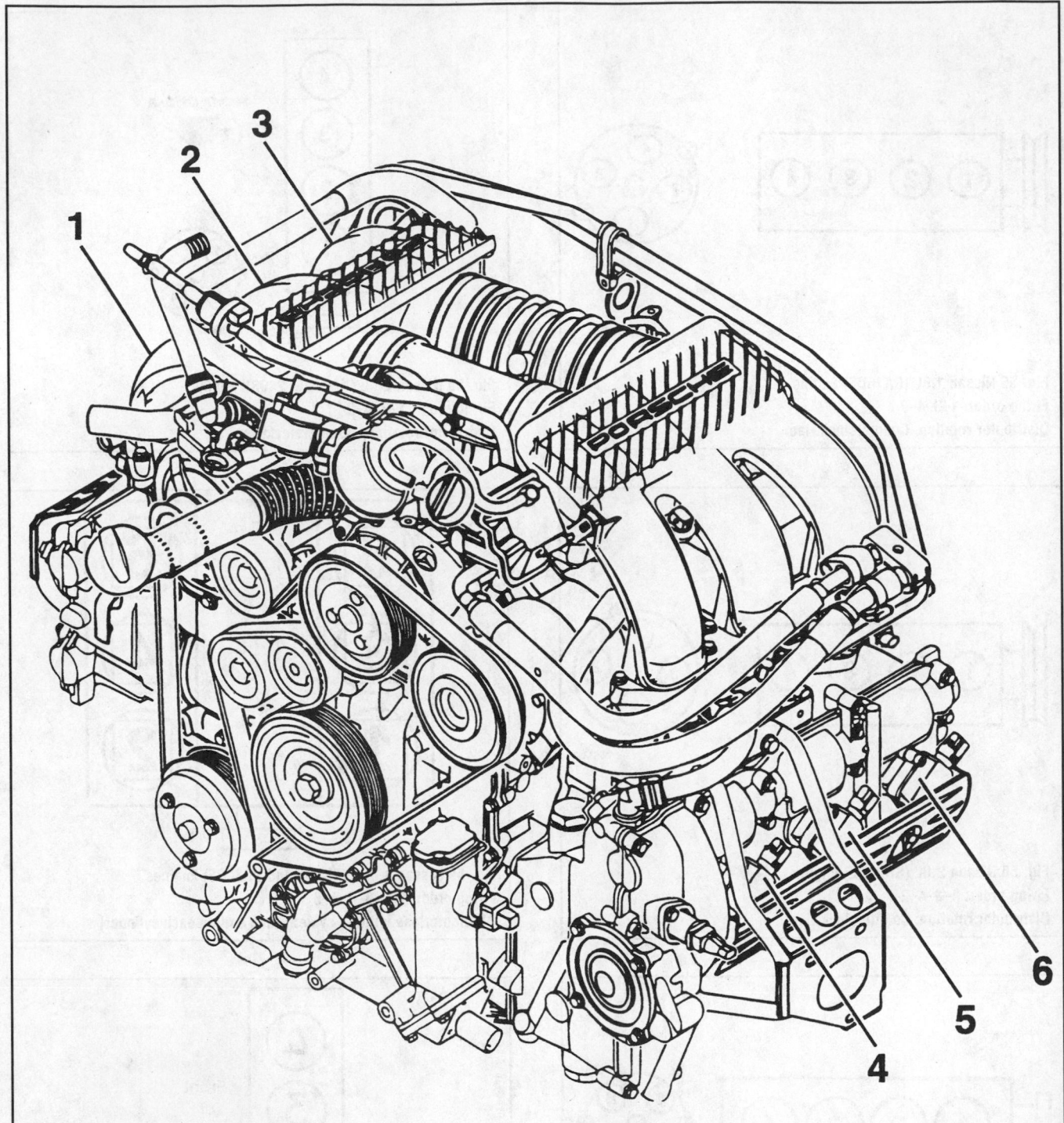

FIRING ORDER 1 - 6 - 2 - 4 - 3 - 5

79233G80

Fig. 61 Porsche 3.6L Engine
Firing order: 1–6–2–4–3–5
Distributorless ignition system (one coil on each cylinder)

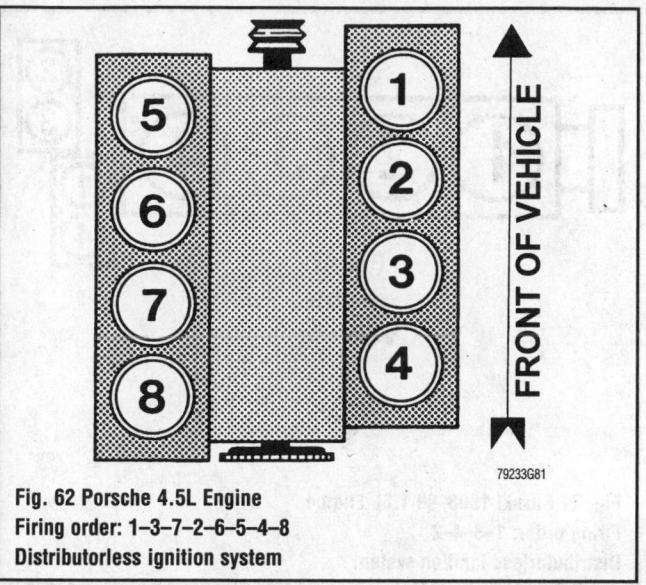

Fig. 62 Porsche 4.5L Engine
Firing order: 1–3–7–2–6–5–4–8
Distributorless ignition system

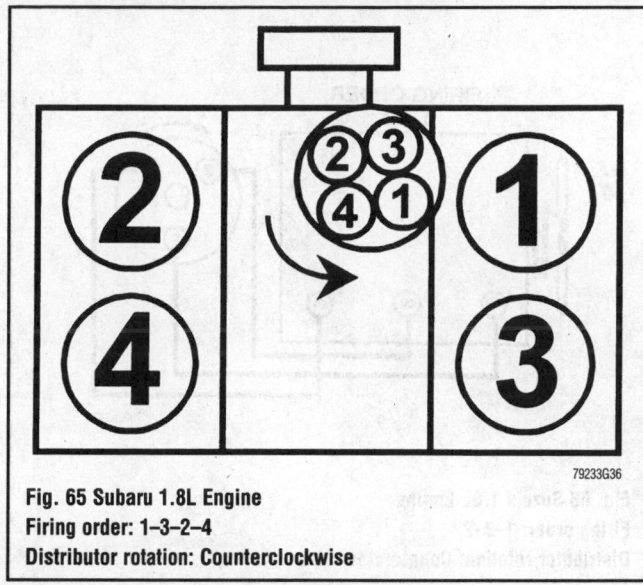

Fig. 65 Subaru 1.8L Engine
Firing order: 1–3–2–4
Distributor rotation: Counterclockwise

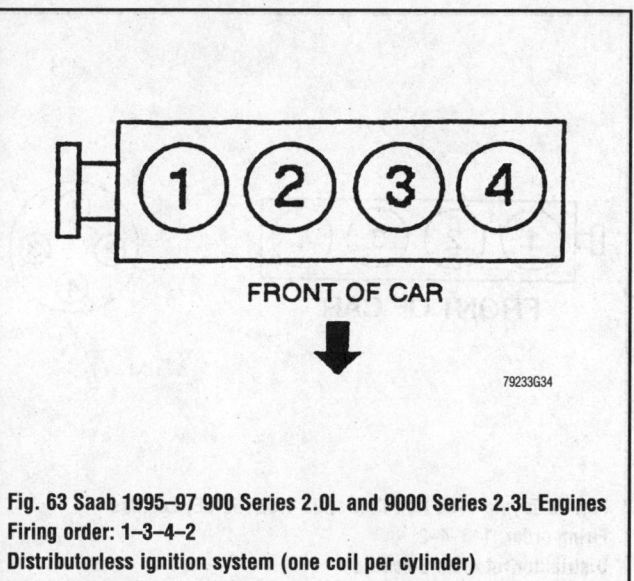

Fig. 63 Saab 1995–97 900 Series 2.0L and 9000 Series 2.3L Engines
Firing order: 1–3–4–2
Distributorless ignition system (one coil per cylinder)

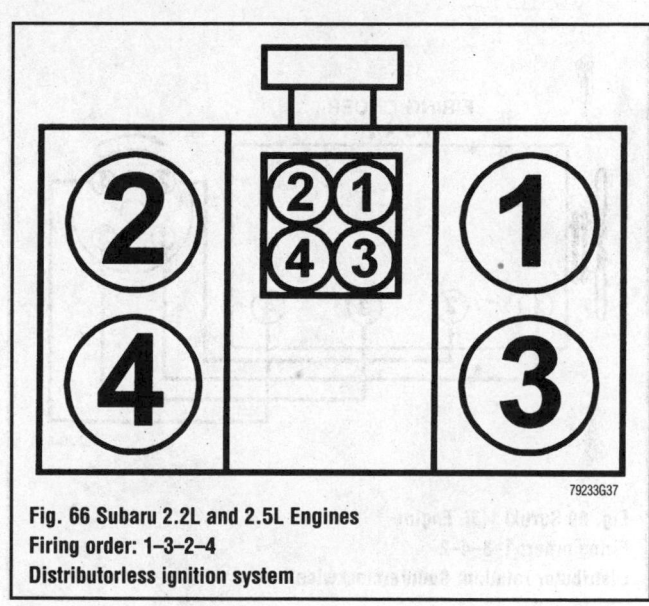

Fig. 66 Subaru 2.2L and 2.5L Engines
Firing order: 1–3–2–4
Distributorless ignition system

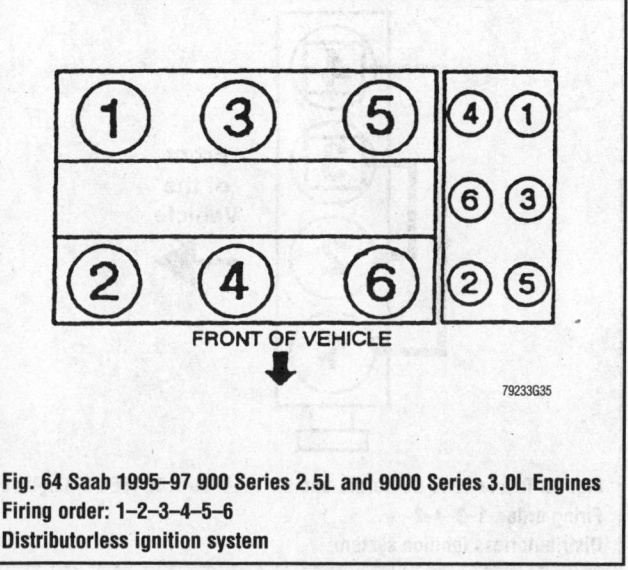

Fig. 64 Saab 1995–97 900 Series 2.5L and 9000 Series 3.0L Engines
Firing order: 1–2–3–4–5–6
Distributorless ignition system

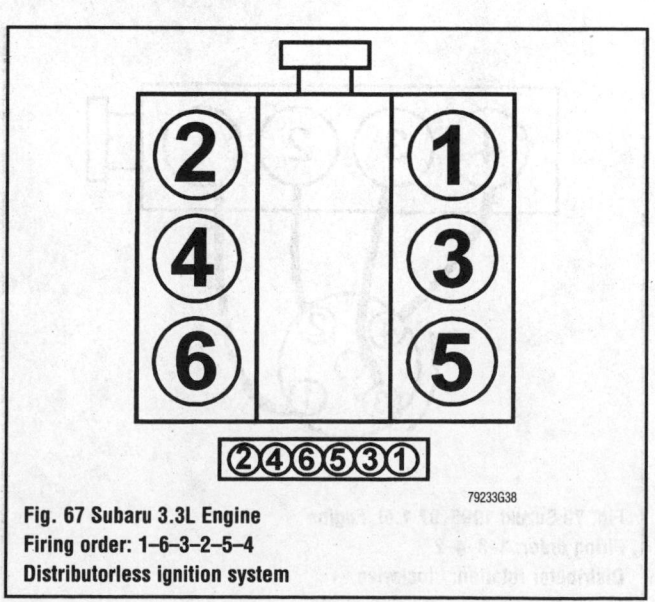

Fig. 67 Subaru 3.3L Engine
Firing order: 1–6–3–2–5–4
Distributorless ignition system

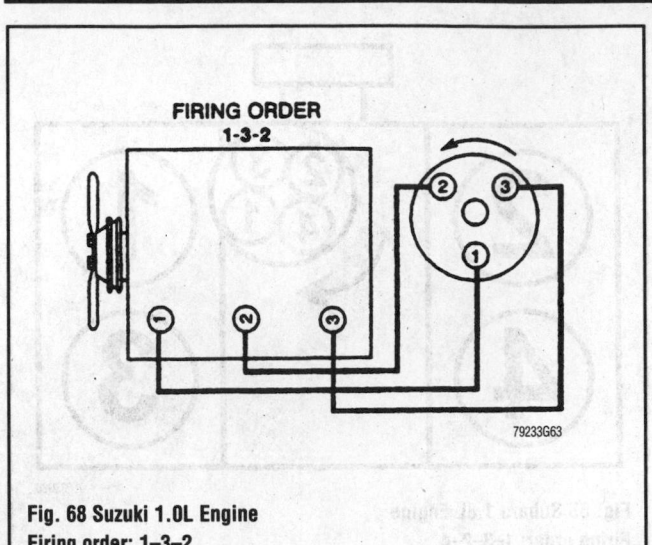

Fig. 68 Suzuki 1.0L Engine
Firing order: 1–3–2
Distributor rotation: Counterclockwise

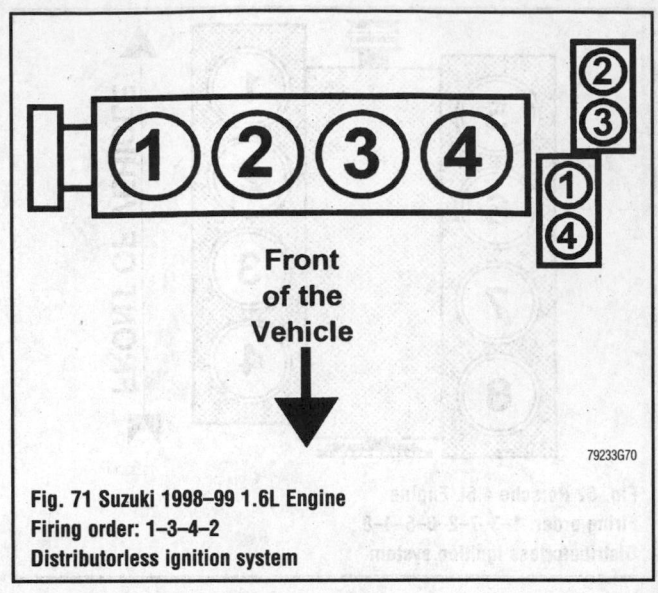

Fig. 71 Suzuki 1998–99 1.6L Engine
Firing order: 1–3–4–2
Distributorless ignition system

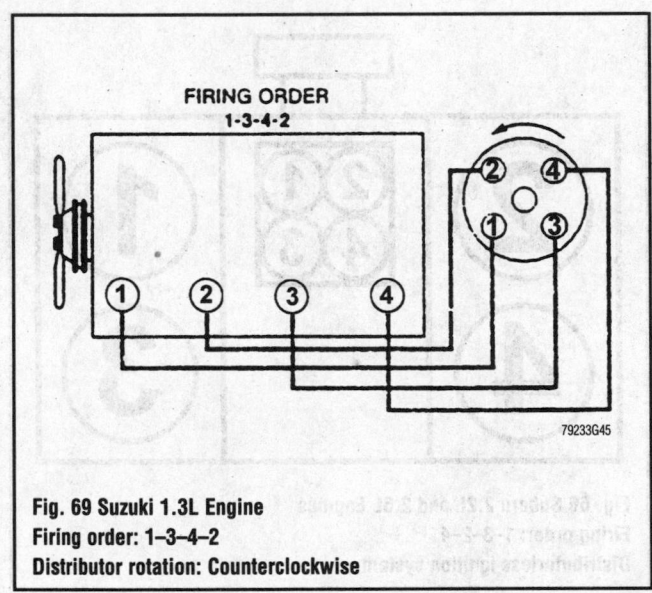

Fig. 69 Suzuki 1.3L Engine
Firing order: 1–3–4–2
Distributor rotation: Counterclockwise

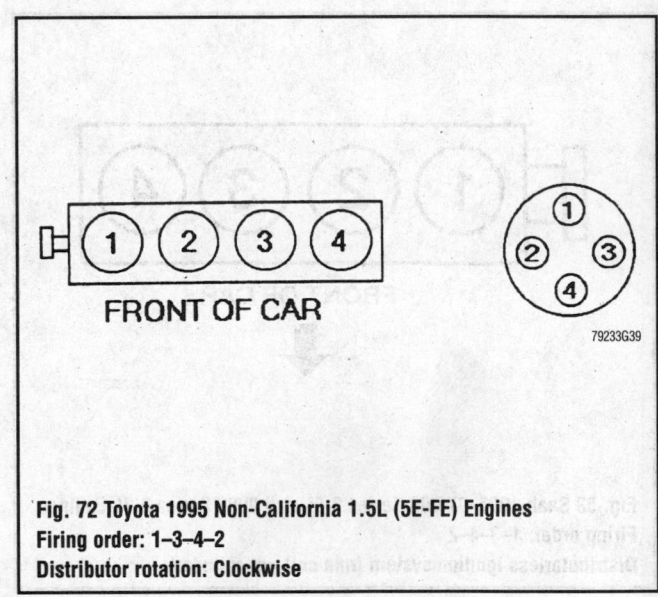

Fig. 72 Toyota 1995 Non-California 1.5L (5E-FE) Engines
Firing order: 1–3–4–2
Distributor rotation: Clockwise

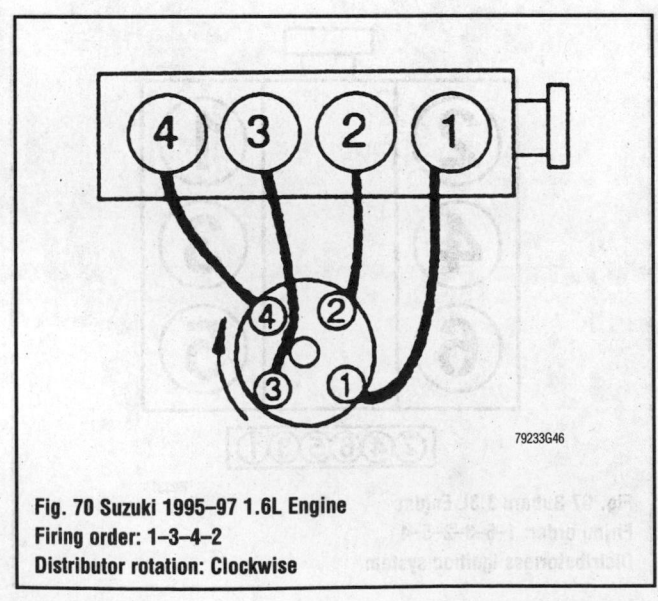

Fig. 70 Suzuki 1995–97 1.6L Engine
Firing order: 1–3–4–2
Distributor rotation: Clockwise

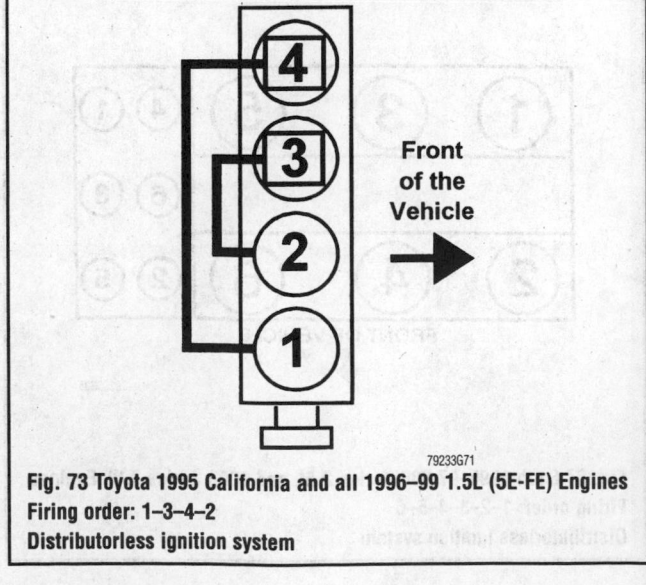

Fig. 73 Toyota 1995 California and all 1996–99 1.5L (5E-FE) Engines
Firing order: 1–3–4–2
Distributorless ignition system

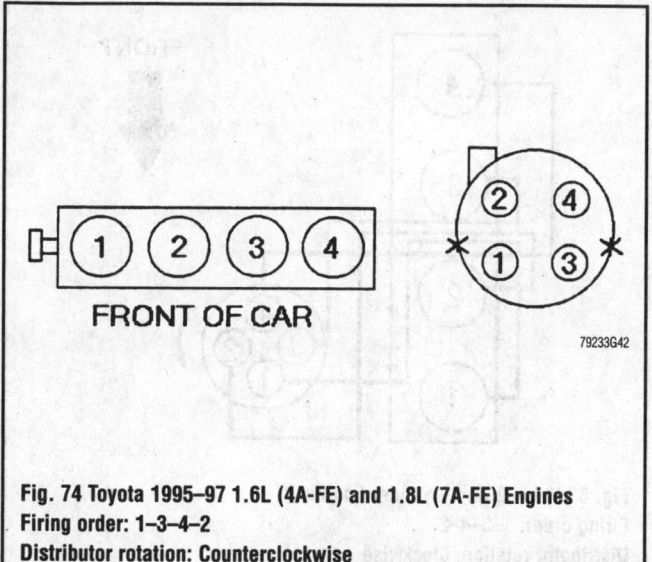

Fig. 74 Toyota 1995–97 1.6L (4A-FE) and 1.8L (7A-FE) Engines
Firing order: 1–3–4–2
Distributor rotation: Counterclockwise

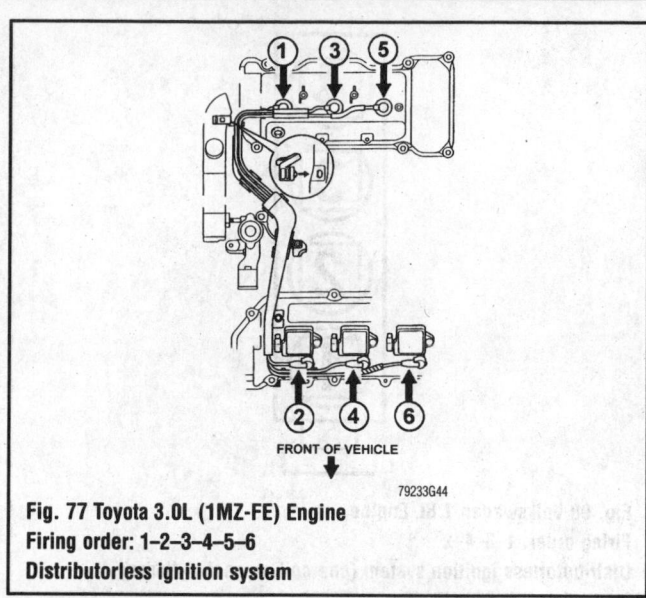

Fig. 77 Toyota 3.0L (1MZ-FE) Engine
Firing order: 1–2–3–4–5–6
Distributorless ignition system

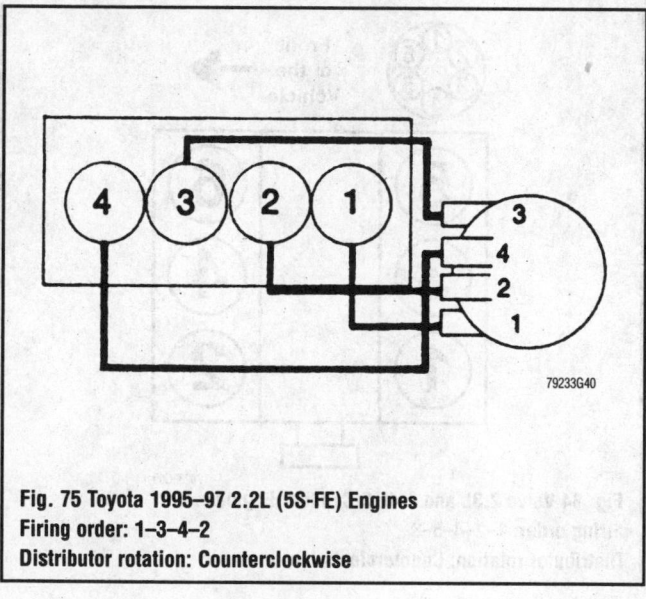

Fig. 75 Toyota 1995–97 2.2L (5S-FE) Engines
Firing order: 1–3–4–2
Distributor rotation: Counterclockwise

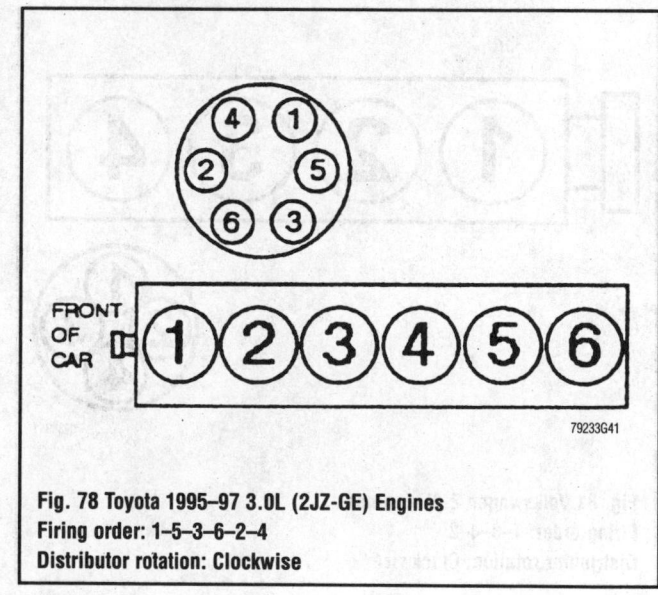

Fig. 78 Toyota 1995–97 3.0L (2JZ-GE) Engines
Firing order: 1–5–3–6–2–4
Distributor rotation: Clockwise

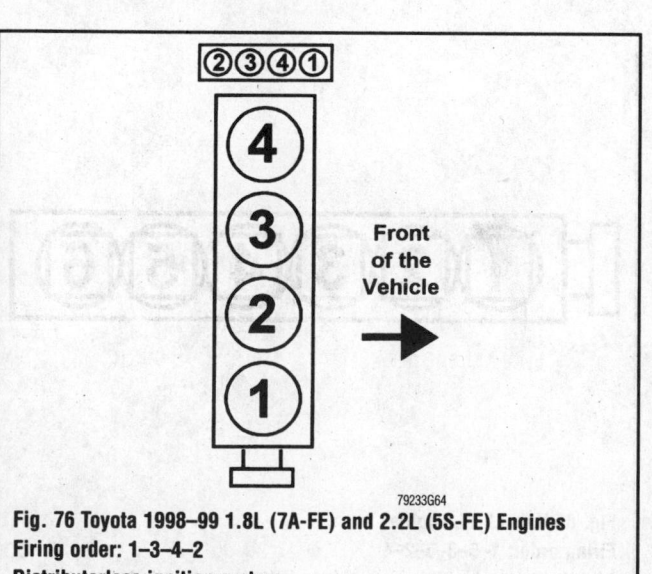

Fig. 76 Toyota 1998–99 1.8L (7A-FE) and 2.2L (5S-FE) Engines
Firing order: 1–3–4–2
Distributorless ignition system

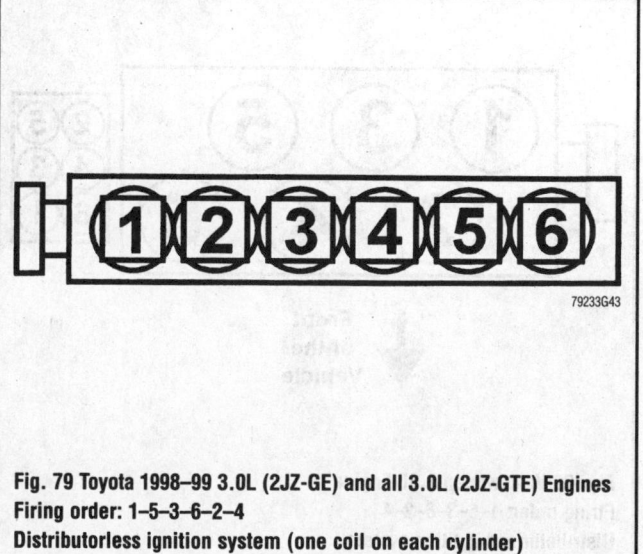

Fig. 79 Toyota 1998–99 3.0L (2JZ-GE) and all 3.0L (2JZ-GTE) Engines
Firing order: 1–5–3–6–2–4
Distributorless ignition system (one coil on each cylinder)

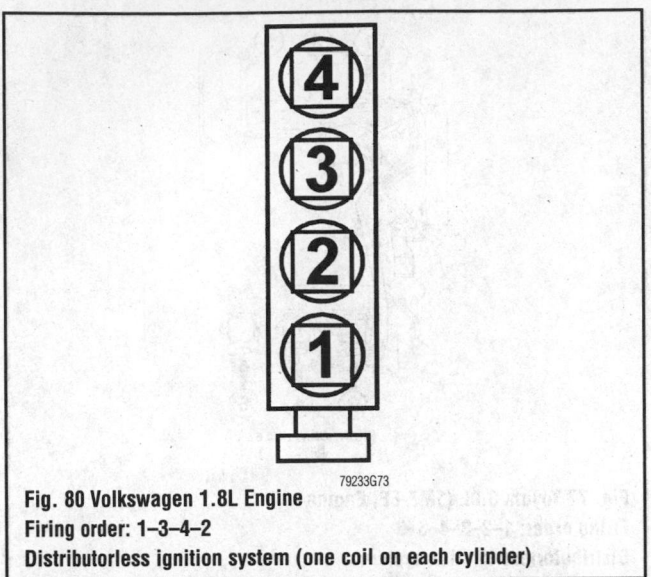

Fig. 80 Volkswagen 1.8L Engine
Firing order: 1–3–4–2
Distributorless ignition system (one coil on each cylinder)

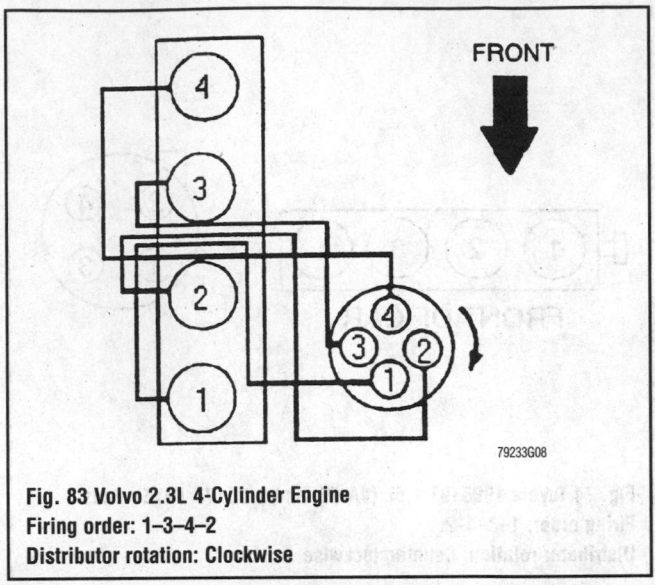

Fig. 83 Volvo 2.3L 4-Cylinder Engine
Firing order: 1–3–4–2
Distributor rotation: Clockwise

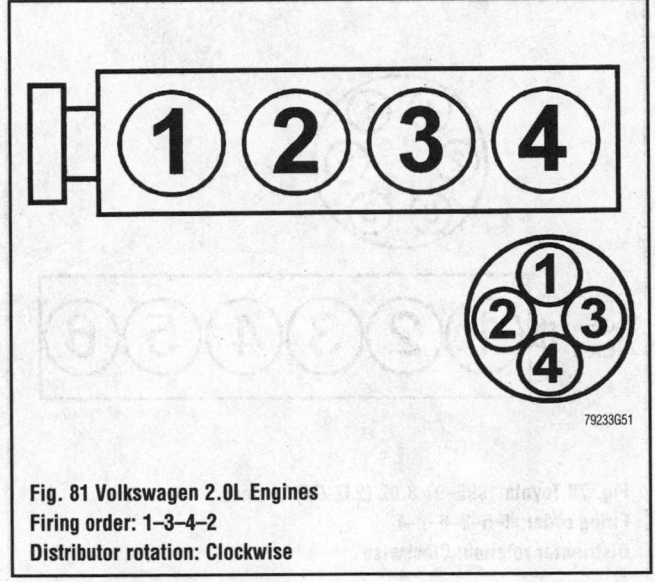

Fig. 81 Volkswagen 2.0L Engines
Firing order: 1–3–4–2
Distributor rotation: Clockwise

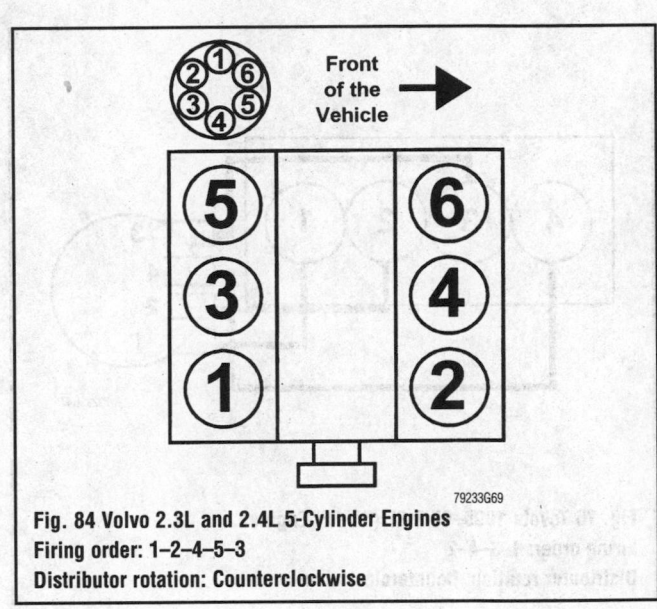

Fig. 84 Volvo 2.3L and 2.4L 5-Cylinder Engines
Firing order: 1–2–4–5–3
Distributor rotation: Counterclockwise

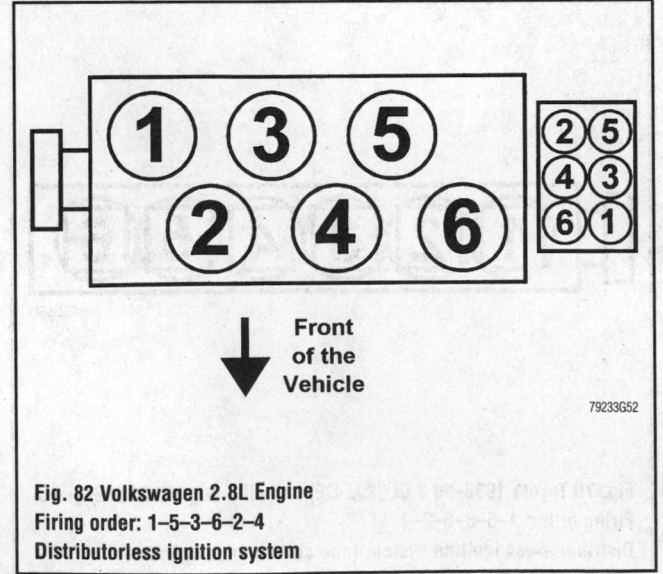

Fig. 82 Volkswagen 2.8L Engine
Firing order: 1–5–3–6–2–4
Distributorless ignition system

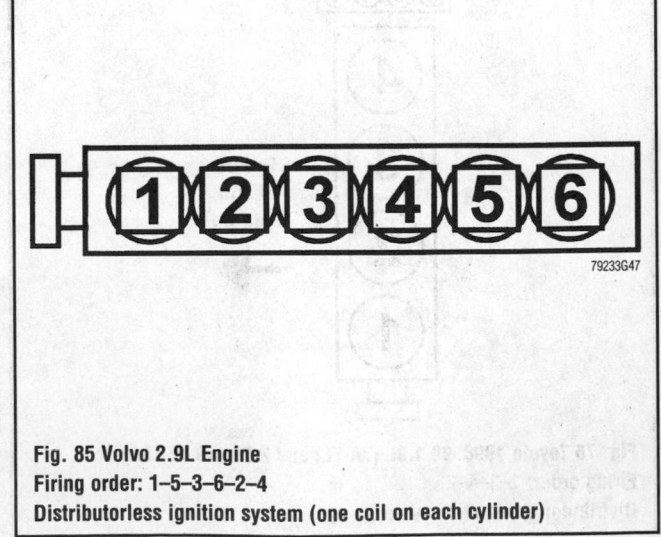

Fig. 85 Volvo 2.9L Engine
Firing order: 1–5–3–6–2–4
Distributorless ignition system (one coil on each cylinder)

ACCESSORY DRIVE BELTS 4

ACCESSORY DRIVE BELTS

Accessory drive belts are usually divided into two basic types: V-belts (conventional, cogged, and flat multi-ribbed) and serpentine (multi-ribbed) belts. The flat multi-ribbed V-belt actually resembles a serpentine belt, however, unlike a serpentine belt, only the inner surface of the belt makes contact with the components' pulleys. (Rarely, the back of multi-ribbed V-belts may ride against an idler or tensioner pulley, however.) V-belts ride in pulleys with V-shaped groove(s) to rotate various accessories, such as the power steering pump, air conditioner compressor, alternator/generator, water pump, and air pump. Only the inside of a V-belt is used, unlike a serpentine belt which utilizes both sides. V-belts typically operate one or two accessories per belt, whereas a single serpentine belt can drive all of the accessories. V-belts and a few serpentine belts require periodic adjustment because the belts are under tension and stretch over time. Most serpentine belts utilize an automatic belt tensioner that constantly provides the proper tension to the belt.

V-Belts

INSPECTION

Although different maintenance intervals are given by each manufacturer, it is a good rule of thumb to inspect the drive belts every 15,000 miles (24,000 km) or 12 months (whichever occurs first). Determine the belt tension at a point half-way between the pulleys by pressing on the belt with moderate thumb pressure. The belt should deflect about ¼ – ½ in. (6–13mm) at this point. Note that

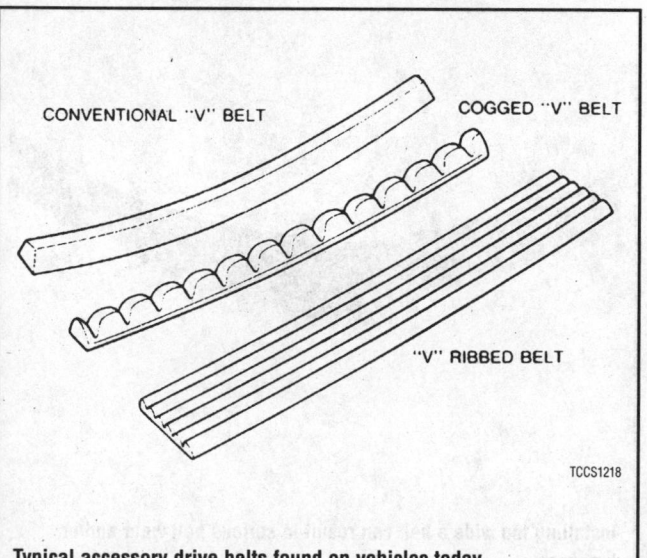

CONVENTIONAL "V" BELT   COGGED "V" BELT

"V" RIBBED BELT

TCCS1218

Typical accessory drive belts found on vehicles today

"deflection" is not play, but the ability of the belt, under actual tension, to stretch slightly and give.

Inspect the belts for the following signs of damage or wear: glazing, cracking, fraying, crumbling or missing chunks. A glazed belt will be perfectly smooth from slippage, while a good belt will have a slight texture of fabric visible. Cracks will usually start at the inner edge of the belt and run outward. A belt that is fraying will have the fabric backing de-laminating its self from the belt. A belt that is crumbling or missing chunks will have voids in the cross-section of the belt, some times the section missing chunks will be in the pulley groove and not easily seen. All worn or damaged drive belts should be replaced immediately. It is best to replace all drive belts at one time, as a preventive maintenance measure.

Although it is generally easier on the component to have the belt too loose than too tight, a very loose belt may place a high impact load on a bearing due to the whipping or snapping action of the belt. A belt that is slightly loose may slip, especially when component loads are high. This slippage may be hard to identify. For example, the generator belt may run okay during the day,, then slip at night when headlights are turned on. Slipping belts wear quickly not only due to the direct effect of slippage but also because of the heat the slippage generates. Extreme slippage may even cause a belt to burn. A very smooth, glazed appearance on the belt's sides, as opposed to the obvious pattern of a fabric cover, indicates that the belt has been slipping.

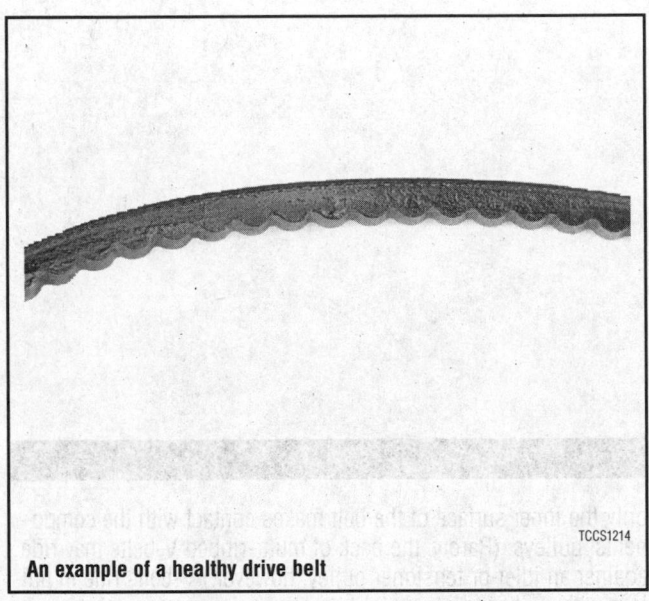

TCCS1214

An example of a healthy drive belt

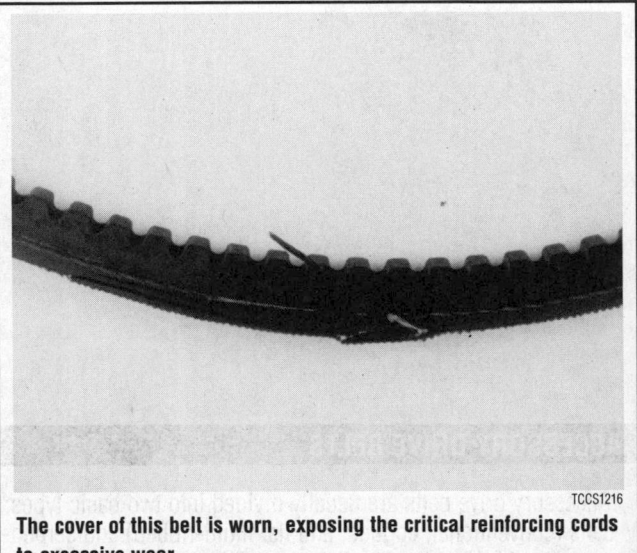

TCCS1216

The cover of this belt is worn, exposing the critical reinforcing cords to excessive wear

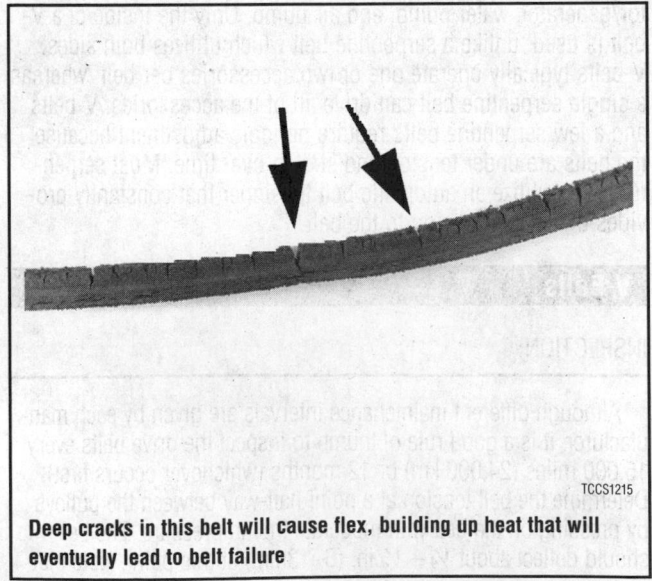

TCCS1215

Deep cracks in this belt will cause flex, building up heat that will eventually lead to belt failure

TCCS1217

Installing too wide a belt can result in serious belt wear and/or breakage

ADJUSTMENT

> ✳✳ **CAUTION**
>
> **On vehicles with an electric cooling fan, disable the power to the fan by disengaging the fan motor wiring connector or removing the negative battery cable before replacing or adjusting the drive belts. Otherwise, the fan may engage even though the ignition is OFF.**

Belt tension can be checked by pressing on the belt at the center point of its longest straight span. The belt should give approximately ¼ – ½ in. (6–13mm). If the belt is loose it will slip, whereas if the belt is too tight it will damage the bearings in the driven unit.

For the purposes of V-belt tensioning, there are generally three types of mounting for the various components driven by the drive belt. The first method, referred to as pivoting type without adjuster, is designed so that the component is secured by at least 2 bolts. One of the bolts is a pivoting bolt and the other is the lockbolt. When both bolts are loosened so that the component may move, the component pivots on the pivoting bolt. The lockbolt passes through the component and a slotted bracket, so that when the lockbolt's nut is tightened the component is held in that position. There are not automatic adjusting mechanisms used with this type of mounting.

The second method of component mounting, referred to as pivoting type with adjuster, is almost identical except for the addition of an adjuster of some sort. Usually the adjuster is composed of a bracket attached to the component and a threaded adjusting bolt. After loosening the pivoting and lockbolts, the adjusting bolt can be tightened or loosened to increase or decrease the drive belt's tension. With this type of mounting, you do not have to hold the component in a tensioned position and tighten the pivoting and lockbolts; the adjusting bolt does the job for you.

A typical pivoting accessory with an adjusting bolt

An accessory that is fixed will have an adjustable pulley—notice the square slot to aid the adjustment

Some versions of this method of mounting use an adjuster which is built into one of the components mounting braces. The brace attaches the component to the engine and incorporates a threaded adjuster in its mid-span, so that when the threaded adjuster is turned the brace shortens or lengthens. This in turn increases or decreases the amount of tension on the component.

The third type of mounting, referred to as stationary type, is designed so that the component is mounted on its brackets. There are no pivoting or lockbolts, and the component is not designed to be moved. Rather, this type of mounting uses an extra tensioner idler pulley assembly. The drive belt is tensioned by adjusting the position of the idler pulley, usually accomplished by turning the adjuster bolt on the idler mechanism.

Pivoting Type

WITHOUT ADJUSTER

1. Disconnect the negative battery cable.
2. Loosen the component's lockbolt and pivoting bolt only enough for the component to move.
3. Using a strong wooden, plastic or metal prytool, move the component either closer to, or farther away from, the engine to provide the correct tension on the belt.

> ✳✳ **WARNING**
>
> **If using a metal prytool, always wrap the end with a rag or towel to prevent accidentally damaging the component from undue stress.**

4. Once the proper amount of tension is applied to the drive belt, hold the prytool with one hand while tightening the lockbolt securely with the other hand.
5. Release the pressure from the prytool and tighten the pivoting bolt securely.
6. Double check the drive belt's tension, in case the component moved slightly while tightening the bolts.
7. Connect the negative battery cable.

WITH ADJUSTER

This type of drive belt is tensioned by a tensioner, which makes precise tension adjustment easy.

1. Disconnect the negative battery cable.
2. Loosen the component's pivot and lockbolts.
3. Inspect the tensioner assembly on the component; the tensioner adjusting bolt may use a locknut or screw to prevent it from loosening over time. On the type of adjuster with a threaded mounting brace, there may be two jam nuts used on either side of the threaded coupling. If such locking fasteners are found, loosen them.
4. Turn the tensioner adjusting bolt or threaded coupling to increase or decrease the amount of tension on the drive belt, as necessary.
5. When the belt tension is correct, tighten the lockbolt and the pivot bolt.
6. If equipped, tighten the tension adjusting bolt locknut or screw to prevent the adjuster from slowly loosening over time. If equipped, tighten the two jam nuts.
7. Connect the negative battery cable.

Stationary Type

IDLER PULLEY WITH ADJUSTING BOLT

1. Loosen the idler bracket pivot bolt and locking bolts.
2. Adjust the belt tension by inserting the proper size ratchet in the square slot of the idler bracket and rotating the bracket until tension is applied.
3. While holding the tension on the belt with the ratchet, tighten the locking bolts, then the pivot bolt.

IDLER PULLEY WITHOUT ADJUSTING BOLT

1. Loosen the mounting/pivot bolt behind the idler pulley.
2. Swivel the idler pulley with a pair of pliers or a wrench on the bearing mounting until the proper tension is achieved.
3. While holding the idler pulley, at the proper tension, tighten the mounting/pivot bolt.

REMOVAL & INSTALLATION

If a belt must be replaced, the driven unit or idler pulley must be loosened and moved to its extreme loosest position, generally by moving it toward the center of the motor. After removing the old belt, check the pulleys for dirt or built-up material which could affect belt contact. Carefully install the new belt, remembering that it is new and unused; it may appear to be just a little too small to fit over the pulley flanges. Fit the belt over the largest pulley (usually the crankshaft pulley at the bottom center of the motor) first, then work on the smaller one(s). Gentle pressure in the direction of rotation is helpful. Some belts run around a third, or idler pulley, which acts as an additional pivot in the belt's path. It may be possible to loosen the idler pulley as well as the main component, making your job much easier. Depending on which belt(s) you are changing, it may be necessary to loosen or remove other interfering belts to get at the one(s) you want.

When buying replacement belts, remember that the fit is critical according to the length of the belt ("diameter"), the width of the belt, the depth of the belt and the angle or profile of the V shape or the ribs. The belt shape should match the shape of the pulley exactly; belts that are not an exact match can cause noise, slippage and premature failure.

After the new belt is installed, draw tension on it by moving the driven unit or idler pulley away from the motor and tighten its mounting bolts. This is sometimes a three or four-handed job; you may find an assistant helpful. Be sure that all the bolts you loosened get retightened and that any other loosened belts also have the correct tension. A new belt can be expected to stretch a bit after installation so be prepared to readjust your new belt, if needed, within the first two hundred miles of use.

Pivoting Type

> **✳✳ CAUTION**
>
> **On vehicles with an electric cooling fan, disable the power to the fan by disengaging the fan motor wiring connector or removing the negative battery cable before replacing or adjusting the drive belts. Otherwise, the fan may engage even though the ignition is OFF.**

WITHOUT ADJUSTER

1. Disconnect the negative battery cable.
2. Loosen the accessory's slotted adjusting bracket bolt. If the hinge bolt is excessively tight, it too will have to be loosened.
3. Push the component toward the engine to provide enough slack in the belt so that it will slide over one of the accessory drive pulleys. Remove the drive belt from the accessory drive pulleys and from the vehicle.

To install:

4. Position the new drive belt over the component pulleys. Be sure that it is routed correctly.
5. Adjust the tension of the belt, as described earlier in this section.
6. Connect the negative battery cable.

WITH ADJUSTER

1. Disconnect the negative battery cable.
2. Loosen the component's pivot and lockbolts.
3. Inspect the tensioner assembly on the component; the tensioner adjusting bolt may use a locknut or screw to prevent it from loosening over time. On the type of adjuster with a threaded mounting brace, there may be two jam nuts used on either side of the threaded coupling. If such locking fasteners are found, loosen them.
4. Turn the tensioner adjusting bolt or threaded coupling to relieve all tension from the drive belt until the most possible slack is gained from the component.
5. Slip the belt off of the accessory pulley, then remove it from the other pulleys. Remove the belt from the vehicle.

To install:

6. Route the new belt on the component pulleys. Make certain that it is routed correctly; incorrect routing could cause a components to spin backward, possibly damaging it.

7. Once the belt is correctly positioned on all of the pulleys, adjust the tension as described earlier in this section.

8. Connect the negative battery cable.

Stationary Type

IDLER PULLEY WITH ADJUSTING BOLT

1. Disconnect the negative battery cable.
2. Loosen the idler bracket pivot bolt and locking bolts.
3. Move the idler pulley until the most amount of slack is gained.
4. Remove the drive belt from the accessory pulley, then from the other applicable pulleys.

To install:

5. Position the new belt over the crankshaft pulley, the idler pulley and the accessory pulley. Make certain that it is correctly routed, otherwise it could cause the accessory to be rotated backward. This could cause damage to the accessory.

6. Adjust the belt tension, as described earlier in this section.

7. While holding the tension on the belt with the ratchet, tighten the locking bolts, then the pivot bolt.

8. Connect the negative battery cable.

IDLER PULLEY WITHOUT ADJUSTING BOLT

1. Disconnect the negative battery cable.
2. Loosen the mounting/pivot bolt behind the idler pulley.
3. Remove the drive belt from the accessory pulley, then from the other applicable pulleys.

To install:

4. Position the new belt over the crankshaft pulley, the idler pulley and the accessory pulley. Make certain that it is correctly routed,

otherwise it could cause the accessory to be rotated backwards. This could cause damage to the accessory.

5. Swivel the idler pulley with a pair of pliers or a wrench on the bearing mounting until the proper tension is achieved.

6. While holding the idler pulley, at the proper tension, tighten the mounting/pivot bolt.

7. Connect the negative battery cable.

Serpentine Belts

INSPECTION

Although many manufacturers recommend that the drive belt(s) be inspected every 30,000 miles (48,000 km) or more, it is really a good idea to check them at least once a year, or at every major fluid change. Whichever interval you choose, the belts should be checked for wear or damage. Obviously, a damaged drive belt can cause problems should it give way while the vehicle is in operation. But, improper length belts (too short or long), as well as excessively worn belts, can also cause problems. Loose accessory drive belts can lead to poor engine cooling and diminished output from the alternator, air conditioning compressor or power steering pump. A belt that is too tight places a severe strain on the driven unit and can wear out bearings quickly.

Serpentine drive belts should be inspected for rib chunking (pieces of the ribs breaking off), severe glazing, frayed cords or other visible damage. Any belt which is missing sections of 2 or more adjacent ribs which are ½ in. (13mm) or longer must be replaced. You might want to note that serpentine belts do tend to form small cracks across the backing. If the only wear you find is in the form of one or more cracks are across the backing and NOT parallel to the ribs, the belt is still good and does not need to be replaced.

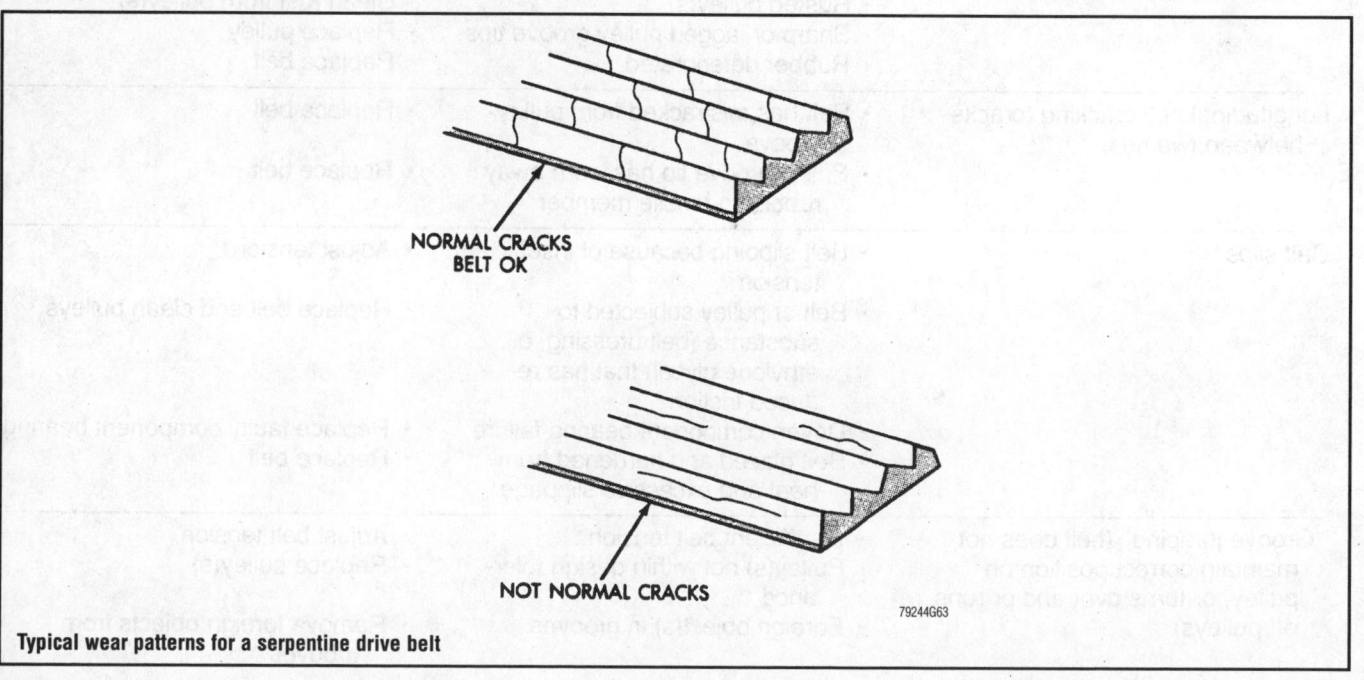

NORMAL CRACKS
BELT OK

NOT NORMAL CRACKS

79244G63

Typical wear patterns for a serpentine drive belt

Troubleshooting the Serpentine Drive Belt

Problem	Cause	Solution
Tension sheeting fabric failure (woven fabric on outside circumference of belt has cracked or separated from body of belt)	• Grooved or backside idler pulley diameters are less than minimum recommended • Tension sheeting contacting (rubbing) stationary object • Excessive heat causing woven fabric to age • Tension sheeting splice has fractured	• Replace pulley(s) not conforming to specification • Correct rubbing condition • Replace belt • Replace belt
Noise (objectional squeal, squeak, or rumble is heard or felt while drive belt is in operation)	• Belt slippage • Bearing noise • Belt misalignment • Belt-to-pulley mismatch • Driven component inducing vibration • System resonant frequency inducing vibration	• Adjust belt • Locate and repair • Align belt/pulley(s) • Install correct belt • Locate defective driven component and repair • Vary belt tension within specifications. Replace belt.
Rib chunking (one or more ribs has separated from belt body)	• Foreign objects imbedded in pulley grooves • Installation damage • Drive loads in excess of design specifications • Insufficient internal belt adhesion	• Remove foreign objects from pulley grooves • Replace belt • Adjust belt tension • Replace belt
Rib or belt wear (belt ribs contact bottom of pulley grooves)	• Pulley(s) misaligned • Mismatch of belt and pulley groove widths • Abrasive environment • Rusted pulley(s) • Sharp or jagged pulley groove tips • Rubber deteriorated	• Align pulley(s) • Replace belt • Replace belt • Clean rust from pulley(s) • Replace pulley • Replace belt
Longitudinal belt cracking (cracks between two ribs)	• Belt has mistracked from pulley groove • Pulley groove tip has worn away rubber-to-tensile member	• Replace belt • Replace belt
Belt slips	• Belt slipping because of insufficient tension • Belt or pulley subjected to substance (belt dressing, oil, ethylene glycol) that has reduced friction • Driven component bearing failure • Belt glazed and hardened from heat and excessive slippage	• Adjust tension • Replace belt and clean pulleys • Replace faulty component bearing • Replace belt
"Groove jumping" (belt does not maintain correct position on pulley, or turns over and/or runs off pulleys)	• Insufficient belt tension • Pulley(s) not within design tolerance • Foreign object(s) in grooves	• Adjust belt tension • Replace pulley(s) • Remove foreign objects from grooves

TCCS3C09

Troubleshooting the Serpentine Drive Belt

Problem	Cause	Solution
"Groove jumping" (belt does not maintain correct position on pulley, or turns over and/or runs off pulleys)	• Excessive belt speed • Pulley misalignment • Belt-to-pulley profile mismatched • Belt cordline is distorted	• Avoid excessive engine acceleration • Align pulley(s) • Install correct belt • Replace belt
Belt broken (Note: identify and correct problem before replacement belt is installed)	• Excessive tension • Tensile members damaged during belt installation • Belt turnover • Severe pulley misalignment • Bracket, pulley, or bearing failure	• Replace belt and adjust tension to specification • Replace belt • Replace belt • Align pulley(s) • Replace defective component and belt
Cord edge failure (tensile member exposed at edges of belt or separated from belt body)	• Excessive tension • Drive pulley misalignment • Belt contacting stationary object • Pulley irregularities • Improper pulley construction • Insufficient adhesion between tensile member and rubber matrix	• Adjust belt tension • Align pulley • Correct as necessary • Replace pulley • Replace pulley • Replace belt and adjust tension to specifications
Sporadic rib cracking (multiple cracks in belt ribs at random intervals)	• Ribbed pulley(s) diameter less than minimum specification • Backside bend flat pulley(s) diameter less than minimum • Excessive heat condition causing rubber to harden • Excessive belt thickness • Belt overcured • Excessive tension	• Replace pulley(s) • Replace pulley(s) • Correct heat condition as necessary • Replace belt • Replace belt • Adjust belt tension

TCCS3C10

ADJUSTMENT

Periodic drive belt tensioning is not necessary, because an automatic spring-loaded tensioner is used with these belts to maintain proper adjustment at all times. The tensioner is also useful as a wear indicator. When the belt is properly installed, the arrow on the tensioner housing must point within the acceptable range lines on the tensioner's face. If the arrow falls outside the range, either an improper belt has been installed or the belt is worn beyond its useful life span. In either case, a new belt must be installed immediately to assure proper engine operation and to prevent possible accessory damage.

REMOVAL & INSTALLATION

Because serpentine belts use a spring loaded tensioner for adjustment, belt replacement tends to be somewhat easier than it used to be on engines where accessories were pivoted and bolted in place for tension adjustment. Basically, all belt replacement involves is to pivot the tensioner to loosen the belt, then slide the belt off of the pulleys. The two most important points are to pay CLOSE attention to the proper belt routing (since serpentine belts tend to be "snaked" all different ways through the pulleys) and to be sure the V-ribs are properly seated in all the pulleys.

Although belt routing diagrams have been included in this section, the first places you should check for proper belt routing are the labels in your engine compartment. These should include a belt routing diagram which may reflect changes made during a production run.

1. Disconnect the negative battery cable for safety. This will help assure that no one mistakenly cranks the engine over with your hands between the pulleys, and that the cooling fan cannot activate while servicing the belt(s).

➡Take a good look at the installed belt and make a note of the routing. Before removing the belt, be sure the routing matches that of the belt routing label or one of the diagrams in this book. If for some reason a diagram does not match (you may not have the original engine or it may have been modified), carefully note the changes on a piece of paper.

2. For tensioners equipped with a ½ in. (13mm) square hole, insert the drive end of a large breaker bar into the hole. Use the breaker bar to pivot the tensioner away from the drive belt. For tensioners not equipped with this hole, use the proper-sized socket and breaker bar (or a large handled wrench) on the tensioner idler pulley center bolt to pivot the tensioner away from the belt. This will loosen the belt sufficiently that it can be pulled off of one or more of the pulleys. It is usually easiest to carefully pull the belt out from underneath the tensioner pulley itself.

3. Once the belt is off one of the pulleys, gently pivot the tensioner back into position. DO NOT allow the tensioner to snap back, as this could damage the tensioner's internal parts.

4. Now finish removing the belt from the other pulleys and remove it from the engine.

To install:

5. While referring to the proper routing diagram (which you identified earlier), begin to route the belt over the pulleys, leaving whichever pulley you first released it from for last.

6. Once the belt is mostly in place, carefully pivot the tensioner and position the belt over the final pulley. As you begin to allow the tensioner back into contact with the belt, run your hand around the pulleys and be sure the belt is properly seated in the ribs. If not, release the tension and seat the belt.

7. Once the belt is installed, take another look at all the pulleys to double check your installation.

8. Connect the negative battery cable, then start and run the engine to check belt operation.

9. Once the engine has reached normal operating temperature, turn the ignition **OFF** and check that the belt tensioner arrow is within the proper adjustment range.

Often the underhood label will display the serpentine drive belt routing

Relieve the belt tension by pivoting the automatic tensioner away from the belt, then remove the belt

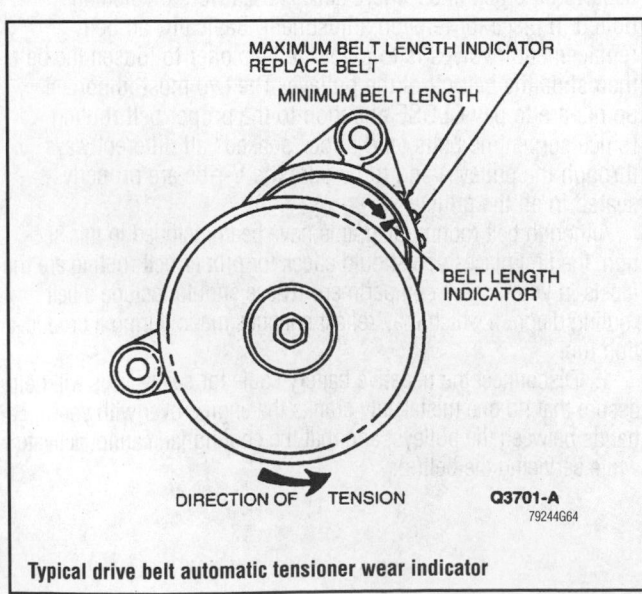

Typical drive belt automatic tensioner wear indicator

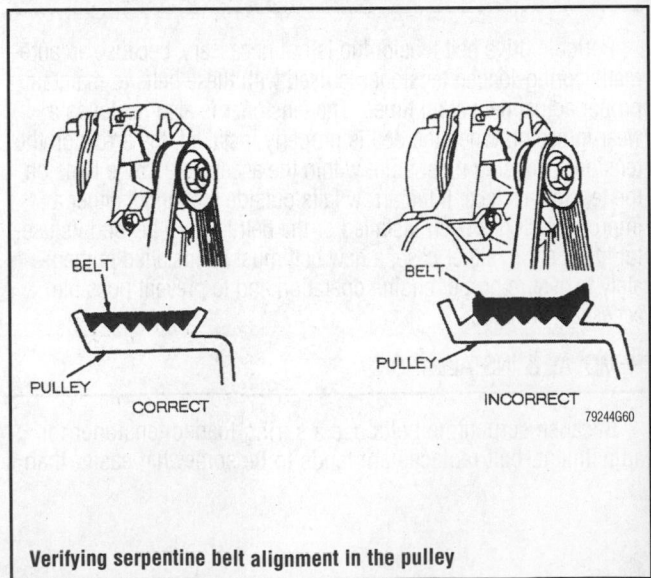

Verifying serpentine belt alignment in the pulley

ACCESSORY DRIVE BELT ROUTING INDEX

MANUFACTURER

ENGINES	DESCRIPTION	FIGURE
Acura		
1.8L engines	Accessory drive belt routing	1
2.2L engines	Accessory drive belt routing	2
2.5L engines	Accessory drive belt routing	3
3.0L engines except NSX	Accessory drive belt routing	4
3.0L NSX engines	Accessory drive belt routing	5
3.5L and 3.2L engines	Accessory drive belt routing	6
Audi		
AAN engines without A/C	Serpentine drive belt routing	7
AAN engines with A/C	Serpentine drive belt routing	8
A4 AFC engines	Serpentine drive belt routing	9
AFC, AAH engines	Serpentine drive belt routing	10
Chrysler Imports		
1.5L and 2.4L engines	Accessory drive belt routing	11
1.8L engines	Accessory drive belt routing	12
Honda		
4-cylinder engines (except Civic/Del Sol) without A/C except Civic	Accessory drive belt routing	13
4-cylinder engines (except Civic/Del Sol) with A/C except Civic	Accessory drive belt routing	14
Civic/Del Sol 4-cylinder engines	Accessory drive belt routing	15
V6 engines	Accessory drive belt routing	16
Hyundai		
Accent 1.5L engines	Accessory drive belt routing	17
1.5L engines except Accent	Accessory drive belt routing	18
1995 1.8L and 2.0L engines	Accessory drive belt routing	19
1996–99 1.8L and 2.0L engines	Accessory drive belt routing	20
3.0L engines	Accessory drive belt routing	21
Infiniti		
2.0L engines	Accessory drive belt routing	22
3.0L (VG30DE) engines	Accessory drive belt routing	23
3.0L (VQ30DE) engines	Accessory drive belt routing	24
4.1L engines	Accessory drive belt routing	25
4.5L engines	Accessory drive belt routing	26
Jaguar		
4.0L non-supercharged engines	Serpentine drive belt routing	27
4.0L supercharged engines	Serpentine drive belt routing	28
6.0L engines	Serpentine drive belt routing	29
KIA		
1.6L and 1.8L engines	Accessory drive belt routing	30
Lexus		
3.0L (2JZ-GE) engines	Serpentine drive belt routing	31
3.0L (1MZ-FE) engines	Accessory drive belt routing	32
4.0L engines	Serpentine drive belt routing	33
Mazda		
4-cylinder engines	Serpentine drive belt routing	34
2.3L (KJ) engines	Serpentine drive belt routing	35
V-6 engines except 2.3L (KJ) engines	Serpentine drive belt routing	36

79234C01

ACCESSORY DRIVE BELT ROUTING INDEX

MANUFACTURER

ENGINES	DESCRIPTION	FIGURE
Mercedes-Benz		
Except 2.2L, 2.3L and Diesel engines	Serpentine drive belt routing	37
2.2L, 2.3L (non-supercharged) and Diesel engines	Serpentine drive belt routing	38
2.3L supercharged engines	Serpentine drive belt routing	39
Mitsubishi		
1.5L, 2.0L (turbo) and 2.4L engines	Accessory drive belt routing	40
2.0L (non-turbo) and 1.8L engines	Accessory drive belt routing	41
3.0L engines	Accessory drive belt routing	42
3.5L engines	Accessory drive belt routing	43
Nissan		
1.6L engines	Accessory drive belt routing	44
2.0L engines	Accessory drive belt routing	45
2.4L engines except 240 SX	Accessory drive belt routing	46
240 SX 2.4L engines	Accessory drive belt routing	47
3.0L engines	Accessory drive belt routing	48
Porsche		
3.6L engines	Accessory drive belt routing	49
4.5L engines	Accessory drive belt routing	50
Saab		
4-cylinder engines	Serpentine drive belt routing	51
6-cylinder engines	Serpentine drive belt routing	52
Subaru		
4-cylinder engines	Accessory drive belt routing	53
6-cylinder engines	Accessory drive belt routing	54
Suzuki		
All engines	Accessory drive belt routing	55
Toyota		
1.5L (5EFE) engines	Accessory drive belt routing	56
1.8L (1ZZFE) engines	Serpentine drive belt routing	57
1.8L (7AFE) and 1.6L (4AFE) engines	Accessory drive belt routing	58
2.2L (5SFE) engines	Accessory drive belt routing	59
3.0L (2JZGTE and 2JZGE) engines	Serpentine drive belt routing	60
3.0L (1MZFE) engines	Accessory drive belt routing	61
Volkswagen		
2.8L engines	Accessory drive belt routing	62
Except 2.8L engines without A/C	Accessory drive belt routing	63
Except 2.8L engines with A/C	Accessory drive belt routing	64
Volvo		
All engines	Serpentine drive belt routing	65

79234C02

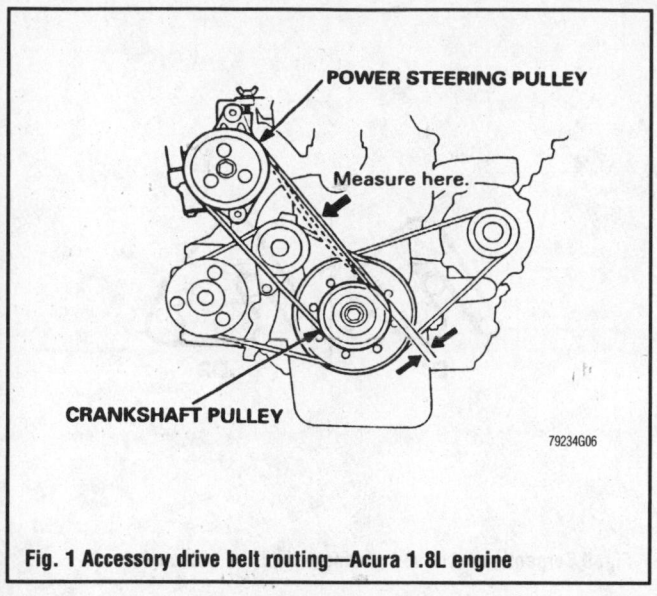

Fig. 1 Accessory drive belt routing—Acura 1.8L engine

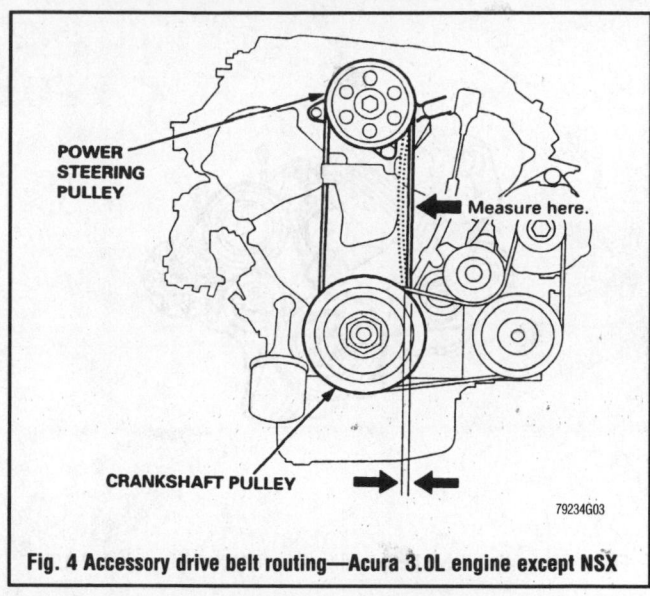

Fig. 4 Accessory drive belt routing—Acura 3.0L engine except NSX

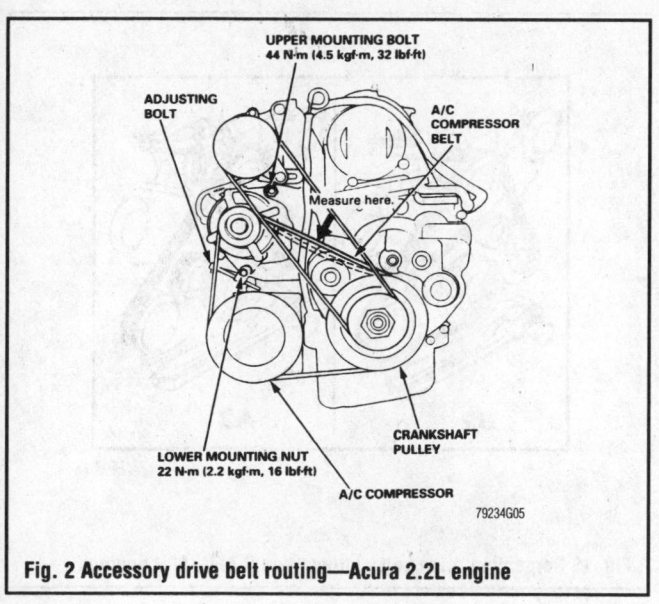

Fig. 2 Accessory drive belt routing—Acura 2.2L engine

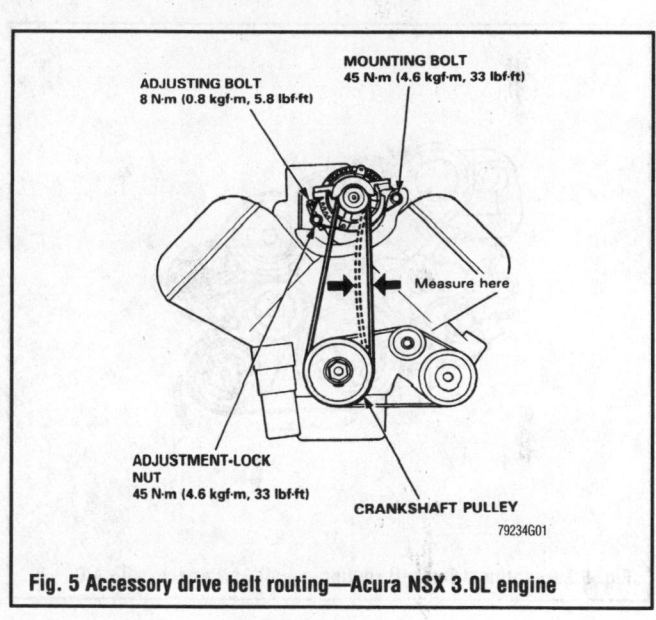

Fig. 5 Accessory drive belt routing—Acura NSX 3.0L engine

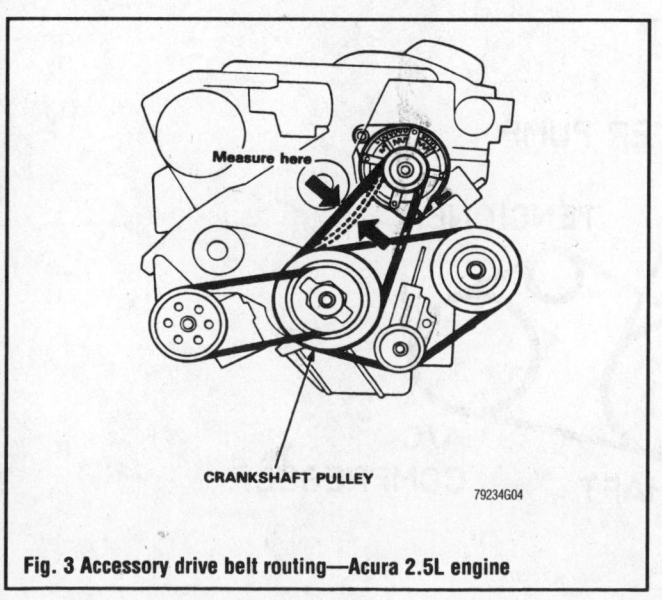

Fig. 3 Accessory drive belt routing—Acura 2.5L engine

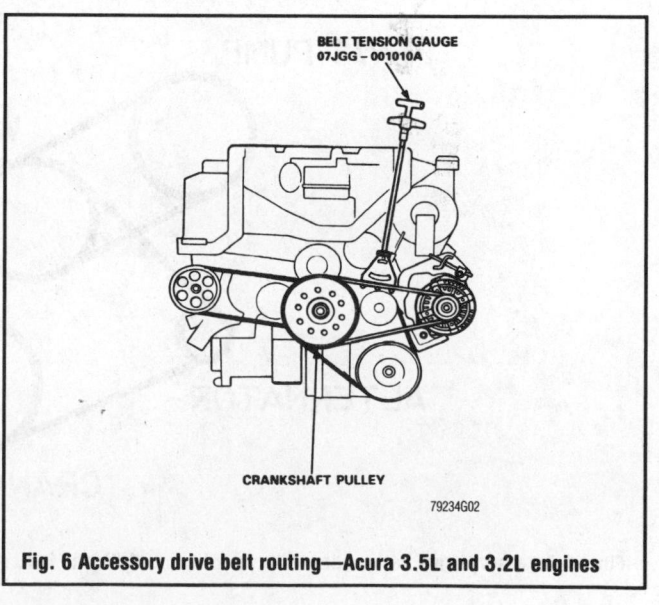

Fig. 6 Accessory drive belt routing—Acura 3.5L and 3.2L engines

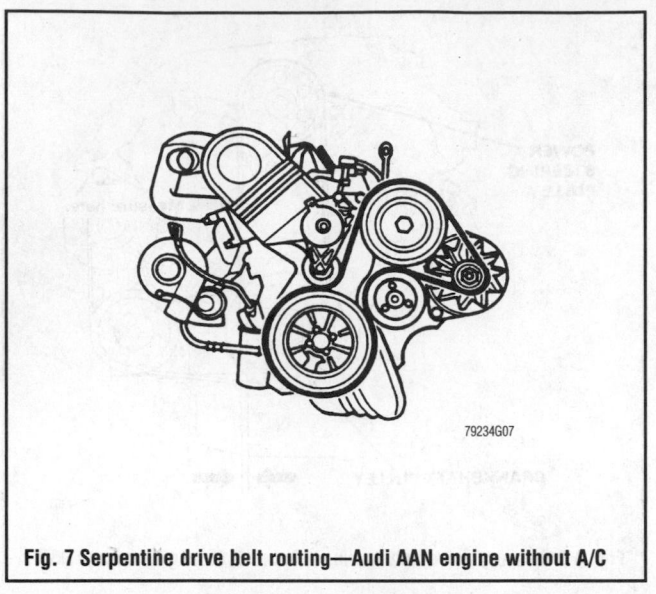

Fig. 7 Serpentine drive belt routing—Audi AAN engine without A/C

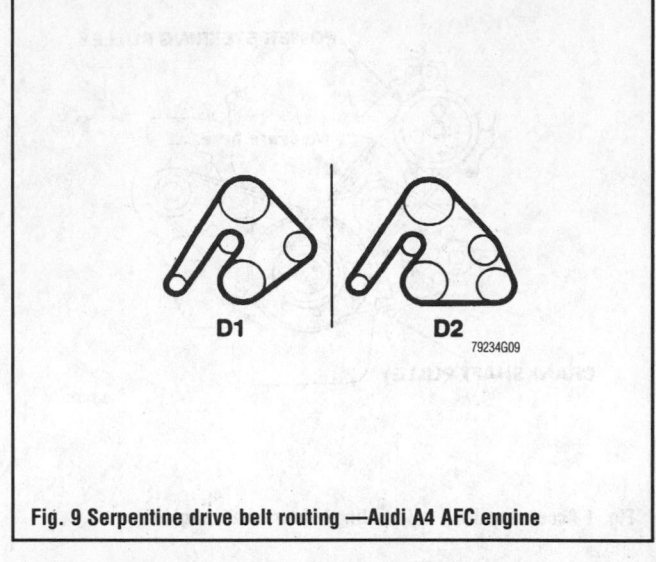

Fig. 9 Serpentine drive belt routing —Audi A4 AFC engine

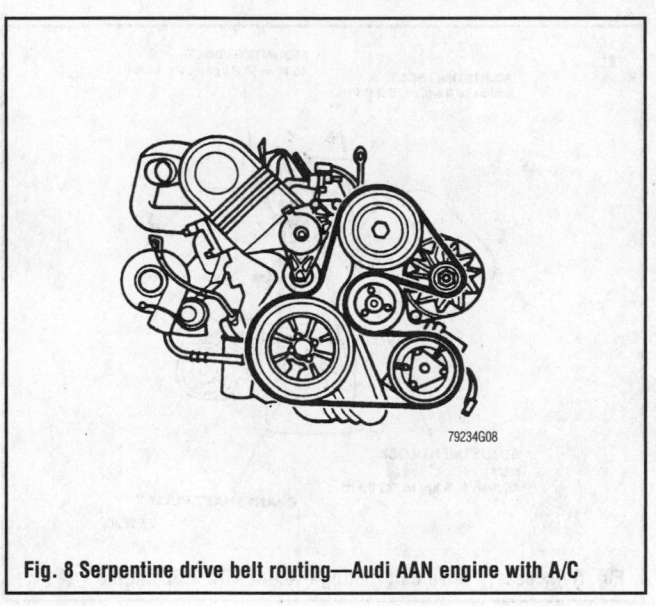

Fig. 8 Serpentine drive belt routing—Audi AAN engine with A/C

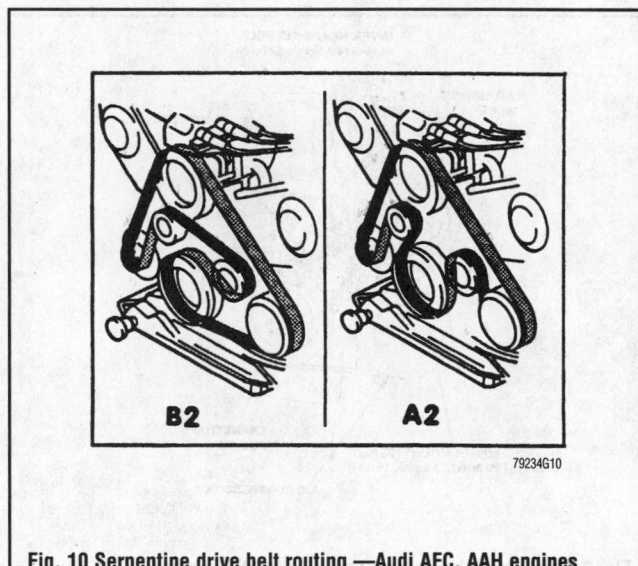

Fig. 10 Serpentine drive belt routing —Audi AFC, AAH engines

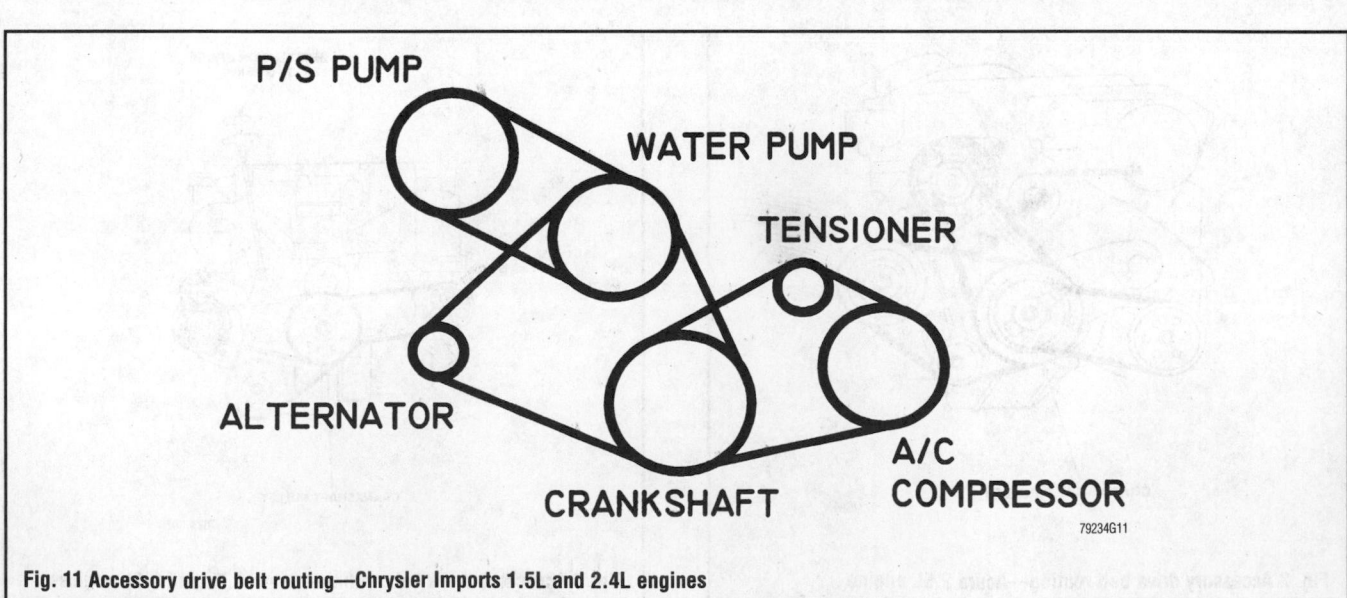

Fig. 11 Accessory drive belt routing—Chrysler Imports 1.5L and 2.4L engines

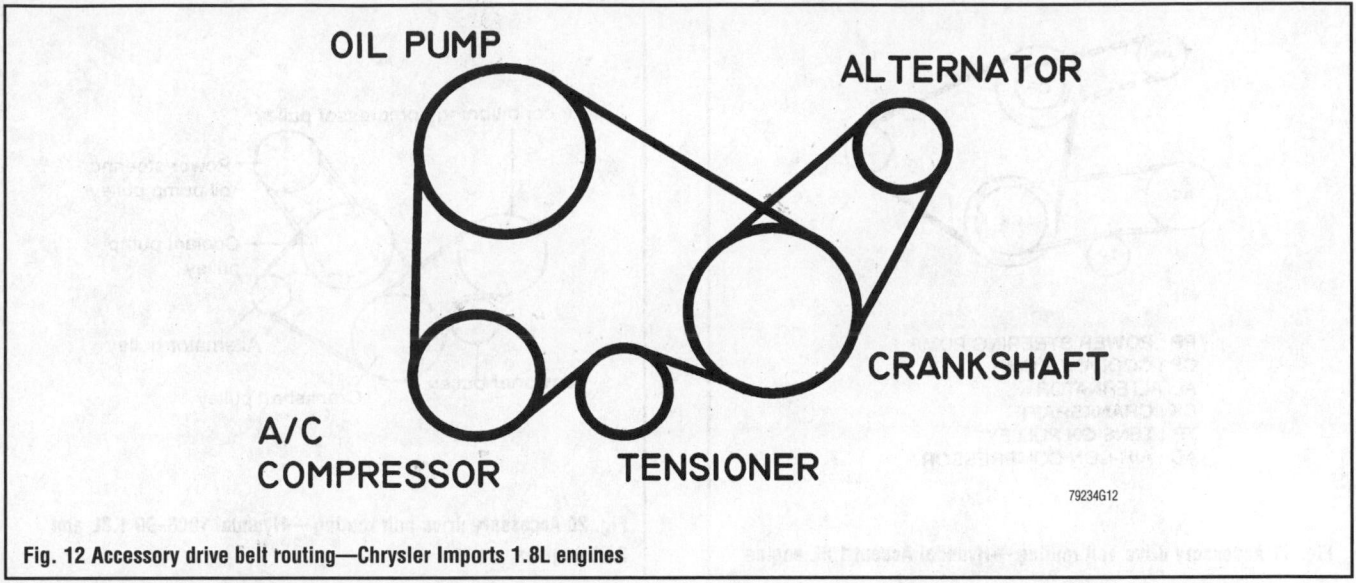

Fig. 12 Accessory drive belt routing—Chrysler Imports 1.8L engines

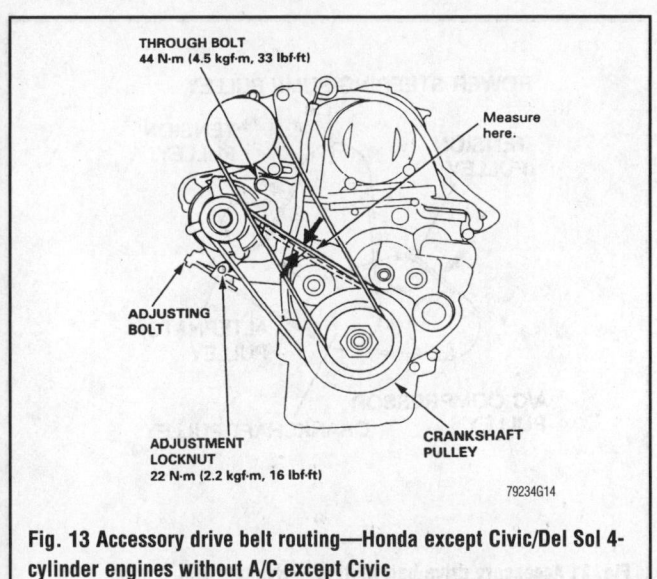

Fig. 13 Accessory drive belt routing—Honda except Civic/Del Sol 4-cylinder engines without A/C except Civic

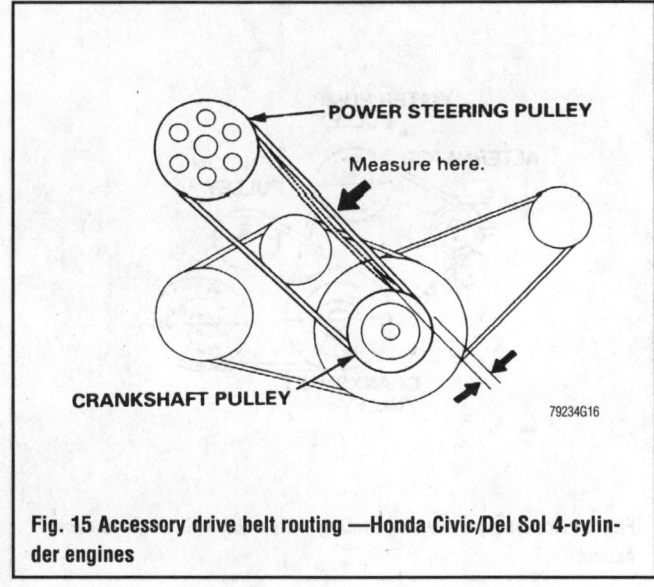

Fig. 15 Accessory drive belt routing —Honda Civic/Del Sol 4-cylinder engines

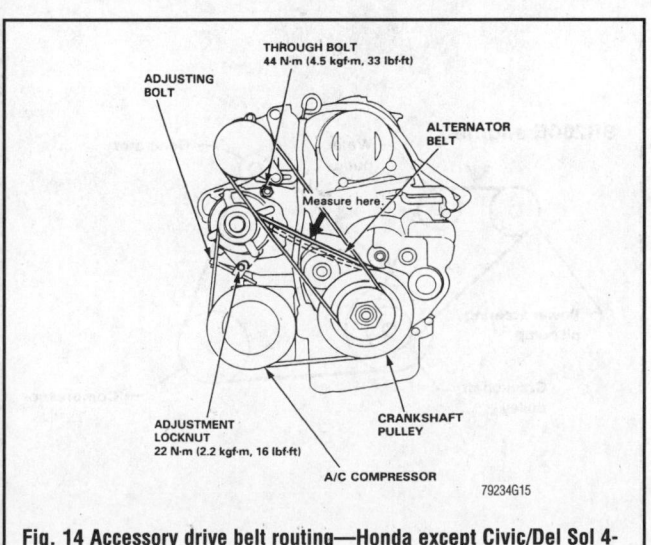

Fig. 14 Accessory drive belt routing—Honda except Civic/Del Sol 4-cylinder engines with A/C except Civic

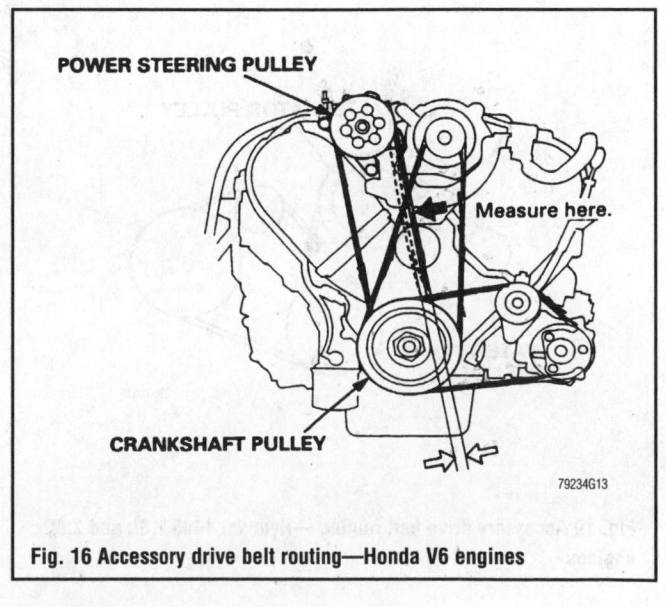

Fig. 16 Accessory drive belt routing—Honda V6 engines

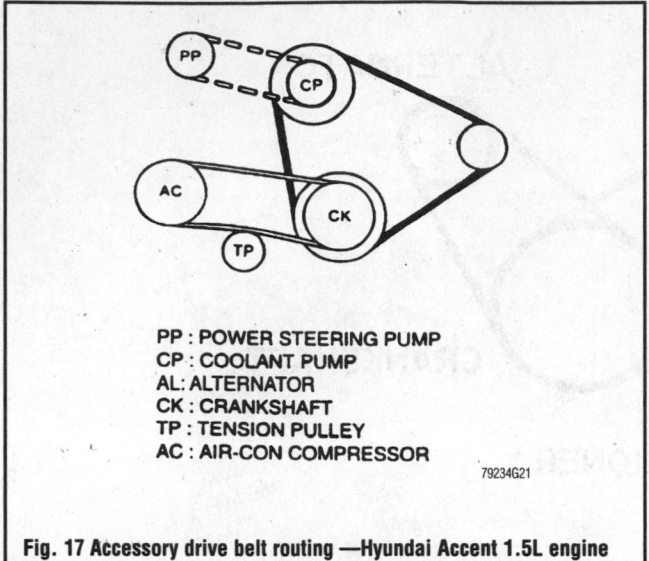

PP : POWER STEERING PUMP
CP : COOLANT PUMP
AL : ALTERNATOR
CK : CRANKSHAFT
TP : TENSION PULLEY
AC : AIR-CON COMPRESSOR

79234G21

Fig. 17 Accessory drive belt routing —Hyundai Accent 1.5L engine

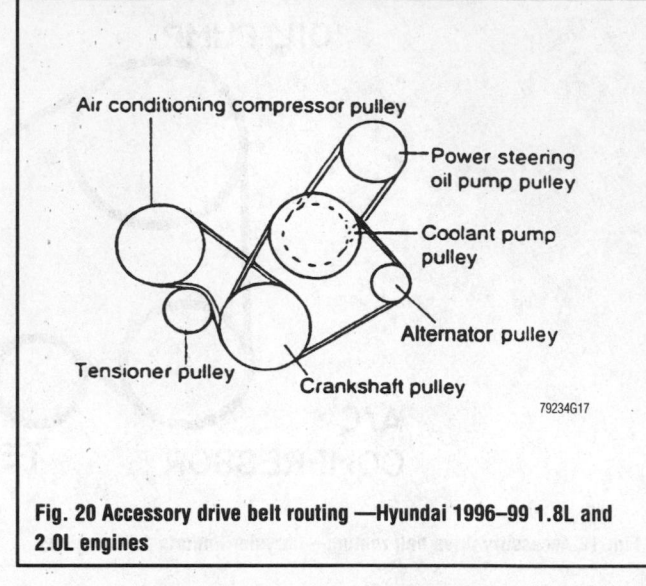

Fig. 20 Accessory drive belt routing —Hyundai 1996–99 1.8L and 2.0L engines

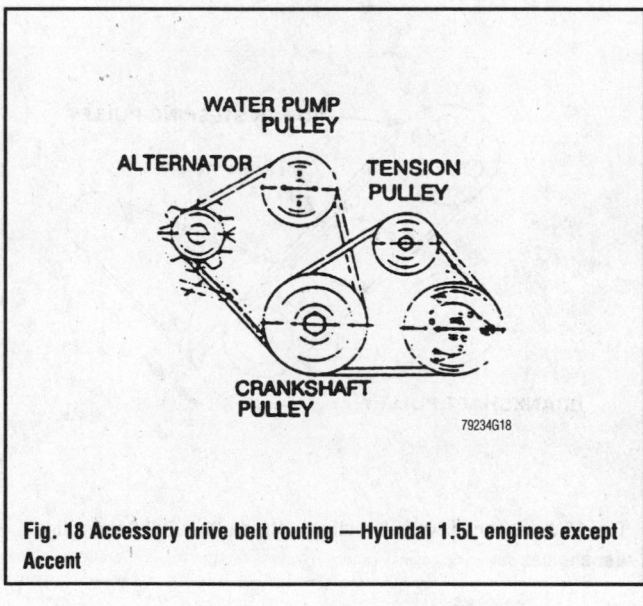

Fig. 18 Accessory drive belt routing —Hyundai 1.5L engines except Accent

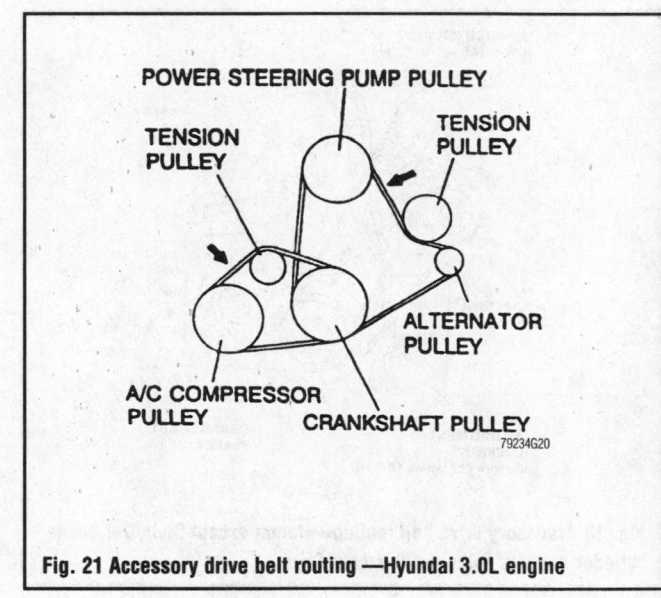

Fig. 21 Accessory drive belt routing —Hyundai 3.0L engine

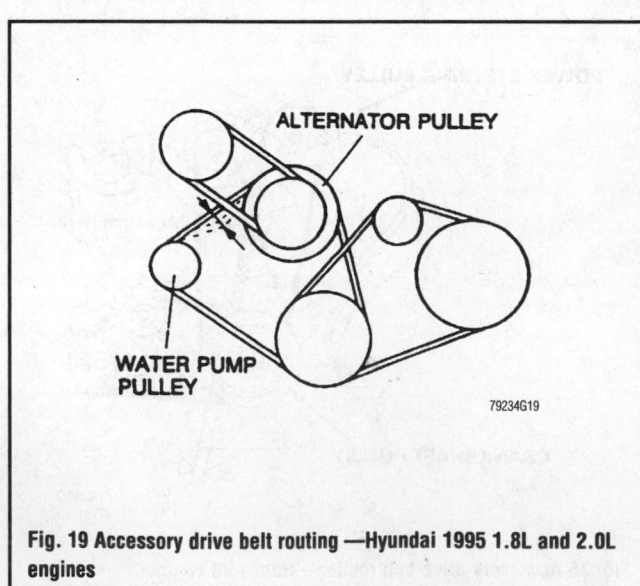

Fig. 19 Accessory drive belt routing —Hyundai 1995 1.8L and 2.0L engines

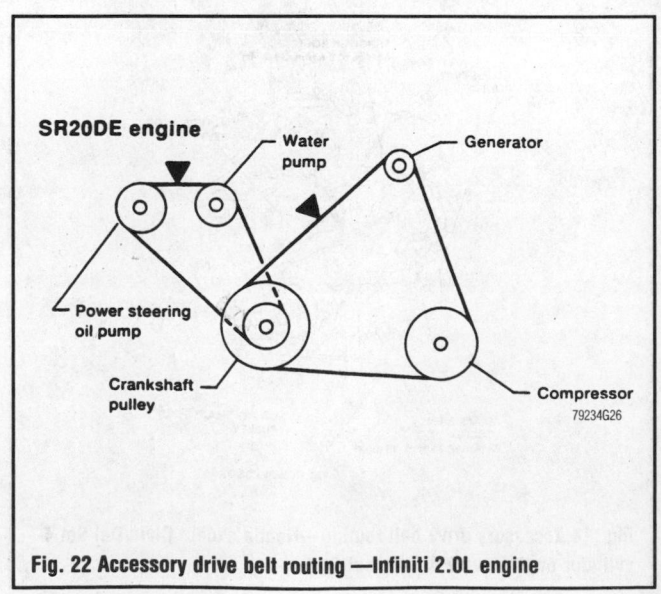

Fig. 22 Accessory drive belt routing —Infiniti 2.0L engine

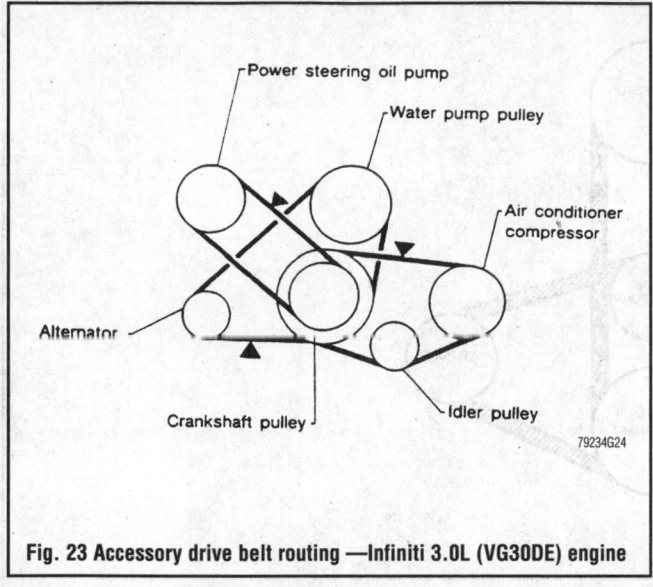

Fig. 23 Accessory drive belt routing —Infiniti 3.0L (VG30DE) engine

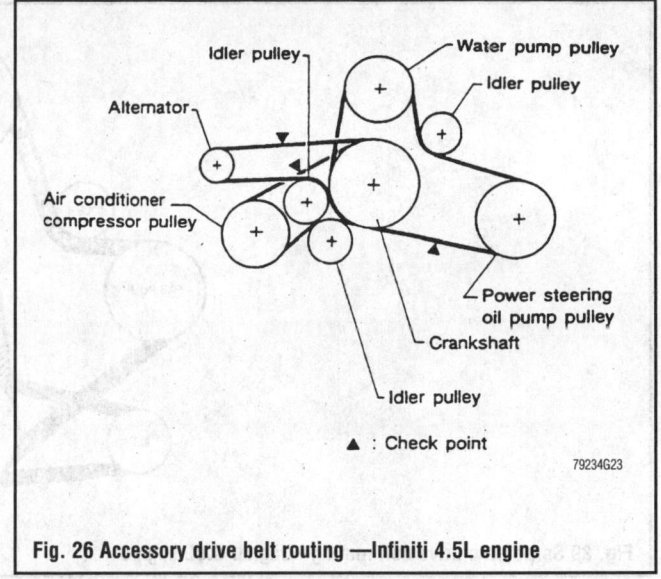

Fig. 26 Accessory drive belt routing —Infiniti 4.5L engine

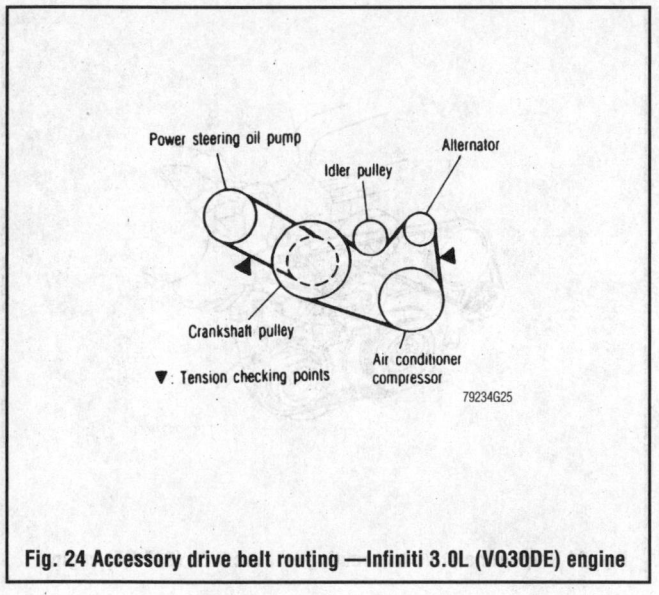

Fig. 24 Accessory drive belt routing —Infiniti 3.0L (VQ30DE) engine

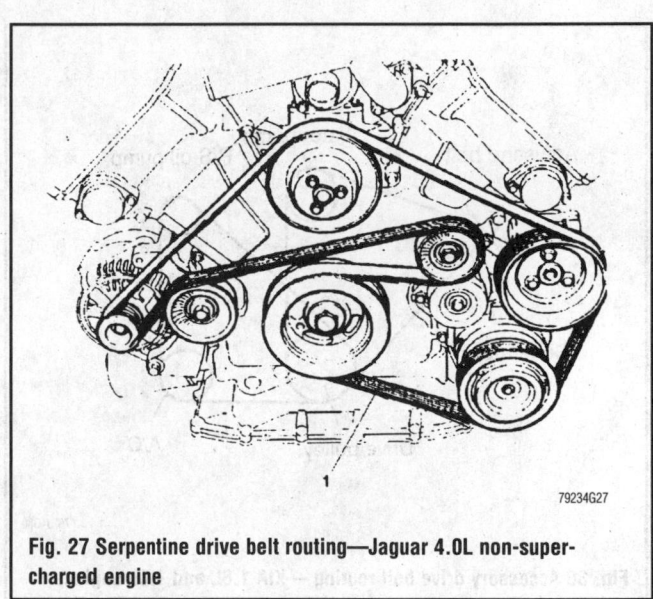

Fig. 27 Serpentine drive belt routing—Jaguar 4.0L non-super-charged engine

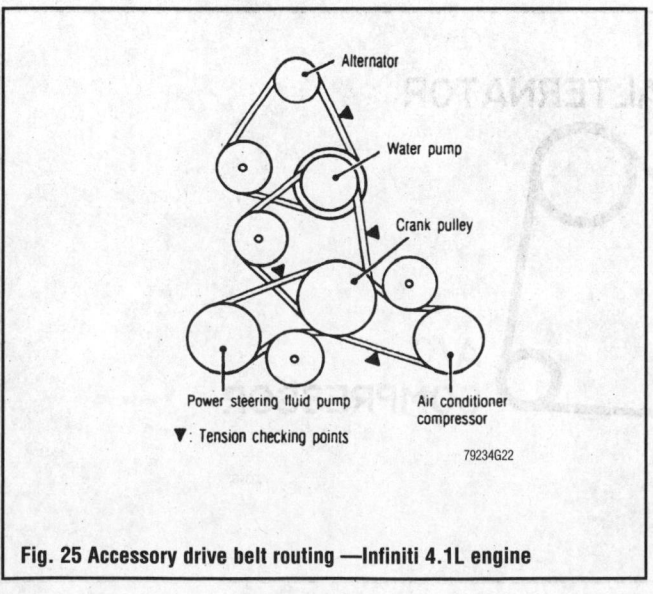

Fig. 25 Accessory drive belt routing —Infiniti 4.1L engine

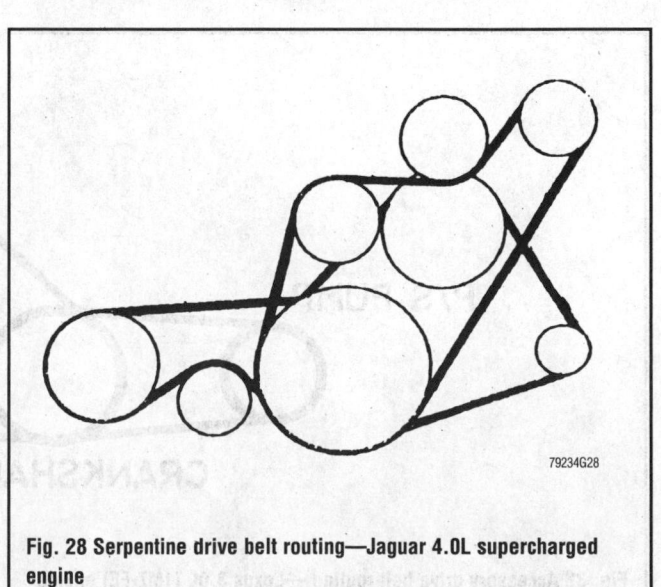

Fig. 28 Serpentine drive belt routing—Jaguar 4.0L supercharged engine

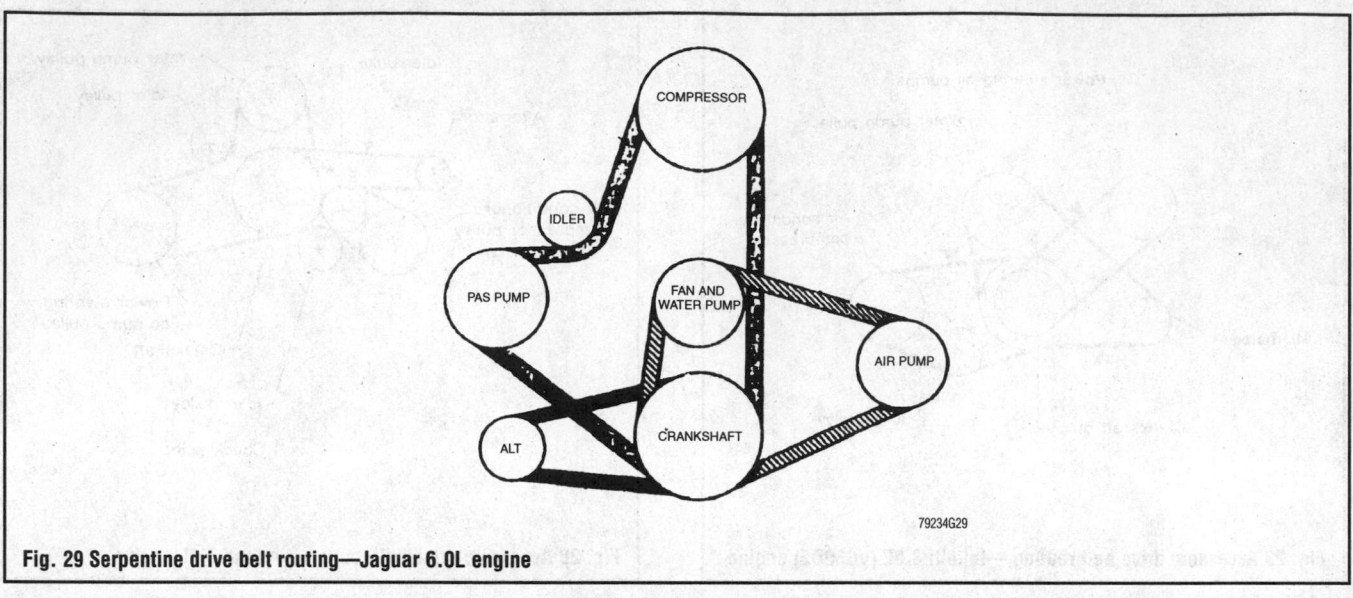

Fig. 29 Serpentine drive belt routing—Jaguar 6.0L engine

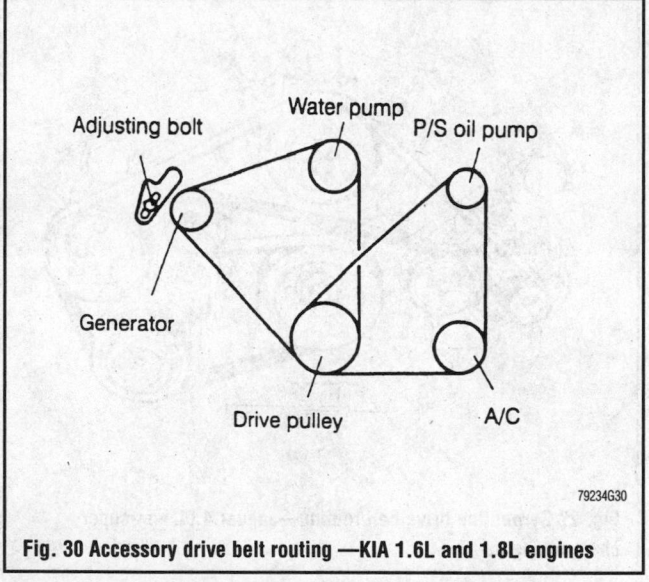

Fig. 30 Accessory drive belt routing —KIA 1.6L and 1.8L engines

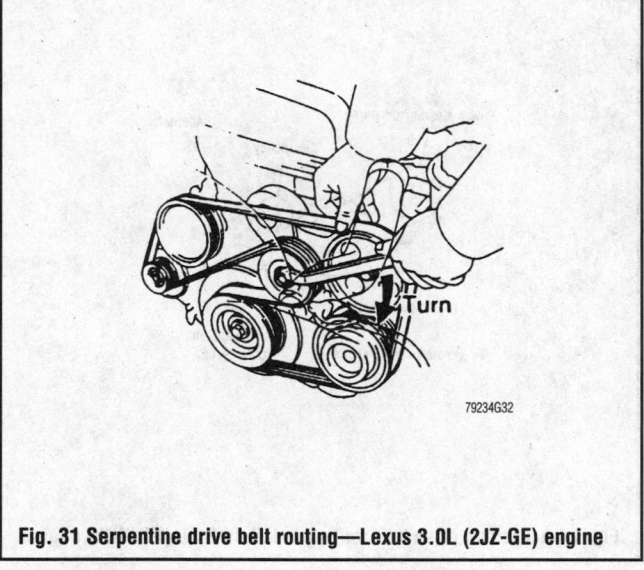

Fig. 31 Serpentine drive belt routing—Lexus 3.0L (2JZ-GE) engine

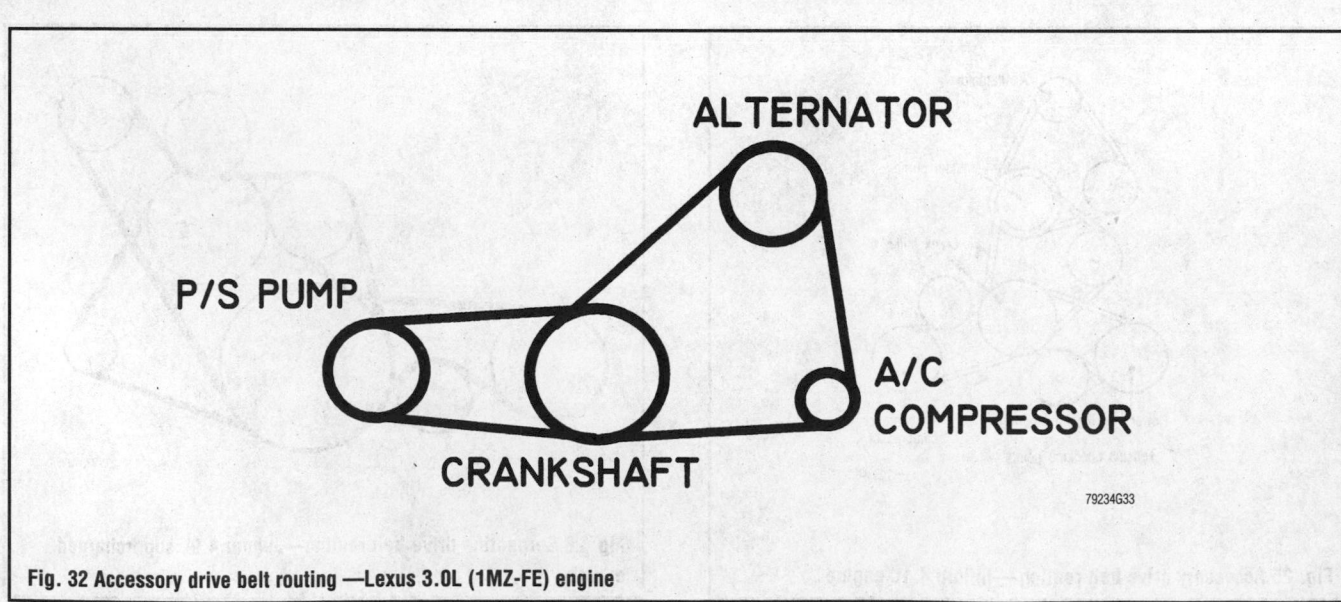

Fig. 32 Accessory drive belt routing —Lexus 3.0L (1MZ-FE) engine

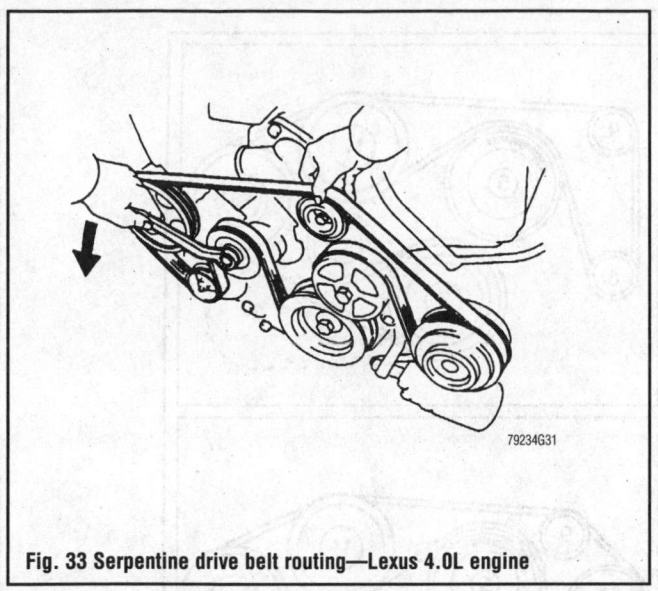

Fig. 33 Serpentine drive belt routing—Lexus 4.0L engine

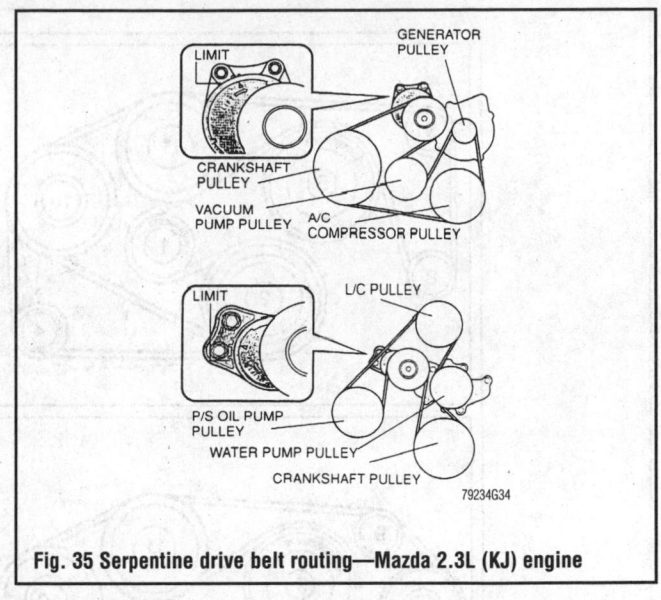

Fig. 35 Serpentine drive belt routing—Mazda 2.3L (KJ) engine

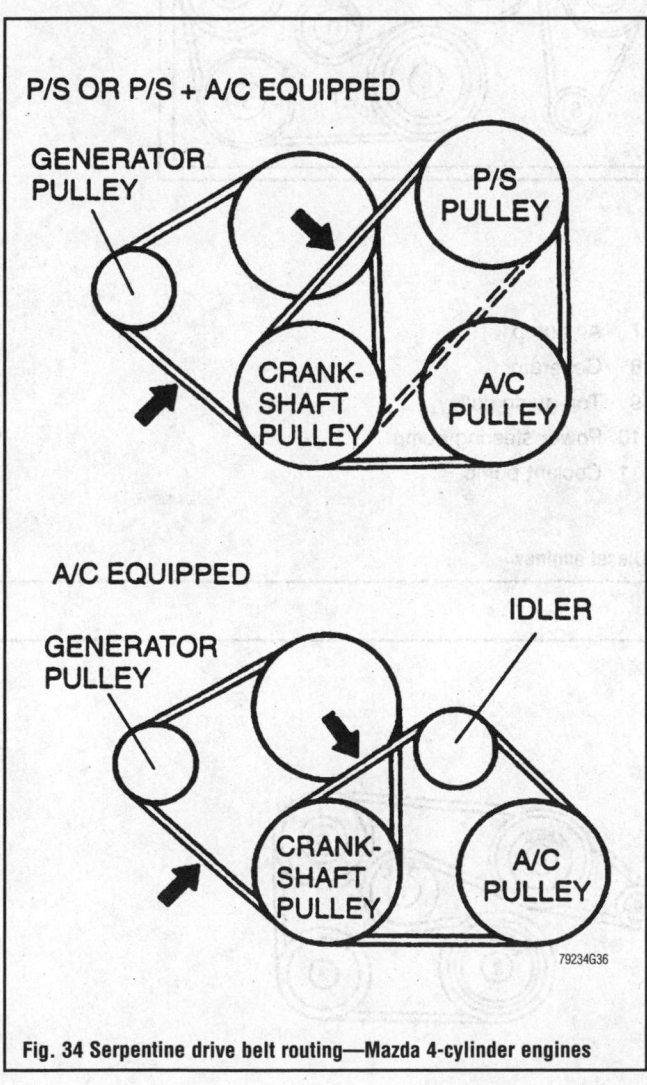

Fig. 34 Serpentine drive belt routing—Mazda 4-cylinder engines

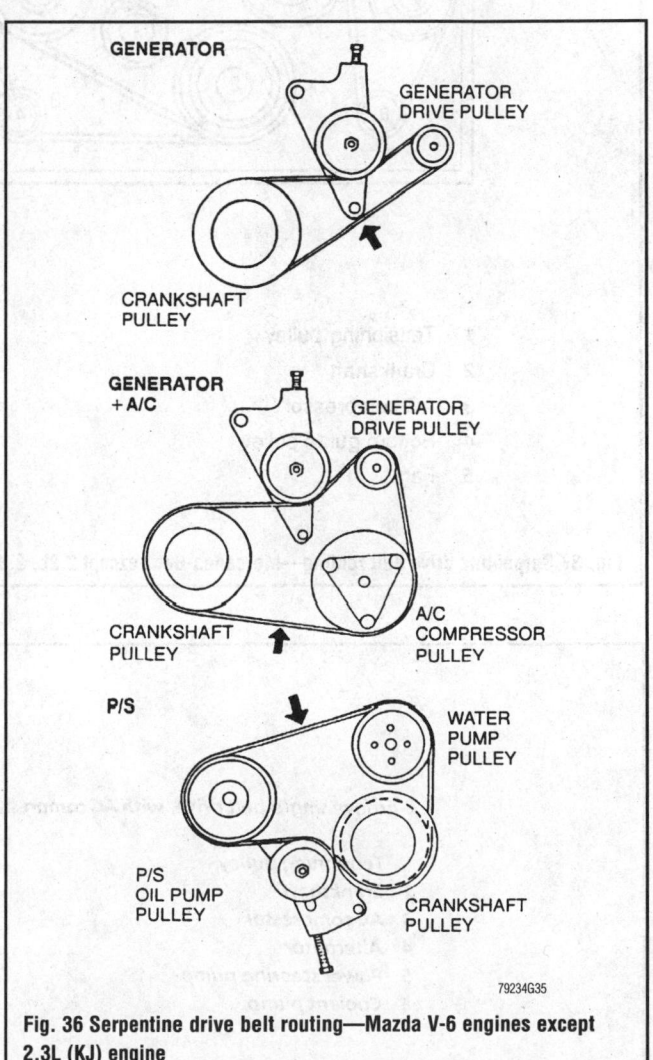

Fig. 36 Serpentine drive belt routing—Mazda V-6 engines except 2.3L (KJ) engine

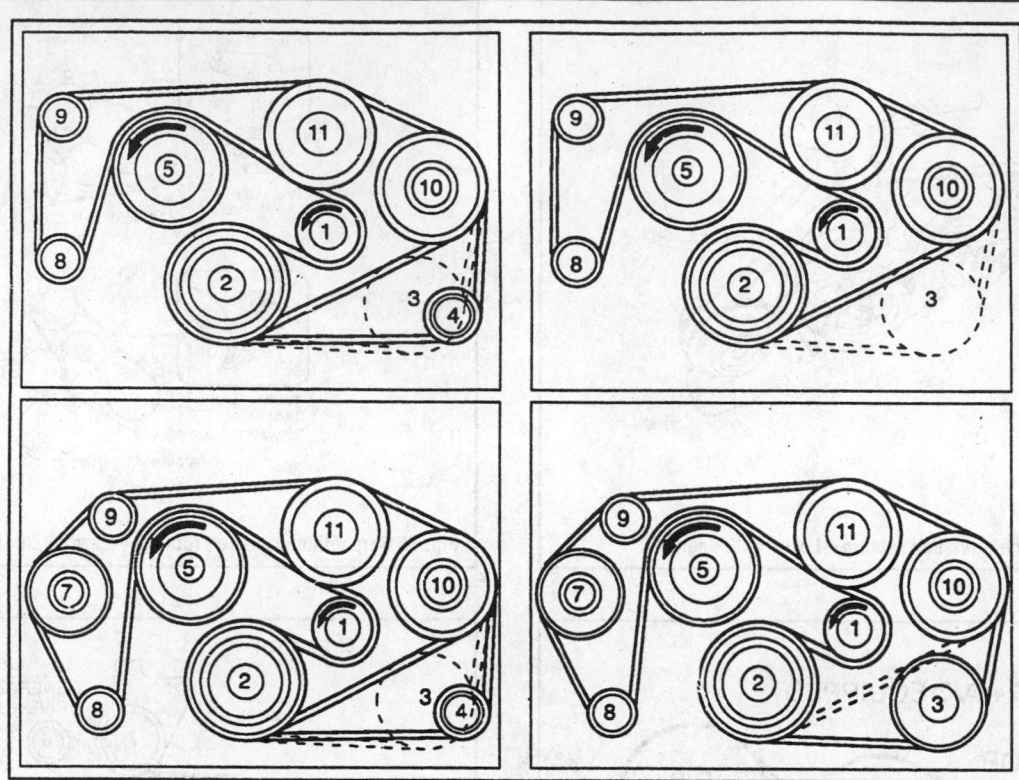

1 Tensioning pulley
2 Crankshaft
3 AC compressor
4 Bottom guide pulley
5 Fan

7 Air pump
8 Generator
9 Top guide pulley
10 Power steering pump
11 Coolant pump

79234G67

Fig. 37 Serpentine drive belt routing—Mercedes-Benz except 2.2L, 2.3L and Diesel engines

6-groove single-belt drive, with AC compressor

1 *Tensioning pulley*
2 *Crankshaft*
3 *AC compressor*
4 *Alternator*
5 *Power steering pump*
6 *Coolant pump*

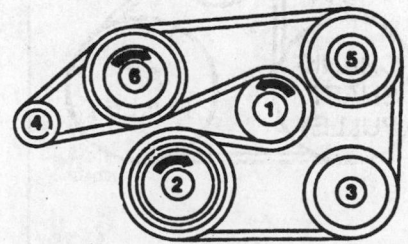

79234G68

Fig. 38 Serpentine drive belt routing—Mercedes-Benz 2.2L, 2.3L (non-supercharged) and Diesel engines

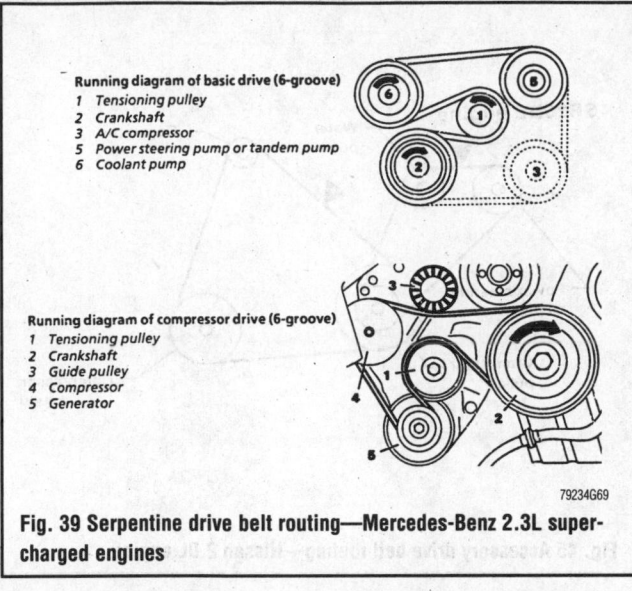

Running diagram of basic drive (6-groove)
1 Tensioning pulley
2 Crankshaft
3 A/C compressor
5 Power steering pump or tandem pump
6 Coolant pump

Running diagram of compressor drive (6-groove)
1 Tensioning pulley
2 Crankshaft
3 Guide pulley
4 Compressor
5 Generator

79234G69

Fig. 39 Serpentine drive belt routing—Mercedes-Benz 2.3L supercharged engines

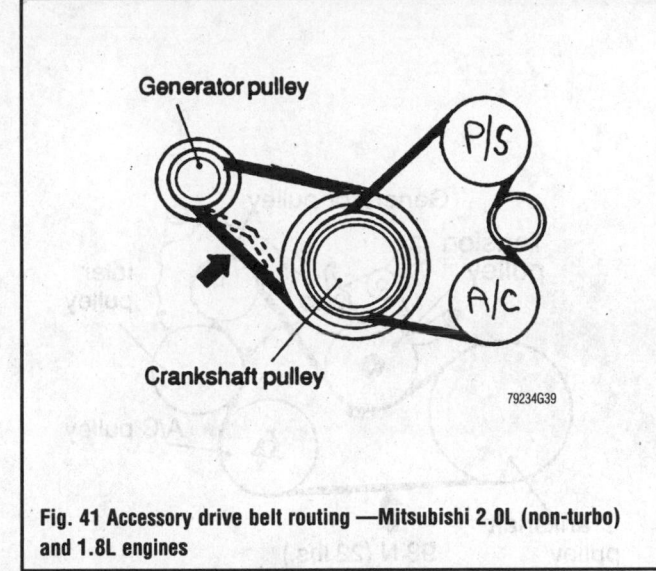

Fig. 41 Accessory drive belt routing —Mitsubishi 2.0L (non-turbo) and 1.8L engines

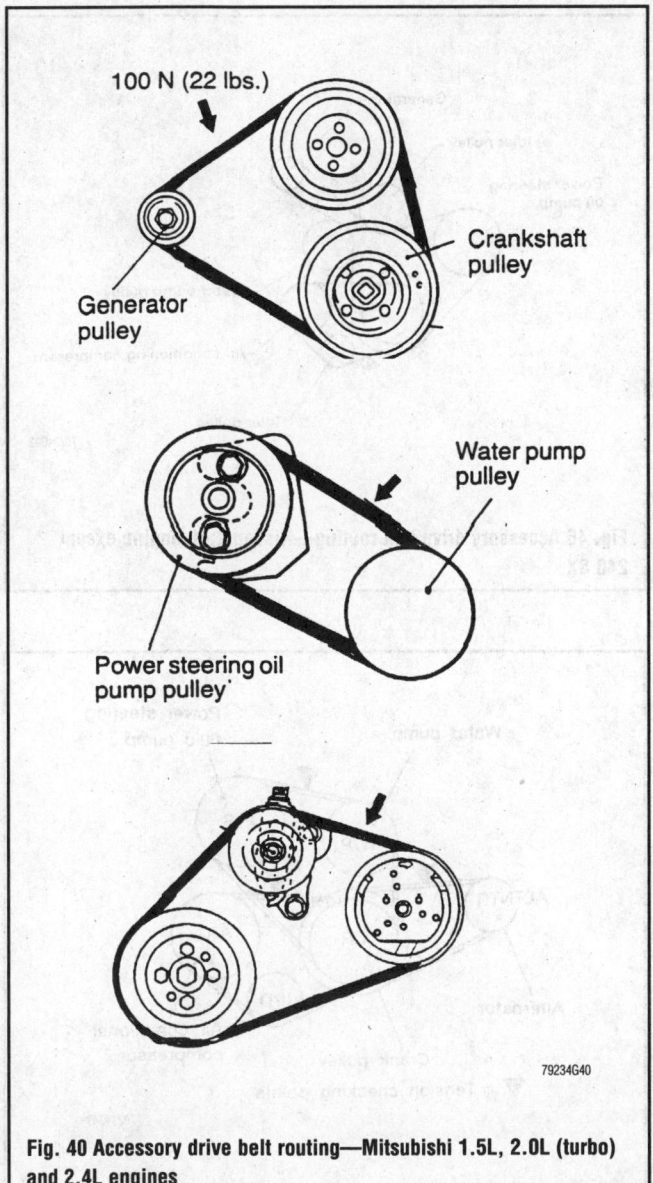

Fig. 40 Accessory drive belt routing—Mitsubishi 1.5L, 2.0L (turbo) and 2.4L engines

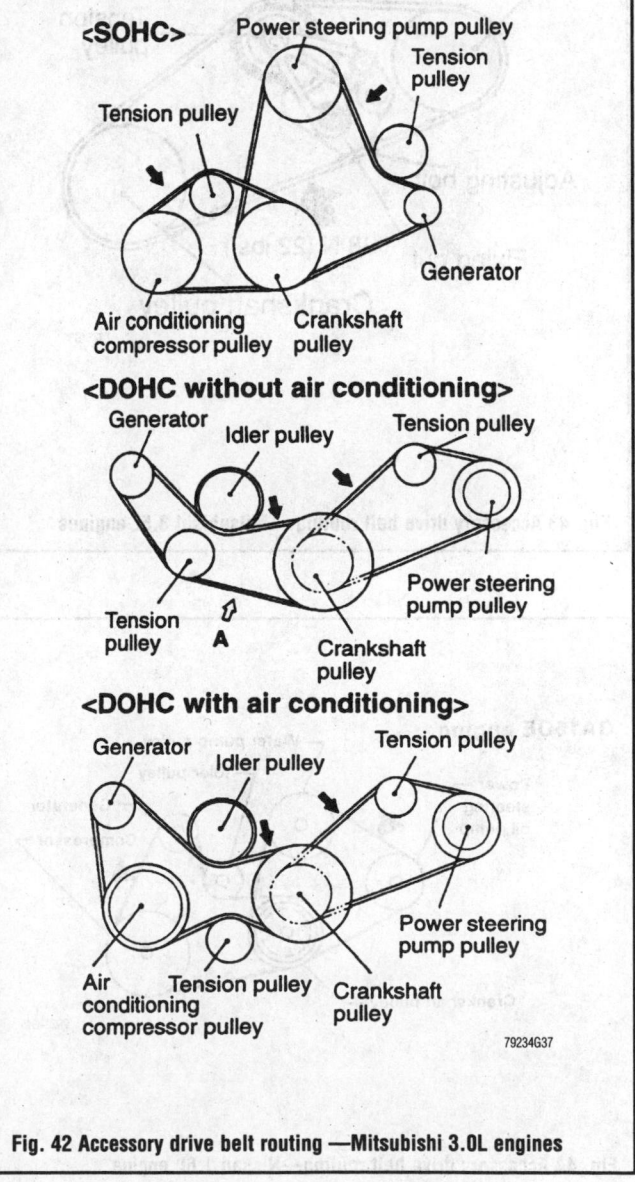

Fig. 42 Accessory drive belt routing —Mitsubishi 3.0L engines

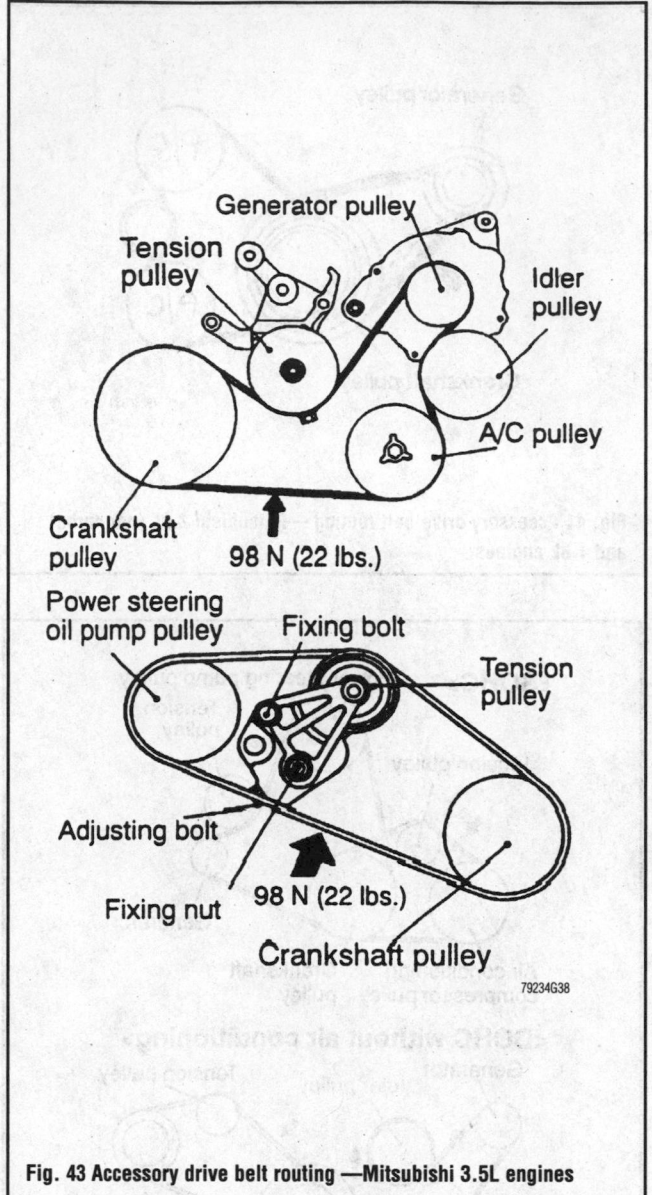

98 N (22 lbs.)

98 N (22 lbs.)

Fig. 43 Accessory drive belt routing —Mitsubishi 3.5L engines

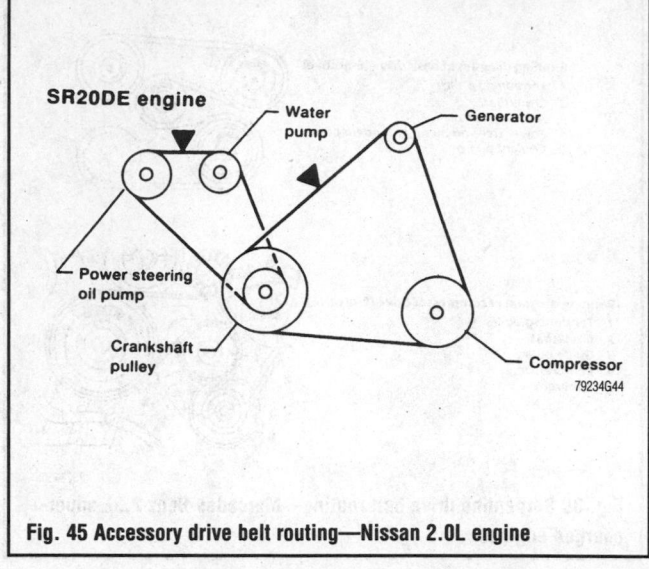

Fig. 45 Accessory drive belt routing—Nissan 2.0L engine

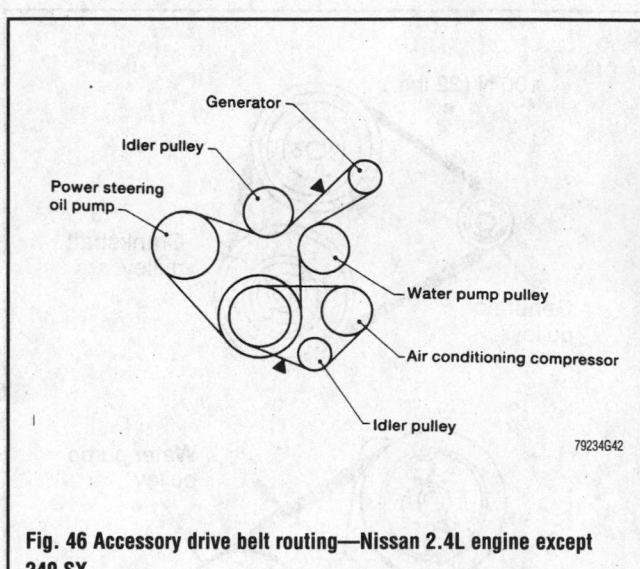

Fig. 46 Accessory drive belt routing—Nissan 2.4L engine except 240 SX

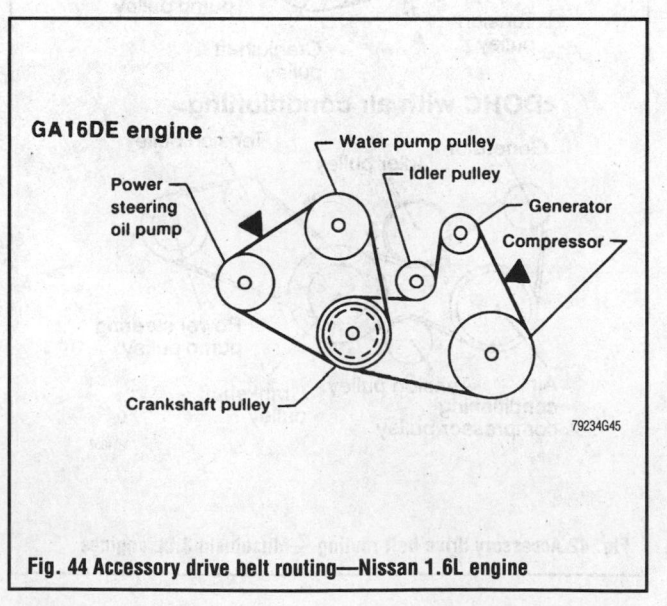

Fig. 44 Accessory drive belt routing—Nissan 1.6L engine

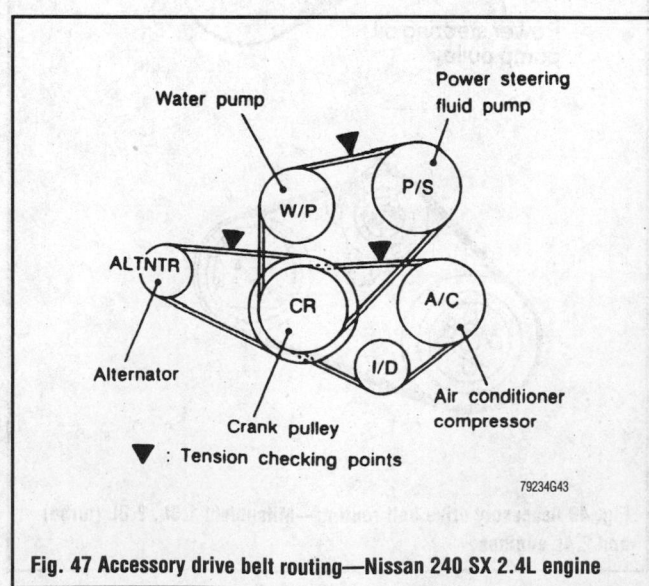

Fig. 47 Accessory drive belt routing—Nissan 240 SX 2.4L engine

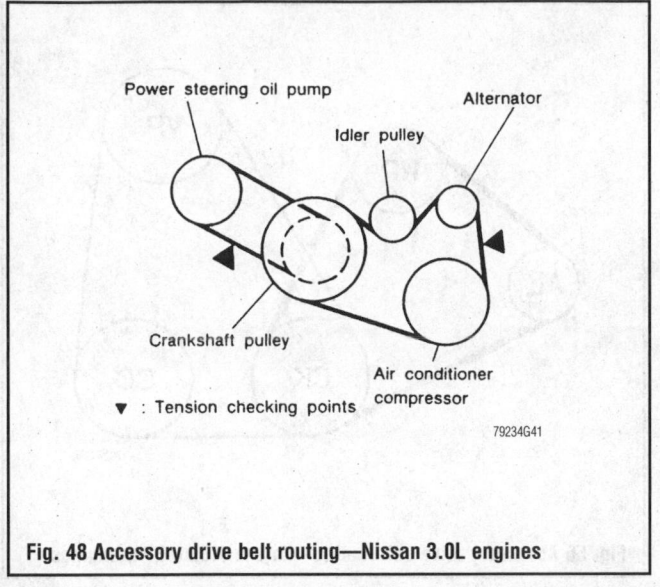

Fig. 48 Accessory drive belt routing—Nissan 3.0L engines

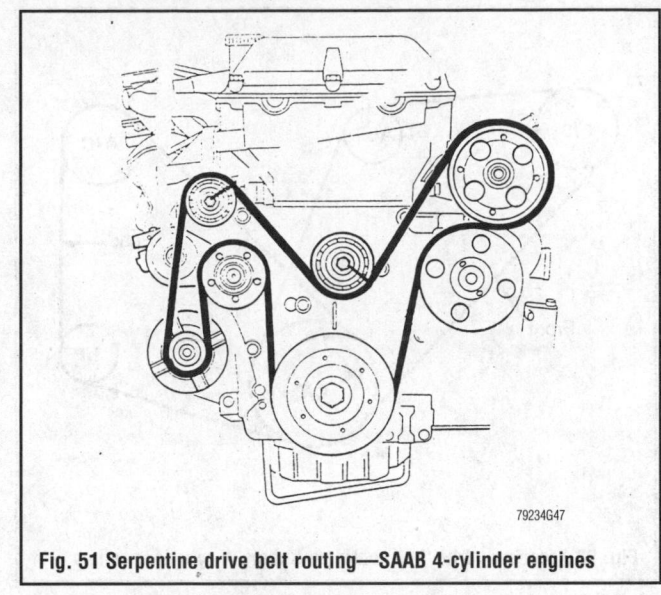

Fig. 51 Serpentine drive belt routing—SAAB 4-cylinder engines

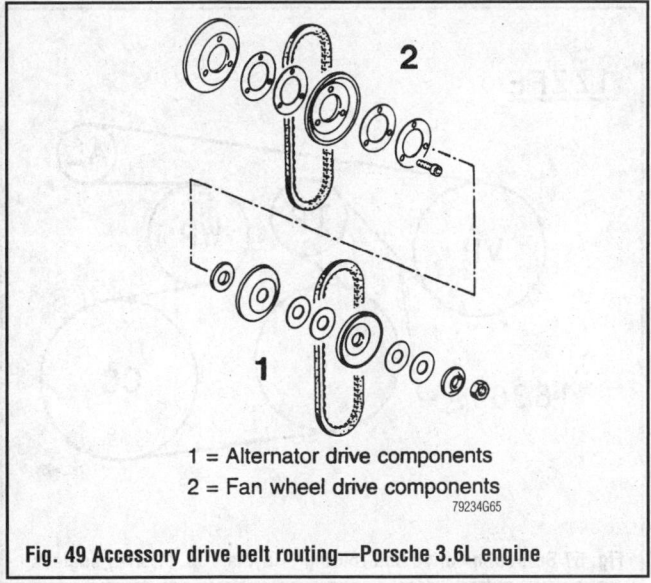

1 = Alternator drive components
2 = Fan wheel drive components

Fig. 49 Accessory drive belt routing—Porsche 3.6L engine

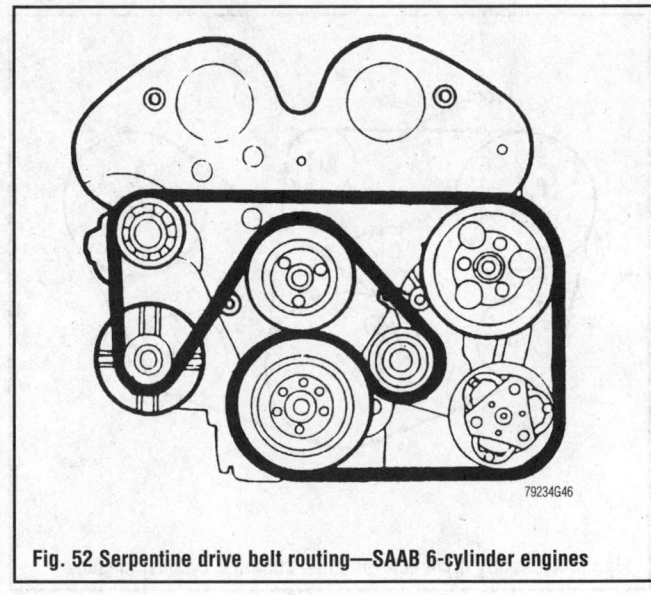

Fig. 52 Serpentine drive belt routing—SAAB 6-cylinder engines

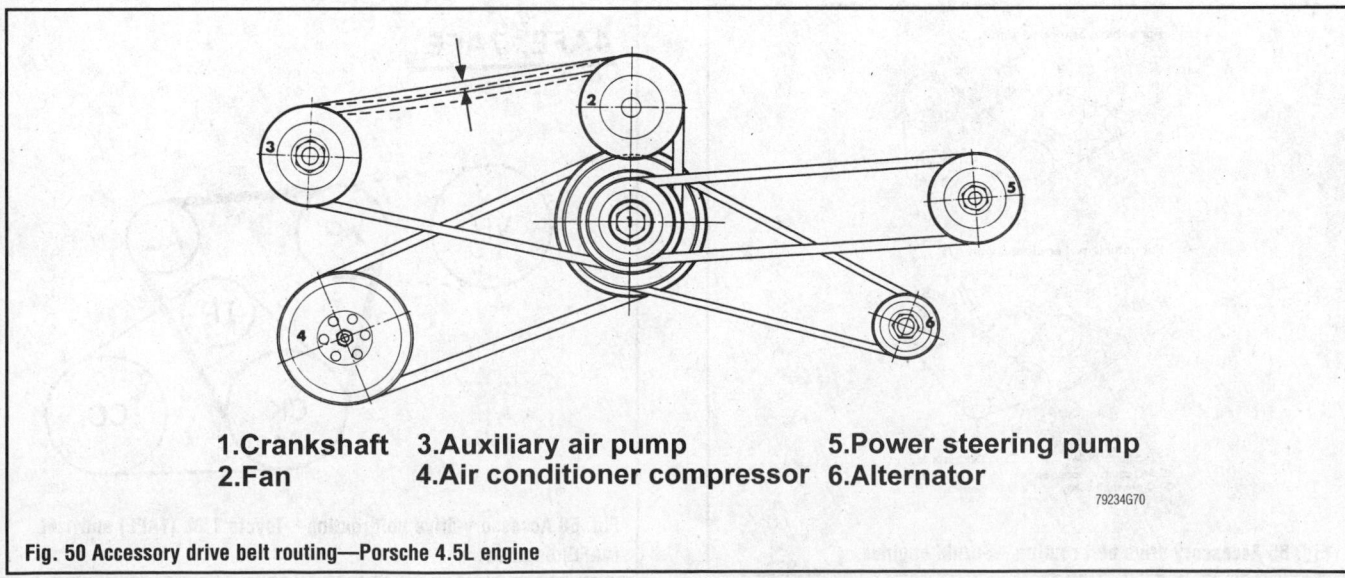

1.Crankshaft   3.Auxiliary air pump   5.Power steering pump
2.Fan   4.Air conditioner compressor   6.Alternator

Fig. 50 Accessory drive belt routing—Porsche 4.5L engine

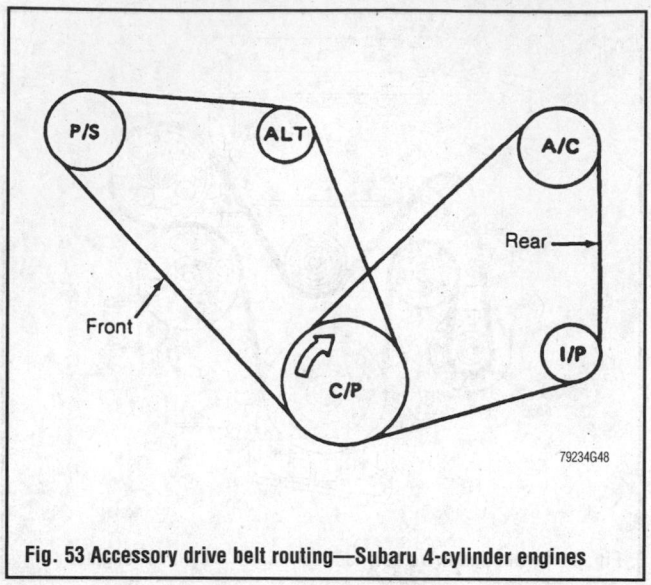

Fig. 53 Accessory drive belt routing—Subaru 4-cylinder engines

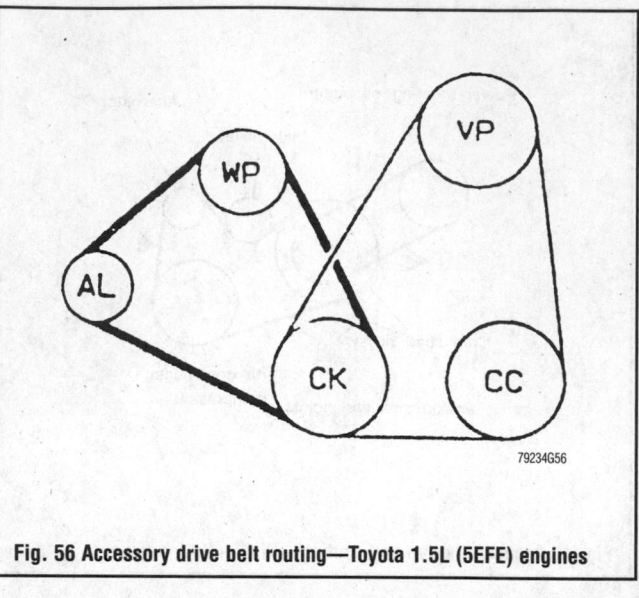

Fig. 56 Accessory drive belt routing—Toyota 1.5L (5EFE) engines

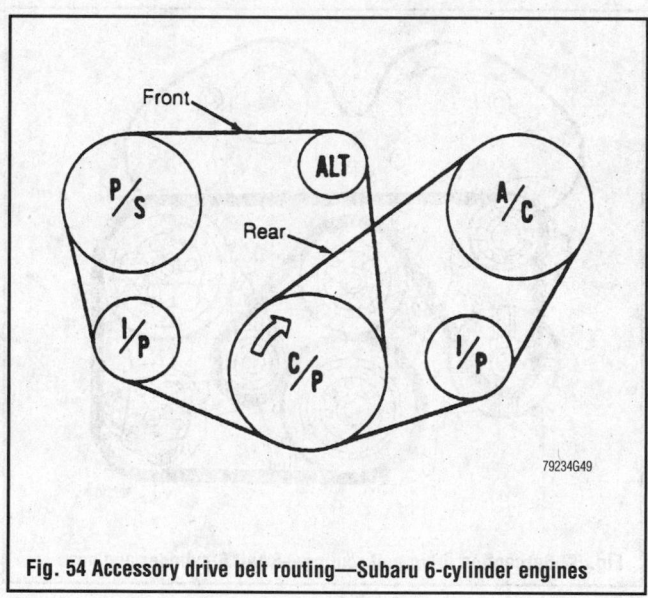

Fig. 54 Accessory drive belt routing—Subaru 6-cylinder engines

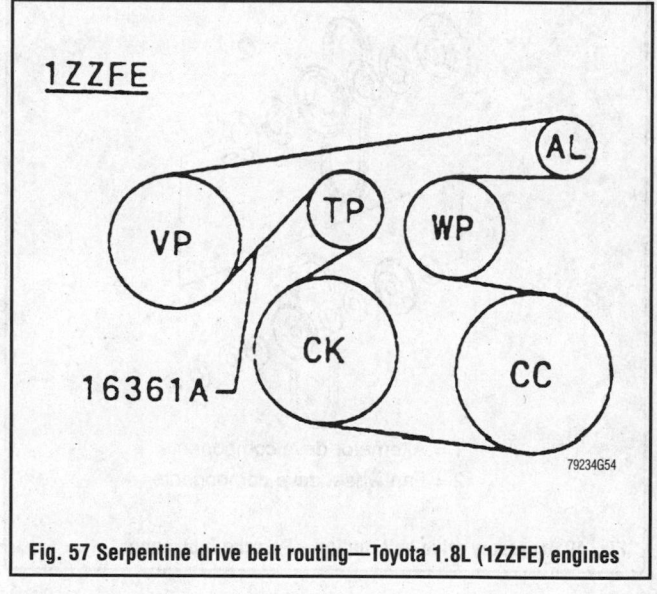

Fig. 57 Serpentine drive belt routing—Toyota 1.8L (1ZZFE) engines

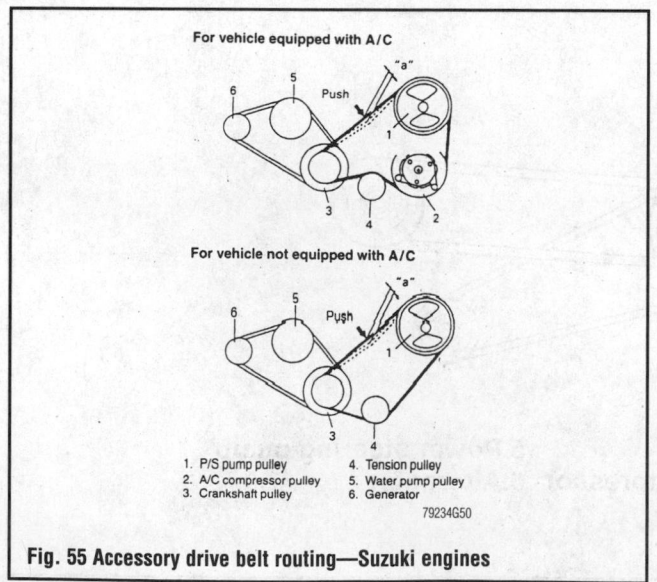

Fig. 55 Accessory drive belt routing—Suzuki engines

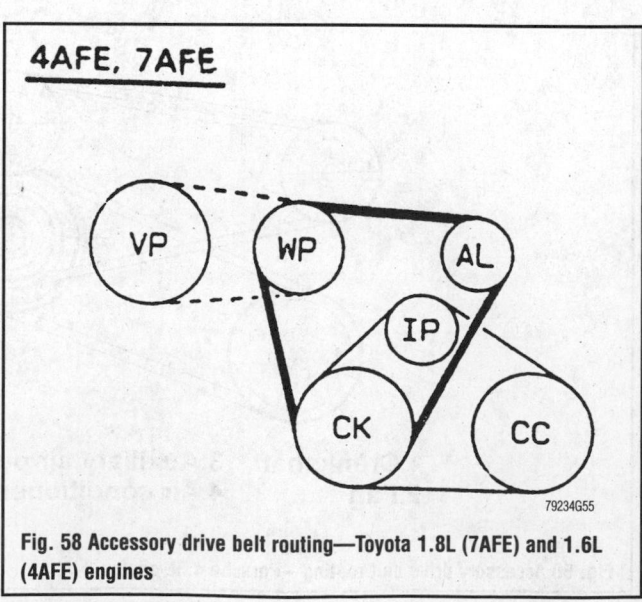

Fig. 58 Accessory drive belt routing—Toyota 1.8L (7AFE) and 1.6L (4AFE) engines

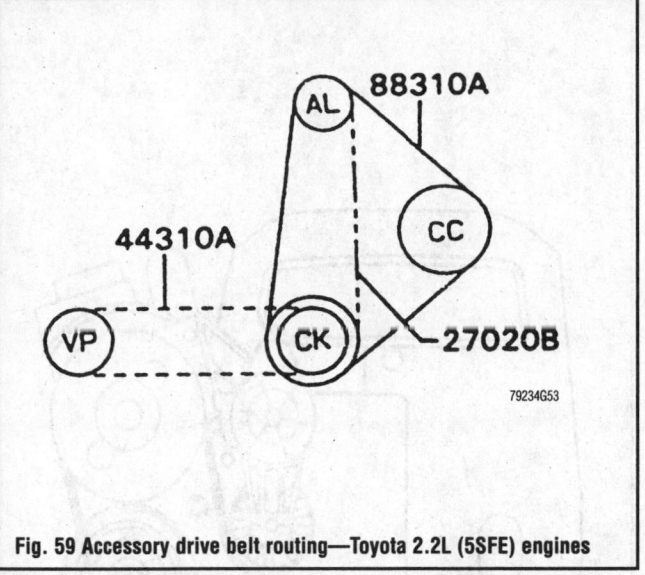

Fig. 59 Accessory drive belt routing—Toyota 2.2L (5SFE) engines

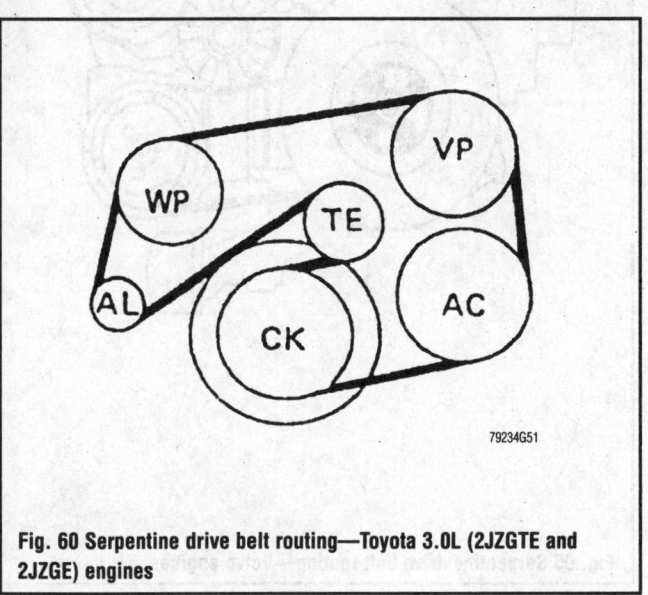

Fig. 60 Serpentine drive belt routing—Toyota 3.0L (2JZGTE and 2JZGE) engines

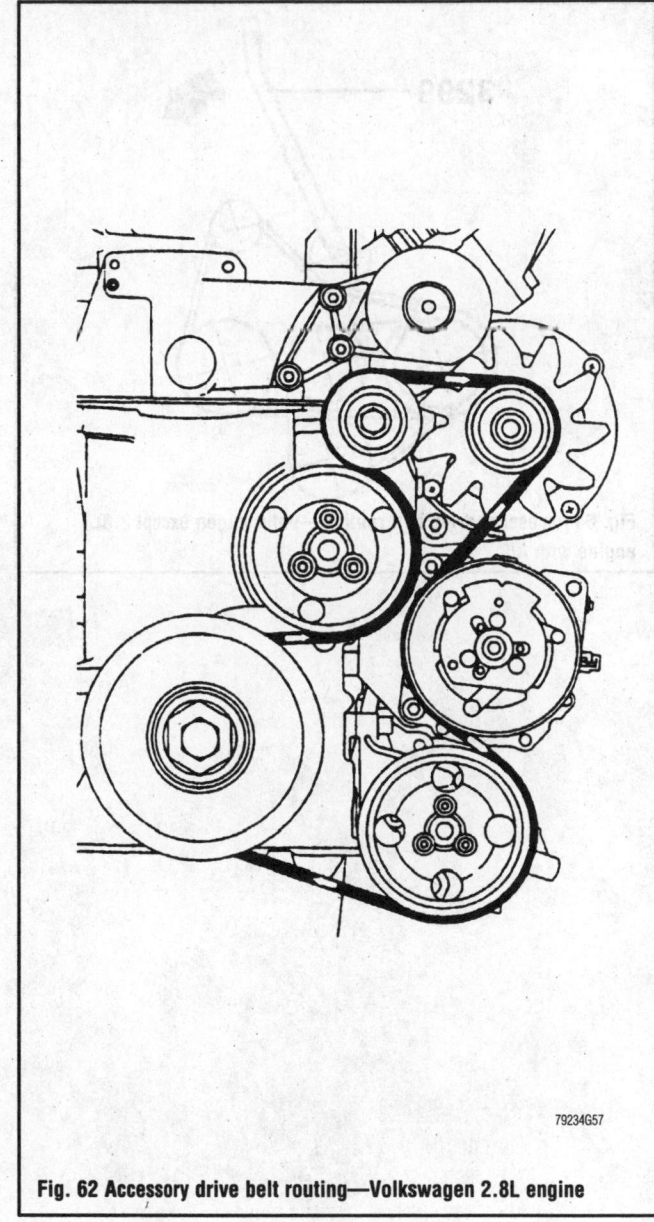

Fig. 62 Accessory drive belt routing—Volkswagen 2.8L engine

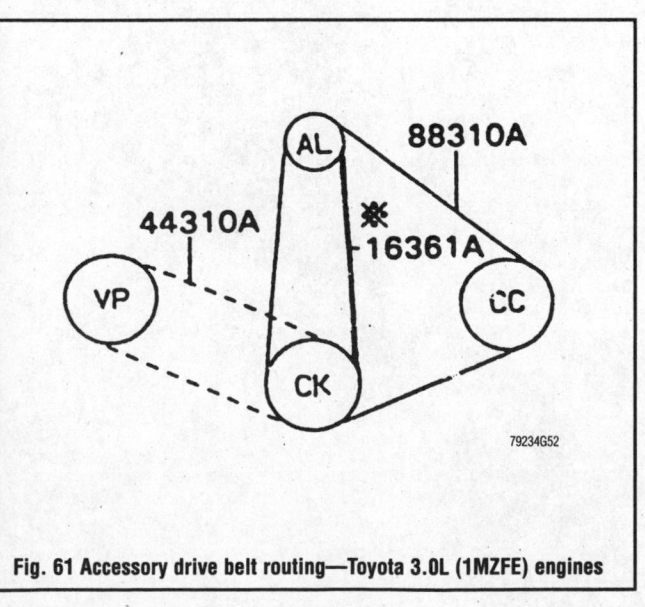

Fig. 61 Accessory drive belt routing—Toyota 3.0L (1MZFE) engines

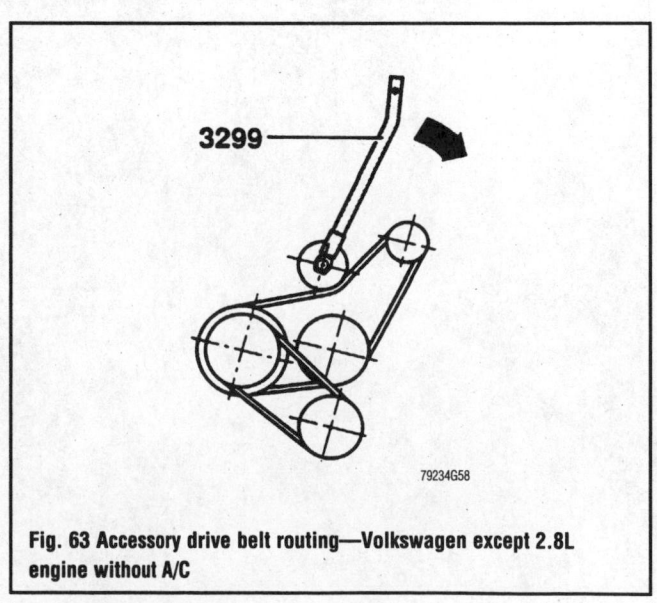

Fig. 63 Accessory drive belt routing—Volkswagen except 2.8L engine without A/C

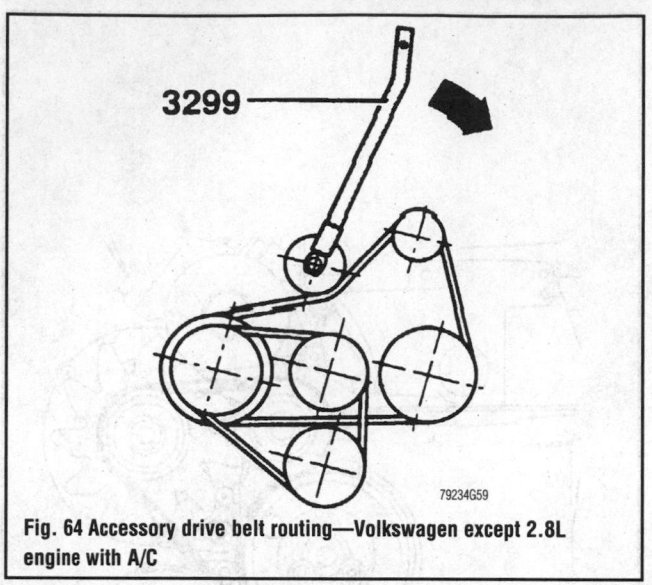

Fig. 64 Accessory drive belt routing—Volkswagen except 2.8L engine with A/C

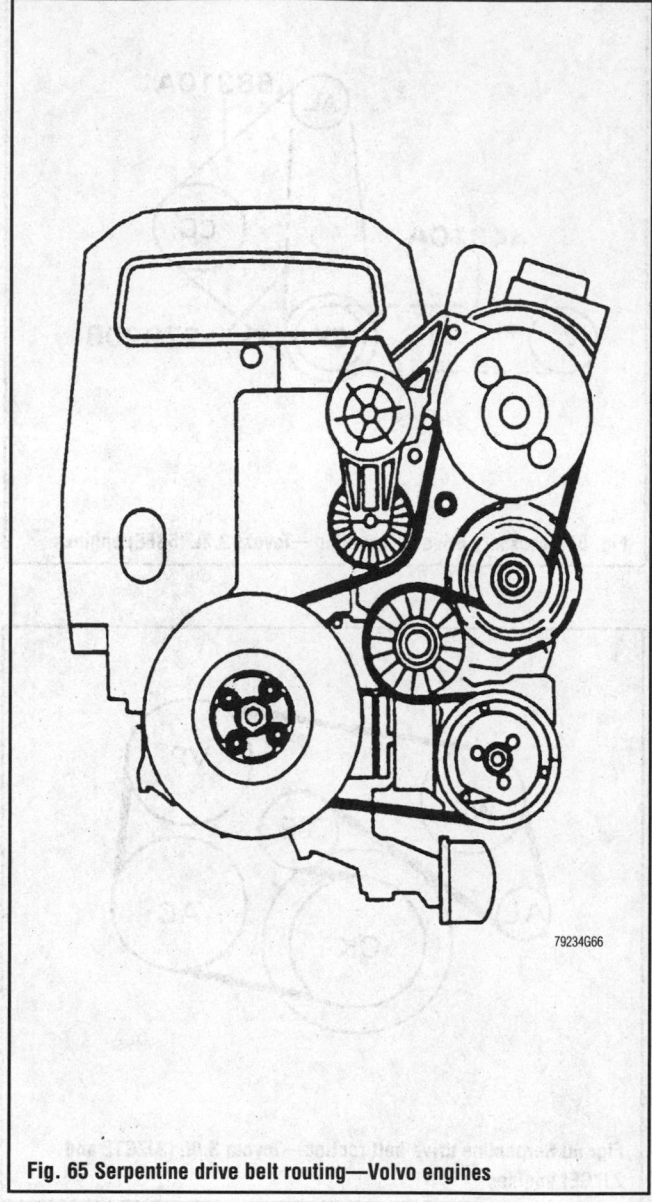

Fig. 65 Serpentine drive belt routing—Volvo engines

TIMING BELTS

5

GENERAL INFORMATION

Timing belts are typically only used on overhead camshaft engines. Timing belts are used to synchronize the crankshaft with the camshaft, similar to a timing chain on a overhead valve (pushrod) engine. Unlike a timing belt, a timing chain will normally last the life of the engine without needing service or replacement. Timing belts use raised teeth to mesh with sprockets to operate the valvetrain of an overhead camshaft engine.

Whenever a vehicle with an unknown service history comes into your repair facility or is recently purchased, here are some points that should be asked to help prevent costly engine damage:

• Does the owner know if, or when the belt was replaced?

• If the vehicle purchased is used, or the condition and mileage of the last timing belt replacement are unknown, it is recommended to inspect, replace, or at least inform the owner that the vehicle is equipped with a timing belt.

• Note the mileage of the vehicle. The average replacement interval for a timing belt is approximately 60,000 miles (96,000 km).

Interference Engines

Engines, chain- or belt-driven, can be classified as either free-running or interference, depending on what would happen if the piston-to-valve timing is disrupted. A free-running engine is designed with enough clearance between the pistons and valves to allow the crankshaft to rotate (pistons still moving) while the camshaft stays in one position (several valves fully open). If this condition occurs normally, no internal engine damage will result. In an interference engine, there is not enough clearance between the pistons and valves to allow the crankshaft to turn without the camshaft being in time.

An interference engine can suffer extensive internal damage if a timing belt fails. The piston design does not allow clearance for the valve to be fully open and the piston to be at the top of its stroke. If the belt fails, the piston will collide with the valve and will bend or break the valve, damage the piston, and/or bend a connecting rod. When this type of failure occurs, the engine will need to be replaced or disassembled for further internal inspection; either choice costing many times that of replacing the timing belt.

TIMING BELT SERVICE

Inspection

→For manufacturer's recommended service interval, refer to the maintenance interval chart located in this manual.

The average replacement interval for a timing belt is approximately 60,000 miles (96,000km). If, however, the timing belt is inspected earlier or more frequently than suggested, and shows signs of wear or defects, the belt should be replaced at that time.

✳✳ WARNING

Never allow antifreeze, oil or solvents to come into with a timing belt. If this occurs immediately wash the solution from the timing belt. Also, never excessive bend or twist the timing belt; this can damage the belt so that its lifetime is severely shortened.

Inspect both sides of the timing belt. Replace the belt with a new one if any of the following conditions exist:
- Hardening of the rubber—back side is glossy without resilience and leaves no indentation when pressed with a fingernail
- Cracks on the rubber backing
- Cracks or peeling of the canvas backing
- Cracks on rib root
- Cracks on belt sides
- Missing teeth or chunks of teeth
- Abnormal wear of belt sides—the sides are normal if they are sharp, as if cut by a knife.

If none of these conditions exist, the belt does not need replacement unless it is at the recommended interval. The belt MUST be replaced at the recommended interval.

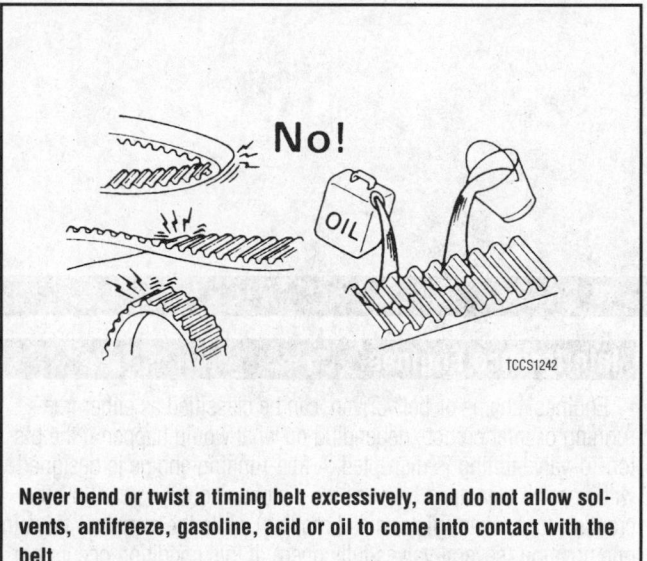

Never bend or twist a timing belt excessively, and do not allow solvents, antifreeze, gasoline, acid or oil to come into contact with the belt

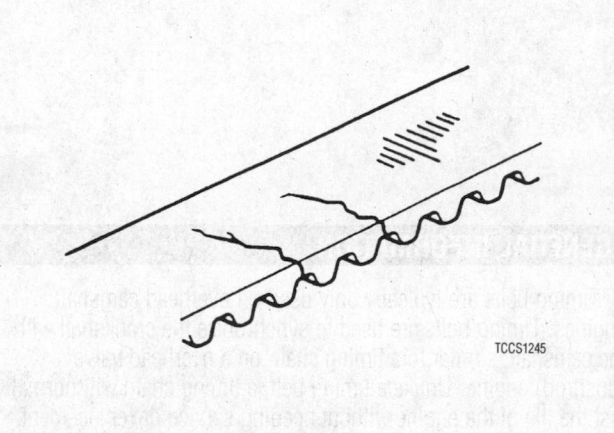

Back surface worn or cracked from a possible overheated engine or interference with the belt cover

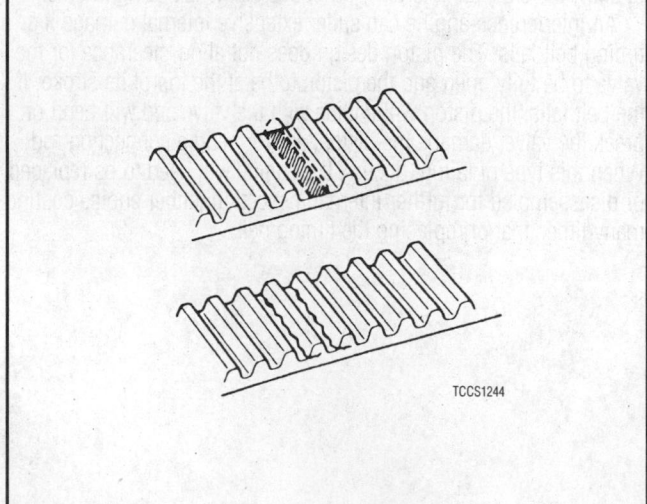

Broken tooth may be due to a damaged pulley

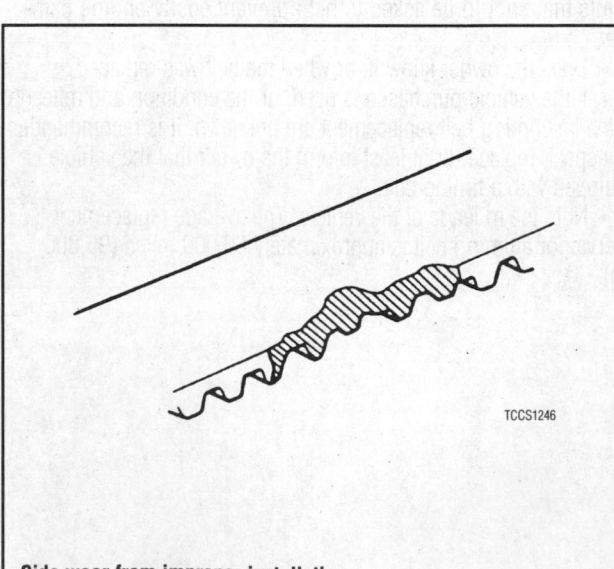

Side wear from improper installation

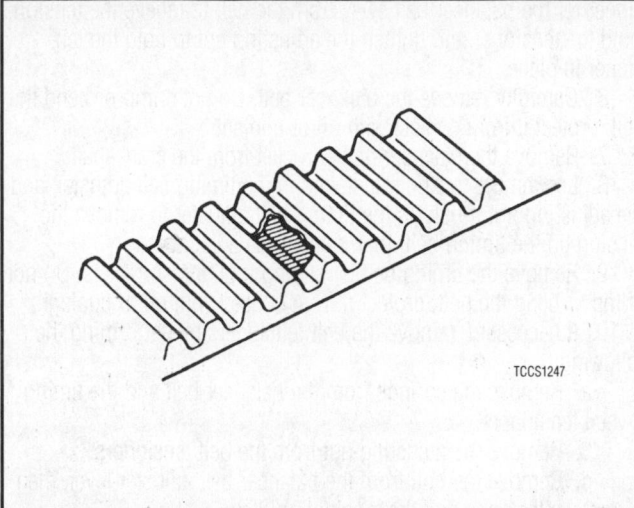

Worn teeth from excessive belt tension, camshaft or distributor not turning properly, or fluid leaking on the belt

✳✳ WARNING

On interference engines, it is very important to replace the timing belt at the recommended intervals, otherwise expensive engine damage will likely result if the belt fails.

Removal & Installation

ACURA MODELS

➡**The radio may have a coded theft protection circuit. Obtain the code before disconnecting the battery, removing the radio fuse, or removing the radio.**

1.8L (B18B1 and B18C1) Engines

1. Turn the crankshaft pulley until cylinder No. 1 is set to Top Dead Center (TDC) on the compression stroke. The white crankshaft pulley mark should be aligned with the pointer on the lower timing belt cover.

2. Remove all necessary components to gain access to the cylinder head and timing belt covers.

3. Remove the cylinder head and timing belt covers.

4. Remove the crankshaft pulley bolt and the crankshaft pulley. To remove the crankshaft pulley bolt a pulley holder (holder attachment tool part No. 07MAB-PY3010A and holder handle tool part No. 07JAB-001020A, or equivalent) will be needed to keep the crankshaft from turning.

✳✳ WARNING

Do not use the timing belt covers to store small parts. Grease or oil can transfer from the parts to the cover, then to the belt. Clean the covers thoroughly before installation.

5. Recheck that the No. 1 piston is at TDC on its compression stroke. Align the groove on the toothed side of the crankshaft timing belt drive sprocket to the arrow pointer on the oil pump.

6. To set the camshafts to top dead center for the No. 1 cylinder, align the hole in each camshaft with the holes in the No. 1

camshaft holders, then push 5.0mm pin punches into the holes. Be sure that the **UP** arrows are pointing up and that the TDC marks on the intake and exhaust sprockets are aligned.

7. Loosen the tensioner adjusting bolt 180 degrees (½ turn). Push on the tensioner to remove the tension from the timing belt, then retighten the bolt. If the timing belt is to be reinstalled, mark the direction of rotation on the belt with a crayon or white paint.

8. Remove the timing belt from the sprockets.

➡**Be sure the water pump pulley turns counterclockwise freely. Check for signs of seal leakage; a small amount of weeping from the bleed hole is normal.**

9. If necessary, remove the timing belt tensioner by performing the following:

a. Remove the timing belt tensioner spring.

b. Remove the bolt from the timing belt tensioner and remove the tensioner.

To install:

10. If the timing belt tensioner was removed, perform the following:

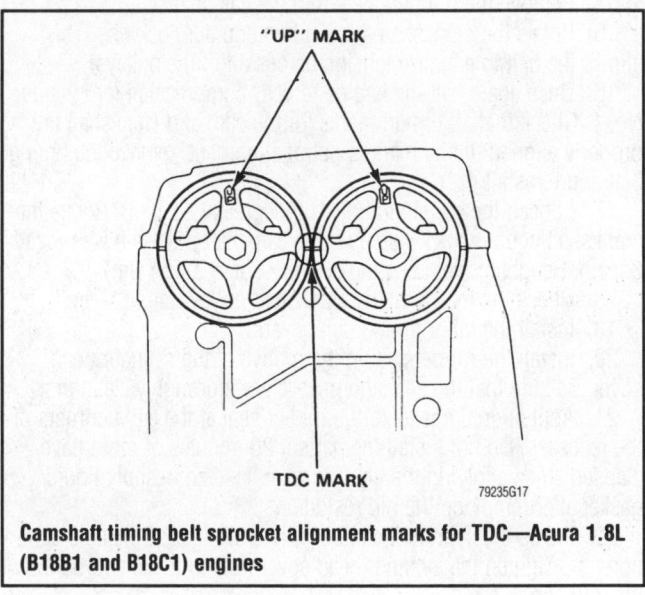

Camshaft timing belt sprocket alignment marks for TDC—Acura 1.8L (B18B1 and B18C1) engines

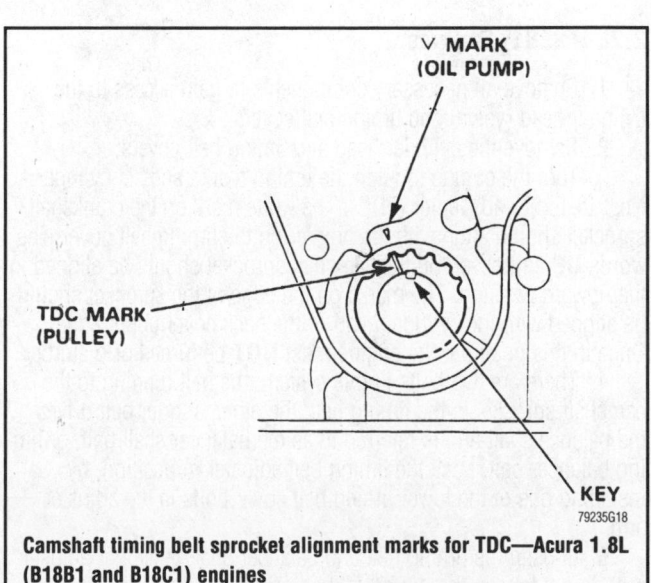

Camshaft timing belt sprocket alignment marks for TDC—Acura 1.8L (B18B1 and B18C1) engines

a. Position the timing belt tensioner on the engine and install the attaching bolt loosely.

b. Install the timing belt tensioner spring.

c. Push the tensioner down, then snug the tensioner bolt to hold this position.

➡ **Before reinstallation, check every component for cleanliness. All covers, pulleys, shields, etc. must be completely free of grease and oil.**

➡ **Install the timing belt in the correct sequence. Also, if installing the old belt, be sure it is turning the same direction.**

11. Install the timing belt first to the crankshaft pulley, then to the adjuster, then to the water pump pulley, the exhaust camshaft and finally to the intake camshaft pulley.

12. Install the lower belt cover. Install the crankshaft pulley, tightening the bolt to 130 ft. lbs. (177 Nm). Lubricate the threads and the flange of the bolt with engine oil before installation.

13. Loosen the adjusting bolt, allowing the adjuster to tension the belt. Retighten the bolt to 40 ft. lbs. (54 Nm).

14. Remove the pin punches from the camshafts.

15. Rotate the crankshaft 4 to 6 turns counterclockwise. This allows the belt to equalize tension across all of the pulleys.

16. Once again, set the engine to TDC compression for cylinder No. 1. Check that all timing marks for the cam and crankshaft are properly aligned. If any mark is out of alignment, remove the timing belt and reinstall it.

17. Loosen the adjusting bolt 180 degrees (½ turn). Rotate the crankshaft counterclockwise until the camshaft pulleys have moved 3 teeth. Retighten the adjusting bolt to 40 ft. lbs. (54 Nm).

18. Check the torque of the crankshaft pulley bolt.

19. Install the other timing belt covers.

20. Install the rubber seal in the groove of the cylinder head cover. Be sure that the seal and groove are thoroughly clean first.

21. Apply liquid gasket to the rubber seal at the eight corners of the recesses. Do not install the parts if 20 minutes or more have elapsed since applying the liquid gasket. Instead, reapply liquid gasket after removing the old residue.

22. Install the cylinder head cover and all other applicable components. Tighten the cylinder head cover nuts in 2 steps to 88 inch lbs. (10 Nm).

2.2L (F22B1) Engines

1. Remove all necessary components to gain access to the cylinder head (valve) and timing belt covers.

2. Remove the cylinder head and timing belt covers.

3. Turn the engine to align the timing marks and set cylinder No.1 to Top Dead Center (TDC). The white mark on the crankshaft sprocket should align with the pointer on the timing belt cover. The words **UP** embossed on the camshaft sprocket should be aligned in the upward position. The marks on the edge of the sprocket should be aligned with the cylinder head or the back cover upper edge. Once in this position, the engine must **NOT** be turned or disturbed.

4. There are two belts in this system; the belt running to the camshaft sprocket is the timing belt, the other, shorter belt drives the balance shaft and is referred to as the balancer shaft belt, or timing balancer belt. Lock the timing belt adjuster in position, by installing one of the lower timing belt cover bolts in the adjuster arm.

5. Loosen the timing belt and balancer shaft tensioner adjuster nut, do not loosen the nut more than one revolution. Push the ten-

sioner for the balancer belt away from the belt to relieve the tension. Hold the tensioner and tighten the adjusting nut to hold the tensioner in place.

6. Carefully remove the balancer belt. Do not crimp or bend the belt; protect it from contact with oil or coolant.

7. Remove the balancer belt sprocket from the crankshaft.

8. Loosen the lockbolt, installed in the timing belt adjuster, and the adjusting nut. Push on the timing belt adjuster to remove the tension on the timing belt, then tighten the adjuster nut.

9. Remove the timing belt by sliding it off the sprockets. Do not crimp or bend the belt; protect it from contact with oil or coolant.

10. If necessary, remove the belt tensioners by performing the following:

a. Remove the springs from the balancer belt and the timing belt tensioners.

b. Remove the adjusting nut from the belt tensioners.

c. Remove the bolt from the balancer belt adjuster lever, then remove the lever and the tensioner pulley.

d. Remove the lockbolt from the timing belt tensioner lever, then remove the tensioner pulley and lever from the engine.

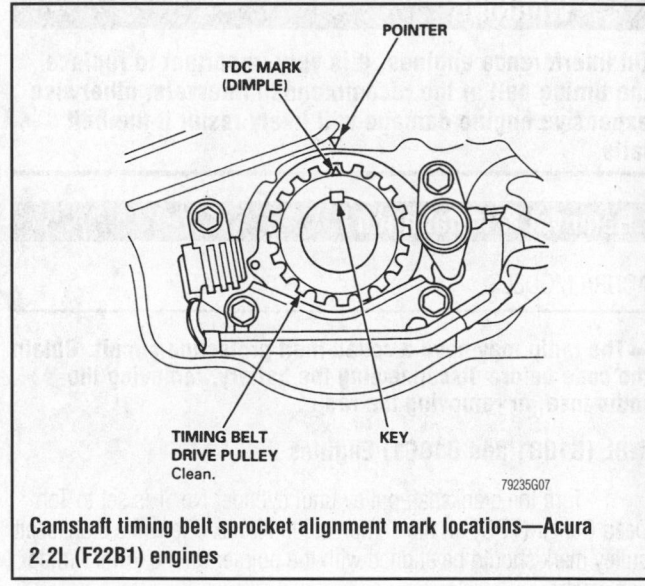

Camshaft timing belt sprocket alignment mark locations—Acura 2.2L (F22B1) engines

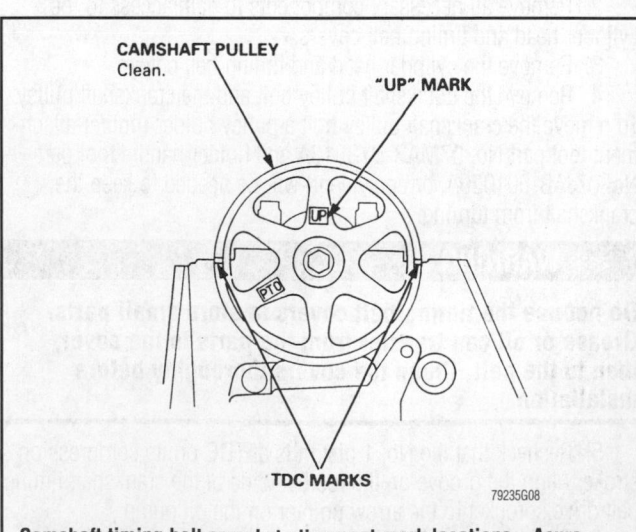

Camshaft timing belt sprocket alignment mark locations—Acura 2.2L (F22B1) engines

11. This is an excellent time to check or replace the water pump. Even if the timing belt is only being replaced as part of a good maintenance schedule, consider replacing the pump at the same time.

To install:

12. If the water pump is to be replaced, install a new O-ring on the pump and make certain it is properly seated. Install the water pump and tighten the mounting bolts to 106 inch lbs. (12 Nm).

13. If the tensioners were removed, perform the following to install them:

   a. Install the timing belt tensioner lever and the tensioner pulley.

   b. Install the balancer belt pulley and adjuster lever.

   c. Install the adjusting nut and bolt to the balancer belt adjuster lever.

   d. Install the springs to the tensioners.

   e. Install the lockbolt to the timing belt tensioner, then move the tensioner it's full deflection and tighten the lockbolt.

   f. Move the balancer belt tensioner it's full deflection and tighten the adjusting nut to hold its position.

14. The pointer on the crankshaft sprocket should be aligned with the pointer on the oil pump; the camshaft sprocket must be aligned so that the word **UP** is at the top of the sprocket and the marks on the edge of the sprocket are aligned with the surfaces of the head, or the back cover upper edge.

15. Install the timing belt in the following sequence:

   a. Start the belt on the crankshaft sprocket.

   b. Then, around the tensioner sprocket.

   c. On the water pump sprocket.

   d. Finally, around the camshaft sprocket.

16. Check the timing marks to be sure that they did not move.

17. Loosen, then retighten the timing belt adjusting nut.

18. Install the timing/balancer belt drive sprocket and the lower timing belt cover.

19. Install the crankshaft pulley and bolt, tighten the bolt to 181 ft. lbs. (245 Nm). Rotate the crankshaft sprocket five or six revolutions to properly position the timing belt on the sprockets.

20. Set the No. 1 cylinder to TDC and loosen the timing belt adjusting nut one turn. Turn the crankshaft counterclockwise until the cam sprocket has moved 3 teeth; this creates the proper tension on the timing belt.

21. Tighten the timing belt adjusting nut.

22. Set the crankshaft sprocket and the camshaft sprocket to TDC. If the sprockets do not align, remove the belt to realign the marks, then install the belt.

23. Remove the crankshaft pulley and the lower cover.

24. With the timing marks aligned, lock the timing belt adjuster in place with one of the lower cover mounting bolts.

25. Loosen the adjusting nut and ensure the timing balancer belt adjuster moves freely.

26. Align the rear timing balancer sprocket using a 6x100mm bolt or rod. Mark the bolt or rod at a point 2.9 in. (74mm) from the end. Remove the bolt from the maintenance hole on the side of the block; insert the bolt/rod into the hole and align the 2.9 in. (74mm) mark with the face of the hole. This will hold the shaft in place during installation.

27. Align the groove on the front balancer shaft sprocket with the pointer on the oil pump.

28. Install the balancer belt. Once the belts are in place, be sure that all the engine alignment marks are still correct. If not, remove the belts, realign the engine and reinstall the belts. Once the belts are properly installed, slowly loosen the adjusting nut, allowing the tensioner to move against the belt. Remove the bolt from the maintenance hole and reinstall the bolt and washer.

29. Install the crankshaft pulley, then turn the crankshaft sprocket one turn counterclockwise and tighten the timing belt adjusting nut to 33 ft. lbs. (45 Nm).

30. Remove the crankshaft pulley and the bolt locking the timing belt adjuster in place.

31. Install the lower timing belt cover and tighten the bolts to 106 inch lbs. (12 Nm).

32. Install a new seal around the adjusting nut. Do not loosen the adjusting nut.

33. Install the crankshaft pulley. Coat the threads and seating face of the pulley bolt with engine oil, then install and tighten the bolt to 181 ft. lbs. (250 Nm).

34. Install a the dipstick tube, then install the side engine mount. Tighten the bolt and nut attaching the mount to the engine to 40 ft. lbs. (55 Nm). Tighten the through-bolt and nut to 47 ft. lbs. (65 Nm), then remove the jack from under the engine.

35. Install the upper timing belt cover. Tighten the bolt toward the exhaust manifold to 89 inch lbs. (10 Nm) and the bolt toward the intake manifold to 106 inch lbs. (12 Nm).

36. Install the cylinder head cover gasket in the groove of the cylinder head cover. Before installing the gasket, thoroughly clean the seal and the groove. Seat the recesses for the camshaft first, then work it into the groove around the outside edges. Be sure the gasket is seated securely in the corners of the recesses.

37. Apply liquid gasket to the four corners of the recesses in the cylinder head cover gasket. Do not install the parts if five minutes or more have elapsed since applying liquid gasket. After assembly, wait at least 20 minutes before filling the engine with oil.

38. Install the cylinder head (valve) cover and all other applicable components. Tighten the bolts attaching the cylinder head cover in two steps to the proper sequence of 89 inch lbs. (10 Nm).

2.5L (G25A4) Engines

1. Set the No. 1 piston at TDC on the compression stroke. The white TDC mark on the crankshaft pulley must line up with the pointers on the lower belt cover.

2. Remove all necessary components to gain access to the timing belt covers.

3. Remove the timing belt upper cover.

4. Be sure the **UP** mark and the TDC marks on the camshaft sprocket are correctly positioned.

5. Rotate the crankshaft to align the white timing mark on the crankshaft pulley with the pointer on the lower cover. There are similar alignment marks on the crankshaft sprocket and oil pump housing

6. Remove the timing belt lower cover.

7. Loosen the adjusting bolt 180 degrees (½ turn), then push the tensioner down to relieve the belt tension. Retighten the adjusting bolt to 33 ft. lbs. (44 Nm).

➡**Do not remove the adjusting bolt and tensioner pulley unless they are to be replaced. The bolt is only loosened and tightened in this procedure to tension the timing belt.**

8. Remove the timing belt.

9. Inspect the timing belt tensioner pulley and tension spring for signs of wear. Remove and replace the tensioner assembly as necessary.

To install:

➡**Replace the timing belt if it shows any signs of wear, damage, or contamination from oil or coolant. The source of any oil or coolant contamination must be determined and corrected before the new timing belt may be installed.**

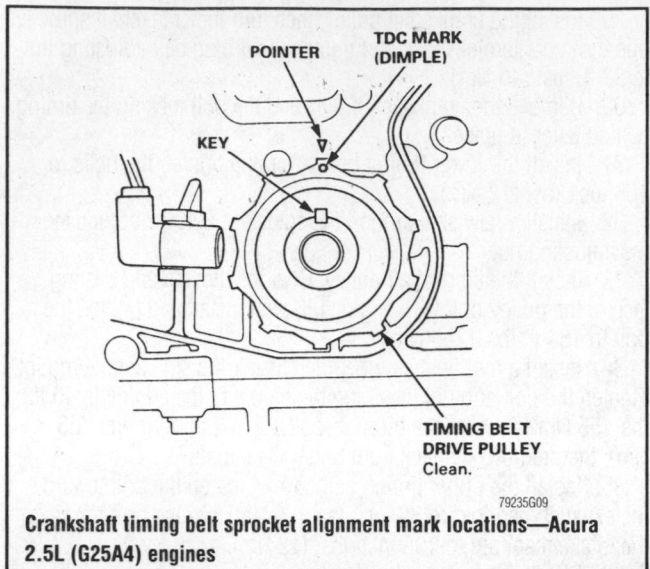

Crankshaft timing belt sprocket alignment mark locations—Acura 2.5L (G25A4) engines

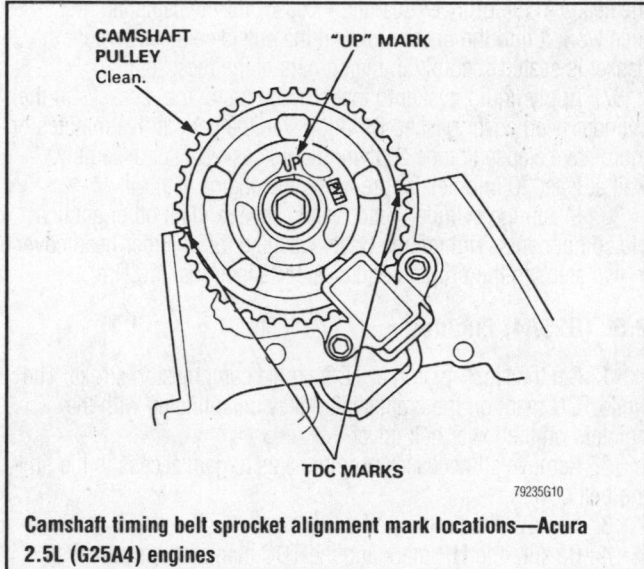

Camshaft timing belt sprocket alignment mark locations—Acura 2.5L (G25A4) engines

10. Verify that the crankshaft and camshaft sprocket matchmarks are properly aligned.

11. Install the timing belt in the following order: first onto the crankshaft sprocket, then the tensioner pulley, the water pump sprocket, and finally the camshaft sprocket.

12. Loosen the tensioner adjusting bolt to allow the spring to set the tension. Then, tighten the bolt to 33 ft. lbs. (44 Nm). Rotate the crankshaft six full turns to seat the belt and verify that the timing marks align properly.

13. Install the lower and upper timing belt covers, and tighten the bolts to 106 inch lbs. (12 Nm).

14. Oil only the threads on the crankshaft pulley bolt. Install the crankshaft pulley and tighten the bolt to 181 ft. lbs. (245–250 Nm).

15. Install the dipstick tube.

16. Install the cylinder head cover with new gaskets and washers. Tighten the cap nuts on Vigor models to 84 inch lbs. (9.5 Nm), or to 106 inch lbs. (12 Nm) on 2.5TL models.

3.2L (C32A1) Engines—Legend

1. Remove all necessary components for access to the timing belt covers.

2. Remove the upper timing belt covers.

3. Rotate the crankshaft to Top Dead Center (TDC) on the compression stroke for the No. 1 piston. The white mark on the crankshaft pulley will be aligned with the pointer on the lower cover, and the camshaft sprocket marks will be aligned with the yellow marks on the rear covers.

> ✻✻ **WARNING**
>
> **Do not rotate the crankshaft or camshafts with the belt removed. The pistons will contact the valves and cause engine damage. If it is necessary to move the camshaft(s) for proper alignment, first advance the crankshaft by 15 degrees from TDC. Adjust the camshaft position as needed, then return the crankshaft 15 degrees to TDC.**

4. Remove the lower timing belt cover.

5. Loosen the timing belt tensioner pulley bolt approximately ½ turn (180 degrees) and push the pulley to slacken the belt tension. Tighten the bolt and remove the belt. If the belt is to be reinstalled, mark the direction of rotation on it with a crayon or with white paint.

To install:

➡ **Replace the belt if it is worn, cracked, or oil soaked. Find and repair the source of the oil leak before installing a new belt. Inspect the water pump. If there is any doubt about the condition of the water pump, it should be replaced now that the timing belt is removed.**

6. Install the belt in the following sequence around the sprockets: crankshaft, adjuster pulley, left camshaft, water pump, right camshaft.

7. To adjust the timing belt tension, loosen the tensioner pulley bolt approximately ½ turn (180 degrees). The spring will automatically set the proper tension.

8. Install the lower cover and the crankshaft pulley. Apply oil to the pulley bolt threads and washer, then tighten the pulley bolt to 174 ft. lbs. (240 Nm).

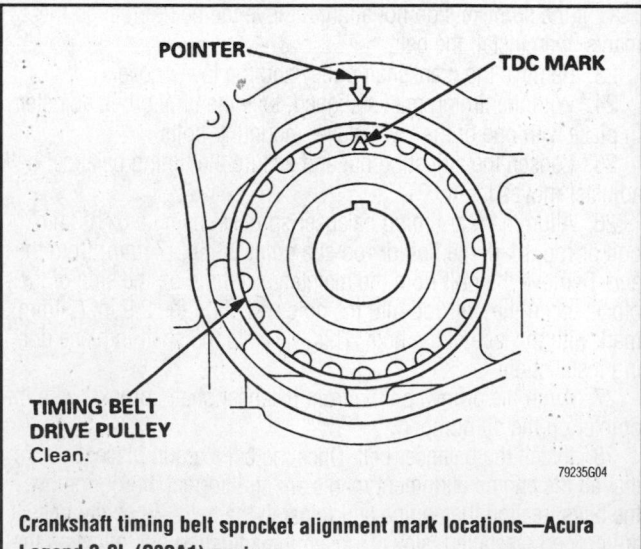

Crankshaft timing belt sprocket alignment mark locations—Acura Legend 3.2L (C32A1) engines

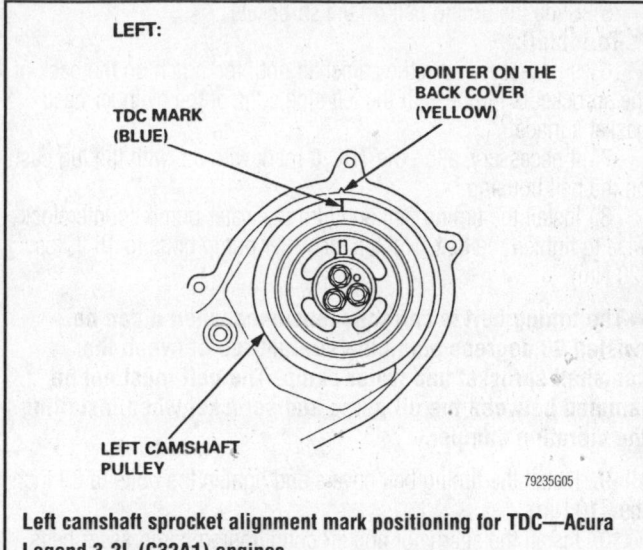

Left camshaft sprocket alignment mark positioning for TDC—Acura Legend 3.2L (C32A1) engines

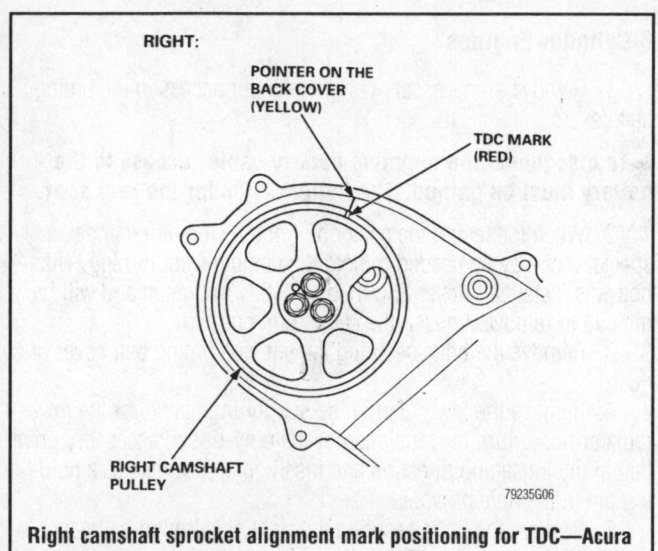

Right camshaft sprocket alignment mark positioning for TDC—Acura Legend 3.2L (C32A1) engines

9. Rotate the crankshaft five or six turns clockwise and check that the timing marks on the crankshaft and camshafts align properly. Adjust the timing belt tension again by rotating the crankshaft to align the blue mark on the pulley with the pointer.

10. Loosen the tensioner pulley bolt so that the tensioner can automatically adjust the belt tension, then retighten the bolt to 31 ft. lbs. (43 Nm).

11. Install the dipstick pipe, then the upper covers. Install all applicable components. When installing the center mount, center bracket, and damper, tighten the center mount and bracket bolts to 28 ft. lbs. (39 Nm) and the damper bolts to 16 ft. lbs. (22 Nm).

3.2L (C32A1) Engines—3.2TL

1. Turn the engine to align the timing marks and set cylinder No.1 to Top Dead Center (TDC) on the compression stroke. The white mark on the crankshaft pulley should align with the pointer on the timing belt cover. Remove the inspection caps on the upper tim-

ing belt covers to check the alignment of the timing marks. The pointers for the camshafts should align with the marks on the camshaft sprockets.

2. Remove all necessary component for access to the timing belt covers.

3. Remove the upper and lower timing belt covers. Clean any dirt, oil or grease from the covers. Do not use the covers for storing removed items.

4. Loosen the timing belt tensioner adjusting bolt 180 degrees (½ turn). Push on the tensioner to remove tension from the timing belt, then tighten the adjusting bolt.

5. Remove the timing belt.

6. If necessary, remove the timing belt tensioner by performing the following:

a. Remove the spring from the tensioner.

b. Remove the bolt mounting the tensioner, then remove the tensioner.

To install:

✳✳ CAUTION

Do not rotate the crankshaft pulley or camshaft pulleys with the timing belt removed. The pistons may hit the valves and cause damage.

7. If necessary, install the timing belt tensioner by performing the following:

a. Install the tensioner and the attaching bolt.

b. Move the tensioner its full deflection to the left and tighten the bolt.

c. Install the spring to the tensioner.

8. Remove the spark plugs.

9. Set the timing belt drive (crankshaft) sprocket so that the No. 1 piston is at top dead center (TDC). Align the TDC mark on the tooth of the timing belt drive sprocket with the pointer on the oil pump.

10. Set the camshaft pulleys so that the No. 1 piston is at TDC. Align the TDC mark on the camshaft pulleys to the pointers on the back covers.

11. Install the timing belt on the sprockets in the following sequence: drive sprocket (crankshaft), tensioner pulley, left camshaft sprocket, water pump pulley, right camshaft sprocket.

12. Loosen, then retighten the timing belt adjuster bolt to tension the timing belt.

13. Install the lower timing belt cover.

14. Install the crankshaft sprocket and pulley bolt. Tighten the bolt to 174 ft. lbs. (235 Nm) with the aid of the crank pulley holder.

15. Rotate the crankshaft five or six turns clockwise so that the timing belt positions itself properly on the sprockets.

16. Set cylinder No. 1 to TDC by aligning the timing marks. If the timing marks do not align, remove the timing belt, then adjust the components and reinstall the timing belt.

17. Rotate the crankshaft clockwise enough to move the camshaft pulley nine teeth (the blue mark on the crankshaft pulley should line up with the pointer on the lower cover).

18. Loosen the timing belt adjusting bolt 180 degrees (½ turn), then tighten the bolt to 31 ft. lbs. (42 Nm).

19. Install the upper timing belt covers, then install all applicable components. When installing the center bracket, tighten the bolts attaching the brackets to 40 ft. lbs. (54 Nm), then the mount through-bolt to 40 ft. lbs. (54 Nm).

AUDI MODELS

5-Cylinder Engines

1. Disconnect the negative battery cable. Using the large bolt on the crankshaft sprocket, rotate the engine until the No. 1 cylinder is at TDC of the compression stroke. Align the TDC mark **0** with the cast mark on the bell housing. If the belt hasn't jumped teeth, the timing mark on the rear face of the camshaft sprocket should be aligned with the upper left edge of the valve cover.

2. Remove the alternator and air conditioner compressor drive belts.

3. Remove the upper and lower timing belt covers.

4. Loosen the water pump bolts only enough to turn the pump clockwise.

➡**By loosening the water pump bolts, the coolant may drain from the engine at the water pump. If necessary, drain the cooling system, remove the water pump and reinstall it with a new O-ring.**

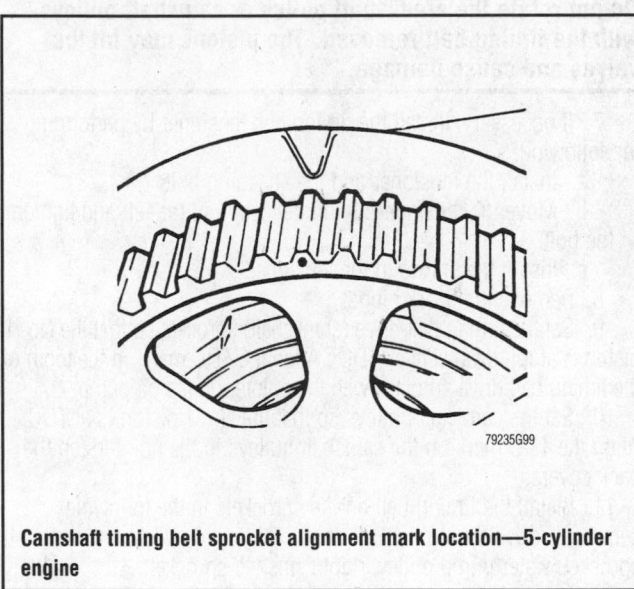

Camshaft timing belt sprocket alignment mark location—5-cylinder engine

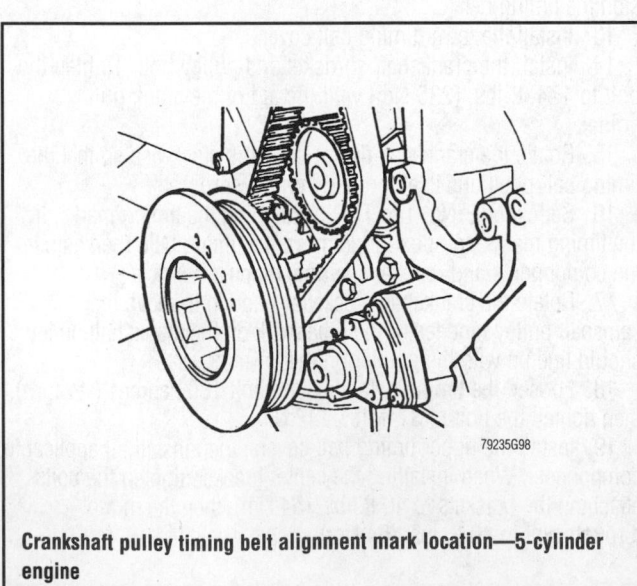

Crankshaft pulley timing belt alignment mark location—5-cylinder engine

5. Slide the timing belt off the sprockets.

To install:

6. If necessary, turn the camshaft until the notch on the back of the sprocket is in line with the left side edge of the cylinder head gasket surface.

7. If necessary, align the TDC **0** mark with the with the lug cast on the bell housing.

8. Install the timing belt and turn the water pump counterclockwise to tighten the belt. Tighten the water pump bolts to 15 ft. lbs. (20 Nm).

➡**The timing belt is correctly tensioned when it can be twisted 90 degrees along the straight run between the camshaft sprocket and water pump. The belt must not be jammed between the oil pump and sprocket when installing the vibration damper.**

9. Install the timing belt covers and tighten the bolts to 89 inch lbs. (10 Nm).

10. Install the alternator and air conditioning compressor belts. These belts are correctly tensioned when they can be depressed ⅜ in. (9.5mm) along their longest straight run.

8-Cylinder Engines

1. Remove all necessary components for access to the timing belt covers.

➡**To disconnect the negative battery cable, access to the battery must be gained. The battery is under the rear seat.**

2. When loosening the center bolt of the vibration damper, a special tool may be needed to hold the damper from turning. This bolt was installed to over 250 ft. lbs. (340 Nm) tighten and will be difficult to remove. Loosen the center bolt one turn.

3. Remove the bolts securing the left side timing belt cover or guard.

4. Remove the lower part of the supporting clamp for the lower radiator hose. Turn the tensioner for the poly-ribbed accessory drive belt in the loosening direction and insert an appropriate size holding pin in the hole provided.

5. Remove the bolts securing the right side timing belt cover, with the exception of the top bolt. Remove the tensioner holding pin. Remove the belt cover top screw and remove the cover or guard. Carefully lift the guard from the bottom to avoid damaging the radiator.

6. Disconnect the ignition cables at the coils. Remove both distributor caps. Turn the engine to TDC. It may be necessary to temporarily install the damper to align the timing marks. Check that the distributor rotor is pointing to the mark on the housing. If not, turn the crankshaft 1 additional turn. Remove both distributors.

7. Remove the stop plate at the timing belt tensioner. Disconnect the shock-absorber shaped damper at the top bolt.

8. Remove the belt from the tensioning idler pulley (with eccentric). Pulley is on right side of engine. Take the belt off both camshaft sprockets.

9. A special holding tool is available that is installed on the back of the camshafts. It fits the locating pin on the distributor flanges. If necessary, use a special hook wrench tool to turn the camshaft until the pins latch into the special holding tool. Secure the special tool with the distributor mounting bolts.

10. At the camshaft sprocket end, loosen the mounting bolts 2 turns. Using a plastic hammer, tap the edge of the camshaft sprockets loose.

➡After the sprockets are removed, note the grooves machined in the camshaft ends. Woodruff keys must not be installed in the camshaft sprocket/camshaft connection. Unlike the other engines, this engine does not use cam keys.

11. Remove the vibration damper which was previously temporarily reinstalled to line up TDC timing marks. A puller can be used. Unscrew 2 opposing bolts of the 4 bolts connecting the vibration damper and toothed belt sprocket. Use a puller in these holes.

12. Remove the timing belt.

To install:

13. Before installing a new belt, the rollers and tensioners must be checked for dirt, rough running and ease of rotation. Clean or replace rollers and tensioners, as necessary.

14. Fit the toothed belt at the crankshaft and install the vibration damper with the belt on the crankshaft.

15. Apply thread locking compound to the center bolt and tighten to 332 ft. lbs. (450 Nm) using hand wrench. A holding tool may be required to keep the crankshaft from turning. It is a must to ensure that the TDC mark on the vibration damper is aligned with the TDC pointer before and after tightening the camshaft sprockets.

16. Guide the toothed belt into position and install the idler pulley with eccentric. Snug the nut enough to hold the eccentric in place but do not fully tighten it yet.

17. Tighten the damper top bolt to 18 ft. lbs. (24 Nm). Engage the damper to the tensioner lever by pressing the lever downward.

18. Perform the basic setting of the toothed belt tensioner by turning the idler pulley eccentric clockwise.

19. A special turning tool may be needed. Turn the eccentric and measure the damper length. Measure the overall length of the barrel not counting the mounting eyes. Turn the eccentric until the damper barrel length is 5.11–5.23 in. (130–133mm).

20. Tighten the idler pulley eccentric to 18 ft. lbs. (24 Nm). Tighten the camshaft sprockets to 33 ft. lbs. (45 Nm).

21. Remove any camshaft and crankshaft locking tool previously installed.

22. Turn engine at least 2 turns. Check the damper length and if necessary, readjust the idler pulley.

23. Reattach the ignition cables at the coils. Install both distributor caps.

24. Install the right side cover or guard.

25. Install the left side cover or guard, and all remaining components.

CHRYSLER IMPORTS

1.5L and 1.8L Engines

1. Rotate the crankshaft clockwise and position the engine at Top Dead Center (TDC) on the compression stroke for the No. 1 cylinder.

2. Remove all necessary components for access to the timing belt upper and lower covers, then remove the covers.

3. If the original timing belt is going to be reused, draw an arrow on the back of the timing belt indicating the direction of rotation so it may be installed in the same direction.

4. For the 1.5L engine, loosen the timing belt tensioner and move the tensioner to provide slack to the timing belt. Tighten the tensioner in this position.

5. For the 1.8L engine, loosen the timing belt tensioner, insert a thin prytool into the tensioner and release tension by prying against the spring tension. Temporarily tighten the tensioner bolt to provide slack.

6. Remove the timing belt.

Coolant and engine oil will damage the rubber in the timing belt, drastically reducing its life. Do not allow engine oil or coolant to contact the timing belt, the sprockets or tensioner assembly.

7. If defective, remove the tensioner spacer, tensioner spring and tensioner assembly.

➡It is recommended that the timing belt be replaced at least every 60,000 miles (96,000 km).

8. Inspect the timing belt for cracks or wear. Check the tensioner pulley for smooth rotation.

To install:

9. If removed, position the tensioner, tensioner spring and tensioner spacer on the engine block.

10. Align the timing marks on the camshaft sprocket and crankshaft sprocket. This will position the No. 1 piston at TDC on its compression stroke.

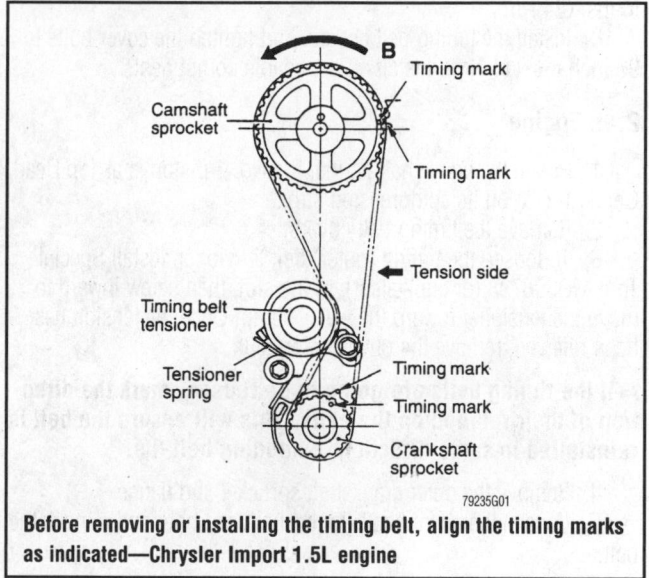

Before removing or installing the timing belt, align the timing marks as indicated—Chrysler Import 1.5L engine

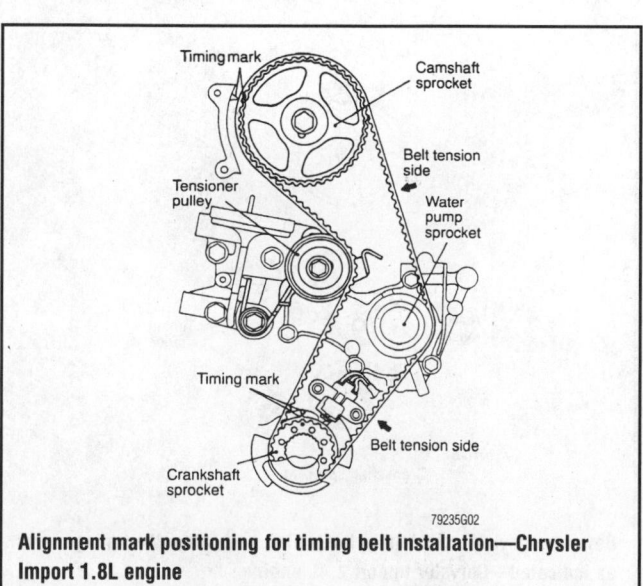

Alignment mark positioning for timing belt installation—Chrysler Import 1.8L engine

11. For the 1.5L engine, position the timing belt on the crankshaft sprocket, keeping the tension side of the belt tight, set it on the camshaft sprocket, then the tensioner sprocket.

12. For the 1.8L engine, position the timing belt on the crankshaft sprocket, water pump sprocket, camshaft sprocket and the tensioner sprocket, keeping the tension side of the belt tight.

13. Apply a slight counterclockwise force to the camshaft sprocket to give tension to the belt, and be sure all timing marks are aligned.

14. Loosen the pivot side tensioner bolt and the slot side bolt. Allow the spring to remove any slack in the timing belt.

15. For the 1.5L engine, tighten the slot side tensioner bolt, then the pivot side bolt. If the pivot side bolt is tightened first, the tensioner could turn with the bolt, causing over tension.

16. For the 1.8L engine, turn the crankshaft clockwise two rotations, then tighten the adjuster bolt to 18 ft. lbs. (24 Nm) and the pivot (spring) bolt to 35 ft. lbs. (45 Nm).

17. Turn the crankshaft clockwise. Loosen the pivot side tensioner bolt, then the slot side bolt to allow the spring to take up any remaining slack. Tighten the slot bolt, then the pivot side bolt to 17 ft. lbs. (24 Nm).

18. Install the timing belt covers, and tighten the cover bolts to 96 inch lbs. (11 Nm). Install all applicable components.

2.4L Engine

1. Rotate the crankshaft so that the No. 1 piston is at Top Dead Center (TDC) on its compression stroke.

2. Remove the timing belt covers.

3. To loosen the timing (outer) belt tensioner, install Special Tool MD998738 (or equivalent), to the slot, then screw inward to move the tensioner toward the water pump. Once the tension has been relieved, remove the outer timing belt.

➡**If the timing belts are going to be reused, mark the direction of their rotation on the belts. This will ensure the belt is reinstalled in same direction, extending belt life.**

4. Remove the outer crankshaft sprocket and flange.

5. Loosen the silent shaft (inner) belt tensioner and remove the belt.

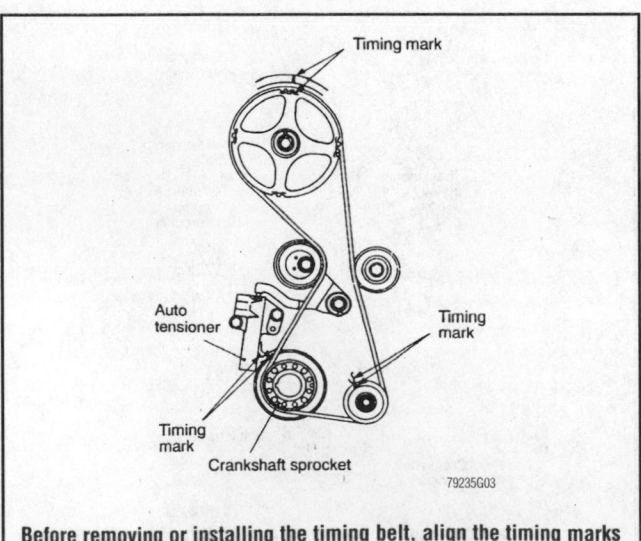

Timing mark

Auto
tensioner

Timing
mark

Timing
mark

Crankshaft sprocket

79235G03

Before removing or installing the timing belt, align the timing marks as indicated—Chrysler Import 2.4L engine

To install:

6. Turn both tensioner pulleys and check for any signs of bearing wear.

7. Align the timing marks of the silent shaft sprockets and the crankshaft sprocket with the timing marks on the front case. Route the timing belt around the sprockets so there is no slack in the upper span of the belt and the timing marks are still aligned.

8. Install the tensioner pulley and move the pulley by hand so the long side of the belt deflects approximately ¼ in. (6mm).

9. Hold the pulley tightly so the pulley cannot rotate when the bolt is tightened. Tighten the bolt to 14 ft. lbs. (19 Nm) and recheck the deflection amount.

10. Align the timing marks of the camshaft, crankshaft and oil pump sprockets with their corresponding marks on the front case or rear cover.

➡**There is a possibility to align all timing marks and have the oil pump sprocket and silent shaft out of time, causing an engine vibration during operation. If the following step is not followed exactly, there is a 50 percent chance that the silent shaft alignment will be 180 degrees (½ turn) off.**

11. Before installing the timing belt, ensure that the left side (rear) silent shaft (oil pump sprocket) is in the correct position as follows:

  a. Remove the plug from the rear side of the block and insert a tool with an outer shaft diameter of 0.31 in. (8mm) into the hole.

  b. With the timing marks still aligned, the shaft of the tool must be able to go in at least 2 ½ (63.5mm). If the tool can only go in approximately 1 in. (25mm), the silent shaft is not in the correct orientation and will cause a vibration during engine operation. Remove the tool from the hole and turn the oil pump sprocket 1 complete revolution. Realign the timing marks and insert the tool. The shaft of the tool should now go in at least 2 ½ (63.5mm)

  c. Recheck and realign the timing marks.

  d. Leave the tool in place to hold the silent shaft while continuing.

12. Install the belt on the crankshaft sprocket, the oil pump sprocket, then the camshaft sprocket—in that order. While doing so, be sure there is no slack between the sprockets, except where the tensioner is installed.

13. To adjust the timing (outer) belt perform the following steps:

  a. Turn the crankshaft ¼ turn counterclockwise, then turn it clockwise to move the No. 1 cylinder to TDC.

  b. Loosen the center bolt. Using tool MD998752 (or equivalent) and a torque wrench, apply 2.6 ft. lbs. (3.6 Nm) to the tensioner, then tighten the center bolt.

  c. Thread the special tool into the engine left support bracket until its end makes contact with the tensioner arm. At this point, thread the special tool in some more and remove the set wire attached to the auto-tensioner, if the wire was not previously removed. Remove the tool.

  d. Rotate the crankshaft two complete turns clockwise and let it sit for approximately 15 minutes. Then, measure the auto-tensioner protrusion (the distance between the tensioner arm and auto-tensioner body) to ensure that it is within 0.15–0.18 in. (3.8–4.5mm). If out of specification, repeat Substeps a through d until the specified value is obtained.

➡**Do not manually overtighten the belt or it will howl.**

14. Install the timing belt covers and all related items.

HONDA MODELS

1995–97 1.5L and 1.6L Engines

1. Rotate the crankshaft to set the engine at Top Dead Center (TDC) on the compression stroke for the No. 1 piston. The white mark on the crankshaft pulley should align with the pointers on the timing cover. Once the engine is in this position, it must not be disturbed.

2. Remove all necessary components for access to the cylinder head and timing belt covers. Cover the rocker arm and shaft assemblies with a towel or sheet of plastic to keep out dust and foreign objects.

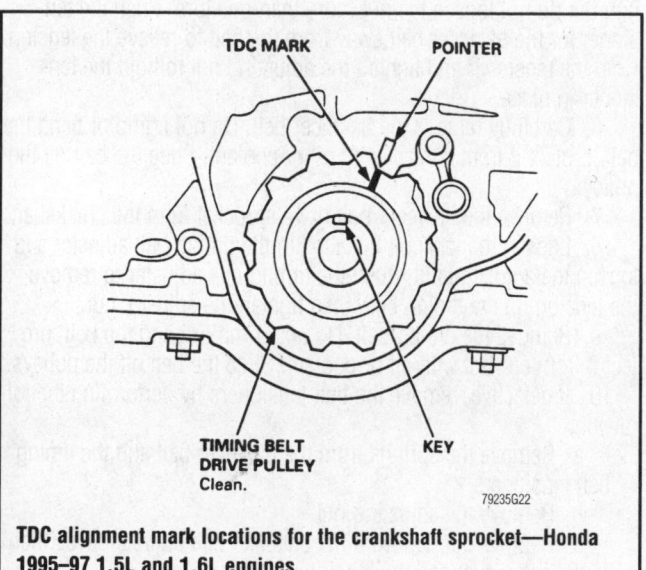

TDC alignment mark locations for the crankshaft sprocket—Honda 1995–97 1.5L and 1.6L engines

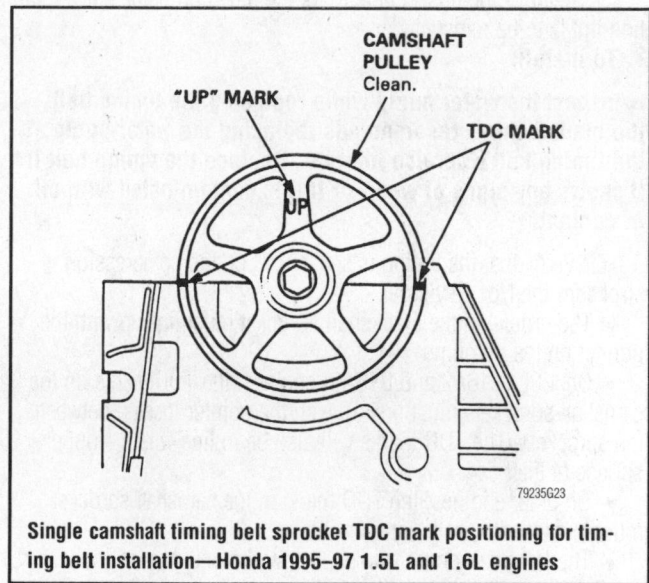

Single camshaft timing belt sprocket TDC mark positioning for timing belt installation—Honda 1995–97 1.5L and 1.6L engines

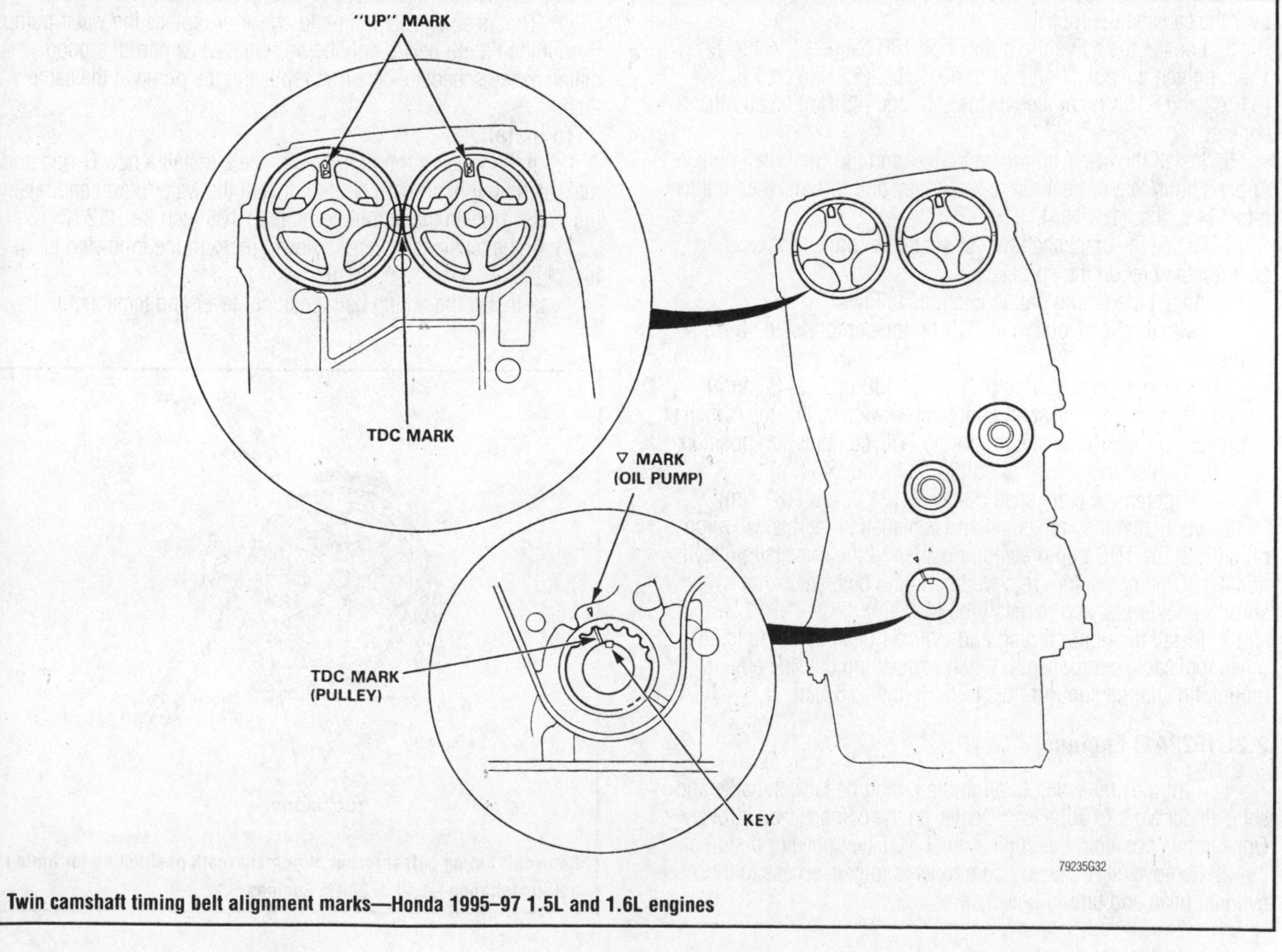

Twin camshaft timing belt alignment marks—Honda 1995–97 1.5L and 1.6L engines

3. Remove the timing belt covers.

4. Loosen the timing belt adjusting bolt 180 degrees (½ turn). Push the tensioner pulley down to release the belt tension. After releasing the tension, retighten the tensioner pulley bolt until snug.

➡**Do not remove the tensioner pulley unless it is to be replaced.**

5. Remove the timing belt. Mark the direction of the belt's rotation if it is to be reinstalled.

To install:

➡**Inspect the water pump when replacing the timing belt; the manufacturer recommends replacing the water pump at the timing belt's service interval. Replace the timing belt if it shows any signs of wear, or if it is contaminated with oil or coolant.**

6. Verify that the timing is set at TDC on the compression stroke for the No. 1 cylinder.

• The groove in the crankshaft sprocket must line up with the pointer on the oil pump.

• On 1.6L (B16A2 and B16A3) engines, the TDC marks on the camshaft sprockets must line up with the pointer located between the sprockets. The TDC marks will also be in line with the upper surface of the head.

• On other engines, the TDC mark on the camshaft sprocket must line up with the pointer on the back cover.

• The **UP** mark on the camshaft sprocket must point up.

7. Install the timing belt onto the crankshaft sprocket, then around the adjusting pulley and water pump sprocket, and finally over the camshaft sprocket.

8. Loosen the adjusting pulley bolt 180 degrees (½ turn). Then, tighten the adjusting bolt to 40 ft. lbs. (55 Nm) on 1.6L (B16A2 and B16A3) engines or to 33 ft. lbs. (45 Nm) on all other engines.

9. Install the lower timing belt cover and the crankshaft pulley. Apply a light coat of fresh oil to the pulley bolt threads, then tighten it to 134 ft. lbs. (181 Nm).

10. Rotate the crankshaft five or six turns counterclockwise to position the belt on the sprockets.

11. Adjust the timing belt tension, as follows:

a. Set the No. 1 piston at TDC on the compression stroke for the No. 1 cylinder.

b. Loosen the adjusting pulley bolt 180 degrees (½ turn).

c. Rotate the crankshaft counterclockwise so that the camshaft sprocket moves three teeth from the TDC/compression position.

d. Tighten the adjusting bolt to 33 ft. lbs. (45 Nm).

e. Tighten the crankshaft pulley to 134 ft. lbs. (181 Nm).

12. Verify that the crankshaft and camshaft sprockets will align properly at the TDC/compression position. If the camshaft pulley is not at TDC/compression, remove the timing belt, adjust the sprocket positions, and reinstall the belt.

13. Install the upper timing and cylinder head covers, and all other applicable components. When reattaching the side engine mount, tighten the support nuts to 54 ft. lbs. (75 Nm).

2.2L (F22A1) Engines

1. Turn the crankshaft to align the timing belt matchmarks and set cylinder No.1 to Top Dead Center on the compression stroke. Once in this position, the engine must NOT be turned or disturbed.

2. Remove all necessary components to gain access to the cylinder head and timing belt covers.

3. Remove the cylinder head and timing belt covers.

4. There are two belts in this system; the one running to the camshaft pulley is the timing belt. The other, shorter one drives the balance shafts and is referred to as the balancer belt or timing balancer belt. Lock the timing belt adjuster in position by installing one of the lower timing belt cover bolts to the adjuster arm.

5. Loosen the timing belt and balancer shafts tensioner adjuster nut, but do not loosen the nut more than one turn. Push the tensioner for the balancer belt away from the belt to relieve the tension. Hold the tensioner and tighten the adjusting nut to hold the tensioner in place.

6. Carefully remove the balancer belt. Do not crimp or bend the belt; protect it from contact with oil or coolant. Slide the belt off the pulleys.

7. Remove the balancer belt drive sprocket from the crankshaft.

8. Loosen the lockbolt installed to the timing belt adjuster and loosen the adjusting nut. Push the timing belt adjuster to remove the tension on the timing belt, then tighten the adjuster nut.

9. Remove the timing belt. Do not crimp or bend the belt; protect it from contact with oil or coolant. Slide the belt off the pulleys.

10. If defective, remove the belt tensioners by performing the following:

a. Remove the springs from the balancer belt and the timing belt tensioners.

b. Remove the adjusting nut.

c. Remove the bolt from the balancer belt adjuster lever, then remove the lever and the tensioner pulley.

d. Remove the lockbolt from the timing belt tensioner lever, then remove the tensioner pulley and lever from the engine.

11. This is an excellent time to check or replace the water pump. Even if the timing belt is only being replaced as part of a good maintenance schedule, consider replacing the pump at the same time.

To install:

12. If the water pump is to be replaced, install a new O-ring and make certain it is properly seated. Install the water pump and retaining bolts. Tighten the mounting bolts to 106 inch lbs. (12 Nm).

13. If the tensioners were removed perform the following to install them:

a. Install the timing belt tensioner lever and tensioner pulley.

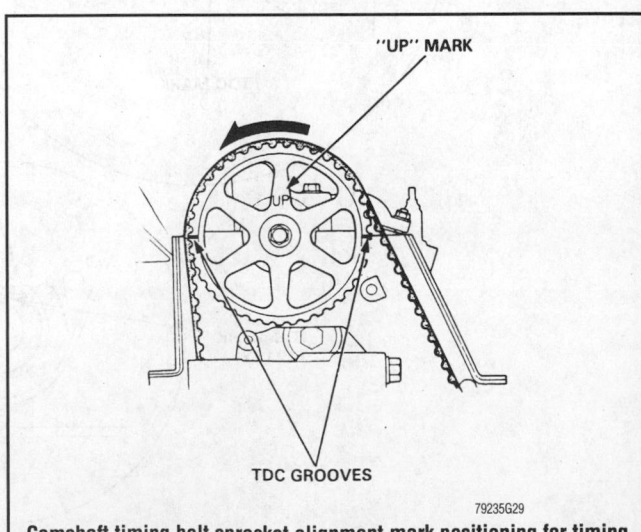

79235G29

Camshaft timing belt sprocket alignment mark positioning for timing belt installation—2.2L (F22A1) Engines

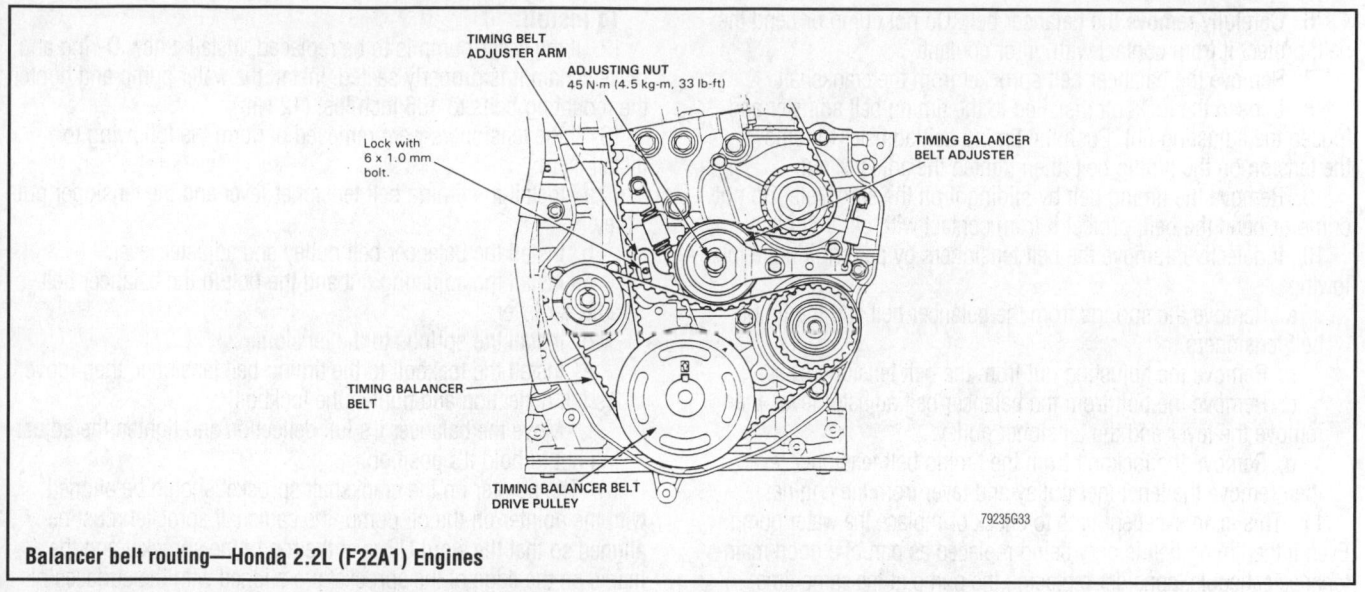

TIMING BELT ADJUSTER ARM
ADJUSTING NUT 45 N·m (4.5 kg-m, 33 lb-ft)
Lock with 6 x 1.0 mm bolt.
TIMING BALANCER BELT ADJUSTER
TIMING BALANCER BELT
TIMING BALANCER BELT DRIVE PULLEY

79235G33

Balancer belt routing—Honda 2.2L (F22A1) Engines

➠**The tensioner lever must be properly positioned on it's pivot pin located on the oil pump. Be sure that the timing belt lever and tensioner moves freely and does not bind.**

   b. Install the lockbolt to the timing belt tensioner, do not tighten the lockbolt at this time.

   c. Install the balancer belt pulley and adjuster lever.

   d. Install the adjusting nut and the bolt to the balancer belt adjuster lever. Do not tighten the adjuster nut or bolt at this time.

➠**Be sure that the balancer lever and tensioner moves freely and does not bind.**

   e. Install the springs to the tensioners.

   f. Move the timing belt tensioner its full deflection and tighten the lockbolt.

   g. Move the balancer its full deflection and tighten the adjusting nut.

14. The crankshaft timing pointer must be perfectly aligned with the white mark on the flywheel or flex-plate; the camshaft pulley must be aligned so that the word UP is at the top of the pulley and the marks on the edge of the pulley are aligned with the surfaces of the head.

15. Install the timing belt over the pulleys and tensioners.

16. Loosen the bolt used to lock the timing belt tensioner. Loosen, then tighten the timing belt adjusting nut.

17. Turn the crankshaft counterclockwise until the cam pulley has moved 3 teeth; this creates tension on the timing belt. Loosen, then tighten the adjusting nut and tighten it to 33 ft. lbs. (45 Nm). Tighten the bolt used to lock the timing belt tensioner.

18. Realign the timing belt marks, then install the balancer belt drive sprocket on the crankshaft.

19. Align the front balancer pulley; the face of the front timing balancer pulley has a mark which must be aligned with the notch on the oil pump body. This pulley is the one at 10 o'clock to the crank pulley when viewed from the pulley end.

20. Align the rear timing balancer pulley (2 o'clock from the crank pulley) using a 6 x 100mm bolt or rod. Mark the bolt or rod at a point 2.9 in. (74mm) from the end. Remove the bolt from the maintenance hole on the side of the block; insert the bolt or rod into the hole. Align the 2.9 in. (74mm) mark with the face of the hole. This pin will hold the shaft in place during installation.

21. Install the balancer belt. Once the belts are in place, be sure that all the engine alignment marks are still correct. If not, remove the belts, realign the engine and reinstall the belts. Once the belts are properly installed, slowly loosen the adjusting nut, allowing the tensioner to move against the belt. Remove the pin from the maintenance hole and reinstall the bolt and washer.

22. Turn the crankshaft one full turn, then tighten the adjuster nut to 33 ft. lbs. (45 Nm). Remove the bolt used to lock the timing belt tensioner.

23. Install the lower cover, ensuring the rubber seals are in place. Install a new seal around the adjusting nut, DO NOT loosen the adjusting nut.

24. Install the key on the crankshaft and install the crankshaft pulley. Apply oil to the bolt threads and tighten it to 181 ft. lbs. (250 Nm).

25. Install the upper timing belt cover and all applicable components. When installing the side engine mount, tighten the bolt and nut attaching the mount to the engine to 40 ft. lbs. (55 Nm) and the through-bolt and nut to 47 ft. lbs. (65 Nm).

2.2L (F22B1 and F22B2) Engines

1. Remove the cylinder head (valve) and upper timing belt covers.

2. Turn the engine to align the timing marks and set cylinder No.1 to TDC. The white mark on the crankshaft sprocket should align with the pointer on the timing belt cover. The words **UP** embossed on the camshaft sprocket should be aligned in the upward position. The marks on the edge of the sprocket should be aligned with the cylinder head or the back cover upper edge. Once in this position, the engine must NOT be turned or disturbed.

3. Remove all necessary components for access to the lower timing belt cover, then remove the cover.

4. There are two belts in this system; the one running to the camshaft sprocket is the timing belt. The other, shorter one drives the balance shaft and is referred to as the balancer shaft belt or timing balancer belt. Lock the timing belt adjuster in position, by installing one of the lower timing belt cover bolts to the adjuster arm.

5. Loosen the timing belt and balancer shafts tensioner adjuster nut, do not loosen the nut more than one turn. Push the tensioner for the balancer belt away from the belt to relieve the tension. Hold the tensioner and tighten the adjusting nut to hold the tensioner in place.

6. Carefully remove the balancer belt. Do not crimp or bend the belt; protect it from contact with oil or coolant.

7. Remove the balancer belt sprocket from the crankshaft.

8. Loosen the lockbolt installed to the timing belt adjuster and loosen the adjusting nut. Push the timing belt adjuster to remove the tension on the timing belt, then tighten the adjuster nut.

9. Remove the timing belt by sliding it off the sprockets. Do not crimp or bend the belt; protect it from contact with oil or coolant.

10. If defective, remove the belt tensioners by performing the following:

   a. Remove the springs from the balancer belt and the timing belt tensioners.

   b. Remove the adjusting nut from the belt tensioners.

   c. Remove the bolt from the balancer belt adjuster lever, then remove the lever and the tensioner pulley.

   d. Remove the lockbolt from the timing belt tensioner lever, then remove the tensioner pulley and lever from the engine.

11. This is an excellent time to check or replace the water pump. Even if the timing belt is only being replaced as part of a good maintenance schedule, consider replacing the pump at the same time.

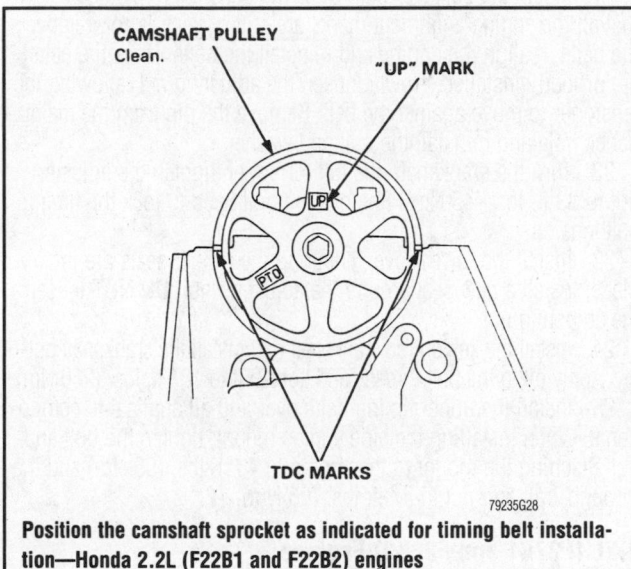

CAMSHAFT PULLEY Clean.

"UP" MARK

PTO

TDC MARKS

79235G28

Position the camshaft sprocket as indicated for timing belt installation—Honda 2.2L (F22B1 and F22B2) engines

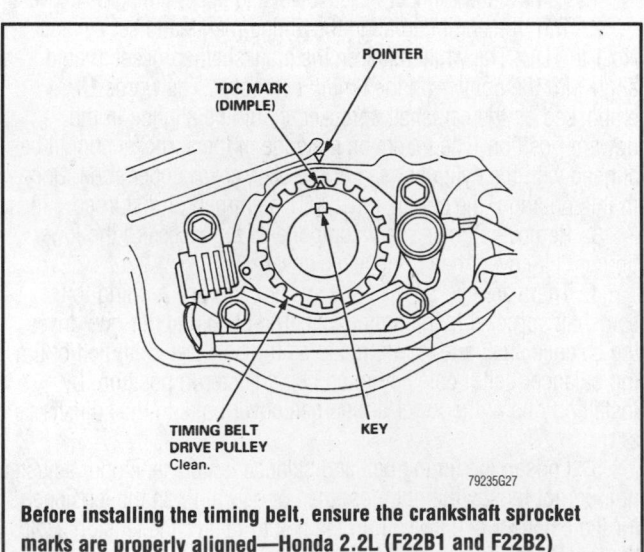

POINTER

TDC MARK (DIMPLE)

TIMING BELT DRIVE PULLEY Clean.

KEY

79235G27

Before installing the timing belt, ensure the crankshaft sprocket marks are properly aligned—Honda 2.2L (F22B1 and F22B2) engines

To install:

12. If the water pump is to be replaced, install a new O-ring and make certain it is properly seated. Install the water pump and tighten the mounting bolts to 106 inch lbs. (12 Nm).

13. If the tensioners were removed perform the following to install them:

   a. Install the timing belt tensioner lever and the tensioner pulley.

   b. Install the balancer belt pulley and adjuster lever.

   c. Install the adjusting nut and the bolt to the balancer belt adjuster lever.

   d. Install the springs to the tensioners.

   e. Install the lockbolt to the timing belt tensioner, then move it it's full deflection and tighten the lockbolt.

   f. Move the balancer it's full deflection and tighten the adjusting nut to hold it's position.

14. The pointer on the crankshaft sprocket should be aligned with the pointer on the oil pump; the camshaft sprocket must be aligned so that the word UP is at the top of the sprocket and the marks on the edge of the sprocket are aligned with the surfaces of the head or the back cover upper edge.

15. Install the timing belt on the sprockets in the following sequence: crankshaft sprocket, tensioner sprocket, water pump sprocket, camshaft sprocket.

16. Check the timing marks to be sure that they did not move.

17. Loosen, then retighten the timing belt adjusting nut; this will apply the proper amount of tension to the timing belt.

18. Install the timing balancer belt drive sprocket and the lower timing belt cover.

19. Install the crankshaft pulley and bolt, tighten the bolt to 181 ft. lbs. (245 Nm). Rotate the crankshaft sprocket five or six turns to position the timing belt on the sprockets.

20. Set the No. 1 cylinder to TDC and loosen the timing belt adjusting nut one turn. Turn the crankshaft counterclockwise until the cam sprocket has moved 3 teeth; this creates tension on the timing belt.

21. Tighten the timing belt adjusting nut.

22. Set the crankshaft sprocket and the camshaft sprocket to TDC. If the sprockets do not align, remove the belt to realign the marks, then install the belt.

23. Remove the crankshaft pulley and the lower cover.

24. With the timing marks aligned, lock the timing belt adjuster in place with one of the lower cover mounting bolts.

25. Loosen the adjusting nut and ensure the timing balancer belt adjuster moves freely.

26. Align the rear timing balancer sprocket using a 6 x 100mm bolt or rod. Mark the bolt or rod at a point 2.9 in. (74mm) from the end. Remove the bolt from the maintenance hole on the side of the block; insert the bolt/rod into the hole and align the 2.9 in. (74mm) mark with the face of the hole. This will hold the shaft in place during installation.

27. Align the groove on the front balancer shaft sprocket with the pointer on the oil pump.

28. Install the balancer belt. Once the belts are in place, be sure that all the engine alignment marks are still correct. If not, remove the belts, realign the engine and reinstall the belts. Once the belts are properly installed, slowly loosen the adjusting nut, allowing the tensioner to move against the belt. Remove the bolt from the maintenance hole and reinstall the bolt and washer.

29. Install the crankshaft pulley, then turn the crankshaft sprocket 1 turn counterclockwise and tighten the timing belt adjusting nut to 33 ft. lbs. (45 Nm).

30. Remove the crankshaft pulley and the bolt locking the timing belt adjuster in place.

31. Install the lower and upper timing belt covers, and all applicable components. When installing the crankshaft pulley, coat the threads and seating face of the pulley bolt with engine oil, then install and tighten the bolt to 181 ft. lbs. (250 Nm).

32. Install the cylinder head cover gasket cover to the groove of the cylinder head cover. Before installing the gasket thoroughly clean the seal and the groove. Seat the recesses for the camshaft first, then work it into the groove around the outside edges. Be sure the gasket is seated securely in the corners of the recesses.

33. Apply liquid gasket to the four corners of the recesses of the cylinder head cover gasket. Do not install the parts if 5 minutes or more have elapsed since applying liquid gasket. After assembly, wait at least 20 minutes before filling the engine with oil.

34. Install the cylinder head (valve) cover and all other applicable components.

2.2L (H22A1) Engine

1. Turn the crankshaft so the No. 1 piston is at top dead center. The No. 1 piston is at top dead center when the pointer on the block aligns with the white painted mark on the driveplate.

2. Remove all necessary components for access to the cylinder head and upper timing belt covers. Then, remove the covers.

3. Ensure the words UP embossed on the camshaft pulleys are aligned in the upward position.

4. Support the engine with a floor jack below the center of the center beam. Tension the jack so that it is just supporting the beam but not lifting it. Remove the 2 rear bolts from the center beam to allow the engine to drop down for clearance to remove the lower cover.

5. Remove and discard the rubber seal from the timing belt adjuster. Do not loosen the adjusting nut.

6. Remove the lock pin from the maintenance bolt.

7. Remove the lower timing belt cover.

8. There are two belts in this system; the one running to the camshaft pulley is the timing belt. The other, shorter one drives the balance shafts and is referred to as the balancer belt or timing balancer belt.

9. Loosen the balancer shafts tensioner adjusting nut, do not loosen the nut more than one turn. Push the tensioner for the balancer belt away from the belt to relieve the tension. Hold the tensioner and tighten the adjusting nut to hold the tensioner in place.

10. Carefully remove the balancer belt by sliding it off of the pulleys. Do not crimp or bend the belt; protect it from contact with oil or coolant.

11. Remove the balancer belt drive sprocket from the crankshaft.

12. Remove the bolts attaching the Crankshaft Position/Top Dead Center (CKP/TDC) sensor and remove the sensor.

13. Remove the timing belt by sliding it off of the pulleys. Do not crimp or bend the belt; protect it from contact with oil or coolant.

14. If defective, remove the two bolts mounting the timing belt auto-tensioner and remove the tensioner from the vehicle.

15. If defective, remove the balancer belt tensioner by performing the following:

   a. Remove the spring from the balancer belt tensioner.

   b. Remove the adjusting nut.

   c. Remove the bolt from the balancer belt adjuster lever, then remove the lever and the tensioner pulley.

16. This is an excellent time to check or replace the water pump. Even if the timing belt is only being replaced as part of a good maintenance schedule, consider replacing the pump at the same time.

To install:

17. If the water pump is to be replaced, install a new O-ring and make certain it is properly seated. Install the water pump and retaining bolts. Tighten the mounting bolts to 106 inch lbs. (12 Nm).

18. If the balancer tensioner was removed, perform the following to install it:

   a. Install the balancer belt pulley and adjuster lever.

   b. Install the adjusting nut and the bolt to the balancer belt adjuster lever. Do not tighten the adjuster nut or bolt at this time.

➡ **Be sure that the balancer lever and tensioner moves freely and does not bind.**

   c. Install the spring to the tensioner.

   d. Move the balancer its full deflection and tighten the adjusting nut.

19. Hold the auto-tensioner with the maintenance bolt pointing up. Remove the maintenance bolt and discard the gasket.

➡ **Handle the tensioner carefully so the oil inside does not spill or leak. Replenish the auto-tensioner with oil if any spills or leaks out. The auto-tensioner total capacity is 1/4 oz. (8 ml).**

20. Clamp the mounting boss of the auto-tensioner in a vise. Use pieces of wood or a cloth to protect the mounting boss.

✳✳ WARNING

Do not clamp the housing of the auto-tensioner, component damage may occur.

21. Insert a flat-bladed prytool into the maintenance hole. Place the stopper (part No. 14540-P13–003) on the auto-tensioner while turning the prytool clockwise to compress the tensioner. Take care not to damage the threads or the gasket contact surface with the prytool.

22. Remove the prytool and install the maintenance bolt with a new gasket. Tighten the maintenance bolt to 71 inch lbs. (8 Nm).

23. Be sure no oil is leaking from the maintenance bolt and install the auto-tensioner to the engine. Tighten the auto-tensioner mounting bolts to 16 ft. lbs. (22 Nm).

24. The pointer on the crankshaft pulley should be aligned with the pointer on the oil pump; the camshaft pulley must be aligned so that the word **UP** is at the top of the pulley and the marks on the edge of the pulley are aligned with the surfaces of the head.

25. Install the timing belt.

26. Remove the stopper from the timing belt adjuster.

27. Install the CKP/TDC sensors and tighten the bolts to 106 inch lbs. (12 Nm). Connect the CKP/TDC sensors connector.

28. Install the balancer belt drive sprocket to the crankshaft.

29. Align the groove on the front balancer shaft pulley with the pointer on the oil pump.

30. Align the rear timing balancer pulley using a 6 x 100mm bolt or rod. Mark the bolt or rod at a point 2.9 in. (74mm) from the end. Remove the bolt from the maintenance hole on the side of the block; insert the bolt/rod into the hole and align the 2.9 in. (74mm) mark with the face of the hole. This pin will hold the shaft in place during installation.

31. Ensure the timing balancer belt adjuster moves freely.

32. Install the balancer belt. Once the belts are in place, be sure that all the engine alignment marks are still correct. If not, remove the belts, realign the engine and reinstall the belts. Once the belts are properly installed, slowly loosen the adjusting nut, allowing the tensioner to move against the belt. Remove the pin from the maintenance hole and reinstall the bolt and washer.

33. Turn the crankshaft pulley one full turn, then tighten the adjusting nut to 33 ft. lbs. (45 Nm).

✳✳ WARNING

Do not apply extra pressure to the pulleys or tensioners while performing the adjustment.

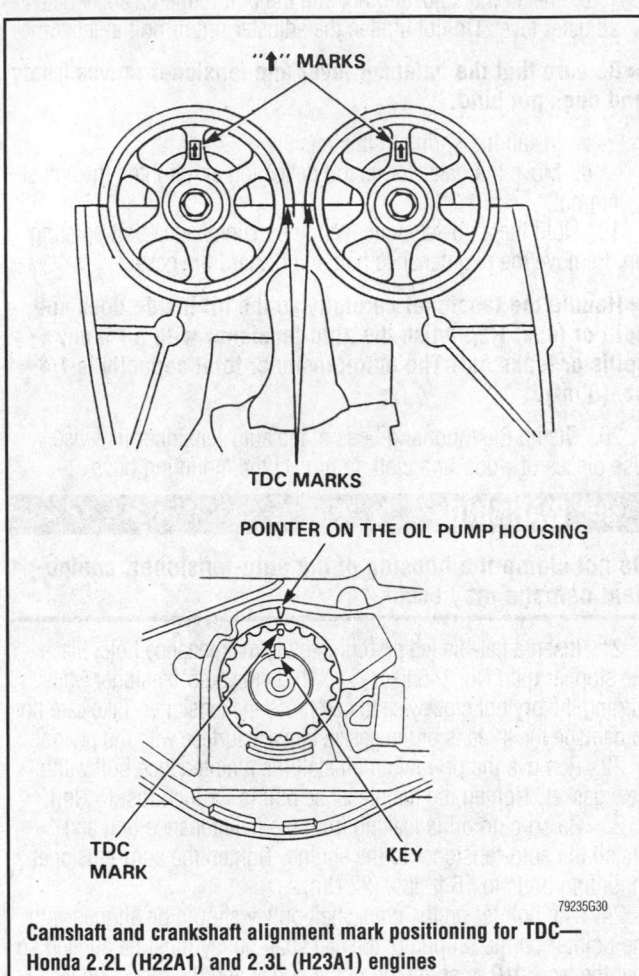

Camshaft and crankshaft alignment mark positioning for TDC—Honda 2.2L (H22A1) and 2.3L (H23A1) engines

34. Install the timing belt and cylinder head covers, and any other applicable components. When installing the side engine mount, tighten the bolt and nut attaching the mount to the engine to 40 ft. lbs. (55 Nm), and the through-bolt and nut to 47 ft. lbs. (65 Nm).

2.3L (H23A1) Engine

1. Turn the crankshaft so the No. 1 piston is at top dead center.

➡ **The No. 1 piston is at top dead center when the pointer on the block aligns with the white painted mark on the flywheel (manual transaxle) or driveplate (automatic transaxle).**

2. Remove all necessary components for access to the cylinder head cover, then remove the cylinder head cover.
3. Ensure the words **UP** embossed on the camshaft pulleys are aligned in the upward position.
4. Insert a 5.0mm pin punch in each of the camshaft caps, nearest to the pulleys, through the holes provided.
5. Remove the upper and middle timing belt covers.
6. Support the engine with a floor jack below the center of the center beam. Tension the jack so that it is just supporting the beam but not lifting it. Remove the 2 rear bolts from the center beam to allow the engine to drop down for clearance to remove the lower cover.
7. Remove and discard the rubber seal from the timing belt adjuster. Do not loosen the adjusting nut.
8. Remove the lower timing belt cover.
9. There are two belts in this system; the one running to the camshaft pulley is the timing belt. The other, shorter one drives the balance shaft and is referred to as the balancer belt or timing balancer belt. Lock the timing belt adjuster in position, by installing one of the lower timing belt cover bolts to the adjuster arm.
10. Loosen the timing belt and balancer shaft tensioner adjuster nut(s), do not loosen the nut(s) more than one turn. Push the tensioner for the balancer belt away from the belt to relieve the tension. Hold the tensioner and tighten the adjusting nut to hold the tensioner in place.
11. Carefully remove the balancer belt by sliding it off of the pulleys. Do not crimp or bend the belt; protect it from contact with oil or coolant.

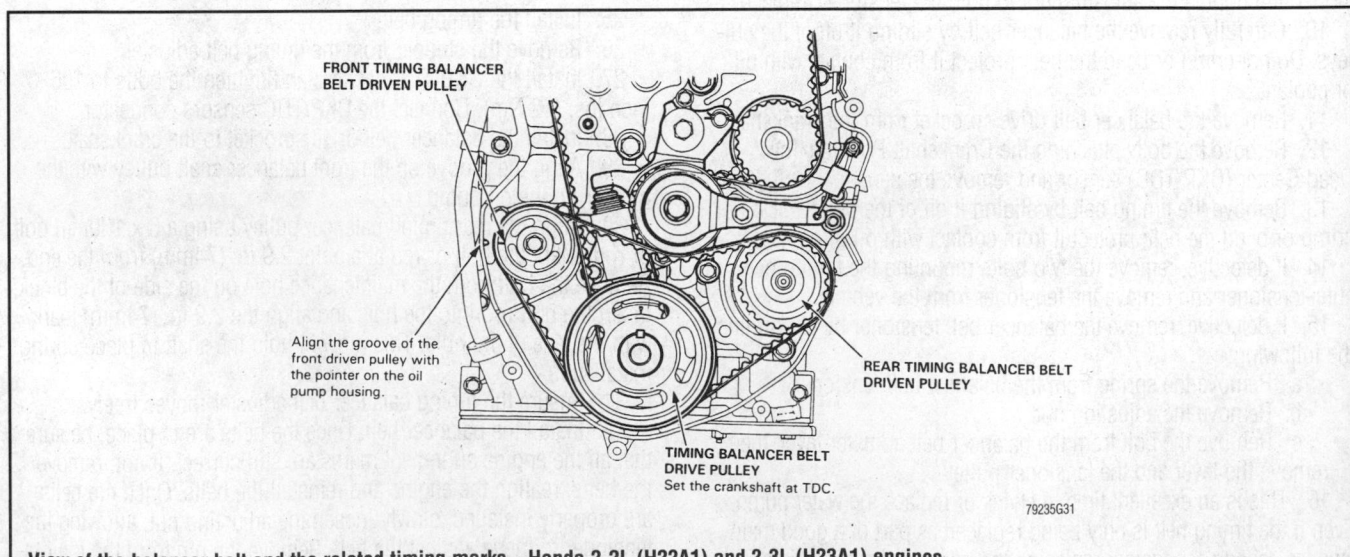

View of the balancer belt and the related timing marks—Honda 2.2L (H22A1) and 2.3L (H23A1) engines

12. Remove the balancer belt drive sprocket from the crankshaft.

13. Loosen the lockbolt installed in the timing belt adjuster and loosen the adjusting nut. Push the timing belt adjuster to remove the tension on the timing belt, then tighten the adjuster nut.

14. Remove the timing belt by sliding it off of the pulleys. Do not crimp or bend the belt; protect it from contact with oil or coolant.

15. If defective, remove the belt tensioners by performing the following:

a. Remove the springs from the balancer belt and the timing belt tensioners.

b. Remove the adjusting nut.

c. Remove the bolt from the balancer belt adjuster lever, then remove the lever and the tensioner pulley.

d. Remove the lockbolt from the timing belt tensioner lever, then remove the tensioner pulley and lever from the engine.

16. This is an excellent time to check or replace the water pump. Even if the timing belt is only being replaced as part of a good maintenance schedule, consider replacing the pump at the same time.

To install:

17. If the water pump is to be replaced, install a new O-ring and make certain it is properly seated. Install the water pump and retaining bolts. Tighten the mounting bolts to 106 inch lbs. (12 Nm).

18. If the tensioners were removed perform the following to install them:

a. Install the timing belt tensioner lever and tensioner pulley.

➡**The tensioner lever must be properly positioned on it's pivot pin located on the oil pump. Be sure that the timing belt lever and tensioner moves freely and does not bind.**

b. Install the lockbolt to the timing belt tensioner, do not tighten the lockbolt at this time.

c. Install the balancer belt pulley and adjuster lever.

d. Install the adjusting nut and the bolt to the balancer belt adjuster lever. Do not tighten the adjuster nut or bolt at this time.

➡**Be sure that the balancer lever and tensioner moves freely and does not bind.**

e. Install the springs to the tensioners.

f. Move the timing belt tensioner it's full deflection and tighten the lockbolt.

g. Move the balancer it's full deflection and tighten the adjusting nut.

19. The crankshaft timing pointer must be perfectly aligned with the white mark on the flywheel or flex-plate; the camshaft pulley must be aligned so that the word UP is at the top of the pulley and the marks on the edge of the pulley are aligned with the surfaces of the head.

20. Install the timing belt.

21. Install the balancer belt drive sprocket to the crankshaft.

22. Remove the two 5.0mm pin punches from the camshaft bearing caps.

23. Loosen the bolt used to lock the timing belt tensioner. Loosen, then tighten the timing belt adjuster nut.

24. Turn the crankshaft counterclockwise until the cam pulley has moved 3 teeth; this creates tension on the timing belt. Loosen, then tighten the adjusting nut and tighten it to 33 ft. lbs. (45 Nm). Tighten the bolt used to lock the timing belt tensioner.

25. Realign the timing belt timing marks.

26. Align the groove on the front balancer shaft pulley with the pointer on the oil pump.

27. Align the rear timing balancer pulley using a 6 x 100mm bolt or rod. Mark the bolt or rod at a point 2.913 in. (74mm) from the end. Remove the bolt from the maintenance hole on the side of the block; insert the bolt/rod into the hole and align the 74mm mark with the face of the hole. This pin will hold the shaft in place during installation.

28. Loosen the adjusting nut and ensure the timing balancer belt adjuster moves freely.

29. Install the balancer belt. Once the belts are in place, be sure that all the engine alignment marks are still correct. If not, remove the belts, realign the engine and reinstall the belts. Once the belts are properly installed, slowly loosen the adjusting nut, allowing the tensioner to move against the belt. Remove the pin from the maintenance hole and reinstall the bolt and washer.

30. Turn the crankshaft pulley one full turn and tighten the adjusting nut to 33 ft. lbs. (45 Nm).

➡**Both belt adjusters are spring loaded to properly tension the belts. Do not apply extra pressure to the pulleys or tensioners while performing the adjustment.**

31. Remove the 6 x 100mm bolt from the timing belt adjuster arm.

32. Install the timing belt and cylinder head covers. Reinstall all applicable components. When installing the crankshaft pulley, coat the threads and seating face of the pulley bolt with engine oil, then install and tighten the bolt to 181 ft. lbs. (250 Nm). When installing the side engine mount, tighten the bolt and nut attaching the mount to the engine to 40 ft. lbs. (55 Nm), and the through-bolt and nut to 47 ft. lbs. (65 Nm). Remove the jack from under the center beam.

2.7L Engines

1. Turn the engine to align the timing marks and set cylinder No.1 to TDC. The white mark on the crankshaft pulley should align with the pointer on the timing belt cover. Remove the inspection caps on the upper timing belt covers to check the alignment of the timing marks. The pointers for the camshafts should align with the green marks on the camshaft sprockets.

2. Remove all necessary components for access to the timing belt covers, then remove the covers.

➡**Do not use the covers to store removed items.**

3. Loosen the timing belt adjuster bolt 180 degrees (½ turn). Push the tensioner to remove the tension from the timing belt, then retighten the adjusting bolt.

4. Remove the timing belt. Do not crimp or bend the belt; protect it from contact with oil or coolant. Slide the belt off the sprockets.

5. Remove the bolts attaching the camshaft sprockets to the camshafts, then remove the sprockets.

6. If the timing belt tensioner is defective, remove the spring from the timing belt tensioner. Remove the tensioner pulley adjusting bolt and the adjuster assembly from the engine.

➡**This is an excellent time to check or replace the water pump. Even if the timing belt is only being replaced as part of a good maintenance schedule, consider replacing the pump at the same time.**

To install:

7. If the water pump is to be replaced, install a new O-ring and make certain it is properly seated. Install the water pump and retaining bolts. Tighten the mounting bolts to 16 ft. lbs. (22 Nm).

8. If removed, install the tensioner pulley and the adjusting bolt, be sure the tensioner is properly positioned on its pivot pin. Install the spring to the tensioner, then push the tensioner to it's full deflection and tighten the adjusting bolt.

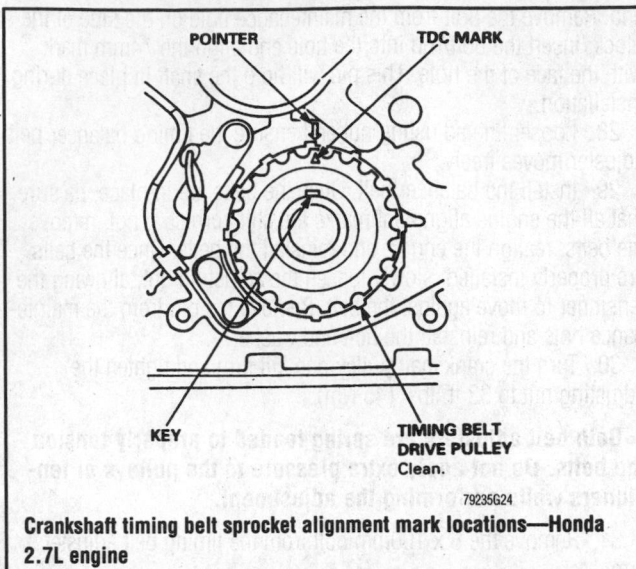

Crankshaft timing belt sprocket alignment mark locations—Honda 2.7L engine

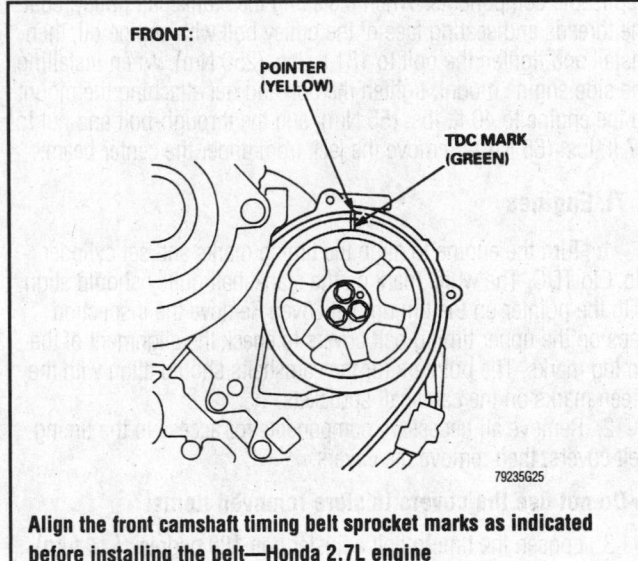

Align the front camshaft timing belt sprocket marks as indicated before installing the belt—Honda 2.7L engine

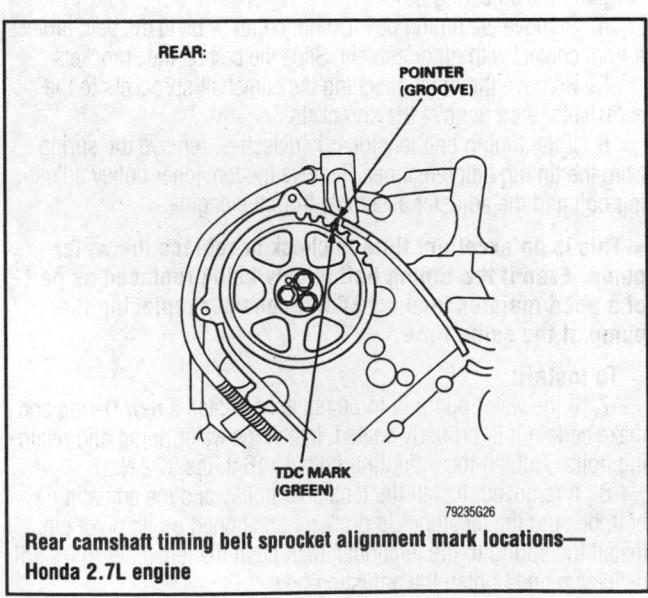

Rear camshaft timing belt sprocket alignment mark locations— Honda 2.7L engine

9. Set the timing belt drive sprocket so that the No. 1 piston is at top dead center (TDC). Align the TDC mark on the tooth of the timing belt drive sprocket with the pointer on the oil pump.

10. Set the camshaft sprockets so that the No. 1 piston is at TDC. Align the TDC marks (green mark) on the camshaft sprockets to the pointers on the back covers.

11. Install the timing belt onto the sprockets in the following sequence: crankshaft sprocket, tensioner pulley, front camshaft sprocket, water pump pulley, rear camshaft sprocket.

12. Loosen, then retighten the timing belt adjuster bolt to tension the timing belt.

13. Install the lower timing belt cover.

14. Install the crankshaft sprocket and the crankshaft pulley bolt. Tighten the bolt to 181 ft. lbs. (245 Nm) with the aid of the crank pulley holder.

15. Rotate the crankshaft five or six turns clockwise so that the timing belt positions on the sprockets.

16. Set cylinder No. 1 to TDC by aligning the timing marks. If the timing marks do not align, remove the timing belt, then adjust the components and reinstall the timing belt.

17. Loosen the timing belt adjusting bolt 180 degrees (½ turn) and retighten the adjusting bolt. Tighten the adjusting bolt to 31 ft. lbs. (42 Nm).

18. Install the upper timing belt cover and all other applicable components. When installing the side engine mount to the engine, use three new attaching bolts. Tighten the new bolts to 40 ft. lbs. (54 Nm).

HYUNDAI MODELS

> **✳✳ CAUTION**
>
> **Timing belt maintenance is extremely important. All Hyundai models use interference-type non-freewheeling engines. Should the timing belt break in these engines, the valves in the cylinder head will come in contact with the pistons, causing major engine damage. The recommended replacement interval for timing belts is 60,000 miles.**

1.5L Engines

1. Remove all necessary components for access to the timing belt cover, then remove the cover.

2. Rotate the crankshaft clockwise and align the timing marks so No. 1 piston will be at TDC of the compression stroke.

3. Loosen the tensioning bolt and the pivot bolt on the timing belt tensioner. Move the tensioner as far as it will go toward the water pump. Tighten the adjusting bolt.

4. Mark the timing belt with an arrow showing direction of rotation.

5. Remove the timing belt.

6. If defective, remove the timing belt tensioner.

To install:

7. Align the timing marks of the camshaft sprocket and check that the crankshaft timing marks are still in alignment.

8. If removed, install the timing belt tensioner, spring and spacer with the bottom end of the spring free. Tighten the adjusting bolt slightly with the tensioner moved as far as possible away from the water pump.

9. Install the free end of the spring into the locating tang on the front case.

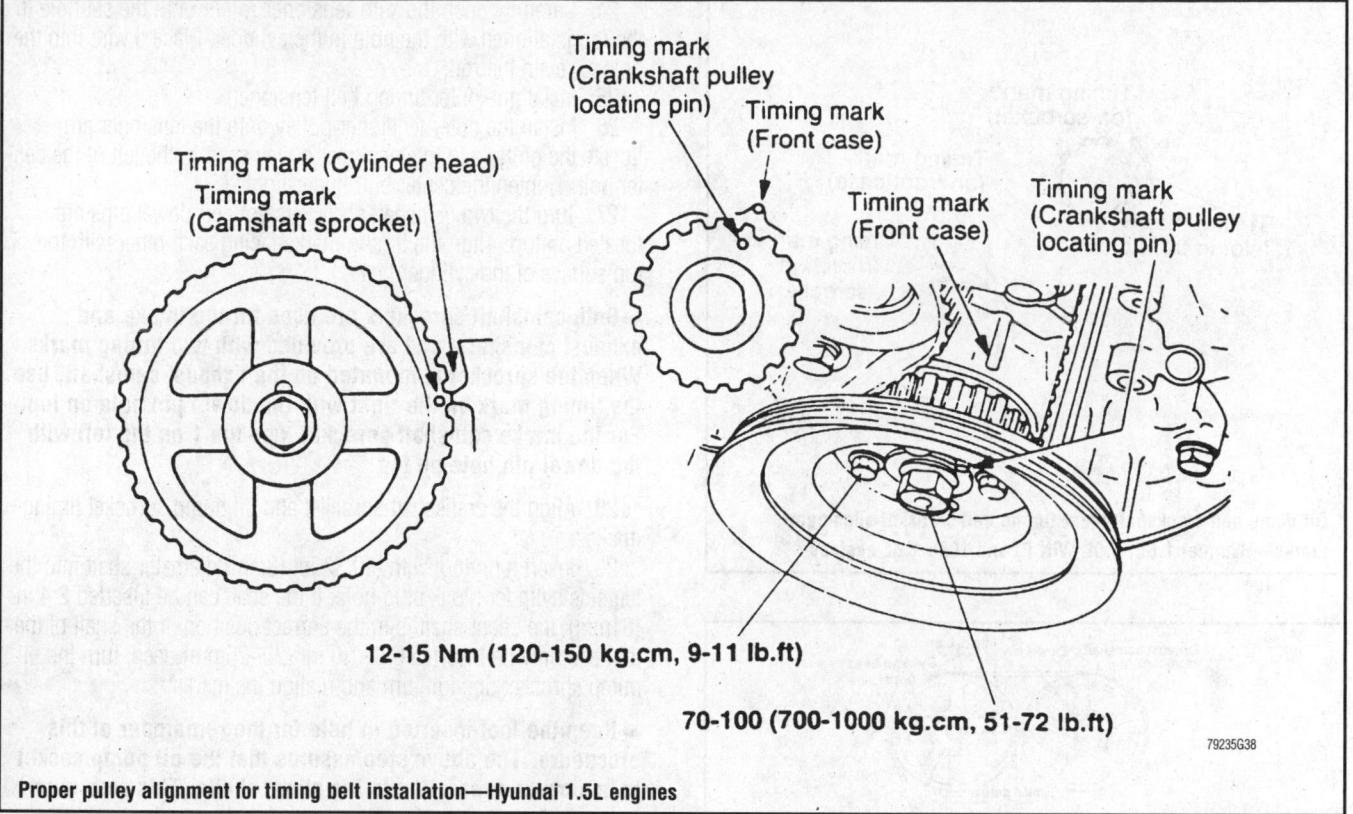

12-15 Nm (120-150 kg.cm, 9-11 lb.ft)

70-100 (700-1000 kg.cm, 51-72 lb.ft)

79235G38

Proper pulley alignment for timing belt installation—Hyundai 1.5L engines

10. Position the timing belt over the crankshaft sprocket, then over the camshaft sprocket. Slip the back of the belt over the tensioner wheel.

11. Turn the camshaft sprocket in the opposite of its normal direction of rotation until the straight side of the belt is tight and be sure the timing marks align.

➡If the timing marks are not properly aligned, shift the belt 1 tooth at a time in the appropriate direction until they are lined up.

12. Loosen the tensioner mounting bolts so the tensioner works, without the interference of any friction, under spring pressure. Be sure the belt follows the curve of the camshaft pulley so the teeth are engaged all the way around. Correct the path of the belt, if necessary.

13. Tighten the tensioner adjusting bolt, then the tensioner pivot bolt to 15–18 ft. lbs. (20–26 Nm).

➡Bolts must be tightened in the stated order or tension won't be correct.

14. Turn the crankshaft 1 turn clockwise until timing marks again align to seat the belt.

15. Loosen both tensioner attaching bolts and let the tensioner position itself under spring tension. Retighten the bolts.

16. Check belt tension by putting a finger on the water pump side of the tensioner wheel and pull the belt toward the water pump. The belt should move toward the pump until the teeth are approximately ½ of the way across the head of the tensioner adjusting bolt. Re-tension the belt, if necessary.

17. Install the timing belt covers and all other related components.

1.6L, 2.0L (VIN P) and 1995 1.8L Engines

1. Remove all necessary components for access to the timing belt covers, then remove the covers.

➡Always rotate the crankshaft in a clockwise direction.

2. Rotate the crankshaft clockwise and align the timing marks so No. 1 piston will be at TDC of the compression stroke. At this time the timing marks on the camshaft sprocket and the upper surface of the cylinder head should coincide, and the dowel pin of the camshaft sprocket should be at the upper side.

3. Remove the outer timing belt tensioner.

4. Mark the timing belts, indicating the direction of rotation.

5. Remove the outer timing belt.

6. Remove the camshaft sprockets.

7. Insert a prytool with a 0.32 in. (8mm) diameter shaft into the left side cylinder block plug hole. The prytool will hold the counterbalance shaft stable while removing the oil pump sprocket retaining nut.

8. Remove the oil pump sprocket.

9. Loosen the right counterbalance shaft sprocket bolt.

10. Remove the inner timing belt tensioner.

11. Remove the inner timing belt.

To install:

12. Install the counterbalance shaft sprocket and tighten the flange bolt finger-tight.

13. Align the timing mark on each sprocket with the corresponding timing mark on the front case.

14. Install the inner timing belt.

➡When installing the inner timing belt, ensure that the tension side has no slack.

15. Install the inner timing belt tensioner with the center of the pulley on the left side of the mounting bolt and with the pulley flange facing the front of the engine.

16. Lift the inner timing belt tensioner to tighten the inner timing belt so that its tension side will be pulled tight.

17. Tighten the bolt to secure the inner tensioner.

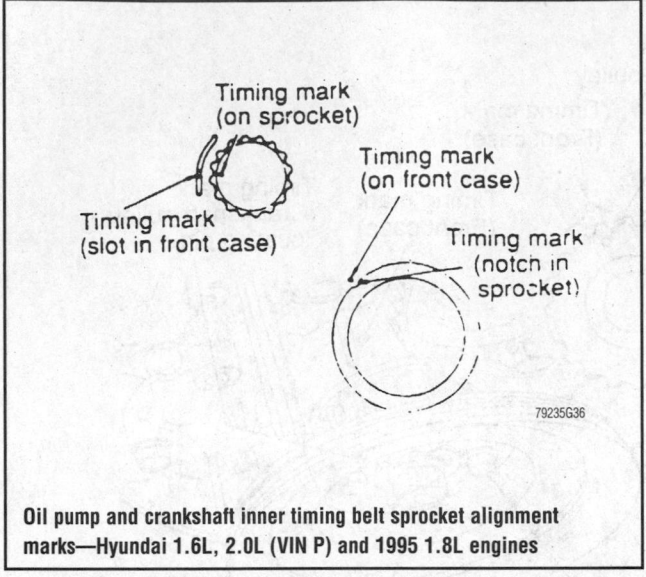

Oil pump and crankshaft inner timing belt sprocket alignment marks—Hyundai 1.6L, 2.0L (VIN P) and 1995 1.8L engines

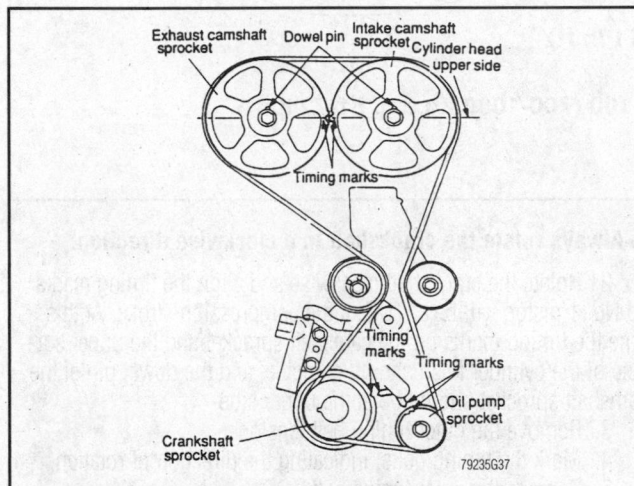

Timing belt sprocket alignment mark locations and positioning for belt removal and installation—Hyundai 1.6L, 2.0L (VIN P) and 1995 1.8L engines

➡ **When tightening the bolt of the tensioner, ensure that the tensioner pulley shaft does not rotate with the bolt. Allowing it to rotate with the bolt can cause excessive tension on the belt.**

18. Ensure the timing marks are in alignment.

19. Check the belt for proper tension by depressing the belt on its long side with your finger and noting the belt deflection. The desired deflection should be 0.20–0.28 in. (5–7mm).

20. Install the flange, crankshaft sprocket and washer on the crankshaft. The flange on the crankshaft sprocket must be installed towards the inner timing belt sprocket. Tighten the bolt to 80–94 ft. lbs. (110–130 Nm).

21. Insert a prytool with a 0.32 in. (8mm) diameter shaft into the left side cylinder block plug hole. The prytool will hold the counterbalance shaft stable while removing the oil pump sprocket retaining nut.

22. Install the oil pump sprocket and tighten the nut to 36–43 ft. lbs. (50–60 Nm).

23. Install the camshaft sprocket and tighten the bolt to 56–72 ft. lbs. (80–100 Nm).

24. Carefully push the auto-tensioner rod in until the set hole in the rod is aligned with the hole in the cylinder. Place a wire into the hole to retain the rod.

25. Install the outer timing belt tensioner.

26. Install the outer tensioner pulley onto the tensioner arm. Locate the pinhole in the tensioner pulley shaft to the left of the center bolt. Tighten the center bolt finger-tight.

27. Turn the two camshaft sprockets so their dowel pins are located on top. Align the timing marks facing each other with the top surface of the cylinder head.

➡ **Both camshaft sprockets are used for the intake and exhaust camshafts and are provided with two timing marks. When the sprocket is mounted on the exhaust camshaft, use the timing mark on the right with the dowel pin hole on top. For the intake camshaft sprocket, use the 1 on the left with the dowel pin hole on top.**

28. Align the crankshaft sprocket and oil pump sprocket timing marks.

29. Insert a prytool with a 0.32 in. (8mm) diameter shaft into the left side cylinder block plug hole. If the shaft can be inserted 2.4 in. (61mm), the silent shaft is in the correct position. If the shaft of the tool can only be inserted 0.8–1.0 in. (20–25mm) deep, turn the oil pump sprocket one full turn and realign the marks.

➡ **Keep the tool inserted in hole for the remainder of this procedure. The above step assures that the oil pump socket is in correct orientation to the silent shafts. This step must not be skipped or a vibration may develop during engine operation.**

30. Install the timing belt around the tensioner pulley and crankshaft sprocket. Hold the belt with your left-hand.

31. Pulling the belt with your right-hand, install it around the oil pump sprocket.

32. Install the belt around the idler pulley and intake camshaft sprocket.

33. Turn the exhaust camshaft sprocket one tooth clockwise to align its timing mark with the cylinder head top surface. Pulling the belt with both hands, install it around the exhaust camshaft sprocket.

34. Gently raise the tensioner pulley so that the belt does not sag and temporarily tighten the center bolt.

35. Turn the crankshaft ¼ turn counterclockwise. Turn the crankshaft clockwise to move the No. 1 cylinder to TDC.

36. Loosen the center bolt and attach special tool (PN 09244–28100) or equivalent to a torque wrench. Apply a torque of 23–25 inch lbs. (2.6–2.8 Nm). Tighten the center bolt.

37. Screw the special tool (PN 09244–28000) or equivalent into the engine left support bracket until its end makes contact with the tensioner arm. At this point, screw the special tool in some more and remove the set wire attached to the auto-tensioner, if the wire was not previously removed. Remove the special tool.

38. Rotate the crankshaft 2 complete turns clockwise and let it sit for approximately 15 minutes. Then, measure the auto-tensioner protrusion (the distance between the tensioner arm and auto-tensioner body) to ensure that it is within 0.15–0.18 in. (3.8–4.5mm).

39. If the timing belt tension adjustment is being performed with the engine mounted in the vehicle, and clearance between the tensioner arm and the auto-tensioner body cannot be measured, the following alternative method can be used:

    a. Screw in special tool (PN 09244–28000) or equivalent, until its end makes contact with the tensioner arm.

b. After the tool makes contact with the arm, screw it in some more to retract the auto-tensioner pushrod while counting the number of turns the tool makes until the tensioner arm is brought into contact with the auto-tensioner body. Be sure the number of turns the special tool makes conforms with the standard value of 2½ –3 turns.

c. Install the rubber plug to the timing belt rear cover.

40. Install the timing belt covers.

1996–99 1.8L (VIN M) and 2.0L (VIN F) Engines

1. Remove all necessary components for access to the timing belt cover, then remove the cover.

2. Rotate the crankshaft clockwise and align the timing marks so No. 1 piston will be at TDC of the compression stroke.

3. Remove the timing belt tensioner and idler pulley.

4. Mark the timing belt with an arrow showing direction of rotation.

5. Remove the timing belt.

To install:

6. Align the timing marks of the camshaft sprocket and check that the crankshaft timing marks are still in alignment.

7. Install the timing belt tensioner.

8. Install the idler pulley, if equipped. Tighten bolt to 32–41 ft. lbs. (43–55 Nm).

9. Position the timing belt over the camshaft sprocket, then over the crankshaft sprocket.

10. Tension the timing belt and tighten the tensioner pulley bolt to 32–41 ft. lbs. (43–55 Nm). When properly tensioned, the timing belt should deflect 0.16–0.24 in. (4–6mm) when a force of 5 lbs. (2.2kg) is placed on the longest span of the belt.

11. Turn the crankshaft sprocket one turn clockwise and realign the crankshaft sprocket timing mark.

12. Recheck the belt tension and adjust as necessary.

13. Install the timing belt cover and all other applicable components.

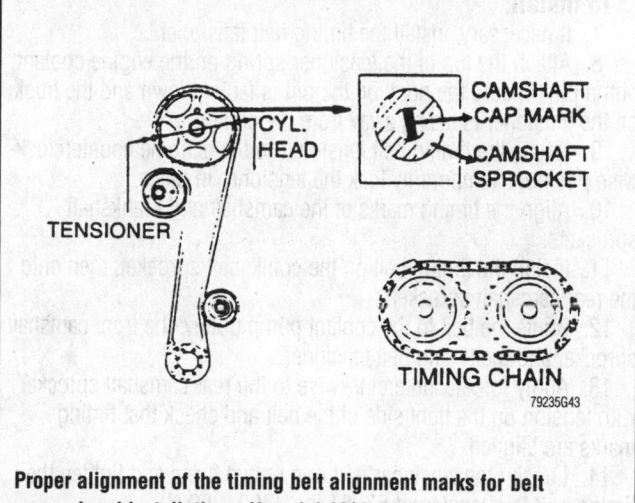

Proper alignment of the timing belt alignment marks for belt removal and installation—Hyundai 1996–98 1.8L and all 2.0L (VIN F) engines

3.0L Engine

1. Remove all necessary components for access to the timing belt covers, then remove the covers.

2. Turn the crankshaft until the timing marks on the camshaft sprocket and cylinder head are aligned.

3. Loosen the timing belt tensioner bolt and turn the tensioner counterclockwise as far as it will go. Tighten the adjusting bolt.

4. Mark the timing belt with an arrow showing direction of rotation.

5. Remove the timing belt.

6. If defective, remove the timing belt tensioner.

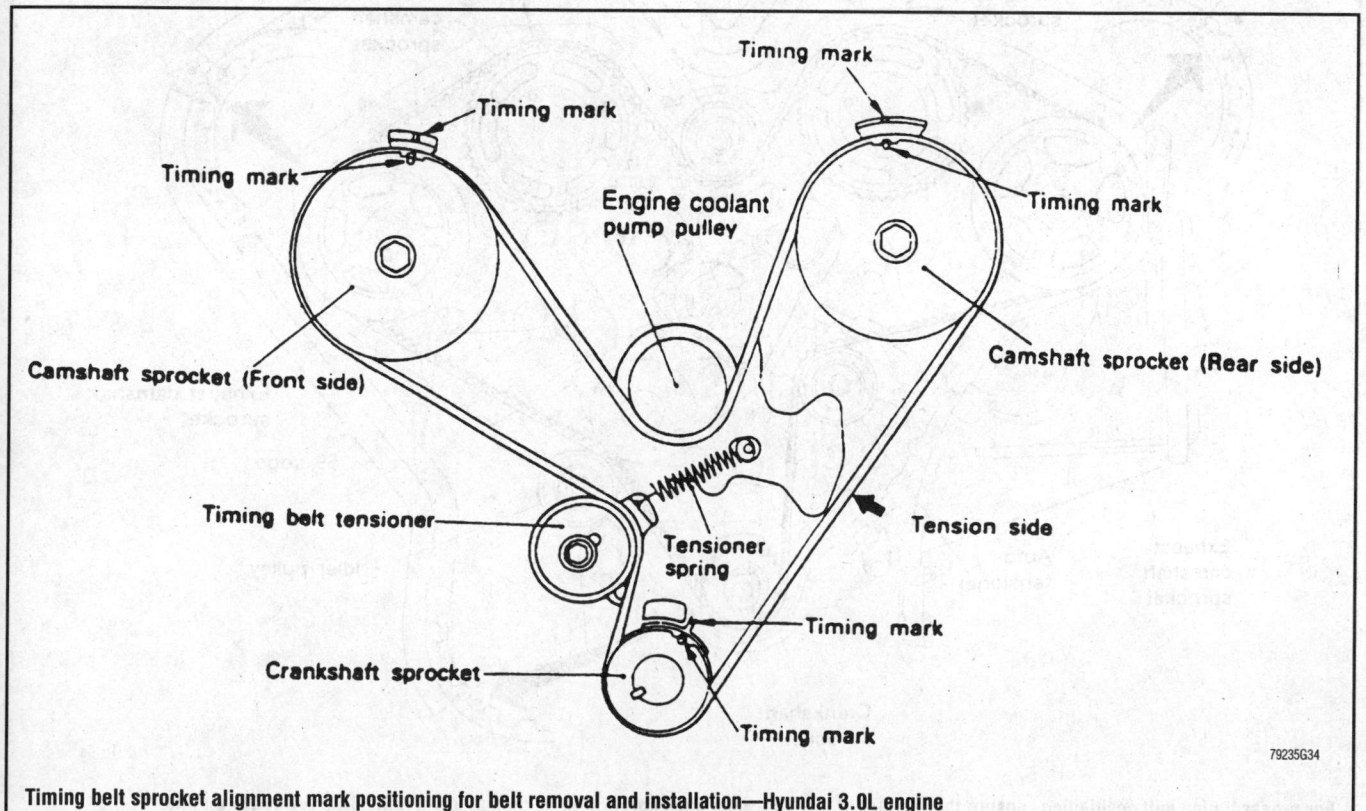

Timing belt sprocket alignment mark positioning for belt removal and installation—Hyundai 3.0L engine

To install:

7. If necessary, install the timing belt tensioner.

8. Attach the top of the tensioner spring on the engine coolant pump pin. Ensure the hook on the pin is facing down and the hook on the tensioner is facing away from the engine

9. Rotate the timing belt tensioner to the extreme counterclockwise position. Temporarily lock the tensioner in place.

10. Align the timing marks of the camshaft and crankshaft sprockets.

11. Install the timing belt on the crankshaft sprocket, then onto the rear camshaft sprocket.

12. Route the belt to the coolant pump pulley, the front camshaft sprocket and the timing belt tensioner.

13. Apply force counterclockwise to the rear camshaft sprocket with tension on the tight side of the belt and check that timing marks are aligned.

14. Loosen the tensioner bolt one or two turns and tighten the timing belt to a tension of 57–84 lbs. (260–380 N).

15. Turn the crankshaft two turns clockwise.

16. Readjust the sprocket timing marks and tighten the tensioner bolts.

17. Install the timing belt covers.

18. Install the crankshaft pulley and tighten to 108-116 ft. lbs. (150-160 Nm).

19. Install all applicable components.

INFINITI MODELS

3.0L (VG30DE) Engine

1. Remove all necessary components for access to the front timing covers, then remove the covers.

2. Set the No. 1 cylinder on TDC of the compression stroke.

3. The automatic belt tensioner is oil damped and spring operated. Install a 6mm bolt to hold the tensioner back against the spring and release tension on the belt.

S0>Remove the auto-tensioner and timing belt.

✳✳ WARNING

Do not rotate the crankshaft or camshaft separately because the pistons will strike the valves causing engine damage.

To install:

4. Confirm that the No. 1 cylinder is at TDC of the compression stroke.

5. Align the marks on the camshaft and crankshaft sprockets with the marks on the rear belt cover and oil pump housing.

6. With the arrows on the timing belt pointing towards the front, align the white lines on the timing belt with the marks on the sprockets and install the belt.

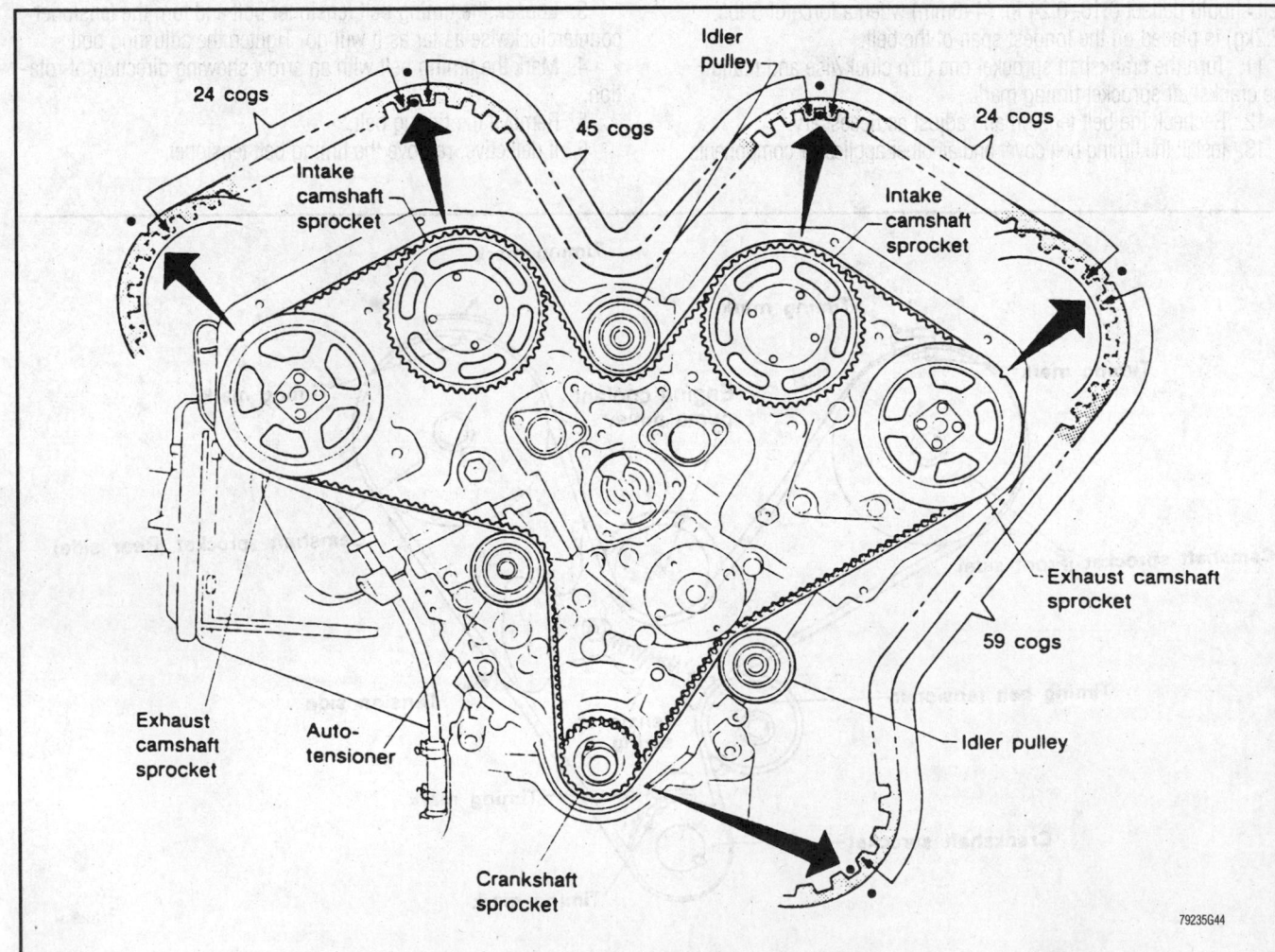

For proper timing belt positioning, ensure the number of teeth between each sprocket is as indicated—Infinity 3.0L (VG30DE) engine

79235G44

7. To prepare the auto-tensioner for installation, perform the following:

a. Remove the bolt holding the tensioner in position.

b. Use a vise to adjust the gap between the tensioner arm and pusher body to 0.160 in. (4mm).

c. Install the bolt again to hold the arm in this position. Do not try to use the bolt to adjust the gap or the threads will be damaged.

8. Install the auto-tensioner, push it towards the belt to just take up the slack, then tighten the bolts finger-tight.

9. Before adjusting the timing belt tension, the slack must be properly distributed:

a. Turn the crankshaft 10 degrees clockwise and tighten the tensioner bolts and nut to 12–15 ft. lbs. (16–21 Nm). Do not push the auto-tensioner hard or the belt will be adjusted too tight.

b. Turn the crankshaft 120 degrees (⅓ turn) counterclockwise.

c. Loosen the tensioner bolts and nut ½ turn and move the tensioner body away from the timing belt as far as it will move.

d. Turn the crankshaft clockwise to TDC again.

e. Push the tensioner against the belt with a force of 13 lbs. (59 N) using a spring scale or similar tool and tighten the bolts again to 12–15 ft. lbs. (16–21 Nm). The pressure specification is important and a special spring scale tool, J-38387, is available to measure the tensioner force.

10. To check the timing belt tension:

a. Turn the crankshaft 120 degrees (⅓ turn) clockwise, then turn counterclockwise and return the engine to TDC.

b. Prepare a steel plate that is approximately ⅜ in. (10mm) wide and longer than the width of the belt.

c. Set the plate on the timing belt between two camshaft sprockets and push against the plate with a force of 11 lbs. (49 N). Note the belt deflection.

d. Repeat the procedure between the other camshaft sprockets and between the exhaust sprockets and idler/tensioner pulleys. There will be a total of four measurements.

e. Add the deflection measurements and divide by four. The average deflection must be 0.240–0.280 in. (6–7mm). If belt tension is not correct, start the entire adjustment procedure again.

11. Confirm the auto-tensioner mounting nuts are tightened to 12–15 ft. lbs. (16–21 Nm) and remove the auto-tensioner stopper bolt.

12. After 5 minutes, measure the clearance between the tensioner arm and the pusher. It should be 0.138–0.205 in. (3.5–5.2mm).

13. Be sure all the sprocket timing marks are correctly aligned. Install the timing belt covers and tighten the bolts to 24–38 inch lbs. (3–5 Nm).

14. Install all applicable components.

LEXUS MODELS

> ✺✺ **CAUTION**
>
> **On models with an air bag, wait at least 90 seconds from the time that the ignition switch is turned to the LOCK position and the battery is disconnected before performing any further work.**

3.0L (1MZ-FE) Engine

1. Remove all necessary components for access to the upper timing belt cover. Remove the 8 bolts and lift off the upper (No. 2) cover.

2. Paint matchmarks on the timing belt at all points where it meshes with the pulleys and the lower timing cover.

3. Set the No. 1 cylinder to TDC of the compression stroke and check that the timing marks on the camshaft timing pulleys are aligned with those on the No. 3 timing cover. If not, turn the engine 1 complete revolution (360 degrees) and check again.

4. Remove the timing belt tensioner and the dust boot.

5. Turn the right camshaft pulley clockwise slightly to release tension, then remove the timing belt from the pulleys.

6. Remove the upper (No. 3) and lower (No. 1) timing belt covers.

7. Remove the timing belt guide.

8. Remove the timing belt from the engine.

➡**If the timing belt is to be reused, draw a directional arrow on the timing belt in the direction of engine rotation (clockwise) and place matchmarks on the timing belt and crankshaft gear to match the drilled mark on the pulley.**

To install:

➡**If the old timing belt is being reinstalled, be sure the directional arrow is facing in the original direction and that the belt and crankshaft gear matchmarks are properly aligned.**

9. Install the lower (No. 1) timing cover and tighten the bolts.

10. Set the No. 1 cylinder to TDC again. Turn the right camshaft until the knock pin hole is aligned with the timing mark on the No. 3 belt cover. Turn the left pulley until the marks on the pulley are aligned with the mark on the No. 3 timing cover.

11. Check that the mark on the belt matches with the edge of the lower cover. If not, shift it on the crank pulley until it does. Turn the left pulley clockwise a bit and align the mark on the timing belt with

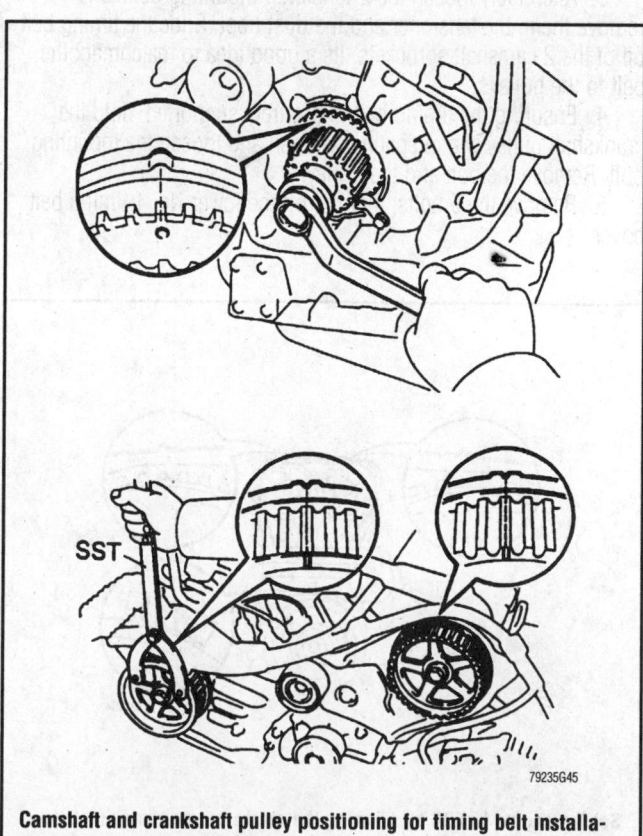

Camshaft and crankshaft pulley positioning for timing belt installation—Lexus 3.0L (1MZ-FE) engine

79235G45

the timing mark on the pulley. Slide the belt over the left pulley. Now move the pulley until the marks on it align with the one on the No. 3 cover. There should be tension on the belt between the crankshaft pulley and the left camshaft pulley.

12. Align the installation mark on the timing belt with the mark on the right side camshaft pulley. Hang the belt over the pulley with the flange facing inward. Align the timing marks on the right pulley with the one on the No. 3 cover and slide the pulley onto the end of the camshaft. Move the pulley until the camshaft knock pin hole is aligned with the groove in the pulley, then install the knock pin. Tighten the bolt to 55 ft. lbs. (75 Nm).

13. Position a plate washer between the timing belt tensioner and the block, then press in the pushrod until the holes are aligned between it and the housing. Slide a 0.05 in. Allen wrench through the hole to keep the pushrod set. Install the dust boot, then install the tensioner. Tighten the bolts to 20 ft. lbs. (26 Nm). Don't forget to pull out the Allen wrench.

14. Turn the crankshaft clockwise 2 complete revolutions and check that all marks are still in alignment. If they aren't, remove the timing belt and start over again.

15. Install the remaining components.

3.0L (2JZ-GE) Engine

1. Remove all necessary components for access to the upper timing belt covers. Using a 5mm Allen wrench, remove the 9 bolts and lift off the two upper (No. 2 and No. 3) timing belt covers.

2. Rotate the crankshaft pulley clockwise so its groove is aligned with the **0** mark in the No. 1 (lower) timing cover. Check that the timing marks on the camshaft timing sprockets are aligned with the marks on the No. 4 (inner) cover. If not, rotate the crankshaft 1 complete revolution (360 degrees).

3. Alternately loosen the 2 tensioner mounting bolts and remove them, the tensioner and the dust boot. Slide the timing belt off of the 2 camshaft sprockets. Its a good idea to matchmark the belt to the pulleys.

4. Ensuring the timing belt is securely supported, hold the crankshaft pulley with a spanner wrench and loosen the mounting bolt. Remove the bolt and the pulley.

5. Remove the 5 bolts, then lift off the lower No. 1 timing belt cover.

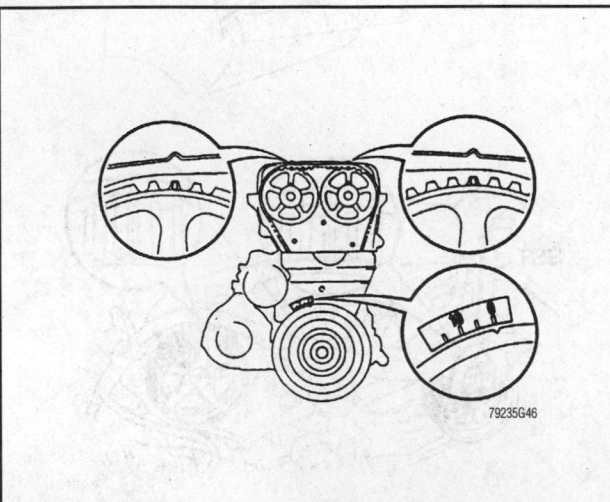

Set the engine to TDC by aligning the marks before removing the lower timing cover—Lexus 3.0L (2JZ-GE) engine

6. Remove the timing belt guide.
7. Remove the timing belt.

➡**If the timing belt is to be reused, draw a directional arrow on the timing belt in the direction of engine rotation (clockwise) and place matchmarks on the timing belt and crankshaft gear to match the drilled mark on the pulley.**

To install:

8. Install the timing belt on the crankshaft timing pulley and the idler pulleys.

➡**If the old timing belt is being reinstalled, be sure the directional arrow is facing in the original direction and that the belt and crankshaft gear matchmarks are properly aligned.**

9. Install the timing belt guide. Install the lower (No. 1) timing cover and tighten the bolts.

10. Align the crankshaft pulley set key with the key groove on the pulley and slide the pulley on. Tighten the bolt to 239 ft. lbs. (324 Nm).

11. Set the No. 1 cylinder to TDC again. Turn the camshaft until the sprocket timing marks are aligned with the timing marks on the No. 4 belt cover.

12. Check that the marks on the belt matches with those on the sprockets, then slide it over the sprockets. If not, shift it on the crank pulley until it does.

13. Position a plate washer between the timing belt tensioner and the a block, then press in the pushrod until the holes are aligned between it and the housing. Slide a 1.5mm Allen wrench through the hole to keep the pushrod set. Install the dust boot, then install the tensioner. Tighten the bolts to 20 ft. lbs. (26 Nm). Don't forget to pull out the Allen wrench.

14. Turn the crankshaft clockwise two complete revolutions and check that all marks are still in alignment. If they aren't, remove the timing belt and start over again.

15. Position new gaskets, then install the upper (No. 2 and No. 3) timing covers.

4.0L (1UZ-FE) Engine

1. Remove all necessary components for access to the right-hand side No. 3 and No. 2, and left-hand side No. 2 timing belt covers, then remove the covers.

2. Turn the crankshaft pulley and align it's groove with the timing mark **0** of the No. 1 timing cover. Check that the timing marks of the camshaft timing pulleys and timing belt rear plates are aligned. If not, turn the crankshaft 1 full revolution (360 degrees).

3. Remove the timing belt tensioner. Using the proper tool, loosen the tension between the left side and right side timing pulleys by slightly turning the left side camshaft clockwise.

4. Disconnect the timing belt from the camshaft timing pulleys. Using the proper tool, remove the bolt and the timing pulleys.

5. Remove the bolt and the crankshaft pulley with the proper tool. Remove the fan bracket. Remove the hydraulic pump on the SC400 model.

6. Remove the mounting bolts and the No. 1 timing belt cover.

7. Remove the 2 upper and lower timing belt covers.

8. Remove the timing belt guide (No. 1 crank position sensor plate).

9. Remove the timing belt.

➡**If the timing belt is to be reused, draw a directional arrow on the timing belt in the direction of engine rotation (clock-**

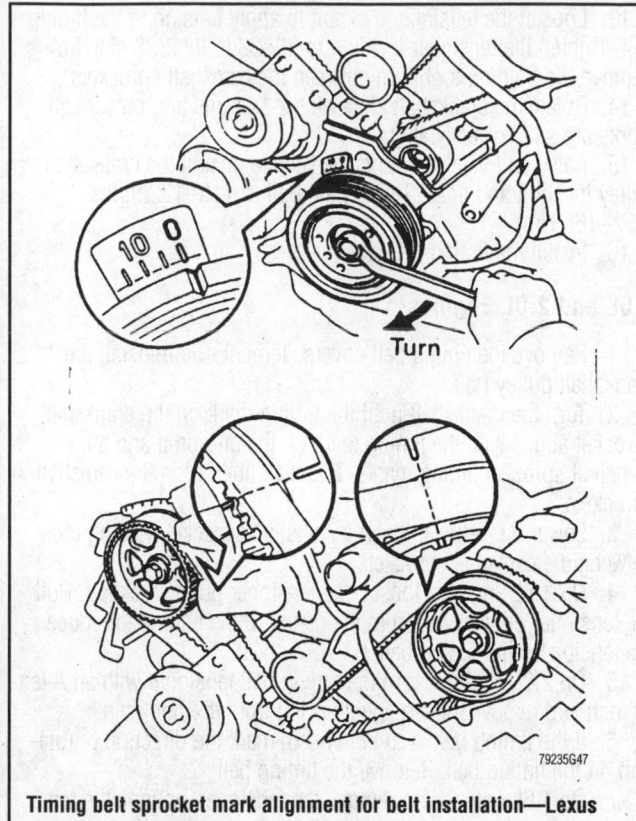

Turn

Timing belt sprocket mark alignment for belt installation—Lexus 4.0L (1UZ-FE) engine

79235G47

wise) and place matchmarks on the timing belt and crankshaft gear to match the drilled mark on the pulley.

To install:

10. Align the installation mark on the timing belt with the drilled mark of the crankshaft timing pulley. Install the timing belt on the crankshaft timing pulley, No. 1 idler pulley and the No. 2 idler pulley.

➡ If the old timing belt is being reinstalled, be sure the directional arrow is facing in the original direction and that the belt and crankshaft gear matchmarks are properly aligned.

11. Install the timing belt guide (No. 1 crank angle sensor plate) with the cup side facing forward. Replace the timing belt cover spacer.

12. Install the No. 1 timing belt cover and tighten the mounting bolts. Install the hydraulic pump on SC400 models. Install the fan bracket.

13. Align the pulley set key on the crankshaft with the key groove of the pulley. Install the pulley, using the proper tool to tap in the pulley. Tighten the pulley bolt to 181 ft. lbs. (245 Nm).

14. Align the knock pin on the right side camshaft with the knock pin of the timing pulley. Slide on the timing pulley with the right side mark facing forward. Tighten the bolt to 80 ft. lbs. (108 Nm).

15. Align the knock pin on the left side camshaft with the knock pin of the timing pulley. Slide on the timing pulley with the left side mark facing forward. Tighten the bolt to 80 ft. lbs. (108 Nm).

16. Turn the crankshaft pulley and align its groove with the **0** timing mark on the No. 1 timing belt cover. Using the proper tool, turn the crankshaft timing pulley and align the timing marks of the camshaft timing pulley and the timing belt rear plate.

17. Install the timing belt to the left side camshaft timing pulley by:

a. Using the proper tool, slightly turn the left side timing pulley clockwise. Align the installation mark of the timing belt with the timing mark of the camshaft timing pulley and hang the timing belt on the left side camshaft pulley.

b. Using the proper tool, align the timing marks of the left side camshaft pulley and the timing belt rear plate.

c. Check that the timing belt has tension between crankshaft timing pulley and the left side camshaft pulley.

18. Install the timing belt to the right side camshaft timing pulley by:

a. Using the proper tool, slightly turn the right side timing pulley clockwise. Align the installation mark of the timing belt with the timing mark of the camshaft timing pulley and hang the timing belt on the right side camshaft pulley.

b. Using the proper tool, align the timing marks of the right side camshaft pulley and the timing belt rear plate.

c. Check that the timing belt has tension between the crankshaft timing pulley and the right side camshaft pulley.

19. The timing belt tensioner must be set prior to installation. The tensioner can be set as follows:

a. Place a plate washer between the tensioner and a block. Using a suitable press, press in the pushrod using 220–2205 lbs. (100–1000kg) of pressure.

b. Align the holes of the pushrod and housing, pass the proper tool (0.05 in. Allen wrench) through the holes to keep the setting position of the pushrod.

c. Release the press and install the dust boot on the tensioner.

20. Install the tensioner and tighten the bolts to 20 ft. lbs. (26 Nm). Remove the tool from the tensioner.

21. Turn the crankshaft pulley two complete revolutions from TDC-to-TDC. Always turn the crankshaft clockwise. Check that each pulley aligns with the timing marks.

22. Install all remaining components in the reverse order of removal.

MAZDA MODELS

Protégé 1.5L (Z5D) and 1.8L (BP) Engines

1. Remove all necessary components for access to the timing belt covers, then remove the covers.

2. Turn the crankshaft until the timing mark on the crankshaft sprocket aligns with the timing mark on the oil pump and the camshaft sprocket timing marks line up on the camshaft sprockets.

3. Remove the crankshaft pulley lockbolt and pulley boss.

4. Lower the vehicle. Insert a camshaft sprocket holding tool between the camshaft sprockets.

5. Loosen the tensioner pulley lockbolt. Pull the tensioner pulley away from the center of the engine to reduce the tension on the timing belt.

6. If the timing belt is to be reused, mark the direction of rotation on the timing belt. Remove the timing belt.

7. To remove the tensioner, unhook the tensioner spring, and remove the pulley lockbolt and tensioner.

To install:

8. Install the crankshaft sprocket bolt. Install the flywheel locking tool, if equipped with automatic transaxle, or place the shift lever in **4th** gear and apply the parking brake, if equipped with manual transaxle. Tighten the bolt to 108–116 ft. lbs. (147–157 Nm).

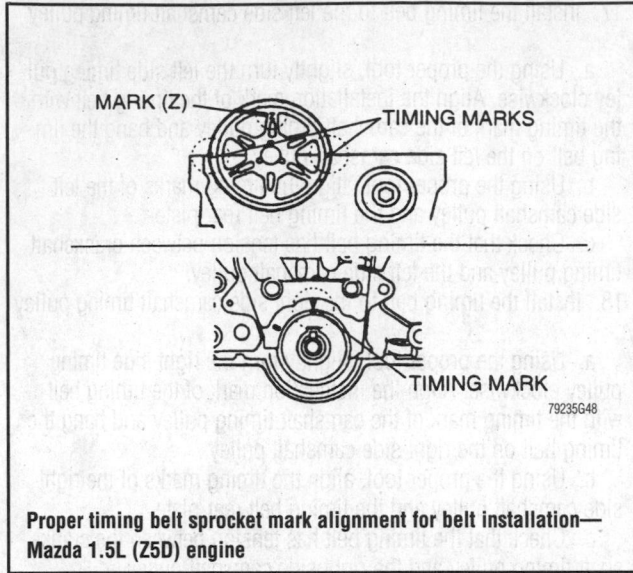

MARK (Z) — TIMING MARKS

TIMING MARK

79235G48

Proper timing belt sprocket mark alignment for belt installation—Mazda 1.5L (Z5D) engine

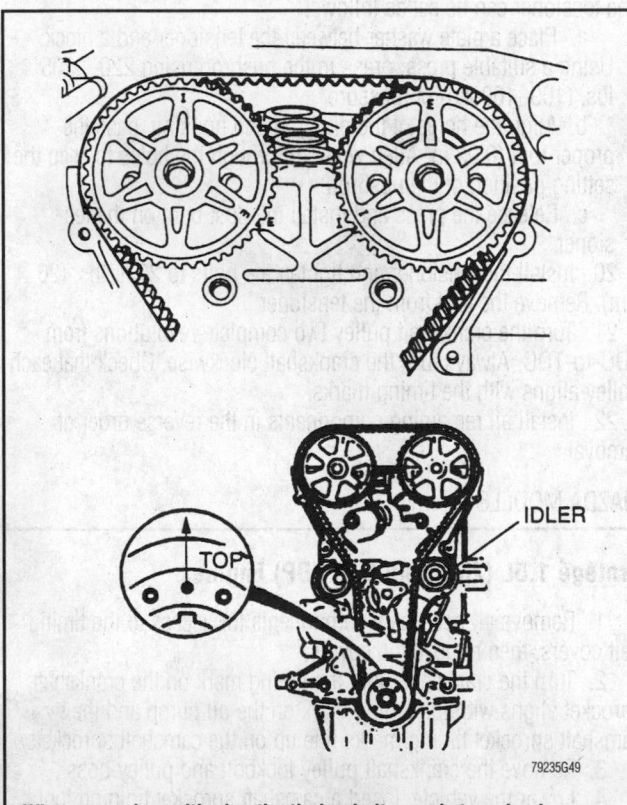

IDLER

TOP

79235G49

When properly positioning the timing belt sprocket marks, be sure the two I's are aligned and the two E's are aligned as indicated—Mazda 1.6L (B6) and Protégé and Miata 1.8L (BP) engines

9. Be sure the timing marks on the camshaft and crankshaft sprockets are still aligned.

10. If removed, position the tensioner with the spring fully extended, and install the lockbolt tightening the mounting bolt to 28–38 ft. lbs. (38–51 Nm).

11. Install the timing belt. If reusing the original timing belt, be sure it is installed in the same direction of rotation.

12. Rotate the crankshaft clockwise 1 ⅚ turns and align the timing marks. Be sure all marks are still correctly aligned.

13. Loosen the tensioner lockbolt to apply tension to the timing belt. Tighten the tensioner lockbolt to 28–38 ft. lbs. (38–51 Nm). Remove the holding tool from between the camshaft sprockets.

14. Rotate the crankshaft clockwise 2 ⅙ turns and be sure all marks are still correctly aligned.

15. Raise and safely support the vehicle. Install the crankshaft pulley lockbolt and boss. Tighten the bolt to 116–122 ft. lbs. (157–166 Nm).

16. Install the timing belt covers.

1.6L and 2.0L Engines

1. Remove the timing belt covers. Temporarily reinstall the crankshaft pulley bolt.

2. Turn the crankshaft until the timing mark on the crankshaft sprocket aligns with the timing mark on the oil pump and the camshaft sprocket timing marks, **E** and **I**, line up on the camshaft sprockets.

3. Lower the vehicle. Insert a camshaft sprocket holding tool between the camshaft sprockets.

4. On 1.6L engines, loosen the tensioner pulley lockbolt. Pull the tensioner pulley away from the center of the engine to reduce the tension on the timing belt.

5. On 2.0L engines, turn the timing belt tensioner with an Allen wrench and remove the tensioner spring from the hook pin.

6. If the timing belt is to be reused, mark the direction of rotation on the timing belt. Remove the timing belt.

7. On 1.6L engines, to remove the tensioner, unhook the tensioner spring, and remove the pulley lockbolt and tensioner.

To install:

8. Install the crankshaft sprocket bolt. Install the flywheel locking tool, if equipped with automatic transaxle, or place the shift lever in **4th** gear and apply the parking brake, if equipped with manual transaxle. Tighten the bolt to 108–116 ft. lbs. (147–157 Nm).

9. Be sure the timing marks on the camshaft and crankshaft sprockets are still aligned.

10. On 1.6L engines, if removed, position the tensioner with the spring fully extended, and install the lockbolt tightening the mounting bolt to 28–38 ft. lbs. (38–51 Nm).

11. Install the timing belt. If reusing the original timing belt, be sure it is installed in the same direction of rotation.

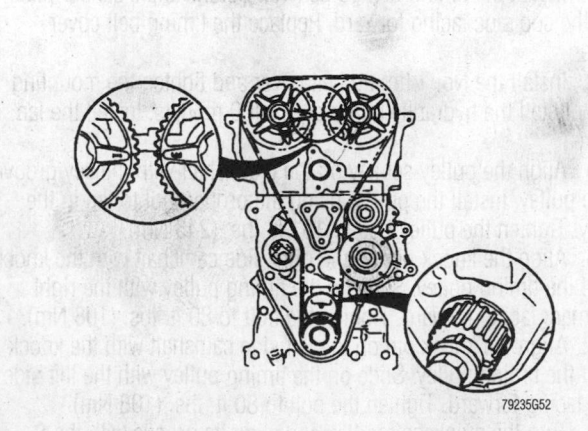

79235G52

When properly aligned for belt removal, the cam gear marks should face each other—Mazda 2.0L (FS) engines

12. On 1.6L engines, loosen the tensioner lockbolt to apply tension to the timing belt. Tighten the tensioner lockbolt. Remove the holding tool from between the camshaft sprockets.

13. On 2.0L engines, turn the tensioner clockwise with an Allen wrench and install the tensioner spring. Remove the holding tool from between the camshaft sprockets.

14. Rotate the crankshaft 2 turns in the normal direction of rotation and align the timing marks. Be sure all marks are still correctly aligned.

15. Raise and safely support the vehicle. Remove the crankshaft pulley bolt and install the timing belt covers.

Miata 1.8L (BP) Engine

1. Remove all necessary components for access to the valve cover, then remove the cover.

2. Remove the spark plugs.

➡**Spark plugs are removed to make it easier to rotate the engine.**

3. Remove the upper, middle and lower timing belt covers.

4. Turn the crankshaft until the timing marks on the crankshaft and camshaft sprockets are aligned. The pin on the pulley boss must face upward. Hold the crankshaft pulley boss with a suitable tool and remove the pulley lockbolt, being careful not to rotate the crankshaft. Remove the crankshaft pulley boss.

5. Mark the direction of rotation on the timing belt. Loosen the tensioner lockbolt and pry the tensioner outward. Tighten the lockbolt with the tensioner spring fully extended. Remove the timing belt.

➡**Protect the tensioner with a shop towel before prying on it. Do not rotate the crankshaft after the timing belt has been removed.**

6. Remove the tensioner and spring. If necessary, remove the idler pulley.

7. Inspect the belt for wear, peeling, cracking, hardening or signs of oil contamination. Inspect the tensioner for free and smooth rotation. Check the tensioner spring free length; it should not exceed 2.331 in. (59.2mm) on the 1995–97 1.8L engine and 2.315 in. (58.8mm) on all others. Inspect the sprocket teeth for wear or damage. Replace parts, as necessary.

To install:

8. Install the crankshaft sprocket bolt. Install the flywheel locking tool, if equipped with automatic transaxle, or place the shift lever in **4th** gear and apply the parking brake, if equipped with manual transaxle. Tighten the bolt to 108–116 ft. lbs. (147–157 Nm).

9. If removed, install the idler pulley and tighten the bolt to 38 ft. lbs. (52 Nm).

10. Install the tensioner and tensioner spring. Pry the tensioner outward and temporarily tighten the tensioner lockbolt with the tensioner spring fully extended.

11. Be sure the crankshaft sprocket timing mark is aligned with the mark on the oil pump housing. Be sure the camshaft sprocket timing marks are aligned with the marks on the seal plate.

12. Install the timing belt so there is no looseness at the idler pulley side or between the camshaft sprockets. If reusing the old belt, be sure it is installed in the same direction of rotation.

13. Temporarily install the pulley boss and lockbolt.

14. Turn the crankshaft 2 turns clockwise and align the crankshaft sprocket timing mark. Face the pin on the pulley boss upright.

Be sure the camshaft sprocket timing marks are aligned. If they are not, repeat the alignment steps.

15. Turn the crankshaft 1 ⅝ turns clockwise and align the crankshaft sprocket timing mark with the tension set mark for proper belt tension adjustment. Remove the lockbolt and pulley boss.

16. Be sure the crankshaft sprocket timing mark is aligned with the tension set mark. Loosen the tensioner lockbolt, and allow the spring to apply tension to the belt. Tighten the tensioner lockbolt to 28–38 ft. lbs. (38–52 Nm).

17. Install the pulley boss and lockbolt.

18. Turn the crankshaft 2 ⅙ turns clockwise and be sure the timing marks are correctly aligned.

19. Apply approximately 22 lbs. (10kg) pressure to the timing belt at a point midway between the camshaft sprockets. The belt should deflect 0.35–0.45 in. (9.0–11.5mm). If the deflection is not correct, repeat the alignment and tensioning procedure.

20. Hold the pulley boss with a suitable tool, and tighten the lockbolt to 123 ft. lbs. (167 Nm).

21. Install the timing belt covers and tighten the bolts to 95 inch lbs. (11 Nm).

22. Install the valve cover and spark plugs, along with all other applicable components.

1.8L (K8) and 2.5L (KL) Engines

1. Remove the timing belt covers. Temporarily reinstall the crankshaft pulley bolt.

2. On Millenia models, support the engine, and remove the nuts and through-bolt from the right side (number three) engine mount sub bracket. Remove the sub bracket.

3. Turn the crankshaft until the timing mark on the crankshaft sprocket aligns with the timing mark on the oil pump and the camshaft

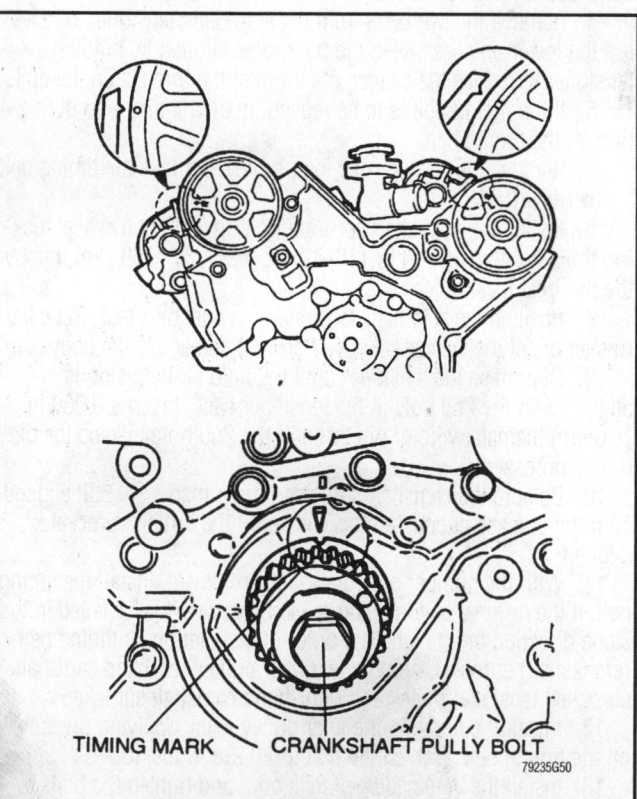

TIMING MARK   **CRANKSHAFT PULLY BOLT**

79235G50

Timing belt sprocket positioning for proper timing belt installation— Mazda 1.8L (K8), and 626/MX6 2.5L (KL) engines

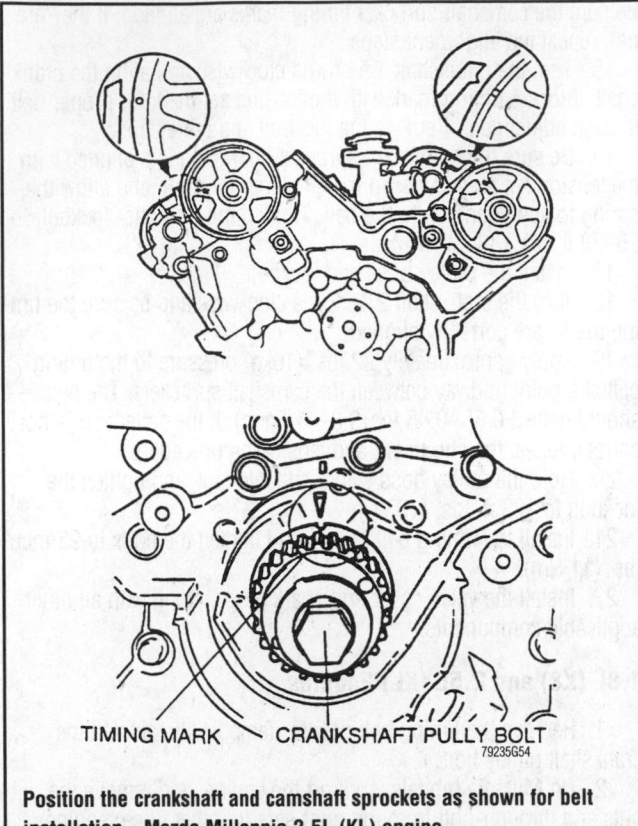

Position the crankshaft and camshaft sprockets as shown for belt installation—Mazda Millennia 2.5L (KL) engine

sprocket timing marks align with the marks on the cylinder head. The number one piston should be at TDC of the compression stroke.

4. Remove the two bolts from the automatic tensioner, removing the lower one first. Keep the bolt holes aligned by holding the tensioner to reduce the chance of stripping the threads on the bolts.

5. If the timing belt is to be reused, mark the direction of rotation on the timing belt.

6. Remove the number one idler pulley. Remove the timing belt.

To install:

7. Install the crankshaft sprocket bolt. Install the flywheel locking tool. Tighten the bolt to 116–122 ft. lbs. (157–166 Nm). remove the flywheel locking tool.

8. Position the automatic tensioner in a suitable press. Set a flat washer under the tensioner body to prevent damage to the body plug.

9. Compress the tensioner until the hole in the piston is aligned with the 2nd hole in the tensioner case. Insert a 0.060 in. (1.6mm) diameter wire or pin through the 2nd hole to keep the piston compressed.

10. Be sure the camshaft sprocket timing marks are still aligned. Turn the crankshaft counterclockwise until the timing sprocket is aligned.

11. With the number one idler pulley removed, install the timing belt. If the original belt is being reused, be sure it is installed in the same direction of rotation. The order of installation is: timing belt (crankshaft) sprocket, number two idler pulley, left-hand camshaft sprocket, tensioner pulley and right-hand camshaft sprocket.

12. Install the number one idler pulley while applying pressure on the timing belt. Tighten the bolt to 28–38 ft. lbs. (38–51 Nm).

13. Install the automatic belt tensioner and tighten the bolts to 14–18 ft. lbs. (19–25 Nm). Remove the wire or pin from the tensioner.

14. Turn the crankshaft clockwise, until the crankshaft sprocket timing mark is again at TDC. This should place all of the belt slack in the automatic tensioner portion of the belt.

15. Rotate the crankshaft 2 turns in the normal direction of rotation and align the timing marks. Be sure all marks are still correctly aligned.

16. Inspect timing belt deflection, 0.24–0.31 in. (6–8mm), between the crankshaft sprocket and the tensioner pulley. If it is out of specification, replace the auto-tensioner.

17. On Millenia models, install the right side (number three) engine mount sub bracket. Tighten the nuts to 55–77 ft. lbs. (75–104 Nm) and the through-bolt to 63–86 ft. lbs. (86–116 Nm). Remove the engine support.

18. Remove the crankshaft damper bolt and install the timing belt covers.

2.3L (KJ) Engine

1. Remove the timing belt covers. Temporarily reinstall the crankshaft pulley bolt.

2. Remove the power steering auto-tensioner and pulley.

3. Turn the crankshaft until the timing mark on the crankshaft sprocket aligns with the timing mark on the oil pump and the camshaft sprocket timing marks align with the marks on the cylinder head. The number one piston should be at TDC of the compression stroke.

4. Remove the two bolts from the automatic tensioner, removing the lower one first. Keep the bolt holes aligned by holding the tensioner to reduce the chance of stripping the threads on the bolts.

5. If the timing belt is to be reused, mark the direction of rotation on the timing belt.

6. Remove the timing belt.

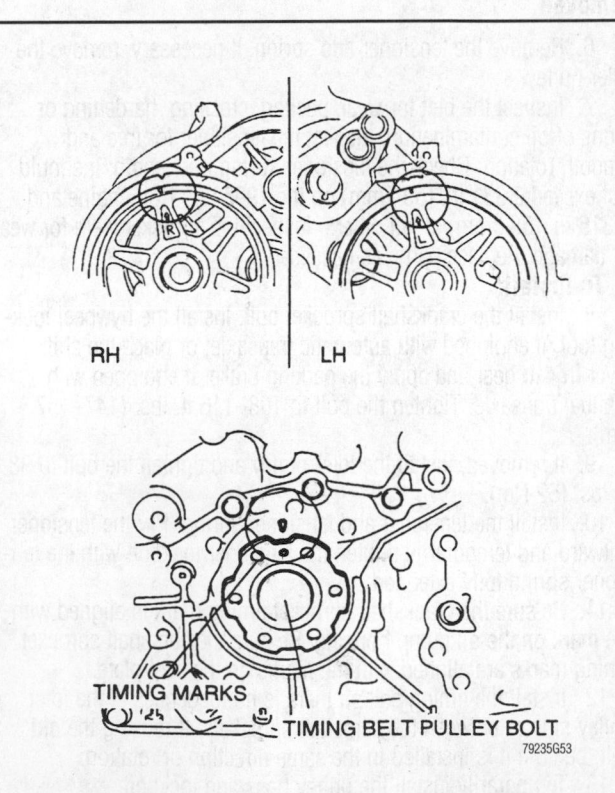

Proper crankshaft and camshaft timing belt sprocket alignment mark positioning—Mazda 2.3L (KJ) engines

To install:

7. Install the crankshaft sprocket bolt. Install the flywheel locking tool. Tighten the bolt to 116–122 ft. lbs. (157–166 Nm). remove the flywheel locking tool.

8. Position the automatic tensioner in a press. Set a flat washer under the tensioner body to prevent damage to the body plug.

9. Compress the tensioner until the hole in the piston is aligned with the 2nd hole in the tensioner case. Insert a 0.063 in. (1.6mm) diameter wire or pin through the 2nd hole to keep the piston compressed.

10. Be sure the camshaft sprocket timing marks are still aligned. Turn the crankshaft clockwise until the timing sprocket is aligned.

11. Install the timing belt. If the original belt is being reused, be sure it is installed in the same direction of rotation. The order of installation is: timing belt (crankshaft) sprocket, number two idler pulley, LEFT-HAND camshaft sprocket, both number one idler pulleys, right-hand camshaft sprocket and the tensioner pulley.

12. Install the automatic belt tensioner and tighten the bolts to 14–18 ft. lbs. (19–25 Nm). Remove the wire or pin from the tensioner.

13. Turn the crankshaft clockwise, until the crankshaft sprocket timing mark is again at TDC. This should place all of the belt slack in the automatic tensioner portion of the belt.

14. Rotate the crankshaft two turns in the normal direction of rotation and align the timing marks. Be sure all marks are still correctly aligned.

15. Inspect timing belt deflection, 0.24–0.31 in. (6–8mm), between the crankshaft sprocket and the tensioner pulley. If it is out of specification, replace the auto-tensioner.

16. Install the power steering auto-tensioner and tighten the bolts to 14–18 ft. lbs. (19–25 Nm). Install the pulley, and tighten the bolt to 29–34 ft. lbs. (40–47 Nm).

17. Remove the crankshaft damper bolt and install the timing belt covers.

MITSUBISHI

1.5L and 1.8L Engines

1. Remove the timing belt upper and lower covers.

2. Make a mark on the back of the timing belt indicating the direction of rotation so it may be reassembled in the same direction if it is to be reused. Loosen the timing belt tensioner and move the tensioner to provide slack to the timing belt. Tighten the tensioner in this position.

3. Remove the timing belt.

❄❄ WARNING

Coolant and engine oil will damage the rubber in the timing belt, drastically reducing its life. Do not allow engine oil or coolant to contact the timing belt, the sprockets or tensioner assembly.

4. If defective, remove the tensioner spacer, tensioner spring and tensioner assembly.

To install:

5. Position the tensioner, tensioner spring and tensioner spacer on engine block.

6. Align the timing marks on the camshaft sprocket and crankshaft sprocket. This will position No. 1 piston on TDC on the compression stroke.

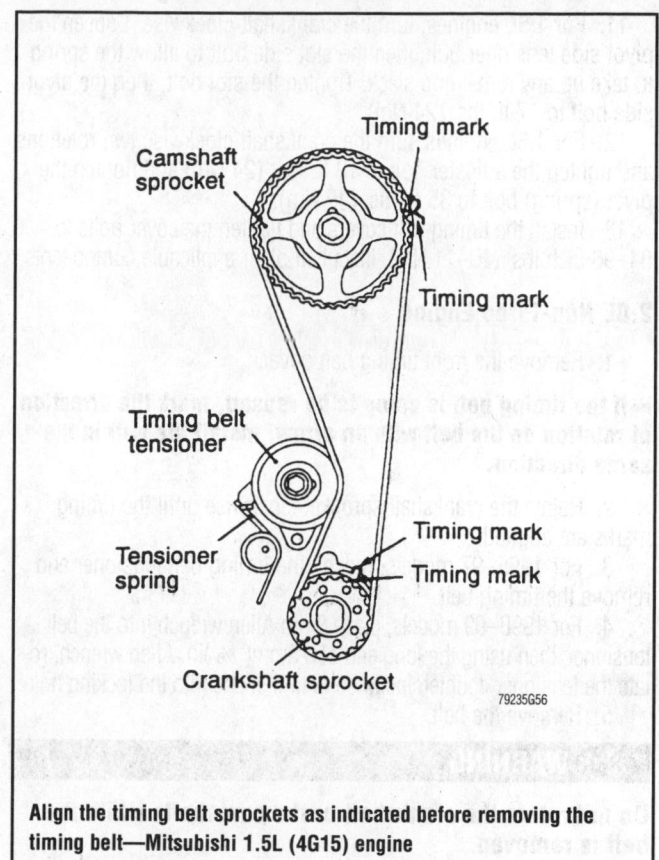

Align the timing belt sprockets as indicated before removing the timing belt—Mitsubishi 1.5L (4G15) engine

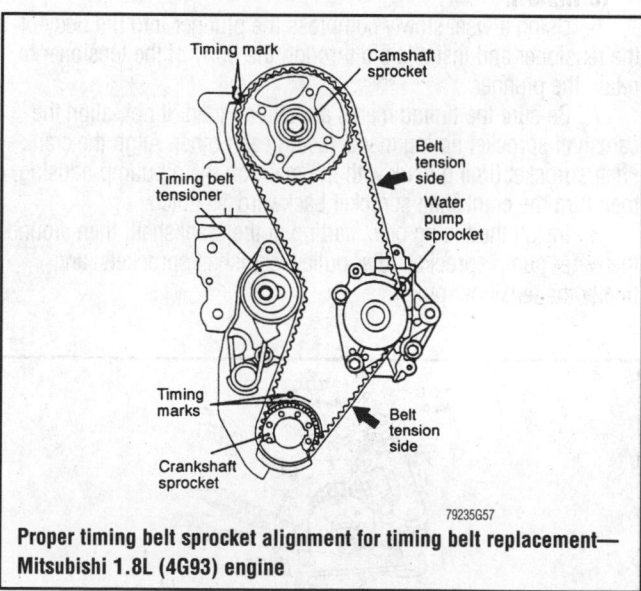

Proper timing belt sprocket alignment for timing belt replacement—Mitsubishi 1.8L (4G93) engine

7. Position the timing belt on the crankshaft sprocket and keeping the tension side of the belt tight, set it on the camshaft sprocket, then the tensioner.

8. Apply slight counterclockwise force to the camshaft sprocket to give tension to the belt and be sure all timing marks are aligned.

9. Loosen the pivot side tensioner bolt and the slot side bolt. Allow the spring to remove the slack.

10. Tighten the slot side tensioner bolt, then the pivot side bolt. If the pivot side bolt is tightened first, the tensioner could turn with bolt, causing over tension.

11. For 1.5L engines, turn the crankshaft clockwise. Loosen the pivot side tensioner bolt, then the slot side bolt to allow the spring to take up any remaining slack. Tighten the slot bolt, then the pivot side bolt to 17 ft. lbs. (24 Nm).

12. For 1.8L engines, turn the crankshaft clockwise two rotations and tighten the adjuster bolt to 18 ft. lbs. (24 Nm) and tighten the pivot (spring) bolt to 35 ft. lbs. (45 Nm).

13. Install the timing belt covers and tighten the cover bolts to 84–96 inch lbs. (10–11 Nm). Install all other applicable components.

2.0L Non-Turbo Engine

1. Remove the front timing belt cover.

➡If the timing belt is going to be reused, mark the direction of rotation on the belt with an arrow. Install the belt in the same direction.

2. Rotate the crankshaft sprocket clockwise until the timing marks are aligned.

3. For 1995–97 models, loosen the timing belt tensioner and remove the timing belt.

4. For 1998–99 models, place 8mm Allen wrench into the belt tensioner, then using the long end of a 3mm(⅛)in. Allen wrench, rotate the tensioner counterclockwise until it slides into the locking hole.

5. Remove the belt.

❊❊❊ WARNING

Do not rotate the crankshaft or the camshafts while the belt is removed.

To install:

6. Using a vise, slowly compress the plunger into the body of the tensioner and install a pin through the body of the tensioner to retain the plunger.

7. Be sure the timing marks are still aligned, if not, align the camshaft sprocket timing marks facing each other. Align the crankshaft sprocket timing mark with the mark on the oil pump housing, then turn the crankshaft sprocket backward ½ notch.

8. Install the timing belt, starting at the crankshaft, then around the water pump sprocket, idler pulley, camshaft sprockets, and finally the tensioner pulley.

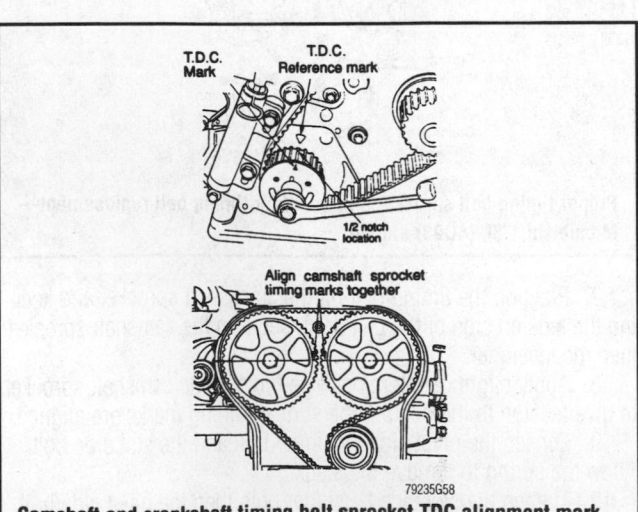

Camshaft and crankshaft timing belt sprocket TDC alignment mark positioning for timing belt removal and installation—Mitsubishi 1995–97 2.0L non-turbo engine

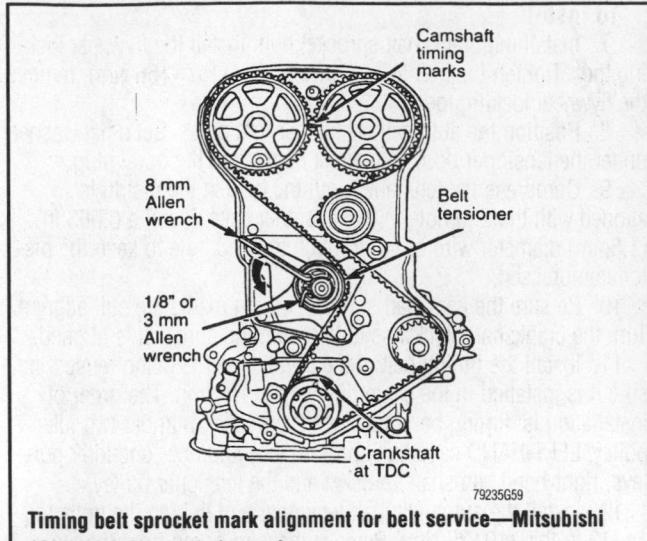

Timing belt sprocket mark alignment for belt service—Mitsubishi 1998–99 2.0L non-turbo engines

9. Turn the crankshaft sprocket ½ notch to TDC to take up the slack in the belt.

10. Install the tensioner on the engine, but do not tighten the bolts.

11. Place a torque wrench on the tensioner pulley and apply 21 ft. lbs. (28 Nm) of torque in the direction of the water pump. Push the tensioner up against the tensioner pulley and tighten the mounting bolts to 23 ft. lbs. (31 Nm).

12. Pull the pin out of the tensioner. Belt tension is correct when the pin can be removed and installed.

13. Rotate the crankshaft two revolutions and check the timing marks for alignment. Repeat the previous steps, if necessary.

14. Install the timing belt cover and all other applicable components.

2.0L Turbo Engine

1. Remove all necessary components for access to the timing belt covers.

2. Remove the stud bolt from the engine support bracket and remove the timing belt covers.

3. Rotate the crankshaft clockwise to line up the camshaft timing marks. Always turn the crankshaft in the normal direction of rotation only.

4. Loosen the tension pulley center bolt.

➡If the timing belt is to be reused, mark the direction of rotation on the flat side of the belt with an arrow.

5. Move the tension pulley towards the water pump and remove the timing belt.

6. Remove the crankshaft sprocket center bolt using special tool MB990767 to hold the crankshaft sprocket while removing the center bolt. Then, use MB998778 or equivalent puller to remove the sprocket.

7. Mark the direction of rotation on the timing belt B with a arrow.

8. Loosen the center bolt on the tensioner and remove the belt.

❊❊❊ WARNING

Do not rotate the camshafts or the crankshaft while the timing belt is removed.

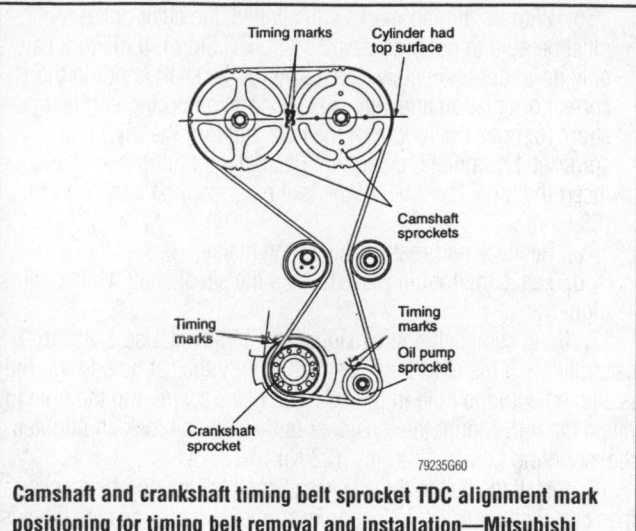

Camshaft and crankshaft timing belt sprocket TDC alignment mark positioning for timing belt removal and installation—Mitsubishi 2.0L turbo engine

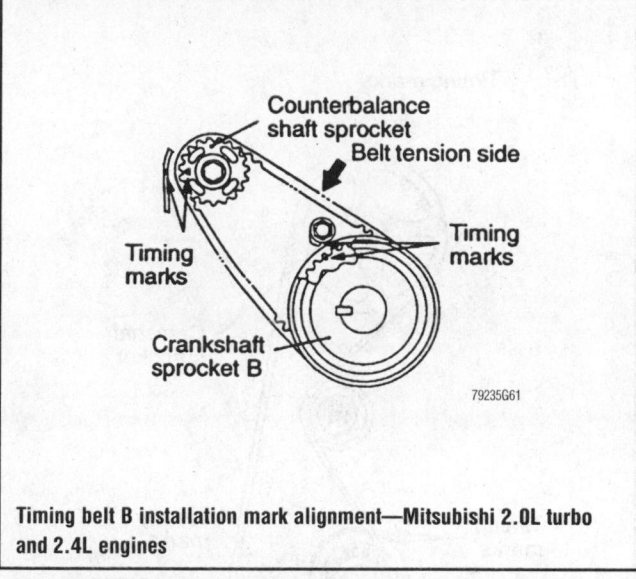

Timing belt B installation mark alignment—Mitsubishi 2.0L turbo and 2.4L engines

To install:

9. Place the crankshaft sprocket on the crankshaft. Use tool MB990767 or equivalent to hold the crankshaft sprocket while tightening the center bolt. Tighten the center bolt to 80–94 ft. lbs. (108–127 Nm).

10. Align the timing marks on the crankshaft sprocket B and the balance shaft.

11. Install timing belt B on the sprockets. Position the center of the tensioner pulley to the left and above the center of the mounting bolt.

12. Push the pulley clockwise toward the crankshaft to apply tension to the belt and tighten the mounting bolt to 14 ft. lbs. (19 Nm). Do not let the pulley turn when tightening the bolt because it will cause excessive tension on the belt. The belt should deflect 0.20–0.28 in. (5–7mm) when finger pressure is applied between the pulleys.

13. Install the crankshaft sensing blade and the crankshaft sprocket. Apply engine oil to the mounting bolt and tighten the bolt to 80–94 ft. lbs. (108–127 Nm).

14. Use a press or vise to compress the auto-tensioner pushrod. Insert a set pin when the holes are lined up.

✳✳ WARNING

Do not compress the pushrod too quickly, damage to the pushrod can occur.

15. Install the auto-tensioner on the engine.

16. Align the timing marks on the camshaft sprocket, crankshaft sprocket and the oil pump sprocket.

17. After aligning the mark on the oil pump sprocket, remove the cylinder block plug and insert a prytool in the hole to check the position of the counterbalance shaft. The prytool should go in at least 2.36 in. (60mm) or more, if not, rotate the oil pump sprocket once and realign the timing mark so the prytool goes in. Do not remove the prytool until the timing belt is installed.

18. Install the timing belt on the intake camshaft and secure it with a clip.

19. Install the timing belt on the exhaust camshaft. Align the timing marks with the cylinder head top surface using two wrenches. Secure the belt with another clip.

20. Install the belt around the idler pulley, oil pump sprocket, crankshaft sprocket and the tensioner pulley.

21. Turn the tensioner pulley so the pinholes are at the bottom. Press the pulley lightly against the timing belt.

22. Screw the special tool into the left engine support bracket until it contacts the tensioner arm, then screw the tool in a little more and remove the pushrod pin from the auto-tensioner. Remove the special tool and tighten the center bolt to 35 ft. lbs. (48 Nm).

23. Turn the crankshaft ¼ turn counterclockwise, then clockwise until the timing marks are aligned.

24. Loosen the center bolt. Install Mitsubishi Special Tool MD998767, or equivalent, on the tensioner pulley. Turn the tensioner pulley counterclockwise with a torque of 2.6 ft. lbs. (3.5 Nm) and tighten the center bolt to 35 ft. lbs. (48 Nm). Do not let the tensioner pulley turn when tightening the bolt.

25. Turn the crankshaft clockwise two revolutions and align the timing marks. After 15 minutes, measure the protrusion of the pushrod on the auto-tensioner. The standard measurement is 0.150–0.177 in (3.8–4.5mm). If the protrusion is out of specification, loosen the tensioner pulley, apply the proper torque to the belt and retighten the center bolt.

26. Install the timing belt covers and all applicable components.

2.4L Engine

1. Position the engine so that the No. 1 piston is at Top Dead Center (TDC).

2. Remove the timing belt covers.

➥If the timing belts are going to be reused, mark the direction of rotation on the belt. This will ensure the belt is reinstalled in same direction, extending belt life.

3. To loosen the timing (outer) belt tensioner, install Mitsubishi Special Tool MD998738 or equivalent, to the slot and screw inward to move the tensioner toward the water pump. Once the tension has been relieved, remove the outer timing belt.

4. If tensioner replacement is required, align the pin hole in the tensioner rod to the hole in the tensioner cylinder. Insert a 0.055 in. (1.4mm) wire in the hole and remove the special tool from the slot. With the cylinder tension relieved, remove the auto-tensioner cylinder assembly two mounting bolts.

5. Remove the outer crankshaft sprocket and flange.

6. Loosen the silent shaft (inner) belt tensioner and remove the belt.

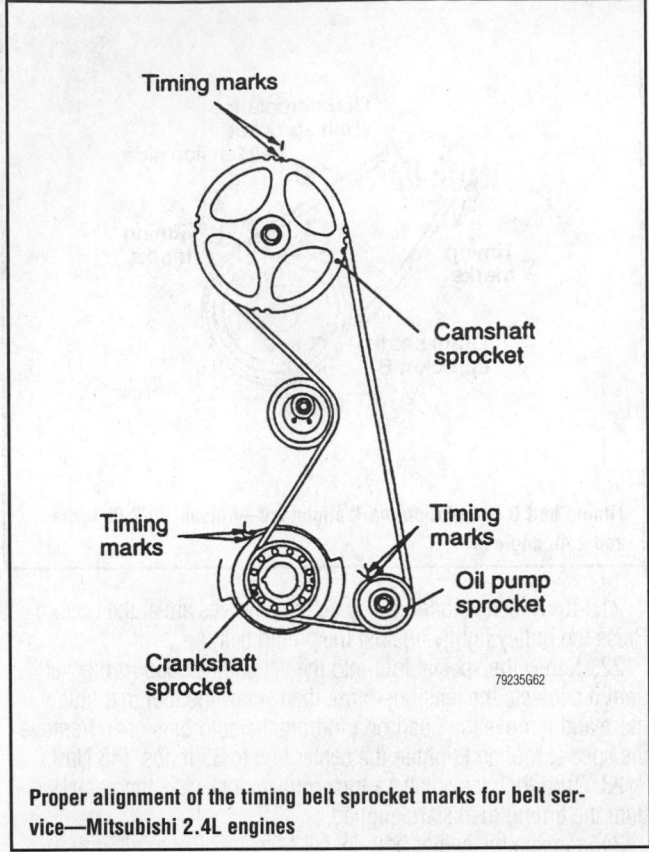

Timing marks

Camshaft sprocket

Timing marks

Timing marks

Oil pump sprocket

Crankshaft sprocket

79235G62

Proper alignment of the timing belt sprocket marks for belt service—Mitsubishi 2.4L engines

To install:

✳✳ WARNING

Do not spray or immerse the sprockets or tensioners in cleaning solvent. The sprocket may absorb the solvent and transfer it to the belt. The tensioners are internally lubricated and the solvent will dilute or dissolve the lubricant.

7. Align the timing marks of the silent shaft sprockets and the crankshaft sprocket with the timing marks on the front case. Route the timing belt around the sprockets so there is no slack in the upper span of the belt and the timing marks are still aligned.

8. Install the tensioner pulley and move the pulley by hand so the long side of the belt deflects approximately ¼ in. (6mm).

9. Hold the pulley tightly so the pulley cannot rotate when the bolt is tightened. Tighten the bolt to 14 ft. lbs. (19 Nm) and recheck the deflection.

10. Align the timing marks of the camshaft, crankshaft and oil pump sprockets with their corresponding marks on the front case or rear cover.

➡**There is a possibility to align all timing marks and have the oil pump sprocket and silent shaft out of time, causing an engine vibration during operation. If the following step is not followed exactly, there is a 50 percent chance that the silent shaft alignment will be 180 degrees (½ turn) off.**

11. Before installing the timing belt, ensure that the left side (rear) silent shaft (oil pump sprocket) is in the correct position as follows:

a. Remove the plug from the rear side of the block and insert a tool with shaft diameter of 0.31 in. (8mm) into the hole.

b. With the timing marks still aligned, the shaft of the tool must be able to go in at least 2 ½ in. (63.5mm). If the tool can only go in approximately 1 in. (25mm), the shaft is not in the correct orientation and will cause a vibration during engine operation. Remove the tool from the hole and turn the oil pump sprocket 1 complete revolution. Realign the timing marks and insert the tool. The shaft of the tool must go in at least 2 ¼ in. (63.5mm).

c. Recheck and realign the timing marks.

d. Leave the tool in place to hold the silent shaft while continuing.

12. If the camshaft belt tensioner was removed, use a vise to carefully push the auto-tensioner rod in until the set hole in the rod is aligned with the hole in the cylinder. Place a wire into the hole to retain the rod. Mount the tensioner to the engine block and tighten the mounting bolt to 17 ft. lbs. (23 Nm).

13. Install the belt to the crankshaft sprocket, oil pump sprocket, then camshaft sprocket, in that order. While doing so, be sure there is no slack between the sprocket except where the tensioner is installed.

14. To adjust the timing (outer) belt perform the following steps:

a. Turn the crankshaft ¼ turn counterclockwise, then turn it clockwise to move No. 1 cylinder to TDC.

b. Loosen the center bolt. Using tool MD998752 (or equivalent) and a torque wrench, apply a torque of 2.6 ft. lbs. (3.6 Nm) to the tensioner. Tighten the center bolt.

c. Screw the special tool into the engine left support bracket until its end makes contact with the tensioner arm. At this point, screw the special tool in some more and remove the set wire attached to the auto-tensioner, if the wire was not previously removed. Then, remove the special tool.

d. Rotate the crankshaft two complete turns clockwise and let it sit for approximately 15 minutes. Then, measure the auto-tensioner protrusion (the distance between the tensioner arm and auto-tensioner body) to ensure that it is within 0.15–0.18 in. (3.8–4.5mm). If out of specification, repeat Substeps a through d until the specified value is obtained.

➡**Do not manually overtighten the belt or it will howl.**

15. Install the upper and lower timing belt covers.

3.0L (6G72) SOHC Engine

1. Position the engine so the No. 1 cylinder is at TDC of its compression stroke.

✳✳ CAUTION

Wait at least 90 seconds after the negative battery cable is disconnected to prevent possible deployment of the air bag.

2. Remove all necessary components for access to the timing belt covers, then remove the covers from the engine.

3. If the same timing belt will be reused, mark the direction of the timing belt's rotation for installation in the same direction. Be sure the engine is positioned so the No. 1 cylinder is at the TDC of its compression stroke and the timing marks are aligned with the engine's timing mark indicators.

4. Loosen the timing belt tensioner bolt and remove the belt. If the tensioner is not being removed, position it as far away from the center of the engine as possible and tighten the bolt.

5. If the tensioner is being removed, mark the outside of the spring to ensure that it is not installed backwards. Unbolt the tensioner and remove it along with the spring.

✳✳ WARNING

Do not rotate the camshafts when the timing belt is removed from the engine. Turning the camshaft when the timing belt is removed could cause the valves to interfere with the pistons thus causing severe internal engine damage.

To install:

6. Install the tensioner, if removed, and hook the upper end of the spring to the water pump pin and the lower end to the tensioner in exactly the same position as originally installed.

7. Ensure both camshafts are still positioned so the timing marks align with those on the rear timing covers. Rotate the crankshaft so the timing mark aligns with the mark on the front cover.

8. Install the timing belt on the crankshaft sprocket and while keeping the belt tight on the tension side, install the belt on the front (left) camshaft sprocket.

9. Install the belt on the water pump pulley, then the rear (right) camshaft sprocket and the tensioner.

10. Loosen the bolt that secures the adjustment of the tensioner and lightly press the tensioner against the timing belt.

11. Check that the timing marks are in alignment.

12. Rotate the crankshaft 2 full turns in the clockwise direction only, then realign the timing marks.

13. Tighten the bolt that secures the tensioner to 19 ft. lbs. (26 Nm).

14. Install the lower and the upper timing belt covers, along with all other applicable components.

3.0L (6G72) DOHC Engine

1. Position the engine so the No. 1 cylinder is at TDC of its compression stroke.

2. Remove all necessary components for access to the timing belt covers, then remove the covers from the engine.

✳✳ CAUTION

Be sure to disconnect the negative battery cable. Wait at least 90 seconds after the negative battery cable is disconnected to prevent possible deployment of the air bag.

3. If the same timing belt will be reused, mark the direction of the timing belt's rotation for installation in the same direction. Be sure the engine is positioned so the No. 1 cylinder is at the TDC of its compression stroke and the timing marks are aligned with the engine's timing mark indicators on the rear timing covers.

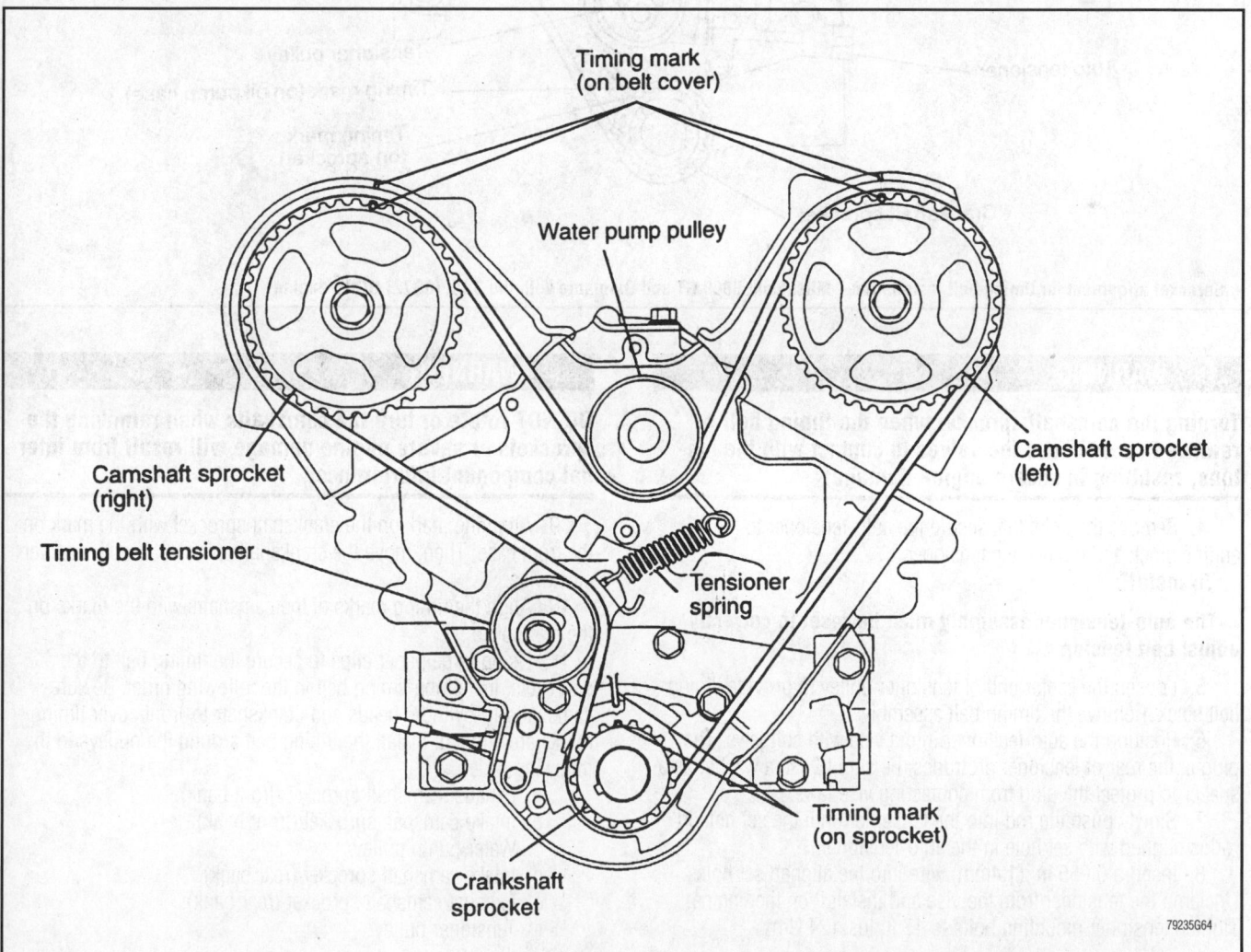

Align the sprockets properly before removing or installing the timing belt—Mitsubishi 3000 GT and Diamante with the 3.0L (6G72) SOHC engine

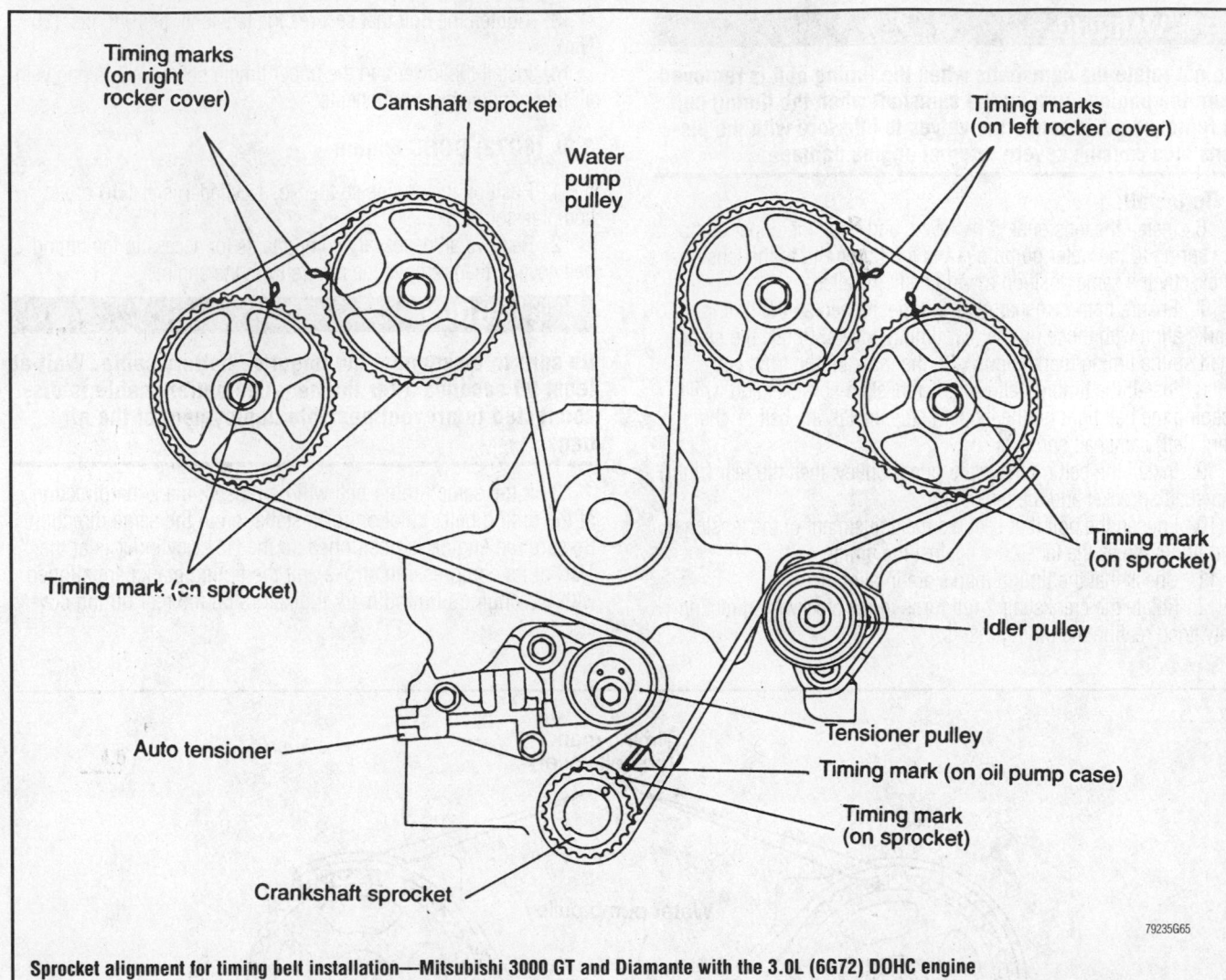

Sprocket alignment for timing belt installation—Mitsubishi 3000 GT and Diamante with the 3.0L (6G72) DOHC engine

---

✻✻ **WARNING**

Turning the camshaft sprocket when the timing belt is removed could cause the valves to contact with the pistons, resulting in severe engine damage.

4. Remove the bolts that secure the auto-tensioner to the engine block and remove the tensioner.

To install:

➡The auto-tensioner assembly must be reset to correctly adjust belt tension.

5. Loosen the center bolt of tensioner pulley to provide timing belt slack. Remove the timing belt assembly.

6. Position the auto-tensioner into a vise with soft jaws. The plug at the rear of tensioner protrudes, be sure to use a washer as a spacer to protect the plug from contacting vise jaws.

7. Slowly push the rod into the tensioner until the set hole in rod is aligned with set hole in the auto-tensioner.

8. Insert a 0.055 in. (1.4mm) wire into the aligned set holes. Unclamp the tensioner from the vise and install it on the engine. Tighten tensioner mounting bolts to 17 ft. lbs. (24 Nm).

✻✻ **WARNING**

DO NOT rotate or turn the camshafts when removing the sprockets or severe engine damage will result from internal component interference.

9. Align the mark on the crankshaft sprocket with the mark on the front case. Then, move the crankshaft sprocket 1 tooth counterclockwise.

10. Align the timing marks of the camshafts with the marks on the rear covers.

11. Using large paper clips to secure the timing belt to the sprockets, install the timing belt in the following order. Be sure camshafts-to-cylinder heads and crankshaft-to-front cover timing marks are aligned. Install the timing belt around the pulleys in the following order:

   a. Exhaust camshaft sprocket (front bank).
   b. Intake camshaft sprocket (front bank).
   c. Water pump pulley.
   d. Intake camshaft sprocket (rear bank).
   e. Exhaust camshaft sprocket (rear bank).
   f. Tensioner pulley.

g. Crankshaft pulley.

h. Idler pulley.

➡**Since the camshaft sprockets turn easily, secure them with box wrenches when installing the timing belt.**

12. Align all timing mark on the crankshaft and raise tensioner pulley against belt to remove slack, snug tensioner bolt.

13. Check the alignment of all the timing marks and remove the clips that secure the timing belt to the camshaft sprockets.

14. Rotate the engine ¼ turn counterclockwise, then rotate the engine clockwise to align the timing marks. Check that all the timing marks are in alignment.

15. Loosen the center bolt on the tensioner pulley. Using tool MD998752 or equivalent and a torque wrench, apply 84 inch lbs. (10 Nm) to the tool on the tensioner. Tighten the tensioner bolt to 35 ft. lbs. (49 Nm) and be sure the tensioner does not rotate with the bolt.

16. Rotate the crankshaft two complete turns clockwise and let it sit for approximately five minutes. Then, check that the set pin can easily be inserted and removed from the hole in the auto-tensioner.

17. Remove the set wire attached to the auto-tensioner.

18. Measure the auto-tensioner protrusion (the distance between the tensioner arm and auto-tensioner body) to ensure that it is within 0.15–0.18 in. (3.8–4.5mm). If out of specification, repeat adjustment procedure until the specified value is obtained.

19. Check again that the timing marks on all sprockets are in proper alignment.

20. Install the timing belt covers and all other applicable components.

NISSAN MODELS

3.0L (VG30DE) Engine

1. Remove the spark plugs and position the engine so that No. 1 piston is at TDC of the compression stroke.

2. Remove all necessary components for access to the timing belt covers, then remove the covers and gaskets.

❊❊❊ WARNING

After the timing belt has been removed, DO NOT rotate the camshafts or the crankshaft. Severe internal engine damage will result from piston and valve contact.

3. Install a suitable 6mm stopper bolt in the tensioner arm of the auto-tensioner so the length of the pusher does not change.

4. Remove the automatic tensioner and the timing belt.

5. Check the automatic tensioner for oil leaks in the pusher rod and diaphragm. If oil is evident, replace the automatic tensioner assembly.

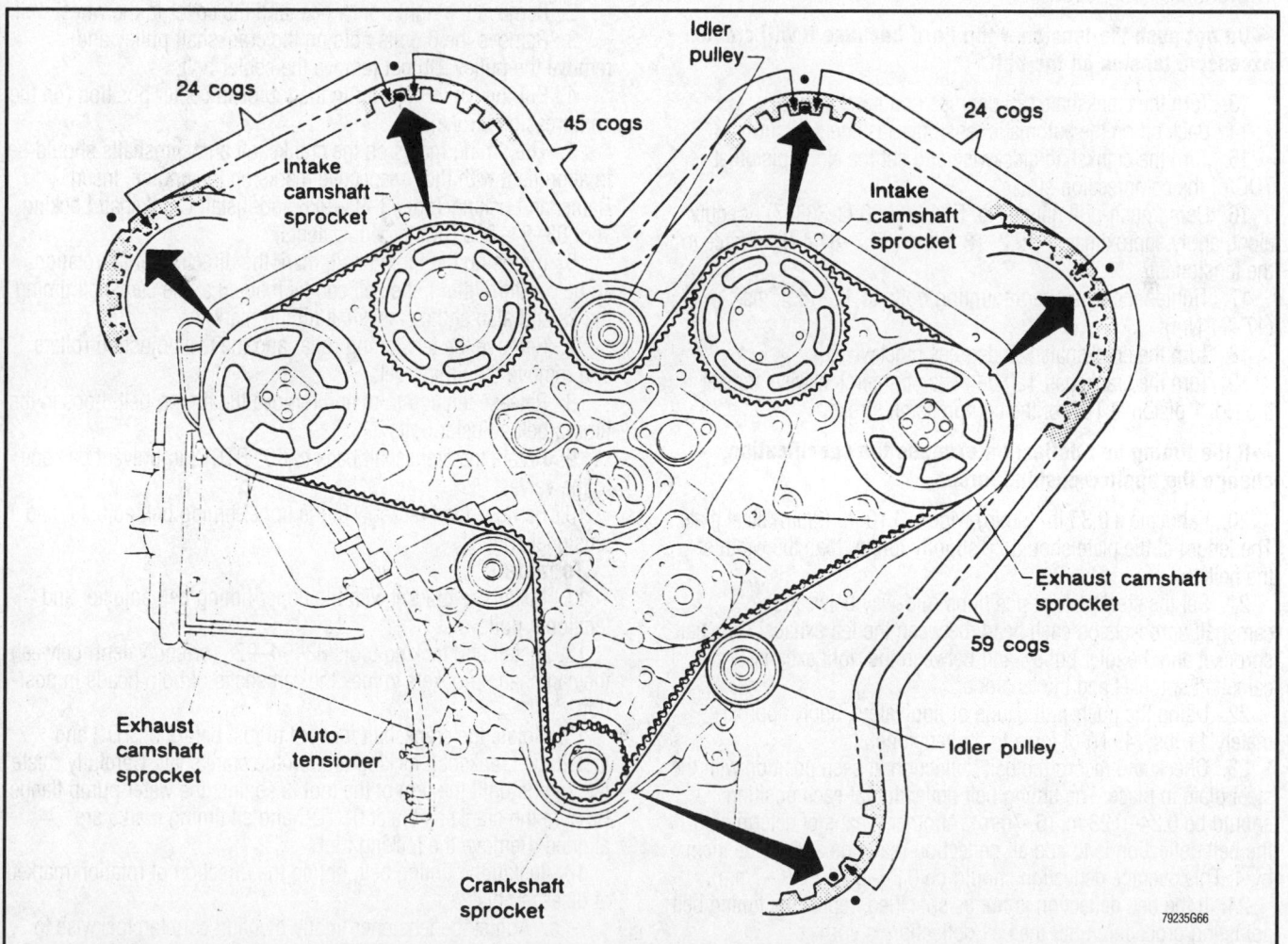

To ensure proper installation of the timing belt, be sure the proper number of belt teeth are between each sprocket, as indicated—Nissan 3.0L (VG30DE) engines

6. Inspect the timing gear teeth for wear.

To install:

7. Verify that the No. 1 piston is at TDC of the compression stroke.

8. Align the timing marks on the camshaft and crankshaft sprockets with the timing marks on the rear timing belt cover and the oil pump housing.

9. With a feeler gauge, check the clearance between the tensioner arm and the pusher of the automatic tensioner. The clearance should be 0.16 in. (4mm) with a slight drag on the feeler gauge. If the clearance is not as specified, mount the tensioner in a vise and adjust the clearance. When the clearance is set, insert the stopper bolt into the tensioner arm to retain the adjustment.

➡**When adjusting the clearance, do not push the tensioner arm with the stopper bolt fitted, because damage to the threaded portion of the bolt will result.**

10. Mount the automatic tensioner and tighten the nuts and bolts by hand.

11. Install the timing belt. Ensure the timing sprockets are free of oil and water. Do not bend or twist the timing belt. Align the white lines on the belt with the timing marks on the camshaft and crankshaft sprockets. Point the arrow on the belt towards the front.

12. Push the automatic tensioner slightly towards the timing belt to prevent the belt from slipping. At the same time, turn the crankshaft 10 degrees clockwise and tighten the tensioner fasteners to 12–15 ft. lbs. (16–21 Nm).

➡**Do not push the tensioner too hard because it will create excessive tension on the belt.**

13. Turn the crankshaft 120 degrees counterclockwise.

14. Back off on the automatic tensioner fasteners ½ turn.

15. Turn the crankshaft clockwise and set the No. 1 piston at TDC of the compression stroke.

16. Using push-pull gauge No. EG1486000 (J-38387) or equivalent, apply approximately 15.2–18.3 lbs. (67.7–81.4 N) of force to the tensioner.

17. Tighten the tensioner mounting bolts to 12–15 ft. lbs. (17–21 Nm).

18. Turn the crankshaft 120 degrees clockwise.

19. Turn the crankshaft 120 degrees counterclockwise and set the No. 1 piston at TDC of the compression stroke.

➡**If the timing belt deflection exceeds the specification, change the applied pushing force.**

20. Fabricate a 0.35 in. (9mm) wide x 0.10 in. (2mm) steel plate. The length of the plate should be slightly longer than the width of the belt.

21. Set the steel plate at positions mid-way between the camshaft sprockets on each head, between the left exhaust camshaft sprocket and the idler pulley, and between the right exhaust camshaft sprocket and the tensioner.

22. Using the push-pull gauge or equivalent, apply approximately 11 lbs. (49 N) of force to the tensioner.

23. Check and record the belt deflection at each position with the steel plate in place. The timing belt deflection at each position should be 0.24–0.28 in. (6–7mm). Another means of determining the belt deflection is to add all deflection readings and divide them by 4. This average deflection should be 0.24–0.28 in. (6–7mm).

24. If the belt deflection is not as specified, repeat the timing belt adjusting procedure until the belt deflection is correct.

25. Once the belt is properly bensioned, tighten the automatic tensioner fasteners to 12–15 ft. lbs. (16–21 Nm).

26. Remove the stopper bolt from the tensioner and wait 5 minutes. After 5 minutes, check the clearance between the tensioner arm and the pusher of the automatic tensioner. The clearance should remain at 0.138–0.205 in. (3.5–5.2mm).

27. Be sure the belt is installed and aligned properly on each pulley and the timing sprocket. There must be no slippage or misalignment.

28. Install the timing belt covers with new gaskets. Tighten the covers bolts to 2–4 ft. lbs. (3–5 Nm).

29. Install the remaining components in the reverse order of removal.

SAAB MODELS

2.5L (B258I) Engine

✳✳ WARNING

To avoid damage to the valves, DO NOT rotate the camshafts once the timing belt is removed. The crankshaft may only be turned between 0° and 60° BTDC when the camshafts are locked in position with the appropriate locking tool.

1. Remove the necessary components for access to the timing cover, then remove the cover.

2. Remove the right front wheel and the cover in the wheel well.

3. Remove the 6 bolts holding the crankshaft pulley and remove the pulley. Do not remove the center bolt.

4. Put the No. 1 cylinder in the top dead center position (on the compression stroke).

5. The timing marks on the crankshaft and camshafts should be in alignment with their respective marks on the engine. Insert Camshaft Locking Tool 83–94–926 and install Crankshaft Locking Tool 83–94–868 (or their equivalents).

6. If reusing the timing belt, mark the direction of its rotation. To help with refitting, the belt can be marked at the camshaft timing marks and also at the crankshaft timing mark.

7. Remove the tensioning roller and the two adjusting rollers and remove the timing belt.

8. Release tension from and remove the timing belt. Loosen the timing belt adjuster bolts.

9. Rotate the crankshaft back to 60° BTDC, to prevent damage to the valves.

10. Remove the bracket with the upper timing belt adjuster and tensioner rollers.

To install:

11. Install the bracket with the upper timing belt adjuster and tensioner pulleys.

12. Install and locking tools 83–94–926 (or equivalent) between the camshaft sprockets to lock the camshafts of both heads in position.

13. Rotate the crankshaft forward to just before 0° TDC and install the crankshaft locking tool on the crankshaft. Carefully rotate the engine until the arm of the tool is against the water pump flange. Be sure the crankshaft is at 0° TDC and all timing marks are aligned. Remove the locking tool.

14. Install the timing belt, noting the direction of rotation marked at disassembly.

 a. Adjust the tensioner lightly by hand counterclockwise to keep the timing belt from falling off. Be sure the crankshaft is a 0° TDC and all timing marks are aligned.

b. Install the camshaft locking tool.

c. Install tool (83–93–985) with a cut piece from an old timing belt, to measure belt tension.

d. Snug the center bolts of the adjusting rollers. Turn the lower adjusting roller counterclockwise, until a belt tension of 202–221 ft. lbs. (275–300 Nm) is registered on the torque wrench.

e. Tighten the adjusting roller center bolts to 30 ft. lbs. (40 Nm).

➡**This adjustment of the timing belt is only preparatory and should, therefore, not be used as a final check.**

15. Adjust the tensioner pulley until the marks are aligned. Tighten the tensioner pulley to 15 ft. lbs. (20 Nm). Remove the camshaft locking tool on camshaft sprockets 1 and 2. Adjust the upper adjusting roller until sprocket No. 2 moves 0.04–0.08 in. (1–2mm) clockwise. Tighten the upper adjusting roller to 30 ft. lbs. (40 Nm) and remove the upper locking tool.

16. Rotate the engine two complete revolutions to just before 0° TDC and install the locking tool on the crankshaft. Carefully turn the crankshaft until the arm of the locking tool is against the water pump flange and tighten the locking tool. Set the camshaft locking tool into position on the front of the camshaft sprockets. Be sure that the timing marks on the camshaft sprockets are aligned with the marks on the tool and that the edge of the timing belt is flush with the edge of the camshaft sprockets.

➡**Also check that the alignment marks on the tensioner pulley are still aligned.**

17. Install the timing belt cover. Tighten the bolts to 6 ft. lbs. (8 Nm). Install all of the remaining components.

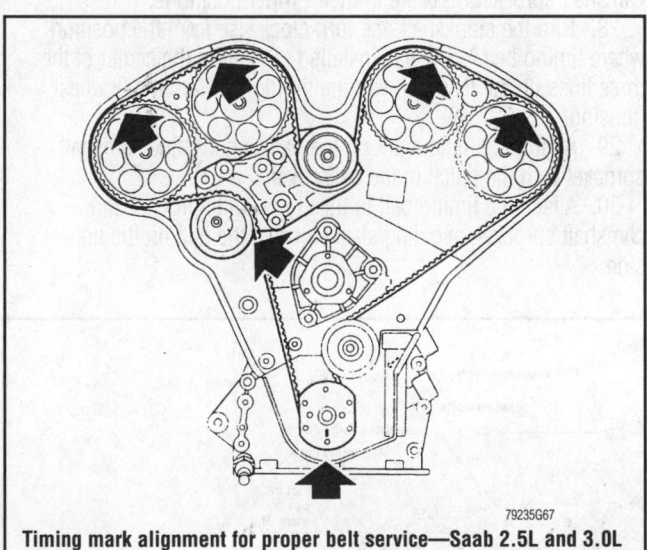

Timing mark alignment for proper belt service—Saab 2.5L and 3.0L engines

79235G67

3.0L (B308I) Engine

✳✳ WARNING

To avoid damage to the valves, DO NOT rotate the camshafts once the timing belt is removed from the engine. The crankshaft may only be turned between 0° and 60° BTDC when the camshafts are locked in position with the appropriate locking tool.

1. Remove all necessary components for access to the timing cover, then remove the cover.

➡**When removing the crankshaft pulley, remove the 6 outer bolts only, DO NOT remove the center bolt.**

2. Remove the crankshaft pulley.

3. Rotate the crankshaft to TDC of No. 1 cylinder.

4. The timing marks on the crankshaft and camshafts should be in alignment with their respective marks on the engine. Install camshaft locking tools (such as Saab tools KM-800–1 for camshaft sprockets 1 and 2 and KM-800–2 for sprockets 3 and 4) and a flywheel locking tool (such as Saab tool 83–94–868).

5. Mark the direction of rotation of the timing belt for reassembly.

6. Release tension from and remove the timing belt. Loosen the timing belt adjuster bolts.

7. Rotate the crankshaft back to 60° BTDC, to prevent damage to the valves.

8. Remove the bracket with the upper timing belt adjuster and tensioner rollers.

To install:

9. Remove the flywheel locking tool and install the flywheel inspection cover.

10. Install the bracket with the upper timing belt adjuster and tensioner pulleys.

11. Install both camshaft locking tools.

12. Rotate the crankshaft forward to just before 0° TDC and install the crankshaft locking tool on the crankshaft. Carefully rotate the engine until the arm of the tool is against the water pump flange. Be sure the crankshaft is at 0° TDC and all timing marks are aligned. Remove the locking tool.

13. If reusing the belt, fit the timing belt according to its marked direction of rotation and timing marks. Adjust the tensioning roller loosely by hand to prevent the belt from slipping out of the cogs. Always adjust counterclockwise.

14. Measure the belt tension with Saab tool 83–93–985, or equivalent.

15. Snug the center bolts of the adjusting rollers. Turn the lower adjusting roller counterclockwise, until a belt tension of 202–220 ft. lbs. (275–300 Nm) is reached. Tighten the adjusting roller center bolts to 30 ft. lbs. (40 Nm).

➡**This is a preliminary adjustment of the belt tension and must not be used as a check when the belt is finally adjusted.**

16. Continue to carry out the adjustment by means of the tensioning roller, mark against mark. Remove the locking tool for camshaft sprockets 1 and 2. Carry out the final adjustment with the upper center adjusting roller until camshaft sprocket No. 2 moves 0.04–0.08 in. (1–2mm) forward.

17. Remove the locking tool for camshaft sprockets 3 and 4 and also remove the crankshaft locking tool.

18. Tighten the tensioning roller to 15 ft. lbs. (20 Nm). Tighten the upper adjusting roller to 30 ft. lbs. (40 Nm) and tighten the lower adjusting roller to 15 ft. lbs. (20 Nm).

19. Rotate the engine two complete revolutions to just before 0° TDC and install the locking tool on the crankshaft. Carefully turn the crankshaft until the arm of the locking tool is against the water pump flange and tighten the locking tool. Set Saab tool KM-800–20 (or equivalent) into position. Be sure that the timing marks on the camshaft sprockets are aligned with the marks on the tool and that the edge of the timing belt is flush with the edge of the camshaft sprockets.

➡️**Also check that the alignment marks on the tensioner pulley are still aligned.**

20. If necessary, install the crankshaft pulley and tighten the retaining bolts to 15 ft. lbs. (20 Nm).

21. Install the timing belt cover, and tighten the bolts to 6 ft. lbs. (8 Nm). Install all of the remaining components in the reverse order of the removal procedure.

SUBARU MODELS

1.8L Engine

1. Remove all necessary components for access to the timing belt covers, then remove the covers.

2. Loosen the timing belt tensioner mounting bolts ½ turn and slacken the timing belt. Tighten the mounting bolts.

3. Mark the rotating direction of the No. 1 timing belt, then remove the belt.

4. Perform the same procedure for the No. 2 timing belt. Remove the crankshaft sprockets.

5. Remove both tensioners together with the tensioner springs.

6. Remove the belt idler. Remove the camshaft sprockets.

7. Remove the No. 2 belt covers.

To install:

8. Inspect the timing belt for breaks, cracks and wear. Replace as required.

9. Check the belt tensioner and idler for smooth rotation. Replace if noisy or excessive play is noticed.

10. Install the driver's side belt cover seal No. 3 to the cylinder block.

11. Install the driver's side belt cover seal, driver's side belt cover seal No. 4, and belt cover mount to the right rear belt cover, then install the assembly on the cylinder block. Tighten to 34 ft. lbs. (45 Nm).

12. Install the driver's side belt cover seal No. 2 and belt cover mounts to driver's side belt cover No. 2, then install to the cylinder head and camshaft case. Tighten to 34 ft. lbs. (45 Nm).

13. Install the passenger's side belt cover seal, belt cover seal No. 2 and belt cover mounts to the passenger's side belt cover No. 2, then install to the cylinder head and camshaft case. Tighten to 34 ft. lbs. (45 Nm).

14. Install the camshaft sprockets to the right and left camshafts. Tighten the bolts gradually in two or three steps to 67 ft. lbs. (91 Nm).

15. Attach the tensioner spring to the tensioner, then install to the right side of the cylinder block. Tighten the bolts temporarily by hand.

16. Attach the tensioner spring to the bolt, tighten the right side bolt, then loosen it ½ turn.

17. Push down the tensioner until it stops, then temporarily tighten the left bolt.

18. Install the left side tensioner in the same manner.

19. Install the belt idler to the cylinder block using care not to turn the seal. Tighten to 29–35 ft. lbs. (39–47 Nm).

20. Install the sprockets on the crankshaft. Install the crankshaft pulley and tighten the bolt temporarily.

21. Align the center of the three lines scribed on the flywheel with the timing mark on the flywheel housing.

22. Align the timing mark on the driver's side camshaft sprocket with the notch on the belt cover.

23. Attach timing belt No. 2 to the crankshaft sprocket No. 2, oil pump sprocket, belt idler, and camshaft sprocket in that order. Avoid downward slackening of the belt.

24. Loosen tensioner No. 2 lower bolt ½ turn to apply tension. Push timing belt by hand to ensure smooth movement of tensioner.

25. Apply 25 ft. lbs. (new belt) or 18 ft. lbs. (used belt) torque to the camshaft sprocket in counterclockwise direction. While applying torque tighten tensioner No. 2 lower bolt temporarily, then tighten upper bolt temporarily.

26. Tighten the lower bolt, then the upper bolt to 13–15 ft. lbs. (17–20Nm) in that order.

27. Check that the flywheel timing mark and driver's side camshaft sprocket mark are in their proper positions.

28. Turn the crankshaft one turn clockwise from the position where timing belt No. 2 was installed, and align the center of the three lines on the flywheel with the timing mark on the flywheel housing.

29. Align the timing mark on the passenger's side camshaft sprocket with the notch in the belt cover.

30. Attach the timing belt to the crankshaft sprocket and camshaft sprocket, avoiding slackening of the belt on the upper side.

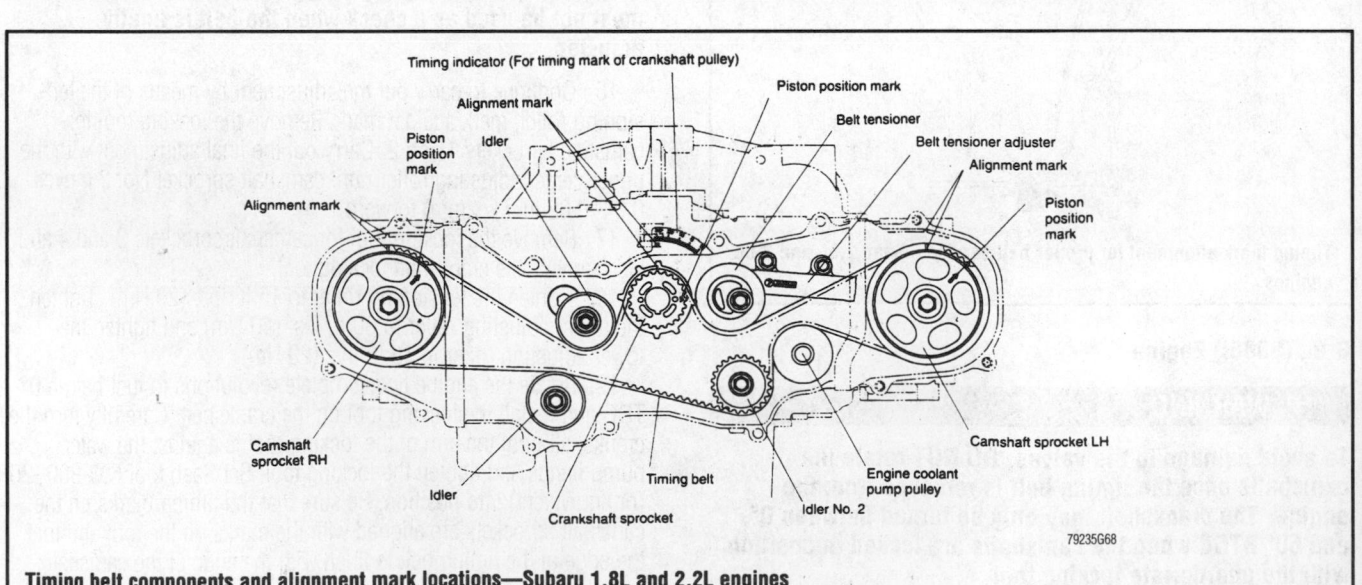

Timing belt components and alignment mark locations—Subaru 1.8L and 2.2L engines

31. Loosen the tensioner ½ turn to apply tension to the belt. Push the belt by hand to ensure smooth operation.

32. Apply 25 ft. lbs. (34 Nm) for new belts, or 18 ft. lbs. (24 Nm) for used belts, tighten to the camshaft sprocket in counterclockwise direction. While applying torque, tighten the tensioner left bolt temporarily, then tighten right bolt temporarily.

33. Tighten the left bolt, then the right bolt to 13–15 ft. lbs. (17–20 Nm) in that order.

34. Check that the flywheel timing mark and driver's side camshaft sprocket mark are in their proper positions.

35. Remove the crankshaft pulley.

36. Install the right front bolt cover seals and bolt cover plug. Install the belt covers to the cylinder block.

37. On turbo-charged engines, install the belt cover plate.

38. Install the crankshaft pulley and tighten to 66–79 ft. lbs. (89–107 Nm).

39. Install the water pump pulley and tighten to 67 ft. lbs. (91 Nm). Install the pulley cover, oil level guide and gauge and oil pressure switch connector.

40. Install and properly tension the accessory drive belt.

2.2L Engine

The engine uses a single cam belt drive system with a serpentine type belt. The left side of the engine uses a hydraulic cam belt tensioner which is self-adjusting.

➡**It is recommended that the timing belt be replaced at least every 60,000 miles (96,618 km).**

1. Disconnect the negative battery cable.
2. Position the No. 1 piston to TDC of its compression stroke.
3. Remove the engine drive belts.
4. Remove the timing belt covers.
5. Align the camshaft sprockets so each sprocket notch aligns with the cam cover notches. Align the crankshaft sprocket top tooth notch, located at the rear of the tooth, with the notch on the crank angle sensor boss. Mark the three alignment points as well as the direction of cam belt rotation.
6. Loosen the tensioner adjusting bolts. Remove the bottom three idlers, the cam belt and the cam belt tensioner. The cam sprockets can, then be removed with a modified camshaft sprocket wrench tool.
7. Remove the sprockets, if necessary. Note the reference sensor at the rear of the left cam sprocket.

To install:

8. Install the sprockets, if removed and tighten the retaining bolts to 47–54 ft. lbs. (64–74 Nm).
9. Install the crankshaft sprocket and the non-adjustable right side idler. Do not install the tensioner idler at this time.
10. Compress the hydraulic tensioner in a vise slowly and temporarily secure the plunger with a pin or suitable Allen wrench. Install the tensioner and the pulley with the adjustable idler pulley. Temporarily tighten the tensioner while the tensioner is pushed to the right.
11. Align the crankshaft sprocket notch on the rear sprocket tooth with the crank angle sensor boss. This places the sprocket notch in the 12 o'clock position.
12. Align the camshaft sprockets with the notches in the cam rear belt cover. This places the sprocket notch in the 12 o'clock position for each camshaft.
13. Install the timing belt with the directional mark and alignment marks properly positioned (if the belt was reused).

14. Loosen the tensioner retaining bolts and slide the tensioner to the left. Tighten the mounting bolts.

15. After verifying the timing marks are correct, remove the stopper pin from the tensioner.

16. Verify the correctness of the timing by noting that the notches on the 2 cam pulleys and the notch on the crankshaft pulley all point to the 12 o'clock position when the belt is properly installed.

17. Complete the engine component assembly by installing the cam belt covers, the crankshaft pulley bolt and pulley and the remaining components.

18. Connect the negative battery cable.

2.5L Engine

The engine uses a single cam belt drive system with a serpentine type belt. The left side of the engine uses a hydraulic cam belt tensioner which is self-adjusting.

➡**It is recommended that the timing belt be replaced at least every 60,000 miles (96,618 km).**

1. Remove all necessary components for access to the left, right and center timing belt covers, then remove the covers.
2. Align the camshaft sprockets so each sprocket notch aligns with the rear cover notches. Align the crankshaft sprocket top tooth notch, located at the rear of the tooth, with the notch. The crankshaft notch will be at 12 o'clock and the key way will be at 6 o'clock.

➡**Mark the sprocket alignment points as well as the direction of cam belt rotation for reinstallation purposes if the belt is to be reused.**

3. Loosen the tensioner adjusting bolts.
4. Remove the lower timing belt idler.
5. Remove the timing belt from the pulleys.

❊❊ WARNING

After the timing belt is removed, DO NOT rotate the camshaft sprockets or the crankshaft. Severe internal damage will result from the valve and/or piston contact.

6. Remove the timing belt tensioner and the timing belt tension adjuster.

To install:

➡**Inspect the timing belt and tensioner for wear or damage and replace as necessary.**

7. Inspect the timing belt tensioner as follows:
   a. When compressing the pushrod of the tensioner with a force of 33 lbs. (147 N), the tensioner should not sink.
   b. When compressing the pushrod of the tensioner with a force of 33–110 lbs. (147–490 N), the tensioner should not sink within 8.5 seconds.
   c. Measure the extension of the rod beyond the body of the tensioner for a length of 0.606–0.646 in. (15.4–16.4mm). If not within specifications, replace the tensioner.

➡**Check the idler sprockets for smooth operation. Replace as necessary.**

8. Using a press, compress the tensioner gradually, taking three minutes or more, and insert a 0.059 in. (1.5mm) pin to secure the rod.
9. Install the tensioner and the pulley with the adjustable idler pulley. Temporarily tighten the tensioner while the tensioner is pushed to the right.

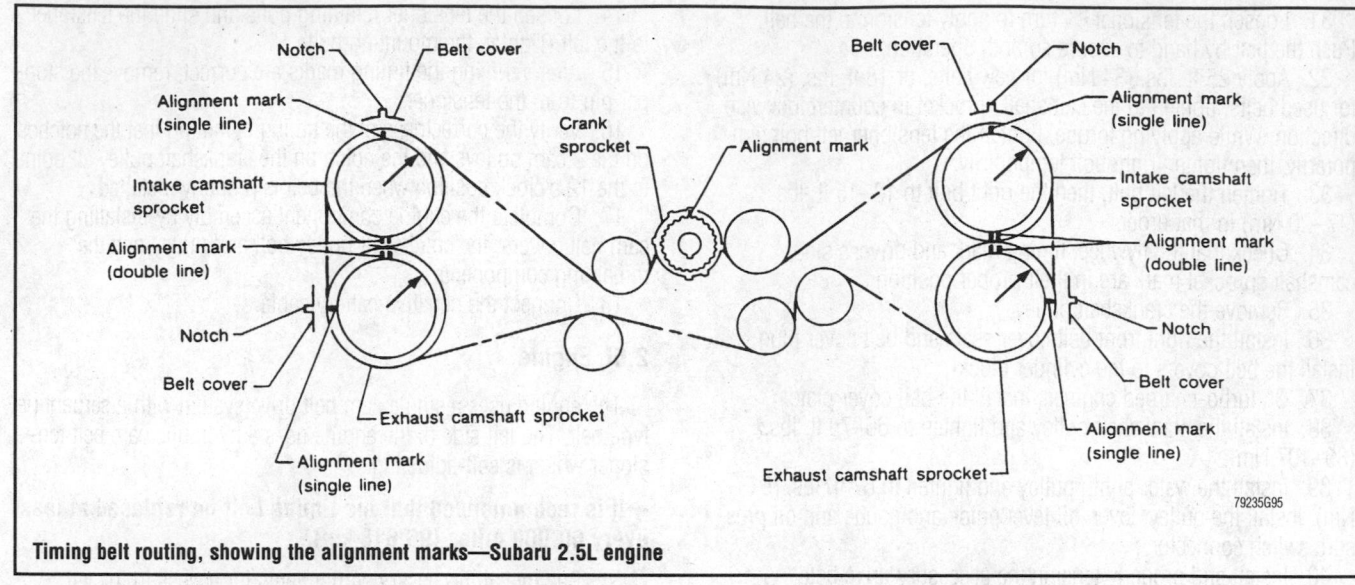

Timing belt routing, showing the alignment marks—Subaru 2.5L engine

10. Align the crankshaft sprocket notch on the rear sprocket tooth with the crank angle sensor boss. This places the sprocket notch in the 12 o'clock position and the key way at the 6 o'clock position.

11. Align the camshaft sprockets with the notches in the cam rear belt cover. This places the sprocket notch in the 12 o'clock position for each camshaft.

12. Install the timing belt in a clockwise direction starting at the crankshaft with the directional mark and alignment marks properly positioned (if the belt was reused).

13. Install the lower timing belt idler and tighten the mounting bolt to 29 ft. lbs. (39 Nm).

✳✳ WARNING

Be sure all the timing marks are properly aligned.

14. Loosen the tensioner retaining bolts and slide the tensioner to the left. Tighten the mounting bolts to 18 ft. lbs. (25 Nm).

15. After verifying the timing marks are correct, remove the stopper pin from the tensioner and recheck the timing marks.

16. Install the center, right, then the center timing belt covers. Tighten the bolts to 44 inch lbs. (5 Nm).

17. Install the remaining components in the reverse order of the removal procedure. When installing the crankshaft sprocket, be sure to tighten the mounting bolt to 94 ft. lbs. (127 Nm).

3.3L Engine

1. Disconnect the negative battery cable.
2. Remove the timing belt covers.
3. Matchmark the timing belt to the sprocket, cover and block marks as follows:

   a. Turn the crankshaft to align the timing marks on the crankshaft sprocket with the mark on the block.

   b. With the crankshaft marks aligned be sure the left and right camshaft sprocket marks are lined up with marks on the timing covers.

   c. If all the marks are in line, use white paint to mark the direction of rotation of the belt as well as mark the spots on the belt where it crosses over the timing marks on the pulleys.

4. Loosen the belt tensioner bolts.

5. Remove belt idler pulley No. 1.
6. Remove belt idler pulley No. 2.
7. Remove the timing belt.
8. Remove the tensioner pulley bolt and remove the tensioner pulley.
9. Remove the two bolts and the tensioner assembly.

To install:

10. Insert a 0.059 in. (1.5mm) diameter stopper pin into place while pushing the tension adjuster rod into the tensioner body.

11. Install the tensioner and tighten the bolts to 18 ft. lbs. (24 Nm), while the tensioner is pushed all the way to the right.

12. Install the tensioner pulley and mounting bolt. DO NOT tighten the idler pulley bolt completely.

13. Be sure the crankshaft and both camshaft sprockets are still lined up with their respective timing marks.

14. Install the timing belt onto the sprockets with the direction of rotation arrow in the correct direction and the timing marks on the belt in line with the marks on the sprockets.

15. Install the number 1 and 2 idler pulleys and tighten the mounting bolts to 29 ft. lbs. (39 Nm).

16. Loosen the tensioner pulley bolt and the tensioner assembly mounting bolts. Slide the tensioner assembly all the way to the left and tighten the bolts to 18 ft. lbs. (24 Nm).

17. Check again that all the timing marks are still in alignment. If they are remove the stopper pin from the tensioner assembly.

18. Install the timing belt covers.
19. Connect the negative battery cable.

SUZUKI MODELS

1.3L Engine

1. Remove all necessary components for access to the upper and lower timing belt outside covers, then remove the covers.

2. Align the camshaft timing belt pulley(s) with their timing marks. The crankshaft mark and camshaft marks are straight up.

3. On single overhead cam models, remove the resonator and the timing belt outside cover.

4. Remove the tensioner stud and loosen the tensioner bolt.

5. Remove the tensioner spring and damper, then remove the timing belt.

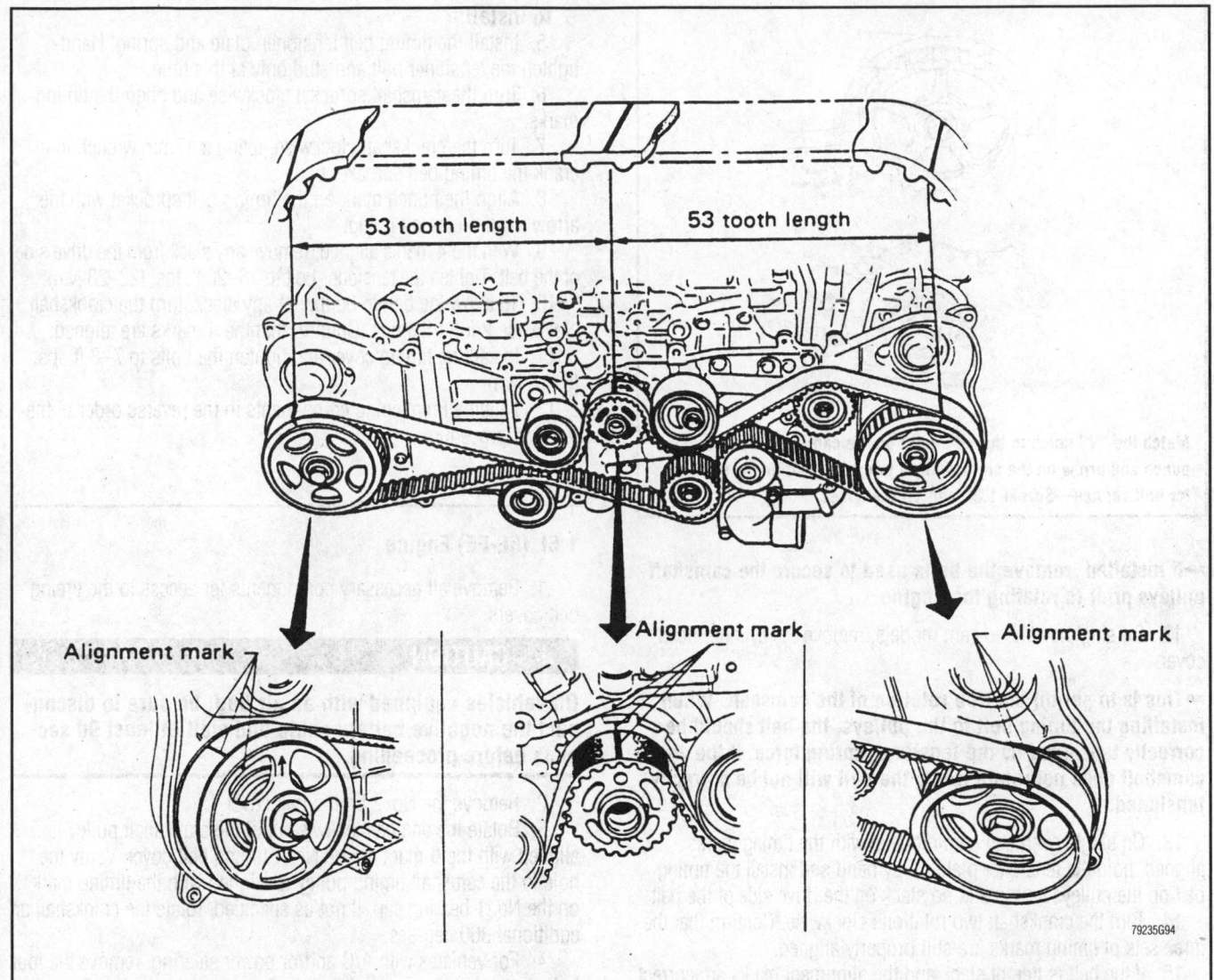

53 tooth length

53 tooth length

Alignment mark

Alignment mark

Alignment mark

79235G94

To ensure proper installation of the timing belt, be sure the proper number of belt teeth are between each sprockets, as indicated—Subaru 3.3L engine

✳✳✳ WARNING

After the timing belt is removed never turn the camshafts or the crankshaft. Interference may occur between the pistons and the valves causing component damage.

6. Remove the tensioner and the tensioner plate.

To install:

7. Install the timing belt tensioner plate and tensioner. Only hand-tighten the tensioner bolt.

➡**Be sure that the lug on the tensioner plate is inserted into the hole on the tensioner.**

8. Be sure the tensioner plate and the tensioner move uniformly. If they do not move together remove the tensioner and the tensioner plate and reinsert the plate lug into the tensioner hole.

9. Check the camshaft sprockets to verify that they have not moved.

➡On dual overhead cam models, if the sprockets will not stay aligned due to valve spring tensions they can be positioned by using four 8mm bolts and two flanged nuts. Install two bolts into the holes on the head in between the camshafts. Take and install the nuts with the flanges on two other bolts, the flange must face away from the head of the bolt. Position the head of the second pair of bolts on the threaded section of the first bolts, the nut should be facing up. The nut can be positioned into a groove on the appropriate camshaft sprocket so the sprocket alignment can be adjusted. Turn the flanged nut without having the sprocket resting on the nut so the sprocket is not damaged.

10. Check the crankshaft alignment by verifying that the punch mark on the timing belt pulley(s) is aligned with the arrow on the oil pump case.

11. On dual overhead cam models, install the timing belt on the three pulleys in such a way that there is no slack in the belt. Install the spring and the spring damper and hand-tighten the tensioner stud.

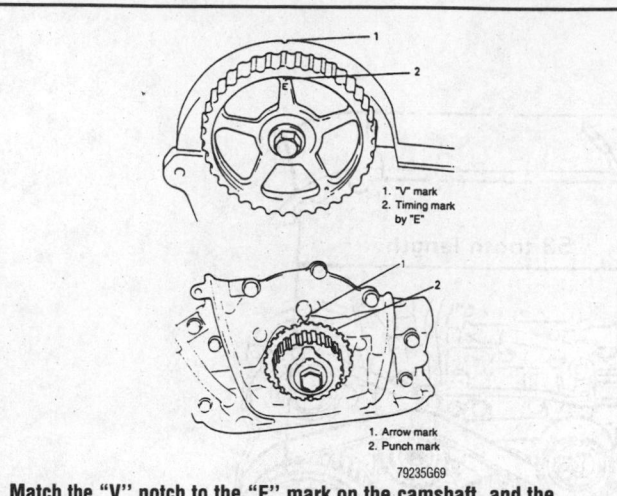

1. "V" mark
2. Timing mark by "E"

1. Arrow mark
2. Punch mark

79235G69

Match the "V" notch to the "E" mark on the camshaft, and the punch and arrow on the crankshaft to properly position the engine for belt service—Suzuki 1.3L and 1.6L engines

➡ **If installed, remove the bolts used to secure the camshaft pulleys prior to rotating the engine.**

12. On single overhead cam models, remove the cylinder head cover.

➡ **This is to permit the free rotation of the camshaft. When installing the timing belt to the pulleys, the belt should be correctly tensioned by the tensioner spring force. If the camshaft does not rotate freely the belt will not be correctly tensioned.**

13. On single overhead cam models, with the timing marks aligned, hold the tensioner plate up by hand and install the timing belt on the pulleys so there is no slack on the drive side of the belt.

14. Turn the crankshaft two rotations clockwise. Confirm that the three sets of timing marks are still properly aligned.

15. If the belt is free of slack and the alignment marks are correct tighten the tensioner stud to 7–8 ft. lbs. (9–12 Nm). Tighten the tensioner bolt to 17–21 ft. lbs. (24–30 Nm).

16. Install the timing belt upper and lower outside covers. Tighten the timing cover bolts to 7–8 ft. lbs. (9–12 Nm).

17. Install all remaining components in the reverse order of the removal procedure.

1.6L Engine

1. Remove all necessary components for access to the timing belt covers, then remove the covers. 2. Loosen but do not remove the tensioner bolt.

❊❊ CAUTION

After the timing belt is removed, never turn the camshaft and crankshaft independently. This engine is an interference engine and if the camshaft or crankshaft is turned beyond a certain point, damage to the valves could occur.

3. Loosen the timing belt tensioner adjusting bolt and pivot nut. Hold pressure on the tensioner to loosen the timing belt and remove the timing belt from the camshaft and crankshaft sprockets.

4. Remove the timing belt tensioner, tensioner plate and tensioner spring.

To install:

5. Install the timing belt tensioner, plate and spring. Hand-tighten the tensioner bolt and stud only at this time.

6. Turn the camshaft sprocket clockwise and align the timing marks.

7. Turn the crankshaft clockwise, using a 17mm wrench to crank the timing belt sprocket bolt.

8. Align the punch mark on the timing belt sprocket with the arrow mark on the oil pump.

9. With the 4 marks aligned, remove any slack from the drive side of the belt. Tighten the tensioner bolt to 16–20 ft. lbs. (22–28 Nm).

10. To allow the belt to be free of any slack, turn the crankshaft clockwise 2 full rotations. Confirm that the 4 marks are aligned.

11. Install the timing cover and tighten the bolts to 7–8 ft. lbs. (9–12 Nm).

12. Install all remaining components in the reverse order of the removal procedure.

TOYOTA MODELS

1.5L (5E-FE) Engine

1. Remove all necessary components for access to the timing belt covers.

❊❊ CAUTION

On vehicles equipped with an air bag, be sure to disconnect the negative battery cable and wait at least 90 seconds before proceeding.

2. Remove the No. 2 timing belt cover.

3. Rotate the engine clockwise until the crankshaft pulley is aligned with the **0** mark on the No. 1 timing belt cover. Verify the hole in the camshaft timing pulley is aligned with the timing mark on the No. 1 bearing cap. If not as specified, rotate the crankshaft an additional 360 degrees.

4. For vehicles with A/C and/or power steering, remove the four bolts to the No. 2 crankshaft pulley. Then, remove the No. 2 crankshaft pulley.

5. Using Toyota Tools SST 09213–14010 and SST 09330–00021 (or their equivalents), remove the No. 1 crankshaft pulley bolt.

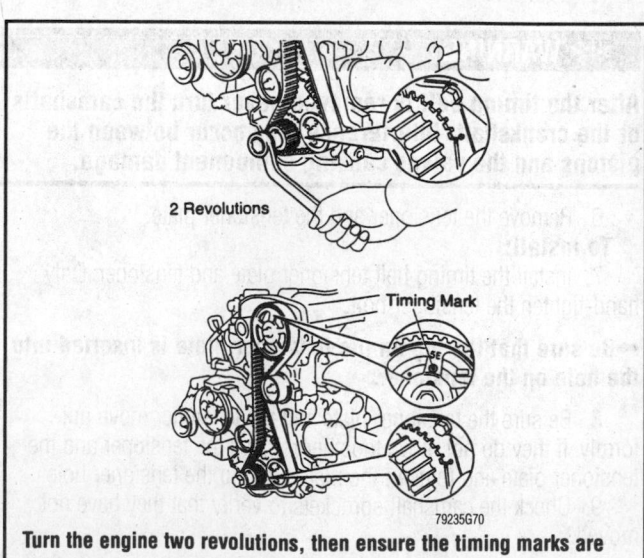

2 Revolutions

Timing Mark

79235G70

Turn the engine two revolutions, then ensure the timing marks are still aligned—Toyota 1.5L (5E-FE) engine

6. Using a crankshaft pulley/damper puller (such as Toyota Tool SST 09950–50010), remove the No. 1 crankshaft pulley from the crankshaft.

7. Remove the No. 3 timing belt cover (plug).

8. Remove the No. 1 timing belt cover and timing belt guide.

9. Place matchmarks on the timing belt on both sides of the cam and crankshaft gear timing marks. Also, place an arrow on the top surface of the belt to indicate the direction of travel.

10. Using pliers, remove the tension spring.

11. Loosen the No. 1 idler pulley bolt and push the pulley to the left as far as it will go, then temporarily tighten the bolt.

12. Remove the belt.

To install:

13. For vehicle with a distributor (distributor ignition), use the crankshaft bolt to turn the crankshaft until the timing marks on the sprocket and oil pump body align. This is method is used to set the piston at TDC before the marks on the belt cover can be seen.

14. For vehicles with a crankshaft position sensor (distributor-less ignition), use the crankshaft bolt to turn the crankshaft until the rotor side of the crankshaft position sensor faces inward.

15. Turn the camshaft and align the hole of the camshaft timing pulley with the timing mark of the bearing cap. The matchmarks, if using the old belt should line up. Place the belt over the crankshaft pulley and the idler pulleys.

16. Install the belt on the crankshaft gear (using the matchmarks if reinstalling the old belt) and install the timing belt guide with flange out.

17. Loosen the No. 1 idler pulley bolt until the pulley is moved slightly by the spring tension.

18. Turn the crankshaft pulley two revolutions from TDC to TDC.

➡**Always rotate the crankshaft clockwise.**

19. Check that the pulleys align with the reference marks. If not, reinstall the belt.

20. When the timing is verified, tighten the adjuster pulley (No. 1 idler pulley) to 13 ft. lbs. (18 Nm).

21. Install No. 1 and No. 3 lower timing belt covers.

22. Install the No. 1 crankshaft pulley and tighten the pulley bolt to 112 ft. lbs. (152 Nm).

23. Tighten the four No. 2 crankshaft pulley bolts to 14 ft. lbs. (19 Nm).

24. Install the No. 2 timing belt cover with the four bolts.

25. Install the remaining components in the reverse order of the removal procedure. When installing the right-hand engine mounting insulator, tighten the bracket bolt to 47 ft. lbs. (64 Nm) and the through-bolt to 54 ft. lbs. (73 Nm).

1.6L (4A-FE) and 1.8L (7A-FE) Corolla engines

1. Remove all necessary components for access to the timing belt covers.

> ❊❊ **CAUTION**
>
> **On vehicles equipped with an air bag, be sure to disconnect the negative battery cable and wait at least 90 seconds before proceeding.**

2. Turn the crankshaft to align the timing mark on crankshaft pulley at **0** , setting the piston in No. 1 cylinder at Top Dead Center (TDC) on the compression stroke. Check that the valve lifters on the No. 1 cylinder are loose. If not, turn crankshaft pulley 1 complete revolution (360 degrees).

3. Remove the nine bolts and timing belt covers from the engine.

4. Slide the timing belt guide from crankshaft.

5. Set the camshaft and crankshaft timing sprockets to align the marks.

➡**Do not turn crankshaft or camshaft independently after removal of timing belt. Binding or damage to engine components could result. If the timing belt is to be reused, mark timing belt with arrow showing direction of engine revolution. Put matchmarks where timing belt meets with crankshaft timing sprocket and camshaft timing sprocket to ensure installation in the same position.**

6. Remove the timing belt tensioner bolt, tensioner, and the tension spring.

7. Remove the timing belt from camshaft and crankshaft timing sprockets.

> ❊❊ **WARNING**
>
> **Do not bend, twist or turn the timing belt inside out. Do not allow the belt to come in contact with oil, coolant or steam.**

To install:

➡**Inspect the camshaft timing sprocket to ensure mark is still aligned as indicated.**

8. Reinstall the timing belt tensioner and the tension spring. Pry the tensioner to the left as far as it will go and temporarily tighten the retaining bolt.

9. Install the timing belt. If reinstalling the old belt, observe the matchmarks made during removal. Be sure the belt is fully and squarely seated on the sprockets.

10. Loosen the retaining bolt for the timing belt tensioner and allow it to tension the belt.

11. Temporarily install the crankshaft pulley bolt and turn the crank clockwise 2 full revolutions from TDC to TDC. Insure that each timing mark realigns exactly.

12. Tighten the timing belt tensioner bolt to 27 ft. lbs. (37 Nm).

13. Measure the timing belt deflection at the **SIDE** point, looking for ¼ in. (5–6mm) of deflection at 4.4 lbs. (2 kg) of pressure. If the deflection is not correct, adjust with the timing belt tensioner.

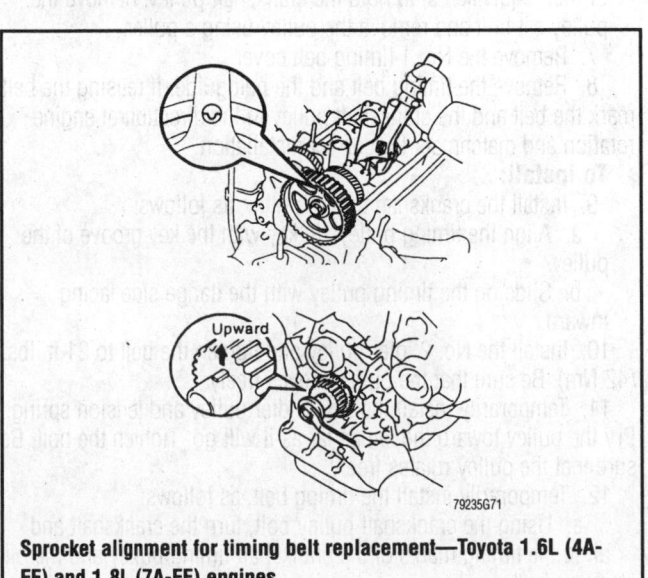

Sprocket alignment for timing belt replacement—Toyota 1.6L (4A-FE) and 1.8L (7A-FE) engines

14. Install the timing belt guide, with the cup side facing outward, onto the crankshaft and install the timing belt covers from the lowest to the highest. Tighten the nine cover bolts to 62 inch lbs. (7 Nm).

15. Install all applicable remaining components. During assembly, be sure to tighten the crankshaft pulley bolt to 87 ft. lbs. (118 Nm), the mounting bracket-to-engine mount bolt to 47 ft. lbs. (64 Nm), the mounting bracket-to-engine mount nuts to 38 ft. lbs. (52 Nm), the engine mount-to-body bolt **A** to 19 ft. lbs. (25 Nm), the engine mount-to-body bolts **B** to 19 ft. lbs. (25 Nm), the engine mount to body bolt **C** to 19 ft. lbs. (25 Nm), if equipped with cruise control.

2.2L (5S-FE) Engines

1. Remove all necessary components for access to the timing belt covers.

❊❊ CAUTION

On vehicles equipped with an air bag, be sure to disconnect the negative battery cable and wait at least 90 seconds before proceeding.

2. Remove the No. 2 timing cover.
3. Position the No. 1 cylinder to TDC on the compression stroke by turning the crankshaft pulley and aligning its groove with the timing mark **0** of the No. 1 timing belt cover. Check that the hole of the camshaft timing pulley is aligned with the alignment mark of the bearing cap. If not, turn the crankshaft one revolution (360 degrees).
4. Remove the timing belt from the camshaft timing pulley, as follows:

   a. If reusing the belt, place matchmarks on the timing belt and the camshaft pulley. Loosen the mount bolt of the No. 1 idler pulley and position the pulley toward the left as far as it will go. Tighten the bolt. Remove the belt from the camshaft pulley.
5. Remove the camshaft timing pulley.

   a. Using Toyota tools Nos. 09249–63010 and 09960–10010, or their equivalents, remove the bolt and the camshaft pulley.
6. Remove the crankshaft pulley.

   a. Using Toyota tools Nos. 09213–54015 and 09330–00021, or their equivalents, to hold the crankshaft pulley. Remove the pulley set bolt and remove the pulley using a puller.
7. Remove the No. 1 timing belt cover.
8. Remove the timing belt and the belt guide. If reusing the belt mark the belt and the crankshaft pulley in the direction of engine rotation and matchmark for correct installation.

   To install:
9. Install the crankshaft timing pulley, as follows:

   a. Align the timing pulley set key with the key groove of the pulley.

   b. Slide on the timing pulley with the flange side facing inward.
10. Install the No. 2 idler pulley and tighten the bolt to 31 ft. lbs. (42 Nm). Be sure that the pulley moves freely.
11. Temporarily install the No. 1 idler pulley and tension spring. Pry the pulley toward the left as far as it will go. Tighten the bolt. Be sure that the pulley rotates freely.
12. Temporarily install the timing belt, as follows:

   a. Using the crankshaft pulley bolt, turn the crankshaft and align the timing marks of the crankshaft timing pulley and the oil pump body.

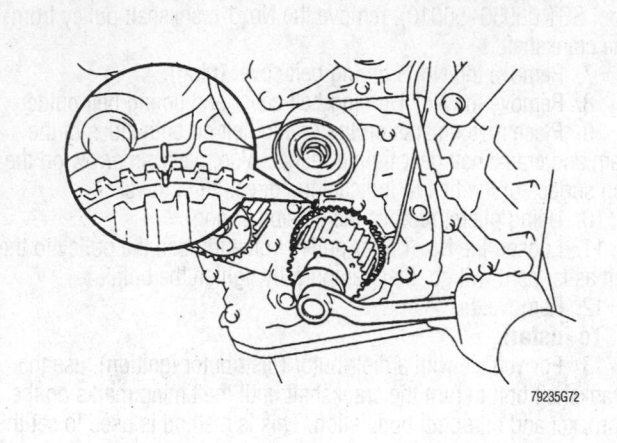

Crankshaft positioning for timing belt removal and installation— Toyota 2.2L (5S-FE) engine

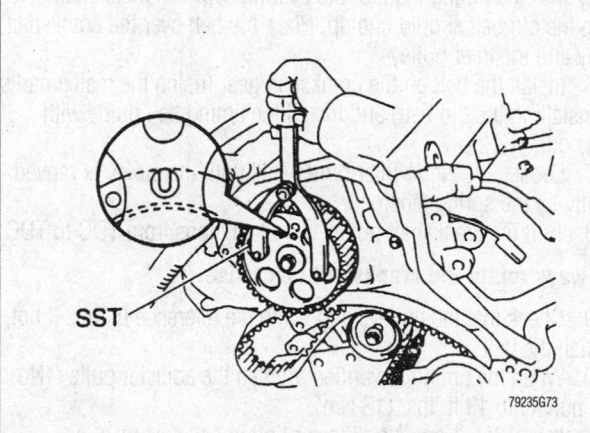

Using a spanner wrench, turn the camshaft into position so that the alignment mark is visible through the hole in the sprocket—Toyota 2.2L (5S-FE) engine

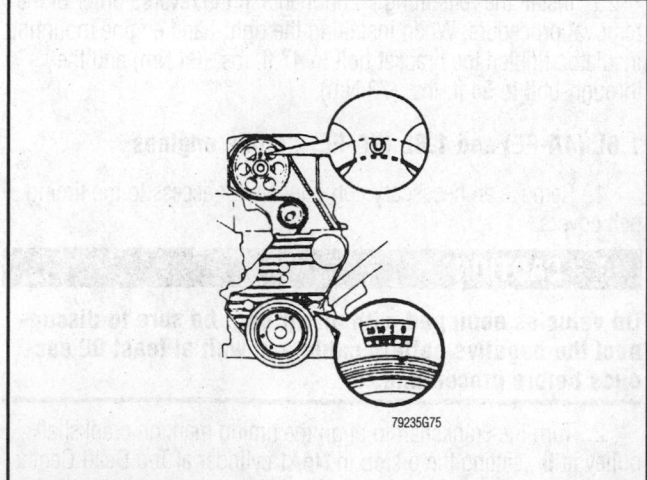

Sprocket alignment for timing belt replacement—Toyota 2.2L (5S-FE) engine

b. If reusing the old belt, align the marks made during removal, and install the belt with the arrow pointing in the direction of the engine revolution.

13. Install the timing belt guide with the cup side facing outward.

14. Install the No. 1 timing belt cover.

15. Install the crankshaft pulley. Align the pulley set key with the key groove of the pulley and slide on the pulley. Tighten the bolt to 80 ft. lbs. (108 Nm).

16. Install the camshaft timing pulley.

a. Align the camshaft knock pin with the knock pin groove of the pulley and slide on the timing pulley. Tighten the bolt to 40 ft. lbs. (54 Nm).

17. With the No. 1 cylinder set at TDC on the compression stroke install, the timing belt (all timing marks aligned). If reusing the belt, align with the marks made during the removal procedure.

a. Turn the crankshaft pulley, and align its groove with the timing mark **0** of the No. 1 timing belt cover. Be sure the camshaft sprocket hole is aligned with the mark on the bearing cap.

18. Connect the timing belt to the camshaft timing pulley.

19. Check that the matchmark on the timing belt matches the end of the No. 1 timing belt cover.

20. Once the belt is installed be sure that there is tension between the crankshaft timing pulley and the camshaft pulley.

21. Check the valve timing.

a. Loosen the No. 1 idler pulley mount bolt ½ turn. Turn the crankshaft pulley two revolutions from TDC in the clockwise direction. Always turn the crankshaft pulley clockwise.

b. Be sure that the all the timing marks are aligned.

c. Slowly turn the crankshaft pulley 1⅞ revolutions. Align its groove with the mark at 45° BTDC on the No. 1 timing belt cover for the No. 1 cylinder.

d. Tighten the No. 1 idler pulley mount bolt to 31 ft. lbs. (42 Nm).

22. Install the No. 2 timing belt cover.

a. Install the upper gasket to the No. 1 timing belt cover.

b. Disconnect the engine wire protector between the cylinder head cover and the No. 3 timing belt cover.

c. Install the gasket to the timing belt cover.

d. Install the belt covers and all remaining components. During assembly, tighten the right engine mount bracket bolts to 38 ft. lbs. (52 Nm), the engine mount insulator bolt to 47 ft. lbs. (64 Nm), the through-bolt to 54 ft. lbs. (78 Nm), the power steering reservoir bracket bolt to 21 ft. lbs. (28 Nm), the power steering reservoir-to-bracket bolt to 27 ft. lbs. (37 Nm) and the nut to 38 ft. lbs. (52 Nm).

3.0L (1MZ-FE) Engine

1. Remove all necessary components for access to the timing belt covers.

> ✳✳ **CAUTION**

On vehicles equipped with an air bag, be sure to disconnect the negative battery cable and wait at least 90 seconds before proceeding.

2. Remove the lower timing belt cover by removing the four bolts.

3. Remove the No. 2 timing belt cover as follows:

a. Remove the bolt and disconnect the engine wire protector from the No. 3 (rear) timing belt cover.

b. Disconnect the engine wire protector clamp from the No. 3 timing belt cover.

c. Remove the five bolts from the No. 2 timing belt cover.

d. Remove the No. 2 cover from the engine.

4. Remove the right engine mounting bracket by removing the nut and two bolts.

5. Remove the crankshaft timing belt guide.

6. Temporarily install the crankshaft pulley bolt.

7. Turn the crankshaft and align the crankshaft timing pulley groove with the oil pump alignment mark. Always turn the engine clockwise.

8. Ensure the timing mark of the camshaft timing pulleys and rear timing belt cover are aligned. If not, turn the engine over an additional 360 degrees (one revolution).

9. Remove the crankshaft pulley bolt.

➡ **If the belt is to be reused, align the installation marks on the belt to the marks on the pulleys. If the marks have worn off, make new ones.**

10. Alternately loosen the two timing belt tensioner bolts. Remove the tensioner and dust boot.

11. Remove the timing belt.

To install:

12. Remove any oil or water from the pulleys.

13. Align the front mark of the timing belt with the dot mark of the crankshaft timing pulley.

14. Align the installation marks on the timing belt with the timing marks of the camshaft pulleys.

15. Install the timing belt in the following order:

a. Crankshaft pulley.

b. Water pump pulley.

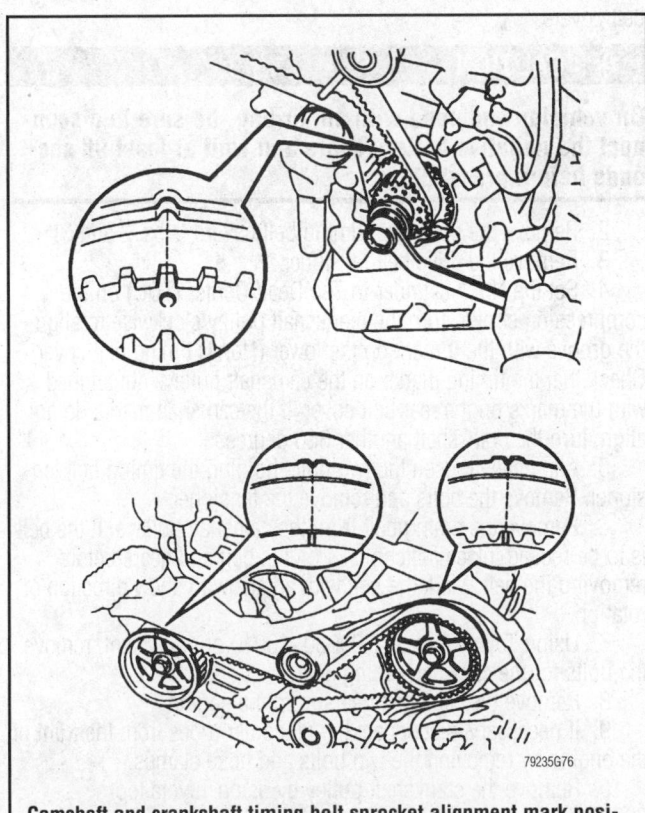

Camshaft and crankshaft timing belt sprocket alignment mark positioning for belt service—Toyota 3.0L (1MZ-FE) engine

79235G76

   c. Left camshaft pulley.
   d. No. 2 idler pulley.
   e. Right camshaft pulley.
   f. No. 1 idler pulley.

16. Using a press, slowly press the timing belt tensioner until the holes of the pushrod and housing align. Insert a 0.05 in. (1.27mm) hexagonal Allen wrench through the holes to preserve the setting position.

17. Install the dust boot to the tensioner.

18. Install the tensioner with the two bolts. Alternately tighten the bolts to 20 ft. lbs. (27 Nm). Remove the Allen wrench.

19. Turn the crankshaft clockwise and align the crankshaft timing pulley groove with the oil pump alignment mark.

20. Ensure the camshaft timing marks align with the timing marks on the rear timing belt cover.

21. Install the timing belt guide.

22. Install the right engine mounting bracket and tighten the bolts to 21 ft. lbs. (28 Nm).

23. Install the upper timing belt cover with the five bolts. Tighten the bolts to 74 inch lbs. (8 Nm).

24. Install the engine wire protector clamp to the No. 3 timing belt cover.

25. Install the engine wire protector to the No. 3 timing belt cover with the bolt.

26. Install the lower timing belt cover by installing the four bolts. Tighten the bolts to 74 inch lbs. (8 Nm).

27. Install the remaining components. During installation be sure to tighten the crankshaft pulley bolt to 159 ft. lbs. (215 Nm) and the No. 2 alternator bracket nut to 21 ft. lbs. (28 Nm).

3.0L (2JZ-GTE and 2JZ-GE) Engines

1. Remove all necessary components for access to the timing belt covers.

✳✳ CAUTION

On vehicles equipped with an air bag, be sure to disconnect the negative battery cable and wait at least 90 seconds before proceeding.

2. Remove the upper two timing belt covers (Nos. 2 and 3).

3. Remove the drive belt tensioner.

4. Set the No. 1 cylinder to Top Dead Center (TDC) on the compression stroke. Turn the crankshaft pulley clockwise to align the groove with the **0** mark on the lower (No. 1) timing belt cover. Check that the timing marks on the camshaft pulleys are aligned with the marks on the rear belt cover. If the camshaft marks do not align, turn the crankshaft another 360 degrees.

5. Alternately loosen the two bolts holding the timing belt tensioner. Remove the bolts and remove the tensioner.

6. Remove the timing belt from the camshaft pulleys. If the belt is to be reused, place matchmarks on the belt and gears before removing the belt. Mark the belt with an arrow to show direction of rotation.

7. Using Toyota tool SST 09960–10010 or equivalent, remove the bolts for the camshaft timing gears.

8. Remove the camshaft gears from the engine.

9. If necessary, disconnect the oil cooler tubes from the front of the engine by removing the two bolts and hose clamps.

10. Remove the crankshaft pulley by using Toyota tool 09330–0021, or equivalent, to hold the pulley and using tool 09213–70010, or equivalent, to remove the pulley bolt.

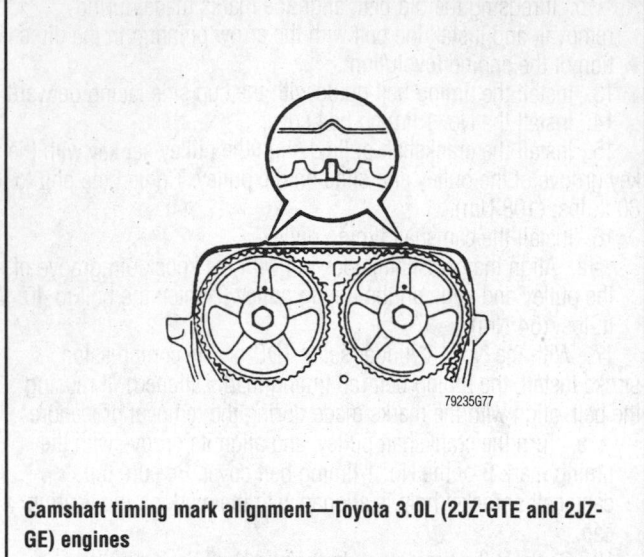

Camshaft timing mark alignment—Toyota 3.0L (2JZ-GTE and 2JZ-GE) engines

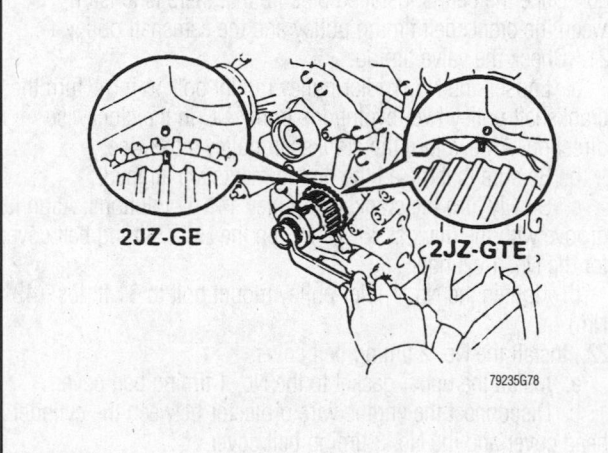

Notice the timing mark difference between the two 3.0L engines—Toyota 3.0L (2JZ-GTE and 2JZ-GE) engines

11. Remove the lower (No. 1) timing belt cover and the timing belt guide.

12. Remove the timing belt. If the belt is to be reused, protect it from contact with oil, grease, or fluids.

To install:

13. Use the crankshaft pulley bolt to turn the crankshaft (clockwise) until the mark on the gear aligns with the oil pump body. Check all the pulleys and gears for cleanliness; remove any grease, oil or coolant. Install the timing belt onto the crankshaft gear and idler pulleys.

14. Install the timing belt guide with the cupped side facing outward.

15. Install the No. 1 timing belt cover.

16. Install the crankshaft pulley. Align the set key with the groove. Hold the pulley with the proper tool and tighten the pulley bolt to 239 ft. lbs. (324 Nm).

17. If equipped with automatic transmission, connect the oil cooler tubes with the clamps and two bolts.

18. Install the camshaft gears as follows:

a. Align the camshaft knock pin with the groove on the gear and slide on the timing gear.

b. Temporarily install the timing gear bolt.

c. Using the same tools as removal, tighten the camshaft gear bolts to 59 ft. lbs. (79 Nm).

d. Turn the crankshaft pulley and align its groove with the timing mark, **0** on the No. 1 timing belt cover.

e. Align the timing marks on the camshaft timing gears and the No. 4 timing belt cover.

19. Finish installing the timing belt.

20. Double check that all the timing marks for the crankshaft pulley and the camshaft gears are aligned as they were during disassembly.

21. Set the timing belt tensioner:

a. Use a press to slowly push in the pushrod on the tensioner. This will require between 220–2200 lbs. (100–1000 kg) of pressure.

b. Align the holes of the pushrod and housing. Place a 0.06 in. (1.5mm) hex wrench through the holes to keep the pushrod retracted.

c. Release the press and install the dust boot onto the tensioner.

22. Install the tensioner; alternately tighten the bolts to 20 ft. lbs. (26 Nm).

23. Remove the hex wrench from the tensioner with a pair of pliers.

24. Turn the crankshaft pulley two full turns clockwise. Check that each pulley's timing marks align correctly after the two turns. If any mark does not align, remove the timing belt and reinstall it.

25. Install the drive belt tensioner and tighten the bolts to 15 ft. lbs. (21 Nm).

26. Install the Nos. 2 and 3 timing belt covers.

27. Install all remaining components. During assembly be sure to tighten the drive belt tensioner damper nuts to 14 ft. lbs. (20 Nm).

VOLKSWAGEN MODELS

1.9L Diesel Engine

Some special tools are required to perform this procedure properly. A flat bar, VW tool 2065A or equivalent, is used to secure the camshaft in position. A pin, VW tool 2064 or equivalent, is used to fix the pump position while the timing belt is removed. The camshaft and pump work against spring pressure and will move out of position when the timing belt is removed. It is not difficult to find substitutes but do not remove the timing belt without these tools.

❊❊ WARNING

Do not turn the engine or camshaft with the timing belt removed. The pistons will contact the valves and cause internal engine damage.

1. Disconnect the negative battery cable and remove the accessory drive belts, crankshaft pulley and the timing belt cover(s). Remove the camshaft cover and rubber plug at the back end of the camshaft.

2. Temporarily reinstall the crankshaft pulley bolt and turn the crankshaft to TDC of No. 1 piston. The mark on the camshaft sprocket should be aligned with the mark on the inner timing belt cover or the edge of the cylinder head.

3. With the engine at TDC, insert the bar into the slot at the back of the camshaft. The bar rests on the cylinder head to will hold the camshaft in position.

4. Insert the pin into the injection pump drive sprocket to hold the pump in position.

5. Loosen the locknut on the tensioner pulley and turn the tensioner counterclockwise to relieve the tension on the timing belt. Slide the timing belt from the sprockets.

To install:

6. Install the new timing belt and adjust the tension so the belt can be twisted 45 degrees at the half-way point between the camshaft and pump sprockets. Tighten the tensioner nut to 33 ft. lbs. (45 Nm).

7. Remove the holding tools.

8. Turn the engine 2 full revolutions to return to TDC for the No. 1 cylinder. Recheck belt tension and timing mark alignment, readjust as required.

9. Install the belt cover and accessory drive belts.

➡**If the belt is too tight, there will be a growling noise that rises and falls with engine speed.**

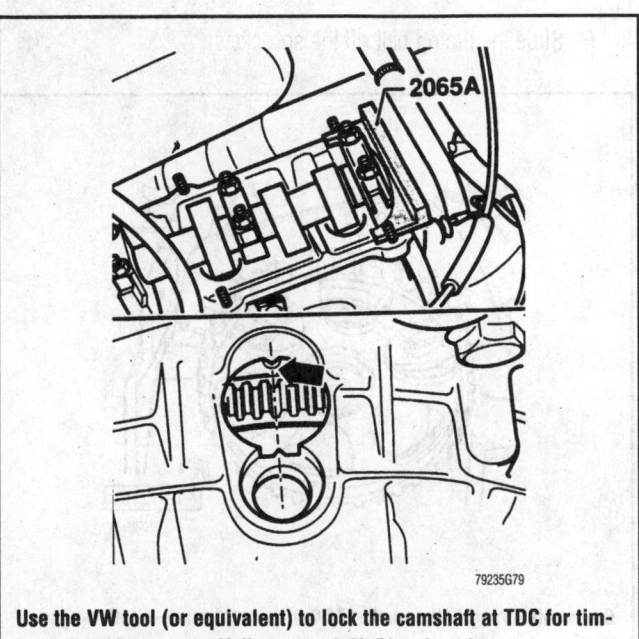

Use the VW tool (or equivalent) to lock the camshaft at TDC for timing belt replacement—Volkswagen 1.9L Diesel engines

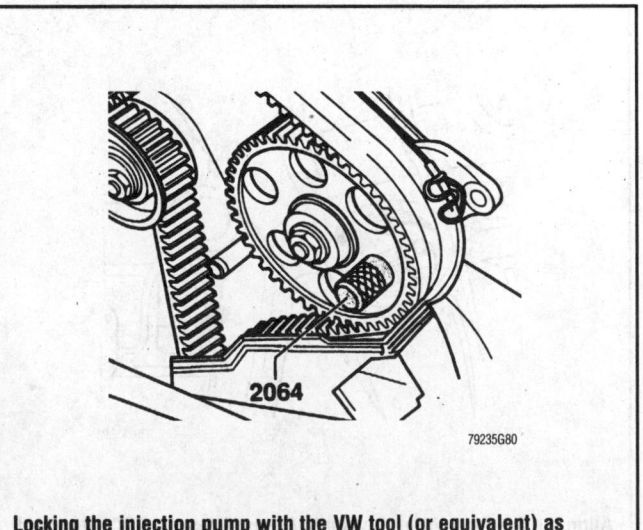

Locking the injection pump with the VW tool (or equivalent) as shown—Volkswagen 1.9L Diesel engines

2.0L Engine

➡️ **Do not turn the engine or camshaft with the camshaft drive belt removed. The pistons will contact the valves and cause internal engine damage.**

1. Disconnect the negative battery cable and remove the accessory drive belts, crankshaft pulley and the timing belt cover(s).

2. Temporarily reinstall the crankshaft pulley bolt, if removed, and turn the crankshaft to TDC of No. 1 piston. The mark on the camshaft sprocket should be aligned with the mark on the inner drive belt cover, if equipped, or the edge of the cylinder head.

3. On 8-valve engines, the notch on the crankshaft pulley should align with the dot on the intermediate shaft sprocket. With the distributor cap removed, the rotor should be pointing toward the No. 1 mark on the rim of the distributor housing.

4. Loosen the locknut on the tensioner pulley and turn the tensioner counterclockwise to relieve the tension on the timing belt.

5. Slide the timing belt off the sprockets.

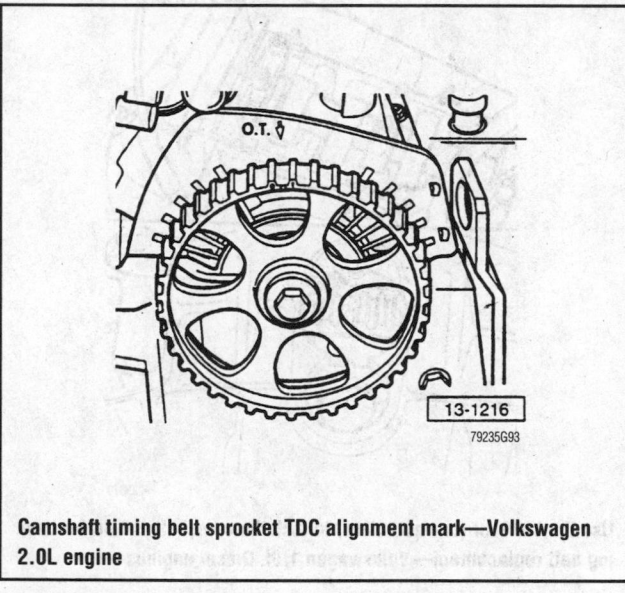

Camshaft timing belt sprocket TDC alignment mark—Volkswagen 2.0L engine

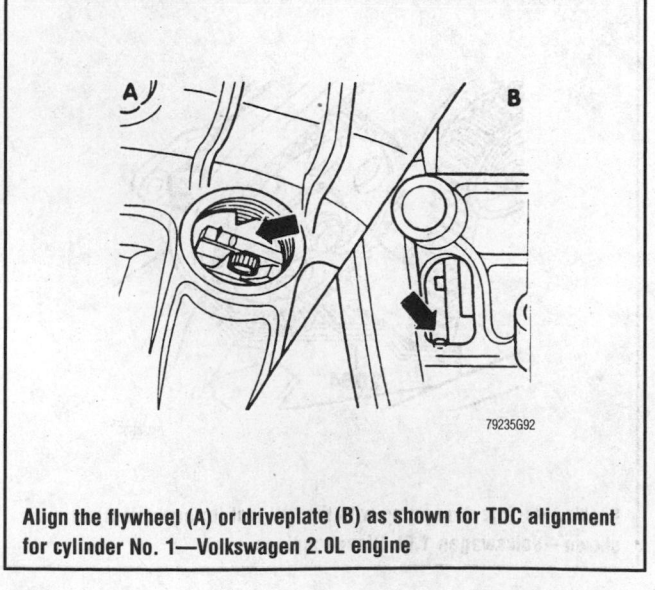

Align the flywheel (A) or driveplate (B) as shown for TDC alignment for cylinder No. 1—Volkswagen 2.0L engine

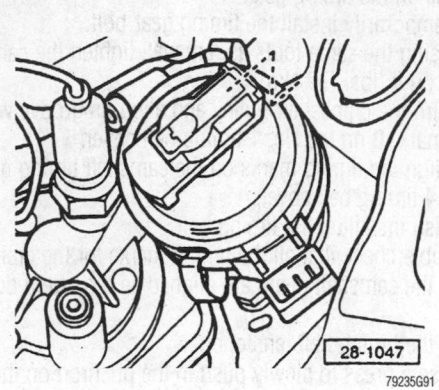

When the No. 1 cylinder is at TDC, the ignition rotor should face the notch in the distributor housing—Volkswagen 2.0L engine

To install:

6. Install the new timing belt and tension the belt so that it can be twisted 90 degrees at the middle of its longest section, between the camshaft and intermediate sprockets.

7. Recheck the alignment of the timing marks, if correct, turn the engine 2 full revolutions to return to TDC of No. 1 piston. Recheck belt tension and timing marks. Readjust as required. Tighten the tensioner nut to 33 ft. lbs. (45 Nm).

8. Reinstall the belt cover and accessory drive belts.

➡️ **When running the engine, there will be a growling noise that rises and falls with engine speed if the belt is too tight.**

VOLVO MODELS

2.3L 4-Cylinder Engine

1. Disconnect the negative battery cable.
2. Remove the drive belts.
3. Remove the cooling fan and shroud.
4. Remove the timing belt cover.

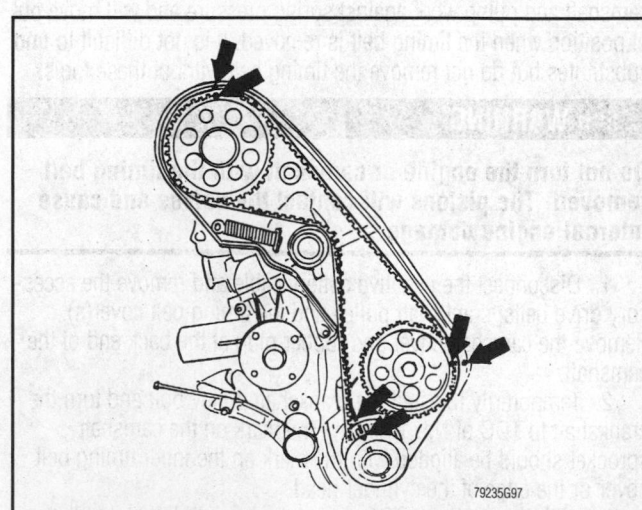

Before removing or installing the timing belt, align the timing marks as indicated—Volvo 2.3L engine

5. Remove the 6 retaining bolts and the crankshaft pulley.

6. To remove the tension from the belt, loosen the nut for the tensioner and press the idler roller back. The tension spring can be locked in this position by inserting the shank end of a 3mm drill through the pusher rod.

7. Remove the belt, taking care not to bend it at any sharp angles.

To install:

8. If the crankshaft, idler shaft or camshaft were disturbed while the belt was out, align each shaft as follows with its corresponding index mark to assure proper valve timing and ignition timing:

   a. Rotate the crankshaft so the notch in the convex crankshaft gear belt guide aligns with the embossed mark on the front cover (12 o'clock position).

   b. Rotate the idler shaft so the dot on the idler shaft drive sprocket aligns with the notch on the timing belt rear cover (4 o'clock position).

   c. Rotate the camshaft so the notch in the camshaft sprocket inner belt guide aligns with the notch in the forward edge of the valve cover (12 o'clock position).

9. Install the timing belt over the sprockets, then over the tensioner roller. Do not use any sharp tools. New belts have yellow marks. The 2 lines on the drive belt should fit toward the crankshaft marks. The next mark should fit toward the intermediate shaft marks, etc.

10. Loosen the tensioner nut and let the spring tension automatically take up the slack. Tighten the tensioner nut to 37 ft. lbs. (51 Nm).

11. Rotate the crankshaft two full revolutions clockwise and be sure the timing marks align.

12. Install the drive belts, radiator fan and shroud. Connect the negative battery cable.

2.3L 5-Cylinder and 2.4L Engines

1. Disconnect the negative battery cable.

2. Remove the coolant expansion tank and place it on top of the engine.

3. Remove the spark plug cover and drive belts.

4. Remove the timing belt cover.

5. Wait five minutes after lining up marks, then install Volvo Gauge 998–8500 (or equivalent) between the exhaust camshaft and water pump. Read the gauge using a mirror, while still installed. For 23mm belts, the tension should be 2.7–4.0 units.

➡**If the belt tension is incorrect, the tensioner must be replaced.**

6. Remove the upper tensioner bolt and loosen the lower bolt, turning the tensioner to free up the pulley.

7. Remove the lower bolt and the tensioner. Remove the timing belt.

To install:

8. Turn all the pulleys listening for bearing noise. Check to see that the contact surfaces are clean and smooth. Remove the tensioner pulley lever and idler pulley, lubricate the contact surfaces and bearing with grease. If the tensioner pulley lever or idler is seized replace it.

9. Install the tensioner pulley lever and idler pulley and tighten to 18 ft. lbs. (25 Nm).

10. Compress the tensioner with Volvo Tool 999–5456 (or equivalent) and insert a 0.079 in. (2.0mm) lockpin in the piston. If the tensioner leaks, has no resistance, or will not compress, replace it. Install the tensioner and tighten to 18 ft. lbs. (25 Nm).

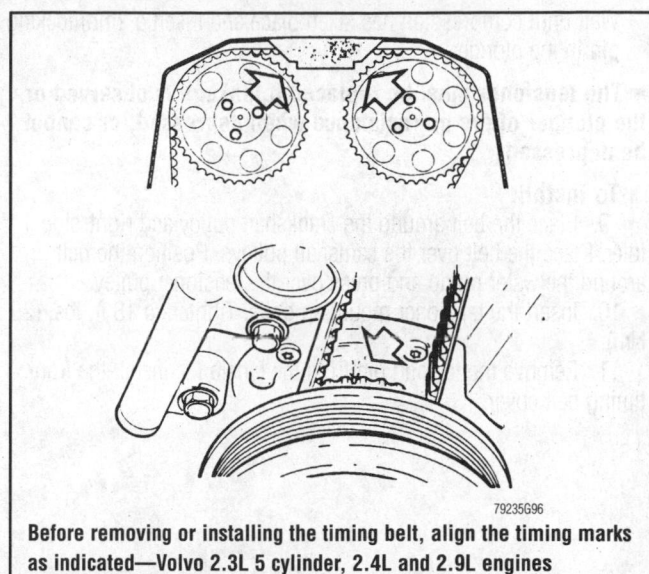

Before removing or installing the timing belt, align the timing marks as indicated—Volvo 2.3L 5 cylinder, 2.4L and 2.9L engines

11. Install the timing belt in the following order:

   a. Around the crankshaft sprocket.

   b. Around the right idler pulley

   c. Around the camshaft sprockets

   d. Around the water pump

   e. Onto the tensioner pulley

12. Pull the lockpin out from the tensioner and install the upper timing cover. Turn the crankshaft two complete revolutions and check to see that the timing marks on the crankshaft and camshaft pulleys are lined up.

13. Install the timing belt covers and the fuel line clips.

14. Install the accessory belts.

15. Install the vibration damper guard and the inner fender well.

16. Install the spark plug cover and any remaining components.

2.9L 6-Cylinder Engines

1. Disconnect the negative battery cable.

2. Remove the drive belts.

3. Remove the timing belt cover.

4. Remove the splash guard, vibration damper guard and ignition coil cover.

5. Rotate the crankshaft clockwise, until the timing marks on the camshaft pulleys and timing belt mounting plate and crankshaft pulley/oil pump housing are aligned.

6. Remove the tensioner upper mounting bolts. Loosen the tensioner lower mounting bolt and twist the tensioner to free the plunger. Remove the lower mounting bolt and remove the tensioner.

7. Remove the timing belt.

➡**Do not rotate the crankshaft while the timing belt is removed.**

8. Check the tensioner and idler pulleys, as follows:

   a. Spin the pulleys and listen for bearing noise.

   b. Check that the pulley surfaces in contact with the belt are clean and smooth.

   c. Check the tensioner pulley arm and idler pulley mountings.

   d. Tighten the tensioner pulley arm to 30 ft. lbs. (40 Nm) and the idler pulley to 18 ft. lbs. (25 Nm).

   e. Compress the tensioner using tool 5456 or equivalent. Mount the tensioner in the tool and tighten the center nut fully.

Wait until compression has taken place and insert a 2mm locking pin in the plunger.

➡ **The tensioner must be replaced if leakage is observed or the plunger offers no resistance when depressed, or cannot be depressed.**

To install:

9. Place the belt around the crankshaft pulley and right-side idler. Place the belt over the camshaft pulleys. Position the belt around the water pump and press over the tensioner pulley.

10. Insert the tensioner mounting bolts. Tighten to 18 ft. lbs. (25 Nm).

11. Remove the locking pin from the tensioner. Install the front timing belt cover.

➡ **The lever bushing must be greased every time the belt is replace or the pulley is removed. Service the bushing, using the following procedure:**

   a. Remove the lever mounting bolt, tensioner pulley and sleeve.

   b. Grease the surfaces of the bushing, bolt and sleeve, using Volvo Part No. 1161246–2 or equivalent.

   c. Install the sleeve, tensioner pulley and lever mounting bolt.

   d. Tighten the bolt to 30 ft. lbs. (40 Nm).

12. Turn the crankshaft two revolutions and check that the timing marks on the crankshaft and camshaft pulleys are correctly aligned.

13. Install the remaining components.

14. Connect the negative battery lead, start and check the engine operation.

BRAKES

BRAKE OPERATING SYSTEM

Basic Operating Principles

Hydraulic systems are used to actuate the brakes of all modern automobiles. The system transports the power required to force the frictional surfaces of the braking system together from the pedal to the individual brake units at each wheel. A hydraulic system is used for two reasons.

First, fluid under pressure can be carried to all parts of an automobile by small pipes and flexible hoses without taking up a significant amount of room or posing routing problems.

Second, a great mechanical advantage can be given to the brake pedal end of the system, and the foot pressure required to actuate the brakes can be reduced by making the surface area of the master cylinder pistons smaller than that of any of the pistons in the wheel cylinders or calipers.

The master cylinder consists of a fluid reservoir along with a double cylinder and piston assembly. Double type master cylinders are designed to separate the front and rear braking systems hydraulically in case of a leak. The master cylinder coverts mechanical motion from the pedal into hydraulic pressure within the lines. This pressure is translated back into mechanical motion at the wheels by either the wheel cylinder (drum brakes) or the caliper (disc brakes).

Steel lines carry the brake fluid to a point on the vehicle's frame near each of the vehicle's wheels. The fluid is, then carried to the calipers and wheel cylinders by flexible tubes in order to allow for suspension and steering movements.

In drum brake systems, each wheel cylinder contains two pistons, one at either end, which push outward in opposite directions and force the brake shoe into contact with the drum.

In disc brake systems, the cylinders are part of the calipers. At least one cylinder in each caliper is used to force the brake pads against the disc.

All pistons employ some type of seal, usually made of rubber, to minimize fluid leakage. A rubber dust boot seals the outer end of the cylinder against dust and dirt. The boot fits around the outer end of the piston on disc brake calipers, and around the brake actuating rod on wheel cylinders.

The hydraulic system operates as follows: When at rest, the entire system, from the piston(s) in the master cylinder to those in the wheel cylinders or calipers, is full of brake fluid. Upon application of the brake pedal, fluid trapped in front of the master cylinder piston(s) is forced through the lines to the wheel cylinders. Here, it forces the pistons outward, in the case of drum brakes, and inward toward the disc, in the case of disc brakes. The motion of the pistons is opposed by return springs mounted outside the cylinders in drum brakes, and by spring seals, in disc brakes.

Upon release of the brake pedal, a spring located inside the master cylinder immediately returns the master cylinder pistons to the normal position. The pistons contain check valves and the master cylinder has compensating ports drilled in it. These are uncovered as the pistons reach their normal position. The piston check valves allow fluid to flow toward the wheel cylinders or calipers as the pistons withdraw. Then, as the return springs force the brake pads or shoes into the released position, the excess fluid reservoir through the compensating ports. It is during the time the pedal is in the released position that any fluid that has leaked out of the system will be replaced through the compensating ports.

Dual circuit master cylinders employ two pistons, located one behind the other, in the same cylinder. The primary piston is actuated directly by mechanical linkage from the brake pedal through the power booster. The secondary piston is actuated by fluid trapped between the two pistons. If a leak develops in front of the secondary piston, it moves forward until it bottoms against the front of the master cylinder, and the fluid trapped between the pistons will operate the rear brakes. If the rear brakes develop a leak, the primary piston will move forward until direct contact with the secondary piston takes place, and it will force the secondary piston to actuate the front brakes. In either case, the brake pedal moves farther when the brakes are applied, and less braking power is available.

All dual circuit systems use a switch to warn the driver when only half of the brake system is operational. This switch is usually located in a valve body which is mounted on the firewall or the frame below the master cylinder. A hydraulic piston receives pressure from both circuits, each circuit's pressure being applied to one end of the piston. When the pressures are in balance, the piston remains stationary. When one circuit has a leak, however, the greater pressure in that circuit during application of the brakes will push the piston to one side, closing the switch and activating the brake warning light.

In disc brake systems, this valve body also contains a metering valve, in some cases, a proportioning valve. The metering valve keeps pressure from traveling to the disc brakes on the front wheels until the brake shoes on the rear wheels have contacted the drums, ensuring that the front brakes will never be used alone. The proportioning valve controls the pressure to the rear brakes to lessen the chance of rear wheel lock-up during very hard braking.

Warning lights may be tested by depressing the brake pedal and holding it while opening one of the wheel cylinder bleeder screws. If this does not cause the light to go on, substitute a new lamp, make continuity checks, finally, replace the switch as necessary.

The hydraulic system may be checked for leaks by applying pressure to the pedal gradually and steadily. If the pedal sinks very slowly to the floor, the system has a leak. This is not to be confused with a springy or spongy feel due to the compression of air within the lines. If the system leaks, there will be a gradual change in the position of the pedal with a constant pressure.

Check for leaks along all lines and at wheel cylinders. If no external leaks are apparent, the problem is inside the master cylinder.

DISC BRAKES

Instead of the traditional expanding brakes that press outward against a circular drum, disc brake systems utilize a disc (rotor) with brake pads positioned on either side of it. An easily-seen analogy is the hand brake arrangement on a bicycle. The pads squeeze onto the rim of the bike wheel, slowing its motion. Automobile disc brakes use the identical principle but apply the braking effort to a separate disc instead of the wheel.

The disc (rotor) is a casting, usually equipped with cooling fins between the two braking surfaces. This enables air to circulate between the braking surfaces making them less sensitive to heat buildup and more resistant to fade. Dirt and water do not drastically affect braking action since contaminants are thrown off by the centrifugal action of the rotor or scraped off the by the pads. Also, the equal clamping action of the two brake pads tends to ensure uniform, straight line stops. Disc brakes are inherently self-adjusting. There are three general types of disc brake:

1. Fixed calipers.
2. Floating calipers.
3. Sliding calipers.

The fixed caliper design uses one or two pistons mounted on each side of the rotor (in each side of the caliper). The caliper is mounted rigidly and does not move.

The sliding and floating designs are quite similar. In fact, these two types are often lumped together. In both designs, the pad on the inside of the rotor is moved into contact with the rotor by hydraulic force. The caliper, which is not held in a fixed position, moves slightly, bringing the outside pad into contact with the rotor.

Floating calipers use threaded guide pins and bushings, or sleeves to allow the caliper to slide and apply the brake pads.

There are typically three methods of securing a sliding caliper to its mounting bracket: with a retaining pin, with a key and bolt, or with a wedge and pin. On calipers which use the retaining pin method, you will find pins driven into the slot between the caliper and the caliper mount. On calipers which use the bolt and key method, a key is used between the caliper and the mounting bracket to allow the caliper to slide. The key is held in position by a lockbolt. On calipers which use the pin and wedge method, a wedge, retained by a pin, is used between the caliper and the mounting bracket.

For pad removal purposes, fixed calipers are usually not removed, floating calipers are either removed or flipped (hinged up or down on one pin), and sliding calipers are removed.

DRUM BRAKES

Drum brakes employ two brake shoes mounted on a stationary backing plate. These shoes are positioned inside a circular drum which rotates with the wheel assembly. The shoes are held in place by springs. This allows them to slide toward the drums (when they are applied) while keeping the linings and drums in alignment. The shoes are actuated by a wheel cylinder which is mounted at the top of the backing plate. When the brakes are applied, hydraulic pressure forces the wheel cylinder's actuating links outward. Since these links bear directly against the top of the brake shoes, the tops of the shoes are, then forced against the inner side of the drum. This action forces the bottoms of the two shoes to contact the brake drum by rotating the entire assembly slightly (known as servo action). When pressure within the wheel cylinder is relaxed, return springs pull the shoes back away from the drum.

Most modern drum brakes are designed to self-adjust themselves during application when the vehicle is moving in reverse. This motion causes both shoes to rotate very slightly with the drum, rocking an adjusting lever, thereby causing rotation of the adjusting screw. Some drum brake systems are designed to self-adjust during application whenever the brakes are applied. This on-board adjustment system reduces the need for maintenance adjustments and keeps both the brake function and pedal feel satisfactory.

POWER BOOSTERS

Virtually all modern vehicles use a power assisted brake system to multiply the braking force and reduce pedal effort. There are two types of power assist used. The most widely used, by far, is the vacuum assist booster. The other is the hydraulically assisted booster.

Vacuum-Assisted Boosters

Most modern vehicles use a vacuum assisted power brake. This system was likely developed, since on all internal combustion engines, except diesels, vacuum is always available when the engine is operating, making the system is simple and efficient.

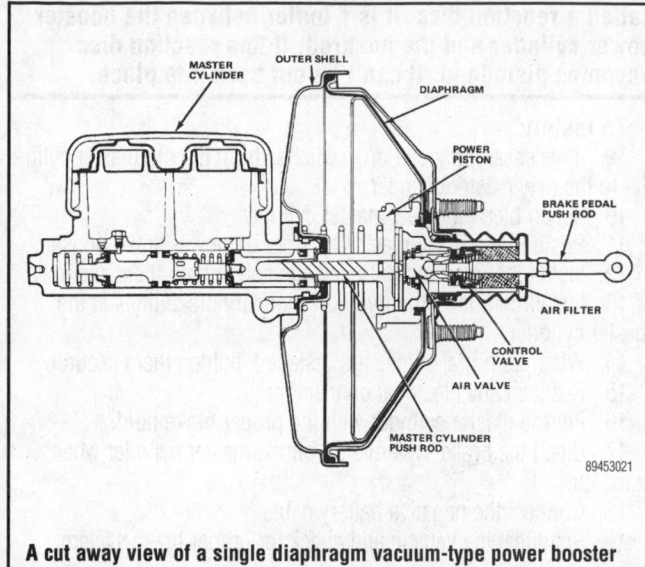

A cut away view of a single diaphragm vacuum-type power booster

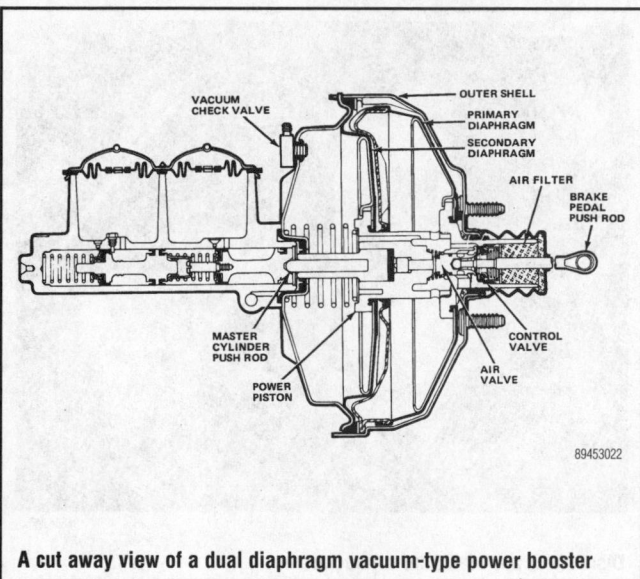

A cut away view of a dual diaphragm vacuum-type power booster

With diesel engines, vacuum is created and stored by way of a belt-driven vacuum pump and reservoir. In either case, the operation of the vacuum assist is the same.

A vacuum diaphragm is located on the front of the master cylinder and assists the driver in applying the brakes, reducing both the effort and travel one must put into moving the brake pedal. The vacuum diaphragm housing is normally connected to the intake manifold by a vacuum hose. A check valve is placed at the point where the hose enters the diaphragm housing, so that during periods of low manifold vacuum brake assist will not be lost.

Depressing the brake pedal closes off the vacuum source and allows atmospheric pressure to enter on one side of the diaphragm. This causes the master cylinder pistons to move and apply the brakes. When the brake pedal is released, vacuum is applied to both sides of the diaphragm and springs return the diaphragm and master cylinder pistons to the released position.

If the vacuum supply fails, the brake pedal rod will contact the end of the master cylinder actuator rod and the system will apply the brakes without any power assistance. The driver will notice that much higher pedal effort is needed to stop the vehicle and that the pedal feels harder than usual.

If you think this is the case you can check it as follows:

VACUUM LEAK TEST

1. Operate the engine at idle without touching the brake pedal for at least one minute
2. Turn off the engine and wait one minute.
3. Test for the presence of assist vacuum by depressing the brake pedal and releasing it several times. If vacuum is present in the system, light application will produce less and less pedal travel. If there is no vacuum, air is leaking into the system.

SYSTEM OPERATION TEST

1. With the engine **OFF**, pump the brake pedal until the supply vacuum is entirely gone.
2. Apply light, steady pressure to the brake pedal.
3. Start the engine and let it idle. If the system is operating correctly, the brake pedal should fall toward the floor if constant pressure is maintained.

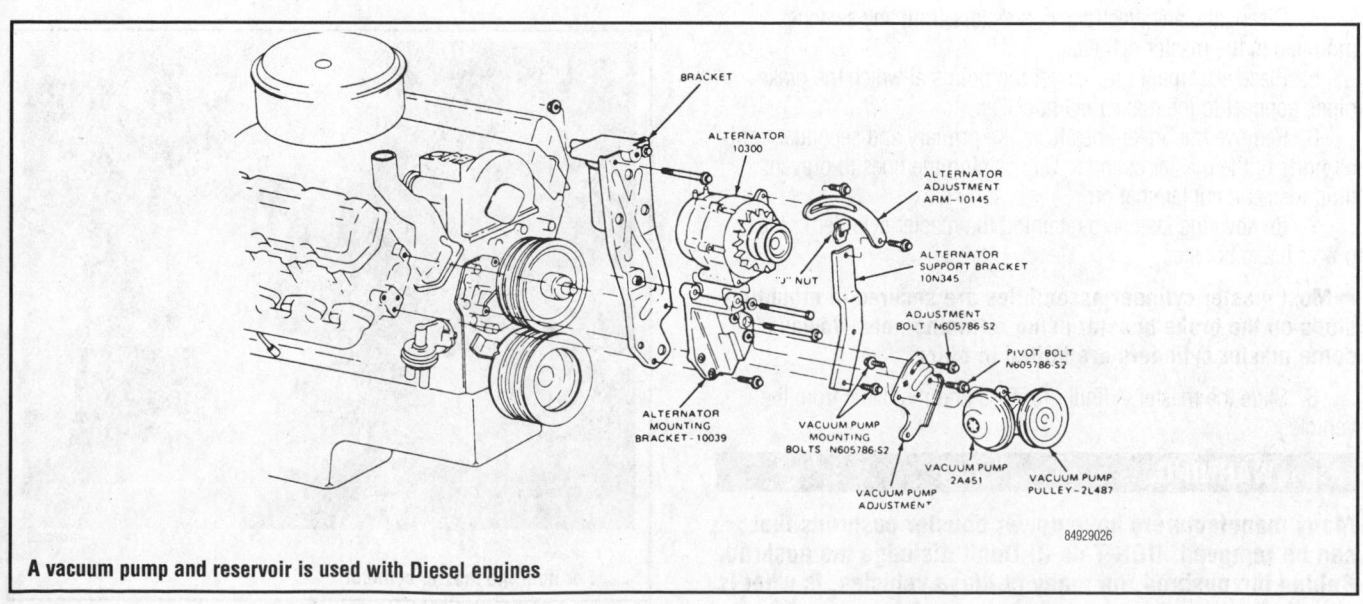

A vacuum pump and reservoir is used with Diesel engines

Power brake systems may be tested for hydraulic leaks just as ordinary systems are tested.

❋❋ WARNING

Clean, high quality brake fluid is essential to the safe and proper operation of the brake system. You should always buy the highest quality brake fluid that is available. If the brake fluid becomes contaminated, drain and flush the system, then refill the master cylinder with new fluid. Never reuse any brake fluid. Any brake fluid that is removed from the system should be discarded.

Hydraulically-Assisted Boosters

Used on some light vehicles, the unit is fed hydraulic fluid through the power steering system. The booster assembly, sometimes known generically by the brand name Hydro-Boost, contains a valve which controls pump pressure while braking, a lever to control the position of the valve and a boost piston to provide the force to operate the master cylinder attached to the front of the booster. The unit has a reserve system designed to store pressurized fluid to provide at least 2 brake applications in the event of hydraulic supply system failure, such as a broken power steering belt. The brakes can also be applied unassisted in the event of system depletion.

Master Cylinder

➡️**The following procedures apply to non-ABS systems and ABS system master cylinders that are separate from other ABS system components. ABS systems with integral master cylinder components often require special tools and model-specific procedures.**

REMOVAL & INSTALLATION

With Power-Assisted Brakes

1. Disconnect the negative battery cable.
2. If applicable, apply the brake pedal several times to exhaust all vacuum from the power boost system.
3. Remove any components in the engine compartment which may interfere with master cylinder removal.
4. Disengage any electrical connectors from any switches mounted in the master cylinder.
5. Place absorbent rags under the points at which the brake pipes connect to the master cylinder.
6. Remove the brake lines from the primary and secondary outlet ports of the master cylinder. Cap or plug the lines to prevent fluid loss and contamination.
7. Remove the fasteners retaining the master cylinder to the power brake booster.

➡️**Most master cylinder assemblies are secured to mounting studs on the brake booster using retaining nuts. However, some master cylinders are bolted in place.**

8. Slide the master cylinder forward and remove it from the vehicle.

❋❋ WARNING

Many manufacturers have power booster pushrods that can be removed. DON'T do it! Don't dislodge the pushrod. Behind the pushrod, on many of these vehicles, is what is called a reaction disc. It is a buffer between the booster power cylinder and the pushrod. If this reaction disc becomes dislodged, it can't be put back into place.

To install:
9. If necessary, transfer any switches from the old master cylinder to the new master cylinder.
10. Bench bleed the new master cylinder.
11. Position the brake master cylinder on power brake booster.
12. Install the retaining nuts or bolts and tighten them securely.
13. Install both the primary and secondary brake lines at the master cylinder.
14. When both brake lines are installed, tighten them securely.
15. Reattach any electrical connectors.
16. Fill the master cylinder with the proper brake fluid.
17. Bleed the brake system. Top off the master cylinder when complete.
18. Connect the negative battery cable.
19. Road test the vehicle and check for proper brake system operation.

Disconnect any electrical connectors at . . .

. . . or near the master cylinder

Use an open-end wrench (a line wrench is preferable) to loosen the brake pipe fittings . . .

. . ., then slide the master cylinder assembly from the mount

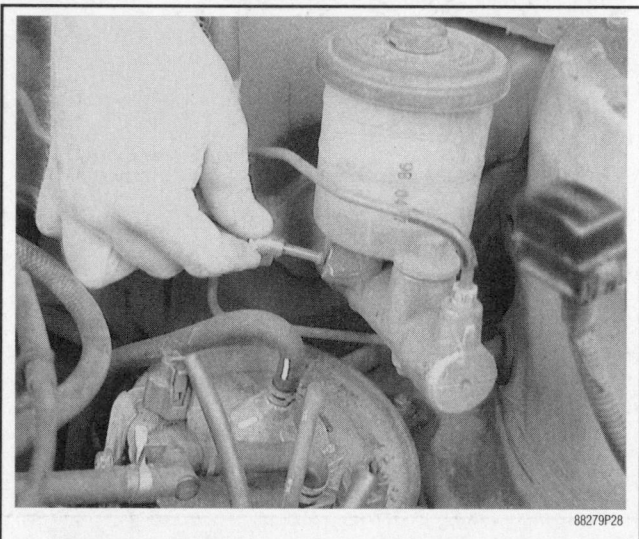

. . . the disconnect the pipes from the master cylinder assembly

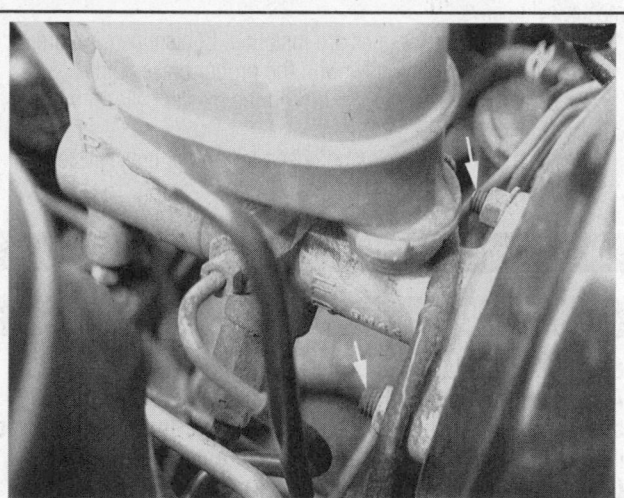

Most master cylinders are secured to the brake booster using 2 retaining nuts

Loosen the master cylinder retainers . . .

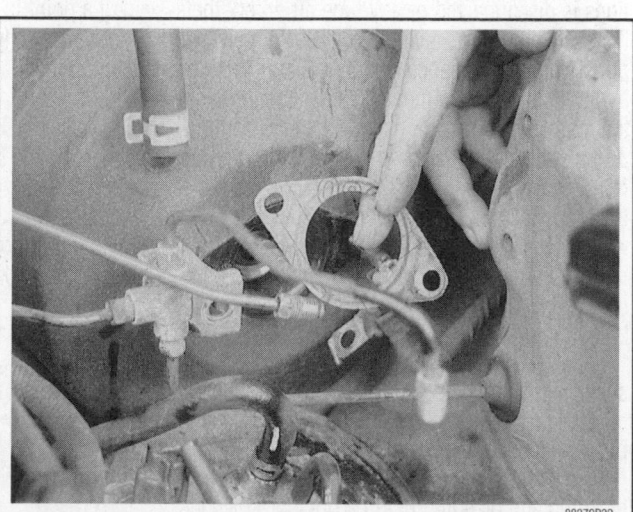

Some vehicles have gaskets between the master cylinder and booster

Without Power-Assisted Brakes

1. Disconnect the master cylinder pushrod from the brake pedal linkage. This connection can be either inside the passenger compartment or inside the engine compartment. The connection is usually by way of a rod fitting over a stud on the pedal arm, retained by washers and a cotter pin or clip.

2. Place absorbent rags under the points at which the brake pipes connect to the master cylinder.

3. Remove the brake lines from the primary and secondary outlet ports of the master cylinder. Cap or plug the lines to prevent fluid loss and contamination.

4. Remove the fasteners retaining the master cylinder to the firewall.

5. Slide the master cylinder forward and remove it from the vehicle.

To install:

6. Bench bleed the new master cylinder.

7. Place the brake master cylinder onto the firewall.

8. Install the fasteners and tighten them securely.

9. Install both the primary and secondary brake lines at the master cylinder.

10. When both brake lines are installed, tighten them securely.

11. Fill the master cylinder with the proper brake fluid.

12. Bleed the brake system. Top off the master cylinder when complete.

13. Road test the vehicle and check for proper brake system operation.

Brake System Bleeding

✳✳ CAUTION

Brake fluid contains polyglycol ethers and polyglycols. Avoid contact with the eyes and wash your hands thoroughly after handling brake fluid. If you do get brake fluid in your eyes, flush your eyes with clean, running water for 15 minutes. If eye irritation persists, or if you have taken brake fluid internally, IMMEDIATELY seek medical assistance.

The hydraulic brake system must be bled any time any of the lines is disconnected or any time air enters the system. If a point in the system, such as a wheel cylinder or caliper brake line is the only point which was opened, the bleeder screws down stream in the hydraulic system are the only ones which must be bled. If however, the master cylinder fittings are opened, or if the reservoir level drops sufficiently that air is drawn into the system, air must be bled from the entire hydraulic system. If the brake pedal feels spongy upon application and travels almost to the floor but regains height when pumped, air has entered the system. It must be bled out. If no fittings were recently opened for service, check for leaks that would have allowed the entry of air and repair them before attempting to bleed the system.

As a general rule, once the master cylinder (and the brake pressure modulator valve or combination valve on ABS systems) is bled, the remainder of the hydraulic system should be bled in the proper sequence.

The hydraulic system can be bled in one of two ways: manual bleeding and bleeding using a pressure bleeder.

Manual Bleeding

MASTER CYLINDER

If the unit is removed from the vehicle, there are 2 ways to "bench-bleed" a master cylinder.

One method is with a large, clear plastic syringe made for the purpose. They are usually available at auto parts stores. In this procedure, the master cylinder is clamped in a soft-jawed vise and filled with fluid. The outlet ports are capped or plugged. Then, uncap each port, place the syringe securely in the outlet port and draw fluid into the syringe until no air is left in the master cylinder, capping the ports when done.

The other is with 2 lengths of hose or pipe (to use as bleeder tubes). Plastic hoses, made for the purpose, are available at most auto parts stores. These hoses have threaded ends for attachment to

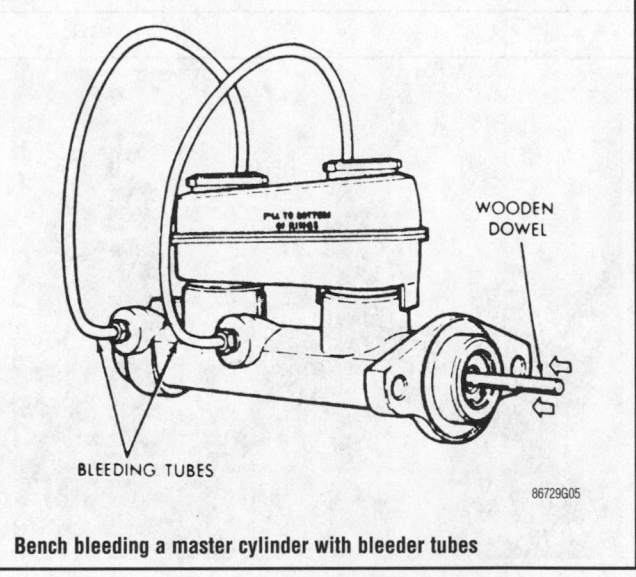

Bench bleeding a master cylinder with bleeder tubes

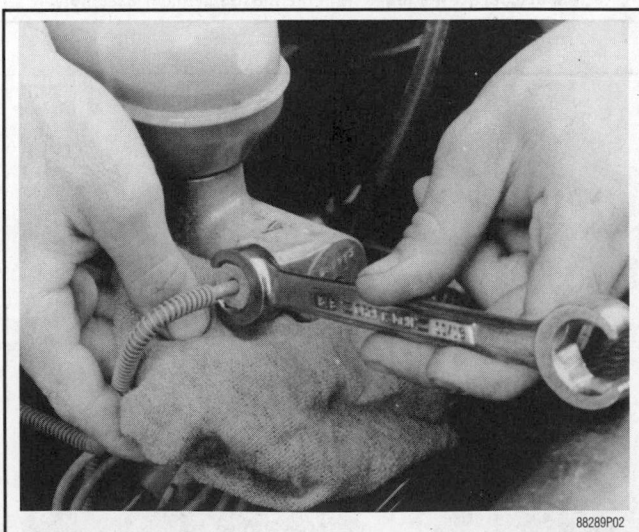

Bleeding the master cylinder by cracking open the fittings

the outlet ports. Otherwise, you'll have to make your own bleeder pipes from 2 lengths of brake pipe equipped with threaded ends. Try to get the plastic ones. In this procedure, clamp the master cylinder in a soft-jawed vise. Connect the pieces of brake pipe or the plastic hoses to the outlet fittings, bend them until the free end is the master cylinder reservoir. Fill the reservoir with fresh DOT 3, or equivalent, brake fluid from a closed container, completely covering the tube ends. Pump the piston slowly until no more air bubbles appear in the reservoir. Remove the tubes, refill the brake master cylinder and securely install the caps or plugs in the ports.

If the brake master cylinder is on the vehicle, place a large, absorbent rag under the fittings. Open the brake lines slightly with the flare nut wrench while pressure is applied to the brake pedal by a helper inside the vehicle. Be sure to tighten the line before the brake pedal is released. Repeat the process with both lines until no air bubbles come out.

In both cases, the rest of the brake system must be bled to assure that all trapped has been removed and that the system will operate properly.

CALIPERS AND WHEEL CYLINDERS

We recommend that the brake system be bled using the jar and tube method. We know some people just let the fluid spray all over the place from the nipple. This is not only unprofessional, but it's messy and potentially dangerous. Brake fluid damages paint, concrete, your clothes, your skin, most importantly, your eyes.

➡**Hydraulic brake systems must be totally flushed if the fluid becomes contaminated with water, dirt or other corrosive chemicals. Also, many manufacturers recommend that the system be flushed routinely, every 2 years or so. To flush, bleed the entire system until all fluid has been replaced and the new brake fluid runs clear.**

The hydraulic system on vehicles with a split system—a 2-chambered master cylinder—can be split either into front/rear or diagonally. In the diagonally split system there is one front and one rear component in each circuit. If you are in doubt as to the design of your vehicle's system, you can check the brake lines. Follow them to each wheel and see which are paired.

➡**If, during the bleeding procedure, you can't get a good flow of fluid from the front brakes, the problem is with the**

Bleeding the calipers

88489P48

Bleeding the wheel cylinders

metering part of the combination valve. Check the valve and you'll see a small stem sticking out of one end. You'll have to fabricate a little clip to hold the stem out as far as it will go. This will allow a full flow to the front brakes. Also, when using this clip on vehicles with power brakes, try bleeding with the engine running. The greater pressure allowed by the power booster will aid in purging the system.

1. Fill the brake master cylinder with the fluid recommended for your vehicle. Check the level often during the procedure. Never let the master cylinder go dry or the procedure will have to be performed again.

2. Raise and support the vehicle safely.

3. If necessary for better access, remove the wheels.

4. On vehicles with a single chamber system or dual chambered systems split front/rear, you can bleed the system in the following order:
   - Right rear
   - Left rear
   - Right front
   - Left front

5. On vehicles with a dual chambered system split diagonally, the usual bleeding order is:
   - Right rear
   - Left front
   - Left rear
   - Right front

6. Find a wrench, a box wrench if possible, of the right size for the bleeder screw and place it on the nipple of the first cylinder to be bled.

7. Connect a clear, vinyl tube to the bleeder nipple. Place the other end of the tube in a clear glass jar of at least 8 oz. (237mL) capacity. The jar should be about ½ full of clean brake fluid. Submerge the end of the tube in the brake fluid.

8. Have an assistant pump the brake pedal, then hold it down. Slowly open the bleeder screw. When the brake pedal reaches the floor, close the bleeder and have the helper slowly release the pedal. Wait 15 seconds, then repeat the procedure until no more air comes out of the bleeder.

9. Repeat the procedure on the remaining calipers or wheel cylinders in the appropriate order.

10. If the brake pedal has a spongy feel, the brake system must be bled again to remove air still trapped in the system.

11. Install the bleeder caps to keep dirt out.

12. If removed for access, install the wheels.

13. Lower the vehicle.

14. Road test the vehicle and check for proper brake system operation.

Pressure Bleeding

A pressure bleeder is a device that uses compressed air and a series of adapters to forcibly expel air from the hydraulic system. When using a pressure bleeder, always follow the manufacturer's instructions. What we've given you here are general instructions.

When using pressure bleeding equipment, it's best to use a bladder-type bleeder tank. In this type of bleeder, the brake fluid is separated from the air by a rubber diaphragm. The bleeder tank must contain enough brake fluid to complete the bleeding operation and should be charged with only 10–30 psi (69–207 kPa). Never exceed 50 psi (345 kPa).

1. Clean all dirt from the master cylinder fluid reservoir filler cap.

➡ **The reservoir must be at least ¾-full during the bleeding procedure. Fill the reservoir as necessary. Use only clean, fresh brake fluid from a sealed container. Fill to the MAX level line on the reservoir.**

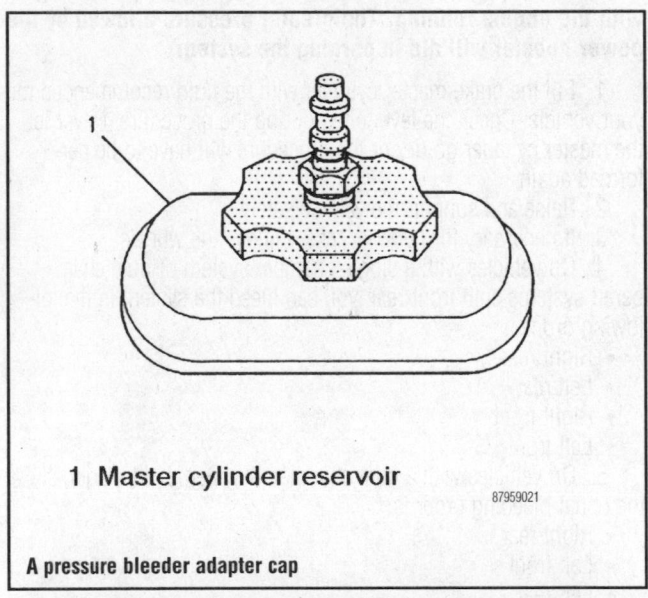

1 Master cylinder reservoir

87959021

A pressure bleeder adapter cap

2. Install the bleeder adapter tool on the master cylinder and attach the hose from the bleeder tank to the fitting on the adapter. Follow the manufacturer's instructions when installing and connecting the master cylinder adapter.

3. Open the valve on the bleeder tank.

MASTER CYLINDER

If the master cylinder is known or suspected to contain air, it must be bled before the wheel cylinders or calipers. Place a large, absorbent rag under the pipe fittings. Beginning at the front of the master cylinder, alternately loosen and tighten the brake line flare nuts. Allow the fluid to flow for several seconds before tightening the flare nut. Repeat this operation several times to be sure all air has been removed from the master cylinder.

CALIPERS AND WHEEL CYLINDERS

Pressure system bleeding must be performed in the correct order. Refer to the manual bleeding procedure for proper bleeding sequences.

1. Raise and safely support the vehicle.

2. Remove the protective bleeder screw cap from the caliper or wheel cylinder and clean the nipple.

3. Place a wrench, preferably a box end wrench, on the bleeder screw.

4. Attach a length of clear vinyl hose onto the bleeder nipple. The hose must fit tightly around the bleeder screw.

5. Submerge the free end of the hose in a large (approximately 16 oz./475mL) clean glass jar about half filled with clean brake fluid.

6. Loosen the bleeder screw approximately ¾ of a turn. When the fluid entering the jar is completely free of bubbles, tighten the bleeder screw.

7. Remove the bleeder hose and attach the protective screw cap.

8. Repeat the bleeding procedure at each brake.

9. Close the valve at the bleeder tank, disconnect the hose from the master cylinder adapter and remove the master cylinder adapter.

10. Check the fluid level in the remote reservoir, refilling with clean, fresh brake fluid, as necessary.

11. Check the brake pedal feel. If spongy, repeat the bleeding process and/or look for defective system components.

MODELS WITH ABS

There are 2 potential problems with attempting to bleed an ABS system. The first is that many use control valves and pressure modulators which might trap air if they are not opened and closed during the procedure using a scan tool. The second potential problem is that some ABS systems operate under extremely high pressure (making bleeding dangerous at worst or messy at best).

With this said, there are still many systems which can be bled with common tools. Many of the control valves have pressure relief knobs at one end of the valve which can be held open using a small tool (or pair of locking pliers)., just about all systems can be bled at the wheels provided that the openings are capped immediately during service. The caps keep enough fluid in the lines to prevent air from working its way back to the control or modulator valves.

Before starting, remember that many manufacturers require the use of special scan tools to bleed any part of the system other than the caliper or wheel cylinders. Some manufacturers recommend the scan tool be used when bleeding any part of the system on some of their models. All manufacturers recommend the use of pressure bleeding equipment for ABS systems, especially when bleeding the rear brakes even though manual bleeding can be done successfully in most cases.

If you decide to attempt bleeding the calipers or wheel cylinders, and you are sure that any residual high pressure is depleted, use the same procedure for bleeding as described for non-ABS systems. During the bleeding procedure, wait 10–15 seconds after closing the bleeder screw before reopening it each time. This is recommended by most manufacturers due to the number of valved components in the system.

Once the procedure is complete, start the engine and allow it to run for 15–30 seconds. Depress the brake pedal. The ABS light should not be **ON**. If the light is **ON**, there is a system problem, probably air still trapped somewhere. At this point, you can try the bleeding procedure again or have the vehicle towed to a dealer or repair shop for system bleeding.

As in all bleeding procedures, DO NOT attempt to move the vehicle unless a firm brake pedal feel has been obtained.

DISC BRAKES

Brake Pads

INSPECTION

To inspect the brake pads, remove the wheel. It is usually possible to view the pad thickness through a large hole in the caliper, or by looking at the side of the pad. However, on a few models, it may be necessary to remove the pads for inspection.

As a rule of thumb, the brake pad lining material should be worn no more than 1/8 in. (3mm). On brake pads glued to the backing material, the pad material can be measured from the edge of the backing material. However, on pads which are riveted to the backing material, the lining should be measured from the rivet heads (in the holes in the lining material)

The brake lining material should not exhibit any dampness, crumbling or cracking. If any such damage is evident the pads must be replaced. If the pads showed evidence of dampness, locate the source of the fluid leak and repair it before installing the new pads. If the brake pads exhibit uneven wear, (such as, one pair of pads is worn more on one side of the vehicle than the other pair of pads on the other side; the inner pad is worn more than the outer pad, or vice versa, on one wheel; the pad lining material is worn more on the front edge of a pad, or more on the rear edge of a pad) the disc brake caliper is either defective or mounted improperly.

※※ WARNING

Never polish the pad lining with sandpaper, because hard particles from the sandpaper will become imbedded in the lining, which will damage the brake rotor. If the pad lining is damaged or worn excessively or unevenly, replace the pads with new ones.

REMOVAL & INSTALLATION

※※ CAUTION

Brake dust may contain asbestos! Asbestos is harmful to your health. Never use compressed air to clean any brake component. A filtering mask should be worn during any brake repair.

Brake pad replacement should always be performed on both front or rear wheels at the same time. Never replace pads on only one wheel. When servicing any brakes use only OEM or better quality pads and parts. When the caliper is removed some brake pads stay with the caliper, others remain on the caliper mounting bracket. Use new pad mounting hardware (springs, anti-rattle clips, or shims) whenever possible to ensure a better repair.

Sliding and Floating Calipers

➡**On certain floating calipers it may be possible to remove one of the guide pins and pivot the caliper up or down to gain access to the brake pads. If you decide to do this, be sure that pivoting the caliper will not damage the flexible brake hose.**

1. Open the hood and locate the master brake cylinder fluid reservoir. Clean the area surrounding the reservoir cap, then remove the cap. Remove some of the brake fluid from the reservoir.

2. Loosen the lug nuts on the applicable wheels.
3. Raise and safely support the vehicle.
4. Remove the wheels.
5. Disconnect any electrical brake pad wear sensors.

➡**It is not necessary, and actually discouraged, to detach the brake hose from the caliper during this procedure. If you decide to detach the hose, it will be necessary for you to bleed your brake system.**

6. Remove and suspend the caliper with a piece of wire, cord or strong string. Be sure that it is not placing any stress on the brake hose.
7. For caliper bracket-mounted pads, perform the following:
   a. If present, remove any anti-squeal shims, noting their positions.
   b. Also, remove any anti-rattle springs that may be present. If these springs don't provide good tension, then replace them.

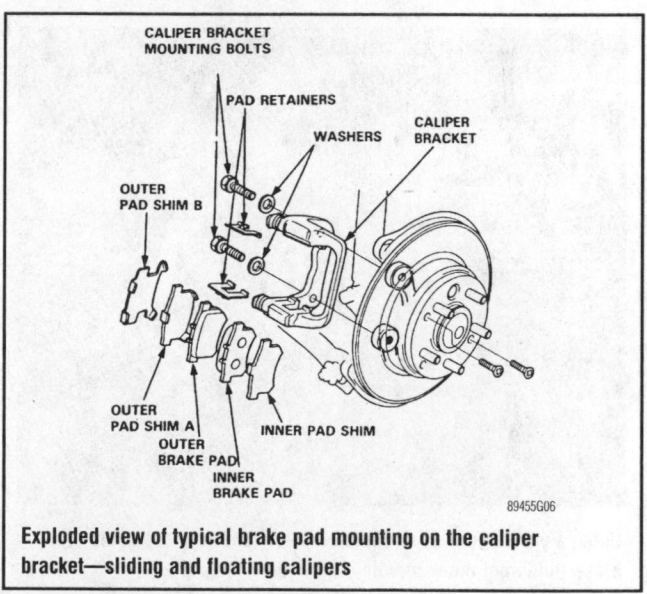

Exploded view of typical brake pad mounting on the caliper bracket—sliding and floating calipers

To remove the brake pads, first clean the brake master cylinder reservoir cap . . .

. . . , then remove it

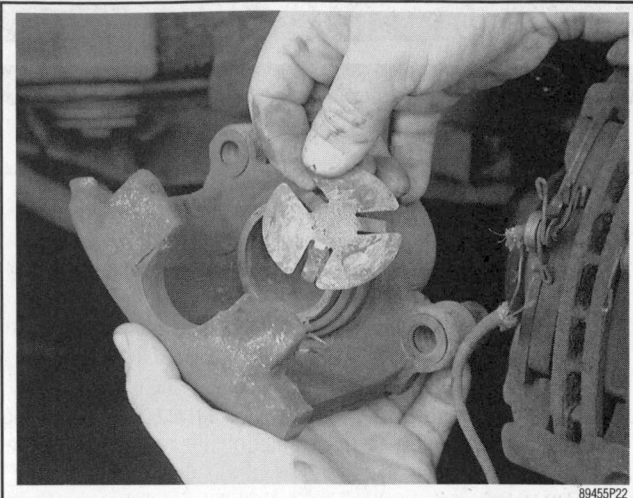

Be sure to note the positions of any clips or springs on the caliper—sliding and floating calipers

Using a vacuum pump, or some other method, remove some of the brake fluid from the reservoir

Remove the outboard pad from the mounting bracket . . .

Remove the disc brake caliper from the rotor—sliding and floating calipers

. . . , then remove the inboard pad—sliding and floating calipers

c. Remove the brake pads from the caliper bracket by lifting the pad out by hand or with a slight tap of a hammer to help.

8. For caliper mounted pads, perform the following:

a. Some outer pads have tabs that are bent over the edge of the caliper, which hold the pads tight in the caliper. Straighten the tabs with pliers before trying to remove the brake pad from the caliper.

b. Then, remove the outer brake pad with a slight tap to the back of the pad with a hammer.

c. Other outer pads use a spring-clip to mount to the caliper. To remove this type of pad, press the pad towards the center of the caliper and slide it off. It may be helpful to use a small pry-bar.

d. Remove the inner pad by pulling it out of the piston.

To install:

9. Clean the caliper sliding area using a wire brush and spray brake cleaner.

10. Lubricate the sliding area of the caliper and the pins with high temperature brake grease.

11. Apply anti-squeal compound to the back side of both brake pads. Allow the compound to set-up according to the instructions on the package.

12. Install one of the old brake pads against the caliper piston, then use a large C-clamp to press the piston back into its bore.

13. Install any new hardware provided with the new pads.

14. For bracket-mounted pads, perform the following steps:

a. Install the pads onto the caliper bracket. Some pads are marked for position.

b. Be sure that the notches or ears of the brake pads are properly engaged on the bracket.

c. Place the caliper over the pads and onto the caliper mounting bracket.

d. Install the caliper mounting hardware and anti-rattle clips. Tighten the guide pins or lockbolt to the proper specification.

➡**It is a good idea to use some thread-locking compound (removable type) to the threaded fasteners of the caliper.**

15. For caliper mounted pads, perform the following:

a. Install the inner pad by pushing the retaining fingers of the pad into the piston of the caliper.

Apply a thin coat of high-temperature brake grease to the sliding surfaces of the bracket and caliper

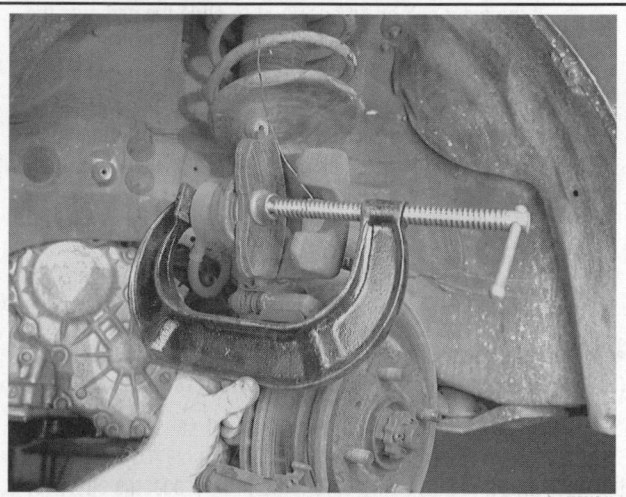

On calipers without integral parking brake mechanisms, a C-clamp can seat the piston in the caliper bore

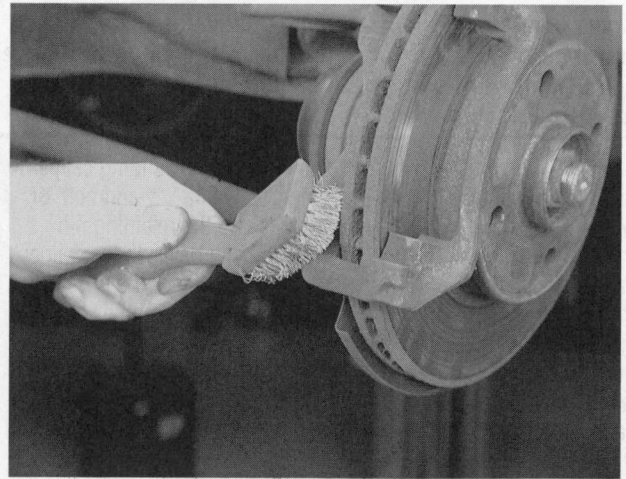

Clean the caliper and mounting bracket with spray brake solvent and a wire brush

Install all of the springs and clips in their original positions—sliding and floating calipers

89455P06

When installing the caliper and pads, be sure not to pinch the sensor wire (if equipped)—sliding and floating calipers

b. If the outer pad has a spring-clip, slide the pad over the edge of the caliper into the caliper frame.

c. If you have the bent-tab style outer brake pad, then test fit the pad; it should fit tight. If the tabs do not secure the pad snugly in the caliper, place the pad on a piece of wood and tap the tab with a hammer to adjust it. It may take a few tries to get it right.

d. Place the caliper with the pads onto the rotor, if equipped, caliper bracket.

e. Install the caliper mounting hardware and anti-rattle clips. Tighten the guide pins or lockbolt(s) securely.

➡**It is a good idea to use some thread-locking compound (removable type) on the threaded fasteners of the caliper.**

16. Connect any electrical brake pad wear sensors.

17. Seat the brake pads, otherwise the vehicle may coast out of the work area and into traffic before the brakes become effective. It will take several pumps of the brake pedal to seat the pads against the rotor.

18. If a firm pedal is not achieved, it may be necessary to bleed the brakes.

19. Check the brake fluid level in the reservoir and top off as needed.

20. Install the wheels and tighten the lug nuts.

21. Road test the vehicle.

Fixed Calipers

➡**It is usually not necessary to remove the caliper to replace the brake pads on a fixed caliper.**

1. Loosen the lug nuts on the applicable wheels.

2. Raise and safely support the vehicle.

3. Remove the wheels.

4. Disconnect any electrical brake pad ware sensors.

5. Remove the pad retaining pins by pulling out the spring-clip or cotter pin, then use a punch and hammer to drive the pin out. Pins without a spring-clip or cotter pin, may be equipped with a spring steel collar on the head of the pin. To remove this style pin, just drive the pin out with a punch and hammer.

6. On calipers with hold-down clips, remove the bolt that holds the clip down.

7. Remove the pads from the caliper with a pair of pliers.

8. To seat the pistons of a fixed caliper, use a piece of wood or a prybar with a rag wrapped around the end, then wedge it between the rotor and the piston and slide the piston into its seat.

➡**It is helpful to replace one pad at a time, to reduce the risk of a piston coming out of its bore, which would lead to the caliper needing to be rebuilt.**

9. Lubricate the sliding area of the caliper and the brake pads with high temperature brake grease.

10. Apply anti-squeal compound to the back side of both brake pads. Allow the compound to set-up according to the instructions on the product.

11. Insert the new pads into the caliper.

12. If equipped, install the anti-rattle clip or retaining pin spring-clip or cotter pin. On pins with a spring steel collar, you must knock them in until seated against the shoulder in the caliper.

➡**It is a good idea to use some thread-locking compound (removable type) to the threaded fasteners of the caliper.**

13. Connect any electrical brake pad wear sensors.

14. Seat the brake pads, otherwise the vehicle may coast out of the work area and into traffic before the brakes become effective. It will take several pumps of the brake pedal to seat the pads against the rotor.

➡**If a firm pedal is not achieved, it may be necessary to bleed the brakes.**

15. Check the brake fluid level in the reservoir and top off as needed.

16. Install the wheels and tighten the lug nuts.

17. Road test the vehicle.

Brake Calipers

REMOVAL & INSTALLATION

Calipers without Integral Parking Brake Mechanisms

SLIDING CALIPERS

✳✳ CAUTION

Brake dust may contain asbestos! Asbestos is harmful to your health. Never use compressed air to clean any brake component. A filtering mask should be worn during any brake repair.

There are typically three methods of securing a sliding caliper to its mounting bracket: with a retaining pin, with a key and bolt, or with a wedge and pin. On calipers which use the retaining pin method, you will find pins driven into the slot between the caliper and the caliper mount. On calipers which use the bolt and key method, a key (small piece of metal) is used between the caliper and the mounting bracket to allow the caliper to slide. The key is held in position by a lockbolt. On calipers which use the pin and wedge method, a wedge, retained by a pin, is used between the caliper and the mounting bracket in much the same manner as with the key and bolt method.

1. Loosen the lug nuts on the applicable wheels.

2. Raise and safely support the vehicle.

3. Remove the wheels.

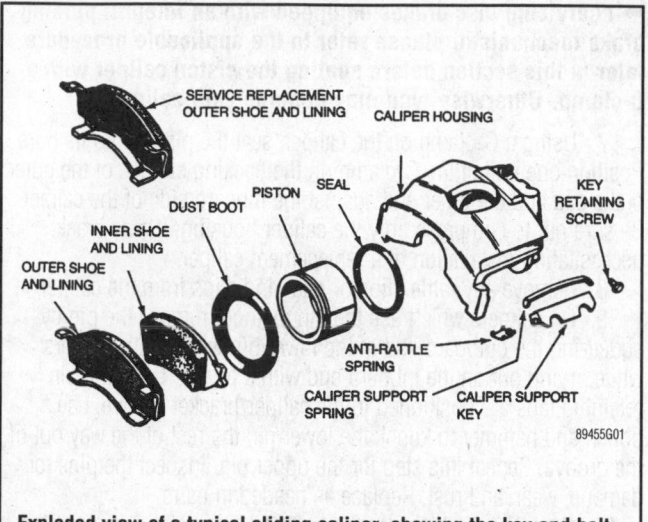

Exploded view of a typical sliding caliper, showing the key and bolt (retaining screw)

✳✳ CAUTION

Any brake fluid that is removed from the system should be discarded. Also, do not allow any brake fluid to come in contact with a painted surface; it will damage the paint. Also, brake fluid contains polyglycol ethers and polyglycols. Avoid contact with the eyes and wash your hands thoroughly after handling brake fluid. If you do get brake fluid in your eyes, flush your eyes with clean, running water for 15 minutes. If eye irritation persists, or if you have taken brake fluid internally, IMMEDIATELY seek medical assistance.

4. Remove some brake fluid from the brake fluid reservoir. Use a clean suction pump, a turkey baster, or an absorbent pad to do so. Never reuse any brake fluid.

5. Place a drain pan under the work area. Clean the brake pad and rotor area with spray brake cleaner.

6. Disconnect any electrical brake pad wear sensor.

Compress the upper pin tabs and pry on the inner end of the pin . . .

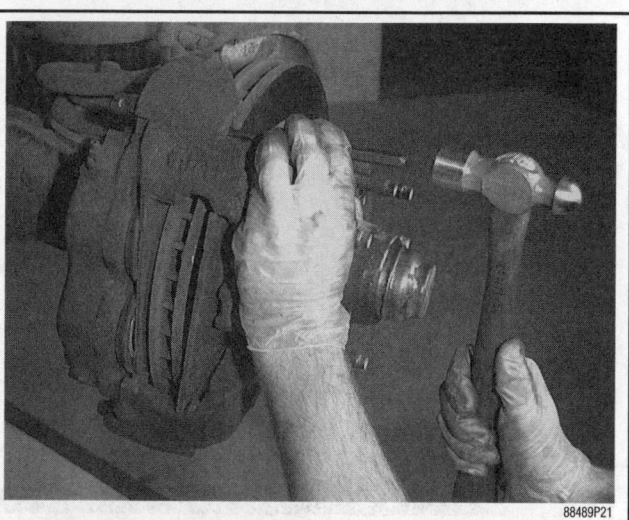

. . ., then use a hammer and punch to drive the pin out of the groove . . .

To remove a typical sliding caliper, remove the anti-rattle clips (if equipped)

. . . until it can be removed by hand

Perform the same for the lower pin as well . . .

➡️If servicing disc brakes equipped with an integral parking brake mechanism, please refer to the applicable procedure later in this section before seating the piston caliper with a C-clamp. Otherwise, you may damage your caliper.

7. Using a C-clamp on the caliper, seat the piston into its bore. Position one end of the C-clamp on the backing surface of the outer brake pad and the other end against the inboard side of the caliper. Be sure not to compress only the caliper housing; it may crack, necessitating installation of a replacement caliper.

8. Remove any rattle clips or retaining clips from the caliper.

9. On calipers which use the pin method, remove the pin by squeezing the outboard end of the lower pin with a pair of pliers while prying out on the inboard end with a prybar. Once the pin retaining tabs are positioned in the caliper/bracket groove, use a punch and hammer to knock the lower pin the rest of the way out of the groove. Repeat this step for the upper pin. Inspect the pins for damage, wear, and rust. Replace as needed in pairs.

10. On calipers which use the bolt and key method, remove the retaining bolt, then use a hammer and punch to drive the key out.

. . ., then pull the caliper off of the rotor and bracket

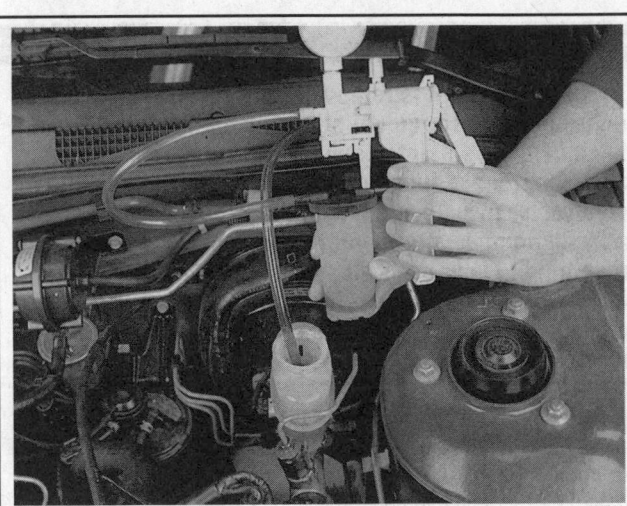

A vacuum pump setup can be used to draw brake fluid from the reservoir

Once the caliper is removed, the brake pads can be removed

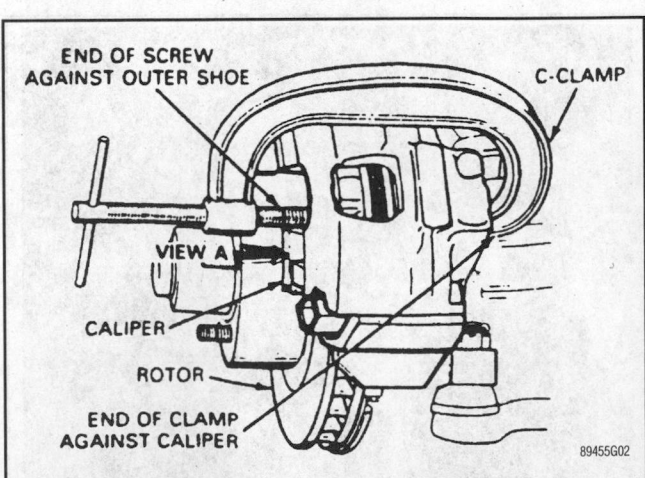

Use a large C-clamp to seat the caliper piston, be sure that one end of the clamp is positioned against the outer shoe—calipers without integral parking brake mechanisms

(Be careful not to lose the caliper support spring, if equipped.) Check parts for wear and replace as necessary.

11. On calipers which use the wedge and pin method, remove the retaining pin from the guide plate, then use a punch and hammer to tap out the guide plate. Inspect parts for wear and replace as necessary.

12. If the caliper is going to be removed for overhaul or replacement, loosen the brake hose, lift off the caliper and remove the brake hose completely. Immediately plug the open end of the rubber brake hose to prevent contamination of the brake fluid. If the brake hose was attached to the caliper with a banjo connection, be sure to remove and discard the two copper washers.

13. If the caliper does not require overhaul or replacement, prepare a length of wire (a coat hanger works well), cord, or a length of strong string to support the caliper. DO NOT let the caliper hang from the brake hose; it may be damaged.

14. Remove the caliper and suspend it from the wire.

15. If the brake pads came off the rotor with the caliper, remove them by prying the pads out of the caliper piston.

16. Inspect the caliper for fluid leakage, torn dust boots, or missing parts. Rebuild or replace the caliper if a problem is found.

17. Inspect the rubber brake hose for cracks or signs of rubbing against the body or steering components. Also, it is a good idea to replace them if they are over 10 years old to maintain proper brake operation.

18. Inspect metal lines for corrosion and kinks from road debris kicked up under the vehicle. If a problem is found, replace the line.

19. Inspect the rotor for non-machine grooves, heat stress cracks, glazing, minimum wear thickness, and disk run-out. Replace the rotor or have it machined to repair the damage.

20. Inspect the brake pads for minimum thickness, loose rivets, or glazing. Install new brake pads if any such problems exist.

To install:

21. Clean the sliding surfaces of the caliper and mounting bracket with spray brake cleaner and a small wire brush, then lubricate them with high temperature brake grease.

22. If necessary, place the pad(s) back onto the caliper or mounting bracket.

23. If the brake hose was removed, reattach it to the caliper. If so equipped, use two new copper washers for the banjo fitting.

24. Install the caliper onto its mounting bracket.

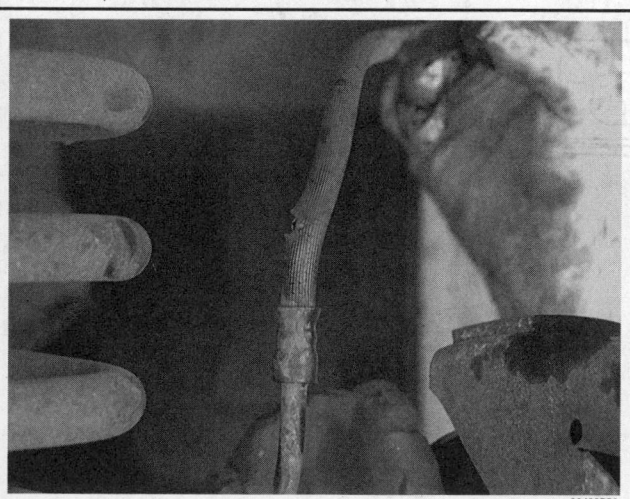

When inspecting the flexible brake hoses, check for rips (as shown), tears and cracks

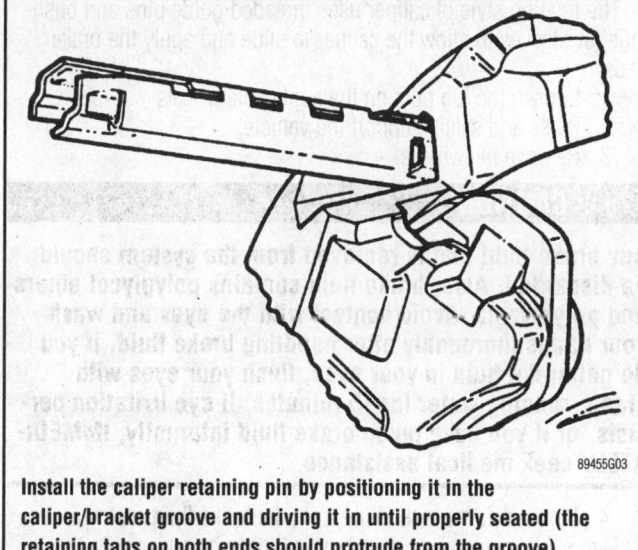

Install the caliper retaining pin by positioning it in the caliper/bracket groove and driving it in until properly seated (the retaining tabs on both ends should protrude from the groove)

25. For calipers which use the pin retaining method, use a hammer to tap the pins back into position, then install any anti-rattle clips.

26. For calipers which use the bolt and key method, use a prybar to lift the caliper up to create a gap into which the key and spring can slide. Tap the key and spring into position, then install the locking bolt and any anti-rattle clips. Tighten the locking bolt securely.

27. For calipers which use the wedge and pin method, slide the guide plates (wedge) between the gaps of the caliper and mounting bracket, then install the retaining pin. Tighten the retaining pin securely.

28. Reattach any electrical brake pad sensors.

✸✸ WARNING

Clean, high quality brake fluid is essential to the safe and proper operation of the brake system. You should always buy the highest quality brake fluid that is available. If the brake fluid becomes contaminated, drain and flush the system, then refill the master cylinder with new fluid. Never reuse any brake fluid. Any brake fluid that is removed from the system should be discarded. Also, do not allow any brake fluid to come in contact with a painted surface; it will damage the paint.

29. Bleed the brakes if a brake line was replaced, or the caliper was detached from a brake line.

30. Seat the brake pads, otherwise the vehicle may coast out of the work area and into traffic before the brakes become effective. It will take several pumps of the brake pedal to seat the pads against the rotor.

31. Check the brake fluid level in the reservoir and top off as needed.

32. Install the wheels and tighten the lug nuts.

33. Road test the vehicle.

FLOATING CALIPERS

✸✸ CAUTION

Brake dust may contain asbestos! Asbestos is harmful to your health. Never use compressed air to clean any brake component. A filtering mask should be worn during any brake repair.

The floating style of caliper uses threaded guide pins and bushings, or sleeves to allow the caliper to slide and apply the brake pads.

1. Loosen the lug nuts on the applicable wheels.
2. Raise and safely support the vehicle.
3. Remove the wheels.

❋❋ CAUTION

Any brake fluid that is removed from the system should be discarded. Also, brake fluid contains polyglycol ethers and polyglycols. Avoid contact with the eyes and wash your hands thoroughly after handling brake fluid. If you do get brake fluid in your eyes, flush your eyes with clean, running water for 15 minutes. If eye irritation persists, or if you have taken brake fluid internally, IMMEDIATELY seek medical assistance.

4. Remove some brake fluid from the brake fluid reservoir. Use a clean suction pump, a turkey baster (not to be returned to the kitchen), or an absorbent pad to do so. Never reuse any brake fluid.
5. Place a drain pan under the work area. Clean the brake pad and rotor area with spray brake cleaner.
6. Disconnect any electrical brake pad wear sensor.
7. If an anti-rattle spring is used and is not part of the brake pad, it can usually be pried off or pulled out.

➡ **If servicing disc brakes equipped with an integral parking brake mechanism, please refer to the applicable procedure later in this section before seating the piston caliper with a C-clamp. Otherwise, you may damage the caliper.**

8. Using a C-clamp on the caliper, seat the piston into its bore. Position one end of the C-clamp on the backing surface of the outer brake pad and the other end against the inboard side of the caliper. Be sure not to compress only the caliper; it may crack, necessitating installation of a replacement caliper.
9. Loosen and remove the guide pins from the caliper.

10. If the caliper is going to be removed for overhaul or replacement, loosen the brake hose, lift off the caliper and remove the brake hose completely. Immediately plug the open end of the rubber brake hose to prevent contamination of the brake fluid. If the brake hose was attached to the caliper via a banjo connection, be sure to remove and discard the two copper washers.
11. If the caliper does not require overhaul or replacement, prepare a length of wire (a coat hanger works well), cord, or a length of strong string to support the caliper. DO NOT let the caliper hang from the brake hose; it may be damaged.
12. Remove the caliper from the rotor, if equipped, the mounting bracket.

➡ **The pads may or may not come off with the caliper; this is normal.**

13. If the brake pads stay on the caliper, they can usually be tapped off with a hammer, or pried out by hand or with prytool.
14. If the brake pads remain on the bracket, when applicable, they can be removed from the bracket by hand.
15. Inspect the caliper for fluid leakage, torn dust boot, or missing parts. Rebuild or replace if a problem is found.
16. Inspect the rubber brake hose for cracks or signs of rubbing against the body or steering components. Install a new rubber hose if any such conditions exist. Also, it is a good idea to replace them if over 10 years old to maintain proper brake operation.
17. Inspect the metal lines for corrosion and kinks from road debris kicked up under the vehicle. If a problem is found, replace the line.
18. Inspect the rotor for non-machine grooves, heat stress cracks, glazing, minimum wear thickness, and disk run-out. Replace the rotor or have it machined to repair the damage.
19. Inspect the brake pads for minimum thickness, loose rivets, or glazing. If such a problem is found, new pads must be installed.
To install:
20. If equipped with a mounting bracket, clean the sliding surfaces of the caliper and mounting bracket with spray brake cleaner

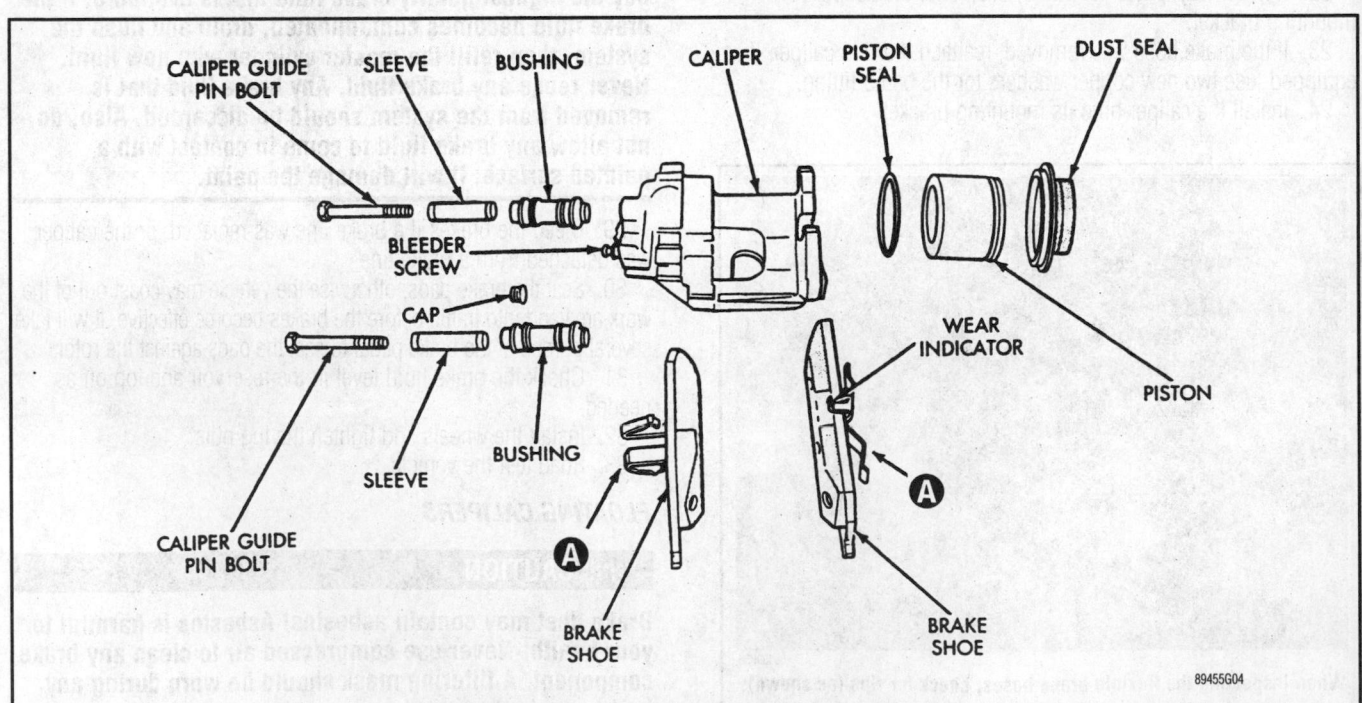

Exploded view of a typical floating caliper—when installing the brake pads, ensure that the retaining clips (A) are properly engaged in the caliper

89455G04

and a small wire brush, then lubricate them with high temperature brake grease.

21. If the brake hose was removed, reattach it to the caliper. If so equipped, use two new copper washers for the banjo fitting.

22. Transfer old pad hardware to the new pads, or install new hardware.

23. Clean and inspect the caliper guide pins, if they are okay, then lubricate them with high temperature brake grease.

24. On caliper mounted pads, position the pads on the caliper, then install the caliper on the rotor.

25. On bracket mounted pads, install the pads on the mounting bracket. Install the caliper on the rotor.

26. Tighten the caliper guide pins securely, and replace any anti rattle clips.

27. Connect any electrical brake pad sensors.

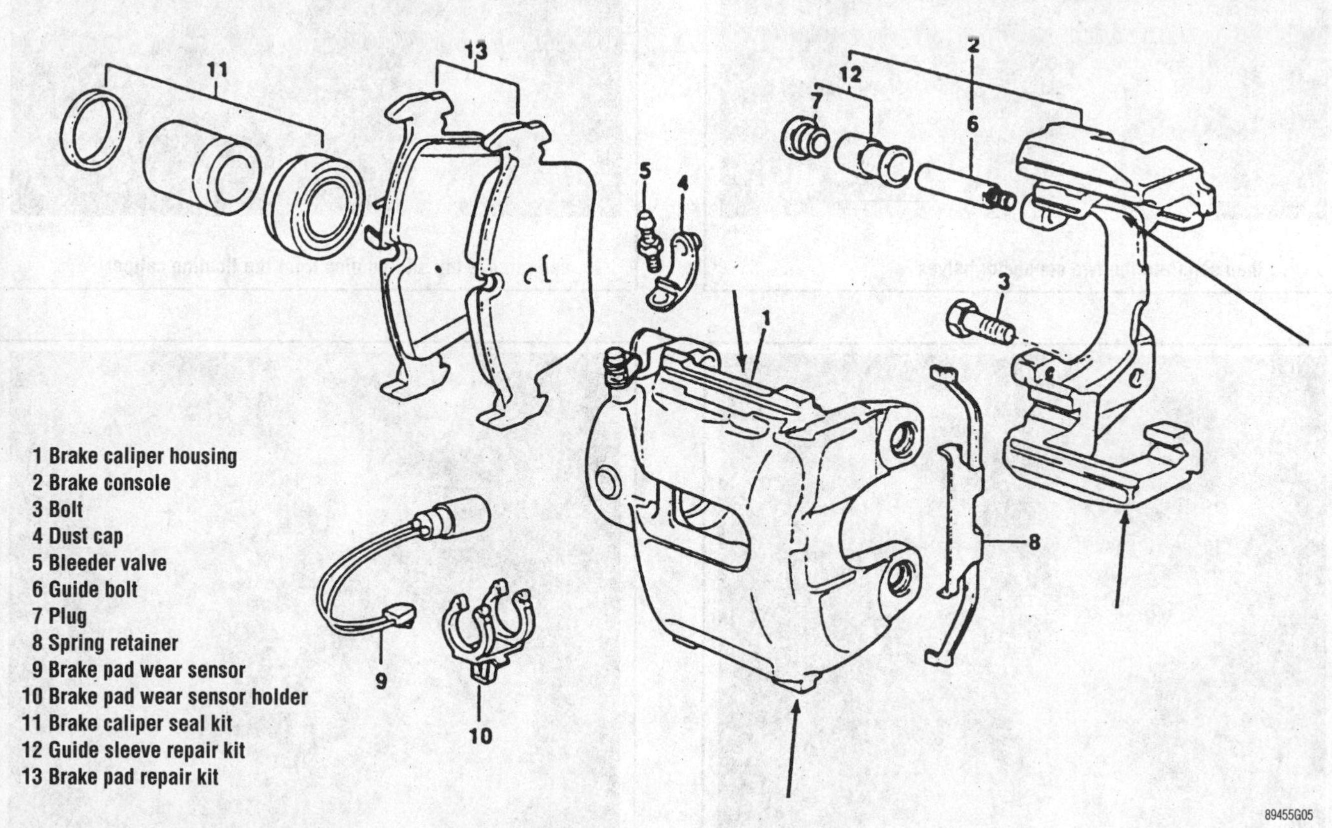

1 Brake caliper housing
2 Brake console
3 Bolt
4 Dust cap
5 Bleeder valve
6 Guide bolt
7 Plug
8 Spring retainer
9 Brake pad wear sensor
10 Brake pad wear sensor holder
11 Brake caliper seal kit
12 Guide sleeve repair kit
13 Brake pad repair kit

89455G05

Exploded view of another floating caliper—note that this vehicle is equipped with a pad wear sensor

89455P04

Some vehicles are equipped with brake pad wear sensors, indicated by the wire leading to the pad

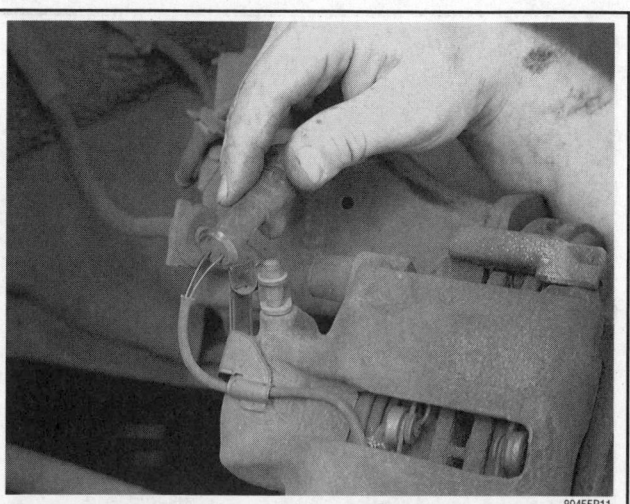

89455P11

To remove a typical floating caliper, disengage the sensor wire connector from its mounting clip (if equipped) . . .

. . . , then separate the two connector halves

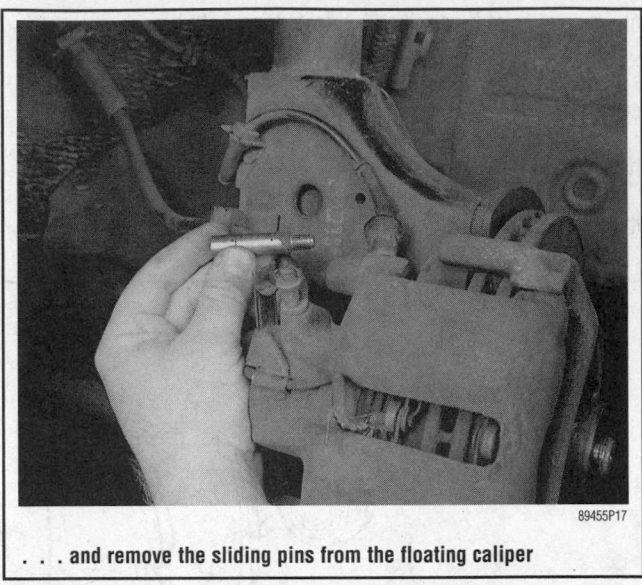

. . . and remove the sliding pins from the floating caliper

If equipped, remove the sliding pin covers . . .

If equipped, unfasten the sensor wire from its retaining clip . . .

. . . , then loosen . . .

. . . , then lift the caliper up and off of the mounting bracket

89455P22

Be sure to note the positions of any clips or springs on the caliper

⚹⚹ WARNING

Clean, high quality brake fluid is essential to the safe and proper operation of the brake system. You should always buy the highest quality brake fluid that is available. If the brake fluid becomes contaminated, drain and flush the system, then refill the master cylinder with new fluid. Never reuse any brake fluid. Any brake fluid that is removed from the system should be discarded. Also, do not allow any brake fluid to come in contact with a painted surface; it will damage the paint.

28. Bleed the brakes, if a brake line was replaced or the caliper was detached from a brake line.

29. Seat the brake pads, otherwise the vehicle may coast out of the work area and into traffic before the brakes become effective. It will take several pumps of the brake pedal to seat the pads against the rotor.

30. Check the brake fluid level in the reservoir and top off as needed.

31. Install the wheels and tighten the lug nuts.

32. Road test the vehicle.

FIXED CALIPERS

⚹⚹ CAUTION

Brake dust may contain asbestos! Asbestos is harmful to your health. Never use compressed air to clean any brake component. A filtering mask should be worn during any brake repair.

The fixed type caliper is bolted to the steering knuckle. The brake pads on this style of caliper are typically held in place by one or two retaining pins. Some other pads use hold down clips. It may not be necessary to remove the brake pads in order to remove the caliper.

1. Loosen the lug nuts on the applicable wheels.

2. Raise and safely support the vehicle.

⚹⚹ WARNING

Any brake fluid that is removed from the system should be discarded. Also, do not allow any brake fluid to come in contact with a painted surface; it will damage the paint.

3. Remove the wheels.

⚹⚹ CAUTION

Brake fluid contains polyglycol ethers and polyglycols. Avoid contact with the eyes and wash your hands thoroughly after handling brake fluid. If you do get brake fluid in your eyes, flush your eyes with clean, running water for 15 minutes. If eye irritation persists, or if you have taken brake fluid internally, IMMEDIATELY seek medical assistance.

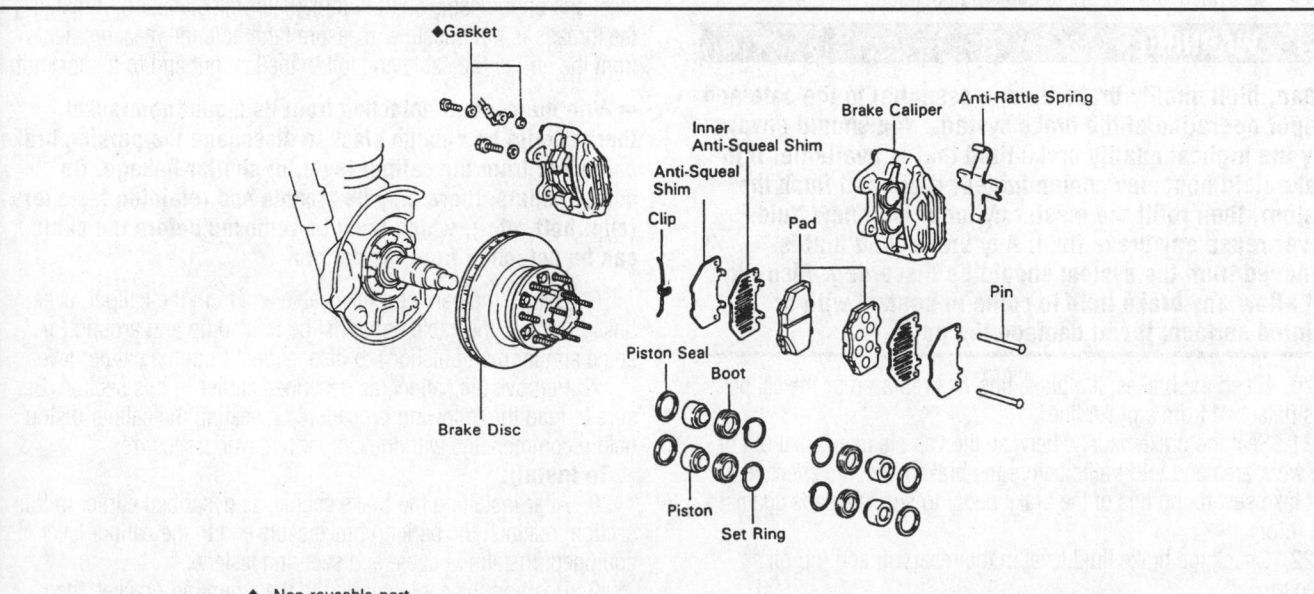

89455G08

Exploded view of a common four piston fixed caliper

4. Remove some brake fluid from the brake fluid reservoir. Use a clean suction pump, a turkey baster, or an absorbent pad to do so. Never reuse any brake fluid.

5. Place a drain pan under the work area. Clean the brake pad and rotor area with spray brake cleaner.

6. If equipped, disconnect any electrical brake pad wear sensor.

7. Although not necessary for caliper removal, the brake pads can now be removed from the caliper.

8. Loosen the caliper mounting bolts.

9. If the caliper is going to be removed for overhaul or replacement purposes, loosen the brake hose, remove the caliper bolts, and disconnect the brake line.

10. If the caliper does not require overhaul or replacement (in other words, you only need to remove it for access to some other component), prepare a length of wire (coat hanger), cord, or a length of strong string from which the caliper can be hung. DO NOT let the caliper hang from the brake hose; it may be damaged and need to be replaced. Remove the caliper and hang it from the wire.

11. Inspect the caliper for fluid leakage, torn dust boot, or missing parts. Rebuild or replace if a problem is found.

12. Inspect the rubber brake hose for cracks or signs of rubbing against the body or steering components. Install a new brake hose if any such damage is evident. Also, it is a good idea to replace them if over 10 years old to maintain proper brake operation.

13. Inspect the metal brake lines for corrosion and kinks from road debris kicked up under the vehicle. If a problem is found replace the line.

14. Inspect the rotor for non-machine grooves, heat stress cracks, glazing, minimum wear thickness, and disk run-out. Replace the rotor or have it machined to repair the damage.

15. Inspect the brake pads for minimum thickness, loose rivets, or glazing. If any such problem is found, new pads must be installed.

To install:

16. Install the caliper and tighten the mounting bolts securely.

17. If the brake hose was removed, reattach it to the caliper. If so equipped, use two new copper washers for the banjo fitting.

18. If removed, install the brake pads.

19. Reconnect any electrical brake pad sensors.

❊❊ WARNING

Clean, high quality brake fluid is essential to the safe and proper operation of the brake system. You should always buy the highest quality brake fluid that is available. If the brake fluid becomes contaminated, drain and flush the system, then refill the master cylinder with new fluid. Never reuse any brake fluid. Any brake fluid that is removed from the system should be discarded. Also, do not allow any brake fluid to come in contact with a painted surface; it will damage the paint.

20. Bleed the brakes, if a brake line was replaced or the caliper was detached from a brake line.

21. Seat the brake pads, otherwise the vehicle may coast out of the work area and into traffic before the brakes become effective. It will take several pumps of the brake pedal to seat the pads against the rotor.

22. Check the brake fluid level in the reservoir and top off as needed.

23. Install the wheels and tighten the lug nuts.

24. Road test the vehicle.

Calipers with Integral Parking Brake Mechanisms

The procedure to remove or replace the caliper and/or pads on vehicles equipped with disc brakes designed with integral parking brake mechanisms is essentially the same as disc brake calipers without integral parking brakes. There are usually two major differences between these two disc brake caliper designs.

➡ **For the actual caliper removal and installation process, refer to the applicable procedure earlier in this section. Read the following two procedures, and perform them in conjunction with the caliper procedures.**

REMOVING THE PARKING BRAKE CABLE

The first, and most obvious, difference is that, in one fashion or another, the parking brake cable is attached to the caliper. Before removing the caliper from the rotor, you must first disengage the parking brake cable from the caliper. To detach the parking brake cable from the caliper, perform the following:

➡ **This is a general procedure and may need slight alteration to apply fully to your specific vehicle. The most important thing to remember is to carefully inspect your caliper to identify the applicable parking brake cable components before disconnecting anything.**

1. Loosen the lug nuts on the applicable wheels.
2. Raise and safely support the vehicle.

➡ **Some vehicles, in fact, may be designed with front parking brake assemblies.**

3. Remove the wheels for easier access to the brake assembly.
4. Relieve the parking brake cable tension.
5. Carefully inspect the parking brake cable mounting and attaching (to the caliper) points. Most parking brake cable conduits are retained to a mounting bracket either by a jam nut and locknut setup, or by a retaining clip. Either remove the jam and locknuts, or pull the retaining clip off of the bracket, then disengage the cable conduit from the mounting bracket. If your vehicle utilizes jam and locknuts to secure the conduit onto the bracket, matchmark the nuts' locations on the cable conduit threads for reinstallation; if marking the threads is not possible, measure (and note the measurements) from the end of the cable conduit to the jam nut and to the locknut.

➡ **With the conduit detached from its mounting bracket, there should be enough slack to disengage the parking brake cable end from the caliper lever, or similar linkage. On some models, there may be a cable end retaining fastener (clip, bolt, etc.), which must be removed before the cable can be detached from the caliper.**

6. Detach the parking brake cable end from the caliper lever, or linkage. Often, the cable end must be twisted up and around (or some similar manipulation) to disengage it from the caliper lever.

7. Remove the caliper, as described earlier in this section. Be sure to read the following procedure on seating the caliper piston before commencing with the caliper removal procedure.

To install:

8. After installing the brake caliper, as described earlier in this section, reattach the parking brake cable end to the caliper lever. If equipped, install the cable end securing fastener.

9. Position the cable conduit in the mounting bracket, then either install the retaining clip, or the jam and locknuts. If equipped with jam and locknuts, position the nuts on the cable conduit so

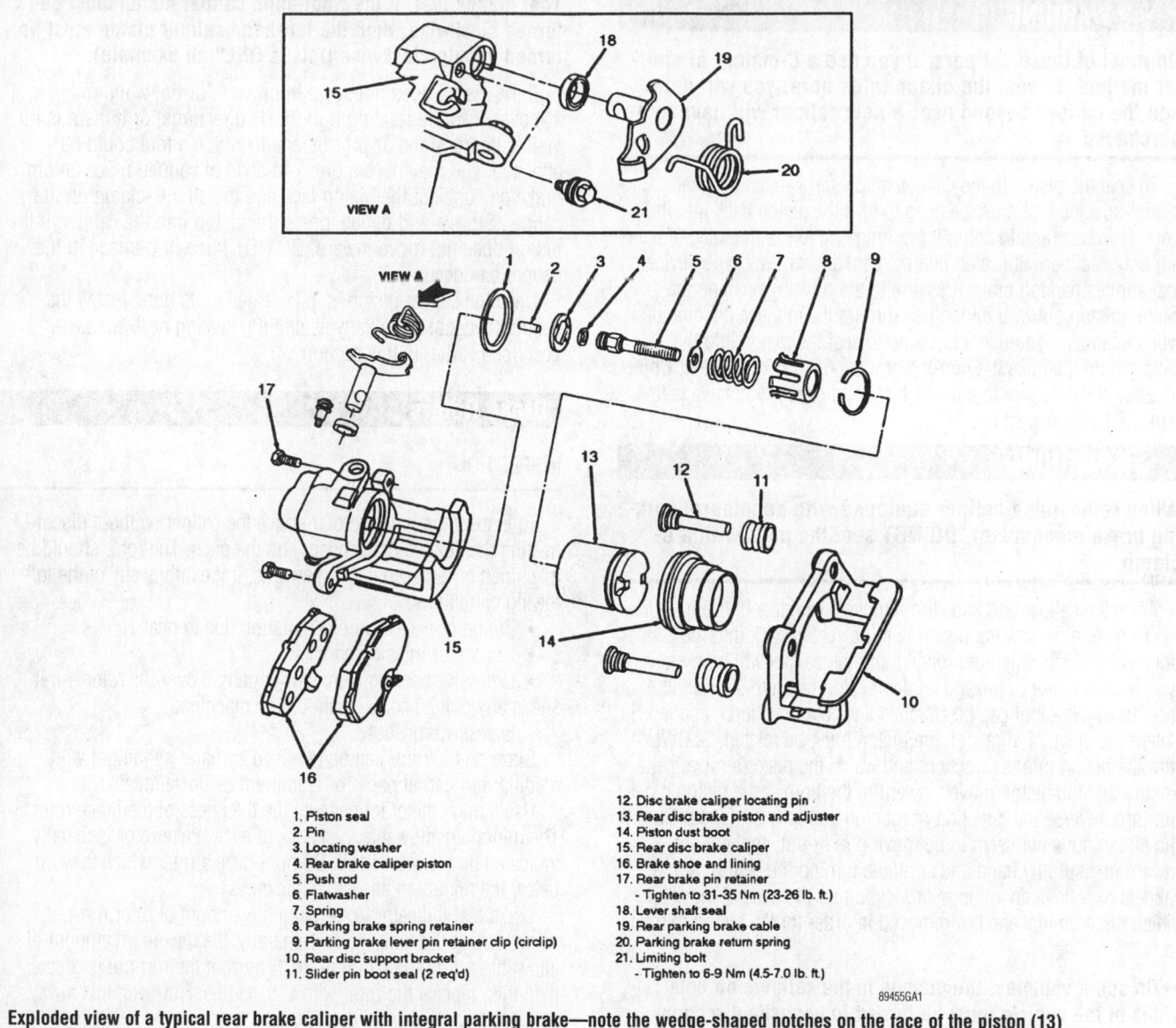

1. Piston seal
2. Pin
3. Locating washer
4. Rear brake caliper piston
5. Push rod
6. Flatwasher
7. Spring
8. Parking brake spring retainer
9. Parking brake lever pin retainer clip (circlip)
10. Rear disc support bracket
11. Slider pin boot seal (2 req'd)

12. Disc brake caliper locating pin
13. Rear disc brake piston and adjuster
14. Piston dust boot
15. Rear disc brake caliper
16. Brake shoe and lining
17. Rear brake pin retainer
    - Tighten to 31-35 Nm (23-26 lb. ft.)
18. Lever shaft seal
19. Rear parking brake cable
20. Parking brake return spring
21. Limiting bolt
    - Tighten to 6-9 Nm (4.5-7.0 lb. ft.)

89455GA1

Exploded view of a typical rear brake caliper with integral parking brake—note the wedge-shaped notches on the face of the piston (13)

that the nuts are positioned as before (using the marks on the threads or a ruler).

10. Adjust the parking brake cable tension, as described in Section 3.

11. Install the wheels and snug the lug nuts.

12. Lower the vehicle.

13. Tighten the lug nuts fully.

14. Depress the brake pedal a few times to ensure that the brake pads are fully seated.

✳✳ CAUTION

If you do not seat the pads before driving the vehicle, the first few times you apply the brake pedal the vehicle may not stop as anticipated; this could lead to an accident with a telephone pole or one of your neighbors' cars.

SEATING THE CALIPER PISTON

➡ **Be sure to read this entirely before commencing with caliper service.**

The second difference between brake calipers with and without integral parking brake mechanisms is in how the caliper pistons should be seated into their bores.

Whereas most pistons on calipers which are not equipped with integral parking brake mechanisms can be seated by using a large C-clamp, this is USUALLY not the case with calipers designed with integral parking brake mechanisms. Most integral parking brake calipers apply parking brake pressure to the rotor as follows: when the parking brake is applied, the cable pulls on the caliper lever. The lever, in turn, applies a rotational (spinning) movement to the caliper piston. The piston is designed much like an ordinary screw, so that when a rotational movement is applied to the piston, it slowly presses in against the rotor. To prevent having to constantly adjust the parking brake cable tension as the brake pads slowly wear down, the internal parking brake mechanism is designed with a ratcheting apparatus, which automatically readjusts the parking brake tension.

Since the caliper is designed to protrude from its bore when turned, usually, it cannot be seated in its bore in a conventional manner (with a large C-clamp).

✳ WARNING

On most of these calipers, if you use a C-clamp, or similar method, to seat the piston in its bore, you will damage the caliper beyond use. A new caliper will have to be purchased.

To seat the piston in the caliper, a spanner wrench or other model-specific tool must be used to turn the piston back into its bore. However (and to complicate things), a few of the integral parking brake calipers utilize an internal cam and/or lever type device that applies parking brake pressure to the rotor by pushing the caliper piston outward rather than turning it. On these uncommon type of calipers, you use a C-clamp to seat the piston into the caliper bore, just like the non-integral parking brake calipers. Unfortunately, the only way to tell which style of caliper you have is to remove it and inspect it.

✳ WARNING

When removing a caliper equipped with an integral parking brake mechanism, DO NOT seat the pads with a C-clamp.

Once the caliper and pads are removed, examine the caliper piston to determine how the piston is to be seated back into the caliper bore. All pistons which are rotated into the caliper will have some type of notch, slot or hexagonal depression or protrusion on its face, to which a tool can be attached and rotational force applied. To determine in which direction the piston must be rotated, SLOWLY turn the piston in one direction, and watch the piston's movement. Ensure that the piston moves inward in the bore. If the piston moves outward, reverse the direction of rotation and fully seat the piston. If the piston does not seem to be moving in or out, apply slight inward pressure by hand and continue turning the piston. Some models may have an adjuster or lockbolt on the back of the caliper which must be loosened or removed in order for the piston to rotate in.

➡ **On some vehicles, the pistons in the calipers on both sides of the vehicle must be turned in opposite directions.**

That means that, if the right-hand caliper piston must be turned clockwise, then the left-hand caliper piston must be turned counterclockwise (this is ONLY an example).

If the piston does not seem to move in or out while rotating, if it moves in while rotating it in BOTH directions, or if there is no visible depressions or protrusions to which a tool could be attached, you may have a press-in style of caliper. Place an old brake pad against the piston face and install a C-clamp on the caliper. Slowly, and gently, press the piston into the caliper. If the piston does not move inward, DO NOT force it! Damage to the caliper can occur.

Once the caliper piston is fully seated in its bore, install the caliper (depending on its type: sliding, floating or fixed) as described earlier in this section.

Brake Rotors

INSPECTION

To inspect the brake rotor, remove the caliper (without disconnecting the flexible brake hose) and the pads. The rotor should be machined or replaced with a new one, if it exhibits any of the following conditions:

- Bluing or excessive discoloration due to heat
- Cracks, or missing chunks
- Excessive scoring (run your fingernail over the rotor—if it snags any of the scores, it should be machined)
- Excessive run-out

Glaze on the rotor can be removed by hand-sanding it with medium grit garnet paper or aluminum oxide sandpaper.

Use a micrometer to measure the thickness of the brake rotor. The minimum allowable thickness of each brake rotor is usually indicated on the rotor itself. Do not utilize a rotor which is worn below the minimum allowable thickness

Use a dial indicator to measure the amount of rotor run-out, while turning the brake rotor. Generally, the maximum amount of allowable run-out is 0.006 in. (0.15mm); if the run-out is greater than this, replace the rotor with a good one. However, it is always better to have less rotor run-out.

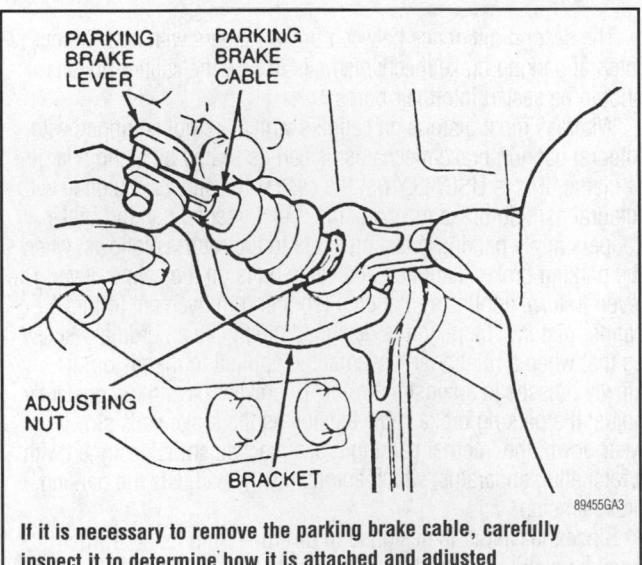

If it is necessary to remove the parking brake cable, carefully inspect it to determine how it is attached and adjusted

Use a micrometer to measure the rotor thickness, and replace it if it is below specifications

REMOVAL & INSTALLATION

Rotors mount in one of 2 ways: either directly on the hub (held in place by the wheels or small fasteners), which are referred to as non-integral (they are not one piece with the hub), or are integral with the hub.

➡ **On some vehicles, the manufacturer installs retaining clips over one or two of the wheel lugs to hold the rotor in place during assembly. Although it is generally thought that these retainers are not necessary and may be discarded, it is a good idea to reinstall them anyway (better safe than sorry). Other manufacturers use one or two small machine screws to hold the rotor in place on the hub; these screws MUST be reinstalled.**

Non-Integral Rotors

1. Loosen the lug nuts on the applicable wheels.
2. Raise and safely support the vehicle.
3. Remove the wheels.
4. Clean the brake assembly thoroughly with spray brake cleaner.

If the rotor is equipped with holes and is difficult to remove, it can be loosened using two small bolts . . .

To remove the rotor, remove the disc brake caliper . . .

. . . , then removed by hand—non-integral rotors

. . . , then pull the rotor off by hand—non-integral rotors

5. Remove the caliper.
6. If any rotor retainers are present, remove them. The push-nut type of retainer is usually damaged during removal; discard the old ones and purchase new ones.
7. Remove the rotor. On some vehicles, the rotor simply slides off the wheel studs. However, some rotors are pressed into place and must be removed by screwing bolts in the threaded holes provided, thereby forcing the rotor off the hub. Other rotors, not equipped with the threaded holes for press-off bolts, may require the use of a puller to dislodge them from the hub.

➡ **The rotor may be rusted in place. Spray the area liberally with WD-40® , Liquid Wrench® or equivalent and tap the rotor loose.**

To install:

➡ **New rotors come with an oily, rust-preventive coating on the braking surface. This coating can be removed with brake parts cleaner or most cleaners which are good for oil removal. Be sure that all traces of the coating are removed. Allow the rotor to dry before installation.**

8. Position the rotor on the hub and install any retainers.
9. Install the caliper.
10. Install the wheels.
11. Lower the vehicle.
12. Seat the brake pads, otherwise the vehicle may coast out of the work area and into traffic before the brakes become effective. It will take several pumps of the brake pedal to seat the pads against the rotor.
13. Check the brake system for proper operation.

Integral Rotor/Hub Assemblies

NON-SEALED HUB/BEARING ASSEMBLIES

1. Loosen the lug nuts on the applicable wheels.
2. Raise and safely support the vehicle.
3. Remove the wheels.
4. Clean the brake assembly thoroughly with spray brake cleaner.
5. Remove the caliper and suspend it out of the way with wire.
6. Remove the hub grease cap.

Remove the cotter pin . . .

Pry loose the grease cap . . .

. . ., then pull the locking cap off of the end of the axle shaft

. . ., then remove it from the rotor hub

Remove the adjusting nut . . .

. . . , then pull the outer bearing out of the hub

Remove the hub/rotor assembly—integral non-sealed hub/bearing (except 4WD vehicle front rotors)

7. Remove the cotter pin and wheel bearing nut locking cap. Discard the cotter pin.

8. Remove the wheel bearing nut.

➡ On some vehicles, a left-hand threaded nut is used on the right wheel spindle. Turn this locknut clockwise to loosen.

9. Remove the brake rotor/hub, washer and bearings as an assembly. Be careful not to let the outer wheel bearing fall out of the hub during removal.

10. If the brake rotor is to be machined or replaced, remove the wheel bearings and grease seal.

To install:

➡ New rotors come with an oily, rust-preventive coating on the braking surface. This coating can be removed with brake parts cleaner or most cleaners which are good for oil removal. Be sure that all traces of the coating are removed. Allow the rotor to dry before installation.

11. If removed, install the inner wheel bearing and a new grease seal.

12. Be sure the bearings and hub contain an adequate amount of clean wheel bearing grease.

13. Position the rotor/hub assembly on the spindle. Keep the hub centered on the spindle to prevent damage to the grease seal and spindle threads.

14. Install the outer wheel bearing, washer and wheel bearing nut.

15. Properly adjust the wheel bearing. On most vehicles (those with tapered roller bearing) this is done by tightening the adjusting nut until drag is felt on the bearing while rotating the rotor;, then, back off the nut about ¼ turn (90°). The rotor/hub should turn freely with no end-play. If you are in any doubt about the proper adjustment procedure, refer to the appropriate model-specific section in this manual.

➡ On some vehicles (those with ball bearings), the nut is not so much an adjuster as a locknut. Tighten this nut to the manufacturer's specifications.

16. Install the wheel bearing nut cover and a new cotter pin.
17. Install the caliper.
18. Install the wheels and snug the lug nuts.
19. Lower the vehicle.
20. Tighten the lug nuts fully.
21. Seat the brake pads, otherwise the vehicle may coast out of the work area and into traffic before the brakes become effective. It will take several pumps of the brake pedal to seat the pads against the rotor.
22. Check the brake system for proper operation.

SEALED HUB/BEARING ASSEMBLIES

These are unitized hubs that contain the bearing assembly. The hub/bearing unit is replaced as an assembly.

1. Loosen the lug nuts on the applicable wheels.
2. Raise and safely support the vehicle.
3. Remove the wheels.
4. Clean the brake assembly thoroughly with spray brake cleaner.
5. Remove the caliper and suspend it out of the way with wire.
6. On models so equipped, disconnect the ABS sensor wire.
7. Working through the hole provided in the rotor, or working from behind the rotor, remove the hub retaining bolts or nuts.
8. Remove the hub assembly.

To install:

➡ New rotors come with an oily, rust-preventive coating on the braking surface. This coating can be removed with brake parts cleaner or most cleaners which are good for oil removal. Be sure that all traces of the coating are removed. Allow the rotor to dry before installation.

9. Clean the mounting surfaces of the hub and spindle.
10. Install the hub assembly and tighten the bolts/nuts securely.
11. Connect the ABS wire on models so equipped.
12. Install the caliper.
13. Install the wheels and snug the lug nuts.
14. Lower the vehicle.
15. Tighten the lug nuts fully.
16. Seat the brake pads, otherwise the vehicle may coast out of the work area and into traffic before the brakes become effective. It will take several pumps of the brake pedal to seat the pads against the rotor.
17. Check the brake system for proper operation.

DRUM BRAKES

Brake Drums

➡Most vehicles have rubber plugs in the backing plates that are removed to access the brake adjusters. However, some vehicles are built with what are called knock-out plugs. These are areas in the backing plate that are made to be knocked out with a hammer and punch. Once the drum is off, the knock-out plug is removed and a rubber plug used in its place.

INSPECTION

➡While the brake drum is removed from the vehicle, inspect the wheel cylinder for damage and leakage.

1. Remove the brake drum from the vehicle.

❊❊ CAUTION

Older brake pads or shoes may contain asbestos, which has been determined to be a cancer causing agent. Never clean the brake surfaces with compressed air! Avoid inhaling any dust from any brake surface! When cleaning brake surfaces, use a commercially available brake cleaning fluid.

2. Thoroughly clean the brake drum.
3. Inspect the brake drum for cracks, scores deep grooves, etc. A damaged drum is unsafe for use, and should be replaced immediately. Do not attempt to weld a cracked drum. If the drum exhibits scoring, and there is enough metal left on the inside diameter of the drum, have the drum cut by a qualified automotive machine shop. Slight scoring can be smoothed using emery cloth.
4. Inspect the drum for excessive wear by measuring the inside diameter of the brake drum with a caliper gauge. The maximum inside drum diameter allowable should be imprinted in the drum itself.
5. If the brake drum exhibits damage, or if the inside diameter is larger than specified, replace it with a new one.

REMOVAL & INSTALLATION

Brake drums are either separate components or an integral part of the hub assembly. Non-integral brake drums are held onto the axle flange or hub by the wheel and lug nuts; once the wheel is removed, the brake drum can be pulled off of the axle flange. Integral (with the hub assembly) brake drums are combined with the bearing hub to comprise one piece, which means that the wheel bearings must be disturbed (loosened or removed) in one way or another to remove the drum/hub assembly.

❊❊ WARNING

If the drum is excessively difficult to remove, loosen the brake pads by adjusting their position with a brake spoon. Access for adjusting the brake pads is often gained through a small hole in the backing plate. If a brake drum is forced off of an axle flange without loosening the brake pads, damage can occur to the brake or axle components.

Non-integral drums (those that are not part of the hub) are usually fairly easy to remove. There are always exceptions to the rule, how-

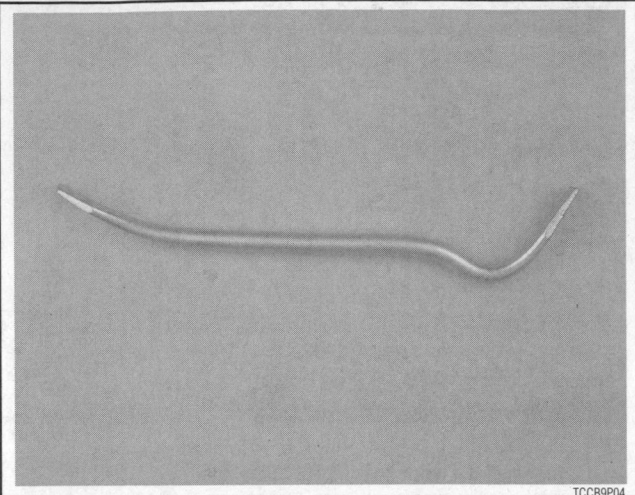

TCCB9P04

A brake spoon can be used to back off the shoe adjustment to allow drum removal

ever. There are drums that are retained to the hub with one or two small bolts. Some drums can be drawn off of the hub by installing two small bolts into threaded holes in the drum; as these bolts are tightened, they slowly press the drum off of the hub. Occasionally a drum is difficult to remove because it binds on the hub flange; these must be worked off by prying gently between the drum and backing plate while applying penetrating oil to the drum/flange contact point. Some older vehicles have a drum assembly that fits over splines on the end of the axle shaft. Others just rust in place. If this occurs, just spray the area around each lug stud and the hub flange with a penetrant such as WD-40®, Liquid Wrench® or equivalent. Let the stuff work for a while, then try pulling or prying the drum off.

Non-Integral Drums

❊❊ CAUTION

It is always a good idea to wear eye protection when working on brake components, especially drum brakes. Drum brakes often use powerful springs which could cause severe eye injury if they accidentally break.

FREE-MOUNTED TYPE

❊❊ CAUTION

Brake shoes may contain asbestos, which is a known cancer-causing agent. As soon as the drum is removed, generously spray the entire brake assembly with brake parts cleaner. Let it dry before proceeding. It's a good idea to wear a filter mask when doing brake work.

➡Some vehicles are built with retainers threaded over 2 or more lug studs to hold the drum in place during assembly. Although these retainers may not be necessary (according to the manufacturer), it may be a good idea to reinstall new retainers anyhow.

1. Loosen the lug nuts on the applicable wheels.
2. Raise and safely support the vehicle.

To remove a free-mounted brake drum, first safely raise the vehicle and remove the wheel . . .

. . . , then grasp hold of the drum and pull it from the axle flange and brake shoes

3. Remove the wheels.

4. If necessary, remove and discard the retainers holding the drum to the hub.

5. If applicable, back off the parking brake adjustment.

6. Back off the brake adjustment until the wheels rotate freely, as follows:

a. On vehicles with a starwheel-type adjuster: Remove the plug on the backing plate, then insert a thin prytool and a brake spoon into the slot. Hold the adjuster lever away from the adjuster wheel with the thin prytool and back of the adjuster wheel with the brake spoon.

b. On vehicles with an expanding-type adjuster, remove the plug and rotate the adjuster screw (usually in an upward motion).

c. On vehicles with ratcheting-type adjusters, remove the plug and insert a thin punch in the hole until it contacts the adjuster assembly pivot. Apply side pressure on this pivot point to allow the adjuster quadrant to ratchet and release the brake adjustment.

d. Some vehicles, notably with manual adjusters, use adjusting cams. On these vehicles, the cam can be turned back from behind the backing plate.

7. Grasp the drum and pull it off the hub.

➡On some vehicles, the drum won't come off even with the shoes completely backed off. This is due to the drum binding on the hub boss. The safest way to remove the drum when this happens, is to spray the binding point with lubricant and to carefully pry between the hub and backing plate. Use a small prybar and pry at various points while rotating the drum. It helps to occasionally tap the hub with a deadblow, or brass mallet.

8. Spray the brake shoe assembly thoroughly with brake parts cleaner and let it dry. Similarly, spray the inside of the drum.

9. Inspect the drum for wear and/or damage, such as deep grooves, excessive thinness, cracks, etc. Machine or replace the drum as necessary. When machining, observe the maximum diameter specification. The maximum machining diameter is stamped into the drum. If the drum braking surface shows signs of blue discoloration, overheating is indicated. If the bluing is extensive the drum must be replaced. Extensive bluing indicates a weakening of the metal.

To install:

➡New brake drums come with an oily, rust-preventive coating on the braking surface. This coating can be removed with brake parts cleaner or most cleaners which are good for oil removal. Be sure that all traces of the coating are removed. Allow the drum to dry before installation.

10. If a new brake drum is being installed, remove the protective coating from the inner braking surface.

11. Adjust the brake shoes to just smaller than the inside diameter of the brake drum.

12. Slide the brake drum onto the hub. Be sure that the brake shoes are not dragging on the brake drum. Install new brake drum retainers.

13. Install the wheels and tighten the lug nuts in a star pattern until tight.

14. Adjust the brakes shoes.

15. Adjust the parking brake.

16. Install the rubber plug in the access hole.

17. Lower the vehicle. To activate the adjusters, some vehicles require you to make several quick pulls on the parking brake lever. On most, however, several short back-ups, about 10 ft. (3m) each, should do it.

18. Road test the vehicle and check for proper brake operation.

FORCE-FIT TYPE

✳✳ CAUTION

Brake shoes may contain asbestos, which is a known cancer-causing agent. As soon as the drum is removed, generously spray the entire brake assembly with brake parts cleaner. Let it dry before proceeding. It's a good idea to wear a filter mask when doing brake work.

1. Loosen the lug nuts on the applicable wheels.

2. Raise and safely support the vehicle.

3. Remove the wheels.

4. If necessary, remove and discard the retainers holding the drum to the hub.

5. If applicable, back off the parking brake adjustment.

6. Back off the brake adjustment until the wheels rotate freely, as follows:

If equipped with threaded holes, it is possible to press a force-fit type drum off of the hub using bolts, as shown

a. On vehicles with a starwheel-type adjuster: Remove the plug on the backing plate, then insert a thin prytool and a brake spoon into the slot. Hold the adjuster lever away from the adjuster wheel with the thin prytool and back of the adjuster wheel with the brake spoon.

b. On vehicles with an expanding-type adjuster, remove the plug and rotate the adjuster screw (usually in an upward motion).

c. On vehicles with ratcheting-type adjusters, remove the plug and insert a thin punch in the hole until it contacts the adjuster assembly pivot. Apply side pressure on this pivot point to allow the adjuster quadrant to ratchet and release the brake adjustment.

d. Some vehicles, notably with manual adjusters, use adjusting cams. On these vehicles, the cam can be turned back from behind the backing plate.

7. Thread the proper size bolts into the holes provided in the drum until each contacts the hub. Turn the bolts evenly, a little at a time, until the drum slides free.

8. Grasp the drum and remove it from the axle flange or hub assembly. Remove the forcing bolts.

9. Spray the brake assembly thoroughly with brake parts cleaner and let it dry. Similarly, spray the inside of the drum.

10. Inspect the drum for wear and/or damage, such as deep grooves, excessive thinness, cracks, etc. Machine or replace the drum as necessary. When machining, observe the maximum diameter specification. The maximum machining diameter is stamped into the drum. If the drum braking surface shows signs of blue discoloration, overheating is indicated. If the bluing is extensive the drum must be replaced. Extensive bluing indicates a weakening of the metal.

To install:

➡ **New brake drums come with an oily, rust-preventive coating on the braking surface. This coating can be removed with brake parts cleaner or most cleaners which are good for oil removal. Be sure that all traces of the coating are removed. Allow the drum to dry before installation.**

11. If a new brake drum is being installed, remove the protective coating from the inner braking surface.

12. Adjust the brake shoes to match the inside diameter of the brake drum.

13. Slide the brake drum onto the hub. Install 2 wheel lug nuts and tighten them, forcing the drum into place on the hub. Remove the lug nuts, then, if equipped, install new drum retainers.

14. Install the wheels.

15. Adjust the brake shoes.

16. Adjust the parking brake.

17. Install the rubber plug in the access hole.

18. Lower the vehicle. To activate the adjusters, some vehicles require you to make several quick pulls on the parking brake lever. On most, however, several short back-ups, about 10 ft. (3m) each, should do it.

19. Road test the vehicle and check for proper brake operation.

BOLTED-IN-PLACE TYPE

> ✳✳ **CAUTION**
>
> **Brake shoes may contain asbestos, which is a known cancer-causing agent. As soon as the drum is removed, generously spray the entire brake assembly with brake parts cleaner. Let it dry before proceeding. It's a good idea to wear a filter mask when doing brake work.**

1. Loosen the lug nuts on the applicable wheels.

2. Raise and safely support the vehicle.

3. Remove the wheels.

4. If applicable, back off the parking brake adjustment.

5. Back off the brake adjustment until the wheels rotate freely, as follows:

a. On vehicles with a starwheel-type adjuster: Remove the plug on the backing plate, then insert a thin prytool and a brake spoon into the slot. Hold the adjuster lever away from the adjuster wheel with the thin prytool and back of the adjuster wheel with the brake spoon.

b. On vehicles with an expanding-type adjuster, remove the plug and rotate the adjuster screw (usually in an upward motion).

c. On vehicles with ratcheting-type adjusters, remove the plug and insert a thin punch in the hole until it contacts the adjuster assembly pivot. Apply side pressure on this pivot point to allow the adjuster quadrant to ratchet and release the brake adjustment.

d. Some vehicles, notably with manual adjusters, use adjusting cams. On these vehicles, the cam can be turned back from behind the backing plate.

6. Remove the drum-to-hub attaching bolts.

7. Grasp the drum and remove it from the axle flange or hub assembly.

8. Spray the brake assembly thoroughly with brake parts cleaner and let it dry. Similarly, spray the inside of the drum.

➡ **On some vehicles, the drum won't come off even with the shoes completely backed off. This is due to the drum binding on the hub boss. The safest way to remove the drum when this happens, is to spray the binding point with lubricant and pry, carefully between the hub and backing plate. Use a small prybar and pry at various points while rotating the drum. It helps to occasionally rap the hub with a deadblow, or brass mallet.**

9. Inspect the drum for wear and/or damage, such as deep grooves, excessive thinness, cracks, etc. Machine or replace the drum as necessary. When machining, observe the maximum diameter specification. The maximum machining diameter is stamped into the drum. If the drum braking surface shows signs of blue discol-

oration, overheating is indicated. If the bluing is extensive the drum must be replaced. Extensive bluing indicates a weakening of the metal.

To install:

→**New brake drums come with an oily, rust-preventive coating on the braking surface. This coating can be removed with brake parts cleaner or most cleaners which are good for oil removal. Be sure that all traces of the coating are removed. Allow the drum to dry before installation.**

10. If a new brake drum is being installed, remove the protective coating from the inner braking surface.

11. Adjust the brake shoes to match the inside diameter of the brake drum.

12. Slide the brake drum onto the hub. Be sure that the brake shoes are not dragging on the brake drum.

13. Install the drum-to-hub attaching bolts and tighten them securely.

14. Install the wheels.

15. Adjust the brakes as follows:

   a. Adjust the brake shoes so that you can feel a slight drag on, or hear a scraping noise coming from the wheel when you spin it.

   b. Back the shoes off just until the drag is no longer felt, or the rasping noise is no longer heard.

16. Adjust the parking brake.

17. Install the rubber plug in the access hole.

18. Lower the vehicle. To activate the adjusters, some vehicles require you to make several quick pulls on the parking brake lever. On most, however, several short back-ups, about 10 ft. (3m) each, should do it.

19. Road test the vehicle and check for proper brake operation.

Integral Drum/Hub Assemblies

❋❋ CAUTION

It is always a good idea to wear eye protection when working on brake components, especially drum brakes. Drum brakes often use powerful springs which could cause severe eye injury if they accidentally break.

Some Rear Wheel Drive (RWD) front drums and some Front Wheel Drive (FWD) rear drums are designed with the bearing hub as an integral assembly with the drum.

❋❋ CAUTION

Brake shoes may contain asbestos, which is a known cancer-causing agent. As soon as the drum is removed, generously spray the entire brake assembly with brake parts cleaner. Let it dry before proceeding. It's a good idea to wear a filter mask when doing brake work.

1. Raise and safely support the vehicle.

2. Remove the wheels.

3. If applicable, back off the parking brake adjustment.

4. Back off the brake adjustment until the wheels rotate freely.

   a. On vehicles with a starwheel-type adjuster: Remove the plug on the backing plate, then insert a thin prytool and a brake spoon into the slot. Hold the adjuster lever away from the adjuster wheel with the thin prytool and back of the adjuster wheel with the brake spoon.

   b. On vehicles with an expanding-type adjuster, remove the plug and rotate the adjuster screw (usually in an upward motion).

   c. On vehicles with ratcheting-type adjusters, remove the plug and insert a thin punch in the hole until it contacts the adjuster assembly pivot. Apply side pressure on this pivot point to allow the adjuster quadrant to ratchet and release the brake adjustment.

   d. Some vehicles, notably with manual adjusters, use adjusting cams. On these vehicles, the cam can be turned back from behind the backing plate.

5. Remove the hub grease cap.

6. Remove the cotter pin and wheel bearing adjusting nut cover. Discard the cotter pin.

7. Remove the wheel bearing nut.

❋❋ WARNING

On some vehicles, a left-hand threaded nut is used on the right wheel spindle. Turn this locknut clockwise to loosen, otherwise damage to the spindle threads will occur.

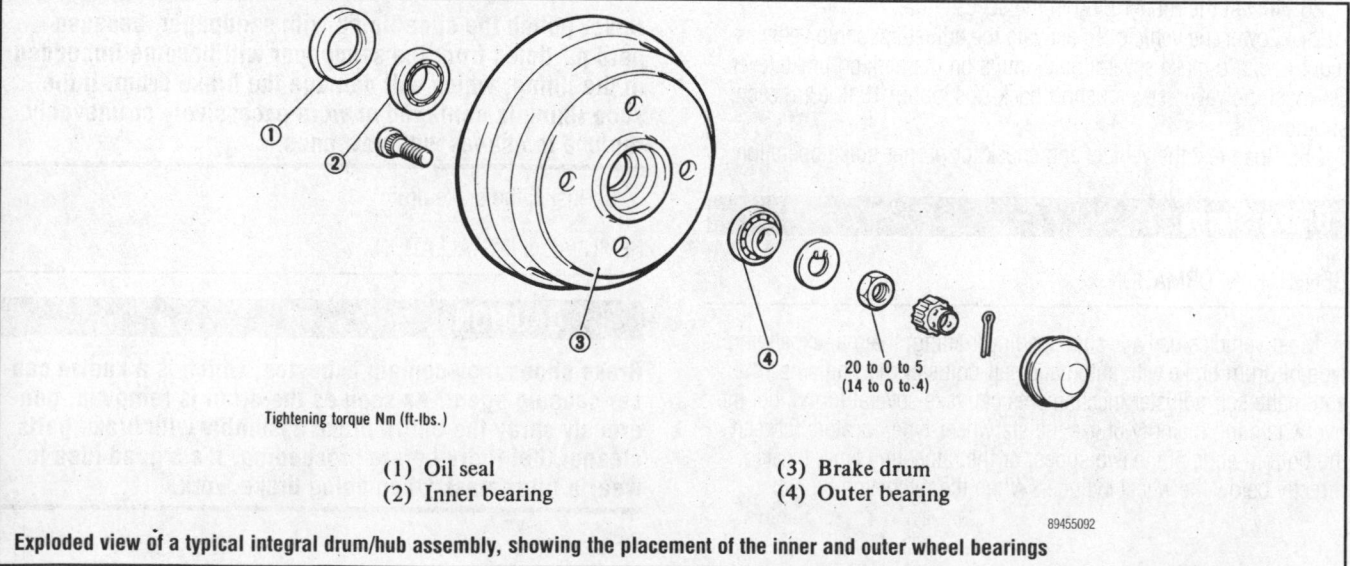

Tightening torque Nm (ft-lbs.)

20 to 0 to 5
(14 to 0 to 4)

(1) Oil seal
(2) Inner bearing

(3) Brake drum
(4) Outer bearing

89455092

Exploded view of a typical integral drum/hub assembly, showing the placement of the inner and outer wheel bearings

8. Remove the brake drum, washer and bearings as an assembly. Be careful not to let the outer wheel bearing fall out of the hub during removal.

9. Spray the brake assembly thoroughly with brake parts cleaner and let it dry. Similarly, spray the inside of the drum.

10. Remove the brake drum/hub assembly. Inspect the drum for wear and/or damage. Machine or replace as necessary. When machining, observe the maximum diameter specification. The maximum machining diameter is stamped into the drum. If the drum braking surface shows signs of blue discoloration, overheating is indicated. If the bluing is extensive the drum/hub assembly must be replaced. Extensive bluing indicates a weakening of the metal.

➡ If the brake drum is to be machined or replaced, remove the inner wheel bearing and grease seal.

To install:

➡ New brake drums come with an oily, rust-preventive coating on the braking surface. This coating can be removed with brake parts cleaner or most cleaners which are good for oil removal. Be sure that all traces of the coating are removed. Allow the drum to dry before installation.

11. If a new brake drum is being installed, remove the protective coating from the inner braking surface.

12. If removed, install the inner wheel bearing and a new grease seal.

13. Be sure the bearings and hub contain an adequate amount of clean wheel bearing grease.

14. Adjust the distance between the brake shoes to match the inner diameter of the brake drum.

15. Position the brake drum on the spindle. Keep the drum centered on the spindle to prevent damage to the grease seal and spindle threads.

16. Install the outer wheel bearing, washer and wheel bearing nut.

17. Properly adjust the wheel bearing; refer to the appropriate model-specific section in this manual.

18. Install the wheel bearing nut cover and a new cotter pin.

19. Install the hub grease cap.

20. Install the wheels.

21. Adjust the brake shoes.

22. Adjust the parking brake.

23. Install the rubber plug in the access hole.

24. Lower the vehicle. To activate the adjusters, some vehicles require you to make several quick pulls on the parking brake lever. On most, however, several short back-ups, about 10 ft. (3m) each, should do it.

25. Road test the vehicle and check for proper brake operation.

Brake Shoes

GENERAL INFORMATION

Most vehicles use a 2-shoe leading/trailing, internal expanding type of drum brake with automatic self-adjuster mechanisms. The automatic self-adjuster mechanisms can take several forms, but the overwhelming majority utilize the starwheel-type, located between the bottom ends of the two shoes, or the ratcheting type, located directly below the wheel cylinder. When the ratcheting type of adjuster is used, the lower ends of the brake shoes usually rest on an anchor plate.

➡ On some vehicles, notably those with unitized rear hubs, and some vehicles with full-floating axles, not only does the brake drum have to be removed, but the hub assembly must be removed as well.

✳✳ CAUTION

Brake shoes must always be replaced as an axle set. That is, do not just replace the shoes on one side of the vehicle. Replace them on both sides. Replacing shoes on only one side will result in poor braking performance. Besides, if the shoes wore out on one side faster than the other side, there is a malfunction in the brake system. Inspect the brake system, if necessary, repair the problem before proceeding.

➡ It is not a good idea to disassemble the brakes on both sides at the same time. There are a lot of parts involved which must be replaced in a certain way. Work on one side at a time, only. If you become confused as to the particular position of the various brake parts during the brake shoe replacement, refer to the other side. Remember, however, the other side is a mirror image (everything is reversed).

INSPECTION

1. Remove the brake drum.

2. Inspect the brake shoe lining material for cracks, crumbling or evidence of wetness. Replace the shoes with new ones if any such damage is found. If evidence of wetness is evident, repair the leaking component prior to installing the new shoes.

3. Measure the thickness of the brake shoe lining (not including the shoe backing). Generally, the minimum allowable lining thickness is either $\frac{1}{16}$ in. (1.6mm) above the head of the rivet (for rivet mounted linings), or $\frac{3}{32}$ in. (2.4mm) from the shoe backing (for glued linings).

4. If one of the brake linings is worn to or beyond the allowable limit, all four of the rear brake shoes must be replaced.

✳✳ WARNING

Never polish the shoe lining with sandpaper, because hard particles from the sandpaper will become imbedded in the lining, which will damage the brake drum. If the shoe lining is damaged or worn excessively or unevenly, replace the shoes with new ones.

5. Install the brake drum.

REMOVAL & INSTALLATION

✳✳ CAUTION

Brake shoes may contain asbestos, which is a known cancer-causing agent. As soon as the drum is removed, generously spray the entire brake assembly with brake parts cleaner. Let it dry before proceeding. It's a good idea to wear a filter mask when doing brake work.

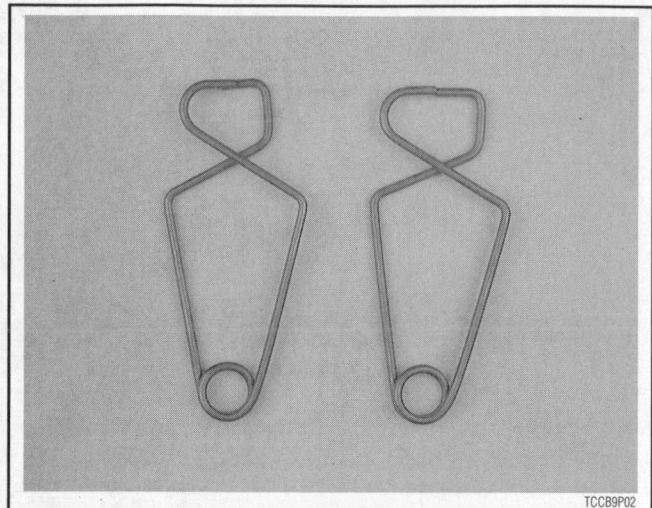

Spring clamp tools, such as those shown, can hold the wheel cylinder pistons in while servicing the shoes

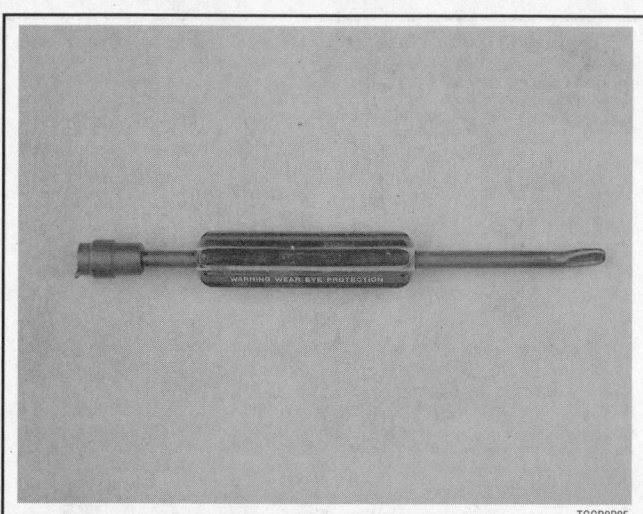

There are several varieties of spring removal and installation tools available, such as this straight one . . .

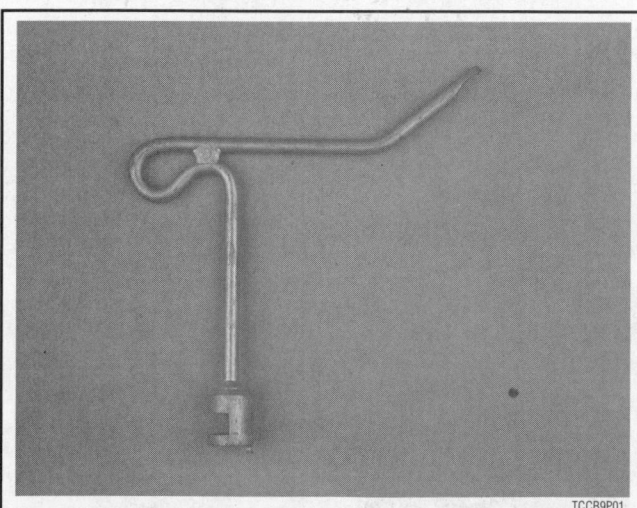

. . . and this curved one—The shape of this tool is designed to provide more leverage during use

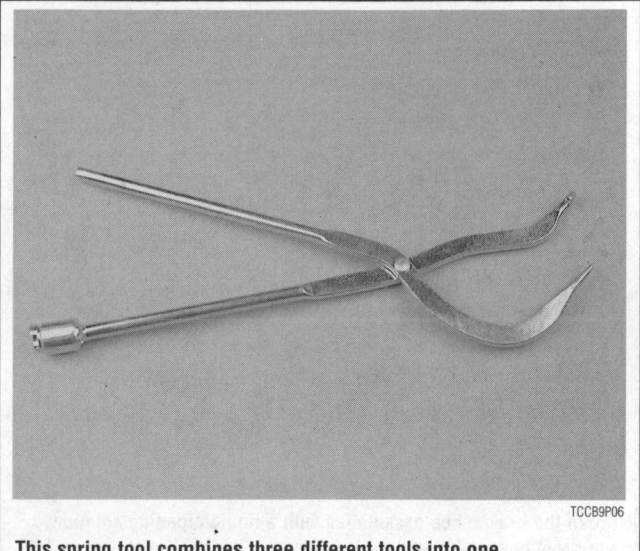

This spring tool combines three different tools into one

Models with Dual Return Springs and a Starwheel-type Adjuster

> ✳✳ **CAUTION**
>
> It is always a good idea to wear eye protection when working on brake components, especially drum brakes. Drum brakes often use powerful springs which could cause severe eye injury if they accidentally break.

1. Remove the brake drum.
2. Spray the brake assembly thoroughly with brake parts cleaner and let it dry. Similarly, spray the inside of the drum.
3. Inspect the drum for wear and/or damage. Machine or replace as necessary. When machining, observe the maximum diameter specification. The maximum machining diameter is stamped into the drum. If the drum braking surface shows signs of blue discoloration, overheating is indicated. If the bluing is extensive the drum/hub assembly must be replaced. Extensive bluing indicates a weakening of the metal.

➡ **Note the location of all springs and clips for proper assembly. If an instant camera is handy, it may be a good idea to take a picture of the brake assembly with the brake drum removed. This will make reassembly much easier.**

4. Completely retract the adjuster by rotating the starwheel to relieve tension on the lower spring.
5. Remove the starwheel assembly and adjuster lever from between the two brake shoes.
6. Using a brake spring tool, remove the 2 upper return springs.
7. Remove the adjuster cable and cable guide.
8. Remove the anchor block plate.
9. Using a hold-down spring tool or pliers, while holding the back of the spring mounting pin with one hand, press inward on the hold-down spring plate, turn it slightly to align the notches and pin ears, then remove the hold-down spring assembly with your other hand. Remove the other hold-down spring in the same manner.
10. Lift the shoes off the pins and remove the pins from the backing plate.
11. Remove the parking brake link.

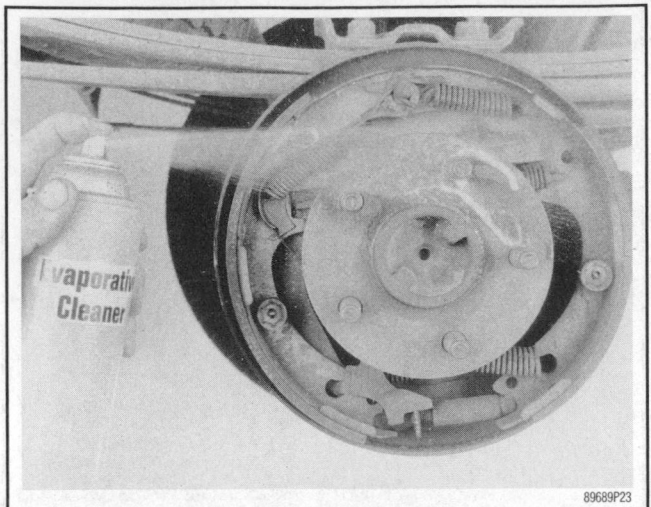

Clean the brake shoe assemblies with a liquid cleaning solution, NEVER with compressed air

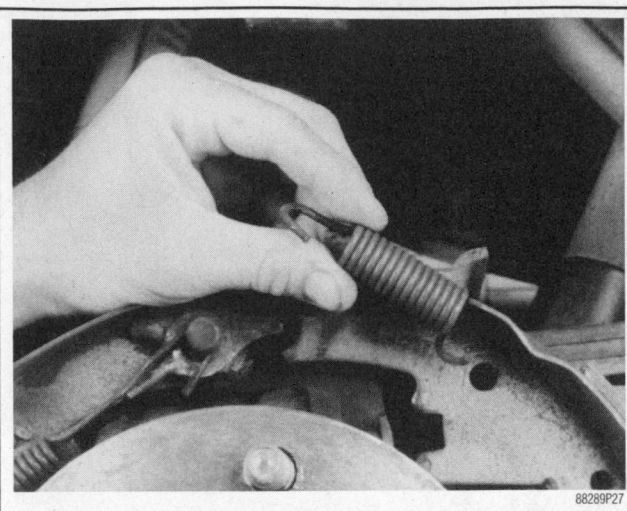

Detach the upper return springs first from the anchor bolt, then from the brake shoes . . .

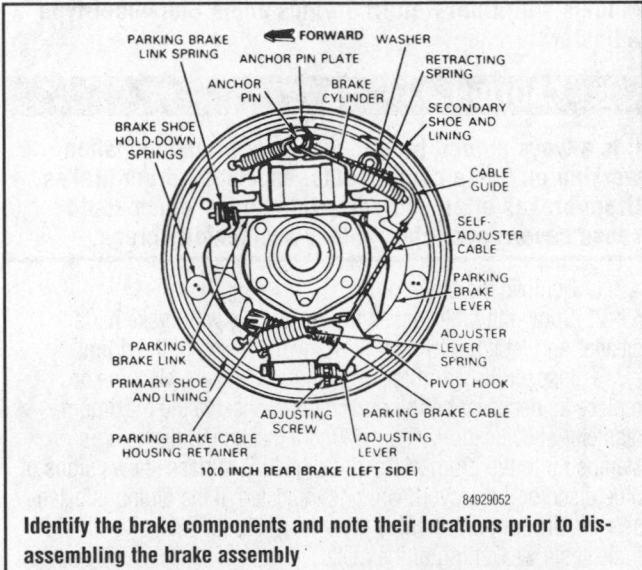

10.0 INCH REAR BRAKE (LEFT SIDE)

Identify the brake components and note their locations prior to disassembling the brake assembly

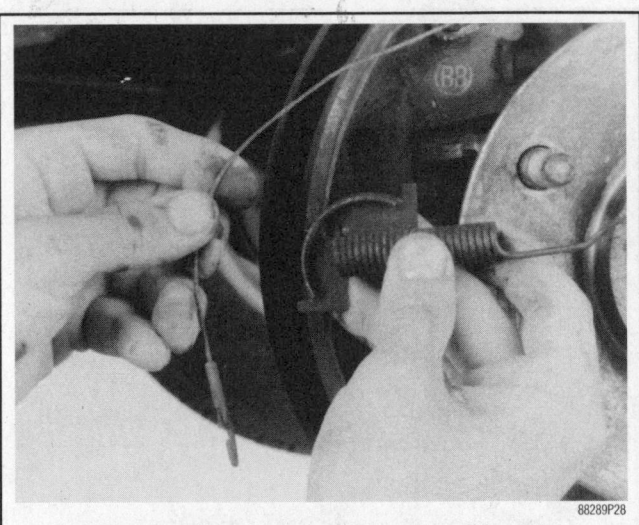

. . ., then remove the adjusting cable from the guide, and the guide from the brake shoe

A specially-designed brake tool can make disconnecting the upper return springs much easier—dual return spring and a starwheel-type adjuster type

Remove the anchor block plate . . .

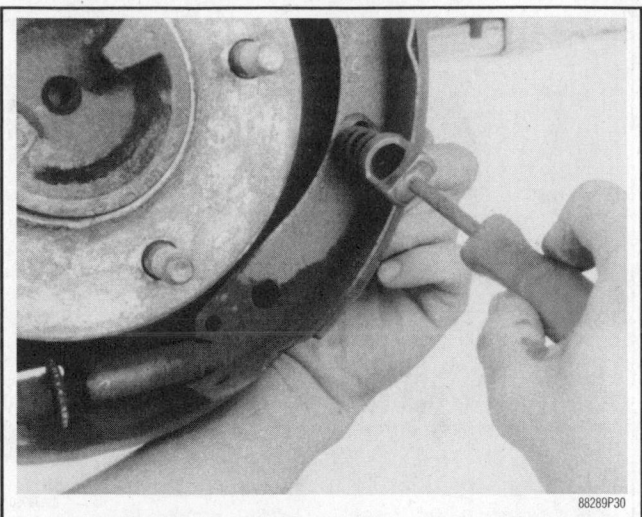

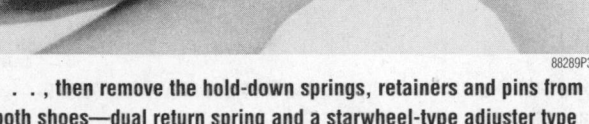

. . . , then remove the hold-down springs, retainers and pins from both shoes—dual return spring and a starwheel-type adjuster type

Another way to remove the shoes for a dual spring setup is to pull the adjuster cable toward the shoe . . .

Lift the brake shoes off of the backing plate . . .

. . . and disconnect the pivot hook from the adjusting lever. Wind the starwheel all the way in

. . . , then detach the parking brake cable from the lever—dual return spring and a starwheel-type adjuster type

Disconnect the adjuster lever return spring from the lever . . .

. . . and remove the spring and the lever—dual return spring and a starwheel-type adjuster type

Repeat the procedure and remove the secondary return spring, adjuster cable and its guide

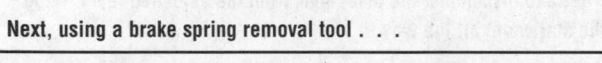

Next, using a brake spring removal tool . . .

Also remove the anchor pin plate—dual return spring and a starwheel-type adjuster type

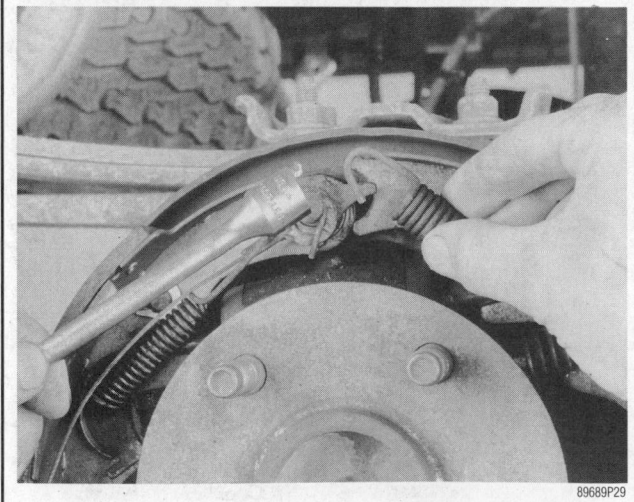

. . . disconnect the primary brake shoe return spring from the anchor pin

Pull the bottoms of the shoes apart and remove the adjuster screw assembly

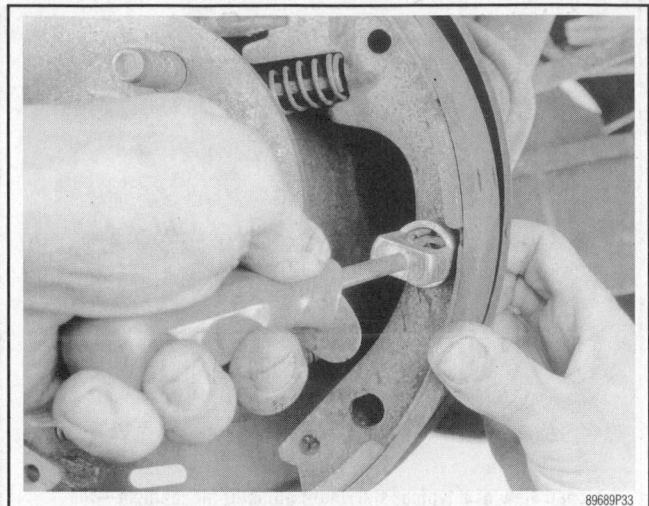

Press in the hold-down springs while holding in on the nail from behind, then turn the cup 90° . . .

. . . and the parking brake strut as well—dual return spring and a starwheel-type adjuster type

. . . and release to remove the hold-down spring. Pull the nail out from the backing plate

Remove the secondary shoe hold-down, pull the shoe out, then press up on the cable spring . . .

Remove the primary (front) brake shoe from the backing plate . . .

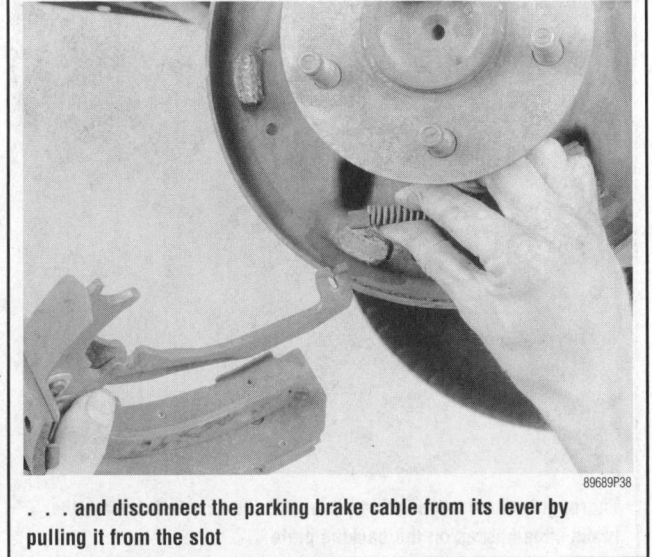

. . . and disconnect the parking brake cable from its lever by pulling it from the slot

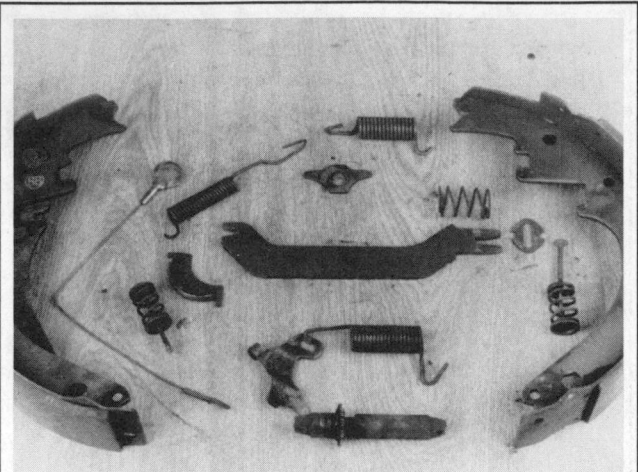

It's a good idea to arrange all the parts in their approximate installed positions on a clean work surface

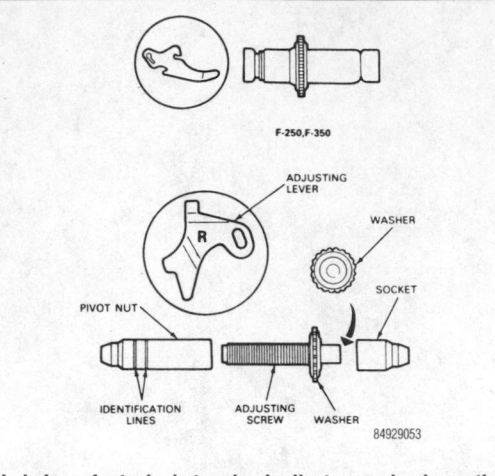

Exploded view of a typical starwheel adjuster mechanism—the adjusting levers may be stamped for left side and right side applications

12. Pull back on the parking brake cable spring and twist the cable out of the parking brake lever.

13. The parking brake lever is held onto the rear shoe with a horseshoe clip. Spread the clip and remove the lever and washer.

To install:

14. Thoroughly clean and dry the backing plate and starwheel assembly.

15. Lubricate the backing plate bosses, anchor plate surfaces, and starwheel threads and contact points with silicone grease. High-temperature wheel bearing grease or synthetic brake grease also work well for this application.

✳✳ CAUTION

When applying lubricant to the backing plate and other components, do not use' so much grease that it may get spread onto the new brake shoes' friction material; this can adversely affect the performance of the new brake shoes, therefore, increase vehicle stopping distance.

Thoroughly clean the backing plate, then be sure to lubricate the brake shoe bosses on the backing plate

16. Insert the parking brake lever pivot stud through the applicable hole in the rear shoe, then install a new wave washer and horseshoe clip. Squeeze the clip ends until the clip cannot be pulled from the lever pivot stud.

17. Connect the parking brake cable to the lever.

18. Position the rear shoe assembly on the backing plate and install the hold-down pin and spring assembly.

19. Install the front shoe and secure it with the hold-down spring assembly.

20. Position the parking brake link and spring between the front shoe and parking brake lever.

21. Position the adjuster cable on the anchor plate pin, install the cable guide and lay the cable across the guide.

22. Be sure that the notch in the upper end of the shoe is engaging the wheel cylinder piston or piston pin.

23. Position the rear shoe return spring into the guide and shoe hole, using a brake spring tool, stretch the spring onto the anchor plate pin. Be sure that the cable guide remained in place.

24. Position the front shoe return spring in its hole in the shoe.

25. Be sure that the parking brake link is properly positioned and that the upper end of the shoe will enter the wheel cylinder or engage the wheel cylinder piston.

26. Using the spring tool, stretch the spring into position on the anchor plate pin.

➡️**If the shoe doesn't properly engage the link or wheel cylinder piston, try again by removing the spring.**

27. Position the adjuster lever in its hole in the rear shoe and hook the cable to it.

28. Position the lower spring in its hole in the front shoe. Now comes the hard part. Clamp a pair of locking pliers, like Vise Grips® on the spring and stretch it to engage the hole in the adjuster lever. Be sure that the cable stays in place on the guide.

29. Check that the shoes are evenly positioned on the backing plate.

30. Turn the starwheel to spread the shoes to the point at which the drum can be installed with very slight drag.

31. Install the drum and adjust the starwheel until the drum can't be turned. Then, back off the adjustment until the drum can just be turned without drag.

32. Install the wheels, lower the vehicle and check brake action. A firm pedal should be felt.

33. To activate the adjusters, some vehicles require you to make several quick pulls on the parking brake lever. On most, however, several short back-ups, about 10 ft. (3m) each, should do it.

Models with a Single Upper Shoe-to-Shoe Return Spring

❊❊ CAUTION

It is always a good idea to wear eye protection when working on brake components, especially drum brakes. Drum brakes often use powerful springs which could cause severe eye injury if they accidentally break. Also, Brake shoes may contain asbestos, which is a known cancer-causing agent. As soon as the drum is removed, generously spray the entire brake assembly with brake parts cleaner. Let it dry before proceeding. It's a good idea to wear a filter mask when doing brake work.

WITH LOWER ANCHOR PLATE

1. Remove the brake drum.

Clean the brake assembly and drum thoroughly with brake parts cleaner and let it dry.

Inspect the drum for wear and/or damage. Machine or replace as necessary. When machining, observe the maximum diameter specification. The maximum machining diameter is stamped into the drum. If the drum braking surface shows signs of blue discoloration, overheating is indicated. If the bluing is extensive the drum/hub assembly must be replaced. Extensive bluing indicates a weakening of the metal.

➡**Note the location of all springs and clips for proper assembly. If you own an instant camera, to make installation easier it may be a good idea to take a picture of your brake assembly with the brake drum removed.**

2. Remove the shoe-to-lever spring and remove the adjuster lever.
3. Remove the auto-adjuster assembly.
4. Remove the retainer spring.
5. Using a hold-down spring tool or pliers, while holding the back of the spring mounting pin with one hand, press inward on the

. . . , then remove the retainer, spring and pin from the shoe and backing plate—models with a single upper shoe-to-shoe return spring and lower anchor plate

hold-down spring plate, turn it slightly to align the notches and pin ears, then remove the hold-down spring assemblies with your other hand.

6. Remove the shoe-to-shoe spring.
7. Remove the brake shoes from the backing plate.
8. Using a flat-tipped tool, pry open the parking brake lever retaining clip. Remove the clip and washer from the pin on the shoe assembly and remove the shoe from the lever assembly.

➡**On some vehicles, the parking brake actuating lever is permanently attached to the trailing brake shoe assembly. Do not attempt to remove it from the original brake shoe assembly or reuse the original actuating lever on a replacement brake shoe assembly. All replacement brake shoe assemblies for these vehicles must come with the actuating lever as part of the trailing brake shoe assembly.**

To install:

9. Thoroughly clean all parts.
10. On vehicles with the ratcheting upper mounted adjuster, clean and inspect the brake support plate and the automatic adjuster

Pliers can be used to disengage the hold-down spring retainer by rotating it until aligned with the pin tabs . . .

Use a pair of needlenose pliers, or similar tool, to detach the upper return spring from both shoes . . .

. . ., then remove the brake shoes from the backing plate . . .

. . . and detach the parking brake cable from the applicable brake shoe—models with a single upper shoe-to-shoe return spring and lower anchor plate

mechanism. Be sure the quadrant (toothed part) of the adjuster is free to rotate throughout its entire tooth contact range and is free to slide the full length of its mounting slot. Check the knurled pin. It should be securely attached to the adjuster mechanism and its teeth should be in good condition. If the adjuster is worn or damaged, replace it. If the adjuster is serviceable, lubricate lightly with high-temperature grease between the strut and the quadrant.

✳✳ CAUTION

The trailing brake shoe assemblies used on the rear brakes of these vehicles are different for the left and right side of the vehicle. Care must be taken to ensure the brake shoes are properly installed in their correct side of the vehicle. Otherwise the brakes will probably malfunction, thereby creating a very dangerous condition. When the trailing shoes are properly installed on their correct side of the vehicle, the park brake actuating lever will be positioned under the brake shoe web.

11. Thoroughly clean and dry the backing plate. Lubricate the backing plate at the brake shoe contact points. Also, lubricate backing plate bosses, anchor pin, and parking brake actuating mechanism with silicone grease. High-temperature wheel bearing grease or synthetic brake grease also work well for this application.

12. Install the parking brake lever assembly on the lever pin. Install the wave washer and a new retaining clip. Use pliers, or the like, to install the retainer on the pin. If removed, connect the parking brake lever to the parking brake cable and verify that the cable is properly routed.

13. Clean and lubricate the adjuster assembly. Be sure the nut-adjuster is drawn all the way to the stop, but the nut must NOT lock firmly at the end of the assembly.

14. Install the brake shoes on the backing plate with the hold-down springs, washers and pins.

15. Install the shoe-to-shoe spring.

16. Install the retainer spring.

17. Install the auto-adjuster assembly and install the adjuster lever and the shoe-to-lever spring.

18. Pre-adjust the shoes so the drum slides on with a light drag and install the brake drum.

19. Adjust the brake shoes.

20. Install the rear wheels.

21. To activate the adjusters, some vehicles require you to make several quick pulls on the parking brake lever. On most, however, several short back-ups, about 10 ft. (3m) each, should do it.

22. Adjust the parking brake cable.

23. Lower the vehicle and check for proper brake operation.

WITH LOWER STARWHEEL-TYPE ADJUSTER

1. Loosen the lug nuts on the applicable wheels.

2. If servicing the front brakes, apply the parking brake, block the rear wheels, then raise and safely support the front of the vehicle securely.

3. If servicing the rear brakes, block the front wheels, then raise and safely support the rear of the vehicle securely.

4. Remove the wheels.

5. Remove the drums.

6. Spray the brake assembly thoroughly with brake parts cleaner and let it dry. Similarly, spray the inside of the drum.

7. Inspect the drum for wear and/or damage. Machine or replace as necessary. When machining, observe the maximum diameter specification. The maximum machining diameter is stamped into the drum. If the drum braking surface shows signs of blue discoloration, overheating is indicated. If the bluing is extensive the drum/hub assembly must be replaced. Extensive bluing indicates a weakening of the metal.

8. Remove the parking brake lever assembly from the backing plate.

9. Remove the adjusting cable assembly from the anchor pin, cable guide and adjusting lever.

10. Remove the brake shoe retracting springs.

11. Remove the brake shoe hold-down spring from each shoe.

12. Remove the brake shoes and adjusting screw assembly.

13. Disassemble the adjusting screw assembly.

➡It's a good idea to arrange all the parts in the approximate installed positions as a guide for reassembly.

To install:

14. Clean the ledge pads on the backing plate. Apply a light coat of silicone grease to the ledge pads (where the brake shoes rub the

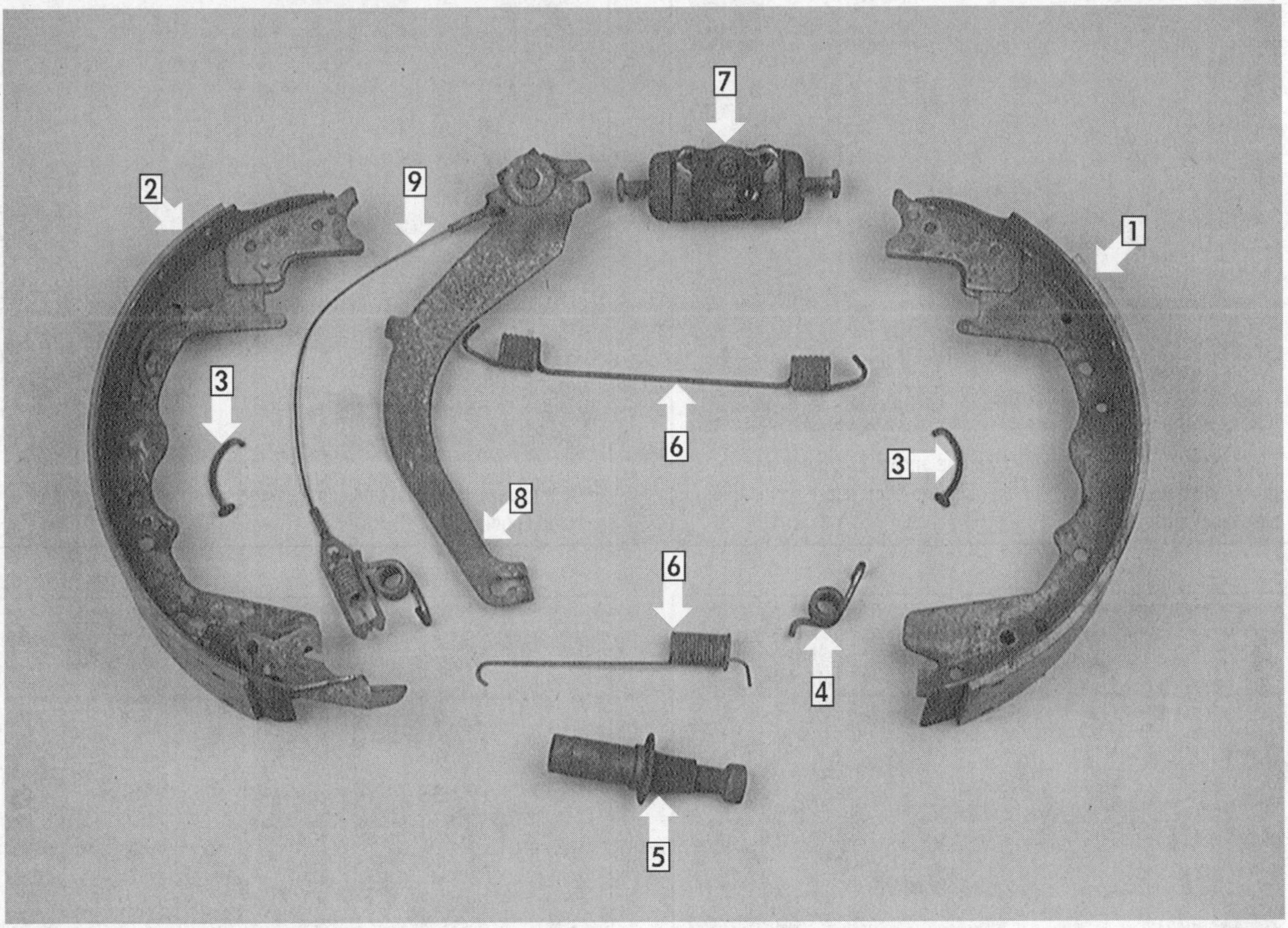

1. Front brake shoe
2. Rear brake shoe
3. Hold-down pin
4. Shoe hold-down spring
5. Adjuster
6. Return spring
7. Wheel cylinder
8. Parking brake lever
9. Parking brake adjuster cable

88489P44

It is a good idea to lay the brake parts out in their positions on a clean work surface as they are removed

88489P31

Remove the brake drum from the rear axle

88489P32

Remove the parking brake lever retaining nut which is located behind the backing plate

Disconnect the adjusting cable from the anchor pin, guide and lever—models with a single upper shoe-to-shoe return spring and starwheel adjuster

Use an appropriate tool to disconnect the return springs from their retaining holes

Slide the parking brake lever out from its mounting—models with a single upper shoe-to-shoe return spring and starwheel adjuster

Disengage the hold-down springs from the retaining clips on the backing plate—models with a single upper shoe-to-shoe return spring and starwheel adjuster

Disconnect the parking brake cable from the lever

Back off the adjusting screw and remove it from the brake assembly

Spread the shoes apart and remove them from the backing plate

backing plate). High-temperature wheel bearing grease or synthetic brake grease (designed specifically for this) also work well. Also, apply grease to the adjusting screw assembly and the hold-down and retracting spring contacts on the brake shoes.

15. Install the upper retracting spring on the primary and secondary shoes, then position the shoe assembly on the backing plate with the wheel cylinder pistons engaged with the shoes.

16. Install the brake shoe hold-down springs.

17. Install the brake shoe adjustment screw assembly so that the slot in the head of the adjusting screw is toward the primary (leading) shoe, along with the lower retracting spring, adjusting lever spring, adjusting lever assembly and connect the adjusting cable to the adjusting lever. Position the cable in the cable guide and install the cable anchor fitting on the anchor pin.

18. Install the adjusting screw assemblies in the same locations from which they were removed.

✳ CAUTION

Interchanging the brake shoe adjusting screws from one side of the vehicle to the other will cause the brake shoes

Connecting the lower retracting spring can often be difficult—be careful and have patience

This is how everything should look after assembly—models with a single upper shoe-to-shoe return spring and starwheel adjuster

to retract rather than expand each time the automatic adjusting mechanism is operated; this will create an extremely dangerous condition when driving the vehicle. To prevent incorrect installation, the socket end of each adjusting screw is usually stamped with an R or an L to indicate their installation on the right or left side of the vehicle. In some cases, the adjusting pivot nuts can be distinguished by the number of lines machined around the body of the nut. Two lines indicate a nut which should be installed on the right side of the vehicle; one line indicates a nut that must be installed on the left side of the vehicle.

19. Install the parking brake assembly in the anchor pin and secure with the retaining nut behind the backing plate.

20. Adjust the brakes before installing the brake drums and wheels. Install the brake drums and wheels.

21. To activate the adjusters, some vehicles require you to make several quick pulls on the parking brake lever. On most, however, several short back-ups, about 10 ft. (3m) each, should do it.

22. Lower the vehicle and road test the brakes. New brakes may pull to one side or the other before they are seated. Continued pulling or erratic braking should not occur.

Models with a Single U-Shaped Return Spring

✳ CAUTION

It is always a good idea to wear eye protection when working on brake components, especially drum brakes. Drum brakes often use powerful springs which could cause severe eye injury if they accidentally break. Also, brake shoes may contain asbestos, which is a known cancer-causing agent. As soon as the drum is removed, generously spray the entire brake assembly with brake parts cleaner. Let it dry before proceeding. It's a good idea to wear a filter mask when doing brake work.

1. Loosen the lug nuts on the applicable wheels.

2. If servicing the front brakes, apply the parking brake, block the rear wheels, then raise and safely support the front of the vehicle securely.

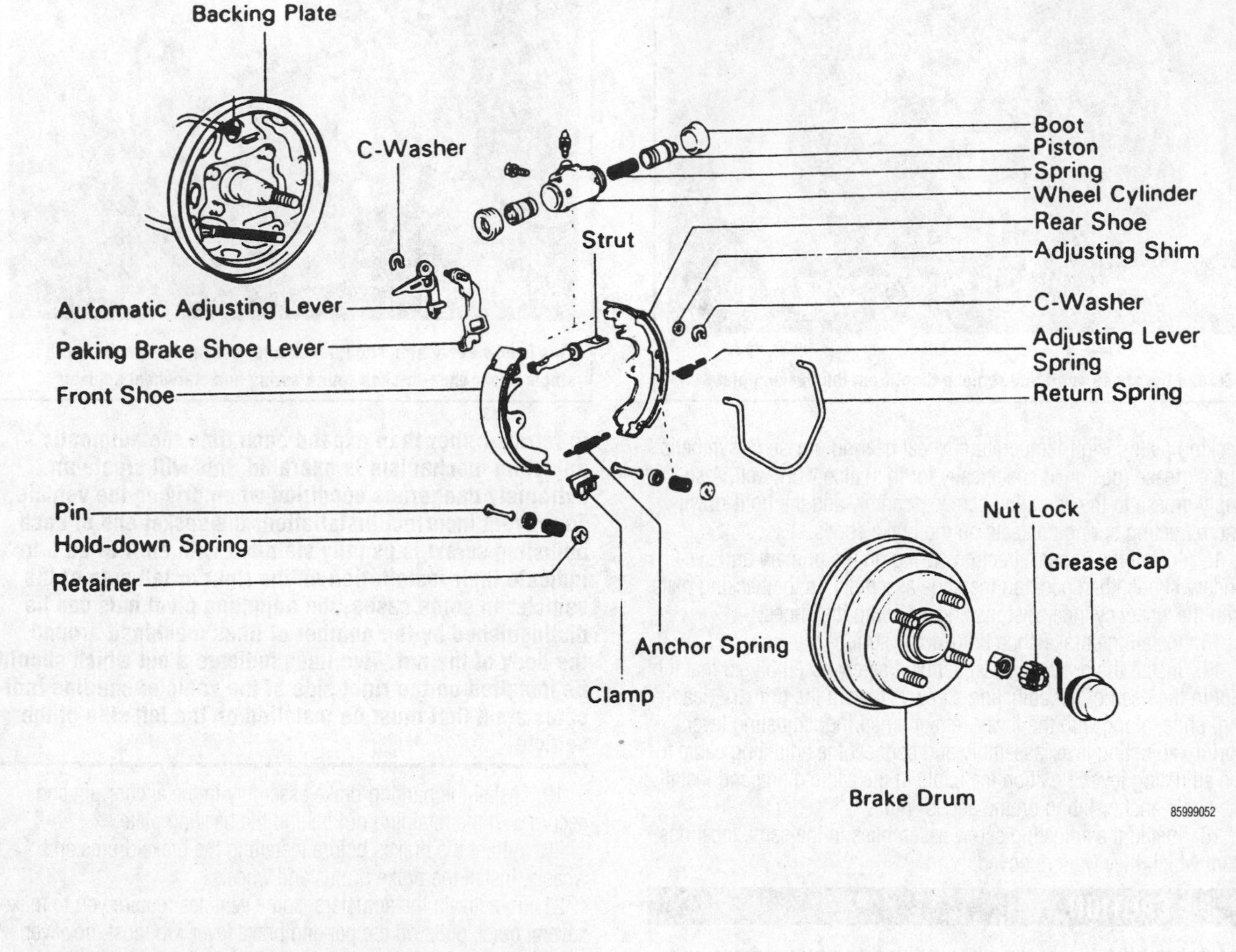

Backing Plate

C-Washer

Boot
Piston
Spring
Wheel Cylinder
Rear Shoe
Adjusting Shim

Strut

Automatic Adjusting Lever

Paking Brake Shoe Lever

Front Shoe

C-Washer

Adjusting Lever
Spring

Return Spring

Pin
Hold-down Spring
Retainer

Nut Lock

Grease Cap

Anchor Spring

Clamp

Brake Drum

85999052

Exploded view of a typical single U-shaped return spring drum brake setup

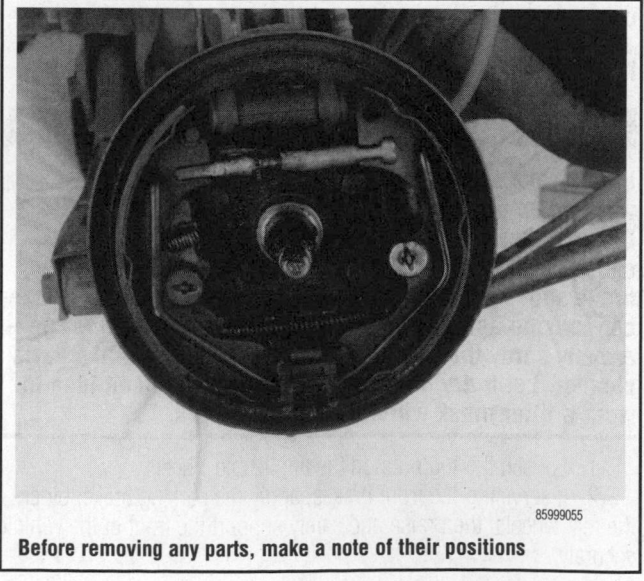

85999055

Before removing any parts, make a note of their positions

3. If servicing the rear brakes, block the front wheels, then raise and safely support the rear of the vehicle securely.

4. Remove the wheels.

5. Remove the brake drum.

6. Spray the brake assembly thoroughly with brake parts cleaner and let it dry. Similarly, spray the inside of the drum.

7. Inspect the drum for wear and/or damage. Machine or replace as necessary. When machining, observe the maximum diameter specification. The maximum machining diameter is stamped into the drum. If the drum braking surface shows signs of blue discoloration, overheating is indicated. If the bluing is extensive the drum/hub assembly must be replaced. Extensive bluing indicates a weakening of the metal.

8. Remove the return spring clip from the lower anchor block.

9. Squeeze the upper ends of the return spring slightly and remove it from the shoes.

10. Using a hold-down spring tool or pliers, remove the hold-down springs. While holding the back of the spring mounting pin with one hand, press inward on the hold-down spring plate, turn it

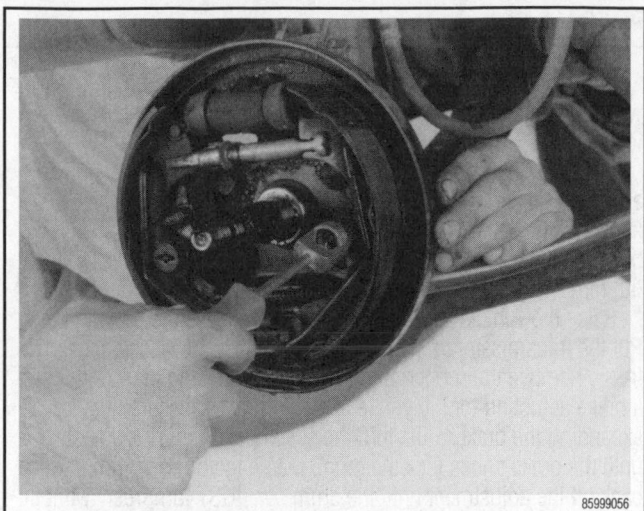

For models with a single U-shaped return spring, depress and rotate the hold-down spring retainer . . .

. . . , then separate the shoes from the backing plate

. . . , then remove the spring, retainer and pin from the backing plate and shoes

A large pair of pliers can be used to disconnect the parking brake cable from the lever

Remove the return spring from both brake shoes . . .

slightly to align the notches and pin ears, then remove the hold-down spring assemblies with your other hand.

11. Lift the shoes off of the pins, then remove the pins from the backing plate.

12. Remove the shoes and adjuster as an assembly.

13. Pull back on the parking brake cable spring and twist the cable out of the parking brake lever.

14. The parking brake lever is held onto the rear shoe with a horseshoe clip. Spread the clip and detach the lever and washer from the shoe.

To install:

15. Thoroughly clean and dry the backing plate assembly.

16. Lubricate the backing plate bosses, anchor plate surfaces, and all contact points with silicone grease. High-temperature wheel bearing grease or synthetic brake grease (designed specifically for this) also work well.

17. Lubricate the parking brake lever pivot stud, then insert the pivot stud through the applicable hole in the rear shoe, then install a new wave washer and horseshoe clip. Squeeze the clip ends until the clip cannot be pulled from the lever pivot stud.

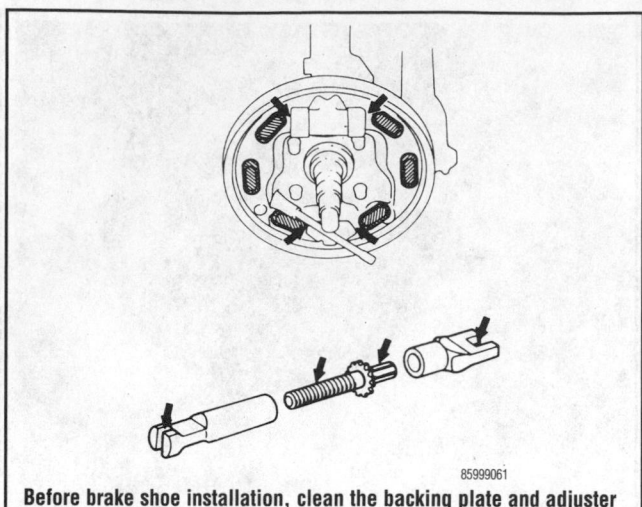

Before brake shoe installation, clean the backing plate and adjuster mechanism, then apply high temperature grease at all shoe-to-backing plate points (arrows)

18. Connect the parking brake cable to the lever.
19. Position the front and rear shoe assemblies and adjuster on the backing plate, then install the hold-down pin and spring assemblies.
20. Position the return spring in the shoes, rotate it down into position on the anchor block, and install the retaining clip.
21. Turn the strut adjusting screw to spread the shoes to the point at which the drum can just be installed without drag.
22. Install the drum.
23. Adjust the brake shoes.
24. Install the wheels, lower the vehicle and check brake action. A firm pedal should be felt.
25. To activate the adjusters, some vehicles require you to make several quick pulls on the parking brake lever. On most, however, several short back-ups, about 10 ft. (3m) each, should do it.

ADJUSTMENT

Drum brakes on all modern vehicles are self-adjusting, however, when the shoes are replaced, a preliminary adjustment makes the job easier.

On most vehicles, the adjustment is made with an expanding adjuster that is a threaded sleeve/stud assembly. Turning the knurled nut or starwheel expands or contracts the spring-loaded brake shoes. On most vehicles, this adjuster can be accessed without removing the drum, or, for that matter, the wheel.

Raise the vehicle and support it safely. Release the parking brake. Put the transmission in neutral. All this allows the wheels to turn freely. Remove the rubber plug in the brake backing plate and insert a brake adjusting tool. If you're applying brake pressure, that is, expanding the brakes, just turn the starwheel or knurled adjuster until the brake shoes lock the drum; meaning you can't turn it. Then, back off the adjustment until the drum can JUST turn freely without any drag. Some manufacturers even say it's okay to have a SLIGHT amount of drag. If the vehicle at hand is equipped with self-adjusters, you'll find that the adjuster can't be backed off. That's because the adjusting lever is holding it in place. You'll have to insert a thin punch or similar device in the hole with the brake adjusting tool. Just push slightly on the adjusting lever. That'll free the adjuster.

There are a few vehicle models that use cam-type adjusters. With these, a hex or square headed stud protrudes through the backing plate. Turning this stud rotates an eccentric cam that contacts the brake shoe. Turning it one way pushes the shoe outward; turning it the other way rotates the cam away from the shoe allowing the springs to pull the shoe away from the drum.

Wheel Cylinders

REMOVAL & INSTALLATION

Wheel cylinders are held in place on the backing plate with either bolts or spring clips. A first glance, this looks like a fairly easy job, and it can be. However, a lot can go wrong. If the wheel cylinder has

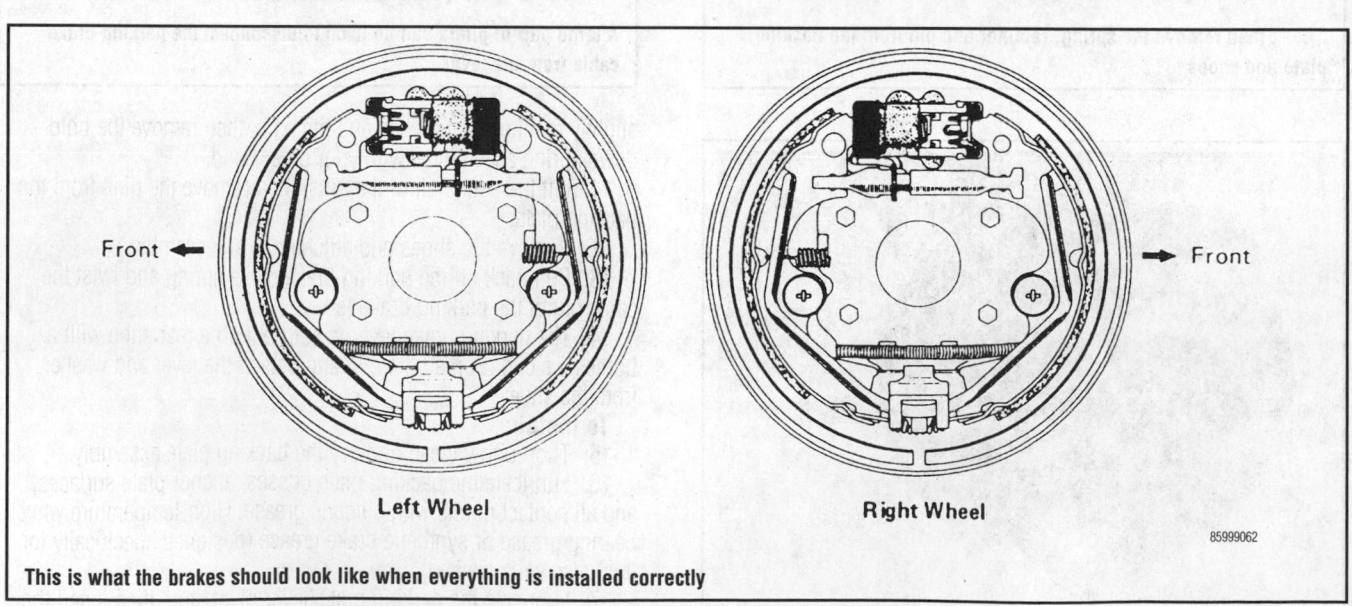

Left Wheel   **Right Wheel**

This is what the brakes should look like when everything is installed correctly

been there a long time, the bolts or clips can be rusted in place. Worse, the brake line flare nut may be rusted in place. The flats on the nut are easily rounded off. Also, the flare nut can be rusted to the line, meaning the line will twist when the nut is turned. So, before starting, it's best to thoroughly soak the area with penetrating oil where the brake line threads into the wheel cylinder. Also, apply penetrating oil to the mounting bolts or clips.

If you run into problems, here are some general tips:
• Use a flare nut wrench on the flare nuts. Sounds logical, does-n't it? Flare nut wrenches are designed to reduce the possibility of rounding-off.
• Use a box end wrench, or, if room permits, a socket on the bolts. The better grip of a box end wrench or socket will help prevent rounding off the bolt head(s).
• If you round off a bolt head, you'll have to try using Vise-Grips® (or equivalent), one of those wrenches designed for rounded-off bolts (space permitting), a nut splitter (again, space permitting), or grind off the bolt head.
• If the brake line won't budge, you fear kinking or twisting the line, or you rounded off the flare nut, try this: remove the wheel cylinder bolts or clips and pull the wheel cylinder, line attached, away from the backing plate. Usually, there is enough play in the brake line. Hold the flare nut with Vise-Grips® or equivalent, and try turning the wheel cylinder. The wheel cylinder gives you greater mechanical advantage than the flare nut. If nothing works, disconnect the line at the junction box. You'll have to install a new line.

Bolt-on Type

✳✳ CAUTION

It is always a good idea to wear eye protection when working on brake components, especially drum brakes. Drum brakes often use powerful springs which could cause severe eye injury if they accidentally break. Also, brake shoes may contain asbestos, which is a known can-cer-causing agent. As soon as the drum is removed, gen-erously spray the entire brake assembly with brake parts cleaner. Let it dry before proceeding. It's a good idea to wear a filter mask when doing brake work.

1. Loosen the lug nuts on the applicable wheels.
2. Raise and safely support the vehicle.
3. Remove the wheels.
4. Remove the drum.
5. Remove the brake shoes.

➡**On some vehicles, it may be possible to just remove the return springs and pull the shoes apart far enough for wheel cylinder removal. We do not recommend this for two rea-sons: wheel cylinder removal involves spilling some brake fluid—brake fluid can contaminate brake shoe friction mate-rial—and leaving the brake shoes on the backing plate can reduce working space and interfere with the job.**

6. Loosen the brake fluid line fitting, then separate the line from the wheel cylinder.

✳✳ CAUTION

Plug the line immediately to prevent contamination of the brake fluid, because brake fluid absorbs water from the atmosphere very quickly. Water reduces the effectiveness of brake fluid, leading to increased brake fade.

Use a flare nut wrench to loosen the brake line fitting from the inboard side of the wheel cylinder

When the brake line is disconnected there will be some fluid leak-age—plug the line to avoid contamination

Remove the wheel cylinder retaining bolts, then separate the cylin-der from the backing plate—bolt-on type

7. Remove the wheel cylinder bolts, and separate the cylinder from the backing plate.

To install:

8. Clean the backing plate thoroughly.

9. Apply a very thin coating of RTV silicone sealer to the cylinder mounting surface. This will aid in keeping moisture and dirt out of the brakes.

10. Position the cylinder on the backing plate, then install the retaining bolts.

11. Reattach the brake line to the wheel cylinder.

12. Install the brake shoes.

13. Install the drum.

14. Bleed the brake system.

15. Adjust the brake shoes.

16. Install the wheels and tighten the lug nuts.

Spring Clip Type

> ### ❋❋ CAUTION
>
> It is always a good idea to wear eye protection when working on brake components, especially drum brakes. Drum brakes often use powerful springs which could cause severe eye injury if they accidentally break. Also, brake shoes may contain asbestos, which is a known cancer-causing agent. As soon as the drum is removed, generously spray the entire brake assembly with brake parts cleaner. Let it dry before proceeding. It's a good idea to wear a filter mask when doing brake work.

1. Loosen the lug nuts on the applicable wheels.
2. Raise and safely support the vehicle.
3. Remove the wheels.
4. Remove the brake drum.
5. Remove the brake shoes.

➡ **On some vehicles, it may be possible to just remove the return springs and pull the shoes apart far enough for wheel cylinder removal. We do not recommend this for two reasons: wheel cylinder removal involves spilling some brake fluid—brake fluid can contaminate brake shoe friction material—and leaving the brake shoes on the backing plate can reduce working space and interfere with the job.**

6. Disconnect and cap the brake line at the wheel cylinder.

> ### ❋❋ CAUTION
>
> Plug the line immediately to prevent contamination of the brake fluid, because brake fluid absorbs water from the atmosphere very quickly. Water reduces the effectiveness of brake fluid, leading to increased brake fade.

7. Using two awls, release the spring clip securing the wheel cylinder to the backing plate.

8. Remove the wheel cylinder from the vehicle.

To install:

9. If you are installing a new wheel cylinder, remove the bleeder screw from the wheel cylinder, then position the cylinder in the backing plate. Removing the bleeder screw will keep it out of harm's way when installing the retaining clip.

10. Hold the wheel cylinder in place with a small prybar, using a socket (usually 1 ⅛ in./28.5mm on domestic vehicles) on the end

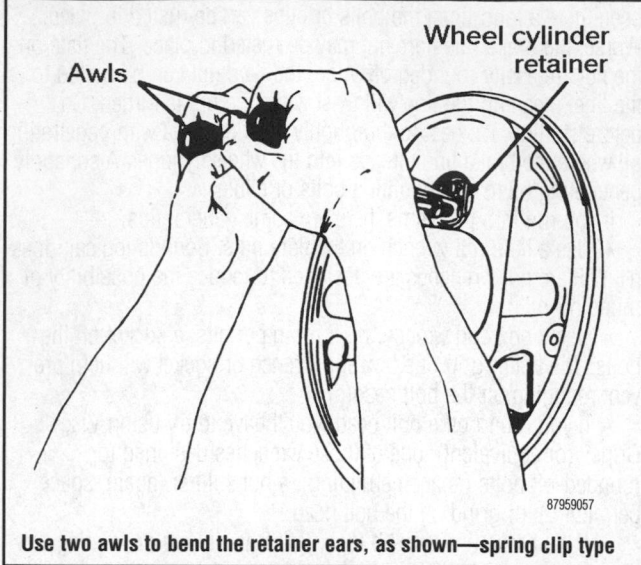

Use two awls to bend the retainer ears, as shown—spring clip type

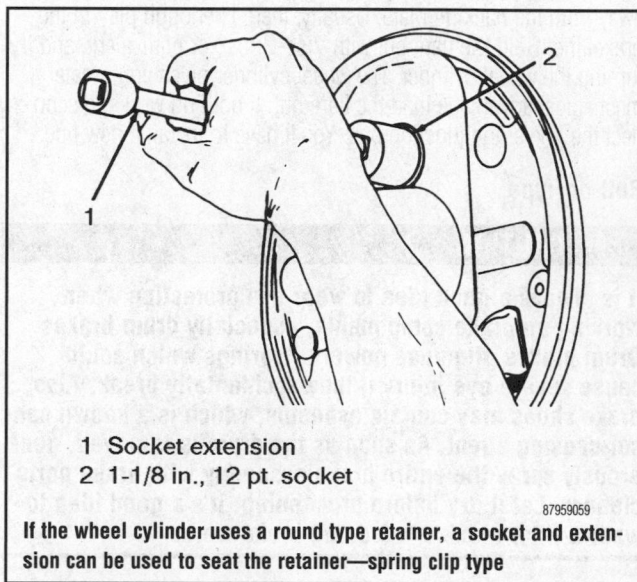

1 Socket extension
2 1-1/8 in., 12 pt. socket

If the wheel cylinder uses a round type retainer, a socket and extension can be used to seat the retainer—spring clip type

of an extension, push the spring clip into place. Be sure both spring clip ears are seated correctly.

11. Connect the brake line to the wheel cylinder.

12. Install the bleeder screw and temporarily tighten it.

13. Install the brake shoes.

14. Install the brake drum.

15. Bleed the brake system.

16. Adjust the brake shoes.

17. Install the wheels and tighten the lug nuts.

OVERHAUL

Wheel cylinders can be overhauled, although most people do not bother. Replacing the wheel cylinder is much easier and requires no special tools or experience. If the cost difference between a rebuilding kit and new cylinder is not great, it's much safer to install the new cylinder.

If you decide to overhaul your wheel cylinder(s), you will need a wheel cylinder hone and a rebuild parts kit.

→It is possible to rebuild the wheel cylinder while still in place on the backing plate. There is no good reason to do so other than that, for some reason, you can't remove the cylinder. If you choose to do this, it is of the UTMOST importance that all material be flushed out of the bore before installing new parts. We DO NOT recommend rebuilding a wheel cylinder while it is installed on the backing plate.

1. Remove the old wheel cylinder.
2. Thoroughly clean the outside of the unit with brake parts cleaner.
3. Place the cylinder on a clean work surface.
4. Remove the boots, then use a finger to push the pistons, cups and spring out of the bore.
5. Inspect the inner bore surface. If it is not badly pitted, rusted or scored, it can be rebuilt.
6. Remove the bleeder screw.
7. Install a wheel cylinder hone into a low-speed drill, and coat the inside of the cylinder with clean brake fluid.
8. Make several passes through the cylinder bore with the hone, never stopping in one place or passing completely through the bore.

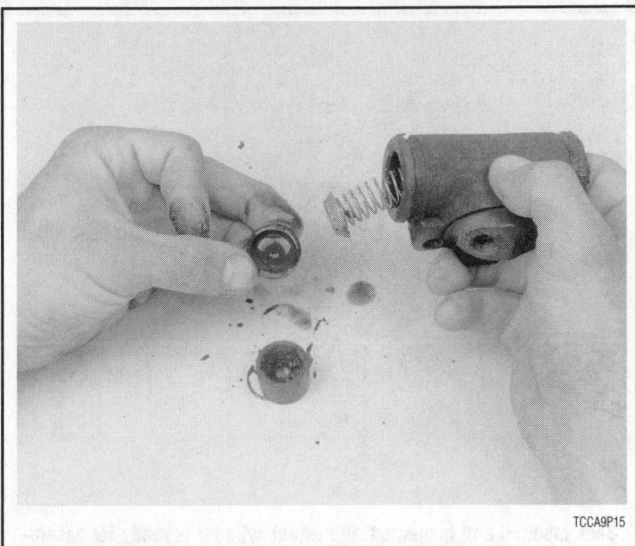

TCCA9P15

Remove the pistons, cup seals and spring from the cylinder

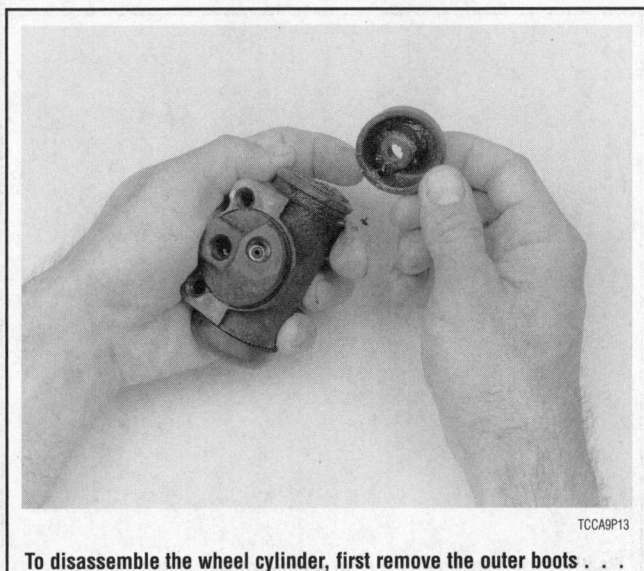

TCCA9P13

To disassemble the wheel cylinder, first remove the outer boots . . .

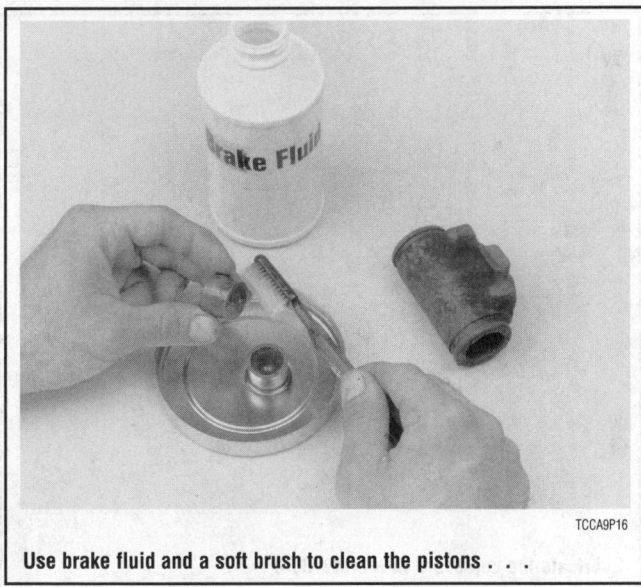

TCCA9P16

Use brake fluid and a soft brush to clean the pistons . . .

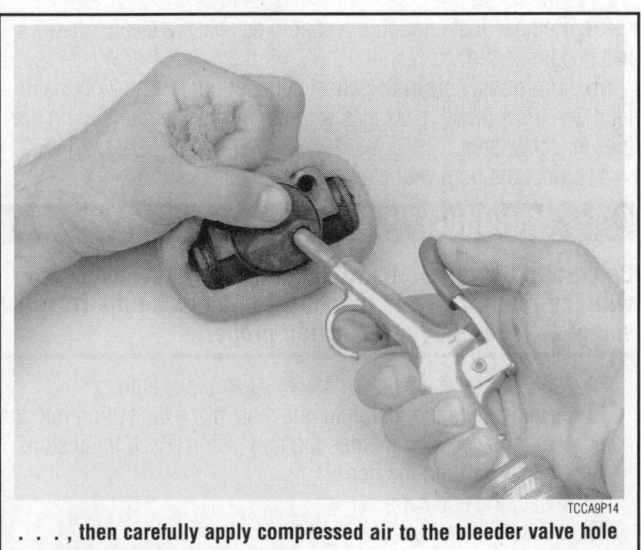

TCCA9P14

. . . , then carefully apply compressed air to the bleeder valve hole to extract the pistons and seals

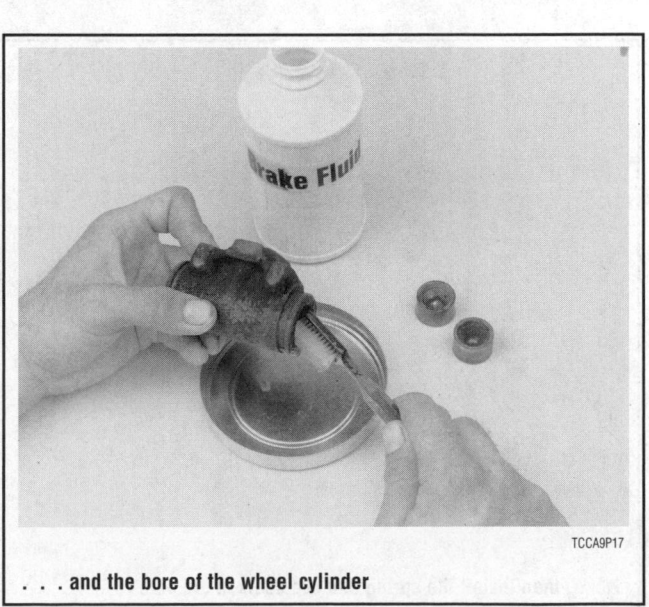

TCCA9P17

. . . and the bore of the wheel cylinder

Once cleaned and inspected, the wheel cylinder is ready for assembly

TCCA9P18

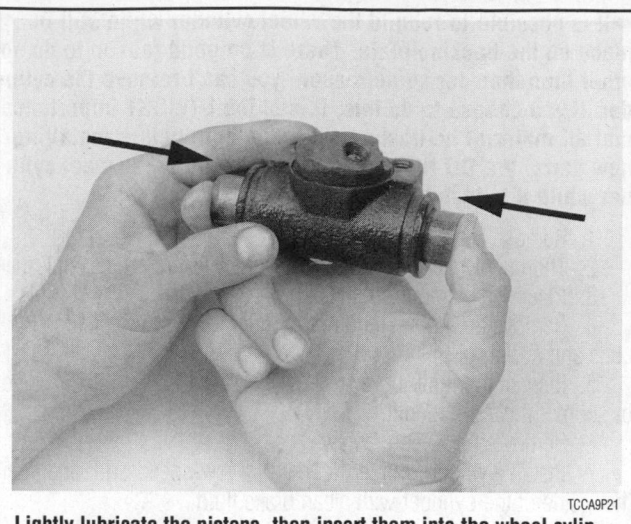

Lightly lubricate the pistons, then insert them into the wheel cylinder bore

TCCA9P21

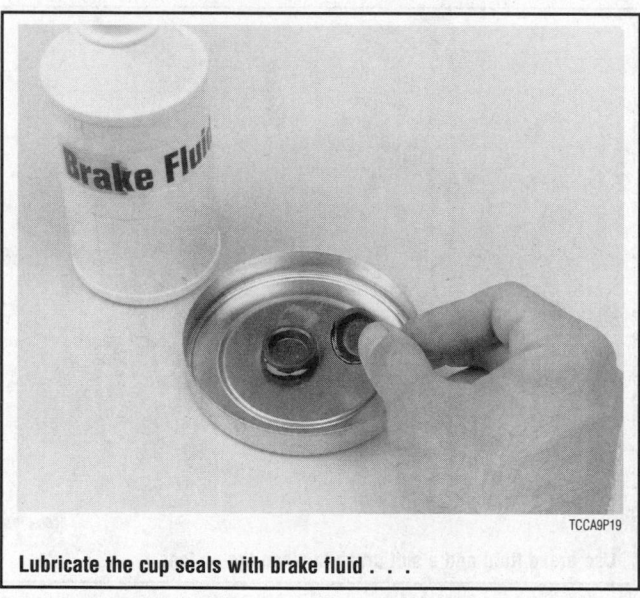

Lubricate the cup seals with brake fluid . . .

TCCA9P19

Finally, install the boots over the wheel cylinder piston ends

TCCA9P22

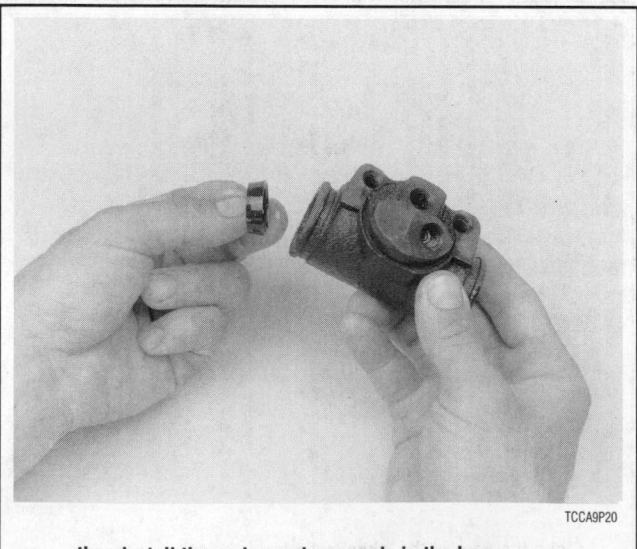

. . . , then install the spring and cup seals in the bore

TCCA9P20

9. Remove just enough material to establish a clean, crosshatched inner surface.

10. Thoroughly clean the wheel cylinder bore with alcohol and let it dry. Blow out all passages with compressed air, including the bleeder screw area.

11. Coat the bore with clean brake fluid.

❈❈ WARNING

Be sure to use all of the replacement parts which come with the rebuild kit you purchased, otherwise the rebuilt wheel cylinder may not function properly.

12. Coat all replacement parts with clean brake fluid.

13. Install a cup and piston in one side, place the spring into the other side, followed by the other cup and piston. Push the pistons in until both are within the bore.

14. Install the end caps.

15. Loosely install the bleeder screw.

16. Install the rebuilt wheel cylinder.

ANTI-LOCK BRAKE SYSTEMS

General Information

The purpose of the Anti-lock Brake System (ABS) is to prevent wheel lock-up under hard braking conditions. This is especially critical on wet or slippery surfaces. ABS is desirable because a vehicle that is stopped without locking one or more wheels, can stop with more control and in a shorter distance than a vehicle with locked wheels.

Under normal braking conditions, the ABS system operates just like a standard system. When one or more wheels shows a tendency to lock during braking, the ABS computer detects this and puts the system into the anti-lock mode. In this mode, hydraulic pressure is modulated to each wheel, preventing any one wheel from locking. The system can hold or reduce pressure at each wheel as necessary, depending on the signal received by the computer.

The effect is sort of like pumping your brakes, although it's done hundreds of time faster. In fact, when driving an ABS vehicle on ice or snow, a driver must overcome the urge to pump the brake during a stop. Let the ABS system work. Pumping the pedal on an ABS equipped vehicle will defeat the system.

PRECAUTIONS

• Do not use rubber hoses or other parts not specifically designed for the ABS system used by your vehicle. When using repair kits, replace all parts included in the kit. Partial or incorrect repair may lead to functional problems and require the replacement of components. NEVER fabricate your own replacement parts!

• Lubricate rubber parts with clean, fresh brake fluid to ease assembly. Do not use lubricated shop air to clean parts; damage to rubber components may result.

• Use only specified brake fluid from an unopened container.

• If any hydraulic component or line is removed or replaced, it may be necessary to bleed the entire system. This is always true when any upper end component (master cylinder, accumulator, control unit, etc.) is opened. It is also true when any lower end component (caliper or wheel cylinder) is opened and too much brake fluid has been lost; this does not happen often. If simply servicing a brake caliper, wheel cylinder, etc. and the line was adequately plugged after it was disconnected, the entire system will not need bleeding; only the component which was serviced. However, when in doubt, play it safe and bleed the entire system.

• A clean repair area is essential. Always clean the reservoir and cap thoroughly before removing the cap. The slightest amount of dirt in the fluid may plug an orifice and impair system function. Perform repairs after components have been thoroughly cleaned; use only denatured alcohol to clean components. Do not allow ABS components to come into contact with any substance containing mineral oil; this includes used shop rags.

• The anti-lock control unit is a microprocessor similar to other computer units in the vehicle. Ensure that the ignition switch is **OFF** before removing or installing controller wiring harnesses. Avoid static electricity discharge at or near the controller.

• If any arc welding is to be done on the vehicle, the control unit should be unplugged before welding operations begin.

ABS DEPRESSURIZING

Some ABS systems store the brake fluid at high pressures, which must be released before any service is attempted.

On these systems, the hydraulic accumulator contains brake fluid and nitrogen gas at extremely high pressures. Certain other system components may also contain brake fluid at high pressure. It is mandatory that the system pressure is relieved before disconnecting any hoses, lines or fittings, otherwise personal injury may result.

❋❋ CAUTION

On ABS systems designed to store brake fluid at high pressures, it is necessary to depressurize the system before disconnecting any hoses, lines or fittings. Otherwise, personal injury may result.

On most vehicles, ABS pressure can be depleted simply by pumping the brake pedal 20–30 times with the ignition switch **OFF**.

On some systems, pressure should be bled using a specific, expensive scan tool. For this reason, we recommend that when in doubt, all ABS system service be referred to a professional, qualified technician.

DIAGNOSTIC TROUBLE CODES

The on-board computer system receives input from sensors all over the vehicle. The sensors signal the operating condition of every controlled component from the engine on down to the wheels.

Part of this overall system is the brake system. When any fault or problem in the brake system is detected by a sensor, a signal is sent to the computer and recorded in its memory in the form of a trouble code. The trouble codes can be accessed, in most cases, through the use of a scan tool. Each ABS equipped vehicle has a connector designed to receive the scan tools wiring harness plug(s).

Vehicle computer systems vary from manufacturer-to-manufacturer and from model-to-model. Because of the large number of different ABS systems, code retrieval information is not included.

➡**The most important thing to remember about ABS trouble codes is that the code does not only implicate the component as defective, but also the component's circuit, possibly, the diagnostic monitor computer. Always check the circuit for faults when diagnosing the brake system based on a trouble code.**

BRAKE SPECIFICATIONS
ACURA 2.2CL, 2.3CL, 2.5TL, 3.0CL, 3.2TL, 3.5RL, INTEGRA, LEGEND, NSX
All measurements in inches unless noted

Year	Model		Master Cylinder Bore	Brake Disc Original Thickness	Brake Disc Minimum Thickness	Brake Disc Maximum Runout	Brake Drum Diameter Original Inside Diameter	Brake Drum Diameter Max. Wear Limit	Brake Drum Diameter Maximum Machine Diameter	Minimum Lining Thickness Front	Minimum Lining Thickness Rear	Brake Caliper Bracket Bolts (ft. lbs.)	Brake Caliper Mounting Bolts (ft. lbs.)
1995	Integra	F	NA	0.830	0.750	0.004	—	—	—	0.06	—	—	24
		R	NA	0.350	0.310	0.006	—	—	—	—	0.06	—	24
	Legend	F	NA	1.100	1.020	0.004	—	—	—	0.06	—	—	36
		R	NA	0.350	0.300	0.006	—	—	—	—	0.06	—	17
	NSX	F	NA	1.100	1.020	0.004	—	—	—	0.06	—	80	36
		R	NA	0.830	0.750	0.004	—	—	—	—	0.06	80	36
	Vigor	F	NA	0.910	0.830	0.004	—	—	—	0.06	—	—	36
		R	NA	0.390	0.310	0.004	—	—	—	—	0.06	—	17
1996	2.5TL	F	NA	0.910	0.830	0.004	—	—	—	0.06	—	—	36
		R	NA	0.350	0.300	0.004	—	—	—	—	0.06	28	17
	3.2TL	F	NA	0.910	0.830	0.004	—	—	—	0.06	—	—	36
		R	NA	0.350	0.300	0.004	—	—	—	—	0.06	28	17
	Integra	F	NA	0.830	0.750	0.004	—	—	—	0.06	—	—	24
		R	NA	0.350	0.310	0.004	—	—	—	—	0.06	—	24
	NSX	F	NA	1.100	1.020	0.004	—	—	—	0.06	—	80	36
		R	NA	0.830	0.750	0.004	—	—	—	—	0.06	80	36
1997	2.2CL	F	NA	0.910	0.830	0.004	—	—	—	0.06	—	—	36
		R	NA	0.350	0.300	0.004	—	—	—	—	0.06	—	17
	3.0CL	F	NA	0.910	0.830	0.004	—	—	—	0.06	—	—	36
		R	NA	0.350	0.300	0.004	—	—	—	—	0.06	28	17
	3.5RL	F	NA	0.910	0.830	0.004	—	—	—	0.06	—	—	36
		R	NA	0.350	0.300	0.004	—	—	—	—	0.06	28	17
	2.5TL	F	NA	0.910	0.830	0.004	—	—	—	0.06	—	—	36
		R	NA	0.350	0.300	0.004	—	—	—	—	0.06	28	17
	3.2TL	F	NA	0.910	0.830	0.004	—	—	—	0.06	—	—	36
		R	NA	0.350	0.300	0.004	—	—	—	—	0.06	28	17
	Integra	F	NA	0.830	0.750	0.004	—	—	—	0.06	—	—	24
		R	NA	0.350	0.310	0.004	—	—	—	—	0.06	—	24
	NSX	F	NA	1.100	1.020	0.004	—	—	—	0.06	—	80	36
		R	NA	0.830	0.750	0.004	—	—	—	—	0.06	80	36
1998-99	2.3CL	F	NA	0.910	0.830	0.004	—	—	—	0.06	—	—	18
		R	NA	0.350	0.300	0.004	—	—	—	—	0.06	28	18
	3.0CL	F	NA	0.910	0.830	0.004	—	—	—	0.06	—	—	36
		R	NA	0.350	0.300	0.004	—	—	—	—	0.06	28	17
	3.5RL	F	NA	0.910	0.830	0.004	—	—	—	0.06	—	—	36
		R	NA	0.350	0.300	0.004	—	—	—	—	0.06	28	17
	2.5TL	F	NA	0.910	0.830	0.004	—	—	—	0.06	—	—	36
		R	NA	0.350	0.300	0.004	—	—	—	—	0.06	28	17
	3.2TL	F	NA	0.910	0.830	0.004	—	—	—	0.06	—	—	36
		R	NA	0.350	0.300	0.004	—	—	—	—	0.06	28	17
	Integra	F	NA	0.830	0.750	0.004	—	—	—	0.06	—	—	24
		R	NA	0.350	0.310	0.004	—	—	—	—	0.06	—	24
	NSX	F	NA	1.100	1.020	0.004	—	—	—	0.06	—	80	36
		R	NA	0.830	0.750	0.004	—	—	—	—	0.06	80	36

NA - Not Available
F - Front
R - Rear

79236C50

BRAKE SPECIFICATIONS
AUDI 90, 90 QUATTRO, A6, A6 QUATTRO, A8, A8 QUATTRO, CABRIOLET, S6
All measurements in inches unless noted

Year	Model		Master Cylinder Bore	Brake Disc Original Thickness	Brake Disc Minimum Thickness	Maximum Runout	Brake Drum Diameter Original Inside Diameter	Max. Wear Limit	Maximum Machine Diameter	Minimum Lining Thickness Front	Minimum Lining Thickness Rear	Brake Caliper Bracket Bolts (ft. lbs.)	Brake Caliper Mounting Bolts (ft. lbs.)
1995	90	F	0.937	0.984	0.866	0.002	—	—	—	①	—	—	②
		R	—	0.394	0.315	0.002	—	—	—	—	①	—	②
	90 Quattro	F	0.937	0.984	0.905	0.002	—	—	—	①	—	—	②
		R	—	0.984	0.905	0.002	—	—	—	—	①	—	②
	A6	F	0.937	0.984	0.905	0.002	—	—	—	①	—	89	18
		R	—	0.394	0.315	0.002	—	—	—	—	①	70	26
	A6 Quattro	F	③	0.984	0.905	0.002	—	—	—	①	—	89	18
		R	—	0.394	0.315	0.002	—	—	—	—	①	70	26
	Cabriolet	F	0.937	0.984	0.866	0.002	—	—	—	①	—	89	18
		R	—	0.394	0.315	0.002	—	—	—	—	①	70	26
	S6	F	③	1.118	1.10	0.002	—	—	—	①	—	—	②
		R	—	0.790	0.71	0.002	—	—	—	—	0.08	—	②
1996	A4	F	0.937	④	⑤	0.002	—	—	—	①	—	—	18
		R	—	④	⑤	0.002	—	—	—	—	①	70	26
	A4 Quattro	F	0.937	④	⑤	0.002	—	—	—	①	—	—	18
		R	—	④	⑤	0.002	—	—	—	—	①	70	44
	A6	F	0.937	0.984	0.905	0.002	—	—	—	①	—	89	18
		R	—	0.394	0.315	0.002	—	—	—	—	①	70	26
	A6 Quattro	F	③	0.984	0.905	0.002	—	—	—	①	—	89	18
		R	—	0.394	0.315	0.002	—	—	—	—	①	70	26
	Cabriolet	F	0.937	0.984	0.866	0.002	—	—	—	①	—	89	18
		R	—	0.394	0.315	0.002	—	—	—	—	①	70	26
1997	A4	F	0.937	④	⑤	0.002	—	—	—	①	—	—	18
		R	—	④	⑤	0.002	—	—	—	—	①	70	26
	A4 Quattro	F	0.937	④	⑤	0.002	—	—	—	①	—	—	18
		R	—	④	⑤	0.002	—	—	—	—	①	70	44
	A6	F	0.937	0.984	0.905	0.002	—	—	—	①	—	89	18
		R	—	0.394	0.315	0.002	—	—	—	—	①	70	26
	A6 Quattro	F	③	0.984	0.905	0.002	—	—	—	①	—	89	18
		R	—	0.394	0.315	0.002	—	—	—	—	①	70	26
	A8	F	1.000	0.984	1.000	0.002	—	—	—	①	—	140	22
		R	—	1.000	0.906	0.002	—	—	—	—	①	44	26
	A8 Quattro	F	1.000	0.984	1.000	0.002	—	—	—	①	—	140	22
		R	—	1.000	0.906	0.002	—	—	—	—	①	44	26
	Cabriolet	F	0.937	0.984	0.866	0.002	—	—	—	①	—	89	18
		R	—	0.394	0.315	0.002	—	—	—	—	①	70	26
1998-99	A4	F	0.937	④	⑤	0.002	—	—	—	①	—	—	18
		R	—	④	⑤	0.002	—	—	—	—	①	70	26
	A4 Quattro	F	0.937	④	⑤	0.002	—	—	—	①	—	—	18
		R	—	④	⑤	0.002	—	—	—	—	①	70	44
	A6	F	0.937	0.984	0.905	0.002	—	—	—	①	—	89	18
		R	—	0.394	0.315	0.002	—	—	—	—	①	70	26
	A6 Quattro	F	③	0.984	0.905	0.002	—	—	—	①	—	89	18
		R	—	0.394	0.315	0.002	—	—	—	—	①	70	26

79236C51

BRAKE SPECIFICATIONS
AUDI 90, 90 QUATTRO, A6, A6 QUATTRO, A8, A8 QUATTRO, CABRIOLET, S6
All measurements in inches unless noted

| Year | Model | | Master Cylinder Bore | Brake Disc | | | Brake Drum Diameter | | | Minimum Lining Thickness | | Brake Caliper | |
				Original Thickness	Minimum Thickness	Maximum Runout	Original Inside Diameter	Max. Wear Limit	Maximum Machine Diameter	Front	Rear	Bracket Bolts (ft. lbs.)	Mounting Bolts (ft. lbs.)
1998-99 (cont.)	A8	F	1.000	0.984	1.000	0.002	—	—	—	①	—	140	22
		R	—	1.000	0.906	0.002	—	—	—	—	①	44	26
	A8 Quattro	F	1.000	0.984	1.000	0.002	—	—	—	①	—	140	22
		R	—	1.000	0.906	0.002	—	—	—	—	①	44	26
	Cabriolet	F	0.937	0.984	0.866	0.002	—	—	—	①	—	89	18
		R	—	0.394	0.315	0.002	—	—	—	—	①	70	26

① 0.28 including backing plate
② Except Teves 25 ft. lbs.
   Teves 18 ft. lbs.
③ Vacuum Booster: 0.937
   Hydraulic Booster: 1.00
④ Front:
   Teves/Ate Calipers:
     -Venetilated Disc: 0.984
     -Non-ventilated Disc: 0.590
   Lucas Calipers: 0.510
   Rear: 0.394
⑤ Front:
   Teves/Ate Calipers:
     -Venetilated Disc: 0.905
     -Non-ventilated Disc: 0.510
   Lucas Calipers: 0.430
   Rear: 0.315

79236C52

BRAKE SPECIFICATIONS
BMW 3 SERIES, 5 SERIES, 7 SERIES, 8 SERIES, M3, Z3
All measurements in inches unless noted

Year	Model	Master Cylinder Bore	Front Brake Disc			Rear Brake Disc			Minimum Lining Thickness		Brake Caliper	
			Original Thickness	Minimum Thickness	Maximum Runout	Original Thickness	Minimum Thickness	Maximum Runout	Front	Rear	Bracket Bolts (ft. lbs.)	Mounting Bolts (ft. lbs.)
1995	318i	NA	NA	0.409	0.008	NA	①	0.008	0.079	0.079	②	22
	318iS	NA	NA	0.409	0.008	NA	①	0.008	0.079	0.079	②	22
	325i	NA	NA	0.803	0.008	NA	①	0.008	0.079	0.079	②	22
	325iS	NA	NA	0.803	0.008	NA	①	0.008	0.079	0.079	②	22
	M3	NA	NA	0.803	0.008	NA	0.724	0.008	0.079	0.079	②	22
	525i	NA	NA	0.803	0.008	NA	0.331	0.008	0.079	0.079	②	22
	530i	NA	NA	0.803	0.008	NA	0.331	0.008	0.079	0.079	②	22
	540i	NA	NA	1.039	0.008	NA	0.724	0.008	0.079	0.079	②	22
	740i	NA	NA	1.118	0.008	NA	0.410	0.008	0.079	0.079	②	22
	740iL	NA	NA	1.118	0.008	NA	0.724	0.008	0.079	0.079	②	22
	750iL	NA	NA	1.118	0.008	NA	0.724	0.008	0.079	0.079	②	22
	840Ci	NA	NA	1.118	0.008	NA	0.724	0.008	0.079	0.079	②	22
	850Ci	NA	NA	1.118	0.008	NA	0.724	0.008	0.079	0.079	②	22
	850CSi	NA	NA	1.118	0.008	NA	0.724	0.008	0.079	0.079	②	22
1996	318i	NA	NA	0.409	0.008	NA	①	0.008	0.079	0.079	②	22
	318iC	NA	NA	0.409	0.008	NA	①	0.008	0.079	0.079	②	22
	318is	NA	NA	0.409	0.008	NA	①	0.008	0.079	0.079	②	22
	318ti	NA	NA	0.409	0.008	NA	①	0.008	0.079	0.079	②	22
	Z3	NA	NA	0.803	0.008	NA	①	0.008	0.079	0.079	②	22
	328i	NA	NA	0.803	0.008	NA	①	0.008	0.079	0.079	②	22
	328is	NA	NA	0.803	0.008	NA	①	0.008	0.079	0.079	②	22
	328iC	NA	NA	0.803	0.008	NA	①	0.008	0.079	0.079	②	22
	M3	NA	NA	1.039	0.008	NA	0.724	0.008	0.079	0.079	②	22
	740iL	NA	NA	1.118	0.008	NA	0.724	0.008	0.079	0.079	②	22
	750iL	NA	NA	1.118	0.008	NA	0.724	0.008	0.079	0.079	②	22
	840Ci	NA	NA	1.118	0.008	NA	0.724	0.008	0.079	0.079	②	22
	850Ci	NA	NA	1.118	0.008	NA	0.724	0.008	0.079	0.079	②	22
1997	318i	NA	NA	0.409	0.008	NA	①	0.008	0.079	0.079	②	22
	318iC	NA	NA	0.409	0.008	NA	①	0.008	0.079	0.079	②	22
	318is	NA	NA	0.409	0.008	NA	①	0.008	0.079	0.079	②	22
	318ti	NA	NA	0.409	0.008	NA	①	0.008	0.079	0.079	②	22
	Z3	NA	NA	0.803	0.008	NA	①	0.008	0.079	0.079	②	22
	328i	NA	NA	0.803	0.008	NA	①	0.008	0.079	0.079	②	22
	328is	NA	NA	0.803	0.008	NA	①	0.008	0.079	0.079	②	22
	328iC	NA	NA	0.803	0.008	NA	①	0.008	0.079	0.079	②	22
	M3	NA	NA	1.039	0.008	NA	0.724	Runout	0.079	0.079	②	22
	740iL	NA	NA	1.118	0.008	NA	0.724	0.008	0.079	0.079	②	22
	750iL	NA	NA	1.118	0.008	NA	0.724	0.008	0.079	0.079	②	22
	840Ci	NA	NA	1.118	0.008	NA	0.724	0.008	0.079	0.079	②	22
	850Ci	NA	NA	1.118	0.008	NA	0.724	0.008	0.079	0.079	②	22

79236C53

BRAKE SPECIFICATIONS
BMW 3 SERIES, 5 SERIES, 7 SERIES, 8 SERIES, M3, Z3
All measurements in inches unless noted

Year	Model	Master Cylinder Bore	Front Brake Disc Original Thickness	Front Brake Disc Minimum Thickness	Front Brake Disc Maximum Runout	Rear Brake Disc Original Thickness	Rear Brake Disc Minimum Thickness	Rear Brake Disc Maximum Runout	Minimum Lining Thickness Front	Minimum Lining Thickness Rear	Brake Caliper Bracket Bolts (ft. lbs.)	Brake Caliper Mounting Bolts (ft. lbs.)
1998-99	318ti	NA	NA	0.409	0.008	NA	①	0.008	0.079	0.079	②	22
	318i	NA	NA	0.409	0.008	NA	①	0.008	0.079	0.079	②	22
	323is	NA	NA	0.803	0.008	NA	①	0.008	0.079	0.079	②	22
	323iC	NA	NA	0.803	0.008	NA	①	0.008	0.079	0.079	②	22
	Z3	NA	NA	0.803	0.008	NA	①	0.008	0.079	0.079	②	22
	328i	NA	NA	0.803	0.008	NA	①	0.008	0.079	0.079	②	22
	328is	NA	NA	0.803	0.008	NA	①	0.008	0.079	0.079	②	22
	328iC	NA	NA	0.803	0.008	NA	①	0.008	0.079	0.079	②	22
	M3 coupe	NA	NA	1.039	0.008	NA	0.724	0.008	0.079	0.079	②	22
	M3 sedan	NA	NA	1.039	0.008	NA	0.724	0.008	0.079	0.079	②	22
	Z3	NA	NA	0.803	0.008	NA	①	0.008	0.079	0.079	②	22
	Z3	NA	NA	0.803	0.008	NA	①	0.008	0.079	0.079	②	22
	528i	NA	NA	0.803	0.008	NA	0.331	0.008	0.079	0.079	②	22
	540i	NA	NA	1.118	0.008	NA	0.724	0.008	0.079	0.079	②	22
	740i	NA	NA	1.118	0.008	NA	0.410	0.008	0.079	0.079	②	22
	740iL	NA	NA	1.118	0.008	NA	0.724	0.008	0.079	0.079	②	22
	750iL	NA	NA	1.118	0.008	NA	0.724	0.008	0.079	0.079	②	22

NA - Not Available
① Solid: 0.331; Vented: 0.685
② Front: 81 ft. lbs; Rear: 50 ft. lbs.

79236C54

BRAKE SPECIFICATIONS
CHRYSLER IMPORTS SUMMIT, SUMMIT WAGON
All measurements in inches unless noted

Year	Model		Master Cylinder Bore	Brake Disc Original Thickness	Brake Disc Minimum Thickness	Brake Disc Maximum Runout	Brake Drum Diameter Original Inside Diameter	Brake Drum Diameter Max. Wear Limit	Brake Drum Diameter Maximum Machine Diameter	Minimum Lining Thickness Front	Minimum Lining Thickness Rear
1995	Summit	①	0.813	0.510	0.449	0.003	7.10	7.20	NA	0.080	0.040
	Summit	②	⑪	0.710 ⑫	0.646 ⑬	0.003	8.00	8.10	NA	0.080	0.040 ⑩
	Summit Wagon	③	③	⑧	⑨	0.003	④	⑤	NA	0.080	0.040 ⑩
1996	Summit	⑥	0.813	0.710 ⑭	0.449	0.003	8.00	8.10	NA	0.080	0.040
	Summit	②	⑦	0.710 ⑫	0.646 ⑬	0.003	8.00	8.10	NA	0.080	0.040 ⑩
	Summit Wagon	③	③	⑧	⑨	0.003	④	⑤	NA	0.080	0.040 ⑩

NA - Not Available
① Hatchback
② Sedan
③ With ABS: 1.000; Without ABS: 0.9375
④ 8" drum: 7.992; 9" drum: 9.000
⑤ 8" drum: 8.071; 9" drum: 9.079

⑥ Coupe
⑦ With ABS: 0.937; Without ABS: 0.812
⑧ Front: 0.945
Rear: 0.394
⑨ Front: 0.882
Rear: 0.331

⑩ Rear disc brakes: 0.080
⑪ Master cylinder bore: 0.813
With ABS: 0.938
⑫ Rear: 0.390
⑬ Rear: 0.330
⑭ Solid front disc: 0.510

79236C55

BRAKE SPECIFICATIONS
HONDA ACCORD, CIVIC, DEL SOL, PRELUDE
All measurements in inches unless noted

Year	Model		Master Cylinder Bore	Brake Disc Original Thickness	Brake Disc Minimum Thickness	Brake Disc Maximum Runout	Brake Drum Diameter Original Inside Diameter	Brake Drum Diameter Max. Wear Limit	Brake Drum Diameter Maximum Machine Diameter	Minimum Lining Thickness Front	Minimum Lining Thickness Rear	Brake Caliper Bracket Bolts (ft. lbs.)	Brake Caliper Mounting Bolts (ft. lbs.)
1995	Accord	F	NA	0.910 ①	0.830 ②	0.004	—	—	—	0.060	—	—	③
		R	—	0.400	0.310	0.004	8.66	8.70	8.70	—	0.080	—	17
	Civic	F	NA	0.830	0.750	0.004	—	—	—	0.060	—	—	③
		R	—	0.350	0.310	0.004	④	⑤	⑤	—	0.080	—	17
	del Sol	F	NA	0.830	0.750	0.004	—	—	—	0.060	—	—	③
		R	—	0.350	0.310	0.004	7.09	7.13	7.13	—	0.080	—	17
	Prelude	F	NA	0.910	0.830	0.004	—	—	—	0.060	—	—	③
		R	—	0.390	0.320	0.004	—	—	—	—	0.060	—	17
1996	Accord	F	NA	0.910 ①	0.830 ②	0.004	—	—	—	0.060	—	—	③
		R	—	0.400	0.310	0.004	8.66	8.70	8.70	—	0.080	—	17
	Civic	F	NA	0.840	0.750	0.004	—	—	—	0.060	—	—	③
		R	—	—	—	—	7.87	7.91	7.91	—	0.080	—	—
	del Sol	F	NA	0.830	0.750	0.004	—	—	—	0.060	—	—	③
		R	—	0.350	0.310	0.004	7.09	7.13	7.13	—	0.080	—	17
	Prelude	F	NA	0.910	0.830	0.004	—	—	—	0.060	—	—	③
		R	—	0.390	0.320	0.004	—	—	—	—	0.060	—	17
1997	Accord	F	NA	0.910 ①	0.830 ②	0.004	—	—	—	0.060	—	—	③
		R	—	0.400	0.310	0.004	8.66	8.70	8.70	—	0.080	—	17
	Civic	F	NA	0.840	0.750	0.004	—	—	—	0.060	—	—	③
		R	—	—	—	—	7.87	7.91	7.91	—	0.080	—	—
	del Sol	F	NA	0.830	0.750	0.004	—	—	—	0.060	—	—	③
		R	—	0.350	0.310	0.004	7.09	7.13	7.13	—	0.080	—	17
	Prelude	F	NA	0.910	0.830	0.004	—	—	—	0.060	—	—	③
		R	—	0.390	0.320	0.004	—	—	—	—	0.060	—	17
1998-99	Accord	F	NA	0.910	0.830	0.004	—	—	—	0.060	—	—	③
		R	—	0.400	0.310	0.004	8.66	8.70	8.70	—	0.080	—	17
	Civic	F	NA	0.840	0.750	0.004	—	—	—	0.060	—	—	③
		R	—	—	—	—	7.87	7.91	7.91	—	0.080	—	—
	Prelude	F	NA	0.910	0.830	0.004	—	—	—	0.060	—	83	③
		R	—	0.390	0.320	0.004	—	—	—	—	0.060	—	17

NA - Not Available

F - Front

R - Rear

① Wagon and V6 models: 0.990

② Wagon and V6 models: 0.910

③ Calipers with long pins beyond bolt threads, 54 ft. lbs.
Calipers with no pin beyond threads, 20 ft. lbs.

④ 7.09: Cars with manual trans.; Except Coupe 1.6L manual trans.
7.87: Cars with automatic trans., Coupe 1.6L manual trans.

⑤ 7.13: Cars with manual trans.; Except Coupe 1.6L manual trans.
7.91: Cars with automatic trans.; Coupe 1.6L manual trans.

79236C56

BRAKE SPECIFICATIONS
HYUNDAI ACCENT, ELANTRA, SCOUPE, SONATA, TIBURON
All measurements in inches unless noted

Year	Model		Brake Disc Original Thickness	Brake Disc Minimum Thickness	Brake Disc Maximum Run-out	Brake Drum Diameter Original Inside Diameter	Brake Drum Diameter Max. Wear Limit	Maximum Machine Diameter	Minimum Lining Thickness Front	Minimum Lining Thickness Rear	Brake Caliper Bracket Bolts (ft. lbs.)	Brake Caliper Mounting Bolts (ft. lbs.)
1995	Accent		0.750	0.669	0.002	7.090	7.165	—	0.039	0.039	48-55	④
	Elantra	①	0.866	0.787	0.002	8.000	8.079	—	0.079	0.059	44-63	16-24
	Elantra	②	0.354	NA	NA	—	—	—	0.079	0.031	44-63	16-24
	Scoupe		0.750	0.669	0.002	7.100	7.165	—	0.039	0.031	48-55	④
	Sonata	①	0.866	0.787	0.004	9.000	9.079	—	0.079	③	51-63	16-24
	Sonata	②	0.472	0.413	0.005	—	—	—	0.079	0.079	51-63	16-24
1996	Accent		0.750	0.669	0.002	7.090	7.165	—	0.039	0.039	48-55	④
	Elantra	①	0.866	0.787	0.002	8.000	8.079	—	0.079	0.059	44-63	16-24
	Elantra	②	0.354	NA	NA	—	—	—	0.079	0.031	44-63	16-24
	Sonata	①	0.866	0.787	0.004	9.000	9.079	—	0.079	③	51-63	16-24
	Sonata	②	0.472	0.413	0.005	—	—	—	0.079	0.079	51-63	16-24
1997	Accent		0.750	0.669	0.002	7.090	7.165	—	0.039	0.039	48-55	④
	Elantra	①	0.866	0.787	0.002	8.000	8.079	—	0.079	0.059	44-63	16-24
	Elantra	②	0.354	NA	NA	—	—	—	0.079	0.031	44-63	16-24
	Sonata	①	0.866	0.787	0.004	9.000	9.079	—	0.079	③	51-63	16-24
	Sonata	②	0.472	0.413	0.005	—	—	—	0.079	0.031	51-63	16-24
	Tiburon	①	0.866	0.787	0.002	8.000	8.079	—	0.079	0.059	44-63	16-24
	Tiburon	②	0.354	NA	NA	—	—	—	0.079	0.031	44-63	16-24
1998-99	Accent		0.750	0.669	0.002	7.090	7.165	—	0.039	0.039	48-55	④
	Elantra	①	0.866	0.787	0.002	8.000	8.079	—	0.079	0.059	44-63	16-24
	Elantra	②	0.354	NA	NA	—	—	—	0.079	0.031	44-63	16-24
	Sonata	①	0.866	0.787	0.004	9.000	9.079	—	0.079	③	51-63	16-24
	Sonata	②	0.472	0.413	0.005	—	—	—	0.079	0.031	51-63	16-24
	Tiburon	①	0.866	0.787	0.002	8.000	8.079	—	0.079	0.059	44-63	16-24
	Tiburon	②	0.354	NA	NA	—	—	—	0.079	0.031	44-63	16-24

NA: Not Available
① With rear drum brakes
② With rear disc brakes
③ With 4-cyl. engine: 0.031 in.
  With 6-cyl. engine: 0.059 in.
④ Lower bolt: 16-24 ft. lbs.
  Upper bolt: 26-33 ft. lbs.

79236C57

BRAKE SPECIFICATIONS
INFINITI G20, I30, J30, Q45
All measurements in inches unless noted

Year	Model		Master Cylinder Bore	Brake Disc Original Thickness	Brake Disc Minimum Thickness	Brake Disc Maximum Runout	Minimum Lining Thickness Front	Minimum Lining Thickness Rear	Brake Caliper Bracket Bolts (ft. lbs.)	Brake Caliper Mounting Bolts (ft. lbs.)
1995	G20	F	0.937	0.870	0.787	0.003	0.079	—	53-72	16-23
		R	—	0.350	0.310	0.003	—	0.059	28-38	16-23
	J30	F	1.000	1.100	1.024	0.003	0.079	—	53-72	16-23
		R	—	0.630	0.551	0.006	—	0.079	28-38	23-30
	Q45	F	1.625	1.000	1.024	0.003	0.079	—	103-118	24-31
		R	—	0.350	0.315	0.003	—	0.079	28-38	23-30
1996	G20	F	0.937	0.870	0.787	0.003	0.079	—	53-72	16-23
		R	—	0.350	0.310	0.003	—	0.059	28-38	16-23
	I30	F	0.937	0.870	0.787	0.003	0.079	—	53-72	16-23
		R	—	0.350	0.310	0.003	—	0.059	16-23	16-23
	J30	F	1.000	1.100	1.024	0.003	0.079	—	53-72	16-23
		R	—	0.630	0.551	0.006	—	0.079	28-38	23-30
	Q45	F	1.000	1.100	1.024	0.003	0.079	—	103-118	24-31
		R	—	0.350	0.315	0.003	—	0.079	28-38	23-30
1997	I30	F	0.937	0.870	0.787	0.003	0.079	—	53-72	16-23
		R	—	0.350	0.310	0.003	—	0.059	28-38	23-30
	J30	F	1.000	1.100	1.024	0.003	0.079	—	53-72	16-23
		R	—	0.630	0.551	0.006	—	0.079	28-38	23-30
	Q45	F	1.000	1.100	1.024	0.003	0.079	—	103-118	24-31
		R	—	0.350	0.315	0.003	—	0.079	28-38	23-30
1998-99	I30	F	0.937	0.870	0.787	0.003	0.079	—	53-72	16-23
		R	—	0.350	0.310	0.003	—	0.059	28-38	23-30
	Q45	F	1.000	1.100	1.024	0.003	0.079	—	103-118	24-31
		R	—	0.350	0.315	0.003	—	0.079	28-38	23-30

F - Front

R - Rear

79236C58

BRAKE SPECIFICATIONS
KIA SEPHIA
All measurements in inches unless noted

Year	Model		Master Cylinder Bore	Brake Disc Original Thickness	Brake Disc Minimum Thickness	Maximum Runout	Brake Drum Diameter Original Inside Diameter	Max. Wear Limit	Maximum Machine Diameter	Minimum Lining Thickness Front	Minimum Lining Thickness Rear	Brake Caliper Bracket Bolts (ft. lbs.)	Brake Caliper Mounting Bolts (ft. lbs.)
1995	Sephia	F	0.875	0.750	0.710	0.004	—	—	—	0.080	—	20	18
		R	—	0.400	0.320	0.004	7.87	7.91	7.91	—	①	—	20
1996	Sephia	F	0.875	0.750	0.710	0.004	—	—	—	0.080	—	20	18
		R	—	0.400	0.320	0.004	7.87	7.91	7.91	—	①	—	20
1997	Sephia	F	0.875	0.750	0.710	0.004	—	—	—	0.080	—	20	18
		R	—	0.400	0.320	0.004	7.87	7.91	7.91	—	①	—	20
1998-99	Sephia	F	②	0.940	0.710	0.004	—	—	—	0.080	—	33-49	19-21
		R	—	0.400	0.320	0.0039	7.87	7.91	7.91	—	0.079	33-49	22-29

F - Front
R - Rear
① Rear disc pad: 0.060 in.
   Rear drum shoe: 0.040 in.
② With ABS: 0.937 in.
   Without ABS: 0.874 in.

79236C59

BRAKE SPECIFICATIONS
LEXUS ES300, GS300, GS400, LS400, SC300, SC400
All measurements in inches unless noted

Year	Model		Master Cylinder Bore	Brake Disc Original Thickness	Brake Disc Minimum Thickness	Brake Disc Maximum Runout	Brake Drum Diameter Original Inside Diameter	Brake Drum Diameter Max. Wear Limit	Brake Drum Diameter Maximum Machine Diameter	Minimum Lining Thickness Front	Minimum Lining Thickness Rear	Brake Caliper Bracket Bolts (ft. lbs.)	Brake Caliper Mounting Bolts (ft. lbs.)
1995	ES300	F	NA	1.102	1.024	0.0020	—	—	—	0.039	—	79	25
		R	NA	0.394	0.354	0.0059	—	—	—	—	0.039	34	14
	GS300	F	NA	1.260	1.181	0.0020	—	—	—	0.039	—	87	25
		R	NA	0.630	0.591	0.0020	—	—	—	—	0.039	—	25
	LS400	F	NA	1.102	1.024	0.0020	—	—	—	0.012	—	87	25
		R	NA	0.630	0.591	0.0020	—	—	—	—	0.098	—	77
	SC300	F	NA	1.102	1.024	0.0020	—	—	—	0.039	—	—	87
		R	NA	0.630	0.591	0.0020	—	—	—	—	0.039	77	25
	SC400	F	NA	1.260	1.181	0.0020	—	—	—	0.039	—	87	25
		R	NA	0.630	0.591	0.0020	—	—	—	—	0.039	—	25
1996	ES300	F	NA	1.102	1.024	0.0020	—	—	—	0.039	—	79	25
		R	NA	0.394	0.354	0.0059	—	—	—	—	0.039	34	14
	SC300	F	NA	1.102	1.024	0.0020	—	—	—	0.039	—	87	25
		R	NA	0.630	0.591	0.0020	—	—	—	—	0.039	—	25
	GS300	F	NA	1.260	1.181	0.0020	—	—	—	0.039	—	87	25
		R	NA	0.630	0.591	0.0020	—	—	—	—	0.039	—	77
	LS400	F	NA	1.102	1.024	0.0020	—	—	—	0.118	—	—	87
		R	NA	0.630	0.591	0.0020	—	—	—	—	0.098	77	25
	SC400	F	NA	1.260	1.181	0.0020	—	—	—	0.039	—	87	25
		R	NA	0.630	0.591	0.0020	—	—	—	—	0.039	—	25
1997	ES300	F	NA	1.102	1.024	0.0020	—	—	—	0.039	—	79	25
		R	NA	0.394	0.354	0.0059	—	—	—	—	0.039	34	14
	SC300	F	NA	1.102	1.024	0.0020	—	—	—	0.039	—	87	25
		R	NA	0.630	0.591	0.0020	—	—	—	—	0.039	—	25
	GS300	F	NA	1.260	1.181	0.0020	—	—	—	0.039	—	87	25
		R	NA	0.630	0.591	0.0020	—	—	—	—	0.039	—	77
	LS400	F	NA	1.102	1.024	0.0020	—	—	—	0.118	—	—	87
		R	NA	0.630	0.591	0.0020	—	—	—	—	0.098	77	25
	SC400	F	NA	1.260	1.181	0.0020	—	—	—	0.039	—	87	25
		R	NA	0.630	0.591	0.0020	—	—	—	—	0.039	—	25
1998-99	ES300	F	NA	1.102	1.024	0.0020	—	—	—	0.039	—	79	25
		R	NA	0.394	0.354	0.0059	—	—	—	—	0.039	34	14
	SC300	F	NA	1.102	1.024	0.0020	—	—	—	0.039	—	87	25
		R	NA	0.630	0.591	0.0020	—	—	—	—	0.039	—	25
	GS300	F	NA	1.260	1.181	0.0020	—	—	—	0.039	—	87	25
		R	NA	0.472	0.413	0.0020	—	—	—	—	0.039	—	77
	LS400	F	NA	1.102	1.024	0.0020	—	—	—	0.039	—	—	87
		R	NA	0.630	0.591	0.0020	—	—	—	—	0.039	77	25
	SC400	F	NA	1.260	1.181	0.0020	—	—	—	0.039	—	87	25
		R	NA	0.630	0.591	0.0020	—	—	—	—	0.039	—	25
	GS400	F	NA	1.260	1.181	0.0020	—	—	—	0.039	—	87	25
		R	NA	0.472	0.413	0.0020	—	—	—	—	0.039	—	25

NA - Not Available

F - Front

R - Rear

79236C60

BRAKE SPECIFICATIONS
MAZDA 626, 929, MIATA, MILLENIA, MX3, MX6, PROTEGE, RX7
All measurements in inches unless noted

Year	Model		Master Cylinder Bore	Brake Disc Original Thickness	Brake Disc Minimum Thickness	Brake Disc Maximum Runout	Brake Drum Original Inside Diameter	Brake Drum Max. Wear Limit	Brake Drum Maximum Machine Diameter	Min Lining Front	Min Lining Rear	Bracket Bolts (ft. lbs.)	Mounting Bolts (ft. lbs.)
1995	RX7	F	NA	0.870	0.790	0.004	-	-	-	0.040	-	58-72	79-98
		R	-	0.790	0.710	0.004	-	-	-	-	0.040	47-62	63-84
	Protege	F	0.937	0.870	0.790	0.002	-	-	-	0.080	-	29-36	40-49
		R	-	0.354	0.276	0.002	7.87	7.91	NA	-	0.040	33-44	46-60
	MX3	F	0.875	0.870	0.790	0.004	-	-	-	0.080	-	29-36	40-49
		R	-	0.354	0.310	0.004	7.87	7.93	NA	-	0.040	12-17	16-24
	Miata	F	0.875	0.790	0.710	0.004	-	-	-	0.040	-	58-65	78-88
		R	-	0.350	0.310	0.004	-	-	-	-	0.040	33-36	45-49
	626	F	0.937	0.940	0.870	0.004	-	-	-	0.080	-	33-36	45-49
		R	-	0.390	0.310	0.004	9.00	NA	NA	-	0.040	22-28	30-41
	MX6	F	0.937	0.940	0.870	0.004	-	-	-	0.080	-	33-36	45-49
		R	-	0.390	0.310	0.004	9.00	NA	NA	-	0.040	26-28	35-39
	929	F	NA	0.950	0.870	0.004	-	-	-	0.040	-	47-62	63-84
		R	-	0.710	0.630	0.004	-	-	-	-	0.040	28-36	38-49
	Millenia	F	1.000	1.100	1.020	0.004	-	-	-	0.080	-	47-62	63-84
		R	-	0.370	0.290	0.004	-	-	-	-	0.080	12-17	16-23
1996	Protege	F	0.937	0.870	0.790	0.002	-	-	-	0.040	-	29-36	40-49
		R	-	0.354	0.276	0.002	7.87	7.91	NA	-	0.040	33-44	46-60
	Miata	F	0.875	0.790	0.710	0.004	-	-	-	0.040	-	58-65	78-88
		R	-	0.350	0.310	0.004	-	-	-	-	0.040	33-36	45-49
	626	F	0.937	0.940	0.870	0.002	-	-	-	0.080	-	33-36	45-49
		R	-	0.390	0.310	0.002	9.00	NA	9.06	-	0.040	22-28	30-41
	MX6	F	0.937	0.940	0.870	0.002	-	-	-	0.080	-	33-36	45-49
		R	-	0.390	0.310	0.002	9.00	NA	9.06	-	0.040	26-28	35-39
	Millenia	F	1.000	1.100	1.020	0.002	-	-	-	0.080	-	47-62	63-84
		R	-	0.370	0.290	0.002	-	-	-	-	0.080	12-17	16-23
1997	Protege	F	0.937	0.870	0.790	0.002	-	-	-	0.040	-	29-36	40-49
		R	-	0.354	0.276	0.002	7.87	7.91	NA	-	0.040	33-44	46-60
	Miata	F	0.875	0.790	0.710	0.004	-	-	-	0.040	-	58-65	78-88
		R	-	0.350	0.310	0.004	-	-	-	-	0.040	33-36	45-49
	626	F	0.937	0.940	0.870	0.002	-	-	-	0.080	-	33-36	45-49
		R	-	0.390	0.310	0.002	9.00	NA	9.06	-	0.040	22-28	30-41
	MX6	F	0.937	0.940	0.870	0.002	-	-	-	0.080	-	33-36	45-49
		R	-	0.390	0.310	0.002	9.00	NA	9.06	-	0.040	26-28	35-39
	Millenia	F	1.000	1.100	1.020	0.002	-	-	-	0.080	-	47-62	63-84
		R	-	0.370	0.290	0.002	-	-	-	-	0.080	12-17	16-23
1998-99	Protege	F	0.937	0.870	0.790	0.002	-	-	-	0.040	-	29-36	40-49
		R	-	0.354	0.276	0.002	7.87	7.91	NA	-	0.040	33-44	46-60
	Miata	F	0.875	0.790	0.710	0.004	-	-	-	0.040	-	58-65	78-88
		R	-	0.350	0.310	0.004	-	-	-	-	0.040	33-36	45-49
	626	F	0.937	0.940	0.870	0.002	-	-	-	0.080	-	33-36	45-49
		R	-	0.390	0.310	0.002	9.00	NA	9.06	-	0.040	22-28	30-41
	Millenia	F	1.000	1.100	1.020	0.002	-	-	-	0.080	-	47-62	63-84
		R	-	0.370	0.290	0.002	-	-	-	-	0.080	12-17	16-23

F - Front
R - Rear

79236C61

BRAKE SPECIFICATIONS
MERCEDES-BENZ C, CLK, E, S, SL, SLK CLASSES
All measurements in inches unless noted

Year	Model	Master Cyl. Bore	Front Brake Disc Original Thickness	Front Brake Disc Minimum Thickness	Front Brake Disc Max. Runout	Rear Brake Disc Original Thickness	Rear Brake Disc Minimum Thickness	Rear Brake Disc Max. Runout	Minimum Lining Thickness Front	Minimum Lining Thickness Rear	Brake Caliper Bracket Bolts (ft. lbs.)	Brake Caliper Mounting Bolts (ft. lbs.)
1995	C220	①	0.866	0.763	0.0047	0.354	0.287	0.0059	0.078	0.078	②	22-30
	C280	①	0.866	0.763	0.0047	0.354	0.287	0.0059	0.078	0.078	②	22-30
	E320	①	0.984 ③	0.881 ③	0.0047	0.354	0.287	0.0059	0.078	0.078	②	22-30
	E420	④	0.984	0.881	0.0047	0.944	0.842	0.0059	0.078	0.078	②	22-30
	E500	④	1.181	1.102	0.0031	0.944	0.842	0.0059	0.078	0.078	②	22-30
	S320	④	1.102 ⑤	0.999 ⑥	0.0031	0.472	0.385	0.0039	0.078	0.078	②	22-30
	S350	④	1.102 ⑤	0.999 ⑥	0.0031	0.472	0.385	0.0039	0.078	0.078	②	22-30
	S420	⑦	1.102 ⑤	0.999 ⑥	0.0031	0.866	0.763	0.0039	0.078	0.078	②	22-30
	S500	⑦	1.102 ⑤	0.999 ⑥	0.0031	0.866	0.763	0.0039	0.078	0.078	②	22-30
	S600	⑦	1.102 ⑤	0.999 ⑥	0.0031	0.866	0.763	0.0039	0.078	0.078	②	22-30
	SL320	④	1.102	0.999	0.0047	0.354	0.287	0.0059	0.078	0.078	②	22-30
	SL500	④	1.102	0.999	0.0047	0.354	0.287	0.0059	0.078	0.078	②	22-30
	SL600	⑦	1.181	1.102	0.0031	0.866	0.763	0.0039	0.078	0.078	②	22-30
1996	C220	①	0.866	0.763	0.0047	0.354	0.287	0.0059	0.078	0.078	②	22-30
	C280	①	0.866	0.763	0.0047	0.354	0.287	0.0059	0.078	NA	②	22-30
	C36	①	1.181	1.102	0.0031	0.945	0.842	0.0059	0.078	0.078	②	22-30
	E300	NA	NA	NA	NA	NA	NA	NA	NA	NA	②	22-30
	E320	NA	NA	NA	NA	NA	NA	NA	NA	NA	②	22-30
	S320	④	1.102 ⑤	0.999 ⑥	0.0031	0.472	0.385	0.0039	0.078	0.078	②	22-30
	SL320	④	1.102	0.999	0.0047	0.866	0.287	0.0059	0.078	0.078	②	22-30
	S420	⑦	1.102 ⑤	0.999 ⑥	0.0031	0.866	0.763	0.0039	0.078	0.078	②	22-30
	S500	⑦	1.102 ⑤	0.999 ⑥	0.0031	0.866	0.763	0.0039	0.078	0.078	②	22-30
	SL500	④	1.102	0.999	0.0047	0.354	0.287	0.0059	0.078	0.078	②	22-30
	S600	⑦	1.102 ⑤	0.999 ⑥	0.0031	0.866	0.763	0.0039	0.078	0.078	②	22-30
	SL600	⑦	1.181	1.102	0.0031	0.866	0.763	0.0039	0.078	0.078	②	22-30
1997	C220	①	0.866	0.763	0.0047	0.354	0.287	0.0059	0.078	0.078	②	22-30
	C280	①	0.866	0.763	0.0047	0.354	0.287	0.0059	0.078	NA	②	22-30
	C360	①	1.181	1.102	0.0031	0.945	0.842	0.0059	0.078	0.078	②	22-30
	E300	NA	NA	NA	NA	NA	NA	NA	NA	NA	②	22-30
	E320	NA	NA	NA	NA	NA	NA	NA	NA	NA	②	22-30
	S320	④	1.102 ⑤	0.999 ⑥	0.0031	0.472	0.385	0.0039	0.078	0.078	②	22-30
	SL320	④	1.102	0.999	0.0047	0.866	0.287	0.0059	0.078	0.078	②	22-30
	S420	⑦	1.102 ⑤	0.999 ⑥	0.0031	0.866	0.763	0.0039	0.078	0.078	②	22-30
	S500	⑦	1.102 ⑤	0.999 ⑥	0.0031	0.866	0.763	0.0039	0.078	0.078	②	22-30
	SL500	④	1.102	0.999	0.0047	0.354	0.287	0.0059	0.078	0.078	②	22-30
	S600	④	1.102 ⑤	0.999 ⑥	0.0031	0.866	0.763	0.0039	0.078	0.078	②	22-30
	SL600	④	1.181	1.102	0.0031	0.866	0.763	0.0039	0.078	0.078	②	22-30
1998-99	C230	①	0.470	0.763	0.0047	0.354	0.287	0.0059	0.078	0.078	②	22-30
	C280	①	0.870	0.763	0.0047	0.35	0.287	0.0059	0.078	0.078	②	22-30
	E300	①	1.035	0.763	0.0047	0.40	0.337	0.0039	0.078	0.078	②	22-30
	E320	①	1.035	0.763	0.0047	0.40	0.337	0.0039	0.078	0.078	②	22-30
	E430	①	1.200	0.763	0.0047	0.48	0.385	0.0039	0.078	0.078	②	22-30

79236C62

BRAKE SPECIFICATIONS
MERCEDES-BENZ C, CLK, E, S, SL, SLK CLASSES
All measurements in inches unless noted

Year	Model	Master Cyl. Bore	Front Brake Disc			Rear Brake Disc			Minimum Lining Thickness		Brake Caliper	
			Original Thickness	Minimum Thickness	Max. Runout	Original Thickness	Minimum Thickness	Max. Runout	Front	Rear	Bracket Bolts (ft. lbs.)	Mounting Bolts (ft. lbs.)
1998-99 (cont.)	S320	⑦	1.200	0.999 ⑥	0.0031	0.48	0.385	0.0039	0.078	0.078	②	22-30
	S420	⑦	1.200	0.999 ⑥	0.0031	0.48	0.763	0.0039	0.078	0.078	②	22-30
	S500	⑦	1.200	0.999 ⑥	0.0031	0.48	0.763	0.0039	0.078	0.078	②	22-30
	S600	⑦	1.200	0.999 ⑥	0.0031	0.48	0.763	0.0039	0.078	0.078	②	22-30
	CL500	NA	1.200	0.999 ⑥	0.0031	0.48	0.763	0.0039	0.078	0.078	②	22-30
	CL600	NA	1.200	0.999 ⑥	0.0031	0.48	0.763	0.0039	0.078	0.078	②	22-30
	SL500	④	1.102	0.999	0.0047	0.354	0.287	0.0059	0.078	0.078	②	22-30
	SL600	⑦	1.200	0.999 ⑥	0.0031	0.87	0.763	0.0039	0.078	0.078	②	22-30
	CLK320	NA	1.100	0.999	0.0047	0.39	0.327	0.0039	0.078	0.078	②	22-30
	SLK230	NA	0.980	0.763	0.0047	0.40	0.337	0.0039	0.078	0.078	②	22-30

NA - Not Available

① Stepped Stepped bore:
　Step 1　Step 1: 0.938
　Step 2　Step 2: 0.750

② Front caliper: 85 ft. lbs.
　Rear caliper: 38 ft. lbs.

③ Wagon:
　Original Thickness: 0.984 - 0.787
　Minimum Thickness 0.881 - 0.686

④ Stepped bore:
　Step 1: 1.000
　Step 2: 0.750

⑤ With 2 piston fixed caliper: 1.102
　With 4 piston fixed caliper: 1.181

⑥ With 2 piston fixed caliper: .999
　With 4 piston fixed caliper: 1.078

⑦ Stepped bore:
　Step 1: 1.063
　Step 2: 1.000

79236C63

BRAKE SPECIFICATIONS
MITSUBISHI DIAMANTE, ECLIPSE, ECLIPSE SPYDER, GALANT, MIRAGE, 3000GT
All measurements in inches unless noted

Year	Model		Master Cylinder Bore	Brake Disc			Brake Drum Diameter			Minimum Lining Thickness		Brake Caliper	
				Original Thickness	Minimum Thickness	Maximum Runout	Original Inside Diameter	Max. Wear Limit	Maximum Machine Diameter	Front	Rear	Caliper-to-Adapter Bolts (ft. lbs.)	Adapter-to-Hub Bolts (ft. lbs.)
1995	Diamante	F	1.000	0.940	0.880	0.003	—	—	—	0.079	—	54	65
		R	—	0.710	0.650	0.003	—	—	—	—	0.079	24	36-43
	Eclipse	F	0.938 ①	0.940	0.880	0.003	—	—	—	0.079	—	54	65
		R	—	②	③	0.003	8.00	—	8.10	—	0.079	54	36-43
	Galant ④	F	0.938 ⑤	0.940	0.880	0.003	—	—	—	0.079	—	54	65
		R	—	0.390	0.330	0.003	—	9.00	—	—	⑥	—	—
	Mirage	F	0.813 ⑦	⑧	⑨	0.003	—	—	—	0.079	—	36	67-81
		R	—	0.390	0.330	0.003	8.00	—	8.10	—	⑥	—	—
	3000GT ⑩	F	1.000 ⑪	0.940	0.880	0.003	—	—	—	0.079	—	54	65
		R	—	0.710	0.650	0.003	—	—	—	—	0.079	20	36-43
	3000GT ⑫	F	1.060	1.180	1.120	0.003	—	—	—	0.079	—	54	65
		R	—	0.790	0.720	0.003	—	—	—	—	0.079	20	36-43
1996	Diamante	F	1.000	0.940	0.880	0.003	—	—	—	0.079	—	54	65
		R	—	0.710	0.650	0.003	—	—	—	—	0.079	24	36-43
	Eclipse	F	0.938 ①	0.940	0.880	0.003	—	—	—	0.079	—	54	65
		R	—	②	③	0.003	9.00	—	9.10	—	0.040	54	36-43
	Galant ④	F	0.938 ⑤	0.940	0.880	0.003	—	—	—	0.079	—	54	65
		R	—	0.390	0.330	0.003	—	9.00	—	—	⑥	—	—
	Mirage	F	0.813 ⑦	⑧	⑨	0.003	—	—	—	0.079	—	36	67-81
		R	—	0.390	0.330	0.003	8.00	—	8.10	—	0.040	—	—
	3000GT ⑩	F	1.000 ⑪	0.940	0.880	0.003	—	—	—	0.079	—	54	65
		R	—	0.710	0.650	0.003	—	—	—	—	0.079	20	36-43
	3000GT ⑬	F	1.060	1.180	1.120	0.003	—	—	—	0.079	—	54	65
		R	—	0.790	0.720	0.003	—	—	—	—	0.079	20	36-43
1997	Diamante	F	1.000	0.940	0.880	0.003	—	—	—	0.079	—	54	65
		R	—	0.710	0.650	0.003	—	—	—	—	0.079	24	36-43
	Eclipse	F	0.938 ①	0.940	0.880	0.003	—	—	—	0.079	—	54	65
		R	—	②	③	0.003	9.00	—	9.10	—	⑥	54	36-43
	Galant ④	F	0.938 ⑤	0.940	0.880	0.003	—	—	—	0.079	—	54	65
		R	—	0.390	0.330	0.003	—	9.00	—	—	⑥	—	—
	Mirage	F	0.813 ⑦	⑧	⑨	0.003	—	—	—	0.079	—	36	67-81
		R	—	0.390	0.330	0.003	8.00	—	8.10	—	⑥	—	—
	3000GT ⑩	F	1.000 ⑪	0.940	0.880	0.003	—	—	—	0.079	—	54	65
		R	—	0.710	0.650	0.003	—	—	—	—	0.079	20	36-43
	3000GT ⑬	F	1.060	1.180	1.120	0.003	—	—	—	0.079	—	54	65
		R	—	0.790	0.720	0.003	—	—	—	—	0.079	20	36-43
1998-99	Diamante	F	1.000	0.940	0.880	0.002	—	—	—	0.080	—	54	65
		R	—	0.410	0.330	0.0023	—	—	—	—	0.039	24	36-43
	3000GT ⑩	F	1.000	0.940	0.880	0.002	—	—	—	0.080	—	54	65
		R	—	0.710	0.650	0.0031	6.60	6.70	6.70	—	0.040	20	36-43
	3000GT ⑬	F	1.500	1.180	1.120	0.002	—	—	—	0.080	—	54	65
		R	—	0.790	0.720	0.0031	6.60	6.70	6.70	—	0.040	20	36-43

79236C64

BRAKE SPECIFICATIONS
MITSUBISHI DIAMANTE, ECLIPSE, ECLIPSE SPYDER, GALANT, MIRAGE, 3000GT
All measurements in inches unless noted

Year	Model		Master Cylinder Bore	Brake Disc Original Thickness	Brake Disc Minimum Thickness	Maximum Runout	Brake Drum Diameter Original Inside Diameter	Brake Drum Diameter Max. Wear Limit	Brake Drum Diameter Maximum Machine Diameter	Minimum Lining Thickness Front	Minimum Lining Thickness Rear	Brake Caliper Caliper-to-Adapter Bolts (ft. lbs.)	Brake Caliper Adapter-to-Hub Bolts (ft. lbs.)
1998-99 (cont.)	Galant	F	0.9375 ⑤	0.940	0.880	0.0031	—	—	—	0.080	—	54	65
		R	—	—	—	—	8.976	9.078	9.078	—	0.040	—	—
	Mirage	F	0.870	0.710	0.650	0.0024	—	—	—	0.080	—	36	67-81
		R	—	—	—	—	8.00	8.10	8.10	—	0.039	—	—
	Eclipse	F	⑭	0.940	0.880	0.0031	—	—	—	0.080	—	54	65
		R	—	0.390	0.330	0.0031	NA	9.10	9.10	—	⑮	54	36-43
	Eclipse Spyder	F	⑭	0.940	0.880	0.0031	—	—	—	0.080	—	54	65
		R	—	0.390	0.330	0.0031	NA	9.10	9.10	—	⑮	54	36-43

① With ABS or AWD: 1.000
② AWD: 0.790
   FWD: 0.390
③ AWD: 0.720
   FWD: 0.330
④ Double overhead camshaft

⑤ With ABS: 1.000
⑥ Drum shoe: 0.040
   Disc pad: 0.079
⑦ 4 door models with ABS: 0.938
⑧ Front solid: 0.510
   Front ventilated: 0.710

⑨ Front solid: 0.450
   Front ventilated: 0.650
⑩ FWD
⑪ With ABS: 1.060
⑫ 7 inch drum: 7.20
   8 inch drum: 8.10

⑬ AWD
⑭ Disc pad: 0.080
   Drum shoe: 0.039
⑮ Models with ABS: 1.000
   Models with AWD: 1.000
   All others: 0.9375

79236C65

BRAKE SPECIFICATIONS
NISSAN 200SX, 240SX, 300ZX, ALTIMA, MAXIMA, SENTRA
All measurements in inches unless noted

Year	Model	Master Cylinder Bore	Brake Disc Original Thickness	Brake Disc Minimum Thickness	Maximum Runout	Brake Drum Diameter Original Inside Diameter	Brake Drum Diameter Max. Wear Limit	Brake Drum Diameter Maximum Machine Diameter	Minimum Lining Thickness Front	Minimum Lining Thickness Rear	Brake Caliper Bracket Bolts (ft. lbs.)	Brake Caliper Mounting Bolts (ft. lbs.)
1995	240SX	①	NA	②	0.003	—	—	—	0.079	0.059	40-47	—
	300ZX	③	NA	④	0.003	—	—	⑤	0.079	0.079	72-87	—
	Altima	⑥	NA	⑦	0.003	9.000	—	9.060	0.079	⑧	53-72	16-23
	Maxima	⑥	NA	⑦	0.003	9.000	—	9.060	0.079	⑧	53-72	16-23
	Sentra/200SX	⑨	NA	⑩	0.003	7.090	—	7.130	0.079	⑧	40-47	—
1996	240SX	①	NA	②	0.003	—	—	—	0.079	0.059	40-47	—
	300ZX	③	NA	④	0.003	—	—	⑤	0.079	0.079	72-87	—
	Altima	⑥	NA	⑦	0.003	9.000	—	9.060	0.079	⑧	53-72	16-23
	Maxima	⑥	NA	⑦	0.003	9.000	—	9.060	0.079	⑧	53-72	16-23
	Sentra/200SX	⑨	NA	⑩	0.003	7.090	—	7.130	0.079	⑧	40-47	—
1997	240SX	①	NA	②	0.003	—	—	—	0.079	0.059	40-47	—
	Altima	⑥	NA	⑦	0.003	9.000	—	9.060	0.079	⑧	53-72	16-23
	Maxima	⑥	NA	⑦	0.003	9.000	—	9.060	0.079	⑧	53-72	16-23
	Sentra/200SX	⑨	⑪	⑩	0.003	7.090	—	7.130	0.079	⑧	40-47	—
1998-99	240SX	⑫	⑬	⑭	0.003	—	—	—	0.079	0.079	40-47	—
	Altima	⑥	⑮	⑦	0.003	9.00	NA	9.06	0.079	0.059	53-72	16-23
	Maxima	0.937	⑯	⑦	0.003	—	—	—	0.079	0.059	53-72	16-23
	Sentra/200SX	⑨	⑪	⑰	0.003	7.09	7.13	7.13	0.079	0.059	40-47	—

NA - Not Available

① With CL22VB front brake and M23 booster: 0.875
   With CL25VA front brake and M195T booster: 0.9375

② Front, without ABS: 0.709
   Front, with ABS: 0.787
   Rear: 0.315

③ Without ABS: 0.9375
   With ABS: 1.0625

④ Front: 1.102
   Rear: 0.630

⑤ Parking brake drum: 6.81

⑥ With ABS: 1.000
   Without ABS: 0.9375

⑦ Front: 0.787
   Rear: 0.315

⑧ Disc brake: 0.079
   Drum brake: 0.059

⑨ With ABS: 0.875
   Without ABS: 0.8125

⑩ Front: AD22VF 0.945
   Front: AD18VE 0.630
   Rear: 0.236

⑪ Front: 0.710
   Rear: 0.280

⑫ Manual transaxle without ABS: 0.875
   Automatic transaxle or all with ABS: 0.937

⑬ Front: 0.790
   Rear: 0.350

⑭ Front: 0.710
   Rear: 0.310

⑮ Front: 0.870
   Rear: 0.390

⑯ Front: 0.870
   Rear: 0.350

⑰ Front: 0.630
   Rear: 0.236

79236C66

BRAKE SPECIFICATIONS
PORSCHE 911 CARRERA, 911 CARRERA 4, 911 TURBO, 928, 968, BOXSTER
All measurements in inches unless noted

Year	Model		Master Cylinder Bore	Brake Disc Original Thickness	Brake Disc Minimum Thickness	Brake Disc Maximum Runout	Minimum Lining Thickness Front	Minimum Lining Thickness Rear	Brake Caliper Bracket Bolts (ft. lbs.)	Brake Caliper Mounting Bolts (ft. lbs.)
1995	911 Carrera	F	0.940	1.102	1.024	0.004	0.079	—	—	63
		R	0.940	0.950	0.866	0.004	—	0.079	—	63
	911 Carrera 4	F	0.940	1.102	1.024	0.004	0.079	—	—	63
		R	0.940	0.950	0.866	0.004	—	0.079	—	63
	928 GTS	F	0.937	1.260	1.181	0.004	0.079	—	—	63
		R	0.937	0.945	0.866	0.004	—	0.079	—	63
	968	F	0.937	1.260	1.181	0.004	0.079	—	—	63
		R	0.937	0.945	0.866	0.004	—	0.079	—	63
1996	911 Carrera	F	0.940	1.102	1.024	0.004	0.079	—	—	63
		R	0.940	0.950	0.866	0.004	—	0.079	—	63
	911 Carrera 4	F	0.940	1.102	1.024	0.004	0.079	—	—	63
		R	0.940	0.950	0.866	0.004	—	0.079	—	63
	911 Turbo	F	1.000	1.102	1.024	0.004	0.079	—	—	63
		R	1.000	0.950	0.866	0.004	—	0.079	—	63
1997	911 Carrera	F	0.940	1.102	1.024	0.004	0.079	—	—	63
		R	0.940	0.950	0.866	0.004	—	0.079	—	63
	911 Carrera 4	F	0.940	1.102	1.024	0.004	0.079	—	—	63
		R	0.940	0.950	0.866	0.004	—	0.079	—	63
	911 Turbo	F	1.000	1.102	1.024	0.004	0.079	—	—	63
		R	1.000	0.950	0.866	0.004	—	0.079	—	63
1998-99	911 Carrera	F	0.940	1.102	1.024	0.004	0.079	—	—	63
		R	0.940	0.950	0.866	0.004	—	0.079	—	63
	911 Carrera 4	F	0.940	1.102	1.024	0.004	0.079	—	—	63
		R	0.940	0.950	0.866	0.004	—	0.079	—	63
	Boxster	F	0.940	0.950	0.870	0.001	0.079	—	—	63
		R	0.940	0.790	0.730	0.001	—	0.079	—	63

F - Front
R - Rear

79236C67

BRAKE SPECIFICATIONS
SAAB 900, 9000, 9-3
All measurements in inches unless noted

Year	Model		Master Cylinder Bore	Brake Disc Original Thickness	Brake Disc Minimum Thickness	Maximum Runout	Minimum Lining Thickness Front	Minimum Lining Thickness Rear	Brake Caliper Bracket Bolts (ft. lbs.)	Brake Caliper Mounting Bolts (ft. lbs.)
1995	900	F	0.940	0.940	0.870	0.002	0.200	—	78	19
		R	—	0.390	0.310	0.003	—	0.200	59	—
	9000	F	0.870	0.980	0.900	0.003	0.160	—	62	21
		R	—	0.350	0.290	0.003	—	0.160	35	21
1996	900	F	0.940	0.940	0.870	0.002	0.200	—	78	19
		R	—	0.390	0.310	0.003	—	0.200	59	—
	9000	F	0.870	0.980	0.900	0.003	0.160	—	62	21
		R	—	0.350	0.290	0.003	—	0.160	35	21
1997	900	F	0.940	0.940	0.870	0.002	0.200	—	78	19
		R	—	0.390	0.310	0.003	—	0.200	59	—
	9000	F	0.870	0.980	0.900	0.003	0.160	—	62	21
		R	—	0.350	0.290	0.003	—	0.160	35	21
1998-99	900	F	0.940	0.940	0.870	0.002	0.200	—	78	19
		R	—	0.390	0.310	0.003	—	0.200	59	—
	9000	F	0.870	0.980	0.900	0.003	0.160	—	62	21
		R	—	0.350	0.290	0.003	—	0.160	35	21
	9-3	F	0.940	0.940	0.870	0.002	0.200	—	78	19
		R	—	0.390	0.310	0.003	—	0.200	59	—

79236C68

BRAKE SPECIFICATIONS
SUBARU IMPREZA, LEGACY, SVX
All measurements in inches unless noted

Year	Model		Master Cylinder Bore	Brake Disc Original Thickness	Brake Disc Minimum Thickness	Brake Disc Maximum Runout	Brake Drum Diameter Original Inside Diameter	Brake Drum Diameter Max. Wear Limit	Brake Drum Diameter Maximum Machine Diameter	Minimum Lining Thickness Front	Minimum Lining Thickness Rear	Brake Caliper Bracket Bolts (ft. lbs.)	Brake Caliper Mounting Bolts (ft. lbs.)
1995	Impreza	F	1.0000 ①	0.940 ②	0.870 ③	0.003	—	—	—	0.059	—	51-65	23-30
		R	—	0.390	0.335	0.004	9.00 ④	9.079 ⑥	NA	—	0.059	—	23-30
	Legacy	F	1.0625 ①	0.940	0.870	0.003	—	—	—	0.059	—	51-65	23-30
		R	—	0.390	0.335	0.004	9.00 ④	9.079 ⑥	NA	—	0.059	—	23-30
	SVX	F	1.0630	1.100	1.020	0.004	—	—	—	0.059	—	51-65	23-30
		R	—	0.390	0.335	0.004	7.48 ⑤	7.52 ⑤	NA	—	0.059	—	23-30
1996	Impreza	F	1.0000 ①	0.940 ②	0.870 ③	0.003	—	—	—	0.059	—	51-65	23-30
		R	—	0.390	0.335	0.004	9.00 ④	9.079 ⑥	NA	—	0.059	—	23-30
	Legacy	F	1.0625 ①	0.940	0.870	0.003	—	—	—	0.059	—	51-65	23-30
		R	—	0.390	0.335	0.004	9.00 ④	9.079 ⑥	NA	—	0.059	—	23-30
	SVX	F	1.0630	1.100	1.020	0.004	—	—	—	0.059	—	51-65	23-30
		R	—	0.390	0.335	0.004	7.48 ⑤	7.52 ⑤	NA	—	0.059	—	23-30
1997	Impreza	F	1.0000 ①	0.940	0.870	0.003	—	—	—	0.059	—	51-65	24-31
		R	—	—	—	—	9.00	9.079	NA	—	0.059	—	24-31
	Legacy	F	1.0625 ①	0.940	0.870	0.003	—	—	—	0.059	—	51-65	24-31
		R	—	0.390	0.335	0.004	9.00 ④	9.079 ⑥	NA	—	0.059	—	24-31
	SVX	F	1.0630	1.100	1.020	0.004	—	—	—	0.059	—	51-65	24-31
		R	—	0.390	0.335	0.004	7.48 ⑤	7.52 ⑤	NA	—	0.059	—	24-31
1998-99	Impreza	F	1.0625 ①	0.940	0.870	0.003	—	—	—	0.059	—	51-65	24-31
		R	—	0.390	0.335	0.004	9.00 ④	9.079 ⑥	NA	—	0.059	—	24-31
	Legacy	F	1.0625 ①	0.940	0.870	0.003	—	—	—	0.059	—	51-65	24-31
		R	—	0.390	0.335	0.004	9.00 ④	9.079 ⑥	NA	—	0.059	—	24-31

NA - Information Not Available

① Models w/out ABS: 0.9374 in.

② Specification is for AWD models. FWD models use 0.710 in. thick disc

③ Specification is for AWD models. Discs on FWD models have a 0.630 in. wear limit/max. diameter

④ Parking brake drum on vehicles with rear disc brakes: 6.69 in.

⑤ Specification is for the parking brake drum

⑥ Parking brake drum on vehicles with rear disc brakes: 6.73 in.

79236C69

BRAKE SPECIFICATIONS
SUZUKI ESTEEM, SWIFT
All measurements in inches unless noted

Year	Model	Master Cylinder Bore	Brake Disc Original Thickness	Brake Disc Minimum Thickness	Brake Disc Maximum Runout	Brake Drum Diameter Original Inside Diameter	Brake Drum Diameter Max. Wear Limit	Brake Drum Diameter Maximum Machine Diameter	Minimum Lining Thickness Front	Minimum Lining Thickness Rear	Brake Caliper Bracket bolts ft lbs.	Brake Caliper Mounting bolts ft lbs.
1995	Swift ①	NA	0.670	0.620	0.004	7.09	7.87	7.87	0.236 ②	0.110 ②	36	27
	Swift	NA	0.670	0.620	0.004	7.16	7.95	7.95	0.236 ②	0.110 ②	36	27
	Esteem	NA	0.790	0.710	0.004	7.87	7.95	7.95	0.240 ②	0.110 ②	62	16
1996	Swift ①	NA	0.670	0.620	0.004	7.09	7.87	7.87	0.236 ②	0.110 ②	36	27
	Swift ①	NA	0.670	0.620	0.004	7.16	7.95	7.95	0.236 ②	0.110 ②	36	27
	Esteem	NA	0.790	0.710	0.004	7.87	7.95	7.95	0.240 ②	0.110 ②	62	16
1997	Swift ①	NA	0.670	0.620	0.004	7.09	7.87	7.87	0.236 ②	0.110 ②	36	27
	Swift	NA	0.670	0.620	0.004	7.16	7.95	7.95	0.236 ②	0.110 ②	36	27
	Esteem	NA	0.790	0.710	0.004	7.87	7.95	7.95	0.240 ②	0.110 ②	62	16
1998-99	Swift ①	NA	0.670	0.620	0.004	7.09	7.87	7.87	0.236 ②	0.110 ②	36	27
	Swift	NA	0.670	0.620	0.004	7.16	7.95	7.95	0.236 ②	0.110 ②	36	27
	Esteem	NA	0.790	0.710	0.004	7.87	7.95	7.95	0.240 ②	0.110 ②	62	16

NA - Not Available
① Hatchback
② Measurement is for lining and backing together.

79236C70

BRAKE SPECIFICATIONS
TOYOTA AVALON, CAMRY, CELICA, COROLLA, MR2, SUPRA, PASEO, TERCEL
All measurements in inches unless noted

Year	Model		Master Cylinder Bore	Brake Disc Original Thickness	Brake Disc Minimum Thickness	Brake Disc Maximum Runout	Brake Drum Diameter Original Inside Diameter	Brake Drum Diameter Max. Wear Limit	Brake Drum Diameter Maximum Machine Diameter	Minimum Lining Thickness Front	Minimum Lining Thickness Rear	Brake Caliper Bracket Bolts (ft. lbs.)	Brake Caliper Mounting Bolts (ft. lbs.)
1995	Avalon	F	NA	1.102	1.024	0.0020	-	-	-	0.039	-	25	79
		R	NA	0.354	0.315	0.0059	-	-	6.73	-	0.039	25	34
	Camry	F	NA	1.102	1.024	0.0020	-	-	9.08	0.039	-	25	79
		R	NA	0.354	0.315	0.0059	9.00	-	6.73	-	0.039	14	20
	Celica	F	NA	⑥	⑦	0.0020	-	-	7.91	0.039	-	25	65
		R	NA	0.394	0.354	0.0059	7.87	-	6.73	-	0.039	-	34
	Corolla		NA	0.866	0.787	0.0020	7.87	-	7.91	0.039	0.039	25	65
	MR2	F	NA	⑧	①	0.0020	-	-	-	0.039	-	65	⑨
		R	NA	②	③	0.0039	-	-	-	-	0.039	43	14
	Supra	F	NA	④	⑤	0.0020	-	-	-	0.039	-	25	87
		R	NA	0.630	0.591	-	7.48	-	7.52	-	0.039	25	77
	Paseo		NA	0.709	0.669	0.0035	7.09	-	7.13	0.039	0.039	18	65
	Tercel		NA	0.709	0.669	0.0035	7.09	-	7.13	0.039	0.039	18	65
1996	Avalon	F	NA	1.102	1.024	0.0020	-	-	-	0.039	-	25	79
		R	NA	0.354	0.315	0.0059	-	-	-	-	0.039	25	34
	Camry	F	NA	1.102	1.024	0.0020	-	-	-	0.039	-	25	79
		R	NA	0.354	0.315	0.0059	9.00	-	9.08	-	0.039	14	20
	Celica	F	NA	⑦	⑧	0.0020	-	-	-	0.039	-	25	79
		R	NA	0.394	0.354	0.0059	7.87	-	7.91	-	0.039	19	14
	Corolla		NA	0.866	0.787	0.0020	7.87	-	7.91	0.039	0.039	25	65
	Paseo		NA	0.709	0.669	0.0035	7.09	-	7.13	0.039	0.039	25	87
	Supra	F	NA	⑤	⑥	0.0020	-	-	-	0.039	-	25	87
		R	NA	0.630	0.591	-	7.48	-	7.52	-	0.039	25	77
	Tercel		NA	0.709	0.669	0.0035	7.09	-	7.13	0.039	0.039	18	65
1997	Avalon	F	NA	1.102	1.024	0.0020	-	-	-	0.039	-	25	79
		R	NA	0.354	0.315	0.0059	-	-	-	-	0.039	25	34
	Camry	F	NA	1.102	1.024	0.0020	-	-	-	0.039	-	25	79
		R	NA	0.354	0.315	0.0059	9.00	-	9.08	-	0.039	14	20
	Celica	F	NA	⑦	⑧	0.0020	-	-	-	0.039	-	25	79
		R	NA	0.394	0.354	0.0059	7.87	-	7.91	-	0.039	19	14
	Corolla		NA	0.866	0.787	0.0020	7.87	-	7.91	0.039	0.039	25	65
	Paseo		NA	0.709	0.669	0.0035	7.09	-	7.13	0.039	0.039	18	65
	Supra	F	NA	⑤	⑥	0.0020	-	-	-	0.039	-	25	87
		R	NA	0.630	0.591	-	7.48	-	7.52	-	0.039	25	77
	Tercel		NA	0.709	0.669	0.0035	7.09	-	7.13	0.039	0.039	18	65
1998-99	Avalon	F	NA	1.102	1.024	0.0020	-	-	-	0.039	-	25	79
		R	NA	0.354	0.315	0.0059	-	-	-	-	0.039	25	34
	Camry	F	NA	1.102	1.024	0.0020	-	-	-	0.039	-	25	79
		R	NA	0.394	0.354	0.0059	9.00	-	9.08	-	0.039	14	20
	Celica	F	NA	⑦	⑧	0.0020	-	-	-	0.039	-	25	79
		R	NA	0.394	0.354	0.0059	7.87	-	7.91	-	0.039	19	34

79236C71

BRAKE SPECIFICATIONS

TOYOTA AVALON, CAMRY, CELICA, COROLLA, MR2, SUPRA, PASEO, TERCEL

All measurements in inches unless noted

Year	Model		Master Cylinder Bore	Brake Disc Original Thickness	Brake Disc Minimum Thickness	Brake Disc Maximum Runout	Brake Drum Diameter Original Inside Diameter	Brake Drum Diameter Max. Wear Limit	Brake Drum Diameter Maximum Machine Diameter	Minimum Lining Thickness Front	Minimum Lining Thickness Rear	Brake Caliper Bracket Bolts (ft. lbs.)	Brake Caliper Mounting Bolts (ft. lbs.)
1998-99 (cont.)	Corolla		NA	0.866	0.787	0.0020	7.87	-	7.91	0.039	0.039	25	65
	Supra	F	NA	⑤	⑥	0.0020	-	-	-	0.039	-	25	87
		R	NA	0.630	0.591	-	7.48	-	7.52	-	0.039	25	77
	Tercel		NA	0.709	0.669	0.0035	7.09	-	7.13	0.039	0.039	18	65

NA - Not Available
F - Front
R - Rear

① 3S-GTE engine: 1.102
5S-FE engine: 0.945

② 3S-GTE engine: 0.866
5S-FE engine: 0.630

③ 3S-GTE engine: 0.8287
5S-FE engine: 0.591

④ 2JZ-GTE engine: 1.181
2JZ-GE engine: 1.260

⑤ 2JZ-GTE engine: 1.102
2JZ-GE engine: 1.181

⑥ 7A-FE engine: 0.984
5S-FE engine: 1.102

⑦ 7A-FE engine: 0.906
5S-FE engine: 1.024

⑧ 3S-GTE engine: 1.181
5S-FE engine: 0.984

⑨ Single piston: 18
Dual piston: 25

79236C72

BRAKE SPECIFICATIONS
VOLKSWAGEN BEETLE, CABRIO, GOLF, GTI, JETTA, PASSAT
All measurements in inches unless noted

Year	Model		Master Cylinder Bore	Brake Disc Original Thickness	Brake Disc Minimum Thickness	Brake Disc Maximum Runout	Drum Diameter Original Inside Diameter	Drum Diameter Maximum Machine Diameter	Minimum Lining Thickness Front	Minimum Lining Thickness Rear	Brake Caliper Bracket Bolts (ft. lbs.)	Brake Caliper Mounting Bolts (ft. lbs.)
1995	Cabrio	F	0.874 ⑦	0.790	0.709	0.002	—	—	0.28	—	18-26	92
		R	—	0.390	0.315	0.002	7.87	7.91	—	0.27 ⑧	22	41
	Golf III ①	F	0.874 ⑦	0.790	0.709	0.002	—	—	0.28	—	18-26	92
		R	—	0.390	0.315	0.002	7.87	7.91	—	0.27 ⑧	22	41
	Golf III ②	F	0.874 ⑦	0.870	0.787	0.002	—	—	0.28	—	18-26	92
		R	—	0.390	0.315	0.002	7.87	7.91	—	0.27	22	41
	GTI	F	0.874 ⑦	0.870	0.787	0.002	—	—	0.28	—	18-26	92
		R	—	0.390	0.315	0.002	7.87	7.91	—	0.27	22	41
	Jetta III ①	F	0.874 ⑦	0.790	0.709	0.002	—	—	0.28	—	18-26	92
		R	—	0.390	0.315	0.002	7.87	7.91	—	0.27 ⑧	22	41
	Jetta III ③	F	0.874 ⑦	0.870	0.787	0.002	—	—	0.28	—	18-26	92
		R	—	0.390	0.315	0.002	7.87	7.91	—	0.27	22	41
	Passat	F	0.874 ⑦	0.870	0.787	0.002	—	—	0.28	—	26	92
		R	—	0.390	0.315	0.002	9.06	9.11	—	0.27 ⑧	26	48
1996	Cabrio	F	0.874 ⑦	0.790	0.709	0.002	—	—	0.28	—	18-26	92
		R	—	0.390	0.315	0.002	7.87	7.91	—	0.27 ⑧	22	41
	Golf ④	F	0.874 ⑦	0.790	0.709	0.002	—	—	0.28	—	18-26	92
		R	—	0.390	0.315	0.002	7.87	7.91	—	0.27 ⑧	22	41
	Golf ②	F	0.874 ⑦	0.870	0.787	0.002	—	—	0.28	—	18-26	92
		R	—	0.390	0.315	0.002	7.87	7.91	—	0.27	22	41
	GTI ⑤	F	0.874 ⑦	0.790	0.709	0.002	—	—	0.28	—	18-26	92
		R	—	0.390	0.315	0.002	7.87	7.91	—	0.27	22	41
	GTI ⑥	F	0.874 ⑦	0.870	0.787	0.002	—	—	0.28	—	18-26	92
		R	—	0.390	0.315	0.002	7.87	7.91	—	0.27	22	41
	Jetta ①	F	0.874 ⑦	0.790	0.709	0.002	—	—	0.28	—	18-26	92
		R	—	0.390	0.315	0.002	7.87	7.91	—	0.27 ⑧	22	41
	Jetta ③	F	0.874 ⑦	0.870	0.787	0.002	—	—	0.28	—	18-26	92
		R	—	0.390	0.315	0.002	7.87	7.91	—	0.27	22	41
	Passat	F	0.874 ⑦	0.870	0.787	0.002	—	—	0.28	—	26	92
		R	—	0.390	0.315	0.002	9.06	9.11	—	0.27 ⑧	26	48
1997	Cabrio	F	0.874 ⑦	0.790	0.709	0.002	—	—	0.28	—	18-26	92
		R	—	0.390	0.315	0.002	7.87	7.91	—	0.27 ⑧	22	41
	Golf ④	F	0.874 ⑦	0.790	0.709	0.002	—	—	0.28	—	18-26	92
		R	—	0.390	0.315	0.002	7.87	7.91	—	0.27 ⑧	22	41
	Golf ②	F	0.874 ⑦	0.870	0.787	0.002	—	—	0.28	—	18-26	92
		R	—	0.390	0.315	0.002	7.87	7.91	—	0.27	22	41
	GTI ⑤	F	0.874 ⑦	0.790	0.709	0.002	—	—	0.28	—	18-26	92
		R	—	0.390	0.315	0.002	7.87	7.91	—	0.27	22	41
	GTI ⑥	F	0.874 ⑦	0.870	0.787	0.002	—	—	0.28	—	18-26	92
		R	—	0.390	0.315	0.002	7.87	7.91	—	0.27	22	41
	Jetta ①	F	0.874 ⑦	0.790	0.709	0.002	—	—	0.28	—	18-26	92
		R	—	0.390	0.315	0.002	7.87	7.91	—	0.27 ⑧	22	41
	Jetta ③	F	0.874 ⑦	0.870	0.787	0.002	—	—	0.28	—	18-26	92
		R	—	0.390	0.315	0.002	7.87	7.91	—	0.28	22	41
	Passat	F	0.874 ⑦	0.870	0.787	0.002	—	—	0.28	—	26	92
		R	—	0.390	0.315	0.002	9.06	9.11	—	0.28	26	48

79236C73

BRAKE SPECIFICATIONS
VOLKSWAGEN BEETLE, CABRIO, GOLF, GTI, JETTA, PASSAT
All measurements in inches unless noted

Year	Model		Master Cylinder Bore	Brake Disc Original Thickness	Brake Disc Minimum Thickness	Brake Disc Maximum Runout	Drum Diameter Original Inside Diameter	Drum Diameter Maximum Machine Diameter	Minimum Lining Thickness Front	Minimum Lining Thickness Rear	Brake Caliper Bracket Bolts (ft. lbs.)	Brake Caliper Mounting Bolts (ft. lbs.)
1998-99	Beetle	F	0.937	0.790	0.950	0.002	—	—	0.27	—	18	92
		R	—	0.390 ⑬	0.315	0.002	9.06	9.09	—	0.27 ⑧	26	48
	Cabrio	F	0.874 ⑦	0.790	0.709	0.002	—	—	0.28	—	18-26	92
		R	—	0.390	0.315	0.002	7.87	7.91	—	0.27 ⑧	22	41
	Golf ④	F	0.874 ⑦	0.790	0.709	0.002	—	—	0.28	—	18-26	92
		R	—	0.390	0.315	0.002	7.87	7.91	—	0.27 ⑧	22	41
	Golf ②	F	0.874 ⑦	0.870	0.787	0.002	—	—	0.28	—	18-26	92
		R	—	0.390	0.315	0.002	7.87	7.91	—	0.28	22	41
	GTI ⑤	F	0.874 ⑦	0.790	0.709	0.002	—	—	0.28	—	18-26	92
		R	—	0.390	0.315	0.002	7.87	7.91	—	0.28	22	41
	GTI ⑥	F	0.874 ⑦	0.870	0.787	0.002	—	—	0.28	—	18-26	92
		R	—	0.390	0.315	0.002	7.87	7.91	—	0.28	22	41
	Jetta ①	F	0.874 ⑦	0.790	0.709	0.002	—	—	0.28	—	18-26	92
		R	—	0.390	0.315	0.002	7.87	7.91	—	0.27 ⑧	22	41
	Jetta ③	F	0.874 ⑦	0.870	0.787	0.002	—	—	0.28	—	18-26	92
		R	—	0.390	0.315	0.002	7.87	7.91	—	0.28	22	41
	Passat	F	0.937	0.980 ⑨	0.900 ⑩	0.002	—	—	0.28	—	22	89
		R	—	0.393	0.314	0.002	N/A	N/A	—	0.27	22	70

NA - Not Available
① GL and GLS models
② VR6 Model
③ GLX Model
④ GL model
⑤ 2.0 L Engine
⑥ 2.8 L Engine
⑦ ABS equipped models have a diameter of .937 in.
⑧ Models equipped with drum brakes have a lining limit of 0.098 in.
⑨ Lucas caliper: 0.87 in.
⑩ Lucas caliper: 0.78 in.

79236C74

BRAKE SPECIFICATIONS
VOLVO 850, 940, 960, C70, S70, S90, V70, V90
All measurements in inches unless noted

Year	Model		Master Cylinder Bore	Brake Disc Original Thickness	Brake Disc Minimum Thickness	Brake Disc Maximum Runout	Minimum Lining Thickness	Brake Caliper Bracket bolts (ft. lbs.)	Brake Caliper Mounting bolts (ft. lbs.)
1995	850	F	0.937	1.024	0.906	0.0024	0.120	74	22
		R	—	0.378	0.330	0.003	0.075	37	22
	940	F	0.937	①	②	0.0024	0.120	78	22
		R	—	③	④	0.003	0.075	44	22
	960	F	0.937	①	②	0.0024	0.120	78	22
		R	—	③	④	0.003	0.075	44	22
1996	850	F	0.937	1.024	0.906	0.0024	0.120	74	22
		R	—	0.378	0.330	0.003	0.075	37	22
	960	F	0.937	①	②	0.0024	0.120	78	22
		R	—	③	④	0.003	0.075	44	22
1997	850	F	0.937	1.024	0.906	0.0024	0.120	74	22
		R	—	0.378	0.330	0.003	0.075	37	22
	960	F	0.937	①	②	0.0024	0.120	78	22
		R	—	③	④	0.003	0.075	44	22
1998-99	C70	F	0.937	1.024	0.906	0.0024	0.120	74	22
		R	—	0.378	0.330	0.003	0.075	37	22
	S70	F	0.937	1.024	0.906	0.0024	0.120	74	22
		R	—	0.378	0.330	0.003	0.075	37	22
	S90	F	0.937	①	②	0.0024	0.120	78	22
		R	—	③	④	0.003	0.075	44	22
	V70	F	0.937	1.024	0.906	0.0024	0.120	74	22
		R	—	0.378	0.330	0.003	0.075	37	22
	V90	F	0.937	①	②	0.0024	0.120	78	22
		R	—	③	④	0.003	0.075	44	22

① With standard vented rotor: 0.870
   With heavy duty rotor: 1.024
② With standard vented rotor: 0.790
   With heavy duty rotor: 0.906
③ With standard rear axle: 0.378
   With multi-link rear axle: 0.393
④ With standard rear axle: 0.330
   With multi-link rear axle: 0.314

79236C75

DRIVESHAFTS, U-JOINTS AND CV-JOINT BOOTS

DRIVESHAFTS, U-JOINTS AND CV-JOINT BOOTS

Driveshafts

➡**The term driveshaft does not refer to halfshaft (often termed driveshaft by various manufacturers), which are used on front wheel drive vehicles.**

GENERAL INFORMATION

The driveshaft is a long steel tube used to transmit power from the transmission to the rear differential. Located at either end of the driveshaft is a universal joint (U-joint), which is designed to transmit torsional power at many different angles (within designed limits) in order to match the motion of the rear axle. A slip joint is attached to the U-joint closest to the transmission or transfer case. The shaft is designed with yokes at each end that are inline with each other in order to produce the smoothest possible running shaft.

Since the vehicles can be obtained in either 2WD or 4WD, and in various combinations (two and four door models), various types of driveshafts may be employed. Some models will be equipped with a one-piece rear driveshaft, while others have a two-piece rear driveshaft which uses a center support bearing.

Because some vehicles covered by this manual utilize 2-piece shafts and splined yokes, it may be possible to reinstall the shaft incorrectly or "out of phase" which would cause vibration. Many of these vehicles utilize a keyed slip yoke to prevent this, but DO NOT risk improper installation. ALWAYS matchmark the shaft ends to the yokes before removal.

At the front of the one or two-piece driveshafts, the U-joint connects the driveshaft to a slip-jointed yoke. This yoke is internally splined and allows the driveshaft to move in and out on the transmission splines (one-piece) or the shaft splines (two-piece). The rear of the one or two-piece driveshaft is attached to the differential, on some vehicles, the U-joint may be clamped or bolted to the yoke on the pinion shaft, on other vehicles the driveshaft terminates with a flange that gets bolted to the pinion flange on the differential.

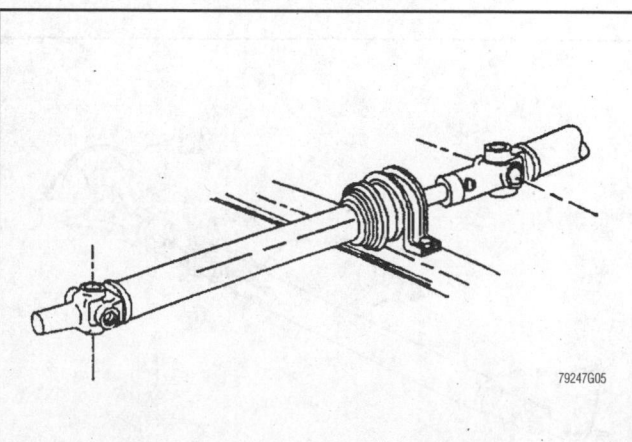

The yokes on either end of the driveshaft should be in phase to prevent vibration

Make alignment marks on the U-joint and shaft before disassembly to prevent possible vibration when assembled

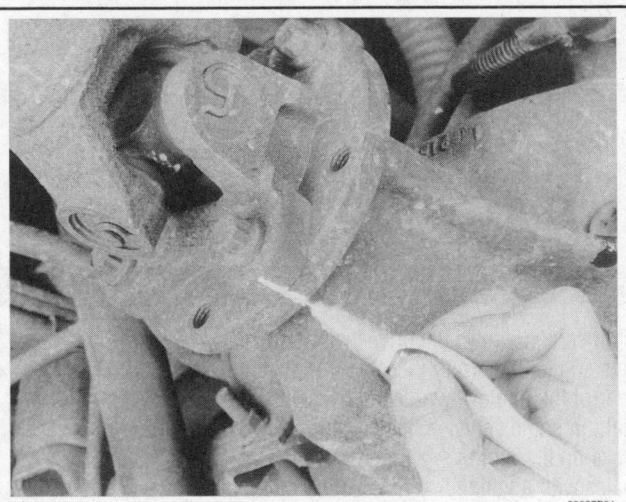

On this type of driveshaft, make alignment marks across the two flanges so it can be installed in the same position

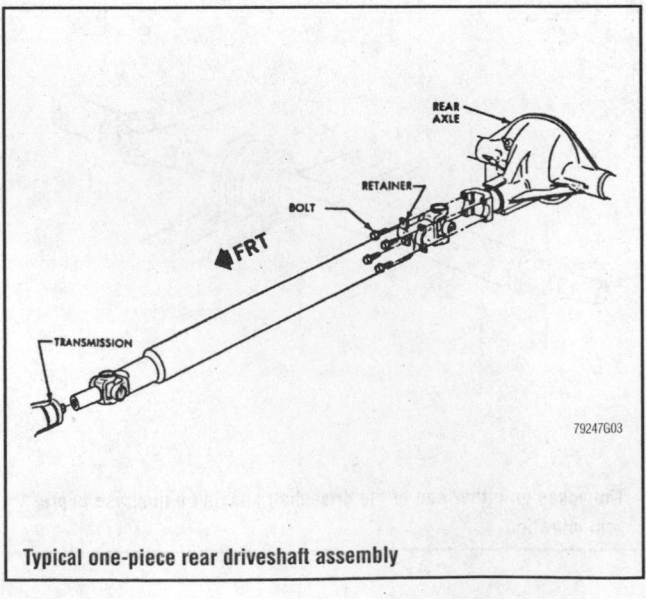

Typical one-piece rear driveshaft assembly

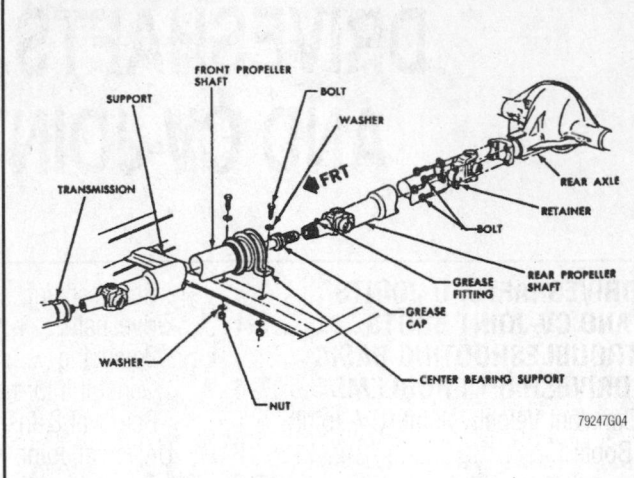

Typical two-piece rear driveshaft assembly with center support bearing

REMOVAL & INSTALLATION

1. Mark the relationship of the driveshaft-to-pinion flange or yoke and disconnect the rear universal joint from the differential by removing the bolts. If the bearing cups are loose, wrap tape around the universal joint to prevent the cups from falling off.

2. If equipped with a one-piece driveshaft, perform the following steps:

   a. Slide the rear driveshaft forward to disengage it from the rear axle flange or yoke.

   b. Move the driveshaft rearward to disengage it from the transmission slip-joint, passing it under the axle housing.

3. If equipped with a two-piece driveshaft, perform the following steps:

   a. Slide the driveshaft forward to disengage it from the rear axle flange or yoke.

   b. Slide the driveshaft rearward to disengage it from the slip-joint of the front driveshaft, passing it under the axle housing.

   c. Remove the center bearing-to-support nuts and bolts.

   d. Slide the front driveshaft rearward to disengage it from the transfer case slip-joint.

➡**DO NOT allow the driveshaft to hang by the U-joint or bend to extreme angles, as damage to the U-joint may occur. Support the driveshaft with wire as needed during removal.**

To install:

4. Inspect the slip-joint for damage, burrs or wear, for these can damage the transmission seal. Apply engine oil to all splined driveshaft joints.

❈❈ WARNING

DO NOT use a hammer to force the driveshaft into place. Check for burrs on the transmission output shaft spline, twisted slip yoke splines or possibly the wrong U-joint. Be sure the splines agree in number and fit. To prevent trunnion seal damage, DO NOT place any tool between the yoke and splines.

5. If installing a one-piece driveshaft, perform the following steps:
   a. Attach the driveshaft to the transmission.

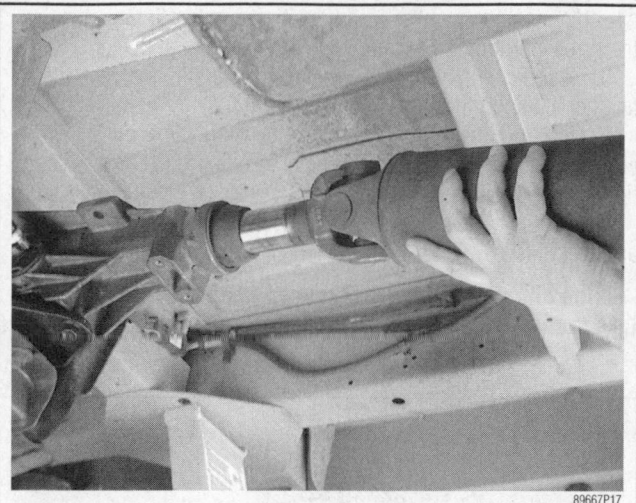

This type of driveshaft is removed from the transmission by simply sliding it out

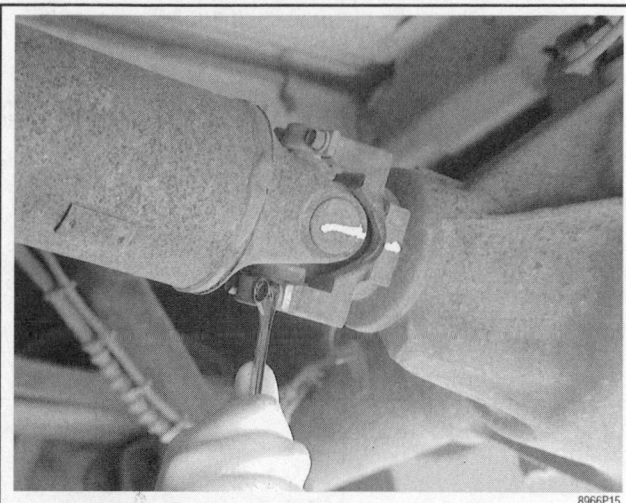

The U-joint on this driveshaft is attached to the pinion yoke by two small brackets and four bolts

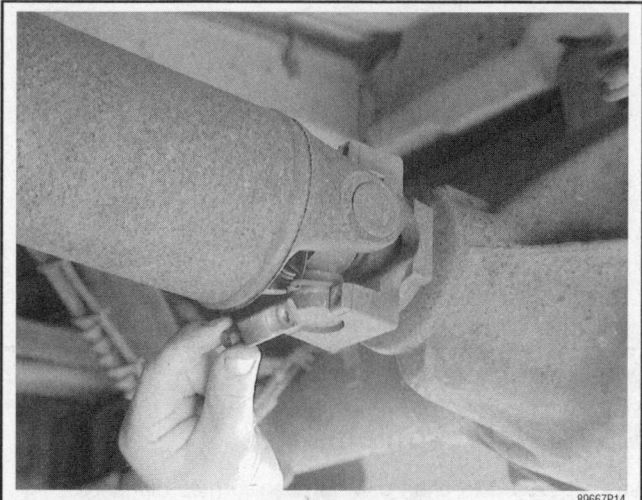

The driveshaft can be detached from the pinion shaft yoke after removing the four bolts and two brackets

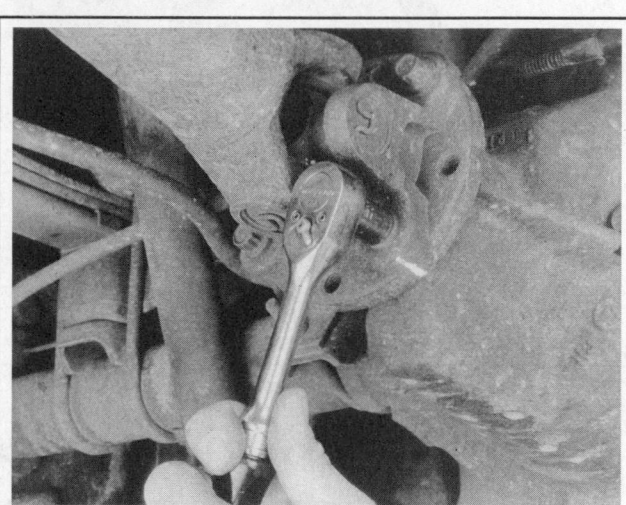

This type of driveshaft is attached to the pinion flange with four bolts and nuts

b. Align the rear universal joint-to-rear axle pinion flange, making sure the bearings are properly seated in the pinion flange yoke.

c. Install the rear driveshaft-to-pinion flange. Tighten the fasteners securely.

6. If installing a two-piece driveshaft, perform the following steps:

a. Install the front driveshaft to the transmission and bolt the center bearing-to-support. Tighten the nuts and bolts securely.

➡**The front driveshaft yoke must be bottomed out in the transmission (fully forward) before being installed to the support.**

b. Rotate the shaft so that the front U-joint trunnion is in the correct position.

➡**Before installing the rear driveshaft, align the U-joint trunnions (a "key" in the output spline of the front driveshaft will align with a missing spline in the rear yoke).**

c. Attach the rear U-joint to the axle flange. Tighten the retainers securely.

7. Road test the vehicle.

BALANCING

The following procedure is used to help eliminate minor driveshaft vibration of an otherwise good driveshaft.

Before attempting this, carefully examine the driveshaft for damage such as dents and deformations. Driveshafts are subjected to large amounts of twisting force which can literally twist the driveshaft. Also check for missing weights that may have been knocked off of the shaft. If the driveshaft is deformed, replace it. If any weights appear to be missing, take the driveshaft to a machine shop that is equipped to balance the shaft and have it repaired. Driveshafts typically turn at speeds 2 ½ to 4 or more times faster than the rear axle; don't use a damaged driveshaft.

This type of balancing is performed by installing one or two hose clamps near the end of the driveshaft closest to the drive axle. The trial and error method is used to determine the best position of the clamp(s).

➡**Removing and turning the driveshaft 180° relative to the yoke may reduce some vibration. This should be done prior to the hose clamp method.**

1. Mark the rear of the driveshaft in four equal sections. Number the marks 1 through 4.

2. Install a hose clamp with the screw portion of the clamp on the No. 1 mark.

3. Test drive the vehicle to see if the vibration condition has improved.

4. Recheck the vibration with the clamp positioned at the remaining three positions. If the vibration is equally reduced at, for example, position number 2 and position number 3, then position the screw portion of the clamp halfway between the marks.

5. Test drive the vehicle. If the vibration is still apparent, install another clamp in the same position as the first.

6. Test drive the vehicle. If the vibration is the same, move both clamps an equal distance from the point determined to be the best position. At first, position the clamps approximately ½ in. (12mm) apart.

7. Continue the process until the vibration is reduced to an acceptable level.

➡**If the vibration cannot be reduced to an acceptable level, take the driveshaft to a qualified machine shop for balancing.**

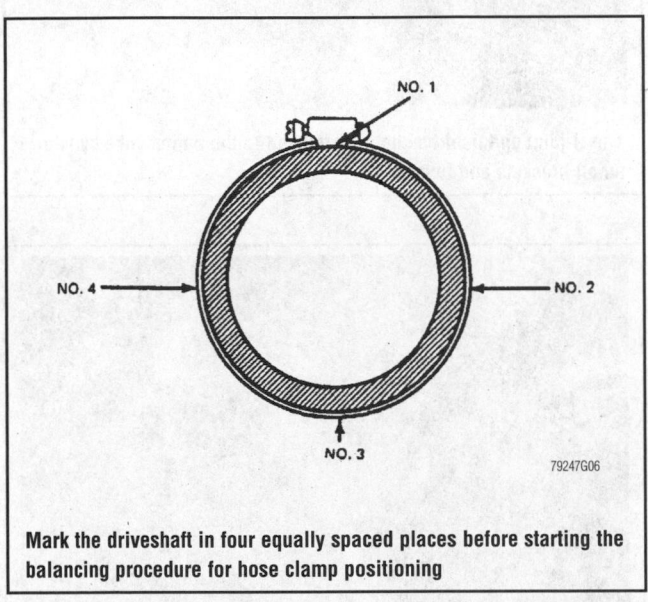

Mark the driveshaft in four equally spaced places before starting the balancing procedure for hose clamp positioning

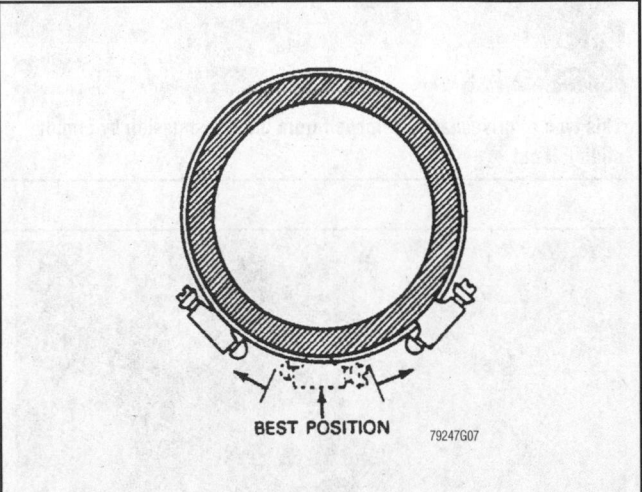

Move the hose clamp heads an equal distance from the best position a little at a time until the vibration is reduced to an acceptable level

Troubleshooting Basic Driveshaft Problems

When abnormal vibrations or noises are detected in the driveshaft area, this chart can be used to help diagnose possible causes. Remember that other components such as wheels, tires, rear axle and suspension can also produce similar conditions.

BASIC DRIVESHAFT PROBLEMS

Problem	Cause	Solution
Shudder as car accelerates from stop or low speed	• Loose U-joint • Defective center bearing	• Replace U-joint • Replace center bearing
Loud clunk in driveshaft when shifting gears	• Worn U-joints	• Replace U-joints
Roughness or vibration at any speed	• Out-of-balance, bent or dented driveshaft • Worn U-joints • U-joint clamp bolts loose	• Balance or replace driveshaft • Replace U-joints • Tighten U-joint clamp bolts
Squeaking noise at low speeds	• Lack of U-joint lubrication	• Lubricate U-joint; if problem persists, replace U-joint
Knock or clicking noise	• U-joint or driveshaft hitting frame tunnel • Worn CV joint	• Correct overloaded condition • Replace CV joint

79247C01

Universal Joints (U-Joints)

GENERAL INFORMATION

The universal joint (U-Joint) is used to provide a strong and flexible connection between the driveshaft and axle assembly. A flexible joint is necessary because of the constant movement of the axle assembly relative to the body of the vehicle. A U-Joint consists of the spider (trunnion), needle (roller) bearings, bearing cups, seals and snaprings. In most cases, U-Joints will last the life of the vehicle. The life of the U-Joint may decrease significantly if the operating angle has been changed or exceeded. This occurs when the vehicle ride height is changed. Vehicles that have been lifted will benefit by using a Double Cardon type joint. The Double Cardon type joint has a greater operating angle than the single U-joint.

When two components are connected by a conventional U-joint, the bend that is formed is called the operating angle. The larger the angle, the larger the amount of angular acceleration and deceleration of the joint. In other words, when the driveshaft is turning at a steady speed, the pinion gear in the differential will actually speed up and slow down. This takes place as long as the driveshaft and pinion gear shaft are at different angles (not in the same plane). The speeding up and slowing down must be canceled out to ensure a smooth flow of power. This is why both yokes on the driveshaft are in line with each other. For example, whereas the transmission output is at a steady speed, the angle at the U-joint causes the driveshaft speed to vary. In such a case, the rear U-joint cancels the fluctuations caused by the front U-joint.

➡**The operating angle is the difference in degrees between the centerline of the driveshaft and the centerline of the transmission and/or axle assembly. The maximum allowable operating angle is determined by engine speed.**

Bad U-joints, requiring replacement, will produce a clunking sound when the vehicle is put into gear and when the transmission shifts from gear-to-gear. This is due to worn needle bearings or scored trunnion ends. Most U-joints are permanently lubricated at the factory and require no periodic maintenance. Those that do have grease fittings should be lubricated at every oil change. Clean the

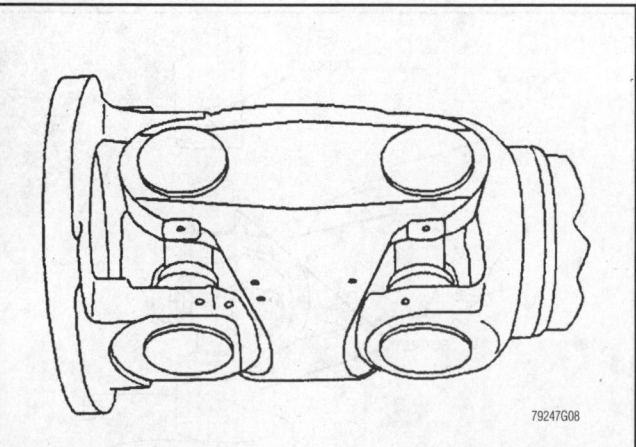

A Double Cardon universal joint has a greater operating angle than a single joint. This joint has been punch marked before disassembly so the components can be reassembled in their original positions

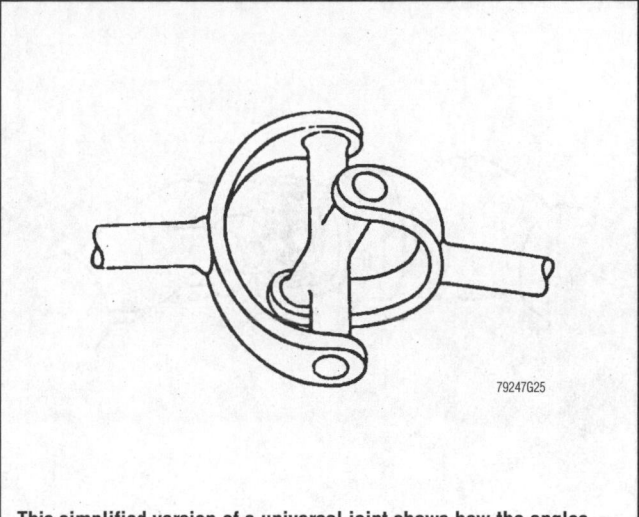

This simplified version of a universal joint shows how the angles can change while still transmitting power

PROPELLER SHAFT R.P.M.	MAX. NORMAL OPERATING ANGLES
5000	3°
4500	3°
4000	4°
3500	5°
3000	5°
2500	7°
2000	8°
1500	11°

Maximum normal operating angle between the driveshaft and transmission and/or axle assembly

fitting with a shop rag before pumping grease to avoid forcing dirt into the joint.

On some production U-joints, nylon is injected through a small hole in the yoke during manufacture and flows along a circular groove between the U-joint and the yoke, creating a non-metallic snapring.

➡**Since plastic retaining rings must be sheared for removal and no snapring grooves are supplied, the production joints must be replaced with service U-joints with a snapring groove whenever they are removed from the shaft.**

INSPECTION

Remove and replace the U-joint if any of the following conditions are present:
• Knocking or clunking noise from the driveshaft when the vehicle is put into gear, or when coasting at 10 mph (16 km/h) in Neutral.
• Squeaking noise from the U-joint that increases in frequency as the speed of the vehicle increases.
• Roughness in the U-joint bearing when felt by hand. The U-joint should turn smoothly.

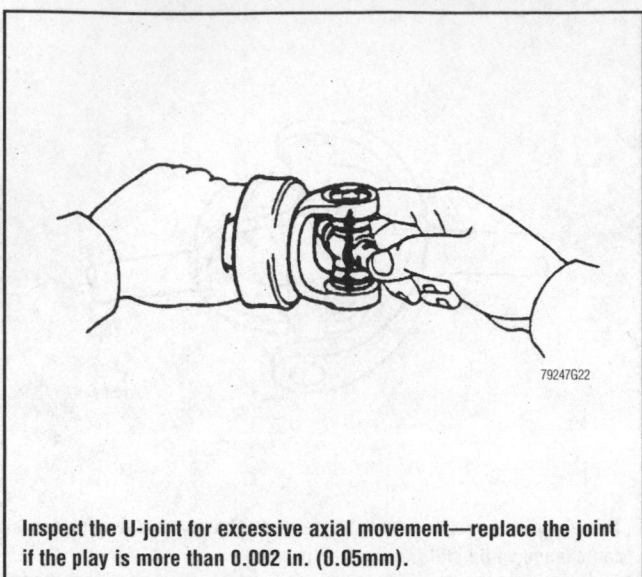

Inspect the U-joint for excessive axial movement—replace the joint if the play is more than 0.002 in. (0.05mm).

• Axial play (up and down movement). Replace the U-joint if the axial play is more than 0.002 in. (0.05mm).

OVERHAUL

1. Position the driveshaft assembly in a sturdy soft-jawed vise, BUT DO NOT place a significant clamp load on the shaft or you will risk deforming and ruining it.

➡Some original equipment U-joints are secured in the yoke by nylon (plastic) that has been injected at the factory. To remove this type of U-joint from the yoke, press the bearing cup until the plastic retaining ring breaks. The replacement U-joint will have a snapring groove like a conventional joint.

2. If applicable, remove the snaprings which retain the bearings in the yoke.

➡A U-joint removal and installation tool (which looks like a large C-clamp) is available to significantly ease the task, but it is very possible to replace the U-joints using an arbor press or a large vise and a variety of sockets.

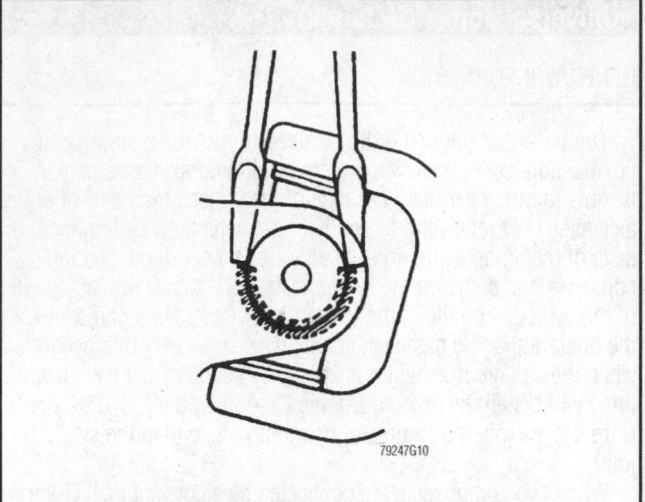

Using two thin prytools is one method for removing the inner snaprings

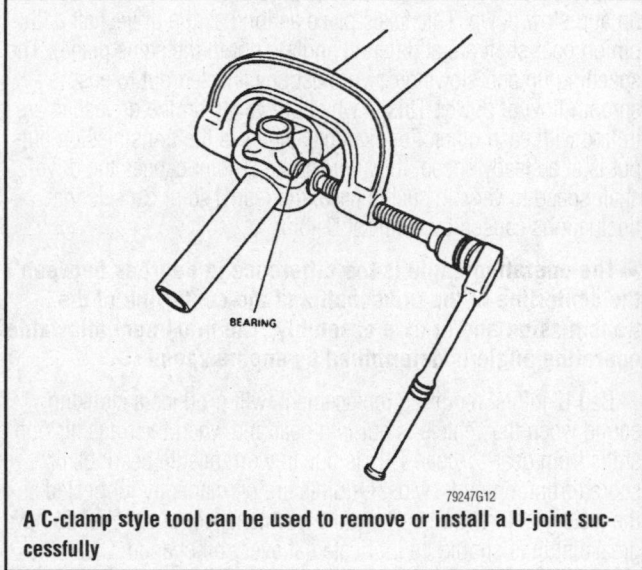

A C-clamp style tool can be used to remove or install a U-joint successfully

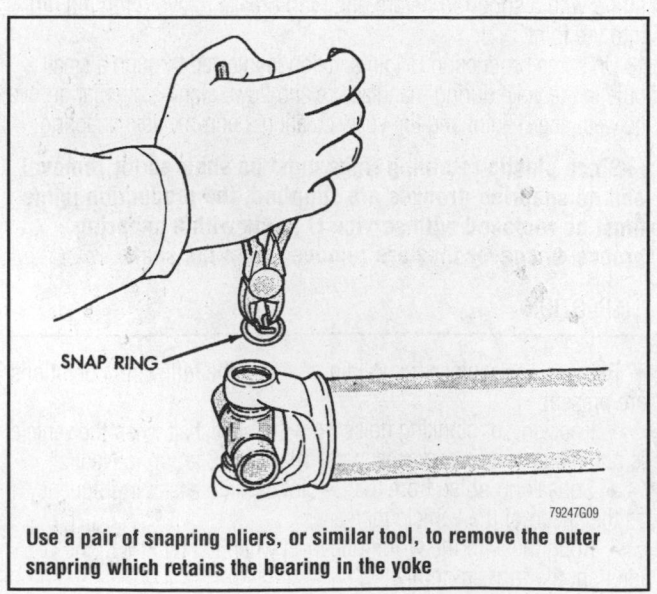

Use a pair of snapring pliers, or similar tool, to remove the outer snapring which retains the bearing in the yoke

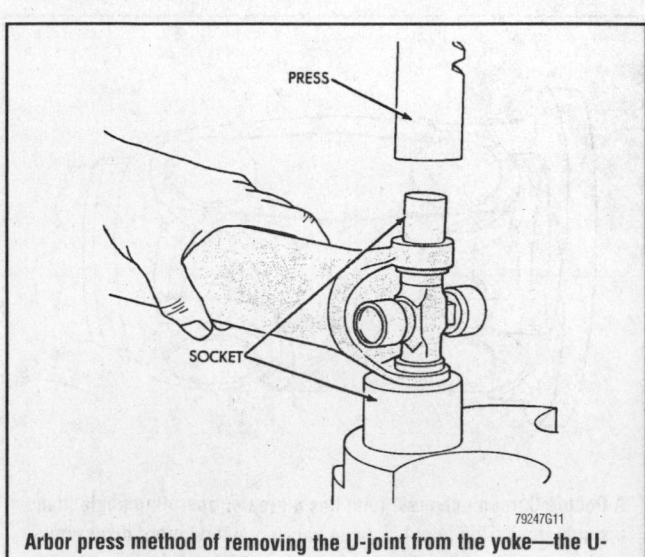

Arbor press method of removing the U-joint from the yoke—the U-joint can also be installed in a similar fashion

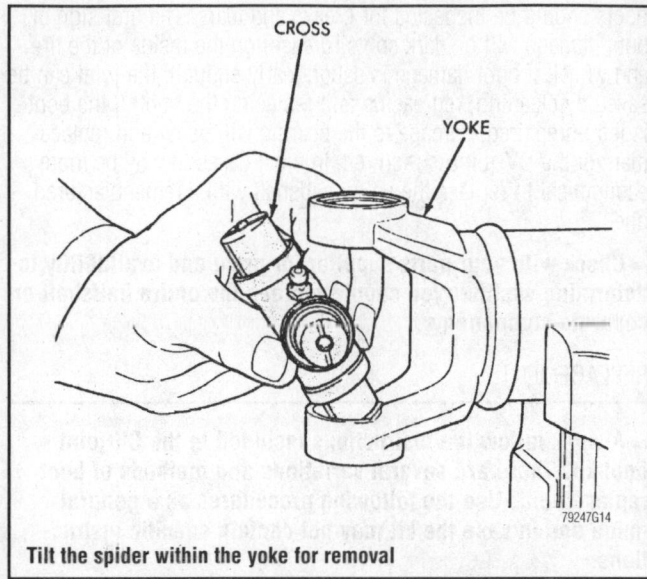

Tilt the spider within the yoke for removal

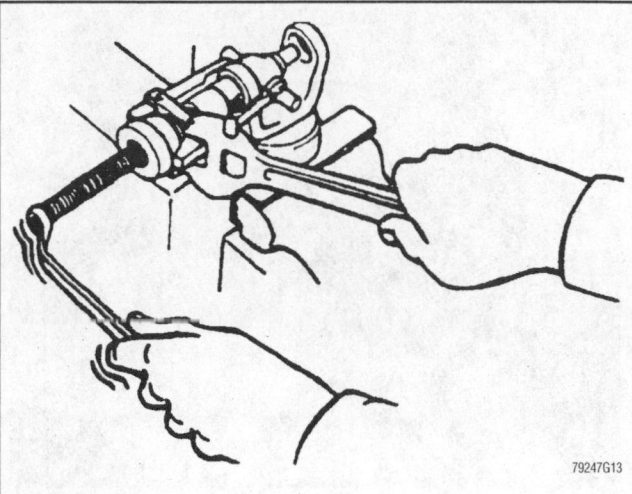

The 2-jawed puller method can also be used to remove or install U-joints

3. Using a large C-clamp, vise or an arbor press, along with a socket smaller than the bearing cap (on one side) and a socket larger than the bearing cap (on the other side), drive one of the bearings in toward the center of the universal joint, which will force the opposite bearing out.

➥The smaller socket is used as a driver here, as it can pass through the opening of the U-joint or slip yoke flange. The larger socket is used to support the other side of the flange so that the bearing cap has room to exit the flange (into the socket).

4. As each bearing is forced far enough out of the universal joint to be accessible, grip it with a pair of pliers and pull it from the driveshaft yoke. Drive the spider in the opposite direction in order to make the opposite bearing accessible and pull it free with a pair of pliers. Use this procedure to remove all the bearings from both universal joints.

5. After removing the bearings, lift the spider from the yoke.

6. Thoroughly clean all dirt and foreign matter from the yokes on both ends of the driveshaft.

To assemble:

❊❊ WARNING

When installing new bearings in the yokes, it is advisable to use an arbor press or the special C-clamp tool. If this tool is not available, the bearings should be pressed into position with extreme care, as a heavy jolt on the needle bearings can easily damage or misalign them. This will greatly shorten their life and hamper their efficiency.

7. Start a new bearing into the yoke at the rear of the driveshaft.

8. Position a new spider in the rear yoke and press the new bearing ¼ in. (6mm) below the outer surface of the yoke.

9. With the bearing in position, install a new snapring.

10. Start a new bearing into the opposite side of the yoke. Press the bearing until the opposite bearing, which you have just installed, contacts the inner surface of the snapring.

11. Install a new snapring on the second bearing. It may be necessary to grind the surface of the second snapring for it to fit in it's groove.

12. Reposition the driveshaft in the vise, so that the front universal joint is accessible.

13. Install the new bearings, new spider and new snaprings in the same manner as for the previously assembled rear joint.

14. Position the slip yoke on the spider. Install new bearings, nylon thrust bearings (if applicable) and snaprings.

15. Check both reassembled joints for freedom of movement. If misalignment of any part is causing a bind, a sharp rap on the side of the yoke with a brass hammer should seat the needle bearings and provide the desired freedom of movement. Care should be exercised to firmly support the shaft end during this operation, as well as to prevent blows to the bearings themselves. Under no circumstance should the driveshaft be installed in a vehicle if there is any binding in the universal joints.

16. Grease the U-joint fittings, if equipped.

Constant Velocity Joint (CV-Joint) Boots

INSPECTION

Whenever undercarriage work is performed, such as brakes, exhaust or suspension work, the Constant Velocity (CV) joint

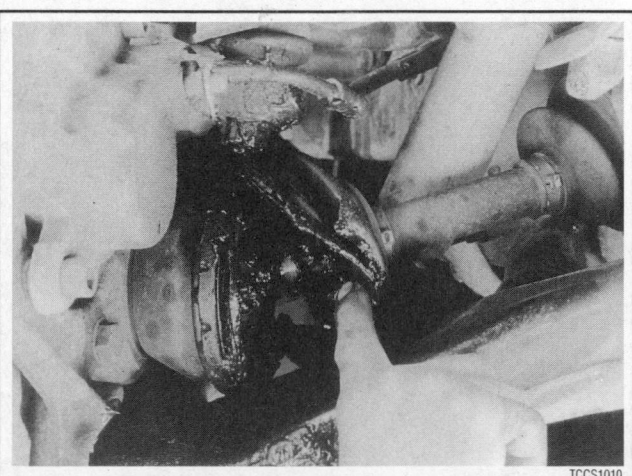

Remove, clean and inspect this CV joint—it may be possible to save the joint by replacing the boot as long as the joint is not beyond repair

Push apart the bellows to inspect for tears or cracks that may developing

boots should be inspected for breaks and tears. The first sign of boot damage will be dark spots (grease) on the inside of the tire and wheel. If boot damage is caught early enough, the joint can be saved by cleaning, regreasing and replacing the boot. If the boot is left unrepaired, damage to the bearing will occur and replacement of the CV-joint is required. In most cases, it may be more economical to replace the entire halfshaft with a remanufactured one.

➡ Check with your parts supplier for price and availability to determine weather you should replace the entire halfshaft or separate components.

REPLACEMENT

➡ Always follow the instructions included in the CV-joint boot kit. There are several variations and methods of boot replacement. Use the following procedures as a general guide and in case the kit may not contain specific instructions.

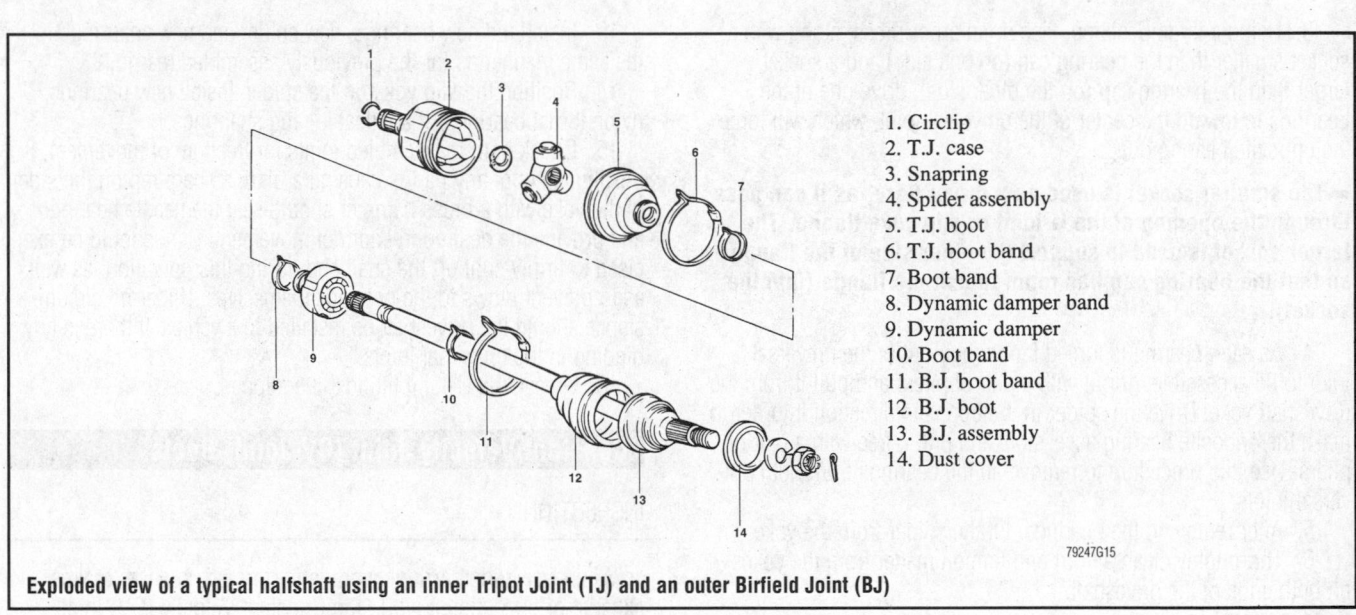

1. Circlip
2. T.J. case
3. Snapring
4. Spider assembly
5. T.J. boot
6. T.J. boot band
7. Boot band
8. Dynamic damper band
9. Dynamic damper
10. Boot band
11. B.J. boot band
12. B.J. boot
13. B.J. assembly
14. Dust cover

Exploded view of a typical halfshaft using an inner Tripot Joint (TJ) and an outer Birfield Joint (BJ)

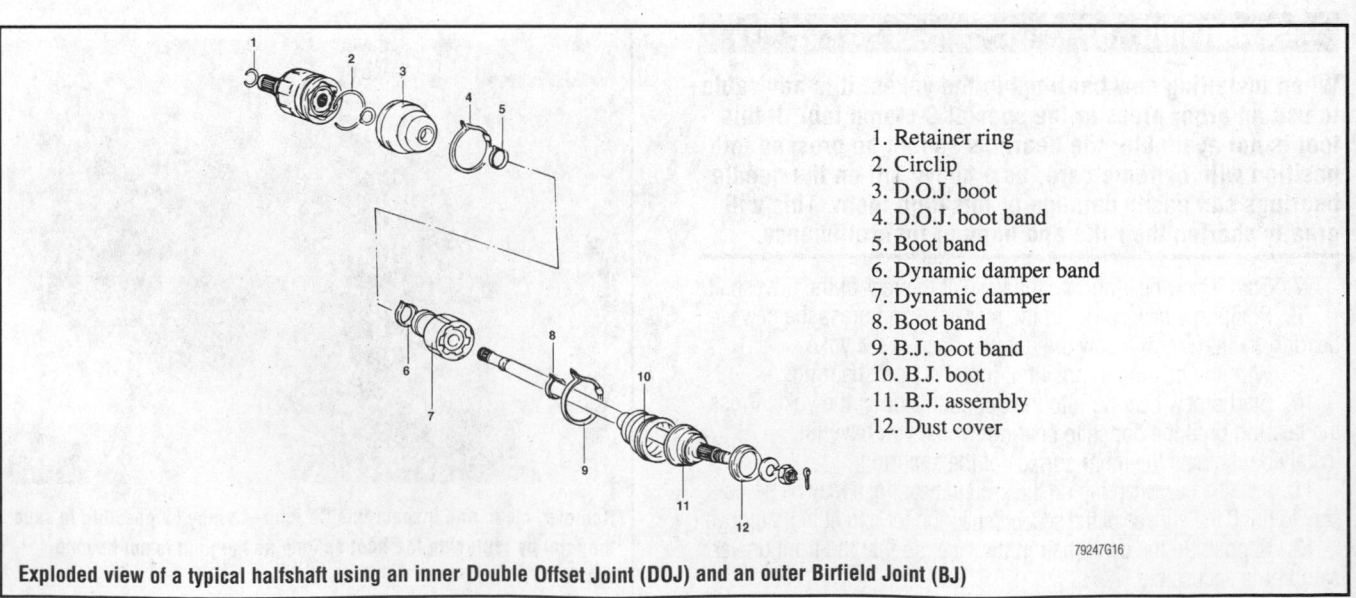

1. Retainer ring
2. Circlip
3. D.O.J. boot
4. D.O.J. boot band
5. Boot band
6. Dynamic damper band
7. Dynamic damper
8. Boot band
9. B.J. boot band
10. B.J. boot
11. B.J. assembly
12. Dust cover

Exploded view of a typical halfshaft using an inner Double Offset Joint (DOJ) and an outer Birfield Joint (BJ)

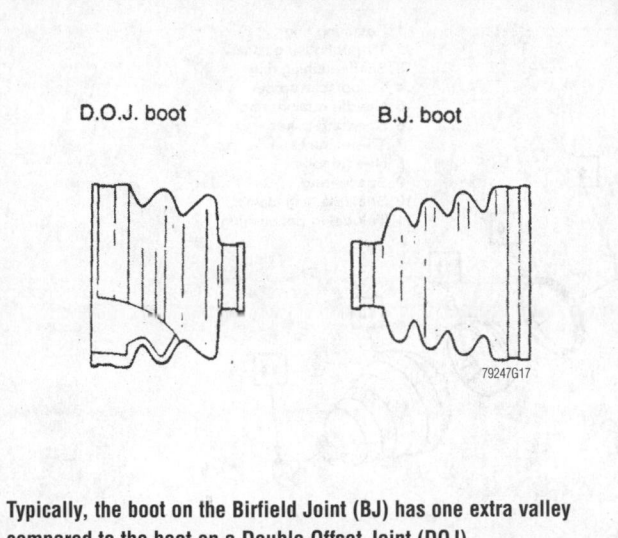

D.O.J. boot   B.J. boot

79247G17

Typically, the boot on the Birfield Joint (BJ) has one extra valley compared to the boot on a Double Offset Joint (DOJ)

TCCS7031

Pry under the hook to remove this type of band from the CV-joint

Most outer CV joints on Asian vehicles, including Chrysler imports, use a Birfield joint, which should not be disassembled. To replace the outer boot, disassemble the inner joint, then slide the outer boot off the inner end of the shaft.

Outer Boot

➡**Generally the Double Offset Joint (DOJ) is used as the outer CV-joint.**

1. Remove the halfshaft and carefully place it in a vise using a protective covering on the vise jaws.

✳✳ WARNING

Some halfshafts may utilize hollow shafts between the CV-joints. Do not tighten the vise more than necessary.

2. Cut the large and small CV-joint boot band clamps and discard them.
3. Slide the boot down the shaft uncovering the outer joint.
4. Clean the grease from the joint to uncover the snapring.

➡**A Tripot Joint (TJ) may also be referred to as a Tulip Joint because of the physical shape of it which resembles a tulip.**

5. Using snapring pliers, open the snapring and slide the outer joint off the shaft.
6. Remove the boot from the shaft.
7. Clean the joint thoroughly using parts cleaner, then dry it completely with compressed air. Inspect the inner bearing and race assembly. If the joint is worn or damaged, replace it.
   To install:
8. Wrap the splines on the end of the halfshaft with tape to prevent damage to the boot during installation.
9. Slide the small CV-joint boot clamp onto the halfshaft and push the boot down several in. past the seal mounting area. Remove the tape from the halfshaft splines.

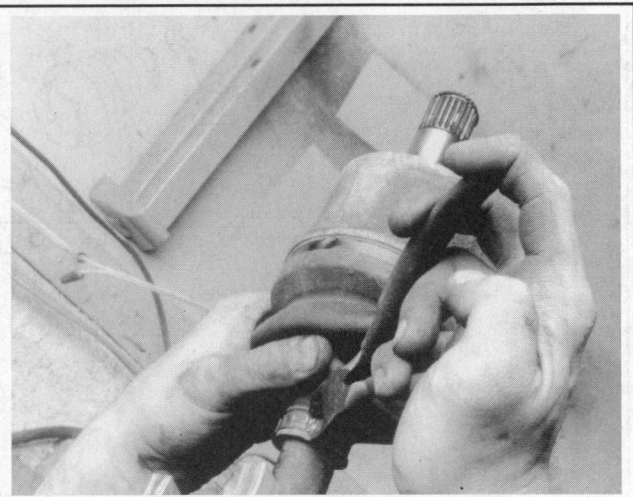

TCCS7032

This type of band is crimped and must be cut before it can be removed

10. Check the snapring in the outer joint for damage or excessive wear and replace as necessary. Pack the joint with half of the grease supplied in the boot kit and install it on the shaft.
11. Insert the shaft into the joint until the splines engage. With a brass drift, lightly tap the joint down until the snapring engages.
12. Pack the remaining grease from the kit into the boot, then pull the large side of the boot over the CV-joint. Seat the small end of the boot on the seal mounting area.

➡**Some CV-joint boot bands require the use of special pliers that are designed to grip the band and allow it to be tightened.**

13. Slide the small clamp into position and secure it.
14. Install the large clamp in the proper position. Slide a small, dull tool under the lip of the boot to equalize the air pressure, then secure the band.

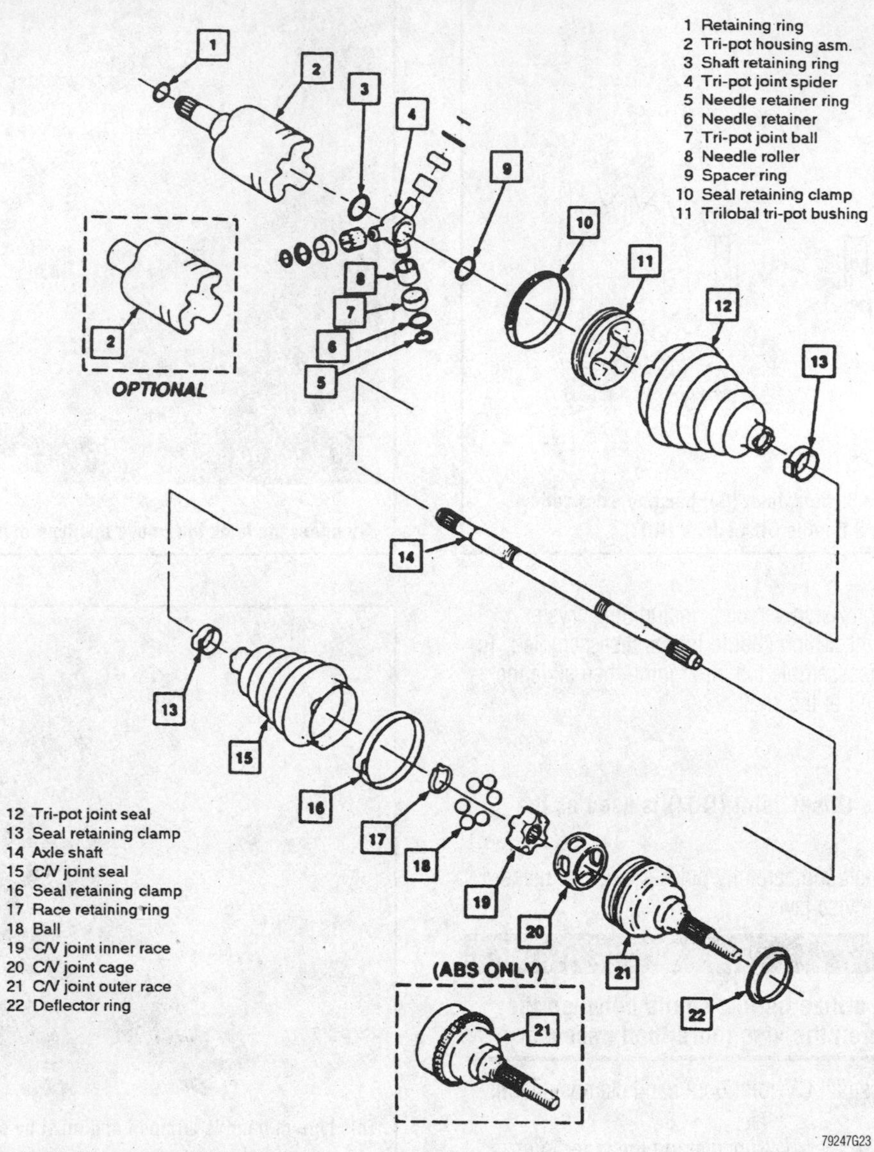

1 Retaining ring
2 Tri-pot housing asm.
3 Shaft retaining ring
4 Tri-pot joint spider
5 Needle retainer ring
6 Needle retainer
7 Tri-pot joint ball
8 Needle roller
9 Spacer ring
10 Seal retaining clamp
11 Trilobal tri-pot bushing

OPTIONAL

12 Tri-pot joint seal
13 Seal retaining clamp
14 Axle shaft
15 C/V joint seal
16 Seal retaining clamp
17 Race retaining ring
18 Ball
19 C/V joint inner race
20 C/V joint cage
21 C/V joint outer race
22 Deflector ring

(ABS ONLY)

79247G23

Exploded view of a halfshaft with TJ inner and DOJ outer joints

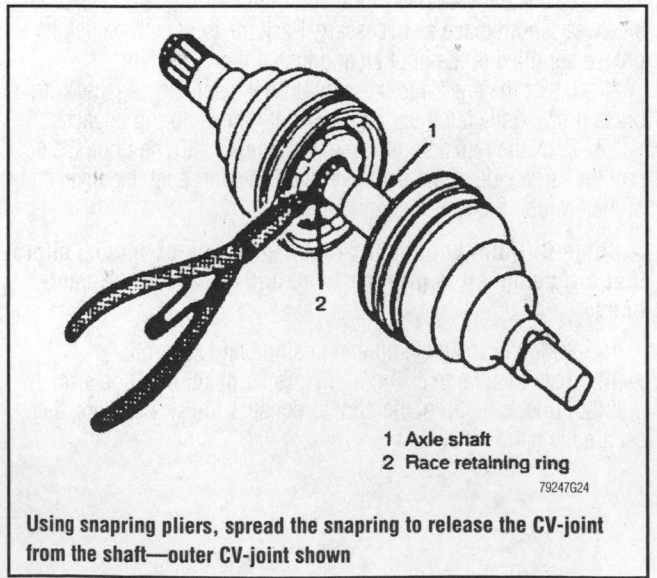

1 Axle shaft
2 Race retaining ring
79247G24

Using snapring pliers, spread the snapring to release the CV-joint from the shaft—outer CV-joint shown

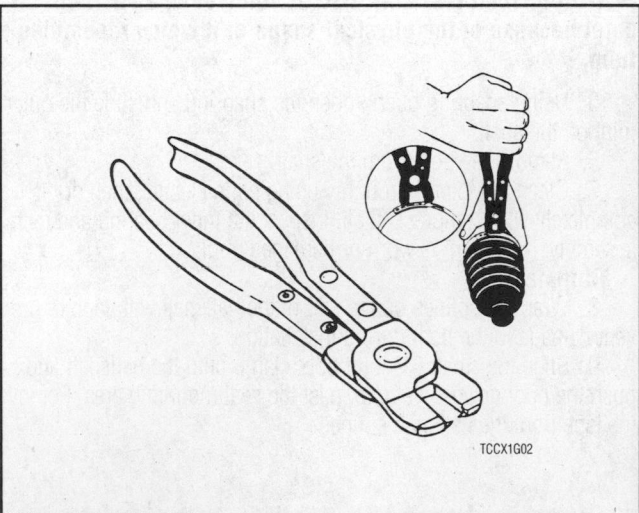

TCCX1G02

The jaws of this tool are designed fit into the small holes on the band and allow it to be tightened

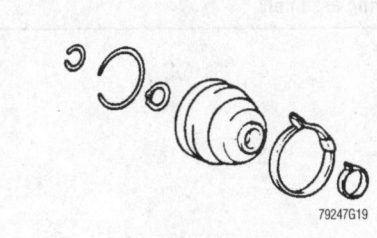

This tool allows a torque wrench to be used when the manufacturer specifies that a certain pressure is required to crimp the band

The typical boot replacement kit contains a new boot, two clamps and special grease—new circlips may also be included is some kits

The boot kit for the Birfield joint should contain an extra set of bands, because the inner joint must be removed in order to install a new boot on the outer (Birfield) joint

✱✱ WARNING

Incorrect CV-joint boot installation may lead to early failure of the boot. The boot must not be dimpled, stretched or out of shape in any way when installed. If the boot is not shaped correctly, carefully insert a thin, blunt tool under the large end of the boot to equalize air pressure. Shape the boot properly by hand, then remove the tool.

15. Install the halfshaft. Road test the vehicle to check for abnormal noise or vibration.

Inner Boot

1. Remove the halfshaft and carefully place it in a vise using a protective covering on the vise jaws.
2. Cut the large and small boot clamps and discard.

✱✱ WARNING

Do not cut through the boot and damage the sealing surface of the CV-joint housing.

3. Pull the boot down the shaft to expose the joint.
4. If equipped, remove the large circlip from the inner edge of the outer bearing race.
5. Matchmark the bearing and outer case so they can be installed in their original positions.
6. Remove the housing from the spider and axle. Clean and dry all components thoroughly. Replace any parts that show signs of wear.
7. Push the spider assembly down the shaft to uncover the snapring on the end of the shaft. Remove the snapring and slide the spider assembly off the end of the shaft.
8. If equipped, remove the spacer ring from the shaft.
9. Remove the remaining circlip and slide the boot off the shaft.
10. Clean and dry all components thoroughly. Replace any parts that show signs of wear.

To install:

11. Slide the small clamp onto the halfshaft.
12. Slide the boot onto the shaft until the small end of the boot is in the original groove that it was removed from.
13. Install the applicable circlip.
14. If equipped, install the spacer ring on the shaft, several in. below the second spacer ring groove.
15. Install the spider assembly far enough down the shaft to expose the top snapring groove. Be sure the counterbored face of the spider faces the end of the shaft.
16. Install the top snapring and pull the spider assembly back up into position.
17. If equipped, lock the spacer ring in the spacer ring groove.
18. Pack the housing with half of the grease supplied in the kit and put the rest of the grease in the boot.
19. Slide the larger clamp over the boot.
20. Push the housing over the spider assembly.
21. If equipped, install the large circlip in the outer race.
22. Slide the larger diameter of the boot into position. Slide a small dull tool under the lip of the boot to equalize the air pressure, then secure the band in position.

✳✳ WARNING

The boot must not be dimpled, stretched or out of shape in any way. If boot is not shaped correctly, carefully insert a thin flat blunt tool at the large end of the boot to equal-

ize pressure. Shape the boot properly by hand, then remove the tool.

23. Install the halfshaft and road test the vehicle.

INNER CV JOINT

ADAPTER

SLIDE HAMMER

79247G26

A special adapter for a slide hammer is available to assist in removing the halfshaft from the transaxle

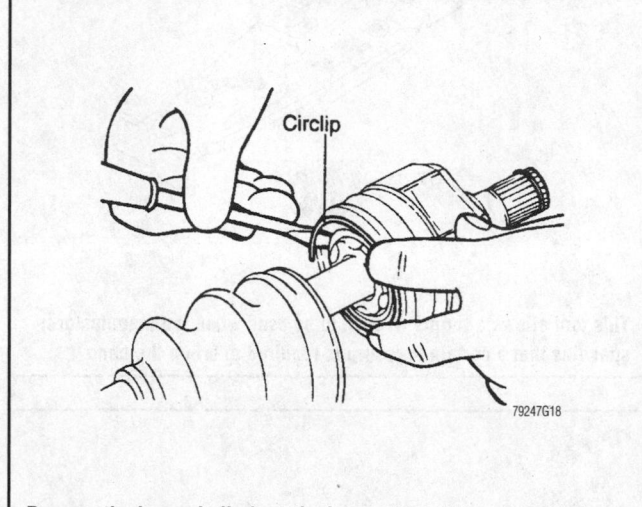

Circlip

79247G18

Remove the large circlip from the inner edge of the outer race to release the bearing assembly

OXYGEN (O₂) SENSORS

8

OXYGEN (O2) SENSORS

General Information

An Oxygen (O₂) sensor is an input device used by the engine control computer to monitor the amount of oxygen in the exhaust gas stream. This information is used by the computer, along with other inputs, to fine-tune the air/fuel mixture so that the engine can run with the greatest efficiency in all conditions. The O₂ sensor sends this information to the computer in the form of a 100–900 millivolt (mV) reference signal, which is actually created by the O₂ sensor itself through chemical interactions between the sensor tip material (zirconium dioxide in almost all cases) and the oxygen levels in the exhaust gas stream and ambient atmosphere gas. At operating temperatures, approximately 1100° F (600° C), the element becomes a semiconductor. Essentially, through the differing levels of oxygen in the exhaust gas stream and in the surrounding atmosphere, the sensor creates a voltage signal which is directly and consistently related to the concentration of oxygen in the exhaust stream. Typically, a higher than normal amount of oxygen in the exhaust stream indicates that not all of the available oxygen was used in the combustion process, because there was not enough fuel (lean condition) present. Inversely, a lower than normal concentration of oxygen in the exhaust stream indicates that a large amount was used in the combustion process, because a larger than necessary amount of fuel was present (rich condition). Thus, the engine control computer can correct the amount of fuel introduced into the combustion chambers.

Since the control computer uses the O₂ sensor output voltage as an indication of the oxygen concentration, and the oxygen concentration directly affects O₂ sensor output, the signal voltage from the sensor to the computer fluctuates constantly. This fluctuation is caused by the nature of the interaction between the computer and the O₂ sensor, which follows a general pattern: detect, compare, compensate, detect, compare, compensate, etc. This means that when the computer detects a lean signal from the O₂ sensor, it compares the reading with known parameters stored within its memory. It calculates that there is too much oxygen present in the exhaust gases, so it compensates by adding more fuel to the air/fuel mixture. This, in turn, causes the O₂ sensor to send a rich signal to the computer, which, then compares this new signal, and adjusts the air/fuel mixture again. This pattern constantly repeats itself: detect rich, compare, compensate lean, detect lean, compare, compensate rich, etc. Since the O₂ sensor fluctuates between rich and lean, and because the lean limit for sensor output is 100 mV and the rich limit is 900 mV, the proper voltage signal from a normally functioning O₂ sensor consistently fluctuates between 100–300 and 700–900 mV.

➡ **The sensor voltage may never quite reach 100 or 900 mV, but it should fluctuate from at least below 300 mV to above 700 mV, and the mid-point of the fluctuations should be centered around 500 mV.**

To improve O₂ sensor efficiency, newer O₂ sensors were designed with a built-in heating element, and were called Heated O₂ (HO₂) sensors. This heating element was incorporated into the sensor so that the sensor would reach optimal operating temperature quicker, meaning that the O₂ sensor output signal could be used by the engine control computer sooner. Because the sensor reaches optimal temperature quicker, modern vehicles enjoy improved driveability and fuel economy even before the engine reaches normal operating temperature.

Although a few manufacturers changed earlier, in 1995 all vehicles were required to implement a new set of engine control pa-

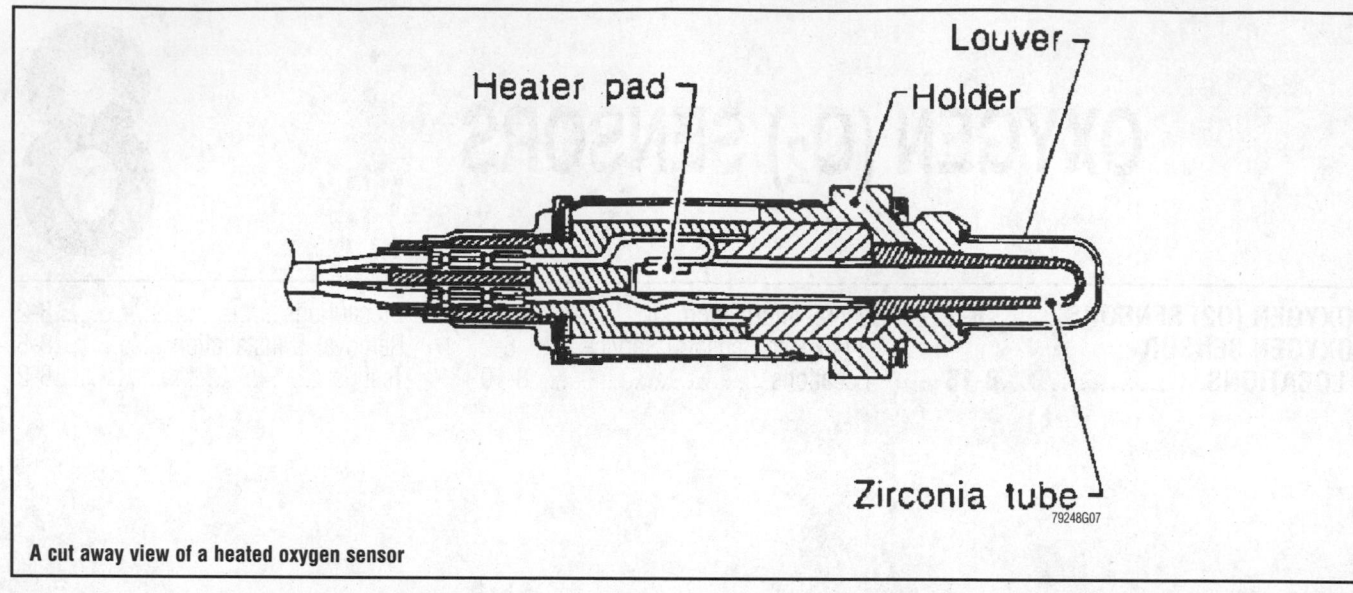

A cut away view of a heated oxygen sensor

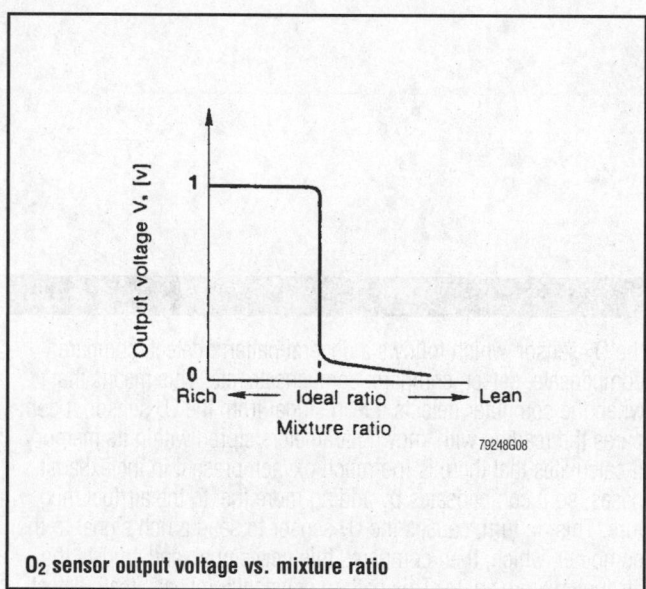

O₂ sensor output voltage vs. mixture ratio

rameters, referred to as On-Board Diagnostics second generation (OBD-II). This updated system (based on the former OBD-I), called for additional O₂ sensors to be used after the catalytic converter, so that catalytic converter efficiency could be measured by the vehicle's engine control computer. The O₂ sensors mounted in the exhaust system after the catalytic converters are not used to affect air/fuel mixture; they are used solely to monitor catalytic converter efficiency.

O₂ (Oxygen Sensors) Service

PRECAUTIONS

When testing or servicing an O₂ sensor you will need to start and warm the engine to operating temperature in order to either perform the necessary testing procedures or to easily remove the sensor from its fitting. This will create a situation in which you will be working around a **HOT** exhaust system. The following is a list of precautions to consider during this service:

- Do not pierce any wires when testing an O₂ sensor, as this can lead to wiring harness damage. Backprobe the connector, when necessary.
- While testing the sensor, be sure to keep out of the way of moving engine components, such as the cooling fan. Refrain from wearing loose clothing which may become tangled in moving engine components.
- Safety glasses must be worn at all times when working on, or near, the exhaust system. Older exhaust systems may be covered with loose rust particles which can shower you when disturbed. These particles are more than a nuisance and can injure your eye.
- Be cautious when working on and around the hot exhaust system. Painful burns will result if skin is exposed to the exhaust system pipes or manifolds.
- The O₂ sensor may be difficult to remove when the engine temperature is below 120° F (48° C). Excessive force may damage the threads in the exhaust manifold or pipe, therefore always start the engine and allow it to reach normal operating temperature prior to removal.
- Since O₂ sensors are usually designed with a permanently-attached wiring pigtail (this allows the wiring harness and sensor connectors to be positioned away from the hot exhaust system), it may be necessary to use a socket or wrench that is designed specifically for this purpose. Before purchasing such a socket, be sure that you can't save some money by using a box end wrench for sensor removal.

TESTING

The best, and most accurate method to test the operation of an O₂ sensor is with the use of either an oscilloscope or a Diagnostic Scan Tool (DST), following their specific instructions for testing. It is possible, however, to test whether the O₂ sensor is functioning properly within general parameters using a Digital Volt-Ohmmeter (DVOM), also referred to as a Digital Multi-Meter (DMM). Newer DMM's are often designed to perform many advanced diagnostic functions, and some are even constructed to be used as an oscilloscope. Two in-vehicle testing procedures, and one bench test procedure, will be provided for the common zirconium dioxide oxygen sensor. The first in-vehicle test makes use of a standard DVOM with a 10 megohm impedance, whereas the second in-vehicle test presented necessitates the usage of an advanced DMM with MIN/MAX/Average functions. Both of these in-vehicle test procedures are likely to set Diagnostic Trouble Codes (DTC's) in the engine control computer. Therefore, after testing, be sure to clear all DTC's before retesting the sensor, if necessary.

These are some of the common DTC's which may be set during testing:

- Open in the O_2 sensor circuit
- Constant low voltage in the O_2 sensor circuit
- Constant high voltage in the O_2 sensor circuit
- Other fuel system problems could set a O_2 sensor code

➡ **Because an improperly functioning fuel delivery and/or control system can adversely affect the O₂ sensor voltage output signal, testing only the O₂ sensor is an inaccurate method for diagnosing an engine driveability problem.**

If after testing the sensor, the sensor is thought to be defective because of high or low readings, be sure to check that the fuel delivery and engine management system is working properly before condemning the O_2 sensor. Otherwise, the new O_2 sensor may continue to register the same high or low readings.

Often, by testing the O_2 sensor, another problem in the engine control management system can be diagnosed. If the sensor

appears to be defective while installed in the vehicle, perform the bench test. If the sensor functions properly during the bench test, chances are that there may be a larger problem in the vehicle's fuel delivery and/or control system.

Many things can cause an O_2 sensor to fail, including old age, antifreeze contamination, physical damage, prolonged exposure to overly-rich exhaust gases, and exposure to silicone sealant fumes. Be sure to remedy any such condition prior to installing a new sensor, otherwise the new sensor may be damaged as well.

➡ **Perform a visual inspection of the sensor. Black sooty deposits may indicate a rich air/fuel mixture, brown deposits may indicate an oil consumption problem, and while gritty deposits may indicate an internal coolant leak. All of these conditions can destroy a new sensor if not corrected before installation.**

O₂ Sensor Terminal Identification

The easiest method for determining sensor terminal identification is to use a wiring diagram for the vehicle and engine in question. However, if a wiring diagram is not available there is a method for determining terminal identification. Throughout the testing procedures, the following terms will be used for clarity:

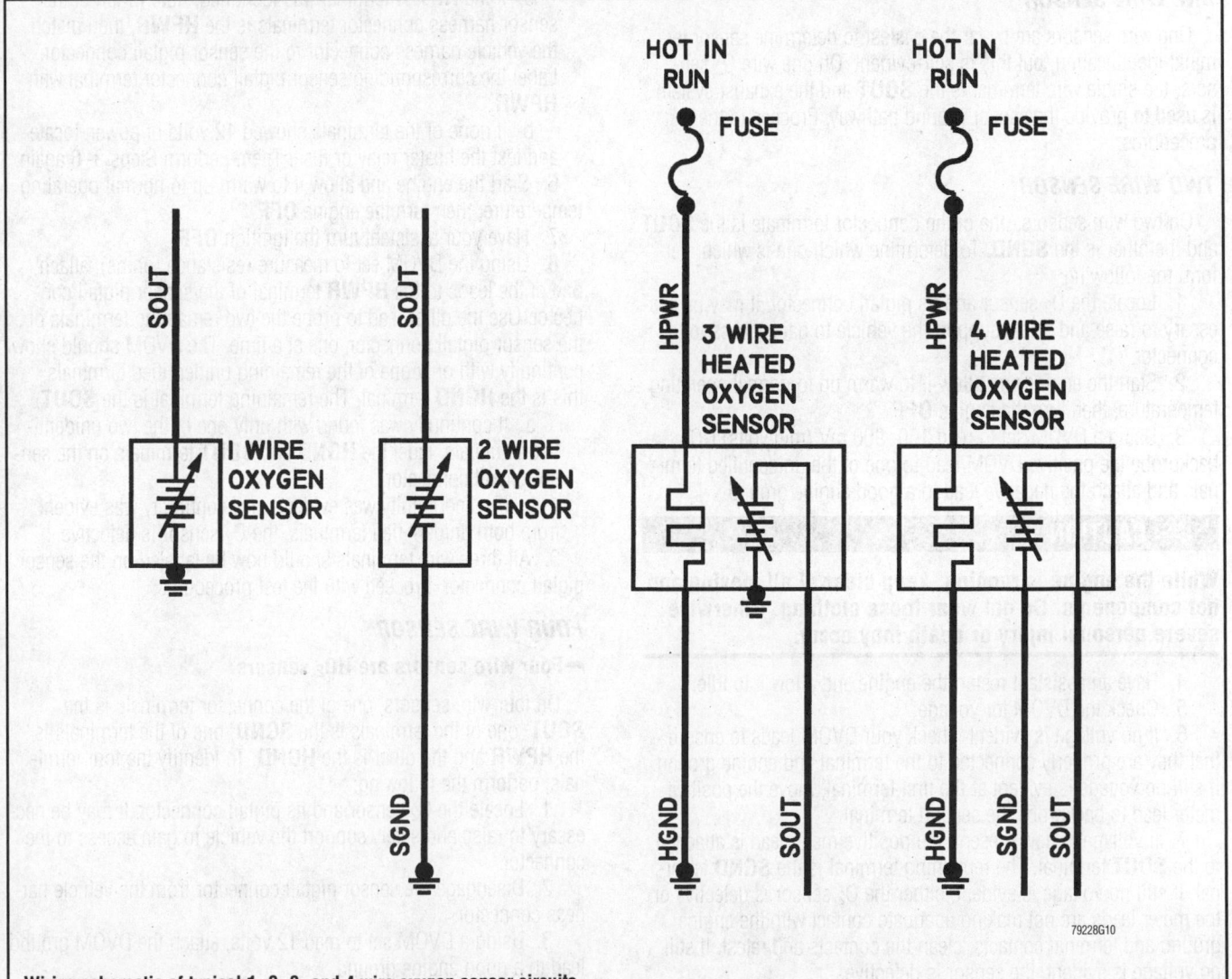

Wiring schematic of typical 1, 2, 3, and 4 wire oxygen sensor circuits

• Vehicle harness connector—this refers to the connector on the wires which are attached to the vehicle; NOT the connector at the end of the sensor pigtail.

• Sensor pigtail connector—this refers to the connector attached to the sensor itself.

• O$_2$ circuit—this refers to the circuit in a Heated O$_2$ (HO$_2$) sensor which corresponds to the oxygen-sensing function of the sensor; NOT the heating element circuit.

• Heating circuit—this refers to the circuit in a HO$_2$ sensor which is designed to warm the HO$_2$ sensor quickly to improve driveability.

• Sensor Output (**SOUT**) terminal—this is the terminal which corresponds to the O$_2$ circuit output. This is the terminal which will register the millivolt signals created by the sensor based upon the amount of oxygen in the exhaust gas stream.

• Sensor Ground (**SGND**) terminal—when a sensor is so equipped, this refers to the O$_2$ circuit ground terminal. Many O$_2$ sensors are not equipped with a ground wire, rather they utilize the exhaust system for the ground circuit.

• Heating Power (**HPWR**) terminal—this terminal corresponds to the circuit which provides the O$_2$ sensor heating circuit with power when the ignition key is turned to the **ON** or **RUN** positions.

• Heating Ground (**HGND**) terminal—this is the terminal connected to the heating circuit ground wire.

ONE WIRE SENSOR

One wire sensors are by far the easiest to determine sensor terminal identification, but this is self-evident. On one wire O$_2$ sensors, the single wire terminal is the **SOUT** and the exhaust system is used to provide the sensor ground pathway. Proceed to the test procedures.

TWO WIRE SENSOR

On two wire sensors, one of the connector terminals is the **SOUT** and the other is the **SGND**. To determine which one is which, perform the following:

1. Locate the O$_2$ sensor and its pigtail connector. It may be necessary to raise and safely support the vehicle to gain access to the connector.

2. Start the engine and allow it to warm up to normal operating temperature, then turn the engine **OFF**.

3. Using a DVOM set to read 100–900 mV (millivolts) DC, backprobe the positive DVOM lead to one of the unidentified terminals and attach the negative lead to a good engine ground.

✳✳ CAUTION

While the engine is running, keep clear of all moving and hot components. Do not wear loose clothing. Otherwise severe personal injury or death may occur.

4. Have an assistant restart the engine and allow it to idle.

5. Check the DVOM for voltage.

6. If no voltage is evident, check your DVOM leads to ensure that they are properly connected to the terminal and engine ground. If still no voltage is evident at the first terminal, move the positive meter lead to backprobe the second terminal.

7. If voltage is now present, the positive meter lead is attached to the **SOUT** terminal. The remaining terminal is the **SGND** terminal. If still no voltage is evident, either the O$_2$ sensor is defective or the meter leads are not making adequate contact with the engine ground and terminal contacts; clean the contacts and retest. If still no voltage is evident, the sensor is defective.

8. Have your assistant turn the engine **OFF**.

9. Label the sensor pigtail **SOUT** and **SGND** terminals.

10. Proceed to the test procedures.

THREE WIRE SENSOR

➡**Three wire sensors are HO$_2$ sensors.**

On three wire sensors, one of the connector terminals is the **SOUT**, one of the terminals is the **HPWR** and the other is the **HGND**. The **SGND** is achieved through the exhaust system, as with the one wire O$_2$ sensor. To identify the three terminals, perform the following:

1. Locate the O$_2$ sensor and its pigtail connector. It may be necessary to raise and safely support the vehicle to gain access to the connector.

2. Disengage the sensor pigtail connector from the vehicle harness connector.

3. Using a DVOM set to read 12 volts, attach the DVOM ground lead to a good engine ground.

4. Have an assistant turn the ignition switch **ON** without actually starting the engine.

5. Probe all three terminals in the vehicle harness connector. One of the terminals should exhibit 12 volts of power with the ignition key **ON**; this is the **HPWR** terminal.

  a. If the **HPWR** terminal was identified, note which of the sensor harness connector terminals is the **HPWR**, then match the vehicle harness connector to the sensor pigtail connector. Label the corresponding sensor pigtail connector terminal with **HPWR**.

  b. If none of the terminals showed 12 volts of power, locate and test the heater relay or fuse. Then, perform Steps 3–6 again.

6. Start the engine and allow it to warm up to normal operating temperature, then turn the engine **OFF**.

7. Have your assistant turn the ignition **OFF**.

8. Using the DVOM set to measure resistance (ohms), attach one of the leads to the **HPWR** terminal of the sensor pigtail connector. Use the other lead to probe the two remaining terminals of the sensor pigtail connector, one at a time. The DVOM should show continuity with only one of the remaining unidentified terminals; this is the **HGND** terminal. The remaining terminal is the **SOUT**.

  a. If continuity was found with only one of the two unidentified terminals, label the **HGND** and **SOUT** terminals on the sensor pigtail connector.

  b. If no continuity was evident, or if continuity was evident from both unidentified terminals, the O$_2$ sensor is defective.

9. All three wire terminals should now be labeled on the sensor pigtail connector. Proceed with the test procedures.

FOUR WIRE SENSOR

➡**Four wire sensors are HO$_2$ sensors.**

On four wire sensors, one of the connector terminals is the **SOUT**, one of the terminals is the **SGND**, one of the terminals is the **HPWR** and the other is the **HGND**. To identify the four terminals, perform the following:

1. Locate the O$_2$ sensor and its pigtail connector. It may be necessary to raise and safely support the vehicle to gain access to the connector.

2. Disengage the sensor pigtail connector from the vehicle harness connector.

3. Using a DVOM set to read 12 volts, attach the DVOM ground lead to a good engine ground.

4. Have an assistant turn the ignition switch **ON** without actually starting the engine.

5. Probe all four terminals in the vehicle harness connector. One of the terminals should exhibit 12 volts of power with the ignition key **ON** ; this is the **HPWR** terminal.

    a. If the **HPWR** terminal was identified, note which of the sensor harness connector terminals is the **HPWR**, then match the vehicle harness connector to the sensor pigtail connector. Label the corresponding sensor pigtail connector terminal with **HPWR**.

    b. If none of the terminals showed 12 volts of power, locate and test the heater relay or fuse. Then, perform Steps 2–6 again.

6. Have your assistant turn the ignition **OFF**.

7. Using the DVOM set to measure resistance (ohms), attach one of the leads to the **HPWR** terminal of the sensor pigtail connector. Use the other lead to probe the three remaining terminals of the sensor pigtail connector, one at a time. The DVOM should show continuity with only one of the remaining unidentified terminals; this is the **HGND** terminal.

    a. If continuity was found with only one of the two unidentified terminals, label the **HGND** terminal on the sensor pigtail connector.

    b. If no continuity was evident, or if continuity was evident from all unidentified terminals, the O₂ sensor is defective.

    c. If continuity was found at two of the other terminals, the sensor is probably defective. However, the sensor may not necessarily be defective, because it may have been designed with the two ground wires joined inside the sensor in case one of the ground wires is damaged; the other circuit could still function properly. Though, this is highly unlikely. A wiring diagram is necessary in this particular case to know whether the sensor was so designed.

8. Reattach the sensor pigtail connector to the vehicle harness connector.

9. Start the engine and allow it to warm up to normal operating temperature, then turn the engine **OFF**.

10. Using a DVOM set to read 100–900 mV (millivolts) DC, backprobe the negative DVOM lead to one of the unidentified terminals and the positive lead to the other unidentified terminal.

✳✳ CAUTION

While the engine is running, keep clear of all moving and hot components. Do not wear loose clothing. Otherwise severe personal injury or death may occur.

11. Have an assistant restart the engine and allow it to idle.

12. Check the DVOM for voltage.

    a. If no voltage is evident, check your DVOM leads to ensure that they are properly connected to the terminals. If still no voltage is evident at either of the terminals, either the terminals were accidentally marked incorrectly or the sensor is defective.

    b. If voltage is present, but the polarity is reversed (the DVOM will show a negative voltage amount), turn the engine **OFF** and swap the two DVOM leads on the terminals. Start the engine and ensure that the voltage now shows the proper polarity.

    c. If voltage is evident and is the proper polarity, the positive DVOM lead is attached to the **SOUT** and the negative lead to the **SGND** terminals.

13. Have your assistant turn the engine **OFF**.

14. Label the sensor pigtail **SOUT** and **SGND** terminals.

In-Vehicle Tests

✳✳ WARNING

Never apply voltage to the O₂ circuit of the sensor, otherwise it may be damaged. Also, never connect an ohmmeter (or a DVOM set on the ohm function) to both of the O₂ circuit terminals (SOUT and SGND) of the sensor pigtail connector; it may damage the sensor.

Test 1 makes use of a standard DVOM with a 10 megohm impedance, whereas Test 2 necessitates the usage of an advanced Digital Multi-Meter (DMM) with MIN/MAX/Average functions or a sliding bar graph function. Both of these in-vehicle test procedures are likely to set Diagnostic Trouble Codes (DTC's) in the engine control computer. Therefore, after testing, be sure to clear all DTC's before retesting the sensor, if necessary. The third in-vehicle test is designed for the use of a scan tool or oscilloscope. The fourth test (Heating Circuit Test) is designed to check the function of the heating circuit in a HO₂ sensor.

➡**If the O₂ sensor being tested is designed to use the exhaust system for the SGND , excessive corrosion between the exhaust and the O₂ sensor may affect sensor functioning.**

The in-vehicle tests may be performed for O₂ sensors located in the exhaust system after the catalytic converter. However, the O₂ sensors located behind the catalytic converter will not fluctuate like the sensors mounted before the converter, because the converter, when functioning properly, emits a steady amount of oxygen. If the O₂ sensor mounted after the catalytic converter exhibits a fluctuating signal (like other O₂ sensors), the catalytic converter is most likely defective.

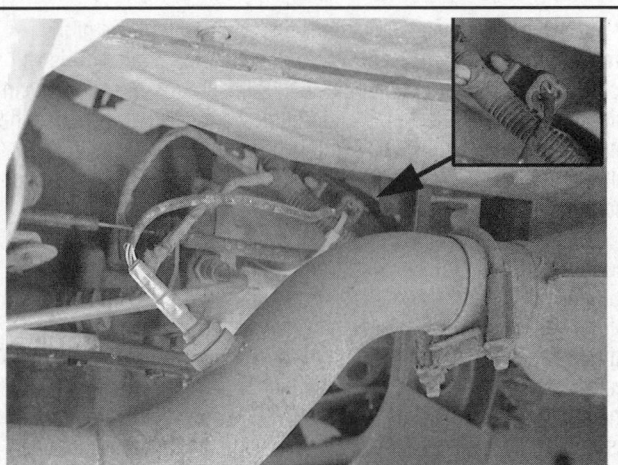

89664P30

To test the O₂ sensor, locate it and its connector (inset), which should be positioned away from the exhaust system to prevent heat damage

TEST 1—DIGITAL VOLT-OHMMETER

This test will not only verify proper sensor functioning, but is also designed to ensure the engine control computer and associated wiring is functioning properly as well.

1. Start the engine and allow it to warm up to normal operating temperature.

➡**If you are using the opening of the thermostat to gauge normal operating temperature, be forewarned: a defective thermostat can open too early and prevent the engine from reaching normal operating temperature. This can cause a slightly rich condition in the exhaust, which can throw the O$_2$ sensor readings off slightly.**

2. Turn the ignition switch **OFF**, then locate the O$_2$ sensor pigtail connector.

3. Perform a visual inspection of the connector to ensure it is properly engaged and all terminals are straight, tight and free from corrosion or damage.

4. Disengage the sensor pigtail connector from the vehicle harness connector.

5. On sensors equipped with a **SGND** terminal (sensors which do not use the exhaust system for the sensor ground pathway), connect a jumper wire to the **SGND** terminal and to a good, clean engine ground (preferably the negative terminal of the battery).

6. Using a DVOM set to read DC voltage, attach the positive lead to the **SOUT** terminal of the sensor pigtail connector, and the DVOM negative lead to a good engine ground.

✳✳ CAUTION

While the engine is running, keep clear of all moving and hot components. Do not wear loose clothing. Otherwise severe personal injury or death may occur.

7. Have an assistant start the engine and hold it at approximately 2,000 rpm. Wait at least 1 minute before commencing with the test to allow the O$_2$ sensor to sufficiently warm up.

➡**Some carbureted Asian models may not switch into closed loop operation until engine speed is above 2,500 rpm.**

8. Using a jumper wire, connect the **SOUT** terminal of the **vehicle harness connector** to a good engine ground. This will fool the engine control computer into thinking it is receiving a lean signal from the O$_2$ sensor, therefore, the computer will enrichen the air/fuel ratio. With the **SOUT** terminal so grounded, the DVOM should register at least 800 mV, as the control computer adds additional fuel to the air/fuel ratio.

9. While observing the DVOM, disconnect the vehicle harness connector **SOUT** jumper wire from the engine ground. Use the jumper wire to apply slightly less than 1 volt to the **SOUT** terminal of the vehicle harness connector. One method to do this is by grasping and squeezing the end of the jumper between your forefinger and thumb of one hand while touching the positive terminal of the battery post with your other hand. This allows your body to act as a resistor for the battery positive voltage, and fools the engine control computer into thinking it is receiving a rich signal. Or, use a mostly-drained AA battery by connecting the positive terminal of the AA battery to the jumper wire and the negative terminal of the battery to a good engine ground. (Another jumper wire may be necessary to do this.) The computer should lean the air/fuel mixture out. This lean mixture should register as 150 mV or less on the DVOM.

10. If the DVOM did not register millivoltages as indicated, the problem may be either the sensor, the engine control computer or the associated wiring. Perform the following to determine which is the defective component:

    a. Remove the vehicle harness connector **SOUT** jumper wire.

    b. While observing the DVOM, artificially enrich the air/fuel charge using propane. The DVOM reading should register higher than normal millivoltages. (Normal voltage for an ideal air/fuel

mixture is approximately 450–550 mV DC). Then, lean the air/fuel intake charger by either disconnecting one of the fuel injector wiring harness connectors (to prevent the injector from delivering fuel) or by detaching one or two vacuum lines (to add additional non-metered air into the engine). The DVOM should now register lower than normal millivoltages. If the DVOM functioned as indicated, the problem lies elsewhere in the fuel delivery and control system. If the DVOM readings were still unresponsive, the O$_2$ sensor is defective; replace the sensor and retest.

➡**Poor wire connections and/or ground circuits may shift a normal O$_2$ sensor's millivoltage readings up into the rich range or down into the lean range. It is a good idea to check the wire condition and continuity before replacing a component which will not fix the problem. A voltage drop test between the sensor case and ground which reveals 14–16 mV, or more, indicates a probable bad ground.**

11. Turn the engine **OFF**, remove the DVOM and all associated jumper wires. Reattach the vehicle harness connector to the sensor pigtail connector. If applicable, reattach the fuel injector wiring connector and/or the vacuum line(s).

12. Clear any DTC's present in the engine control computer memory, as necessary.

TEST 2—DIGITAL MULTI-METER

This test method is a more straight forward O$_2$ sensor test, and does not test the engine control computer's response to the O$_2$ sensor signal. The use of a DMM with the MIN/MAX/Average function or sliding bar graph/wave function is necessary for this test. Don't forget that the O$_2$ sensor mounted after the catalytic converter (if equipped) will not fluctuate like the other O$_2$ sensor(s) will.

1. Start the engine and allow it to warm up to normal operating temperature.

➡**If you are using the opening of the thermostat to gauge normal operating temperature, be forewarned: a defective thermostat can open too early and prevent the engine from reaching normal operating temperature. This can cause a slightly rich condition in the exhaust, which can throw the O$_2$ sensor readings off slightly.**

2. Turn the ignition switch **OFF**, then locate the O$_2$ sensor pigtail connector.

3. Perform a visual inspection of the connector to ensure it is properly engaged and all terminals are straight, tight and free from corrosion or damage.

4. Backprobe the O$_2$ sensor connector terminals. Attach the DMM positive test lead to the **SOUT** terminal of the sensor pigtail connector and the negative lead to either the **SGND** terminal of the sensor pigtail connector (if equipped—refer to the terminal identification procedures earlier in this section for clarification) or to a good, clean engine ground.

5. Activate the MIN/MAX/Average or sliding bar graph/wave function on the DMM.

✳✳ CAUTION

While the engine is running, keep clear of all moving and hot components. Do not wear loose clothing. Otherwise severe personal injury or death may occur.

6. Have an assistant start the engine and wait a few minutes before commencing with the test to allow the O$_2$ sensor to sufficiently warm up.

7. Read the minimum, maximum and average readings exhibited by the O₂ sensor, or observe the bar graph/wave form. The average reading for a properly functioning O₂ sensor is be approximately 450–550 mV DC. The minimum and maximum readings should vary more than 300–600 mV. A typical O₂ sensor can fluctuate from as low as 100 mV to as high as 900 mV; if the sensor range of fluctuation is not large enough, the sensor is defective. Also, if the fluctuation range is biased up or down in the scale. For example, if the fluctuation range is 400 mV to 900 mV the sensor is defective, because the readings are pushed up into the rich range (as long as the fuel delivery system is functioning properly). The same goes for a fluctuation range pushed down into the lean range. The mid-point of the fluctuation range should be around 400–500 mV. Finally, if the O₂ sensor voltage fluctuates too slowly (usually the voltage wave should oscillate past the mid-way point of 500 mV several times per second) the sensor is defective. (When an O₂ sensor fluctuates too slowly, it is referred to as being "lazy.")

➡**Poor wire connections and/or ground circuits may shift a normal O₂ sensor's millivoltage readings up into the rich range or down into the lean range. It is a good idea to check the wire condition and continuity before replacing a component which will not fix the problem. A voltage drop test between the sensor case and ground which reveals 14–16 mV, or more, indicates a probable bad ground.**

8. Using the propane method, enrichen the air/fuel mixture and observe the DMM readings. The average O₂ sensor output signal voltage should rise into the rich range.

9. Lean the air/fuel mixture by either disconnecting a fuel injector wiring harness connector or by disconnecting a vacuum line. The O₂ sensor average output signal voltage should drop into the lean range.

10. If the O₂ sensor did not react as indicated, the sensor is defective and should be replaced.

11. Turn the engine **OFF**, remove the DMM and all associated jumper wires. Reattach the vehicle harness connector to the sensor pigtail connector. If applicable, reattach the fuel injector wiring connector and/or the vacuum line(s).

12. Clear any DTC's present in the engine control computer memory, as necessary.

TEST 3—OSCILLOSCOPE

This test is designed for the use of an oscilloscope to test the functioning of an O₂ sensor.

➡**This test is only applicable for O₂ sensors mounted in the exhaust system before the catalytic converter.**

1. Start the engine and allow it to reach normal operating temperature.

2. Turn the engine **OFF**, and locate the O₂ sensor connector. Backprobe the scope lead to the O₂ sensor connector **SOUT** terminal. Refer to the manufacturer's instructions for more information on attaching the scope to the vehicle.

3. Turn the scope ON.

4. Set the oscilloscope amplitude to 200 mV per division, and the time to 1 second per division. Use the 1:1 setting of the probe, and be sure to connect the scope's ground lead to a good, clean engine ground. Set the signal function to automatic or internal triggering.

5. Start the engine and run it at 2,000 rpm.

6. The oscilloscope should display a wave form, representative of the O₂ sensor switching between lean (100–300 mV) and rich

(700–900 mV). The sensor should switch between rich and lean, or lean and rich (crossing the mid-point of 500 mV) several times per second. Also, the range of each wave should reach at least above 700 mV and below 300 mV. However, an occasional low peak is acceptable.

7. Force the air/fuel mixture rich by introducing propane into the engine, then observe the oscilloscope readings. The fluctuating range of the O₂ sensor should climb into the rich range.

8. Lean the air/fuel mixture out by either detaching a vacuum line or by disengaging one of the fuel injector's wiring connectors. Watch the scope readings; the O₂ sensor wave form should drop toward the lean range.

9. If the O₂ sensor's wave form does not fluctuate adequately, is not centered around 500 mV during normal engine operation, does not climb toward the rich range when propane is added to the engine, or does not drop toward the lean range when a vacuum hose or fuel injector connector is detached, the sensor is defective.

10. Reattach the fuel injector connector or vacuum hose.

11. Disconnect the oscilloscope from the vehicle.

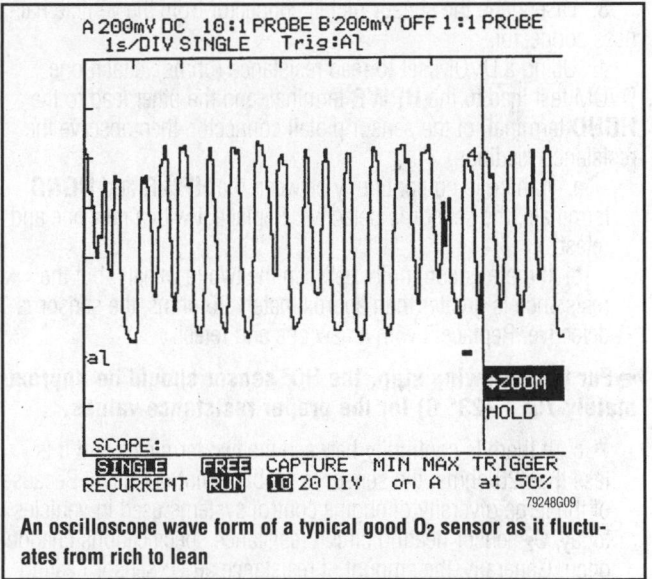

An oscilloscope wave form of a typical good O₂ sensor as it fluctuates from rich to lean

HEATING CIRCUIT TEST

The heating circuit in an O₂ sensor is designed only to heat the sensor quicker than a non-heated sensor. This provides an advantage of increased engine driveability and fuel economy while the engine temperature is still below normal operating temperature, because the fuel management system can enter closed loop operation (more efficient than open loop operation) sooner.

Therefore, if the heating element goes bad, the O₂ sensor may still function properly once the sensor warms up to its normal temperature. This will take longer than normal and may cause mild driveability-related problems while the engine has not reached normal operating temperature.

If the heating element is found to be defective, replace the O₂ sensor without wasting your time testing the O₂ circuit; if necessary, you can perform the O₂ circuit test with the new O₂ sensor and save yourself some time.

1. Locate the O₂ sensor pigtail connector.

2. Perform a visual inspection of the connector to ensure it is properly engaged and all terminals are straight, tight and free from corrosion or damage.

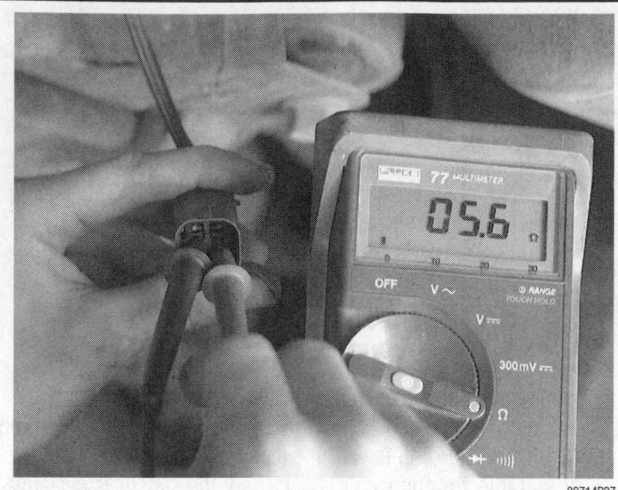

The heating circuit of the O₂ sensor can be tested with a DMM set to measure resistance

3. Disengage the sensor pigtail connector from the vehicle harness connector.

4. Using a DVOM set to read resistance (ohms), attach one DVOM test lead to the **HPWR** terminal, and the other lead to the **HGND** terminal, of the sensor pigtail connector, then observe the resistance readings.

    a. If there is no continuity between the **HPWR** and **HGND** terminals, the sensor is defective. Replace it with a new one and retest.

    b. If there is continuity between the two terminals, but the resistance is greater than approximately 20 ohms, the sensor is defective. Replace it with a new one and retest.

➡**For the following step, the HO² sensor should be approximately 75° F (23° C) for the proper resistance values.**

    c. If there is continuity between the two terminals and it is less than 20 ohms, the sensor is probably not defective. Because of the large diversity of engine control systems used in vehicles today, O₂ sensor heating circuit resistance specifications change often. Generally, the amount of resistance an O₂ sensor heating circuit should exhibit is between 2–9 ohms. However, some manufacturer's O₂ sensors may show resistance as high as 15–20 ohms. As a rule of thumb, 20 ohms of resistance is the upper limit allowable.

5. Turn the engine **OFF**, remove the DVOM and all associated jumper wires. Reattach the vehicle harness connector to the sensor pigtail connector.

6. Clear any DTC's present in the engine control computer memory, as necessary.

Bench Test

➡**Utilize one of the in-vehicle tests before performing this test.**

This test is designed to test an O₂ sensor which does not seem to fluctuate fully beyond 400–700 mV. The sensor is to be secured in a table-mounted vise.

❊❊ **CAUTION**

This test can be very dangerous. Take the necessary precautions when working with a propane torch. Ensure that all combustible substances are removed from the work area and have a fire extinguisher ready at all times. Be sure to wear the appropriate protective clothing as well.

1. Remove the O₂ sensor.

➡**Perform a visual inspection of the sensor. Black sooty deposits may indicate a rich air/fuel mixture, brown deposits may indicate an oil consumption problem, and white gritty deposits may indicate an internal coolant leak. All of these conditions can destroy a new sensor if not corrected before installation.**

2. Position the sensor in a vise so that the vise holds the sensor by the hex portion of its case.

3. Attach one lead of a DVOM set to read DC millivoltages to the sensor case and the other lead to the **SOUT** terminal of the sensor pigtail connector.

4. Carefully use a propane torch to heat the tip (and ONLY the tip) of the sensor. Once the sensor reaches close to normal operating temperature range, alternately heat the sensor up and allow it to cool down; the sensor output voltage signal should change with the temperature change.

➡**This may also clean a sensor covered with a heavy coat of carbon.**

5. If the sensor voltage does not change with the fluctuation in temperature, replace the sensor with a new one. Install the new sensor and perform one of the in-vehicle tests to rule out additional fuel management system faults.

REMOVAL & INSTALLATION

1. Start the engine and allow it to reach normal operating temperature, then turn the ignition switch **OFF**.

2. Disconnect the negative battery cable.

3. Open the hood and locate the O₂ sensor connector. It may be necessary to raise and safely support the vehicle for access to the sensor and its connector.

➡**On a few models, it may be necessary to remove the passenger seat and lift the carpeting in order to access the connector for a downstream O₂ sensor.**

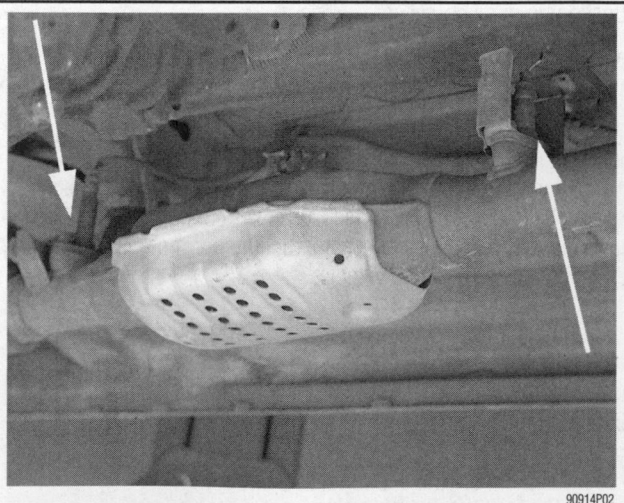

Since sensor locations vary between vehicles, the first step in removal is to locate the O₂ sensors (arrows) . . .

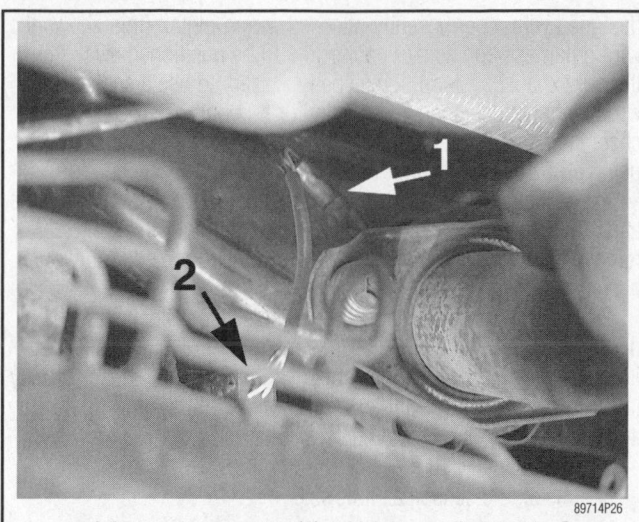

. . . and the sensor connector (2), which is usually near the O₂ sensor (1), but removed enough from the heat of the exhaust system

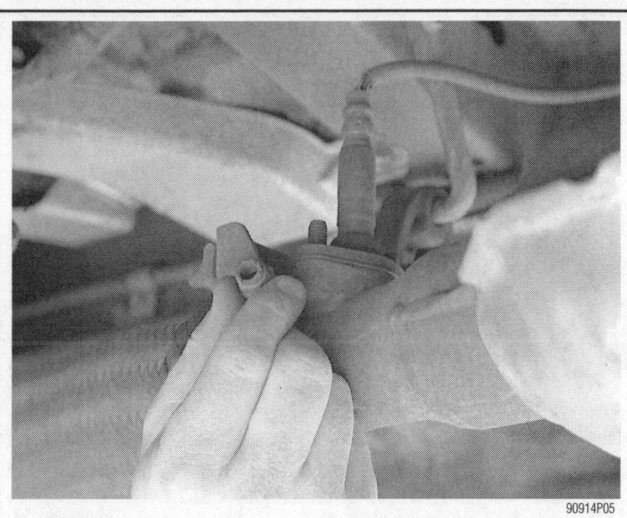

. . . which happen to be nuts in this particular case—some models may use bolts rather than nuts

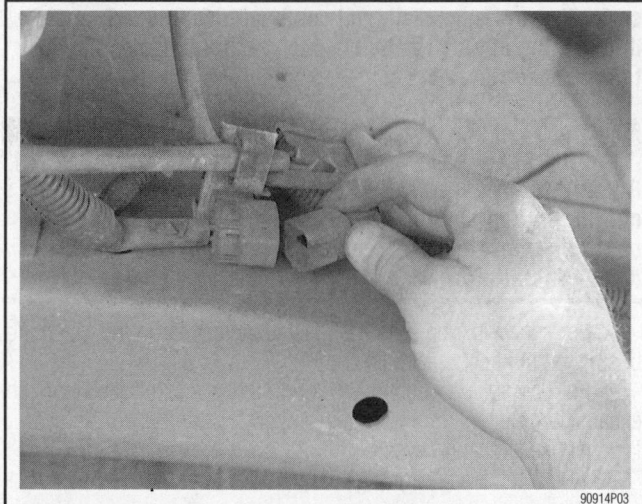

Disengage the sensor pigtail connector half from the vehicle harness connector half

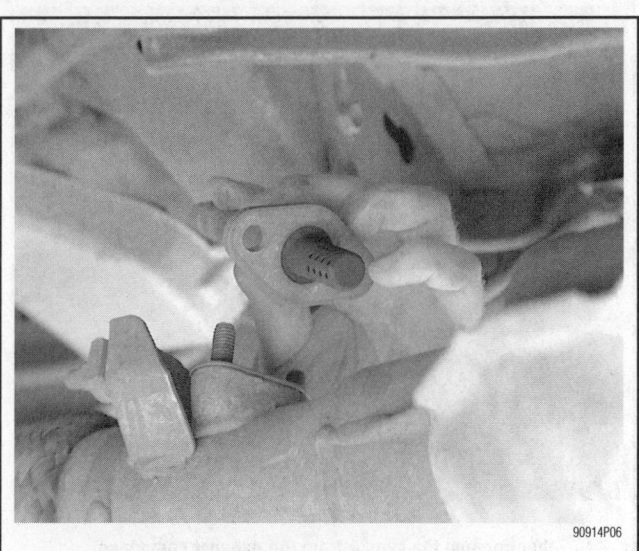

Then, pull the sensor out of the exhaust component

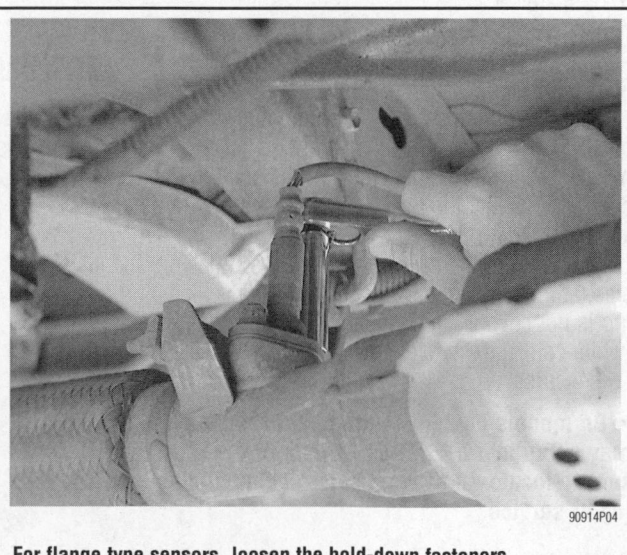

For flange type sensors, loosen the hold-down fasteners . . .

For screw-in type sensors (arrow) . . .

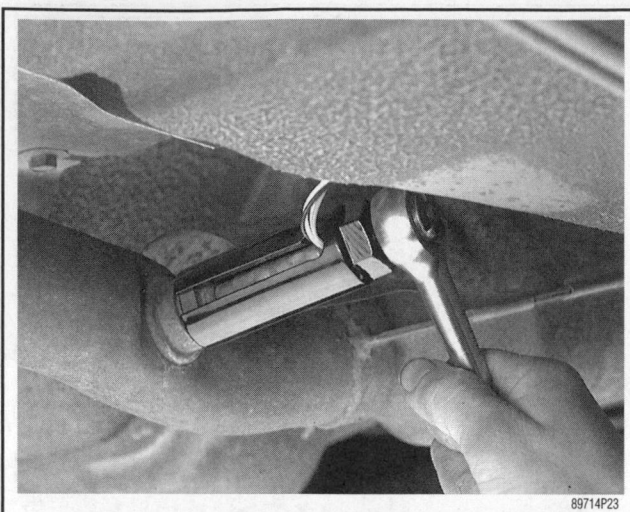

. . . either use a box end wrench to loosen the sensor or a socket designed expressly for this purpose . . .

. . . , then remove the sensor from the exhaust component

4. Disengage the O_2 sensor pigtail connector from the vehicle harness connector.

➡There are generally two methods used to mount an O_2 sensor in the exhaust system: either the O_2 sensor is threaded directly into the exhaust component (screw-in type), or the O_2 sensor is retained by a flange and two nuts or bolts (flange type).

❊❊ WARNING

To prevent damaging a screw-in type O_2 sensor, if excessive force is needed to remove the sensor lubricate it with penetrating oil prior to removal. Also, be sure to protect the tip of the sensor; O_2 sensor tips are very sensitive and may be easily damaged if allowed to strike or come in contact with other objects.

5. Remove the sensor, as follows:
• For screw-in type sensors—Since O_2 sensors are usually

designed with a permanently-attached wiring pigtail (this allows the wiring harness and sensor connectors to be positioned away from the hot exhaust system), it may be necessary to use a socket or wrench that is designed specifically for this purpose. Before purchasing such a socket, be sure that you can't save some money by using a box end wrench for sensor removal.

• For flange type sensors—Loosen the hold-down nuts or bolts and pull the sensor out of the exhaust component. Be sure to remove and discard the old sensor gasket, if equipped. You will need a new gasket for installation.

6. Perform a visual inspection of the sensor. Black sooty deposits may indicate a rich air/fuel mixture, brown deposits may indicate an oil consumption problem, and white gritty deposits may indicate an internal coolant leak. All of these conditions can destroy a new sensor if not corrected before installation.

To install:
7. Install the sensor, as follows:

➡A special anti-seize compound is used on most screw-in type O_2 sensor threads, and is designed to ease O_2 sensor removal. New sensors usually have the compound already applied to the threads. However, if installing the old O_2 sensor or the new sensor did not come with compound, apply a thin coating of electrically-conductive anti-seize compound to the sensor threads.

❊❊ WARNING

Be sure to prevent any of the anti-seize compound from coming in contact with the O_2 sensor tip. Also, take precautions to protect the sensor tip from physical damage during installation.

• For screw-in type sensors—Install the sensor in the mounting boss, then tighten it securely.
• For flange type sensors—Position a new sensor gasket on the exhaust component and insert the sensor. Tighten the hold-down fasteners securely and evenly.

8. Reattach the sensor pigtail connector to the vehicle harness connector.
9. Lower the vehicle.
10. Connect the negative battery cable.
11. Start the engine and ensure no Diagnostic Trouble Codes (DTC's) are set.

LOCATIONS

Generally, there are only five different locations in the exhaust system where O_2 sensors are positioned. The five locations have been given numbers and will be used in the accompanying charts to identify the positions of O_2 sensors in most 1995–99 vehicles.

Due to mid-year production changes or factory inconsistencies, all models may not be covered. If a vehicle you are servicing is not covered in the charts, inspect the exhaust system (while cold!) in the five general locations to find the applicable O_2 sensors.

➡On models equipped with dual exhaust systems, there may be up to 4 or 5 O_2 sensors in the exhaust system. Be sure to locate all of them before commencing with any testing or service.

The five locations are as follows:

- Location No.1—exhaust manifold or down pipe.
- Location No.2—both exhaust manifolds or down pipes of a V-type engine.
- Location No.3—exhaust collector.
- Location No.4—outlet of the catalytic converter.
- Location No.5—both the inlet and outlet of catalytic converter. This location is used to monitor the efficiency of the catalytic converter.

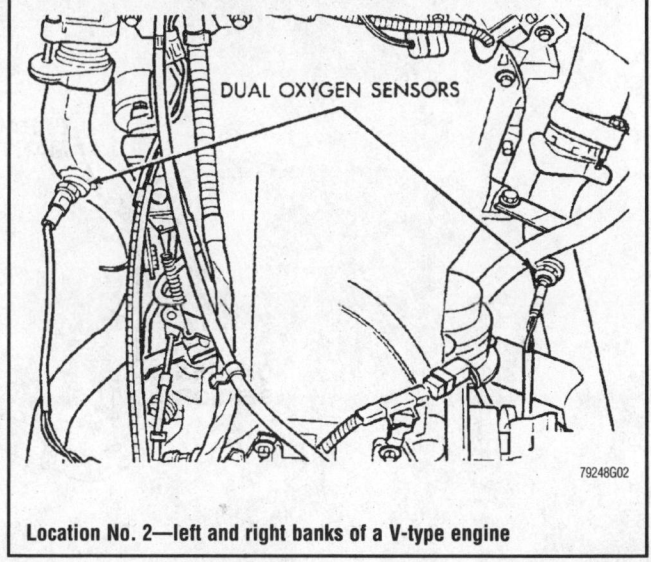

Location No. 2—left and right banks of a V-type engine

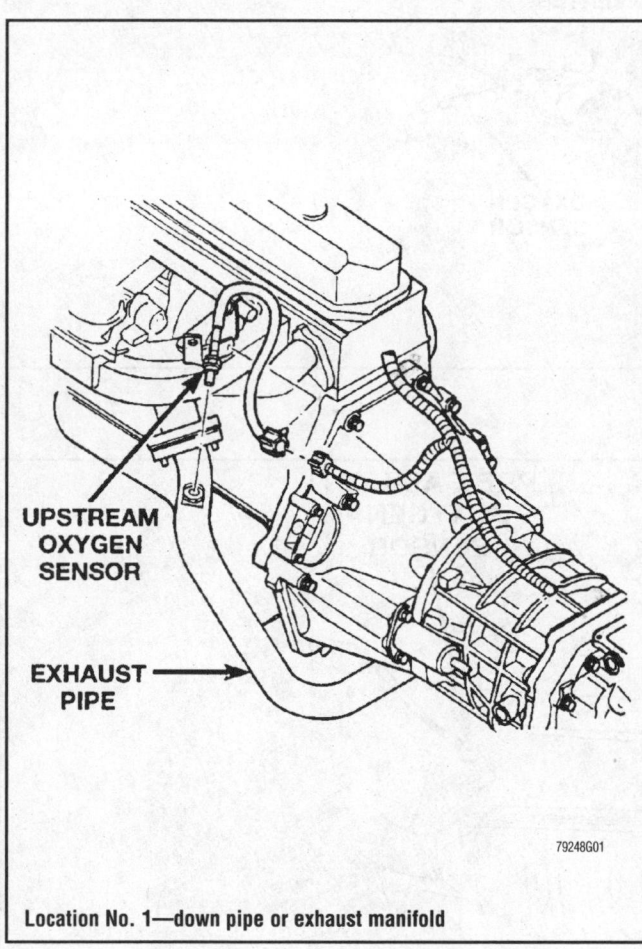

Location No. 1—down pipe or exhaust manifold

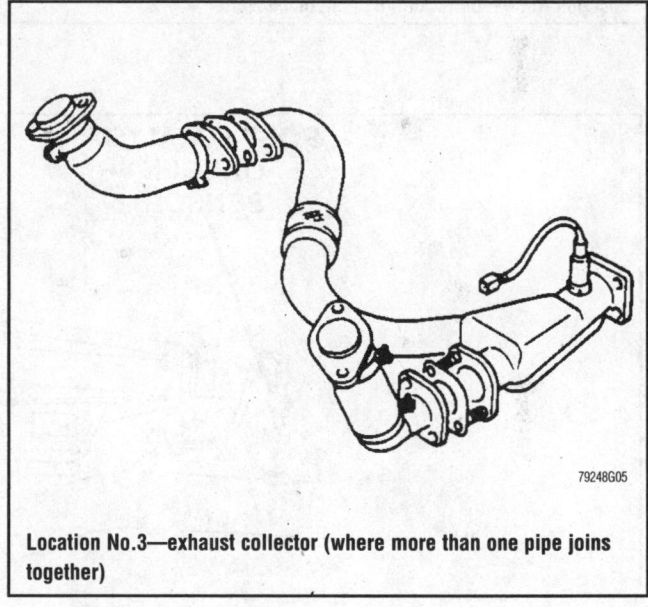

Location No.3—exhaust collector (where more than one pipe joins together)

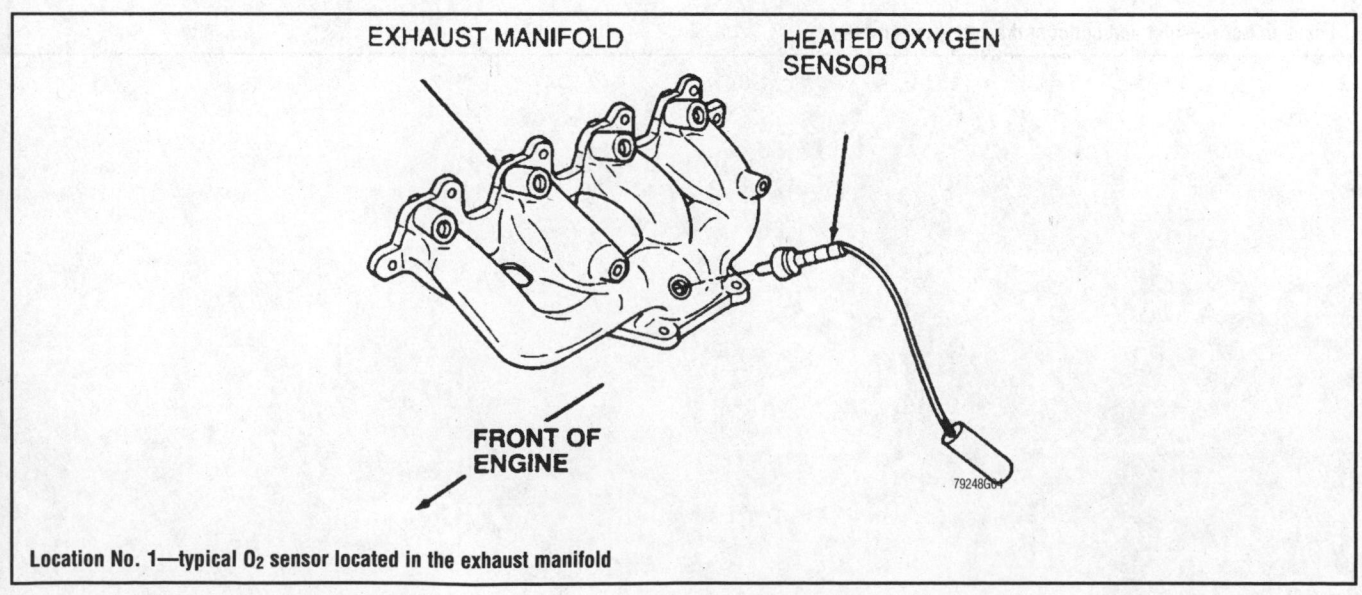

Location No. 1—typical O₂ sensor located in the exhaust manifold

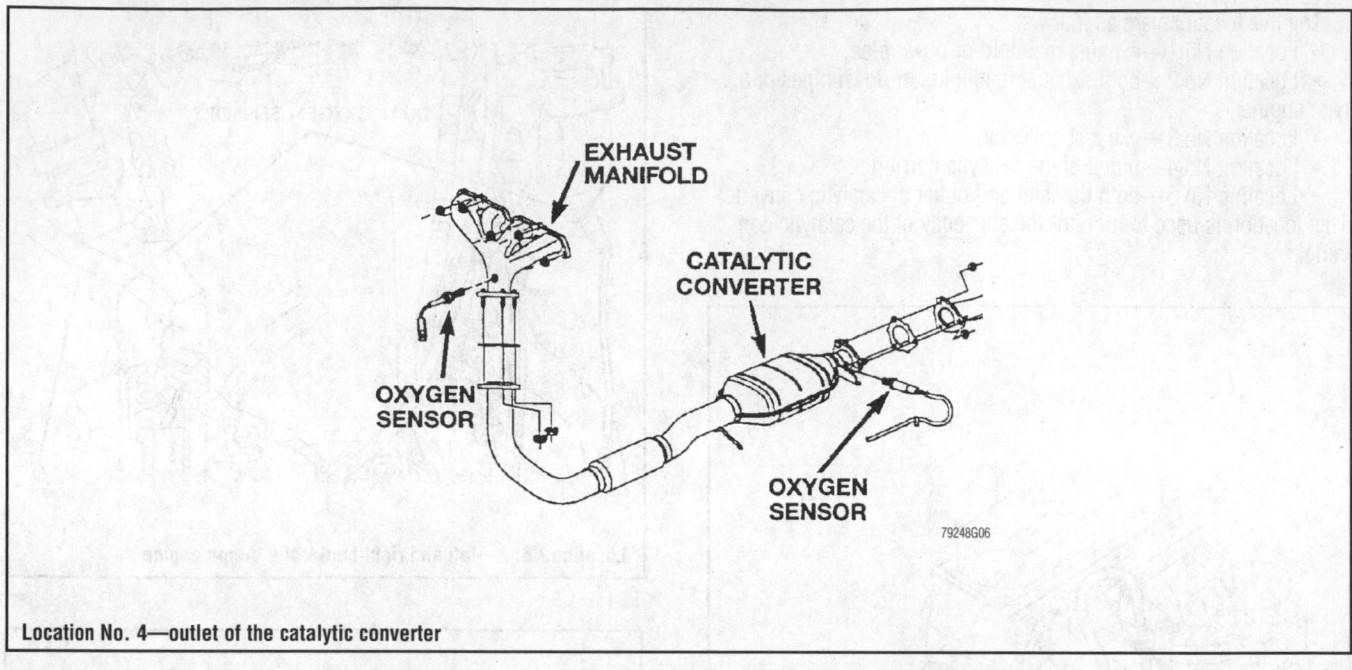

EXHAUST MANIFOLD

CATALYTIC CONVERTER

OXYGEN SENSOR

OXYGEN SENSOR

79248G06

Location No. 4—outlet of the catalytic converter

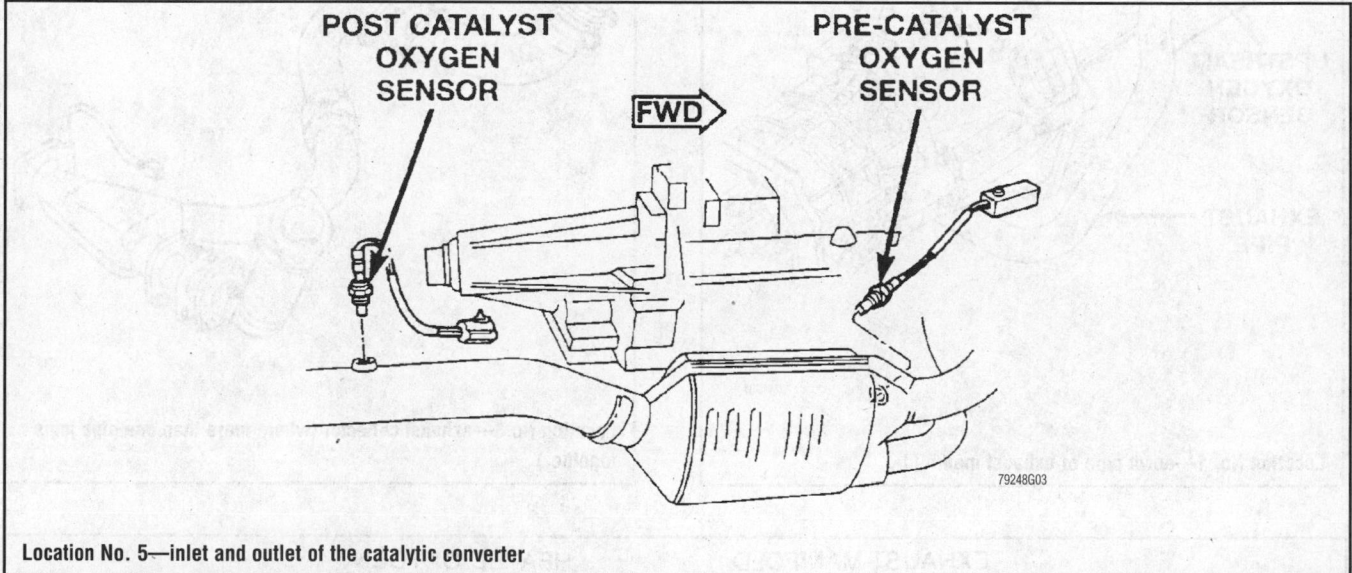

POST CATALYST OXYGEN SENSOR

PRE-CATALYST OXYGEN SENSOR

FWD

79248G03

Location No. 5—inlet and outlet of the catalytic converter

OXYGEN SENSOR LOCATIONS

Manufacturer Model Year	Engine	No. of Sensors	Locations
Acura			
Integra			
1995	1.8L	1	2
1996-99	1.8L	2	5
Legend			
	3.2L	2	2
NSX			
	3.0L	4	2, 4
2.2CL			
	2.2L	2	5
2.5TL			
	2.5L	2	3, 4
3.0CL			
	3.0L	2	3, 4
3.2TL			
	3.2L	3	2, 4
3.5RL			
	3.5L	3	2, 4
Audi			
All Models			
	1.8L	2	1, 4
	2.8L	2	2, 4
	3.7L	4	5
	4.2L	4	5
BMW			
All Models			
	1.8L	1	1
	1.9L	2	2
	2.5L	2	5
	2.8L	2	5
	3.0L	4	5
	3.2L	4	5
	4.0L	4	5
	4.4L	4	5
	5.4L	4	5
	5.6L	4	5
Chrysler Imports			
All Models			
1995	1.5L	1	1
	1.8L	1	1
	2.4L	1	1
1996	1.5L	2	1, 4
	1.8L	2	1, 4
	2.4L	2	3, 4

79238C01

OXYGEN SENSOR LOCATIONS

Manufacturer Model Year	Engine	No. of Sensors	Locations
Honda			
Accord			
1995	2.2L	1	1
1996-99	2.2L	2	1, 4
	2.7L	2	5
Civc			
1995	All	1	1
1996-99	All	2	1, 4
Del Sol			
1995	All	1	1
1996-97	All	2	1, 4
Prelude			
1995	All	1	1
1996-99	All	2	1, 4
Hyundai			
All Models			
	1.5L	2	1, 4
	1.8L	1	1
	2.0L	1 (2 CA)	1 (4 CA)
	3.0L	1 (2 CA)	1 (4 CA)
Infiniti			
G20			
1995	2.0L	1	1
1996	2.0L	2	1, 4
I30			
1996	3.0L	2	1, 4
1997-99	3.0L	3	2, 4
J30			
1995	3.0L	2	2
1996-97	3.0L	4	2, 4
Q45			
	4.5L	4	2, 4
Jaguar			
All Models			
	4.0L	2	5
	6.0L	4	2, 4
Kia			
Sephia			
1995	1.8L	1	1
1996-99	1.8L	2	1, 4
Lexus			
ES 300			
	3.0L	3	2, 4

79238C02

OXYGEN SENSOR LOCATIONS

Manufacturer Model Year	Engine	No. of Sensors	Locations
Lexus (cont.)			
GS/SC 300			
3.0L	3	2, 4
GS/LS/SC 400			
4.0L	4	2, 4
Mazda			
929			
3.0L	2	2
MX3			
1.6L	2, 4 CA	2, (4 CA)
MX6/626			
1995	2.0L	1	1
	2.5L	2	2
1996-97	2.0L	2	1, 4
	2.5L	3	2, 4
RX-7			
1.3L	1	1
Miata			
1.8L	1	1, 4
Millenia			
2.0L	4	2, 4
2.3L	4	2, 4
Protégé			
1.5L	2	1, 4
1.8L	2	1, 4
Mercedes-Benz			
All Models			
2.2L (111)	2	5
2.3L (111)	2	5
2.8L (104)	2	5
3.2L (104)	2	5
3.2L (112)	4	5
4.2L (119)	4	5
4.3L (113)	4	5
5.0L (119)	4	5
Mitsubishi			
Eclipse			
2.0L	2	1, 4
Diamante			
3.0L/3.5L	2	3, 4
Mirage			
1.8L	2	3, 4

79238C03

OXYGEN SENSOR LOCATIONS

Manufacturer Model Year	Engine	No. of Sensors	Locations
Mitsubishi (cont.)			
Galant			
	2.4L	2	3, 4
3000GT			
	3.0L	3	2, 4
Nissan			
300 ZX			
	3.0L	4	2, 4
240 SX			
	2.4L	2	1, 4
Sentra/200 SX			
	1.6L	2	1, 4
	2.0L	2	1, 4
Altima			
	2.4L	2	1, 4
Maxima			
1995-96	3.0L	2	1, 4
1997-99	3.0L	3	2, 4
Porsche			
911			
	3.6L	4	5
928			
	5.4L	1	3
968			
	3.0L	2	5
Boxster			
	2.5L	4	5
Saab			
900 and 9000			
	4 Cyl.	2	1, 4
	6 Cyl.	3	2, 4
9-3 and 9-5			
	4 Cyl.	2	1, 4
	6 Cyl.	3	2, 4
Subaru			
All Models			
1995	1.8L	1	3
	2.2L	2	3, 4
	3.3L	3	2, 4
1996-99	1.8L	2	3, 4
	2.2L	2	3, 4
	2.5L	2	3, 4
	3.3L	3	2, 4

79238C04

OXYGEN SENSOR LOCATIONS

Manufacturer Model Year	Engine	No. of Sensors	Locations
Suzuki			
All Models			
	All	2	1, 4
Toyota			
MR2			
	2.0L	1, 2 CA	1, (4 CA)
	2.2L	1, 2 CA	1, (4 CA)
Avalon			
	All	3	2, 4
Camry			
	2.2L	2	1, 4
	3.0L	3	2, 4
Celica			
	All	2	1, 4
Corolla			
1995	All	1 (2 CA)	1 (4 CA)
1996-99	All	2	1, 4
Paseo			
1995	1.5L	1, 2 CA	1, (4 CA)
1996-98	1.5L	2	1, 4
Supra			
	All	3	2, 4
Tercel			
	1.5L	2	1, 4
Volkswagen			
Beetle			
	2.0L	2	1, 4
Cabrio			
	2.0L	2	1, 4
Golf			
	2.0L	2	1, 4
GTI			
	2.0L	2	1, 4
	2.8L	2	1, 4
Jetta			
	1.8L	2	1, 4
	2.0L		
	2.8L	2	1, 4
Passat			
	1.8L	2	1, 4
	2.8L	2	1, 4
Volvo			
940			
	All	2	5

79238C05

OXYGEN SENSOR LOCATIONS

Manufacturer Model Year	Engine	No. of Sensors	Locations
Volvo (cont.)			
960			
	All	2	5
850			
	All	2	5
C70/S70/V70			
	All	2	5
S90/V90			
	All	2	5

79238C06

ELECTRIC COOLING FANS

9

ELECTRIC COOLING FANS

General Information

A basic vehicle cooling system consists of a radiator, water pump, thermostat, electric or engine-driven cooling fan, and hoses. Electric cooling fans are common on today's vehicles due to engine compartment space limitations or engine layout. Electric cooling fans operate in either a pusher or a puller capacity. A pusher type fan is typically mounted on the front of the radiator assembly and forces air through the radiator, whereas a puller type fan is mounted on the engine side of the radiator and draws air through the grill and radiator assembly. Vehicles that utilize a transversely-mounted engine will always be equipped with at least one electric cooling fan (most having two), because none of the engine pulleys are inline with the radiator air-flow.

There are generally two types of electric cooling fans: primary cooling fans and secondary cooling fans. Primary cooling fans are typically of the puller style. Vehicles that do not incorporate an engine-driven mechanical cooling fan will utilize a primary cooling fan. The secondary cooling fan, also known as a A/C condenser fan or auxiliary cooling fan by certain manufacturers, could be of either a pusher or a puller style. Vehicles equipped with A/C will either utilize the radiator cooling fan or a separate fan as the A/C condenser cooling fan (which performs the same function as an auxiliary cooling fan on vehicles with a primary mechanical fan). The engine control computer that receives inputs from various sensors in the engine compartment commonly controls electric cooling fans. The engine control computer receives inputs from the engine coolant temperature sensors and A/C system pressure switches, then actuates the necessary cooling fan relays to engage the applicable cooling fan for the condition. On models equipped with only one electric primary

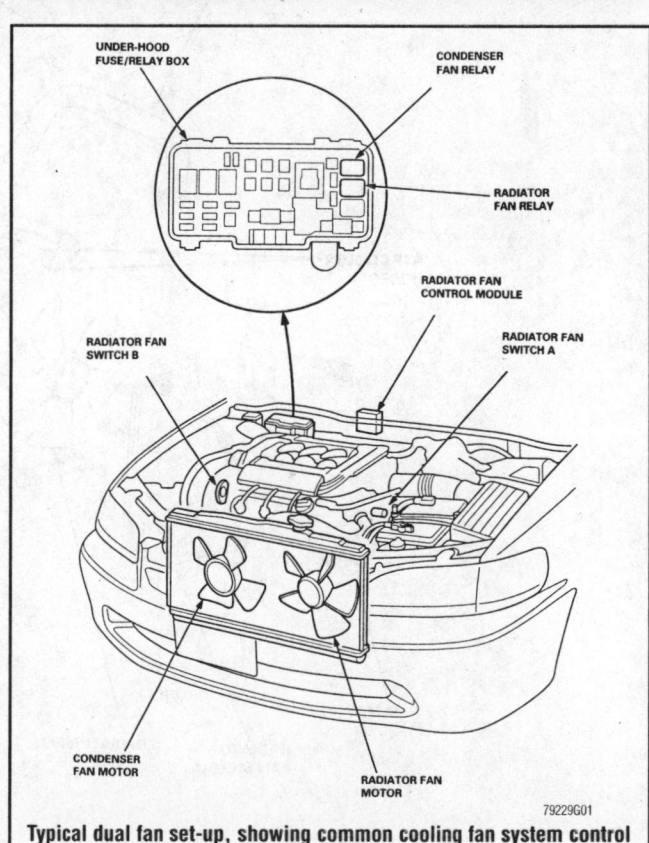

Typical dual fan set-up, showing common cooling fan system control components used on many vehicles with A/C

cooling fan, the fan can operate at two speeds: low speed and high speed. The low speed condition is enabled when the engine begins to heat up or when the A/C is engaged. As the engine demands more cooling, the cooling fan will be stepped-up to high speed.

Electric Cooling Fan Service

Due to the wide variety of vehicle manufacturers and suppliers of electric cooling fans it is almost impossible to cover every specific combination of cooling fan and model. The following procedures will cover the most common types of mountings and troubleshooting techniques.

REMOVAL & INSTALLATION

Puller Type

➡It may be simpler to remove the cooling fan(s) with the radiator as an assembly.

1. Disconnect the negative battery cable.
2. Inspect the cooling fan and take note of any wires, hoses or A/C lines which may hamper fan removal. Also at this time, decide whether it is necessary to remove the fan along with the radiator or not.

3. Position aside all wires, hoses and A/C lines for fan removal. It may not always be possible to create enough clearance for fan removal by simply moving these obstructions aside; often they must be disconnected. If any cooling system lines must be disconnected, drain and recycle the engine coolant. If any of the A/C lines must be disconnected, the A/C system will need to be discharged and evacuated by a MVAC-trained technician using an approved recovery machine.

4. Disengage the cooling fan wiring harness connector.
5. If the fan can be removed without the radiator, perform the following:

  a. Loosen the mounting fasteners. Usually there are two nuts or bolts along the top edge of the cooling fan shroud and either two retaining clips or bolts along the bottom edge.

  b. Carefully lift the fan up and out of the engine compartment, making sure that no wires or hoses get hung up on it.

6. If it is necessary to remove the radiator for fan removal, perform the following:

  a. Disconnect all cooling system hoses from it after draining the cooling system.

  b. Locate all of the radiator mounting fasteners (usually two or more nuts or bolts along the top, possibly two along the bottom).

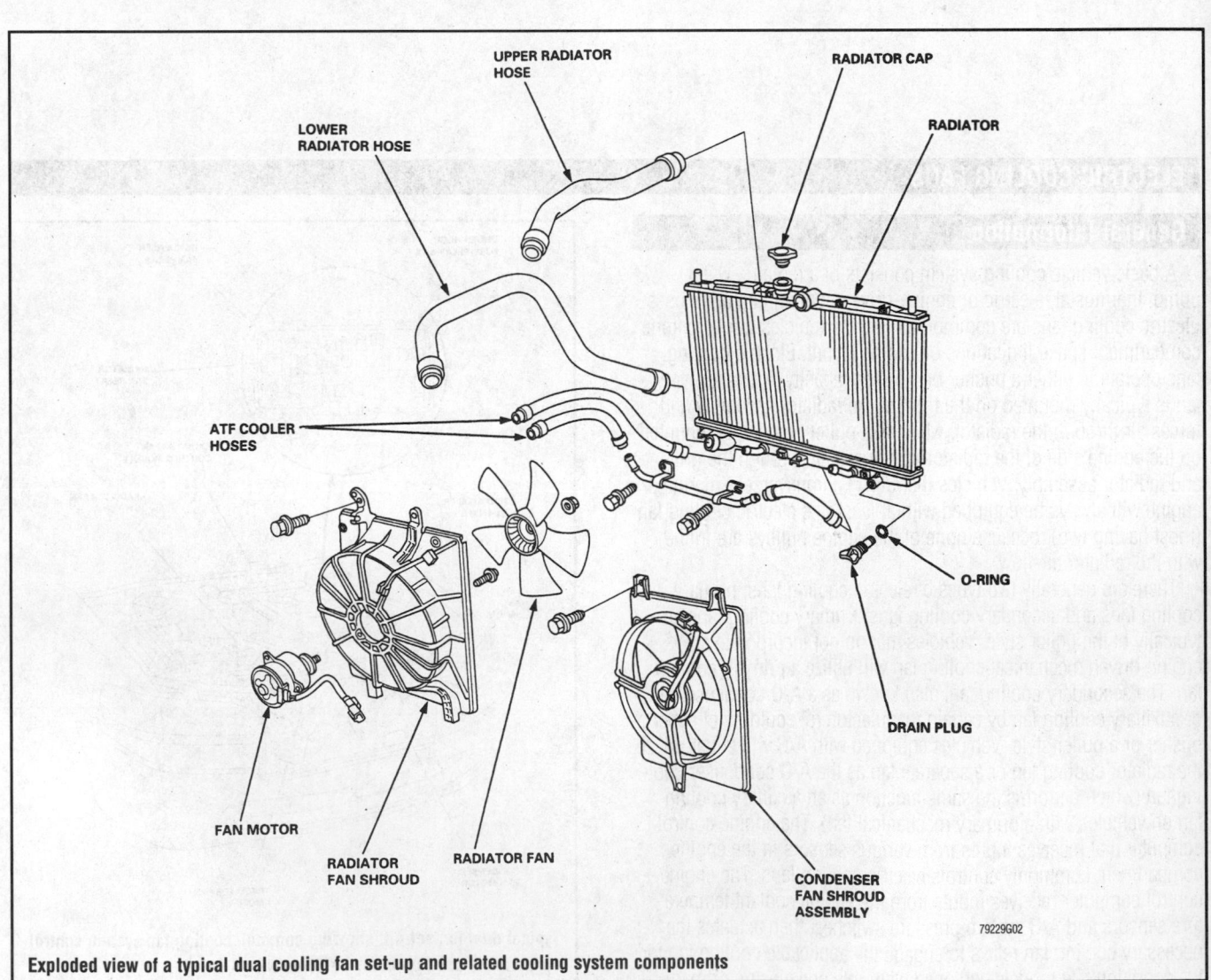

Exploded view of a typical dual cooling fan set-up and related cooling system components

79229G02

To remove a common puller type cooling fan, first detach any braces (1), wires (2) or other obstructions . . .

. . . and loosen all fan mounting fasteners

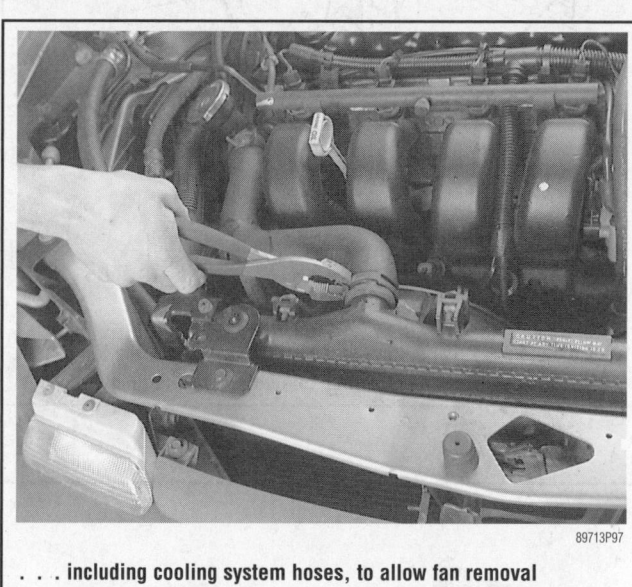

. . . including cooling system hoses, to allow fan removal

Separate the fan from the radiator . . .

Disengage the fan wiring connector(s) . . .

. . ., then lift the fan up and out of the engine compartment

➡**Quite a few radiators are secured along the bottom by two posts which fit into rubber grommets. The rubber grommets help isolate the radiator from harsh vibrations in the frame. If no nuts or bolts can be located along the bottom of the radiator, chances are that the radiator is secured with the posts and grommets.**

   c. Lift the radiator and cooling fan up and out of the engine compartment together.

   d. Separate the cooling fan from the radiator by removing the attaching fasteners.

To install:

7. If applicable, install the cooling fan on the radiator.

8. Install the cooling fan and shroud assembly (also the radiator if necessary). Tighten the fan shroud mounting bolts.

9. Reattach all wires, hoses and A/C lines as applicable. If the A/C lines were detached, the system must be evacuated and recharged by a MVAC-trained technician.

10. If drained, refill and bleed the cooling system.

11. Reattach the cooling fan electrical harness connector.

12. Connect the negative battery cable.

13. Start the engine and check for leaks.

14. Verify the operation of the cooling fan(s).

Pusher Type

Vehicles that utilize the pusher type of electric cooling fan, may require the removal of the grilles and/or upper radiator shroud in order to gain access the fasteners that mount the fan assembly in the vehicle.

1. Disconnect the negative battery cable.

2. Access the cooling fan.

3. Label and disconnect the cooling fan electrical harness.

➡**It may be necessary to loosen the mounting bolts for the A/C condenser to the body**

4. Remove the fasteners that mount the cooling fan to the A/C condenser or radiator.

5. Lift the cooling fan out of the vehicle.

To install:

6. Insert the cooling fan into the vehicle.

7. Mount the cooling fan to the A/C condenser or radiator

8. Connect the cooling fan electrical harness.

9. If removed, install any shrouding or grills.

10. Connect the negative battery cable.

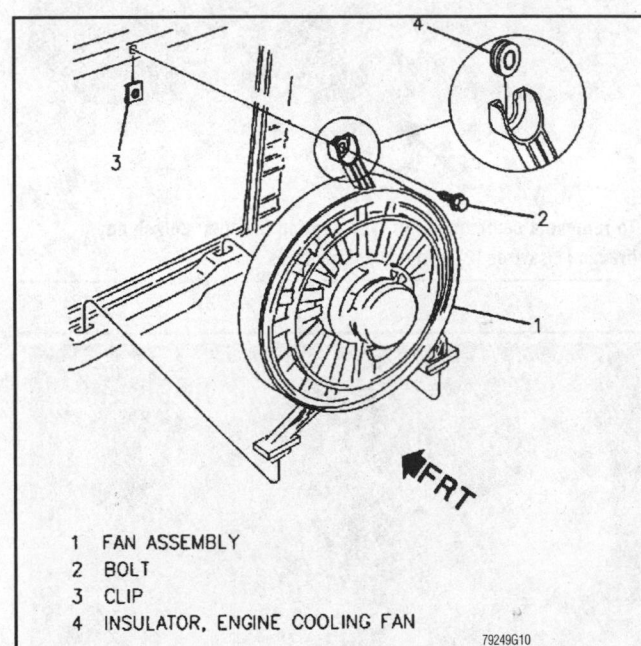

1   FAN ASSEMBLY
2   BOLT
3   CLIP
4   INSULATOR, ENGINE COOLING FAN

79249G10

Notice the slots in the bottom of the radiator, in which the fan housing posts rest—common mounting of a puller type cooling fan

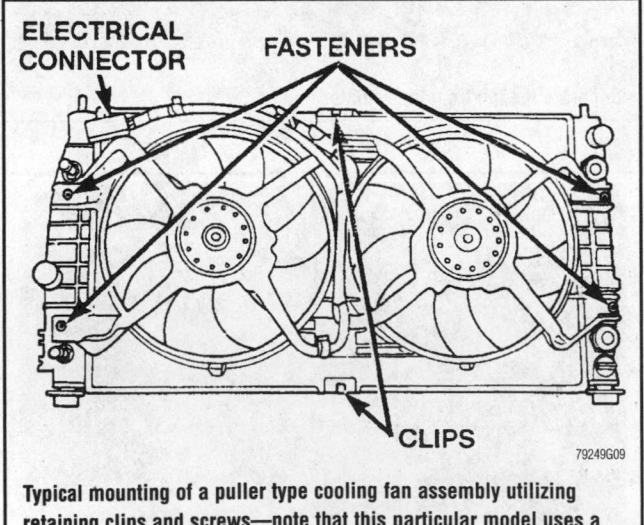

79249G09

Typical mounting of a puller type cooling fan assembly utilizing retaining clips and screws—note that this particular model uses a dual puller fan setup

79249G11

This fan mounts to the fan shroud, then the shroud mounts to the radiator—molded clips in the radiator hold the bottom in place and screws at the top

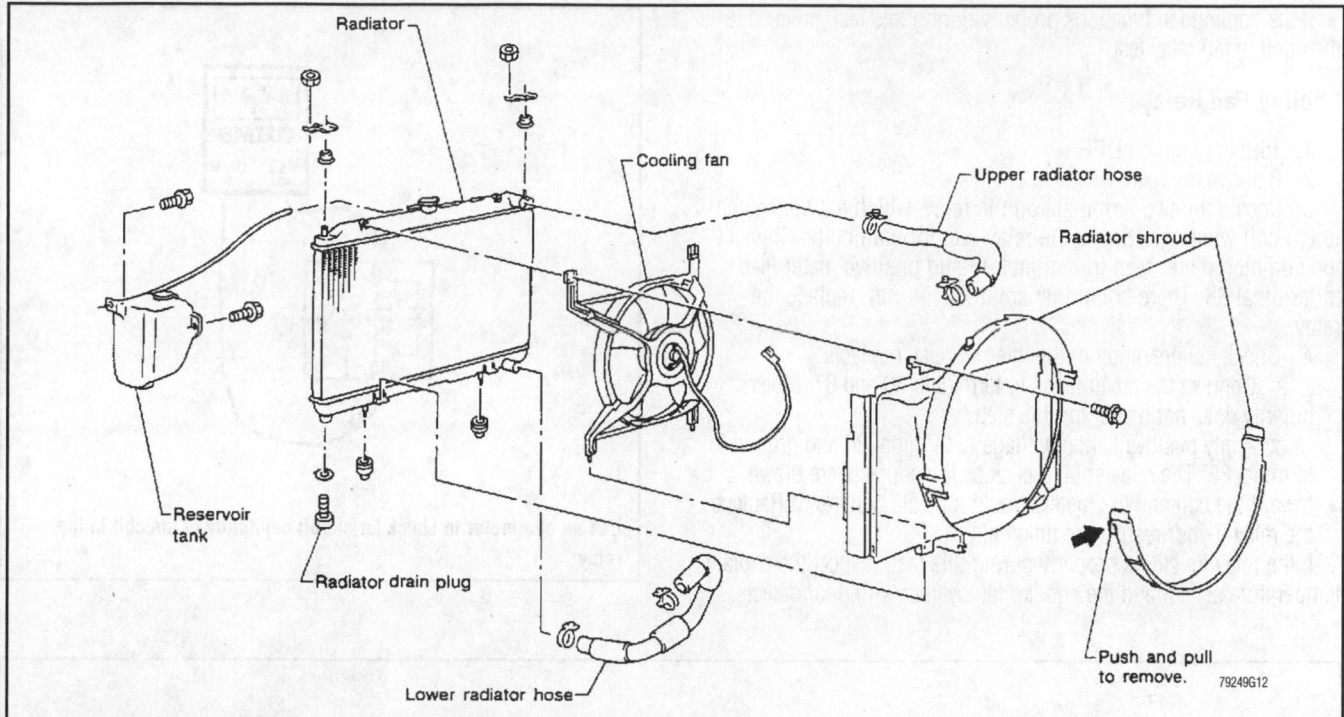

Typically the cooling fan is rubber mounted to isolate vibration and noise—usually the rubber grommets are located at the mount, verify their position before installation

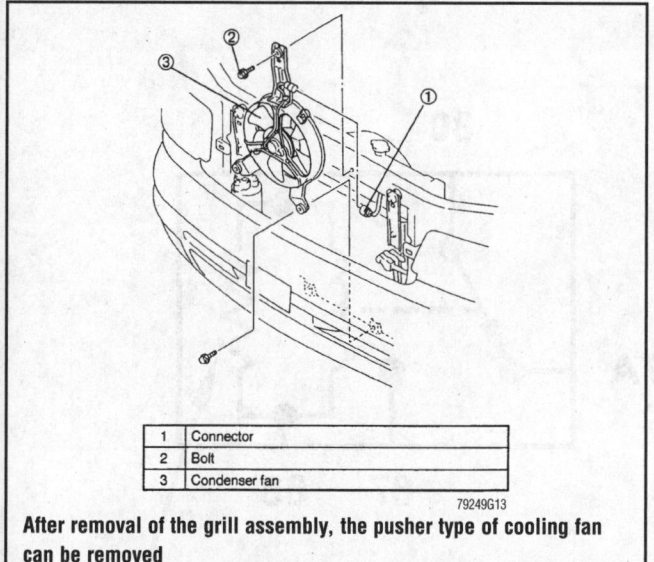

1	Connector
2	Bolt
3	Condenser fan

79249G13

After removal of the grill assembly, the pusher type of cooling fan can be removed

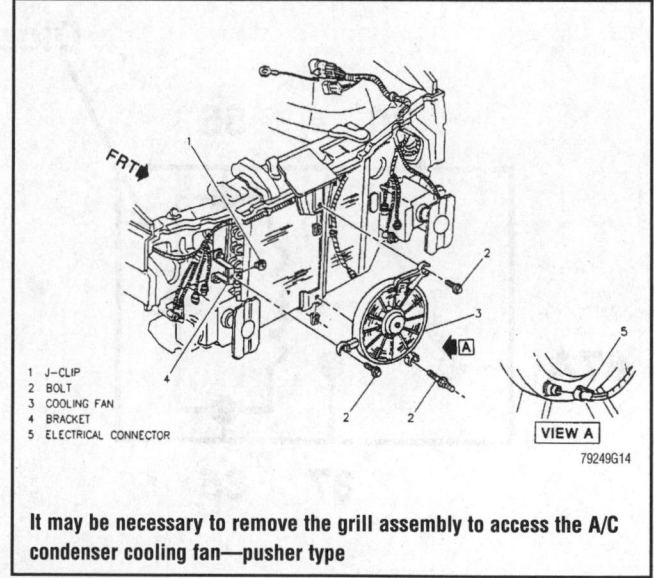

1 J-CLIP
2 BOLT
3 COOLING FAN
4 BRACKET
5 ELECTRICAL CONNECTOR

VIEW A

79249G14

It may be necessary to remove the grill assembly to access the A/C condenser cooling fan—pusher type

TROUBLESHOOTING

When diagnosing an inoperative cooling fan it may be necessary to use a diagnostic scan tool to monitor engine coolant temperature and the engine control computer.

1. Perform a visual inspection of the cooling fan. If the fan does not turn with ease, the fan motor is seized and needs to be replaced.

2. Check all the fuses and fusible links related to the cooling fan circuit.

3. Check the integrity of the electrical connections related to the cooling fan circuit.

4. Check the cooling fan motor.

5. Check the relays associated with the cooling fan circuit.

6. Using a scan tool, determine if the engine control computer is calling for the fan to activate.

Cooling Fan Motor

1. Disconnect the negative battery cable.

2. Disengage the cooling fan motor connector.

3. Identify and label the ground and the power terminals of the cooling fan connector using the wiring diagrams provided.

4. Using jumper leads with a fuse in series, apply battery voltage to the appropriate terminals of the cooling fan.

5. The cooling fan should operate. If not, replace the cooling fan.

If the cooling fan functions properly during this test, proceed to the cooling fan relay test.

Cooling Fan Relay

1. Turn the ignition **OFF**.
2. Remove the relay.
3. Locate the two terminals on the relay, which are connected to the coil windings. Check the relay coil for continuity. Connect the common meter lead to terminal 85 and positive meter lead to terminal 86. There should be continuity. If not, replace the relay.
4. Check the operation of the internal relay contacts.

   a. Connect the meter leads to terminals 30 and 87. Meter polarity does not matter for this step.

   b. Apply positive battery voltage to terminal 86 and ground to terminal 85. The relay should click as the contacts are drawn toward the coil and the meter should indicate continuity. Replace the relay if your results are different.

If the relay functions properly during this test, inspect the coolant temperature sensor and the cooling fan system wiring for defects.

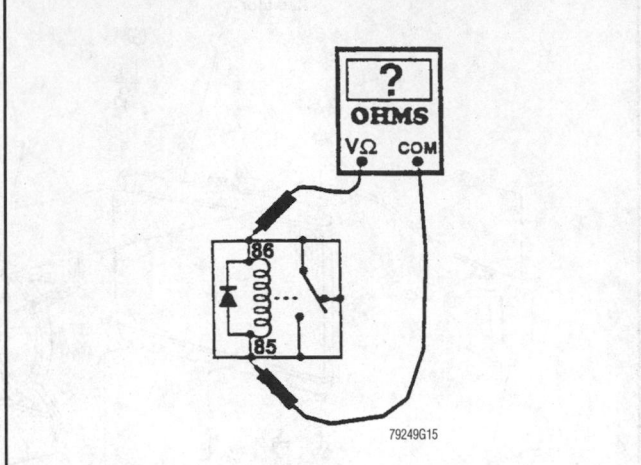

Use an ohmmeter to check for circuit continuity of the coil in the relay

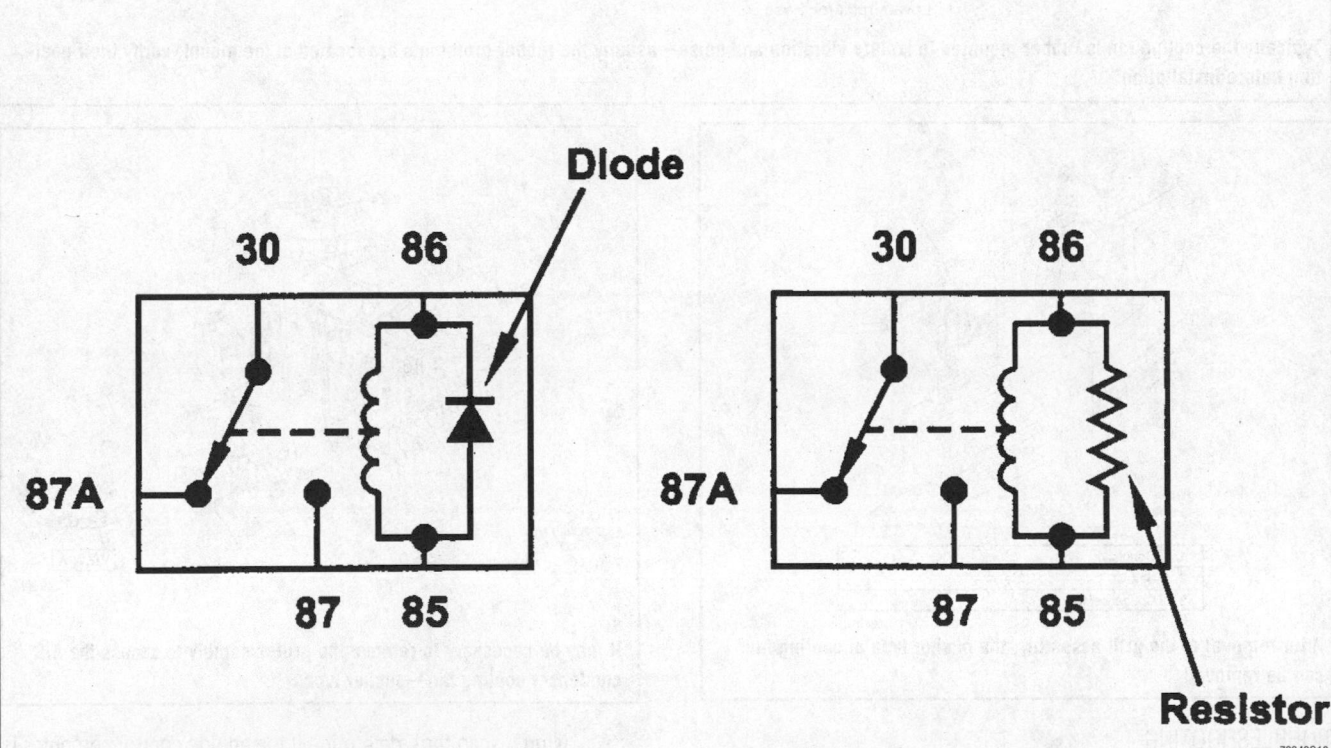

Terminal identification of the most common types of relays. Diodes and resistors in the relay prevent voltage spikes induced when the current is removed from the coil from damaging electronic components

COOLING FAN DIAGRAM INDEX

MANUFACTURER MODEL AND ENGINE		DIAGRAM
Acura		
Integra, GSR 1.8L		1
2.5TL		2
Legend Coupe 3.2L		3
3.0L		4
3.5RL		5
2.2CL		6
NSX 3.0L / 3.2L		7
Audi		
90 2.8L	1995 with automatic A/C and A/T	11
	1995 with automatic A/C and M/T	10
	1995 with manual A/C and M/T	8
	1995 with manual A/C and A/T	9
A4 1.8L		12
A4 2.8L		13
A6 2.8L	1995 with automatic A/C and A/T	11
	1995 with automatic A/C and M/T	10
	1995 with manual A/C and M/T	8
	1995 with manual A/C and A/T	9
	1996-99	13
Cabriolet 2.8L	1995-96	9
	1997-99	13
S6 2.8L		13
BMW		
318i 1.8L / 2.5L	1995	14
318is-c, 320i, 325i-c, M3 2.8L	1996-99	15
525i 2.5L / 3.0L / 4.0L		16
318is-c, 320i, 325i-c, M3 1.9L / 2.8L	1996-99	17
318ts 1.9L / 2.8L	1996-99	18
Chrysler Imports		
Summit Wagon, Expo 1.8L / 2.4L		19
Summit 1.5L		20
Colt, Summit 1.8L		21
Honda		
Accord 2.2L / 2.7L		24
Civic 1.5L / 1.6L	1995	22
	1996-99	23
Del Sol 1.6L		23
Prelude 2.2L / 2.3L	1995	25
	1996	26
	1997-99	27
Hyundai		
Accent 1.5L	1995	28
	1996-99	31

79239C01

COOLING FAN DIAGRAM INDEX

79239C02

COOLING FAN DIAGRAM INDEX

MANUFACTURER MODEL AND ENGINE		DIAGRAM
Mitsubishi (cont.)		
Eclipse 2.0L Turbo A/T		66
Eclipse 2.4L M/T		65
Eclipse 2.4L A/T		66
Galant 2.4L		67
Mirage 1.5L/1.8L	1995-96 1.5L	68
	1995-96 1.8L	69
	1997-99	70
Nissan		
200SX, Sentra 1.6L / 2.0L M/T		71
200SX, Sentra 1.6L / 2.0L A/T		72
240SX 2.4L		73
300ZX 3.0L		74
Altima 2.4L		75
Maxima 3.0L		76
Porsche		
928 GTS		77
968		77
911 Carrera, Carrera 4, Turbo		78
Boxster		79
Saab		
900 2.0L / 2.3L / 2.5L		80
9000 2.3L		81
9000, 9-5 3.0L		82
Subaru		
Impreza 1.8L / 2.2L	1995-96	83
Impreza 2.2L / 2.5L	1997-99	84
Legacy 2.0L	1995	85
Legacy 2.2L / 2.5L	1995-97	86
	1998-99	87
Legacy Brighton, Outback 2.2L / 2.5L		88
SVX 3.3L		89
Suzuki		
Esteem 1.6L		90
Swift 1.0L / 1.3L		91
Toyota		
Avalon 3.0L		92
Camry 2.2L/3.0L		
	1995-96 2.2L	93
	1995-96 3.0L	94
	1997-99	95
Celica 1.8L / 2.2L		96
Corolla 1.8L		97
Paseo, Tercel 1.5L		98
Supra 3.0L		99

79239C03

COOLING FAN DIAGRAM INDEX

MANUFACTURER MODEL AND ENGINE	DIAGRAM
Volkswagen	
Beetle, Cabrio, Golf, Jetta 2.0L	100
Beetle, Golf, Jetta 1.9L Turbo Diesel	101
GTI 2.0L	100
GTI, Jetta 2.8L	102
Passat 1.9L Turbo Diesel	103
Passat 2.0L	104
Passat 2.8L	105
Volvo	
850, C70, V70 2.3L Turbo / 2.4L Diesel	106
940 2.3L	107
960, S90, V90 2.9L	108

79239C04

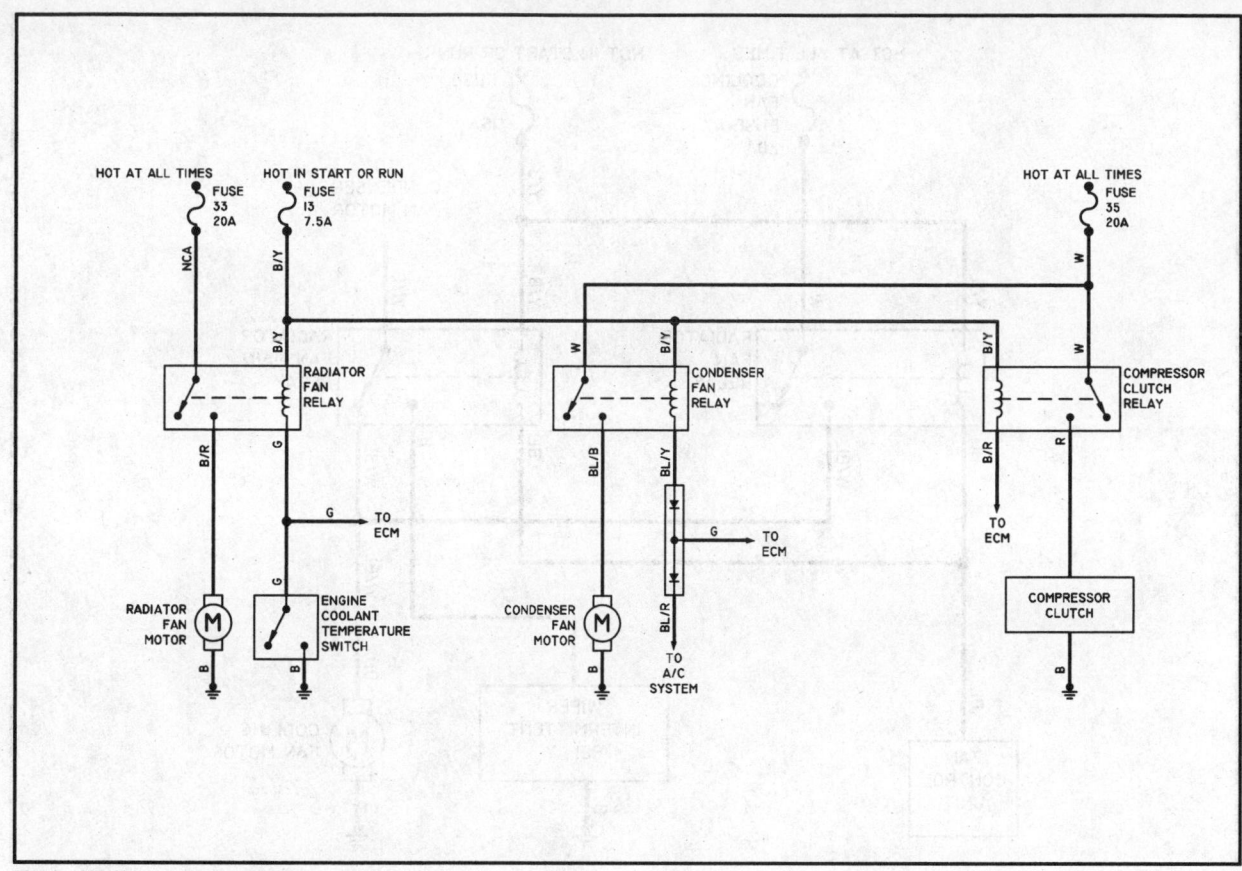

DIA. 1- 1995-99 Acura Integra, GSR 1.8L

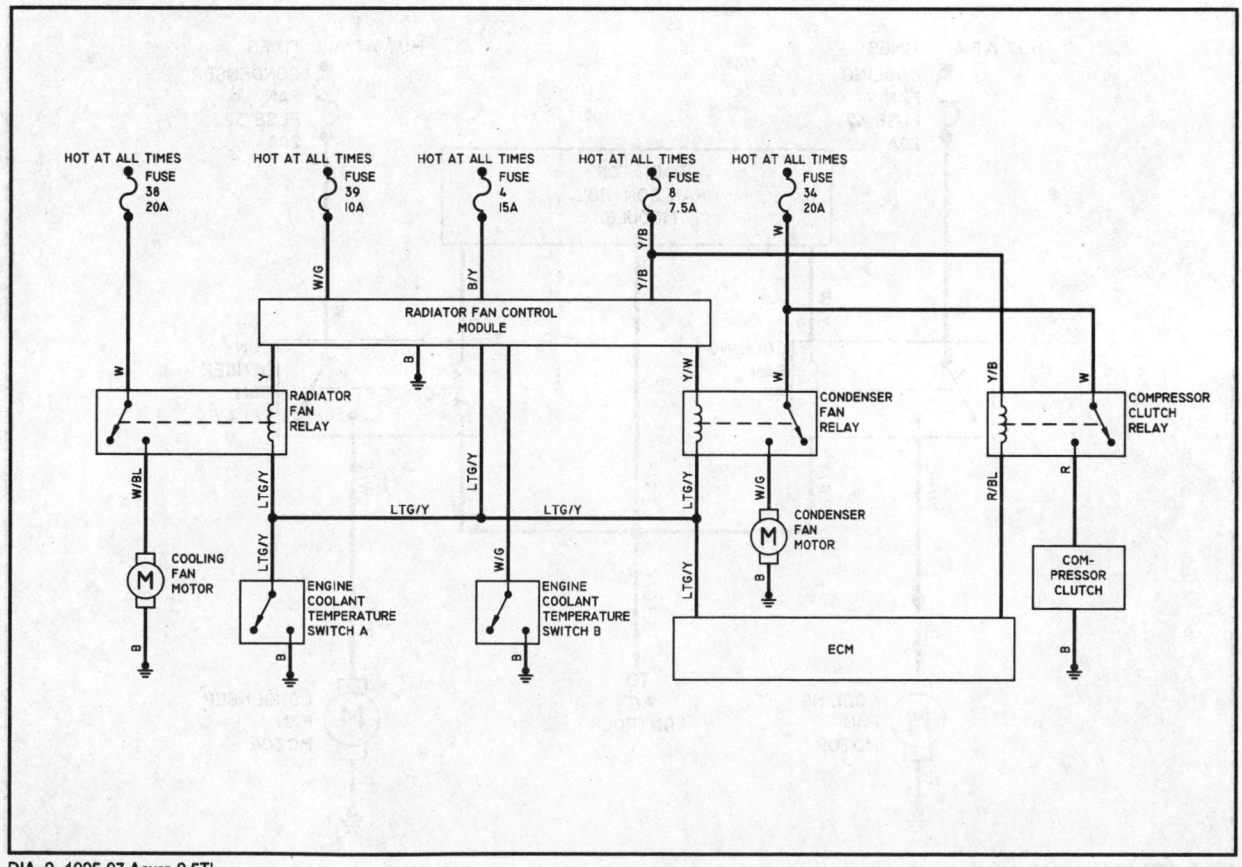

DIA. 2- 1995-97 Acura 2.5TL

79239W01

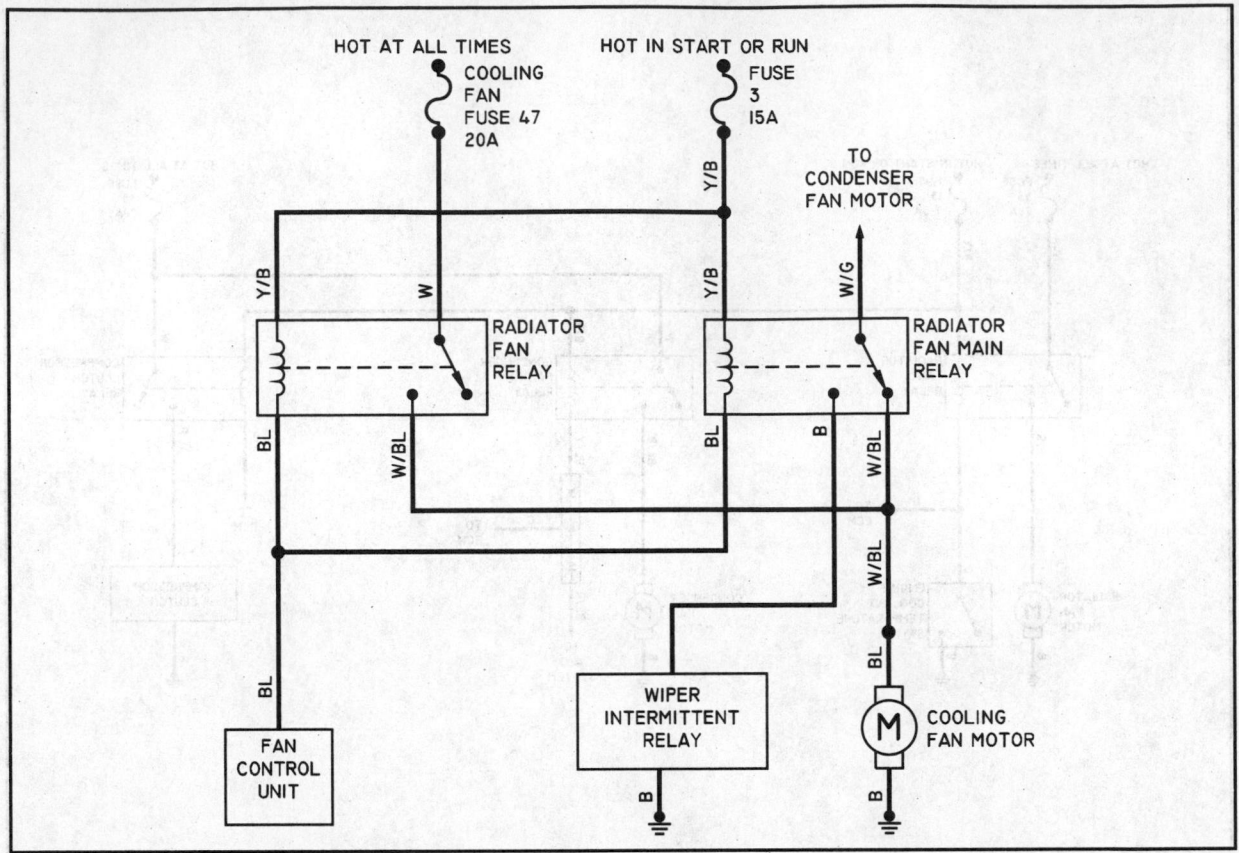

DIA. 3- 1995 Acura Legend Coupe 3.2L

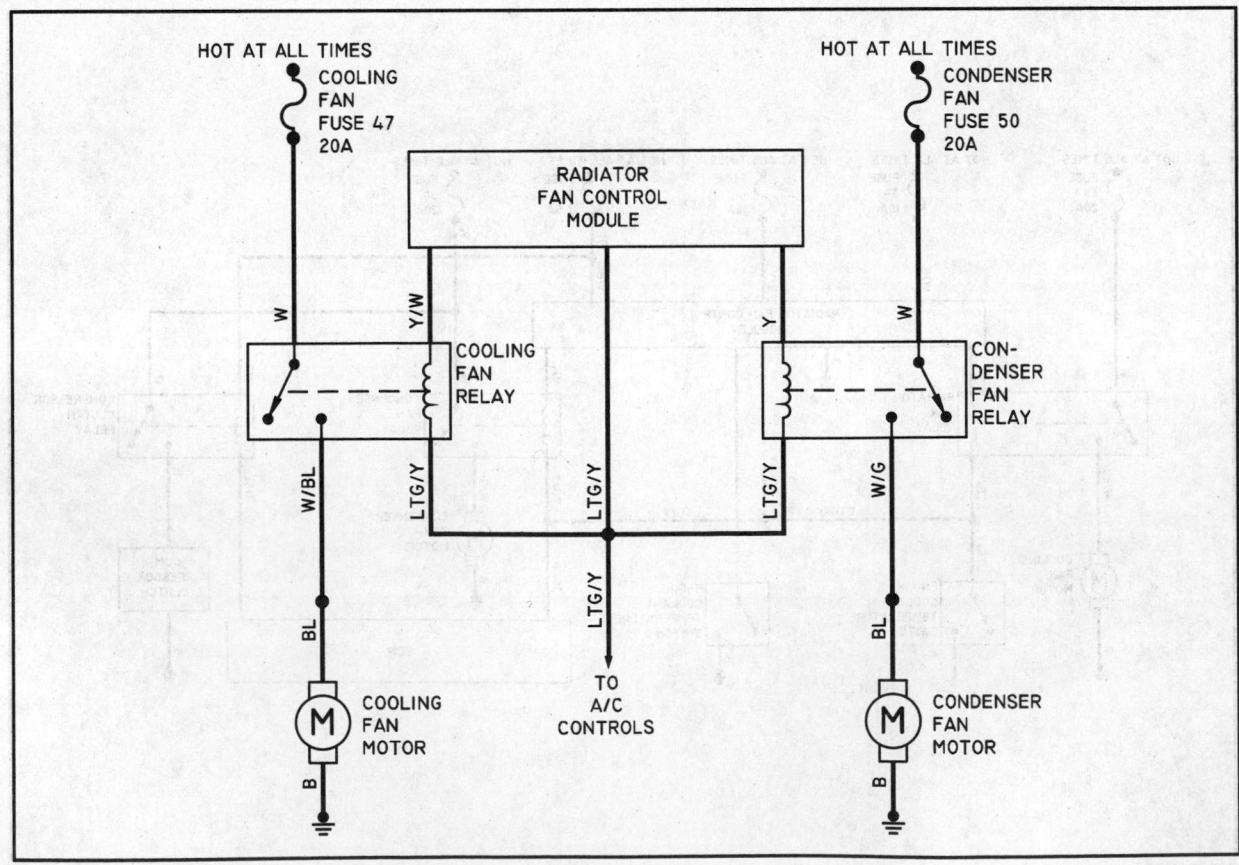

DIA. 4- 1995-97 Acura 3.0L

79239W02

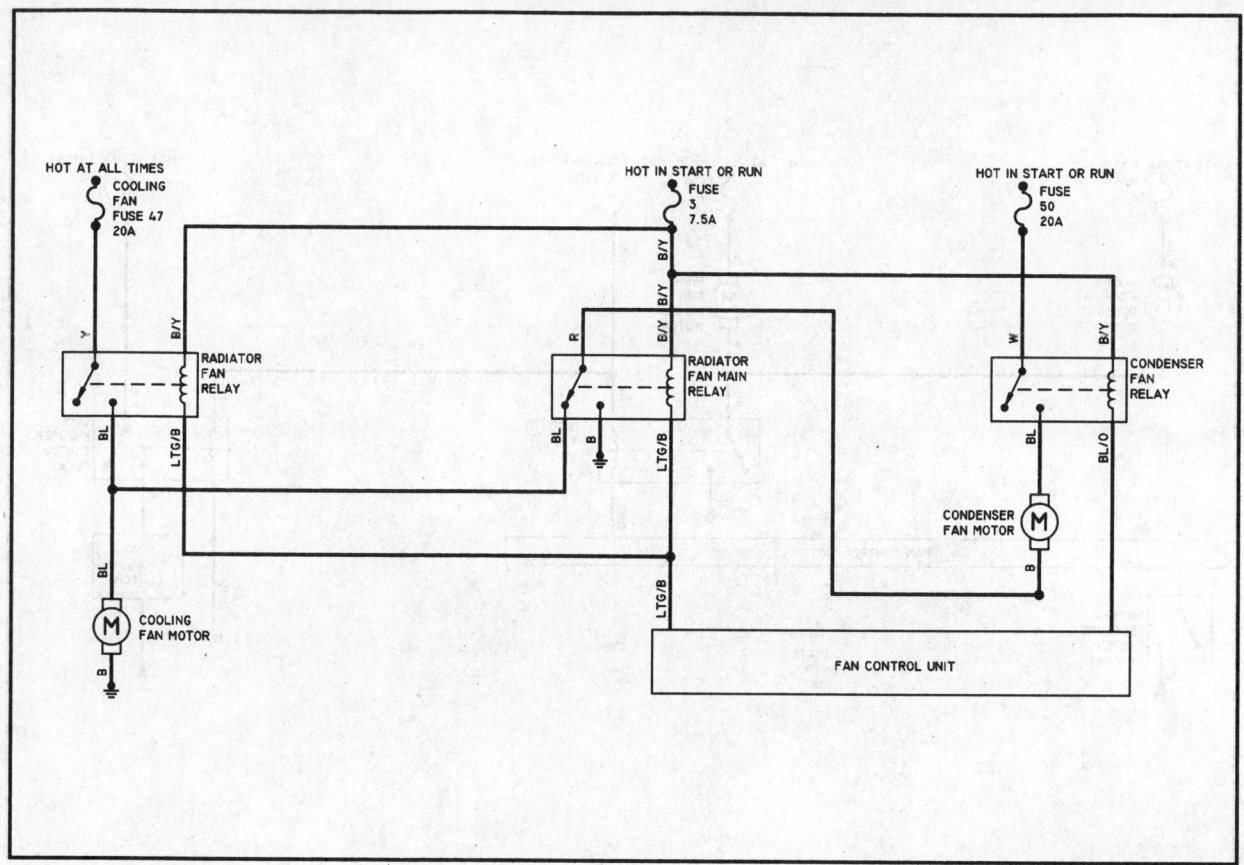

DIA. 5- 1996-99 Acura 3.5RL

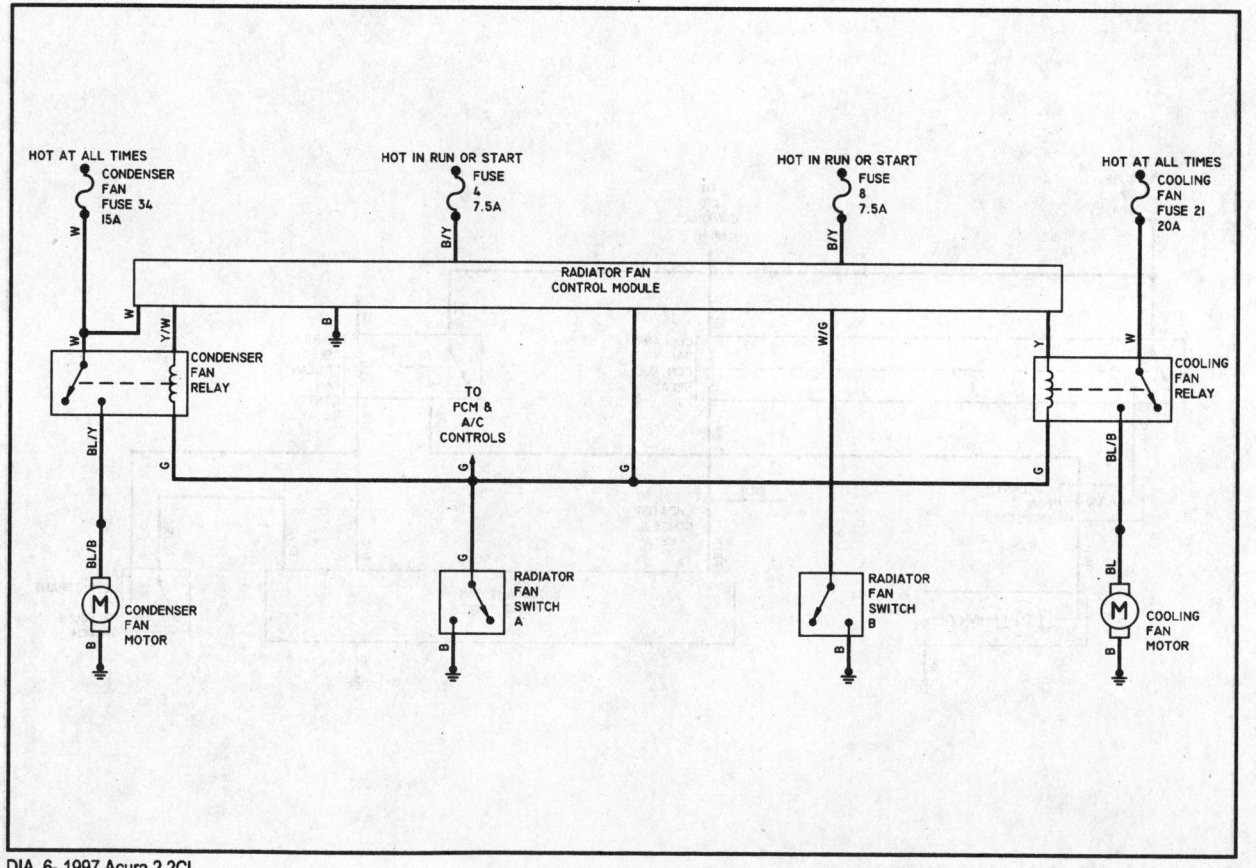

DIA. 6- 1997 Acura 2.2CL

79239W03

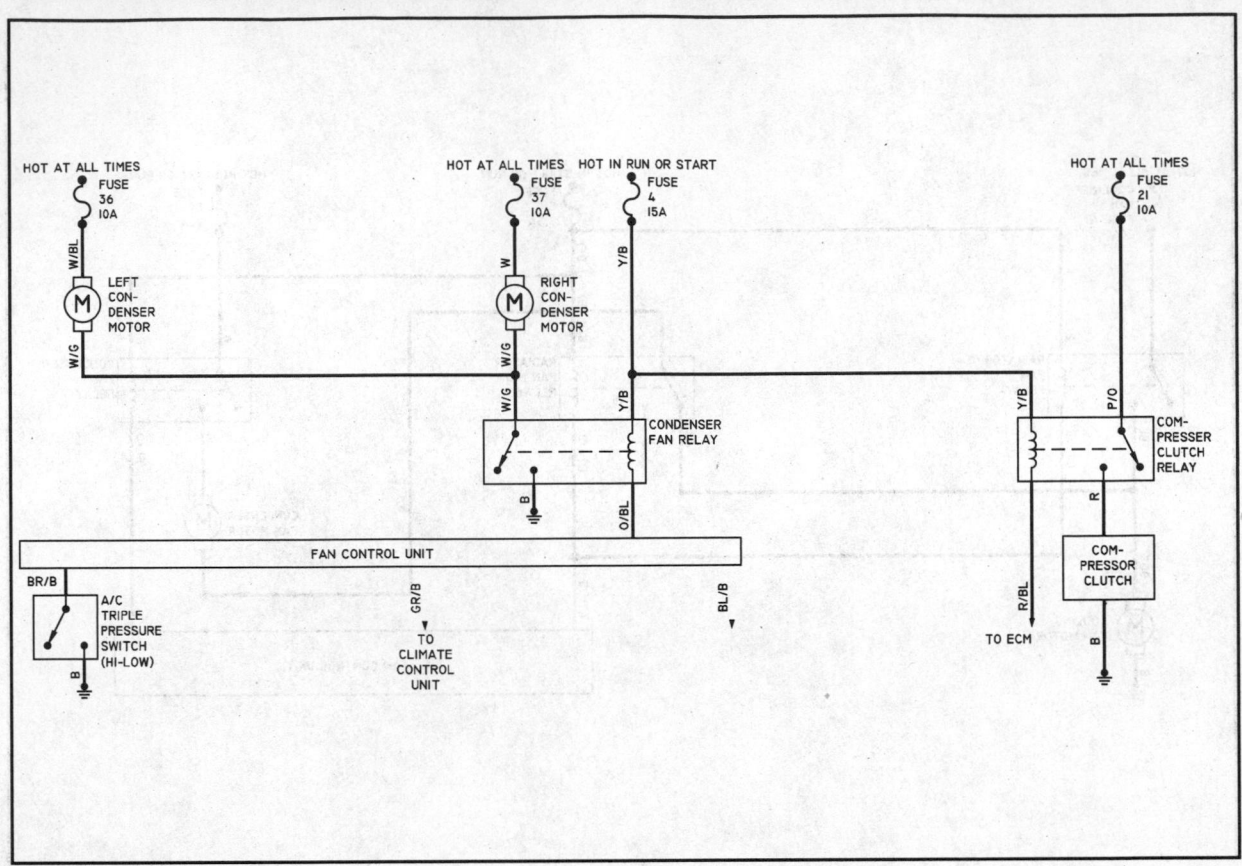

DIA. 7- 1995-99 Acura NSX 3.0L / 3.2L

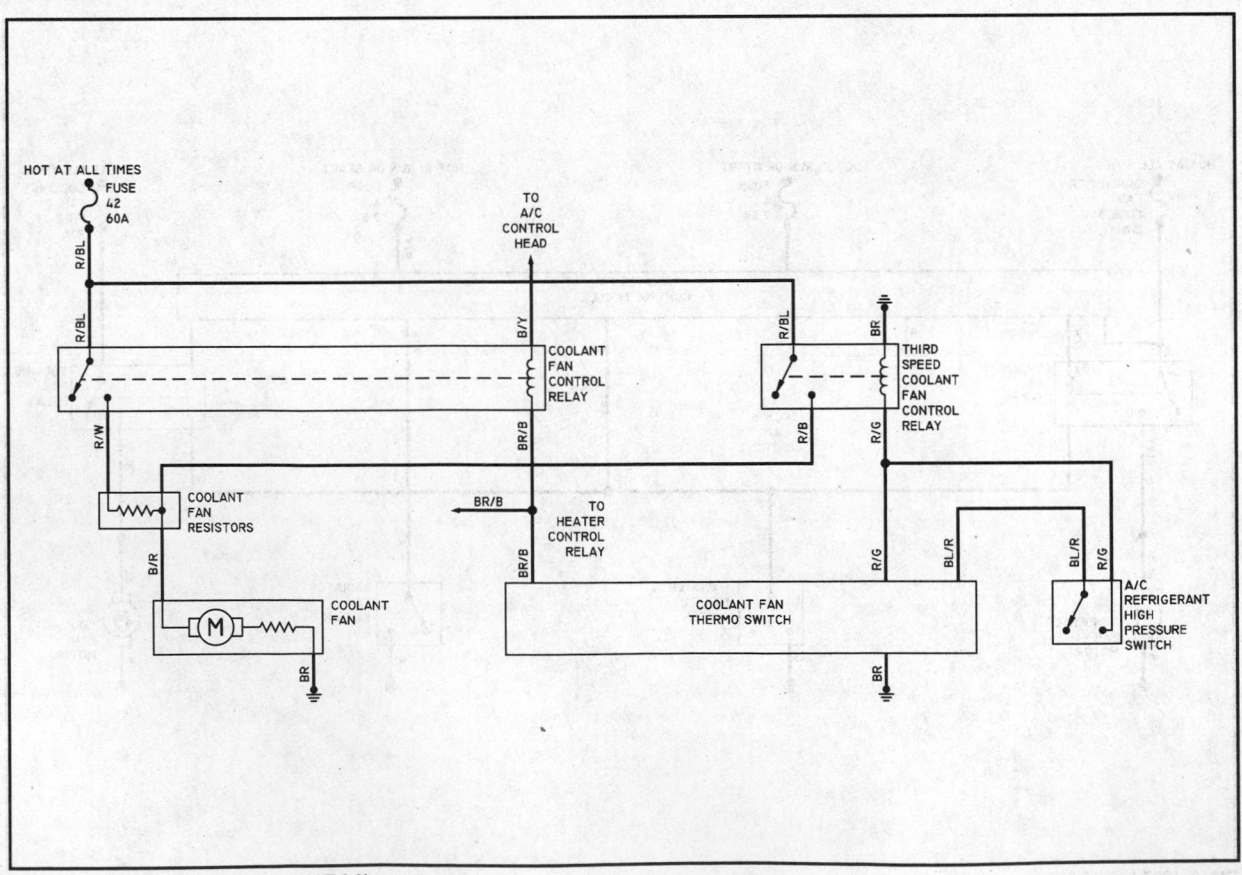

DIA. 8- 1995 Audi 90, A6 Manual A/C w/ M/T 2.8L

79239W04

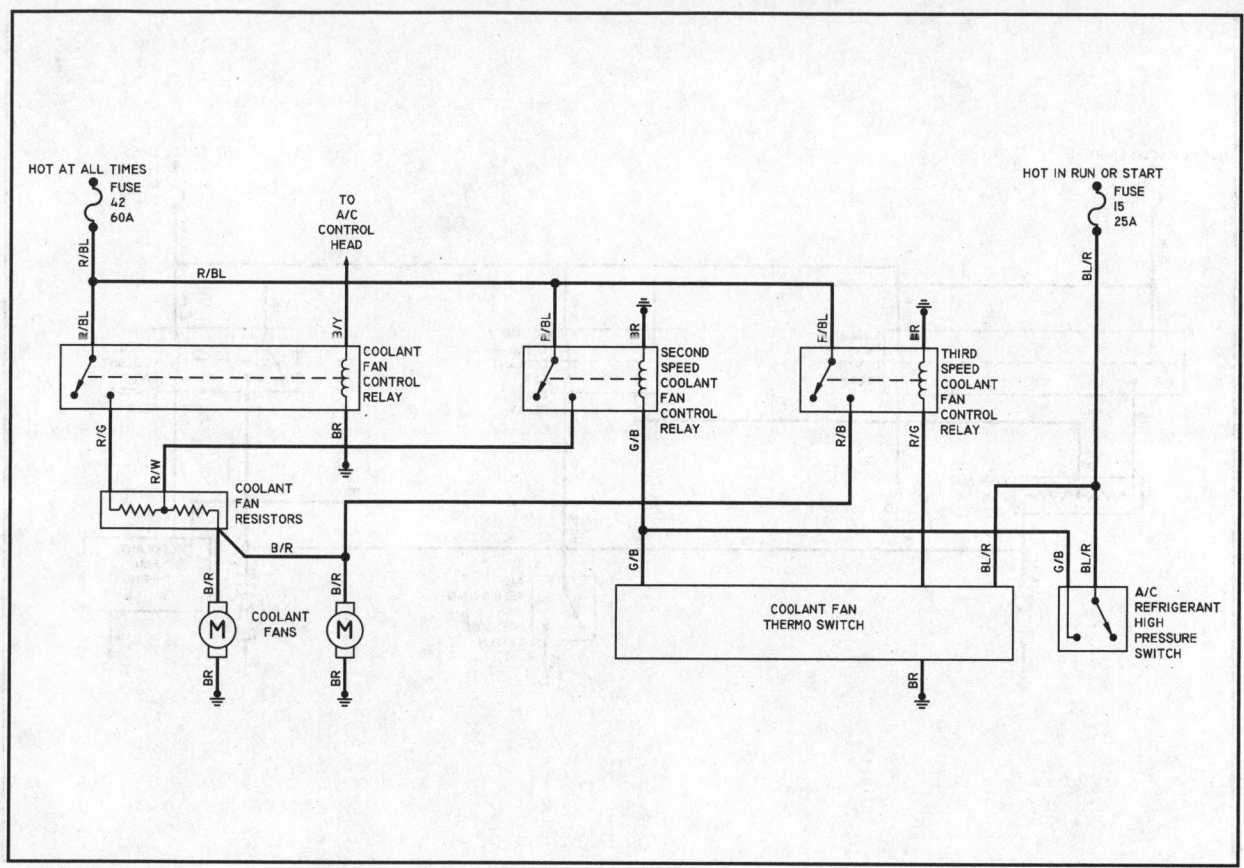

DIA. 9- 1995 Audi 90, A6 Manual A/C w/ A/T 2.8L
1995-96 Cabriolet 2.8L

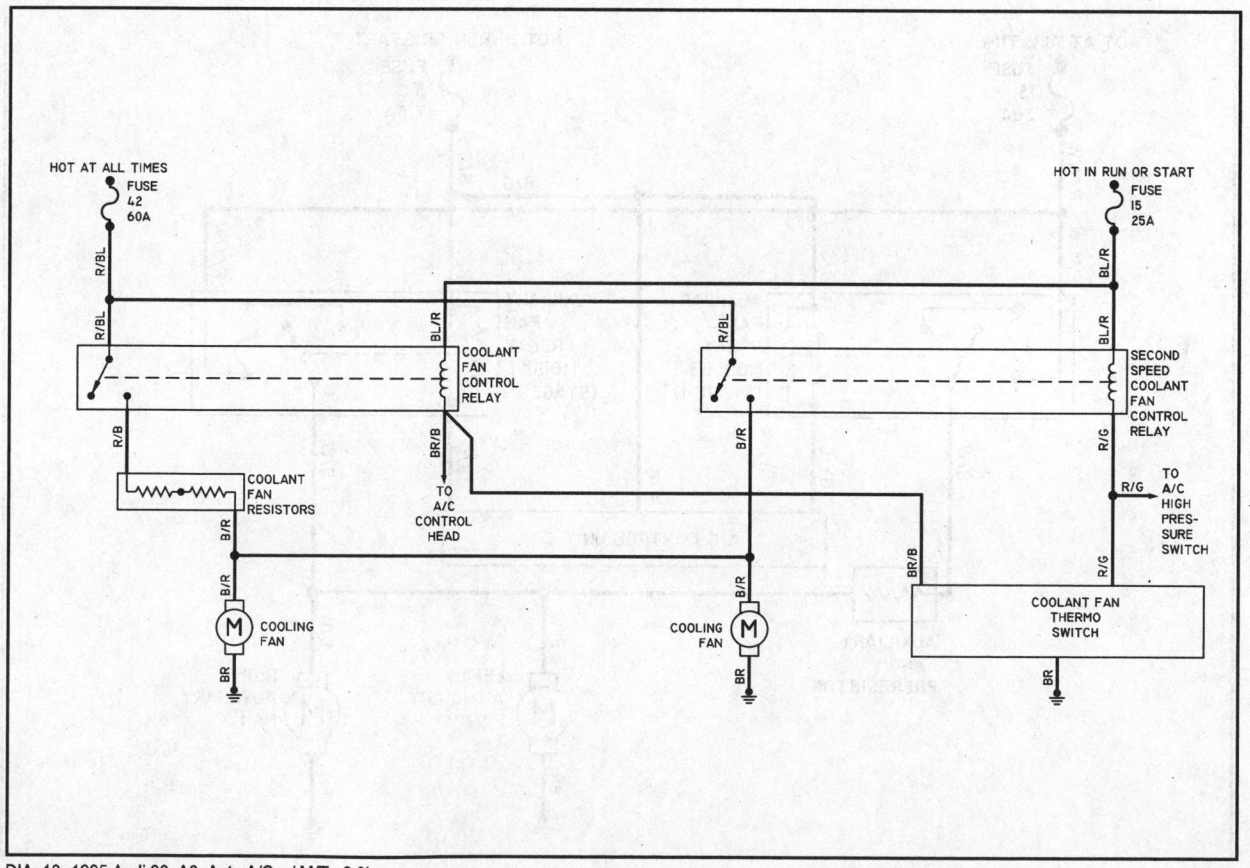

DIA. 10- 1995 Audi 90, A6 Auto A/C w/ M/T 2.8L

79239W05

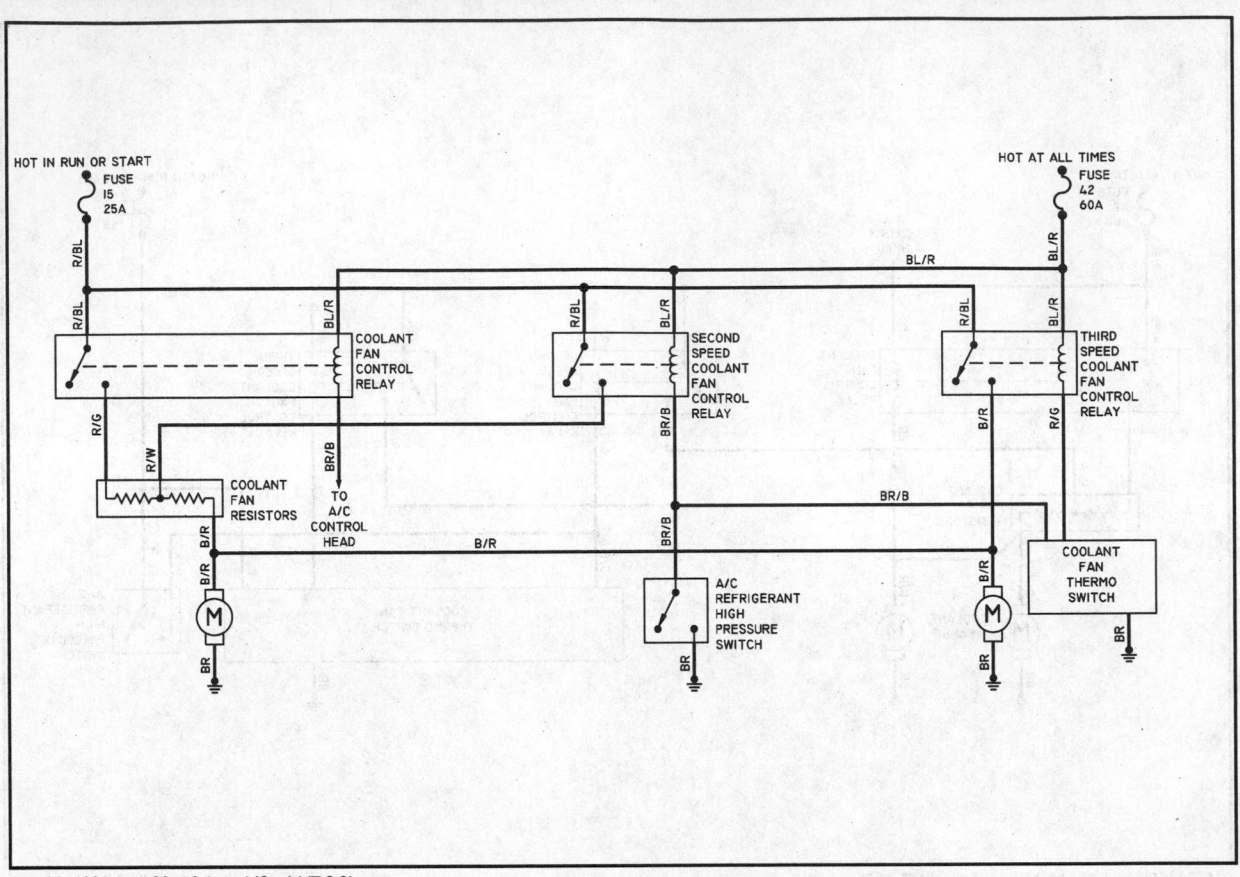

DIA. 11- 1995 Audi 90, A6 Auto A/C w/ A/T 2.8L

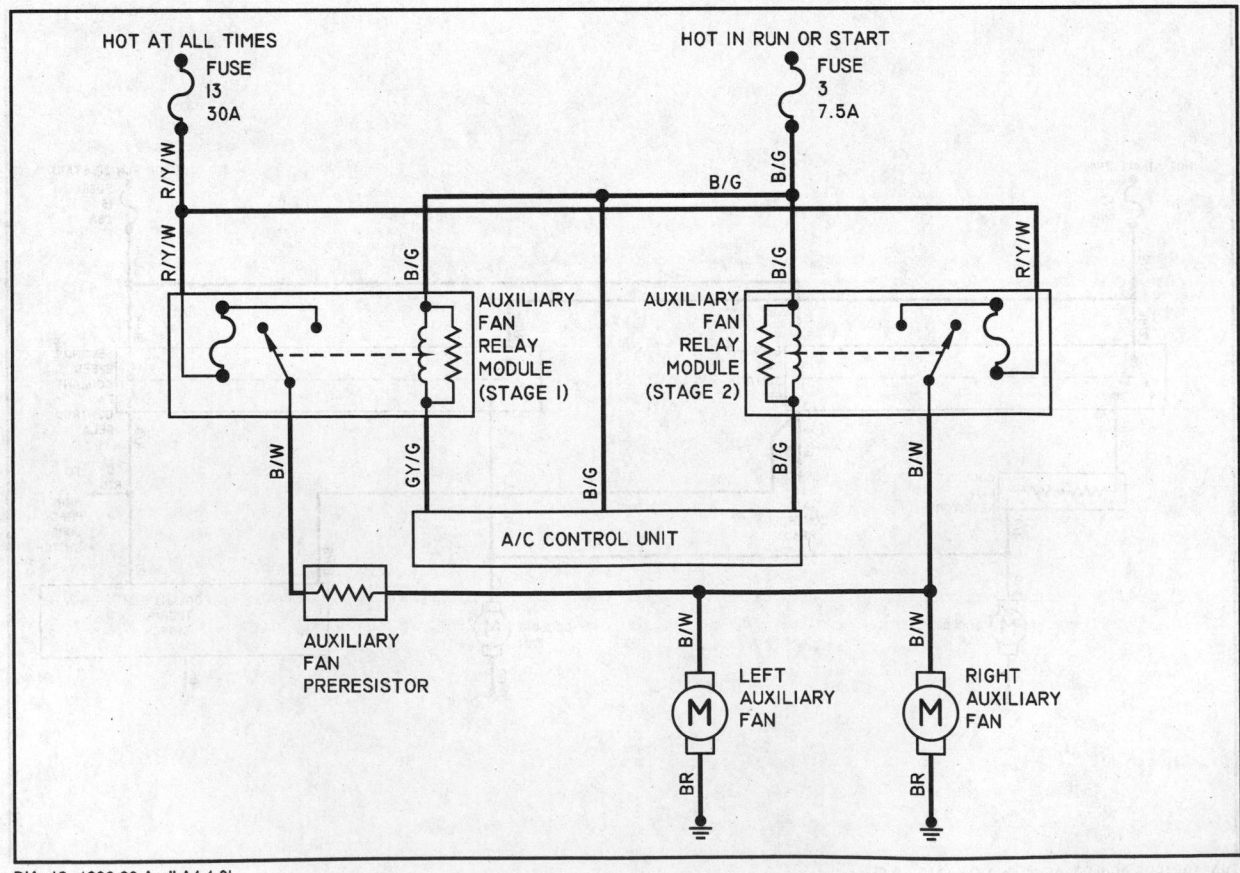

DIA. 12- 1996-99 Audi A4 1.8L

79239W06

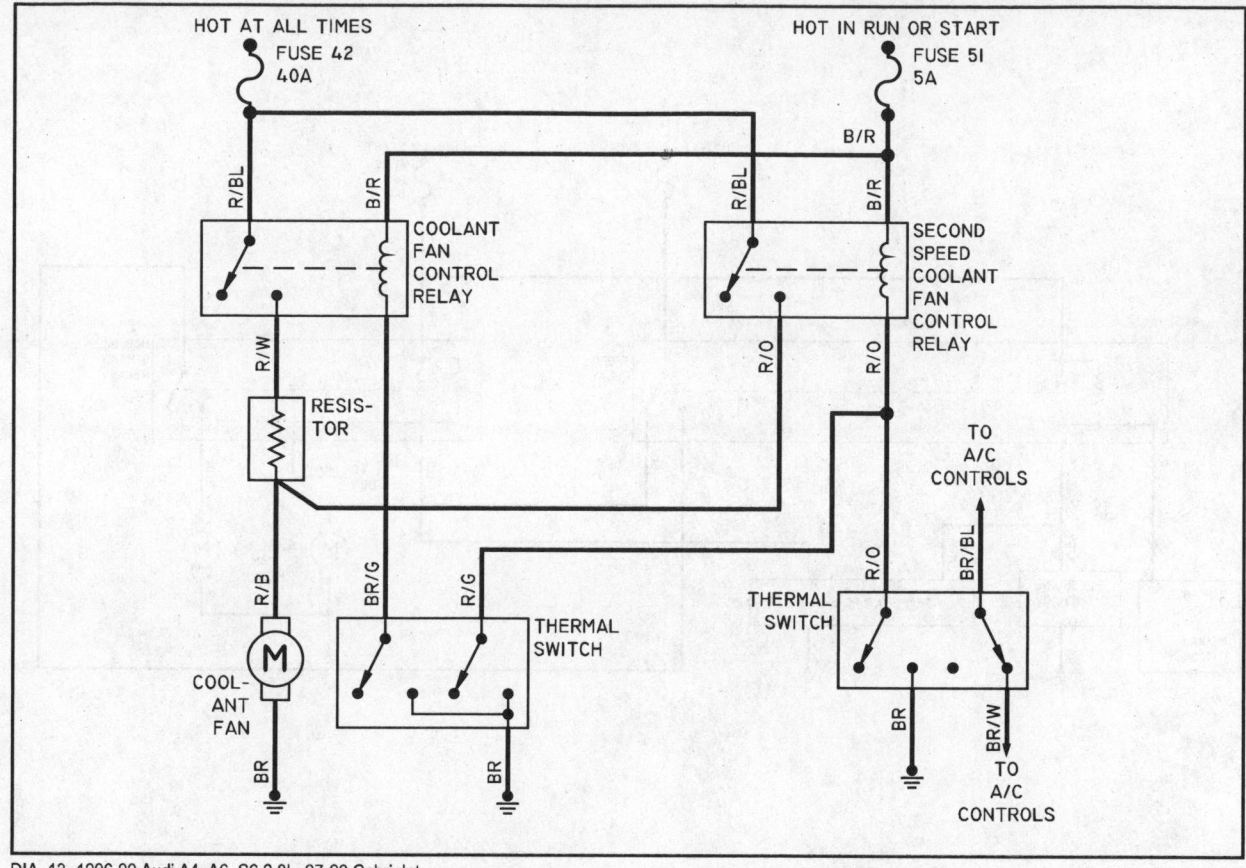

DIA. 13- 1996-99 Audi A4, A6, S6 2.8L, 97-99 Cabriolet

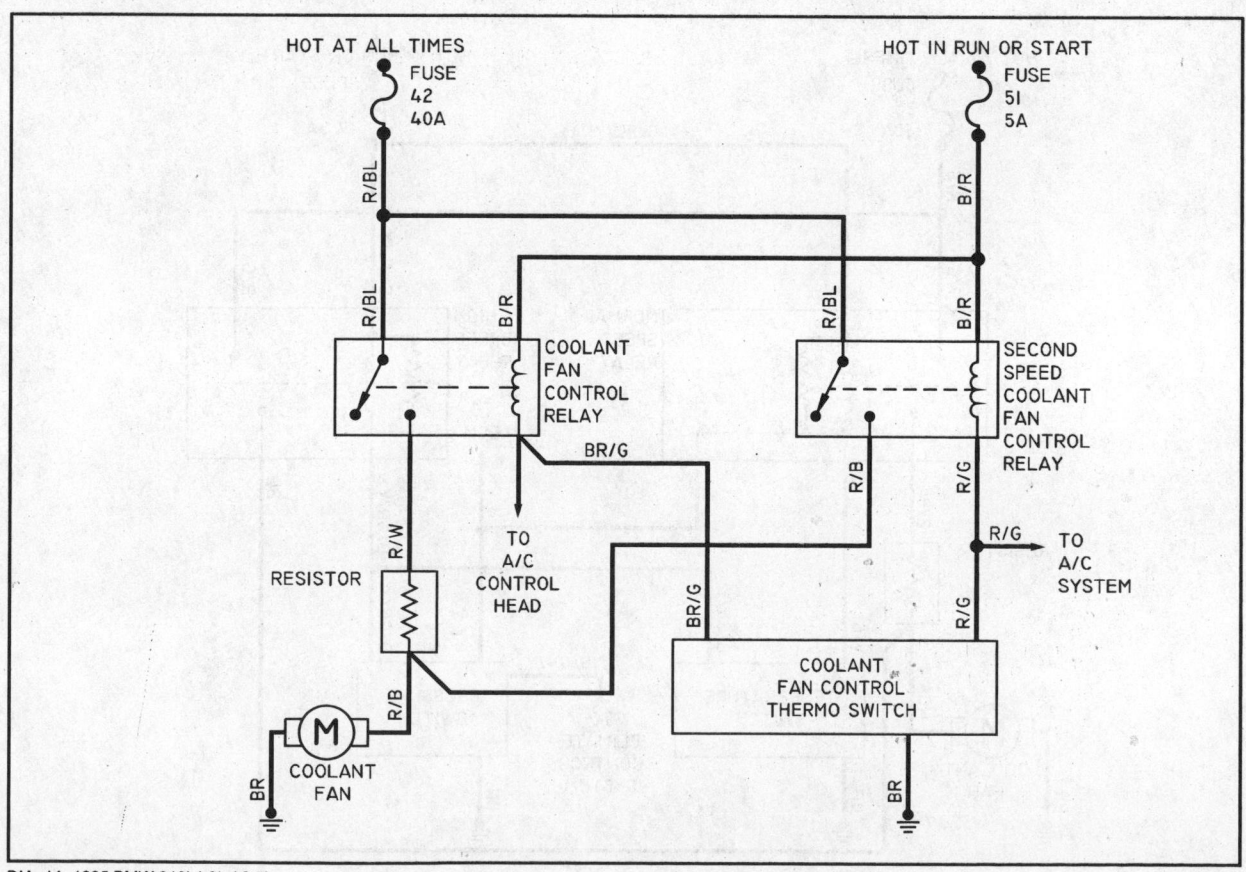

DIA. 14- 1995 BMW 318i 1.8L / 2.5L

79239W07

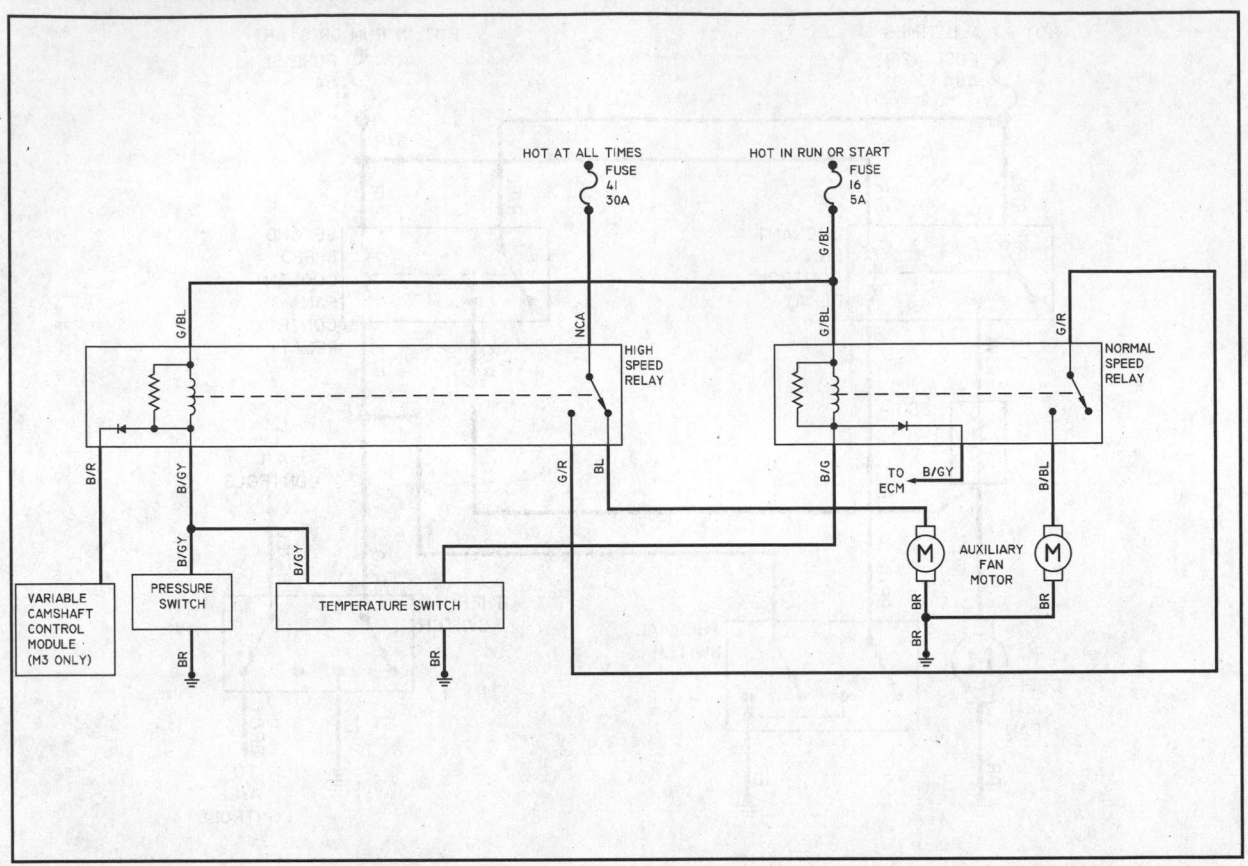

DIA. 15- 1996-99 BMW 318is-c, 320i, 325i-c, M3 2.8L

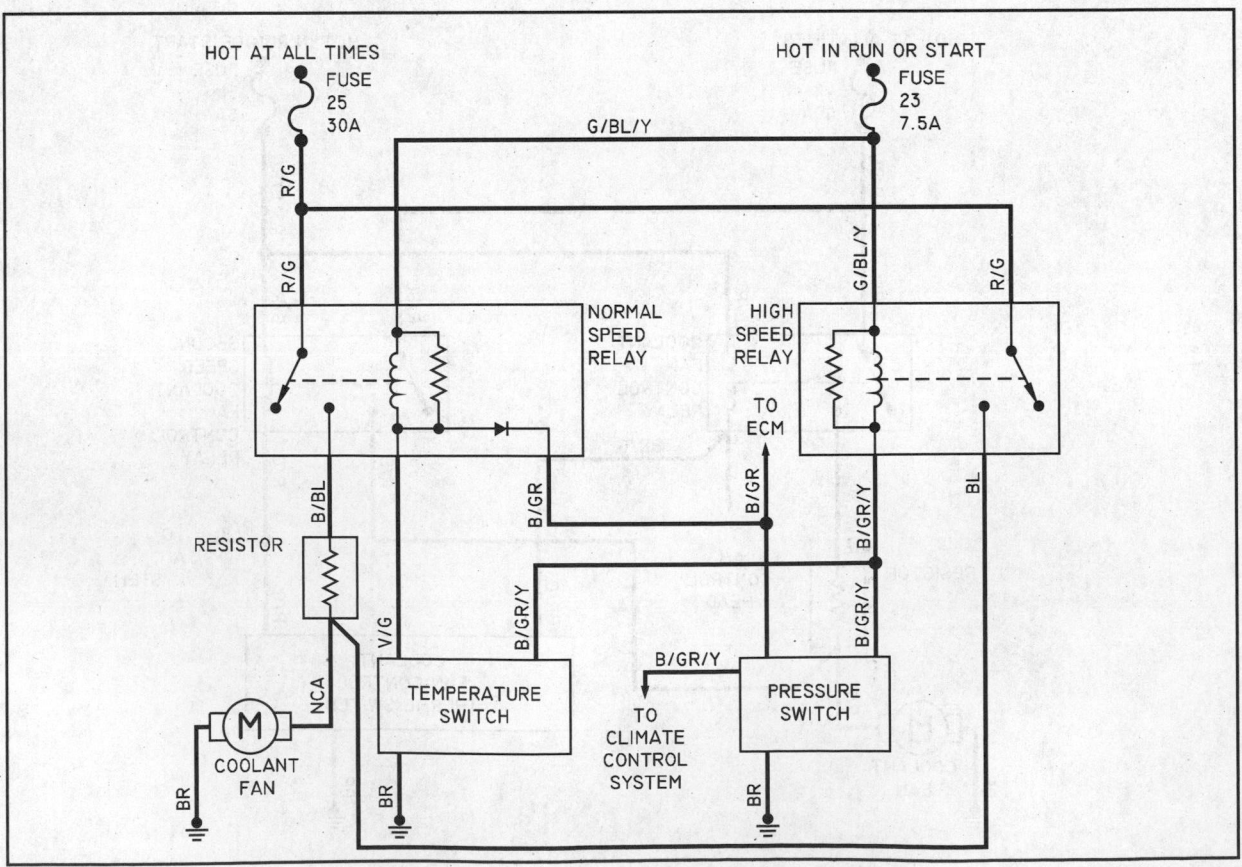

DIA. 16- 1995-99 BMW 525i 2.5L / 3.0L / 4.0L

79239W08

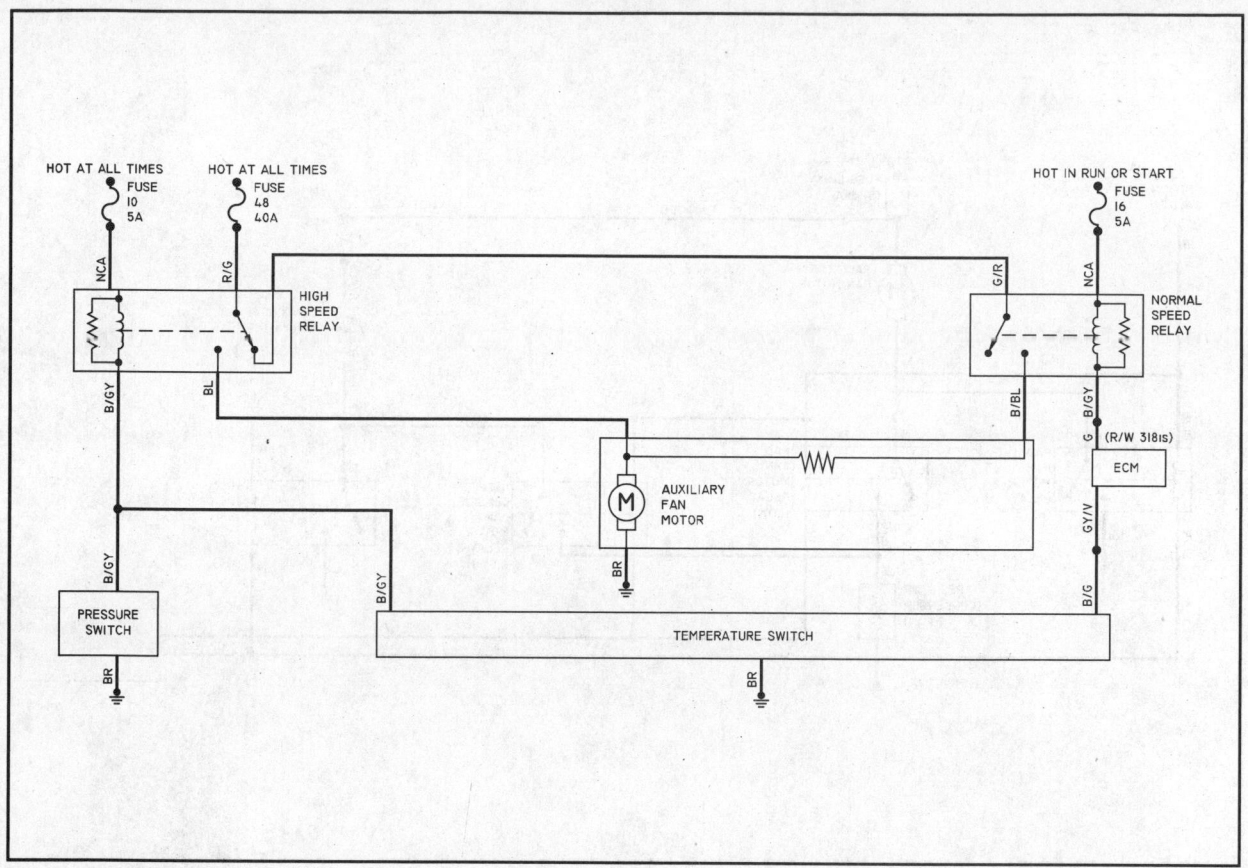

DIA. 17- 1996-99 BMW 318is-c, 320i, 325i-c, M3 1.9L / 2.8L

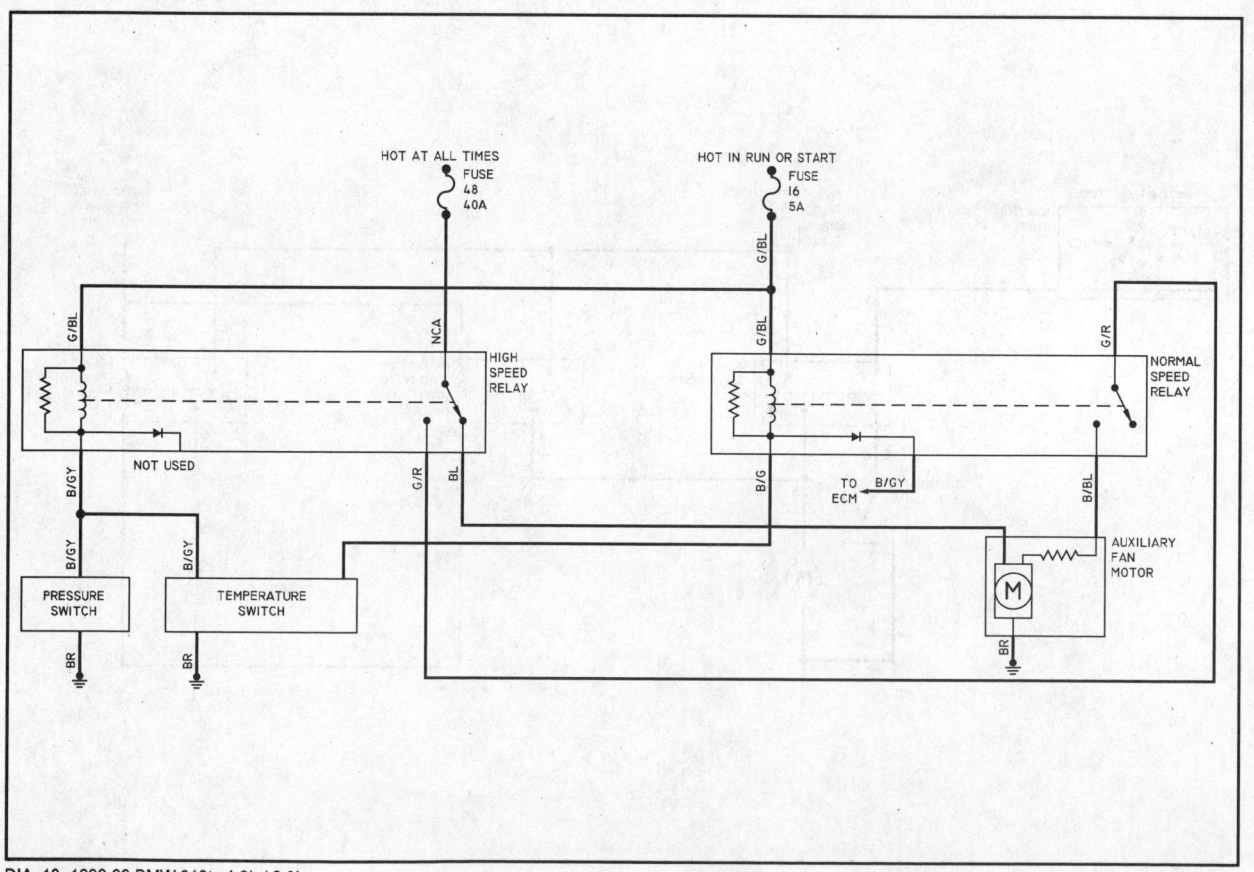

DIA. 18- 1996-99 BMW 318ts 1.9L / 2.8L

79239W09

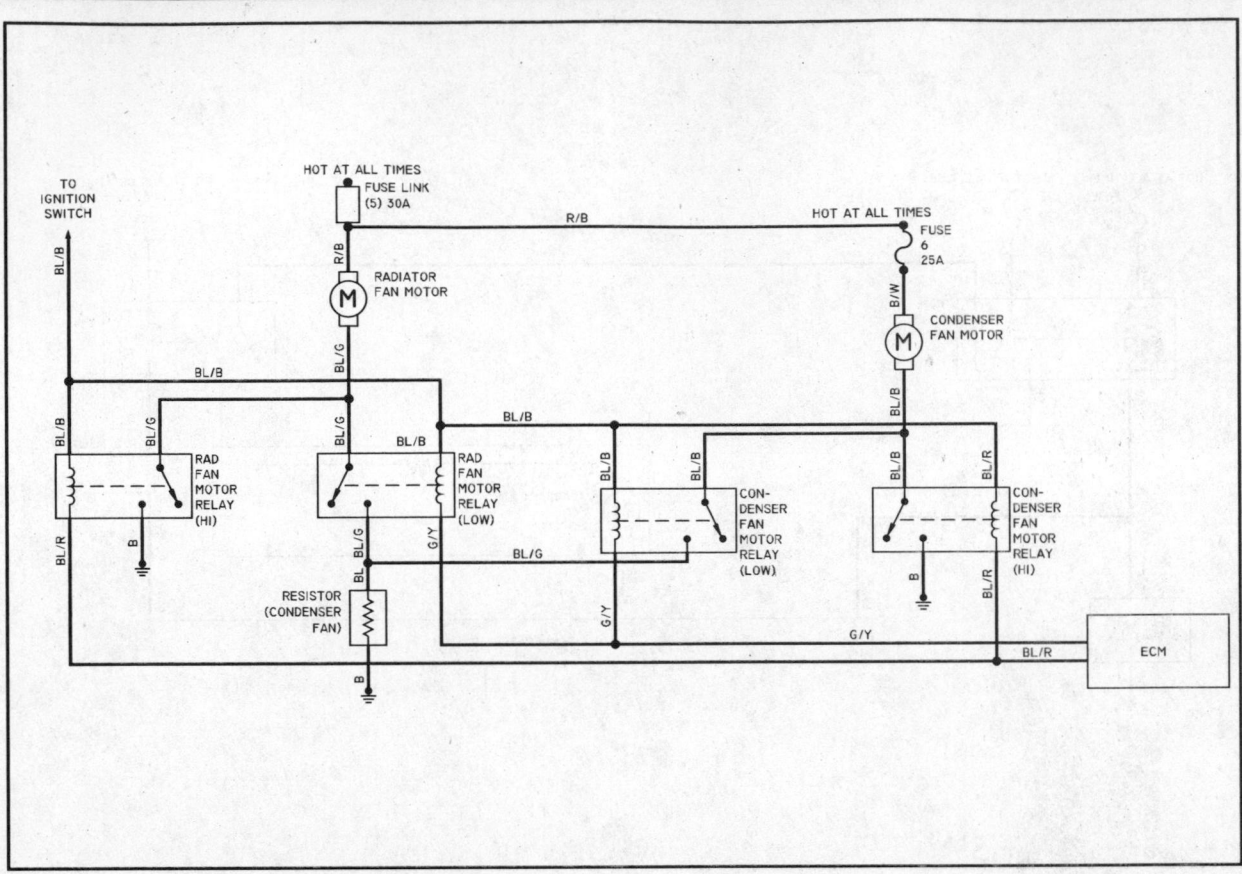

DIA. 19- 1995-96 Chrysler Imports Colt, Summit Wagon, Expo 1.8L / 2.4L

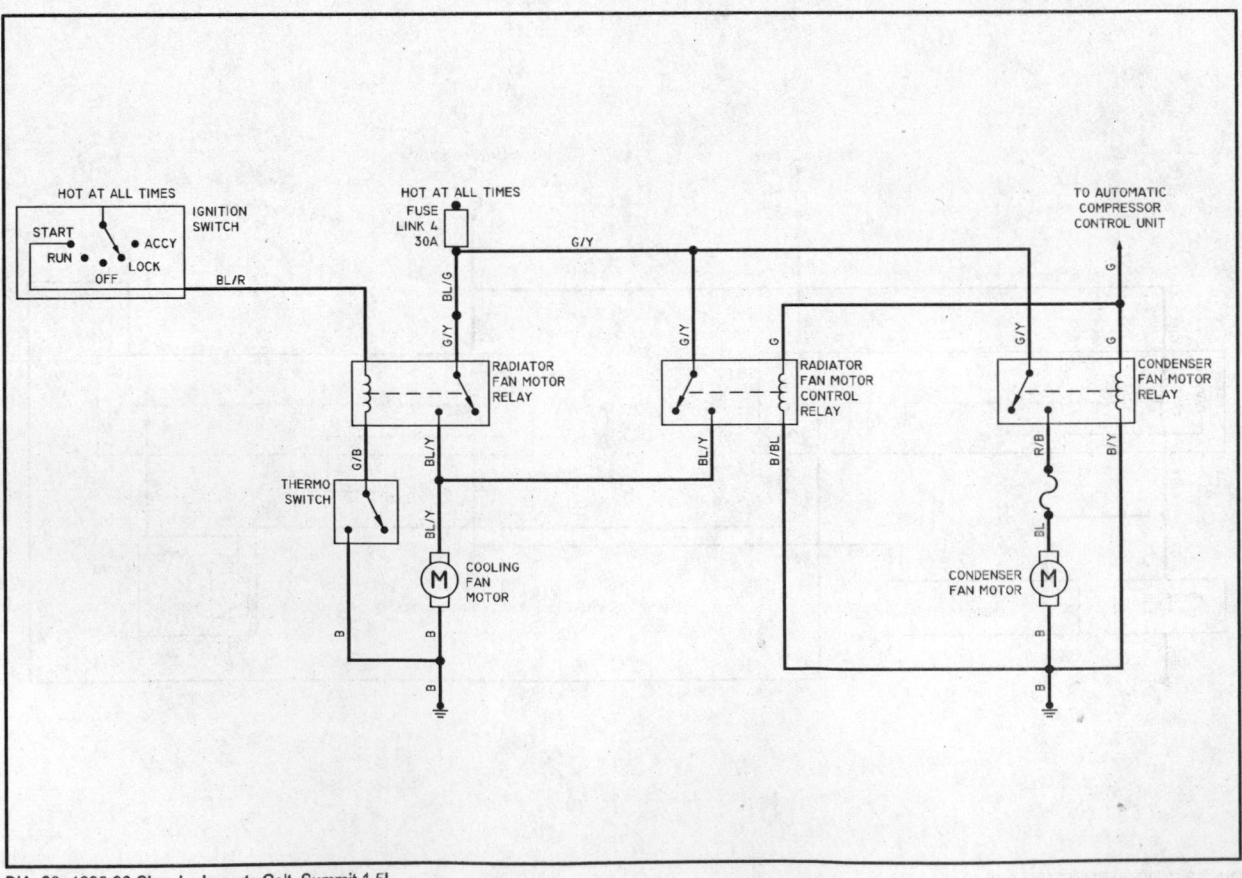

DIA. 20- 1995-96 Chrysler Imports Colt, Summit 1.5L

79239W10

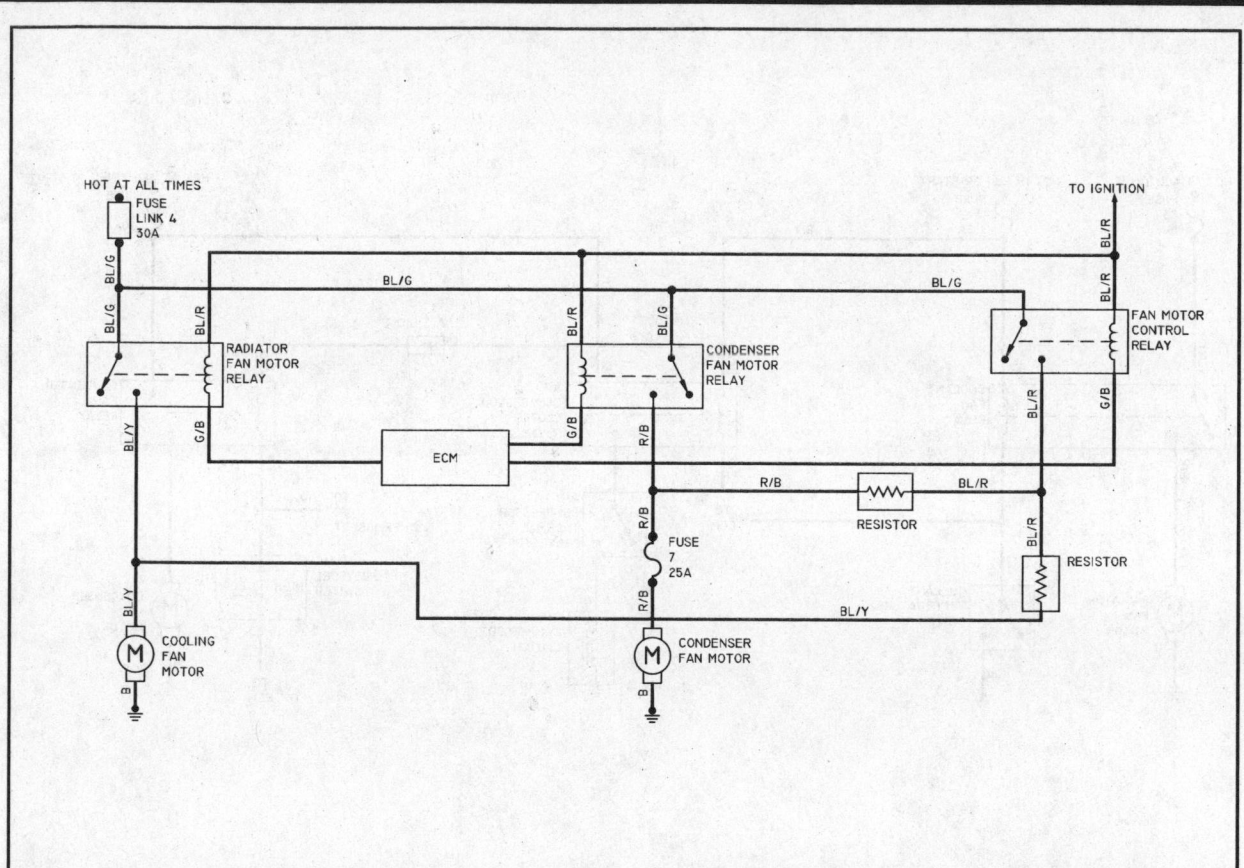

DIA. 21- 1995-96 Chrysler Imports Colt, Summit 1.8L

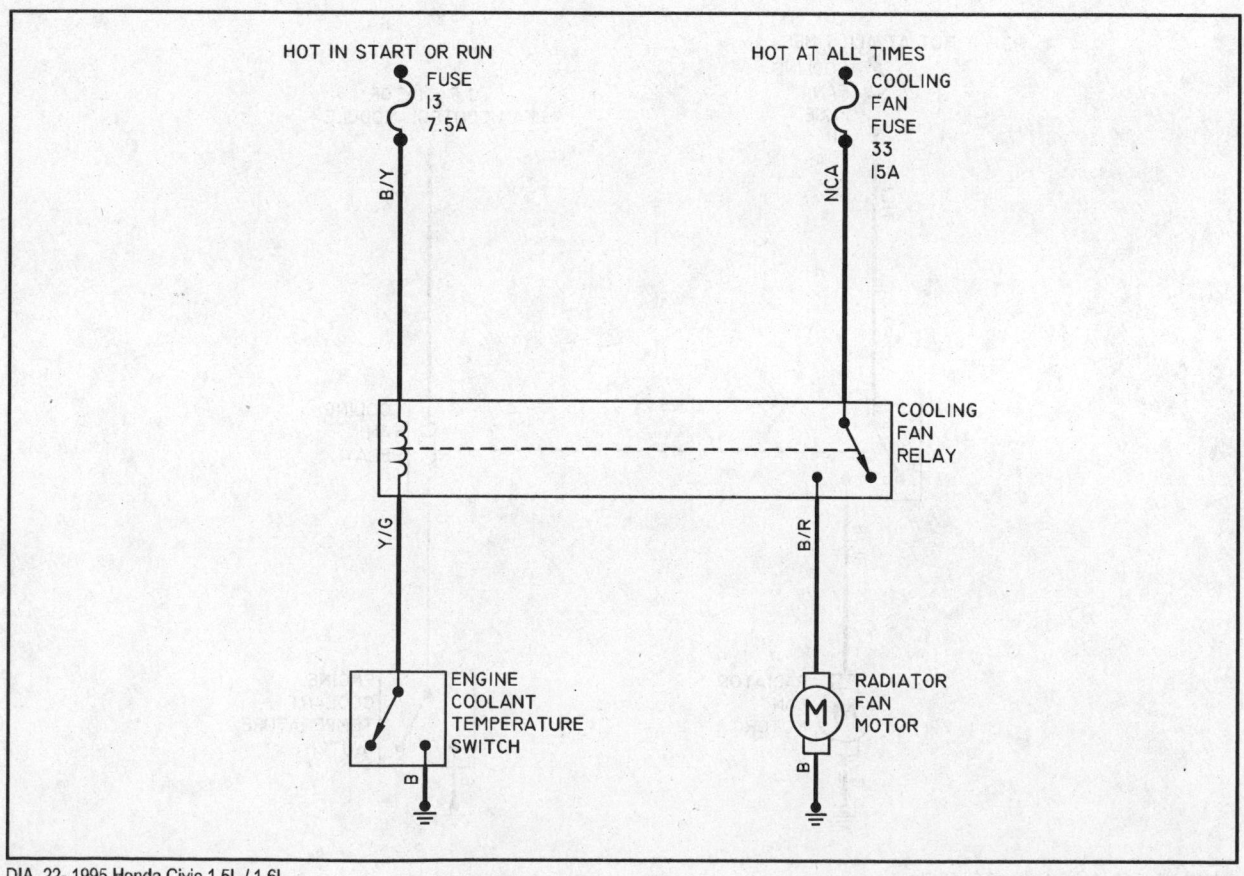

DIA. 22- 1995 Honda Civic 1.5L / 1.6L

79239W11

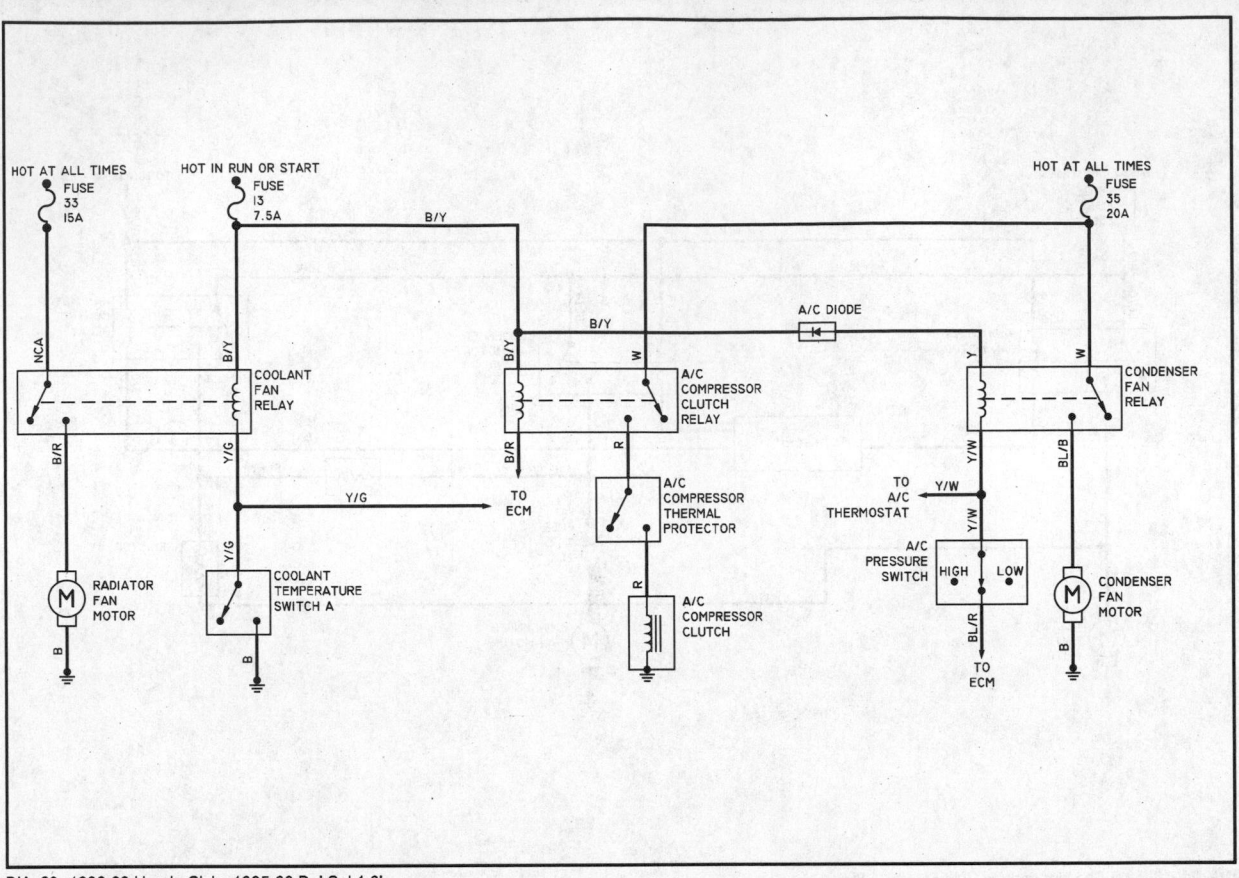

DIA. 23- 1996-99 Honda Civic, 1995-99 Del Sol 1.6L

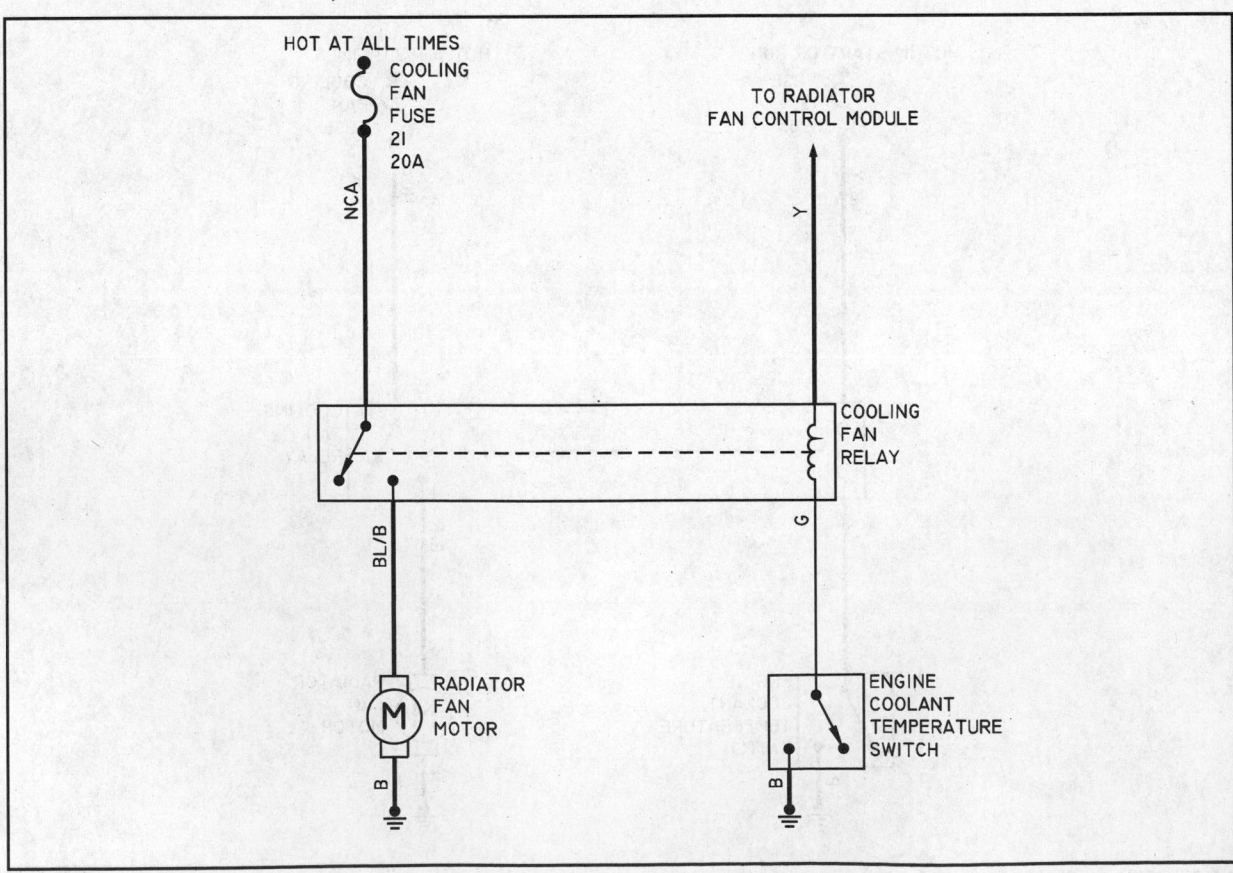

DIA. 24- 1995-99 Honda Accord 2.2L / 2.7L

79239W12

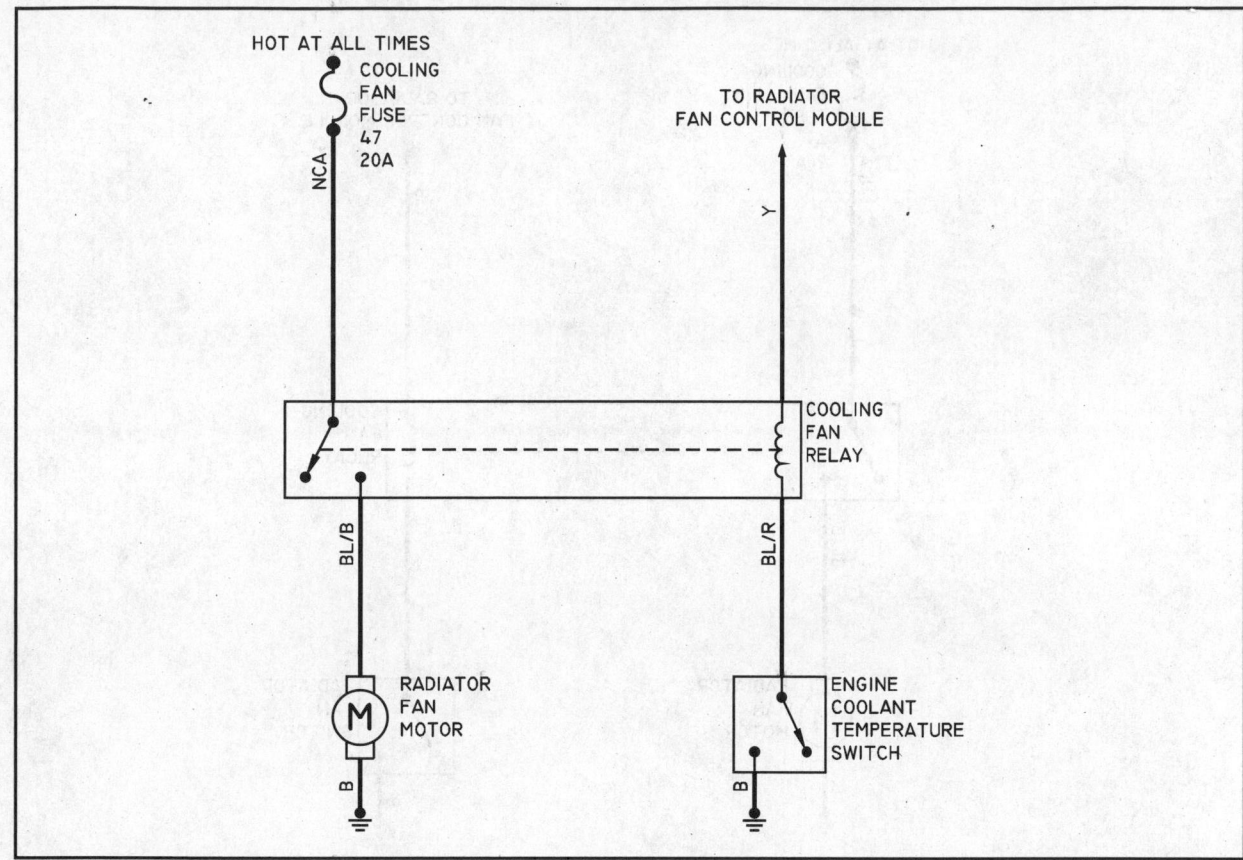

DIA. 25- 1995 Honda Prelude 2.2L / 2.3L

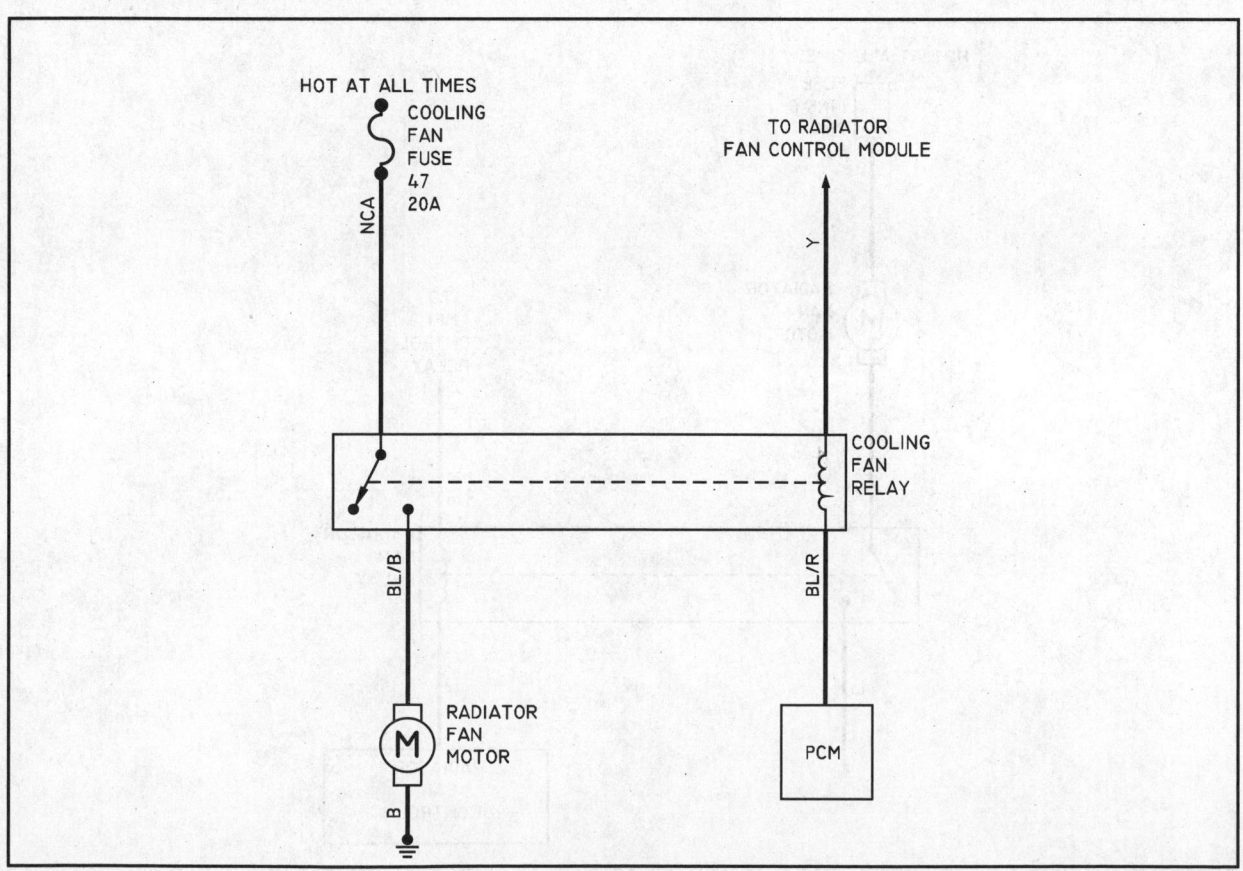

DIA. 26- 1996 Honda Prelude 2.2L / 2.3L

79239W13

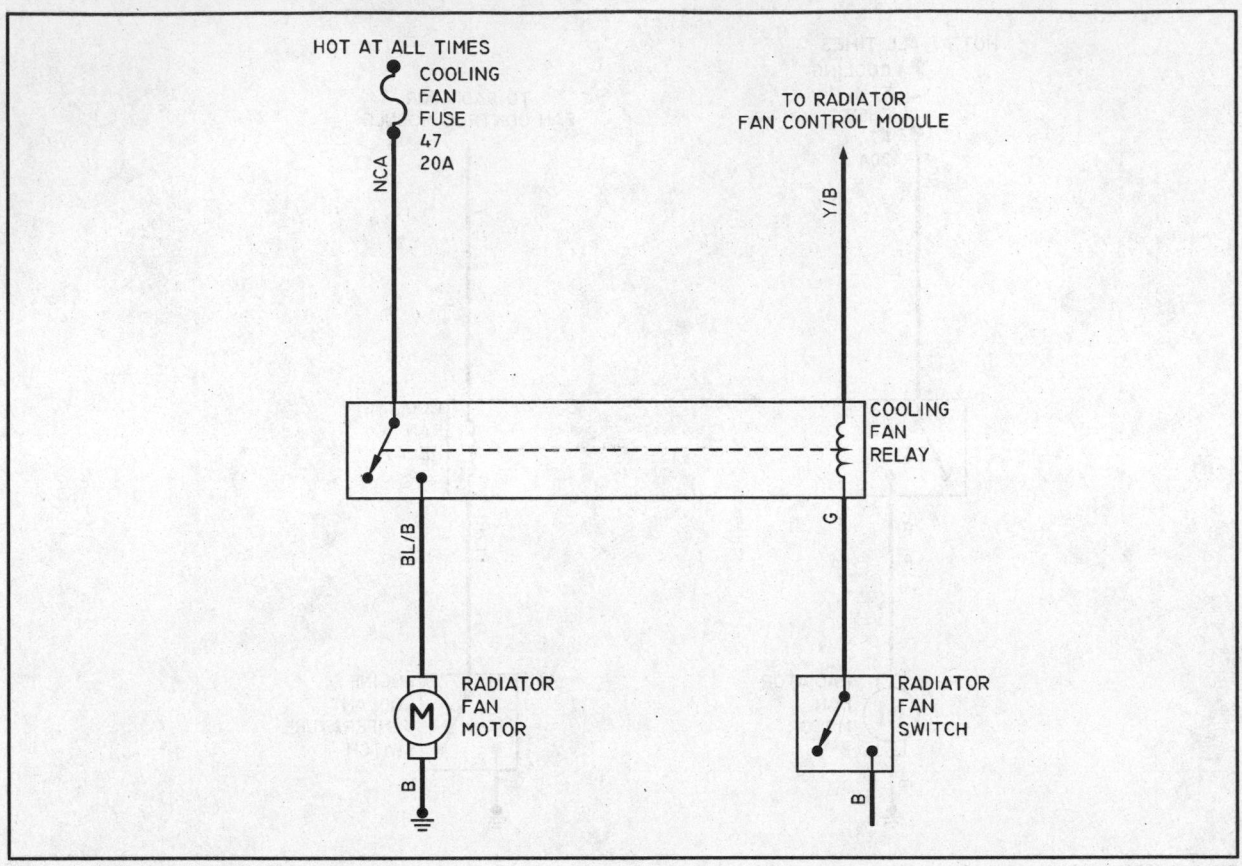

DIA. 27- 1997-99 Honda Prelude 2.2L

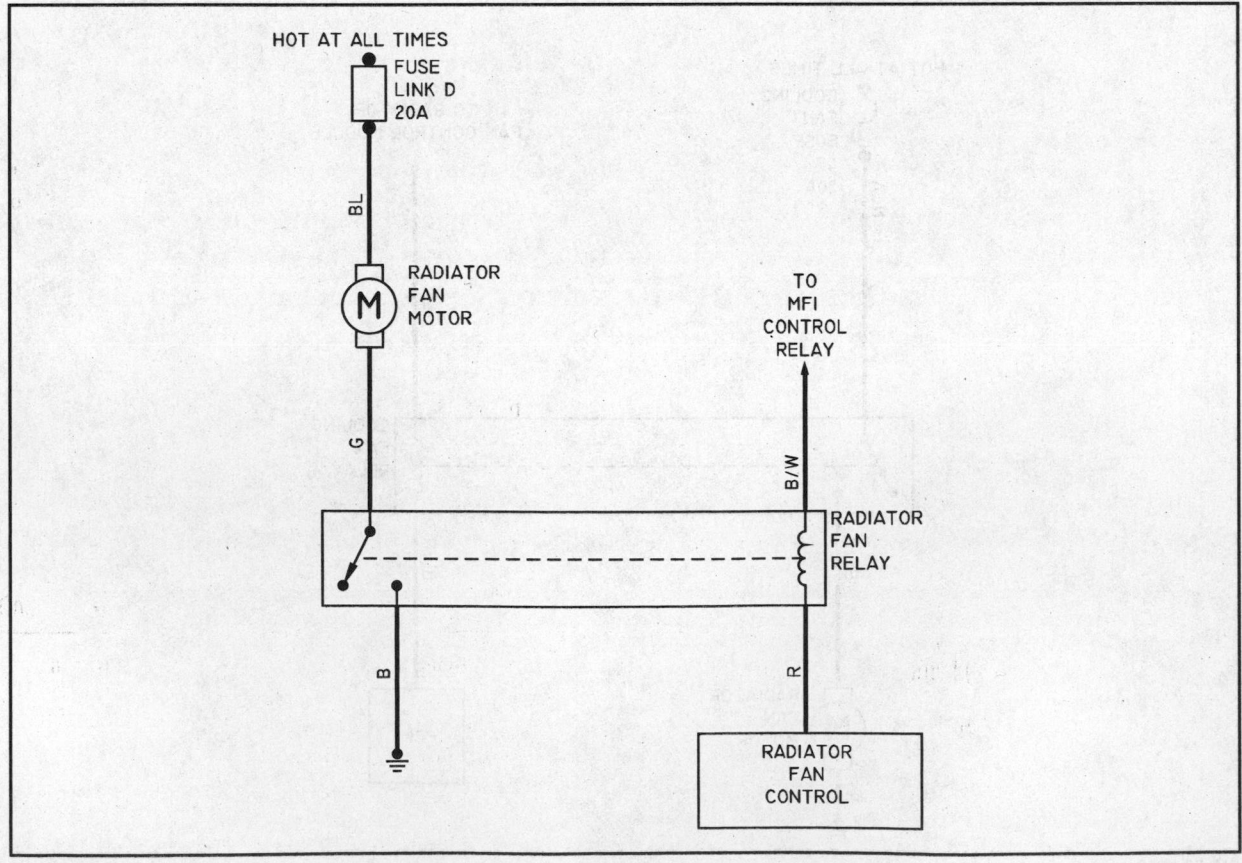

DIA. 28- 1995 Hyundai Accent 1.5L

79239W14

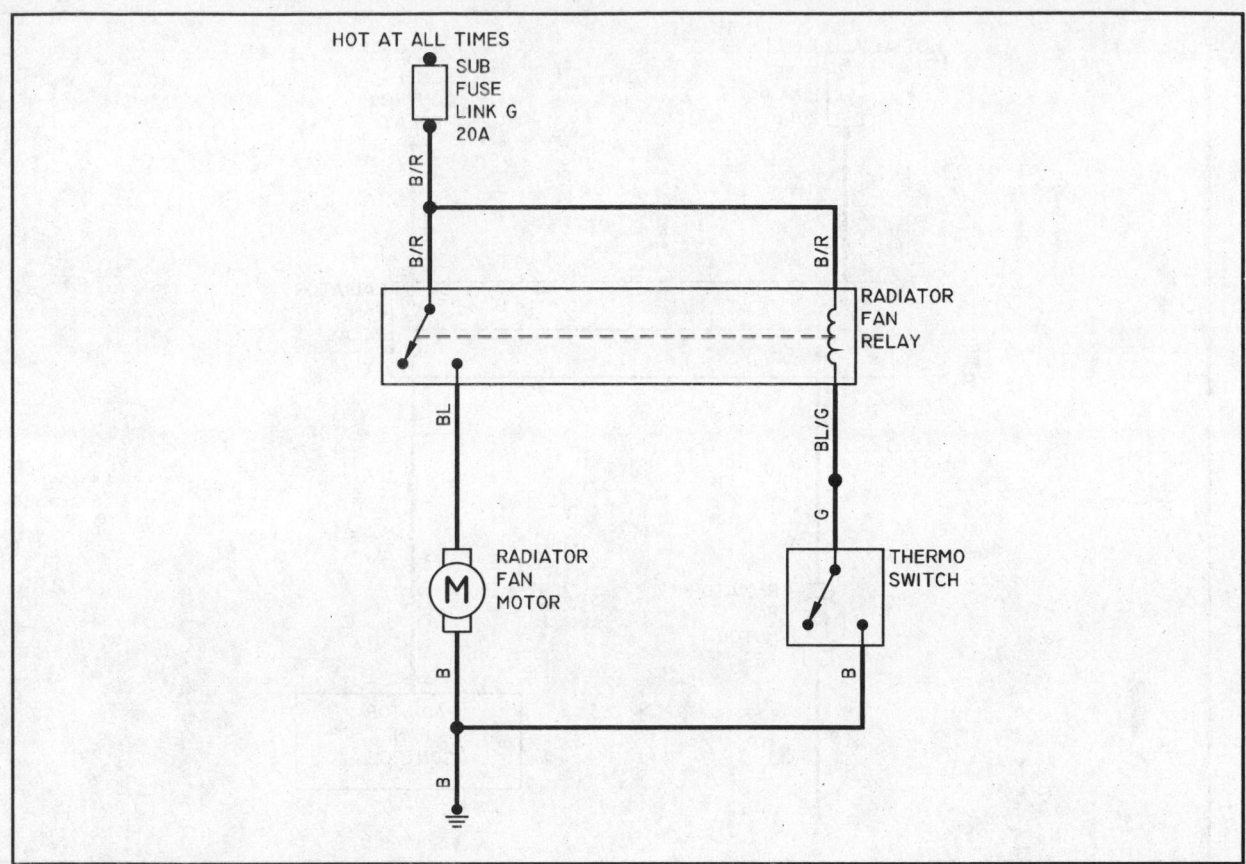

DIA. 29- 1995 Hyundai Elantra 1.5L / 1.6L / 1.8L

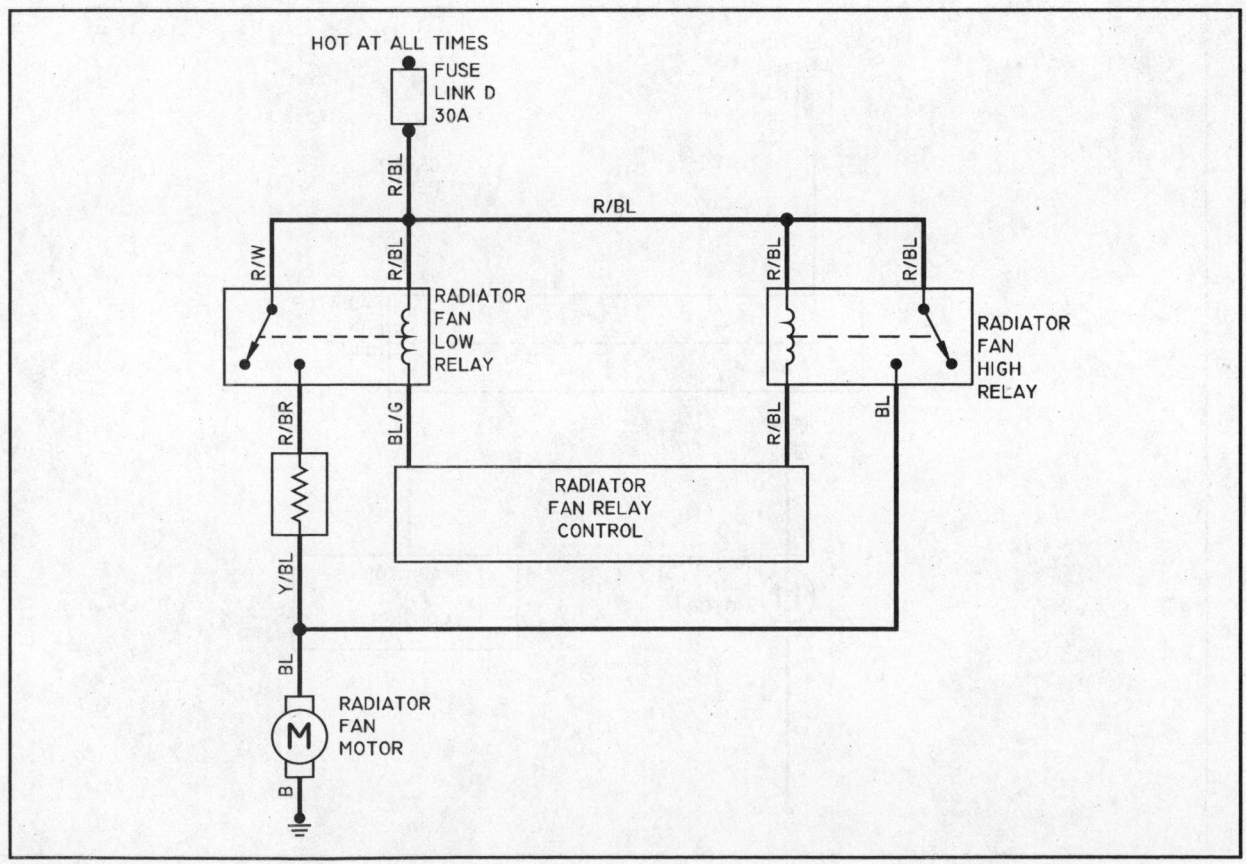

DIA. 30- 1995 Hyundai Sonata 2.0L / 3.0L

79239W15

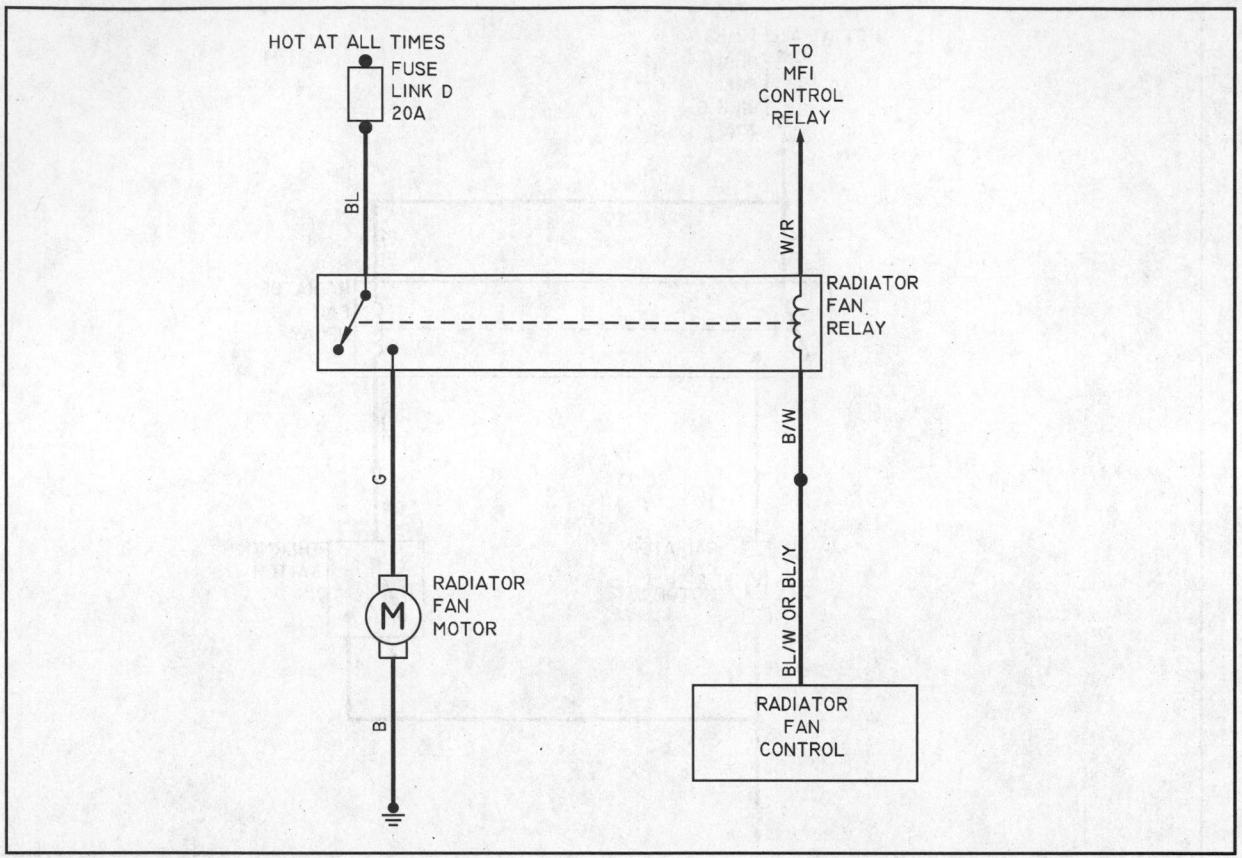

DIA. 31- 1996-99 Hyundai Accent 1.5L

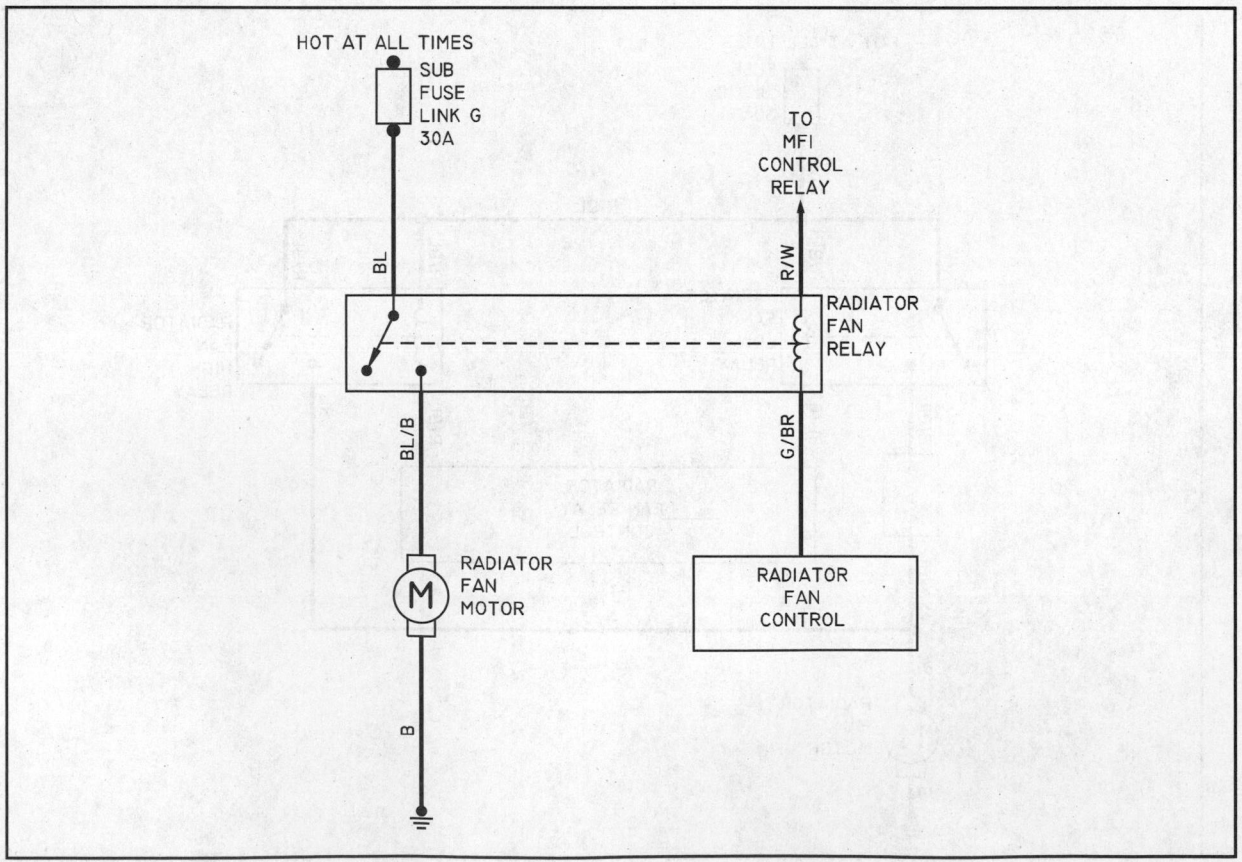

DIA. 32- 1996-99 Hyundai Elantra 1.8L / 2.0L

79239W16

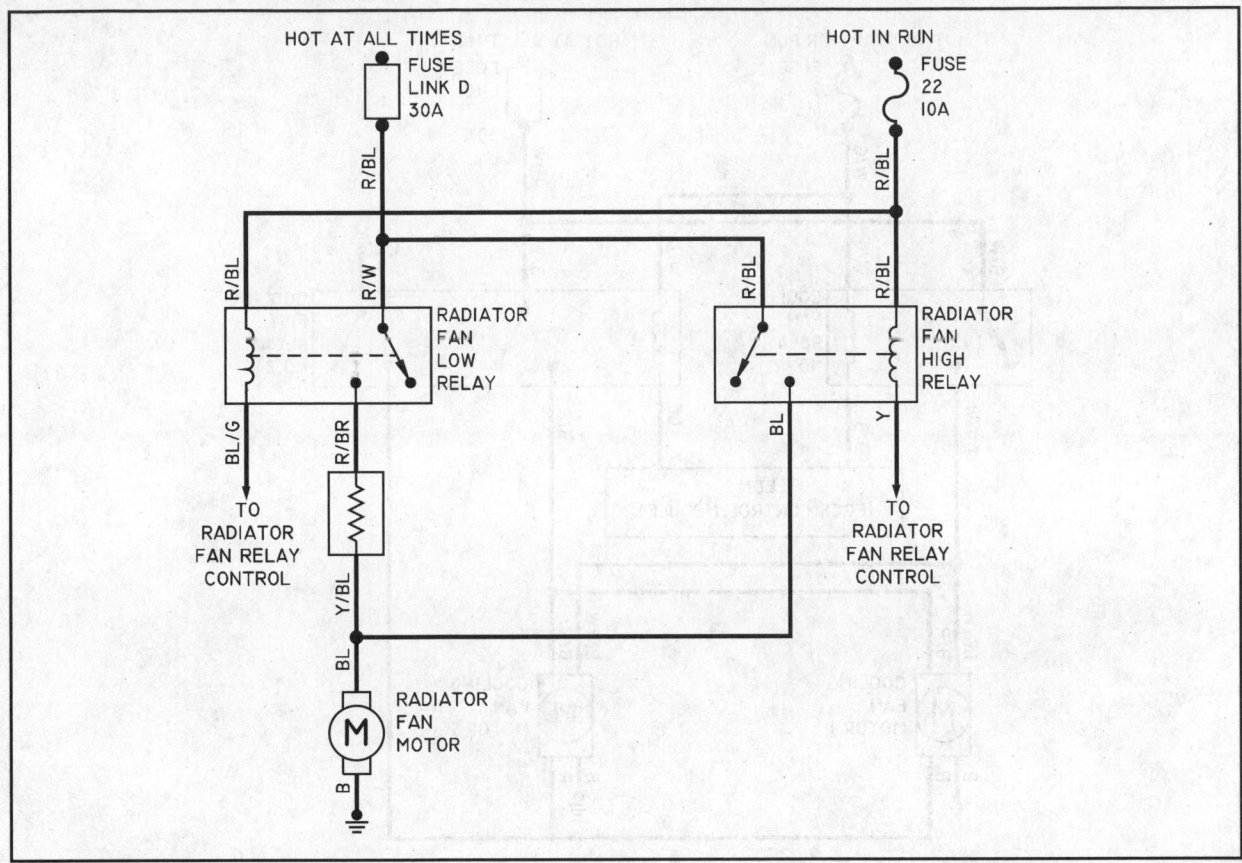

DIA. 33- 1996-99 Hyundai Sonata 2.0L / 3.0L

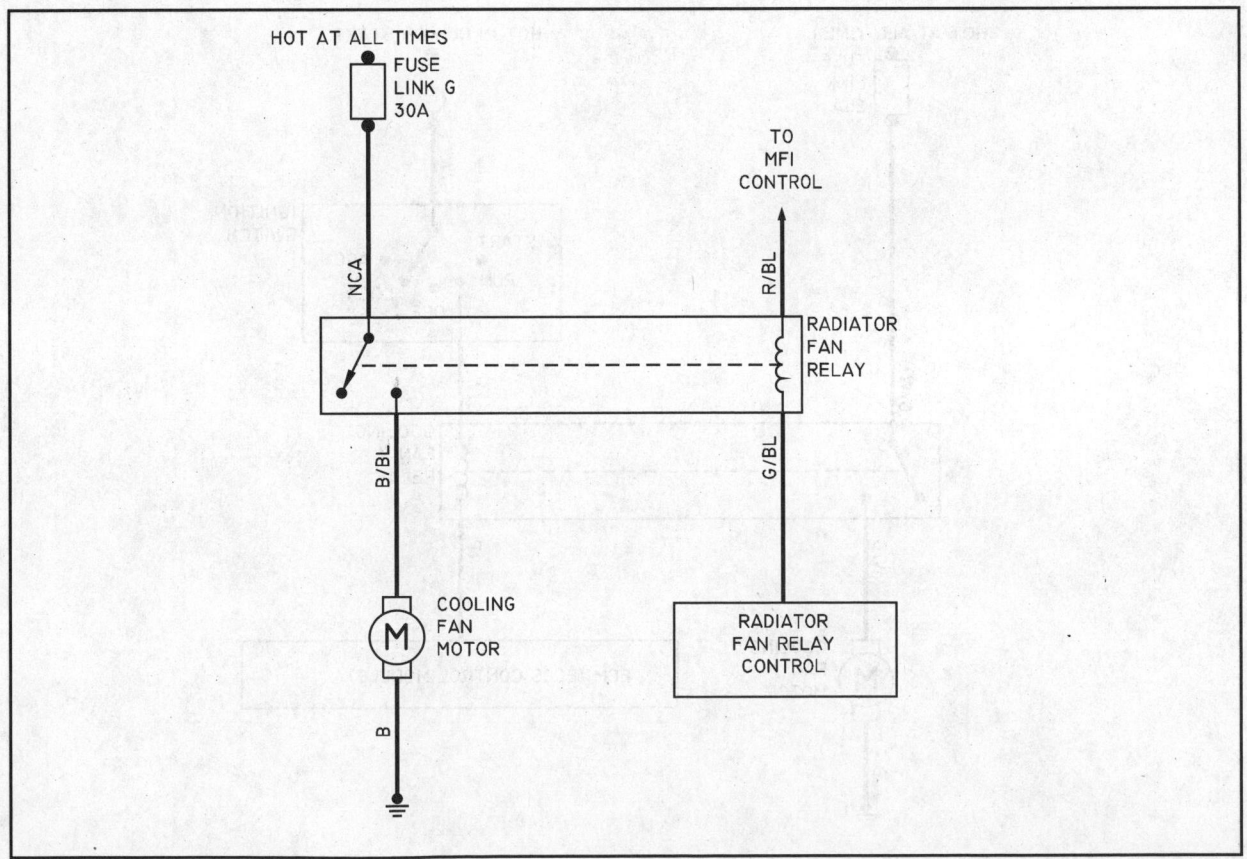

DIA. 34- 1997-99 Hyundai Tiburon 1.8L / 2.0L

79239W17

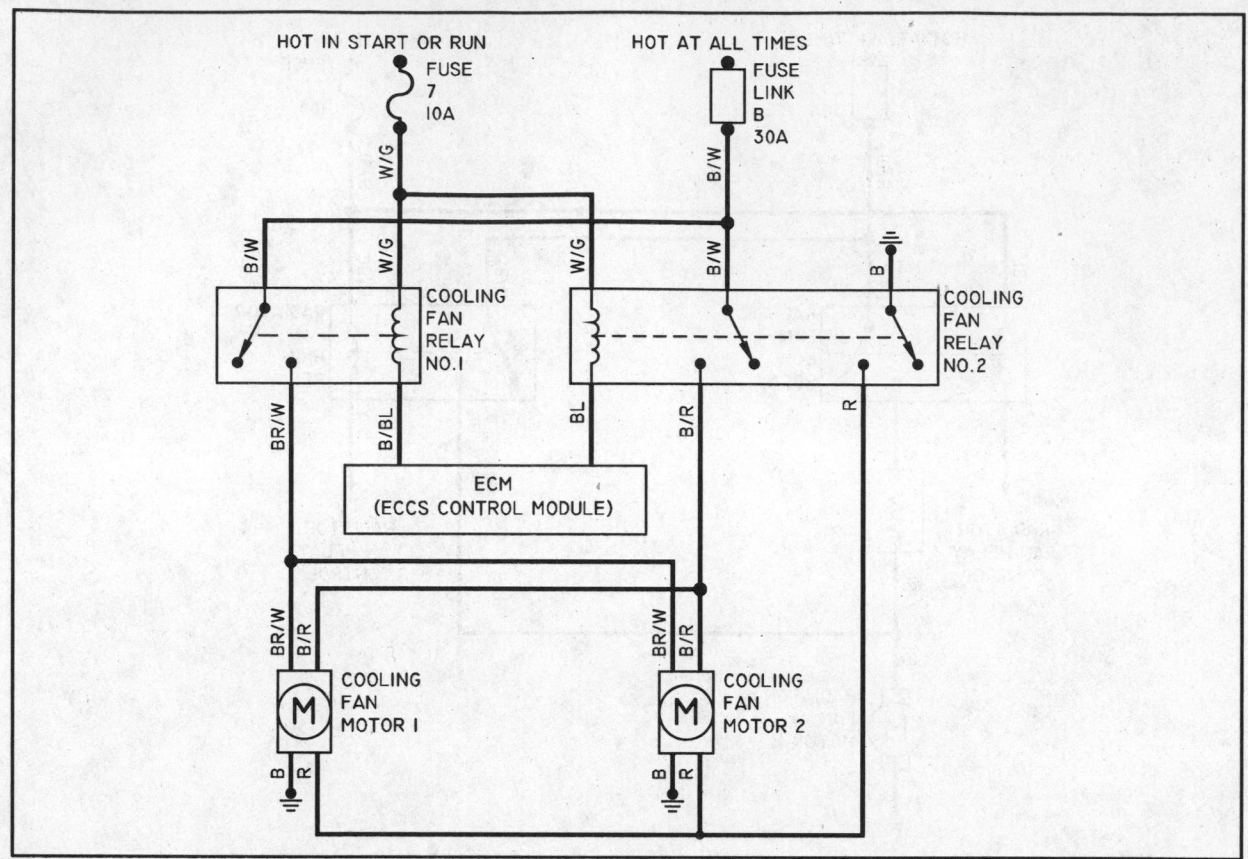

DIA. 35- 1995-97 Infiniti G20 2.0L

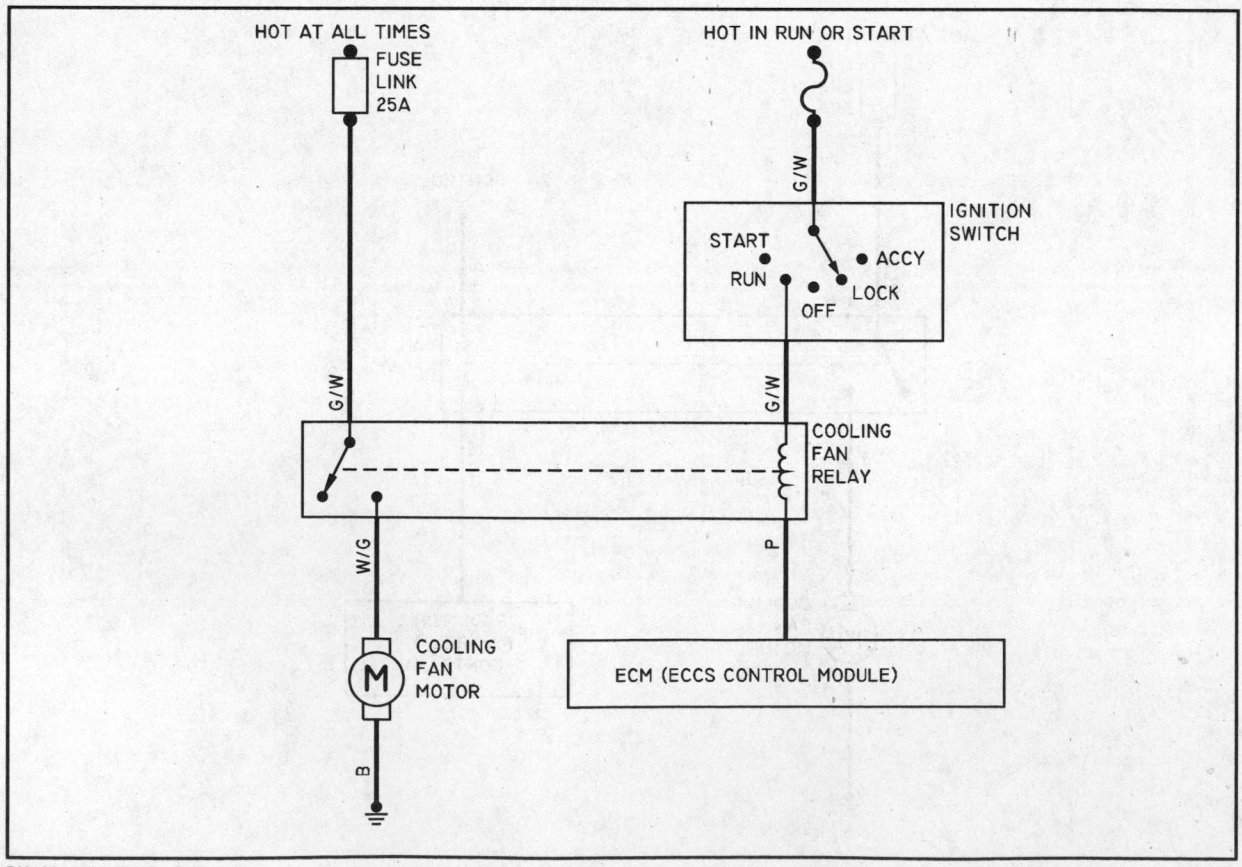

DIA. 36- 1995-97 Infiniti J30 3.0L

79239W18

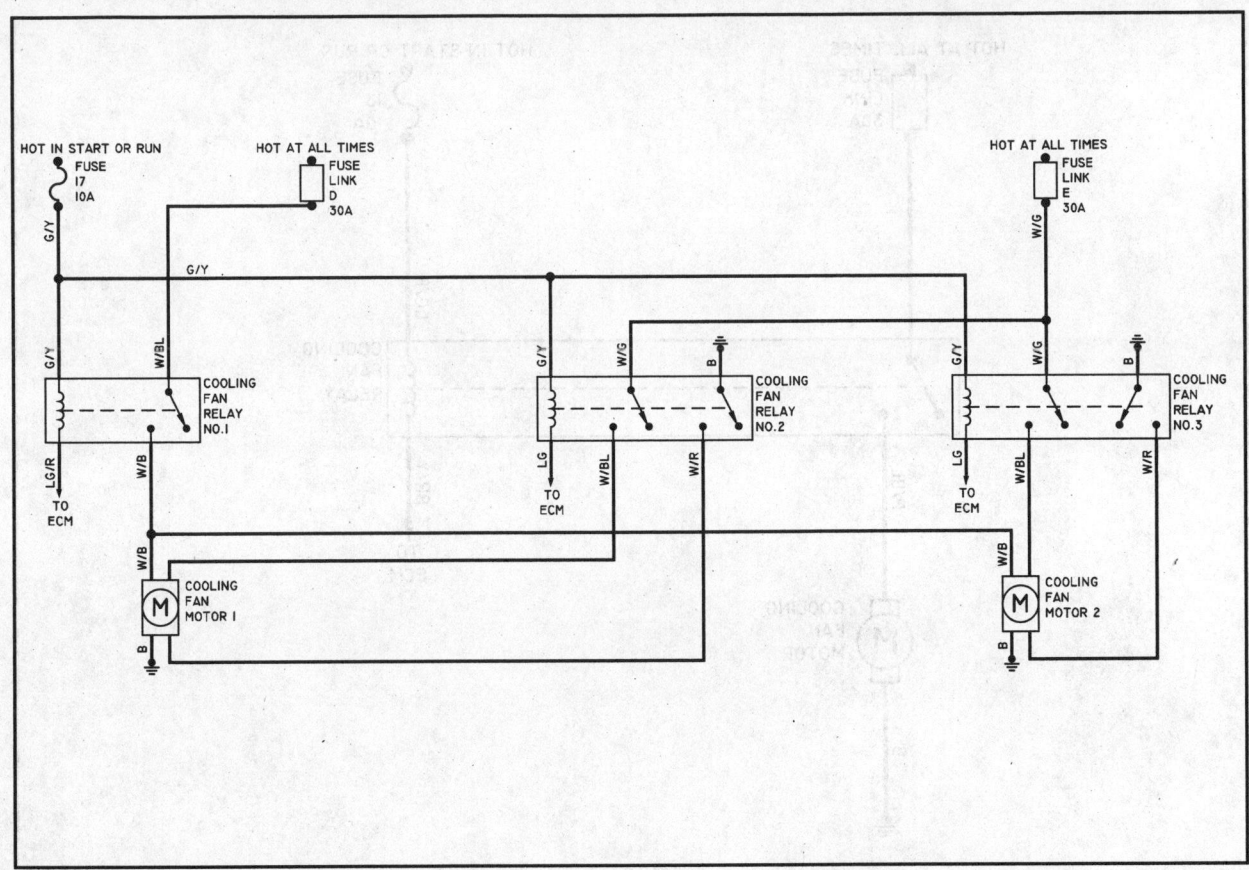

DIA. 37- 1996-99 Infiniti I30 3.0L

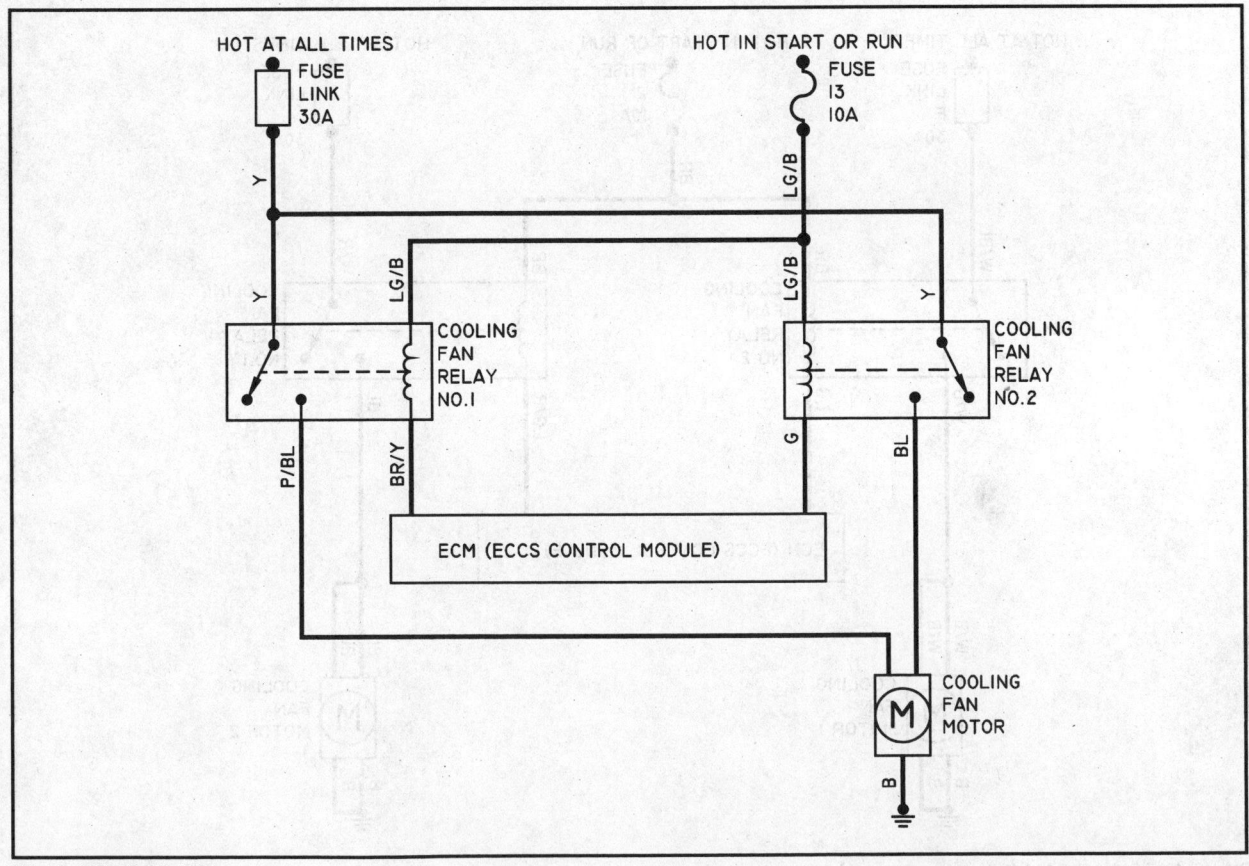

DIA. 38- 1995-96 Q45 (US)

79239W19

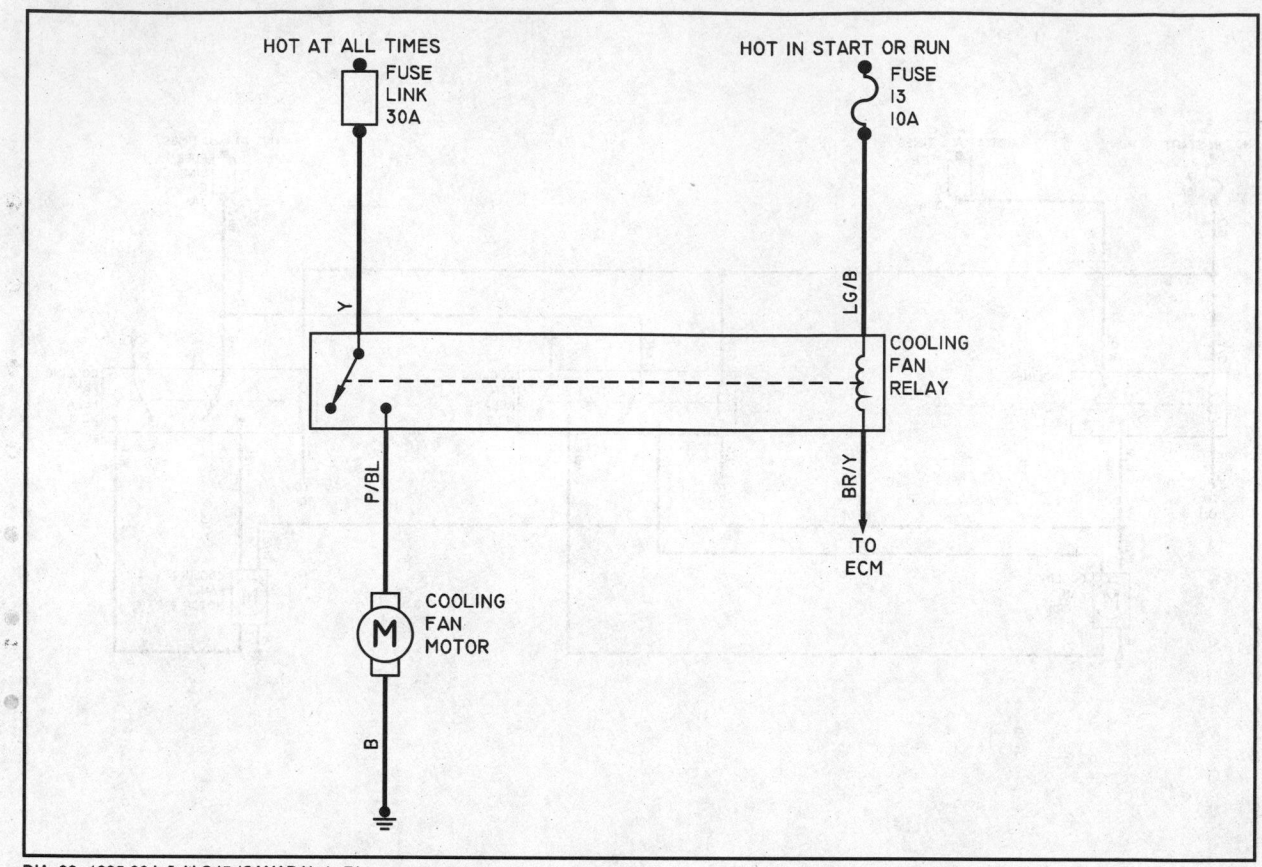

DIA. 39- 1995-96 Infiniti Q45 (CANADA) 4.5L

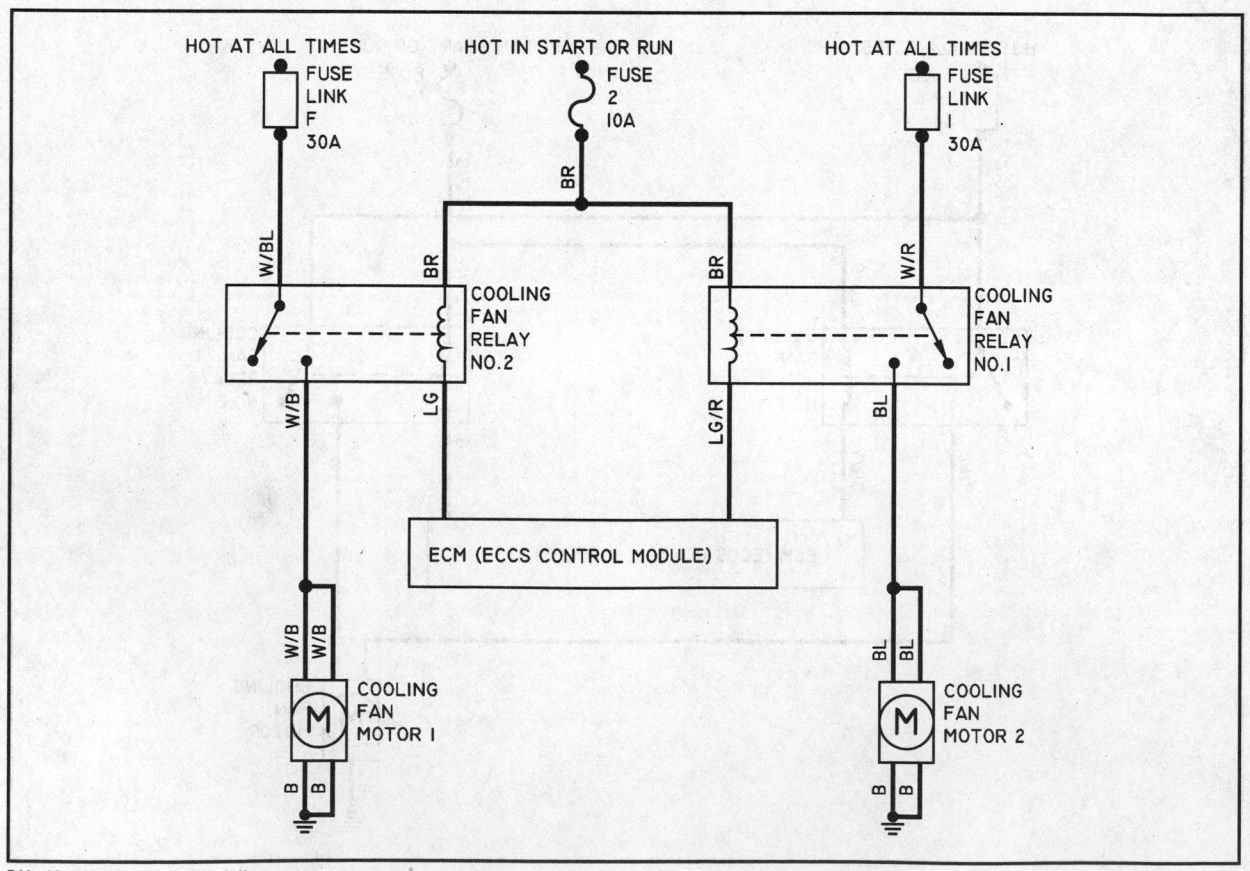

DIA. 40- 1997-99 Infiniti Q45 4.1L

79239W20

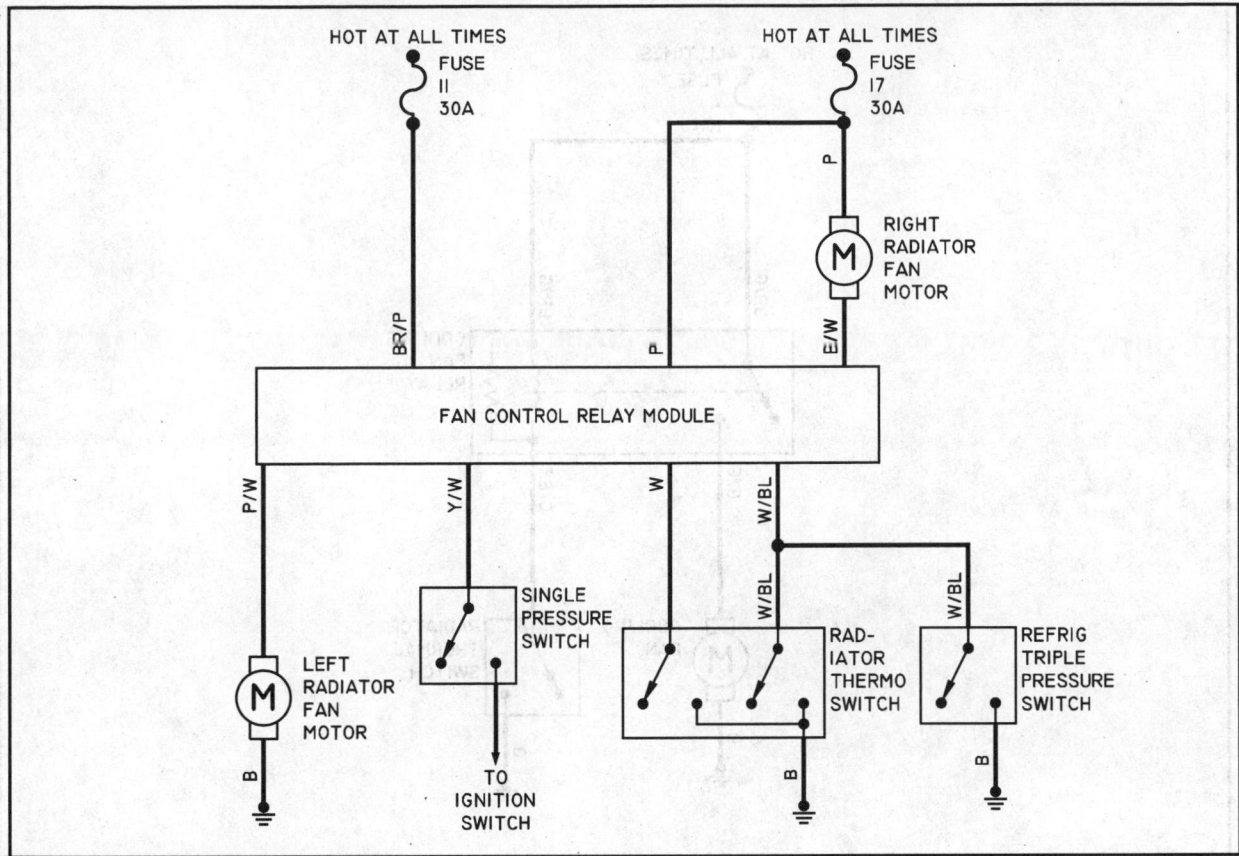

DIA. 41- 1995-99 Jaguar XJ6, XJ12, XJR 4.0L

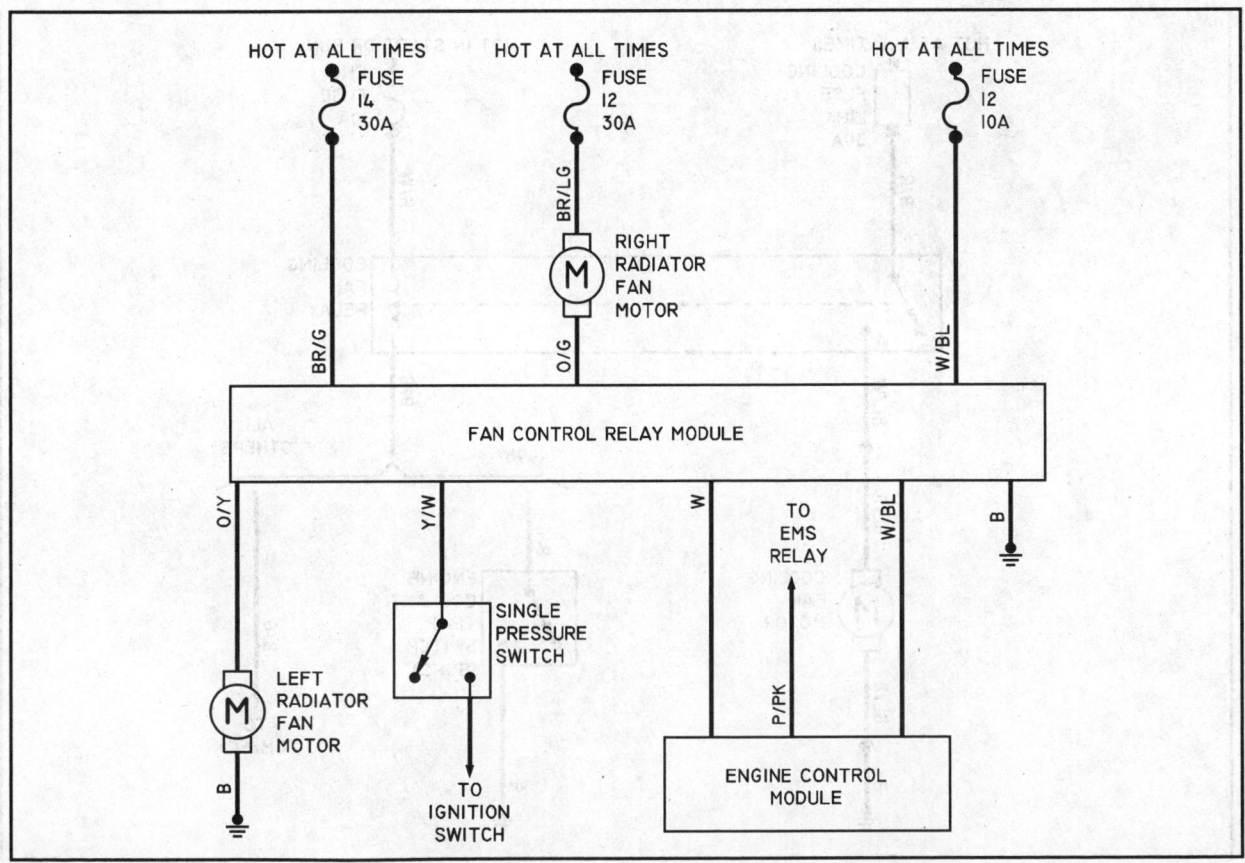

DIA. 42- 1997-99 Jaguar XK8, XJ8, 4.0L

79239W21

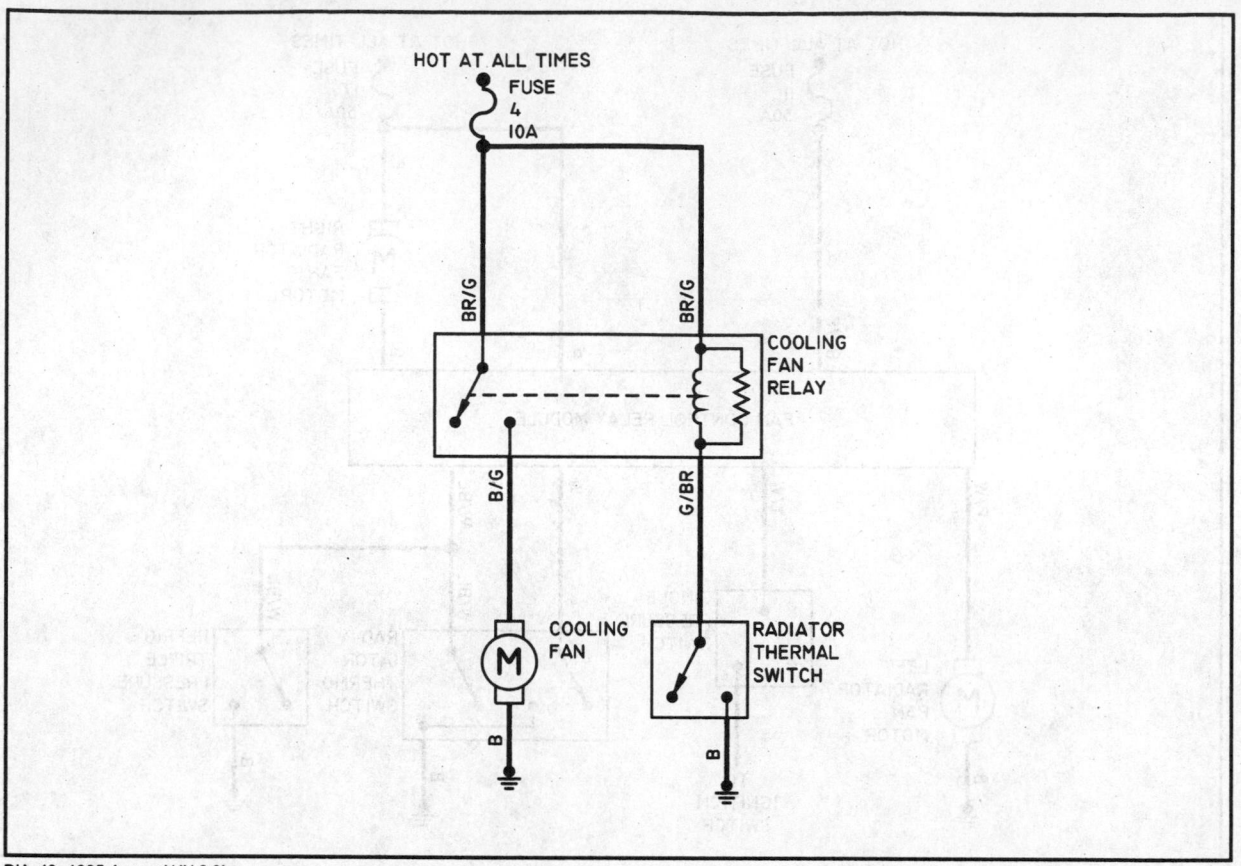

DIA. 43- 1995 Jaguar XJX 6.0L

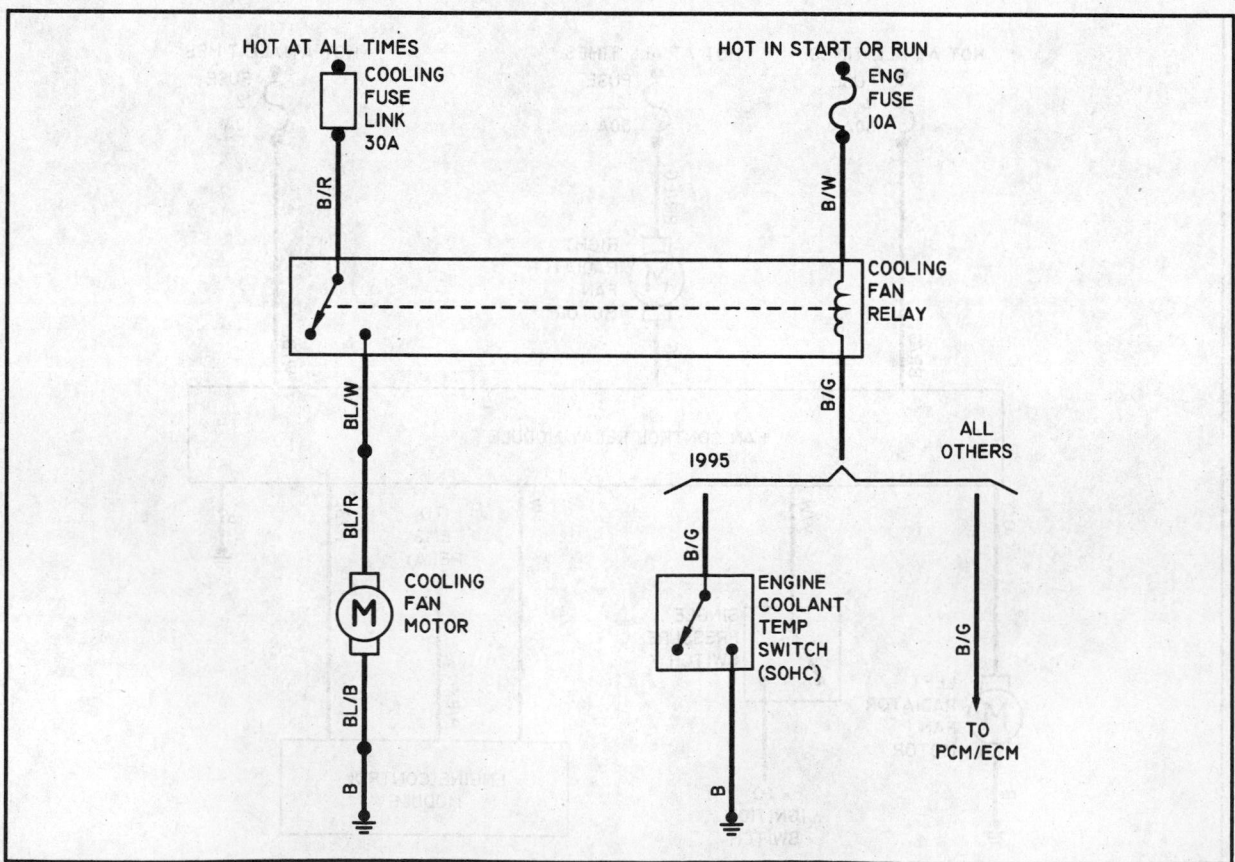

DIA. 44- 1995-99 Kia Sephia 1.6L / 1.8L

79239W22

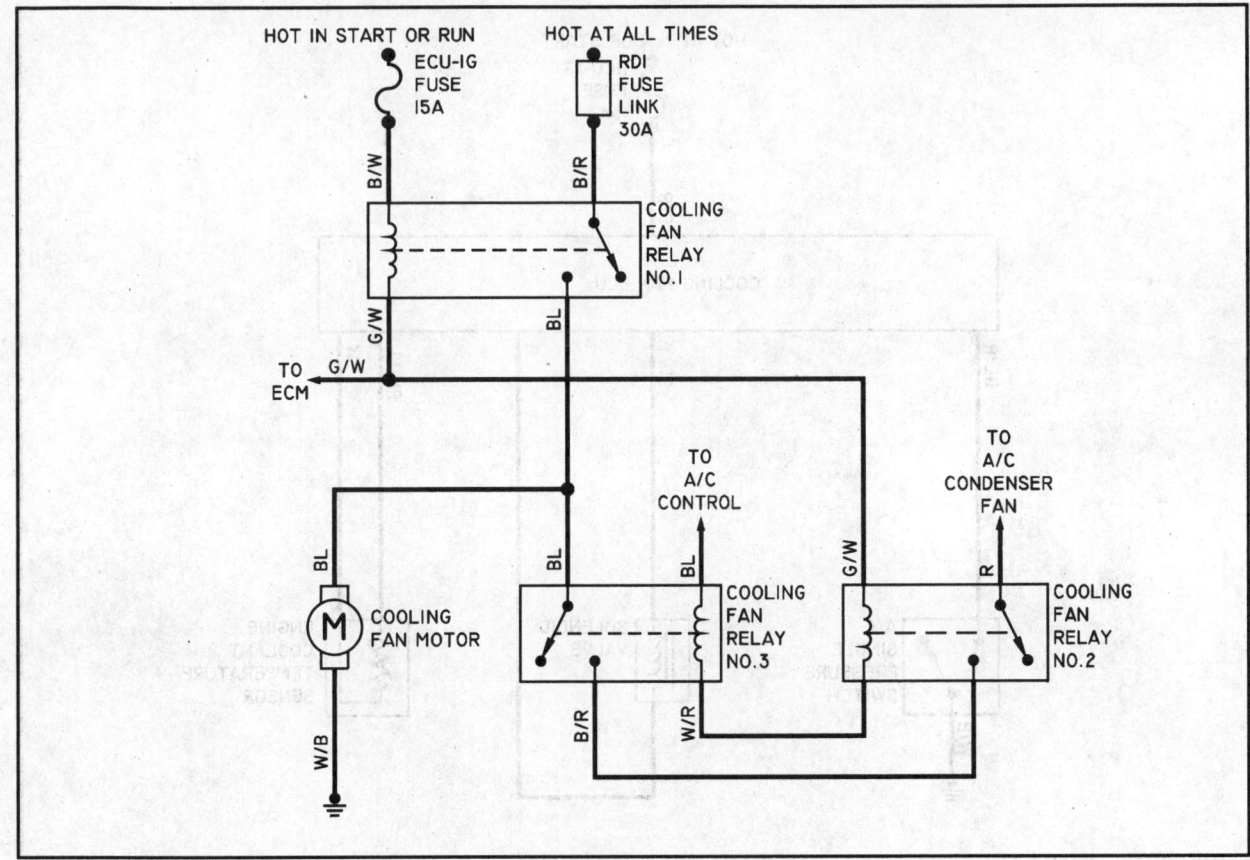

DIA. 45- 1995-99 Lexus ES300 3.0L

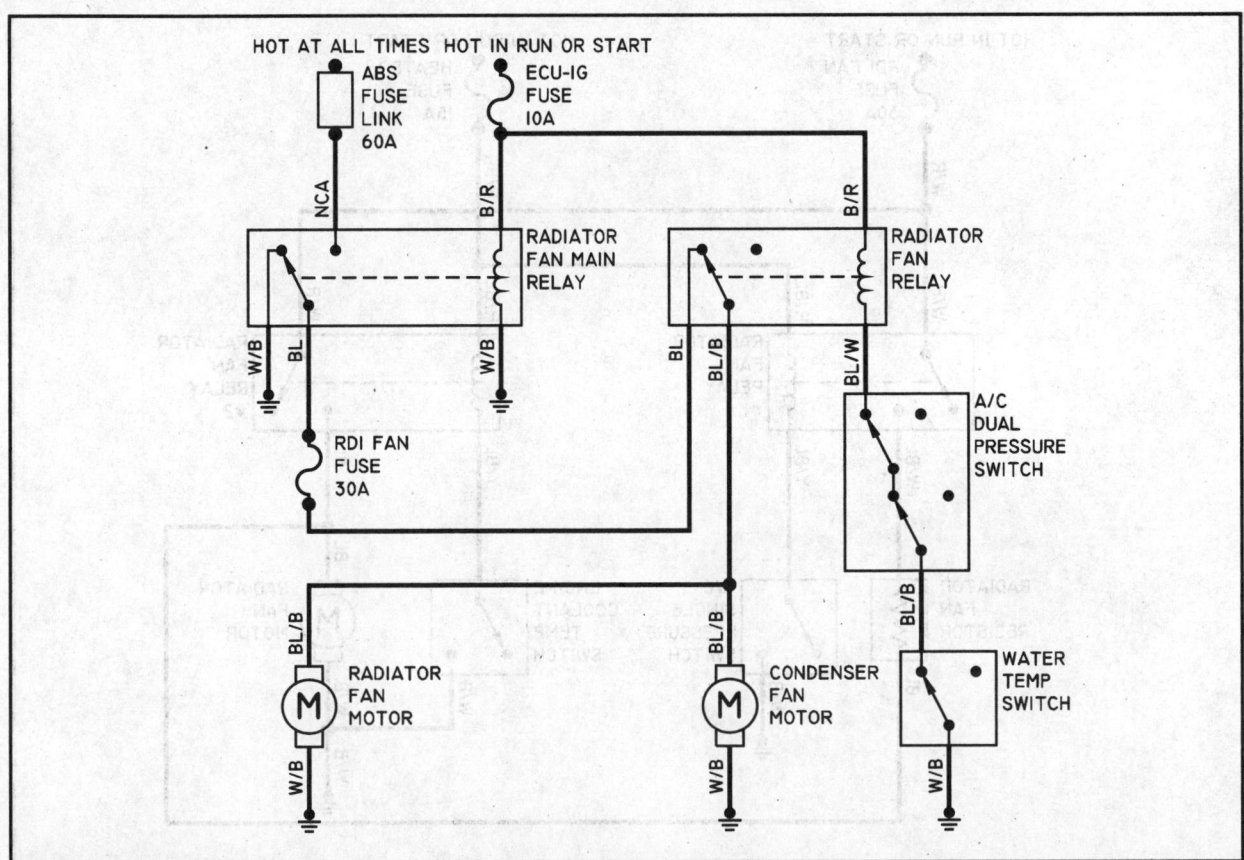

DIA. 46- 1995-99 Lexus GS 300, SC 300 3.0L

79239W23

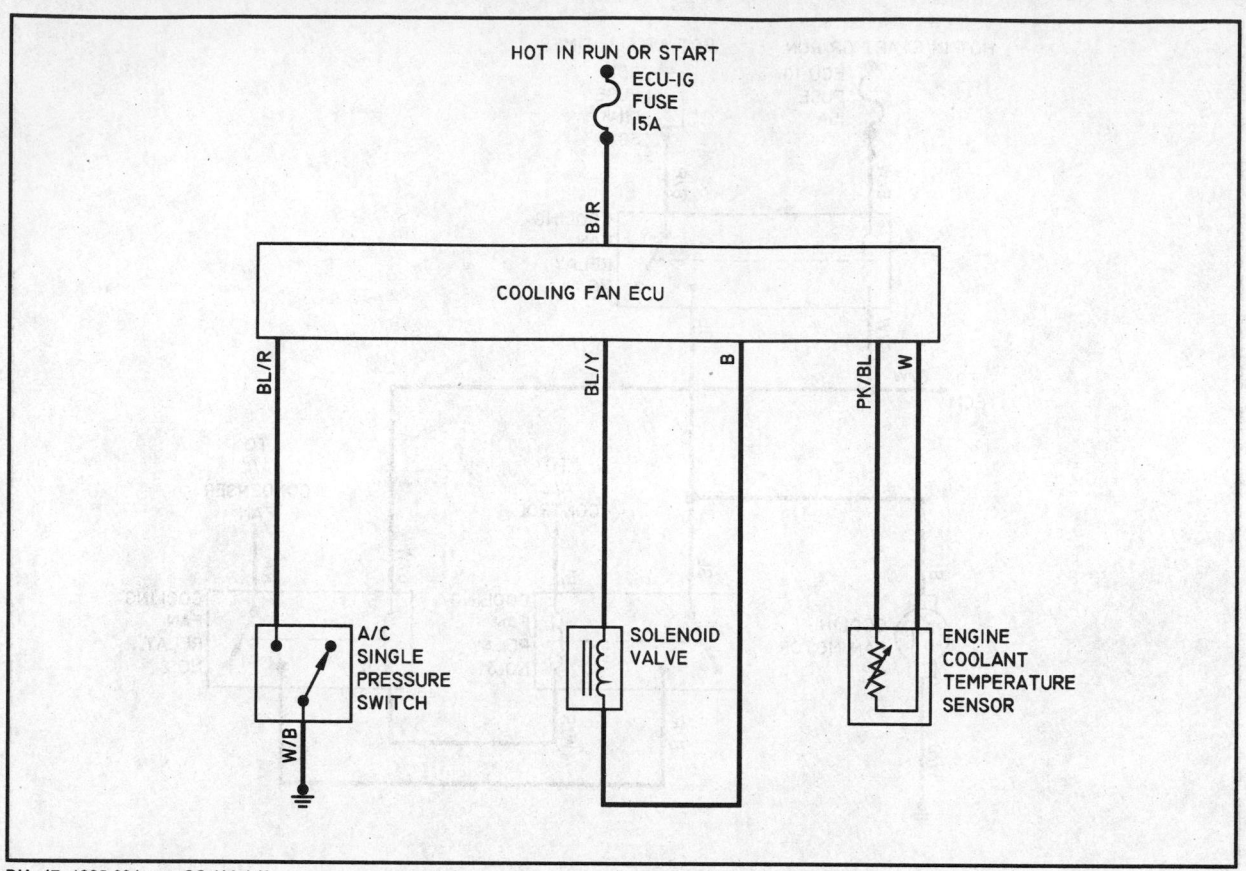

DIA. 47- 1995-99 Lexus SC 400 4.0L

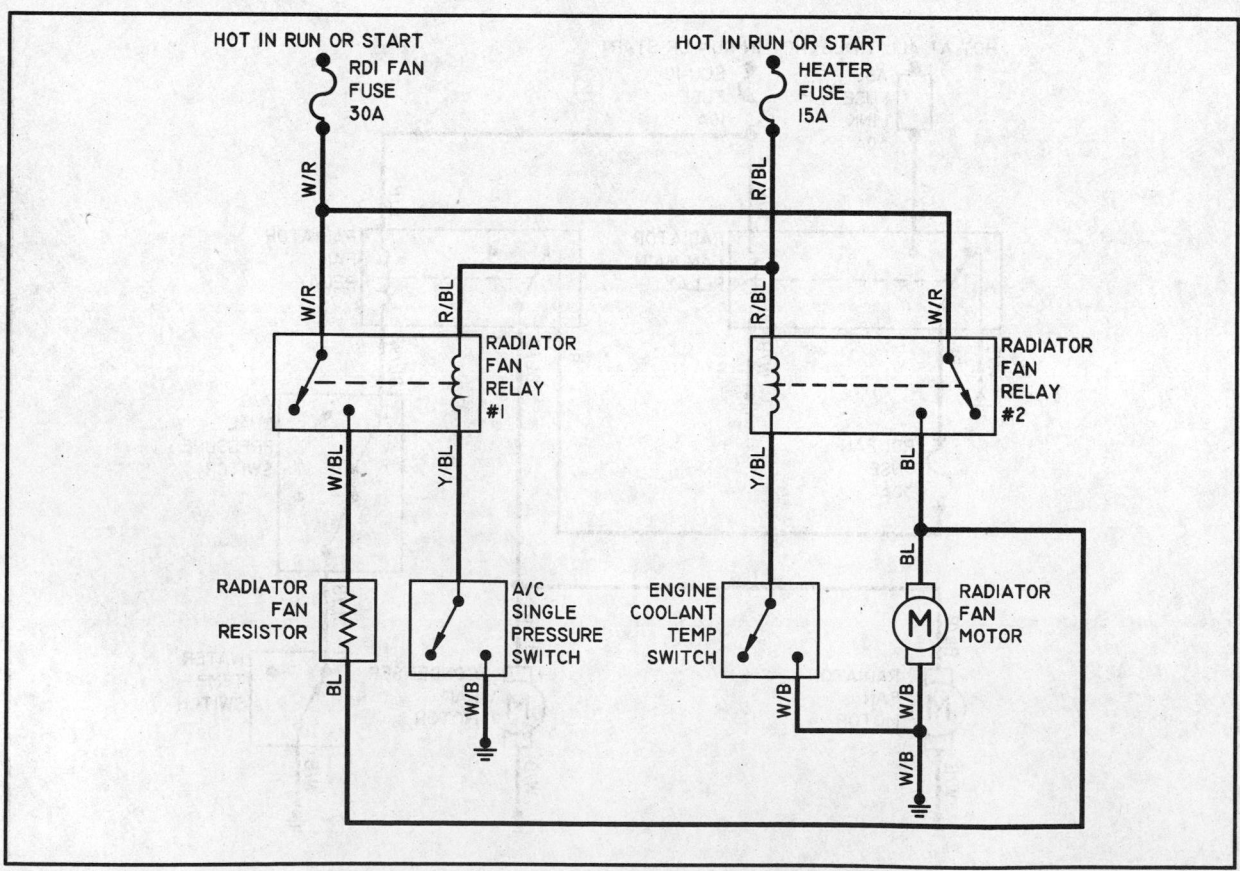

DIA. 48- 1995-99 Lexus LS 400 4.0L

79239W24

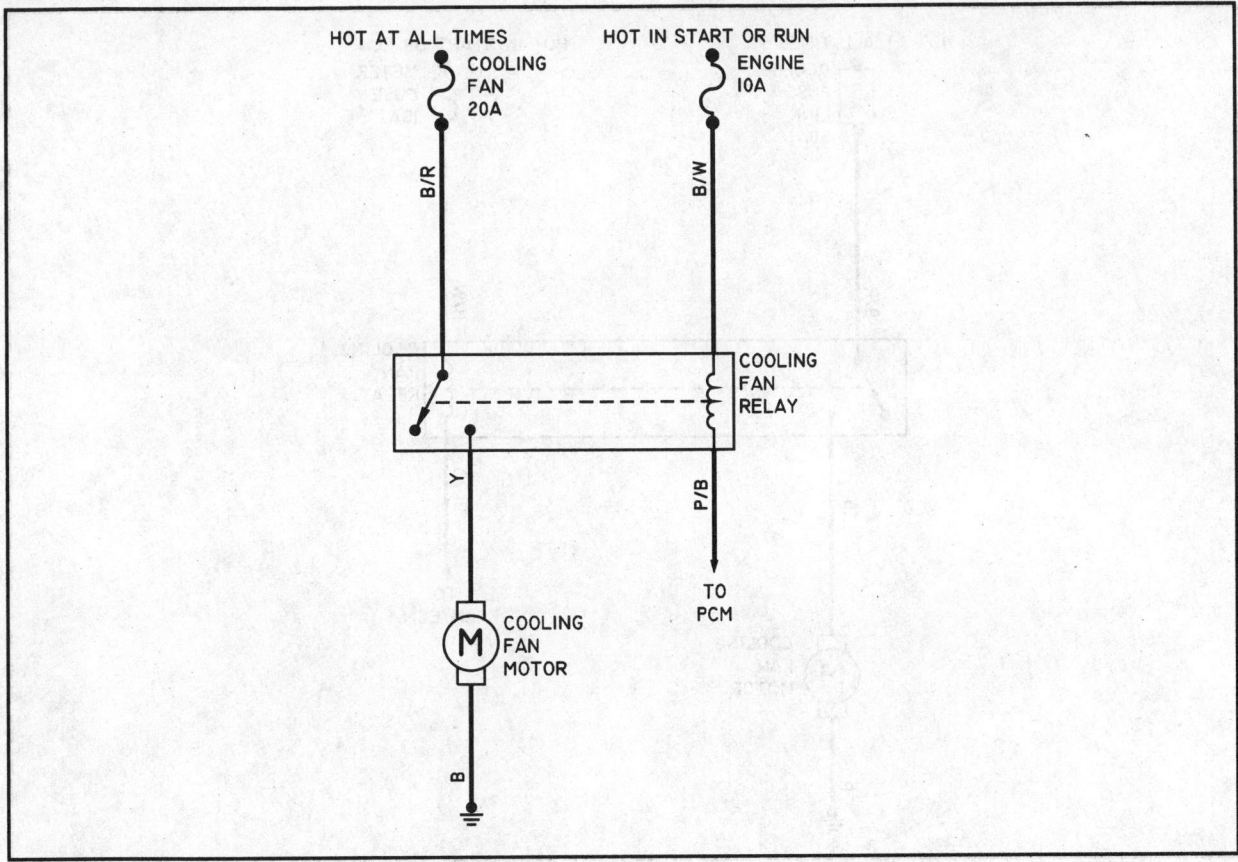

DIA. 49- 1995-99 Mazda 323, Protege 1.5L / 1.8L

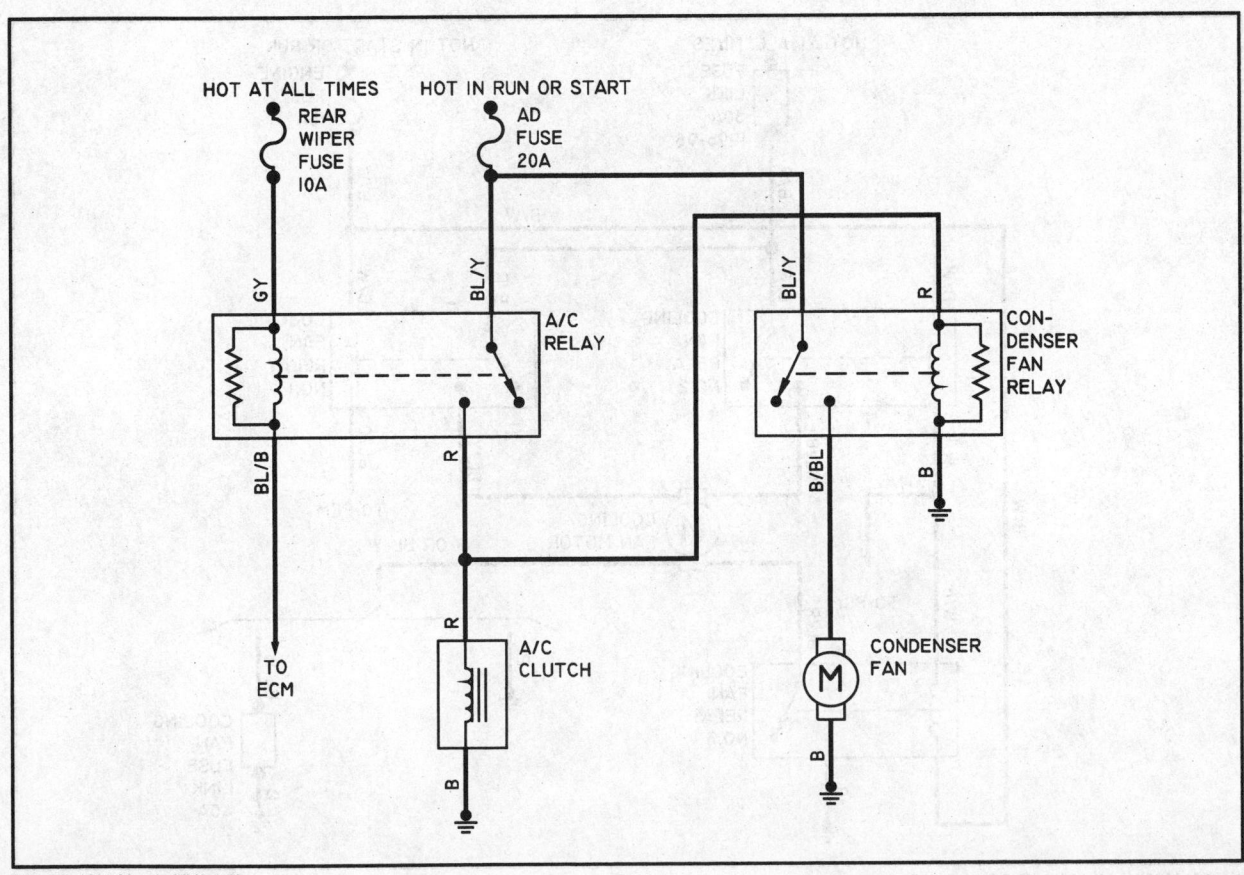

DIA. 50- 1995 Mazda MX3 1.6L

79239W25

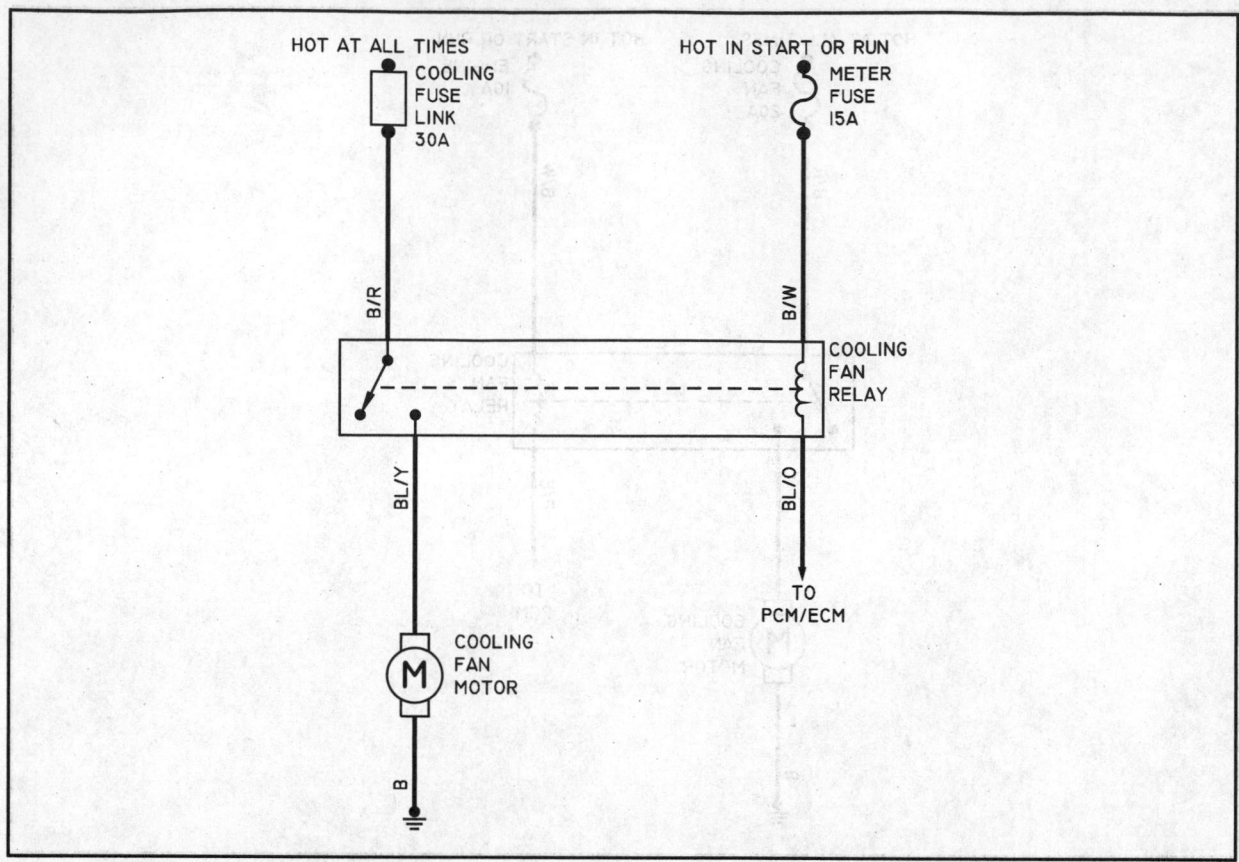

DIA. 51- 1995 Mazda 626, MX6 2.0L M/T, 1996-99 2.0L

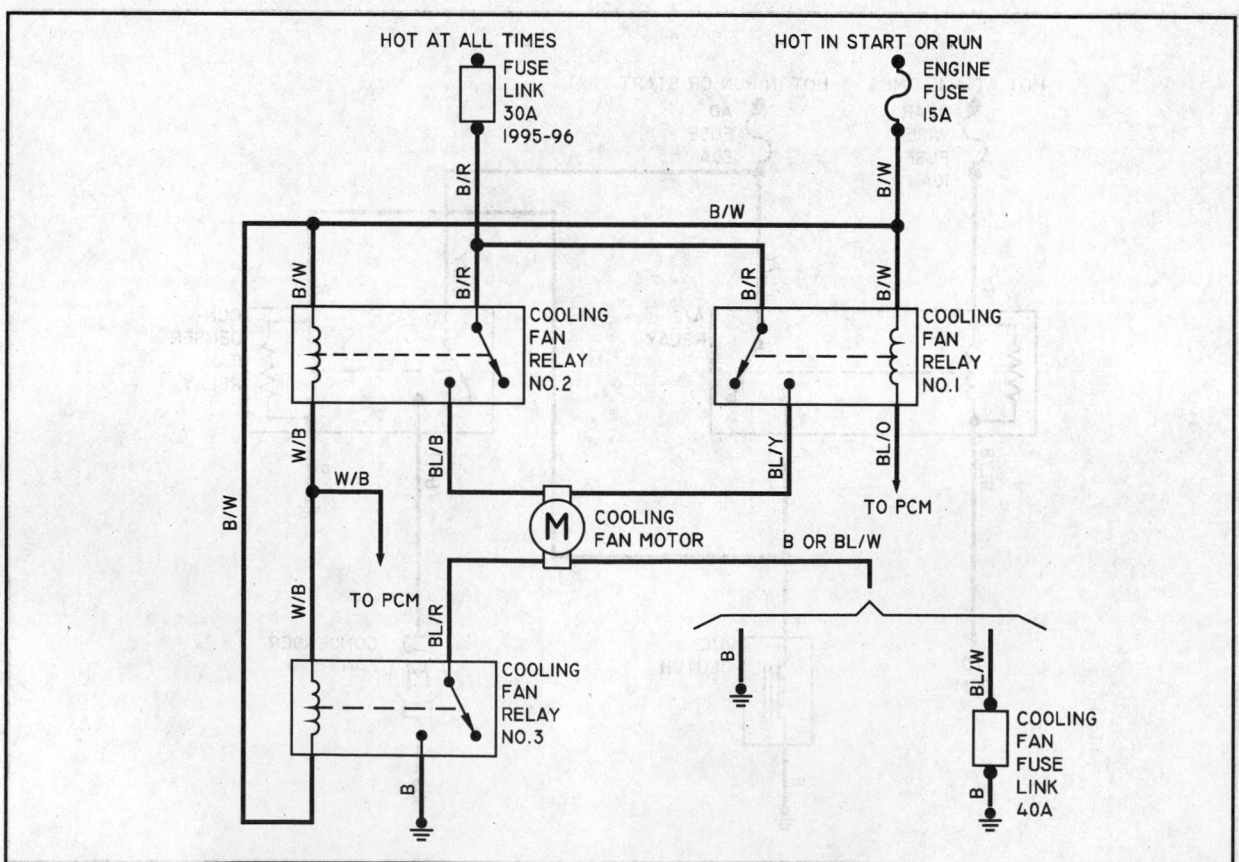

DIA. 52- 1995 Mazda 626, MX6 2.0L A/T, 1996-99 2.5L

79239W26

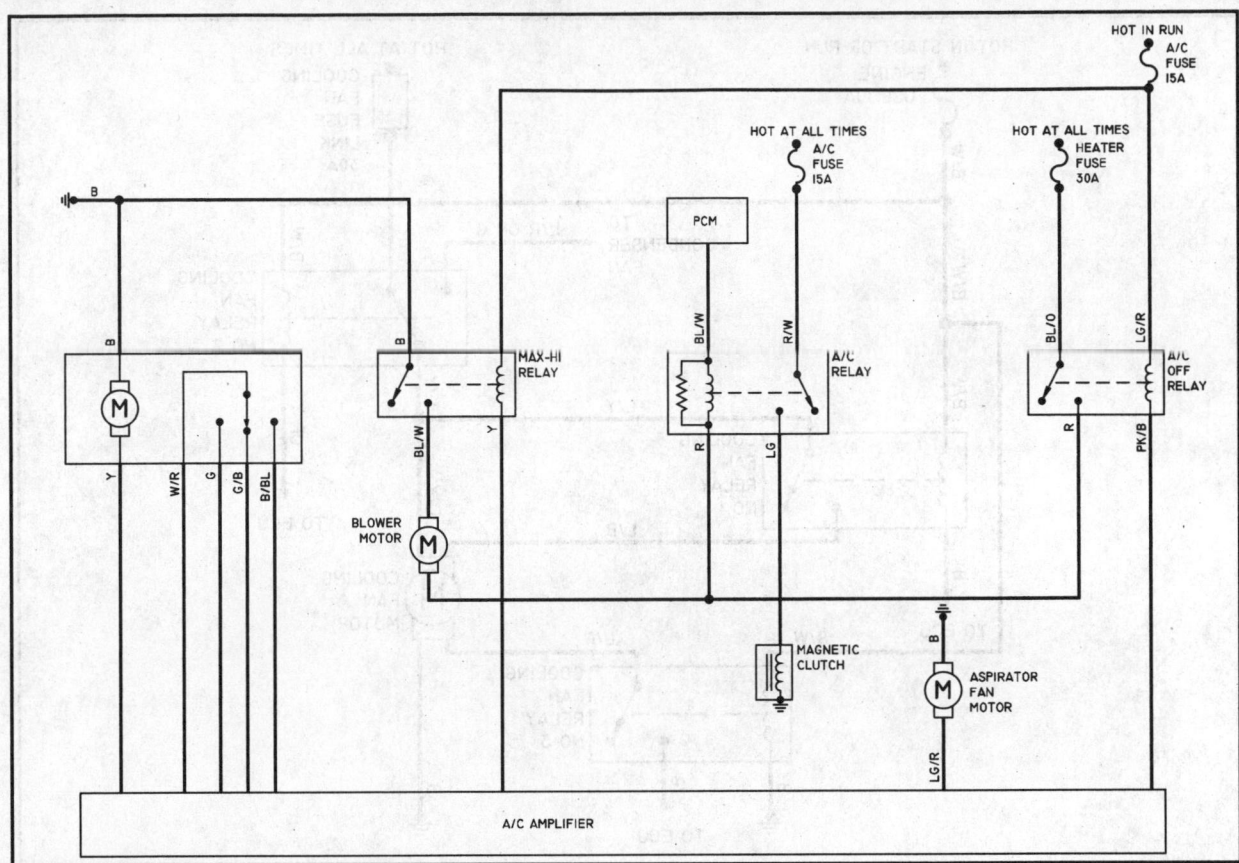

DIA. 53- 1995 Mazda 929 3.0L

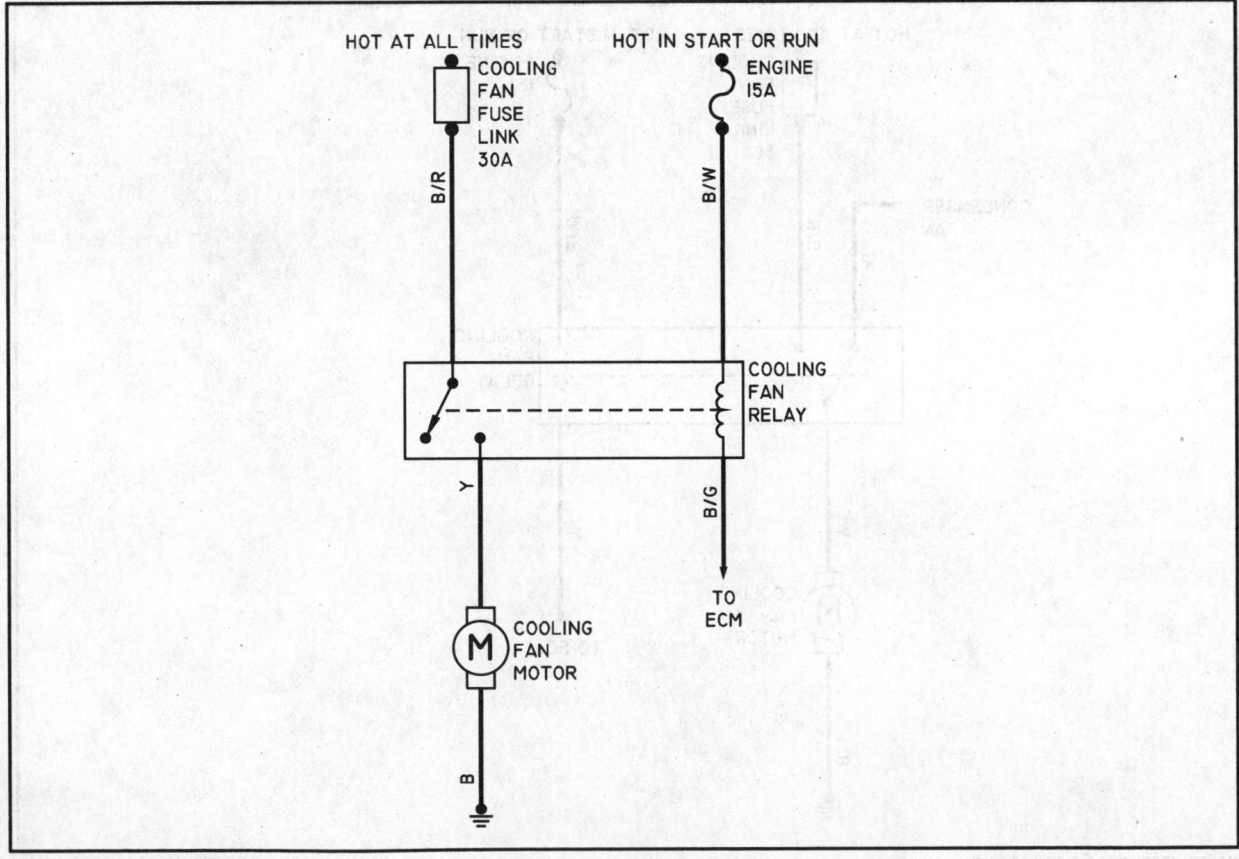

DIA. 54- 1995-99 Mazda Miata 1.8L

79239W27

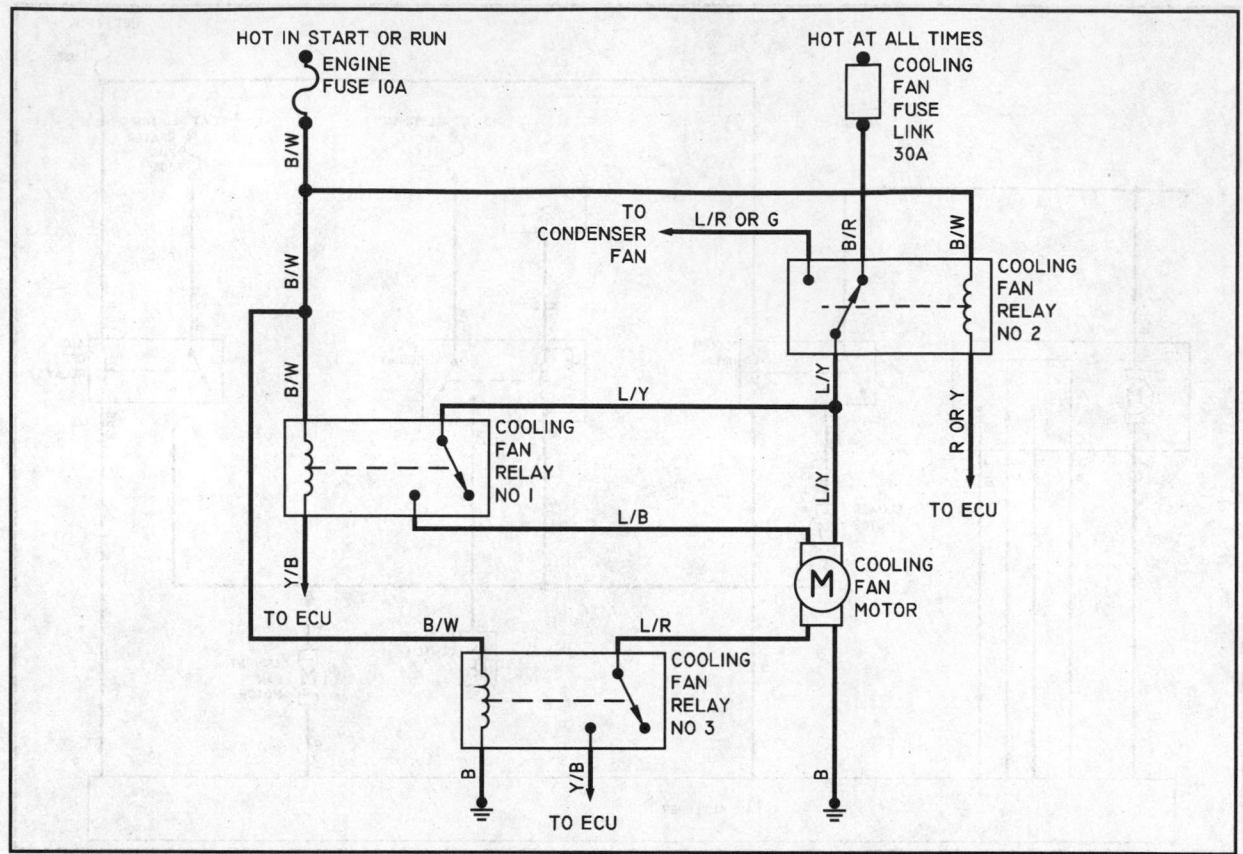

DIA. 55- 1995-99 Mazda Millenia 2.3L

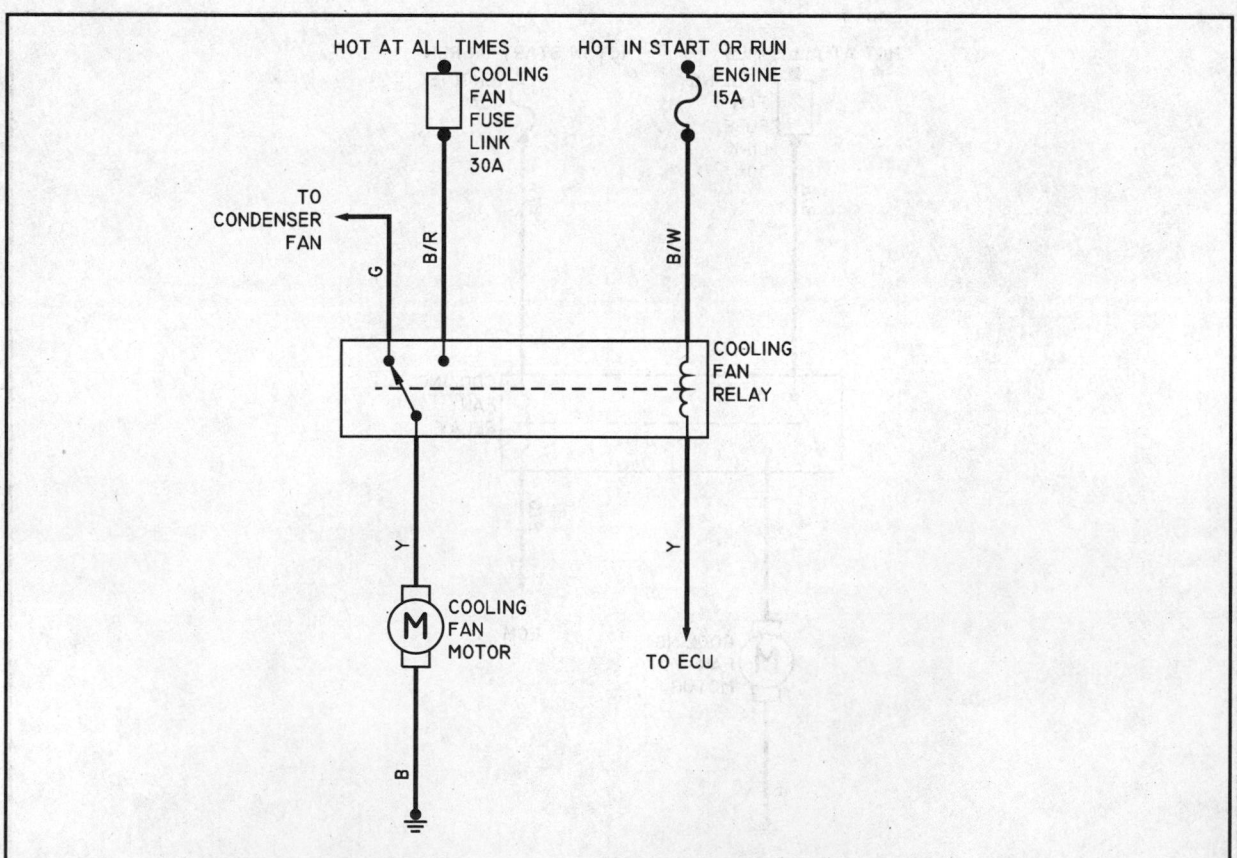

DIA. 56- 1995-99 Mazda Millenia 2.5L

79239W28

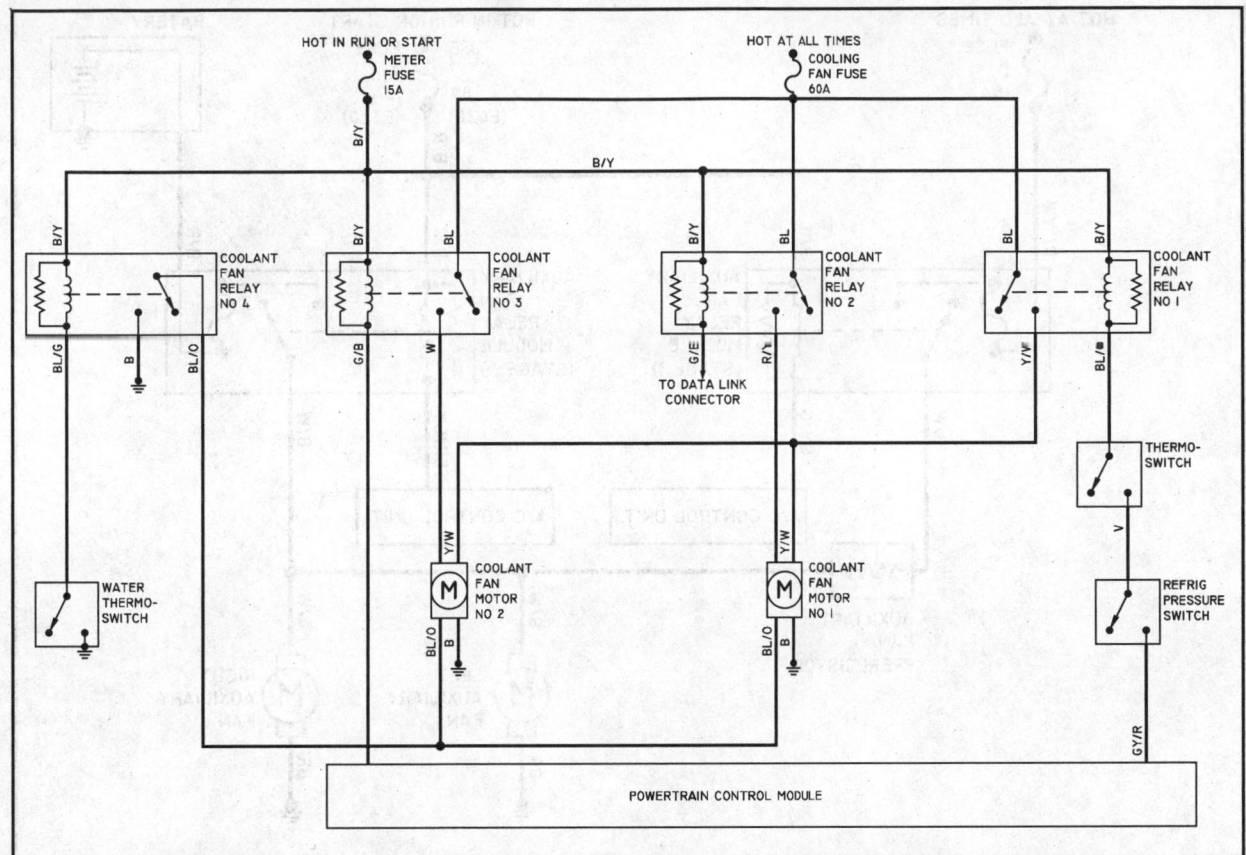

DIA. 57- 1995 Mazda RX7 1.3L

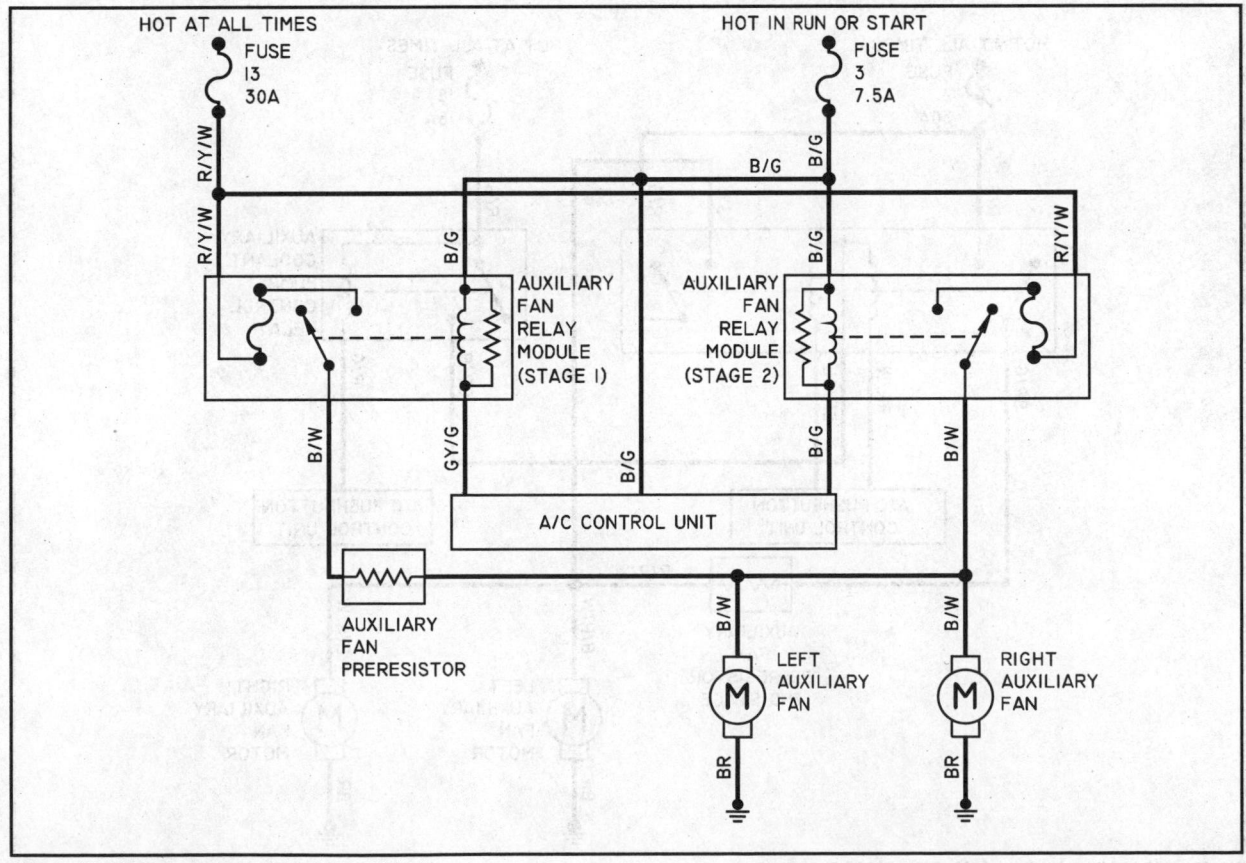

DIA. 58- 1995-99 Mercedes-Benz C220, C280 2.2L / 2.8L

79239W29

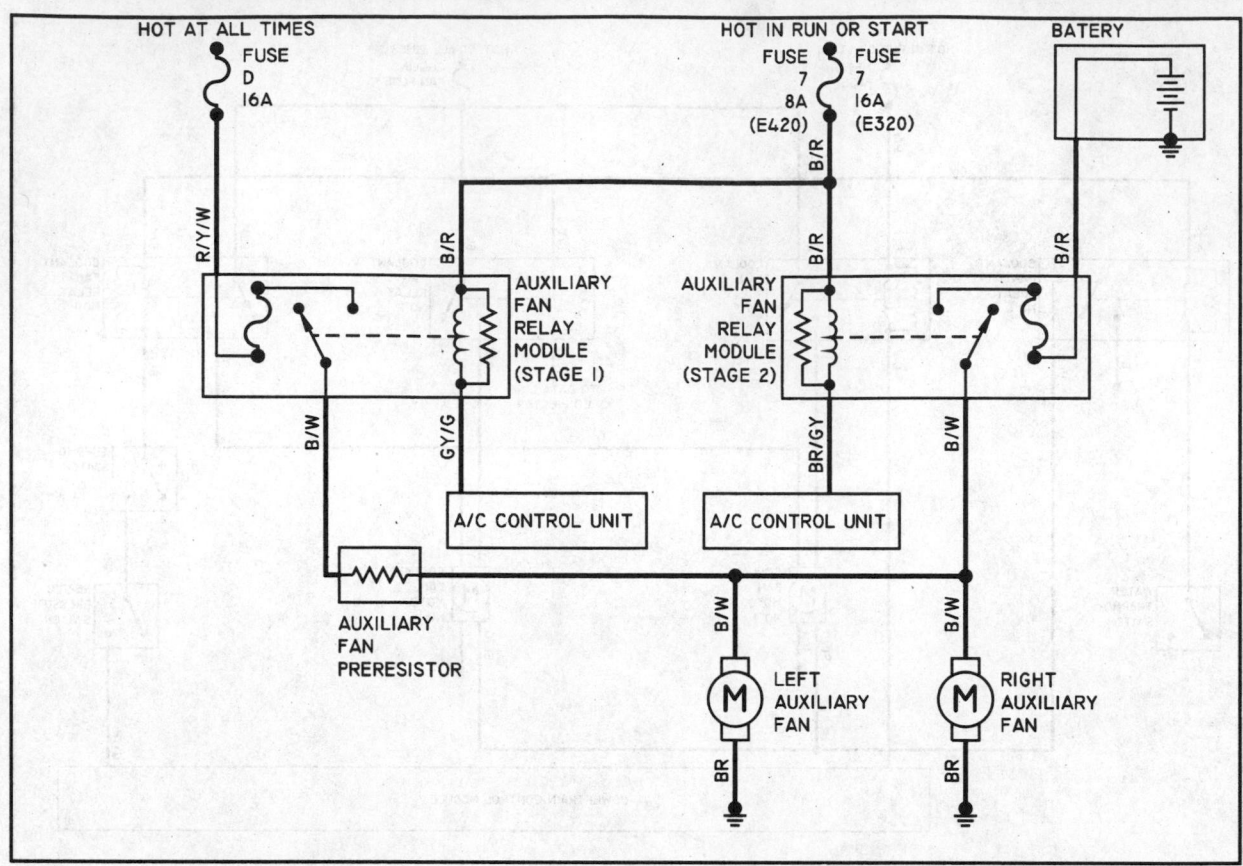

DIA. 59- 1995-99 Mercedes-Benz E320, E420 3.2L / 4.2L

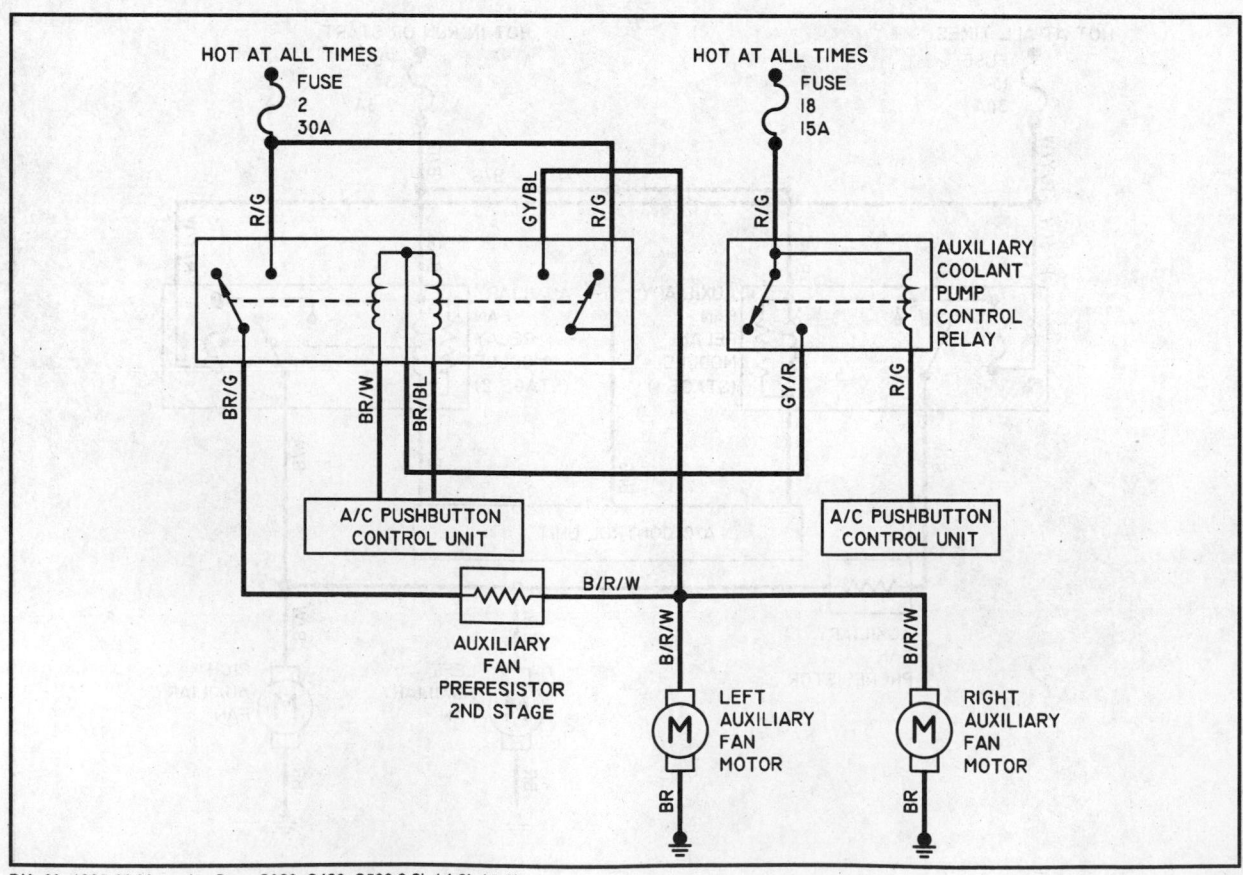

DIA. 60- 1995-99 Mercedes-Benz S320, S420, S500 3.2L / 4.2L / 5.0L

79239W30

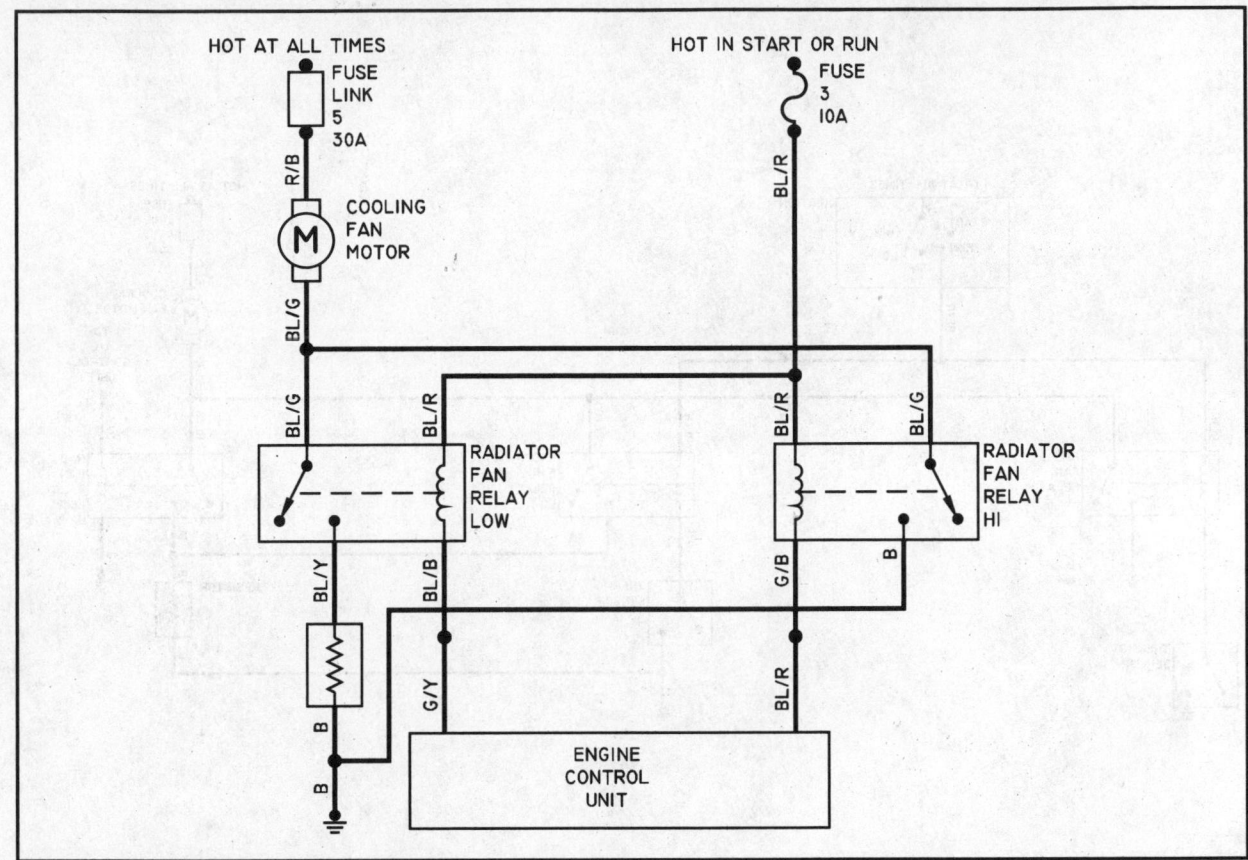

DIA. 61- 1995-99 Mitsubishi 3000GT 3.0L

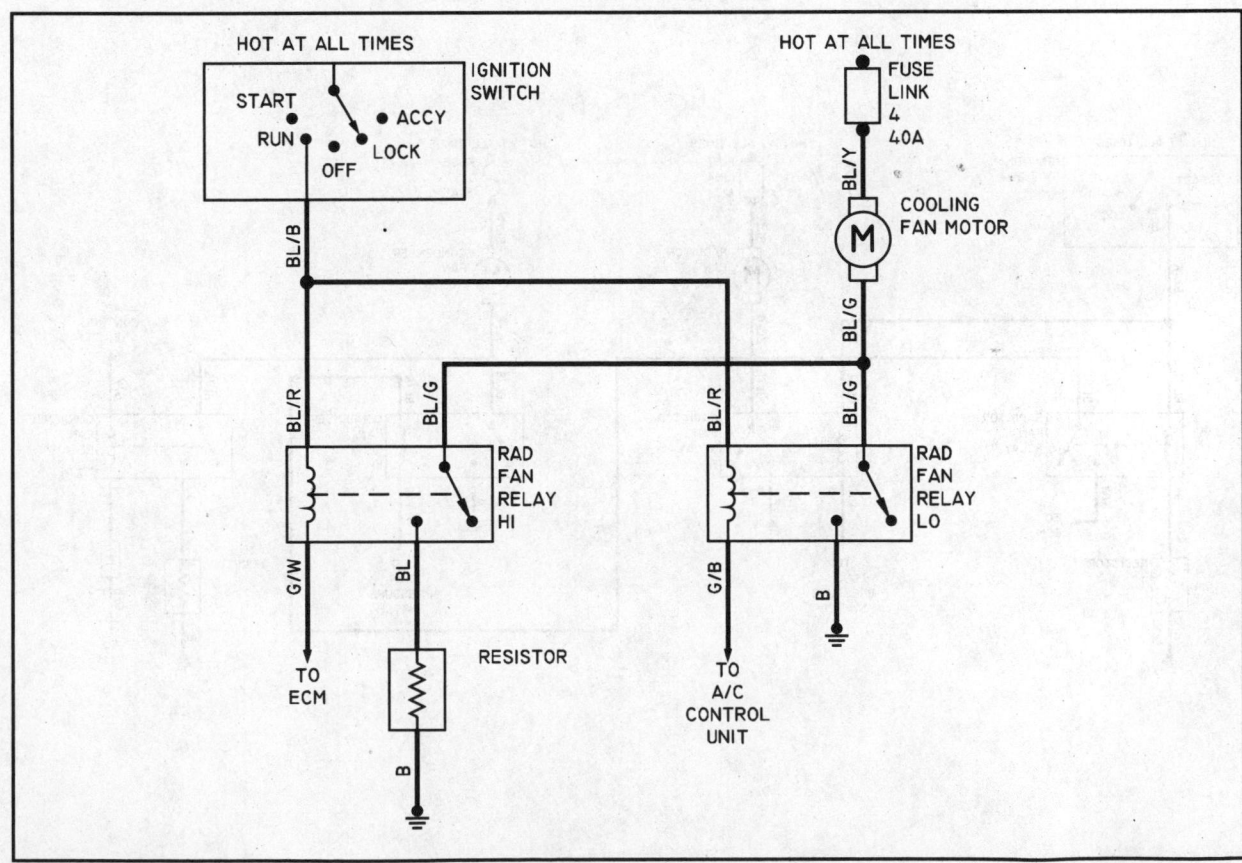

DIA. 62- 1995 Mitsubishi Diamante 3.0L (Except wagon)

79239W31

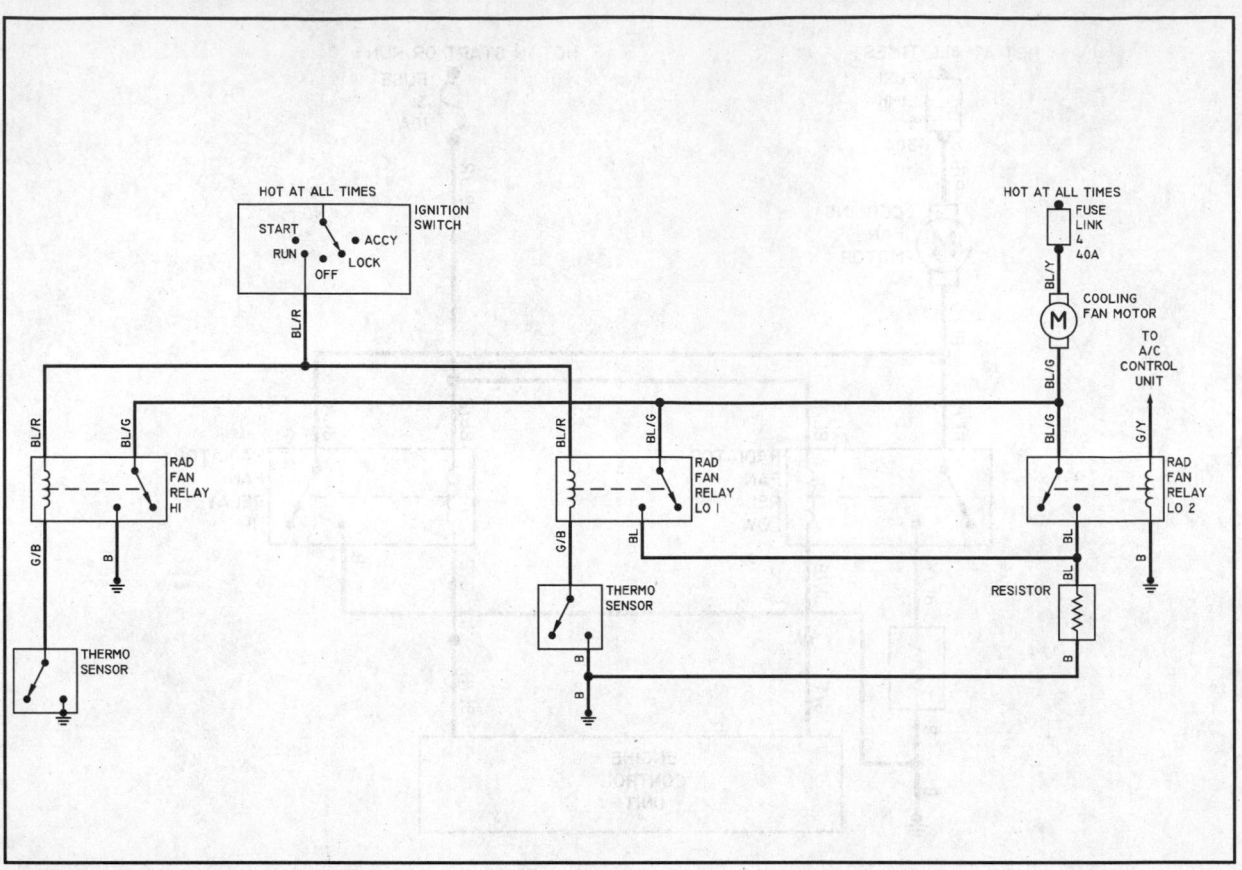

DIA. 63- 1995 Mitsubishi Diamante 3.0L (WAGON)

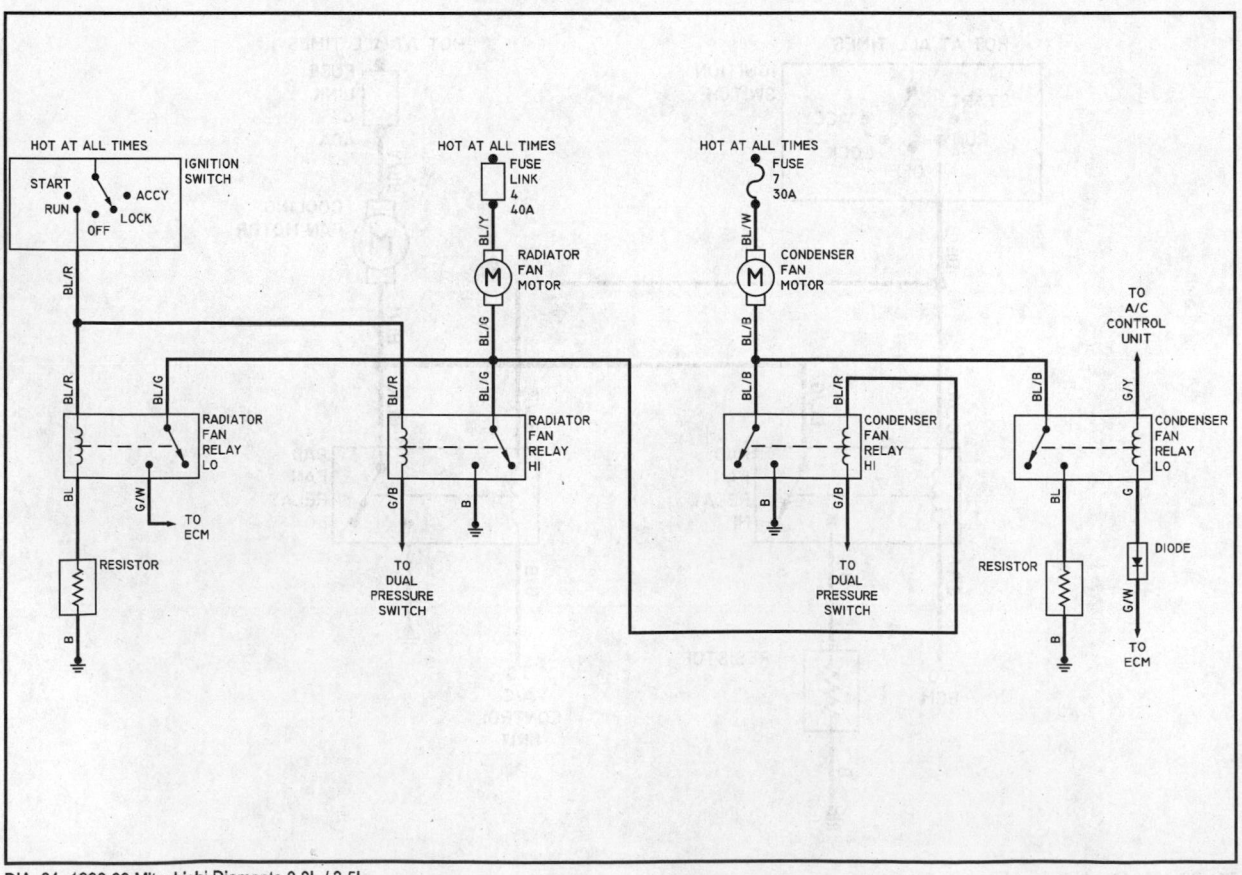

DIA. 64- 1996-99 Mitsubishi Diamante 3.0L / 3.5L

79239W32

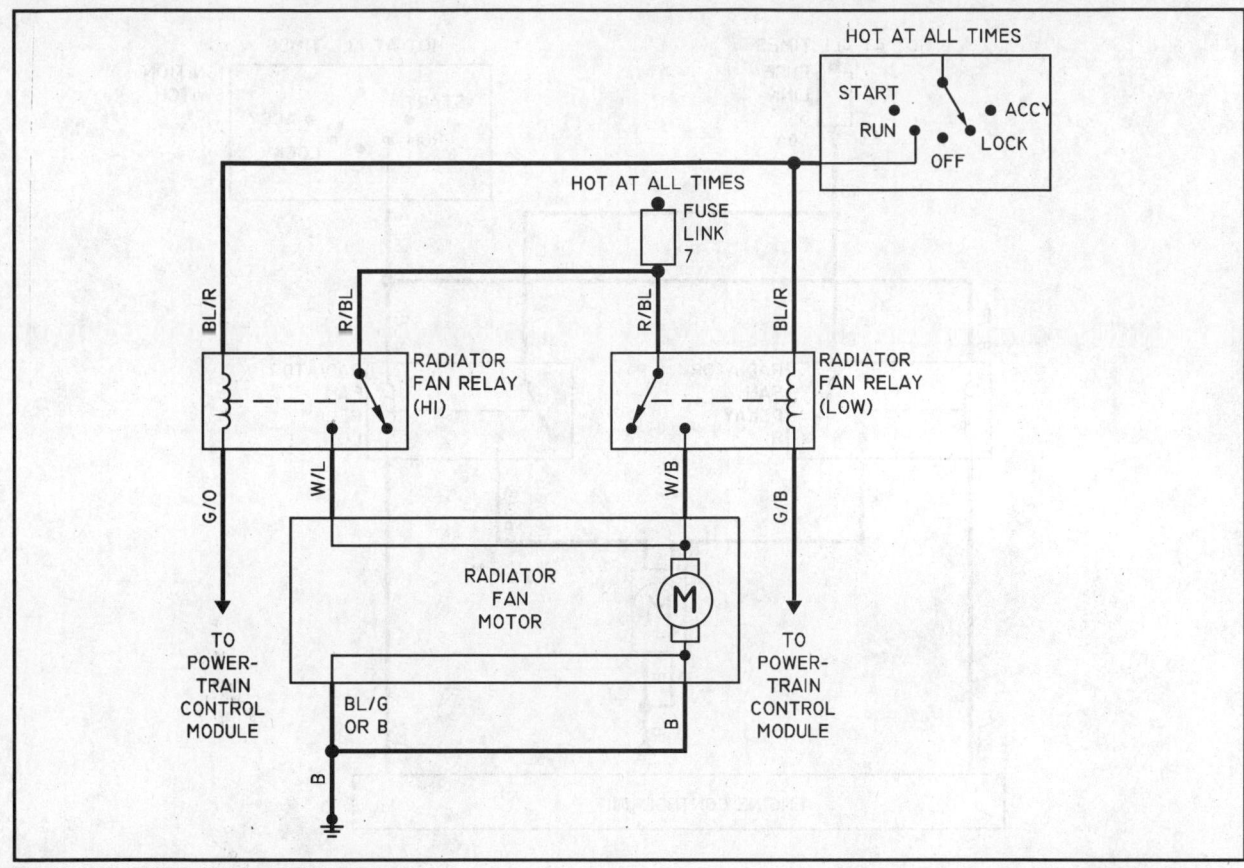

DIA. 65- 1995-99 Mitsubishi Eclipse 2.0L Non-turbo, 2.0L
Turbo M/T, 2.4L M/T

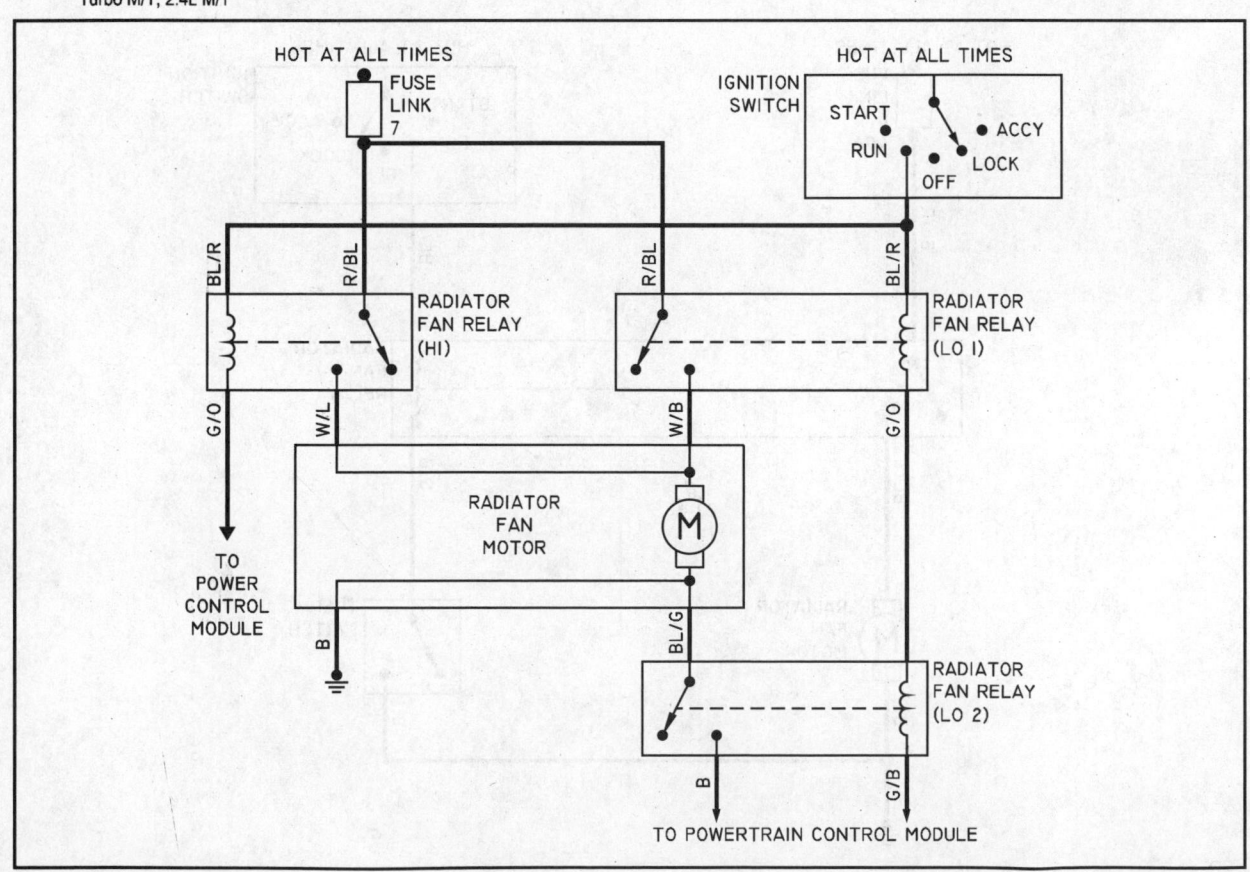

DIA. 66- 1996-99 Mitsubishi Eclipse 2.0L Turbo A/T, 2.4L A/T

79239W33

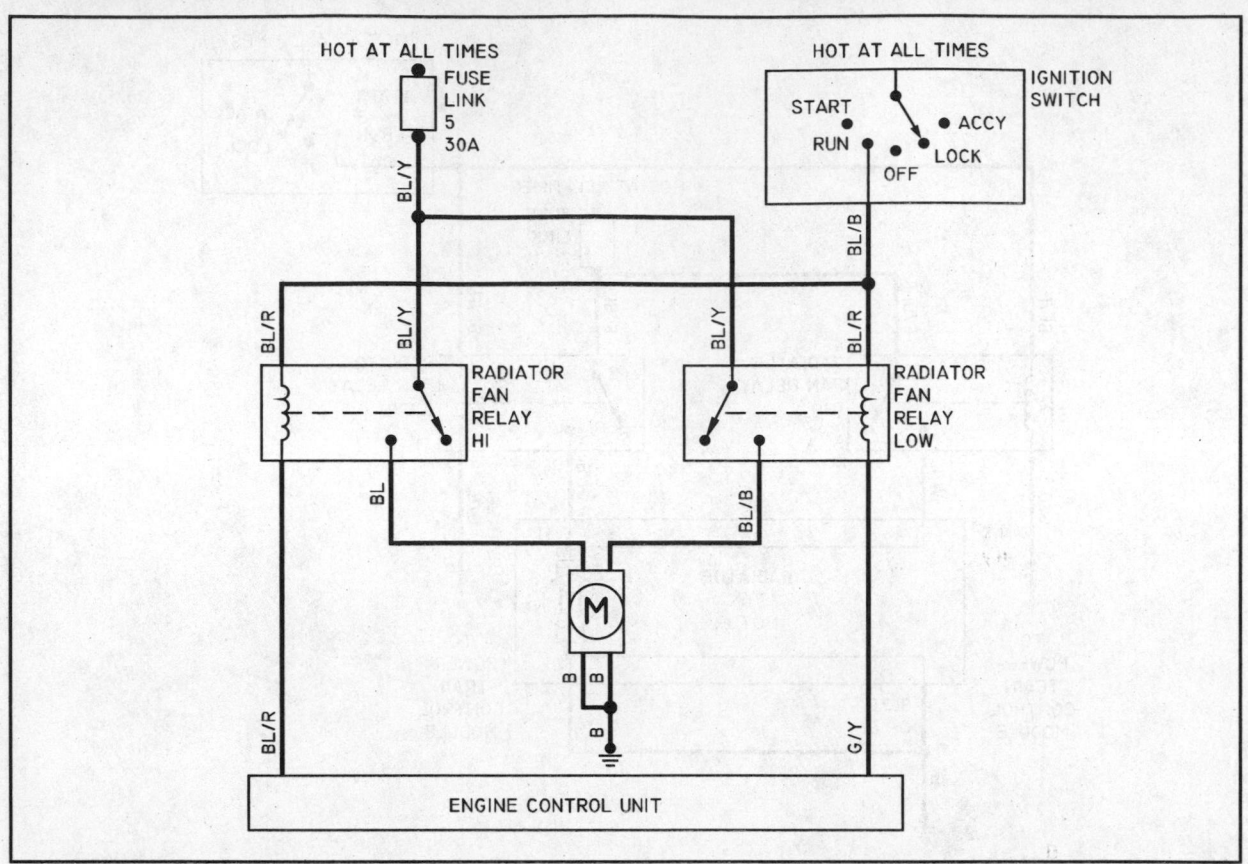

DIA. 67- 1995-99 Mitsubishi Galant 2.4L

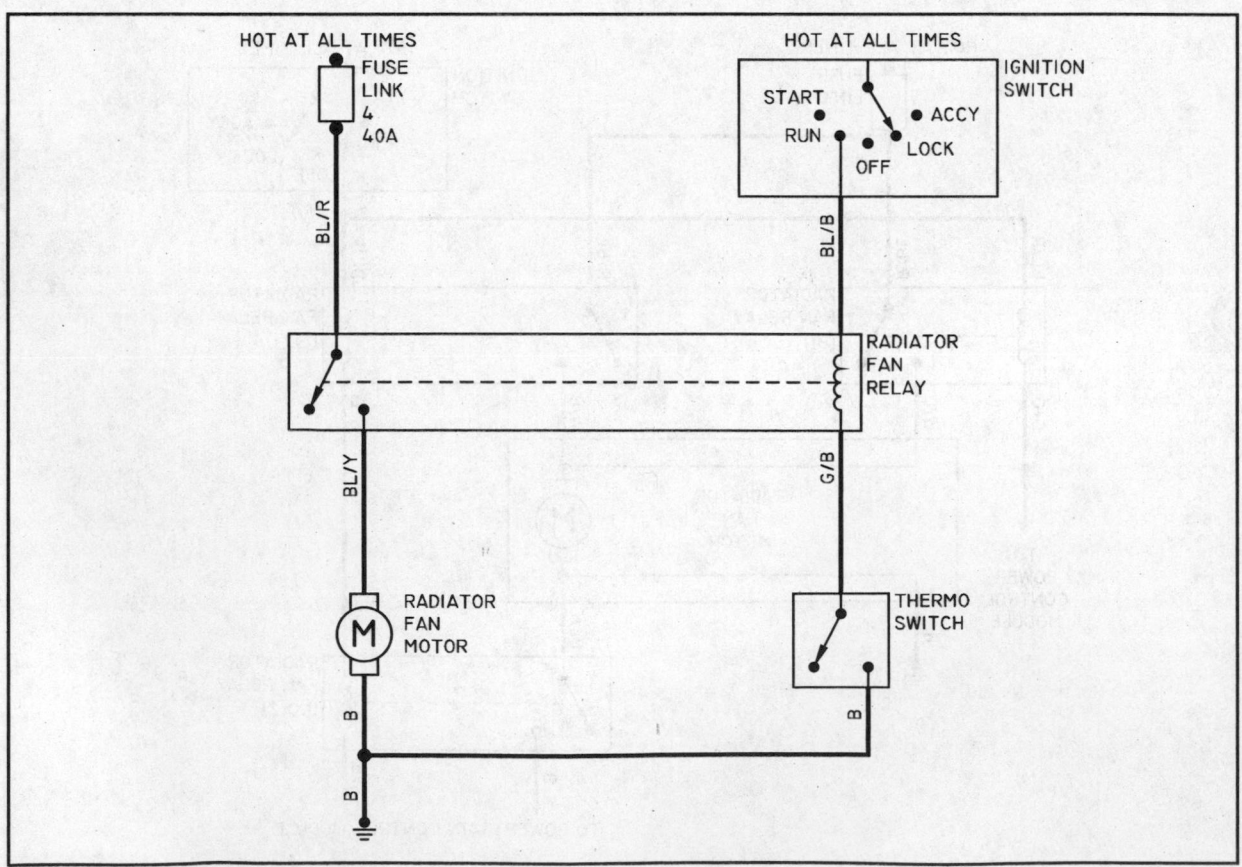

DIA. 68- 1995-96 Mitsubishi Mirage 1.5L

79239W34

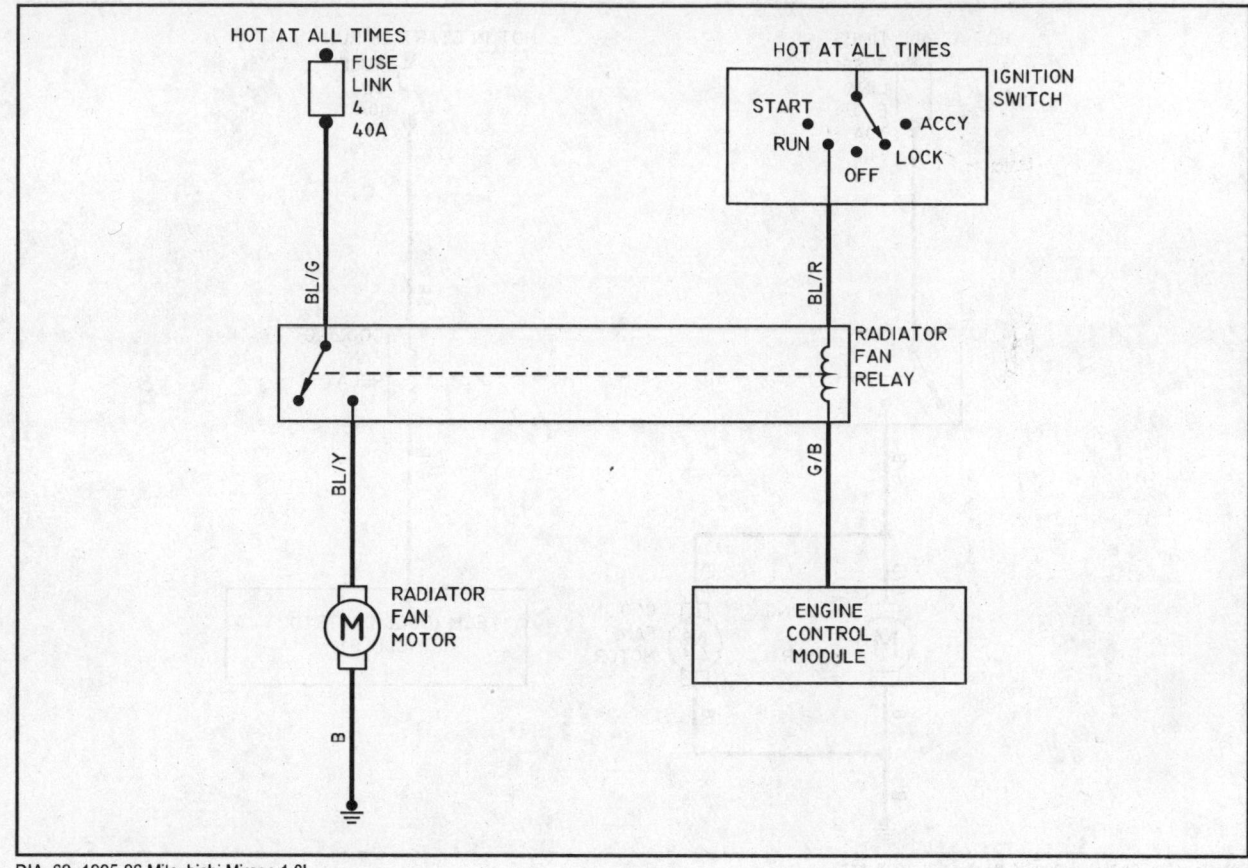

DIA. 69- 1995-96 Mitsubishi Mirage 1.8L

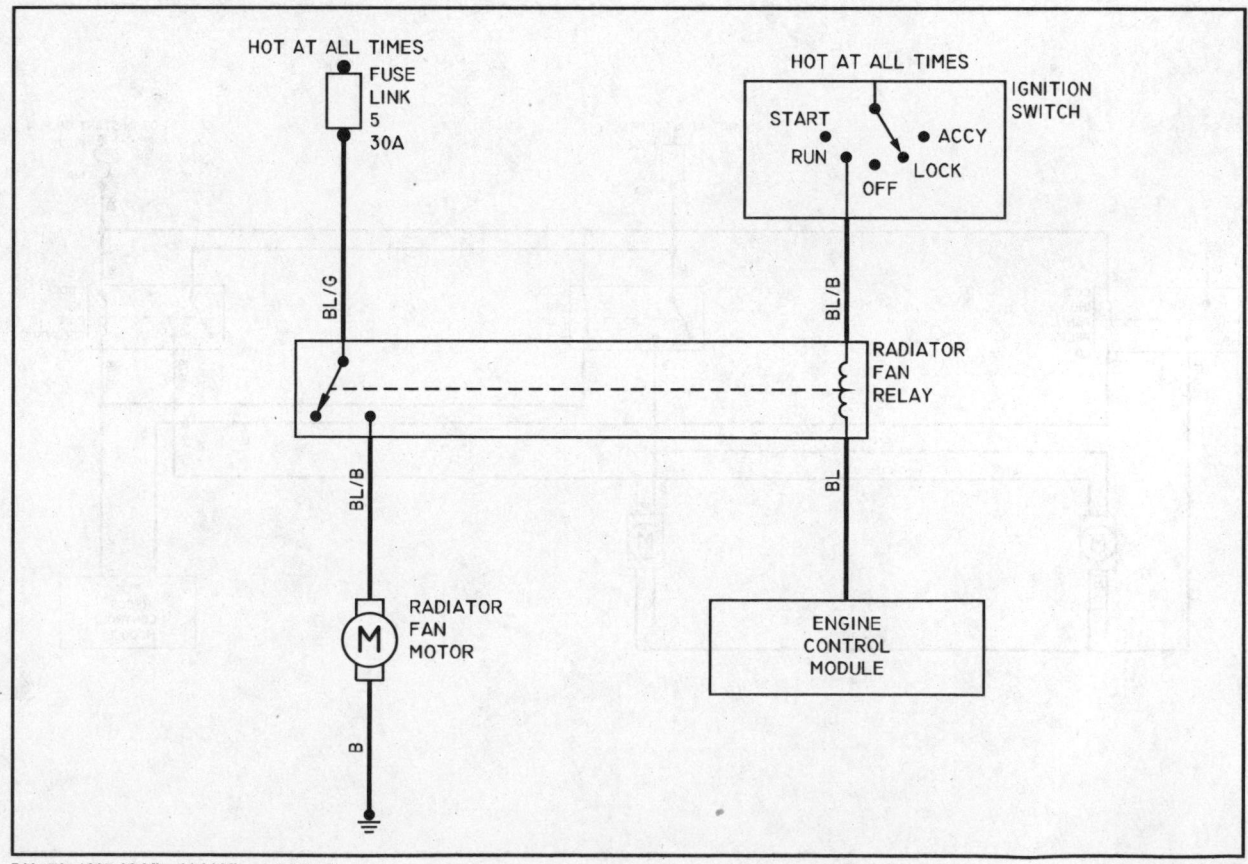

DIA. 70- 1997-99 Mitsubishi Mirage 1.5L/1.8L

79239W35

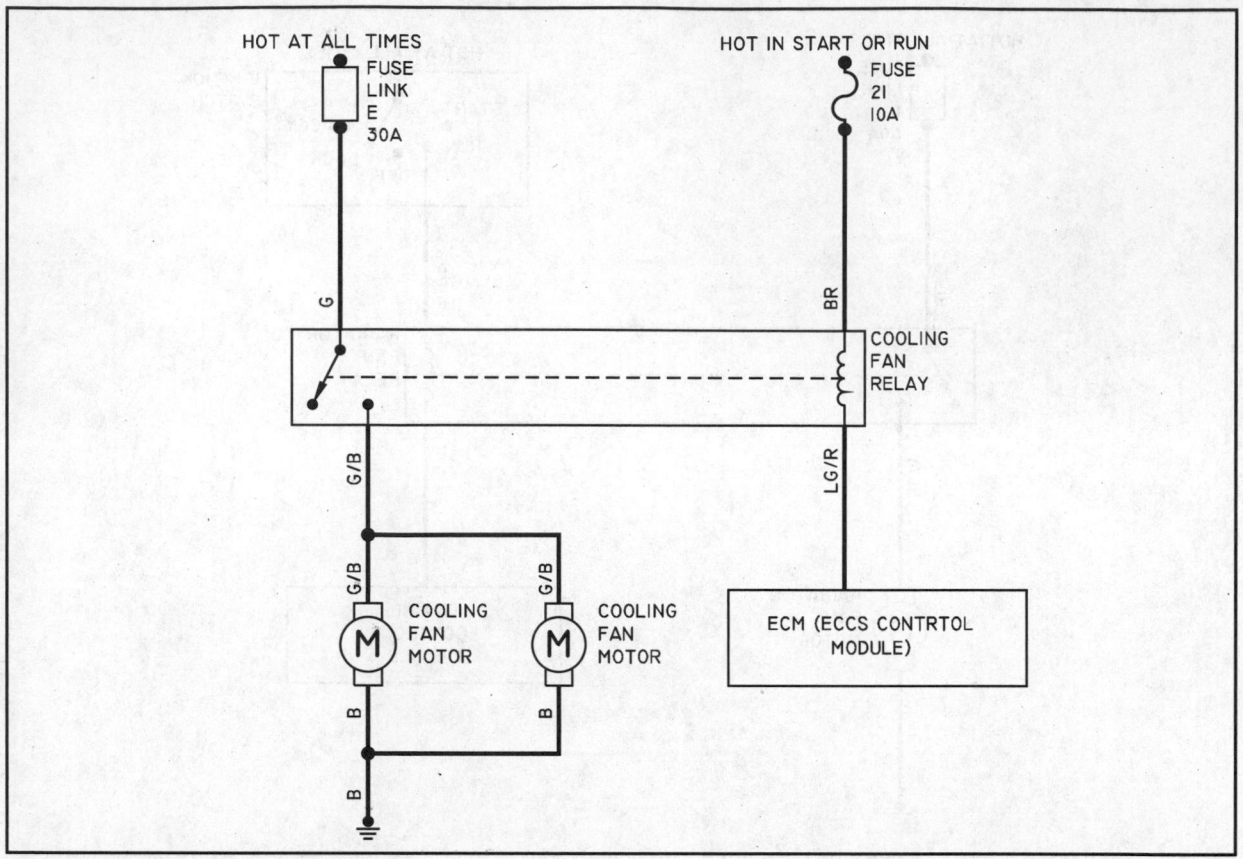

DIA. 71- 1995-99 Nissan 200SX, Sentra 1.6L / 2.0L M/T

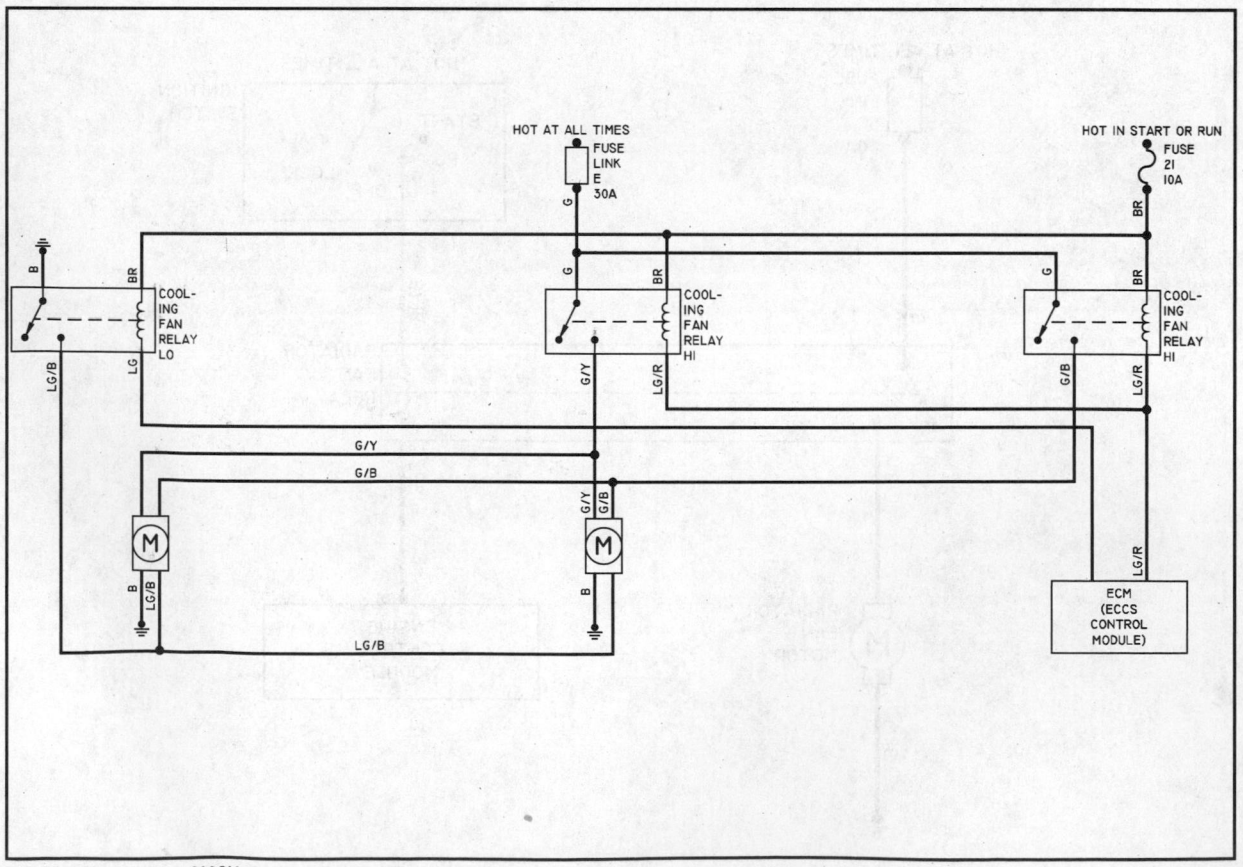

DIA. 72- 1995-99 Nissan 200SX, Sentra 1.6L/2.0L A/T

79239W36

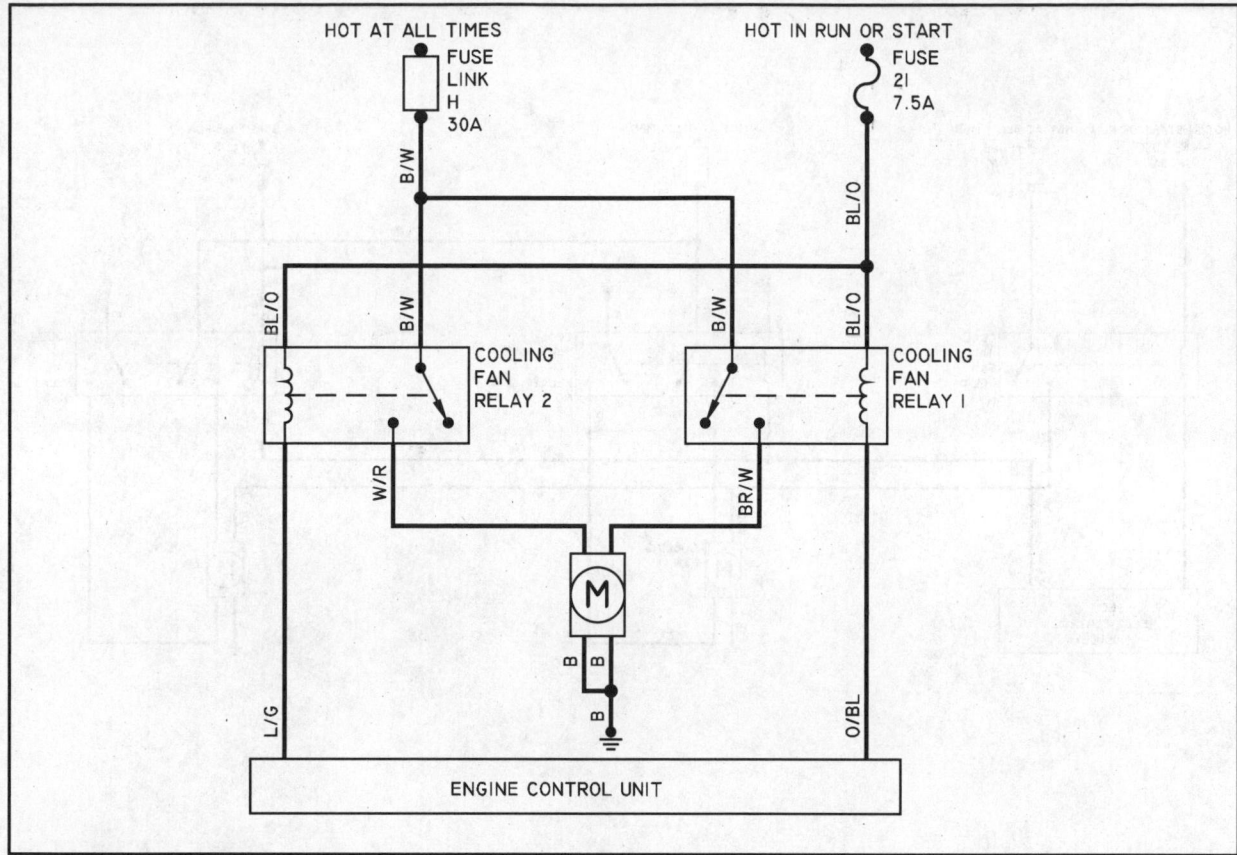

DIA. 73- 1995-99 Nissan 240SX 2.4L

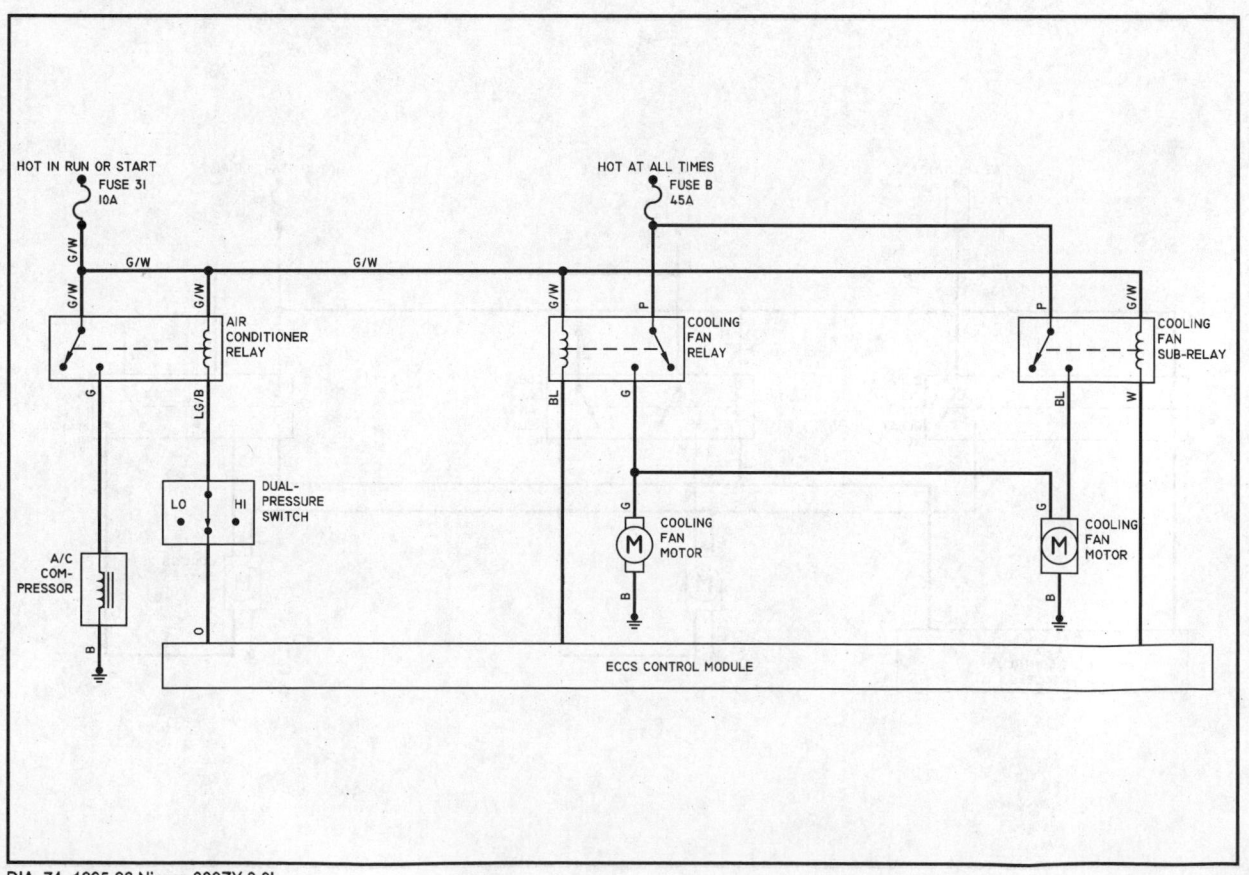

DIA. 74- 1995-96 Nissan 300ZX 3.0L

79239W37

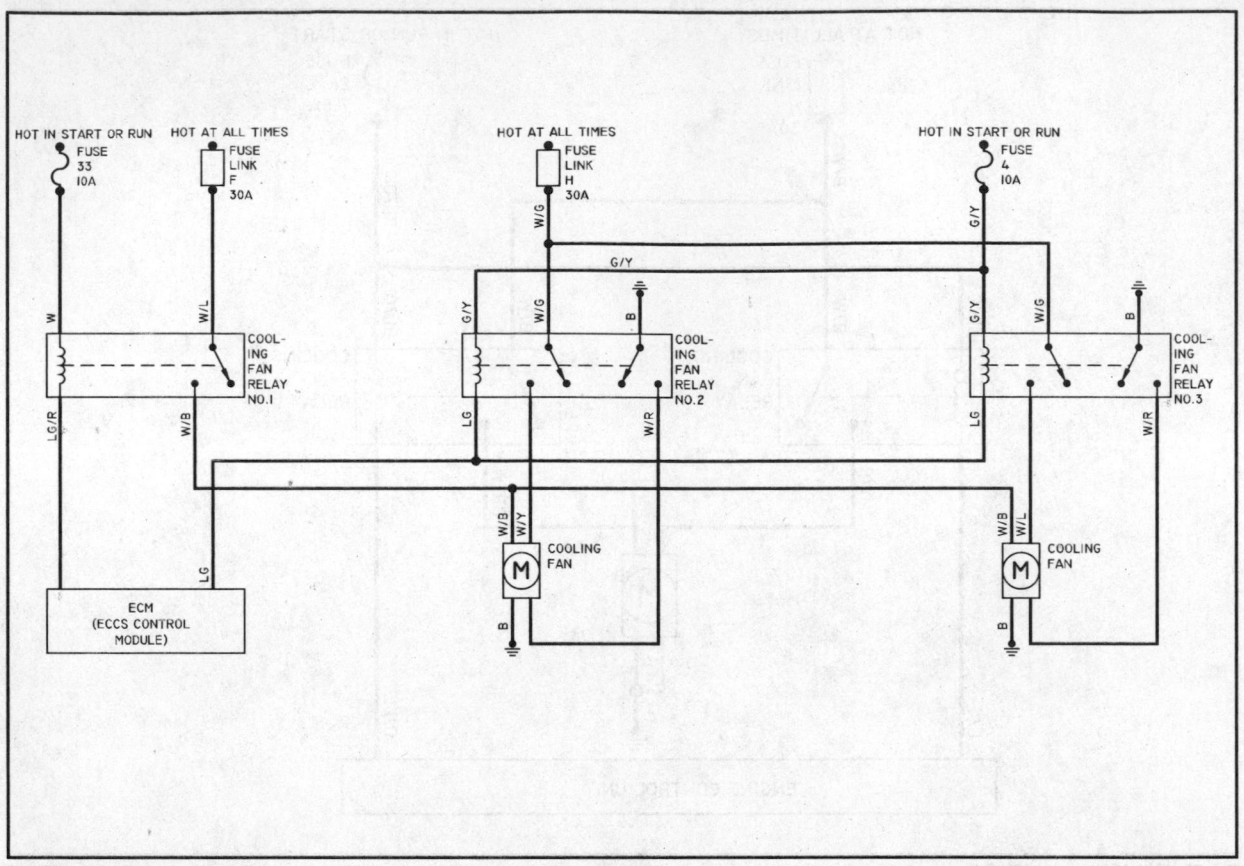

DIA. 75- 1995-99 Nissan Altima 2.4L

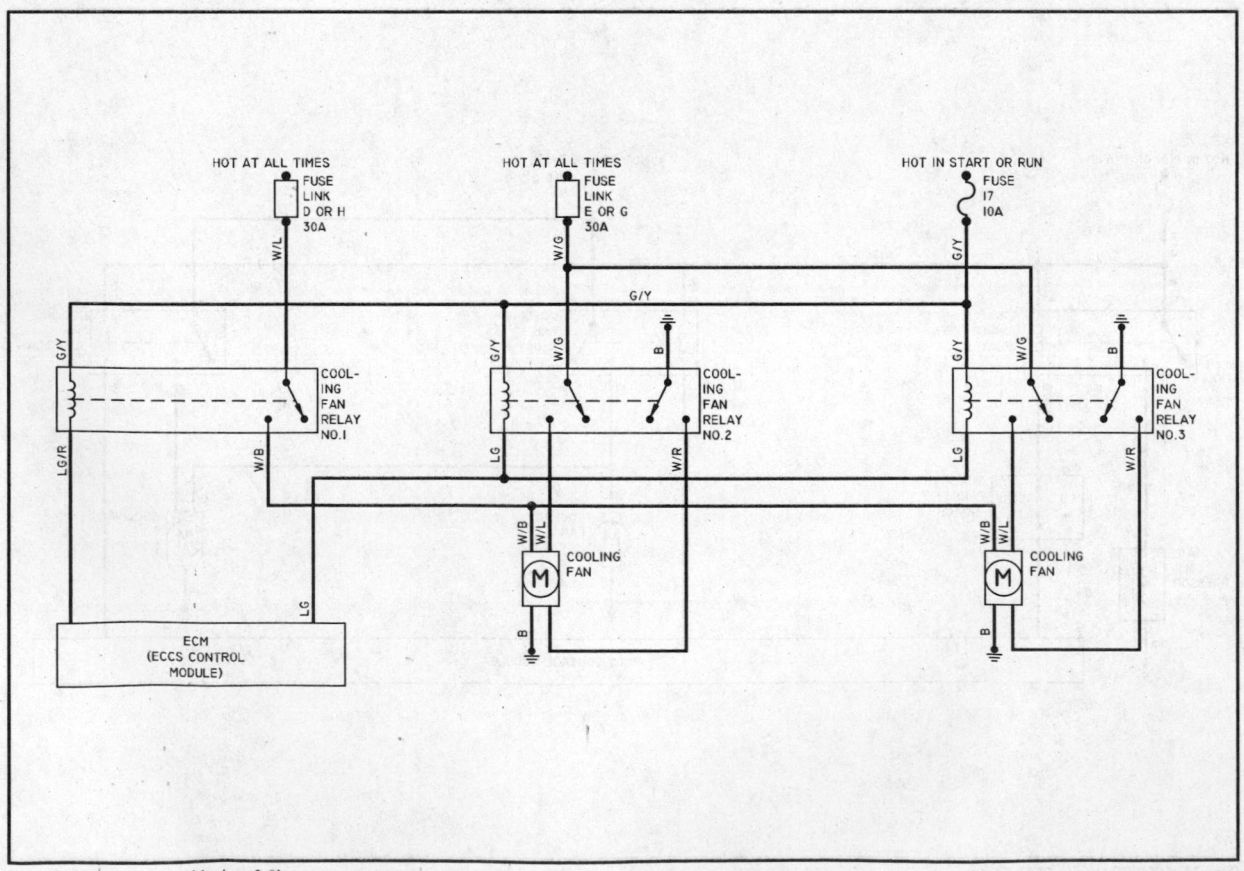

DIA. 76- 1995-99 Nissan Maxima 3.0L

79239W38

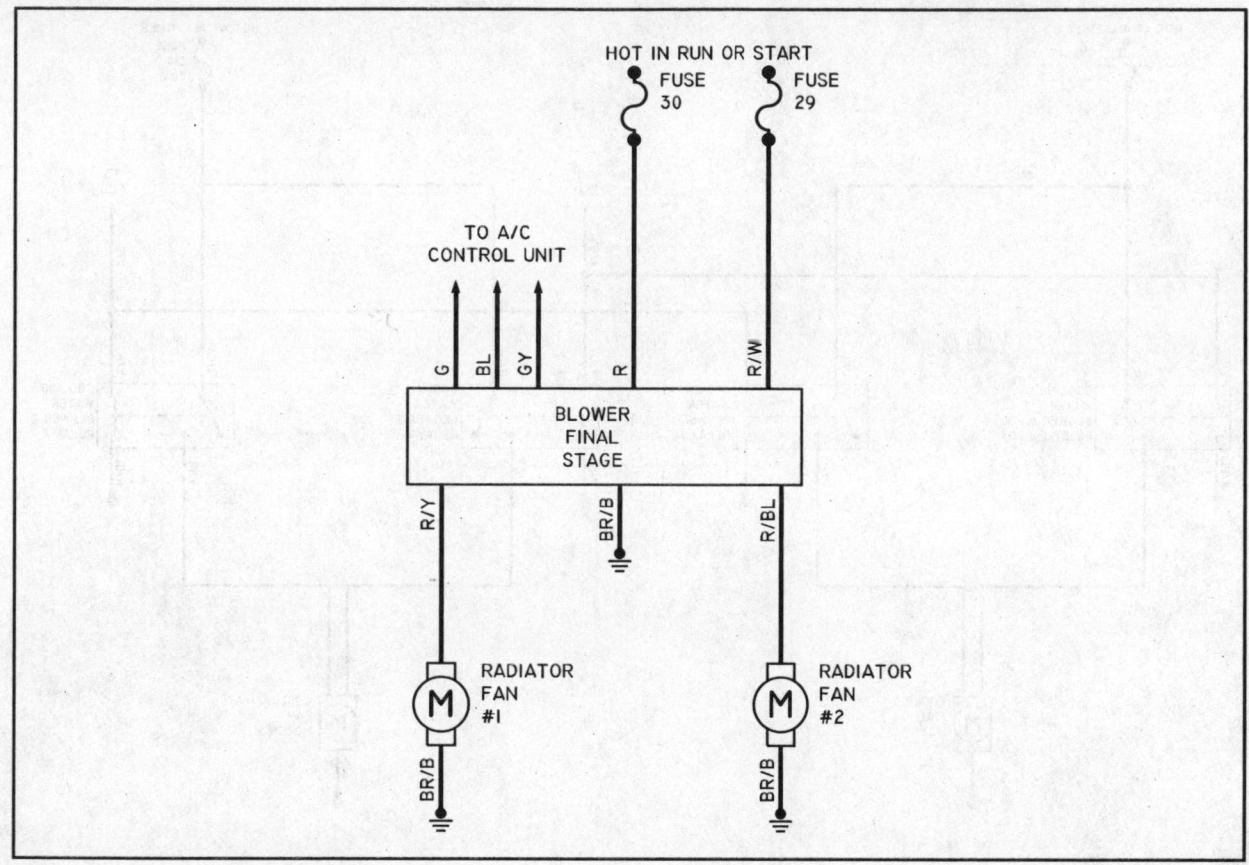

DIA. 77- 1995-99 Porsche 928 GTS, 1995 968

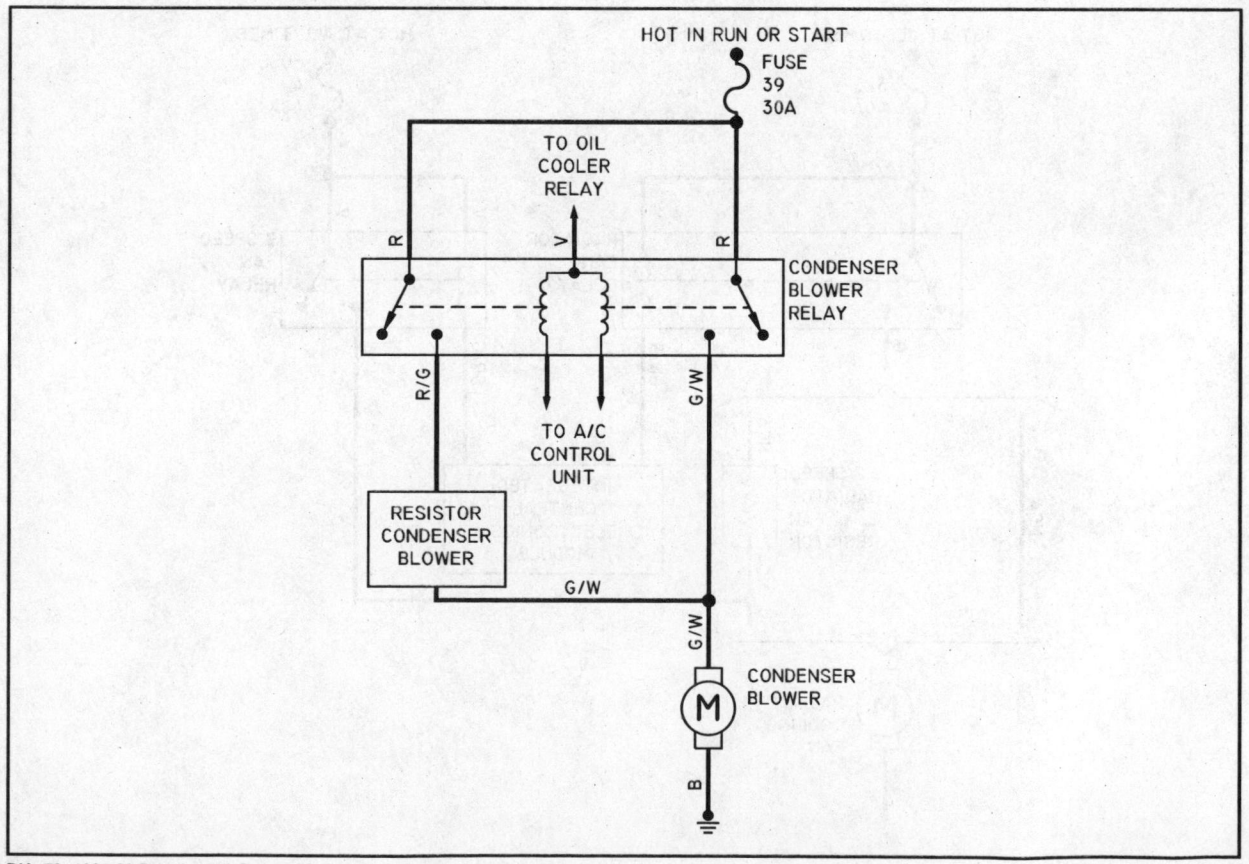

DIA. 78- 1995-99 Porsche 911 Carrera, Carrera 4, Turbo

79239W39

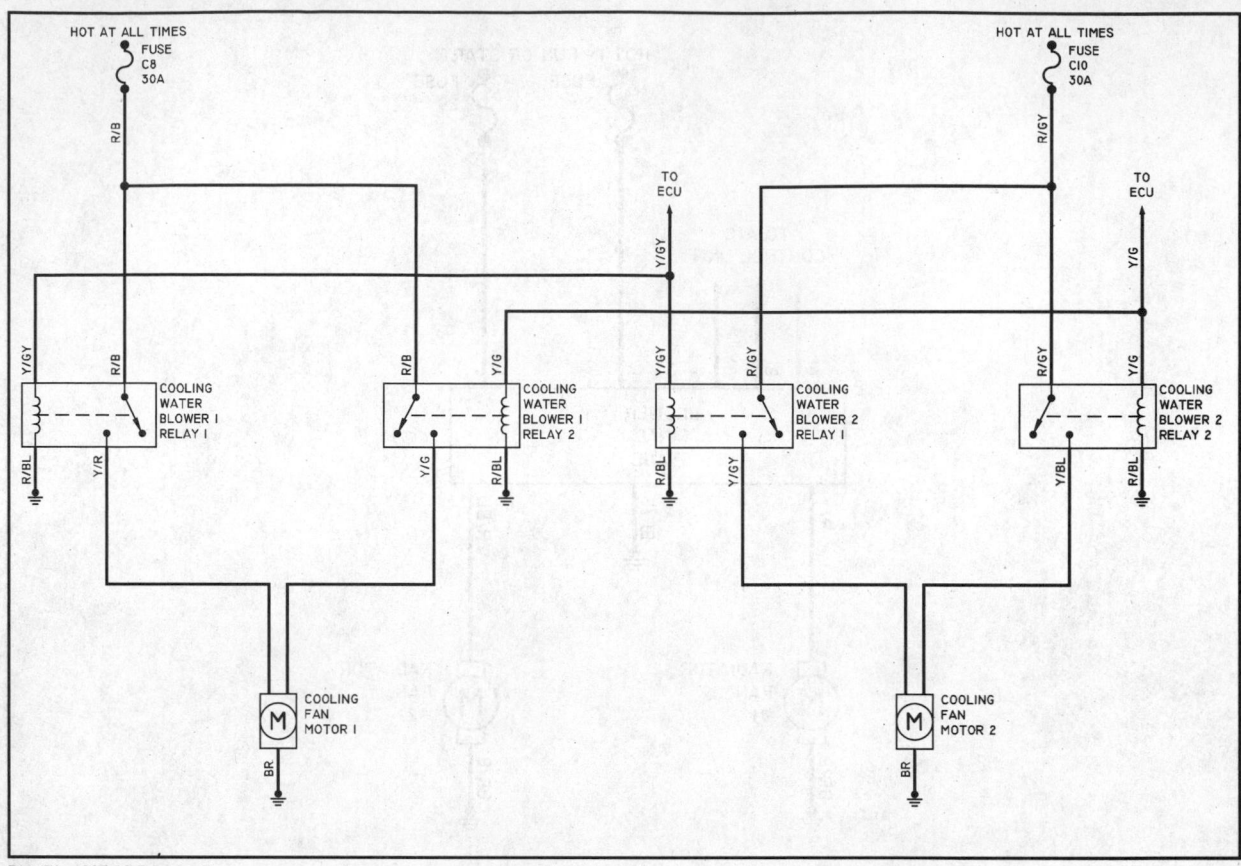

DIA. 79- 1997-99 Porsche Boxter

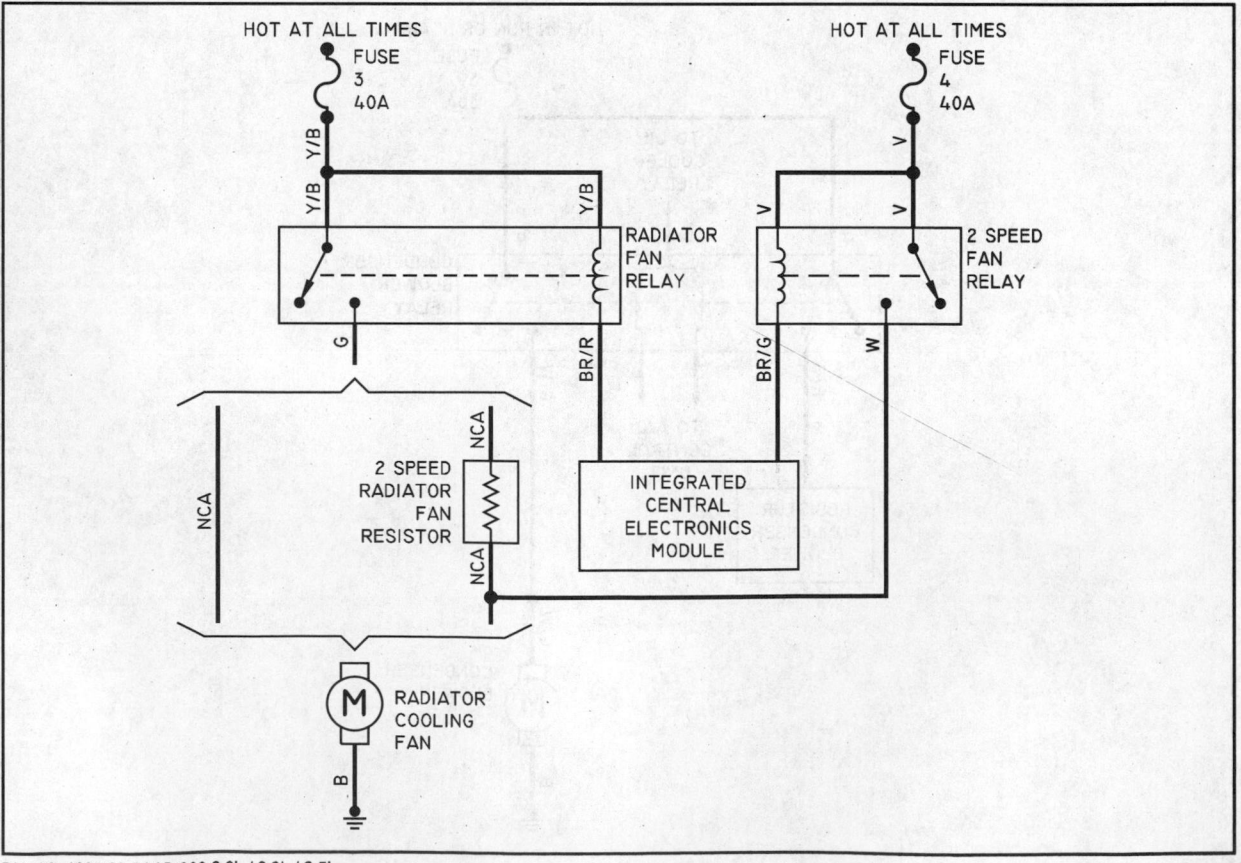

DIA. 80- 1995-99 SAAB 900 2.0L / 2.3L / 2.5L

79239W40

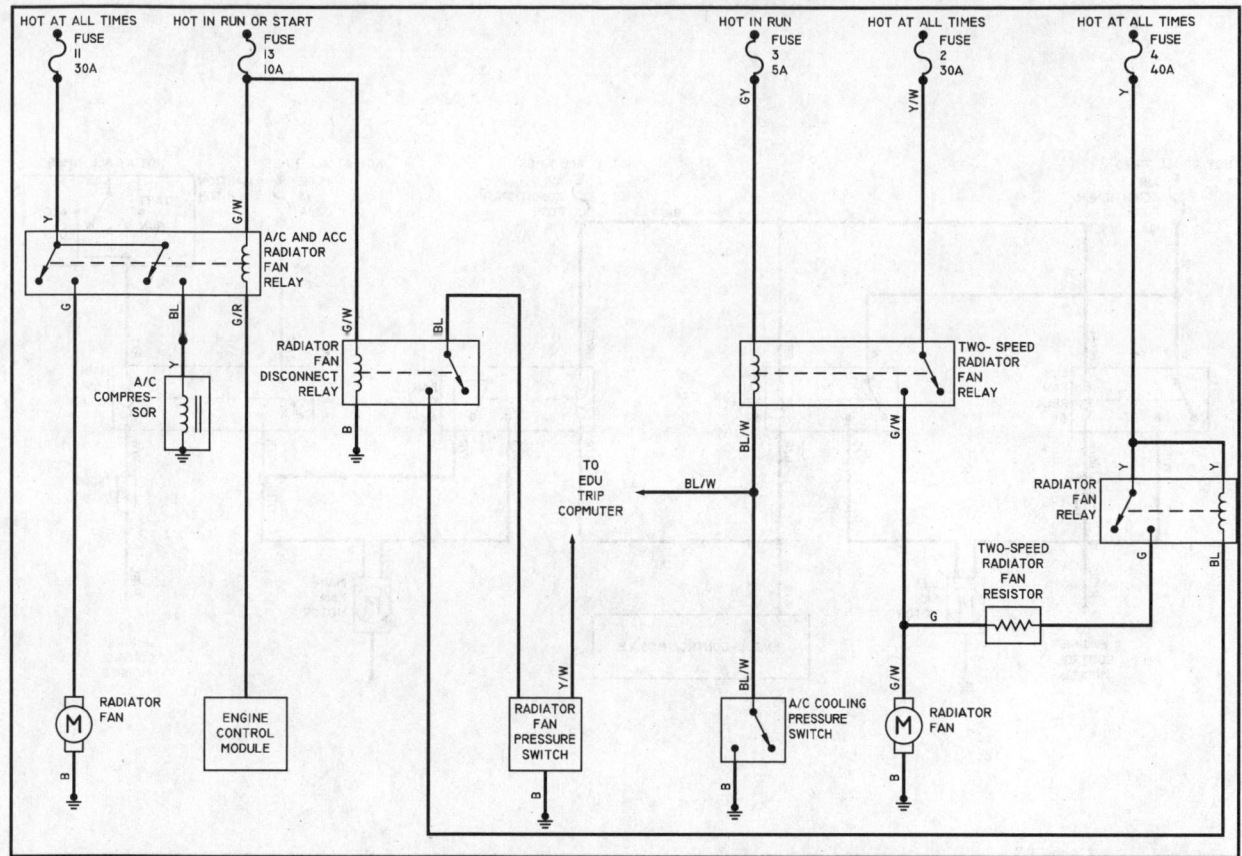

DIA. 81- 1995-99 SAAB 9000 2.3L

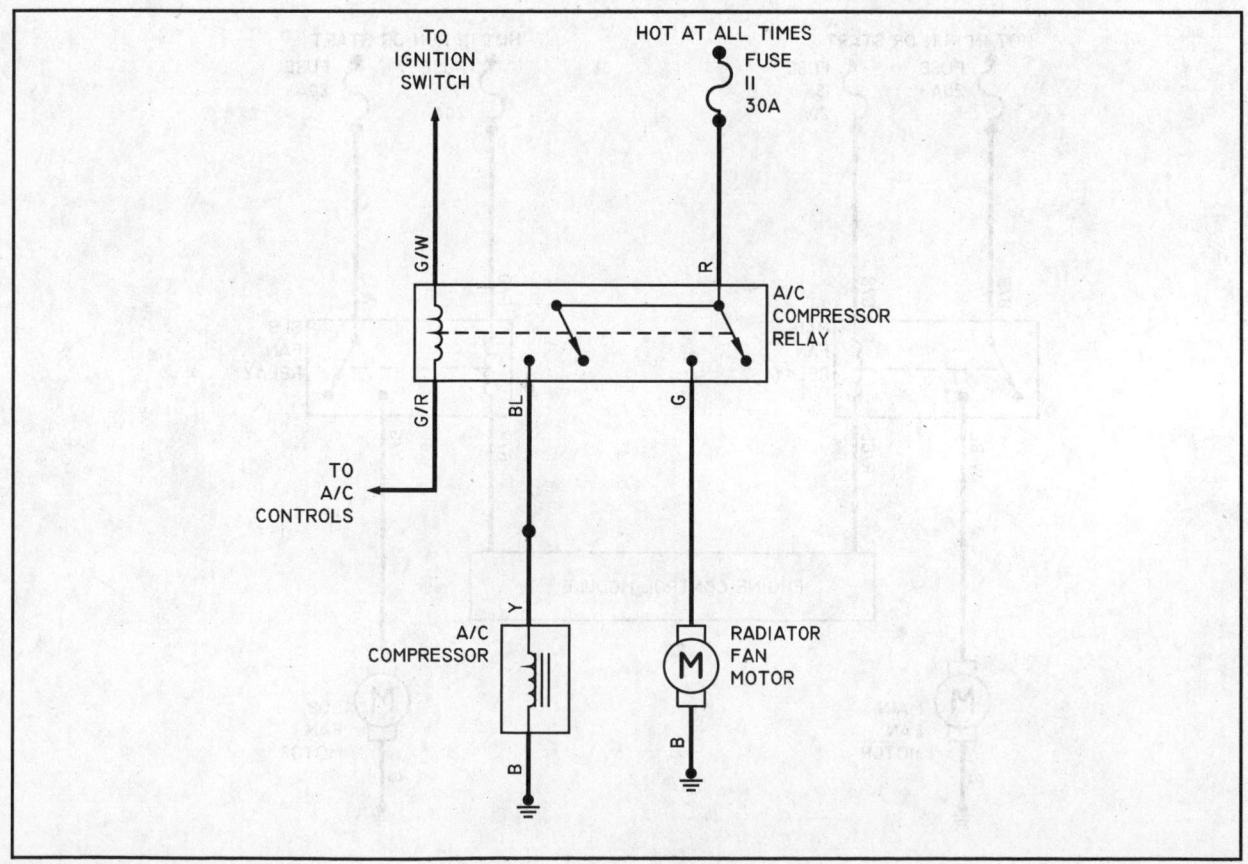

DIA. 82- 1996-99 SAAB 9000, 9-5 3.0L

79239W41

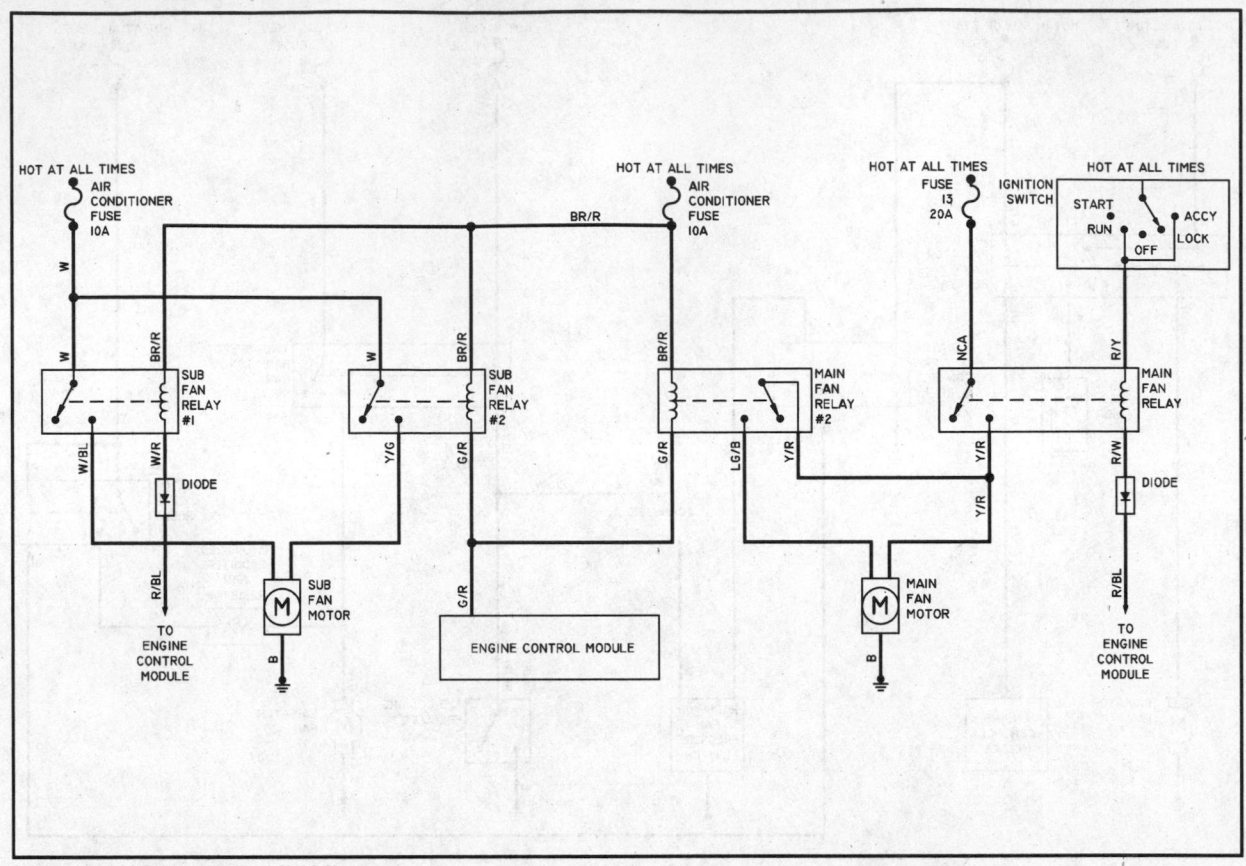

DIA. 83- 1995-96 Subaru Impreza 1.8L / 2.2L

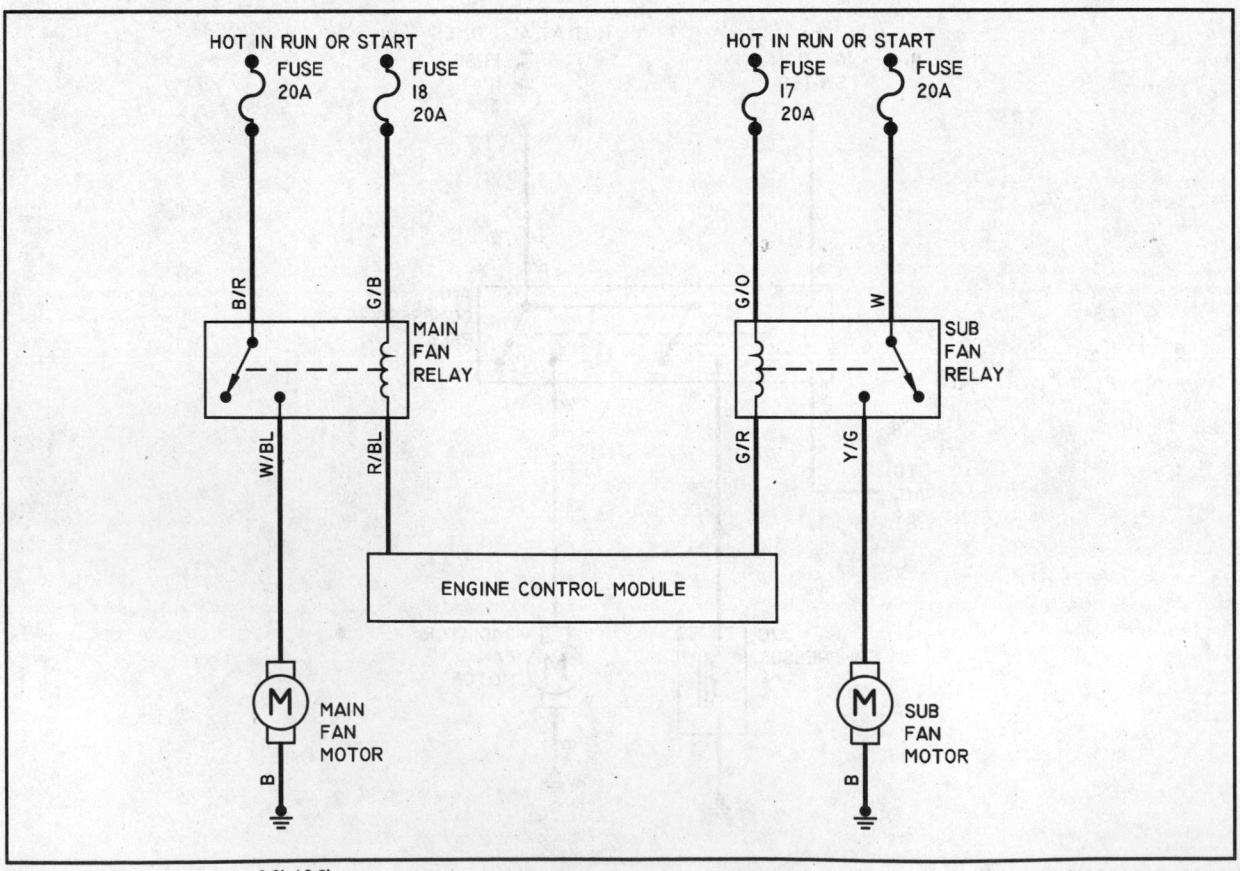

DIA. 84- 1997-99 Subaru Impreza 2.2L / 2.5L

79239W42

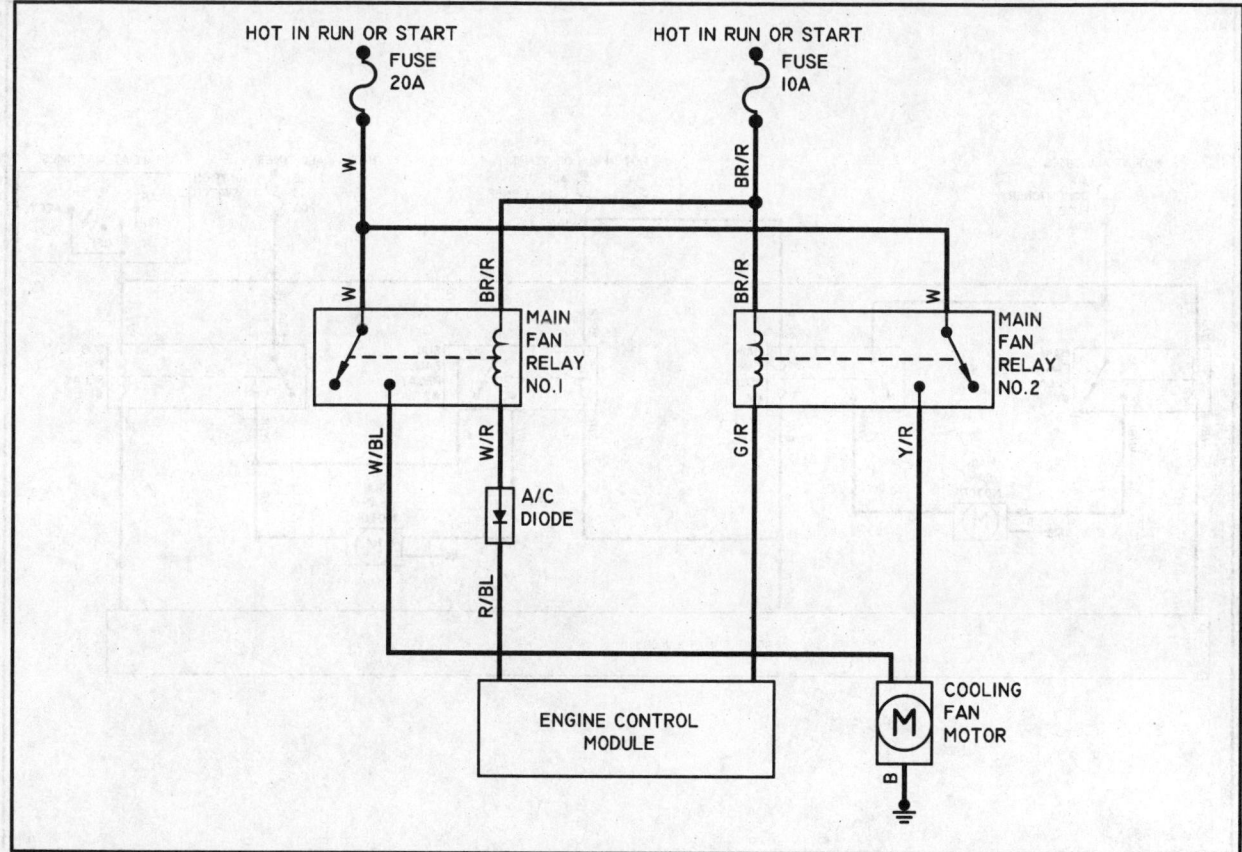

DIA. 85- 1995 Subaru Legacy 2.0L

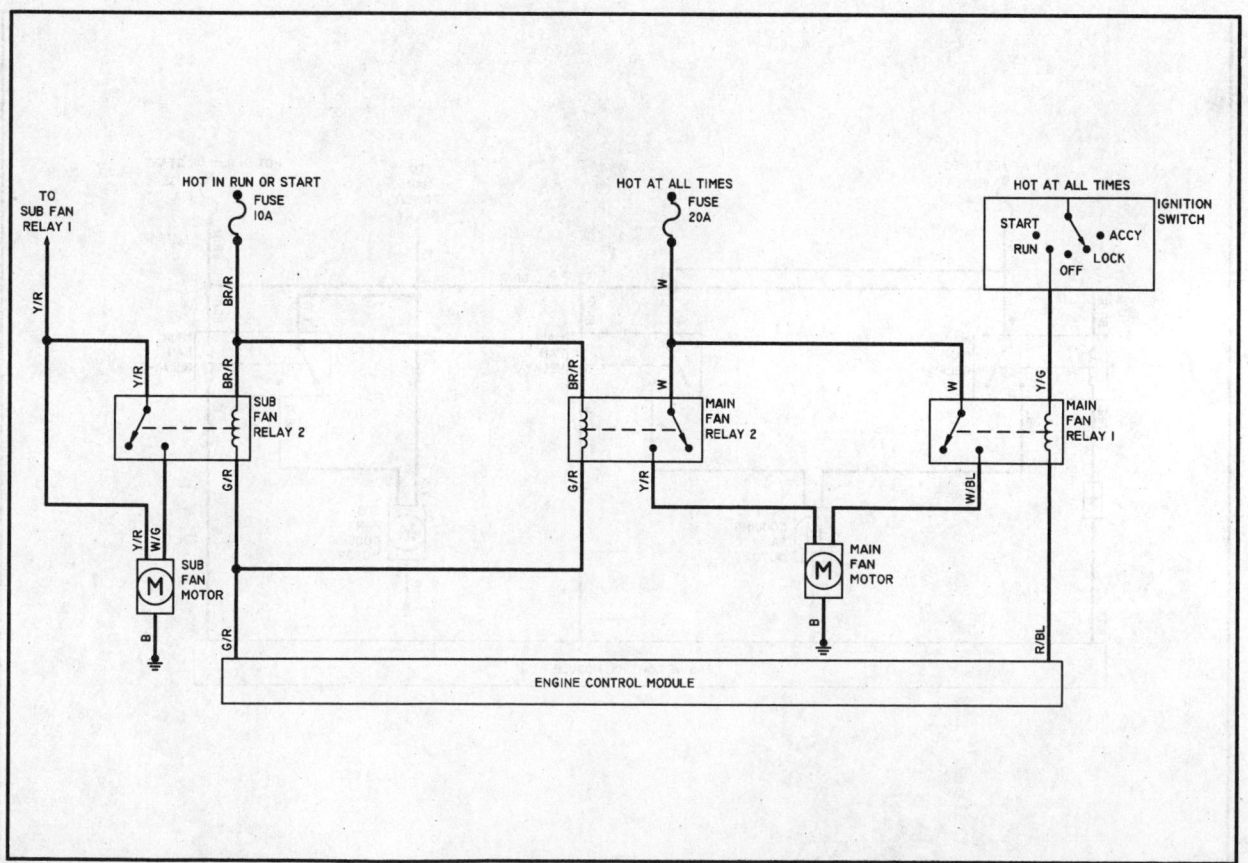

DIA. 86- 1995-97 Subaru Legacy 2.2L / 2.5L

79239W43

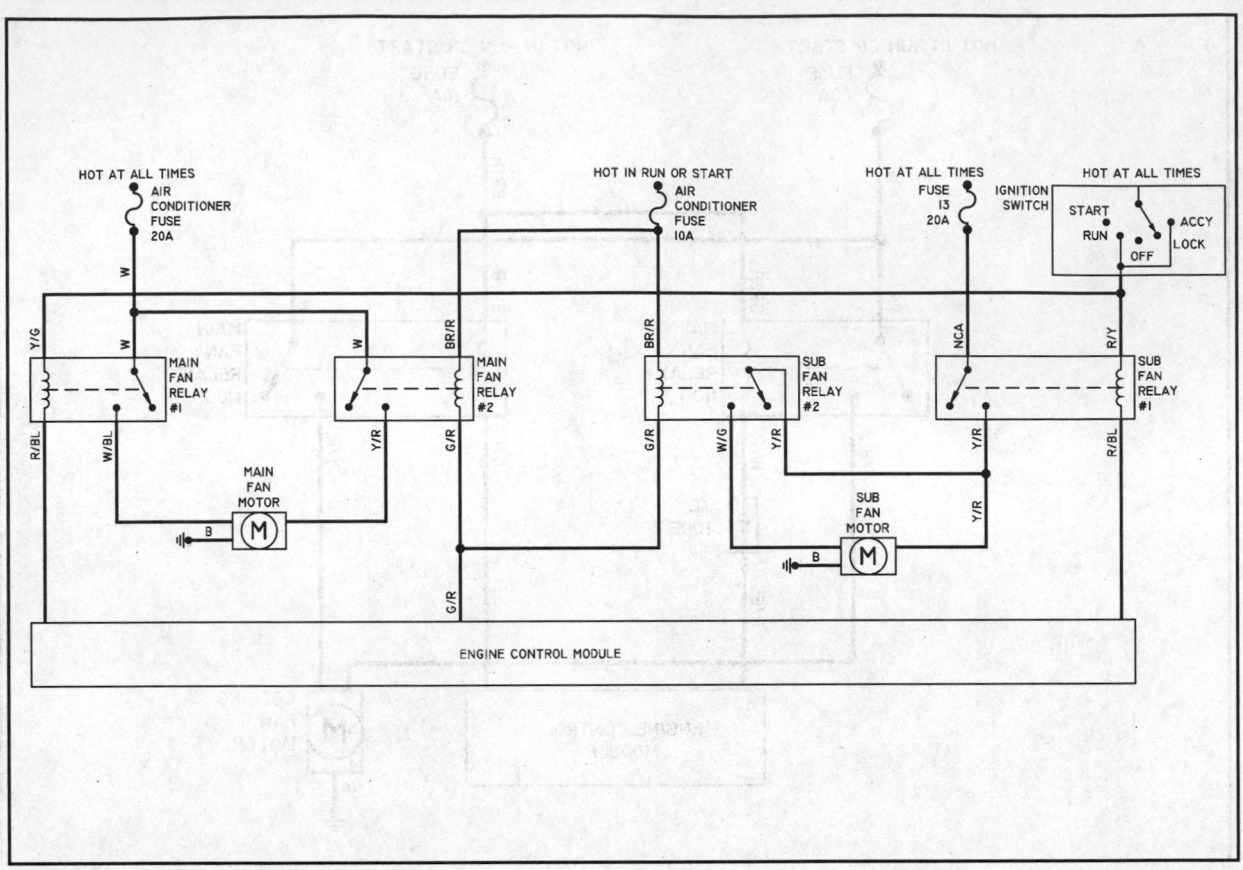

DIA. 87- 1998-99 Subaru Legacy 2.2L / 2.5L

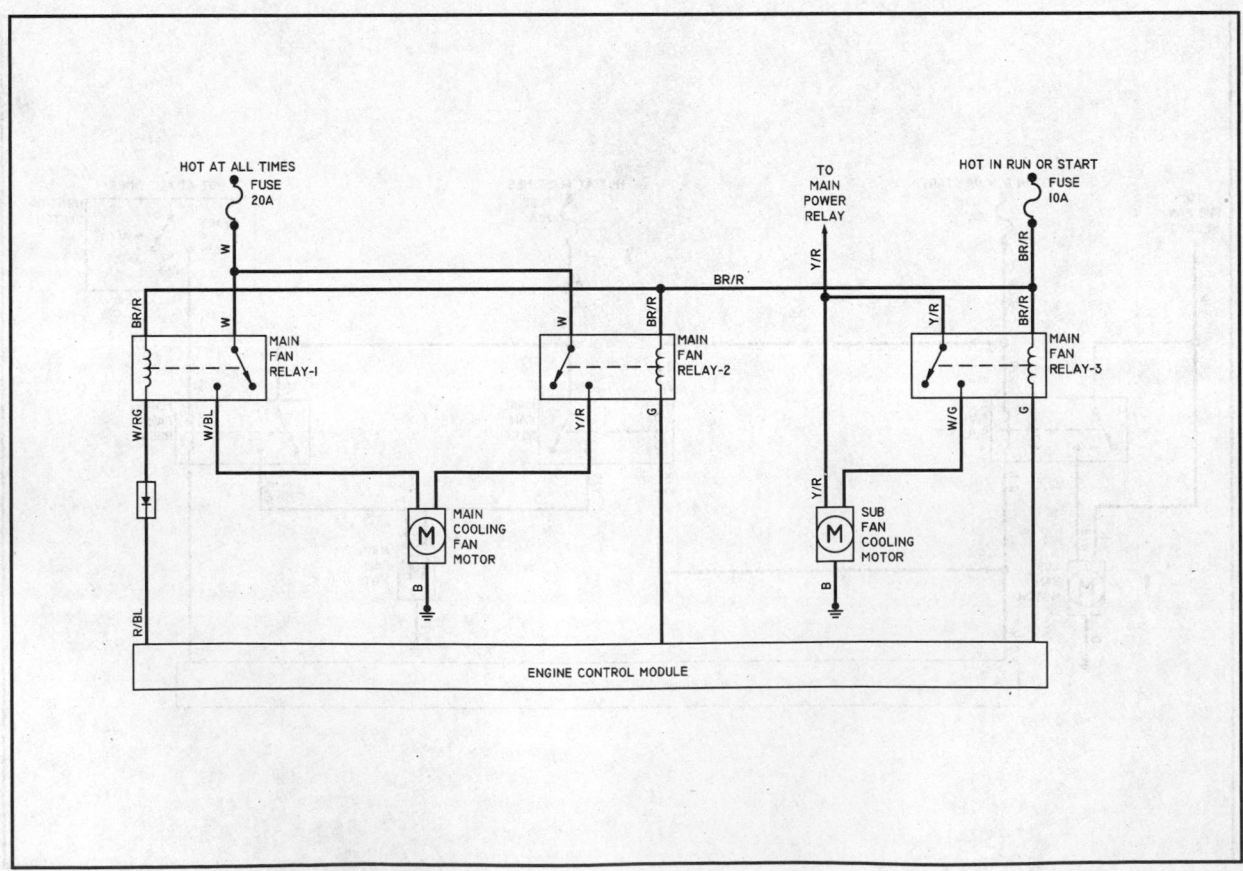

DIA. 88- 1995-99 Subaru Legacy Brighton, Outback 2.2L, 2.5L

79239W44

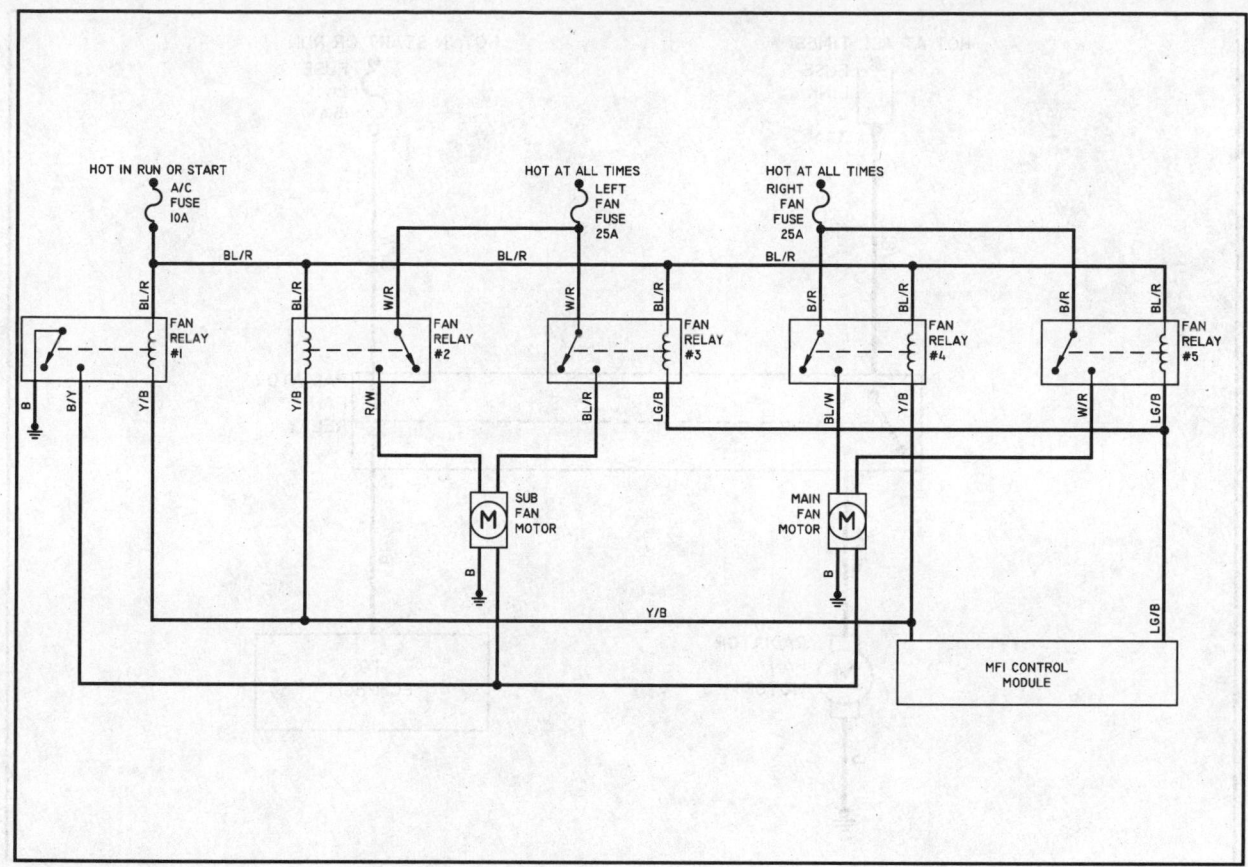

DIA. 89- 1995-97 Subaru SVX 3.3L

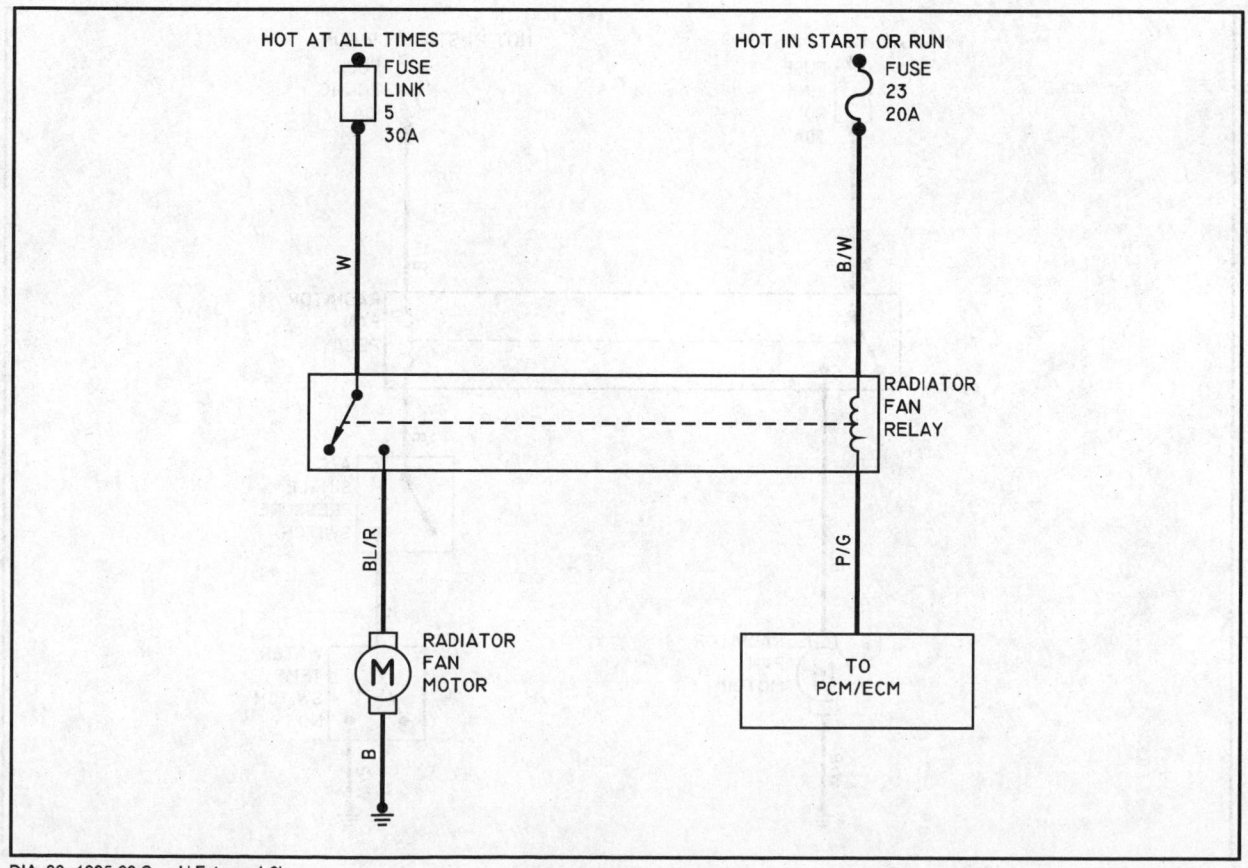

DIA. 90- 1995-99 Suzuki Esteem 1.6L

79239W45

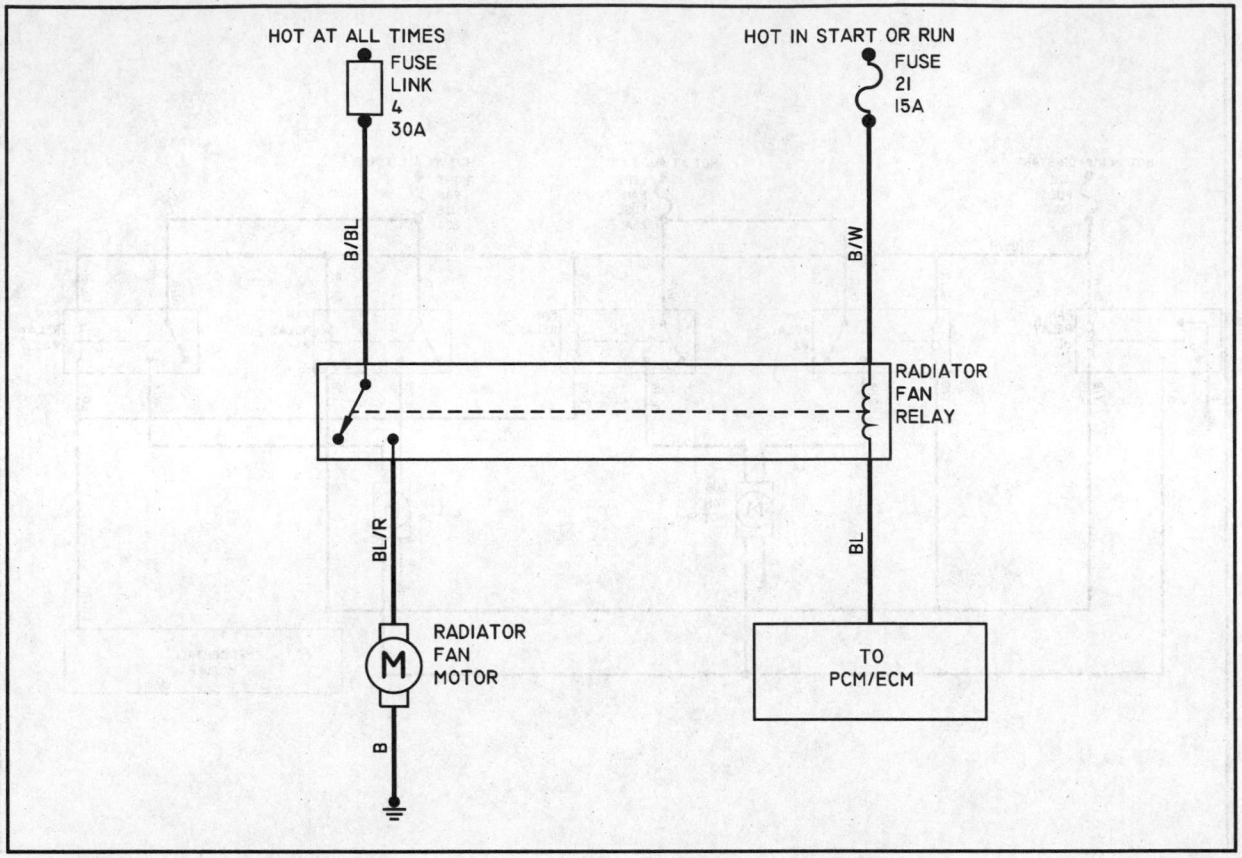

DIA. 91- 1995-99 Suzuki Swift 1.0L / 1.3L

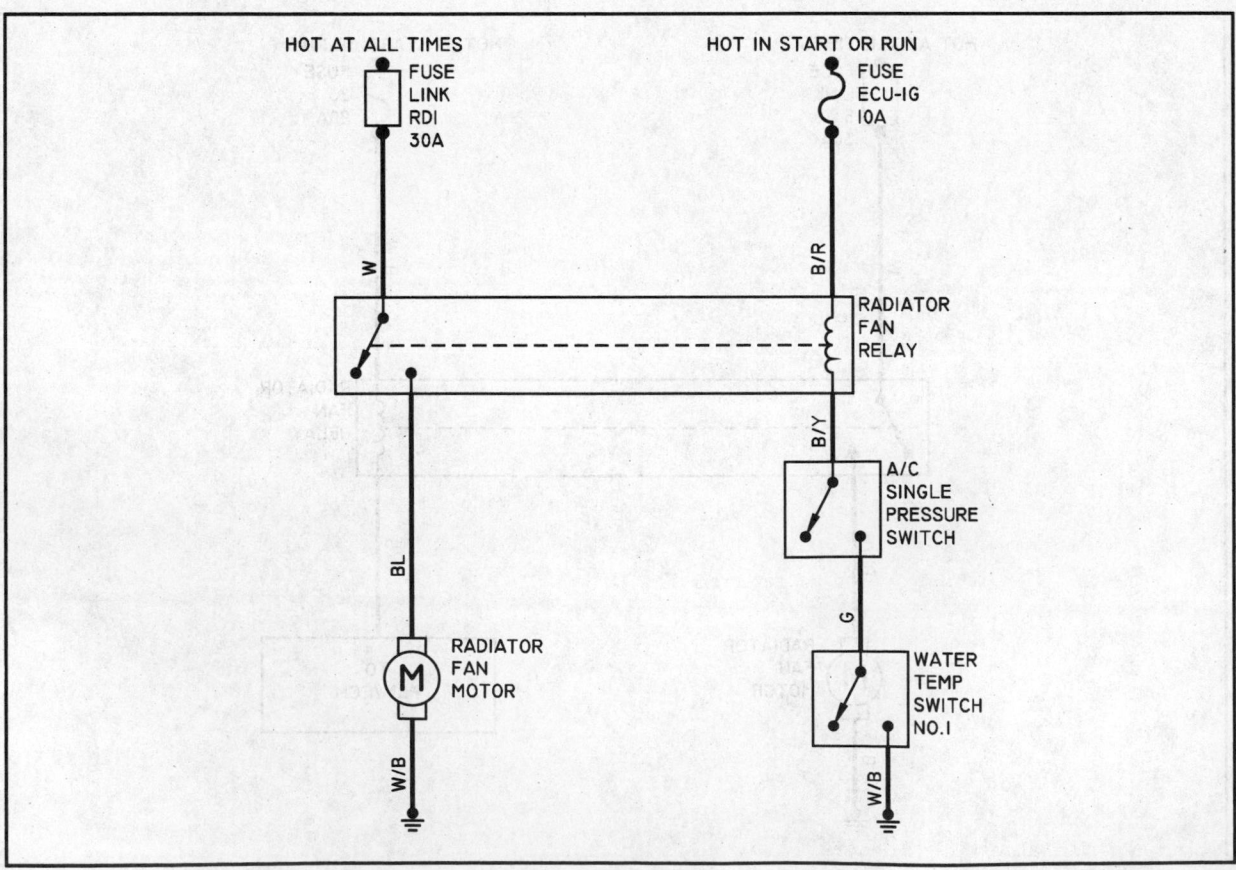

DIA. 92- 1995-99 Toyota Avalon 3.0L

79239W46

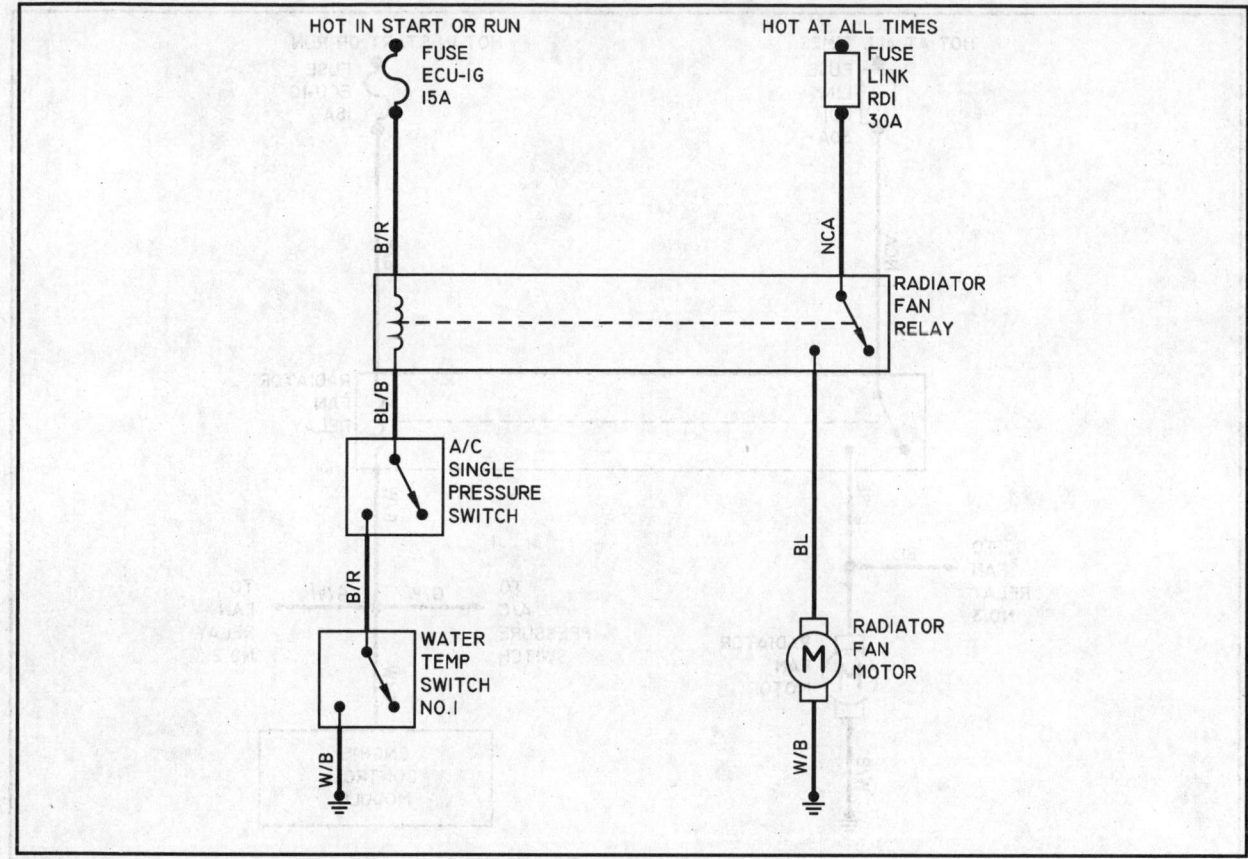

DIA. 93- 1995-96 Toyota Camry 2.2L

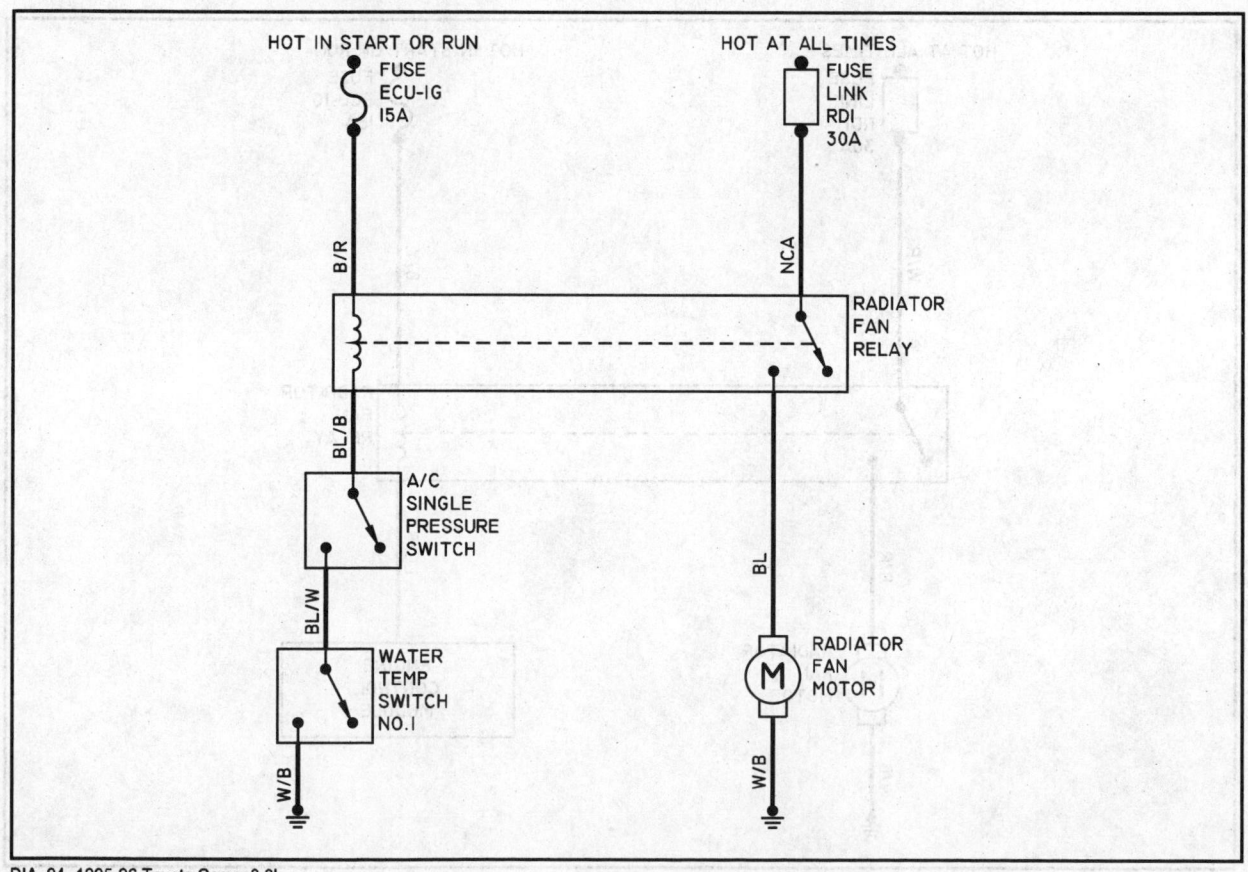

DIA. 94- 1995-96 Toyota Camry 3.0L

79239W47

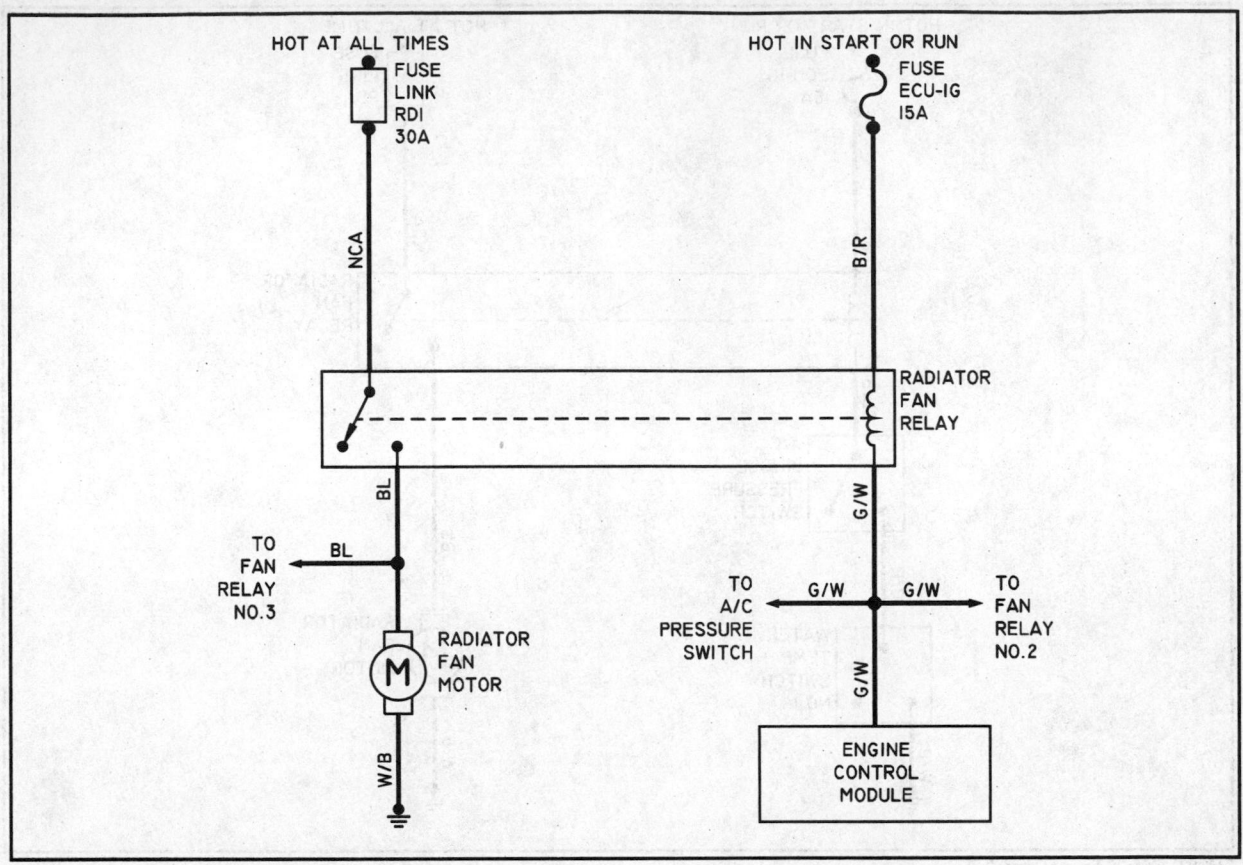

DIA. 95- 1997-99 Toyota Camry 2.2L / 3.0L

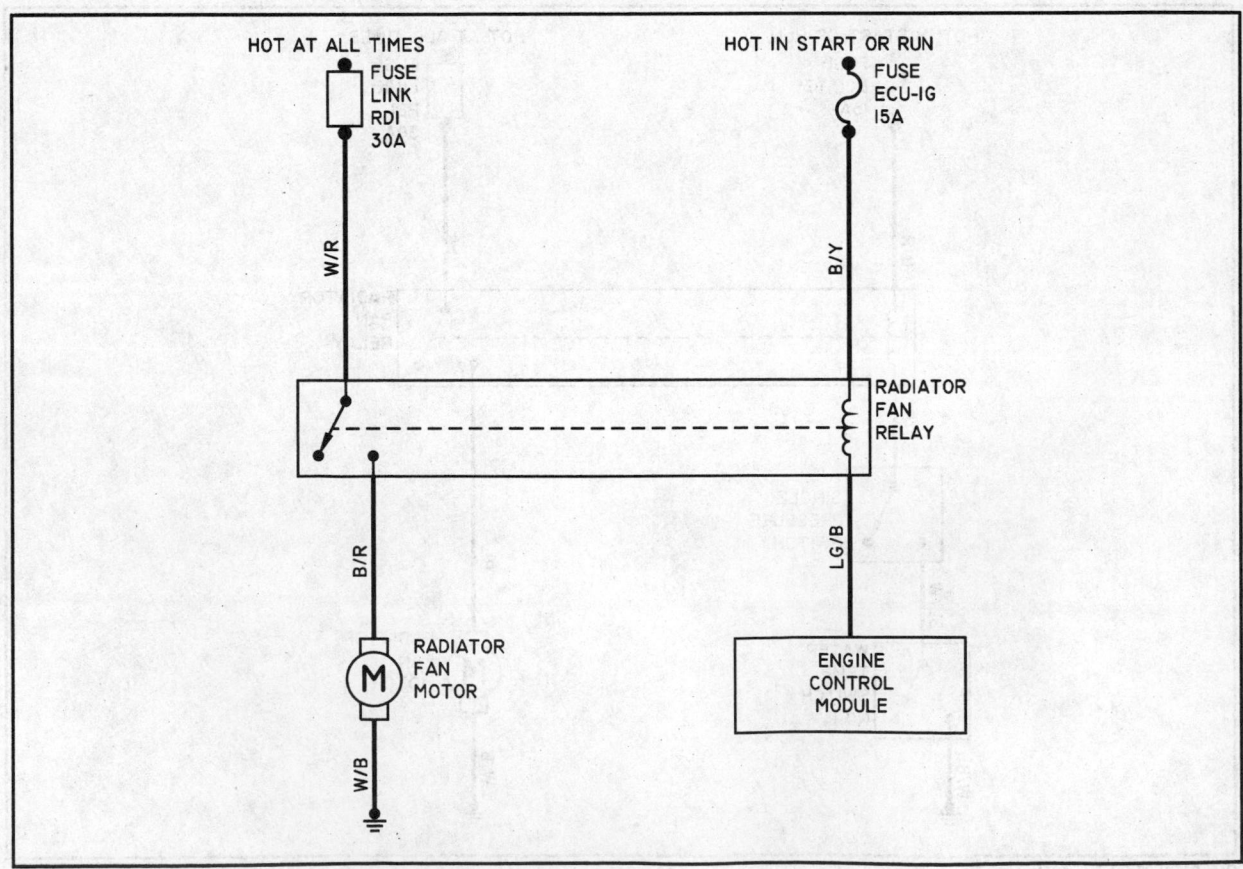

DIA. 96- 1995-99 Toyota Celica 1.8L / 2.2L

79239W48

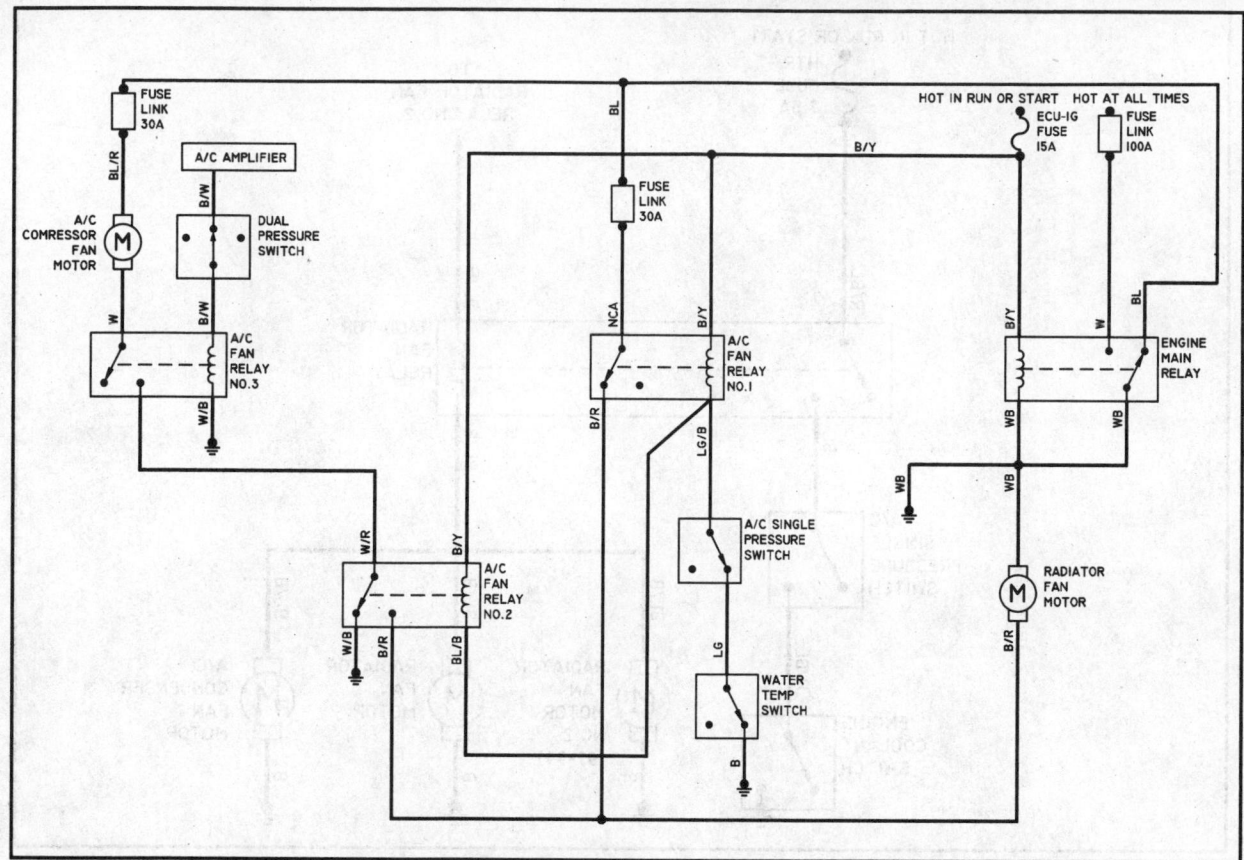

DIA. 97- 1995-99 Toyota Corolla 1.6L

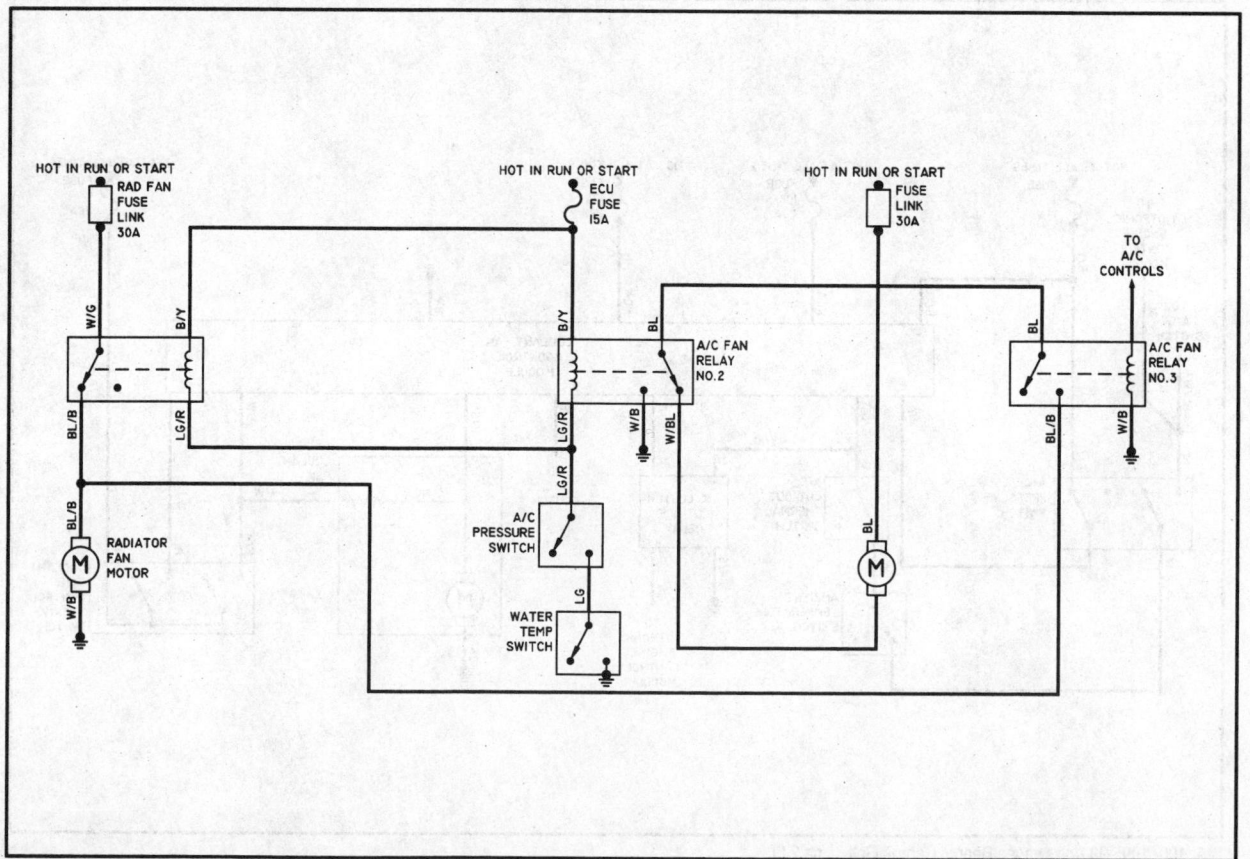

DIA. 98- 1995-99 Toyota Paseo, Tercel 1.5L

79239W49

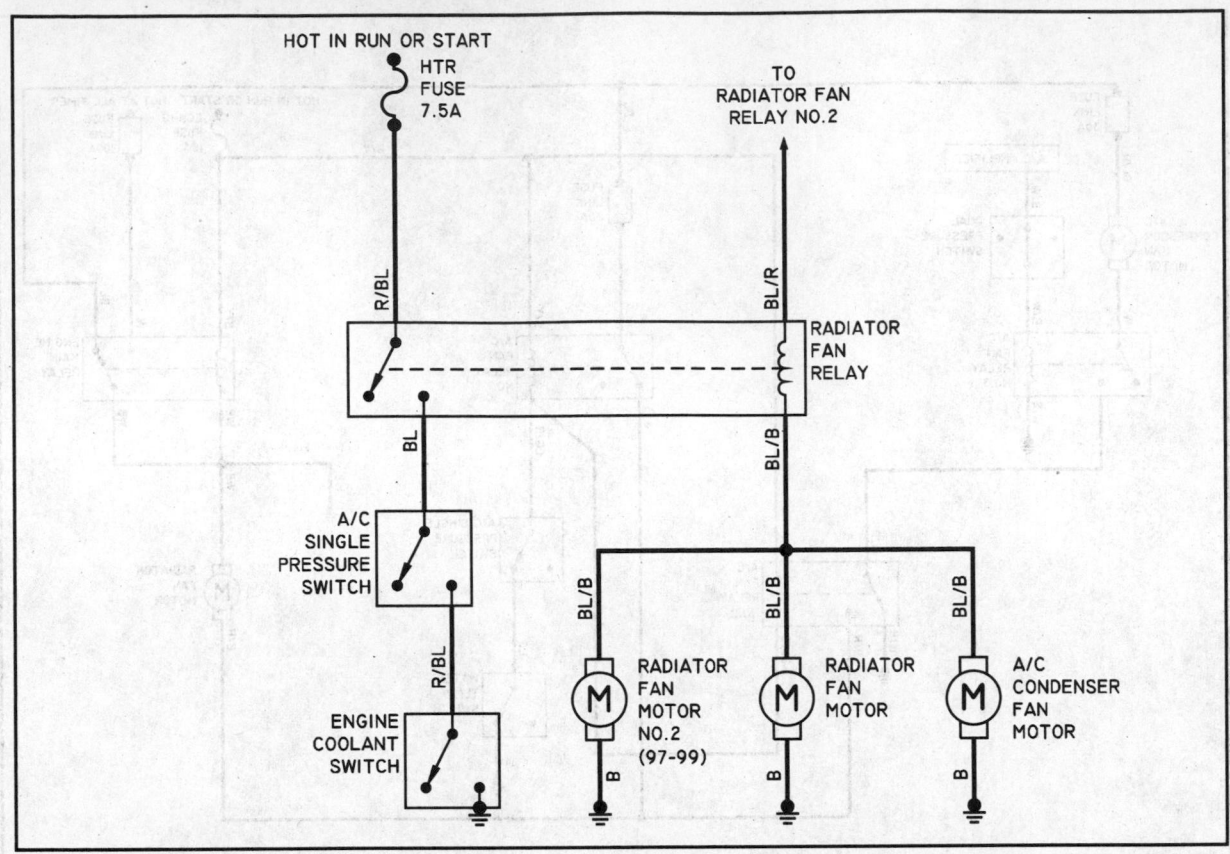

DIA. 99- 1995-99 Toyota Supra 3.0L

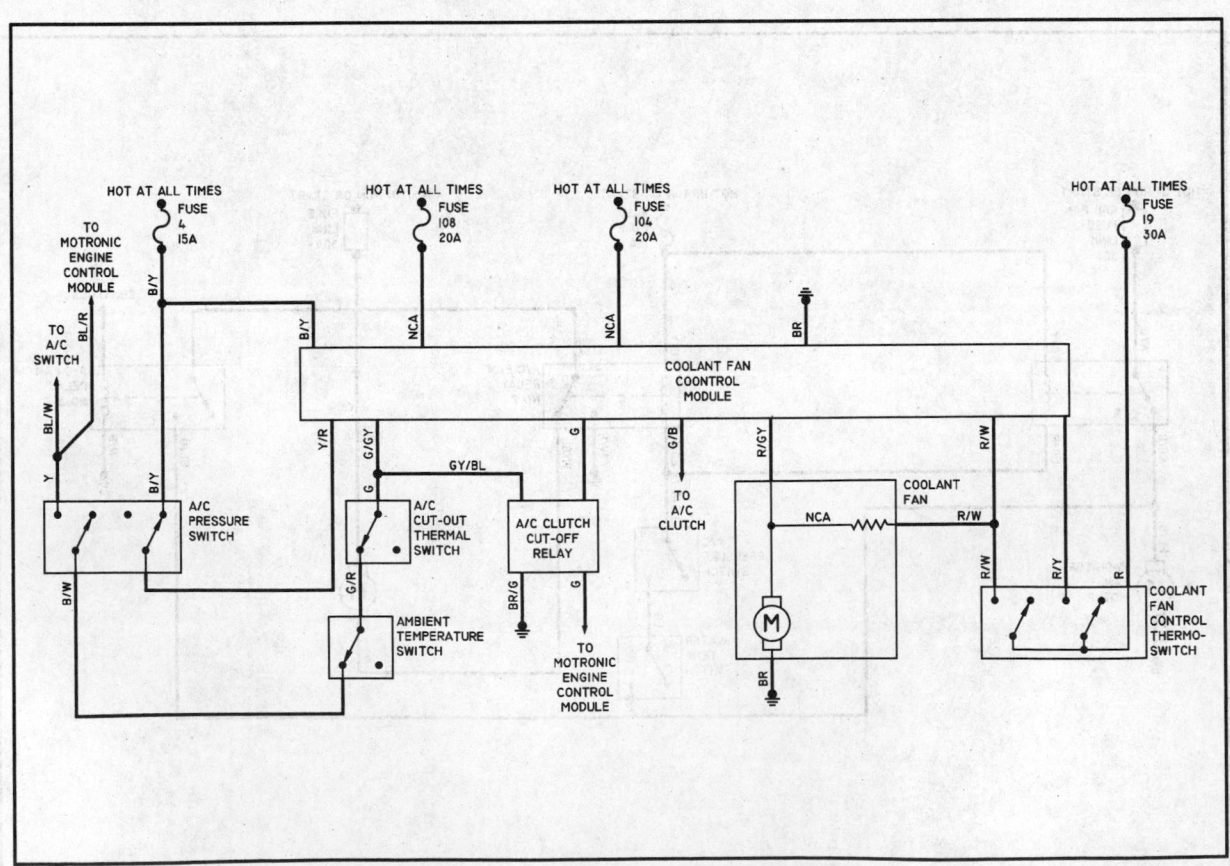

DIA. 100- 1995-99 Volkswagen Beetle, Cabroi, Golf, Jetta 2.0L
1997-99 GTI 2.0L

79239W50

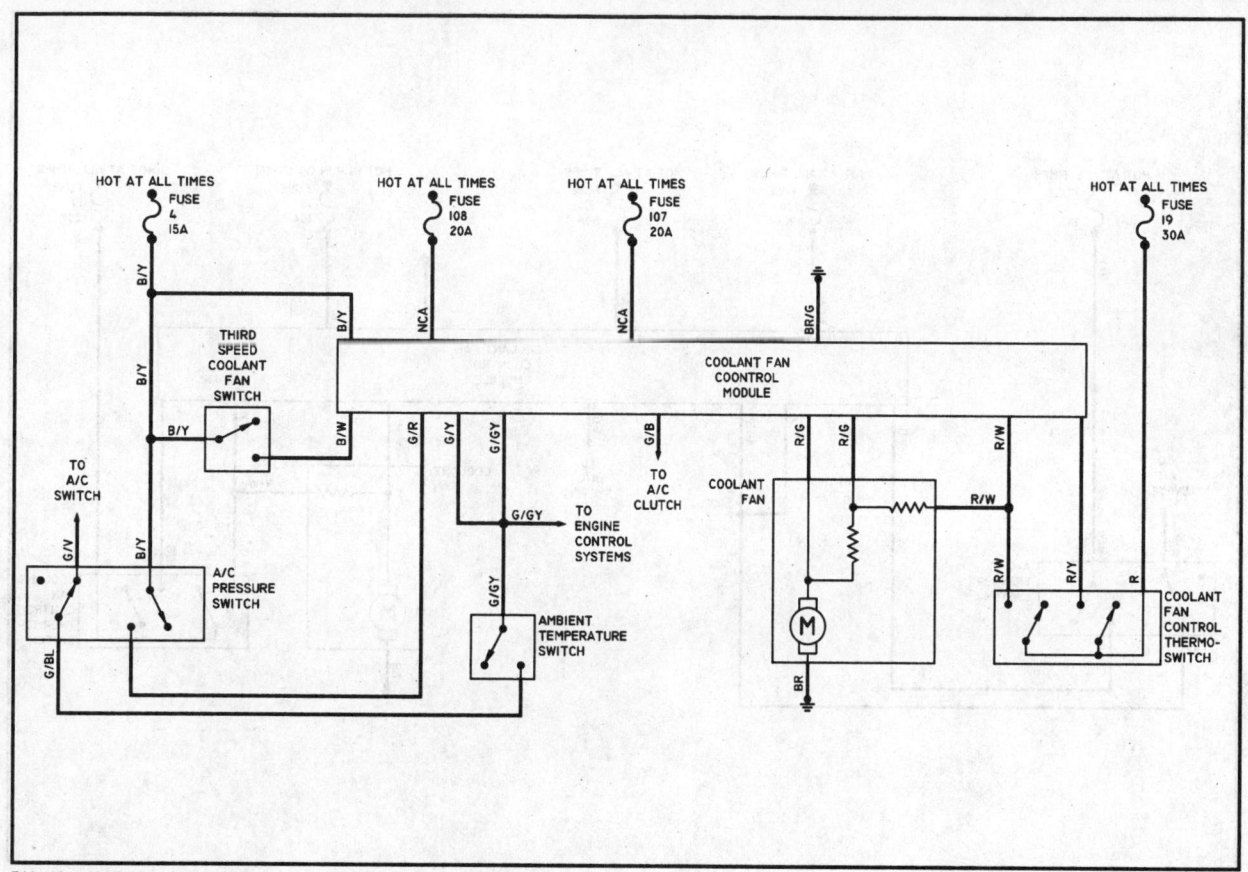

DIA. 101- 1997-99 Volkswagen Beetle, Golf, Jetta 1.9L Turbo Diesel

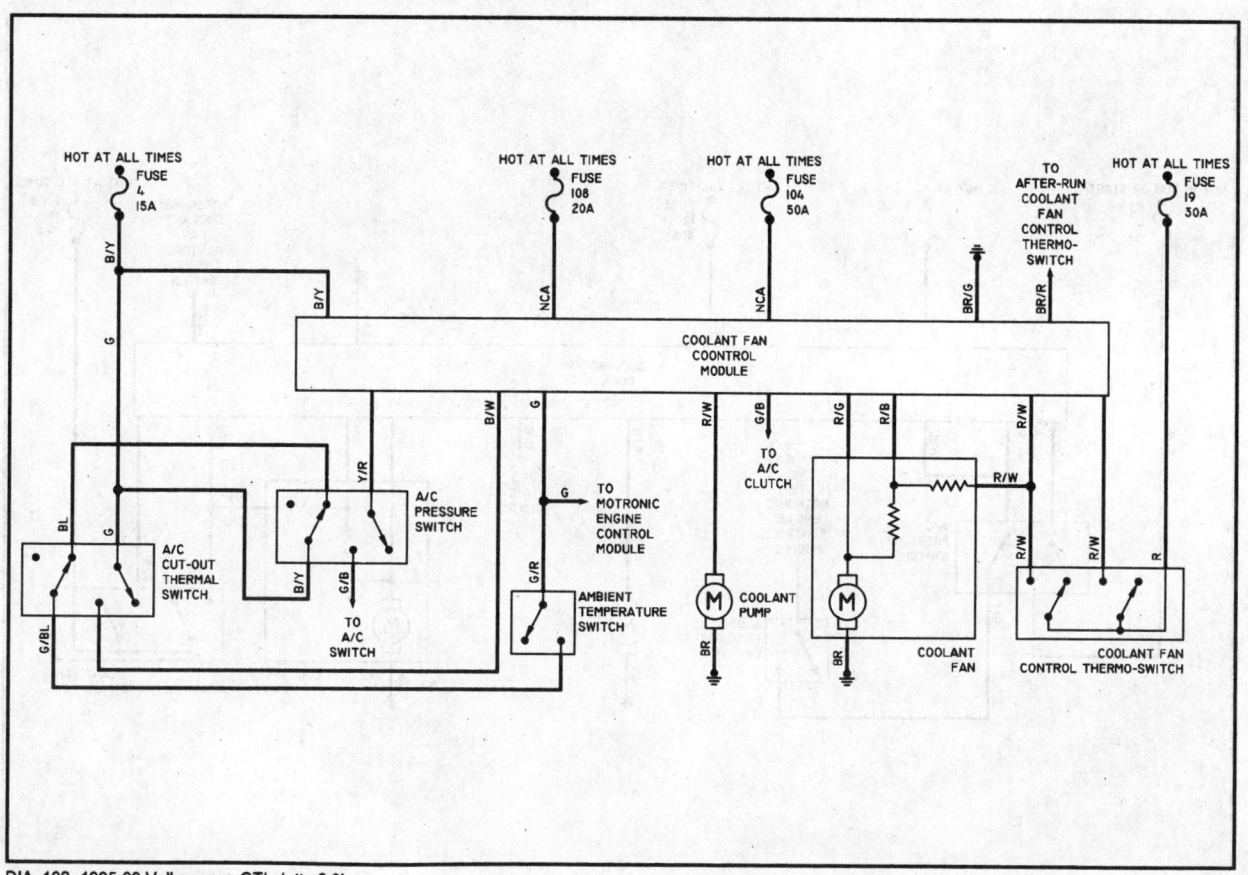

DIA. 102- 1995-99 Volkswagen GTI, Jetta 2.8L

79239W51

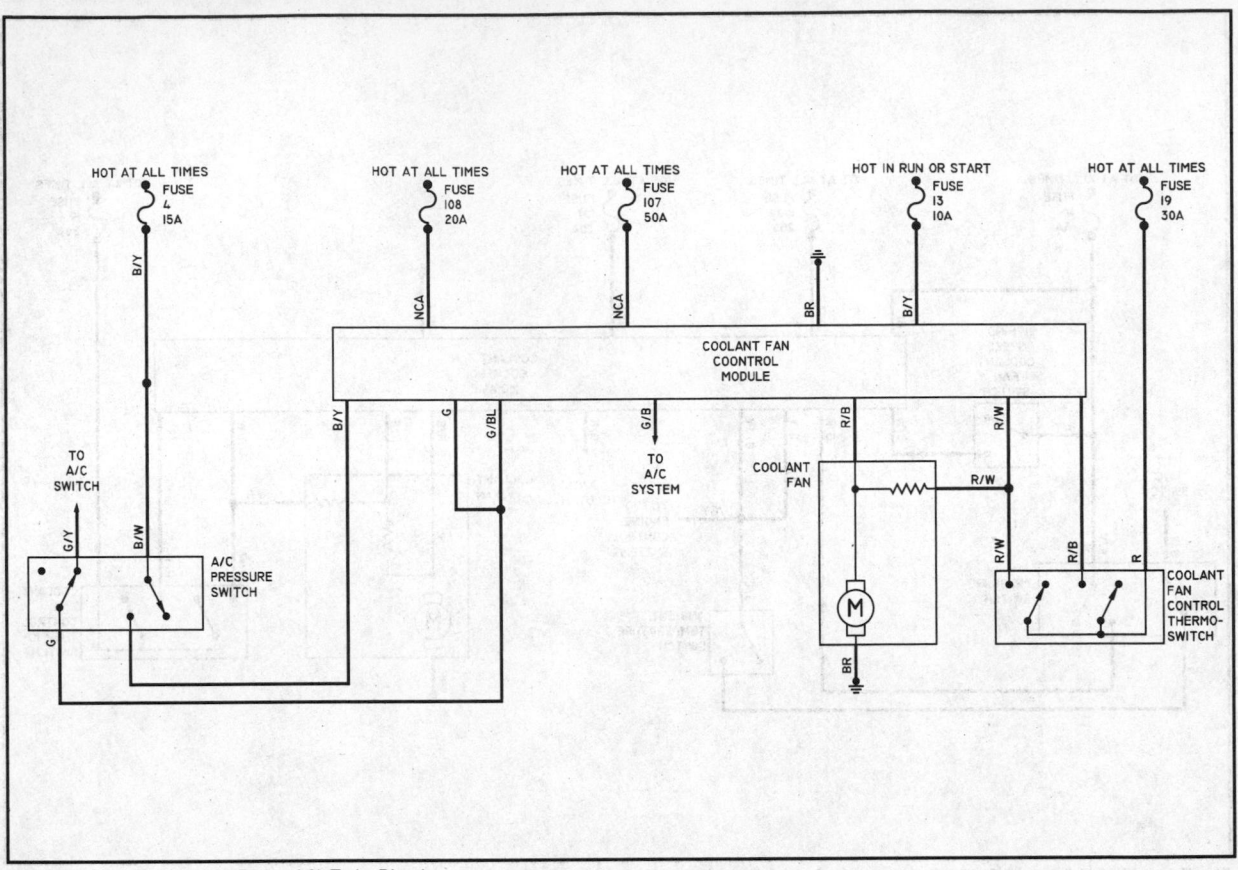

DIA. 103- 1997-99 Volkswagen Passat 1.9L Turbo Diesel

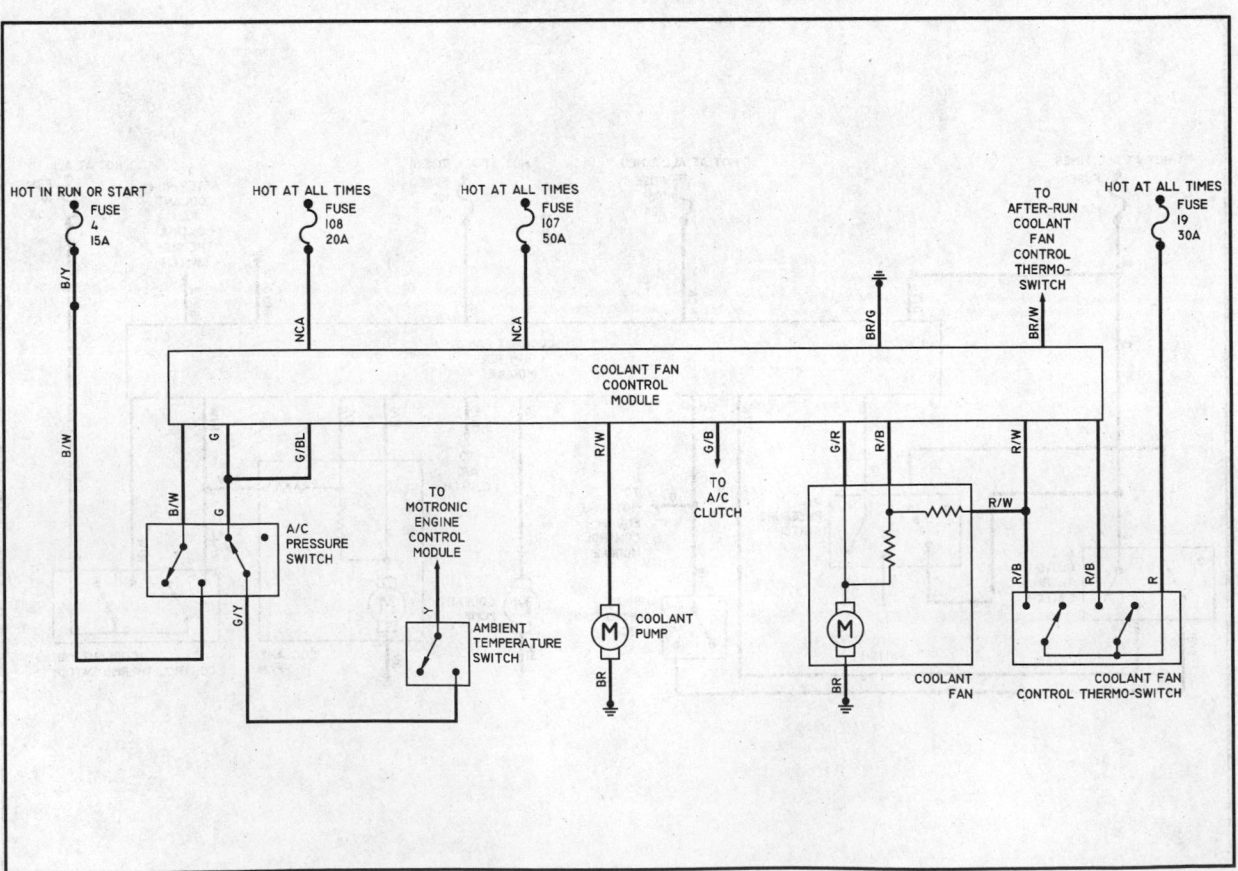

DIA. 104- 1995-99 Volkswagen Passat 2.0L

79239W52

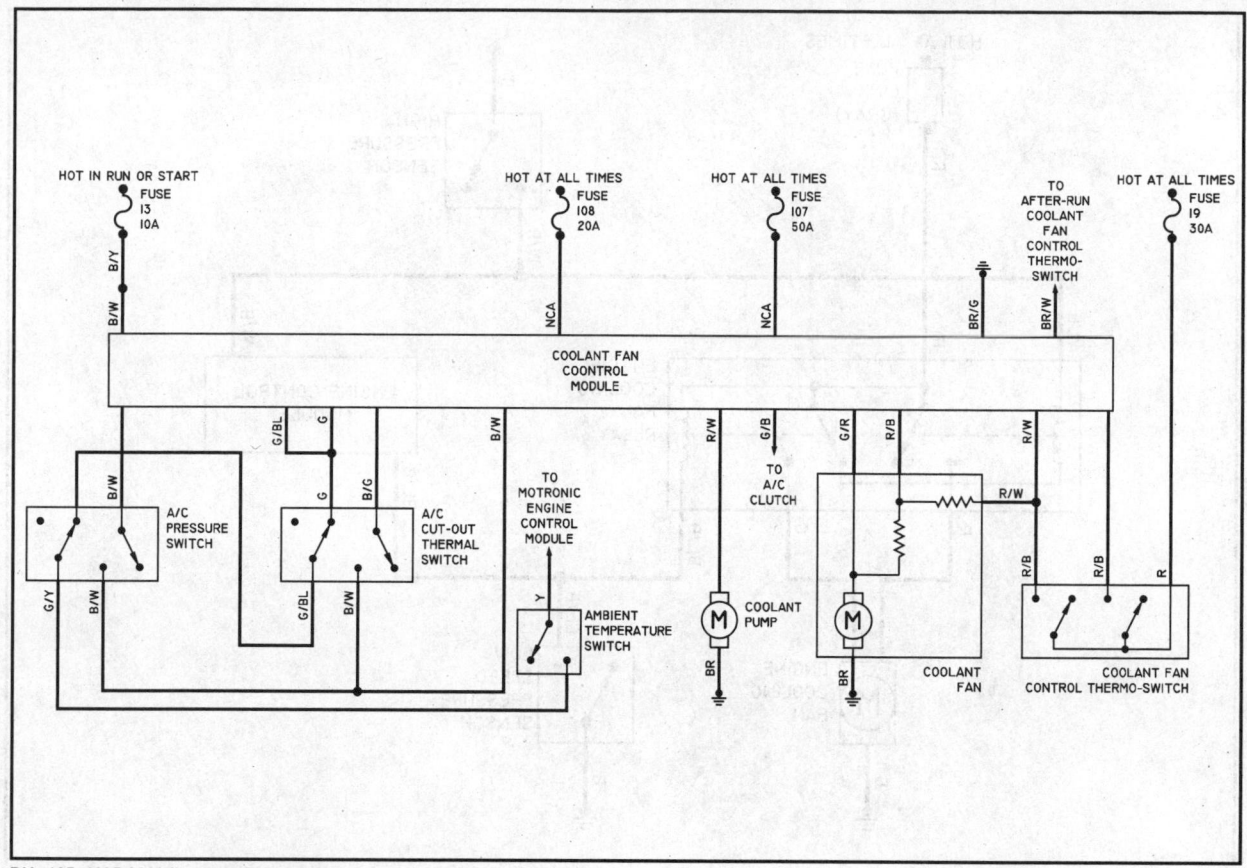

DIA. 105- 1995-99 Volkswagen Passat 2.8L

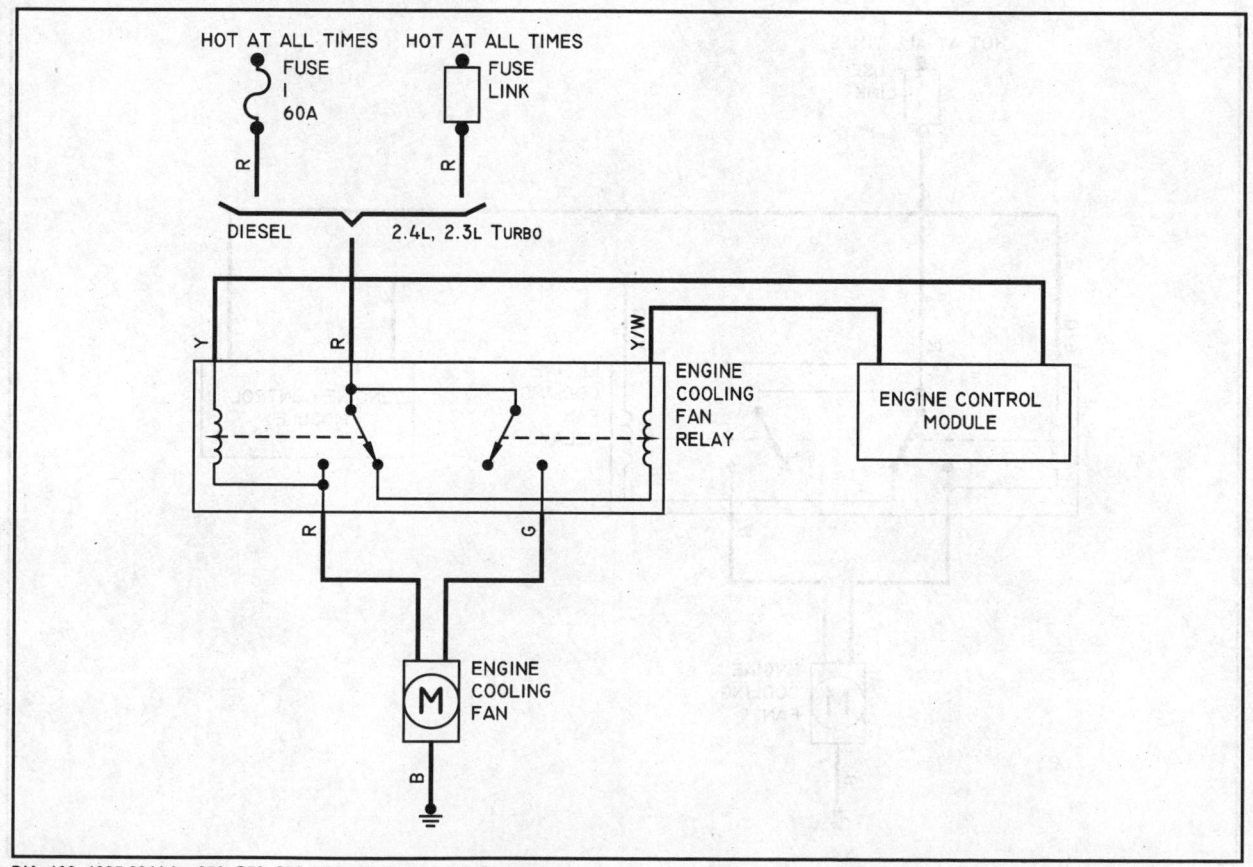

DIA. 106- 1995-99 Volvo 850, C70, S70, V70 2.3L Turbo / 2.4L Diesel

79239W53

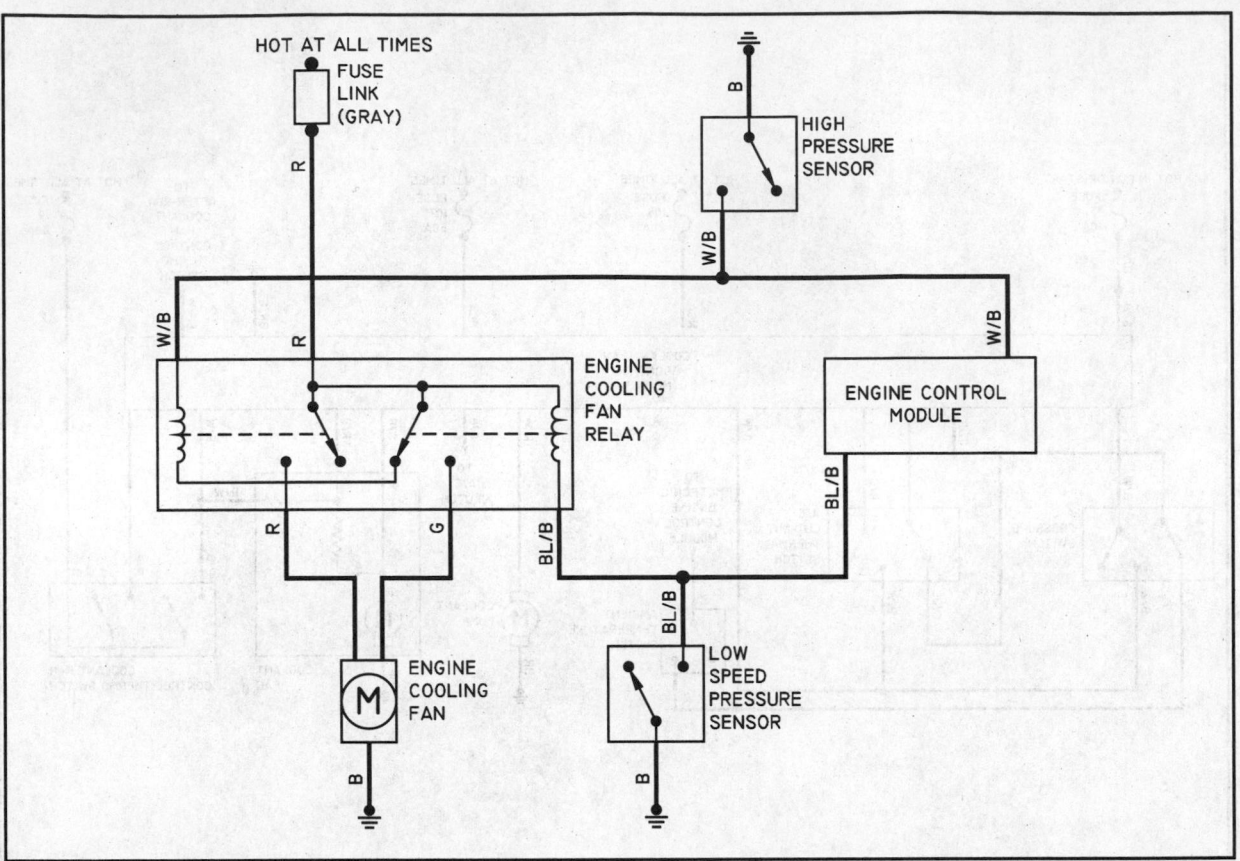

DIA. 107- 1995-99 Volvo 940 2.3L

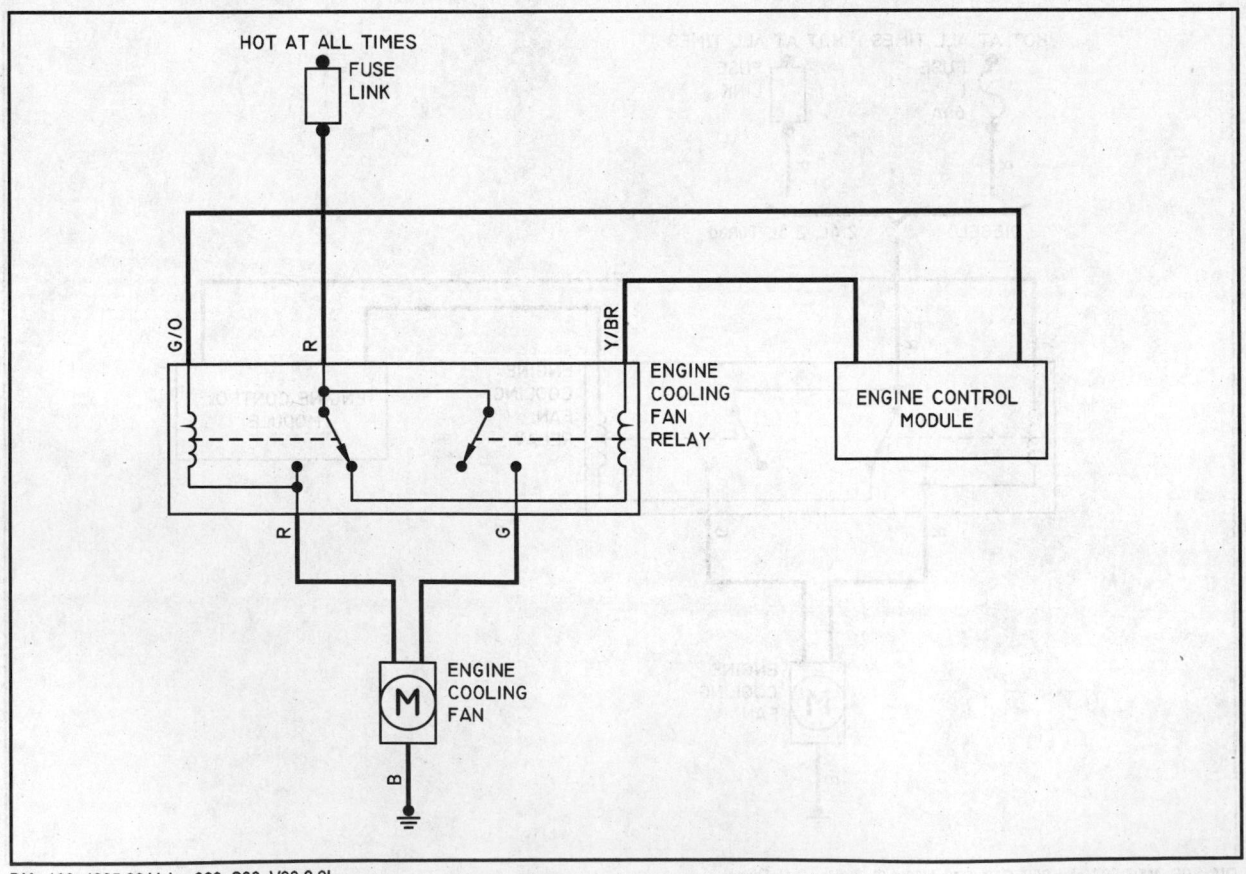

DIA. 108- 1995-99 Volvo 960, S90, V90 2.9L

79239W54

STARTING AND CHARGING SYSTEMS

10

STARTING SYSTEM

General Information

The typical starting system includes the battery, starter motor, solenoid, ignition switch, and in some cases, a starter relay. An inhibitor (neutral safety) switch is included in the starting system circuit to prevent the vehicle from being started while in gear.

When the ignition key is turned to the **START** position, current flows and energizes the starter's solenoid coil. The energized coil becomes an electromagnet which pulls the plunger into the coil, the plunger closes a set of contacts which allow high current to reach the starter motor. On models where the solenoid is mounted on the starter, the plunger also serves to push the starter pinion to mesh with the teeth on the flywheel/flexplate.

To prevent damage to the starter motor when the engine starts, the pinion gear incorporates an over-running (one-way) clutch which is splined to the starter armature shaft. The rotation of the running engine may speed the rotation of the pinion but not the starter motor itself.

Some starting systems employ a starter relay in addition to the solenoid. This relay may be located under the instrument panel, in the kickpanel or in the fuse/relay center under the hood. This relay is used to reduce the amount of current the starting (ignition) switch must carry.

PRECAUTIONS

To prevent damage to the on-board computer, alternator and regulator, the following precautionary measures must be taken when working with the electrical system.

• Always disconnect the negative battery cable before servicing the starter motor. Battery voltage is always present at the large (**B**) terminal on the solenoid. When removing the starter motor, be prepared to support its weight after the last bolt is removed because the starter motor is a fairly heavy component.

• Never operate the starter motor for more than 30 seconds at a time. Too much cranking will cause the starter motor to overheat,

TCCA1P02

Before servicing the electrical system always disconnect the negative battery cable to prevent system damage

causing permanent damage. Allow the starter motor to cool for at least two minutes between starting attempts.

• Wear safety glasses when working on or near the battery.

• Don't wear a watch with a metal band when servicing the battery. Serious burns can result if the band completes the circuit between the positive battery terminal and ground.

• Be absolutely sure of the polarity of a booster battery before making connections. Connect the cables positive to positive, and negative to negative. Connect positive cables first, then make the last connection to ground on the body of the booster vehicle so that arcing cannot ignite hydrogen gas that may have accumulated near the battery. Even momentary connection of a booster battery with the polarity reversed will damage the alternator diodes.

• Disconnect both vehicle battery cables before attempting to charge a battery.

• Be cautious when using metal tools around a battery to avoid creating a short circuit between the terminals.

• When installing a battery, be sure that the positive and negative cables are not reversed.

• When jump-starting the car, be sure that like terminals are connected. This also applies to using a battery charger. Reversed polarity will burn out the alternator and regulator in a matter of seconds.

• Always disconnect the battery (negative cable first) when charging it.

System Testing

➡A good quality digital multimeter with at least 10 megohm/volt impedance should be used when testing modern automotive circuits. These meters can accurately detect very small amounts of voltage, current and resistance. This type of meter also has a high internal resistance that will not load the circuit being tested. Loading the circuit causes inaccurate readings, and may cause damage to sensitive computer circuits. Although we are not testing computer circuits in this section, accuracy is very important.

WITH STARTER MOUNTED SOLENOID

1. Check the battery and clean the connections as follows:

a. If the battery cells have removable caps, check the water level. Add distilled water if low. Load test the battery and charge if necessary. See Battery Testing in this section for the procedure.

b. Remove the cables and clean them with a wire brush. Reconnect the cables.

2. Check the starter motor ground circuit with a voltage drop test as follows:

a. Set the meter to read DC voltage on the lowest possible scale.

b. Connect the negative lead of your multimeter to the negative terminal of the battery.

c. Connect the positive lead to the body of the starter. Be sure the starter mounting bolts are tight. The meter should read 0.2 volts or less. If the voltage reading is greater, remove and clean the negative battery connection on the engine block. The voltage reading should now be within specification: if not, replace the negative battery cable.

3. Check the motor feed circuit with a voltage drop test as follows:

a. Disconnect the coil wire or the fuel injector harness to prevent the engine from starting.

b. Connect the positive lead of your meter to the positive terminal of the battery.

c. Connect the negative meter lead to the motor feed terminal. The motor feed terminal comes out of the body of the starter motor and connects to the solenoid.

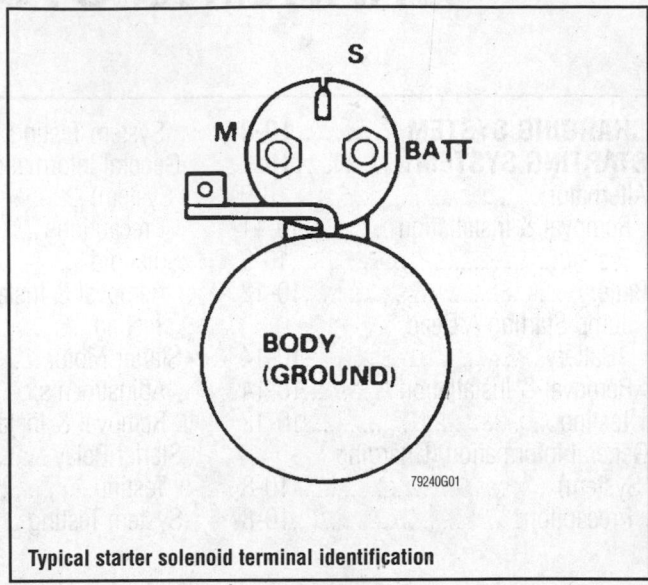

Typical starter solenoid terminal identification

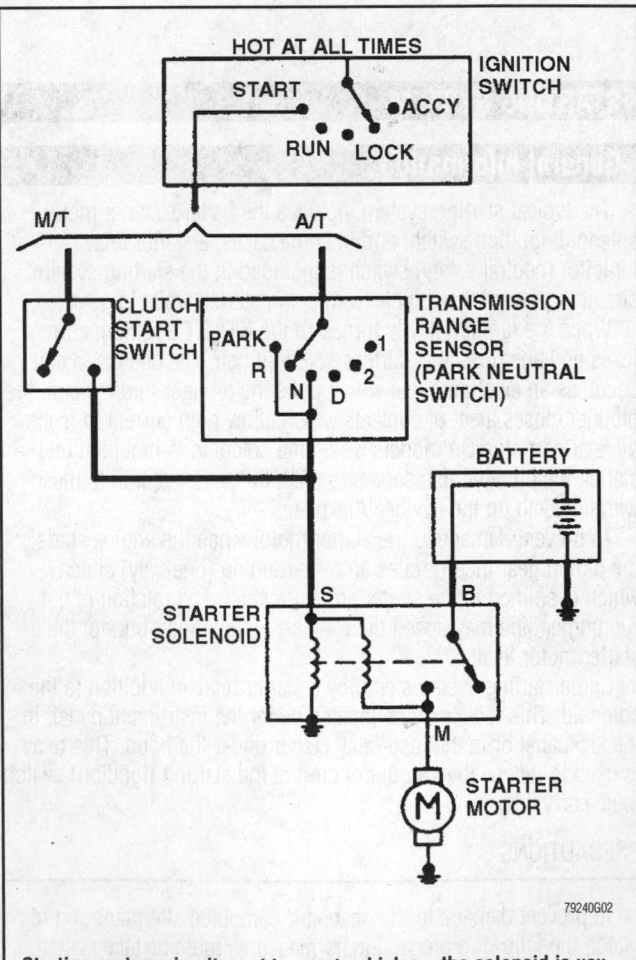

Starting system circuit used by most vehicles—the solenoid is usually mounted on the starter as indicated

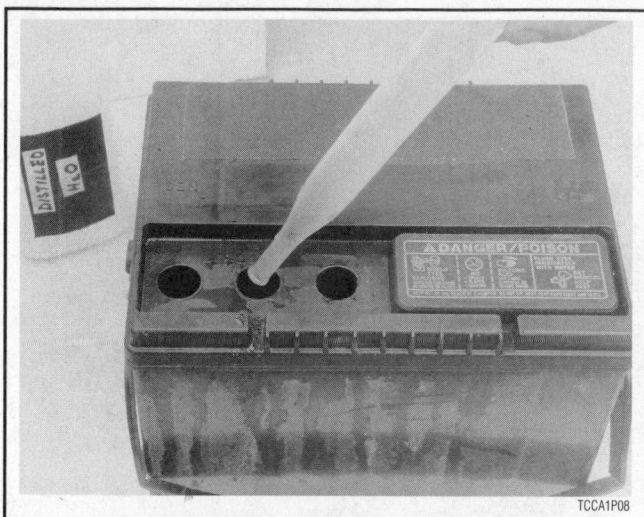

Before testing the system, be sure that the battery is in good shape, which includes ensuring the cells are full

Also, disconnect both battery cables (negative first) . . .

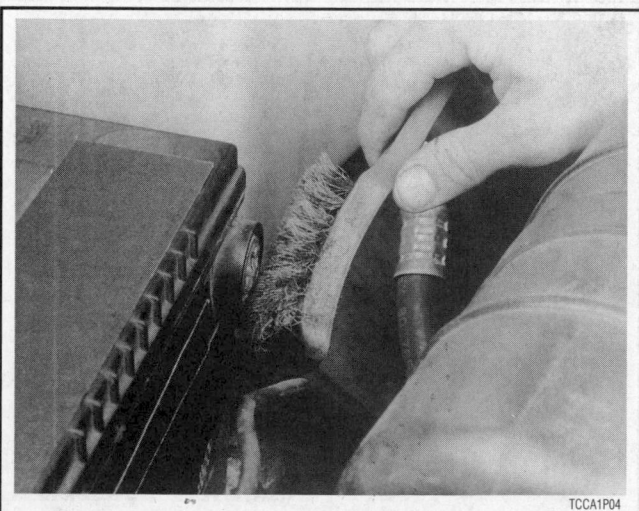

. . . clean the cable terminals of all dirt and corrosion with a wire brush . . .

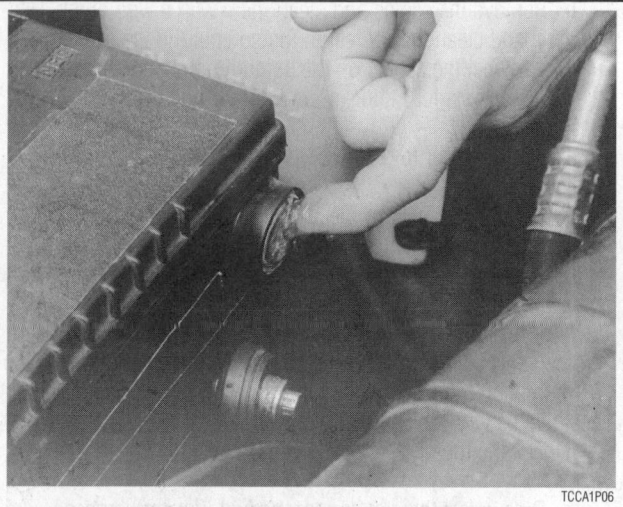

. . . and apply petroleum jelly or multi-purpose grease to the terminals before reattaching the cables

d. Turn the ignition key to the **START** position. The meter should read 0.2 volts or less. If the voltage reading is greater, remove and clean the positive battery connection on the starter solenoid. The voltage reading should now be within specification, if not replace the positive battery cable.

e. Connect the coil wire or fuel injector harness.

4. Check for battery voltage at the **S** terminal on the starter solenoid as follows:

a. Disconnect the coil wire or the fuel injector harness to prevent the engine from starting.

b. Set the meter to read battery voltage. Move it to the next higher range if set on the 2 volt scale.

c. Connect the positive lead to the **S** terminal on the starter solenoid and the negative lead to a good ground.

d. Turn the ignition key to the **START** position and crank the engine. The meter should read battery voltage. If battery voltage is not present, check the inhibitor (neutral safety) switch, fuse(s) and wiring between the ignition switch and starter solenoid. If battery voltage is present at the **S** terminal on the solenoid and the starter does not operate, replace the starter and solenoid assembly.

e. Connect the coil wire or fuel injector harness.

WITH EXTERNAL SOLENOID

➡ Not all solenoids are mounted on the starter motor. Some models use a solenoid (relay) mounted on the inner fender or firewall. Both types of solenoids serve to make the connection between the battery and starter motor. Trace the wires for positive identification. The small wire comes from the ignition switch, one large cable from the battery and the other large cable to the starter. The terminals are S , B and M respectively.

1. Check the battery and clean the connections as follows:

a. If the battery cells have removable caps, check the water level. Add distilled water if low. Load test the battery and charge if necessary. See Battery Testing in this section for the procedure.

✳✳ CAUTION

Alway remove the negative battery cable first, and install it last.

b. Remove the cables and clean them with a wire brush. Disconnect and clean the cables on the solenoid in the same manner. Reconnect the cables on the solenoid, then the battery.

2. Check the starter motor ground circuit with a voltage drop test as follows:

a. Set the meter to read DC voltage on the lowest possible scale.

b. Connect the negative lead of your multimeter to the negative terminal of the battery.

c. Connect the positive lead to the body of the starter. Be sure the starter mounting bolts are tight. The meter should read 0.2 volts or less. If the voltage reading is greater, remove and clean the negative battery connection on the engine block. The voltage reading should now be within specification; if not, replace the negative battery cable.

3. Check the motor feed circuit with a voltage drop test as follows:

a. Disconnect the coil wire or the fuel injector harness to prevent the engine from starting.

b. Connect the positive lead of your meter to the positive terminal of the battery.

c. Connect the negative meter lead to the motor feed terminal at the starter. This is the heavy cable on the starter. Turn the ignition key to the **START** position and crank the engine. The meter should read 0.2 volts or less. If the voltage reading is greater, remove and clean the positive battery connections on the starter and solenoid. The voltage reading should now be within specification; if not, replace the positive battery cable.

d. Connect the coil wire or fuel injector harness.

4. Check for battery voltage at the **S** terminal on the starter solenoid as follows:

a. Disconnect the coil wire or the fuel injector harness to prevent the engine from starting.

b. Set the meter to read battery voltage. Move it to next higher range, if previously set on the 2 volt scale.

c. Connect the positive lead to the **S** terminal on the starter solenoid and the negative lead to a good ground.

d. Turn the ignition key to the **START** position. The meter should read battery voltage. If battery voltage is not present, check the inhibitor (neutral safety) switch, fuse(s) and wiring between the ignition switch and starter solenoid. If battery voltage is present at the **S** and **B** terminals but not at the motor feed terminal, replace the solenoid. If battery voltage is present at all three terminals and the starter does not operate, replace the starter motor.

e. Connect the coil wire or fuel injector harness.

Starter Motor

REMOVAL & INSTALLATION

1. Disconnect the negative battery cable.

2. Remove all components necessary to gain access to the starter motor (such as exhaust pipes, air intake ducts, hoses, brackets and heat shields).

3. Disconnect the wiring from the starter. In some cases, the wiring may be more accessible after removing the mounting bolts and moving the starter.

4. Remove the starter mounting bolts, if not already done.

5. Remove the starter assembly from the vehicle. In some cases, the starter will have to be turned to a different angle to clear

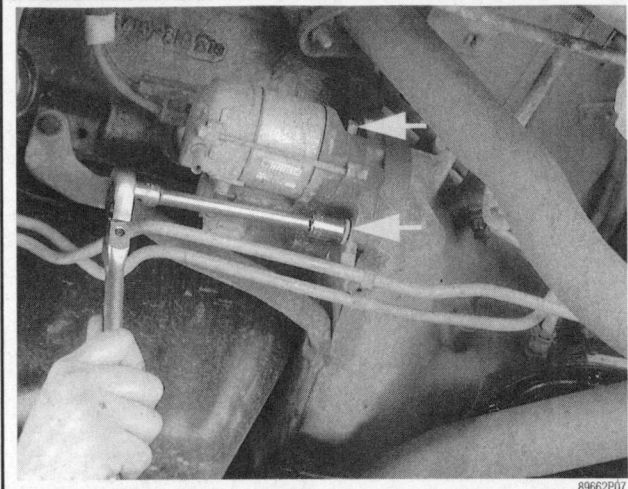

To remove the starter, raise the vehicle if needed and loosen the starter mounting fasteners (arrows) . . .

. . ., then pull the starter out of the transmission bell housing . . .

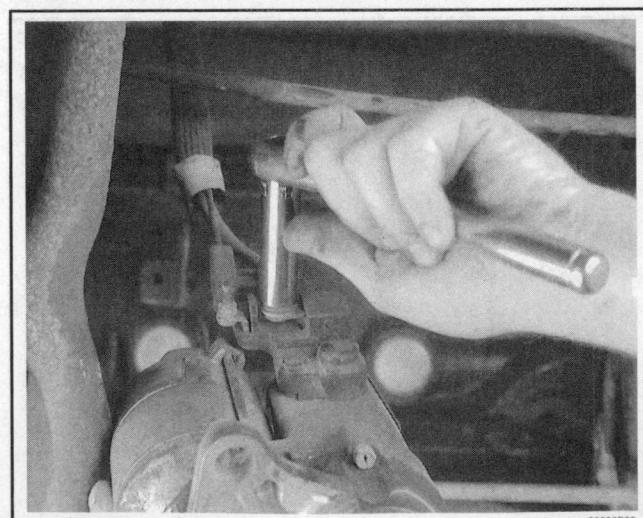

. . . and disconnect the starter motor wires, if not already done

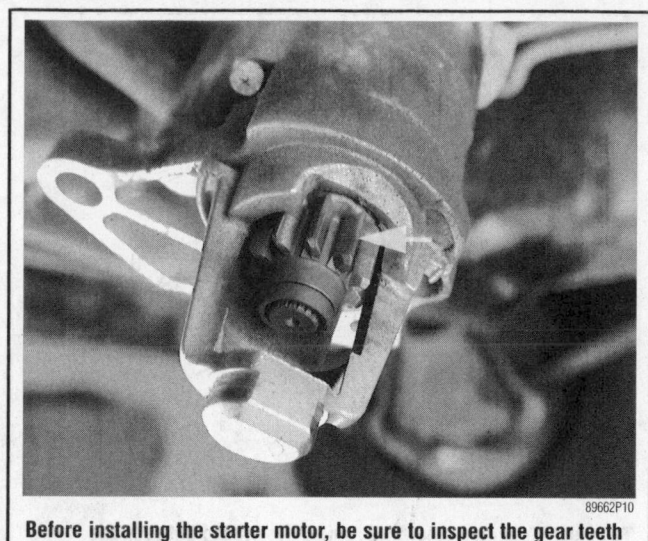

Before installing the starter motor, be sure to inspect the gear teeth (arrow) . . .

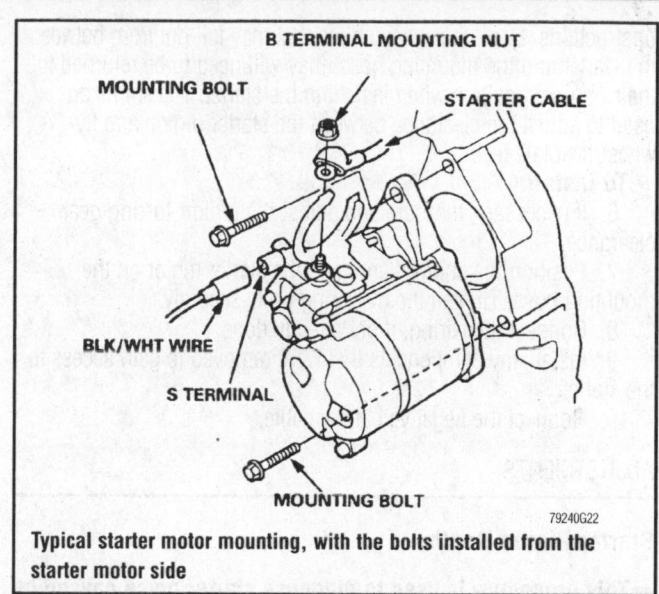

Typical starter motor mounting, with the bolts installed from the starter motor side

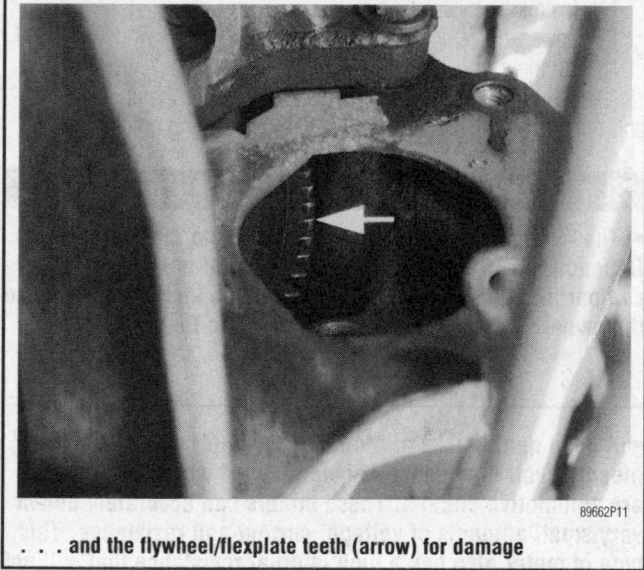

. . . and the flywheel/flexplate teeth (arrow) for damage

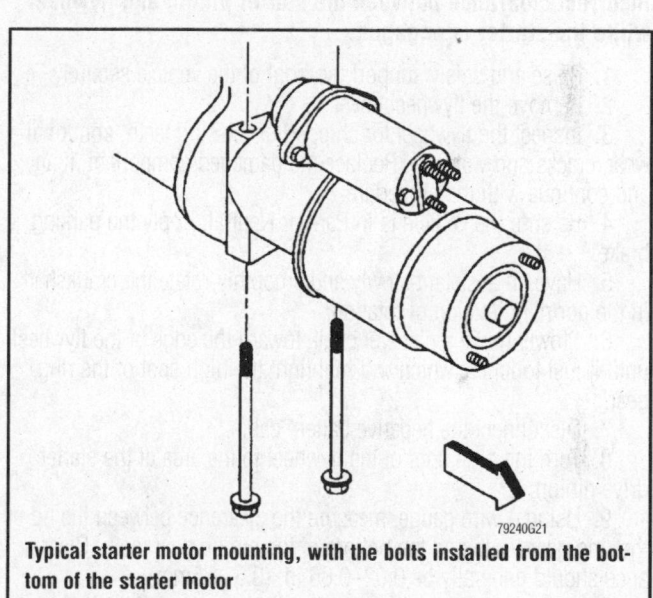

Typical starter motor mounting, with the bolts installed from the bottom of the starter motor

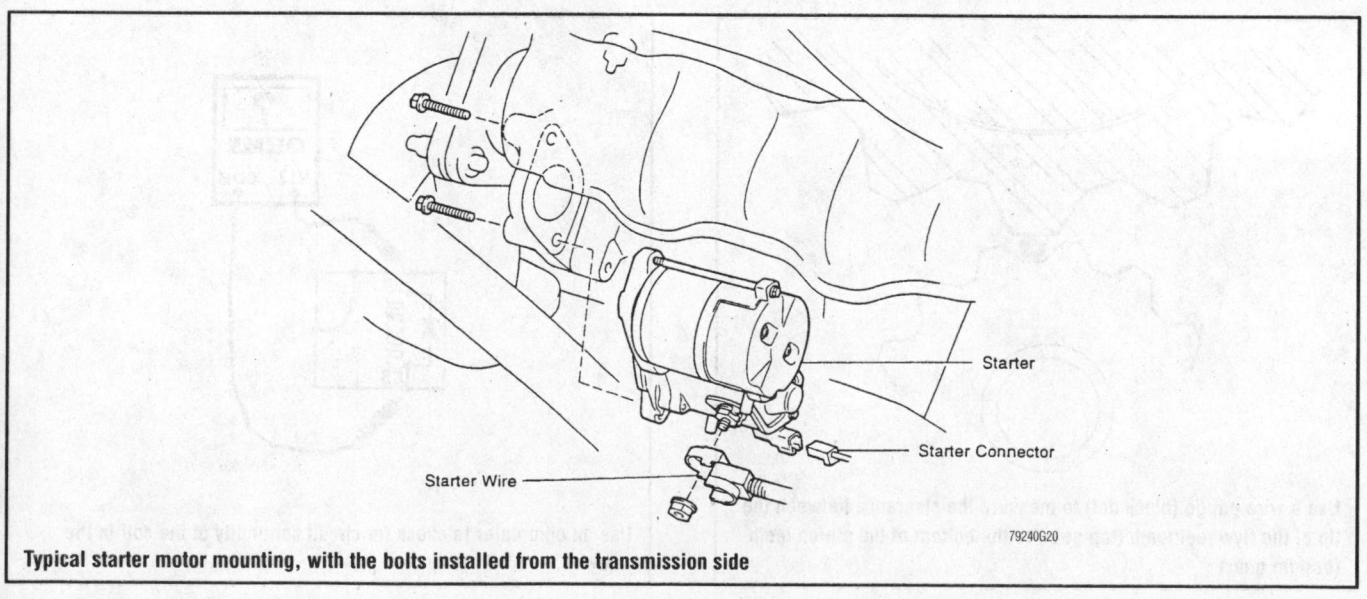

Typical starter motor mounting, with the bolts installed from the transmission side

obstructions. Don't loose any shims that may fall out from between the starter and the mounting boss, they will need to be returned to their original position when installing the starter. The shims are used to adjust the clearance between the starter pinion and fly-wheel/flexplate teeth.

To install:

6. If necessary, measure and adjust the pinion-to-ring gear clearance.

7. Position the shim (if any) and the starter motor on the mounting boss. Tighten the mounting bolts securely.

8. Connect the wiring, if not already done.

9. Install any components that were removed to gain access to the starter.

10. Connect the negative battery cable.

ADJUSTMENTS

Starter Pinion Depth

➡**This procedure is used to diagnose starter noise caused by incorrect clearance between the starter pinion and flywheel while the starter is engaged.**

1. Raise and safely support the front of the vehicle securely.

2. Remove the flywheel cover.

3. Inspect the flywheel for chipped or missing teeth, abnormal wear, cracks and warpage. Replace the damaged component, if any, and continue with the procedure.

4. Be sure the vehicle is in Park or Neutral. Apply the parking brake.

5. Have an assistant slowly and smoothly rotate the crankshaft in the normal direction of rotation.

6. Slowly move a piece of chalk toward the edge of the flywheel until it just touches, which will highlight the high spot of the ring gear.

7. Disconnect the negative battery cable.

8. Turn the high spot of the flywheel to the area of the starter drive pinion.

9. Using a wire gauge, measure the clearance between the tip of the ring gear tooth and the bottom of the pinion gear teeth. Clearance should generally be 0.02–0.06 in. (0.5–1.5mm).

10. Add or remove shims to adjust the clearance, if needed.

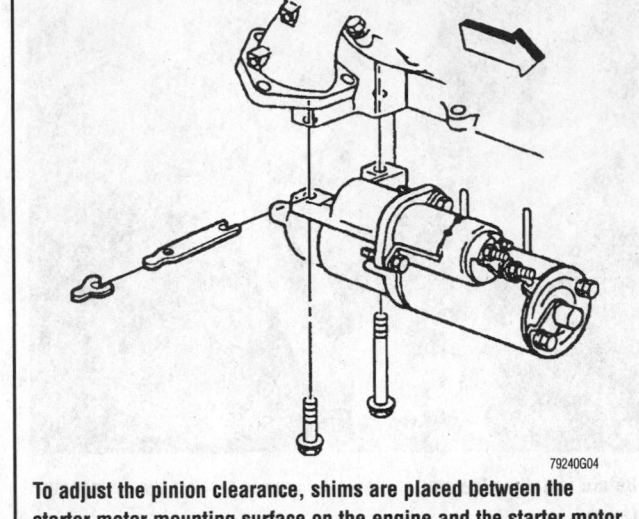

To adjust the pinion clearance, shims are placed between the starter motor mounting surface on the engine and the starter motor

11. Install the flywheel cover.

12. Lower the vehicle to the floor.

13. Connect the negative battery cable.

Generally, add shims if the starter whines after the engine starts, and remove shims if the starter whines only during cranking.

Starter Relay

➡**The starter relay is usually located in the fuse/relay panel. Depending on the manufacturer, it may be in the engine compartment, under the dash or behind a kickpanel. Refer to the owner's manual for the location of the fuse/relay box.**

TESTING

➡**A good quality Digital Multimeter (DMM) with at least 10 megohm/volt impedance should be used when testing modern automotive circuits. These meters can accurately detect very small amounts of voltage, current and resistance. This type of meter also has a high internal resistance that will not**

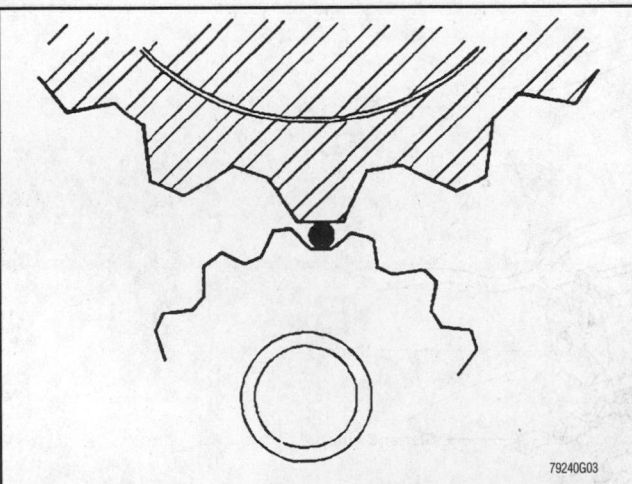

Use a wire gauge (black dot) to measure the clearance between the tip of the flywheel tooth (top gear) to the bottom of pinion teeth (bottom gear)

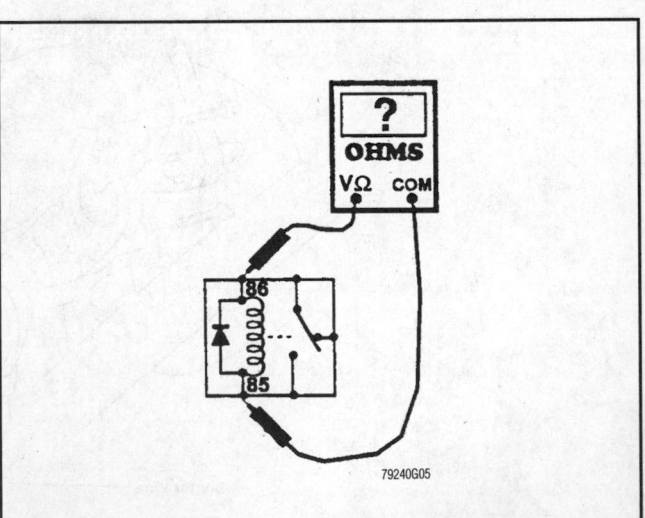

Use an ohmmeter to check for circuit continuity of the coil in the relay

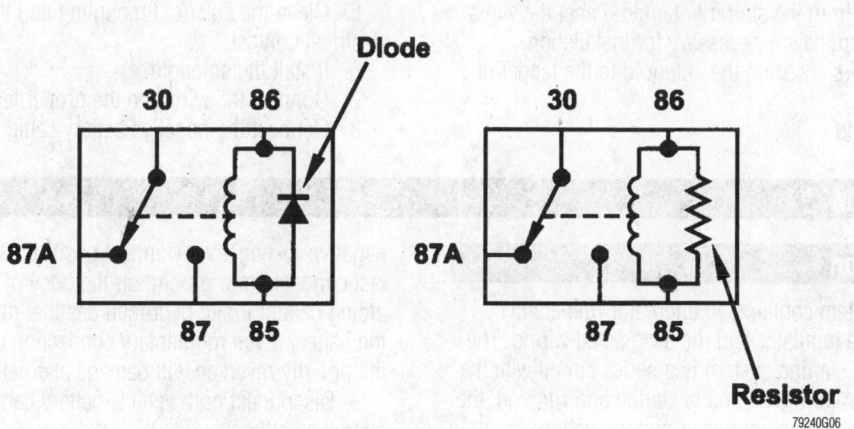

Terminal identification of the most common types of relays. Diodes and resistors in the relay prevent voltage spikes, induced when the current is removed from the coil, from damaging electronic components

load the circuit being tested. Loading the circuit gives inaccurate readings and may cause damage to sensitive computer circuits.

1. Turn the ignition **OFF**.
2. Remove the relay.
3. Locate the two terminals on the relay which are connected to the coil windings. Check the relay coil for continuity. Connect the negative meter lead to terminal **85** and positive meter lead to terminal **86**. There should be continuity. If not, replace the relay.
4. Check the operation of the internal relay contacts, as follows:

   a. Connect the meter leads to terminals **30** and **87**. Meter polarity does not matter for this step.

   b. Apply positive battery voltage to terminal **86** and ground to terminal **85**. The relay should click as the contacts are drawn toward the coil and the meter should indicate continuity. Replace the relay if your results are different.

Solenoid

TESTING

1. Disconnect the negative battery cable.
2. Remove the wire connections from the starter solenoid.
3. Using a self-powered test light or ohmmeter, check for continuity between the following:
   • Solenoid **B** terminal and solenoid case or ground terminal—no continuity
   • **S** terminal and solenoid case or ground terminal—continuity
   • **S** terminal and **M** terminal—continuity
   • **M** terminal and solenoid case or ground terminal—continuity
4. If the actual results of the test are different than indicated, replace the starter solenoid.

REMOVAL & INSTALLATION

➡This procedure is for externally mounted starter solenoids only. For solenoids mounted on the starter, we recommend replacing the complete assembly.

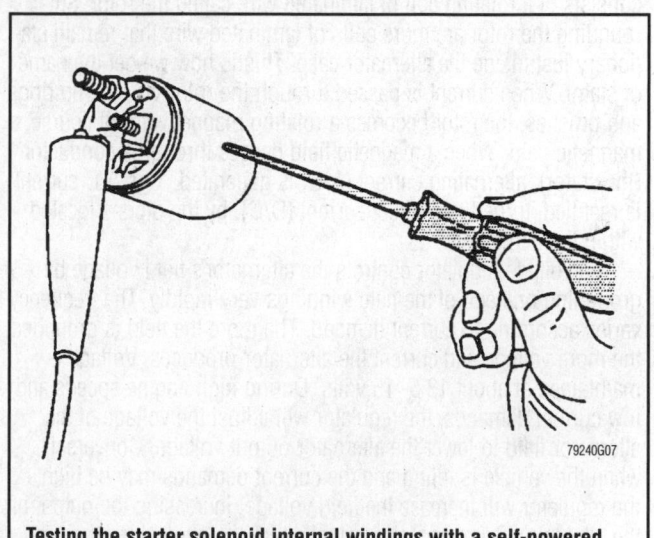

Testing the starter solenoid internal windings with a self-powered test light—starter mounted solenoid shown

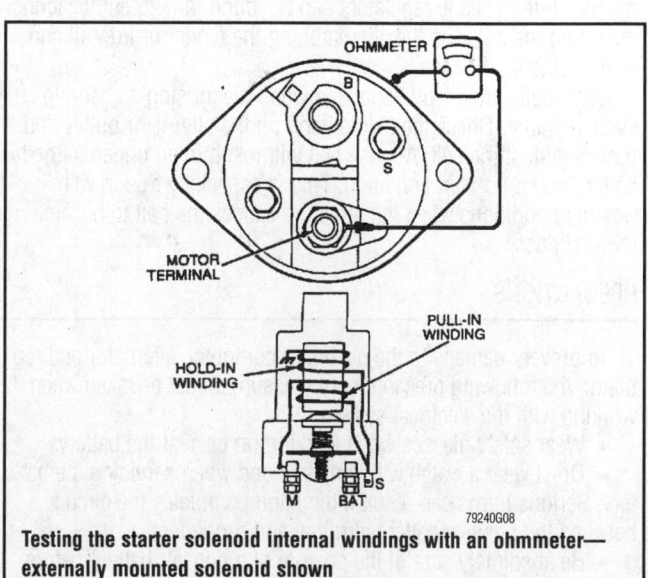

Testing the starter solenoid internal windings with an ohmmeter—externally mounted solenoid shown

1. Disconnect the negative battery cable.
2. Remove the wiring from the starter solenoid. Label the wires and the corresponding terminals if necessary for installation.
3. Remove the fasteners securing the solenoid to the fender or firewall.
4. Remove the solenoid.

To install:

5. Clean the solenoid mounting and the solenoid to ensure good electrical contact.
6. Install the solenoid.
7. Connect the wiring to the proper terminals.
8. Connect the negative battery cable.

CHARGING SYSTEM

General Information

A typical charging system contains an alternator (generator), drive belt, battery, voltage regulator and the associated wiring. The charging system, like the starting system is a series circuit with the battery wired in parallel. After the engine is started and running, the alternator takes over as the source of power and the battery, then becomes part of the load on the charging system.

Some vehicle manufacturers use the term generator instead of alternator. Many years ago there used to be a difference, now they are one and the same. The alternator which is driven by the belt, consists of a rotating coil of laminated wire called the rotor. Surrounding the rotor are more coils of laminated wire that remain stationary just inside the alternator case. This is how we get the name of stator. When current is passed through the rotor via the slip rings and brushes, the rotor becomes a rotating magnet with, of course, a magnetic field. When a magnetic field passes through a conductor (the stator), alternating current (A/C) is generated. This A/C current is rectified, turned into direct current (D/C), by the diodes located within the alternator.

The voltage regulator controls the alternator's field voltage by grounding one end of the field windings very rapidly. The frequency varies according to current demand. The more the field is grounded, the more voltage and current the alternator produces. Voltage is maintained at about 13.5–15 volts. During high engine speeds and low current demands, the regulator will adjust the voltage of the alternator field to lower the alternator output voltage. Conversely, when the vehicle is idling and the current demands may be high, the regulator will increase the field voltage, increasing the output of the alternator. Some vehicles actually turn the alternator off during periods of no load and/or wide open throttle. This was designed to reduce fuel consumption and increase power. Depending on the manufacturer, voltage regulators can be found in different locations, including inside or on the alternator, on the fender or firewall and even inside the PCM.

Drive belts are often overlooked when diagnosing a charging system failure. Check the belt tension on the alternator pulley and replace/adjust the belt. A loose belt will result in an undercharged battery and a no-start condition. This is especially true in wet weather conditions when the moisture causes the belt to become more slippery.

PRECAUTIONS

To prevent damage to the on-board computer, alternator and regulator, the following precautionary measures must be taken when working with the electrical system:

- Wear safety glasses when working on or near the battery.
- Don't wear a watch with a metal band when servicing the battery. Serious burns can result if the band completes the circuit between the positive battery terminal and ground.
- Be absolutely sure of the polarity of a booster battery before making connections. Connect the cables positive-to-positive, and negative-to-negative. Connect positive cables first,, then make the last connection to ground on the body of the booster vehicle so that arcing cannot ignite hydrogen gas that may have accumulated near the battery. Even momentary connection of a booster battery with the polarity reversed will damage alternator diodes.
- Disconnect both vehicle battery cables before attempting to charge a battery.
- Never ground the alternator or generator output or battery terminal. Be cautious when using metal tools around a battery to avoid creating a short circuit between the terminals.
- Never ground the field circuit between the alternator and regulator.
- Never run an alternator or generator without load unless the field circuit is disconnected.
- Never attempt to polarize an alternator.
- When installing a battery, be sure that the positive and negative cables are not reversed.
- When jump-starting the car, be sure that like terminals are connected. This also applies to using a battery charger. Reversed polarity will burn out the alternator and regulator in a matter of seconds.
- Never operate the alternator with the battery disconnected or on an otherwise uncontrolled open circuit.
- Do not short across or ground any alternator or regulator terminals.
- Do not try to polarize the alternator.
- Do not apply full battery voltage to the field (brown) connector.
- Always disconnect the battery ground cable before disconnecting the alternator lead.
- Always disconnect the battery (negative cable first) when charging it.
- Never subject the alternator to excessive heat or dampness. If you are steam cleaning the engine, cover the alternator.
- Never use arc-welding equipment on the vehicle with the alternator connected.

SYSTEM TESTING

The charging system should be inspected if:

- A Diagnostic Trouble Code (DTC) is set relating to the charging system
- The charging system warning light is illuminated
- The voltmeter on the instrument panel indicates improper charging (either high or low) voltage
- The battery is overcharged (electrolyte level is low and/or boiling out)
- The battery is undercharged (insufficient power to crank the starter)

The starting point for all charging system problems begins with the inspection of the battery, related wiring and the alternator drive belt. The battery must be in good condition and fully charged before system testing. If a Diagnostic Trouble Code (DTC) is set, diagnose and repair the cause of the trouble code first.

If equipped, the charging system warning light will illuminate if the charging voltage is either too high or too low. The warning light should light when the key is turned to the **ON** position as a bulb check. When the alternator starts producing voltage due to the engine starting, the light should go out. A good sign of voltage that is too high are lights that burn out and/or burn very brightly. Overcharging can also cause damage to the battery and electronic circuits.

Alternator

TESTING

➡Before testing, be sure all connections and mounting bolts are clean and tight. Many charging system problems are related to loose and corroded terminals or bad grounds. Don't overlook the engine ground connection to the body, or the tension of the alternator drive belt.

Voltage Drop Test

➡A good quality Digital Multimeter (DMM) with at least 10 megohm/volt impedance should be used when testing modern automotive circuits. These meters can accurately detect very small amounts of voltage, current and resistance. This type of meter also has a high internal resistance that will not load the circuit being tested. Loading the circuit gives inaccurate readings and may cause damage to sensitive computer circuits.

1. Be sure the battery is in good condition and fully charged.
2. Perform a voltage drop test of the positive side of the circuit as follows:
   • Start the engine and allow it to reach normal operating temperature.
   • Turn the headlamps, heater blower motor and interior lights on.
   • Bring the engine to about 2,500 rpm and hold it there.
   • Connect the negative (-) voltmeter lead directly to the battery positive (+) terminal.
   • Touch the positive (+) voltmeter lead directly to the alternator **B+** output stud, not the nut. The meter should read no higher than about 0.5 volts. If it does, then there is higher than normal resistance between the positive side of the battery and the **B+** output at the alternator.
   • Move the positive (+) meter lead to the nut and compare the voltage reading with the previous measurement. If the voltage reading drops substantially, then there is resistance between the stud and the nut.

➡The theory is to keep moving closer to the battery terminal one connection at a time in order to find the area of high resistance (bad connection).

3. Perform a voltage drop test of the negative side of the circuit as follows:
   a. Start the engine and allow it to reach normal operating temperature.
   b. Turn the headlamps, heater blower motor and interior lights ON.
   c. Bring the engine to about 2,500 rpm and hold it there.
   d. Connect the negative (-) voltmeter lead directly to the negative battery terminal.

e. Touch the positive (+) voltmeter lead directly to the alternator case or ground connection. The meter should read no higher than about 0.3 volts. If it does, then there is higher than normal resistance between the battery ground terminal and the alternator ground.

f. Move the positive (+) meter lead to the alternator mounting bracket, if the voltage reading drops substantially, then you know that there is a bad electrical connection between the alternator and the mounting bracket.

➡The theory is to keep moving closer to the battery terminal one connection at a time in order to find the area of high resistance (bad connection).

Current Output Test

➡A good quality Digital Multimeter (DMM) with at least 10 megohm/volt impedance should be used when testing modern automotive circuits. These meters can accurately detect very small amounts of voltage, current and resistance. This type of meter also has a high internal resistance that will not load the circuit being tested. Loading the circuit gives inaccurate readings and may cause damage to sensitive computer circuits.

1. Perform a current output test as follows:

➡The current output test requires the use of a volt/amp tester with battery load control and an inductive amperage pick-up. Follow the manufacturer's instructions on the use of the equipment.

   a. Start the engine and allow it to reach normal operating temperature.
   b. Apply the parking brake and turn OFF all electrical accessories.
   c. Connect the tester to the battery terminals and cable according to the instructions.
   d. Bring the engine to about 2,500 rpm and hold it there.

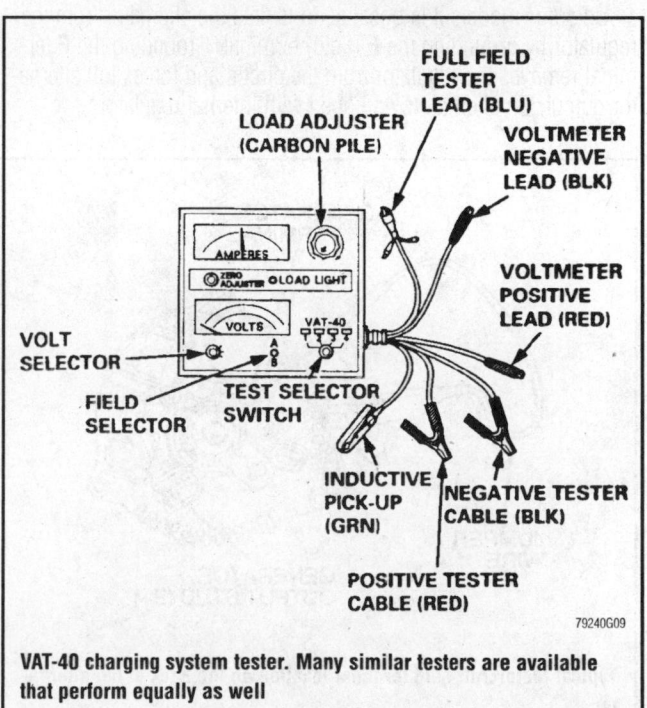

VAT-40 charging system tester. Many similar testers are available that perform equally as well

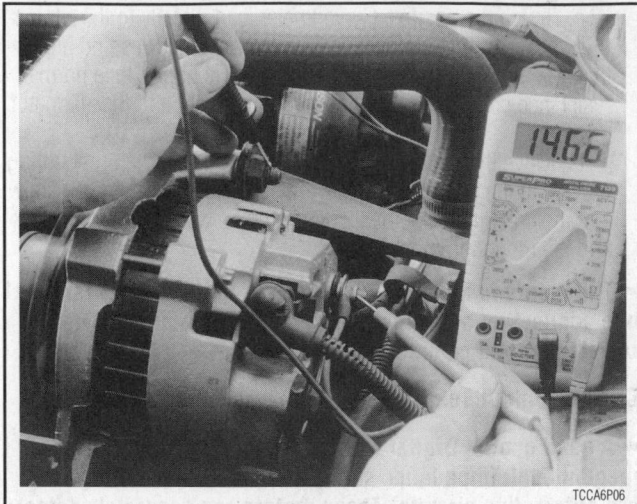

The output voltage of the alternator can be quickly measured by probing between the output terminal and a good ground

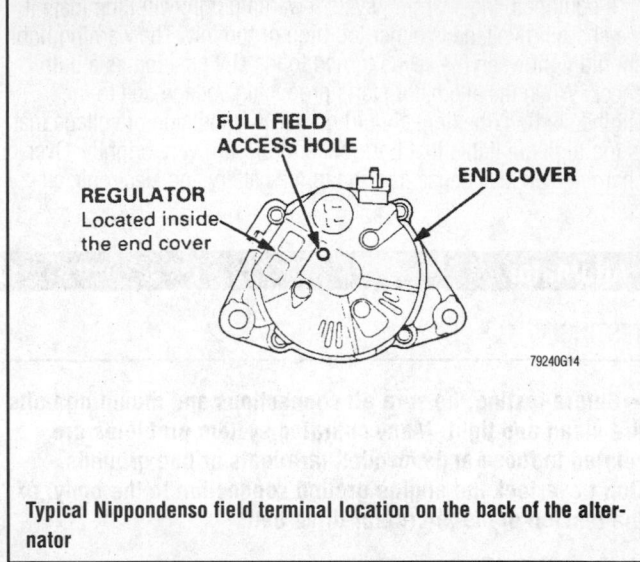

Typical Nippondenso field terminal location on the back of the alternator

e. Apply a load to the charging system with the rheostat on the tester. Do not let the voltage drop below 12 volts.

f. The alternator should deliver to within 10% of the rated output. If the amperage is not within 10% and all other components test good, replace the alternator.

Alternator Isolation Test

➡A good quality Digital Multimeter (DMM) with at least 10 megohm/volt impedance should be used when testing modern automotive circuits. These meters can accurately detect very small amounts of voltage, current and resistance. This type of meter also has a high internal resistance that will not load the circuit being tested. Loading the circuit gives inaccurate readings and may cause damage to sensitive computer circuits.

On some models it is possible to isolate the alternator from the regulator by grounding the **F** (field) terminal. Grounding the **F** terminal removes the regulator from the circuit and forces full alternator output. On alternators equipped with internal regulators, we

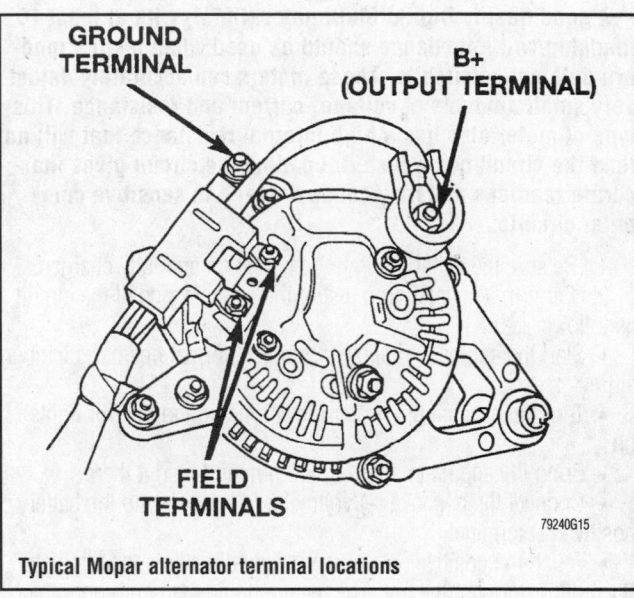

Typical Mopar alternator terminal locations

recommend replacing the complete assembly if either the alternator or regulator is defective.

✷✷ WARNING

Do not allow the voltage to rise above 18 volts. Damage to electrical circuits may occur.

1. Connect a voltmeter across the battery terminals so the voltage can be monitored.

2. Start the engine and allow it reach normal operating temperature.

3. Connect a jumper lead to a good ground.

4. Locate the field terminal (negative) on the back of the alternator.

5. Momentarily connect the grounded jumper to the field terminal. If the alternator is OK, the voltage will climb rapidly. Disconnect the jumper before the output reaches 18 volts. If the voltage does not rise, replace the alternator. If the voltage rises, then the regulator is bad.

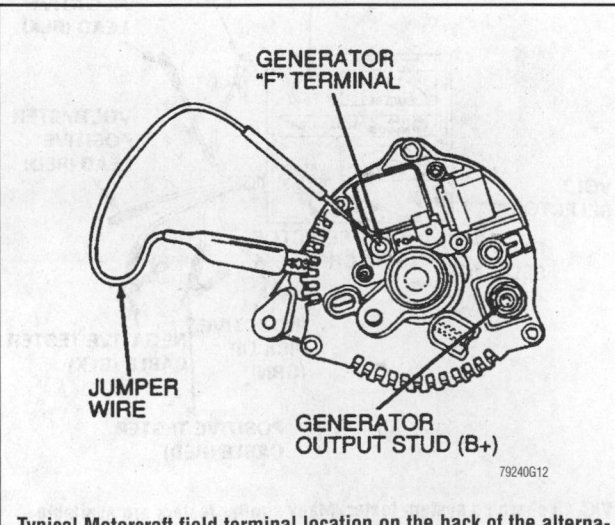

Typical Motorcraft field terminal location on the back of the alternator

➡Chrysler models have two field terminals, one positive and one negative. The positive (+) terminal will have battery voltage present and the negative (-) terminal will have 3–5 volts less. Ground the negative (-) terminal when testing this type of alternator.

REMOVAL & INSTALLATION

1. Disconnect the negative battery cable.
2. Remove the drive belt from the alternator pulley.

➡In some cases, it may be easier to disconnect the wiring after the alternator has been removed. Be sure to support the alternator by hand while removing the wiring.

3. Disconnect the wiring from the alternator.
4. Remove the alternator.

To install:

➡If necessary, attach the wiring to the alternator before installation.

When removing the mounting bolts, be sure to retain any washers, spacers or nuts for reassembly

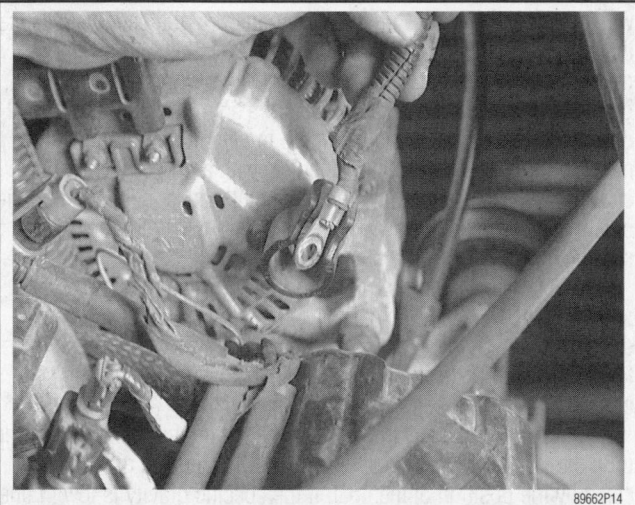

To remove a common alternator, first detach the wiring terminals from it (if possible) . . .

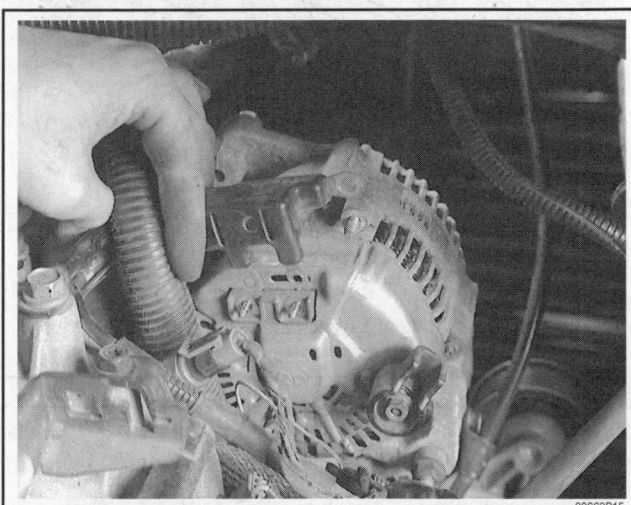

If not possible earlier, now disconnect any applicable wiring from the alternator

. . . then loosen the alternator mounting bolts (arrows)

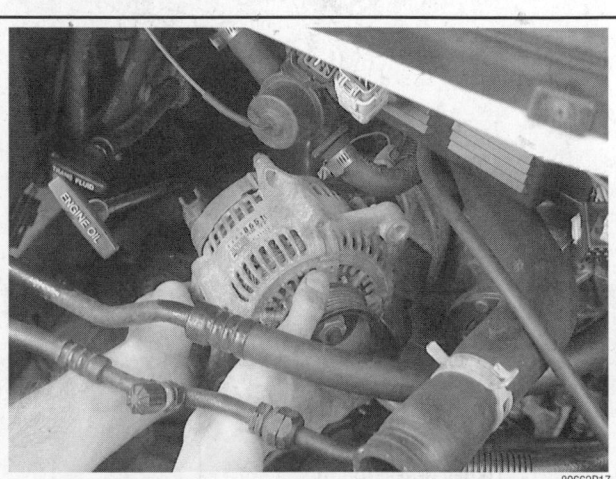

Finally, carefully remove the alternator from the engine compartment—although not shown here some alternators must be dropped out the bottom of the vehicle

5. Install the alternator and attach the wiring if not already done.

6. Install the drive belt on the alternator pulley. Adjust the belt if necessary.

7. Connect the negative battery cable.

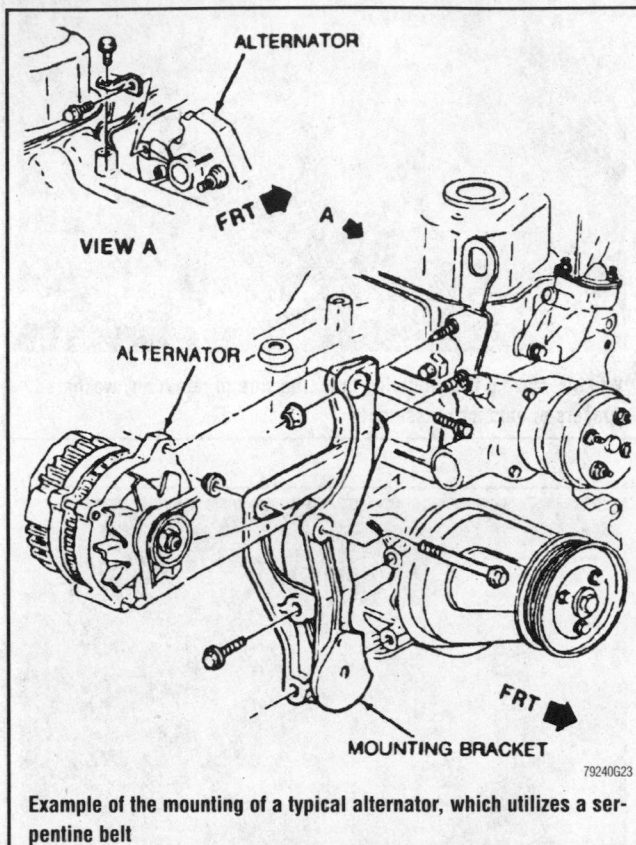

Example of the mounting of a typical alternator, which utilizes a serpentine belt

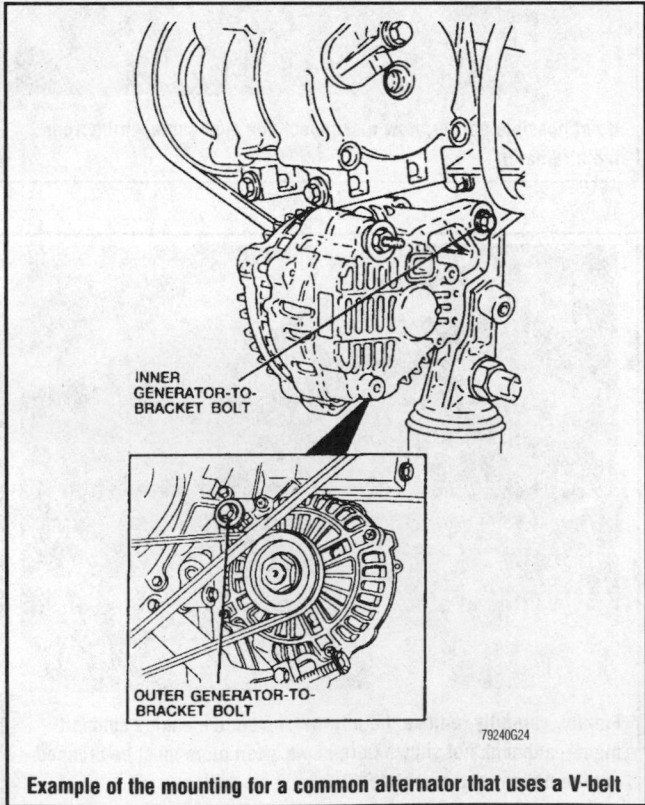

Example of the mounting for a common alternator that uses a V-belt

Battery

TESTING

✳✳ CAUTION

If the battery shows signs of freezing, cracking, leaking, loose posts or low electrolyte level, do not attempt to test, charge or jump start. Internal arcing may occur and cause the battery to explode. Always replace a battery that is physically damaged. If only the water level is low and the battery can be filled, add distilled water to the proper level. When charging, disconnect the battery cables, attach the connections to the battery first, then turn the charger ON. Never disconnect the battery cable(s) while the engine is running. Always wear safety glasses when servicing the battery.

Specific Gravity Test

The fluid (sulfuric acid solution) contained in the battery cells will tell you many things about the condition of the battery. Because the cell plates must be kept submerged below the fluid level in order to operate, maintaining the fluid level is extremely important., because the specific gravity of the acid is an indication of electrical charge, testing the fluid can be an aid in determining if the battery must be replaced. A battery in a vehicle with a properly operating charging system should require little maintenance, but careful, periodic inspection should reveal problems before they leave you stranded.

At least once a year, check the specific gravity of the battery. It should be between 1.20 and 1.26 on the gravity scale. Most auto supply stores carry a variety of inexpensive battery testing hydrometers. These can be used on any non-sealed battery to test the specific gravity in each cell.

Draw some of the electrolyte from the battery into the hydrometer until the float is lifted from its seat. Read the specific gravity indicated by the position of the float. If the specific gravity is low in one or more cells, the battery should be slowly charged and checked again to see if the gravity has come up. Generally, if after charging,

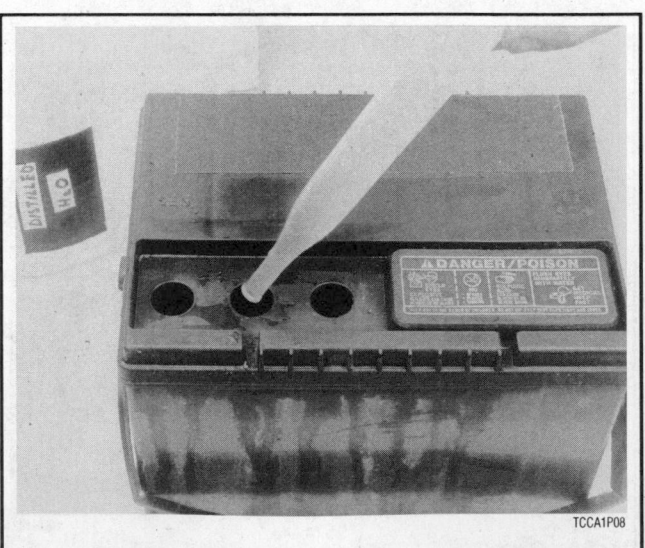

If the battery cells are low on fluid, top them up with distilled water

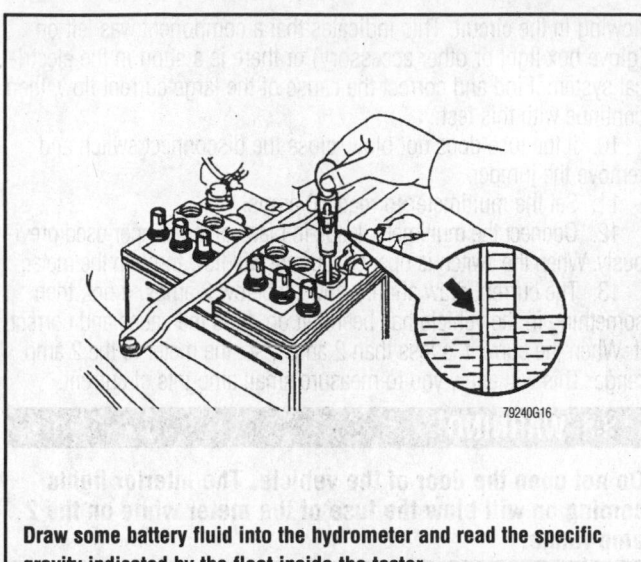

Draw some battery fluid into the hydrometer and read the specific gravity indicated by the float inside the tester

Open Circuit Voltage	
Open Circuit Volts	**Charge Percentage**
11.7 volts or less	0%
12.0 volts	25%
12.2 volts	50%
12.4 volts	75%
12.6 volts or more	100%

Compare the actual voltage measured with these values to determine the percent of charge based on no load test results

the specific gravity between any two cells varies more than 50 points (0.50), replace the battery, as it can no longer produce sufficient voltage to guarantee proper operation.

No Load Voltage Test

➡A good quality Digital Multimeter (DMM) with at least 10 megohm/volt impedance should be used when testing modern automotive circuits. These meters can accurately detect very small amounts of voltage, current and resistance. This type of meter also has a high internal resistance that will not load the circuit being tested. Loading the circuit gives inaccurate readings and may cause damage to sensitive computer circuits.

1. Perform a no load voltage test to determine the state of charge by doing the following:

   a. If the battery has just been charged, remove the surface charge by turning on the headlamps for 15 seconds, then let the voltage stabilize for about 5 minutes before making any measurements.

   b. Disconnect the negative battery cable.

   c. Measure the battery voltage with a DMM.

   d. Compare the readings to the chart to determine the state of charge.

High Capacity Discharge Test

1. Perform a high capacity discharge test to determine the cranking capacity as follows:

   a. Fully charge the battery.

   b. Connect a VAT-40 or equivalent load tester to the battery.

   c. Apply a load equal to ½ of the Cold Cranking Amp (CCA) rating of the battery for 15 seconds. The CCA is usually found on the battery label, if not, apply a load equal to 200 amps.

   d. If the voltmeter reading falls below 9.6 volts at 70° F (21° C) or more, the battery should be replaced. The minimum battery voltage will be lower depending on the ambient temperature. Refer to the chart for testing in temperatures lower than 70° F (21° C).

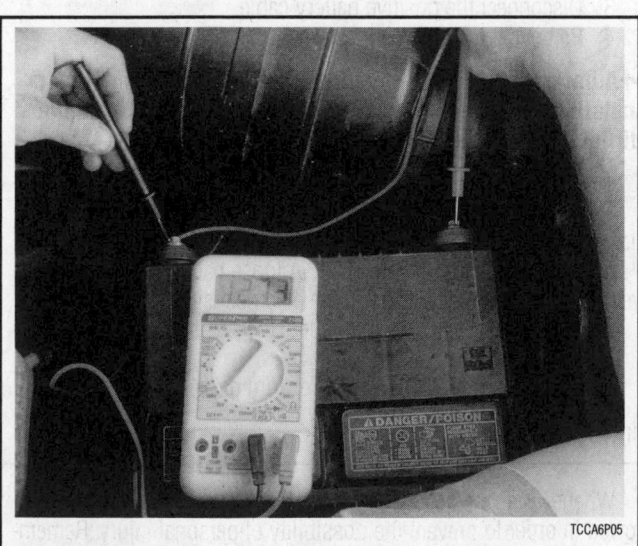

Use a high quality DMM to measure the battery voltage

Load Test Temperature		
Minimum Voltage	**Temperature**	
	°F	**°C**
9.6 volts	70° and above	21° and above
9.5 volts	60°	16°
9.4 volts	50°	10°
9.3 volts	40°	4°
9.1 volts	30°	-1°
8.9 volts	20°	-7°
8.7 volts	10°	-12°
8.5 volts	0°	-18°

High capacity discharge test minimum voltage/temperature chart

Parisitic Draw Test

➡A good quality Digital Multimeter (DMM) with at least 10 megohm/volt impedance should be used when testing modern automotive circuits. These meters can accurately detect very small amounts of voltage, current and resistance. This type of meter also has a high internal resistance that will not load the circuit being tested. Loading the circuit gives inaccurate readings and may cause damage to sensitive computer circuits.

This test measures the amount of current that the vehicle draws while it is parked and not in use. A small amount of current should be flowing for such things as the on-board computer memory, automatic climate control, clock, and radio station presets. If there is a short in the vehicle electrical system or something has been left on, the excess current draw will eventually drain the battery and cause a no-start condition.

1. Be sure all accessories are turned **OFF**. Disconnect the negative battery cable.

2. Install a battery quick-disconnect switch (such as GM Parasitic Draw Test Switch J 38758) between the negative cable and the negative battery terminal. A battery disconnect switch will work in most cases.

3. Road test the vehicle while activating all accessories including the radio and air conditioning. Then, turn all accessories **OFF**.

4. Turn the vehicle **OFF** and open the hood.

5. If equipped, disable the underhood light.

6. Allow approximately 20 minutes for the vehicle computer system(s) to power down.

7. Connect one end of a jumper with a 10 amp fuse to the side of the quick-disconnect switch closest to the negative battery terminal. Be sure the jumper is on the metal part of the switch.

8. Connect the remaining end of the jumper to the other side of the switch closest to the negative battery cable.

✳✳ WARNING

Do not connect the multimeter to the circuit if more than 10 amps are flowing. Damage to the meter may occur.

9. Open the switch so all current flows through the jumper with the 10 amp fuse. If the fuse blows, there is more than 10 amps

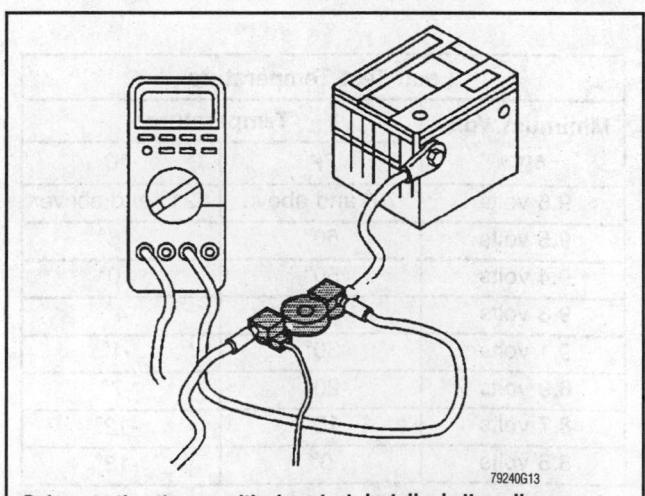

Before starting the parasitic draw test, install a battery disconnect switch between the negative battery cable and the battery terminal, as shown

79240G13

flowing in the circuit. This indicates that a component was left on (glove box light or other accessory) or there is a short in the electrical system. Find and correct the cause of the large current flow, then continue with this test.

10. If the fuse does not blow, close the disconnect switch and remove the jumper.

11. Set the multimeter to read 10 amps.

12. Connect the multimeter leads in place of the jumper used previously. When the switch is opened, current will flow through the meter.

13. The current draw should now be below 2 amps. If not, then something in the vehicle has been left on. Find the cause and correct it. When the current is less than 2 amps, set the meter to the 2 amp range. This will allow you to measure small amounts of current.

✳✳ WARNING

Do not open the door of the vehicle. The interior lights coming on will blow the fuse of the meter while on the 2 amp range.

14. Normal current draw should be less than ¼ of the reserve capacity of the battery. If the reserve capacity is unknown, normal current draw should be somewhere in the range of 0.005–0.040 amps depending on the type and amount of equipment on the vehicle.

➡The reserve capacity is the amount of time, in minutes, it takes for the battery voltage to fall below 10.5 volts at a discharge rate of 25 amps at 80° F (26.7° C). In most cases, this number can be found on the battery label.

15. If the current draw is higher than specified, pull fuses and/or disconnect components until the problem is found. Don't overlook the alternator connection.

REMOVAL & INSTALLATION

➡Disconnecting the negative battery cable on some vehicles may interfere with the functioning of the on-board computer system, and may require the computer to undergo a relearning process once the negative battery cable is reconnected.

1. Turn the ignition key to the **OFF** position.

2. Disconnect the negative battery cable first. On some vehicles, a cover or trim panel may have to be removed first.

3. Disconnect the positive battery cable.

4. Remove the battery hold-down.

➡A battery strap or holding device can make removing or installing the battery much easier. In some cases it can be difficult to get your hands under the battery.

To install:

5. Position the battery in the vehicle. Pay attention to the location of the terminals.

6. Install the battery hold-down. A loose battery may cause a vehicle fire or severe damage to the electrical system.

7. Clean the terminals and connect the positive battery cable first, then the negative cable.

8. If equipped, install the cover or trim panel.

JUMP STARTING A DEAD BATTERY

Whenever a vehicle is jump started, precautions must be followed in order to prevent the possibility of personal injury. Remember that batteries contain a small amount of explosive hydrogen gas which is a by-product of battery charging. Sparks should always be avoided when working around batteries, especially when attaching jumper cables. To minimize the possibility of accidental sparks, follow the procedure carefully.

✳ CAUTION

NEVER hook the batteries up in a series circuit or the entire electrical system will go up in smoke, including the starter!

Vehicles equipped with a diesel engine may utilize two 12 volt batteries. If so, the batteries are connected in a parallel circuit (positive terminal-to-positive terminal, negative terminal-to-negative terminal). Hooking the batteries up in parallel circuit increases battery cranking power without increasing total battery voltage output. Output remains at 12 volts. On the other hand, hooking two 12 volt batteries up in a series circuit (positive terminal-to-negative terminal, positive terminal-to-negative terminal) increases total battery output to 24 volts (12 volts plus 12 volts).

Jump Starting Precautions

To avoid personal injury and/or vehicle damage, please read all of the following precautions prior to jump starting a discharged battery:

• NEVER hook the batteries up in a series circuit or the entire electrical system will go up in smoke, including the starter!

• Be sure that both batteries are of the same voltage. Vehicles covered by this manual and most vehicles on the road today utilize a 12 volt charging system.

• Be sure that both batteries are of the same polarity (have the same terminal, in most cases NEGATIVE grounded).

• Be sure that the vehicles are not touching, otherwise a short could occur.

• On serviceable batteries, be sure the vent cap holes are not obstructed.

• Do not smoke or allow sparks anywhere near the batteries.

• In cold weather, be sure the battery electrolyte is not frozen. This can occur more readily in a battery that has been in a state of discharge.

• Do not allow electrolyte to contact your skin or clothing.

Jump Starting Procedure

1. Be sure that the voltages of the two batteries are the same. Most batteries and charging systems are of the 12 volt variety.

2. Pull the vehicle with the good battery into a position so the jumper cables can reach the dead battery and that vehicle's engine compartment. Be sure that the vehicles DO NOT touch.

➡**Remote power terminals are usually provided on vehicles where the battery is located in the fender or other location that makes connecting jumper cables difficult. These power terminals are located in the engine compartment. If this is the situation, use the remote terminals instead of the terminals on the battery.**

3. Place the transmissions/transaxles of both vehicles in Neutral, manual transmissions, or P (park), automatic transmissions, as applicable, then firmly set their parking brakes.

➡**If necessary for safety reasons, the hazard lights on both vehicles may be operated throughout the entire procedure without significantly increasing the difficulty of jumping the dead battery.**

4. Turn all lights and accessories OFF on both vehicles. Be sure the ignition switches on both vehicles are turned to the **OFF** position.

5. Cover the battery cell caps with a rag, but do not cover the terminals.

6. Be sure the terminals on both batteries are clean and free of corrosion, otherwise proper electrical connection will be impeded. If necessary, clean the battery terminals before proceeding.

7. Identify the positive (+) and negative (-) terminals on both batteries.

8. Connect the first jumper cable to the positive (+) terminal of the dead battery, then attach the other end of that cable to the positive (+) terminal of the booster (good) battery.

9. Connect the clamp of the negative jumper cable to the negative (-) terminal on the good battery and the final cable clamp to an engine bolt head, alternator bracket or other solid, metallic point on the engine with the dead battery. Try to pick a ground on the engine that is positioned away from the battery in order to minimize the possibility of explosion due to the sparks created when the last connection is made. DO NOT connect this clamp to the negative terminal of the bad battery.

✳ WARNING

Be very careful to keep the jumper cables away from moving parts (cooling fan, belts, etc.) on both engines.

10. Ensure the cables are routed away from any moving parts, then start the donor vehicle's engine. Run the engine at moderate speed for several minutes to allow the dead battery a chance to receive some initial charge.

11. With the donor vehicle's engine still running slightly above idle, try to start the vehicle with the dead battery. Crank the engine for no more than 10 seconds at a time and let the starter cool for at least 20 seconds between tries. If the vehicle does not start in 3 tries, it is likely that something else is also wrong, or that the battery needs additional time to charge.

12. Once the vehicle is started, allow it to run at idle for a few seconds to be sure that it is operating properly.

13. Turn ON the headlights, heater blower, if equipped, the rear defroster of both vehicles in order to reduce the severity of voltage spikes and subsequent risk of damage to the vehicles' electrical systems when the cables are disconnected. This step is especially important to any vehicle equipped with computer control modules.

14. Carefully remove the cables in the reverse order of connection. Start with the negative cable that is attached to the engine ground, then the negative cable on the donor battery. Disconnect the positive cable from the donor battery, and finally disconnect the positive cable from the formerly dead battery. Be careful when disconnecting the cables from the positive terminals not to allow the alligator clips to touch any metal on either vehicle or a short and sparks will occur.

Voltage Regulator

TESTING

➡**Most regulators are integral (built in) to the alternator or Powertrain Control Module (PCM). If the regulator is found to be defective on these models, the alternator or PCM should be replaced.**

For voltage regulator testing, refer to the Alternator Isolation test.

REMOVAL & INSTALLATION

➡**The following procedure is only for voltage regulators mounted on the back (outside) of the alternator or elsewhere in the engine compartment.**

1. Disconnect the negative battery cable.
2. If equipped, remove the exterior alternator cover to expose the regulator. Do not disassemble the alternator case that houses the rotor and stator.

3. If equipped, disengage the electrical connector from the regulator.
4. Remove the regulator mounting screws and remove the regulator.

To install:
5. Position the regulator in its original position and install the mounting screws.
6. Connect any wiring that was removed from the regulator.
7. If equipped, install the cover.
8. Connect the negative battery cable.

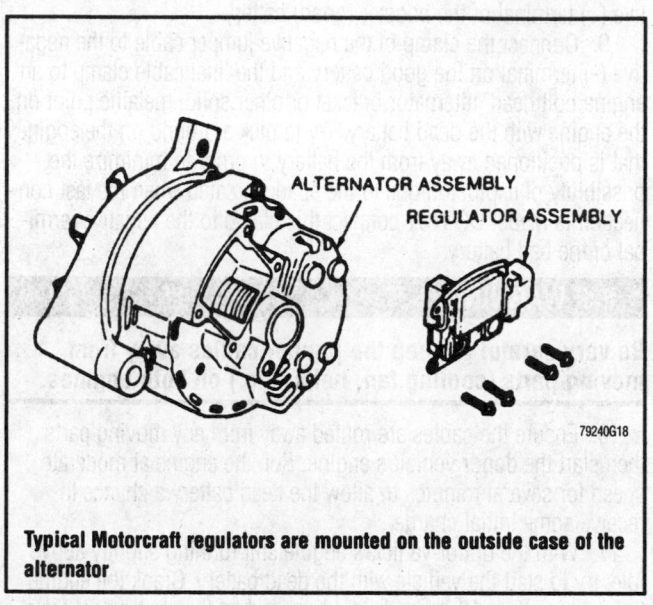

Typical Motorcraft regulators are mounted on the outside case of the alternator

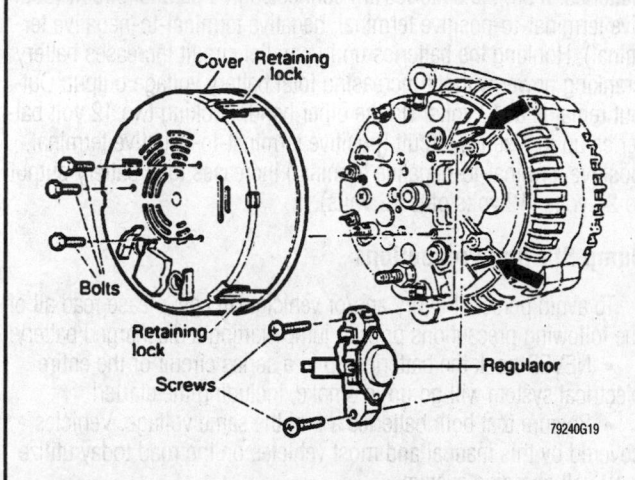

The regulator on a common Bosch alternator is mounted under a cover on the rear of the alternator

PISTON, PISTON RING AND CONNECTING ROD POSITIONING

PISTON, PISTON RING AND CONNECTING ROD POSITIONING

When assembling the pistons, piston rings and connecting rods, and when installing these assemblies into the engine block, it is vitally important to ensure that these three components are properly positioned with respect to each other. Often times the engine block is designed so that if a connecting rod or piston is installed backwards, or in the wrong bank of cylinders, internal engine damage may occur once the engine is started. The piston ring end-gap spacing that is recommended by the engine manufacturer is often with the purpose of increased compression pressures during the engine break-in period. Failure to properly space the piston ring end-gaps may lead to increased oil consumption and extended break-in time. Therefore, always be sure to position the pistons, rings and connecting rods as shown in the accompanying illustrations.

✳ WARNING

Always be sure to matchmark the connecting rods and caps prior to disassembly so that they may be reassembled with their original counterparts. If the caps are not installed on their original connecting rods, the assemblies will most likely need machining to avoid bearing, connecting rod and/or crankshaft damage.

PISTON, PISTON RING AND CONNECTING ROD POSITIONING INDEX

7923AC01

PISTON, PISTON RING AND CONNECTING ROD POSITIONING INDEX

7923AC02

PISTON, PISTON RING AND CONNECTING ROD POSITIONING INDEX

MANUFACTURER		
ENGINE	**DESCRIPTION**	**FIGURE**
Mazda		
All engines	Connecting Rod Bearing Cap Identification	62
	Compression Ring Identification And Positioning	63
	Upper, Spacer And Lower Oil Ring Identification And Positioning	64
	Piston Ring End-Gap Spacing	65
1.5L (Z5) engines	Piston-To-Engine Block Mark Location	66
1.6L (B6) engines	Piston-To-Engine Block Mark Location	66
1.8L (BP) engines	Piston-To-Engine Block Mark Location	66
1.8L (K8) engines	Piston-To-Engine Block Mark Location	67
2.0L (FS) engines	Piston-To-Engine Block Mark Location	67
2.3L (KJ) engines	Piston-To-Engine Positioning Mark Location	68
2.5L (KL) engines	Piston-To-Engine Block Mark Location	67
3.0L (JE) engines	Piston/Connecting Rod Assembly-To-Engine Block Positioning	69
Mitsubishi		
All engines	Connecting Rod Bearing Cap Identification	70
All engines - except 2.0L (420A) engines	Piston Ring End-Gap Spacing	74
1.5L engines	Compression Ring Identification Mark Locations	71
	Piston Positioning Mark Location	76
1.8L engines	Compression Ring Identification Mark Locations	72
	Oil Side And Spacer Ring Positioning	73
	Piston Positioning Mark Location	76
2.0L engines	Piston Positioning Mark Location	76
2.0L (420A) engines	Piston Ring End-Gap Spacing	75
2.4L engines	Compression Ring Identification Mark Locations	71
	Piston Positioning Mark Location	76
3.0L engines	Piston-To-Engine Block Mark Locations	77
3.5L engines	Piston And Connecting Rod Assembly Positioning	78
Nissan		
All engines	Piston And Connecting Rod Assembly Positioning	84
	Piston Ring End-Gap Spacing	83
	Piston Ring Mounting	79
1.6L engines	Piston Ring Positioning	80
2.0L engines	Piston Ring Positioning	80
2.4L engines	Piston Ring Positioning	80
3.0L (VG30DE) engines	Piston Ring Positioning	81
	Piston Ring Positioning	82
Saab		
2.0L and 2.3L engines	Piston And Connecting Rod Assembly Positioning	85
Subaru		
All engines	Piston And Connecting Rod Assembly Positioning	92
1.8L engines	Compression Ring End-Gap Spacing	86
	Upper, Spacer And Lower Oil Ring End-Gap Spacing	87
2.2L engines	Compression Ring End-Gap Spacing	88
	Upper, Spacer And Lower Oil Ring End-Gap Spacing	89
2.5L engines	Compression Ring End-Gap Spacing	90
	Upper, Spacer And Lower Oil Ring End-Gap Spacing	91

7923AC03

PISTON, PISTON RING AND CONNECTING ROD POSITIONING INDEX

7923AC04

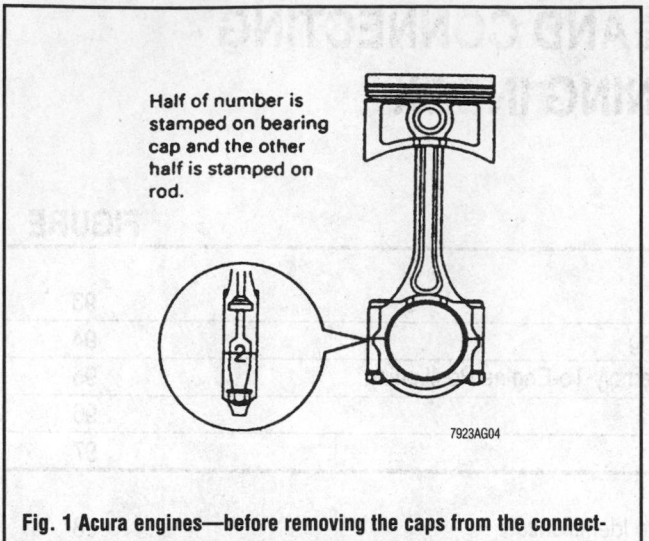

Half of number is stamped on bearing cap and the other half is stamped on rod.

7923AG04

Fig. 1 Acura engines—before removing the caps from the connecting rods, be sure to matchmark them as shown

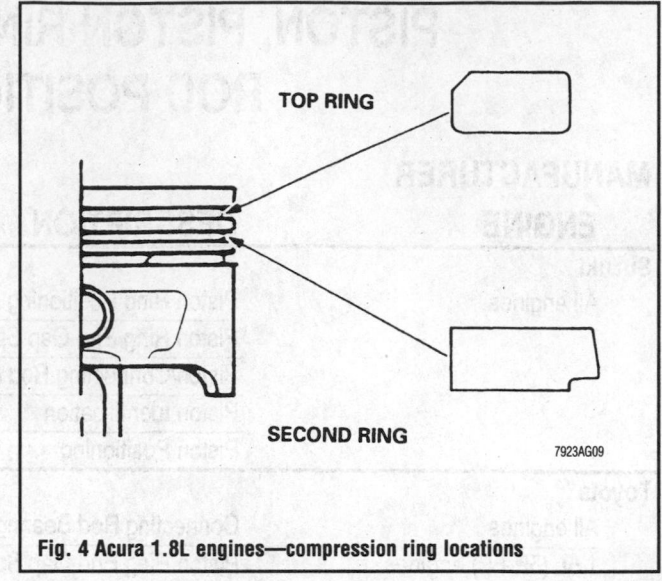

TOP RING

SECOND RING

7923AG09

Fig. 4 Acura 1.8L engines—compression ring locations

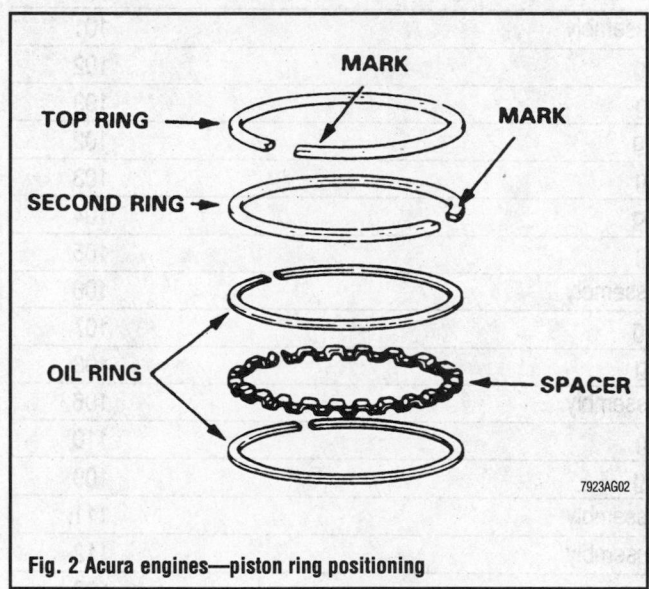

MARK

MARK

TOP RING

SECOND RING

OIL RING

SPACER

7923AG02

Fig. 2 Acura engines—piston ring positioning

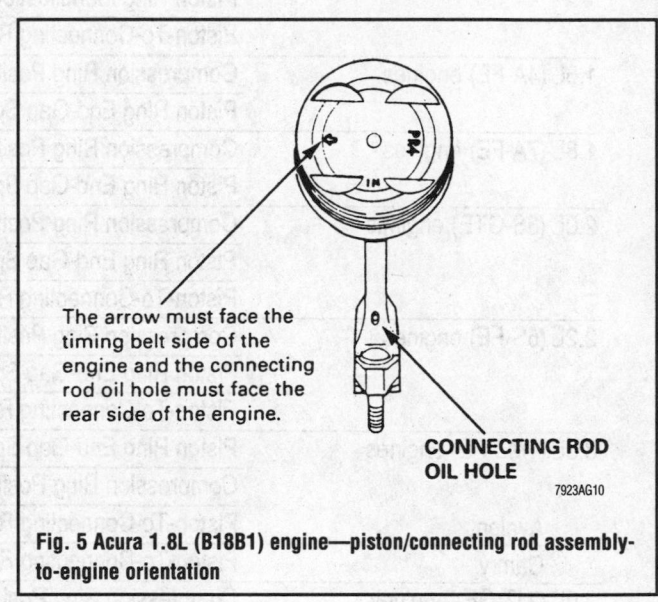

The arrow must face the timing belt side of the engine and the connecting rod oil hole must face the rear side of the engine.

CONNECTING ROD OIL HOLE

7923AG10

Fig. 5 Acura 1.8L (B18B1) engine—piston/connecting rod assembly-to-engine orientation

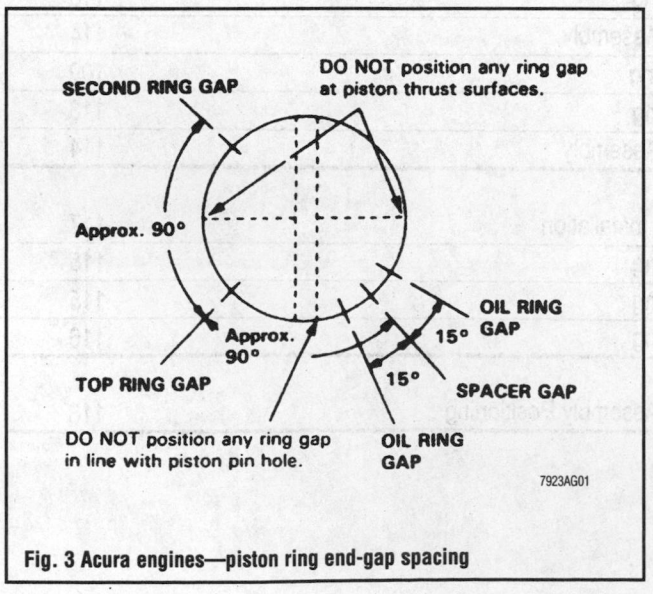

DO NOT position any ring gap at piston thrust surfaces.

SECOND RING GAP

Approx. 90°

Approx. 90°

TOP RING GAP

OIL RING GAP

15°

15°

SPACER GAP

OIL RING GAP

DO NOT position any ring gap in line with piston pin hole.

7923AG01

Fig. 3 Acura engines—piston ring end-gap spacing

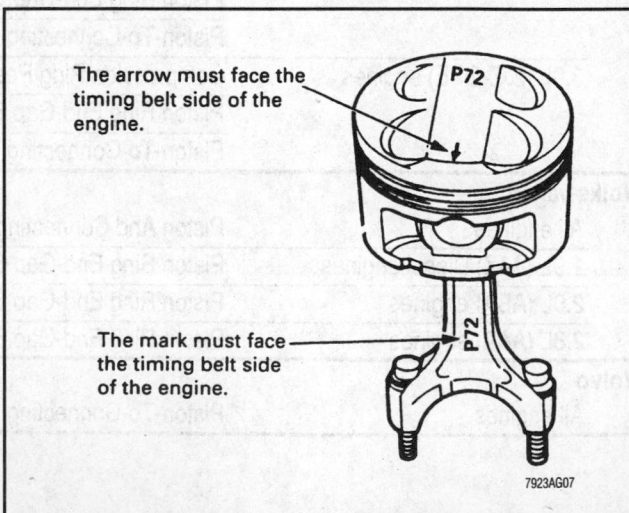

The arrow must face the timing belt side of the engine.

P72

The mark must face the timing belt side of the engine.

7923AG07

Fig. 6 Acura 1.8L (B18C1) engine—piston/connecting rod assembly-to-engine orientation

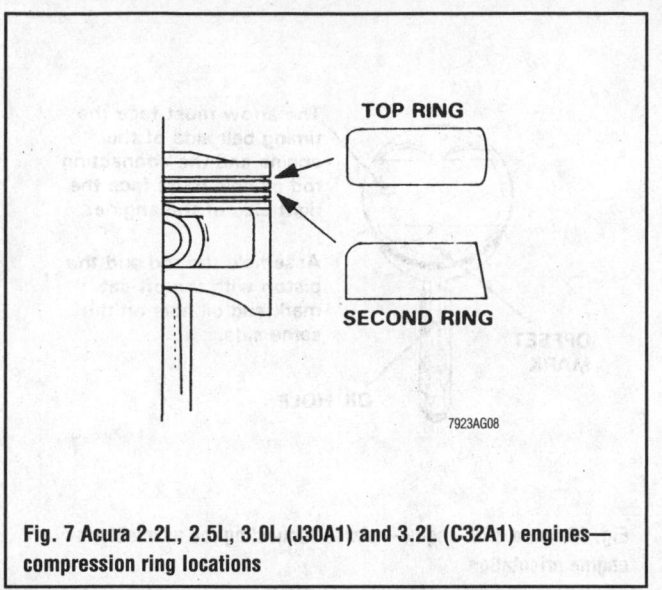

Fig. 7 Acura 2.2L, 2.5L, 3.0L (J30A1) and 3.2L (C32A1) engines—compression ring locations

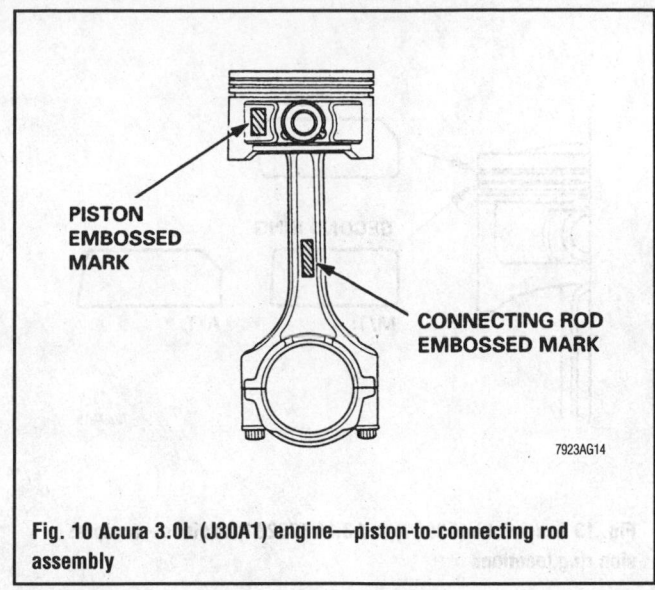

Fig. 10 Acura 3.0L (J30A1) engine—piston-to-connecting rod assembly

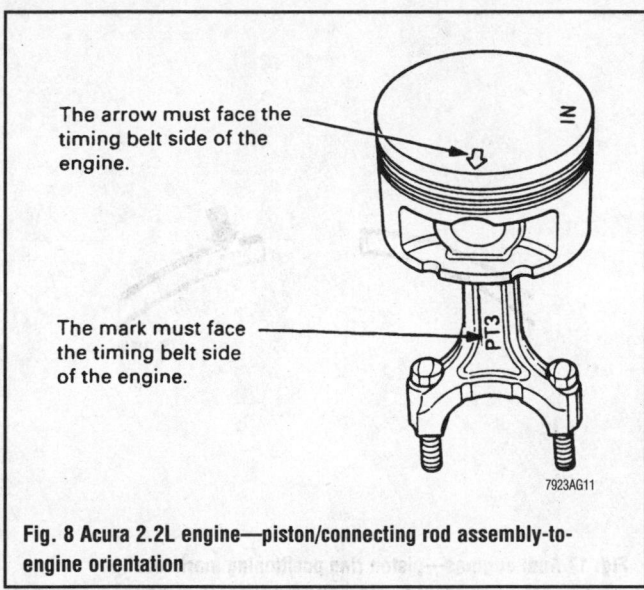

Fig. 8 Acura 2.2L engine—piston/connecting rod assembly-to-engine orientation

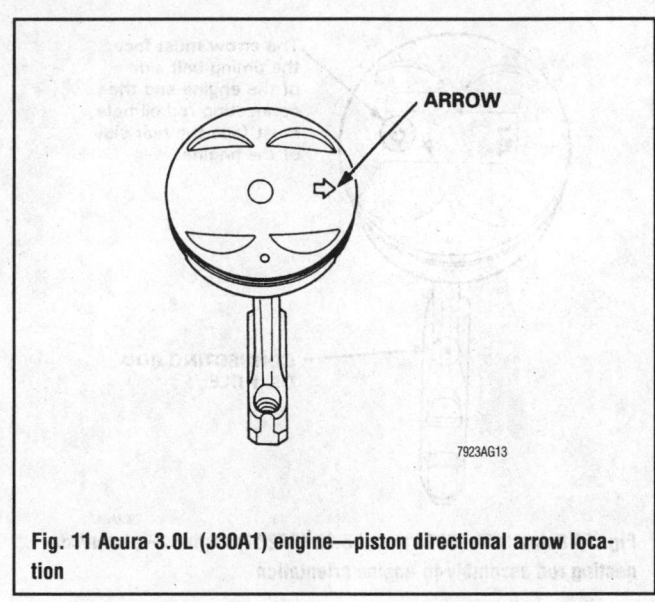

Fig. 11 Acura 3.0L (J30A1) engine—piston directional arrow location

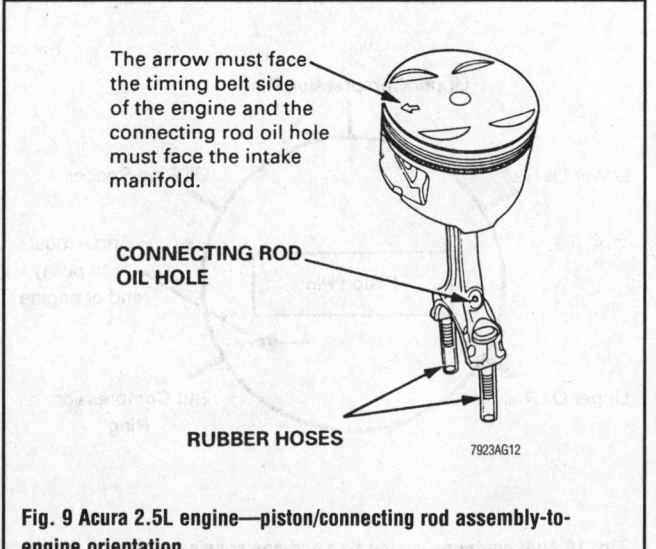

Fig. 9 Acura 2.5L engine—piston/connecting rod assembly-to-engine orientation

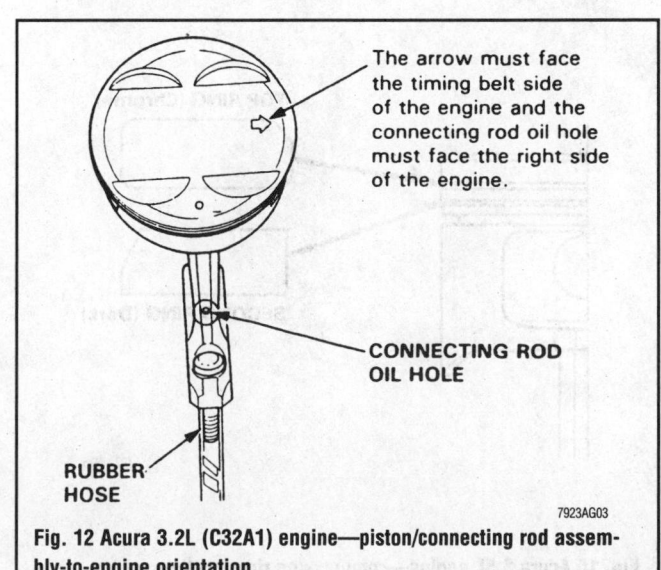

Fig. 12 Acura 3.2L (C32A1) engine—piston/connecting rod assembly-to-engine orientation

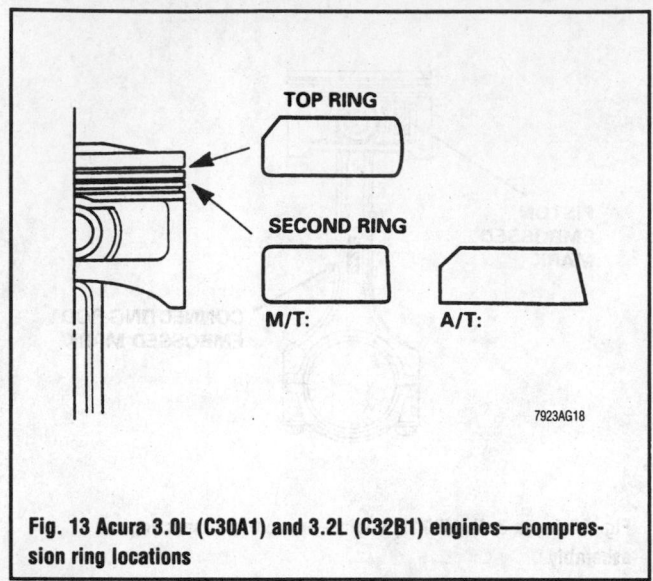

TOP RING

SECOND RING

M/T: **A/T:**

7923AG18

Fig. 13 Acura 3.0L (C30A1) and 3.2L (C32B1) engines—compression ring locations

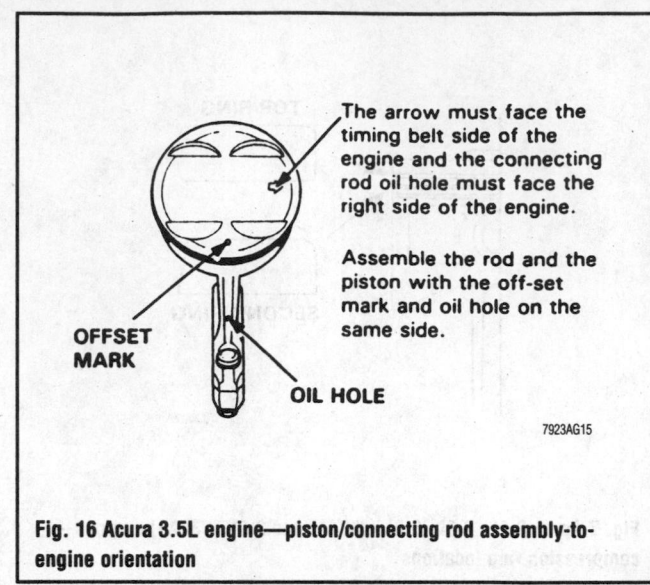

The arrow must face the timing belt side of the engine and the connecting rod oil hole must face the right side of the engine.

Assemble the rod and the piston with the off-set mark and oil hole on the same side.

OFFSET MARK

OIL HOLE

7923AG15

Fig. 16 Acura 3.5L engine—piston/connecting rod assembly-to-engine orientation

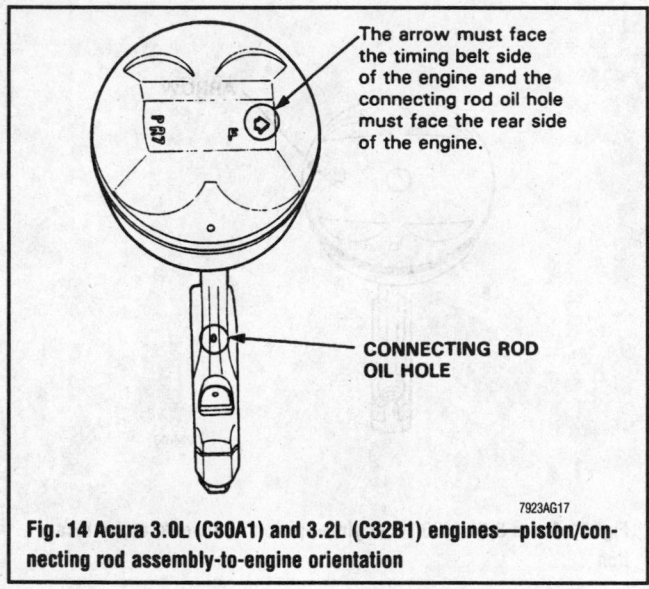

The arrow must face the timing belt side of the engine and the connecting rod oil hole must face the rear side of the engine.

CONNECTING ROD OIL HOLE

7923AG17

Fig. 14 Acura 3.0L (C30A1) and 3.2L (C32B1) engines—piston/connecting rod assembly-to-engine orientation

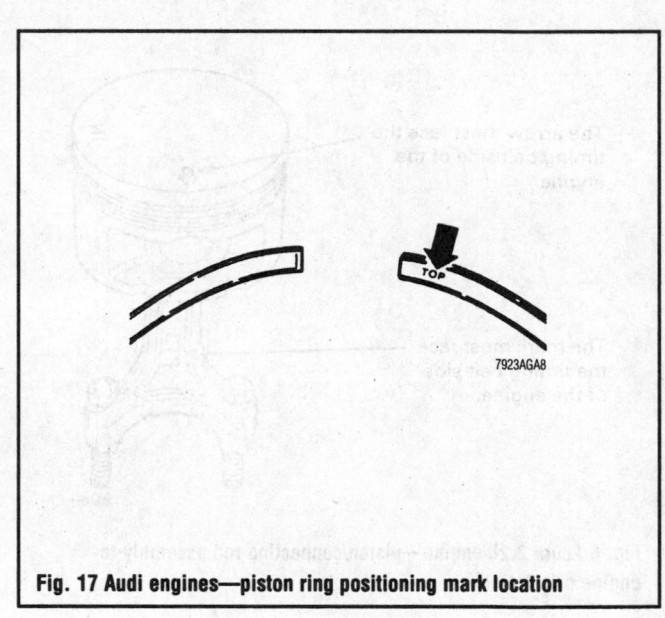

TOP

7923AGA8

Fig. 17 Audi engines—piston ring positioning mark location

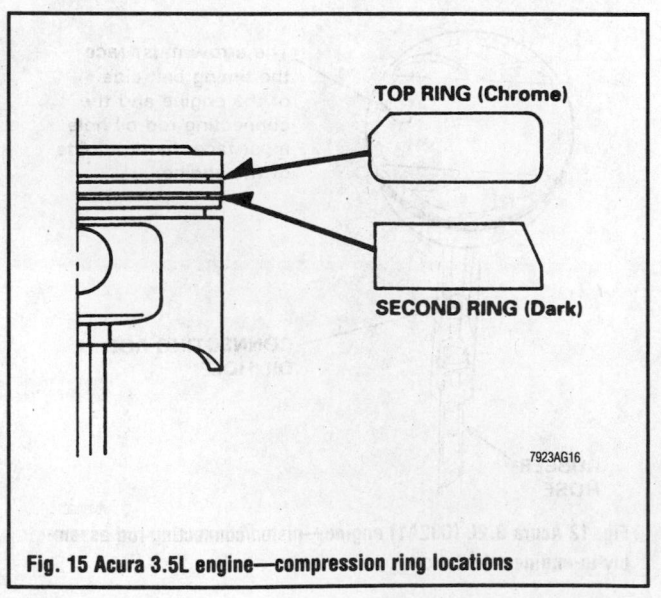

TOP RING (Chrome)

SECOND RING (Dark)

7923AG16

Fig. 15 Acura 3.5L engine—compression ring locations

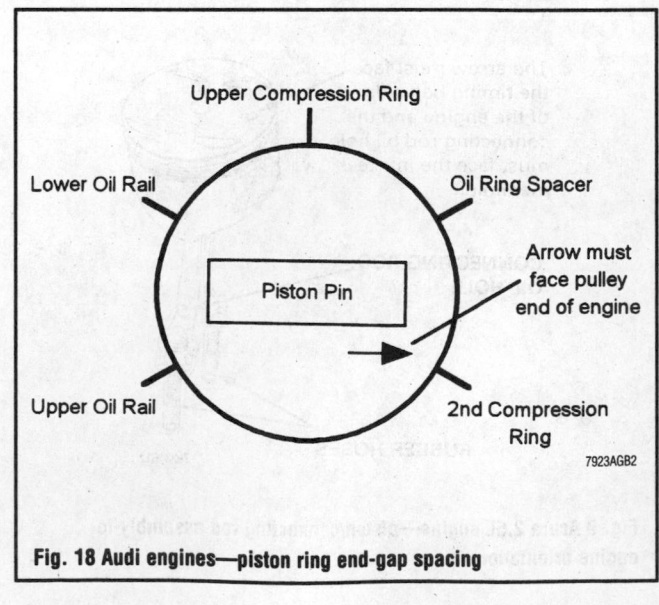

Upper Compression Ring

Lower Oil Rail

Oil Ring Spacer

Piston Pin

Arrow must face pulley end of engine

Upper Oil Rail

2nd Compression Ring

7923AGB2

Fig. 18 Audi engines—piston ring end-gap spacing

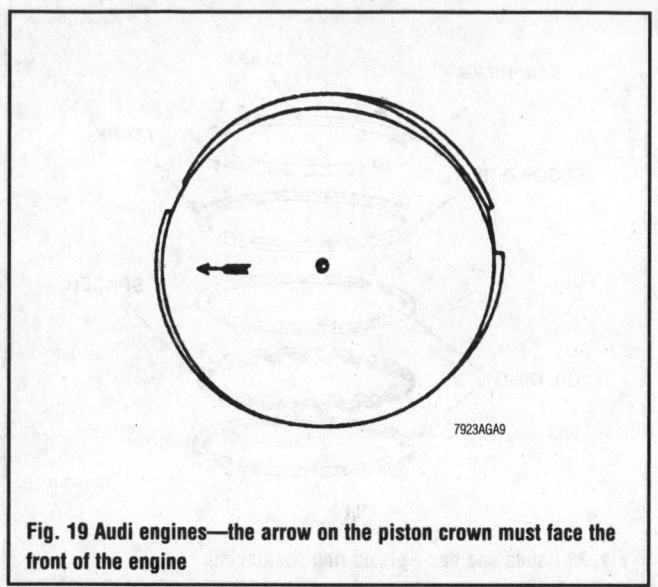

Fig. 19 Audi engines—the arrow on the piston crown must face the front of the engine

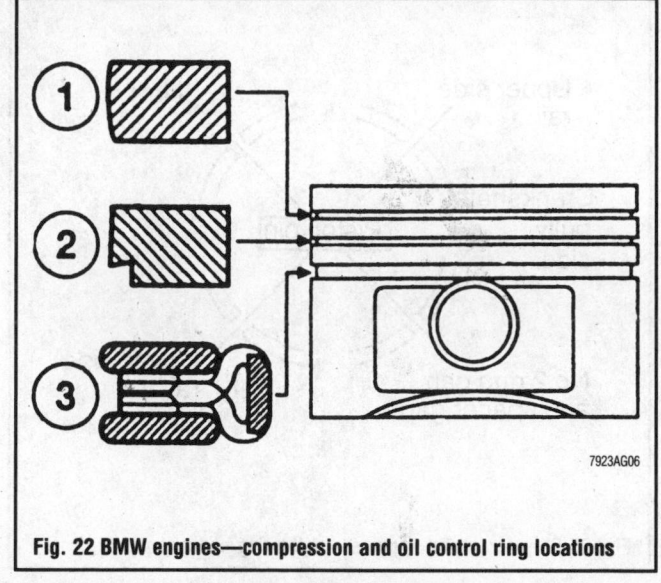

Fig. 22 BMW engines—compression and oil control ring locations

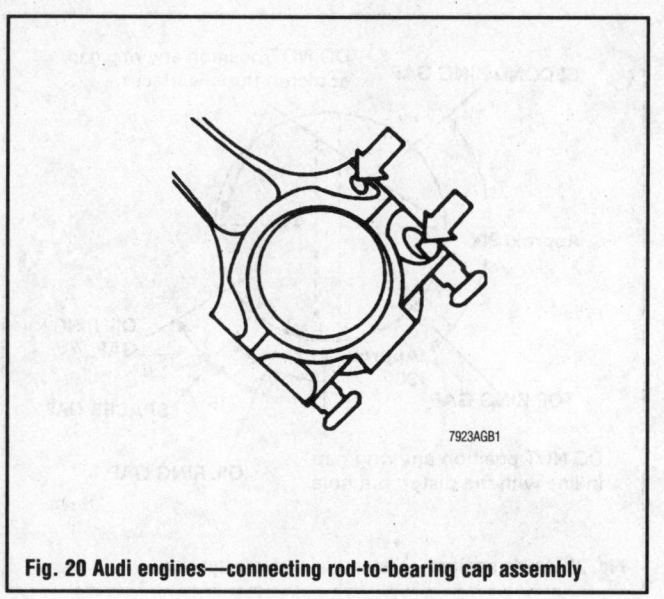

Fig. 20 Audi engines—connecting rod-to-bearing cap assembly

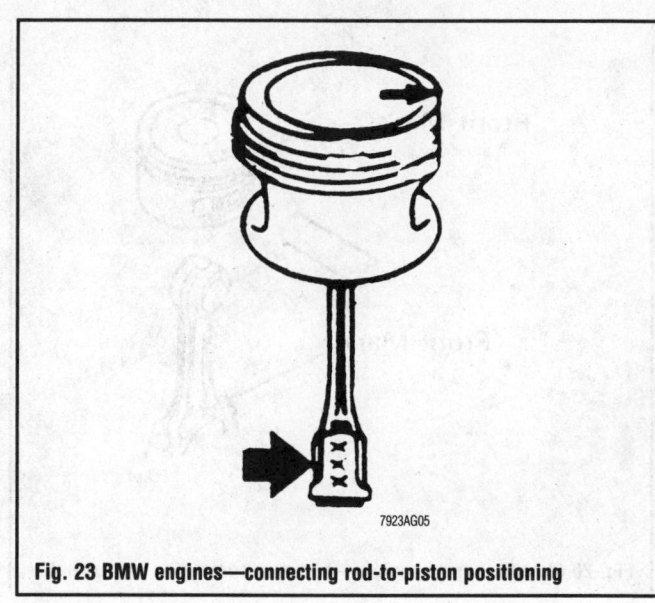

Fig. 23 BMW engines—connecting rod-to-piston positioning

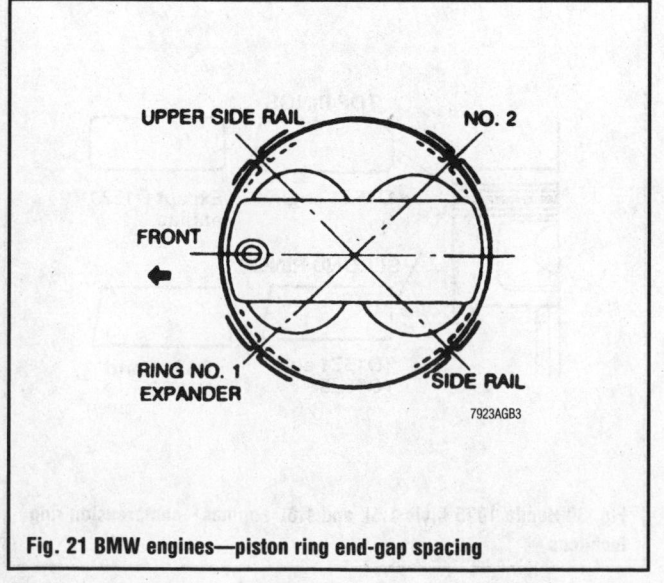

Fig. 21 BMW engines—piston ring end-gap spacing

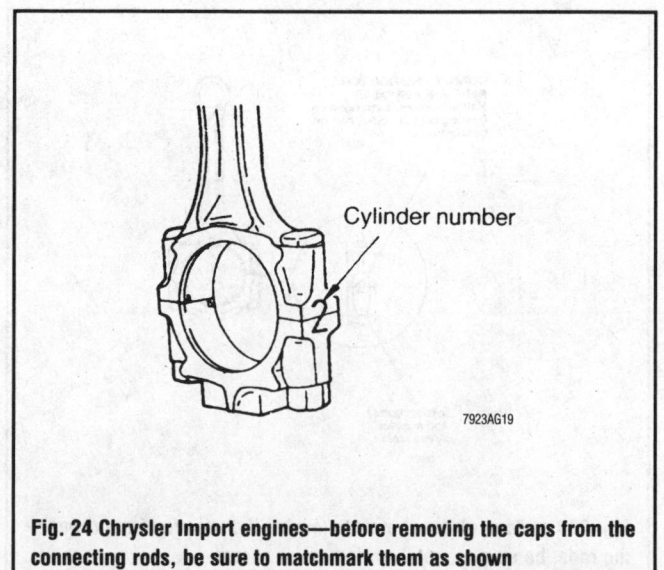

Fig. 24 Chrysler Import engines—before removing the caps from the connecting rods, be sure to matchmark them as shown

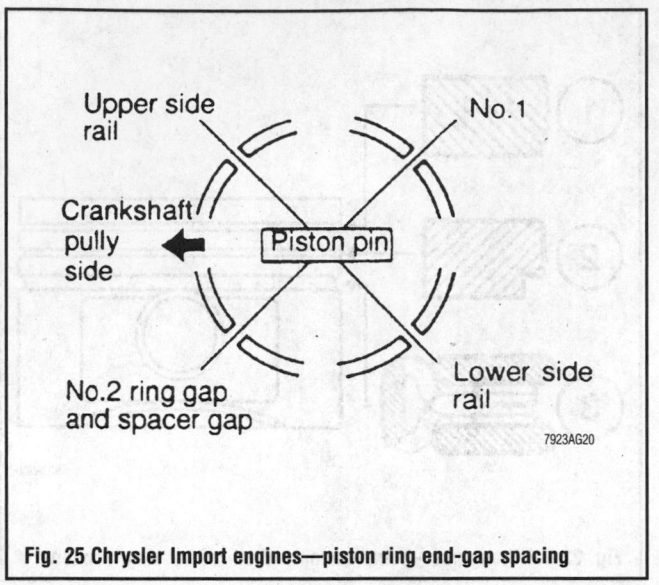

Fig. 25 Chrysler Import engines—piston ring end-gap spacing

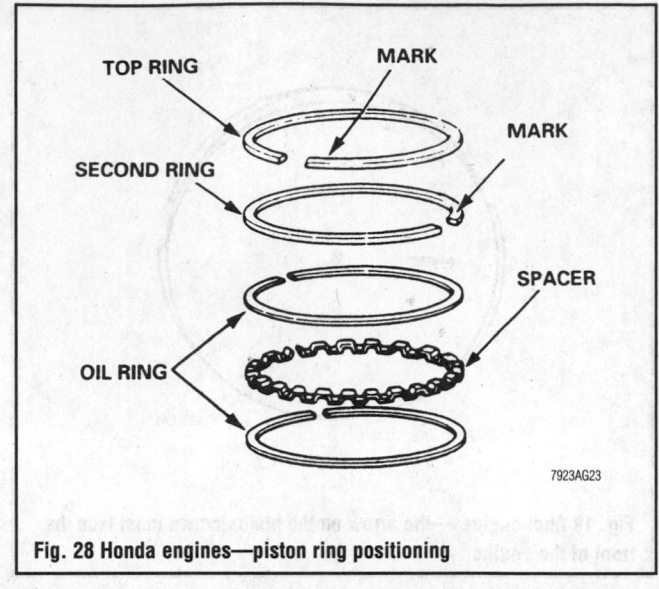

Fig. 28 Honda engines—piston ring positioning

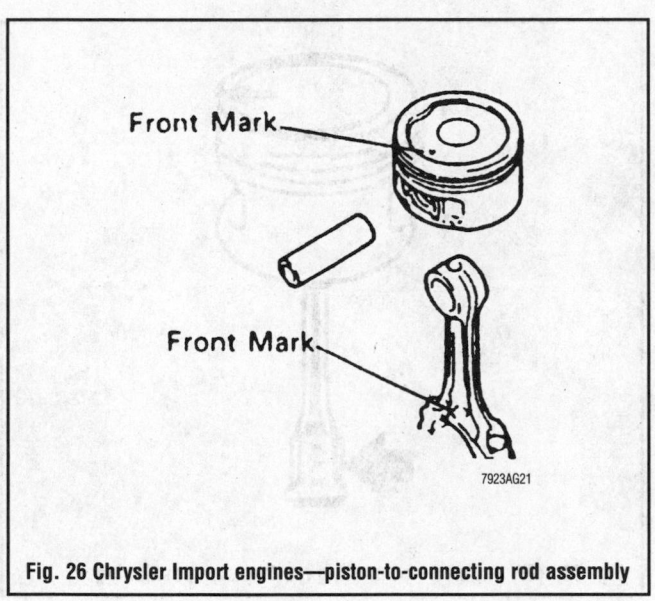

Fig. 26 Chrysler Import engines—piston-to-connecting rod assembly

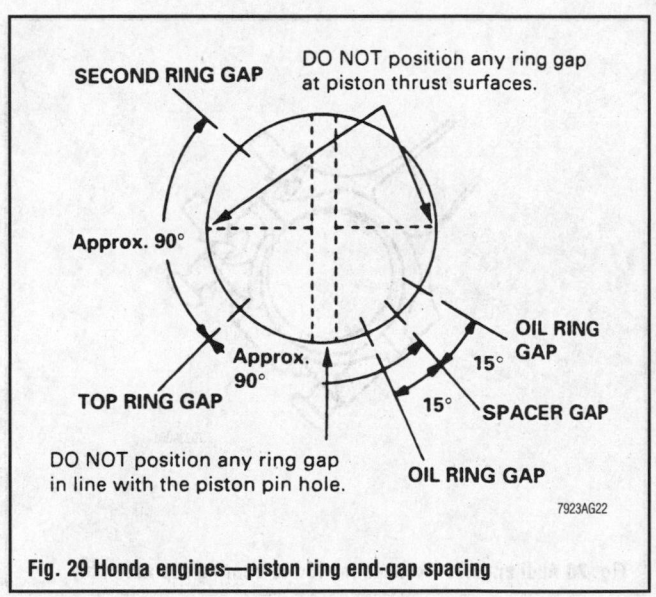

Fig. 29 Honda engines—piston ring end-gap spacing

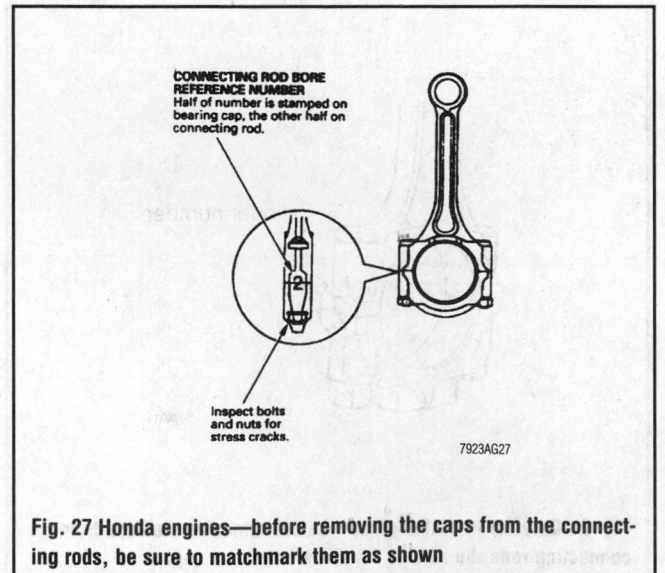

Fig. 27 Honda engines—before removing the caps from the connecting rods, be sure to matchmark them as shown

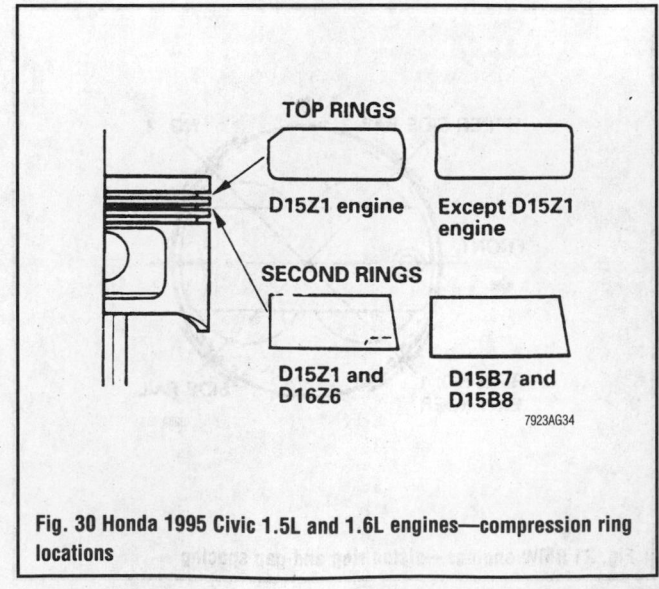

Fig. 30 Honda 1995 Civic 1.5L and 1.6L engines—compression ring locations

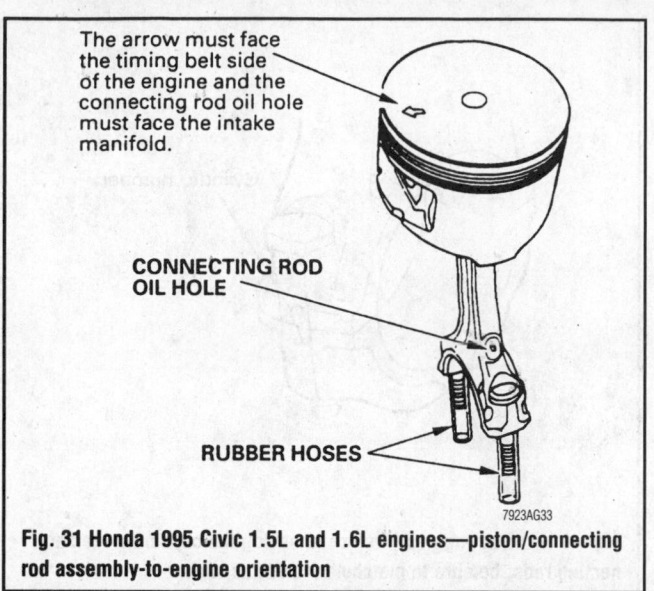

The arrow must face the timing belt side of the engine and the connecting rod oil hole must face the intake manifold.

CONNECTING ROD OIL HOLE

RUBBER HOSES

7923AG33

Fig. 31 Honda 1995 Civic 1.5L and 1.6L engines—piston/connecting rod assembly-to-engine orientation

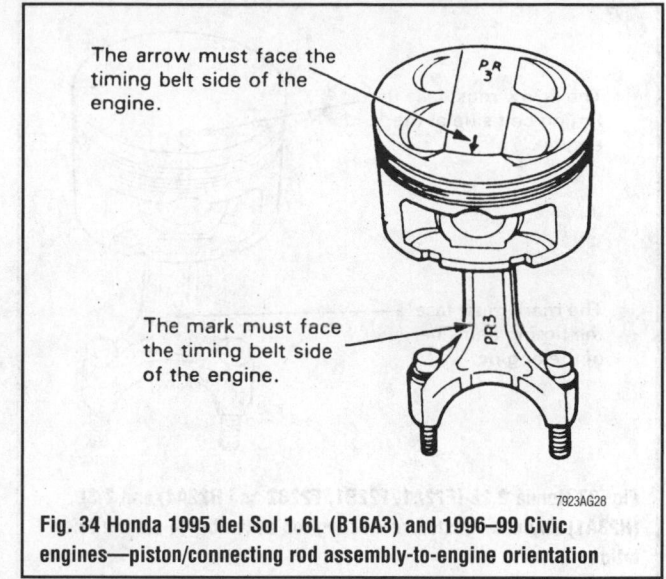

The arrow must face the timing belt side of the engine.

The mark must face the timing belt side of the engine.

PR3

7923AG28

Fig. 34 Honda 1995 del Sol 1.6L (B16A3) and 1996–99 Civic engines—piston/connecting rod assembly-to-engine orientation

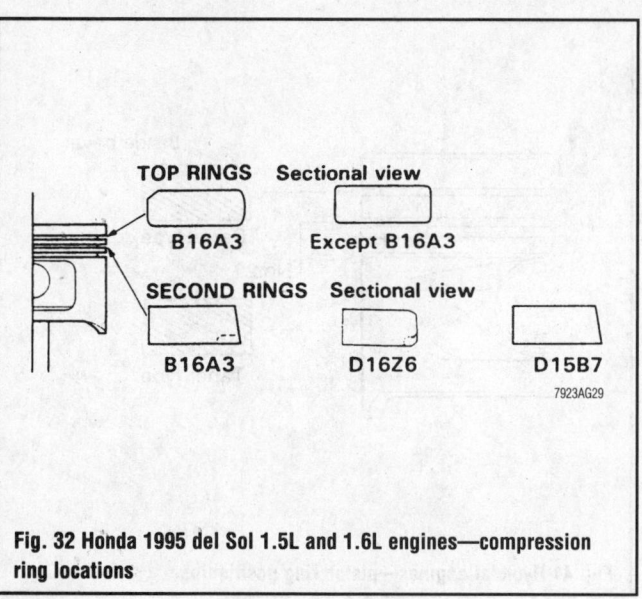

TOP RINGS   Sectional view

B16A3   Except B16A3

SECOND RINGS   Sectional view

B16A3   D16Z6   D15B7

7923AG29

Fig. 32 Honda 1995 del Sol 1.5L and 1.6L engines—compression ring locations

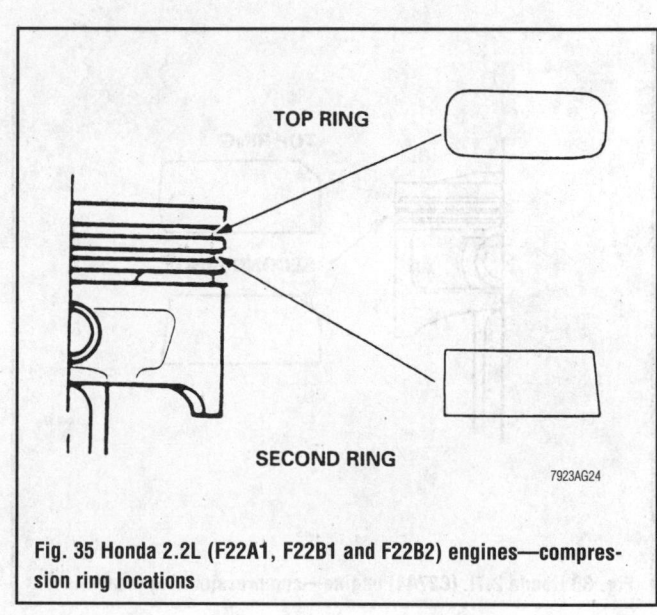

TOP RING

SECOND RING

7923AG24

Fig. 35 Honda 2.2L (F22A1, F22B1 and F22B2) engines—compression ring locations

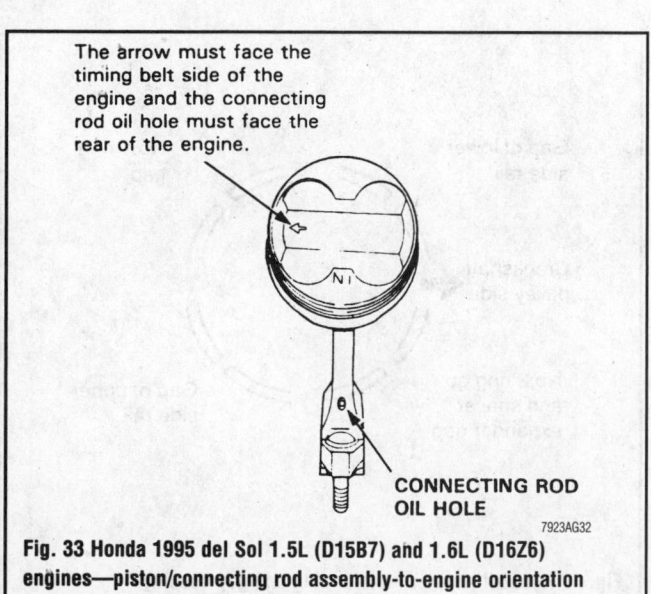

The arrow must face the timing belt side of the engine and the connecting rod oil hole must face the rear of the engine.

CONNECTING ROD OIL HOLE

7923AG32

Fig. 33 Honda 1995 del Sol 1.5L (D15B7) and 1.6L (D16Z6) engines—piston/connecting rod assembly-to-engine orientation

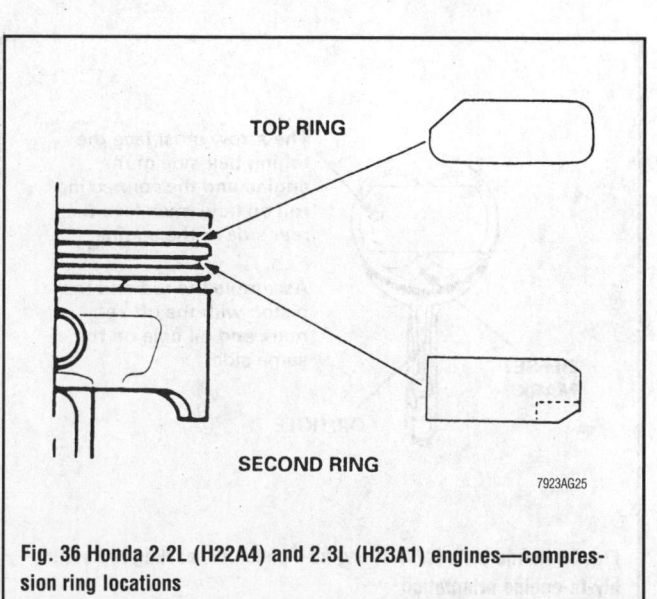

TOP RING

SECOND RING

7923AG25

Fig. 36 Honda 2.2L (H22A4) and 2.3L (H23A1) engines—compression ring locations

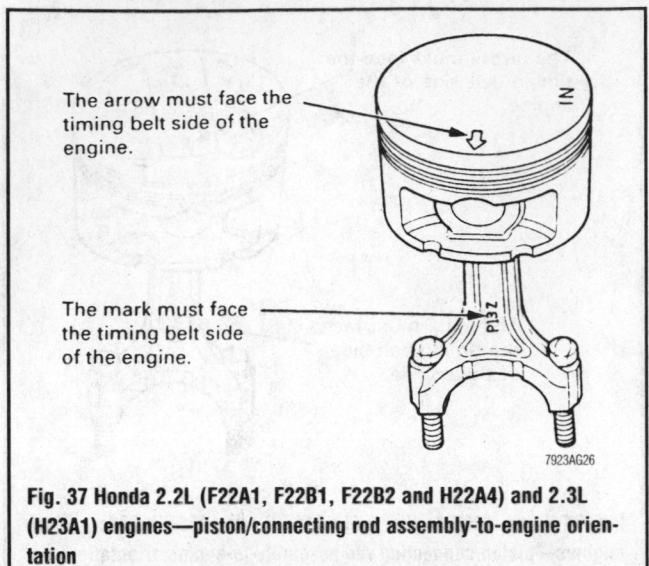

Fig. 37 Honda 2.2L (F22A1, F22B1, F22B2 and H22A4) and 2.3L (H23A1) engines—piston/connecting rod assembly-to-engine orientation

The arrow must face the timing belt side of the engine.

The mark must face the timing belt side of the engine.

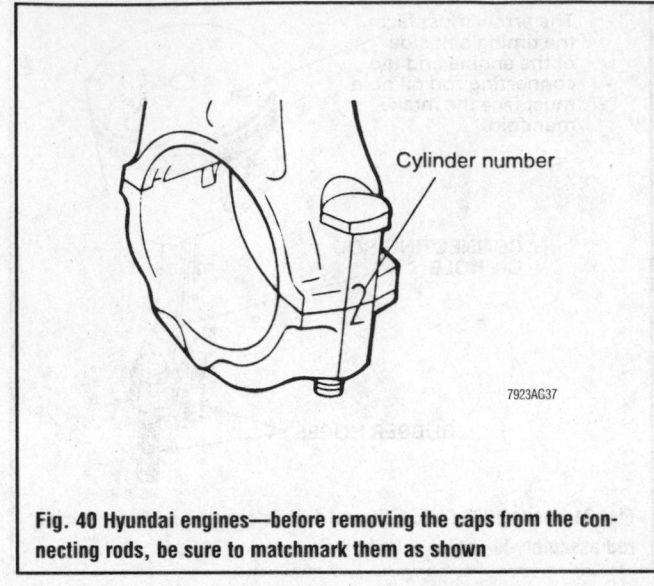

Fig. 40 Hyundai engines—before removing the caps from the connecting rods, be sure to matchmark them as shown

Cylinder number

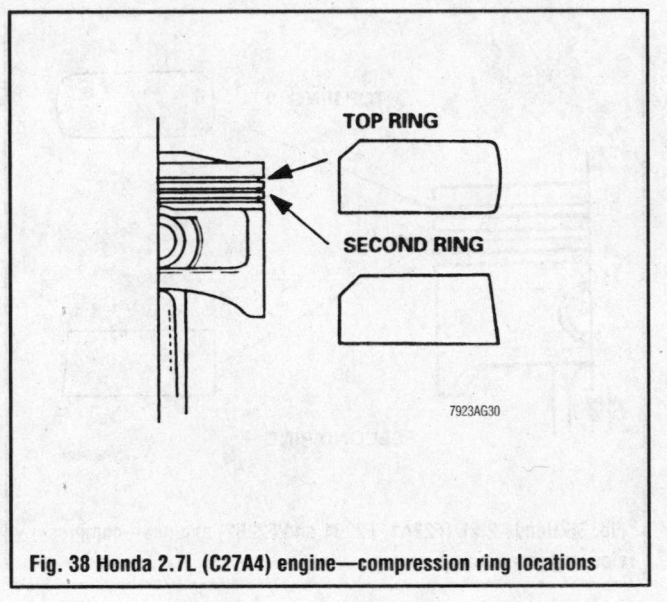

Fig. 38 Honda 2.7L (C27A4) engine—compression ring locations

TOP RING

SECOND RING

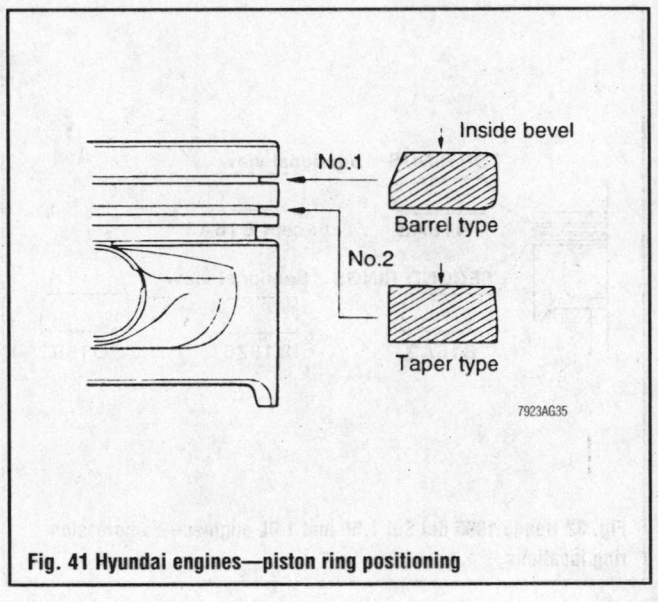

Fig. 41 Hyundai engines—piston ring positioning

Inside bevel

No.1

Barrel type

No.2

Taper type

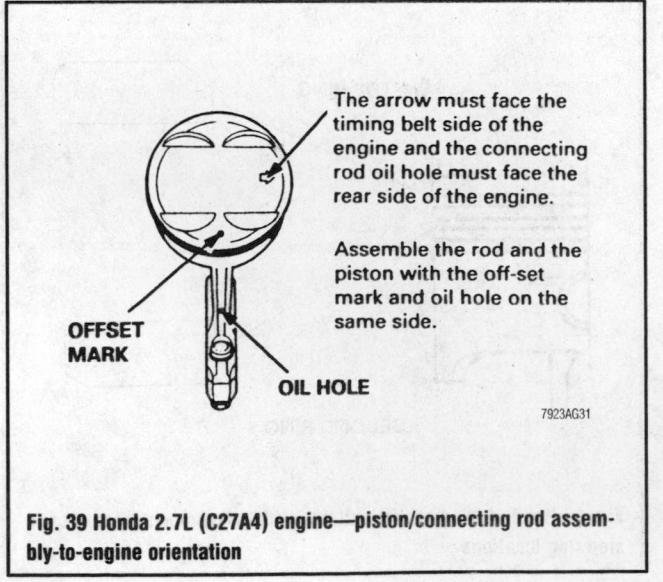

Fig. 39 Honda 2.7L (C27A4) engine—piston/connecting rod assembly-to-engine orientation

The arrow must face the timing belt side of the engine and the connecting rod oil hole must face the rear side of the engine.

Assemble the rod and the piston with the off-set mark and oil hole on the same side.

OFFSET MARK

OIL HOLE

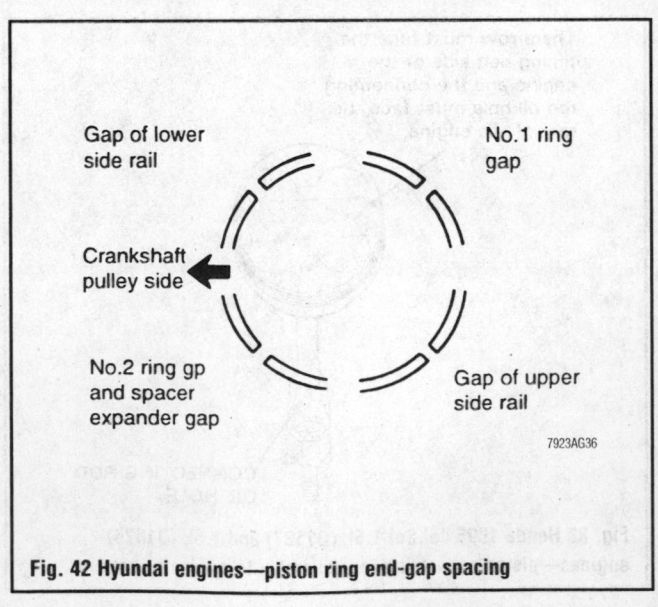

Fig. 42 Hyundai engines—piston ring end-gap spacing

Gap of lower side rail

No.1 ring gap

Crankshaft pulley side

No.2 ring gp and spacer expander gap

Gap of upper side rail

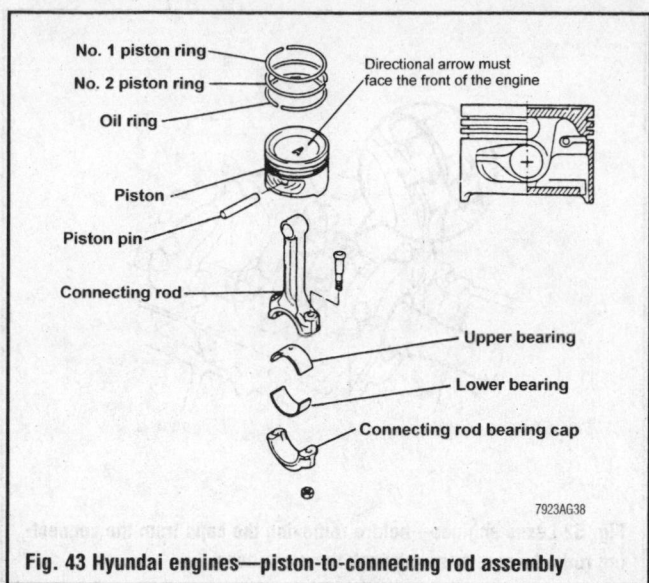

Fig. 43 Hyundai engines—piston-to-connecting rod assembly

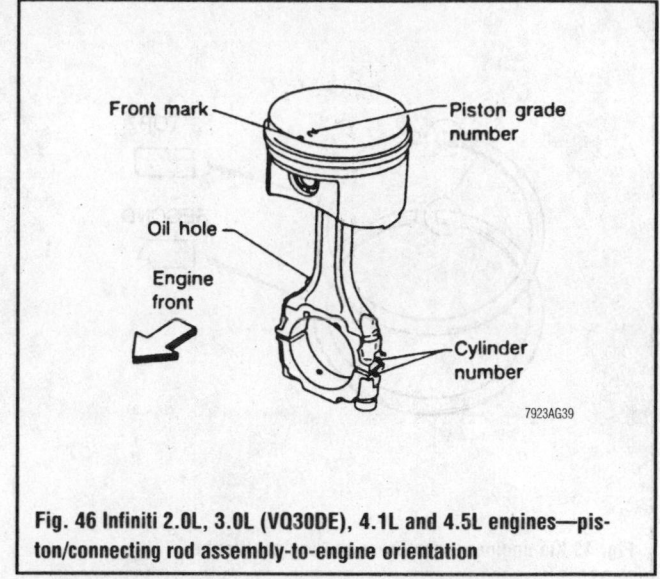

Fig. 46 Infiniti 2.0L, 3.0L (VQ30DE), 4.1L and 4.5L engines—piston/connecting rod assembly-to-engine orientation

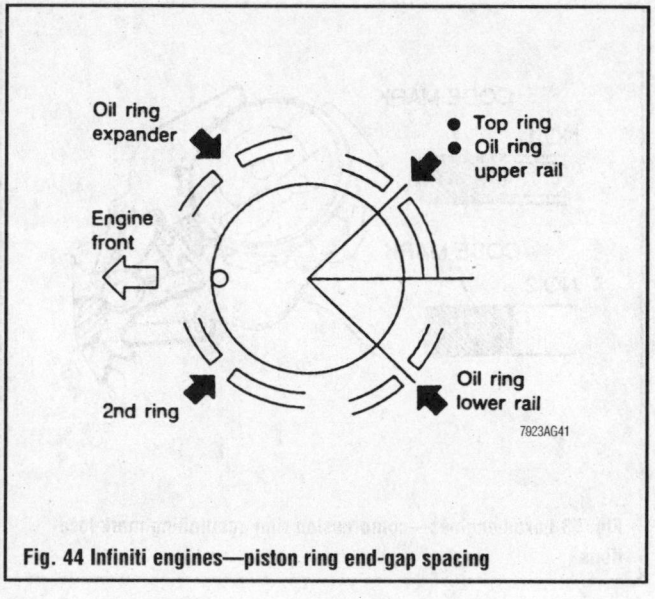

Fig. 44 Infiniti engines—piston ring end-gap spacing

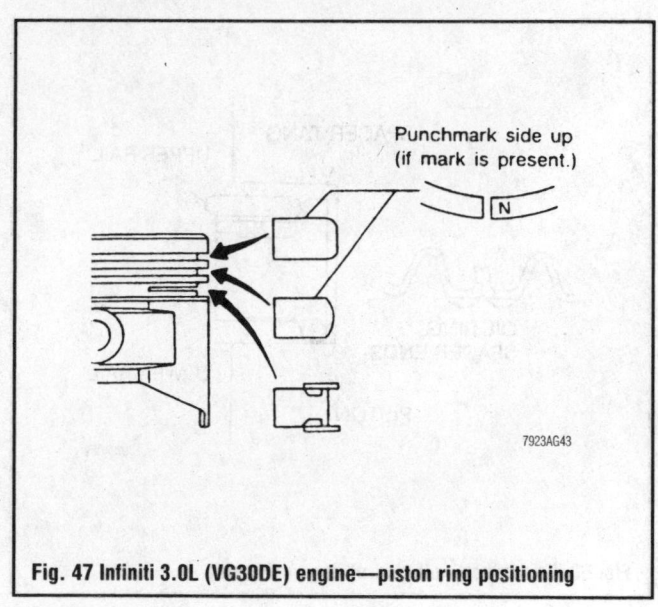

Fig. 47 Infiniti 3.0L (VG30DE) engine—piston ring positioning

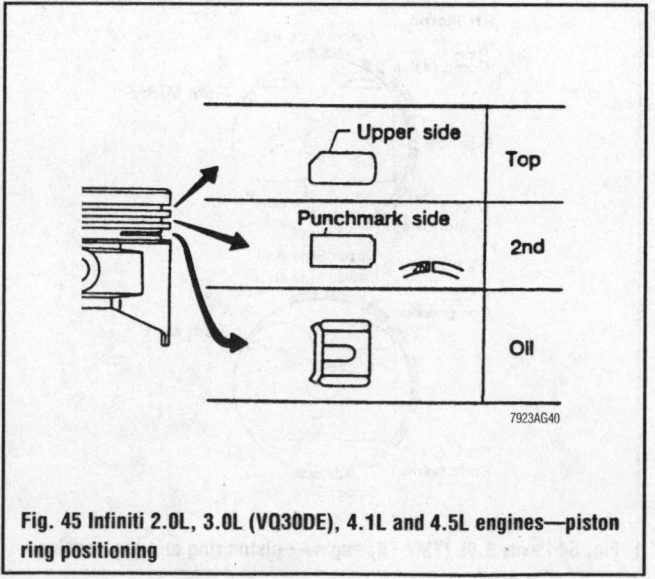

Fig. 45 Infiniti 2.0L, 3.0L (VQ30DE), 4.1L and 4.5L engines—piston ring positioning

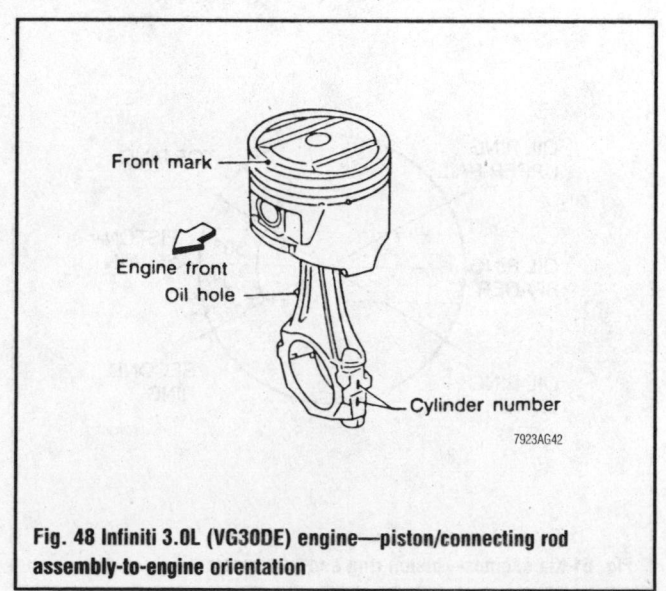

Fig. 48 Infiniti 3.0L (VG30DE) engine—piston/connecting rod assembly-to-engine orientation

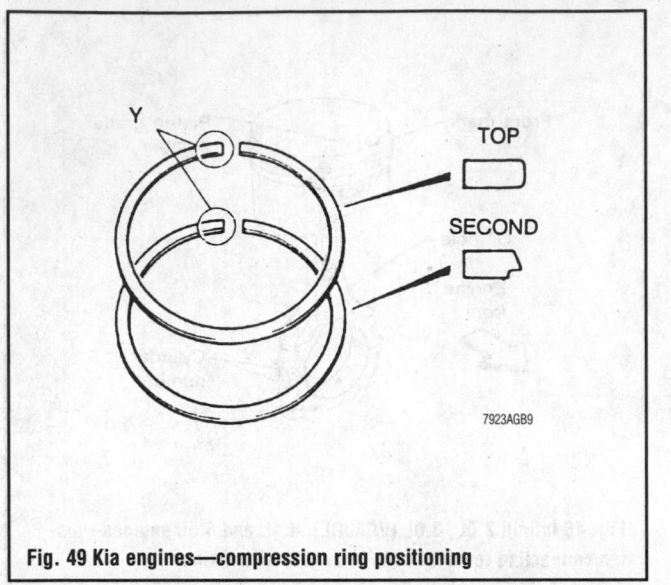

Fig. 49 Kia engines—compression ring positioning

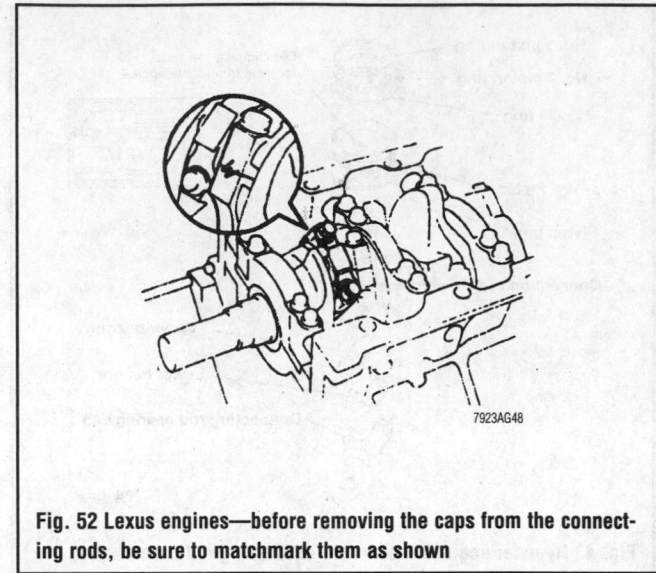

Fig. 52 Lexus engines—before removing the caps from the connecting rods, be sure to matchmark them as shown

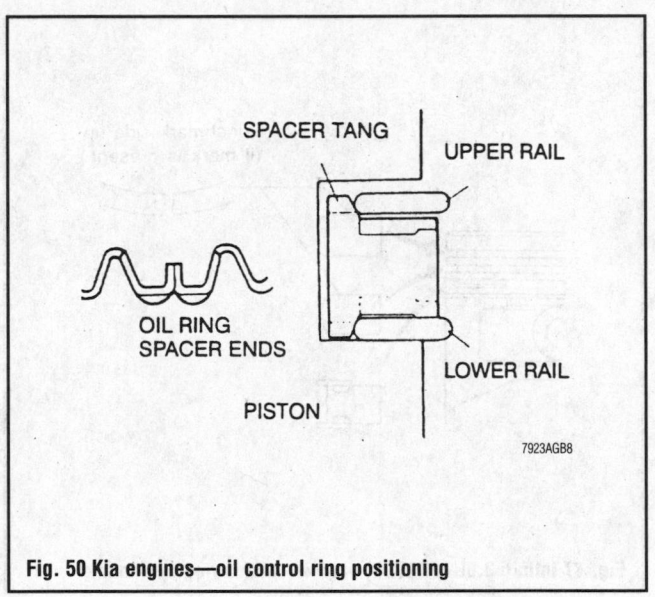

Fig. 50 Kia engines—oil control ring positioning

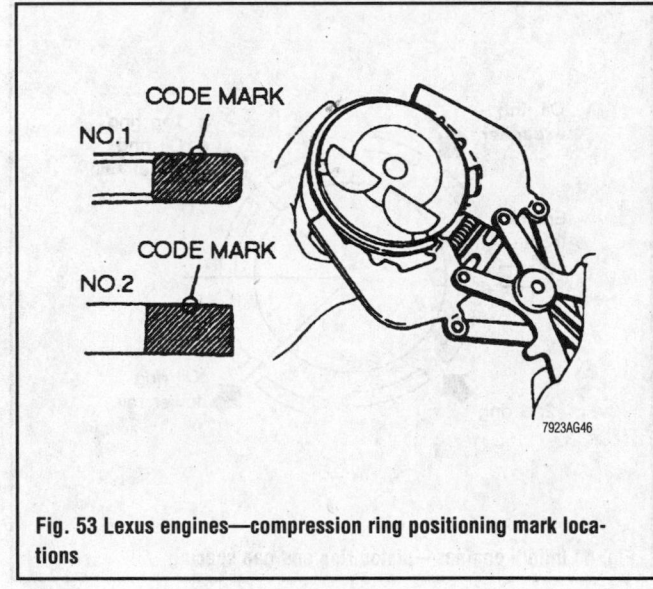

Fig. 53 Lexus engines—compression ring positioning mark locations

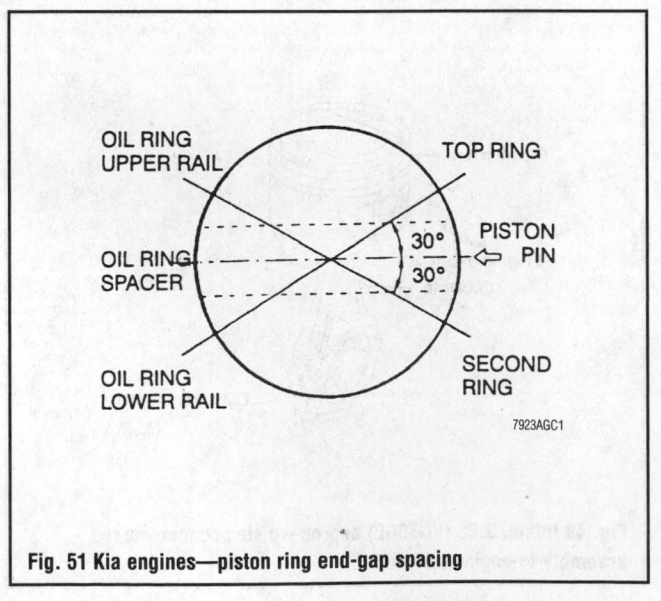

Fig. 51 Kia engines—piston ring end-gap spacing

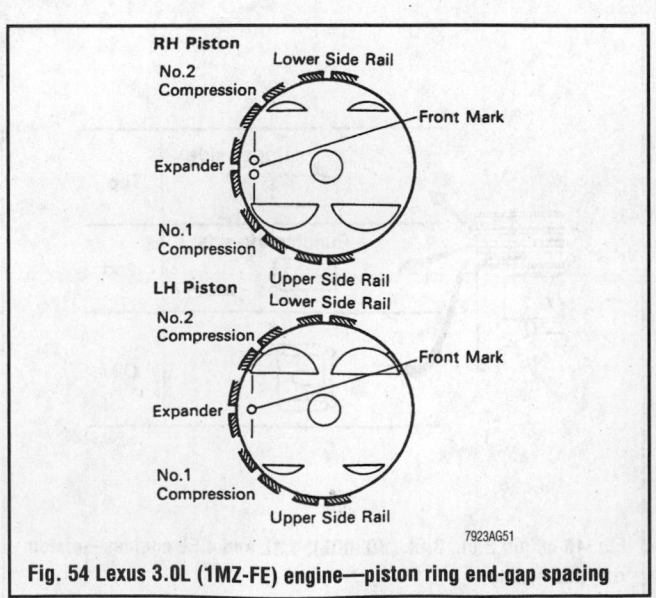

Fig. 54 Lexus 3.0L (1MZ-FE) engine—piston ring end-gap spacing

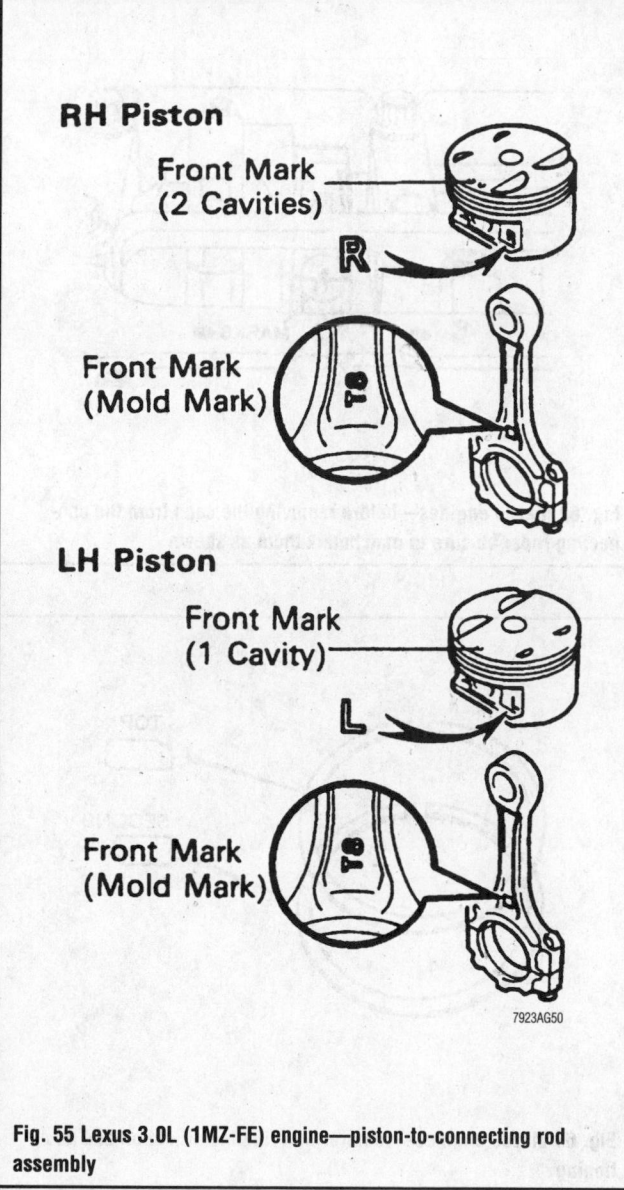

Fig. 55 Lexus 3.0L (1MZ-FE) engine—piston-to-connecting rod assembly

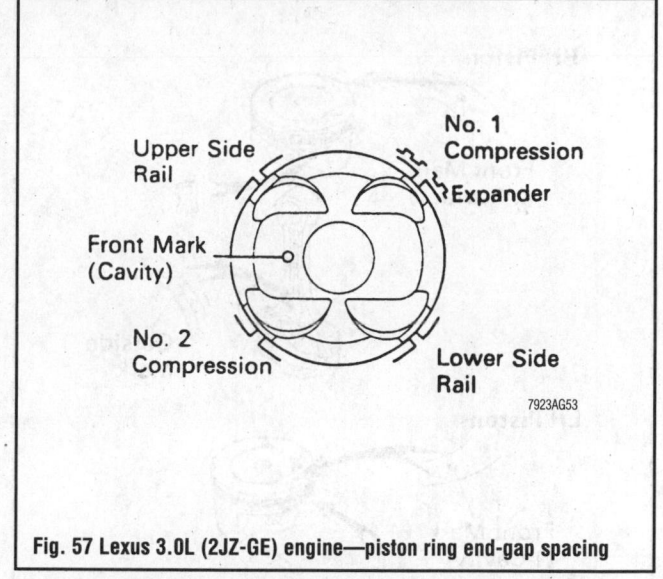

Fig. 57 Lexus 3.0L (2JZ-GE) engine—piston ring end-gap spacing

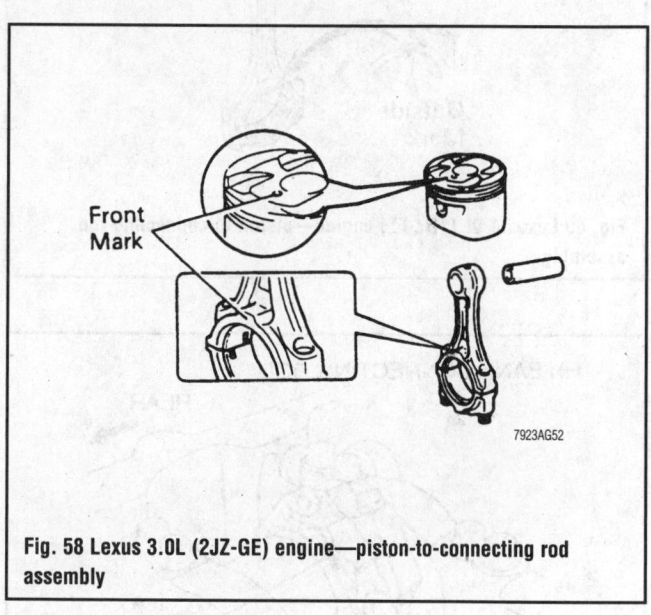

Fig. 58 Lexus 3.0L (2JZ-GE) engine—piston-to-connecting rod assembly

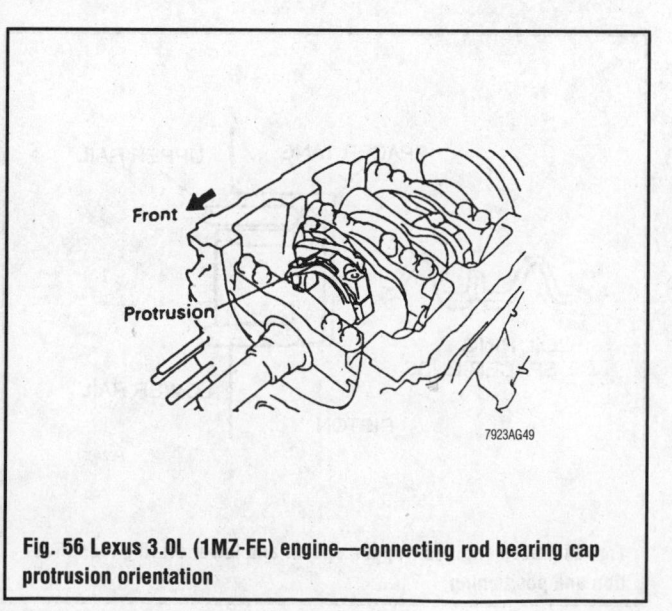

Fig. 56 Lexus 3.0L (1MZ-FE) engine—connecting rod bearing cap protrusion orientation

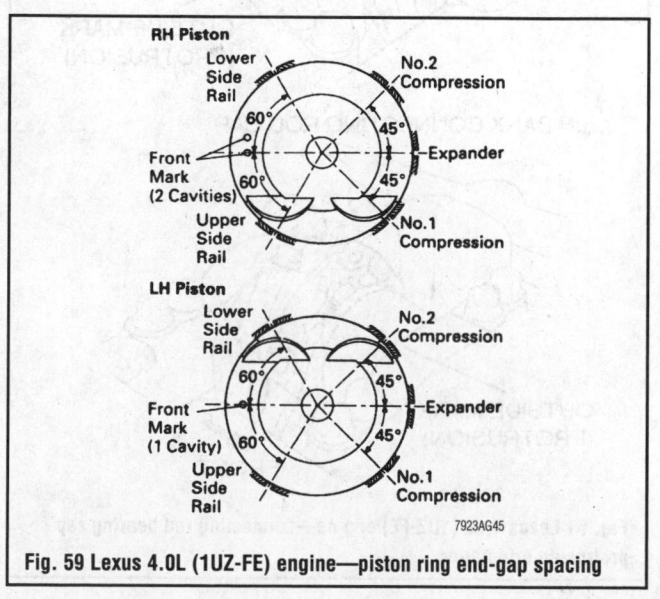

Fig. 59 Lexus 4.0L (1UZ-FE) engine—piston ring end-gap spacing

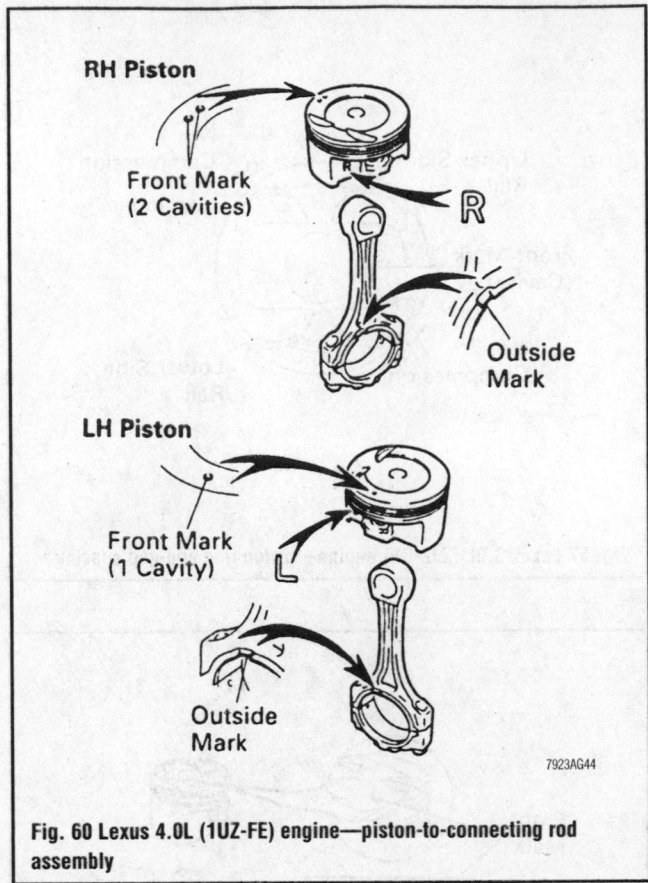

Fig. 60 Lexus 4.0L (1UZ-FE) engine—piston-to-connecting rod assembly

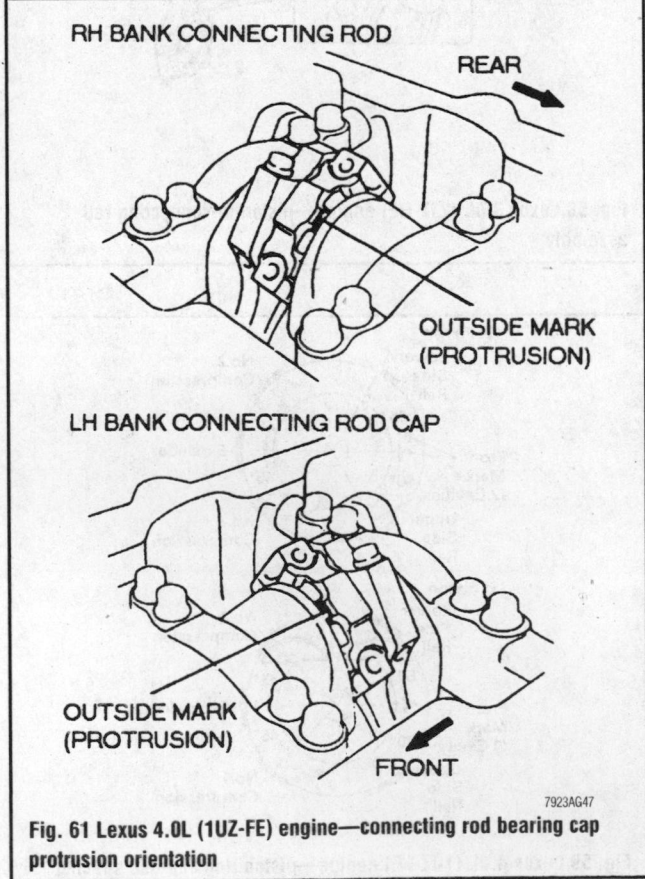

Fig. 61 Lexus 4.0L (1UZ-FE) engine—connecting rod bearing cap protrusion orientation

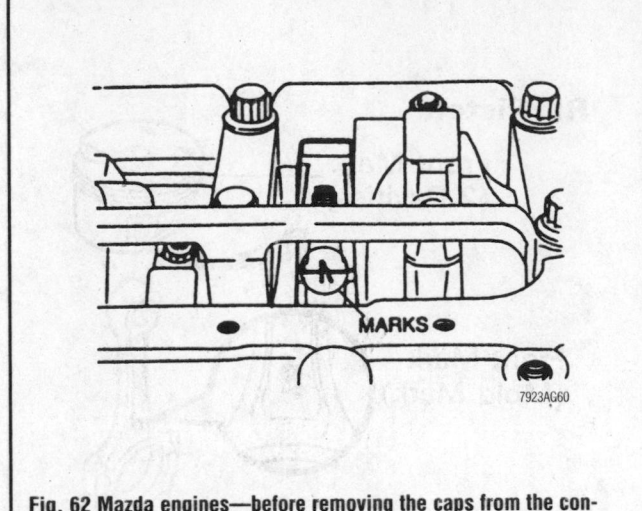

Fig. 62 Mazda engines—before removing the caps from the connecting rods, be sure to matchmark them as shown

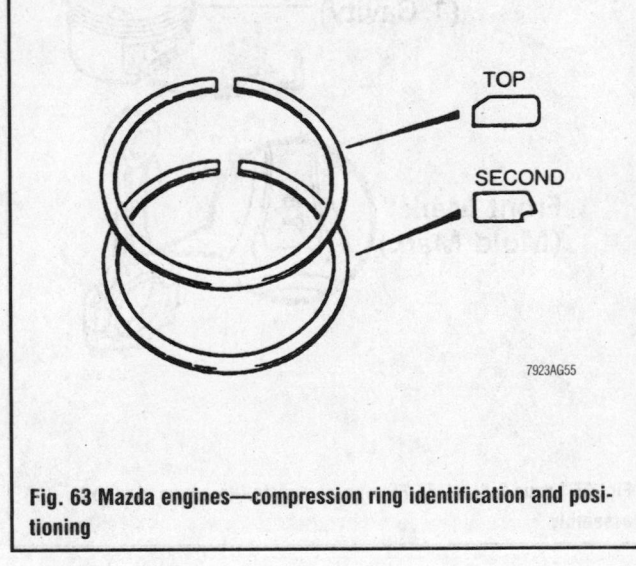

Fig. 63 Mazda engines—compression ring identification and positioning

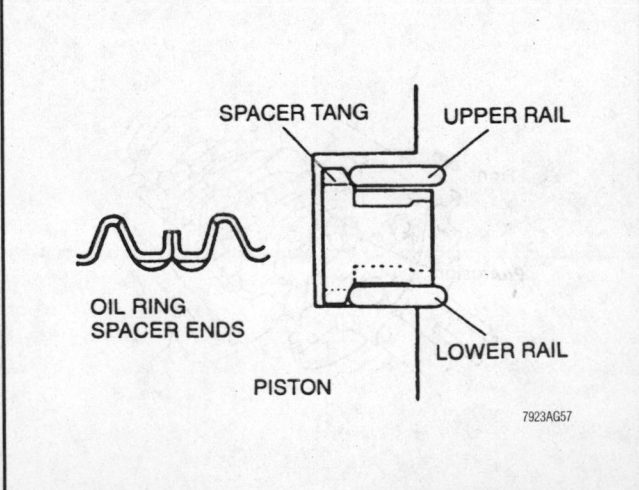

Fig. 64 Mazda engines—upper, spacer and lower oil ring identification and positioning

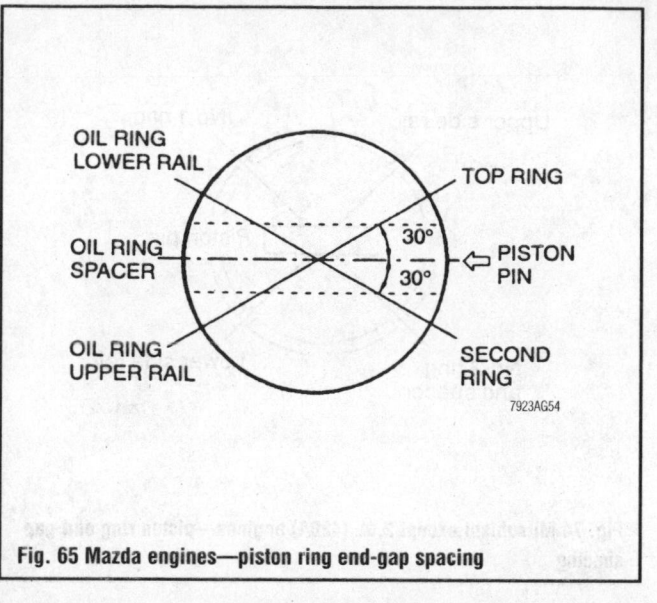

Fig. 65 Mazda engines—piston ring end-gap spacing

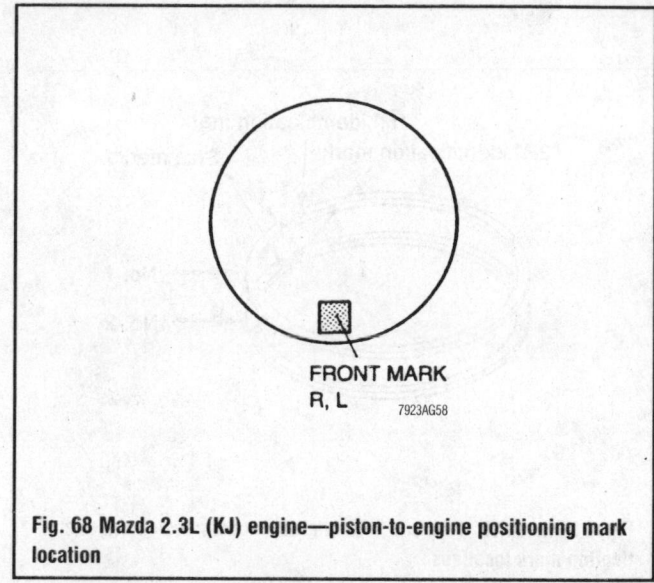

Fig. 68 Mazda 2.3L (KJ) engine—piston-to-engine positioning mark location

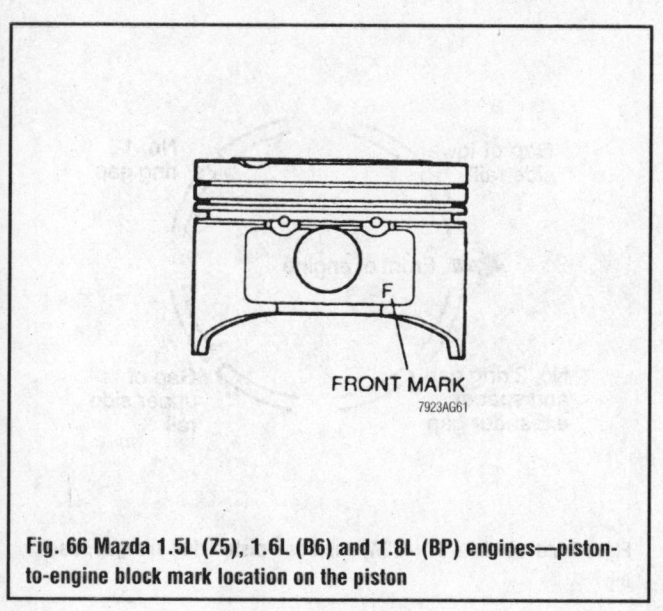

Fig. 66 Mazda 1.5L (Z5), 1.6L (B6) and 1.8L (BP) engines—piston-to-engine block mark location on the piston

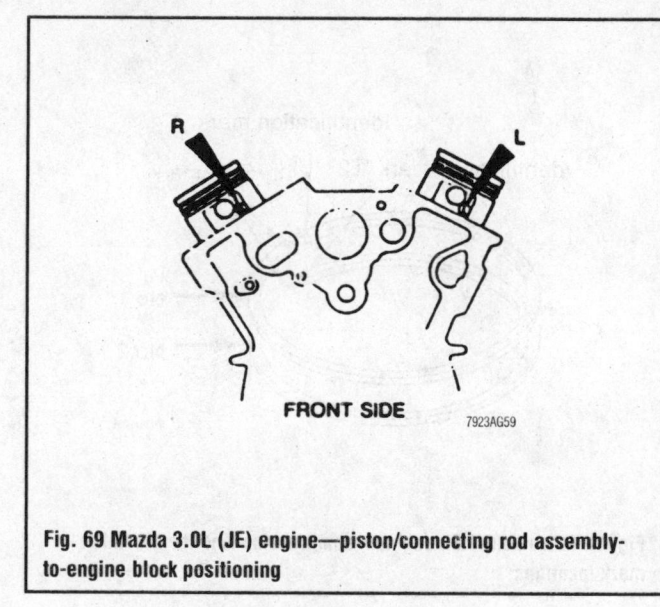

Fig. 69 Mazda 3.0L (JE) engine—piston/connecting rod assembly-to-engine block positioning

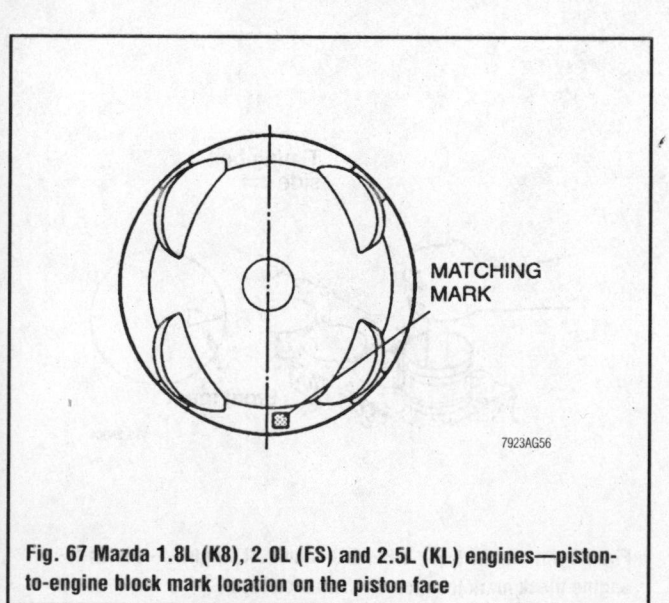

Fig. 67 Mazda 1.8L (K8), 2.0L (FS) and 2.5L (KL) engines—piston-to-engine block mark location on the piston face

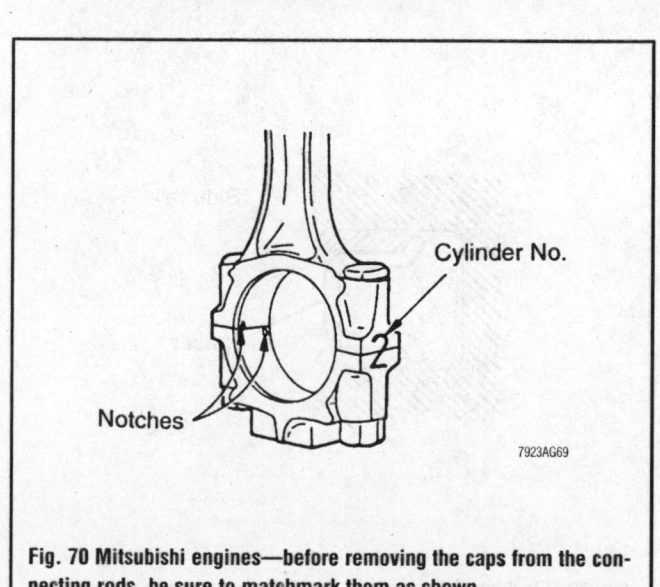

Fig. 70 Mitsubishi engines—before removing the caps from the connecting rods, be sure to matchmark them as shown

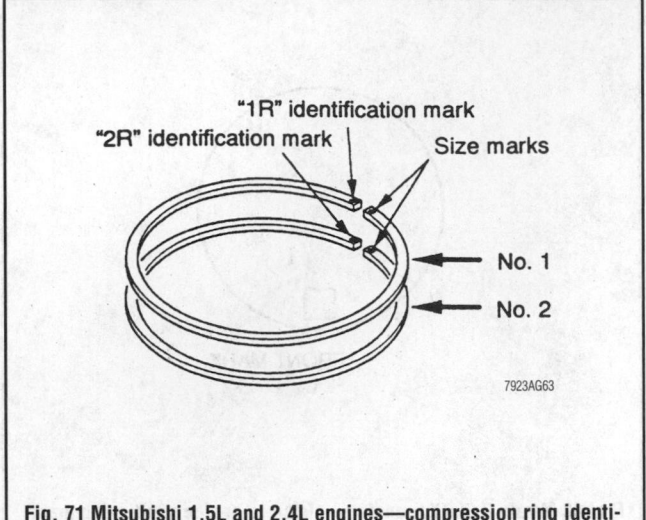

Fig. 71 Mitsubishi 1.5L and 2.4L engines—compression ring identification mark locations

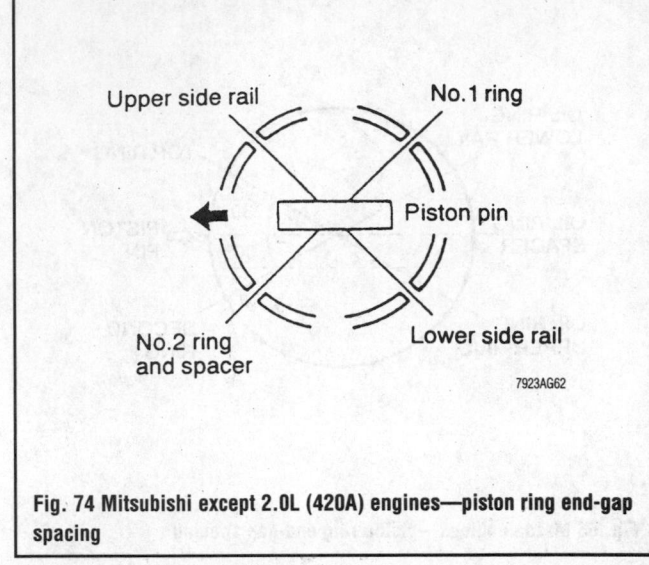

Fig. 74 Mitsubishi except 2.0L (420A) engines—piston ring end-gap spacing

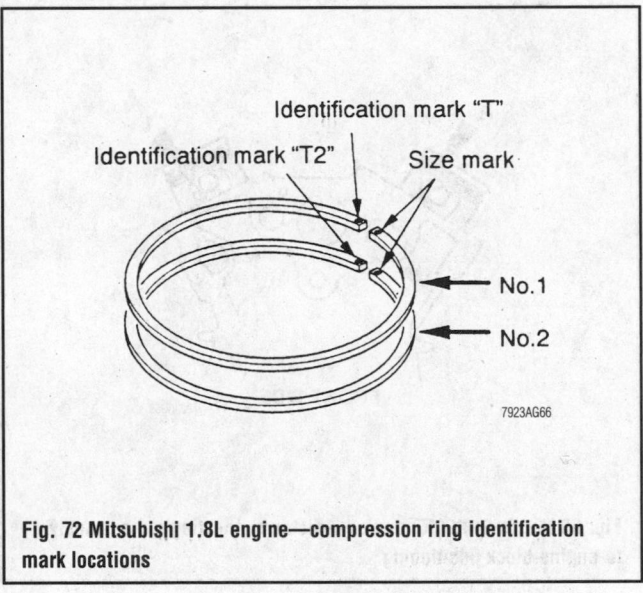

Fig. 72 Mitsubishi 1.8L engine—compression ring identification mark locations

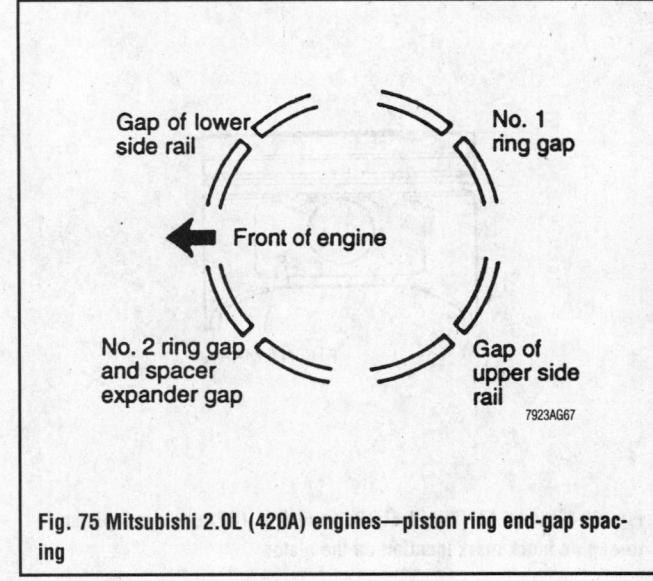

Fig. 75 Mitsubishi 2.0L (420A) engines—piston ring end-gap spacing

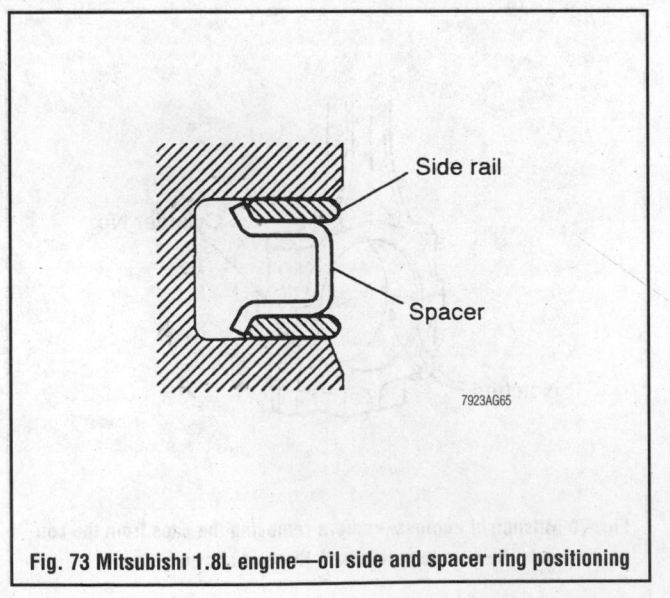

Fig. 73 Mitsubishi 1.8L engine—oil side and spacer ring positioning

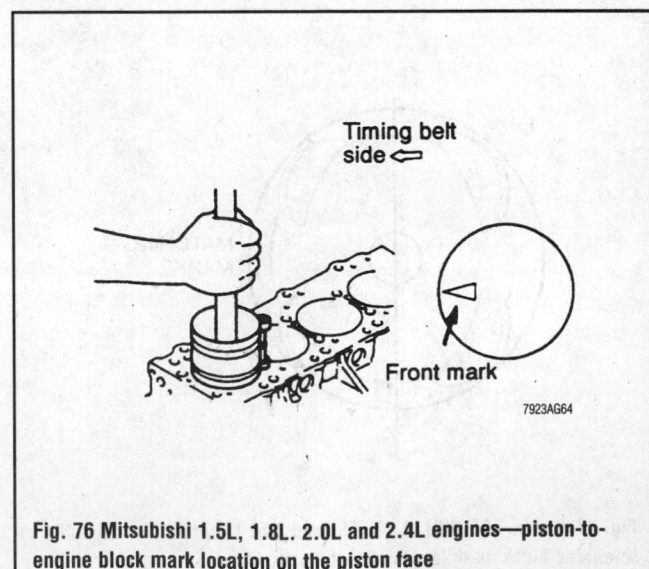

Fig. 76 Mitsubishi 1.5L, 1.8L. 2.0L and 2.4L engines—piston-to-engine block mark location on the piston face

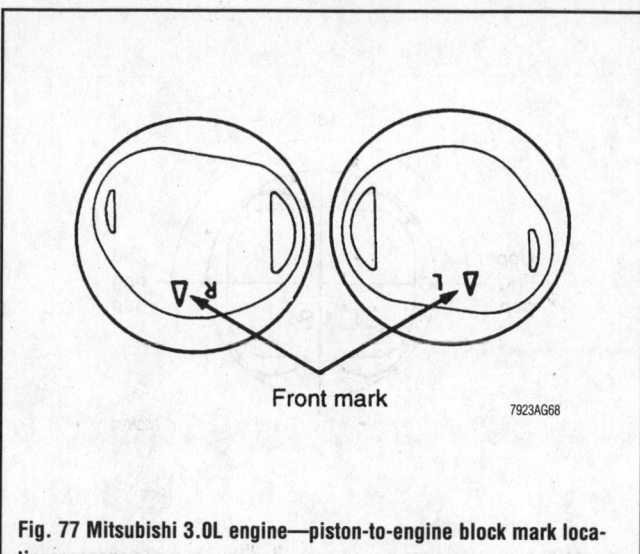

Fig. 77 Mitsubishi 3.0L engine—piston-to-engine block mark locations

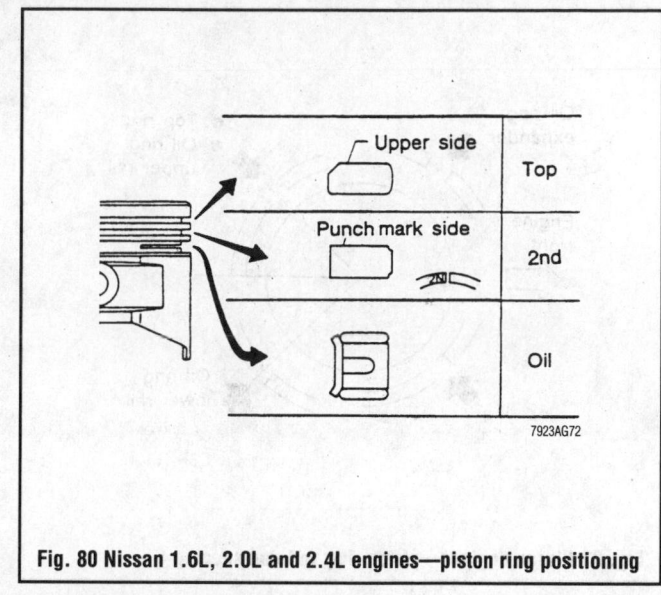

Fig. 80 Nissan 1.6L, 2.0L and 2.4L engines—piston ring positioning

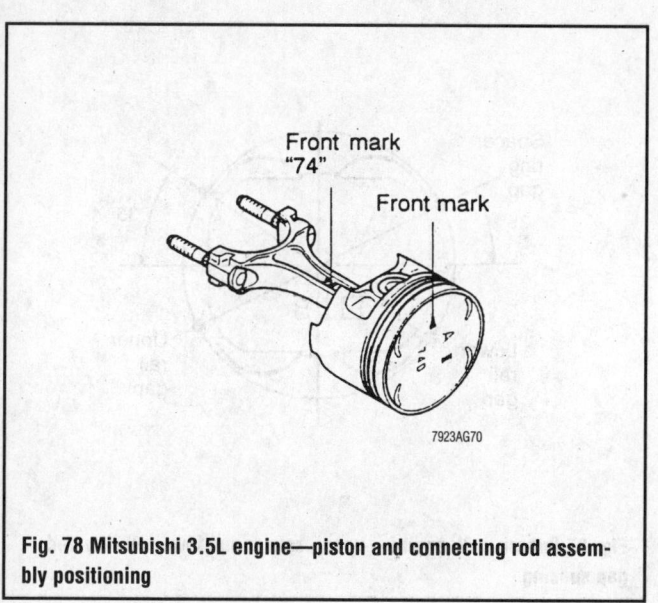

Fig. 78 Mitsubishi 3.5L engine—piston and connecting rod assembly positioning

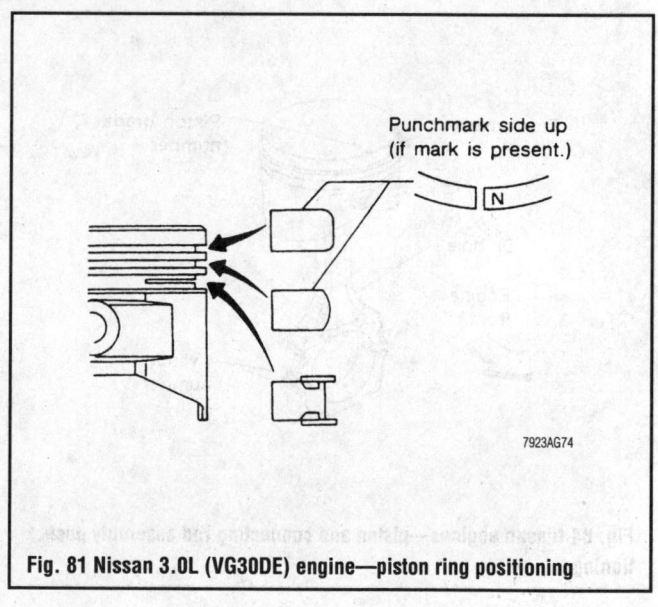

Fig. 81 Nissan 3.0L (VG30DE) engine—piston ring positioning

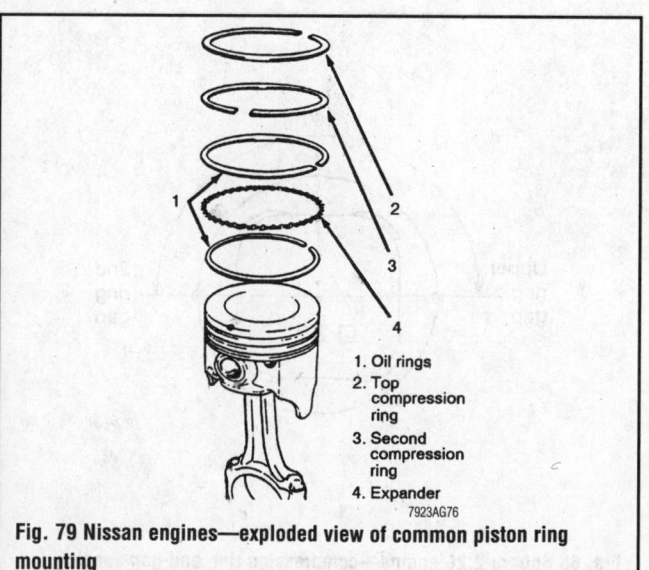

1. Oil rings
2. Top compression ring
3. Second compression ring
4. Expander

Fig. 79 Nissan engines—exploded view of common piston ring mounting

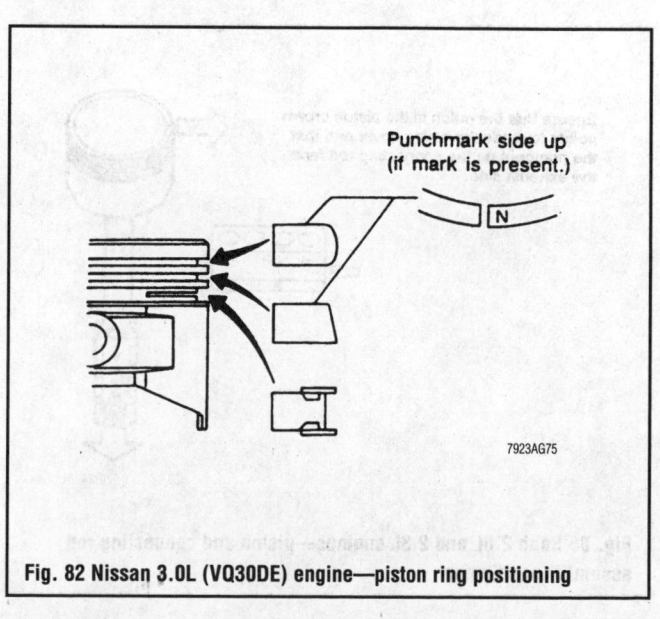

Fig. 82 Nissan 3.0L (VQ30DE) engine—piston ring positioning

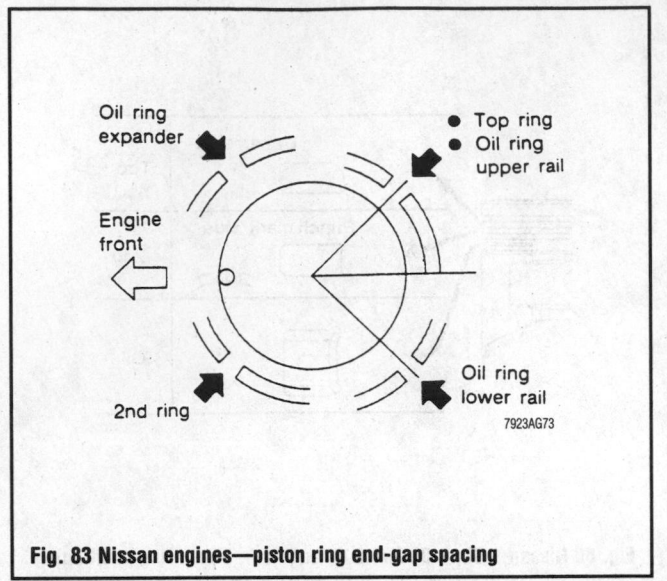

Fig. 83 Nissan engines—piston ring end-gap spacing

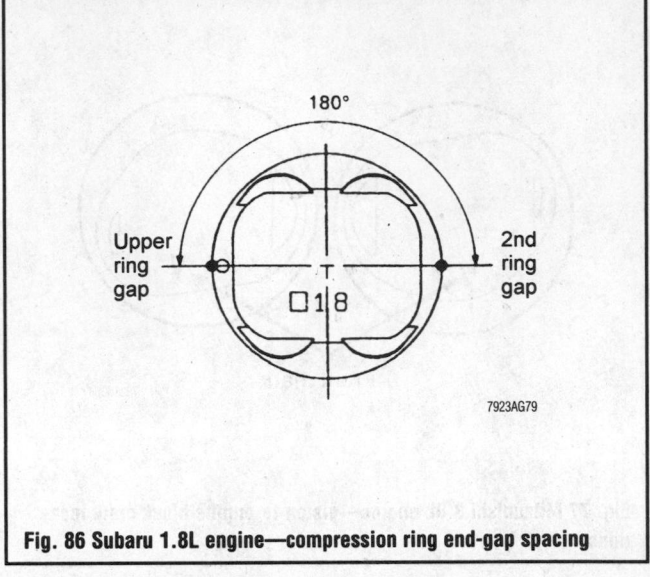

Fig. 86 Subaru 1.8L engine—compression ring end-gap spacing

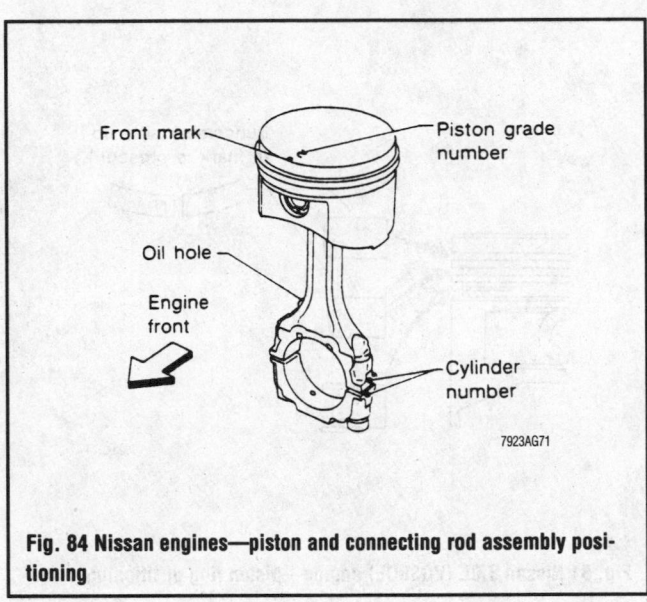

Fig. 84 Nissan engines—piston and connecting rod assembly positioning

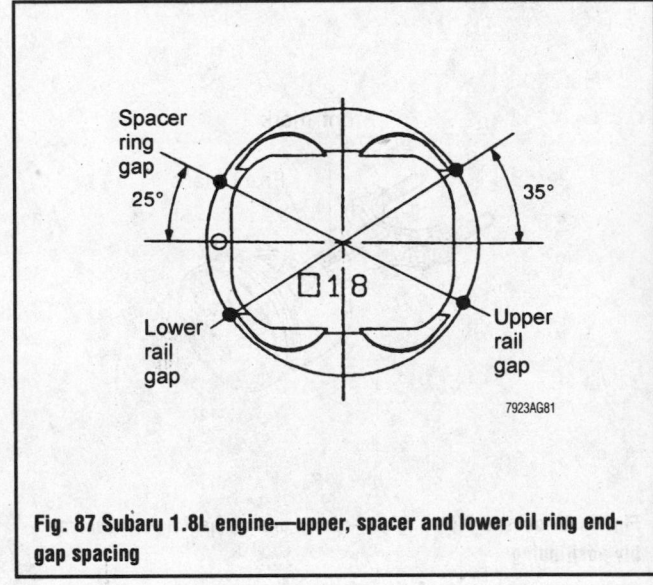

Fig. 87 Subaru 1.8L engine—upper, spacer and lower oil ring end-gap spacing

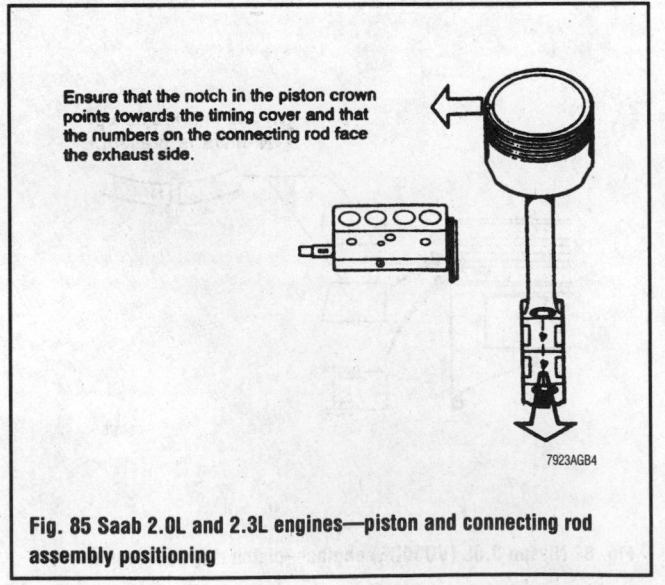

Fig. 85 Saab 2.0L and 2.3L engines—piston and connecting rod assembly positioning

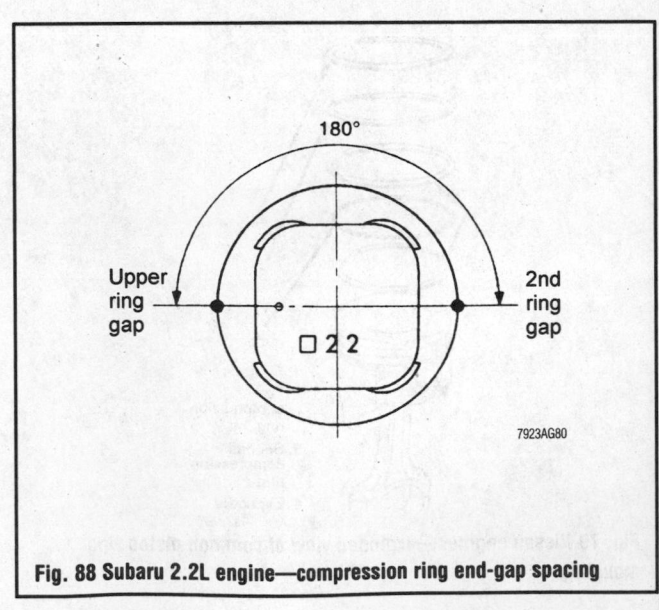

Fig. 88 Subaru 2.2L engine—compression ring end-gap spacing

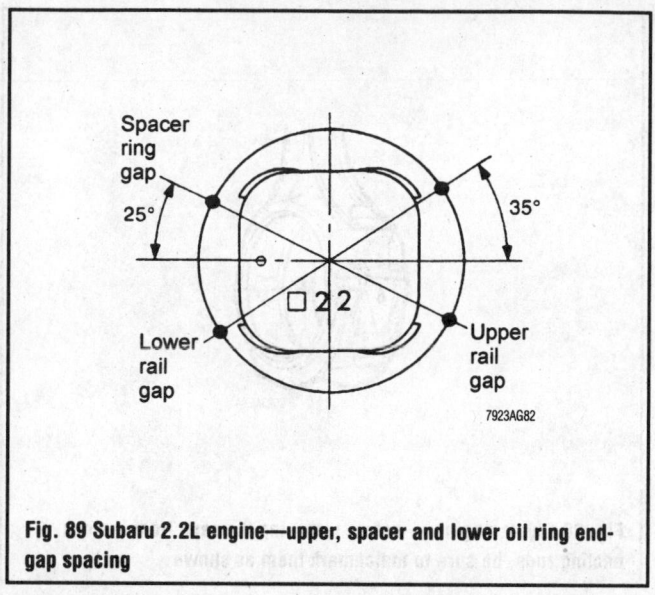

Fig. 89 Subaru 2.2L engine—upper, spacer and lower oil ring end-gap spacing

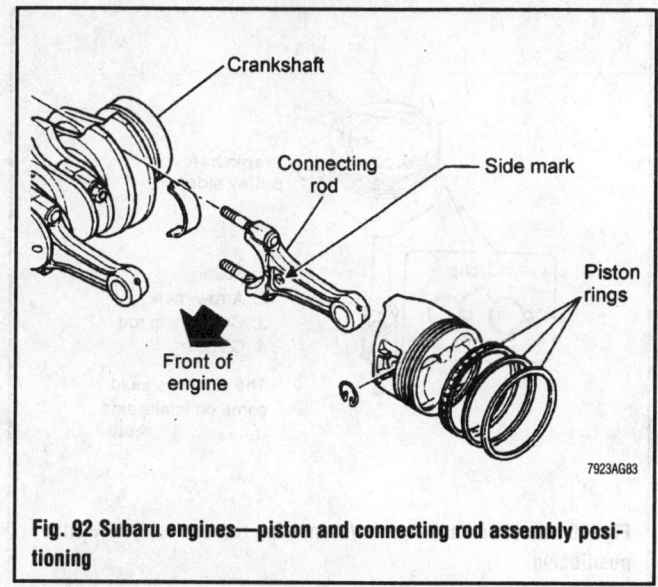

Fig. 92 Subaru engines—piston and connecting rod assembly positioning

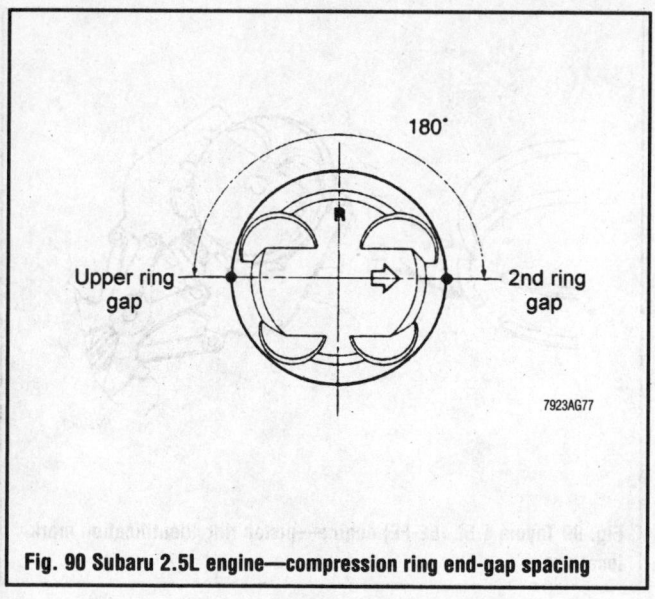

Fig. 90 Subaru 2.5L engine—compression ring end-gap spacing

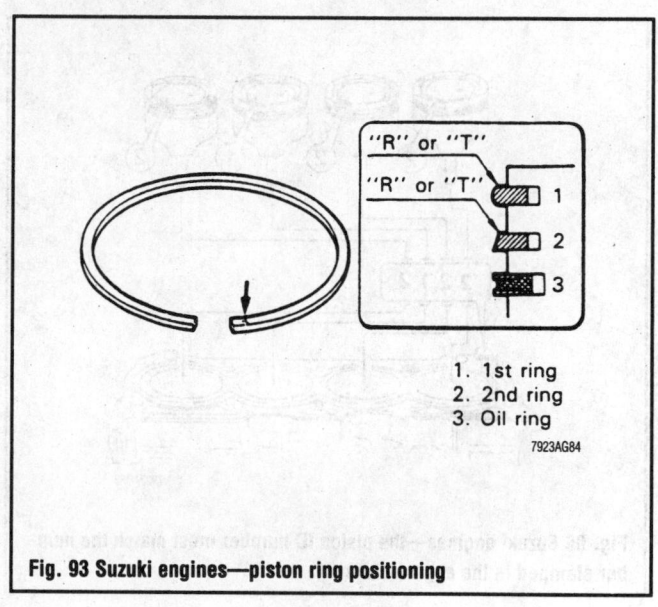

Fig. 93 Suzuki engines—piston ring positioning

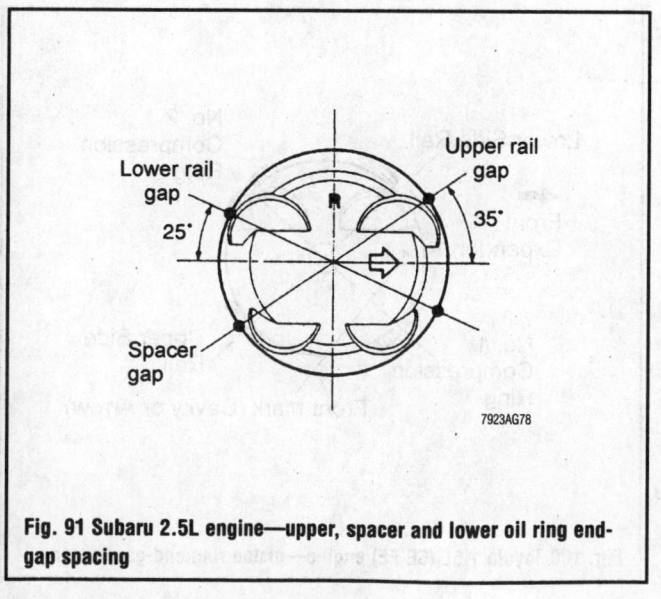

Fig. 91 Subaru 2.5L engine—upper, spacer and lower oil ring end-gap spacing

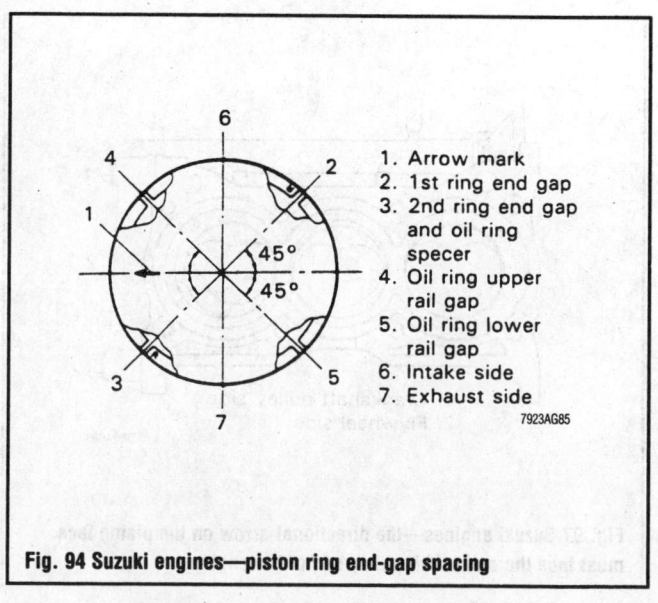

Fig. 94 Suzuki engines—piston ring end-gap spacing

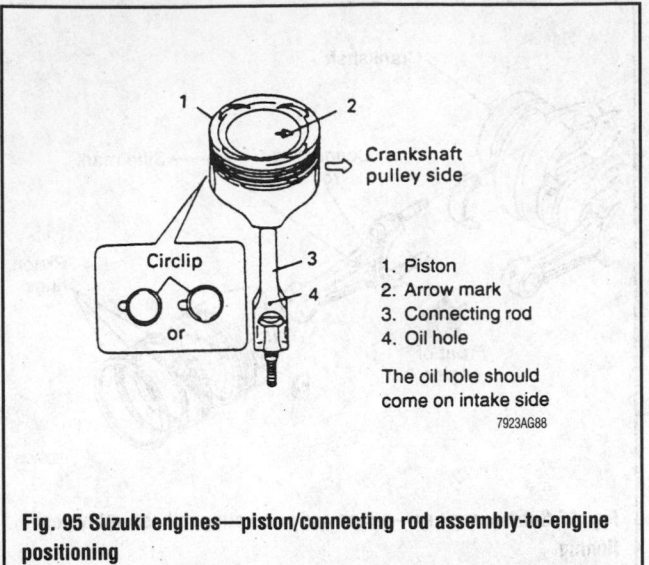

1. Piston
2. Arrow mark
3. Connecting rod
4. Oil hole

The oil hole should come on intake side

Fig. 95 Suzuki engines—piston/connecting rod assembly-to-engine positioning

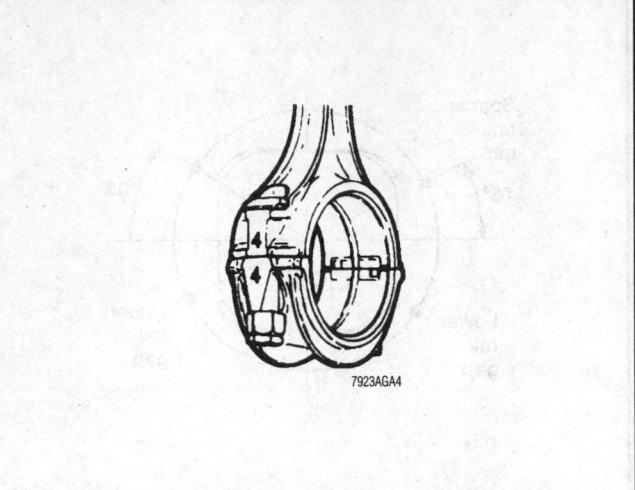

Fig. 98 Toyota engines—before removing the caps from the connecting rods, be sure to matchmark them as shown

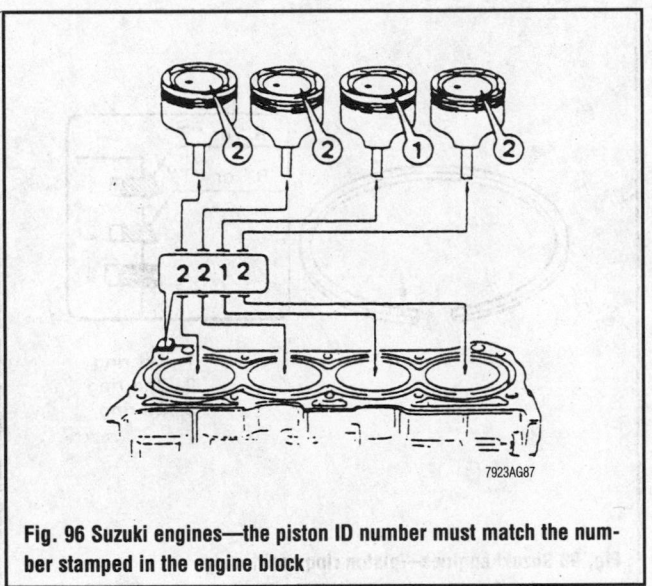

Fig. 96 Suzuki engines—the piston ID number must match the number stamped in the engine block

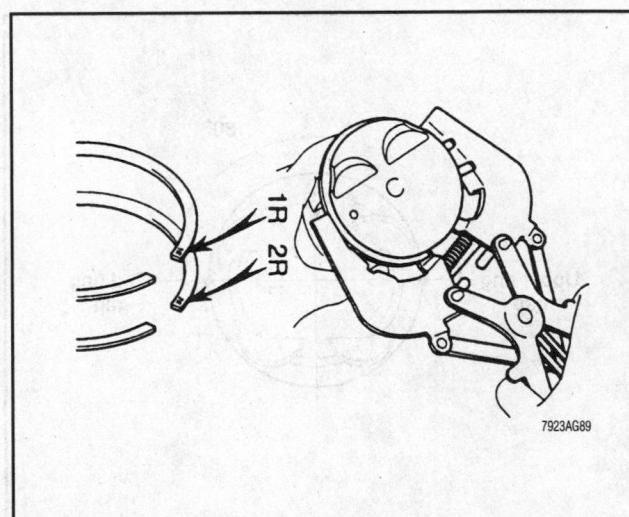

Fig. 99 Toyota 1.5L (5E-FE) engine—piston ring identification mark locations

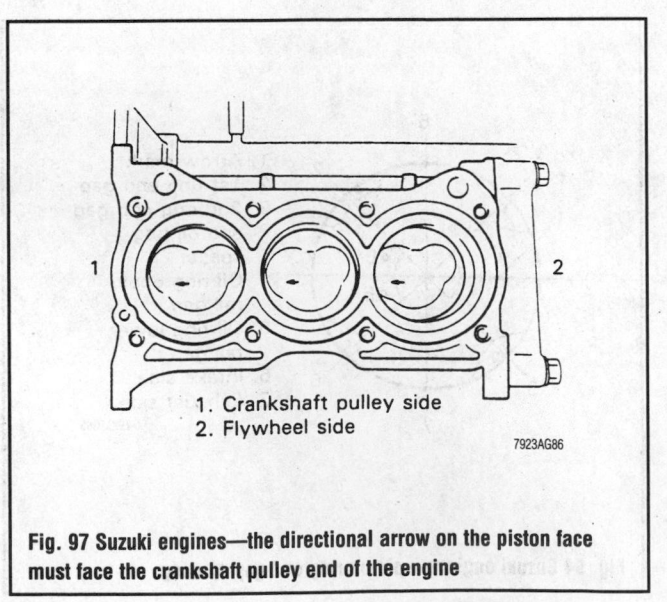

1. Crankshaft pulley side
2. Flywheel side

Fig. 97 Suzuki engines—the directional arrow on the piston face must face the crankshaft pulley end of the engine

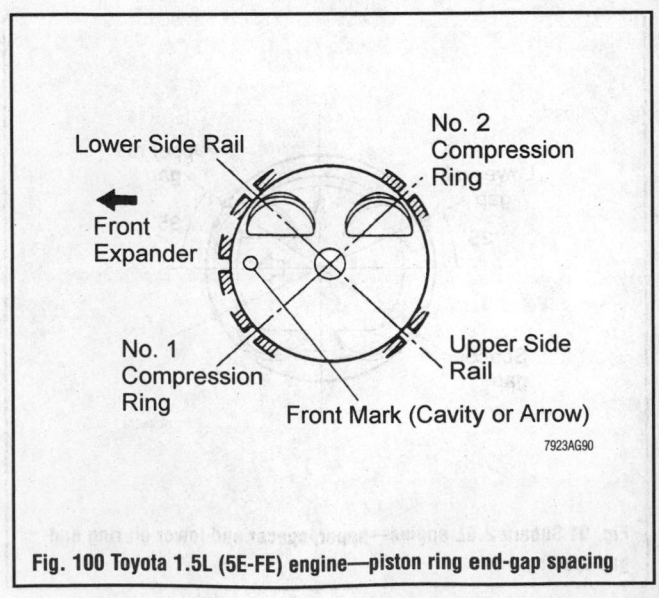

Fig. 100 Toyota 1.5L (5E-FE) engine—piston ring end-gap spacing

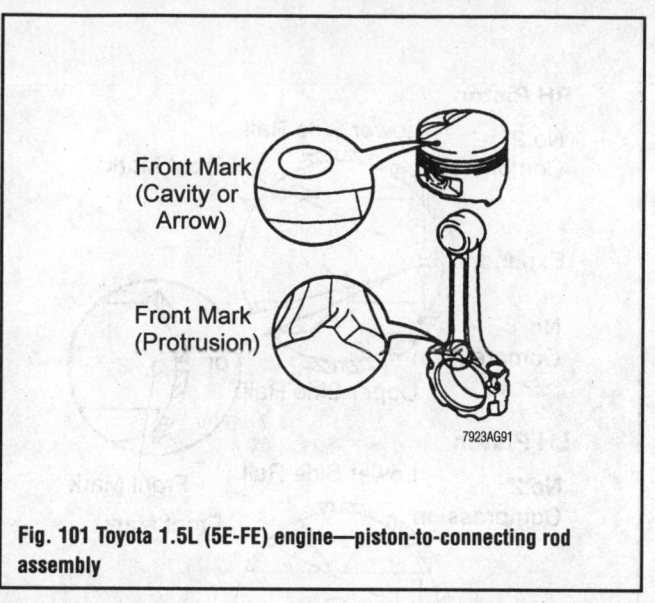

Fig. 101 Toyota 1.5L (5E-FE) engine—piston-to-connecting rod assembly

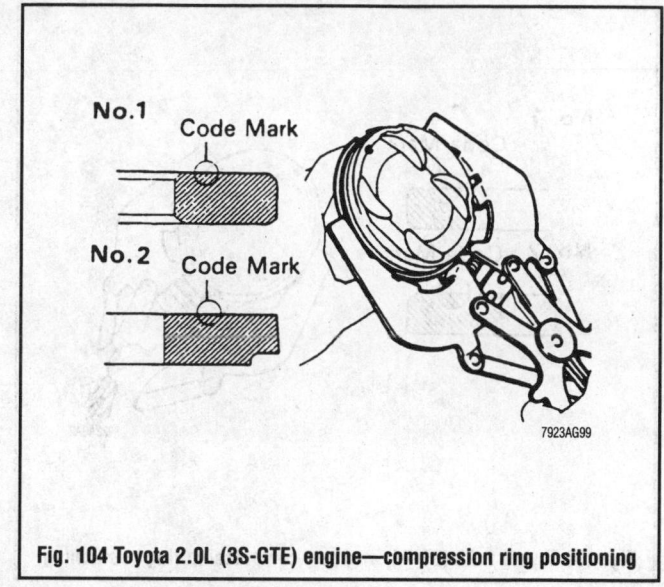

Fig. 104 Toyota 2.0L (3S-GTE) engine—compression ring positioning

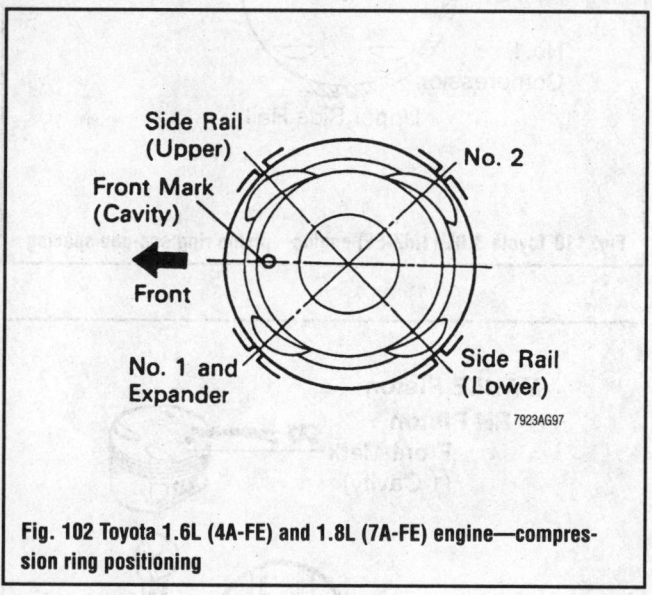

Fig. 102 Toyota 1.6L (4A-FE) and 1.8L (7A-FE) engine—compression ring positioning

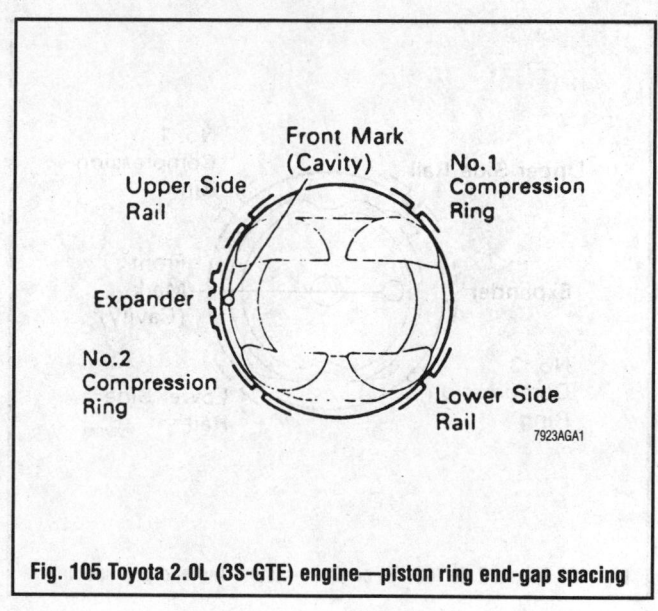

Fig. 105 Toyota 2.0L (3S-GTE) engine—piston ring end-gap spacing

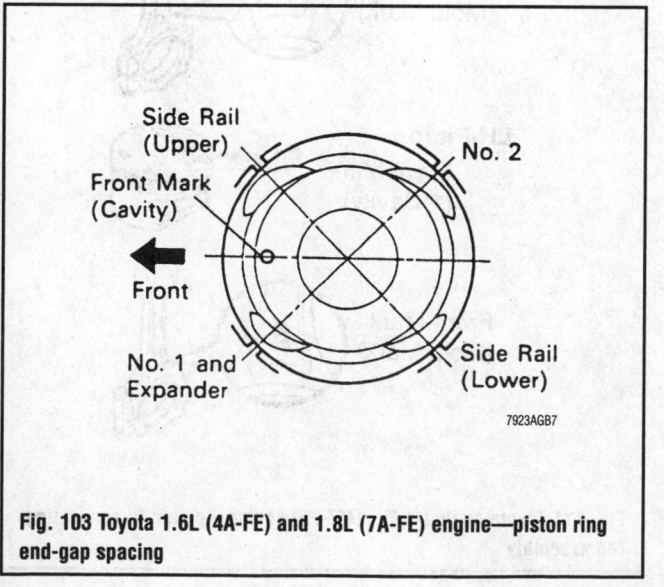

Fig. 103 Toyota 1.6L (4A-FE) and 1.8L (7A-FE) engine—piston ring end-gap spacing

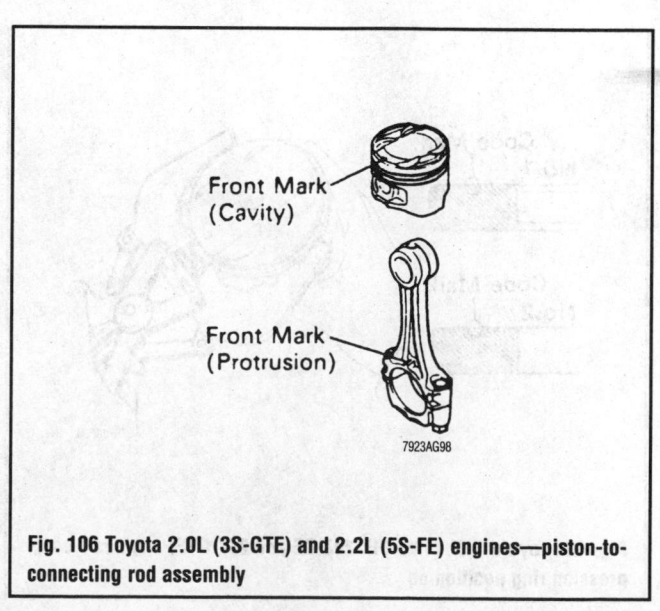

Fig. 106 Toyota 2.0L (3S-GTE) and 2.2L (5S-FE) engines—piston-to-connecting rod assembly

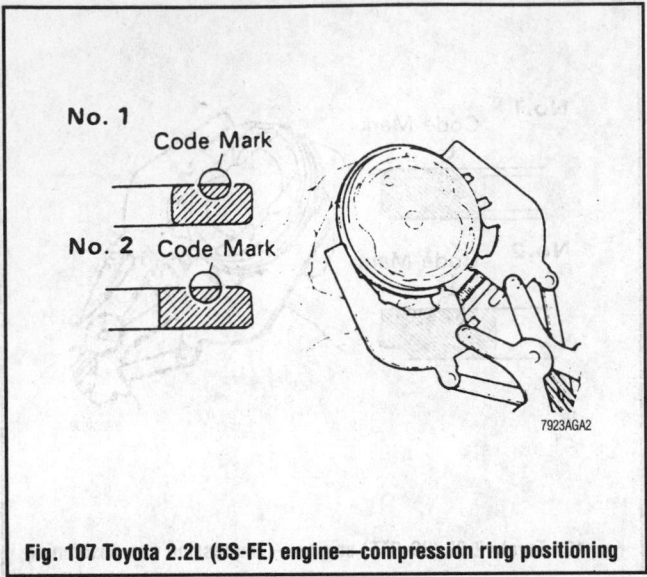

Fig. 107 Toyota 2.2L (5S-FE) engine—compression ring positioning

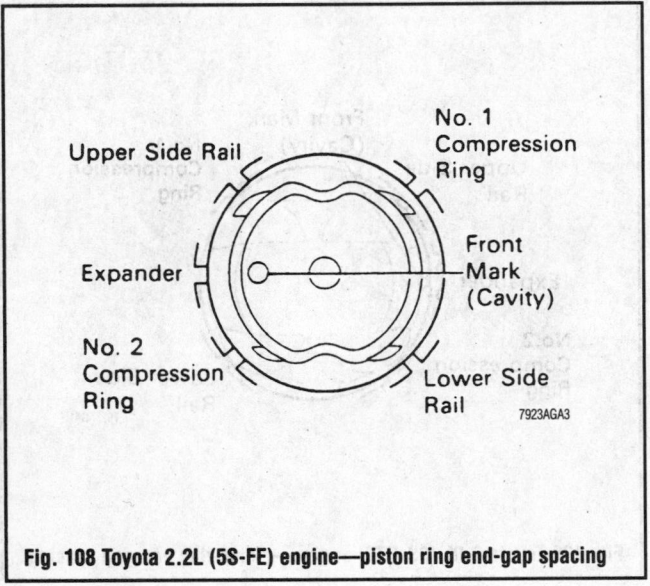

Fig. 108 Toyota 2.2L (5S-FE) engine—piston ring end-gap spacing

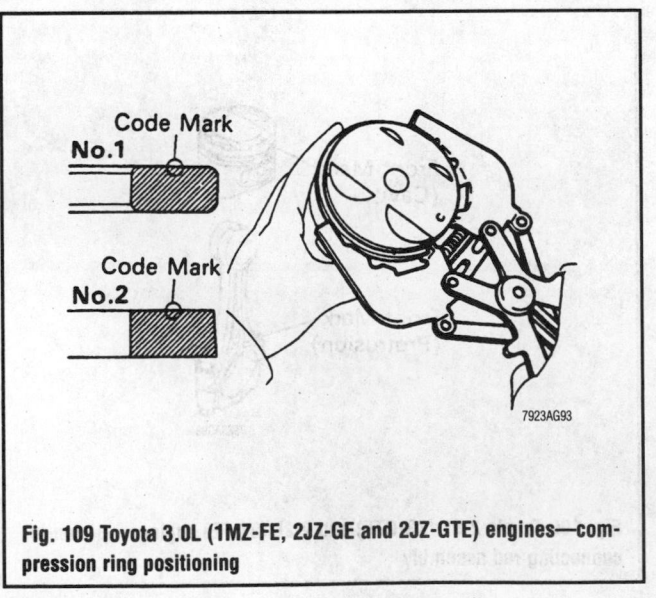

Fig. 109 Toyota 3.0L (1MZ-FE, 2JZ-GE and 2JZ-GTE) engines—compression ring positioning

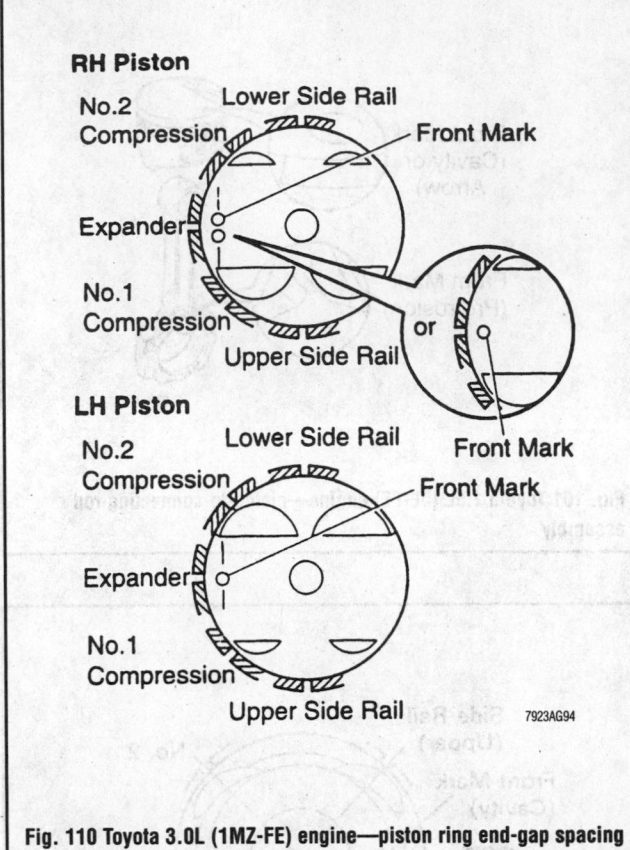

Fig. 110 Toyota 3.0L (1MZ-FE) engine—piston ring end-gap spacing

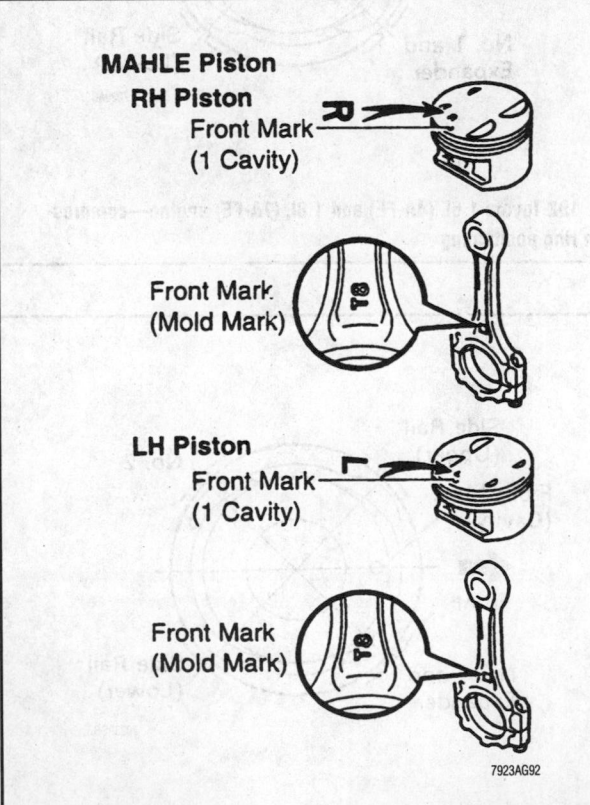

Fig. 111 Toyota Avalon 3.0L (1MZ-FE) engine—piston-to-connecting rod assembly

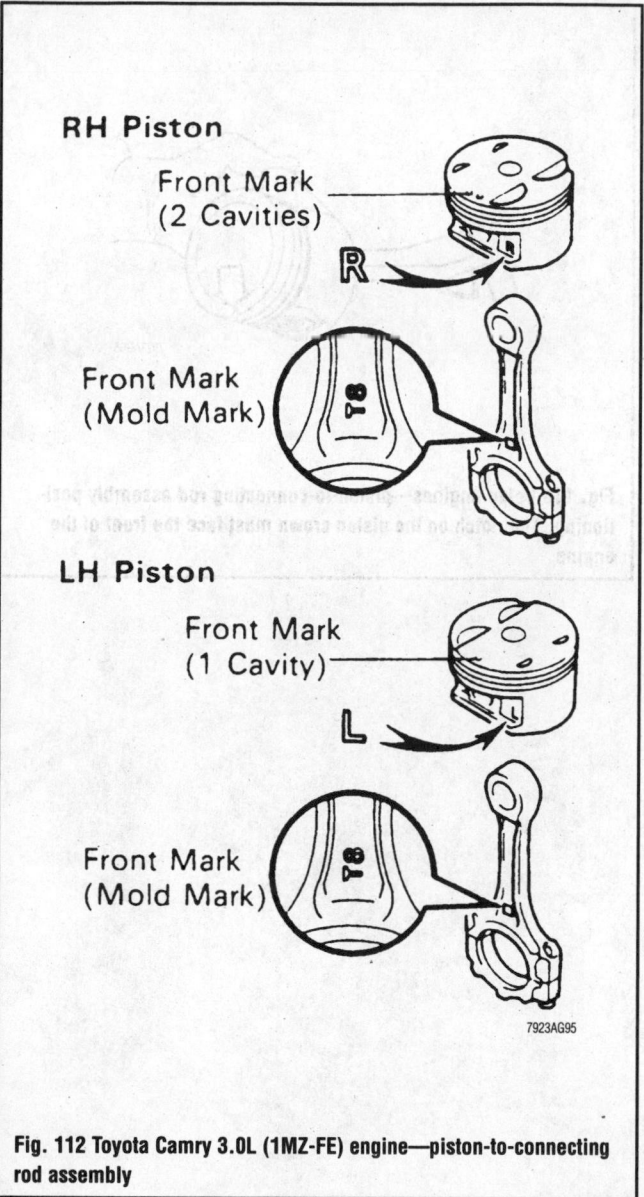

RH Piston

Front Mark
(2 Cavities)

R

Front Mark
(Mold Mark)

LH Piston

Front Mark
(1 Cavity)

L

Front Mark
(Mold Mark)

7923AG95

Fig. 112 Toyota Camry 3.0L (1MZ-FE) engine—piston-to-connecting rod assembly

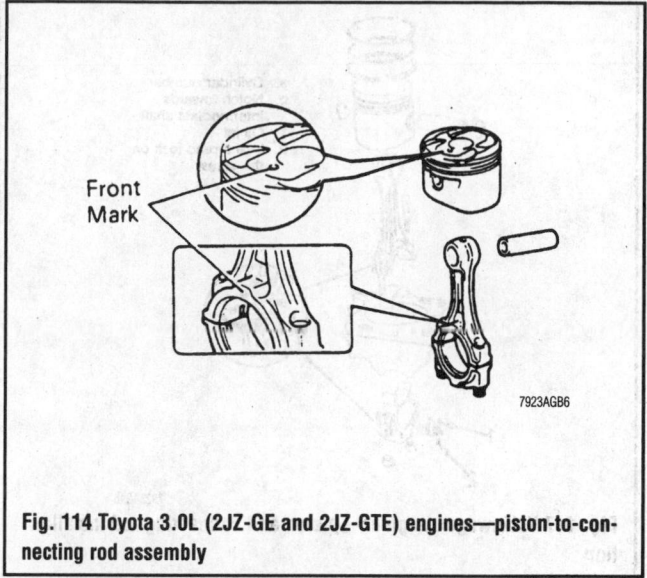

Front Mark

7923AGB6

Fig. 114 Toyota 3.0L (2JZ-GE and 2JZ-GTE) engines—piston-to-connecting rod assembly

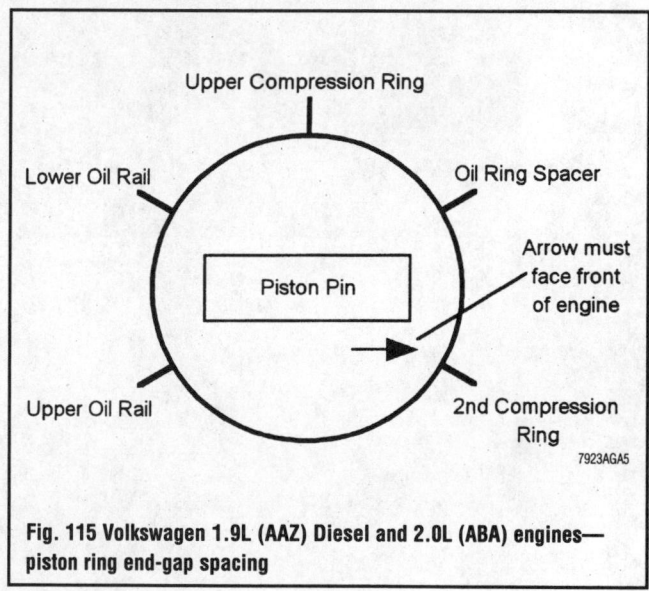

Upper Compression Ring

Lower Oil Rail

Oil Ring Spacer

Piston Pin

Arrow must face front of engine

Upper Oil Rail

2nd Compression Ring

7923AGA5

Fig. 115 Volkswagen 1.9L (AAZ) Diesel and 2.0L (ABA) engines—piston ring end-gap spacing

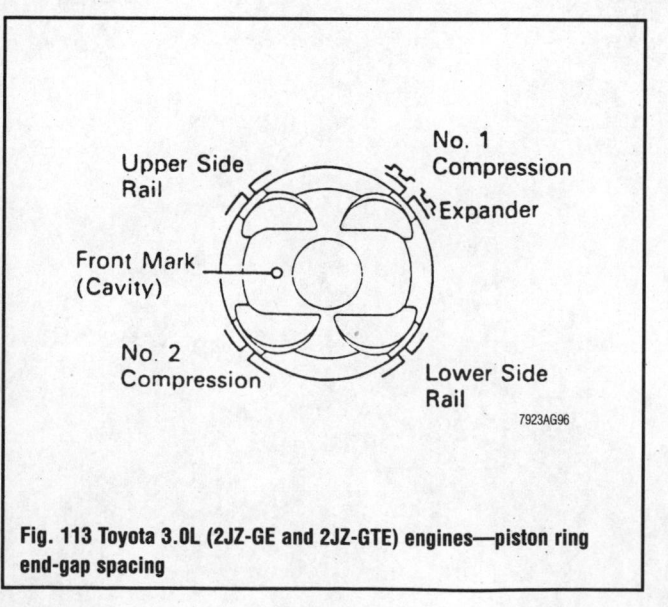

Upper Side Rail

No. 1 Compression

Expander

Front Mark
(Cavity)

No. 2 Compression

Lower Side Rail

7923AG96

Fig. 113 Toyota 3.0L (2JZ-GE and 2JZ-GTE) engines—piston ring end-gap spacing

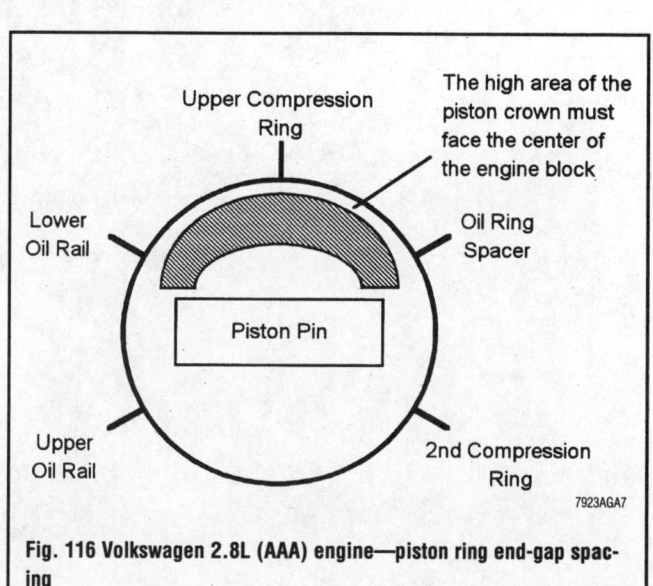

Upper Compression Ring

The high area of the piston crown must face the center of the engine block

Lower Oil Rail

Oil Ring Spacer

Piston Pin

Upper Oil Rail

2nd Compression Ring

7923AGA7

Fig. 116 Volkswagen 2.8L (AAA) engine—piston ring end-gap spacing

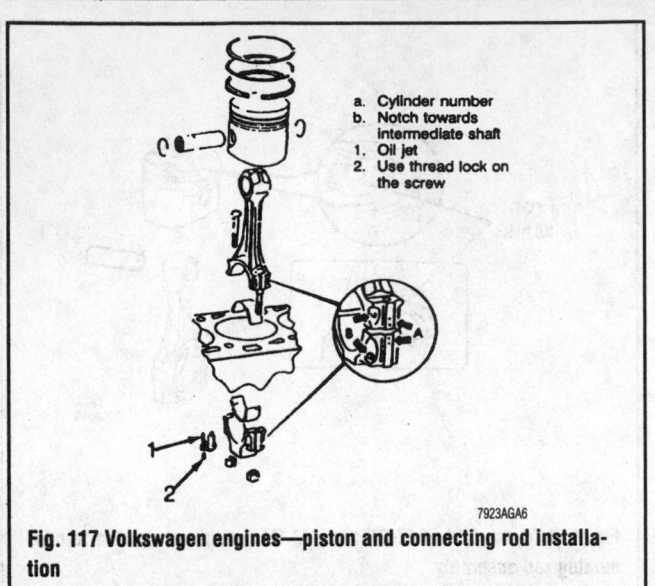

a. Cylinder number
b. Notch towards intermediate shaft
1. Oil jet
2. Use thread lock on the screw

7923AGA6

Fig. 117 Volkswagen engines—piston and connecting rod installation

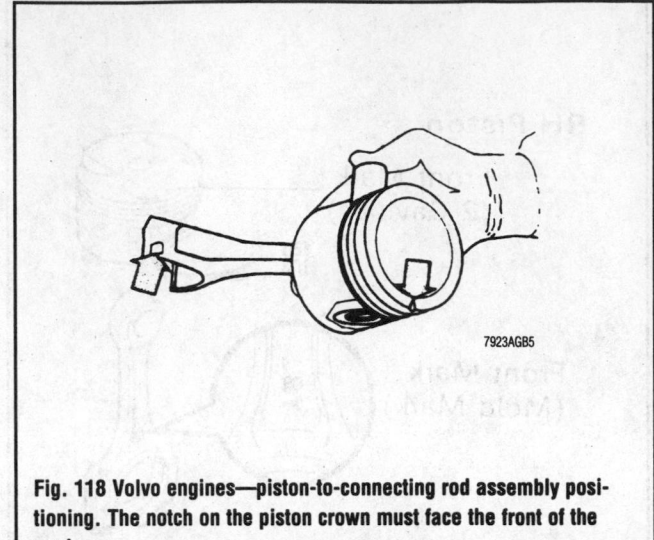

7923AGB5

Fig. 118 Volvo engines—piston-to-connecting rod assembly positioning. The notch on the piston crown must face the front of the engine

ACURA

2.2CL • 2.3CL • 2.5TL • 3.0CL • 3.2TL • 3.5RL • Integra • Integra GSR • Legend

ENGINE REPAIR

➡The PCM idle memory must be reset after reconnecting the battery. Start the engine and hold it at 3000 rpm until the cooling fan comes on. Then allow the engine to idle for about five minutes with all accessories OFF and with the transmission in Park or Neutral.

Distributor

REMOVAL & INSTALLATION

1.8L Engines

➡The radio may contain a coded anti-theft circuit. Be sure you have the security code number before disconnecting the battery.

1. Disconnect the negative battery cable.
2. Disconnect engine wiring harness and connectors from distributor.
3. Disconnect the spark plug wires from distributor cap.
4. If removing the ignition coil, remove the distributor cap, rotor, and cap seal, then remove the leak cover.
5. Remove the 2 screws to disconnect the wires from the coil.
6. Remove the 2 screws and slide the ignition coil out of the distributor housing.
7. Remove distributor hold-down bolts, and remove distributor from cylinder head.

To install:

8. Use new O-ring on distributor housing. Coat new O-ring with engine oil before installation.
9. Slip the distributor into position.

➡Lugs on the end of the distributor and the matching grooves in the camshaft end are offset to eliminate any possibility of installing the distributor 180 degrees out of time.

10. Install hold-down bolts, hand-tighten.
11. Slide the ignition coil into the distributor housing and install the 2 mounting screws.
12. Reconnect the 2 wires to the coil and install the 2 screws. Install the leak cover, rotor, cap seal, and cap.
13. Connect engine wiring harness and connector to distributor.
14. Connect the spark plug wires. Reconnect the negative battery cable.
15. Set timing, using a timing light, then

tighten the hold-down bolts to 16 ft. lbs. (22 Nm).

2.3, 2.5L and 3.0L Engines

➡The radio contains a coded anti-theft circuit. Be sure you have the security code number before disconnecting the battery.

1. Disconnect the negative battery cable.
2. Disconnect the spark plug and coil wires from the distributor cap and mark their positions.
3. Detach the harness connector(s) from the distributor.
4. Remove the distributor mounting bolts. Remove the distributor from the cylinder head.

To install:

5. Install a new O-ring on the distributor housing. Coat the O-ring with engine oil before installation.
6. Install the distributor into position, verifying that the lugs on the distributor shaft end fit into the grooves on the camshaft end.
7. Install the mounting bolts. Tighten the bolt(s) to 13 ft. lbs. (18 Nm).
8. Connect the spark plug and coil wires. Connect the negative battery cable.
9. Check the ignition timing with a timing light. The timing marks are located on the crankshaft pulley and lower timing cover.

Ignition Timing

ADJUSTMENT

1.8L Engines

1. If equipped with an automatic transaxle, place the shifter in Park. If equipped with a manual transaxle place the shifter in Neutral. Set the parking brake and block the drive wheels.
2. Start the engine and hold the engine speed at 3000 rpm, until the radiator fan comes on. The engine will be at normal operating temperature. Be sure all electrical systems (radio, air conditioning, lights, etc.,) are shut OFF.
3. Pull the service check connector located behind the right kick panel. Connect the BRN/WHT and BLK terminals with the SCS service connector or equivalent device.
4. Connect a timing light to No. 1 ignition wire and point the light toward the pointer on the timing belt cover.
5. The red mark on the crankshaft pulley should be aligned with the pointer no the

timing belt cover. The ignition timing specification should be: 14–18° BTDC (red mark) at 700–800 rpm.

➡The red mark on the crank pulley is 16° BTDC.

6. Adjust the ignition timing by: loosening the distributor mounting bolts, turn the distributor housing counterclockwise to advance the timing, turn the distributor housing clockwise to retard the timing.
7. Tighten the distributor bolts to 17 ft. lbs. (24 Nm) and recheck the timing.
8. Remove the SCS service connector from the service check connector.

2.3L Engine

1. Connect a PGM tester (scan tool) to the data link connector.
2. Connect a timing light to the No. 1 ignition cable.
3. Start the engine and allow it to warm up until the electric fan comes on.
4. Be sure to turn off all accessories.
5. Verify the idle speed is 650–750 rpm.
6. Point the light at the timing belt cover near the crankshaft pulley and read the timing. Correct timing is 10–14° BTDC for both automatic and manual transmissions. If necessary, loosen the distributor hold-down bolt and rotate the distributor slightly to adjust the timing. Turn it counterclockwise to advance and clockwise to retard the timing.
7. Tighten the hold-down bolt to 16 ft. lbs. (22 Nm). Recheck the timing after the bolt is tight to confirm the correct timing.
8. Disconnect the PGM tester.

2.5L Engine

❊❊ CAUTION

All Supplemental Restraint System (SRS, air bag system) electrical wiring harnesses are covered with yellow outer insulation for easy identification. To avoid the possibility of personal injury, if any SRS component or wiring harness must be disconnected, the air bags must first be disabled. Replace the entire affected SRS harness assembly if there is an open circuit or damage to the wiring.

➡This vehicle's ignition timing is not adjustable. The timing is controlled by the PCM.

1. Start the engine and allow it to idle at 3000 rpm with all electrical accessories off and the transmission in **N** or **P**. Allow the

engine to warm up until the cooling fan comes on.

2. Pull the service check connector out from under the glove box. Connect the WHT/BLU and ORN terminals with a service connector, No. 07PAZ-0010100.

3. Check the idle speed and adjust if necessary:

    a. Connect a test tachometer to the test tachometer connector on the right side of the engine compartment to the rear of the right shock tower.

    b. Idle speed must be 650–750 rpm with the transmission in **N** or **P** and all electrical accessories off.

4. Connect a timing light to the No. 1 plug wire. While engine idles, point the light toward the pointer on the timing belt cover.

5. Inspect ignition timing at idle. Timing should be 13–17 BTDC, using the red timing mark, at 650–750 rpm in **N** or **P**.

6. If the ignition timing is incorrect, it cannot be adjusted. The PCM must be replaced.

➡**All mechanical and electrical systems should checked for proper operation before replacing the PCM. Only replace the PCM as a last resort.**

7. Remove the timing light.

8. Remove the service connector. Reconnect the check connector, and tuck it under the glove box.

3.0L Engine

The ignition timing is only adjustable by the PCM, but the ignition base timing can be checked by performing the following:

1. Connect a PGM tester (scan tool) to the data link connector.

2. Connect a timing light to the No. 1 ignition cable.

3. Start the engine and allow it to warm up until the electric fan comes on.

4. Be sure to turn off all accessories.

5. Verify that the idle speed is 630–730 rpm.

6. Point the light at the timing belt cover near the crankshaft pulley and read the timing. Correct timing is 8–12° BTDC. If the ignition timing is different from the specification, replace the PCM.

3.2L Engine

LEGEND

1. Start the engine and allow it to warm up until the cooling fan comes on.

2. Pull up the upper right edge of the

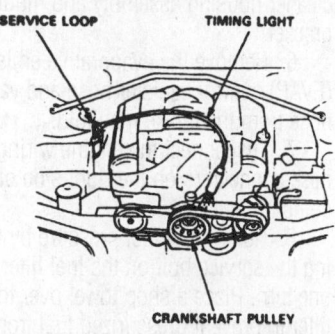

Ignition timing check—Legend

front passenger's floor carpeting. Pull the 2-P service check connector out from under the right side of the dash.

3. Connect the WHT and BLK terminals with a jumper wire.

4. Connect an inductive timing light to the service loop located on the right shock tower. Point light toward the pointer on timing belt cover while the engine is idling.

5. Check the idle speed by connecting a test tachometer to the test connector on the right shock tower. Adjust the idle speed if necessary.

6. Ignition timing should be 13–17 BTDC, using the RED timing mark, at 550–650 rpm in **N** or **P** for vehicles with automatic transmissions or 600–700 rpm in **N** for vehicles with manual transmissions.

7. If timing adjustment is necessary, follow this procedure:

    a. Remove the control box cover from the firewall. Be careful not to damage the vacuum hoses when removing the cover.

    b. Drill the two cover rivets off with a ⅛ in. (3.2mm) drill bit. Then, remove the cover from the adjuster.

✳✳ CAUTION

To avoid personal injury, always wear eye protection when drilling. Do not damage the adjuster body when removing rivets.

---

    c. Turn the adjusting screw counter-clockwise to retard the timing, or clockwise to advance the timing.

    d. After the adjustment, install the cover to the ignition timing adjuster with new rivets.

    e. Reinstall the control box cover.

    f. Remove the jumper wire from the service check connector.

8. Recheck the ignition timing and remove the timing light.

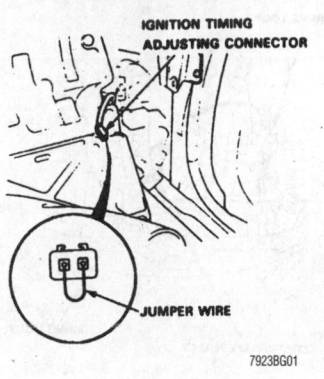

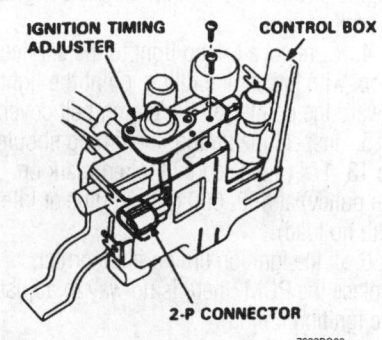

Control box and timing adjuster—Legend

3.2TL AND 3.5RL

This procedure is for inspection only. The ignition timing is not adjustable on these vehicles.

1. Start the engine and hold the engine at 3000 rpm with no load (shift lever in **P** or **N**) until the radiator fan comes on, then let the engine idle.

2. Locate the service connector under the glove box. Connect the wire terminals with the special tool (SCS service connector Part # 07PAZ-0010100).

3. Check the idle speed, the engine should idle at 590–690 rpm on the 3.2L

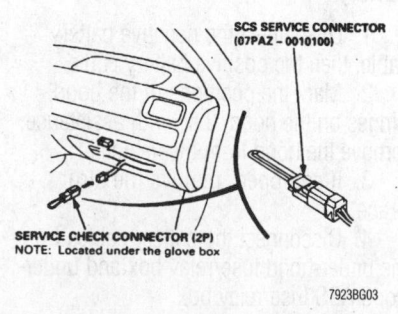

Service check connector—3.2TL and

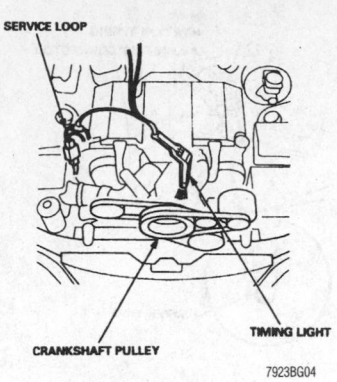

Timing light attachment—3.2TL

engine or 600–700 rpm on the 3.5L engine in park or neutral.

4. Connect a timing light to the service loop, with the engine idling, point the light toward the pointer on the timing belt cover.

5. Inspect the timing: the timing should be 13–17° (indicated by the red mark on the pulley) at 590–690 rpm (engine at idle with no load).

6. If the ignition timing is incorrect, replace the PCM (there is no way to adjust the ignition timing).

➡**All mechanical and electrical systems should checked for proper operation before replacing the PCM. Only replace the PCM as a last resort.**

7. Remove the timing light.

8. Disconnect the special tool (SCS service connector) from the service check connector.

Engine Assembly

REMOVAL & INSTALLATION

1.8L Engines

➡**The radio may contain a coded theft protection circuit. Always obtain the code number before disconnecting the battery.**

1. Disconnect the negative battery cable, then the positive battery cable.

2. Mark the positions of the hood hinges on the hood, then with assistance, remove the hood from the vehicle.

3. If equipped, remove the strut brace.

4. Disconnect the battery cables from the under-hood fuse/relay box and under-hood ABS fuse/relay box.

5. Remove the intake air duct, air cleaner housing assembly and mounting bracket.

6. Remove the evaporative emission (EVAP) control canister hose and vacuum hose from the intake manifold.

7. Disconnect the engine wiring harness connectors on the right side of engine compartment.

8. Relieve the fuel pressure by loosening the service bolt on the fuel filter about one turn. Place a shop towel over the fuel filter to prevent pressurized fuel from spraying over the engine.

✳✳ CAUTION

The fuel injection system remains under pressure after the engine has been turned OFF. Properly relieve fuel pressure before disconnecting any fuel lines. Failure to do so may result in fire or personal injury.

9. Disconnect the fuel feed hose, brake booster vacuum hose, and fuel return hose.

10. Remove the throttle cable by loosening the locknut, then slip the cable end out

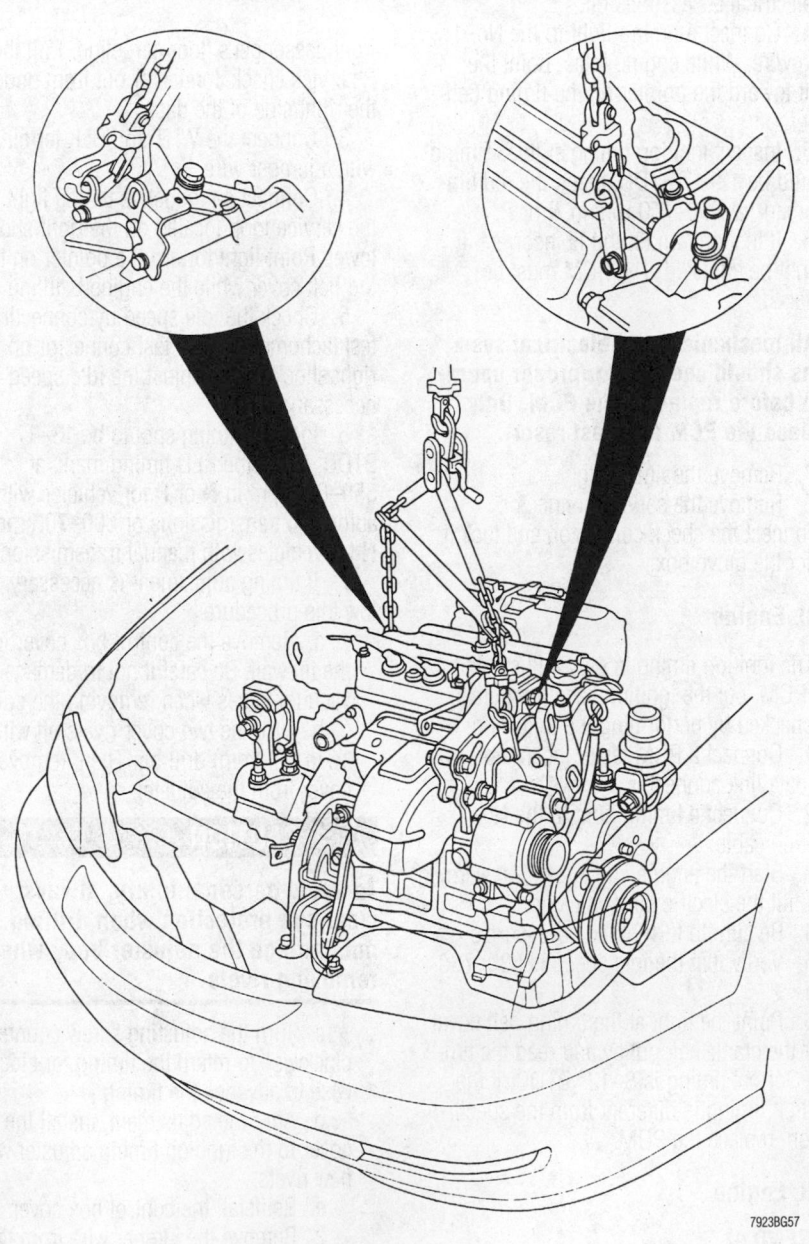

Engine lifting points—2.3CL

of the accelerator linkage. Be careful not to bend the cable when removing it. Replace the cable if it gets kinked.

11. Remove the engine wiring harness connectors, terminal, and clamps on the left side of engine compartment.

12. Remove the cruise control actuator, and engine ground cable at the body end.

13. Remove the adjusting bolt and mounting bolt, then remove the power steering belt and pump. Do not disconnect the power steering hoses.

14. Loosen the idler pulley bolt and adjusting bolt, then remove the air conditioning compressor belt.

15. On manual transmission only, remove the clutch slave cylinder and pipe/hose assembly. Do not disconnect the pipe/hose assembly.

16. Remove the transmission ground cable and hose clamp. Remove the radiator cap.

17. Safely raise and support the vehicle, then remove the front wheels and splash shield.

18. Drain the engine coolant, engine oil, and transmission fluid. Reinstall the drain plugs using new washers. Be careful not to overtighten the drain plugs.

19. Disconnect the upper and lower radiator hoses and the heater hoses from the engine.

20. If equipped with a automatic transmissions, disconnect the ATF (automatic transmission fluid) cooler hoses.

21. Remove the radiator assembly.

22. Remove the air conditioning compressor mounting bolts and position the compressor out of the way. Suspend the compressor on a wire, do not let it hang by its hoses. Do not disconnect the hoses.

23. Disconnect the heated oxygen sensor (HO2S) connector.

24. Remove the nuts and bolts connecting exhaust pipe A to the catalytic converter. Discard the gasket and the locknuts.

25. Remove and discard the nuts attaching exhaust pipe A to the exhaust hanger.

26. Remove and discard the locknuts attaching exhaust pipe A to the exhaust manifold, then remove exhaust pipe A from the vehicle. discard the exhaust gaskets.

27. If equipped with a manual transaxle, disconnect the shift rod and extension rod from the transaxle.

28. If equipped with a automatic transaxle, remove the shift cable cover, then disconnect the shift cable from the transaxle.

29. Remove the right strut fork bolt, discard the nut.

30. Remove the right strut pinch bolt, then remove the strut fork.

31. Disconnect the suspension lower arm ball joints using a ball joint remover.

32. Carefully pry the inner CV-joint away from the transaxle to force the set ring at the inner end past the groove. Remove the other CV-joint out of the intermediate shaft. Do not let the halfshafts hang down. Support the halfshafts or hang them from the body with wire and cover the halfshaft ends with plastic bags.

33. Attach a hoist to the engine.

34. Remove the left and right front mounts and brackets, then remove the rear mount bracket.

35. Remove the side engine mount, then remove the transmission mount.

36. Check that the engine is completely clear of vacuum hoses, fuel and engine coolant hoses, and electrical wiring.

37. Slowly raise the engine approximately 6 in. (150mm). Check one more time to be sure that all hoses and wires are disconnected from the engine.

38. Raise the engine and transaxle assembly all the way and remove it from the vehicle.

39. Separate the engine and transaxle.

To install:

40. Install the transaxle to the engine assembly. If equipped with a manual transaxle, tighten the transaxle housing mounting bolts to 47 ft. lbs. (64 Nm), the two bolts and new washers to the rear mounting bracket to 87 ft. lbs. (118 Nm), and tighten the upper mounting bolts to 47 ft. lbs. (64 Nm). If equipped with a automatic transaxle, tighten the transaxle housing mounting bolts to 43 ft. lbs. (59 Nm), the two bolts and new washers to the rear mounting bracket to 87 ft. lbs. (118 Nm), and tighten

the upper mounting bolts to 54 ft. lbs. (74 Nm).

41. If equipped with a automatic transaxle, tighten the bolts attaching the

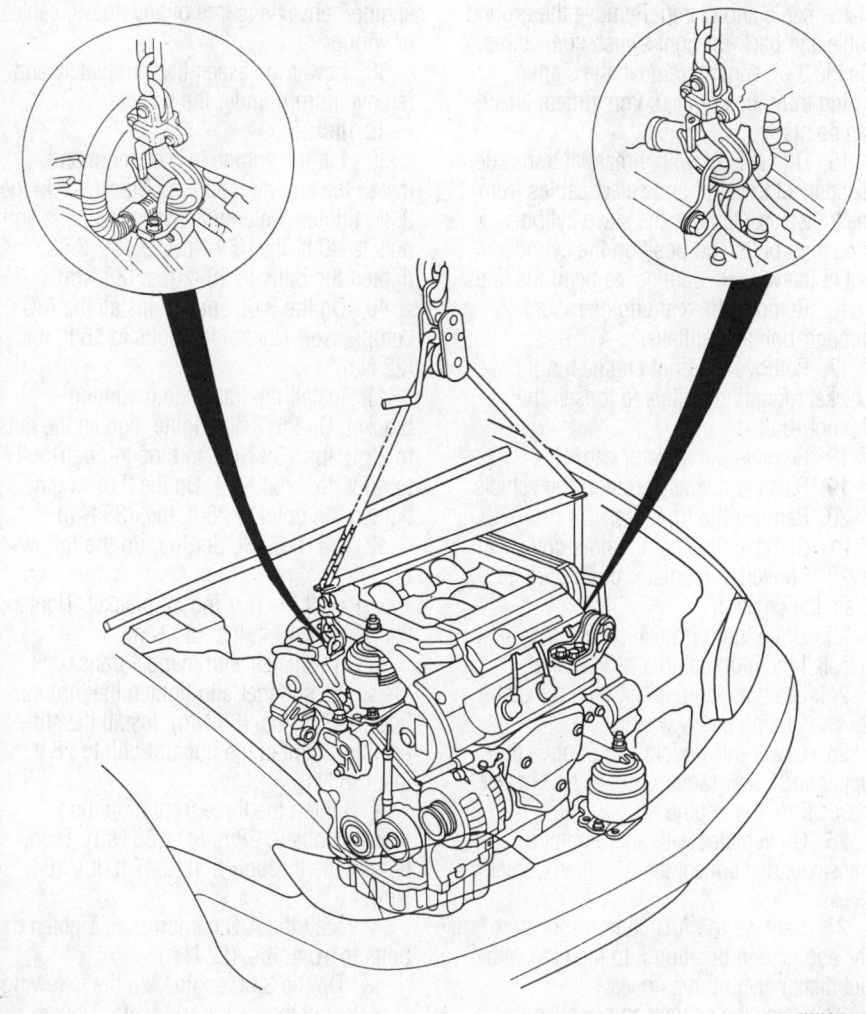

7923BG58

Engine lifting points—3.0CL

torque converter to the drive plate to 8.7 ft. lbs. (12 Nm).

42. Install the torque converter/clutch cover.

43. Install the rear engine stiffener, tighten the bolts attaching the stiffener to the engine to 17 ft. lbs. (24 Nm). If equipped with a manual transaxle, tighten the bolts attaching the stiffener to the transaxle to 42 ft. lbs. (57 Nm). If equipped with a automatic transaxle, tighten the bolts attaching the stiffener to the transaxle to 32 ft. lbs. (43 Nm).

44. If equipped with the VTEC (B18C1) engine, install the front engine stiffener. Tighten the bolt attaching the stiffener to the engine to 17 ft. lbs. (24 Nm), tighten the bolts attaching the stiffener to the transaxle to 42 ft. lbs. (57 Nm).

45. Install the engine and transaxle into the engine compartment. Install the transmission mount, then tighten the bolt/nuts on the transmission side. Leave the mount bolt loose.

46. Install the engine side mount, then tighten the bolt/nuts on the engine side. Leave the mount bolt loose. Tighten the mount bolt on the transmission mount, then tighten the mount bolt on the side engine mount.

47. Install the rear mount bracket, then tighten the bolts in the proper sequence.

48. Install the right front mount/bracket, then tighten the bolts in the proper sequence.

49. Install the left front mount, then tighten the bolts in the proper sequence.

50. The remaining components are installed in the reverse order of removal.

51. Reconnect the battery cables to the fuse/relay boxes and the battery. Connect the positive cable, then the negative cable to the battery.

52. Turn on the ignition switch (do not operate the starter) so that the fuel pump operates for approximately 2 seconds and the fuel line pressurizes. Repeat this operation 2 or 3 times and check for fuel leakage.

53. Enter the radio security code, then test drive the vehicle.

2.3L and 3.0L Engines

1. Obtain the anti-theft code for the radio, then disconnect the battery cables. Be sure to disconnect the negative cable first.

2. Remove the air intake duct.

3. Secure the hood in the open position with a long prop rod such as P/N 74145-S84-A00.

4. Detach both battery cables and the connector from the underhood relay box. On the 3.0L engine, remove the battery and tray.

5. Remove the bolt securing the relay box to the body.

6. Remove the accelerator and cruise control cables from the throttle body and bracket.

7. Properly relieve the fuel system pressure.

8. Detach the fuel hoses from the fuel rail.

9. Remove the following hoses:
- Brake booster vacuum
- EVAP canister
- Vacuum hose from the canister

10. Remove the hose securing the power steering hose on the engine.

11. Remove the power steering pump belt, then remove the pump and position it out of the way. Use wire if necessary.

12. Detach the ECM/PCM connectors from the control module. Remove the grommet and pull the connectors through.

13. Detach the wiring harness connectors at the right side of the engine compartment. Left side on the 3.0L engine.

14. On the 2.3L engine, remove the starter cable and clamp. Remove the ground cable and back-up light switch connectors. On the 3.0L engine, remove the starter wiring from the engine compartment attaching points.

15. On vehicles with a manual transaxle, disconnect the shift and select cables from the transaxle. Remove the slave cylinder mounting bolts and position the cylinder out of the way. Be sure not to bend the line.

16. Remove the rear engine mount through-bolt and stiffener.

17. Remove the front engine mount bracket mounting bolts and loosen the through-bolt.

18. Remove the radiator cap.

19. Raise and safely support the vehicle.

20. Remove the front tires.

21. Remove the engine under cover.

22. Loosen the radiator drain plug and drain the coolant.

23. Drain the transaxle oil or fluid, then reinstall the plug using a new washer.

24. Drain the engine oil, then reinstall the plug using a new washer.

25. Lower the vehicle and remove the upper and lower radiator hoses and heater hoses from the engine.

26. On vehicles with an automatic transaxle, disconnect the ATF fluid cooler lines.

27. Remove the A/C compressor from the engine and position it to the side without disconnecting the hoses.

28. Raise the vehicle and remove the front exhaust pipe.

29. Remove the two bolts for the shift cable holder, then remove the shift cable cover. To prevent damage to the linkage, be sure to remove the shift cable holder before removing the bolts for the cover.

30. Remove the lockbolt from the control lever, then remove the shift cable with the control lever.

31. Remove the through-bolt securing the bottom of the shock absorber to the control arm.

32. Remove the halfshafts.

33. Remove the rear engine mounting bracket.

34. Mark the location of the front beams on the rear beams. Remove the four bolts and the subframe.

35. Lower the vehicle about half way and attach a chain hoist to the engine lifting points as shown. Apply slight upward pressure to the engine/transaxle assembly.

36. Remove the remaining engine and transaxle mounting brackets.

37. Lower the engine about 6 in. (150mm) and check that the engine/transaxle is free of any hoses, cables or wiring.

38. Lower the assembly completely and remove it from under the vehicle.

To install:

39. Lift the engine into position and install the engine mounting brackets. On the 2.3L, tighten the engine mounting bolts and nuts to 40 ft. lbs. (54 Nm). On the 3.0L, tighten the bolts to 28 ft. lbs. (38 Nm).

40. On the 3.0L engine, install the A/C compressor. Tighten the bolts to 16 ft. lbs. (22 Nm).

41. Install the transaxle mounting bracket. On the 2.3L engine, tighten the nuts to 28 ft. lbs. (38 Nm) and the through-bolt to 40 ft. lbs. (54 Nm). On the 3.0L engine, tighten the bolts to 28 ft. lbs. (38 Nm).

42. On the 2.3L engine, do the following:
- Install the rear mount bracket. Tighten the bolts to 40 ft. lbs. (54 Nm).
- On vehicles with manual transaxles, install the stiffener and tighten the through-bolt to 47 ft. lbs. (64 Nm). Install the stiffener and tighten the nut and bolt to 28 ft. lbs. (38 Nm)
- Tighten the three front mounting bracket bolts to 28 ft. lbs. (38 Nm). Then, tighten the through-bolt to 47 ft. lbs. (64 Nm).
- Install the A/C compressor. Tighten the bolts to 16 ft. lbs. (22 Nm).

43. On the 3.0L engine, do the following:
- Install the radius rod bolts. Tighten them to 119 ft. lbs. (162 Nm).

- Install the front mounting bracket support nut. Tighten it to 40 ft. lbs. (54 Nm).
- Install the rear mounting bracket nut and bolt. Tighten the nut to 40 ft. lbs. (54 Nm) and the bolt to 28 ft. lbs. (38 Nm).
- Install the side mounting bracket. Tighten the bolts to 40 ft. lbs. (54 Nm) and the through-bolt to 40 ft. lbs. (54 Nm).

44. Assemble the exhaust system.

45. If equipped with an automatic transaxle, connect the shift linkage.

46. The remainder of the installation is the reverse of the removal.

47. Refill and bleed the cooling system.

❊❊ WARNING

Operating the engine without the proper amount and type of engine oil will result in severe engine damage.

48. Fill the engine with the correct amount of oil.

49. Install the battery if removed. Start the engine and check for leaks.

2.5L Engine

➡**A hydraulic lift is very helpful for removing this engine. The recommended engine removal and installation procedure requires the vehicle to be raised and lowered to unbolt the transmission mounts before the engine may be lifted out.**

1. On vehicles with automatic transmissions, be sure the selector is in **P**. On manual transmissions, put the selector into first gear.

2. Open the hood and secure it in its fully opened position (vertical). The hood may be removed if more clearance and working room is desired.

3. Disconnect the battery and remove it. Remove the battery tray.

4. Remove the splash shield.

5. Drain the engine oil and the differential and transmission oil.

6. Remove the engine ground cables and ignition coil wire.

7. Unbolt the ABS relay and move it out of the way. Remove the battery heat shield.

8. Remove the battery cables from the under-hood fuse/relay box.

9. Disconnect the engine harness connectors on the right side of the engine compartment.

10. Remove the intake air cleaner duct and the air cleaner housing.

11. Loosen the component adjusting and mounting bolts and remove the power steering pump and air conditioning compressor belts.

12. Without disconnecting the hoses, remove the power steering pump and the air conditioner compressor and secure them out of the way.

➡**Do not loosen or disconnect the air conditioning refrigerant lines.**

13. Loosen the service bolt on the fuel filter to relieve the fuel system pressure. Remove the banjo bolt to remove the fuel feed hose from the fuel filter. Remove the fuel return hose from the pressure regulator.

❊❊ CAUTION

The fuel injection system remains under pressure after the engine has been turned OFF. Properly relieve fuel pressure before disconnecting any fuel lines. Failure to do so may result in fire or personal injury.

14. Remove the throttle cable by loosening the locknut, then slip the cable end out of the throttle bracket and accelerator linkage. Do not bend the cable when removing it. Unbolt the throttle cable clamp and move the cable aside.

15. Label and disconnect the engine wiring harness connectors on the left side of the engine compartment.

16. Disconnect the charcoal canister hoses, fuel return hose, brake booster hose and the emission control vacuum hoses.

17. Disconnect the transmission wiring connector that is located near the firewall.

18. The distributor may be removed for extra access to the upper transmission case bolts. Disconnect the wiring and remove the two bolts to remove the distributor. Do not lose the collar that fits on the distributor shaft.

19. Remove the upper transmission bolts and the 26mm differential shim. Note the location of the shim for installation.

20. Remove the power steering speed sensor from the differential case without disconnecting the hydraulic hoses. Disconnect the wiring and secure the sensor out of the way.

21. Drain the engine coolant.

22. Disconnect the heater hoses.

23. Disconnect the transmission cooler hoses from the radiator tank. Disconnect the upper and lower radiator hoses and the fan wiring.

24. Remove the two upper brackets and lift the radiator and cooling fan assembly from the engine compartment.

25. Raise and support the vehicle safely.

26. On vehicles with automatic transmissions, remove the small torque converter cover. Rotate the crankshaft to remove the torque converter bolts.

27. Disconnect the halfshafts. The lower strut fork and ball joint must be disconnected so that the inner CV-joint can be separated from the intermediate shaft.

  a. Remove the front wheels. Remove the strut fork nut and pinch bolt. Remove the strut fork from the lower arm.

  b. Remove the lower ball joint castle nut. Press the ball joint out of the lower arm using a suitable puller.

  c. Carefully pry the inner CV-joint away from the transaxle to force the set ring at the inner end past the groove.

  d. Pull the inner CV-joint, not the halfshaft, and remove the CV-joint from the intermediate shaft.

➡**Do not pull on the halfshaft, the CV-joint may come apart. Use care when prying out the assembly and pull it straight to avoid damaging the intermediate shaft seals.**

  e. Hang the halfshafts from the body with wire. Do not let them hang from the outer CV-joint or it will be damaged.

28. Disconnect the oxygen sensor lead from the exhaust system and remove the front exhaust pipe and its brackets.

29. Remove the transmission side mount and bracket.

30. Be sure the transmission is in **P** or first gear. Remove the extension shaft sealing cap on the lower left side of transmission housing.

31. Use an extension shaft puller tool, Acura tool number 07LAC-PW50101, to disengage the extension shaft from the differential. The differential is removed with the engine.

32. Remove the transmission mid-mounts and the mid-mount spacer.

33. Support the transmission with a jack and remove the transmission case bolts.

34. Install the transmission mid-mounts and spacer to hold the transmission in the vehicle. Be sure the engine will separate from the transmission, and that the transmission will be supported by the mounts after the engine has been removed.

35. Lower the vehicle and install a chain hoist onto the engine lifting hooks.

36. Raise the hoist just enough to take up the weight of the engine.

37. Unbolt the front engine mounts.

38. If only the engine mounts need to be removed, it is not necessary to remove the engine. However, the weight of the engine and transaxle must be safely supported by a floor jack or engine hoist before the mounts may be removed.

39. Slowly raise the engine a few in. to separate it from the transmission. On man-

ual transmission vehicles, be sure the engine clears the mainshaft. Verify that all wiring harnesses, fuel and coolant lines, and vacuum hoses are disconnected.

40. Lift the engine out of the vehicle. Be sure the engine clears the mounts, the transmission case, and the differential extension shaft.

To install:

➡**Use new mounting bolts when installing the transmission side mount and bracket.**

41. Install the front engine mounts into the engine compartment.

42. Install the engine in the vehicle. Keep the lifting chain attached to hold the weight of the engine. Install the mounting nuts to hold the engine in place. Do not tighten the nuts and bolts at this time. Be sure the differential lines up with the extension shaft. Be sure the mainshaft is aligned in the clutch pressure plate, or the torque converter is flush against the drive plate and mounted on the mainshaft.

43. Raise the vehicle.

44. Install new snap and set rings onto the extension shaft. Apply high-temperature molybdenum grease to its splines and the shaft mating surface in the differential.

45. Support the transmission with a jack and remove the mid-mounts. Carefully fit the engine into position and start the

upper engine bolts. Slowly tighten two bolts on opposite sides just enough to draw the engine and transmission together. Install all the remaining bolts except for the differential bolt and its shim. Do not fully tighten the bolts yet.

46. If either the engine, transmission or differential is being replaced, the space between the differential and transmission housings must be measured and the correct shim installed. Shims are available in increments of 0.004 in. (0.1mm). Measure the space between the housings with a feeler gauge. Install the largest shim possible that does not exceed the measurement. If the wrong shim is installed, the differential or transmission housing will be out of alignment and could develop cracks.

47. Install the shim and tighten the bolts to 54 ft. lbs. (75 Nm). Tighten the transmission case bolts to 54 ft. lbs. (75 Nm).

48. Install the mid-mount and spacer. Loosely install the nuts and bolts for all the mounts and set the engine into place. Remove the engine lifting equipment.

49. Tighten the mounting nuts and bolts in the to the correct torque in the proper sequence. This step is important to preload the engine and transmission mounts. Fol-

lowing the proper sequence will minimize engine vibration and premature mount failure. Be sure the rubber strut mounting surface is not contaminated with oil.

- Left and right front mount nut: 54 ft. lbs. (75 Nm).
- Left front strut bolt: 28 ft. lbs. (39 Nm).
- Left front strut bracket bolt: 40 ft. lbs. (55 Nm).
- 2.5TL front mount-to-subframe bolts: 28 ft. lbs. (38 Nm).
- Mid-mount nuts: 32 ft. lbs. (43 Nm)
- Mid-mount bolts: 28 ft. lbs. (38 Nm)

50. After the engine and transmission have been bolted together, install the extension shaft. Be sure the set ring snaps firmly into place. Coat the threads of the 33mm sealing cap with a sealing compound, install the cap and tighten it to 58 ft. lbs. (80 Nm).

51. On vehicles with automatic transmissions, install the torque converter bolts and tighten them to 9 ft. lbs. (12 Nm). Do not over tighten these bolts or the drive plate will warp. Install the torque converter cover.

52. Install the transmission side mount and bracket. Tighten the bracket bolts to 40 ft. lbs. (54 Nm). Tighten the mount bolts to 47 ft. lbs. (64 Nm).

53. The balance of the installation is the reverse of the removal procedure.

54. Fill the engine with fresh oil. Refill the differential and transmission. Fill the radiator with a coolant mixture containing no more than 50–60 percent antifreeze.

55. Verify that all components have been reinstalled and connected properly. Check for loose or disconnected wires and lines.

56. Connect the negative and positive battery cables.

57. Bleed the cooling system. Check fluid levels. Run the engine and check its operation.

3.2L Engine

LEGEND

➡**The engine and transaxle are removed as an assembly.**

1. Do not remove the hood. Disconnect the hood stay strut and reconnect it to hold the hood in a vertical position.

2. Disconnect the negative battery cable, then the positive battery cable. Remove the battery and the battery box.

3. Remove the radiator cap.

4. Remove the strut bar and its bracket from the bulkhead.

5. Working underneath the vehicle,

remove the splash shield. Drain the engine coolant and oil, the transaxle fluid, and the differential oil.

6. Label and disconnect the starter and battery wiring from the main fuse/relay box. Remove the ground cable from the engine block. Label and disconnect the main engine wiring harness.

7. Remove the throttle cable cover. Without turning the adjusting nut, loosen the locknut, which is closer to the throttle, and disconnect the throttle cable from the throttle and bracket.

8. Remove the air cleaner assembly and intake duct.

9. Disconnect the igniter unit located on the right shock tower and remove the wiring harness clamp. Disconnect the engine ground cable.

10. On the firewall behind the right cylinder head is a control box containing emission control equipment. Unplug the electrical connectors and remove the control box from the firewall without disconnecting the vacuum lines. Lay the box on top of the engine.

11. Label and disconnect the main engine wiring harness connectors and remove the bracket.

12. Relieve the fuel system pressure by slowly loosening the service bolt one turn.

✳✳ CAUTION

The fuel injection system remains under pressure after the engine has been turned OFF. Properly relieve fuel pressure before disconnecting any fuel lines. Failure to do so may result in fire or personal injury.

13. Remove the fuel supply hose and disconnect the return hose from the pressure regulator.

14. Disconnect the vacuum hose to the brake booster at the check valve.

15. The transaxle wiring harness is located at the left rear of the engine compartment. Disconnect the harness and remove the clamp.

16. Disconnect and plug the transaxle cooling hoses at the radiator. Disconnect the upper and lower hoses and the fan and sensor wiring. Remove the radiator and fans as an assembly.

17. Remove the bypass solenoid valve assembly, vacuum pipes, and air tank. The valve assembly is located near the power steering reservoir.

18. On GS models with traction control, remove the TCS control valve and its mounting bracket. The valve assembly is

located on the right side of the engine compartment behind the air intake duct.

19. Remove the power steering pump without disconnecting the hoses, and secure it out of the way.

20. Raise and safely support the vehicle and remove the front wheels.

21. Remove the lower strut forks and remove the nut from the lower ball joint. Use a suitable ball joint removal tool to disconnect the ball joint from the lower control arm.

22. Carefully pry the inner CV-joints from the differential. Hang the halfshafts out of the way with wire to avoid damaging the outer CV-joint. Cover the inner joints with plastic bags to protect the splines.

23. Remove the steering rack lower cover plate from the rear beam.

24. Without disconnecting any hoses, remove the power steering speed sensor from the differential housing.

25. Disconnect the front exhaust pipe from the catalyst and remove it from the manifolds. On GS models, remove the twin warm-up catalytic converters

26. Leaving the lines connected, remove the air conditioning compressor and hang it from the body with wire.

➡**Do not loosen or disconnect the air conditioning refrigerant lines. Do not vent refrigerant into the air.**

27. On vehicles with manual transaxles, remove the clutch slave cylinder without disconnecting the hydraulic line. Disconnect the shift lever torque rod and disconnect the shift linkage by driving out the 8mm roll pin.

28. On vehicles with automatic transaxles, remove the converter heat shields. Unbolt the shift cable from its mounting bracket. Disconnect the shift cable from the input shaft and hang it from the underbody of the vehicle.

29. Remove the engine mid-mounts and the transaxle rear mount and bracket.

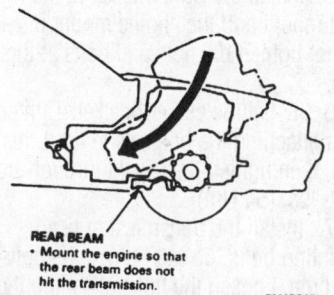

REAR BEAM
• Mount the engine so that the rear beam does not hit the transmission.

7923BG05

Tilt the rear of the transmission downward while installing it in the engine compartment—Legend

30. Working from above, remove one of the EGR valve passage bolts and install a lifting hook. Attach a chain hoist and take up the slack.

31. Remove all engine and transaxle mounting nuts and bolts. Raise the engine/transaxle slightly and check to be sure that all hoses and wires have been disconnected. As the unit is raised, allow it to tilt up so the transaxle will clear the rear beam.

32. The engine and transaxle mounts can be unbolted, then removed from the engine compartment and engine assembly at this time.

33. If removing and replacing the engine mounts, it is not necessary to remove the engine. However, the weight of the engine must be off of the mount and firmly supported by an engine hoist or floor jack before performing this procedure.

To install:

➡**Portions of the sub-frame are made of aluminum alloy. Using normal steel bolts will cause corrosion and looseness. Only use bolts that are specifically designed for this application. These parts are available from the dealer.**

34. When installing the engine/transaxle, rotate the unit up in front to avoid hitting the rear beam with the transaxle.

35. Check carefully to be sure the rubber mounts are not twisted or offset. Start all the mounting nuts and bolts before tightening them. This step is important to help minimize engine vibrations.

36. Tighten the front engine mounts to 28 ft. lbs. (39 Nm). The strut mounting bolts are tightened to 16 ft. lbs. (22 Nm).

37. Tighten the mid mount to subframe nut to 35 ft. lbs. (49 Nm).

38. Tighten the rear transaxle mount and bracket bolts to 28 ft. lbs. (39 Nm).

39. Remove the lifting hook and install the EGR valve passage bolt. Be sure the rubber strut mounting surface is not contaminated with oil.

40. On vehicles with manual transaxles, reconnect the shift linkage and install the clutch slave cylinder.

41. On vehicles with automatic transaxles, reconnect the shift cable and its mounting bracket. Install the heat shield.

42. When installing the exhaust pipe and catalytic converter, use new gaskets and self-locking nuts. Tighten the manifold nuts to 40 ft. lbs. (55 Nm) and the catalyst flange nuts to 16 ft. lbs. (22 Nm).

43. The balance of the removal is the reverse of the installation. Refer to the appropriate portions of this section for tightening torque specifications needed for the various components to be installed.

44. Verify that all engine wiring, fluid lines, and vacuum lines have been connected properly.

45. When the battery is connected, turn the ignition switch **ON** and **OFF** a number of times to pressurize the fuel system and check for leaks.

46. After checking carefully and making sure everything is properly connected, start the engine and bleed the cooling system. When the engine is at operating temperature, stop the engine and adjust the throttle cable.

47. To adjust the throttle cable, loosen both nuts and take up the slack in the cable. Back the adjusting nut away from the bracket so there is 0.120 in. (3.0mm) gap between the nut and bracket. Be sure the throttle opens and closes fully with pedal movement.

48. Start the engine and check for leaks and proper operation.

3.2TL

➡**The engine and transaxle are removed as an assembly.**

1. Move the front passenger's seat forward.

2. Do not remove the hood. Disconnect the hood support strut and reconnect it to hold the hood in a vertical position.

➡**The radio may contain a coded theft protection circuit. Always obtain the code number before disconnecting the battery.**

3. Disconnect the negative battery cable, then the positive battery cable. Remove the battery and the battery box.

4. Remove the engine cover.

5. Remove the air cleaner assembly and intake duct.

6. Remove the throttle cable cover. Without turning the adjusting nut, loosen the locknut, which is closer to the throttle, and disconnect the throttle cable and cruise control cable from the throttle and bracket.

7. Disconnect the engine wiring harness connector on the left side of the engine compartment.

8. Remove the engine ground cable and engine wiring harness clamps.

9. Disconnect the vacuum hoses, then remove the clamp from the under-hood fuse/relay box.

10. Label, then disconnect the battery cables from the under-hood fuse/relay box,

then remove the under-hood fuse/relay box.

11. Disconnect the engine wiring harness connector, located by the under-hood fuse/relay box.

12. Raise the power steering fluid reservoir, then disconnect the vacuum hoses and remove the vacuum pipe and vacuum tank.

➡ **Do not disconnect the power steering hoses.**

13. Disconnect the ignition control module (igniter) located on the right shock tower and remove the wiring harness clamp.
Disconnect the engine ground cable.

14. Disconnect the engine wiring harness connectors on the right side of the engine compartment.

15. Remove the ground cable and wiring harness clamp.

16. Disconnect the wiring harness from the control box and solenoid valve, then remove the control box.

17. Disconnect the brake booster vacuum hose.

18. Remove the two bolts mounting the heater valve.

✳✳ CAUTION

The fuel injection system remains under pressure after the engine has been turned OFF. Properly relieve fuel pressure before disconnecting any fuel lines. Failure to do so may result in fire or personal injury. Do not allow fuel spray or fuel vapors to come in contact with a spark or open flame. Keep a dry chemical fire extinguisher nearby. Never store fuel in an open container due to risk of fire or explosion.

19. Properly relieve the fuel pressure.

20. Disconnect the engine fuel feed hose from the fuel filter and disconnect the fuel return hose from the fuel regulator.

21. Disconnect the evaporative emissions (EVAP) control canister hose and the vacuum hose.

22. Disconnect the transaxle sub-harness connector, and remove the wiring harness clamp.

23. Loosen the alternator mounting bolt, lockbolt and adjusting rod, then remove the drive belt.

24. Loosen the A/C idler pulley center nut and adjusting bolt, then remove the drive belt.

25. Disconnect the power steering pressure switch connector.

26. Remove the power steering pump

adjusting bolt, locknut and mounting bolt, then remove the drive belt and pump.

27. Pull the carpet back under the passenger seat to expose the secondary heated oxygen sensor (HO2S) connector, then disconnect the connector.

28. Remove the radiator cap.

29. Raise and safely support the vehicle.

30. Remove the front wheels and the splash guard.

31. Drain the engine coolant into a sealable container.

32. Drain the transaxle fluid into a proper container, then install the drain plug with a new washer.

33. Drain the oil from the differential, then install the drain plug with a new washer.

34. Drain the engine oil into a proper container, then install the drain plug with a new washer.

35. Remove the front suspension strut forks.

36. Disconnect the lower ball joints from the steering knuckles.

37. Disconnect the halfshafts from the differential and the intermediate shaft. Support the halfshafts with wire out of the way and cover the inner CV-joints with plastic bags.

38. Disconnect the A/C compressor clutch connector. Remove the compressor, without disconnecting the hoses.

39. Disconnect the vehicle speed sensor connector, then remove the VSS/power steering sensor. Do not disconnect the fluid hoses.

40. Remove the heat shields from the front exhaust pipe.

41. Remove the nuts attaching the front exhaust pipe to the exhaust manifolds and the catalytic converter. Remove the front exhaust pipe.

42. Remove the oxygen sensor wiring harness cover and grommet, then remove the catalytic converter. Discard the nuts and gasket.

43. Remove the exhaust heat shield from the floor of the vehicle.

44. Disconnect the transaxle cooler hoses, then plug the hoses and pipes.

45. Remove the shift cable cover mounting bolts and remove the wiring harness clamps from the cover. Remove the shift cable cover from the transaxle.

46. Remove the shift cable holder from the holder base, do not lose the washers.

47. Remove the locknut attaching the shift cable to the control lever, then remove the shift cable.

48. Lower the vehicle.

49. Remove the upper and lower radiator hoses.

50. Remove the radiator assembly.

51. Remove the heater hoses.

52. Attach a hoist to the engine lifting points.

53. Remove the center bracket from the front engine mounts.

54. Remove the center mount from the front beam.

55. Remove the nuts and bolts attaching the left and right engine mount brackets to the left and right brackets.

56. Working under the vehicle, Remove the shift cable guide bracket.

57. Remove the bolts attaching the transmission beam to the body, and loosen the three bolts on the transmission beam.

58. Remove the stop holder, the mid mount stops and the mid mounts.

59. Verify that the engine and transaxle assembly is completely free of vacuum hoses, fuel and coolant lines and electrical wiring.

60. Slowly raise the engine and transaxle. Remove the left end right brackets from the front engine mounts.

61. Raise the engine all of the way and remove it from the vehicle. Separate, then engine and the transaxle assembly.

To install:

62. Carefully install the engine and transaxle into the engine compartment, take care to not hit the rear beam.

➡ **Check carefully to be sure that the rubber mounts are not twisted or offset. Start all of the mount nuts and bolts before tightening them. This is important to help minimize engine vibrations.**

63. Install the center mount to the front beam and tighten the bolts to 40 ft. lbs. (54 Nm).

64. Install the left bracket to the left front mount and the engine mount bracket. Do not tighten the nuts and bolts at this time.

65. Install the right bracket to the right front mount and the engine mount bracket. Do not tighten the nuts and bolts at this time.

66. Install the center bracket. Tighten the bolts attaching the brackets to 40 ft. lbs. (54 Nm), then tighten the mount through-bolt to 40 ft. lbs. (54 Nm).

67. Install the transmission beam mounting bolts, do not tighten the bolts at this time. Loosen the bolts attaching the mount to the beam and the mount through-bolt.

68. Install the mid mounts, then the mid

mount stops. Tighten the 8mm bolts loosely, then install the stop holder.

69. Tighten the mid mount 10mm bolts to 28 ft. lbs. (38 Nm) and tighten the 8mm bolts to 16 ft. lbs. (22 Nm). Tighten the new nuts attaching the mid-mounts to 35 ft. lbs. (48 Nm) and tighten the nuts attaching the stop holder to 40 ft. lbs. (54 Nm).

70. Tighten the nut and bolt attaching the left bracket and the left engine mount bracket to 28 ft. lbs. (38 Nm). Tighten the left bracket through-bolt to 40 ft. lbs. (54 Nm).

71. Tighten the nut and bolt attaching the right bracket and the right engine mount bracket to 28 ft. lbs. (38 Nm). Tighten the right bracket through-bolt to 40 ft. lbs. (54 Nm).

72. Tighten the bolts attaching the transmission beam to the vehicle to 28 ft. lbs. (38 Nm), then tighten the bolts attaching the mount to the beam to 40 ft. lbs. (54 Nm). Tighten the three bolts attaching the bracket to the transaxle to 28 ft. lbs. (38 Nm), then tighten the mount through-bolt to 40 ft. lbs. (54 Nm).

73. Remove the engine hoist.

74. Connect the heater hoses to the heater core at the bulkhead.

75. Install the radiator assembly.

76. Install the upper and lower radiator hoses.

77. Raise and safely support the vehicle.

78. Connect the shift cable control lever to the control shaft, then install the washer and nut. Tighten the nut to 8.7 ft. lbs. (12 Nm).

79. Install the shift cable holder to the shift cable holder base with the mounting washers. Install the attaching bolts and tighten the bolts to 8.7 ft. lbs. (12 Nm).

80. Install the shift cable cover and tighten the mounting bolts to 8.7 ft. lbs. (12 Nm).

81. The balance of the removal is the reverse of the installation.

82. Fill the engine with engine oil.

83. Fill the transmission with transmission oil.

84. Connect the positive, then the negative battery cables and enter the radio security code.

85. Switch the ignition **ON** but do not engage the starter. The fuel pump should run for approximately 2 seconds, building pressure within the lines. Switch the ignition **OFF**, then **ON** 2 or 3 more times to build full system pressure. Check for fuel leaks.

86. Fill the cooling system and bleed the air from the cooling system.

87. Run the engine and check for leaks, check the transmission oil level and add if necessary.

3.5L Engine

➡**The engine and transaxle are removed as an assembly.**

1. Move the front passenger's seat forward.

2. Do not remove the hood. Disconnect the hood support strut and reconnect it to hold the hood in a vertical position.

➡**The radio may contain a coded theft protection circuit. Always obtain the code number before disconnecting the battery.**

3. Disconnect the negative battery cable, then the positive battery cable.

4. Remove the strut brace.

5. Remove the engine cover.

6. Remove the air cleaner assembly and intake duct.

7. Remove the throttle cable cover. Without turning the adjusting nut, loosen the locknut, which is closer to the throttle, and disconnect the throttle cable and cruise control cable from the throttle and bracket.

8. Raise the coolant reservoir, then remove the battery and tray.

9. Remove the relay box, ground cable and wiring harness clips from the firewall.

10. Disconnect the vacuum hoses, then remove the clamp from the under-hood fuse/relay box.

11. Disconnect the engine wiring harness connector on the left side of the engine compartment.

12. Properly relieve the fuel system pressure.

13. Remove the fuel feed hose from the fuel filter. Disconnect the fuel return hose.

14. Remove the EVAP canister and brake booster hoses.

15. Disconnect the transmission sub-harness connector.

16. Detach the connector and remove the control box.

17. Detach the wiring harness connectors on the right side of the engine compartment.

18. Disconnect the spark plug voltage detection module and remove the engine ground cables.

19. Remove the accessory drive belts.

20. Pull the carpet back under the front passengers seat and detach the secondary heated oxygen sensor connector.

✳✳ CAUTION

Never open, service or drain the radiator or cooling system when hot; serious burns can occur from the steam and hot coolant.

21. Remove the radiator cap.

22. Raise and safely support the vehicle.

23. Drain the engine coolant, transmission fluid, differential and engine oil.

24. Disconnect the lower ball joints and remove the halfshafts.

25. Disconnect the A/C compressor clutch connector. Remove the compressor, without disconnecting the hoses.

26. Disconnect and remove the VSS.

27. Remove the transmission stop collars.

28. Remove the nuts attaching the front exhaust pipe to the exhaust manifolds and the catalytic converter. Remove the front exhaust pipe.

29. Remove the heat shield.

30. Disconnect the ATF cooler hoses

31. Remove the shift cable cover, shift control solenoid valve/linear solenoid harness connector from the shift cable cover.

32. Disconnect the shift cable from the transmissions.

33. Remove the control lever from the control shaft.

34. Lower the vehicle to the floor.

35. Remove the radiator hoses and the radiator.

36. Disconnect the heater hoses.

37. Attach a chain hoist to the engine.

38. Raise and safely support the vehicle.

39. Remove the shift cable guide and the transmission beam.

40. Remove the transmission mount and bracket.

41. Lower the vehicle.

42. Separate the left and right front mount brackets from the mounts.

43. Remove the nuts from the right and left engine mounts.

44. Raise the engine slightly, be sure all connections have be removed.

45. Remove the engine/transmission from the vehicle.

To install:

46. Position the engine/transmission in the vehicle.

47. Install the transmission mount bracket. Tighten the bolts to 28 ft. lbs. (38 Nm).

48. Install the nuts on the right and left engine mounts. Tighten them to 47 ft. lbs. (64 Nm).

49. Install the transmission beam.

Bracket Bolts Torque Specifications:

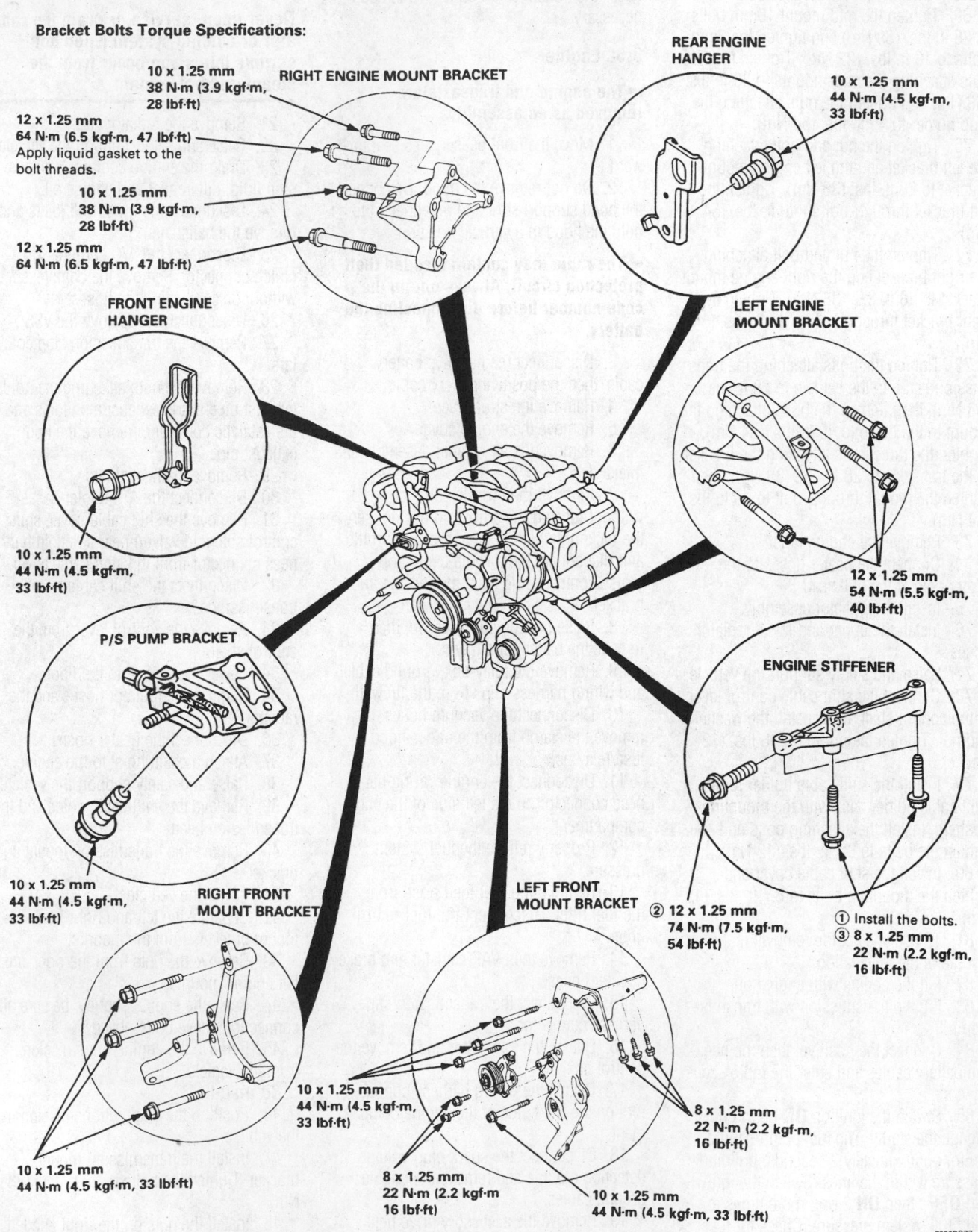

RIGHT ENGINE MOUNT BRACKET

10 x 1.25 mm
38 N·m (3.9 kgf·m, 28 lbf·ft)

12 x 1.25 mm
64 N·m (6.5 kgf·m, 47 lbf·ft)
Apply liquid gasket to the bolt threads.

10 x 1.25 mm
38 N·m (3.9 kgf·m, 28 lbf·ft)

12 x 1.25 mm
64 N·m (6.5 kgf·m, 47 lbf·ft)

REAR ENGINE HANGER

10 x 1.25 mm
44 N·m (4.5 kgf·m, 33 lbf·ft)

FRONT ENGINE HANGER

10 x 1.25 mm
44 N·m (4.5 kgf·m, 33 lbf·ft)

LEFT ENGINE MOUNT BRACKET

12 x 1.25 mm
54 N·m (5.5 kgf·m, 40 lbf·ft)

P/S PUMP BRACKET

10 x 1.25 mm
44 N·m (4.5 kgf·m, 33 lbf·ft)

ENGINE STIFFENER

② 12 x 1.25 mm
74 N·m (7.5 kgf·m, 54 lbf·ft)

① Install the bolts.
③ 8 x 1.25 mm
22 N·m (2.2 kgf·m, 16 lbf·ft)

RIGHT FRONT MOUNT BRACKET

10 x 1.25 mm
44 N·m (4.5 kgf·m, 33 lbf·ft)

10 x 1.25 mm
44 N·m (4.5 kgf·m, 33 lbf·ft)

LEFT FRONT MOUNT BRACKET

10 x 1.25 mm
44 N·m (4.5 kgf·m, 33 lbf·ft)

8 x 1.25 mm
22 N·m (2.2 kgf·m, 16 lbf·ft)

8 x 1.25 mm
22 N·m (2.2 kgf·m, 16 lbf·ft)

10 x 1.25 mm
44 N·m (4.5 kgf·m, 33 lbf·ft)

7923BG77

View of the engine mounting bracket showing torque specifications—3.5RL

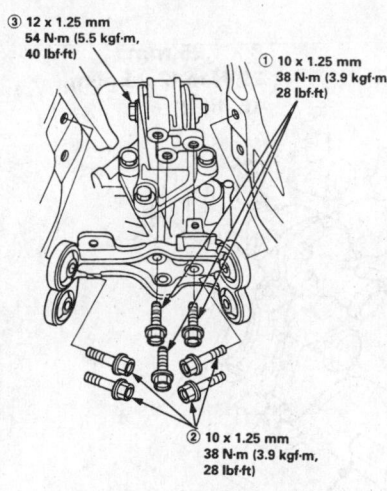

③ 12 x 1.25 mm
54 N·m (5.5 kgf·m,
40 lbf·ft)

① 10 x 1.25 mm
38 N·m (3.9 kgf·m,
28 lbf·ft)

② 10 x 1.25 mm
38 N·m (3.9 kgf·m,
28 lbf·ft)

7923BG78

Transmission beam bolt tightening sequence and torque specifications— 3.5RL

Tighten the bolts in the sequence shown to the correct specification.

50. Install the bolts in the left and right front mounts. Tighten them to 52 ft. lbs. (74 Nm).

51. Install and connect the remaining components.

✳✳ WARNING

Operating the engine without the proper amount and type of engine oil will result in severe engine damage.

52. Refill all fluids.
53. Start the engine and check for leaks.

Water Pump

REMOVAL & INSTALLATION

➡ The radio may have a coded theft protection circuit. Obtain the code from the owner before disconnecting the battery, removing the radio fuse, or removing the radio.

1.8L Engines

1. Disconnect the negative battery cable.
2. If applicable, remove the front under panel.
3. Gradually release the system pressure by slowly and carefully removing the radiator cap. Be sure to protect your hands with gloves or a shop rag.
4. Drain the engine coolant into a sealable container.
5. Remove the timing belt from the engine.

6. Remove the camshaft pulleys and remove the back cover.
7. Remove the five water pump mounting bolts and remove the water pump.
8. Remove and discard the old O-ring.
9. Remove the dowel pins from the oil water pump.
10. Clean the O-ring groove and the water pump mounting surface on the engine.

To install:
11. Install the dowel pins to the new water pump.
12. Position a new O-ring to the new water pump. Apply a small amount of sealant to the O-ring to hold it in position.
13. Place the new water pump on the engine and install the mounting bolts. Tighten the mounting bolts to 8.7 ft. lbs. (12 Nm).
14. Install the back cover and the camshaft pulleys.
15. Install the timing belt.
16. Fill the engine with coolant and bleed the air from the cooling system.
17. Connect the negative battery cable and enter the radio security code.
18. Run the engine and check for cooling system leaks.

2.5L and 3.2L Engines

➡ **Perform this service operation with the engine cold.**

1. Disconnect the negative battery cable.
2. Remove the front splash panel and release the system pressure by slowly removing the radiator cap.
3. Drain the cooling system.
4. Remove the timing belt. Inspect the timing belt for any signs of damage or oil and coolant contamination. Replace the timing belt if there is any doubt about its condition.
5. If extra clearance is required, remove the camshaft pulleys and the timing belt rear cover.
6. Remove the two mounting bolts from the thermostat housing.
7. Remove the water pump bolts. Then, remove the water pump and sprocket assembly from the engine block. Remove the O-rings from the water passage.

To install:
8. Before installation, be sure all gasket and O-ring groove surfaces are clean.
9. Install the water pump with a new O-ring. Use liquid gasket if it was present on

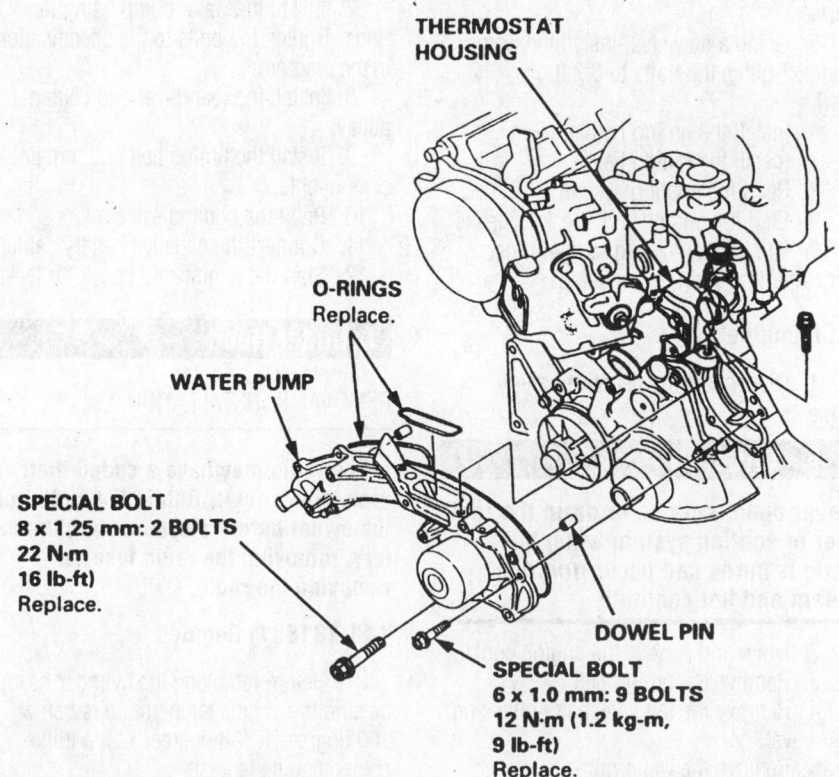

THERMOSTAT HOUSING

O-RINGS
Replace.

WATER PUMP

SPECIAL BOLT
8 x 1.25 mm: 2 BOLTS
22 N·m
16 lb-ft)
Replace.

DOWEL PIN

SPECIAL BOLT
6 x 1.0 mm: 9 BOLTS
12 N·m (1.2 kg-m,
9 lb-ft)
Replace.

7923BG06

Water pump—Legend

the sealing surfaces of the water pump that was removed. Use new 6mm mounting bolts and evenly tighten them to 9 ft. lbs. (12 Nm). Use new 8mm bolts and tighten them to 16 ft. lbs. (22 Nm).

10. Install the timing belt rear cover and camshaft pulleys. Tighten the pulley bolts for 2.5L (G25A1, G25A4) engines to 33 ft. lbs. (44 Nm). For 3.2L (C32A1) engines, tighten the pulley bolts to 23 ft. lbs. (32 Nm).

11. Install the thermostat housing and the two mounting bolts. Use a new O-ring.

12. Install the timing belt and timing belt covers.

13. Install and adjust the tension of the accessory drive-belts.

14. Close the cooling system drain plug. Refill and bleed the cooling system.

15. Connect the negative battery cable.

16. Start the engine, allow it to reach normal operating temperature, and check for leaks.

3.0L Engine

1. Remove the timing belt.
2. Remove the timing belt tensioner.
3. Remove the five water pump mounting bolts, then remove the pump and seal.

To install:

4. Clean the seal groove and mating surfaces.

5. Using a new seal, install the water pump. Tighten the bolts to 8.7 ft. lbs. (12 Nm).

6. Install the timing belt tensioner.
7. Install the timing belt.
8. Refill the cooling system.
9. Start the engine and check for leaks.
10. Top off the cooling system if necessary after the engine has cooled.

3.5L Engine

1. Disconnect the negative battery cable.

❊❊ CAUTION

Never open, service or drain the radiator or cooling system when hot; serious burns can occur from the steam and hot coolant.

2. Drain and recycle the engine coolant.
3. Remove the timing belt.
4. Remove the left camshaft pulley and rear cover.
5. Remove the water pump.

To install:

6. Clean the mounting surface and the O-ring grooves.

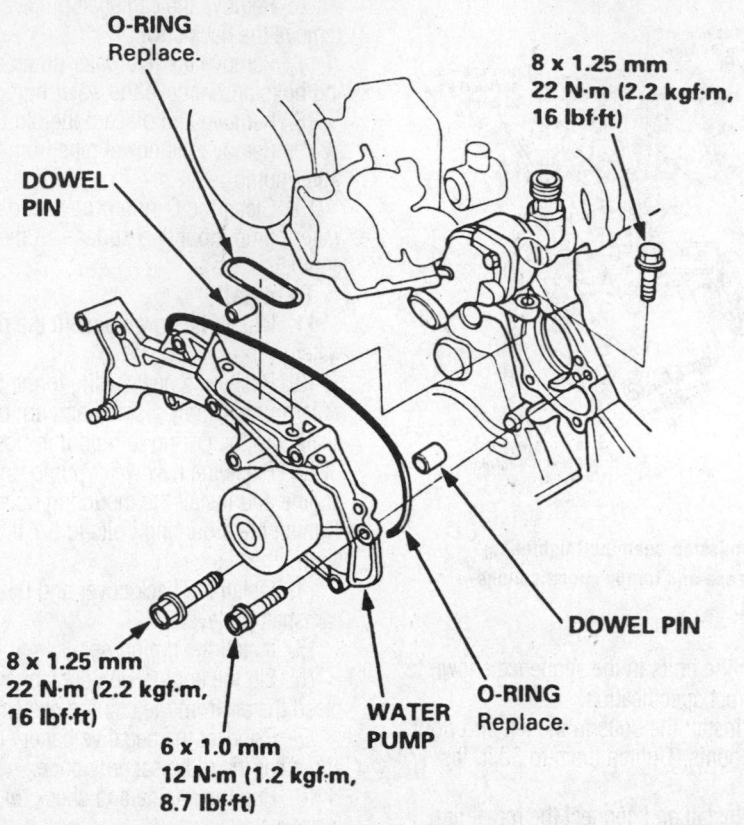

Water pump mounting and bolt torque specifications—3.5L engine

7. Install the water pump using new O-rings. Tighten the bolts to the specification in the diagram.

8. Install the rear cover and camshaft pulley.

9. Install the timing belt and remaining components.

10. Refill the cooling system.
11. Connect the negative battery cable.
12. Start the engine and check for leaks.

Cylinder Head

REMOVAL & INSTALLATION

➡**The radio may have a coded theft protection circuit. Obtain the code from the owner before disconnecting the battery, removing the radio fuse, or removing the radio.**

1.8L (B18B1) Engines

1. Before removing the cylinder head, be sure the engine temperature is below 100 degrees F. (38 degrees C.); a fully cooled engine is best.

2. Disconnect the negative battery cable.

3. Be sure the crankshaft is at TDC on

No. 1 cylinder by aligning the white mark on the crankshaft pulley with the pointer on the lower timing belt cover.

4. Drain the engine coolant. Remove the radiator cap to speed draining.

5. Remove the intake air duct.

6. Relieve the fuel pressure. Exercise proper safety precautions.

❊❊ CAUTION

Do not allow fuel spray or vapors to come in contact with a spark or open flame. Never store fuel in an open container due to risk of fire or explosion.

7. Disconnect the fuel feed hose. Be sure to mark all connectors and vacuum hoses before disconnecting them.

8. Disconnect the breather hose, water bypass hose and the evaporative emission control canister hose.

9. Remove the PCV hose and the fuel return hose.

10. Remove the brake booster vacuum hose, water bypass and EVAP (Evaporative emissions) purge control solenoid vacuum hose.

11. Remove the throttle cable. Remove the throttle control cable (automatic

transaxle only). Be careful not to bend the cables when removing them.

12. Remove the wiring harness clamps and disconnect the following:
- Four fuel injector connectors
- Intake Air Temperature (IAT) sensor connector
- Engine Coolant Temperature (ECT) sensor connector
- TDC/CKP/CYP sensor connector
- Ignition coil connector
- ECT gauge sending unit connector
- Throttle Position (TP) sensor connector
- Manifold Absolute Pressure (MAP) sensor connector
- Idle Air Control (IAC) valve connector
- EVAP purge control solenoid valve connector
- Crankshaft Speed Fluctuation (CKF) sensor connector, if equipped

13. Remove the upper radiator hose, heater hose and water bypass hose.

14. Remove the splash shield.

15. Remove the power steering adjusting and mounting bolts, then remove the power steering pump and belt. Do not disconnect the power steering hoses.

16. Remove the air conditioning compressor and alternator belts, then remove the cruise control actuator.

17. Remove the engine side mount.

18. Remove the cylinder head cover, timing belt cover, and timing belt.

19. Remove the camshaft pulleys and back cover.

20. Remove the exhaust manifold cover, bracket, and exhaust manifold.

21. Remove the bolts attaching the intake manifold to the support bracket.

22. Remove the nuts attaching the intake manifold to the cylinder head. Remove the nuts in a crisscross pattern, beginning from the center and moving out to both ends.

23. Remove the manifold and the old gasket.

24. Loosen the locknuts and adjusting screws, then remove the camshaft holder bolts. Remove the camshaft holders, camshafts and rocker arms.

25. Remove the cylinder head bolts, then remove the cylinder head. To prevent warpage, unscrew the bolts in the reverse of the torque sequence, ⅓ turn at a time. Repeat the sequence until all bolts are loosened.

To install:

26. Install the cylinder head onto the engine block, after making sure the mating surface was cleaned and a new gasket was installed. Be sure to pay attention to the following points:

- Be sure the No. 1 cylinder is at top dead center and the camshaft pulley UP mark is on the top before positioning the head in place.
- The cylinder head dowel pins and oil control orifice must be cleaned and aligned.
- Replace the washer when damaged or deteriorated.
- Apply engine oil to the cylinder head bolts and the washers.
- Use the longer cylinder head bolts at the No. 1 and No. 2 positions.

27. Tighten the cylinder head bolts in two steps. In the first step tighten all bolts in sequence to 22 ft. lbs. (29 Nm), then in the second step tighten all bolts in the same sequence to 63 ft. lbs. (85 Nm).

28. Use new gaskets and install the intake manifold onto the cylinder head and tighten the nuts in a crisscross pattern in 2–3 steps, beginning in the middle. Tighten the nut nuts to 17 ft. lbs. (23 Nm).

29. Install and tighten the intake manifold bracket bolts to 17 ft. lbs. (24 Nm).

30. Install the exhaust manifold and tighten the new self-locking nuts in a crisscross pattern in 2–3 steps, beginning with the inner nuts. Tighten the nuts to 23 ft. lbs. (31 Nm). Install a new exhaust pipe gasket and tighten the new nuts to 40 ft. lbs. (54 Nm).

31. Be sure that the keyways on the camshafts are facing up and that the rocker arms are in their original position. The valve locknuts should be loosened and the adjusting screw backed off before installation.

32. Place the rocker arms on the pivot bolts and the valve stems.

33. Install the camshafts, then install the camshaft seals with the open side facing in. Install the rubber cap with liquid gasket applied. If the rubber cap has two horizontal marks, align the marks with the cylinder head upper surface.

34. Apply liquid gasket to the head of the

mating surfaces of the No. 1 and No. 6 camshaft holders, then install them, along with No. 2, 3, 4, and 5. Be sure to pay attention to the following points:

- "I" or "E" marks are stamped on the camshaft holders.
- Do not apply oil to the holder mating surface of camshaft seals.
- The arrows marked on the camshaft holders should point to the timing belt.

35. Tighten the camshaft holders temporarily. Be sure that the rocker arms are properly positioned on the valve stems.

36. Tighten each bolt in 2 steps to ensure that the rockers do not bind on the valves. Tighten the bolts to 7 ft. lbs. (10 Nm) working from the middle outward.

37. Install the keys into the camshaft grooves. To set the No. 1 piston at TDC, align the holes on the camshaft with the holes in the No. 1 camshaft holders and insert 5.0mm pin punches into the holes.

38. Install the back cover and push the camshaft pulleys onto the camshafts, then tighten the retaining bolts to 27 ft. lbs. (37 Nm). Install the timing belt and adjust the tension, then install the timing belt covers.

39. Adjust the valve clearance.

40. Install the cylinder head cover.

41. Install the engine side mount, tighten the two new nuts and new bolt to the engine to 38 ft. lbs. (52 Nm) and tighten the bolt attaching the mount to the vehicle to 54 ft. lbs. (74 Nm).

42. The balance of the removal is the reverse of the installation. If necessary, refer to the appropriate portion of this section for information on the various components to be installed.

43. Connect the negative battery cable and enter the radio security code.

44. After installation, check to see that all hoses and wires are installed correctly.

45. Fill and bleed the air from the cooling system.

1.8L (B18C1) Engines

1. Before removing the cylinder head, be sure the engine temperature is below 100° F degrees; a fully cooled engine is best.

2. Disconnect the negative battery cable.

3. Be sure the crankshaft is at TDC/compression on No. 1 cylinder. Align the white mark on the crankshaft pulley with the pointer on the lower timing belt cover.

4. Drain the engine coolant into a sealable container. Remove the radiator cap to speed draining.

5. Remove the strut brace.

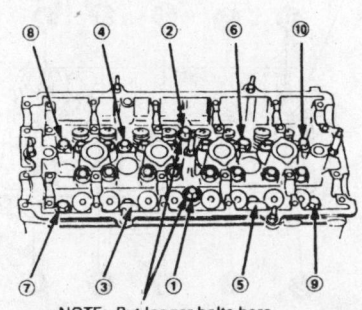

NOTE: Put longer bolts here.

7923BG07

Cylinder head bolt torque sequence—1.8L (B18B1) engines

6. Remove the intake air duct.
7. Relieve the fuel pressure. Exercise proper safety precautions.

CAUTION

Do not allow fuel spray or vapors to come in contact with a spark or open flame. Never store fuel in an open container due to risk of fire or explosion.

8. Disconnect the fuel feed hose.
9. Be sure to mark all connectors and vacuum hoses before disconnecting them. Disconnect the EVAP (Evaporative emissions) purge control solenoid vacuum hose and the EVAP control canister hose.
10. Remove the PCV hose and the water bypass hose.
11. Remove the brake booster vacuum hose, and the fuel return hose.
12. Remove the throttle cable. Remove the throttle control cable (automatic transaxle only). Be careful not to bend the cables when removing them.
13. Remove the wiring harness clamps and disconnect the following:
   - Four fuel injector connectors
   - Intake Air Temperature (IAT) sensor connector
   - Engine Coolant Temperature (ECT) sensor connector
   - TDC/CKP/CYP sensor connector
   - Ignition coil connector
   - ECT gauge sending unit connector
   - Throttle Position (TP) sensor connector
   - VTEC solenoid connector
   - VTEC pressure switch connector
   - Manifold Absolute Pressure (MAP) sensor connector
   - Idle Air Control (IAC) valve connector
   - EVAP purge control solenoid valve connector
   - Intake Air Bypass (IAB) control solenoid valve connector
   - Crankshaft Speed Fluctuation (CKF) sensor connector, if equipped
14. Remove the spark plug wires and distributor from the cylinder head.
15. Disconnect the upper radiator hose, heater hose and water bypass hose.
16. Remove the splash shield.
17. Remove the engine ground cable.
18. Remove the power steering adjusting and mounting bolts, then remove the power steering pump and belt. Do not disconnect the power steering hoses.
19. Remove the heat shield from the power steering bracket.
20. Remove the air conditioning compressor and alternator belts.
21. Remove the cruise control actuator.
22. Remove the engine side mount.
23. Remove the cylinder head cover, timing belt cover, and timing belt.
24. Remove the camshaft sprockets and back cover.
25. Remove the exhaust manifold cover, bracket, and exhaust manifold.
26. Remove the bolts attaching the intake manifold to the support bracket.
27. Remove the nuts attaching the intake manifold to the cylinder head. Remove the nuts in a crisscross pattern, beginning from the center and moving out to both ends.
28. Remove the manifold and the old gasket.
29. Remove the VTEC solenoid from the cylinder head.
30. Loosen the rocker arm locknuts and adjusting screws.
31. Remove the camshaft holder bolts, then remove the camshaft holder plates, the camshaft holders, and camshafts.
32. Remove the cylinder head bolts, then remove the cylinder head. To prevent warpage, loosen the bolts in the reverse of the torque sequence 1/3 turn at a time. Repeat this sequence until all bolts are loosened.

To install:

33. Install the cylinder head onto the engine block, after making sure the mating surface was cleaned and a new gasket was installed. Be sure to pay attention to the following points:
   - Be sure the No. 1 cylinder is at top dead center and the camshaft pulley UP mark is on the top before positioning the head in place.
   - The cylinder head dowel pins and oil control orifice must be cleaned and aligned.
   - Replace the washer when damaged or deteriorated.
   - Apply engine oil to the cylinder head bolts and the washers.
   - Use the longer cylinder head bolts at the No. 1 and No. 2 positions.
34. Tighten the cylinder head bolts in two steps. In the first step tighten all bolts in sequence to 22 ft. lbs. (29 Nm). In the second step tighten all the bolts in the same sequence to 60 ft. lbs. (81 Nm) for 1995 models, and to 63 ft. lbs. (85 Nm) for 1996–99 models.
35. Use new gaskets and install the intake manifold onto the cylinder head; tighten the nuts in a crisscross pattern in 2–3 steps, beginning in the middle. Tighten the nuts to 17 ft. lbs. (23 Nm).
36. Install and tighten the intake manifold bracket bolts to 17 ft. lbs. (24 Nm).
37. Install the VTEC solenoid with a new filter, tighten the attaching bolts to 17 ft. lbs. (24 Nm).
38. Install the exhaust manifold and tighten the new self-locking nuts in a crisscross pattern in 2–3 steps, beginning with the inner nuts. Tighten the nuts to 23 ft. lbs. (31 Nm). Install a new exhaust pipe gasket and tighten the new nuts to 40 ft. lbs. (54 Nm).
39. Install the exhaust manifold bracket and cover. Tighten the bracket attaching bolts to 33 ft. lbs. (44 Nm) and the cover bolts to 17 ft. lbs. (24 Nm).
40. Be sure that the keyways on the camshafts are facing up and that the rocker arms are in their original position. The valve locknuts should be loosened and the adjusting screw backed off.

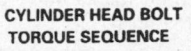

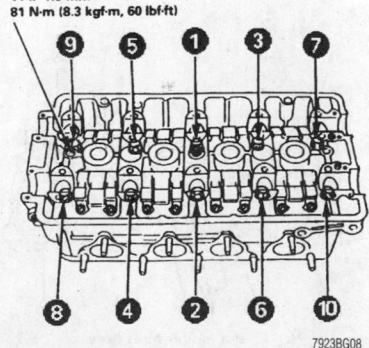

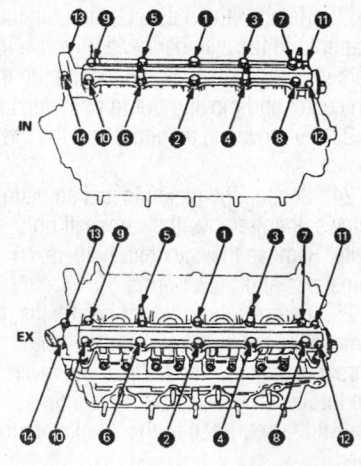

CYLINDER HEAD BOLT
TORQUE SEQUENCE
11 x 1.5 mm
81 N·m (8.3 kgf·m, 60 lbf·ft)
9 5 1 3 7
8 4 2 6 10
7923BG08

Cylinder head bolt torque sequence—1.8L (B18C1) engine

13 9 5 1 3 7 11
IN
14 10 6 2 4 8 12
13 9 5 1 3 7 11
EX
14 10 6 2 4 8 12
1 - 10: 8 x 1.25 mm 27 N·m (2.8 kgf·m, 20 lbf·ft)
11 - 14: 6 x 1.0 mm 9.8 N·m (1.0 kgf·m, 7.2 lbf·ft)
7923BG09

Camshaft holder plate torque sequence—1.8L (B18C1) engine

41. Install the camshafts, then install the camshaft seals with the open side facing in. Install the rubber cap with liquid gasket applied. If the rubber cap has two horizontal marks, align the marks with the cylinder head upper surface.

42. Install a new O-ring and the dowel pin to the oil passage of the No. 3 camshaft holder.

43. Apply liquid gasket to the head of the mating surfaces of the No. 1 and No. 5 camshaft holders, then install them, along with No. 2, 3, and 4. Be sure to pay attention to the following points:

• Do not apply oil to the holder mating surface of camshaft seals.

• The arrows marked on the camshaft holders should point to the timing belt.

44. Tighten the camshaft holders temporarily. Be sure that the rocker arms are properly positioned on the valve stems.

45. Tighten each bolt in 2 steps to ensure that the rockers do not bind on the valves. Tighten the 8x1.25mm bolts to 20 ft. lbs. (27 Nm), and the 6x1.0mm bolts to 7.2 ft. lbs. (9.8 Nm).

46. Install the back cover and tighten the attaching bolt to 7.2 ft. lbs. (9.8 Nm). Install the keys into the camshaft grooves, then push the camshaft pulleys onto the camshafts, then tighten the retaining bolts to 41 ft. lbs. (56 Nm).

47. Install the timing belt and adjust the tension, then install the timing belt covers.

48. Adjust the valve clearance.

49. Install the rubber seal in the groove of the cylinder head cover. Be sure that the seal and groove are thoroughly clean first.

50. Apply liquid gasket to the rubber seal at the eight corners of the recesses. Do not install the parts if 20 minutes or more have elapsed since applying the liquid gasket. Instead, reapply liquid gasket after removing old residue.

51. Install the cylinder head cover and engine ground cable. Be sure the contact surfaces are clean and do not touch surfaces where liquid gasket has been applied.

52. Tighten the cylinder head cover nuts in 2–3 steps. In the final step, tighten all nuts in sequence, to 7 ft. lbs. (10 Nm).

53. Install the engine side mount, tighten the two new nuts and new bolt to the engine to 38 ft. lbs. (52 Nm) and tighten the bolt attaching the mount to the vehicle to 54 ft. lbs. (74 Nm).

54. The balance of the removal is the reverse of the installation. If necessary, refer to the appropriate portion of this section for information on the various components to be installed.

55. Connect the negative battery cable and enter the radio security code.

56. After installation, check to see that all hoses and wires are installed correctly.

57. Fill and bleed the air from the cooling system.

58. Change the engine oil. Wait at least 20 minutes for the sealant to cure before filling the engine with oil.

2.3L Engine

1. Disconnect the negative battery cable.

2. Turn the crankshaft so the No. 1 piston is at Top Dead Center (TDC).

➡**The No. 1 piston is at top dead center when the pointer on the block aligns with the white painted mark on the flywheel (manual transaxle) or driveplate (automatic transaxle).**

3. Drain the engine coolant into a sealable container.

4. Relieve the fuel system pressure.

5. Remove the air intake duct.

6. Remove the evaporative emissions (EVAP) control canister hose from the intake manifold.

7. Remove the throttle cable from the throttle body. On automatic transaxle equipped vehicles, remove the throttle control cable at the throttle body.

➡**Be careful not to bend the cable when removing. Always replace a kinked cable with a new one.**

8. Disconnect the fuel feed and return hose.

9. Remove the brake booster vacuum hose from the intake manifold.

10. Disconnect the following engine wiring harness connectors:

   a. Fuel injector connectors

   b. Intake Air Temperature (IAT) sensor connector

   c. Idle Air Control (IAC) valve connector

   d. Throttle Position (TP) sensor connector

   e. Exhaust Gas Recirculation (EGR) valve lift sensor

   f. Ground cable terminals

   g. Engine Coolant Temperature (ECT) switch B connector

   h. Heated oxygen sensor (HO2S) connector

   i. ECT sensor

   j. ECT gauge sending unit connector

   k. Ignition Control Module (ICM) connector

   l. CKP/TDC/CYP sensor connector

   m. Vehicle Speed Sensor (VSS) connector

   n. Ignition coil connector

   o. Intake air bypass solenoid valve connector

   p. ECT switch A connector

   q. Knock sensor connector

11. Remove the engine ground cable from the cylinder head cover.

12. Remove the connector and the terminal from the alternator, then remove the engine wiring harness from the valve cover.

13. Remove the mounting bolts and drive belt from the power steering pump. Pull the pump away from the mounting bracket, without disconnecting the hoses. Support the pump out of the way.

14. Remove the ignition coil.

15. Tag, then disconnect the emissions vacuum hoses from the intake manifold assembly.

16. Remove the bypass hose from the intake manifold.

17. Remove the upper radiator hose and the heater hose from the cylinder head.

18. Remove the lower radiator hose and bypass hose from the thermostat housing.

19. Remove the thermostat housing mounting bolts. Remove the thermostat housing from the intake manifold and the connecting pipe, by pulling and twisting the housing. Discard the O-rings.

20. Raise and safely support the vehicle.

21. Remove the front wheel and tire assemblies.

22. Remove the splash shield.

23. Remove the intake manifold bracket bolts.

24. Remove the intake manifold.

25. Disconnect the exhaust pipe from the exhaust manifold.

26. Remove the exhaust manifold and the exhaust manifold heat insulator.

27. Label, then disconnect the electrical connectors from the distributor and the spark plug wires from the spark plugs. Mark the position of the distributor and remove it from the cylinder head. Disconnect the ignition coil wire from the distributor.

28. Remove the Positive Crankcase Ventilation (PCV) hose, then remove the cylinder head cover. Replace the rubber seals if damaged or deteriorated.

29. Remove the timing belt.

30. Insert a 5.0mm pin punch in each of the camshaft caps, nearest to the sprockets, through the holes provided. Remove the camshaft sprocket attaching bolts, then remove the sprockets. Do not lose the sprocket keys.

31. Loosen all of the rocker arm adjusting screws, then remove the pin punches from the camshaft caps.

32. Remove the camshaft holders, note the holders locations for ease of installation.

33. Remove the rubber cap from the head, located at the end of the intake camshaft.

34. Remove the rocker arms from the cylinder head. Note the locations of the rocker arms.

➡**The rocker arms have to be installed to their original locations if being reused.**

35. Remove the side engine mount bracket B, then the back cover from behind the camshaft sprockets.

36. Remove the cylinder head bolts in the proper sequence.

➡**To prevent warpage, unscrew the bolts in sequence ⅓ turn at a time. Repeat the sequence until all bolts are loosened.**

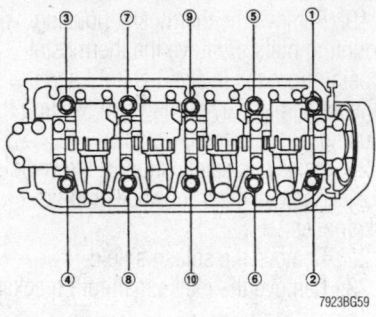

Cylinder head bolt removal sequence— 2.3L engine

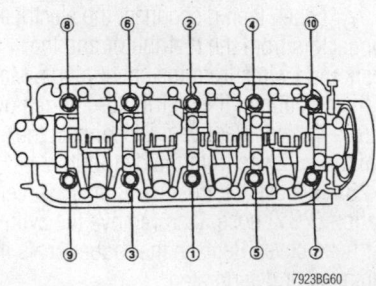

Cylinder head bolt torque sequence—2.3L engine

37. Separate the cylinder head from the engine block with a suitable flat bladed pry-tool.

To install:

38. Be sure all cylinder head and block gasket surfaces are clean. Check the cylinder head for warpage. If warpage is less than 0.002 in. (0.05mm), cylinder head resurfacing is not required. Maximum resurface limit is 0.008 in. (0.2mm) based on a cylinder head height of 5.20 in. (132.0mm).

39. Always use a new head gasket.

40. Be sure the No. 1 cylinder is at TDC.

41. Clean the oil control orifice and install a new O-ring. The cylinder head dowel pins and oil control jet must be aligned.

42. Install the bolts that secure the intake manifold to it's bracket but do not tighten them.

43. Install the cylinder head, then tighten the cylinder head bolts sequentially in 3 steps:

- Step 1: 29 ft. lbs. (40 Nm).
- Step 2: 51 ft. lbs. (70 Nm).
- Step 3: 72 ft. lbs. (100 Nm).

➡**A beam type torque wrench is recommended. If a bolt makes any noise while being tightened, loosen the bolt and retighten it.**

44. Install the intake manifold with a new gasket.

45. Install the exhaust manifold with a new gasket.

46. Install the exhaust manifold bracket, then install the exhaust pipe, bracket and upper shroud.

47. Install the camshafts and rocker arms.

48. Install the timing belt back cover.

49. Install the side engine mount bracket B. Tighten the bolt attaching the bracket to the cylinder head to 33 ft. lbs. (45 Nm). Tighten the bolts attaching the bracket to the side engine mount to 16 ft. lbs. (22 Nm).

50. Install the camshaft sprockets onto the camshafts.

51. Install the timing belt.

52. Adjust the valves.

53. Tighten the crankshaft pulley bolt to 181 ft. lbs. (250 Nm).

54. Install the splash shield and the front wheels.

55. Lower the vehicle.

56. Install the remaining components in the reverse order of removal.

57. Drain the oil from the engine into a sealable container. Install the drain plug and refill the engine with clean oil.

58. Fill and bleed the air from the cooling system.

59. Connect the negative battery cable and enter the radio security code.

60. Start the engine, checking carefully for any leaks.

61. Check and adjust the ignition timing. Tighten the distributor bolts to 13 ft. lbs. (18 Nm).

62. If equipped with 4WS, start the engine and turn the steering wheel lock-to-lock to reset the 4WS control unit.

2.5L Engine

1. Disconnect the negative battery cable and drain the coolant.

2. Disconnect the wiring from the ignition coil and the ground wire.

3. Remove the ABS motor relay box and the battery heat shield.

✳✳ **CAUTION**

The fuel injection system remains under pressure after the engine has been turned OFF. Properly relieve fuel pressure before disconnecting any fuel lines. Failure to do so may result in fire or personal injury.

4. Remove the fuel filler cap and loosen the service bolt on the fuel filter banjo bolt to relieve the fuel system pressure. Remove the banjo bolt to disconnect the fuel feed hose from the fuel filter. Disconnect the fuel return hose from the pressure regulator.

5. Remove the intake air duct and air cleaner assembly.

6. Loosen the air conditioner compressor and alternator adjustment bolts. Remove the drive belts.

7. Remove the throttle cable by loosening the locknut, then slip the cable end out of the throttle bracket and accelerator linkage. Do not bend the cable when removing it. Unbolt the throttle cable clamp and move the cable aside.

8. Label and disconnect the fuel and vacuum hoses from the intake manifold. Be sure to mark all electrical connectors and vacuum hoses before disconnecting them.

9. Disconnect the upper radiator hose, the heater hoses, and the water bypass hoses and unbolt the wiring harness clips.

10. Disconnect the brake booster hose, canister hose, and the two vacuum hoses from the rear of the cylinder head.

11. Remove the two distributor mounting bolts. Remove the distributor, ignition wires, and ground cables from the cylinder head.

12. Label and disconnect the wiring harness holder that is routed across the front of the cylinder head.

13. Disconnect and label the fuel injector

leads; throttle position sensor; IAC, EGR, EVAP, and air bypass solenoid valves, and the ECT connections. Disconnect the TDC, crankshaft, and camshaft position sensors.

14. Unbolt the intake manifold support brackets. The manifold may be removed after removing the cylinder head.

15. Disconnect the oxygen sensor wire.

16. Remove the exhaust manifold heat shields and disconnect the exhaust pipe from the manifold.

17. Remove the support bracket and remove the exhaust manifold.

18. Remove the cylinder head cover and upper timing belt cover.

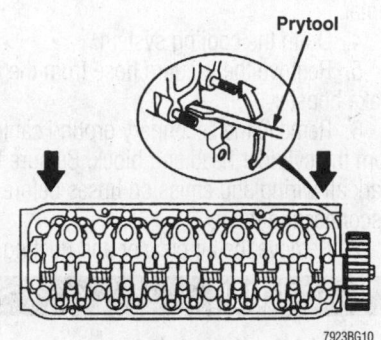

Cylinder head prying points—2.5L engine

19. Remove the timing belt. Replace the belt if it shows any signs of stress or damage.

20. Remove the camshaft position sensor and the camshaft sprocket.

21. Remove the back cover and unbolt the TDC/CKP sensor.

22. Loosen each cylinder head bolt about ⅓ turn at a time. Follow the reverse of the installation sequence to prevent warping the head. Repeat until all bolts are loose and can be removed.

23. If the cylinder head is stuck to the block, there are pry points at each end of the cylinder head. Do not pry against the gasket surfaces.

24. Carefully remove the cylinder head from the vehicle. If the intake and exhaust manifolds are still attached, have an assistant help in removal since the head assembly will be very heavy.

25. After removing the cylinder head, the intake manifold and exhaust manifolds maybe removed.

To install:

26. Be sure the cylinder head and the engine block sealing surfaces are flat and clean. Resurface the head if it is warped. Clean all gasket surfaces and run a tap through the bolt holes in the block to clean the threads. Be sure the bolt holes are clean

and dry so the head can be tightened properly.

27. Install a new O-ring onto the oil control orifice and install the orifice and dowel pins onto the block. Lay the new head gasket in place.

28. Install the intake manifold onto the cylinder head with a new gasket. Tighten the nuts in a crisscross pattern in 2 steps to 16 ft. lbs. (22 Nm).

29. Verify that the crankshaft and camshaft are both at TDC for number one piston.

30. Carefully fit the cylinder head to the block. Be sure the oil control orifice is properly aligned.

31. Lightly oil the threads and washer surfaces of the cylinder head bolts and install them. Tighten the bolts in 3 steps to 72 ft. lbs. (100 Nm) in the correct sequence.

32. Install the intake manifold brackets.

33. Loosely install the exhaust manifold bracket onto the manifold. Install the exhaust manifold with a new gasket and new self-locking nuts and tighten the nuts to 23 ft. lbs. (32 Nm).

34. Connect the exhaust pipe and install the manifold shields.

35. Reconnect the oxygen sensor.

36. Install the timing belt and covers.

37. Adjust the valves.

38. Apply silicone sealant to the ends of the cylinder head near the camshaft holders. Install the cylinder head cover with new rubber seals as required.

39. The balance of the removal is the reverse of the installation. If necessary, refer to the appropriate portion of this section for information on the various components to be installed.

40. Add fresh engine oil and a new filter.

41. Verify that all wiring, grounds, hoses, and cables are properly connected.

42. Connect the battery cable. Run the engine to bleed the cooling system and check for leaks. Check for proper cooling system and engine operation.

3.0L Engine

1. Obtain the security code for the radio.

2. Disconnect the negative battery cable.

3. Drain the coolant.

4. Remove the EVAP canister hose from the throttle body.

5. Remove the air intake duct.

6. Remove the upper engine covers.

7. Disconnect the accelerator and cruise control cables from the throttle body.

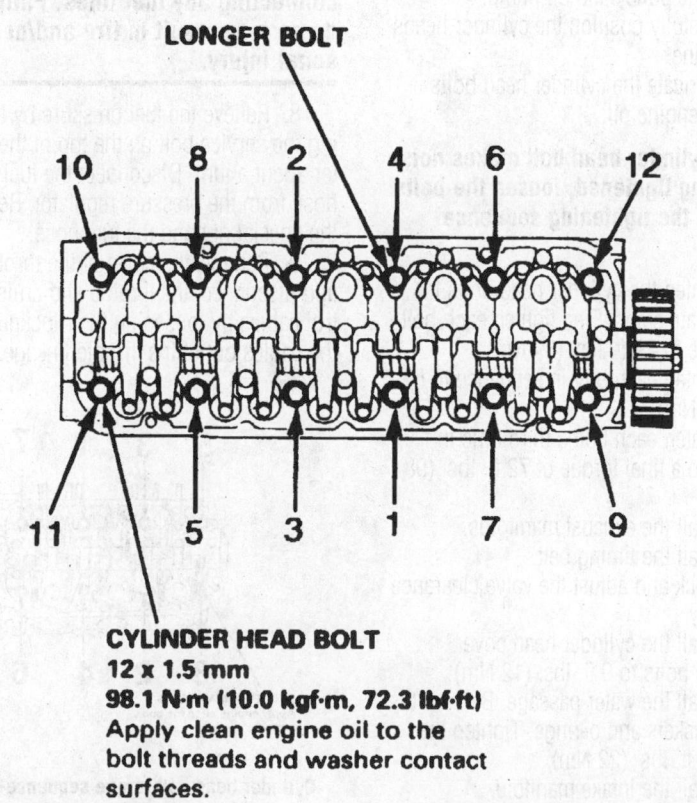

LONGER BOLT

CYLINDER HEAD BOLT
12 x 1.5 mm
98.1 N·m (10.0 kgf·m, 72.3 lbf·ft)
Apply clean engine oil to the bolt threads and washer contact surfaces.

Tighten the cylinder head bolts in the correct sequence as shown—2.5TL

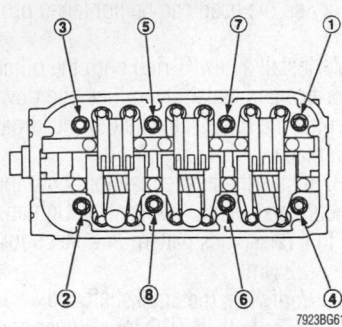

Loosen the cylinder head bolts in the sequence shown to prevent damage to the head—3.0L engine

8. Remove the spark plug wire holder, cover and intake manifold covers.

9. Properly relieve the fuel system pressure.

10. Disconnect the fuel hoses from the supply rail.

11. Disconnect the following hoses and lines:

- Brake booster vacuum hose
- PCV hose
- Breather hose
- Water bypass hose
- Vacuum hose from the throttle body

12. Remove the ground cable from the engine.

13. Remove the alternator belt.

14. Support the engine with a jack and a block of wood and remove the side engine mounting bracket.

15. Remove the power steering pump without disconnecting the hoses.

16. Remove the alternator.

17. Detach the wiring harness connectors from the components on the engine that may interfere with removing the cylinder head.

18. Remove the distributor and spark plug wires.

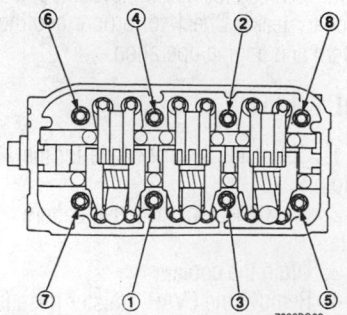

Tighten the cylinder head bolts in the sequence shown to prevent damage to the head—3.0L engine

19. Remove the intake manifold.

20. Detach the connectors from the fuel injectors.

21. Remove the fuel supply rails.

22. Remove the vacuum hoses from the fuel control valve.

23. Set the engine to TDC by aligning the marks on the crankshaft and camshaft pulleys.

24. Remove the timing belt.

25. Remove the upper and lower radiator hoses.

26. Disconnect the heater hoses.

27. Remove both exhaust manifolds.

28. Remove the water passage assembly.

29. Remove the camshaft pulleys and rear timing belt covers.

30. Loosen each cylinder head bolt ⅓ turn at a time in the correct sequence. This will take several passes.

31. Remove the cylinder heads.

To install:

32. Clean the cylinder head and the surface of the cylinder block.

33. Install the oil control orifices and install them using new o-rings.

34. If removed, install the dowel pins.

35. Position new cylinder head gaskets on the cylinder block.

36. If moved, set the crankshaft and camshaft pulleys to TDC by aligning the marks on the pulley and oil pump.

37. Carefully position the cylinder heads on the engine.

38. Lubricate the cylinder head bolts with clean engine oil.

➡ **If any cylinder head bolt makes noise while being tightened, loosen the bolts and begin the tightening sequence again.**

39. Tighten the cylinder head bolts in three separate steps. First tighten each bolt in sequence to 29 ft. lbs. (39 Nm).

40. Tighten each bolt in sequence to 51 ft. lbs. (69 Nm).

41. Tighten each bolt a third time in sequence to a final torque of 72 ft. lbs. (98 Nm).

42. Install the exhaust manifolds.

43. Install the timing belt.

44. Check and adjust the valve clearance if necessary.

45. Install the cylinder head cover. Tighten the bolts to 9 ft. lbs. (12 Nm).

46. Install the water passage. Be sure to use new gaskets and o-rings. Tighten the bolts to 16 ft. lbs. (22 Nm).

47. Install the intake manifold.

48. Install all of the remaining hoses,

tubes and connectors are installed correctly.

49. Connect the negative battery cable.

50. Enter the security code for the radio.

3.2L and 3.5L Engines

LEGEND

1. Disconnect the negative, then the positive battery cables.

2. Remove the battery and battery tray.

3. Engine temperature must be below 100 degrees F. (38 degrees C.) before performing this procedure. Turn the crankshaft so that the No. 1 cylinder is at top dead center.

4. Drain the cooling system.

5. Remove the vacuum hose from the brake booster.

6. Remove the secondary ground cable from the cylinder head and block. Be sure to mark all wiring and emission hoses before disconnecting them.

7. Remove the air cleaner and ducting.

✳✳ CAUTION

The fuel injection system remains under pressure, even after the engine has been turned OFF. The fuel system pressure must be relieved before disconnecting any fuel lines. Failure to do so may result in fire and/or personal injury.

8. Relieve the fuel pressure by loosening the service bolt on the top of the fuel filter about a turn. Disconnect the fuel return hose from the pressure regulator. Remove the special nut and the fuel hose.

9. Remove the cover at the throttle body and disconnect the throttle and cruise control cables by loosening their locknuts. Slip the cables out of the throttle linkage.

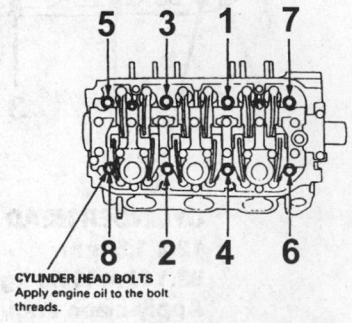

CYLINDER HEAD BOLTS
Apply engine oil to the bolt threads.

Cylinder head bolt torque sequence—3.2L engines

10. Disconnect the charcoal canister hose from the throttle body.

11. Remove the intake manifold.

12. Remove the upper timing belt covers.

13. Do not remove the timing belt adjuster bolt. Loosen it ½ turn, relieve the belt tension and tighten the bolt.

14. Remove the camshaft timing belt and the camshaft pulleys.

15. Remove the timing belt cover plates from the heads and remove the crank angle/cylinder sensor from the left head.

16. Remove the cylinder head covers.

17. Remove the bolts from the alternator and power steering pump brackets as required.

18. Working under the vehicle, unbolt the twin three-way catalytic converters and disconnect them from the exhaust manifolds.

19. Loosen each head bolt about ½ turn in the opposite of the installation sequence. This is important to prevent warping the heads. Repeat until all bolts are loose and the head can be removed.

20. Remove the heads,, then remove the exhaust manifolds from the heads.

To install:

21. It is easier to install the exhaust manifolds and their covers onto the heads before installing the heads to the engine. Use new gaskets and self-locking nuts. Tighten the nuts to 22 ft. lbs. (31 Nm).

22. Install the heads with new gaskets and O-rings, making sure the dowel pins and control orifices are properly positioned. Oil the threads and washers on the head bolts and tighten in two steps in the sequence shown to 56 ft. lbs. (78 Nm).

23. Apply liquid gasket to the corners of the camshaft holders and install the cylinder head covers.

24. Install the crankshaft/cylinder sensor to the left cylinder head, then install both timing belt cover plates.

25. Install the camshaft pulleys and tighten the bolts to 23 ft. lbs. (32 Nm). The left and right pulleys are different; the left one goes with the crankshaft/cylinder sensor.

26. Align the timing marks on the crankshaft and camshaft pulleys, install the timing belt and adjust the belt tension. Install the timing belt covers.

27. The balance of the removal is the reverse of the installation. If necessary, refer to the appropriate portion of this section for information on the various components to be installed.

28. Refill the cooling system. Whenever cylinder heads are removed and installed, it is always recommended to change the oil.

29. Verify that all wires and hoses are properly connected and install the battery. Before starting the engine, turn the ignition switch **ON** and **OFF** a number of times to pressurize the fuel system. Check for leaks.

30. After starting the engine, bleed the cooling system.

3.2TL AND 3.5RL

➥**The PCM idle memory must be reset after reconnecting the battery. Start the engine and hold it at 3000 rpm until the cooling fan comes on. Then allow the engine to idle for about five minutes with all accessories OFF and with the transmission in Park or Neutral.**

1. Disconnect the negative, then the positive battery cables.

2. Drain the engine coolant into a sealable container.

3. Remove the engine cover.

4. Remove the air intake dust and the air cleaner housing.

5. Remove the throttle cable cover. Without turning the adjusting nut, loosen the locknut, which is closer to the throttle, and disconnect the throttle cable and cruise control cable from the throttle and bracket.

➥**Take care to not bend the throttle cables, replace the cables if they become bent or kinked.**

6. Remove the upper and lower radiator hoses.

7. Remove the battery and the battery base.

8. Disconnect the vacuum hoses, then remove the clamp from the under-hood fuse/relay box.

9. Label, then disconnect the battery cables from the under-hood fuse/relay box, then remove the under-hood fuse/relay box.

10. Disconnect the engine wiring harness connector, located by the under-hood fuse/relay box.

11. Disconnect the wiring harness from the control box and solenoid valve, then remove the control box.

12. Properly relieve the fuel pressure.

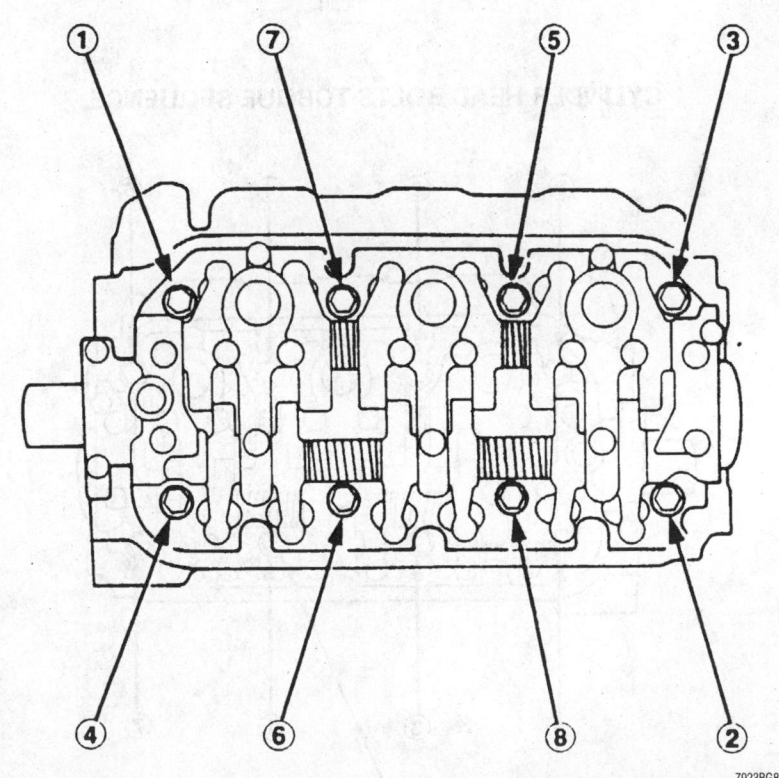

Loosen the cylinder head bolts in the sequence shown—3.5L engine

✳✳ CAUTION

Do not allow fuel spray or fuel vapors to come in contact with a spark or open flame. Keep a dry chemical fire extinguisher nearby. Never store fuel in an open container due to risk of fire or explosion. The fuel injection system remains under pressure after the engine has been turned OFF. Properly relieve fuel pressure before disconnecting any fuel lines. Failure to do so may result in fire or personal injury.

13. Disconnect the engine fuel feed hose from the fuel filter and disconnect the fuel return hose from the fuel regulator.

14. Disconnect the evaporative emissions (EVAP) control canister hose and the vacuum hose.

15. Disconnect the bypass hoses and the heater hose.

16. Loosen the alternator mounting bolt, lockbolt and adjusting rod, then remove the drive belt.

17. Loosen the A/C idler pulley center nut and adjusting bolt, then remove the drive belt.

18. Remove the power steering pump drive belt.

19. Remove the bolt attaching the engine ground cable to the intake manifold. Remove the engine wiring harness clamps.

20. Remove the Traction Control System (TCS) control valve assembly upper and lower brackets.

21. Disconnect the TCS throttle sensor connector and the TCS throttle actuator connector, then remove the TCS control valve assembly.

22. Remove the oil pressure switch connector, engine ground cable and the engine wiring harness cover.

23. Remove the EGR pipe, then remove the intake manifold and water passage.

24. Remove the covers from the exhaust manifolds, then remove the exhaust manifolds.

25. Remove the timing belt from the engine.

26. Remove the left and right camshaft sprockets, then remove the timing belt back covers.

27. Remove the left and right cylinder head covers.

28. Remove the three bolts attaching the alternator bracket to the cylinder head.

29. Remove the two bolts attaching the power steering pump bracket to the cylinder head.

✳✳ WARNING

To prevent cylinder head damage, loosen each cylinder head bolt in sequence 1/3 turn until all of the bolts are loose.

30. Remove the cylinder head bolts 1/3 at turn at a time in the reverse of the torque sequence. Repeat until all the bolts are loosened.

31. Remove the cylinder heads from the cylinder block.

To install:

32. Remove the oil control orifices, then clean them and install new O-rings.

33. Clean the cylinder block and the head mating surfaces.

34. Install the oil control orifices and the dowel pins to the cylinder block.

35. Install new gaskets to the cylinder block, then install the cylinder heads.

36. Tighten the cylinder head bolts in two or three steps following the proper sequence to 56 ft. lbs. (76 Nm). The manufacturer recommends using a beam-type torque wrench.

37. Clean the groove in the cylinder head cover and install the gasket into the groove. Seat the recesses for the camshaft first, then work it into the groove around the outside edges. Apply liquid gasket to the cylinder head cover gasket at the four corners of the recesses.

38. Install the cylinder head cover, slide the cover back and forth to seat the cylinder head cover gasket. Replace the washer when damaged or deteriorated. Tighten the nuts in two or three steps following the proper sequence, to 8.7 ft. lbs. (12 Nm).

39. Install the two bolts used to attach the power steering pump bracket to the cylinder head, then tighten the bolts to 33 ft. lbs. (44 Nm).

40. Install the three bolts used to attach the alternator bracket to the cylinder head, then tighten the bolts to 16 ft. lbs. (22 Nm).

41. Install the timing belt back covers and tighten the bolts to 8.7 ft. lbs. (12 Nm).

42. Install the left and right camshaft sprockets and tighten the attaching bolts to 23 ft. lbs. (31 Nm).

43. Install the timing belt and timing belt covers.

44. The balance of the removal is the reverse of the installation.

45. Install the engine cover and tighten the nuts to 8.7 ft. lbs. (12 Nm).

46. Install the battery base and tighten the bolts to 16 ft. lbs. (22 Nm), then install the battery.

CYLINDER HEAD BOLTS TORQUE SEQUENCE

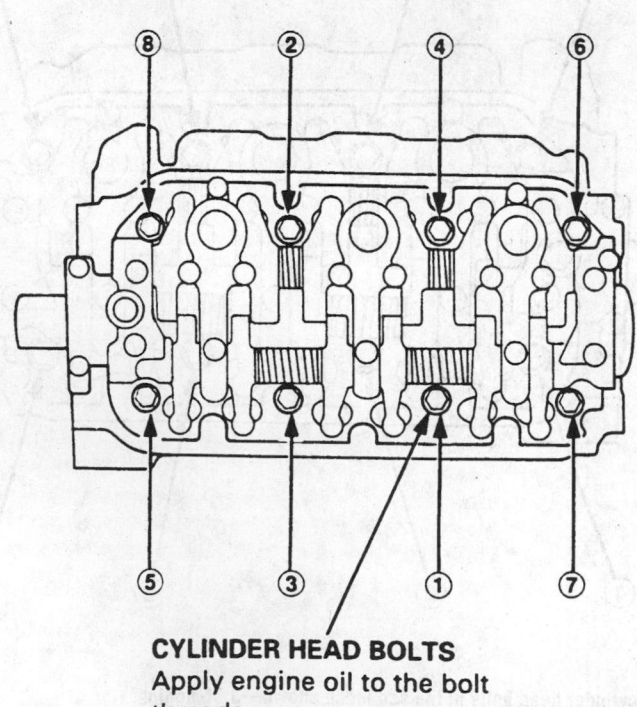

CYLINDER HEAD BOLTS
Apply engine oil to the bolt threads.

7923BG82

Cylinder head bolt tightening sequence—3.5L engine

NOTE: Clean the oil control orifice when installing.

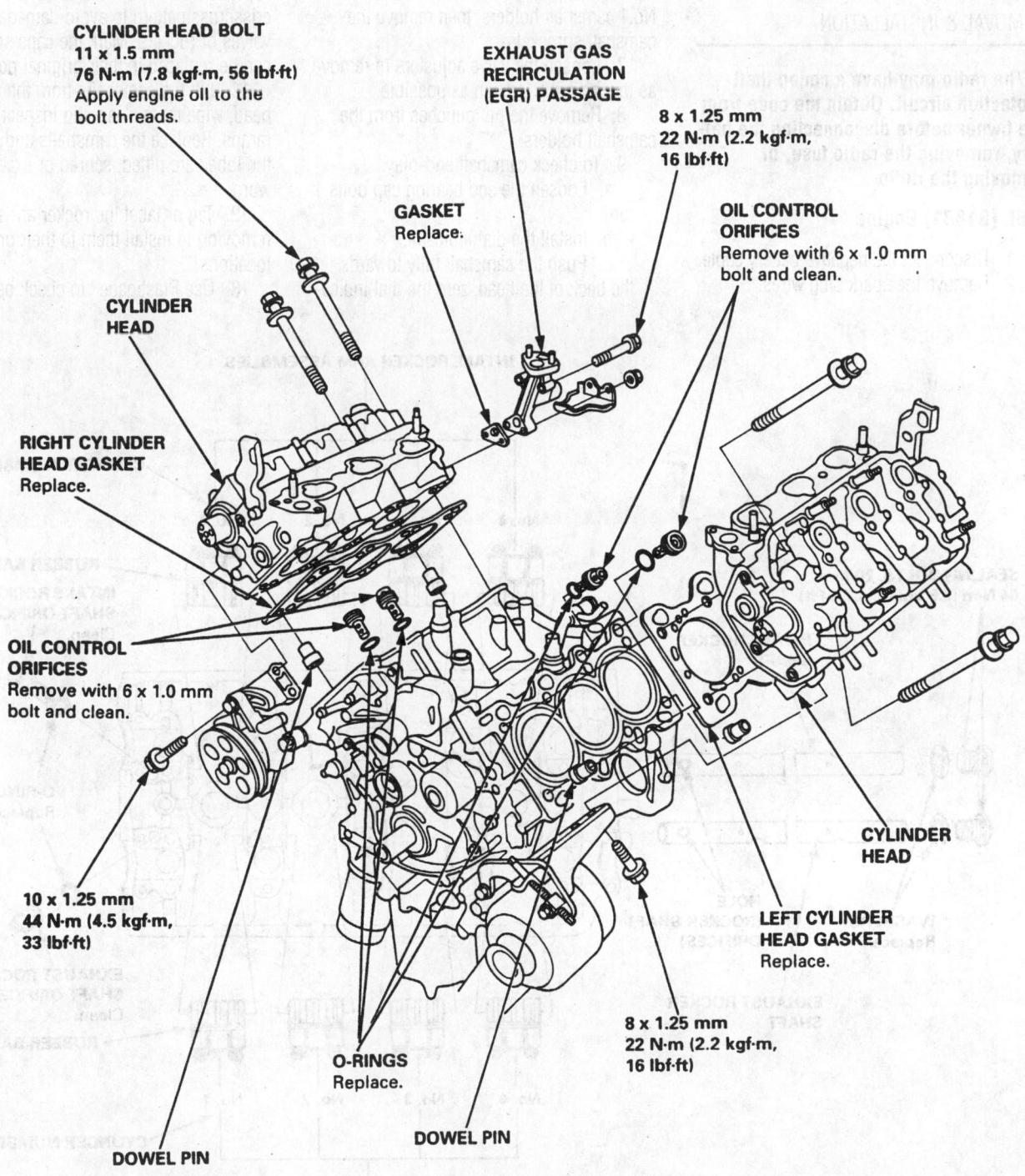

CYLINDER HEAD BOLT
11 x 1.5 mm
76 N·m (7.8 kgf·m, 56 lbf·ft)
Apply engine oil to the
bolt threads.

**CYLINDER
HEAD**

**RIGHT CYLINDER
HEAD GASKET**
Replace.

**OIL CONTROL
ORIFICES**
Remove with 6 x 1.0 mm
bolt and clean.

10 x 1.25 mm
44 N·m (4.5 kgf·m,
33 lbf·ft)

DOWEL PIN

O-RINGS
Replace.

DOWEL PIN

GASKET
Replace.

**EXHAUST GAS
RECIRCULATION
(EGR) PASSAGE**

8 x 1.25 mm
22 N·m (2.2 kgf·m,
16 lbf·ft)

**OIL CONTROL
ORIFICES**
Remove with 6 x 1.0 mm
bolt and clean.

**CYLINDER
HEAD**

**LEFT CYLINDER
HEAD GASKET**
Replace.

8 x 1.25 mm
22 N·m (2.2 kgf·m,
16 lbf·ft)

7923BG80

Exploded view of the cylinder heads and related components—3.2L and 3.5L engines

47. Fill and bleed the air from the cooling system.

48. Connect the positive, then the negative battery cables. Enter the radio security code.

Rocker Arms/Shafts

REMOVAL & INSTALLATION

➡ **The radio may have a coded theft protection circuit. Obtain the code from the owner before disconnecting the battery, removing the radio fuse, or removing the radio.**

1.8L (B18B1) Engine

1. Disconnect the negative battery cable.
2. Remove the spark plug wires.

3. Remove the cylinder head cover and timing belt cover.

4. Rotate the crankshaft to TDC, compression of No. 1 piston and remove the timing belt.

5. Remove the distributor from the cylinder head.

6. Install 5.0mm pin punches to the No.1 camshaft holders, then remove the camshaft sprockets.

7. Loosen the valve adjusters to remove as much spring tension as possible.

8. Remove the pin punches from the camshaft holders.

9. To check camshaft end-play:
   a. Loosen the end bearing cap bolts 1 turn.
   b. Install the dial indicator.
   c. Push the camshaft fully towards the back of the head, zero the dial indica-

tor and push the camshaft fully the other way to read end-play.
   d. End-play on a new camshaft should be 0.002–0.006 in. (0.05–0.15mm), 0.020 in. (0.5mm) is the service limit.

10. To remove the camshaft bearing caps, loosen each bolt 2 turns at a time in a crisscross pattern to avoid damage to the valves or rockers. Mark the caps so they can be replaced in their original position.

11. Lift the camshafts from the cylinder head, wipe them clean and inspect the lift ramps. Replace the camshafts and rockers if the lobes are pitted, scored or excessively worn.

12. Tag or label the rocker arms before removing to install them to their original locations.

13. Use Plastigage® to check bearing

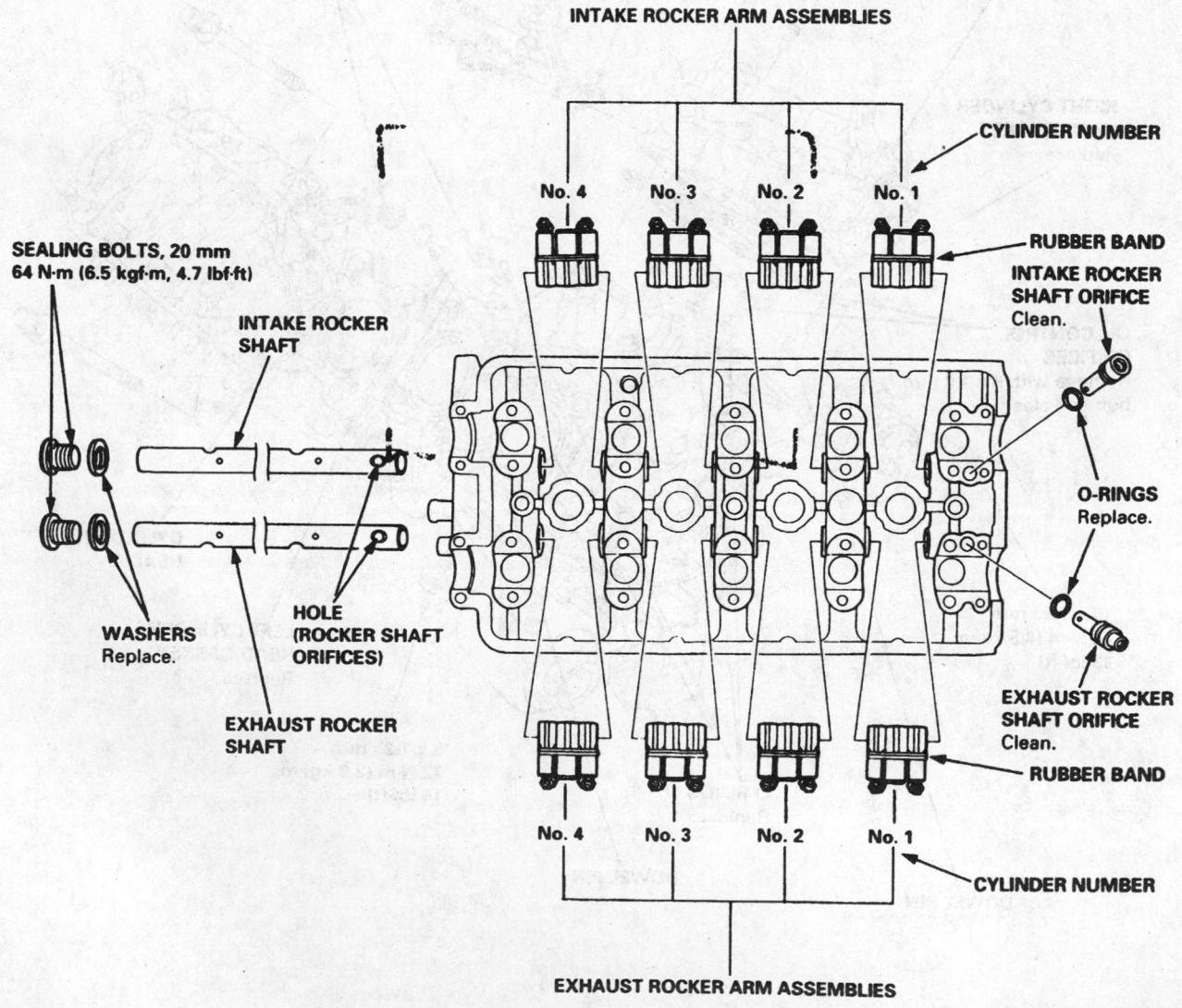

Rocker arms and shafts—1.8L (B18B1 and B18C1) engines

7923BG13

clearance. The standard clearance is 0.0012–0.0027 in. (0.030–0.069mm), and absolute service limit is 0.006 in. (0.15mm).

To install:

14. Check the following before installing the camshafts:

    a. Be certain the keyways on the camshafts are facing UP (No. 1 cylinder at TDC).

    b. The valve adjuster locknuts should be loosened and the adjusting screws backed off before installation.

15. Lubricate the rocker arms and camshafts with clean oil.

16. Place the rocker arms on the pivot bolts and the valve stems, making sure that the rocker arms are in their original positions.

17. Install the camshaft seals with the open side (spring) facing in. Lubricate the lip of the seal.

18. Be sure the keyways on the camshafts are facing up and install the camshafts to the cylinder head.

19. Apply liquid gasket to the head mating surfaces of the No. 1 and No. 6 camshaft holders, then install them along with Nos. 2, 3, 4 and 5 camshaft holders. The arrows stamped on the holders should point toward the timing belt. Do not apply oil to the holder mating surface where the camshaft seals are housed.

20. Tighten the camshaft holders temporarily and be sure that the rocker arms are properly positioned.

21. Press the oil seals into the No.1 camshaft holders with a seal driver.

22. Tighten the bolts in a crisscross pattern to 7 ft. lbs. (10 Nm). Check that the rockers do not bind on the valves.

23. Install the cylinder head plug to the end of the cylinder head. If the plug has alignment marks, align the marks with the cylinder head upper surface.

24. If equipped with a timing belt back

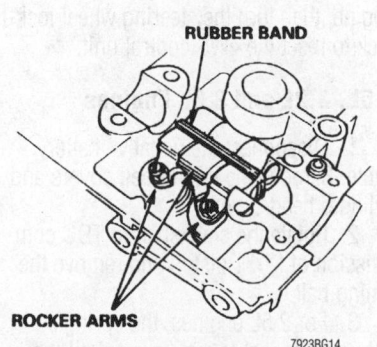

RUBBER BAND

ROCKER ARMS

7923BG14

Rocker arms with rubber band installed— 1.8L (B18B1 and B18C1) engines

cover, install the cover and tighten the bolts to 7.2 ft. lbs. (9.8 Nm).

25. Install 5.0mm pin punches to the No.1 camshaft holders, then install the camshaft pulley keys onto the grooves in the camshafts.

26. Push the camshaft pulleys onto the camshafts, then tighten the retaining bolts to 27 ft. lbs. (38 Nm).

27. Install the timing belt and timing belt covers. Remove the pin punches from the camshaft holders.

28. Adjust the valves and pour oil over the camshafts and rocker arms.

29. Apply liquid gasket to the rubber seal at the eight corners of the recesses. Do not install the parts if 20 minutes or more have elapsed since applying the liquid gasket. Instead, reapply liquid gasket after removing old residue.

30. Install the cylinder head cover and engine ground cable. Be sure the contact surfaces are clean and do not touch surfaces where liquid gasket has been applied.

31. Tighten the cylinder head cover nuts in 2–3 steps. In the final step, tighten all nuts in sequence, to 7 ft. lbs. (10 Nm).

32. Install the distributor to the cylinder head and reconnect the spark plug wires to the spark plugs.

33. Connect the negative battery cable and enter the radio security code.

34. Change the engine oil. Wait at least 20 minutes for the sealant to cure before filling the engine with oil.

1.8L (B18C1) Engine

1. Disconnect the negative battery cable.

2. Remove the cylinder head from the vehicle.

3. Hold each rocker arm assembly together with a rubber band to prevent them from separating.

4. Remove the intake and exhaust rocker shaft orifices from the cylinder head. The rocker shaft orifices are different and should be identified when removed. Discard the O-rings on the orifices.

5. Remove the VTEC solenoid from the cylinder head and discard the filter.

6. Remove the rocker arm shaft sealing bolts, discard the washers.

7. Insert 12mm bolts into the rocker arm shafts. Remove each rocker arm set while slowly pulling out the rocker arm shaft.

➡**Tag each rocker arm set to assure installation in their original locations.**

8. Inspect the rocker arm pistons. If they do not move smoothly, replace the rocker arm assembly.

9. Remove the lost motion assembly from the cylinder head. Inspect the lost motion assembly by pushing the plunger with your finger. Replace the lost motion assembly if it does not move smoothly.

To install:

10. Install the lost motion assembly to the cylinder head.

11. Apply engine oil to the rocker arm pistons, then bundle the rocker arms with a rubber band. Apply a light coat of clean engine oil to the rocker arms.

12. Position the rocker arms in their original locations, if they are being reused. If new assembles are being used place them in the cylinder head.

13. Lightly coat the rocker arm shafts with clean engine oil, then install the rocker arm shafts into the cylinder head. A 12mm bolt can be installed into the end of the rocker arm shafts to aid in their installation. Be sure to install the shafts in the proper positions. Remove the 12mm bolts from the rocker arm shafts, if used.

14. Clean and install the rocker arm shaft orifices with new O-rings. If the holes in the rocker arm shafts are not aligned screw a 12mm bolt into the end of the shaft to position the shaft.

15. Install the sealing bolts with new washers, tighten the bolts to 47 ft. lbs. (64 Nm).

16. Install the cylinder head into the vehicle.

2.3L Engine

1. Disconnect the negative battery cable.

2. Turn the crankshaft so the No. 1 piston is at top dead center.

➡**The No. 1 piston is at top dead center when the pointer on the block aligns with the white painted mark on the flywheel (manual transaxle) or driveplate (automatic transaxle).**

3. Remove the air intake duct.

4. Remove the engine ground cable from the cylinder head cover.

5. Remove the connector and the terminal from the alternator, then remove the engine wiring harness from the valve cover.

6. Remove the ignition coil.

7. Label, then disconnect the electrical connectors from the distributor and the spark plug wires from the spark plugs. Mark the position of the distributor and remove it from the cylinder head. Disconnect the ignition coil wire from the distributor.

8. Remove the Positive Crankcase Ventilation (PCV) hose, then remove the cylin-

der head cover. Replace the rubber seals if damaged or deteriorated.

9. Remove the timing belt middle cover.

10. Ensure the words **UP** embossed on the camshaft pulleys are aligned in the upward position.

11. Mark the rotation of the timing belt if it is to be used again. Loosen the timing belt adjusting nut 1/2 turn, then release the tension on the timing belt. Push the tensioner to release tension from the belt, then tighten the adjusting nut.

12. Remove the timing belt from the camshaft sprockets.

❊❊ WARNING

Do not crimp or bend the timing belt more than 90° or less, then 1 in. (25mm) in diameter

13. Insert a 5.0mm pin punch in each of the camshaft caps, nearest to the sprockets, through the holes provided. Remove the camshaft sprocket attaching bolts, then remove the sprockets. Do not lose the sprocket keys.

14. Remove the side engine mount bracket B, then the timing belt back cover from behind the camshaft sprockets.

15. Loosen all of the rocker arm adjusting screws, then remove the pin punches from the camshaft caps.

16. Remove the camshaft holders, note the holders locations for ease of installation. Loosen the bolts in the reverse order of the holder bolts torque sequence.

17. Remove the camshafts from the cylinder head, then discard the camshaft seals.

18. Remove the rubber cap from the head, located at the end of the intake camshaft.

19. Remove the rocker arms from the cylinder head. Note the locations of the rocker arms.

➡ **The rocker arms have to be installed to their original locations if being reused.**

To install:

20. Lubricate the rocker arms with clean oil, then install the rocker arms on the pivot bolts and the valve stems. If the rocker arms are being reused, install them to their original locations. The locknuts and adjustment screws should be loosened before installing the rocker arms.

21. Lubricate the camshafts with clean oil.

22. Install the camshaft seals to the end of the camshafts that the timing belt sprockets attach to. The open side (spring) should be facing into the cylinder head when installed.

23. Be sure the keyways on the camshafts are facing up and install the camshafts to the cylinder head.

24. Install the rubber plug to the cylinder head at the end of the intake camshaft.

25. Apply liquid gasket to the head mating surfaces of the No. 1 and No. 6 camshaft holders, then install them along with No. 2, 3, 4 and 5. **I** or **E** marks are stamped on the camshaft holders to identify them as Intake or Exhaust side holders. The arrows stamped on the holders should point toward the timing belt.

26. Snug the camshaft holders in place.

27. Press the camshaft seals securely into place.

28. Tighten the camshaft holder bolts in two steps, following the proper sequence, to ensure that the rockers do not bind on the valves. Tighten all the bolts, except the four studs, to 7 ft. lbs. (10 Nm). Tighten the studs (number 5 and 7 bolts in the correct sequence) to 9 ft. lbs. (12 Nm).

29. Install the timing belt back cover.

30. Install the side engine mount bracket B. Tighten the bolt attaching the bracket to the cylinder head to 33 ft. lbs. (45 Nm). Tighten the bolts attaching the bracket to the side engine mount to 16 ft. lbs. (22 Nm).

31. Insert a 5.0mm pin punch in each of the camshaft caps, nearest to the pulleys, through the holes provided. Install the keys into the camshaft grooves.

32. Push the camshaft sprockets onto the camshafts, then tighten the retaining bolts to 27 ft. lbs. (38 Nm).

33. Ensure the words **UP** embossed on the camshaft pulleys are aligned in the

upward position. Install the timing belt to the camshaft sprockets, then remove the two 5.0mm pin punches from the camshaft bearing caps.

34. Loosen, then tighten the timing belt adjuster nut.

35. Turn the crankshaft counterclockwise until the cam pulley has moved 3 teeth; this creates tension on the timing belt. Loosen, then tighten the adjusting nut and tighten it to 33 ft. lbs. (45 Nm).

36. Adjust the valves.

37. Tighten the crankshaft pulley bolt to 181 ft. lbs. (250 Nm).

38. Install the middle timing belt cover and tighten the attaching bolts to 9 ft. lbs. (12 Nm).

39. Install the cylinder head cover and tighten the cap nuts to 7 ft. lbs. (10 Nm). Install the PCV hose to the cylinder head cover.

40. Install the distributor to the cylinder head, snug the attaching bolts until the timing has been checked and adjusted.

41. Connect the spark plug wires to the correct spark plugs, then connect the distributor electrical connectors. Install the ignition coil wire to the distributor.

42. Install the ignition coil.

43. Install the alternator wiring harness to the cylinder head cover, then connect the terminal and connector to the alternator.

44. Connect the engine ground cable to the cylinder head cover.

45. Install the air intake duct.

46. Drain the oil from the engine into a sealable container. Install the drain plug and refill the engine with clean oil.

47. Connect the negative battery cable and enter the radio security code.

48. Start the engine, checking carefully for any leaks.

49. Check and adjust the ignition timing. Tighten the distributor bolts to 13 ft. lbs. (18 Nm).

50. If equipped with 4WS, start the engine, then turn the steering wheel lock-to-lock to reset the 4WS control unit.

2.5L, 3.2L and 3.5L Engines

1. Disconnect the negative battery cable. Remove the timing belt covers and cylinder head covers.

2. Rotate the crankshaft to TDC compression of No. 1 piston and remove the timing belt.

3. For 2.5L engines, the springs between the rocker arms are not all the

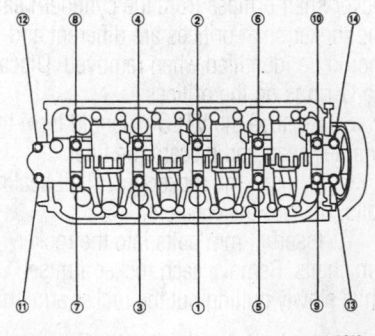

Camshaft holders torque sequence—2.3L engine

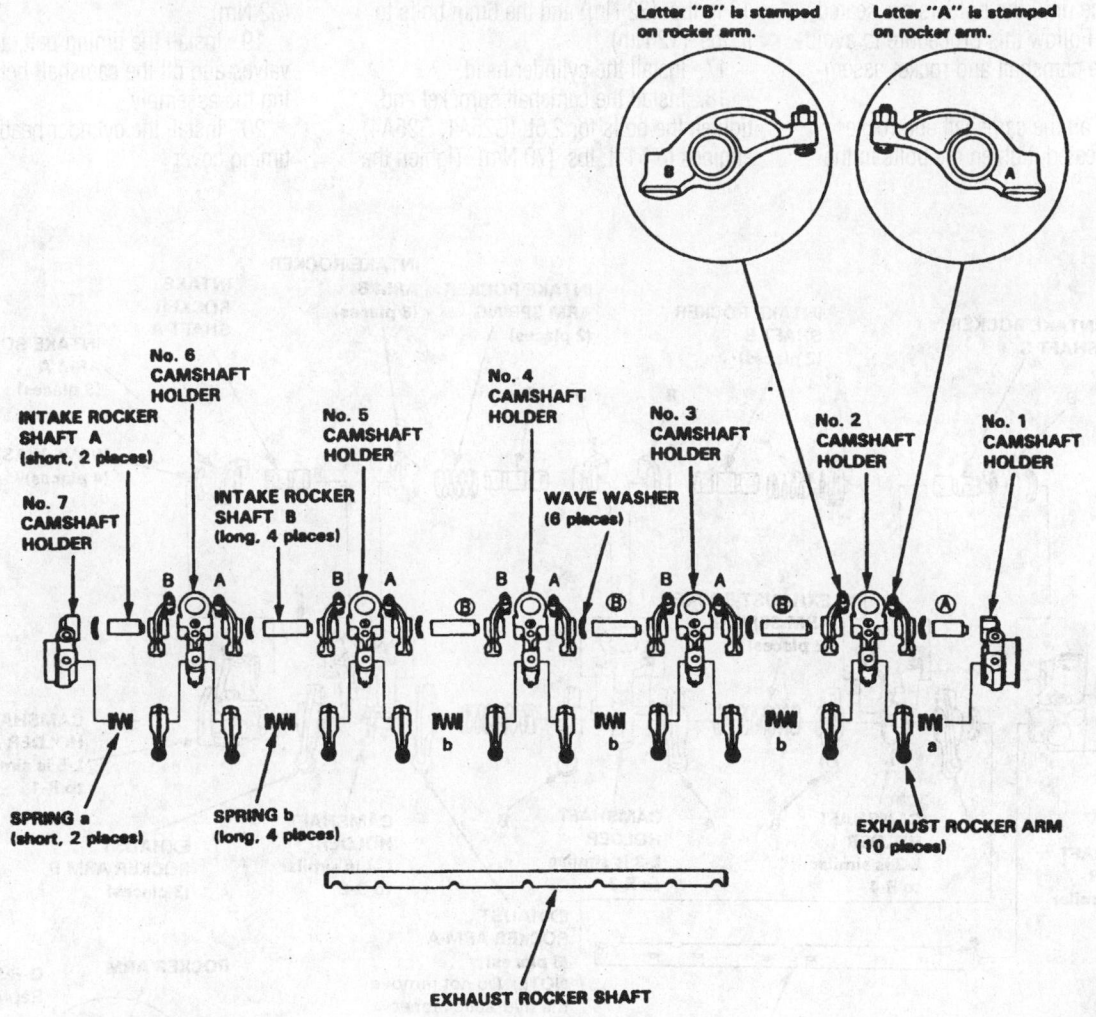

Letter "B" is stamped on rocker arm.

Letter "A" is stamped on rocker arm.

No. 6 CAMSHAFT HOLDER

No. 5 CAMSHAFT HOLDER

No. 4 CAMSHAFT HOLDER

No. 3 CAMSHAFT HOLDER

No. 2 CAMSHAFT HOLDER

No. 1 CAMSHAFT HOLDER

INTAKE ROCKER SHAFT A (short, 2 places)

No. 7 CAMSHAFT HOLDER

INTAKE ROCKER SHAFT B (long, 4 places)

WAVE WASHER (6 places)

SPRING a (short, 2 places)

SPRING b (long, 4 places)

EXHAUST ROCKER ARM (10 places)

EXHAUST ROCKER SHAFT

7923BG15

Rocker arm and shaft assembly—2.5L (G25A1, G25A4) engines

Specified torque:
8 mm bolts: 22 N·m (2.2 kg-m, 16 lb-ft)
6 mm bolts: 12 N·m (1.2 kg-m, 9 lb-ft)

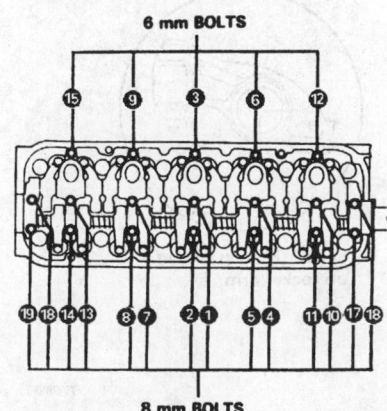

6 mm BOLTS

8 mm BOLTS

7923BG16

Rocker arm assembly holder bolt torque sequence—2.5L (G25A1, G25A4) engines

same length. Carefully note their positions during disassembly.

4. Remove the camshaft sprocket.

5. Remove the cylinder head from the vehicle.

6. Loosen the rocker shaft holder bolts 1 turn at a time in the opposite of the installation sequence. Following this procedure will prevent the camshafts and rocker assemblies from warping.

7. After all bolts are loose, remove the rocker arm shafts as an assembly with the bolts still in the holders.

8. If the rocker shafts are to be disassembled, note that each rocker arm has a letter **A** or **B** stamped into the side. Before disassembling the rocker arms, make a note of the position of each letter so the arms can be reassembled the same way.

9. For 3.2L (C32A6) and 3.5L engines, do not remove the hydraulic tappets from the rocker arms unless they are

to be replaced. Handle the rocker arms carefully so the oil does not drain out of the tappets.

10. Lift the camshafts from the cylinder head, wipe them clean and inspect the lift ramps. Replace the camshafts and rockers if the lobes are pitted, scored, or excessively worn.

To install:

11. Lubricate the camshaft and its journals with fresh engine oil.

12. Place a new camshaft seal on the end of the camshaft. The spring side of the seal must face in. Lubricate the journals and set the camshaft in place on the head.

13. Install the camshaft onto the cylinder head with the keyway pointed up.

14. Apply liquid gasket to the mounting surfaces of the camshaft end holders.

15. Set the rocker arm assemblies in place and start all the cam holder bolts. Be sure the rocker arms are properly posi-

tioned and turn each bolt in sequence two turns at a time until the holders are seated on the head. Follow this procedure to avoid damaging the camshaft and rocker assemblies.

16. When all the camshaft and rocker holders are seated, tighten the bolts in the same sequence. Tighten the 8mm bolts to 16 ft. lbs. (22 Nm) and the 6mm bolts to 9 ft. lbs. (12 Nm).

17. Install the cylinder head.

18. Install the camshaft sprocket and tighten the bolts for 2.5L (G25A1, G25A4) engines to 51 ft. lbs. (70 Nm). Tighten the bolts for 3.2L (C32A6) engine to 23 ft. lbs. (32 Nm).

19. Install the timing belt, adjust the valves and oil the camshaft before completing the assembly.

20. Install the cylinder head cover and timing cover.

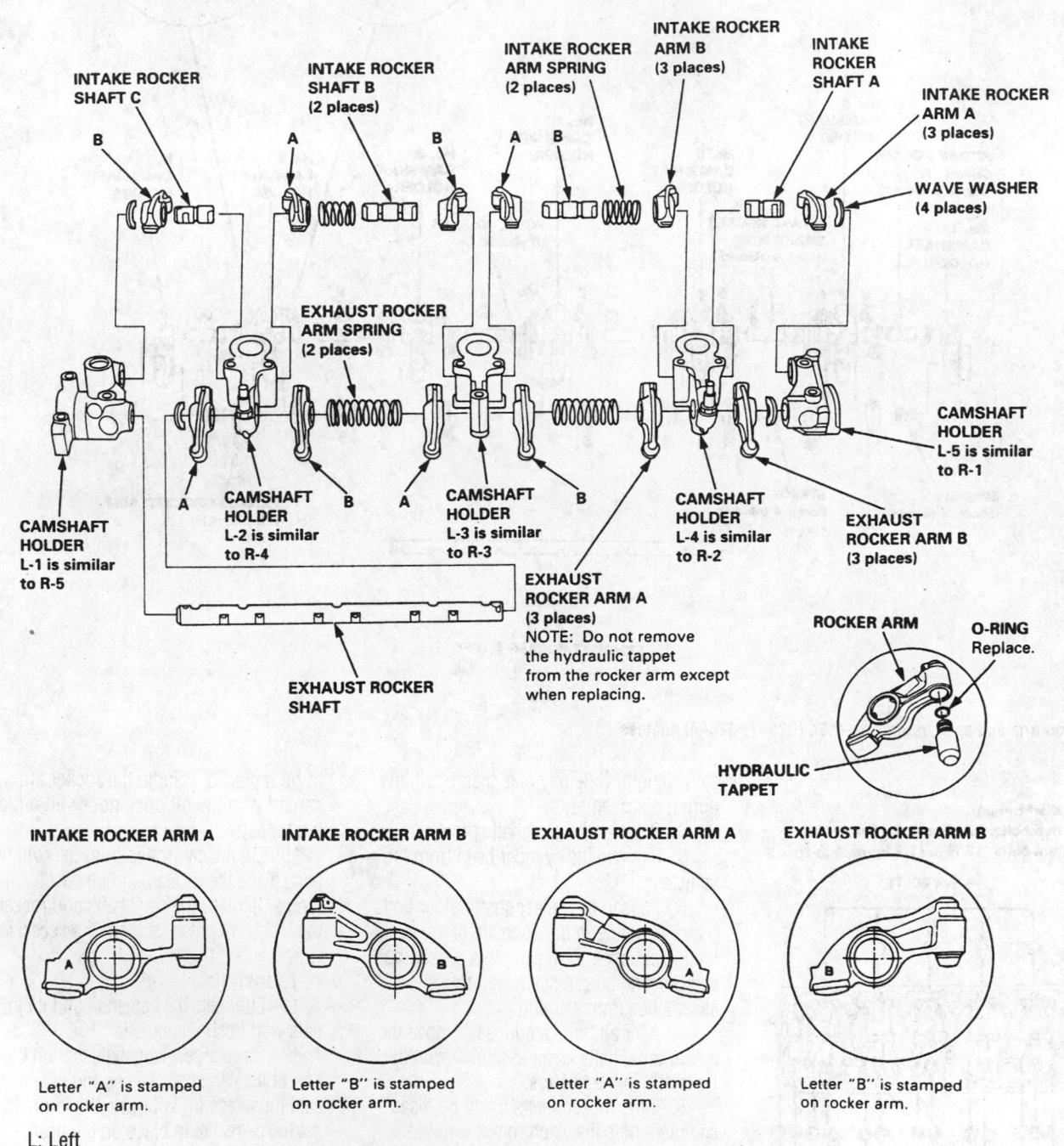

Exploded view of the rocker arms and related components—3.5L engine

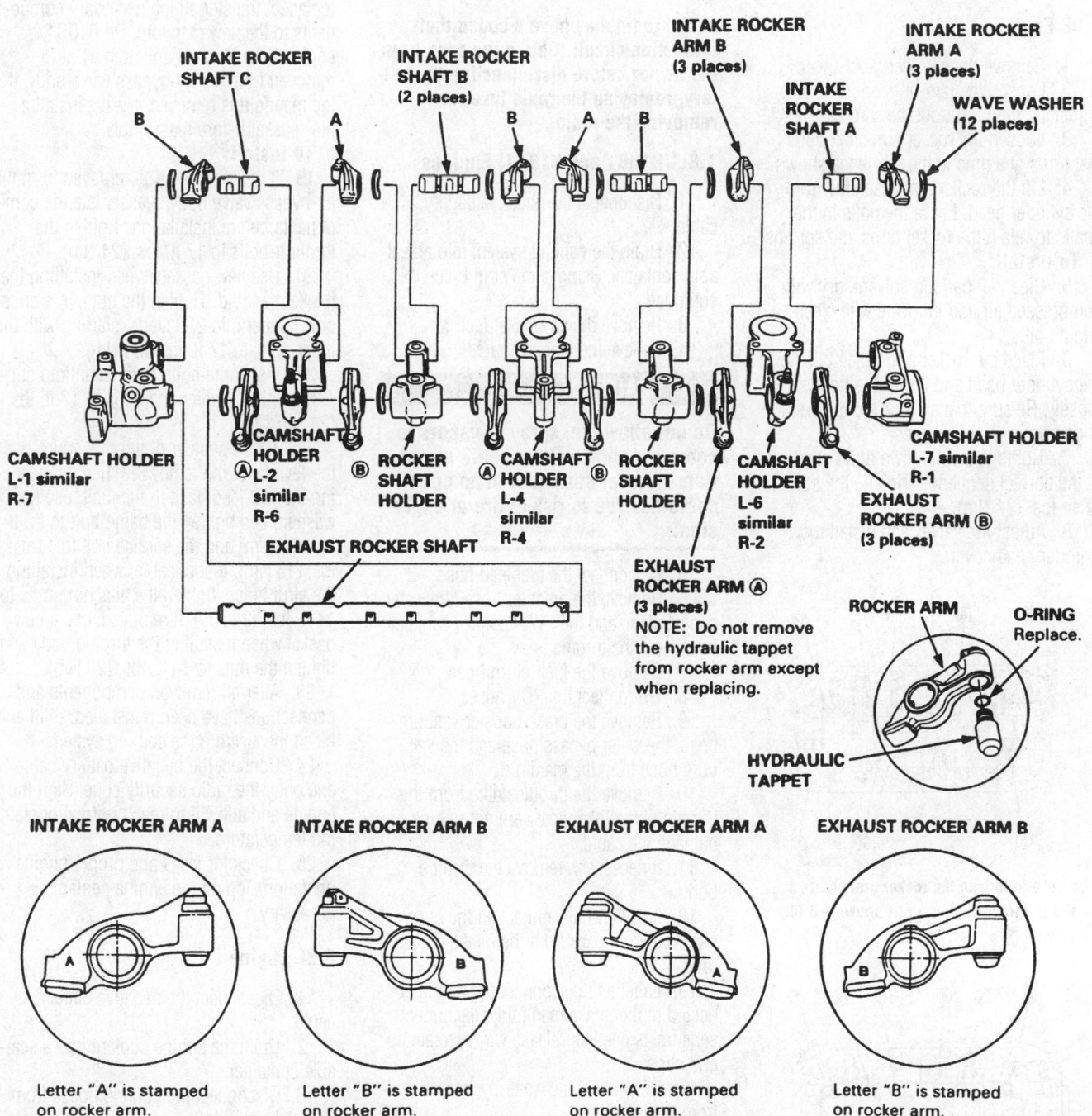

INTAKE ROCKER SHAFT C

INTAKE ROCKER SHAFT B (2 places)

INTAKE ROCKER ARM B (3 places)

INTAKE ROCKER ARM A (3 places)

INTAKE ROCKER SHAFT A

INTAKE ROCKER ARM A (3 places)

WAVE WASHER (12 places)

CAMSHAFT HOLDER L-1 similar R-7

CAMSHAFT HOLDER Ⓐ L-2 similar R-6

ⒷROCKER SHAFT HOLDER

ⒶCAMSHAFT HOLDER L-4 similar R-4

ⒷROCKER SHAFT HOLDER

CAMSHAFT HOLDER L-6 similar R-2

CAMSHAFT HOLDER L-7 similar R-1

EXHAUST ROCKER ARM Ⓑ (3 places)

EXHAUST ROCKER SHAFT

EXHAUST ROCKER ARM Ⓐ (3 places)
NOTE: Do not remove the hydraulic tappet from rocker arm except when replacing.

ROCKER ARM

O-RING Replace.

HYDRAULIC TAPPET

INTAKE ROCKER ARM A

Letter "A" is stamped on rocker arm.

INTAKE ROCKER ARM B

Letter "B" is stamped on rocker arm.

EXHAUST ROCKER ARM A

Letter "A" is stamped on rocker arm.

EXHAUST ROCKER ARM B

Letter "B" is stamped on rocker arm.

L: Left
R: Right

7923BG84

Exploded view of the rocker arms and related components—3.2L (C32A6) engine

21. Install the distributor.
22. Reconnect the negative battery cable.
23. Check for proper engine and valve-train operation.

3.0L Engine

1. Remove the cylinder head cover.
2. Loosen the jam-nuts on the adjusters, then back out the screws.
3. Loosen the rocker arm shaft bolts two turns at a time in the sequence shown.
4. Lift the rocker arm assembly from the cylinder head. Leave the bolts in the shafts to retain the rocker arms and springs.

To install:

5. Clean all parts in solvent, dry with compressed air and lubricate with clean engine oil.
6. Place the rocker arm assemblies on the cylinder head and install the bolts loosely. Be sure that all rocker arms are in alignment with their valves.
7. Tighten each bolt two turns at a time in the correct sequence. Tighten the bolts to 17 ft. lbs. (24 Nm).
8. Adjust the valves and install the cylinder head covers.

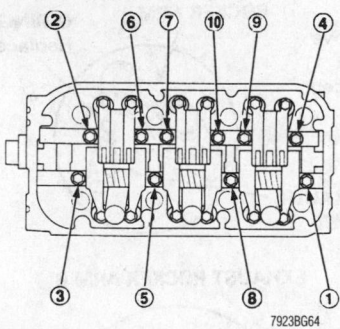

Be sure to loosen the rocker arm shaft bolts in the correct order as shown—3.0L engine

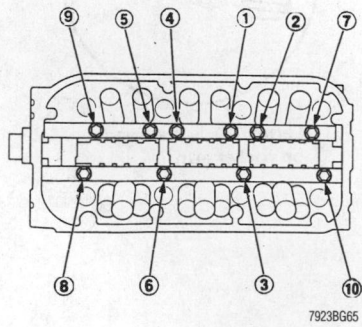

Tighten the bolts two turns at a time in the sequence shown—3.0L engine

Intake Manifold

REMOVAL & INSTALLATION

➡The radio may have a coded theft protection circuit. Obtain the code from the owner before disconnecting the battery, removing the radio fuse, or removing the radio.

1.8L (B18B1 and B18C1) Engines

1. Disconnect the negative battery cable.
2. Drain the cooling system into a sealable container. Remove the strut brace if equipped.
3. Remove the air intake duct.
4. Relieve the fuel pressure.

✻✻ CAUTION

Do not allow fuel spray or vapors to come in contact with a spark or open flame. Never store fuel in an open container due to risk of fire or explosion.

5. Disconnect the fuel feed hose.
6. Remove the breather hose, the water bypass hose and the EVAP control canister hose from the throttle body.
7. Remove the fuel return hose.
8. Disconnect the PCV hose.
9. Remove the brake booster vacuum hose, the water bypass hose and the vacuum hose from the manifold.
10. Remove the throttle cable from the throttle body. Take great care not to kink or damage the cable.
11. If necessary, remove the throttle body.
12. Label and disconnect all the emission vacuum hoses from the intake manifold.
13. Label and disconnect the wiring connected to the intake manifold. Disconnect sensors as needed; release wiring retainers and clips.
14. Disconnect the water bypass hoses from the manifold.
15. Remove the bolts attaching the intake manifold to the support bracket.
16. Remove the nuts attaching the intake manifold to the cylinder head. Remove the nuts in a crisscross pattern, beginning from the center and moving out to both ends.
17. Remove the manifold and the old gasket.

18. Clean the intake manifold mating surfaces. Inspect the manifold for cracks, flatness and/or damage; replace the parts, if necessary. If the intake manifold is to be replaced, transfer all the necessary components to the new manifold. On B18C1 engines, the intake manifold may be removed from the air bypass valve body. If the manifold is removed, always install a new gasket before reassembly.

To install:

19. If the manifold was removed from the air bypass valve body, reassemble the components before installation. Tighten the through-bolts to 17 ft. lbs. (24 Nm).
20. Use new gaskets when installing the intake manifold. Tighten the nuts, in a crisscross pattern, in 2–3 steps, starting with the inner nuts, to 17 ft. lbs. (23 Nm).
21. Install the bolts to the manifold support bracket. Tighten the bolts to 17 ft. lbs. (24 Nm).
22. The remainder of the procedure is the reverse of the removal. When connecting the fuel feed hose to the filter, use new washers and tighten the banjo bolt to 25 ft. lbs. (33 Nm) and the service bolt to 11 ft. lbs. (15 Nm). If applicable, when installing the strut brace, tighten the attaching nuts to 17 ft. lbs. (24 Nm). If removed, use a new gasket when installing the throttle body and tighten the nuts to 14 ft. lbs. (20 Nm).
23. After all removed components and connections have been reinstalled, refill and bleed the air from the cooling system.
24. Connect the negative battery cable and enter the radio security code. Start the engine and allow it to reach normal operating temperature.
25. Check for leaks and proper engine operation. Top off the engine coolant as necessary.

2.3L Engine

1. Disconnect the negative battery cable.
2. Drain the engine coolant into a sealable container.
3. Disconnect the cooling hoses from the intake manifold.
4. Label and unplug the vacuum hoses and electrical connectors on the manifold and throttle body. Unplug the connector from the Exhaust Gas Recirculation (EGR) valve. Position the wiring harnesses out of the way.
5. Disconnect the throttle cable from the throttle body.

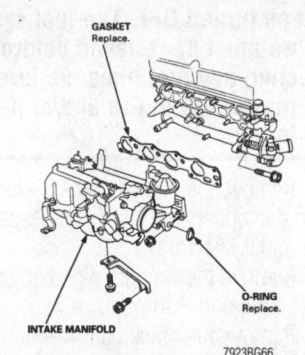

**Intake manifold and related components—
2.3L engine**

6. Relieve the fuel pressure.

7. Remove the fuel rail and fuel injectors.

8. Remove the thermostat housing mounting bolts. Remove the thermostat housing from the intake manifold and the connecting pipe, by pulling and twisting the housing. Discard the O-rings.

9. It may be necessary to remove the upper intake manifold plenum and throttle body assembly in order to access the nuts securing the manifold to the head.

10. Remove the intake manifold support bracket bolts and the bracket. It may be necessary to access it from under the vehicle; raise and support the vehicle safely.

11. While supporting the intake manifold, remove the nuts attaching the intake manifold to the cylinder head, then remove the manifold. Remove the old gasket from the cylinder head.

12. Clean any old gasket material from the cylinder head and the intake manifold. check and clean the FIA chamber on the cylinder head.

To install:

13. Using a new gasket, place the manifold into position and support.

14. Install the support bracket to the manifold. Tighten the bolt holding the bracket to the manifold to 16 ft. lbs. (22 Nm).

15. Starting with the inner or center nuts, tighten the nuts, in a crisscross pattern, to the correct torque. The tension must be even across the entire face of the manifold if leaks are to be prevented. Correct torque is 16 ft. lbs. (22 Nm).

16. Using a new gasket, install the upper intake manifold and throttle body assembly, if removed as a separate unit. Tighten the nuts and bolts holding the chamber to 16 ft. lbs. (22 Nm).

17. Install a new O-ring to the coolant connecting pipe, and to the thermostat housing. Install the housing to the coolant

pipe and the intake manifold. Tighten the mounting bolts to 16 ft. lbs. (22 Nm).

18. Connect and adjust the throttle cable.

19. Install the fuel rail/injector assembly. Connect the fuel lines.

20. Properly position the wiring harnesses and connect the electrical connectors.

21. Connect the vacuum hoses.

22. Fill and bleed the air from the cooling system.

23. Connect the negative battery cable and enter the radio security code.

24. If equipped with 4WS, turn the steering wheel lock-to-lock to reset the 4WS control unit.

25. Start the engine, checking carefully for any leaks of fuel, coolant or vacuum. Check the manifold gasket areas carefully for any leakage of vacuum.

2.5L Engine

1. Disconnect the negative battery cable.

2. Remove the fuel filler cap and loosen the service bolt on the fuel filter to relieve the fuel system pressure. Remove the banjo bolt to remove the fuel feed hose from the fuel filter. Remove the fuel return hose from the pressure regulator.

> ❋❋ **CAUTION**

The fuel injection system remains under pressure after the engine has been turned OFF. Properly relieve fuel pressure before disconnecting any fuel lines. Failure to do so may result in fire or personal injury. Do not allow fuel spray or fuel vapors to come in contact with a spark or open flame. Keep a dry chemical fire extinguisher nearby. Never store fuel in an open container due to risk of fire or explosion.

3. Remove the intake air duct and air cleaner assembly.

4. Remove the throttle cable by loosening the locknut, then slip the cable end out of the throttle bracket and throttle linkage. Take care not to bend the cable when removing it. Move the cable aside.

5. Remove the engine harness cover. Label and disconnect the vacuum hoses and all wiring from the intake manifold.

6. To avoid having to drain the cooling system, remove the fast idle valve and the IAC valve without disconnecting the coolant hoses. Move these components out of the work area so that they will not be damaged.

7. Remove the EGR pipe and the vacuum pipe.

8. Remove the fuel rail. Remove the fuel injectors from the manifold. Handle the injectors and fuel rail carefully to avoid damaging them or contaminating them with dirt.

9. Unbolt the top bolts on the intake manifold brackets.

10. Remove the nuts that secure the manifold to the head. Remove the intake manifold from the engine.

To install:

➡**Use new O-rings when installing the IAC and fast idle valves.**

11. Inspect the manifold and its components for any signs of damage.

12. Fit the manifold to the engine with a new gasket and tighten the nuts to 16 ft. lbs. (22 Nm). Tighten the manifold bracket bolts to 16 ft. lbs. (22 Nm).

13. Install the fuel injectors into the rail and install the assembly onto the manifold with new sealing rings and cushion rings to prevent noise and leakage.

14. Connect the fuel injector harnesses and install the harness cover.

15. Install the IAC, fast idle, EGR, and EVAP valves. Use new O-rings.

16. Connect the wiring, vacuum hoses, and fuel lines. Use new sealing washers when connecting the fuel lines.

17. Install the throttle cable into its bracket and linkage. The throttle cable deflection is the measured by pressing down on the cable between the rubber boot and the linkage. The deflection must be 0.39–0.47 in. (10–12mm). Adjust the throttle cable as required.

18. Verify that all wiring and vacuum hoses are installed correctly.

19. Connect the negative battery cable. Run the engine and check for leaks.

3.0L (J30A1) Engine

1. Obtain the security code for the radio.

2. Disconnect the negative battery cable.

3. Drain the coolant.

4. Remove the EVAP canister hose from the throttle body.

5. Remove the air intake duct.

6. Remove the upper engine covers.

7. Disconnect the accelerator and cruise control cables from the throttle body.

8. Ensure that all components have be removed from the intake manifold.

9. Remove the intake manifold.

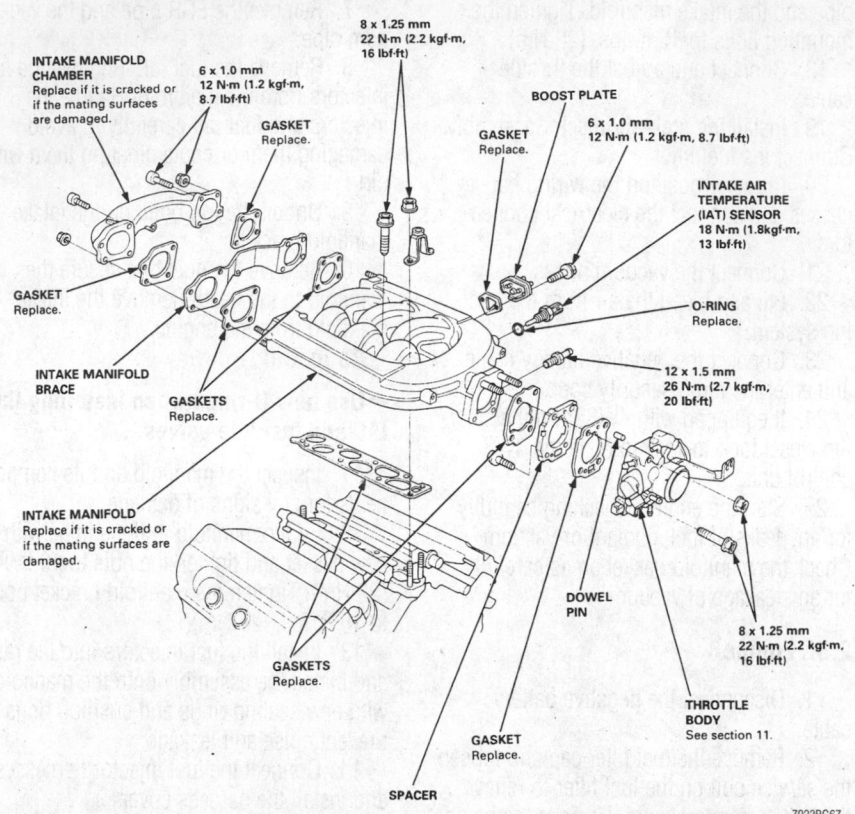

Exploded view of the intake manifold and related components—3.0L engine

7923BG67

To install:

10. Clean the mounting surfaces.

11. Install a new gasket on the engine and install the manifold. Tighten the bolts to 16 ft. lbs. (22 Nm).

12. Install all removed hoses and wiring on the intake manifold and throttle body.

13. Install the engine covers.

14. Install the intake air duct.

15. Refill the cooling system.

16. Connect the negative battery cable, start the engine and check for leaks.

3.2L and 3.5L Engines

1. Disconnect the negative battery cable.

2. Drain the cooling system.

3. Remove the air intake duct from the throttle body.

4. On GS models, remove the TCS control valve assembly and its brackets from the throttle body.

5. Relieve the fuel system pressure by loosening the service bolt on the fuel filter about 1 turn, then disconnect the fuel supply and return lines from the manifold.

✳✳ CAUTION

The fuel injection system remains under pressure, even after the engine

has been turned OFF. The fuel system pressure must be relieved before disconnecting any fuel lines. Failure to do so may result in fire and/or personal injury.

6. Remove the engine harness covers, tag and disconnect the wiring harnesses from the fuel injectors.

7. Remove the vacuum pipe harness, air inlet pipe, and EGR pipe.

8. Remove the pulsed air injection pipe and valve.

9. Remove the intake manifold nuts and bolts in a crisscross pattern, beginning from the center and moving out to both ends of the manifold.

10. Verify that all vacuum lines are disconnected and remove the intake manifold and throttle body as a unit.

11. Remove the water passage and clean the gasket mounting surfaces.

12. Inspect the manifold for cracks, flatness, or other damage; replace any damaged parts. If the intake manifold is to be replaced, transfer all the necessary components to the new manifold.

To install:

➡Always use new gaskets and O-rings during installation.

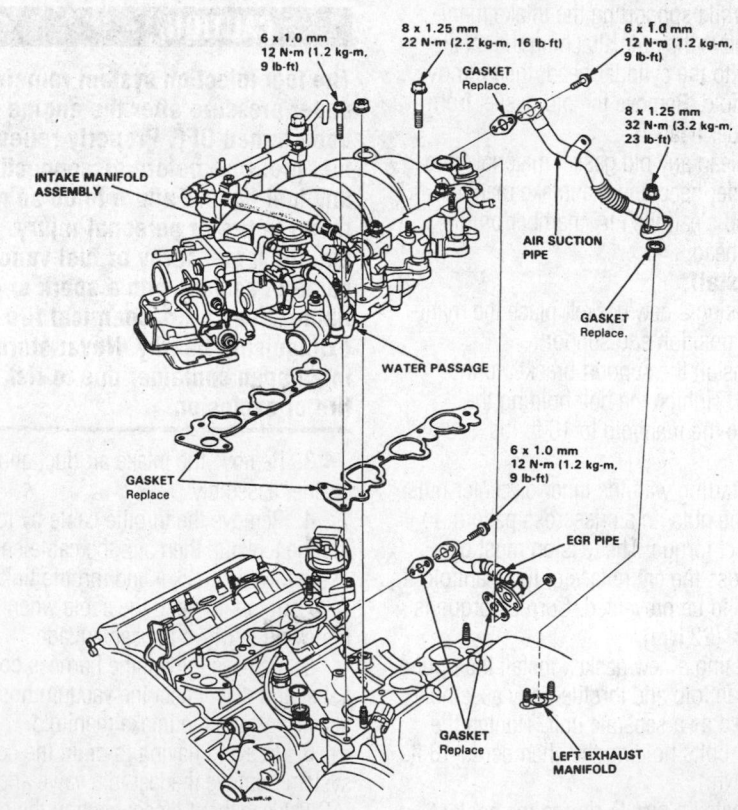

Intake manifold assembly—3.2L (C32A1) Engines (Legend shown, 3.2TL similar)

7923BG17

13. Install the water passage.

14. Install the intake manifold and tighten the nuts/bolts, in a crisscross pattern in 2–3 steps, starting with the inner nuts. Tighten the 8mm bolts to 16 ft. lbs. (22 Nm) and the 6mm bolts to 9 ft. lbs. (12 Nm).

15. Reconnect the vacuum lines and the air inlet, and the EGR pipes.

16. Reconnect the fuel supply and return lines to the manifold. Tighten the fuel system service bolt.

17. Reconnect all intake manifold wiring connectors.

18. Install the TCS control valve assembly and brackets.

19. Install the air duct to the throttle body. Refill the cooling system.

20. Reconnect the negative battery cable. Start the engine, allow it to reach normal operating temperature and check for leaks and proper engine operation.

Exhaust Manifold

REMOVAL & INSTALLATION

➡The radio may have a coded theft protection circuit. Obtain the code from the owner before disconnecting the battery, removing the radio fuse, or removing the radio.

1.8L Engines

✳✳ WARNING

This procedure should only be performed on a cold engine.

1. Disconnect the negative battery cable.
2. Remove the exhaust manifold cover.
3. Raise and safely support the vehicle.
4. Remove the three nuts attaching the front exhaust pipe to the exhaust manifold. Discard the nuts. Separate the exhaust pipe from the manifold and discard the gaskets.
5. Lower the vehicle.
6. Disconnect the bracket attaching the exhaust manifold to the engine.
7. Remove the exhaust manifold attaching nuts and discard the nuts.
8. Remove the exhaust manifold from the engine. Clean any old gasket material from the engine and the exhaust manifold mating surfaces.
9. Remove the rear cover from the exhaust manifold.
   To install:
10. Install the rear cover to the exhaust manifold and tighten the mounting bolts to 17 ft. lbs. (24 Nm).

11. Install a new exhaust manifold gasket to the cylinder head.

12. Install the exhaust manifold to the engine and install new attaching nuts. Tighten the nuts to 23 ft. lbs. (31 Nm).

13. Install the bracket to the exhaust manifold and the engine. Tighten the bolts to 33 ft. lbs. (44 Nm).

14. Raise and safely support the vehicle.

15. Install new gaskets to the front exhaust pipe where it connects to the exhaust manifold.

16. Connect the front exhaust pipe to the exhaust manifold. Install new nuts and tighten the nuts to 40 ft. lbs. (54 Nm).

17. Lower the vehicle.

18. Install the exhaust manifold cover and tighten the bolts to 17 ft. lbs. (24 Nm).

19. Connect the negative battery cable and enter the radio security code.

20. Run the engine and check for exhaust leaks.

2.3L Engine

1. Disconnect the negative battery cable.
2. Safely raise and support the vehicle.
3. If the oxygen sensor is located in the exhaust manifold, disconnect the oxygen sensor connector.

4. Remove the exhaust manifold upper cover.

5. If equipped with air conditioning, remove the heat insulator from the manifold.

6. Remove the nuts attaching the exhaust manifold to the front exhaust pipe. Separate the pipe from the manifold and discard the gasket. Support the pipe with wire; do not allow it to hang by itself.

7. Remove the exhaust manifold bracket(s) bolts and remove the bracket(s).

8. Using a crisscross pattern (starting from the center), remove the exhaust manifold attaching nuts.

9. Remove the manifold and discard the gasket. Clean the manifold and cylinder head mating surfaces.

10. If equipped, remove the lower manifold cover from the manifold.
    To install:
11. If equipped, install the lower manifold cover, tighten the attaching bolts to 16 ft. lbs. (22 Nm).

12. Using a new gasket and nuts, place the manifold into position and support it. Install the nuts snug on the studs.

13. Install the support bracket(s) below the manifold. Tighten the bracket(s) mounting bolts to 33 ft. lbs. (44 Nm).

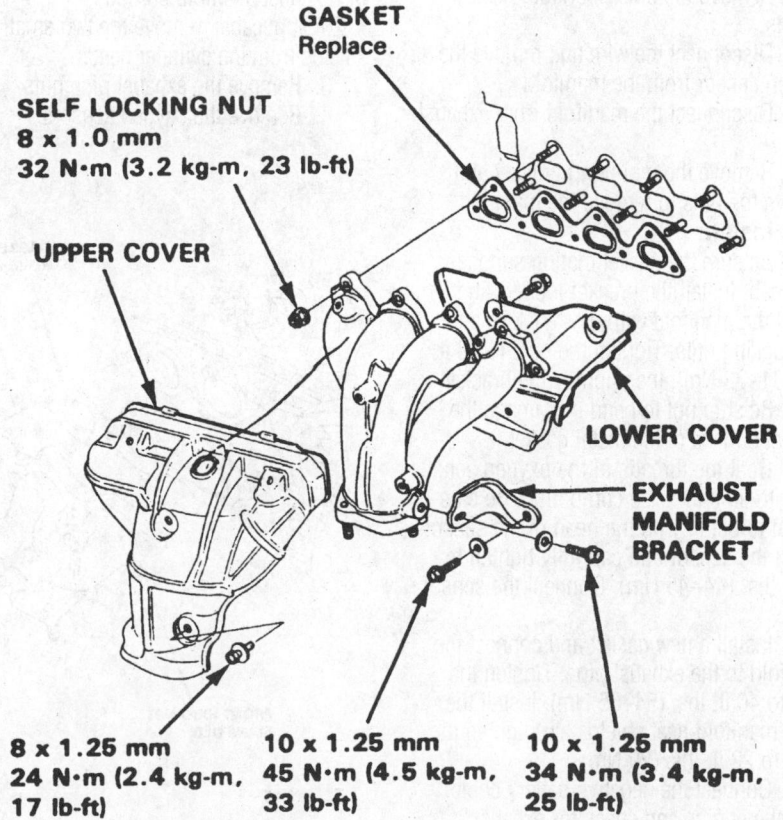

GASKET
Replace.

SELF LOCKING NUT
8 x 1.0 mm
32 N·m (3.2 kg-m, 23 lb-ft)

UPPER COVER

LOWER COVER

EXHAUST MANIFOLD BRACKET

8 x 1.25 mm
24 N·m (2.4 kg-m, 17 lb-ft)

10 x 1.25 mm
45 N·m (4.5 kg-m, 33 lb-ft)

10 x 1.25 mm
34 N·m (3.4 kg-m, 25 lb-ft)

7923BG18

Exhaust manifold—1.8L engines

14. Starting with the manifold inner or center nuts, tighten the nuts in a criss-cross pattern to the correct torque. The tension must be even across the entire face of the manifold if leaks are to be prevented. Tighten the nuts to 23 ft. lbs. (31 Nm).

15. If equipped with air conditioning, install the heat insulator to the manifold. Tighten the attaching bolts to 7 ft. lbs. (10 Nm) on Prelude models and 9 ft. lbs. (12 Nm) on Accord models.

16. Install the upper manifold cover, tighten the bolts to 16 ft. lbs. (22 Nm).

17. If disconnected, connect the oxygen sensor connector.

18. Connect the front exhaust pipe, using new gaskets and nuts. Tighten the exhaust pipe attaching nuts to 40 ft. lbs. (55 Nm).

19. Connect the negative battery cable and enter the radio security code.

20. Start the engine and check for exhaust leaks.

21. If equipped with 4WS, turn the steering wheel lock-to-lock to reset the 4WS control unit.

2.5L Engine

1. Disconnect the negative battery cable.
2. Remove the outer manifold heat shields.
3. Disconnect the wire and remove the oxygen sensor from the manifold.
4. Disconnect the manifold from exhaust pipe.
5. Remove the mounting bracket and remove the nuts to remove the manifold.

To install:

6. Be sure the gasket mating surfaces are clean. Install the bracket loosely and install the manifold with new gaskets and self-locking nuts. Tighten the nuts to 23 ft. lbs. (31–32 Nm), then tighten the bracket bolts. Be sure not to bend or damage the contact surface of the metal gasket.

7. Coat the threads of the oxygen sensor with an anti-seize compound. Be careful not to get any on the head of the sensor. Install the sensor and carefully tighten to 33 ft. lbs. (44–45 Nm). Connect the sensor wire.

8. Install a new gasket and connect the manifold to the exhaust pipe. Tighten the nuts to 40 ft. lbs. (54–55 Nm). Install the outer manifold heat shields and tighten the bolts to 22 ft. lbs. (29 Nm).

9. Connect the negative battery cable. Start the engine and check for exhaust leaks.

3.0L Engine

1. Raise and safely support the vehicle.
2. Remove the engine undercover.
3. Disconnect the exhaust pipe from the manifold to be removed.
4. Lower the vehicle.
5. Remove the exhaust manifold heat shield.
6. Remove the mounting nuts and the exhaust manifold.

To install:

7. Clean the mounting surfaces.
8. Position a new gasket on the cylinder head.
9. Install the exhaust manifold. Tighten the nuts to 23 ft. lbs. (31 Nm).
10. Install the heat shield. Tighten the bolts to 16 ft. lbs. (22 Nm).
11. Raise the vehicle and connect the exhaust pipe to the manifold using a new gasket. Tighten the nuts to 40 ft. lbs. (54 Nm).

3.2L and 3.5L Engines

➡ **This operation should be performed with the engine and exhaust cold.**

1. Disconnect the negative battery cable. Be sure the engine is not hot or warm before performing this operation. Remove the exhaust manifold shrouds.

2. If applicable, remove the two small heat shields from the cylinder heads.

3. Remove the exhaust pipe nuts.

4. Remove the oxygen sensors.

5. Remove the air suction tube.

6. Remove the exhaust attaching nuts in a crisscross pattern starting from the center of the manifold.

7. Clean the gasket mounting surfaces. Inspect the manifold for cracks, flatness and/or damage; replace the parts, if necessary.

To install:

8. To install, use new gaskets and self-locking nuts. Be sure all mating surfaces are clean before installing exhaust manifold. Tighten the manifold nuts in a crisscross pattern starting from the center, for Legend models to 25 ft. lbs. (34 Nm). For 3.2TL, tighten the manifold nuts to 22 ft. lbs. (30 Nm).

9. If applicable, install the two small heat shields and tighten the attaching bolts to 16 ft. lbs. (22 Nm).

10. Use new gaskets when installing the exhaust pipe to the manifold and tighten the nuts to 40 ft. lbs. (55 Nm).

➡ **On GS series Legends, the exhaust manifold is bolted to twin three-way catalytic converters.**

11. Install the air suction tube, then install the oxygen sensors. Tighten the oxygen sensors to 33 ft. lbs. (45 Nm).

12. Install the manifold shrouds, tightening the bolts to 16 ft. lbs. (22 Nm).

13. Verify that all vacuum lines and wiring are properly connected.

14. Reconnect the negative battery cable.

15. Start the engine and check for leaks.

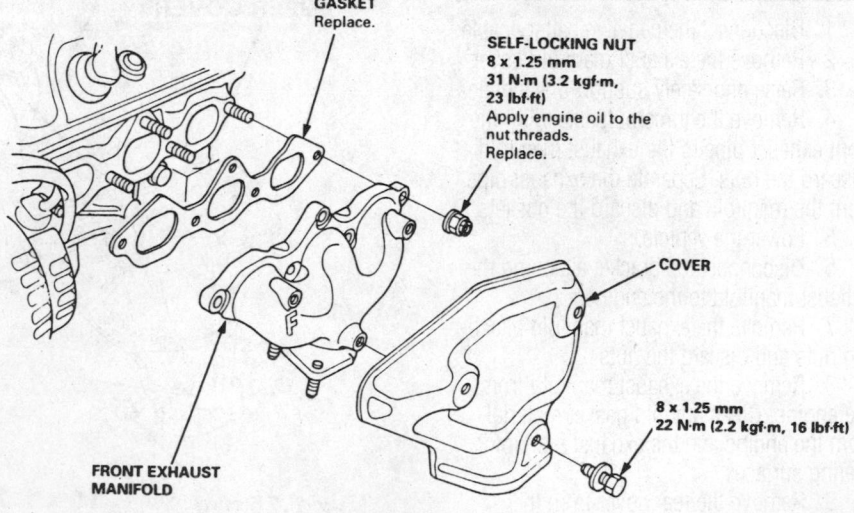

GASKET
Replace.

SELF-LOCKING NUT
8 x 1.25 mm
31 N·m (3.2 kgf·m, 23 lbf·ft)
Apply engine oil to the nut threads.
Replace.

COVER

8 x 1.25 mm
22 N·m (2.2 kgf·m, 16 lbf·ft)

FRONT EXHAUST MANIFOLD

7923BG68

Exploded view of the front exhaust manifold mounting—3.0L engine

Front Crankshaft Seal

REMOVAL & INSTALLATION

➡**The radio may have a coded theft protection circuit. Obtain the code from the owner before disconnecting the battery, removing the radio fuse, or removing the radio.**

1. Disconnect negative cable at the battery.

2. Raise and safely support the vehicle. Drain the engine oil and properly dispose of it.

3. Be sure the crankshaft is at TDC on No. 1 cylinder by aligning the white mark on the crankshaft pulley with the pointer on the lower timing belt cover.

4. Remove the cylinder head cover, timing belt cover and timing belt.

5. If equipped with a Crankshaft Speed Fluctuation (CKF) sensor, remove the sensor form the oil pump.

6. Remove the timing belt gear from the crankshaft.

7. Using a small prying tool, remove the seal from the oil pump.

To install:

8. Apply a light coat of oil to the seal lip.

9. Position the seal on the oil pump, then using a seal driver, install the seal into the oil pump.

10. Install the timing belt pulley to the crankshaft. If equipped with a CKF sensor, install the sensor to the oil pump and tighten the attaching bolts to 8 ft. lbs. (11 Nm).

11. Install the timing belt, timing belt covers and cylinder head cover.

12. Lower the vehicle and fill the engine with oil.

13. Connect the negative battery cable and enter the radio security code.

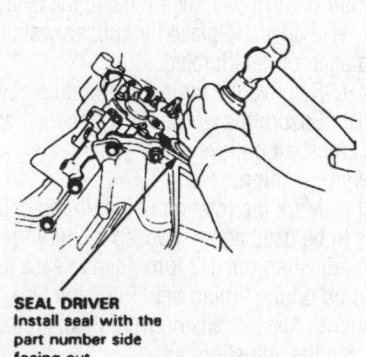

SEAL DRIVER
Install seal with the part number side facing out.

7923BG19

Installing the seal

14. Run the engine and check for leaks.

15. Turn off engine and check the oil level. Top off the oil level if necessary.

Camshaft and Valve Lifters

REMOVAL & INSTALLATION

➡**The radio may have a coded theft protection circuit. Obtain the code from the owner before disconnecting the battery, removing the radio fuse, or removing the radio.**

1.8L (B18C1) Engines

1. Disconnect the negative battery cable.

2. Be sure the crankshaft is at TDC/compression on No. 1 cylinder by aligning the white mark on the crankshaft pulley with the pointer on the lower timing belt cover.

3. Remove the strut brace.

4. Remove the cylinder head cover, timing belt cover, and timing belt.

5. Remove the camshaft pulleys and back cover.

6. Loosen the rocker arm locknuts and adjusting screws.

7. Remove the camshaft holder bolts, then remove the camshaft holder plates, the camshaft holders, and camshafts.

To install:

8. Be sure that the keyways on the camshafts are facing up and that the rocker arms are in their original position. The valve locknuts should be loosened and the adjusting screw backed off before installation

9. Install the camshafts, then install the camshaft seals with the open side facing in. Install the rubber cap with liquid gasket applied.

10. Install a new O-ring and the dowel pin to the oil passage of the No. 3 camshaft holder.

11. Apply liquid gasket to the head of the mating surfaces of the No. 1 and No. 5 camshaft holders, then install them, along with No. 2, 3, and 4. Be sure to pay attention to the following points:

• Do not apply oil to the holder mating surface of camshaft seals.

• The arrows marked on the camshaft holders should point to the timing belt.

12. Tighten the camshaft holders temporarily. Be sure that the rocker arms are properly positioned on the valve stems.

13. Tighten the camshaft holder bolts in two steps, following the proper sequence, to ensure that the rockers do not bind on the

valves. Tighten the 8x1.25mm bolts to 20 ft. lbs. (27 Nm). Tighten the 6x1.0mm bolts to 7 ft. lbs. (10 Nm).

14. Install the keys into the camshaft grooves. To set the No. 1 piston at TDC, align the holes on the camshaft with the holes in the No. 1 camshaft holders and insert 5.0mm pin punches into the holes.

15. Install the back cover and push the camshaft pulleys onto the camshafts, then tighten the retaining bolts to 27 ft. lbs. (37 Nm). Install the timing belt and adjust the tension, then install the timing belt covers.

16. Adjust the valve clearance.

17. Install the cylinder head cover. Be sure that the seal and groove are thoroughly clean first.

18. Install the engine side mount, tighten the two new nuts and new bolt to the engine to 38 ft. lbs. (52 Nm) and tighten the bolt attaching the mount to the vehicle to 54 ft. lbs. (74 Nm).

19. Install the distributor to the cylinder head and reconnect the spark plug wires to the spark plugs.

20. Install the intake air duct.

21. Install the strut brace, tighten the nuts to 17 ft. lbs. (24 Nm).

22. Connect the negative battery cable and enter the radio security code.

23. Drain the engine oil. Wait at least 20 minutes before filling the engine with oil; the time delay allows the sealant to cure.

1.8L (B18B1) Engines

1. Disconnect the negative battery cable.

2. Remove the spark plug wires.

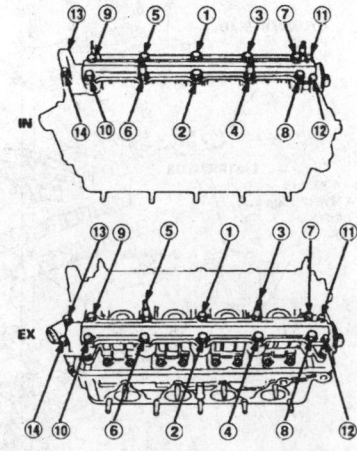

1-10: 8 x 1.25 mm 27 N·m (2.8 kgf·m, 20 lbf·ft)
11-14: 6 x 1.0 mm 9.8 N·m (1.0 kgf·m, 7.2 lbf·ft)

7923BG20

Camshaft holder plates torque sequence—1.8L (B18C1) engines

3. Remove the cylinder head cover and timing belt cover.

4. Rotate the crankshaft to TDC, compression of No. 1 piston and remove the timing belt.

5. Remove the distributor from the cylinder head.

6. Install 5.0mm pin punches to the No.1 camshaft holders, then remove the camshaft sprockets.

7. Loosen the valve adjusters to remove as much spring tension as possible.

8. Remove the pin punches from the camshaft holders.

To install:

9. Check the following before installing the camshafts:

   a. Be certain the keyways on the camshafts are facing UP (No. 1 cylinder at TDC).

   b. The valve adjuster locknuts should be loosened and the adjusting screws backed off before installation.

10. Lubricate the rocker arms and camshafts with clean oil.

11. Place the rocker arms on the pivot bolts and the valve stems, making sure that the rocker arms are in their original positions.

12. Install the camshaft seals with the open side (spring) facing in. Lubricate the lip of the seal.

13. Be sure the keyways on the camshafts are facing up and install the camshafts to the cylinder head.

14. Apply liquid gasket to the head mating surfaces of the No. 1 and No. 6 camshaft holders, then install them along with No. 2, 3, 4 and 5 camshaft holders. The arrows stamped on the holders should point toward the timing belt. Do not apply oil to the holder mating surface where the camshaft seals are housed.

15. Tighten the camshaft holders temporarily and be sure that the rocker arms are properly positioned.

16. Press the oil seals into the No.1 camshaft holders with a seal driver.

17. Tighten the bolts in a crisscross pattern to 7 ft. lbs. (10 Nm). Check that the rockers do not bind on the valves.

18. Install the cylinder head plug to the end of the cylinder head. If the plug has alignment marks, align the marks with the cylinder head upper surface.

19. If equipped with a timing belt back cover, install the cover and tighten the bolts to 7.2 ft. lbs. (9.8 Nm).

20. Install 5.0mm pin punches to the No.1 camshaft holders, then install the camshaft pulley keys onto the grooves in the camshafts.

21. Push the camshaft pulleys onto the camshafts, then tighten the retaining bolts to 27 ft. lbs. (38 Nm).

22. Install the timing belt and timing belt covers. Remove the pin punches from the camshaft holders.

23. Adjust the valves and pour oil over the camshafts and rocker arms.

24. Install the cylinder head cover and engine ground cable.

25. Install the distributor to the cylinder head and reconnect the spark plug wires to the spark plugs.

26. Connect the negative battery cable and enter the radio security code.

27. Change the engine oil. Wait at least 20 minutes for the sealant to cure before filling the engine with oil.

2.3L Engine

1. Disconnect the negative battery cable.

2. Turn the crankshaft so the No. 1 piston is at top dead center.

➡The No. 1 piston is at top dead center when the pointer on the block aligns with the white painted mark on the flywheel (manual transaxle) or driveplate (automatic transaxle).

3. Remove the air intake duct.

4. Remove the engine ground cable from the cylinder head cover.

5. Remove the connector and the terminal from the alternator, then remove the engine wiring harness from the valve cover.

6. Remove the ignition coil.

7. Label, then disconnect the electrical connectors from the distributor and the spark plug wires from the spark plugs. Mark the position of the distributor and remove it from the cylinder head. Disconnect the ignition coil wire from the distributor.

8. Remove the Positive Crankcase Ventilation (PCV) hose, then remove the cylinder head cover. Replace the rubber seals if damaged or deteriorated.

9. Remove the timing belt middle cover.

10. Ensure the words **UP** embossed on the camshaft pulleys are aligned in the upward position.

11. Mark the rotation of the timing belt if it is to be used again. Loosen the timing belt adjusting nut 1/2 turn, then release the tension on the timing belt. Push the tensioner to release tension from the belt, then tighten the adjusting nut.

12. Remove the timing belt from the camshaft sprockets.

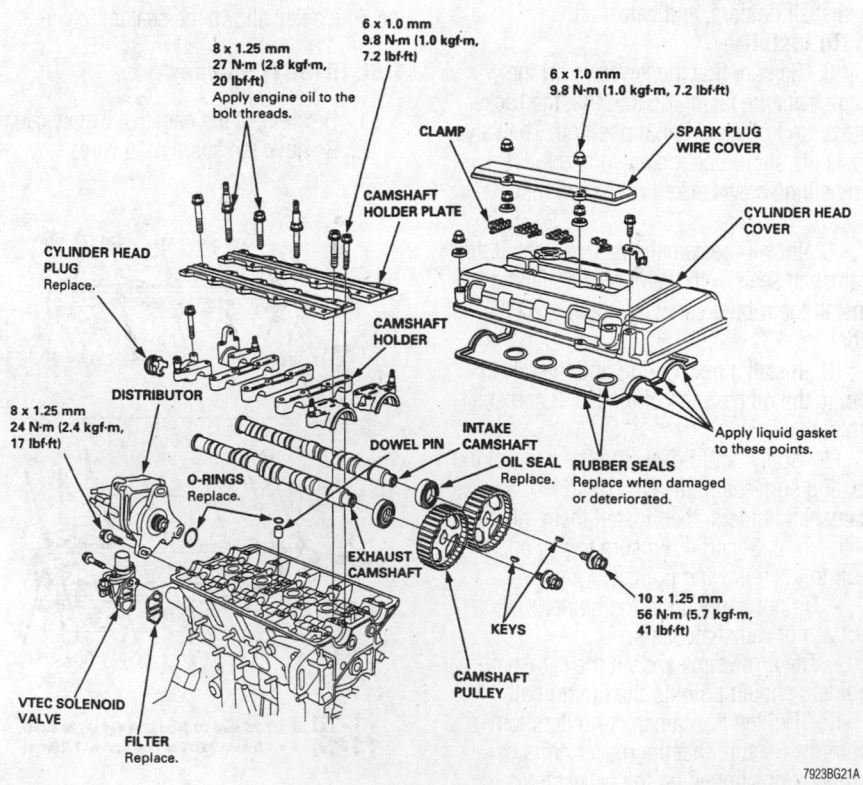

Exploded view of the cylinder head—1.8L (B18B1) engine

7923BG21A

❊❊ WARNING

Do not crimp or bend the timing belt more than 90° or less, then 1 in. (25mm) in diameter

13. Insert a 5.0mm pin punch in each of the camshaft caps, nearest to the sprockets, through the holes provided. Remove the camshaft sprocket attaching bolts, then remove the sprockets. Do not lose the sprocket keys.

14. Remove the side engine mount bracket, then the timing belt back cover from behind the camshaft sprockets.

15. Loosen all of the rocker arm adjusting screws, then remove the pin punches from the camshaft caps.

16. Remove the camshaft holders, note the holders locations for ease of installation. Loosen the bolts in the reverse order of the installation.

17. Remove the camshafts from the cylinder head, then discard the camshaft seals.

18. Remove the rubber cap from the head, located at the end of the intake camshaft.

19. Remove the rocker arms from the cylinder head. Note the locations of the rocker arms.

➡The rocker arms have to be installed to their original locations if being reused.

To install:

20. Lubricate the rocker arms with clean oil, then install the rocker arms on the pivot bolts and the valve stems. If the rocker arms are being reused, install them to their original locations. The locknuts and adjustment screws should be loosened before installing the rocker arms.

21. Lubricate the camshafts with clean oil.

22. Install the camshaft seals to the end

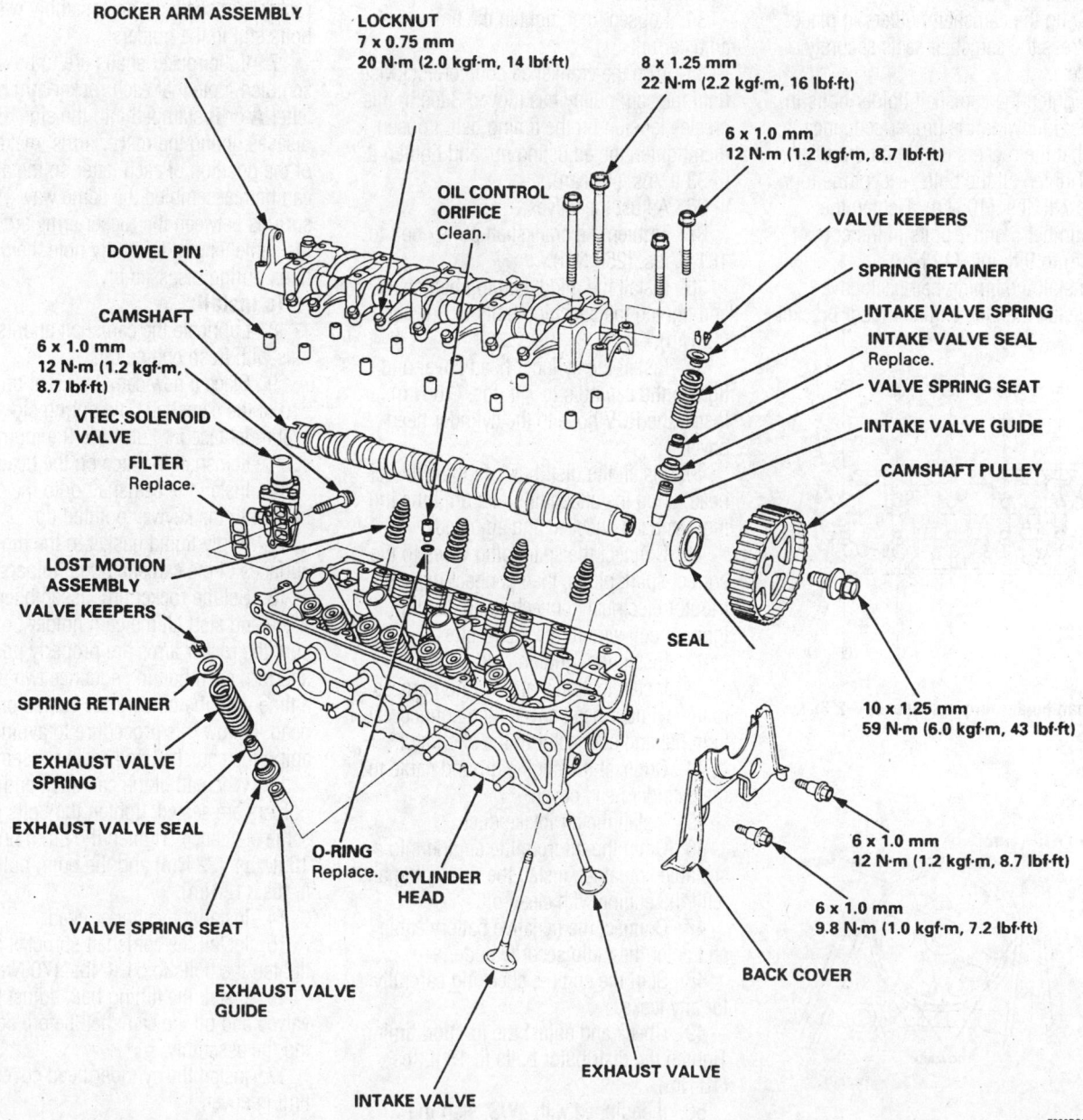

Exploded view of the cylinder head and related components—2.3L engine

7923BG70

of the camshafts that the timing belt sprockets attach to. The open side (spring) should be facing into the cylinder head when installed.

23. Be sure the keyways on the camshafts are facing up and install the camshafts to the cylinder head.

24. Install the rubber plug to the cylinder head at the end of the intake camshaft.

25. Apply liquid gasket to the head mating surfaces of the No. 1 and No. 6 camshaft holders, then install them along with No. 2, 3, 4 and 5. **I** or **E** marks are stamped on the camshaft holders to identify them as Intake or Exhaust side holders. The arrows stamped on the holders should point toward the timing belt.

26. Snug the camshaft holders in place.

27. Press the camshaft seals securely into place.

28. Tighten the camshaft holder bolts in two steps, following the proper sequence, to ensure that the rockers do not bind on the valves. Tighten all the bolts, except the four studs, to 7 ft. lbs. (10 Nm). Tighten the studs (number 5 and 7 bolts in the correct sequence) to 9 ft. lbs. (12 Nm).

29. Install the timing belt back cover.

30. Install the side engine mount bracket

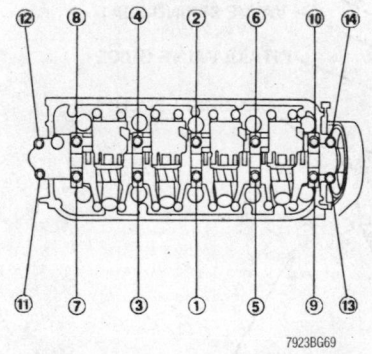

Camshaft holder torque sequence—2.3L engine

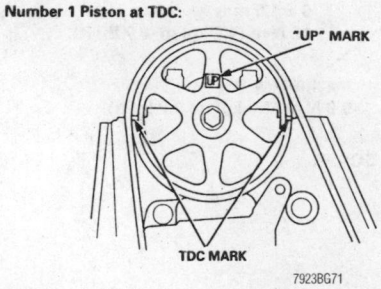

Camshaft sprocket alignment—2.3L

B. Tighten the bolt attaching the bracket to the cylinder head to 33 ft. lbs. (45 Nm). Tighten the bolts attaching the bracket to the side engine mount to 16 ft. lbs. (22 Nm).

31. Insert a 5.0mm pin punch in each of the camshaft caps, nearest to the pulleys, through the holes provided. Install the keys into the camshaft grooves.

32. Push the camshaft sprocket onto the camshaft, then tighten the retaining bolt to 27 ft. lbs. (38 Nm).

33. Ensure the words **UP** embossed on the camshaft pulleys are aligned in the upward position. Install the timing belt to the camshaft sprocket, then remove the two 5.0mm pin punches from the camshaft bearing caps.

34. Loosen, then tighten the timing belt adjuster nut.

35. Turn the crankshaft counterclockwise until the cam pulley has moved 3 teeth; this creates tension on the timing belt. Loosen, then tighten the adjusting nut and tighten it to 33 ft. lbs. (45 Nm).

36. Adjust the valves.

37. Tighten the crankshaft pulley bolt to 181 ft. lbs. (250 Nm).

38. Install the middle timing belt cover and tighten the attaching bolts to 9 ft. lbs. (12 Nm).

39. Install the cylinder head cover and tighten the cap nuts to 7 ft. lbs. (10 Nm). Install the PCV hose to the cylinder head cover.

40. Install the distributor to the cylinder head, snug the attaching bolts until the timing has been checked and adjusted.

41. Connect the spark plug wires to the correct spark plugs, then connect the distributor electrical connectors. Install the ignition coil wire to the distributor.

42. Install the ignition coil.

43. Install the alternator wiring harness to the cylinder head cover, then connect the terminal and connector to the alternator.

44. Connect the engine ground cable to the cylinder head cover.

45. Install the air intake duct.

46. Drain the oil from the engine into a sealable container. Install the drain plug and refill the engine with clean oil.

47. Connect the negative battery cable and enter the radio security code.

48. Start the engine, checking carefully for any leaks.

49. Check and adjust the ignition timing. Tighten the distributor bolts to 13 ft. lbs. (18 Nm).

50. If equipped with 4WS, start the engine, then turn the steering wheel lock-to-lock to reset the 4WS control unit.

2.5L Engine

1. Disconnect the negative battery cable. Remove the timing belt covers and cylinder head covers.

2. Rotate the crankshaft to TDC compression of No. 1 piston and remove the timing belt.

3. Remove the camshaft sprocket.

4. Remove the cylinder head from the vehicle.

5. Loosen the rocker shaft holder bolts 1 turn at a time in the opposite of the installation sequence. Following this procedure will prevent the camshafts and rocker assemblies from warping.

6. After all bolts are loose, remove the rocker arm shafts as an assembly with the bolts still in the holders.

7. If the rocker shafts are to be disassembled, note that each rocker arm has a letter **A** or **B** stamped into the side. Before disassembling the rocker arms, make a note of the position of each letter so the arms can be reassembled the same way. The springs between the rocker arms are not all the same length. Carefully note their positions during disassembly.

To install:

8. Lubricate the camshaft and its journals with fresh engine oil.

9. Place a new camshaft seal on the end of the camshaft. The spring side of the seal must face in. Lubricate the journals and set the camshaft in place on the head.

10. Install the camshaft onto the cylinder head with the keyway pointed up.

11. Apply liquid gasket to the mounting surfaces of the camshaft end holders.

12. Set the rocker arm assemblies in place and start all the cam holder bolts. Be sure the rocker arms are properly positioned and turn each bolt in sequence two turns at a time until the holders are seated on the head. Follow this procedure to avoid damaging the camshaft and rocker assemblies.

13. When all of the camshaft and rocker holders are seated, tighten the bolts in the same sequence. Tighten the 8mm bolts to 16 ft. lbs. (22 Nm) and the 6mm bolts to 9 ft. lbs. (12 Nm).

14. Install the cylinder head.

15. Install the camshaft sprocket and tighten the bolts to 51 ft. lbs. (70 Nm).

16. Install the timing belt, adjust the valves and oil the camshaft before completing the assembly.

17. Install the cylinder head cover and timing cover.

18. Install the distributor.

19. Reconnect the negative battery cable.

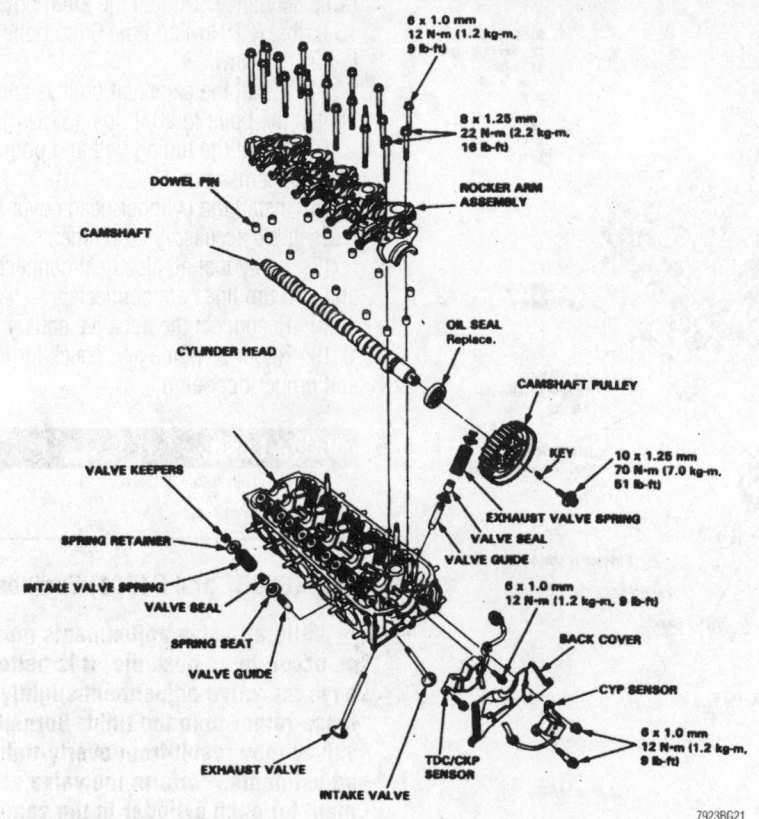

Camshaft and rocker arm assembly—2.5L engine

20. Check for proper engine and valve-train operation.

3.0L Engine

1. Disconnect the negative battery cable.
2. Remove the timing belt.
3. Remove the cylinder head.
4. Remove the camshaft sprocket and rear cover.
5. Remove the rocker arm/shaft assembly.
6. Remove the camshaft thrust cover and O-ring.

7. Pull out the camshaft.

To install:

8. Lubricate the camshaft with clean engine oil.
9. Slide the camshaft into position.
10. Install the thrust plate using a new O-ring. Tighten the bolts to 16 ft. lbs. (22 Nm).
11. Install the rocker arm/shaft assembly.

12. Install the cylinder head.
13. Install the rear cover and camshaft sprocket. Tighten the bolt to 67 ft. lbs. (90 Nm).
14. Install the timing belt.
15. Adjust the valves.

3.2L and 3.5L Engines

1. Disconnect the negative battery cable.
2. Remove the timing belt covers and cylinder head covers.
3. Rotate the crankshaft to TDC for the No. 1 piston and remove the timing belt.
4. Remove the camshaft sprockets.
5. Loosen the rocker shaft holder bolts one turn at a time in the reverse of the torque sequence to avoid damaging the valves, camshafts, or rocker assemblies.
6. After all bolts are loose, remove the rocker arm shafts as an assembly with the bolts still in the holders.
7. If the rocker shafts are to be disassembled, note that each rocker arm has a letter **A** or

B stamped into the side. Before disassembling the rocker arms, make a note of the position of each letter so that the arms can be reassembled in the same position.

8. Do not remove the hydraulic tappets from the rocker arms unless they are to be replaced. Handle the rocker arms carefully so the oil does not drain out of the tappets.

9. Lift the camshafts from the cylinder head, wipe them clean and inspect the lift ramps. Replace the camshafts and rockers if the lobes are pitted, scored, or excessively worn.

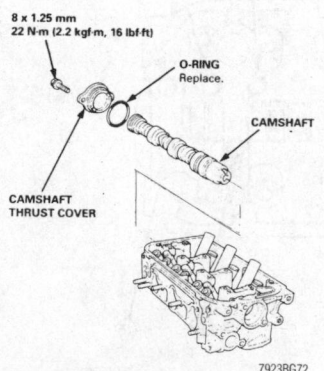

Camshaft installation—3.0L engine

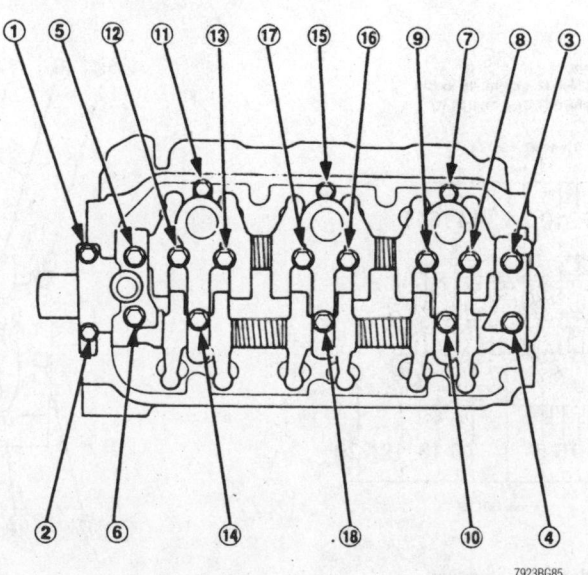

Loosen the camshaft holder bolts in the specified sequence—3.5L engine

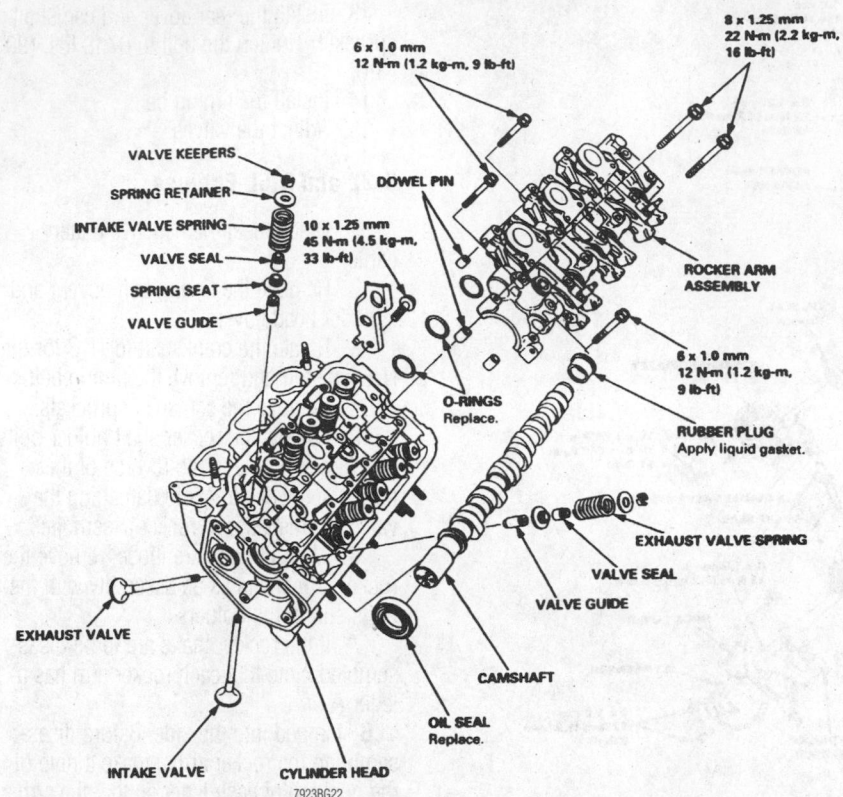

Camshaft and rocker arm assembly—3.2L engines

To install:

10. Place a new seal on the end of the camshaft, lubricate the journals and set the camshaft in place on the head.

➡**The pin hole in the front of the camshaft designates the top position.**

11. Apply liquid gasket to the mounting surfaces of the camshaft end holders.

12. Set the rocker arm assemblies in place and start all of the camshaft holder bolts. Be sure the rocker arms are properly positioned and turn each bolt in sequence two turns at a time until the holders are seated on the head to avoid damaging the valves or rocker assemblies.

13. When all the camshaft and rocker holders are seated, tighten the bolts in the same sequence. Tighten the 8mm bolts to 16 ft. lbs. (22 Nm) and the 6mm bolts to 9 ft. lbs. (12 Nm).

14. Install the camshaft pulleys and tighten the bolts to 23 ft. lbs. (32 Nm).

15. Install the timing belt and pour oil over the camshafts.

16. Install the cylinder head cover and reassemble accessory components.

17. Verify that all electrical connections and vacuum lines are connected.

18. Reconnect the negative battery cable.

19. Run the engine and check for leaks and proper operation.

Valve Lash

ADJUSTMENT

1.8L (B18B1 and B18C1) Engines

➡**While all valve adjustments must be as accurate as possible, it is better to have the valve adjustment slightly loose rather than too tight. Burned valves may result from overly-tight adjustments. Perform the valve adjustment for each cylinder in the same sequence as the firing order: 1–3–4–2.**

1. Be sure the engine is cold; cylinder head temperature must be below 100° F (38° C). Overnight cold is best.

2. Remove the cylinder head cover and the upper timing belt cover.

3. Set the No. 1 cylinder to Top Dead Center (TDC). The word **UP** should appear at the top and the TDC grooves on the pulley should align with the cylinder head surface or the mark on the rear belt cover.

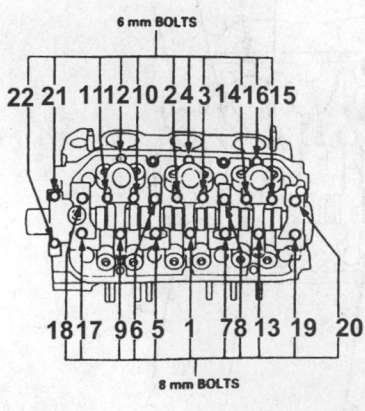

Specified torque:
8 mm bolts: 22 N·m (2.2 kg-m, 16 lb-ft)
6 mm bolts: 12 N·m (1.2 kg-m, 9 lb-ft)

Camshaft holder bolt tightening
sequence—3.2L engine

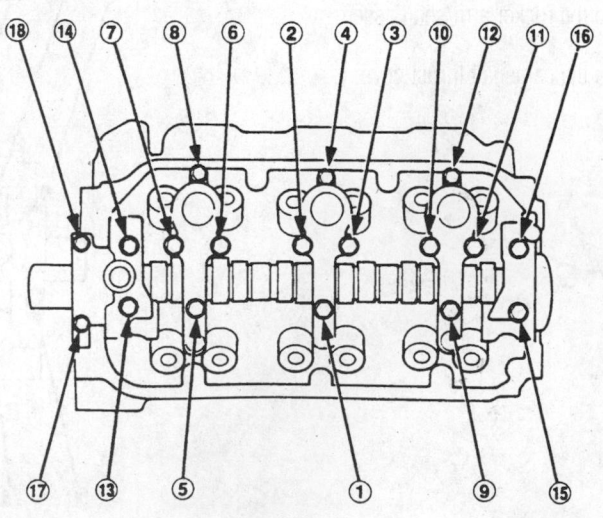

Tighten the camshaft holder bolts in the specified sequence—3.5L engine

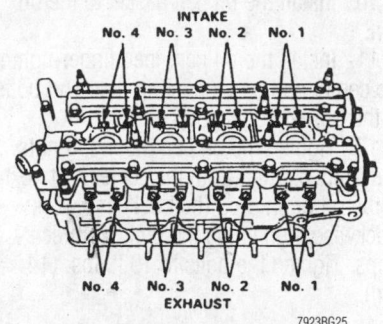

Valve arrangement—1.8L (B18B1 and B18C1) engines

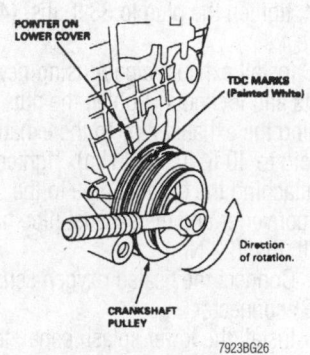

Crankshaft pulley timing mark—1.8L engines

4. Valve clearances are:

a. B18B1 engine: Intake— 0.003–0.005 in. (0.08–0.12mm) Exhaust—0.006–0.008 in. (0.16–0.20mm)

b. B18C1 (VTEC) engine: Intake— 0.006–0.007 in. (0.15–0.19mm) Exhaust—0.007–0.008 in. (0.17–0.20mm)

5. With the No. 1 cylinder at TDC, adjust the valves of the No. 1 cylinder by performing the following procedures:

a. Hold the rocker arm against the valve and place the feeler gauge between the rocker arm and the camshaft lobe. There should be a slight drag on the feeler gauge.

b. If adjustment is required, loosen the valve adjusting the screw locknut.

c. Turn the adjusting screw to obtain the proper clearance.

d. Hold the adjusting screw and tighten the locknut(s) to 18 ft. lbs.

e. Recheck the clearance.

6. Turn the crankshaft 180 degrees counterclockwise; the cam pulley will turn 90 degrees. With the No. 3 cylinder at TDC, the **UP** marks should be at the exhaust side. Adjust the valves on the No. 3 cylinder.

7. Turn the crankshaft 180 degrees

counterclockwise; the cam pulley will turn 90 degrees. With the No. 4 cylinder at TDC, both **UP** marks should be at the bottom. Adjust the valves on the No. 4 cylinder.

8. Turn the crankshaft 180 degrees counterclockwise. The No. 2 cylinder will now be on TDC and the **UP** marks should be at the intake side. Adjust the valves on the No. 2 cylinder.

9. Install the cylinder head cover and upper timing belt cover.

2.3L Engine

➡ **The valve should be adjusted only when the engine temperature is below 100°F (38°C). Retighten the crankshaft pulley bolt to 181 ft. lbs. (245 Nm) after adjusting the valves.**

1. Turn the crankshaft so the No. 1 piston is at TDC. Be sure the UP mark on the camshaft pulley is at the 12 o'clock position.

2. Adjust the valves on the No. 1 cylinder. To the following specifications:
- Intake—0.010 in. (0.26mm)
- Exhaust—0.012 in. (0.30mm)

3. Tighten the locknut to 14 ft. lbs. (20 Nm).

4. Turn the crankshaft counterclockwise 180°. Be sure the UP mark on the camshaft pulley is at the 9 o'clock position.

5. Adjust the valves on the No. 3 cylinder. Tighten the locknut to 14 ft. lbs. (20 Nm).

6. Turn the crankshaft counterclockwise 180°. Be sure the UP mark on the camshaft pulley is at the 6 o'clock position.

7. Adjust the valves on the No. 4 cylinder. Tighten the locknut to 14 ft. lbs. (20 Nm).

8. Turn the crankshaft counterclockwise 180°. Be sure the UP mark on the camshaft pulley is at the 3 o'clock position.

9. Adjust the valves on the No. 2 cylinder. Tighten the locknut to 14 ft. lbs. (20 Nm).

10. Retighten the crankshaft pulley bolt to 181 ft. lbs. (245 Nm) after adjusting the valves.

2.5L Engines

1. Disconnect the negative battery cable.

2. Remove the cylinder head cover and the upper timing belt cover.

3. Rotate the crankshaft to align the white TDC on the crankshaft pulley with the pointer on the cover. Be sure the **UP** mark on the camshaft sprocket is up and the TDC marks align with the edge of the cylinder head.

4. Align the No. 1 mark on the back of the camshaft sprocket with the notch in the camshaft holder.

5. Hold a No. 1 cylinder rocker arm against the camshaft and use a feeler gauge to check the clearance at the valve stem. Intake valve clearance should be 0.010 in. (026mm), exhaust valve clearance should be 0.012 in. (0.30mm). The service limit for both intake and exhaust valves is plus or minus 0.0008 in. (0.02mm). Loosen the locknut and turn the adjusting screw to adjust the clearance. Tighten the locknut and recheck the clearance.

6. Rotate the crankshaft counterclockwise to align the TDC marks for each piston with the notch. Adjust the valves of each cylinder. The adjustment order is 1, 2, 4, 5 and 3.

7. Install the cylinder head and timing belt covers.

8. Reconnect the negative battery cable.

3.0L Engine

1. Remove the cylinder head cover.

2. Remove the upper front timing belt cover.

3. Rotate the crankshaft so the No. 1 piston is at TDC on compression to adjust the valves for the No. 1 cylinder.

4. Loosen the locknuts and adjust the screws until a slight drag can be felt with the feeler gage when the gage is placed between the valve and rocker arm tip as shown. The specifications are as follows:
- Intake—0.008–0.009 in. (0.20–0.24mm)
- Exhaust—0.011–0.013 in. (0.28–0.32mm)

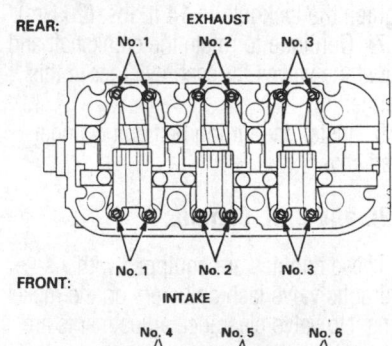

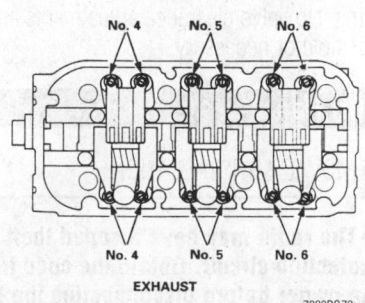

Adjusting screw locations for valve lash adjustment—3.0L engine

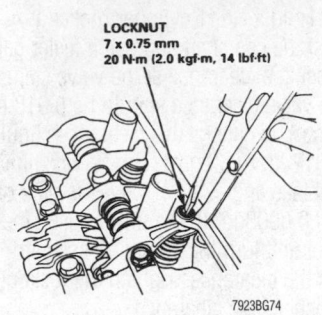

**LOCKNUT
7 x 0.75 mm
20 N·m (2.0 kgf·m, 14 lbf·ft)**

7923BG74

Slide the feeler gauge between the valve and rocker arm while turning the adjusting screw—3.0L engine

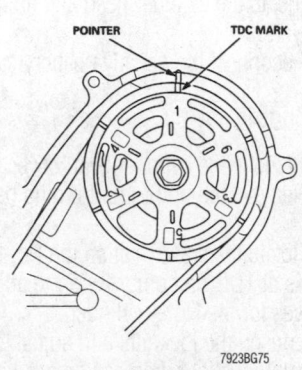

POINTER   TDC MARK

7923BG75

Camshaft sprocket position when No. 1 piston is at TDC—3.0L engine

5. Rotate the crankshaft clockwise until the No. 4 on the camshaft sprocket is near the pointer on the rear cover. This is the No. 4 cylinder firing position.

6. Adjust the valves for the No. 4 cylinder while the sprocket is in this position. Tighten the locknuts to 14 ft. lbs. (20 Nm).

7. Continue to rotate the crankshaft and adjust the valves for each cylinder in this manner.

8. Install the timing belt and cylinder head covers.

3.2L and 3.5L Engines

These engines are equipped with hydraulic valve lash adjusters on the rocker arms. No valve clearance adjustments are possible or necessary.

Oil Pan

REMOVAL & INSTALLATION

➡**The radio may have a coded theft protection circuit. Obtain the code from the owner before disconnecting the battery, removing the radio fuse, or removing the radio.**

1.8L and 2.3L Engines

1. Disconnect negative cable at the battery.

2. Raise and safely support the vehicle. Drain the oil and remove the lower splash panel.

3. If equipped, disconnect the heated oxygen sensor (HO2S) connector.

4. Remove the nuts and bolts connecting exhaust pipe A to the catalytic converter. Discard the gasket and the locknuts.

5. Remove and discard the nuts attaching exhaust pipe A to the exhaust hanger.

6. If applicable, remove the mounting bolts from the center beam. Remove the center beam from the subframe.

7. Remove and discard the locknuts attaching exhaust pipe A to the exhaust manifold, then remove exhaust pipe A from the vehicle. discard the exhaust gaskets.

8. Loosen the oil pan bolts in a criss-cross pattern. To remove the oil pan, lightly tap the corners of the oil pan with a rubber or plastic faced mallet. Clean off all the old gasket material.

To install:

9. Apply liquid gasket to the oil pan mating surface where the oil pump and the right side cover meet the engine block.

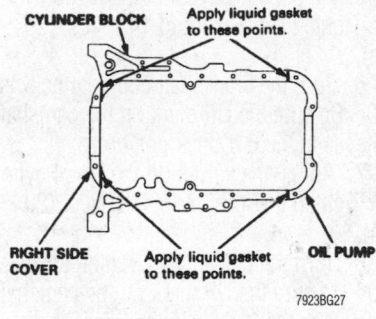

CYLINDER BLOCK   Apply liquid gasket to these points.

RIGHT SIDE COVER   Apply liquid gasket to these points.   OIL PUMP

7923BG27

Apply liquid gasket to the oil pan as shown—1.8L (B18B1 and B18C1) engines

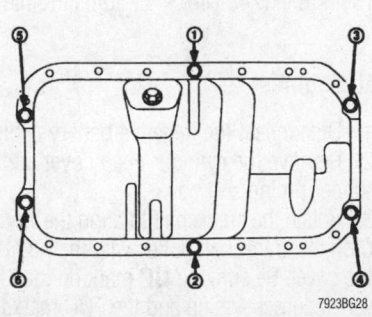

7923BG28

Oil pan bolt tightening sequence—1.8L (B18B1 and B18C1) engines

10. Install the oil pan gasket to the oil pan.

11. Install the oil pan, then finger-tighten the center and end mounting nuts and bolts in the proper sequence.

12. Tighten the oil pan mounting nuts and bolts starting from the center bolt next to the oil drain plug (bolt # 1) and work clockwise, tightening the bolts in three steps. Tighten the bolts to 10 ft. lbs. (14 Nm).

➡**Excessive tightening can cause distortion of the oil pan gasket and oil leakage.**

13. Install the oil drain plug with a new gasket, tighten the plug to 33 ft. lbs. (44 Nm).

14. Install exhaust pipe A using new gaskets and locknuts. Tighten the nuts attaching the exhaust pipe to the exhaust manifold to 40 ft. lbs. (54 Nm). Tighten the nuts attaching the exhaust pipe to the catalytic converter and the exhaust pipe hanger to 16 ft. lbs. (22 Nm).

15. Connect the heated oxygen sensor (HO2S) connector.

16. Install the lower splash panel, then lower the vehicle.

17. Fill the engine with oil.

18. Connect the negative battery cable and enter the radio security code.

19. Run the engine and check for leaks.

20. Turn off engine and check the oil level. Top off the oil level if necessary.

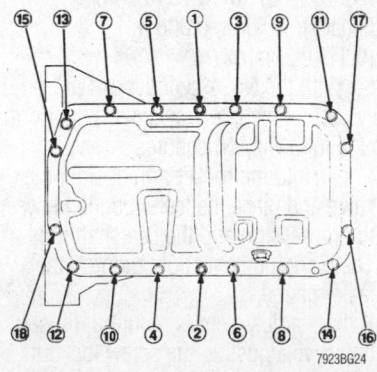

7923BG24

Oil pan bolt tightening sequence—2.3L engine

2.5L Engine

1. Shift the manual transmission to 1st gear or automatic transmission to the **P** position.

2. Disconnect the negative battery cable.

3. Remove the air cleaner housing.

4. Raise and safely support the vehicle and remove the front wheels.

5. Drain the engine oil and coolant.

6. Remove the strut forks.

7. Remove the lower ball joint nut. Use a ball joint remover to disconnect the ball joint from the control arm.

8. Carefully pry the inner CV-joints out of their sockets. Wrap them in plastic to keep them clean. Do not let the driveshafts hang by the outer CV-joint.

9. Drain the differential oil.

10. Disconnect the differential oil cooler hoses.

11. Install the shaft puller, Acura tool number 07LAC-PW50101, and disengage the extension shaft from the differential.

12. Remove the side splash shield.

13. Attach a chain hoist to the lifting hooks and take up the engine's weight.

14. Remove the transmission side mount and bracket.

15. Unbolt and remove the left front engine mount bracket.

16. Remove the power steering speed sensor from the differential. Do not disconnect the hoses.

17. Remove the differential mounting bolts and the 26mm shim. Remove the differential from the vehicle.

18. Unbolt the intermediate shaft bearing housing from the oil pan and pull the intermediate shaft assembly from the oil pan pipe.

19. Remove the A/C compressor, then its mounting bracket. Leave the A/C lines connected to the compressor. Support the compressor with a piece of wire to move it out of the work area and take the weight off the A/C lines.

20. Remove the set plate that holds the oil pan inner pipe from the right side of the engine.

21. Unbolt the oil pan and remove it from the vehicle.

To install:

22. Clean the oil pan and engine block mating surfaces. Apply an even bead of liquid gasket to the engine block sealing surface. Apply some liquid gasket to the inner threads of the bolt holes.

23. Install the oil pan and tighten the bolts in the correct sequence to 16–17 ft. lbs. (22–24 Nm).

24. Install new O-rings on the oil pan inner pipe. Install the pipe and tighten the set plate bolts to 9 ft. lbs. (12 Nm).

25. Install the differential, making sure the original shim is in the proper position.

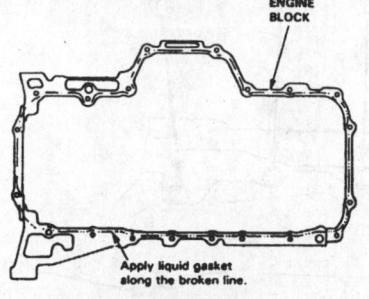

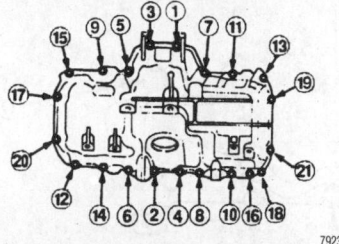

Apply liquid gasket and tighten the oil pan bolts as shown—2.5L engine

Tighten the bolts to 54 ft. lbs. (75 Nm). Connect the cooling hoses.

26. Install new set and snaprings on the extension shaft. Coat the splines and their mating surfaces with high temperature grease. Thread the special installation tool into the transmission case to install the extension shaft.

27. Pack the extension shaft cavity with high temperature grease and install the 33mm sealing bolt. Tighten the bolt to 58 ft. lbs. (80 Nm) and install the secondary cover.

28. Install the intermediate shaft, tighten the bolts to 16 ft. lbs. (22 Nm).

29. Install the left front engine mount bracket and tighten the bolts to 40 ft. lbs. (54 Nm). Tighten the mounting bolt to 54 ft. lbs. (74 Nm).

30. Install the transmission side bracket and mount. Tighten the bracket mounting bolts and through-bolt to 40 ft. lbs. (54 Nm). Install new mount bolts and tighten them to 47 ft. lbs. (64 Nm).

31. Install the A/C compressor mount. Tighten the mounting bolts on the oil pan, then the mounting bolts on the engine block, 36 ft. lbs. (49 Nm). Install the A/C compressor onto the mount and tighten the bolts to 16 ft. lbs. (22 Nm).

32. Install the speed sensor and tighten the mounting bolt to 7 ft. lbs. (10 Nm).

33. Install new set rings on the CV-joints and press them into their sockets.

34. Refill the differential.

35. Connect the lower ball joints to the control arms and install the nuts, tighten them to 36–43 ft. lbs. (49–59 Nm) and install a new cotter pin. Tighten the strut

fork bolts to 47 ft. lbs. (64 Nm). Install the front wheels.

36. Lower the vehicle.

37. Remove the chain hoist.

38. Refill the engine oil and cooling system.

39. Install the air cleaner and intake duct.

40. Bleed the cooling system by opening the bleeder on the upper radiator hose inlet when filling the system.

41. Connect the negative battery cable.

3.0L Engine

1. Raise and safely support the vehicle.

2. Drain the engine oil.

3. Remove the exhaust pipe if necessary.

4. Remove the oil pan mounting bolts.

5. Remove the oil pan.

To install:

6. Clean the sealing surface on the engine and oil pan flange.

7. Apply a bead of sealant to the oil pan flange and install the oil pan. Tighten the bolts in the sequence shown to 9 ft. lbs. (12 Nm).

8. If removed, install the exhaust pipe.

❋❋ WARNING

Operating the engine without the proper amount and type of engine oil will result in severe engine damage.

9. Add the correct amount of engine oil to the crankcase.

10. Start the engine and check for leaks.

3.2L and 3.5L Engines

1. Disconnect the negative, then the positive battery cables.

2. Remove the air conditioning compressor drive belt.

3. Raise and safely support the vehicle.

4. Remove the front wheels and splash shield.

5. For Legend models, remove the strut forks. Remove the lower ball joint nut and use a ball joint press tool to disconnect the ball joint from the control arm.

6. Remove the halfshafts from the differential and the intermediate shaft.

7. Remove the intermediate shaft from the oil pan.

8. Drain the oil from the differential into a sealable container the install the drain plug with a new washer.

9. Drain the engine oil into a sealable container.

10. If equipped, disconnect the Vehicle speed Sensor (VSS) harness, then remove

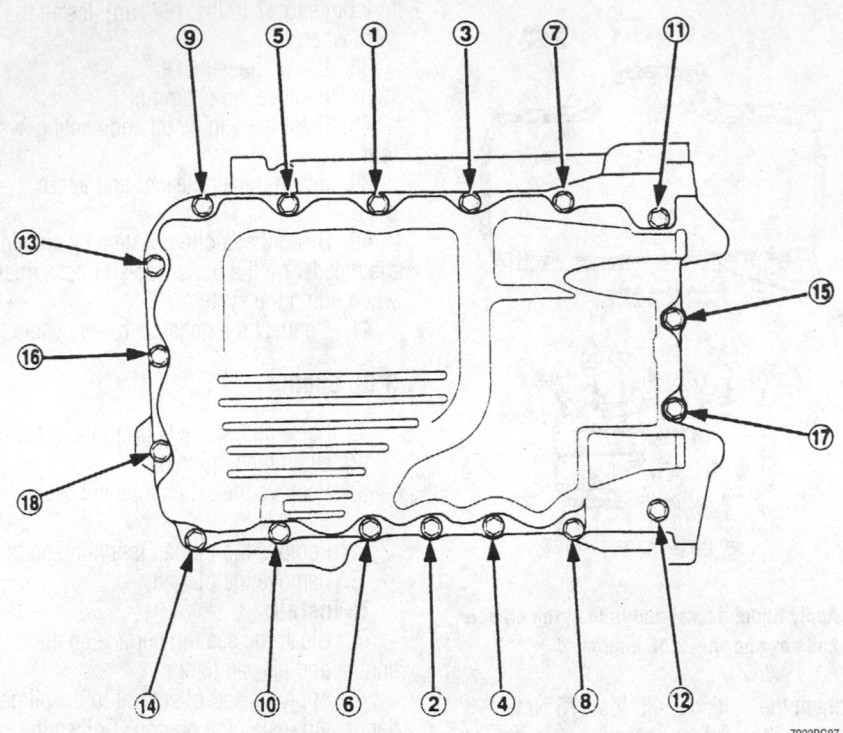

Oil pan bolt tightening sequence—3.0L engine

the VSS/power steering speed sensor. Do not disconnect the fluid hoses, support the sensor out of the way.

11. Remove the right front beam bridge.

12. Remove the lower plate from the rack and pinion, install the two rack and pinion mounting bolts that were removed.

13. Remove the 36mm sealing bolt on the transaxle. Ensure that the transaxle is in 1st gear (manual) or **P** (automatic).

14. Disconnect the extension shaft from the differential with the extension shaft puller (part # 07LAC-PW50101).

15. Remove the differential mounting bolts and the 26mm shim, then remove the differential.

16. Disconnect the A/C compressor clutch connector, then remove the compressor. Do not disconnect the A/C hoses from the compressor and do not let the compressor hang by the hoses.

17. Remove the rear engine stiffener.

18. Remove the flywheel cover or the torque converter covers.

19. Remove the oil pan. Do not lose the dowel pins from the oil pan.

To install:

20. Clean the oil pan and cylinder block mating surfaces, then apply liquid gasket to the cylinder block. Be sure that the mating surfaces are clean and dry before installing the liquid gasket. Do not apply liquid gasket to the O-ring grooves.

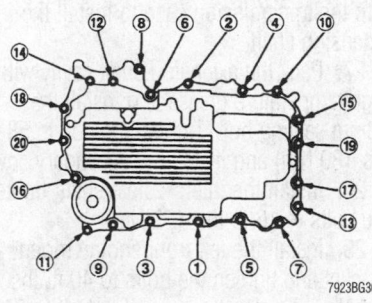

Be sure to tighten the oil pan bolts in the sequence shown—3.2L engines

21. Install the dowel pins to the oil pan and new O-rings coated with clean oil. Install the oil pan to the cylinder block. Coat the oil pan bolts with liquid gasket, then install them. Tighten the bolts in the proper sequence to 16 ft. lbs. (22 Nm).

22. For manual transmission, install the flywheel cover and engine stiffener.

23. For automatic transmission, install the torque converter covers and tighten the mounting bolts to 8.7 ft. lbs. (12 Nm)

24. Install the rear engine stiffener. Tighten the bolt attaching the engine stiffener to the transaxle first, to 47 ft. lbs. (64 Nm), then tighten the bolts to the engine block to 16 ft. lbs. (22 Nm).

25. Install the A/C compressor to the

engine block and tighten the mounting bolts to 16 ft. lbs. (22 Nm).

26. Connect the A/C clutch connector.

27. Install the dowel pins to the differential, then install the differential to the engine. Install the mounting bolts loosely and install the 26mm shim. Tighten all of the mounting bolts to 47 ft. lbs. (64 Nm).

28. Install a new set ring to the extension shaft. Using the extension shaft installer (part # 07MAF-PY40100 or 07MAF-PY40101) install the shaft to the differential.

29. Fill the secondary gear with super high temperature grease (Part # 08798–9002). Applying sealer to the threads of the 36mm sealing bolt, then install the bolt and tighten to 58 ft. lbs.(78 Nm).

30. Remove the two bolts from the rack and pinion necessary to install the lower plate, then install the lower plate and the attaching bolts. Tighten the lower plate attaching bolt to 28 ft. lbs. (38 Nm) and tighten the rack and pinion bolts to 43 ft. lbs. (59 Nm).

31. For Legend models, when installing the lower ball joint nuts, tighten them to 51–58 ft. lbs. (70–80 Nm) and install a new cotter pin. Tighten the strut fork bolts to 51 ft. lbs. (70 Nm).

32. Install the right beam bridge and tighten the attaching bolts to 28 ft. lbs. (38 Nm).

33. Install the VSS and tighten the attaching bolt to 8.7 ft. lbs. (12 Nm). Connect the VSS harness to the VSS sensor.

34. Install the intermediate shaft and the halfshafts.

35. Fill the differential with oil.

36. Install the engine splash shield and tighten the bolts to 7.2 ft. lbs. (9.8 Nm).

37. Install the front wheels.

38. Lower the vehicle.

39. Install the A/C compressor drive belt.

40. Fill the engine with oil.

41. Connect the positive, then the negative battery cables and enter the radio security code.

42. Run the engine and check for leaks.

43. Check the front wheel alignment.

Oil Pump

REMOVAL & INSTALLATION

➡**The radio may have a coded theft protection circuit. Obtain the code from the owner before disconnecting the battery, removing the radio fuse, or removing the radio.**

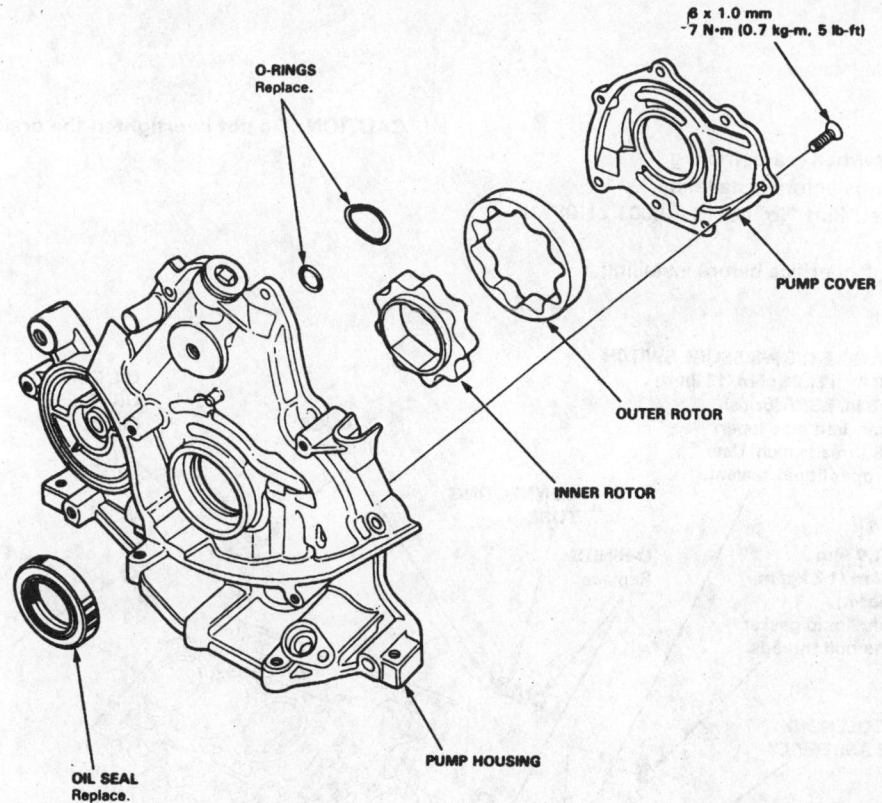

Exploded view of the oil pump—1.8L (B18B1 and B18B2) engines

1. Disconnect the negative battery cable. Raise and safely support the vehicle. Drain the oil and remove the lower splash panel, if necessary.

2. Be sure the crankshaft is at TDC on No. 1 cylinder and remove the timing belt cover, timing belt, and the gear off the crankshaft.

3. For Legend models, remove the oil pump cap.

➡**On GS model Legends, an engine oil cooler is installed in place of the oil pump cap. Remove this assembly and disconnect the oil cooler lines.**

4. Remove the oil pan. Remove the pick-up screen.

5. Remove the oil filter assembly, if necessary.

6. Remove the oil pump from the front of the engine. Any time the oil pump is removed, the front oil seal should be replaced.

To install:

7. Install the oil pump, using new O-rings and liquid gasket applied to a clean pump mounting face. For all engines, except

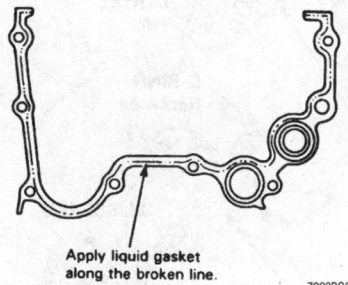

Apply sealant to the oil pump sealing surface as shown—2.5L engine

the 1.8L (B18B1, B18C1) engines, tighten the 6mm bolts to 9 ft. lbs. (12 Nm) and the 8mm bolts to 16 ft. lbs. (22 Nm). For 1.8L (B18B1, B18C1) engines, tighten the 8x1.25mm bolts to 17 ft. lbs. (24 Nm), tighten the 6x1.0mm bolts to 8 ft. lbs. (11 Nm).

8. Install the oil pump cap or oil cooler unit, as applicable. Replace the cooler hoses if they show signs of damage. Tighten the center bolt to 30 ft. lbs. (42 Nm).

❊❊ WARNING

The B18B1 and B18C1 engines use different oil pumps, be sure that you have the correct oil pump. Match the crankshaft timing mark on the new oil pump with the timing mark on the old oil pump, because the timing marks are in different locations. If an oil pump is used with the timing mark in the wrong position the pistons may contact the valves.

9. Install the pick-up screen, then the oil pan. Tighten the oil pan bolts to 9 ft. lbs. (12 Nm).

10. Install the oil filter assembly, exhaust pipe, center beam, and lower splash panel, if necessary.

11. Wait at least 30 minutes after completion of procedure before refilling the engine with oil. The waiting period is to allow a curing period for the silicone sealant. Refill the engine with oil and connect the negative battery cable. Start the engine and check the engine for leaks.

NOTE:
- Use new O-rings when reassembling.
- Apply oil to O-rings before installation.
- Use liquid gasket, Part No. 08718 – 0001 or 08718 – 0003.
- Clean the oil control orifice before installing.

CAUTION: Do not overtighten the drain bolt.

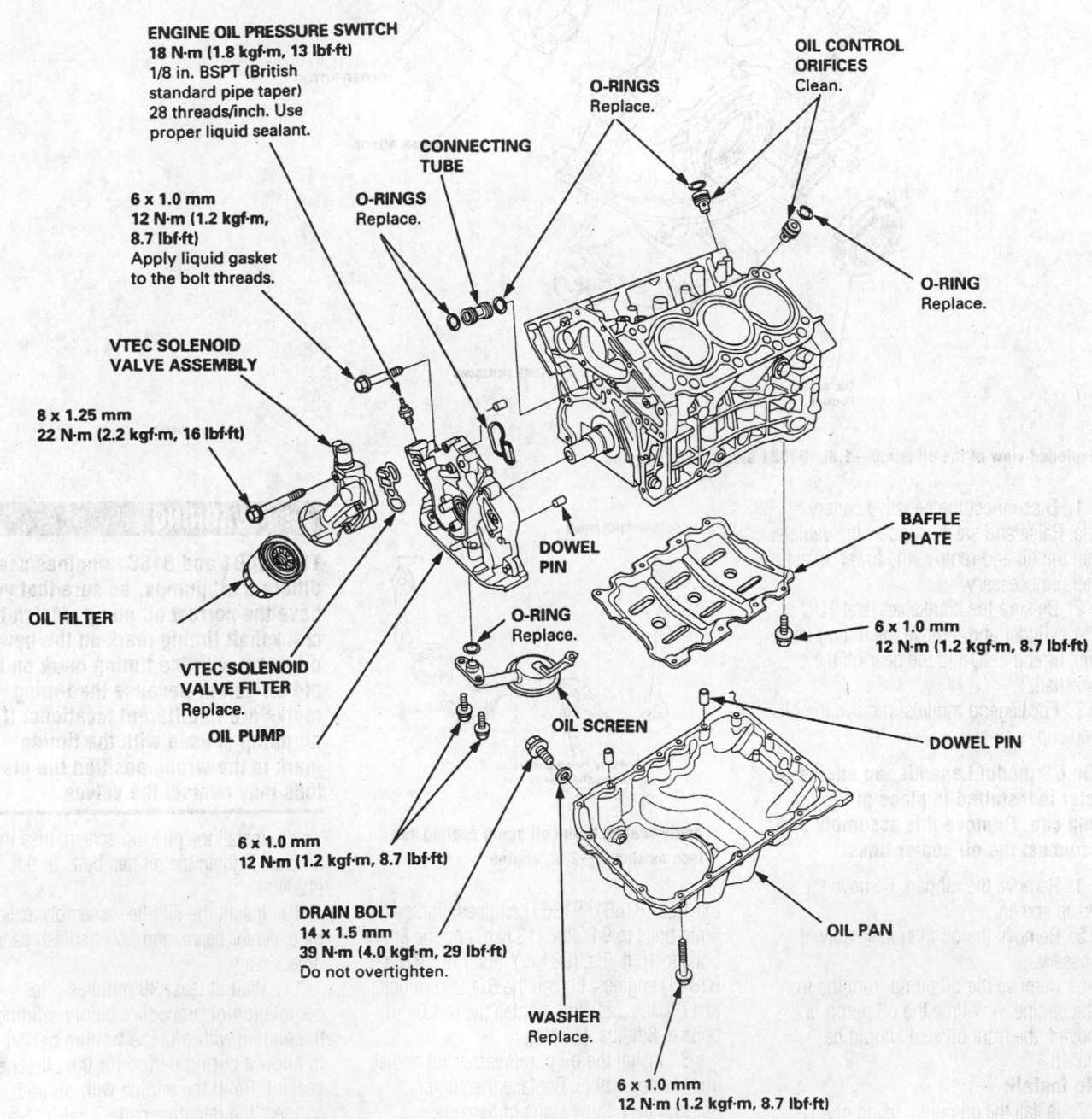

ENGINE OIL PRESSURE SWITCH
18 N·m (1.8 kgf·m, 13 lbf·ft)
1/8 in. BSPT (British standard pipe taper) 28 threads/inch. Use proper liquid sealant.

O-RINGS
Replace.

OIL CONTROL ORIFICES
Clean.

CONNECTING TUBE

O-RINGS
Replace.

O-RING
Replace.

6 x 1.0 mm
12 N·m (1.2 kgf·m, 8.7 lbf·ft)
Apply liquid gasket to the bolt threads.

VTEC SOLENOID VALVE ASSEMBLY

8 x 1.25 mm
22 N·m (2.2 kgf·m, 16 lbf·ft)

DOWEL PIN

BAFFLE PLATE

6 x 1.0 mm
12 N·m (1.2 kgf·m, 8.7 lbf·ft)

OIL FILTER

VTEC SOLENOID VALVE FILTER
Replace.

OIL PUMP

O-RING
Replace.

OIL SCREEN

DOWEL PIN

6 x 1.0 mm
12 N·m (1.2 kgf·m, 8.7 lbf·ft)

DRAIN BOLT
14 x 1.5 mm
39 N·m (4.0 kgf·m, 29 lbf·ft)
Do not overtighten.

OIL PAN

WASHER
Replace.

6 x 1.0 mm
12 N·m (1.2 kgf·m, 8.7 lbf·ft)

7923BG88

Lubrication system—3.0L engine

NOTE:
- Use new O-rings when reassembling.
- Apply oil to O-rings before installation.
- Use liquid gasket, Part No. 08718 – 0001 or 08718 – 0003.
- Clean the oil control orifice before installing.
- Remove the balancer shaft

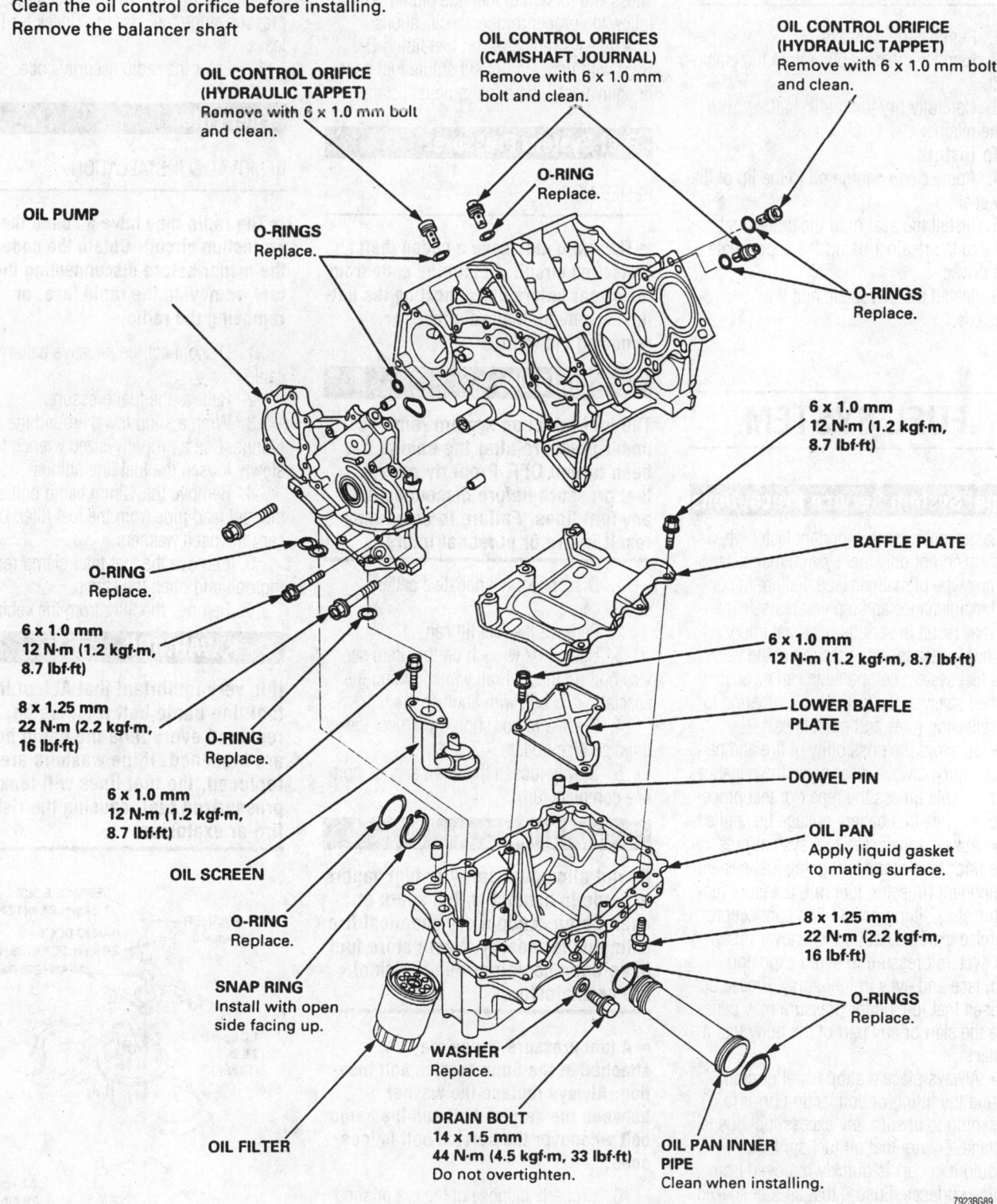

OIL CONTROL ORIFICE (HYDRAULIC TAPPET)
Remove with 6 x 1.0 mm bolt and clean.

OIL CONTROL ORIFICES (CAMSHAFT JOURNAL)
Remove with 6 x 1.0 mm bolt and clean.

OIL CONTROL ORIFICE (HYDRAULIC TAPPET)
Remove with 6 x 1.0 mm bolt and clean.

OIL PUMP

O-RING Replace.

O-RINGS Replace.

O-RINGS Replace.

6 x 1.0 mm
12 N·m (1.2 kgf·m, 8.7 lbf·ft)

BAFFLE PLATE

O-RING Replace.

6 x 1.0 mm
12 N·m (1.2 kgf·m, 8.7 lbf·ft)

8 x 1.25 mm
22 N·m (2.2 kgf·m, 16 lbf·ft)

O-RING Replace.

6 x 1.0 mm
12 N·m (1.2 kgf·m, 8.7 lbf·ft)

6 x 1.0 mm
12 N·m (1.2 kgf·m, 8.7 lbf·ft)

LOWER BAFFLE PLATE

DOWEL PIN

OIL SCREEN

OIL PAN
Apply liquid gasket to mating surface.

8 x 1.25 mm
22 N·m (2.2 kgf·m, 16 lbf·ft)

O-RING Replace.

O-RINGS Replace.

SNAP RING
Install with open side facing up.

WASHER
Replace.

OIL FILTER

DRAIN BOLT
14 x 1.5 mm
44 N·m (4.5 kgf·m, 33 lbf·ft)
Do not overtighten.

OIL PAN INNER PIPE
Clean when installing.

7923BG89

Exploded view of the lubrication system—3.5L engine

12. Turn off engine and check the oil level. Top off the oil level if necessary.

Rear Main Seal

REMOVAL & INSTALLATION

1. Remove the transaxle.
2. Remove the driveplate from the crankshaft.
3. Carefully pry the crankshaft seal out of the retainer.

To install:

4. Apply clean engine oil to the lip of the new seal.
5. Install the seal onto the crankshaft and into the retainer using the appropriate seal driver.
6. Install the driveplate and the transaxle.

FUEL SYSTEM

Fuel System Service Precautions

Safety is the most important factor when performing not only fuel system maintenance but any type of maintenance. Failure to conduct maintenance and repairs in a safe manner may result in serious personal injury or death. Maintenance and testing of the vehicle's fuel system components can be accomplished safely and effectively by adhering to the following rules and guidelines.

• To avoid the possibility of fire and personal injury, always disconnect the negative battery cable unless the repair or test procedure requires that battery voltage be applied.

• Always relieve the fuel system pressure prior to disconnecting any fuel system component (injector, fuel rail, pressure regulator, etc.), fitting or fuel line connection. Exercise extreme caution whenever relieving fuel system pressure to avoid exposing skin, face and eyes to fuel spray. Please be advised that fuel under pressure may penetrate the skin or any part of the body that it contacts.

• Always place a shop towel or cloth around the fitting or connection prior to loosening to absorb any excess fuel due to spillage. Ensure that all fuel spillage (should it occur) is quickly removed from engine surfaces. Ensure that all fuel soaked cloths or towels are deposited into a suitable waste container.

• Always keep a dry chemical (Class B) fire extinguisher near the work area.

• Do not allow fuel spray or fuel vapors to come into contact with a spark or open flame.

• Always use a back-up wrench when loosening and tightening fuel line connection fittings. This will prevent unnecessary stress and torsion to fuel line piping. Always follow the proper torque specifications.

• Always replace worn fuel fitting O-rings with new. Do not substitute fuel hose or equivalent, where fuel pipe is installed.

Fuel System Pressure

RELIEVING

➡**The radio may have a coded theft protection circuit. Obtain the code from the owner before disconnecting the battery, removing the radio fuse, or removing the radio.**

✳ CAUTION

The fuel injection system remains under pressure after the engine has been turned OFF. Properly relieve fuel pressure before disconnecting any fuel lines. Failure to do so may result in fire or personal injury.

1. Disconnect the negative battery cable.
2. Remove the fuel fill cap.
3. Use a box wrench on the 6mm service bolt on the fuel rail while holding the special banjo bolt with another wrench.
4. Place a rag or shop towel over the 6mm service bolt.
5. Slowly loosen the 6mm service bolt one complete turn.

✳ CAUTION

Do not allow fuel spray or fuel vapors to come in contact with a spark or open flame. Keep a dry chemical fire extinguisher nearby. Never store fuel in an open container due to risk of fire or explosion.

➡**A fuel pressure gauge may be attached at the 6mm service bolt location. Always replace the washer between the service bolt and the banjo bolt whenever the service bolt is loosened.**

6. Properly dispose of the rag or shop towel.
7. Remove the service bolt and install a new washer. Tighten the 6mm service bolt to 9 ft. lbs. (12 Nm).

8. Clean up any fuel spilled on the engine and intake manifold.
9. Install the fuel fill cap.
10. Reconnect the negative battery cable.
11. After servicing the vehicle, turn the ignition **ON**, but don't start the engine. Repeat this process two or three times to pressurize the fuel system. Check for fuel leaks.
12. Enter the radio security code.

Fuel Filter

REMOVAL & INSTALLATION

➡**The radio may have a coded theft protection circuit. Obtain the code from the owner before disconnecting the battery, removing the radio fuse, or removing the radio.**

1. Disconnect the negative battery cable.
2. Relieve the fuel pressure.
3. Wrap a shop towel around the filter fittings. Use a properly-sized wrench to slowly loosen the fuel line fittings.
4. Remove the 12mm banjo bolt and the fuel feed pipe from the fuel filter. Discard the used washers.
5. Remove the fuel filter clamp retaining bolt and open the clamp.
6. Remove the filter from the vehicle.

✳ WARNING

It is very important that ALL of the fuel line banjo bolt washers be replaced every time the banjo bolts are loosened. If the washers are not replaced, the fuel lines will leak pressurized fuel, causing the risk of fire or explosion.

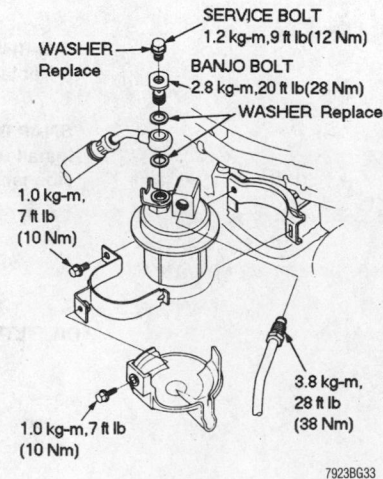

SERVICE BOLT
1.2 kg-m,9 ft lb(12 Nm)
WASHER Replace
BANJO BOLT
2.8 kg-m,20 ft lb(28 Nm)
WASHER Replace
1.0 kg-m, 7 ft lb (10 Nm)
1.0 kg-m,7 ft lb (10 Nm)
3.8 kg-m, 28 ft lb (38 Nm)

7923BG33

Exploded view of the fuel filter—Legend

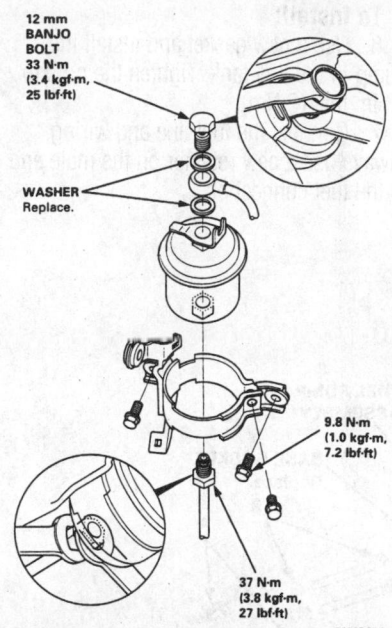

12 mm
BANJO
BOLT
33 N·m
(3.4 kgf-m,
25 lbf-ft)

WASHER
Replace.

9.8 N·m
(1.0 kgf-m,
7.2 lbf-ft)

37 N·m
(3.8 kgf-m,
27 lbf-ft)

7923BG90

Fuel filter assembly—3.0CL shown, others are similar

To install:

7. Install the new filter in position and tighten the clamp mounting bolt to 7 ft. lbs. (10 Nm).

8. Install the banjo bolts with new washers. Tighten the bolt to 25 ft. lbs. (33 Nm).

9. Connect the fuel feed line and tighten the fitting to 27 ft. lbs. (37 Nm).

10. Connect the negative battery cable and enter the radio security code.

11. Start the vehicle and check for leaks.

Fuel Pump

REMOVAL & INSTALLATION

➡The radio may have a coded theft protection circuit. Obtain the code from the owner before disconnecting the battery, removing the radio fuse, or removing the radio.

1.8L Engines

❋❋ CAUTION

Do not smoke while working on the fuel system. Keep open flames away from your work area.

1. Disconnect the negative battery cable.
2. Relieve the fuel system pressure.
3. Remove the rear seat to gain access to the fuel pump access panel. Remove the maintenance access cover.

4. Disconnect the electrical connector from the fuel pump.

5. If equipped with quick connect fittings, hold the fuel line connector with one hand and press down the retainer tabs with the other hand, then pull the connector off. Check the contact area of the pipe for dirt or damage, clean or replace the pipe or pump as required. Remove the old retainer from the pipe and discard. Cover the connector and pipe with plastic bags to prevent damage and keep foreign material out.

6. Remove the fuel pump mounting nuts, then remove the fuel pump from the fuel tank.

To install:

7. Install the fuel pump into the fuel tank. Tighten the mounting nuts to 4 ft. lbs. (6 Nm).

8. Reconnect the fuel lines and electrical connectors. Use new sealing washers when connecting the fuel pressure hose. Tighten the fuel line banjo bolt to 21 ft. lbs. (27 Nm).

9. Install a new retainer into the fuel tube connector and press the tube on to the pipe. The retainer pawls should lock with a clicking sound.

10. Install fuel pump access cover.
11. Install the rear seat.
12. Reconnect the negative battery cable and enter the radio security code.

➡The PCM idle memory must be reset after reconnecting the battery. Start the engine and hold it at 3000 rpm until the cooling fan comes on. Then allow the engine to idle for about five minutes with all accessories OFF and with the transmission in Park or Neutral.

13. Turn the ignition switch on and off two or three times to pressurize the system and check for leaks.

2.3L and 3.0L Engines

1. Properly relieve the fuel system pressure.

2. Lower the fuel tank and detach the connector and fuel lines from the pump assembly.

3. Remove the nuts and the fuel pump from the tank.

To install:

4. Use a new gasket and install the fuel pump assembly. Tighten the nuts to 48 in. lbs. (6 Nm).

5. Position the tank near the vehicle and connect the fuel line and electrical connector. Always use a new retainer on the male end of the fuel connection.

6. Install the fuel tank.

7. Pressurize the fuel system and check for leaks.

2.5L and 3.2L Engines

❋❋ CAUTION

Observe all applicable fuel precautions when working around fuel. Do not allow fuel spray or fuel vapors to come in contact with a spark or open flame. Keep a dry chemical (Class B) fire extinguisher near the work area. Never drain or store fuel in an open container due to the possibility of fire or explosion.

1. Disconnect the negative battery cable. Raise and safely support the vehicle. Remove the left rear wheel.

2. Remove the tank drain bolt and drain the fuel into an approved container.

3. Disconnect the pump and float wiring connectors located under the trunk floor.

4. Remove the fuel hose and pipe covers from the inside of the quarter panel.

5. Support the tank with a transmission jack, remove the straps and lower the tank out of the vehicle. If it sticks on the undercoating, carefully pry it free using a blunt or wooden instrument as a lever.

6. Disconnect the fuel line by removing the banjo bolt or uncoupling the quick-connect fittings.

7. Remove the fuel pump mounting nuts. Remove the fuel pump from the fuel tank.

To install:

8. Install fuel pump and mounting nuts. Tighten the mounting nuts to 4 ft. lbs. (6 Nm).

9. Use a jack to place the fuel tank in position. Reconnect the fuel hoses with new washers, or reconnect the quick-connect fittings with new retainers. Reconnect the pump and float wiring connectors. Tighten the fuel line to 28 ft. lbs. (38 Nm).

10. Tighten the strap bolts on 2.5TL and 3.2TL to 28 ft. lbs. (38 Nm).

11. Use a new sealing washer on the drain plug and tighten the plug to 36 ft. lbs. (49–50 Nm).

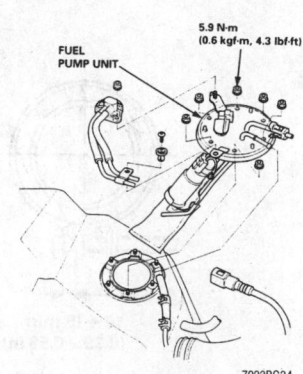

5.9 N·m
(0.6 kgf-m, 4.3 lbf-ft)

FUEL
PUMP UNIT

7923BG34

Fuel pump mounting—2.5L engine

12. Refill the tank. Check for leaks.

13. Connect the negative battery cable and enter the radio security code.

3.5L Engine

1. Properly relieve the fuel system pressure.

NOTE: Check all hose clamps and retighten if necessary.

2. Remove the rear seat cushion.

3. Remove the access panel from the floor.

4. Detach the fuel line and wiring from the fuel pump assembly.

5. Remove the mounting nuts and the fuel pump from the fuel tank.

To install:

6. Use a new gasket and install the pump in the fuel tank. Tighten the nuts to 48 in. lbs. (6 Nm).

7. Connect the fuel line and wiring. Always use a new retainer on the male end of the fuel connection.

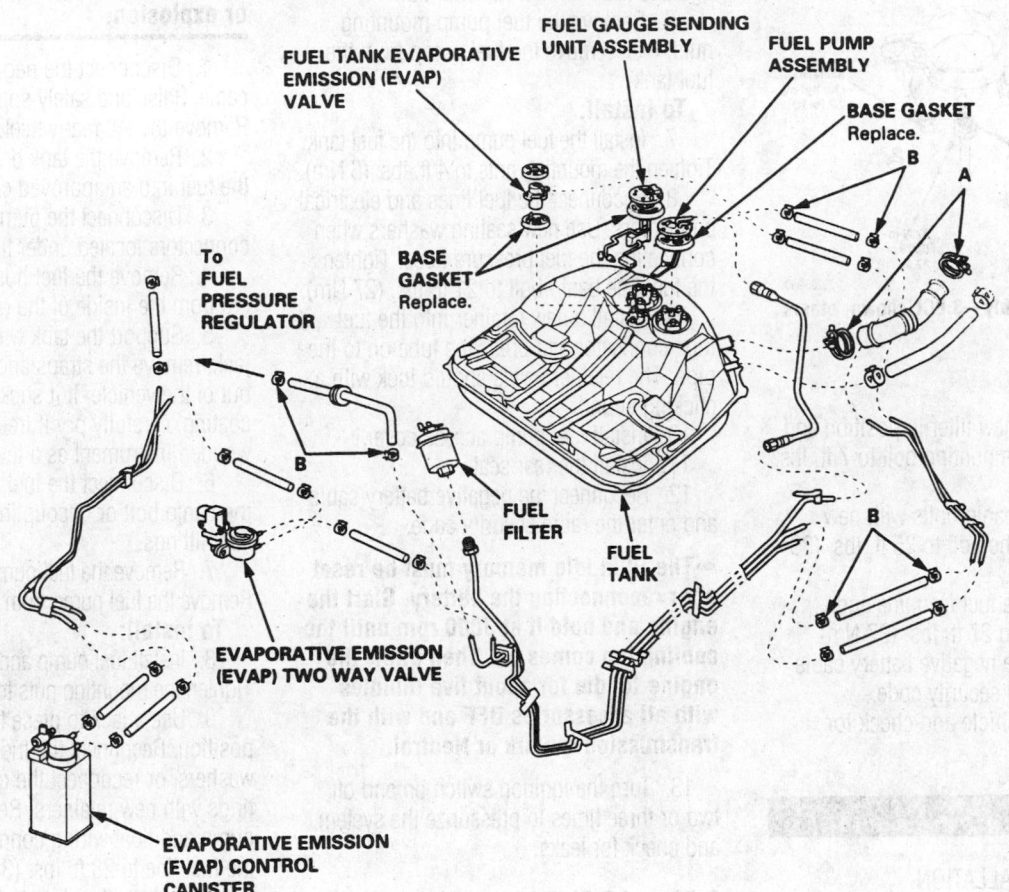

FUEL TANK EVAPORATIVE EMISSION (EVAP) VALVE

FUEL GAUGE SENDING UNIT ASSEMBLY

FUEL PUMP ASSEMBLY

BASE GASKET Replace.

To FUEL PRESSURE REGULATOR

BASE GASKET Replace.

FUEL FILTER

FUEL TANK

EVAPORATIVE EMISSION (EVAP) TWO WAY VALVE

EVAPORATIVE EMISSION (EVAP) CONTROL CANISTER

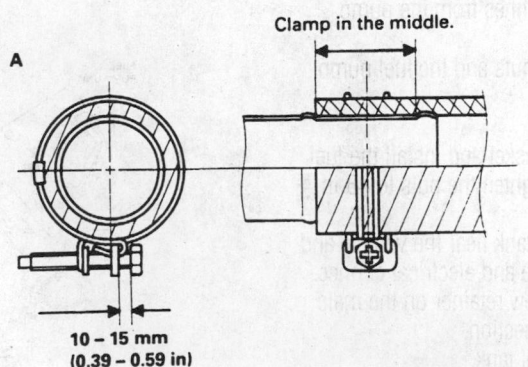

Clamp in the middle.

A

10 – 15 mm
(0.39 – 0.59 in)

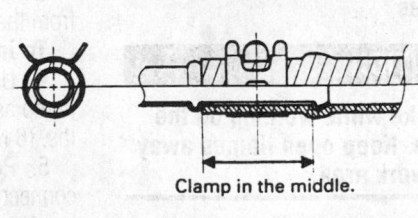

B

Clamp in the middle.

7923BG91

Exploded view of the fuel line routing—3.5L engine

8. Pressurize the fuel system and check for leaks.

9. Install the access cover and the rear seat cushion.

DRIVE TRAIN

Transaxle Assembly

REMOVAL & INSTALLATION

Manual

➡ **The radio may have a coded theft protection circuit. Obtain the code from the owner before disconnecting the battery, removing the radio fuse, or removing the radio.**

INTEGRA AND INTEGRA GSR

1. Disconnect the negative battery cable, then the positive battery cable.

2. Drain the transaxle oil. Install the drain plug with a new washer.

3. Remove the air cleaner case and the air intake tube.

4. Disconnect the back-up light switch connector and the transaxle ground wire.

5. Remove the lower radiator hose clamp from the transaxle hanger.

6. Remove the wiring harness clips.

7. Disconnect the starter motor cables and the Vehicle Speed Sensor (VSS) connector.

8. Remove the clutch pipe bracket and slave cylinder. Do not operate the clutch pedal once the slave cylinder has been removed.

9. Remove the three upper transaxle mounting bolts and the lower starter mounting bolt.

10. Safely raise and support the vehicle.

11. Remove the engine splash shield.

12. Disconnect the Heated Oxygen Sensor (HO2S) connector.

13. Remove and discard the two nuts attaching exhaust pipe A to the hanger bracket.

14. Remove and discard the nuts attaching exhaust pipe A to the exhaust manifold, discard the exhaust gaskets.

15. Remove and discard the three nuts attaching the exhaust system to the catalytic converter, discard the exhaust gasket. Remove exhaust pipe A from the vehicle.

16. Remove the cotter pins and castle nuts from the front lower ball joints. Separate the ball joints from the lower control arms.

17. Remove the right strut fork pinch bolt and lower nut and bolt, then remove the strut fork from the vehicle.

18. Pry the right halfshaft out of the differential, discard the set ring on the inner joint.

19. Pry the left halfshaft out of the intermediate shaft, discard the set ring on the inner joint.

20. Tie plastic bags over the halfshaft joints to keep the splines of the joints clean.

21. Remove the intermediate shaft mounting bolts, and remove the intermediate shaft.

22. Remove the set ring from the intermediate shaft and install a new set ring.

23. Disconnect the extension rod and the shift rod:

    a. If equipped with a VTEC (B18C1) engine, remove the heat shield.

    b. Disconnect the shift extension rod from the transaxle case.

    c. Slide the boot on the shift rod back to expose the clip and spring pin. Remove the clip from the shift rod.

    d. Drive out the spring pin with a punch and disconnect the shift control rod. Note that on reassembly, install the clip back into place after driving the spring pin in.

24. Remove the rear engine stiffener, and if equipped with a VTEC (B18C1) engine remove the front engine stiffener.

25. Remove the clutch cover.

26. Remove the right front mount/bracket. Discard the long self-locking bolt.

27. Place a transmission jack under the transaxle and a jack stand under the engine.

28. Remove the transaxle mount.

29. Remove the transaxle mounting bolts and the bolts from the rear mounting bracket. Discard the self-locking bolts from the rear mounting bracket.

30. Pull the transaxle assembly away from the engine until it clears the main shaft, then lower it on the transmission jack.

To install:

31. Install the dowel pins to the clutch housing.

32. Apply super high temperature grease to the following components:

    a. The release fork bolt.

    b. The spline of the transaxle input shaft.

    c. The inside of the release bearing and the sleeve on the transaxle input shaft where the release bearing rides.

    d. The tips of the release fork and where the slave cylinder pin rides on the release fork.

33. Install the release fork boot.

34. Place the transaxle assembly on the transmission jack and raise it to engine level.

35. Install the transaxle mounting bolts and new rear mount bracket bolts. Tighten the transaxle mounting bolts to 47 ft. lbs. (64 Nm) and the rear mount bolts to 87 ft. lbs. (118 Nm).

36. Raise the transaxle and install the transaxle mount. First tighten the mounting nuts and bolt to the transaxle to 47 ft. lbs. (64 Nm), then tighten the mounting bolt to 54 ft. lbs. (74 Nm).

37. Install the three upper transaxle mounting bolts and the lower starter bolt. Tighten the upper transaxle mounting bolts to 47 ft. lbs. (54 Nm), tighten the lower starter bolt to 33 ft. lbs. (44 Nm).

38. Install the right front mount/bracket. Tighten the new self locking bolt to 61 ft. lbs. (83 Nm), and tighten the other mounting bolts to 33 ft. lbs. (44 Nm).

39. Install the clutch cover. Tighten the 6x1mm bolts to 9 ft. lbs. (12 Nm), tighten the 8x1.25mm bolts to 17 ft. lbs. (24 Nm). If equipped with a NON-VTEC (B18B1) engine, tighten the 12x1.25mm bolt to 42 ft. lbs. (57 Nm).

40. Install the rear engine stiffener and the front engine stiffener, if equipped. Tighten the bolts attaching the stiffener(s) to the transaxle to 42 ft. lbs. (57 Nm). Tighten the bolts attaching the stiffener(s) to the engine to 17 ft. lbs. (24 Nm).

41. Remove the transmission jack and the jack stand from the engine.

42. Install the shift rod to the transaxle and install the spring pin and clip. Install the shift rod boot to the shift rod, making sure the drain hole is facing down.

43. Install the extension rod and tighten the attaching bolt to 16 ft. lbs. (22 Nm).

44. If equipped with a VTEC (B18C1) engine, install the heat shield. Tighten the mounting bolts to 7 ft. lbs. (9.8 Nm).

45. Install new set ring to the intermediate shaft and the halfshafts.

46. Install the intermediate shaft, tighten the mounting bolts to 29 ft. lbs. (39 Nm).

47. Turn the right steering knuckle fully outward and slide the halfshaft into the differential unit until you feel the spring clip engage. Turn the left steering knuckle fully outward and slide the halfshaft onto the intermediate shaft until you feel the spring clip engage.

48. Install the right strut fork. Tighten the pinch bolt to 32 ft. lbs. (43 Nm) and tighten the fork lower nut and bolt to 47 ft. lbs. (64 Nm).

49. Connect the lower ball joints to the lower control arms, tighten the castle nuts

to 36–43 ft. lbs. (49–59 Nm). Install new cotter pins.

50. Install exhaust pipe A using new gaskets and locknuts. Tighten the nuts attaching the exhaust pipe to the exhaust manifold to 40 ft. lbs. (54 Nm). Tighten the nuts attaching the exhaust system to the catalytic converter to 25 ft. lbs. (33 Nm). Tighten the exhaust pipe hanger nuts to 12 ft. lbs. (16 Nm).

51. Install the engine splash shield.

52. Apply super high temperature grease to the tip of the slave cylinder and install the slave cylinder to the transaxle. Tighten the mounting bolts to 16 ft. lbs. (22 Nm). Install the clutch bracket pipe and tighten the mounting bolts to 7 ft. lbs. (9.8 Nm).

53. Install the front wheels and lower the vehicle.

54. Connect the VSS sensor and the starter motor connectors.

55. Connect the lower radiator hose clamp to the transaxle hanger.

56. Connect the transaxle ground cable and the back-up light switch connector.

57. Install the air cleaner housing assembly with the air intake tube.

58. Refill the transaxle with 4.6 pints of 10W-30 or 10W-40, SH or SJ grade oil.

59. Connect the positive, then the negative battery cables.

60. Check the operation of the clutch and smooth operation of the shifter.

61. Check the front wheel alignment and road test the vehicle.

LEGEND

1. Disconnect both battery cables.

2. Unbolt the strut bar from the bulkhead and shock towers.

3. Drain the transaxle fluid and replace the plug with a new washer.

4. Remove the control box, but do not disconnect the vacuum lines.

5. Disconnect the neutral switch, back-up light switch, and the reverse lockout solenoid connector.

6. Remove the transaxle housing bolts and the clutch hose bracket from the rear engine hanger.

7. Remove the front exhaust pipe and catalytic converter. On Legends with 6–speed transaxle, remove the twin three-way catalytic converters and their brackets.

8. Remove the converter heat shield and bracket.

➡ **An extension shaft puller, Acura #07LAC-PW50100 or equivalent, should be used to remove the extension shaft before removing the transaxle.**

9. Lock the transaxle by shifting it into first gear.

10. Remove the extension shaft secondary cover and the 36mm sealing bolt. Use an extension shaft puller to remove the extension shaft from the rear of the transaxle case. The differential is not removed with the transmission.

11. Disconnect the transaxle linkage extension and the shift rod.

12. Disconnect the oil cooler lines at the oil pump pipes.

13. Remove the release fork cover and the clutch slave cylinder.

14. Support the steering rack with a jack, and remove the rack cover plate. Reinstall the steering rack bolts to hold the rack in the vehicle.

15. Remove the exhaust pipe bracket.

16. Remove the transaxle rear mount and bracket assembly.

17. Pull out on the clutch fork to release it from the throw-out bearing. Don't remove it from the clutch housing.

18. Use a jack to take up the transaxle's weight. Remove the transaxle mid mounts.

19. Remove the transaxle housing mounting bolts.

20. Remove the engine stiffener and the clutch housing cover.

21. Remove the transaxle mounting bolts and the 26mm shim under the transaxle.

22. Verify that all linkages, vacuum lines, and wiring harnesses have been disconnected.

23. Slide the transaxle back and off the input shaft, and lower it from the vehicle.

To install:

24. Make sure that the transaxle mounting dowel pins are seated in the clutch housing.

25. Clean and lightly lubricate the input shaft and release fork contact points with molybdenum grease and install the fork.

26. Install the extension shaft in place. Use a new set ring on the shaft and lightly lubricate the splines with molybdenum grease. Install the sealing bolt and secondary cover.

27. Install the transaxle and start all of the bolts. Install the 26mm transaxle shim. Torque the 12mm bolts to 55 ft. lbs. (75 Nm).

28. Install the clutch cover and engine stiffener bolts. Torque to stiffener bolts to 16 ft. lbs. (22 Nm), and the clutch cover bolts to 9 ft. lbs. (12 Nm).

29. Install the mid mounts and the exhaust bracket. Torque the 10mm bolts to 29 ft. lbs. (39 Nm), 10mm nuts to 36 ft. lbs. (49 Nm).

30. Install the transaxle rear mount and bracket assembly. Torque the mount bolts to 29 ft. lbs. (39 Nm).

31. Install the shift linkage and extension rod. Make sure the hole in the shift rod boot is facing down.

32. Install the release fork and the slave cylinder.

33. Install the release fork cover and connect the oil cooler hoses.

34. With the transaxle in gear, install the extension shaft using the special tool. Coat the extension shaft with molybdenum grease and use a new set ring. Make sure the shaft snaps into place on the set ring.

35. Pack the shaft area with molybdenum grease, but keep the thread area clean. Apply liquid gasket to the sealing bolt threads and install the bolt and cover.

36. Support the steering rack with a jack and remove the bolts. Install the rack cover plate and torque the bolts to 28 ft. lbs. (39 Nm). These bolts thread into aluminum and must have the special Dacro® coating to avoid corrosion.

37. On six-speed Legends, install the twin three-way catalytic converters and brackets.

38. Install the heat shield and the exhaust pipe and catalytic converter. Use new locking nuts and gaskets. Torque the exhaust flange nuts to 40 ft. lbs. (55 Nm) and the catalyst flange nuts to 26 ft. lbs. (34 Nm).

39. Install the clutch hose bracket.

40. Install the upper transaxle mounting bolts and torque to 55 ft. lbs. (75 Nm).

41. Reconnect the neutral position switch, backup light switch, and the reverse lockout solenoid connectors.

42. Install the control box.

43. Install the strut bar.

44. Verify that all linkages, vacuum lines, and wiring harnesses have been reconnected.

45. Refill the transaxle fluid.

46. Connect the battery cables.

47. Check the clutch operation and adjust if necessary.

48. Shift the transaxle through the gear range and check for smooth operation.

2.2CL AND 2.3CL

1. Disconnect the negative battery cable, then the positive cable. Remove the battery.

2. Drain the transaxle fluid.

3. Remove the air cleaner, air duct and resonator assembly.

4. Remove the starter.

5. Tag and remove all wiring from the transaxle.

6. Shift the transaxle into reverse.

7. Remove the cable bracket and the cables from the transaxle.

8. Remove the bolts and raise the clutch damper bracket.

9. Remove the slave cylinder.

10. Remove the two upper transmission mounting bolts.

11. Remove the engine splash shield.

12. Disconnect the shock absorber from the lower arm.

13. Separate the lower ball joint from the knuckle assembly.

14. Swing the hub assemblies outward and remove the halfshafts and intermediate shaft from the transaxle..

15. Remove the center beam from the sub-frame.

16. Remove the clutch cover.

17. Remove the intake manifold bracket.

18. Remove the three bolts from the rear bracket.

19. Jack up the transaxle slightly and remove the mounting bracket.

20. Remove the three lower transaxle-to-engine mounting bolts.

21. Pull the transaxle away from the engine. Do not damage the clutch hydraulic lines.

To install:

22. Be sure the two dowels are installed in the clutch housing.

23. Grease the release fork and bearing.

24. Install the transaxle in the reverse of the removal. Tighten the transaxle-to-engine bolts to 47 ft. lbs. (64 Nm). Tighten the rear mount bracket bolts to 40 ft. lbs. (54 Nm). Tighten the intake manifold bracket bolts to 16 ft. lbs. (22 Nm). Tighten the center beam bolts to 37 ft. lbs. (50 Nm).

25. Refill the transaxle with Honda manual transmission fluid.

Automatic

INTEGRA

1. Disconnect the negative battery cable, then the positive battery cable.

2. Remove the air cleaner housing assembly with intake air tube.

3. Disconnect the starter cables and remove the cable holder from the starter.

4. Disconnect the transaxle ground cable from the transaxle hanger.

5. Disconnect the lock-up control solenoid valve connector and the shift control solenoid valve connector. Remove the harness clamp on the lock-up control solenoid harness from the harness stay.

6. Disconnect the vehicle speed sensor (VSS), main shaft speed sensor and the counter shaft speed sensor connectors.

7. Remove the upper transaxle mounting bolts.

8. Remove the drain plug from the transaxle and drain the used fluid into a sealable container. Properly dispose of the used fluid. Reinstall the drain plug with a new sealing washer.

9. Remove the splash shield.

10. Remove the front wheels.

11. Remove the cotter pins and castle nuts from the front lower ball joints. Separate the ball joints from the lower control arms.

12. Remove the right strut fork bolt, discard the nut.

13. Remove the right strut pinch bolt, then remove the strut fork.

14. Pry the right halfshaft out of the differential, discard the set ring on the inner joint.

15. Pry the left halfshaft out of the intermediate shaft, discard the set ring on the inner joint.

16. Tie plastic bags over the halfshaft joints to keep the splines of the joints clean.

17. Disconnect the heated oxygen sensor (HO2S) connector.

18. Remove the nuts and bolts connecting exhaust pipe A to the catalytic converter. Discard the gasket and the locknuts.

19. Remove and discard the nuts attaching exhaust pipe A to the exhaust hanger.

20. Remove and discard the locknuts attaching exhaust pipe A to the exhaust manifold, then remove exhaust pipe A from the vehicle. discard the exhaust gaskets.

21. Remove the intermediate shaft mounting bolts, and remove the intermediate shaft.

22. Remove the set ring from the intermediate shaft and install a new set ring.

23. Remove the shift cable cover, then remove the shift cable by removing the control lever. Discard the lockwasher.

⁕⁕ WARNING

Do not bend the shift control cable when removing it.

24. Remove the right front mount/bracket and discard the two long attaching bolts.

25. Remove the end of the throttle control cable from the throttle control drum.

26. Disconnect the ATF cooler hoses from the joint pipes. Turn the ends of the cooler hoses up to prevent ATF from flowing out, then plug the joint pipes.

27. Remove the engine stiffener and the torque converter cover.

28. Remove the eight drive plate bolts one at a time while rotating the crankshaft pulley.

29. Place a transmission jack under the transaxle, raise the transaxle just enough to take the weight off of the mounts, then remove the transmission mount.

30. Remove the transaxle mounting bolts and rear engine mounting bolts.

31. Pull the transaxle away from the engine until it clears the 14mm dowel pins, then lower it on the transaxle jack.

32. Remove the starter from the transaxle.

To install:

33. Flush the ATF cooler.

34. Install the starter to the transaxle, tighten the bolts to 33 ft. lbs. (45 Nm). Install the 14mm dowel pins to the torque converter housing.

35. Place the transaxle on a transmission jack, and raise to engine level.

36. Fit the transaxle to the engine, then install the transaxle housing mounting bolts and the two rear engine mounting bolts with new washers. Tighten the transaxle housing mounting bolts to 43 ft. lbs. (59 Nm) and the rear engine mounting bolts to 86.8 ft. lbs. (118 Nm).

37. Install the transmission mount. Tighten the bolt to 54 ft. lbs. (74 Nm) and the nuts to 47 ft. lbs. (64 Nm).

38. Install the three transaxle upper mounting bolts, tighten the bolts to 54 ft. lbs. (74 Nm).

39. Remove the transmission jack.

40. Attach the torque converter to the drive plate with eight bolts, tighten the bolts to 8.7 ft. lbs. (12 Nm). Rotate the crankshaft as necessary to tighten the bolts to 1/2 the specified torque, then the final torque, in a crisscross pattern. After tightening the last bolts, check that the crankshaft rotates freely.

41. Install the torque converter cover, tighten the three 6x1mm bolts, to 8.7 ft. lbs. (12 Nm) and tighten the 10x1.25 bolt to 33 ft. lbs. (44 Nm).

42. Install the engine stiffener. Tighten the bolt attaching the engine stiffener to the transaxle to 32 ft. lbs. (43 Nm). Tighten the bolts attaching the stiffener to the engine to 17 ft. lbs. (24 Nm).

43. Tighten the crankshaft pulley bolt to 130 ft. lbs. (177 Nm).

44. Connect the transaxle cooler inlet hose to the joint pipe. Leave the drain hose on the return line.

45. Connect the throttle control cable to the control drum and install the right front mount/bracket. Tighten the two new bolts (12x1.25mm) to 47 ft. lbs. (64 Nm). Tighten the two 10x1.25 bolts to 33 ft. lbs. (44 Nm).

46. Connect the control lever and shifter cable, using a new lockwasher, then install

the shift cable cover. Tighten the control lever bolt to 10 ft. lbs. (14 Nm), and the shift cable cover bolts to 8.7 ft. lbs. (12 Nm).

47. Install new set ring to the intermediate shaft and the halfshafts.

48. Install the intermediate shaft, tighten the mounting bolts to 29 ft. lbs. (39 Nm).

49. The remaining components are installed in the reverse order from which they were removed.

50. Refill the transaxle to the proper level.

51. Start the engine, with the parking brake set, and shift the transaxle through all gears three times.

52. Check and adjust the shift cable as necessary.

53. Let the engine reach operating temperature (the cooling fan comes on) with the transaxle in **P**

or **N**, then turn the engine off and check the fluid level.

54. Road test the vehicle.

2.5TL

1. Shift the transmission into **P**.

2. Disconnect both battery cables and remove the battery and battery tray.

3. Without disconnecting the wires, remove the ABS relay box and set it aside.

4. Remove the heat shield and sub-ground cable.

➡**The distributor may be removed for better access to the transmission case bolts.**

5. Remove the emission control equipment box from the firewall without disconnecting the vacuum hoses.

6. Remove the torque converter cover and rotate the crankshaft as required to remove the eight torque converter flexplate bolts.

7. Disconnect the transmission wiring harnesses and tag them for reassembly.

8. Remove the transmission ground cable.

9. Remove the upper transmission bolts and the 26mm shim.

10. Raise and support the vehicle.

11. Remove the guard plate and remove the plug to drain the transmission fluid.

12. Remove the transmission left-side mount and bracket.

13. Disconnect the oil cooler hoses.

14. The differential stays on the vehicle. Remove the secondary cover and the 33mm sealing bolt and install an extension shaft removal tool. Disconnect the differential extension shaft from the transmission.

➡**An extension shaft puller and installer, Acura part numbers 07LAC-PW50101 and 07MAF-PY40100, or their equivalents are needed to disconnect the extension shaft from the differential.**

15. Remove the front exhaust pipe and its mounting brackets.

16. Remove the shift cable cover and disconnect the shift cable from the control shaft. Remove the cable mounting bracket and wire the cable up and out of the way.

17. Place a transmission jack securely under the transmission and raise it to take the weight off the mounts.

18. Use an offset wrench to remove the mid-mounts and the mid-mount spacer.

19. Remove the transmission case bolts. Remove the torque converter cover mounting bolt located on the converter housing. Do not remove the torque converter cover.

20. Slide the transmission back and away from the engine. Carefully lower it from the vehicle.

To install:

21. Flush the ATF cooler lines.

22. Install the torque converter onto the mainshaft using a new O-ring. Install the mounting pins into the transmission case.

23. Install a new set ring on the extension shaft and lightly lubricate the splines with high temperature molybdenum grease. Pack the opening in the drive pinion with high temperature molybdenum grease.

24. Install the transmission and start all of the bolts. Don't forget the 26mm shim between the transmission and differential. Tighten the 12mm bolts to 54 ft. lbs. (75 Nm). Tighten the torque converter cover bolt to 9 ft. lbs. (12 Nm).

25. Install the mid-mounts and brackets. Tighten the bolts to 28 ft. lbs. (39 Nm) and the nuts to 32 ft. lbs. (43–44 Nm).

26. Install a new set ring, and install the extension shaft using an extension shaft installer. Be sure the shaft snaps into place.

27. Pack the shaft area with molybdenum grease, but keep the thread area clean. Apply liquid gasket to the sealing bolt threads and tighten the bolt to 58 ft. lbs. (78–80 Nm). Install the cover.

28. Install the transmission left-side mount and bracket. Tighten the mount bolts to 47 ft. lbs. (65 Nm). Tighten the bracket bolts to 40 ft. lbs. (54 Nm).

29. Install the torque converter bolts and tighten them in 2 steps in a crisscross pattern to 9 ft. lbs. (12 Nm). Install the torque converter cover and tighten the bolts to 9 ft. lbs. (12 Nm).

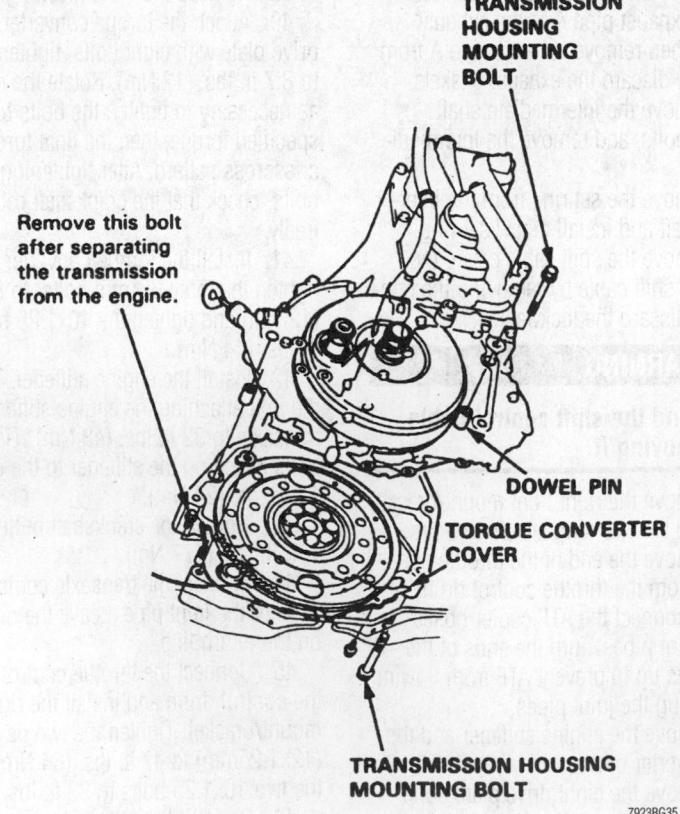

Remove this bolt after separating the transmission from the engine.

TRANSMISSION HOUSING MOUNTING BOLT

DOWEL PIN

TORQUE CONVERTER COVER

TRANSMISSION HOUSING MOUNTING BOLT

7923BG35

Automatic transaxle—2.5L engines

30. The remaining components are installed in the reverse order from which they were removed.

31. Connect all the wiring harnesses and the battery cables. Refill the transmission with fresh fluid.

32. When all parts have been installed, start the engine and shift through all the gears 3 times to fill all the passages with fluid, then check the shift cable and adjust as needed. When the engine is fully warmed up, stop the engine and check the fluid level.

LEGEND AND 3.2TL

1. Disconnect the negative, then the positive battery cables.

2. Shift the transmission into **P** (Park).

3. Remove the control box from the bulkhead without disconnecting the vacuum hoses. Place the control box out of the way.

4. Disconnect the transmission sub-harness connectors, and remove the sub-harness clamp.

5. If applicable, remove the three bolts securing the ATF dipstick pipe bracket.

6. Remove the upper transmission mounting bolts.

7. Drain the fluid from the transmission into a sealable container. Install the drain plug with a new washer and tighten the plug to 36 ft. lbs. (49 Nm).

8. Pull the carpet back under the passenger seat to expose the secondary heated oxygen connector (HO2S sensor 2) connector. Disconnect the connector and push it out from the inside of the vehicle.

9. Remove the heat shields from exhaust pipe A.

10. Remove the nuts attaching exhaust pipe A to the exhaust manifolds and the catalytic converter. Remove the front exhaust pipe and discard the gaskets.

11. Remove the oxygen sensor wiring harness cover and grommet, then remove the catalytic converter. Discard the nuts and gasket.

12. Remove the exhaust heat shield from the floor of the vehicle.

13. Disconnect the transmission cooler hoses, then plug the hoses and pipes.

14. Remove the shift cable cover mounting bolts and remove the wiring harness clamps from the cover. Remove the shift cable cover from the transmission.

15. Remove the shift cable holder from the holder base, do not lose the washers.

16. Remove the locknut attaching the shift cable to the control lever, then remove the shift cable.

17. Remove the ATF dipstick pipe from the torque converter housing.

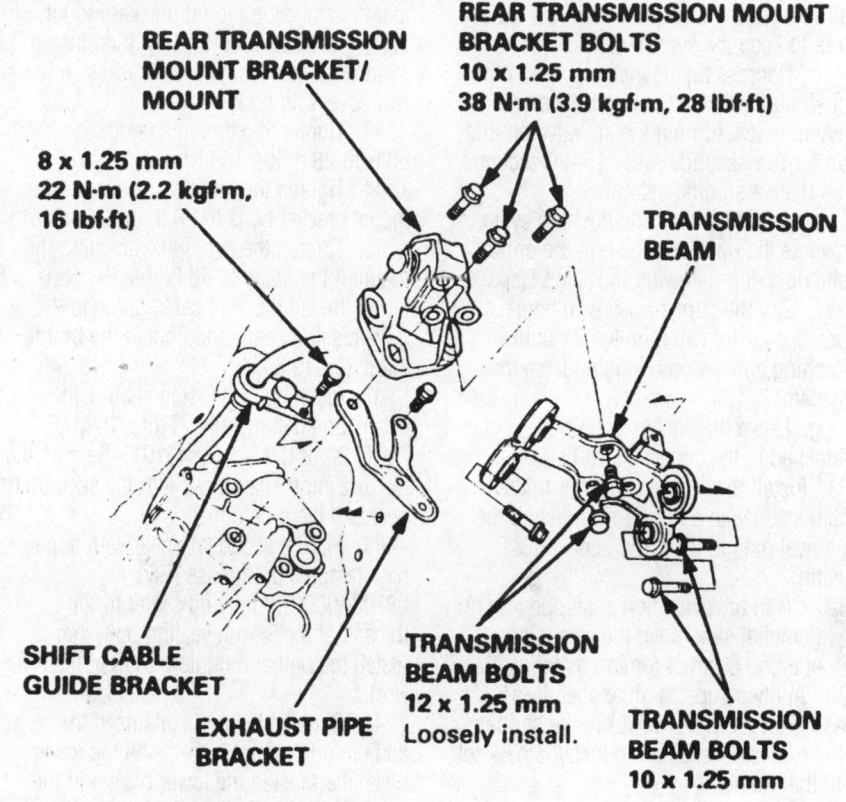

REAR TRANSMISSION MOUNT BRACKET/ MOUNT

REAR TRANSMISSION MOUNT BRACKET BOLTS
10 x 1.25 mm
38 N·m (3.9 kgf·m, 28 lbf·ft)

8 x 1.25 mm
22 N·m (2.2 kgf·m, 16 lbf·ft)

TRANSMISSION BEAM

SHIFT CABLE GUIDE BRACKET

EXHAUST PIPE BRACKET

TRANSMISSION BEAM BOLTS
12 x 1.25 mm
Loosely install.

TRANSMISSION BEAM BOLTS
10 x 1.25 mm

7923BG36

Exploded view of the rear transmission mount—3.2TL

18. Remove the lower plate from under the rack and pinion, then install the two rack and pinion mounting bolts.

19. Remove the shift cable guide bracket from the transmission beam.

20. Remove the transmission beam, rear transmission mount racket/mount and exhaust pipe hanger.

21. Be sure that the transmission is in park, then remove the 36mm sealing bolt from the transmission.

22. Install the extension shaft puller (Part # 07LAC-PW50100 or 07LAC-PW50101) onto the end of the extension shaft. Using the extension shaft puller, disconnect the extension shaft from the differential. Pull the extension shaft out enough to remove the set ring.

➡**Do not try to remove the extension shaft, it cannot be removed from the transmission this way.**

23. Place a transmission jack under the transmission and raise the transmission to take the weight off of the mounts.

24. Remove the stop holder, the mid mount stops and the mid mounts.

25. Remove the engine stiffener.

26. Remove the torque converter covers.

27. Remove the six drive plate bolts one at a time while rotating the crankshaft.

➡**If necessary, remove the spark plugs while removing the drive plate bolts.**

28. Remove the transmission mounting bolts.

29. Pull the transmission away from the engine until it clears the dowel pins, then lower it on the transmission jack.

To install:

30. Flush the transmission cooling lines before installing the transmission. Use a pressurized flushing canister, such as Honda tool No. J38405-a, or its equivalent. Use only biodegradable flushing fluid, Honda part No. J35944–20. Other types of flushing fluid may damage the A/T cooling system.

a. Fill the flusher with 21 ounces of fluid (canister is 2/3 full). Pressurize the flusher to 80–120 PSI (560–845 kPa), following the procedure on the fluid container and flusher.

b. Clamp the discharge hose of the flusher to the cooler return line. Clamp the drain hose to the cooler inlet line and route it into a bucket or drain tank.

c. Connect the flusher to air and water lines. Open the flusher water valve and flush the cooler for ten seconds. The air line should be equipped with a water trap to keep the system dry.

d. Depress the flusher trigger to mix flushing fluid with the water. Flush for two minutes, turning the air valve on and off for five seconds every 15–20 seconds to create a surging action.

e. After finishing one flushing cycle, reverse the hose and flush in the opposite direction following the same steps.

f. Dry the cooler lines with compressed air for two minutes, or until flushing agent stops draining from the system.

g. Leave the flusher drain hose attached to the cooler return line.

31. Install the torque converter to the transmission with a new O-ring. Install the two dowel pins to the torque converter housing.

32. Clean the extension shaft opening on the differential side. Keep the extension shaft opening clean of foreign material.

33. Apply a super high temperature grease (Part # 08798–9002) to the splines on the extension shaft, then install a new set rig to the groove.

34. Raise the transmission into position and attach the transmission to the engine. Install the housing mounting bolt with the 26mm shim. Do not install the transmission housing bolt on the engine stiffener side at this time.

35. Attach the torque converter to the drive plate with the six bolts. Rotate the crankshaft as necessary to tighten the bolts to 10 ft. lbs. (13 Nm), then tighten the bolts in a crisscross pattern to 20 ft. lbs. (26 Nm). After tightening the bolts check that the crankshaft rotates freely.

36. Install the torque converter covers and tighten the bolts to 8.7 ft. lbs. (12 Nm).

37. Install the engine stiffener. Tighten the 8mm bolts loosely and tighten the transmission housing bolt to 16 ft. lbs. (22 Nm), then tighten the rest of the bolts to 16 ft. lbs. (22 Nm).

38. Install the transmission housing mounting bolts from the transmission side. Tighten the bolts to 47 ft. lbs. (64 Nm).

39. Install the mid mounts, then the mid mount stops. Tighten the 8mm bolts loosely, then install the stop holder.

40. Tighten the mid mount 10mm bolts to 28 ft. lbs. (38 Nm) and tighten the 8mm bolts to 16 ft. lbs. (22 Nm). Tighten the new nuts attaching the mid mounts to 35 ft. lbs. (48 Nm) and tighten the nuts attaching the stop holder to 40 ft. lbs. (54 Nm).

41. Remove the transmission jack from the transmission.

42. Install the transmission beam to the rear transmission mount bracket/mount. Tighten the two bolts loosely, then install them with the exhaust pipe bracket on the rear cover and body.

43. Tighten the three transmission beam bolts to 28 ft. lbs. (38 Nm).

44. Tighten the three rear transmission mount bracket bolts to 28 ft. lbs. (38 Nm).

45. Tighten the two bolts attaching the mount it the beam to 40 ft. lbs. (54 Nm).

46. Install the shift cable guide to the transmission beam and tighten the bolt to 7.2 ft. lbs. (9.8 Nm).

47. Install the extension shaft using the extension shaft installer (Part # 07MAF-PY40100 or 07MAF-PY40101). Be sure that the extension shaft locks into the secondary gear and the differential.

48. Fill the secondary gear with super high temperature grease (Part # 08798–9002). Applying sealer to the threads of the 36mm sealing bolt, then install the bolt and tighten to 58 ft. lbs.(78 Nm).

49. Remove the two bolts from the rack and pinion necessary to install the lower plate, then install the lower plate and the attaching bolts. Tighten the lower plate attaching bolt to 28 ft. lbs. (38 Nm) and tighten the rack and pinion bolts to 43 ft. lbs. (59 Nm).

50. Connect the shift cable control lever to the control shaft, then install the washer and nut. Tighten the nut to 8.7 ft. lbs. (12 Nm).

51. Install the shift cable holder to the shift cable holder base with the mounting washers. Install the attaching bolts and tighten the bolts to 8.7 ft. lbs. (12 Nm).

52. Install the shift cable cover and tighten the mounting bolts to 8.7 ft. lbs. (12 Nm).

53. Connect the cooler feed hose to the pipe.

54. Install the ATF dipstick pipe with a new O-ring on the torque converter housing.

55. Install the transmission sub-harness clamp to the harness.

56. Install the exhaust heat shield to the floor of the vehicle and tighten the bolts to 7.2 ft. lbs. (9.8 Nm).

57. Install the catalytic converter with a new gasket. Install the wiring harness and the grommet to the vehicle, then install the harness cover. Tighten the cover bolts to 7.2 ft. lbs. (9.8 Nm).

58. Install exhaust pipe A with new gaskets and new nuts. Tighten the nuts attaching the pipe to the manifolds to 40 ft. lbs. (54 Nm) and tighten the nuts attaching the

exhaust pipe to the catalytic converter to 16 ft. lbs. (22 Nm). Tighten the catalytic converter rear attaching nuts to 24 ft. lbs. (32 Nm).

59. Install the heat shields to exhaust pipe A and tighten the nuts to 8.7 ft. lbs. (12 Nm).

60. Connect the secondary heated oxygen sensor (HO2S) connector, located under the passenger front seat.

61. Install the transmission upper mounting bolts and tighten the bolts to 47 ft. lbs. (64 Nm).

62. Install the ATF dipstick pipe bracket bolts and tighten the bolts to 8.7 ft. lbs. (12 Nm).

63. Connect the transmission sub-harness connectors.

64. Install the control box and tighten the mounting bolts to 8.7 ft. lbs. (12 Nm).

65. Connect the positive, then the negative battery cable.

66. Fill the transmission with ATF. Use only Honda Premium ATF or an equivalent DEXRON® II ATF.

a. Leave the flusher drain hose attached to the cooler return line.

b. With the transmission in park, run the engine for 30 seconds, or until approximately one quart of fluid is discharged. As soon as one quart of fluid drains, shut off the engine. This completes the cooler flushing process.

c. Remove the drain hose and reconnect the cooler return line.

d. Refill the transmission to the proper level with ATF.

67. Let the engine reach proper operating temperature (the radiator fan comes on) with the transmission in **N** (neutral) or **P** (park). Turn the engine **OFF** and check the fluid level.

68. Enter the radio security code.

Clutch

REMOVAL & INSTALLATION

➡**The radio may have a coded theft protection circuit. Obtain the code from the owner before disconnecting the battery, removing the radio fuse, or removing the radio.**

Integra

1. Disconnect the negative battery cable.

2. Remove the manual transaxle assembly from the vehicle.

3. Insert the clutch alignment shaft (part # 07NAF-PR30100) with the clutch

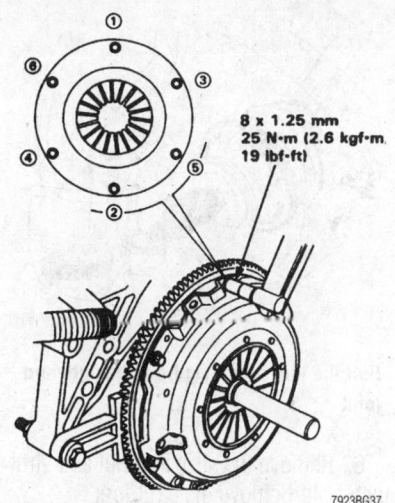

8 x 1.25 mm
25 N·m (2.6 kgf·m,
19 lbf·ft)

Pressure plate bolt torque sequence— Integra

7923BG37

alignment disc (part # 07JAF-PM7011A) and handle (part # 07936–3710100). Use a feeler gauge and measure the clearance between the pressure plate spring fingers and the clutch alignment disc. There should be a maximum of 0.02 in. (0.6mm) of clearance for a new pressure plate with 0.03 in. (0.8mm) limit for a used pressure plate.

4. Remove the clutch alignment disc.

5. Install a flywheel holder (part # 07LAB-PV00100 or 07924-pD20003) to aid in the removal of the pressure plate and clutch disc.

6. Matchmark the flywheel and pressure plate for easy reassembly. Remove the pressure plate bolts in a crisscross pattern 2 turns at a time to prevent warping the plate.

7. Remove the pressure plate, then the clutch disc with the alignment shaft.

To install:

8. If the flywheel was removed, align the hole in the flywheel with the crankshaft dowel pin and install the mounting bolts finger-tight.

9. Install the flywheel holder, then tighten the flywheel mounting bolts in a crisscross pattern in several steps. The mounting bolts final torque should be 76 ft. lbs. (103Nm).

10. Apply high temperature grease (part # 08798 9002) to the spline of the clutch disc, then install the clutch disc using the clutch alignment shaft.

11. Install the pressure plate, tighten the mounting bolts in the proper sequence to 19 ft. lbs. (25 Nm).

12. Remove the flywheel holding tool and the clutch alignment shaft.

13. Insert the clutch alignment shaft with the clutch alignment disc and handle. Use a

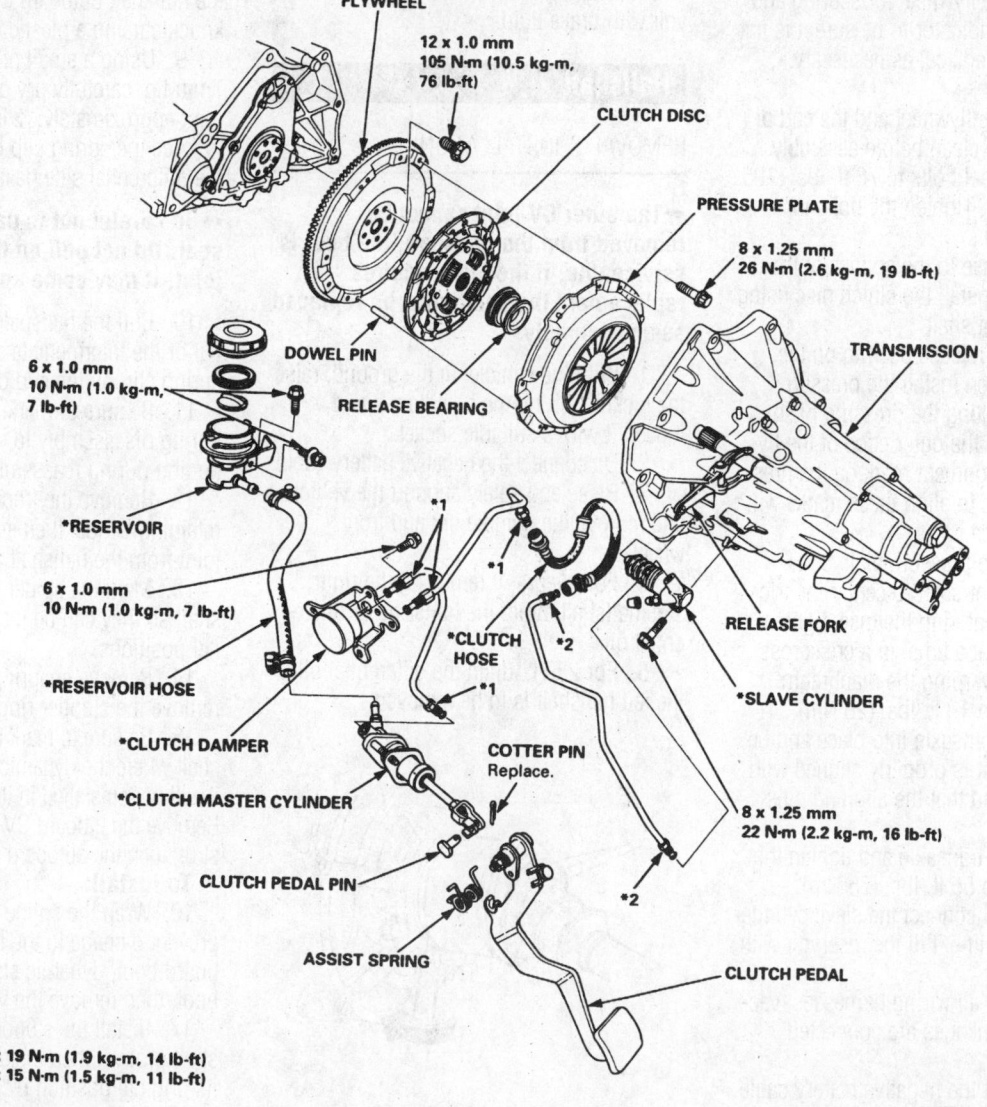

Clutch components—Legend shown, others similar

7923BG38

feeler gauge and measure the clearance between the pressure plate spring fingers and the clutch alignment disc. There should be a maximum of 0.02 in. (0.6mm) of clearance for a new pressure plate with 0.03 in. (0.8mm) limit for a used pressure plate.

14. Install the transaxle assembly.

15. Connect the negative battery cable and enter the radio security code.

Legend

1. Disconnect the negative battery cable.

2. Raise and support the vehicle.

3. Remove the transaxle.

4. Matchmark the flywheel and pressure plate for reassembly. Remove the pressure plate bolts in a crisscross pattern 2 turns at a time to prevent warping the plate.

5. Inspect the pressure plate and clutch disk for signs of wear.

6. Inspect the flywheel for scoring and wear. Use a dial indicator to be sure it is flat and resurface or replace, as necessary.

To install:

7. Be sure the flywheel and the end of the crankshaft are clean before assembly. Tighten the flywheel bolts to 76 ft. lbs. (105 Nm) on all others. Tighten the bolts in a crisscross pattern.

8. Apply grease to the splines of the clutch disc, and install the clutch disc using a clutch alignment shaft.

9. Install the release bearing on the pressure plate, then install the pressure plate. When installing the pressure plate, align the mark on the outer edge of the flywheel with the alignment mark on the pressure plate. Failure to align these marks will result in imbalance.

10. Tighten the pressure plate bolts using an alignment shaft to center the friction disc. After centering the disc, tighten the bolts 2 turns at a time, in a crisscross pattern to avoid warping the diaphragm springs; tighten to 19 ft. lbs. (26 Nm).

11. Jack the transaxle into place and be sure the mainshaft is properly aligned with the disc spline and that the aligning pins are in place.

12. Install the transaxle and tighten the mounting bolts to 55 ft. lbs. (75 Nm).

13. Install and connect the slave cylinder and its hydraulic line. Fill the reservoir with fluid.

14. Verify that all wiring harnesses, vacuum lines, and linkages are connected properly.

15. Reconnect the negative battery cable.

16. Check the clutch adjustment and road test the vehicle.

Hydraulic Clutch System

BLEEDING

➡ **Use DOT 3 or 4 brake fluid in the clutch master and slave cylinders. Brake fluid will damage the vehicle's paint—immediately clean up any spills.**

1. Fit a flare or box-end wrench onto the slave cylinder bleeder screw.

2. Attach a rubber tube to the slave cylinder bleeder screw and suspend it into a clear drain container partially filled with brake fluid.

3. Fill the clutch master cylinder with brake fluid.

4. Open the bleeder screw and pump the clutch pedal until no more bubbles appear in the tube.

5. Close the bleeder screw.

6. Refill the clutch master cylinder reservoir with brake fluid.

Halfshaft

REMOVAL & INSTALLATION

➡ **The outer CV-joint cannot be removed from the halfshaft, the boot is serviceable, if the joint requires replacement the shaft must be replaced as an assembly.**

1. With the vehicle on the ground, raise the locking tab on the spindle nut and loosen it with a suitable socket.

2. Disconnect the negative battery cable.

3. Raise and safely support the vehicle and remove the spindle nut and front wheels.

4. For Integra, if removing the right side halfshaft, drain the transaxle or differential oil.

5. For 2.5TL, drain the differential oil if the left halfshaft is to be removed.

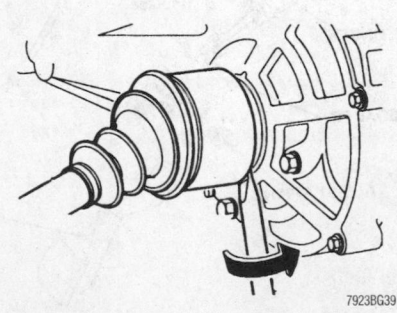

Carefully pry the inboard joint from the transaxle

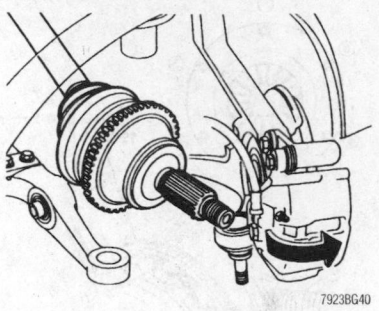

Pull the hub assembly from the outboard joint

6. Remove the strut fork nut and strut pinch bolt. Remove the strut fork.

7. Remove the lower ball joint nut and separate the lower ball joint using a ball joint remover.

8. Pull the knuckle outward and remove the halfshaft outboard CV-joint from the knuckle using a plastic mallet.

9. Using a small prybar with a 3.5 **x** 7mm tip, carefully pry out the inboard CV-joint approximately ½ in. (13mm) in order to force the spring clip out of the groove in the differential side gears.

➡ **Be careful not to damage the oil seal. Do not pull on the inboard CV-joint, it may come apart.**

10. Pull the halfshaft out of the differential or the intermediate shaft. Replace the spring clip on the end of the inboard joint.

11. Be sure to mark the roller grooves during disassembly to ensure proper positioning during reassembly.

12. Remove the front and rear boot retaining bands, then separate the inboard joint from the halfshaft assembly.

13. Mark the spider gear and the driveshaft so they can be installed in their original positions.

14. Remove snapring, spider gear, then remove the stopper ring.

15. Be sure to mark the position on the shaft where the dynamic strut goes, to ensure it will be reinstalled in its original position. Remove the inboard CV-joint boot, dynamic strut, then the outboard CV-joint boot.

To install:

16. Wrap the spline with vinyl tape to prevent damage to the boots. Install the outboard boot, dynamic strut, and inboard boot, then remove the vinyl tape.

17. Install the stopper ring onto the halfshaft groove, then install the spider gear in its original position by aligning the marks.

18. Fit the snapring into the halfshaft groove.

19. Pack the outboard joint boot with CV-joint grease only. Do not use a substitute or mix types of grease.

20. Fit the rollers to the spider gear with their high shoulders facing outward. Reinstall the rollers in their original positions on the spider gear.

21. Pack the inboard joint boot with CV-joint grease.

22. Fit the inboard joint onto the halfshaft. Hold the halfshaft assembly so the inboard joint points up to prevent it from falling off.

23. With the boots installed, adjust the CV-joints in or out to place the inner boot ends in the original positions.

24. Install the new boot bands on the boots and bend both sets of locking tabs. Lightly tap on the locking tabs to ensure a good fit.

25. Always use a new set ring whenever the driveshaft is being installed. Be sure the driveshaft locks in the differential side gear groove and that the CV-joint sub-axle bottoms in the differential or the intermediate shaft.

26. Tighten the ball joint nut to:
- Integra, 2.2CL, 2.3CL, 3.0CL and 2.5TL: 36–43 ft. lbs. (49–59 Nm)
- Legend, 3.2TL and 3.5RL: 54 ft. lbs. (75 Nm)

27. Install and tighten the lower strut nut and bolt for all models except Legend and 3.2TL to 47 ft. lbs. (65 Nm). For Legend and 3.2TL, tighten the lower strut nut and bolt to 50 ft. lbs. (70 Nm).

28. Install and tighten the upper pinch bolt for all models except Legend and 3.2TL to 32 ft. lbs. (44 Nm). For Legend and 3.2TL, tighten the upper pinch bolt to 36 ft. lbs. (49 Nm).

29. With the vehicle on the ground, tighten the spindle nut, then stake the nut.
- Integra, 2.2CL, 2.3CL and 3.0CL: 134 ft. lbs. (182 Nm)
- 2.5TL: 181 ft. lbs. (245 Nm)
- Legend, 2.5TL, 3.2TL and 3.5RL: 242 ft. lbs. (355 Nm)

STEERING AND SUSPENSION

Air Bag

✷✷ CAUTION

Some vehicles are equipped with an air bag system, also known as the

Supplemental Restraint System (SRS). The system must be disabled before performing service on or around system components, steering column, instrument panel components, wiring and sensors. Failure to follow safety and disabling procedures could result in accidental air bag deployment, possible personal injury and unnecessary system repairs.

PRECAUTIONS

Several precautions must be observed when handling the inflator module to avoid accidental deployment and possible personal injury.

- Never carry the inflator module by the wires or connector on the underside of the module.
- When carrying a live inflator module, hold securely with both hands, and ensure that the bag and trim cover are pointed away.
- Place the inflator module on a bench or other surface with the bag and trim cover facing up.
- With the inflator module on the bench, never place anything on or close to the module which may be thrown in the event of an accidental deployment.

DISARMING

➡**The radio may have a coded theft protection circuit. Obtain the code from the owner before disconnecting the battery, removing the radio fuse, or removing the radio.**

Integra and Legend

✷✷ CAUTION

The Supplemental Restraint System (SRS, air bag system) must be disarmed before any of its components are disconnected or removed. Failing to disable the SRS before servicing its components may cause accidental deployment of the air bag, resulting in unnecessary repairs and possible personal injury.

1. Turn the ignition switch **OFF**.
2. Wait 3 minutes to let the capacitor in the back-up circuit discharge.
3. Disconnect the negative battery cable, then disconnect the positive battery cable.
4. For the driver air bag:

a. Remove the access panel lid below the air bag assembly on the steering wheel and remove the red shorting connector.

b. Disconnect the connector between the air bag and cable reel.

c. Connect the red shorting connector to the air bag side of the connector.

5. For the passenger air bag:

a. If necessary, remove the glove box, then remove the red shorting connector from its holder.

b. Disconnect the 3-pin connector between the passenger air bag and the main harness.

c. Connect the shorting connector to the air bag side of the connector.

6. After installing the shorting connectors on the air bags, connect shorting connector 07-mAZ-SP0020A, or equivalent, on the cable reel connector and another on the main harness connector of the passenger air bag to prevent static electricity from setting off the seat belt pre-tensioners before you disconnect them.

7. For the seat belt pre-tensioners, disarm them one side at a time:

a. Remove the B-pillar trim panels.

b. Remove the red shorting connector from the short connector holder.

c. Disconnect the pre-tensioner 3-pin connector, then install the shorting connector to the pre-tensioner side of the connector.

To enable:

8. Enable the seat belt pre-tensioners:

a. Disconnect the shorting connector from the 3-pin connector. Then, reconnect the 3-pin connector.

b. Fit the shorting connector into its holder and reinstall the B-pillar trim panels.

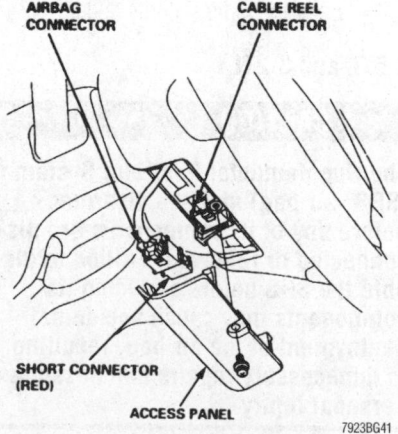

Access panel and air bag connectors—1995 Integra

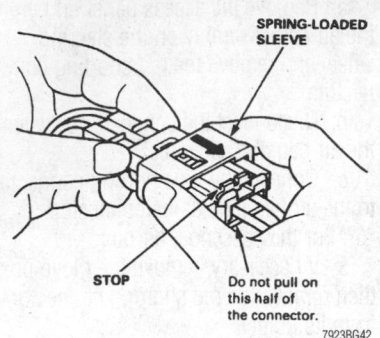

Spring-sleeve connectors

9. Enable the passenger air bag:

a. Disconnect the shorting connectors from the air bag and main harness connectors.

b. Reconnect the air bag connector to the main harness.

c. Fit the short connector into its holder.

d. If removed, install the glove box.

10. Disconnect the shorting connector from the cable reel connection.

11. Enable the driver's air bag:

a. Disconnect the shorting connector from the air bag connector.

b. Reconnect the air bag and cable reel connectors.

c. Fit the shorting connector back into its holder.

d. Install the steering wheel access cover.

12. Reconnect the positive and negative battery cables.

13. Turn the ignition switch to the **ON** position, but don't start the engine. The SRS indicator light should turn on for six seconds, then turn off. If the SRS indicator light doesn't come on, or stays on longer than six seconds, the system fault must be diagnosed.

14. Enter the radio security code.

2.5TL and 3.2TL

❊❊ CAUTION

The Supplemental Restraint System (SRS, air bag) must be disarmed before any of its components are disconnected or removed. Failing to disable the SRS before servicing its components may cause accidental deployment of the air bag, resulting in unnecessary repairs and possible personal injury.

1. Turn the ignition switch to the **LOCK** position. Remove the key.

2. Disconnect the negative and positive battery cables.

3. Always wait at least three minutes after disconnecting the battery before working around the air bags.

4. Remove the steering wheel lower access cover.

5. Remove the clip securing the air bag module/cable reel connection to the steering column.

6. Uncouple the air bag and cable reel connection. Immediately install the red shorting connector onto the air bag module connector.

➡ **The driver's side air bag connection contains a spring-contact self-disabling device. A shorting connector doesn't need to be installed on the driver's air bag connector.**

7. After servicing has been completed, couple the air bag and cable reel connectors.

8. Install the clip securing the air bag/cable reel connection to the steering column.

9. Install the access cover.

10. Reconnect the positive and negative battery cables.

11. Turn the ignition switch to the ON position, but don't start the engine. The SRS indicator light should turn on for six seconds, then turn off. If the SRS indicator light doesn't come on, or stays on longer than six seconds, the system fault must be diagnosed.

12. Enter the radio security code.

2.2CL, 2.3CL, 3.0CL and 3.5RL

1. Disconnect the negative battery cable, then the positive cable.

2. Wait three minutes for the air bag reserve power to discharge before preceding with work.

Rack and Pinion Steering Gear

REMOVAL & INSTALLATION

➡ **The radio may have a coded theft protection circuit. Obtain the code from the owner before disconnecting the battery, removing the radio fuse, or removing the radio.**

Power

2.2CL, 2.3CL, 3.0CL AND INTEGRA

1. Lift the power steering reservoir and disconnect the return hose that goes to the oil cooler.

2. Connect a hose of suitable diameter to the disconnected return hose and place the end of the hose in a container to collect the power steering fluid.

❊❊ CAUTION

Take care not to spill the fluid on the body and engine assembly. Wipe off any spilled fluid at once.

3. Start the engine, let it run at idle, and turn the steering wheel from lock to lock several times. When fluid stops running out of the hose, shut off the engine. Connect the return hose to the reservoir. Properly dispose of the used fluid.

4. Disconnect the negative battery cable.

5. Using cleaning solvent and a brush, wash any oil and dirt off the valve body unit and its lines, and the end of the rack.

6. Remove the steering joint cover.

7. Remove the ignition key and lock the steering wheel in the straight forward position.

8. Remove the steering joint lower bolt and pull the joint toward the column.

9. Raise and safely support the vehicle.

10. Remove the front wheels.

11. Remove the cotter pins and unscrew the tie rod end ball joint nuts halfway.

12. Break the tie rod ball joints loose from the steering knuckles, by using a tie rod end removal tool (part No. 07MAC-SL00200) or equivalent.

13. Remove the nuts and lift the tie rod ends out of the steering knuckles.

14. If equipped with a manual transaxle, disconnect the extension rod and the shift rod by performing the following:

a. If equipped with a VTEC engine, remove the heat shield.

b. Disconnect the shift extension rod from the transaxle case.

c. Slide the boot on the shift rod back to expose the clip and spring pin. Remove the clip from the shift rod.

d. Drive out the spring pin with a punch and disconnect the shift control rod. Note that on reassembly, install the clip back into place after driving the spring pin in.

15. If equipped with an automatic transaxle, disconnect the shift cable from the transaxle by performing the following:

a. Remove the shift cable holder from the vehicle.

b. Remove the shift cable cover from the transaxle.

c. Remove the bolt attaching the control lever to the control shaft, then remove the control lever from the control shaft. Discard the lockwasher.

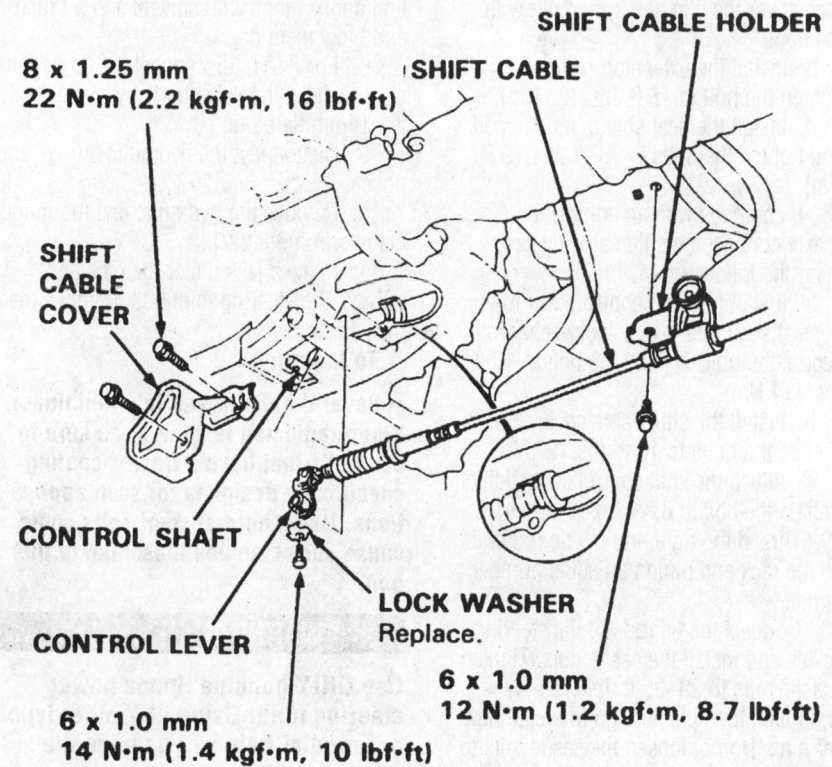

SHIFT CABLE HOLDER

SHIFT CABLE

8 x 1.25 mm
22 N·m (2.2 kgf·m, 16 lbf·ft)

SHIFT CABLE COVER

CONTROL SHAFT

LOCK WASHER
Replace.

CONTROL LEVER

6 x 1.0 mm
12 N·m (1.2 kgf·m, 8.7 lbf·ft)

6 x 1.0 mm
14 N·m (1.4 kgf·m, 10 lbf·ft)

7923BG43

Automatic transaxle shift cable attachment—Integra

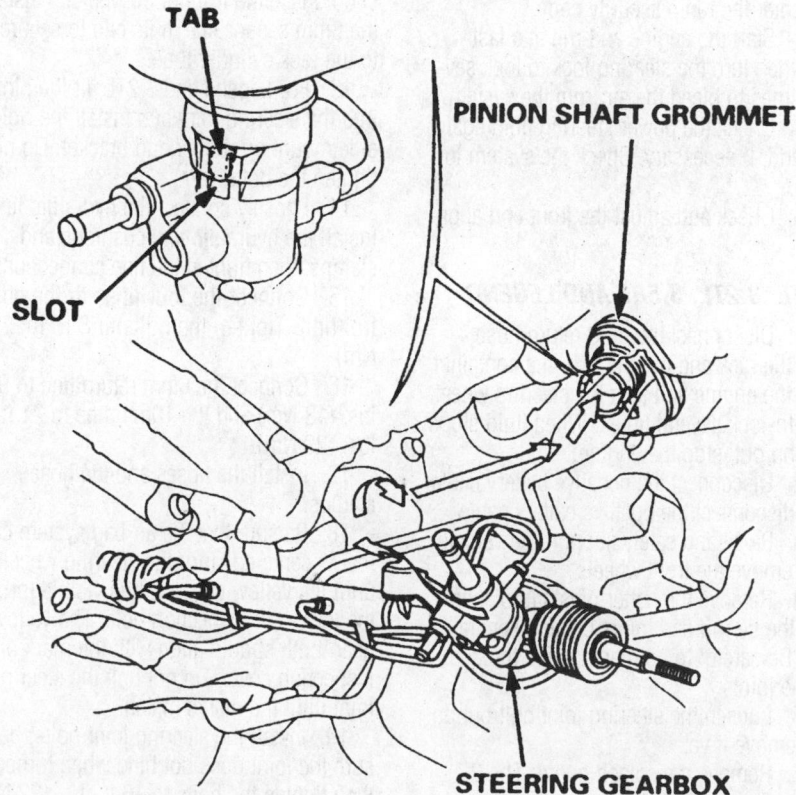

TAB

PINION SHAFT GROMMET

SLOT

STEERING GEARBOX

7923BG44

Installing the rack and pinion—Integra

d. Position the cable out of the way without bending the cable.

16. Disconnect the oxygen sensor electrical connector.

17. Remove the catalytic converter front attaching nuts and bolts, then remove the rear attaching nuts. Remove the catalytic converter from the vehicle. Discard the lock-nuts and old exhaust gaskets.

18. Remove the return clamp from the left side of the rear beam, and move the return pipe above the rack and pinion.

19. Remove the rear beam brace.

20. Remove the left tie rod end, then slide the inner tie rod all of the way to the right.

21. Disconnect the two lines from the valve body unit on the rack and pinion. Place the pipe and hose to the rear side of the rack and pinion, so they do not hinder in the removal of the rack and pinion.

✳✳ CAUTION

After disconnecting the hose and pipe, plug or cap the hose and pipe to prevent foreign materials from entering the valve body unit.

➡**Do not loosen the cylinder pipes between the valve body unit and the cylinder.**

22. Remove the gearbox mounting bolts.

23. Pull the rack and pinion all the way down to clear the pinion shaft from the bulkhead, and remove the pinion shaft grommet.

24. Holding the rack and pinion assembly, slide the rack all of the way to the right, then place the left rack end below the rear beam.

25. Move the rack and pinion assembly to the left, and tilt the left side down to remove it from the vehicle.

To install:

26. Gently push the left inner tie rod end into the rack and pinion assembly until it reaches the end of its travel.

27. Place the right side of the rack and pinion over the rear beam and move the assembly completely to the right. Lift the left side of the rack and pinion over the rear beam.

28. Install the pinion shaft grommet, then slide the rack and pinion to the left and up into position. Be sure the pinion shaft fits the pinion shaft grommet properly by aligning the tab on the grommet with the slot in the valve body.

29. With the rack and pinion properly positioned install the mounting bolts. Tighten the left side mounting bolts to 28 ft.

lbs. (38 Nm), and the right side mounting bolts to 43 ft. lbs. (58 Nm).

➥**After installing the rack and pinion, check the air hose connections for interference with adjacent parts.**

30. Center the steering rack within its stroke.

31. If applicable, be sure that the cable reel in the steering column is cantered by performing the following:

    a. Turn the steering wheel left approximately 150° , to check the cable reel position with the indicator.

    b. If the cable reel is centered, the yellow gear tooth lines up with the alignment mark on the cover.

    c. Return the steering wheel right approximately 150° , to position the steering wheel in the straight ahead position.

✳✳ CAUTION

Do not connect the steering joint to the pinion without the cable reel being centered. Damage to the SRS system components and personal injury may occur.

32. Slip the lower end of the steering joint onto the pinion shaft (line up the bolt hole with the groove around the shaft), and tighten the lower bolt. Tighten the bolt to 16 ft. lbs. (22 Nm).

➥**Be sure that the lower steering joint is securely in the groove in the steering pinion. If the steering wheel and rack and pinion are not centered, reposition the serrations at the lower end of the steering joint.**

33. Install the steering joint cover with the clamps and clips.

34. Connect the feed pipe to the rack and pinion valve body unit, tighten the fitting to 27 ft. lbs. Connect the return hose to the rack and pinion valve body unit and tighten the hose clamp.

35. Install the rear beam brace rod and the return pipe clamp on the rear beam. Tighten the rear beam brace rod bolts to 28 ft. lbs. (38 Nm).

36. Install the catalytic converter with new gaskets and new locknuts. Tighten the rear nuts to 25 ft. lbs. (33 Nm). Tighten the front nuts and bolts to 16 ft. lbs. (22 Nm). Connect the oxygen sensor electrical connector.

37. If equipped with a manual transaxle connect the shift linkage by performing the following:

    a. Install the shifter rod and install the spring pin to attach the shifter rod to the transaxle.

    b. Install the clip over the spring pin, then cover the clip and spring pin with the boot.

    c. Install the extension rod and tighten the bolt to 16 ft. lbs. (22 Nm).

    d. Install the heat shield, if equipped and tighten the bolts to 7.2 ft. lbs. (9.8 Nm).

38. If equipped with an automatic transaxle connect the shift cable by performing the following:

    a. Install the shift control lever to the control shaft, use a new lockwasher to secure the bolt. Tighten the bolt to 10 ft. lbs. (14 Nm).

    b. Install the shift cable cover and tighten the bolts to 16 ft. lbs. (22 Nm).

    c. Attach the shift control cable holder, tighten the bolt to 8.7 ft. lbs. (12 Nm).

39. Thread the right and left tie rod ends on to the rack and pinion an equal number of turns.

40. Connect the tie rods to the steering knuckles and install the castle nuts. Tighten the castle nuts to 29–35 ft. lbs. (39–47 Nm), tighten the nuts enough to install new cotter pins. Do not loosen the castle nuts to install the cotter pins.

41. Install the front wheels.

42. Fill the power steering reservoir to the upper line.

43. Connect the negative battery cable and enter the radio security code.

44. Start the engine and run at a fast idle, then turn the steering lock to lock several times to bleed the air from the system.

45. Check the power steering fluid again and add, if necessary. Check the system for leaks.

46. Check and adjust the front end alignment.

2.5TL, 3.2TL, 3.5RL AND LEGEND

1. Disconnect the fluid return hose from the rack and put the end in a container. Start the engine and turn the steering wheel lock-to-lock several times. When fluid stops coming out, stop the engine.

2. Disconnect the negative battery cable, then disconnect the positive battery cable.

3. Raise and safely support the vehicle and remove the front wheels.

4. Remove the cotter pins and disconnect the tie rod end joints using a separator tool. Be careful to not damage the threads on the joints.

5. Loosen the steering joint bolt but do not remove it yet.

6. Remove the splash guard. The 2 long bolts also hold the rack in place, and the rack will now be partially hanging on the steering joint.

7. Carefully clean all the hydraulic fitting connections with solvent and a brush and blow them dry.

8. For 2.5TL, disconnect the 8mm sensor line from the valve body by removing the 14mm flare nut.

9. Disconnect the hydraulic fittings and hoses.

10. Remove the hydraulic line mounting clamps from the rack.

11. Place a jack under the rack and remove the steering joint bolt. Remove the rack assembly.

To install:

➥**Several bolts thread into aluminum. When replacing fasteners, be sure to use bolts that have a Dacro® coating specifically designed for such applications. Using normal steel bolts could cause corrosion and loosening of the bolt.**

✳✳ WARNING

Use ONLY genuine Honda power steering fluid. Using ANY other type or brand of fluid will damage the power steering system.

12. For 2.5TL, loosely install the right mounting bracket to hold the rack in the vehicle. The arrow stamped on the bracket should face the front of the vehicle. Install the 8mm sensor line in its clip to secure it to the rack cylinder tube.

13. For Legend and 3.2TL, fit the pinion into the steering joint and install the right side mounting rubber and bracket. Do not tighten the bolts yet.

14. Loosely connect the hydraulic lines. Install the hydraulic line cushions and clamps, then tighten the line connections.

15. Connect the four lines to the control unit. Tighten the bolts to 8 ft. lbs. (11 Nm)

16. Connect the 6mm return line to 9 ft. lbs. (13 Nm) and the 10mm line to 21 ft. lbs. (29 Nm)

17. Install the hoses and the hose clamps.

18. Be sure that the air bag system cable reel is centered. Turn the steering wheel left until the yellow gear tooth is visible through the lower left inspection hole. The yellow gear tooth should align with the mark on the inspection cover. Do not bolt the steering joint until the marks match.

19. Install the steering joint bolts, be sure the joint does not bind when turned, then tighten the bolts to 16 ft. lbs. (22 Nm).

20. Tighten the right side mount bolts to 28 ft. lbs. (39 Nm).

21. Install the splash guard and tighten the short bolts to 28 ft. lbs. (39 Nm). Tighten the long bolts to 43 ft. lbs. (60 Nm).

22. For 2.5TL, connect the feed line, the inlet hose, and the outlet hose to the valve body unit. Tighten the bolts to 8 ft. lbs. (11 Nm). Install the 8mm sensor line to the valve body unit. Tighten the line to 18 ft. lbs. (25 Nm).

23. Connect the tie rod ends and tighten the nuts to 36–43 ft. lbs. (50–60 Nm), then tighten them enough to install a new cotter pin.

24. When installation is complete, refill the hydraulic reservoir with new steering fluid, reconnect the battery, start the engine, and turn the steering wheel lock-to-lock several times to bleed the system. Check fluid level again.

25. Check the system for leaks.

26. Check and adjust the front wheel alignment

Shock Absorber

REMOVAL & INSTALLATION

Front

1. Raise and safely support the vehicle and remove the front wheels.

2. Support the lower suspension arm with a jack.

CAUTION:
- Replace the self-locking nuts after removal.
- The vehicle should be on the ground before any bolts or nuts connected to rubber mounts or bushings are tightened.
- Torque the castle nut to the lower torque specification, then tighten it only far enough to align the slot with the pin hole. Do not align the nut by loosening.

NOTE: Wipe off the grease before tightening the nut at the ball joint.

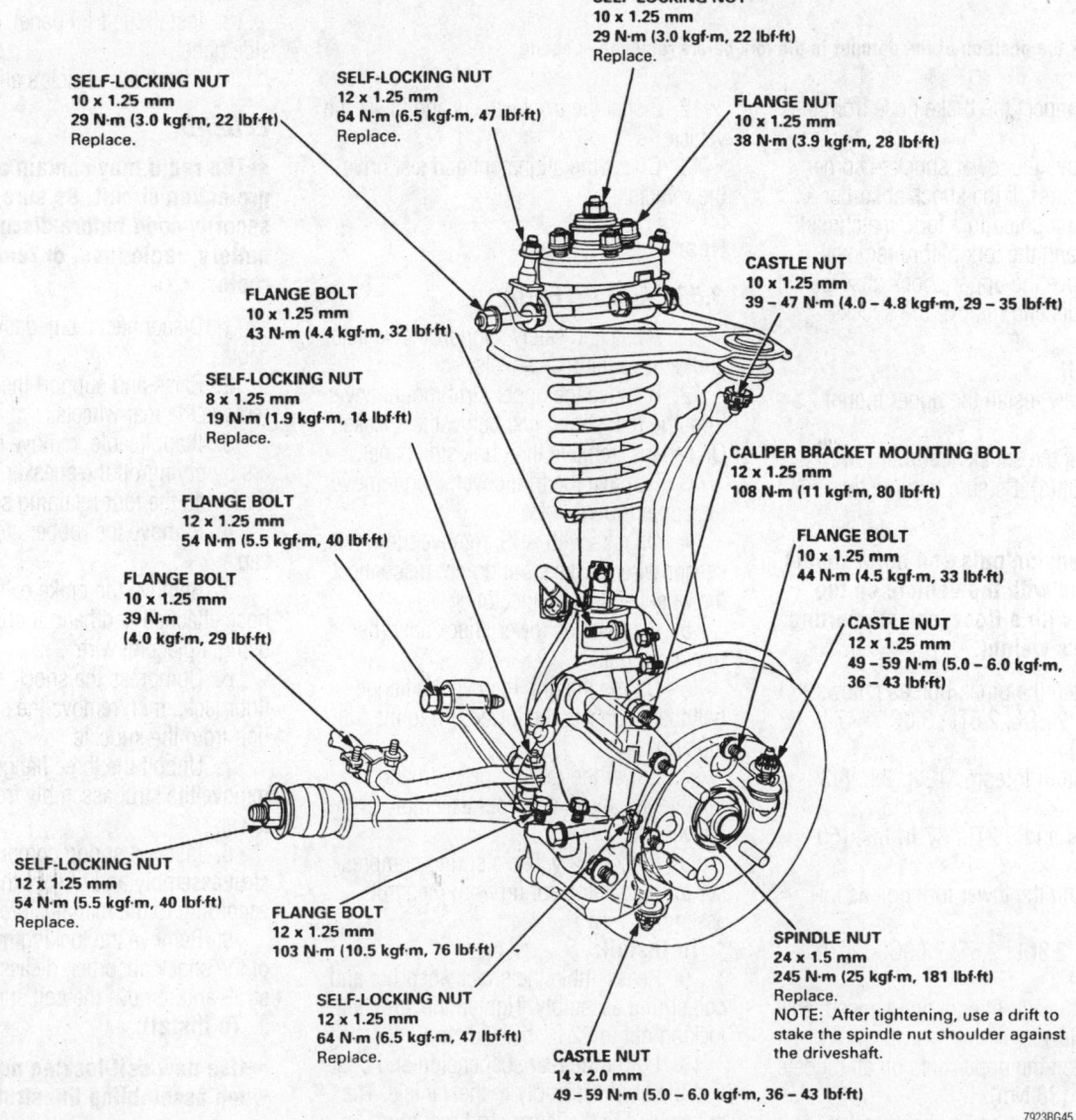

SELF-LOCKING NUT
10 x 1.25 mm
29 N·m (3.0 kgf·m, 22 lbf·ft)
Replace.

SELF-LOCKING NUT
12 x 1.25 mm
64 N·m (6.5 kgf·m, 47 lbf·ft)
Replace.

SELF-LOCKING NUT
10 x 1.25 mm
29 N·m (3.0 kgf·m, 22 lbf·ft)
Replace.

FLANGE NUT
10 x 1.25 mm
38 N·m (3.9 kgf·m, 28 lbf·ft)

CASTLE NUT
10 x 1.25 mm
39 – 47 N·m (4.0 – 4.8 kgf·m, 29 – 35 lbf·ft)

FLANGE BOLT
10 x 1.25 mm
43 N·m (4.4 kgf·m, 32 lbf·ft)

SELF-LOCKING NUT
8 x 1.25 mm
19 N·m (1.9 kgf·m, 14 lbf·ft)
Replace.

CALIPER BRACKET MOUNTING BOLT
12 x 1.25 mm
108 N·m (11 kgf·m, 80 lbf·ft)

FLANGE BOLT
12 x 1.25 mm
54 N·m (5.5 kgf·m, 40 lbf·ft)

FLANGE BOLT
10 x 1.25 mm
44 N·m (4.5 kgf·m, 33 lbf·ft)

FLANGE BOLT
10 x 1.25 mm
39 N·m
(4.0 kgf·m, 29 lbf·ft)

CASTLE NUT
12 x 1.25 mm
49 – 59 N·m (5.0 – 6.0 kgf·m, 36 – 43 lbf·ft)

SELF-LOCKING NUT
12 x 1.25 mm
54 N·m (5.5 kgf·m, 40 lbf·ft)
Replace.

FLANGE BOLT
12 x 1.25 mm
103 N·m (10.5 kgf·m, 76 lbf·ft)

SELF-LOCKING NUT
12 x 1.25 mm
64 N·m (6.5 kgf·m, 47 lbf·ft)
Replace.

CASTLE NUT
14 x 2.0 mm
49 – 59 N·m (5.0 – 6.0 kgf·m, 36 – 43 lbf·ft)

SPINDLE NUT
24 x 1.5 mm
245 N·m (25 kgf·m, 181 lbf·ft)
Replace.
NOTE: After tightening, use a drift to stake the spindle nut shoulder against the driveshaft.

7923BG45

Front suspension showing the torque specifications—2.5TL shown, other vehicles are similar

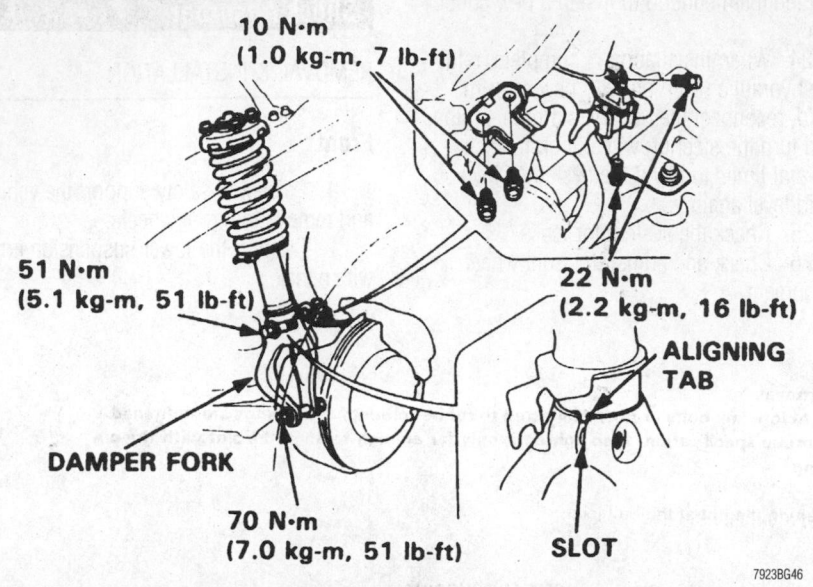

10 N·m
(1.0 kg-m, 7 lb-ft)

51 N·m
(5.1 kg-m, 51 lb-ft)

22 N·m
(2.2 kg-m, 16 lb-ft)

ALIGNING
TAB

DAMPER FORK

70 N·m
(7.0 kg-m, 51 lb-ft)

SLOT

7923BG46

Matchmark the position of the damper in the fork before removal—Legend

3. Disconnect the brake hose from the shock.

4. Remove the lower shock absorber mounting bolt(s). If the shock absorber (damper) is mounted in a fork, matchmark the damper and the fork before removal.

5. Remove the upper shock absorber mounting nuts and remove the shock absorber.

To install:

6. Loosely install the upper mount nuts.

7. Install the shock absorber lower mounting bolt(s). Be sure to align the matchmark.

➡**All suspension nuts and bolts should be tightened with the vehicle on the ground, or with a floor jack supporting the vehicle's weight.**

8. Tighten the pinch bolt as follows:
• 2.2CL, 2.3CL, 2.5TL, 3.0CL : 47 ft. lbs. (65 Nm)
• 3.5RL and Integra: 32 ft. lbs. (43 Nm)
• Legend and 3.2TL: 37 ft. lbs. (50 Nm)

9. Tighten the lower fork bolt as follows:
• 2.2CL, 2.3CL, 2.5TL, 3.0CL : 47 ft. lbs. (65 Nm)
• 3.2TL, 3.5RL, Integra and Legend: 51 ft. lbs. (69 Nm)

10. Tighten the upper nuts on all models to 28 ft. lbs. (38 Nm).

11. Connect the brake hose bracket to the shock absorber.

12. Install the front wheels and lower the vehicle.

13. Check the alignment and test drive the vehicle.

Rear

3.5RL AND INTEGRA

1. Raise and safely support the vehicle and remove the rear wheels.

2. Remove the upper strut mount cover from the rear panel, just below the speaker. On sedans, remove the trunk side panel.

3. Remove the trim cover, then remove the upper mount nuts.

4. On cars with ABS, remove the wheel sensor wire brackets but do not disconnect the wheel sensor connector.

5. Remove the lower shock absorber mounting bolt.

6. On the Integra, remove the flange bolt that connects the lower arm to the trailing arm.

7. Lower the rear suspension and remove the shock absorber assembly from the vehicle.

8. If necessary, use a spring compressor to remove the spring from the strut assembly.

To install:

9. Reassemble the shock absorber and coil spring assembly. Tighten the strut self-locking nut to 22 ft. lbs. (30 Nm).

10. Lower the rear suspension and position the strut assembly in the vehicle. The nut welded to the lower strut mounting should face the front of the vehicle.

11. Loosely install the upper mounting nuts.

12. On the Integra, raise the rear suspension and install the bolt connecting the lower arm to the trailing arm.

13. Install the shock absorber lower mounting bolt.

14. Raise the vehicle until the vehicle just lifts off the safety stand and tighten the lower strut bolt and lower control arm bolt. On the Integra, tighten the lower strut bolt and the control arm bolt to 40 ft. lbs. (54 Nm). In the 3.5RL, tighten the lower shock absorber mounting bolt to 76 ft. lbs. (103 Nm)

15. Install the wheel sensor wire bracket on cars with ABS.

16. Tighten the upper mounting nuts to 36 ft. lbs. (49 Nm).

17. Install the rear wheels, then lower the vehicle.

18. Install the trim panel, or the trunk side panel.

19. Check the vehicle's alignment.

LEGEND

➡**The radio may contain a coded theft protection circuit. Be sure you have the security code before disconnecting the battery, radio fuse, or removing the radio.**

1. Disconnect the negative battery cable.

2. Raise and support the vehicle, remove the rear wheels.

3. If applicable, remove the rear speakers by prying off the speaker grilles and removing the four retaining screws.

4. Remove the rubber strut mounting cap.

5. Remove the brake caliper with the hose attached and hang it out of the way using a piece of wire.

6. Compress the shock slightly with a floor jack;, then remove the shock mounting bolt from the knuckle.

7. Unbolt the three flange bolts and remove the strut assembly from the shock tower.

8. Install a spring compressor onto the strut assembly and tighten the compressor according to the manufacturer's instructions.

9. Remove the locking nut from the top of the shock absorber, disassemble the strut, and remove the coil spring.

To install:

➡**Use new self-locking nuts and bolts when assembling the strut.**

10. Install a spring compressor onto the coil spring.

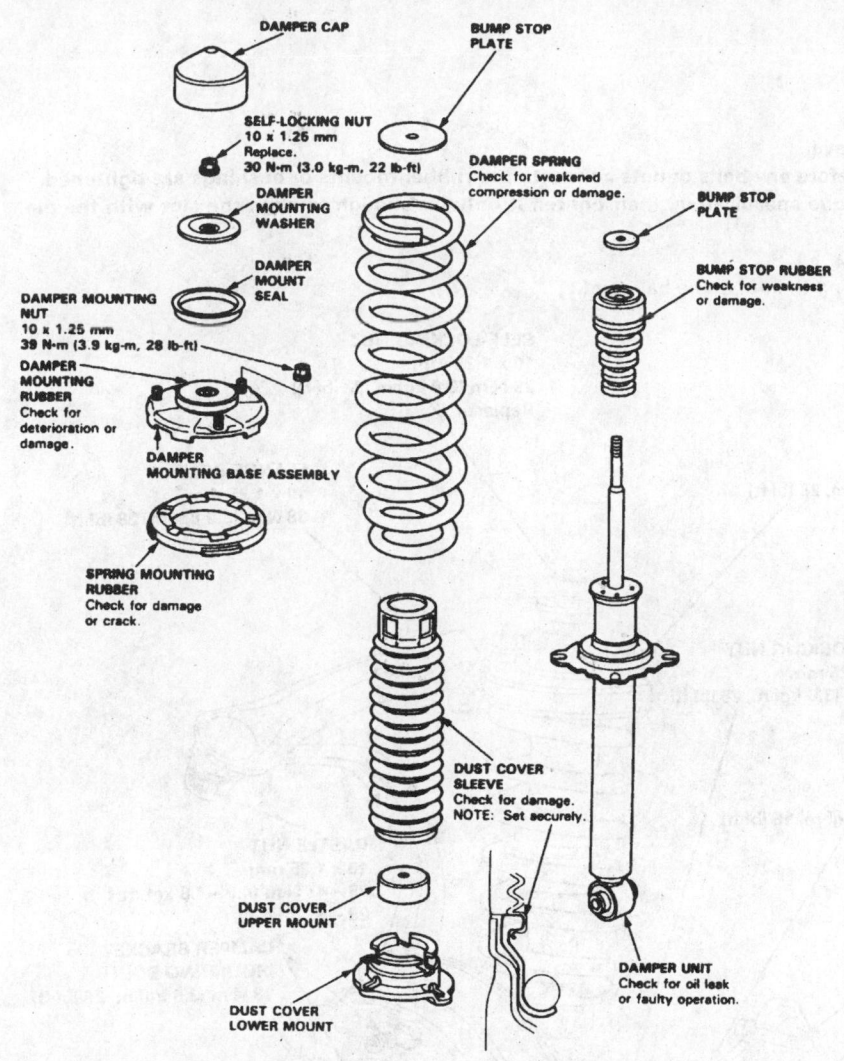

DAMPER CAP

BUMP STOP
PLATE

SELF-LOCKING NUT
10 x 1.25 mm
Replace.
30 N·m (3.0 kg-m, 22 lb-ft)

DAMPER SPRING
Check for weakened
compression or damage.

BUMP STOP
PLATE

DAMPER MOUNTING
WASHER

BUMP STOP
RUBBER
Check for weakness
or damage.

DAMPER
MOUNT
SEAL

DAMPER MOUNTING
NUT
10 x 1.25 mm
39 N·m (3.9 kg-m, 28 lb-ft)

DAMPER
MOUNTING
RUBBER
Check for
deterioration or
damage.

DAMPER
MOUNTING BASE ASSEMBLY

SPRING MOUNTING
RUBBER
Check for damage
or crack.

DUST COVER
SLEEVE
Check for damage.
NOTE: Set securely.

DUST COVER
UPPER MOUNT

DUST COVER
LOWER MOUNT

DAMPER UNIT
Check for oil leak
or faulty operation.

7923BG47

Exploded view of the rear strut—Legend

11. Assemble the upper and lower strut mounts, dust covers, and the shock absorber.

12. Install the mounting washer, and loosely install a new self-locking nut.

13. Hold the shock absorber piston with a hex wrench and tighten the self-locking nut. Tighten the self-locking nut to 22 ft. lbs. (30 Nm).

➡**All suspension nuts and bolts should be tightened with the vehicle on the ground, or with a floor jack supporting the vehicle's weight.**

14. Install the strut assembly into the vehicle. Tighten the upper mounting nuts to 28 ft. lbs. (39 Nm).

15. Install the shock mounting bolt at the knuckle and tighten to 76 ft. lbs. (105 Nm).

16. Install the brake caliper. Tighten the mounting bolts to 28 ft. lbs. (39 Nm).

17. Install the wheel, and tighten the wheel nuts to 80 ft. lbs. (110 Nm).

18. Install the strut mount caps and the rear speakers. Reconnect the negative battery cable.

2.2CL, 2.3CL, 2.5TL, 3.0CL and 3.2TL

1. Raise and safely support the vehicle and remove the rear wheels.

2. Remove the rear seat:

a. Remove the lower cushion bolt located under the armrest.

b. Pull the rear of the lower cushion up and lift it forward to release it from the clips.

c. Pull down the trunk bulkhead trim and release the armrest lid clips.

d. Remove the three bolts from the back cushion,, then lift it up and forward to disengage the securing hooks.

3. Place a floor jack under the lower arm and slightly compress the spring.

4. Remove the upper mounting nuts and the lower flange bolt.

5. Lower the jack to remove the strut. Be sure and mark the right and left struts so they can be reinstalled on the proper sides.

6. Use a spring compressor to remove the spring from the struts.

To install:

7. Reassemble the spring and strut assembly. Tighten the strut self-locking nut to 22 ft. lbs. (30 Nm).

8. Install the struts into the vehicle. Loosely install the mounting nuts and mounting bolt, but do not tighten them until the weight of the vehicle is on the suspension.

9. Raise the rear suspension with a floor jack until the weight of the vehicle is on the strut. Tighten the upper mounting nuts to 28 ft. lbs. (39 Nm), then tighten the lower mounting bolts to 40 ft. lbs. (55 Nm). Be careful not to pinch the ABS speed
sensor wire between the strut and bracket.

10. Install the rear wheels and lower the vehicle.

11. Install the rear seat cushions.

Coil Spring

REMOVAL & INSTALLATION

All Models

1. Raise and support the vehicle, remove the front wheels.

2. Remove the shock absorber (damper).

3. Compress the coil spring with a spring compressor.

4. Remove the locking nut from the top of the shock absorber, and remove the coil spring.

CAUTION:
- Replace the self-locking nuts after removal.
- The vehicle should be on the ground before any bolts or nuts connected to rubber mounts or bushings are tightened.
- Torque the castle nut to the lower torque specification, then tighten it only far enough to align the slot with the pin hole. Do not align the nut by loosening.

NOTE: Wipe off the grease before tightening the nut at the ball joint.

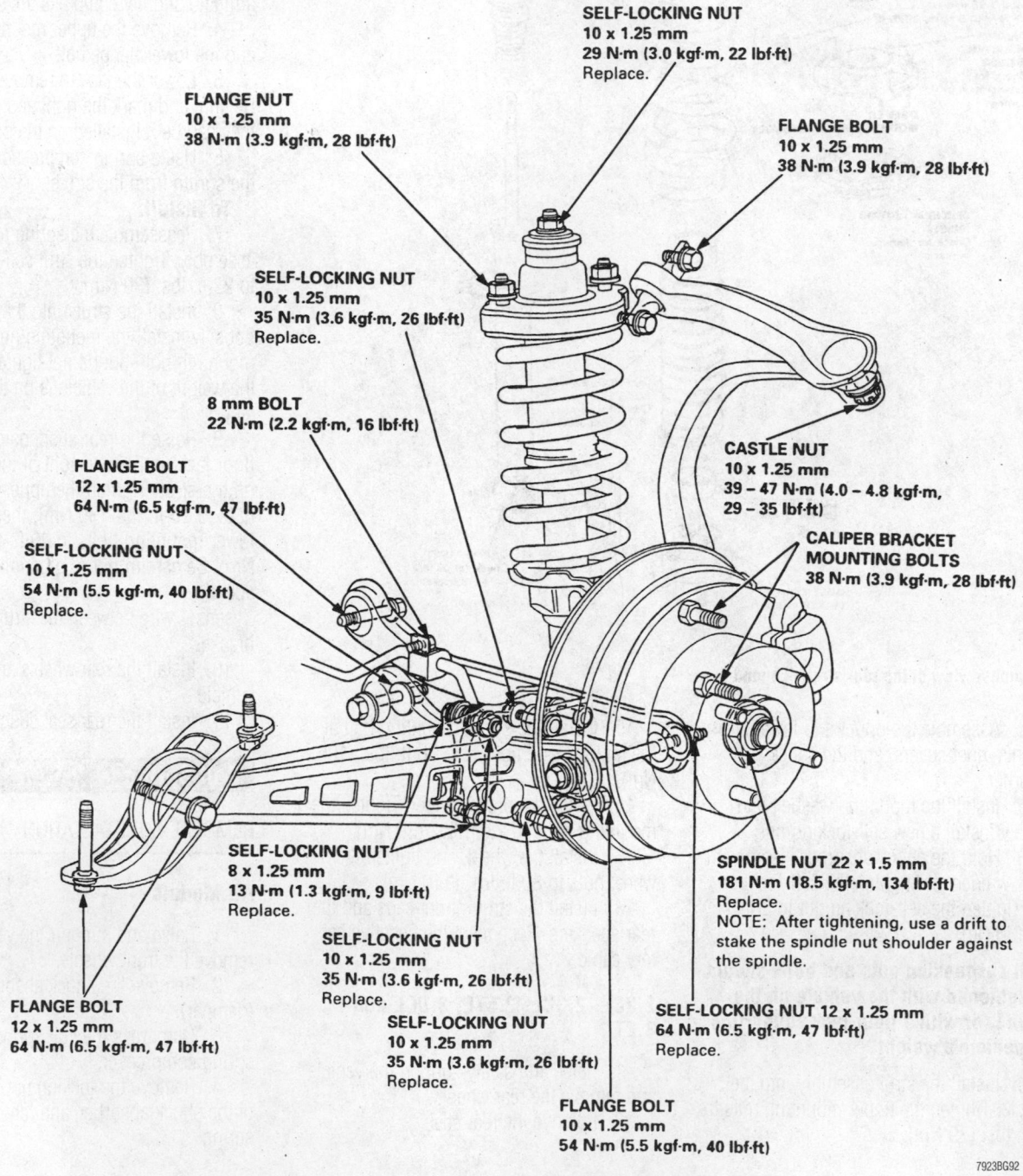

SELF-LOCKING NUT
10 x 1.25 mm
29 N·m (3.0 kgf·m, 22 lbf·ft)
Replace.

FLANGE NUT
10 x 1.25 mm
38 N·m (3.9 kgf·m, 28 lbf·ft)

FLANGE BOLT
10 x 1.25 mm
38 N·m (3.9 kgf·m, 28 lbf·ft)

SELF-LOCKING NUT
10 x 1.25 mm
35 N·m (3.6 kgf·m, 26 lbf·ft)
Replace.

8 mm BOLT
22 N·m (2.2 kgf·m, 16 lbf·ft)

FLANGE BOLT
12 x 1.25 mm
64 N·m (6.5 kgf·m, 47 lbf·ft)

SELF-LOCKING NUT
10 x 1.25 mm
54 N·m (5.5 kgf·m, 40 lbf·ft)
Replace.

CASTLE NUT
10 x 1.25 mm
39 – 47 N·m (4.0 – 4.8 kgf·m,
29 – 35 lbf·ft)

**CALIPER BRACKET
MOUNTING BOLTS**
38 N·m (3.9 kgf·m, 28 lbf·ft)

SELF-LOCKING NUT
8 x 1.25 mm
13 N·m (1.3 kgf·m, 9 lbf·ft)
Replace.

SELF-LOCKING NUT
10 x 1.25 mm
35 N·m (3.6 kgf·m, 26 lbf·ft)
Replace.

SELF-LOCKING NUT
10 x 1.25 mm
35 N·m (3.6 kgf·m, 26 lbf·ft)
Replace.

FLANGE BOLT
12 x 1.25 mm
64 N·m (6.5 kgf·m, 47 lbf·ft)

SPINDLE NUT 22 x 1.5 mm
181 N·m (18.5 kgf·m, 134 lbf·ft)
Replace.
NOTE: After tightening, use a drift to
stake the spindle nut shoulder against
the spindle.

SELF-LOCKING NUT 12 x 1.25 mm
64 N·m (6.5 kgf·m, 47 lbf·ft)
Replace.

FLANGE BOLT
10 x 1.25 mm
54 N·m (5.5 kgf·m, 40 lbf·ft)

7923BG92

Rear suspension showing the torque specifications—2.5TL

To install:

➡**Use new self-locking nuts and bolts when assembling the strut.**

5. Install the compressed coil spring on the shock absorber.

6. Assemble the upper spring seat/bearing and related components.

7. Install the mounting washer, and loosely install a new self-locking nut.

8. Hold the shock absorber piston rod with a hex wrench and tighten the self-locking nut. Tighten the self-locking nut to 22 ft. lbs. (30 Nm).

➡**All suspension nuts and bolts should be tightened with the vehicle on the ground.**

9. Install the shock absorber in the vehicle.

10. Check and adjust the vehicle's front wheel alignment.

Upper Ball Joint

REMOVAL & INSTALLATION

All Models

➡**The upper ball joint cannot be removed from the control arm. If the ball joint is damaged, the upper arm assembly must be replaced.**

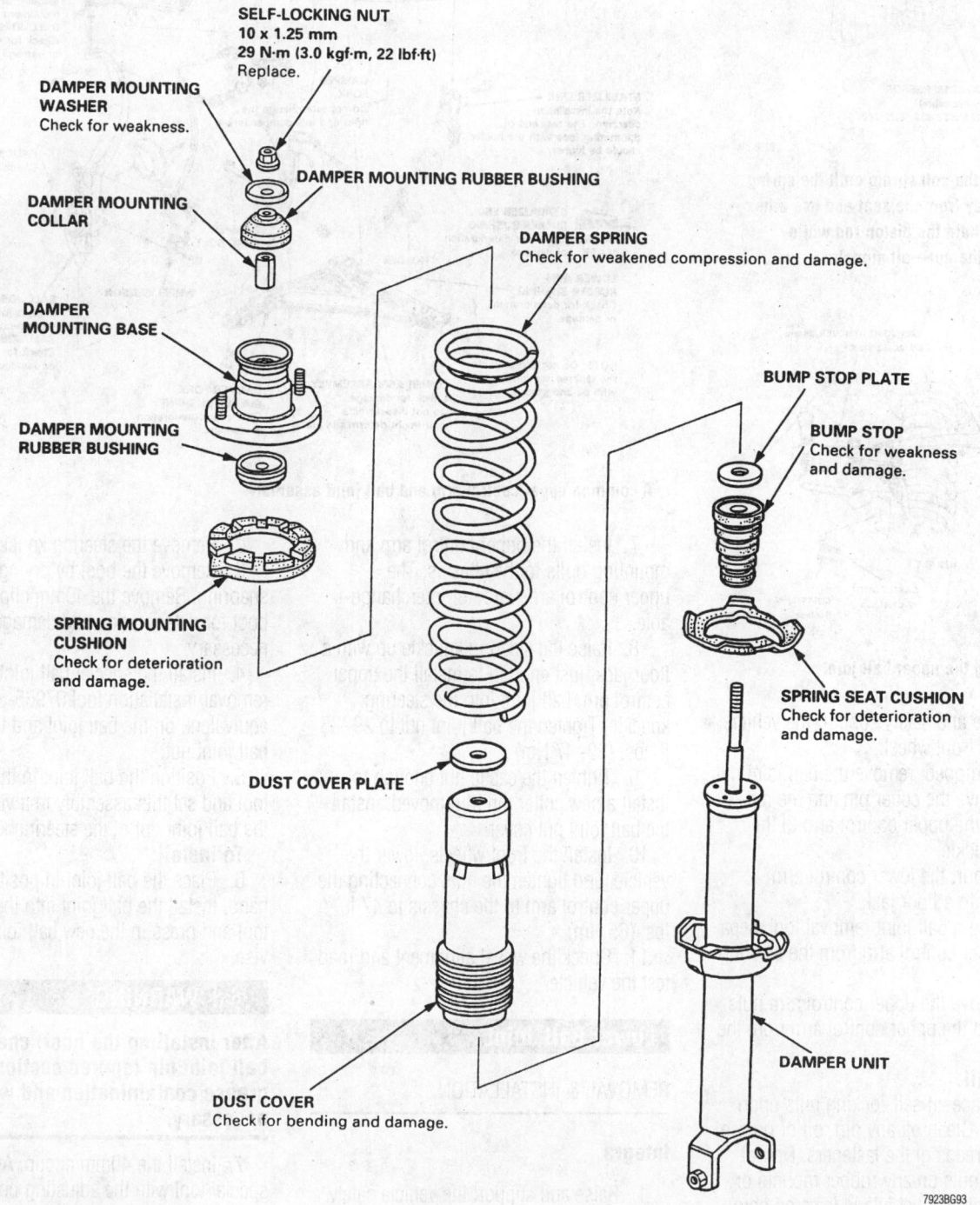

Exploded view of the rear shock absorber (damper)—2.5TL shown, other vehicles are similar

SELF-LOCKING NUT
10 x 1.25 mm
29 N·m (3.0 kgf·m, 22 lbf·ft)
Replace.

DAMPER MOUNTING WASHER
Check for weakness.

DAMPER MOUNTING RUBBER BUSHING

DAMPER MOUNTING COLLAR

DAMPER MOUNTING BASE

DAMPER MOUNTING RUBBER BUSHING

SPRING MOUNTING CUSHION
Check for deterioration and damage.

DAMPER SPRING
Check for weakened compression and damage.

DUST COVER PLATE

DUST COVER
Check for bending and damage.

BUMP STOP PLATE

BUMP STOP
Check for weakness and damage.

SPRING SEAT CUSHION
Check for deterioration and damage.

DAMPER UNIT

7923BG93

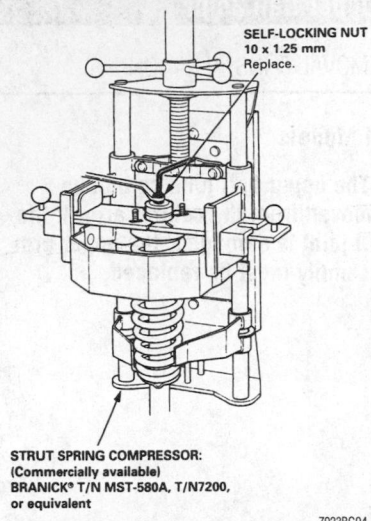

SELF-LOCKING NUT
10 x 1.25 mm
Replace.

STRUT SPRING COMPRESSOR:
(Commercially available)
BRANICK® T/N MST-580A, T/N7200,
or equivalent

7923BG94

Compress the coil spring until the spring moves away from the seat and use a hex wrench to hold the piston rod while removing the nut—all models

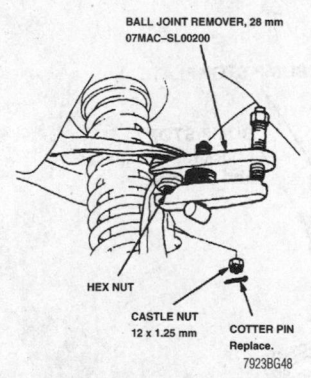

BALL JOINT REMOVER, 28 mm
07MAC–SL00200

HEX NUT

CASTLE NUT
12 x 1.25 mm

COTTER PIN
Replace.

7923BG48

Separating the upper ball joint

1. Raise and safely support the vehicle. Remove the front wheel.
2. If equipped, remove the ball joint nut cover. Remove the cotter pin and the nut connecting the upper control arm to the steering knuckle.
3. Support the lower control arm assembly with a floor jack.
4. Using a ball joint removal tool, separate the upper control arm from the steering knuckle.
5. Remove the upper control arm nuts, washers and the upper control arm from the vehicle.

To install:
6. Replace all self-locking nuts upon installation. Clean off any dirt, oil or grease off of the threads of the fasteners. Do not tighten any nuts on any rubber mounts or bushings until the vehicle is lowered onto the ground.

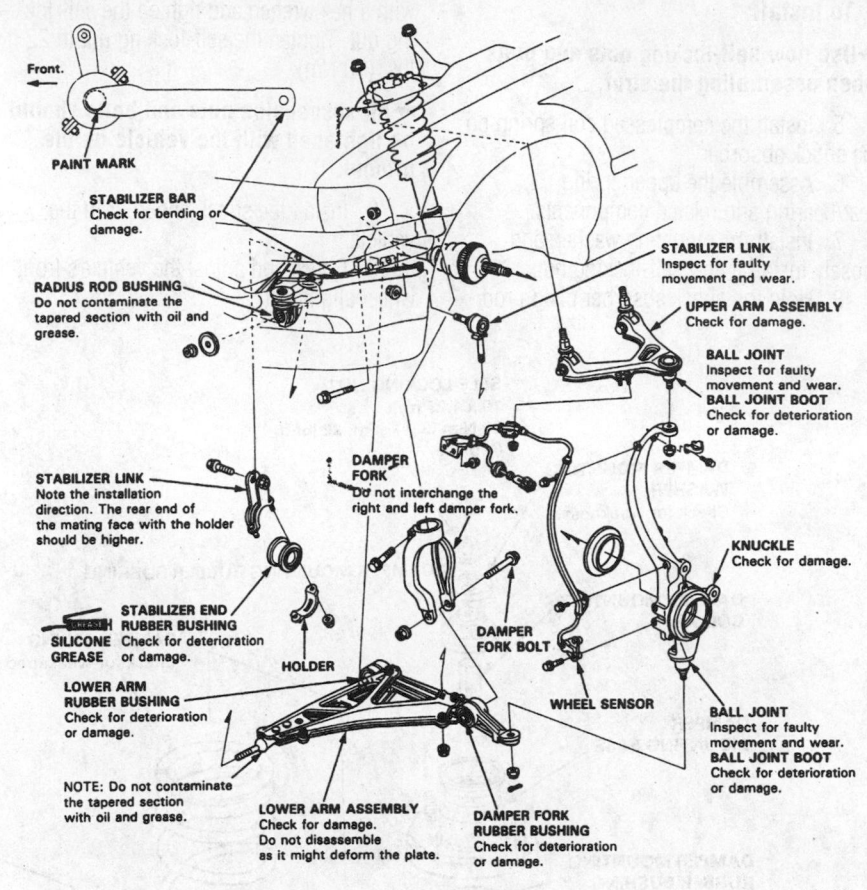

Front.

PAINT MARK

STABILIZER BAR
Check for bending or damage.

RADIUS ROD BUSHING
Do not contaminate the tapered section with oil and grease.

STABILIZER LINK
Note the installation direction. The rear end of the mating face with the holder should be higher.

STABILIZER END RUBBER BUSHING
Check for deterioration or damage.

SILICONE GREASE

LOWER ARM RUBBER BUSHING
Check for deterioration or damage.

HOLDER

NOTE: Do not contaminate the tapered section with oil and grease.

LOWER ARM ASSEMBLY
Check for damage. Do not disassemble as it might deform the plate.

DAMPER FORK
Do not interchange the right and left damper fork.

DAMPER FORK BOLT

DAMPER FORK RUBBER BUSHING
Check for deterioration or damage.

STABILIZER LINK
Inspect for faulty movement and wear.

UPPER ARM ASSEMBLY
Check for damage.

BALL JOINT
Inspect for faulty movement and wear.

BALL JOINT BOOT
Check for deterioration or damage.

KNUCKLE
Check for damage.

WHEEL SENSOR

BALL JOINT
Inspect for faulty movement and wear.

BALL JOINT BOOT
Check for deterioration or damage.

7923BG49

A common upper control arm and ball joint assembly

7. Install the upper control arm and mounting bolts to the chassis. The upper control arms are not interchangeable.
8. Raise the steering knuckle up with a floor jack, just enough to install the upper control arm ball joint into the steering knuckle. Tighten the ball joint nut to 29–35 ft. lbs. (39–47 Nm).
9. Tighten the castle nut enough to install a new cotter pin. If removed, install the ball joint nut cover.
10. Install the front wheels, lower the vehicle, and tighten the nuts connecting the upper control arm to the chassis to 47 ft. lbs. (65 Nm).
11. Check the wheel alignment and road test the vehicle.

Lower Ball Joint

REMOVAL & INSTALLATION

Integra

1. Raise and support the vehicle safely. Remove the front wheel assemblies.

2. Remove the steering knuckle.
3. Remove the boot by prying off the snapring. Remove the 40mm clip. Check the boot for deterioration and damage, replace if necessary.
4. Install the special ball joint with a removal/installation tool 07965-SB00100 or equivalent, on the ball joint and tighten the ball joint nut.
5. Position the ball joint in this special tool and set this assembly in a vise. Press the ball joint out of the steering knuckle.

To install:
6. Place the ball joint in position by hand. Install the ball joint into the special tool and press in the new ball joint in the vise.

✳✳ WARNING

After installing the boot, check the ball joint pin tapered section for grease contamination and wipe it if necessary.

7. Install the 40mm circlip. Adjust the special tool with the adjusting bolt until the end of the tool aligns with the groove on the

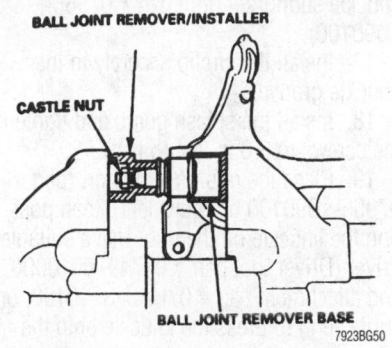

Use a vise or press to remove the ball joint from the steering knuckle

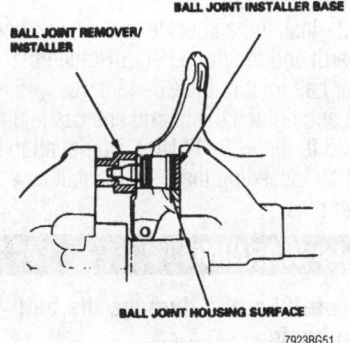

Press the new ball joint into the steering knuckle

boot. Slide the clip over the tool and into position.

8. Install the knuckle.

9. Check the front wheel alignment and adjust if necessary.

2.2CL, 2.3CL, 2.5TL, 3.0CL

➡**The lower ball joint is pressed into the steering knuckle and cannot be removed.**

1. Pry up the lock tab and loosen the spindle nut. Slightly loosen the lug nuts.

2. Raise and safely support the vehicle. Remove the front wheel and spindle nut.

3. Remove the brake caliper mounting bolts and remove the caliper from the knuckle. Hang the caliper out of the way with a length of wire.

4. Remove the ABS speed sensor from the knuckle.

5. Disconnect the tie rod end from the knuckle using a ball joint remover. Be careful not to damage the joint boot.

6. Remove the cotter pin and castle nut from the lower arm ball joint. Separate lower control arm from the knuckle.

7. Remove the cotter pin and castle nut. Separate the upper arm from the knuckle using the ball joint remover.

8. Remove the knuckle and hub by sliding the assembly off of the halfshaft. Tap the end of the halfshaft with a plastic mallet to release it from the knuckle.

9. To remove the hub and rotor assembly from the knuckle, remove the four self-locking bolts from the back of the knuckle.

10. Remove the four bolts from the hub to separate it from the brake disc.

11. The bearing can be pressed off the hub with a hydraulic press. The inner race will stay on the hub and can be removed with a bearing puller. Any time the hub and bearing are separated, the wheel bearing must be replaced with a new one.

To install:

12. Clean all the parts and examine them for wear. A worn or damaged hub will cause premature bearing failure and should be replaced.

➡**When pressing on a new bearing, be sure to press only on the inner race or the bearing will be damaged.**

13. Install the brake disc and tighten the bolts to 40 ft. lbs. (55 Nm).

14. Install the hub assembly and tighten the self-locking bolts to 33 ft. lbs. (45 Nm).

➡**Be sure that all the hub bolts are properly tightened to avoid warpage of the brake disc.**

15. Install the knuckle and hub assembly onto the halfshaft.

16. Install the knuckle on the tie rod, and the upper and lower control arms. Tighten the lower ball joint nut to 40 ft. lbs. (54 Nm) and tighten as required to install a new cotter pin.

17. Tighten the upper ball joint nut to 32 ft. lbs. (44 Nm) and tighten as required to install a new cotter pin. Tighten the tie rod end to 36 ft. lbs. (50 Nm) and tighten as required to install a new cotter pin. Install the knuckle protector.

18. Install the speed sensor, sensor wire, and mounting bolts. Be careful to avoid twisting the wires.

19. Install the brake caliper, brake hoses, and mounting bolts.

20. With the wheel installed and the vehicle on the ground, tighten the spindle nut to 180 ft. lbs. (250 Nm) and stake it in place. Tighten the wheel nuts to 80 ft. lbs. (110 Nm).

1995 Legend, 3.2TL and 3.5RL

➡**These special tools, or their equivalents are recommended by Acura for removing and replacing wheel bearings:**

- #07749–0010000: driver
- #07HAD-SG00100: driver attachment
- #07746–0010500: wheel bearing driver attachment
- #07GAF-SD40700: hub assembly base
- #07GAF-SE00100: hub assembly tool

1. Pry the lock tab away from the spindle and loosen the nut. Slightly loosen the lug nuts.

2. Raise and safely support the vehicle. Remove the front wheel and spindle nut.

3. Remove the wheel sensor from the knuckle, but do not disconnect it.

4. Remove the caliper mounting bolts. Hang the caliper out of the way with a piece of wire.

5. Remove the brake rotor retaining screws. Screw both 12mm bolts into the disc brake removal holes and turn the bolts to press the rotor from the hub. Only turn each bolt 2 turns at a time to prevent cocking the disc.

6. Remove the tie rod from the knuckle using a tie rod end removal tool. Use care not to damage the ball joint seals.

7. Remove the cotter pin from the lower arm ball joint and remove the castle nut.

➡**The lower ball joints cannot be separated from the steering knuckle**

8. Remove the lower control arm from the knuckle using the ball joint removal tool.

9. Remove the cotter pin from the upper arm ball joint and remove the castle nut.

10. Remove the upper arm from the knuckle using the ball joint remover.

11. Remove the knuckle and hub by sliding the assembly off of the halfshaft. Be sure to clean any dirt or grease off of the ball joints.

12. The hub can be removed with a slide hammer. Clamp the knuckle in a vise and secure the slide hammer to the wheel studs.

13. Remove the splash guard and snaprings.

14. Support the knuckle and press the bearing out towards the wheel side.

15. If the inner bearing race stayed on the hub, use a puller to remove it.

To install:

16. Clean all parts and examine for wear and damage.

17. When pressing in a new bearing, install the inner snapring first and press the bearing in from the wheel side. Be sure to press only on the outer race or the bearing will be damaged.

18. Install the outer snapring and the splash guard.

19. Properly support the knuckle and press the hub into the bearing. Do not press

on the wheel studs or they will press out of the hub. Support the knuckle by the inner race or the bearing will be damaged. Be sure to lubricate the bearings.

20. Install the knuckle/hub/bearing assembly onto the halfshaft and reassemble the knuckle to the upper and lower control arms.

21. Tighten the lower ball joint nut to 51–58 ft. lbs. (70–80 Nm) and tighten only enough to install a new cotter pin.

22. Tighten the upper ball joint nut to 29–35 ft. lbs. (40–48 Nm) and tighten only enough to install a new cotter pin. Tighten the tie rod end to 36–43 ft. lbs. (50–60 Nm) and tighten only enough to install a new cotter pin.

23. Install the disc brake rotor, caliper, mounting bolts, and brackets. Reconnect the wheel sensor bracket to the knuckle.

24. With the wheel installed and all 4 wheels on the ground, tighten the spindle nut to specification and stake it in place.

Wheel Bearings

ADJUSTMENT

The front and rear wheel bearings are not adjustable or repairable and should be replaced if found defective.

REMOVAL & INSTALLATION

Front

INTEGRA

1. Raise and safely support the vehicle.
2. Remove the front wheels, then pry the lock tab away and loosen the spindle nut.
3. Remove the brake hose mounting bolts.
4. Remove the brake caliper bolts and remove the caliper from the knuckle. Do not allow the caliper to hang by the brake hose, support it with a length of wire.

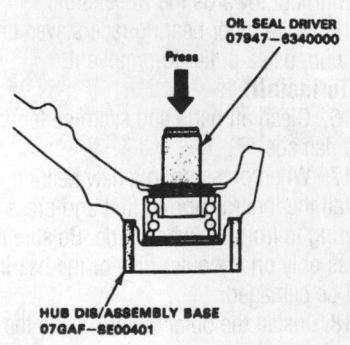

Separating the bearing from the knuckle— Integra

5. Remove the disc brake rotor.
6. If equipped with ABS, remove the wheel sensor wire bracket, then remove the wheel sensor from the knuckle. Do not disconnect the wheel sensor connector.
7. Remove the lower ball joint.
8. Remove the upper ball joint using a suitable ball joint removal tool (tool part # 07MAC-SL00200 or equivalent).
9. Pull the knuckle outward and remove the halfshaft outboard joint from the knuckle using a plastic hammer, then remove the knuckle.
10. Place the knuckle on a base (tool part # 07GAF-SD40700 or equivalent), take care not to distort the splash shield. Insert a disassembly tool into the hub (tool part # 07GAF-SE00100 or equivalent), then using a press, remove the hub from the knuckle. Hold onto the hub to keep it from falling when pressed clear.
11. Remove the knuckle ring from the rear of the knuckle.
12. Remove the circlip from the knuckle, then remove the splash guard.
13. Place the knuckle on the disassembly base and install a driver (driver tool part # 07749–0010000 and attachment tool part # 07746–0010500, or equivalent) to the bearing. Using a press, remove the bearing from the knuckle.
14. Press the wheel bearing inner race from the hub using the hub disassembly tool (part # 07GAF-SE00100) and a bearing separator.

To install:

15. Remove the old grease from the hub and knuckle and thoroughly dry and wipe clean all components.
16. Press a new wheel bearing into the knuckle with a suitable driver (Driver tool part # 07749–0010000 and attachment part # 07HAD-SF10100, or equivalent) and the

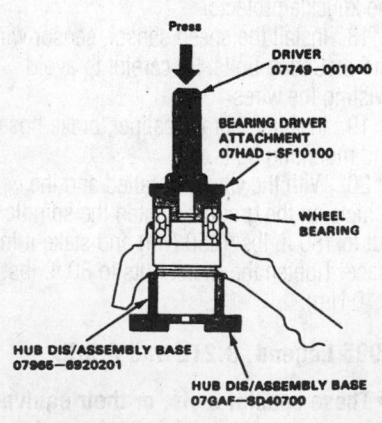

Pressing the bearing into the knuckle— Integra

knuckle supported (tool part # 07965-sD90100).

17. Install the circlip securely in the knuckle groove.
18. Install the splash guard and tighten the screws to 3.6 ft. lbs. (5 Nm).
19. Place the hub on a support (part # 07965-sD90100 or equivalent), then position the knuckle on the hub. Use a suitable driver (Driver tool part # 07749–0010000 and attachment part # 07HAD-SF10100, or equivalent) to press the knuckle onto the hub.
20. Install the knuckle ring to the rear of the knuckle.
21. Install the knuckle/hub assembly onto the halfshaft.
22. Install the knuckle to the lower control arm and the tie rod end. Tighten the lower ball joint nut to 36–43 ft. lbs. (49–59 Nm) and tighten the tie rod end castle nut to 29–35 ft. lbs. (39–47 Nm). Do not align the nuts by loosening them, and install new cotter pins.

❊❊ WARNING

Be careful not to damage the ball joint boots.

23. Install the knuckle to the upper control arm, then tighten the castle nut to 29–35 ft. lbs. (39–47 Nm). Do not align the nut by loosening it, and install a new cotter pin.
24. Install the wheel sensor and wheel sensor wire bracket onto the knuckle (for cars with ABS only).

➡**Be careful not to twist the ABS sensor wires during installation.**

25. Clean the brake rotor mating surfaces, then install the disc brake rotor and its retaining screws. Tighten the screws to 7 ft. lbs. (10 Nm).
26. Install the brake caliper, caliper bracket, and mounting bolts.
27. Install the brake hose mounting bolts, tighten the mounting bolts to 7 ft. lbs. (10 Nm).
28. Install a new spindle nut and tighten the nut to 134 ft. lbs. (181 Nm). Stake the shoulder of the nut against the halfshaft.
29. Lower the vehicle and check the front wheel alignment. Road test the vehicle.

2.2CL, 2.3CL and 3.0CL

1. Remove the steering knuckle.
2. Remove the four bolts securing the hub/rotor assembly to the knuckle.
3. Remove the brake rotor from the hub.

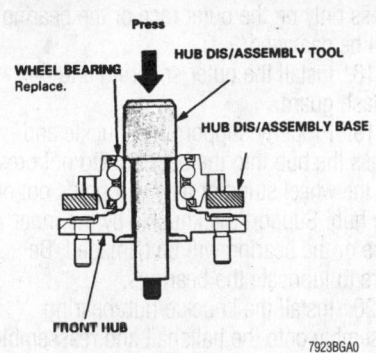

Press the hub/flange out of the bearing assembly—2.2CL and 3.0CL

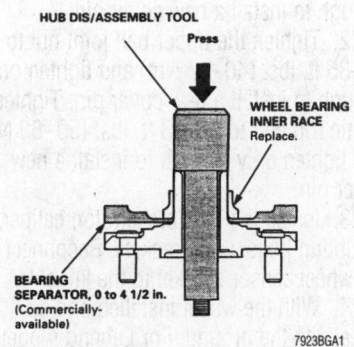

Use a press to remove the inner race from the hub/flange—2.2CL and 3.0CL

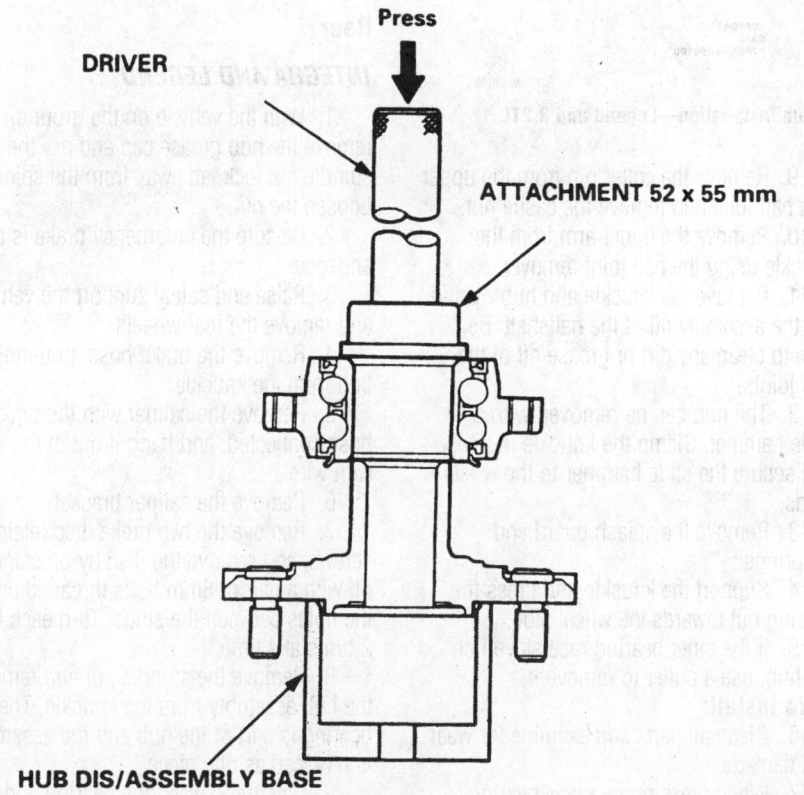

Press the new bearing onto the hub/flange—2.2CL and 3.0CL

➡**Be prepared to catch the hub/flange as it is pressed out of the bearing assembly.**

4. Press the hub/flange out of the bearing assembly.

5. Press the inner race off of the hub/flange.

To install:

6. Press a new wheel bearing onto the hub/flange.

7. Install the brake rotor on the hub. Tighten the bolts to 40 ft. lbs. (54 Nm).

8. Install the hub/rotor assembly on the knuckle. Tighten the four bolts to 33 ft. lbs. (44 Nm).

9. Install the knuckle assembly.

2.5TL

1. Pry up the lock tab and loosen the spindle nut. Slightly loosen the lug nuts.

2. Raise and safely support the vehicle. Remove the front wheel and spindle nut.

3. Remove the brake caliper mounting bolts and remove the caliper from the knuckle. Hang the caliper out of the way with a length of wire.

4. Remove the ABS speed sensor from the knuckle.

5. Disconnect the tie rod end from the knuckle using a ball joint remover. Be careful not to damage the joint boot.

6. Remove the cotter pin and castle nut from the lower arm ball joint. Separate the lower control arm from the knuckle using a ball joint remover.

7. Remove the cotter pin and castle nut. Separate the upper arm from the knuckle using the ball joint remover.

8. Remove the knuckle and hub by sliding the assembly off of the halfshaft. Tap the end of the halfshaft with a plastic mallet to release it from the knuckle.

9. To remove the hub and rotor assembly from the knuckle, remove the four self-locking bolts from the back of the knuckle. Remove the four bolts from the hub to separate it from the brake disc.

10. The bearing can be pressed off the hub with a hydraulic press. The inner race will stay on the hub and can be removed with a bearing puller. Any time the hub and bearing are separated, the wheel bearing must be replaced with a new one.

To install:

11. Clean all the parts and examine them for wear. A worn or damaged hub will cause premature bearing failure and should be replaced.

➡**When pressing on a new bearing, be sure to press only on the inner race or the bearing will be damaged.**

12. Install the brake disc and tighten the bolts to 40 ft. lbs. (55 Nm).

13. Install the hub assembly and tighten the self-locking bolts to 33 ft. lbs. (45 Nm).

➡**Be sure that all the hub bolts are properly tightened to avoid warpage of the brake disc.**

14. Install the knuckle and hub assembly onto the halfshaft.

15. Install the knuckle on the tie rod, and the upper and lower control arms. Tighten the lower ball joint nut to 40 ft. lbs. (54 Nm) and tighten as required to install a new cotter pin.

16. Tighten the upper ball joint nut to 32 ft. lbs. (44 Nm) and tighten as required to install a new cotter pin. Tighten the tie rod end to 36 ft. lbs. (50 Nm) and tighten as required to install a new cotter pin. Install the knuckle protector.

17. Install the speed sensor, sensor wire, and mounting bolts. Be careful to avoid twisting the wires.

18. Install the brake caliper, brake hoses, and mounting bolts.

19. With the wheel installed and the vehicle on the ground, tighten the spindle

nut to 180 ft. lbs. (250 Nm) and stake it in place. Tighten the wheel nuts to 80 ft. lbs. (110 Nm).

Legend, 3.2TL and 3.5RL

➡These special tools, or their equivalents are recommended by Acura for removing and replacing wheel bearings:

- #07749–0010000: driver
- #07HAD–SG00100: driver attachment
- #07746–0010500: wheel bearing driver attachment
- #07GAF–SD40700: hub assembly base
- #07GAF–SE00100: hub assembly tool

1. Pry the lock tab away from the spindle and loosen the nut. Slightly loosen the lug nuts.

2. Raise and safely support the vehicle. Remove the front wheel and spindle nut.

3. Remove the wheel sensor from the knuckle, but do not disconnect it.

4. Remove the caliper mounting bolts. Hang the caliper out of the way with a piece of wire.

5. Remove the brake rotor retaining screws. Screw both 12mm bolts into the disc brake removal holes and turn the bolts to press the rotor from the hub. Only turn each bolt 2 turns at a time to prevent cocking the disc.

6. Remove the tie rod from the knuckle using a tie rod end removal tool. Use care not to damage the ball joint seals.

7. Remove the cotter pin from the lower arm ball joint and remove the castle nut.

➡The lower ball joints cannot be separated from the steering knuckle

8. Remove the lower control arm from the knuckle using the ball joint removal tool.

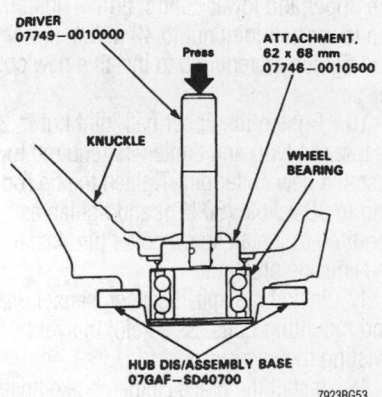

Pressing out the wheel bearing—Legend and 3.2TL

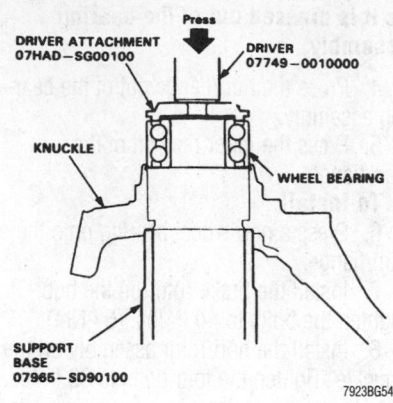

Installing the wheel bearing—Legend and 3.2TL

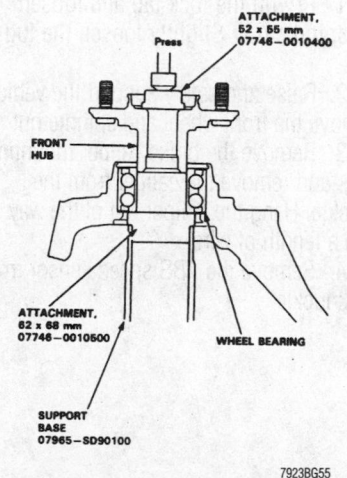

Hub installation—Legend and 3.2TL

9. Remove the cotter pin from the upper arm ball joint and remove the castle nut.

10. Remove the upper arm from the knuckle using the ball joint remover.

11. Remove the knuckle and hub by sliding the assembly off of the halfshaft. Be sure to clean any dirt or grease off of the ball joints.

12. The hub can be removed with a slide hammer. Clamp the knuckle in a vise and secure the slide hammer to the wheel studs.

13. Remove the splash guard and snaprings.

14. Support the knuckle and press the bearing out towards the wheel side.

15. If the inner bearing race stayed on the hub, use a puller to remove it.

To install:

16. Clean all parts and examine for wear and damage.

17. When pressing in a new bearing, install the inner snapring first and press the bearing in from the wheel side. Be sure to

press only on the outer race or the bearing will be damaged.

18. Install the outer snapring and the splash guard.

19. Properly support the knuckle and press the hub into the bearing. Do not press on the wheel studs or they will press out of the hub. Support the knuckle by the inner race or the bearing will be damaged. Be sure to lubricate the bearings.

20. Install the knuckle/hub/bearing assembly onto the halfshaft and reassemble the knuckle to the upper and lower control arms.

21. Tighten the lower ball joint nut to 51–58 ft. lbs. (70–80 Nm) and tighten only enough to install a new cotter pin.

22. Tighten the upper ball joint nut to 29–35 ft. lbs. (40–48 Nm) and tighten only enough to install a new cotter pin. Tighten the tie rod end to 36–43 ft. lbs. (50–60 Nm) and tighten only enough to install a new cotter pin.

23. Install the disc brake rotor, caliper, mounting bolts, and brackets. Reconnect the wheel sensor bracket to the knuckle.

24. With the wheel installed and all 4 wheels on the ground. For Legend models, tighten the spindle nut to 242 ft. lbs. (335 Nm). For 3.2TL, tighten the spindle nut to 181 ft. lbs. (245 Nm). Stake the nut in place.

Rear

INTEGRA AND LEGEND

1. With the vehicle on the ground, remove the hub grease cap and pry the spindle nut lock tab away from the spindle. Loosen the nut.

2. Be sure the emergency brake is disengaged.

3. Raise and safely support the vehicle and remove the rear wheels.

4. Remove the brake hose mounting bolt from the knuckle.

5. Remove the caliper with the brake hose connected, and hang it out of the way with wire.

6. Remove the caliper bracket.

7. Remove the two brake disc retaining screws, and remove the disc by pressing it off with a pair of 8mm bolts threaded into the holes between the studs. Turn each bolt 2 turns at a time.

8. Remove the spindle nut and remove the hub assembly from the knuckle. The bearing is part of the hub and the assembly is replaced as one piece.

9. Be sure to wash the bearing and spindle thoroughly in solvent before reassembly.

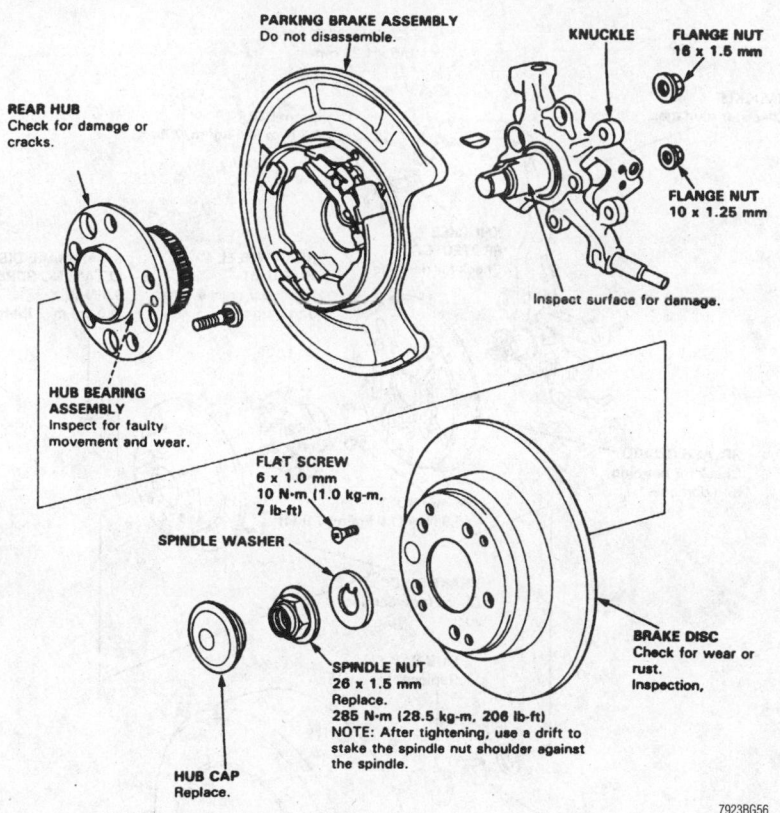

Hub and bearing assembly—Legend

To install:

10. Install the hub and bearing assembly onto the spindle and install the spindle nut. Tighten the spindle nut, but do not tighten it until the vehicle is on the ground.

11. Install the brake disc and tighten the disc retaining screws to 7 ft. lbs. (10 Nm).

12. Install the caliper mounting bracket and tighten the bolts to 28 ft. lbs. (39 Nm).

13. Install the brake caliper and the brake hose mounting bolts. For Integra, tighten the mounting bolts to 28 ft. lbs. (38 Nm). For Legend, tighten the caliper mounting bolts to 17 ft. lbs. (23 Nm).

14. Install the brake hose mounting bolt, tighten the bolt to 16 ft. lbs. (22 Nm).

15. If applicable, install the brake caliper shield. Tighten the mounting bolts to 7 ft. lbs. (10 Nm).

16. Tighten the brake disc retaining screws to 7 ft. lbs. (10 Nm).

17. Install the rear wheels and lower the vehicle to the ground.

18. Install a new hub nut and tighten the nut for Integra models to 134 ft. lbs. (181 Nm). For Legend models, tighten it to 206 ft. lbs. (285 Nm). Stake the spindle nut, and install the grease cap.

2.2CL, 2.3CL and 3.0CL

1. Remove the rear wheel.

2. Apply the parking brake.

3. Remove the grease cap the covers the spindle nut.

4. Raise the locking tab on the nut and remove it from the spindle.

5. Release the parking brake.

6. Remove the bolts securing the brake hose bracket.

7. Remove the brake caliper and hang it out of the way with wire or string.

8. Remove the brake rotor retaining screws. Install two 8 x 1.25mm bolts into the brake rotor. Tighten each screw two turns at a time to push the rotor off the hub.

9. Remove the hub/bearing unit from the spindle.

To install:

10. Install a new hub/bearing unit on the spindle.

11. Remove the two 8 x 1.25mm bolts and install the brake rotor on the hub/bearing. Tighten the two 6mm screws to 7 ft. lbs. (10 Nm).

12. Install the spindle washer and a new nut on the spindle. Tighten the nut to 134 ft. lbs. (181 Nm) on the 2.2CL and 3.0CL or

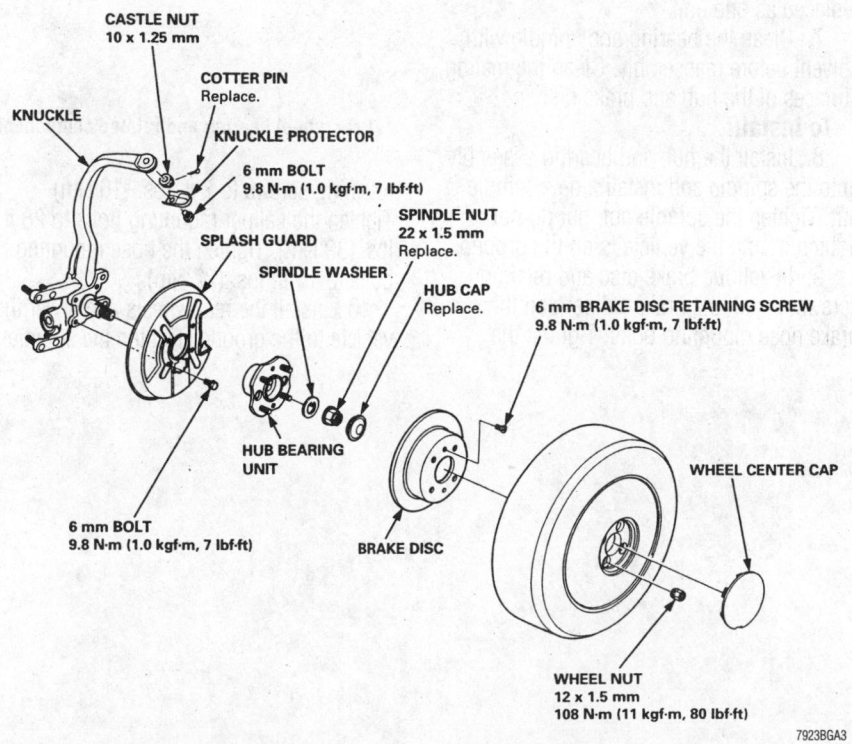

Exploded view of the rear bearing and related components—2.2CL and 3.0CL

181 ft. lbs. (245 Nm) on the 2.3CL. Stake the nut to the spindle.

13. Install the brake caliper and hose bracket.

14. Install the rear wheel. Tighten the nuts to 80 ft. lbs. (101 Nm).

2.5TL, 3.2TL and 3.5RL

1. With the vehicle on the ground, remove the hub cap and pry the spindle nut lock tab away from the spindle. Loosen the spindle nut.

2. Engage the parking brake to provide leverage to help with loosening the brake disc retaining screws.

3. Raise and safely support the vehicle and remove the rear wheels.

4. Remove the brake hose mounting bolt. Remove the caliper without disconnecting the hydraulic hose. Support the caliper with a piece of wire.

5. Remove the two disc retaining screws. Remove the disc by pressing it off with a pair of 8mm bolts threaded into the holes between the studs. Turn each bolt two turns at a time. Release the parking brake after removing the disc brake retaining screws, and before removing the brake disc.

6. Remove the spindle nut and remove the hub and bearing assembly from the knuckle. The wheel bearing is part of the hub assembly and the components are replaced as one unit.

7. Clean the bearing and spindle with solvent before reassembly. Clean the mating surfaces of the hub and brake disc.

To install:

8. Install the hub and bearing assembly onto the spindle and install a new spindle nut. Tighten the spindle nut, but do not tighten it until the vehicle is on the ground.

9. Install the brake disc and retaining screws. Install the brake caliper and the brake hose mounting bolts. Tighten the

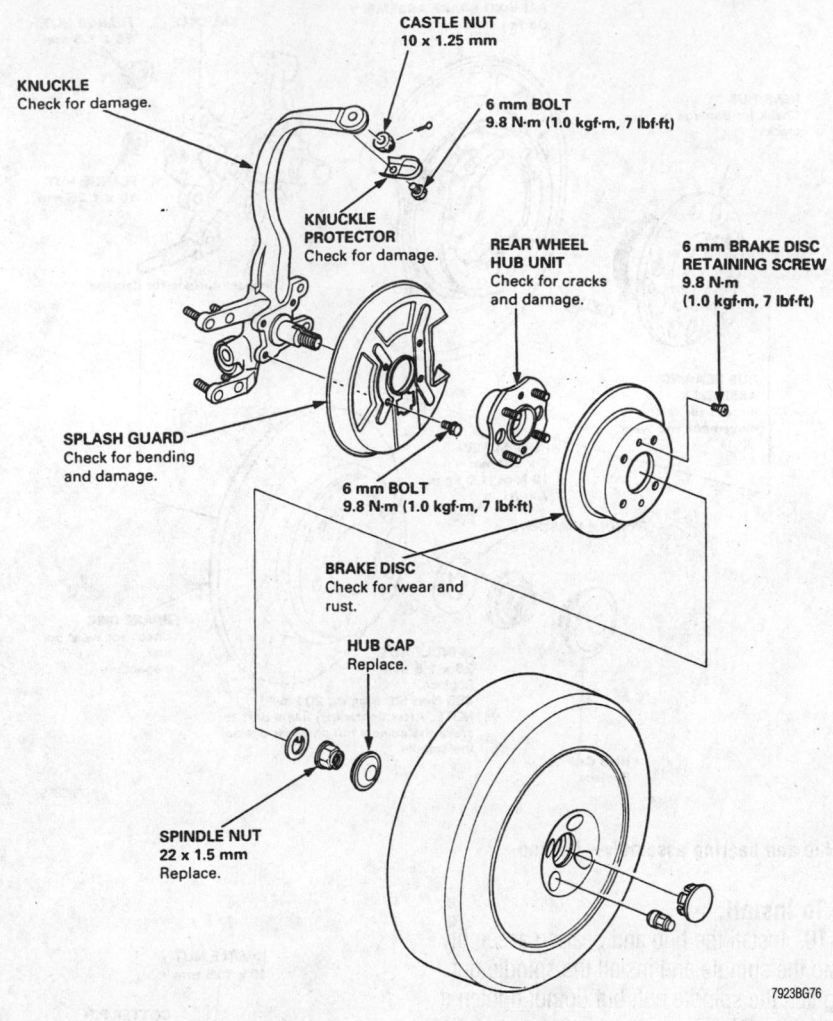

Rear wheel bearing and related components—2.5TL and 3.2TL

retaining screws to 7 ft. lbs. (10 Nm). Tighten the caliper mounting bolts to 28 ft. lbs. (39 Nm). Tighten the hose mounting bolts to 16 ft. lbs. (22 Nm).

10. Install the rear wheels and lower the vehicle to the ground. Tighten the spindle

nut for the Integra to 134 ft. lbs. (181–185 Nm). For the and 2.5TL, 181 ft. lbs. (245 Nm), and for the 3.2TL, tighten to 242 ft. lbs. (355 Nm). Stake the nut to the spindle with a punch. Install the hub cap.

ENGINE REPAIR

➡Disconnecting the negative battery cable on some vehicles may interfere with the functions of the on-board computer systems and may require the computer to undergo a relearning process, once the negative battery cable is disconnected. Most vehicles are equipped with theft protected radios, which cannot be operated if power to the radio is interrupted. Before disconnecting the battery cables, obtain the security code for the radio.

Ignition Timing

Audi vehicles are equipped with distributorless ignition systems; no adjustments are possible.

Engine Assembly

REMOVAL & INSTALLATION

4-Cylinder Engine

➡Tag all hoses and wiring during removal, to use as reference during reassembly.

1. Lock the carrier into service position as follows:
   a. Remove the front bumper.
   b. Tag and remove any wiring or connector that would inhibit locking the carrier.
   c. Remove the three quick-release screws on the front noise insulation panel.
   d. Unbolt the air guide between the lock carrier and the air filter.
   e. If installed, remove the retaining clamps for the wiring harness at the left side of the radiator frame.
   f. Remove the No. 2 bolts and install Support tool 3369 or equivalent.
   g. Remove the remaining bolts and pull the lock carrier out to the stop.
   h. To secure the lock carrier, install the appropriate M6 bolts into the rear of the lock carrier and fender.
2. Position the wipers to the vertical position.
3. Properly relieve the fuel system pressure.
4. Disconnect the negative battery cable.
5. Remove the engine under cover.
6. Drain and recycle the engine coolant.
7. Remove the front bumper.

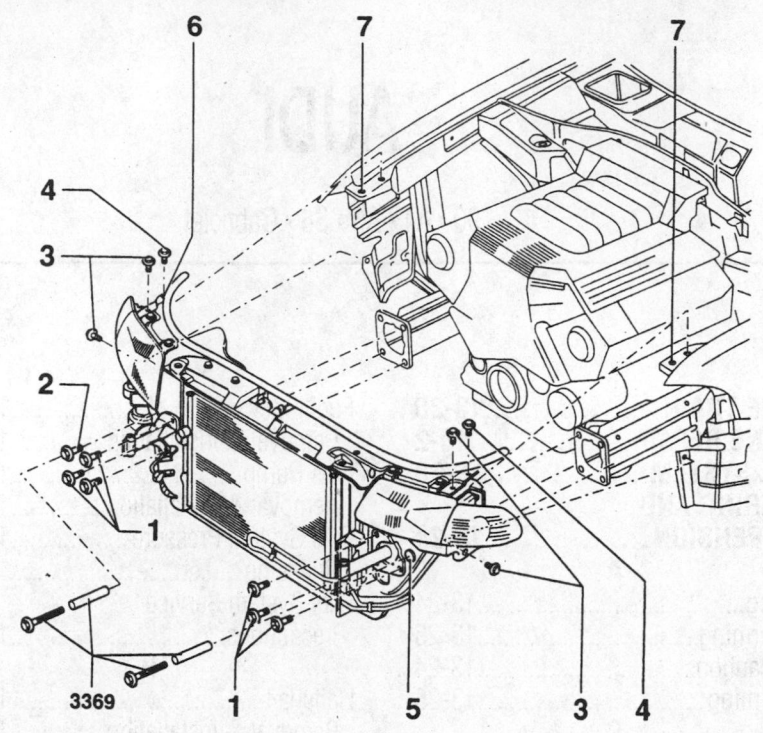

1. Bolts 33 ft. lbs. (45 Nm)
2. Bolts 33 ft. lbs. (45 Nm)
3. Bolts 7 ft. lbs. (10 Nm)
4. Bolts 7 ft. lbs. (10 Nm)
5. Bore for support tool
6. Lock carrier bore
7. Fender bore

7923CG12

Moving the lock carrier into the service position—A4 models

8. Unbolt the power steering cooling coil from the radiator, leaving it connected and hanging.
9. If equipped, disconnect the transaxle oil cooling lines.
10. Detach the electric cooling fan thermal switch at the lower left of the radiator.
11. Unbolt and remove the air intake duct and assembly from the vehicle.
12. Disconnect the headlight height adjuster wiring harness.
13. Remove the turn signal bulbs from the light housing.
14. Remove the coolant hose from the radiator at the upper coolant pipe.
15. Remove the hood release cable at the carrier lock.
16. Remove the power steering fluid reservoir cap/dipstick.

17. Disconnect the wiring harness for the ABS hydraulic unit.
18. Detach the horn electrical connectors.
19. Unbolt the air guides at the left and right sides of the radiator.
20. Unbolt the A/C condenser retaining fasteners.
21. Disconnect the A/C low pressure switch.
22. Cover the right fender, then pull the condenser out of its mounting and position it over the fender.
23. Detach the green harness connector from the A/C compressor magnetic clutch.
24. Remove the engine covers.
25. Detach the wiring harness connectors for the wastegate bypass regulator valve, the Evaporative Emissions (EVAP)

canister purge regulator valve, the power output stage, and the Mass Air Flow (MAF) sensor.

26. Detach the Engine Coolant Level (ECL) warning switch.

27. Disconnect the coolant hoses at the expansion tank, then remove the tank and position it aside.

28. If equipped with cruise control, detach the actuating rod from the throttle valve control module, then remove the vacuum hose from the vacuum unit.

29. Detach the accelerator pedal cable from the throttle valve control module.

30. Remove the hose for the Leak Detection Pump (LDP).

31. Disconnect the fuel supply and return lines.

32. Disconnect the brake booster vacuum hose.

33. Remove the vacuum hose for the EVAP canister purge regulator valve.

34. Uncover the E-box.

35. Unclip the Motronic Engine Control Module (ECM) retaining bracket.

36. Detach the wiring harness to the ECM.

37. If equipped with an automatic transaxle, detach the kickdown switch connector.

38. Detach the heated oxygen sensor wiring harness.

39. Detach the ground connection at the plenum chamber.

40. Disconnect the heater hoses from the heater core.

41. Detach the Vehicle Speed Sensor (VSS) from the transaxle and position it aside.

42. If equipped with a manual transaxle, detach the back-up light switch from the transaxle.

43. Remove the engine driven cooling fan.

44. Remove the accessory drive belts.

45. Unbolt the A/C compressor from the mounting bracket and position it aside using wire.

46. Disconnect the P/S pump and position it aside leaving the hoses attached.

➡ **The flexpipe at the front exhaust pipe must not be bent more than 10°, otherwise it may be damaged.**

47. Disconnect the catalytic converter from the turbocharger.

48. Remove the starter, and the ground strap at the right engine mount.

49. If equipped with an automatic transaxle, remove the three torque converter-to-driveplate mounting bolts through the opening left by the starter.

50. Loosen the upper nuts for the left and right engine mounts.

51. Place matchmarks on the threaded bolt and centering sleeves at the bottom of the left and right engine mounts, then remove the mounting nuts.

52. Remove the lower engine-to-transaxle mounting bolts.

53. If equipped with an automatic transaxle, remove the ATF cooler line bracket form the left side of the engine.

54. Remove the upper nuts from the engine mounts.

55. Position an Engine Support Bridge 10–222A or equivalent to the bolted flanges of the fenders with the spindle facing forward.

56. Attach the Engine Support Adapter 3147 or equivalent to a bolt hole in the transaxle bell housing.

57. Connect the Engine Support Adapter 3147 and the Engine Support Bridge 10–222A using Adapter 2024A/1 and Extension 2024A/2 or their equivalents.

58. Attach an engine sling between the engine and the hoist.

59. Remove the upper engine-to-transaxle mounting bolts.

60. Separate the engine from the transaxle, then slowly lift the engine up and out the front of the engine compartment.

61. If equipped with an automatic transaxle, secure the torque converter to prevent it from falling out.

To install:

➡ **Be sure that the centering sleeves for the engine-to-transaxle are correctly installed in the cylinder block.**

62. Verify that the intermediate plate is over the centering sleeves.

63. Install the engine into the vehicle.

64. Install the upper engine-to-transaxle mounting bolts.

65. Lower the engine into position, then remove the engine sling and hoist.

66. Remove the transaxle support apparatus from the vehicle.

67. Install the engine mounting fasteners without any tension or pre-load.

68. If equipped with an automatic transaxle, install the ATF cooler line bracket to the left side of the engine.

69. Install the lower engine-to-transaxle mounting bolts and tighten the mounting bolts as follows: M12 bolts to 48 ft. lbs. (65 Nm) and M10 bolts to 33 ft. lbs. (45 Nm).

70. Tighten the engine mounting nuts/bolts to 18 ft. lbs. (25 Nm).

71. If equipped with an automatic transaxle, install the driveplate-to-torque converter mounting bolts through the starter opening and tighten to 63 ft. lbs. (85 Nm).

72. Install the starter and attach the ground strap to the right engine mount.

73. Attach the catalytic converter to the turbocharger and tighten the mounting bolts to 22 ft. lbs. (30 Nm).

74. Install the P/S pump, the A/C compressor, and the engine cooling fan, then the accessory drive belts.

75. If equipped with a manual transaxle, connect the back-up light switch to the transaxle.

76. Connect the Vehicle Speed Sensor (VSS) to the transaxle.

77. Connect the heater hoses to the heater core.

78. Attach the ground connection at the plenum chamber.

79. Attach the heated oxygen sensor wiring harness.

80. If equipped with an automatic transaxle, attach the kickdown switch connector.

81. Connect the wiring harness to the ECM.

82. Install the Motronic Engine Control Module (ECM) retaining bracket and cover the E-box.

83. Connect the vacuum hose for the EVAP canister purge regulator valve.

84. Connect the fuel supply and return lines.

85. Connect the brake booster vacuum hose.

86. The completion of the installation procedure is the reverse of the removal, keeping in mind the following items.

• If equipped with an automatic transaxle, check the ATF level

• Fill the engine with coolant

• Fully close all power windows to stop, operate all window switches for at least one second in the close direction to activate the one-touch opening/closing function

• Check the oil level before starting the engine

87. Connect the negative battery cable.

88. Set the clock to the correct time.

➡ **Diagnostic Trouble Codes (DTCs) are stored when harness connectors are detached.**

89. Read the DTCs and clear the fault codes.

90. Adjust the headlights.

5-Cylinder Engine

➡ **Tag all hoses and wiring during removal, to use as reference during reassembly.**

1. Disconnect the negative battery cable. Relieve fuel pressure.

2. Open the heater control valve all the way and drain the cooling system.

3. Remove the fuel injector cooling fan blower motor and intake hose from the engine.

4. If necessary, remove the upper radiator cover, grille, bumper strip. Disconnect the wiring harness in bumper for turn signals and headlights. Remove the bumper.

5. Disconnect the electrical connector from the coolant fan. Remove the upper radiator hose from the engine. Remove the radiator-to-expansion tank hose from the tank and the bleeder hose from the auxiliary radiator.

6. Disconnect the wire from the thermo-switch. Remove the radiator mounting bolts, right-side radiator cover and bottom radiator cover.

7. Remove the windshield washer reservoir from the mount and support it aside.

➡If your vehicle is equipped with air conditioning. Only a MVAC-trained, EPA-certified, automotive technician should service the A/C system or its components.

8. Properly discharge the air conditioning system and disconnect the refrigerant hoses from the air conditioning condenser.

9. Remove the radiator and air conditioning condenser together. Remove the air conditioning compressor and mounting bracket from the engine.

10. Remove the power steering pump drive belt from the pump. Remove the pump from the mounts. Leaving the hoses attached, support it aside.

11. Disconnect the coolant hose from the thermostat housing and disconnect the wires from the oil pressure switch and temperature sender. Disconnect the wire plugs from the control pressure regulator.

12. Remove the control pressure regulator from the engine but leave the fuel lines connected. Support it aside.

13. Remove the throttle rod clips and remove the rod from the engine. Remove the injector line holder and remove the fuel injectors from the cylinder head.

14. Disconnect the electrical connector from the cold start valve and remove the valve from the intake manifold. Leave the fuel line connected.

15. At the throttle body, disconnect the electrical connectors from the throttle valve switches and intake air temperature switch.

16. Disconnect the air intake hose. Disconnect the wire from the auxiliary air regulator, pull off the vacuum hoses and disconnect the breaker hose from the engine.

17. At the 2-way valve, remove and tag the vacuum hoses. Remove the thermo-pneumatic valve. Leave the vacuum lines connected and remove the rpm sensor.

18. Disconnect the speedometer cable from the transaxle.

19. Remove the distributor from the engine.

20. Disconnect and tag the thermo-time switch and overheating warning lamp connectors. Disconnect the heater hoses from the engine.

21. At the left engine mount, disconnect the brake booster and reservoir from the firewall and leave the lines connected. On Quattro vehicles, disconnect the differential lock control lights connector. Disconnect the back-up light switch wires.

22. Disconnect the tie rods from the steering rack. Disconnect the steering linkage.

23. If equipped with a manual transaxle, remove the clutch slave cylinder from the bell housing. Leave the line attached. Remove the bracket and pin under the transaxle bracket.

24. Disconnect the left engine mount ground strap. Disconnect the vacuum hose from the auxiliary air valve.

25. Remove the air duct from the intercooler and remove the intercooler.

26. Disconnect and tag the electrical connectors from the alternator. Remove the oil cooler. Leave the lines attached. Disconnect and tag the starter wiring.

27. Disconnect the exhaust pipe at the turbocharger. Remove the transaxle cover plates and the right side transaxle mount. Disconnect the halfshafts from the transaxle. On Quattro vehicles, disconnect the driveshaft from the rear of the transaxle.

28. On Quattro vehicles at the transaxle, disconnect the differential lock, remove the front and rear circlips and push back the boot. Disconnect the cable.

29. Remove the left-side transaxle mounting bolt and mounts from both sides.

30. At both front wheels, remove the ball joint pinch bolts. At the subframe, remove the mounting bolts and subframe. Separate the ball joints from the steering knuckle.

31. Install an engine lifting device on the engine. Raise the engine slightly and remove the left and right engine mounts. Lower the engine/transaxle assembly from the vehicle.

32. Raise the front of the vehicle and slide the engine/transaxle assembly from under the vehicle.

33. Separate the engine from the transaxle.

To install:

34. Install the engine assembly and temporarily secure the engine mounts.

35. Install the steering joints to the steering knuckles and tighten to 22 ft. lbs. (30 Nm). Install the ball joints with the pinch bolts and tighten to 44 ft. lbs. (65 Nm).

36. Install the exhaust and 4WD differential lock clips, if equipped.

37. Reconnect the exhaust pipe at the turbocharger.

38. Install the halfshafts to the transaxle.

39. Install the right side transaxle mount and transaxle cover plates.

40. On Quattro vehicles, reconnect the driveshaft to the rear of the transaxle.

41. Lower the vehicle and install all electrical connections.

42. Install the oil cooler.

43. Install the intercooler assembly and air duct.

44. Reconnect the left engine mount ground strap.

45. Install the clutch slave cylinder, if equipped.

46. Install the tie rods to the steering rack assembly.

47. Install the brake booster and reservoir. On Quattro vehicles, connect the differential lock control lights connector. Reconnect the back-up light switch wires.

48. Install the heater hoses.

49. Install the distributor.

50. Install the speedometer cable and all vacuum lines.

51. Install the fuel injectors from the cylinder head.

52. Connect the throttle body connectors, throttle linkage and fuel injection pressure regulator.

53. Install the power steering pump and all drive belts.

54. Install the radiator and air conditioning condenser. Connect the hoses.

55. Install the windshield washer reservoir.

56. If removed, install the upper radiator cover and grille. Reconnect the wiring harness in bumper for turn signals and headlights. Install the bumper.

57. Install the cooling fan blower motor and intake hose to the engine.

58. Refill and bleed the cooling system. Connect the battery negative cable.

59. To minimize vibration, loosen all the engine and subframe mounting bolts, then tighten to 25 ft. lbs. (34 Nm) while the engine is running at idle.

60. Check all fluid levels, road test for proper operation.

6-Cylinder Engine

➡The engine is removed without the transaxle, through the top of the engine compartment.

1. Disconnect the negative battery cable. On some vehicles the battery is located under the rear seat.

2. Drain and recycle the engine coolant.

3. Remove the engine undercover, then the soundproofing material holder from the engine mount.

4. Remove the engine cooling fan.

5. If equipped, unbolt the power steering cooling coil from the bottom left of the radiator.

6. Remove the stabilizer brace from the right rear of the engine compartment.

7. Remove the air intake hose and air cleaner assembly from the vehicle.

8. Label and detach all hoses, wiring harnesses, lines and cables as necessary for engine removal.

9. Remove the accessory drive belt guard, then the belt.

10. Detach the front engine mount at the crossmember.

11. Disconnect the power steering pressure hose from the P/S pump.

12. Detach the ground strap from the right side engine support.

13. Remove the air intake for the generator.

14. Disconnect the exhaust system from the manifolds.

15. Remove the crossmember.

16. Disconnect the front exhaust pipes with the catalytic converters.

17. Unbolt the starter and pull out rearward, leaving the wires attached, then tie the starter to the engine at one side.

18. Remove the oil filter, then loosen and remove the oil cooler from under the filter.

19. Unbolt the A/C compressor and tie it aside leaving the hoses attached.

20. If equipped with A/T, remove the flexplate-to-torque converter attaching bolts through the starter opening.

21. Remove the engine-to-transmission mounting bolts.

22. Support the weight of the transmission using the engine support bridge 10–222A and hook 3147 or their equivalents.

23. Attach the engine sling 2024A or equivalent to the right-rear and left-front of the engine.

24. Attach an engine hoist to the sling.

25. Lift the engine slightly.

26. Remove the front engine mount from the side member.

27. Carefully lift the engine in conjunction with the transmission to clear the right engine mount.

28. Carefully pull the engine forward until it is separated from the transmission.

29. Lift the starter out after the engine has been separated from the transmission,

then lift the engine up and out of the vehicle.

To install:

➡ **Before installing the engine, inspect the clutch components for wear or damage, and replace if necessary. Short blocks are supplied without the sleeve in the crankshaft. For vehicles equipped with an automatic transmission, tap the sleeve into place before installing the flexplate. If equipped with a manual transmission, install a pilot bearing. Lightly lube the clutch release bearing and the transmission input shaft with G 000 100 or equivalent grease. Check that the alignment sleeves for centering the engine and transmission are properly installed in the engine block. Install sleeves if missing or damaged. Replace all self locking nuts.**

30. Slowly lower the engine into the vehicle.

31. Position the starter, then engage the engine to the transmission.

32. Simultaneously lower the engine and transmission into position.

➡ **Install the engine mounts without any pre-load and without tension. Ensure proper alignment by rocking the engine before tightening the engine mounts.**

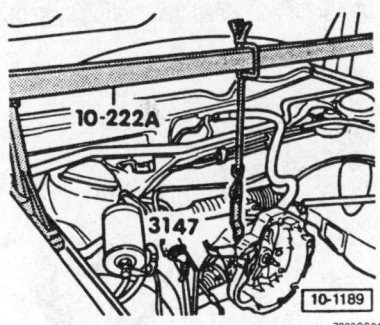

Using the engine support bridge and hook to support the weight of the transmission—shown with the engine removed

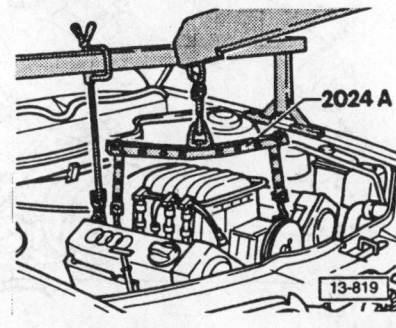

Be sure to attach the engine sling properly

33. Install the engine mounts and tighten the bolts to 33 ft. lbs. (45 Nm).

34. The completion of the installation is the reverse of the removal procedure. Note the following items.

• Tighten the flexplate mounting bolts to 44 ft. lbs. (60 Nm) plus 90°, flywheel mounting bolts to 30 ft. lbs. (40 Nm) plus 90°, then an additional 90°

• Tighten the flexplate-to-torque converter mounting bolts to 26 ft. lbs. (35 Nm)

• Tighten the engine-to-transmission attaching bolts to 18 ft. lbs. (24 Nm) for the M8 bolts, 33 ft. lbs. (45 Nm) for the M10 bolts, and 48 ft. lbs. (65 Nm) for the M12 bolts

Water Pump

REMOVAL & INSTALLATION

4-Cylinder Engine

➡ **The coolant pump is bolted to the brackets for the generator, power steering pump, and cooling fan.**

1. Lock the carrier into service position as follows:

a. Remove the front bumper.

b. Tag and remove any wiring or connector that would inhibit locking the carrier.

c. Remove the three quick-release screws on the front noise insulation panel.

d. Unbolt the air guide between the lock carrier and the air filter.

e. If installed, remove the retaining clamps for the wiring harness at the left side of the radiator frame.

f. Remove the No. 2 bolts and install Support tool 3369 or equivalent.

g. Remove the remaining bolts and pull the lock carrier out to the stop.

h. To secure the lock carrier, install the appropriate M6 bolts into the rear of the lock carrier and fender.

2. Turn the ignition switch to the **OFF** position, then disconnect the negative battery cable.

3. Remove the accessory drive belt, then the engine driven cooling fan.

4. Drain and recycle the engine coolant.

5. Loosen the clamps for the coolant hoses at the water pump.

6. Remove the intake air duct between the intake manifold and the charge air cooler.

7. Remove the generator mounting bolts and slide it forward.

8. Disconnect the wiring from the generator once it is removed.

9. Unbolt the following supports and brackets for the generator, power steering pump, and engine cooling fan:
- Intake manifold support
- Support for the engine torque bracket
- Brace to the cylinder block (remove completely)

10. Remove the brackets for the generator, power steering pump, and engine cooling fan.

11. Position the brackets for the generator, power steering pump, and engine cooling fan to the left side using a piece of wire.

12. Pull off the coolant hoses from the pump and thermostat housing.

13. Unbolt the coolant pump housing from the timing belt cover.

14. Remove the coolant pump mounting bolts, then the pump.

15. Unbolt the impeller housing from the pump housing.

16. Clean all gasket and O-ring sealing surfaces.

To install:

17. Using a new gasket, mount the new coolant pump to the pump housing and tighten the mounting bolts to 7 ft. lbs. (10 Nm).

18. Using a new gasket and O-ring, install the coolant pump and tighten the mounting bolts in numerical sequence to 18 ft. lbs. (25 Nm).

19. Tighten the coolant pump housing to the timing belt cover to 7 ft. lbs. (10 Nm).

20. Install the coolant hoses to the pump and thermostat housing.

21. Install the brackets that were removed and tighten the mounting bolts to 18 ft. lbs. (25 Nm).

22. Connect the wires to the generator, then install the generator.

23. Install the air intake duct between the intake manifold and the charge air cooler.

24. The remaining steps are the reverse of the removal procedure noting the following items:

- Fill the engine with coolant
- Verify that the key is in the **OFF** position before connecting the battery
- Fully close all power windows to stop, operate all window switches for at least one second in the close direction to activate the one-touch opening/closing function
- Set the clock to the correct time
- After installing the lock carrier, check the wiring for proper routing near the cooling fan

Except 4-Cylinder Engine

1. Drain the cooling system.

2. Remove the V-belts and the timing belt covers. Remove the timing belt from the water pump.

3. Remove the water pump mounting bolts, then the pump.

➡**Always replace the old gasket or O-ring.**

To install:

4. Installation is the reverse of the removal procedure. Tighten the water pump retaining bolts to 15 ft. lbs. (20 Nm) on the 5-Cylinder engines and 7 ft. lbs. (10 Nm) for the 6-Cylinder engines.

5. Reinstall the timing belt and properly tension the belt with the water pump. Refer to the necessary service procedures.

6. Refill and bleed the cooling system.

1. Thermostat
2. Seal
3. Thermostat housing

Cylinder Head

➡**Before removing or installing the cylinder head, align the engine timing marks at TDC. Rotate the crankshaft mark away about ¼ turn (BTDC). This will prevent the valves from hitting the piston heads. Be sure to turn the crankshaft to the proper position after cylinder head installation.**

REMOVAL & INSTALLATION

➡**Cylinder head removal should not be attempted unless the engine is cold.**

4-Cylinder Engine

1. Place the lock carrier into the service position as follows:
   a. Remove the front bumper.
   b. Tag and remove any wiring or connector that would inhibit locking the carrier.
   c. Remove the three quick-release screws on the front noise insulation panel.
   d. Unbolt the air guide between the lock carrier and the air filter.
   e. If installed, remove the retaining clamps for the wiring harness at the left side of the radiator frame.
   f. Remove the No. 2 bolts and install Support tool 3369 or equivalent.

4. Bolt   7. Bolt
5. Gasket
6. Coolant pump

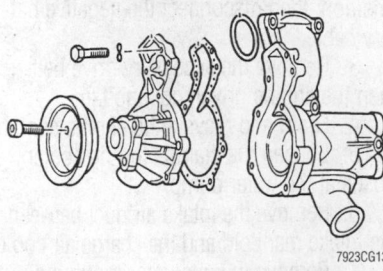

Exploded view of the water pump, housing and related components—4-cylinder engine

7923CG13

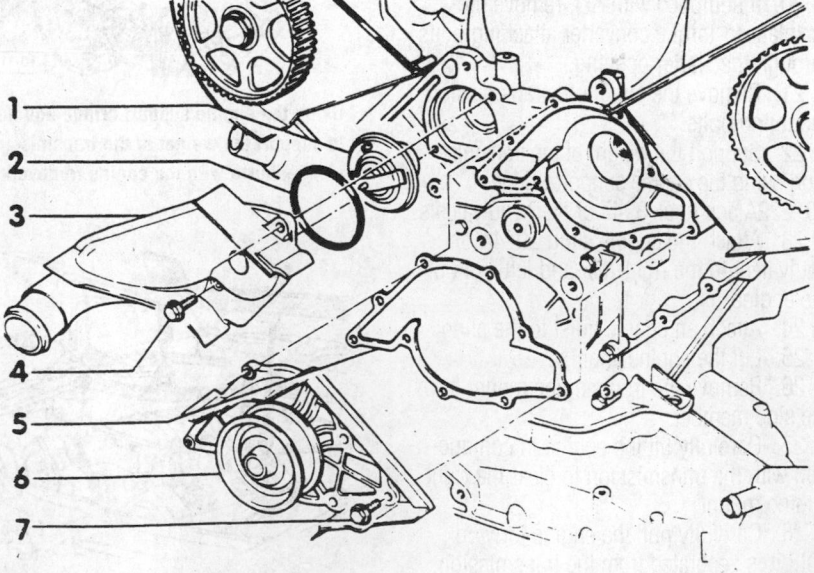

Exploded view of the water pump and related components—6-cylinder engine

7923CG03

g. Remove the remaining bolts and pull the lock carrier out to the stop.

h. To secure the lock carrier, install the appropriate M6 bolts into the rear of the lock carrier and fender.

2. Turn the ignition switch to the **OFF** position, then disconnect the negative battery cable.

3. Remove the accessory drive belt, then the engine driven cooling fan.

4. Drain and recycle the engine coolant.

5. Remove the intake manifold.

6. Remove the accessory drive belts.

7. Label and detach the following lines and electrical connectors:

- Wastegate bypass regulator valve
- Evaporative Emission (EVAP) canister purge regulator valve
- Power outage stage
- Mass Air Flow (MAF) sensor

8. Remove the air cleaner housing.

9. Detach the Engine Coolant Temperature (ETC) and the temperature II sensor harness connector.

10. Label and detach all connections from the cylinder head and position them aside.

11. Remove the crankcase breather line.

12. Disconnect the oil supply line at the cylinder head.

13. Remove the exhaust manifold heat shield.

14. Unbolt the turbocharger from the exhaust manifold.

15. Disconnect the coolant hose to the heat exchanger at the rear of the cylinder head.

16. Remove the upper timing belt cover.

17. Turn the crankshaft, in the direction of rotation (clockwise), until the No. 1 cylinder is at Top Dead Center (TDC).

18. Using Torx® wrench T45, loosen the timing belt tensioner.

19. Push down on the tensioner, and remove the belt from the camshaft gear.

20. Remove the Torx® bolt and swing the tensioner assembly bracket forward.

21. Remove the valve cover.

22. Remove the cylinder head bolts in sequence, as shown.

23. Remove the cylinder head, then clean the gasket mating surfaces.

24. Clean and dry out the cylinder head bolt holes.

To install:

➡**Always replace the cylinder head bolts. Always replace self-locking nuts, bolts, gaskets and O-rings. Do not remove the new head gasket from the package until immediately before installing.**

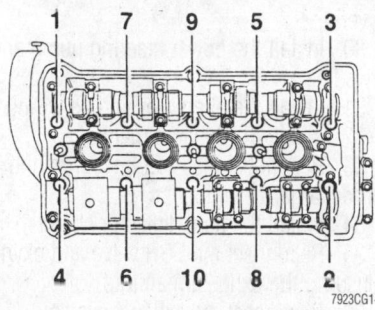

Cylinder head bolt removal sequence—4-cylinder engine

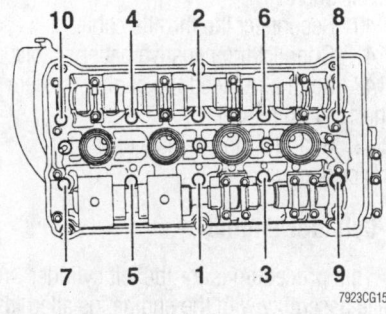

Cylinder head bolt tightening sequence—4-cylinder engine

25. Before installing the cylinder head, set the crankshaft and camshaft to TDC for the No. 1 cylinder.

26. Loosen the turbocharger support bracket to reduce the likelihood of any tension while installing the cylinder head.

27. Install the head gasket with the part number visible from the intake side.

28. Install the cylinder head.

29. Install the new cylinder head bolts and tighten by hand.

30. Tighten the new cylinder head bolts in sequence in two steps:

- Step 1 –44 ft. lbs. (60 Nm)
- Step 2 -additional ½ turn (180°)

➡**It is not necessary to retighten the cylinder head bolts.**

31. Using new gaskets, install the turbocharger to the exhaust manifold, coat the bolts with Hot Bolt Paste G 052 112 A3 (or equivalent), then tighten the mounting bolts to 26 ft. lbs. (35 Nm). Tighten the turbo support bracket mounting bolts to 33 ft. lbs. (40 Nm).

32. Install the valve cover.

33. Install the timing belt.

34. Install the accessory drive belts.

35. Install the exhaust manifold heat shield.

36. Connect the oil supply lines to the cylinder head and tighten the retaining straps to 15 ft. lbs. (20 Nm).

37. Install the crankcase breather.

38. Attach any items removed during disassembly.

39. Connect the coolant temperature sensors, and install the air cleaner housing.

40. Fill the engine with coolant.

41. Connect the negative battery cable.

42. Fully close all power windows to stop, operate all window switches for at least one second In the close direction to activate the one-touch opening/closing function

43. Check the oil level before starting the engine

44. Set the clock to the correct time.

➡**Diagnostic Trouble Codes (DTCs) are stored when harness connectors are detached.**

45. Read the DTCs and clear the fault codes.

46. Adjust the headlights.

5-Cylinder Engine

1. Disconnect the negative battery cable.

2. Drain the cooling system.

3. Disconnect the air duct from the throttle valve assembly on all vehicles except the Turbo and Quattro. On the Turbo and Quattro, remove the hose which runs between the air duct and the turbocharger.

4. Disconnect the throttle cable from the throttle valve assembly.

5. Remove the air duct for the injector cooling fan on the Turbo and Quattro.

6. Clean and remove the fuel injectors and all other fuel lines.

➡**Protect the fuel injectors and the cold start valve with caps.**

7. Tag and disconnect all vacuum and PCV lines.

8. Remove the hose which runs from the intake manifold to the turbocharger on the Turbo and Quattro.

9. Tag and disconnect all electrical lines leading to the cylinder head.

10. Remove the intake manifold.

11. Disconnect all radiator and heater hoses where they are attached to the cylinder head. Position them aside.

12. Separate the exhaust manifold from the exhaust pipe.

➡**Exhaust pipe detachment differs slightly on the Turbo and Quattro. First the exhaust pipe must be unbolted from the turbocharger. Second, it must be unbolted from the wastegate at the rear of the engine.**

13. Disconnect the EGR valve and oxygen sensor from the exhaust manifold.

14. Remove the heat deflector shield.

15. Remove the oil lines (2) from the turbocharger.

16. Remove the exhaust manifold.

➡**When removing the exhaust manifold on the Turbo and Quattro, the manifold, turbocharger and wastegate should all be removed as a unit.**

17. Remove the air hose cover from the back of the alternator.

18. Tag and disconnect all wires coming from the back of the alternator and remove the alternator from the engine.

19. Disconnect and plug the hoses coming from the power steering pump.

20. Remove the power steering pump and the V-belt.

21. Remove the timing belt cover and belt.

22. Remove the valve cover.

23. Loosen the cylinder head bolts in the reverse order of the tightening sequence.

24. Remove the bolts and lift the cylinder head off the engine.

To install:

25. Clean the cylinder head and engine block mating surfaces thoroughly and install the new gasket without any sealing compound. Be sure the words **TOP** or **OBEN** are facing up, when the gasket is installed.

26. Place the cylinder head on the engine block and install bolts No. 8 and 10 first. These holes are smaller and will properly locate the gasket and the head on the engine block.

27. Install the remaining bolts. Tighten them in sequence in 3 stages as follows:
- Step 1—30 ft. lbs. (40 Nm)
- Step 2—44 ft. lbs. (60 Nm)
- Step 3—Tighten ½ turn more (180 degrees).

28. Install the valve cover.

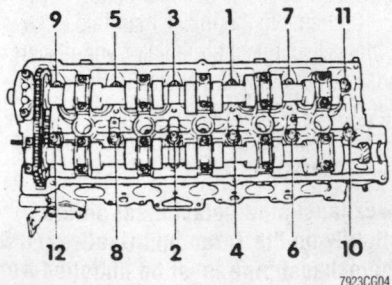

Cylinder head bolt tightening sequence—5-cylinder engine

29. Install timing belt and timing belt cover.

30. Install the power steering pump and drive belt.

31. Install the alternator and all wiring.

32. Install the exhaust manifold.

33. Reconnect the 2 oil lines from the turbocharger assembly.

34. Install the heat deflector shield.

35. Reconnect the EGR valve and oxygen sensor to the exhaust manifold.

36. Reconnect the exhaust system.

37. Install the intake manifold.

38. Install the radiator and heater hoses.

39. Install the fuel injectors. Reconnect the air duct.

40. Reconnect the throttle cable.

41. Connect the negative battery cable.

42. Refill and bleed the cooling system. Check all fluid levels.

43. Road test the vehicle check for proper operation.

6-Cylinder Engine

This procedure is for the left cylinder head assembly with the engine installed in the vehicle. Modify the service procedure as necessary for the right side.

1. Disconnect the negative battery cable. Drain engine coolant. Relieve fuel pressure.

2. Remove the ribbed V-belt.

3. Remove the timing belt. The vibration damper should remain installed.

4. Remove the exhaust pipe from the manifold.

5. Remove the EGR valve hose at manifold.

6. Remove the air guide hose between air mass sensor and intake manifold.

7. Remove and tag all spark plug wires and injector connectors.

8. Remove the crankcase breathers on the left and right cylinder head covers.

9. Remove the fuel feed and return lines. Remove the silencer.

10. Remove the left side cover for fuel line.

11. Disconnect the throttle cable.

12. Mark and remove all vacuum hoses from vacuum pump and intake manifold.

13. Disconnect connectors on idling stabilization valve and throttle valve potentiometer.

14. Disconnect vacuum hose on vacuum control unit.

15. Disconnect connectors on oil pressure sender and oil pressure switch.

16. Disconnect connector for Hall sender sensor.

17. Remove the EGR valve from the intake manifold.

18. Remove the intake manifold assembly.

19. Remove the coolant pipe at the rear of the cylinder head.

20. Remove the oxygen sensor.

21. Remove the heatshield on the exhaust manifold.

22. Remove the cylinder head cover.

23. Remove the timing belt rear belt guard. Remove the hydraulic line from reservoir to pump.

24. Reverse the installation torque sequence and remove the cylinder head assembly from the engine.

To install:

25. Clean all sealing surfaces. Check cylinder head for distortion. Measure at several locations. The maximum permissible distortion is 0.1mm.

26. Install cylinder head gasket. The lettering must face upwards.

27. Install the cylinder head assembly, check centering pins in the cylinder block.

28. Install cylinder head bolts by hand.

29. Tighten the cylinder head bolts in sequence in 2 steps. Step 1—44 ft. lbs. and Step 2—½ turn (180 degrees). It is not necessary to retighten the cylinder head bolts after repairs or as part of inspection service.

30. Install the timing belt rear belt guard.

31. Install the cylinder head cover.

32. Install the oxygen sensor. Install the heatshield on the exhaust manifold.

33. Install the intake manifold assembly.

34. Install the EGR valve to the intake manifold.

35. Reconnect the hall sender sensor, oil pressure sender and oil pressure switch.

36. Reconnect connectors on idling stabilization valve and throttle valve potentiometer.

37. Install all vacuum hoses to vacuum pump and intake manifold.

38. Reconnect the throttle cable.

39. Reconnect the fuel feed and return lines. Install the silencer.

40. Install the crankcase breather on the cylinder head cover.

41. Install all spark plug wires and injector connectors.

42. Install the exhaust manifold.

43. Install the timing belt and V-belt.

44. Refill and bleed the cooling system.

45. Check all fluid levels. Operate the engine at normal operating temperatures and check for leaks.

46. Road test the vehicle for proper operation.

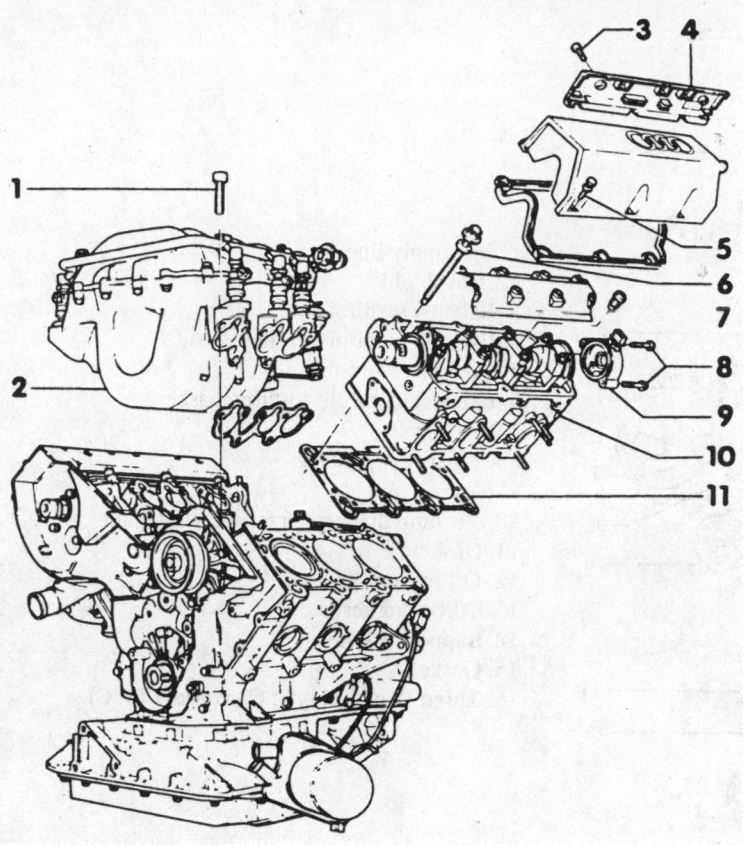

1. Bolt 15 ft. lbs. (20 Nm)
2. Intake manifold
3. Bolt 7 ft. lbs. (10 Nm)
4. Cover
5. Bolt 7 ft. lbs. (10 Nm)
6. Cylinder head bolts

7. Pressure relief valve
8. Bolt 7 ft. lbs. (10 Nm)
9. Camshaft Position (CMP) sensor
10. Cylinder head
11. Cylinder head gasket

7923CG06

Exploded view of the cylinder head mounting and related components—6-cylinder engine

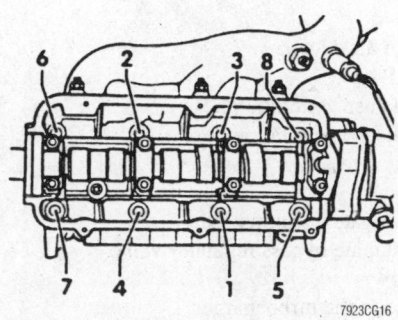

7923CG16

Cylinder head mounting bolt torque sequence—6-cylinder engine

Turbocharger

REMOVAL & INSTALLATION

4-Cylinder Engine

1. Disconnect the negative battery cable.

2. Remove the engine undercover, and unbolt the A/C compressor.

3. Unbolt the turbocharger support bracket.

4. Disconnect the oil return line at the turbocharger.

5. Remove the air hoses from the turbocharger.

6. Disconnect the oil feed line at the turbocharger.

7. Disconnect the hose for the boost pressure regulation valve vacuum diaphragm.

8. Unbolt the bracket for the coolant supply line at the boost pressure regulation valve vacuum diaphragm.

9. Using Clamp 3094 or equivalent, pinch off the coolant supply hose.

10. Remove the intake air duct between the cowl and the air cleaner housing.

11. Remove the air cleaner housing cover.

12. Label and detach the following lines and electrical connectors:
   • Wastegate bypass regulator valve
   • Evaporative Emission (EVAP) canister purge regulator valve
   • Power outage stage
   • Mass Air Flow (MAF) sensor

13. Remove the air cleaner housing and the engine cover.

14. Disconnect the crankcase breather hose at the valve cover.

15. Disconnect the oil supply line at the turbocharger.

16. Remove the heat shield, and sleeve from the coolant return hose.

17. Using Clamp 3094 or equivalent, pinch off the coolant return hose, then remove.

➡**The exhaust flexpipe may be damaged if bent more than 10°.**

18. Disconnect the Three Way Catalytic Converter (TWC) from the turbo.

19. Unbolt the turbo from the exhaust manifold.

20. Move the turbo aside to disconnect the coolant supply banjo fitting.

21. Remove the turbo.

To install:

22. Install the turbo.

23. Connect the coolant supply banjo fitting and tighten to 18 ft. lbs. (25 Nm).

24. Using new gaskets, install the turbocharger to the exhaust manifold, coat the bolts with Hot Bolt Paste G 052 112 A3 (or equivalent), then tighten the mounting bolts to 26 ft. lbs. (35 Nm). Tighten the turbo support bracket mounting bolts to 33 ft. lbs. (45 Nm).

25. Attach the Three Way Catalytic Converter (TWC) to the turbo.

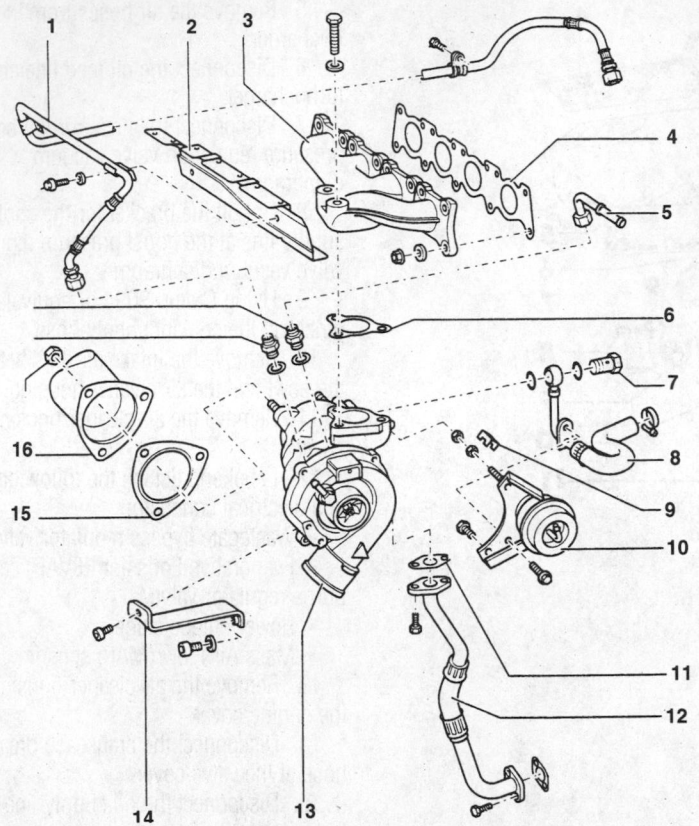

1. Oil supply line
2. Heat shield
3. Exhaust manifold
4. Exhaust manifold gasket
5. Coolant return line
6. Exhaust manifold-to-turbo gasket
7. Banjo bolt
8. Coolant supply hose
9. Fuse
10. Vacuum diaphragm for the wastegate
11. Gasket
12. Oil return line
13. Turbocharger
14. Support bracket
15. Gasket
16. Three Way Catalytic Converter (TWC)

7923CG17

Exploded view of the turbocharger and related components—4-cylinder engine

1. Vacuum hose
2. Boost pressure recirculation valve
3. Hose
4. Intake air duct
5. EVAP hose
6. Crankcase ventilation hose
7. Crankcase ventilation hose
8. PCV valve
9. Hose
10. Wastegate vacuum hose
11. Wastegate bypass regulator valve
12. Elbow
13. Hose to the turbocharger

7923CG18

Exploded view of the vacuum hoses related to the turbocharger—4-cylinder engine

26. Connect the coolant supply hose.

27. Install the sleeve and heat shield to the return hose.

28. Connect the oil return hose.

29. Add oil to the turbo through the oil feed line.

30. Connect the oil supply line to the turbo and tighten to 18 ft. lbs. (25 Nm).

31. Connect the crankcase breather, and install the engine cover and air cleaner housing.

32. Attach the following lines and electrical connectors:
- Mass Air Flow (MAF) sensor
- Power outage stage
- Wastegate bypass regulator valve

33. Connect the hoses and brackets for the boost pressure regulation valve vacuum diaphragm.

34. Connect the air hoses to the air cleaner assembly and the turbo.

35. Install the A/C compressor and engine undercovers.

36. Refill the coolant system and check the oil level.

37. Connect the negative battery cable.

38. Start the vehicle and check for leaks, then let the engine idle for approx. 1 minute without increasing the engine speed. This ensures adequate oil supply to the turbo.

5-Cylinder Engine

1. Disconnect the negative battery cable. Spray all mounting bolts with a lubricant.

2. Remove the vacuum tube between the intake air boot and turbocharger.

3. Remove the intake boot and crankcase ventilation hose. Remove the hose assembly between the intake manifold and throttle housing.

4. Remove the air box cover and remove the filter element.

5. Remove the right side engine mount heatshield.

6. Remove the oil supply pipe from the turbocharger. Remove the exhaust pipe from the corrugated pipe. Loosen the exhaust pipe at the transaxle mount and catalytic converter.

7. Remove the retaining clamp from the starter housing and sensor air hose.

8. Remove the exhaust pipe from the turbocharger.

9. Remove the alternator support bolt and position the alternator to the side.

10. Remove the oil return pipe from the turbocharger. Remove mounting bolts and turbocharger.

To install:

11. Install the turbocharger with new gaskets and tighten the mounting nuts to 43 ft. lbs. (60 Nm).

12. Connect the oil supply and return lines using new gaskets.

13. Install the alternator assembly.

14. Connect the exhaust pipe and tighten the nuts to 25 ft. lbs. (34 Nm). Install all exhaust brackets.

➡**Bolts and nuts exposed to high temperatures should receive a light coating of anti-seize compound on the threads before assembly. After servicing the turbocharger, always replace the engine oil along with the turbocharger filter and engine oil filter.**

15. Connect the outlet air hose and breather hoses.

16. Install heatshield, filter element and air box cover.

17. Reconnect the battery. Operate the engine at normal operating temperatures and check for leaks.

Intake Manifold

REMOVAL & INSTALLATION

4-Cylinder Engine

1. Turn the ignition switch to the **OFF**, then disconnect the negative battery cable.

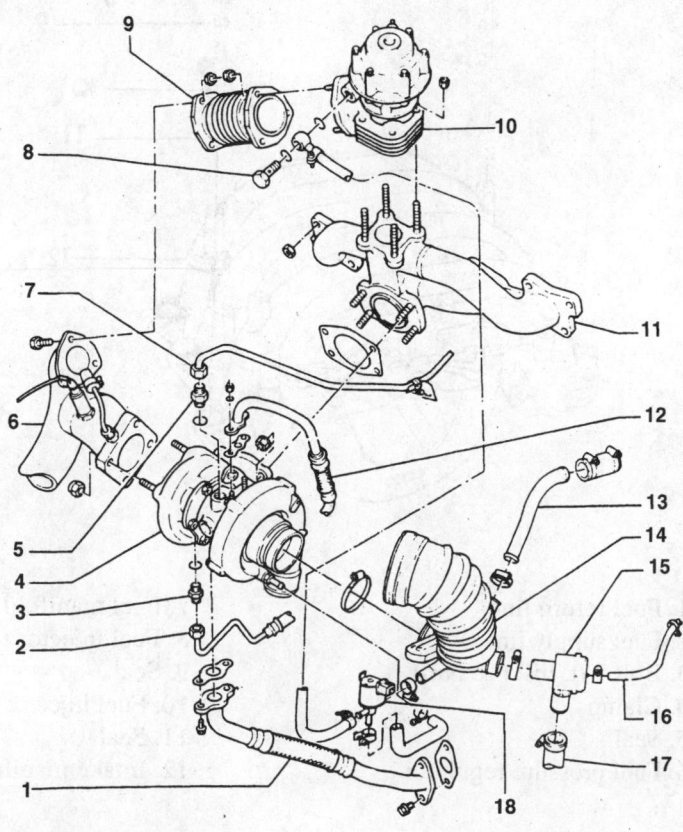

1. Oil return line
2. Coolant return line
3. Fitting
4. Turbocharger
5. Fitting
6. Exhaust pipe
7. Coolant supply line
8. Banjo bolt
9. Corrugated pipe
10. Wastegate
11. Exhaust manifold
12. Oil supply line
13. Hose
14. Air hose
15. Bypass valve
16. Vacuum hose
17. Hose
18. Wastegate bypass regulator valve

Exploded view of the turbocharger mounting and related components—5-cylinder engine

7923CG19

2. Remove the engine covers.

3. Drain and recycle the engine coolant.

4. Disconnect the hose for the Leak Detection Pump (LDP).

5. Disconnect the accelerator pedal cable from the throttle valve control module.

6. Remove the air guide hose from the throttle valve control module.

7. Disconnect the vacuum line from the Evaporative Emissions (EVAP) canister.

8. Detach the brake booster vacuum hose.

9. Disconnect the Intake Air Temperature (IAT) sensor and the throttle valve control module.

10. Detach the Camshaft Position (CMP) sensor wiring harness connector.

11. Disconnect the fuel rail with the injectors and position it aside in the engine compartment on a clean cloth.

12. Disconnect the coolant hoses attached to the intake manifold.

13. Disconnect the crankcase breather hose at the intake manifold.

14. Remove the intake manifold brace and the oil dipstick.

15. Unbolt the manifold at the mounting flange, then remove.

16. Stuff clean shop rags into the cylinder head ports to prevent debris from entering the engine.

17. Clean all gasket surfaces of any old gasket material.

To install:

18. Remove the rags from the cylinder head ports.

19. Install the intake manifold and tighten the mounting fasteners to 7 ft. lbs. (10 Nm).

20. Connect the manifold brace and tighten the bracket mounting bolts to 15 ft. lbs. (20 Nm).

21. Install the dipstick.

22. Connect the crankcase breather and coolant hoses to the intake manifold.

23. Replace the fuel injector sealing O-ring, then install the fuel rail with the injectors and tighten the retaining bolts to 7 ft. lbs. (10 Nm).

24. Connect the CMP, then the IAT sensor to the throttle valve control module.

25. Connect the brake booster and EVAP canister vacuum hoses.

26. Connect the air guide hose and the accelerator pedal cable to the throttle valve control module.

27. Connect the hose for the LDP.

28. Top off the engine coolant.

29. Turn the ignition switch **OFF**, then reconnect the negative battery cable.

30. Fully close all power windows to stop, operate all window switches for at least one second in the close direction to activate the one-touch opening/closing function

31. Check the oil level before starting the engine

32. Set the clock to the correct time.

➡**Diagnostic Trouble Codes (DTCs) are stored when harness connectors are detached.**

33. Read the DTCs and clear the fault codes.

5-Cylinder Engine

1. Disconnect the negative battery cable.
2. Relieve the fuel system pressure.

3. On non-turbocharged engines, disconnect the air duct from the throttle valve assembly. On turbocharged engines, remove the hose between the air duct and turbocharger.

4. Disconnect the throttle cable and rod from the throttle valve assembly.

5. On turbocharged engines, remove the air duct for the injector cooling fan.

6. Remove the fuel injectors from the cylinder head, with the fuel lines attached.

7. Disconnect the cold start valve wiring and remove the fuel line from the valve.

➡**Protect the fuel injectors and cold start valve with caps.**

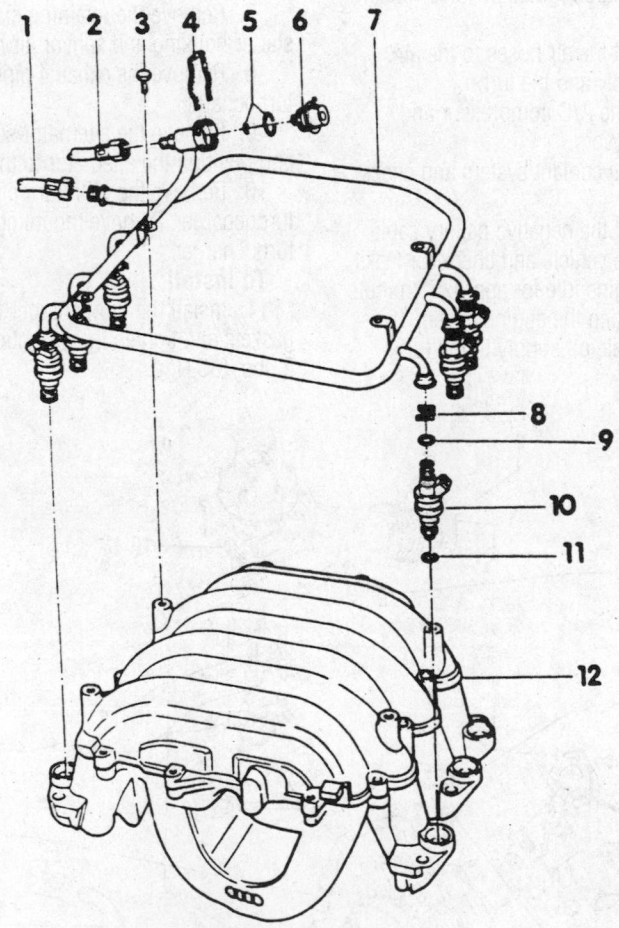

1. Fuel return line
2. Fuel supply line
3. Bolt 7 ft. lbs. (10 Nm)
4. Clamp
5. Seal
6. Fuel pressure regulator
7. Fuel manifold
8. Fuel injector retainer
9. Seal
10. Fuel injector
11. Seal
12. Intake manifold

7923CG07

Exploded view of the fuel injector assembly—6-cylinder engine

8. Tag and disconnect all vacuum and PCV lines.

9. Tag and disconnect all electrical lines leading to the cylinder head.

10. On turbocharged engines, remove the hose which runs from the intake manifold to the turbocharger. On the Quattro vehicle, remove the hose which runs from the intake manifold to the intercooler.

11. Remove the auxiliary air regulator. Remove the air box cover and filter element.

12. Remove the intake manifold mounting nuts and remove the manifold from the engine.

To install:

13. Clean the gasket mating surfaces on the manifold and engine.

14. Using a new gasket, install the manifold on the cylinder head and tighten the nuts to 15 ft. lbs. (20 Nm).

15. Install the auxiliary air regulator.

16. On turbocharged engines, install the hose which runs from the intake manifold to the turbocharger. On the Quattro vehicle, install the hose which runs from the intake manifold to the intercooler.

17. Reconnect all vacuum and electrical connections.

18. Install the fuel line to the cold start valve and reconnect the cold start valve wiring.

19. Install the fuel injectors.

20. Connect the throttle cable and rod to the throttle valve assembly. Install all air ducts.

21. Check all fluid levels. Operate the engine at normal operating temperatures and check for leaks.

22. Road test the vehicle for proper operation.

6-Cylinder Engine

1. Disconnect the negative battery cable.
2. Properly relieve the fuel system pressure.
3. Disconnect the fuel supply and return lines.
4. Detach the fuel injector electrical connectors and the retainers for the fuel injectors.
5. Remove the fuel pressure regulator clamp, then the regulator.
6. Unbolt and remove the fuel manifold, then the injectors.
7. Label and detach any hoses or connectors associated with the intake manifold.
8. Remove the upper intake manifold mounting bolts, then the manifold and gasket.
9. Remove the Idle Air Control (IAC) valve.
10. Disconnect the EGR valve.

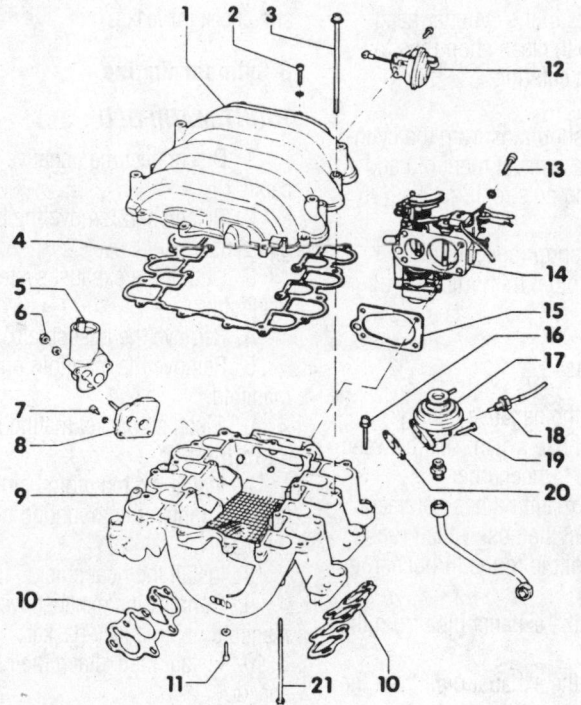

1. Upper intake manifold
2. Bolt 7 ft. lbs. (10 Nm)
3. Bolt 15 ft. lbs. (20 Nm)
4. Gasket
5. Idle Air Control (IAC) valve
6. Bolt 7 ft. lbs. (10 Nm)
7. Bolt 53 inch lbs. (6 Nm)
8. Flange
9. Lower intake manifold
10. Gasket
11. Bolt 7 ft. lbs. (10 Nm)
12. Vacuum unit
13. Bolt 15 ft. lbs. (20 Nm)
14. Throttle body
15. Gasket
16. Bolt 15 ft. lbs. (20 Nm)
17. EGR valve
18. EGR temp sensor
19. Bolt 7 ft. lbs. (10 Nm)
20. Gasket
21. Bolt 7 ft. lbs. (10 Nm)

7923CG08

Exploded view of the upper/lower intake manifold assembly and related components—6-cylinder engine

11. Disconnect the accelerator cable form the throttle body.

12. Remove the lower intake manifold mounting bolts, then the manifold.

13. Clean all gasket mating surfaces of old gasket material.

14. Place shop rag in the cylinder head openings to prevent dirt and debris from entering the open port.

To install:

15. Remove the shop rags from the cylinder head openings.

16. Install the lower intake manifold using a new gasket, and tighten the mounting bolts to 15 ft. lbs. (20 Nm).

17. Connect the EGR valve.

18. Attach the accelerator cable to the throttle body.

19. Install the IAC valve and tighten the mounting bolts to 7 ft. lbs. (10 Nm).

20. Install the upper intake manifold and tighten the short mounting bolts to 7 ft. lbs. (10 Nm) and the long mounting bolts to 15 ft. lbs. (20 Nm).

21. Attach any hoses or connectors that were removed.

22. Install the fuel injectors using new O-rings, then the fuel manifold and tighten the mounting bolts to 7 ft. lbs. (10 Nm).

23. Install the pressure regulator and clamp.

24. Install the fuel injector retainers, then attach the electrical connectors.

25. Attach the fuel return and supply lines.

26. Connect the negative battery cable.

Exhaust Manifold

REMOVAL & INSTALLATION

4-Cylinder Engine

1. Remove the turbocharger.
2. Seal any openings in the turbo with clean shop rags to prevent debris from entering.

3. Remove the exhaust manifold mounting nuts, then remove the manifold.

4. Seal openings in the cylinder head and exhaust pipe with clean shop rags to prevent debris from entering.

To install:

5. Remove the shop rags from the cylinder head, install the exhaust manifold and tighten the mounting nuts to 18 ft. lbs. (25 Nm).

6. Remove the shop rags from the turbo and exhaust pipe, then install the turbo.

5-Cylinder Engine

1. Disconnect the negative battery cable. Remove the hose which runs between the air duct and the turbocharger.

2. If the intake manifold has not been removed, disconnect the hose which runs from the intake manifold to the turbocharger or intercooler.

3. Disconnect the exhaust pipe from the turbocharger.

4. Disconnect the exhaust pipe from the wastegate on the rear of the manifold.

5. Disconnect the EGR valve and the oxygen sensor, if necessary, from the manifold.

6. Remove the oil lines from the turbocharger.

7. Remove the line from the bottom of the turbocharger to the intercooler, if equipped.

➡ **The manifold, turbocharger and wastegate are removed as a unit.**

8. Remove the manifold assembly.

To install:

9. Clean the gasket mating surfaces of the manifolds and engine.

10. Install the exhaust manifold. Tighten the exhaust manifold mounting nuts to 26 ft. lbs. (35 Nm). Always use new gaskets and O-rings where necessary.

➡ **The oxygen sensor, EGR tube, bolts and nuts exposed to high temperatures should receive a light coating of anti-seize compound on the threads before assembly.**

11. Install the line from the bottom of the turbocharger to the intercooler, if equipped.

12. Reconnect the oil lines from the turbocharger.

13. Install the EGR valve and oxygen sensor, if necessary.

14. Install the exhaust system.

15. Install all hoses or air ducts.

16. Reconnect the battery. Operate the engine at normal operating temperatures and check for leaks.

6-Cylinder Engine

RIGHT MANIFOLD

1. Disconnect the negative battery cable.

2. Detach and remove the heated oxygen sensor.

3. Unbolt the exhaust system from the manifold.

4. Remove the heat shield.

5. Remove the manifold nuts, then the manifold.

6. Clean all gasket mating surfaces.

To install:

7. Install the manifold using a new gasket, and tighten the mounting nuts to 18 ft. lbs. (24 Nm).

8. Install the heat shield.

9. Connect the exhaust system to the manifold using a new gasket.

10. Install, then attach the heated oxygen sensor.

11. Connect the negative battery cable.

12. Start the vehicle and check for leaks.

LEFT MANIFOLD

1. Disconnect the negative battery cable.

2. Drain and recycle the engine coolant.

3. Detach and remove the heated oxygen sensor.

4. Unbolt the exhaust system from the manifold.

5. Unbolt, and if necessary, remove the coolant tube from the cylinder head.

6. Remove the heat shield.

7. Disconnect the EGR tube from the rear of the manifold.

8. Remove the manifold nuts, then the manifold.

9. Clean all gasket mating surfaces.

To install:

10. Install the manifold using a new gasket, and tighten the mounting nuts to 18 ft. lbs. (24 Nm).

11. Connect the exhaust system to the manifold using a new gasket.

12. Install, then attach the heated oxygen sensor.

13. Connect the EGR tube to the rear of the manifold.

14. Install the heat shield.

15. Attach the coolant tube to the cylinder head.

16. Fill the engine with coolant.

17. Connect the negative battery cable.

Front Crankshaft Seal

REMOVAL & INSTALLATION

4-Cylinder Engine

1. Place the lock carrier into the service position as follows:

   a. Remove the front bumper.

   b. Tag and remove any wiring or connector that would inhibit locking the carrier.

   c. Remove the three quick-release screws on the front noise insulation panel.

   d. Unbolt the air guide between the lock carrier and the air filter.

   e. If installed, remove the retaining clamps for the wiring harness at the left side of the radiator frame.

   f. Remove the No. 2 bolts and install Support tool 3369 or equivalent.

   g. Remove the remaining bolts and pull the lock carrier out to the stop.

   h. To secure the lock carrier, install the appropriate M6 bolts into the rear of the lock carrier and fender.

2. Turn the ignition switch to the **OFF** position, then disconnect the negative battery cable.

3. Remove the accessory drive belts.

4. Remove the timing belt.

5. Remove the torque arm mounting bracket.

6. Remove the crankshaft timing belt gear retaining bolt.

7. Using the appropriate gear puller, remove the crankshaft timing belt gear.

8. Wrap a shop rag around the end of the crankshaft, then use a seal puller to remove the crankshaft oil seal.

9. Remove the shop rag, then the oil seal.

To install:

10. Lightly oil the lip of the seal.

11. Install the oil seal over the end of the crankshaft.

12. Using the appropriate seal driver, install the seal flush.

13. Clean the crankshaft of oil, then install the timing belt gear and tighten the new retaining bolt to 66 ft. lbs. (90 Nm) plus ¼ turn (90°).

14. Install the torque arm mounting bracket and tighten the bolts to 18 ft. lbs. (25 Nm).

15. Install the timing belt, then the accessory drive belts.

16. Turn the ignition switch **OFF**, then reconnect the negative battery cable.

17. Fully close all power windows to stop, operate all window switches for at least one second in the close direction to activate the one-touch opening/closing function

18. Check the oil level before starting the engine

19. Set the clock to the correct time.

➡**Diagnostic Trouble Codes (DTCs) are stored when harness connectors are detached.**

20. Read the DTCs and clear the fault codes.

5-Cylinder Engine

The oil seal is a part of the oil pump.

1. Disconnect the negative battery cable. Remove the timing belt.

2. Using a small prybar, pry the oil seal from the oil pump.

3. Clean out the seal seat. Using a new seal, lubricate the lip with engine oil. Using a suitable socket, drive the seal in to the seal seat.

➡**When installing a new seal, be careful not to damage the lip of the seal.**

4. Reinstall the timing belt. Tighten the crankshaft pulley bolt to 253 ft. lbs. (343 Nm).

6-Cylinder Engine

1. Remove the timing belt.

2. Remove the timing belt sprocket from the crankshaft.

3. Remove the seal with Seal remover 3203 or equivalent.

4. Clean the running and sealing surfaces.

To install:

5. Slide the seal over the Installing Sleeve 3202 or equivalent.

6. Press the oil seal flush with Seal Installer 3265, or equivalent, and the center crankshaft bolt.

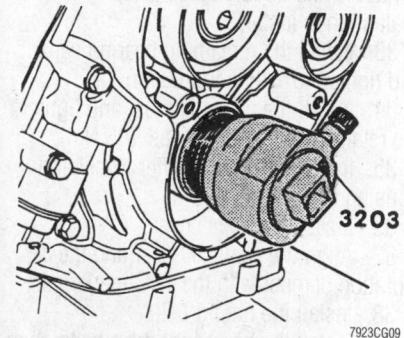

Removing the seal using the seal remover—6-cylinder engine

7923CG09

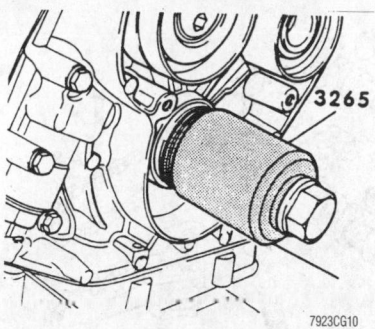

7923CG10

Installing the seal using the seal installer and the crankshaft center bolt—6-cylinder engine

7. Install the timing belt sprocket and tighten the center crankshaft bolt to 148 ft. lbs. (200 Nm), plus ½ turn.

8. Install the timing belt.

Camshaft

REMOVAL & INSTALLATION

4-Cylinder Engine

1. Place the lock carrier into the service position as follows:

a. Remove the front bumper.

b. Tag and remove any wiring or connector that would inhibit locking the carrier.

c. Remove the three quick-release screws on the front noise insulation panel.

d. Unbolt the air guide between the lock carrier and the air filter.

e. If installed, remove the retaining clamps for the wiring harness at the left side of the radiator frame.

f. Remove the No. 2 bolts and install Support tool 3369 or equivalent.

g. Remove the remaining bolts and pull the lock carrier out to the stop.

h. To secure the lock carrier, install the appropriate M6 bolts into the rear of the lock carrier and fender.

2. Turn the ignition switch to the **OFF** position, then disconnect the negative battery cable.

3. Remove the accessory drive belts.

4. Remove the engine covers.

5. Remove timing belt upper cover.

6. Turn the crankshaft, in the direction of rotation (clockwise), until the No. 1 cylinder is at Top Dead Center (TDC).

7. Using Torx® wrench T45, loosen the timing belt tensioner.

8. Push down on the tensioner, and remove the belt from the camshaft gear.

9. Remove the Torx® bolt and swing the tensioner assembly bracket forward.

10. Remove the valve cover.

11. Using Retainer tool 3036 or equivalent, loosen the cam gear retaining bolt.

12. Remove the camshaft gear.

13. Remove the housing for Camshaft Position (CMP) sensor and shutter wheel.

14. Secure the hydraulic chain tensioner with Bracket-Tensioner 3366 or equivalent.

15. Verify that the camshafts are at Top Dead Center (TDC) for the No. 1 cylinder. Both camshaft markings must align with arrows on the bearing caps.

16. Clean the drive chain and the cam chain gears opposite both arrows on the bearing caps. Matchmark the installed position using paint.

➡**The distance between the two arrows/paint marks is equivalent to 16 drive chain rollers, and the notch on the exhaust camshaft is slightly offset inward toward the drive chain roller.**

17. Remove the bearing caps 3 and 5 from the intake and exhaust camshafts.

18. Remove the double bearing cap.

19. Remove both bearing caps from the chain gears on the intake and exhaust camshafts.

20. Remove the hydraulic chain tensioner retaining bolts.

21. In an alternating and diagonal sequence, loosen the bearing caps 2 and 4 of the intake and exhaust manifold, then remove.

22. Remove the camshafts with the hydraulic chain tensioner.

To install:

✳✳ CAUTION

After installing the lifters or the camshaft(s), the engine must NOT be started for at least 30 minutes. Otherwise the valves could strike the pistons. Rotate the engine by hand, at least two revolutions, to ensure that the valves do not strike the pistons.

23. Replace the rubber/metal chain tensioner gasket and apply sealant to the hatched area, as shown.

24. Install the drive chain on the camshaft as follows:

a. If installing the old chain, align the paint marks with the camshaft marks.

b. If installing a new chain, the distance between the notches A and B on the camshafts must equal the distance between 16 drive chain rollers.

25. Slide the hydraulic chain tensioner between the drive chain.

26. Install the camshafts with the chain tensioner into the cylinder head.

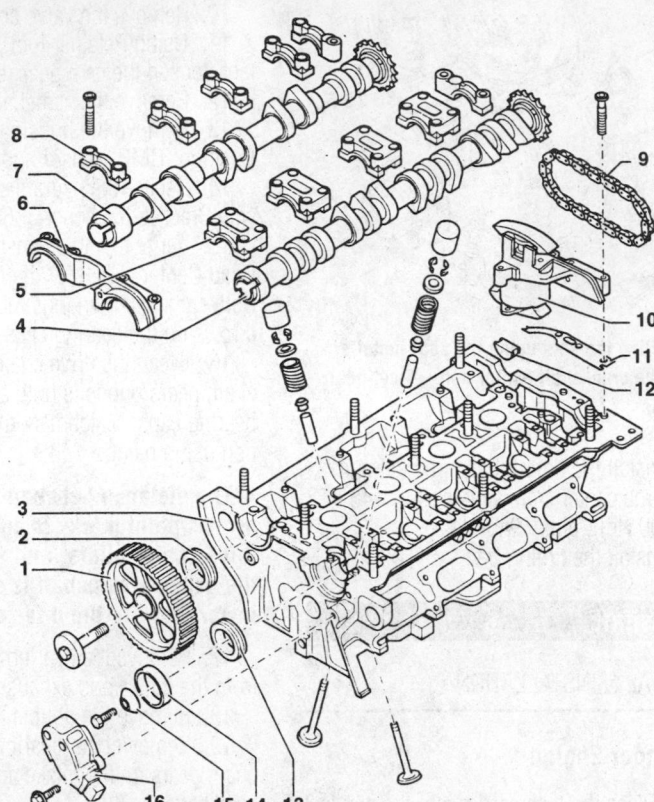

1. Camshaft gear
2. Oil seal
3. Cylinder head
4. Intake camshaft
5. Intake camshaft bearing cap
6. Double bearing cap
7. Exhaust camshaft
8. Exhaust camshaft bearing cap
9. Drive chain
10. Hydraulic chain tensioner
11. Rubber/metal seal
12. Gasket
13. Oil seal
14. Shutter wheel for the CMP
15. Washer
16. CMP sensor housing

Exploded view of the camshaft mounting and related components—4-cylinder engine

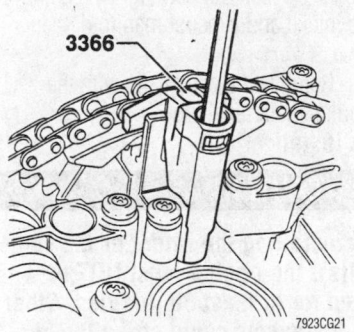

Be sure not to overtighten the chain tensioner (3366), it can be damaged—4-cylinder engine

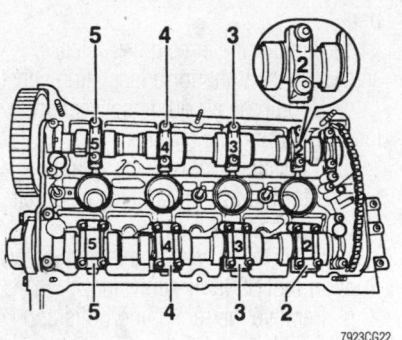

Camshaft bearing cap identification—4-cylinder engine

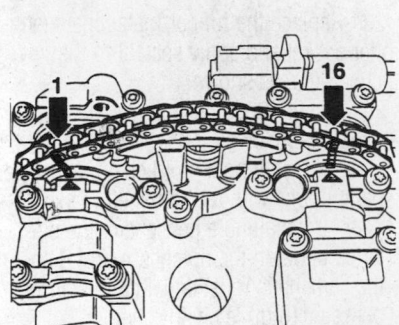

To ensure proper installation, matchmark the chain-to-camshaft position—4-cylinder engine

To ensure a proper seal, be sure to apply sealant to the hatched area—4-cylinder engine

27. Oil the camshaft contact surfaces.

➡ When installing the bearing caps, verify the markings on the caps are readable from the intake side of the cylinder head.

28. Tighten the bearing caps 2 and 4 of the intake and exhaust camshafts in an alternating diagonal sequence to 7 ft. lbs. (10 Nm).

29. Install both bearing caps on the chain sprockets of the intake and exhaust camshafts and tighten to 7 ft. lbs. (10 Nm).

30. Verify the correct positions of the camshafts.

31. Remove the bracket-tensioner 3366.

32. Lightly coat the cylinder head mating surface of the double bearing cap with sealant, then install.

33. Install the remaining bearing caps and tighten to 7 ft. lbs. (10 Nm).

34. Install the camshaft gear and tighten the retaining bolt to 48 ft. lbs. (65 Nm).

35. Install the CMP shutter wheel and housing cover.

36. Install the valve cover.

37. Align the camshaft gear and the vibration damper with the TDC markings.

38. Install the timing belt.

39. Install the accessory drive belts, then the engine cover.

40. Reinstall the lock carrier.
41. Connect the negative battery cable.
42. Fully close all power windows to stop, operate all window switches for at least one second in the close direction to activate the one-touch opening/closing function
43. Check the oil level before starting the engine
44. Set the clock to the correct time.

➡**Diagnostic Trouble Codes (DTCs) are stored when harness connectors are detached.**

45. Read the DTCs and clear the fault codes.
46. Adjust the headlights.

5-Cylinder Engines

1. Disconnect the negative battery cable. Remove the upper drive belt cover, valve cover and upper part of intake manifold, if necessary.
2. Using the large bolt on the crankshaft sprocket, rotate the engine until the No. 1 cylinder is at TDC of the compression stroke. Align the TDC mark **0** with the cast mark on the bell housing. If the belt hasn't jumped, the timing mark on the rear face of the camshaft sprocket should be aligned with the upper left edge of the valve cover.
3. Remove the timing belt from the camshaft sprocket. Remove the camshaft sprocket.
4. Diagonally loosen bearing caps No. 2 and 4 and remove the bearing caps. Bearing caps are marked on the top.
5. Diagonally loosen bearing caps No. 1 and 3 and remove the bearing caps.
6. Lift the camshaft out of the cylinder head.
    To install:

❊❊ CAUTION

After installing the lifters or the camshaft(s), the engine must NOT be started for at least 30 minutes. Otherwise the valves could strike the pistons. Rotate the engine by hand, at least two revolutions, to ensure that the valves do not strike the pistons.

7. When installing, lightly oil the camshaft and bearing journals with clean engine oil.
8. Position the caps on the same journals from which they were removed.
9. Tighten the nuts of caps 2 and 4 until snug.
10. Tighten all nuts to 15 ft. lbs. (20 Nm).

11. Install the camshaft sprocket and timing belt. Install the valve cover. The camshaft sprocket bolt is tightened to 58 ft. lbs. (79 Nm).

6-Cylinder Engine

1. Remove the timing belt.
2. Remove the valve cover(s).
3. On the left cylinder head, remove the Camshaft Position Sensor (CPS).
4. On the right cylinder head, remove the plug/cover on the head.
5. Remove the camshaft timing belt sprocket.
6. Identify the bearing caps.

➡**DO NOT allow the bearing caps to become mixed up.**

7. Remove the camshaft bearing caps 2 and 3.
8. Gradually and evenly, loosen the nuts for the camshaft bearing caps 1 and 4, in a diagonal sequence.
9. Remove the cam and lift out the valve lifter. If it is to be reused, it must go in the bore from which it was removed.
    To install:
10. Install the lifters into their respective bore.
11. Install bearing caps 1 and 4 in a alternating and diagonal sequence.
12. Install bearing caps 2 and 3, then tighten all bearing caps to 15 ft. lbs. (20 Nm).

❊❊ CAUTION

After installing the lifters or the camshaft(s), the engine must NOT be started for at least 30 minutes. Otherwise the valves could strike the pistons. Rotate the engine by hand, at least two revolutions, to ensure that the valves do not strike the pistons.

13. Install the camshaft timing belt sprocket and tighten the mounting bolt to 52 ft. lbs. (71 Nm).

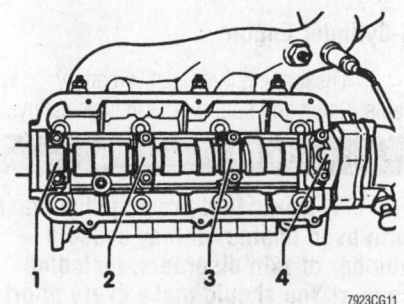

Camshaft bearing cap identification—6-cylinder engine

14. On the right cylinder head, install the plug/cover on the head.
15. On the left cylinder head, install the Camshaft Position Sensor (CPS) and tighten the mounting bolts to 89 inch lbs. (10 Nm).
16. Install the valve cover.
17. Install the timing belt.

Valve Lash

ADJUSTMENT

Audi engine's are equipped with hydraulic lash adjusters. No adjustment is necessary.

Oil Pan

REMOVAL & INSTALLATION

4-Cylinder Engine

1. Disconnect the negative battery cable.
2. Raise and support the vehicle safely.
3. Remove the engine undercover

❊❊ CAUTION

The EPA warns that prolonged contact with used engine oil may cause a number of skin disorders, including cancer! You should make every effort to minimize your exposure to used engine oil. Protective gloves should be worn when changing the oil. Wash your hands and any other exposed skin areas as soon as possible after exposure to used engine oil. Soap and water, or waterless hand cleaner should be used.

4. Drain the engine oil into a suitable container.
5. Remove the accessory drive belts and the A/C belt tension pulley.
6. Remove the torque support stop and side brace.
7. Disconnect the starter wiring.
8. Remove the hose from the turbocharger at the air guide tube in the lock carrier.
9. Remove the bottom nuts from the lower engine mount.
10. Remove the top engine cover.
11. Set the Engine Support Bridge 10–222A or equivalent across the fender mounting edges.
12. Install the Engine Sling 2024A or equivalent, and lift the engine up as far as possible.

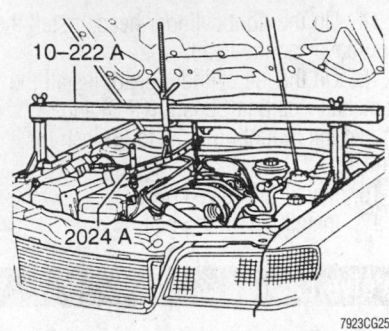

The engine must be supported, because the subframe mounting bolts must be loosened—4-cylinder engine

13. Use an engine hoist under the vehicle to support the subframe.

14. Remove the front bolts 2 and 3, then bolt 1 from the subframe.

15. Slowly lower the subframe with the engine hoist.

16. If equipped with a manual transaxle, loosen the left transaxle mount nut until it is aligned with the lower edge of the bolt (approx. 4 turns).

17. If equipped with an automatic transaxle, loosen the rear bolt for the left transaxle mount several turns, then remove the front bolt for the transaxle mount.

18. Loosen the rear bolt for the right transaxle mount several turns, then remove the front bolt for the right transaxle mount.

➡**If equipped with a manual transaxle, both of the rear bolts on the oil pan can be accessed through the opening on the flywheel. Turn the flywheel as needed.**

19. Unbolt the oil pan, then remove. It may be necessary to tap lightly on the pan using a rubber mallet to loosen it from the engine block.

20. Clean all gasket mating surfaces.

To install:

21. Apply sealant to the front and rear contact areas of the oil seal carriers.

For vehicles equipped with a manual transaxle, align the flywheel as shown to remove the rear oil pan bolts—4-cylinder engine

22. Install the oil pan and tighten the oil pan-to-engine block mounting bolts to 44 inch lbs. (5 Nm).

23. Tighten the bolts between the oil pan and the transmission to 33 ft. lbs. (45 Nm).

24. Tighten the M10 bolts between the oil pan and engine block to 33 ft. lbs. (45 Nm).

25. Tighten the M6 bolts between the oil pan and engine block to 7 ft. lbs. (10 Nm).

26. Tighten the subframe bolts/nuts using the illustration as follows:

• Tighten bolts 2 and 5 to 81 ft. lbs. (110 Nm) plus ¼ turn (90°)

• Tighten bolts 6 to 55 ft. lbs. (75 Nm)

• Tighten bolts 1 to 17 ft. lbs. (23 Nm)

• Tighten nuts 3 and 4 to 30 ft. lbs. (40 Nm)

27. Tighten the transaxle-to-subframe nuts to 17 ft. lbs. (23 Nm).

28. Lower the engine into position and tighten the engine mount-to-subframe nuts to 18 ft. lbs. (25 Nm).

29. Install the turbocharger air hose.

30. Connect the starter wiring.

31. Install the torque support stop and brace, then tighten the mounting hardware to 18 ft. lbs. (25 Nm).

32. Install the A/C belt tensioner, then the accessory drive belts.

✳ WARNING

Operating the engine without the proper amount and type of engine oil will result in severe engine damage.

33. Fill engine with oil and check the level.

34. Connect the negative battery cable.

35. Fully close all power windows to stop, operate all window switches for at least one second in the close direction to activate the one-touch opening/closing function. Reset the clock time.

36. Start the vehicle and check for leaks, then recheck the engine oil level.

37. Install the engine cover and undercover.

5-Cylinder Engine

1. Disconnect the negative battery cable. Raise and support the vehicle safely.

✳ CAUTION

The EPA warns that prolonged contact with used engine oil may cause a number of skin disorders, including cancer! You should make every effort to minimize your exposure to used engine oil. Protective gloves should be worn when changing the oil. Wash your hands and any other exposed skin areas as soon as possible after exposure to used engine oil. Soap and water, or waterless hand cleaner should be used.

2. Drain the oil from the crankcase. Remove the cover plate from under the engine, if equipped.

3. If necessary, remove the 4 bolts from the subframe and lower the subframe. Remove the oil pan bolts while supporting the pan.

4. Lower the pan from the engine. Discard the gasket. Note that some engine uses a 2 piece oil pan as well as a honeycomb baffle insert. Both an upper and lower pan gasket will be required.

To install:

5. Coat both sides of a new gasket with sealer and install the gasket and oil pan.

6. Tighten the pan bolts to 15 ft. lbs. (20 Nm).

7. If removed, install the 4 subframe bolts and tighten to 81 ft. lbs. (110 Nm).

8. Fill the crankcase with oil.

9. If equipped, install the engine under cover.

10. Connect the negative battery cable.

6-Cylinder Engine

1. Raise and safely support the vehicle.

✳ CAUTION

The EPA warns that prolonged contact with used engine oil may cause a number of skin disorders, including cancer! You should make every effort to minimize your exposure to used engine oil. Protective gloves should be worn when changing the oil. Wash your hands and any other exposed skin areas as soon as possible after exposure to used engine oil. Soap and water, or waterless hand cleaner should be used.

2. Drain the engine oil.

3. Loosen and remove the bolts retaining the oil pan.

4. Lower the pan from the engine.

To install:

5. Be sure the gasket surface is flat and install the pan with a new gasket.

6. Tighten the retaining bolts in a crisscross pattern to 11 ft. lbs. (15 Nm).

✳ WARNING

Operating the engine without the proper amount and type of engine oil will result in severe engine damage.

7. Refill the engine with oil. Start the engine and check for leaks.

Oil Pump

REMOVAL & INSTALLATION

4-Cylinder Engine

1. Remove the oil pan.
2. Unhinge the baffle plate, then remove.
3. Remove the two oil pump-to-engine mounting bolts.
4. Press down on the subframe and remove the bolts.

To install:

5. Press down on the subframe and install the oil pump.
6. Install the oil pump-to-engine mounting bolts and tighten to 18 ft. lbs. (25 Nm).
7. Install the oil pan.

5-cylinder Engine

1. Disconnect the negative battery cable. Loosen and remove the crankshaft pulley bolt.
2. Remove the timing belt covers.
3. Loosen the water pump bolts and turn the pump body clockwise.
4. Remove the timing belt and V-belt pulley with the timing belt sprocket.

5. Remove the dipstick. Raise and safely support vehicle. Drain the engine oil.
6. Remove the front bolts on the subframe and remove the oil pan.
7. Remove the oil suction pipe from the base of the oil pump and bracket to the engine block.
8. Remove the oil pump bolts and remove the oil pump from the front of the engine.
9. Installation is the reverse of the removal procedure. Refill with oil to correct level.

6-Cylinder Engine

The oil pump is part of the engine front cover, and the cooling system does not have to be opened during this procedure.

1. Remove the timing belt.
2. Remove the engine under cover and drain the engine oil.
3. Remove the engine oil dipstick.
4. Unscrew the engine oil dipstick tube, then remove the tube from the crankcase.
5. Remove the crankshaft vibration damper and sprocket.
6. Unbolt the timing belt idler and tensioner pulleys.
7. Label and detach any wiring, hoses, lines and cables that interfere with oil pump removal.

8. Remove the oil filter, then disconnect the oil cooler line from the filter housing.
9. Slowly, remove the front bolts for the engine support, being prepared for the engine support to drop ⅜ inch.
10. Remove the starter bolts.
11. Remove the upper and lower oil pans, then the oil suction tube.
12. Unbolt, then remove the oil pump.

To install:

13. Clean the gasket mating surfaces for the oil pan and pump.
14. Install the oil pump and verify the oil pump drive is properly engaged.
15. Install the upper and lower oil pans, then the starter bolts.
16. Install the engine support bolts.
17. Install the oil cooler line to the oil filter housing, then a new oil filter.
18. Attach any wiring, hoses, lines and cables that were removed.
19. Install the timing belt idler pulley and tighten the mounting bolt to 15 ft. lbs. (20 Nm), then the tensioner pulleys and tighten after adjusting the timing belt.
20. Install the crankshaft sprocket and vibration damper, then tighten the center bolt to 148 ft. lbs. (200 Nm), plus ½ turn.
21. Screw the dipstick tube into the crankcase, and install the engine oil dipstick.

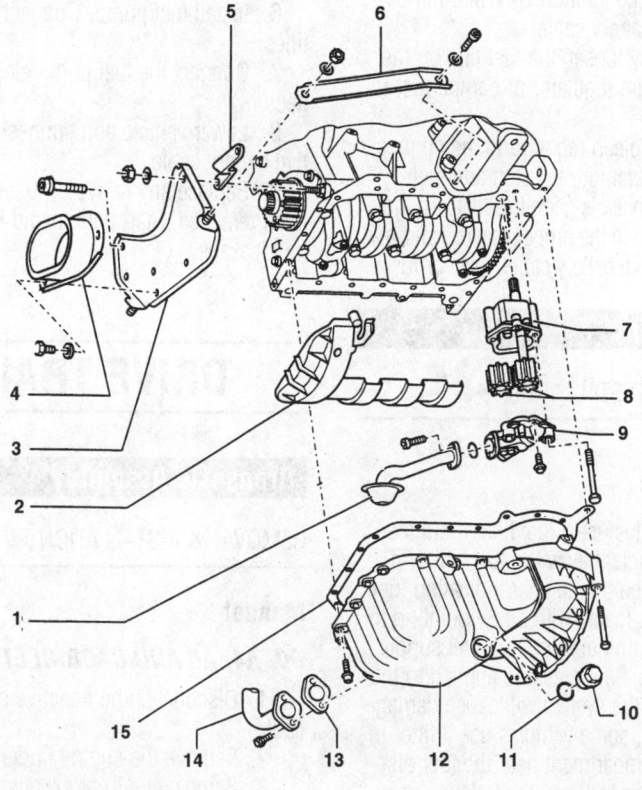

1. Suction pipe
2. Baffle plate
3. Bracket
4. Stop for torque support
5. Brace
6. Side brace
7. Oil pump housing
8. Gears
9. Oil pump cover with pressure relief valve
10. Oil drain plug
11. Sealing washer
12. Oil pan
13. Gasket
14. Oil return line
15. Gasket

Exploded view of the oil pan and pump—4-cylinder engine

7923CG27

22. Install the engine under cover and fill the engine with oil.

23. Install the timing belt.

Rear Main Seal

REMOVAL & INSTALLATION

All Engines

The rear main oil seal is located at the rear of the engine block. It can be found in a housing or flange behind the flywheel/flexplate. To replace the seal, remove the transaxle or pull the engine.

1. Disconnect the negative battery cable. Remove the transaxle.

2. Remove the flywheel/flexplate.

3. Using a suitable tool, pry the old seal out of its housing.

4. To install, lightly oil the replacement seal and press it into place.

➡**Be careful not to damage the seal or score the crankshaft.**

5. Install the flywheel/flexplate and the transaxle.

FUEL SYSTEM

Fuel System Service Precautions

Safety is the most important factor when performing not only fuel system maintenance but any type of maintenance. Failure to conduct maintenance and repairs in a safe manner may result in serious personal injury or death. Maintenance and testing of the vehicle's fuel system components can be accomplished safely and effectively by adhering to the following rules and guidelines.

• To avoid the possibility of fire and personal injury, always disconnect the negative battery cable unless the repair or test procedure requires that battery voltage be applied.

• Always relieve the fuel system pressure prior to disconnecting any fuel system component (injector, fuel rail, pressure regulator, etc.), fitting or fuel line connection. Exercise extreme caution whenever relieving fuel system pressure to avoid exposing skin, face and eyes to fuel spray. Please be advised that fuel under pressure may penetrate the skin or any part of the body that it contacts.

• Always place a shop towel or cloth around the fitting or connection prior to loosening to absorb any excess fuel due to spillage. Ensure that all fuel spillage is

quickly removed from engine surfaces. Ensure that all fuel soaked cloths or towels are deposited into a suitable waste container.

• Always keep a dry chemical (Class B) fire extinguisher near the work area.

• Do not allow fuel spray or fuel vapors to come into contact with a spark or open flame.

• Always use a back-up wrench when loosening and tightening fuel line connection fittings. This will prevent unnecessary stress and torsion to fuel line piping. Always follow the proper torque specifications.

• Always replace worn fuel fitting O-rings with new. Do not substitute fuel hose or equivalent where fuel pipe is installed.

Fuel System Pressure

RELIEVING

The fuel injection system operates under high pressure. This makes it necessary to first relieve the system of pressure before servicing. The pressurized fuel, when released, may ignite or cause personal injury.

1. Disconnect the power to the fuel pump by removing the relay or the fuel pump fuse. Check the list on the fuse box lid to be sure. The fuse can be removed to stop the fuel pump from running. With the engine operating at idle, wait until the engine stalls from fuel starvation.

2. Switch the ignition **OFF** and remove the negative battery cable.

3. Carefully loosen the fuel line on the control pressure regulator or component to be serviced.

4. Wrap a clean rag around the connection, while loosening, to catch any fuel.

5. After service is complete, discard the fuel soaked rag in the proper manner and reconnect negative battery cable, relay or fuses.

Fuel Filter

REMOVAL & INSTALLATION

All Vehicles

Most vehicles use a fuel filter mounted under the vehicle, below the fuel tank. An arrow should be on the filter indicating fuel flow direction. Install with arrow pointing to engine. Use care not to mix up fuel supply or return lines. Fuel pressure applied to the return side of the system will cause damage.

In addition, some vehicles use a filter in the engine compartment near the fuel distributor. If equipped, use the following procedure:

1. Make certain to follow precautions and relieve fuel pressure.

2. Disconnect the fuel lines leading into and out of the fuel filter.

3. Unscrew the filter retaining bracket and remove the filter.

4. Install a new filter in the bracket and reattach the bracket. Be sure the arrows are pointing in the direction of the fuel flow.

5. Reconnect the fuel lines, start the engine and check for leaks.

Fuel Pump

REMOVAL & INSTALLATION

All Vehicles

The fuel pump is located under the vehicle on a bracket in front of the fuel tank. The fuel pump assembly is located on the right side of front wheel drive vehicles and on the left side on Quattro vehicles.

1. Make certain to follow precautions and relieve fuel pressure.

2. Disconnect the negative battery cable.

3. Raise and safely support vehicle.

4. Carefully loosen fuel line at fuel pump. Catch excess fuel in a container.

5. Remove fuel pump electrical connectors and remove the fuel pump.

To install:

6. Install fuel pump. Connect the fuel lines.

7. Connect the fuel pump electrical connectors.

8. Lower vehicle and connect the negative battery cable.

9. Replace any relays or fuses, that had been removed. Start engine and inspect for fuel leakage.

DRIVE TRAIN

Transaxle Assembly

REMOVAL & INSTALLATION

Manual

90, A4, A6 AND CABRIOLET MODELS

1. Disconnect the negative battery cable.

2. Remove the engine undercover.

3. Remove the front exhaust pipe with the catalytic converter.

4. On All-Wheel drive models, match-mark and remove the driveshaft.

5. Disconnect and remove the starter.

6. Remove the securing bolt for the shift rod and joint at the transaxle, then separate from the rear of the shift rod.

7. Disconnect the shift rod.

8. Unplug the connector for the Vehicle Speed Sensor (VSS) and back-up light switch.

9. Using a transmission/transaxle jack, support the transaxle.

10. Remove the transaxle mount heat shield.

11. Remove the right mount at the transaxle.

12. Remove the left mount with the bushings.

13. Disconnect the left and right half-shafts and heat shield.

14. Remove the remaining engine-to-transaxle mounting bolts.

15. Pry the transaxle off the dowel sleeves and carefully lower the transaxle until the slave cylinder is accessible (approx. 6 in. (15 cm)).

16. Remove the clutch slave cylinder from the transaxle, leaving the hydraulic line attached.

17. Carefully lower transaxle from the vehicle.

To install:

18. Clean and lubricate, with grease, the splines of the input shaft and clutch hub.

19. Clean the threaded hole for attaching the slave cylinder and the shift lever with a threaded tap to remove any residual thread locking compound.

20. Raise the transaxle into the vehicle.

21. Install the slave cylinder and tighten the mounting bolts to 18 ft. lbs. (25 Nm).

22. Be sure to align the dowel pins, then install the transaxle to the engine.

23. Install the engine-to-transaxle mounting bolts. Tighten the M12 bolts to 48 ft. lbs. (65 Nm), the M10 bolts to 33 ft. lbs. (45 Nm), and the M8 bolts to 18 ft. lbs. (25 Nm).

24. Install the left and right halfshafts and heat shield.

25. Install the transaxle mounts and tighten the mounting bolts to 30 ft. lbs. (40 Nm).

26. Install the transaxle heat shield.

27. Remove the transmission/transaxle jack.

28. Connect the VSS and the back-up light switch.

29. Connect the shift rod and tighten to 15 ft. lbs. (20 Nm).

30. Install the starter.

31. On All-Wheel drive models, align the matchmarks and install the driveshaft.

32. Connect the front exhaust pipe.

33. Install the engine undercover.

34. Connect the negative battery cable.

S6 MODELS

1. Disconnect the negative battery cable.

2. Remove the upper engine to transaxle bolts.

3. Remove the connector for the speedometer sender by pressing in the clips.

4. Remove the clip from the clutch slave cylinder and drive out spring pin, if equipped. Remove the bolt securing the clutch slave cylinder to the transaxle and remove the cylinder. Leave the hydraulic line connected.

5. Support the engine. Tie up coolant hoses and cables, as needed.

6. Remove the right side guard plate.

7. Disconnect the halfshafts from the flanges and rest both halfshafts on top of the subframe.

8. Tag and disconnect the wire from the back-up light switch. Disconnect vacuum hoses at the servo if so equipped.

9. Pry off the shift and adjusting rods.

10. Remove the lower engine-to-transaxle bolts.

11. Remove the starter.

12. Remove the guard plate from the subframe.

13. With suitable jack, lift transaxle slightly.

14. Remove both rear subframe mounting bolts.

15. Remove both transaxle support bolts from the subframe.

16. Remove the bracket from the transaxle, push tension system cable and bracket off the retainer on transaxle. The retainer can only be removed with the transaxle out of the vehicle.

17. Remove the right side transaxle bracket.

18. Pull transaxle off dowel sleeves.

19. Lower transaxle and take out from below.

To install:

20. Installation is the reverse of the removal procedure. Before installing the transaxle, rest both halfshafts on top of the subframe.

21. Lubricate mainshaft splines.

22. Install transaxle onto dowels and install the lower bolts.

23. Install the tensioning system bracket and cable to the transaxle.

24. Tighten the transaxle bracket and subframe upper bolts to 29 ft. lbs. (39 Nm).

25. Check alignment of transaxle and tighten transaxle-to-engine bolts to 40 ft. lbs. (54 Nm).

26. Tighten subframe-to-body bolts to 80 ft. lbs. (108 Nm).

27. Tighten halfshaft-to-drive flange bolts to 58 ft. lbs. (79 Nm).

Automatic

1. Disconnect the negative battery cable.

2. Remove the upper engine-to-transaxle bolts. Raise and support the vehicle safely.

3. Using a suitable engine support tool, secure it to the engine and the vehicle.

4. At the front of the engine, remove both top bolts. Remove the starter.

5. Through the starter opening, remove the torque converter to drive plate bolts and remove torque converter cover plate.

6. Clamp off the coolant hoses at the ATF cooler and remove the hoses from the cooler.

7. Remove the speedometer cable from the transaxle.

8. Remove the inner halfshaft-to-transaxle bolts. Using a wire, tie up the halfshafts.

9. At the left control arm, mark the position of the ball joint and remove the ball joint and the support.

10. Place an oil catch pan under the transaxle, remove the oil filler tube from the oil pan and drain the fluid.

11. Remove the exhaust pipe-to-transaxle bracket.

12. Remove the selector cable bracket from the transaxle. At the transaxle shift lever, remove the selector cable circlip and the cable.

13. At the transaxle, remove the accelerator cable bracket and the cable from the operating lever.

14. Remove the center bolt, from the transaxle mount. Using the engine support tool, lift the engine slightly.

15. Remove the throttle cable bracket bolts and the bracket.

16. Support the transaxle and lift it slightly. Remove the lower transaxle-to-engine bolts.

17. Separate the engine from the transaxle and lower it from the vehicle. Be sure to secure the torque converter.

To install:

18. When installing the transaxle, should the torque converter slip off the one-way clutch support, the oil pump shaft could be pulled from the oil pump. This may cause severe damage when bolting the transaxle to the engine. Be sure the torque converter is properly positioned before installing the bolts.

19. Tighten the engine-to-transaxle bolts to 41 ft. lbs. (56 Nm), the subframe bolts to 52 ft. lbs. (71 Nm) and the transaxle mount center bolt to 30 ft. lbs. (40 Nm).

20. Install the torque converter-to-drive-plate bolts and tighten to 22 ft. lbs. (30 Nm). Install the cover plate.

21. Install the halfshaft-to-transaxle bolts and tighten to 33 ft. lbs. (45 Nm).

22. Connect the ball joint to the control arm and tighten the bolts to 48 ft. lbs. (65 Nm).

23. Install the exhaust system. When installing the exhaust, tighten the clamps after making sure everything is properly positioned to minimize vibration.

24. Connect the hoses to the oil cooler. Install the oil filler tube and refill the transaxle.

25. Connect and adjust the selector cable as required.

26. Connect and adjust the accelerator linkage and align the engine-to-transaxle mounts, if necessary.

Clutch

ADJUSTMENT

All vehicles use a hydraulic clutch release mechanism. No free-play adjustment is required or possible. If the clutch does not release or engage properly, try bleeding the system before moving on to more extensive repairs.

REMOVAL & INSTALLATION

1. Disconnect the negative battery cable. Raise and safely support the vehicle and remove the transaxle.

2. If the pressure plate is to be reused, mark its relationship to the flywheel.

3. Using a suitable tool, lock the flywheel. Unbolt the pressure plate from the flywheel, loosening the bolts alternately, a little at a time, to prevent warpage.

To install:

4. Install the clutch with the driven plate on the pressure plate so the spring cage is facing the pressure plate.

5. Hold the clutch assembly against the flywheel, aligning the marks made in Step 2 and the dowel pins on the flywheel with the pressure plate. Insert an alignment shaft tool through the pressure plate and the driven plate into the crankshaft pilot bearing.

6. Install the pressure plate bolts finger-tight. Tighten the bolts evenly, in a diagonal pattern, to avoid distortion. Tighten the bolts to 18 ft. lbs. (24 Nm). Remove the alignment shaft.

7. The clutch release bearing in the front of the transaxle should be checked before reassembly. It is retained by 2 springs.

8. Replace the transaxle.

Hydraulic Clutch System

BLEEDING

The clutch system should be bled using a pressure bleeder. Follow the instructions that come with the bleeder tank, for the proper bleeding procedure. The maximum line pressure must not exceed 36 psi (248 kPa).

Halfshaft

REMOVAL & INSTALLATION

Front

A6, S6 AND CABRIOLET MODELS

➡**When loosening or tightening axle nuts, be sure the vehicle is on the ground. Axle nut torque is high enough that attempting to loosen it may cause the vehicle to fall off the support.**

1. Remove the halfshaft end bolt.

2. Raise and support the vehicle safely and remove the wheels. If equipped with ABS, slide the speed sensor partly out of its mount.

3. Disconnect the halfshaft from the transaxle flange.

4. Press the halfshaft upward toward the front of the vehicle.

5. Turn the steering to full lock and remove the halfshaft.

To install:

6. If equipped, replace the inner CV joint gasket.

7. Slide the halfshaft into the wheel hub.

8. Attach the halfshaft-to-transaxle flange, and tighten to 33 ft. lbs. (45 Nm) for the M8 bolts and 59 ft. lbs. (80 Nm) for the M10 bolts.

9. If equipped with ABS, install the speed sensor.

10. With the wheels installed and the vehicle on the ground, tighten the axle nut to 148 ft. lbs. (200 Nm) plus an additional ¼ turn.

90 MODELS

➡**When loosening or tightening axle nut or bolt, be sure the vehicle is on the ground. Axle nut torque is high enough that attempting to loosen it may cause the vehicle to fall off the support.**

1. Loosen the axle nut or bolt. Raise and safely support the vehicle.

2. Unbolt and remove the halfshaft-to-transaxle drive flange bolts.

3. Mark the position of the ball joint on the control arm, remove the 2 retaining nuts and disconnect the ball joint.

4. Remove the ball joint-to-steering knuckle bolt and separate the knuckle from the ball joint. Remove the mounting bolts for the control arm/stabilizer and push control arm downward, if necessary.

5. Pivot the strut outward and remove the halfshaft.

To install:

6. When installing the right halfshaft, take care not to damage the boot on the cover plate.

7. Tighten the ball joint-to-control arm/knuckle nuts/bolt to 47 ft. lbs. (64 Nm). Tighten the halfshaft flange bolts to 33 ft. lbs. (45 Nm).

8. Install the wheel and snug the axle nut or bolt. Place the vehicle on the ground and tighten the axle nut or bolt to 200 ft. lbs. (270 Nm).

9. Check and adjust wheel alignment when finished.

A4 MODELS

➡**When loosening or tightening axle nut or bolt, be sure the vehicle is on the ground. Axle nut torque is high enough that attempting to loosen it may cause the vehicle to fall off the support.**

1. Remove the hub cap or center cap.

2. Loosen the hex collar bolt.

3. Raise and safely support the vehicle, then remove the front wheels.

4. Remove the halfshaft-to-transaxle flange bolts, then the hex collar bolt.

5. Remove the ABS wheel speed sensor cable from the brake caliper bracket.

6. Slide the ABS speed sensor partly out of its mount.

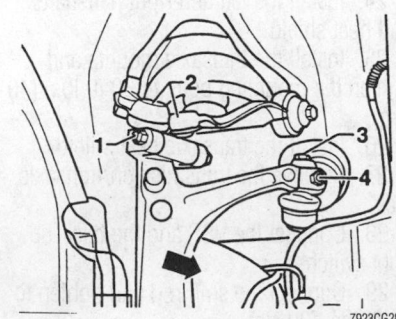

Loosen nut (1), remove the hex bolt, and pull both arms (2) upward and out—A4 models

7. Remove nut/bolt No. 1, as shown, then pull both arms up and out of the swing arm.

➡**The slots in the swing arm must not be widened. Do not loosen the bolts No. 3 and 4, otherwise the axle geometry must be checked.**

8. Tilt the swing arm out and to the rear of the vehicle, then remove the halfshaft.

To install:

9. Install the halfshaft into the wheel hub.

10. Install the swing arm bolt and tighten to 30 ft. lbs. (40 Nm).

11. Attach the halfshaft-to-transaxle flange, and tighten to 30 ft. lbs. (40 Nm) for the M8 bolts and 57 ft. lbs. (77 Nm) for the M10 bolts.

12. Install the ABS wheel speed sensor, and the sensor cable into the caliper bracket.

13. With the wheels installed and the vehicle on the ground, tighten the axle bolt as follows: M14 bolt 85 ft. lbs. (115 Nm) plus an additional ¼ (90°) turn, M16 bolt 140 ft. lbs. (190 Nm) plus an additional ¼ (90°) turn.

Rear

AWD 90 MODELS

➡**When loosening or tightening axle nut/bolt, be sure the vehicle is on the ground. Axle nut/bolt torque is high enough that attempting to loosen it may cause the vehicle to fall off the support.**

1. With the vehicle resting on the ground, loosen the halfshaft nut.

2. Raise and support the vehicle safely.

3. Remove the halfshaft nut, wheel bolts and wheel assembly.

4. Remove the ball joint nut. Using a ball joint removal tool, separate the ball joint from the strut.

5. Using a suitable tool, pry downward on the lower control arm to remove the ball joint from the control arm. If necessary, loosen lower control arm mounting bolts.

6. Pull the brake hose and parking brake cable, with grommets, from the holding fixture.

7. Remove the inner halfshaft flange bolts. Separate the shaft from the flange and support it.

8. Using a halfshaft pulling tool, attach it to the wheel hub and press the halfshaft out of the hub.

To install:

9. Clean the halfshaft splines of any grease, dirt or locking compound. Using the locking compound D-6, or equivalent, apply

a ¼ in. (3mm) bead around the outer edge of the splines. Allow the locking compound to dry for an hour after installation.

10. When installing, use a new inner flange gasket and reverse the removal procedures. Tighten the ball joint nut to 47 ft. lbs. (64 Nm).

11. Tighten the inner halfshaft flange bolts to 58 ft. lbs. (79 Nm) and install the wheel.

12. With the vehicle on the ground, tighten the halfshaft to hub nut to 238 ft. lbs. (322 Nm).

AWD A6 AND S6 MODELS

➡**When loosening or tightening axle nuts/bolt, be sure the vehicle is on the ground. Axle nut/bolt torque is high enough that attempting to loosen it may cause the vehicle to fall off the support.**

1. With the vehicle weight on the ground, loosen the halfshaft end bolt.

2. Raise and support the vehicle safely.

3. Remove the halfshaft bolt, wheel bolts and wheel assembly.

4. Remove the brake caliper to strut retaining bolts and remove the caliper, without disconnecting the hydraulic line. Using wire, support the caliper.

5. Remove the brake rotor. Remove the inner halfshaft flange bolts and support the halfshaft.

6. If necessary, remove the fuel tank cover plate and/or inner CV joint heat shield.

7. Slide the ABS speed sensor partly out of its mount.

8. Remove the lower mounting bolt for suspension strut.

9. Remove the transverse link-to-wheel bearing housing nut and remove the link.

10. Press down on wheel bearing housing and remove the halfshaft.

11. Clean the halfshaft splines of any grease, dirt or locking compound.

To install:

12. Use a new inner flange gasket and reverse the removal procedures.

13. Tighten the halfshaft flange bolts as follows: M8 bolts to 33 ft. lbs. (45 Nm) and the M10 bolts to 59 ft. lbs. (80 Nm).

14. Install caliper and tighten the bolts to 48 ft. lbs. (65 Nm). Adjustment of parking brake may be necessary.

15. Install the halfshaft bolt and washer assembly, tighten until just snug.

16. Make certain speed sensor sleeve is in place and install speed sensor, by hand, until seated. Install wheels.

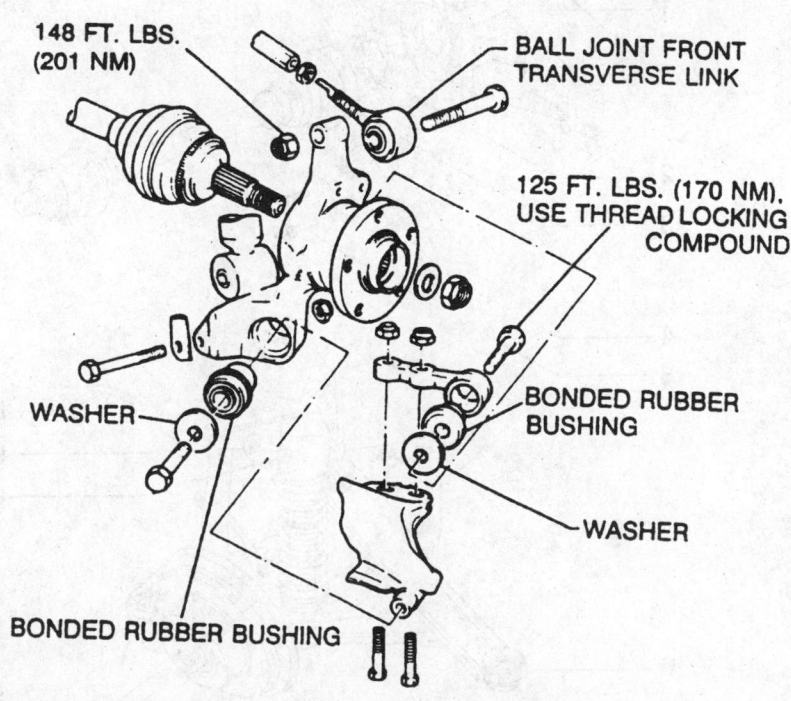

148 FT. LBS. (201 NM)

BALL JOINT FRONT TRANSVERSE LINK

125 FT. LBS. (170 NM), USE THREAD LOCKING COMPOUND

BONDED RUBBER BUSHING

WASHER

WASHER

BONDED RUBBER BUSHING

7923CG29

Exploded view of the rear suspension and halfshaft—AWD A6 and S6 models

17. Lower vehicle and tighten halfshaft bolts to 148 ft. lbs. (200 Nm) plus an additional ¼ turn.

AWD A4 MODELS

➡ When loosening or tightening axle nuts/bolts, be sure the vehicle is on the ground. Axle nut/bolt torque is high enough that attempting to loosen it may cause the vehicle to fall off the support.

1. With the vehicle weight on the ground, loosen the halfshaft end bolt.
2. Raise and support the vehicle safely.
3. Remove the halfshaft bolt, wheel bolts and wheel assembly.
4. Slide the ABS speed sensor partly out of its mount.
5. Unbolt the CV joint from the final drive.
6. Loosen the sway bar link mounting bolt at the wheel bearing housing.
7. Loosen the upper control arm mounting bolt at the wheel bearing housing.
8. If servicing the left halfshaft, remove the center and rear mufflers.
9. Lower the halfshaft at the final drive, then remove from the wheel bearing housing.

To install:

10. Install the halfshaft into the wheel bearing, then the final drive.

11. Tighten the upper control arm mounting bolt to 52 ft. lbs. (70 Nm) plus ¼ (90°) turn.
12. Tighten the sway bar link mounting bolt to 37 ft. lbs. (50 Nm).
13. If removed, install the center and rear mufflers.
14. Tighten the CV joint-to-final drive bolts to 30 ft. lbs. (40 Nm).
15. Install the ABS wheel speed sensor.
16. Install the halfshaft mounting bolt, tire/wheel assembly, lower the vehicle, then and tighten to 85 ft. lbs. (115 Nm) plus ¼ (90°) turn.

STEERING AND SUSPENSION

Air Bag

✳✳ CAUTION

Some vehicles are equipped with an air bag system, also known as the Supplemental Inflatable Restraint (SIR) or Supplemental Restraint Sys- tem (SRS). The system must be disabled before performing service on or around system components, steering column, instrument panel components, wiring and sensors. Failure to follow safety and disabling procedures could result in accidental air bag deployment, possible personal injury and unnecessary system repairs.

PRECAUTIONS

Several precautions must be observed when handling the inflator module to avoid accidental deployment and possible personal injury.

• Never carry the inflator module by the wires or connector on the underside of the module.

• When carrying a live inflator module, hold securely with both hands, and ensure that the bag and trim cover are pointed away from you.

• Place the inflator module on a bench or other surface with the bag and trim cover facing up.

• With the inflator module on the bench, never place anything on or close to the module which may be thrown in the event of an accidental deployment.

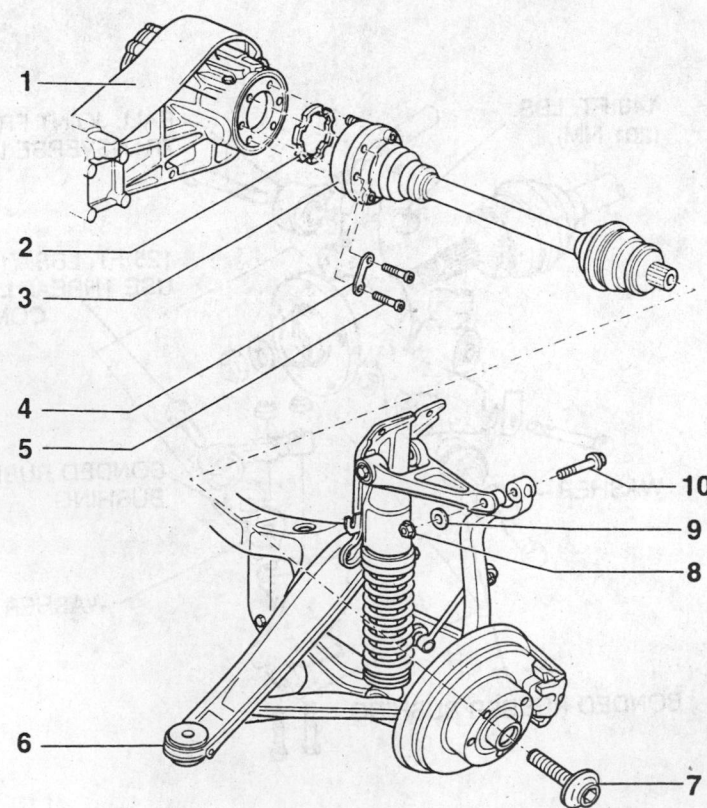

1. Rear final drive
2. Gasket
3. Halfshaft
4. Spacer plate
5. Halfshaft retaining bolts
6. Subframe
7. Collar bolt
8. Self-locking nut
9. Washer
10. Halfshaft retaining bolt

Exploded view of the rear halfshaft and related component mounting—AWD A4 models

7923CG30

• Before installing a computer memory saver on vehicles with electronic radio lock, detach the air bag voltage connector.

• DO NOT use air bag components that have been dropped from heights of 18 in. (45cm) or higher.

• Disable the SRS before performing electric welding on the vehicle.

• SRS can only be tested using Diagnostic Tester (VAG 1551) and Adapter Test Harness (VAG 1551/1) or their equivalents. DO NOT use Air Bag Tester (VAG 1619) or equivalent. Never use a test light, ohmmeter, or volt meter to test the air bag system, except when testing the clockspring.

DISARMING

1995 Models

1. Disconnect and shield the negative battery cable.
2. Disconnect the voltage supply RED connector, located behind the inspection cover, on the driver-side lower instrument panel cover.

1996–99 Models

1. Disconnect and shield the negative battery cable and wait at least 5 minutes before servicing the vehicle.

REARMING

1995 Models

1. After performing the required procedure, arm the system as follows:
2. Attach the voltage supply connector and the negative battery cable.

1996–99 Models

1. After performing the required procedure, arm the system as follows:
2. Remove the shield, then connect the negative battery cable.

Power Rack and Pinion Steering Gear

REMOVAL & INSTALLATION

90 Models

1. Raise and safely support the vehicle.
2. Remove the lower left instrument panel cover, the steering column-to-steering rack clamp bolt and the steering column-to-dash bolts. Remove the steering column from the vehicle.
3. Using a pair of locking pliers, clamp off the fluid return line to the reservoir. Dis-connect the fluid pressure line from the steering gear.
4. At the steering column boot, press in on the clips and remove the boot from the panel. From inside the vehicle, remove the fluid return line from the control valve body. On 5-cylinder vehicles, push off the dash panel boot and push the boot into the passenger compartment to access the pressure and return line.
5. At the left wheel housing, disconnect the steering rack from the frame.
6. At the steering rack, remove the tie rod coupling locknuts/bolts and the tie rods from the rack. Push the rack back into the steering housing.
7. Disconnect the steering assembly from the firewall. Turn the wheels to the right. Remove the assembly between the left wheel housing and the control arm.

To install:

8. Install the rack assembly and tighten the left side bolts and nuts to 14 ft. lbs. (20 Nm). Use new self-locking nuts to secure the rack to the firewall and tighten to 35 ft. lbs. (45 Nm).
9. Use new self-locking nuts and secure the tie rod coupling to the rack. Tighten the nuts to 35 ft. lbs. (45 Nm).
10. Install the steering column, connect the hydraulic lines and bleed the system.

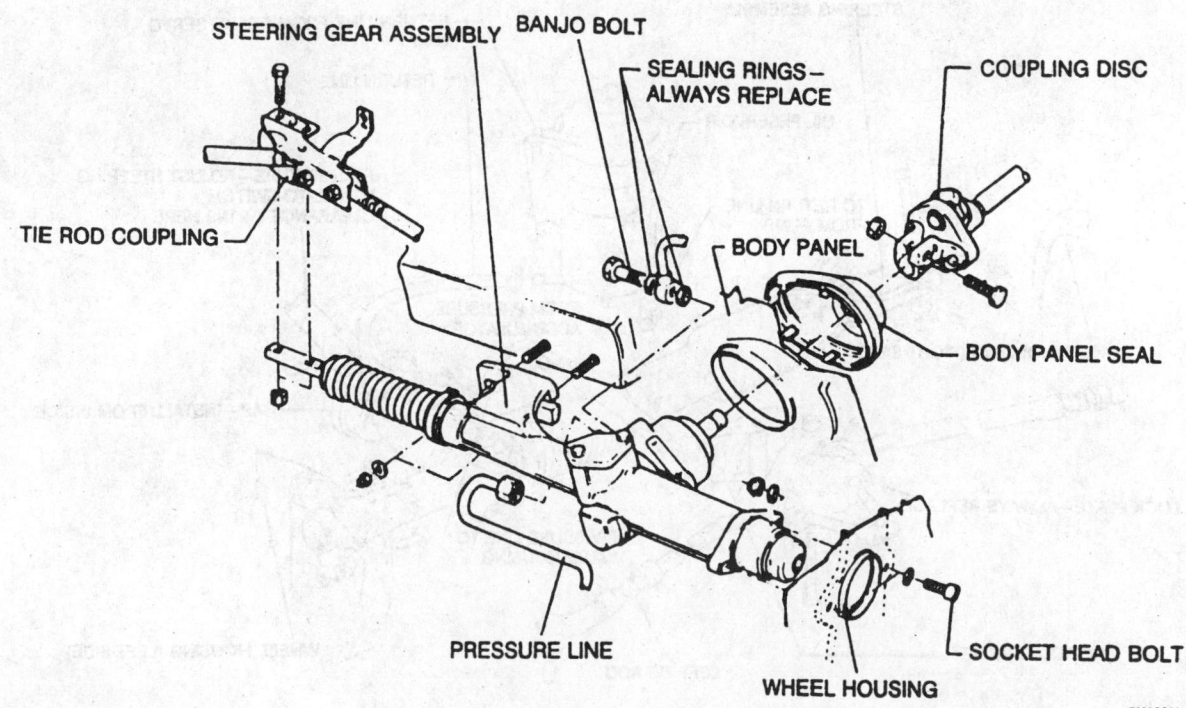

Exploded view of the steering rack assembly—90 models

A6, S6 and Cabriolet Models

1. Disconnect the negative battery cable.
2. Pry off the sound cover with a screwdriver.
3. Disconnect the crankcase breather hose.
4. Lift the sound cover, disconnect the vacuum hose, then remove the sound cover.
5. Drain the brake fluid from the reservoir.
6. Disconnect the hoses and lines from the brake fluid reservoir and master cylinder.
7. Remove the check valve from the brake booster.
8. If equipped, disconnect the vacuum unit from the left valve cover and set it aside.
9. Remove the left-side storage shelf and footwell under the instrument panel.
10. Dismount the center electrical panel and/or relay panel and set it aside.
11. Remove the brake booster clevis pin.
12. Remove the pedal bracket mounting nuts.
13. Unclip the fuel line from the left side of the instrument panel bracket and move towards the front of the vehicle.
14. Carefully remove the brake booster from the vehicle, being careful not to bend any fuel or brake lines.
15. Remove the mounting bolt for the steering gear pinion U-joint-to-steering column.
16. Disconnect the U-joint.
17. Clamp the pressure and return hydraulic hoses, using Camp 3094 or equivalent.
18. Remove the hoses from the steering gear.
19. Remove both tie rods with the carrier.
20. Cut the cable tie attaching the right wheel housing wiring harness.
21. Remove the right-side steering gear mounting bolts.
22. Detach the left front ABS wheel speed sensor connector from the mounting bracket.
23. Remove the front wheels.
24. Remove the steering gear mounting bolts on the left and right wheelhousing.
25. Pull the steering gear toward the front of the vehicle, to clear the instrument panel seal.
26. With the aid of an assistant, pull the steering gear to the left in the left wheel housing hole so that the wiring harness can be pushed toward the back on the right side of the steering gear and the steering gear can be removed from the opening in the right wheelhousing.

To install:

27. Slowly, install the steering gear into the vehicle through the right wheelhousing.
28. Ensure the proper routing of the wiring harness around the steering gear.
29. Install the steering gear mounting bolts and tighten to 37 ft. lbs. (50 Nm).
30. Connect the tie rods and tighten the attaching bolts to 44 ft. lbs. (60 Nm) when the vehicle is on its wheels.
31. Connect the hydraulic hoses to the steering gear and tighten the banjo bolts to 30 ft. lbs. (40 Nm).
32. Attach the left front wheel ABS wheel speed sensor bracket.
33. The completion of installation is the reverse of the removal, with following these additional points:

• Remove the hose clamps and check the fluid levels
• Ensure the brake lines are connected and bled properly
• Tighten the U-joint attaching bolts to 18 ft. lbs. (25 Nm)
• Tighten the pedal bracket mounting bolts to 18 ft. lbs. (25 Nm)
• Install the brake booster clevis pin

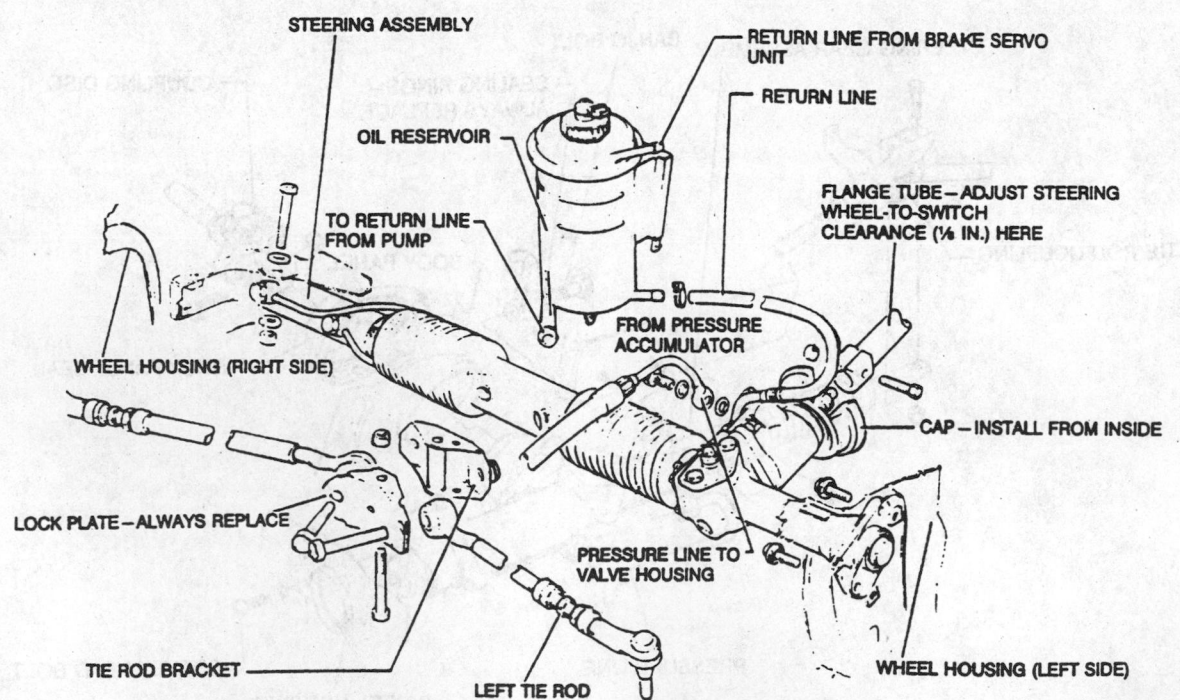

STEERING ASSEMBLY
RETURN LINE FROM BRAKE SERVO UNIT
RETURN LINE
OIL RESERVOIR
TO RETURN LINE FROM PUMP
FLANGE TUBE—ADJUST STEERING WHEEL-TO-SWITCH CLEARANCE (⅛ IN.) HERE
FROM PRESSURE ACCUMULATOR
WHEEL HOUSING (RIGHT SIDE)
CAP—INSTALL FROM INSIDE
LOCK PLATE—ALWAYS REPLACE
PRESSURE LINE TO VALVE HOUSING
TIE ROD BRACKET
LEFT TIE ROD
WHEEL HOUSING (LEFT SIDE)
7923CG32

Exploded view of the steering rack assembly—A6, S6 and Cabriolet models

A4 Models

1. Disconnect and remove the battery, then the battery box.

2. Remove the bolt at the steering column U-joint.

3. Release the eccentric by turning the Torx® T50 bolt clockwise, then remove the bolt.

4. Before removing the steering column form the steering gear, secure the steering column with safety wire.

※※ CAUTION

Be sure to lock the steering wheel, otherwise the air bag unit coil spring may be damaged.

5. Lock the steering wheel in the center position and do not move during the repairs.

➡**The splines between the top and bottom part of the steering column must not be separated.**

6. Move the U-joint down and out of the way.

7. Using Hose Clamps 3094 or equivalent, pinch off the suction and return lines to the steering gear.

8. Raise and safely support the vehicle, and remove the front wheels.

9. Disconnect the left and right tie rods.

10. Remove the tie rod opening cover.

➡**Place a drip tray under the vehicle to catch any residual power steering fluid.**

11. Remove the banjo bolts for the steering gear suction and return hydraulic hoses.

12. Remove the steering gear mounting bolts.

13. With the aid of an assistant, remove the steering gear through the left side wheel opening.

To install:

14. Remove the screw plug to lock the steering gear in the center position with Locking tool VAG 1907 or equivalent, and tighten to 13 ft. lbs. (18 Nm).

15. Insert the steering gear into the vehicle through the left side wheel opening.

16. Hand-tighten mounting bolts 1 and 2.

17. Install bolt 3 and tighten to 48 ft. lbs. (65 Nm), then tighten bolts 1 and 2 to 48 ft. lbs. (65 Nm).

18. Using new sealing gaskets, install the return hose and tighten the banjo bolt to 37 ft. lbs. (50 Nm), then the suction hose banjo bolt to 30 ft. lbs. (40 Nm).

19. Connect the left and right tie rods, and tighten the mounting through-bolt to 33 ft. lbs. (45 Nm).

20. Install the tie rod opening cover.

21. Fit the J-joint to the steering gear, then insert the Torx® adjusting bolt by turning it clockwise.

22. Remove the Locking tool VAG 1907, then reinstall the screw plug and tighten to 13 ft. lbs. (18 Nm).

23. Tighten the adjusting bolt nut to 30 ft. lbs. (40 Nm).

24. Remove the steering wheel lock.

25. Remove the hose clamps 3094, and check the hydraulic fluid.

26. Install the battery tray, then connect the battery.

27. Start the vehicle, check for leaks, then have the alignment set.

Strut

REMOVAL & INSTALLATION

Front

90, A6, S6 AND CABRIOLET MODELS

1. Disconnect the negative battery cable. With the vehicle on the ground,

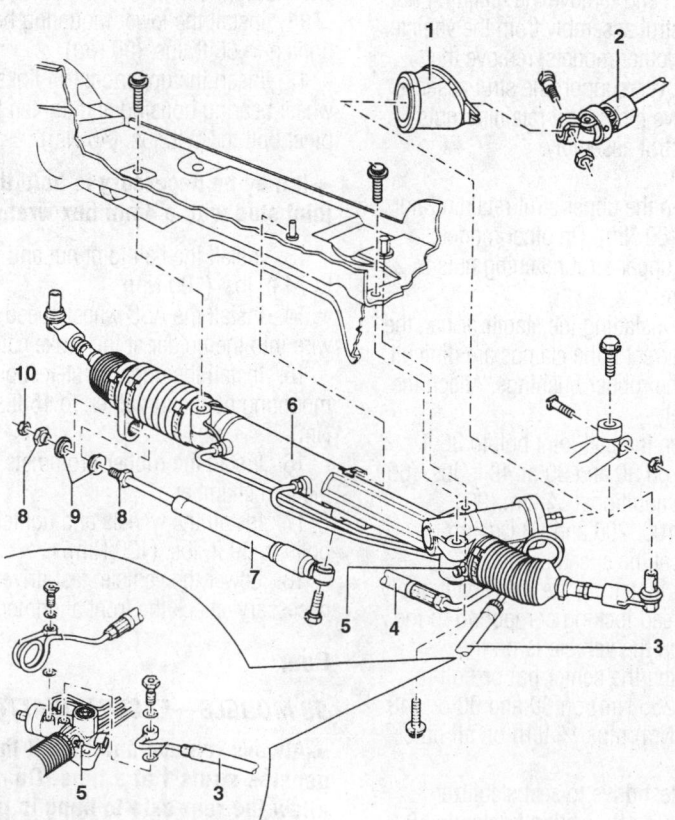

1. Boot seal
2. Steering column
3. Return hose
4. Flexible hose
5. Screw plug for centering the steering wheel
6. Rack and pinion steering gear
7. Steering damper
8. Bushing
9. Two-piece rubber bushing
10. Nut

Exploded view of the steering gear mounting—A4 models

7923CG33

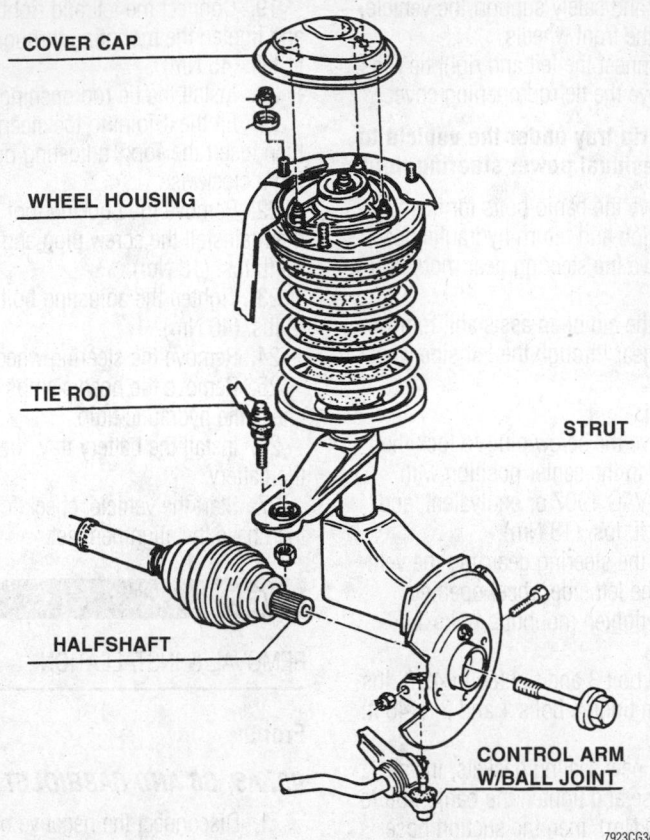

COVER CAP

WHEEL HOUSING

TIE ROD

STRUT

HALFSHAFT

CONTROL ARM
W/BALL JOINT

7923CG34

Exploded view of the front strut mounting—90, A6, S6 and Cabriolet models

remove the front axle nut or bolt and loosen the wheel bolts.

2. Raise and support the vehicle safely. Remove the wheel assembly.

3. Remove the brake caliper mounting bolts and disconnect the brake line from the bracket without disconnecting the line from the caliper. Remove speed sensor, if equipped.

4. Remove the brake caliper with the line still attached and support it aside.

5. Remove disc brake rotor.

6. Remove the ball joint clamp bolt and nut.

7. Remove the tie rod end nut and separate the tie rod end from the strut.

8. If equipped with a stabilizer bar, remove the retaining bolt and remove the stabilizer bar end clamps. Pivot the stabilizer bar downward.

9. Remove the 2 center stabilizer bar clamps and unbolt stabilizer bar from the lower control arm.

10. Remove the pinch bolt from the steering knuckle and separate the lower ball joint from the knuckle by pushing down on the control arm.

11. Using a suitable hub puller, press the halfshaft out of the hub.

12. Support the strut assembly, hold the shock absorber piston rod with an internal

socket wrench and remove the retaining nut. Remove the strut assembly from the vehicle.

13. On all other models, remove the upper strut cover, support the strut assembly and remove the 3 strut retaining nuts. Remove the strut assembly.

To install:

14. Tighten the upper strut retaining nut to 44 ft. lbs. (60 Nm). On other models, tighten the 3 upper strut retaining nuts to 22 ft. lbs. (30 Nm).

15. When installing the stabilizer bar, the position is correct if the clamps are difficult to install in the rubber bushings. Attach the clamps loosely.

16. Tighten the ball joint bolt to 36 ft. lbs. (49 Nm) on 80 and 90 or 48 ft. lbs. (65 Nm), plus an additional ¼ turn (90 degrees) on 100, 200 and V8 Quattro. Install and seat the speed sensor.

17. When installing the axle shaft, apply a bead of thread locking compound to the splines. When the vehicle is on the ground, tighten the center nut or bolt to 195 ft. lbs. (265 Nm) on 80 and 90 or 148 ft. lbs. (200 Nm) plus ¼ turn on all other models.

18. After test drive to seat stabilizer bushings in correct position tighten to 18 ft. lbs. (24 Nm).

A4 MODELS

1. Raise and safely support the vehicle.

2. Remove the front wheels.

3. Raise the hood and remove the rubber grommets from the plenum chamber.

4. Remove the upper strut-to-body mounting nuts.

5. Detach the ABS wheel speed sensor wire from the bracket at the brake caliper.

6. Remove the upper control arm pinchbolt, then lift out both upper control links upward and out.

7. Swivel the wheel bearing housing aside to disconnect the guide link ball joint.

8. Remove the lower strut mounting bolt.

➡**When removing the strut be sure not to damage the CV joint boot.**

9. Remove the strut downward.

To install:

➡**The bonded rubber bushing can only turned to a limited extent. The bolted connections between the suspension strut and the lower track control links should therefore only be tightened when the vehicle is standing on the ground.**

10. Install the strut into the vehicle and position it so that the hole in the spring plate faces the middle of the vehicle.

11. Install the lower mounting bolt and tighten to 66 ft. lbs. (90 Nm).

12. Insert the upper control links to the wheel bearing housing and tighten the pinchbolt to 30 ft. lbs. (40 Nm).

➡**It may be necessary to hold the ball joint stud with a 4mm hex wrench.**

13. Install the ball joint nut and tighten to 74 ft. lbs. (100 Nm).

14. Install the ABS wheel speed sensor wire into the holder at the brake caliper.

15. Install the upper strut-to-body mounting nuts and tighten to 15 ft. lbs. (20 Nm).

16. Install the rubber grommets into the plenum chamber.

17. Install the wheels and tighten the lug bolts to 89 ft. lbs. (120 Nm).

18. Lower the vehicle, test drive and if necessary, check the front alignment.

Rear

90 MODELS—EXCEPT QUATTRO

➡**Always remove and install the suspension struts 1 at a time. Do not allow the rear axle to hang in place as this may cause damage to the brake lines.**

1. With the vehicle at ground level, open the trunk and remove the trim from around the shock tower.

2. Remove the rubber cap.

3. Hold the strut rod and remove the strut mounting nut.

4. Raise and support the vehicle safely.

5. Remove the lower strut mounting bolt from the axle beam and remove the strut.

6. Installation is the reverse of removal. Tighten the upper strut mounting bolt to 14 ft. lbs. (19 Nm) and the lower strut mounting bolt to 43 ft. lbs. (58 Nm).

90 MODELS—QUATTRO WITH SINGLE PIECE STRUT

1. With the vehicle on the ground, remove the axle nut.

2. Raise and safely support the vehicle and remove the rear wheels.

3. Unbolt the halfshaft from the differential flange.

4. Remove the self-locking lower ball joint nut and press the ball joint out of the wheel bearing housing. Use the nut to protect the threads.

5. Loosen the control arm mounting bolts and allow the arm to pivot down out of the way.

6. Remove the tie rod nut and press the tie rod joint out of the bearing housing arm. Use the nut to protect the threads.

7. Pull the parking brake cable out of the strut bracket. Remove the brake caliper with its bracket without disconnecting the hydraulic line. Hang the caliper from the body with wire.

8. Remove the brake disc and slide the wheel speed sensor out of its mounting, if equipped. Press the halfshaft out of the hub.

9. Support the strut from below. In the luggage compartment, remove the trim and the rubber cap at the top of the strut.

10. Hold the top of the strut with an Allen wrench and remove the self-locking nut. Lower the strut from the vehicle.

To install:

11. Replace all self-locking nuts. Install the strut into the upper mount and tighten the new upper nut to 48 ft. lbs. (60 Nm).

12. Be sure the axle shaft splines are clean and apply a bead of threat locking compound to the outer end of the splines. Install the axle shaft but do not tighten the center nut until the vehicle is on its wheels. Tighten the axle flange bolts to 35 ft. lbs. (45 Nm).

13. Install the brake disc and caliper. Tighten the caliper bracket mounting bolts to 92 ft. lbs. (125 Nm).

14. Attach the ball joint to the control arm and the tie rod to the bearing housing

and install new self-locking nuts. Tighten the tie rod nut to 29 ft. lbs. (40 Nm) and the ball joint nut to 54 ft. lbs. (75 Nm).

15. When installation is complete and the vehicle is on the ground, tighten the center axle nut to 236 ft. lbs. (320 Nm).

90 MODELS—QUATTRO WITH 2-PIECE STRUT

1. In the luggage compartment, remove the trim and the cap from the top of the strut. With the vehicle on its wheels, hold the strut rod from turning and remove the upper strut nut.

2. Place a block of wood between the axle shaft and the frame. Carefully raise and safely support the vehicle and remove the rear wheels.

3. Remove the bolts securing the lower strut to the wheel bearing housing and remove the strut. These are stretch bolts that cannot be reused and must be replaced with the newer type.

To install:

4. Install the strut to the bearing housing with new bolts. Tighten the new stretch bolts to 59 ft. lbs. (80 Nm), plus an additional ½ turn.

5. Install the wheel and lower the vehicle until the wood block can be removed. Be sure to carefully guide the strut into the upper mount.

6. Install a new self-locking upper strut mounting nut and tighten to 44 ft. lbs. (60 Nm). Check the wheel alignment.

A6, S6 AND CABRIOLET MODELS—EXCEPT QUATTRO

➡The struts must be removed with the weight of the vehicle on the rear wheels. If not, a spring compressor must be used on the rear springs.

1. If the vehicle is not on its wheels, install the spring compressor and compress the spring. Do not attempt to remove the shock with the rear wheels raised without a compressor.

2. Remove the upper strut mounting nut.

3. Remove the lower strut mounting nut.

4. Remove the shock absorber.

5. Installation is the reverse of removal. Tighten the lower mounts to 66 ft. lbs. (89 Nm) and the upper to 14 ft. lbs. (19 Nm).

A6, S6 AND CABRIOLET MODELS—QUATTRO

1. Raise and support the vehicle safely. Remove the wheel assembly.

2. Open the trunk and remove the shock absorber covers, the remove the shock absorber-to-body nuts/bolts.

3. Remove the shock absorber-to-rear wheel knuckle assembly. Remove the shock absorber from the vehicle.

4. To install, reverse the removal procedures. Tighten the shock absorber-to-body nuts/bolts to 15 ft. lbs. (20 Nm) and the shock absorber-to-rear wheel knuckle assembly bolt to 66 ft. lbs. (89 Nm).

A4 MODELS

1. Raise and safely support the vehicle.

2. Support the trailing arms.

3. Remove the lower strut mounting bolt.

4. Remove the rear seat backrest side bolster cover or backrest to access the upper mounting.

5. Unbolt the upper strut-to-body mounting nuts.

➡In addition to the bolted connection, the strut is also attached to the body by four retaining lugs.

6. Turn the strut until the retaining lugs are positioned above the recesses, then pull the strut downward out of its mount.

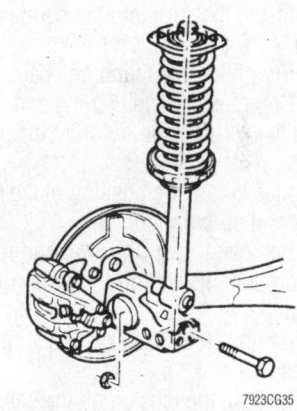

7923CG35

Be sure to support the trailing arm before removing the lower strut mounting bolt—A4 models

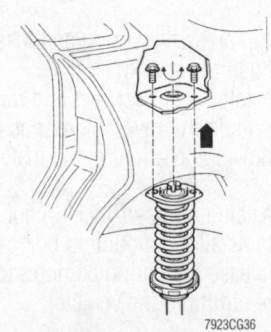

7923CG36

After loosening the two attaching bolts, rotate the upper strut mount to disengage the strut from the vehicle—A4 models

To install:

➡ **The bonded rubber bushing can only turned to a limited extent. The bolted connections between the suspension strut and the rear axle should therefore only be tightened when the vehicle is standing on the ground.**

7. Install the strut into the vehicle and be sure to engage the retaining lugs.

8. Install the upper strut-to-body mounting nuts and tighten to 18 ft. lbs. (25 Nm).

9. Install the lower strut mounting and tighten to 37 ft. lbs. (50 Nm) plus ¼ turn (90°) with the suspension loaded.

10. Install the rear seat backrest side bolster cover or backrest.

11. Lower the vehicle.

Coil Spring

REMOVAL & INSTALLATION

All Models

1. Remove the strut from the vehicle.

2. Clamp the Spring Compressor VAG 1752/2 or equivalent in a vise.

3. Install the strut into the spring compressor.

4. Pry off the mounting bolt cap.

5. Compress the coil spring and remove the self-locking nut from the piston rod.

6. Matchmark the position of the spring retainer and spring mount.

7. Remove the spring seat and related components noting the order of removal.

8. Remove the strut from the spring compressor.

9. Release the tension on the coil spring, and remove the spring out of the compressor.

To install:

10. Install the new spring into the compressor.

11. Compress the spring and insert the strut through the spring.

12. Install the spring seat and related components in the reverse order as they were removed and aligning the matchmarks.

13. Install a new self-locking nut.

14. Reinstall the mounting bolt cap.

15. Release the spring compressor and install the strut into the vehicle.

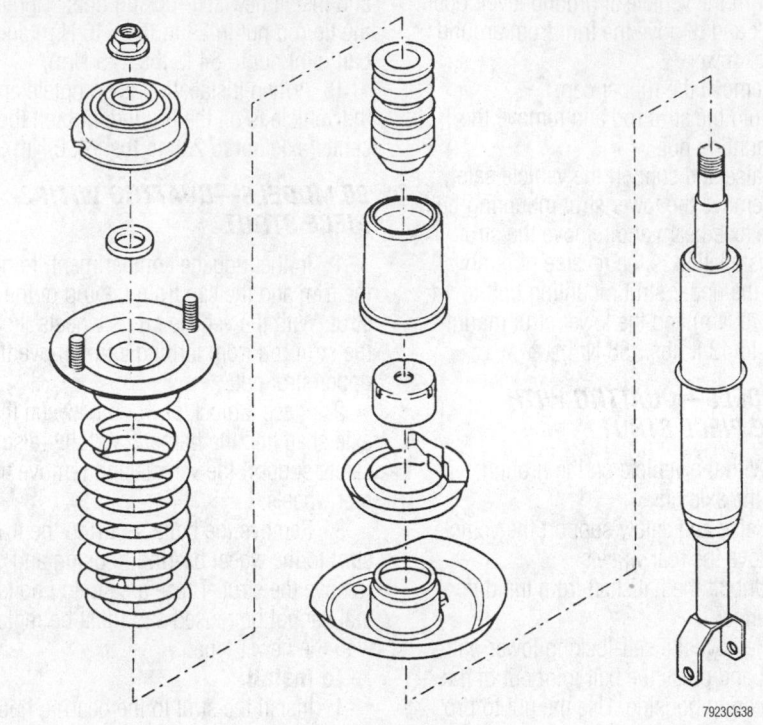

Exploded view of the front strut—A4 model

7923CG38

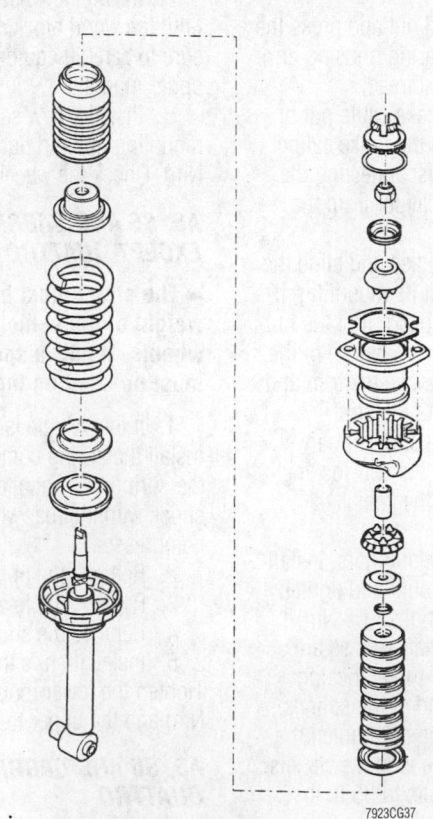

Exploded view of the rear strut—A4 model

7923CG37

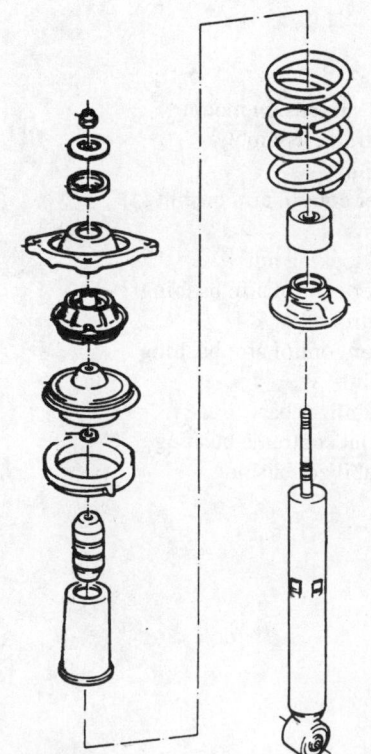

Exploded view of the front strut—except A4 model

7923CG40

Exploded view of the rear strut—except A4 model

7923CG39

Upper Ball Joint

REMOVAL & INSTALLATION

A4 Models

The Audi A4 front suspension is equipped with two separate upper ball joints that are not replaceable, the upper link (front or rear) must be replaced. To remove this link, perform the following:

1. Raise and safely support the vehicle, and remove the front wheels.

2. Remove clip No. 1 as shown. The clip does not have to be replaced.

3. Remove the pinchbolt and pull both control arms upward and out.

4. Cover the steering gear boot.

5. Remove the guide link ball joint and press off the joint.

6. Detach the ABS wheel speed sensor wire front the bracket on the brake caliper.

7. Support the suspension from excessive rebound travel.

8. Remove the lower strut mounting bolt.

9. Swing the wheel bearing housing aside.

10. Raise the hood and remove the rubber grommets from the plenum chamber.

11. Remove the upper strut-to-body mounting nuts.

12. Remove the strut together with the mounting bracket.

13. Clamp the strut in a vise with the protective jaw covers.

14. Remove the upper link bolts and detach both of the links.

15. Remove the bracket-to-strut mounting nuts, then separate.

To install:

16. Position the brackets and links as shown, and tighten the bracket-to-strut mounting nuts to 15 ft. lbs. (20 Nm).

17. Align the links as shown, then tighten to 37 ft. lbs. (50 Nm) plus ¼ turn (90°).

18. Install the strut with mounting bracket into the vehicle and tighten the upper strut-to-body mounting nuts to 48 ft. lbs. (75 Nm).

19. Install the lower strut mounting bolt and tighten to 66 ft. lbs. (90 Nm).

20. Install the nut on the ball joint and tighten to 74 ft. lbs. (100 Nm).

21. Install the upper links to the wheel bearing housing and tighten the pinchbolt to 30 ft. lbs. (40 Nm).

22. Attach the ABS wiring to the brake caliper bracket.

23. Install the wheel, lower the vehicle, and check the front suspension alignment.

Lower Ball Joint

REMOVAL & INSTALLATION

90 Models

1. Raise and safely support the vehicle. Remove the wheel assembly.

2. Remove the lower ball joint clamp nut and bolt and pry the control arm down to disengage the ball joint from the steering knuckle.

3. Remove the ball joint to control arm mounting bolts and remove the ball joint.

To install:

4. Install the new ball joint on the control arm and tighten the mounting bolts to 46 ft. lbs. (62 Nm).

5. Slowly allow the lower control arm and ball joint to fit into the strut assembly. Install the bolt and tighten to 48 ft. lbs. (65 Nm).

6. Install the wheel assembly and lower the vehicle.

7. Reset the front end alignment when finished.

A6, S6 and Cabriolet Models

1. Raise and support the vehicle safely. Remove the wheel assembly.

2. Remove the ball joint to strut bolt and nut. Pry and hold the control arm down.

3. Disconnect the end of the stabilizer bar and note the position of the washers and bushings.

4. Remove the control arm-to-subframe bolts and remove the control arm.

To install:

5. Installation is in the reverse order of removal. Check control arm bushings for cracking or undue wear. Tighten the control arm-to-subframe bolts to 81 ft. lbs. (110 Nm) plus ¼ turn (90°); the ball joint-to-strut bolt to 48 ft. lbs. (65 Nm) and the stabilizer bar nut to 89 ft. lbs. (120 Nm) plus ¼ turn (90°).

A4 Models

The A4 is equipped with two lower ball joints that are not serviceable. The control arms must be replace if a joint is worn. The lower track control link ball joint stud faces down, and the guide link ball joint stud faces up.

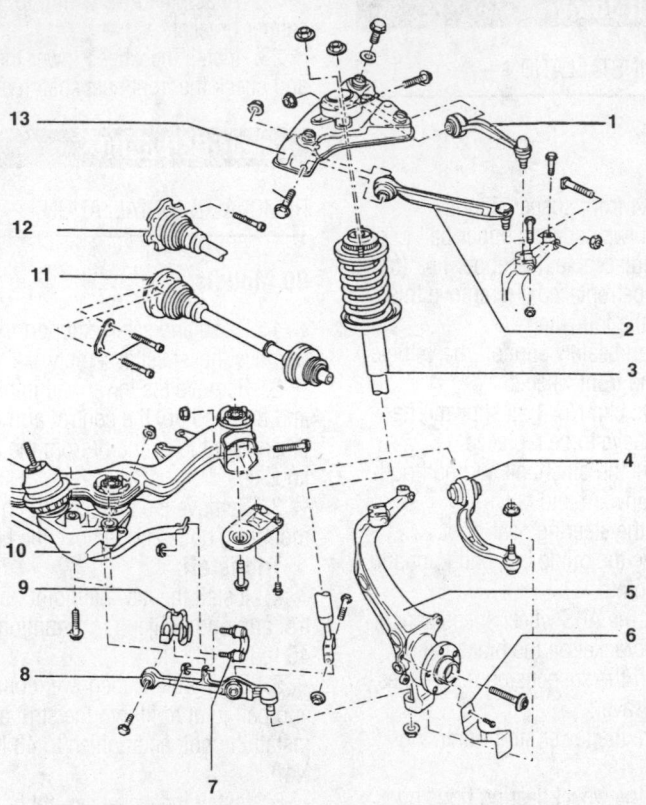

1. Upper link, rear
2. Upper link, front
3. Suspension strut
4. Guide link
5. Wheel bearing housing
6. Splash shield
7. Connecting link
8. Lower track control link
9. Clamp
10. Subframe
11. Halfshaft w/CV joint
12. Halfshaft w/triple-rotor joint
13. Mounting bracket

Exploded view of the front suspension—A4 models

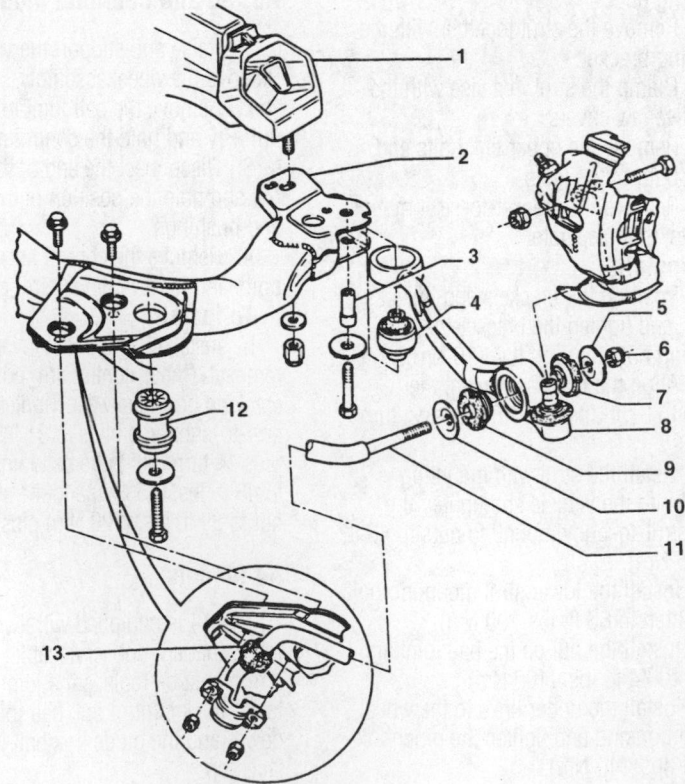

1. Left suspension mount
2. Subframe assembly
3. Bushing
4. Inner control arm bushing
5. Shims
6. Self-locking nut
7. Outer control arm bushing
8. Control arm
9. Outer control arm bushing
10. Shims
11. Stabilizer bar
12. Front subframe bushing
13. Stabilizer bushing

Exploded view of the front suspension—A6, S6 and Cabriolet models

LOWER TRACK CONTROL LINK

1. Raise and safely support the vehicle and remove the front wheels.

2. Remove the nut from the lower track control link; then press the ball joint out of the tapered seat.

3. Support the wheel bearing housing to prevent excessive rebound travel in the suspension.

4. Remove the stabilizer link and lower strut mounting bolt.

5. Remove the lower track control link-to-subframe attaching bolt.

6. Remove the lower track control link from the vehicle.

To install:

7. Install the lower track control link into the vehicle and install the subframe attaching bolt.

8. Tighten the lower strut mounting bolt to 74 ft. lbs. (100 Nm).

9. Install the stabilizer link and tighten the upper attaching bolt to 30 ft. lbs. (40 Nm) plus ¼ turn (90°), then the lower attaching bolt to 74 ft. lbs. (100 Nm).

10. Load the suspension and tighten the subframe attaching bolt to 59 ft. lbs. (80 Nm) plus ¼ turn (90°).

11. Install the wheels and lower vehicle.

12. Check the front suspension alignment.

LOWER GUIDE LINK

1. Raise and safely support the vehicle.

2. Remove the front wheels.

3. Remove the nut from the lower guide link joint and press out the joint from the wheel bearing housing.

4. Loosen the lower guide link-to-subframe attaching bolt.

➡**The subframe must be lowered at the rear to remove the lower guide link-to-subframe attaching bolt.**

5. Loosen the rear subframe support plate bolts and subframe bolts.

6. Remove the lower guide link-to-subframe attaching bolt.

7. Remove the link from the vehicle.

To install:

8. Install the link into the vehicle.

9. Install the guide link-to-subframe mounting bolt.

10. Tighten the support plate bolts as follows:

• If equipped with bolt type "A", tighten to 18 ft. lbs. (25 Nm)

• If equipped with bolt type "B", tighten to 55 ft. lbs. (75 Nm)

11. Install new subframe bolts and tighten to 81 ft. lbs. (110 Nm) plus ¼ turn (90°).

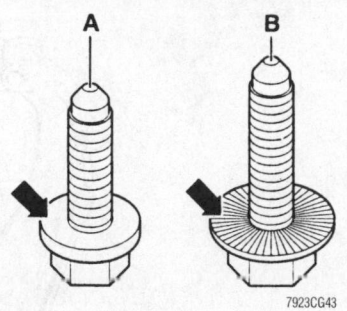

Subframe support bracket bolt identification—A4 model

12. Install the joint end into the wheel bearing housing and tighten the nut to 74 ft. lbs. (100 Nm).

13. Load the suspension and tighten the lower guide link-to-subframe attaching bolt to 66 ft. lbs. (90 Nm) plus ¼ turn (90°).

14. Install the front wheels and lower the vehicle.

15. Check the front suspension alignment.

Wheel Bearings

ADJUSTMENT

Front

The front wheel bearings are sealed, no adjustment is necessary or possible.

Rear

FWD VEHICLES

1. Raise and support the vehicle safely.

2. Remove the grease cap.

3. Remove the cotter pin and the locking nut.

4. While turning the wheel, so the wheel bearing does not jam, tighten the adjusting nut firmly.

5. Back the nut off slightly. The nut is properly adjusted when it is possible to pry the thrust washer side to side with some drag but using light pressure on the tool.

6. Install the locking nut and a new cotter pin. When installing the cap, be sure it is securely in place.

AWD VEHICLES

The wheel bearings are sealed; no adjustment is necessary or possible.

Front

90 MODELS

➡**90 vehicles use 2 types of front wheel bearing housing assemblies. They are a single-piece unit that cannot be separated from the strut and a bearing housing that is removable from the strut for service. The repair procedures are similar for both with the exception that the single piece unit housing, if defective, must be replaced as a strut assembly.**

1. Raise and safely support vehicle. Remove the halfshafts.

2. Remove the strut housing to body nuts and remove the strut/hub assembly from the vehicle.

3. Remove the brake disc and splash shield.

4. Using an arbor press with suitable drivers, press the wheel hub from the strut housing.

5. Using snapring pliers, remove the snaprings from both sides of the wheel bearing.

6. Using an arbor press with suitable drivers, press the wheel bearing from the strut housing.

7. Using a wheel puller, pull the wheel bearing race from the wheel hub.

To install:

8. Lightly grease inside the strut housing before installing the new bearing.

9. Be sure to press only on the outer race when pressing the new bearing into the hub. When installing the snaprings, be sure they are properly seated.

10. Install the brake plate onto the bearing housing. Be sure to press only on the inner bearing race when pressing the bearing over the hub.

11. To install the strut and wheel hub assembly, reverse the removal procedures. Tighten the upper strut to body nut to 44 ft. lbs. (60 Nm). On all other models, tighten the 3 upper strut nuts to 22 ft. lbs. (30 Nm). Check front wheel alignment.

A6, S6 AND CABRIOLET MODELS

1. Raise and safely support vehicle. Remove the halfshafts.

2. Remove the strut to vehicle nuts and remove the strut from the vehicle.

3. Remove the disc brake rotor and splash shield.

4. Using an arbor press with suitable drivers, press the wheel hub from the strut housing.

5. Using snapring pliers, remove the snaprings from both sides of the wheel bearing.

6. Using an arbor press with suitable driver, press the wheel bearing from strut housing.

7. Using a suitable puller, remove the bearing race from the wheel hub.

To install:

8. Install the outer snapring into the strut housing.

9. Be sure to press only on the outer race when pressing the new bearing into the hub. When installing the snapring, be sure it is properly seated.

10. Install the brake plate onto the bearing housing. Be sure to press only on the inner bearing race when pressing the bearing over the hub.

11. To install the strut and wheel hub assembly, reverse the removal procedures. Tighten the strut to body nuts to 44 ft. lbs. (60 Nm). Check front wheel alignment.

A4 MODELS

1. Loosen the halfshaft retaining bolt.
2. Raise and safely support the vehicle.
3. Remove the front wheel.
4. Remove the ABS wheel speed sensor.
5. Remove the caliper bracket mounting bolts, then the rotor.
6. Remove the brake splash guard.
7. Loosen the mounting nuts for the lower guide and track links.
8. Disconnect the tie rod end from the wheel bearing housing.
9. Remove the mounting nuts for the lower guide and track links and press out the joints.
10. Remove the upper control arm pinchbolt and disconnect the arms.
11. Remove the wheel bearing housing.
12. Place the wheel bearing housing on a press.
13. Drive out the hub with the wheel bearing.
14. Using a bearing separator and press, drive hub out of the bearing.

To install:

15. Press the new wheel bearing into the bearing housing using the appropriate bearing driver.
16. Press the hub into the wheel bearing using the appropriate bearing driver.
17. Install the wheel bearing housing.
18. Slide the CV-joint through the wheel hub and hand-tighten the new nut.
19. Connect the lower track control and guide link, then tighten the new self-locking nut to 74 ft. lbs. (100 Nm).

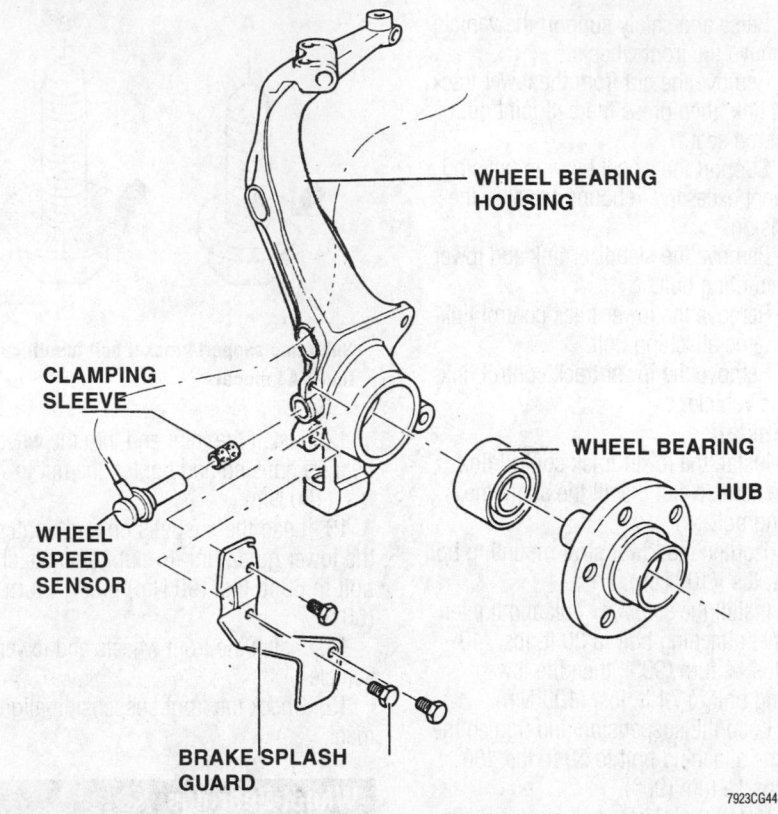

Exploded view of the front wheel bearing housing—A4 models

20. Install both of the upper link ball joints into the wheel bearing and tighten the pinchbolt to 30 ft. lbs. (40 Nm).
21. Install the tie rod end and tighten the new self-locking nut to 37 ft. lbs. (50 Nm), then the bolt to 44 inch lbs. (5 Nm).
22. Install the ABS wheel speed sensor.
23. Install the brake splash guard and tighten the retaining bolts to 7 ft. lbs. (10 Nm).
24. Install the brake rotor.
25. Install the brake caliper and tighten the retaining bolt to 89 ft. lbs. (120 Nm).
26. Install the wheels and the lug bolts to 89 ft. lbs. (120 Nm).
27. Tighten the halfshaft retaining bolt as follows:
• If equipped with a M14 bolt, tighten to 85 ft. lbs. (115 Nm) plus ½ turn (180°)
• If equipped with a M16 bolt, tighten to 140 ft. lbs. (190 Nm) plus ½ turn (180°)
28. Check the front suspension alignment, if necessary, adjust.

Rear

FWD MODELS

1. Raise and support the vehicle safely. Remove the wheel assembly.
2. Without disconnecting the hydraulic line, remove caliper assembly from rotor.

Suspend caliper from the body with wire, do not let it hang by brake hose.

3. Pry off the grease cap and remove the cotter pin, nut and washer.
4. Remove the outer bearing.
5. Remove the brake rotor.
6. Remove the bearing inner bearing and seal from the rotor hub, using a soft drift or press.
7. Remove the bearing inner and outer race(s) from the rotor, using a soft drift or press.

To install:

8. Clean and inspect mating surfaces for bearing races.
9. Install new races, using soft drift or press.
10. Pack the new bearing with grease and set it into the inner race.
11. Install seal, making sure it is square in the rotor hub.
12. Install rotor, outer bearing, washer, nut and adjust bearing play.
13. Install cotter pin and dust cap.
14. Install caliper assembly.
15. If hydraulic lines had be removed, install and bleed brakes.
16. If parking brake cable has been remove, install and adjust as necessary.
17. Install the wheel assembly.

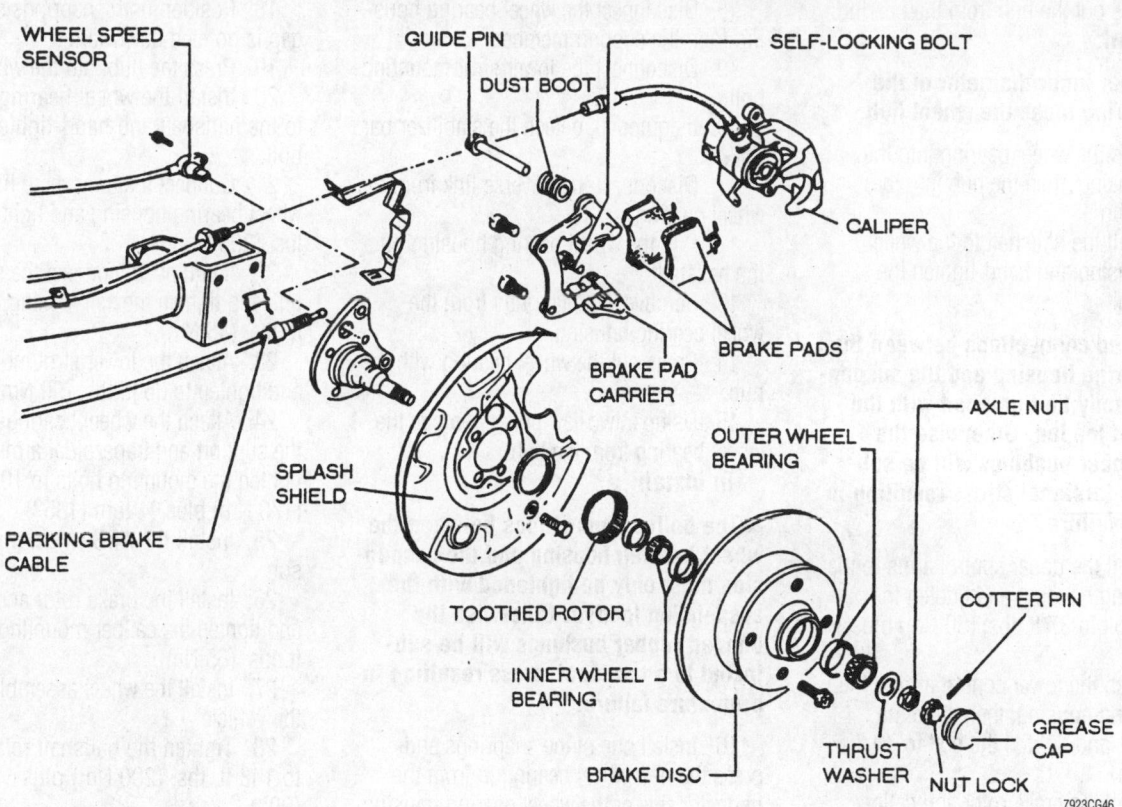

Exploded view of the rear wheel bearing—front wheel drive vehicles

18. Lower vehicle and check brakes for proper operation.

AWD A4 MODELS

1. Raise and safely support the vehicle.
2. Remove the wheel assembly.
3. Loosen, then remove the halfshaft retaining bolt.
4. Remove the ABS wheel speed sensor from the wheel bearing housing.
5. Remove the nut for the stabilizer bar connecting link to the wheel bearing housing.
6. Separate the track rod from the wheel bearing housing.
7. Remove the caliper mounting bolts and suspend the caliper from the body.
8. Remove the brake rotor.
9. Matchmark the position of the eccentric washer for the lower control arm-to-wheel bearing housing bolt and remove.
10. Remove the wheel bearing housing-to-upper control arm mounting bolt.
11. Remove the halfshaft from the wheel bearing housing.
12. Clean any dirt and debris from around the machined area for the wheel bearing path.
13. Place the wheel bearing housing in a press, then using an appropriate bearing driver, press out the bearing with the hub.

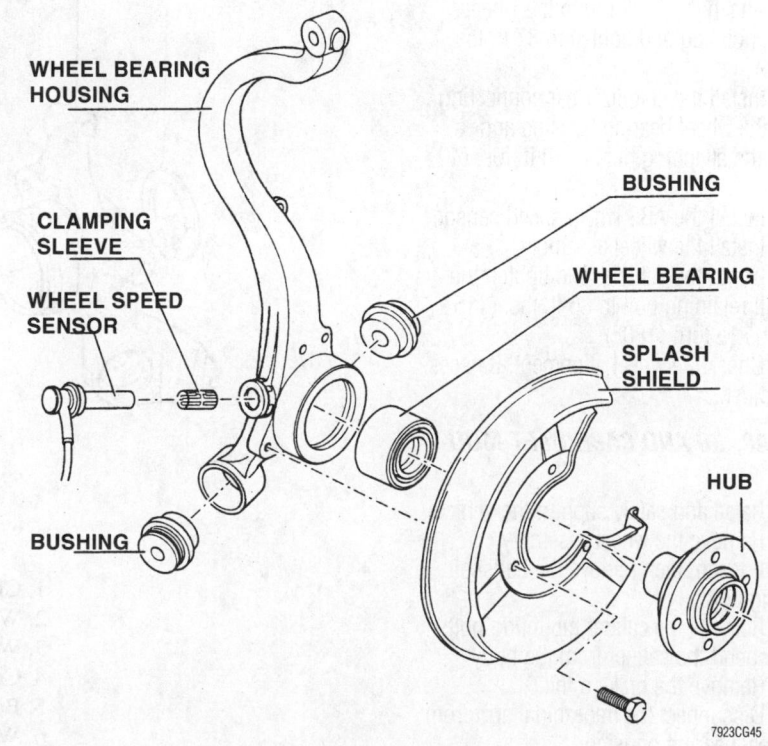

Exploded view of the rear wheel bearing housing—AWD A4 models

14. Press out the hub from the bearing.

To install:

➡**The larger inner diameter of the wheel bearing faces the wheel hub.**

15. Press the wheel bearing into the bearing housing, then the hub into the wheel bearing.

16. Install the halfshaft to the wheel bearing housing and hand-tighten the retaining bolt.

➡**The bolted connections between the wheel bearing housing and the suspension must only be tightened with the suspension loaded. Otherwise the bonded rubber bushings will be subjected to a torsional stress resulting in premature failure.**

17. Install the upper control arms to the wheel bearing housing and tighten the attaching bolt to 37 ft. lbs. (50 Nm) plus ¼ turn (90°).

18. Attach the lower control arm to the wheel bearing housing, then align the matchmarks and tighten the bolt to 70 ft. lbs. (95 Nm).

19. Install the brake rotor and caliper, and tighten the caliper mounting bolts as follows:
   • If equipped with socket-head bolts, tighten to 44 ft. lbs. (60 Nm)
   • If equipped with ribbed bolts, tighten to 70 ft. lbs. (95 Nm)

20. Attach the track rod to the wheel bearing housing and tighten to 37 ft. lbs. (50 Nm).

21. Install the stabilizer bar connecting link to the wheel bearing housing and tighten the attaching nuts to 30 ft. lbs. (40 Nm).

22. Install the ABS wheel speed sensor.

23. Install the wheel assembly.

24. Lower the vehicle, and tighten the halfshaft retaining bolt to 85 ft. lbs. (115 Nm) plus ½ turn (180°).

25. Check the wheel alignment, if necessary, adjust.

AWD A6, S6 AND CABRIOLET MODELS

1. Raise and safely support the vehicle.

2. Remove the wheel assembly.

3. Loosen, then remove the halfshaft retaining bolt.

4. Remove the caliper mounting bolts and suspend the caliper from the body.

5. Remove the brake rotor.

6. Disconnect the trapezoidal arm from the wheel bearing housing.

7. Remove the ABS wheel speed sensor from the wheel bearing housing.

8. Disconnect the wheel bearing housing from the support member.

9. Disconnect the lower strut mounting bolt.

10. If equipped, detach the stabilizer bar link rod.

11. Disconnect the traverse link from the wheel bearing.

12. Pull the wheel bearing housing off the halfshaft.

13. Remove the snaprings from the wheel bearing housing.

14. Press out the wheel bearing with the hub.

15. Using a two-jaw puller, remove the wheel bearing from the hub.

To install:

➡**The bolted connections between the wheel bearing housing and the suspension must only be tightened with the suspension loaded. Otherwise the bonded rubber bushings will be subjected to a torsional stress resulting in premature failure.**

16. Install one of the snaprings and press the new wheel bearing in from the opposite side of the wheel bearing housing.

17. After the bearing is seated against the snapring, install the other snapring.

18. Position both snaprings so that the gap is pointed downward.

19. Press the hub into the wheel bearing.

20. Install the wheel bearing housing to the halfshaft and hand-tighten the bolt.

21. Connect the transverse link to the wheel bearing housing and tighten to 148 ft. lbs. (200 Nm).

22. If equipped, connect the stabilizer link and tighten the self locking nuts to 33 ft. lbs. (45 Nm).

23. Attach the lower strut mounting bolt and tighten to 66 ft. lbs. (90 Nm).

24. Attach the wheel bearing housing to the support and trapezoidal arms, then tighten the mounting bolts to 104 ft. lbs. (170 Nm) plus ¾ turn (135°).

25. Install the ABS wheel speed sensor.

26. Install the brake rotor and caliper, and tighten the caliper mounting bolts to 48 ft. lbs. (65 Nm).

27. Install the wheel assembly, and lower the vehicle.

28. Tighten the halfshaft retaining bolt to 148 ft. lbs. (200 Nm) plus ¼ turn (90°).

29. Check the wheel alignment, if necessary, adjust.

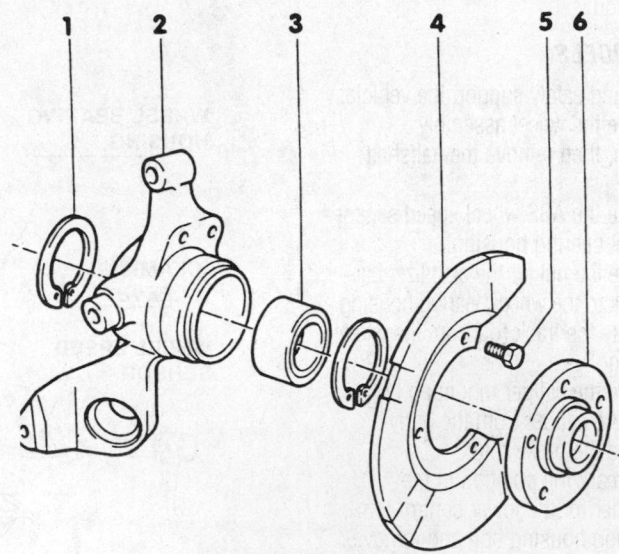

1. Circlip
2. Wheel bearing housing
3. Wheel bearing
4. Cover plate
5. Bolt
6. Wheel hub

7923CG47

Exploded view of the wheel bearing and related components—AWD A6 and S6 models

BMW

M3 • Z3 • 3 Series • 5 Series • 7 Series • 8 Series

14

ENGINE REPAIR

➡ **Disconnecting the negative battery cable on some vehicles may interfere with the functions of the on-board computer systems and may require the computer to undergo a relearning process.**

Ignition Timing

ADJUSTMENT

All ignition and fuel injection functions are controlled by the Digital Motor Electronics (DME) control unit. Ignition timing is fully electronically controlled; there is no vacuum advance or manual adjustment. Ignition functions are calculated from internal maps and from the same sensors used for the fuel injection system. Vehicles with an automatic transmission, the control unit will retard ignition timing briefly when the transmission is about to shift up or down. For this reason, there is a data link between the DME control unit and the transmission control unit.

Variations of this ignition system are used on different engines. The M42B18 engine uses distributorless ignition with a coil pack mounted on the inner fender. The M50B25 engine uses distributorless ignition with a coil pack mounted above each spark plug.

Since the ignition timing is controlled by the DME, checking and adjusting the timing is impossible. There is no method of setting dynamic or static timing.

Engine Assembly

REMOVAL & INSTALLATION

M42/M44 Engines

1. Disconnect the battery ground cable. Remove the transmission and remove the engine splash guard. Disconnect the gas spring and prop rod and support hood safely in the fully open position.

2. Remove the fan cowl by turning the expansion rivets on the left and right sides. Lift the cowl up and out of the engine compartment.

3. Hold the fan pulley while unscrewing the fan nut from the shaft. The shaft uses left-hand threads; turn the nut counterclockwise to unscrew.

4. Drain the coolant from the engine block. Disconnect the bottom hose from the radiator expansion tank, the engine coolant hoses and the heater hoses from the splash wall. Drain all coolant into clean containers for reuse or proper disposal.

5. Disconnect the air flow meter electrical plug and loosen the hose clamp and mounting screws. Lift the air sensor with the air cleaner up and out of the engine compartment.

6. Unclip the throttle cable and pull the cable out with the rubber holder.

7. Disconnect the fuel lines taking note of their positions. Pull off the vent hose to the filter for tank venting.

8. Disconnect the vacuum fitting at the brake booster.

9. Remove the ignition leads from the coil. Unscrew the connections at the alternator and starter. Disconnect the 2 plugs from the electrical duct. Remove the plug from the throttle valve potentiometer located at the throttle neck. Pull off the tank venting valve plug located next to the air cleaner. Disconnect the fuel injector plug located at the end of the electrical duct near to the fuel pipes. Pull off the idle speed control connector at the rear of the intake manifold. Disconnect the oil pressure switch electrical connection.

10. Unscrew the front and rear intake manifold supports.

11. Remove the electrical duct from the engine. Disconnect the coolant temperature senders for the gauge and the DME.

12. Disconnect the electrical duct and wiring harness on the engine and lay it off to the side of the engine.

13. Use a suitable lifting yoke to attach to the engine lifting eyes. Unscrew the motor mounts and the engine ground strap. Lift out the engine.

To install:

14. Lower engine into engine compartment. Fasten the motor mounts and the ground strap.

15. The balance of installation is the reverse of the removal procedure.

16. Install the fan using tool 11 5 040 or equivalent. Tighten the nut to 29 ft. lbs. (40 Nm). If using the fan tool, set the torque wrench to 22 ft. lbs. (30 Nm); the additional length of the tool multiplies the torque to achieve 29 ft. lbs. (40 Nm) at the nut.

17. Add the proper coolant mixture and bleed the cooling system.

18. Connect the battery leads and check all fluid levels before starting the engine.

M50/M52/S50 Engines

3 SERIES

1. Disconnect the battery ground cable.
2. Remove the transmission from the vehicle.

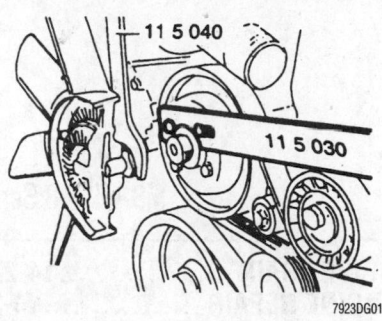

Cooling fan removal tools

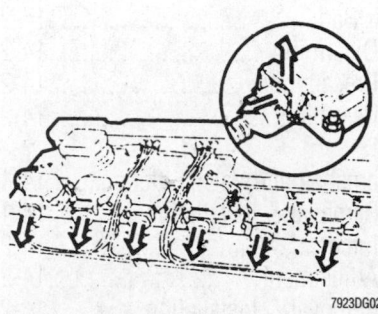

Ignition leads on coil—3 Series with M50/M52/S50 engines

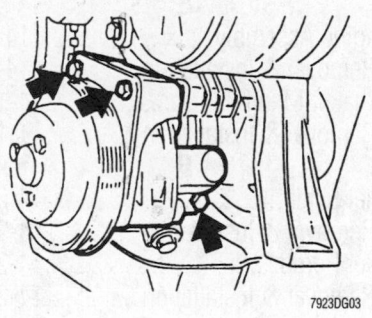

Power steering pump mounting bolts—3 Series with M50/M52/S50 engines

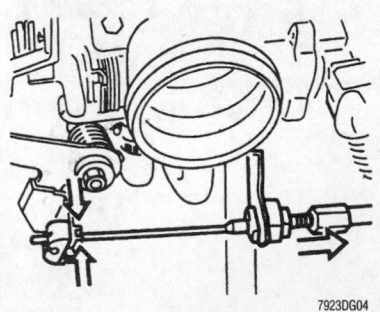

Throttle cable—3 Series with M50/M52/S50 engines

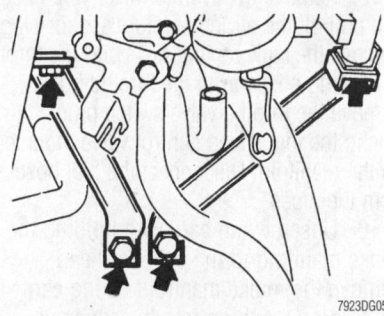

Front and rear intake manifold supports—3 Series with M50/M52/S50 engines

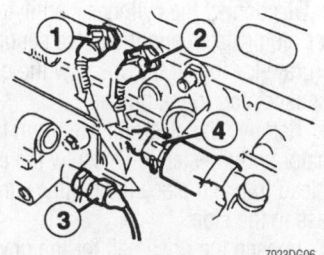

Temperature sensor (1), temperature gauge (2), oil pressure switch (3) and idle speed control valve (4) electrical connector locations—3 Series with M50/M52/S50 engines

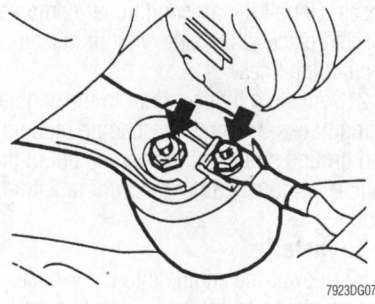

Right engine mount, showing the ground cable attachment—3 Series with M50/M52/S50 engines

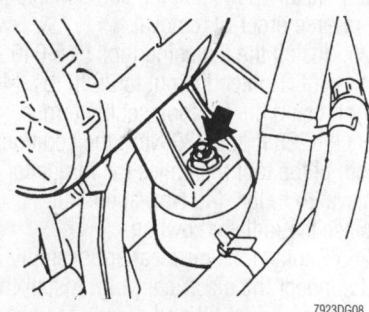

Left engine mount—3 Series with M50/M52/S50 engines

Air cleaner and air mass sensor—3 Series with M50/M52/S50 engines

3. Remove the engine splash guard.

4. Press the hinge so it goes over center and support hood safely in the fully open position.

5. Disconnect the air mass sensor plug and loosen the air intake duct hose clamp. Remove the air cleaner assembly.

6. Disconnect the hoses for the idle speed control and the crankcase breather.

7. Unscrew and remove the ducting for the alternator. Pull out the fan cowl expansion rivets and remove the cowl upwards.

8. Hold the pulley with tool 11–5–030 or equivalent, and unscrew the fan clockwise. Remove the fan and keep upright.

9. Drain the cooling system. The engine block drain plug is accessible through the exhaust manifold.

10. Remove the upper and lower radiator hoses from the radiator. Disconnect the coolant level switch and the automatic transmission cooler lines, if equipped. Plug the cooler lines.

11. Disconnect the right side hose and the temperature sensor.

12. Insert a tool into the radiator support clips and press down on the tab. Pull back on the radiator to release it. Remove the radiator.

13. Disconnect the heater hoses from the heater and heater valve.

14. Remove the grill from the air intake cowl at the base of the windshield.

   a. Remove the electrical lead tray.

   b. Remove the screws on the right side cowl holder bracket and the screw on the left side.

   c. Remove the cowl from the engine compartment.

15. Unscrew the fastener from the throttle cable cover and pull the cover forward and off. Unclip the cable and pull the cable out with the rubber holder.

16. Pull the vacuum fitting from the brake booster and plug the openings.

17. Remove the engine and intake manifold covers. Unscrew the bolt holding the

ground strap on the front lifting eye. Replace the bolt before lifting the engine.

18. Unscrew the 2 bolts holding the plug plate and pull off the plug plate. Be careful not to damage the rubber seals. Take off the ignition coil electrical plugs. Remove the plug plate complete with the electrical leads.

19. Remove the cylinder head vent hose and pull off the air temperature sensor plug.

20. Remove the tank venting hose and the throttle heating hoses from the throttle body.

21. Remove the throttle valve switch plug.

22. Unclip the idle speed control valve mounted on the manifold.

23. Disconnect the fuel hoses from the pipes.

24. Unscrew the hardware holding the intake manifold to the cylinder head. Remove the intake manifold taking care not to drop anything into the exposed ports.

25. Disconnect the plugs from the temperature sensor, temperature gauge, the oil pressure switch and the idle speed control valve.

26. Disconnect the cylinder identifying sender plug (black) and the pulse sender plug (gray) for the DME. Unscrew the oxygen sensor plug in the holder.

27. Remove the electric leads from the alternator and the starter. Unscrew the electrical lead tray and place the engine wiring harness to the side.

28. Loosen the drive belt for the power steering pump and the air conditioner compressor by turning their respective tensioners clockwise. This will release the tension on the belt and allow the belt to be removed.

29. Unbolt the power steering pump and place to the side without disconnecting the hoses.

30. Unbolt the air conditioner compressor and place to the side without disconnecting the lines.

31. Attach a lifting fixture to the engine lifting hooks.

32. Unscrew the engine mounts and ground strap.

33. Lift the engine out of the vehicle being careful of the front radiator mount.

To install:

34. Lower the engine into the vehicle and attach the motor mounts and ground strap. Tighten the engine mount: 8mm bolts to 16 ft. lbs. (22 Nm), and the 10mm bolts to 31 ft. lbs. (42 Nm).

35. Install the power steering pump and the air conditioner compressor. Install the drive belts.

36. Reposition the wiring harness and electrical lead tray on the engine. Connect the leads to the starter and alternator.

37. Screw in the plug for the oxygen sensor holder.

38. Connect the leads for the cylinder identifying sender, DME pulse sender, temperature sensor, temperature gauge sender, oil pressure switch and idle speed control valve.

39. Install the intake manifold making sure that the intake seals are intact. Replace the intake seals if any signs of deterioration are evident.

40. Attach the fuel lines. The upper line is the return and the lower is the feed.

41. Attach the idle speed control valve hose located on the manifold.

42. Connect the throttle valve switch plug, the throttle valve heating lines and the tank vent line.

43. Connect the air temperature sensor plug and attach the cylinder head venting hoses.

44. Reconnect the plugs for the ignition coils and mount the plug plate. Attach the ground strap to the front lifting eye.

45. Replace the engine and manifold covers. Connect the line to the brake booster.

46. Reconnect the throttle cable and cover. Install the air intake cowl at the base of the windshield.

47. Connect the heater hoses to the valve and inlet.

48. Remount the radiator by pressing down on the mounting clips to fasten. Check that the lower mounts are in place.

49. Connect the temperature switch plug for the air conditioner and replace the trim panel.

50. Connect the cooling system hoses and the automatic transmission lines. Use new seals on the transmission lines and tighten to 13–15 ft. lbs. (18–21 Nm).

51. Install and tighten the engine block drain plug.

52. Install the fan using tool 11–5–040 or equivalent wrench and holding tool 11–5–030 or equivalent. Tighten the nut to 29 ft. lbs. (40 Nm). If using the tool set the torque wrench to 22 ft. lbs. (30 Nm): the additional length of the tool multiplies the torque to achieve 29 ft. lbs. (40 Nm) at the nut.

53. Replace the radiator cowling into its mounting slots and press in the rivets.

54. Install the alternator air ducting.

55. Connect the idle speed and crankcase breather hose to the air intake duct. Replace the air cleaner assembly and connect the electrical plug.

56. Install the transmission to the vehicle.

57. Refill and bleed the cooling system.

58. Install the splash shield.

59. Connect the negative battery cable.

60. Check all fluids before starting engine.

5 SERIES

1. Disconnect the battery terminals, negative side first and remove the battery. Unscrew and remove the battery tray. Remove the transmission.

2. Loosen the clamp on the cooling duct to the alternator and remove the duct.

3. Disconnect the plug to the air flow meter and loosen the clamps to the air cleaner duct. Unscrew the mounting bolts and remove the air cleaner assembly.

4. Pull out the expansion rivets that hold the fan cowl. Remove the cowl by pulling up out of the engine compartment.

5. Hold the fan pulley while unscrewing the fan nut from the shaft. The shaft uses left-hand threads; turn the nut counterclockwise to unscrew.

6. Drain the coolant from the block. The drain plug is located between the exhaust manifolds. Disconnect the coolant hoses from the radiator and remove the coolant level switch plug. On automatic transmission equipped vehicles, remove the oil lines to the radiator and plug.

7. Disconnect the bottom radiator hose and remove the trim panel from the right side of the engine compartment to expose the side of the radiator and the air conditioner condenser.

8. Pull the plug off of the air conditioner temperature switch.

9. Remove the radiator supporting clips by inserting a small prybar down from above into the slot and pulling back. Pull the radiator free from the clip. Remove the radiator from the vehicle.

10. Disconnect the heater hoses from the heater valve and the heater.

11. Unscrew the fastener from the throttle cable cover and pull the cover forward and off. Unclip the cable and pull the cable out with the rubber holder.

12. Pull the vacuum fitting from the brake booster and plug the openings.

13. Remove the engine and intake manifold covers. Unscrew the bolt holding the ground strap on the front lifting eye. Replace the bolt before lifting the engine.

14. Unscrew the 2 bolts holding the plug plate and pull off the plug plate. Be careful not to damage the rubber seals. Take off the ignition coil electrical plugs. Remove the plug plate complete with the electrical leads.

15. Remove the cylinder head vent hose and pull off the air temperature sensor plug. Remove the tank venting hose and the throttle heating hoses from the throttle body. Remove the throttle valve switch plug. Unclip the idle speed control valve mounted on the manifold. Disconnect the fuel hoses from the pipes.

16. Unscrew the hardware holding the intake manifold to the cylinder head. Remove the intake manifold taking care not to drop anything into the exposed ports.

17. Disconnect the plugs from the temperature sensor, temperature gauge, the oil pressure switch and the idle speed control valve. Disconnect the cylinder identifying sender plug (black) and the pulse sender plug (gray) for the DME. Unscrew the oxygen sensor plug in the holder.

18. Remove the electric leads from the alternator and the starter. Unscrew the electrical lead tray and place the engine wiring harness to the side.

19. Loosen the drive belt for the power steering pump and the air conditioner compressor by turning their respective tensioners clockwise. This will release the tension on the belt and allow the belt to be removed.

20. Unbolt the power steering pump and place to the side without disconnecting the hoses. Unbolt the air conditioner compressor and place to the side without disconnecting the lines.

21. Attach a lifting fixture to the engine lifting hooks. Unscrew the engine mounts and ground strap. Lift the engine out of the vehicle being careful of the front radiator mount.

To install:

22. Lower the engine into the vehicle and attach the motor mounts and ground strap.

23. Install the power steering pump and the air conditioner compressor. Install the drive belts.

24. Install the remaining components in the reverse order of removal.

25. Install the fan using tool 11 5 040 or equivalent. Tighten the nut to 29 ft. lbs. (40 Nm). If using the fan tool, set the torque wrench to 22 ft. lbs. (30 Nm); the additional length of the tool multiplies the torque to achieve 29 ft. lbs. (40 Nm) at the nut. Replace the radiator cowling.

26. Replace the air cleaner assembly and connect the electrical plug. Install the transmission and fill and bleed the cooling system. Install the battery tray and battery. Check all fluids before starting engine.

M60/M62 and M70/M73/S70 Engines

5 SERIES

1. Allow the engine to cool. Disconnect the battery cable from the battery, negative side first. Open the hood to the widest position possible and secure in place.

2. Remove the splash shield from under the vehicle. Drain the coolant and remove the radiator.

3. Remove the transmission.

4. Remove the nut holding the transmission oil cooler lines to the engine oil pan.

5. Loosen and remove the drive belts to the power steering pump and the air conditioner compressor. Remove the bolts holding the pump and compressor to the engine and remove them from the engine, keeping the lines connected. Wire the pump and compressor out of the way and without any tension on the hoses.

6. Disconnect the hoses from the coolant expansion tank. remove the screws on the side of the expansion tank and remove the expansion tank from the engine compartment.

7. Disconnect the heater hoses from the heater control valve and the heater inlet pipe.

8. Pull off the connections to the ignition coil. Remove the air cleaner assembly. Disconnect and remove the idle speed control from the intake duct.

9. Disconnect the harness to the air flow meter. Disconnect the ducting to the air flow meter and remove along with the crankcase breather vacuum line.

10. On non-ASC equipped cars, disconnect the cruise control cable and the throttle cable at the throttle. Remove the cable mounting bracket. If equipped with ASC, remove the connector to the throttle control unit as there will be no throttle cable to disconnect.

11. Disconnect the leads to the starter. Detach the 2 electrical connectors in the starter area. Detach the oil level sender leads and the alternator connections. Remove the air duct to the alternator.

12. Disconnect the tank venting valve and the hose to the carbon canister.

13. Mark the feed and return fuel lines. Disconnect the lines and catch any spilled fuel.

14. Disconnect the vacuum line from the brake booster and plug the opening. Disconnect the harness connections to the temperature sensors and DME sensors.

15. Disconnect the ground strap and make a check for any remaining lines or electrical leads still attached.

16. Attach a lifting sling to the engine. Remove the engine mount nuts and bolts and lift the engine from the engine bay.

To install:

17. Install the engine into the engine compartment. Tighten the engine mounts to 32.5 ft. lbs. (45 Nm). Connect the ground strap.

18. Connect the brake booster vacuum line, the temperature sensors and the DME sensors.

19. Connect the fuel lines to the proper locations as previously marked. Connect the tank venting valve and the carbon canister line. Attach the alternator leads and the cooling duct.

20. The remaining components are installed in the reverse order of removal.

21. Install the transmission and the radiator. Fill the cooling system and check the engine fluids. Install the splash shield and connect the battery.

22. Run the engine and check for leaks. Bleed the cooling system.

7 SERIES

1. Disconnect the negative battery cable, then the positive. Remove the transmission. Scribe hinge locations and remove the hood, or remove support struts and prop it securely all the way up.

2. Remove the splash guard from underneath the engine. Then, with the engine cool, remove the drain plugs in the radiator and block and drain the engine coolant.

3. Loosen the power steering pump bolts from underneath. Turn the adjusting pinion to loosen the belt and remove the belt. Then, remove the mounting bolts and remove the power steering pump without disconnecting the hoses. Support the pump out of the way so as to avoid stressing the hoses.

4. Do the same with the air conditioner compressor, this unit does not have the belt adjusting pinion—it is necessary only to loosen all the bolts and push the compressor toward the engine to remove the belt.

5. Loosen the air intake hose clamp and disconnect the hose. Remove the mounting nut, then remove the air cleaner(s).

6. Unscrew the oil filter cover bolt and disconnect the oil cooler lines and the plug from the oil pressure switch on 750iL.

7. The unit on the opposite side of the intake hose from the air cleaner contains the idle speed control valve, which must be removed next. Loosen the hose clamps and pull off the hoses. Disconnect the electrical connector. Remove the mounting nut, then pull the idle speed control out of the air intake hose.

8. Pull off the retainers for the air flow sensor, then pull the unit off its mounts,

disconnecting the vacuum hose from the PCV system at the same time.

9. Working on the coolant expansion tank, detach the electrical connector. Remove the nuts on both sides. Loosen their clamps, then disconnect the hoses and remove the tank.

10. Disconnect the heater hoses at both the control valve and at the heater core.

11. Disconnect the throttle and cruise control cables at the throttle lever. Unbolt the cable housing retainer and remove the housing and cables.

12. Pull off the low amperage starter connectors and detach the high amperage connector coming from the battery.

13. Loosen its clamp, then disconnect the coolant hose that runs to the alternator.

14. Disconnect the connecting plug for the oxygen sensor, as well as the other plugs.

15. Loosen the clamps, then disconnect the fuel supply and return pipes.

16. Disconnect the fuel pipe at the injector supply manifold. Disconnect the plug. Detach the electrical connector at the throttle body. Lift off the protective caps, then remove the attaching nuts for the protective cover for the wiring harness for the injectors and remove it.

17. Disconnect the ground strap at the block. Remove the engine mount nut from the top on both sides.

18. Attach a lifting sling to the engine and support the assembly. Disconnect the ground lead. Carefully lift the engine out of the compartment, tilting the front of the engine upward for clearance.

To install:

19. Keep these points in mind during installation:

   a. Tighten the engine mounting bolts to 32.5 ft. lbs. (43 Nm).

   b. Adjust the belt tension for the air conditioning compressor and power steering pump drive belts to give 1/2 – 3/4 in. deflection.

   c. Tighten the oil cooler line flare nuts to 25 ft. lbs. (34 Nm).

   d. When reconnecting the intake manifold to the throttle necks, inspect, if necessary, replace the O-rings. Tighten the mounting nuts to 6.5 ft. lbs. (8 Nm).

20. Lower the engine into the engine compartment. When the engine is positioned, the guide pin must fit in the bore of the axle carrier. Tighten the mounting bolts on the front axle carrier (small bolt) to 18–20 ft. lbs. (25–27 Nm); the larger bolt to 31–35 ft. lbs. (40–47 Nm). The mount-to-bracket bolts are tightened to 31–35 ft. lbs. (40–47 Nm).

21. Connect the fuel lines, use new hose clamps to connect the fuel lines to the fuel filter. Connect all of the multi-prong plugs and all vacuum hoses.

22. The balance of installation is the reverse of the removal procedure.

23. Be sure all fluid levels are correct before starting the engine. Bleed air from the cooling system.

8 SERIES

1. Disconnect the negative battery cable, then the positive. Remove the transmission. Scribe hinge locations and remove the hood, or remove support struts and prop it securely all the way up.

2. Remove the mass air flow sensor from the engine. Remove the windshield washer tank.

3. Loosen the oil filter cap to permit the oil to drain back into the oil pan. Then, remove the oil filter lines.

4. Remove the radiator and expansion tank.

5. Unclip the diagnostic clip. Disconnect the wires at the left and right coils. Unscrew the right ignition coil.

6. Disconnect the D+ (thin lead) from the alternator. Disconnect the oil sender.

7. Disconnect the hoses and wire connections from the tank venting valves.

8. Remove the oil catching tray. Detach the wire connectors from the sensors, injectors and throttle body.

9. Disconnect the wiring cover at the rear of the engine.

10. Disconnect hoses at pressure regulator, noting there arrangement.

11. Disconnect the temperature sensors and remove the injector wiring harness.

12. Disconnect the alternator main feed wires, at the B+ connection point. Disconnect the starter wire connections.

13. Remove the air conditioning compressor, leaving the hoses connected.

14. Remove the cool air duct for the alternator.

15. Disconnect and plug the fuel lines.

16. Remove the heat shields located under the vehicle on the thrust struts.

17. Remove the heat shields on the right exhaust manifold and remove the right exhaust pipe from the manifold.

18. Drain the power steering fluid and remove the hose from the supply tank.

19. Remove the starter assembly. Disconnect any necessary connections.

20. Disconnect the heater hoses.

21. Attach a suitable lifting device to the engine and unscrew the ground strap and engine mounts.

22. Remove the guide tube for the oil dipstick.

23. Lift the engine slightly and remove the right engine bracket. Turn the rear of the engine to the right to clear the left exhaust pipe past the steering spindle.

24. Remove the engine and place on a suitable holding fixture.

To install:

25. Keep these points in mind during installation:

   a. Tighten the engine mounting bolts to 32.5 ft. lbs. (43 Nm).

   b. Adjust the belt tension for the air conditioning compressor and power steering pump drive belts to give ½ – ¾ in. deflection.

   c. Ensure all hose and wires are connected as prior to removal.

26. Lower the engine into the engine compartment. The mount-to-bracket bolts are tightened to 31–35 ft. lbs. (40–47 Nm).

27. Install the remaining components in the reverse order of removal.

28. Be sure all fluid levels are correct before starting the engine. Bleed air from the cooling system.

Water Pump

REMOVAL & INSTALLATION

M42/M44 Engines

1. Disconnect the negative battery cable. Drain the cooling system.

2. Remove the drive belt and the water pump pulley.

3. Remove the pump mounting bolts.

4. Screw 2 M6 bolts into the tapped bores and press the water pump out of the cover uniformly.

To install:

5. Lubricate and install a new O-ring.

6. Install the water pump and tighten the bolts to 6 ft. lbs. (9 Nm).

7. The remaining components are installed in the reverse order from which they were removed.

8. Start the engine and check for coolant leaks.

M50/M52/S50 and M70/M73/S70 Engines

1. Disconnect the negative battery cable. Drain the cooling system.

2. Remove the fan cowl and fan, if necessary.

3. Remove the drive belt and the pulley. Disconnect the bracket, if necessary.

4. Remove the air cleaner with the air flow sensor, if needed.

5. Disconnect the cooling hoses and remove the water pump.

6. The installation is the reverse of the removal procedure. Tighten the M8 bolts to 16 ft. lbs. (22 Nm) and the M6 bolts to 6.5 ft. lbs. (9 Nm).

M60/M62 Engines

1. Disconnect the negative battery cable.

2. Drain the cooling system.

3. Remove the heat shields at the left and right-hand sides of the front axle carrier.

4. Remove the front and rear engine splash guards.

5. Remove the fan. The fan must be held stationary with tool 11–5–030 or some sort of flat blade cut to fit over the hub and drilled to fit over 2 of the studs on the front of the pulley. Remove the fan coupling nut; left-hand thread-turn clockwise to remove.

6. Remove the drive belt tensioner, and the serpentine drive belt.

7. Remove the vibration damper and hub. There are eight mounting bolts, and a central bolt. Use a suitable holding tool 11–2–230 or equivalent for the central bolt.

8. Disconnect the coolant hose at the cover of the thermostat housing. Remove the thermostat.

9. Remove the water pump pulley by counterholding the pulley with the drive belt and removing the four pulley mounting bolts.

10. Disconnect the hoses from the water pump. Remove the 6 mounting bolts and remove the water pump.

To install:

11. Clean the gasket surfaces and use a new gasket.

12. Check for the correct seating of the dowel sleeves. Install the water pump in position. Tighten the M8 mounting bolts to 16 ft. lbs. (22 Nm), and the M6 mounting bolts to 7 ft. lbs. (10 Nm). Connect the hoses.

13. Install the vibration damper and hub. Tighten the eight mounting bolts to 17 ft. lbs. (24 Nm). Tighten the central mounting bolt in four steps as follows:

   • An initial torque of 74–81 ft. lbs. (100–110 Nm)
   • Add an additional 60 degrees torque
   • Add an additional 60 degrees torque
   • Then, add an additional 30 degrees torque.

14. Install the thermostat. Connect the coolant hose.

15. Install the drive belt tensioner and the drive belt.

16. Install the pulley and tighten the bolts to 6–7 ft. lbs. (8–10 Nm). Install the belt and tighten. Install the fan.

17. Install the heat shields and splash guards.

18. Connect the negative battery terminal.

19. Refill and bleed the cooling system.

Cylinder Head

REMOVAL & INSTALLATION

M42/M44 Engines

1. Disconnect the negative battery cable.

2. Remove the ignition coil cover and pull off the spark plug connectors.

3. Remove the complete ignition tackle. Remove the cylinder head cover.

4. Disconnect the coolant hoses and unscrew the temperature sensor.

5. Remove the thermostat housing and thermostat. Unscrew the upper timing case cover.

6. Rotate the engine in the direction of the rotation until the camshaft peaks of the intake and exhaust camshafts for cylinder No. 1 face each other. The arrows on the sprocket face up.

7. Remove the chain tensioner. Remove the upper chain guide, chain guide bolt on the right side and the sprockets.

8. Remove the cylinder head bolts from the outside to the inside in several steps using the proper tool.

9. Remove the cylinder head. Clean the sealing surfaces on the cylinder head and the crankcase.

To install:

10. Install the cylinder head onto the engine with a new gasket.

11. Tighten the cylinder head in 3 steps by following the sequence shown as follows:

    a. Step 1—24 ft. lbs. (32.5 Nm)

    b. Step 2—90–95 degree turn

    c. Step 3—plus an additional 90–95 degree turn.

12. The balance of installation is the reverse of the removal procedure.

13. Connect the negative battery cable, then start the engine and inspect for any fuel, vacuum or coolant leaks.

M50/M52/S50 Engines

1. If engine is not already removed from the vehicle, disconnect the negative battery cable and drain the engine coolant. Remove the intake manifold and throttle valve. Disconnect the exhaust pipes and the oxygen sensor wires. Remove the exhaust manifolds. Remove the thermostat housing and engine lifting eye.

2. Pull off the connectors for the ignition coils and remove the coils. Unscrew the cylinder heads cover and remove. Remove the sender from the head and the electrical lead duct.

3. Remove the upper timing case cover and the camshaft cover. Crank the engine in the direction of rotation so the intake and exhaust camshaft peaks for cylinder No. 1 face each other. Hold the camshafts in place with tool 11 3 240 or equivalent. With the camshafts in this alignment, the arrows on the sprockets will be facing up. Remove the valve cover mounting studs. Lock the flywheel in place to prevent movement of the crankshaft.

4. Unscrew the chain tensioner and carefully remove. There is a spring contained within the tensioner and may eject if care is not taken.

5. Press down on the upper chain tensioner and lock it into place using tool 11 3 290 or equivalent. Unscrew the transfer timing chain sprockets and pull the 2 off together with the chain. Remove the upper chain tensioner and the lower chain guide. Pull off the main timing chain sprocket along with the chain. Use a bent piece of wire to hold the chain from falling down into the engine. Do not rotate the engine

after this point or the valve timing will be disturbed when the engine is reassembled.

6. Unscrew the bolts on the head at the ends of the cams. Using a proper sized Torx® bit or tool 11 2 250, loosen the cylinder head bolts in several steps. Use an outside to inside pattern to prevent warpage. On production heads the bolt washers are locked into place while on replacement heads the washers are loose. Keep track of the bolt washers.

To Install:

7. If the camshafts have been removed and reinstalled a waiting period dependent on the ambient temperature is necessary before mounting the cylinder head on the engine. At room temperature wait 4 minutes to allow the lifters to compress fully. At temperatures down to 50° F (10° C) wait 11 minutes. At temperatures lower than 50° F (10° C) wait 30 minutes. This is to prevent contact between the valves and the piston tops. The engine may not be cranked under the same condition for a period of 10 minutes at room temperature; 30 minutes for temperatures down to 50° F (10° C); 75 minutes for temperatures below 50° F (10° C).

8. Clean all mounting surfaces and check the head for warpage. Take care not to drop any pieces of gasket or dirt into the oil or coolant passages. Check the condition of the head locating dowel sleeves.

9. Place a new head gasket on the engine block over the locating dowels and gently place the head on the engine. Align the head with the dowel sleeves and check that the head sits flat on the engine.

10. Cylinder head bolts may only be used once. Lightly oil the threads of the new cylinder head bolts. Check that the head bolt washers are in place and install the bolts. Tighten the head bolts as follows.

11. If equipped with a cast iron block tighten in 3 steps; Step 1 to 22 ft. lbs. (30 Nm), Step 2 and 3 to 90 degree torque angle. Tighten the center bolts first and go out in a diagonal pattern.

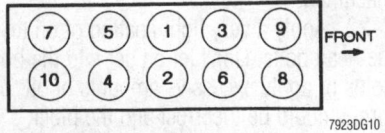

Cylinder head tightening sequence—
M42/M44 engines

7923DG10

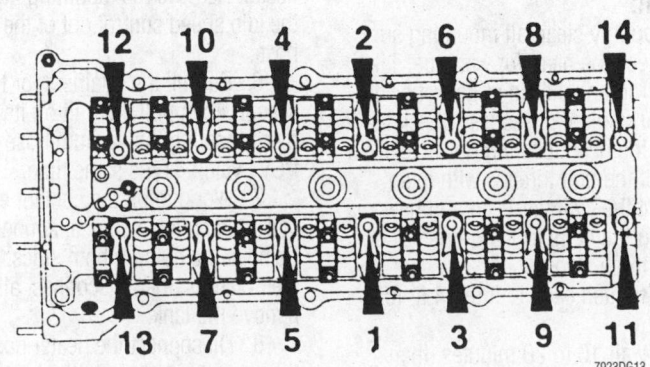

Cylinder head mounting bolt tightening sequence—M50/M52/S50 engines

7923DG13

12. If equipped with a cast aluminum block tighten in 3 steps; Step 1 to 29.5 ft. lbs. (40 Nm), Step 2 and 3 to 90 degree torque angle. Tighten the center bolts first and go out in a diagonal pattern.

13. Align main timing chain and sprocket on the can so the arrow faces up. The bolt holes in the camshaft should be on the left sides of the sprocket slots. This will allow the tensioner to take up the slack in the chain and rotate the gear to the counter-clockwise position.

14. The balance of installation is the reverse of the removal procedure.

15. Connect the negative battery cable, start the engine and check for any leaks.

M60/M62 Engines

1. Disconnect negative battery cable.

2. Remove both exhaust manifolds from each side of the engine. Remove the heat shields from the front axle carrier.

3. Drain the engine coolant, and remove the coolant expansion tank.

4. Remove the upper timing case cover.

5. Remove the oil pipes on the cylinder head.

6. Remove the intake manifold.

7. Remove the cylinder head cover.

8. Remove the engine vent pipe together with the O-ring. Disconnect all the coolant hoses on the coolant collector. Remove the coolant collector mounting bolts, and remove the coolant collector.

9. Remove all eight spark plugs.

10. Remove the camshaft sprockets, and the timing chain tensioner.

11. Remove the bolts retaining the guide rail on the cylinder head's left-hand side.

12. Remove the cyclone oil trap.

13. Remove the cylinder head bolts from the outside to inside. Lift off the cylinder head.

➡ **The cylinder head bolts must be replaced.**

To install:

14. Thoroughly clean all mounting surfaces and check the head for warpage. Take care not to drop any pieces of gasket or dirt into the oil or coolant passages. Check the condition of the head locating dowel sleeves and clean out the bolt threads with a tap.

15. Mount the cylinder head and new bolts. Tighten the bolts in the proper sequence in three steps:
- Step 1–tighten each bolt to 24 ft. lbs. (32 Nm)
- Step 2–wait 10 to 20 minutes, then turn each bolt an additional 80°

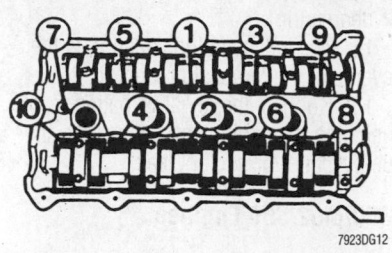

Cylinder head bolt tightening sequence—M60/M62 engines

- Step 3–wait 10 to 20 minutes, then turn each bolt an additional 80°

16. Install the cyclone oil trap.

17. Install the bolts retaining the guide rail on the cylinder head's left-hand side.

18. Install the camshaft sprockets and the timing chain tensioner. Tighten the sprocket mounting bolts to 11 ft. lbs. (15 Nm).

19. The remaining components are installed in the reverse order from which they were removed.

20. Connect negative battery cable.

M70/M73/S70 Engines

1. Unbolt the exhaust pipe connections at the manifold and at the transmission pipe clamp. Disconnect the negative battery cable.

2. Remove the splash shield from under the engine. With the engine cool, remove the drain plugs from the bottom of the radiator and block. Drain the engine oil.

3. Remove the fan. Lift out the expansion rivets on either side and remove the fan shroud.

4. Loosen the hose clamp and disconnect the air inlet hose. Remove the mounting nut and remove the air cleaner.

5. The unit on the opposite side of the intake hose from the air cleaner contains the idle speed control valve, which must be removed next. Loosen the hose clamps and pull off the hoses. Detach the electrical connector. Remove the mounting nut, then pull the idle speed control out of the air intake hose.

6. Pull off the retainers for the air flow sensor, then pull the unit off its mounts, disconnecting the vacuum hose from the PCV system at the same time.

7. Working on the coolant expansion tank, detach the electrical connector. Remove the nuts on both sides. Loosen their clamps, then disconnect all hoses and remove the tank.

8. Disconnect the heater hoses at both the control valve and at the heater core. Remove the valve, if needed.

9. Disconnect the throttle and cruise control cables at the throttle lever. Unbolt the cable housing retainer and remove the housing and cables.

10. Disconnect the plugs near the thermostat housing. Loosen the hose clamps and pull off the coolant hoses.

11. Disconnect the plug in the line leading to the oxygen sensor. Disconnect the other plugs.

12. Disconnect the fuel supply and return lines, collecting fuel in a metal container for safe disposal.

13. Detach the fuel pipe running along the cylinder head, near the manifold. Pull off the electrical connector at the throttle body. Remove the caps, then remove the attaching bolts and remove the wiring harness carrier and harness for the fuel injectors.

14. Disconnect the coil high tension lead. Disconnect the high tension wires at the plugs. Then, remove the mounting nuts and remove the carrier for the high tension wires from the head.

15. Remove the attaching nuts for the camshaft cover and remove it.

16. Turn the engine until the timing marks are at TDC and the No. 6 valves are at overlap, both slightly open.

17. Remove the upper timing case cover. Remove the timing chain tensioner piston.

18. Remove the upper timing chain sprocket bolts and pull the sprocket off, holding it upward, then supporting it securely so the relationship between the chain and sprockets top and bottom will not be lost.

19. Disconnect the upper radiator hose at the thermostat housing. Remove the bolts and remove the support for the intake manifold.

20. Remove the cylinder head bolts in the opposite of numbered order. Then, install 4 special pins part 11 1 063 or equivalent. This is necessary to keep the rocker arm shafts from moving. Then, lift off the head.

To install:

21. Make checks of the lower cylinder head and block deck surface to be sure they are true. Install a new head gasket, making sure all bolt, oil and coolant holes line up. Use a gasket marked M30B35. Use a 0.3mm thicker gasket, if the head has been machined.

22. Apply a very light coating of oil to the head bolts. Don't let oil get into the bolt holes or apply excessive amounts of oil, or torque could be incorrect and the block could crack. Use the type of bolt without a collar. Install the bolts finger-tight.

23. Tighten bolts 1–6 in the correct order to 42–44 ft. lbs. (57–60 Nm). Remove

the pins holding the rocker shafts in place. Now, complete the first stage of torquing by tightening bolts 7–14 in the correct order, to the same specification. Adjust the valves after a 15 minute wait. Tighten the bolts, in the correct order, with a torque angle gauge 30–36 degrees, using special tool 11 2 110 or equivalent.

24. Reinstall the timing sprocket to the camshaft. Be sure the camshaft is in proper time, that new lockplates are used and that nuts are properly tightened.

25. When reinstalling the timing cover, be sure to apply a liquid sealer to the joints between upper and lower timing covers. The remainder of installation is the reverse of removal. Note these points:

a. Adjust throttle, speed control and accelerator cables. Inspect and if necessary replace the exhaust manifold gasket.

b. When reinstalling the cylinder block coolant plug, coat it with sealer. Be sure to refill the cooling system and bleed it. Be sure to refill the oil pan with the correct amount of oil.

c. Install the timing chain so the down pin on the camshaft sprocket is at the 8 o'clock when its tapped bores are at right angles to the engine. Tighten the sprocket bolts to 6.5–7.5 ft. lbs. (8–10 Nm).

d. Check the camshaft cover gasket, replacing, as necessary. Tighten camshaft cover bolts in the order shown. Tighten the bolts to 6.5–7.5 ft. lbs, (8–10 Nm).

e. When reinstalling the fan shroud, be sure all guides are located properly.

f. Coat the tapered portion of the exhaust pipe connection flange with the proper sealant. Tighten the attaching nuts to 4.5 ft. lbs. (6 Nm) and loosen 1 ½ turns.

26. Start the engine and run it until hot (25 minutes). Then, again remove the valve cover and turn the head bolts, in the correct order, 30–40 degrees.

Rocker Arms/Shafts

REMOVAL & INSTALLATION

The BMW models covered do not use rocker arms/shafts. The camshaft directly actuates the valves.

Intake Manifold

REMOVAL & INSTALLATION

M42/M44 Engines

1. Disconnect the negative battery cable. Unscrew the upper manifold section.

2. Disconnect the rear mounting bracket and remove the coolant hose.

3. Loosen the front mounting bracket and disconnect the holder for the preheater.

4. Remove the mounting bolts and lift off the upper manifold section. Pull the hose off the fuel pressure regulator at the same time.

5. Pull the plug plate off the fuel injectors and remove the wire holding clamp.

6. Remove the injection pipe with the fuel injectors attached and remove the lower manifold section.

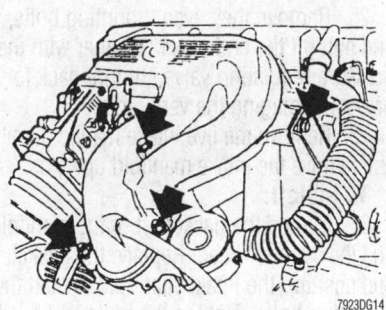

Upper intake manifold and support bracket mounting bolt locations—M42/M44 engines

Be sure the hollow bushings (1) are properly installed—M42/M44 engines

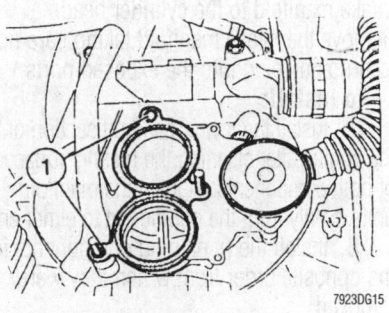

Check the dowel sleeves (1) for damage and correct installation position—M42/M44 engines

To install:

7. Install the lower manifold section onto the engine after cleaning the mating surfaces of both components. Tighten the manifold bolts evenly from the middle out to either end.

8. Clean the fuel injector mounting holes, then install the injection pipe and fuel injectors onto the lower intake manifold.

9. The remaining components are installed in the reverse order from which they were removed.

10. Connect the negative battery cable, cycle the ignition **ON** and **OFF** several times to allow fuel pressure to build. Each time allow the ignition to remain **ON** for 5–7 seconds.

11. Check for fuel leaks.

M50/M52/S50 Engines

1. Disconnect the negative battery cable and drain the coolant to a level below that of the throttle housing. Unscrew the fastener from the throttle cable cover and pull the cover forward and off. Unclip the cable and pull the cable out with the rubber holder.

2. Pull the vacuum fitting from the brake booster and plug the openings.

3. Remove the engine and intake manifold covers. Unscrew the bolt holding the

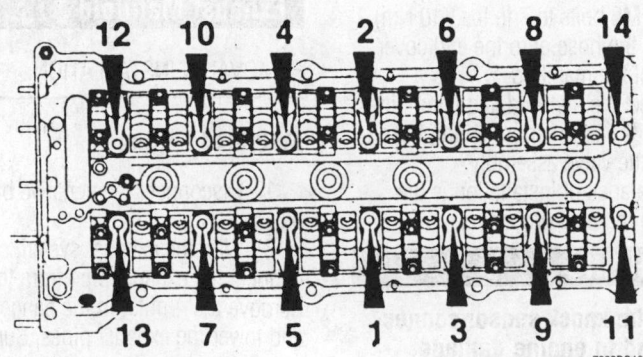

Cylinder head bolt tightening sequence, both cylinder heads are the same—M70/M73/S70 engines

ground strap on the front lifting eye. Replace the bolt before lifting the engine.

4. Unscrew the 2 bolts holding the plug plate and pull off the plug plate. Be careful not to damage the rubber seals. Take off the ignition coil electrical plugs. Remove the plug plate complete with the electrical leads.

5. Remove the cylinder head vent hose and pull off the air temperature sensor plug. Remove the tank venting hose and the throttle heating hoses from the throttle body. Remove the throttle valve switch plug. Unclip the idle speed control valve mounted on the manifold. Disconnect the fuel hoses from the pipes.

6. Unscrew the hardware holding the intake manifold to the cylinder head. Remove the intake manifold taking care not to drop anything into the exposed ports.

To install:

7. Install the lower manifold section onto the engine after cleaning the mating surfaces of both components. Tighten the manifold bolts evenly from the middle out to either end.

8. Install the remaining components in the opposite order from which they were removed.

9. Connect the negative battery cable, cycle the ignition **ON** and **OFF** several times to allow fuel pressure to build. Each time allow the ignition to remain **ON** for 5–7 seconds.

10. Check for fuel leaks.

M60/M62 Engines

1. Relieve the fuel system pressure. Disconnect the negative battery cable.

2. Remove the center cover from the cylinder head cover.

3. Loosen the hose clamps on the idle speed control and the throttle valve assembly.

4. Disconnect the plug on the mass air flow sensor.

5. Unclip and remove the upper section of the air cleaner assembly along with the mass air flow sensor.

6. Remove the right-hand cover from the cylinder head cover.

7. Raise and safely support the vehicle. Disconnect the plug for the oil level switch. Lower the vehicle.

8. Disconnect the plugs for the ignition coils.

9. Disconnect both knock sensors (For cylinders 1 and 2 along with 3 and 4), and the pulse sensor.

10. Disconnect the intake air temperature sensor, throttle valve potentiometer and the idle speed control.

11. Disconnect the diagnosis plug and the engine plug.

12. Unscrew the ignition coil ground wire, located near the rear engine lifting eye. Disconnect the temperature sensor (black) for the temperature gauge, and the temperature sensor (white) for the DME.

13. Remove the four bolts for the holder of the intake manifold cover. Remove the holder.

14. Disconnect and remove the throttle cable.

15. Remove the left-hand cover from the cylinder head cover.

16. Detach the ignition coil electrical connectors.

17. Disconnect both knock sensors (For cylinders 5 and 6 along with 7 and 8), and the camshaft sender.

18. Disconnect the coolant expansion tank plug and spill hose. Remove the two mounting bolts, and move the tank aside.

19. Detach the oil pressure switch electrical connector and remove the wiring.

20. Remove the screws for the wiring ducts on the cylinder heads.

21. Disconnect the vacuum hoses on the radiator, and loosen the hose clamp. Pull the vacuum hose off of the brake booster.

22. Remove the tank vapor venting hose off of the throttle valve assembly.

23. Disconnect the fuel feed and return lines.

24. Remove the hose off of the end cover on the back of the manifold.

25. Remove the seven mounting bolts, and pull off the end cover together with the pressure regulating valve straight back to prevent damaging the vent pipe.

26. Remove the five intake manifold bolts, and remove the intake manifold upwards.

To install:

27. Scrape the gasket off of the manifold and the cylinder head. Replace the gasket, and position the intake manifold. Install the mounting bolts. Tighten the bolts to 14–17 ft. lbs. (20–24 Nm).

28. Check and replace the seal and gasket for the end cover, if necessary. Position the end cover and install the mounting bolts. Tighten the M8 bolts to 14–17 ft. lbs. (20–24 Nm), and the M6 bolts to 7 ft. lbs. (10 Nm).

29. Install the hose onto the end cover on the back of the manifold.

30. Connect the fuel feed and return lines.

31. Install the tank vapor venting hose onto the throttle valve assembly.

32. The balance of installation is the reverse of the removal procedure.

> ### ❊❊ WARNING
>
> **Mixing up the knock sensor connectors will lead to engine damage.**

33. Connect negative battery cable.

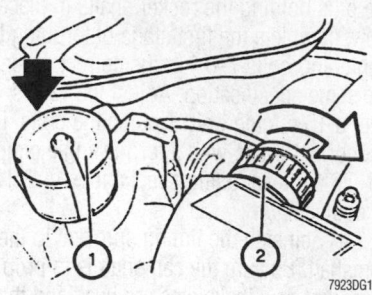

Diagnosis connector (1) and engine connector (2) locations—M60/M62 engines

M70/M73/S70 Engines

1. Disconnect the negative battery cable. Loosen the clamps for the fuel lines.

2. Pull off the vacuum hoses for the pressure regulators. Lift out the injection pipes with the injectors attached.

3. Remove the distributor caps and the throttle valve necks on the manifolds.

4. Disconnect the spark plug wires and remove the ignition lead pipes.

5. Disconnect the crankcase breather hose and loosen the manifold support nuts.

6. Disconnect the nose guard and remove the intake manifold, using the proper tool.

To install:

7. Scrape the gasket off of the manifold and the cylinder head. Replace the gasket, and position the intake manifold. Install the mounting bolts to 14–17 ft. lbs. (19–23 Nm) from the center of the manifold out to the ends.

8. Install the remaining components in the opposite order from which they were removed.

9. Connect the negative battery cable, cycle the ignition **ON** and **OFF** several times to allow fuel pressure to build. Each time allow the ignition to remain **ON** for 5–7 seconds.

10. Check for fuel leaks.

Exhaust Manifold

REMOVAL & INSTALLATION

M42/M44 Engine

1. Disconnect the negative battery terminal.

2. With the exhaust system cool, disconnect the exhaust pipe from the manifold. Remove the 4 nuts on the flange connection and lower the exhaust pipes. Support the exhaust system. Be sure the oxygen sensor wire is not being stretched.

3. Remove the nuts securing the manifold to the cylinder head. Remove the manifolds.

To install:

4. Clean the mounting surfaces on the manifolds and the cylinder head. Check the condition of the studs and replace if necessary.

5. Install the new exhaust manifold gaskets with the graphite side towards the cylinder head and install the manifolds. Tighten the nuts to 16–18 ft. lbs. (22–25 Nm). Use new nuts and anti-seize.

6. Connect the exhaust pipe to the manifolds. Connect the negative battery terminal. Start engine and check for leaks.

7. After 1200 miles, loosen, then tighten each nut to 10 ft. lbs. (12 Nm).

M50/M52/S50 Engine

1. Disconnect the negative battery terminal.

2. Remove the mounting nuts on each flange connection and separate the exhaust pipes from the manifold. Support the exhaust system. Be sure the oxygen sensor wire is not being stretched.

3. Remove the nuts securing the manifold to the cylinder head. Remove the manifolds.

To install:

4. Clean the mounting surfaces on the manifolds and the cylinder head. Check the condition of the studs and replace if necessary.

5. Install the new exhaust manifold gaskets with the graphite side towards the cylinder head and install the manifolds. Tighten the nuts to 14 ft. lbs. (19 Nm). Use new nuts and anti-seize.

6. Install the exhaust pipe to the manifolds using new mounting nuts.

7. Connect the negative battery terminal.

8. Start engine and check for exhaust leaks.

M60/M62 Engine

LEFT EXHAUST MANIFOLD

1. Disconnect negative battery cable.

2. Disconnect the oxygen sensor plug and remove the exhaust assembly.

3. Remove the alternator.

4. Remove the left cylinder head cover.

5. Remove the complete air cleaner upper section along with the mass air flow sensor.

6. Remove the bolts from the left and right engine mounts at the bottom.

7. Remove the rear engine splash guard.

8. Remove the bolts of the center of gravity mount to front axle carrier. Remove the left heat shields on the front axle carrier.

9. Lift the engine, with a suitable tool, at the front eye. Ensure clearance between the engine and the firewall.

10. Remove the manifold bolts and remove the manifolds downwards.

To install:

11. Scrape the old gasket off of the cylinder head and exhaust manifold and replace the gasket. The gasket beads face the exhaust manifolds.

12. Coat the upper row of the exhaust bolts with locking fluid. Position the exhaust manifold and install the bolts. Tighten the mounting bolts to 16 ft. lbs. (22 Nm).

13. Lower the engine to its original position. Install the left heat shields and the center of gravity mount-to-front axle carrier bolts.

14. Install the rear engine splash guard.

15. Install the bolts for the left and right engine mounts at the bottom. Tighten the 10mm bolts to 31 ft. lbs. (42 Nm) and the 8mm bolts to 16 ft. lbs. (22 Nm).

16. Install the complete air cleaner upper section along with the mass air flow sensor.

17. Install the left cylinder head cover and replace the gasket. Tighten the mounting bolts in a crisscross pattern to 11 ft. lbs. (15 Nm)

18. Install the alternator. Connect the oxygen sensor plug, and install the exhaust assembly.

19. Connect negative battery cable.

RIGHT EXHAUST MANIFOLD

1. Disconnect negative battery cable.

2. Disconnect the oxygen sensor plug and remove the exhaust assembly.

3. Remove the right heat shields on the front axle carrier.

4. Remove the rear engine splash guard.

5. Remove the washing fluid tank.

➡**Remove the manifold for cylinders two and four first.**

6. Remove the manifold bolts and remove the manifolds upwards.

To install:

7. Scrape the old gasket off of the cylinder head and exhaust manifold and replace the gasket. The gasket beads face the exhaust manifolds.

8. Coat the upper row of the exhaust bolts with locking fluid. Position the exhaust manifold and install the bolts. Tighten the mounting bolts to 16 ft. lbs. (22 Nm).

9. Install the washing fluid tank.

10. Install the rear engine splash guard.

11. Install the right heat shields.

12. Connect the oxygen sensor plug and install the exhaust assembly.

13. Connect negative battery cable.

M70/M73/S70 Engine

1. Disconnect the negative battery terminal.

2. Remove the left side upper section of the air cleaner assembly along with the air mass sensor.

3. Remove the clamp on the left and right split pipes.

4. Remove the heat shields on the left manifold and on the steering gear.

5. Remove the manifold/split pipe bolts on the left-hand side.

6. On the left-hand side, remove the front and rear manifolds along with the gaskets.

7. Remove the nuts on the stay bolts. Remove the stay bolts in the cylinder head for the left manifold.

8. Remove the right side upper section of the air cleaner assembly along with the air mass sensor.

9. Remove the windshield washing fluid tank.

10. Remove the oil dipstick guide tube.

11. Remove the heat shields on the right manifold.

12. On the right-hand side, remove the front and rear manifolds along with the gaskets.

13. Remove the nuts on the stay bolts. Remove the stay bolts in the cylinder head for the right manifold.

To install:

14. Clean the mounting surfaces on the manifolds and the cylinder head. Check the condition of the studs and replace if necessary.

15. Install the stay bolts in the cylinder head for the right manifold. Install the nuts onto the stay bolts.

16. On the right-hand side, install the new exhaust manifold heat shield gaskets and install the manifolds. Tighten the nuts to 16–18 ft. lbs. (22–25 Nm). Use new self-locking nuts.

17. Install the heat shields on the right manifold.

18. Install the oil dipstick guide tube.

19. Install the windshield washing fluid tank.

20. Install the right side upper section of the air cleaner assembly along with the air mass sensor.

21. Install the stay bolts in the cylinder head for the left manifold. Install the nuts onto the stay bolts.

22. On the left-hand side, install the new exhaust manifold heat shield gaskets and

install the manifolds. Tighten the nuts to 16–18 ft. lbs. (22–25 Nm). Use new self-locking nuts.

23. Install the manifold/split pipe bolts on the left-hand side.

24. Install the heat shields on the left manifold and on the steering gear.

25. Install the clamp on the left and right split pipes.

26. Install the left side upper section of the air cleaner assembly along with the air mass sensor.

27. Connect the negative battery terminal. Start engine and check for leaks.

Camshaft and Valve Lifters

REMOVAL & INSTALLATION

M42/M44 Engines

1. Disconnect the negative battery cable.

2. Unscrew the ignition coil cover and remove the spark plug connectors and spark plugs. Remove mounting nuts and lift out complete ignition assembly.

3. Disconnect and tag all wiring and hoses which may interfere with cylinder head cover removal. Position aside.

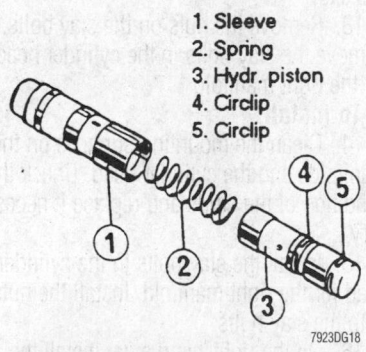

1. Sleeve
2. Spring
3. Hydr. piston
4. Circlip
5. Circlip

Hydraulic lash adjuster components—M42/M44 engines

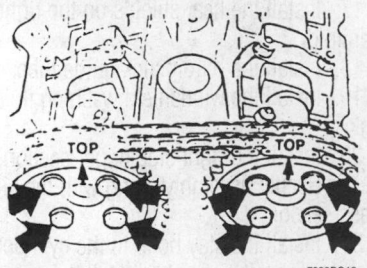

The camshaft sprockets should be positioned as shown when the engine is at TDC on the No. 1 cylinder—M42/M44 engines

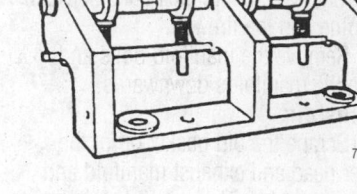

Camshaft removal tool 11–3–260—M42/M44 engines

Bearing cap bolt locations—M42/M44 engines

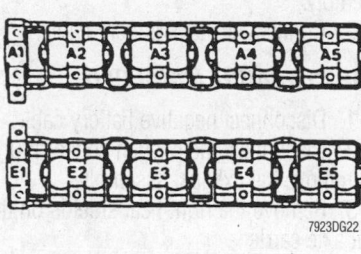

Bearing caps are marked with A1 through A5 for exhaust side and E1 through E5 for intake side—M42/M44 engines

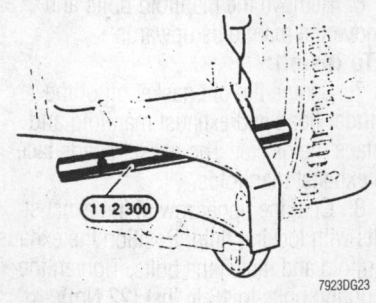

Hold the crankshaft in TDC position with tool 11–2–300 or equivalent—M42/M44 engines

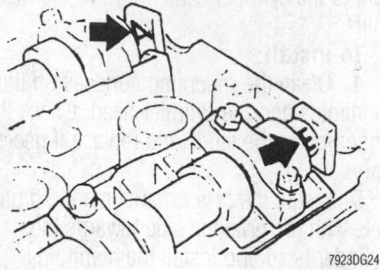

Camshafts are marked with "A" for exhaust and "E" for intake—M42/M44 engines

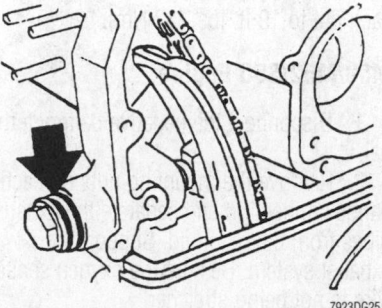

Chain tensioner location—M42/M44 engines

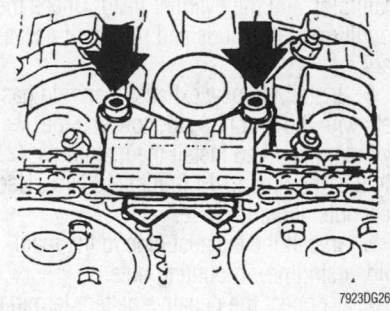

Upper chain guide location—M42/M44 engines

4. Unscrew and remove the cylinder head cover.

5. Rotate the crankshaft in normal direction of rotation until the cam lobes of intake and exhaust cams of No. 1 cylinder faces each other. The arrows on the timing chain sprockets face UP.

➡ **If the camshaft must be rotated with the timing chain and sprockets removed, the crankshaft must first be rotated approximately 90 degrees away from TDC position, in the direction of engine rotation. In this position contact between valves and pistons will be avoided.**

6. To avoid loss of ignition timing, mount the crankshaft holding tool 11–2–300 or equivalent into position.

7. Matchmark the timing chain and sprockets relative to each other.

8. Remove the timing chain tensioner.

9. Unbolt and remove the upper chain guide.

10. Unbolt and remove the timing chain and sprockets. Suspend the timing chain and sprockets after removing.

11. Mount the special fixture 11–3–260 or equivalent in spark plug holes and tighten to 17 ft. lbs. (23 Nm).

✳✳ WARNING

The camshafts can be damage or broken when removing/installing without the fixture.

12. Apply load to the bearing caps by turning the eccentric shaft of the special fixture. Loosen and remove the bearing cap bolts.

13. Unscrew and remove the special fixture. Lift out the bearing caps and camshafts. Camshafts are marked with "A" for exhaust and "E" for intake. Bearing caps are marked with "A1 through A5" for the exhaust side and "E1 through E5" for the intake side.

To install:

14. Fit the camshafts and bearing caps into position. Note the correct location of the intake and exhaust camshafts. Also note the position of each bearing caps. Do not tighten the bearing caps at this time.

15. Align the camshafts so that lobes of the intake and exhaust cams face each other in no. 1 cylinder (see note above). The camshafts can be turned on the hexagon with a 27mm wrench. The camshafts can be held in position using special holding tool 11–3–240 or equivalent.

➡ **The valve clearance compensating elements expand whenever no load is applied by the camshaft (camshaft removed) and require a certain amount of time after installation before they compress again. Always wait approximate 10 minutes (at room temperature 68° (20° C) after installation of the camshaft, before the engine is cranked. Allow a longer waiting time at lower temperature.**

16. Mount the special fixture 11–3–260 or equivalent and apply load to the bearing caps by turning the eccentric shaft of the special fixture. Tighten the bearing cap bolts M6: 7 ft. lbs. (10 Nm), M7: 11 ft. lbs. (15 Nm) or M8: 15 ft. lbs. (20 Nm).

17. Install the camshaft sprockets and timing chain. Tighten the mounting bolts M6: 11 ft. lbs. (15 ft. lbs.), all others 7 ft. lbs. (10 Nm). Ensure that the matchmarks are correctly aligned. The arrows on the timing chain sprockets face UP.

18. Before installing the timing chain tensioner, adjust the basic position:

a. Knock the outside sleeve of the chain tensioner on a hard surface. This will cause the piston to jump of the lock.

b. Reassemble the parts and clamp in a vice fitted with soft jaws.

c. Push the chain tensioner together and fit the circlip in the slip bevel of the sleeve.

d. Push the chain tensioner together even further until the circlip is heard engage.

e. Remove the chain tensioner form the vice and install the chain tensioner to the engine and tighten to 29 ft. lbs. (40 Nm). Using a suitable tool, reach down and push the chain tensioner rail against the hydraulic piston until the tensioning element is released.

➡ **After removing the timing chain tensioner, the hydraulic plunger is locked and cannot be pressed back. With no spring action, the chain or tensioner would break. The plunger of the timing chain tensioner must therefore be brought into basic position before installing.**

19. Install the upper timing chain guide.

20. Check the cylinder head cover gasket and replace if necessary. Install the cylinder head cover and check for correct seating of the gasket at the rear of the cylinder head. Tighten the cover bolts to 7–11 ft. lbs. (10–15 Nm).

21. Attach all wiring and hoses. Install complete ignition assembly, spark plugs, spark plug connectors and ignition coil cover.

✳✳ WARNING

Remove the crankshaft holding tool before operating the engine.

22. Connect the negative battery cable.

23. Start the engine and check for proper operation.

➡ **A new or disassembled timing chain tensioner is without oil. To assure correct functioning, the engine must be operated the first time at 3500 rpm for approximately 20 seconds.**

M50/M52/S50 Engines

1. Disconnect the negative battery cable.

2. Remove the cylinder head.

➡ **Special tools are required to perform this operation. BMW tools 11–3–260/ 270/250 or equivalent are required for proper removal and installation of the camshafts and for retention of the valve lash compensators. Without these tools the camshafts will be damaged during removal or installation.**

3. Remove the spark plugs and attach the 11–3–260 (plus addition 11–3–270) camshaft removal fixture. Tighten the hold down bolts in the spark plug bores to 17 ft. lbs. (23 Nm).

4. Apply load to the bearing caps by rotating the eccentric shaft. This relieves the

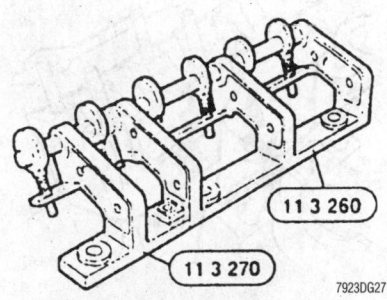

Camshaft removal tools 11 3 260 and 11 3 270—M50/M52/S50 engines

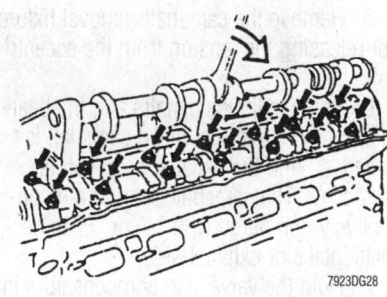

Bearing cap bolt locations—M50/M52/S50 engines

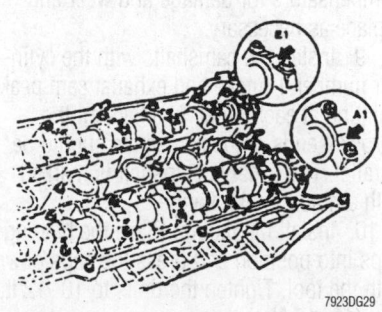

Bearing cap ID markings—M50/M52/S50 engines

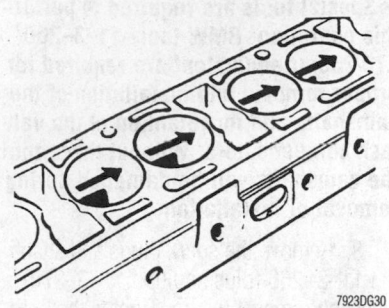

Check bearing surfaces of valve clearance compensators for scoring—M50/M52/S50 engines

Bearing plate markings—M50/M52/S50 engines

tension on the bearing cap bolts. Loosen and remove the bearing cap bolts.

5. Remove the camshaft removal fixture after releasing the tension from the eccentric shaft.

6. Remove the camshafts and the bearing caps. Note that the intake camshaft is marked "E" and the exhaust camshaft is marked "A". The camshaft bearing are consecutively numbered with "A" or "E" to designate intake or exhaust side.

7. Hold the valve lash compensators in place using tool 11–3–250 or equivalent, and remove the bearing plate along with the valve plungers.

To install:

8. Inspect the camshafts and valve lash compensators for damage and wear and replace as necessary.

9. Install the camshafts with the cylinder number 1 intake and exhaust cam peaks pointing at each other. The flats on the sprocket ends of the camshafts should be parallel. The exhaust camshaft is marked with a notch on the flange.

10. Install the fixture. Place the bearing caps into position and press the caps down with the tool. Tighten the bolts to 10–12 ft. lbs. (13–17 Nm).

11. When the camshafts have been removed and reinstalled a waiting period

dependent on the ambient temperature is necessary before mounting the cylinder head on the engine. At room temperature wait 4 minutes to allow the lifters to compress fully. At temperatures down to 50° F(10° C) wait 11 minutes. At temperatures lower than 50° F(10° C) wait 30 minutes. This is to prevent contact between the valves and the piston tops.

12. The engine may not be cranked under the same conditions as above for a period of 10 minutes at room temperature; 30 minutes for temperatures down to 50° F(10° C); 75 minutes for temperatures below 50° F(10° C).

M60/M62 Engines

LEFT CAMSHAFT (CYLINDER BANK 5–8)

1. Disconnect negative battery cable.
2. Remove the left and right cylinder head covers.
3. Remove all spark plugs.
4. Remove the top left timing case cover.
5. Remove the splash guard.
6. Remove all oil lines to the left and right cylinder head.
7. Rotate the crankshaft in direction of rotation until the first cylinder is in TDC position.

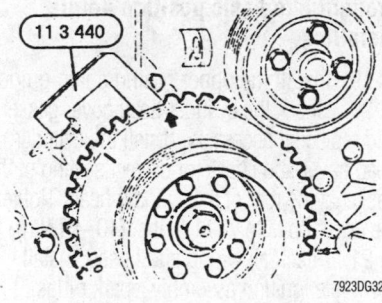

Gap in increment gear must fit in special tool 11 3 440—M60/M62 engines

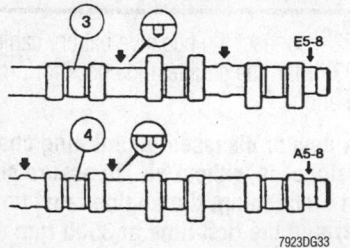

Left side camshaft identification (cylinder bank 5 to 8): hex head (3) on intake camshaft between cylinders 7 and 8, hex head (4) on exhaust camshaft between cylinders 5 and 6—M60/M62 engines

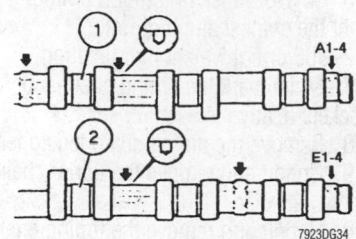

Right side camshaft identification (cylinder bank 1 to 4): hex-head (2) on intake camshaft between cylinders 3 and 4, hex-head (1) on exhaust camshaft between cylinders 1 and 2—M60/M62 engines

Camshaft positioning—M60/M62 engines

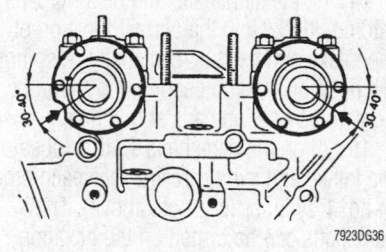

Install left-hand camshafts: Recesses in camshaft point downwards approximately 30–40 degrees from plane of cylinder head—M60/M62 engines

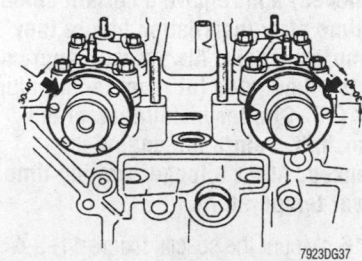

Install right-hand camshafts: Recesses in camshaft point upwards approximately 30–40 degrees from plane of cylinder head—M60/M62 engines

8. Brace the camshaft on the hex head with a suitable open-end wrench and loosen the 3 accessible screws on each right sprocket approximately ½ a turn.

a. Turn the engine over once and loosen the remaining 3 screws on each right sprocket approximately ½ a turn.

b. Unscrew and remove the primary sprocket from the left-hand intake camshaft (cylinder bank 5–8). Secure the chain to prevent it from dropping.

9. Rotate the engine to 45 degrees BTDC setting position. Rotate the crankshaft against direction of rotation until the gap in the increment gear fits in the special tool 11–3–440 or equivalent.

10. Remove the screws on the exhaust camshafts with a spanner tool or equivalent.

11. Remove the screws on the exhaust camshaft sprocket. Do not remove the sprocket.

12. Compress the chain tensioner and install special tool 11–3–420 or equivalent to lock the tensioner in place.

13. Lift off both secondary camshaft sprockets together with the chain.

14. Rotate the intake and exhaust camshafts to the installed position:

a. Using special tool 11–3–430 or equivalent, rotate the camshafts until the recess in both camshaft flange points approximately 30–40 degrees downwards from the plane of the cylinder head.

b. Check the installed position by installing special tool 11–2–430 or equivalent to the camshafts. The cylinder designation of the special tool must point upwards.

15. Loosen the both camshaft bearing caps uniformly from outside to inside ½ turn.

16. Remove all bearing caps. Label each bearing cap to facilitate re-assemble and position aside.

17. Remove the camshafts noting their locations.

18. To remove the hydraulic valve lifters use special tool 11–3–250 to pull them out of the cylinder head. Be sure that no damage occurs to the guides in the head. Inspect the bearing surfaces of the bucket tappets (lifters) for wear and scoring.

To install:

19. If the lifters were removed, install them with special tool 11–32–250 or equivalent.

20. Lubricate and install the camshafts in their correct position.

➡ **The intake camshaft will have a hexagon between cylinders 7 and 8. The exhaust camshaft will have the hexagon between cylinders 5 and 6.**

21. Rotate the intake and exhaust camshafts to the installed position:

a. Using special tool 11–3–430 or equivalent, rotate the camshafts until the recess in both camshaft flange points approximately 30–40 degrees downwards from the plane of the cylinder head.

b. Check the installed position by installing special tool 11–2–430 or equivalent to the camshafts. The cylinder designation of the special tool must point upwards.

22. Install the bearing caps. Tighten the bearing caps from outside to inside in 1/2 turn increments. Tighten the bolts to 9–13 ft. lbs. (12–17 Nm).

➡ **Do not confuse camshaft bearing caps of cylinders No. 1–4 and 5–8. The exhaust camshaft bearing caps are marked with A1-a5 from intake side. The intake camshaft bearing caps are marked with E1-e5 from intake end.**

23. Fit the special tool 11–3–430 or equivalent to the camshaft. Rotate the camshaft until the marker bores face upwards.

a. Install special tools 11–2–442/446 or equivalent to the camshaft on cylinder bank 5–8.

b. Install special tools 11–2–441/445 or equivalent to the camshaft on cylinder bank 1–4.

c. Using a suitable open-end wrench, align all camshafts in such a way that the special tools fit on the cylinder heads without any gaps.

d. Fit special tools 11–2–443 or equivalent to special tools 11–2–441/442/445/446 and secure them with special tools

11–2–444 using spark plug threads.

24. Install the secondary sprockets together with chain to the camshafts on cylinder bank 5–8.

25. Install the screws on the exhaust camshaft sprocket and tighten snug.

26. Remove the special tool used to lock the chain tensioner in position.

27. Rotate the engine from 45 degrees BTDC in direction of rotation as far as TDC setting. Install special tool 11–2–300 at the flywheel to lock the crankshaft in TDC position.

28. Assemble the primary sprocket and chain to the intake camshaft with the arrow pointing upwards (in cylinder axis) and the long bores centrally aligned. Install the screws snug.

29. Install the special tool 11–3–390 or equivalent in the right timing case cover and with a suitable torque wrench, tension the tool to 1.3 Nm.

30. Tighten the sprockets to 11 ft. lbs. (15 Nm) in the following order:
- All screws on the left exhaust camshaft
- 3 screws in the right exhaust camshaft
- All screws on the left intake camshaft
- 3 screws in the right intake camshaft

31. Remove special tools 11–2–444/443/441/445.

32. Remove special tools 11–2–444/443/442/446.

33. Remove special tool 11–2–300 used to locked the crankshaft in TDC position.

34. Turn the engine over once.

35. Tighten the remaining 3 screws on right exhaust camshaft and remaining 3 screws on right intake camshaft to 11 ft. lbs. (15 Nm).

36. Relieve the load and remove the special tool 11–3–390 or equivalent from the right timing case cover.

37. The balance of installation is the reverse of the removal procedure.

38. Start the engine. Check for leaks and proper operation.

RIGHT CAMSHAFT (CYLINDER BANK 1–4)

1. Disconnect negative battery cable.

2. Remove the left and right cylinder head covers.

3. Remove all spark plugs.

4. Remove the fan assembly.

5. Remove the top right timing case cover.

6. Remove the splash guard.

7. Remove all oil lines to the left and right cylinder head.

8. Rotate the crankshaft in direction of rotation until the first cylinder is in TDC position.

9. Brace the camshaft on the hex head with a suitable open-end wrench and loosen the 3 accessible screws on each left sprocket approximately ½ a turn.

a. Turn the engine over once and loosen the remaining 3 screws on each left sprocket approximately ½ a turn.

b. Unscrew and remove the primary sprocket from the right-hand intake camshaft (cylinder bank 1–4). Secure the chain to prevent it from dropping.

10. Rotate the engine to 45 degrees BTDC setting position. Rotate the crankshaft against direction of rotation until the gap in the increment gear fits in the special tool 11–3–440 or equivalent.

11. Remove the screws on the exhaust camshaft sprocket. Do not remove the sprocket.

12. Compress the chain tensioner and install special tool 11–3–420 or equivalent to lock the tensioner in place.

13. Lift off both secondary camshaft sprockets together with the chain.

14. Rotate the intake and exhaust camshafts to the installed position:

a. Using special tool 11–3–430 or equivalent, rotate the camshafts until the recess in both camshaft flange points approximately 30–40 degrees upwards from the plane of the cylinder head.

b. Check the installed position by installing special tool 11–2–430 or equivalent to the camshafts. The cylinder designation of the special tool must point upwards.

15. Loosen the both camshaft bearing caps uniformly from outside to inside ½ turn.

16. Remove all bearing caps. Label each bearing cap to facilitate re-assemble and position aside.

17. Remove the camshafts noting their locations.

18. To remove the hydraulic valve lifters use special tool 11–3–250 to pull them out of the cylinder head. Be sure that no damage occurs to the guides in the head. Inspect the bearing surfaces of the bucket tappets (lifters) for wear and scoring.

To install:

19. If the lifters were removed, lubricate and install them with special tool 11–32–250 or equivalent.

20. Lubricate and install the camshafts in their correct position.

➡**The intake camshaft will have a hexagon between cylinders 3 and 4. The exhaust camshaft will have the hexagon between cylinders 1 and 2.**

21. Rotate the intake and exhaust camshafts to the installed position:

a. Using special tool 11–3–430 or equivalent, rotate the camshafts until the recess in both camshaft flange points approximately 30–40 degrees upwards from the plane of the cylinder head.

b. Check the installed position by installing special tool 11–2–430 or equivalent to the camshafts. The cylinder designation of the special tool must point upwards.

22. Install the bearing caps. Tighten the bearing caps from outside to inside in 1/2 turn increments. Tighten the bolts to 9–13 ft. lbs. (12–17 Nm).

➡**Do not confuse camshaft bearing caps of cylinders No. 1–4 and 5–8. The exhaust camshaft bearing caps are marked with A1-a5 from intake side. The intake camshaft bearing caps are marked with E1-e5 from intake end.**

23. Fit the special tool 11–3–430 or equivalent to the camshaft. Rotate the camshaft until the marker bores face upwards.

a. Install special tools 11–2–442/446 or equivalent to the camshaft on cylinder bank 5–8.

b. Install special tools 11–2–441/445 or equivalent to the camshaft on cylinder bank 1–4.

c. Using a suitable open-end wrench, align all camshafts in such a way that the special tools fit on the cylinder heads without any gaps.

d. Fit special tools 11–2–443 or equivalent to special tools 11–2–441/442/445/446 and secure them with special tools 11–2–444 using spark plug threads.

24. Install the secondary sprockets together with chain to the camshafts on cylinder bank 1–4.

25. Install the screws on the exhaust camshaft sprocket and tighten snug.

26. Remove the special tool used to lock the chain tensioner in position.

27. Rotate the engine from 45 degrees BTDC in direction of rotation as far as TDC setting. Install special tool 11–2–300 at the flywheel to lock the crankshaft in TDC position.

28. Assemble the primary sprocket/chain with sensor pin to the intake camshaft with the arrow pointing upwards (in cylinder axis) and the long bores centrally aligned. Install the screws snug.

29. Install the special tool 11–2–400 or equivalent to the right cylinder head (cylinder bank 1–4). Install the special tool 11–3–390 to special tool 11–2–400. Using a suitable torque wrench, tension the tool to 1.3 Nm.

30. Tighten the sprockets to 11 ft. lbs. (15 Nm) in the following order:

• 3 screws on the left exhaust camshaft
• All screws on the right exhaust camshaft
• 3 screws on the left intake camshaft
• All screws in the right intake camshaft

31. Remove special tools 11–2–444/443/441/445.

32. Remove special tools 11–2–444/443/442/446.

33. Remove special tool 11–2–300 used to locked the crankshaft in TDC position.

34. Turn the engine over once.

35. Tighten the remaining 3 screws on left exhaust camshaft and remaining 3 screws on left intake camshaft to 11 ft. lbs. (15 Nm).

36. Relieve the load and remove the special tool 11–3–390 and 11–2–400.

37. Install the remaining components in the reverse order of removal.

38. Start the engine. Check for leaks and proper operation.

M70/M73/S70 Engines

1. Disconnect the negative battery cable. Drain the cooling system and remove the fan assembly.

2. Remove both intake manifolds and distributor housings.

3. Disconnect the round rubber mounts, bolts and nuts. Remove both cylinder head covers.

4. Remove the mounting bolts and lift out the upper timing cover.

5. Set the engine to TDC. Install a holder in the crankshaft. The valves of cylinders one and seven are closed. The dowel pins in the camshafts should face in.

6. Press off the anti-tamper lock for the chain tensioner with a screwdriver. Loosen the nut, then loosen the adjusting screw several turns,, then unscrew the plug. Remove the timing belt tensioning piston, using care not to lose the spring that is between the plug and the piston.

7. Remove mounting bolts for the timing chain guide, and remove the guide. Remove the tensioning rail.

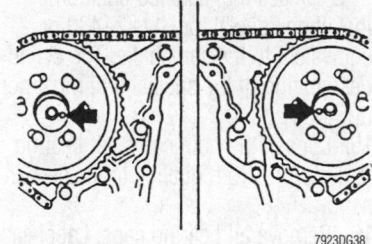

At TDC, the dowel pins in the camshaft sprockets should face each other—M70/M73/S70 engines

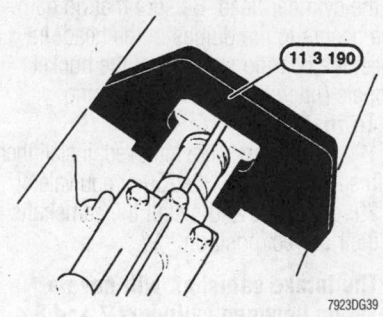

Camshaft alignment gauge—M70/M73/S70 engines

8. Remove the mounting bolts on the camshaft sprockets, and carefully remove the sprockets. Do not allow the timing chain to fall into the engine.

9. Remove the oil pipe mounting bolts from the top of the camshaft bearings.

10. Unbolt the bearing caps and remove the camshaft.

❊❊ WARNING

The bearing caps are matched with the bearings, do not mix up the order of the caps.

To install:

11. With the crankshaft positioned at TDC, install the camshaft with the dowel pin facing the center of the engine. Position the bearing caps and install the mounting bolts from inside to outside. Tighten the bolts to 11 ft. lbs. (15 Nm).

12. Hold both camshafts in position with special tool 11 3 190.

13. Mount the oil pipes with the oil outlet bores facing the camshaft. Install the hollow union bolt in the bearing cover. Install the mounting bolts and tighten to 9 ft. lbs. (12 Nm).

14. Install the camshaft sprockets mounting bolts finger-tight. Position the timing chain on the sprockets in the opposite direction of engine rotation, beginning at the crankshaft. Verify that the timing chain is correctly aligned on all the sprockets, and remove the crankshaft holder.

15. Position and install the timing chain guide, and tensioning rail.

16. Install the timing chain tensioner.

17. Tighten the camshaft sprocket bolts to 7 ft. lbs. (10 Nm).

18. Install the upper timing cover and a new gasket.

19. Install the cylinder head covers.

20. Install both intake manifolds and distributor housings.

21. Install the fan assembly. Fill and bleed the cooling system.

22. Connect negative battery cable.

Valve Lash

ADJUSTMENT

All engines are equipped with hydraulic valve lash adjusters. This design does not require adjustments nor are adjustments possible.

Oil Pan

REMOVAL & INSTALLATION

M42/M44 Engines

1. Disconnect the negative battery cable. Raise and safely support the vehicle.

2. Drain the engine oil.

3. Disconnect the exhaust pipe, if necessary.

4. Remove the lower oil pan mounting bolts and take off the lower oil pan. Remove the upper section oil pan bolts and remove the upper oil pan.

To install:

5. Clean the mounting surfaces and install new gaskets.

6. Position the oil pan against the engine and install the mounting bolts. Tighten the mounting bolts to 6 ft. lbs. (9 Nm).

7. Install the exhaust pipe, if removed.

8. Install and tighten the oil pan drain plug, then fill the engine with the correct viscosity and amount of clean engine oil.

9. Lower the vehicle and connect the negative battery cable.

10. Start the engine and check that oil pressure is present; if the oil pressure lamp does not turn off within 5–7 seconds of starting the engine, turn the engine **OFF**. Check for any oil leaks.

M50/M52/S50 Engines

3 SERIES

1. Disconnect the negative battery cable. Raise the vehicle and support it. Drain the engine oil.

2. Remove the front lower splash guard, if necessary.

3. Disconnect the electrical terminal from the oil sending unit.

4. Remove the power steering gear from the front axle carrier, if necessary.

5. Remove the flywheel cover.

6. Remove the oil pan bolts and lower the oil pan. Remove the oil pump bolts and take out the oil pump and oil pan.

To install:

7. Before installing the oil pan, clean the gasket surfaces and install a new gasket on the oil pan.

8. Coat the joints on the ends of the front engine cover with a universal sealing compound.

9. Install the flywheel cover.

10. If the power steering gear was removed, be sure to refill and bleed this system.

11. Connect the electrical wiring harness to oil sending unit.

12. Install the front lower splash guard, if removed.

13. Install and tighten the oil pan drain plug, then fill the engine with the correct viscosity and amount of clean engine oil.

14. Lower the vehicle and connect the negative battery cable.

15. Start the engine and check that oil pressure is present; if the oil pressure lamp does not turn off within 5–7 seconds of starting the engine, turn the engine **OFF**. Check for any oil leaks.

5 SERIES

1. Disconnect the negative battery terminal and raise and safely support the vehicle. Drain the engine oil.

2. Loosen the holding bolt for the oil dipstick guide pipe and remove the clamp. Pull the guide tube free of the pan.

3. Remove all the oil pan bolts and remove the pan. Raise the engine slightly if needed for clearance.

To install:

4. Apply sealer to the joint between the pan, front cover and block.

5. Install new gaskets and install the pan. Tighten mounting bolts to 6.5–8.0 ft. lbs. (9–11 Nm).

6. Install the dipstick guide tube using a new base seal and tighten the holding bolt.

M60/M62 Engines

1. Disconnect the negative battery cable.

2. Remove the intake manifold cover. Remove the top clips on the radiator.

3. Remove the cooling fan.

4. Remove the guide tube for the oil dipstick.

5. Remove the engine splash guards.

6. Remove the cover for the oil filter so the oil will run back to the pan. Drain the engine oil. Disconnect the plug for the level switch.

7. Remove the lower oil pan bolts and remove the lower oil pan. Remove the gasket and clean the mounting surfaces.

8. Disconnect the left and right engine mounts at the bottom.

9. Unbolt the power steering pump at the holder. On automatic transmissions, remove the oil pipes at the power steering pump.

10. Remove the banjo bolt for the oil return pipe from the oil filter at the oil pan.

11. Remove the mounting bolt on the sprocket for the oil pump and remove the sprocket along with the chain.

12. Remove the three oil pump mounting bolts and remove the oil pump. Remove the oil pipes out of the crankcase.

13. Lift the engine by the front eye hook. Observe the distance between the engine and the firewall while lifting the engine.

14. Unscrew the upper oil pan bolts and remove the upper oil pan.

To install:

15. Clean the mounting surfaces and install a new gasket.

16. Install the upper oil pan and tighten the bolts to 7–8 ft. lbs. (9–11 Nm).

17. Lower the engine.

18. Check the seals on the oil pipes and replace it if necessary. Lubricate the seals with oil and the oil pipes.

19. Check the seal in the oil pump and replace it if necessary. Screw the hexagon adapter back into the oil pump until it stops.

20. Position the oil pump and install the two right side oil pump mounting bolts. Tighten the bolts to 14–17 ft. lbs. (20–24 Nm). Position the chain on the pump and the sprocket and install the sprocket. Tighten the bolt to 35 ft. lbs. (47 Nm). Verify that the chain is positioned correctly.

21. Adjust the chain sag to 0.315–0.472 in. (8–12mm) by turning the hexagon adapter in the oil pump. Install the left side mounting bolt.

22. Install the banjo bolt for the oil return pipe from the oil filter at the oil pan.

23. Install the power steering pump and connect the oil lines (if equipped).

24. Making sure to connect the ground strap. Connect the left and right engine mounts at the bottom and tighten to 32 ft. lbs. (43 Nm).

➡**If replacing the lower oil pan, remove the level switch from the old pan and install it in the new pan with a new O-ring.**

25. Position the lower oil pan with a new gasket. Install the bolts and tighten to 7–8 ft. lbs. (9–11 Nm), beginning in the middle and working to the outside.

26. Connect the plug for the level switch, making sure to replace the O-ring.

27. Install the engine splash guards.

28. Install the oil dipstick guide tube, making sure to replace the O-ring. Install the cooling fan.

29. Install the intake manifold cover. Install the top clips on the radiator.

30. Fill the engine oil. Connect negative battery cable.

M70/M73/S70 Engines

7 SERIES

1. Disconnect the negative battery cable. Raise and safely support the vehicle.

2. Remove the transmission and the oil pump assembly. Lower the vehicle.

3. Disconnect and remove the windshield washer tank and the coolant expansion tank.

4. Remove the guide tube for the oil dipstick. Disconnect the oil pipe on the tandem pump. Remove the mounting bracket.

5. Unscrew the belt tensioner and remove the oil drain hose.

6. Crank the engine to TDC and unscrew the flywheel using the proper tool.

7. Disconnect the left and right engine mounts at the bottom. Pull off the pipe adapter for oil extraction.

8. Remove the oil pump consoles. Unscrew the oil pan bolts and remove the oil pan.

To install:

9. Clean the mounting surfaces and install a new gasket.

10. Install the oil pan and tighten the mounting bolts to 7 ft. lbs. (11 Nm).

11. Connect the left and right engine mounts at the bottom and tighten to 32.5 ft. lbs. (43 Nm).

12. Replace the oil consoles and tighten to 25 ft. lbs. (34 Nm).

13. Install the flywheel and tighten the bolts to 72 ft. lbs. (97 Nm).

14. The remainder of the installation is the reverse of the removal procedure.

8 SERIES

1. Disconnect the negative battery cable. Raise and safely support the vehicle.

2. Remove the transmission and the oil pump assembly. Lower the vehicle.

3. Disconnect and remove the windshield washer tank and the coolant expansion tank.

4. Remove the guide tube for the oil dipstick. Disconnect the oil pipe on the tandem pump. Remove the mounting bracket.

5. Unscrew the belt tensioner and remove the oil drain hose.

6. Crank the engine to TDC and unscrew the flywheel using the proper tool.

7. Disconnect the left and right engine mounts at the bottom. Pull off the pipe adapter for oil extraction.

8. Remove the oil pump consoles. Unscrew the oil pan bolts and remove the oil pan.

To install:

9. Clean the mounting surfaces and install a new gasket.

10. Install the oil pan and tighten the mounting bolts to 7 ft. lbs. (11 Nm).

11. Connect the left and right engine mounts at the bottom and tighten to 32.5 ft. lbs. (43 Nm).

12. Replace the oil consoles and tighten to 25 ft. lbs. (34 Nm).

13. Install the flywheel and tighten the bolts to 72 ft. lbs. (97 Nm).

14. The remainder of the installation is the reverse of the removal procedure.

<table>
<tr><td>**Oil Pump**</td></tr>
</table>

REMOVAL & INSTALLATION

M42/M44 Engines

1. Disconnect the negative battery cable.

2. Raise and safely support the vehicle. Drain the engine oil.

3. Remove the timing case cover.

4. Disconnect the oil pump cover mounting bolts and remove the oil pump assembly.

To install:

5. Clean the oil pump mounting surfaces, then position the oil pump on the engine. Install the oil pump cover mounting bolts.

6. Install the timing case cover.

7. Install and tighten the oil pan drain plug, then fill the engine with the correct viscosity and amount of clean engine oil.

8. Lower the vehicle and connect the negative battery cable.

9. Start the engine and check that oil pressure is present; if the oil pressure lamp does not turn off within 5–7 seconds of starting the engine, turn the engine **OFF**. Check for any oil leaks.

M50/M52/S50 Engines

1. Raise and safely support vehicle. Disconnect the negative battery cable. Drain the oil from the engine. Remove the oil pan to access the oil pump drive sprocket.

2. Remove the oil pump drive sprocket nut. Note that it is a left-hand thread. Remove the oil pump drive sprocket from the oil pump shaft. Check the shaft splines.

3. Unbolt the oil pump body from the block and remove. Check the condition of the dowel sleeves.

To install:

4. Clean the oil pump mounting surfaces, then position the oil pump on the engine. Install the oil pump body mounting bolts to 16 ft. lbs. (22 Nm).

5. Install the oil pump drive sprocket onto the oil pump shaft. Install the oil pump drive sprocket nut to 18 ft. lbs. (25 Nm). The sprocket nut must be tightened in a counterclockwise direction; it has a reverse, or left-hand thread.

6. Install the oil pan.

7. Install and tighten the oil pan drain plug, then fill the engine with the correct viscosity and amount of clean engine oil.

8. Lower the vehicle and connect the negative battery cable.

9. Start the engine and check that oil pressure is present; if the oil pressure lamp does not turn off within 5–7 seconds of starting the engine, turn the engine **OFF**. Check for any oil leaks.

M60/M62 Engines

1. Disconnect the negative battery cable.

2. Remove the lower oil pan.

3. Disconnect the left and right engine mounts at the bottom.

4. Unbolt the power steering pump at the holder. On automatic transmissions, remove the oil pipes at the power steering pump.

5. Remove the banjo bolt for the oil return pipe from the oil filter at the oil pan.

6. Remove the mounting bolt on the sprocket for the oil pump and remove the sprocket along with the chain.

7. Remove the three oil pump mounting bolts and remove the oil pump. Remove the oil pipes out of the crankcase.

To install:

8. Check the seals on the oil pipes and replace it if necessary. Lubricate the seals with oil and the oil pipes.

9. Check the seal in the oil pump and replace it if necessary. Screw the hexagon adapter back into the oil pump until it stops.

10. Position the oil pump and install the two right side oil pump mounting bolts. Tighten the bolts to 14–17 ft. lbs. (20–24 Nm). Position the chain on the pump and the sprocket and install the sprocket. Tighten the bolt to 35 ft. lbs. (47 Nm). Verify that the chain is positioned correctly.

11. Adjust the chain sag to 0.315–0.472 in. (8–12mm) by turning the hexagon adapter in the oil pump. Install the left side mounting bolt.

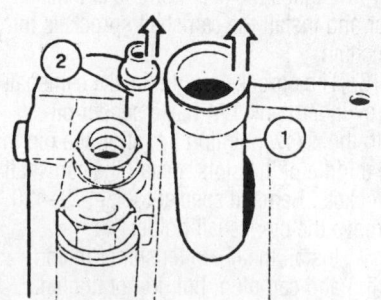

7923DG40

Fresh oil pipe (1), and pure oil pipe (2) locations—M60/M62 engines

12. Install the banjo bolt for the oil return pipe from the oil filter at the oil pan.

13. Install the power steering pump and connect the oil lines (if equipped).

14. Install the lower oil pan.

15. Fill the engine oil. Connect negative battery cable.

M70/M73/S70 Engines

1. Disconnect the negative battery cable and remove the oil pan.

2. Remove the bolts retaining the sprocket to the oil pump shaft and remove the sprocket.

3. Remove the oil pump retaining bolts and lower the oil pump from the engine block. There are 3 bolts at the front and 2 bolts attaching the rear of the oil pick-up to the lower end of a support bracket. It is necessary to remove all 5 bolts.

4. Do not loosen the chain adjusting shims from the 2 mounting locations.

To install:

5. Add or subtract shims between the oil pump body and the engine block to obtain a slight movement of the chain under light thumb pressure.

6. Install the oil pump in position.

➡**When used, the 2 shim thicknesses must be the same. Tighten the pump holder at the pick-up end after shimming is completed to avoid stress on the pump.**

7. After the main pump mounting bolts are tightened, loosen the bolts at the bracket on the rear of the pick-up, allowing the pick-up to assume its most natural position. This will relieve tension on the bracket.

8. The balance of installation is the reverse of the removal procedure.

9. Install the oil pan.

10. Install and tighten the oil pan drain plug, then fill the engine with the correct viscosity and amount of clean engine oil.

11. Lower the vehicle and connect the negative battery cable.

12. Start the engine and check that oil pressure is present; if the oil pressure lamp does not turn off within 5–7 seconds of starting the engine, turn the engine **OFF**. Check for any oil leaks.

Rear Main Seal

REMOVAL & INSTALLATION

The rear main bearing oil seal can be replaced after the transmission and clutch/flywheel or the converter/flywheel has been removed from the engine.

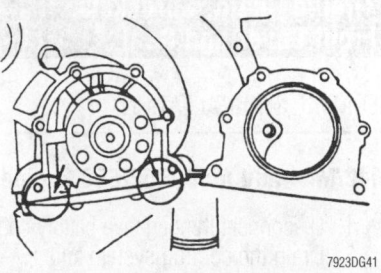

7923DG41

Apply sealer to the joints (as marked) during rear main seal housing installation

1. Raise and safely support the vehicle. Drain the engine oil and loosen the oil pan bolts. Carefully use a sharp bladed tool to separate the oil pan gasket from the lower surface of the end cover housing.

2. Remove the 2 rear oil pan bolts.

3. Remove the bolts around the outside of the cover housing and remove the end cover housing from the engine block. Remove the gasket from the block surface.

4. Remove the seal from the housing.

To install:

5. Coat the sealing lips of the new seal with oil. Install a new seal into the end cover housing with a special seal installer tool. On the 6-cylinder engines, press the seal in until it is about 0.039–0.079 in. (0.991–2.070mm) deeper than the standard seal, which was installed flush.

6. While the cover is off, check the plug in the rear end of the main oil gallery. If the plug shows signs of leakage, replace it with another, coating it with the proper sealant to keep it in place.

➡**Fill the cavity between the sealing lips of the seal with grease before installing.**

7. Coat the mating surface between the oil pan and end cover with sealer. Using a new gasket, install the end cover on the engine block and bolt it into place.

8. Install the remaining components in the opposite order from which they were removed.

9. Install and tighten the oil pan drain plug, then fill the engine with the correct viscosity and amount of clean engine oil.

10. Lower the vehicle and connect the negative battery cable.

11. Start the engine and check that oil pressure is present; if the oil pressure lamp does not turn off within 5–7 seconds of starting the engine, turn the engine **OFF**. Check for any oil leaks.

Timing Chain, Sprockets, Front Cover and Seal

REMOVAL & INSTALLATION

M42/M44 and M50/M52/S50 Engines

1. Disconnect the negative battery cable.

2. Drain the cooling system and remove the radiator and fan assembly.

3. Remove the drive belts and any accessories that block access to the timing cover. Remove the engine splash shield, if necessary.

4. Remove the vibration damper using the proper tool. Unscrew the central bolt and remove the vibration damper hub.

5. Remove the timing case cover bolts and remove the timing cover.

➡ **The timing case cover can be removed without removing the water pump.**

6. Unscrew the upper chain guide and top bolt on the right chain guide.

7. Remove the timing chain sprockets and the lift out the chain. Remove the timing chain guide.

8. Remove the tensioning rail, if necessary. Remove the crankshaft sprocket with the proper tool and lift out the Woodruff key.

9. Remove the reversing roller, if needed.

➡ **The reversing roller can only be replaced complete with bearings.**

To install:

10. Install the Woodruff key into the channel in the crankshaft. Slide the crankshaft sprocket over the end of the crankshaft with the Woodruff key aligning with the channel in the crankshaft. Use the central mounting bolt to draw the sprocket entirely into position.

11. The remaining components are installed in the reverse order from which they were removed. Be sure to tighten the camshaft sprocket bolts to 16 ft. lbs. (22 Nm), the timing cover bolts to 6.5–8.0 ft. lbs. (9–11 Nm) for M6 bolts and 16 ft. lbs. (22 Nm) for M8 bolts, the vibration damper hub bolt to 295 ft. lbs. (410 Nm) and the vibration damper bolts to 17 ft. lbs. (23 Nm). Use sealer at the intersections of the timing cover and the pan.

12. Connect the negative battery cable, start the engine and check for leaks.

M60/M62 Engines

1. Disconnect the negative battery cable.

2. Remove the left upper timing cover as follows:

a. Remove the left cylinder head cover.

b. Remove the alternator.

c. Remove the cover from the oil filter. Unbolt the return pipe at the oil filter housing. Remove the full flow oil filter housing retaining nuts and remove the oil filter housing.

d. Remove the battery positive wire from the alternator. Remove the protective tube mounting screws and move the wire aside.

e. Remove the nine timing case mounting bolts and remove the timing case.

3. Remove the right timing cover and tensioner as follows:

a. Remove the right cylinder head cover.

b. Remove the air cleaner upper section along with the mass air flow sensor.

c. Unscrew the timing chain tensioner mounting element from the side of the cover. Remove the complete mounting element and hydraulic tensioner.

d. Remove the camshaft sender screw.

e. Remove the upper mounting bolt for the oil dipstick guide tube. Remove the lower mounting nut for the tube and remove the tube.

f. Remove the nine right timing case mounting bolts and remove the cover from the cylinder head.

4. Remove the lower timing cover and seal as follows:

a. Remove the intake manifold cover. Remove the intake hose between the throttle body and the air volume meter.

b. Remove the cooling fan and the drive belt.

c. Remove the pulley on the water pump.

d. Remove the eight mounting bolts for the vibration damper and remove the damper.

e. Raise and safely support the vehicle. Remove the front engine splash shield. Remove the cooling air guide for the alternator, located on the engine carrier. Lower the vehicle.

f. Position special tool 11 2 450, or equivalent, on the center bolt for the vibration damper hub. Remove the bolt. Install a suitable puller on the hub and remove the hub.

g. Press out the oil seal using special tool 11 2 380 and 11 2 383, or equivalent.

h. Remove the 15 mounting bolts for the lower timing case cover and remove the cover.

5. Turn the engine in the direction of rotation and set cylinder number 1 to TDC. The arrows on the sprockets should face up in the cylinder axis. Use a crankshaft holder to keep the TDC position.

6. Loosen and remove the camshaft sprocket bolts from both banks of cylinders. Compress the hydraulic tensioning element to loosen the timing chain. Lock the element with special tool 11–3–420, and remove the sprockets with the chain. Do not rotate the engine with the timing chain removed.

7. Guide the chain out of the tensioner rails and off the lower sprocket.

8. To remove the guide rail:

a. On the left side (cylinder bank 1–4), remove the lower mounting bolt and remove the tensioning rail. Pull off the spacer with oil supply for the tensioning rail.

b. On the left side, remove the two mounting bolts for the guide rail. Do not mix up the two bolts, it is important to install the same bolt in the same hole. Remove the sliding rail.

c. On the right side (cylinder bank 5–8), remove the two mounting bolts on the tensioning rail, and the two bolts on the guide rail. Remove the rails.

To install:

9. On the right cylinder bank, position the guide rail and install the mounting bolts. Position the tensioning rail, and install the mounting bolts.

10. On the left cylinder bank, check the seal for the spacer. Position the guide rail, and install the mounting bolts in the correct holes. Install the spacer. Position the tensioning rail, and install the lower mounting bolt.

11. Inspect the sprockets for wear and replace if necessary.

12. Install the chain in position.

13. Be sure No. 1 piston remains at the top of its firing stroke and the key on the crankshaft is in the 12 o'clock position.

14. Position the chain on the guide rail and swing the chain inward and to the left.

15. Engage the chain on the crankshaft gear and install the camshaft sprockets into the chain.

16. The sprocket on the intake camshaft for cylinder bank 1–4 has a sender pin. With the arrow pointing up, align the pin in the middle of the slots. Install the camshaft sprockets. Remove special tool 11–3–420. Remove the crankshaft holder.

17. Install the chain tensioner piston, spring and cap plug, but do not tighten.

18. To bleed the chain tensioner, fill the oil pocket, located on the upper timing housing cover, with engine oil and move the

tensioner back and forth with a suitable pry-bar until oil is expelled at the cap plug. Tighten the cap plug securely.

19. Install the lower timing cover as follows:

   a. Check for the correct seating of the dowel sleeves. Clean the sealing surfaces and remove all pieces of gasket. Position a new gasket on the lower cover.

   b. Cut off the protruding ends of the gasket, making sure the cutting tool is level. Do not allow the pieces to fall into the engine.

   c. Position the lower cover and install the mounting bolts with an even distribution of pressure. Tighten the 6mm bolts to 7 ft. lbs. (10 Nm), 8mm bolts to 16 ft. lbs. (22 Nm) and 10mm bolts to 35 ft. lbs. (47 Nm)

   d. Install the oil seal in the timing case cover using special tool 11 1 220 or an appropriate seal install tool and the crankshaft damper hub bolt. Be sure the seal is flush with the cover.

   e. Install the vibration damper hub and install the bolt. Tighten the hub bolt in three steps:
   • Step 1: 74–81 ft. lbs. (100–110 Nm)
   • Step 2: turn an additional 60 degrees
   • Step 3: turn an additional 60 degrees
   • Step 4: turn an additional 30 degrees

   f. Raise and safely support the vehicle. Install the cooling air guide for the alternator, located on the engine carrier. Install the front engine splash shield. Lower the vehicle.

   g. Position the vibration damper. Install the mounting bolts on the vibration damper.

   h. Install the pulley on the water pump.

   i. Install the drive belt and the cooling fan.

   j. Install the intake hose between the throttle body and the air volume meter. Install the manifold cover.

20. Install the right timing cover as follows:

   a. Replace the hydraulic tensioner oil seal in the timing case cover.

   b. Check for the correct seating of the dowel sleeves. Clean the sealing surfaces to remove any oil or old gasket. Position a new gasket.

   c. Mount the timing case cover. Screw in the vertically mounted bolts until the cover contacts the cylinder head. Do not tighten the bolts yet.

21. Install the horizontally mounted bolts, then tighten the vertically mounted bolts in two steps. After the vertically mounted bolts are tight, tighten the horizon-

tally mounted bolts in two steps. Tighten the 6mm bolts to 7 ft. lbs. (10 Nm), 8mm bolts to 16 ft. lbs. (22 Nm), and 10mm bolts to 35 ft. lbs. (47 Nm).

   a. Install the oil dipstick guide tube, making sure to replace the O-ring.

   b. Install the camshaft sender screw and the chain tensioner.

   c. Install the air cleaner upper section along with the mass air flow sensor.

   d. Install the right cylinder head cover.

22. Install the left timing chain cover as follows:

   a. Check for the correct seating of the dowel sleeves. Clean the sealing surfaces to remove any oil or old gasket. Position a new gasket.

   b. Mount the timing case cover together with the inserted bolt. This bolt cannot be installed with the cover in place. Install the rest of the mounting bolts and screw in the vertically mounted bolts until the cover just contacts the cylinder head. Do not tighten the bolts yet.

   c. Install the horizontally mounted bolts, then tighten the vertically mounted bolts in two steps. After the vertically mounted bolts are tight, tighten the horizontally mounted bolts in two steps. Tighten the 6mm bolts to 7 ft. lbs. (10

Nm), 8mm bolts to 16 ft. lbs. (22 Nm), and 10mm bolts to 35 ft. lbs. (47 Nm).

   d. Position the battery positive wire for the alternator and install the protective tube mounting screws. Connect the wire to the alternator.

   e. Install the oil filter housing. Install the return pipe and replace the housing cover.

   f. Install the alternator and the cylinder head cover.

23. Connect negative battery cable.

M70/M73/S70 Engines

1. Disconnect the negative battery cable. Drain the cooling system and remove the fan assembly.

2. Remove the drive belts and the engine splash shield. Remove the tensioning bolt.

3. Remove both intake manifolds and distributor housings.

4. Disconnect the round rubber mounts, bolts and nuts. Remove both cylinder head covers.

5. Remove the mounting bolts and lift out the timing cover.

6. Unscrew the bolts but do not remove the vibration damper. Remove the central hub bolt with the proper tool.

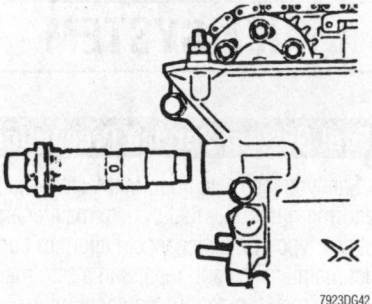

The timing chain tensioner slides out of the side of the front cover—M60/M62 engines

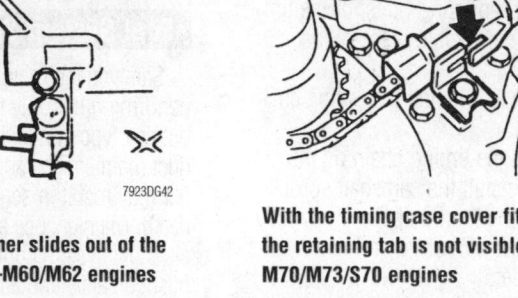

With the timing case cover fitted correctly, the retaining tab is not visible—M70/M73/S70 engines

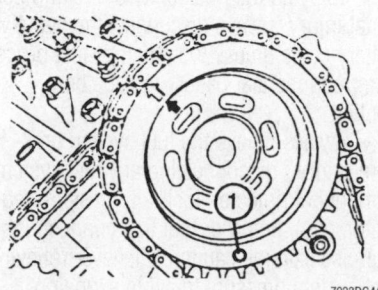

Camshaft sprocket sender pin—M60/M62 engines

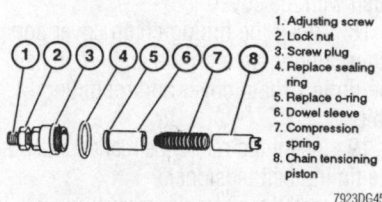

1. Adjusting screw
2. Lock nut
3. Screw plug
4. Replace sealing ring
5. Replace o-ring
6. Dowel sleeve
7. Compression spring
8. Chain tensioning piston

Timing chain tensioner components—M70/M73/S70 engines

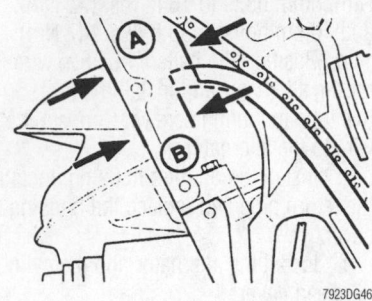

Timing chain tensioner adjustment: dimension "B" from "A" should be 0.216–0.256 in. (5.48–6.5mm)—M70/M73/S70 engines

7. Remove the vibration damper using the proper tool to pull the vibration damper hub from the crankshaft.

8. Drain the engine oil and remove the lower section of the oil pan. Remove the bottom mounting screws from the timing case cover and loosen the adjacent oil pan bolts on both sides.

9. Remove the timing belt tensioner and reference mark sender.

10. Remove the mounting screws and take off the timing case cover.

11. Press out the front cover oil seal using special tool 11 1 210 or equivalent.

12. Set the engine to TDC. Install a holder in the crankshaft. The valves of No. 1 and No. 7 cylinders are closed. The dowel pins in the camshafts should face in.

13. Remove the mounting bolts on the camshaft sprockets and carefully remove the sprockets with the timing chain.

14. Remove mounting bolts for the timing chain guide and remove the guide.

To install:

15. Position and install the timing chain guide.

16. Position the timing chain on the sprockets and install the camshaft sprockets. Verify that the timing chain is correctly aligned on all the sprockets and remove the crankshaft holder.

17. Lubricate the sealing lip of the shaft seal with oil. Install the new seal using a suitable seal installer. The seal should be flush with the cover.

18. Clean the timing chain cover and the mounting area on the engine. Install the timing chain cover and mounting bolts.

19. Install the reference mark sender and the timing belt tensioner.

20. Install the remaining components in the reverse order of removal. Be sure to tighten the central hub bolt to 318 ft. lbs. (430 Nm) and the vibration damper mounting bolts to 17 ft. lbs. (25 Nm).

21. Clean the timing cover and engine block mounting surfaces, then position the timing cover on the engine. Install the mounting bolts.

22. Install both cylinder head covers.

23. The remaining components are installed in the reverse order from which they were removed.

24. Install the engine splash shield. Install the fan assembly.

25. Connect negative battery cable.

26. Before starting the engine, change the engine oil. Refill and bleed the cooling system.

27. Start the engine and check for proper operation.

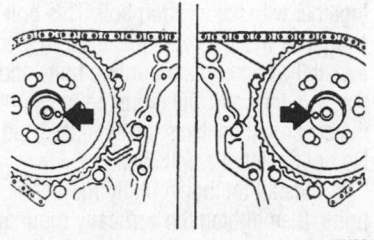

Camshaft dowel pins with the engine at TDC—M70/M73/S70 engines

FUEL SYSTEM

Fuel System Service Precautions

Safety is the most important factor when performing not only fuel system maintenance but any type of maintenance. Failure to conduct maintenance and repairs in a safe manner may result in serious personal injury or death. Maintenance and testing of the vehicle's fuel system components can be accomplished safely and effectively by adhering to the following rules and guidelines.

• To avoid the possibility of fire and personal injury, always disconnect the negative battery cable unless the repair or test procedure requires that battery voltage be applied.

• Always relieve the fuel system pressure prior to disconnecting any fuel system component (injector, fuel rail, pressure regulator, etc.), fitting or fuel line connection. Exercise extreme caution whenever relieving fuel system pressure to avoid exposing skin, face and eyes to fuel spray. Fuel under pressure may penetrate the skin or any part of the body that it contacts.

• Always place a shop towel or cloth around the fitting or connection prior to loosening to absorb any excess fuel due to spillage. Ensure that all fuel spillage (should it occur) is quickly removed from engine surfaces. Ensure that all fuel soaked cloths or towels are deposited into a suitable waste container.

• Always keep a dry chemical (Class B) fire extinguisher near the work area.

• Do not allow fuel spray or fuel vapors to come into contact with a spark or open flame.

• Always use a back-up wrench when loosening and tightening fuel line connection fittings. This will prevent unnecessary stress and torsion to fuel line piping. Always follow the proper torque specifications.

• Always replace worn fuel fitting O-rings with new. Do not substitute fuel hose or equivalent where fuel pipe is installed.

Fuel System Pressure

RELIEVING

To relieve the pressure in the system, first find the fuel pump relay plug, located on the cowl. Unplug the relay, leaving it in a safe position where the connections cannot ground. If necessary, tape the plug in place or tape over the connector prongs with electrical tape. Then, start the engine and operate it until it stalls. Crank the engine for 10 seconds after it stalls to remove any residual pressure.

Fuel Filter

REMOVAL & INSTALLATION

On filters that are located near the fuel tank, it is necessary to clamp the fuel lines closed before disconnecting them, or fuel will run out continuously.

1. Disconnect the negative battery cable. Relieve fuel system pressure. Clamp the lines closed if the filter is mounted low, near the fuel tank. Then, loosen the clamps and disconnect the inlet and outlet hoses. Remove the hose clamps or slide them back, well off the connections to make it easier to pull off the hoses, if necessary.

2. The filters will usually be attached to a frame, floor pan or wheel well by a bracket. Loosen the bracket and remove the filter. Note the direction of flow, then remove the filter.

To install:

3. Situate the new fuel filter, while observing the direction of flow markings on the filter, into position on the frame, floor pan or wheel

well (depending on the particular model). Install the mounting bracket until snug.

4. Install the fuel lines onto the correct fuel filter fittings. Tighten the fuel line clamps until tight, but not to the point where the fuel lines become excessively pinched or damaged.

5. Connect the negative battery cable and cycle the ignition **ON** and **OFF** several times to build fuel pressure.

6. Inspect the fuel filter and fuel lines for any fuel leaks.

Fuel Pump

REMOVAL & INSTALLATION

3 Series

1. Relieve fuel system pressure. Disconnect the negative battery cable. Going to the pump, which is under the vehicle and near the fuel tank, push back any protective caps, note the routing and attach the electrical connector(s).

2. Securely clamp the suction hose (coming from the tank) and plug the discharge hose so no fuel can escape.

3. Open the hose clamp connecting the suction hose to the pump and disconnect it.

4. Remove the attaching nuts which mount the pump and bracket to the floor pan and remove both as an assembly.

5. Remove the bolt passing through the 2 parts of the bracket and also mounting the hose attaching strap to the bracket. Then, pull the pump out of the bracket.

6. Loosen the hose clamp for the discharge hose and disconnect it at the pump. Pull the rubber ring off the pump.

7. Note the code number on the pump and be sure to replace it with one of the same number. Inspect all the rubber mounts on the pump mounting bracket and replace any that are cracked or crushed.

To install:

8. Attach the discharge hose to the fuel pump, then slide the pump into the mounting bracket. Install the hose strap mounting bolt to the pump mounting bracket.

9. Install the remaining components in the opposite order from which they were removed.

10. Connect the negative battery cable. Be sure to remove the clamps from the hoses, then run the engine and check for leaks. Check the fuel system pressure.

5 Series

The fuel pump is an electrical unit, delivering fuel through a pressure regulator, to a fuel distributor or a ring-line for the injection valves. The fuel pump is mounted under the vehicle, in the fuel tank, or in the engine compartment.

1. Relieve fuel system pressure. Detach the negative battery connector. Push back any protective caps and detach the electrical connector(s).

2. If the fuel lines are flexible, pinch them closed with an appropriate tool. Disconnect the fuel lines and plug the ends.

3. Remove the retaining bolts and remove the pump and expansion tank as an assembly.

4. The pump can be separated from the expansion tank after removal.

To install:

5. Install the pump in the correct position, be sure to use similar types of hose clamps, if any need replacing. The wrong type clamp can damage the pressure lines.

6. Run the engine and check the fuel lines for leakage. Check the fuel system pressure.

7 Series

The pump on this vehicle is mounted in the top of the tank along with the fuel level sending unit.

1. Disconnect the negative battery cable. Drain the fuel tank, enough to prevent spillage when removing the pump.

2. Relieve fuel system pressure. Remove the trim panels from the trunk. Then, remove the screws from the cover for the pump/sending unit assembly.

3. Label the fuel hoses connecting at the top of the pump/sending unit assembly.

1. Fuel level transmitter
2. Gasket
3. Inlet line
4. Return line
5. Pressure damper
6. Check valve
7. Fuel pump
8. Pump insulating sleeve
9. Fuel intake filter
10. Pump holder

7923DG48

View of the in-tank fuel pump—7 Series

Unclamp and disconnect the fuel hoses, then plug them.

4. Slide the collar for the electrical connector to one side, then unplug the connector.

5. Remove the mounting screws and remove the pump/sending unit assembly. Replace the gasket.

6. Press the retaining locks for the pump unit inward and slide the pump out of the pump/sending unit assembly.

7. Note the routing of the fuel and electrical lines to the pump from the top of the pump/sending unit assembly. Loosen the hose clamp screws and the screws attaching the electrical connectors to the pump. Detach the hose and connector.

8. Unscrew the pressure regulator from the top of the check valve. Then, unscrew the check valve from the top of the pump.

9. Pull the insulating sleeve off the pump. Then, loosen the retaining screw and slide the filter off the pump.

To install:

10. Slide a new filter onto the pump, then install and tighten the retaining screw.

11. Install the insulating sleeve on the pump.

12. Install the remaining components in the reverse order of removal.

13. Be careful to ensure that the 2 retaining locks fasten the pump in place in a secure manner. Operate the engine and check for leaks.

8 Series

The pump on this vehicle is mounted in the top of the tank along with the fuel level sending unit.

1. Disconnect the negative battery cable. Drain the fuel tank, enough to prevent spillage when removing the pump.

2. Relieve fuel system pressure. Remove the rear seat. Then, remove the screws from the cover for the pump/sending unit assembly.

3. Label the fuel hoses connecting at the top of the pump/sending unit assembly. Unclamp and disconnect the fuel hoses, then plug them.

4. Slide the collar for the electrical connector to one side, then unplug the connector.

5. Remove the coupling nut and remove the pump/sending unit assembly. Replace the gasket.

6. Remove the filter screens and rubber liner.

7. Mark the routing of the fuel and electrical lines to the pump from the top of the pump/sending unit assembly. Loosen the

hose clamp screws and the screws attaching the electrical connectors to the pump. Detach the hose and connector.

To install:

8. Attach the hose and connector to the fuel pump. Tighten the hose clamps and electrical connector retaining screws until snug.

9. Install the filter screens and rubber liner onto the pump.

10. The remaining components are installed in the reverse order from which they were removed.

11. Be careful to ensure that the 2 pumps are connected as previously marked.

12. Connect the negative battery cable. Operate the engine and check for leaks.

DRIVE TRAIN

Transmission Assembly

REMOVAL & INSTALLATION

Manual

3 SERIES

1. Disconnect the negative battery cable. Raise and safely support the vehicle. Remove the exhaust system. Remove the cross brace and heat shield.

2. Hold the nuts on the front with one wrench and remove bolts from the rear with another to disconnect the flexible coupling at the front of the driveshaft. Some vehicles have a vibration damper at this point in the drive train. This damper is mounted on the transmission output flange with bolts that are pressed into the damper. On these vehicles, unscrew and remove the nuts located behind the damper.

3. Loosen the threaded sleeve on the driveshaft. Get a special tool to hold the splined portion of the shaft while turning the sleeve.

4. Remove its mounting bolts and remove the center driveshaft mount. Then, bend the driveshaft down at the center and pull it off the transmission output flange. Keep the sections of the driveshaft from pulling apart and suspend it from the vehicle with wire.

5. Remove the retainer and washer and pull out the shift selector rod.

6. Use a hex-head wrench to remove the self-locking bolts that retain the shift rod bracket at the rear of the transmission, then

remove the bracket. If equipped with a shift arm, use a suitable prybar to pry the spring clip up off the boss on the transmission case and swing it upward. Then, pull out the shift shaft pin.

7. Unscrew and remove the clutch slave cylinder and support it so the hydraulic line can remain connected.

8. The transmission incorporates sending units for flywheel rotating speed and position. Remove the heat shield that protects these from exhaust heat, then remove the retaining bolt for each sending unit. Note that the speed sending unit, which has no identifying ring goes in the bore on the right, and that the reference mark sending unit, which has a marking ring, goes in the bore on the left. If the sending units are installed in reverse positions, the engine will not run at all. Pull these units out of the flywheel housing.

9. Detach the wiring connector going to the back-up light switch and pull the wires out of the harness.

10. Support the transmission from underneath in a secure manner. Remove mounting bolts and remove the crossmember holding the rear of the transmission to the body. Then, lower the transmission onto the front axle carrier.

11. Using the proper tool, remove the bolts holding the transmission flywheel housing to the engine at the front. Be sure to retain the washers with the bolts. Pull the transmission rearward to slide the input shaft out of the clutch disc, then lower the transmission and remove from the vehicle.

To install:

12. Install the transmission in position under the vehicle. Align the input shaft and install the transmission.

13. The remaining components are installed in the reverse order from which they were removed, note the following points:

a. Coat the input shaft splines and flywheel housing guide pins with a light coating of suitable grease.

b. Be sure the front mounting bolts are installed with their washers. Tighten them to 46–58 ft. lbs. (62–80 Nm).

c. Before reinstalling the sending units for flywheel position and speed, be sure their faces are free of either grease or dirt, then coat them with a light coating of a suitable lubricant. Inspect the O-rings and replace them if they are cut, cracked, crushed, or stretched.

d. When installing the shift rod bracket at the rear of the transmission, use new self-locking bolts and be sure the bracket is level before tightening them. Tighten the shift rod bracket bolts

to 16.5 ft. lbs. (22 Nm) except on the M3, which uses an aluminum bracket.

e. Install the clutch slave cylinder with the bleed screw downward.

f. When installing the driveshaft center bearing, preload it forward 0.157–0.236 in. (3.98–5.99mm). Check the driveshaft alignment with an appropriate tool such. Replace the nuts, then tighten the center mount bolts to 16–17 ft. lbs. (21–23 Nm).

g. Tighten the flexible coupling bolts to 83–94 ft. lbs. (114–129 Nm).

14. Install the exhaust system, then lower the vehicle to the ground. Connect the negative battery cable.

5 SERIES

1. Disconnect the negative battery cable. Raise and safely support the vehicle. Disconnect and lower the exhaust system to provide clearance for transmission removal. Remove the heat shield brace and transmission heat shield.

2. Support the driveshaft, then unscrew the driveshaft coupling at the rear of the transmission. Use a wrench on both the nut and the bolt.

3. Working at the front of the driveshaft center bearing, unscrew the screw-on ring type connector which attaches the driveshaft to the center bearing. Then, unbolt the center bearing mount. Bend the driveshaft down and pull it off the centering pin. If equipped with a vibration damper, turn it and pull it back over the output flange before pulling the driveshaft off the guide pin. Suspend it from the vehicle.

4. Pull off the wires for the back-up light switch. Unscrew the passenger compartment console to disconnect it from the top of the transmission by removing the self-locking bolts. Discard and replace.

5. Pull out the locking clip and disconnect the shift rod at the rear of the transmission. Take care to keep all the washers.

6. If the transmission is linked to the shift lever with an arm, use a small prybar to lift the spring out of the holder on the bracket, then raise the arm. Pull out the shift shaft bolt.

7. If equipped with a flywheel housing cover (semi-circular in shape), remove the mounting bolts and remove the cover.

8. The speed sensor and reference mark sensor on the flywheel housing must be disconnected. Note their locations. The speed sensor goes in the upper bore, marked D. The reference mark sensor, which has a ring, goes in the lower bore, marked B. Check the O-rings for the sensors and install new ones if they are damaged.

9. Support the transmission securely. Then, unbolt and remove the rear transmission crossmember.

10. Remove the upper and lower attaching nuts and remove the clutch slave cylinder, supporting it so the hydraulic line need not be disconnected. Disconnect the reverse gear back-up light switch and pull the wires out of the holders.

11. Unscrew the bolts fastening the transmission to the bell housing, using the proper tool. On some vehicles there are Torx® bolts used ; use a Torx® wrench for these. Pull the transmission rearward until the input shaft has disengaged from the clutch disc, then lower and remove it.

To install:

12. Place the transmission in gear. Insert the guide sleeve of the input shaft into the clutch pilot bearing carefully. Turn the output shaft to rotate the front of the input shaft until the splines line up and it engages the clutch disc.

13. Perform the remaining portions of the procedure in reverse of removal, observing the following points:

a. Be sure the arrows on the rear crossmember point forward.

b. Preload the center bearing mount forward of its most natural position 0.079–0.157 in. (2.07–3.99mm).

c. In tightening the driveshaft screw on ring, use tool 26 1 040 or equivalent.

d. When reconnecting the nuts and bolt at the transmission coupling, replace the nuts with new ones and turn only the nut, holding the bolts stationary.

e. Be sure DME sensor faces are clean. Coat the sensor outside diameters with the proper lubricant.

f. If equipped with a shift arm, lubricate the bolt with a light layer of a suitable lubricant.

g. Observe these torque figures:
- Transmission-to-bell housing—52–58 ft. lbs. (70–80 Nm).
- Rear/top transmission Torx® bolts—46–58 ft. lbs. (62–80 Nm).
- Center mount-to-body—16–17 ft. lbs. (21–23 Nm).
- Front joint-to-transmission—83–94 ft. lbs. (114–129 Nm).

14. Install the exhaust system, then lower the vehicle to the ground. Connect the negative battery cable.

7 AND 8 SERIES

1. Disconnect the negative battery cable. Raise and safely support the vehicle. Remove the exhaust system. Remove the attaching bolts and remove the heat shield mounted just to the rear of the transmission on the floorpan.

2. Support the transmission securely from underneath. Then, remove the crossmember that supports it at the rear from the body by removing the mounting bolts on both sides.

3. Using wrenches on both the bolt heads and on the nuts, remove the bolts passing through the vibration damper and front universal joint at the front of the driveshaft.

4. Remove its mounting bolts and remove the center driveshaft mount. Then, bend the driveshaft down at the center and pull it off the transmission output flange. Keep the sections of the driveshaft from pulling apart and suspend it from the vehicle with wire.

5. Pull out the circlip, slide off the washer, then pull the shift selector rod off the transmission shift shaft. Disconnect the back-up light switch.

6. Lower the transmission slightly for access. Then, use a small prybar to lift the spring out of the holder on the bracket, then raise the arm. Pull out the shift shaft bolt.

7. Remove the upper and lower attaching nuts and remove the clutch slave cylinder, supporting it so the hydraulic line need not be disconnected.

8. Unscrew the bolts fastening the transmission to the bell housing. Use a Torx® wrench to remove the bolts. Be sure to retain the washer with each bolt to ensure that they can be readily removed later, if necessary. Pull the transmission rearward until the input shaft has disengaged from the clutch disc, then lower and remove the transmission.

To install:

9. Install the transmission in position under the vehicle. Align the input shaft and install the transmission.

10. Preload the center bearing mount forward of its most natural position 0.157–0.236 in. (3.98–5.99mm).

11. When reconnecting the nuts and bolt at the transmission coupling, replace the nuts with new ones and turn only the nut, holding the bolts stationary. Tighten the center mount-to-body nut to 16–17 ft. lbs. (21–23 Nm). Tighten the front joint-to-transmission nut to 58 ft. lbs. (80 Nm).

12. Install the remaining components in the reverse order of removal.

13. Reconnect the shift arm, if equipped, and lubricate the bolt with a light layer of a suitable lubricant, then check the O-ring for crushing, cracks or cuts, replacing it, if damaged.

14. When installing the clutch slave cylinder, be sure the bleeder screw faces downward.

15. Install the exhaust system. Connect the negative battery cable, then lower the vehicle to the ground.

Automatic

EXCEPT 3 SERIES

➡To perform this operation, the following tools or equivalents are required. Special transmission support tools 24 0 120 and 00 2 020 and driveshaft locking ring tool 26 1 040 or equivalent.

1. Disconnect the battery ground cable. Loosen the throttle cable adjusting nuts, release the cable tension and disconnect the cable at the throttle lever. Then, remove the nuts and pull the cable housing out of the bracket.

2. Disconnect the exhaust system at the manifold and hangers and lower it out of the way. Remove the hanger that runs across under the driveshaft. Remove the exhaust heat shield from under the center of the vehicle.

3. Support the transmission a suitable lifting device. Remove the crossmember that supports the transmission at the rear.

4. Remove the driveshaft coupling through-bolts and nuts or the CV-joint through-bolts and nuts. Either type is located right at the rear of the transmission. Discard used self-locking coupling nuts. Keep the CV-joint clean and replace its gasket.

5. Unscrew the transmission locking ring at the center mount, if equipped. Then, remove the bolts and remove the center mount. Bend the driveshaft downward and pull it off the centering pin. Suspend it with wire from the underside of the vehicle.

6. Drain the transmission oil and discard it. Remove the oil filler neck. Disconnect the oil cooler lines at the transmission by unscrewing the flare nuts and plug the open connections.

7. If equipped, remove the converter cover by removing the Torx® bolts from behind and the regular bolts from underneath.

8. Remove the bolts fastening the torque converter to the driveplate, turning the flywheel as necessary to gain access from below.

9. If equipped, remove the guard for the speed and reference mark sensors. Remove the attaching bolt for each and remove each sensor. Keep the sensors clean.

10. Disconnect the shift cable by loosening the locknut fastening it to the shift lever and disconnecting the cable at the cable housing bracket.

11. If the transmission has an electrical connection, turn the bayonet fastener to the left to release the connection, disconnect it and pull the wire out of the ties.

12. Lower the transmission as far as possible. Then, remove all the Torx® or standard type bolts attaching the transmission to the engine.

13. Remove the small grill from the bottom of the transmission. Then, press the converter off with a large prybar through this opening while sliding the transmission out.

To install:

14. Install the transmission under the vehicle and raise it into position. Slide the torque converter and the transmission together before installing the transmission completely to the engine.

15. Install the small grille onto the transmission.

16. Raise the transmission to install it against the engine block.

17. The remaining components are installed in the reverse order from which they were removed. Be sure the converter is fully installed onto the transmission—so the ring on the front is inside the edge of the case.

a. When reinstalling the driveshaft, tighten the lockring with a special tool. If the driveshaft has a simple coupling, rather than a CV-joint, be sure to replace the self-locking nuts and to hold the bolts still while tightening the nuts to keep from distorting the coupling.

b. When installing the center mount, preload it forward from its most natural position 0.157–0.236 in. (3.98–5.99mm).

18. Connect the negative battery cable, then adjust the throttle cables.

3 SERIES

➡**To perform this operation, a support for the transmission, BMW tool 24 0 120 and 00 2 020 or equivalent and a tool for tightening the driveshaft locking ring, BMW tool 26 1 040 or equivalent, are required.**

1. Disconnect the battery ground cable. Loosen the throttle cable adjusting nuts, release the cable tension and disconnect the cable at the throttle lever. Then, remove (and retain) the nuts and pull the cable housing out of the bracket.

2. Disconnect the exhaust system at the manifold and hangers and lower it aside. Remove the hanger that runs across under the driveshaft. Remove the exhaust heat shield from under the center of the vehicle.

3. Drain the transmission oil and discard it. Remove the oil filler neck. Disconnect the oil cooler lines at the transmission by unscrewing the flare nuts and plug the open connections.

4. Support the transmission with the proper tools. Separate the torque converter housing from the transmission by removing the Torx® bolts with the proper tool from behind and the regular bolts from underneath. Retain the washers used with the Torx® bolts.

5. Remove bolts attaching the torque converter housing to the engine, making sure to retain the spacer used behind one of the bolts. Then, loosen the mounting bolts for the oil level switch just enough so the plate can be removed while pushing the switch mounting bracket to one side.

6. Remove the bolts attaching the torque converter to the driveplate. Turn the flywheel as necessary to gain access to each of the bolts, which are spaced at equal intervals around it. Be sure to re-use the same bolts and retain the washers.

7. To remove the speed and reference mark sensors, remove the attaching bolt for each and remove each sensor. Keep the sensors clean.

8. Turn the bayonet type electrical connector counterclockwise, then pull the plug out of the socket. Then, lift the wiring harness out of the harness bails.

9. Support the transmission using the proper jack. Then, remove the crossmember that supports the transmission at the rear.

10. Disconnect the transmission shift rod. Then, remove the nuts, then the through-bolts from the damper-type U-joint at the front of the transmission.

11. Unscrew the transmission locking ring at the center mount, if equipped, using the special tool designed for this purpose. Then, remove the bolts and remove the center mount. Bend the driveshaft downward and pull it off the centering pin. Suspend it with wire from the underside of the vehicle.

12. Lower the transmission as far as possible. Then, remove all the Torx® or standard type bolts attaching the transmission to the engine.

13. Remove the small grill from the bottom of the transmission. Then, press the converter off with a large prybar passing through this opening while sliding the transmission out.

To install:

14. Install the transmission in position under the vehicle and install the torque converter onto the transmission.

15. Be sure the converter is fully installed onto the transmission—so the

ring on the front is inside the edge of the case. Install the small grille onto the bottom of the transmission.

16. Raise the transmission up and bring it together with the engine, then install the engine-to-transmission attaching bolts until tight.

17. The remaining components are installed in the reverse order from which they were removed. When reinstalling the driveshaft, tighten the lockring with the proper tool. Be sure to replace the self-locking nuts on the driveshaft flexible joint and to hold the bolts still while tightening the nuts to keep from distorting it. When installing the center mount, preload it forward from its most natural position 0.157–0.236 in. (3.98–5.99mm). When reconnecting the bayonet type electrical connector, be sure the alignment marks are aligned after the plug it twisted into its final position. When reinstalling the speed and reference mark sensors, inspect the O-rings used on the sensors and install new ones, if necessary. Be sure to install the speed sensor into the bore marked **D** and the reference mark sensor, which is marked with a ring, into the bore marked **B**. Tighten the crossmember mounting bolts to 16–17 ft. lbs. (21–23 Nm). If O-rings are used with the transmission oil cooler connections, replace them.

18. Install the negative battery cable, then adjust the throttle cables.

SHIFT LINKAGE ADJUSTMENT

1. Move the selector lever to **P** position. Loosen the nut until the cable is free.

2. Push the transmission lever to the **D** or **P** position. Then, push the cable rod in the opposite direction.

3. Clamp down the cable rod without tension.

4. Tighten the nut to 7.0–8.5 ft. lbs. (9–11 Nm).

➡**Do not bend the cable.**

THROTTLE LINKAGE ADJUSTMENT

1. On the injection system throttle body, loosen the 2 locknuts at the end of the throttle cable and adjust the cable until there is a play of 0.010–0.030 in. (0.254–0.762mm).

2. Loosen the locknut and lower the kickdown stop under the accelerator pedal. Have someone depress the accelerator pedal until the transmission detent can be felt. Then, back the kickdown stop back out until it just touches the pedal.

3. Check that the distance from the seal at the throttle body end of the cable housing is at least 1.732 in. (43.9mm) from the rear end of the threaded sleeve. If this dimension checks out, tighten all the locknuts.

Clutch

ADJUSTMENT

These vehicles are equipped with a hydraulic clutch actuating system that is self-adjusting.

REMOVAL & INSTALLATION

1. Disconnect the negative battery cable. Raise and safely support the vehicle. Remove the heat shield, then the mounting bolts. Disconnect the speed and reference mark sensors at the flywheel housing. Mark the plugs for reinstallation.

2. Remove the transmission and clutch housing.

3. On vehicles with 6-cylinder engines, a Torx® socket is required. If equipped with a 265/6 transmission (without an integral clutch housing), remove the clutch housing.

4. Prevent the flywheel from turning, using a locking tool.

5. Loosen the mounting bolts one after another gradually, 1–1 ½ turns at a time, to relieve tension from the clutch.

6. Remove the mounting bolts, clutch and driveplate. Coat the splines of the transmission input shaft with Molykote® Long-term 2, Microlube® GL 2611 or equivalent. Be sure the clutch pilot bearing, located in the center of the crankshaft, turns easily.

7. Check the clutch driven disc for excess wear or cracks. Check the integral torsional damping springs, used with lighter flywheels only, for tight fit. Inspect the rivets to be sure they are all tight. Check the flywheel to be sure it is not scored, cracked, or burned, even at a small spot. Use a straight-edge to be sure the contact surface is true. Replace any defective parts.

To install:

8. To install, fit the new clutch plate and disc in place and install the mounting bolts.

9. When installing the clutch retaining bolts turn them in gradually to evenly tighten the clutch disc and to prevent warpage.

10. Install the transmission and the clutch housing.

11. If equipped, install the speed and reference mark sensors. Install the heat shield.

12. Note that on vehicles with 6-cylinder engines, the clutch pressure plate must fit

over dowel pins. Tighten the clutch mounting bolts to 16–19 ft. lbs. (21–26 Nm).

Hydraulic Clutch System

BLEEDING

1. Fill the reservoir.

2. Connect a bleeder hose from the bleeder screw to a container filled with brake fluid so air cannot be drawn in during bleeding procedures.

3. Pump the clutch pedal about 10 times, then hold it down.

4. Open the bleeder screw and watch the stream of escaping fluid. When no more bubbles escape, close the bleeder screw and tighten it.

5. Release the clutch pedal and repeat the above procedure until no more bubbles can be seen when the screw is opened.

6. If this procedure fails to produce a bubble-free stream:

a. Pull the slave cylinder off the transmission without disconnecting the fluid line.

➡**Do not depress the clutch pedal while the slave cylinder is dismounted.**

b. Depress the pushrod in the cylinder until it hits the internal stop. Then, reinstall the cylinder.

Halfshafts

REMOVAL & INSTALLATION

3 Series

1. Raise and safely support the vehicle.

2. Remove the rear tire and wheel assembly.

3. Disconnect the output shaft at the outer flange and suspend it with wire.

4. Unbolt the caliper and suspend it with the brake line connected. Unbolt and remove the rear disc.

5. Remove the large nut and remove the lock plate. If equipped with ABS, disconnect, then remove the ABS speed sensor by unscrewing it.

6. Unscrew the collar nut. Then, pull off the drive flange with the proper tool(s).

7. Screw on the collar nut until it is just flush with the end of the shaft and use a suitable hammer to knock out the shaft.

8. Remove the snap ring. Pull out the wheel bearings, using the proper tool.

9. Pull the inner bearing race off the axle shaft with special tool 00 7 500 or equivalent.

To install:

10. Install the new bearing assembly using the proper tools. Then, reinstall the snap ring.

11. Install the rear axle shaft with special tools: 23 1 300, 33 4 080 and 33 4 020 or equivalent.

12. Lubricate and install the collar nut, and drive in the lock plate with the proper tool(s). Tighten the collar nut to 148 ft. lbs. (200 Nm).

13. Install and connect the ABS speed sensor.

14. Remount the brake disc and caliper.

15. Reconnect the output shaft.

16. Install the rear tire and wheel assembly.

17. Lower the vehicle.

Remove the snap ring from the hub—3 and 5 Series

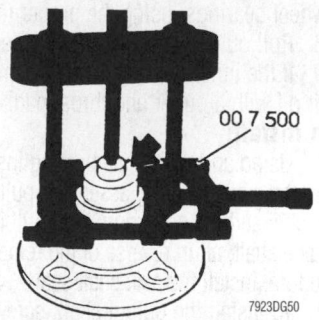

Special Tool 00 7 500, bearing puller—3 and 5 Series

5 and 7 Series

1. Raise and safely support the vehicle. Remove the rear tire and wheel assembly.

2. Lift out the lock plate and if equipped, remove the ABS sensor.

3. Remove the retaining nut from the output flange. Remove the flange.

4. Disconnect the output (half) shaft from the final drive (differential carrier) by pressing out with the proper tool and suspend it.

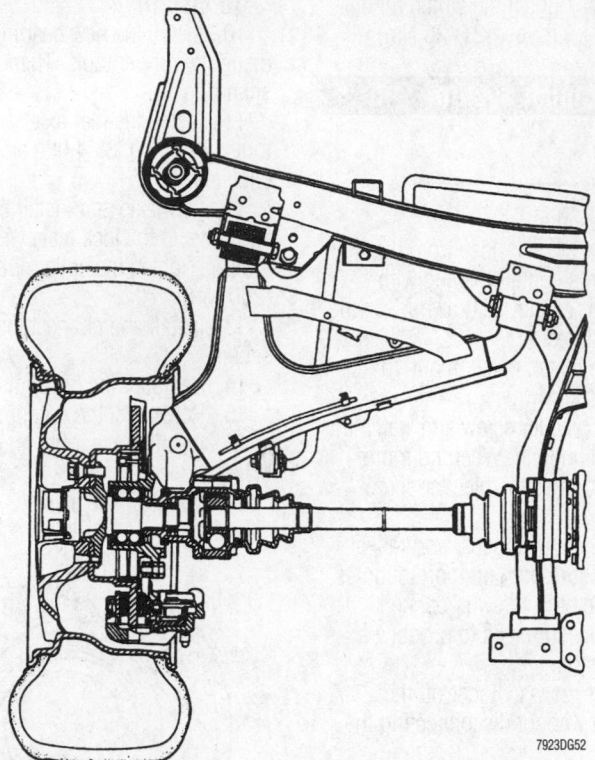

Cutaway schematic of the rear halfshaft and suspension system—5 Series

5. Pull out the output shaft from the drive flange hub with a special tool.

6. Drive out the rear axle shaft with the proper tool.

7. Lift out the snap ring. Then, pull out the wheel bearings, using the proper tool.

8. Pull out the seal with a suitable tool.

9. If the inner bearing shell is damaged, pull it off with a puller and thrust pad.

To install:

10. Using an appropriate bearing installer, pull in the wheel bearing assembly, pull in the seal, insert the snap ring, then pull in the rear axle shaft, all in reverse of the removal procedure. Install the axle shaft seal.

11. To install the output shaft, screw the threaded spindle into the shaft all the way, then use the nut and washer against the outside of the bridge.

12. Reconnect the output shaft to the final drive. Tighten the mounting bolts to 42–46 ft. lbs. (58–63 Nm).

13. Lubricate the bearing surface of the outer nut with oil and install the nut. Tighten the nut to 169–188 ft. lbs. (234–260 Nm).

14. Install the ABS sensor, if removed.

15. Install the rear tire and wheel assembly and lower the vehicle.

8 Series

1. Raise and safely support the vehicle. Remove the rear tire and wheel assembly.

2. Remove the ABS sensor.

3. Remove the retaining nut from the output flange. Remove the drive flange hub.

4. Remove the half-shaft from the vehicle by unscrewing the shaft from the final drive output flange and by pressing the half-shaft out of the drive flange hub with special tools 33–2–111, 116 and 117.

5. Press out the drive flange hub with special tools 33–3–250 and 33–2–105.

6. Lift out the snap-ring. Then, pull out the wheel bearings, using the special tools 33–3–261, 262, and 263.

7. Pull out the seal with a suitable tool.

8. If the bearing inner race is damaged, pull it off of the drive flange hub with special tool 33–3–240.

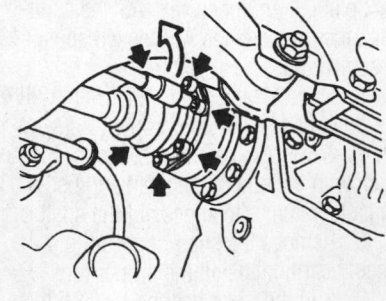

Halfshaft-to-final drive (rear differential) bolt locations—8 Series, others similar

To install:

9. Using an appropriate bearing installer, pull in the wheel bearing assembly, pull in the seal, insert the snap-ring, then pull in the drive flange hub with special tools 33–3–261, 263, and 264. Install the axle shaft seal.

10. To install the output shaft, pull in half-shaft with special tools 33–2–116 and 118.

11. Reconnect the output shaft to the final drive. Tighten the M10 mounting bolts to 61 ft. lbs. (83 Nm).

12. Lubricate the bearing surface of the outer nut with oil and install. Tighten the nut to 221 ft. lbs. (300 Nm).

13. Install the ABS sensor.

14. Install the rear tire and wheel assembly and lower the vehicle.

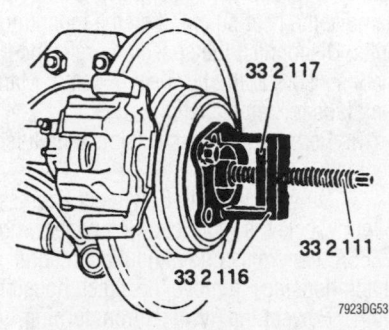

Halfshaft special tools for removal: 33 2 111, 116, and 117—8 Series

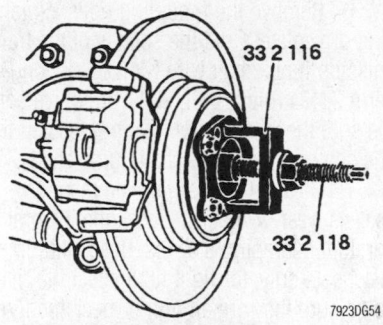

Halfshaft special tools for installation: 33 2 116 and 118—8 Series

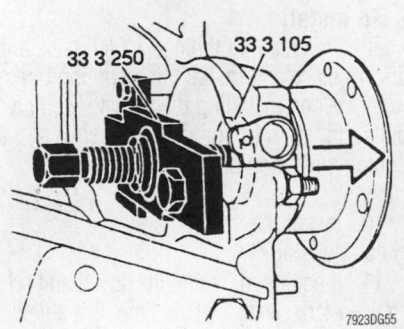

Drive flange hub removal special tools: 33 3 250 and 105—8 Series

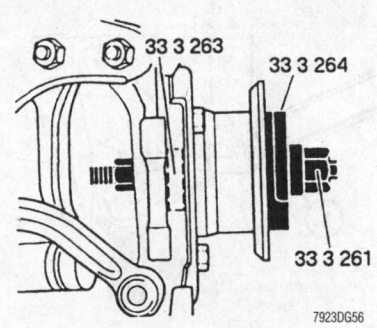

Drive flange hub installation special tools: 33 3 261, 263, and 264—8 Series

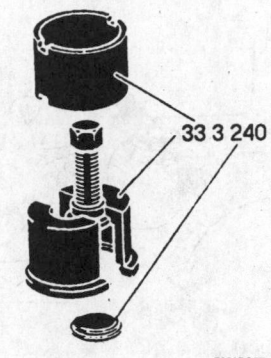

Special tool: 33 3 240—8 Series

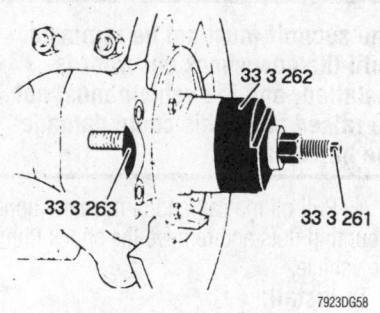

Bearings special tools: 33 3 261, 262 and 263—8 Series

STEERING AND SUSPENSION

Air Bag

✳✳ CAUTION

Some vehicles are equipped with an air bag system, also known as the Supplemental Inflatable Restraint (SIR) or Supplemental Restraint Sys- **tem (SRS). The system must be disarmed before performing service on, or around, system components, the steering column, instrument panel components, wiring and sensors. Failure to follow the safety precautions and the disarming procedure could result in accidental air bag deployment, possible personal injury and unnecessary system repairs.**

PRECAUTIONS

Several precautions must be observed when handling the inflator module to avoid accidental deployment and possible personal injury.

• Never carry the inflator module by the wires or connector on the underside of the module.

• When carrying a live inflator module, hold securely with both hands, and ensure that the bag and trim cover are pointed away.

• Place the inflator module on a bench or other surface with the bag and trim cover facing up.

• With the inflator module on the bench, never place anything on or close to the module which may be thrown in the event of an accidental deployment.

DISARMING

1. Place the ignition switch in the **OFF** position.

2. Disconnect the negative battery terminal and cover the battery terminal to prevent accidental contact.

3. Once the battery has been disconnected, wait for a period of approximately 10 minutes allowing the capacitor in the control unit to discharged. Once the capacitor is discharged, a trigger pulse cannot be generated inadvertently.

REARMING

1. Place the ignition switch in the **OFF** position.

2. Attach the sensors, the steering column connector and the seat belt tensioner connectors.

3. Connect the negative battery terminal.

4. Place the ignition switch in the **ON** position. Check that the SRS light illuminates for 6 seconds and extinguishes. If it illuminates in any other pattern, there is a problem that needs to be rectified by a qualified BMW technician.

Power Rack and Pinion Steering Gear

REMOVAL & INSTALLATION

3 Series

1. Raise and safely support the vehicle and remove front wheels. Remove the pinch bolt and loosen bolt. Press the spindle off the steering gear.

2. Use a syringe to empty the power steering fluid reservoir. Loosen the clamp and pull off the hydraulic fluid return line from the power steering unit. Discard drained fluid.

3. Disconnect and plug the pressure line.

4. Unscrew left and right side nuts and press off the tie rods where they connect to the spring struts.

5. Remove the bolts attaching the steering unit to the front crossmember and remove it.

To install:

6. Install the steering unit to the front crossmember, then install and tighten the mounting bolts.

7. Install in reverse order, keeping the following points in mind:

   a. The steering unit bolts to the rear holes of the axle carrier. Use new self-locking nuts and tighten them to 29–34 ft. lbs. (40–46 Nm).

   b. When reconnecting tie rods to the spring struts, be sure tie rod pins and strut bores are clean. Replace self-locking nut and tighten to 40–48 ft. lbs. (54–66 Nm).

   c. Replace the seals on the power steering pump connection and tighten the bolt to 29–32 ft. lbs. (40–43 Nm).

8. Refill the fluid reservoir with specified fluid. Idle the engine and turn the steering wheel back and forth until it has reached right and left lock 2 times each. Then, turn the engine **OFF** and refill the reservoir.

Power Recirculating Ball Steering Gear

REMOVAL & INSTALLATION

5, 7 and 8 Series

1. Disconnect the negative battery cable.

2. Remove the steering wheel, if equipped with an air bag (SRS).

3. Discharge the pressure reservoir by pushing in on the brake pedal about 10 times. Draw off hydraulic fluid in the supply tank.

4. Unscrew the bolt and press the tie rod off the steering drop arm with the proper tool.

5. Remove the heat shield on the steering gear and disconnect the ride level height control pipes on the 750iL.

6. Remove the bolt and push the U-joint from the steering gear. Disconnect and plug the hydraulic lines.

7. Unscrew the steering gear mounting bolts and remove the steering gear.

➡ **If necessary, move the steering drop arm by turning the steering stub to enable the removal of the gear assembly.**

To install:

8. Install the steering gear and tighten the mounting bolts.

9. Connect the hydraulic lines, using new seals.

10. Turn the steering wheel counterclockwise or clockwise against the stop, then back about 1.7 turns until the marks are aligned.

11. Connect the U-joint to the steering gear making sure the bolt is in the locking groove of the steering stub.

12. Install the tie rod to the steering drop arm and replace the self locking nut.

13. Replace the heat shield on the steering gear and connect the ride level height control pipes on the 750iL.

14. Refill the hydraulic fluid and replace the steering wheel, if equipped with an air bag (SRS).

15. Connect the negative battery cable.

Strut

REMOVAL & INSTALLATION

Front

3 SERIES

1. Disconnect the negative battery cable.
2. Raise and safely support the vehicle. Remove the tire and wheel assembly.
3. Disconnect the brake pad wear indicator plug and ground wire. Pull the wires out of the holder on the strut. Remove the ABS pulse sender, if equipped.
4. Unbolt the caliper and pull it away from the strut, suspending it with a piece of wire from the body. Do not disconnect the brake line.
5. Remove the attaching nut, then detach the pushrod on the stabilizer bar at the strut.

6. Unscrew the attaching nut and press off the guide joint with the proper tool.

7. Unscrew the nut and press off the tie rod joint.

8. Press the bottom of the strut outward and push it over the guide joint pin, using the proper tool. Support the bottom of the strut.

9. Unscrew the nuts at the top of the strut, from inside the engine compartment, then remove the strut.

To install:

10. Position the strut in the vehicle, then tighten the upper strut mounting nuts 16–17 ft. lbs. (21–23 Nm). The upper strut mounting nuts must be replaced with new self-locking nuts.

11. The remaining components are installed in the reverse order from which they were removed. Tie rod and guide joints must have both pins and both bores clean for reassembly. Replace both self-locking nuts. Tighten the control arm to spring strut attaching nut to 43–51 ft. lbs. (59–69 Nm).

12. Install the front tire and wheel assemblies, then lower the front of the vehicle.

13. Connect the negative battery cable.

5, 7 AND 8 SERIES

1. Disconnect the negative battery cable.
2. Raise and safely support the vehicle. Remove the tire and wheel assembly.
3. Disconnect the brake pad wear indicator plug and ground wire. Pull the wires out of the holder on the strut. Remove the ABS pulse sender, if equipped.
4. Disconnect the stabilizer pushrod with the proper tool.
5. Disconnect the lower strut bolts at the control arm.
6. Support the bottom of the strut and unscrew the nuts at the top of the strut, from inside the engine compartment. Remove the strut.
7. The installation is the reverse of the removal procedure.

Shock Absorber

REMOVAL & INSTALLATION

Rear

3 SERIES

1. Remove the trunk trim panel to expose the upper shock mounts.
2. Raise the rear of the vehicle and support safely.
3. Support the trailing arm and remove the lower mounting bolt.

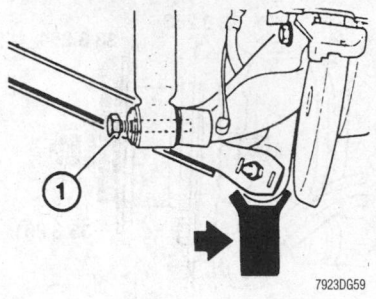

Support the trailing arm and remove the bolt (1)—3 Series

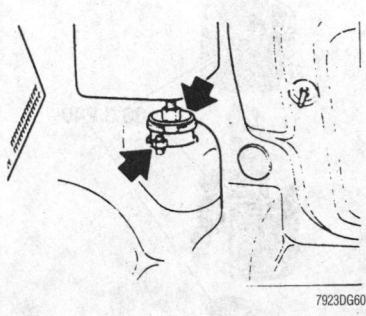

Upper mounting nut locations—3 Series

✳✳ WARNING

The support must not be removed until the new shock absorber is installed, and the vehicle must not be raised since this could damage the halfshafts.

4. Pull off the cap and remove the upper mounting nuts and remove the shock from the vehicle.

To install:

5. Place the shock into position with new seals fitted between the shock absorber and body. Renew the upper self-locking nuts and tighten to 11 ft. lbs. (15 Nm) for the Z3 and 318ti, and to 16 ft. lbs. (22 Nm) for all other 3 Series vehicles (including the M3).

6. Install the trunk trim panel.

7. Install the lower shock mounting to the rear axle assembly. The thrust washer on the rubber mount must face the screw head.

8. Lower the vehicle. With the vehicle resting at standard ride height, tighten the mounting bolt to 63 ft. lbs. (87 Nm), or to 94 ft. lbs. (130 Nm) if marked with 10.9, for the Z3 and 318ti models, or to 74 ft. lbs. (100 Nm) for all other 3 Series models (including the M3).

5 AND 7 SERIES WITH STANDARD SUSPENSION

1. Raise and support the rear of the vehicle.

2. Remove the rear seat cushion and back rest. Remove the trim panel over the strut mount.

3. Support the control arm, pull off rubber cap and remove the nuts at the top of the strut mount.

4. Remove the lower mounting bolt and lower the spring/shock assembly. Remove the assembly from the vehicle.

5. Use a spring compressor and compress the spring. Remove the top nut and pull the top mount off. Remove the spring.

To install:

6. Compress the new spring or replace the old spring on the shock. Install the mount and washers. Use a new locknut and tighten to 18 ft. lbs. (25 Nm). Release the spring.

7. Install the shock and tighten the upper mount nuts to 16 ft. lbs. (21.5 Nm). Loosely install the lower mounting bolt.

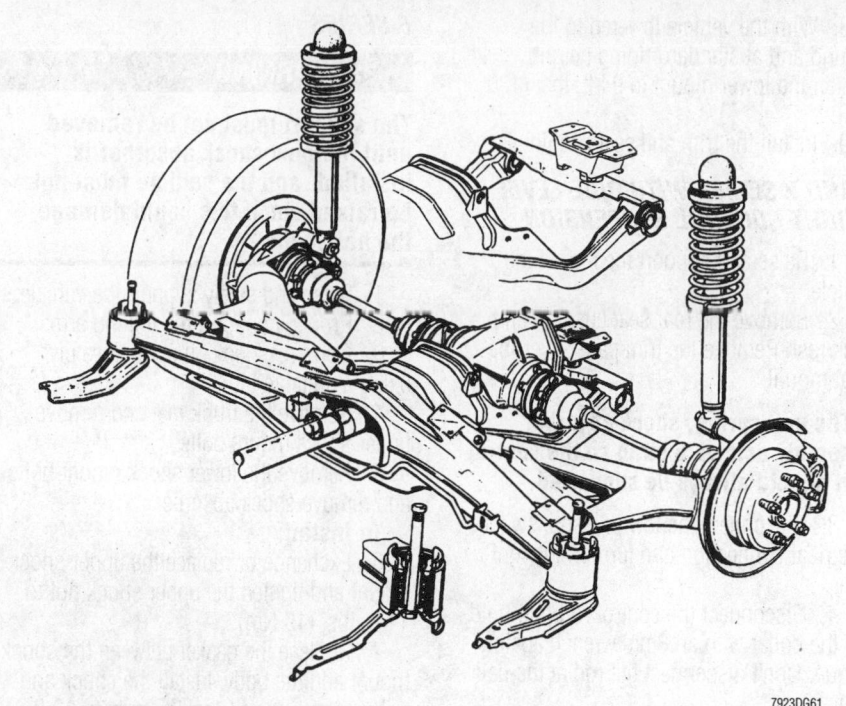

7923DG61

Rear suspension system—5 Series

REAR SPRING STRUT LAYOUT DRAWING

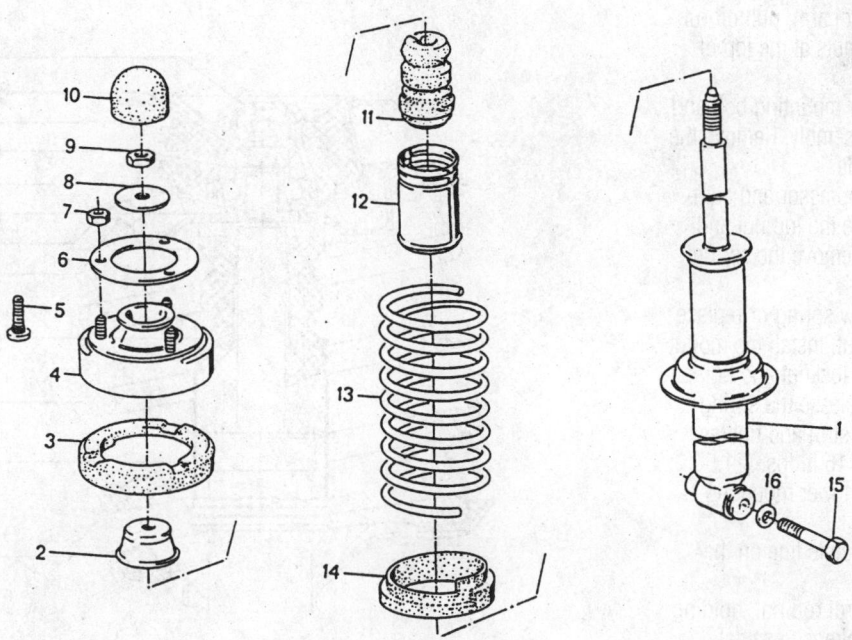

3	Upper spring ring	
4	Mount	
5	Bolt	
6	Insulator	
7	Collar nut M 8	
8	Disc	
9	Hexagon nut M 10 x 1.8 ZN	

12	Protective tube
13	Coil spring
14	Lower spring ring
15	Bolt M 14 x 1.5 x 85
16	Washer

7923DG62

Exploded view of the shock and coil spring assembly—5 and 7 Series

8. With the vehicle lowered to the ground and at standard riding height, tighten the lower mount to 94 ft. lbs. (130 Nm).

9. Install the trim and seat cushions.

5 AND 7 SERIES WITH RIDE LEVEL HEIGHT CONTROL SUSPENSION

1. Raise and support the rear of the vehicle.

2. Remove the rear seat cushion and back rest. Remove the trim panel over the strut mount.

➡ **The coil spring, shock absorber assembly acts as a strap so the control arm should always be supported.**

3. Disconnect the low pressure switch electrical connection and turn on the ignition.

4. Disconnect the control rod nut, holding the collar with an 8mm wrench against torque. Don't disconnect the rod at the ball joint.

5. Operate the lever on the control switch in the "discharge" direction for about 20 seconds to discharge fluid from the lines.

6. Disconnect the hydraulic line on the strut and turn off the ignition.

7. Support the control arm, pull off rubber cap and remove the nuts at the top of the strut mount.

8. Remove the lower mounting bolt and lower the spring strut assembly. Remove the assembly from the vehicle.

9. Use a spring compressor and compress the spring. Remove the top nut and pull the top mount off. Remove the spring.

To install:

10. Compress the new spring or replace the old spring on the strut. Install the mount and washers. Use a new locknut and tighten to 18 ft. lbs. (25 Nm). Release the spring.

11. Install the spring strut and tighten the upper mount nuts to 16 ft. lbs. (21.5 Nm). Loosely install the lower mounting bolt.

12. Connect the hydraulic line on the strut.

13. Connect the control rod nut, holding the collar with an 8mm wrench against torque.

14. Connect the low pressure switch electrical connection.

15. With the vehicle lowered to the ground and at standard riding height, tighten the lower mount to 94 ft. lbs. (130 Nm).

16. Install the trim and seat cushions.

8 SERIES

✳✳ WARNING

The support must not be removed until the new shock absorber is installed, and the vehicle must not be raised since this could damage the halfshafts.

1. Raise and safely support the vehicle.

2. Properly support the trailing arms.

3. Compress the coil springs safely, using a suitable tool.

4. Remove the trunk mat and remove upper shock mount bolts.

5. Remove the lower shock mount bolts and remove shock absorber.

To install:

6. Exchange or replace the upper shock mount and tighten the upper shock nut to 11 ft. lbs. (15 Nm).

7. Replace the gasket between the shock mount and the body. Install the shock and tighten the new self-locking nuts to 16 ft. lbs. (22 Nm). Install the trunk mat.

8. The thrust disk on the rubber mount must face the screw head. Install the lower shock mounting bolt and replace the cap.

9. Release the coil spring, and remove the support for the trailing arm. Lower the vehicle. With the vehicle resting at standard ride height, tighten the mounting bolt to 85 ft. lbs. (115 Nm).

Coil Spring

REMOVAL & INSTALLATION

Front

✳✳ CAUTION

This procedure calls for the spring to be compressed. A compressed spring has high potential energy and if released suddenly can cause severe damage and personal injury. If not comfortable with dealing with a compressed spring, have a professional technician remove the spring from the strut for you.

1. Remove the strut from the vehicle and mount in a vise using a strut holder. This will prevent damage to the strut tube.

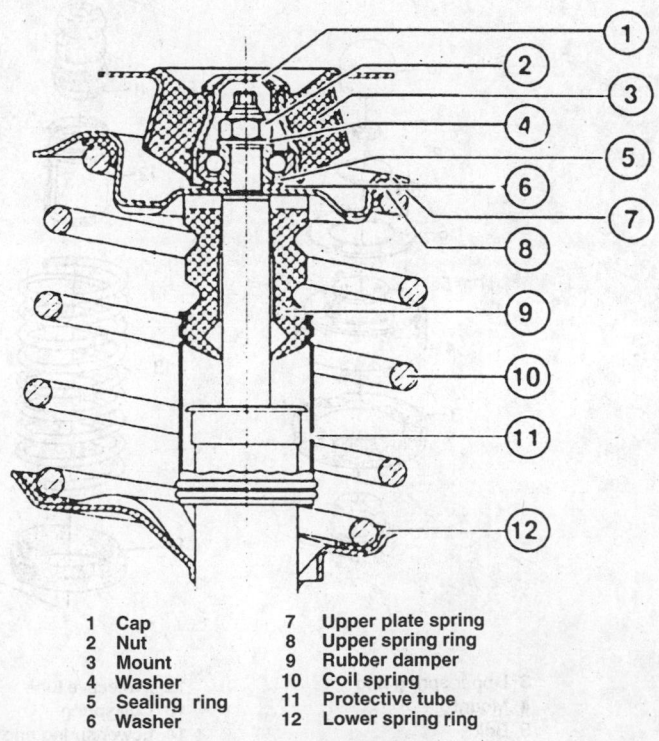

1	Cap	7	Upper plate spring
2	Nut	8	Upper spring ring
3	Mount	9	Rubber damper
4	Washer	10	Coil spring
5	Sealing ring	11	Protective tube
6	Washer	12	Lower spring ring

7923DG65

Cut away view of the strut mount and related components—3 Series with 4-cylinder engine

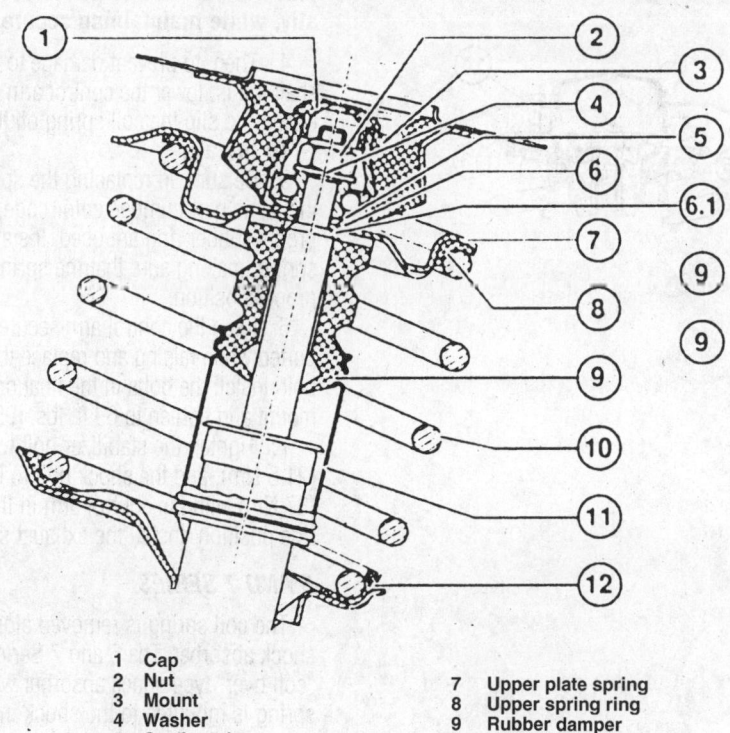

1	Cap	7	Upper plate spring
2	Nut	8	Upper spring ring
3	Mount	9	Rubber damper
4	Washer	10	Coil spring
5	Sealing ring	11	Protective tube
6	Washer	12	Lower spring ring
6.1	Ring for hollow piston rod		

7923DG66

Cut away view of the strut mount and related components—3 Series with 6-cylinder engine

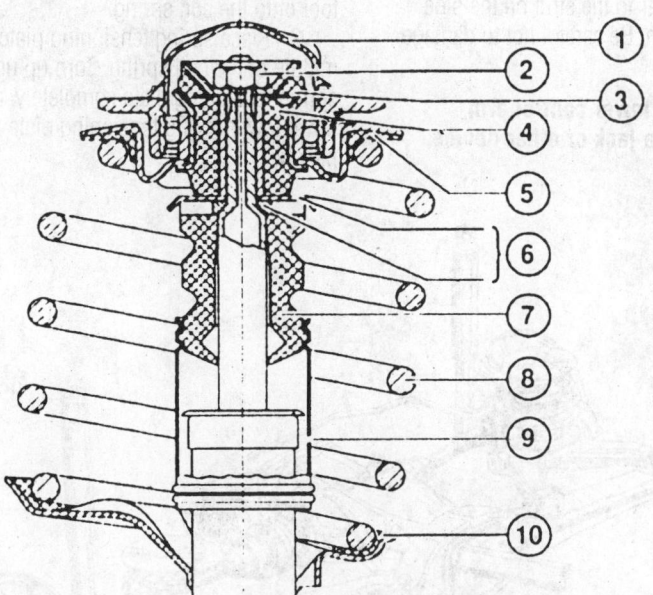

1.	Cap	6.	Support with ring for hollow piston rod
2.	Nut	7.	Rubber damper
3.	Stop washer	8.	Coil spring
4.	Mount	9.	Protective tube
5.	Upper spring ring	10.	Lower spring ring

7923DG67

Cut away view of the strut mounting with a separate support bearing and related components—3 Series with 6-cylinder engine

2. Using a proper spring compressor, compress the spring and lock into place. Use all the safety hooks provided and never point the compressed spring at a person.

3. Remove the top nut of the strut mount. Counterhold the strut rod during removal.

4. Pull the strut mount off the strut rod. Note the positioning of the spacers and washer for replacement.

5. Pull the spring off the strut and place somewhere safe.

6. Slowly release the compression of the spring.

To install:

7. Install the spring in the compressor and compress.

8. Install the spring and strut mount with all the spacers and washers in their original positions. Tighten the new strut rod nut to 47 ft. lbs. (65 Nm).

9. Release the spring slowly and check that it seats in the spring holders. Install the strut in the vehicle.

Rear

3 SERIES—EXCEPT Z3 AND 318TI

1. Raise the rear of the vehicle and support securely. Do not support on the suspension parts.

2. Remove the tire and wheel assembly.

3. Support the lower trailing arm at the hub and disconnect the stabilizer bar at the control arm and the subframe.

4. Remove the shock absorber lower mounting bolt.

5. Lower the trailing arm slowly and remove the spring to the side.

To install:

6. Install the spring with the bushing in place and the top of the upper spring ring lubricated.

7. Raise the trailing arm to a level where the bolt can be replaced in the lower shock mount. Connect the stabilizer bar. Do not tighten any bolts yet.

8. Install the tire and wheel assembly.

9. Lower the rear of the vehicle.

10. Tighten the stabilizer bolt to 16 ft. lbs. (21.5 Nm) and the shock bolt to 63 ft. lbs. (87 Nm) with the control arm in the normal ride position.

Z3 AND 318TI

1. Disconnect the rear portion of the exhaust system and hang it from the body.

2. Disconnect the final drive rubber mount, push it down, and hold it down with a wedge.

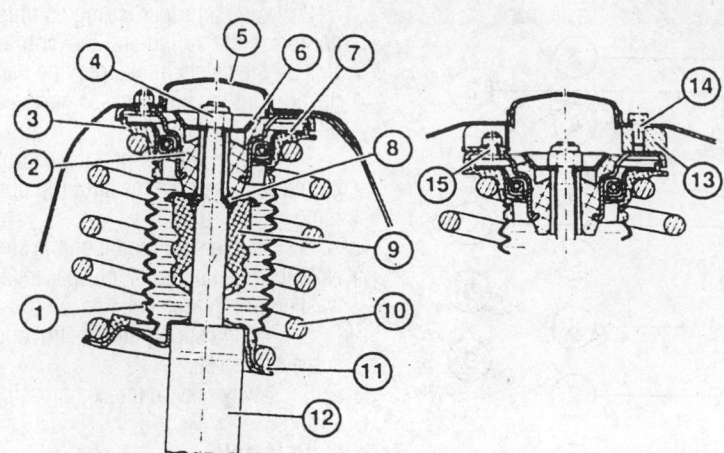

1. Rubber Gaiter
2. Thrust Bearing
3. Top spring support
4. Nut
5. Cap
6. Joint seat
7. Top spring plate
8. Support disc
9. Auxiliary spring
10. Coil spring
11. Bottom spring support
12. Spring strut shock absorber

7923DG68

Cut away view of the strut mount and related components—5, 7 and 8 Series

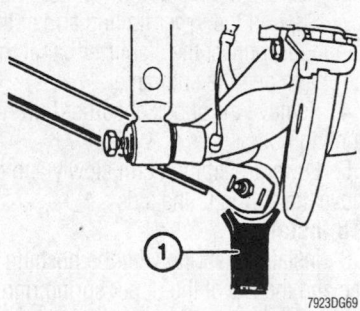

7923DG69

Support the trailing arm (1)—3 Series, except Z3 and 318ti

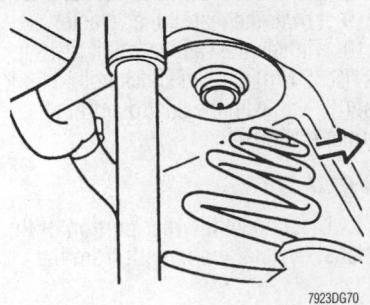

7923DG70

Remove the coil spring—3 Series, except Z3 and 318ti

3. Remove the bolt that connects the rear stabilizer bar to the strut on the side being worked on. Be careful not to damage the brake line.

➡ **Support the lower control arm securely with a jack or other device**

that will permit it to be lowered gradually, while maintaining secure support.

4. Then, to prevent damage to the output shaft joints, lower the control arm only enough to slip the coil spring off the retainer.

To install:

5. Be sure, in replacing the spring, that the same part number, color code, and proper rubber ring are used. Install the spring, making sure that the spring is in proper position.

6. Keep the control arm securely supported while raising and replace the shock bolt. Install the bolts in the final drive rubber mount and tighten to 69 ft. lbs. (95 Nm).

7. Tighten the stabilizer bolt to 16 ft. lbs. (21.5 Nm), and the shock bolt to 63 ft. lbs. (87 Nm) with the control arm in the normal ride position. Install the exhaust system.

5 AND 7 SERIES

The coil spring is removed along with the shock absorber. The 5 and 7 Series use a "coil over" type shock absorber where the spring is mounted to the shock in one compact unit. Once the shock is removed from the vehicle, the spring can be compressed and separated from the shock absorber.

8 SERIES

1. Raise and safely support the vehicle.
2. Position the coil spring compressor tool onto the coil spring.
3. Place spring tensioning plate in the middle of the coil spring. Turn up upper spring tensioning plate completely, and turn down lower spring tensioning plate completely.

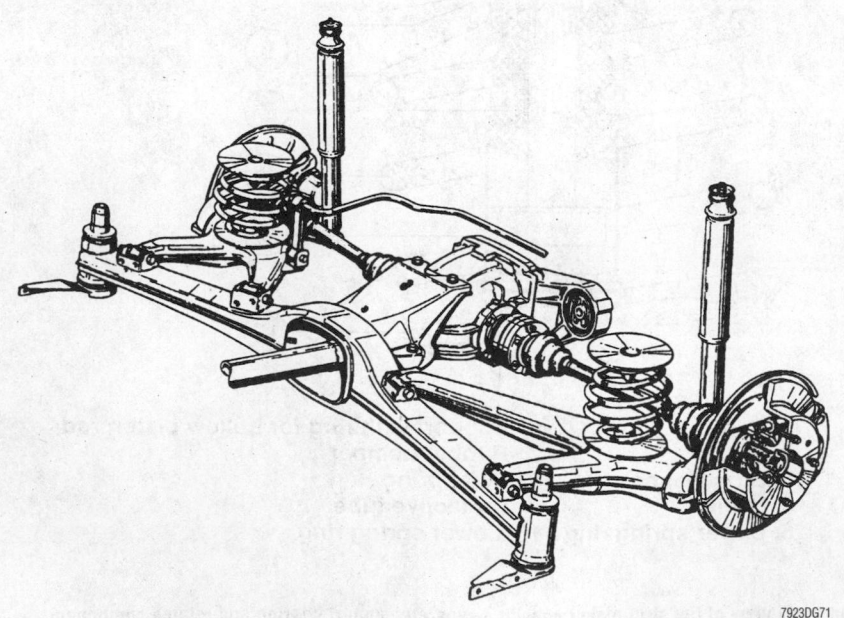

7923DG71

Rear suspension setup on Z3 and 318ti models

4. Slide in and turn the tensioning shaft until cross head is inserted perfectly in the opening of the upper spring retainer.

5. Compress the spring.

6. Properly support the trailing arm.

7. Remove the spring assembly.

To install:

8. Be sure, in replacing the spring, that the same part number, color code, and proper rubber ring are used. Install the spring, making sure that the spring is in proper position.

9. Keep the control arm securely supported. Check for correct position of the rubber liners.

10. Release the spring.

11. Remove the support and lower the vehicle.

Lower Ball Joint

REMOVAL & INSTALLATION

The 5, 7 and 8 Series uses a multi-link type suspension. There is a lower control arm to support the strut housing and a thrust rod to control fore and aft motion. The thrust rod is not used on 3 Series vehicles.

3 Series

1. Raise and safely support the vehicle. Remove the front tire and wheel assembly. Use a piece of wire to prevent the strut from extending to far and damaging the brake hose.

2. Disconnect the rear control arm bushing bracket where it connects to the body by removing the bolts.

3. Remove the nut and disconnect the link on the front stabilizer bar where it connects to the control arm.

4. Unscrew the nut which attaches the control arm to the crossmember and remove the nut from above the crossmember. Then, use a plastic hammer to knock the stud out of the crossmember.

5. Unscrew the nut to the point it contacts the strut housing. Remove the bolts connecting the hub to the struts. Press off the ball joint where the control arm attaches to the lower end of the strut, using the proper tool.

To install:

6. Clean the threaded holes in the hub. Install new micro-encapsulated hub mounting bolts to the strut and tighten to 58 ft. lbs. (80 Nm).

7. Be sure the ball joints studs and the bores in the crossmember and strut are clean before inserting the studs. Replace the

original nuts with replacement nuts and washers. Tighten the ball joint nut to 47 ft. lbs. (65 Nm) for 2 wheel drive. Tighten the control arm to subframe nut to 61 ft. lbs. (85 Nm) for 2 wheel drive.

8. Install the control arm bushing bracket and tighten the bolts to 30 ft. lbs. (42 Nm).

9. Install the stabilizer bar link and tighten to 43 ft. lbs. (59 Nm).

5, 7 and 8 Series

CONTROL ARM

1. Raise and safely support the front end. Do not place the jackstands under any suspension parts. Remove the wheel.

2. Remove the 3 bolts holding the steering knuckle to the bottom of the strut.

3. Remove the ball joint nut and press the stud out of the steering knuckle with a ball joint remover tool.

4. Remove the nut and bolt at the subframe end of the control arm. Remove the control arm.

To install:

5. Install the control arm to the subframe using a new nut and washer on both sides. Do not tighten at this point.

6. Clean the grease and dirt off of the ball joint stud and bore. Install the ball joint stud into the steering knuckle and tighten the new nut to 67 ft. lbs. (93 Nm).

7. Clean the threads and bores of the steering knuckle mounting bolts and the strut housing. Install the bolts, using threadlocker, and tighten to 80 ft. lbs. (110 Nm). There is a groove that will align the strut and knuckle.

8. Install the wheel and lower the vehicle to the ground. Load 150 lbs. into each of the front seats and in the center of the rear seat. Tighten the control arm to subframe bolt to 56 ft. lbs. (77.5 Nm).

THRUST ROD

Always replace the strut rods in pairs. If the strut rods are not replaced in pairs, uneven driving response may result.

1. Raise and safely support the front end. Do not place the jackstands under any suspension parts. Remove the wheel.

2. Remove the thrust rod ball joint nut and press the stud out of the steering knuckle with a ball joint remover tool.

3. Remove the nut and bolt at the subframe end of the strut rod. Remove the strut rod.

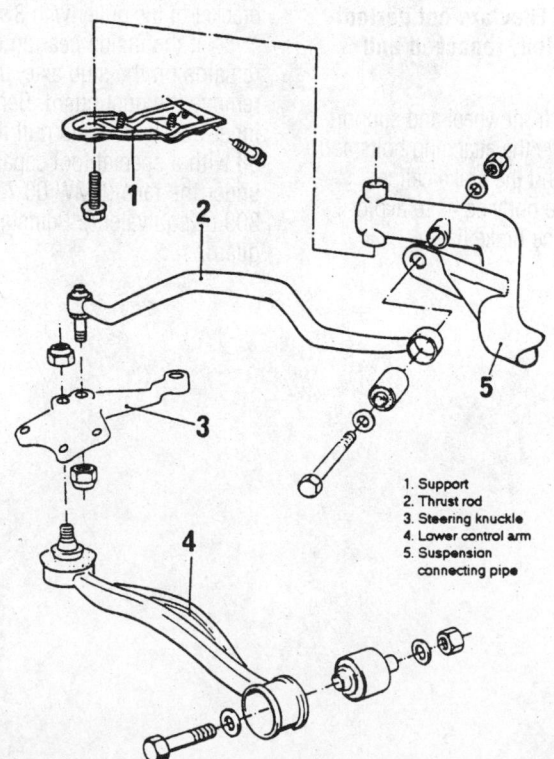

Fig. 8 Front suspension components — E34 5 Series

1. Support
2. Thrust rod
3. Steering knuckle
4. Lower control arm
5. Suspension connecting pipe

7923DG74

Exploded view of the front suspension—5, 7, and 8 Series

To install:

4. Install the strut to the subframe using a new nut and washer on both sides. Do not tighten at this point.

5. Clean the grease and dirt off of the ball joint stud and bore. Install the ball joint stud into the steering knuckle and tighten the new nut to 67 ft. lbs. (93 Nm).

6. Install the wheel and lower the vehicle to the ground. Load 150 lbs. into each of the front seats and in the center of the rear seat. Tighten the control arm to subframe bolt to 92 ft. lbs. (127 Nm).

Wheel Bearings

ADJUSTMENT

Wheel bearings can not be adjusted. The bearings must be replaced as a unit and never be reused once removed from the spindle

REMOVAL & INSTALLATION

Front

➡ **The wheel bearings are only removed if they are worn. They cannot be removed without destroying them (due to side thrust created by the bearing puller). They are not periodically disassembled, repacked and adjusted.**

1. Remove the front wheel and support the vehicle. Remove the attaching bolts and remove and suspend the brake caliper, hanging it from the body so as to avoid putting stress on the brake line.

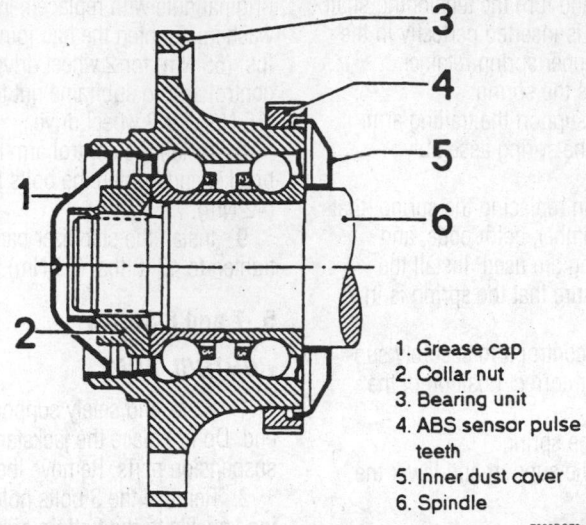

1. Grease cap
2. Collar nut
3. Bearing unit
4. ABS sensor pulse teeth
5. Inner dust cover
6. Spindle

7923DG73

Cut away view of the front wheel bearing

2. Remove the setscrew with an Allen wrench. Pull off the brake disc and pry off the dust cover with a small prybar.

3. Using a chisel, knock the tab on the collar nut away from the shaft. Unscrew and discard the nut.

4. Pull off the bearing with a puller set such as 31 2 101/102/104 and discard it. On the M3, use a puller set such as 31 2 102/105/106. On the M3, install the main bracket of the puller with 3 wheel bolts.

5. If the inside bearing inner race remains on the stub axle, unbolt and remove the dust guard. Bend back the inner dust guard and pull the inner race off with a special tool capable of getting under the race (BMW 00 7 500 and 33 1 309 or equivalent). Reinstall the dust guard.

To install:

6. If the dust guard has been removed, install a new one. Install a special tool (BMW 31 2 120 or equivalent; on M3, use 31 2 110 or equivalent) over the stub axle and screw it in for the entire length of the guide sleeve's threads. Press the bearing on.

7. Reverse the remaining removal procedures to install the disc and caliper. Tighten the wheel hub collar nut to 210 ft. lbs. (290 Nm). Lock the collar nut by bending over the tab.

Rear

Refer to the halfshaft procedure for the removal and installation of the rear wheel bearings.

CHRYSLER IMPORTS

15

Dodge/Plymouth-Colt • Eagle-Summit • Summit Wagon • Mitsubishi-Expo

ENGINE REPAIR

➡**Disconnecting the negative battery cable on some vehicles may interfere with the functions of the on board computer systems and may require the computer to undergo a relearning process, once the negative battery cable is reconnected.**

Distributor

REMOVAL

Before removing the distributor, position No. 1 cylinder at TDC on the compression stroke and align the timing marks.

1. Disconnect the negative battery cable.
2. Disconnect the spark plug wires from the distributor cap.
3. Remove the vacuum hose from the advance unit, if equipped.
4. Remove the cap from the distributor.
5. Verify the rotor points to the No. 1 cylinder position and the timing marks on the crankshaft pulley and the timing tab are aligned at TDC.
6. Mark the distributor body to the exact place the rotor points. Matchmark both the distributor mounting flange and the cylinder head.
7. Loosen and remove the retaining nut from the mounting stud. Lift the distributor from the cylinder head. The rotor may turn slightly from the mark on the distributor body. Make note of how far. When the distributor is reinstalled, this is the point to position the rotor.

INSTALLATION

Engine Not Disturbed

1. Position the distributor into the engine while aligning the matchmarks made during removal.
2. Verify the rotor points to the No. 1 cylinder position and the timing marks on the crankshaft pulley and the timing tab are aligned at TDC.
3. Install the distributor retaining nut on the mounting stud and tighten.
4. Install the cap on the distributor.
5. Connect the spark plug wires to the distributor cap.
6. Connect the negative battery cable to the battery.
7. Start the engine and check the ignition timing whenever the distributor has been removed.

Engine Disturbed

1. With the distributor removed from the engine, turn the crankshaft so the No. 1 piston is on the TDC of the compression stroke and the timing marks are aligned.

➡**With the distributor properly installed, the rotor will be pointed toward the No. 1 terminal of the distributor cap.**

2. Insert the distributor; if resistance is met, slight wiggling of the rotor shaft will help seat the distributor.
3. When the distributor seats against the head, align the matchmarks and install the retaining nut. Do not tighten the retaining nut all the way, as the timing must be checked.
4. Reinstall the rotor, cap, plug wires, primary lead or harness and connect the vacuum hoses.
5. Connect the negative battery cable. Start the engine, allow it to reach operating temperature and check the ignition timing. Adjust the timing by turning the distributor as needed. Tighten the retaining nut.

Ignition Timing

ADJUSTMENT

Colt

1. Attach the timing light according to the manufacturer's instructions.
2. Locate the timing tab line on the front of the engine and the notch on the crankshaft pulley. Mark them so they are easily recognizable with the timing light. Connect a tachometer to the engine.
3. Start the engine and allow it to reach operating temperature.
4. Check that the engine idle is at specifications.
5. Turn the engine OFF.
6. Locate the ignition timing adjustment connector (brown) and connect a jumper wire to from the connector to a good ground connection.

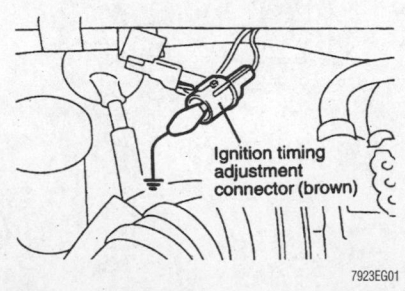

Timing connector identification—Colt models

➡**Grounding the ignition timing adjustment connector will set the engine to basic ignition timing.**

7. Point the timing light at the crankshaft pulley marks and inspect the ignition timing.
8. If the timing is not within specifications, loosen the distributor mounting nut and rotate the distributor slowly, in either direction, to align the timing marks.
9. Tighten the distributor mounting nut when the ignition timing is correct. Stop the engine and remove the timing light.
10. Adjust engine idle if needed.

Summit Wagon and Expo

1. Set the parking brake, start and run the engine until normal operating temperature is obtained. Keep all lights and accessories OFF and the front wheels straight-ahead. Place the transaxle in **P** or automatic transaxle or neutral for manual transaxle.

➡**On Canadian vehicles the lights will remain on when the vehicle is running, this will not be a problem.**

2. Locate the wire connector on the ignition coil connector. Insert a paper clip behind the TACH terminal connector to act as a tachometer adapter. Connect a tachometer to the paper clip. If not at specification, set the idle speed at the correct level.
3. Turn the engine **OFF** and remove the water-proof cover from the ignition timing adjusting connector. This connector is a brown connector located near the center of the firewall. Connect a jumper wire from this terminal to a good ground.
4. Connect a conventional power timing light to the No. 1 cylinder spark plug wire. Start the engine and run at idle.
5. Aim the timing light at the timing scale located near the crankshaft pulley.

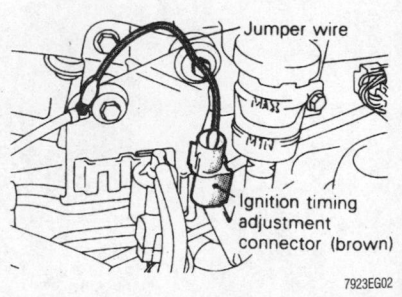

Ignition timing connector location—Summit Wagon and Expo models with 2.4L engines

6. Loosen the distributor hold-down nut just enough so the housing can be rotated.

7. Turn the housing in the proper direction until the specified timing is reached. Tighten the hold-down nut and recheck the timing. Turn the engine **OFF**.

8. Remove the jumper wire from the ignition timing adjusting terminal and install the water-proof cover.

9. Start the engine and check the actual timing without the terminal grounded. This reading should be approximately 5 degrees more than the basic timing. Actual timing may increase according to altitude. Also, actual timing may fluctuate because of slight variation accomplished by the ECU. As long as the basic timing is correct, the engine is timed correctly.

10. Turn the engine **OFF**. Disconnect the timing equipment and tachometer.

Engine Assembly

REMOVAL & INSTALLATION

All Engines

1. Relieve fuel system pressure.
2. Disconnect the negative battery cable. Remove the undercover if equipped.
3. Matchmark the hood and hinges and remove the hood assembly. Remove the air cleaner assembly and all adjoining air intake duct work.
4. Drain the engine coolant and remove the radiator assembly, coolant reservoir and intercooler.
5. Remove the transaxle assembly.
6. Disconnect the ground cable, accelerator cable, breather hose and heater hose connections from the engine.
7. Note the locations and remove the vacuum hoses from engine. Be sure to disconnect the brake booster vacuum supply.
8. For 2.4L engines, disconnect the heater hoses at the cylinder head and coolant inlet pipe.
9. Disconnect fuel feed and return hoses.
10. Disconnect the crankshaft and camshaft sensor wiring.
11. Disconnect oxygen sensor connections, coolant temperature gauge and coolant temperature sensor connections.
12. Disconnect the oil pressure switch connection.
13. On models with automatic transmissions, disconnect the thermo switch.
14. Disconnect harness connections for the idle speed control motor and throttle position sensor.

15. Disconnect harness connections for the intake air temperature sensor.
16. Disconnect EGR temperature sensor (California).
17. Note locations for reassembly and disconnect injector harness plugs.
18. Disconnect power transistor and the ignition coil connections.
19. Disconnect alternator and power steering switch wiring.
20. Remove the air conditioner drive belt and the air conditioning compressor. Leave the hoses attached. Do not discharge the system. Wire the compressor aside.
21. Remove the power steering pump and wire aside.
22. Remove the alternator and starter harness clamps.
23. Remove the exhaust manifold to head pipe nuts. Discard the gasket.
24. Attach a hoist to the engine and support the engine weight. Remove the engine mount bracket. Remove any torque control brackets (roll stoppers).
25. Remove the engine assembly from the vehicle.

To install:
26. Install the engine and secure in position. The front lower mount through-bolt nut should not be tightened until the full weight of the engine is on the mount. Tighten the bolts to the following specifications:
- 1.5L and 1.8L through-bolt—72 ft. lbs. (100 Nm)
- 1.5L and 1.8L bracket mounting bolts—42 ft. lbs. (58 Nm)
- 1.5L and 1.8L bracket mounting nut—38 ft. lbs. (53 Nm)
- 2.4L through-bolt—51 ft. lbs. (69 Nm)
- 2.4L bracket mounting bolts—42 ft. lbs. (58 Nm)

27. Using a new gasket, position exhaust pipe onto the manifold and tighten the flange nuts to 36 ft. lbs. (50 Nm).
28. Install the remaining components in the opposite order from which they were removed.
29. Connect negative battery cable and run engine.
30. Inspect all connections and check all fluid levels.

Water Pump

REMOVAL & INSTALLATION

1.5L and 1.8L Engines

1. Disconnect the negative battery cable.
2. Rotate the engine and position the No. 1 piston to TDC of its compression stroke.

3. Drain the cooling system.
4. Remove the engine undercover.
5. Disconnect the clamp bolt from the power steering hose.
6. Support the engine with the appropriate equipment and remove the engine mount bracket.
7. Remove the timing belt from the front of the engine.
8. Remove the timing belt rear cover.
9. Remove the power steering pump bracket.
10. Remove the alternator brace if necessary.

➡The water pump mounting bolts are different in length, note their positioning for reassembly.

11. Remove the water pump mounting bolts and remove the pump.

To install:
12. Thoroughly clean and dry both mating surfaces of the water pump and block.
13. Apply a 0.09–0.12 in. (2.5–3.0mm) continuous bead of sealant to water pump and install the pump assembly.

➡Install the water pump within 15 minutes of the application of the sealant. Wait 1 hour after installation of the water pump to refill the cooling system or starting the engine.

14. Properly position the bolts and tighten the bolts to 18 ft. lbs. (24 Nm).
15. The remaining components are installed in the reverse order from which they were removed.
16. Fill the system with coolant.
17. Connect the negative battery cable, run the vehicle until the thermostat opens and fill the radiator completely.
18. Once the vehicle has cooled, recheck the coolant level.

2.4L Engine

1. Disconnect the negative battery cable.
2. Drain the cooling system.
3. Remove the engine undercover.
4. Disconnect the clamp bolt from the power steering hose.
5. Support the engine with the appropriate equipment and remove the engine mount bracket.
6. Remove the timing belt from the front of the engine.
7. Disconnect the coolant hoses from the pump, if equipped.
8. Remove the alternator brace.
9. Remove the water pump, gasket and O-ring where the water inlet pipe(s) joins the pump.

To install:

10. Thoroughly clean and dry both gasket surfaces of the water pump and block.

11. Install a new O-ring into the groove on the front end of the water inlet pipe. Do not apply oils or grease to the O-ring. Wet with clean antifreeze only.

12. Install the gasket and pump assembly and tighten the bolts.

13. Connect the hoses to the pump.

14. Reinstall the timing belt and related parts.

15. Install the engine drive belts and adjust.

16. Fill the system with coolant.

17. Connect the negative battery cable, run the vehicle until the thermostat opens and fill the radiator completely.

18. Once the vehicle has cooled, recheck the coolant level.

Cylinder Head

REMOVAL & INSTALLATION

1.5L Engine

1. Relieve the fuel system pressure. Disconnect the negative battery cable.

2. Position the No. 1 piston to TDC of its compression stroke.

3. Drain the cooling system.

4. Remove the air intake hose and the air cleaner assembly.

5. Label, then detach all coolant hoses, cables, vacuum lines, electrical harness connectors and other connections from the cylinder head, intake manifold, exhaust manifold and all related components.

6. Remove the upper radiator hose, throttle body hoses, bypass hose and heater hose connections.

7. Place a shop towel around the high pressure fuel line to absorb any residual fuel remaining in the system. Disconnect the high pressure fuel line.

8. Disconnect and plug the fuel return line.

9. Remove the spark plug cables.

10. Remove the clamp that holds the power steering pressure hose to the engine mounting bracket.

11. Place a jack and wood block under the oil pan and carefully lift just enough to take the weight off the engine mounting bracket, remove the bracket.

12. Remove the valve cover and gasket.

13. Remove the timing belt front upper cover.

14. If not already done, rotate the crankshaft clockwise and align the timing marks.

15. While securing the sprocket, remove the sprocket bolt and remove the sprocket with the timing belt attached. Be sure to keep the sprocket and belt wired or tied together as one unit.

16. Remove the timing belt rear upper cover.

✳✳ CAUTION

The crankshaft must always be rotated clockwise. Do not rotate engine by turning the camshaft.

17. Remove the exhaust pipe self-locking nuts and separate the exhaust pipe from the exhaust manifold. Discard the gasket.

18. Loosen the cylinder head mounting bolts in the reverse sequence as shown in the tightening illustration, using three steps. Lift off the cylinder head assembly and remove the head gasket.

To install:

19. Thoroughly clean and dry the mating surfaces of the head and block. Check the cylinder head for cracks, damage or engine coolant leakage. Remove scale, sealing compound and carbon. Clean oil passages thoroughly. Check the head for flatness. End to end, the head should be within 0.002 in. with 0.008 in. the maximum allowed out of true. The total thickness allowed to be removed from the head and block is 0.008 in. maximum.

20. Place a new head gasket on the cylinder block with the identification marks facing upward. Be sure the gasket has the proper identification mark for the engine. Do not use sealer on the gasket.

➡ **The cylinder head torque specifications are for a cold cylinder head.**

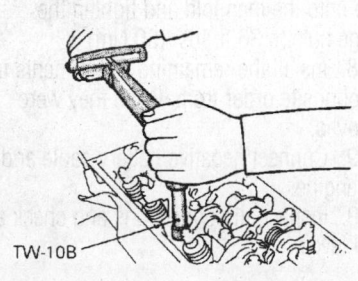

TW-10B

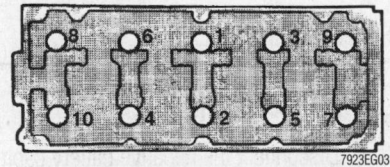

◄ Front of engine

Cylinder head bolt tightening sequence— 1.5L engine

21. Carefully install the cylinder head on the block. Using 3 even steps, tighten the head bolts in sequence to 53 ft. lbs. (73 Nm).

22. The remaining components are installed in the reverse order from which they were removed. Tighten the camshaft sprocket retaining bolt to 51 ft. lbs. (70 Nm) and the valve cover retaining bolts to 16 inch lbs. (2 Nm).

23. Fill the system with coolant.

24. Connect the negative battery cable, run the vehicle until the thermostat opens, fill the radiator completely.

25. Check and adjust the idle speed and ignition timing.

26. Once the vehicle has cooled, recheck the coolant level.

1.8L Engine

1. Relieve fuel system pressure. Disconnect the negative battery cable.

2. Position the No. 1 piston to TDC of its compression stroke.

3. Remove the air cleaner assembly.

4. Drain the cooling system.

5. Label, then detach all coolant hoses, cables, vacuum lines, electrical harness connectors and other connections from the cylinder head, intake manifold, exhaust manifold or related components.

6. Remove the upper radiator hose, overflow tube and the water hose from the thermostat to the throttle body.

7. Wrap the connection with a shop towel and disconnect the high pressure fuel line at the fuel rail. Discard the O-ring.

8. Disconnect the fuel return hose from the fuel pressure regulator.

9. Remove the thermostat housing, thermostat and the thermostat case with O-ring from the engine.

10. Remove the rocker cover.

11. Remove the timing belt upper cover.

12. If not already done, rotate the crankshaft in the clockwise direction to align the camshaft timing marks. Matchmark the camshaft sprocket and the timing belt. Tie the camshaft sprocket and the timing belt together so the sprocket will not move with respect to the timing belt.

13. While holding the camshaft sprocket in position using the appropriate wrench, remove the camshaft sprocket and with the belt attached. Wire the sprocket and belt aside making sure constant tension is maintained on the belt. Do not allow the belt to slacken or engine timing may be altered.

➡ **When removing the camshaft sprocket, do not allow the crankshaft to rotate. Confirm proper engine timing during installation.**

14. Loosen the cylinder head bolts in two or three steps in the reverse of the proper tightening sequence.

15. Remove the cylinder head from the engine.

✳✳ CAUTION

When removing the cylinder head, take care not to bend or damage the plug guide. The plug guide can not be replaced.

16. Remove the cylinder head gasket from the block.

To install:

17. Thoroughly clean and dry the mating surfaces of the head and block. Check the cylinder head for cracks, damage or engine coolant leakage. Remove scale, sealing compound and carbon. Clean oil passages thoroughly. Check the head for flatness end to end, the head should be within 0.002 in. with 0.008 in. the maximum allowed out of true. The total thickness allowed to be removed from the head and block is 0.008 in. maximum.

18. Place a new head gasket on the cylinder block with the identification marks facing upward. Be sure the gasket has the proper identification mark for the engine. Do not use sealer on the gasket.

19. Carefully install the cylinder head on the block.

20. Inspect the cylinder head bolt prior to installation. The length below the head of the bolts should not exceed 3.795 in. (96.4mm). If bolt shank length exceeds limit, bolt must be replaced. New bolts are always recommended.

21. Apply a small amount of engine oil to the thread section of the bolt and install so the chamfer of the washer faces upward.

22. Tighten the cylinder head bolts in the proper order as follows:

 a. In the proper tightening sequence, tighten bolts to 54 ft. lbs. (75 Nm).

 b. In the reverse order of the tightening sequence, fully loosen bolts.

 c. In the proper tightening sequence, tighten bolts to 14 ft. lbs. (20 Nm).

 d. In the proper tightening sequence, tighten bolts an additional ¼ turn (90 degrees).

 e. In the proper tightening sequence, tighten bolts an additional ¼ turn (90 degrees).

23. Install the remaining components in the opposite order from which they were removed. Tighten the camshaft sprocket mounting bolt to 65 ft. lbs. (90 Nm), the valve cover retaining bolts to 29 inch lbs. (3

Nm), the thermostat case assembly bolts to 16 ft. lbs. (22 Nm) and the thermostat housing bolts to 10 ft. lbs. (14 Nm).

24. Fill the system with coolant.

25. Connect the negative battery cable, run the vehicle until the thermostat opens, fill the radiator completely.

26. Check and adjust the idle speed and ignition timing.

27. Check all systems for leaks. Allow the engine to cool and recheck the coolant level.

2.4L Engine

1. Relieve fuel system pressure. Disconnect the negative battery cable.

2. Drain the cooling system.

3. Label, then detach all coolant hoses, cables, vacuum lines, electrical harness connectors and other connections from the cylinder head, intake manifold, exhaust manifold or related components.

4. Remove the radiator.

5. Wrap the connection with a shop towel and disconnect the high pressure fuel line at the fuel rail.

6. Disconnect the fuel return hose and remove the O-ring.

7. Remove the bolt retaining the power steering hose and air conditioner hose clamp.

8. Remove the coolant reservoir. Remove the bolt holding the ground wire to the manifold.

9. Place a jack and wood block under the oil pan and carefully lift just enough to take the weight off the engine mounting bracket. Then, remove the engine mounting bracket taking note of the position of the mount stopper.

10. Remove the valve cover, gasket and half-round seal.

11. Remove the timing belt front upper cover.

12. If possible, rotate the crankshaft clockwise until the timing marks on the cam sprocket and belt align. Matchmark the timing sprocket to the belt. Remove the sprocket bolt and remove the sprocket with the timing belt attached. Attach a flexible cord to the hood and suspend the sprocket so it cannot turn and there is no slack in the belt. Remove the timing belt rear upper cover.

13. Loosen the head bolts in the reverse of the correct tightening sequence in 2 or 3 steps. Remove the cylinder head bolts and head assembly from the block.

To install:

14. Thoroughly clean and dry the mating surfaces of the head and block. Remove

scale, sealing compound and carbon. Clean oil passages thoroughly.

15. Perform the following checks before reassembly:

 • Check the cylinder head for cracks, damage or engine coolant leakage.

 • Check the head for flatness. End to end, the head should be within 0.002 in. (0.051mm) normally with 0.008 in. (0.200mm) the maximum allowed out of true. The total thickness allowed to be removed from the head and block is 0.008 in. (0.200mm) maximum.

 • Check the cylinder head bolts for stretching or necking. The shank length should not exceed 3.91 in. (99.4mm).

16. Place a new head gasket on the cylinder block with the identification marks at the top (upward) position. Be sure the gasket has the proper identification mark for the engine. Do not use sealer on the gasket. Replace the turbo gasket and ring, if equipped.

17. Carefully install the cylinder head on the block. Tighten the cylinder head using the following procedure:

 a. Tighten all bolts in sequence to 58 ft. lbs. (78 Nm).

 b. Using loosening sequence, fully loosen all bolts.

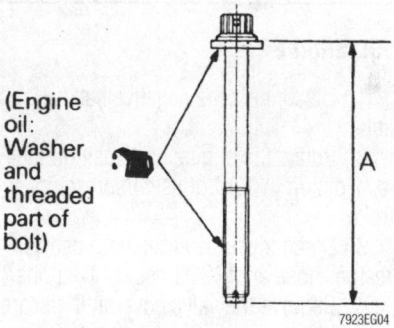

(Engine oil: Washer and threaded part of bolt)

A

Checking a cylinder head bolt for necking—2.4L engines

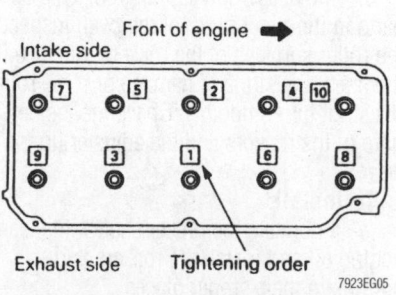

Front of engine ➡

Intake side

7 5 2 4 10

9 3 1 6 8

Exhaust side   Tightening order

To avoid damage to the cylinder head, tighten the bolts in the sequence shown— 1.8L and 2.4L engines

c. Tighten all bolts in sequence to 14 ft. lbs. (20 Nm).

d. Tighten all bolts in sequence ¼ turn (90°).

e. Once again, tighten all bolts in sequence ¼ turn (90°).

➡**Install the head bolt washer so the sagging side made by tapping out the washer is facing upward.**

18. Install the camshaft sprocket and tighten bolt to 65 ft. lbs. (90 Nm), while holding the sprocket in place using the appropriate wrench. Confirm proper timing mark alignment.

19. Apply sealer to the perimeter of the half-round seal and to the lower edges of the half-round portions of the belt-side of the new gasket. Install the valve cover.

20. Install the remaining components in the reverse order of removal.

21. Firmly set the parking brake. Start the engine and allow to idle until the thermostat opens, add coolant as required to fill system to the appropriate level.

22. Check all systems for leaks. Allow the engine to cool and recheck the coolant level.

Rocker Arms/Shafts

REMOVAL & INSTALLATION

1.5L Engine

1. Disconnect the negative battery cable.

2. Rotate the engine and position the No. 1 piston to TDC of its compression stroke.

3. Disconnect the accelerator cable, breather hose and PCV hose connections.

4. Remove the valve cover and discard the gasket.

5. Loosen both rocker arm assemblies gradually and evenly and remove the rocket shafts from the vehicle.

6. If disassembly is required, keep all parts in the exact order of removal. Inspect the roller surfaces of the rockers. Replace if there are any signs of damage or if the roller does not turn smoothly. Check the inside bore of the rockers and the adjuster tip for wear.

To install:

7. Lubricate the rocker shaft with clean engine oil and install the rockers and springs in their proper places.

8. Install the rocker shaft assemblies. Tighten the bolts gradually and evenly to 23 ft. lbs. (32 Nm).

9. Check valve adjustment and install the valve cover with a new gasket. Tighten the valve cover bolts to 16 inch lbs. (1.8 Nm).

10. Connect the accelerator cable, breather hose and PCV hose.

11. Connect the negative battery cable and check the engine adjustments.

1.8L Engine

1. Disconnect the negative battery cable.

2. Rotate the engine and position the No. 1 piston to TDC of its compression stroke.

3. Label and disconnect the spark plug cables.

4. Disconnect the air flow sensor connector and remove the air cleaner case cover.

5. Disconnect the accelerator cable, breather hose and PCV hose connections.

6. Remove the rocker cover and discard the gasket.

7. Loosen both rocker arm shaft assemblies gradually and evenly and remove the rocket shafts from the vehicle. Do not disassembly rocker arms and rocker arm shaft assemblies.

8. If disassembly is required, keep all parts in the exact order of removal. Inspect the roller surfaces of the rockers. Replace if there are any signs of damage or if the roller does not turn smoothly. Check the inside bore of the rockers and the adjuster tip for wear.

To install:

9. Lubricate the rocker shaft with clean engine oil and install the rockers and springs in their proper places.

10. Install the rocker arm and shaft assemblies. Tighten the rocker arm shaft retainer bolts to 23 ft. lbs. (32 Nm).

11. Check valve adjustment and install valve cover with a new gasket. Tighten the valve cover bolts to 29 inch lbs. (3.3 Nm).

12. Connect the spark plug cables.

13. Connect the accelerator cable, breather hose and PCV hose.

14. Connect the air flow sensor connector and install the air cleaner case cover.

15. Connect the negative battery cable. Run the engine at idle until normal operating temperature is reached. Check idle speed and ignition timing and adjust as required.

2.4L Engine

1. Disconnect the negative battery cable.

2. Disconnect the PCV and breather hoses. Remove the valve cover.

3. Install lash adjuster retainer tools MD998443 or equivalent, to prevent the auto-lash adjuster from falling out of the rocker arm.

4. Loosen rocker arm and shaft assembly evenly in several steps. Remove the rocker arm and shaft assembly as a complete unit.

➡**If any parts are to be reused, it is essential that all parts be kept in the same order and orientation for reinstallation. Be sure to mark and separate parts, so parts won't be mixed during reassembly.**

5. Carefully disassemble the shaft assembly. Visually inspect the rocker arm roller and replace if damage or seizure is evident. Check the roller for smooth rotation. Replace if excess play or binding is present. Also, inspect valve contact surface for possible damage or seizure. It is recommended that all rocker arms and lash adjusters be replaced together.

To install:

6. Immerse the lash adjusters in clean diesel fuel. Using a small wire, move the plunger of the lash adjuster up and down 4 or 5 times while pushing down lightly on the check ball in order to bleed out the air. Install the lash adjusters in the rocker arms.

7. Assemble the rocker assembly components in their original locations, lubricate the rocker shaft with heavy engine oil and position on the cylinder head.

8. Install the shaft mounting bolts and tighten all bolts evenly and gradually to 21–25 ft. lbs. (28–34 Nm).

9. Remove the lash adjuster retainers.

10. Install the valve cover, with a new gasket and semi-circular packing in place.

11. Connect the negative battery cable.

Intake Manifold

REMOVAL & INSTALLATION

All Engines

1. Relieve the fuel system pressure.

✳✳ CAUTION

The fuel injection system remains under pressure after the engine has been turned OFF. Properly relieve fuel pressure before disconnecting any fuel lines. Failure to do so may result in fire or personal injury.

2. Disconnect battery negative cable and drain the cooling system.

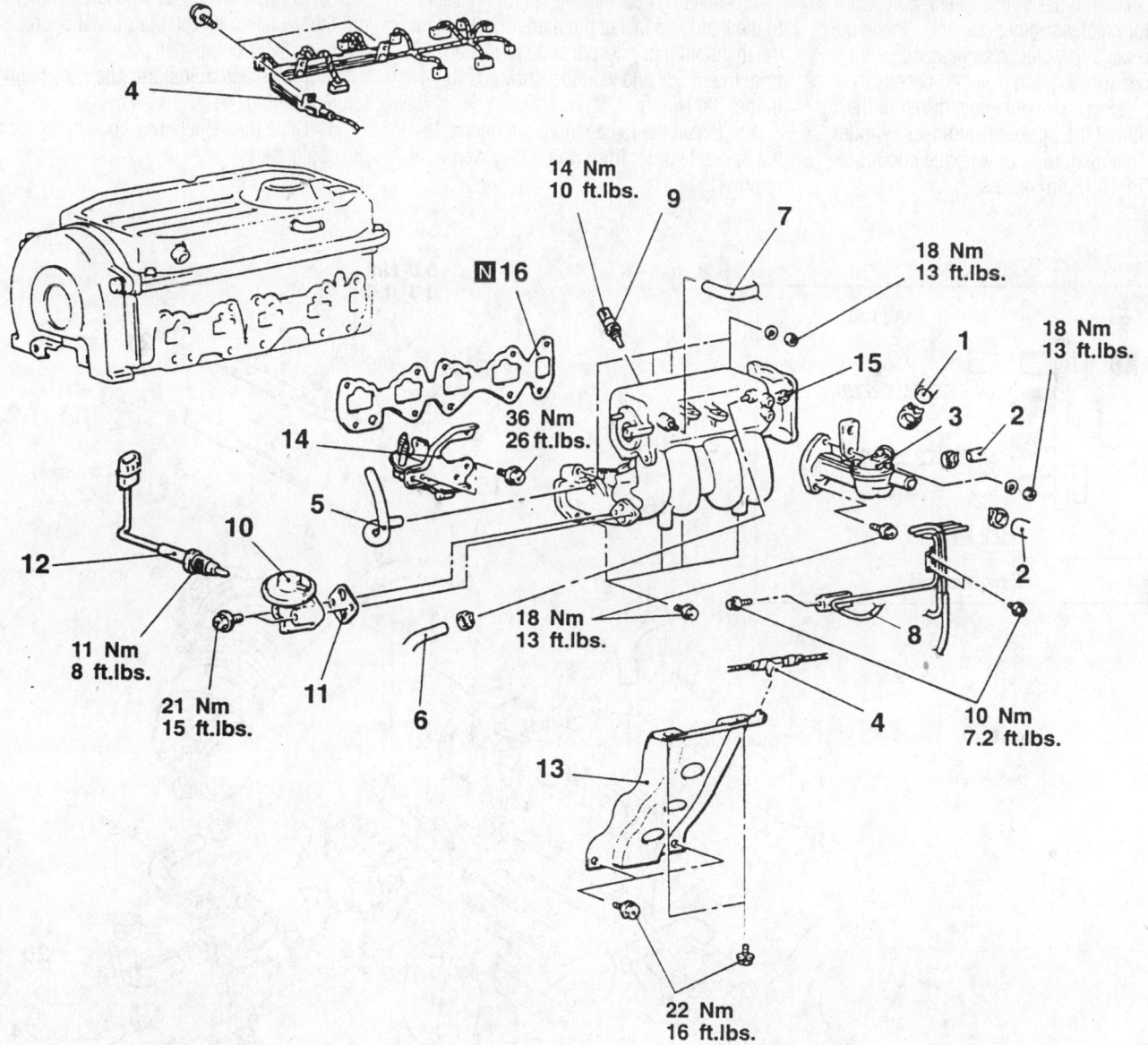

14 Nm
10 ft.lbs.

18 Nm
13 ft.lbs.

18 Nm
13 ft.lbs.

36 Nm
26 ft.lbs.

11 Nm
8 ft.lbs.

21 Nm
15 ft.lbs.

18 Nm
13 ft.lbs.

10 Nm
7.2 ft.lbs.

22 Nm
16 ft.lbs.

7923EG29

1. Radiator upper hose
2. Water hose
3. Water outlet fitting and thermostat housing assembly
4. Control wiring harness
5. PCV hose
6. Brake booster vacuum hose
7. Vacuum hose connection
8. Vacuum hose and pipe assembly
9. Intake air temperature sensor
10. EGR valve <Vehicles for Federal and Canada>
11. EGR gasket <Vehicles for Federal and Canada>
12. EGR temperature sensor <Vehicles for Federal and Canada>
13. Intake manifold stay
14. Engine mount stay
15. Intake manifold
16. Intake manifold gasket

Exploded view of the intake manifold mounting—1.5L engine

3. Label, then detach all coolant hoses, cables, vacuum lines, electrical harness connectors and other connections from the intake manifold and related components.

4. Disconnect the high pressure fuel line and the fuel return hose.

5. Remove the fuel rail, fuel injectors, pressure regulator and insulators.

6. Remove the intake manifold support bracket.

7. If the thermostat housing is preventing removal of the intake manifold, remove it.

8. Remove the intake manifold mounting bolts/nuts and remove the intake manifold assembly.

To install:

9. Clean all gasket material from the cylinder head intake mounting surface and

intake manifold assembly. Check both surfaces for cracks or other damage. Check the intake manifold water passages and jet air passages for clogging. Clean if necessary.

10. Using a straight edge, measure the distortion of the intake manifold-to-cylinder head. Total distortion or warpage should be 0.006 in. (0.15mm or less.

11. Install a new intake manifold gasket to the head and install the manifold. Tighten the manifold in a crisscross pattern, starting from the inside and working outwards to 14 ft. lbs. (20 Nm).

12. Install the remaining components in the opposite order from which they were removed.

13. Connect the negative battery cable, run the vehicle until the thermostat opens, fill the radiator completely.

14. Check and adjust the idle speed and ignition timing.

15. Once the vehicle has cooled, recheck the coolant level.

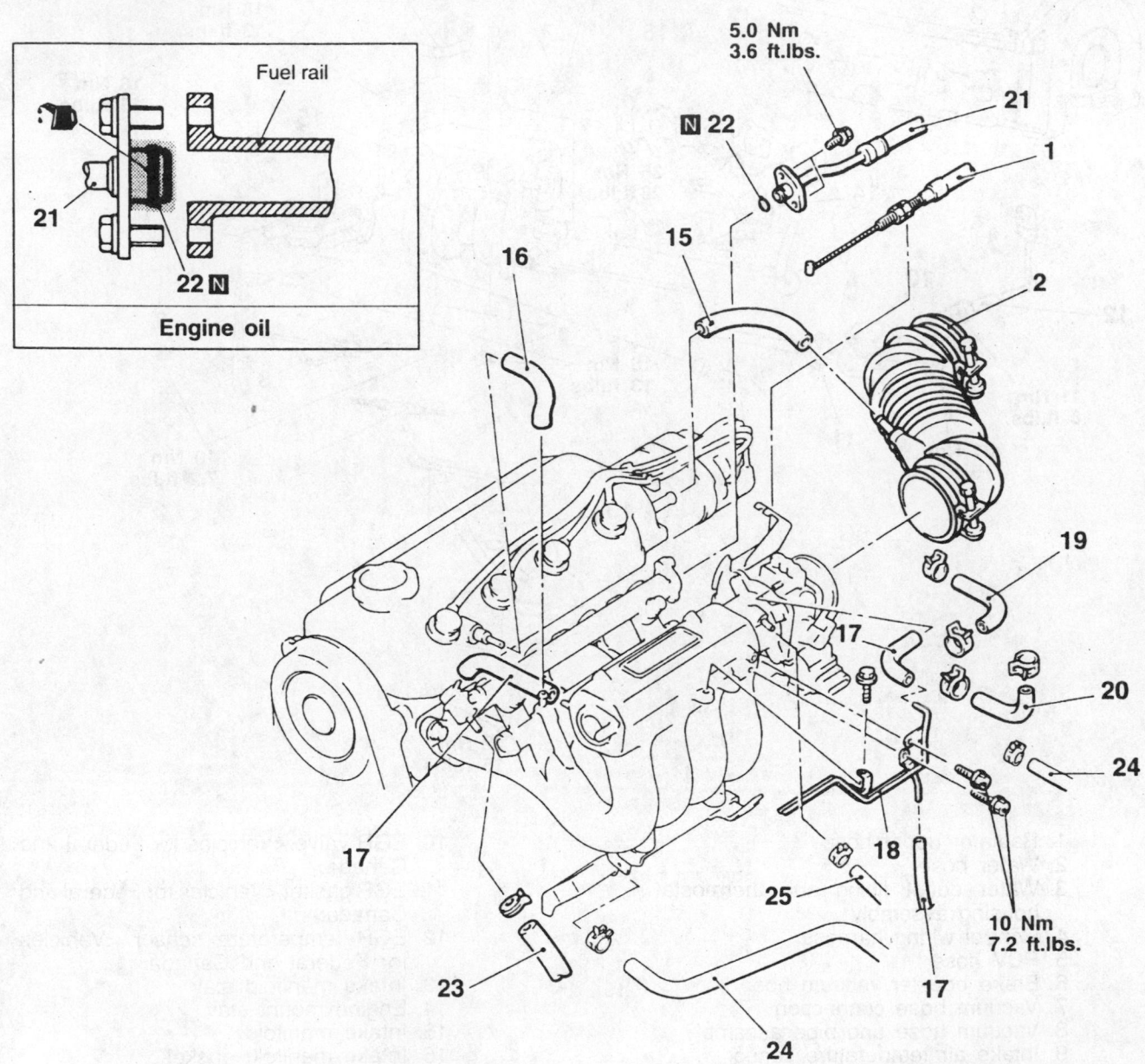

Removal steps

1. Accelerator cable connection
2. Air intake hose

Exploded view of the intake manifold mounting—1.8L engine, 1 of 2

7923EG30

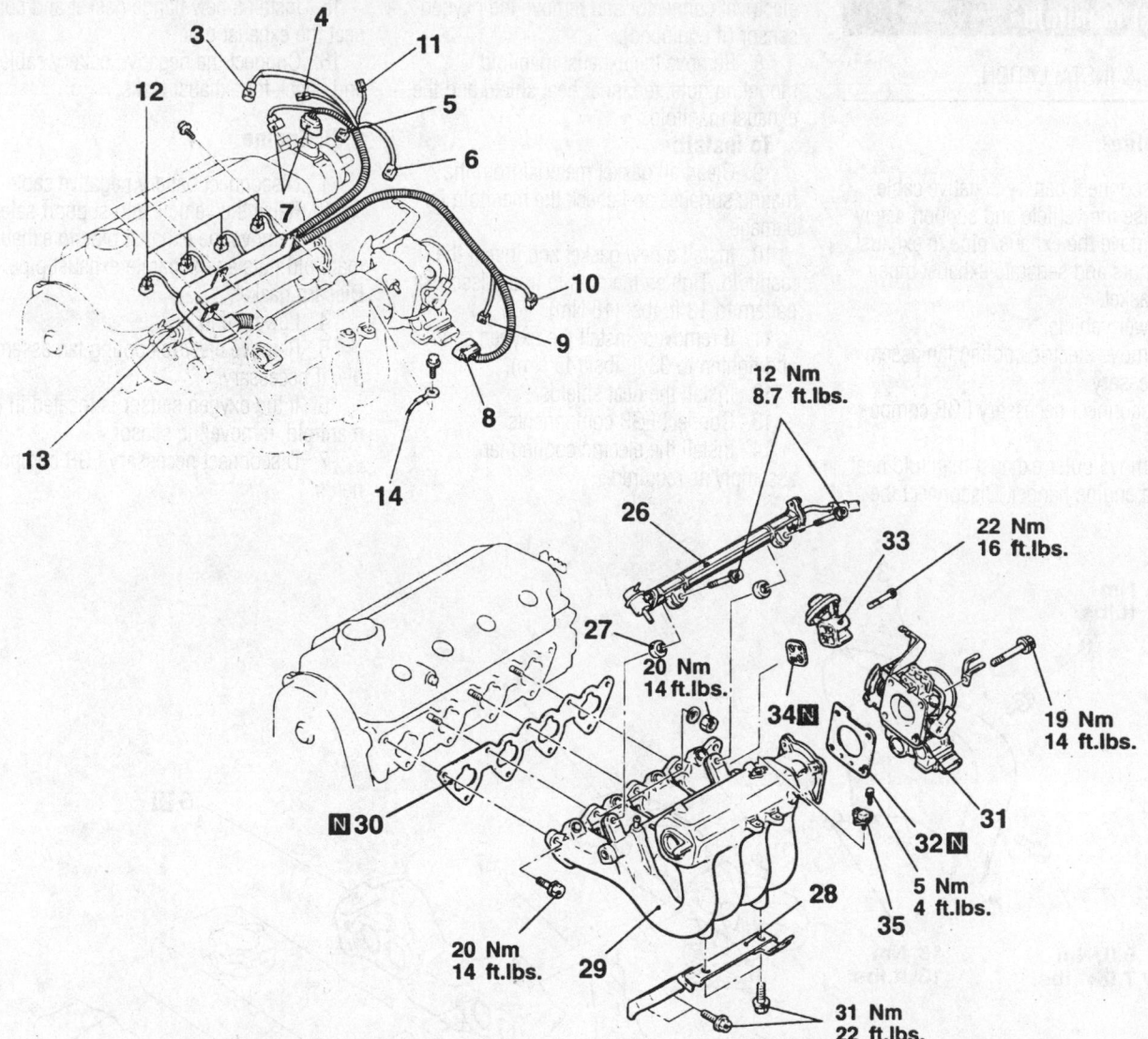

12 Nm
8.7 ft.lbs.

22 Nm
16 ft.lbs.

20 Nm
14 ft.lbs.

19 Nm
14 ft.lbs.

5 Nm
4 ft.lbs.

20 Nm
14 ft.lbs.

31 Nm
22 ft.lbs.

Exploded view of the intake manifold mounting—1.8L engine, 2 of 2

3. Heated oxygen sensor connector
4. Oil pressure switch connector
5. Engine temperature gauge unit connector
6. Engine coolant temperature sensor connector
7. Distributor connector
8. Idle air control motor connector
9. Heated oxygen sensor connector <Vehicles for California>
10. EGR temperature sensor connector <Vehicles for Federal and Canada>
11. Throttle position sensor connector
12. Injector connector
13. Control harness assembly
14. Ground wire
15. Breather hose connection
16. PCV hose connection
17. Vacuum hose connection
18. Vacuum pipe
19. Water hose connection (Thermostat case → Throttle body)
20. Water hose connection (Throttle body → Water inlet fitting)
21. High-pressure fuel hose connection
22. O-ring
23. Fuel return hose connection
24. Heater hose connection
25. Brake booster vacuum hose connection
26. Fuel rail, injector and pressure regulator assembly
27. Insulator
28. Intake manifold stay
29. Intake manifold
30. Intake manifold gasket
31. Throttle body
32. Throttle body gasket
33. EGR valve
34. EGR valve gasket
35. Manifold differential pressure sensor <Vehicles for California>

7923EG31

Exhaust Manifold

REMOVAL & INSTALLATION

1.5L Engine

1. Disconnect battery negative cable.
2. Raise the vehicle and support safely.
3. Remove the exhaust pipe to exhaust manifold nuts and separate exhaust pipe. Discard gasket.
4. Lower vehicle.
5. Remove electric cooling fan assembly, if necessary.
6. Disconnect necessary EGR components.
7. Remove outer exhaust manifold heat shield and engine hanger. Disconnect the electrical connector and remove the oxygen sensor (if equipped).
8. Remove the exhaust manifold mounting nuts, the inner heat shield and the exhaust manifold.

To install:

9. Clean all gasket material from the mating surfaces and check the manifold for damage.
10. Install a new gasket and install the manifold. Tighten the nuts to in a crisscross pattern to 13 ft. lbs. (18 Nm).
11. If removed, install the oxygen sensor and tighten to 33 ft. lbs. (45 Nm).
12. Install the heat shields.
13. Connect EGR components.
14. Install the electric cooling fan assembly as required.
15. Install a new flange gasket and connect the exhaust pipe.
16. Connect the negative battery cable and check for exhaust leaks.

1.8L Engine

1. Disconnect battery negative cable.
2. Raise the vehicle and support safely.
3. Remove the exhaust pipe to exhaust manifold nuts and separate exhaust pipe. Discard gasket.
4. Lower vehicle.
5. Remove electric cooling fan assembly, if necessary.
6. If the oxygen sensor is located in the manifold, remove the sensor.
7. Disconnect necessary EGR components.

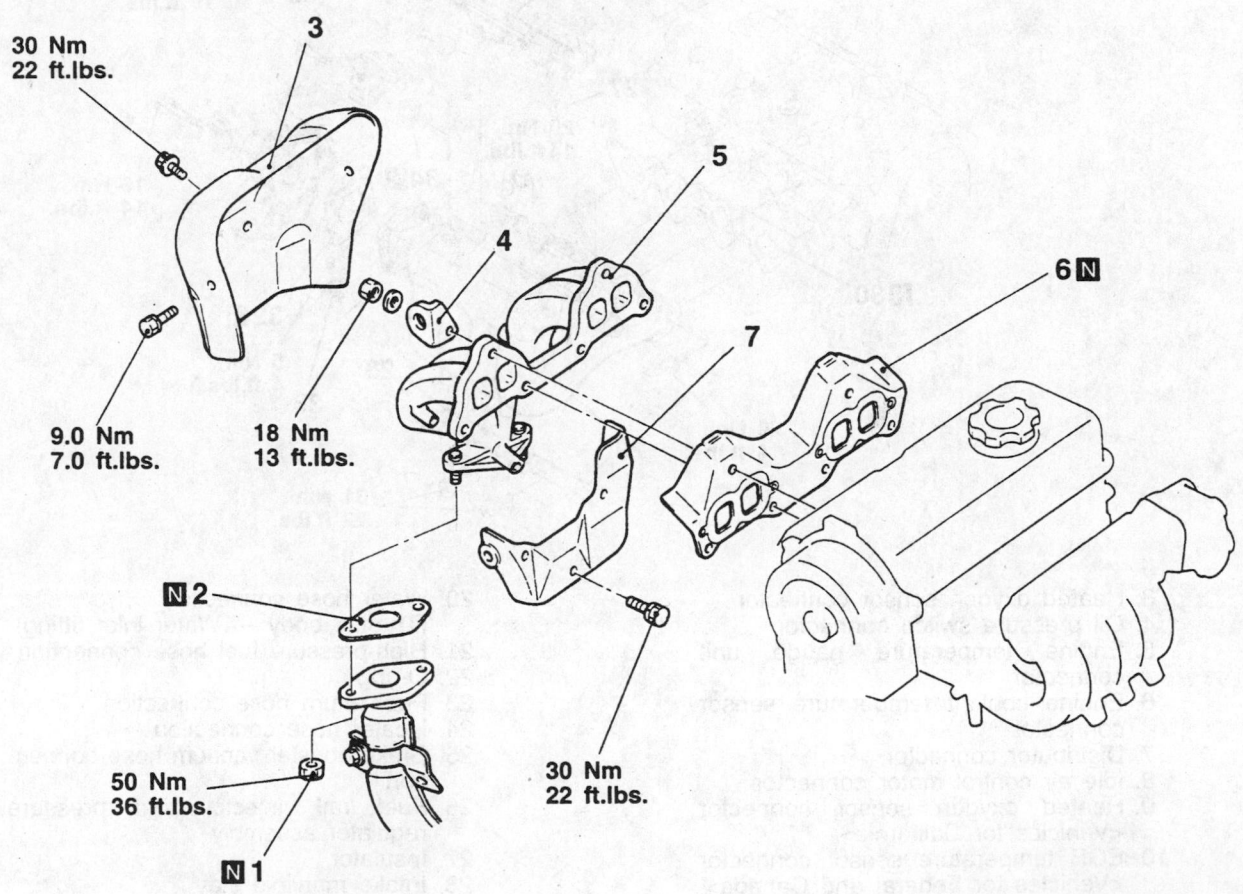

Removal steps

1. Self locking nut
2. Gasket
3. Exhaust manifold cover A
4. Engine hanger
5. Exhaust manifold
6. Exhaust manifold gasket
7. Exhaust manifold cover B

Exploded view of the exhaust manifold mounting—1.5L engine

7923EG32

8. Remove outer exhaust manifold heat shield and engine hanger. Disconnect the electrical connector and remove the oxygen sensor.

9. Remove the exhaust manifold mounting bolts, the inner heat shield and the exhaust manifold.

To install:

10. Clean all gasket material from the mating surfaces and check the manifold for damage.

11. Using a new gasket and install the manifold. Tighten the inner nuts to in a crisscross pattern to 13 ft. lbs. (18 Nm) and tighten the two outer (larger) nuts to 22 ft. lbs. (30 Nm).

12. Install the heat shields.

13. Connect EGR components.

14. If removed, install the oxygen sensor.

15. Install the electric cooling fan assembly as required.

16. Install a new flange gasket and connect the exhaust pipe.

17. Connect the negative battery cable and check for exhaust leaks.

2.4L Engine

1. Disconnect battery negative cable.

2. Raise the vehicle and support safely.

3. Leaving hoses connected, disconnect the power steering pump and position aside.

4. Remove the exhaust pipe to exhaust manifold nuts and separate exhaust pipe. Discard gasket.

5. Lower the vehicle.

6. Remove the electric cooling fan assembly.

7. Remove the outer exhaust manifold heat shield and engine hanger.

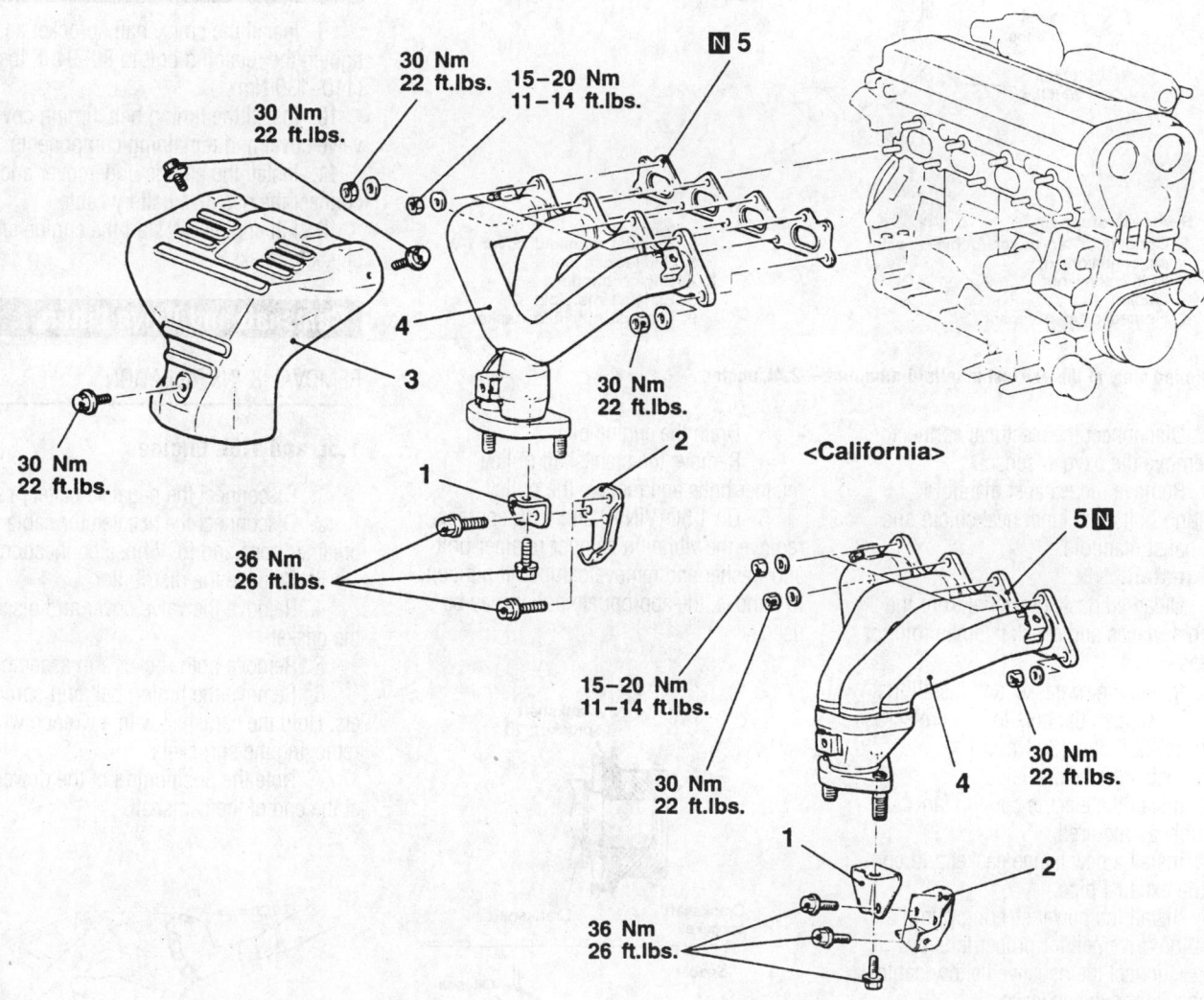

Removal steps

1. Bracket exhaust manifold B
2. Bracket exhaust manifold A
3. Exhaust manifold cover
4. Exhaust manifold
5. Exhaust manifold gasket

Exploded view of the exhaust manifold mounting—1.8L engine

7923EG33

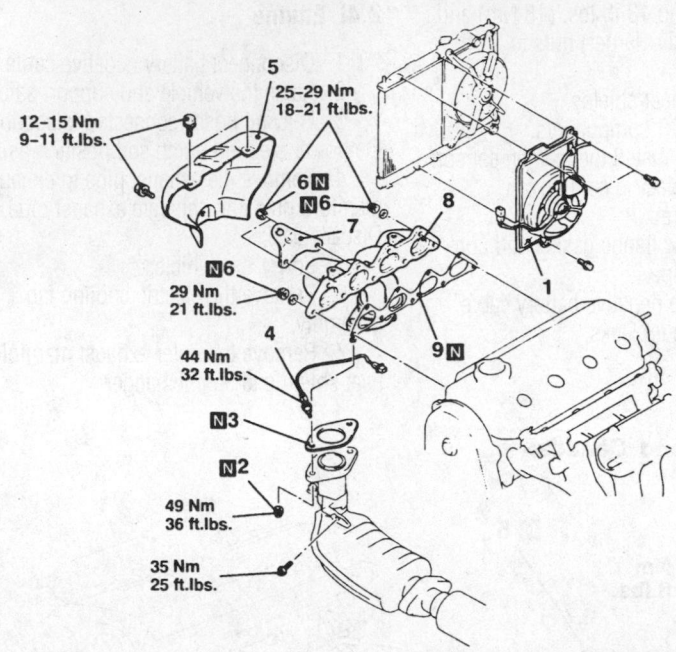

Removal steps
1. Condenser fan motor (Vehicles with air conditioning)
2. Self-locking nut
3. Gasket
4. Heated oxygen sensor
5. Exhaust manifold cover (A)
6. Self-locking nut
7. Engine hanger
8. Exhaust manifold
9. Exhaust manifold gasket

7923EG34

Exploded view of the exhaust manifold mounting—2.4L engine

8. Disconnect the electrical connector and remove the oxygen sensor.

9. Remove the exhaust manifold mounting bolts, the inner heat shield and the exhaust manifold.

To install:

10. Clean all gasket material from the mating surfaces and check the manifold for damage.

11. Install a new gasket and install the manifold. Tighten the nuts to in a crisscross pattern to 22 ft. lbs. (30 Nm).

12. Install the heat shields.

13. Install the electric cooling fan assembly as required.

14. Install a new flange gasket and connect the exhaust pipe.

15. Install the power steering pump and adjust the drive belt for proper tension.

16. Connect the negative battery cable and check for exhaust leaks.

Front Crankshaft Seal

REMOVAL & INSTALLATION

All Engines

1. Disconnect the negative battery cable.
2. Remove the timing belt.

3. Drain the engine oil.

4. Remove the crankshaft pulley retainer bolts and remove the pulley.

5. On 1.5L (VIN A) and 1.8L engines, remove the vibration damper retainer bolt and washer and remove damper. If difficult to remove, the appropriate puller may be used.

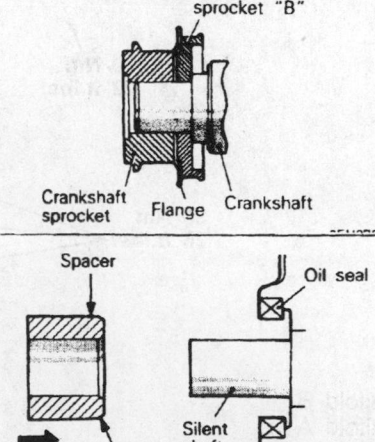

Front crankshaft oil seal installation— 2.4L engine

7923EG06

6. Remove the crankshaft sprocket retainer bolt and washer from the sprocket, if used, and remove the sprocket. If no bolts are used on the sprocket, use the appropriate puller to remove.

7. Pry out the oil seal from front of engine.

To install:

8. Using proper size driver, install new front seal.

❋❋ WARNING

Small nicks and burrs on crankshaft surface will damage oil seal. Use care when installing oil seal not to damage crankshaft surface.

9. Install the crankshaft sprocket and tighten the retaining bolt to 80–94 ft. lbs. (110–130 Nm).

10. Install the timing belt, timing covers, valve cover and remaining components.

11. Install the engine undercover and connect the negative battery cable.

12. Fill engine oil, start the engine and check for leaks.

Camshaft and Valve Lifters

REMOVAL & INSTALLATION

1.5L and 1.8L Engines

1. Disconnect the negative battery cable.
2. Disconnect the accelerator cable, breather hose and PCV hose connections.
3. Remove the distributor.
4. Remove the valve cover and discard the gasket.
5. Remove both rocker arm assemblies.
6. Remove the timing belt and sprockets. Hold the camshaft with a wrench when removing the sprockets.
7. Note the positioning of the dowel pin at the end of the camshaft.

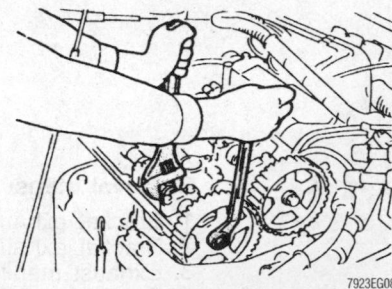

7923EG08

Hold the camshaft with a wrench while removing the sprockets—1.5L and 1.8L engines

8. Remove the camshaft oil seal from the front of the cylinder head.

9. Remove the camshaft from the head.

10. Carefully check all parts for damage and wear.

To install:

11. Lubricate the camshaft with heavy engine oil and slide it into the head. Be sure to position the dowel pin at the 12 o'clock position.

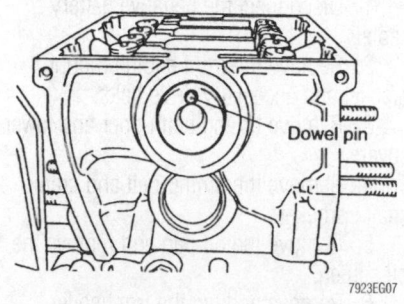

Positioning of the dowel pin during camshaft installation—1.5L engine

12. Check the camshaft end-play between the thrust case and camshaft. The camshaft end-play should be 0.002–0.008 in. (0.05–0.20mm). If the end-play is not within specification, replace the camshaft thrust bearing.

13. Install a new camshaft oil seal. Be sure to lubricate the lips of the seal with clean engine oil.

14. The balance of installation is the reverse of the removal procedure.

15. Connect the negative battery cable and check the ignition timing.

2.4L Engine

1. Disconnect the negative battery cable.

2. Remove the breather hose. Disconnect the PCV hose.

3. Matchmark and remove the distributor.

4. Remove the timing belt.

5. Disconnect and tag the spark plug wires.

6. Remove the rocker cover.

7. Remove the camshaft sprocket retainer bolt while holding shaft stationary with appropriate spanner wrench. Remove the sprocket from the shaft.

8. Remove the camshaft oil seal.

9. Install the lash adjuster holders on the rocker arms to prevent them from falling out.

10. Remove both rocker arm shaft assemblies from the head. Do not disas-

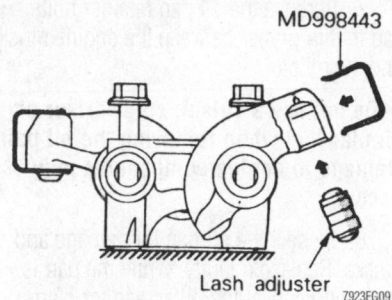

Install the lash adjuster holders on the rocker arms—2.4L engine

semble rocker arms from the rocker arm shaft unless worn or damaged.

11. Remove the camshaft from the cylinder head.

12. Inspect the bearing journals on the camshaft, cylinder head, and bearing caps.

To install:

13. Lubricate the camshaft journals and camshaft with clean engine oil and install the camshaft in the cylinder head.

14. Install the rocker arm and shaft assemblies. Tighten the rocker arm shaft retainer bolts to 21–25 ft. lbs. (29–35 Nm).

15. Install new camshaft oil seal.

16. Install camshaft sprocket and retainer bolt tightening to 65 ft. lbs. (90 Nm).

17. The remaining components are installed in the reverse order from which they were removed.

18. Connect the negative battery cable. Run the engine at idle until normal operating temperature is reached. Check idle speed and ignition timing and adjust as required.

Valve Lash

ADJUSTMENT

1.5L and 1.8L Engines

➡**Incorrect valve clearances will cause unsteady engine operation, excessive noise and reduced engine output. Check the valve clearances and adjust as required while the engine is hot.**

1. Warm the engine to operating temperature, turn **OFF** and disconnect the negative battery cable.

2. Remove all spark plugs so engine can be easily turned by hand.

3. Remove the valve cover.

4. Turn the crankshaft clockwise to position the No. 1 cylinder at TDC of its compression stroke. The notch on the crankshaft pulley will be aligned with the **T** mark on the timing belt lower cover.

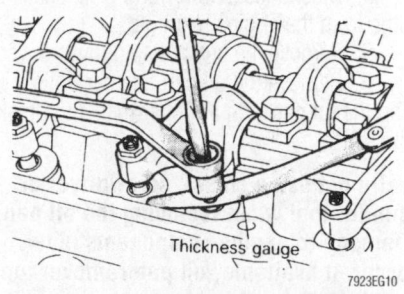

Proper method of adjusting valve clearance, using a wrench, feeler gauge and a screwdriver—1.5L and 1.8L engines

➡**When the No. 1 cylinder at TDC of its compression stroke, the No. 1 cylinder intake and exhaust valves will have valve lash.**

5. Check the valve lash at cylinder No. 1 intake, cylinder No. 1 exhaust, cylinder No. 2 intake and cylinder No. 3 exhaust valves.

6. Rotate the crankshaft clockwise 1 complete turn and align the **T** mark. This will position the No. 4 cylinder at TDC of its compression stroke.

➡**When the No. 4 cylinder at TDC of its compression stroke, the No. 4 cylinder intake and exhaust valves will have valve lash.**

7. Check the valve lash at cylinder No. 2 exhaust, cylinder No. 3 intake and cylinder No. 4 intake and exhaust valves.

8. If the valve clearances are out of specification, loosen the rocker arm locknut and adjust the clearance using a feeler gauge while turning the adjusting screw. Be sure to hold the screw to prevent it from turning when tightening the locknut.

9. After adjusting the valves, install the valve cover with new gasket and spark plugs, and connect the negative battery cable.

2.4L Engine

The 2.4L engines utilize hydraulic lash adjusters, which automatically retain the correct valve lash. These engines do not require manual valve lash adjustments.

Oil Pan

REMOVAL & INSTALLATION

1.5L and 1.8L Engines

1. Disconnect the negative battery cable.

2. Raise the vehicle and support safely.

3. Remove the oil pan drain plug and drain the engine oil.

4. Disconnect and lower the exhaust pipe from the engine manifold.

5. Remove the bell housing lower cover.

6. Remove the oil pan retainer bolts. Tap the oil pan with a rubber mallet to break the seal.

➡ Do not use a chisel, screwdriver or similar tool when removing the oil pan. Damage to engine components may occur. If available, oil pan remover tool MD998727 or equivalent may be used break the seal.

7. Inspect the oil pan for damage and cracks. Replace if faulty. While the pan is removed, inspect the oil screen for clogging, damage and cracks. Replace if faulty.

To install:

8. Using a wire brush or other tool, scrape clean all gasket surfaces of the cylinder block and the oil pan so that all loose material is removed. Clean sealing surfaces of all dirt and oil.

9. Apply sealant around the gasket surfaces of the oil pan in such a manner that all bolt holes are circled and there is a continuous bead of sealer around the entire outside edge of the oil pan.

➡ **The continuous bead of sealer should be applied in a bead approximately 0.16 in. (4mm) in diameter.**

10. Install the oil pan onto the cylinder block within 15 minutes after applying sealant. Install the fasteners and tighten to 60 inch lbs. (7 Nm).

11. Install the bell housing cover.

12. Connect the exhaust pipe from the engine manifold with new gasket in place. Tighten the exhaust pipe to manifold flange nuts to 33 ft. lbs. (45 Nm). Install and tighten the support bolt to 18 ft. lbs. (25 Nm).

13. Install the oil drain plug and tighten to 29 ft. lbs. (40 Nm).

14. Lower the vehicle and fill the crankcase to the proper level with clean engine oil.

15. Connect the negative battery cable. Start the engine and check for leaks.

2.4L Engine

1. Disconnect the negative battery cable.

2. Raise the vehicle and support safely.

3. Remove the oil pan drain plug and drain the engine oil.

4. Disconnect and lower the exhaust pipe from the engine manifold.

5. On FWD models, remove the left side axle shaft.

6. On AWD models, remove the transfer assembly.

7. Remove the oil pan retainer bolts. Tap in thin prybar between the engine block and the oil pan.

➡ **Do not use a chisel, screwdriver or similar tool when removing the oil pan. Damage to engine components may occur.**

8. Inspect the oil pan for damage and cracks. Replace if faulty. While the pan is removed, inspect the oil screen for clogging, damage and cracks. Replace if faulty.

To install:

9. Using a wire brush or other tool, scrape clean all gasket surfaces of the cylinder block and the oil pan so that all loose material is removed. Clean sealing surfaces of all dirt and oil.

10. Apply sealant around the gasket surfaces of the oil pan in such a manner that all bolt holes are circled and there is a continuous bead of sealer around the entire perimeter of the oil pan.

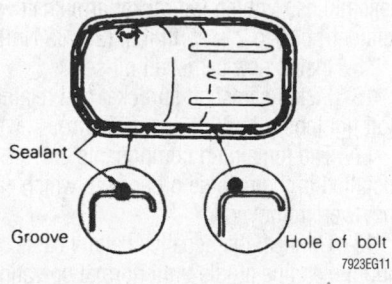

Apply sealant to the oil pan as shown— 1.8L and 2.4L engines

➡ **The continuous bead of sealer should be applied in a bead approximately 0.16 in. (4mm) in diameter.**

11. Install the oil pan onto the cylinder block within 15 minutes after applying sealant. Install the fasteners and tighten to 4–6 ft. lbs. (6–8 Nm).

12. If applicable, install the transfer assembly and check fluid level.

13. Connect the exhaust pipe from the engine manifold with new gasket in place. Tighten the exhaust pipe to manifold flange nuts to 29 ft. lbs. (40 Nm). Install and tighten the support bolt to 29 ft. lbs. (40 Nm).

14. Install the oil drain plug and tighten to 29 ft. lbs. (40 Nm), If not already done.

15. Lower the vehicle and fill the crankcase to the proper level with clean engine oil.

16. Connect the negative battery cable. Start the engine and check for leaks.

REMOVAL & INSTALLATION

1.5L and 1.8L Engines

➡ **Whenever the oil pump is disassembled or the cover removed, it is suggested the gear cavity is filled with petroleum jelly for priming purposes. Do not use grease.**

1. Disconnect the negative battery cable.

2. Remove the front engine mount bracket and accessory drive belts.

3. Remove timing belt upper and lower covers.

4. Remove the timing belt and crankshaft sprocket.

5. Remove the oil pan and remove the oil screen.

6. Remove and tag the front cover mounting bolts. Note the lengths of the mounting bolts as they are removed for proper installation.

7. Remove the front case assembly and oil pump assembly.

8. Remove the oil pump cover.

9. Remove the inner and outer gears from the front case.

➡ **The outer gear has no identifying marks to indicate direction of rotation. Clean the gear and mark it with an indelible marker.**

10. Check the front case for damage or cracks. Replace the front seal. Replace the oil screen O-ring. Clean all parts thoroughly with a safe solvent. Check the pump gears for wear or damage, and ensure that the relief valve can slide freely in the case.

To install

11. Remove all gasket material from the mating surfaces and clean all parts.

12. Thoroughly coat both oil pump gears with clean engine oil and install them in the correct direction of rotation.

13. Install the pump cover and tighten the bolts to 84 inch lbs. (10 Nm).

14. Coat the relief valve and spring with clean engine oil, install them and tighten the plug to 33 ft. lbs. (45 Nm).

15. Install a new front crankshaft seal and coat the lips of the seal with clean engine oil.

16. Install the front case and oil pump assembly to the engine block using a new gasket. Use the noted locations of the mounting bolts for proper positioning and tighten the bolts to 10 ft. lbs. (14 Nm).

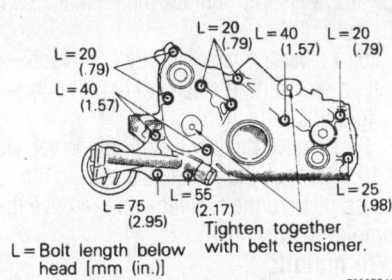

Front cover bolt locations—2.4L engine

17. Install the oil screen with new gasket. Tighten the screen bolts to 14 ft. lbs. (19 Nm).

18. Install the oil pan.

19. Install the sprocket, timing belt and pulley.

20. Connect the battery cable refill the engine oil, run the engine and check for leaks.

FUEL SYSTEM

Fuel System Service Precautions

Safety is the most important factor when performing not only fuel system maintenance but any type of maintenance. Failure to conduct maintenance and repairs in a safe manner may result in serious personal injury or death. Maintenance and testing of the vehicle's fuel system components can be accomplished safely and effectively by adhering to the following rules and guidelines.

• To avoid the possibility of fire and personal injury, always disconnect the negative battery cable unless the repair or test procedure requires that battery voltage be applied.

• Always relieve the fuel system pressure prior to disconnecting any fuel system component (injector, fuel rail, pressure regulator, etc.), fitting or fuel line connection. Exercise extreme caution whenever relieving fuel system pressure to avoid exposing skin, face and eyes to fuel spray. Please be advised that fuel under pressure may penetrate the skin or any part of the body that it contacts.

• Always place a shop towel or cloth around the fitting or connection prior to loosening to absorb any excess fuel due to spillage. Ensure that all fuel spillage (should it occur) is quickly removed from engine surfaces. Ensure that all fuel soaked cloths or towels are deposited into a suitable waste container.

• Always keep a dry chemical (Class B) fire extinguisher near the work area.

• Do not allow fuel spray or fuel vapors to come into contact with a spark or open flame.

• Always use a back-up wrench when loosening and tightening fuel line connection fittings. This will prevent unnecessary stress and torsion to fuel line piping.

• Always replace worn fuel fitting O-rings with new. Do not substitute fuel hose or equivalent, where fuel pipe is installed.

Fuel System Pressure

RELIEVING

✳✳ CAUTION

The fuel injection system remains under pressure after the engine has been turned OFF. Properly relieve the fuel pressure before disconnecting any fuel lines. Failure to do so may result in fire or personal injury.

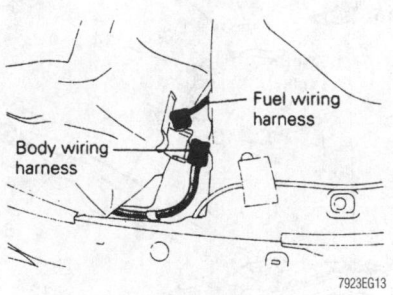

Fuel pump harness location—all models

1. Loosen the fuel filler cap to release fuel tank pressure.

2. Remove rear seat for access and disconnect the fuel pump harness connector.

3. Start the vehicle and allow it to run until it stalls from lack of fuel. Turn the key to the **OFF** position.

4. Disconnect the negative battery cable, then reconnect the fuel pump connector and reinstall the fuel filler cap.

✳✳ WARNING

Always wrap shop towels around a fitting that is being disconnected to absorb residual fuel in the lines.

Fuel Filter

REMOVAL & INSTALLATION

All Engines

A replaceable fuel filter is located in the engine compartment.

✳✳ CAUTION

Do not use conventional fuel filters, hoses or clamps when servicing fuel injection systems. They are not compatible with the injection system and could fail, causing personal injury or damage to the vehicle. Use only hoses and clamps specifically designed for fuel injection.

✳✳ CAUTION

The fuel injection system remains under pressure, after the engine has been turned OFF. Properly relieve fuel pressure before disconnecting any fuel lines. Failure to do so may result in fire or personal injury.

1. Relieve the fuel system pressure.

➡**Wrap shop towels around the fitting that is being disconnected to absorb residual fuel in the lines.**

2. Hold the fuel filter nut securely with a back-up or spanner wrench. Cover the hoses with shop towels and remove the eye bolt. Discard the gaskets.

3. Separate the flare nut connection at the filter. Discard the gaskets.

4. Remove the mounting bolts and the fuel filter from the vehicle.

To install:

5. If equipped with flare fitting, tighten the fitting by hand before installing the filter to the vehicle.

6. Install the filter to its bracket only finger-tight. Movement of the filter will ease attachment of the fuel lines.

7. Install new gaskets and connect the high pressure hose and eye bolt, then the main pipe. While holding the fuel filter nut, tighten the eye bolts to 22 ft. lbs. (30 Nm). Tighten the flare nut to 27 ft. lbs. (37 Nm).

8. Tighten the filter mounting bolts fully.

9. Install the air cleaner assembly, if removed.

10. Connect the negative battery cable, install the fuel filler cap, turn the key to the **ON** position to pressurize the fuel system and check for leaks.

11. Release the fuel pressure and repair leaks as required.

Fuel Pump

REMOVAL & INSTALLATION

Colt

1. Relieve the fuel system pressure using proper procedures. Disconnect negative battery cable.

➡ **Wrap shop towels around the fitting that is being disconnected to absorb residual fuel in the lines.**

✳✳ CAUTION

The fuel injection system remains under pressure after the engine has been turned OFF. Properly relieve fuel pressure before disconnecting

any fuel lines. Failure to do so may result in fire or personal injury.

2. Raise the vehicle and support safely.
3. Drain the fuel from the fuel tank into an approved container.
4. Disconnect the return hose, high pressure hose and vapor hoses from the fuel pump.
5. Disconnect the electrical connectors at the pump/sending unit.

✳✳ CAUTION

Cover all fuel hose connections with a shop towel, prior to disconnecting, to prevent splash of fuel that could be caused by residual pressure remaining in the fuel line.

6. Disconnect the filler and vent hoses.
7. Place a transmission jack under the center of the fuel tank and apply a slight

upward pressure. Remove the fuel tank strap retaining nut.

8. Lower the tank slightly and disconnect any remaining electrical or hose connectors at the fuel tank.
9. Remove the fuel tank from the vehicle.
10. Remove the five nuts securing the access plate to the fuel tank and remove the pump assembly.

To install:

11. Install fuel pump into fuel tank, with new packing gasket, and tighten mounting nuts.
12. Install the fuel tank onto the transmission jack. Raise the tank in position under the vehicle. Leave enough clearance to attach the electrical and hose connections to the top of the fuel pump.
13. Attach all connections to the top of the tank.
14. Raise the tank completely and position the retainer straps around the fuel tank.

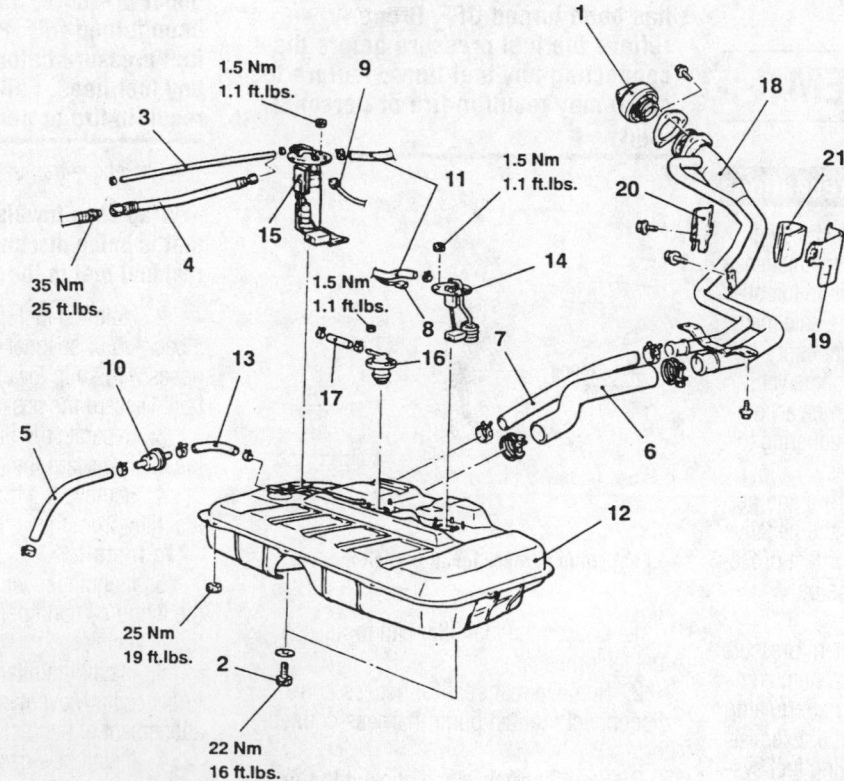

Removal steps
1. Fuel tank filler tube cap
2. Drain plug
3. Return hose
4. High-pressure fuel hose
5. Vapor hose
6. Filler hose
7. Vapor hose
8. Fuel gauge unit connector
9. Fuel gauge and pump assembly connector
10. Fuel tank pressure control valve
11. Suction hose
12. Fuel tank
13. Vapor hose
14. Fuel gauge unit
15. Fuel guage and pump assembly
16. Fuel cut off valve
17. Vapor hose
18. Fuel tank filler tube
19. Protector (A)
20. Protector (B)
21. Insulator

7923EG35

Exploded view of the fuel tank and related components—Colt/Summit Wagon and Expo

Install new fuel tank self-locking nuts and tighten to 22 ft. lbs. (31 Nm).

15. Connect the return hose and high pressure hoses.

16. Install the vapor hose and the filler hose. Install the filler hose retainer screws to the fender, if removed.

17. Lower the vehicle and pour the drained fuel into the gas tank.

18. Connect the negative battery cable. Check the fuel pump for proper pressure and inspect the entire system for leaks.

Colt/Summit Wagon and Expo

1. Relieve fuel system pressure. Remove the fuel filler cap.

❋❋ CAUTION

Fuel injection systems remain under pressure after the engine has been turned OFF. Properly relieve fuel pressure before disconnecting any fuel lines. Failure to do so may result in fire or personal injury.

2. Disconnect the negative battery cable.
3. Raise and safely support the vehicle.
4. The fuel pump is located in the fuel tank. Drain the fuel from the fuel tank.

❋❋ CAUTION

Do not allow fuel spray or fuel vapors to come in contact with a spark or open flame. Keep a dry chemical fire extinguisher nearby. Never store fuel in an open container due to risk of fire or explosion.

5. On vehicles equipped with AWD, remove the rear propeller shaft from the vehicle as follows:

a. Remove the center exhaust pipe bracket.

b. Matchmark the differential companion flange to the propeller flange yoke.

c. Remove the bolts, washers and nuts from the center support. Remove the propeller shaft assembly in a straight and level manner to avoid damage to the boot caused by pinching.

d. Install cover into the rear end of the transfer case to prevent the entry of foreign materials.

6. Disconnect the return hose, high pressure hose and all other hoses and connectors connected to the pump and sending unit.

7. Disconnect the filler and vent hoses. Place a support under the tank and remove the retaining nuts. Lower the tank from vehicle.

8. Remove retaining nuts and remove the fuel pump assembly from tank.

To install:

9. Install the replacement pump using a new gasket. Be certain the pump is installed in the same location, facing the same direction as before.

10. Install the fuel tank and secure the retainer nuts. Connect all electrical harness connectors. Reconnect all vent hoses, fuel supply and fuel return hoses securing with the proper clamps.

11. On vehicles equipped with AWD, install the propeller shaft aligning the matchmarks prior to installation. Tighten the rear yoke nuts to 22–25 ft. lbs. (30–35 Nm) and the center support self-locking nuts to 22 ft. lbs. (30 Nm).

12. Install the exhaust pipe center bracket. Check that electrical connectors are properly installed and all fuel hose connections are tight.

13. Connect the negative battery cable and check the entire fuel system for proper operation and leaks. If repairing of a fuel leak is required, release the fuel system pressure prior to repairing system.

DRIVE TRAIN

Transaxle Assembly

REMOVAL & INSTALLATION

Manual

COLT

➡ **If the vehicle is going to be rolled while the halfshafts are out of the vehicle, obtain 2 outer CV-joints or proper equivalent tools and install to the hubs. If the vehicle is rolled without the proper torque applied to the front wheel bearings, the bearings will no longer be usable.**

➡ **The suspension components should not be tightened until the vehicles weight is resting on the ground.**

1. Disconnect the negative battery cable.

2. Remove the front wheels and the inner wheel panels.

3. Remove the air cleaner assembly and vacuum hoses.

4. Note the locations and disconnect the shifter cables.

5. Disconnect the back-up lamp switch connector, speedometer cable connection and remove the starter motor.

6. Remove the upper transaxle-to-engine mounting bolts.

7. Raise the vehicle and support safely.

8. Remove the undercover and splash pan.

9. Drain the transaxle oil.

10. Support the engine and remove the crossmember.

11. Remove the upper transaxle mounting bolt and bracket.

12. Disconnect the stabilizer bar, tie rod ends and the lower ball joint connections.

13. Remove the clutch release cylinder and clutch oil line bracket. Do not disconnect the fluid lines and secure the slave cylinder with wire. Disconnect the clutch cable, if equipped with cable controlled clutch system.

14. Remove the halfshafts by inserting a prybar between the transaxle case and the driveshaft and prying the shaft from the transaxle. Do not pull on the driveshaft. Doing so damages the inboard joint. Do not insert the prybar so far the oil seal in the case is damaged.

➡ **It is not necessary to disconnect the halfshafts from the steering knuckle. Remove the shaft with the hub and knuckle as an assembly. Tie the shafts aside. Note the circle clip on the end of the inboard shafts should not be reused.**

15. Remove the bell housing lower cover.

16. Remove the transaxle to engine bolts and lower the transaxle from the vehicle.

To install:

➡ **When installing the transaxle, be sure to align the splines of the transaxle with the clutch disc.**

17. Install the transaxle to the engine and install the mounting bolts. Tighten the bolts to specifications.

18. Install the bell housing cover.

19. Install the remaining components in the opposite order from which they were removed.

20. Connect the negative battery cable and check the transaxle for proper operation. Be sure the reverse lights operate when in reverse.

EXPO AND SUMMIT WAGON (FWD MODELS)

1. Disconnect the battery cables, negative cable first. Remove the battery and tray.

2. Remove the coolant reservoir.

3. Remove the air cleaner.

4. Disconnect the clutch cable, speedometer cable and back-up light wiring from the transaxle.

5. Remove the upper engine-to-transaxle bolts.

6. Disconnect the select control lever and switch harness.

7. Remove the starter.

8. Disconnect and tag all wiring from the transaxle.

9. Raise and support the vehicle safely.

10. Remove the front wheels.

11. Drain the transaxle fluid.

12. Remove the extension and shift rod from the engine compartment.

13. Remove the stabilizer and strut bar from the lower control arm.

14. Remove the left and right halfshafts.

15. Support the transaxle with a suitable floor jack, taking care to avoid damaging the pan.

16. Remove the bell housing cover.

17. Remove the remaining transaxle-to-engine bolts.

18. Remove the transaxle mounting bolt.

19. Lower the jack and slide the transaxle from under the vehicle.

To install:

20. Secure the transaxle on a transaxle jack and position it to the engine.

21. Carefully guide the transaxle input shaft into the clutch assembly. Be sure the transaxle is seated properly and is flush to the engine flange. Install two transaxle-to-engine bolts.

22. Install the transaxle mounting bolt. Tighten the bolt to 29–36 ft. lbs. (40–50 Nm).

23. Install the lower transaxle-to-engine bolts. Tighten the bolts to 32–39 ft. lbs. (43–53 Nm).

24. Remove the floor jack from the transaxle.

25. Install the remaining components in the reverse order of removal.

26. Tighten the upper engine-to-transaxle bolts to 32–39 ft. lbs. (43–53 Nm).

27. Install the coolant reservoir and refill with coolant.

28. Install the battery tray and battery. Connect the battery cables, positive cable first.

EXPO AND SUMMIT WAGON (AWD MODELS)

1. Disconnect the battery cables, negative cable first. Remove the battery.

2. Remove the coolant reserve tank.

3. Disconnect the speedometer cable, shift control cable and back-up light harness at the transaxle.

4. Remove the range select control valves and connectors.

5. Tag and disconnect all other wiring attached to the transaxle.

6. Remove the clutch slave cylinder.

7. Remove the vacuum reservoir tank.

8. Disconnect the starter wiring and remove.

9. Remove the upper engine-to-transaxle bolts.

10. Raise and support the vehicle safely.

11. Remove the front wheels, lower engine cover and skid plate.

12. Drain the transaxle and transfer case.

13. Remove the driveshaft.

14. Remove the transfer case extension housing.

15. Remove the left and right halfshafts.

16. Disconnect the right strut from the lower arm.

17. Remove the right fender liner.

18. Take up the weight of the transaxle with a suitable floor jack.

19. Remove the bell housing cover bolts and remove the cover.

20. Remove the remaining engine-to-transaxle bolts.

21. Remove the transaxle mount insulator bolt.

22. Remove the transaxle mounting bracket attaching bolts.

23. Move the transaxle/transfer case assembly to the right. Tilt the right side of the transaxle down, until the transfer case is about level with the upper part of the steering rack tube, then turn it to the left and lower the assembly.

To install:

24. Secure the transaxle/transfer case assembly to a transaxle jack.

25. Raise the assembly in position to the engine. It may be necessary to tilt or angle the assembly in and around the steering rack tube.

26. Once the transaxle/transfer assembly is positioned to the engine, carefully guide the input shaft into the clutch assembly.

27. Install the transaxle mounting bracket attaching bolts. Tighten the bolts to 40–43 ft. lbs. (55–60 Nm).

28. Install the transaxle mount insulator bolt. Tighten to 40–43 ft. lbs. (55–60 Nm).

29. Install the lower engine-to-transaxle bolts. Tighten to 31–40 ft. lbs. (43–55 Nm).

30. Install the bell housing cover bolts and install the cover.

31. The remaining components are installed in the reverse order from which they were removed.

32. Install the battery. Connect the battery cables, positive cable first.

Automatic

COLT

➡ If the vehicle is going to be rolled on its wheels while the halfshafts are out of the vehicle, obtain two outer CV-joints or proper equivalent tools and install to the hubs. If the vehicle is rolled without the proper torque applied to the front wheel bearings, the bearings will no longer be usable.

1. Disconnect the negative battery cable.

2. Remove the battery and battery tray.

3. Remove the air hose and air cleaner assembly.

4. Raise the vehicle and support safely.

5. Remove the under guard pan.

6. Drain the transaxle oil.

7. If equipped with 1.6L engine, remove the tension rod.

8. Disconnect the control cable and cooler lines.

9. Disconnect the shift control solenoid valve connector.

10. Disconnect the inhibitor switch, kick-down servo switch, the pulse generator and oil temperature sensor, if equipped.

11. Disconnect the speedometer cable and remove the starter.

12. Remove the transaxle mounting bolts and bracket.

13. Disconnect the stabilizer bar from the lower control arm.

14. Disconnect the steering tie rod end and the ball joint from the steering arm.

15. Remove the halfshafts at the inboard side from the transaxle. Tie the joint assembly aside.

➡ It is not necessary to disconnect the halfshafts from the wheel hubs.

16. Support the engine and remove the center member.

17. Remove the bell housing cover and remove the driveplate bolts.

18. Remove the transaxle assembly lower connecting bolt, located just over the halfshaft opening.

19. Properly support the transaxle assembly and lower it moving it to the right for clearance.

To install:

20. After the torque converter has been mounted on the transaxle, install the transaxle assembly on the engine. Install the mounting bolts and torque to specifications.

21. Tighten the driveplate bolts to 33–38 ft. lbs. (46–53 Nm). Install the bell housing cover.

22. Install the center member and tighten the bolts to specifications.

23. The balance of installation is the reverse of the removal procedure.

24. Refill with Dexron® II, Mopar ATF Plus type 7176 or equivalent, automatic transaxle fluid.

25. Start the engine and allow to idle for two minutes. Apply parking brake and move selector through each gear position, ending in **N**. Recheck fluid level and add if necessary. Fluid level should be between the marks in the **HOT** range.

EXPO AND SUMMIT WAGON

➡**On both Front Wheel Drive (FWD) and All Wheel Drive (AWD) vehicles, the transaxle and converter must be removed and installed as an assembly.**

1. Disconnect negative battery cable.

2. Remove the air cleaner assembly.

3. Disconnect the transaxle control lever. Disconnect and plug the oil cooler lines.

4. Disconnect the pulse generator connector, oil temperature connector, kickdown servo switch connector, inhibitor switch connector and solenoid valve connection.

5. Disconnect the speedometer cable connection. Remove the oil level dipstick and tube.

6. Install holding fixture to the top of the engine to support engine weight.

7. Remove the top transaxle upper coupling bolts.

8. Raise and safely support the vehicle.

9. Remove the starter motor leaving wiring harness attached.

10. Remove the right side undercover. Drain the transaxle fluid.

11. Disconnect the tie rod ends, stabilizer bar and lower ball joints.

12. Remove the axle shafts from the vehicle.

13. On AWD models, remove the driveshaft from the transfer case, insert a prybar between the driveshaft and the transaxle case and pry the shaft from the transaxle housing. Swing the shafts out of the way keeping the joints straight, and suspend using wire. Turn the right shaft 90 degrees toward the front of the vehicle so it will be out of the way.

➡**Do not pull on the shaft during removal from the transaxle. This will damage the inboard joint. Do not insert the prybar so deep as to damage the oil seal.**

14. Remove the lower bell housing cover. Scribe a mark on the driveplate and transaxle converter face using chalk. Remove the driveplate connecting bolts while turning the crankshaft.

15. Support the transaxle using a transmission jack. Remove the center support.

16. Remove the transaxle mount bolt and bracket.

17. On AWD models, disconnect the front exhaust pipe, then drain and remove the transfer assembly.

18. Remove the lower transaxle case coupling bolts, press the torque converter towards the transfer case to prevent separation during removal and lower the transfer case from the vehicle.

To install:

19. Install the transaxle into the vehicle and secure using the lower case coupling bolts.

20. Install the transaxle mount bolt and bracket, tighten through-bolt nut to 51 ft. lbs. (70 Nm).

21. Align the scribe marks on the converter and the driveplate. Install the driveplate connecting bolts tightening to 33–38 ft. lbs. (46–53 Nm).

22. Install the transfer assembly, if applicable, and the center crossmember. Remove the transmission jack.

23. The remaining components are installed in the reverse order from which they were removed.

24. Refill with DEXRON® II, Mopar ATF Plus type 7176, or equivalent automatic transaxle fluid. Fill the transfer case to proper level GL-4 or higher, SAE 75W-90W.

25. Start the engine and allow to idle for 2 minutes. Apply parking brake and move selector through each gear position, ending in **N**. Recheck fluid level and add if necessary. Fluid level should be between the marks in the **HOT** range. Check operation of all gauges and meters.

Clutch Assembly

ADJUSTMENT

All models

➡**The following procedure if for cable actuated clutch systems. Hydraulic clutch systems are self-adjusting and require no periodic maintenance.**

1. Measure the clutch pedal height (measurement A). The specification is 6.38–6.50 in. (162–165mm) on the Colt or 7.68–7.87 in. (195–200mm) on the Colt/Summit wagon.

➡**The clutch pedal height is not adjustable. If not within specifications, part replacement is required.**

2. Depress clutch pedal several times and check the pedal free-play (measurement B).

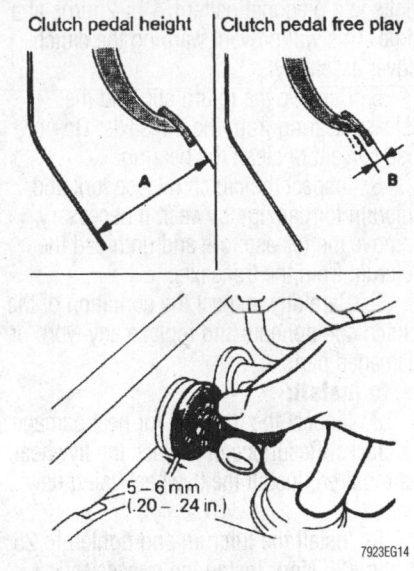

Proper method of adjusting clutch pedal free-play—Colt models

3. If measurement is not 0.67–0.87 in. (17–22mm) on the Colt or 0.04–0.12 in. (1–3mm) on the Colt/Summit wagon, adjustment is required.

4. To adjust the free-play, turn the outer cable adjusting nut, located at the firewall, until free-play is within range.

5. Depress clutch pedal several times and recheck measurement.

REMOVAL & INSTALLATION

All Models

1. Disconnect the negative battery cable. Raise and safely support the vehicle.

2. Remove the transaxle assembly from the vehicle.

3. Install a dummy shaft through the clutch disc to support the clutch during the removal procedure.

4. Remove the pressure plate attaching bolts, pressure plate and clutch disc. If the pressure plate is to be reused, loosen the

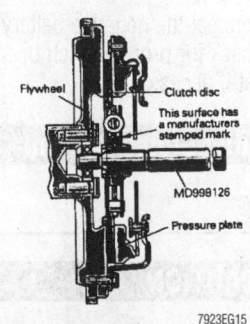

Use the alignment tool to center the clutch disc while tightening the pressure plate mounting bolts—all models

bolts in a diagonal pattern, 1 or 2 turns at a time. This will prevent warping the clutch cover assembly.

5. Remove the return clip and the release bearing from the transaxle. Do not use solvent to clean the bearing.

6. Inspect the clutch release fork and fulcrum for damage or wear. If necessary, remove the release fork and unthread the fulcrum from the transaxle.

7. Carefully inspect the condition of the clutch components and replace any worn or damaged parts.

To install:

8. Inspect the flywheel for heat damage or cracks. Resurface or replace the flywheel as required. Install the flywheel using new bolts.

9. Install the fulcrum and tighten to 25 ft. lbs. (35 Nm). Install the release fork. Apply a coating of multi-purpose grease to the point of contact with the release bearing. Apply a coating of multi-purpose grease to the end of the release cylinder's pushrod and the pushrod hole in the release fork.

➡**When installing the clutch be careful not to apply excessive grease. Excessive grease will cause clutch slippage and shudder.**

10. Apply multi-purpose grease to the clutch release bearing. Pack the bearing inner surface and the groove with grease. Do not apply grease to the resin portion of the bearing. Place the bearing in position and install return clip.

11. Apply a coating of grease to the clutch disc splines, then use a brush to rub it in the grooves. Using a universal clutch disc aligner, position the clutch disc on the flywheel. Install the retainer bolts and tighten gradually in a diagonal sequence. Tighten them to a final torque of 15 ft. lbs. (21 Nm). Remove the aligning tool.

12. Install the transaxle assembly and check fluid level.

13. Connect the negative battery cable.

14. Check for proper clutch operation and adjust if necessary.

Hydraulic Clutch System

BLEEDING

❄❄ CAUTION

The clutch hydraulic system uses brake fluid. Use care; brake fluid is harmful to painted surfaces.

1. Fill the reservoir with clean brake fluid meeting DOT 3 specifications.

2. Press the clutch pedal to the floor, then open the bleeder screw on the slave cylinder.

3. Tighten the bleed screw and release the clutch pedal.

4. Repeat the procedure until the fluid is free of air bubbles.

➡**It is suggested that a hose be attached to the bleeder with the other end immersed in a container at least half full of brake fluid during the bleeding operation. Do not allow the reservoir to run out of fluid during bleeding.**

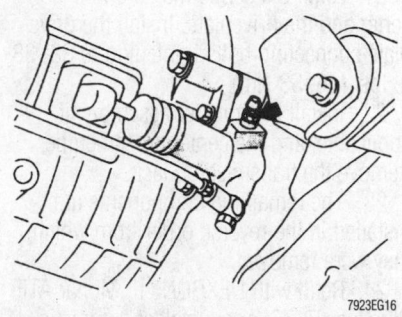

Clutch hydraulic system bleeder screw location—all models

Transfer Case Assembly

REMOVAL & INSTALLATION

1. Disconnect the battery negative cable.

2. Raise and properly support vehicle. Drain the transfer assembly.

3. Disconnect the front exhaust pipe and hanger.

4. Matchmark and remove the driveshaft.

5. Unbolt the transfer case assembly and remove by sliding out from the transaxle. Cover the opening in the transaxle and transfer case to keep oil from dripping and to keep dirt out.

To install:

6. Install the transfer case assembly to the transaxle. Tighten the transfer case to transaxle bolts to proper specification.

7. Lubricate the driveshaft sleeve yoke and oil seal lip on the transfer extension housing. Install the driveshaft.

➡**Use care when installing the rear driveshaft to the transfer case, not to damage the output shaft seal.**

8. Connect the exhaust pipe, using a new gasket.

9. Refill the transfer case with gear oil of correct classification. Check fluid level in transaxle and add as required.

10. Lower the vehicle and connect the negative battery cable.

Halfshaft

REMOVAL & INSTALLATION

Front

EXCEPT LEFT SIDE OF AWD VISTA AND SUMMIT WAGON

➡**If the vehicle is going to be rolled while the halfshafts are out of the vehicle, obtain 2 outer CV-joints or proper equivalent tools and install to the hubs. If the vehicle is rolled without the proper torque applied to the front wheel bearings, the bearings will no longer be usable.**

1. Disconnect the negative battery cable.

2. Remove the cotter pin, halfshaft nut and washer.

3. Raise the vehicle and support safely. Disconnect the lower ball joint and the tie rod end from the steering knuckle.

4. On vehicles with an inner shaft, remove the center support bearing bracket bolts and washers.

5. Remove the halfshaft by setting up a puller on the outside wheel hub and pushing the halfshaft from the front hub. After pressing the outer shaft, insert a prybar between the transaxle case and the halfshaft and pry the shaft from the transaxle. Do not pull on the shaft; doing so damages the inboard joint. Do not insert the prybar too far or the oil seal in the case may be damaged.

To install:

6. Inspect the halfshaft boot for damage or deterioration. Check the ball joints and splines for wear.

7. Replace the circlips on the ends of the halfshafts.

8. Insert the halfshaft into the transaxle. Be sure it is fully seated.

9. Pull the strut assembly out and install the outer end of the shaft to the hub.

10. Install the washer so the chamfered edge faces outward. Install the nut and tighten temporarily.

11. Install the tie rod end and ball joint.

12. Install the wheel and lower the vehicle to the floor. Tighten the axle nut with the brakes applied. Tighten the nut to a maximum torque of 188 ft. lbs. (260 Nm). Install the cotter pin and bend to secure.

LEFT SIDE OF AWD VISTA AND SUMMIT WAGON

1. Remove the center wheel hub. Loosen the wheel lugs and the center half-shaft nut.

2. Raise and safely support the front of the vehicle with the suspension hanging.

3. Remove the front wheels.

4. Drain the transaxle fluid.

5. Disconnect the lower ball joint from the steering knuckle.

6. Remove the strut bar and the stabilizer bar from the lower arm.

7. Remove the center bearing mount snapring/bolts from the bracket.

8. Lightly tap the double-offset joint outer race with a wooden mallet and disconnect the Cardan joint.

9. Disconnect the halfshaft from the center bearing bracket.

10. Use a pusher/puller tool mounted to the wheel studs and press the halfshaft from the drive hub.

11. Unbolt and remove the bearing bracket.

12. Use a wooden mallet and lightly tap the Cardan joint yoke and remove it from the transaxle. DO NOT pry the Cardan joint from the transaxle, damage can be caused to the joint and boot.

To install:

13. Service the halfshaft as required. Install the Cardan joint.

14. Apply grease to the Cardan joint contact surfaces.

15. Attach a new O-ring to the oil seal retainer.

16. Install the center bearing bracket. Tighten the mounting to 40 ft. lbs.

17. Insert the center bearing into the mounting bracket, be sure it is fully seated. Secure with the snapring bolts.

18. Coat the halfshaft splines with grease and slide it into the Cardan joint.

19. Slide the halfshaft into the drive hub. Install the suspension components and wheel. Lower the vehicle and tighten the wheel lugs and the center halfshaft nut. Tighten the nut to 188 ft. lbs.

Rear

AWD VISTA AND SUMMIT WAGON

1. Raise the vehicle and support safely.

2. Remove the bolts that attach the rear halfshaft to the companion flange.

3. Remove the cotter pin and axle nut from the outer shaft.

4. Remove the rear driveshaft from the vehicle.

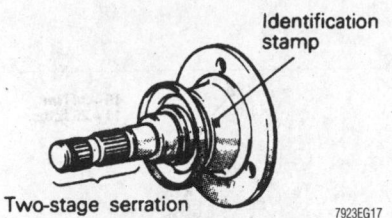

Identification stamp

Two-stage serration

7923EG17

Companion shaft identification—Summit Wagon and Expo models

5. If the differential companion shaft is to be removed, connect a slide hammer to the flange and pull the shaft from the differential.

To install:

➤If the companion shaft is being replaced or both shafts were removed together, it is important to properly identify the companion shaft. The right and left side companion shafts are different, as are limited slip differentials, which use a two stage serration on the companion shaft.

6. Replace the circlip and install the companion flange to the differential case. Be sure it snaps in place. On limited slip differentials, ensure that both serrations are fully engaged to the differential.

7. Install the axle shaft through the hub and install the axle nut. Tighten the nut to 145 to 188 ft. lbs. (200 to 260 Nm) and secure with cotter pin.

8. Install the companion flange bolts and tighten to 40 to 47 ft. lbs. (55 to 65 Nm).

9. Check the fluid level in the rear differential.

STEERING AND SUSPENSION

Air Bag

✳ CAUTION

Some vehicles are equipped with an air bag system, also known as the Supplemental Inflatable Restraint (SIR) or Supplemental Restraint System (SRS). The system must be disabled before performing service on or around system components, steering column, instrument panel components, wiring wand sensors. Failure to follow safety and disabling procedures could result in accidental air bag deployment, possible personal injury and unnecessary system repairs.

PRECAUTIONS

Several precautions must be observed when handling the inflator module to avoid accidental deployment and possible personal injury.

• Never carry the inflator module by the wires or connector on the underside of the module.

• When carrying a live inflator module, hold securely with both hands, and ensure that the bag and trim cover are pointed away.

• Place the inflator module on a bench or other surface with the bag and trim cover facing up.

• With the inflator module on the bench, never place anything on or close to the module which may be thrown in the event of an accidental deployment.

DISARMING

1. Position the front wheels in the straight-ahead position and place the key in the **LOCK** position. Remove the key from the ignition lock cylinder.

2. Disconnect the negative battery cable and insulate the cable end with high-quality electrical tape or similar non-conductive wrapping.

3. Wait at least one minute before working on the vehicle. The air bag system is designed to retain enough voltage to deploy the air bag for a short period of time after the battery has been disconnected.

REARMING

Connect the negative battery cable, turn the ignition switch to the **ON** position and check the SRS warning light for proper operation.

Rack and Pinion Steering Gear

REMOVAL & INSTALLATION

Manual

➤If equipped with air bag, prior to removal of the steering rack, center the front wheels and remove the ignition key. Failure to do so may damage the SRS clockspring and render SRS system inoperative, risking serious driver injury. Be sure to properly disarm the air bag system.

1. Disconnect the battery negative cable. Raise the vehicle and support safely and remove the wheels.

2. Disconnect the oxygen sensor and remove the front exhaust pipe.

3. Properly support the engine. Remove both roll stopper mounting bolts and the four center member installation bolts.

4. Remove the center member.

➡**Matchmark the pinion input shaft of the rack to the lower steering column joint for installation purposes.**

5. Remove the pinch bolt holding the lower steering column joint to the rack and pinion input shaft.

6. Remove the cotter pins and disconnect the tie rod ends from the steering knuckle.

7. Remove the rack and pinion steering assembly and its rubber mounts from the right side of the vehicle.

To install:

8. Align the matchmarks of the input shaft and install the rack to the vehicle.

9. Secure the rack using the retainer clamps and bolts. Tighten the bolts to 51 ft. lbs. (70 Nm).

10. Tighten the steering column pinch bolt to 13 ft. lbs. (18 Nm).

11. Install the center member.

12. Install the front exhaust pipe.

13. Connect the tie rod ends to the steering knuckles and tighten the castle nuts to 25 ft. lbs. (34 Nm). Install new cotter pins.

14. Install the wheels and connect the negative battery cable.

15. Perform a front end alignment.

Power

COLT

➡**If equipped with air bag, prior to removal of the steering gear box, center the front wheels and remove the ignition key. Failure to do so may damage the SRS clockspring and render SRS system inoperative, risking serious driver injury.**

1. Drain power steering system:

a. Disconnect the return hose at the reservoir and place into a suitable container.

b. Disable the ignition system. While cranking the engine, turn the wheels several times, until system has been drained.

2. Disconnect the battery negative cable. Raise the vehicle and support safely.

3. Disconnect the oxygen sensor and remove the front exhaust pipe.

➡**It may be easier on some vehicles, to disconnect all changers and lower the complete exhaust system.**

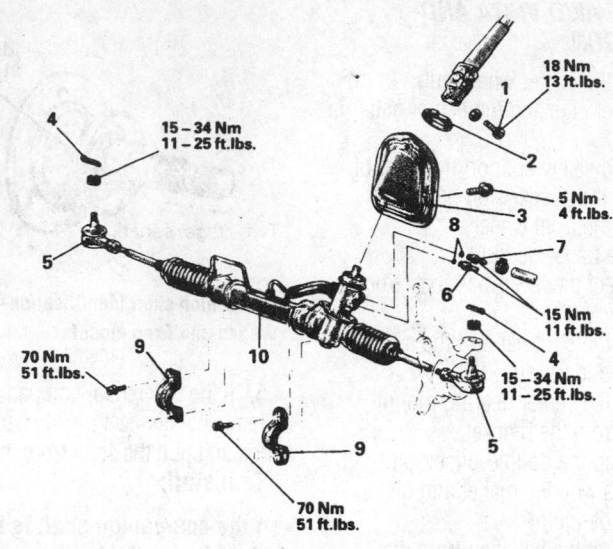

1. Joint assembly and gear box connecting bolt
2. Band
3. Steering cover
4. Cotter pin
5. Connection for tie-rod end and knuckle
6. Pressure pipe
7. Return pipe
8. O-ring
9. Clamp
10. Gear box assembly

7923EG18

Power-assisted steering rack and pinion and related parts—Colt models

4. Properly support the engine. Remove both roll stopper mounting bolts and the four center member installation bolts. Remove the center member.

5. Remove the center member.

➡**Matchmark the pinion input shaft of the rack to the lower steering column joint for installation purposes.**

6. Remove the pinch bolt holding the lower steering column joint to the rack and pinion input shaft.

7. Remove the cotter pins and disconnect the tie rod ends from the steering knuckle.

8. Disconnect the power steering fluid pressure pipe and return hose from the rack fittings.

9. Remove the rack and pinion steering assembly and its rubber mounts from the right side of the vehicle.

To install:

10. Align the matchmarks of the input shaft and install the rack to the vehicle.

11. Secure the rack using the retainer clamps and bolts. Tighten the bolts to 51 ft. lbs. (70 Nm).

12. Tighten the steering column pinch bolt to 13 ft. lbs. (18 Nm).

13. Using new O-rings, connect the power steering fluid lines to the rack fittings.

14. Install the center member.

15. Install the front exhaust pipe.

16. Connect the tie rod ends to the steering knuckles and tighten the castle nuts to 25 ft. lbs. (34 Nm). Install new cotter pins.

17. Install the wheels and connect the negative battery cable.

18. Refill the reservoir and bleed the system.

19. Perform a front end alignment.

EXPO AND SUMMIT WAGON

❋❋ WARNING

If equipped with air bag, prior to removal of the steering gear box, center the front wheels and remove the ignition key. Failure to do so may damage the SRS clockspring and render SRS system inoperative, risking serious driver injury.

1. Drain power steering system:

a. Disconnect the return hose at the reservoir and place into a suitable container.

b. Disable the ignition system. While cranking the engine, turn the wheels several times, until system has been drained.

2. Disconnect the battery negative cable. Raise the vehicle and support safely.

3. On AWD vehicles, properly support the engine and remove the rear engine mount bracket.

4. Remove the pinch bolt holding the lower steering column joint to the rack and pinion input shaft.

5. Remove the cotter pins and disconnect the tie rod ends from the steering knuckle.

6. Disconnect the power steering fluid pressure pipe and return hose from the rack fittings.

7. Remove the rack and pinion steering assembly and its rubber mounts from the right side of the vehicle.

To install:

8. Align steering shaft and install the steering gear into the vehicle. Secure rack assembly using the retainer clamps and bolts.

9. Install the pinch bolt and tighten to 13 ft. lbs. (18 Nm).

10. Install the engine mount bracket, if removed.

11. Connect the power steering fluid lines to the rack fittings.

12. Connect the tie rod ends to the steering knuckles.

13. Connect the negative battery cable. Refill the reservoir and bleed the system.

14. Perform a front end alignment.

Strut and Coil Spring

REMOVAL & INSTALLATION

Front

1. Disconnect the negative battery cable.

2. If applicable, remove the daytime running lamp relay mounting bracket and position relay assembly aside.

3. Raise and safely support the vehicle.

4. Remove the brake hose and tube bracket retainer bolt and bracket from the front strut. Do not pry the brake hose and tube clamp away when removing.

5. If equipped with ABS, disconnect the front speed sensor mounting clamp from the strut.

6. Support the lower arm using floor jack or equivalent. Remove the lower strut to knuckle bolts. Once the mounting bolts have been removed, jack up the lower arm. Use a piece of wire to attach the brake hose, tube and driveshaft to the knuckle and to help keep the weight off. These components are not to be pulled.

7. Before removing the top bolts, make matchmarks on the body and the strut insulator for proper reassembly. If this plate is installed improperly, the wheel alignment will be wrong. Remove the strut upper mounting bolts. Remove the strut assembly from the vehicle.

8. To remove the coil spring from the strut assembly, perform the following:

a. Hold the spring upper seat with a spring compressor.

✳✳ CAUTION

Do not remove the nut unless the spring is held by a spring compressor. Failure to do so may result in personal injury.

b. Compress the spring, then remove the self-locking nut holding the strut insulator.

c. Remove the spring.

To install:

9. To install the spring onto the strut, perform the following:

a. With the spring being held in the spring compressor, align the spring in the grooves in the upper and lower seats.

b. Install the self-locking nut and tighten to 43–51 ft. lbs. (60–70 Nm).

10. Install the strut to the vehicle and install the top mounting bolts. Be sure the insulator is installed so the matchmarks made during disassembly are in alignment. Tighten the mounting bolts to 33 ft. lbs. (45 Nm) for Summit Wagon and Vista models, or to 29 ft. lbs. (40 Nm) for Colt and Summit models.

11. Position the strut on the knuckle and install the mounting bolts. While holding the head of the lower mounting bolt, tighten the nuts to 78 ft. lbs. (108 Nm) for Summit Wagon and Vista models, or to 80–94 ft. lbs. (110–130 Nm) for Colt and Summit models.

12. Install the brake hose bracket and the ABS clamp.

13. Install the relay mounting bracket and connect the negative battery cable.

14. Install the wheel and tire assembly. Perform a front end alignment.

Rear

COLT

➡ **The strut assembly is a load bearing component, therefore the vehicle chassis and axle weight must be supported separately, requiring the use of two separate lifting devices.**

➡ **Matchmark the upper spring plate to the vehicle chassis for reassembly.**

1. Remove the trunk interior trim to gain access to the top mounting nuts.

2. Remove the top cap and upper shock mounting nuts.

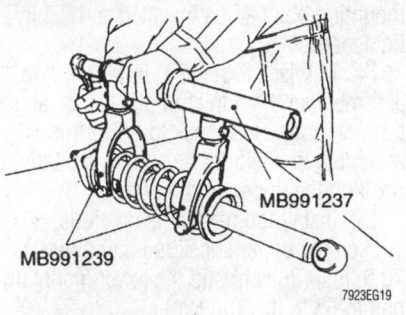

Compress the coil spring using the proper tool—Colt models

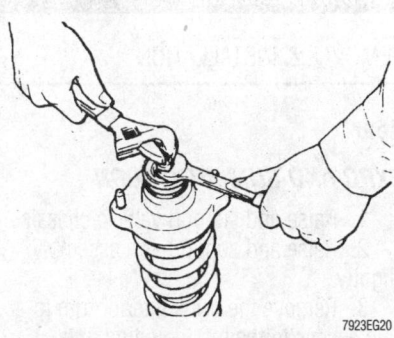

Hold the piston rod and remove the self locking nut—Colt models

3. Raise and support vehicle chassis.

4. Raise and support the trailing arm assembly slightly.

5. Remove the shock lower mounting bolt and remove the assembly from the vehicle.

6. Compress the coil spring using the proper spring compressor.

✳✳ CAUTION

Do not use air tools to tighten the spring compressor.

7. Hold the piston rod with a wrench and remove the self locking nut.

8. Remove the washer, upper bushing A, bracket, spring pad, upper bushing B, collar, cup, dust cover and bump rubber.

➡ **Align the stepped part of the spring pad with the end of the spring.**

9. Remove the coil spring.

To install:

10. Install the coil spring on the strut.

11. Install the bump rubber, dust cover, cup, collar, upper bushing A, spring pad, bracket, upper bushing B and the washer.

12. Temporarily install a new self locking nut, carefully release the spring from the compressor and tighten the self locking nut to specifications.

13. Position strut assembly so that lower mounting bolt can be installed and lightly tightened.

14. Use jack to raise or lower the axle assembly so that top strut plate studs aligns through body. Raise jack to hold strut assembly in position. Be sure to properly position the upper spring plate.

15. Install top plate nuts on studs. Tighten the upper shock mounting nuts to 20 ft. lbs. (28 Nm) and the lower mounting bolt to 65 ft. lbs. (90 Nm).

16. Lower the vehicle. Install top cap and interior trim.

Shock Absorber

REMOVAL & INSTALLATION

Rear

EXPO AND SUMMIT WAGON

1. Raise and support vehicle chassis.
2. Raise and support arm assembly slightly.
3. Remove the trunk interior trim to gain access to the top mounting nuts.
4. Remove the top cap and upper shock mounting nuts.
5. Remove the shock lower retaining nut and remove the assembly from the vehicle.
   To install:
6. Position strut assembly so that lower mounting nut can be installed and lightly tightened.
7. Use jack to raise or lower arm, so that top strut plate studs aligns through body. Raise jack to hold strut assembly in position.
8. Install top plate nuts on studs. Tighten the upper shock mounting nuts to 33 ft. lbs. (45 Nm).
9. With the full weight of the vehicle on the suspension, tighten the lower mounting bolt to 72 ft. lbs. (100 Nm).
10. Install top cap and interior trim.

Coil Spring

REMOVAL & INSTALLATION

Front

➡Refer the front strut procedure for service of the front coil spring.

Rear

COLT

Refer to the rear strut procedure for service of the rear coil spring.

EXPO AND SUMMIT WAGON

1. Raise and properly support the vehicle.
2. Remove the rear stabilizer bar.

➡**Perform the following steps, working on one side at a time.**

3. On AWD models, remove the rear driveshaft mounting bolts at the carrier flange and hang the driveshaft from the vehicle body using wire.
4. Using a jack to support the lower arm, remove the rear shock absorber lower mounting bolt.
5. If equipped with ABS, remove the speed sensor clamp bolt and relocate out of the way. Do not apply tension to the wiring harness of the connector.
6. Scribe mating marks on the lower control arm shaft (inner mounting bolt) and the crossmember. To remove the coil spring, loosen the shaft assembly nut and flange bolt nut (outer mounting bolt), then slowly lower the rear end of the lower arm. It is not necessary to remove the nuts, only to loosen them.
   To install:
7. Install the coil spring into the seats making sure both ends of the spring are correctly aligned with the spring seat groove.
8. Slowly raise the rear the rear end of the lower arm and align the scribe marks made during disassembly. Once the full weight of the vehicle is on the ground, tighten shaft and flange mounting nuts to 69 ft. lbs. (95 Nm).
9. Install the speed sensor clamp to it's original location and secure the wiring harness making.
10. Reconnect the lower portion of the shock and tighten the retaining bolt to 72 ft. lbs. (100 Nm).
11. On AWD models, install the rear driveshaft to the flange and secure tightening mounting bolts to 40–47 ft. lbs. (55–65 Nm).
12. Install the stabilizer bar.
13. Lower the arm and remove the jack.
14. Check rear alignment.

Lower Ball Joint

REMOVAL & INSTALLATION

The lower ball joint is an integral part of the lower control arm assembly, and can not be serviced separately. A worn or damaged ball joint, requires replacement of lower control arm assembly.

Lower Control Arm

REMOVAL & INSTALLATION

All Models

➡**The suspension components should not be tightened until the vehicle's weight is resting on its wheels.**

1. Raise the vehicle and support safely.
2. Remove the wheel and tire assembly.
3. For Colt and Summit models, remove sway bar links or mounting nuts and bolts from lower control arm. Remove the joint cups and bushings.
4. For Vista and Summit Wagon models, disconnect the sway link by holding the ball stud with a hex wrench and removing the self-locking nut with a box wrench.
5. Disconnect the ball joint stud from the steering knuckle.
6. Remove the inner lower arm mounting bolt and nut.
7. Remove the rear mount bolts from the retaining clamp. Remove the rear retainer clamp if equipped.
8. Remove the arm from the vehicle.
   To install:
9. Install the control arm to the vehicle and install the inner mounting bolt. Install new nut and torque to 78 ft. lbs. (108 Nm).
10. Install the rear mount clamp and bolts. Torque the clamp mounting bolts to 65 ft. lbs. (90 Nm) for Colt and Summit models, or to 51 ft. lbs. (70 Nm) for Summit Wagon and Vista models.
11. Connect the ball joint stud to the knuckle. Install a new nut and torque to 49 ft. lbs. (68 Nm).
12. Install the sway bar and links.
13. Lower the vehicle to the floor for the final tightening of the inner frame mount bolt.
14. Inspect all suspension bolts, making sure they all have been fully tightened.
15. Install the wheel and tire assembly.

Wheel Bearings

ADJUSTMENT

Front

1. Remove the hub, knuckle and bearing assembly from the vehicle.
2. Using pressing tool MB990998 or equivalent, mount the front hub assembly into the knuckle. Tighten the nut of the pressing tool to 144–188 ft. lbs. (200–260 Nm). Rotate the hub to seat the bearing.

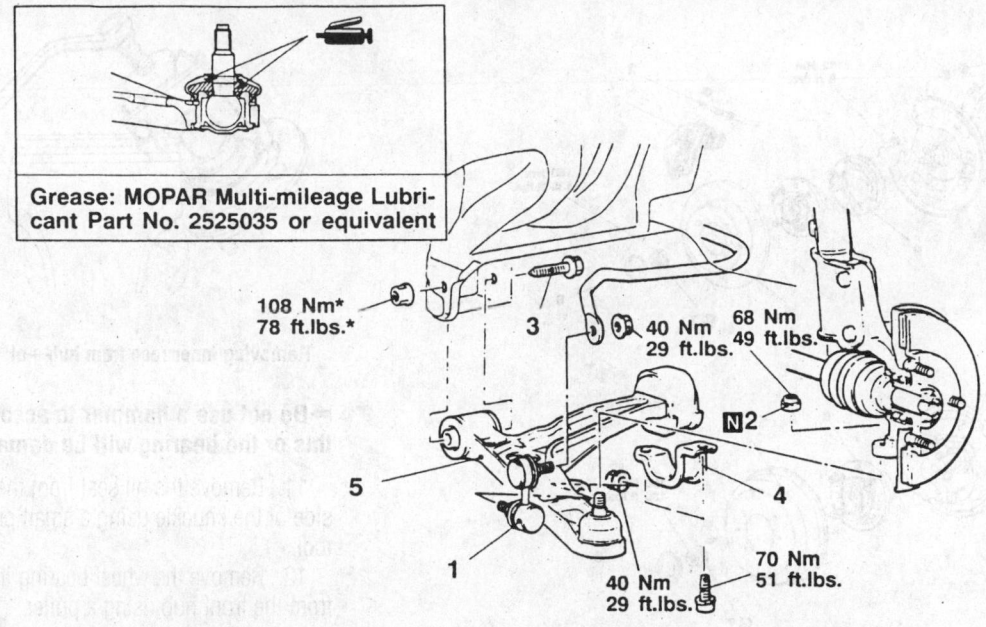

Grease: MOPAR Multi-mileage Lubricant Part No. 2525035 or equivalent

108 Nm*
78 ft.lbs.*

3

40 Nm
29 ft.lbs.

68 Nm
49 ft.lbs.

N2

5

4

1

40 Nm
29 ft.lbs.

70 Nm
51 ft.lbs.

Removal steps

1. Stabilzer link
2. Self-locking nut
3. Bolt
4. Clamp
5. Lower arm

Caution
* : Indicates parts which should be temporarily tightened, and then fully tightened with the vehicle on the ground in the unladen condition.

7923EG36

Exploded view of the lower control arm and related parts—wagon models

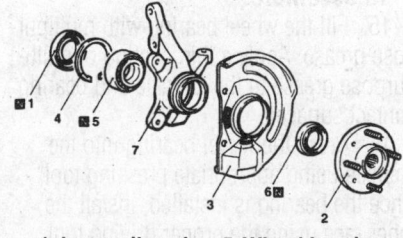

1. Inner oil seal
2. Hub
3. Dust cover
4. Snap ring
5. Wheel bearing
6. Outer oil seal
7. Knuckle

7923EG21

Wheel bearing assembly exploded view— Colt models

3. Mount the knuckle assembly in a vise. Check the hub assembly turning torque and end-play as follows:

· a. Using a torque wrench and socket MB990998 or equivalent, turn the hub in the knuckle assembly. Note the reading on the torque wrench and compare to the desired reading of 16 inch lbs. (1.8 Nm) or less. This is known as the break-away torque.

b. Check for roughness when turning the bearing.

c. Mount a dial indicator on the hub so the pointer contacts the machined surface on the hub.

d. Check the end-play.

e. Compare the reading to the limit of 0.002 in. (0.05mm).

4. If the starting torque or the hub end-play are not within specifications while the nut is tightened to 144–188 ft. lbs. (200–260 Nm), the bearing, hub or knuckle have probably not been installed correctly. Repeat the disassembly and assembly procedure and recheck starting torque and end-play.

5. Install the hub and knuckle assembly onto the vehicle.

Rear

COLT

➡**Never disassemble the rear hub bearing. The wheel bearing is serviced by replacement of the hub.**

1. Raise and safely support the vehicle.
2. Remove the rear wheel.
3. Remove the caliper and brake disc or brake drum.
4. Remove the dust cap and tighten the flange nut to 130 ft. lbs. (180 Nm).
5. Using a dial indicator, measure wheel bearing end-play. The maximum limit for end-play is 0.0020 in. (0.05mm).
6. Using a spring scale and a rope wrapped around the bolts, measure the rotary sliding resistance of the bearing/hub. The maximum limit for resistance is 4 lbs. (19 N).

7. If any of the readings exceed the specifications, replacement of the hub is required.
8. Install the dust cap.
9. Install the brake disc and caliper, or brake drum.

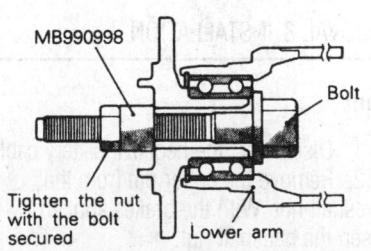

MB990998

Bolt

Tighten the nut
with the bolt
secured

Lower arm

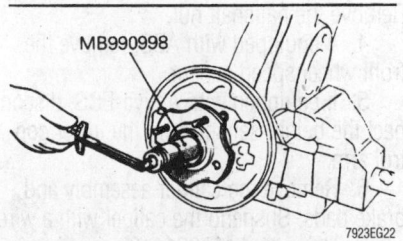

MB990998

7923EG22

Wheel bearing preload adjustment method—Summit Wagon and Expo models

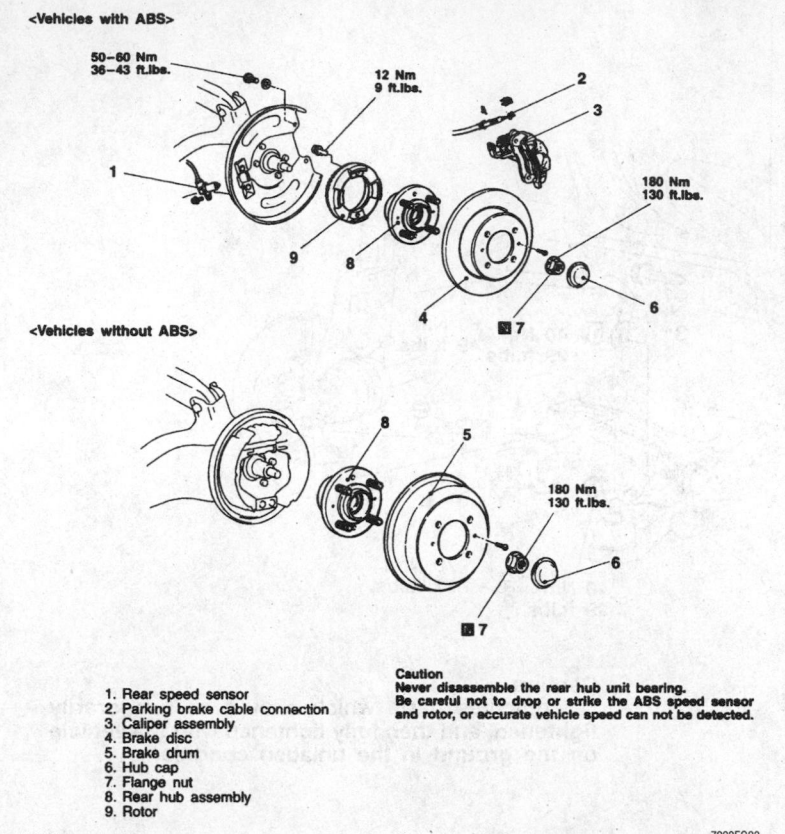

‹Vehicles with ABS›

50—60 Nm
36—43 ft.lbs.

12 Nm
9 ft.lbs.

180 Nm
130 ft.lbs.

‹Vehicles without ABS›

180 Nm
130 ft.lbs.

Caution
Never disassemble the rear hub unit bearing.
Be careful not to drop or strike the ABS speed sensor
and rotor, or accurate vehicle speed can not be detected.

1. Rear speed sensor
2. Parking brake cable connection
3. Caliper assembly
4. Brake disc
5. Brake drum
6. Hub cap
7. Flange nut
8. Rear hub assembly
9. Rotor

7923EG23

Exploded view of the rear axle hub—Colt models

10. Install the rear wheel assembly and lower the vehicle to the floor.

EXPO AND SUMMIT WAGON

The rear wheel bearings on the Vista and Summit Wagon models do not require manual adjustment. If the rear wheel bearings (on FWD models) show damage or excessive looseness, the rear hub must be replaced.

REMOVAL & INSTALLATION

Front

1. Disconnect the negative battery cable.
2. Remove the cotter pin from the driveshaft nut. With the brakes applied, loosen the halfshaft nut.
3. Raise the vehicle and support safely. Remove the halfshaft nut.
4. If equipped with ABS, remove the front wheel speed sensor.
5. If equipped with Active-ECS, disconnect the height sensor from the lower control arm.
6. Remove the caliper assembly and brake pads. Suspend the caliper with a wire.
7. Using tool MB991113 or equivalent, disconnect the ball joint and tie rod end from the steering knuckle.

➡It is important to use proper methods of joint separation. Use of unproved techniques can result in damage to joint and possible failure.

8. Remove the halfshaft by setting up a puller on the outside wheel hub and pushing the halfshaft from the front hub. After pressing the outer shaft, insert a prybar between the transaxle case and the halfshaft and pry the shaft from the transaxle.

9. Unbolt the lower end of the strut and remove the hub and steering knuckle assembly from the vehicle.

10. Install the hub/knuckle assembly in a vise. Using puller MB991056 or equivalent, remove the hub from the knuckle.

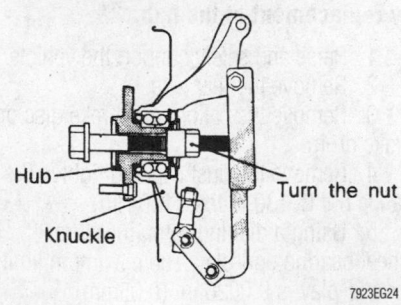

Hub

Knuckle

Turn the nut

7923EG24

Use of press tool for hub removal—all models

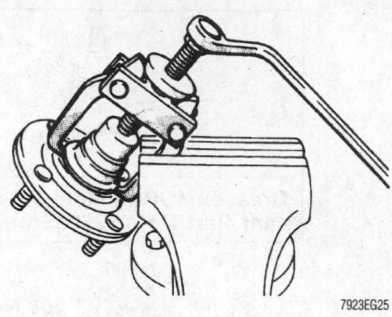

7923EG25

Removing inner race from hub—all models

➡Do not use a hammer to accomplish this or the bearing will be damaged.

11. Remove the oil seal from the axle side of the knuckle using a small prying tool.

12. Remove the wheel bearing inner race from the front hub using a puller.

➡Be careful that the front hub does not fall when the inner race is removed.

13. Remove the snapring from the axle side of the knuckle. Remove the bearing from the knuckle using a puller.

14. Once the bearing is removed, the bearing outer race can be removed by tapping out with a brass drift pin and a hammer.

To assemble:

15. Fill the wheel bearing with multipurpose grease. Apply a thin coating of multipurpose grease to the knuckle and bearing contact surfaces.

16. Press the wheel bearing into the knuckle using appropriate pressing tool. Once the bearing is installed, install the inner race using the proper driving tool.

17. Drive the oil seal into the knuckle by using the proper size driver. Drive seal into knuckle until it is flush with the knuckle end surface.

18. Using pressing tool MB990998 or equivalent, mount the front hub assembly

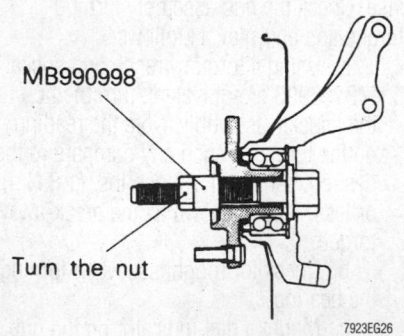

MB990998

Turn the nut

7923EG26

Pressing new bearing assembly into knuckle—all models

into the knuckle. Tighten the nut of the pressing tool to 144–188 ft. lbs. (200–260 Nm). Rotate the hub to seat the bearing.

19. Mount the knuckle assembly in a vise. Check the hub assembly turning torque and end-play as follows:

    a. Using a torque wrench and socket MB990998 or equivalent, turn the hub in the knuckle assembly. Note the reading on the torque wrench and compare to the desired reading of 16 inch lbs. (1.8 Nm) or less. This is known as the break-away torque.

    b. Check for roughness when turning the bearing.

    c. Mount a dial indicator on the hub so the pointer contacts the machined surface on the hub.

    d. Check the end-play.

    e. Compare the reading to the limit of 0.002 in. (0.05mm).

20. If the starting torque or the hub end-play are not within specifications while the nut is tightened to 144–188 ft. lbs. (200–260 Nm), the bearing, hub or knuckle have probably not been installed correctly. Repeat the disassembly and assembly procedure and recheck starting torque and end-play.

21. Install the hub and knuckle assembly onto the vehicle. Install the lower ball joint stud into the steering knuckle and install new nut. Tighten to 52 ft. lbs. (72 Nm).

22. Install the halfshaft into the transaxle extension housing and guide the outer end through the hub/knuckle assembly.

23. Install the two front strut lower mounting bolts and tighten to 80–94 ft. lbs. (110–130 Nm).

24. Install the connection for the tie rod end and tighten nut to 25 ft. lbs. (34 Nm). Install new cotter pin and bend to locknut in position.

25. Install the brake disc and caliper assembly.

26. If equipped with Active-ECS, connect the height sensor and tighten the mounting bolt to 15 ft. lbs. (20 Nm).

27. Install the front speed sensor, if removed.

➡**When installing front speed sensor, be sure harness is routed in the original position and that it is not twisted.**

28. Install the washer and new locknut to the end of the halfshaft. Tighten the locknut snugly.

29. Install the tire and wheel assembly onto the vehicle. Lower the vehicle to the ground.

30. With the weight of the vehicle on the ground and the brakes applied, tighten the

locknut to 144–188 ft. lbs. (200–260 Nm).

31. Install the cotter pin in the first matching holes and bend it securely.

Rear

COLT

➡**Some vehicles may be equipped with a non-serviceable bearing/hub assembly, and are identified as a Type II assembly. If the vehicle has this design, the bearing/hub is serviced as an assembly.**

1. Loosen the lug nuts. Raise the vehicle and support it safely.

2. Remove the wheel and tire assemblies.

3. Remove the grease cap.

4. Remove the nut.

5. Pull the drum off. If equipped with disc brakes, remove the caliper assembly, then remove the disc rotor.

    a. The outer bearing will fall out while the drum is coming off. Do not drop it. Remove the hub assembly.

    b. Pry out the oil seal. Discard it.

    c. Remove the inner bearing.

    d. Check the bearing races. If any scoring, heat checking or damage is noted, they should be replaced.

➡**When bearing or races need replacement, replace them as a set.**

    e. Inspect the bearings. If wear or looseness or heat checking is found, replace them.

    f. If the bearings and races are to be replaced, drive out the race with a brass drift.

To install:

6. Before installing new races, coat them with wheel bearing grease. Drive into place with proper size driver. Be sure they are fully seated.

7. Thoroughly pack the bearings and lubricate the hubs with wheel bearing grease. Install the inner bearing and coat the lip and rim of the grease seal with grease. Drive the seal into place with a seal driver.

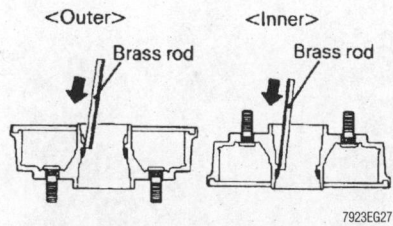

Proper method of removing the bearing races—Colt models

8. To determine if the self-locking nut is reusable:

    a. Screw in the self-locking nut until about 1/10 in. (2.54mm) of the spindle is showing.

    b. Measure the torque required to turn the self-locking nut counterclockwise.

    c. The lowest allowable torque is 48 inch lbs. (5.5 Nm). If the measured torque is less than the specification, replace the nut.

9. Place the drum or rotor on the shaft and install the outer bearing.

10. Tighten the self-locking nut to 108–145 ft. lbs. (150–200 Nm).

11. If brake caliper was removed, reinstall.

12. Install the wheel and lower the vehicle.

EXPO AND SUMMIT WAGON WITH FWD

1. Raise the vehicle and support safely.

2. Remove the tire and wheel assembly.

3. Remove the bolt(s) holding the speed sensor bracket to the knuckle and remove the assembly from the vehicle.

➡**The speed sensor has a pole piece projecting from it. This exposed tip must be protected from impact or scratches. Do not allow the pole piece to contact the toothed wheel during removal or installation.**

4. Remove the brake drums. If equipped with rear disc brakes, remove the caliper from the brake disc and suspend with a wire. Remove the brake rotor.

5. Remove the grease cap, locking nut and tongued washer.

6. Remove the rear hub and bearing assembly.

➡**The rear hub assembly can not be disassembled. If bearing replacement is required, replace the assembly as a unit.**

To install:

7. Install the hub and bearing assembly.

8. Install the tongued washer and a new locking nut. Tighten the locknut to 166 ft. lbs. (230 Nm). Once the locknut has been properly tightened, crimp the nut flange over the slot in the spindle.

9. Install the grease cap and brake parts.

10. Temporarily install the speed sensor to the knuckle; tighten the bolts only finger-tight.

11. Route the speed sensor cable correctly and loosely install the clips and retainers. All clips must be in their original position and the sensor cable must not be twisted. Improper installation may cause cable damage or system failure.

➡**The wiring in the harness is easily damaged by twisting and flexing. Use the white stripe on the outer insulation to keep the sensor harness properly placed.**

12. Use a brass or other non-magnetic feeler gauge to check the air gap between

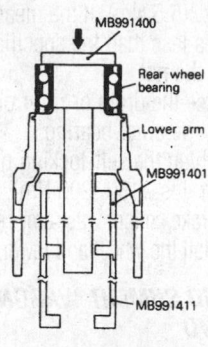

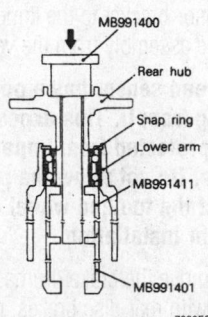

Proper method of assembling the bearing and hub—Summit Wagon and Expo models

the tip of the pole piece and the toothed wheel. Correct gap is 0.012–0.035 in. (0.3–0.9mm). Tighten the 2 sensor bracket bolts to 10 ft. lbs. (14 Nm) with the sensor located so the gap is the same at several points on the toothed wheel. If the gap is incorrect, it is likely that the toothed wheel is worn or improperly installed.

13. Install the tire and wheel assembly.

✳✳ CAUTION

Be sure to pump brake pedal until firm before moving vehicle.

EXPO AND SUMMIT WAGON WITH AWD

1. Raise and safely support the rear of the vehicle, with the suspension hanging free.

2. Remove the rear wheels.

3. Remove the brake drums. If equipped with rear disc brakes, remove the caliper and rotor assemblies.

4. Remove the bolts that attach the rear halfshaft to the rear carrier.

5. Remove the cotter pin, driveshaft nut cover and nut from the rear driveshaft.

✳✳ WARNING

Do not apply the vehicle weight to the wheel bearing while loosening the driveshaft nut or bearing damage may occur.

6. Use a slide hammer puller and proper adapter to remove the hub assembly from the axle-shaft.

7. Remove the lower control arm.

8. Using a hydraulic press and the appropriate adapters, press the inner race from the hub assembly. Remove the outer snapring and press the outer race from the lower control arm.

To install:

9. Press the new bearing into the lower control arm.

10. Using special adapters MB991400, MB991401 & MB991411 to properly support the bearing races, press the hub into the bearing.

11. Install a wheel bearing preload tool MB990998 to the hub and bearing. Tighten the tool nut to 145 to 188 ft. lbs. (200 to 260 Nm). With preload tool in place, use a torque wrench and socket to measure the rotating torque of the bearings. The torque should be 9 inch lbs. (1.1 Nm) or less.

12. Install the lower control arm.

13. Install the rear brake components.

14. Install the axle shaft and tighten the retainers on the rear carrier to 40 to 47 ft. lbs. (55 to 65 Nm) and the shaft end nut to 145 to 188 ft. lbs. (200 to 260 Nm).

15. Install the rear wheel assemblies and lower the vehicle.

➡**Be sure to pump brake pedal until firm before moving vehicle.**

HONDA

Accord • Civic • Del Sol • Prelude

16

ENGINE REPAIR

➡️**Disconnecting the negative battery cable on some vehicles may interfere with the functions of the on board computer systems and may require the computer to undergo a relearning process, once the negative battery cable is reconnected.**

Distributor

REMOVAL & INSTALLATION

➡️**The radio may contain a coded theft protection circuit. Always obtain the code number before disconnecting the battery. If the vehicle is equipped with 4WS, the steering control unit is shut down when the battery is disconnected. After connecting the battery, turn the steering wheel lock-to-lock to reset the steering control unit.**

1. Disconnect the negative battery cable.

2. Rotate the crankshaft to bring No. 1 cylinder to TDC, and align the white mark on the crankshaft pulley with the pointer on the timing belt cover.

3. Remove the distributor cap with the ignition wires attached.

4. Disconnect the electrical connectors from the distributor.

5. Mark the direction the ignition rotor is pointing on the distributor housing to aid in installation.

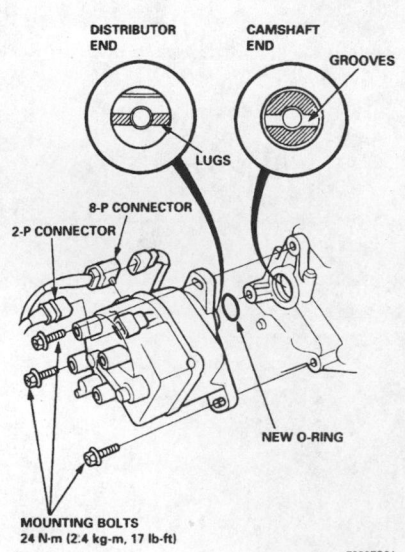

Distributor components—1.5L and 1.6L engines

6. Match mark the distributor housing with the cylinder head to aid in installation.

7. Remove the distributor mounting bolts and remove the distributor.

8. Remove and discard the O-ring from the distributor housing.

To install:

9. Coat a new O-ring with clean engine oil and install it to the distributor housing.

10. Align the ignition rotor with the mark made on the distributor housing. The drive lugs are off-set so the distributor cannot be installed incorrectly. Fit the distributor into place and turn the rotor until the drive lugs engage and the distributor seats in the cylinder head.

➡️**The lugs on the end of the distributor and their mating grooves in the camshaft end, are offset to eliminate the possibility of installing the distributor 180° out of time.**

11. Align the matchmark on the distributor housing and the cylinder head and install the mounting bolts snugly.

12. Install the distributor cap with the ignition wires.

13. Connect the distributor electrical connectors.

14. Connect the negative battery cable and enter the radio security code.

15. If equipped with 4WS, start the engine and turn the steering wheel lock-to-lock to reset the 4WS control unit.

16. Adjust the ignition timing.

17. Tighten the distributor mounting bolts to 16 ft. lbs. (22 Nm), except on 1996–99 Civics. On 1996–99 Civics, tighten the bolt to 13 ft. lbs. (18 Nm).

Ignition Timing

ADJUSTMENT

1.5L and 1.6L Engines

1. Set the parking brake and block the front wheels.

2. Connect a timing light to the No. 1 spark plug wire.

3. Start the engine and allow it to warm up.

4. Pull out the service check connector located behind the right kick panel. On the 2-P connector, connect the WHT/BGN or BRN and BLK terminals with service connector 07PAZ-0010100, or equivalent. Don't connect a jumper wire to the 3-P data link connector.

5. Shift the transaxle to neutral. All electrical accessories must be off. If

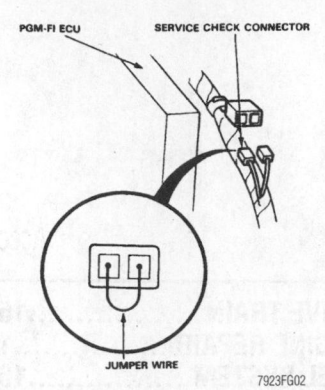

Service check connector—1.6L engines

equipped with Daytime Running Lights (DRL's), turn them off by engaging the parking brake lever.

6. Connect a test tachometer to the test tachometer connector located on the left shock tower. Check the idle speed.

7. While the engine idles, point the timing light at the mark on the timing belt cover.

8. Timing specifications: D15Z1 engine
• M/T: 16° BTDC at 600 rpm (USA) and 700 rpm (Canada)

9. Timing specifications: D15B8 engine
• M/T: 12° BTDC at 650 rpm (USA) and 750 rpm (Canada)

10. Timing specifications: D15B7 and D16Z6 engines
• M/T: 16° BTDC at 650 rpm (USA) and 750 rpm (Canada)
• A/T: 16° BTDC at 700 rpm (USA) and 750 (Canada)

11. Timing specifications: B16A2 and B16A3 engines
• M/T: 16° BTDC at 650–750 (USA) and 700–800 (Canada)

12. Timing specifications: D16Y5 engine
• M/T: 10–14° BTDC at 620–720 rpm (USA only)
• A/T and CVT: 10–14° BTDC at 650–750 rpm (USA only)

13. Timing specifications: D16Y7 and D16Y8 engines

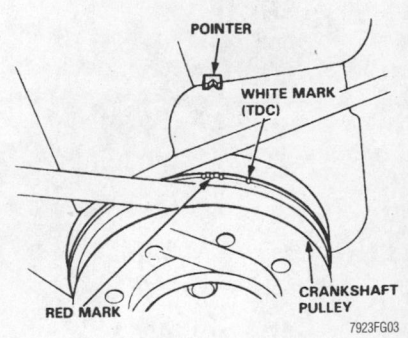

Typical crankshaft pulley timing mark location

- M/T: 10–14° BTDC at 620–730 rpm (USA) or 700–800 rpm (Canada)
- A/T: 10–14° BTDC at 650–750 rpm (USA) or 700–800 rpm (Canada)

14. If adjustment is needed, loosen the distributor adjusting bolts and turn the distributor counterclockwise to advance the timing or clockwise to retard the timing.

15. Tighten the distributor adjusting bolts to 17 ft. lbs. (24 Nm) and recheck the timing and the idle.

16. After everything has been rechecked, remove the service connector from the service check connector. Tuck the service check connector back behind the kick panel.

2.3L Engines

1. Connect a PGM tester (scan tool) to the data link connector.

2. Connect a timing light to the No. 1 ignition cable.

3. Start the engine and allow it to warm up until the electric fan comes on.

4. Be sure to turn off all accessories.

5. Verify the idle speed is 650–750 rpm.

6. Point the light at the timing belt cover near the crankshaft pulley and read the timing. Correct timing is 10–14° BTDC for both automatic and manual transmissions. If necessary, loosen the distributor hold-

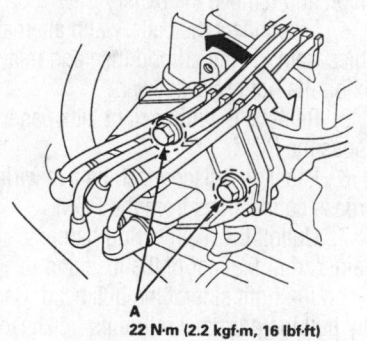

A
22 N·m (2.2 kgf·m, 16 lbf·ft)
7923FG04
Distributor hold-down bolt locations—2.3L (F23A1 and F23A4) engine

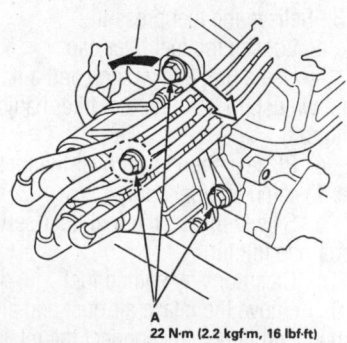

A
22 N·m (2.2 kgf·m, 16 lbf·ft)
7923FG05
Distributor hold-down bolt locations—2.3L (F23A5) engine

down bolt and rotate the distributor slightly to adjust the timing. Turn it counterclockwise to advance and clockwise to retard the timing.

7. Tighten the hold-down bolt to 16 ft. lbs. (22 Nm). Recheck the timing after the bolt is tight to confirm the correct timing.

8. Disconnect the PGM tester.

2.7L Engines

The ignition timing is only adjustable by the PCM, but the ignition base timing can be checked by performing the following:

1. Connect a timing light to the number 1 spark plug wire.

➡ **Set the parking brake and block the front wheels.**

2. Start the engine and allow it to warm up.

3. Pull out the service check connector located behind the glove box. Connect the GRN/BLU and RED terminals with the SCS service connector.

4. While the engine idles, point the timing light toward the pointer on the timing belt cover.

5. Timing specifications:
- 13–17° BTDC (Red mark on the crankshaft pulley) at 650–750 rpm in neutral

6. After everything has been rechecked, remove the SCS service connector from the check connector.

3.0L Engine

The ignition timing is only adjustable by the PCM, but the ignition base timing can be checked by performing the following:

1. Connect a PGM tester (scan tool) to the data link connector.

2. Connect a timing light to the No. 1 ignition cable.

3. Start the engine and allow it to warm up until the electric fan comes on.

4. Be sure to turn off all accessories.

5. Verify that the idle speed is 630–730 rpm.

6. Point the light at the timing belt cover near the crankshaft pulley and read the timing. Correct timing is 8–12° BTDC. If the ignition timing is different from the specification, replace the PCM.

Engine Assembly

REMOVAL & INSTALLATION

➡ **The original radio contains a coded anti-theft circuit. Obtain the security**

code number before disconnecting the battery cables.

1995 Civic and 1995–97 Del Sol

1. Disconnect the negative and positive battery cables.

2. Raise the vehicle and support it safely.

3. Remove the radiator cap.

➡ **The engine and transaxle are removed from the vehicle as one unit.**

4. Remove the engine splash shield.

5. Drain the coolant, transaxle oil and engine oil.

6. Lower the vehicle, Secure the hood as far open as possible.

7. Unbolt the underhood ABS fuse/relay box and move it out of the way.

8. Remove the air intake hose, resonator and air cleaner assembly.

9. Relieve the system fuel pressure by turning the fuel filter service bolt one turn.

10. Remove the fuel feed hose and charcoal canister hose from the intake manifold.

11. Remove the throttle cable by loosening the locknut and slipping the cable end out of the accelerator linkage.

12. Disconnect the engine wiring harness connectors at the left side of the engine compartment.

13. Remove the fuel return hose and brake booster vacuum hose.

14. Disconnect the engine wiring harness connectors, terminal, and clamps on the right side of the engine compartment.

15. Remove the battery/starter cable from the underhood fuse/relay box. Remove the ABS power cable from the battery terminal.

16. Remove the engine ground cable from the cylinder head.

17. Remove the power steering pump and belt, but do not disconnect the power steering hoses.

18. If equipped with A/C, unbolt the left front engine mount bracket from the body.

19. Remove the air conditioning belt and compressor and disconnect the electrical connector, but do not disconnect the air conditioning hoses.

20. Remove the transaxle ground cable. On vehicles with automatic transaxles, disconnect the ATF cooler lines.

21. On vehicles with manual transaxles, remove the slave cylinder without disconnecting the hydraulic line.

22. Raise and support the vehicle safely. Remove the front wheels.

23. Remove the upper and lower radiator hoses and heater hoses.

24. Remove the front exhaust pipe and stay.

25. On automatic transaxle, disconnect the shift cable.

26. On manual transaxle, disconnect the shift rod and extension rod from the transaxle.

27. Remove the strut damper fork. Disconnect the suspension lower arm ball joint using a ball joint separator.

28. Remove the driveshafts. Tie plastic bags over the inboard CV-joints to prevent damage to the boots.

29. Attach a chain hoist to the engine lifting brackets and raise it slightly to take up the weight of the engine/transaxle assembly.

30. At this point the engine mounts can be removed individually if any need to be replaced without the removal of the entire engine.

31. Remove the left and right engine stopper rubbers and brackets.

32. Remove the rear engine mounting bracket. Remove the engine support nuts.

33. Loosen the mount bolt and pivot the engine side mount out of the way.

34. Remove the transaxle mount nuts and pivot the mount out of the way.

35. Raise the chain hoist so it is tight.

36. Verify that all electrical, vacuum, coolant and fuel lines have been disconnected.

37. Remove engine/transaxle assembly from the vehicle.

To install:

➡**Use new self-locking nuts and gaskets when installing the front exhaust pipe and when assembling the front suspension. Use new set rings on the inboard CV-joint shaft.**

38. Lower the engine into the vehicle.

39. Install and connect the engine and transaxle mounts and brackets. At this point only tighten the mounting nuts and bolts by hand.

➡**Failure to tighten the bolts in the proper sequence can cause excessive noise and vibration and reduce bushing life. Be sure to check that the bushings are not twisted or offset.**

40. The engine and transaxle mount and bracket fasteners must be tightened in the proper sequence with the weight of the engine resting upon them. This step is important for engine mount pre-loading. Tighten the engine mount bolts in the following sequence:

   a. Side transaxle mount nuts—47 ft. lbs. (65 Nm).

   b. Engine side mount nuts—47 ft. lbs. (65 Nm).

   c. Side transaxle mount bolt—54 ft. lbs. (75 Nm).

   d. Engine side mount bolt—54 ft. lbs. (75 Nm).

   e. Rear engine mount-to-engine bracket—61 ft. lbs. (85 Nm).

   f. Rear engine mount through-bolt—43 ft. lbs. (60 Nm).

   g. Right front stopper bracket—47 ft. lbs. (65 Nm). Stopper-to-body bolts:33 ft. lbs. (45 Nm).

   h. Left front mount—stud: 61 ft. lbs. (85 Nm), nut:43 ft. lbs. (60 Nm), bolts: 33 ft. lbs. (45 Nm).

41. Check that the spring clip on the end of each driveshaft clicks into place. Be sure to use new spring clips on installation.

42. Install the damper fork and reconnect the lower ball joint. With the vehicle on the ground, tighten the pinch bolt to 32 ft. lbs. (44 Nm), and the fork bolt to 47 ft. lbs. (65 Nm).

43. Install the slave cylinder and connect the shift rod and extension rod to the transaxle on manual transaxle vehicles. On automatic transaxle vehicles, connect the shift cable and install its cover.

44. Install the front exhaust pipe. Use new self-locking nuts. Tighten the converter flange nuts to 16 ft. lbs. (22 Nm). Tighten the exhaust manifold nuts to 40 ft. lbs. (55 Nm).

45. Reconnect the radiator and heater hoses. Reconnect the ATF cooler lines.

46. Install the A/C compressor and tighten its mounting bolts to 16 ft. lbs. (22 Nm).

47. Reconnect the engine and transaxle ground cables. Reconnect the starter cables and the ABS power cable.

48. Reconnect the engine wiring harnesses.

49. Reconnect the throttle cable and adjust its deflection to be 10–12mm (0.39–0.47 in.).

50. Reconnect the fuel lines to the fuel rail and fuel filter. Use new sealing washers. Tighten the fuel filter banjo bolt to 25 ft. lbs. (34 Nm) and the service bolt to 11 ft. lbs. (15 Nm).

51. Install the relay box, resonator, air cleaner, and intake duct.

52. Install the accessory belts and adjust their tensions.

53. Install the splash shield.

54. Refill the cooling system.

55. Refill the engine with fresh oil.

56. Refill the transaxle with the proper fluid.

57. Install and reconnect the battery.

58. Verify that all fuel and vacuum lines and electrical harnesses and ground cables have been reconnected properly.

59. After assembling the fuel line parts, turn the ignition switch to the **ON**, but don't start the engine. Then, turn the ignition switch **OFF**. Repeat this procedure two or three times to pressurize the fuel system and check for leak.

60. Bleed the air from the cooling system at the bleed bolt with the heater valve open.

61. Adjust the clutch cable free-play and check that the transaxle shifts into gear smoothly.

62. Check the ignition timing.

63. Check the and adjust the front wheel alignment.

64. Road test the vehicle.

1996–99 Civic

1. Disconnect the negative and positive battery cables.

Wait at least three minutes before working around the air bags.

➡**The engine and transaxle are removed from the vehicle as one unit.**

2. Support the hood as far open as possible. If the hood is to be removed, first matchmark the hinge plates with a felt-tipped marker.

3. Remove the battery from the vehicle. Unbolt and remove the battery tray.

4. Disconnect the battery and alternator cables from the underhood fuse and relay box on the right shock tower.

5. Remove the lower right kick panel to expose the PCM.

6. Label and disconnect the five wiring harness connections from the PCM.

7. Unbolt the main wiring harness retainer from the rear of the fuse and relay box on the right side of the bulkhead. Carefully pull the grommet out of its bulkhead opening. Next, pull the PCM harness and connectors through the opening. Be careful not to damage the wiring, insulation, or connectors.

8. Relieve the fuel pressure:

   a. Loosen the fuel filler cap.

   b. Use a box-end wrench and a flare nut wrench to hold the fuel filter banjo fitting.

   c. Place a shop towel over the fuel filter to catch the fuel spray.

   d. Slowly loosen the fuel filter service bolt one full turn.

   e. Clean up any spilled fuel.

9. Remove the intake air duct and air cleaner. If equipped, disconnect the intake air temperature (IAT) sensor connector from the air cleaner case.

10. Disconnect the fuel feed hose from the fuel filter. Disconnect the fuel return hose from the fuel rail.

11. Label and disconnect the following vacuum lines:
- Intake manifold/throttle body vacuum hoses
- Brake booster vacuum hose
- EVAP canister vacuum hose

12. Disconnect the power steering pressure switch (PSP) and detach its clamp from the bracket below the brake booster.

13. Disconnect the transaxle ground cable. Remove the radiator hose bracket.

14. Loosen the throttle cable's locknut, then disconnect the cable from the throttle body linkage. Don't kink the cable: move it out of the work area.

15. Loosen the power steering pump mounting bolts. Slip power steering belt off its pulleys. Unbolt the steering pump and move it out of the work area. Don't disconnect the hydraulic hoses.

16. Label and disconnect the engine wiring harness connectors at the left side of the engine compartment.

17. Drain the coolant from the radiator and engine block.

18. Disconnect the upper and lower radiator hoses, then remove them. Disconnect the heater hoses from the cylinder head.

19. If equipped with a CVT transaxle, loosen the shift cable locknut. Remove the spring clip and washers and disconnect the shift cable from its linkage. Be careful not to kink the cable or damage its boot.

20. If equipped with a manual transaxle, unbolt the hydraulic line brackets from the top of the transaxle case.

21. Attach a chain hoist to the engine lifting brackets. Don't raise the hoist to lift the engine yet.

22. Raise the vehicle and support it safely. Remove the front wheels.

23. Remove the engine splash shield.

24. Drain the engine oil.

25. Drain the fluid from the transaxle.

26. If equipped with A/C, unbolt the left front engine mount bracket from the shock tower.

27. Loosen the compressor idler pulley and adjusting bolt. Slip the belt around the engine mount stud to remove it.

28. Unbolt the compressor mounting bolts to separate the compressor from its mounting plate. Move the compressor out of the work area. Do not disconnect the air conditioning refrigerant lines.

29. If equipped with an automatic transaxle, disconnect the ATF cooler lines. Plug the cooler lines to prevent fluid leakage and contamination.

30. If equipped with a manual transaxle, unbolt the slave cylinder from the transaxle case without disconnecting its hydraulic line.

31. Separate the front exhaust pipe from the exhaust manifold and catalytic converter. Unbolt its hanger bracket and remove the exhaust pipe.

32. If equipped with a automatic transaxle, disconnect the shift cable from the transaxle control shaft

33. If equipped with a manual transaxle, disconnect the shift rod and extension rod from the transaxle.

34. Unbolt and remove the strut damper fork. Disconnect the steering knuckle ball joint from the lower control arm using a ball joint separator.

35. Pry the inboard CV-joints from the transaxle. Then, move the halfshafts away from the transaxle and wire them to the undercarriage of the vehicle. Tie plastic bags over the inboard CV-joints to prevent damage to the boots and splined shafts.

36. Raise the hoist slightly to take up the weight of the engine and transaxle assembly.

37. Disconnect the engine mounts in the following order:
   a. Unbolt and remove the left front engine mount.
   b. Unbolt and remove the right front engine mount and bracket assembly.
   c. Remove the rear engine mount through-bolt. Then, unbolt the rear mount bracket from the engine block.

38. If necessary, lower the vehicle slightly to gain access to the side engine and transaxle mounts. Do not release the tension of the chain hoist—the engine must be securely supported.

39. Unbolt the side engine mount bracket from the engine block bracket and mount damper.

40. Unbolt the transaxle mount bracket from the transaxle case. Then, unbolt the mount from the shock tower.

41. Raise the chain hoist to lift the engine a few in. off of its mounts.

42. Verify that all electrical, vacuum, and fuel lines have been disconnected.

43. Raise the engine and transaxle assembly and remove it from the vehicle.

To install:

➡**Use new self-locking nuts and gaskets when installing the front exhaust pipe and when assembling the front suspension. Use new set rings on the inboard CV-joint splined shafts.**

44. Lower the engine and transaxle assembly into the vehicle.

45. Install and connect the engine and transaxle mounts and brackets. Use new self-locking nuts and color-coded bolts. At this point, only tighten the mounting nuts and bolts by hand.

46. Before installing the left front engine mount, fit the A/C compressor back into place and install the compressor belt. Tighten the compressor bolts to 17 ft. lbs. (22 Nm).

➡**Failure to tighten the bolts in the proper sequence can cause excessive noise and vibration and reduce bushing life. Be sure to check that the bushings are not twisted or offset.**

47. The engine and transaxle mount and bracket fasteners must be tightened in the proper sequence with the weight of the engine resting upon them. This step is important for engine mount pre-loading. Tighten the engine mount bolts in the following sequence:
   a. Transaxle mount bolts: 47 ft. lbs. (64 Nm); or 28 ft. lbs. (38 Nm) for CVT-equipped vehicles
   b. Side engine mount bracket nuts: 54 ft. lbs. (74 Nm)
   c. Rear mount bracket bolts: 61 ft. lbs. (83 Nm); or 43 ft. lbs. (59 Nm) for CVT-equipped vehicles
   d. Rear mount through-bolt: 43 ft. lbs. (59 Nm)
   e. Transaxle mount bracket nuts or bolts: 47 ft. lbs. (64 Nm).
   f. Transaxle mount through-bolt: 54 ft. lbs. (74 Nm)
   g. Right front mount bracket bolts: 33 ft. lbs. (44 Nm).
   h. Right front mount carrier bolts: 33 ft. lbs. (44 Nm).
   i. Left front mount: stud: 61 ft. lbs. (85 Nm); carrier bolts: 33 ft. lbs. (44 Nm); nut: 43 ft. lbs. (59 Nm).

48. Remove the chain hoist from the engine lifting hooks.

49. Install new set rings on the inboard splined shafts of each halfshaft. Check that the set ring on each inboard CV-joint clicks into place when the halfshafts are installed into the transaxle.

50. Install the damper fork and reconnect the lower ball joint. When the weight of the vehicle is resting on its suspension, tighten the pinch bolt to 32 ft. lbs. (44 Nm) and the fork bolt to 47 ft. lbs. (65 Nm). Tighten the ball joint castle nut to 36–43 ft. lbs. (50–60 Nm). Next, tighten the castle nut only enough to install a new cotter pin.

51. If equipped, install the slave cylinder. Tighten the slave cylinder mounting bolts to 16 ft. lbs. (22 Nm). If the clutch hydraulic line was disconnected, the fluid must be bled.

52. If equipped, reconnect the transaxle shift and extension rods to the linkage at the transaxle case. Install a new 8mm spring pin into the shift rod linkage. Then, install the retainer clip and boot. Tighten the extension rod bolt to 16 ft. lbs. (22 Nm).

53. If equipped with an automatic transaxle, connect the shift cable to the control shaft. Use a new lockwasher and tighten the lockbolt to 10 ft. lbs. (14 Nm). Tighten the shift cable cover bolts to 16 ft. lbs. (22 Nm). Install the shift cable cover and tighten its bolts to 16 ft. lbs. (22 Nm).

54. Install the front exhaust pipe using new self-locking nuts.

• If equipped with the D16Y8 engine, tighten the converter flange nuts to 16 ft. lbs. (22 Nm). Tighten the exhaust manifold nuts to 40 ft. lbs. (55 Nm).

• If equipped with the D16Y5 or D16Y7 engine, tighten the converter flange nuts to 25 ft. lbs. (33 Nm). Tighten the exhaust flange bolts to 16 ft. lbs. (22 Nm).

55. Reconnect the ATF cooler lines. If the rubber cooler lines are cracked or stressed, they must be replaced.

56. Install the engine splash shield.

57. Refill the engine with fresh oil.

58. Refill the transaxle with the proper fluid.

59. Lower the vehicle.

60. If equipped, fit the clutch hydraulic line brackets back into place. Tighten the 8mm bolts to 17 ft. lbs. (24 Nm). Tighten the 6mm bolts to 8 ft. lbs. (11 Nm).

61. If equipped with a CVT transaxle, Reconnect the shift cable to the linkage. Use new plastic washers and a new spring clip. Tighten the locknut to 22 ft. lbs. (29 Nm).

62. Adjust the alternator and A/C compressor belt tensions.

63. Install and reconnect the upper and lower radiator hoses. Reconnect the heater hoses.

64. Install the power steering pump into its mounts. Adjust the pump belt's tension, then tighten the mounting bolts to 17 ft. lbs. (24 Nm).

65. Reconnect the PSP switch connector and attach its harness clamp.

66. Reconnect the following vacuum lines:

• Intake manifold/throttle body vacuum hoses
• Brake booster vacuum hose
• EVAP canister vacuum hose

67. Reconnect the fuel line fittings to the fuel filter and fuel rail. Use new sealing washers. Tighten the banjo fittings to 25 ft. lbs. (33 Nm), and the service bolts to 11 ft. lbs. (15 Nm). Don't overtighten the fittings.

68. Reconnect the throttle cable and adjust its deflection to 10–12mm (0.39–0.47 in.).

69. Feed the PCM harness through the hole in the bulkhead. Apply sealant to the grommet, then install the retainer.

70. Reconnect any engine wiring harness and ground cables that were disconnected during engine removal. Be sure the grounds are free of corrosion to ensure good contact.

71. Fit the fuse and relay box back into position. Reconnect the battery and alternator cables.

72. Install the air cleaner case and air intake duct. Reconnect the IAT connector.

73. Reconnect the five PCM connectors. Install the kick panel.

74. Install the battery tray and the battery.

75. Verify that all wiring harnesses and grounds, vacuum lines, fuel lines have been reconnected.

76. Refill the radiator with fresh coolant.

77. If it was removed, install the hood. Reconnect the windshield washer tubing. After installation, check to be sure that the hood, fender, and grille panel gaps are equal.

78. Reconnect the positive and negative battery cables.

79. Turn the ignition switch to the **ON** position, but don't start the engine. Then, turn the ignition **OFF**. Repeat this procedure two or three times and check for a any indications of fuel leaks.

80. Start the engine and allow it to warm up to its normal operating temperature.

81. Bleed the air from the cooling system with the heater valve open.

82. Check the throttle cable deflection and operation.

83. Check the and adjust the ignition timing.

84. Shut the engine off and check the drive belt adjustments.

85. Check all fluid levels and top up as necessary.

86. Check the and adjust the front wheel alignment.

87. Road test the vehicle.

Prelude

1. Secure the hood as far open as possible.

2. Disconnect the negative battery cable, then the positive battery cable.

3. Remove the radiator cap.

4. Raise and safely support the vehicle. Remove the front wheels and the engine splash shield.

5. Drain the engine coolant into a sealable container.

6. Drain the transaxle oil/fluid into a sealable container. Install the drain plug with a new gasket.

7. Lower the vehicle to a working level.

8. Remove the air intake duct and the air cleaner case.

9. Remove the pulsed secondary air Injection (PAIR) vacuum tank and bracket.

10. Remove the battery and the battery base. Disconnect the battery cable and starter cable harnesses from the body.

11. Relieve the pressure from the fuel system.

✻✻ CAUTION

The fuel injection system remains under pressure after the engine has been turned OFF. Properly relieve fuel pressure before disconnecting any fuel lines. Failure to do so may result in fire or personal injury.

12. Disconnect the fuel feed hose from the fuel rail and disconnect the fuel return line from the fuel pressure regulator.

13. Disconnect the injector resistor connector on the left side of the engine compartment.

14. Remove the throttle cable by loosening the locknut, then slip the cable end out of the throttle linkage. Take care not to bend the cable when removing it. Always replace any kinked cable with a new one.

15. Disconnect the engine wiring harness connectors, terminal and clamps on the right side of the engine.

16. Remove the power cable from the under-hood fuse/relay box.

17. Disconnect the brake booster vacuum hose and emissions control vacuum tubes from the intake manifold.

18. Disconnect the cruise control actuator electrical connector and vacuum tube, then remove the actuator.

19. Remove the engine ground cable from the body side.

20. Remove the power steering pump drive belt, then remove the pump.

21. Remove the air conditioning (A/C) condenser fan, then install a protector plate on the radiator.

22. Loosen the alternator mounting bolt, nut and adjusting nut, then remove the alternator drive belt.

23. Disconnect the A/C compressor electrical connector and loosen the compressor mounting bolts. Remove the compressor without disconnecting the A/C hoses. Support the compressor with a strong wire out of the way.

24. Remove the upper and lower radiator hoses, then disconnect the heater hoses from the engine.

25. Remove the transaxle ground cable.

26. If equipped with a automatic transaxle, disconnect the cooler hoses.

27. If equipped with a manual transaxle perform the following:

a. Disconnect the shift cable and the select cable from the transaxle. Do not bend the cables when removing them. Replace any kinked cable with a new one.

b. Remove the clutch slave cylinder and the pipe/hose assembly. Do not operate the clutch once the slave cylinder has been removed.

c. Remove the clutch damper assembly.

28. Remove the Vehicle Speed Sensor (VSS)/Power Steering (P/S) speed sensor assembly. Do not disconnect the hoses.

29. Remove the nuts attaching exhaust pipe A to the exhaust manifold and the catalytic converter. Remove the bolts from the exhaust pipe hanger, then remove the exhaust pipe and discard the gaskets.

30. If equipped with a automatic transaxle, remove the shift cable cover, then disconnect the shift cable. Do not bend the cable and replace the cable if it becomes kinked.

31. Remove the left and the right side damper forks.

32. Disconnect the lower ball joints from the lower control arms.

33. Pry the halfshafts from the transaxle. Cover the inner CV joints with plastic bags to protect them.

34. Swing the halfshafts under the fender out of the way.

35. Attach an engine hoist to the engine lifting points and raise the hoist to remove all slack from the chain.

36. Remove the rear engine mount bracket.

37. Remove the front engine mount bracket.

38. Remove the left side engine mount.

39. Remove the transaxle mount and the mount bracket.

40. Check that the engine is completely free of vacuum hoses, fuel and coolant hoses, and electrical wiring.

41. Slowly raise the engine approximately 6 in. (150mm). Check once again that all hoses and wires have been disconnected from the engine.

42. Raise the engine all the way and remove it from the vehicle.

43. Remove the transaxle.

44. If equipped with a manual transaxle, remove the clutch cover (pressure plate) and clutch disc.

45. Mount the engine on an engine stand, making sure the mounting bolts are tight. If an engine stand is not available,

support the engine in an upright position with blocks. Never leave an engine hanging from a lift or hoist.

To install:

46. Assemble the clutch disc and pressure plate to the flywheel for manual transaxle vehicles.

47. Install the transaxle.

48. Lift the engine into position and lower it into the car, aligning the mounts and bushings.

➡**When installing the engine mounts and vibration dampers in the following steps, they must be tightened to the correct tension in the correct order if they are to damp vibration properly.**

49. Install the side engine mount and the through-bolt. Do not tighten the through-bolt at this time. Install the nut and bolt attaching the mount to the engine. Tighten the nut and bolt attaching the side mount to the engine to 40 ft. lbs. (55 Nm).

50. Install the transaxle mount and through-bolt. Do not tighten the through-bolt at this time.

51. Install the rear engine mount and new bolts attaching the mount to the engine. Tighten the three new bolts attaching the mount to the engine assembly to 40 ft. lbs. (55 Nm). Install a new rear engine mount through-bolt and tighten the new through-bolt to 47 ft. lbs. (65 Nm).

52. Install the front mount and the three bolts attaching the mount to the engine assembly, only snug the bolts in place. Install a new through-bolt to the front mount and tighten the new through-bolt to 47 ft. lbs. (65 Nm).

53. Install the nuts to the transaxle mount. Tighten the nuts to 28 ft. lbs. (39 Nm).

54. Tighten the side engine mount through-bolt to 47 ft. lbs. (65 Nm).

55. Tighten the transaxle mount through-bolt to 47 ft. lbs. (65 Nm).

56. Tighten the three bolts attaching the front mount to the engine to 28 ft. lbs. (39 Nm).

57. Remove the hoist equipment from the engine.

58. Install new spring clips to the inner CV-joints. Install the halfshafts into the transaxle, be sure that the spring clips on the inner joints click into place.

59. Connect the lower ball joints to the lower control arms. Tighten the nuts to 36–43 ft. lbs. (50–60 Nm). Install a new cotter pin to the ball joint stud.

60. Install the damper forks. Tighten a new self-locking bolt attaching the damper

fork to the strut to 32 ft. lbs. (44 Nm). Tighten the new nut and bolt attaching the damper fork to the lower control arm to 47 ft. lbs. (65 Nm).

61. If equipped with a automatic transaxle, connect the shift cable to the transaxle. Install a new lockwasher and tighten the attaching bolt to 7 ft. lbs. (10 Nm). Install the shift cable cover. Tighten the shift cable cover attaching bolts to 13 ft. lbs. (18 Nm).

62. Install exhaust pipe A with new gaskets. Tighten the new nuts attaching the exhaust pipe to the exhaust manifold to 40 ft. lbs. (55 Nm). Tighten the new nuts attaching exhaust pipe A to the catalytic converter to 25 ft. lbs. (34 Nm). Install new attaching bolts to the exhaust pipe hanger and tighten the bolts to 13 ft. lbs. (18 Nm).

63. Install the VSS and connect the electrical connector and tighten the mounting bolt to 13 ft. lbs. (18 Nm).

64. If equipped with a manual transaxle perform the following:

a. Install the clutch damper assembly and tighten the attaching bolts to 16 ft. lbs. (22 Nm).

b. Install the clutch slave cylinder and the pipe/hose assembly and tighten the slave cylinder mounting bolts to 16 ft. lbs. (22 Nm).

c. Connect the shift cable and the select cable to the transaxle. Adjust the shift cable and select cable.

65. If equipped with a automatic transaxle, connect the cooler hoses.

66. Install the transaxle ground cable.

67. Install the upper and lower radiator hoses and connect the heater hoses to the engine.

68. Install the A/C compressor and connect the electrical connector. Tighten the mounting bolts to 16 ft. lbs. (22 Nm).

69. Install and adjust the alternator drive belt.

70. Remove the protector plate from the radiator and install the A/C condenser fan.

71. Install and the power steering pump and drive belt. Adjust the drive belt tension, then tighten the attaching nuts and bolts to 16 ft. lbs. (22 Nm).

72. Attach the engine ground cable to the body.

73. Install the cruise control actuator, then connect the electrical connector and vacuum tube. Tighten the mounting bolts to 7 ft. lbs. (10 Nm).

74. Connect the brake booster vacuum hose and the emissions control vacuum tubes to the intake manifold.

75. Connect the engine wiring harness connectors, terminal and clamps.

76. Install and adjust the throttle cable.

77. Connect the injector resistor connector on the left of the engine compartment.

78. Connect the fuel return hose to the regulator. Connect the fuel feed hose to the fuel rail with new washers. Tighten the cap nut to 16 ft. lbs. (22 Nm).

79. Connect the battery cable and the starter cable to the body. Install the battery base and the battery. Tighten the battery base attaching bolts to 16 ft. lbs. (22 Nm).

80. Install the PAIR vacuum tank and bracket. Tighten the mounting bolts to 8 ft. lbs. (10 Nm).

81. Install the air cleaner duct and housing.

82. Install the engine splash shield and the front wheels.

83. Lower the vehicle.

84. Fill the engine with oil and the transaxle with oil/fluid.

85. Fill and bleed the air from the cooling system.

86. Connect the positive, then the negative battery cable and enter the radio security code.

87. Switch the ignition **ON** but do not engage the starter. The fuel pump should run for approximately 2 seconds, building pressure within the lines. Switch the ignition **OFF**, then **ON** 2 or 3 more times to build full system pressure. Check for fuel leaks.

88. Disconnect the coil wire from the distributor. Insulate or protect the end of the cable so it does not arc to the engine or surrounding metal. Without touching the accelerator, turn the ignition switch to the START position and crank the engine for about 5–10 seconds; this will develop some oil pressure within the motor. Do not exceed 10 seconds cranking.

89. Switch the ignition OFF and reconnect the coil.

90. Start the engine, allowing it to idle. Check the hoses and lines carefully for any sign of leakage.

91. Check the timing and idle speed.

92. After the engine has warmed up fully and the fan(s) have come on at least once, recheck the engine for fluid leaks. Switch the engine OFF.

93. Adjust the belts and throttle cable as necessary.

94. If equipped with 4WS, start the engine and turn the steering wheel lock-to-lock to reset the 4WS control unit.

95. Road test the vehicle, then loosen and retighten the three bolts attaching the front engine mount to the engine. Tighten the bolts to 28 ft. lbs. (39 Nm).

1995–97 Accord

2.2L ENGINES

1. Secure the hood as far open as possible.

2. Disconnect the negative battery cable, then the positive battery cable.

3. Remove the battery and the battery base. Disconnect the engine ground cable.

4. Remove the throttle cable and the cruise control cable, by loosening the locknuts, then slip the cable ends out of the throttle linkage. Take care not to bend the cables when removing them. Always replace any kinked cable with a new one.

5. Remove the intake air duct B and intake air duct/air cleaner housing.

6. Disconnect the Intake Air Resonator (IAR) control solenoid valve connector, then remove the IAR from the vehicle.

7. Remove the battery cables from the under-hood fuse/relay box and under-hood ABS fuse/relay box.

8. Disconnect the engine wiring harness connectors on the right side of the engine compartment.

9. Remove the brake booster vacuum hose, then label and disconnect the other vacuum hoses from the intake manifold.

10. Relieve the pressure from the fuel system.

✳✳ CAUTION

The fuel injection system remains under pressure after the engine has been turned OFF. Properly relieve fuel pressure before disconnecting any fuel lines. Failure to do so may result in fire or personal injury.

11. Disconnect the fuel feed hose from the fuel rail and disconnect the fuel return line from the fuel pressure regulator.

12. Remove the engine wiring harness connectors, terminal and clamps on the left side of the engine compartment.

13. Disconnect the injector resistor connector on the left side of the engine compartment.

14. Remove the power steering hose clamp.

15. Remove the power steering pump mounting nuts and adjusting bolt, then remove the power steering pump drive belt and the pump.

16. Loosen the alternator mounting bolt, nut and adjusting bolt, then remove the alternator belt.

17. If equipped with a manual transaxle perform the following:

　a. Disconnect the shift cable and the select cable from the transaxle. Do not bend the cables when removing them. Replace any kinked cable with a new one.

　b. Disconnect the back-up light switch connectors and starter motor cable.

　c. Remove the clutch slave cylinder and the pipe/hose assembly. Do not operate the clutch once the slave cylinder has been removed.

18. Disconnect the vehicle speed sensor (VSS).

19. Remove the radiator cap.

20. Raise and safely support the vehicle. Remove the front wheels and the engine splash shield.

21. Drain the engine coolant into a sealable container.

22. Drain the transaxle oil/fluid into a sealable container. Install the drain plug with a new gasket.

23. Drain the engine oil into a sealable container.

24. Lower the vehicle to a working level.

25. Remove the upper and lower radiator hoses, then disconnect the heater hoses from the engine.

26. If equipped with a automatic transaxle, disconnect the ATF cooler hoses.

27. Remove the radiator assembly from the vehicle.

28. Loosen the A/C mounting bolts, then remove the compressor. Do not disconnect the A/C hoses. Disconnect the compressor electrical connector and support the compressor with a strong wire.

29. Remove the center beam from under the engine.

30. Remove the nuts attaching exhaust pipe A to the exhaust manifold and the catalytic converter. Remove the nuts from the exhaust pipe hanger, then remove the exhaust pipe and discard the gaskets.

31. If equipped with a automatic transaxle, remove the shift cable cover, then disconnect the shift cable. Do not bend the cable and replace the cable if it becomes kinked.

32. Remove the left and the right side damper forks.

33. Disconnect the lower ball joints from the lower control arms.

34. Pry the halfshafts from the transaxle. Cover the inner CV joints with plastic bags to protect them.

35. Swing the halfshaft under the fender out of the way.

36. Attach an engine hoist to the engine lifting points and raise the hoist to remove all slack from the chain.

37. Remove the rear engine mount bracket.

38. Remove the front engine mount bracket.

39. Remove the side engine mount.

40. Remove the transaxle mount and the mount bracket.

41. Check that the engine is completely free of vacuum hoses, fuel and coolant hoses, and electrical wiring.

42. Slowly raise the engine approximately 6 in. (150mm). Check once again that all hoses and wires have been disconnected from the engine.

43. Raise the engine all the way and remove it from the vehicle.

44. Remove the transaxle from the engine.

45. If manual transaxle, remove the clutch cover (pressure plate) and clutch disc.

46. Mount the engine on an engine stand, making sure the mounting bolts are tight. If an engine stand is not available, support the engine in an upright position with blocks. Never leave an engine hanging from a lift or hoist.

To install:

47. Assemble the clutch disc and pressure plate to the flywheel, if equipped with a manual transaxle.

48. Install the transaxle to the engine.

49. Lift the engine into position and lower it into the car, aligning the mounts and bushings.

➡ **When installing the engine mounts and vibration dampers in the following steps, they must be tightened to the correct tension in the correct order if they are to damp vibration properly.**

50. Install the side engine mount. Install a 6 x 100mm bolt to the mount to properly position the mount. Do not tighten the nut and bolt attaching the mount to the engine. Tighten the through-bolt to 47 ft. lbs. (64 Nm), then remove the 6 x 100mm bolt from the mount.

51. Install the transaxle mount. Install a 6 x 100mm bolt to the mount to properly position the mount. Do not tighten the nuts attaching the mount to the transaxle at this time. Tighten the through-bolt to 47 ft. lbs. (64 Nm), then remove the 6 x 100mm bolt from the mount.

52. Install the rear engine mount bracket using new bolts. Tighten the new bolts attaching the mount to the engine assembly to 40 ft. lbs. (54 Nm). Install a new rear engine mount through-bolt. Tighten the new through-bolt to 47 ft. lbs. (64 Nm).

➡ **Tighten the bolts attaching the mount to the engine assembly first, then the through-bolt. If this order is not followed excessive engine vibration may be felt and mount damage may occur.**

53. Install the front mount bracket. Do not tighten the nuts attaching the mount to the engine assembly, only snug the nuts in place. Install a new through-bolt to the front mount. Tighten the new through-bolt to 47 ft. lbs. (64 Nm).

54. Tighten the side engine mount nut and bolt to 47 ft. lbs. (64 Nm).

55. Tighten the nuts attaching the transaxle mount to the transaxle to 28 ft. lbs. (38 Nm).

56. Tighten the three bolts attaching the front mount bracket to the engine to 28 ft. lbs. (38 Nm).

57. Remove the hoist equipment from the engine.

58. Install new spring clips to the inner CV-joints. Install the halfshafts into the transaxle, be sure that the spring clips on the inner joints click into place.

59. Connect the lower ball joints to the lower control arms, tighten the nuts to 36–43 ft. lbs. (49–59 Nm).

Install a new cotter pin to the ball joint stud.

60. Install the damper forks, tighten a new self-locking bolt attaching the damper fork to the strut to 32 ft. lbs. (43 Nm). Tighten the new nut and bolt attaching the damper fork to the lower control arm to 47 ft. lbs. (64 Nm).

61. If equipped with a automatic transaxle, connect the shift cable to the transaxle. Install a new lockwasher and tighten the attaching bolt to 10 ft. lbs. (14 Nm). Install the shift cable cover and tighten the shift cable cover attaching bolts to 13 ft. lbs. (18 Nm).

62. Install exhaust pipe A with new gaskets. Tighten new nuts attaching the exhaust pipe to the exhaust manifold to 40 ft. lbs. (54 Nm). Tighten new nuts attaching exhaust pipe A to the catalytic converter to 16 ft. lbs. (22 Nm). Install new attaching nuts to the exhaust pipe hanger and tighten the nuts to 13 ft. lbs. (18 Nm).

63. Install the center beam and tighten the bolts attaching the center beam to 37 ft. lbs. (50 Nm).

64. Install the A/C compressor and connect the electrical connector and tighten the mounting bolts to 16 ft. lbs. (22 Nm).

65. Install the radiator assembly.

66. If equipped with a automatic transaxle, connect the ATF cooler hoses.

67. Install the upper and lower radiator hoses and connect the heater hoses to the engine.

68. Install the engine splash shield and the front wheels.

69. Connect the VSS (vehicle speed sensor) electrical connector.

70. If equipped with a manual transaxle perform the following:

a. Install the clutch slave cylinder and the pipe/hose assembly. Tighten the slave cylinder mounting bolts to 16 ft. lbs. (22 Nm).

b. Connect the starter motor cable and the back-up light switch connectors.

c. Connect the shift cable and the select cable to the transaxle. Adjust the shift cable and select cable.

71. Install and adjust the alternator drive belt.

72. Install the power steering pump and drive belt. Adjust the drive belt and tighten the attaching nuts and bolts to 16 ft. lbs. (22 Nm).

73. Install the power steering hose clamp.

74. Connect the injector resistor connector on the left of the engine compartment.

75. Connect the engine wiring harness connectors, terminal and clamps on the left side of the engine compartment.

76. Connect the fuel return hose to the regulator. Connect the fuel feed hose to the fuel rail with new washers and tighten the banjo nut to 16 ft. lbs. (22 Nm).

77. Connect the vacuum hoses to the intake manifold.

78. Connect the engine wiring harness connectors on the right side of the engine compartment.

79. Connect the battery cables to the under-hood fuse/relay box and under-hood ABS fuse/relay box.

80. Install the vacuum hose and IAR, then connect the IAR control solenoid valve connector.

81. Install the intake air duct/air cleaner housing, then intake air duct B.

82. Install and adjust the throttle cable.

83. Connect the engine ground cable and install the battery base. Tighten the base mounting bolts to 16 ft. lbs. (22 Nm).

84. Install the battery and connect the positive, then the negative battery cables. Enter the radio security code.

85. Fill the engine with oil and the transaxle with oil/fluid.

86. Fill and bleed the air from the cooling system.

87. Switch the ignition **ON** but do not engage the starter. The fuel pump should run for approximately 2 seconds, building pressure within the lines. Switch the ignition **OFF**, then **ON** 2 or 3 more times to build full system pressure. Check for fuel leaks.

88. Disconnect the coil wire from the distributor. Insulate or protect the end of the cable so it does not arc to the engine or surrounding metal. Without touching the

accelerator, turn the ignition switch to the START position and crank the engine for about 5–10 seconds; this will develop some oil pressure within the motor. Do not exceed 10 seconds cranking.

89. Switch the ignition **OFF** and reconnect the coil.

90. Start the engine, allowing it to idle. Check the hoses and lines carefully for any sign of leakage.

91. Check the timing and idle speed.

92. After the engine has warmed up fully and the fan(s) have come on at least once, recheck the engine for fluid leaks. Switch the engine **OFF**.

93. Adjust the belts, clutch and throttle cable as necessary.

2.7L ENGINES

1. Disconnect the support struts from the engine hood, then fix the hood in a vertical position.

2. Disconnect the negative, then the positive battery cables.

3. Remove the battery, battery base and bracket. Disconnect the engine ground cable, located next to the battery.

4. Remove the intake air duct.

5. Remove intake manifold cover B, then disconnect the throttle cable and cruise control cables from the throttle linkage. Loosen the cable locknuts, then slip the cables ends out of the accelerator linkage. Take care to not bend the cables, always replaced a kinked cable.

6. Remove the starter cable from the strut brace, then remove the strut brace.

7. Disconnect the engine wiring harness connectors on the left side of the engine compartment.

8. Disconnect the injector resistor connector on the left side of the engine compartment.

9. Relieve the pressure from the fuel system.

✳✳ CAUTION

The fuel injection system remains under pressure after the engine has been turned OFF. Properly relieve fuel pressure before disconnecting any fuel lines. Failure to do so may result in fire or personal injury.

10. Disconnect the fuel feed hose from the fuel filter. Disconnect the fuel return hose from the regulator.

11. Disconnect the brake booster vacuum hose and the evaporative emissions (EVAP) control canister hose.

12. Label, then disconnect the vacuum hoses from the engine.

13. Remove the battery cables from the under-hood fuse/relay box and under-hood ABS fuse/relay box.

14. Disconnect the engine wiring harness connectors on the right side of the engine compartment.

15. Remove the side engine mount, discard the bolts attaching the mount to the engine.

16. Loosen the air conditioning idler pulley center nut and adjusting bolt, then remove the drive belt.

17. Disconnect the ground cable from the body, located toward the front of the vehicle by the drive belts.

18. Loosen the alternator mounting nut, bolt and adjusting bolt, then remove the drive belt.

19. Loosen the power steering pump mounting and adjusting nuts, then remove the drive belt.

20. Disconnect the power steering inlet hose from the pump, plug or cap the connections.

21. Remove the power steering pump mounting nuts and adjusting bolt, then remove the pump.

22. Remove the radiator cap.

23. Raise and safely support the vehicle.

24. Remove the front wheels and the splash shield.

25. Drain the engine coolant into a sealable container.

26. Drain the transaxle fluid into a sealable container. Install the drain plug with a new gasket.

27. Drain the engine oil into a sealable container. Install the drain plug with a new gasket.

28. Remove the center beam from under the engine.

29. Disconnect the oxygen sensor electrical connector. Remove the nuts attaching exhaust pipe A to the exhaust manifolds. Remove the nuts attaching exhaust pipe A to the catalytic converter and remove the exhaust pipe. Discard the locknuts and gaskets.

30. Remove the crankshaft pulley bolt and remove the crankshaft pulley. Use a crank pulley holder (part # 07MAB-PY3010A) and holder handle (part # 07JAB-001020A) to hold the crankshaft pulley in place while removing the bolt.

31. Remove the oil filter.

32. Disconnect the oil pressure switch terminal, then remove the oil filter base attaching bolts.

33. Remove the oil filter base and discard the O-rings.

34. Remove the shift cable cover, then remove the shift cable attaching bolts. Discard the lockwasher.

35. Remove the left and right damper forks.

36. Disconnect the lower ball joints from the lower control arms.

37. Remove the halfshafts from the vehicle.

38. Remove the upper and lower radiator hoses, then disconnect the heater hoses from the engine.

39. Disconnect the transaxle cooler hoses, then plug the pipes and hoses.

40. Remove the radiator from the vehicle.

41. Disconnect the air conditioning (A/C) compressor electrical connector. Remove the A/C compressor mounting bolts, position the compressor out of the way and support it with a strong wire. Do not disconnect the A/C hoses from the compressor.

42. Attach an engine hoist to the engine lifting points and raise the hoist to remove all slack from the chain.

43. Remove the transaxle mount.

44. Remove the bolts attaching the front mount to the beam.

45. Disconnect the vacuum hose from the rear engine mount control solenoid valve.

46. Remove the bolts attaching the rear mount to the beam.

47. Check that the engine is completely free of vacuum hoses, fuel and coolant hoses, and electrical wiring.

48. Slowly raise the engine approximately 6 in. (150mm). Check once again that all hoses and wires have been disconnected from the engine.

49. Raise the engine all the way and remove it from the vehicle.

50. Remove the transaxle from the engine.

51. Mount the engine on an engine stand, making sure the mounting bolts are tight. If an engine stand is not available, support the engine in an upright position with blocks. Never leave an engine hanging from a lift or hoist.

To install:

52. Install the transaxle.

53. Install the front and rear engine mounts to their mounting brackets and tighten the attaching nuts to 40 ft. lbs. (54 Nm).

54. Lift the engine into position and lower it into the car, aligning the mounts and bushings.

55. Place the power steering pump drive belt over the bracket the side engine mount attaches to.

➡**When installing the engine mounts and vibration dampers in the following** steps, they must be tightened to the correct tension in the correct order if they are to damp vibration properly.

56. Install the rear mount and tighten the bolts attaching the mount to the beam to 43 ft. lbs. (59 Nm).

57. Install the front mount and tighten the bolts attaching the mount to the beam to 43 ft. lbs. (59 Nm).

58. Install the side engine mount. Use three new bolts to attach the mount to the engine, tighten the bolts to 40 ft. lbs. (54 Nm). Do not tighten the through-bolt at this time.

59. Install the transaxle mount and tighten the three nuts attaching the mount to the transaxle to 28 ft. lbs. (38 Nm). Do not tighten the through-bolt at this time.

60. Tighten the side engine mount through-bolt to 47 ft. lbs. (64 Nm).

61. Tighten the transaxle mount through-bolt to 47 ft. lbs. (64 Nm).

62. Remove the engine hoist.

63. Position the A/C compressor on the engine and install the mounting bolts. Tighten the mounting bolts to 16 ft. lbs. (22 Nm), then connect the A/C compressor electrical connector.

64. Install the radiator assembly.

65. Connect the transaxle cooler hoses to the transaxle cooler lines.

66. Install the upper and lower radiator hoses and connect the heater hoses to the engine.

67. Install the halfshafts with new snaprings, be sure the snaprings click into place.

68. Connect the lower ball joints to the lower control arms. Tighten the ball joint nuts to 36–43 ft. lbs. (49–59 Nm), then install a new cotter pin.

69. Install the damper forks, tighten the flange bolt to 32 ft. lbs. (43 Nm). Install the lower bolt and a new locknut, tighten the nut to 47 ft. lbs. (64 Nm).

70. Connect the shift cable to the transaxle and install a new lockwasher to the attaching bolt. Tighten the bolt attaching the shift cable end to the transaxle to 10 ft. lbs. (14 Nm). Install the shift cable cover and tighten the cover attaching bolts to 20 ft. lbs. (26 Nm). Tighten the two bolts attaching the cable housing to the transaxle to 9 ft. lbs. (12 Nm).

71. Install new O-rings to the oil filter base and install the oil filter base to the engine. Tighten the mounting bolts to 16 ft. lbs. (22 Nm).

72. Connect the oil pressure switch terminal and tighten the attaching bolt to 1.8 ft. lbs. (2.5 Nm).

73. Install the oil filter.

74. Install the crankshaft pulley, use the pulley holder when installing and tightening the pulley bolt. Tighten the bolt to 181 ft. lbs. (245 Nm).

75. Install exhaust pipe A with new gaskets and new locknuts. Tighten the nuts attaching the exhaust pipe to the catalytic converter to 25 ft. lbs. (33 Nm). Tighten the nuts attaching the exhaust pipe to the exhaust manifolds to 40 ft. lbs. (54 Nm). Connect the oxygen sensor electrical connector.

76. Install the center beam and tighten the center beam attaching bolts to 37 ft. lbs. (50 Nm).

77. Install the splash shield and the front wheels.

78. Install the power steering pump and connect the inlet hose. Do not tighten the power steering pump mounting bolts and nuts at this time.

79. Install and adjust the power steering pump belt and tighten the mounting nuts to 16 ft. lbs. (22 Nm).

80. Install and adjust the alternator belt, then tighten the alternator mounting nut and bolt to 16 ft. lbs. (22 Nm).

81. Install the A/C belt and adjust the belt tension. Tighten the idler center nut to 33 ft. lbs. (44 Nm).

82. Connect the engine wiring harness connectors on the right side of the engine compartment.

83. Connect the battery cables to the under-hood fuse/relay box and under-hood ABS fuse/relay box.

84. Connect the vacuum hoses to the intake manifold.

85. Connect the EVAP control canister hose and the power brake booster hose to the engine assembly.

86. Connect the fuel return hose to the regulator. Connect the fuel feed hose to the fuel filter with new washers. Tighten the banjo nut to 16 ft. lbs. (22 Nm) and tighten the service bolt to 9 ft. lbs. (12 Nm).

87. Connect the injector resistor connector on the left of the engine compartment.

88. Connect the engine wiring harness connectors on the left side of the engine compartment.

89. Install the strut brace and tighten the strut brace bolts to 16 ft. lbs. (22 Nm). Install the starter cable to the strut brace.

90. Connect the throttle cable and cruise control cables to the throttle linkage, adjust the cable as necessary. Install intake manifold cover B, tighten the attaching bolt to 9 ft. lbs. (12 Nm).

91. Install the intake air duct.

92. Connect the engine ground cable and install the battery base and bracket.

Tighten the base mounting bolts to 16 ft. lbs. (22 Nm).

93. Install the battery and connect the positive, then the negative battery cables. Enter the radio security code.

94. Fill the engine with oil and the transaxle with fluid.

95. Fill and bleed the air from the cooling system.

96. Switch the ignition **ON** but do not engage the starter. The fuel pump should run for approximately 2 seconds, building pressure within the lines. Switch the ignition **OFF**, then **ON** 2 or 3 more times to build full system pressure. Check for fuel leaks.

97. Disconnect the coil wire from the distributor. Insulate or protect the end of the cable so it does not arc to the engine or surrounding metal. Without touching the accelerator, turn the ignition switch to the START position and crank the engine for about 5–10 seconds; this will develop some oil pressure within the motor. Do not exceed 10 seconds cranking.

98. Switch the ignition **OFF** and reconnect the coil.

99. Start the engine, allowing it to idle. Check the hoses and lines carefully for any sign of leakage.

100. Check the timing and idle speed.

101. After the engine has warmed up fully and the fan(s) have come on at least once, recheck the engine for fluid leaks. Switch the engine **OFF**.

1998–99 Accord

2.3L AND 3.0L ENGINES

1. Obtain the anti-theft code for the radio, then disconnect the battery cables. Be sure to disconnect the negative cable first.

2. Remove the air intake duct.

3. Secure the hood in the open position with a long prop rod such as P/N 74145-S84-A00.

4. Detach both battery cables and the connector from the underhood relay box. On the 3.0L engine, remove the battery and tray.

5. Remove the bolt securing the relay box to the body.

6. Remove the accelerator and cruise control cables from the throttle body and bracket.

7. Properly relieve the fuel system pressure.

8. Detach the fuel hoses from the fuel rail.

9. Remove the following hoses:
- Brake booster vacuum
- EVAP canister

- Vacuum hose from the canister
10. Remove the hose securing the power steering hose on the engine.
11. Remove the power steering pump belt, then remove the pump and position it out of the way. Use wire if necessary.
12. Detach the ECM/PCM connectors from the control module. Remove the grommet and pull the connectors through.
13. Detach the wiring harness connectors at the right side of the engine compartment. Left side on the 3.0L engine.
14. On the 2.3L engine, remove the starter cable (A) and clamp (B). Remove the ground cable (C) and back-up light switch connectors (D). On the 3.0L engine, remove the starter wiring from the engine compartment attaching points.
15. On vehicles with a manual transaxle, disconnect the shift and select cables from the transaxle. Remove the slave cylinder mounting bolts and position the cylinder out of the way. Be sure not to bend the line.
16. Remove the rear engine mount through-bolt and stiffener.
17. Remove the front engine mount bracket mounting bolts and loosen the through-bolt.
18. Remove the radiator cap.
19. Raise and safely support the vehicle.
20. Remove the front tires.

21. Remove the engine under cover.
22. Loosen the radiator drain plug and drain the coolant.
23. Drain the transaxle oil or fluid, then reinstall the plug using a new washer.
24. Drain the engine oil, then reinstall the plug using a new washer.
25. Lower the vehicle and remove the upper and lower radiator hoses and heater hoses from the engine.
26. On vehicles with an automatic transaxle, disconnect the ATF fluid cooler lines.
27. Remove the A/C compressor from the engine and position it to the side without disconnecting the hoses.
28. Raise the vehicle and remove the front exhaust pipe.
29. On vehicles with A/T, remove the two bolts (A) for the shift cable holder (B), then remove the shift cable cover (C). To prevent damage to the linkage, be sure to remove the shift cable holder before removing the bolts for the cover.
30. Remove the lockbolt (D) from the control lever (E), then remove the shift cable (F) with the control lever.
31. Remove the through-bolt securing the bottom of the shock absorber to the control arm.

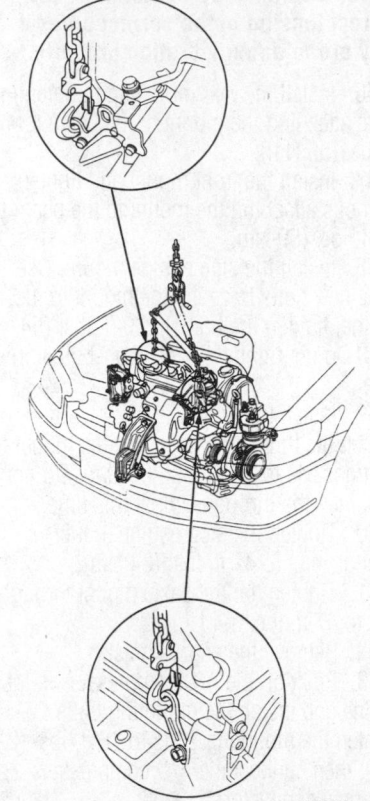
Engine lifting points—1998–99 2.3L Accord

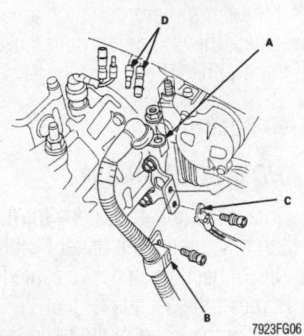
Starter cable, clamp, ground cable and back-up light switch connector locations—1998–99 2.3L Accord

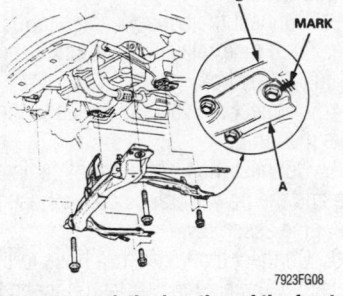
Be sure to mark the location of the front beams (A) on the rear beams (B) before removing the subframe—1998–99 2.3L Accord

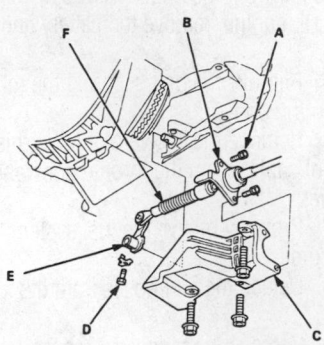
Automatic transaxle linkage components—1998–99 2.3L Accord

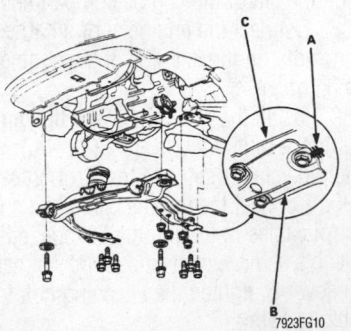
Mark the location of the front beams (A) on the rear beams (B) before removing the subframe—1998–99 3.0L Accord

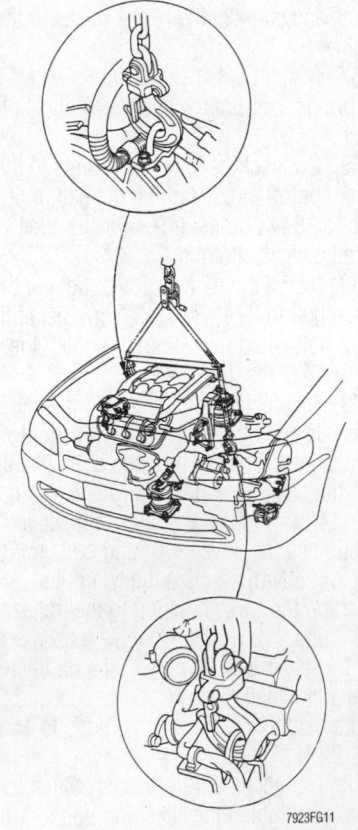
Engine lifting points—1998–99 3.0L Accord

32. Remove the halfshafts.
33. Remove the rear engine mounting bracket.
34. Remove the two flange bolts from each of the radius rods.
35. Mark the location of the front beams (A) on the rear beams (B). Remove the four bolts and the subframe.
36. Lower the vehicle about half way and attach a chain hoist to the engine lifting points as shown. Apply slight upward pressure to the engine/transaxle assembly.
37. Remove the remaining engine and transaxle mounting brackets.
38. Lower the engine about 6 in. (150mm) and check that the engine/transaxle is free of any hoses, cables or wiring.
39. Lower the assembly completely and remove it from under the vehicle.

To install:
40. Lift the engine into position and install the engine mounting brackets. On the 2.3L, tighten the engine mounting bolts and nuts to 40 ft. lbs. (54 Nm). On the 3.0L, tighten the bolts to 28 ft. lbs. (38 Nm).
41. On the 3.0L engine, install the A/C compressor. Tighten the bolts to 16 ft. lbs. (22 Nm).
42. Install the transaxle mounting bracket. On the 2.3L engine, tighten the nuts to 28 ft. lbs. (38 Nm) and the through-bolt to 40 ft. lbs. (54 Nm). On the 3.0L engine, tighten the bolts to 28 ft. lbs. (38 Nm).
43. Install the sub-frame in its original position. On the 2.3L engine, tighten the rear bolts to 47 ft. lbs. (64 Nm) and the front bolts to 76 ft. lbs. (103 Nm). On the 3.0L engine, tighten the rear bolts to 40 ft. lbs. (54 Nm), front bolts to 76 ft. lbs. (103 Nm) and the nuts to 28 ft. lbs. (38 Nm).
44. On the 2.3L engine, do the following:
• Install the radius rod bolts. Tighten them to 119 ft. lbs. (162 Nm).
• Install the rear mount bracket. Tighten the bolts to 40 ft. lbs. (54 Nm).
• On vehicles with manual transaxles, install the stiffener and tighten the through-bolt to 47 ft. lbs. (64 Nm). On vehicles with automatic transaxles, install the stiffener and tighten the nut and bolt to 28 ft. lbs. (38 Nm)
• Tighten the three front mounting bracket bolts to 28 ft. lbs. (38 Nm). Then, tighten the through-bolt to 47 ft. lbs. (64 Nm).
• Install the A/C compressor. Tighten the bolts to 16 ft. lbs. (22 Nm).
45. On the 3.0L engine, do the following:
• Install the radius rod bolts. Tighten them to 119 ft. lbs. (162 Nm).
• Install the front mounting bracket support nut. Tighten it to 40 ft. lbs. (54 Nm).

• Install the rear mounting bracket nut and bolt. Tighten the nut to 40 ft. lbs. (54 Nm) and the bolt to 28 ft. lbs. (38 Nm).
• Install the side mounting bracket. Tighten the bolts to 40 ft. lbs. (54 Nm) and the through-bolt to 40 ft. lbs. (54 Nm).
46. Assemble the exhaust system.
47. If equipped with an automatic transaxle, connect the shift linkage.
48. The remainder of the installation is the reverse of the removal.
49. Refill and bleed the cooling system.

✷✷ WARNING

Operating the engine without the proper amount and type of engine oil will result in severe engine damage.

50. Fill the engine with the correct amount of oil.
51. Install the battery if removed. Start the engine and check for leaks.

Water Pump

REMOVAL & INSTALLATION

I.5L, 1.6L, 2.2L and 2.3L Engines

➡**The original radio contains a coded anti-theft circuit. Obtain the security code number before disconnecting the battery cables.**

1. Disconnect the negative battery cable.
2. Drain the cooling system.
3. Remove the accessory drive belts, the valve cover, and the upper timing belt cover.
4. Set the timing at TDC/compression for No. 1 piston.
5. Remove the crankshaft pulley and lower timing belt cover.
6. Remove the timing belt. Replace the timing belt if it is contaminated with oil or coolant or shows any signs of wear and damage.

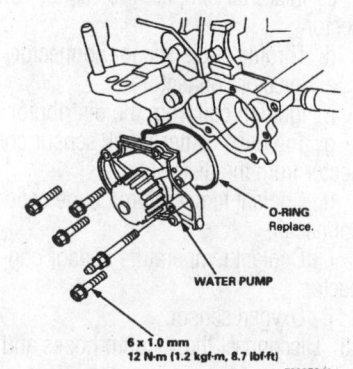

O-RING
Replace.

WATER PUMP

6 x 1.0 mm
12 N·m (1.2 kgf-m, 8.7 lbf-ft)

7923FG12

Water pump—2.2L and 2.3L engines

7. If equipped with a Crankshaft Speed Fluctuation (CKF) sensor at the crankshaft sprocket, unbolt the sensor bracket and move the sensor out of the way. Cover the sensor with a shop towel to keep coolant off of it.
8. Unbolt the water pump and remove it from the engine block. On 1.5L and 1.6L engines, the top right water pump mounting bolt also secures the alternator adjusting bracket. Leave the bracket attached to the alternator.

To install:
9. Clean the water pump and O-ring mating surfaces before installation.
10. Install the water pump with a new O-ring. Coat only the bolt threads with liquid gasket and tighten them to 9 ft. lbs. (12 Nm). On 1.5L and 1.6L engines, tighten the bracket bolt to 33 ft. lbs. (44 Nm).
11. Install the timing belt. Be sure it is fitted and adjusted properly.
12. If equipped, install the CKF sensor and tighten the bracket bolts to 9 ft. lbs. (12 Nm).
13. Install the lower belt cover and crankshaft pulley.
14. Install the upper timing belt cover, the valve cover, and the accessory drive belts.
15. Be sure the cooling system drain plug is closed. Refill and bleed the cooling system.
16. Connect the negative battery cable and enter the radio security code.
17. Start the engine, allow it to reach normal operating temperature, and check for coolant leaks Check the tensions of the accessory belts.
18. If equipped with 4WS, turn the steering wheel lock-to-lock to reset the 4WS control unit.

2.7L Engine

1. Disconnect the negative battery cable.
2. Drain the coolant into a sealable container.
3. Remove the timing belt covers and the timing belt.
4. Remove the timing belt tensioner.
5. Remove the nine water pump bolts, take note of their locations for reinstallation.
6. Remove the water pump from the engine and discard the O-ring. Remove the dowel pins.
To install:
7. Clean the water pump mounting surface and O-ring groove, then install the dowel pins to the engine.

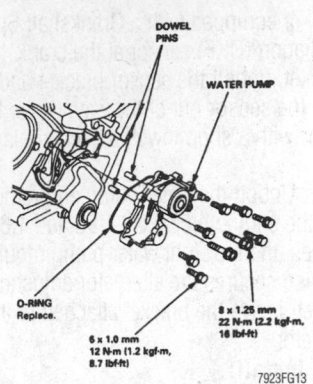

Water pump—2.7L engine

8. Install a new O-ring to the engine, then install the water pump, be careful not to pinch the O-ring. Install the mounting bolts to their original locations and when tightening the bolts, be sure that the O-ring does not bulge out of the groove. Tighten the six 1.0mm bolts to 9 ft. lbs. (12 Nm), and tighten the eight 1.25mm bolts to 16 ft. lbs. (22 Nm).

9. Inspect the water pump, making sure that the pump turns freely.

10. Install the timing belt tensioner.

11. Install the timing belt and timing belt covers.

12. Refill and bleed the air from the cooling system.

13. Connect the negative battery cable and enter the radio security code.

3.0L Engine

1. Remove the timing belt.

2. Remove the timing belt tensioner.

3. Remove the five water pump mounting bolts, then remove the pump and seal.

To install:

4. Clean the seal groove and mating surfaces.

5. Using a new seal, install the water pump. Tighten the bolts to 8.7 ft. lbs. (12 Nm).

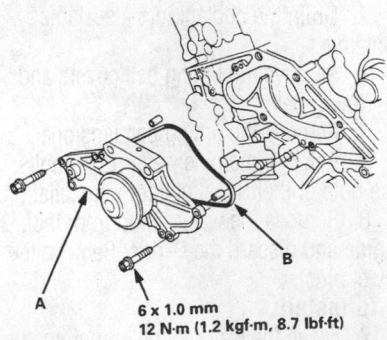

Exploded view of the water pump mounting—3.0L engine

6. Install the timing belt tensioner.

7. Install the timing belt.

8. Refill the cooling system.

9. Start the engine and check for leaks.

10. Top off the cooling system if necessary after the engine has cooled.

Cylinder Head

REMOVAL & INSTALLATION

➥**The radio may contain a coded theft protection circuit. Always obtain the code number before disconnecting the battery. If the vehicle is equipped with 4WS, the steering control unit is shut down when the battery is disconnected. After connecting the battery, turn the steering wheel lock-to-lock to reset the steering control unit.**

1995 Civic and Del Sol

1. The engine should be cold before the cylinder head is removed.

2. Disconnect the negative battery cable. Drain the cooling system.

3. Remove the brake booster vacuum hose from the brake master cylinder power booster. Remove the engine secondary ground cable from the valve cover.

4. Remove the air intake hose and the air chamber. Relieve the fuel pressure. Disconnect the fuel hoses and fuel return hose.

5. Remove the air intake hose and resonator hose. Disconnect the throttle cable at the throttle body. On vehicles equipped with automatic transaxles, disconnect the throttle control cable at the throttle body.

6. Disconnect the charcoal canister hose at the throttle valve.

7. Disconnect the following engine wire connectors from the cylinder head and the intake manifold:

  a. 14 prong connector from the main wiring harness

  b. EACV connector

  c. Intake air temperature sensor connector

  d. Throttle angle sensor connector

  e. Injector connectors

  f. Ignition coil from the distributor

  g. Top dead center/crank sensor connector from the distributor.

  h. Coolant temperature gauge sender connector.

  i. Coolant temperature sensor connector.

  j. Oxygen sensor.

8. Disconnect the vacuum hoses and the water bypass hoses from the intake manifold and throttle body.

9. Remove the upper radiator hose and the heater hoses from the cylinder head.

10. Remove the PCV hose, charcoal canister hose and vacuum hose from the intake manifold, and remove the vacuum hose from the brake master cylinder power booster.

11. Loosen the air conditioning idler pulley and remove the air conditioning belt. Remove the alternator belt. If equipped with power steering, remove the power steering belt and pump bracket.

12. Remove the intake manifold bracket and the exhaust manifold bracket.

13. Remove the exhaust manifold shroud, then remove the exhaust manifold.

14. Mark the position of the distributor in relation to the engine block, remove and tag the spark plug wires and remove the distributor assembly.

15. Remove the valve cover. Remove the timing belt cover.

16. Mark the direction of rotation on the timing belt. Loosen the timing belt adjuster bolt 180°, then remove the timing belt from the camshaft pulley. Retighten the adjuster bolt to 33 ft. lbs. (45 Nm).

➥**Do not crimp or bend the timing belt more than 90 degrees or less than 1 in. (25mm) in diameter (width).**

17. Remove the cylinder head bolts in the reverse order of the tightening sequence. Once the bolts are all removed, remove the cylinder head along with the intake manifold from the engine. Remove the intake manifold from the cylinder head. If the head sticks to the engine block, tap it with a plastic or wooden mallet, or pry it loose with a large, flat screwdriver. Leverage slots are located at each end of the back side of the head.

To install:

➥**Use new O-ring, seals, and gaskets when installing the cylinder head and its components.**

18. Be sure the cylinder head and the engine block surfaces are clean, level, and straight.

19. Be sure the cylinder head dowel pins and control jet are aligned. Clean the oil control orifice and reinstall it with a new O-ring.

20. Install the cylinder head onto the engine with new head gasket.

21. Be sure the **UP** mark on the timing belt pulley is at the top.

22. Install the intake manifold and tighten the nuts in a crisscross pattern in 2–3 steps to 17 ft. lbs. starting with the inner nuts.

23. Install the bolts that secure the intake manifold to its bracket but do not tighten them at this point.

24. Position the cam so the TDC marks align, and install the cylinder head bolts.

25. Tighten the cylinder head bolts in 2 steps. On the first step tighten all the bolts, in sequence, to 22 ft. lbs. (30 Nm). On the final step, using the same sequence, tighten the bolts to 47 ft. lbs. (65 Nm) for D15B7 and D15B8 engines. On D16Z6 and D15Z1 engine, the final torque for the head bolts is 53 ft. lbs. (73 Nm).

26. Install the exhaust manifold and tighten the nuts in a crisscross pattern in two or three steps to 25 ft. lbs. (34 Nm) starting with the inner nuts.

27. Install the exhaust manifold to the head pipe. Tighten the bolts to the intake manifold bracket. Install the header pipe on to its bracket.

28. Install the timing belt, timing cover, and crankshaft pulley. Install the upper timing cover and valve cover. Coat the valve cover spark plug seals with oil before installation.

29. Reconnect the throttle cable. Adjust its tension so the cable has a deflection of 10–12mm (0.39–0.47 in.).

30. Install the power steering pump bracket and tighten the bolts to 33 ft. lbs. (45 Nm). Tighten the power steering pump mounting bolt to 17 ft. lbs. (24 Nm).

31. Install the distributor. Install new spark plugs and reconnect the spark plug wires.

32. After the installation procedure is complete, check that all tubes, hoses, and connectors are installed correctly. Check the tension of the accessory drive belts.

33. Adjust the valve clearance and ignition timing.

34. Change the engine oil and filter.

35. Refill the cooling system with fresh coolant and bleed the system.

36. Reconnect the negative battery cable.

37. Run the engine and road test the vehicle.

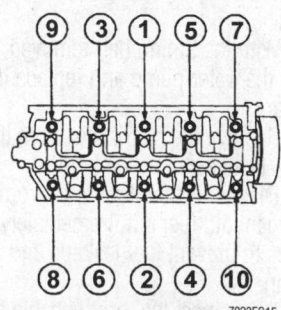

CYLINDER HEAD BOLTS TORQUE SEQUENCE:

7923FG15

Cylinder head bolt tightening sequence— 1995–97 Civic and Del Sol

1996–97 Civic and Del Sol

1.6L (D16Y5, D16Y7 AND D16Y8) ENGINES

1. Be sure the cylinder head is cool to the touch before beginning the removal procedure. The coolant temperature must be below 100° F (38° C).

2. Disconnect the negative battery cable.

3. Drain the cooling system.

4. Label and disconnect the ignition wires.

5. Remove the air intake duct and the air cleaner assembly.

6. Relieve the fuel pressure.

7. Clean up any fuel that may have spilled on the engine or intake manifold.

8. Disconnect the upper radiator hose from the coolant inlet.

9. Disconnect the coolant bypass hoses and the heater hose from the intake manifold.

10. Loosen the power steering pump mounting bolts to release the belt tension. Remove the power steering pump belt.

11. Remove the power steering pump from its mounting bracket and lift the power steering reservoir from its mount. Move the pump and reservoir out of the work area and secure them. Don't disconnect the hydraulic lines.

12. Place a block of wood on the pad of a floor jack. Place the floor jack under the engine for support.

13. If equipped with A/C: unbolt the left-front engine mount bracket.

14. Loosen the A/C compressor idler pulley bolt. Then, loosen the adjusting bolt to release the belt's tension. Slip the A/C compressor belt around the engine mount to remove it.

15. Loosen the alternator mounts;, then, remove the alternator belt.

16. Be sure the engine is supported with the padded floor jack. Loosen the nuts from left side engine mount. Remove the engine mount bracket.

17. Remove the valve cover and the upper timing belt cover.

18. Remove the crankshaft pulley and the lower timing belt cover. Separate the dipstick tube from its catches on the timing cover. Remove the timing belt.

19. With the timing belt removed, inspect the water pump and replace it if necessary.

20. Remove the distributor from the cylinder head as an assembly.

21. Unbolt and remove the camshaft sprocket.

22. Disconnect the fuel lines from the intake manifold fuel rail. Immediately plug the lines to prevent fuel leakage and contamination.

23. Disconnect the throttle cable from the linkage by first loosening its locknut, then slipping it out of its holder.

24. Label and disconnect the following engine harness connectors from the cylinder head and the intake manifold:

   a. Fuel injector wiring harness connectors

   b. VTEC solenoid valve and pressure switch connectors (D16Y5, D16Y8 engines only)

   c. Idle air control valve (IAC) connector

   d. Throttle position sensor (TPS) connector

   e. EGR valve lift sensor connectors (D16Y5 engine only)

   f. Engine coolant temperature sensor, switch, and gauge sender (ECT) connectors

   g. Manifold absolute pressure sensor (MAP) connector

   h. Primary and secondary (D16Y5, D16Y7 engines only) oxygen sensor (HO2S) connectors

25. Label and disconnect the vacuum hoses and PCV hose from the intake manifold and throttle body.

26. Disconnect the charcoal canister (EVAP) and breather hoses from the intake manifold.

27. Remove the intake manifold together with the throttle body and plenum.

28. Remove the exhaust manifold.

29. Remove the power steering pump bracket.

30. Loosen the cylinder head bolts in a three-step crisscross pattern in the reverse order of the tightening sequence. Start with the outermost bolts and work toward the middle of the cylinder head. Loosen the bolts in the reverse order of installation.

31. Remove the cylinder head. If the head sticks to the engine block, tap it with a plastic or wooden mallet.

32. Inspect the cylinder head for warpage and cracking. Repair, machine, or replace as necessary. The warpage limit is 0.002 in. (0.05mm). Standard cylinder head height is 3.659–3.663 in. (92.95–93.05mm).

33. Remove the old cylinder head gasket and thoroughly clean the mating surfaces.

34. Cover the engine block with a sheet of plastic to keep out dust and foreign objects.

To install:

➡**Use new O-ring, seals, and gaskets when installing the cylinder head and its components.**

35. Be sure the cylinder head and the engine block surfaces are clean, level, and straight.

36. Be sure the cylinder head dowel pins and control orifice are aligned. Clean the oil control orifice and reinstall it with a new O-ring.

37. Install a new head gasket onto the engine block.

38. If the camshaft was removed, reinstall it with the keyway facing up so that the engine will remain at TDC/compression for the No. 1 cylinder. Lubricate and install a new camshaft seal.

39. Use new cylinder head bolts and washers. Used or previously-tightened bolts may be stretched, and therefore they have reduced clamping and sealing power under compression. Apply clean engine oil to the threads of each head bolt.

40. Fit the cylinder head into place. Hand-tighten all the cylinder head bolts.

41. Tighten the cylinder head bolts to their final torque specification in four steps. Use a crisscross sequence starting with the bolts at the middle of the head and working toward the outer bolts.

   a. Step 1: Tighten each bolt to 14 ft. lbs. (20 Nm).

   b. Step 2: Tighten each bolt to 36 ft. lbs. (49 Nm).

   c. Step 3: Tighten each bolt to 49 ft. lbs. (67 Nm).

   d. Step 4: Tighten only the two center bolts to an additional 49 ft. lbs. (67 Nm).

42. Apply oil to the camshaft sprocket bolt. Install the sprocket with the UP mark and the keyway pointing straight up. Tighten the sprocket bolt to 27 ft. lbs. (37 Nm).

43. Install the intake manifold with a new gasket, and tighten the nuts in a crisscross pattern in 2–3 steps to 17 ft. lbs. (24 Nm) starting with the inner nuts.

44. Install the bolts that secure the intake manifold to its bracket and tighten them to 17 ft. lbs. (24 Nm).

45. Install the power steering pump bracket and tighten its bolts to 33 ft. lbs. (44 Nm).

46. Install the exhaust manifold with a new gasket. Apply anti-seize paste to the studs, and tighten the nuts to 23 ft. lbs. (31 Nm) in a crisscross sequence.

47. Connect the exhaust manifold to the front exhaust pipe. Tighten the self-locking nuts to 25 ft. lbs. (33 Nm). On vehicles with the D16Y8 engine, tighten the nuts to 40 ft. lbs. (55 Nm).

48. Verify that the engine is at TDC/compression for the No. 1 cylinder.

49. Install the timing belt. After the timing belt has been properly tensioned, Tighten the adjusting bolt to 33 ft. lbs. (44 Nm).

50. Install the lower timing belt cover. Install the crankshaft pulley and tighten its bolt to 134 ft. lbs. (181 Nm). Fit the dipstick tube back into its catches.

51. Adjust the valves. If equipped with a VTEC engine, also check the rocker arms for free and smooth motion.

52. If equipped with a VTEC engine, remove the VTEC solenoid valve and its filter. Install a new filter, then reinstall the VTEC solenoid valve and tighten its bolt to 9 ft. lbs. (12 Nm).

53. Install the distributor. The lugs on the distributor drive fit into the groove on the end of the camshaft. Don't fully tighten the distributor mounting bolts yet.

54. Be sure all the spark plug tube sealing gaskets are fully seated.

55. Install a new gasket onto to the valve cover. Apply liquid gasket to the corner recesses of the gasket. Don't let the sealant to cure before installing the valve cover onto the cylinder head.

56. Install the valve cover. Gently wiggle the valve cover to be sure it is fully seated. Tighten the valve cover bolts in a crisscross pattern to 7 ft. lbs. (10 Nm).

57. Install new spark plugs.

58. Reconnect the ignition wires.

59. Reconnect the upper radiator hose, heater hoses, and intake manifold coolant bypass hoses.

60. Reconnect the intake manifold vacuum lines, PCV, EVAP canister, and breather hoses.

61. Connect the fuel lines to the fuel rail. Use new sealing washers on the banjo fitting. Carefully tighten the banjo fitting to 21 ft. lbs. (28 Nm) for the D16Y5 engine, or to 16 ft. lbs. (22 Nm) for all other engines. Tighten the service bolt to 9–11 ft. lbs. (12–15 Nm).

62. Reconnect the throttle cable. Adjust its tension so the cable has a deflection of 10–12mm (0.39–0.47 in.).

63. Installation of the remaining components is the reverse of removal.

1.6L (B16A2 AND B16A3) ENGINES

1. Before beginning the cylinder head removal procedure, be sure the engine temperature is below 100° F (38° C). To prevent warping, the cylinder head should be removed when the engine is cold.

2. Disconnect the negative battery cable.

3. Label and disconnect the ignition wires.

4. Drain the engine coolant. Remove the radiator cap to speed draining.

5. Remove the strut brace.

6. Remove the intake air duct and disconnect the breather hose.

7. Relieve the fuel pressure:

   a. Loosen the fuel filler cap.

   b. Hold the fuel filter banjo bolt with a back-up wrench. Hold the fuel filter service bolt with a box end wrench.

   c. Place a shop rag over the fuel filter to absorb fuel spray.

   d. Slowly loosen the fuel filter service bolt one complete turn.

8. Clean up any fuel that may have spilled on the engine or intake manifold.

9. Disconnect the upper radiator hose from the coolant inlet.

10. Disconnect the coolant bypass hoses and the heater hose from the intake manifold.

11. Loosen the power steering pump mounting bolts to release the belt tension. Remove the power steering pump belt.

12. Remove the power steering pump from its mounting bracket and lift the power steering reservoir from its mount. Move the pump and reservoir out of the work area and secure them. Don't disconnect the hydraulic lines.

13. Place a block of wood on the pad of a floor jack. Place the floor jack under the engine for support.

14. If equipped with A/C: unbolt the left-front engine mount bracket.

15. Loosen the A/C compressor idler pulley bolt. Then, loosen the adjusting bolt to release the belt's tension. Slip the A/C compressor belt around the engine mount to remove it.

16. Loosen the alternator mounts;, then, remove the alternator belt.

17. Be sure the engine is supported with the padded floor jack. Loosen the left side engine mount nuts. Remove the engine mount bracket.

18. Remove the valve cover and the upper timing belt cover.

19. Remove the crankshaft pulley and the lower timing belt cover. Remove the timing belt.

20. With the timing belt removed, inspect the water pump and replace it if necessary.

21. Remove the distributor from the cylinder head as an assembly.

22. Disconnect the fuel lines from the intake manifold fuel rail. Immediately plug the lines to prevent fuel leakage and contamination.

23. Disconnect the throttle cable from the linkage by first loosening its locknut, then slipping it out of its holder.

24. Label and disconnect the following engine harness connectors from the cylinder head and the intake manifold:

    a. Fuel injector wiring harness connectors

    b. VTEC solenoid valve and pressure switch connectors

    c. Idle air control valve (IAC) connector

    d. Throttle position sensor (TPS) connector

    e. Engine coolant temperature sensor, switch, and gauge sender (ECT) connectors

    f. Manifold absolute pressure sensor (MAP) connector

    g. Primary oxygen sensor (HO2S) connector

25. Label and disconnect the vacuum hoses and PCV hose from the intake manifold and throttle body.

26. Disconnect the charcoal canister (EVAP) and breather hoses from the intake manifold.

27. Loosen the intake manifold nuts in a crisscross sequence. Then, remove the intake manifold together with the throttle body and plenum.

28. Remove the exhaust manifold heat shield. Then, loosen the exhaust manifold nuts in a crisscross sequence. Remove the exhaust manifold. Be careful not to damage the oxygen sensors when removing the manifold. Cover the front exhaust pipe flange with a shop towel to keep dirt out.

29. Remove the power steering pump bracket.

30. Remove the camshaft pulleys and back cover.

31. Loosen the camshaft holder plate bolts in a crisscross sequence working toward the middle of the cylinder head.

32. Loosen the valve adjusting screws.

33. Lift the camshaft holder plates and holders from the cylinder head. The holder bolts will keep the components together. Note the positions of each camshaft holder for reassembly.

34. Lift the camshafts from the cylinder head. Mark the exhaust and intake camshafts so that they will not be confused.

35. Loosen the cylinder head bolts in a three-step crisscross pattern. Start with the outermost bolts and work toward the middle of the cylinder head.

36. Remove the cylinder head. If the head sticks to the engine block, tap it with a plastic-faced or wooden mallet.

37. Inspect the cylinder head for warpage and cracking. Repair, machine, or replace as necessary. The warpage limit is 0.002 in. (0.05mm). Standard cylinder head height is 5.589–5.593 in. (141.95–142.05mm).

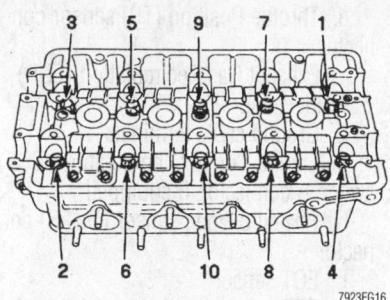

Cylinder head bolt loosening sequence—1.6L (B16A2 and B16A3) engines

To install:

➡ **Use new O-ring, seals, and gaskets when installing the cylinder head and its components.**

38. Be sure the cylinder head and the engine block surfaces are clean, level, and straight.

39. Be sure the cylinder head dowel pins and oil control orifice are aligned. Clean the oil control orifice and reinstall it with a new O-ring.

40. Install a new head gasket onto the engine block.

41. Use new cylinder head bolts and washers. Used or previously-tightened bolts may be stretched; and therefore, they have reduced clamping and sealing power under compression. Apply clean engine oil to the threads of each head bolt.

42. Fit the cylinder head into place. Hand-tighten all the cylinder head bolts.

43. Tighten the cylinder head bolts to their final torque specification in two steps. Use a crisscross sequence starting with the bolts at the middle of the head and working toward the outer bolts.

    a. Step 1: Tighten each bolt to 22 ft. lbs. (30 Nm).

    b. Step 2: Tighten each bolt to 61 ft. lbs. (85 Nm).

44. Install the dowel pin in the No. 3 cylinder head camshaft holder with a new O-ring.

45. Thoroughly clean the intake and exhaust camshaft oil control orifices. Reinstall them with new O-rings.

46. Install the camshafts.

47. Install the intake manifold with a new gasket, and tighten the nuts in a crisscross pattern in 2–3 steps to 17 ft. lbs. (24 Nm) starting with the inner nuts.

48. Install the bolts that secure the intake manifold to its bracket and tighten them to 17 ft. lbs. (24 Nm).

49. Install the power steering pump bracket and tighten its bolts to 33 ft. lbs. (44 Nm).

50. Install the exhaust manifold with a new gasket. Apply anti-seize paste to the studs, and tighten the nuts to 23 ft. lbs. (31 Nm) in a crisscross sequence. Tighten the exhaust manifold bracket bolts to 17 ft. lbs. (24 Nm).

51. Connect the exhaust manifold to the front exhaust pipe. Tighten the self-locking nuts to 40 ft. lbs. (55 Nm).

52. Verify that the engine is at TDC/compression for the No. 1 cylinder.

53. Install the timing belt. After the timing belt has been properly tensioned, Tighten the adjusting bolt to 40 ft. lbs. (55 Nm).

54. Install the lower timing belt cover. Install the crankshaft pulley and tighten its bolt to 130 ft. lbs. (180 Nm).

55. Adjust the valves.

56. Inspect the VTEC rocker arms for free and smooth motion.

57. Remove the VTEC solenoid valve and its filter. Install a new filter, then reinstall the VTEC solenoid valve and tighten its bolts to 9 ft. lbs. (12 Nm).

58. Install the distributor. The lugs on the distributor drive fit into the groove on the end of the intake camshaft. Don't fully-tighten the distributor mounting bolts yet.

59. Be sure all the spark plug tube sealing gaskets are fully seated.

60. Install a new gasket onto to the valve cover. Apply liquid gasket to the corners of the gasket that meet the camshaft holders. Don't let the sealant cure before installing the valve cover onto the cylinder head.

61. Install the valve cover. Gently wiggle the valve cover to be sure it is fully seated. Tighten the valve cover bolts in a crisscross pattern to 7 ft. lbs. (10 Nm).

62. Install new spark plugs.

63. Reconnect the ignition wires.

64. Reconnect the upper radiator hose, heater hoses, and intake manifold coolant bypass hoses.

65. Reconnect the intake manifold vacuum lines, PCV, EVAP canister, and breather hoses.

66. Connect the fuel lines to the fuel rail. Use new sealing washers on the banjo fitting. Carefully tighten the banjo fitting to 25 ft. lbs. (33 Nm). Tighten the service bolt to 11 ft. lbs. (15 Nm).

67. Reconnect the throttle cable. Adjust its tension so the cable has a deflection of 10–12mm (0.39–0.47 in.).

68. Installation of the remaining components is the reverse of removal.

69. After the installation procedure is complete, check that all tubes, hoses, and connectors are installed correctly.

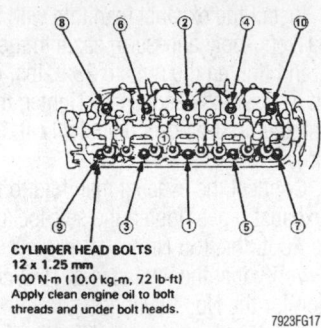

CYLINDER HEAD BOLTS
12 x 1.25 mm
100 N·m (10.0 kg-m, 72 lb-ft)
Apply clean engine oil to bolt
threads and under bolt heads.

7923FG17

Cylinder head bolt tightening sequence— 1.6L (B16A2 and B16A3) and 2.2L (F22A1) engines

2.2L (F22A1) ENGINES

1. Disconnect the negative battery cable.
2. Bring the No. 1 cylinder to TDC.
3. Drain the engine coolant into a sealable container.
4. Relieve the fuel system pressure.
5. Remove the vacuum hose, breather hose and air intake duct.
6. Remove the water bypass hose from the cylinder head.
7. Disconnect the fuel feed and return hose from the fuel rail.
8. Remove the evaporative emissions (EVAP) control canister hose from the intake manifold.
9. Remove the brake booster vacuum hose from the intake manifold. On automatic transaxle equipped vehicles, remove the vacuum hose mount.
10. Remove the throttle cable from the throttle body. On automatic transaxle equipped vehicles, remove the throttle control cable at the throttle body.

➡**Be careful not to bend the cable when removing. Do not use pliers to remove the cable from the linkage. Always replace a kinked cable with a new one.**

11. Remove the ignition coil.
12. Label, then disconnect the electrical connectors from the distributor and the spark plug wires from the spark plugs. Mark the position of the distributor and remove it from the cylinder head. Disconnect the ignition coil wire from the distributor.
13. Remove the connector and the terminal from the alternator, then remove the engine wiring harness from the valve cover.
14. Disconnect the following engine wiring harness connectors:
   a. Fuel injector connectors
   b. Intake Air Temperature (IAT) sensor connector, if equipped
   c. Idle Air Control (IAC) valve connector

   d. Throttle Position (TP) sensor connector
   e. Exhaust Gas Recirculation (EGR) valve lift sensor
   f. Ground cable terminals
   g. Engine Coolant Temperature (ECT) switch B connector, if equipped
   h. Heated oxygen sensor (HO2S) connector
   i. ECT sensor
   j. ECT gauge sending unit connector
   k. CKP/TDC/CYP sensor connector, if equipped
   l. Vehicle Speed Sensor (VSS) connector
   m. ECT switch A connector
15. Remove the upper radiator hose and the heater inlet hose from the cylinder head.
16. Remove the lower radiator hose and heater outlet hose from the intake manifold.
17. Remove the bypass hose from the thermostat housing and intake manifold.
18. Remove the thermostat housing mounting bolts. Remove the thermostat housing from the intake manifold and the connecting pipe, by pulling and twisting the housing. Discard the O-rings.
19. Tag, then disconnect the emissions vacuum hoses from the intake manifold assembly.
20. Disconnect the cruise control actuator electrical connector and the vacuum tube, then remove the cruise control actuator.
21. Remove the engine ground cable from the body.
22. Remove the mounting bolts and drive belt from the power steering pump. Pull the pump away from the mounting bracket, without disconnecting the hoses. Support the pump out of the way.
23. Raise and safely support the vehicle.
24. Remove the front wheel and tire assemblies.
25. Remove the splash shield.
26. Remove the intake manifold bracket bolts.
27. Remove the intake manifold.
28. Disconnect the exhaust pipe from the exhaust manifold.
29. Remove the exhaust manifold and the exhaust manifold heat insulator.
30. Remove the power steering pump mounting bracket.
31. Remove the Positive Crankcase Ventilation (PCV) hose, then remove the cylinder head cover. Replace the rubber seals if damages or deteriorated.
32. Remove the timing belt.
33. Remove the cylinder head bolts in the reverse order of installation.

➡**To prevent warpage, unscrew the bolts in sequence ⅓ turn at a time. Repeat the sequence until all bolts are loosened.**

34. Separate the cylinder head from the engine block with a suitable flat bladed pry-tool.

To install:

35. Be sure all cylinder head and block gasket surfaces are clean. Check the cylinder head for warpage. If warpage is less than 0.002 in. (0.05mm), cylinder head resurfacing is not required. Maximum resurface limit is 0.008 in. (0.2mm) based on a cylinder head height of 3.94 in. (100mm).
36. Always use a new head gasket.
37. The **UP** mark on the camshaft pulley should be at the top.
38. Be sure the No. 1 cylinder is at TDC.
39. Clean the oil control orifice and install a new O-ring. Install and align the cylinder head dowel pins and oil control jet.
40. Install the bolts that secure the intake manifold to it's bracket but do not tighten them.
41. Position the camshaft correctly.
42. Install the cylinder head, then tighten the cylinder head bolts sequentially in 3 steps:
   • Step 1: 29 ft. lbs. (40 Nm).
   • Step 2: 51 ft. lbs. (70 Nm).
   • Step 3: 72 ft. lbs. (100 Nm).
43. Install the intake manifold and tighten the nuts in a crisscross pattern, in 2–3 steps, beginning with the inner nuts. Final torque should be 16 ft. lbs. (22 Nm). Always use a new intake manifold gasket.
44. Connect the intake manifold bracket to the intake manifold. Tighten the bolt to 16 ft. lbs. (22 Nm).
45. Install the heat insulator to the cylinder head and the block.
46. Install the power steering pump mounting bracket to the cylinder head. Tighten the two 10x1.25mm bolts to 36 ft. lbs. (50 Nm). Torque the 8x1.25 bolt to 16 ft. lbs. (22 Nm).
47. Install the exhaust manifold and tighten the nuts in a crisscross pattern in 2–3 steps, beginning with the inner nut. Final torque should be 23 ft. lbs. (32 Nm). Always use a new exhaust manifold gasket.
48. Install the exhaust manifold bracket, then install the exhaust pipe, bracket and upper shroud.
49. Be sure the camshaft sprocket and the crankshaft pulleys are aligned to TDC. Install the timing belt.
50. Install the splash shield and the front wheels.

51. Lower the vehicle.
52. Check and adjust the valves, as necessary.
53. Tighten the crankshaft pulley bolt to 181 ft. lbs. (250 Nm).
54. Installation of the remaining components is the reverse of removal.
55. Connect the negative battery cable and enter the radio security code.
56. Start the engine, checking carefully for any leaks.
57. Check the ignition timing and tighten the distributor bolts to 13 ft. lbs. (18 Nm).
58. If equipped with 4WS, turn the steering wheel lock-to-lock to reset the 4WS control unit.

2.3L (H23A1) ENGINE

1. Disconnect the negative battery cable.
2. Turn the crankshaft so the No. 1 piston is at Top Dead Center (TDC).

➡**The No. 1 piston is at top dead center when the pointer on the block aligns with the white painted mark on the flywheel (manual transaxle) or driveplate (automatic transaxle).**

3. Drain the engine coolant into a sealable container.
4. Relieve the fuel system pressure.
5. Remove the air intake duct.
6. Remove the evaporative emissions (EVAP) control canister hose from the intake manifold.
7. Remove the throttle cable from the throttle body. On automatic transaxle equipped vehicles, remove the throttle control cable at the throttle body.

➡**Be careful not to bend the cable when removing. Always replace a kinked cable with a new one.**

8. Disconnect the fuel feed and return hose.
9. Remove the brake booster vacuum hose from the intake manifold.
10. Disconnect the following engine wiring harness connectors:
    a. Fuel injector connectors
    b. Intake Air Temperature (IAT) sensor connector
    c. Idle Air Control (IAC) valve connector
    d. Throttle Position (TP) sensor connector
    e. Exhaust Gas Recirculation (EGR) valve lift sensor
    f. Ground cable terminals
    g. Engine Coolant Temperature (ECT) switch B connector
    h. Heated oxygen sensor (HO2S) connector

i. ECT sensor
j. ECT gauge sending unit connector
k. Ignition Control Module (ICM) connector
l. CKP/TDC/CYP sensor connector
m. Vehicle Speed Sensor (VSS) connector
n. Ignition coil connector
o. Intake air bypass solenoid valve connector
p. ECT switch A connector
q. Knock sensor connector
11. Remove the engine ground cable from the cylinder head cover.
12. Remove the connector and the terminal from the alternator, then remove the engine wiring harness from the valve cover.
13. Remove the mounting bolts and drive belt from the power steering pump. Pull the pump away from the mounting bracket, without disconnecting the hoses. Support the pump out of the way.
14. Remove the ignition coil.
15. Tag, then disconnect the emissions vacuum hoses from the intake manifold assembly.
16. Remove the bypass hose from the intake manifold.
17. Remove the upper radiator hose and the heater hose from the cylinder head.
18. Remove the lower radiator hose and bypass hose from the thermostat housing.
19. Remove the thermostat housing mounting bolts. Remove the thermostat housing from the intake manifold and the connecting pipe, by pulling and twisting the housing. Discard the O-rings.
20. Raise and safely support the vehicle.
21. Remove the front wheel and tire assemblies.
22. Remove the splash shield.
23. Remove the intake manifold bracket bolts.
24. Remove the intake manifold.
25. Disconnect the exhaust pipe from the exhaust manifold.
26. Remove the exhaust manifold and the exhaust manifold heat insulator.
27. Label, then disconnect the electrical connectors from the distributor and the spark plug wires from the spark plugs. Mark the position of the distributor and remove it from the cylinder head. Disconnect the ignition coil wire from the distributor.
28. Remove the Positive Crankcase Ventilation (PCV) hose, then remove the cylinder head cover. Replace the rubber seals if damaged or deteriorated.
29. Remove the timing belt.
30. Insert a 5.0mm pin punch in each of the camshaft caps, nearest to the sprockets, through the holes provided. Remove the

camshaft sprocket attaching bolts, then remove the sprockets. Do not lose the sprocket keys.
31. Loosen all of the rocker arm adjusting screws, then remove the pin punches from the camshaft caps.
32. Remove the camshaft holders, note the holders locations for ease of installation.
33. Remove the rubber cap from the head, located at the end of the intake camshaft.
34. Remove the rocker arms from the cylinder head. Note the locations of the rocker arms.

➡**The rocker arms have to be installed to their original locations if being reused.**

35. Remove the side engine mount bracket B, then the back cover from behind the camshaft sprockets.
36. Remove the cylinder head bolts in the proper sequence.

➡**To prevent warpage, unscrew the bolts in sequence ⅓ turn at a time. Repeat the sequence until all bolts are loosened.**

37. Separate the cylinder head from the engine block with a suitable flat bladed pry-tool.

To install:

38. Be sure all cylinder head and block gasket surfaces are clean. Check the cylinder head for warpage. If warpage is less than 0.002 in. (0.05mm), cylinder head resurfacing is not required. Maximum resurface limit is 0.008 in. (0.2mm) based on a cylinder head height of 5.20 in. (132.0mm).
39. Always use a new head gasket.
40. Be sure the No. 1 cylinder is at TDC.
41. Clean the oil control orifice and install a new O-ring. The cylinder head dowel pins and oil control jet must be aligned.

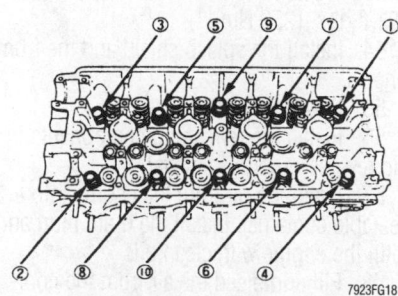

Cylinder head bolt removal sequence— 2.3L (H23A1) engine

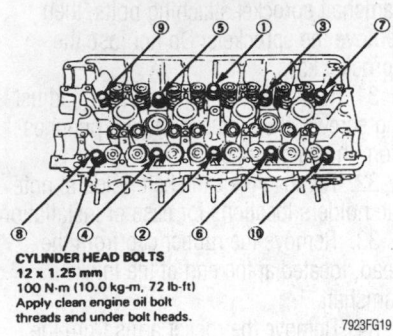

CYLINDER HEAD BOLTS
12 x 1.25 mm
100 N·m (10.0 kg-m, 72 lb-ft)
Apply clean engine oil bolt
threads and under bolt heads.

7923FG19

Cylinder head bolt torque sequence—2.3L (H23A1) engine

42. Install the bolts that secure the intake manifold to it's bracket but do not tighten them.

43. Install the cylinder head, then tighten the cylinder head bolts sequentially in 3 steps:
- Step 1: 29 ft. lbs. (40 Nm).
- Step 2: 51 ft. lbs. (70 Nm).
- Step 3: 72 ft. lbs. (100 Nm).

➡**A beam type torque wrench is recommended. If a bolt makes any noise while being tightened, loosen the bolt and retighten it.**

44. Install the intake manifold with a new gasket.

45. Install the exhaust manifold with a new gasket.

46. Install the exhaust manifold bracket, then install the exhaust pipe, bracket and upper shroud.

47. Install the camshafts and rocker arms.

48. Install the timing belt back cover.

49. Install the side engine mount bracket B. Tighten the bolt attaching the bracket to the cylinder head to 33 ft. lbs. (45 Nm). Tighten the bolts attaching the bracket to the side engine mount to 16 ft. lbs. (22 Nm).

50. Install the camshaft sprockets onto the camshafts.

51. Install the timing belt.

52. Adjust the valves.

53. Tighten the crankshaft pulley bolt to 181 ft. lbs. (250 Nm).

54. Install the splash shield and the front wheels.

55. Lower the vehicle.

56. Install the remaining components in the reverse order of removal.

57. Drain the oil from the engine into a sealable container. Install the drain plug and refill the engine with clean oil.

58. Fill and bleed the air from the cooling system.

59. Connect the negative battery cable and enter the radio security code.

60. Start the engine, checking carefully for any leaks.

61. Check and adjust the ignition timing. Tighten the distributor bolts to 13 ft. lbs. (18 Nm).

62. If equipped with 4WS, start the engine and turn the steering wheel lock-to-lock to reset the 4WS control unit.

1995–97 Accord

2.2L (F22B1, F22B2) ENGINES

1. Disconnect the negative battery cable, then the positive battery cable.

2. Turn the engine to align the timing marks and set cylinder No.1 to TDC. The white mark on the crankshaft pulley should align with the pointer on the timing belt cover.

3. Raise and safely support the vehicle.

4. Drain the engine coolant into a sealable container.

5. Remove the front wheel and tire assemblies.

6. Remove the splash shield.

7. Disconnect the exhaust pipe from the exhaust manifold.

8. Remove the intake manifold bracket bolts.

9. Lower the vehicle to a working level without placing it on the floor.

10. Remove the throttle cable from the throttle body. On automatic transaxle equipped vehicles, remove the throttle control cable. If equipped with cruise control, remove the cruise control cable.

➡**Be careful not to bend the cable when removing it. Always replace a kinked cable with a new one. Do not use pliers to remove the cable from the linkage.**

11. Remove the intake air duct.

12. Remove the breather hose, Positive Crankcase Ventilation (PCV) hose and evaporative emissions (EVAP) control canister hose.

13. Relieve the fuel system pressure.

14. Remove the fuel feed and return hose from the fuel rail.

15. Disconnect the vacuum hoses attached to the engine located near the fuel feed and return hoses.

16. Remove the brake booster vacuum hose from the intake manifold. Label and remove the other vacuum hoses from the intake manifold.

17. Remove the clamp holding the power steering hose to the strut tower.

18. Remove the wiring harness clamp and the ground cable from the intake manifold.

19. Remove the connector and the terminal from the alternator, then remove the engine wiring harness from the valve cover.

20. Remove the mounting bolts and drive belt from the power steering pump. Pull the pump away from the mounting bracket, without disconnecting the hoses. Support the pump out of the way.

21. Loosen the adjusting and mounting bolts for the alternator and remove the drive belt.

22. Remove the engine wiring harness and bypass hose from the lower side of the intake manifold.

23. Disconnect the following engine wiring harness connectors:
   a. Fuel injector connectors
   b. Intake Air Temperature (IAT) sensor connector
   c. Idle Air Control (IAC) valve connector
   d. Throttle Position (TP) sensor connector
   e. Manifold Absolute Pressure (MAP) sensor connector
   f. Heated oxygen sensor (HO2S) connector
   g. Engine Coolant Temperature (ECT) sensor connector
   h. ECT switch connector
   i. ECT gauge sending unit connector
   j. VTEC solenoid valve connector (VTEC engine)
   k. VTEC pressure switch connector (VTEC engine)
   l. Exhaust Gas Recirculation (EGR) valve lift sensor
   m. CKP/TDC/CYP sensor connector
   n. Ignition coil connector (Non VTEC engine)
   o. Fuel Injection Air (FIA) control solenoid valve connector (VTEC engine)

24. Label, then disconnect the electrical connectors from the distributor and the spark plug wires from the spark plugs. Mark the position of the distributor and remove it from the cylinder head. Disconnect the ignition coil wire from the distributor.

25. Remove the upper radiator hose and the heater inlet hose from the cylinder head.

26. Remove the lower radiator hose from the thermostat housing.

27. Remove the coolant bypass hoses.

28. Use a jack to support the engine, be sure to place a cushion between the oil pan and the jack. Remove the through-bolt from the side engine mount and remove the mount.

29. Remove the cylinder head cover. Replace the rubber seals if damaged or deteriorated.

30. Remove the timing belt covers and the timing belt.

31. Remove the camshaft sprocket and the back cover. Do not lose the sprocket key.

32. Remove the exhaust manifold heat insulator and the exhaust manifold.

33. Remove the thermostat housing mounting bolts. Remove the thermostat housing from the intake manifold and the connecting pipe, by pulling and twisting the housing. Discard the O-rings.

34. Remove the fuel rail and fuel injectors.

35. Remove the intake manifold.

36. Remove the cylinder head bolts in the reverse order proper sequence, then remove the cylinder head.

➡To prevent warpage, unscrew the bolts in sequence ⅓ turn at a time. Repeat the sequence until all bolts are loosened.

To install:

37. Be sure all cylinder head and block gasket surfaces are clean. Check the cylinder head for warpage. If warpage is less than 0.002 in. (0.05mm), cylinder head resurfacing is not required. Maximum resurface limit is 0.008 in. (0.2mm) based on a cylinder head height of 3.94 in. (100mm).

38. Always use a new head gasket.

39. Be sure the No. 1 cylinder is at TDC.

40. Clean the oil control orifice and install a new O-ring (VTEC engine only).

41. Install the dowel pins to the engine block.

42. Install the bolts that secure the intake manifold to it's bracket but do not tighten them.

43. Position the camshaft correctly.

44. Install the cylinder head, then tighten the cylinder head bolts sequentially in 3 steps:
- Step 1: 29 ft. lbs. (39 Nm).
- Step 2: 51 ft. lbs. (69 Nm).
- Step 3: 72 ft. lbs. (98 Nm).

45. Install the intake manifold with a new gasket.

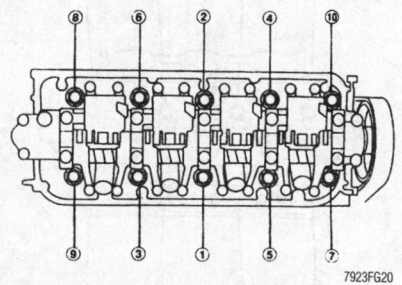

Cylinder head bolt torque sequence—2.2L (F22B1, F22B2) engines

46. Connect the intake manifold bracket to the intake manifold and tighten the bolt to 16 ft. lbs. (22 Nm).

47. Install the fuel rail with the fuel injectors.

48. Install the exhaust manifold with a new gasket.

49. Install the exhaust manifold bracket.

50. Install the timing belt back cover to the cylinder head. Tighten the cover bolt on the non VTEC engine to 9 ft. lbs. (12 Nm). On the VTEC engine tighten the bolt on the intake side of the head to 9 ft. lbs. (12 Nm) and tighten the bolt on the exhaust side of the head to 7 ft. lbs. (10 Nm).

51. Install the key to the camshaft, then install the camshaft sprocket. Tighten the sprocket bolt to 27 ft. lbs. (37 Nm).

52. Be sure the camshaft sprocket and the crankshaft pulleys are aligned to TDC and install the timing belt.

53. Install the lower timing belt cover and tighten the bolts to 9 ft. lbs. (12 Nm).

54. Install a new seal around the adjusting nut. Do not loosen the adjusting nut.

55. Install the crankshaft pulley. Coat the threads and seating face of the pulley bolt with engine oil. Install and tighten the bolt to 181 ft. lbs. (250 Nm).

56. Install the side engine mount. Tighten the bolt and nut attaching the mount to the engine to 40 ft. lbs. (55 Nm). Tighten the through nut and bolt to 47 ft. lbs. (65 Nm), remove the jack from under the center beam.

57. Adjust the valves.

58. Install the upper timing belt cover. Tighten the bolt on the intake side of the head to 9 ft. lbs. (12 Nm) and tighten the bolt on the exhaust side of the head to 7 ft. lbs. (10 Nm).

59. Raise and safely support the vehicle.

60. Connect the exhaust pipe to the exhaust manifold with new gaskets. Tighten the nuts 40 ft. lbs. (54 Nm).

61. Install the splash shield and the front wheels.

62. Lower the vehicle.

63. Install the cylinder head cover gasket cover to the groove of the cylinder head cover. Before installing the gasket thoroughly clean the seal and the groove. Seat the recesses for the camshaft first, then work it into the groove around the outside edges. Be sure the gasket is seated securely in the corners of the recesses.

64. Apply liquid gasket to the four corners of the recesses of the cylinder head cover gasket. Do not install the parts if 5 minutes or more have elapsed since applying liquid gasket. After assembly, wait at least 20 minutes before filling the engine with oil.

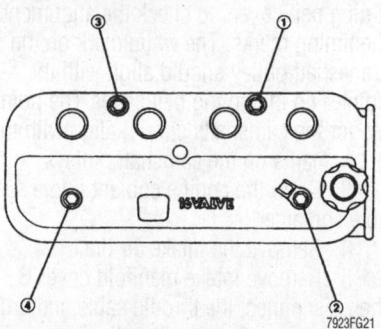

Non-VTEC engine cylinder head cover torque sequence—2.2L (F22B1, F22B2) engines

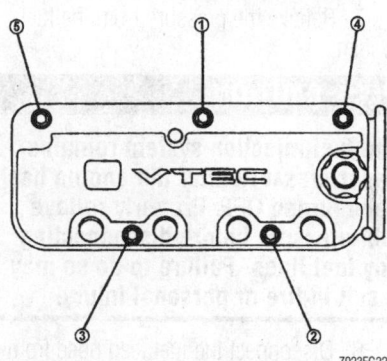

VTEC engine cylinder head cover torque sequence—2.2L (F22B1, F22B2) engines

65. If equipped with a VTEC engine, install the spark plug seals on the spark plug pipes. Take care not to damage the spark plug seals when installing the cylinder head cover.

66. Clean the cylinder head cover contacting surface with a shop towel. Install the cylinder head cover, tighten the cylinder head cover bolts in two or three steps. Tighten the cap nuts in the proper sequence to 7 ft. lbs. (10 Nm).

67. Install the remaining components in the reverse order of removal.

68. Drain the oil from the engine into a sealable container. Install the drain plug and refill the engine with clean oil.

69. Fill and bleed the air from the cooling system.

70. Connect the positive, then the negative battery cable. Enter the radio security code.

71. Start the engine, checking carefully for any leaks.

72. Check the ignition timing.

2.7L ENGINES

1. Disconnect the negative battery cable.

2. Turn the engine to align the timing marks and set cylinder No.1 to TDC. Remove the inspection caps on the upper

timing belt covers to check the alignment of the timing marks. The white mark on the crankshaft pulley should align with the pointer on the timing belt cover. The pointers for the camshafts should align with the green marks on the camshaft pulleys.

3. Drain the engine coolant into a sealable container.

4. Remove the intake air duct.

5. Remove intake manifold cover **B**, then disconnect the throttle cable and cruise control cables from the throttle linkage. Take care to not bend the cables, always replaced a kinked cable.

6. Remove the starter cable from the strut brace, then remove the strut brace.

7. Relieve the pressure from the fuel system.

❉❉ CAUTION

The fuel injection system remains under pressure after the engine has been turned OFF. Properly relieve fuel pressure before disconnecting any fuel lines. Failure to do so may result in fire or personal injury.

8. Disconnect the fuel feed hose from the fuel filter. Disconnect the fuel return hose from the regulator.

9. Disconnect the brake booster vacuum hose and the evaporative emissions (EVAP) control canister hose.

10. Label and disconnect all vacuum hoses from the throttle body, intake manifold and cylinder head.

11. Disconnect the coolant hoses from the idle air control valve, the fast idle thermo valve and the water passage.

12. Remove the intake manifold cover.

13. Disconnect the breather hose from the cylinder head cover.

14. Remove the PCV hose from the cylinder head cover.

15. Loosen the idler pulley center nut and adjusting bolt, then remove the air conditioning compressor belt.

16. Disconnect the engine ground cable from the body, located near the drive belts.

17. Remove the vacuum pipe assembly.

18. Loosen the alternator mounting bolt, nut and adjusting bolt, then remove the alternator drive belt.

19. Support the engine with a floor jack on the oil pan (use a cushion between the jack and pan). Tension the jack so that it is just supporting the engine but not lifting it.

20. Remove the three bolts from the side engine mount, then loosen the through-bolt. Pivot the side engine mount out of the way.

21. Loosen the power steering pump mounting nuts and adjusting nut, then remove the power steering drive belt.

22. Disconnect the inlet hose from the power steering pump, then plug the hose and pump. Remove the power steering pump and place it out of the way.

23. Remove the wiring harness cover and the ground cable from the water passage inlet.

24. Disconnect the following engine harness connectors from the cylinder head and the intake manifold:

   a. Six injector connectors.

   b. Intake Air Temperature (IAT) sensor connector.

   c. TDC/CYP sensor connector.

   d. Intake Air Control (IAC) valve connector.

   e. Manifold Absolute Pressure (MAP) sensor connector.

   f. Engine Coolant Temperature (ECT) sensor connector.

   g. ECT gauge sending unit connector.

   h. ECT switch connector.

   i. Exhaust Gas Recirculation (EGR) valve lift sensor connector.

   j. Throttle position sensor connector.

   k. Intake Air Bypass (IAB) control solenoid valve connector.

   l. Engine oil temperature sensor connector.

   m. Evaporative emissions (EVAP) purge control solenoid valve connector.

   n. Alternator connector.

25. Label, then disconnect the electrical connectors from the distributor and the spark plug wires from the spark plugs. Remove the distributor from the cylinder head.

26. Remove the upper and lower radiator hoses, then disconnect the heater hoses from the engine.

27. Remove the intake air bypass vacuum tank.

28. Remove the bolts attaching the water passage to the cylinder heads, then remove it from the engine. Discard the O-rings.

29. Remove the engine wiring harness covers from the intake manifold.

30. Remove the nuts attaching the EGR pipe to the intake manifold and discard the gasket. Loosen the nut attaching the EGR pipe to the exhaust manifold and remove the pipe from the vehicle.

31. Remove the intake manifold.

32. Remove the exhaust manifolds from the engine.

33. Remove the cylinder head covers and the side covers.

34. Remove the timing belt covers and the timing belt.

35. Remove the camshaft sprockets and the timing belt back covers.

36. Remove the bolts attaching the camshaft holder plates and camshaft holders in the reverse order of installation.

37. Remove the camshaft holder plates, camshaft holders and the dowel pins from the cylinder head.

38. Remove the camshafts from the cylinder heads and the rubber cap from the rear cylinder head. Discard the camshaft seals.

39. Remove the intake rocker arms, exhaust inside rocker arms and the pushrods. Identify the location of the parts as they are removed, to ensure reinstallation to the original locations.

40. Remove the cylinder head bolts in the proper sequence.

➡ **To prevent warpage, loosen the bolts in sequence ⅓ turn at a time. Repeat the sequence until all bolts are removed.**

41. Remove the cylinder heads from the engine block.

42. Remove and clean the oil control orifices, then install new O-rings to the orifices.

To install:

43. Be sure all cylinder head and block gasket surfaces are clean. Check the cylinder head for warpage. If warpage is less than 0.002 in. (0.05mm), cylinder head resurfacing is not required. Maximum resurface limit is 0.008 in. (0.2mm) based on a cylinder head height of 5.24 in. (133mm).

44. Install new head gaskets.

45. Be sure the No. 1 cylinder is at TDC.

46. Install and align the cylinder head dowel pins and oil control orifices.

47. Install the cylinder heads. Apply clean oil to the threads of the cylinder head bolts and washers, then install and tighten

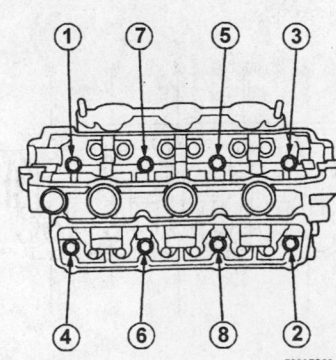

Cylinder head bolt removal sequence— 2.7L engine

7923FG23

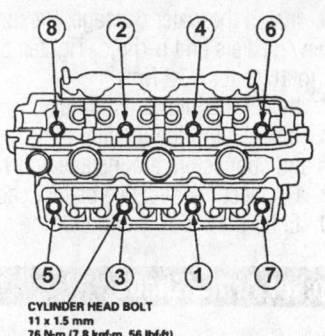

CYLINDER HEAD BOLT
11 x 1.5 mm
76 N·m (7.8 kgf·m, 56 lbf·ft)

7923FG24

Cylinder head bolt torque sequence—2.7L engine

the cylinder head bolts sequentially in 2 steps:

- Step 1: 29 ft. lbs. (39 Nm).
- Step 2: 56 ft. lbs. (76 Nm).

48. Fill the hydraulic tappet mounting hole and the oil fillers with clean engine oil.

49. Install the hydraulic tappets.

✳✳ WARNING

Do not rotate the hydraulic tappets while installing them.

50. Apply clean engine oil to the rocker arms, pushrods and the camshafts.

51. Loosen the exhaust rocker arm adjusting screws and locknuts, then install the pushrods, exhaust inside rocker arms and the intake rocker arms. Install the parts to their original locations.

52. Be sure the rocker arms are properly positioned on the valve stems. Advance the crankshaft 30° from TDC to prevent interference between the pistons and valves, then install the camshafts. Position the rear camshaft on the cylinder head so the cam is not pushing on any valves.

53. Install the timing belt back covers and tighten the attaching bolts to 9 ft. lbs. (12 Nm).

54. Install the camshaft sprockets and tighten the attaching bolts to 23 ft. lbs. (31 Nm).

55. Set the camshaft sprockets so that the No. 1 piston is at TDC. Align the TDC marks (green mark) on the camshaft pulleys to the pointers on the back covers.

56. Turn the crankshaft counterclockwise to set it at TDC. Align the TDC mark on the tooth of the timing belt drive pulley with the pointer on the oil pump.

57. Install the timing belt and timing belt covers.

58. Set No. 1 cylinder to TDC.

59. Tighten the adjusting screws for No. 1, No. 2 and No. 4 cylinders. Tighten the screw until it contacts the valve, then tighten

the screw 1 1/8 turns. Hold the screw in place and tighten the locknut to 14 ft. lbs. (20 Nm).

60. Rotate the crankshaft pulley one turn clockwise, then tighten the adjusting screws for No. 3, No. 5 and No. 6 cylinders. Tighten the screw until it contacts the valve, then tighten the screw 1 1/8 turns. Hold the screw in place and tighten the locknut to 14 ft. lbs. (20 Nm).

61. Install the cylinder head cover gasket into the groove of the cylinder head cover. Seat the recesses for the camshaft first, then work it into the groove around the outside edges.

➡ **Before installing the cylinder head cover gasket, thoroughly clean the seal groove.**

62. Apply liquid gasket to the cylinder head cover gasket at the four corners of the recesses. Use a shop towel and wipe the cylinder heads where the cylinder head covers will come in contact.

63. Install the cylinder head covers, hold the gasket in the groove by placing your fingers on the camshaft contacting surfaces. With the cylinder head cover on the cylinder heads, slide the covers slightly back and forth to seat the cylinder head cover gaskets. Replace the washers if damaged or deteriorated.

64. Tighten the cylinder head cover bolts in two or three steps. In the final step, tighten all the bolts, in sequence, to 11 ft. lbs. (15 Nm).

65. Install the cylinder head side covers with new O-rings and tighten the bolts to 9 ft. lbs. (12 Nm).

66. Install the intake manifold.

67. Install the exhaust manifolds.

68. Install new O-rings to the water passage and install the water passage to the engine, tighten the mounting bolts to 16 ft. lbs. (22 Nm).

69. Install the EGR pipe with a new gasket at the intake manifold. Tighten the nuts attaching the pipe to the intake manifold to 9 ft. lbs. (12 Nm) and tighten the exhaust manifold fitting to 43 ft. lbs. (59 Nm).

70. Install the side engine mount. Use three new bolts to attach the mount to the engine, tighten the bolts to 40 ft. lbs. (54 Nm).

71. Tighten the side engine mount through-bolt to 47 ft. lbs. (64 Nm).

72. Install the remaining components in the reverse of removal.

73. Drain the engine oil into a sealable container, then refill the engine with clean oil.

74. Connect the negative battery cable and enter the radio security code.

75. Fill and bleed the air from the cooling system.

76. Switch the ignition **ON** but do not engage the starter. The fuel pump should run for approximately 2 seconds, building pressure within the lines. Switch the ignition **OFF**, then **ON** 2 or 3 more times to build full system pressure. Check for fuel leaks.

77. Start the engine, allow it to idle and check for any signs of leakage.

1998–99 Accord

3.0L ENGINE

1. Obtain the security code for the radio.

2. Disconnect the negative battery cable.

3. Drain the coolant.

4. Remove the EVAP canister hose from the throttle body.

5. Remove the air intake duct.

6. Remove the upper engine covers.

7. Disconnect the accelerator and cruise control cables from the throttle body.

8. Remove the spark plug wire holder, cover and intake manifold covers.

9. Properly relieve the fuel system pressure.

10. Disconnect the fuel hoses from the supply rail.

11. Disconnect the following hoses and lines:

- Brake booster vacuum hose
- PCV hose
- Breather hose
- Water bypass hose
- Vacuum hose from the throttle body

12. Remove the ground cable from the engine.

13. Remove the alternator belt.

14. Support the engine with a jack and a block of wood and remove the side engine mounting bracket.

15. Remove the power steering pump without disconnecting the hoses.

16. Remove the alternator.

17. Detach the wiring harness connectors from the components on the engine that may interfere with removing the cylinder head.

18. Remove the distributor and spark plug wires.

19. Remove the intake manifold.

20. Detach the connectors from the fuel injectors.

21. Remove the fuel supply rails.

22. Remove the vacuum hoses from the fuel control valve.

23. Set the engine to TDC by aligning the marks on the crankshaft and camshaft pulleys.

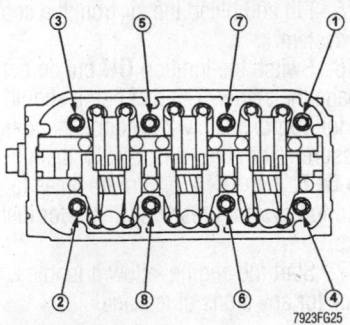

Loosen the cylinder head bolts in the sequence shown to prevent damage to the head—3.0L engine

24. Remove the timing belt.
25. Remove the upper and lower radiator hoses.
26. Disconnect the heater hoses.
27. Remove both exhaust manifolds.
28. Remove the water passage assembly.
29. Remove the camshaft pulleys and rear timing belt covers.
30. Loosen each cylinder head bolt ⅓ turn at a time in the correct sequence. This will take several passes.
31. Remove the cylinder heads.

To install:
32. Clean the cylinder head and the surface of the cylinder block.
33. Install the oil control orifices and install them using new o-rings.
34. If removed, install the dowel pins.
35. Position new cylinder head gaskets on the cylinder block.
36. If moved, set the crankshaft and camshaft pulleys to TDC by aligning the marks on the pulley and oil pump.
37. Carefully position the cylinder heads on the engine.
38. Lubricate the cylinder head bolts with clean engine oil.

➡ **If any cylinder head bolt makes noise while being tightened, loosen the bolts and begin the tightening sequence again.**

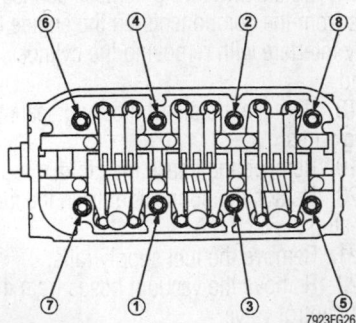

Tighten the cylinder head bolts in the sequence shown to prevent damage to the head—3.0L engine

39. Tighten the cylinder head bolts in three separate steps. First tighten each bolt in sequence to 29 ft. lbs. (39 Nm).
40. Tighten each bolt in sequence to 51 ft. lbs. (69 Nm).
41. Tighten each bolt a third time in sequence to a final torque of 72 ft. lbs. (98 Nm).
42. Install the exhaust manifolds.
43. Install the timing belt.
44. Check and adjust the valve clearance if necessary.
45. Install the cylinder head cover. Tighten the bolts in sequence to 9 ft. lbs. (12 Nm).

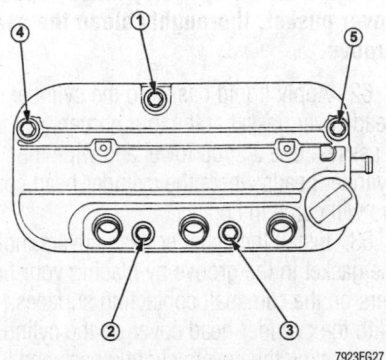

Tighten the cylinder head cover bolts in the sequence shown—3.0L engine

46. Install the water passage. Be sure to use new gaskets and o-rings. Tighten the bolts to 16 ft. lbs. (22 Nm).
47. Install the intake manifold.
48. Install all of the remaining hoses, tubes and connectors are installed correctly.
49. Connect the negative battery cable.
50. Enter the security code for the radio.

Rocker Arms/Shafts

REMOVAL & INSTALLATION

➡ **The radio may contain a coded theft protection circuit. Always obtain the code number before disconnecting the battery. If the vehicle is equipped with 4WS, the steering control unit is shut down when the battery is disconnected. After connecting the battery, turn the steering wheel lock-to-lock to reset the steering control unit.**

Civic and Del Sol

1.5L (D15B7, D15B8) ENGINE

1. Disconnect the negative battery cable.
2. Remove the valve cover and bring the No. 1 cylinder to TDC for the compression stroke.

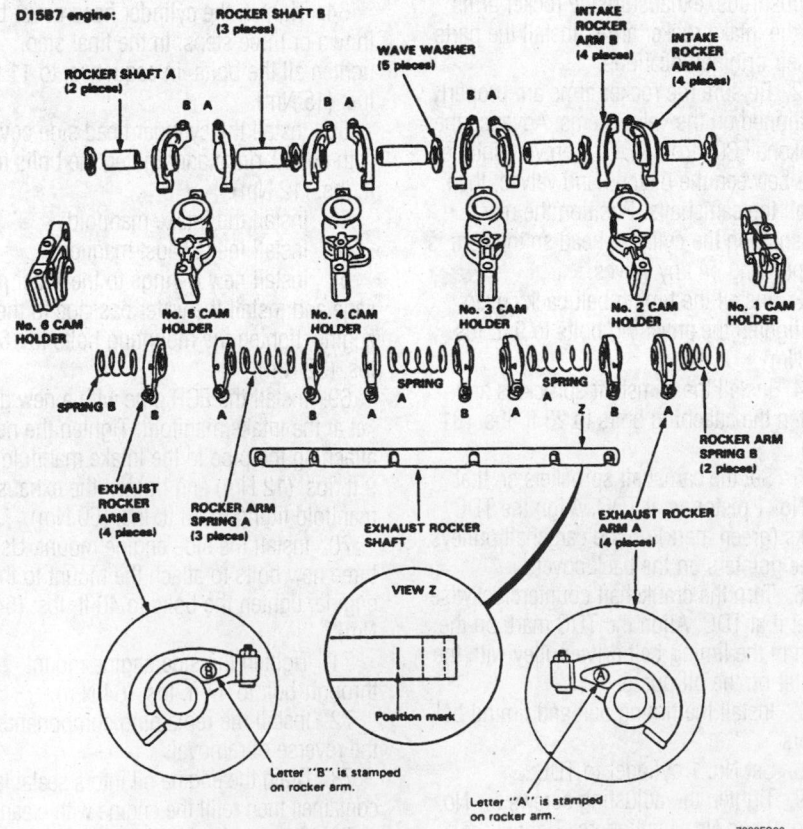

Rocker arm and shaft assembly—1.5L (D15B7) engine

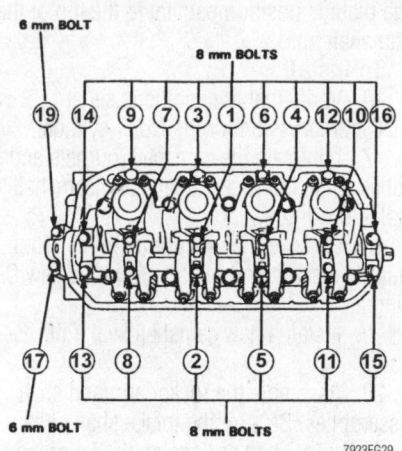

Rocker arm shaft bolt torque sequence—
1.5L (D15B7) engine

3. Loosen the valve adjusting screws.
4. Remove the rocker arm bolts. Unscrew the bolts two turns at a time, in a crisscross pattern, to prevent damaging the valves or rocker assembly.

➡The rocker arms and shafts are an assembly; they must be removed from the engine as a unit. Always follow the torque sequence carefully when installing the rocker shaft assembly.

5. Remove the rocker arm/shaft assemblies. Do not remove the camshaft holder bolts. The bolts keep the camshaft bearing caps, springs, and rocker arms in place on the shafts.
6. If the rocker arms or shafts are to be replaced, identify the parts as they are removed from the shafts to ensure reinstallation in the original location.

To install:
7. Lubricate the camshaft journals and lobes.

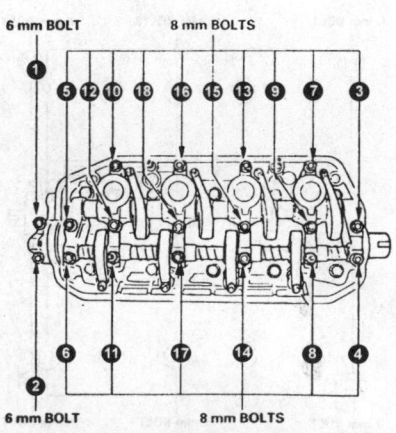

Rocker arm shaft bolt loosening
sequence—1.5L (D15B8) engine

8. Set the rocker arm assembly in place and loosely install the bolts. Tighten each bolt two turns at a time in the proper sequence to ensure that the rockers do not bind on the valves. Tighten the 8mm rocker arm bolts to 16 ft. lbs. (22 Nm). Tighten the 6mm bolts to 9 ft. lbs. (12 Nm).
9. Adjust the valves and tighten the locknuts to 10 ft. lbs. (14 Nm).
10. Replace the valve cover and connect the negative battery cable.

1.5L (D15Z1) AND 1.6L (D16Z6) ENGINE

1. Disconnect the negative battery cable.
2. Label and disconnect the ignition wires. Remove the spark plugs and note their positions.
3. Remove the valve cover and bring the No. 1 cylinder to TDC for the compression stroke.
4. Remove the distributor as an assembly.

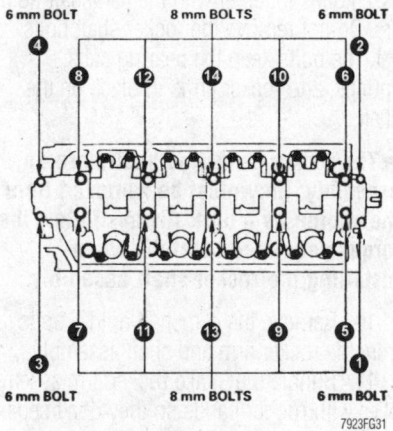

Rocker arm/shaft bolt loosening
sequence—1.5L (D15Z1) engine

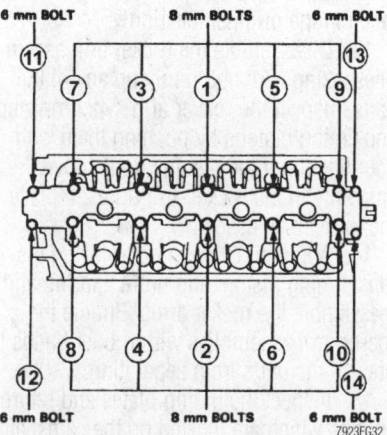

Rocker arm/shaft bolt torque sequence—
1.5L (D15Z1) engine

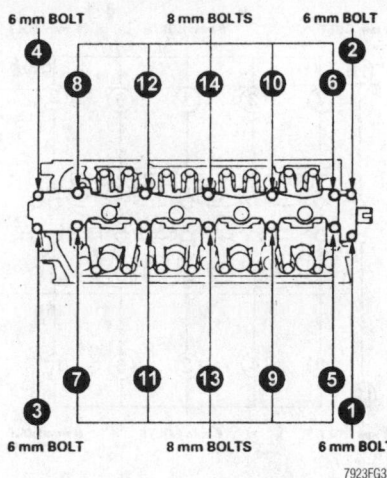

Rocker arm/shaft bolt loosening
sequence—1.6L (D16Z6) engine

5. Loosen the valve adjusting screws.
6. Remove the VTEC solenoid valve.
7. Remove the rocker arm bolts. Unscrew the bolts two turns at a time, in a crisscross pattern, to prevent damaging the valves or rocker assembly.
8. Remove the rocker arm/shaft assemblies. Do not remove the rocker shaft bolts yet. The bolts keep the bearing caps, springs and rocker arms in place on the shafts.

➡The rocker arms and shafts are an assembly; they must be removed from the engine as a unit. Always follow the torque sequence carefully when installing the rocker shaft assembly.

9. Disassemble the rocker arm/shaft assemblies. Identify the parts as they are removed from the shafts to ensure reinstallation in the original location.
10. Disassemble the rocker arm assemblies. Inspect the rocker arm piston by pushing it. If the piston doesn't move smoothly, replace the rocker arm assembly.
11. Apply oil to the pistons and reassemble the rocker arms. Bundle the rocker arm assemblies with rubber bands to prevent the parts from separating.
12. Remove the lost motion assembly from its holder in cylinder head. Inspect the lost motion assembly by pushing down on its piston. If the piston doesn't move smoothly, replace the assembly.
To install:
13. Lubricate the camshaft journals and lobes. Coat the rocker shafts and camshaft holders with oil.
14. Install the lost motion assemblies into the lost motion assembly holder.
15. Reassemble the rocker arms assemblies onto the rocker shafts and camshaft

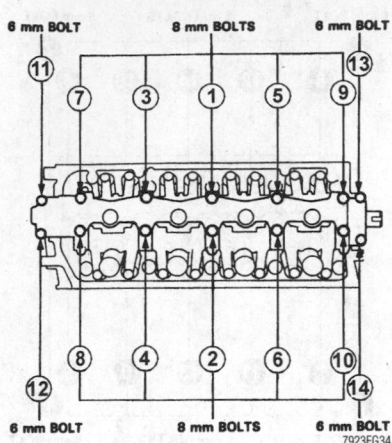

Rocker arm/shaft bolt torque sequence—
1.6L (D16Z6) engine

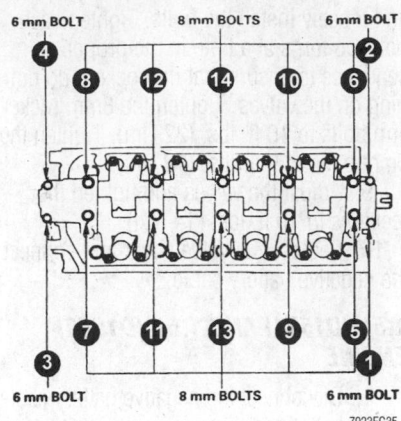

Rocker arm/shaft bolt loosening
sequence—1.6L (D16Y5) engine

holders. After the rocker arms and shafts are reassembled, remove the rubber bands

16. Set the rocker arm assembly in place and loosely install the bolts. Tighten each bolt two turns at a time in the proper sequence to ensure that the rockers do not bind on the valves. Tighten the 8mm rocker arm bolts to 14 ft. lbs. (20 Nm). Tighten the 6mm bolts to 9 ft. lbs. (12 Nm).

17. Install the VTEC solenoid valve with a new filter. Tighten the bolts to 9 ft. lbs. (12 Nm).

18. Install the distributor, but don't tighten the bolts yet.

19. Adjust the valves. Tighten the locknuts to 7 ft. lbs. (10 Nm).

20. Install the valve cover and connect the negative battery cable.

21. Install the spark plugs and reconnect the ignition wires.

22. Warm the engine up to normal operating temperature. Check the ignition timing, and adjust if necessary. Tighten the distributor mounting bolts to 17 ft. lbs. (24 Nm).

1.6L (D16Y5) ENGINE

1. Disconnect the negative battery cable.

2. Label and disconnect the ignition wires. Remove the spark plugs and note their cylinder assignments.

3. Remove the valve cover.

4. Rotate the crankshaft to set the No. 1 cylinder to TDC for the compression stroke. The white TDC mark on the crankshaft pulley aligns with the pointers on the lower timing cover.

5. Remove the distributor.

6. Loosen the valve adjusting screws.

7. Label and disconnect the VTEC solenoid valve connector.

8. Loosen the camshaft holder bolts two turns at a time in a crisscross pattern to prevent damaging the valves or rocker assembly.

9. Remove the rocker arm and shaft assemblies together with the camshaft holders. Do not remove the rocker shaft bolts yet. The bolts keep the bearing caps, springs, and rocker arms in place on the shafts.

➡ **The rocker arms and shafts are an assembly; they must be removed from the engine as a unit. Always follow the torque sequence carefully when installing the rocker shaft assembly.**

10. Remove the camshaft holder bolts from the rocker arm and shaft assembly.

11. Bundle the intake rocker arm assemblies with rubber bands so they don't separate when the intake rocker shaft is removed.

12. Disassemble the rocker arm and shaft assemblies. Label the parts as they are removed from the shafts to ensure reinstallation in the original location.

13. Disassemble the rocker arm assemblies taking care not to mix up any of the parts. Inspect the rocker arm synchronizing and timing pistons by pushing them with your fingers. If the pistons don't move smoothly in the rocker arm bores, replace the rocker arm assembly.

14. Apply oil to the synchronizing pistons, timing piston, and timing spring and reassemble the rocker arms. Bundle the rocker arm assemblies with rubber bands to prevent the parts from separating.

15. Inspect the timing plates and return springs which are located on the camshaft holders. Set each timing plate and return spring so that the C-shaped upper arm of

the plate is position parallel to the top of the camshaft holder.

To install:

16. Verify that the engine is set at TDC/compression for the No. 1 cylinder.

17. Lubricate the camshaft journals and lobes. Coat the rocker shafts and camshaft holders with oil.

18. Remove the oil control orifice. Thoroughly clean it and reinstall it with a new O-ring.

19. Install a new camshaft seal if necessary.

20. Assemble the rocker arm and shaft assemblies. Be sure the intake shaft collars and exhaust shaft springs are in the proper locations.

21. After the rocker arms and shafts are assembled, cut the rubber bands and remove them from the intake rockers. Be sure that no rubber band fragments are left in the engine.

22. Apply fresh oil to the threads of camshaft holder bolts, then install them.

23. Apply liquid gasket to the cylinder head mating surfaces of the No. 1 and No. 5 camshaft holders. Do not allow the sealant to cure before installation.

24. Set the rocker arm and shaft assembly in place. Install and hand-tighten the bolts. Tighten each bolt two turns at a time in the crisscross sequence so that the rockers are evenly tightened and don't bind on the valves. Tighten the 8mm rocker arm bolts to 14 ft. lbs. (20 Nm). Tighten the 6mm bolts to 9 ft. lbs. (12 Nm).

25. Starting with the No. 1 cylinder at TDC/compression, adjust the valve clearances. After the clearance has been reached, tighten the locknuts to 14 ft. lbs. (20 Nm). Set the No. 3, No. 4, and No. 2 cylinders at TDC/compression, and adjust their valve clearances.

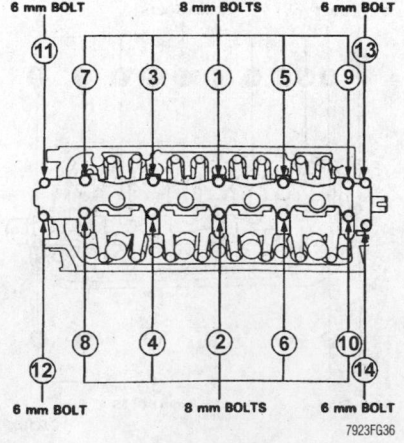

Rocker arm/shaft bolt torque sequence—
1.6L (D16Y5) engine

- Intake: 0.007–0.009 in. (0.18–0.22mm)
- Exhaust: 0.009–0.011 in. (0.23–0.27mm)

26. Remove the VTEC solenoid valve, then remove the valve's filter. Install a new VTEC solenoid valve filter. Tighten the solenoid valve bolts to 9 ft. lbs. (12 Nm) and reconnect the solenoid valve connector.

27. Rotate the crankshaft to set the No. 1 cylinder at TDC/compression. Then, manually inspect the operation of each of the VTEC intake rocker arms:

a. Move the No. 1 cylinder's secondary intake rocker arm up and down.

b. Verify that the secondary intake rocker arm moves independently of the primary intake rocker arm.

c. Repeat the rocker arm inspection for the other three cylinders with each cylinder set at TDC/compression.

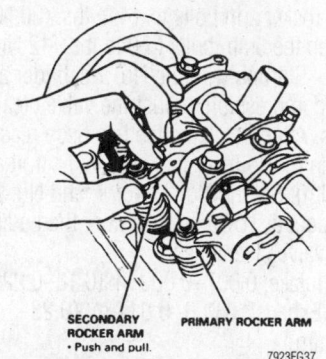

VTEC rocker arm inspection—1.6L (D16Y5) engine

28. Rotate the crankshaft back to TDC/compression for the No. 1 cylinder. Install the distributor, but do not tighten the mounting bolts yet.

29. Tighten the crankshaft pulley bolt to 134 ft. lbs. (181 Nm).

30. Install the valve cover. Be sure the gasket is in good condition, and apply sealant to the corners where the gasket meets the camshaft holders.

31. Install the spark plugs and reconnect the ignition wires.

32. Drain the engine oil and remove the oil filter. Install a new oil filter and refill the engine with fresh oil.

33. Connect the negative battery cable.

34. Warm the engine up to normal operating temperature.

35. Check the ignition timing and adjust it if necessary. Then, tighten the distributor mounting bolts to 17 ft. lbs. (24 Nm).

36. Check all fluid levels. Test drive the vehicle and observe the engine RPM changes at various speeds.

1.6L (D16Y7) ENGINE

1. Disconnect the negative battery cable.

2. Label and disconnect the ignition wires. Remove the spark plugs and note their cylinder assignments.

3. Remove the valve cover and the upper timing belt cover.

4. Set the No. 1 cylinder to TDC for the compression stroke. Verify that the TDC marks are correctly aligned. Once the engine is set in this position, it must not be disturbed.

5. Remove the distributor as an assembly.

6. Loosen the valve adjusting screws.

7. Cover the timing belt with a clean shop towel to protect it from engine oil. If the belt is contaminated with oil, it must be replaced.

8. Remove the camshaft holder bolts. Unscrew the bolts two turns at a time in a crisscross pattern to prevent damaging the valves, camshaft, or rocker arm assembly.

➡ **The rocker arms and shafts are an assembly; they must be removed from the engine as a unit. To prevent warpage, always follow the torque sequence carefully when removing or installing the rocker shaft assembly.**

9. Remove the rocker arm and shaft assemblies. Do not remove the camshaft holder bolts. The bolts keep the camshaft bearing caps, springs, and rocker arms in place on the shafts.

10. If the rocker arms or shafts are to be replaced, identify the parts as they are removed from the shafts to ensure reinstallation in the original location.

To install:

11. Verify that the engine is set to TDC/compression for the No. 1 cylinder.

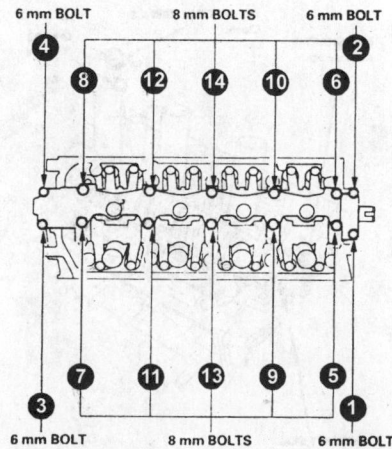

Rocker arm/shaft bolt loosening sequence—1.6L (D16Y7) engine

The camshaft keyway faces up when the engine is at TDC/compression.

12. Lubricate the camshaft journals and lobes with clean engine oil. Install a new camshaft seal if necessary.

13. Remove the oil control orifice. Thoroughly clean it and install it with a new O-ring.

14. Assemble the rocker arms, shafts, and camshaft bearing caps.

15. Apply sealant to the mating surfaces of the No. 1 and No. 5 camshaft bearing caps. Do not allow the sealant to cure before the rocker arm assembly is installed.

16. Set the rocker arm assembly in place. Apply engine oil to the holder bolt threads, then loosely install the bolts. Tighten each bolt in a two step crisscross pattern to ensure that the rockers do not bind on the valves. Tighten the 8mm bolts to 14 ft. lbs. (20 Nm). Tighten the 6mm bolts to 8.7 ft. lbs. (12 Nm).

17. Verify that the engine is at TDC/compression for the No. 1 piston, and install the distributor.

18. Adjust the valves and tighten the locknuts to 14 ft. lbs. (20 Nm).

19. Install the valve cover and upper timing belt cover.

20. Reconnect the negative battery cable.

21. Check the ignition timing and adjust if necessary. Tighten the distributor mounting bolts to 17 ft. lbs. (24 Nm).

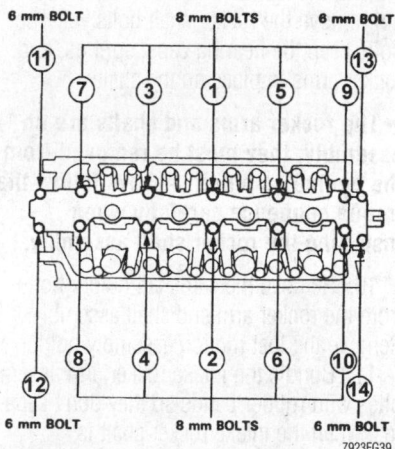

Rocker arm/shaft bolt tightening sequence—1.6L (D16Y7) engine

1.6L (D16Y8) ENGINE

1. Disconnect the negative battery cable.

2. Label and disconnect the ignition wires. Remove the spark plugs and note their cylinder assignments.

3. Remove the valve cover.

4. Rotate the crankshaft to set the No. 1 cylinder to TDC for the compression stroke.

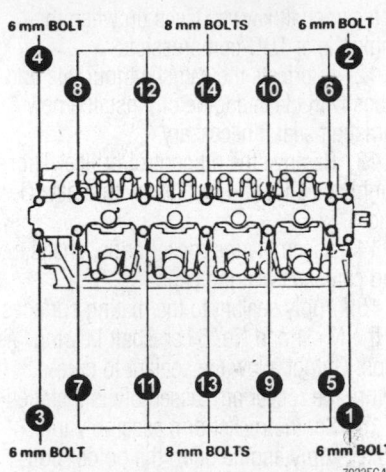

Rocker arm/shaft bolt loosening sequence—1.6L (D16Y8) engine

The white TDC mark on the crankshaft pulley aligns with the TDC pointers on the lower timing belt cover.

5. Remove the distributor from the cylinder head.

6. Loosen the valve adjusting screws.

7. Label and disconnect the VTEC solenoid valve connector.

8. Loosen the camshaft holder bolts two turns at a time in a crisscross pattern to prevent damaging the valves or rocker assembly.

9. Remove the rocker arm and shaft assemblies together with the camshaft holders and the lost motion assembly holder. Do not remove the rocker shaft bolts yet. The bolts keep the bearing caps, springs, and rocker arms in place on the shafts.

➡**The rocker arms and shafts are an assembly; they must be removed from the engine as a unit. Always follow the torque sequence carefully when installing the rocker shaft assembly.**

10. Remove the camshaft holder bolts from the rocker arm and shaft assembly. Remove the lost motion assembly holder.

11. Bundle the intake rocker arm assemblies with rubber bands so they don't separate when the intake rocker shaft is removed.

12. Disassemble the rocker arm and shaft assemblies. Label the parts as they are removed from the shafts to ensure reinstallation in the original location.

13. Disassemble the rocker arm assemblies taking care not to mix up any of the parts. Inspect the rocker arm synchronizing pistons by pushing them with your fingers. If the pistons don't move smoothly in their rocker arm bores, replace the rocker arm assembly.

14. Apply oil to the synchronizing pistons and reassemble the rocker arms. Bundle the rocker arm assemblies with rubber bands to prevent the parts from separating.

15. Remove each lost motion assembly from its port in the lost motion assembly holder. Inspect each lost motion assembly by pushing down on its piston. If the piston doesn't move smoothly, replace the lost motion assembly. Lost motion assemblies cannot be bled like hydraulic lash adjusters.

16. Install the lost motion assemblies back into the lost motion assembly holder.

To install:

17. Verify that the engine is set at TDC/compression for the No. 1 cylinder.

18. Lubricate the camshaft journals and lobes. Coat the rocker shafts and camshaft holders with fresh oil.

19. Remove the oil control orifice. Thoroughly clean it and reinstall it with a new O-ring.

20. Install a new camshaft seal if necessary.

21. Assemble the rocker arm and shaft assemblies. Be sure the intake shaft collars and exhaust shaft springs are in the proper locations.

22. After the rocker arms and shafts are assembled, cut the rubber bands and remove them from the intake rockers. Be sure that no rubber band fragments are left in the engine.

23. Install the lost motion assembly holder onto the camshaft holder. Apply fresh oil to the threads of camshaft holder bolts, then install them.

24. Apply liquid gasket to the cylinder head mating surfaces of the No. 1 and No. 5 camshaft holders. Don't allow the sealant to cure before installation.

25. Set the rocker arm and shaft assembly in place. Install and hand-tighten the bolts.

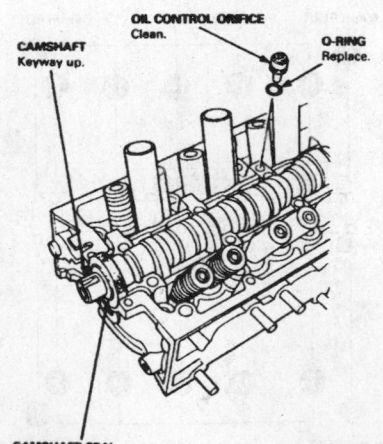

Camshaft seal and oil control orifice— 1.6L (D16Y8) engine

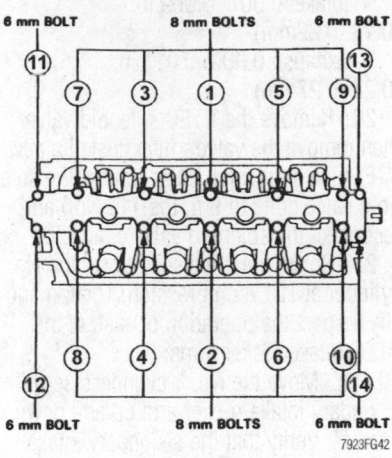

Rocker arm/shaft bolt tightening sequence—1.6L (D16Y8) engine

Tighten each bolt two turns at a time in the crisscross sequence to ensure that the rockers do not bind on the valves. Tighten the 8mm rocker arm bolts to 14 ft. lbs. (20 Nm). Tighten the 6mm bolts to 9 ft. lbs. (12 Nm).

26. Starting with the No. 1 cylinder at TDC/compression, adjust the valve clearances. After the clearance has been reached, tighten the adjuster locknuts to 14 ft. lbs. (20 Nm). Set the No. 3, No. 4, and No. 2 cylinders at TDC/compression, then adjust their valve clearances.

- Intake: 0.007–0.009 in. (0.18–0.22mm)
- Exhaust: 0.009–0.011 in. (0.23–0.27mm)

27. Remove the VTEC solenoid valve, then remove the valve's filter. Install a new VTEC solenoid valve filter. Tighten the solenoid valve bolts to 9 ft. lbs. (12 Nm) and reconnect the solenoid valve connectors.

28. Rotate the crankshaft to set the No. 1 cylinder at TDC/compression. Then, manually inspect the operation of each of the VTEC intake rocker arms:

a. Push the in on the No. 1 cylinder's mid-intake rocker arm.

b. Verify that the mid-intake rocker arm moves independently of the primary and secondary intake rocker arms.

c. Repeat the rocker arm inspection for the other three cylinders with each cylinder set at TDC/compression.

29. Rotate the crankshaft back to TDC/compression for the No. 1 cylinder. Install the distributor, but do not tighten the mounting bolts yet.

30. Tighten the crankshaft pulley to 134 ft. lbs. (181 Nm).

31. Install the valve cover. Be sure the gasket is in good condition, and apply sealant to the corners where the gasket meets the camshaft holders.

32. Install the spark plugs and reconnect the ignition wires.

33. Drain the engine oil and remove the oil filter. Install a new oil filter and refill the engine with fresh oil.

34. Connect the negative battery cable.

35. Warm the engine up to normal operating temperature.

36. Check the ignition timing and adjust it if necessary. Then, tighten the distributor mounting bolts to 17 ft. lbs. (24 Nm).

37. Check all fluid levels. Test drive the vehicle and observe the engine RPM changes at various speeds.

1.6L (B16A2, B16A3) ENGINES

1. Disconnect the negative battery cable.

2. Label and disconnect the ignition wires.

3. Rotate the crankshaft to set the engine at TDC for the compression stroke of the No. 1 cylinder. The white TDC mark on the crankshaft pulley should align with the pointer on the lower timing belt cover.

4. Remove the strut brace.

5. Remove the intake air duct.

6. Loosen the power steering pump adjusting bolt to release the belt tension. Slip the belt off the pulleys. Loosen the air conditioner and alternator adjusting bolts, and slip their belts off the crankshaft pulley.

7. Use a floor jack padded with a block of wood to support the engine.

8. Remove the engine ground cable.

9. Unbolt and remove the engine side mount.

10. Remove the valve cover and the upper timing belt cover.

11. Verify that the engine is set at TDC/compression. Loosen the timing belt tensioner bolt 180°. Then, remove the crankshaft pulley, the lower timing cover, and timing belt.

❊❊ WARNING

Inspect the timing belt for signs of cracked and broken teeth, as well as oil or coolant contamination. If the timing belt is damaged, or has been in contact with oil or coolant, it must be replaced to avoid potential failure.

12. Remove the distributor.

13. Disconnect and remove the VTEC solenoid valve. Remove the solenoid valve's filter and inspect it for clogging.

14. Remove the camshaft sprockets and back cover.

15. Loosen the camshaft holder plate bolts in a crisscross sequence working toward the middle of the cylinder head.

16. Loosen the valve adjusting screws.

17. Lift the camshaft holder plates and holder from the cylinder head. The holder bolts will keep the components together. Note the positions of each camshaft holder for reassembly.

18. Lift the camshafts from the cylinder head. Mark the exhaust and intake camshafts so that they will not be confused.

19. Hold each rocker arm assembly together with a rubber band to prevent them from separating.

20. Remove the intake and exhaust rocker shaft orifices from the cylinder head. The rocker shaft orifices are different and should be identified when removed. Thoroughly clean the orifices and reinstall them with new O-rings.

21. Remove the rocker arm shaft sealing bolts, discard the washers.

22. Insert 12mm bolts into the rocker arm shafts. Remove each rocker arm set while slowly pulling out the rocker arm shaft.

➡**Tag each rocker arm set to assure installation in their original locations.**

23. Inspect the rocker arm pistons. If they do not move smoothly, replace the rocker arm assembly.

24. Remove the two lost motion assemblies from the cylinder head. Inspect each lost motion assembly by pushing the plunger with your finger. Replace the lost motion assembly if it does not move smoothly.

To install:

25. Install the two lost motion assemblies to the cylinder head.

26. Apply engine oil to the rocker arm pistons, then bundle the rocker arms with a rubber band. Apply a light coat of clean engine oil to the rocker arms.

27. Position the rocker arms in their original locations, if they are being reused. If new assembles are being used place them in the cylinder head.

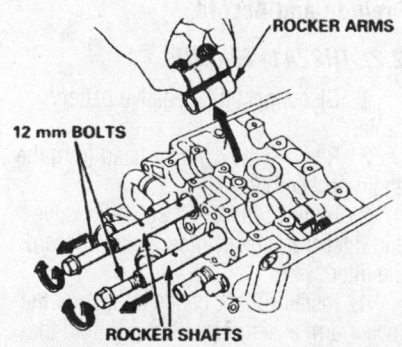

Removing the rocker arms—1.6L (B16A2, B16A3) engines

28. Lightly coat the rocker arm shafts with clean engine oil, then install the rocker arm shafts into the cylinder head. A 12mm bolt can be installed into the end of the rocker arm shafts to aid in their installation. Be sure to install the shafts in the proper positions. Remove the 12mm bolts from the rocker arm shafts, if used.

29. Clean and install the rocker arm shaft orifices with new O-rings. If the holes in the rocker arm shafts are not aligned screw a 12mm bolt into the end of the shaft to position the shaft.

30. Install the sealing bolts with new washers, tighten the bolts to 47 ft. lbs. (64 Nm).

31. Lubricate the camshaft lobes and journals with clean engine oil.

32. Set the camshafts into the cylinder head. Both the intake and exhaust camshafts should be installed with their keyways pointing straight up.

33. Lubricate and install new camshaft seals. Apply liquid gasket to a new camshaft end-plug and install it. If the end-plug has is marked, the mark should be aligned with the cylinder head surface.

34. Apply liquid gasket to the cylinder head mating surfaces of the No. 1 and No. 5 camshaft holders, then install them, along with No. 2, 3, and 4 holders. Be sure to pay attention to the following points:

• Do not apply oil to the holder mating surface of camshaft seals.

• The arrows marked on the camshaft holders should point to the timing belt.

35. Install the camshaft holder plates.

36. Lubricate the threads of the 10mm holder bolts. Then, install all the camshaft holder bolts, but don't tighten them yet.

37. Evenly hand-tighten the camshaft holders. Be sure that the rocker arms are properly positioned on the valve stems.

38. Use a two-step crisscross pattern to tighten the camshaft holder bolts. Begin tightening with the bolts in the middle of the cylinder head, and work toward the outer edges. Final torque specifications are as follows:

• B16A2 engine: 8mm bolts 20 ft. lbs. (28 Nm).

• B16A3 engines: 8mm bolts to 16 ft. lbs. (22 Nm)

• 6mm bolts to 7–8 ft. lbs. (10–11 Nm)

39. Verify that the camshaft keyways are pointing straight up and that the engine is at TDC/compression for the No. 1 cylinder. Fit the camshaft sprocket keys into their keyways.

40. Install the back cover and push the camshaft pulleys onto the camshafts. Then, tighten the sprocket retaining bolts to 37 ft.

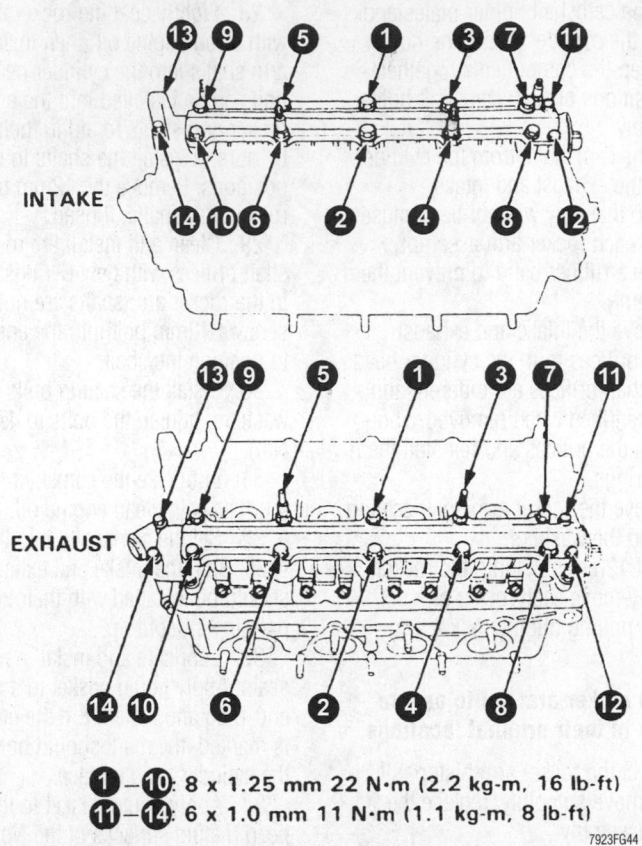

Camshaft holder bolt torque sequences—1.6L (B16A2, B16A3) engines

lbs. (51 Nm) for 1995 vehicles, or 41 ft. lbs. (57 Nm) for 1996–99vehicles.

41. Install and tension the timing belt.

42. Install the lower timing cover and the crankshaft pulley. Tighten the pulley bolt to 130 ft. lbs. (180 Nm).

43. Adjust the valve clearance.

44. Inspect the VTEC rocker arms for smooth and independent movement.

45. Install the VTEC solenoid valve with a new filter. Tighten the valve mounting bolts to 9 ft. lbs. (12 Nm).

46. Install the distributor.

47. Clean the valve cover gasket surfaces. Fit the gasket into the groove of the valve cover.

48. Apply liquid gasket to the rubber seal at the eight corners where the gaskets meet the camshaft holders. Don't allow the sealant to cure before installing the cylinder head cover.

49. Install the cylinder head cover and engine ground cable. Be sure the contact surfaces are clean and do not touch surfaces where liquid gasket has been applied.

50. Tighten the valve cover nuts in to 7 ft. lbs. (10 Nm) in a crisscross sequence.

51. Install the upper timing cover.

52. Install the accessory drive belts and adjust their tensions.

53. Install the engine side mount, tighten the two new nuts to 54 ft. lbs. (75 Nm) and tighten the bolt attaching the mount to the vehicle to 54 ft. lbs. (74 Nm).

54. Install the strut bar and tighten the mounting bolts to 16 ft. lbs. (22 Nm).

55. Reconnect the ignition wires.

56. Drain the engine oil. Install a new oil filter and refill the engine with fresh oil.

57. Reconnect the negative battery cable.

58. Warm the engine up to its normal operating temperature. Then, check and adjust the ignition timing. Tighten the distributor mounting bolts to 17 ft. lbs. (24 Nm).

Prelude and Accord

2.2L (H22A1) ENGINE

1. Disconnect the negative battery cable.

2. Remove the cylinder head from the engine assembly.

3. Remove the VTEC solenoid valve and filter from the cylinder head. Discard the filter.

4. Install rubber bands to each of the rocker arm assemblies, this will hold the rocker arms together and prevent them from separating.

5. Remove the intake and exhaust rocker shaft orifices. The intake and exhaust shaft orifices are different, note their locations for installation.

6. Remove the rocker arm shaft sealing bolts from the cylinder head, discard the washers.

7. Install 12mm bolts into the rocker arm shafts. Remove each rocker arm while slowly pulling out the intake and exhaust rocker arm shafts.

8. Remove the lost motion assemblies from the cylinder head and inspect it. Pushing it gently with a finger will cause it to sink slightly. Increasing the force on it will cause it to sink deeper.

To install:

➡ Clean the rocker shaft orifices and install new O-rings. Clean the rocker arms and the shafts in solvent, dry them and apply clean oil to any contact surfaces.

9. Install the lost motion assemblies to the cylinder head.

10. Install the rocker arms to their original locations while passing the rocker arm shaft through the cylinder head.

11. Install the rocker arm shaft orifices. If the holes in the rocker arm shaft and cylinder head are not in line with each other, thread a 12mm bolt into the rocker arm shaft and rotate the shaft. Be sure that the orifices are installed in the correct locations, the intake and exhaust orifices are different. The rocker shafts should not turn if the orifices are installed correctly.

12. Install the rocker arm sealing bolts with new washers, tighten the bolts to 43 ft. lbs. (60 Nm). Remove the rubber bands from the rocker arm assemblies.

13. Install the VTEC solenoid valve with a new filter, tighten the mounting bolts to 9 ft. lbs. (12 Nm).

14. Install the cylinder head onto the engine block.

15. Adjust the valves and ignition timing.

16. Connect the negative battery cable and enter the radio security code.

17. If equipped with 4WS, start the engine and turn the steering wheel lock-to-lock to reset the 4WS control unit.

18. Run the engine and check for leaks, then road test the vehicle.

2.2L (F22A1, F22A6, F22B1, F22B2) ENGINES

1. Disconnect the negative battery cable.

2. Remove the air intake duct.

3. Remove the positive crankcase ventilation (PCV) hose, then remove the cylinder

head cover. Replace the rubber seals if damaged or deteriorated.

4. Remove the timing belt upper cover.

5. Bring the No. 1 cylinder to TDC. The white mark on the crankshaft pulley should align with the pointer on the timing belt cover. The words **UP** embossed on the camshaft pulley should be aligned in the upward position. The marks on the edge of the pulley should be aligned with the cylinder head or the back cover upper edge. Once in this position, the engine must NOT be turned or disturbed.

6. Label, then disconnect the electrical connectors from the distributor and the spark plug wires from the spark plugs. Mark the position of the distributor and remove it from the cylinder head.

7. Loosen the power steering mounting bolts and remove drive belt from the pump.

8. Mark the rotation of the timing belt if it is to be used again. Loosen the timing belt adjusting bolt 3/4 to one turn, then release the tension on the timing belt. Push the tensioner to release tension from the belt, then tighten the adjusting bolt.

9. Remove the timing belt from the camshaft sprocket.

❊❊ WARNING

Do not crimp or bend the timing belt more than 90° or less, then 1 in. (25mm) in diameter

10. Ensure the words **UP** embossed on the camshaft pulley is aligned in the upward position, then remove the camshaft sprocket bolt. Pull the sprocket from the camshaft and remove the sprocket key.

11. Remove the timing belt back cover.

12. Loosen the valve adjusting screws.

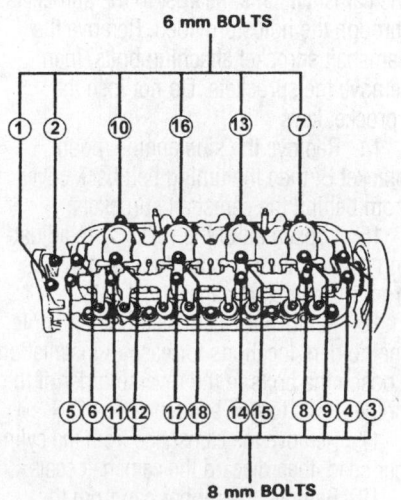

Rocker arm assembly bolt removal sequence—2.2L (F22A1, F22A6) engines

7923FG46

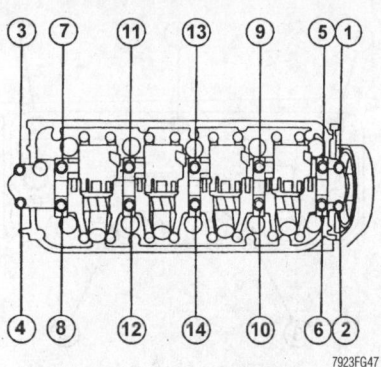

Rocker arm assembly bolt loosening sequence—2.2L (F22B1) engine

7923FG47

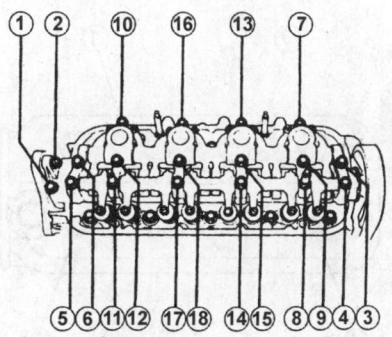

Rocker arm assembly bolt loosening sequence—2.2L (F22B2) engine

7923FG48

13. Loosen the camshaft holder attaching bolts two turns at a time, in the proper sequence to prevent damaging the valves or rocker arm assemblies.

➡**When removing the rocker arm assembly, do not remove the camshaft holder bolts. The bolts will keep the camshaft holders, springs, and the rocker arms on the shafts.**

14. Carefully remove the camshaft holders and rocker arm assembly. If the rocker arm and shaft assembly needs to be disassembled for service, note the location of the components as they are removed. Install a rubber band around the VTEC rocker arm assemblies, to keep them from coming apart during disassembly of the rocker arm assembly. The rocker arms must be installed in the same position if reused.

15. Remove the camshaft from the cylinder head and discard the seal.

16. Remove the oil control orifice.

To install:

17. Wipe the camshaft and the camshaft journals clean, then lubricate both surfaces and install the camshaft.

18. Turn the camshaft so that its keyway is facing up (No. 1 cylinder will be at TDC).

19. Clean the oil control orifice and install a new O-ring, then install the oil control orifice.

20. Reassemble the rocker arm and shaft assembly, if it was disassembled. Lubricate the rocker arm and shaft assembly with clean oil, then apply liquid gasket to the head mating surfaces of the No. 1 and No. 6 camshaft holders.

21. Set the camshaft holders and rocker arm assembly in place, then loosely install the attaching bolts.

22. Apply clean oil to the camshaft oil seal lip and the seal guide (part # 07NAG-PT0010A), then install the seal to the seal guide. Install the seal guide to the camshaft, then the installer cup (part # 07NAF-PT0010A) and the installer shaft (part # 07NAF-PT0020A). Tighten the nut on the installer shaft to press the seal into the cylinder head.

23. Tighten the camshaft holder bolts two turns at a time in the proper sequence. The final torque for the 8mm bolts is 16 ft. lbs. (22 Nm) and the final torque for the 6mm bolts is 9 ft. lbs. (12 Nm).

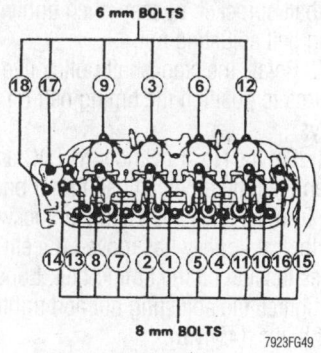

Rocker arm assembly torque sequence—2.2L (F22A1, F22A6) engines

7923FG49

Specified torque:
8 mm bolts: 22 N·m (2.2 kgf·m, 16 lbf·t)
6 mm bolts: 12 N·m (1.2 kgf·m, 8.7 lbf·t)
6 mm bolts: ⑪ ⑫ ⑬ ⑭

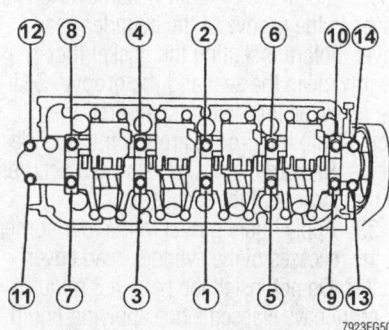

Rocker arm assembly torque sequence—2.2L (F22B1) engine

7923FG50

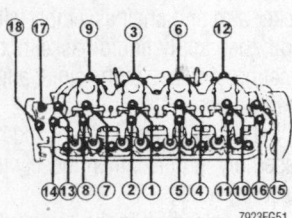

Specified torque:
8 mm bolts: 22 N·m (2.2 kgf·m, 16 lbf·t)
6 mm bolts: 12 N·m (1.2 kgf·m, 8.7 lbf·t)

6 mm bolts: ③, ⑥, ⑨, ⑫, ⑰, ⑱

7923FG51

Rocker arm assembly torque sequence—2.2L (F22B2) engine

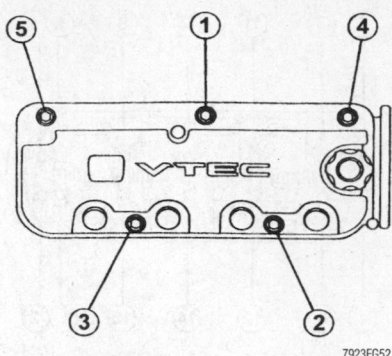

7923FG52

Cylinder head cover torque sequence—2.2L (F22B1) engine

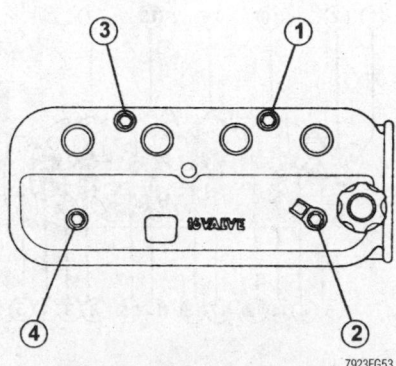

7923FG53

Cylinder head cover torque sequence—2.2L (F22B2) engine

24. Install the timing belt back cover and a new gasket, if necessary. Tighten the bolt toward the exhaust manifold to 7 ft. lbs. (10 Nm) and tighten the bolt toward the intake manifold to 9 ft. lbs. (12 Nm).

25. Install the camshaft sprocket key to the camshaft, then install the camshaft sprocket. Install the bolt and tighten it to 27 ft. lbs. (37 Nm).

26. Ensure the words **UP** embossed on the camshaft pulley is aligned in the upward position, then install the timing belt onto the camshaft sprocket. Loosen, then tighten the timing belt adjusting nut.

27. Rotate the crankshaft pulley five or six turns to position the timing belt on the pulleys.

28. Set the No. 1 cylinder to TDC and loosen the timing belt adjusting nut one turn. Turn the crankshaft counterclockwise until the cam pulley has moved 3 teeth; this creates tension on the timing belt. Loosen, then tighten the adjusting nut and tighten it to 33 ft. lbs. (45 Nm).

29. Adjust the valves.

30. Tighten the crankshaft pulley bolt to 181 ft. lbs. (245 Nm) on (F22B1) engines and 159 ft. lbs. (220 Nm) on all other engines.

31. Install the upper timing belt cover. Tighten the bolt toward the exhaust manifold to 7 ft. lbs. (10 Nm) and tighten the bolt toward the intake manifold to 9 ft. lbs. (12 Nm).

32. Install the cylinder head cover gasket cover to the groove of the cylinder head cover. Before installing the gasket thoroughly clean the seal and the groove. Seat the recesses for the camshaft first, then work it into the groove around the outside edges. Be sure the gasket is seated securely in the corners of the recesses.

33. Apply liquid gasket to the four corners of the recesses of the cylinder head cover gasket. Do not install the parts if 5 minutes or more have elapsed since applying liquid gasket. After assembly, wait at least 20 minutes before filling the engine with oil.

34. Install the cylinder head (valve) cover. Tighten the bolts attaching the cylinder head cover in the proper sequence to 7 ft. lbs. (10 Nm).

35. Install the PCV hose to the cylinder head cover.

36. Install and adjust the power steering belt.

37. Install the distributor to the cylinder head, snug the attaching bolts until the timing has been checked and adjusted.

38. Connect the spark plug wires to the correct spark plugs, then connect the distributor electrical connectors.

39. Install the air intake duct.

40. Drain the oil from the engine into a sealable container. Install the drain plug and refill the engine with clean oil.

41. Connect the negative battery cable and enter the radio security code.

42. Start the engine, checking carefully for any leaks.

43. Check and adjust the ignition timing as necessary, then tighten the distributor bolts to 13 ft. lbs. (18 Nm).

2.3L (H23A1) ENGINE

1. Disconnect the negative battery cable.
2. Turn the crankshaft so the No. 1 piston is at top dead center.

➡The No. 1 piston is at top dead center when the pointer on the block aligns with the white painted mark on the flywheel (manual transaxle) or driveplate (automatic transaxle).

3. Remove the air intake duct.

4. Remove the engine ground cable from the cylinder head cover.

5. Remove the connector and the terminal from the alternator, then remove the engine wiring harness from the valve cover.

6. Remove the ignition coil.

7. Label, then disconnect the electrical connectors from the distributor and the spark plug wires from the spark plugs. Mark the position of the distributor and remove it from the cylinder head. Disconnect the ignition coil wire from the distributor.

8. Remove the Positive Crankcase Ventilation (PCV) hose, then remove the cylinder head cover. Replace the rubber seals if damaged or deteriorated.

9. Remove the timing belt middle cover.

10. Ensure the words **UP** embossed on the camshaft pulleys are aligned in the upward position.

11. Mark the rotation of the timing belt if it is to be used again. Loosen the timing belt adjusting nut 1/2 turn, then release the tension on the timing belt. Push the tensioner to release tension from the belt, then tighten the adjusting nut.

12. Remove the timing belt from the camshaft sprockets.

✳✳ WARNING

Do not crimp or bend the timing belt more than 90° or less, then 1 in. (25mm) in diameter

13. Insert a 5.0mm pin punch in each of the camshaft caps, nearest to the sprockets, through the holes provided. Remove the camshaft sprocket attaching bolts, then remove the sprockets. Do not lose the sprocket keys.

14. Remove the side engine mount bracket B, then the timing belt back cover from behind the camshaft sprockets.

15. Loosen all of the rocker arm adjusting screws, then remove the pin punches from the camshaft caps.

16. Remove the camshaft holders, note the holders locations for ease of installation. Loosen the bolts in the reverse order of the holder bolts torque sequence.

17. Remove the camshafts from the cylinder head, then discard the camshaft seals.

18. Remove the rubber cap from the head, located at the end of the intake camshaft.

19. Remove the rocker arms from the cylinder head. Note the locations of the rocker arms.

➡ **The rocker arms have to be installed to their original locations if being reused.**

To install:

20. Lubricate the rocker arms with clean oil, then install the rocker arms on the pivot bolts and the valve stems. If the rocker arms are being reused, install them to their original locations. The locknuts and adjustment screws should be loosened before installing the rocker arms.

21. Lubricate the camshafts with clean oil.

22. Install the camshaft seals to the end of the camshafts that the timing belt sprockets attach to. The open side (spring) should be facing into the cylinder head when installed.

23. Be sure the keyways on the camshafts are facing up and install the camshafts to the cylinder head.

24. Install the rubber plug to the cylinder head at the end of the intake camshaft.

25. Apply liquid gasket to the head mating surfaces of the No. 1 and No. 6 camshaft holders, then install them along with No. 2, 3, 4 and 5. **I** or **E** marks are stamped on the camshaft holders to identify them as Intake or Exhaust side holders. The arrows stamped on the holders should point toward the timing belt.

26. Snug the camshaft holders in place.

27. Press the camshaft seals securely into place.

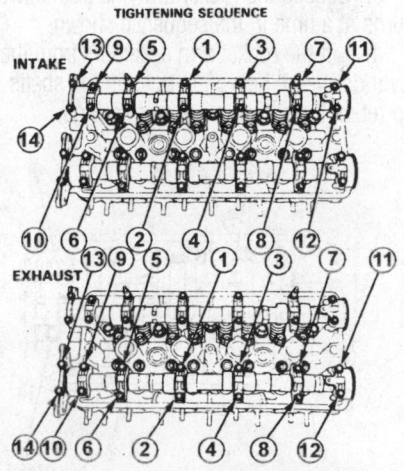

Specified torque:
Except Intake ⑤, ⑦, Exhaust ⑥, ⑧:
 10 N·m (1.0 kg-m, 7 lb-ft)
Intake ⑤, ⑦, Exhaust ⑥, ⑧:
 12 N·m (1.2 kg-m, 9 lb-ft)

TIGHTENING SEQUENCE

Camshaft holders torque sequence—2.3L (H23A1) engine

28. Tighten the camshaft holder bolts in two steps, following the proper sequence, to ensure that the rockers do not bind on the valves. Tighten all the bolts, except the four studs, to 7 ft. lbs. (10 Nm). Tighten the studs (number 5 and 7 bolts in the correct sequence) to 9 ft. lbs. (12 Nm).

29. Install the timing belt back cover.

30. Install the side engine mount bracket B. Tighten the bolt attaching the bracket to the cylinder head to 33 ft. lbs. (45 Nm). Tighten the bolts attaching the bracket to the side engine mount to 16 ft. lbs. (22 Nm).

31. Insert a 5.0mm pin punch in each of the camshaft caps, nearest to the pulleys, through the holes provided. Install the keys into the camshaft grooves.

32. Push the camshaft sprockets onto the camshafts, then tighten the retaining bolts to 27 ft. lbs. (38 Nm).

33. Ensure the words **UP** embossed on the camshaft pulleys are aligned in the upward position. Install the timing belt to the camshaft sprockets, then remove the two 5.0mm pin punches from the camshaft bearing caps.

34. Loosen, then tighten the timing belt adjuster nut.

35. Turn the crankshaft counterclockwise until the cam pulley has moved 3 teeth; this creates tension on the timing belt. Loosen, then tighten the adjusting nut and tighten it to 33 ft. lbs. (45 Nm).

36. Adjust the valves.

37. Tighten the crankshaft pulley bolt to 181 ft. lbs. (250 Nm).

38. Install the middle timing belt cover and tighten the attaching bolts to 9 ft. lbs. (12 Nm).

39. Install the cylinder head cover and tighten the cap nuts to 7 ft. lbs. (10 Nm). Install the PCV hose to the cylinder head cover.

40. Install the distributor to the cylinder head, snug the attaching bolts until the timing has been checked and adjusted.

41. Connect the spark plug wires to the correct spark plugs, then connect the distributor electrical connectors. Install the ignition coil wire to the distributor.

42. Install the ignition coil.

43. Install the alternator wiring harness to the cylinder head cover, then connect the terminal and connector to the alternator.

44. Connect the engine ground cable to the cylinder head cover.

45. Install the air intake duct.

46. Drain the oil from the engine into a sealable container. Install the drain plug and refill the engine with clean oil.

47. Connect the negative battery cable and enter the radio security code.

48. Start the engine, checking carefully for any leaks.

49. Check and adjust the ignition timing. Tighten the distributor bolts to 13 ft. lbs. (18 Nm).

50. If equipped with 4WS, start the engine, then turn the steering wheel lock-to-lock to reset the 4WS control unit.

2.7L ENGINE

1. Disconnect the negative battery cable.

2. Remove the timing belt covers and the timing belt.

3. Remove the camshafts from the cylinder heads.

4. Remove the intake rocker arms, exhaust inside rocker arms and the pushrods. Identify the location of the parts as they are removed, to ensure reinstallation to the original locations.

5. Remove the valve lifters (hydraulic tappets) from the cylinder heads.

6. Remove the intake manifold, then the cylinder heads from the vehicle.

7. Remove the rocker arm shaft sealing bolt from the cylinder head and discard the washer.

8. Install a 12 x 1.25mm bolt into the rocker arm shaft. Slowly remove the shaft from the cylinder head and remove the exhaust rocker arms and washers. Identify the location of the parts as they are removed, to ensure reinstallation to the original locations.

To install:

9. Clean the rocker arms and rocker arm shafts in solvent, dry them, then oil the contact surfaces of the parts.

10. Install a 12 x 1.25mm bolt into the rocker arm shaft. Install the rocker arms to their original locations while passing the rocker arm shaft through the cylinder head.

11. Remove the bolt from the rocker arm shaft and install the sealing bolt with a new washer. Tighten the sealing bolt to 33 ft. lbs. (44 Nm).

12. Install the cylinder heads and the intake manifold.

13. Fill the valve lifter (hydraulic tappets) mounting hole and the oil fillers with clean engine oil.

14. Install the valve lifters.

❊❊ WARNING

Do not rotate the valve lifters while installing them.

15. Apply clean engine oil to the rocker arms, pushrods and the camshafts.

16. Loosen the exhaust rocker arm adjusting screws and locknuts, then install

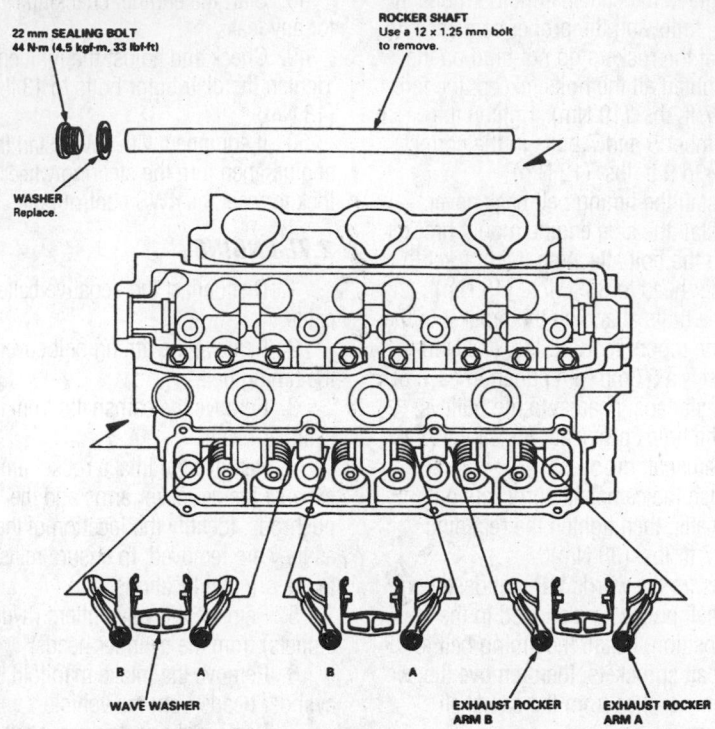

Exhaust rocker arms and rocker arm shaft component locations—2.7L engine

NOTE: The wave washer should be firmly fitted to the cylinder head groove.

7923FG54

the pushrods, exhaust inside rocker arms and the intake rocker arms. Install the parts to their original locations.

17. Be sure the rocker arms are properly positioned on the valve stems. Advance the crankshaft 30° from TDC to prevent interference between the pistons and valves when the camshafts are installed.

18. Install the camshafts and camshaft holders.

19. Install the timing belt and set No. 1 cylinder to TDC.

20. Tighten the adjusting screws for No. 1, No. 2 and No. 4 cylinders. Tighten the screw until it contacts the valve, then tighten the screw 1 turn. Hold the screw in place and tighten the locknut to 14 ft. lbs. (20 Nm).

21. Rotate the crankshaft pulley one turn clockwise, then tighten the adjusting screws for No. 3, No. 5 and No. 6 cylinders. Tighten the screw until it contacts the valve, then tighten the screw 1 1/8 turns. Hold the screw in place and tighten the locknut to 14 ft. lbs. (20 Nm).

22. Install the cylinder head gasket into the groove of the cylinder head cover. Seat the recesses for the camshaft first, then work it into the groove around the outside edges.

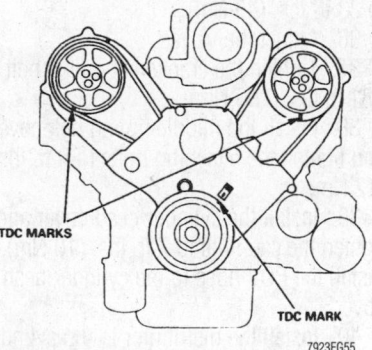

Timing marks for valve adjustment for cylinders 3, 5 and 6—2.7L engine

➡Before installing the cylinder head cover gasket, thoroughly clean the seal groove.

23. Apply liquid gasket to the cylinder head cover gasket at the four corners of the recesses. Use a shop towel and wipe the cylinder heads where the cylinder head covers will come in contact.

24. Install the cylinder head covers, hold the gasket in the groove by placing your fingers on the camshaft contacting surfaces. With the cylinder head cover on the cylinder heads, slide the covers slightly back and forth to seat the cylinder head cover gas-

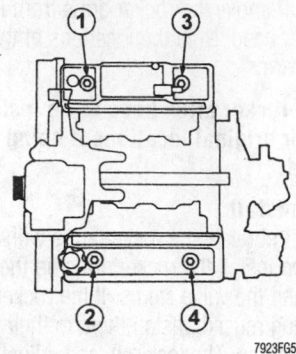

Cylinder head cover torque sequence— 2.7L engine

kets. Replace the washers if damaged or deteriorated

25. Install the cylinder head side covers with new O-rings and tighten the bolts to 9 ft. lbs. (12 Nm).

26. Tighten the cylinder head cover bolts in two or three steps. In the final step, tighten all the bolts, in sequence, to 11 ft. lbs. (15 Nm).

27. Install the distributor to the cylinder head and tighten the mounting bolt to 16 ft. lbs. (22 Nm).

28. Connect the spark plug wires to the correct spark plugs, then connect the distributor electrical connectors.

29. Drain the engine oil into a sealable container, then refill the engine with clean oil.

30. Connect the negative battery cable and enter the radio security code.

31. Start the engine, allowing it to idle and check for any signs of leakage.

3.0L Engine

1. Remove the cylinder head cover.
2. Loosen the jam nuts on the adjusters, then back out the screws.
3. Loosen the rocker arm shaft bolts two turns at a time in the sequence shown.
4. Lift the rocker arm assembly from the cylinder head. Leave the bolts in the shafts to retain the rocker arms and springs.

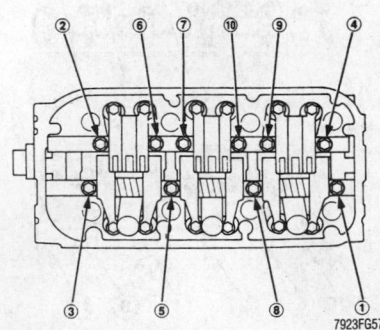

Be sure to loosen the rocker arm shaft bolts in the correct order as shown—3.0L engine

To install:

5. Clean all parts in solvent, dry with compressed air and lubricate with clean engine oil.

6. Place the rocker arm assemblies on the cylinder head and install the bolts loosely. Be sure that all rocker arms are in alignment with their valves.

7. Tighten each bolt two turns at a time in the correct sequence. Tighten the bolts to 17 ft. lbs. (24 Nm).

8. Adjust the valves and install the cylinder head covers.

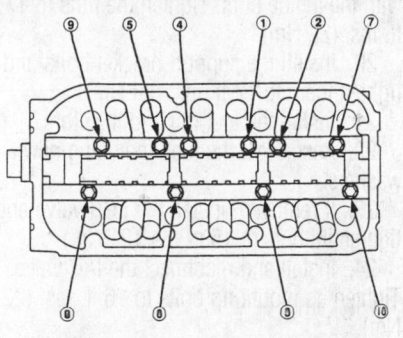

7923FG59

Tighten the bolts two turns at a time in the sequence shown—3.0L engine

Intake Manifold

REMOVAL & INSTALLATION

➡The radio may contain a coded theft protection circuit. Always obtain the code number before disconnecting the battery. If the vehicle is equipped with 4WS, the steering control unit is shut down when the battery is disconnected. After connecting the battery, turn the steering wheel lock-to-lock to reset the steering control unit.

INTAKE ROCKER SHAFT

INTAKE ROCKER ARM ASSEMBLY

A   B   B   A

EXHAUST ROCKER ARM B

SPRING

EXHAUST ROCKER ARM A

EXHAUST ROCKER SHAFT

Letter "B" is stamped on rocker arm.

Letter "A" is stamped on rocker arm.

7923FG58

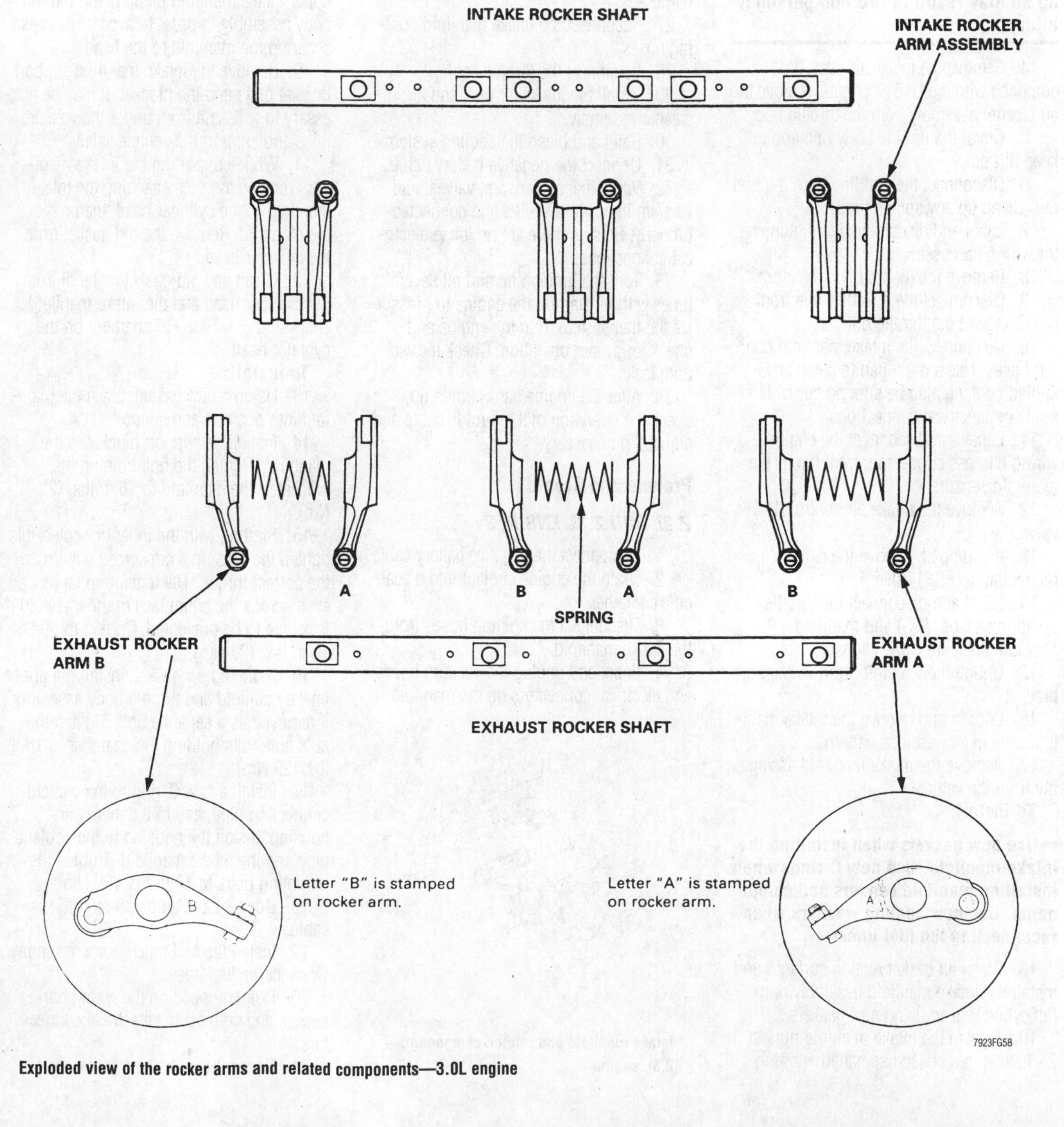

Exploded view of the rocker arms and related components—3.0L engine

Civic and Del Sol

1. Disconnect the negative battery cable.

2. Drain the cooling system to a level below the upper radiator hose.

3. Relieve the fuel system pressure by loosening the fuel filter service bolt.

✳✳ CAUTION

The fuel injection system remains under pressure even after the engine has been turned off. The fuel system pressure must be relieved before disconnecting any fuel lines. Failure to do so may result in fire and personal injury.

4. Remove the intake air duct. If equipped with the D16Y7 engine, remove the air cleaner assembly from the throttle body.

5. Cover the throttle body opening to keep dirt out.

6. Disconnect the fuel line from the fuel rail. Clean up any spilled fuel.

7. Label and disconnect the fuel injector wiring harnesses.

8. Remove the fuel rail and injectors.

9. Disconnect the throttle cable from the linkage at the throttle body.

10. Disconnect the intake manifold cooling hoses. Use a drain pan to catch any spilled coolant, also be sure no coolant spills on electrical connections.

11. Label and disconnect the engine wiring harness connectors from the intake manifold sensors.

12. Remove the intake air control (IAC) valve.

13. If equipped, remove the exhaust gas recirculation (EGR) valve.

14. Label and disconnect the throttle position sensor (TPS) and manifold absolute pressure (MAP) sensor.

15. Unbolt the manifold from its support bracket.

16. Loosen and remove the intake manifold nuts in a crisscross pattern.

17. Remove the intake manifold assembly from the vehicle.

To install:

➡Use new gaskets when installing the intake manifold. Use new O-rings when installing manifold sensors and components. Use new sealing washers when reconnecting the fuel lines.

18. Clean all gasket mating surfaces and install the intake manifold assembly onto the cylinder head using new gaskets.

19. Tighten the intake manifold nuts in 2–3 steps in a crisscross pattern starting with the inside nuts. Tighten the nuts to 17 ft. lbs. (23 Nm).

20. Install the support bracket bolts and tighten them to 17 ft. lbs. (24 Nm).

21. Install the fuel rail and injectors.

22. Reconnect the fuel line using new washers.

23. If removed, install the EGR valve and tighten its nuts to 15 ft. lbs. (21 Nm).

24. Install and reconnect the IAC valve. Tighten its mounting bolts to 16 ft. lbs. (22 Nm).

25. Reconnect the fuel injector wiring harnesses.

26. Reconnect the intake manifold wiring harnesses.

27. Reconnect the intake manifold cooling hoses.

28. Reconnect the throttle cable.

29. Install the intake air duct and air cleaner assembly.

30. Refill and bleed the cooling system.

31. Connect the negative battery cable.

32. Verify that all sensors, valves, and vacuum lines are installed and connected properly. Be sure there are no loose electrical connections.

33. Turn the ignition on and off several times without starting the engine to pressurize the fuel system. Run the engine and check for proper operation. Check for vacuum leaks.

34. After the engine has warmed up, check the operation of the throttle cable and adjust it if necessary.

Prelude and Accord

2.2L AND 2.3L ENGINES

1. Disconnect the negative battery cable.

2. Drain the engine coolant into a sealable container.

3. Disconnect the cooling hoses from the intake manifold.

4. Label and unplug the vacuum hoses and electrical connectors on the manifold

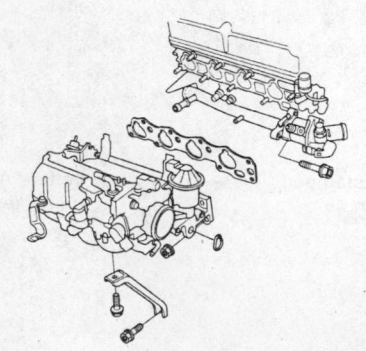

Intake manifold and related components— 2.3L engine

7923FG60

and throttle body. Unplug the connector from the Exhaust Gas Recirculation (EGR) valve. Position the wiring harnesses out of the way.

5. Disconnect the throttle cable from the throttle body.

6. Relieve the fuel pressure.

7. Remove the fuel rail and fuel injectors.

8. Remove the thermostat housing mounting bolts. Remove the thermostat housing from the intake manifold and the connecting pipe, by pulling and twisting the housing. Discard the O-rings.

9. It may be necessary to remove the upper intake manifold plenum and throttle body assembly in order to access the nuts securing the manifold to the head.

10. Remove the intake manifold support bracket bolts and the bracket. It may be necessary to access it from under the vehicle; raise and support the vehicle safely.

11. While supporting the intake manifold, remove the nuts attaching the intake manifold to the cylinder head, then remove the manifold. Remove the old gasket from the cylinder head.

12. Clean any old gasket material from the cylinder head and the intake manifold. check and clean the FIA chamber on the cylinder head.

To install:

13. Using a new gasket, place the manifold into position and support.

14. Install the support bracket to the manifold. Tighten the bolt holding the bracket to the manifold to 16 ft. lbs. (22 Nm).

15. Starting with the inner or center nuts, tighten the nuts, in a crisscross pattern, to the correct torque. The tension must be even across the entire face of the manifold if leaks are to be prevented. Correct torque is 16 ft. lbs. (22 Nm).

16. Using a new gasket, install the upper intake manifold and throttle body assembly, if removed as a separate unit. Tighten the nuts and bolts holding the chamber to 16 ft. lbs. (22 Nm).

17. Install a new O-ring to the coolant connecting pipe, and to the thermostat housing. Install the housing to the coolant pipe and the intake manifold. Tighten the mounting bolts to 16 ft. lbs. (22 Nm).

18. Connect and adjust the throttle cable.

19. Install the fuel rail/injector assembly. Connect the fuel lines.

20. Properly position the wiring harnesses and connect the electrical connectors.

21. Connect the vacuum hoses.

22. Fill and bleed the air from the cooling system.

23. Connect the negative battery cable and enter the radio security code.

24. If equipped with 4WS, turn the steering wheel lock-to-lock to reset the 4WS control unit.

25. Start the engine, checking carefully for any leaks of fuel, coolant or vacuum. Check the manifold gasket areas carefully for any leakage of vacuum.

2.7L ENGINES

1. Disconnect the negative battery cable.

2. Drain the engine coolant into a sealable container.

3. Relieve the fuel pressure.

4. Remove the feed hose from the fuel filter.

5. Remove the PCV valve from the cylinder head cover.

6. Remove the air intake duct.

7. Remove the intake manifold covers.

8. Remove the throttle cable and cruise control cable by loosening the locknut, then slip the cable end out of the accelerator linkage.

9. Label and unplug all electrical connections on the manifold and throttle body.

10. Disconnect the hoses from the brake booster and the evaporative canister.

11. Remove the wiring harness holders and position them out of the way.

12. Disconnect the vacuum hose from the fuel pressure regulator.

13. Remove the fuel rail attaching nuts.

14. Remove the fuel rail from the injectors, leaving the injectors in the manifold.

15. Remove each injector, noting its position, and remove the seal ring from each manifold port.

16. Remove the cushion ring and O-ring from each injector.

17. Label and disconnect all vacuum and coolant hoses from the intake manifold and throttle body. If necessary, remove the vacuum pipe assembly.

18. Remove the Exhaust Gas Recirculation (EGR) crossover pipe. Discard the gasket.

19. Remove the bolts and nuts securing the intake manifold to the engine. Be sure all vacuum and electrical connections are unplugged. Carefully lift the manifold from the engine. Discard the gaskets.

To install:

20. Using new gaskets, place the manifold into position. Install the nuts/bolts until just snug.

21. Starting with the inner/center bolts, tighten the nuts and bolts in a crisscross pattern to the correct torque. The tension must be even across the entire face of the manifold if leaks are to be prevented. Correct torque is 16 ft. lbs. (22 Nm).

22. Install the EGR crossover pipe using a new gasket. Tighten the nuts (to the intake manifold) to 9 ft. lbs. (12 Nm) and the pipe (to the exhaust manifold) to 43 ft. lbs. (59 Nm).

23. If removed, install the vacuum pipe assembly. Tighten the bolts to 9 ft. lbs. (12 Nm).

24. Connect the coolant hoses to the intake manifold and the throttle body.

25. Install new cushion rings on each injector.

26. Coat new O-rings with clean engine oil and install them on the fuel injectors.

27. Install the injectors into the fuel rail. Make certain the O-rings seat properly and are not distorted.

➡**Assembling each injector into the fuel rail prevents damage to the O-rings. Handle the rail and injector assembly carefully when reinstalling it to the manifold. Don't drop the injectors or bang their tips.**

28. Coat new seal rings with a light coat of clean, thin oil and install them into the manifold.

29. Install the fuel rail and injectors to the intake manifold.

30. With all injectors seated in the manifold, be sure each injector is positioned properly.

31. Install the fuel rail retaining nuts and tighten them evenly to 9 ft. lbs. (12 Nm).

32. Connect the vacuum hose to the regulator.

33. Install the wiring harness to the fuel rail, tighten the mounting bolts to 9 ft. lbs (12 Nm).

34. Connect the electrical harness to the injectors.

35. Connect all electrical and vacuum connections to the throttle body and manifold.

36. Connect the brake booster, evaporative canister, and fuel return hoses.

37. Connect the feed hose to the fuel filter with new gaskets. Tighten the union bolt to 16 ft. lbs. (22 Nm) and the service bolt to 9 ft. lbs. (12 Nm).

38. Install and adjust the throttle and cruise control cables.

39. Install the air intake duct.

40. Install the PCV valve to the cylinder head cover.

41. Refill and bleed the air from the cooling system.

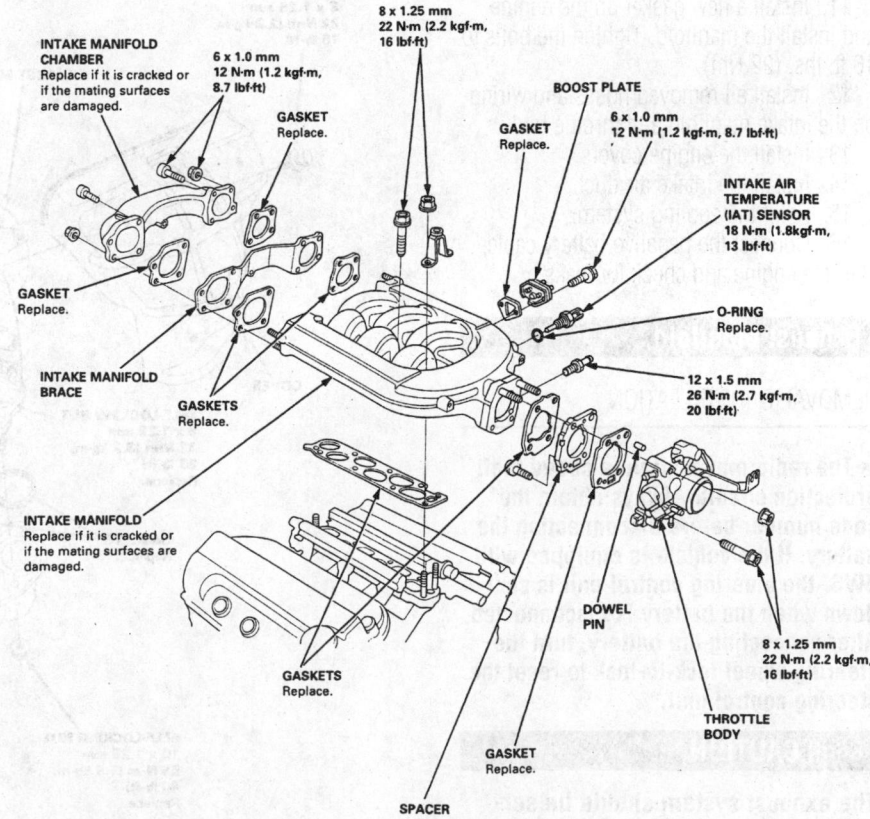

Exploded view of the intake manifold and related components—3.0L engine

7923FGC4

42. Connect the negative battery cable and enter the radio security code.

43. Switch the ignition **ON** but do not engage the starter. The fuel pump should run for approximately 2 seconds, building pressure within the lines. Switch the ignition **OFF**, then **ON** 2 or 3 more times to build full system pressure. Check for fuel leaks.

44. Start the engine, checking carefully for any leaks of fuel, coolant or vacuum. Check the manifold gasket areas carefully for any leakage of vacuum.

45. Install the intake manifold covers, tighten the attaching bolts to 9 ft. lbs. (12 Nm).

3.0L ENGINE

1. Obtain the security code for the radio.
2. Disconnect the negative battery cable.
3. Drain the coolant.
4. Remove the EVAP canister hose from the throttle body.
5. Remove the air intake duct.
6. Remove the upper engine covers.
7. Disconnect the accelerator and cruise control cables from the throttle body.
8. Ensure that all components have been removed from the intake manifold.
9. Remove the intake manifold.

To install:

10. Clean the mounting surfaces.
11. Install a new gasket on the engine and install the manifold. Tighten the bolts to 16 ft. lbs. (22 Nm).
12. Install all removed hoses and wiring on the intake manifold and throttle body.
13. Install the engine covers.
14. Install the intake air duct.
15. Refill the cooling system.
16. Connect the negative battery cable, start the engine and check for leaks.

Exhaust Manifold

REMOVAL & INSTALLATION

➡ **The radio may contain a coded theft protection circuit. Always obtain the code number before disconnecting the battery. If the vehicle is equipped with 4WS, the steering control unit is shut down when the battery is disconnected. After connecting the battery, turn the steering wheel lock-to-lock to reset the steering control unit.**

❈❈ CAUTION

The exhaust system should be serviced with the engine cold.

Civic and Del Sol

1. Disconnect the negative battery cable.
2. Raise and support the front of the vehicle and block the rear wheels.
3. Unbolt the front exhaust pipe from the exhaust manifold/catalytic converter. Unbolt the exhaust manifold support brackets if their bolts are accessible from this angle. The splash shield may be removed for better access.
4. Lower the vehicle.

➡ **Remove any rust or dirt from the exhaust manifold before removal. This will prevent dirt from entering the exhaust pipes.**

5. Unbolt and remove the manifold heat shield.
6. Disconnect the oxygen sensor (HO2S) harness. Use an oxygen sensor socket or box end wrench to unscrew the sensor from the manifold. Handle the sensor carefully.
7. Unbolt the exhaust manifold brackets.
8. Unbolt the exhaust manifold and separate it from the cylinder head. Remove the exhaust manifold and its gasket.

To install:

➡ **Use new gaskets and self-locking nuts when installing the exhaust manifold.**

9. Clean the gasket mating surfaces of the manifold and cylinder head ports. Install the new gasket onto the cylinder head. Install new gaskets onto the exhaust pipe flange.

10. Install the exhaust manifold. Apply anti-seize paste to the studs. Tighten the self-locking nuts to 23 ft. lbs. (32 Nm) in a crisscross pattern starting in the center of the manifold and working outward.

11. Install the manifold brackets and tighten their bolts to 17 ft. lbs. (24 Nm) for the B16A2, D15B8 and D15Z1 engines and 33 ft. lbs. (45 Nm) for all other engines.

12. Carefully coat only the threads of the oxygen sensor body with anti-seize paste. Don't get any anti-seize on the sensor probe. Install the oxygen sensor and carefully tighten it to 33 ft. lbs. (45 Nm).

13. Install the heat shield and tighten the bolts to 16 ft. lbs. (22 Nm).

14. Reconnect the oxygen sensor connector.

15. Raise and support the front of the vehicle and block the rear wheels.

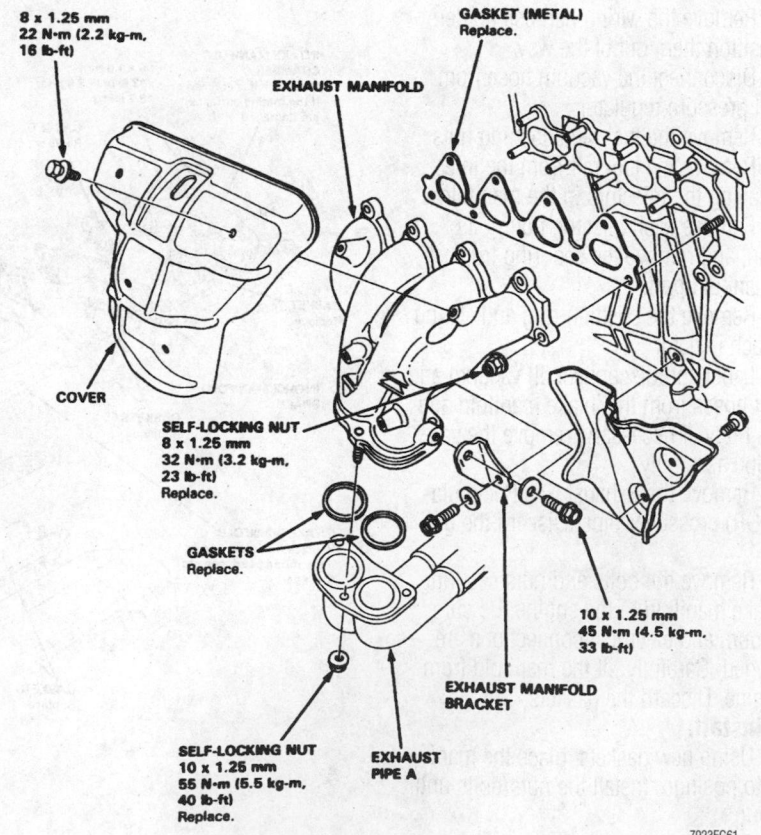

8 x 1.25 mm
22 N·m (2.2 kg-m,
16 lb-ft)

GASKET (METAL)
Replace.

EXHAUST MANIFOLD

COVER

SELF-LOCKING NUT
8 x 1.25 mm
32 N·m (3.2 kg-m,
23 lb-ft)
Replace.

GASKETS
Replace.

10 x 1.25 mm
45 N·m (4.5 kg-m,
33 lb-ft)

EXHAUST MANIFOLD
BRACKET

SELF-LOCKING NUT
10 x 1.25 mm
55 N·m (5.5 kg-m,
40 lb-ft)
Replace.

EXHAUST
PIPE A

7923FG61

Exhaust manifold components—1.6L engines

16. Reconnect the front exhaust pipe and the exhaust manifold/catalytic converter. Tighten the self-locking nuts to 40 ft. lbs. (55 Nm), if the converter is not attached to the manifold. If the converter is attached, tighten to 25 ft. lbs., (34 Nm). Install any manifold brackets and tighten them to 33 ft. lbs. (45 Nm). Install the splash shield if it was removed.

17. Lower the vehicle and connect the negative battery cable.

18. Run the engine and check for exhaust leaks.

Prelude and Accord

2.2L AND 2.3L ENGINES

1. Disconnect the negative battery cable.

2. Safely raise and support the vehicle.

3. If the oxygen sensor is located in the exhaust manifold, disconnect the oxygen sensor connector.

4. Remove the exhaust manifold upper cover.

5. If equipped with air conditioning, remove the heat insulator from the manifold.

6. Remove the nuts attaching the exhaust manifold to the front exhaust pipe. Separate the pipe from the manifold and discard the gasket. Support the pipe with wire; do not allow it to hang by itself.

7. Remove the exhaust manifold bracket(s) bolts and remove the bracket(s).

8. Using a crisscross pattern (starting from the center), remove the exhaust manifold attaching nuts.

9. Remove the manifold and discard the gasket. Clean the manifold and cylinder head mating surfaces.

10. If equipped, remove the lower manifold cover from the manifold.

To install:

11. If equipped, install the lower manifold cover, tighten the attaching bolts to 16 ft. lbs. (22 Nm).

12. Using a new gasket and nuts, place the manifold into position and support it. Install the nuts snug on the studs.

13. Install the support bracket(s) below the manifold. Tighten the bracket(s) mounting bolts to 33 ft. lbs. (44 Nm).

14. Starting with the manifold inner or center nuts, tighten the nuts in a crisscross pattern to the correct torque. The tension must be even across the entire face of the manifold if leaks are to be prevented. Tighten the nuts to 23 ft. lbs. (31 Nm).

15. If equipped with air conditioning, install the heat insulator to the manifold. Tighten the attaching bolts to 7 ft. lbs. (10 Nm) on Prelude models and 9 ft. lbs. (12 Nm) on Accord models.

16. Install the upper manifold cover, tighten the bolts to 16 ft. lbs. (22 Nm).

17. If disconnected, connect the oxygen sensor connector.

18. Connect the front exhaust pipe, using new gaskets and nuts. Tighten the exhaust pipe attaching nuts to 40 ft. lbs. (55 Nm).

19. Connect the negative battery cable and enter the radio security code.

20. Start the engine and check for exhaust leaks.

21. If equipped with 4WS, turn the steering wheel lock-to-lock to reset the 4WS control unit.

2.7L ENGINES

1. Disconnect the negative battery cable.

2. Remove the radiator cap and drain the cooling system into a sealable container.

3. Remove the radiator.

4. Detach the starter cable from the strut tower brace. Remove the strut tower brace.

5. If necessary for additional clearance, remove the vacuum control box on the bulkhead. Position it aside with the vacuum hoses attached.

6. Raise and safely support the vehicle.

7. Remove the front wheels, then remove the splash shield from under the engine.

8. Remove the center beam.

9. Disconnect the front oxygen sensor electrical connector.

10. Disconnect the exhaust pipe from the exhaust manifolds and the catalytic converter. Discard the locknuts attaching the downpipe to the manifolds and the catalytic converter. Remove the exhaust pipe from the vehicle and discard the gaskets.

11. Remove the bolts securing the heat shields on the exhaust manifolds.

12. Disconnect the EGR crossover pipe from the engine.

13. Remove the nuts securing the manifolds to the cylinder heads. Remove the manifolds and gaskets from the engine. Remove any old gasket material from the cylinder heads.

To install:

14. Using new gaskets and nuts, place the manifolds into position. Lightly oil the threads, then install the nuts snug on the studs.

15. Starting with the center nuts, tighten the nuts in a crisscross pattern to the correct torque. Tighten the nuts to 22 ft. lbs. (30 Nm).

16. Install the EGR crossover pipe with a new gasket. Tighten the nuts to 9 ft. lbs. (12 Nm) and the exhaust manifold connection to 43 ft. lbs. (59 Nm).

17. Install the heat shields to the manifolds. Tighten the bolts to 16 ft. lbs. (22 Nm).

18. Install the exhaust pipe with new nuts and gaskets. Tighten the exhaust manifold connections to 40 ft. lbs. (54 Nm) and the catalytic converter to 25 ft. lbs. (33 Nm).

19. Connect the front oxygen sensor electrical connector.

20. Install the center beam. Tighten the bolts to 37 ft. lbs. (50 Nm).

21. Install the splash shield and front wheels.

22. Install the vacuum control box and the radiator.

23. Install the strut tower brace and secure the starter cable. Tighten the strut tower bolts to 16 ft. lbs. (22 Nm).

24. Fill and bleed the air from the cooling system.

25. Connect the negative battery cable and enter the radio security code.

26. Start the engine and allow it to reach normal operating temperature. Check for leaks.

3.0L ENGINE

1. Raise and safely support the vehicle.

2. Remove the engine undercover.

3. Disconnect the exhaust pipe from the manifold to be removed.

4. Lower the vehicle.

5. Remove the exhaust manifold heat shield.

6. Remove the mounting nuts and the exhaust manifold.

To install:

7. Clean the mounting surfaces.

8. Position a new gasket on the cylinder head.

9. Install the exhaust manifold. Tighten the nuts to 23 ft. lbs. (31 Nm).

10. Install the heat shield. Tighten the bolts to 16 ft. lbs. (22 Nm).

11. Raise the vehicle and connect the exhaust pipe to the manifold using a new gasket. Tighten the nuts to 40 ft. lbs. (54 Nm).

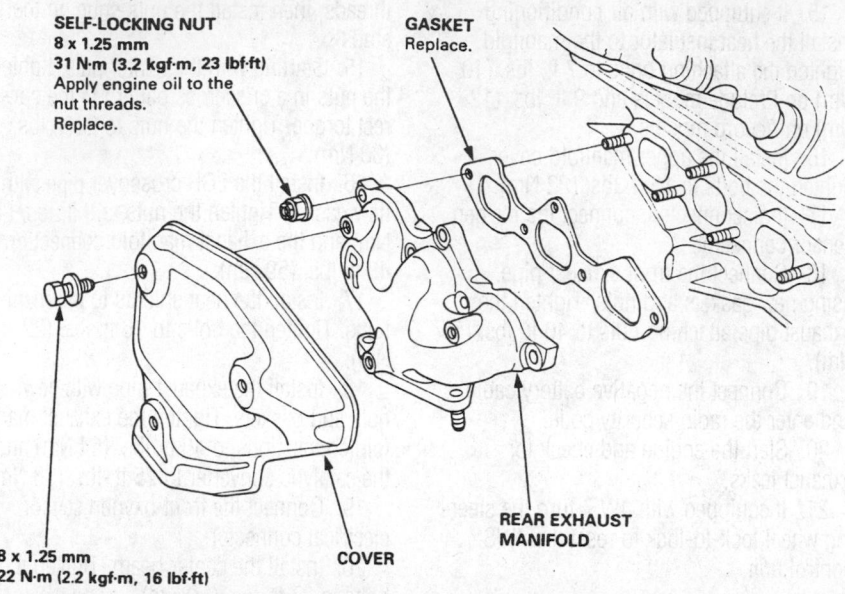

SELF-LOCKING NUT
8 x 1.25 mm
31 N·m (3.2 kgf·m, 23 lbf·ft)
Apply engine oil to the
nut threads.
Replace.

GASKET
Replace.

8 x 1.25 mm
22 N·m (2.2 kgf·m, 16 lbf·ft)

COVER

REAR EXHAUST MANIFOLD

7923FG93

Exploded view of the rear exhaust manifold mounting—3.0L engine

Front Crankshaft Seal

REMOVAL & INSTALLATION

➡ **The original radio may contain a coded anti-theft circuit. Obtain the security code number before disconnecting the battery cables.**

1. Disconnect the negative battery cable.
2. Safely raise and support the vehicle.
3. Remove the splash shield.
4. Remove the engine accessory drive belts.
5. Turn the engine to align the timing marks and set cylinder No.1 to TDC. The white mark on the crankshaft pulley should align with the pointer on the timing belt cover. Remove the inspection caps on the upper timing belt covers to check the alignment of the timing marks. The pointers for the camshafts should align with the green marks on the camshaft sprockets.
6. Remove the upper timing belt covers and crankshaft pulley. Remove the lower timing belt cover

➡ **Mark the direction of the timing belt's rotation if it is to be reinstalled.**

7. Remove the timing belt.
8. If equipped, remove the Crankshaft Position (CKP) sensor from the oil pump, then remove the stopper plate.
9. Remove the timing belt sprocket from the crankshaft, do not lose the sprocket key.

10. Using a suitable seal removal tool, remove the seal from the front of the engine.

To install:

11. Clean the seal mounting surfaces on the engine block.
12. Apply a thin coat of grease on the crankshaft and seal lips.
13. Install the seal with the part number facing out. Use a seal driver to seat the seal against the oil pump. Clean any excess grease off the crankshaft and be sure the seal lip is not distorted.
14. Install the timing belt sprocket and key to the crankshaft.
15. Install the stopper plate and if equipped, the CKP sensor to the oil pump, tighten the stopper plate and sensor mounting bolts to 9 ft. lbs. (12 Nm).
16. Verify that the engine is at TDC for the no. 1 cylinder on the compression stroke. Install and tension the timing belt.
17. Install the timing belt covers and crankshaft pulley. Tighten the crankshaft pulley bolt to 181 ft. lbs. (245 Nm), with the aid of a crank pulley holder.
18. Install and adjust the accessory drive belts.
19. Verify that all engine components that may have been removed have been reinstalled correctly.
20. Install the splash shield and lower the vehicle.
21. Connect the negative battery cable.
22. Top up the engine oil if necessary.
23. Run the engine and check for leaks.

Camshaft

➡ **The radio may contain a coded theft protection circuit. Always obtain the code number before disconnecting the battery. If the vehicle is equipped with 4WS, the steering control unit is shut down when the battery is disconnected. After connecting the battery, turn the steering wheel lock-to-lock to reset the steering control unit.**

REMOVAL & INSTALLATION

Civic and Del Sol

1.5L (D15B7, D15B8 AND D15Z1) AND 1.6L (D16Z6, D16Y5, D16Y7, D16Y8) ENGINES

1. Disconnect the negative battery cable.
2. Label and disconnect the ignition wires.
3. Remove the valve cover and the upper timing belt cover.
4. Rotate the crankshaft to set the No. 1 cylinder at TDC for the compression stroke. Once the engine is in this position, it shouldn't be disturbed.
5. Remove the timing belt. If the timing belt is contaminated with oil or coolant, it must be replaced. If the timing belt is to be reused, mark its direction of rotation.
6. Remove the distributor as an assembly.
7. Unbolt and remove the camshaft sprocket and its key. Remove the upper back cover timing cover.
8. Loosen the rocker arm locknuts and back off the valve adjusting screws.
9. Loosen the camshaft holder bolts in a two-step crisscross sequence, starting at the edges and working toward the center of the cylinder head.
10. Remove the rocker arm and shaft assembly. Leave the camshaft holder bolts in the camshaft holders to hold the rocker arm and shaft assembly together.
11. Wrap rubber bands around the VTEC rocker arm assemblies so that they do not separate.
12. Store the rocker arm and shaft assembly away from your work area. Cover the assembly with shop towels or a sheet of plastic to protect it from dust.
13. Lift the camshaft from the cylinder head. Remove the camshaft seal.
14. Inspect the camshaft journals and lobes for signs of scoring or other damage.

To install:

15. Remove the oil control orifice. Thoroughly clean it and reinstall it with a new O-ring.

16. Clean and inspect the camshaft bearing caps in the cylinder head.

17. Lubricate the lobes and journals of the camshaft prior to installation. Install the camshaft with the keyway facing up so that the camshaft will be at TDC/compression for the No. 1 cylinder.

18. Lightly lubricate a new camshaft seal with engine oil and install it.

19. Install the rocker arm and shaft assembly as follows:

   a. Remove the rubber bands from the VTEC rocker arms.

   b. Lubricate the rocker arm contact surfaces.

   c. Apply liquid gasket to the head mating surfaces of the No. 1 and No. 5 camshaft holders. Don't allow the sealant to cure before installing the rocker arm assembly.

   d. Set the rocker arm and shaft assembly in place. If equipped, install the lost motion assembly holder.

   e. Coat the threads of the camshaft holder bolts with clean oil and loosely install them.

   f. Tighten each bolt two turns at a time in the crisscross sequence to ensure that the rockers and camshaft holder do not bind on the camshaft journals.

   g. Tighten the 8mm camshaft holder bolts to 14 ft. lbs. (20 Nm), and the 6mm camshaft holder bolts to 9 ft. lbs. (12 Nm).

20. Install the camshaft sprocket and key. Tighten the retaining bolt to 27 ft. lbs. (38 Nm).

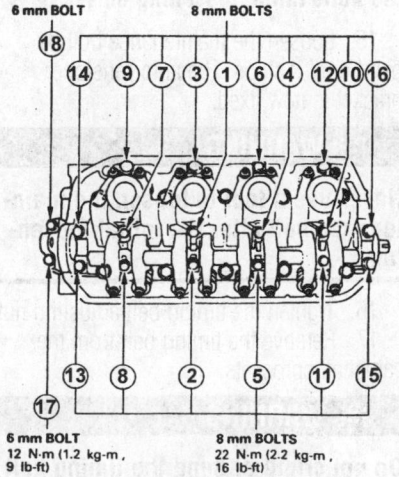

6 mm BOLT   **8 mm BOLTS**

6 mm BOLT
12 N·m (1.2 kg-m,
9 lb-ft)

8 mm BOLTS
22 N·m (2.2 kg-m,
16 lb-ft)

7923FG62

Camshaft holder bolt tightening sequence—1.5L (D15B7, D15B8) engines

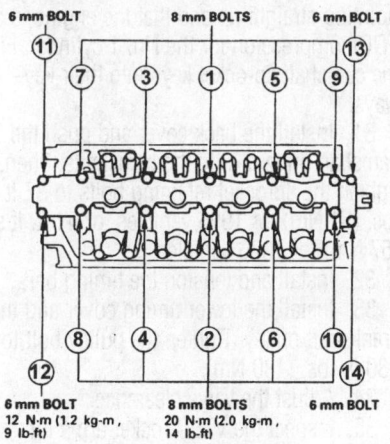

6 mm BOLT   **8 mm BOLTS**   **6 mm BOLT**

6 mm BOLT
12 N·m (1.2 kg-m,
9 lb-ft)

8 mm BOLTS
20 N·m (2.0 kg-m,
14 lb-ft)

7923FG63

Camshaft holder bolt tightening sequence—1.5L (D15Z1) and 1.6L (D16Z6, D16Y5, D16Y7, D16Y8) engines

21. Verify that the engine remains at TDC/compression for the No. 1 cylinder.

22. Install the distributor. The lugs on the distributor drive fit into the groove on the end of the camshaft. Don't fully tighten the distributor mounting bolts yet.

23. Install the timing belt. Tighten the tensioner bolt to 33 ft. lbs. (44 Nm) once the belt has been properly tensioned.

24. Install the lower timing cover. Tighten the crankshaft pulley bolt to 134 ft. lbs. (181 Nm).

25. Adjust the valves.

26. Manually inspect the VTEC rocker arms for smooth motion.

27. Be sure all the spark plug tube sealing gaskets are fully seated.

28. Apply liquid gasket to the corner recesses of a new valve cover gasket. Then, install the gasket to the valve cover. Don't allow the sealant to cure before installation.

29. Install the valve cover. Gently wiggle the valve cover to be sure it is fully seated. Tighten the valve cover bolts in a crisscross pattern to 7 ft. lbs. (10 Nm).

30. Reconnect the ignition wires.

31. Refill the engine with fresh oil and install a new filter.

32. Reconnect the battery cable.

33. Warm the engine up to normal operating temperature. Check for oil leaks.

34. Check the ignition timing and adjust it if necessary. Then, tighten the distributor mounting bolts to 17 ft. lbs. (24 Nm).

1.6L (B16A2, B16A3) ENGINES

1. Disconnect the negative battery cable.

2. Label and disconnect the ignition wires.

3. Rotate the crankshaft to set the engine at TDC for the compression stroke of the No. 1 cylinder. The white TDC mark on the crankshaft pulley should align with the pointer on the lower timing belt cover.

4. Remove the strut brace.

5. Remove the intake air duct.

6. Loosen the power steering pump adjusting bolt to release the belt tension. Slip the belt off the pulleys. Loosen the air conditioner and alternator adjusting bolts, and slip their belts off the crankshaft pulley.

7. Use a floor jack padded with a block of wood to support the engine.

8. Remove the engine ground cable.

9. Unbolt and remove the engine side mount.

10. Remove the valve cover and the upper timing belt cover.

11. Verify that the engine is set at TDC/compression. Loosen the timing belt tensioner bolt 180°. Then, remove the crankshaft pulley, the lower timing cover, and timing belt.

✷✷ WARNING

Inspect the timing belt for signs of cracked and broken teeth, as well as oil or coolant contamination. If the timing belt is damaged, or has been in contact with oil or coolant, it must be replaced to avoid potential failure.

12. Remove the distributor.

13. Disconnect and remove the VTEC solenoid valve. Remove the solenoid valve's filter and inspect it for clogging.

14. Remove the camshaft sprockets and back cover.

15. Loosen the camshaft holder plate bolts in a crisscross sequence working toward the middle of the cylinder head.

16. Loosen the valve adjusting screws.

17. Lift the camshaft holder plates and holders from the cylinder head. The holder bolts will keep the components together. Note the positions of each camshaft holder for reassembly.

18. Lift the camshafts from the cylinder head. Mark the exhaust and intake camshafts so that they will not be confused.

19. Remove the intake and exhaust oil control orifices. Thoroughly clean each, and reinstall them with new O-rings.

20. Inspect the camshaft lobes and journals for any signs of damage.

To install:

21. Install a new O-ring and the dowel pin to the oil passage of the No. 3 camshaft holder.

22. Lubricate the camshaft lobes and journals with clean engine oil.

23. Set the camshafts into the cylinder head. Both the intake and exhaust camshafts should be installed with their keyways pointing straight up.

24. Install new camshaft seals. Apply liquid gasket to a new camshaft end-plug and install it. If the end-plug has is marked, the mark should be aligned with the cylinder head surface.

25. Apply liquid gasket to the cylinder head mating surfaces of the No. 1 and No. 5 camshaft holders, then install them, along with No. 2, 3, and 4 holders. Be sure to pay attention to the following points:

• Do not apply oil to the holder mating surface of camshaft seals.

• The arrows marked on the camshaft holders should point to the timing belt.

26. Install the camshaft holder plates.

27. Lubricate the threads of the 8mm holder bolts. Then, install all the camshaft holder bolts, but don't tighten them yet.

28. Evenly hand-tighten the camshaft holders. Be sure that the rocker arms are properly positioned on the valve stems.

29. Use a two-step crisscross pattern to tighten the camshaft holder bolts. Begin tightening with the bolts in the middle of the cylinder head, and work toward the outer edges. Final torque specifications are as follows:

• 1996–97 B16A2 engine: 8mm bolts 20 ft. lbs. (28 Nm).

• 1995 B16A3 engines: 8mm bolts to 16 ft. lbs. (22 Nm)

• 1995–97: 6mm bolts to 7–8 ft. lbs. (10–11 Nm)

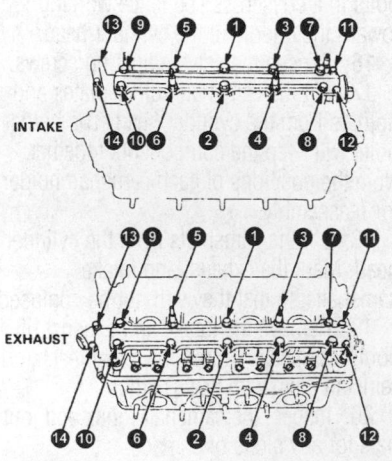

● – ⑩ 8 x 1.25 mm 22 N·m (2.2 kg-m, 16 lb-ft)
⑪ – ⑭ 6 x 1.0 mm 11 N·m (1.1 kg-m, 8 lb-ft)

7923FG64

Camshaft holder bolt tightening sequences—1.6L (B16A2, B16A3) engines

30. Verify that the camshaft keyways are pointing straight up and that the engine is at TDC/compression for the No. 1 cylinder. Fit the camshaft sprocket keys into their keyways.

31. Install the back cover and push the camshaft pulleys onto the camshafts. Then, tighten the sprocket retaining bolts to 37 ft. lbs. (51 Nm) for 1995 vehicles, or 41 ft. lbs. (57 Nm) for 1996 vehicles.

32. Install and tension the timing belt.

33. Install the lower timing cover and the crankshaft pulley. Tighten the pulley bolt to 130 ft. lbs. (180 Nm).

34. Adjust the valve clearance.

35. Inspect the VTEC rocker arms for smooth and independent movement.

36. Install the VTEC solenoid valve with a new filter. Tighten the valve mounting bolts to 9 ft. lbs. (12 Nm).

37. Install the distributor.

38. Clean the valve cover gasket surfaces. Fit the gasket into the groove of the valve cover.

39. Apply liquid gasket to the rubber seal at the eight corners where the gasket meets the camshaft holders. Don't allow the sealant to cure before installing the cylinder head cover.

40. Install the cylinder head cover and engine ground cable. Be sure the contact surfaces are clean and do not touch surfaces where liquid gasket has been applied.

41. Tighten the valve cover nuts in to 7 ft. lbs. (10 Nm) in a crisscross sequence.

42. Install the upper timing cover.

43. Install the accessory drive belts and adjust their tensions.

44. Install the engine side mount, tighten the two new nuts to 54 ft. lbs. (75 Nm) and tighten the bolt attaching the mount to the vehicle to 54 ft. lbs. (74 Nm).

45. Install the strut bar and tighten the mounting bolts to 16 ft. lbs. (22 Nm).

46. Reconnect the ignition wires.

47. Drain the engine oil. Install a new oil filter and refill the engine with fresh oil.

48. Reconnect the negative battery cable.

49. Warm the engine up to its normal operating temperature. Then, check and adjust the ignition timing. Tighten the distributor mounting bolts to 17 ft. lbs. (24 Nm).

Accord and Prelude

2.2L (H22A1) ENGINE

1. Disconnect the negative battery cable.

2. Turn the crankshaft so the No. 1 piston is at top dead center.

➡ **The No. 1 piston is at top dead center when the pointer on the block aligns with the white painted mark on the flywheel (manual transaxle) or driveplate (automatic transaxle).**

3. Remove the air intake duct.

4. Remove the engine ground cable from the cylinder head cover.

5. Remove the connector and the terminal from the alternator, then remove the engine wiring harness from the valve cover.

6. Loosen the mounting bolts and remove drive belt from the power steering pump.

7. Remove the ignition coil.

8. Remove the plug wire cover from the cylinder head cover. Label, then disconnect the electrical connectors from the distributor and the spark plug wires from the spark plugs. Mark the position of the distributor and remove it from the cylinder head.

9. Remove the Positive Crankcase Ventilation (PCV) hose, then remove the cylinder head cover. Replace the rubber seals if damaged or deteriorated.

10. Remove the timing belt middle cover.

11. Ensure the arrows embossed on the camshaft pulleys are aligned in the upward position (12 o'clock).

12. Mark the rotation of the timing belt if it is to be used again.

13. Loosen the timing belt adjusting nut, do not loosen the nut more than one turn.

14. Use a open end wrench to loosen the maintenance bolt for the timing belt tensioner. If the maintenance bolt cannot be loosen with an open end wrench, a box wrench can be used after removing the lock pin.

➡ **The use of a wrench to loosen the maintenance bolt should be limited to the bolts initial loosening only.**

15. Loosen the maintenance bolt by hand until it stops. The auto-tensioner bracket is now fixed.

✳✳ WARNING

Never use a tool to loosen the maintenance bolt after the initial loosening.

16. Tighten the timing belt adjusting nut.

17. Remove the timing belt from the camshaft sprockets.

✳✳ WARNING

Do not crimp or bend the timing belt more than 90° or less, then 1 in. (25mm) in diameter

18. Remove the camshaft sprocket attaching bolts, then remove the sprockets. Do not lose the sprocket keys.

19. Remove the side engine mount bracket B, then the timing belt back cover.

20. Loosen all of the rocker arm adjusting screws.

21. Remove the camshaft holder bolts two turns at a time, in the reverse of the tightening sequence, to prevent damaging the valves or the rocker arm assemblies. Remove the camshaft holder plates and the camshaft holders, note the holders locations for ease of installation.

22. Remove the rubber cap from the head, located at the end of the intake camshaft. Replace the rubber cap if damaged or deteriorated.

23. Remove the camshafts and discard the camshaft seals.

To install:

24. Lubricate the rocker arms and camshafts with clean oil.

25. Install the camshaft seals to the end of the camshafts that the timing belt sprockets attach to. The open side (spring) should be facing into the cylinder head when installed.

26. Be sure the keyways on the camshafts are facing up and install the camshafts to the cylinder head.

27. Install the rubber cap to the cylinder head at the end of the intake camshaft.

28. Clean the oil control orifice and install a new O-ring. Install the oil control orifice into the oil passage of the No. 3 camshaft holder.

29. Apply liquid gasket to the head mating surfaces of the No. 1 and No. 5 camshaft holders, then install them along with No. 2, 3 and 4 camshaft holders. The arrows stamped on the holders should point toward the timing belt.

30. Install the camshaft holder plates and attaching bolts, then snug the camshaft holders in place.

31. Press the camshaft seals in securely against the base of the camshaft holders.

32. Tighten the camshaft holder bolts in two steps, following the proper sequence, to ensure that the rockers do not bind on the valves. Tighten the 8X 23mm bolts to 19 ft. lbs. (26 Nm). Tighten the 6X1.0mm bolts to 9 ft. lbs. (12 Nm).

33. Install the timing belt back cover and tighten the attaching bolts to 9 ft. lbs. (12 Nm).

34. Install the side engine mount bracket B. Tighten the bolt attaching the bracket to the cylinder head to 33 ft. lbs. (45 Nm) and tighten the bolts attaching the bracket to the side engine mount to 16 ft. lbs. (22 Nm).

35. Install the keys into the camshaft grooves. Push the camshaft sprockets onto the camshafts, then tighten the retaining bolts to 37 ft. lbs. (51 Nm).

36. Ensure the arrows embossed on the camshaft pulleys are aligned in the upward position (12 o'clock). Ensure that the TDC mark on the flywheel is aligned with the pointer on the engine block.

37. Install the timing belt slider (part # 07NAG-P130100) to the intake camshaft sprocket, then install the timing belt to the camshaft sprockets.

➡ **If the auto—tensioner has been extended and the timing belt cannot be installed, remove the auto tensioner, compress it and reinstall it.**

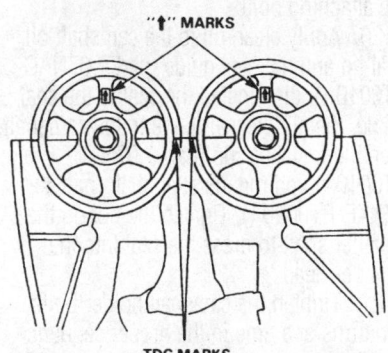

"↑" MARKS

TDC MARKS

CRANKSHAFT TDC POSITION:

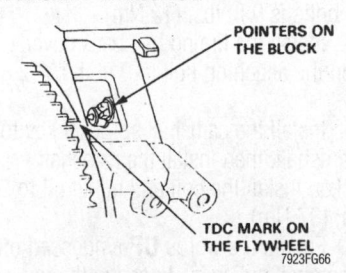

POINTERS ON THE BLOCK

TDC MARK ON THE FLYWHEEL
7923FG66

Timing marks—2.2L (H22A1) engine

38. Tighten the auto-tensioner maintenance bolt to 16 ft. lbs. (22 Nm), this will make the auto-tensioner functional.

39. Loosen the timing belt adjusting nut.

40. Turn the crankshaft pulley 1 turn, then tighten the adjusting nut to 33 ft. lbs. (45 Nm).

41. Adjust the valves.

42. Tighten the crankshaft pulley bolt to 181 ft. lbs. (250 Nm).

43. Install the middle timing belt cover and tighten the cover attaching bolts to 9 ft. lbs. (12 Nm).

44. Install the cylinder head cover, tighten the cap nuts to 7 ft. lbs. (10 Nm). Install the PCV hose to the cylinder head cover.

45. Install the distributor to the cylinder head, snug the attaching bolts until the timing has been checked and adjusted.

46. Connect the spark plug wires to the correct spark plugs, then connect the distributor electrical connectors.

47. Install the spark plug wire cover to the cylinder head cover and tighten the cap nuts to 7 ft. lbs. (10 Nm).

48. Install the ignition coil.

49. Install and adjust the power steering belt.

50. Install the alternator wiring harness to the cylinder head cover, then connect the terminal and connector to the alternator.

51. Connect the engine ground cable to the cylinder head cover.

52. Install the air intake duct.

53. Drain the oil from the engine into a sealable container. Install the drain plug and refill the engine with clean oil.

54. Connect the negative battery cable and enter the radio security code.

55. Start the engine, checking carefully for any leaks.

56. Check and adjust the ignition timing, then tighten the distributor mounting bolts to 13 ft. lbs. (18 Nm).

57. If equipped with 4WS, turn the steering wheel lock-to-lock to reset the 4WS control unit.

2.2L (F22A1, F22A6, F22B1 AND F22B2) ENGINES

1. Disconnect the negative battery cable.

2. Remove the air intake duct.

3. Remove the cylinder head cover and replace the rubber seals if damaged or deteriorated.

4. Remove the timing belt upper cover.

5. Turn the engine to align the timing marks and set cylinder No.1 to TDC, compression. Once in this position, the engine must NOT be turned or disturbed.

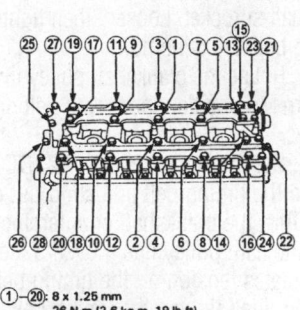

25 27 19 17 11 9 3 1 7 5 13 23 21
⑮

26 28 20 18 10 12 2 4 6 8 14 16 24 22

① – ㉑ 8 x 1.25 mm
    26 N·m (2.6 kg-m, 19 lb-ft)
㉑ – ㉘ 6 x 1.0 mm
    12 N·m (1.2 kg-m, 9 lb-ft)
7923FG65

Camshaft holders torque sequence—2.2L (H22A1) engine

6. Label, then disconnect the electrical connectors from the distributor and the spark plug wires from the spark plugs. Mark the position of the distributor and remove it from the cylinder head.

7. Loosen the mounting bolts and remove drive belt from the power steering pump.

8. Mark the rotation of the timing belt if it is to be used again. Loosen the timing belt adjusting bolt 3/4 to one turn, then release the tension on the timing belt. Push the tensioner to release tension from the belt, then tighten the adjusting bolt.

9. Remove the timing belt from the camshaft sprocket.

✴✴ WARNING

Do not crimp or bend the timing belt more than 90° or less, then 1 in. (25mm) in diameter

10. Ensure the words **UP** embossed on the camshaft pulley is aligned in the upward position, then remove the camshaft sprocket bolt. Pull the sprocket from the camshaft and remove the sprocket key.

11. Remove the timing belt back cover.

12. Loosen the valve adjusting screws.

13. Loosen the camshaft holder attaching bolts two turns at a time, in the proper sequence to prevent damaging the valves or rocker arm assemblies.

➡ **When removing the rocker arm assembly, do not remove the camshaft holder bolts. The bolts will keep the camshaft holders, springs, and the rocker arms on the shafts.**

14. Carefully remove the camshaft holders and rocker arm assembly. If the rocker

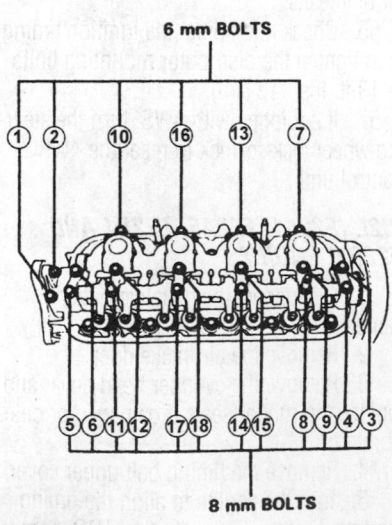

Rocker arm assembly bolt removal sequence—2.2L (F22A1, F22A6) engines

arm and shaft assembly needs to be disassembled for service, note the location of the components as they are removed. The rocker arms must be installed in the same position if reused.

15. Remove the camshaft from the cylinder head and discard the seal.

To install:

16. Wipe the camshaft and the camshaft journals clean, then lubricate both surfaces and install the camshaft.

17. Turn the camshaft so that its keyway is facing up (No. 1 cylinder will be at TDC).

18. Reassemble the rocker arm and shaft assembly, if it was disassembled. Lubricate the rocker arm and shaft assembly with clean oil, then apply liquid gasket to the head mating surfaces of the No. 1 and No. 6 camshaft holders.

19. Set the camshaft holders and rocker arm assembly in place, then loosely install the attaching bolts.

20. Apply clean oil to the camshaft oil seal lip and the seal guide (part # 07NAG-PT0010A), then install the seal to the seal guide. Install the seal guide to the camshaft, then the installer cup (part # 07NAF-PT0010A) and the installer shaft (part # 07NAF-PT0020A). Tighten the nut on the installer shaft to press the seal into the cylinder head.

21. Tighten the camshaft holder bolts two turns at a time in the proper sequence. The final torque for the 8mm bolts is 16 ft. lbs. (22 Nm) and the final torque for the 6mm bolts is 9 ft. lbs. (12 Nm).

22. Install the timing belt back cover, tighten the attaching bolt to 9 ft. lbs. (12 Nm).

23. Install the camshaft sprocket key to the camshaft, then install the camshaft sprocket. Install the bolt and tighten it to 27 ft. lbs. (37 Nm).

24. Ensure the words **UP** embossed on the camshaft pulley is aligned in the upward position, then install the timing belt onto the

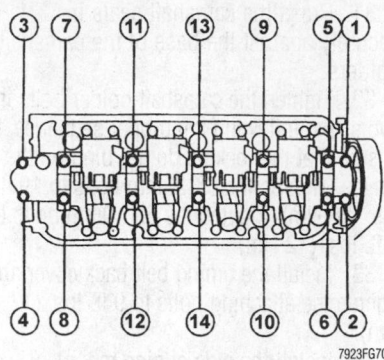

Rocker arm assembly bolt loosening sequence—2.2L (F22B1) engine

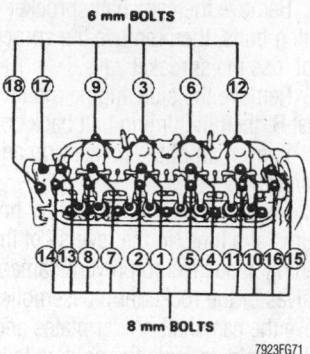

Rocker arm assembly torque sequence—2.2L (F22A1, F22A6) engines

Specified torque:
8 mm bolts: 22 N·m (2.2 kgf·m, 16 lbf·t)
6 mm bolts: 12 N·m (1.2 kgf·m, 8.7 lbf·t)
6 mm bolts: ⑪,⑫,⑬,⑭

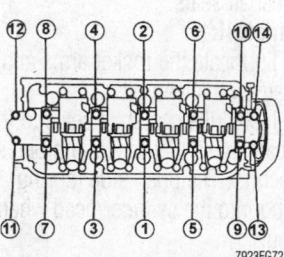

Rocker arm assembly torque sequence—2.2L (F22B1) engine

Specified torque:
8 mm bolts: 22 N·m (2.2 kgf·m, 16 lbf·t)
6 mm bolts: 12 N·m (1.2 kgf·m, 8.7 lbf·t)

6 mm bolts: ③,⑥,⑨,⑫,⑰,⑱

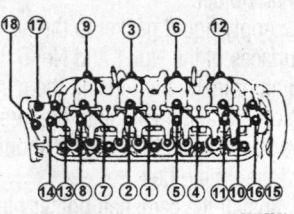

Rocker arm assembly torque sequence—2.2L (F22B2) engine

camshaft sprocket. Loosen, then tighten the timing belt adjusting nut.

25. Rotate the crankshaft pulley five or six turns to position the timing belt on the pulleys.

26. Set the No. 1 cylinder to TDC and loosen the timing belt adjusting nut one turn. Turn the crankshaft counterclockwise until the cam pulley has moved 3 teeth; this creates tension on the timing belt. Loosen, then tighten the timing belt adjusting nut and tighten it to 33 ft. lbs. (45 Nm).

27. Adjust the valves.

28. Tighten the crankshaft pulley bolt to 181 ft. lbs. (245 Nm) on F22B1 and F22B2 models and 159 ft. lbs. (220 Nm) on all other models.

29. Install the upper timing belt cover and tighten the bolt to 9 ft. lbs. (12 Nm).

30. Install the cylinder head cover gasket cover to the groove of the cylinder head cover. Before installing the gasket thoroughly clean the seal and the groove. Seat the recesses for the camshaft first, then work it into the groove around the outside edges. Be sure the gasket is seated securely in the corners of the recesses.

31. Install the cylinder head (valve) cover and tighten the cap nuts to 7 ft. lbs. (10 Nm).

32. Install and adjust the power steering belt.

33. Install the distributor to the cylinder head, snug the attaching bolts until the timing has been checked and adjusted.

34. Connect the spark plug wires to the correct spark plugs, then connect the distributor electrical connectors. Install the ignition coil wire to the distributor.

35. Install the air intake duct.

36. Drain the oil from the engine into a sealable container. Install the drain plug and refill the engine with clean oil.

37. Connect the negative battery cable and enter the radio security code.

38. Start the engine, checking carefully for any leaks.

39. Check and adjust the ignition timing as necessary, then tighten the distributor bolts to 16 ft. lbs. (22 Nm).

2.3L (H23A1) ENGINE

1. Disconnect the negative battery cable.

2. Turn the crankshaft so the No. 1 piston is at top dead center.

➡The No. 1 piston is at top dead center when the pointer on the block aligns with the white painted mark on the flywheel (manual transaxle) or driveplate (automatic transaxle).

3. Remove the air intake duct.

4. Remove the engine ground cable from the cylinder head cover.

5. Remove the connector and the terminal from the alternator, then remove the engine wiring harness from the valve cover.

6. Remove the ignition coil.

7. Label, then disconnect the electrical connectors from the distributor and the spark plug wires from the spark plugs. Mark the position of the distributor and remove it from the cylinder head. Disconnect the ignition coil wire from the distributor.

8. Remove the Positive Crankcase Ventilation (PCV) hose, then remove the cylinder head cover. Replace the rubber seals if damaged or deteriorated.

9. Remove the timing belt middle cover.

10. Ensure the words **UP** embossed on the camshaft pulleys are aligned in the upward position.

11. Mark the rotation of the timing belt if it is to be used again. Loosen the timing belt adjusting nut 1/2 turn, then release the tension on the timing belt. Push the tensioner to release tension from the belt, then tighten the adjusting nut.

12. Remove the timing belt from the camshaft sprockets.

✳✳ WARNING

Do not crimp or bend the timing belt more than 90° or less, then 1 in. (25mm) in diameter

13. Insert a 5.0mm pin punch in each of the camshaft caps, nearest to the sprockets, through the holes provided. Remove the camshaft sprocket attaching bolts, then remove the sprockets. Do not lose the sprocket keys.

14. Remove the side engine mount bracket B, then the timing belt back cover from behind the camshaft sprockets.

15. Loosen all of the rocker arm adjusting screws, then remove the pin punches from the camshaft caps.

16. Remove the camshaft holders, note the holders locations for ease of installation. Loosen the bolts in the reverse order of the installation.

17. Remove the camshafts from the cylinder head, then discard the camshaft seals.

18. Remove the rubber cap from the head, located at the end of the intake camshaft.

19. Remove the rocker arms from the cylinder head. Note the locations of the rocker arms.

➡The rocker arms have to be installed to their original locations if being reused.

To install:

20. Lubricate the rocker arms with clean oil, then install the rocker arms on the pivot bolts and the valve stems. If the rocker arms are being reused, install them to their original locations. The locknuts and adjustment screws should be loosened before installing the rocker arms.

21. Lubricate the camshafts with clean oil.

22. Install the camshaft seals to the end of the camshafts that the timing belt sprockets attach to. The open side (spring) should

be facing into the cylinder head when installed.

23. Be sure the keyways on the camshafts are facing up and install the camshafts to the cylinder head.

24. Install the rubber plug to the cylinder head at the end of the intake camshaft.

25. Apply liquid gasket to the head mating surfaces of the No. 1 and No. 6 camshaft holders, then install them along with No. 2, 3, 4 and 5. **I** or **F** marks are stamped on the camshaft holders to identify them as Intake or Exhaust side holders. The arrows stamped on the holders should point toward the timing belt.

26. Snug the camshaft holders in place.

27. Press the camshaft seals securely into place.

28. Tighten the camshaft holder bolts in two steps, following the proper sequence, to ensure that the rockers do not bind on the valves. Tighten all the bolts, except the four studs, to 7 ft. lbs. (10 Nm). Tighten the studs (number 5 and 7 bolts in the correct sequence) to 9 ft. lbs. (12 Nm).

29. Install the timing belt back cover.

30. Install the side engine mount bracket B. Tighten the bolt attaching the bracket to the cylinder head to 33 ft. lbs. (45 Nm). Tighten the bolts attaching the bracket to the side engine mount to 16 ft. lbs. (22 Nm).

31. Insert a 5.0mm pin punch in each of the camshaft caps, nearest to the pulleys, through the holes provided. Install the keys into the camshaft grooves.

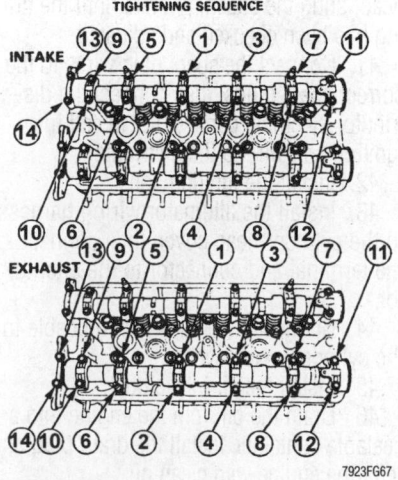

Specified torque:
Except Intake ⑤, ⑦. Exhaust ⑥, ⑧:
      10 N·m (1.0 kg-m, 7 lb-ft)
Intake ⑤, ⑦. Exhaust ⑥, ⑧:
      12 N·m (1.2 kg-m, 9 lb-ft)

TIGHTENING SEQUENCE

7923FG67

Camshaft holders torque sequence—2.3L (H23A1) engine

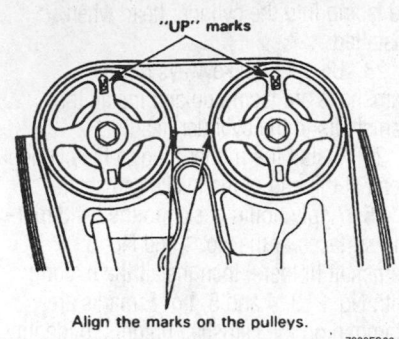

Align the marks on the pulleys.

7923FG68

Camshaft sprockets alignment—2.3L (H23A1) engine

32. Push the camshaft sprockets onto the camshafts, then tighten the retaining bolts to 27 ft. lbs. (38 Nm).

33. Ensure the words **UP** embossed on the camshaft pulleys are aligned in the upward position. Install the timing belt to the camshaft sprockets, then remove the two 5.0mm pin punches from the camshaft bearing caps.

34. Loosen, then tighten the timing belt adjuster nut.

35. Turn the crankshaft counterclockwise until the cam pulley has moved 3 teeth; this creates tension on the timing belt. Loosen, then tighten the adjusting nut and tighten it to 33 ft. lbs. (45 Nm).

36. Adjust the valves.

37. Tighten the crankshaft pulley bolt to 181 ft. lbs. (250 Nm).

38. Install the middle timing belt cover and tighten the attaching bolts to 9 ft. lbs. (12 Nm).

39. Install the cylinder head cover and tighten the cap nuts to 7 ft. lbs. (10 Nm). Install the PCV hose to the cylinder head cover.

40. Install the distributor to the cylinder head, snug the attaching bolts until the timing has been checked and adjusted.

41. Connect the spark plug wires to the correct spark plugs, then connect the distributor electrical connectors. Install the ignition coil wire to the distributor.

42. Install the ignition coil.

43. Install the alternator wiring harness to the cylinder head cover, then connect the terminal and connector to the alternator.

44. Connect the engine ground cable to the cylinder head cover.

45. Install the air intake duct.

46. Drain the oil from the engine into a sealable container. Install the drain plug and refill the engine with clean oil.

47. Connect the negative battery cable and enter the radio security code.

48. Start the engine, checking carefully for any leaks.

49. Check and adjust the ignition timing. Tighten the distributor bolts to 13 ft. lbs. (18 Nm).

50. If equipped with 4WS, start the engine, then turn the steering wheel lock-to-lock to reset the 4WS control unit.

2.7L ENGINE

1. Disconnect the negative battery cable.

2. Turn the engine to align the timing marks and set cylinder No.1 to TDC. The white mark on the crankshaft pulley should align with the pointer on the timing belt cover. Remove the inspection caps on the upper timing belt covers to check the alignment of the timing marks. The pointers for the camshafts should align with the green marks on the camshaft pulleys.

3. Remove the intake air duct.

4. Remove the starter cable from the strut brace, then remove the strut brace.

5. Remove intake manifold covers.

6. Disconnect the breather hose from the cylinder head cover.

7. Remove the PCV hose from the cylinder head cover.

8. Loosen the idler pulley center nut and adjusting bolt, then remove the air conditioning compressor belt.

9. Loosen the alternator mounting bolt, nut and adjusting bolt, then remove the alternator drive belt.

10. Loosen the power steering mounting nuts and adjusting nut, then remove the power steering drive belt.

11. Label, then disconnect the electrical connectors from the distributor and the spark plug wires from the spark plugs. Remove the distributor from the cylinder head.

12. Remove the cylinder head covers and the side covers.

13. Remove the timing belt covers and the timing belt.

14. Remove the camshaft sprockets and the timing belt back covers.

15. Remove the bolts attaching the camshaft holder plates and camshaft holders in the opposite order of the installation sequence.

16. Remove the camshaft holder plates, camshaft holders and the dowel pins. Discard the O-rings.

17. Remove the camshafts from the cylinder heads and the rubber cap from the rear cylinder head. Discard the camshaft seals.

To install:

18. Apply clean engine oil to the rocker arms and the camshafts.

19. Loosen the exhaust rocker arm adjusting screws and locknuts.

20. Be sure the rocker arms are properly positioned on the valve stems. Advance the crankshaft 30° from TDC to prevent interference between the pistons and valves, then install the camshafts. Position the rear camshaft on the cylinder head so the cam is not pushing on any valves.

21. Apply liquid gasket around the rubber cap, then install it to the cylinder head.

22. Install the camshaft seals to the camshafts with the open side (spring) facing in.

23. Apply liquid gasket the cylinder head and camshaft holder mating surfaces, then install the camshaft holders and the camshaft plates with the dowel pins. Install new O-rings to the camshaft holder plates

24. Apply clean oil to the camshaft holder bolts, then install the bolts and tighten them in the proper sequence. Tighten the 8mm bolts to 20 ft. lbs. (27 Nm), and the 6mm bolts to 8.7 ft. lbs. (12 Nm).

25. Install the timing belt back covers, tighten the attaching bolts to 9 ft. lbs. (12 Nm).

26. Install the camshaft sprockets and tighten the attaching bolts to 23 ft. lbs. (31 Nm).

27. Set the camshaft sprockets so that the No. 1 piston is at TDC. Align the TDC marks (green mark) on the camshaft pulleys to the pointers on the back covers.

28. Turn the crankshaft counterclockwise to set it at TDC. Align the TDC mark on the

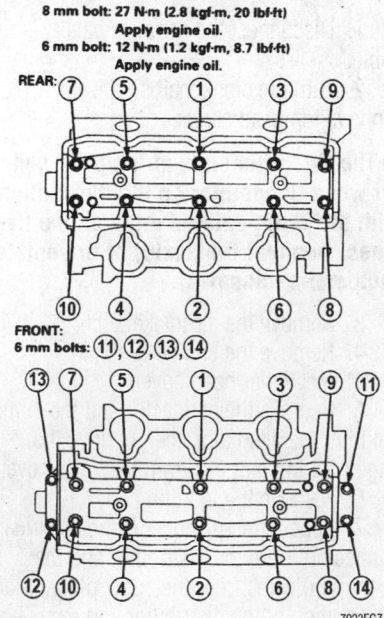

8 mm bolt: 27 N-m (2.8 kgf-m, 20 lbf-ft)
Apply engine oil.
6 mm bolt: 12 N-m (1.2 kgf-m, 8.7 lbf-ft)
Apply engine oil.

7923FG74

Camshaft holders torque sequence—2.7L engine

tooth of the timing belt drive pulley with the pointer on the oil pump.

29. Install the timing belt and timing belt covers.

30. Set No. 1 cylinder to TDC.

31. Tighten the valve adjusting screws for No. 1, No. 2 and No. 4 cylinders. Tighten the screw until it contacts the valve, then tighten the screw 1 1/8 turns. Hold the screw in place and tighten the locknut to 14 ft. lbs. (20 Nm).

32. Rotate the crankshaft pulley one turn clockwise, then tighten the adjusting screws for No. 3, No. 5 and No. 6 cylinders. Tighten the screw until it contacts the valve, then tighten the screw 1 1/8 turns. Hold the screw in place and tighten the locknut to 14 ft. lbs. (20 Nm).

33. Install the cylinder head cover gasket into the groove of the cylinder head cover. Seat the recesses for the camshaft first, then work it into the groove around the outside edges.

➡**Before installing the cylinder head cover gasket, thoroughly clean the seal groove.**

34. Apply liquid gasket to the cylinder head cover gasket at the four corners of the recesses. Use a shop towel and wipe the

cylinder heads where the cylinder head covers will come in contact.

35. Install the cylinder head covers, hold the gasket in the groove by placing your fingers on the camshaft contacting surfaces. With the cylinder head cover on the cylinder heads, slide the covers slightly back and forth to seat the cylinder head cover gaskets. Replace the washers if damaged or deteriorated

36. Tighten the cylinder head cover bolts in two or three steps. In the final step, tighten all the bolts, in sequence, to 11 ft. lbs. (15 Nm).

37. Install the cylinder head side covers with new O-rings and tighten the bolts to 9 ft. lbs. (12 Nm).

38. Install the distributor to the cylinder head and tighten the mounting bolt to 16 ft. lbs. (22 Nm).

39. Connect the spark plug wires to the correct spark plugs, then connect the distributor electrical connectors.

40. Install and adjust the power steering belt.

41. Install and adjust the alternator belt. Tighten the alternator mounting nut and bolt to 16 ft. lbs. (22 Nm).

42. Install the A/C belt and adjust the belt tension. Tighten the idler center nut to 33 ft. lbs. (44 Nm).

43. Install the PCV hose to the cylinder head cover.

44. Connect the breather hose to the cylinder head cover.

45. Install the intake manifold cover and tighten the bolts to 9 ft. lbs. (12 Nm).

46. Install the intake air duct.

47. Install the strut brace and tighten the mounting bolts to 16 ft. lbs. (22 Nm). Install the starter cable to the strut brace.

48. Drain the engine oil into a sealable container, then refill the engine with clean oil.

49. Connect the negative battery cable and enter the radio security code.

50. Start the engine, allowing it to idle and check for any signs of leakage.

Valve Clearance

ADJUSTMENT

➡**The radio may contain a coded theft protection circuit. Always obtain the code number before disconnecting the battery. If the vehicle is equipped with 4WS, the steering control unit is shut down when the battery is disconnected. After connecting the battery, turn the steering wheel lock-to-lock to reset the steering control unit.**

Del Sol and Civic

1. Disconnect the negative battery cable.

2. Remove the cylinder head cover and the upper timing belt cover.

3. Rotate the crankshaft to align the white TDC mark on the crankshaft pulley with the pointer on the cover for the No. 1 cylinder compression stroke. Be sure the **UP** mark on the camshaft sprocket is up and the TDC marks align with the edge of the cylinder head.

4. Hold a No. 1 cylinder rocker arm against the camshaft and use a feeler gauge to check the clearance at the valve stem. Except on B16A2 and B16A3 engines, intake valve clearance should be 0.007–0.009 in. (0.18–0.26mm), exhaust valve clearance should be 0.009–0.011 in. (0.23–0.27mm). On B16A2 and B16A3 engines, the intake valve clearance should be 0.006–0.007 in. (0.15–0.19mm), exhaust valve clearance should be 0.007–0.008 in. (0.17–0.21mm). Loosen the locknut and turn the adjusting screw to adjust the clearance. Tighten the locknut to 10 ft. lbs. (14 Nm) on D15B7 and D15B8 engines and 14 ft. lbs. (20 Nm) on all other models and recheck the clearance. Don't

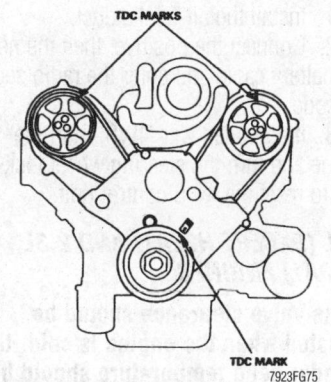

Timing marks—2.7L engine

7923FG75

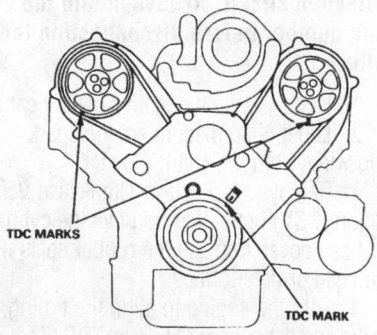

Timing marks for valve adjustment on cylinders 3, 5 and 6—2.7L engine

7923FG76

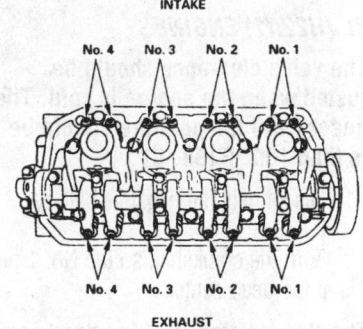

7923FG77

Valve adjuster locations—except 1.5L (D15B8) engine

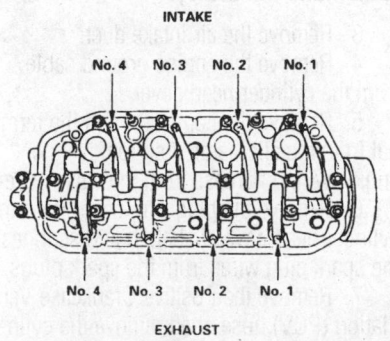

7923FG78

Valve adjuster locations—1.5L (D15B8) engine

overtighten the locknut, the aluminum rockers will strip easily.

5. The adjustment order is 1–3–4–2. Rotate the crankshaft counterclockwise 180° (the camshaft sprocket will rotate 90°) to bring each cylinder to TDC/compression. Adjust each set of valves.

a. At TDC for the No. 3 cylinder, the UP mark is pointed to the exhaust side of the cylinder head.

b. At TDC for the No. 4 cylinder, the UP mark is pointed down, and the TDC marks align with the edge of the cylinder head.

c. At TDC for the No. 2 cylinder, the UP mark is pointed to the intake side of the cylinder head.

6. After adjusting the valves of a VTEC engine, inspect its intake rocker arms for smooth and independent motion.

7. Apply sealant to the edges of the valve cover gasket where it meets the camshaft holders. Be sure the spark plug tube seals are properly seated.

8. Install the cylinder head and timing belt covers.

9. Tighten the crankshaft pulley bolt to 134 ft. lbs. (185 Nm).

10. Reconnect the negative battery cable. Enter the radio security code.

Prelude and Accord

2.2L (H22A1) ENGINE

➡**The valve clearance should be adjusted when the engine is cold. The cylinder head temperature should be less than 100° F (38° C)**

1. Disconnect the negative battery cable.

2. Turn the crankshaft so the No. 1 piston is at top dead center.

➡**The No. 1 piston is at top dead center when the pointer on the block aligns with the white painted mark on the flywheel (manual transaxle) or driveplate (automatic transaxle).**

3. Remove the air intake duct.

4. Remove the engine ground cable from the cylinder head cover.

5. Remove the connector and the terminal from the alternator, then remove the engine wiring harness from the valve cover.

6. Remove the plug wire cover from the cylinder head cover. Label, then disconnect the spark plug wires from the spark plugs.

7. Remove the Positive Crankcase Ventilation (PCV) hose, then remove the cylinder head cover. Replace the rubber seals if damaged or deteriorated.

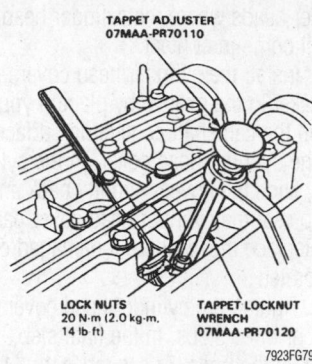

Checking and adjusting the valve clearance—2.2L (H22A1) engine

8. Ensure the arrows embossed on the camshaft pulleys are aligned in the upward position (12 o'clock).

9. Adjust the valves on cylinder No. 1:

a. Insert a feeler gauge in between the camshaft lobe and the rocker arm.

➡**The intake valve clearance specification is 0.006–0.007 in (0.15–0.19mm) and the exhaust valve clearance specification is 0.007–0.008 in. (0.17–0.21mm).**

b. Loosen the locknut and turn the adjusting screw until the feeler gauge slides back and forth with a slight amount of drag.

c. Tighten the locknut and recheck the valve clearance. Repeat the valve adjustment if necessary.

10. Rotate the crankshaft 180° counterclockwise (the camshaft pulleys will turn 90°) The arrow marks should be pointing to the exhaust side of the cylinder head.

11. Adjust the valves on cylinder No. 3:

a. Insert a feeler gauge in between the camshaft lobe and the rocker arm.

b. Loosen the locknut and turn the adjusting screw until the feeler gauge slides back and forth with a slight amount of drag.

c. Tighten the locknut and recheck the valve clearance. Repeat the valve adjustment if necessary.

12. Rotate the crankshaft 180° counterclockwise (the camshaft pulleys will turn 90°) to bring No. 4 piston to TDC. The arrow marks should be pointing down, toward the crankshaft.

13. Adjust the valves on cylinder No. 4:

a. Insert a feeler gauge in between the camshaft lobe and the rocker arm.

b. Loosen the locknut and turn the adjusting screw until the feeler gauge slides back and forth with a slight amount of drag.

c. Tighten the locknut and recheck the valve clearance. Repeat the valve adjustment if necessary.

14. Rotate the crankshaft 180° counterclockwise (the camshaft pulleys will turn 90°) to bring piston No. 2 to TDC. The arrow marks should be pointing to the intake side of the cylinder head.

15. Adjust the valves on cylinder No. 2:

a. Insert a feeler gauge in between the camshaft lobe and the rocker arm.

b. Loosen the locknut and turn the adjusting screw until the feeler gauge slides back and forth with a slight amount of drag.

c. Tighten the locknut and recheck the valve clearance. Repeat the valve adjustment if necessary.

16. Install the cylinder head cover, tighten the cap nuts to 7 ft. lbs. (10 Nm). Install the PCV hose to the cylinder head cover.

17. Connect the spark plug wires to the correct spark plugs.

18. Install the spark plug wire cover to the cylinder head cover and tighten the cap nuts to 7 ft. lbs. (10 Nm).

19. Install the alternator wiring harness to the cylinder head cover, then connect the terminal and connector to the alternator.

20. Connect the engine ground cable to the cylinder head cover.

21. Install the air intake duct.

22. Connect the positive, then the negative battery cable and enter the radio security code.

23. If equipped with 4WS, Start the engine and turn the steering wheel lock-to-lock to reset the 4WS control unit.

2.2L (EXCEPT H22A1) AND 2.3L (H23A1) ENGINES

➡**The valve clearance should be adjusted when the engine is cold, the cylinder head temperature should be less than 100° F (38° C).**

➡**The radio may contain a coded theft protection circuit. Always obtain the code number before disconnecting the battery.**

1. Disconnect the negative battery cable.

2. Label, then disconnect the spark plug wires from the spark plugs.

3. Remove the Positive Crankcase Ventilation (PCV) hose, then remove the cylinder head cover. Replace the rubber seals if damaged or deteriorated.

4. Turn the engine to align the timing marks and set cylinder No.1 to TDC. The white mark on the crankshaft pulley should align with the pointer on the timing belt

cover. The words **UP** embossed on the camshaft pulley should be aligned in the upward position. The marks on the edge of the pulley should be aligned with the cylinder head or the back cover upper edge.

5. Adjust the valves on cylinder No. 1 by performing the following:

a. Insert a feeler gauge in between the camshaft lobe and the rocker arm.

➡**The intake valve clearance specification is 0.009–0.011 in (0.24–0.28mm) and the exhaust valve clearance specification is 0.011–0.013 in. (0.27–0.32mm).**

b. Loosen the locknut and turn the adjusting screw until the feeler gauge slides back and forth with a slight amount of drag.

c. Tighten the locknut to 14 ft. lbs. (20 Nm) and recheck the valve clearance. Repeat the valve adjustment if necessary.

6. Rotate the crankshaft 180° counterclockwise (the camshaft pulleys will turn 90°) The **UP** arrow marks should be pointing to the exhaust side of the cylinder head.

7. Adjust the valves on cylinder No. 3 by performing the following:

a. Insert a feeler gauge in between the camshaft lobe and the rocker arm.

b. Loosen the locknut and turn the adjusting screw until the feeler gauge slides back and forth with a slight amount of drag.

c. Tighten the locknut to 14 ft. lbs. (20 Nm) and recheck the valve clearance. Repeat the valve adjustment if necessary.

8. Rotate the crankshaft 180° counterclockwise (the camshaft pulleys will turn 90°) to bring No. 4 piston to TDC. The **UP** arrow marks should be pointing down, toward the crankshaft.

9. Adjust the valves on cylinder No. 4 by performing the following:

a. Insert a feeler gauge in between the camshaft lobe and the rocker arm.

b. Loosen the locknut and turn the adjusting screw until the feeler gauge slides back and forth with a slight amount of drag.

c. Tighten the locknut to 14 ft. lbs. (20 Nm) and recheck the valve clearance. Repeat the valve adjustment if necessary.

10. Rotate the crankshaft 180° counterclockwise (the camshaft pulleys will turn 90°) to bring piston No. 2 to TDC. The **UP** arrow marks should be pointing to the intake side of the cylinder head.

11. Adjust the valves on cylinder No. 2 by performing the following:

a. Insert a feeler gauge in between the camshaft lobe and the rocker arm.

b. Loosen the locknut and turn the adjusting screw until the feeler gauge slides back and forth with a slight amount of drag.

c. Tighten the locknut to 14 ft. lbs. (20 Nm) and recheck the valve clearance. Repeat the valve adjustment if necessary.

12. Install the cylinder head cover gasket cover to the groove of the cylinder head cover. Before installing the gasket thoroughly clean the seal and the groove. Seat the recesses for the camshaft first, then work it into the groove around the outside edges. Be sure the gasket is seated securely in the corners of the recesses.

13. Apply liquid gasket to the four corners of the recesses of the cylinder head cover gasket. Do not install the parts if 5 minutes or more have elapsed since applying liquid gasket. After assembly, wait at least 20 minutes before filling the engine with oil.

14. Install the cylinder head (valve) cover. Tighten the bolts attaching to 7 ft. lbs. (10 Nm).

15. Connect the spark plug wires to the correct spark plugs.

16. Connect the positive, then the negative battery cable and enter the radio security code.

17. If equipped with 4WS, turn the steering wheel lock-to-lock to reset the 4WS control unit.

2.7L ENGINE

The V-6 engine in the Accord has adjustments on the exhaust rocker arm assemblies that are located under the cylinder head side covers. The exhaust inside rocker arms and intake rocker arms operate with valve lifters (hydraulic tappets) and are not adjustable.

1. Disconnect the negative battery cable.

2. Turn the engine to align the timing marks and set cylinder No.1 to TDC. The white mark on the crankshaft pulley should align with the pointer on the timing belt cover. Remove the inspection caps on the upper timing belt covers to check the alignment of the timing marks. The green TDC marks on the camshaft pulleys should align with the yellow pointer mark on the timing belt back covers.

3. Remove the timing belt upper covers.

4. Remove the side covers from the cylinder heads.

5. Loosen the exhaust rocker arm adjusting screws and locknuts, and be sure the rocker arms are properly positioned on the valve stems.

6. Tighten the adjusting screws for No. 1, No. 2 and No. 4 cylinders. Tighten the screw until it contacts the valve, then tighten the screw 1 1/8 turns. Hold the screw in place and tighten the locknut to 14 ft. lbs. (20 Nm).

7. Rotate the crankshaft pulley one turn clockwise, then tighten the adjusting screws for No. 3, No. 5 and No. 6 cylinders. Tighten the screw until it contacts the valve, then tighten the screw 1 1/8 turns. Hold the screw in place and tighten the locknut to 14 ft. lbs. (20 Nm).

8. Install the cylinder head side covers with new O-rings and tighten the bolts to 9 ft. lbs. (12 Nm).

9. Install the timing belt upper covers and tighten the bolts to 9 ft. lbs. (12 Nm).

10. Connect the negative battery cable.

Oil Pan

REMOVAL & INSTALLATION

➡**The radio may contain a coded theft protection circuit. Always obtain the code number before disconnecting the battery. If the vehicle is equipped with 4WS, the steering control unit is shut down when the battery is disconnected. After connecting the battery, turn the steering wheel lock-to-lock to reset the steering control unit.**

Civic and Del Sol

1. Disconnect the negative battery cable.

2. Raise and safely support the vehicle.

3. Drain the oil and remove the lower splash panel.

4. Remove the nuts and bolts connecting exhaust pipe A to the catalytic converter. Discard the gasket and the locknuts.

5. Remove the nuts attaching exhaust pipe A to the exhaust hanger.

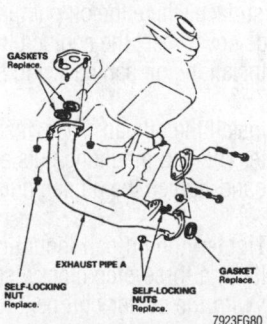

Exhaust pipe A—1.5L (D15B7 D15B8, D15Z1) and 1.6L (D16Z6, B16A2, B16A3 engines

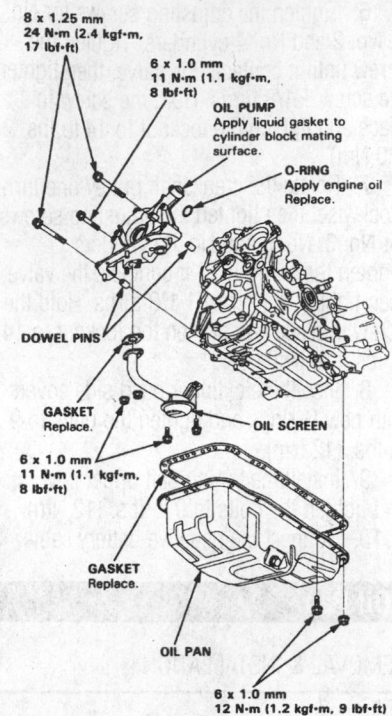

Oil pan and oil screen—1.6L (B16A2, B16A3) engines

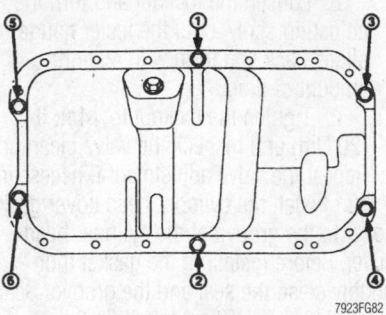

Oil pan bolt tightening sequence—1.5L (D15B7 D15B8, D15Z1) and 1.6L (D16Z6, B16A2, B16A3 engines

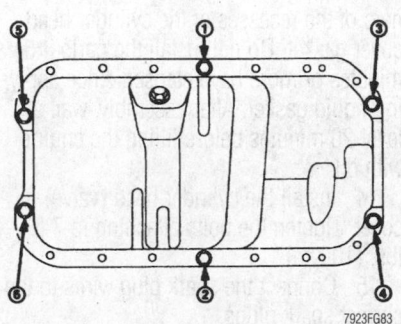

Oil pan bolt tightening sequence—1.6L (D16Y5, D16Y7, D16Y8 engines

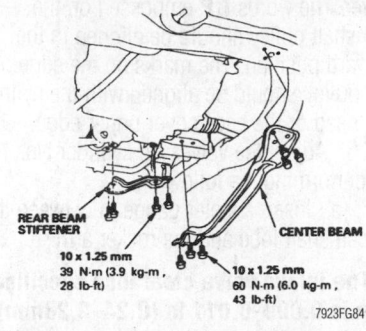

To gain access to the oil pan, remove the center beam—2.2L (F22A1, H22A1) and 2.3L (H23A1) engines

6. Remove and discard the locknuts attaching exhaust pipe A to the exhaust manifold, then remove exhaust pipe A from the vehicle. Discard the exhaust gaskets.

7. Loosen the oil pan bolts in a crisscross pattern. To remove the oil pan, lightly tap the corners of the oil pan with a rubber or plastic faced mallet. Clean off all the old gasket material.

8. Inspect the oil screen and pick-up tube for damaged and clogging. If the screen and tube are clogged with oil residue, they should be thoroughly cleaned or replaced.

To install:

9. If removed, install the oil screen and tube with a new gasket. Tighten the mounting nuts and bolts to 8 ft. lbs. (11 Nm).

10. Apply liquid gasket to the oil pan mating surface where the oil pump and the right side cover meet the engine block.

11. Install the oil pan gasket to the oil pan.

12. Install the oil pan. Then, install all the center and end mounting nuts and bolts. Evenly hand-tighten the oil pan nuts and bolts.

13. Tighten the oil pan mounting nuts and bolts in a three-step clockwise pattern starting with the center bolt next to the oil drain plug. The final torque value for the nuts and bolts is 9–10 ft. lbs. (12–14 Nm).

➡**Excessive tightening can cause distortion of the oil pan gasket and oil leakage.**

14. Install the oil drain plug with a new crush washer, tighten the plug to 33 ft. lbs. (44 Nm).

15. Install exhaust pipe A using new gaskets and locknuts. Tighten the nuts attaching the exhaust pipe to the exhaust manifold to 40 ft. lbs. (54 Nm). Tighten the nuts attaching the exhaust pipe to the catalytic converter and the exhaust pipe hanger to 16 ft. lbs. (22 Nm).

16. Install the lower splash panel. Then, lower the vehicle.

17. Refill the engine with clean oil.

18. Connect the negative battery cable and enter the radio security code.

19. Run the engine and check for leaks.

20. Turn off the engine and check the oil level. Top off the oil level if necessary.

Prelude and Accord

2.2L, 2.3L AND 2.7L ENGINES

1. Disconnect the negative battery cable.

2. Raise and safely support the vehicle.

3. Drain the engine oil into a sealable container.

4. Install the drain bolt with a new gasket, tighten the bolt to 33 ft. lbs. (44 Nm).

5. Remove the front wheels and the splash shield.

6. Remove the center beam.

7. Disconnect the oxygen sensor electrical connector.

8. Remove the bolts from the support bracket on exhaust pipe A.

9. Remove the nuts attaching exhaust pipe A to the exhaust manifold and the catalytic converter. Remove exhaust pipe A and discard the gaskets.

10. If equipped with a automatic transaxle, remove the tighten converter cover.

11. If equipped with a manual transaxle, remove the clutch cover.

12. Remove the oil pan nuts and bolts (in a crisscross pattern) and the oil pan; if necessary, use a mallet to tap the corners of the oil pan. Do NOT pry on the pan to get it loose.

13. Clean the oil pan mounting surface of old gasket material and engine oil.

To install:

14. Install a new oil pan gasket to the oil pan. Apply liquid gasket to the corners of the curved section of the gasket.

15. Install the oil pan to the engine.

16. Install the oil pan nuts and bolts, tighten the nuts and bolts in sequence. Tighten the nuts and bolts in two steps to 10 ft. lbs. (14 Nm).

17. If equipped with a automatic transaxle, install the torque converter cover. Tighten the bolts to 9 ft. lbs. (12 Nm).

18. If equipped with a manual transaxle, install the clutch cover. Tighten the bolts to 9 ft. lbs. (12 Nm).

19. Install exhaust pipe A with new gaskets and new locknuts. Tighten the nuts attaching exhaust pipe A to the manifold to 40 ft. lbs. (54 Nm), tighten the nuts attaching exhaust pipe A to the catalytic converter to 25 ft. lbs. (33 Nm). Install the bolts to the exhaust pipe support bracket and tighten the bolts to 13 ft. lbs. (18 Nm).

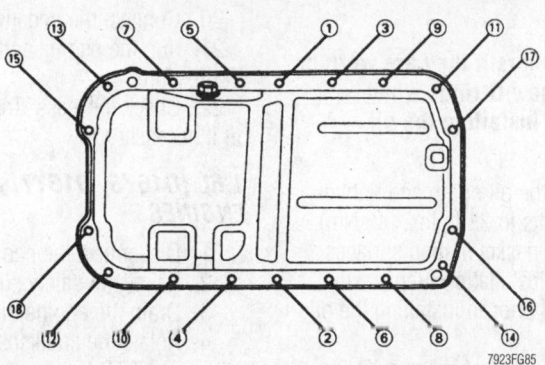

Oil pan mounting bolt tightening sequence—2.2L (F22A1, F22B1, F22B2, H22A1 and F22A6) and 2.3L H23A1 engines

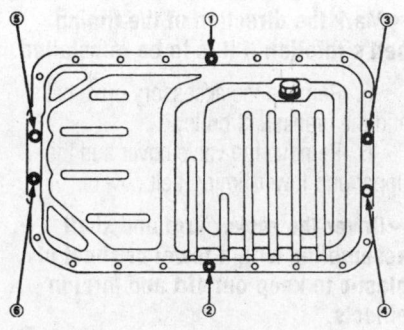

Oil pan tightening sequence—2.7L engines

20. Connect the oxygen sensor electrical connector.

21. Install the center beam, tighten the mounting bolts as follows;
- Prelude: 43 ft. lbs. (60 Nm)
- 1995–97 Accord and 2.7L engines: 37 ft. lbs. (50 Nm)

22. Install the splash shield, tighten the mounting bolts to 7 ft. lbs. (10 Nm).

23. Install the front wheels.

24. Lower the vehicle and fill the engine with oil.

25. Connect the negative battery cable and enter the radio security code.

26. Start the engine and check for leaks.

27. If equipped with 4WS, turn the steering wheel lock-to-lock to reset the 4WS control unit.

28. If the oxygen sensor is located in the mid exhaust pipe, disconnect the oxygen sensor electrical connector.

3.0L ENGINE

1. Disconnect the negative battery cable.

2. Raise and safely support the vehicle.

3. Remove the undercover.

4. Drain the engine oil and replace the drain plug.

5. Remove the front exhaust pipe.

6. Remove the oil pan mounting bolts.

7. Hammer a seal cutter between the engine block and oil pan to break the seal.

8. Remove the oil pan.

To install:

9. Clean the oil pan flange and engine block mounting surface.

10. Apply sealant to the oil pan flange. Be sure to apply sealant toward the inside of the bolt holes.

11. Install the oil pan on the engine. Tighten the bolts in sequence to 10 ft. lbs. (14 Nm).

12. Install the exhaust pipe.

13. Install the undercover.

14. Lower the vehicle.

❋❋ WARNING

Operating the engine without the proper amount and type of engine oil will result in severe engine damage.

15. Refill the engine with the correct amount of oil.

16. Connect the negative battery cable.

17. Start the engine and check for leaks.

Oil Pump

REMOVAL & INSTALLATION

➡The original radio may contain a coded anti-theft circuit. Always obtain the security code number before disconnecting the battery cables.

Civic and Del Sol

1.5L (D15B7 D15B8, D15Z1) AND 1.6L (D16Z6) ENGINE

1. Disconnect the negative battery cable.

2. Raise and safely support the vehicle.

3. Drain the engine oil.

4. Set the No. 1 cylinder to TDC for the compression stroke. The mark on the crank-

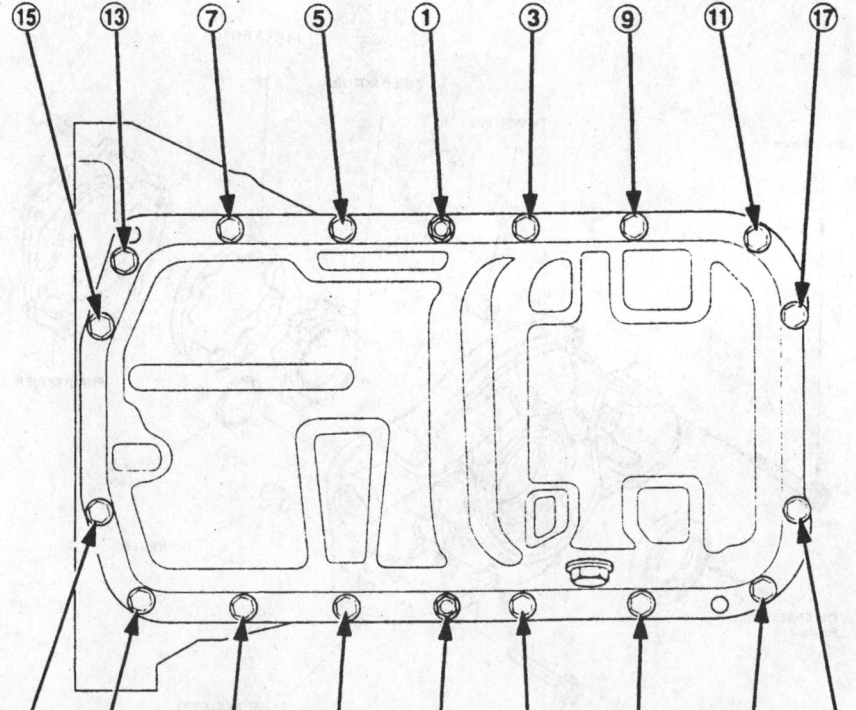

Oil pan mounting bolt tightening sequence—3.0L engine

shaft pulley should align with the index mark on the timing cover.

➡**Mark the direction of the timing belt's rotation if it is to be reinstalled.**

5. Remove the accessory drive belts and the crankshaft pulley.

6. Remove the valve cover and the timing belt covers.

7. Remove the following components:
   a. Timing belt tensioner
   b. Timing belt
   c. Timing belt crankshaft sprocket

8. Remove the oil pan and oil screen.

9. Remove the oil pump mount bolts and the oil pump assembly.

10. Disassemble the oil pump and inspect the rotors:
   a. Inner-to-outer rotor clearance is 0.02–0.14mm (0.001–0.006 in.). The service limit is 0.20mm (0.008 in.).
   b. Rotor-to-pump housing axial clearance is 0.03–0.08mm (0.001–0.003 in.). The service limit is 0.15mm (0.006 in.). Check the clearance using a steel bar and feeler gauge.
   c. Outer rotor-to-pump housing clearance is 0.10–0.18mm (0.004–0.007 in.). The service limit is 0.20mm (0.008 in.).

To install:

➡**Replace the rotors if they are worn or damaged. Use new O-rings when assembling and installing the oil pump.**

11. Assemble the oil pump and tighten the rotor cover bolts to 33 ft. lbs. (45 Nm).

12. Be sure all gasket mating surfaces are clean prior to installation. Replace the crankshaft oil seal prior to installing the oil pump.

13. Apply liquid gasket to the cylinder block mating surface of the block. Apply a light coat of oil to the crankshaft seal lip. Install a new O-ring on the cylinder block and install the oil pump. Apply liquid gasket to the threads of the oil pump mounting bolts and tighten them to 8–9 ft. lbs. (11–12 Nm).

14. Install the oil screen.

15. Install the oil pan.

16. Tighten the crankshaft pulley bolt to specification.

17. Install and tension the timing belt components after installation. Install the valve cover.

18. Install and adjust the accessory drive belts.

19. Refill the engine with oil.

20. Connect the negative battery cable.

21. Run the engine and check for proper oil pressure.

22. Check for leaks. Top up the engine oil if necessary.

1.6L (D16Y5, D16Y7, D16Y8) ENGINES

1. Disconnect the negative battery cable.

2. Raise and safely support the vehicle.

3. Drain the engine oil.

4. Rotate the crankshaft to set the No. 1 cylinder to TDC for the compression stroke. The white TDC mark on the crankshaft pulley should align with the TDC pointers on the lower timing cover.

➡**Mark the direction of the timing belt's rotation if it is to be reinstalled.**

5. Remove the accessory drive belts and the crankshaft pulley.

6. Remove the valve cover and the upper and lower timing belt covers.

➡**Cover the rocker arm and shaft assemblies with a towel or sheet of plastic to keep out dirt and foreign objects.**

7. Remove the dipstick and its tube from the oil pump housing.

8. Release the timing belt's tension. Then, remove the timing belt.

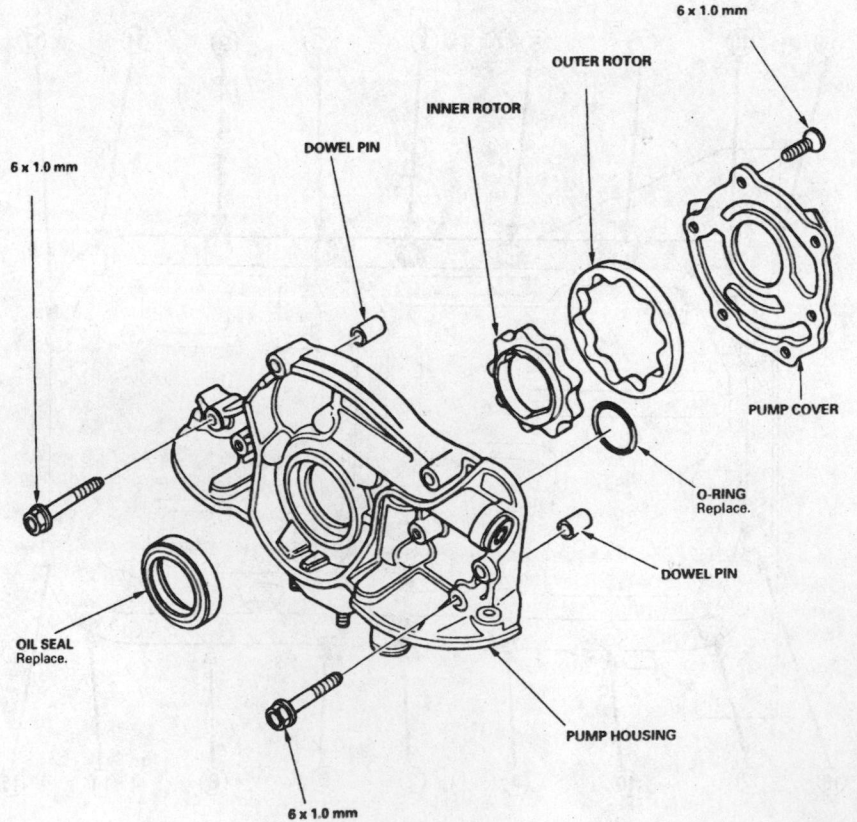

Oil pump exploded view—1.5L (D15B7 D15B8, D15Z1) and 1.6L (D16Z6 engines

7923FG87

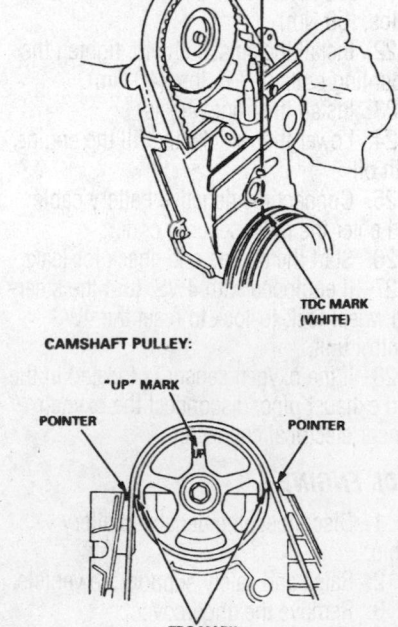

Crankshaft and camshaft TDC marks—
1.6L (D16Y5, D16Y7, D16Y8) engines

7923FG88

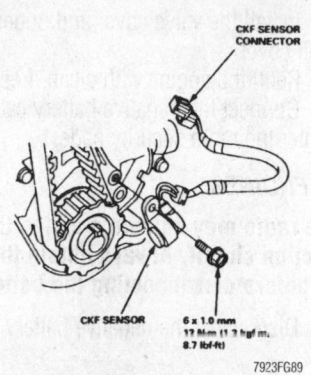

CKF sensor location—1.6L (D16Y5, D16Y7, D16Y8) engines

9. Unbolt the Crankshaft Speed Fluctuation (CKF) sensor from the oil pump cover. Disconnect the sensor and remove it so that it will not come in contact with oil or become damaged.

10. Remove the crankshaft sprocket.

11. Remove the oil pan.

12. Unbolt the oil screen and pick-up tube from the oil pump housing and crankshaft buttress. If the screen and pick-up tube are blocked with oil residue, clean or replace them as necessary.

13. Unbolt, then remove the oil pump assembly.

➡If the rotors are to be reused, match-mark them with a felt-tipped marker for assembly.

To install:

➡Replace the rotors if they are worn or damaged. Use new O-rings when assembling and installing the oil pump.

14. Install the rotors back into their original positions. Be sure they move without binding. Pack the rotor cavity with petroleum jelly to prevent oil starvation damage when the engine is initially started.

15. Assemble the oil pump and tighten the rotor cover bolts to 5 ft. lbs. (7 Nm).

16. Be sure all gasket mating surfaces are clean prior to installation.

17. Install a new crankshaft oil seal into the oil pump housing.

18. Apply liquid gasket to the cylinder block mating surface of the block. Apply a light coat of oil to the crankshaft seal lip. Install a new O-ring on the cylinder block and install the oil pump. Apply liquid gasket to the threads of the oil pump mounting bolts and tighten them to 8 ft. lbs. (11 Nm).

19. Lightly lubricate the relief valve piston and spring, then install them. Install the sealing bolt with a new crush washer, and tighten it to 29 ft. lbs. (39 Nm).

20. Install the oil screen. Tighten the fastening nuts and bolts to 8 ft. lbs. (11 Nm).

21. Install the oil pan. Tighten the oil pan nuts and bolts to 9 ft. lbs. (12 Nm).

22. Install the crankshaft sprocket. The concave surface of the spacer must face the engine block.

23. Verify that the engine is at TDC/compression for the No. 1 cylinder.

24. Install and tension the timing belt. Tighten the tensioner adjusting bolt to 33 ft. lbs. (44 Nm).

25. Install and reconnect the CKF sensor. Tighten the sensor mounting bolt to 9 ft. lbs. (12 Nm).

26. Install the upper and lower timing belt covers. Install the valve cover. Be sure all rubber seals and gaskets are properly seated.

27. Install the dipstick tube with a new O-ring.

28. Tighten the crankshaft pulley bolt to 134 ft. lbs. (181 Nm).

29. Install a new oil filter. Refill the engine with fresh oil.

30. Slowly rotate the engine several times by hand to prime the oil pump and verify that the timing belt has been installed and tensioned correctly.

31. Install and adjust the accessory drive belts.

32. Connect the negative battery cable.

33. Run the engine and check for proper oil pressure.

34. Check for leaks. Top up the engine oil if necessary.

1.6L (B16A2, B16A3) ENGINES

1. Disconnect the negative battery cable.

2. Raise and safely support the vehicle.

3. Drain the engine oil.

4. Label and disconnect the ignition wires.

5. Set the No. 1 cylinder to TDC for the compression stroke. The mark on the crankshaft pulley should align with the index mark on the timing cover.

➡Mark the direction of the timing belt's rotation if it is to be reinstalled.

6. Release the tensions of the accessory drive belts and slip them off their pulleys. Remove the crankshaft pulley.

7. Remove the valve cover and the upper and lower timing belt covers.

➡Cover the rocker arm and shaft assemblies with a towel or sheet of plastic to keep out dirt and foreign objects.

8. Release the timing belt's tension. Then, remove the timing belt.

9. If equipped with a Crankshaft Speed Fluctuation (CKF) sensor at the crankshaft sprocket, unbolt the sensor's bracket. Then, disconnect the sensor and remove it.

10. Remove the crankshaft sprocket.

11. Remove the oil pan and oil screen.

12. Unbolt, then remove the oil pump assembly.

To install:

➡Replace the rotors if they are worn or damaged. Use new O-rings when assembling and installing the oil pump.

13. Install the rotors back into their original positions. Be sure they rotate without binding. Pack the rotor cavity with petroleum jelly to prevent oil starvation damage when the engine is initially started.

14. Assemble the oil pump and tighten the rotor cover bolts to 5 ft. lbs. (7 Nm).

15. Be sure all gasket mating surfaces are clean prior to installation. Replace the crankshaft oil seal prior to installing the oil pump.

16. Install a new crankshaft oil seal into the oil pump housing.

17. Apply liquid gasket to the oil pump mating surface of the cylinder block. Apply a light coat of oil to the crankshaft seal lip. Install a new oil passage O-ring on the cylinder block, then install the oil pump. Apply liquid gasket to the threads of the oil pump mounting bolts and tighten them the 6mm bolts to 8 ft. lbs. (11 Nm). Tighten the 8mm bolts to 17 ft. lbs. (24 Nm).

18. Lightly lubricate the relief valve piston and spring, then install them. Install the sealing bolt with a new crush washer and tighten it to 29 ft. lbs. (39 Nm).

19. Install the oil screen.

20. Install the oil pan. Wait for the sealant to cure before refilling the engine with oil.

21. Install the CKF sensor. Tighten the sensor mounting bolts to 8 ft. lbs. (11 Nm).

22. Install the crankshaft sprocket. The concave surface of the spacer faces out.

23. Install and tension the timing belt. Tighten the tensioner adjusting bolt to 40 ft. lbs. (55 Nm).

24. Install and reconnect the CKF sensor. Tighten the sensor mounting bolt to 9 ft. lbs. (12 Nm).

25. Install the upper and lower timing belt covers. Install the valve cover. Be sure all rubber seals and gaskets are properly seated.

26. Tighten the crankshaft pulley bolt to 130 ft. lbs. (180 Nm).

27. Install a new oil filter. Refill the engine with fresh oil.

28. Slowly rotate the engine several times by hand to prime the oil pump and to verify that the timing belt has been installed and tensioned correctly.

29. Install and adjust the accessory drive belts.

30. Connect the negative battery cable.

31. Run the engine and check for proper oil pressure.

32. Check for leaks. Top up the engine oil if necessary.

Accord and Prelude

2.2L AND 2.3L ENGINES

1. Disconnect the negative battery cable.

2. Drain the engine oil into a sealable container.

3. Turn the engine to align the timing marks and set cylinder No.1 to TDC. The white mark on the crankshaft pulley should align with the pointer on the timing belt cover.

4. Remove the valve cover and upper timing belt cover.

5. Remove the power steering pump belt and the alternator belt, also the air conditioning belt if so equipped.

6. Remove the crankshaft pulley and the lower timing belt cover.

7. Remove the balancer belt and the timing belt, be sure to mark the rotation of the timing belt if it is going to be reused.

8. Remove the timing belt and balancer belt tensioners.

9. If equipped, remove the bolts mounting the CKP/TDC sensor and carefully remove the CKP/TDC sensor from the oil pump. Disconnect the CKP/TDC sensor connector and remove it from the vehicle.

10. Remove the timing belt drive pulley and key from the crankshaft.

11. Insert a suitable tool into the maintenance hole in the front balancer shaft and remove the balancer driven pulley.

12. Align the rear timing balancer pulley using a 6 x 100mm bolt or rod. Mark the bolt or rod at a point 2.9 in. (74mm) from the end. Remove the bolt from the maintenance hole on the side of the block; insert the bolt/rod into the hole. Align the 74mm mark with the face of the hole. This pin will hold the shaft in place.

13. Remove the balancer gear case and the dowel pins. Discard the O-ring.

14. Remove the balancer driven gear attaching bolt and the balancer driven gear.

15. Remove the oil pan and the oil screen. Discard the screen gasket.

16. Remove the oil pump mounting bolts and remove the oil pump assembly. Remove the dowel pins from the engine and clean the oil pump mating surfaces of old gasket material and oil. Discard the O-rings.

To install:

17. Install the two dowel pins and new O-rings to the cylinder block.

18. Be sure that the mating surfaces are clean and dry. Apply a liquid gasket evenly in a narrow bead, centered on the mating surface. Once the sealant is applied, do not wait longer than 20 minutes to install the parts; the sealant will become ineffective. After final assembly, wait at least 30 minutes before adding oil to the engine, giving the sealant time to set. To prevent leakage of oil, apply a suitable thread sealer to the inner threads of the bolt holes.

19. Install the oil pump to the engine block. Tighten the mounting bolts to 9 ft. lbs. (12 Nm).

20. Install the oil screen. Tighten the screen mounting bolts and nuts to 9 ft. lbs. (12 Nm).

21. Install the oil pan.

22. Install the balancer driven pulley to the front balancer belt, hold the balancer shaft in place with a suitable tool. Tighten the attaching bolt to 22 ft. lbs. (29 Nm).

23. Install the balancer driven gear to the rear balancer shaft. Tighten the bolt to 18 ft. lbs. (25 Nm).

24. Before installing the balancer driven gear and the gear case, apply molybdenum disulfide (lithium grease) to the thrust surfaces of the balancer gears.

25. Align the groove on the pulley edge to the pointer on the balancer gear case.

26. Install the balancer gear case to the engine and install the mounting bolts and nut. The rear balancer shaft is being held in place with a 6x100mm bolt. Tighten the mounting bolts and nut to 18 ft. lbs. (25 Nm).

27. Check the alignment of the pointer on the balancer pulley to the pointer on the oil pump.

28. Install the drive pulley to the crankshaft.

29. If equipped, connect the CKP/TDC sensor connector to the wiring harness on the engine. Install the CKP/TDC sensor to the oil pump and install the mounting bolts. Tighten the mounting bolts to 9 ft. lbs. (12 Nm).

30. Install the timing belt tensioners.

31. Install the timing belt and the balancer belt.

32. Install the crankshaft pulley and the lower timing belt cover.

33. Install the drive belts for the alternator, power steering and A/C compressor; adjust the tension.

34. Install the valve cover and upper timing belt cover.

35. Refill the engine with clean, fresh oil.

36. Connect the negative battery cable and enter the radio security code.

2.7L ENGINES

➡The radio may contain a coded theft protection circuit. Always obtain the code before disconnecting the battery.

1. Disconnect the negative battery cable.

2. Turn the engine to align the timing marks and set cylinder No.1 to TDC. The white mark on the crankshaft pulley should align with the pointer on the timing belt cover. Remove the inspection caps on the upper timing belt covers to check the alignment of the timing marks. The pointers for the camshafts should align with the green marks on the camshaft pulleys.

3. Raise and safely support the vehicle.

4. Drain the engine oil into a sealable container.

5. Remove the front wheels and the engine splash shield.

6. Remove the timing belt covers and the timing belt.

7. Disconnect the CKP sensor electrical connector, then remove the CKP sensor from the oil pump.

8. Remove the timing belt tensioner.

9. Remove the stopper plate from the oil pump, then remove the timing belt drive pulley.

10. Remove the oil filter.

11. Disconnect the oil pressure switch terminal, then remove the oil filter base attaching bolts.

12. Remove the oil filter base and discard the O-rings.

13. Remove the oil pan.

14. Remove the oil screen and baffle plate.

15. Remove the oil pass pipe attaching bolts, then remove the pipe from the oil pump. Discard the O-rings.

16. Remove the bolts from the oil pump, note the location of the bolts.

17. Remove the oil pump and dowel pins from the engine.

To install:

18. Install the two dowel pins to the cylinder block.

19. Be sure that the mating surfaces are clean and dry. Apply a liquid gasket evenly in a narrow bead, centered on the mating surface. Once the sealant is applied, do not wait longer than 20 minutes to install the parts; the sealant will become ineffective.

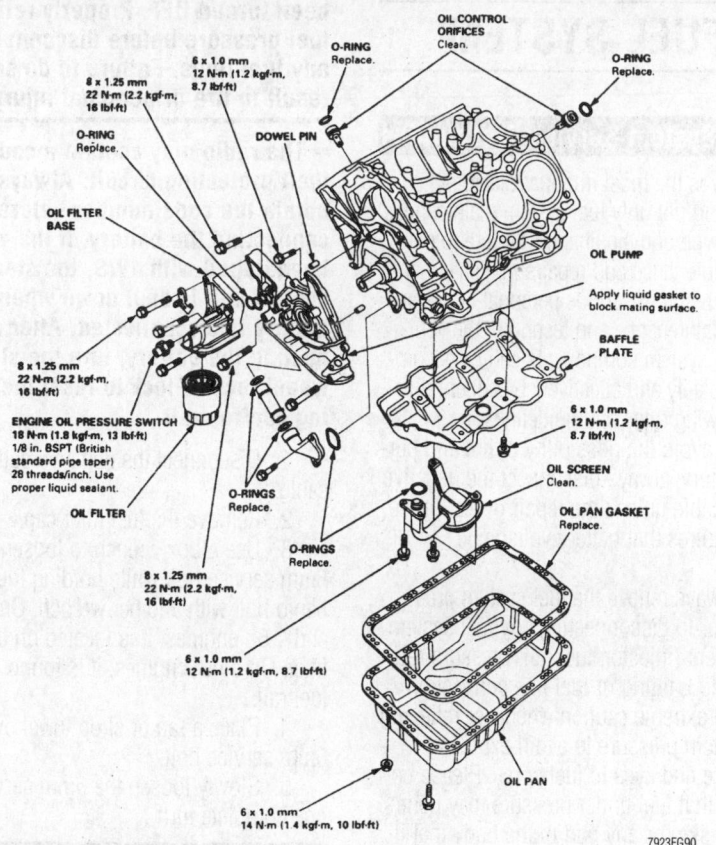

Oil pump and related components—2.7L engines

After final assembly, wait at least 30 minutes before adding oil to the engine, giving the sealant time to set. To prevent leakage of oil, apply a suitable thread sealer to the inner threads of the bolt holes.

20. Apply grease to the lips of the crankshaft oil seal. Install the oil pump to the engine block. Tighten the 6x1.0mm mounting bolts to 9 ft. lbs. (12 Nm) and tighten the 8x1.25mm mounting bolts to 16 ft. lbs. (22 Nm). Clean any excess grease from the crankshaft, then be sure that the seal lips are not distorted.

21. Install new O-rings to the oil pass pipe, then install the pass pipe to the oil pump and the crankshaft bridge. Tighten the attaching bolts to 9 ft. lbs. (12 Nm).

22. Install the baffle plate. Tighten the baffle plate mounting bolts to 9 ft. lbs. (12 Nm).

23. Install the oil screen with a new O-ring. Tighten the screen mounting bolts to 9 ft. lbs. (12 Nm).

24. Install the oil pan.

25. Install new O-rings to the oil filter base and install the oil filter base to the engine. Tighten the mounting bolts to 16 ft. lbs. (22 Nm).

26. Connect the oil pressure switch terminal. Tighten the attaching bolt to 1.8 ft. lbs. (2.5 Nm).

27. Install the oil filter.

28. Install the timing belt drive pulley and the stopper plate. Tighten the stopper plate mounting bolts to 9 ft. lbs. (12 Nm).

29. Install the timing belt tensioner.

30. Install the CKP sensor to the oil pump and connect the electrical connector. TIGHTEN the mounting bolts to 9 ft. lbs. (12 Nm).

31. Install the timing belt and the timing belt covers.

32. Install the engine splash shield and the front wheels.

33. Lower the vehicle and fill the engine with oil.

34. Connect the negative battery cable and enter the radio security code.

35. Run the engine and check for leaks.

3.0L ENGINE

1. Drain the engine oil.

2. Turn the crankshaft to position the No. 1 piston at TDC on the compression stroke.

3. Remove the timing belt.

4. Remove the idler pulley.

5. Remove the crankshaft position sensor.

6. Remove the VTEC solenoid valve and the oil filter.

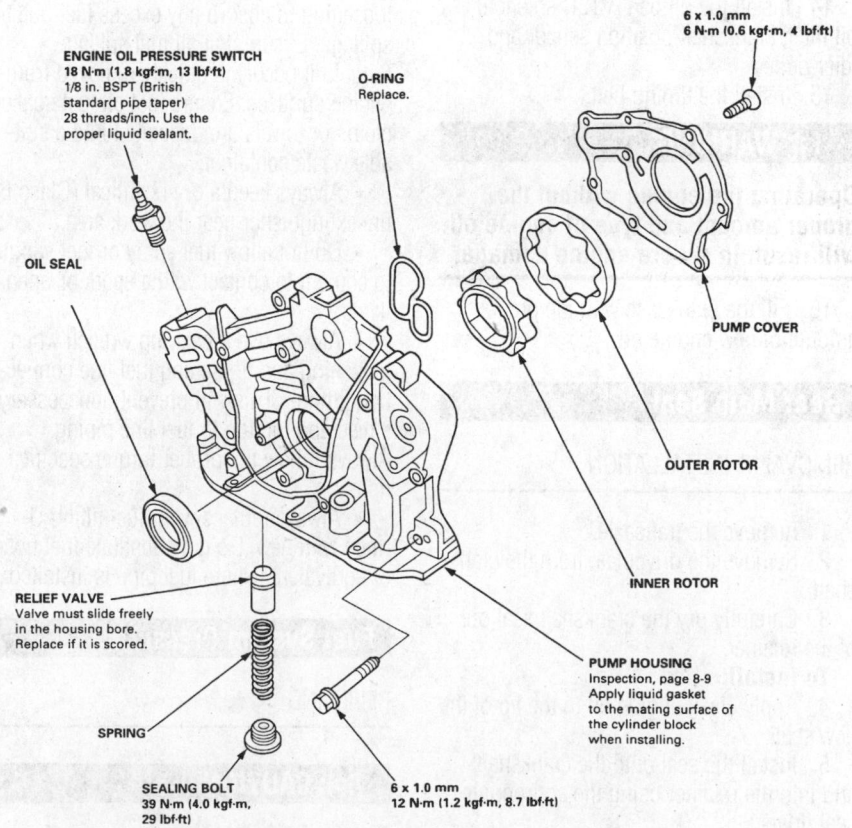

Exploded view of the oil pump—3.0L engine

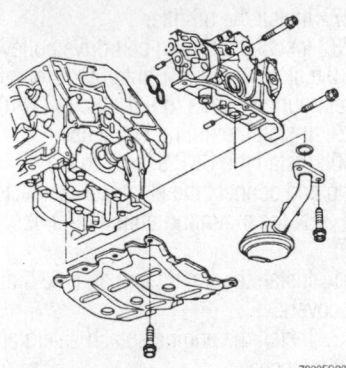

7923FG92

Oil pump mounting—3.0L engine

7. Remove the oil pan and pick-up.

8. Remove the oil pump assembly.

To install:

9. Install a new crankshaft seal in the oil pump.

10. Apply sealant to the oil pump mounting surface and bolt holes on the engine block.

11. Grease the lip of the new seal and apply engine oil to the O-ring.

12. Install the dowel pin and oil pump while aligning the inner rotor with the crankshaft. Tighten the bolts to 9 ft. lbs. (12 Nm).

13. Install the oil pump pick-up. Tighten the mounting bolts to 9 ft. lbs. (12 Nm).

14. Install the oil pan, VTEC solenoid, oil filter, crankshaft position sensor and idler pulley.

15. Install the timing belt.

✳ WARNING

Operating the engine without the proper amount and type of engine oil will result in severe engine damage.

16. Fill the crankcase with the proper amount of new engine oil.

Rear Main Seal

REMOVAL & INSTALLATION

1. Remove the transaxle.

2. Remove the driveplate from the crankshaft.

3. Carefully pry the crankshaft seal out of the retainer.

To install:

4. Apply clean engine oil to the lip of the new seal.

5. Install the seal onto the crankshaft and into the retainer using the appropriate seal driver.

6. Install the driveplate and the transaxle.

FUEL SYSTEM

Fuel System Service Precautions

Safety is the most important factor when performing not only fuel system maintenance but any type of maintenance. Failure to conduct maintenance and repairs in a safe manner may result in serious personal injury or death. Maintenance and testing of the vehicle's fuel system components can be accomplished safely and effectively by adhering to the following rules and guidelines.

• To avoid the possibility of fire and personal injury, always disconnect the negative battery cable unless the repair or test procedure requires that battery voltage be applied.

• Always relieve the fuel system pressure prior to disconnecting any fuel system component (injector, fuel rail, pressure regulator, etc.), fitting or fuel line connection. Exercise extreme caution whenever relieving fuel system pressure to avoid exposing skin, face and eyes to fuel spray. Please be advised that fuel under pressure may penetrate the skin or any part of the body that it contacts.

• Always place a shop towel or cloth around the fitting or connection prior to loosening to absorb any excess fuel due to spillage. Ensure that all fuel spillage (should it occur) is quickly removed from engine surfaces. Ensure that all fuel soaked cloths or towels are deposited into a suitable waste container.

• Always keep a dry chemical (Class B) fire extinguisher near the work area.

• Do not allow fuel spray or fuel vapors to come into contact with a spark or open flame.

• Always use a back-up wrench when loosening and tightening fuel line connection fittings. This will prevent unnecessary stress and torsion to fuel line piping. Always follow the proper torque specifications.

• Always replace worn fuel fitting O-rings with new. Do not substitute fuel hose or equivalent, where fuel pipe is installed.

Fuel System Pressure

RELIEVING

✳ CAUTION

The fuel injection system remains under pressure after the engine has

been turned OFF. Properly relieve fuel pressure before disconnecting any fuel lines. Failure to do so may result in fire or personal injury.

➡**The radio may contain a coded theft protection circuit. Always obtain the code number before disconnecting the battery. If the vehicle is equipped with 4WS, the steering control unit is shut down when the battery is disconnected. After connecting the battery, turn the steering wheel lock-to-lock to reset the steering control unit.**

1. Disconnect the negative battery cable.

2. Remove the fuel filler cap.

3. Use a box wrench to loosen the 6mm service bolt while holding the special banjo bolt with another wrench. On 1.5L and 1.6L engines, it is located on the fuel filter. On other engines, it is found on the fuel rail.

4. Place a rag or shop towel over the 6mm service bolt.

5. Slowly loosen the 6mm service bolt one complete turn.

✳ CAUTION

Do not allow fuel spray or fuel vapors to come in contact with a spark or open flame. Keep a dry chemical fire extinguisher nearby. Never store fuel in an open container due to risk of fire or explosion.

➡**A fuel pressure gauge may be attached at the 6mm service bolt location. Always replace the washer between the service bolt and the banjo bolt whenever the service bolt is loosened.**

6. Remove the service bolt and install a new washer. Tighten the 6mm service bolt to 9 ft. lbs. (12 Nm). Don't overtighten the service bolts, their threads may strip and cause leaks.

7. Clean up any fuel spilled on the engine and intake manifold.

8. Install the fuel filler cap.

9. Reconnect the negative battery cable.

10. Turn the ignition **ON**, but don't start the engine. Repeat this two or three times to pressurize the fuel system. Check for fuel leaks.

11. Enter the radio security code.

12. If equipped with 4WS, turn the steering wheel lock-to-lock to reset the 4WS control unit.

Fuel Filter

REMOVAL & INSTALLATION

Civic and Del Sol

✷✷ CAUTION

The fuel injection system remains under pressure, even after the engine has been turned OFF. The fuel system pressure must be relieved before disconnecting any fuel lines. Failure to follow this procedure may result in fire or explosion.

➡The original radio contains a coded anti-theft circuit. Obtain the security code number before disconnecting the battery.

1. Disconnect the negative battery cable.
2. Place a rag under the fuel filter to catch fuel spray.
3. Relieve the fuel pressure by first loosening the fuel filler cap. Use a 6mm flare nut wrench to hold the banjo bolt. Then, loosen the service bolt one complete turn with a box-end wrench or socket.
4. Use two flare nut wrenches to disconnect the fuel inlet line from the bottom of the filter. Plug the fuel line to keep out dirt.
5. Unbolt and remove the fuel filter clamp. Remove the filter from its bracket.

To install:

➡Use new sealing washers when installing the fuel filter to prevent fuel leaks and the possibility of fire.

6. Clean the fuel line fittings before installing the filter.
7. Install the fuel filter and its clamp. Tighten the clamp bolt to 7 ft. lbs. (10 Nm).
8. Connect the fuel inlet line and carefully tighten its fitting to 27 ft. lbs. (38 Nm).

9. Connect the fuel line with new washers and install the banjo bolt. Tighten the banjo bolt to 16 ft. lbs. (22 Nm). Install the service bolt and tighten it to 9 ft. lbs. (12 Nm).
10. Connect the battery cable. Tighten the fuel filler cap.
11. Turn the ignition on and off several times to pressurize the fuel system. Start and run the engine and check for fuel leaks.

Prelude and 1995–97 Accord

➡The radio may contain a coded theft protection circuit. Always obtain the code number before disconnecting the battery. If the vehicle is equipped with 4WS, the steering control unit is shut down when the battery is disconnected. After connecting the battery, turn the steering wheel lock-to-lock to reset the steering control unit.

1. Disconnect the negative battery cable.
2. Place a shop towel under and around the fuel rail, then relieve the fuel pressure.

✷✷ CAUTION

Do not allow fuel spray or fuel vapors to come in contact with a spark or open flame. Keep a dry chemical fire extinguisher nearby. Never store fuel in an open container due to risk of fire or explosion.

3. Remove the 12mm banjo bolt and the fuel feed pipe from the fuel filter. Discard the washers.
4. Remove the fuel filter clamp and the fuel filter.

To install:

5. Position the fuel filter on the bracket and install the filter clamp. Tighten the clamp bolts to 7 ft. lbs. (10 Nm).

➡Clean the fuel fittings thoroughly before reconnecting them.

6. Connect the fuel feed pipe to the filter, tighten the fitting to 28 ft. lbs. (38 Nm).
7. Connect the fuel outlet pipe to the filter using new gaskets around the fitting. Tighten the banjo bolt to 20 ft. lbs. (28 Nm) on all Prelude models and 16 ft. lbs. (22 Nm) on all other models.
8. Connect the negative battery cable and enter the radio security code.
9. If equipped with 4WS, turn the steering wheel lock-to-lock to reset the 4WS control unit.
10. Turn the ignition **ON** and check for fuel leaks.

1998–99 Accord

1. Remove the fuel pump from the tank.
2. Remove the fuel filter from the pump module.

To install:

3. Install the new filter on the pump module.
4. Install the fuel pump in the tank.

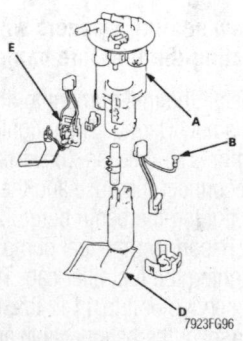

Fuel filter (A), wiring harness (B), suction filter (D) and sending unit (E)—1998–99 Accord

Fuel Pump

REMOVAL & INSTALLATION

➡The radio may contain a coded theft protection circuit. Always obtain the code number before disconnecting the battery. If the vehicle is equipped with 4WS, the steering control unit is shut down when the battery is disconnected. After connecting the battery, turn the steering wheel lock-to-lock to reset the steering control unit.

Civic and Del Sol

✷✷ CAUTION

The fuel injection system remains under pressure, even after the engine has been turned OFF. The fuel system

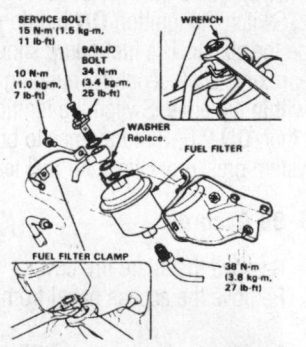

Fuel filter components—1.5L and 1.6L engines

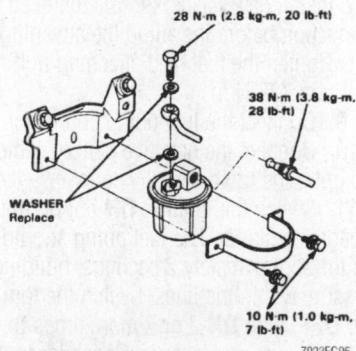

Fuel filter mounting—Prelude with 2.2L (F22A1) and 2.3L (H23A1) engines

pressure must be relieved before disconnecting any fuel lines. Failure to follow this procedure may result in fire, explosion, or personal injury.

1. Disconnect the negative battery cable.
2. Loosen the fuel filler cap. Then, loosen the fuel filter service bolt to relieve the fuel pressure.
3. Remove the rear seat cushions (Civic), or the rear compartment trim (Del Sol).
4. Remove the fuel pump access panel.
5. Disconnect the two-wire fuel pump harness.
6. Clean the fuel line fittings before disconnecting them.
7. Disconnect the fuel line and the hose from the fuel pump.
8. Unbolt the fuel pump and lift it out the fuel tank. Allow the fuel in the pump drain into the tank before removing the pump from the vehicle.
9. Disconnect and remove the fuel pump motor from its bracket.

To install:

➡️ Use new sealing washers when reconnecting the fuel line banjo bolt.

10. Install the fuel pump into the fuel tank with a new O-ring. Then, tighten the mounting nuts to 4 ft. lbs. (6 Nm).
11. Reconnect the hose and the fuel line. Carefully tighten the banjo bolt to 20 ft. lbs. (28 Nm). Reconnect the fuel pump harness.
12. Tighten the fuel filler cap. Tighten the fuel filter service bolt to 11 ft. lbs. (15 Nm).
13. Connect the battery cable and turn the ignition switch **ON** and **OFF** several times to pressurize the fuel system.

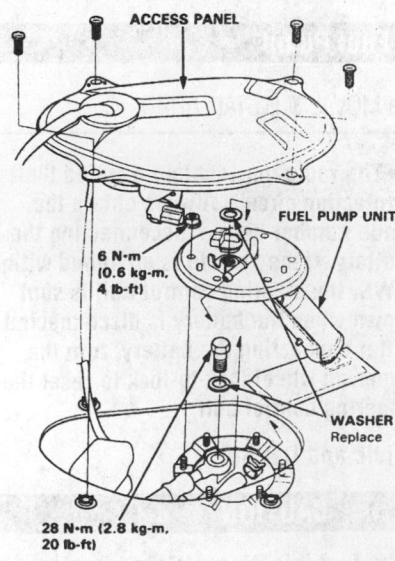

Fuel pump components-1995 Civic and 1995–97 Del Sol

14. Check the connections at the fuel pump for any leaks. Check the fuel filter service bolt for leaks.
15. Install the fuel pump access cover.
16. Install the rear seat cushions or rear compartment trim. Be sure the clips are properly seated.

Prelude

❄❄ CAUTION

The fuel injection system remains under pressure after the engine has been turned OFF. Properly relieve fuel pressure before disconnecting any fuel lines. Failure to do so may result in fire or personal injury.

1. Disconnect the negative battery terminal.
2. Relieve the fuel pressure.
3. Lift or reposition the carpet in the luggage area. Remove the fuel pump maintenance access cover in the floor.
4. Disconnect the electrical connector at the pump unit.
5. Label and disconnect the fuel lines. Discard the washers from the fuel feed connection.
6. Carefully remove the retaining nuts holding the pump. When all are removed, lift the pump up and out of the tank.

➡️ The pump sits on an angle and may require some manipulation to remove. If the pump still won't come out, loosen the fuel tank mounting nuts under the car, slide the tank downward a bit to give more clearance at the top.

To install:

7. Using a new sealing ring, reinstall the pump making certain it is correctly seated and not wedged or jammed. Install the retaining nuts, tightening them evenly and alternately to 4 ft. lbs. (6 Nm).
8. Install the fuel lines. Make certain the clamp is secure; use new ones if necessary. Install new washers to the fuel feed connection before installing the attaching bolt. Tighten the fuel feed attaching bolt to 20 ft. lbs. (28 Nm).
9. Connect the fuel pump connector.
10. Connect the negative battery cable and enter the radio security code.
11. Switch the ignition **ON** but do not engage the starter. The fuel pump should run for approximately 2 seconds, building pressure within the lines. Switch the ignition **OFF**, then **ON** 2 or 3 more times to build full system pressure. Check for fuel leaks.

12. If equipped with 4WS, turn the steering wheel lock-to-lock to reset the 4WS control unit.
13. Install the maintenance access cover and seal or gasket, if used.
14. Reposition the carpeting in the luggage compartment.

1995–97 Accord

1. Disconnect the negative battery cable.
2. Relieve the fuel pressure.

❄❄ CAUTION

The fuel injection system remains under pressure after the engine has been turned OFF. Properly relieve fuel pressure before disconnecting any fuel lines. Failure to do so may result in fire or personal injury.

3. Remove the fuel tank from the vehicle.
4. Disconnect the electrical connector from the fuel pump.
5. Remove the fuel feed line attaching bolt, and discard the washers. Disconnect the fuel return line from the fuel pump.
6. Remove the fuel pump mounting nuts.
7. Carefully remove the fuel pump from the fuel tank.

To install:

8. Clean the fuel pump mounting surface and install a new gasket.
9. Install the fuel pump into the tank, being careful not to damage the pick-up screen.
10. Install the mounting nuts; tighten the nut s to 4 ft. lbs. (6 Nm).
11. Connect the fuel return hose to the pump and be sure that the clamp is secure.
12. Connect the fuel feed line to the pump with new washers, tighten the bolt to 21 ft. lbs. (27 Nm).
13. Connect the fuel pump electrical connector.
14. Install the fuel tank into the vehicle.
15. Connect the negative battery cable and enter the radio security code.
16. Switch the ignition **ON** but do not engage the starter. The fuel pump should run for approximately 2 seconds, building pressure within the lines. Switch the ignition **OFF**, then **ON** 2 or 3 more times to build full system pressure. Check for fuel leaks.

1998–99 Accord

1. Remove the spare tire cover.
2. Remove the access panel from the floor.
3. Turn the ignition switch **OFF** and detach the five pin connector from the pump assembly.

4. Remove the fuel cap and relieve the fuel system pressure.

5. Detach the quick-connect connections from the pump assembly.

6. Remove the mounting bolts an the pump assembly from the tank.

To install:

7. Install the pump using a new gasket.

8. Attach the quick-connect fuel lines to the pump assembly.

9. Install the fuel cap.

10. Attach the five pin connector to the pump.

11. Switch the ignition **ON** but do not engage the starter. The fuel pump should run for approximately 2 seconds, building pressure within the lines. Switch the ignition **OFF**, then **ON** 2 or 3 more times to build full system pressure. Check for fuel leaks.

12. If there are no leaks, install the access panel cover and the tire cover.

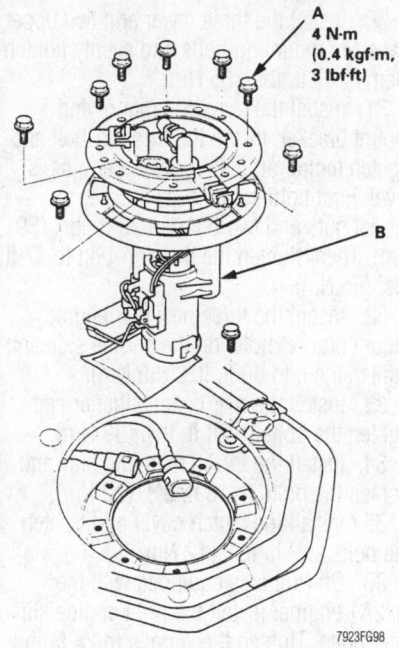

Exploded view of the fuel pump assembly—1998–99 Accord

DRIVE TRAIN

Transaxle Assembly

REMOVAL & INSTALLATION

Manual

➡**The radio may contain a coded theft protection circuit. Always obtain the**

code number before disconnecting the battery. If the vehicle is equipped with 4WS, the steering control unit is shut down when the battery is disconnected. After connecting the battery, turn the steering wheel lock-to-lock to reset the steering control unit.

CIVIC AND DEL SOL

✳✳ WARNING

Use only genuine Honda manual transaxle fluid (MTF)-it is specially formulated for use in Honda transaxles. If Honda MTF is not available, API SG/SJ 10W-30 or 10W-40 motor oil may be used as a temporary lubricant. However, motor oil will cause increased transaxle wear and shifting effort. Refill the transaxle with Honda MTF as soon as possible.

1. Disconnect the negative and positive battery cables.

2. Drain the transaxle fluid.

3. Remove the resonator, the air cleaner box, and the air intake duct.

4. Disconnect the starter cables and the transaxle ground cable.

5. Disconnect the back-up light switch connection.

6. Move the upper radiator hose out of its clamp.

7. Disconnect the vehicle speed sensor (VSS) connector.

8. Unbolt the clutch fluid line bracket. Unbolt and remove the slave cylinder. It isn't necessary to disconnect the clutch fluid line.

9. Raise and safely support the vehicle. Remove the front wheels.

10. Remove the strut pinch bolt and fork bolt. Disconnect the lower ball joint from the steering knuckle using a ball joint remover.

11. Pry the halfshaft inboard joints out of the transaxle case. Swing the steering knuckles out to free the halfshafts from the transaxle.

12. Tie the halfshafts up and out of the way with wire so that the joints will not be stressed. Tie plastic bags over the inboard joints to prevent damage to the CV-boots and splined shafts.

13. Disconnect the shift rod and extension rod from the transaxle case. Drive the shift rod retaining pin out with a pin punch.

14. Disconnect and remove the front exhaust pipe.

15. Remove the engine-to-transaxle stiffener brackets and the clutch cover plate.

16. Attach a lifting chain to the engine and lift slightly to ease the tension on the mounts.

17. Remove the splash shield from underneath the vehicle.

18. Unbolt and remove the right-front mount/bracket assembly.

19. Place a jack under the transaxle to support its weight.

20. Remove the transaxle side mount and its bracket.

21. Remove the starter's lower mounting bolt. Remove the upper three transaxle case bolts.

22. Remove the three rear transaxle mount bracket bolts. Next, remove the lower three transaxle case bolts.

23. Pull the transaxle away from the engine until it clears the mainshaft. Lower the transaxle out of the vehicle Be careful not to bend the clutch hydraulic line.

To install:

➡**Use new self-locking nuts and color-coded bolts when installing the transaxle and suspension components.**

24. Apply high temperature grease to the mainshaft splines, release fork contact points, and throw-out bearing. The manufacturer recommends part No. 08798–9002, Honda Super High temp Urea Grease.

25. Place the transaxle on a transaxle jack and raise it to the level of the engine.

26. Align the transaxle and engine. Be sure the transaxle case dowel pins are securely seated, and fit the transaxle onto the engine. Install the upper and lower transaxle case bolts and the 14mm rear mount bolts and washers, only hand-tighten them at this time.

27. Raise the transaxle and install the side mount. Tighten the upper and lower transaxle case bolts to 47 ft. lbs. (64 Nm). Tighten the 14mm rear mount bracket bolts to 61 ft. lbs. (84 Nm).

28. First, tighten the transaxle side mount bracket nuts and bolt to 47 ft. lbs. (64 Nm). each. Next, tighten the mount bushing bolts to 47 ft. lbs. (64 Nm). Finally, tighten the through-bolt to 54 ft. lbs. (74 Nm).

29. Install the right-front mount/bracket assembly. Use three new 12mm bolt and washers, and tighten them to 47 ft. lbs. (64 Nm). Tighten the two 10mm bolts to 33 ft. lbs. (45 Nm).

30. Install the clutch cover.

31. Install the engine-to-transaxle stiffener brackets and tighten the 8mm bolts to 17 ft. lbs. (24 Nm). Tighten the 10mm bolts to 33 ft. lbs. (44 Nm).

32. Once the transaxle is bolted to the engine, and the transaxle mounts are

securely tightened, the engine lifting chain may be removed.

33. Connect the shift rod with a new spring pin and clip. Then, fit the shift rod boot back into place. Connect the torque rod and tighten its bolt to 16 ft. lbs. (22 Nm).

34. Install the front exhaust pipe. Use new self-locking nuts and gaskets. Tighten the rear flange nuts to 16 ft. lbs. (22 Nm). Tighten the front flange nuts to 25 ft. lbs. (33 Nm), or 40 ft. lbs. (54 Nm) for D16Y8 and D16Y7 engines.

35. Install new set rings on the inboard CV joint splines. Install the halfshafts into the transaxle case and intermediate shaft. The inboard joints must snap into place.

36. Connect the lower ball joint and damper fork. Install the front wheels.

37. Install the slave cylinder and clutch pipe stay. Coat the slave cylinder's tip with high temperature grease. Be sure it snaps into the release fork. Tighten the slave cylinder mounting bolts to 16 ft. lbs. (22 Nm).

38. Connect the VSS connector and back-up light switch connectors.

39. Install the wiring harness clamps and starter cables. Install the resonator, air cleaner box and air intake duct. Fit the upper radiator hose back into its bracket.

40. Lower the vehicle and tighten the strut pinch bolts to 32 ft. lbs. (44 Nm). Tighten the fork bolts to 47 ft. lbs. (65 Nm). Tighten the ball joint castle nuts to 40 ft. lbs. (55 Nm), then tighten them only enough to install new cotter pins.

41. Turn the breather cap so that the arrow with the **F** mark points toward the front of the vehicle.

42. Refill the transaxle with the Honda MTF fluid.

43. Reconnect the positive and negative battery cables.

44. Bleed the clutch hydraulic system.

45. Check the clutch and transaxle for smooth operation.

46. Check and adjust the front wheel alignment.

PRELUDE

1. Shift the transaxle to **R**.

2. Disconnect the negative and positive battery cables. Remove the battery.

3. Remove the intake duct and air cleaner case. Remove the battery base.

4. Remove the vacuum tank and bracket. Do not disconnect the hoses.

5. Disconnect the starter wires and remove the starter.

6. Loosen, but do not remove the two upper transaxle mounting bolts.

7. Disconnect the transaxle ground cable and the back-up light switch wire. Unbolt the engine harness clamp.

8. Leave both shift cables attached their bracket. Remove the shift cables from the transaxle case and wire them safely out of the work area.

9. Disconnect the vehicle speed sensor connector. Leave the sensor hoses connected and remove the sensor from the transaxle case.

10. Remove the slave cylinder mounting bolts. Leave the hydraulic line connected to the slave cylinder. Remove the slave cylinder from the release fork, and move it out of the work area.

➡**Do not operate the clutch pedal once the slave cylinder has been removed. Be careful not to kink the metal hydraulic line.**

11. Raise and safely support the vehicle. Drain the transaxle fluid.

12. Remove the clutch damper mounting bolts and raise the clutch damper.

13. If equipped, remove the rear engine mount bracket stay.

14. Remove the front wheels.

15. Remove the cotter pins and lower arm ball joint nuts. Separate the ball joints and lower arms using a press-type ball joint tool.

16. Remove the damper fork bolt and the radius rod on the right side of the vehicle only.

17. Use a suitable tool to pry the right and left halfshafts out of the differential and the intermediate shaft. Pull on the inboard joint and remove the right and left halfshafts.

18. Unbolt and remove the intermediate shaft from the differential. Tie plastic bags over the halfshaft inboard joints to prevent damage to the boots and splines. Wire the halfshafts to the underbody of the vehicle so that their weight doesn't hang on their outboard joints.

19. Remove the center beam and remove the clutch cover. On vehicles with the H22A1 engine, remove the front engine stiffener plate.

20. Remove the rear beam stiffener and the intake manifold stay.

21. Remove the three rear engine mount bracket bolts.

22. Place a transaxle jack under the transaxle and raise the transaxle just enough to take its weight off the of mounts.

23. Remove the transaxle mount and mount bracket.

24. Remove the two upper transaxle housing mounting bolts and the three lower transaxle housing bolts.

25. Pull the transaxle away from the engine to clear the mainshaft.

26. Lower the transaxle from the vehicle.
To install:

➡**Use new self-locking nuts and set rings when assembling the front suspension components and halfshafts. Use new self-locking bolts when installing the center beam and rear engine mount bracket. These fasteners can be purchased from a Honda dealer.**

27. Be sure the dowel pins are installed into the transaxle case.

28. Apply heavy duty high temperature grease (use Honda part number 08798–9002) to the mainshaft splines, release fork bolt and paws, and the throwout bearing. Install the bearing and release fork. Be sure the release fork snaps into place.

29. Raise the transaxle into position with a transaxle jack.

30. Install the three lower and two upper transaxle mounting bolts and evenly tighten them to 47 ft. lbs. (65 Nm).

31. Install the transaxle mount and mount bracket. Install the through-bolt and tighten temporarily. Be sure the engine is level. First tighten the three bracket-to-mount nuts and two bolts to 28 ft. lbs. (39 Nm). Then, tighten the through-bolt to 47 ft. lbs. (65 Nm).

32. Install the three new rear engine mount bracket bolts on the engine side and tighten them to 40 ft. lbs. (55 Nm).

33. Install the rear beam stiffener and tighten the bolts to 28 ft. lbs. (39 Nm).

34. Install the intake manifold stay and tighten the bolts to 16 ft. lbs. (22 Nm).

35. Install the clutch cover and tighten the bolts to 9 ft. lbs. (12 Nm).

36. On Preludes equipped with the H22A1 engine, install the front engine stiffener bolts. Tighten these bolts to 28 ft. lbs. (39 Nm). First tighten the one bolt threaded into the transaxle case, then tighten the two bolts to the engine block.

37. Install the center beam and tighten the bolts to 43 ft. lbs. (60 Nm).

38. Install the intermediate shaft. Tighten its mounting bolts to 28 ft. lbs. (39 Nm).

39. Install new set rings onto the halfshaft inboard joint splines. Install the halfshafts, making sure that they lock into place.

40. Install the radius rod and damper fork. Only hand-tighten their fasteners at this time.

41. Install the ball joint to the lower arm. Tighten the castle nut to 36–43 ft. lbs. (50–60 Nm). Then, only tighten the nut enough to install a new cotter pin.

42. If equipped, install the rear engine mount bracket stay. Tighten the nut to 15 ft. lbs. (21 Nm) and the bolt to 28 ft. lbs. (39 Nm).

43. Install the clutch damper and tighten the bolts to 16 ft. lbs. (22 Nm).

44. Install the front wheels.

45. Lower the vehicle.

46. Use a floor jack placed under the right front control arm to raise the vehicle enough so that its weight is supported by the jack. Tighten the radius rod mounting bolts to 76 ft. lbs. (105 Nm) and the radius rod nut to 32 ft. lbs. (44 Nm). Tighten the damper pinch bolt to 32 ft. lbs. (44 Nm). Tighten the damper fork bolt to 47 ft. lbs. (65 Nm). After pre-loading the suspension, lower the vehicle and remove the floor jack.

47. Coat the tip of the slave cylinder with heavy duty high temperature grease. Install the clutch hose pipe and clutch slave cylinder to the transaxle housing. Be sure the slave cylinder snaps into the release fork. Tighten the slave cylinder mounting bolts to 16 ft. lbs. (22 Nm).

48. Install the speed sensor. Tighten the mounting bolt to 14 ft. lbs. (19 Nm).

49. Install the shift cable and select cable to the shift arm lever. Install the shift cable assembly onto the transaxle case. Tighten the cable bracket mounting bolts to 16 ft. lbs. (22 Nm). Install new cotter pins.

50. Connect the back-up light switch coupler and the transaxle ground cable. Install the harness clamp.

51. Install the starter. Tighten the 10x1.25mm bolt to 32 ft. lbs. (45 Nm) and the 12x1.25mm bolt to 54 ft. lbs. (75 Nm). Connect the starter wires.

52. Loosen the three front engine mount bracket bolts. Tighten them to 28 ft. lbs. (39 Nm).

53. Install the vacuum tank and its bracket. Install the air cleaner case and intake duct.

54. Fill the transaxle with the proper type and quantity of oil.

55. Install the battery base stay and the battery base. Tighten the battery base bolts to 16 ft. lbs. (22 Nm). Install the battery and connect the battery cables.

56. Check the clutch pedal free-play.

57. Start the vehicle and check the transaxle and clutch for smooth operation.

58. On Preludes equipped with 4WS, start the engine and turn the steering wheel lock-to-lock to reset the steering control unit.

59. Check and adjust the front wheel alignment.

60. Enter the radio security code.

ACCORD

1. Shift the transaxle into **R**.

2. Disconnect the negative and positive battery cables. Remove the battery.

3. Disconnect the IAC solenoid connector. Remove the intake duct, resonator, and air cleaner case, and battery base.

4. Disconnect the starter wires and remove the starter.

5. Disconnect the transaxle ground cable and the back-up light switch wire.

6. Remove the cable stay, then disconnect the cables from the top housing of the transaxle. Remove both cables and the stay together.

7. Disconnect the vehicle speed sensor connector and remove the speed sensor. Leave the speed sensor hoses connected.

8. Remove the shift cable bracket. Then, disconnect the shift and select cables from the top of the transaxle case. Leave the cables and bracket together, and wire them out of the work area.

9. Remove the mounting bolts and clutch slave cylinder with the clutch pipe and pushrod.

10. Remove the mounting bolt and clutch hose joint with the clutch pipe and clutch hose.

➡**Do not operate the clutch pedal once the slave cylinder has been removed. Be careful not to kink the metal hydraulic lines.**

11. Remove the two upper transaxle case bolts.

12. Raise and safely support the vehicle.

13. Remove the front wheels.

14. Remove the engine splash shield.

15. Drain the transaxle fluid.

16. Remove the clutch damper bracket and raise it out of the way.

17. Remove the subframe center beam.

18. Remove the cotter pins and lower arm ball joint nuts. Separate the ball joints and lower arms using a press type ball joint tool.

19. Remove the right damper fork bolt. Remove the right damper pinch bolt, then separate the damper fork and damper. Remove the radius rod bolts and nut;, then, remove the right radius rod.

20. Use a suitable prytool to separate the right and left halfshafts from the differential and the intermediate shaft. Remove the left halfshaft.

21. Remove the intermediate shaft from the differential by removing its three bearing shaft mounting bolts.

22. Swing the right halfshaft out and wire it up inside the right fender well. Tie plastic bags over the inboard CV-joints to protect the boots and splines from damage.

23. Remove the engine stiffener and the clutch cover.

24. Remove the intake manifold bracket.

25. Remove the rear engine mount bracket. Remove and discard the three rear engine mount bracket mounting bolts.

26. Place a transaxle jack under the transaxle. Raise the transaxle just enough to take the weight off the its mounts.

➡**A chain hoist may be attached the transaxle lifting hooks to steady it and aid in lowering it from the vehicle.**

27. Remove the transaxle housing mounting bolt on the engine side.

28. Remove the transaxle mount bolt and loosen the mount bracket nuts.

29. Remove the three transaxle housing mounting bolts.

30. Remove the transaxle from the vehicle.

To install:

➡**Use new self-locking nuts when assembling the front suspension. Install new set rings onto the inboard CV-joints. Use new self-locking bolts when installing transaxle rear mount bracket (the bolts are color coded by type). New fasteners are available from a Honda dealer.**

31. Be sure the two dowel pins are installed into the transaxle case.

32. Apply heavy duty high temperature grease (use Honda part No. 08798–9002) to the release bearing, mainshaft splines, and the release fork pawls. Install the release fork and release bearing.

33. Raise the transaxle into position.

34. Install the three lower transaxle case bolts and tighten to 47 ft. lbs. (65 Nm).

35. Install the transaxle mount and mount bracket. Install the through-bolt and tighten temporarily. Be sure the engine is level and tighten the three mount bracket nuts to 40 ft. lbs. (55 Nm), or 28 ft. lbs. (38 Nm) for 1995 Accords. Tighten the through-bolt to 47 ft. lbs. (65 Nm).

36. Install the upper transaxle case bolts on the engine side and tighten to 47 ft. lbs. (65 Nm).

37. Install the three new rear engine bracket mounting bolts and tighten to 40 ft. lbs. (55 Nm).

38. Install the intake manifold bracket and tighten the bolts to 16 ft. lbs. (22 Nm).

39. Install the clutch cover and tighten the bolts to 9 ft. lbs. (12 Nm).

40. Install the subframe center beam with new self-locking bolts. Evenly tighten the bolts to 37 ft. lbs. (50 Nm).

41. If equipped, install the engine stiffener plate and loosely install the mounting

bolts. Tighten the stiffener-to-transaxle case mounting bolt to 28 ft. lbs. (39 Nm), then tighten the two stiffener-to-engine block mounting bolts to 28 ft. lbs. (39 Nm) beginning with the bolt closest to the transaxle.

42. Install the radius rod and the damper fork. Install all the fasteners, but only hand-tighten them at this time.

43. Install the intermediate shaft, tighten its mounting bolts to 28 ft. lbs. (39 Nm).

44. Install a new set ring on the end of each halfshaft. Install the right and left half-shafts. Turn the right and left steering knuckle fully outward and slide the axle into the differential, until the set ring is felt engaging the differential side gear.

45. Reconnect the lower control arm ball joints. Tighten the castle nuts to 40 ft. lbs. (50 Nm). Then, tighten them only enough to install a new cotter pin.

46. Install the clutch damper and tighten its mounting bolts to 16 ft. lbs. (22 Nm).

47. Install the front wheels. Lower the vehicle.

48. Place a floor jack under the right front knuckle, and raise the jack until it is supporting the vehicle's weight.

49. Tighten the radius rod mounting bolts to 76 ft. lbs. (105 Nm) and the radius rod nut to 32 ft. lbs. (44 Nm). Tighten the damper fork nut while holding the damper fork bolt to 40 ft. lbs. (55 Nm). Tighten the damper pinch bolt to 32 ft. lbs. (44 Nm).

50. Coat the tip of the slave cylinder with high temperature grease. Install the clutch hose joint and clutch slave cylinder to the transaxle housing. Be sure the slave cylinder's tip snaps into the release fork. Tighten the slave cylinder mounting bolts to 16 ft. lbs. (22 Nm).

51. Install the speed sensor. Tighten the mounting bolt to 13 ft. lbs. (18 Nm).

52. Install the shift cable and select cable to the shift arm lever. Tighten the cable bracket mounting bolts to 20 ft. lbs. (27 Nm). Install new cotter pins.

53. Connect the back-up light switch.

54. Install the starter. Tighten the 10x1.25mm bolt to 32 ft. lbs. (45 Nm) and the 12x1.25mm bolt to 54 ft. lbs. (75 Nm). Connect the starter wires.

55. Install the transaxle ground cable.

56. Fill the transaxle with the proper type and quantity of oil.

57. Install the air cleaner case and the resonator, then the intake duct.

58. Install the battery tray bracket and battery tray, tighten the bolts to 16 ft. lbs. (22 Nm).

59. Install the battery and connect the battery cables.

60. Check the clutch pedal free-play.

61. Check and adjust the front wheel alignment.

62. Road test the vehicle and check the transaxle for smooth operation.

63. Loosen the three front engine mount bracket mounting bolts, then retighten them to 28 ft. lbs. (38 Nm).

64. Enter the radio security code.

Automatic

CIVIC AND DEL SOL

1. Disconnect the negative and positive battery cables.

2. Remove the resonator, the air cleaner box, and the air intake duct.

3. Disconnect the starter cables and the transaxle ground cable. Remove the engine wiring harness clip. Label and disconnect the lock-up control solenoid connector.

4. Label and disconnect the vehicle speed sensor (VSS) and countershaft speed sensor connectors.

5. Loosen the upper transaxle case bolts and the rear engine mounting bolt.

6. Raise and safely support the vehicle. Remove the front wheels.

7. Drain the automatic transaxle fluid. Then, install the drain plug with a new crush washer. Note the color, consistency, and odor of the drained fluid.

8. Remove the front splash shield.

9. Label and disconnect the shift control and linear solenoid connectors. Disconnect the mainshaft speed sensor connector.

10. Remove the strut pinch bolt and fork bolt. Disconnect the lower ball joint using a ball joint remover.

11. Pry the halfshaft inboard joints out of the transaxle case and intermediate shaft. Swing the steering knuckles out to free the halfshafts from the transaxle.

12. Tie the halfshafts up and out of the way with wire. Tie plastic bags over the inboard joints to prevent damage to the CV-boots and splined shafts.

13. Disconnect and remove the front exhaust pipe.

14. Remove the shift cable cover. Disconnect the shift cable from the transaxle control shaft. Move the shift cable out of the way, and tie it up with wire.

15. Disconnect the ATF cooler hoses from the cooler lines. Cap the lines hoses to prevent fluid lose and contamination.

16. Remove the right-front mount and bracket assembly.

17. Remove the engine stiffener and the torque converter cover plate.

18. Remove the eight torque converter-to-driveplate bolts one at a time by rotating the crankshaft pulley. There are no gear

teeth on the driveplate; the starter motor engages a ring gear on the inner edge of the torque converter.

19. After unbolting the torque converter from the driveplate, Rotate the crankshaft to set the engine at TDC/compression for the No. 1 cylinder.

20. Label and disconnect the ignition wires. Unbolt and remove the distributor assembly so that an engine lifting chain hook can be bolted to the distributor mount.

21. Attach a lifting chain to the engine and lift slightly to ease the tension on the mounts.

22. Place a transaxle jack under the transaxle and remove the transaxle side mount and bracket.

23. With the transaxle supported, remove the transaxle rear mount bracket bolts and transaxle case bolts.

24. Pull the transaxle away from the engine until it clears the locating dowel pins. Carefully lower the transaxle from the vehicle with the torque converter angled upward so it doesn't drop out of the transaxle.

25. Remove the torque converter from the transaxle. Inspect the ring gear teeth for breakage and inspect the converter's hub for burrs and scoring. Check the condition of the converter's fluid. Replace the torque converter if necessary.

26. Inspect the transaxle's front oil pump bearing and seal for signs of leakage and scoring. Inspect the mainshaft for burrs, scoring, and roughness.

27. With the transaxle removed, carefully inspect the driveplate for stress cracks, enlarged bolt holes, and other defects. Replace it if necessary.

To install:

➡**Use new self-locking nuts and color-coded bolts when installing the transaxle and suspension components.**

28. Flush the transaxle cooler lines to remove any contaminated fluid and residual clutch material:

a. Use a pressurized flusher (Honda J38405-A or equivalent). Use only Honda flushing fluid (Honda J35944–20); other fluids may damage the system.

b. Fill the flusher with 21 ounces of fluid. Pressurize the flusher to 80–120 PSI, following the procedure on the fluid container and flusher.

c. Clamp the discharge hose of the flusher to the cooler return line. Clamp the drain hose to the cooler inlet line and route it into a bucket or drain tank.

d. Connect the flusher to air and water lines. The air line use a water trap to keep excess moisture out.

e. Open the flusher water valve and flush the cooler for ten seconds.

f. Depress the flusher trigger to mix flushing fluid with the water. Flush for two minutes, turning the air valve on and off for five seconds every 15–20 seconds to create a surging action.

g. After finishing one flushing cycle, reverse the hose and flush in the opposite direction.

h. Dry the cooler lines with compressed air for two minutes or longer to remove all excess moisture from the system.

29. If removed, install the starter motor onto the transaxle case and tighten its mounting bolts to 33 ft. lbs. (45 Nm). Install the torque converter with a new hub O-ring.

30. Place the transaxle on a transaxle jack and raise to the level of the engine.

31. Align the transaxle and engine. Install the transaxle case bolts. Install new 14mm rear mount bolts and washers.

32. Raise the transaxle and install the side mount. Tighten the case bolts to 47 ft. lbs. (64 Nm). Tighten all of the 14mm rear mount bolts to 61 ft. lbs. (85 Nm).

33. Install the transaxle side mount and bracket. Tighten the bracket nuts to 47 ft. lbs. (64 Nm). Tighten the mount through-bolt to 54 ft. lbs. (74 Nm).

34. Remove the transaxle jack.

35. Rotate the crankshaft and install the torque converter-to-driveplate bolts. Tighten the bolts to 9 ft. lbs. (12 Nm) is a crisscross pattern. Tighten the bolts to the specification in two steps.

36. Rotate the crankshaft to reset the engine at TDC/compression for the No. 1 cylinder. After the engine is set at TDC, it must not be disturbed until the distributor has been reinstalled.

37. Install the torque converter cover and tighten the bolts to 9 ft. lbs. (12 Nm).

38. Install the engine stiffener and tighten the 8mm bolts to 17 ft. lbs. (24 Nm). Tighten the 10mm bolts to 33 ft. lbs. (45 Nm).

39. Install the right-front mount and bracket assembly. Tighten the 10mm bolt to 33 ft. lbs. (44 Nm), and the 12mm bolts to 47 ft. lbs. (64 Nm).

40. Remove the lifting chain and chain hooks.

41. Verify that the engine is at TDC/compression for the No. 1 cylinder. Align the tabs on the distributor drive with the grooves on the end of the camshaft. Install the distributor and hand-tighten the mounting bolts. Reconnect the ignition wires.

42. Tighten the crankshaft pulley to 134 ft. lbs. (181 Nm).

43. Reconnect the transaxle cooler lines.

44. Install new set rings on the inboard CV-joint splines. Install the halfshafts into the transaxle case and intermediate shaft. The inboard joints must snap into place.

45. Connect the lower ball joint and damper fork. Install the front wheels.

46. Connect the shift cable linkage to the transaxle control shaft. Install a new lock-washer and tighten the linkage bolt to 10 ft. lbs. (14 Nm). Install the shift cable cover and tighten its bolt to 16 ft. lbs. (22 Nm).

47. Install the front exhaust pipe. Use new self-locking nuts and gaskets. Tighten the rear flange nuts to 16 ft. lbs. (22 Nm), and the front flange nuts to 47 ft. lbs. (64 Nm).

48. Connect the vehicle speed sensor (VSS) and countershaft speed sensor connectors. Connect the lock-up control solenoid connector.

49. Connect the shift control and linear solenoid connectors. Connect the mainshaft speed sensor connector.

50. Install the wiring harness clamps and starter cables. Install the resonator, air cleaner box and air intake duct.

51. Install the front splash shield.

52. Lower the vehicle and tighten the strut pinch bolts to 32 ft. lbs. (44 Nm). Tighten the fork bolts to 47 ft. lbs. (65 Nm). Tighten the ball joint castle nuts to 40 ft. lbs. (55 Nm), then tighten them only enough to install new cotter pins.

53. Refill the transaxle with fresh ATF. Use only Honda Premium ATF or an equivalent DEXRON® II or III ATF. Reconnect the positive and negative battery cables.

a. Leave the flusher drain hose attached to the cooler return line.

b. With the transaxle in park, run the engine for 30 seconds, or until approximately one quart of fluid is discharged. Immediately shut the engine off. This completes the cooler flushing process.

c. Remove the drain hose and reconnect the cooler return line.

d. Refill the transaxle to the proper level.

54. Check shift cable and throttle cable adjustments.

55. Check the ignition timing. Rotate the distributor counterclockwise to advance the timing, or clockwise to retard the timing. When the timing has been set, tighten the distributor mounting bolts to 13 ft. lbs. (18 Nm).

56. Start the engine and shift through all the gears three times.

57. Let the engine warm up to operating temperature and check the fluid level with the transaxle in the **P** or **N** position.

58. Check and adjust the front wheel alignment.

59. Road test the vehicle. Recheck the transaxle fluid level.

PRELUDE

1. Disconnect both cables from the battery.

2. Shift the transaxle into **N**.

3. Remove the battery hold-down and remove the battery.

4. Drain the transaxle fluid and reinstall the drain plug with a new crush washer.

5. Remove the air intake duct, air cleaner case, and resonator.

6. Disconnect the connector from the vacuum tank and remove the vacuum tank and tank bracket. Do not remove the vacuum tube from the vacuum tank.

7. Disconnect the transaxle-to-body ground cable.

8. Remove the battery base with the ground cable and remove the battery base stay.

9. Disconnect the lock-up control solenoid valve and shift control solenoid valve connectors.

10. Disconnect the throttle control cable from the throttle control lever.

11. Disconnect the countershaft speed sensor connector.

12. Disconnect the vehicle speed sensor connector.

13. Remove the rear stiffener, then remove the vehicle speed sensor and power steering speed sensor.

➡**Do not disconnect the power steering pressure hoses from the vehicle speed sensor and power steering speed sensor.**

14. Disconnect the ATF cooler hoses at the joint pipes. Turn the ends of the cooler hoses upward to prevent fluid loss. Plug the joint pipes.

15. Remove the starter motor.

16. Remove the upper transaxle housing mounting bolts.

17. Loosen the front engine mount bracket bolts.

18. Remove the transaxle mount.

19. Raise and support the vehicle safely. Remove the front wheels.

20. Remove the splash shield and remove the subframe center beam and rear beam stiffener.

21. Remove the cotter pins and castle nuts from the lower ball joints. Use a press-type ball joint tool to separate the ball joints from the lower arm.

22. Remove the damper fork bolts and separate the damper fork and the damper.

23. Use a suitable prytool to separate the right and left halfshafts from the differential.

24. Pull on the inboard joint and remove the right and left halfshafts. Tie plastic bags over the halfshaft ends to protect the boots and splined shafts from damage.

25. Remove the right damper pinch bolt and separate the right damper fork from the strut.

26. Remove the right radius rod bolts and nut. Remove the radius rod.

27. Remove the torque converter cover and the shift cable cover.

28. Remove the control lever lockbolt and remove the shift cable with the lever. Do not bend the shift control cable during removal. Wire the cable to the underbody of the vehicle our of the work area.

29. Remove the driveplate bolts while rotating the crankshaft.

30. Place a transaxle jack below the transaxle and raise it enough to take the weight off the mounts.

31. Remove the intake manifold bracket.

32. Remove the lower transaxle housing mounting bolts and lower rear engine mounting bolts.

33. Pull the transaxle away from the engine until it clears the dowel pins. Lower the transaxle out of the vehicle.

To install:

➡**Use new self-locking nuts when assembling the front suspension components. Use new set rings on the halfshaft inboard joints. Use new self-locking bolts for the subframe beams. These fasteners are available from a Honda dealer.**

34. Flush the transaxle cooling lines before installing the transaxle. Use a pressurized flushing canister, such as Honda tool No. J38405-A, or its equivalent. Use only biodegradable flushing fluid, Honda part No. J35944–20.

a. Fill the flusher with 21 ounces of fluid. Pressurize the flusher to 80–120 PSI, following the procedure on the fluid container and flusher.

b. Clamp the discharge hose of the flusher to the cooler return line. Clamp the drain hose to the cooler inlet line and route it into a bucket or drain tank.

c. Connect the flusher to air and water lines. Open the flusher water valve and flush the cooler for ten seconds.

d. Depress the flusher trigger to mix flushing fluid with the water. Flush for two minutes, turning the air valve on and off for five seconds every 15–20 seconds.

e. After finishing one flushing cycle, reverse the hose and flush in the opposite direction.

f. Dry the cooler lines with compressed air so that no moisture remains in the cooler lines.

35. Install the starter motor onto the transaxle case. Install the torque converter with a new hub O-ring. Tighten the starter bolts to 33 ft. lbs. (45 Nm).

36. Place the transaxle on a transaxle jack and raise it to the level of the engine.

37. Align the transaxle to the engine and install the transaxle housing mounting bolts and lower rear engine mounting bolts. Tighten the rear engine mounting bolts to 40 ft. lbs. (55 Nm) and the transaxle mounting bolts to 47 ft. lbs. (65 Nm). Install the intake manifold bracket and tighten the bolts to 16 ft. lbs. (22 Nm).

38. Tighten the front engine mount bracket bolts to 28 ft. lbs. (39 Nm).

39. Install the transaxle mount. Tighten the bolt to 47 ft. lbs. (65 Nm) and the nuts to 28 ft. lbs. (39 Nm).

40. Remove the transaxle jack.

41. Attach the torque converter to the driveplate and install the mounting bolts. Turn the crankshaft to rotate the driveplate. Tighten the bolts in 2 steps, first to 4.5 ft. lbs. (6 Nm) in a crisscross pattern and finally to 9 ft. lbs. (12 Nm) for 1995. (75 Nm). Check for free rotation after tightening the last bolt.

42. Install the shift cable onto the control shaft and tighten the lockbolt to 10 ft. lbs. (14 Nm).

43. Install the torque converter cover and the shift cable cover.

44. Install a new set ring onto the inboard joint of each halfshaft.

45. Install the damper fork bolts and ball joint nuts to the lower arms. Tighten the ball joint nut to 47 ft. lbs. (65 Nm) and install a new cotter pin. Install the radius rod and connect the damper fork. Only hand-tighten the radius rod and damper fork fasteners at this point.

46. Turn the right steering knuckle fully outward and slide the axle into the differential until the spring clip is felt engaging the differential side gear. Repeat the procedure on the left side.

47. Install the subframe rear beam stiffener and the center beam. Tighten the stiffener bolts to 28 ft. lbs. (39 Nm). Tighten the subframe center beam bolts to 43 ft. lbs. (60 Nm).

48. Install the front wheels and lower the vehicle.

49. Use a floor jack to place the weight of the vehicle onto the right front knuckle.

Tighten the radius rod bolts to 76 ft. lbs. (105 Nm) and the nut to 40 ft. lbs. (55 Nm). Tighten the damper pinch bolt to 32 ft. lbs. (44 Nm). Tighten the nut to 47 ft. lbs. (65 Nm) while holding the damper fork bolt.

50. Install the speedometer sensor. Tighten the sensor bolt to 9 ft. lbs. (12 Nm). On 1993 Preludes only, install the rear stiffener and tighten the bolt to 28 ft. lbs. (39 Nm), tighten the nut to 15 ft. lbs. (21 Nm).

51. Connect the ATF cooler hoses to the joint pipes.

52. Connect the lock-up control solenoid and shift control solenoid valve connectors.

53. Connect the vehicle speed sensor and power steering speed sensor connectors.

54. Connect the starter motor cables and install the battery base and base stay.

55. Connect the ground cables on the body and on the transaxle.

56. Install the vacuum tank, tank bracket, and connect the tank connector.

57. Install the resonator, air cleaner case, and air intake duct.

58. Refill the transaxle with ATF. Use only Honda Premium ATF or an equivalent DEXRON® II ATF. Connect the negative and positive battery cables.

a. Leave the flusher drain hose attached to the cooler return line.

b. With the transaxle in park, run the engine for 30 seconds, or until approximately one quart of fluid is discharged. This completes the cooler flushing process.

c. Remove the drain hose and reconnect the cooler return line.

d. Refill the transaxle to the proper level with ATF.

59. Start the engine, set the parking brake and shift the transaxle through all gears 3 times. Check for proper control cable adjustment.

60. On Preludes equipped with 4WS, start the engine and turn the steering wheel lock-to-lock to reset the steering control unit.

61. Check and adjust the front wheel alignment.

62. Let the engine reach operating temperature with the transaxle in **N** or **P**, then turn the engine OFF and check the fluid level.

63. After road testing the vehicle, loosen the front engine mount bolts, and tighten them to 28 ft. lbs. (39 Nm).

64. Enter the radio security code.

Accord

2.2L ENGINES

1. Disconnect the negative, then the negative battery cables.

2. Remove the battery from the vehicle.

3. Shift the transaxle into **N**.

4. Remove the air intake hose, air cleaner housing and the resonator assembly.

5. Remove the battery base and the base stay.

6. Disconnect the throttle cable from the throttle control lever.

7. Disconnect the transaxle ground cable and the speed sensor connectors. Disconnect the solenoid valve connectors.

8. Disconnect the lock-up control solenoid valve and shift control solenoid valve connectors.

9. Disconnect the transaxle cooler hoses from the joint pipes and plug the hoses.

10. Disconnect the starter cables and remove the starter.

11. Disconnect the countershaft speed sensor connector.

12. Disconnect the vehicle speed sensor connector.

13. Install a hoist to the engine.

14. Remove the four upper bolts attaching the transaxle to the engine block.

15. Loosen the three bolts attaching the front engine mount bracket to the engine.

16. Remove the transaxle mount.

17. Raise and safely support the vehicle. Remove the front wheels.

18. Drain the transaxle fluid and reinstall the drain plug with a new washer.

19. Remove the splash shield.

20. Remove the subframe center beam.

21. Remove the cotter pins and lower arm ball joint nuts, then separate the ball joints from the lower arms using a suitable tool.

22. Remove the right damper pinch bolt, then separate the damper fork and damper.

23. Remove the bolts and nut, then remove the right radius rod.

24. Using a small prying device, carefully pry the right and left halfshafts out of the differential. Remove the right and left halfshafts. Tie plastic bags over the halfshaft ends to prevent damage to the CV boots and splines.

25. Remove the bolts mounting the intermediate shaft, then remove the intermediate shaft from the differential.

26. Remove the torque converter cover and shift cable cover.

27. Remove the shift control cable by removing the lockbolt. Remove the shift cable lever from the control shaft. Don't disconnect the control lever from the shift cable. Wire the shift cable out of the work area and be careful not to kink it.

28. Remove the eight drive plate bolts one at a time while rotating the crankshaft pulley.

29. Place a suitable jack under the transaxle and raise the jack just enough to take weight off of the mounts.

30. Remove the intake manifold bracket.

31. Remove the transaxle housing mounting bolts.

32. Remove the mounting bolts from the rear engine mount bracket.

33. Remove the four transaxle housing mounting bolts and three mount bracket nuts.

34. Pull the transaxle away from the engine until it clears the 14mm dowel pins, then lower it using the jack.

To install:

➡ **Use new self-locking nuts when assembling the front suspension components. Install new set rings onto the halfshaft inboard joint splines. Replace any color-coded self-locking bolts.**

35. Flush the transaxle cooler lines before installing the transaxle. Use a pressurized flushing unit such as Honda J38405-A or equivalent. Use only Honda biodegradable flushing fluid, Honda J35944-20. Other fluids will damage the A/T cooling system.

  a. Fill the flusher with 21 ounces of fluid. Pressurize the flusher to 80-120 PSI, following the procedure on the fluid container and flusher.

  b. Clamp the discharge hose of the flusher to the cooler return line. Clamp the drain hose to the cooler inlet line and route it into a bucket or drain tank.

  c. Connect the flusher to air and water lines. Open the flusher water valve and flush the cooler for ten seconds. The air line should be equipped with a water trap to keep the system dry.

  d. Depress the flusher trigger to mix flushing fluid with the water. Flush for two minutes, turning the air valve on and off for five seconds every 15-20 seconds to create a surging action.

  e. After finishing one flushing cycle, reverse the hose and flush in the opposite direction following the same steps.

  f. Dry the cooler lines with compressed air so that no moisture is left in the cooler system.

36. Be sure the two 14mm dowel pins are installed into the torque converter housing.

37. Install the torque converter onto the transaxle mainshaft with a new hub O-ring. Install the starter motor onto the transaxle case and tighten the mounting bolts to 33 ft. lbs. (44 Nm).

38. Raise the transaxle into position and install the transaxle housing mounting bolts. Tighten the bolts to 47 ft. lbs. (65 Nm).

39. Install the rear engine mounting bolts and tighten to 40 ft. lbs. (54 Nm).

40. Install the intake manifold bracket and tighten the bolts to 16 ft. lbs. (22 Nm).

41. Install the upper bolts attaching the transaxle to the engine and tighten the bolts to 47 ft. lbs. (64 Nm).

42. Tighten the front engine mount bracket bolts to 28 ft. lbs. (38 Nm).

43. Install the transaxle mount and loosely install the nuts and bolt that attach the mount. Tighten the nuts first to 28 ft. lbs. (38 Nm), then tighten the bolt to 47 ft. lbs. (64 Nm).

44. Remove the jack from the transaxle.

45. Attach the torque converter to the drive plate with the eight bolts. Tighten the bolts in two steps in a crisscross pattern: first to 4.5 ft. lbs. (6 Nm), and finally to 9 ft. lbs. (12 Nm). Check for free rotation after tightening the last bolt.

46. Install the shift control cable and control cable holder. Tighten the shift cable lockbolt to 10 ft. lbs. (14 Nm). Tighten the shift cable cover bolts to 13 ft. lbs. (18 Nm).

47. Install the torque converter cover and tighten the bolts to 9 ft. lbs. (12 Nm).

48. Remove the engine hoist.

49. Install the radius rod and damper fork.

50. Install the intermediate shaft into the differential and tighten the mounting bolts to 28 ft. lbs. (38 Nm).

51. Install a new set ring on the end of each halfshaft.

52. Turn the right steering knuckle fully outward and slide the axle into the differential until the set ring snaps into the differential side gear. Repeat the procedure on the left side.

53. Install the damper fork bolts and ball joint nuts to the lower arms. Tighten the ball joint nut to 40 ft. lbs. (55 Nm) and install a new cotter pin.

54. Install the subframe center beam and tighten the center beam bolts to 28 ft. lbs. (39 Nm).

55. Install the splash shield.

56. Install the front wheels and lower the vehicle.

57. Reconnect the speed sensor connector.

58. Support the right front knuckle with a floor jack until the weight of the vehicle is held by the jack. Tighten the damper fork pinch bolt to 32 ft. lbs. (44 Nm). Tighten the radius rod bolts to 76 ft. lbs. (105 Nm), and the radius rod nut to 32 ft. lbs. (44 Nm). Hold the damper fork bolt with a wrench, and tighten the nut to 40 ft. lbs. (55 Nm).

59. Connect the cables to the starter.

60. Reconnect the throttle control cable.

61. Connect the lock-up control solenoid valve and shift control solenoid valve connectors.

62. Connect the speed sensor connectors and the transaxle ground cable.

63. Connect the transaxle cooler inlet hose to the joint pipe. Attach a drain hose to the return line.

64. Install the battery base stay and the battery base.

65. Install the resonator assembly, the air cleaner assembly and the air intake hose.

66. Install the battery and connect the positive, then the negative battery cables to the battery.

67. Refill the transaxle with ATF. Use only Honda Premium ATF or an equivalent DEXRON® II ATF.

   a. With the flusher drain hose attached to the cooler return line.

   b. Place the transaxle in **P**, run the engine for 30 seconds, or until approximately one quart of fluid is discharged. Immediately shut off the engine. This completes the cooler flushing process.

   c. Remove the drain hose and reconnect the cooler return line.

   d. Refill the transaxle to the proper level with ATF.

68. Start the engine, set the parking brake and shift the transaxle through all gears 3 times. Check for proper shift cable adjustment.

69. Let the engine reach operating temperature with the transaxle in **P** or **N**. Then, shut off the engine and check the fluid level.

70. Road test the vehicle.

71. After road testing the vehicle, loosen the front engine mount bracket bolts, then retighten them to 28 ft. lbs. (39 Nm).

72. Check and adjust the vehicle's front end alignment.

73. Enter the radio security code.

2.7L ENGINE

➡**Several of the engine mounts must be removed during the transaxle removal procedure. The objective of this procedure is to allow the engine to tilt so that the transaxle will clear the left shock tower. A chain host is necessary for this operation.**

1. Disconnect the negative and positive battery cables.

2. Prop the hood open and remove the support struts. Secure the hood open in a vertical position.

3. Remove the battery. Remove the battery tray and bracket. Unbolt the cable holder from the battery tray.

4. Remove the intake air duct.

5. Detach the clips securing the starter cable to the strut brace. Remove the strut brace.

6. Drain the transaxle fluid. Install the drain plug with a new crush washer.

7. Uncouple the ATF cooler lines. Plug the lines, and turn them upward to lessen fluid spillage.

8. Remove the starter cables. Unbolt the starter cable clamp and the transaxle ground cable. Unbolt the engine sub-harness clamp from the transaxle case, and move the harness holder out of the way.

9. Disconnect the following electrical couplings:

   • Shift control solenoid valve
   • Mainshaft speed sensor connector
   • Lock-up control solenoid valve
   • Vehicle speed sensor connector
   • Linear solenoid connector
   • Counter shaft speed sensor

10. Remove the intake air bypass (IAB) vacuum tank from the transaxle. Do not disconnect its vacuum hoses.

11. Loosen the top two upper transaxle case bolts, they may not be easily accessible when the vehicle is raised.

12. Raise and safely support the vehicle. Remove the front wheels.

13. Remove the splash shield.

14. Remove the subframe center beam.

15. Remove the lower ball joint castle nuts. Separate the ball joints from the lower control arms using a suitable press type tool.

16. Remove the damper fork bolts, and separate the damper forks from the lower control arm.

17. Use a flat-bladed tool to carefully pry the halfshafts out of the differential and intermediate shaft.

18. Pull the inboard joints away from the differential and intermediate shaft. Tie plastic bags over the inboard joints to protect the boots and splines.

19. Remove the left damper pinch bolt;, then, separate the damper fork from the strut. Unbolt and remove the left radius rod. Use wire to support the left control arm and halfshaft out of the work area.

20. Remove the intermediate shaft.

21. Remove the shift cable holder and cover. Remove the shift cable lockbolt, and slide the cable lever off of the control rod. Be careful not to kink the shift cable.

22. Remove the torque converter cover. Remove the eight driveplate bolts one at a time while rotating the crankshaft pulley.

23. Attach a chain hoist to the transaxle. Place a jack under the engine for support.

24. Remove the rear mount stiffener;, then, unbolt and remove the rear mount.

25. Remove the front mount bracket;, then, remove the front mount.

26. Remove the four upper transaxle case bolts.

27. Remove the side transaxle mount.

28. Remove the two lower transaxle case bolts.

✳✳ CAUTION

Be sure the transaxle is securely supported before removing the rear mounting bracket bolts. Place a transaxle jack under the transaxle for additional support.

29. Remove the rear mount bolts. Raise the engine and transaxle slightly;, then, remove the rear mount bracket from the rear mount.

30. Carefully lower the transaxle from the vehicle. Tilt the engine just enough for the transaxle to clear the left shock tower.

➡**Do not allow the A/C compressor to hit the right shock tower when the engine is being tilted.**

31. Pull the transaxle away from the engine until it clears the dowel pins, then lower it from the vehicle.

To install:

➡**Use new self-locking nuts when assembling the front suspension components. Install a new set ring onto the inboard halfshafts. Replace any color-coded self-locking nuts and bolts when installing the engine and transaxle mounts. These fasteners are available from a Honda dealer.**

32. Flush the transaxle cooler lines before installing the transaxle. Use a pressurized flushing unit such as Honda J38405-A or equivalent. Use only Honda biodegradable flushing fluid, Honda J35944–20. Other fluids will damage the A/T cooling system.

   a. Fill the flusher with 21 ounces of fluid. Pressurize the flusher to 80–120 PSI, following the procedure on the fluid container and flusher.

   b. Clamp the discharge hose of the flusher to the cooler return line. Clamp the drain hose to the cooler inlet line and route it into a bucket or drain tank.

   c. Connect the flusher to air and water lines. Open the flusher water valve and flush the cooler for ten seconds. The air line should be equipped with a water trap to keep the system dry.

   d. Depress the flusher trigger to mix flushing fluid with the water. Flush for two minutes, turning the air valve on and off for five seconds every 15–20 seconds to create a surging action.

e. After finishing one flushing cycle, reverse the hose and flush in the opposite direction following the same steps.

f. Dry the cooler lines with compressed air so that no moisture is left in the cooler system.

33. Be sure the two 14mm dowel pins are installed into the torque converter housing.

34. Install the torque converter onto the transaxle mainshaft with a new O-ring. Install the starter motor onto the transaxle case and tighten the mounting bolts to 33 ft. lbs. (44 Nm).

35. Position the transaxle under the vehicle and attach the chain hoist to it.

36. Jack or hoist the engine into place;, then, lift the transaxle into place.

37. Mate the transaxle to the engine. Install the transaxle case bolts, and tighten them to 47 ft. lbs. (65 Nm). The one long bolt fits behind the starter.

38. Install the rear mount bracket onto the rear mount and hand-tighten the nut and upper bolt. Install two new rear mount bracket-to-transaxle case bolts, and tighten them to 40 ft. lbs. (55 Nm).

39. Verify that all of the transaxle case bolts have been installed.

40. Install the front mount, tighten the three bolts to 43 ft. lbs. (59 Nm). Install the front mount bracket onto the transaxle and tighten the bolts to 28 ft. lbs. (39 Nm). Install and hand-tighten the mount nut.

41. Install the side transaxle mount. Install and hand-tighten the three nuts and the through-bolt. Tighten the nuts to 28 ft. lbs. (39 Nm), then tighten the bolt to 47 ft. lbs. (65 Nm).

42. With all the engine and transaxle mounts installed, tighten the mount nuts to 40 ft. lbs. (55 Nm). Install the rear mount stiffener. Tighten the rear stiffener bolt and upper rear mount bracket bolt to 28 ft. lbs. (39 Nm).

43. Remove the chain hoist and jacks from the engine and transaxle.

44. Install the eight driveplate bolts and tighten them to their final torque specification in two steps in a crisscross pattern. The first tightening step is 4.3–4.5 ft. lbs. (6 Nm). The second tightening step is 8.7 ft. lbs. (12 Nm). After tightening the last bolt, check the crankshaft for free rotation.

45. Install the torque converter cover and tighten its bolts to 8.7 ft. lbs. (12 Nm).

46. Use a holder tool and a torque wrench to retighten the crankshaft pulley to 181 ft. lbs. (245 Nm).

47. Reconnect the shift cable lever to the control shaft with a new lockwasher, and

tighten the bolt to 10 ft. lbs. (14 Nm). Install the shift cable holder and cover. Tighten the cover bolts to 20 ft. lbs. (26 Nm).

48. Install the intermediate shaft and tighten the bolts to 16 ft. lbs. (22 Nm).

49. Install the left radius rod and damper fork. Hand-tighten the bolts.

50. Install new set rings on the halfshaft inboard CV-joints. Install the halfshafts into the differential and intermediate shaft. Be sure the set rings snap into place.

51. Reassemble the damper forks and lower control arm ball joints. Tighten the ball joint castle nuts to 36–43 ft. lbs. (49–59 Nm), then tighten them only enough to install new cotter pins.

52. Install the center beam and tighten the bolts to 37 ft. lbs. (50 Nm). Install the splash shield.

53. Install the front wheels and lower the vehicle.

54. Use a floor jack to raise the left front knuckle until it is supporting the weight of the vehicle. Tighten the damper pinch bolt to 32 ft. lbs. (43 Nm). Tighten the radius rod bolts to 76 ft. lbs. (101 Nm), and the nuts to 32 ft. lbs. (43 Nm). Tighten the damper fork nuts to 47 ft. lbs. (65 Nm).

55. Install the IAB vacuum tank and the engine wiring harness clamp.

56. Reconnect the following electrical couplings:

- Shift control solenoid valve
- Mainshaft speed sensor connector
- Lock-up control solenoid valve
- Vehicle speed sensor connector
- Linear solenoid connector
- Counter shaft speed sensor

57. Install and connect the starter cables and the transaxle ground cable.

58. Reconnect the ATF cooler hoses.

59. Install the strut brace and tighten its mounting bolts to 16 ft. lbs. (22 Nm). Attach the starter cable clips to the strut brace.

60. Install the intake air duct. Install the battery tray and tighten its mounting bolts to 16 ft. lbs. (22 Nm). Reconnect the cable holder and ground terminal.

61. Install the battery and reconnect the battery cables.

62. Prop the hood and install the support struts.

63. Refill the transaxle with ATF. Use only Honda Premium ATF or an equivalent DEXRON® II ATF. Connect the battery cables.

a. Leave the flusher drain hose attached to the cooler return line.

b. With the transaxle in **P**, run the engine for 30 seconds, or until approximately one quart of fluid is discharged. Immediately shut off the engine. This completes the cooler flushing process.

c. Remove the drain hose and reconnect the cooler return line.

d. Refill the transaxle to the proper level with ATF.

64. Start the engine, set the parking brake and shift the transaxle through all gears three times. Check for proper shift cable adjustment.

65. Let the engine reach operating temperature with the transaxle in **P** or **N**. Then, shut off the engine and check the fluid level.

66. Road test the vehicle.

67. After road testing the vehicle, loosen the front and rear engine mount nuts, then retighten them to 40 ft. lbs. (54 Nm).

68. Check and adjust the vehicle's front end alignment.

69. Enter the radio security code.

3.0L ENGINE

1. Disconnect the negative battery cable, then the positive cable.

2. Remove the battery and tray.

3. Remove the clamps securing the battery cables to the base.

4. Remove the intake air duct and the air cleaner assembly.

5. Raise the vehicle and drain the transaxle fluid. Replace the drain plug with a new washer.

6. Remove the starter wiring and harness clamps, remove the breather and radiator hoses from the retainer.

7. Detach the wiring connectors from the transaxle assembly.

8. Disconnect the cooler lines and point them up to prevent fluid drainage.

9. Remove the bolt and nut securing the rear stiffener and remove the stiffener.

10. Remove the bolts securing the transaxle to the engine.

11. Remove the front mounting bracket bolts.

12. Remove the engine under cover.

13. Disconnect the lower shock absorber mounting and the lower ball joints from the control arms.

14. Remove the bolts securing the radius rods to the lower arms.

15. Remove the halfshafts. Keep the splined ends of the shafts clean.

16. Mark the position of the sub-frame on the main-frame and remove it.

17. Remove the engine brace from the rear of the engine.

18. Remove the shift cable cover, bracket and cable.

19. Remove the eight bolts securing the drive plate to the torque converter.

20. Attach a chain hoist to the engine and raise it slightly.

21. Place a jack under the transaxle.

22. Remove the transaxle mount bracket.

23. Remove the intake manifold support bracket.

24. Remove the rear mount bracket.

25. Pull the transaxle back slightly until it comes off the dowels and lower it from the vehicle. Do not let the torque converter fall out of the transaxle.

To install:

26. If removed, install the torque converter using a new O-ring.

27. Install the dowel pins in the torque converter housing.

28. Raise the transaxle to the engine and install the rear mount bracket. Tighten the 8x1.25 bolt to 16 ft. lbs. (22 Nm) and the 12x1.25 bolts to 40 ft. lbs. (54 Nm).

29. Install the transaxle-to-engine bolts. Tighten the bolts to 47 ft. lbs. (64 Nm).

30. Connect the breather tube with the dot facing up and install the transaxle mount bracket. Tighten the nuts to 28 ft. lbs. (38 Nm) and the through-bolt to 40 ft. lbs. (54 Nm).

31. Install the driveplate-to-torque converter bolts. Tighten them to 9 ft. lbs. (12 Nm) in a crisscross pattern.

32. Install the shift cable, bracket and cover.

33. Install the engine brace on the rear of the engine.

34. Install the halfshafts.

35. Install the sub-frame after aligning the matchmarks. Tighten the rear bolts to 47 ft. lbs. (64 Nm) and the front bolts to 76 ft. lbs. (103 Nm).

36. Install the front mount. Tighten the bolts to 28 ft. lbs. (38 Nm).

37. Connect the shock absorbers and the radius rods to the lower control arms.

38. Install the engine under cover.

39. Attach all the wiring connectors.

40. Connect the starter wiring and install the harness clamps.

41. Install the battery.

42. Install the air cleaner assembly and intake duct.

43. Refill the transaxle with Genuine Honda® premium automatic transmission fluid.

Clutch

REMOVAL & INSTALLATION

➡️**The original radio contains a coded anti-theft circuit. Obtain the security code number before disconnecting the battery cables. On vehicles equipped with 4WS, the steering control unit is shut down when the battery is disconnected. After reconnecting the battery cables, turn the steering wheel lock-to-lock to reset the control unit.**

1. Disconnect the negative battery cable.

2. Raise and safely support the vehicle.

3. Remove the transaxle from the vehicle. Matchmark the flywheel and clutch for reassembly.

4. Use a flywheel ring-gear holder to lock the flywheel in position.

5. Loosen the pressure plate bolts two turns at a time working in a crisscross pattern to prevent warping the pressure plate. Remove the pressure plate and clutch disc.

6. Inspect the flywheel, disc, and pressure plate for wear, cracks, and warpage. Light scoring of the flywheel may be polished out; gouges, warpage, burn marks, cracks, or chipped teeth require replacement of the flywheel.

➡️**If the flywheel is to be removed, but is going to be reused, matchmark it to the engine block prior to removal. Aligning the matchmarks upon reassembly will preserve driveline balance.**

7. Inspect the flywheel's ball bearing: turn the inner race of the bearing with your finger, and be sure it turns smoothly and quietly. If the bearing is loose or noisy, or exhibits rough motion, replace it.

8. Remove the release fork boot, Squeeze the release fork retaining spring to disengage the fork from its pivot. Remove the release fork from the clutch housing.

9. Remove the release bearing. Spin the bearing by hand to check its degree of play. Replace the release bearing if it has excessive play or is leaking grease.

10. Inspect the rear main bearing oil seal for signs of leakage. If necessary, replace the seal to prevent oil leakage onto the clutch's friction surfaces.

To install:

11. If necessary, drive out the flywheel bearing, then use a suitably-sized bearing driver to install a new one. Use a crisscross pattern to tighten the flywheel mounting bolts in several steps to 87 ft. lbs. (118 Nm) for vehicles with SOHC engines. If equipped with the B16A2 or B16A3 engine, tighten the flywheel bolts to 76 ft. lbs. (105 Nm).

12. Install the clutch disc and pressure plate by aligning the dowels on the flywheel with the dowel holes in the pressure plate. If a new pressure plate is not being installed, align the matchmarks that were made during removal. Install and hand-tighten the pressure plate bolts.

13. Insert a suitable clutch disc alignment tool into the splined hole in the clutch disc. Align the clutch and pressure plate.

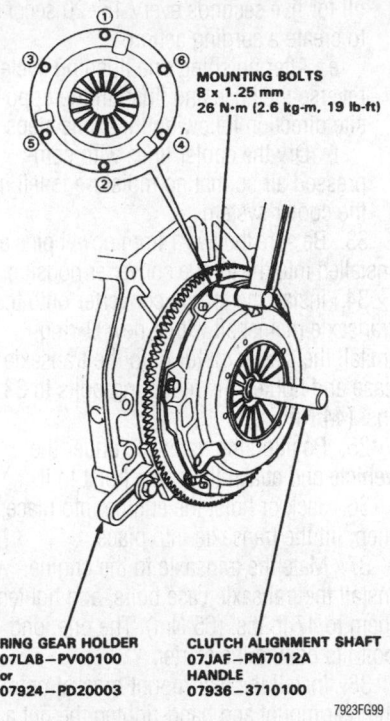

MOUNTING BOLTS
8 x 1.25 mm
26 N·m (2.6 kg-m, 19 lb-ft)

RING GEAR HOLDER
07LAB—PV00100
or
07924—PD20003

CLUTCH ALIGNMENT SHAFT
07JAF—PM7012A
HANDLE
07936—3710100

7923FG99

Clutch alignment tools and pressure plate torque sequence

14. Tighten the pressure plate bolts in a crisscross pattern two turns at a time to prevent warping the pressure plate. The final torque is 19 ft. lbs. (26 Nm).

15. Remove the alignment tool and ring gear holder.

16. Coat the mainshaft with heavy-duty high-temperature grease. The manufacturer recommends part No. 08798–9002, Honda super high-temp urea grease.

17. Coat the release fork pawls and the inner race of the release bearing with high temperature grease and install them into the clutch housing. Be sure the release fork retainer spring snaps into place on the pivot. The bearing and fork must fit together properly and slide back and forth smoothly.

18. Coat the tip of the slave cylinder with grease. Install the release fork boot.

19. Install the transaxle, making sure the mainshaft is properly aligned with the clutch disc splines and the transaxle case dowels are properly aligned with the engine block.

20. Install the transaxle case bolts and sequentially tighten them to 47 ft. lbs. (65 Nm).

21. Bleed the clutch hydraulic system.

22. Adjust the clutch pedal free-play.

23. Verify that all engine and transaxle components are installed and connected properly.

24. Reconnect the negative battery cable.

25. Road test the vehicle.

Hydraulic Clutch System

BLEEDING

1. Fill the clutch master cylinder reservoir with clean DOT 3 or 4 brake fluid.
2. Attach a rubber tube to the clutch slave cylinder bleed screw. Route the tube into a container of clean brake fluid.
3. Loosen the bleed screw.
4. Slowly pump the clutch pedal until the fluid draining from the slave cylinder is free of air bubbles.
5. Tighten the bleed screw to 6–7 ft. lbs. (8–10 Nm).
6. Refill the clutch master cylinder reservoir with brake fluid.

Halfshaft

REMOVAL & INSTALLATION

1. Loosen the front spindle nut.
2. Raise and safely support the vehicle.
3. Remove the front wheels and the spindle nut.
4. Drain the transaxle fluid and install the drain plug with a new washer. If the halfshaft to be removed is installed into the intermediate shaft, the transaxle fluid does not need to be drained.
5. Remove the damper fork nut and damper pinch bolt.
6. Remove the damper fork.
7. Remove the cotter pin and castle nut from the lower arm ball joint. Install a hex nut flush onto the ball joint stud to prevent the ball joint tool from damaging the stud threads.
8. Using a ball joint tool, separate the lower arm from the knuckle.
9. Pull the knuckle outward. Remove the halfshaft outboard joint from the hub by tapping it with a plastic hammer.
10. Carefully pry the inner CV-joint away from the transaxle case to force the halfshaft set ring out of the groove.
11. Pull on the inboard CV-joint and remove the halfshaft from the differential case or intermediate shaft.

➡ **Do not pull on the halfshaft as the CV-joint may come apart. Use care when prying out the assembly and pull it straight to avoid damaging the differential oil seal or intermediate shaft oil or dust seals.**

To install:

12. Replace the differential oil seal or intermediate shaft seal if either were damaged during removal.

13. Install new set rings on the ends of the halfshafts.
14. Install the halfshafts and be sure the set ring locks in the differential gear groove and the halfshaft bottoms in the differential or intermediate shaft.
15. Install the outboard joint into the hub. Be sure the splines mesh together and the joint is fully seated into the hub.
16. Fit the ball joint stud into the lower control arm. Install the damper fork into position. Tighten the upper damper pinch bolt to 32 ft. lbs. (44 Nm) and the fork nut to 47 ft. lbs. (65 Nm).
17. Tighten the ball joint castle nut to 40 ft. lbs. (55 Nm);, then, tighten the nut just enough to install a new cotter pin.
18. Install the front wheels. Install a new spindle nut, but don't tighten it yet.
19. Lower the vehicle.
20. Tighten the spindle nut to 181 ft. lbs. (245 Nm) and stake its tab. Tighten the wheel nuts to 80 ft. lbs. (110 Nm).
21. Fill the transaxle with the proper type and quantity of fluid.
22. Warm the engine up, check the transaxle fluid level, and road test the vehicle.

STEERING AND SUSPENSION

Air Bag

The Supplemental Restraint System (SRS/air bag) is a passive safety device designed to reduce the risk and severity of injuries to the front seat passengers of a vehicle involved in a frontal collision. The SRS is designed to be used in conjunction with the vehicle's seat belts. Airbags are less effective in an accident if the passengers are not wearing seat belts. The SRS includes the driver's and passenger's air bag modules, a cable reel in the steering column, crash sensors, and a dedicated control module.

✳✳ CAUTION

The SRS is designed to operate on extremely low power levels. A back-up power circuit energizes the SRS if the vehicle's battery power is disconnected or interrupted in an accident. Due to the sensitive nature of the SRS, the air bag modules must be disabled if they, or any other part of the SRS, must be serviced or disconnected. Failing to disable the SRS

before servicing its components may cause accidental air bag deployment and possible personal injury.

PRECAUTIONS

Several precautions must be observed when handling the inflator module to avoid accidental deployment and possible personal injury.

• Never carry the inflator module by the wires or connector on the underside of the module.

• When carrying a live inflator module, hold securely with both hands, and ensure that the bag and trim cover are pointed away.

• Place the inflator module on a bench or other surface with the bag and trim cover facing up.

• With the inflator module on the bench, never place anything on or close to the module which may be thrown in the event of an accidental deployment.

DISARMING

➡ **The radio may contain a coded theft protection circuit. Always obtain the code number before disconnecting the battery.**

Driver's Side

1. Disconnect the negative and positive battery cables.
2. Always wait at least three minutes after disconnecting the battery before working around the air bag.
3. Remove the steering wheel lower access cover.
4. Remove the clip securing the air bag module/cable reel connection to the steering column.

➡ **Spring-loaded air bag connectors contain a self-disabling contact. A shorting connector doesn't need to be installed on the driver's air bag connector.**

5. Uncouple the spring-loaded connectors:
   a. Hold the connector body, not the wiring.
   b. Pull the spring-loaded locking sleeve toward its stop while holding the opposite half of the connector.
   c. After releasing the locking sleeve, uncouple the connectors.
6. After servicing has been completed, couple the air bag and cable reel connectors. Press the sleeve side of the connector into the pawl side until the sleeve locks the connectors together.

7. Install the clip securing the air bag/cable reel connection to the steering column.

8. Install the access cover.

9. Reconnect the positive and negative battery cables.

10. Turn the ignition switch to the **ON** position, but don't start the engine. The air bag indicator light should turn on for six seconds, then turn off. If the air bag indicator light doesn't come on, or stays on longer than six seconds, the system fault must be diagnosed.

11. Enter the radio security code.

Passenger's Side

1. Disconnect the negative and positive battery cables.

2. Always wait at least three minutes after disconnecting the battery before working around the air bag.

3. Remove the glove box door and frame.

4. If equipped, remove any lower mounting brackets that may cover the air bag connection.

5. Disconnect the passenger's air bag connector. Pull the spring loaded sleeve toward the stop while holding the opposite half of the connector and pull the connector apart.

REARMING

1. After servicing has been completed, immediately couple the air bag and cable reel connectors.

2. If equipped, install any lower mounting brackets that may have been removed.

3. Install the glove box frame and glove box door.

4. Reconnect the positive and negative battery cables.

5. Turn the ignition switch to the **ON** position, but don't start the engine. The air bag indicator light should turn on for six seconds, then turn off. If the air bag indicator light doesn't come on, or stays on longer than six seconds, the system fault must be diagnosed.

6. Enter the radio security code.

Rack and Pinion Steering

REMOVAL & INSTALLATION

Manual

CIVIC AND DEL SOL

✳✳ CAUTION

The supplemental restraint system (air bag) must be disabled before

removing the steering wheel to center the cable reel. Failure to disarm the air bag system may cause accidental air bag deployment, resulting in unnecessary air bag system repairs and the risk of personal injury.

1. Position the front wheels straight ahead. Lock the steering column and remove the ignition key.

2. Disconnect the negative battery cable and positive battery cables.

3. Disable the Supplemental Restraint System (air bag).

4. Remove the steering joint cover. Remove the upper and lower steering joint bolts.

5. Raise and support the vehicle safely.

6. Remove the front wheels.

7. Remove the tie-rod end cotter pins and castle nuts. Using a ball joint tool, disconnect the tie rod ends from the steering knuckles.

8. Remove the left tie-rod end and slide the rack all the way to the right.

9. Remove the self-locking nuts and separate the catalytic converter or front exhaust pipe from the rear exhaust pipes. Remove the catalytic converter or front exhaust pipe.

10. Manual transaxle: Disconnect the shift lever extension rod from the clutch housing. Slide the pin retainer out of the way, drive out the spring pin, and disconnect the shift rod.

11. Automatic transaxle: Unbolt the shift cable bracket and holder. Disconnect the shift cable from the control shaft. Suspend the cable from the underbody with a piece of wire.

12. Unbolt and remove the steering rack stiffener plate.

13. Remove the steering rack mounting bracket.

14. Pull the steering rack down to release it from the pinion shaft.

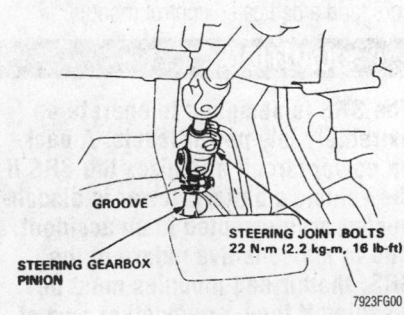

GROOVE

STEERING GEARBOX PINION

STEERING JOINT BOLTS 22 N·m (2.2 kg-m, 16 lb-ft)

7923FG00

Steering joint bolts—1995 Civic and Del Sol

15. Drop the steering rack far enough to permit the end of the pinion shaft and the grommet to come out of the hole in the bulkhead.

16. Slide the gearbox to the right until the left tie rod clears the subframe, then drop it down and out of the vehicle to the left.

To install:

➡**Use new self-locking nuts and gaskets when installing the catalytic converter.**

17. Install the steering rack into position. Install the pinion shaft grommet and insert the pinion through the hole in the bulkhead.

18. Install the steering rack mounting cushion, bracket, and bolts. The arrow on the bracket faces the front of the vehicle. Tighten the bracket bolts to 28 ft. lbs. (39 Nm).

19. Install the steering rack stiffener plate. Tighten the steering rack mounting bolts to 43 ft. lbs. (59 Nm). Tighten the stiffener plate bolts to 28 ft. lbs. (39 Nm).

20. Center the rack ends within their steering strokes.

21. Install the tie rod ends onto the rack ends. Connect the tie rod ends to the steering knuckles and install the castle nuts. Install the front wheels.

22. Install the catalytic converter using new gaskets and self-locking nuts. Tighten the front nuts to 16 ft. lbs. (22 Nm), and the rear nuts to 25 ft. lbs. (34 Nm).

23. Manual transaxle: Reconnect the shift linkage by installing a new spring pin and clip. Install the extension rod and tighten its bolt to 16 ft. lbs. (22 Nm).

24. Automatic transaxle: Reconnect the shift cable and brackets. Tighten the bracket bolts to 9 ft. lbs. (12 Nm). Tighten the cable lockbolt to 10 ft. lbs. (14 Nm). Tighten the cable holder bolts to 16 ft. lbs. (22 Nm).

25. Verify that the rack is centered within its strokes. Lower the vehicle.

26. Center the air bag cable reel:

• Remove the steering wheel.

• Turn the cable reel clockwise until it stops.

• Turn the steering wheel counterclockwise (approximately two turns) until the arrow on the label points straight up.

• Install the steering wheel.

27. During steering wheel installation, verify that the slot on the steering wheel shaft engages with the tabs on the turn signal canceling sleeve. The pins on the cable reel fit into the holes on the steering wheel body. Install a new steering wheel nut and tighten it to 36 ft. lbs. (50 Nm).

28. Line up the bolt hole in the steering joint with the groove in the pinion shaft.

Slip the joint onto the pinion shaft. Pull the joint up and down to be sure the splines are fully seated. Tighten the joint bolts to 16 ft. lbs. (22 Nm).

➡**Connect the steering joint and pinion shaft with the cable reel and steering rack centered. Verify that the lower joint bolt is securely seated in the pinion shaft groove. If the steering wheel and rack are not centered, reposition the serrations at the lower end of the steering joint.**

29. Install the steering joint cover.
30. Tighten the ball joint castle nuts to 29–35 ft. lbs. (40–48 Nm). Then, tighten them only enough to install new cotter pins.
31. Enable the Supplemental Restraint System (air bag).
32. Install the steering wheel's lower access cover.
33. Reconnect the negative and positive battery cables.
34. Turn the ignition switch to the **ON** position. The air bag indicator light should come on for six seconds, then turn off. This light sequence indicates that the air bag system is enabled and functioning normally. If the air bag light stays on longer, or doesn't turn on, the system must be diagnosed.
35. Check the front wheel alignment and steering wheel spoke angle. Make adjustments by turning the left and right tie-rod ends equally.
36. Road test the vehicle.

Power

➡**The radio may contain a coded theft protection circuit. Always obtain the code number before disconnecting the battery. If the vehicle is equipped with 4WS, the steering control unit is shut down when the battery is disconnected. After connecting the battery, turn the steering wheel lock-to-lock to reset the steering control unit.**

CIVIC AND DEL SOL

✳✳ CAUTION

The supplemental restraint system (air bag) must be disabled before removing the steering wheel to center the cable reel. Failure to disarm the air bag system may cause accidental air bag deployment, resulting in unnecessary air bag system repairs and the risk of personal injury.

1. Lift the power steering reservoir off of its mount and disconnect the inlet hose.

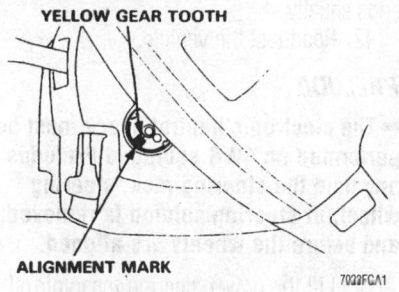

YELLOW GEAR TOOTH

ALIGNMENT MARK

7022FGA1

Cable reel alignment—1995 Civic and Del Sol

2. Insert a length of tubing into the inlet hose and route the tubing into a drain container.
3. With the engine running at idle, turn the steering wheel lock-to-lock several times until fluid stops running out of the hose.
4. Position the front wheels straight ahead. Shut off the engine and lock the steering column and remove the ignition key. Reconnect the reservoir inlet hose.
5. Disconnect the negative battery cable and positive battery cables. Wait three minutes before working around the air bags.
6. Remove the steering wheel's lower access cover.
7. Uncouple the air bag connector from the cable reel connector:
   a. Hold the cable reel connector. With your other hand, slide the spring-loaded sleeve toward the stop tab on the air bag connector.
   b. Separate the two connectors. There is no need to install a shorting connector, as the connectors are automatically grounded when they are uncoupled.
8. Remove the steering joint cover and remove the upper and lower steering joint bolts.
9. Raise and support the vehicle safely.
10. Remove the front wheels.
11. Remove the tie rod end cotter pins and castle nuts. Using a ball joint tool, disconnect the tie rod ends from the steering knuckles.
12. Manual transaxle: Disconnect the shift lever torque rod from the clutch housing. Slide the pin retainer out of the way, drive out the spring pin and disconnect the shift rod.
13. Automatic transaxle: Unbolt the shift cable bracket and holder. Disconnect the shift cable from the control shaft. Suspend the cable from the underbody with a piece of wire.
14. Remove the self-locking nuts and separate the catalytic converter from the

exhaust pipes. Remove the catalytic converter.
15. Use a flare nut wrench to disconnect the hydraulic line and hose from the rack valve body.
16. Remove the left tie rod end and slide the rack all the way to the right.
17. Remove the steering rack mounting bolts.
18. Pull the steering rack down to release it from the pinion shaft.
19. Drop the gearbox far enough to permit the end of the pinion shaft to come out of the hole in the frame channel.
20. Slide the gearbox to the right until the left tie rod clears the subframe, then drop it down and out of the vehicle to the left.

To install:

➡**Use new self-locking nuts when installing the catalytic converter.**

✳✳ WARNING

Use only genuine Honda power steering fluid. Any other type or brand of fluid will damage the power steering pump.

21. Install the steering rack into position. Install the pinion shaft grommet and insert the pinion through the hole in the bulkhead.
22. Install the rack mounting bolts. Tighten the bracket bolts to 28 ft. lbs. (39 Nm). Tighten the mounting bolt under the valve body to 43 ft. lbs. (59 Nm).
23. Reconnect the two hydraulic lines to the rack valve body. Carefully tighten the hydraulic line fitting to 28 ft. lbs. (38 Nm). Securely tighten the return hose clamp.
24. Center the rack ends within their steering strokes.
25. Install the tie rod ends onto the rack ends. Connect the tie rod ends to the steering knuckles and install the castle nuts. Install the front wheels.
26. Install the catalytic converter using new gaskets and self-locking nuts. Tighten the front nuts to 16 ft. lbs. (22 Nm), and the rear nuts to 25 ft. lbs. (34 Nm).
27. Manual transaxle: Reconnect the shift linkage by installing a new spring pin and clip. Install the extension rod and tighten its bolt to 16 ft. lbs. (22 Nm).
28. Automatic transaxle: Reconnect the shift cable and brackets. Tighten the bracket bolts to 9 ft. lbs. (12 Nm). Tighten the cable lockbolt to 10 ft. lbs. (14 Nm). Tighten the cable holder bolts to 16 ft. lbs. (22 Nm).
29. Verify that the rack is centered within its strokes. Lower the vehicle.
30. Center the air bag cable reel:

a. Remove the steering wheel.

b. Turn the cable reel clockwise until it stops.

c. Turn the steering wheel counter-clockwise (approximately two turns) until the arrow on the label points straight up.

d. Install the steering wheel.

31. During steering wheel installation, verify that the slot on the steering wheel shaft engages with the tabs on the turn signal canceling sleeve. The pins on the cable reel fit into the holes on the steering wheel body. Install a new steering wheel nut and tighten it to 36 ft. lbs. (50 Nm).

32. Line up the bolt hole in the steering joint with the groove in the pinion shaft. Slip the joint onto the pinion shaft. Pull the joint up and down to be sure the splines are fully seated. Tighten the joint bolts to 16 ft. lbs. (22 Nm).

➡**Connect the steering joint and pinion shaft with the cable reel and steering rack centered. Verify that the lower joint bolt is securely seated in the pinion shaft groove. If the steering wheel and rack are not centered, reposition the serrations at the lower end of the steering joint.**

33. Install the steering joint cover.

34. Tighten the ball joint castle nuts to 29–35 ft. lbs. (40–48 Nm). Then, tighten them only enough to install new cotter pins.

35. Reconnect the air bag and cable reel connectors: Be sure the connectors fit squarely together. Then, press the connectors to couple them. The spring-loaded sleeve will lock into place as the two connectors are coupled.

36. Install the steering wheel's lower access cover.

37. Reconnect the negative and positive battery cables.

38. Turn the ignition switch to the **ON** position. The air bag indicator light should come on for six seconds, then turn off. This light sequence indicates that the air bag system is enabled and functioning normally. If the air bag light stays on longer, or doesn't turn on, the system must be diagnosed.

39. Be sure the reservoir inlet line has been reconnected. Fill the reservoir to the upper line with Honda power steering fluid. Run the engine at idle and turn the steering wheel lock-to-lock several times to bleed air from the system and fill the rack valve body. Recheck the fluid level and add more if necessary.

40. Check the power steering system for leaks.

41. Check the front wheel alignment and steering wheel spoke angle. Make adjust-

ments by turning the left and right tie rod ends equally.

42. Road test the vehicle.

PRELUDE

➡**The electronic neutral check must be performed on 4WS equipped Preludes any time the steering rack, steering wheel, or steering column is removed, and before the wheels are aligned.**

1. Lift the power steering reservoir off of its mount and disconnect the inlet hose.

2. Insert a length of tubing into the inlet hose and route the tubing into a drain container.

3. With the engine running at idle, turn the steering wheel lock-to-lock several times until fluid stops running out of the hose. Shut off the engine.

4. Position the front wheels straight ahead. Lock the steering column with the ignition key. Reconnect the reservoir inlet hose.

5. Disconnect the negative battery cable.

6. Remove the steering joint cover and remove the upper and lower steering joint bolts.

7. Raise and support the vehicle safely.

8. Remove the front wheels.

9. Remove the tie rod end cotter pins and castle nuts. Install a 12mm nut onto the end of the ball joint stud to protect the threads from damage. Using a ball joint tool, disconnect the tie rod ends from the steering knuckles.

10. Disconnect the heated oxygen sensor.

11. Remove the self-locking nuts and separate the catalytic converter from exhaust pipe A. Unbolt exhaust pipe A from the intake manifold, and remove it from the vehicle.

12. On vehicles equipped with automatic transaxles, remove the shift cable cover, disconnect the shift cable, and wire it up and out of the way.

13. Clean any oil or dirt off of the valve body with solvent.

14. Remove the center beam from the subframe.

15. Remove the valve body shield.

16. Use a flare nut wrench to disconnect the four hydraulic lines from the rack valve body. Plug the lines to keep dirt and moisture out.

17. On models with 4WS, carefully cut the wire tie securing the cover to the front sub-steering angle sensor. Remove the cover.

18. Remove the sensor wiring harness from the two securing clamps, then, discon-

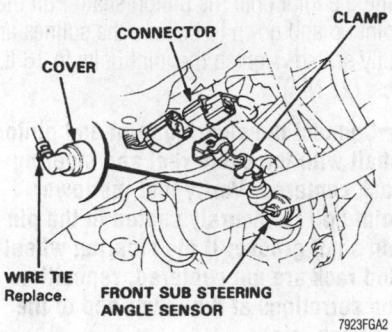

4WS front sub-steering angle sensor— Prelude

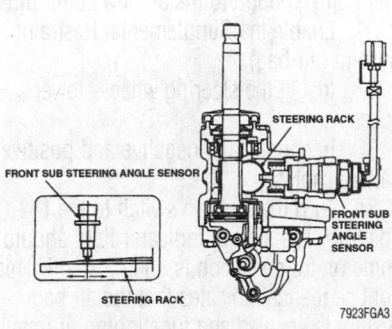

Front sub-steering angle sensor, harness, and steering rack—Prelude

nect sensor connector from the 4WS steering main wiring harness.

19. Remove the steering joint bolt and slide the pinion shaft out of the joint.

20. Remove the left mounting bracket;, then remove the right mounting brackets.

21. Remove the left tie rod end and slide the rack all the way to the right.

22. Pull the steering rack down to release it from the pinion shaft.

23. Slide the steering rack to the right until the left tie rod clears the subframe, then drop it down and out of the vehicle to the left.

To install:

➡**Use new gaskets and self-locking nuts when installing exhaust pipe A.**

❊❊ WARNING

Use only genuine Honda power steering fluid. Any other type or brand of fluid will damage the power steering pump.

24. Install the steering rack into position. Install the pinion shaft grommet and insert the pinion through the hole in the firewall.

25. Install the right and left mounting brackets. Tighten the short bolts to 28 ft.

lbs. (39 Nm), and the long bolts to 32 ft. lbs. (44 Nm).

26. Center the rack ends within their steering strokes.

27. Center the air bag cable reel:

• Turn the steering wheel clockwise until it stops.

• Turn the steering wheel counterclockwise until the yellow gear tooth lines up with the alignment mark on the lower column cover.

20. Line up the bolt hole in the steering joint with the groove in the pinion shaft. Slip the joint onto the pinion shaft. Pull the joint up and down to be sure the splines are fully seated. Tighten the joint bolts to 16 ft. lbs. (22 Nm).

➡**Connect the steering joint and pinion shaft with the cable reel and steering rack centered. Verify that the lower joint bolt is securely seated in the pinion shaft groove. If the steering wheel and rack are not centered, reposition the serrations at the lower end of the steering joint.**

29. Reconnect the four hydraulic lines to the rack valve body. Carefully tighten the 12mm fittings to 9 ft. lbs. (13 Nm), the 14mm inlet fitting to 28 ft. lbs. (37 Nm), and the 17mm oil cooler fitting to 21 ft. lbs. (29 Nm).

30. Connect the front sub-steering angle sensor to the 4WS harness. Place the wire back into its clamps, making sure that it doesn't interfere with the stabilizer bar. Install the sensor cover with a new wire tie.

31. Install the valve body shield.

32. Install the center beam. Use new self-locking bolts and tighten them to 43 ft. lbs. (60 Nm).

33. On automatic transaxle equipped vehicles, reconnect the shift cable and tighten the locknut to 10 ft. lbs. (14 Nm). Install the cable holder and tighten its bolts to 13 ft. lbs. (18 Nm).

34. Install the catalytic converter using new gaskets and self-locking nuts. Tighten the exhaust manifold nuts to 40 ft. lbs. (55 Nm), and the rear nuts to 25 ft. lbs. (34 Nm).

35. Reconnect the heated oxygen sensor.

36. Install the tie rod ends onto the rack ends. Connect the tie rod ends to the steering knuckles and install the castle nuts. Install the front wheels.

37. Verify that the rack is centered within its strokes. Lower the vehicle.

38. Install the steering joint cover.

39. Tighten the ball joint castle nuts to 36–43 ft. lbs. (50–60 Nm). Then, tighten them only enough to install new cotter pins.

40. Reconnect the negative battery cable.

41. Be sure the reservoir inlet line has been reconnected. Fill the reservoir to the upper line with Honda power steering fluid. Run the engine at idle and turn the steering wheel lock-to-lock several times to bleed air from the system and fill the rack valve body. Recheck the fluid level and add more if necessary.

42. Check the power steering system for leaks.

43. On Preludes without 4WS, check and adjust the front wheel alignment. On Preludes with 4WS, the electronic neutral check must be performed on the 4WS system.

ACCORD

1. Lift the power steering reservoir off of its mount and disconnect the inlet hose.

2. Insert a length of tubing into the inlet hose and route the tubing into a drain container.

3. With the engine running at idle, turn the steering wheel lock-to-lock several times until fluid stops running out of the hose. Immediately shut off the engine.

4. Position the front wheels straight ahead. Lock the steering column with the ignition key. Reconnect the reservoir inlet hose.

5. Disconnect the negative battery cable.

6. Remove the steering joint cover and remove the upper and lower steering joint bolts.

7. Raise and support the vehicle safely.

8. Remove the front wheels.

9. Remove the tie rod end cotter pins and castle nuts. Using a ball joint tool, disconnect the tie rod ends from the steering knuckles.

10. Remove the left tie rod end and slide the rack all the way to the right.

11. Disconnect the heated oxygen sensor.

12. Remove the self-locking nuts and separate the catalytic converter from exhaust pipe A. Remove the catalytic converter.

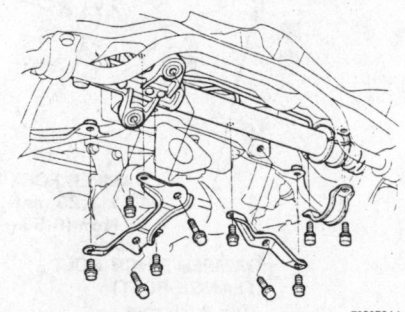

Power rack and pinion steering gear mounting—Accord

13. On vehicles equipped with manual transaxles, disconnect the shift linkage from the transaxle case.

14. On vehicles equipped with automatic transaxles, remove the shift cable cover, disconnect the cable, and wire it up and out of the way.

15. Use a flare nut wrench to disconnect the two hydraulic lines from the rack valve body. Plug the lines to keep dirt and moisture out. Carefully move the disconnected lines to the rear of the rack assembly so that they are not damaged when the rack is removed.

16. Remove the rack stiffener plate;, then, remove the steering rack mounting bolts.

17. Pull the steering rack down to release it from the pinion shaft.

18. Drop the steering rack far enough to permit the end of the pinion shaft to come out of the hole in the frame channel.

19. Slide the steering rack to the right until the left tie rod clears the subframe, then drop it down and out of the vehicle to the left.

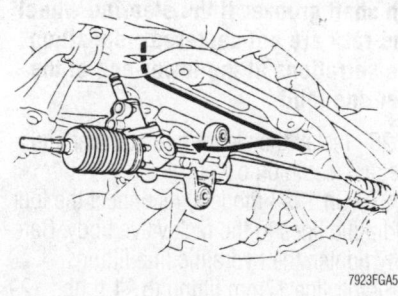

Move the steering rack to the right, then down and out of the vehicle—Accord

To install:

➡**Use new gaskets and self-locking nuts when installing the catalytic converter.**

✳✳ WARNING

Use only genuine Honda power steering fluid. Any other type or brand of fluid will damage the power steering pump.

20. Before installing the rack & pinion, slide the ends all the way to the right. Install the pinion shaft grommet. The lug on the pinion shaft grommet aligns with the slot on the valve body.

21. Install the steering rack into position. Install the pinion shaft grommet and insert the pinion through the hole in the bulkhead.

22. Install the rack mounting bolts. Tighten the bracket bolts to 28 ft. lbs. (39 Nm). Tighten the stiffener plate mounting bolts to 32 ft. lbs. (43 Nm).

23. Center the rack ends within their steering strokes.

24. Center the air bag cable reel, as follows:

a. Turn the steering wheel left approximately 150°, to check the cable reel position with the indicator.

b. If the cable reel is centered, the yellow gear tooth lines up with the alignment mark on the cover.

c. Return the steering wheel right approximately 150° to position the steering wheel in the straight ahead position.

25. Line up the bolt hole in the steering joint with the groove in the pinion shaft. Slip the joint onto the pinion shaft. Pull the joint up and down to be sure the splines are fully seated. Tighten the joint bolts to 16 ft. lbs. (22 Nm).

➡**Connect the steering joint and pinion shaft with the cable reel and steering rack centered. Verify that the lower joint bolt is securely seated in the pinion shaft groove. If the steering wheel and rack are not centered, reposition the serrations at the lower end of the steering joint.**

26. Install the steering joint cover and the rack & pinion cover.

27. On 1993 models, reconnect the four hydraulic lines to the rack valve body. Carefully tighten the hydraulic line fittings: reservoir line 17mm fitting to 21 ft. lbs. (29 Nm), oil cooler line 12mm fitting to 9 ft. lbs. (13 Nm), pump outlet line 14mm fitting to 28 ft. lbs. (38 Nm), vehicle speed sensor 12mm fitting to 9 ft. lbs. (13 Nm).

28. On 1995–97 models, reconnect the two hydraulic lines to the rack valve body. Carefully tighten the 14mm inlet fitting to 27 ft. lbs. (37 Nm) and the 16mm outlet fitting to 21 ft. lbs. (28 Nm).

29. If equipped with a manual transaxle, connect the shift cable and the select cable to the transaxle with new cotter pins.

30. If equipped with a automatic transaxle, connect the shift cable to the transaxle using a new lockwasher. Tighten the lockbolt to 10 ft. lbs. (14 Nm).

31. Install the catalytic converter using new gaskets and self-locking nuts. Tighten the front nuts to 16 ft. lbs. (22 Nm), and the rear nuts to 25 ft. lbs. (34 Nm).

32. Reconnect the heated oxygen sensor.

33. Install the tie rod ends onto the rack ends. Connect the tie rod ends to the steering knuckles and install the castle nuts.

34. Tighten the ball joint castle nuts to 29–35 ft. lbs. (40–48 Nm). Then, tighten them only enough to install new cotter pins.

35. Install the front wheels.

36. Lower the vehicle.

37. Reconnect the negative battery cable.

38. Be sure the reservoir inlet line has been reconnected. Fill the reservoir to the upper line with Honda power steering fluid. Run the engine at idle and turn the steering wheel lock-to-lock several times to bleed air from the system and fill the rack valve body. Recheck the fluid level and add more if necessary.

39. Check the power steering system for leaks.

40. Check the front wheel alignment and steering wheel spoke angle. Make adjustments by turning the left and right tie rod ends equally.

41. Road test the vehicle.

Strut

REMOVAL & INSTALLATION

Front

CIVIC AND DEL SOL

1. Raise and safely support the vehicle. Remove the front wheels.

2. Unbolt the brake hose brackets from the bottom of the strut tube. Do not disconnect the brake hoses.

➡**Some Civic models may not have brake hose brackets on their struts. In these cases, there is no need to unbolt the brackets.**

3. Remove the damper pinch bolt.

4. Remove the damper fork nut and bolt. Remove the damper fork.

5. Remove the two strut mounting bolts from the shock tower and remove the strut from the vehicle.

6. Install a spring compressor onto the strut assembly and tighten the compressor according to the manufacturer's instructions.

7. Remove the locking nut from the top of the shock absorber piston. Disassemble the strut and remove the coil spring.

To install:

➡**Use new self-locking nuts when installing the strut.**

8. Install a spring compressor onto the coil spring.

9. Assemble the lower strut mounts, dust covers, coil spring, and upper strut mount onto the shock absorber. Position the strut bearing mounting studs so that they will line up with the mounting holes in the shock tower.

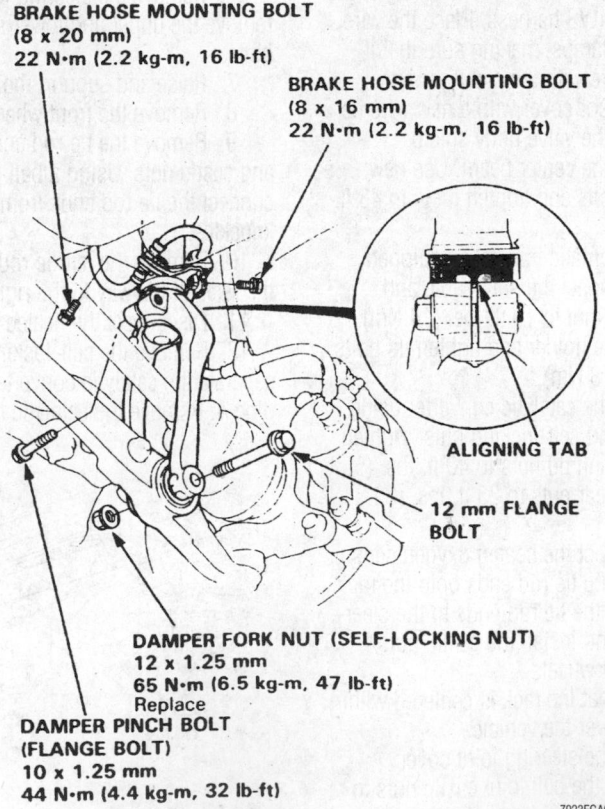

BRAKE HOSE MOUNTING BOLT
(8 x 20 mm)
22 N·m (2.2 kg-m, 16 lb-ft)

BRAKE HOSE MOUNTING BOLT
(8 x 16 mm)
22 N·m (2.2 kg-m, 16 lb-ft)

ALIGNING TAB

12 mm FLANGE BOLT

DAMPER FORK NUT (SELF-LOCKING NUT)
12 x 1.25 mm
65 N·m (6.5 kg-m, 47 lb-ft)
Replace

DAMPER PINCH BOLT
(FLANGE BOLT)
10 x 1.25 mm
44 N·m (4.4 kg-m, 32 lb-ft)

7923FGA6

Damper fork components—Civic and Del Sol

10. Install the mounting washer, and loosely install a new self-locking nut.

11. Hold the shock absorber piston with a hex wrench and tighten the self-locking nut. Tighten the self-locking nut to 22 ft. lbs. (30 Nm).

12. Install the strut into the vehicle. Hand-tighten the strut mounting bolts. The alignment mark on the strut tube faces away from the wheel.

13. Install the damper fork onto the strut and lower control arm. Install the pinch and fork bolts.

14. Connect the brake hose brackets to the strut tube and tighten them to 16 ft. lbs. (22 Nm).

15. Install the front wheels and lower the vehicle.

16. Tighten the strut mount bolts to 36 ft. lbs. (50 Nm).

17. Tighten the pinch bolt to 32 ft. lbs. (44 Nm). Tighten the damper fork nut to 47 ft. lbs. (65 Nm).

18. Tighten the wheel nuts to 80 ft. lbs. (110 Nm).

19. Check the vehicle's front end alignment and adjust it if necessary.

PRELUDE AND ACCORD

1. Raise and safely support the vehicle.
2. Remove the front wheels.
3. Remove the brake hose clamp bolts from the strut.
4. Remove the damper fork bolts and remove the damper fork.
5. Remove the three strut mounting nuts. Remove the strut from the vehicle.

To install:

➡ **Use new self-locking bolts when installing the struts and assembling the damper forks.**

6. Install the strut into the vehicle. Hand-tighten the mounting nuts.

7. Install the strut into the damper fork. The alignment mark on the strut tube fits into the groove on the damper fork.

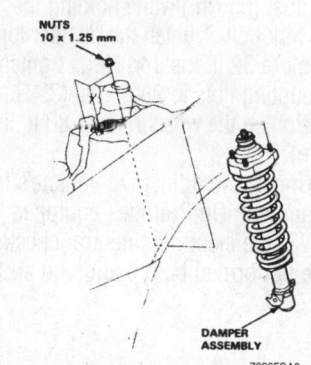

NUTS
10 x 1.25 mm

DAMPER
ASSEMBLY

7923FGA8

Front strut and strut mount—Prelude and Accord

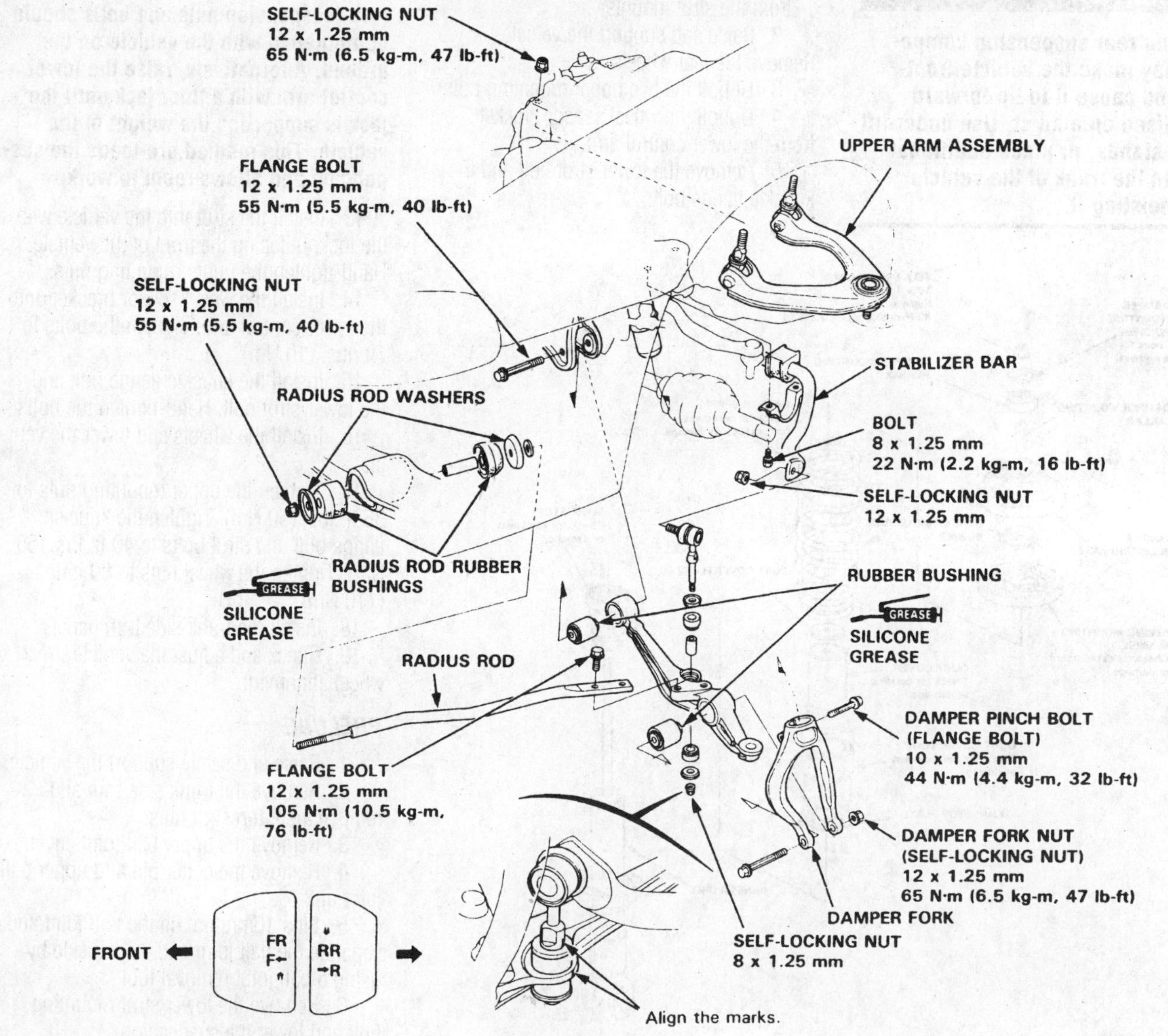

SELF-LOCKING NUT
12 x 1.25 mm
65 N·m (6.5 kg-m, 47 lb-ft)

FLANGE BOLT
12 x 1.25 mm
55 N·m (5.5 kg-m, 40 lb-ft)

SELF-LOCKING NUT
12 x 1.25 mm
55 N·m (5.5 kg-m, 40 lb-ft)

RADIUS ROD WASHERS

RADIUS ROD RUBBER BUSHINGS

GREASE
SILICONE GREASE

RADIUS ROD

FLANGE BOLT
12 x 1.25 mm
105 N·m (10.5 kg-m, 76 lb-ft)

UPPER ARM ASSEMBLY

STABILIZER BAR

BOLT
8 x 1.25 mm
22 N·m (2.2 kg-m, 16 lb-ft)

SELF-LOCKING NUT
12 x 1.25 mm

RUBBER BUSHING

GREASE
SILICONE GREASE

DAMPER PINCH BOLT (FLANGE BOLT)
10 x 1.25 mm
44 N·m (4.4 kg-m, 32 lb-ft)

DAMPER FORK NUT (SELF-LOCKING NUT)
12 x 1.25 mm
65 N·m (6.5 kg-m, 47 lb-ft)

DAMPER FORK

SELF-LOCKING NUT
8 x 1.25 mm

FRONT ⬅   FR / F← / ↑   ↓ / RR / →R   ➡

Align the marks.

7923FGA7

Front suspension components—Prelude and Accord

8. Install the pinch bolt and damper fork bolt. Only hand-tighten these bolts.

9. Install the front wheels and lower the vehicle.

10. With all four of the vehicle's wheels on the ground, tighten the damper fork nut to 47 ft. lbs. (65 Nm) while holding the damper fork bolt. Tighten the damper fork pinch bolt to 32 ft. lbs. (44 Nm). Tighten the strut mounting nuts to 28 ft. lbs. (39 Nm).

11. Tighten the wheel nuts to 80 ft. lbs. (110 Nm).

12. Check and adjust the vehicle's front end alignment. On Preludes equipped with 4WS, the electronic neutral check must be performed before aligning all four wheels.

Rear

CIVIC AND DEL SOL

✷✷ CAUTION

Removing rear suspension components may make the vehicle front-heavy and cause it to tip forward when raised on a hoist. Use under-lift support stands, or place additional weight in the trunk of the vehicle before hoisting it.

1. Remove the interior or trunk trim pieces that cover the strut mount:

   a. **Sedan and coupe models:** Fold down the upper rear seat cushion. Carefully pry out the clips that secure the trunk and shock tower trim to the body. Remove the trunk trim to expose the strut mounts.

   b. **Hatchback models:** Fold down the rear seat. Unbolt and remove the rear side shelf/speaker grille assemblies. Disconnect and remove the speaker. Carefully pry out the clips and remove the screws to remove shock tower trim panel.

   c. **Del Sol models:** Support the trunk lid in a fully-open position. Remove the trunk lid support struts. Lift the roof storage frame up to get it out of the work area. It is not necessary to remove the roof storage frame. Carefully loosen and remove the screw clips. Then, position the trunk side trim panels away from the strut mounts.

2. Raise and support the vehicle. Remove the rear wheels.

3. Unbolt the two upper mounting nuts.

4. Unbolt the wheel sensor bracket from the lower control arm.

5. Remove the lower strut bolt and the knuckle flange bolt.

6. Remove the strut from the vehicle.

7. Install a spring compressor onto the strut assembly and tighten the compressor according to the manufacturer's instructions.

8. Remove the locking nut from the top of the shock absorber. Disassemble the strut and remove the coil spring.

To install:

➡**Use new self-locking nuts when installing the strut.**

9. Install a spring compressor onto the coil spring.

10. Assemble the upper and lower strut mounts, dust covers, and coil spring onto the shock absorber.

11. Install the mounting washer, and loosely install a new self-locking nut.

12. Hold the shock absorber piston with a hex wrench and tighten the self-locking nut. Tighten the self-locking nut to 22 ft. lbs. (30 Nm).

➡**All suspension nuts and bolts should be tightened with the vehicle on the ground. Alternatively, raise the lower control arm with a floor jack until the jack is supporting the weight of the vehicle. This method pre-loads the suspension and allows room to work.**

13. Install the strut into the vehicle with the locknut facing the front of the vehicle. Hand-tighten the upper mounting nuts.

14. Install the wheel sensor bracket onto the lower control arm. Tighten the bolts to 7 ft. lbs. (10 Nm).

15. Install the knuckle flange bolt and the lower strut bolt. Hand-tighten the bolts.

16. Install the wheels and lower the vehicle.

17. Tighten the upper mounting nuts to 36 ft. lbs. (50 Nm). Tighten the knuckle flange bolt and strut bolts to 40 ft. lbs. (55 Nm). Tighten the wheel nuts to 80 ft. lbs. (110 Nm).

18. Install the trunk side trim panels.

19. Check and adjust the vehicle's rear wheel alignment.

PRELUDE

1. Raise and safely support the vehicle.

2. Remove the trunk side trim and remove the 2 top strut nuts.

3. Remove the upper ball joint cover.

4. Remove the cotter pin and upper ball joint nut.

5. Fit a 10mm nut on the ball joint and separate the ball joint and the knuckle by using a ball joint removal tool.

6. Remove the lower strut mounting bolt and lower the suspension.

7. Remove the strut from the vehicle.

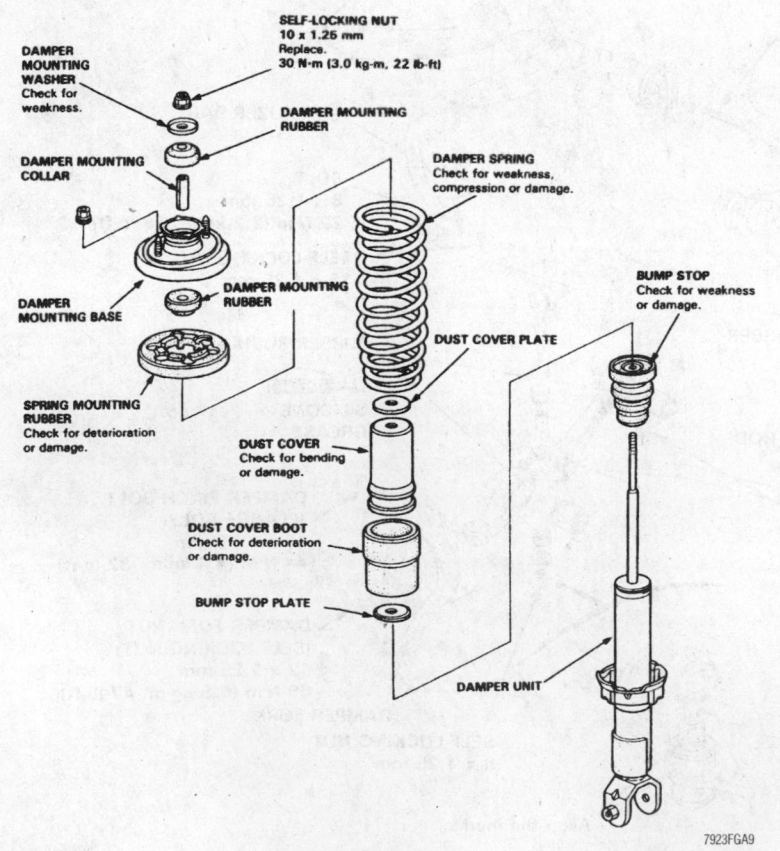

DAMPER MOUNTING WASHER Check for weakness.

SELF-LOCKING NUT 10 x 1.25 mm Replace. 30 N·m (3.0 kg-m, 22 lb-ft)

DAMPER MOUNTING RUBBER

DAMPER MOUNTING COLLAR

DAMPER SPRING Check for weakness, compression or damage.

DAMPER MOUNTING RUBBER

BUMP STOP Check for weakness or damage.

DAMPER MOUNTING BASE

DUST COVER PLATE

SPRING MOUNTING RUBBER Check for deterioration or damage.

DUST COVER Check for bending or damage.

DUST COVER BOOT Check for deterioration or damage.

BUMP STOP PLATE

DAMPER UNIT

7923FGA9

Exploded view of the rear suspension strut—Civic and Del Sol

To install:

➡️**Use new self-locking nuts when installing the rear struts.**

8. Install the strut and loosely install the lower mounting bolt. Do not tighten.

9. Install the upper strut mounting bolts. Tighten the bolts to 28 ft. lbs. (39 Nm).

10. Connect the upper arm and knuckle and tighten the castle nut to 29–35 ft. lbs. (40–48 Nm).

11. Install the upper ball joint cover.

12. Raise the rear suspension with a floor jack until the weight is on the strut.

13. Tighten the lower strut mounting bolt to 47 ft. lbs. (65 Nm).

14. Install the rear wheels and lower the vehicle.

15. Tighten the rear wheel nuts to 80 ft. lbs. (110 Nm).

16. Check and adjust the vehicle's rear wheel alignment.

ACCORD

1. Fold the rear seat forward and remove the side bolster cushions. The side bolster cushions are secured by a screw at the bottom and two clips at the top.

2. Remove the strut mount cap. Remove the upper strut mounting nuts.

3. Raise and safely support the vehicle.

4. Remove the rear wheels.

5. Support the knuckle with a floor jack.

6. Remove the strut mounting bolt, lower the jack, and remove the strut.

To install:

➡️**Use new self-locking nuts when installing the strut.**

7. Fit the strut into the upper mount. Only hand-tighten the upper mounting nuts.

8. Fit the strut into position on the knuckle. Install the mounting bolt.

9. Place a jack under the lower strut mount. Raise the jack until the weight of the vehicle is on the jack.

10. With the suspension under load, tighten the lower mount bolt to 40 ft. lbs. (55 Nm). Tighten the upper nuts to 28 ft. lbs.(39 Nm).

11. Install the rear wheel. Lower the vehicle to the ground.

12. Tighten the wheel nuts to 80 ft. lbs. (110 Nm).

13. Install the rear seat side bolsters and fold the seat back into place.

14. Check and adjust the vehicle's rear wheel alignment.

Coil Spring

REMOVAL & INSTALLATION

Front

CIVIC AND DEL SOL

1. Raise and safely support the vehicle. Remove the front wheels.

2. Unbolt the brake hose brackets from the bottom of the strut tube. Do not disconnect the brake hoses.

➡️**Some Civic models may not have brake hose brackets on their struts.**

3. Remove the damper fork pinch bolt and flange bolt;, then, remove the damper fork.

4. Remove the strut's upper mounting nuts. Remove the strut assembly from the vehicle.

5. Install a spring compressor onto the strut assembly and tighten the compressor according to the manufacturer's instructions.

6. Remove the locking nut from the top of the shock absorber piston. Disassemble the strut and remove the coil spring.

To install:

➡️**Use new self-locking nuts when assembling the strut.**

7. Install a spring compressor onto the coil spring.

8. Assemble the lower strut mounts, dust covers, coil spring, and upper strut mount onto the shock absorber. Position the strut bearing mounting studs so that they will line up with the mounting holes in the shock tower.

9. Install the mounting washer, and loosely install a new self-locking nut.

10. Hold the shock absorber piston with a hex wrench and tighten the self-locking nut. Tighten the self-locking nut to 22 ft. lbs. (30 Nm).

➡️**All suspension nuts and bolts should be tightened with the vehicle on the ground.**

11. Install the strut assembly into the vehicle. Tighten the upper mounting nuts to 36 ft. lbs. (50 Nm).

12. Install the damper fork. Tighten the pinch bolt to 32 ft. lbs. (44 Nm), and the fork bolt to 47 ft. lbs. (64 Nm).

13. Install the brake hose clamps. Tighten them to 16 ft. lbs. (22 Nm).

14. Install the wheel, and tighten the wheel nuts to 80 ft. lbs. (110 Nm).

15. Check and adjust the vehicle's front wheel alignment.

ACCORD AND PRELUDE

1. Raise and safely support the vehicle.

2. Remove the front wheels.

3. Unbolt the brake hose clamp from the strut.

4. Remove the damper fork bolts and remove the damper fork.

5. Remove the three strut mounting nuts. Remove the strut from the vehicle.

6. Place the strut in vice and install a spring compressor onto the coil spring. Follow the spring compressor manufacturer's instructions.

7. Compress the spring and remove the self-locking nut from the top of the strut. Disassemble the strut mounts and remove the coil spring.

➡️**The left and right front coil springs on 1995–97 Accords equipped with the F22B1 (2.2L VTEC) engine are not interchangeable. Remember this when ordering parts or reassembling the strut.**

8. Inspect the strut mounts for wear and damage. Replace any damaged or worn parts.

To install:

➡️**Use new self-locking nuts when assembling and installing the struts.**

9. Install the spring compressor onto the coil spring. Set the spring onto the strut

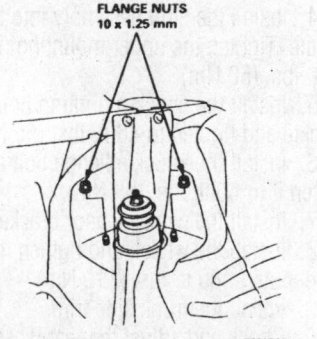

FLANGE NUTS 10 x 1.25 mm

7923FGB1

Rear strut upper mounting nut locations— Accord

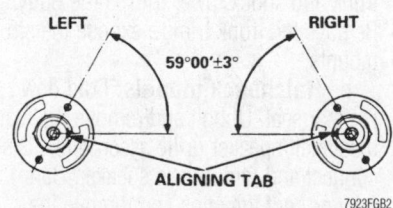

LEFT   RIGHT

59°00'±3°

ALIGNING TAB

7923FGB2

Strut bearing installation direction—Civic and Del Sol

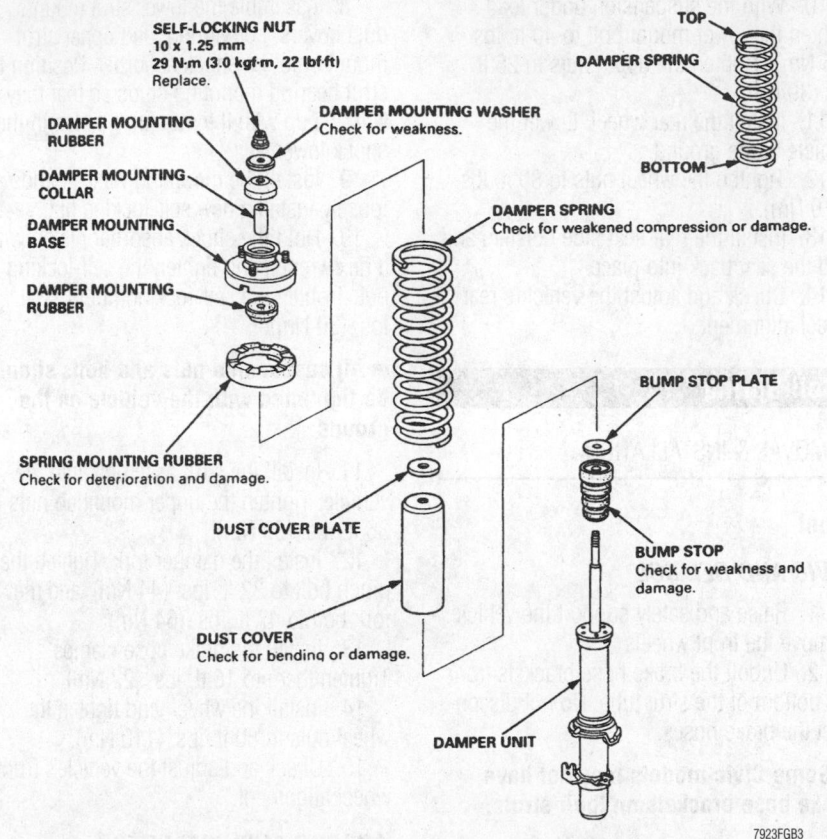

Coil spring, strut cartridge, and strut mount components—Accord and Prelude

cartridge. The flat part of the coil spring is its top.

10. Assemble the strut mount and its washer onto the strut. Tighten the self-locking nut to 22 ft. lbs. (29 Nm). Remove the spring compressor.

11. Install the strut into the vehicle. Hand-tighten the mounting nuts.

12. Install the strut into the damper fork. The alignment mark on the strut tube fits into the groove on the damper fork.

13. Install the pinch bolt and damper fork bolt. Only hand-tighten these bolts.

14. Install the front wheels and lower the vehicle.

15. With all four of the vehicle's wheels on the ground, tighten the damper fork nut to 47 ft. lbs. (65 Nm) while holding the damper fork bolt. Tighten the damper fork pinch bolt to 32 ft. lbs. (44 Nm). Tighten the strut mounting nuts to 28 ft. lbs. (39 Nm).

16. Tighten the wheel nuts to 80 ft. lbs. (110 Nm).

17. Check and adjust the vehicle's front wheel alignment. On Preludes equipped with 4WS, the electronic neutral check must be performed before all four wheels are aligned.

Rear

CIVIC AND DEL SOL

> ※※ **CAUTION**
>
> **Removing rear suspension components may make the vehicle front-heavy and cause it to tip forward when raised on a hoist. Use under-lift support stands, or place additional weight in the trunk of the vehicle before hoisting it.**

1. Remove the interior or trunk trim pieces that cover the strut mount:

a. **Sedan and coupe models:** Fold down the upper rear seat cushion. Carefully pry out the clips that secure the trunk and shock tower trim to the body. Remove the trunk trim to expose the strut mounts.

b. **Hatchback models:** Fold down the rear seat. Unbolt and remove the rear side shelf/speaker grille assemblies. Disconnect and remove the speaker. Carefully pry out the clips and remove the screws to remove shock tower trim panel.

c. **Del Sol models:** Support the trunk lid in a fully-open position.

Remove the trunk lid support struts. Lift the roof storage frame up to get it out of the work area. It is not necessary to remove the roof storage frame. Carefully loosen and remove the screw clips. Then, position the trunk side trim panels away from the strut mounts.

2. Raise and safely support the vehicle.

3. Remove the two strut mounting bolts.

4. Unbolt the wheel sensor brackets from the lower control arm. Do not disconnect the sensor.

5. Support the lower control arm with a floor jack.

6. Remove the strut mounting flange bolt and the knuckle flange bolt.

7. Lower the floor jack and remove the strut from the vehicle.

8. Install a spring compressor onto the strut assembly and tighten the compressor according to the manufacturer's instructions.

9. Remove the locking nut from the top of the shock absorber. Disassemble the strut and remove the coil spring.

To install:

➡**Use new self-locking nuts when assembling the strut.**

10. Install a spring compressor onto the coil spring.

11. Assemble the upper and lower strut mounts, dust covers, and coil spring onto the shock absorber.

12. Install the mounting washer, and loosely install a new self-locking nut.

13. Hold the shock absorber piston with a hex wrench and tighten the self-locking nut. Tighten the self-locking nut to 22 ft. lbs. (30 Nm).

➡**All suspension nuts and bolts should be tightened with the vehicle on the ground. Alternatively, raise the lower control arm with a floor jack until the jack is supporting the weight of the vehicle. This method pre-loads the suspension and allows room to work.**

14. Install the strut assembly into the vehicle. Tighten the upper mounting nuts to 36 ft. lbs. (50 Nm).

15. Install the shock mounting bolt at the knuckle and tighten to 40 ft. lbs. (55 Nm).

16. Install the knuckle flange bolt and tighten it to 40 ft. lbs. (55 Nm).

17. Install the wheel sensor brackets.

18. Install the wheel, and tighten the wheel nuts to 80 ft. lbs. (110 Nm).

19. Install the trunk side trim.

20. Check and adjust the rear wheel alignment.

PRELUDE

1. Raise and safely support the vehicle.
2. Remove the trunk side trim and remove the two strut mounting nuts.
3. Remove the upper ball joint cover.
4. Remove the cotter pin and upper ball joint nut.
5. Fit a 10mm nut on the ball joint and separate the ball joint and the knuckle by using a ball joint removal tool.
6. Remove the lower strut mounting bolt and lower the suspension.
7. Remove the strut from the vehicle.
8. Place the strut in vice and install a spring compressor onto the coil spring. Follow the spring compressor manufacturer's instructions.
9. Compress the spring and remove the self-locking nut from the strut. Disassemble the strut mounts and remove the coil spring.
10. Inspect the strut mounts for wear and damage. Replace any damaged or worn parts.

To install:

➡**Use new self-locking nuts when installing the rear struts.**

11. Install the spring compressor onto the coil spring. Set the spring onto the strut cartridge. The flat part of the coil spring is its top.
12. Assemble the strut mount and its washer onto the strut. Tighten the self-locking nut to 22 ft. lbs. (29 Nm). Remove the spring compressor.
13. Install the strut to the vehicle and loosely install the lower mounting bolt. Do not tighten.
14. Install the upper strut mounting bolts. Tighten the bolts to 28 ft. lbs. (39 Nm).
15. Connect the upper arm and knuckle and tighten the castle nut to 29–35 ft. lbs. (40–48 Nm).
16. Install the upper ball joint cover.
17. Raise the rear suspension with a floor jack until the weight is on the strut.
18. Tighten the lower strut mounting bolt to 47 ft. lbs. (65 Nm).
19. Install the rear wheels and lower the vehicle.
20. Tighten the rear wheel nuts to 80 ft. lbs. (110 Nm).
21. Install the trunk trim.
22. Check and adjust the vehicle's rear wheel alignment. On Preludes equipped with 4WS, the electronic neutral check must be performed before all four wheels are aligned.

ACCORD

1. Remove the strut.
2. Place the strut in vice and install a spring compressor onto the coil spring.

Follow the spring compressor manufacturer's instructions.

3. Compress the spring and remove the self-locking nut from the strut. Disassemble the strut mounts and remove the coil spring.
4. Inspect the strut mounts for wear and damage. Replace any damaged or worn parts.

To install:

➡**Use new self-locking nuts when assembling and installing the struts.**

5. Install the spring compressor onto the coil spring. Set the spring onto the strut cartridge. The flat part of the coil spring is its top.
6. Assemble the strut mount and its washer onto the strut. Tighten the self-locking nut to 22 ft. lbs. (29 Nm). Remove the spring compressor.
7. Install the strut into the vehicle. Hand-tighten the mounting nuts.
8. Fit the strut into position on the knuckle. Install the mounting bolt.
9. Place a jack under the lower strut mount. Raise the jack until the weight of the vehicle is on the jack.
10. With the suspension under load, tighten the lower mount bolt to 40 ft. lbs.

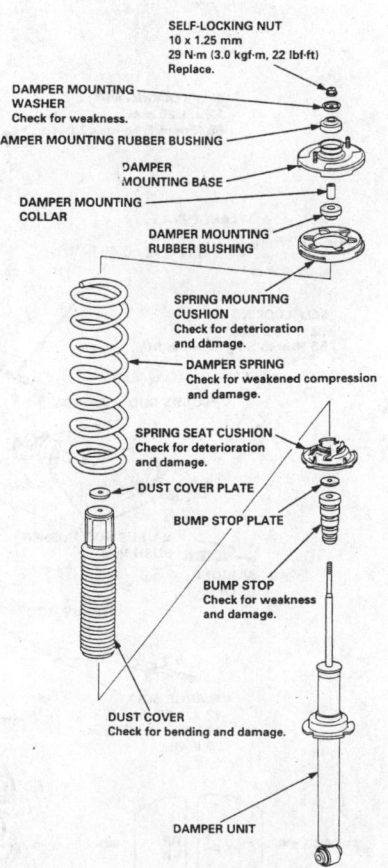

Exploded view of the rear suspension strut assembly—Accord

(55 Nm). Tighten the upper nuts to 28 ft. lbs. (39 Nm).

11. Install the rear wheel. Lower the vehicle to the ground.
12. Tighten the wheel nuts to 80 ft. lbs. (110 Nm).
13. Install the rear seat side bolsters and fold the seat back into place.
14. Check and adjust the vehicle's rear wheel alignment.

Upper Ball Joint

REMOVAL & INSTALLATION

Front and Rear

ALL MODELS

The upper ball joint cannot be removed from the upper control arm. If the ball joint is faulty or worn, the entire control arm must be replaced. If the upper ball joint boot is damaged and the ball joint itself is still usable, the boot can be replaced.

Upper Control Arm

REMOVAL & INSTALLATION

Front

CIVIC AND DEL SOL

1. Raise and support the vehicle safely.
2. Remove the front wheels.
3. Unbolt the damper fork from the lower control arm.
4. Unbolt the strut mounting nuts and remove the strut from the vehicle.
5. Separate the upper ball joint from the steering knuckle using a suitable ball joint remover.
6. Remove the self-locking nuts and remove the upper arm from the vehicle.
7. Remove the upper arm bolts to separate the control arm from its anchor bolt assembly. Inspect the bushings for signs of deterioration and replace them if they are damaged.
8. Place the upper control arm anchor bolt assembly into a vice and drive out the upper arm bushings.

To install:

➡**Use new self-locking nuts when assembling the anchor bolts and when installing the control arm into the vehicle.**

9. Drive the new upper arm bushings into the upper arm anchor bolts. Center the

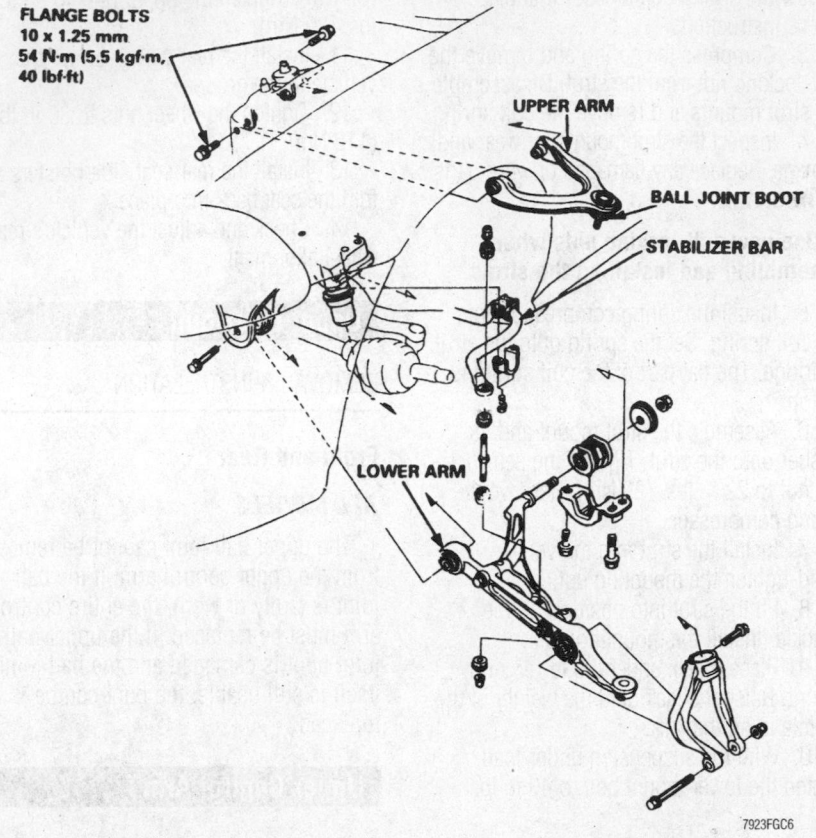

FLANGE BOLTS
10 x 1.25 mm
54 N·m (5.5 kgf·m,
40 lbf·ft)

UPPER ARM

BALL JOINT BOOT

STABILIZER BAR

LOWER ARM

7923FGC6

Front suspension components—Civic

bushing in the anchor bolt so that equal amounts of the bushing sleeve protrude on either side.

10. Install the anchor bolt assembly onto the control arm. Align the marks on the arm and anchor assembly. Tighten the nuts to 22 ft. lbs. (30 Nm).

11. Install the upper control arm assembly into the shock tower.

12. Install the strut into the vehicle. Install the damper fork bolt and nut.

13. Connect the steering arm and upper ball joint.

14. Install the front wheels. Lower the vehicle to the ground.

15. Torque the strut mounting nuts to 36 ft. lbs. (50 Nm).

16. Torque the upper control arm mounting nuts to 47 ft. lbs. (65 Nm).

17. Torque the damper fork nut to 47 ft. lbs. (65 Nm).

18. Torque the upper ball joint castle nut to 29–35 ft. lbs. (40–48 Nm). Then, tighten the nut only enough to Install a new cotter pin.

19. Tighten the wheel nuts to 80 ft. lbs. (108 Nm).

20. Check the vehicle's front end alignment and adjust it if necessary. Road test the vehicle.

ACCORD AND PRELUDE

➡Do not disassemble the upper arm. If the ball joint or bushings are faulty, or the upper arm is damaged, the entire upper arm must be replaced.

1. Raise and support the vehicle safely.

2. Remove the front wheels. Support the lower control arm assembly with a floor jack.

3. Separate the upper ball joint from the steering knuckle using a ball joint separator tool.

4. Remove the self-locking nuts from the upper arm anchor bolts. Remove the upper arm from the vehicle.

➡Do not disassemble the upper arm. If the ball joint or bushings are faulty, or the upper arm is damaged, the entire upper arm must be replaced.

To install:
➡Use new self-locking nuts when installing the upper arm and strut.

5. Install the upper control arm assembly into the strut tower.

6. Connect the upper ball joint.

7. Install the front wheels and lower the vehicle.

8. With all four of the vehicle's wheels on the ground, torque the upper control arm

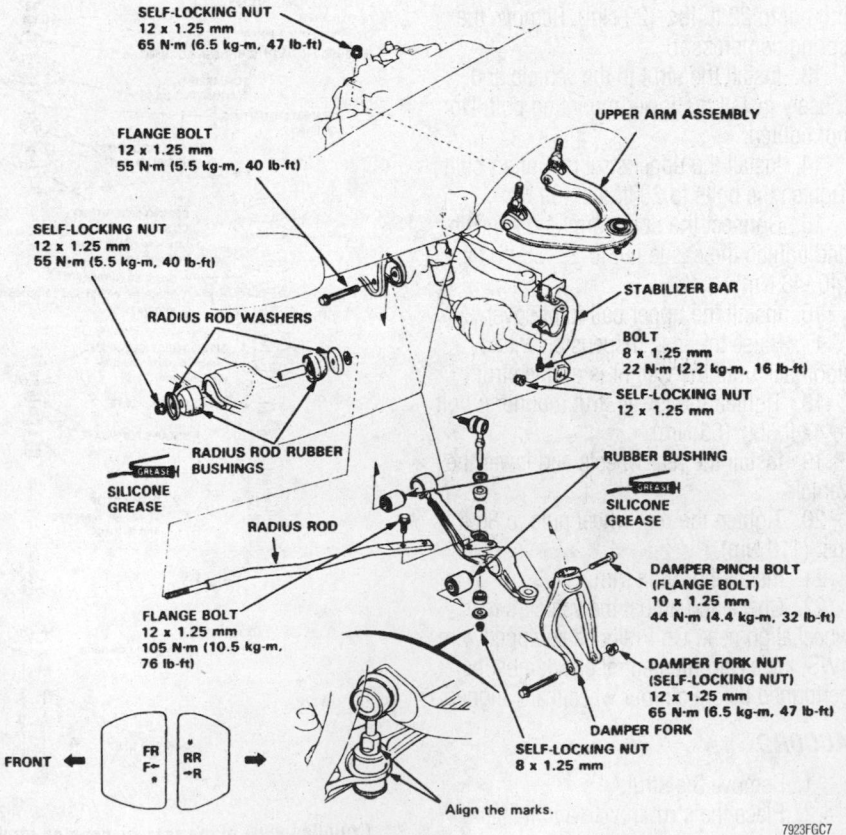

SELF-LOCKING NUT
12 x 1.25 mm
65 N·m (6.5 kg·m, 47 lb·ft)

FLANGE BOLT
12 x 1.25 mm
55 N·m (5.5 kg·m, 40 lb·ft)

SELF-LOCKING NUT
12 x 1.25 mm
55 N·m (5.5 kg·m, 40 lb·ft)

RADIUS ROD WASHERS

RADIUS ROD RUBBER BUSHINGS

SILICONE GREASE

RADIUS ROD

FLANGE BOLT
12 x 1.25 mm
105 N·m (10.5 kg·m, 76 lb·ft)

UPPER ARM ASSEMBLY

STABILIZER BAR

BOLT
8 x 1.25 mm
22 N·m (2.2 kg·m, 16 lb·ft)

SELF-LOCKING NUT
12 x 1.25 mm

RUBBER BUSHING

SILICONE GREASE

DAMPER PINCH BOLT (FLANGE BOLT)
10 x 1.25 mm
44 N·m (4.4 kg·m, 32 lb·ft)

DAMPER FORK NUT (SELF-LOCKING NUT)
12 x 1.25 mm
65 N·m (6.5 kg·m, 47 lb·ft)

DAMPER FORK

SELF-LOCKING NUT
8 x 1.25 mm

FRONT ←

FR F→

*RR →R

Align the marks.

7923FGC7

Front suspension components—Prelude and Accord

nuts to 47 ft. lbs. (65 Nm). Torque the castle nut to 32 ft. lbs. (44 Nm); then, only tighten it only enough to install a new cotter pin.

9. Tighten the wheel nuts to 80 ft. lbs. (110 Nm).

10. Check and adjust the vehicle's front end alignment. On Preludes equipped with 4WS, the electronic neutral check must be performed before all four wheels are aligned.

Rear

CIVIC AND DEL SOL

✵ CAUTION

Removing rear suspension components may make the vehicle front-heavy and cause it to tip forward when raised on a hoist. Use under-lift support stands, or place additional weight in the trunk of the vehicle before hoisting it.

1. Raise and safely support the vehicle.
2. Remove the rear wheels.
3. Support the lower control arm with a floor jack.
4. Unbolt the upper control arm from the trailing arm.
5. Unbolt the upper control arm flange bar from its vehicle body mount. Remove the upper control arm.
6. Inspect the upper control and its bushings for signs of wear and distortion. The bushings are replaceable:

   a. Press the bushings out of the upper control arm using suitably sized press fixtures.

   b. Matchmark the bolt flange bar to the body of the upper control arm.

   c. Lubricate the new bushings with silicon grease before installation.

   d. Press the new bushings into the control arm. Make sure the bolt flange bar matchmarks align. The leading edges of the control arm bushings must be flush with the edges of the control arm body.

To install:

➡ **Use new self-locking nuts and color-coded bolts when assembling suspension components.**

7. Install the control arm to its body mount. Hand-tighten the flange bolts.
8. Install the control arm to the trailing arm. Hand-tighten the flange bolt.
9. Install the rear wheel and lower the vehicle.
10. Torque the bolts with the vehicle on the ground. Tighten the control arm bolts-

to-body to 29 ft. lbs. (40 Nm). Tighten the control arm-to-trailing arm bolt to 40 ft. lbs. (55 Nm).

11. Check and adjust the vehicle's rear wheel alignment.

12. Tighten the wheel nuts to 80 ft. lbs. (110 Nm).

PRELUDE

1. Raise and support the vehicle safely.
2. Remove the rear wheels. Support the knuckle and lower control arm assembly with a jack.
3. Separate the upper ball joint from the knuckle using a ball joint separator tool.
4. Pull back the trunk side trim and remove the two strut mounting nuts.
5. Remove the self-locking nuts from the upper arm anchor bolts. Remove the upper arm from the vehicle.

➡ **Do not disassemble the upper arm. If the ball joint or bushings are faulty, or the upper arm is damaged, the entire upper arm must be replaced.**

To install:
➡ **Use new self-locking nuts when installing the upper arm and strut.**

6. Install the upper control arm assembly into the strut tower.

7. Connect the upper ball joint.
8. Install the rear wheels and lower the vehicle.
9. With all four of the vehicle's wheels on the ground, torque the upper control arm nuts to 47 ft. lbs. (65 Nm). Torque the castle nut to 32 ft. lbs. (44 Nm); then, only tighten it only enough to install a new cotter pin.
10. Tighten the wheel nuts to 80 ft. lbs. (110 Nm).
11. Put the trunk side trim back into position.
12. Check and adjust the vehicle's rear end wheel alignment. On Preludes equipped with 4WS, the electronic neutral check must be performed before all four wheels are aligned.

ACCORD

1. Raise and safely support the vehicle.
2. Remove the rear wheels.
3. Support the knuckle and lower control arm with a floor jack to compress the strut.
4. Remove the castle nut cap, cotter pin and castle nut from the upper ball joint. Use a ball joint separator tool to separate the ball joint from the knuckle.
5. Unbolt and remove the upper control arm.

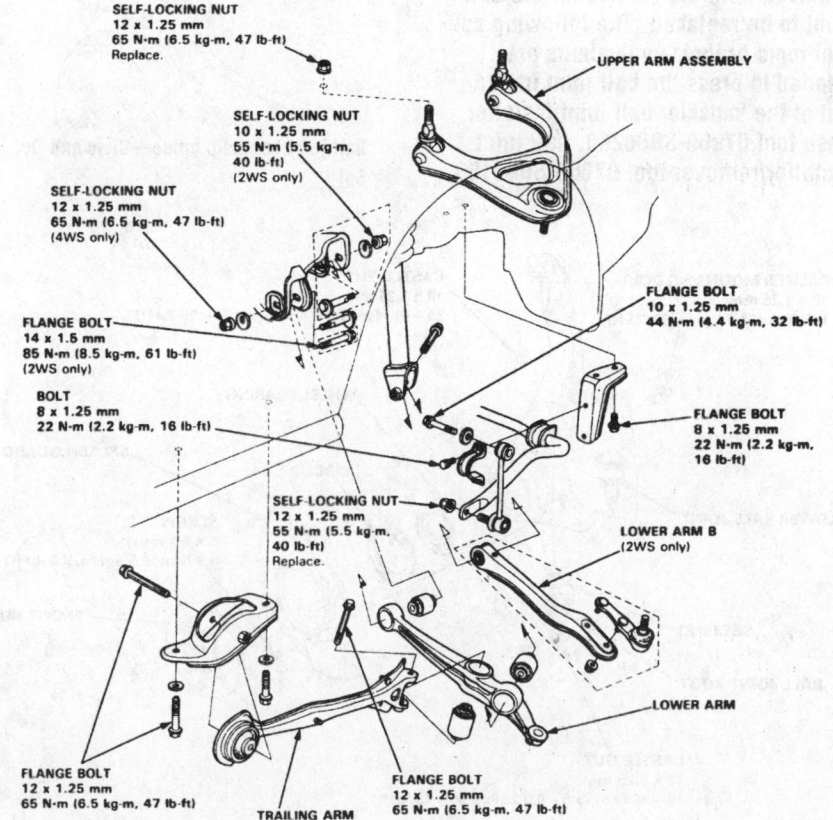

SELF-LOCKING NUT
12 x 1.25 mm
65 N·m (6.5 kg-m, 47 lb-ft)
Replace.

SELF-LOCKING NUT
10 x 1.25 mm
55 N·m (5.5 kg-m,
40 lb-ft)
(2WS only)

SELF-LOCKING NUT
12 x 1.25 mm
65 N·m (6.5 kg-m, 47 lb-ft)
(4WS only)

FLANGE BOLT
14 x 1.5 mm
85 N·m (8.5 kg-m, 61 lb-ft)
(2WS only)

BOLT
8 x 1.25 mm
22 N·m (2.2 kg-m, 16 lb-ft)

SELF-LOCKING NUT
12 x 1.25 mm
55 N·m (5.5 kg-m,
40 lb-ft)
Replace.

UPPER ARM ASSEMBLY

FLANGE BOLT
10 x 1.25 mm
44 N·m (4.4 kg-m, 32 lb-ft)

FLANGE BOLT
8 x 1.25 mm
22 N·m (2.2 kg-m,
16 lb-ft)

LOWER ARM B
(2WS only)

LOWER ARM

FLANGE BOLT
12 x 1.25 mm
65 N·m (6.5 kg-m, 47 lb-ft)

FLANGE BOLT
12 x 1.25 mm
65 N·m (6.5 kg-m, 47 lb-ft)

TRAILING ARM

Rear suspension components—Prelude

7923FGC8

6. Check upper arm and its bushing for signs of wear and damage. Replace the upper arm if the ball joint is faulty.

To install:

→Use new self-locking nuts when assembling suspension components.

7. Install the upper arm into the vehicle. Install the mounting bolts and only hand–tighten them. Reconnect the upper arm to the knuckle.

8. Tighten the castle nut at the ball joint to 32 ft. lbs. (44 Nm). Tighten the castle nut only enough to install a new cotter pin. Install the castle nut cap.

9. Install the rear wheels and lower the vehicle.

10. Tighten the upper mounting bolts to 28 ft. lbs. (39 Nm).

11. Tighten the wheel nuts to 80 ft. lbs. (110 Nm).

12. Check and adjust the vehicle's rear wheel alignment.

Lower Ball Joint

REMOVAL & INSTALLATION

Civic and Del Sol

→The steering knuckle must be removed from the vehicle for the ball joint to be replaced. The following special tools or their equivalents are needed to press the ball joint in and out of the knuckle: ball joint installer base tool 07965-SB00200, ball joint installer/remover tool 07965-SB00100,

and ball joint remover base tool 07965-SH20200. A large vise will be required to hold the knuckle and the press tools. A ball joint clip guide tool 07974-SA50700 or 07GAG-SD40700 is used to install the retaining clip on the joint boot.

BALL JOINT REMOVER/INSTALLER 07965–SB00100

CASTLE NUT

BALL JOINT REMOVER BASE 07JAF–SH20200

7923FGB5

Ball joint removal tools—Civic and Del Sol

ADJUSTING BOLT Adjust the depth by turning the bolt.

BALL JOINT BOOT CLIP GUIDE 07974–SA50700

SET RING

BOOT

7923FGB6

Ball joint boot clip guide—Civic and Del Sol

1. Remove the steering knuckle assembly from the vehicle. Remove the ball joint boot snapring and the boot.

2. Pry the snapring out of the groove in the ball joint body.

3. Install the ball joint removal tool onto the ball joint with the large end facing out. Install the ball joint nut to attach the tool to the joint.

4. Position the removal base tool on the ball joint and set the assembly in a large vise. Press the ball joint out of the steering knuckle.

To install:

5. Position the new ball joint into the hole of the steering knuckle.

6. Install the ball joint installer tool over the ball joint with the small end facing out.

7. Position the installation base tool on the ball joint and set the assembly in a large vise. Press the ball joint into the steering knuckle.

8. Seat the snapring in the groove of the ball joint.

9. Adjust the boot clip tool with the adjusting bolt until the end of the tool aligns with the groove on the boot. Slide the clip over the tool and into position.

10. Install the ball joint stud in the steering knuckle. Tighten the nut to 44 ft. lbs. (60 Nm).

Wheel Bearings

ADJUSTMENT

Front

CIVIC AND DEL SOL

1. Raise and support the vehicle safely.

2. Remove the front and/or rear wheels.

3. Install the lug nuts and tighten them to 80 ft. lbs. (110 Nm).

4. Use a dial gauge to measure front bearing end-play at the hub flange.

5. Use a dial gauge to measure rear bearing end-play at the center of the hub's grease cap.

6. Move the rotor or drum assembly in and out to measure the play. Then, compare the dial gauge readings.

7. The standard bearing end-play for both front and rear wheels is 0–0.002 in. (0–0.05mm). If the end-play measurement exceeds the standard, the wheel bearings must be replaced. The wheel bearings cannot be adjusted.

PRELUDE AND ACCORD

The wheel bearings are not adjustable or repairable and should be replaced if found defective.

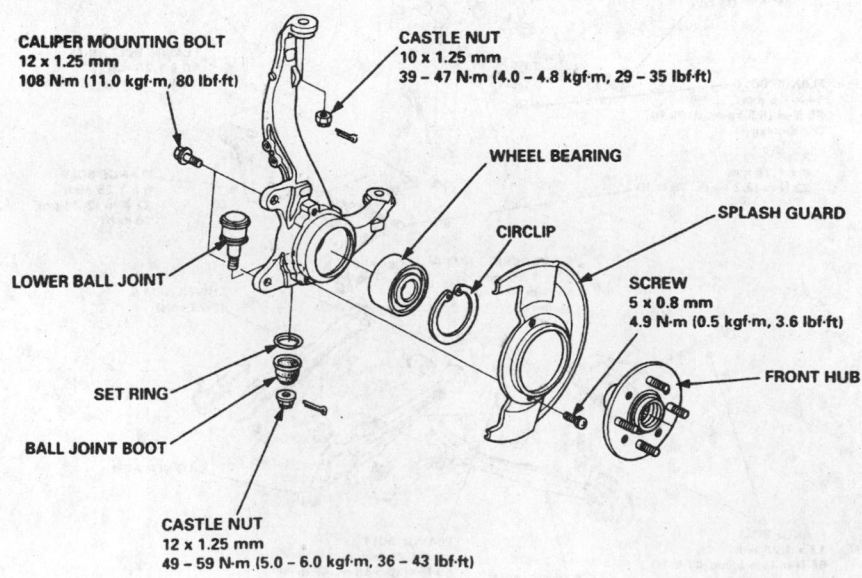

CALIPER MOUNTING BOLT
12 x 1.25 mm
108 N·m (11.0 kgf·m, 80 lbf·ft)

CASTLE NUT
10 x 1.25 mm
39 – 47 N·m (4.0 – 4.8 kgf·m, 29 – 35 lbf·ft)

WHEEL BEARING

CIRCLIP

SPLASH GUARD

LOWER BALL JOINT

SCREW
5 x 0.8 mm
4.9 N·m (0.5 kgf·m, 3.6 lbf·ft)

SET RING

FRONT HUB

BALL JOINT BOOT

CASTLE NUT
12 x 1.25 mm
49 – 59 N·m (5.0 – 6.0 kgf·m, 36 – 43 lbf·ft)

7923FGB4

Knuckle components—Civic and Del Sol

Rear

CIVIC AND DEL SOL

1. Raise and support the vehicle safely.
2. Remove the front and/or rear wheels.
3. Install the lug nuts and tighten them to 80 ft. lbs. (110 Nm).
4. Use a dial gauge to measure front bearing end-play at the hub flange.
5. Use a dial gauge to measure rear bearing end-play at the center of the hub's grease cap.
6. Move the rotor or drum assembly in and out to measure the play. Then, compare the dial gauge readings.
7. The standard bearing end-play for both front and rear wheels is 0–0.002 in. (0–0.05mm). If the end-play measurement exceeds the standard, the wheel bearings must be replaced. The wheel bearings cannot be adjusted.

PRELUDE AND ACCORD

The wheel bearings are not adjustable or repairable and should be replaced if found defective.

REMOVAL & INSTALLATION

Front

CIVIC AND DEL SOL

➡**A hydraulic press and several bearing drivers and attachments are needed to remove and install the hub and bearing.**

1. Pry the spindle nut stake away from the spindle, then loosen the nut.
2. Raise and safely support the vehicle.
3. Remove the front wheel and the spindle nut.
4. Remove the wheel sensor wire bracket from the knuckle, but don't disconnect it.
5. Remove the caliper mounting bolts and the caliper. Support the caliper out of

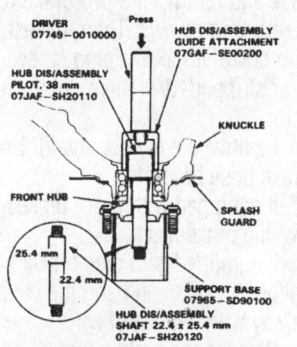

Hub installation driver, guide, and base—1995 Civic and 1995–97 Del Sol

the way with a length of wire. Do not let the caliper hang from the brake hose.

6. Remove the 6mm brake disc retaining screws. Screw two 12mm bolts into the disc to push it away from the hub.
7. Remove the tie rod castle nut. Disconnect the tie rod ball joint using a suitable ball joint remover.
8. Remove the cotter pin and loosen the lower arm ball joint nut half the length of the joint threads.
9. Separate the ball joint and lower arm using a suitable puller with the pawls applied to the lower arm.

➡**Avoid damaging the ball joint boot. If necessary, apply penetrating type lubricant to loosen the ball joint.**

10. Remove the ball joint nut cover. Remove the cotter pin and remove the upper ball joint nut.
11. Separate the upper ball joint and knuckle using a ball joint remover.
12. Use a plastic mallet to free the halfshaft from the knuckle. Pull the knuckle out to remove it.

➡**A new wheel bearing must be used when the hub is removed.**

13. Place the knuckle in a press and use a base and pilot to press the hub assembly out of the wheel bearing.
14. Remove the knuckle ring seal and circlip. Remove the splash guard from the knuckle.
15. Press the wheel bearing out of the knuckle using a driving attachment.

To install:
16. Clean the knuckle and hub assembly and inspect them for damage.
17. Press a new wheel bearing into the hub using a driving tool.
18. Install the circlip in the outer groove of the knuckle.
19. Install the splash guard.
20. Press the hub assembly into the steering knuckle using a base and a driving and guide tool.
21. Install the knuckle ring seal.
22. Install the knuckle onto the spindle.
23. Install the knuckle onto the upper and lower ball joints and tighten the castle nuts. Install the tie rod ball joint onto the steering knuckle.
24. Tighten the upper ball joint nut and tie rod nut to 29–35 ft. lbs. (40–48 Nm) and the lower ball joint castle nut to 36–43 ft. lbs. (50–60 Nm).
25. Install the ABS wheel sensor wire brackets onto the knuckle. Tighten the mounting bolts to 7 ft. lbs. (10 Nm).

26. Install the brake disc and use two lug nuts to evenly draw the disc onto the hub. Install the retainer screws and tighten them to 7 ft. lbs. (10 Nm). Install the spindle washer and nut. Don't tighten the nut until the vehicle is on the ground.
27. Install the brake caliper and tighten the bolts to 80 ft. lbs. (110 Nm).
28. Install the front wheels and lower the vehicle.
29. Tighten the spindle nut to 134 ft. lbs. (185 Nm), stake the nut, and install the grease cap.
30. Check and adjust the vehicle's front wheel alignment.

➡**Avoid damaging the ball joint boot. If necessary, apply penetrating-type lubricant to loosen the ball joint.**

PRELUDE AND ACCORD

➡**Once the hub has been removed, the wheel bearings must be replaced. A hydraulic press and bearing drivers must be used to remove and install the bearing.**

1. Pry the spindle nut stake away from the spindle and loosen the nut. Do not tighten or loosen a spindle nut unless the vehicle is sitting on all four wheels. The torque required is high enough to cause the vehicle to fall off the stands even when properly supported.
2. Raise and safely support the vehicle.
3. Remove the wheel and the spindle nut.
4. Remove the caliper mounting bolts and the caliper. Support the caliper out of the way with a length of wire. Do not let the caliper hang from the brake hose.
5. Remove the 6mm brake disc retaining screws. Screw two 8 x 1.25 12mm bolts into the disc to push it away from the hub.

➡**Turn each bolt two turns at a time to prevent cocking the brake disc.**

6. Remove the cotter pin from the tie rod castle nut, then remove the nut. Separate the tie rod ball joint using a ball joint remover, then lift the tie rod out of the knuckle.
7. Remove the cotter pin and loosen the lower arm ball joint nut half the length of the joint threads. The nut will retain the arm when the joint comes loose.
8. Separate the ball joint and lower arm using a puller with the pawls applied to the lower arm. Avoid damaging the ball joint boot. If necessary, apply penetrating lubricant to loosen the ball joint.
9. Remove the upper ball joint shield, if equipped.

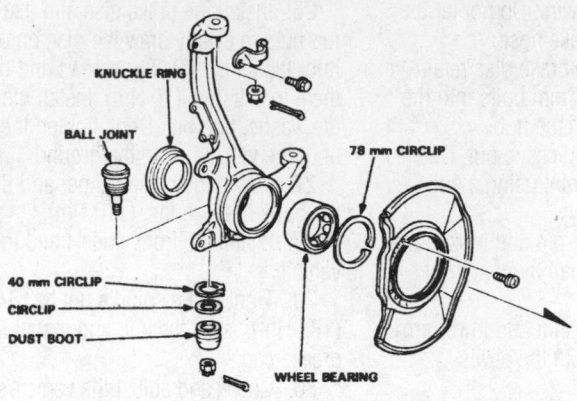

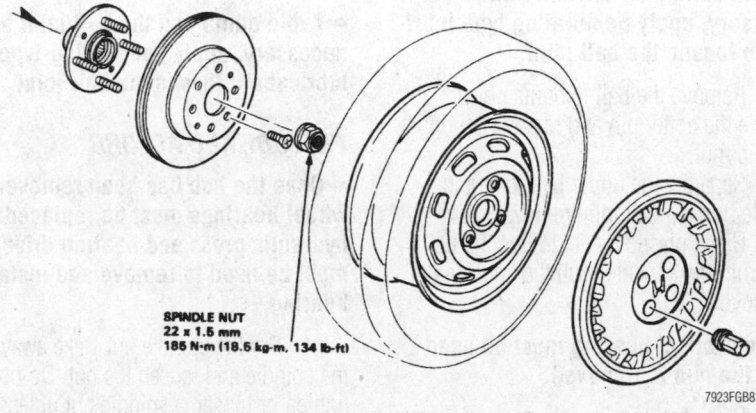

Hub and steering knuckle components—Prelude and Accord

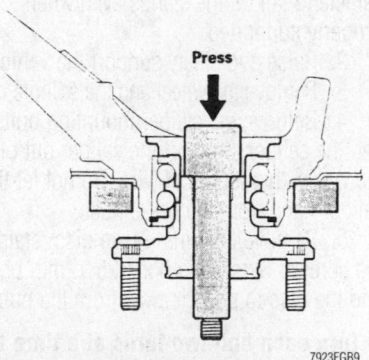

Press the hub out of the knuckle—Prelude and Accord

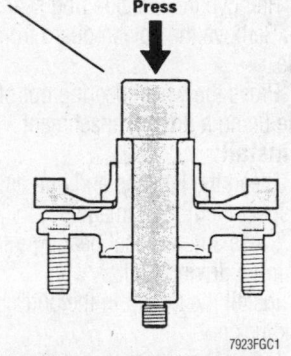

Use a press to remove the inner bearing race from the hub—Prelude and Accord

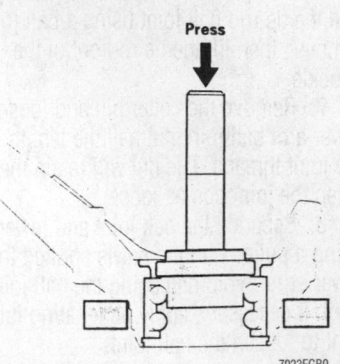

Press the bearing out of the knuckle—Prelude and Accord

10. Pry off the cotter pin and remove the upper ball joint nut.

11. Separate the upper ball joint and knuckle.

12. Remove the knuckle and hub by sliding them off the halfshaft.

13. Remove the splash guard screws from the knuckle.

14. Position the knuckle/hub assembly in a hydraulic press. Press the hub from the knuckle using a driver of the proper diameter while supporting the knuckle. The inner bearing race may stay on the hub.

15. Remove the splash guard and snapring from the knuckle.

16. Press the wheel bearing out of the knuckle while supporting the knuckle.

17. If necessary, remove the outboard bearing inner race from the hub using a bearing puller.

To install:

18. Clean the knuckle and hub thoroughly.

19. Press a new wheel bearing into the knuckle. Be sure the press tool contacts only the outer bearing race and properly support the knuckle so it is stable.

20. Install the snapring.

21. Install the splash shield. Don't overtighten the screws.

22. Place the hub on the press table and press the knuckle onto the hub. Be sure the press tool contacts only the inner bearing race.

23. Install the front knuckle ring on the knuckle.

24. Install the knuckle/hub assembly on the vehicle. Tighten the upper ball joint nut and tie rod end nut to 32 ft. lbs. (44 Nm). Install new cotter pins. Tighten the lower ball joint nut to 40 ft. lbs. (55 Nm) and install a new cotter pin.

25. Install the brake disc and caliper. Tighten the caliper bracket bolts to 80 ft. lbs. (110 Nm).

26. Install the front wheels and lower the vehicle.

27. Tighten the spindle nut to 180 ft. lbs. (250 Nm). Tighten the wheel nuts to 80 ft. lbs. (110 Nm).

28. Check and adjust the vehicle's front wheel alignment.

Rear

CIVIC AND DEL SOL

1. Remove the hub dust cap and loosen the spindle nut.

2. Raise and safely support the vehicle. Remove the rear wheels.

3. Engage the parking brake for added leverage and remove the two brake rotor or drum retaining screws. Then, release the parking brake. If the retaining screws are stuck or stripped, drill them out, or use an extractor.

4. Remove the caliper shield. Unbolt the brake hose bracket.

5. If equipped with drum brakes, remove the brake drum.

6. If equipped with disc brakes, unbolt the caliper bracket and hang the caliper out of the way with a piece of wire.

7. Remove the brake rotor.

8. Remove the hub assembly from the spindle.

9. Clean the hub assembly in solvent.

10. Inspect the hub assembly for any signs of wear or damage. If the wheel bearings are damaged, the hub assembly must be replaced.

To install:

11. Clean the spindle and the brake rotor/drum mounting surfaces.

12. Install the hub assembly onto the spindle. Install the spindle washer.

13. Install the brake rotor or brake drum. Apply anti-seize paste to the retaining screws and tighten them to 7 ft. lbs. (10 Nm). Don't overtighten the retaining screws.

14. Install the brake caliper and tighten the mounting bolts to 28 ft. lbs. (39 Nm). Install the brake hose bracket onto its mount. Install the caliper dust shield and tighten the bolts to 7 ft. lbs. (10 Nm).

15. Install a new spindle nut. Install the wheel and lower the vehicle.

16. Tighten the spindle nut to 134 ft. lbs. (185 Nm). Tighten the wheel nuts to 80 ft. lbs. (110 Nm). Stake the spindle nut with a punch. If the dust cap was bent during removal—install a new one.

PRELUDE AND ACCORD

➡ **The rear wheel bearing and hub unit are replaced as a unit.**

1. Set the parking brake, then loosen the rear wheel nuts and the spindle nut.

2. Raise the vehicle and support it safely.

3. Remove the rear wheels.

4. Remove the brake disc retaining screws.

5. Release the parking brake.

6. Unbolt the brake hose brackets from the knuckle.

7. Remove the caliper bracket mounting bolts and hang the caliper out of the way with a piece of wire.

8. Remove the brake disc. If the disc is frozen on the hub, screw two 8 x 1.25mm bolts evenly into the disc to push it away from the hub.

9. Remove the spindle nut and pull the hub unit off of the spindle.

➡ **Clean the backing plate and the mating surfaces of the brake disc and hub** with brake cleaner. Clean the spindle, washer, and hub with solvent.

To install:

10. Inspect the hub unit for signs of damage or wear. If the bearings are worn, the entire unit must be replaced.

11. Install the hub unit and spindle washer onto the spindle. Install the spindle nut but do not tighten it.

12. Install the brake disc and tighten the retaining screws to 7 ft. lbs. (10 Nm).

13. Install the brake caliper and tighten the mounting bolts to 28 ft. lbs. (39 Nm). Install the brake hose brackets onto the knuckle and tighten the bolts to 16 ft. lbs. (22 Nm).

14. Install the rear wheels and lower the vehicle.

15. With the vehicle on the ground, tighten the new spindle nut to 185 Nm (134 ft. lbs.), then stake the nut with a punch.

16. Tighten the wheel nuts to 80 ft. lbs. (110 Nm).

17. Test the operation of the brakes.

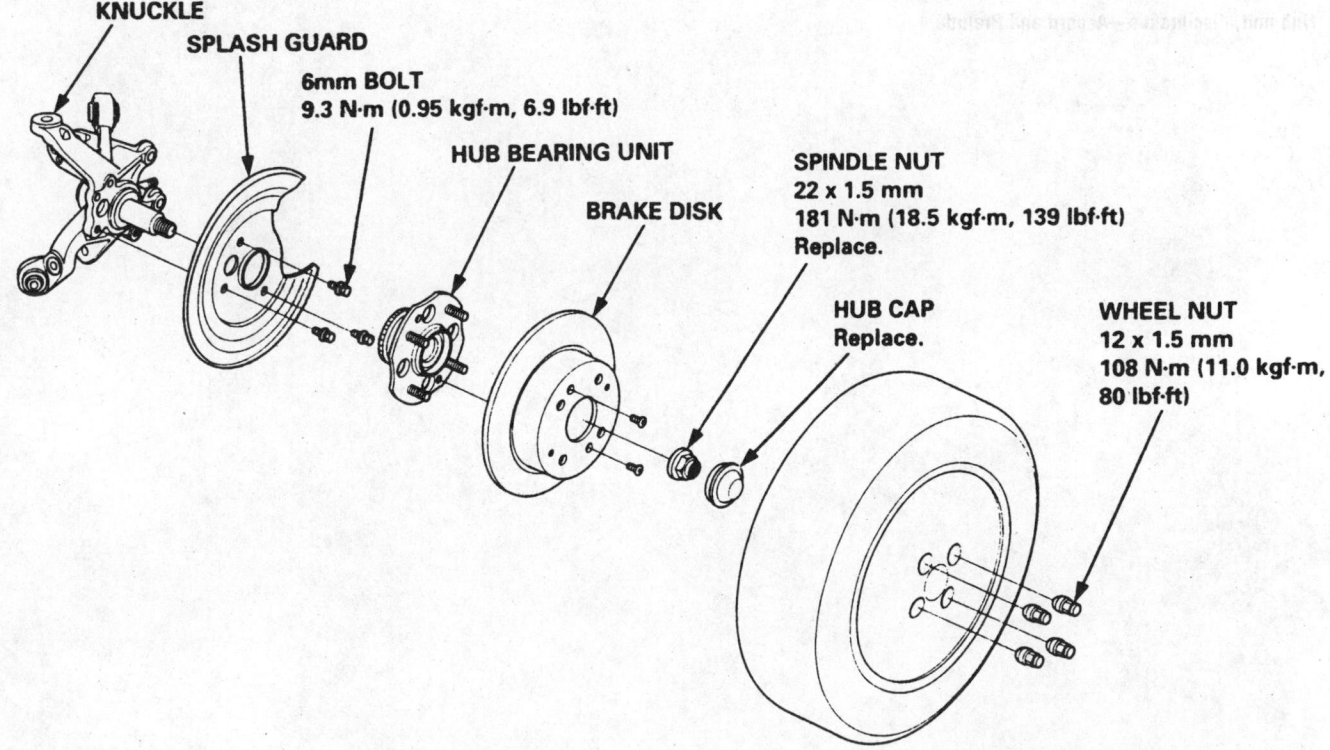

KNUCKLE

SPLASH GUARD

6mm BOLT
9.3 N·m (0.95 kgf·m, 6.9 lbf·ft)

HUB BEARING UNIT

BRAKE DISK

SPINDLE NUT
22 x 1.5 mm
181 N·m (18.5 kgf·m, 139 lbf·ft)
Replace.

HUB CAP
Replace.

WHEEL NUT
12 x 1.5 mm
108 N·m (11.0 kgf·m, 80 lbf·ft)

7923FGC2

Exploded view of the hub unit, drum brakes—Accord

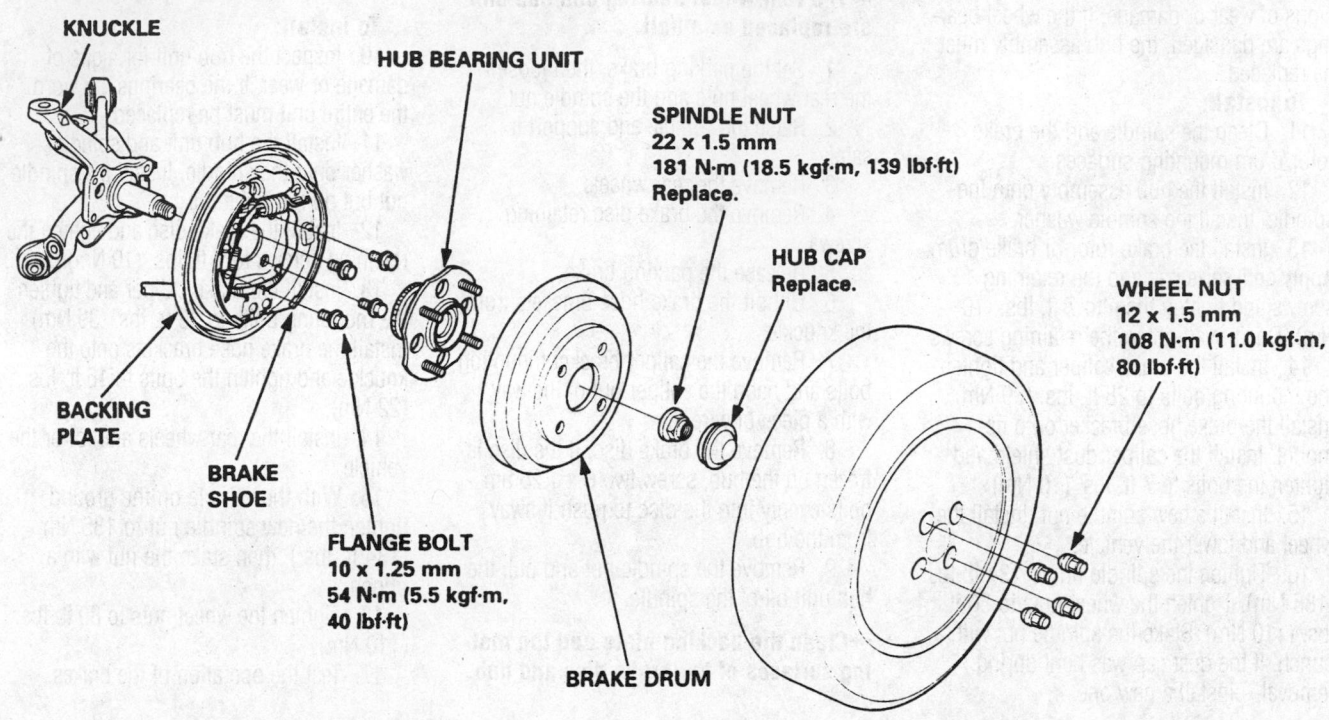

KNUCKLE

HUB BEARING UNIT

SPINDLE NUT
22 x 1.5 mm
181 N·m (18.5 kgf·m, 139 lbf·ft)
Replace.

HUB CAP
Replace.

WHEEL NUT
12 x 1.5 mm
108 N·m (11.0 kgf·m,
80 lbf·ft)

**BACKING
PLATE**

**BRAKE
SHOE**

FLANGE BOLT
10 x 1.25 mm
54 N·m (5.5 kgf·m,
40 lbf·ft)

BRAKE DRUM

7923FGC3

Hub unit, disc brakes—Accord and Prelude

HYUNDAI

Accent • Elantra • Scoupe • Sonata • Tiburon

ENGINE REPAIR

➡Disconnecting the negative battery cable on some vehicles may interfere with the functions of the on board computer systems and may require the computer to undergo a relearning process, once the negative battery cable is reconnected.

Distributor

REMOVAL & INSTALLATION

1. Rotate the engine and bring the No.1 piston to TDC of its compression stroke.
2. Disconnect the negative battery cable.
3. Label and disconnect the electrical harness from the distributor.
4. Remove the distributor cap and lay it aside with the spark plug wires still attached.
5. Matchmark the rotor to the distributor, and the distributor to the engine.

➡Do not rotate the engine after removing the distributor.

6. Remove the distributor mounting nut and remove the distributor assembly.

➡If the distributor is hard to remove, the O-rings on the distributor shaft are probably damaged.

To install:

7. Carefully inspect the O-rings on the distributor shaft and replace as necessary.

➡If the engine was disturbed while the distributor was removed, it will be necessary to remove the No. 1 spark plug and rotate the engine clockwise until No. 1 piston is on the compression stroke. Align the timing pointer with TDC on the crankshaft damper.

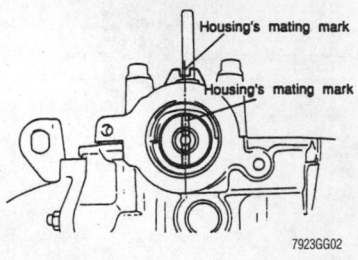

Aligning the distributor housing and gear mating marks—3.0L Sonata

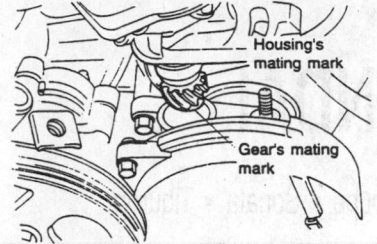

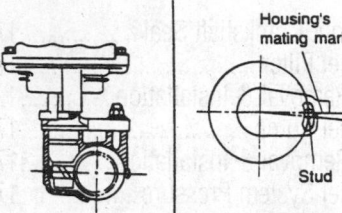

Aligning the distributor housing and gear mating marks—Scoupe

8. Align the distributor housing mating mark with the gear mating mark, as illustrated.
9. Lubricate the O-rings on the distributor shaft.
10. Install the distributor into the engine and ensure all matchmarks made during disassembly are aligned.
11. Snug the distributor hold down nut.
12. Start the engine and adjust the ignition timing.

Ignition Timing

➡No periodic adjustment of the ignition timing is necessary for any of the vehicles covered. However, the ignition system used on the 1.6L (VIN R), 1.8L (VIN M) AND 3.0L (VIN T) engines does allow for adjustment, should the distributor be removed or otherwise disturbed.

ADJUSTMENT

1.6L (VIN R), 1.8L (VIN M) and 3.0L (VIN T) Engines

➡Do not use a scan tool to check ignition timing. A scan tool connected to the data link connector reads the ordinary ignition timing, not the basic the ignition timing (with timing connector grounded) that is necessary to set timing properly.

1. Place the vehicle in **P** or **N** with the emergency brake applied and the drive wheels blocked.
2. Start the engine and let it reach normal operating temperature. Be sure all accessories are off.

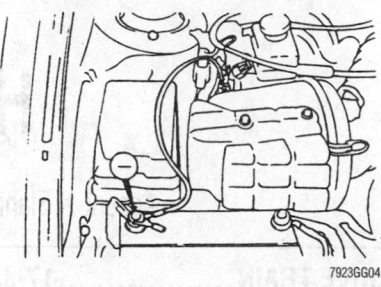

Connect the jumper wire from ground to the ignition timing terminal—3.0L (VIN R) engine

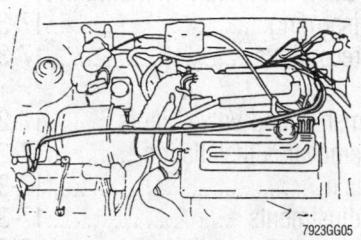

Connect the jumper wire from ground to the ignition timing terminal—1.6L (VIN R) and 1.8L (VIN M) engines

Aim the timing light at the timing marks or scale, found on the rim of the crankshaft pulley and timing cover

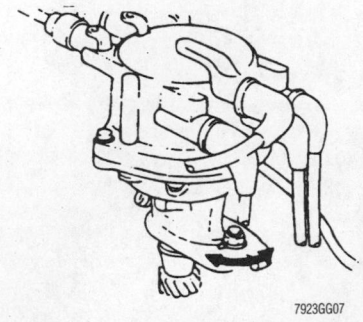

Loosen the distributor hold-down nut and turn the distributor to adjust the timing—3.0L (VIN R) engine

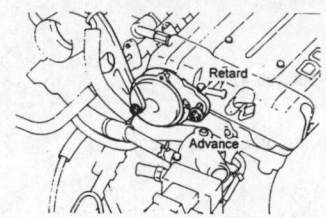

7923GG08

Loosen the crankshaft position sensor hold-down nut and turn the sensor counterclockwise to retard or clockwise to advance the timing —1.6L (VIN R) and 1.8L (VIN M) engines

3. Connect a suitable tachometer and timing light to the engine, as per the manufacturers' instructions.

4. On 1.6L (VIN R) and 1.8L (VIN M) engines, increase engine speed to 2,000–3,000 rpm for 5 seconds, then allow engine to idle for 2 minutes

5. Check that the idle speed is within the specified rpm range and adjust as necessary.

6. Stop the engine and connect a jumper wire from the ignition timing adjustment connector (located at the rear of the engine compartment) to ground.

7. Start engine and allow it to idle.

8. Following the manufacturer's instructions, aim the timing light and check the basic ignition timing. As the light flashes, note the position of the mark on the crankshaft pulley against the scale on the timing cover. Basic timing (with the connector grounded) should be 3–7 degrees BTDC.

9. If timing is not within specification, loosen the distributor hold-down nut and turn the distributor as needed to obtain a proper basic timing.

10. Tighten the distributor hold-down nut to 7–9 ft. lbs. (10–13 Nm).

11. Recheck the basic timing and readjust as necessary.

12. Stop the engine and remove the jumper wire. Be sure to remove the tachometer and timing light.

Engine Assembly

REMOVAL & INSTALLATION

The most important part of engine removal is the labeling of components, wires and hoses to be removed or disconnected from the engine. In most cases, the engine will be removed one day and installed several days later. This lapse in time makes it very difficult (even for professional mechanics) to remember where each and every connection must be made. A little time spent labeling and taking photographs of the engine compartment will pay big dividends once the engine is ready for installation.

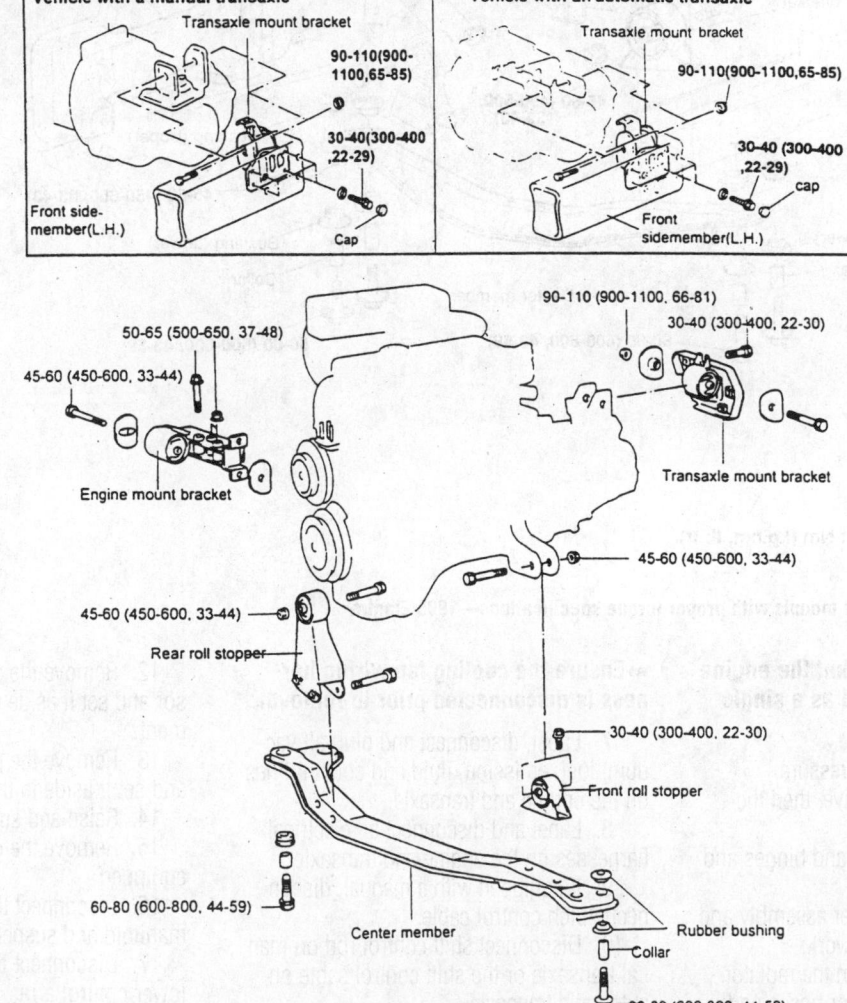

Exploded view of the engine mounts with proper torque specifications—Accent

7923GG09

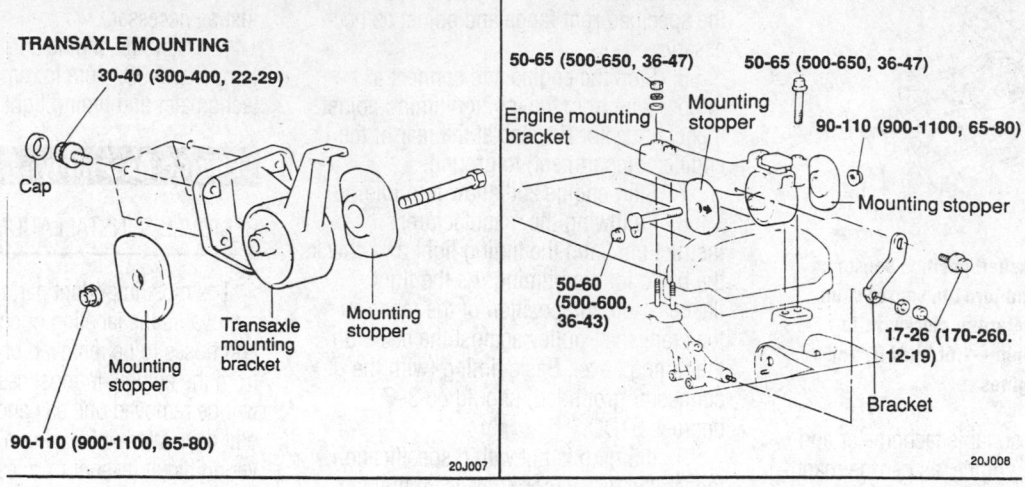

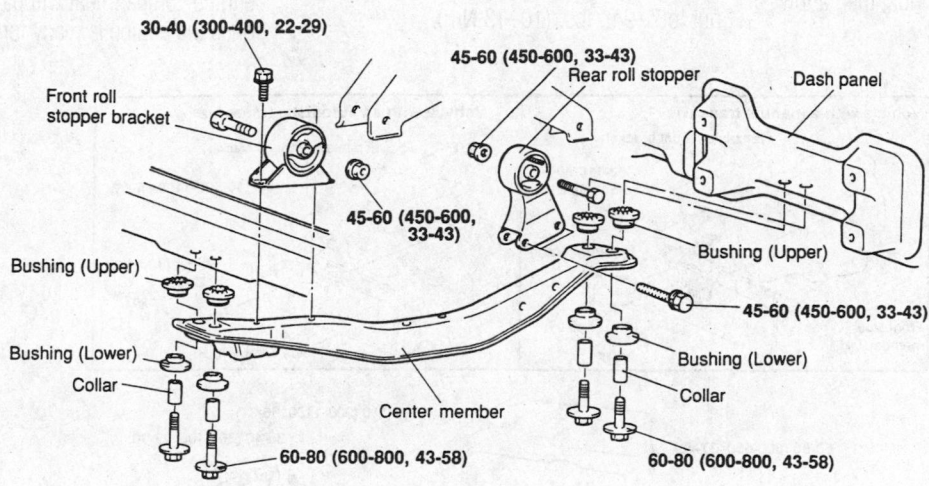

TORQUE : Nm (kg.cm, lb.ft)

7923GG12

Exploded view of the engine mounts with proper torque specifications—1995 Elantra

➡**Hyundai recommends that the engine and transaxle be removed as a single unit on all models.**

1. Relieve fuel system pressure.

2. Disconnect the negative, then the positive battery cable.

3. Matchmark the hood and hinges and remove the hood assembly.

4. Remove the air cleaner assembly and all adjoining air intake duct work.

5. Drain the coolant from the radiator.

6. Disconnect the radiator hoses and remove the radiator assembly with the electric cooling fan attached.

➡**Ensure the cooling fan wiring harness is disconnected prior to removal.**

7. Label, disconnect and plug all vacuum, fuel, emission, fluid and coolant lines on the engine and transaxle.

8. Label and disconnect all electrical harnesses on the engine and transaxle.

9. If equipped with a manual, disconnect clutch control cable.

10. Disconnect shift control rod on manual transaxle or the shift control cable on automatic transaxle.

11. Disconnect the accelerator and speedometer cables.

12. Remove the air conditioner compressor and set it aside in the engine compartment.

13. Remove the power steering pump and set it aside in the engine compartment.

14. Raise and support the vehicle safely.

15. Remove the engine under cover, if equipped.

16. Disconnect the exhaust pipe at the manifold and suspend the pipe with wire.

17. Disconnect the stabilizer bar at both lower control arms. Remove the bolts that attach the lower control arms to the body on either side. Support the arms from the body.

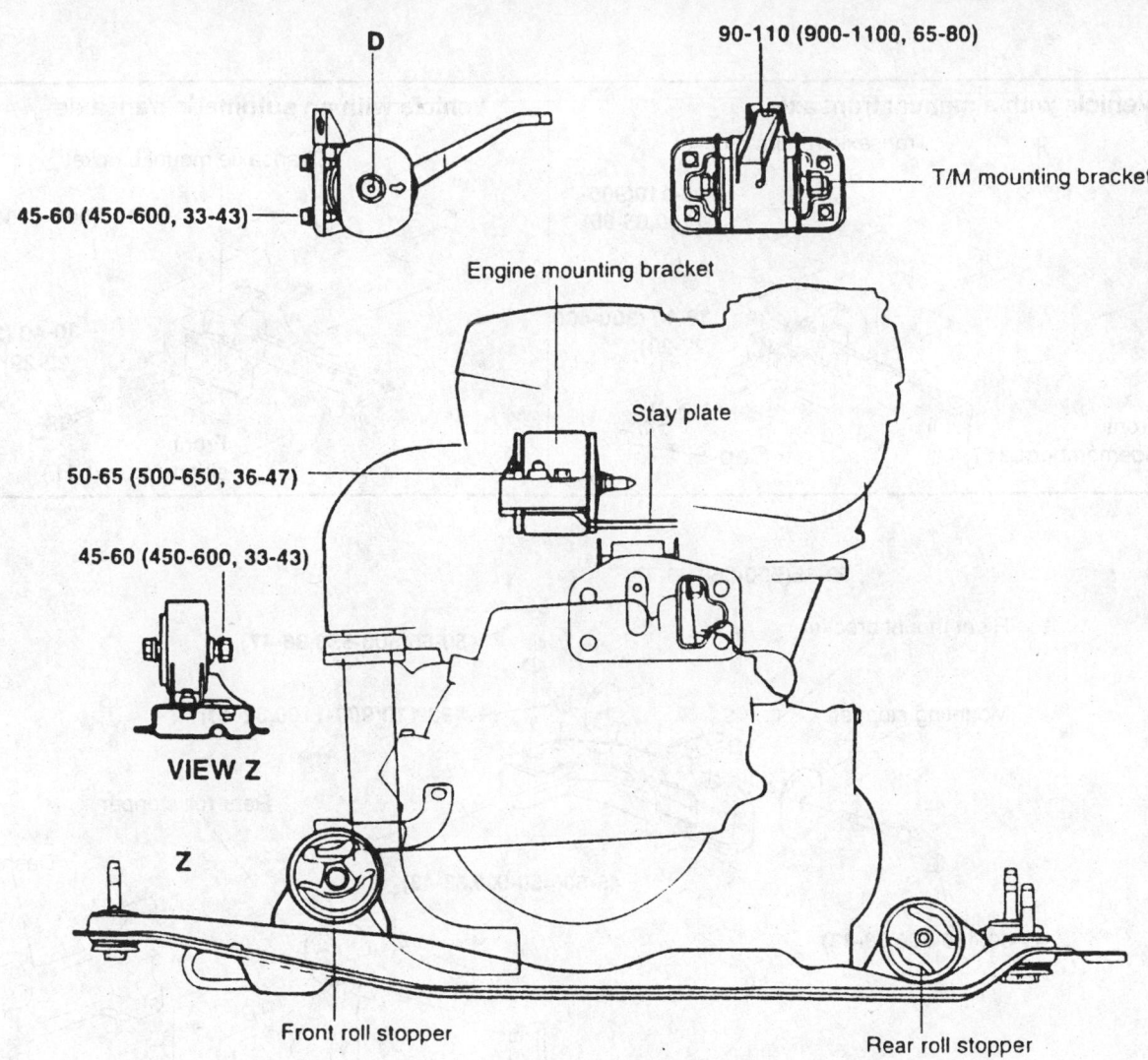

D

90-110 (900-1100, 65-80)

45-60 (450-600, 33-43)

T/M mounting bracket

Engine mounting bracket

Stay plate

50-65 (500-650, 36-47)

45-60 (450-600, 33-43)

VIEW Z

Z

Front roll stopper

Rear roll stopper

TORQUE: Nm (kg.cm, lb.ft)

7923GG13

Exploded view of the engine mounts with proper torque specifications—Tiburon and 1996–99 Elantra

18. Remove the front halfshafts and seal off the openings to prevent the entry of dirt.

19. Lower the vehicle.

20. Attach an engine lift, via chains or cables, to both the engine lifting hooks. Put just a little tension on the cables. Then, remove the nut and bolt from the front roll stopper; unbolt the brace from the top of the engine damper.

21. Separate the rear roll stopper from the crossmember.

22. Remove the nut from the left mount

insulator bolt, but do not remove the bolt.

23. Raise the engine just enough that the lifting device is supporting its weight.

24. Remove the transaxle mounting bracket bolts.

➡**Prior to lifting the engine/transaxle assembly, check that everything is disconnected.**

25. Remove the left mount insulator bolt. Then, press downward on the transaxle while lifting the engine/transaxle assembly to guide it up and out of the vehicle.

➡**Ensure the engine/transaxle assembly does not hit anything in the engine compartment during removal.**

To install:

26. Installation is the reverse of removal. Please note the following important steps:

27. After installing the engine/transaxle assembly, temporarily tighten the front roll stopper.

28. The front and rear center member rubber bushings and collar are different.

29. After the weight of the engine/transaxle assembly has been put on each

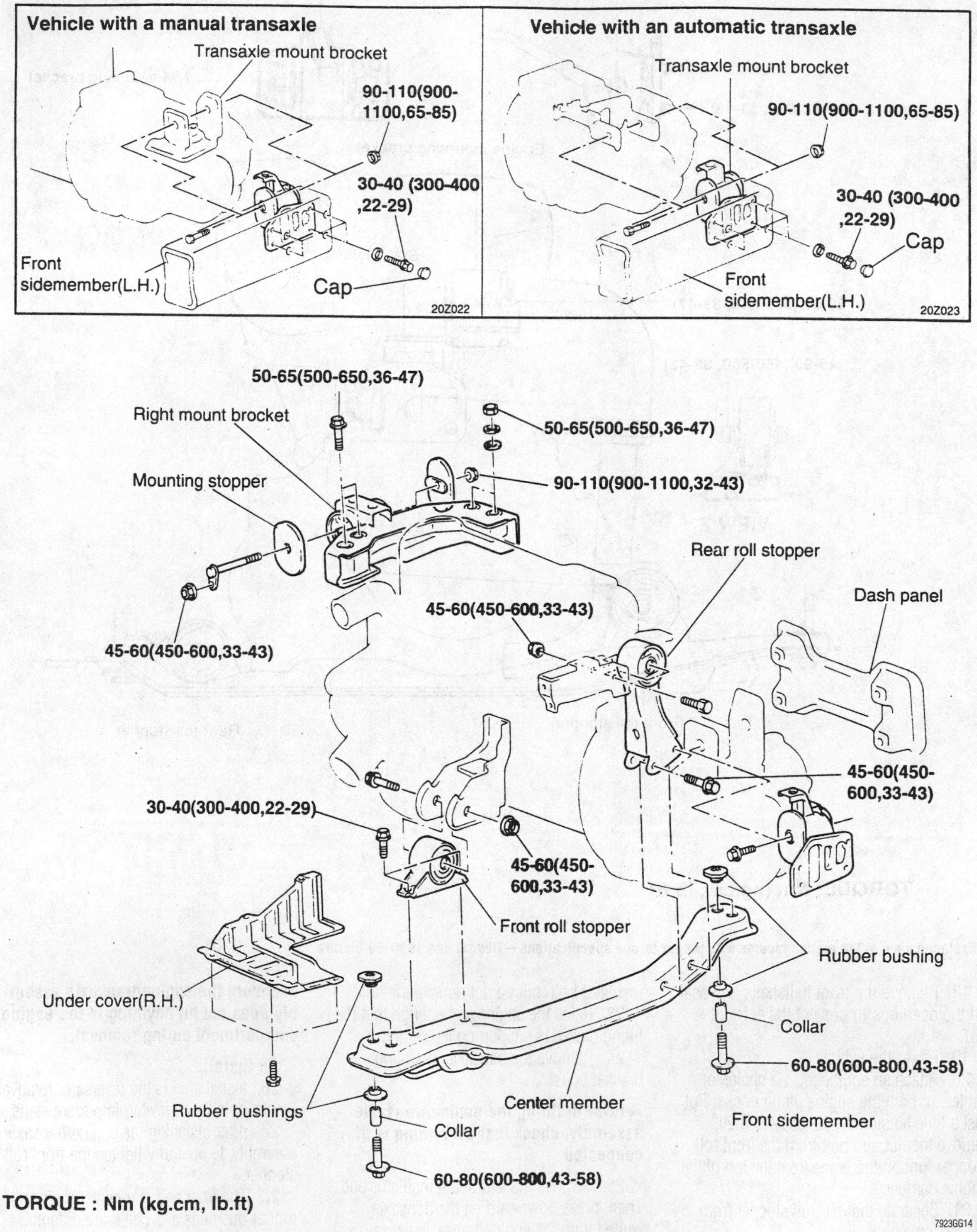

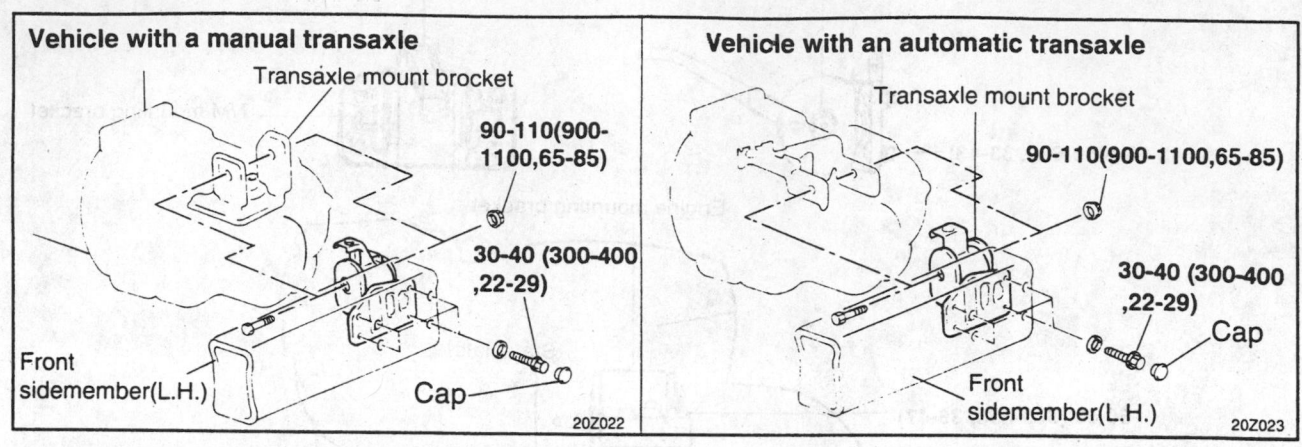

Vehicle with a manual transaxle

Transaxle mount brocket

90-110(900-1100,65-85)

30-40 (300-400,22-29)

Front sidemember(L.H.)

Cap

20Z022

Vehicle with an automatic transaxle

Transaxle mount brocket

90-110(900-1100,65-85)

30-40 (300-400,22-29)

Cap

Front sidemember(L.H.)

20Z023

50-65(500-650,36-47)

Right mount brocket

50-65(500-650,36-47)

Mounting stopper

90-110(900-1100,32-43)

Rear roll stopper

Dash panel

45-60(450-600,33-43)

45-60(450-600,33-43)

45-60(450-600,33-43)

30-40(300-400,22-29)

45-60(450-600,33-43)

Front roll stopper

Rubber bushing

Under cover(R.H.)

Collar

60-80(600-800,43-58)

Rubber bushings

Collar

Center member

Front sidemember

60-80(600-800,43-58)

TORQUE : Nm (kg.cm, lb.ft)

7923GG14

Exploded view of the engine mounts with proper torque specifications—Scoupe

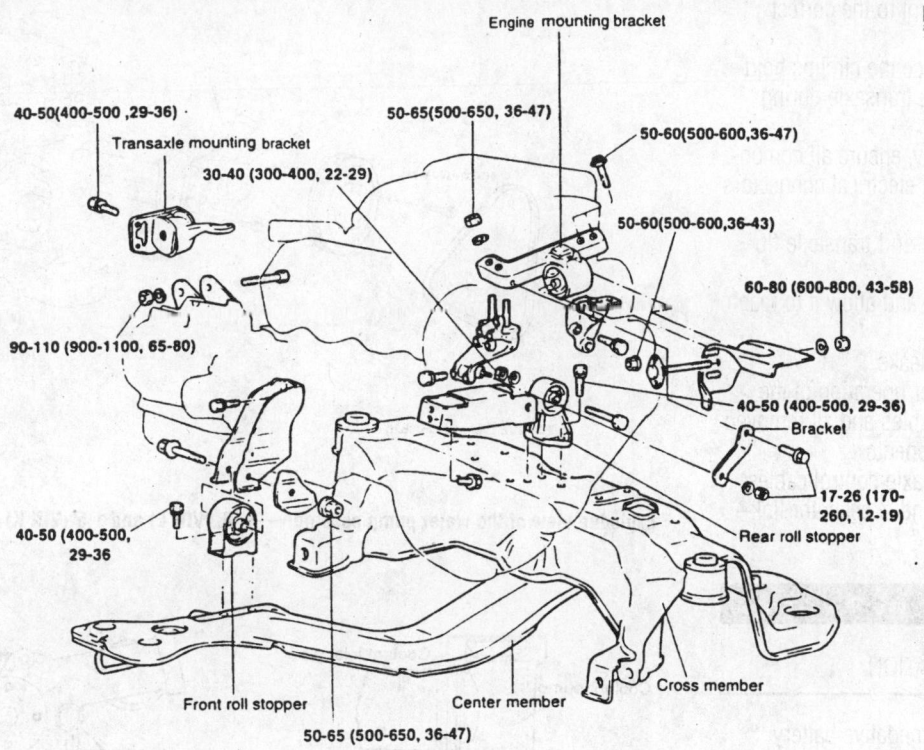

Engine mounting bracket

40-50(400-500 ,29-36)
Transaxle mounting bracket

30-40 (300-400, 22-29)

50-65(500-650, 36-47)

50-60(500-600,36-47)

50-60(500-600,36-43)

60-80 (600-800, 43-58)

90-110 (900-1100, 65-80)

40-50 (400-500, 29-36)

Bracket

17-26 (170-260, 12-19)

Rear roll stopper

40-50 (400-500, 29-36)

Front roll stopper   Center member   Cross member

50-65 (500-650, 36-47)

TORQUE : Nm (kg.cm, lb.ft)

7923GG15

Exploded view of the engine mounts with proper torque specifications—Sonata

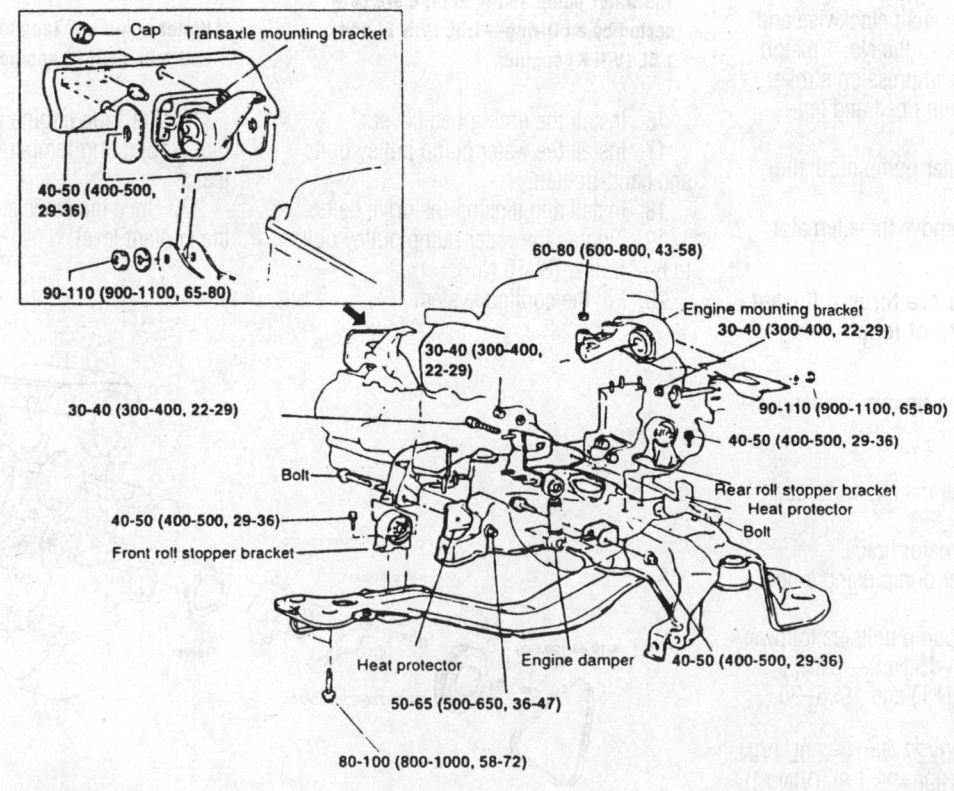

Cap   Transaxle mounting bracket

40-50 (400-500, 29-36)

90-110 (900-1100, 65-80)

60-80 (600-800, 43-58)

Engine mounting bracket

30-40 (300-400, 22-29)

30-40 (300-400, 22-29)

90-110 (900-1100, 65-80)

30-40 (300-400, 22-29)

40-50 (400-500, 29-36)

Bolt

Rear roll stopper bracket
Heat protector
Bolt

40-50 (400-500, 29-36)

Front roll stopper bracket

Heat protector   Engine damper   40-50 (400-500, 29-36)

50-65 (500-650, 36-47)

80-100 (800-1000, 58-72)

TORQUE : Nm (kg.cm, lb.ft)

7923GG16

Exploded view of the engine mounts with proper torque specifications—3.0L Sonata

mount, tighten the mount to the correct torque.

30. Be sure to replace the circlips holding the halfshafts in the transaxle during assembly.

31. During assembly, ensure all components, hoses, lines and electrical connectors are installed securely.

32. Refill all engine and transaxle fluids.

33. Start the engine and allow it to reach operating temperature.

34. Check for fluid leaks.

35. Check for proper operation of the transaxle, all control cables and all removed or disconnected components.

36. Adjust the transaxle control cables, accessory drive belts and accelerator linkages as required.

Water Pump

REMOVAL & INSTALLATION

1. Disconnect the negative battery cable

2. Remove the water pump pulley bolts.

3. Remove the drive belt.

4. Drain the engine coolant.

5. Remove the timing belt covers.

6. Rotate the crankshaft clockwise and align the timing marks so the No. 1 piston will be at TDC of the compression stroke.

7. Remove the timing belt and tensioner.

8. Remove the water pump mounting bolts.

9. As required, remove the alternator brace.

➡**Water pump bolts are three different lengths. Make a note of length and location.**

10. Remove the water pump, disconnecting the water outlet pipe.

To install:

11. Clean all gasket mating surfaces thoroughly.

12. Install the alternator brace.

13. Install the water pump using a new O-ring and gaskets.

14. Tighten water pump bolts as follows:

• 9–11 ft. lbs. (12–15 Nm)—except 2.0L (VIN F), 3.0L (VIN T) and 1996–98 1.8L (VIN M) engines

• 14–20 ft. lbs. (20–27 Nm)—2.0L (VIN F), 3.0L (VIN T) and 1996–98 1.8L (VIN M) engines

15. Install the timing belt and tensioner. Properly tension the timing belt.

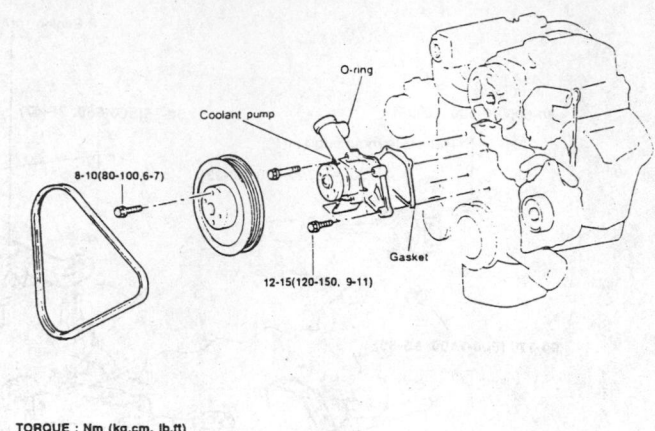

TORQUE : Nm (kg.cm, lb.ft)

7923GG17

Exploded view of the water pump assembly—1.5L (VIN E) and 1.5 (VIN K) engines

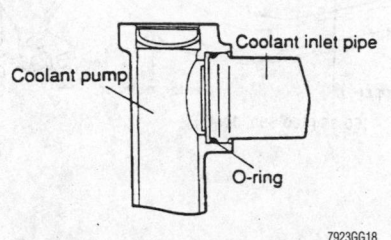

7923GG18

The water pump and inlet pipe are connected by an O-ring—1.5L (VIN E) and 1.5L (VIN K) engines

16. Install the timing belt covers.

17. Install the water pump pulley bolts and hand-tighten.

18. Install and tension the drive belts.

19. Tighten the water pump pulley bolts to 6–7 ft. lbs. (8–10 Nm).

20. Fill the cooling system.

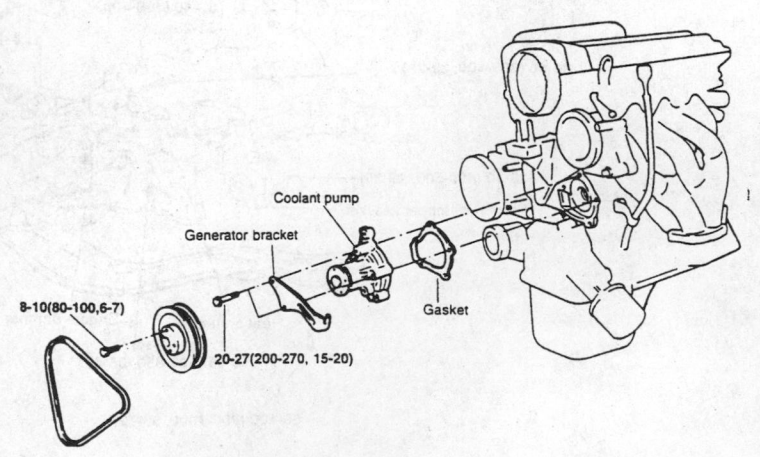

7923GG19

Water pump bolt lengths—1.5L (VIN E) and 1.5L (VIN K) engines

21. Start the engine and allow it to reach operating temperature. Check for leaks.

22. Once the vehicle has cooled, recheck the coolant level.

7923GG20

Water pump assembly—2.0L (VIN F) and 1996–99 1.8L (VIN M) engines

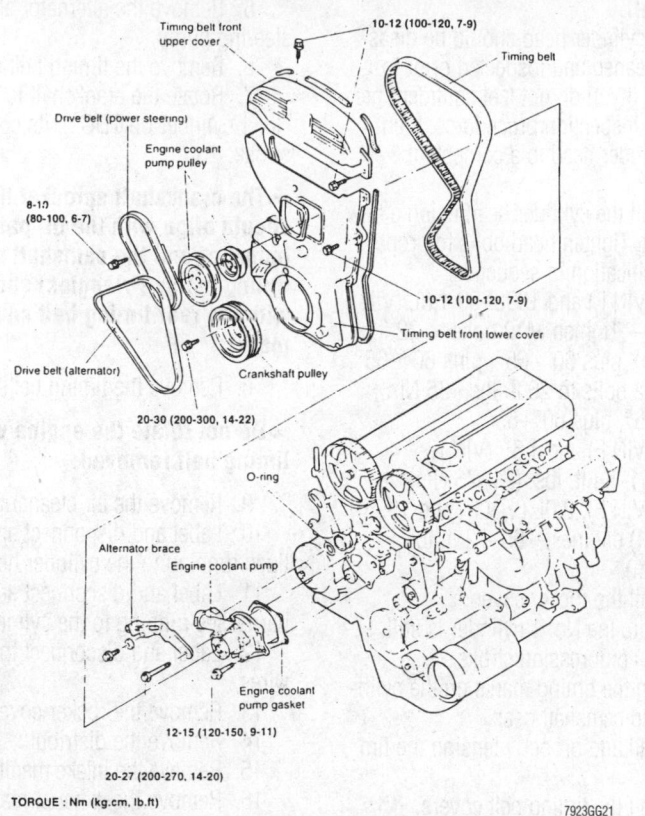

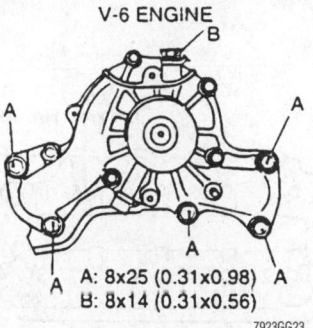

A: 8x25 (0.31x0.98)
B: 8x14 (0.31x0.56)

Water pump bolt lengths—3.0L (VIN T) engines

TORQUE : Nm (kg.cm, lb.ft)

Exploded view of the Water pump assembly and related components—1.6L (VIN R), 2.0L (VIN P) and 1995 1.8L (VIN M) enginespump assembly—3.0L (VIN T) engines

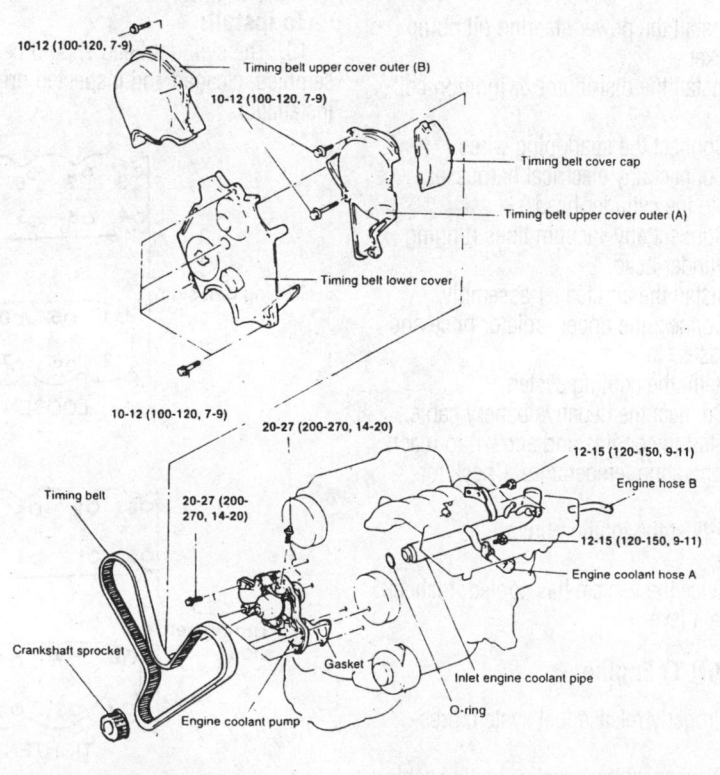

TORQUE : Nm (kg.cm, lb.ft)

Water pump assembly—3.0L (VIN T) engines

Cylinder Head

REMOVAL & INSTALLATION

➡The cylinder heads are made of aluminum. Do not remove the cylinder head unless the engine is cold. A hot cylinder head will warp once removed from the engine.

Except 3.0L (VIN T) Engine

1. Properly release fuel system pressure.
2. Disconnect the negative battery cable.
3. Drain the cooling system.
4. Disconnect the upper radiator hose and heater hoses.
5. Remove the air cleaner assembly.
6. Label and disconnect any vacuum lines running to the cylinder head.
7. Label and disconnect any electrical harnesses running to the cylinder head.
8. Label and disconnect the spark plug wires.
9. Turn the crankshaft until the No. 1 cylinder is at TDC on the compression stroke.
10. Remove the distributor or ignition coil pack.
11. Remove the power steering oil pump and bracket.
12. Remove the intake and exhaust manifolds.
13. Remove the water pump and crankshaft pulleys.
14. Remove the timing belt covers.
15. Remove the timing belt.

➡Do not rotate the engine with the timing belt removed.

16. Remove the rocker cover.
17. Loosen the cylinder head bolts in proper sequence.
18. Carefully remove the cylinder head from the engine.

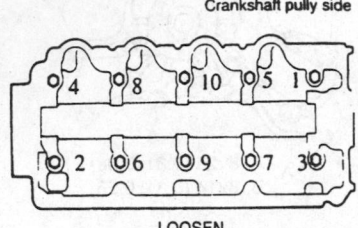

Cylinder head bolt loosen and tighten sequence—1.5L (VIN E) engine

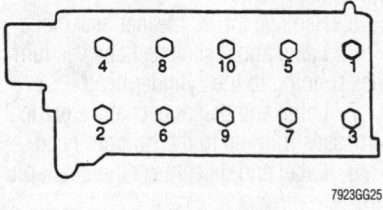

Cylinder head bolt loosen and tighten sequence—1.5L (VIN K), 1.6L (VIN R), 2.0L (VIN F) and 1996–99 1.8L (VIN M) engines

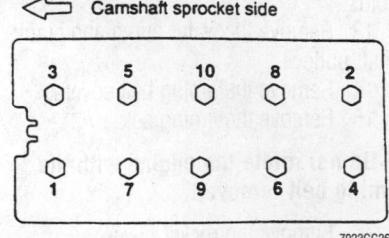

Cylinder head bolt loosen and tighten sequence—2.0L (VIN P) and 1995 1.6L (VIN R) and 1.8L (VIN M) engines

To install:

19. The cylinder head should be disassembled, cleaned and inspected prior to installation. If you do not feel confident performing the inspection procedures, then take the cylinder head to a competent machinist.

20. Install the cylinder head using a new head gasket. Tighten head bolts to proper torque specification in sequence.

- 2.0L (VIN F) and 1996–99 1.8L (VIN M) engines—Tighten M10 bolts to 22 ft. lbs. (30 Nm), plus 60°–65°, plus 60°–65°. Tighten M12 bolts to 26 ft. lbs. (35 Nm), plus 60°–65°, plus 60°–65°.
- 1.5L (VIN E) and 1.5L (VIN K) engines—51–54 ft. lbs. (71–75 Nm).
- 1.6L (VIN F), 2.0L (VIN P) and 1995 1.8L (VIN M) engines—65–72 ft. lbs. (90–100 Nm).

21. Install the rocker cover.

22. Ensure the No. 1 cylinder is still TDC on the compression stroke.

23. Align the timing marks on the cylinder head and camshaft gear.

24. Install and properly tension the timing belt.

25. Install the timing belt covers.

26. Install the water pump and crankshaft pulleys.

27. Install the intake and exhaust manifolds.

28. Install the power steering oil pump and bracket.

29. Install the distributor or ignition coil pack.

30. Connect the spark plug wires.

31. Connect any electrical harnesses running to the cylinder head.

32. Connect any vacuum lines running to the cylinder head.

33. Install the air cleaner assembly.

34. Connect the upper radiator hose and heater hoses.

35. Refill the cooling system.

36. Connect the negative battery cable.

37. Start the engine and allow it to reach normal operating temperature. Check for leaks.

38. Adjust the ignition timing, as required.

39. Once the vehicle has cooled, recheck the coolant level.

3.0L (VIN T) Engine

1. Properly release fuel system pressure.

2. Disconnect the negative battery cable.

3. Drain the cooling system.

4. Remove the air conditioning compressor.

5. Remove the alternator and power steering pump.

6. Remove the timing belt covers.

7. Rotate the crankshaft to position the No. 1 cylinder on TDC of its compression stroke.

➡**The crankshaft sprocket timing mark should align with the oil pan timing indicator and the camshaft sprocket timing marks (triangles) should align with the rear timing belt cover timing marks.**

8. Remove the timing belts.

➡**Do not rotate the engine with the timing belt removed.**

9. Remove the air cleaner assembly.

10. Label and disconnect any vacuum lines running to the cylinder head.

11. Label and disconnect any electrical harnesses running to the cylinder head.

12. Label and disconnect the spark plug wires.

13. Remove the rocker covers.

14. Remove the distributor.

15. Remove the intake manifold assembly.

16. Remove the exhaust manifolds.

17. Loosen the cylinder head bolts in proper sequence.

18. Carefully remove the cylinder head from the engine.

To install:

19. The cylinder head should be disassembled, cleaned and inspected prior to installation.

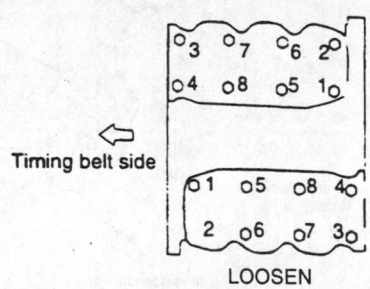

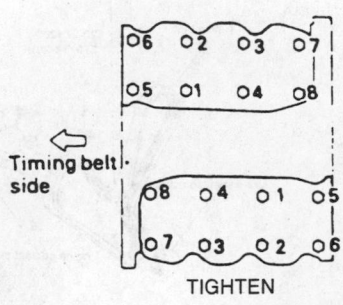

Cylinder head bolt loosen and tighten sequence—3.0L (VIN T) engine

20. Install the cylinder head using a new head gasket. Tighten head bolts in sequence to 76–83 ft. lbs. (105–115 Nm).

21. Install the exhaust manifolds.

22. Install the intake manifold assembly.

23. Install the distributor.

24. Install the rocker covers.

25. Connect the spark plug wires.

26. Connect any electrical harnesses running to the cylinder head.

27. Connect any vacuum lines running to the cylinder head.

28. Install the air cleaner assembly.

➡**Prior to installing the timing belt, check the timing marks for proper alignment. The crankshaft sprocket timing mark should align with the oil pan timing indicator and the camshaft sprocket timing marks (triangles) should align with the rear timing belt cover timing marks.**

29. Install and tension the timing belts.

30. Install the timing belt covers.

31. Install the alternator and power steering pump.

32. Install the air conditioning compressor.

33. Refill the cooling system.

34. Connect the negative battery cable.

35. Start the engine and allow it to reach normal operating temperature. Check for leaks.

36. Adjust the ignition timing.

37. Once the vehicle has cooled, recheck the coolant level.

Rocker Arms/Shafts

REMOVAL & INSTALLATION

1.5L (VIN E) and 1.5L (VIN K) Engines

1. Remove the rocker cover.

2. Loosen the rocker shaft bolts evenly and remove the rocker shaft assembly from the cylinder head with the bolts still in place.

3. Disassemble the rocker shaft by progressively removing each bolt, then the associated springs and rockers, keeping all parts in the exact order of disassembly.

➡**It is important to keep all parts in the exact order of removal. Two types of rocker arms are used, and "A" type and a "B" type. Do not mix them up.**

To install:

4. Inspect the components as follows:

   a. On engines with roller rockers, inspect the roller on the rocker arm.

Replace any rockers where the rollers are dented, damaged or show evidence of seizure. Ensure the oil hole on the bottom of the rocker arm near the roller is not clogged.

   b. On engines with standard rockers, inspect the rocker face contacting the cam lobe and the adjusting screw that contacts the valve stem for excess wear. If badly worn or damaged, replace the rocker.

   c. Inspect the fit of the rockers on the shaft. Replace rockers or the shaft as necessary.

   d. On engines with hydraulic lash adjusters, inspect the lash adjuster face that contacts the valve stem. Replace the lash adjuster if worn or damaged.

5. Reassemble the rocker shaft making sure all components are installed in their original positions.

6. Install the rocker shaft on the cylinder head and tighten bolts evenly to 14–20 ft. lbs. (20–26 Nm).

7. Install the rocker cover using a new gasket.

8. Start the engine and allow it to reach normal operating temperature. Check for leaks.

9. On 1.5L (VIN E) engine, adjust the valve lash.

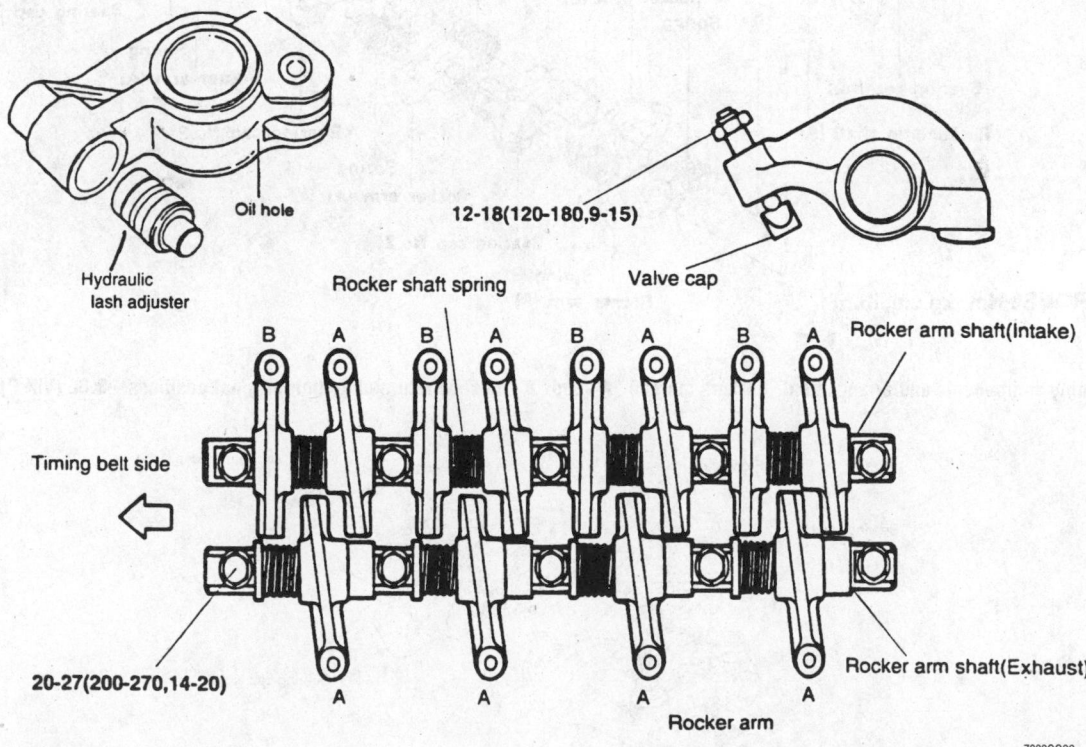

Rocker assembly components and arrangement. Rockers marked "A" and "B" must be returned to their original positions—1.5L (VIN E) and 1.5L (VIN K) engines

3.0L (VIN T) Engines

➡ The 3.0L (VIN T) engine is equipped with hydraulic lash adjusters. To prevent the lash adjusters from falling out during removal of the rocker arms, special holding clips (PN 09426 32000) are used to hold the adjusters in place. The lash adjuster is filled with diesel fuel. Store the adjusters in the upright position or cover with masking tape to prevent the diesel fuel from spilling out. If the fuel spills out the adjusters must be bled.

1. Remove the rocker cover.
2. Install special holding clip (PN 09246 3200) on each rocker arm to prevent the hydraulic lash adjusters from falling out during removal (and installation).
3. Loosen the bearing cap bolts evenly and remove the rocker shaft assembly from the cylinder head with the bolts still in place.
4. Disassemble the rocker shaft by removing the bearing cap bolts and sliding the rocker shafts from the bearing caps.

➡ It is important to keep all parts in the exact order of removal. Two types of rocker arms are used, and "A" type and a "B" type. Do not mix them up.

To install:

5. Inspect the components as follows:
   a. On engines with roller rockers, inspect the roller on the rocker arm. Roller should not bind or have excessive play. If eccentric rotation or backlash is evident, or if rollers are dented, damaged or show evidence of seizure, replace the rocker. Ensure the oil hole on the bottom of the rocker arm near the roller is not clogged.

   b. On engines with standard rockers, inspect the rocker face contacting the cam lobe and the adjusting screw that contacts the valve stem for excess wear. If badly worn or damaged, replace the rocker.

   c. Inspect the fit of the rockers on the shaft. Replace rockers or the shaft as necessary.

   d. On engines with hydraulic lash adjusters, inspect the lash adjuster face that contacts the valve stem. Replace the lash adjuster if worn or damaged.

6. Lubricate the rocker arm and shafts with clean engine oil.
7. Observe the mating marks and reassemble the rocker arm shafts, springs, rockers and bearing caps in the reverse order of removal.

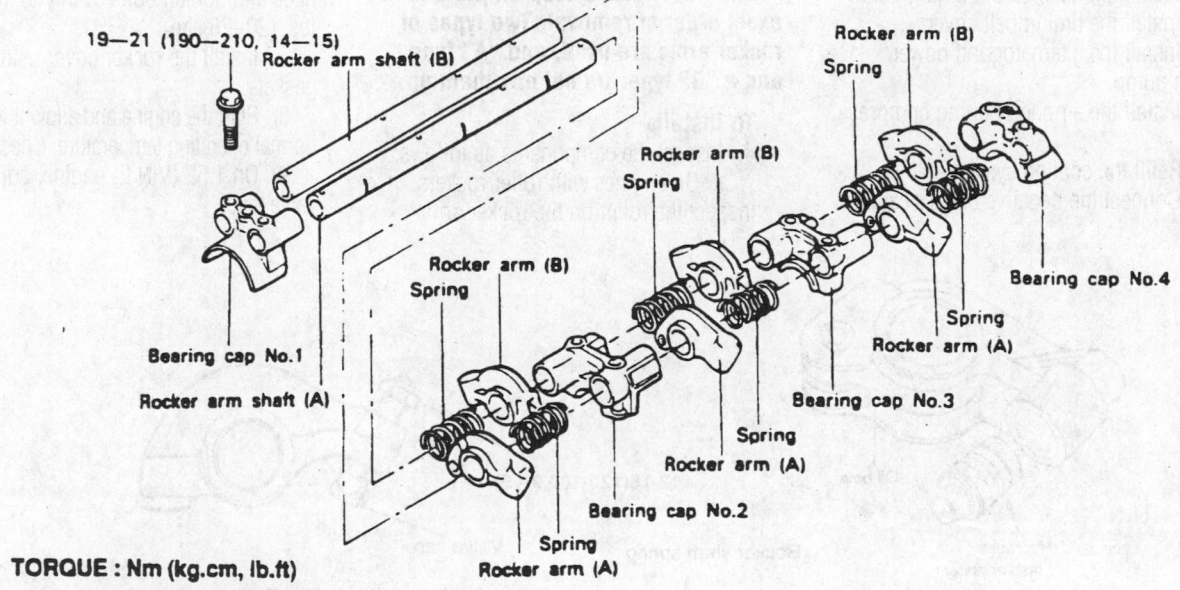

19—21 (190—210, 14—15)
Rocker arm shaft (B)

Rocker arm (B)
Spring

Rocker arm (B)
Spring

Rocker arm (B)
Spring

Bearing cap No.4

Bearing cap No.1

Rocker arm shaft (A)

Rocker arm (B)
Spring

Spring
Rocker arm (A)

Bearing cap No.3

Bearing cap No.2

Spring
Rocker arm (A)

Spring
Rocker arm (A)

TORQUE : Nm (kg.cm, lb.ft)

7923GG29

Rocker assembly components and arrangement. Rockers marked "A" and "B" must be returned to their original positions—3.0L (VIN P) engine

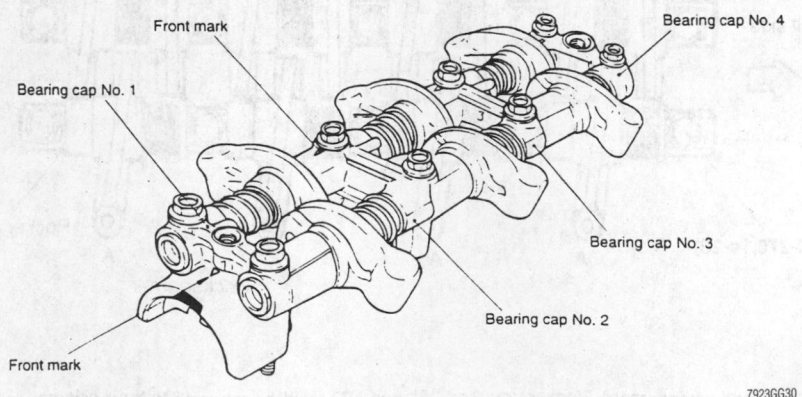

Front mark

Bearing cap No. 4

Bearing cap No. 1

Bearing cap No. 3

Bearing cap No. 2

Front mark

7923GG30

Rocker assembly identification marks—3.0L (VIN P) engine

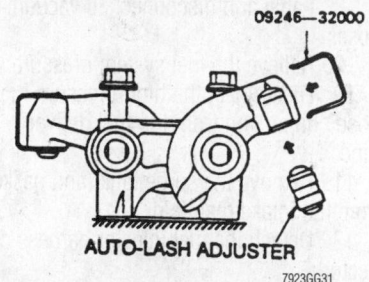

Install special holding clip (PN 09240 3200) on each rocker arm to prevent the hydraulic lash adjusters from falling out during servicing—3.0L (VIN P) engine

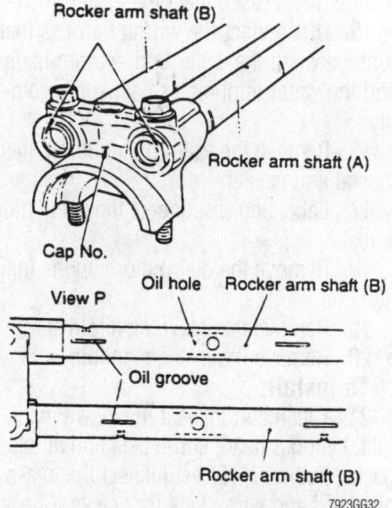

Insert bearing cap No. 1 so the notch on the end of the shaft faces in the direction shown. Ensure the oil groove faces downward and the oil port is located on the shaft "A" side—3.0L (VIN P) engine

➡Insert the rocker arm shaft into the front bearing cap so the notch on the end of the shaft is facing up. The No. 2, 3 and 4 bearing caps have roughly the same shape and are stamped with identification marks. When assembling the caps on the rocker shafts, be sure that they are installed in their original positions.

8. Install the rocker arm shaft assemblies on the cylinder head and tighten the bearing cap bolts 14–15 ft. lbs. (19–21 Nm) starting from the center and working out.

9. Remove the special holding tools from the auto lash adjusters.

10. Install the rocker cover using a new gasket.

11. Start the engine and allow it to reach normal operating temperature. Check for leaks.

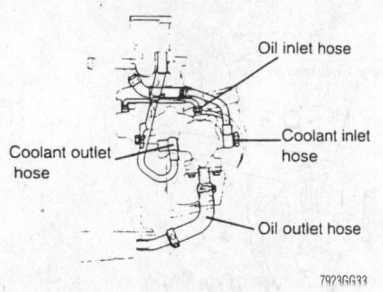

Turbocharger oil and coolant hose locations

Turbocharger

REMOVAL & INSTALLATION

1.5L (VIN E) Turbo Engine

1. Disconnect the negative battery cable.

2. Remove the turbocharger air intake pipe and the air intake hose.

3. Disconnect the water return and feed hoses.

4. Detach the oil feed and drain pipe connectors from the turbo housing.

5. Remove the turbocharger discharge pipe and bracket from the outlet (exhaust) side of the turbo.

6. Remove the turbo mounting bolts and remove the turbo.

To install:

7. Install the turbo in position on the manifold, using a new gasket. Tighten the turbo mounting bolts to 18–25 ft. lbs. (25–35 Nm).

8. Connect the discharge pipe and bracket to the turbo, tighten the bolts to 18–25 ft. lbs. (25–35 Nm).

9. Connect the oil lines to the turbo housing.

10. Connect the water return and feed hoses.

11. Connect the air intake pipe and hose.

12. Connect the negative battery cable.

13. Start the engine and allow it to reach operating temperature. Check for leaks.

Intake Manifold

REMOVAL & INSTALLATION

Except 3.0L (VIN T) Engine

1. Disconnect the negative battery cable.

2. Disconnect the air intake hose.

3. Disconnect the accelerator cable.

4. Drain the cooling system and disconnect the upper radiator hose.

5. Label and disconnect all wiring harnesses.

6. Remove the throttle body and gasket.

7. Disconnect the PCV hose from the rocker cover and disconnect the brake vacuum hoses.

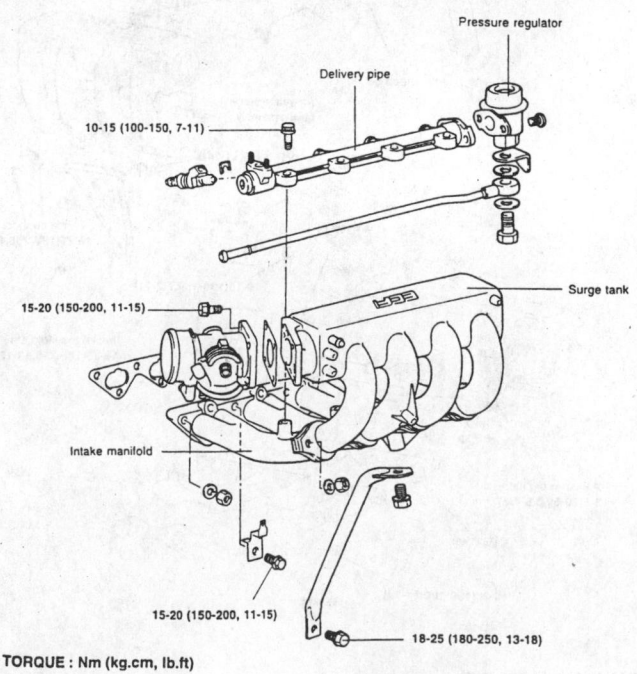

TORQUE : Nm (kg.cm, lb.ft)

Surge tank and intake manifold components—1.5L (VIN K) engine

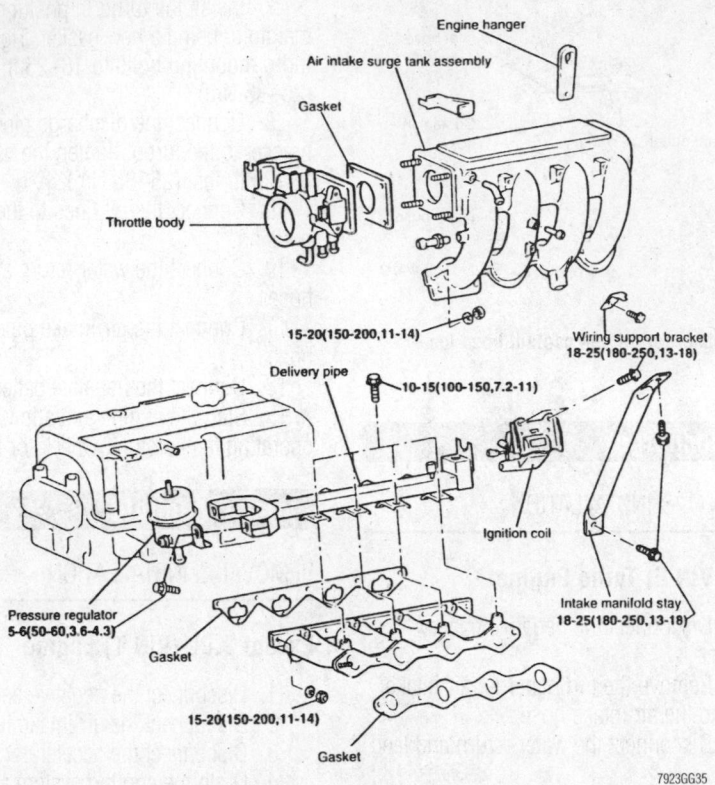

Surge tank and intake manifold components—1.5L (VIN E) engine

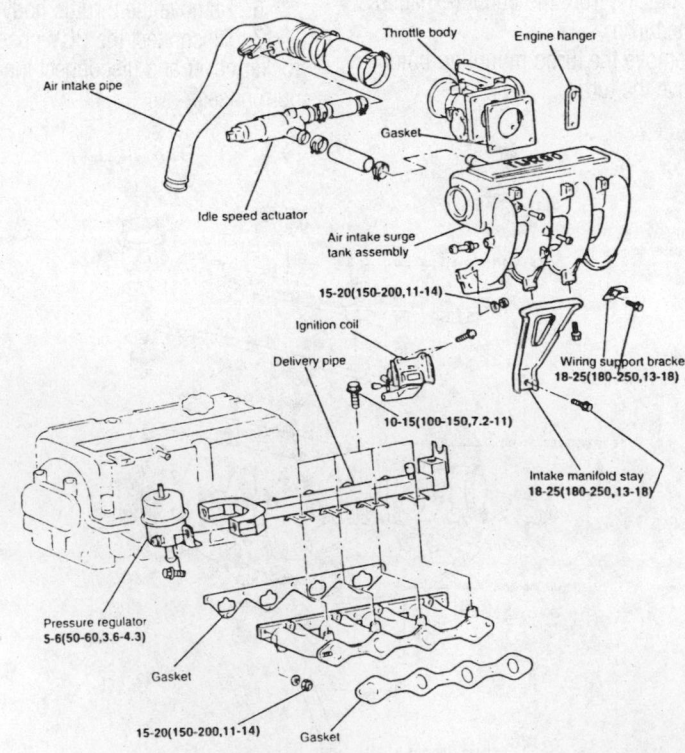

TORQUE : Nm (kg.cm, lb.ft)

7923GG36

Surge tank and intake manifold components—1.5L (VIN E) Turbo engine

8. Label and disconnect all vacuum hoses.

9. Relieve the fuel system pressure.

10. Disconnect the high pressure fuel hose connection from the fuel delivery pipe.

11. Remove the surge tank (and gasket) from the intake manifold.

12. Detach the fuel injector harness connectors.

13. Remove the fuel delivery pipe with the pressure regulator attached.

▶**Take care not to drop the injectors when removing the delivery pipe.**

14. Remove the insulator from the intake manifold and disconnect the heater hose.

15. Disconnect the wiring harness that runs between the water temperature gauge and the water temperature sensor assembly.

16. Remove the water outlet fitting, thermostat and gasket.

17. Label and disconnect the spark plug wires.

18. Remove the distributor and the ignition coil.

19. Remove the intake manifold stay.

20. Remove the intake manifold.

To install:

21. Clean and inspect the intake manifold, cylinder head, surge tank and all other gasket mating surfaces. Inspect the intake manifold and surge tank for cracks. Check the coolant passages for restrictions.

22. Install the intake manifold using a new gasket. Tighten intake manifold nuts to 11–14 ft. lbs. (15–20 Nm). starting from the center and working outwards.

23. Install the intake manifold stay and tighten bolts to 13–18 ft. lbs. (18–25 Nm).

24. Install the ignition coil, distributor, and spark plug wires.

25. Install the thermostat, gasket and water outlet housing. Tighten bolts to 12–14 ft. lbs. (17–20 Nm).

26. Connect the wiring harness that runs between the temperature sensor and temperature gauge.

27. Install the fuel delivery pipe onto the intake manifold. Torque the retaining bolts to 7–9 ft. lbs. (10–13 Nm).

28. Attach the fuel injector harness connectors.

29. Install the surge tank using a new gasket. Tighten bolts to 11–14 ft. lbs. (15–20 Nm).

30. Connect the high pressure hose to the fuel delivery pipe.

31. Connect the intake manifold, brake and PCV vacuum hoses.

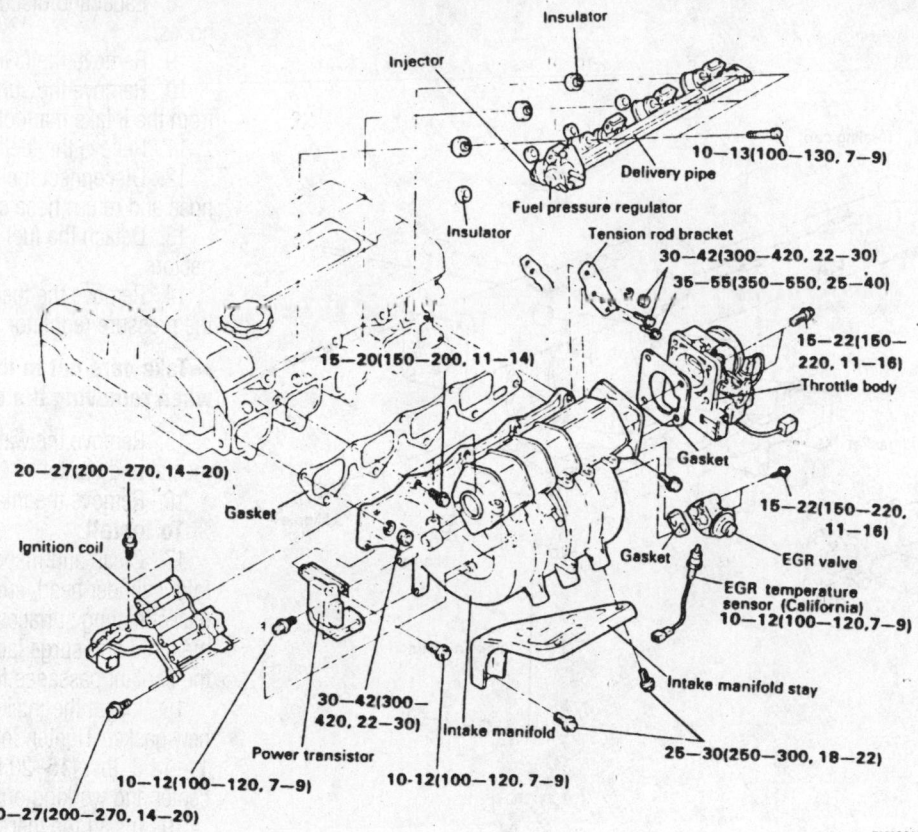

Insulator

Injector

10—13(100—130, 7—9)

Delivery pipe

Fuel pressure regulator

Insulator

Tension rod bracket

30—42(300—420, 22—30)

35—55(350—550, 25—40)

15—22(150—220, 11—16)

Throttle body

15—20(150—200, 11—14)

Gasket

20—27(200—270, 14—20)

Gasket

15—22(150—220, 11—16)

Ignition coil

EGR valve

Gasket

EGR temperature sensor (California)
10—12(100—120,7—9)

Intake manifold stay

30—42(300—420, 22—30)

Intake manifold

25—30(250—300, 18—22)

Power transistor

10—12(100—120, 7—9)

10—12(100—120, 7—9)

20—27(200—270, 14—20)

7923GG37

Surge tank and intake manifold components—1.6L (VIN R) and 1.8L (VIN M) engines

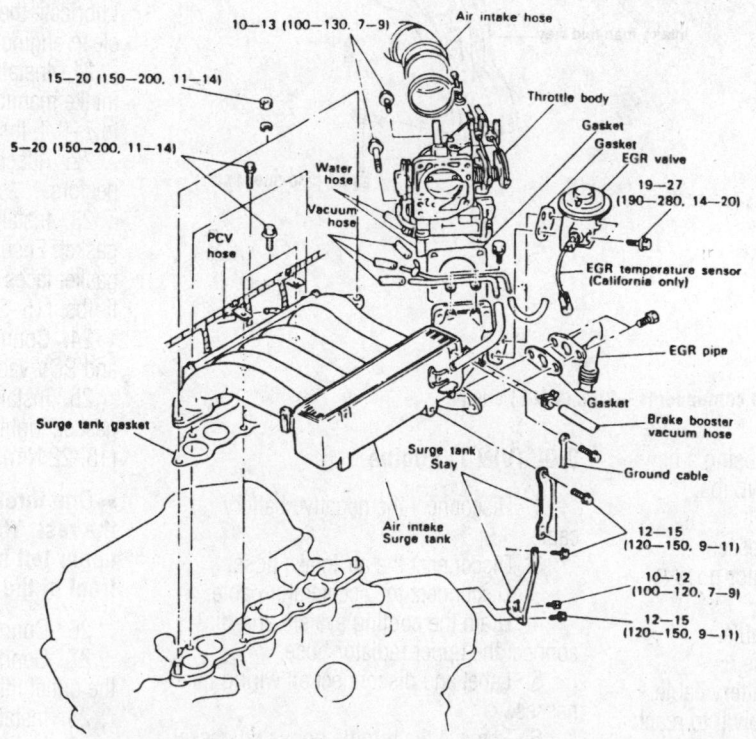

10—13 (100—130, 7—9)

Air intake hose

15—20 (150—200, 11—14)

Throttle body

5—20 (150—200, 11—14)

Gasket
Gasket

EGR valve

Water hose

19—27 (190—280, 14—20)

Vacuum hose

EGR temperature sensor (California only)

PCV hose

EGR pipe

Gasket

Surge tank gasket

Brake booster vacuum hose

Surge tank Stay

Ground cable

Air intake Surge tank

12—15 (120—150, 9—11)

10—12 (100—120, 7—9)

12—15 (120—150, 9—11)

TORQUE : Nm (kg.cm, lb.ft)

7923GG38

Surge tank and intake manifold components—2.0L (VIN P) engine

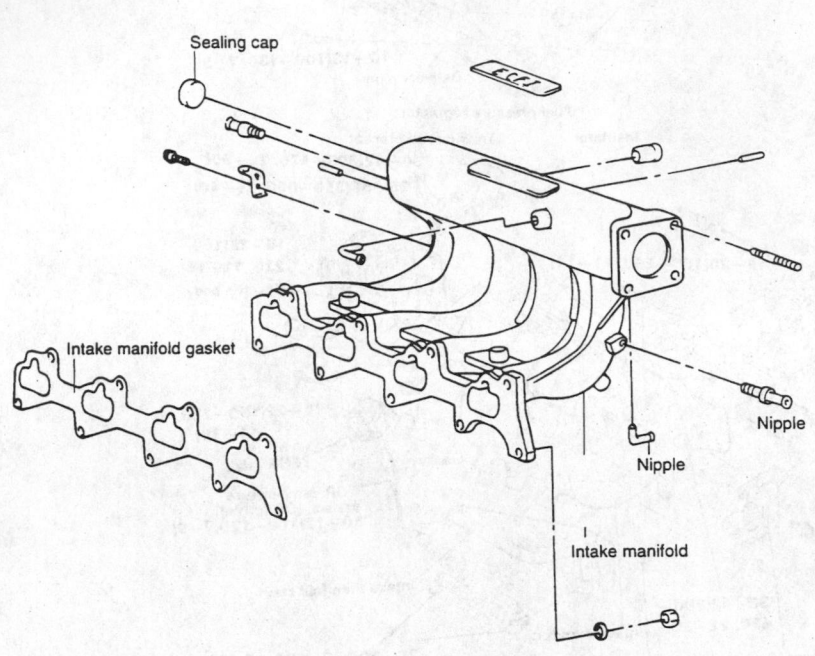

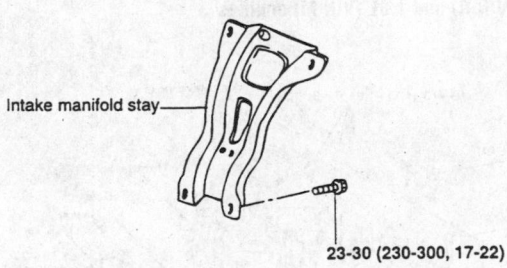

23-30 (230-300, 17-22)

TORQUE : Nm (kg.cm, lb.ft)

7923GG39

Surge tank and intake manifold components—2.0L (VIN F) engine

32. Install the throttle body using a new gasket. Tighten bolts to 11–16 ft. lbs. (15–22 Nm).

33. Connect all wiring harnesses.

34. Connect the upper radiator hose to the outlet fitting.

35. Install the accelerator cable.

36. Install the air intake hose.

37. Connect the negative battery cable.

38. Start the engine and allow it to reach operating temperature.

39. Check for fuel and coolant leaks.

40. Adjust ignition timing, idle speed and accelerator cable.

3.0L (VIN T) Engine

1. Disconnect the negative battery cable.

2. Disconnect the air intake hose.

3. Disconnect the accelerator cable.

4. Drain the cooling system and disconnect the upper radiator hose.

5. Label and disconnect all wiring harnesses.

6. Remove the throttle body and gasket.

7. Disconnect the PCV hose from the rocker cover and disconnect the brake vacuum hoses.

8. Label and disconnect all vacuum hoses.

9. Remove the EGR pipe and gasket.

10. Remove the surge tank (and gasket) from the intake manifold.

11. Relieve the fuel system pressure.

12. Disconnect the high pressure fuel hose and return hose connections.

13. Detach the fuel injector harness connectors.

14. Remove the fuel delivery pipe with the pressure regulator attached.

➡**Take care not to drop the injectors when removing the delivery pipe.**

15. Remove the water outlet fitting, thermostat and gasket.

16. Remove the intake manifold.

To install:

17. Clean and inspect the intake manifold, cylinder head, surge tank and all other gasket mating surfaces. Inspect the intake manifold and surge tank for cracks. Check the coolant passages for restrictions.

18. Install the intake manifold using a new gasket. Tighten intake manifold nuts to 11–14 ft. lbs. (15–20 Nm). starting from the center and working outwards.

19. Install the thermostat, gasket and water outlet housing. Tighten bolts to 12–14 ft. lbs. (17–20 Nm).

20. Be sure the injector holes are clean. Lubricate the injector O-rings with a drop of clean engine oil.

21. Install the fuel delivery pipe onto the intake manifold. Torque the retaining bolts to 7–9 ft. lbs. (10–13 Nm).

22. Attach the fuel injector harness connectors.

23. Install the surge tank using a new gasket. Ensure the printed surface of the gasket faces upward. Tighten bolts to 11–14 ft. lbs. (15–20 Nm).

24. Connect the intake manifold, brake and PCV vacuum hoses.

25. Install the throttle body using a new gasket. Tighten bolts to 11–16 ft. lbs. (15–22 Nm).

➡**One throttle body bolt is shorter than the rest. This bolt is installed in the upper left hole when viewed from the front of the throttle body.**

26. Connect all wiring harnesses.

27. Connect the upper radiator hose to the outlet fitting.

28. Install the accelerator cable.

29. Install the air intake hose.

30. Connect the negative battery cable.

31. Start the engine and allow it to reach operating temperature.

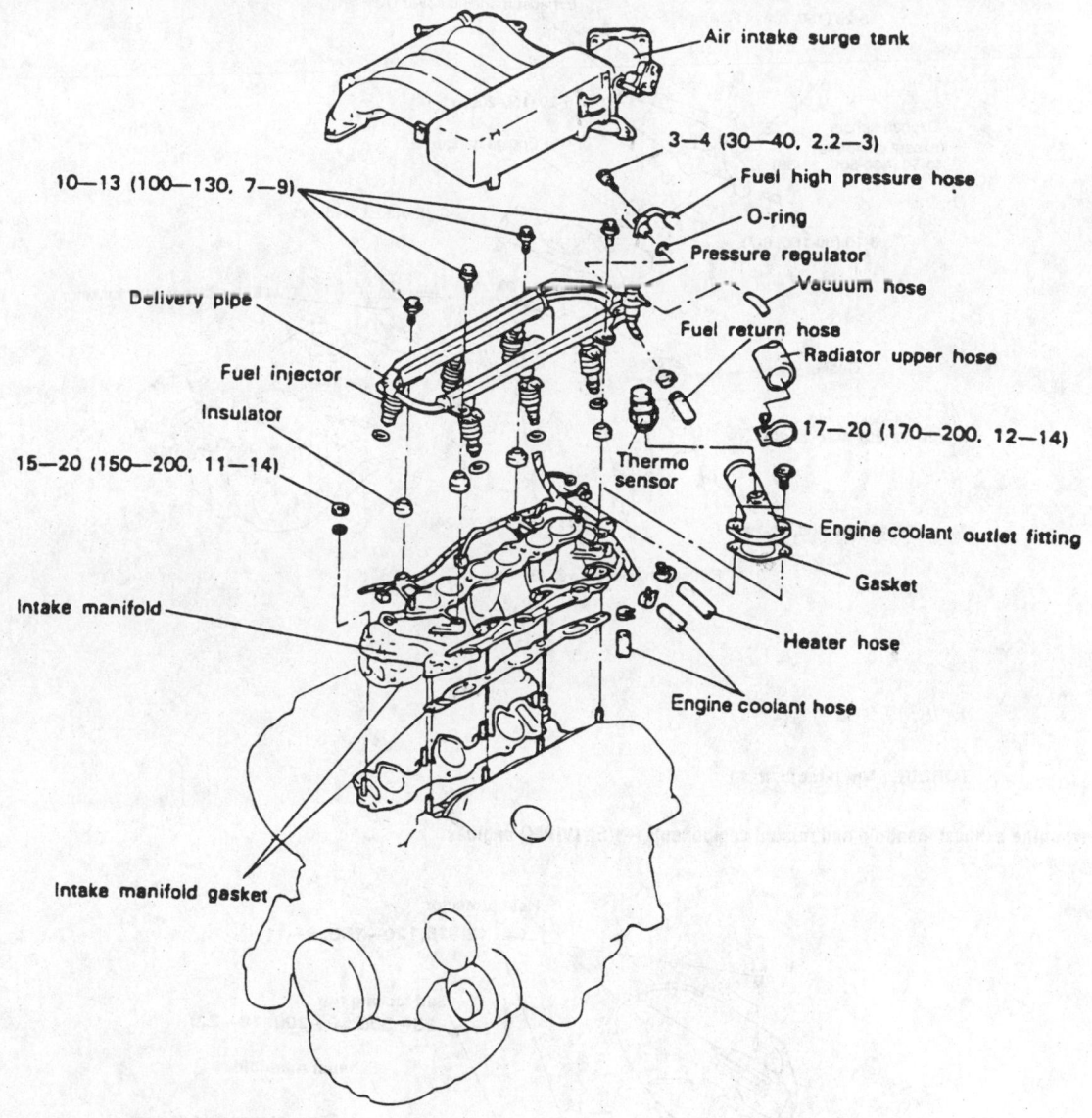

Air intake surge tank

10—13 (100—130, 7—9)

3—4 (30—40, 2.2—3)

Fuel high pressure hose

O-ring

Pressure regulator

Vacuum hose

Delivery pipe

Fuel return hose

Radiator upper hose

Fuel injector

17—20 (170—200, 12—14)

Insulator

Thermo sensor

15—20 (150—200, 11—14)

Engine coolant outlet fitting

Gasket

Intake manifold

Heater hose

Engine coolant hose

Intake manifold gasket

TORQUE : Nm (kg.cm, lb.ft)

7923GG40

Surge tank and intake manifold components—3.0L (VIN T) engine

32. Check for fuel and coolant leaks.
33. Adjust the accelerator and cruise control cables, as necessary.

Exhaust Manifold

Exhaust system fasteners are notorious for rusting which makes removing them without snapping or rounding an almost impossible task. Before working on any exhaust system component, identify which flanges, brackets, U-bolts, manifold, etc. have to be removed and inspect the mater-

ial condition of the fasteners. If necessary, wire brush any rusted fastener to remove loose rust particles, then spray the area with penetrating oil and allow it to soak overnight.

REMOVAL & INSTALLATION

Except 1.5L (VIN E) Turbo and 3.0L (VIN T) Engines

1. With the exhaust manifold cool, soak all nuts, studs and bolts with a liquid pene-

trant. Allow the penetrant to soak in overnight.
2. Disconnect the negative battery cable.
3. Label and disconnect the oxygen sensor electrical harness.
4. Remove the oxygen sensor.
5. Remove the heat shield on the exhaust manifold.
6. Disconnect the exhaust pipe at the exhaust manifold.
7. Support the exhaust manifold and remove all attaching nuts and washers.

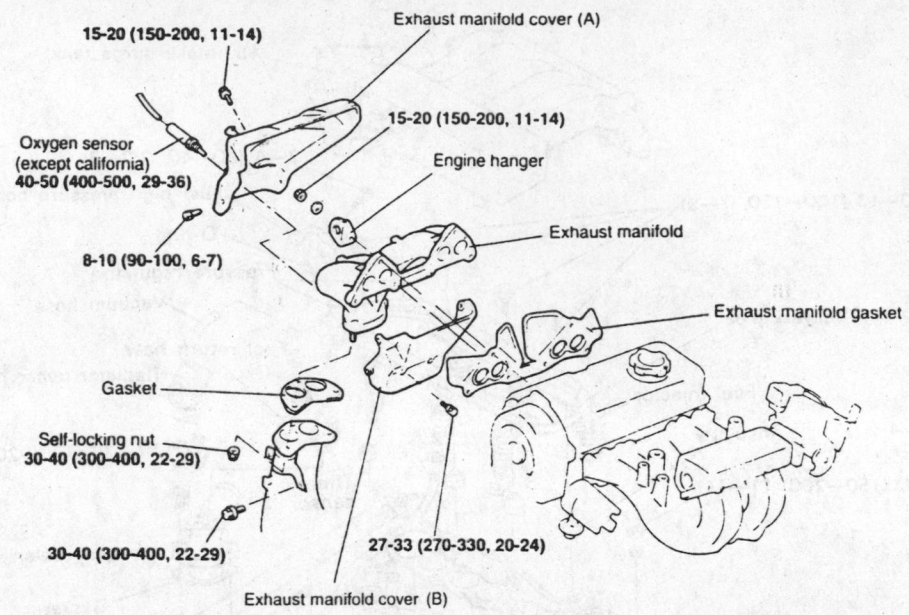

15-20 (150-200, 11-14)
Exhaust manifold cover (A)

15-20 (150-200, 11-14)
Engine hanger

Oxygen sensor
(except california)
40-50 (400-500, 29-36)

Exhaust manifold

Exhaust manifold gasket

8-10 (90-100, 6-7)

Gasket

Self-locking nut
30-40 (300-400, 22-29)

30-40 (300-400, 22-29)

27-33 (270-330, 20-24)

Exhaust manifold cover (B)

TORQUE : Nm (kg.cm, lb.ft)

7923GG41

Exploded view of the exhaust manifold and related components—1.5L (VIN K) engines

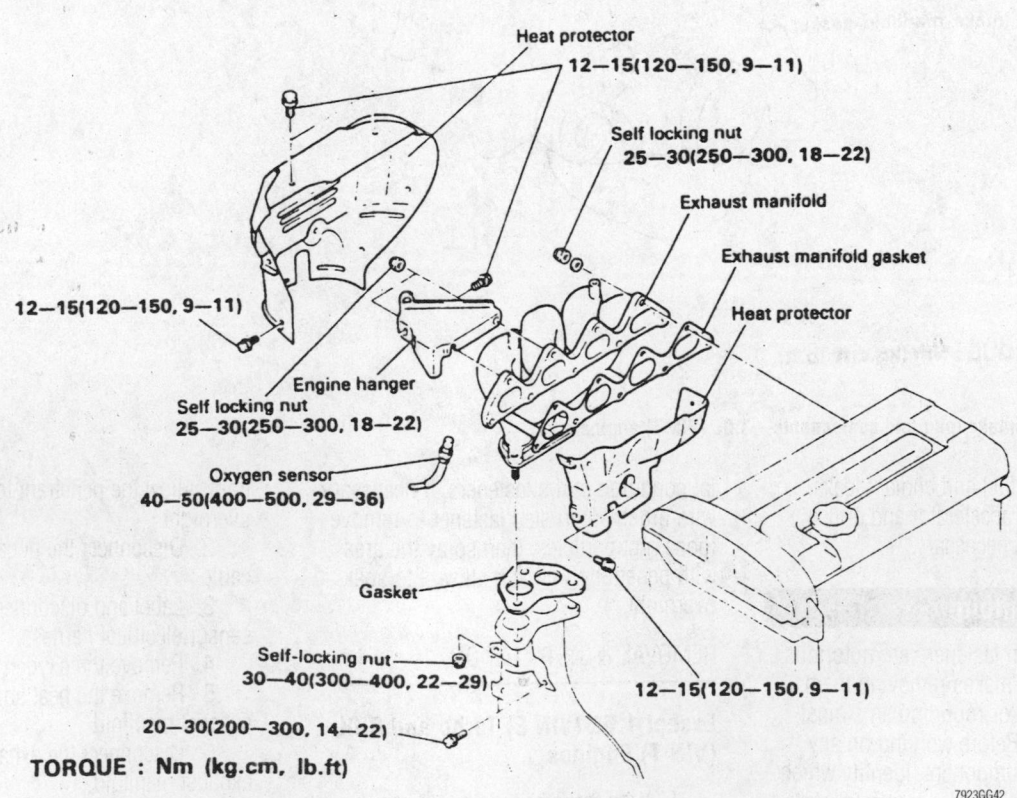

Heat protector
12—15(120—150, 9—11)

Self locking nut
25—30(250—300, 18—22)

Exhaust manifold

Exhaust manifold gasket

Heat protector

12—15(120—150, 9—11)

Engine hanger

Self locking nut
25—30(250—300, 18—22)

Oxygen sensor
40—50(400—500, 29—36)

Gasket

12—15(120—150, 9—11)

Self-locking nut
30—40(300—400, 22—29)

20—30(200—300, 14—22)

TORQUE : Nm (kg.cm, lb.ft)

7923GG42

Exploded view of the exhaust manifold components—1.6L (VIN R), 1.8L (VIN M), 2.0L (VIN P) engines

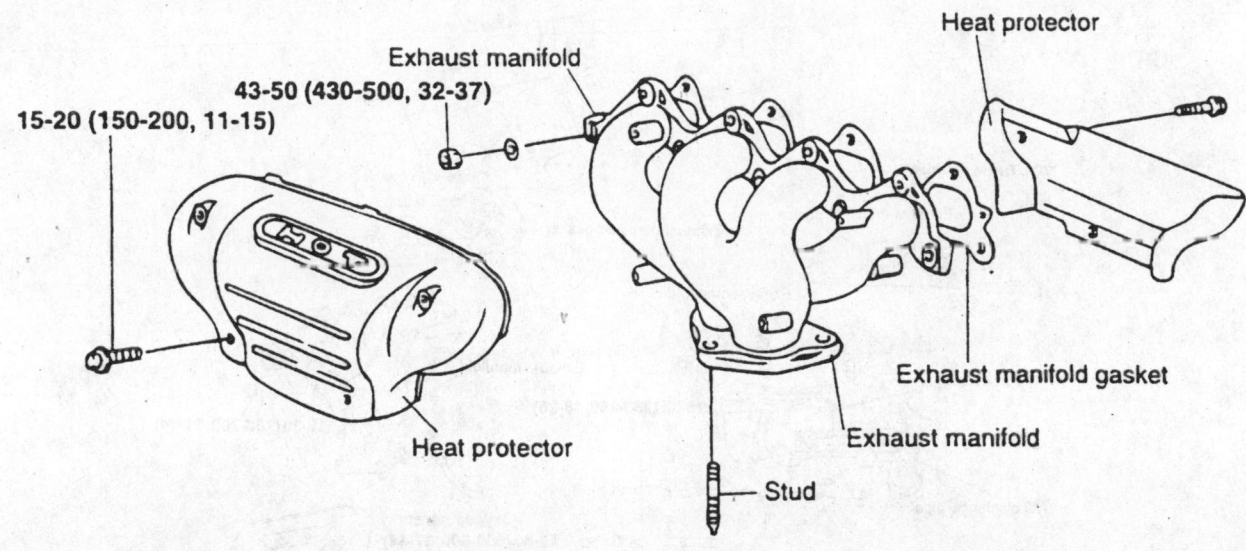

15-20 (150-200, 11-15)

43-50 (430-500, 32-37)

Exhaust manifold

Heat protector

Heat protector

Exhaust manifold gasket

Exhaust manifold

Stud

TORQUE: Nm (kg.cm, lb.ft)

7923GG43

Exploded view of the exhaust manifold components—2.0L (VIN F) engine

8. Remove the exhaust manifold and old gasket from the cylinder head.

To install:

9. Thoroughly clean the sealing surfaces on the cylinder head and manifold.

10. Ensure all the nuts and bolts turn freely, oiling them lightly, if necessary. Also, ensure all studs are properly installed in the cylinder head. Replace any nuts, washers or studs that are excessively rusted or may have been damaged during removal.

11. Use a straightedge to check the manifold sealing surfaces for flatness. If distortion is greater than 0.006 in. (0.15mm) machine the surface or replace the exhaust manifold.

12. Install new gaskets so all bolt holes and ports are aligned.

13. Place the manifold in position and install all washers and nuts hand tight.

14. Exhaust manifold nuts should be tightened alternately and in several stages. Tighten nuts as follows:

• 1.5L (VIN E) and 1.5L (VIN K): 11–15 ft. lbs. (15–20 Nm)

• 1.6L (VIN R), 1.8L (VIN M) and 2.0L (VIN P): 18–22 ft. lbs. (25–30 Nm)

• 1.6L (VIN R), 2.0L (VIN P) and 1995 1.8L (VIN M): 18–22 ft. lbs. (25–30 Nm)

• 2.0L (VIN F) and 1996–99 1.8L (VIN M): 32–41 ft. lbs. (43–50 Nm)

15. Connect exhaust pipe and tighten nuts to 22–29 ft. lbs. (30–40 Nm).

16. Install the heat shield and tighten bolts to 11–15 ft. lbs. (15–20 Nm).

17. Coat the oxygen sensor threads with anti-seize compound and install the oxygen sensor. Tighten to 29–36 ft. lbs. (30–40 Nm).

18. Connect the oxygen sensor electrical harness.

19. Connect the negative battery cable.

20. Start the engine and allow it to reach operating temperature. Check for leaks.

1.5L (VIN E) Turbo Engine

1. With the exhaust manifold cool, soak all nuts, studs and bolts with a liquid penetrant. Allow the penetrant to soak in overnight.

2. Disconnect the negative battery cable.

3. Label and disconnect the oxygen sensor electrical harness.

4. Remove the oxygen sensor.

5. Remove the heat shield on the exhaust manifold.

6. Remove the air intake pipe assembly.

7. Disconnect the exhaust pipe at the turbocharger discharge.

8. Support the exhaust manifold and remove all attaching nuts and washers.

9. Carefully remove the exhaust manifold (with turbocharger attached).

10. Remove the old gasket from the cylinder head.

To install:

11. Thoroughly clean the sealing surfaces on the cylinder head and manifold.

12. Ensure all the nuts and bolts turn freely, oiling them lightly, if necessary. Also, ensure all studs are properly installed in the cylinder head. Replace any nuts, washers or studs that are excessively rusted or may have been damaged during removal.

13. Use a straightedge to check the manifold sealing surfaces for flatness. If distortion is greater than 0.006 in. (0.15mm) machine the surface or replace the exhaust manifold.

14. Install new gaskets so all bolt holes and ports are aligned.

15. Place the manifold in position and install all washers and nuts hand tight.

16. Exhaust manifold nuts should be tightened alternately and in several stages. Tighten nuts to 18–25 ft. lbs. (25–35 Nm).

17. Connect exhaust pipe and tighten nuts to 22–29 ft. lbs. (30–40 Nm).

18. Install the air intake pipe assembly.

19. Install the heat shield and tighten bolts to 11–15 ft. lbs. (15–20 Nm).

20. Coat the oxygen sensor threads with

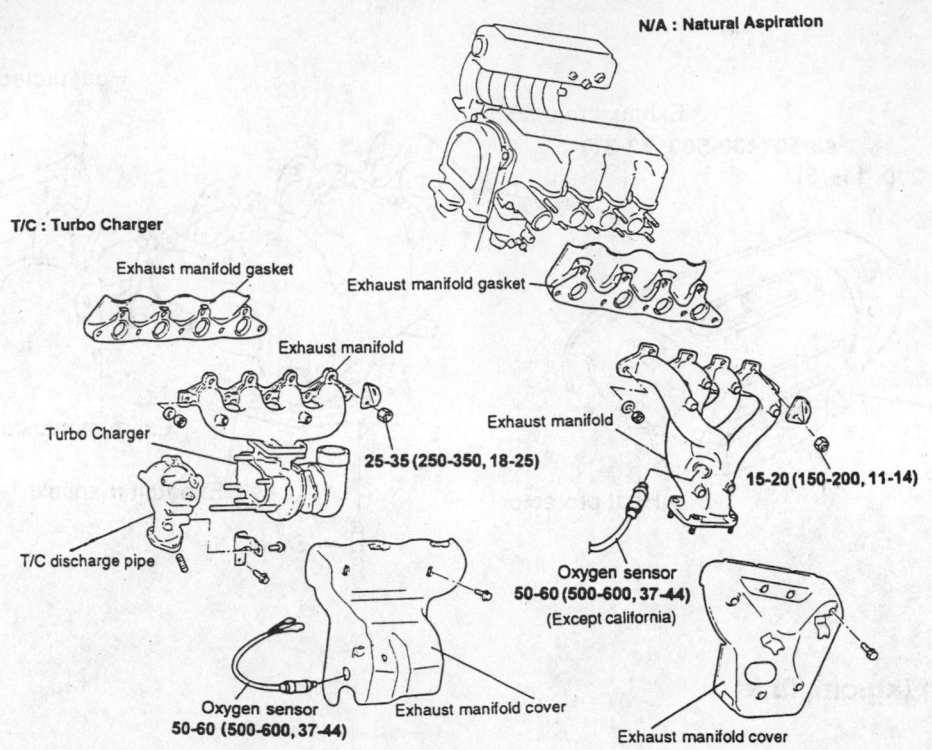

N/A : Natural Aspiration

T/C : Turbo Charger

Exhaust manifold gasket

Exhaust manifold gasket

Exhaust manifold

Turbo Charger

25-35 (250-350, 18-25)

Exhaust manifold

15-20 (150-200, 11-14)

T/C discharge pipe

Oxygen sensor
50-60 (500-600, 37-44)
(Except california)

Oxygen sensor
50-60 (500-600, 37-44)

Exhaust manifold cover

Exhaust manifold cover

TORQUE : Nm (kg.cm, lb.ft)

7923GG44

Exhaust manifold components—1.5L (VIN E) engine

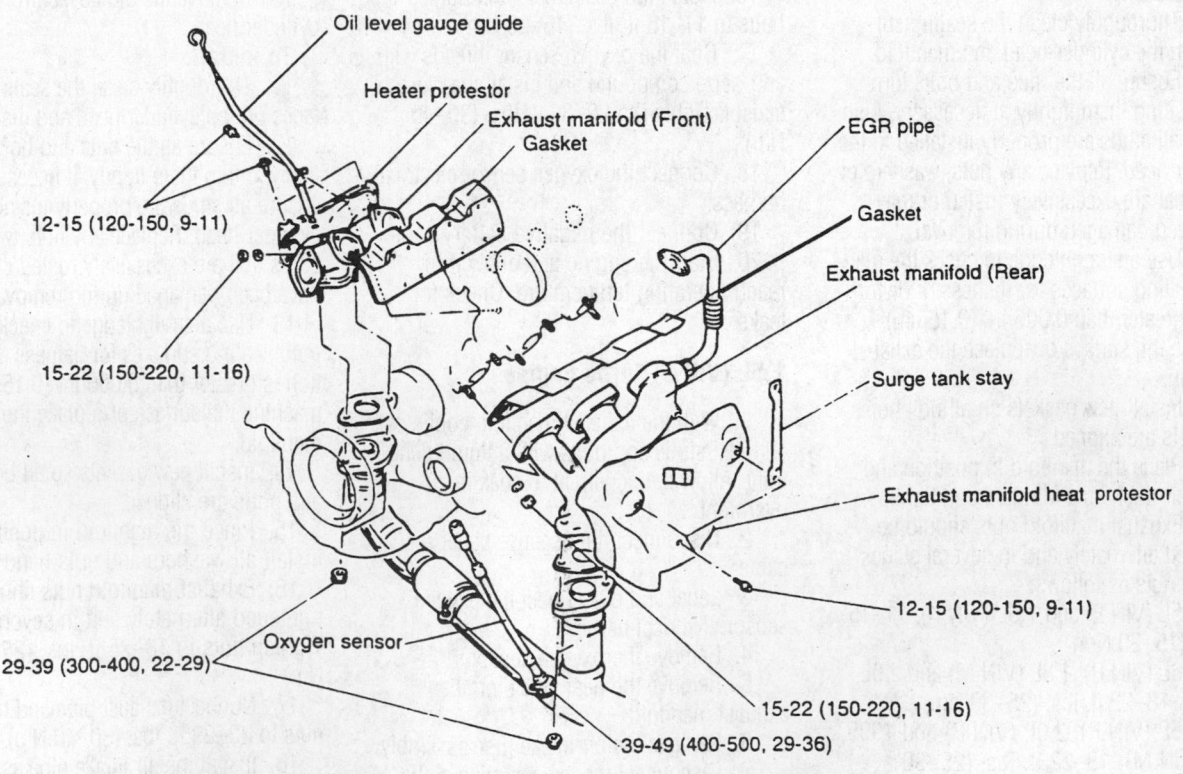

Oil level gauge guide

Heater protestor

Exhaust manifold (Front)
Gasket

EGR pipe

Gasket

12-15 (120-150, 9-11)

Exhaust manifold (Rear)

15-22 (150-220, 11-16)

Surge tank stay

Exhaust manifold heat protestor

12-15 (120-150, 9-11)

Oxygen sensor

29-39 (300-400, 22-29)

15-22 (150-220, 11-16)

39-49 (400-500, 29-36)

TORQUE : Nm (kg.cm, lb.ft)

7923GG45

Exhaust manifold components—3.0L (VIN T) engines

anti-seize compound and install the oxygen sensor. Tighten to 29–36 ft. lbs. (30–40 Nm).

21. Connect the oxygen sensor electrical harness.

22. Connect the negative battery cable.

23. Start the engine and allow it to reach operating temperature. Check for leaks.

3.0L (VIN T) Engine

REAR MANIFOLD

1. Disconnect the negative battery cable.

2. Disconnect the crossover pipe at the exhaust manifold.

3. Disconnect the EGR tube.

4. Support the exhaust manifold and remove all attaching nuts and washers.

5. Carefully remove the exhaust manifold.

6. Remove the old gasket from the cylinder head.

To install:

7. Thoroughly clean the sealing surfaces on the cylinder head and manifold.

8. Ensure all the nuts and bolts turn freely, oiling them lightly, if necessary. Also, ensure all studs are properly installed in the cylinder head. Replace any nuts, washers or studs that are excessively rusted or may have been damaged during removal.

9. Use a straightedge to check the manifold sealing surfaces for flatness. If distortion is greater than 0.006 in. (0.15mm) machine the surface or replace the exhaust manifold.

10. Install new gaskets so all bolt holes and ports are aligned.

➡️ **When installing, the numbers 1–3–5 on the gaskets are used with the rear cylinders and 2–4–6 are on the gasket for the front cylinders.**

11. Place the manifold in position and install all washers and nuts hand tight.

12. Exhaust manifold nuts should be tightened alternately and in several stages. Tighten nuts to 11–16 ft. lbs. (15–22 Nm).

13. Connect the EGR tube.

14. Connect the crossover pipe and tighten nuts to 22–29 ft. lbs. (30–40 Nm).

15. Connect the negative battery cable.

16. Start the engine and allow it to reach operating temperature. Check for leaks.

FRONT MANIFOLD

1. Disconnect the negative battery cable.

2. Disconnect the crossover pipe at the exhaust manifold.

3. Remove the oil level dipstick.

4. Support the exhaust manifold and remove all attaching nuts and washers.

5. Carefully remove the exhaust manifold.

6. Remove the old gasket from the cylinder head.

To install:

7. Thoroughly clean the sealing surfaces on the cylinder head and manifold.

8. Ensure all the nuts and bolts turn freely, oiling them lightly, if necessary. Also, ensure all studs are properly installed in the cylinder head. Replace any nuts, washers or studs that are excessively rusted or may have been damaged during removal.

9. Use a straightedge to check the manifold sealing surfaces for flatness. If distortion is greater than 0.006 in. (0.15mm) machine the surface or replace the exhaust manifold.

10. Install new gaskets so all bolt holes and ports are aligned.

➡️ **When installing, the numbers 1–3–5 on the gaskets are used with the rear cylinders and 2–4–6 are on the gasket for the front cylinders.**

11. Place the manifold in position and install all washers and nuts hand tight.

12. Exhaust manifold nuts should be

tightened alternately and in several stages. Tighten nuts to 11–16 ft. lbs. (15–22 Nm).

13. Install the oil level dipstick.

14. Connect the crossover pipe and tighten nuts to 22–29 ft. lbs. (30–40 Nm).

15. Connect the negative battery cable.

16. Start the engine and allow it to reach operating temperature. Check for leaks.

Front Crankshaft Seal

The front crankshaft seal replacement is covered under the oil pump removal and installation.

Camshaft and Valve Lifters

REMOVAL & INSTALLATION

1.5L (VIN K) SOHC and 1.5L (VIN E) Engines

1. Disconnect the negative battery cable.

2. Remove the timing belt cover.

3. Loosen the 2 bolts and move the timing belt tensioner toward the water pump as far as it will go, then retighten the timing belt tensioner adjusting bolt.

4. Remove the timing belt from the camshaft sprocket

➡️ **The timing belt may be left engaged with the crankshaft sprocket and tensioner.**

5. Remove the camshaft sprocket.

6. Remove the rocker cover.

7. Remove the rocker shaft assembly.

8. Remove the cylinder head rear cover (distributorless ignition) or distributor (distributor ignition).

➡️ **On distributorless ignition equipped engines, a cylinder head rear cover is used to block off the space formerly occupied by the distributor.**

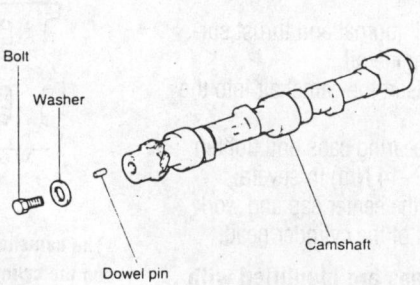

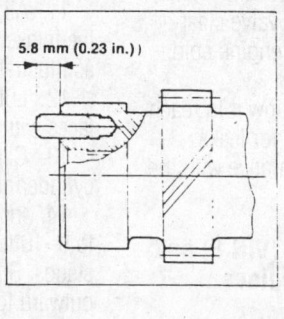

7923GG46

The camshaft sprocket pin should protrude from the camshaft as illustrated—1.5L (VIN K) SOHC and 1.5L (VIN E) engines

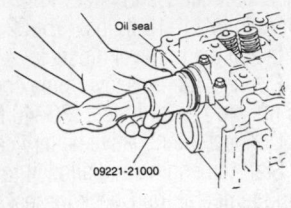

7923GG47

Use special tool (PN 09221–21000) or equivalent, drive a new front camshaft oil seal into the clearance between the cam and head, making sure the seal seats fully—1.5L (VIN K) SOHC and 1.5L (VIN E) engines

9. Remove the camshaft thrust case tightening bolt.

10. Carefully, slide the camshaft out of the head, being careful that the cam lobes do not strike the bearing bores in the head.

To install:

11. Lubricate all journal and thrust surfaces with clean engine oil.

12. Carefully insert the camshaft into the engine. Be sure the camshaft goes in with the threaded hole in the top of the thrust case straight upward.

13. Align the bolt hole in the thrust case and the cylinder head surface.

14. Install the thrust case bolt and tighten securely.

15. Install the cylinder head rear cover (distributorless ignition) with a new gasket. Tighten bolts to 6–7 ft. lbs. (8–10 Nm). Install the distributor (distributor ignition).

16. Coat the external surface of the front oil seal with engine oil.

17. Using special installer tool MD 998306–01 or equivalent, drive a new front camshaft oil seal into the clearance between the cam and head, making sure the seal seats fully.

18. Install the rocker shaft assembly.

19. Install the camshaft sprocket and torque the bolt to 47–54 ft. lbs. (64–74 Nm).

20. Install and tension the timing belt.

21. Install the timing belt covers

22. Temporarily adjust the valve clearance to specification with the engine cold.

23. Install the rocker cover.

24. Start the engine and allow it to reach operating temperature. Check for leaks.

25. Readjust the valve clearance with the engine warm.

1.5L (VIN K) DOHC, 2.0L (VIN F) and 1996–99 1.8L (VIN M) Engines

1. Disconnect the negative battery cable.
2. Remove the timing belt cover.
3. Remove the timing belt tensioner and idler.

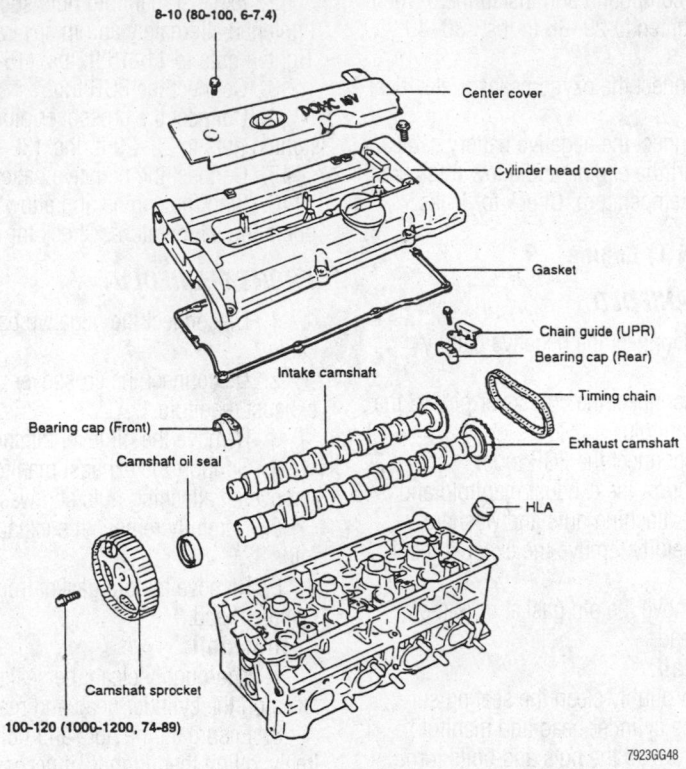

8-10 (80-100, 6-7.4)

100-120 (1000-1200, 74-89)

7923GG48

Camshaft assembly components—1.5L (VIN K) DOHC, 2.0L (VIN F) and 1996–99 1.8L (VIN M) engines

4. Remove the timing belt from the camshaft sprocket.

➡**The timing belt may be left engaged with the crankshaft sprocket and tensioner.**

5. Remove the camshaft sprocket.
6. Remove the rocker cover.
7. Remove the camshaft bearing caps and timing chain.
8. Remove the camshaft from the cylinder head, being careful that the cam lobes do not strike the bearing bores in the head.
9. Remove the hydraulic lash adjusters.

To install:

10. Install the hydraulic lash adjusters.
11. Align the camshaft timing chain with the intake and exhaust camshaft sprockets as illustrated
12. Lubricate all journal and thrust surfaces with clean engine oil.
13. Carefully insert the camshaft into the cylinder head.
14. Install the bearing caps and tighten to 9–10 ft. lbs. (12–14 Nm) in several stages. Start from the center cap and work outward to the end of the cylinder head.

➡**The bearing caps are identified with a letter and number stamp. The letter indicates either intake or exhaust and**

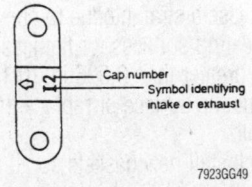

7923GG49

The camshaft bearing caps are identified with a letter and number stamp. The letter indicates either intake or exhaust and the number is sequential from the cylinder head end opposite the timing chain—1.5L (VIN K) DOHC, 2.0L (VIN F) and 1996–99 1.8L (VIN M) engines

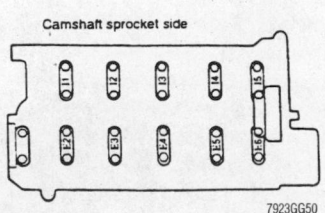

7923GG50

The camshaft bearing caps are arranged on the cylinder head as illustrated—1.5L (VIN K) DOHC, 2.0L (VIN F) and 1996–99 1.8L (VIN M) engines

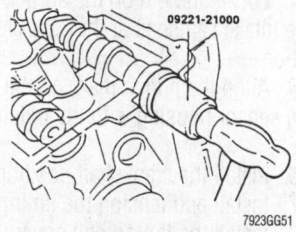

7923GG51

Using special tool (PN 09221–21000) or equivalent, install a new oil seal—1.5L (VIN K) DOHC, 2.0L (VIN F) and 1996–99 1.8L (VIN M) engines

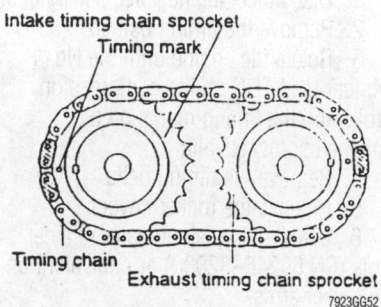

Intake timing chain sprocket
Timing mark
Timing chain
Exhaust timing chain sprocket

7923GG52

Align the timing chain and camshaft sprockets as illustrated—1.5L (VIN K) DOHC engine

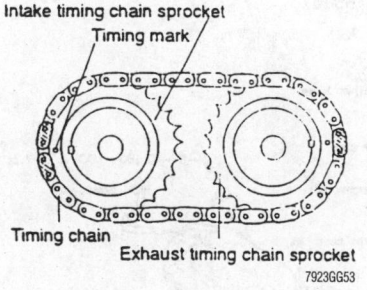

Intake timing chain sprocket
Timing mark
Timing chain
Exhaust timing chain sprocket

7923GG53

Align the timing chain and camshaft sprockets as illustrated—2.0L (VIN F) and 1996–99 1.8L (VIN M) engines

the number is sequential from the cylinder head end opposite the timing chain.

15. Using special tool (PN 09221–21000) or equivalent, install a new oil seal.

16. Install the camshaft sprocket and torque the bolt to 60–74 ft. lbs. (80–100 Nm).

17. Install and tension the timing belt.

18. Install the timing belt covers

19. Install the rocker cover.

20. Start the engine and allow it to reach operating temperature. Check for leaks.

1.6L (VIN R), 1.8L (VIN M) and 2.0L (VIN P) Engines

1. Disconnect battery negative cable.
2. Remove the timing belt cover.
3. Remove the rocker cover.
4. Remove the crankshaft position sensor.
5. Loosen the bearing cap bolts in 2–3 steps.
6. Label and remove all camshaft bearing caps.

➡ **If the bearing caps are difficult to remove, use a plastic hammer to gently tap the rear part of the camshaft.**

7. Remove the intake and exhaust camshafts.

➡ **The hydraulic lash adjusters can be removed without disassembling the cylinder head by using special tool (PN 09246–34000), or equivalent.**

8. Remove the rocker arms and lash adjusters.

To install:

9. Lubricate the components with heavy engine oil.

➡ **Do not confuse the intake camshaft with the exhaust camshaft. The intake camshaft has a split on its rear end for driving the crank angle sensor.**

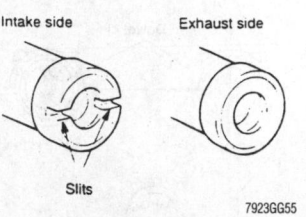

Intake side   Exhaust side
Slits

7923GG55

The intake camshaft can be identified by the slits on its rear end. The exhaust camshaft does not have these slits—1.6L (VIN R), 2.0L (VIN P) and 1995 1.8L (VIN M) engines

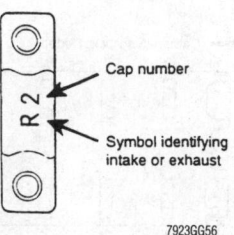

Cap number
Symbol identifying intake or exhaust

7923GG56

The camshaft bearing caps are identified by a letter indicating intake or exhaust and a number—1.6L (VIN R), 2.0L (VIN P) and 1995 1.8L (VIN M) engines

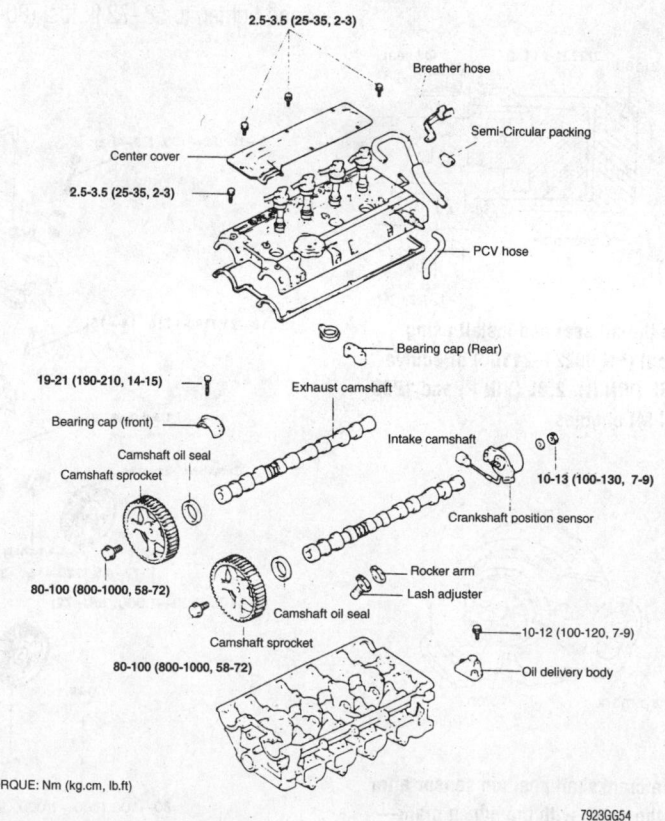

2.5-3.5 (25-35, 2-3)
Breather hose
Center cover
Semi-Circular packing
2.5-3.5 (25-35, 2-3)
PCV hose
Bearing cap (Rear)
19-21 (190-210, 14-15)
Exhaust camshaft
Bearing cap (front)
Camshaft oil seal
Intake camshaft
Camshaft sprocket
10-13 (100-130, 7-9)
Crankshaft position sensor
80-100 (800-1000, 58-72)
Rocker arm
Lash adjuster
Camshaft oil seal
10-12 (100-120, 7-9)
Camshaft sprocket
80-100 (800-1000, 58-72)
Oil delivery body

TORQUE: Nm (kg.cm, lb.ft)

7923GG54

Exploded view camshaft and rocker arm assembly components—1.6L (VIN R), 2.0L (VIN P) and 1995 1.8L (VIN M) engines

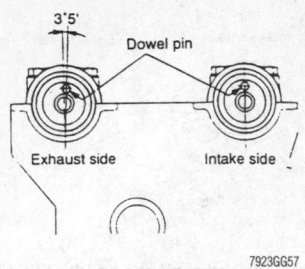

Ensure the camshaft dowel pins are located at the 12 o'clock position—1.6L (VIN R), 2.0L (VIN P) and 1995 1.8L (VIN M) engines

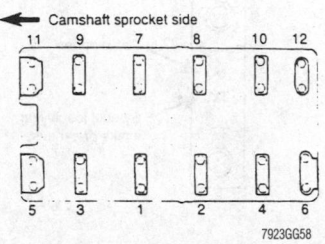

Tighten the camshaft bearing caps in the sequence specified—1.6L (VIN R), 2.0L (VIN P) and 1995 1.8L (VIN M) engines

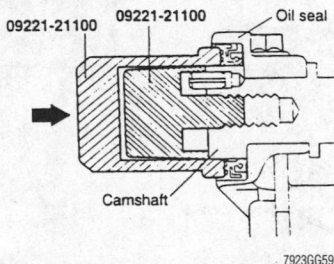

Lubricate the oil seal and install using special tool (PN 09221–21100) or equivalent—1.6L (VIN R), 2.0L (VIN P) and 1995 1.8L (VIN M) engines

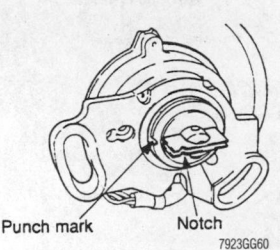

Install the crankshaft position sensor after aligning the notch with the punch mark—1.6L (VIN R), 2.0L (VIN P) and 1995 1.8L (VIN M) engines

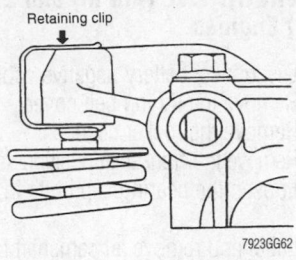

Before removing the rocker arm assemblies, use the lash adjuster retaining clips (PN 09246–32000) or equivalent to retain the lash adjusters—3.0L (VIN T) engine

10. Install the camshafts.

11. Install the bearing caps. Ensure the rocker arm is correctly mounted on the lash adjuster and the valve stem end. Tighten in sequence using 2 or 3 steps to 14–15 ft. lbs. (19–21 Nm).

➡ **Number 2 and 5 caps are of the same shape. Check the markings on the caps to identify the cap number and intake/exhaust symbol. Only L (intake) or R (exhaust) is stamped on No. 1 bearing cap.**

12. Lubricate the oil seal and install using special tool (PN 09221–21100) or equivalent. Install the camshaft sprockets and tighten to 58–72 ft. lbs. (80–100 Nm).

13. Install the rocker covers.

14. Locate the pin on the sprocket side of the intake camshaft at the 12 o'clock position.

15. Align the punch mark on the crank angle sensor housing with the notch in the plate.

16. Install the crankshaft position sensor.

17. Install and tension the timing belt.

18. Install the timing belt covers.

19. Connect the negative battery cable.

20. Start the engine and allow it to reach normal operating temperature. Check for leaks.

3.0L (VIN T) Engine

1. Disconnect the negative battery cable.

2. Remove the timing belt covers.

3. Rotate the engine until the No. 1 cylinder is at TDC on the compression stroke and the timing marks on the camshaft sprockets align.

4. Remove the timing belts.

5. Remove the rocker covers.

6. Install auto lash adjuster retainer tools (PN 09246–32000) or equivalent on the rocker arms.

7. If removing the right side (front) camshaft, remove the distributor extension.

8. Remove the rocker arm and shaft assembly.

9. Remove the camshaft from the cylinder head.

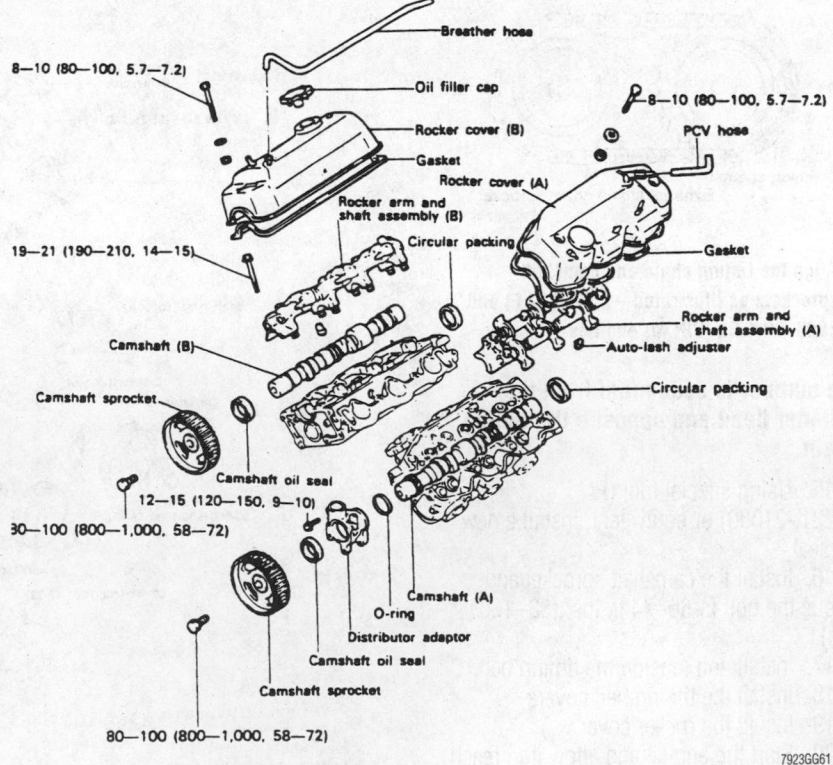

Exploded view of the camshaft and rocker arm assembly components—3.0L (VIN T) engine

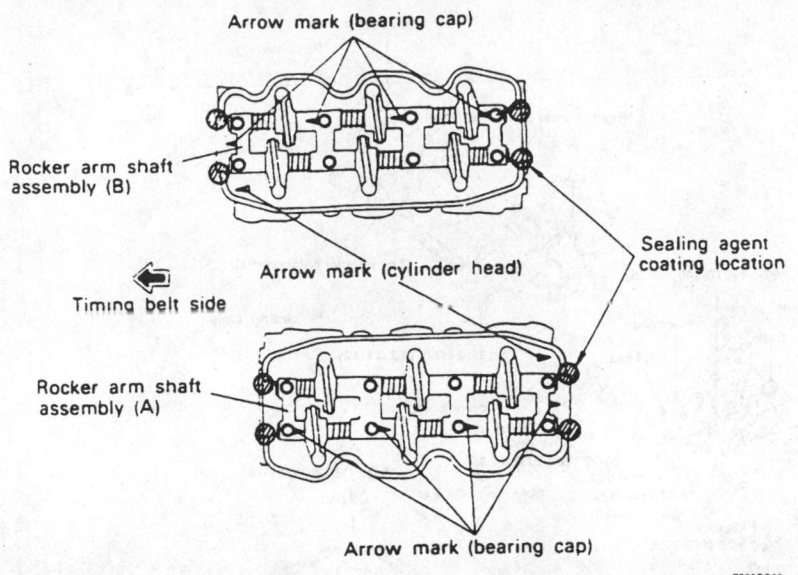

Install the rocker arm assemblies with the arrows as indicated. Place sealer at the corners—3.0L (VIN T) engine

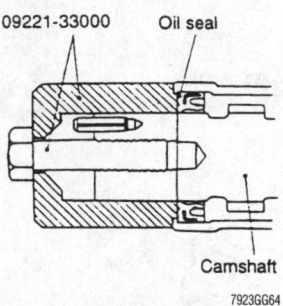

Lubricate the oil seal and install using special tool (PN 09221–33000) or equivalent—3.0L (VIN T) engine

To install:

10. Lubricate the camshaft journals and camshaft with clean engine oil and install the camshaft in the cylinder head.

11. Align the rocker arm and shaft assemblies. Apply sealer (Threebond No. 1324 or equivalent) at the ends of the bearing caps and install the assembly. Tighten bolts to 14–15 ft. lbs. (19–21 Nm).

12. Lubricate the oil seal and install using special tool (PN 09221–33000) or equivalent.

13. Install the distributor extension, if removed.

14. Remove auto lash adjuster retainer tools.

15. Install the rocker covers.

16. Install and tension the timing belts.

17. Install the timing belt covers.

18. Connect the negative battery cable.

19. Start the engine and allow it to reach operating temperature. Check for leaks.

Valve Lash

All engines, except the 1.5L (VIN E), use hydraulic valve lash adjusters. No periodic valve lash adjustments are necessary or possible on these engines. If the engine is determined to have a valve tap, a complete inspection of the valvetrain must be made to determine the faulty components.

The 1.5L (VIN E) engines use manual valve lifters which require periodic adjustment.

ADJUSTMENT

1.5L (VIN J) and 1.5L (VIN E) Engines

1. Run the engine until it reaches normal operating temperature, then turn it OFF.

2. Remove the spark plug wires from their clips on the rocker cover.

3. Remove the rocker cover bolts and carefully lift the rocker cover off the cylinder head.

4. Using a torque wrench, ensure all cylinder head bolts are all tightened to specification.

5. Remove the spark plugs to make it easier to turn the engine manually.

6. Turn the crankshaft pulley to bring the No. 1 piston to TDC of the compression stroke.

7. Adjust the valves marked "A" in the illustration, using a feeler gauge.

8. Valve clearance should be as follows:

   a. 1.5L (VIN E) engine—0.010 in. (0.25mm) Intake, 0.012 in. (0.30mm) Exhaust

9. A feeler gauge of the proper size

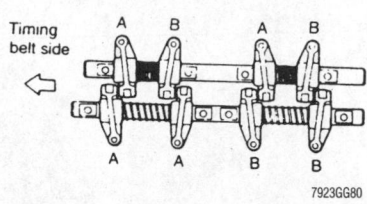

Adjust the valves marked "A" when the No. 1 cylinder is at TDC. Adjust the valves marked "B" when the No. 4 cylinder is at TDC

should fit between the rocker arm and the tip of the valve with a slight drag. If the clearance is not correct, loosen the locknut and turn the adjusting screw to obtain the proper clearance. Tighten the locknut securely.

10. Turn the crankshaft pulley to bring the No. 4 piston to TDC of the compression stroke.

11. Adjust the valves marked "B" in the illustration, using a feeler gauge.

12. Using a new gasket and the proper adhesive, install the rocker cover and tighten bolts to 48–60 inch lbs. (5.4–6.8 Nm)

13. Start the engine and check idle speed. Readjust as necessary.

Oil Pan

REMOVAL & INSTALLATION

1. Disconnect the negative battery cable.

2. Raise the vehicle and support it safely.

3. Remove the underbody splash shield.

4. Drain the engine oil.

5. Remove the oil pan bolts and slide the oil pan out from under the vehicle.

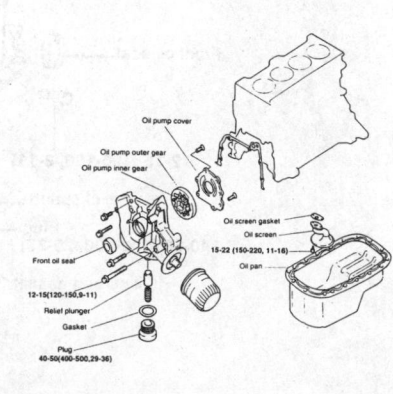

Exploded view of oil pump and pan—1.5L (VIN E) engine

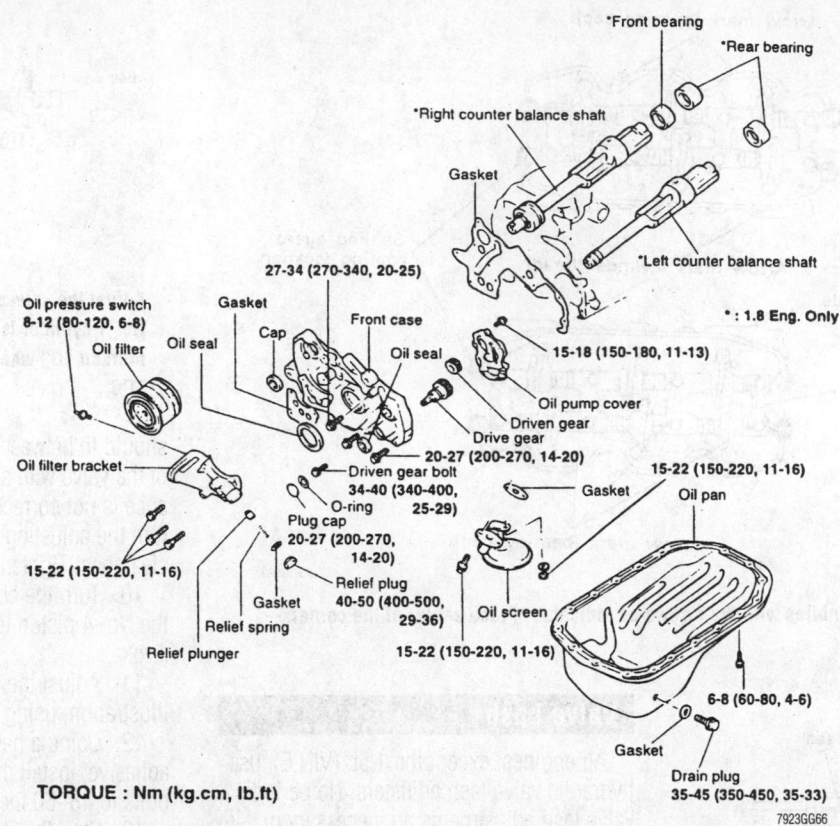

TORQUE : Nm (kg.cm, lb.ft)

7923GG66

Exploded view of oil pump and pan—2.0L (VIN P), 1995 1.6L (VIN R) and 1.8L (VIN M) engine

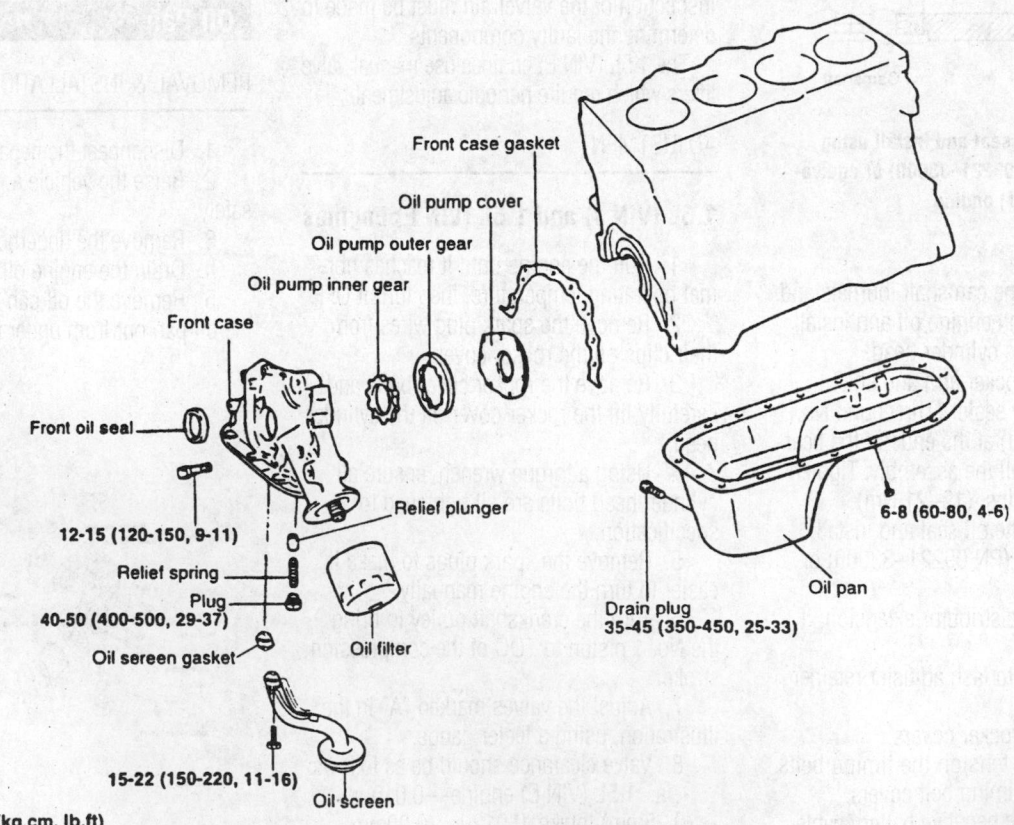

TORQUE : Nm (kg.cm, lb.ft)

7923GG67

Exploded view of oil pump and pan—1.5L (VIN K), 2.0L (VIN F) and 1996–99 1.8L (VIN M) engine

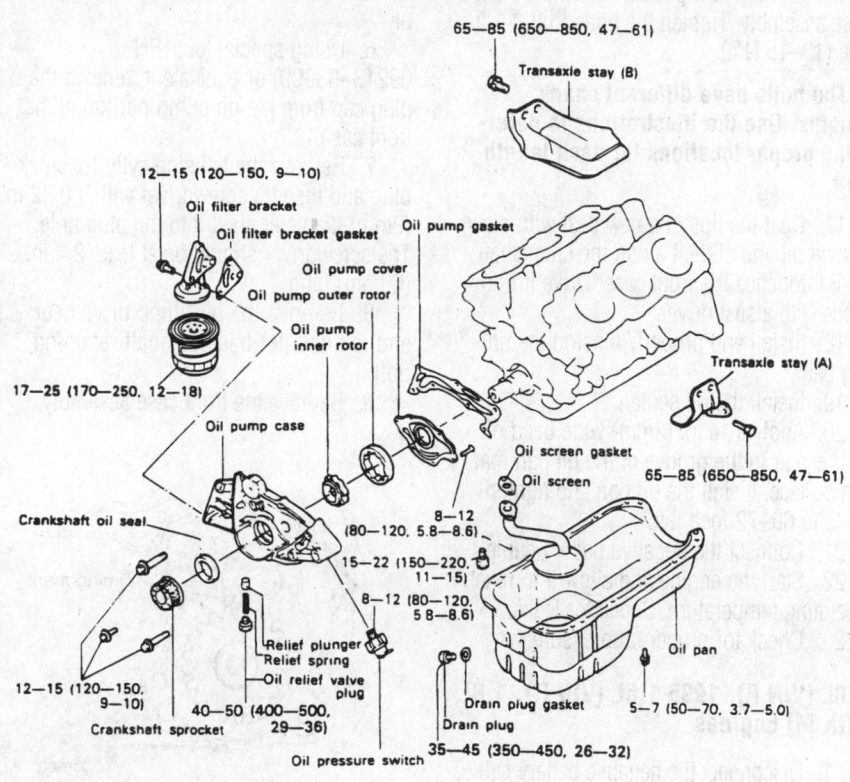

65—85 (650—850, 47—61)
Transaxle stay (B)

12—15 (120—150, 9—10)
Oil filter bracket
Oil filter bracket gasket
Oil pump cover
Oil pump outer rotor
Oil pump inner rotor
Oil pump gasket

17—25 (170—250, 12—18)
Oil pump case

Transaxle stay (A)
65—85 (650—850, 47—61)

Crankshaft oil seal
Oil screen gasket
Oil screen

8—12 (80—120, 5.8—8.6)
15—22 (150—220, 11—15)
8—12 (80—120, 5.8—8.6)

Oil pan

12—15 (120—150, 9—10)
Relief plunger
Relief spring
Oil relief valve plug
Drain plug gasket
Drain plug
Oil pressure switch
5—7 (50—70, 3.7—5.0)

Crankshaft sprocket
40—50 (400—500, 29—36)
35—45 (350—450, 26—32)

TORQUE : Nm (kg.cm, lb.ft)

7923GG68

Exploded view of oil pump and pan—3.0L (VIN T) engine

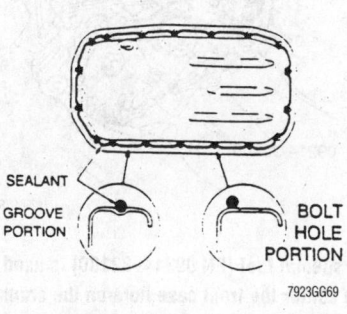

SEALANT
GROOVE PORTION
BOLT HOLE PORTION

7923GG69

Oil pan sealant applications points—except 3.0L (VIN T) engine

To install:

6. Clean the mating surfaces of the oil pan and the engine block.

7. Apply a ⅛ in. (3mm) bead of RTV sealer along the groove in the oil pan.

8. Install the oil pan and tighten bolts to 4–6 ft. lbs. (6–8 Nm) on all engines except 1.5L (VIN E) and 11–16 ft. lbs. (15–22 Nm) on 1.5L (VIN E).

9. Install the oil pan drain plug and tighten to 25–33 ft. lbs. (35–45 Nm).

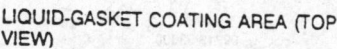

LIQUID-GASKET COATING AREA (TOP VIEW)

BOLT HOLE AREAS

TIMING BELT SIDE

SEALANT
GROOVE AREA

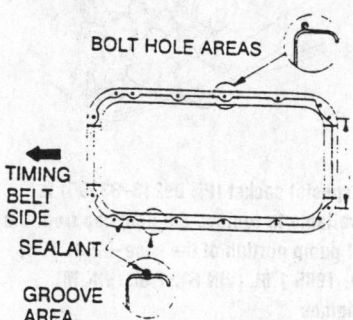

TIGHTENING SEQUENCE OF FLANGE BOLTS

TIMING BELT SIDE

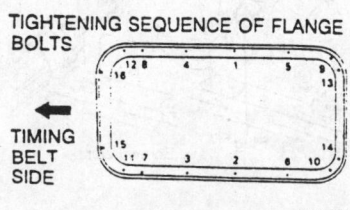

7923GG70

Oil pan sealant applications points and tightening sequence—3.0L (VIN T) engine

10. Install the splash shield
11. Lower the vehicle.
12. Refill the crankcase with oil.
13. Start the engine and allow it to reach operating temperature. Check for leaks.

Oil Pump

REMOVAL & INSTALLATION

➡ **Whenever the oil pump is disassembled or the cover removed, the gear cavity must be filled with petroleum jelly for priming purposes. Do not use grease.**

Except 2.0L (VIN P), 1995 1.6L (VIN R), 1.8L (VIN M) Engines

1. Disconnect the negative battery cable.
2. Remove the timing belt.
3. Remove the oil pan.
4. Remove the oil screen.
5. Remove the front case assembly.
6. Remove the oil pump cover.
7. Remove the inner and outer gears from the front case.

➡ **The inner and outer gears indicate the direction of installation.**

8. Remove the plug, relief valve spring and relief valve from the case.

To install:
9. Check the front case for damage or cracks. Replace the front seal. Replace the oil screen O-ring. Clean all parts thoroughly with a safe solvent.

10. Check the pump gears for wear or damage. Clean the gears thoroughly and place them in position in the case to check the clearances.

11. Check that the relief valve can slide freely in the case.

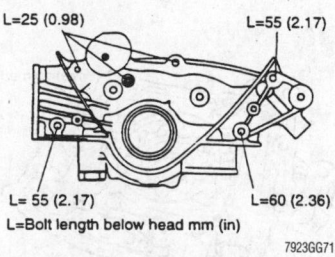

L=25 (0.98)   L=55 (2.17)

L=55 (2.17)   L=60 (2.36)
L=Bolt length below head mm (in)

7923GG71

Oil pump cover bolt lengths and locations—3.0L (VIN T) engine

12. Check the relief valve spring for damage.

13. Thoroughly coat both oil pump gears with clean engine oil and install them in the correct direction of rotation.

14. Install the pump cover and torque the bolts as follows:

    a. 6–8 ft. lbs. (8–12 Nm)—1.5L (VIN E) engine

    b. 4–6 ft. lbs. (6–9 Nm)—except 1.5L (VIN E) engine

15. Coat the relief valve and spring with clean engine oil, install them and tighten the plug to 30–36 ft. lbs. (39–49 Nm).

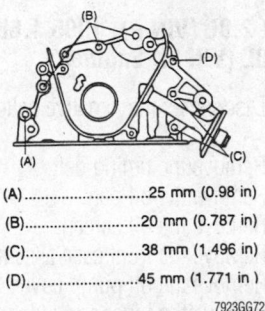

(A)	25 mm (0.98 in)
(B)	20 mm (0.787 in)
(C)	38 mm (1.496 in)
(D)	45 mm (1.771 in)

7923GG72

Oil pump cover bolt lengths and locations—1.8L (VIN M) and 2.0L (VIN F) engines

(A)	25 mm (0.98 in.)
(B)	30 mm (1.18 in.)
(C)	45 mm (1.77 in.)
(D)	60 mm (2.36 in.)

7923GG73

Oil pump cover bolt lengths and locations—1.5L (VIN E) and 1.5L (VIN K) engines

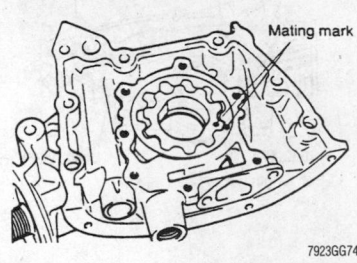

7923GG74

Align the mating marks on the oil pump gears—1.5L (VIN E) and 1.5L (VIN K) engines

16. Using a new gasket, install the front case assembly. Tighten the bolts to 8–11 ft. lbs. (12–15 Nm).

➡ **The bolts have different shank lengths. Use the illustrations to determine proper locations for each length bolt.**

17. Coat the lips of a new seal with clean engine oil and slide it along the crankshaft until it touches the front case. Drive it into place with a seal driver.

18. Install and properly tension the timing belt.

19. Install the oil screen.

20. Apply a ⅛ in. (3mm) wide bead of RTV sealer in the groove of the oil pan mating surface. Install the oil pan and tighten bolts to 60–72 inch lbs.

21. Connect the negative battery cable.

22. Start the engine and allow it to reach operating temperature. Check for leaks.

23. Check for proper oil pressure.

2.0L (VIN P), 1995 1.6L (VIN R), 1.8L (VIN M) Engines

1. Disconnect the negative battery cable.
2. Remove the timing belt.
3. Remove the oil pan.
4. Remove the oil screen.

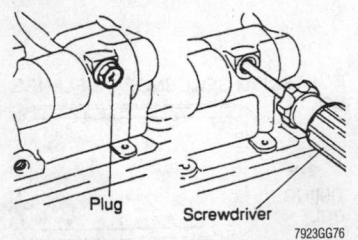

7923GG75

A special socket (PN 09213–33000) is available to remove the plug cap from the oil pump portion of the case—2.0L (VIN P), 1995 1.6L (VIN R), 1.8L (VIN M) engines

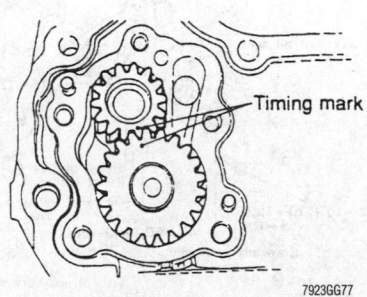

7923GG76

Remove the left side cylinder block plug and insert a screwdriver into the hole to hold the balance shaft from turning—2.0L (VIN P), 1995 1.6L (VIN R), 1.8L (VIN M) engines

5. Remove the oil filter bracket assembly.

6. Using special tool (PN 09213–33000) or equivalent, remove the plug cap from the oil pump portion of the front case.

7. Remove the left side cylinder block plug and insert a screwdriver with a 0.32 in. (8mm) diameter shaft into the plug hole. The screwdriver should be at least 2.4 in. (60mm) long.

8. Remove the oil pump drive gear and left counter balance shaft retaining bolt.

9. Remove the front case assembly.

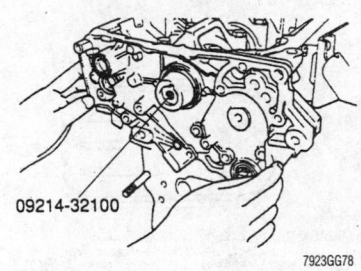

7923GG77

Align the timing marks on the gears during assembly—2.0L (VIN P), 1995 1.6L (VIN R), 1.8L (VIN M) engines

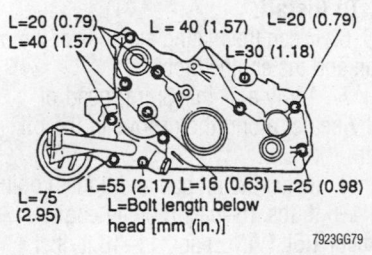

09214-32100

7923GG78

A special tool (PN 09214–32100) is used to center the front case hole on the crankshaft—2.0L (VIN P), 1995 1.6L (VIN R), 1.8L (VIN M) engines

L=20 (0.79)
L=40 (1.57)
L = 40 (1.57)
L=20 (0.79)
L=30 (1.18)
L=75 (2.95)
L=55 (2.17)
L=16 (0.63)
L=25 (0.98)
L=Bolt length below head [mm (in.)]

7923GG79

Oil pump cover bolt lengths and locations—2.0L (VIN P), 1995 1.6L (VIN R), 1.8L (VIN M) engines

10. Remove the oil pump cover from the front case.

11. Remove the oil pump gears.

To install:

12. Check the oil holes for clogging and clean as necessary.

13. Check counter balance shaft front bearing section for wear, damage or seizure. If there is any damage, replace the front case.

14. Check front case for cracks and other damage. Replace the front case as necessary.

15. Lubricate, align and install the oil pump gears.

16. Using a seal driver, install a new front case oil seal.

17. Lubricate and install special tool (PN 09214–32100) or equivalent, on the crankshaft.

18. Install the front case assembly using a new gasket. Install all bolts except those for the filter bracket and tighten finger-tight.

19. Tighten front case bolts to specification, as per the illustration.

20. Insert a screwdriver with a 0.32 in. (8mm) diameter shaft into the left side cylinder block plug hole. The screwdriver will hold the shaft stable while tightening the oil pump drive gear retaining bolt.

21. Install the oil pump drive gear and left counter balance shaft retaining bolt. Tighten to 25–29 ft. lbs. (34–40 Nm).

22. Install a new O-ring to the groove on the front case.

23. Using special tool (PN 09213–33000) or equivalent, install the plug cap on the oil pump portion of the front case. Tighten to 14–20 ft. lbs. (20–27 Nm).

24. Install the oil filter bracket assembly.

25. Install the oil screen.

26. Install the oil pan.

27. Install and properly tension the timing belt.

28. Connect the negative battery cable.

29. Start the engine and allow it to reach operating temperature. Check for leaks.

30. Check for proper oil pressure.

OIL SEAL REPLACEMENT

→**The front case oil seal is removed and installed from the rear of the case. The case assembly must be removed from the engine to replace the oil seal.**

1. Remove the front case assembly.

2. Using a seal remover or brass drift, drive the seal from the front case.

To install:

3. Clean the seal bore and inspect for damage. If damage is evident, replace the front case assembly.

4. Lubricate the seal bore and install the seal.

5. Using a seal driver, drive the seal into the front case.

Rear Main Seal

REMOVAL & INSTALLATION

1. Remove the transaxle.

2. Remove the flywheel.

3. Remove the oil seal case.

4. Remove the separator from the case, as required.

5. Remove the seal from the case.

To install:

6. Clean the seal case and separator.

7. Lubricate and install the new seal using a seal driver.

8. Install the separator into the housing so that the oil hole faces down.

9. Lubricate the lips of the seal.

10. Apply sealer or install a new gasket on the seal case and install. Tighten case bolts to 7–9 ft. lbs. (8–10 Nm).

11. Install the flywheel.

12. Install the transaxle.

FUEL SYSTEM

Fuel System Service Precautions

Safety is the most important factor when performing not only fuel system maintenance but any type of maintenance. Failure to conduct maintenance and repairs in a safe manner may result in serious personal injury or death. Maintenance and testing of the vehicle's fuel system components can be accomplished safely and effectively by adhering to the following rules and guidelines.

• To avoid the possibility of fire and personal injury, always disconnect the negative battery cable unless the repair or test procedure requires that battery voltage be applied.

• Always relieve the fuel system pressure prior to disconnecting any fuel system component (injector, fuel rail, pressure regulator, etc.), fitting or fuel line connection. Exercise extreme caution whenever relieving fuel system pressure to avoid exposing skin, face and eyes to fuel spray. Please be advised that fuel under pressure may penetrate the skin or any part of the body that it contacts.

• Always place a shop towel or cloth around the fitting or connection prior to loosening to absorb any excess fuel due to spillage. Ensure that all fuel spillage (should it occur) is quickly removed from engine surfaces. Ensure that all fuel soaked cloths or towels are deposited into a suitable waste container.

• Always keep a dry chemical (Class B) fire extinguisher near the work area.

• Do not allow fuel spray or fuel vapors to come into contact with a spark or open flame.

• Always use a back-up wrench when loosening and tightening fuel line connection fittings. This will prevent unnecessary stress and torsion to fuel line piping. Always follow the proper torque specifications.

• Always replace worn fuel fitting O-rings with new. Do not substitute fuel hose or equivalent, where fuel pipe is installed.

Fuel System Pressure

RELIEVING

Accent, Elantra, Sonata and Tiburon

The fuel pump connector is located at the fuel tank sending unit on top of the fuel tank. The connector is accessible through a door located under the rear seat.

1. Remove the rear seat cushion.

2. Disengage the fuel pump harness connector at the fuel tank sending unit.

3. Start the engine and allow it to run until it stalls.

4. Turn the ignition switch to the **OFF** position.

5. Disconnect the negative battery cable.

6. Attach the fuel pump harness connector.

Scoupe

The fuel pump connector is located to the right of the fuel tank.

1. Disengage the fuel pump harness connector at the rear of the fuel tank.

2. Start the engine and allow it to run until it stalls.

3. Turn the ignition switch to the **OFF** position.

4. Disconnect the negative battery cable.

5. Attach the fuel pump harness connector.

Fuel Filter

REMOVAL & INSTALLATION

All Engines

✳✳ CAUTION

The fuel injection system remains under pressure after the engine has been turned OFF. Properly relieve fuel pressure before disconnecting any fuel lines. Failure to do so may result in fire or personal injury. Do not allow fuel spray or fuel vapors to come in contact with a spark or open flame. Keep a dry chemical fire extinguisher nearby. Never store fuel in an open container due to risk of fire or explosion.

1. Relieve the fuel pressure.

2. Disconnect the negative battery cable.

3. Remove the fuel line inlet and outlet union bolts while holding the fuel filter stationary with a back-up wrench.

4. Remove the fuel filter mounting bolts.

5. Pull the old filter from the mounting bracket.

To install:

6. Place the new fuel filter into the mounting bracket.

7. Install and tighten the fuel filter mounting bolts to 18–25 ft. lbs. (25–35 Nm).

8. Using new gaskets, connect the fuel inlet and outlet lines to the filter. Carefully tighten the union bolts to avoid stripping the threads.

9. Connect the negative battery cable.

10. Start the engine and check the filter connections for leaks by running the tip of your finger around each union bolt connection.

Fuel Pump

REMOVAL & INSTALLATION

1. Relieve the fuel system pressure.

2. Raise and support the vehicle safely.

3. Remove the fuel tank drain plug and drain the fuel into an approved container.

4. Remove the fuel tank from the vehicle.

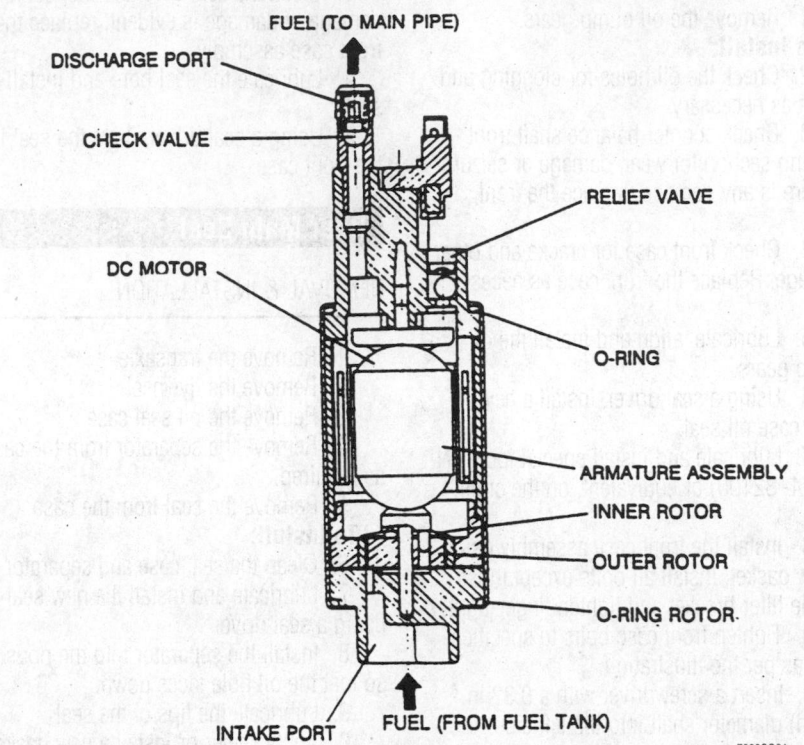

Cut away view of the electric fuel pump

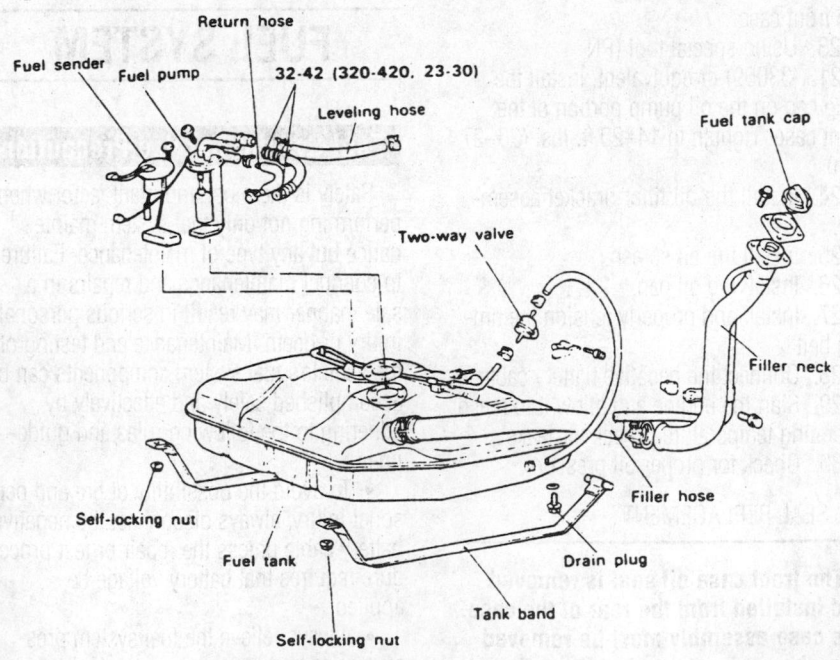

TORQUE ; Nm (kg.cm, lb.ft)

Exploded view of the fuel tank and fuel pump assembly—1995 Elantra

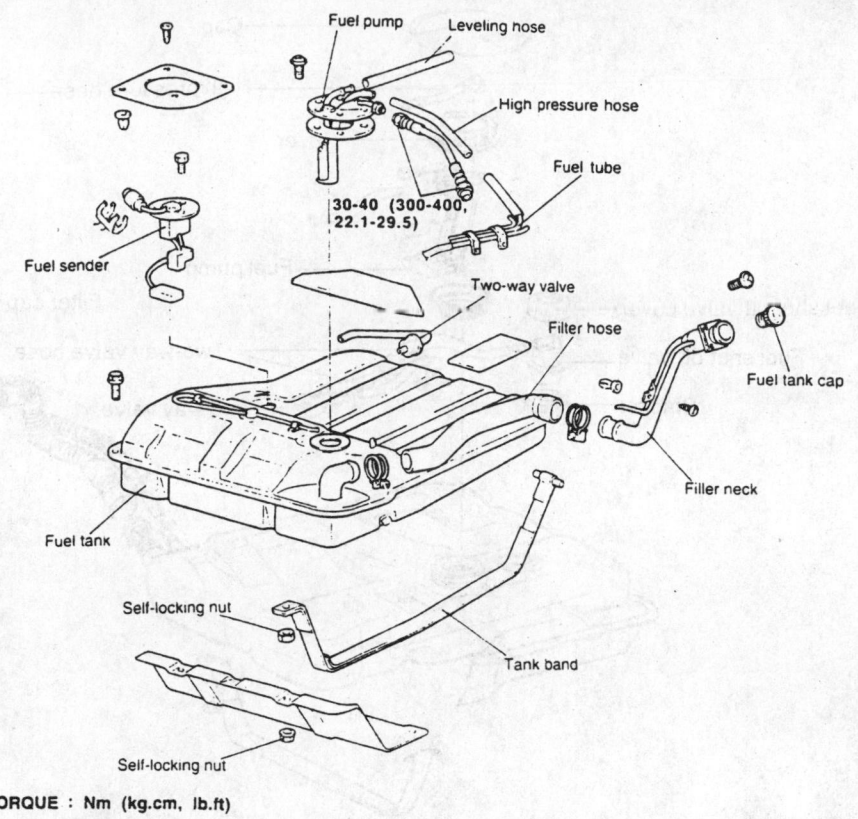

Exploded view of the fuel tank and fuel pump assembly—Scoupe

7923GG83

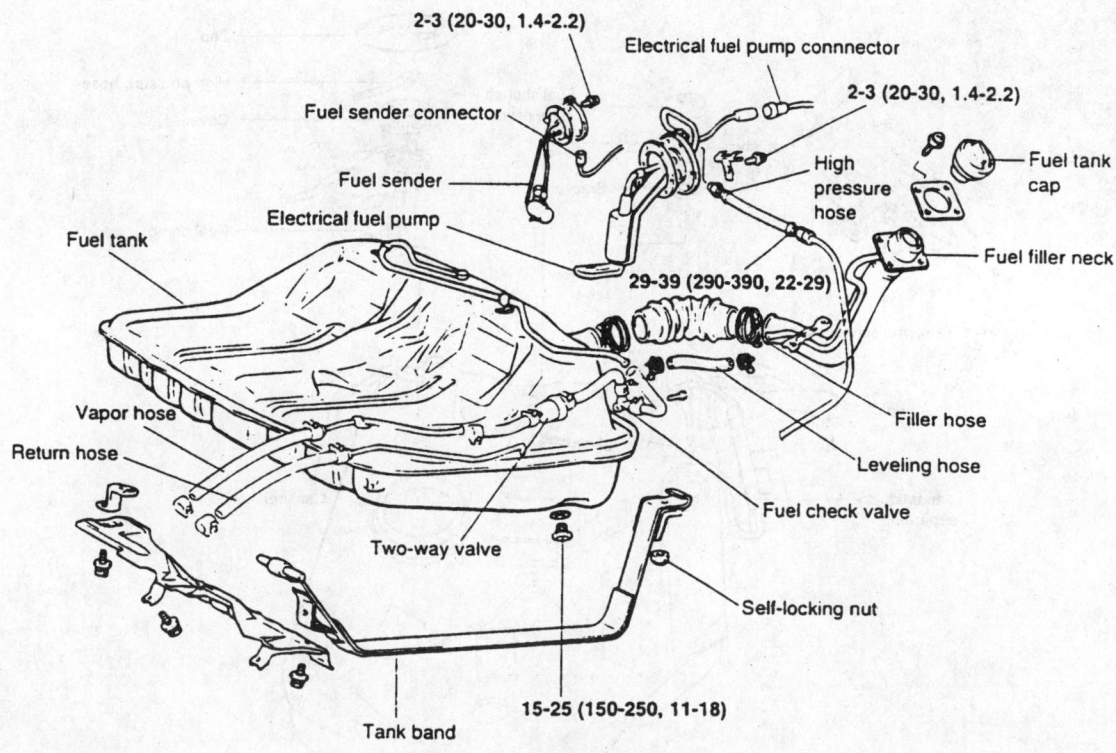

Exploded view of the fuel tank and fuel pump assembly—Sonata

7923GG84

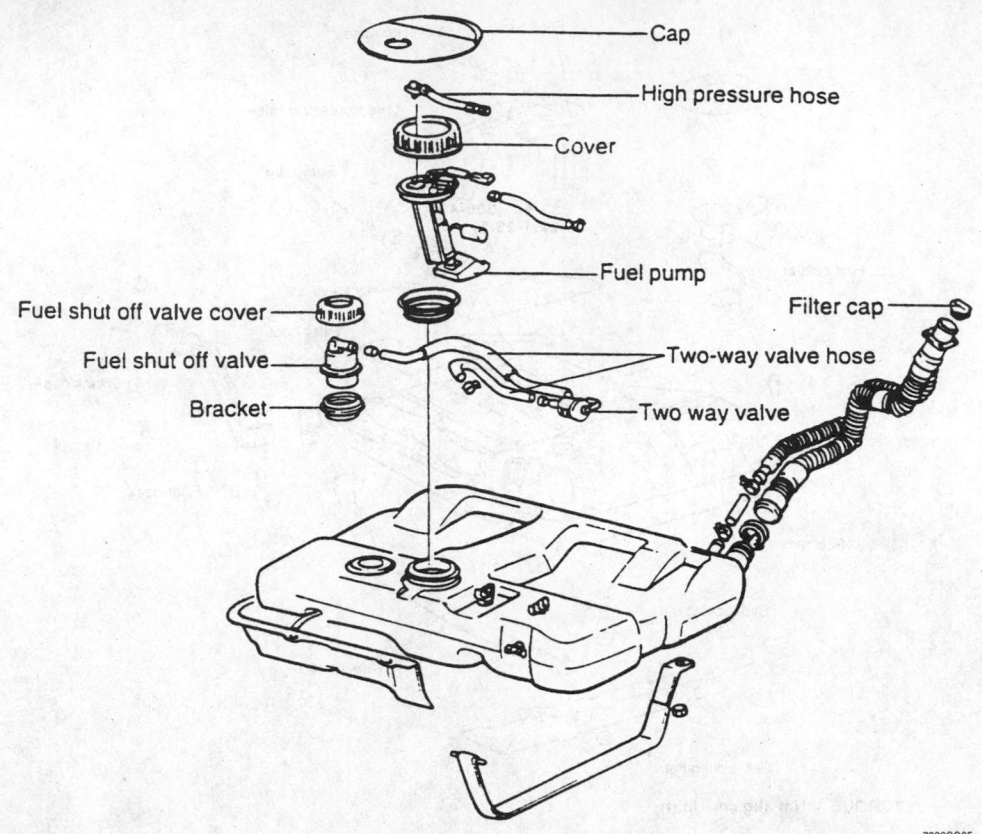

Exploded view of the fuel tank and fuel pump assembly—Accent

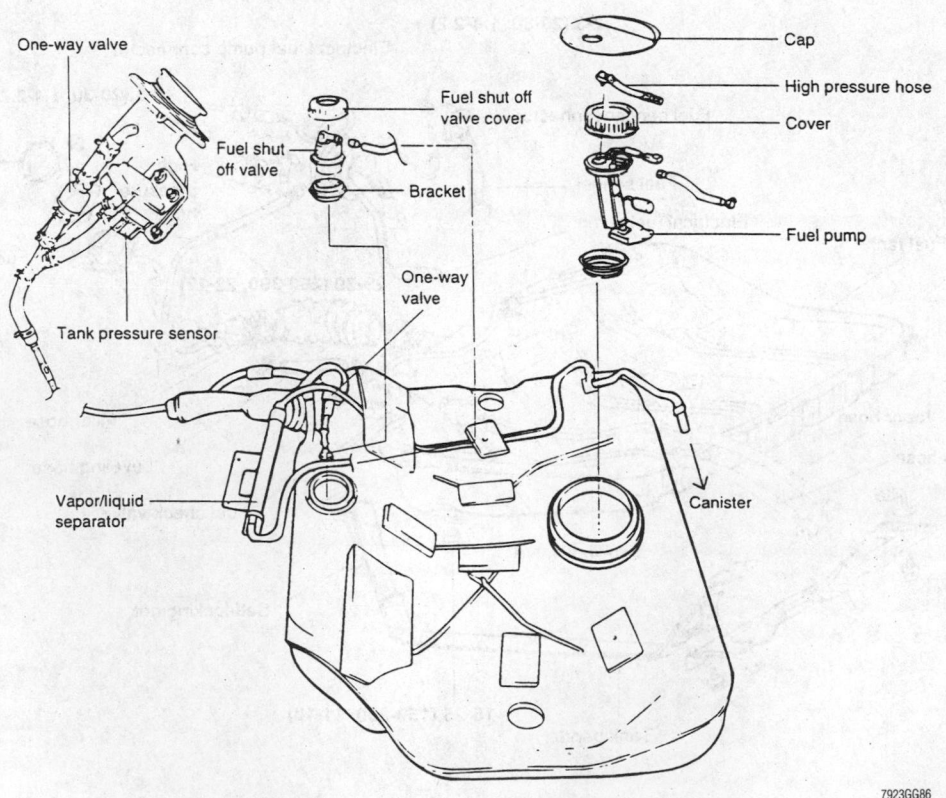

Fuel tank and fuel pump assembly—Tiburon and 1996–99 Elantra

5. Label, disconnect and plug the vapor and fuel hoses at the pump.

6. Label and disconnect the electrical harness at the pump.

7. Loosen the fuel pump mounting bolts.

8. Remove the fuel pump from the tank and discard the gasket.

To install:

9. Install the fuel pump using a new gasket. Tighten mounting bolts to 1–2 ft. lbs. (3–4 Nm).

10. Connect the electrical harness at the pump.

11. Connect the vapor and fuel hoses at the pump. Tighten fuel hose connections to 22–29 ft. lbs. (29–39 Nm).

12. Install the fuel tank and fuel drain plug.

13. Lower the vehicle.

14. Fill tank with fuel and check for proper fuel pump operation.

DRIVE TRAIN

Transaxle Assembly

REMOVAL & INSTALLATION

Manual

1. Disconnect the negative battery cable.
2. Drain the transaxle oil.
3. Remove the air duct and air cleaner assembly, as required.
4. Disconnect the back-up light switch.
5. Disconnect the clutch release mechanism and remove the clutch release cylinder.
6. Disconnect the speedometer cable.
7. Remove the pin clips and cotter pins and disconnect the select and shift cables from the control levers.
8. Label and disconnect the starter wiring harness. Remove the starter.
9. Raise and support the vehicle safely.
10. Disconnect the halfshafts.
11. Unbolt and remove the bell housing cover.
12. Support the bottom of the transaxle with a transmission jack.
13. Support the engine by the engine lifting tabs and remove the transaxle mounting brackets and insulator.
14. Remove the transaxle-to-engine bolts.

15. Slide the transaxle back, then lower it away from the engine.

To install:

16. Raise the transaxle into position on the engine.

17. Install the transaxle-to-engine bolts. Tighten M8 bolts to 6–7 ft. lbs. (8–10 Nm), M10 bolts to 22–25 ft. lbs. (30–35 Nm) and M12 bolts to 32–39 ft. lbs. (43–55 Nm).

18. Install the transaxle mounting brackets and tighten bolts to 65–80 ft. lbs. (90–110 Nm).

19. Install the bell housing cover and tighten bolts to 6–7 ft. lbs. (8–10 Nm).

20. Connect the halfshafts.

21. Lower the vehicle.

22. Install the starter and connect the starter wiring harness.

23. Connect the select and shift cables. Install new pin clips and cotter pins.

24. Connect the speedometer cable.

25. Install the clutch release cylinder and connect the clutch release mechanism.

26. Connect the back-up light switch.

27. Install the air duct and air cleaner assembly, as required.

28. Fill the transaxle with oil.

29. Connect the negative battery cable.

Automatic

1. Disconnect the negative battery cable.
2. Drain the transaxle oil.
3. Remove the air duct and air cleaner assembly, as required.
4. Disconnect and plug the transaxle cooler lines.
5. Disconnect the control cable.
6. Disconnect the speedometer cable.
7. Label and disconnect the pulse generator, inhibitor, kickdown servo, solenoid valve and oil temperature sensor electrical harnesses.
8. Label and disconnect the starter wiring harness. Remove the starter.
9. Raise and support the vehicle safely.
10. Disconnect the halfshafts.
11. Unbolt and remove the bell housing cover.
12. Remove the torque converter bolts.
13. Support the bottom of the transaxle with a transmission jack.
14. Remove the transaxle mounting brackets and center member.
15. Remove the transaxle-to-engine bolts.
16. Slide the transaxle back, then lower it away from the engine.

To install:

17. Raise the transaxle into position on the engine.

18. Install the transaxle-to-engine bolts. Tighten M8 bolts to 6–7 ft. lbs. (8–10 Nm), M10 bolts to 22–25 ft. lbs. (30–35 Nm) and M12 bolts to 32–39 ft. lbs. (43–55 Nm).

19. Install the transaxle mounting brackets and tighten bolts to 65–80 ft. lbs. (90–110 Nm).

20. Install the torque converter bolts and tighten to 34–39 ft. lbs. (46–53 Nm).

21. Install the bell housing cover and tighten bolts to 6–7 ft. lbs. (8–10 Nm).

22. Connect the halfshafts.

23. Lower the vehicle.

24. Install the starter and connect the starter wiring harness.

25. Connect and adjust the control cables.

26. Connect the speedometer cable.

27. Connect the pulse generator, inhibitor, kickdown servo, solenoid valve and oil temperature sensor electrical harnesses.

28. Install the air duct and air cleaner assembly, as required.

29. Fill the transaxle with oil.

30. Connect the negative battery cable.

Clutch

ADJUSTMENTS

Pedal Height and Free-Play

1. Measure the clutch pedal height (from the face of the pedal pad to the floorboard). Distance should be 7.0 in. (178mm).

2. Measure the clutch pedal clevis pin play (measured at the face of the pedal pad). Distance should be 0.04–0.11 in. (1–3mm).

3. If either measurement is not within specification, adjust as follows:

4. Turn and adjust the stop bolt so that the pedal height is within specification. and secure with the locknut.

5. When the pedal height is lower than specification, loosen the bolt or clutch pedal position switch and turn the pushrod to make the adjustment. Turn the bolt or clutch pedal position switch until it reaches the pedal stopper, then lock with the locknut.

➡**When adjusting clutch pedal height or pedal clevis pin play, be careful not to push the clutch master cylinder rod toward the master cylinder.**

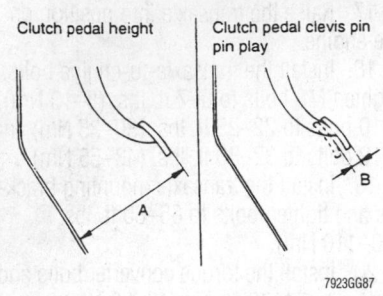

Measuring clutch pedal height (A) and clutch pedal clevis pin play (B)

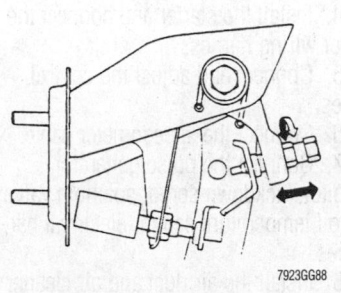

Loosen the stop bolt or clutch pedal position switch and turn the pushrod to make the adjustment

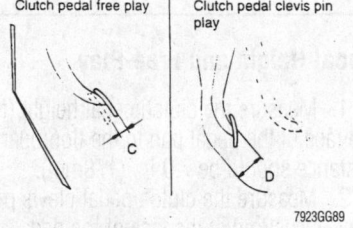

Measuring clutch pedal free-play (C) and clutch pedal clevis pin play (D)

6. Measure the clutch pedal free-play (measured at the face of the pedal pad). Distance should be 0.2–0.5 in. (6–13mm).

7. Measure the clutch pedal clevis pin play (measured at the face of the pedal pad with the clutch pedal depressed). Distance should be 2.8 in. (90mm).

8. If the clutch pedal free-play and the distance between the clutch pedal and the firewall when the clutch is disengaged are not within specification, it may be the result of air in the hydraulic line or a faulty hydraulic component.

9. Bleed the hydraulic clutch system.

REMOVAL & INSTALLATION

1. Remove the transaxle.

2. Insert the forward end of an old transaxle input shaft or a clutch disc guide tool into the splined center of the clutch disc, pressure plate and the pilot bearing in the crankshaft. This will keep the disc from dropping when the pressure plate is removed from the flywheel.

3. Loosen the clutch mounting bolts alternately and diagonally in very small increments, no more than 2 turns at a time, so as to avoid warping the cover flange.

4. Remove the pressure plate and disc.

5. Remove the return clip and the clutch release bearing.

6. On early model clutches, insert tool 09414–24000, or equivalent, in the spring pin and attach the round nut to the end of the tool. While holding the shaft of the special tool, rotate the sleeve with a wrench to force the spring pin out.

7. Remove the clutch release shaft, packings, return spring and the release fork.

To install:

8. Apply a light coating of high temperature grease to the release fork shaft and the clutch release bearing contact surfaces.

9. On early model clutches, align the lock pin holes of the release fork and shaft and drive 2 new spring pins into the holes. Be sure the spring pin slot is at right angles to the centerline of the control shaft.

10. Apply grease into the groove in the release bearing and install bearing into the front bearing retainer in the transaxle. Install the return clip to the release bearing and fork.

11. Be sure the surfaces of the pressure plate and flywheel are wiped clean of grease and lightly sand them with crocus cloth. Lightly grease the clutch disc and transaxle input shaft splines making sure not to allow any grease to contact the clutch disc material or clutch slip may result.

12. Locate the clutch disc on the flywheel with the stamped mark facing outward. Use a clutch disc guide or old input shaft to center the disc on the flywheel, then install the pressure plate over it. Install the bolts and tighten them evenly. Tighten them in increments of 2 turns or less to avoid warping the pressure plate. Torque to 11–15 ft. lbs. (15–21 Nm).

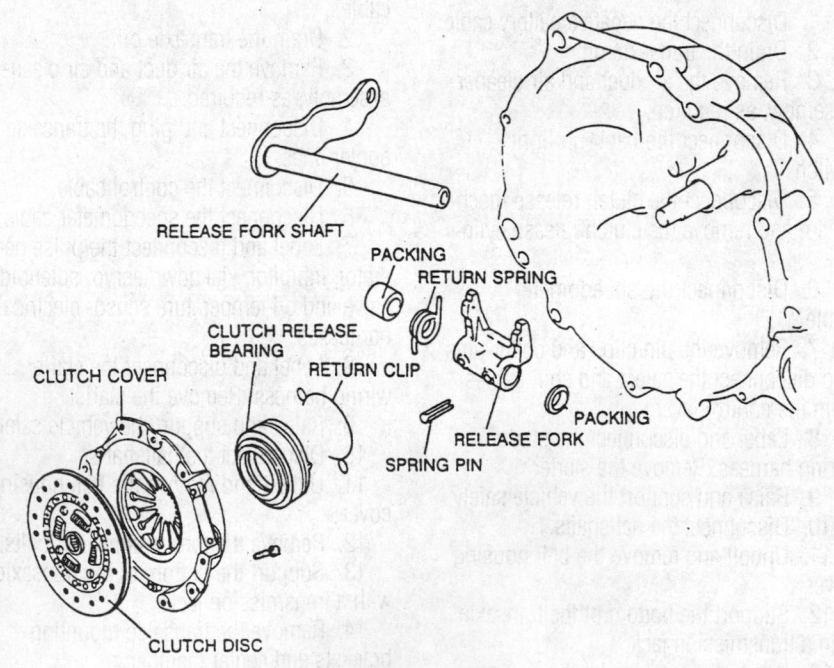

Exploded view of the clutch disk and pressure plate components—early models

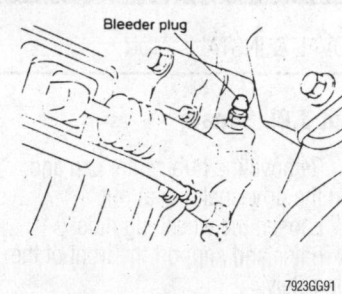

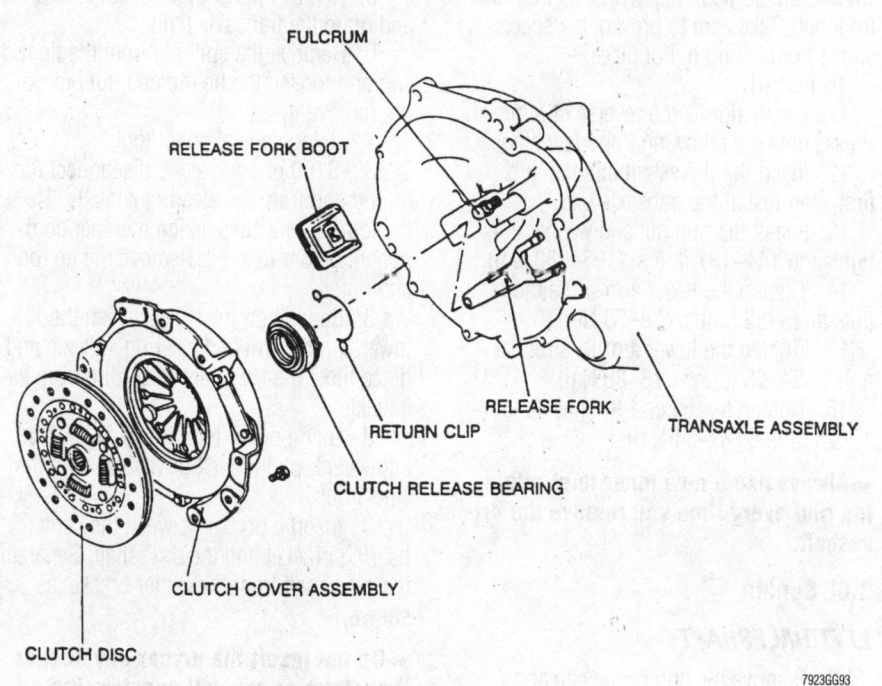

Exploded view of the clutch disk and pressure plate components—late models

13. Remove the clutch disc centering tool.
14. Install the transaxle.
15. Adjust the clutch free-play.

Hydraulic Clutch System

BLEEDING

Whenever a clutch system hydraulic component is removed, a hydraulic line disconnected or the system is opened for any reason, the system must be bled to remove any air that may be trapped inside the system.

To bleed the system you will need: a good supply of brake fluid that meets or exceeds DOT 3 specifications, a small transparent plastic container, a wrench to loosen and tighten the bleeder screw, a small length of clear plastic hose to attach to the bleeder screw and an assistant to work the clutch pedal.

1. Unscrew the clutch fluid reservoir cap.
2. Loosen the bleeder screw just so fluid starts to leak out of the bleeder hole.
3. Fill the plastic container halfway with brake fluid.

The hydraulic bleeder is located on the release cylinder

4. Connect the hose to the bleeder screw and place the other end of hose into the container of brake fluid.
5. Fill the reservoir to the MAX fill line and have the assistant pump the clutch pedal slowly. You will notice air bubbles rising to the top of the fluid container. Keep the reservoir full at all times.
6. Repeat the previous step until all the air bubbles are gone.
7. Tighten the bleeder screw and cheek the fluid level. Add fluid as necessary until the proper level is reached.
8. Install the reservoir cap.

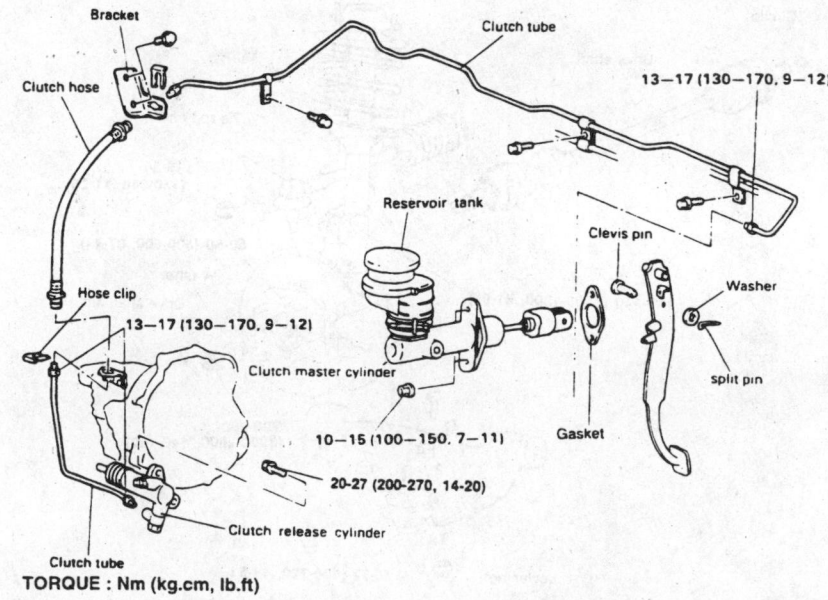

TORQUE : Nm (kg.cm, lb.ft)

Exploded view of the clutch hydraulic system

Halfshafts

REMOVAL & INSTALLATION

Except 3.0L Sonata

1. Remove the hub center cap and loosen the driveshaft (axle) nut.
2. Loosen the wheel lug nuts.
3. Raise and support the front of the vehicle safely.
4. Remove the front wheels.
5. Remove the engine splash shield.
6. Remove the lower ball joint and strut bar from the lower control arm.

➡ **Place the lower arm ball joint on the lower arm to prevent damage to the ball joint dust boot.**

7. Drain the transaxle fluid into a suitable waste container.
8. Insert a prybar between the transaxle case (on the raised rib) and the driveshaft inner joint case. Move the bar to the right to withdraw the left driveshaft; left, to remove the right driveshaft.

➡ **Do not insert the prybar too deeply (7mm) or you will damage the oil seal.**

9. Plug the transaxle case with a clean rag to prevent dirt from entering the case.

10. Use a puller/driver mounted on the wheel studs to push the driveshaft from the front hub. Take care to prevent the spacer shims from falling out of place.

To install:

11. Installation is the reverse of removal. Please note the following important steps.
12. Insert the driveshaft into the hub first, then install the transaxle end.
13. Install the hub nut and washer and tighten to 144–187 ft. lbs. (195–253 Nm).
14. Tighten the lower arm-to-ball joint nuts to 43–52 ft. lbs. (58–70 Nm).
15. Tighten the lower arm-to-strut bar nuts to 54–65 ft. lbs. (73–88 Nm).
16. Tighten tie rod end-to-knuckle to 17–25 ft. lbs. (23–34 Nm).

➡ **Always use a new inner joint retaining ring every time you remove the driveshaft.**

3.0L Sonata

LEFT HALFSHAFT

1. Remove the hub center cap and remove the split pin, driveshaft (axle) nut and washer. Make a mental note of how the washer is installed.
2. Loosen the wheel lug nuts.
3. Raise and safely support the front of the vehicle.

4. Remove the front wheels.
5. Remove the engine splash shield and drain the transaxle fluid.
6. Remove the split pin from the tie rod end and loosen the tie rod end nut but do not remove it.
7. Using special puller tool 09568–3100 or equivalent, disconnect the tie rod end from the steering knuckle. Tie the tool off to a suspension member component before using it. Remove the tie rod end nut.
8. Reposition the tool between the lower control arm and steering knuckle and disconnect the lower arm ball joint from the knuckle.
9. Using puller tool 09526–11001 or equivalent, pull the left driveshaft from the wheel hub.
10. Insert a prybar between the center bearing bracket and the driveshaft. Separate the driveshaft from the center bracket as shown.

➡ **Do not insert the prybar any deeper than 7mm or you will puncture the oil seal and also damage the joint. When separating the driveshaft, do not allow the full weight of the vehicle to be placed on the wheel bearing. If the weight of the vehicle must be applied for any reason, support the wheel bearing with holding tool 09517–21500 or equivalent.**

11. Remove the oxygen sensor connector from the center bearing bracket. One screw holds the connector to the bracket.
12. Remove the two center bracket mounting bolts. Insert a prybar between the center bearing bracket, inner shaft and cylinder block. Then, pull the center bracket and inner shaft assembly from the transaxle case. Plug the transaxle case with a clean rag to prevent dirt from entering the case.

To install:

➡ **Always use a new inner joint retaining ring every time you remove the driveshaft.**

13. Insert the inner shaft and bracket assembly into the transaxle case and install the center bracket mounting bolts. Tighten the bolts to 26–33 ft. lbs. (35–45 Nm).
14. Connect the oxygen sensor to the center mounting bracket with the mounting screw.
15. Insert the driveshaft into the center bearing, then into the wheel hub.
16. Connect the lower ball joint to the steering knuckle and tighten the nut to 43–52 ft. lbs. (58–71 Nm).

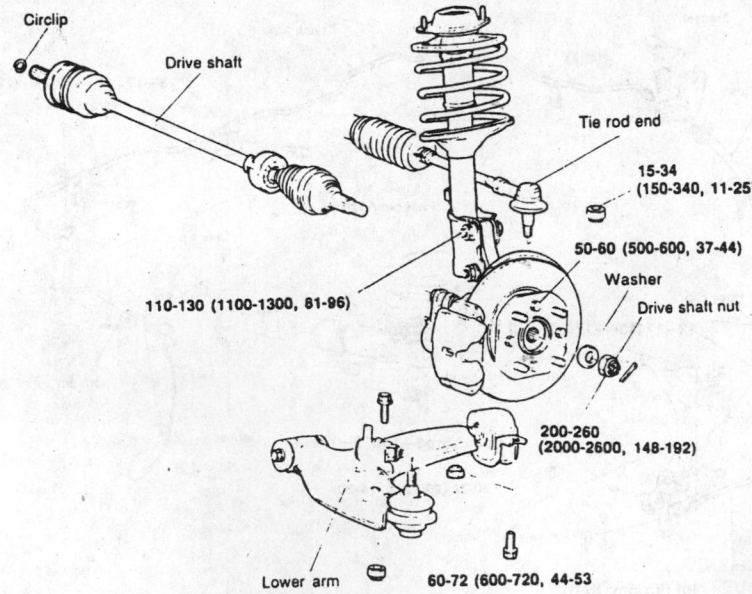

Circlip
Drive shaft
Tie rod end
15-34 (150-340, 11-25)
50-60 (500-600, 37-44)
Washer
Drive shaft nut
110-130 (1100-1300, 81-96)
200-260 (2000-2600, 148-192)
Lower arm
60-72 (600-720, 44-53)

TORQUE : Nm (kg.cm, lb.ft)

7923GG94

Halfshaft components—except 3.0L Sonata

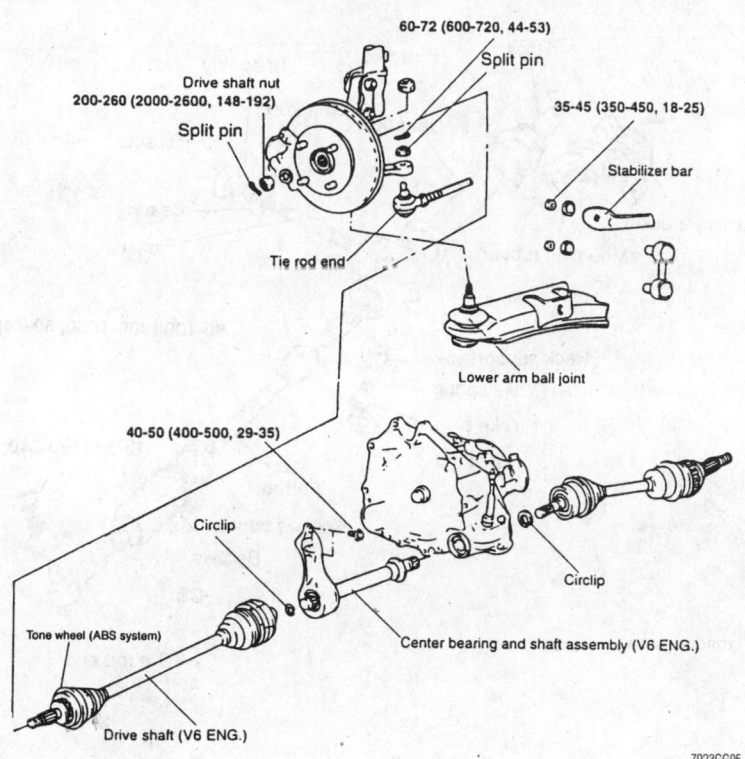

60-72 (600-720, 44-53)
Split pin
Drive shaft nut
200-260 (2000-2600, 148-192)
Split pin
35-45 (350-450, 18-25)
Stabilizer bar
Tie rod end
Lower arm ball joint
40-50 (400-500, 29-35)
Circlip
Circlip
Tone wheel (ABS system)
Center bearing and shaft assembly (V6 ENG.)
Drive shaft (V6 ENG.)

7923GG95

Halfshaft components—3.0L Sonata

17. Connect the tie rod end to the steering knuckle and torque the tie rod end nut to 17–25 ft. lbs. (23–34 Nm).

18. Install the splash shield.

19. Mount the front wheels, tighten the lug nuts, and lower the vehicle to the ground.

20. Install the axle washer and nut. Be sure the washer is installed properly. Torque the axle nut to 145–188 ft. lbs. (196–254 Nm) and secure the nut with a new split pin.

21. Fill the transaxle to the proper level with the specified fluid.

RIGHT HALFSHAFT

1. Remove the hub center cap and loosen the driveshaft (axle) nut.

2. Loosen the wheel lug nuts.

3. Raise and safely support the front of the vehicle.

4. Remove the front wheels.

5. Remove the engine splash shield and drain the transaxle fluid.

6. Remove the split pin from the tie rod end and loosen the tie rod end nut but do not remove it.

7. Using special puller tool 09568–3100 or equivalent, disconnect the tie rod end from the steering knuckle. Tie the tool off to a suspension member component before using it.

8. Remove the tie rod end nut.

9. Reposition the tool between the lower control arm and steering knuckle and disconnect the lower arm ball joint from the knuckle.

10. Insert a prybar between the transaxle case (on the raised rib) and the driveshaft inner joint case, Move the bar to the left to remove the right driveshaft.

➡**Do not insert the prybar any deeper than 7mm or you will puncture the oil seal.**

11. Plug the transaxle case with a clean rag to prevent dirt from entering the case.

12. Use a puller/driver mounted on the wheel studs to push the driveshaft from the front hub. Take care to prevent the spacer shims from falling out of place.

To install:

13. Installation is the reverse of removal. Please note the following important steps.

14. Insert the driveshaft into the hub, first, then install the transaxle end.

15. Install the hub nut washer as illustrated and tighten the axle shaft hub nut 145–188 ft. lbs. (196–254 Nm).

16. Tighten the lower arm ball joint-to-knuckle to 42–50 ft. lbs. (57–68 Nm)

17. Tighten the tie rod end-to-knuckle to 17–25 ft. lbs. (23–34 Nm).

➡**Always use a new inner joint retaining ring every time you remove the driveshaft.**

STEERING AND SUSPENSION

Air Bag

❊❊ CAUTION

Some vehicles are equipped with an air bag system, also known as the Supplemental Restraint System (SRS). The system must be disabled before performing service on or around system components, steering column, instrument panel components, wiring and sensors. Failure to follow safety and disabling procedures could result in accidental air bag deployment, possible personal injury and unnecessary system repairs.

PRECAUTIONS

Several precautions must be observed when handling the inflator module to avoid accidental deployment and possible personal injury.

• Never carry the inflator module by the wires or connector on the underside of the module.

• When carrying a live inflator module, hold securely with both hands, and ensure that the bag and trim cover are pointed away.

• Place the inflator module on a bench or other surface with the bag and trim cover facing up.

• With the inflator module on the bench, never place anything on or close to the module which may be thrown in the event of an accidental deployment.

DISARMING

❊❊ CAUTION

The air bag system must be disarmed before removing the steering wheel or air bag module. Failure to do so may cause accidental deployment, property damage or personal injury.

1. Disconnect the negative battery cable.
2. Wait 30 seconds.
3. Remove the air bag module mounting nuts and lift the module from the steering wheel.
4. Locate and unplug the connector. Protect this connector from accidental electrical contact, including static electricity.
5. The air bag is now disarmed and the battery can be connected to perform electrical testing on other systems.
6. When ready to reconnect the air bag; disconnect the battery, connect the air bag connector, and install the cover or panel.
7. Install the mounting nuts. Be sure no one is in the vehicle when connecting the battery.

Rack and Pinion Steering Gear

REMOVAL & INSTALLATION

Manual

1. Raise and support the vehicle safely.
2. Turn the wheels to the straight ahead position.
3. Remove the wheels.
4. Remove all components necessary to gain access to the rack.
5. Matchmark the steering shaft to the pinion.
6. Remove the steering shaft-to-pinion coupling bolt.
7. Disconnect the tie rod ends from the steering knuckles.
8. Remove the clamps securing the rack to the crossmember.
9. Remove the rack from the vehicle.
   To install:
10. Install the rubber mount for the rack and pinion with the slit on the downside.
11. Install the rack from the vehicle.
12. Install the clamps securing the rack to the crossmember. Tighten bolts to 44–59 ft. lbs. (60–80 Nm).
13. Install the steering shaft-to-pinion coupling bolt and tighten to 11–14 ft. lbs. (15–19 Nm).

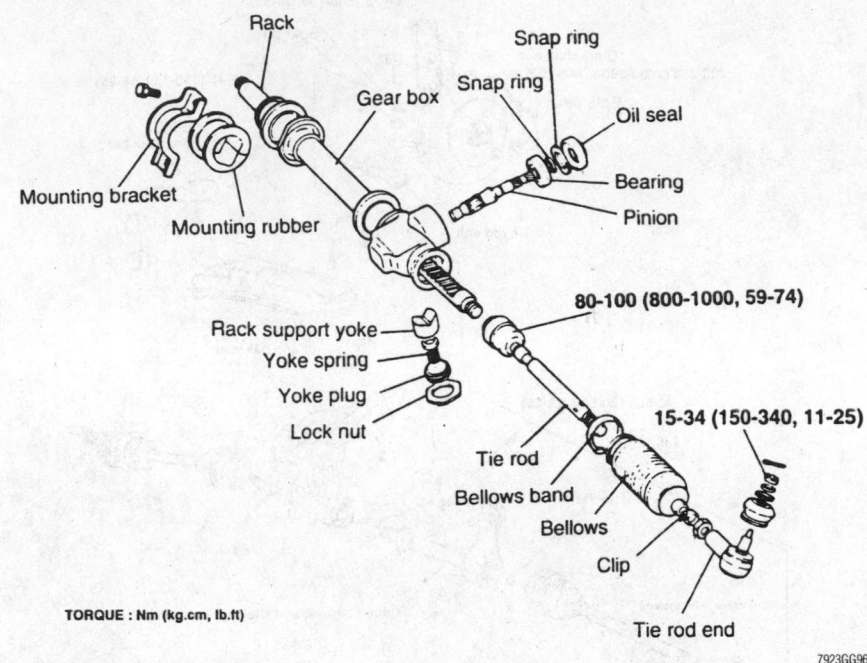

TORQUE : Nm (kg.cm, lb.ft)

80-100 (800-1000, 59-74)

15-34 (150-340, 11-25)

Exploded view of the manual rack and pinion assembly

Manual rack and pinion mounting bolt locations

14. Connect the tie rod ends to the steering knuckles.
15. Install all components previously removed to gain access to the rack.
16. Install the wheels.
17. Lower the vehicle.

Power

1. Raise and support the vehicle safely.
2. Turn the wheels to the straight ahead position.
3. Remove the wheels.

4. Remove all components necessary to gain access to the rack.
5. Drain the fluid from the power steering system.
6. Disconnect and plug the fluid hoses.
7. Matchmark the steering shaft to the pinion.
8. Remove the steering shaft-to-pinion coupling bolt.
9. Disconnect the tie rod ends from the steering knuckles.
10. Remove the clamps securing the rack to the crossmember.
11. Remove the rack from the vehicle.
   To install:
12. Install the rubber mount for the rack and pinion with the slit on the downside.
13. Install the rack from the vehicle.
14. Install the clamps securing the rack to the crossmember. Tighten bolts to 44–59 ft. lbs. (60–80 Nm).
15. Install the steering shaft-to-pinion coupling bolt and tighten to 11–14 ft. lbs. (15–19 Nm).
16. Connect the tie rod ends to the steering knuckles.

COMPONENTS

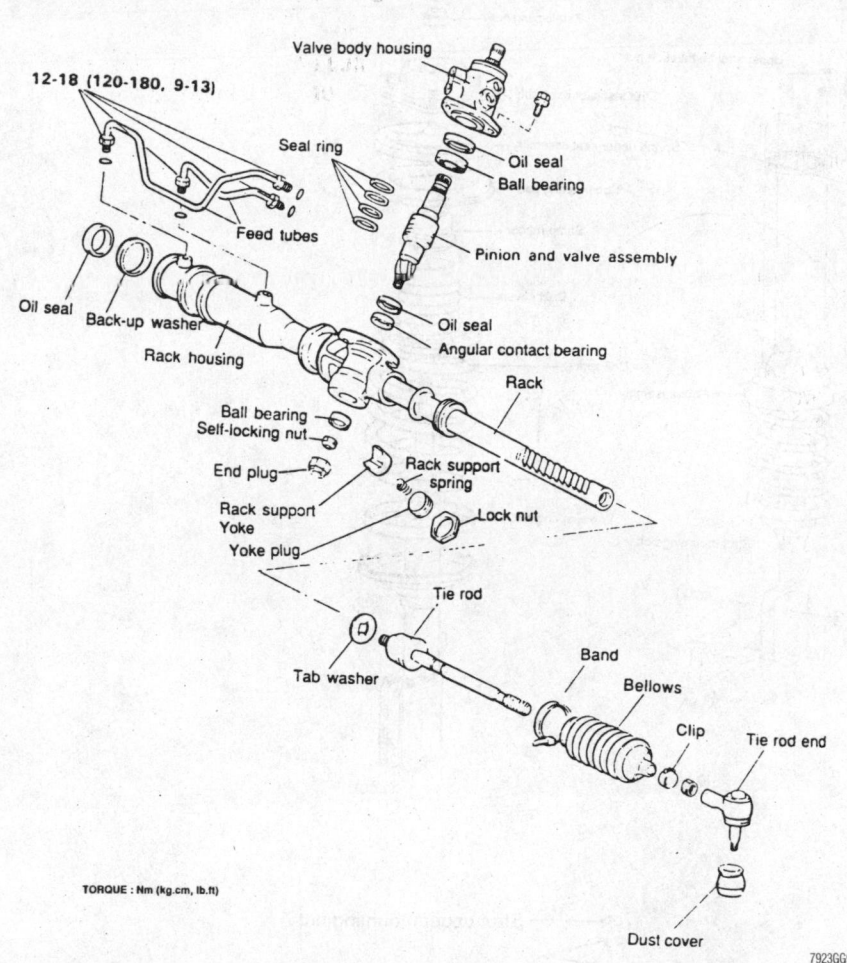

12-18 (120-180, 9-13)

Valve body housing

Seal ring

Oil seal
Ball bearing

Feed tubes

Pinion and valve assembly

Oil seal

Back-up washer

Oil seal

Rack housing

Angular contact bearing

Ball bearing
Self-locking nut

Rack

End plug

Rack support
spring

Rack support
Yoke

Lock nut

Yoke plug

Tie rod

Tab washer

Band

Bellows

Clip

Tie rod end

Dust cover

TORQUE : Nm (kg.cm, lb.ft)

7923GG98

Exploded view of the power rack and pinion assembly

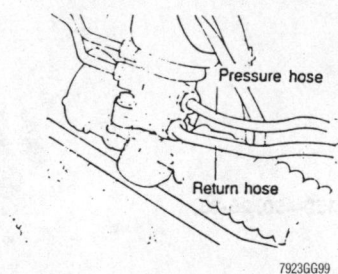

Pressure hose

Return hose

7923GG99

Pressure and return hose location on the rack

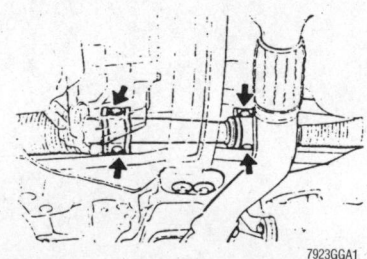

7923GGA1

Power rack and pinion mounting bolt locations

17. Connect the fluid hoses and tighten fittings to 9–13 ft. lbs. (12–18 Nm).

18. Install all components previously removed to gain access to the rack.

19. Install the wheels.

20. Lower the vehicle.

21. Fill the power steering system with fluid.

22. Bleed the power steering system.

Struts

REMOVAL & INSTALLATION

Front

1. Raise and support the vehicle safely.

2. Remove the front wheels.

3. Detach the brake hose from the clip on the strut.

4. Unbolt the strut from the knuckle.

5. Remove the four strut-to-fender nuts.

6. Pull the strut away from the steering knuckle and wheelhouse and out from the vehicle.

➡On some models it is helpful to raise the lower arm with a jack and attach the brake hose, brake line, front speed sensor harness and drive shaft to the steering knuckle with a piece of rope or wire after the strut is separated from the knuckle.

To install:

7. Before installing the strut, be sure the surface where the strut attaches to the knuckle is clean. This ensures a good connection.

8. Position the strut onto the knuckle and inside fender apron and install the upper and lower attaching hardware.

9. Observe the following torque specifications:

• Strut-to-knuckle bolts—65–76 ft. lbs. (95–105 Nm) for Accent, Scoupe and Sonata.

• Strut-to-knuckle bolts—80–94 ft. lbs. (110–130 Nm) for Elantra and Tiburon.

• Strut-to-fender nuts—11–14 ft. lbs. (15–20 Nm) for Scoupe.

• Strut-to-fender nuts—14–22 ft. lbs. (20–30 Nm) for Accent.

• Strut-to-fender nuts—25–33 ft. lbs. (35–45 Nm) for Elantra and Tiburon.

• Strut-to-fender nuts—18–25 ft. lbs. (25–34 Nm) for Sonata.

10. Install the tire and wheel and lower the vehicle.

Rear

EXCEPT SCOUPE

1. Remove the access panel and locate the strut upper mounting nuts.

2. Remove the strut upper mounting nuts.

3. Raise and support the vehicle safely.

4. Remove the rear wheels.

5. If equipped with ABS, disconnect the wheel sensor electrical harness.

6. Disconnect the stabilizer link from the strut body.

7. Support the rear suspension assembly with a floor jack.

8. Unbolt the strut from the knuckle.

9. Remove the strut from the vehicle.

To install:

10. Before installing the strut, be sure the surface where the strut attaches to the knuckle is clean. This ensures a good connection.

11. Install the strut assembly and tighten mounting bolts/nuts as follows:

• Strut-to-body—14–22 ft. lbs. (20–30 Nm)

• Strut-to-knuckle bolts—80–90 ft. lbs. (110–130 Nm)

• Stabilizer link-to-knuckle bolt—25–33 ft. lbs. (35–45 Nm)

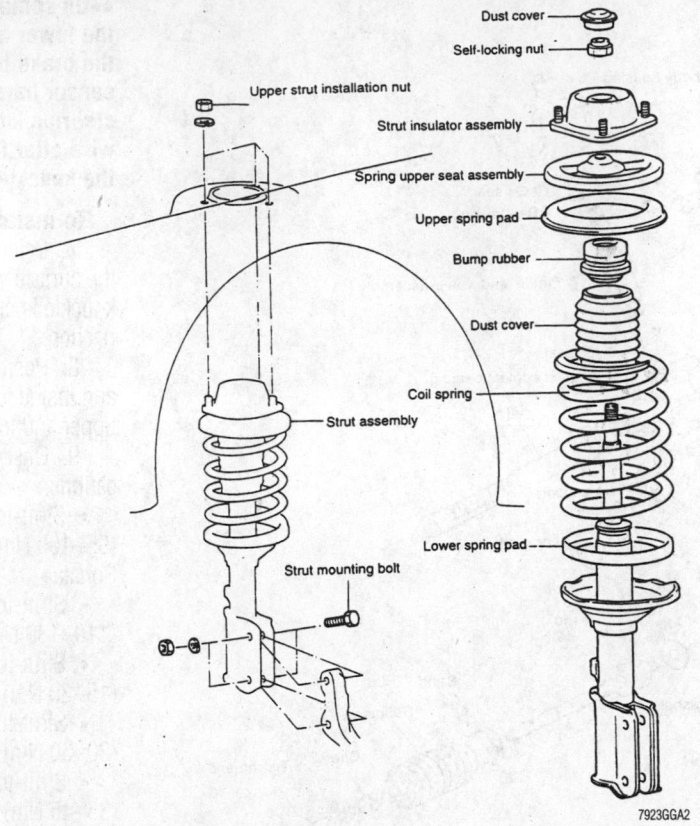

Dust cover

Self-locking nut

Upper strut installation nut

Strut insulator assembly

Spring upper seat assembly

Upper spring pad

Bump rubber

Dust cover

Coil spring

Strut assembly

Lower spring pad

Strut mounting bolt

7923GGA2

Exploded view of the strut assembly components

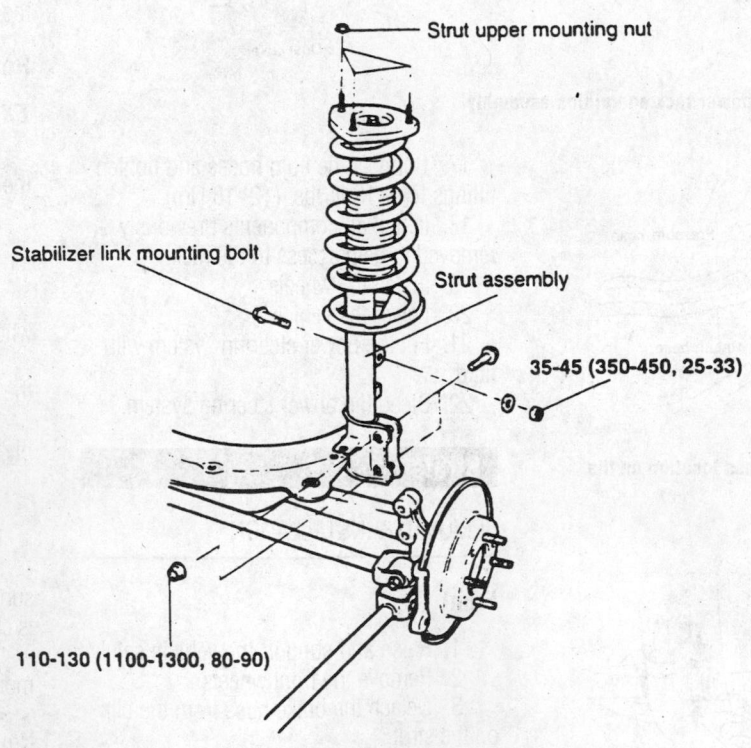

Strut upper mounting nut

Stabilizer link mounting bolt

Strut assembly

35–45 (350–450, 25-33)

110–130 (1100–1300, 80-90)

TORQUE : Nm (kg·cm, lb·ft)

7923GGA3

Exploded view of the rear strut assembly used in multi-link rear suspension

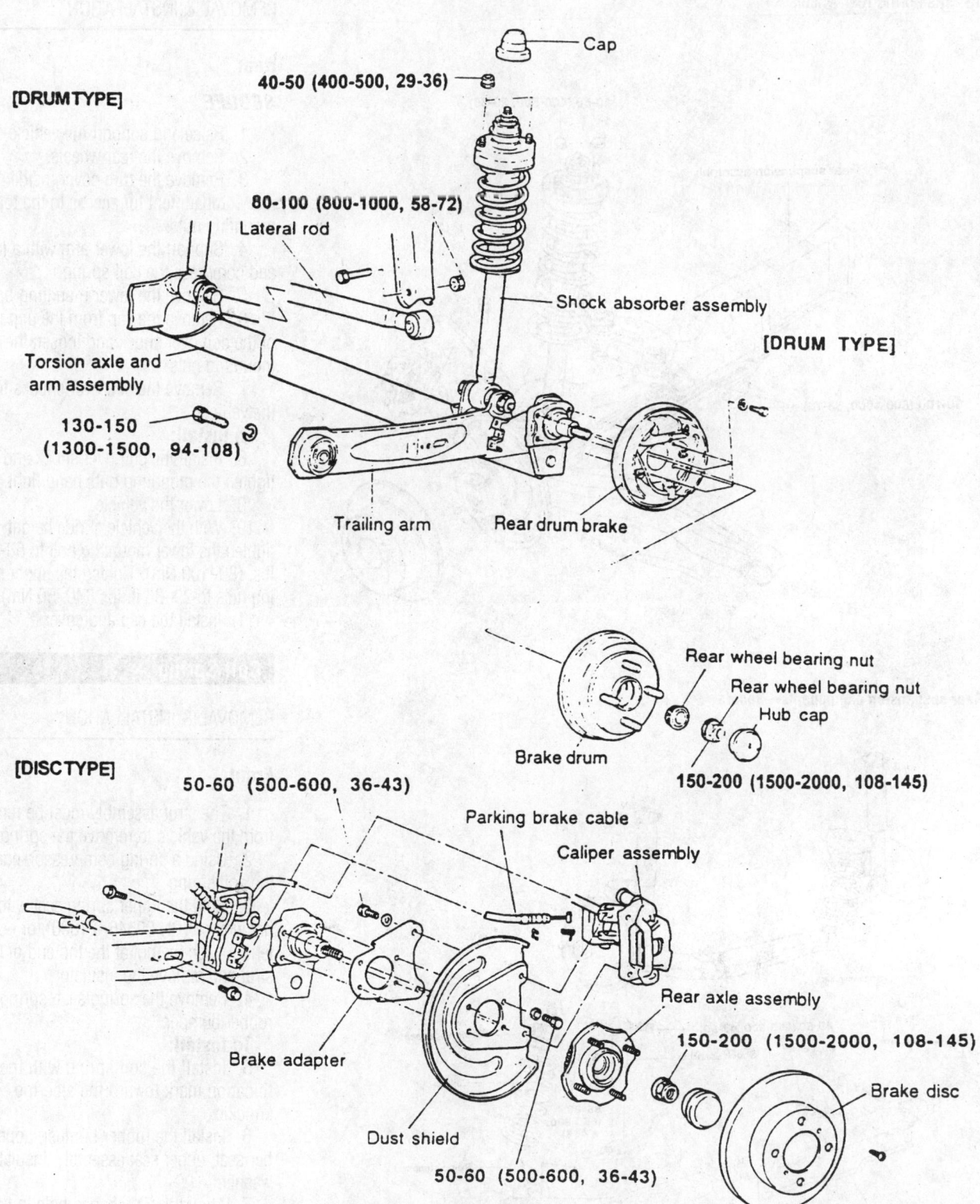

[DRUMTYPE]

Cap

40-50 (400-500, 29-36)

80-100 (800-1000, 58-72)
Lateral rod

Shock absorber assembly

[DRUM TYPE]

Torsion axle and
arm assembly

130-150
(1300-1500, 94-108)

Trailing arm

Rear drum brake

Rear wheel bearing nut
Rear wheel bearing nut
Hub cap

Brake drum

150-200 (1500-2000, 108-145)

[DISCTYPE]

50-60 (500-600, 36-43)

Parking brake cable

Caliper assembly

Brake adapter

Rear axle assembly

150-200 (1500-2000, 108-145)

Brake disc

Dust shield

50-60 (500-600, 36-43)

TORQUE : Nm (kg.cm, lb.ft)

7923GGA8

Rear suspension components—1995 Elantra

12. If equipped with ABS, connect the wheel sensor electrical harness.
13. Install the rear wheels.

14. Lower the vehicle.
15. Install the access panel.

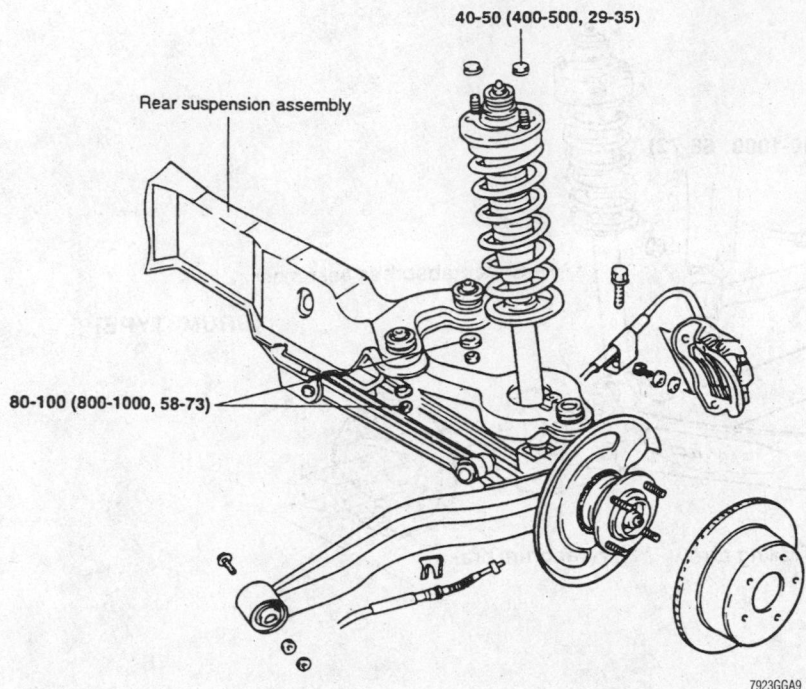

40-50 (400-500, 29-35)

Rear suspension assembly

80-100 (800-1000, 58-73)

7923GGA9

Rear suspension components—Sonata

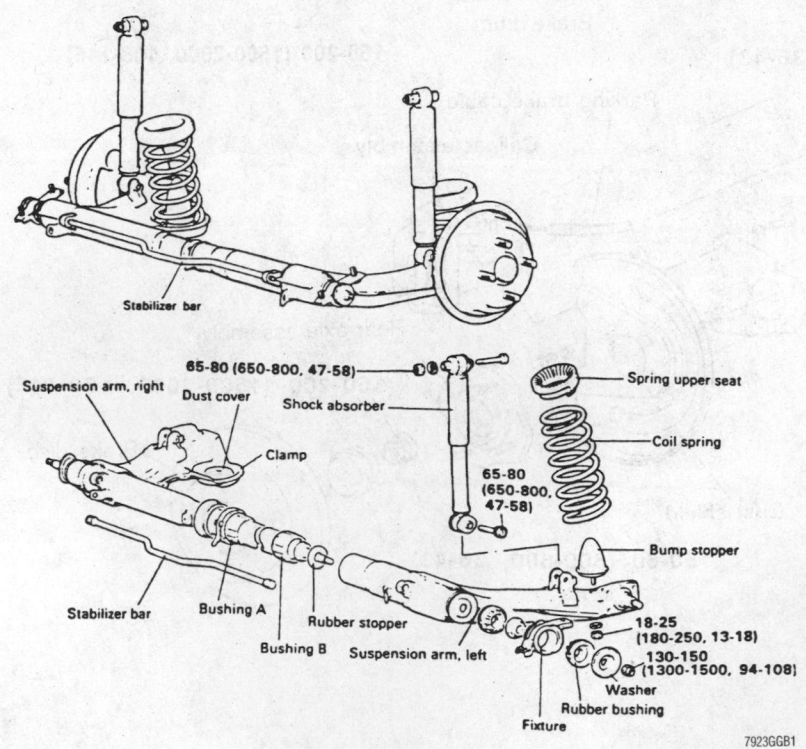

Stabilizer bar

65-80 (650-800, 47-58)

Spring upper seat

Suspension arm, right

Dust cover

Shock absorber

Clamp

Coil spring

65-80 (650-800, 47-58)

Bump stopper

Stabilizer bar

Bushing A

Rubber stopper

Bushing B

Suspension arm, left

18-25 (180-250, 13-18)

130-150 (1300-1500, 94-108)

Washer

Rubber bushing

Fixture

7923GGB1

Exploded view of the rear suspension—Scoupe models

Shock Absorber

REMOVAL & INSTALLATION

Rear

SCOUPE

1. Raise and support the vehicle safely.
2. Remove the rear wheels.
3. Remove the trim cover inside the rear compartment for access to the top mounting nuts.
4. Support the lower arm with a jack and compress the coil spring.
5. Remove the lower mounting bolt.
6. Remove the cap from the upper end of the coil over shock and loosen the mounting nuts.
7. Remove the coil over shocks from the vehicle.

To install:

8. Install the coil over shock and tighten the mounting nuts hand tight.
9. Lower the vehicle.
10. With the vehicle at ride height, tighten the lower mounting bolt to 58–72 ft. lbs. (80–100 Nm). Tighten the upper mounting nuts to 29–36 ft. lbs. (40–50 Nm).
11. Install the cap and cover.

Coil Spring

REMOVAL & INSTALLATION

Front

1. The strut assembly must be removed from the vehicle to remove the spring
2. Using a spring compressor, compress the coil spring.
3. Hold the upper spring seat with spanner wrench (PN 09546–21000), or equivalent, loosen the nut at the top end of the strut and remove the insulator.
4. Remove the spring seat, spring and rubber bumper.

To install:

5. Install the coil spring with the identification mark toward the steering knuckle.
6. Install the rubber bumper, upper rubber seat, upper seat assembly, insulator and washer.
7. Align the "D" shaped hole in the spring seat upper assembly with the indentation o the piston rod.
8. After seating the upper and lower ends of the coil spring in the upper and lower spring seat groves, tighten locknut to 29–36 ft. lbs. (40–50 Nm).

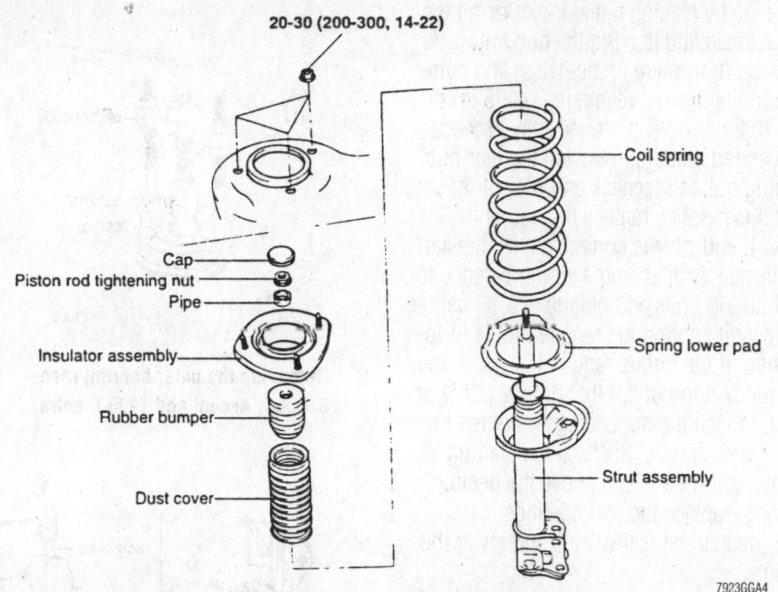

20-30 (200-300, 14-22)

Cap
Piston rod tightening nut
Pipe
Insulator assembly
Rubber bumper
Dust cover

Coil spring
Spring lower pad
Strut assembly

7923GGA4

TORQUE : Nm (kg·cm, lb·ft)

Rear strut components

9. Pack grease in the strut upper bearing and install the cap.

➡**Ensure grease does not contact the insulator rubber.**

Rear

WITH STRUT TYPE SUSPENSION

1. The strut assembly must be removed from the vehicle to remove the spring
2. Using a spring compressor, compress the coil spring.
3. Hold the upper spring seat with spanner wrench (PN 09546–11000), or equivalent, loosen the nut at the top end of the strut and remove the insulator.
4. Remove the spring seat, spring and rubber bumper.

To install:

5. Install the coil spring with the identification mark toward the steering knuckle.
6. Install the rubber bumper, upper rubber seat, upper seat assembly, insulator and washer.
7. Align the "D" shaped hole in the spring seat upper assembly with the indentation o the piston rod.
8. After seating the upper and lower ends of the coil spring in the upper and lower spring seat groves, tighten locknut to 29–36 ft. lbs. (40–50 Nm).
9. Pack grease in the strut upper bearing and install the cap.

➡**Ensure grease does not contact the insulator rubber.**

WITH SHOCK ABSORBER TYPE SUSPENSION

1. Raise the support the vehicle safely.
2. Remove the rear wheels.
3. Support the rear suspension arm with a floor jack.
4. Remove the lower shock absorber attaching bolt, nut and lockwasher.
5. Slowly, lower the jack just to the point where the spring can be removed

➡**If the spring is being replaced, transfer the spring seat to the new spring.**

To install:

6. Install the spring in the reverse order of removal.

➡**Ensure the smaller diameter of the spring is installed upward and the spring identification and load markings match up.**

7. Torque the lower shock mounting bolt to 47–58 ft. lbs. (64–78 Nm).

Lower Ball Joint

REMOVAL & INSTALLATION

Bolt-On Type

1. Raise the vehicle and support it safely.
2. Remove the front wheel.
3. Disconnect the stabilizer bar from the lower arm.

4. Remove the ball joint-to-steering knuckle nut and separate the ball joint from the knuckle.
5. Remove the ball joint-to-lower arm mounting bolts and remove the joint from the arm.
6. Remove the dust cover from the ball joint.

To install:

7. Lubricate the ball joint with grease and install dust cover.
8. Install the joint to the lower arm. Tighten the ball joint mounting bolts to 69–87 ft. lbs. (95–120 Nm).
9. Connect the ball joint to the steering knuckle and tighten nut to 43–52 ft. lbs. (60–72 Nm).
10. Connect the stabilizer bar from the lower arm.
11. Install the front wheel.
12. Lower the vehicle

Press-In Type

1. Raise the vehicle and support it safely.
2. Remove the front wheel.
3. Remove the lower arm from the vehicle.

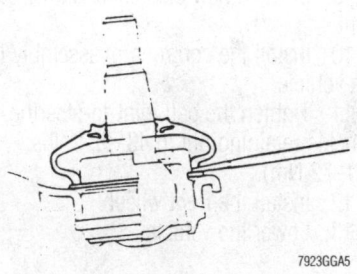

7923GGA5

Use a small prybar to remove the ball joint boot

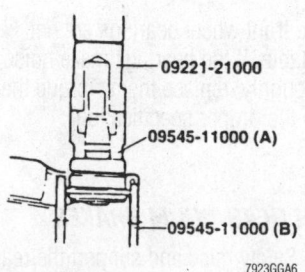

09221-21000
09545-11000 (A)
09545-11000 (B)

7923GGA6

The ball joint is removed from the lower arm using the special tools illustrated and a hydraulic press

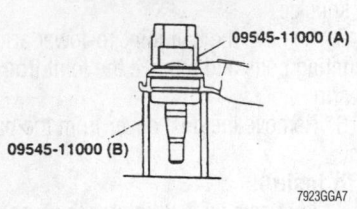

The ball joint is installed in the lower arm using the special tools illustrated and a hydraulic press

4. Remove the ball joint dust cover.

5. Using a special tools (PN 09221–21000, 09545–11000A and 09545–11000B), press the ball joint from the control arm.

To install:

6. Apply grease to the lip of the control arm and to the ball joint contact surfaces.

7. Place the ball joint in the control arm.

8. Using a special tools (PN 09545–11000A and 09545–11000B), press the ball joint into the control arm.

➡**The ball joint must be pressed evenly into the control arm.**

9. Install a new dust cover on the ball joint.

10. Install the control arm assembly into the vehicle.

11. Tighten the ball joint-to-steering knuckle retaining nut to 43–52 ft. lbs. (60–72 Nm).

12. Install the front wheel.

13. Lower the vehicle.

Wheel Bearings

ADJUSTMENT

Front

The front wheel bearings are not adjustable. If the bearings make noise or turn roughly, replace them. Torque the axle nut to the proper specification.

Rear

WITH REAR DRUM BRAKES

1. Safely raise and support the rear of the vehicle.

2. Remove the rear wheels.

3. Remove the grease cap and loosen the axle shaft nut.

4. Tighten the nut to 108–145 ft. lbs. (150–200 Nm). Check for correct bearing

end-play by placing a dial indicator on the hub surface and moving the hub outward. Note the movement of the gauge and compare to the desired reading of 0.008 in. or less (0.2mm or less). If end-play exceeds the desired reading, retighten the rear hub bearing nut and recheck end-play. If reading is still excessive, replace the hub unit.

5. If end-play is correct, check the starting torque by attaching a spring balance to the hub lug bolts and pulling at a 90 degree angle while noting the required force to turn the hub. If the torque required is above the desired reading of 4.9 lbs. or less (22 N or less), loosen the nut and again tighten to the desired torque. Recheck the starting torque. If torque is still above the desired reading, replace the rear bearings.

6. Install the rear wheels and lower the vehicle.

7. Prior to moving the vehicle, pump the brakes until a firm pedal is obtained.

WITH REAR DISC BRAKES

The rear wheel bearing is an integral part of the rear hub and is not serviceable. If the wheel bearing is defective, replace the wheel hub as an assembly.

REMOVAL & INSTALLATION

Front

SCOUPE, ACCENT AND 1995 ELANTRA

➡**The following procedure requires the use of several special tools.**

1. Raise and support the vehicle safely.

2. Remove the steering knuckle assembly.

3. Install first the arm, then the body of special tool (PN 09517–21600) on the knuckle and tighten the nut.

4. Using special tool (PN 09517–21500), separate the hub from the knuckle.

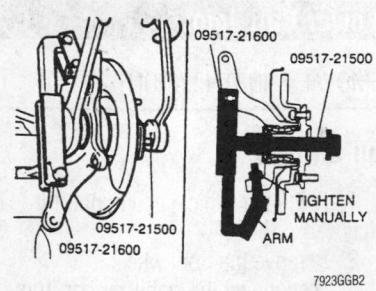

Separating the hub and knuckle—Scoupe, Accent and 1995 Elantra

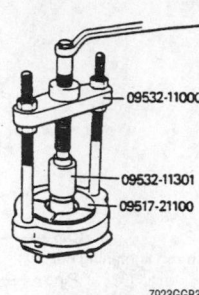

Removing the outer bearing race— Scoupe, Accent and 1995 Elantra

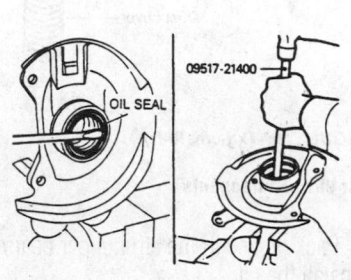

Removing the oil seal and inner bearing race—Scoupe, Accent and 1995 Elantra

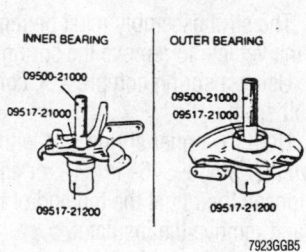

Installing the outer races—Scoupe, Accent and 1995 Elantra

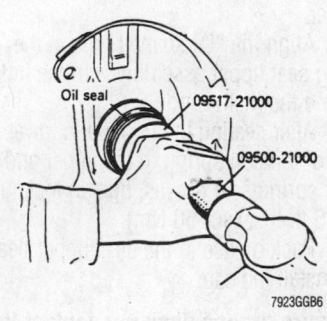

Installing the outer bearing inner race— Scoupe, Accent and 1995 Elantra

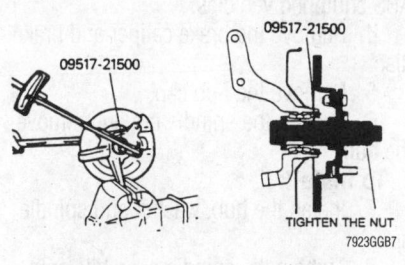

Assembling the hub and knuckle—Scoupe, Accent and 1995 Elantra

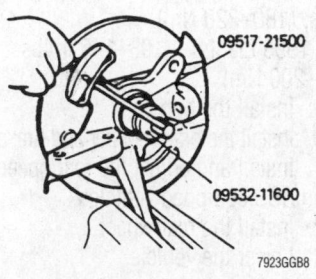

Measuring bearing starting torque—Scoupe, Accent and 1995 Elantra

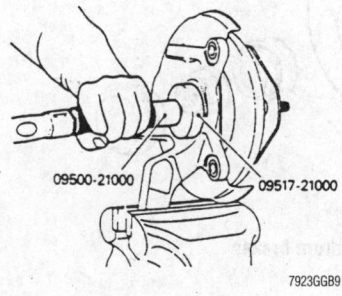

Halfshaft side seal installation—Scoupe, Accent and 1995 Elantra

→Prying or hammering will damage the bearing. Use these special tools, or their equivalent to separate the hub and knuckle.

5. Place the knuckle in a protected jaw vise and separate the rotor from the hub by removing the four attaching bolts.

6. Using special tools (PN 09532–11000, 0953211301 and 09517–21100), remove the outer bearing inner race.

7. Drive the oil seal and inner bearing inner race from the knuckle with a brass drift.

8. Drive out both outer races in a similar fashion.

→Always replace bearings and races as a set. Never replace just an inner or outer bearing. If either is in need of replacement, both sets must be replaced.

9. Thoroughly clean and inspect all parts. Any suspect part should be replaced.

To install:

10. Pack the wheel bearings with lithium based wheel bearing grease. Coat the inside of the knuckle with similar grease and pack the cavities in the knuckle.

→Apply a thin coating of grease to the outer surface of the race before installation.

11. Using special tools (PN 09500–21000, 09517- 21300, and 09517–21200), install the outer races.

12. Install the rotor on the hub and torque the bolts to 36–43 ft. lbs. (50–60 Nm).

13. Drive the outer bearing inner race into position.

14. Coat the out ring and lip of the oil seal and drive the hub side oil seal into place, using a seal driver.

15. Place the inner bearing in the knuckle.

16. Mount the knuckle in a vise. Position the hub and knuckle together. Install tool (PN 09517–21500) and tighten the tool to 145–188 ft. lbs. (200–260 Nm). Rotate the hub to seat the bearing.

17. With the knuckle still in the vise, measure the hub starting torque with an inch lbs. torque wrench and tool (PN 09517–215000). Starting torque should be 11.5 inch lbs. If the starting torque is 0, measure the hub bearing axial play with a dial indicator. If axial play exceeds 0.11mm, while the nut is tightened to specification, the assembly has not been done correctly. Disassemble the knuckle and hub and start again.

18. Remove the special tool.

19. Place the outer bearing in the hub and drive the seal into place.

20. Install the steering knuckle assembly.

21. Lower the vehicle.

SONATA, TIBURON AND 1996–99 ELANTRA

→The following procedure requires the use of several special tools.

1. Raise and support the vehicle safely.

2. Remove the steering knuckle assembly.

3. Remove the snapring from the axle side of the hub.

4. Secure knuckle in a vise and separate the hub and knuckle using special tools (PN 09517–21500, 09517–29000 and 09517–33000), or equivalent.

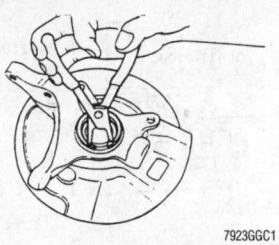

Removing the snapring—Sonata, Tiburon and 1996–99 Elantra

5. Using special tool (PN 09455–21000), or equivalent, remove the outer wheel bearing inner race from the hub.

To install:

6. Install the outer wheel bearing inner race using special tool (PN 09517–21000), or equivalent.

7. Fill the wheel bearing with multipurpose grease.

8. Apply a thin coating of grease to the knuckle and bearing contact surfaces.

9. Press the wheel bearing into the knuckle using special tool (PN 09517–215000, or equivalent.

10. Install the snap ring.

11. Measure the wheel bearing starting rotation torque using a torque wrench and special tools (PN 09517–21500 and 09532–11600), or equivalent. Starting torque should be 11.5 inch lbs. (1.3 Nm).

12. Install the steering knuckle.

13. Lower the vehicle.

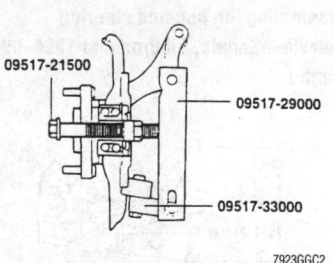

Separating the hub and knuckle—Sonata, Tiburon and 1996–99 Elantra

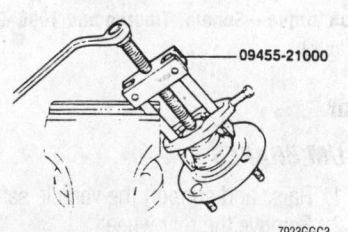

Removing the outer wheel bearing inner race—Sonata, Tiburon and 1996–99 Elantra

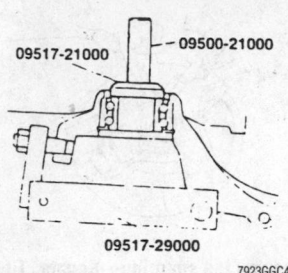

Installing the outer wheel bearing inner race—Sonata, Tiburon and 1996–99 Elantra

Pressing the bearing into the steering knuckle—Sonata, Tiburon and 1996–99 Elantra

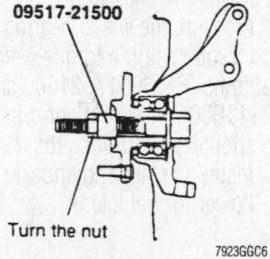

Assembling the hub and steering knuckle—Sonata, Tiburon and 1996–99 Elantra

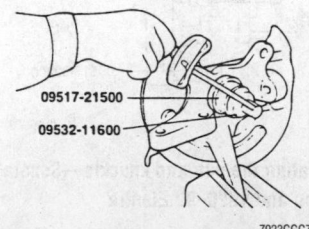

Measuring the wheel bearing starting rotation torque—Sonata, Tiburon and 1996–99 Elantra

Rear

DRUM BRAKES

1. Raise and support the vehicle safely.
2. Remove the rear wheel.
3. Remove the dust cap.
4. Loosen the spindle nut and remove the brake drum.

5. Remove the outer bearing from the drum.
6. Using a prybar, remove the oil seal from inside the drum.
7. Remove the inner bearing.
8. Using a brass drift, remove the inner and outer bearing races.

To install:

9. Using a race driver, install the inner and outer bearing races.
10. Lubricate and install the inner bearing.
11. Using a seal driver, install the oil seal on the inside of the drum.
12. Lubricate and install the outer bearing.
13. Install the brake drum and tighten the spindle nut to 108–145 ft. lbs. (150–200 Nm). Turn the brake drum while tightening the spindle nut to seat the wheel bearings.
14. Install the dust cap.
15. Install the rear wheel.
16. Lower the vehicle.

DISC BRAKES

1. Raise and support the vehicle safely.
2. Remove the rear wheel.

3. Remove the rear speed sensor on ABS equipped vehicles.
4. Remove the brake caliper and brake disc.
5. Remove the hub cap.
6. Loosen the spindle nut and remove the hub.

To install:

7. Install the hub, washer and spindle nut.
8. Tighten the spindle nut while spinning the hub to the following torque:
  • Tiburon—143–164 ft. lbs. (200–230 Nm)
  • Sonata—146–189 ft. lbs. (200–260 Nm)
  • Accent and 1996–99 Elantra-130–159 ft. lbs. (180–220 Nm)
  • 1995 Elantra—108–145 ft. lbs. (150–200 Nm)
9. Install the hub cap.
10. Install the brake caliper and brake disc.
11. Install and adjust the rear speed sensor on ABS equipped vehicles.
12. Install the rear wheel.
13. Lower the vehicle.

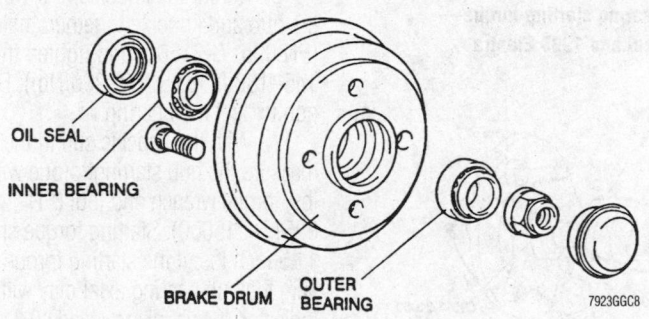

Exploded view of the wheel bearing assembly—with drum brakes

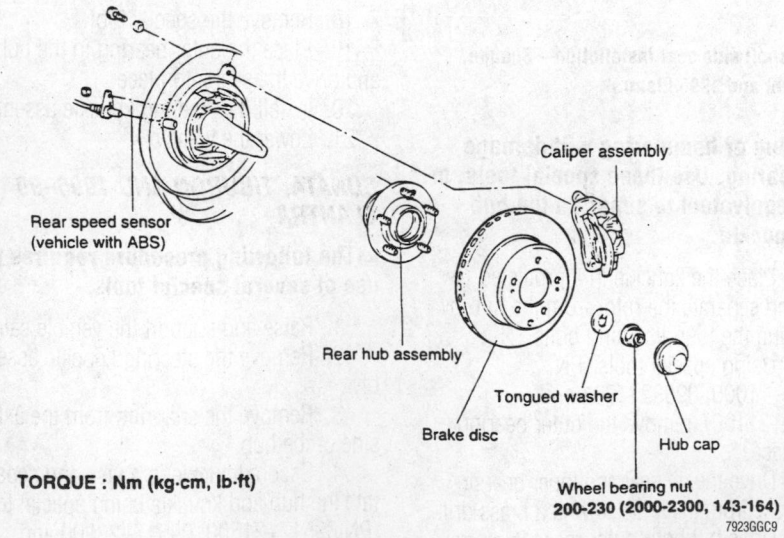

TORQUE : Nm (kg·cm, lb·ft)

Exploded view of the wheel bearing assembly—with disc brakes

INFINITI

G20 • I30 • J30 • Q45

ENGINE REPAIR

➡**Disconnecting the negative battery cable on some vehicles may interfere with the functions of the on board computer systems and may require the computer to undergo a relearning process, once the negative battery cable is reconnected.**

Distributor

REMOVAL

2.0L Engine

1. Disconnect the negative battery cable.
2. Remove the splash shield, if equipped. Unplug the distributor connections but leave the ignition wires in place.
3. Unscrew the distributor cap hold-down screws and lift off the distributor cap with all ignition wires still connected.
4. Matchmark the rotor to the distributor housing and the distributor housing to the engine.

➡**Do not crank the engine during this procedure. If the engine is cranked, the matchmark must be disregarded.**

5. Remove the hold-down bolt.
6. Remove the distributor from the engine.

3.0L (VG30DE and VQ30DE), 4.1L and 4.5L Engines

The J30 and I30 3.0L engine is equipped with a distributorless ignition. As a result, there is no distributor to remove.

INSTALLATION

2.0L Engine

ENGINE NOT DISTURBED

1. If the engine was not disturbed, proceed as follows:
   a. Install a new distributor housing O-ring.
   b. Install the distributor in the engine so the rotor is aligned with the matchmark on the housing and the housing is aligned with the matchmark on the engine. Be sure the distributor is fully seated and the distributor gear is fully engaged.
   c. Install and snug the hold-down bolt.
   d. Connect the distributor pick-up lead wires.

e. Install the distributor cap and tighten the screws. Install the splash shield.
   f. Connect the negative battery cable.
   g. Check and/or adjust the ignition timing and tighten the hold-down bolt.

ENGINE DISTURBED

1. If the engine was disturbed (cranked or turned over with the distributor removed), proceed as follows:
   a. Install a new distributor housing O-ring.
   b. Position the engine so the No. 1 piston is at TDC of its compression stroke and the mark on the vibration damper is aligned with **0** on the timing indicator.
   c. Install the distributor in the engine so the rotor is aligned with the position of the No. 1 ignition wire on the distributor cap. Be sure the distributor is fully seated and that the distributor shaft is fully engaged.

➡**There are distributor cap runners inside the cap on 2.0L engine. Be sure the rotor is pointing to where the No. 1 runner originates inside the cap.**

   d. Install and snug the hold-down bolt.
   e. Connect the distributor pick-up lead wires.
   f. Install the distributor cap and tighten the screws. Install the splash shield, if equipped.
   g. Connect the negative battery cable.
   h. Check and/or adjust the ignition timing and tighten the hold-down bolt to 9–12 ft. lbs. (13–16 Nm).

3.0L (VG30DE and VQ30DE), 4.1L and 4.5L Engines

The J30 and I30 3.0L engine is equipped with a distributorless ignition. As a result, there is no distributor to remove.

Ignition Timing

ADJUSTMENT

2.0L Engine

➡**The engine should be in good mechanical condition and all electrical connectors and vacuum hoses connected before making this adjustment.**

1. Start the engine and let it warm up to normal operating temperature.
2. Open the hood and run the engine under no load at about 2,000 rpm for about two minutes.

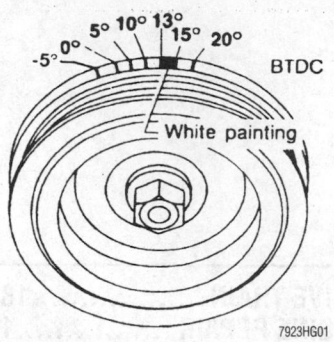

Crankshaft pulley and timing marks—2.0L engine

3. Perform Diagnostic Test Mode II and repair any causes of trouble codes as needed.
4. Run the engine under no load at 2,000 rpm for about two minutes. Rev the engine two or three times and let it idle for one minute.
5. Turn **OFF** the engine and disconnect the throttle position sensor connector. Connect a timing light to the No. 1 spark plug wire. Start the engine.
6. Adjust the timing to 15° ±2° BTDC by loosening the distributor mounting bolts and turning the distributor. When the timing is correct, tighten the mounting bolts and turn the engine **OFF**.
7. Reconnect the throttle position sensor connector. Start the engine and check the ignition timing again.

3.0L (VG30DE and VQ30DE), 4.1 and 4.5 Engines

➡**The engine should be in good mechanical condition and all electrical connectors and vacuum hoses attached before making this adjustment.**

1. Start the engine and let it warm up to normal operating temperature.
2. Open the hood and run the engine under no load at about 2,000 rpm for about two minutes.
3. Perform Diagnostic Test Mode II and repair any causes of trouble codes as needed.
4. Run the engine under no load at 2,000 rpm for about two minutes. Rev the engine two or three times and let it idle for one minute.
5. Turn **OFF** the engine and disconnect the throttle position sensor connector. Remove the No. 1 ignition coil. Connect the coil to the spark plug using a spare piece of high-tension wire so you have a place to connect your timing light. Start the engine.

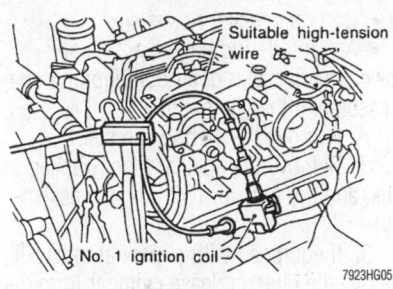

Connect the No. 1 ignition coil to the spark plug with the spare piece of high-tension wire—4.1L engine shown

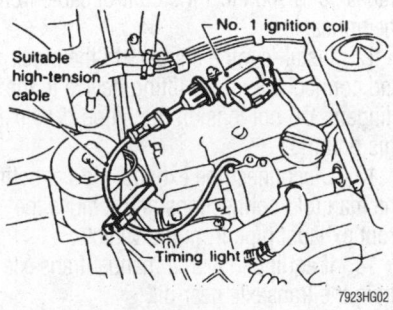

Timing light connection to spark plug cable—3.0L (VG30DE) engine

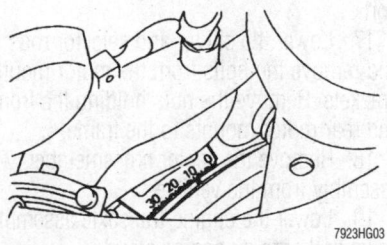

Location of timing marks—3.0L (VQ30DE) engine

6. Run the engine under no load at 2,000 rpm for about two minutes. Rev the engine two or three times and let it idle.

7. Check the ignition timing and adjust if needed.

• Correct ignition timing for 3.0L engines is 8–12° BTDC.

• Correct ignition timing for the 4.1L and 4.5L engines is 13–17° BTDC.

8. Adjustment is made by loosening the screws and turning the camshaft position sensor until the mark on the crankshaft pulley is pointing at 10° BTDC. Tighten the mounting screws and confirm ignition timing has not changed.

9. Turn the engine **OFF** and connect the throttle position sensor connector.

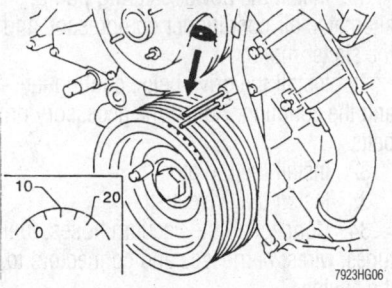

Location of timing marks—4.1L engine

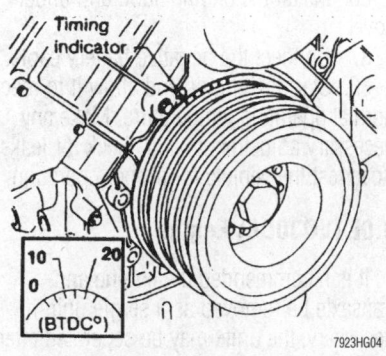

Location of timing marks—4.5L engine

Engine Assembly

REMOVAL & INSTALLATION

2.0L Engine

1. Disconnect the negative battery cable.

2. Raise and support the vehicle safely.

3. Remove the engine undercover.

4. Matchmark the hood with the hood hinges and remove.

5. Drain the coolant from both the cylinder block and radiator.

6. Drain the engine and transaxle oil.

✳ CAUTION

The fuel injection system remains under pressure after the engine has been turned OFF. Properly relieve fuel pressure before disconnecting any fuel lines. Failure to do so may result in fire or personal injury.

7. Release fuel system pressure and remove fuel line.

8. Label and remove all vacuum lines and wiring harness connectors.

9. Remove exhaust tubes, ball joints, and driveshafts.

10. Remove the radiator and fans.

11. Remove the drive belts.

12. Remove the alternator, A/C compressor, and the power steering pump from the engine and lay them aside. Do not disconnect the compressor or power steering pump lines.

13. Support the engine with a hoist and the transaxle with a suitable jack. Raise the engine and transaxle slightly and remove the center member.

14. Remove the engine mounting bolts from both sides and slowly lower the hoist and transaxle jack.

15. Remove the engine and transaxle from beneath the vehicle.

To install:

16. Install the center member bracket (manual transmission) on the engine, if removed. Ensure that all insulators are correctly positioned on the brackets. Torque insulator through-bolts to 32–41 ft. lbs. (43–55 Nm).

17. If equipped with manual transaxle, ensure that the distance between the center of the insulator through-bolt and the center member is 2.28–2.36 in. (58–60mm). Torque through-bolt to 46–58 ft. lbs. (62–78 Nm).

18. Carefully install the engine and tighten the center member-to-frame bolts to 57–72 ft. lbs. (77–98 Nm).

19. Install the alternator, A/C compressor, and power steering pump.

20. Connect all vacuum hoses and wiring harness connectors. Connect the fuel line.

21. Install the exhaust tubes, ball joints, driveshafts, the radiator and fans and drive belts.

22. Fill the cooling system and fill the crankcase and transaxle with the proper oil.

23. Install the engine undercover and hood.

24. Connect the negative battery cable and road test the vehicle for proper operation.

3.0L (VG30DE) Engine

✳ CAUTION

The fuel injection system remains under pressure after the engine has been turned OFF. Properly relieve fuel pressure before disconnecting any fuel lines. Failure to do so may result in fire or personal injury.

1. Relieve the fuel system pressure.

2. Disconnect the negative battery cable.

3. Remove the engine undercover. Remove the hood after matchmarking it to the hinges.

4. Drain the cooling system. Use the cylinder block drain plugs to drain the engine.

5. Label and disconnect all the vacuum hoses, fuel tubes, wires, harnesses and connectors from the engine.

6. Remove the driveshaft.

7. Remove the radiator.

8. Remove the drive belts, cooling fan, and the coupling.

9. Raise and support the vehicle safely.

10. Remove the power steering pump, alternator, air conditioner compressor, and the starter motor from the engine.

11. Remove the exhaust tube front nuts. Remove the front tubes after removing the exhaust tube bracket.

12. Remove the fluid filler pipe from the transmission.

13. Remove the cooler pipes from the transmission.

14. Remove the control linkage from the selector lever.

15. Disconnect the inhibitor switch and solenoid harness connectors.

16. Remove the gusset securing the transmission to the engine.

17. Remove the crankshaft position sensor to prevent it from being damaged.

18. Remove the bolts securing torque converter to the driveplate.

19. Support the engine and separate engine from transmission.

20. Lift the engine slightly and remove the engine mounting bolts from both sides.

21. Carefully remove the engine from the top side of the vehicle.

To install:

22. Lower the engine into the vehicle and tighten the engine mounting bolts to 32–41 ft. lbs. (43–55 Nm).

23. Install the transmission. Torque the converter bolts to 33–43 ft. lbs. (44–59 Nm). Torque the engine-to-transmission bolts as follows:

   a. Torque the upper six bolts to 29–36 ft. lbs. (39–49 Nm).

   b. Torque the lower four bolts to 22–29 ft. lbs. (29–39 Nm).

24. Install the crankshaft position sensor.

25. Connect the inhibitor switch and solenoid harness connectors.

26. Install the control linkage on the selector lever.

27. Install the fluid cooler pipes on the transmission.

28. Install the fluid filler pipe on the transmission.

29. Install the front exhaust tubes and exhaust tube bracket. Torque the nuts to 33–44 ft. lbs. (45–60 Nm).

30. Install the power steering pump, alternator, air conditioner compressor, and the starter motor.

31. Install the drive belts, cooling fan, and the coupling. Adjust the accessory drive belts.

32. Install the radiator.

33. Install the driveshaft.

34. Connect all the vacuum hoses, fuel tubes, wires, harnesses and connectors to the engine.

35. Check and fill the cooling system, engine oil, transmission fluid and power steering fluid.

36. Install the engine hood and undercover.

37. Connect the negative battery cable.

38. Start the engine and allow it to reach normal operating temperature. Make any necessary adjustments and check for leaks. Road test the vehicle for proper operation.

3.0L (VQ30DE) Engine

It is recommended the engine and transaxle be removed as a single unit. If necessary, the units may be separated after removal.

➡**The engine and transaxle assembly must be removed from the under side of the vehicle.**

1. Matchmark the hood hinge relationship and remove the hood.

2. Release the fuel system pressure and disconnect the negative battery cable.

3. Raise and safely support the vehicle.

4. Drain the coolant from the cylinder block and the radiator. Drain the crankcase and the automatic transaxle, if equipped.

5. Remove the air cleaner, the air intake tube, the air flow meter and disconnect the throttle linkage.

6. Disconnect and/or remove the following:

• Drive belts
• Engine ground cable
• Electrical connector from the crank angle sensor
• Engine electrical harness connectors

❊❊ CAUTION

The fuel injection system remains under pressure after the engine has been turned OFF. Properly relieve the fuel pressure before disconnecting any fuel lines. Failure to do so may result in fire or personal injury.

• Fuel feed and fuel return hoses
• Upper and lower radiator hoses
• Heater inlet and outlet hoses

• Engine vacuum hoses
• Carbon canister hoses
• Any interfering engine accessory: power steering pump, air conditioning compressor, and the alternator

7. Remove the carbon canister.

8. Remove the auxiliary fan, washer tank, and the radiator (with the fan assembly).

9. If equipped with a manual transaxle, remove the clutch release cylinder from the clutch housing.

10. If equipped with a manual transaxle, disconnect the shift control rod and disconnect the shift support rod.

11. If equipped with a automatic transaxle, disconnect the control cable from the transaxle.

12. Install engine slingers to the block and connect a suitable lifting device to the slingers. Do not tension the lifting device at this point.

13. Disconnect the exhaust pipe at both the manifold connections and remove the front exhaust pipe from the vehicle.

14. If equipped with a manual transaxle, drain the transaxle gear oil.

15. Support the engine and transaxle assembly with proper jack.

16. Disconnect the right and left side halfshafts from their side flanges and remove the bolt holding the radius link support.

17. Lower the shifter and selector rods and remove the bolts from the motor mount brackets. Remove the nuts holding the front and rear motor mounts to the frame.

18. Remove the center crossmember assembly from the vehicle.

19. Lower the engine/transaxle assembly down and onto an engine stand.

To install:

20. Raise the engine/transaxle assembly into the vehicle. When raising the engine onto the mounts, be sure to keep it as level as possible.

21. Check the clearance between the frame and clutch housing and be sure the engine mount bolts are seated in the groove of the mounting bracket.

22. Remove the transaxle and engine jack assembly.

23. Install the center crossmember and secure the engine mounting bolts.

24. If removed, raise the shifter and selector rods to their normal operating positions and secure them with the mounting bolts.

25. Install the halfshafts.

26. Connect the exhaust pipe assembly.

27. Disconnect and remove the engine slingers.

28. If equipped with a automatic transaxle, connect the control cable to the transaxle.

29. If equipped with a manual transaxle, install the clutch release cylinder to the clutch housing.

30. Install the auxiliary fan, washer tank, and the radiator (with the fan assembly).

31. Install the carbon canister.

32. Connect and/or install the following:
- Any interfering engine accessory: power steering pump, air conditioning compressor, and the alternator
- Carbon canister hoses
- Engine vacuum hoses
- Heater inlet and outlet hoses
- Upper and lower radiator hoses
- Fuel feed and fuel return hoses
- Engine electrical harness connectors
- Electrical connector from the crank angle sensor
- Engine ground cable
- Drive belts

33. Connect the throttle linkage, then install the air cleaner, the air flow meter, and the air intake duct.

34. Fill the transaxle, the engine, and the cooling system to the proper levels with the appropriate fluids.

35. Install the hood and connect the negative battery cable.

36. Make all the necessary engine adjustments. Charge the air conditioning system, if discharged. Road test the vehicle for proper operation.

4.1L and 4.5L Engines

✳✳ CAUTION

The fuel injection system remains under pressure after the engine has been turned OFF. Properly relieve fuel pressure before disconnecting any fuel lines. Failure to do so may result in fire or personal injury.

1. Disconnect the negative battery cable.

2. Relieve the pressure from the fuel system.

3. Mark the relation of the hood to the hinge brackets and remove the hood.

4. Raise and safely support the vehicle.

5. Remove the engine splash shield.

6. Drain the coolant and the engine oil.

7. Disconnect the transmission cooler lines from the radiator.

8. Remove the radiator hoses and remove the radiator and shroud.

9. Tag and disconnect all the vacuum hoses, fuel lines, and electrical connectors.

10. Disconnect the exhaust pipes from the exhaust manifolds.

11. Mark the position of the driveshaft on the flanges and remove the driveshaft.

12. Remove the accessory drive belts.

13. Remove the alternator, air conditioning compressor, and the power steering pump. Do not disconnect the coolant or hydraulic lines from the compressor and pump.

14. Remove the lower steering joint.

15. Remove the sway bar, transverse link, and the tension rod with bracket.

16. Place a suitable jack under the transmission and attach a hoist to the engine.

17. Disconnect the transmission rear mount.

18. Remove the suspension member attaching bolts.

19. Remove the engine mounting bolts.

20. Attach a suitable hoist to the engine. Lower the transmission jack and the hoist and lower the engine and transmission from under the vehicle.

To install:

21. Install the engine and transmission into position. Check the clearance between the frame and transaxle and be sure the engine mount bolts are seated in the groove of the mounting bracket.

22. Torque the engine mounts to the following specifications:

a. Front engine mount bracket-to-engine bolts to 32–41 ft. lbs. (43–55 Nm).

b. Front engine mount-to-frame bolts to 41–49 ft. lbs. (55–67 Nm).

c. Rear engine mount-to-crossmember bolts to 16–21 ft. lbs. (22–28 Nm).

d. Rear crossmember-to-frame bolts to 32–41 ft. lbs. (43–55 Nm).

23. Install the suspension member bolts.

24. Remove the hoist and the transmission jack.

25. Install the sway bar, transverse link, and the tension rod with bracket.

26. Install the lower steering joint.

27. Install the alternator, air conditioning compressor, and the power steering pump.

28. Install and adjust the accessory drive belts.

29. Install the radiator and shroud. Install the radiator hoses.

30. Install the driveshaft, aligning the marks, that were made during the removal procedure.

31. Connect the exhaust pipes to the exhaust manifolds.

32. Connect all the electrical connectors, fuel lines, and vacuum hoses.

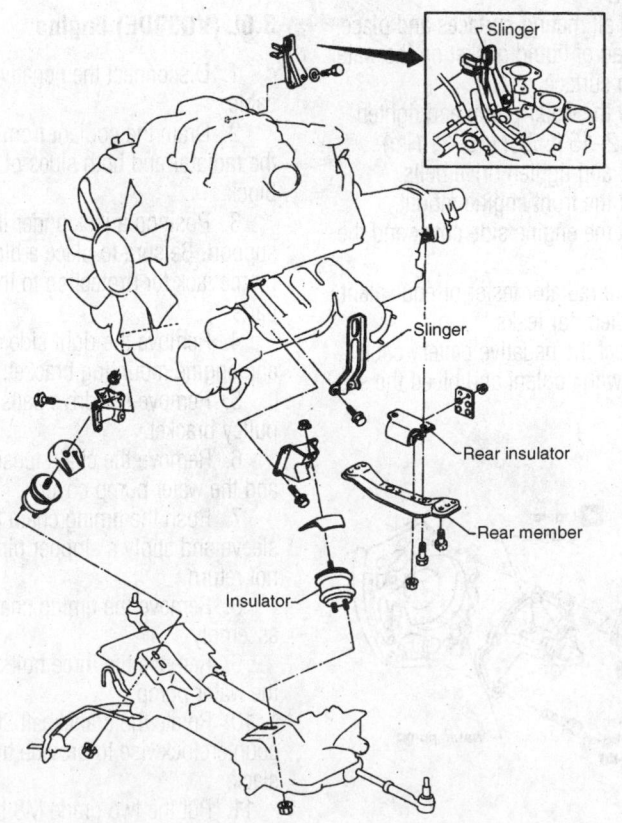

7923HG07

Engine mounts and brackets—4.5L engine shown, 4.1L engine is similar

33. Install the transmission cooling lines and install the engine splash shield.

34. Fill the crankcase with the proper type of engine oil to the required level. Fill the cooling system with the proper type and quantity of coolant.

35. Install the hood, aligning the marks that were made during the removal procedure.

36. Connect the negative battery cable, start the engine and check for leaks. Road test the vehicle for proper operation.

Water Pump

REMOVAL & INSTALLATION

2.0L Engine

1. Disconnect the negative battery cable.
2. Drain the coolant from the radiator and engine block. The drain plug in the engine block is located at the left front of the cylinder block.
3. Remove the right wheel and the engine side cover.
4. Remove the drive belts.
5. Remove the front engine mount.
6. Loosen the water pump attaching bolts and remove the water pump. Take care not to drip coolant on the drive belts.

To install:

7. Clean all mating surfaces and place a 2–3mm bead of liquid gasket on the water pump mating surface.
8. Install the water pump and tighten the bolts to 12–15 ft. lbs. (16–21 Nm).
9. Install and tighten drive belts.
10. Install the front engine mount.
11. Install the engine side cover and the right wheel.
12. Using a radiator tester or equivalent, check the system for leaks.
13. Connect the negative battery cable.
14. Refill with coolant and bleed the system of air.

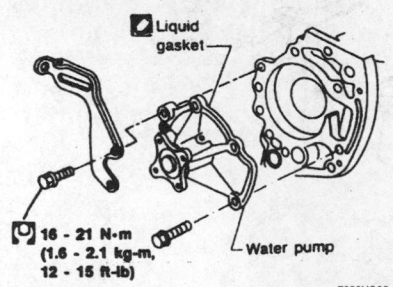

Exploded view of the water pump mounting—2.0L engine

3.0L (VG30DE) Engine

1. Disconnect the negative battery cable.
2. Drain the coolant from the radiator and from the drain plugs on both sides of the cylinder block.
3. Remove the cooling fan assembly. Remove the timing belt covers.

➡ **Use the proper precautions to avoid getting coolant on the timing belt.**

4. Remove the water pump mounting bolts and remove the pump from the engine.

To install:

5. Thoroughly clean and dry the mating surfaces, bolts and bolt holes.
6. Apply liquid gasket to the water pump and install to the engine. Torque the bolts to 12–15 ft. lbs. (16–21 Nm).
7. Open the air release plug, as required. Fill the cooling system and check for leaks using a pressure tester before continuing.
8. Install the timing belt covers and all related parts.
9. Connect the negative battery cable, run the vehicle until the thermostat opens and fill the radiator completely. Recheck for coolant leaks.
10. Once the vehicle has cooled, recheck the coolant level.

3.0L (VQ30DE) Engine

1. Disconnect the negative battery cable.
2. Drain the coolant from the plugs on the radiator and both sides of the engine block.
3. Position a jack under the oil pan for support. Be sure to place a block of wood on the jack for protection to the engine parts.
4. Remove the right side engine mount and engine mounting bracket.
5. Remove the drive belts and the idler pulley bracket.
6. Remove the chain tensioner cover and the water pump cover.
7. Push the timing chain tensioner sleeve and apply a stopper pin so it does not return.
8. Remove the timing chain tensioner assembly.
9. Remove the three bolts that secure the water pump.
10. Rotate the crankshaft 20 degrees counterclockwise to provide timing chain slack.
11. Put the two grade M8 bolts in the two M8 threaded holes of the water pump.

12. Tighten each bolt by turning alternately ½ turn until they reach the timing chain rear case. Be sure to turn each bolt ½ turn at a time to prevent damage.
13. Lift up the water pump and remove it.
14. When removing the water pump, do not allow the water pump gear to hit the timing chain.
15. Remove and discard the O-rings from the water pump.
16. Clean all traces of liquid gasket from the water pump and covers.

To install:

17. Using new O-rings, install the water pump to the engine block.
18. Tighten the three water pump mounting bolts evenly to 62–86 inch lbs. (7–10 Nm).
19. Rotate the crankshaft pulley to its original position by turning it 20 degrees clockwise.
20. Install the timing chain tensioner and tighten the mounting bolts to 75–96 inch lbs. (9–10 Nm).
21. Remove the stopper pin from the timing chain tensioner.
22. Apply a continuous 0.091–0.130 in. (2.3–3.3mm) bead of liquid sealant to the mating surfaces of the timing chain tensioner and water pump covers.
23. Install the timing chain tensioner and water pump covers to the engine block. Tighten the cover mounting bolts to 84–108 inch lbs. (10–13 Nm).
24. Install the drive belts and the idler pulley bracket.
25. Install the right side engine mounting bracket and the engine mount.
26. Remove the jack from under the engine and install the drain plugs to the cylinder block.
27. Connect the negative battery cable and refill the cooling system.
28. Start the engine, bleed the cooling system, and check for leaks.

4.1L and 4.5L Engines

1. Disconnect the negative battery cable.
2. Drain the coolant from the radiator and from the drain cocks on both sides of the cylinder block.
3. Remove the drive belts.
4. Unbolt the fan shroud and move it backward in order to remove the fan and coupling. Remove the fan to water pump bolts and remove the fan, coupling, water pump pulley and shroud.
5. Remove all necessary accessories to gain access to the water pump.

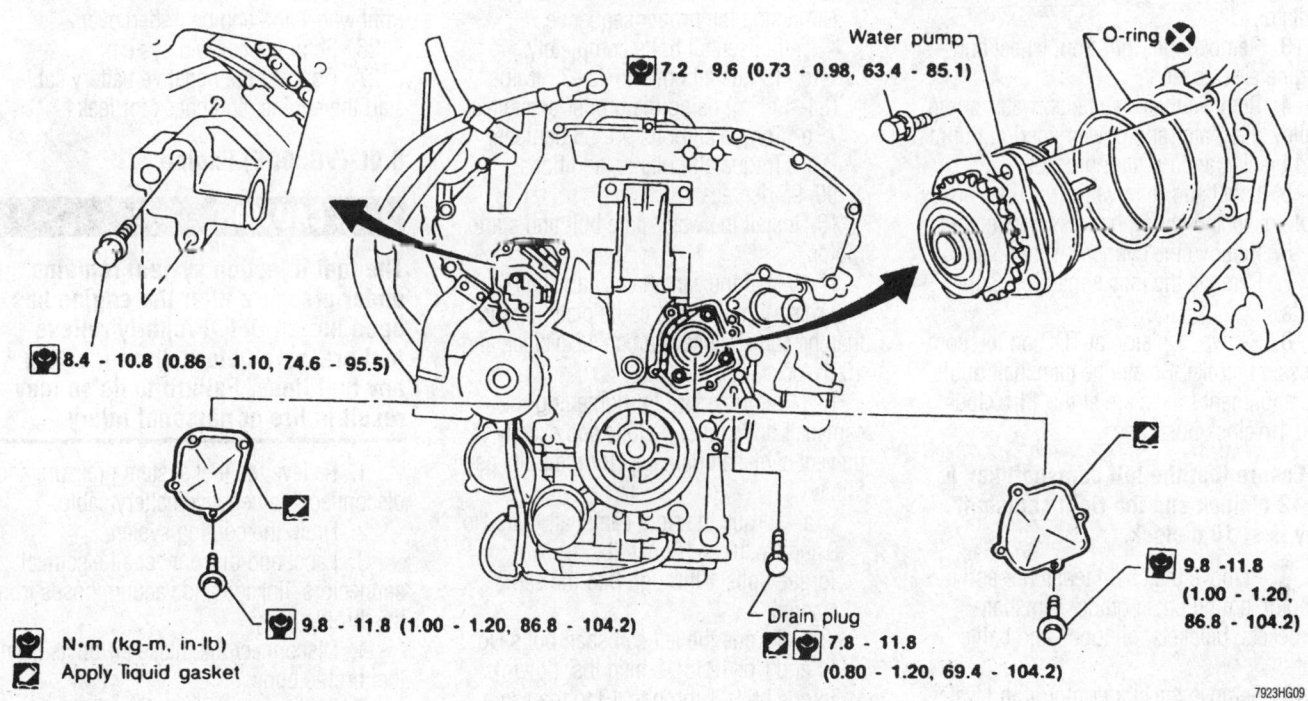

Water pump and timing cover assembly—3.0L (VQ30DE) engine

7.2 - 9.6 (0.73 - 0.98, 63.4 - 85.1)

Water pump

O-ring

8.4 - 10.8 (0.86 - 1.10, 74.6 - 95.5)

9.8 - 11.8 (1.00 - 1.20, 86.8 - 104.2)

N·m (kg-m, in-lb)
Apply liquid gasket

Drain plug
7.8 - 11.8
(0.80 - 1.20, 69.4 - 104.2)

9.8 -11.8
(1.00 - 1.20,
86.8 - 104.2)

7923HG09

6. Note the positioning of the clamp and disconnect the hose from the water pump.

7. Remove the water pump mounting bolts and remove the pump from the engine.

To install:

8. Thoroughly clean and dry the mating surfaces, bolts and bolt holes.

9. Apply liquid gasket to the water pump and install it to the engine. Torque the bolts to 10–13 (14–18 Nm) on the 4.1L or 12–15 ft. lbs. (16–21 Nm) on the 4.5L engine.

10. Connect the hose and install the clamp in the same position as when it was removed. Fill the cooling system and check for leaks using a pressure tester before continuing.

11. Install all removed accessories.

12. Install the shroud, pulley, coupling and fan. Torque the water pump pulley nuts to 7 ft. lbs. (10 Nm). Install and adjust all the belts.

13. Connect the negative battery cable, run the vehicle until the thermostat opens and fill the radiator completely. Recheck for coolant leaks.

14. Once the vehicle has cooled, recheck the coolant level.

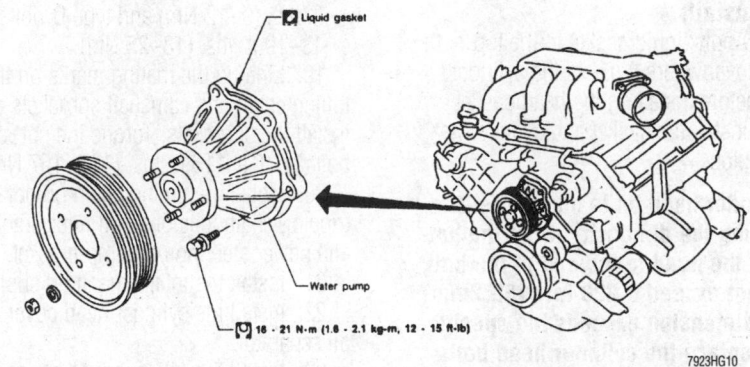

Liquid gasket

Water pump

16 - 21 N·m (1.6 - 2.1 kg-m, 12 - 15 ft-lb)

7923HG10

Water pump assembly—4.5L engine

7923HG11

Apply a continuous bead of RTV sealant to the mounting surface of the water pump assembly—4.1L engine

Cylinder Head

REMOVAL & INSTALLATION

2.0L Engine

✳✳ CAUTION

The fuel injection system remains under pressure after the engine has been turned OFF. Properly relieve fuel pressure before disconnecting any fuel lines. Failure to do so may result in fire or personal injury.

1. Relieve the fuel system pressure and disconnect the negative battery cable.

2. Drain the coolant and remove the radiator.

3. Remove the right front wheel and engine side cover.

4. Remove the drive belts, water pump pulley, alternator and power steering pump and bracket and oil filter bracket.

5. Label and remove the vacuum hoses, fuel hoses and wiring harness connectors.

6. Remove the cylinder head cover.

7. Remove the intake manifold supports.

8. Set No. 1 piston at TDC on the compression stroke. Rotate the camshaft until the alignment marks are at the 11 o'clock and 1 o'clock positions.

➡**Ensure that the left camshaft key is at 12 o'clock and the right camshaft key is at 10 o'clock.**

9. Remove the chain tensioner and distributor, timing chain guide, camshaft sprockets, brackets, oil tubes and baffle plate.

10. Remove the starter motor and water pipe bolt.

11. Loosen the cylinder head bolts in 2–3 steps.

12. Remove the cylinder head with the intake and exhaust manifolds attached.

To install:

13. Apply liquid gasket to the top of the chain cover where it meets the cylinder block before installing the head gasket.

14. Install the gasket and cylinder head on the block.

➡**Cylinder head bolts may be reused providing the dimension from the bottom of the head to the end of the bolt does not exceed 6.228 in. (158.2mm). If the dimension exceeds the specification, replace the cylinder head bolts.**

15. Torque cylinder head bolts as follows:

a. Torque all bolts to 29 ft. lbs. (39 Nm) using the proper sequence.

b. Torque all bolts to 58 ft. lbs. (78 Nm) using the proper sequence.

c. Loosen all bolts completely.

d. Torque all bolts to 25–33 ft. lbs. (34–44 Nm) using the proper sequence.

e. Torque all bolts 90–95 degrees.

f. Torque all bolts an additional 90–95 degrees.

16. Install the water pipe bolt and starter motor.

17. Install the camshafts, camshaft brackets, oil tubes and baffle plate. Ensure that the camshaft keys are at 12 o'clock and 10 o'clock.

18. The procedure for tightening camshaft bolts must be followed exactly to prevent camshaft damage. Torque bolts as follows:

a. Torque the right camshaft bolts No. 9 and No.10 to 18 inch lbs. (2 Nm). Torque bolts 1 through 8 to the same amount.

b. Torque the left camshaft bolts No. 11 and No. 12 to 18 inch lbs. (2 Nm). Torque bolts 1 through 10 to the same amount.

c. Torque all bolts in sequence to 54 inch lbs. (6 Nm).

d. Torque all bolts in sequence again. Torque type A, B, and C bolts to 6.5–8.5 ft. lbs. (9–12 Nm) and type D bolts to 13–19 ft. lbs. (18–25 Nm).

19. Line up the mating marks on the timing chain and camshaft sprockets and install the sprockets. Torque the sprocket bolts to 101–116 ft. lbs. (137–157 Nm).

20. Install the timing chain guide, distributor, chain tensioner, oil filter bracket and power steering oil pump bracket.

21. Install the intake manifold supports.

22. Install the cylinder head cover and oil separator.

23. Install the AIV, spark plugs, power steering pump, alternator water pump pulley and drive belts, air duct to the intake manifold and the radiator.

24. Install all vacuum and fuel hoses and reconnect all electrical connections.

25. Install the engine side cover, right front wheel and engine undercover.

26. Refill the cooling system.

27. Connect the negative battery cable, start the engine, and check for leaks.

3.0L (VG30DE) Engine

✳✳ CAUTION

The fuel injection system remains under pressure after the engine has been turned OFF. Properly relieve fuel pressure before disconnecting any fuel lines. Failure to do so may result in fire or personal injury.

1. Relieve the fuel system pressure and disconnect the negative battery cable.

2. Drain the cooling system.

3. Label and disconnect all electrical connectors, linkage and vacuum hoses from the throttle body.

4. Disconnect the intake air ducts from the throttle body.

5. Disconnect the fuel injector connectors and remove the injector pipe assembly.

6. Remove the cylinder head covers, timing belt, idler pulley and the mounting bolt.

7. Remove the intake manifold.

8. Disconnect the exhaust pipe from the exhaust manifold.

9. Loosen the cylinder head bolts in several steps.

10. Remove the cylinder head with the exhaust manifold attached.

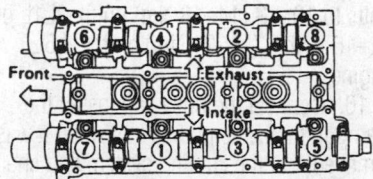

Right cylinder head

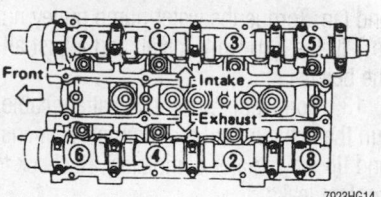

Left cylinder head

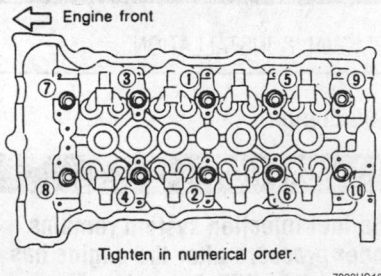

Tighten in numerical order.

7923HG12

Tighten the cylinder head bolts in sequence to prevent leakage and cylinder head damage—2.0L engine

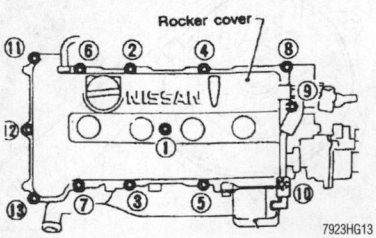

Rocker cover

NISSAN

7923HG13

Tighten the cylinder head cover bolts according to the sequence shown—2.0L engine

7923HG14

Cylinder head bolt tightening sequence— 3.0L (VG30DE) engine

To install:

11. Set the No. 1 piston at TDC on its compression stroke. Ensure that the crankshaft sprocket alignment mark is aligned with the one on the oil pump body and the camshaft sprocket alignment mark is aligned with the one on the timing belt rear cover.

12. Fit the cylinder head into place with a new gasket. Install the cylinder head bolts with washers.

13. Torque the cylinder head bolts in sequence.

14. Install the cylinder head covers, timing belt and intake manifold.

15. Install the exhaust pipe on the exhaust manifold.

16. Install the injector pipe assembly and intake manifold collector.

17. Connect the intake air ducts to the throttle body.

18. Connect all electrical connectors and vacuum hoses.

19. Start the engine, allow it to reach operating temperature and make any necessary adjustments. Check for leaks.

3.0L (VQ30DE) Engine

1. Relieve the fuel system pressure.

✴✴ CAUTION

The fuel injection system remains under pressure after the engine has been turned OFF. Properly relieve fuel pressure before disconnecting any fuel lines. Failure to do so may result in fire or personal injury.

2. Disconnect the negative battery cable.

3. Drain the engine oil and the cooling system.

➡**Before disconnecting any hoses or connectors, note the locations for reassembly.**

4. Remove the air duct to intake manifold hose, collector hose, blow-by hose, and vacuum hoses.

5. Remove the fuel hoses and disconnect the harness connections.

6. Disconnect the canister purge hoses.

7. Disconnect and remove all six ignition coils from the spark plugs.

8. Remove the bolts that secure the EGR tube and remove the tube.

9. Remove the intake manifold collector supports and remove the collector. Remove the manifold from the cylinder head.

10. Remove the cylinder head covers from the cylinder head.

11. Remove the right front wheel and engine side covers.

12. Remove the drive belts and idler pulley followed by the power steering oil pump belt and remove the power steering oil pump assembly.

13. Remove the camshaft position sensor (PHASE) and crankshaft position sensors (REF)/(POS).

14. Set the No. 1 piston to TDC of compression stroke by rotating the crankshaft.

15. Remove the ring gear cover access plate and remove the crankshaft pulley bolt.

16. Remove the timing chain tensioner and slack side chain guide.

17. Remove the engine oil pan.

18. Remove the camshaft sprockets first. Be sure to hold the flats of the camshafts while removing the sprocket bolts.

19. Loosen the camshaft bearing caps in several steps. The bearing caps MUST be loosened in sequence.

➡**Keep all bearing caps and camshafts in proper order for reinstallation.**

20. Remove the cylinder head bolts in sequence.

To install:

21. Turn the crankshaft until the No. 1 piston is set 240 degrees before TDC on compression stroke.

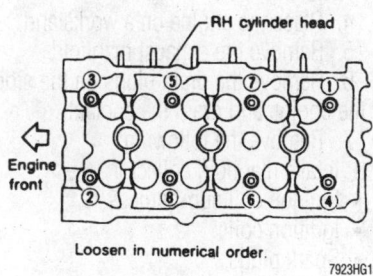

Right cylinder head bolt loosening sequence—3.0L (VQ30DE) engine

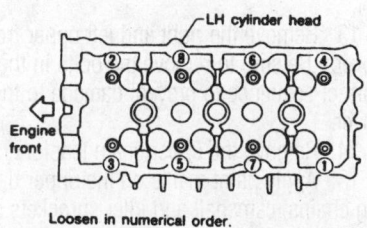

Left cylinder head bolt loosening sequence—3.0L (VQ30DE) engine

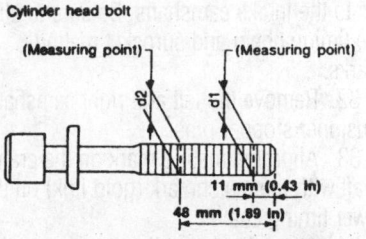

Measuring the cylinder head bolts—3.0L (VQ30DE) engine

22. Using new head gaskets, install the cylinder heads.

➡**If possible, replacement of the head bolts is suggested.**

23. If replacement of the head bolts is not possible, perform the following bolt measurement:

a. Measure the diameter of the head bolt 0.43 in. (11mm) from the bottom of the bolt.

b. Measure the diameter of the head bolt 1.89 in. (48mm) from the bottom of the bolt.

c. Whenever the size difference between the two measurements exceeds 0.0043 in. (0.11mm) the head bolts must be replaced.

24. Install the cylinder head bolts and tighten in sequence.

25. Install the camshaft tensioners. Tighten the tensioner mounting bolts to 75–96 inch lbs. (8.4–10.8 Nm).

➡**The camshafts can be identified by the paint marks on the camshaft. The left cylinder head camshafts have a YELLOW paint mark and the right cylinder head camshafts have a WHITE paint mark.**

➡**When installing the camshafts, position the camshaft keys at the 12 o'clock position in respect to the cylinder head angle.**

26. Install the camshafts and install the bearing caps.

27. Install new O-rings to the front of the engine block.

28. Install the crankshaft sprocket with the mating mark facing out.

29. Rotate the crankshaft clockwise and position the crankshaft to TDC of compression stroke and align the dowels of the camshaft sprockets to the 12 o'clock position.

30. Install the lower chain guide on the dowel pin with the front mark on the guide facing upward.

31. Install the timing chains and sprockets to the intake camshafts. Be sure to align the timing chain and sprocket mating marks.

32. Remove the left and right camshaft tensioner stopper pins.

33. Align the mating mark on the crankshaft with the matchmark (gold link) on the lower timing chain.

34. Attach the lower timing chain to the water pump sprocket.

35. Working counterclockwise, install the lower timing chain camshaft sprockets. Be sure to align the sprocket marks with the blue links of the timing chain during installation.

36. Install the intake sprocket bolts and tighten to 88–95 ft. lbs. (119–128 Nm). Be sure to secure the camshafts while tightening the bolts.

37. Install the timing chain guide, upper timing chain guide, lower timing chain tensioner and slack side timing chain guide.

38. Install the timing cover evenly and gently. Be sure to align the dowel pin holes. Tighten the mounting bolts in sequence.

39. Install the front exhaust pipe and its support.

40. Install the A/C compressor and bracket.

41. Install the crankshaft pulley to the crankshaft and install the mounting bolt.

42. Tighten the mounting bolt to 14–22 ft. lbs. (20–29 Nm). Tighten the crankshaft bolt an additional 60–66 degrees clockwise. This is about the angle from one hexagon bolt head corner to another.

43. Install the ring gear cover plate.

44. Install the camshaft position sensor (PHASE) and crankshaft position sensors (REF)/(POS).

45. Install the power steering pump assembly.

46. Install the drive belts and the idler pulley.

47. Install the right front inner wheel cover and install the right front wheel.

48. Install the engine undercovers.

49. Using new gaskets, install the intake manifold. Tighten the nuts and bolts in sequence.

50. Install the intake manifold collector gasket with the arrow facing forward.

51. Install the intake manifold collector assembly and tighten the mounting bolts to 16–18 ft. lbs. (22–25 Nm).

52. Install the intake manifold collector support brackets.

53. Using new gaskets, install the EGR tube and tighten the mounting bolts to 15–20 ft. lbs. (21–26 Nm) in two progressive steps.

54. Install the spark plugs.

55. Install the ignition coils and tighten the mounting bolts to 27–33 inch lbs. (2.9–3.8 Nm).

56. Install the cylinder head cover ornament on the left side.

57. Install the water hoses to the cylinder head and intake manifold.

58. Connect the canister purge hoses.

59. Connect the fuel hoses and wiring harness connections to the fuel rail.

60. Install and connect the air duct to intake manifold hose, collector hose, blow-by hose, and vacuum hoses.

61. Refill the engine crankcase and cooling system with the proper type and amount of fluid.

62. Connect the negative battery cable.

63. Start the engine and run at 3000 RPM under no load to purge the air from the high pressure chamber. The engine may produce a rattling noise. This indicates that air still remains in the chamber and is not a matter of concern.

64. Verify that there are no leaks.

4.1L Engine

1. Disconnect the negative and positive battery cables.

2. Have the A/C system discharged by an EPA- certified technician.

3. Remove the engine assembly from the vehicle.

4. Place the engine on a workstand.

5. Remove the exhaust manifold.

6. Remove the drain plugs on the sides of the engine and drain the coolant.

7. Remove the following:
- Intake manifold collector
- Ignition coil sub-harness
- Ignition coils
- Spark plugs

8. Remove the fuel rail with injectors. Do not disassembly the fuel hose.

9. Remove the intake manifold.

10. Remove the rocker arm covers.

11. Bring the No. 1 piston to TDC on the compression stroke.

12. Remove the camshaft position sensor.

13. Remove the right and left upper front covers. Be sure to remove the bolts in the correct sequence to prevent damage to the cover.

14. Remove the upper chain tensioners.

15. Apply paint marks on the upper timing chains, camshaft and idler sprockets so they can be installed in their original positions.

16. Remove the camshaft sprocket.

17. Remove the idler sprocket bolt.

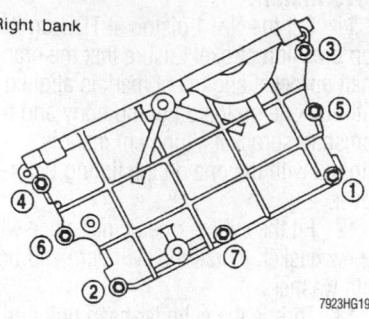

Upper front cover bolt removal sequence—4.1L engine, right bank

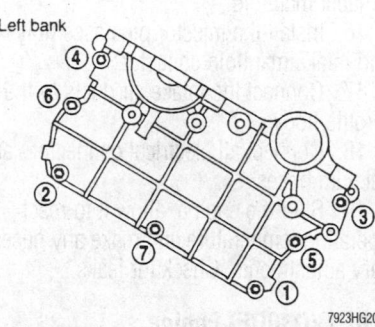

Upper front cover bolt removal sequence—4.1L engine, left bank

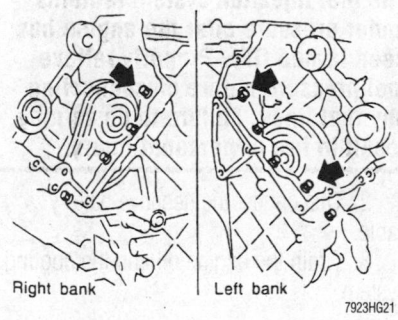

Be sure to install the long bolts in the positions indicated by the arrows—4.1L engine

18. Remove the cylinder head sub-bolts. The bolts are different lengths, note their so they can be installed in their original positions.

19. Loosen the cylinder head bolts gradually in the proper removal sequence, then remove the cylinder head.

To install:

20. Be sure all mating surfaces are clean before installation.

21. Check the cylinder head surface for warpage using a feeler gauge and a suitable straightedge. If the cylinder head is warped more than 0.004 in. (0.1mm), it must be resurfaced or replaced. The total amount

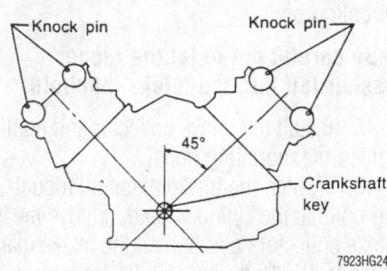

Be sure to position the camshaft knock pins as shown before installing the cylinder heads—4.1L engine

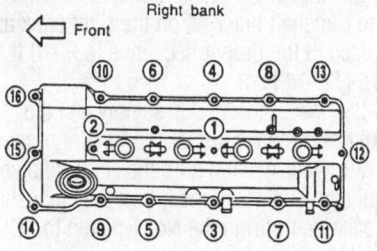

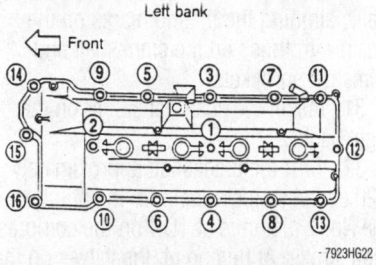

Cylinder head cover bolt tightening sequence—4.1L engine

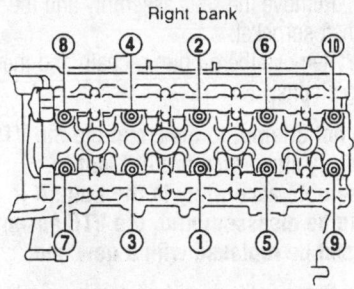

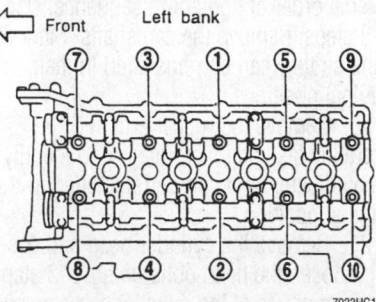

Cylinder head torque sequence—4.1L and 4.5L engines

machined from the head or head and block combined, cannot total more than 0.008 in. (0.2mm).

22. Place new gaskets on the cylinder block.

23. Be sure the knock pins on the camshafts are in the positions shown. Carefully place the cylinder heads on the engine. Do not damage the head gasket.

24. Lubricate the cylinder head bolt with engine oil. Tighten the bolts in sequence using the following sub-steps:

   a. Tighten the bolts to 22 ft. lbs. (29 Nm)

   b. Tighten the bolts to 69 ft. lbs. (93 Nm)

   c. Loosen the bolts completely

   d. Tighten the bolts to 18–25 ft. lbs. (25–35 Nm)

   e. Turn each bolts clockwise 90–95°

25. Install the cylinder head sub-bolts. Tighten the bolts to 56–74 in. lbs. (6.3–8.3 Nm). Be sure the long bolts are returned to the positions shown.

26. Install the idler and camshaft sprockets.

27. Install the chain tensioners.

28. Using a new gaskets, install the cylinder head covers. Be sure to apply RTV silicone sealant to the gasket arch and rubber plugs.

29. Tighten the cylinder head cover bolts in sequence using the following sub-steps:

   a. Bolts Nos. 1 through 16 to 35–52 inch lbs. (4–6 Nm)

   b. Bolts Nos. 1 through 16 to 61–78 inch lbs. (7–9 Nm)

   c. Bolts Nos. 1 and 2 to 61–78 inch lbs. (7–9 Nm)

30. Install the intake valve timing control solenoid with a new O-ring. Tighten the solenoid to 18–25 ft. lbs. (25–34 Nm).

31. Apply a bead of RTV silicone sealant to the upper front covers, then install them. Tighten the bolts to 56–74 in. lbs. (6.3–8.3 Nm).

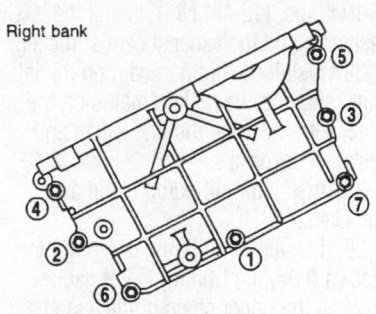

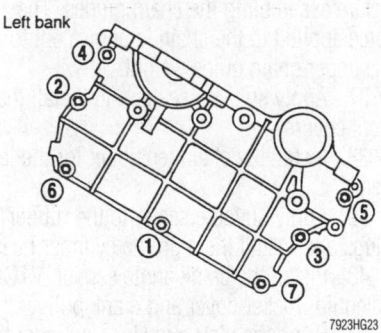

Tighten the upper front cover bolts in the sequence shown—4.1L engine

32. Install the remaining part in the reverse of the removal.

33. Drain the refill the crankcase with new engine oil.

34. Refill and bleed the cooling system.

4.5L Engine

1. Disconnect the negative battery cable.

2. Remove the engine and transmission assembly from the vehicle.

3. Remove the suspension member and engine mounts from the engine.

4. Remove the air compressor bracket and the exhaust manifolds.

5. Remove the cooling fan with coupling and the engine gusset.

6. Separate the engine from the transmission and mount the engine on a suitable workstand.

7. Remove the oil pan. Remove the intake collector.

8. Disconnect the injector harness connector and remove the injector tube assembly with injector. Loosen bolts in opposite sequence of tightening.

➡**Be careful not to let the rubber washer fall into the intake manifold.**

9. Remove the intake manifold.

10. Remove the ornamental cylinder head cover and remove the ignition coils and spark plugs.

11. Bring the No. 1 piston to TDC on the compression stroke.

12. Use a suitable puller to remove the crankshaft pulley.

13. Remove the cylinder heads cover.

14. Remove the crank angle sensor and the Valve Timing Control (VTC) solenoid.

15. Remove the chain tensioners and the upper front covers.

16. Remove the front timing chain cover.

➡**The timing chain will not be disengaged or dislocated from the crankshaft sprocket unless the front cover is removed. The cast portion of the front cover is located on the lower side of the crankshaft sprocket so the timing chain is not disengaged from the sprocket.**

17. Remove the VTC assembly and the camshaft sprocket.

18. Remove the oil pump chain and the timing chains.

➡Do not attempt to disassemble the VTC assembly since they are difficult to reassemble accurately in the field. If it should be disassembled, the VTC assembly must be replaced with a new one.

19. Remove the camshaft brackets in the reverse order of tightening sequence. Use 2–3 steps. Remove the camshafts. Mark the parts so they can be reinstalled in their original positions.

20. Remove the rocker arm and hydraulic lash adjuster. Be sure to identify each adjuster so it can be reinstalled in it's original position.

21. Remove the cylinder head and gasket. Loosen the head bolts using 2–3 steps in the opposite of the installation sequence.

To install:

22. Be sure all mating surfaces are clean before installation.

23. Check the cylinder head surface for warpage using a feeler gauge and a suitable straightedge. If the cylinder head is warped more than 0.004 in. (0.1mm), it must be resurfaced or replaced. The total amount machined from the head or head and block combined, cannot total more than 0.008 in. (0.2mm).

24. Be sure the No. 1 piston is still at TDC of the compression stroke, then turn the crankshaft until the No. 1 piston is at approximately 45 degrees before TDC on the compression stroke. At this point, the No. 3 piston will be at the same height as the No. 1 piston to prevent contact between the valves and pistons.

25. Install the cylinder heads with new gaskets. Be sure to install washers between the bolts and the cylinder heads. Temporarily install the cylinder head bolts to avoid damage to the head gasket.

❈❈ WARNING

Do not rotate the crankshaft or camshaft separately or the valves will hit the pistons.

26. Install the hydraulic lash adjusters and check them as follows:

   a. When the rocker arm can be moved at least 0.04 in. (1.0mm) by pushing at the hydraulic lash adjuster location, it indicates that there is air in the high pressure chamber The adjuster will have to be bled.

➡Air cannot be bled from the lash adjusters by running the engine.

27. Install the rocker arms, camshafts and camshaft brackets on the right bank and tighten in the proper sequence to 9–10 ft. lbs. (12–14 Nm).

28. Install the VTC assembly and the exhaust camshaft sprocket on the right bank.

29. After making sure the camshafts are still correctly positioned, turn the crankshaft clockwise to bring the No. 1 piston to TDC on the compression stroke.

30. Install the timing chain on the right bank, aligning the mating marks on the chain with those on the crankshaft and camshaft sprockets.

31. Install the chain tensioner on the right bank.

32. Turn the crankshaft approximately 120 degrees clockwise from the point where the No. 1 piston is at TDC on the compression stroke. At this point, the valves on the left bank still remain closed.

33. Correctly position the camshafts and tighten brackets in the proper sequence to 9–10 ft. lbs. (12–14 Nm). Install the VTC assembly and the exhaust cam sprocket.

34. Install the timing chain on the left bank, aligning the mating marks on the chain with those on the crankshaft and camshaft sprockets.

35. Install the oil pump chain and sprockets.

36. Install the oil pump chain guides. Place a 0.04 in. (1.0mm) feeler gauge between the upper chain guide and chain before assembling the chain guides. The force applied to the chain is equivalent to the upper chain guide weight.

37. Apply suitable sealer and install the front covers.

38. Install the chain tensioner for the left bank.

39. Apply suitable sealer to the rubber plugs and install them on the cylinder head.

40. Install the crank angle sensor, VTC solenoid, rocker cover and crank pulley.

41. Bring the piston in No. 1 cylinder to TDC on the compression stroke.

42. Tighten the bolts in sequence using the following sub-steps:

   a. Tighten the bolts to 22 ft. lbs. (29 Nm)

   b. Tighten the bolts to 69 ft. lbs. (93 Nm)

   c. Loosen the bolts completely

   d. Tighten the bolts to 18–25 ft. lbs. (25–35 Nm)

   e. Turn each bolts clockwise 90–95°

43. Install the intake manifold bolts in their proper positions on the cylinder head and lightly tighten the mounting bolts.

44. Connect the injector tube assemblies, including the fuel injectors, to the

intake manifolds and lightly tighten the mounting bolts.

➡Be careful not to let the rubber washer fall into the intake manifold.

45. Install the intake collector and lightly tighten the mounting bolts.

46. Tighten the intake manifold mounting bolts at the cylinder head, remove the intake collectors and tighten the intake manifolds to 12–15 ft. lbs. (16–21 Nm).

47. Tighten the sub-fuel tubes, in sequence, first to 3.1–4.3 ft. lbs. (4.2–5.9 Nm), then to 6.2–8.0 ft. lbs. (8.4–10.8 Nm).

48. Tighten the injector tube assemblies, in sequence, first to 6.9–8.0 ft. lbs. (9.3–10.8 Nm), then to 15–20 ft. lbs. (21–26 Nm).

49. Install the intake collectors and tighten to 9–11 ft. lbs. (12–15 Nm).

50. Install the exhaust manifolds.

51. Install the cylinder head covers and tighten in the proper sequence to 5–7 ft. lbs. (7–10 Nm).

52. Install all other remaining components. Join the engine and transmission and install the assembly in the vehicle.

Rocker Arms/Shafts

REMOVAL & INSTALLATION

2.0L Engine

1. Relieve the fuel system pressure and disconnect the negative battery cable.

2. Drain the coolant from the radiator and engine block. Remove the radiator.

3. Raise and support the vehicle safely. Remove the right front wheel and engine side cover.

4. Remove the air duct to the intake manifold.

5. Remove the drive belts, water pump pulley, alternator and power steering pump.

6. Label and remove the vacuum hoses, fuel hoses and wiring harness connectors.

7. Remove all the spark plugs, the AIV valve and resonator.

8. Remove the rocker cover and oil separator. Loosen rocker cover bolts, using 2 to 3 steps, in the opposite sequence of tightening

9. Remove the intake manifold supports.

10. Remove the oil filter bracket and power steering oil pump bracket.

11. Set No. 1 piston at TDC on the compression stroke by rotating the crankshaft.

12. Remove the chain tensioner.

13. Remove the distributor. Do not turn the rotor with the distributor removed.

14. Remove the timing chain guide, camshaft sprockets, camshafts, brackets, oil tubes and baffle plate. The camshaft bracket bolts must be loosened in sequence to prevent damage to the camshafts or the head.

15. Remove rocker arm assembly.

To install:

16. Check the hydraulic lash adjusters to ensure they did not bleed down during disassembly by trying to compress them. If the lash adjuster can be compressed 0.04 in. (1mm), air has entered and it must be bleed.

➡**Air cannot be bled from the lash adjusters by running the engine.**

17. Clean the camshaft end bracket and coat with liquid gasket. Install the camshafts, camshaft brackets, oil tubes and baffle plate. Ensure the left camshaft key is at 12 o'clock and the right camshaft key is at 10 o'clock.

➡**The procedure for tightening camshaft bracket bolts must be followed exactly to prevent camshaft damage.**

18. Line up the mating marks on the timing chain and camshaft sprockets and install the sprockets. Tighten sprocket bolts to 101–116 ft. lbs. (137–157 Nm).

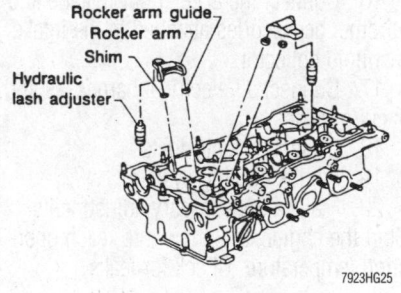

Rocker arm guide
Rocker arm
Shim
Hydraulic lash adjuster

7923HG25

Exploded view of the rocker arms and related components—2.0L engine

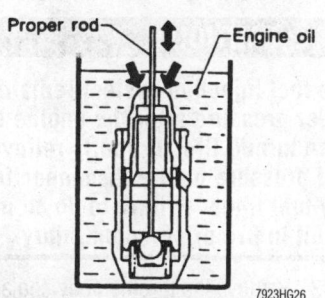

Proper rod — Engine oil

7923HG26

Submerge the lash adjuster in engine oil, lightly unseat the check ball with a thin rod and push on the plunger to release the air

19. Install the timing chain guide, distributor (ensure that rotor head is at 5 o'clock position) and chain tensioner.

20. Install intake manifold supports. Clean the rocker cover and mating surfaces and apply a continuous bead of liquid gasket to the mating surface.

21. Install the rocker cover and oil separator. Tighten the rocker cover bolts in sequence.

22. Install the oil filter bracket and the power steering pump bracket.

23. Install the spark plugs, AIV valve and the resonator.

24. Connect the fuel lines, vacuum hoses and wiring connectors.

25. Install the water pump pulley, alternator and power steering pump.

26. Install the adjust the drive belts.

27. Install the intake manifold air duct, engine side cover and right wheel.

28. Install the radiator and refill the cooling system.

29. Connect the negative battery cable.

3.0L Engines

➡**The valves in the 3.0L engine are directly actuated by the camshaft. No rocker arms are used in this engine.**

4.1 and 4.5L Engines

> ❄❄ **CAUTION**

The fuel injection system remains under pressure after the engine has been turned OFF. Properly relieve fuel pressure before disconnecting any fuel lines. Failure to do so may result in fire or personal injury.

1. Disconnect the negative battery cable.

2. Remove the engine and transmission assembly from the vehicle.

3. Remove the suspension member and engine mounts from the engine.

4. Remove the air compressor bracket.

5. Remove the cooling fan with coupling and the engine gusset.

6. Separate the engine from the transmission and mount the engine on a suitable workstand.

7. Remove the oil pan.

8. Remove the crank angle sensor and the Valve Timing Control (VTC) solenoid.

9. Remove the chain tensioners and the upper front covers.

10. Remove the front timing chain cover.

➡**The timing chain will not be disengaged or dislocated from the crankshaft sprocket unless the front cover is**

removed. The cast portion of the front cover is located on the lower side of the crankshaft sprocket so the timing chain is not disengaged from the sprocket.

11. Remove the VTC assembly and the camshaft sprocket.

12. Remove the oil pump chain and the timing chains.

➡**Do not attempt to disassemble the VTC assembly since they are difficult to reassemble accurately in the field. If it should be disassembled, the VTC assembly must be replaced with a new one.**

13. Remove the camshaft brackets and the camshafts. Mark the parts so they can be reinstalled in their original positions.

14. Remove the rocker arms. Be sure to identify each rocker arm so it can be reinstalled in it's original position.

To install:

15. Be sure all mating surfaces are clean before installation.

16. Install the rocker arms, camshafts and camshaft brackets on the right bank. Properly lubricate the rocker arms and camshafts prior to installation.

17. Install the VTC assembly and the exhaust cam sprocket on the right bank.

18. Be sure the camshafts are still correctly positioned and the piston in the No. 1 cylinder is still at TDC.

19. Install the timing chain on the right bank, aligning the mating marks on the chain with those on the crankshaft and camshaft sprockets.

20. Install the chain tensioner on the right bank.

21. Turn the crankshaft approximately 120 degrees clockwise from the point where the No. 1 piston is at TDC on the compression stroke. At this point, the valves on the left bank still remain closed.

22. Correctly position the camshafts and rocker arms for the left cylinder head. Prop-

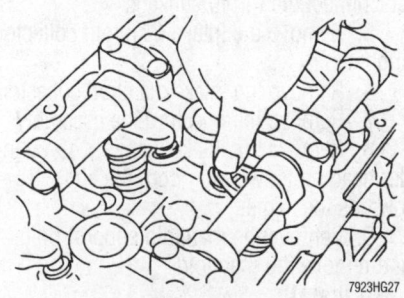

7923HG27

Press down on the lash adjuster to check for bleed-down—4.1L and 4.5L engines

erly lubricate the rocker arms and camshafts prior to installation. Install the VTC assembly and the exhaust cam sprocket.

23. Install the timing chain on the left bank, aligning the mating marks on the chain with those on the crankshaft and camshaft sprockets.

24. Install the oil pump chain and sprockets.

25. Install the oil pump chain guides. Place a 0.04 in. (1.0mm) feeler gauge between the upper chain guide and chain before assembling the chain guides. The force applied to the chain is equivalent to the upper chain guide weight.

26. Apply suitable sealer and install the front covers.

27. Install the chain tensioner for the left bank.

28. Apply suitable sealer to the rubber plugs and install them on the cylinder head.

29. Install the crank angle sensor, VTC solenoid, rocker cover and crank pulley.

30. Install the transmission on the engine and install the engine assembly in the vehicle.

Intake Manifold

REMOVAL & INSTALLATION

2.0L Engine

1. Disconnect the negative battery cable.

✺ CAUTION

The fuel injection system remains under pressure after the engine has been turned OFF. Properly relieve fuel pressure before disconnecting any fuel lines. Failure to do so may result in fire or personal injury.

2. Properly relieve the fuel system pressure.

3. Drain the cooling system.

4. Tag and disconnect the fuel lines, vacuum hoses and electrical connectors. Disconnect the throttle linkage.

5. Remove the intake manifold collector support.

6. Remove the intake manifold collector.

7. Remove the injector tube assembly.

8. Remove the intake manifold-to-cylinder head bolts, working from the ends towards the center.

9. Remove the manifold support bolts and remove the manifold.

To install:

10. Be sure all mating surfaces are clean prior to installation.

11. Fit a new gasket and the manifold into place. Start the support bolts to hold the manifold in place.

12. Install the intake manifold bolts. Torque the bolts in sequence to 13–15 ft. lbs. (18–21 Nm). Tighten the bolts in two steps, starting at the center and working towards the ends.

13. Install the injector tube assembly. Torque the bolts first to 6.9–8.0 ft. lbs. (9.3–10.8 Nm), then to 15–20 ft. lbs. (21–26 Nm).

14. Use a new gasket and install the intake manifold collector. Torque the mounting bolts in sequence to 13–15 ft. lbs. (18–21 Nm).

15. Reconnect the fuel lines, vacuum hoses, electrical connectors and the throttle linkage.

16. Refill the cooling system, connect the negative battery cable, start engine and test for leaks.

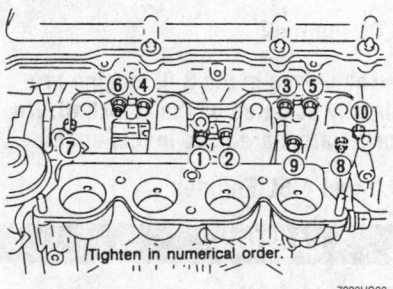

Intake manifold bolt tightening sequence—2.0L engine

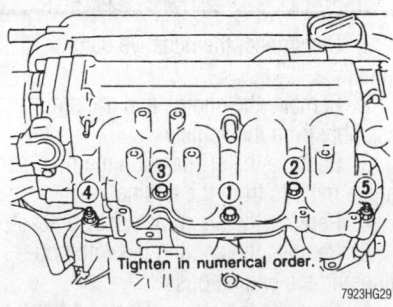

Intake manifold collector bolt tightening sequence—2.0L engine

3.0L (VG30DE) Engine

✺ CAUTION

The fuel injection system remains under pressure, after the engine has been turned OFF. Properly relieve fuel pressure before disconnecting any fuel lines. Failure to do so may result in fire or personal injury.

1. Disconnect the negative battery cable.

2. Remove the timing belt.

3. Remove the air ducts.

4. Label and disconnect all wiring and vacuum hoses from the intake manifold.

5. Disconnect the accelerator linkage and all other accessories attached to the intake manifold collector.

6. Remove the intake manifold collector.

7. Relieve the fuel system pressure. Disconnect the fuel feed and return pipe.

8. Remove the injector pipe assembly.

9. Remove the timing belt idler pulley.

10. Remove the intake manifold.

To install:

11. Install the new gaskets and fit the manifold into place. Tighten the nuts and bolts hand-tight, working from the center towards the ends, then tighten the bolts to 12–14 ft. lbs. (16–20 Nm) and the nuts to 17–20 ft. lbs. (24–27 Nm).

12. Install the stud bolt, idler pulley and timing belt.

13. Install the injector pipe assembly.

14. Connect the fuel feed and return pipes.

15. Install the intake manifold collector and tighten the bolts to 12–15 ft. lbs. (16–21 Nm).

16. Connect the accelerator linkage and all other accessories attached to the intake manifold collector.

17. Connect all electrical harnesses and vacuum hoses.

18. Connect the air ducts.

19. Fill the cooling system.

20. Make any necessary adjustments. Start the engine and allow it to reach operating temperature. Check for leaks.

3.0L (VQ30DE) Engine

1. Disconnect the negative battery cable and drain the cooling system.

2. Release the fuel system pressure.

✺ CAUTION

The fuel injection system remains under pressure after the engine has been turned OFF. Properly relieve the fuel pressure before disconnecting any fuel lines. Failure to do so may result in fire or personal injury.

3. Remove the throttle body coolant hoses.

4. Label and disconnect the electrical connectors from the throttle position sensor.

5. Label and disconnect the hoses from the throttle body, the EGR valve, intake manifold collector, IAC valve, and the fuel pressure regulator.

6. Disconnect the canister purge hose and blow-by hose.

7. Disconnect the EGR guide tube.

8. Disconnect the accelerator cable from the throttle body.

9. Remove the intake manifold collector support brackets.

10. Disconnect the right side electrical connectors from the ignition coils.

11. If necessary, disconnect the electrical connector from the crank angle sensor and the power transistor.

12. Remove the intake manifold collector-to-intake manifold bolts/nuts and remove the intake manifold collector.

13. Remove the fuel injector assembly by performing the following procedures:

a. Disconnect the electrical connectors from the fuel injectors.

b. Disconnect the fuel lines from the fuel injector assembly.

c. Remove the fuel rail-to-cylinder head bolts.

d. Remove the fuel rail assembly from the engine.

14. Remove the intake manifold bolts/nuts in the reverse sequence of the torque procedure.

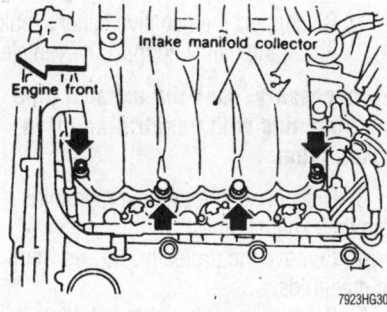

Intake manifold collector mounting bolts— 3.0L (VQ30DE) engine

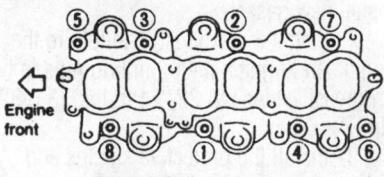

Tighten the intake manifold bolts in the proper order (loosen the bolts in the reverse sequence)—3.0L (VQ30DE) engine

15. Remove the intake manifold from the engine and discard the gaskets.

16. Clean all gasket mounting surfaces.

To install:

17. Using new gaskets, install the intake manifold to the engine.

18. Tighten the bolts/nuts in sequence as follows:

a. Tighten nuts and bolts to 44–86 inch lbs. (5–10 Nm).

b. Tighten nuts and bolts to 20–23 ft. lbs. (26–31 Nm).

19. Install the fuel injector assembly by performing the following procedures:

a. Install the fuel rail assembly to the engine.

b. Install the fuel rail-to-cylinder head bolts and tighten the bolts to 15–20 ft. lbs. (21–26 Nm) in two progressive steps.

c. Connect the fuel lines to the fuel injector assembly.

d. Connect the electrical connectors to the fuel injectors.

20. Using a new gasket, install the intake manifold collector and tighten the intake manifold collector-to-intake manifold bolts/nuts to 13–16 ft. lbs. (18–22 Nm).

21. Install the intake manifold collector supports and tighten the bolts to 14–18 ft. lbs. (20–25 Nm).

22. If disconnected, connect the electrical connector to the crank angle sensor and the power transistor.

23. Connect the electrical connectors to the ignition coils and tighten the mounting bolts to 27–33 inch lbs. (2.9–3.8 Nm)

24. Connect the accelerator cable to the throttle body.

25. Connect the EGR guide tube and tighten the bolts to 15–20 ft. lbs. (21–26 Nm) in two progressive steps.

26. Connect the canister purge hose and blow-by hose.

27. Connect the hoses to the throttle body, EGR valve, intake manifold collector, IAC valve, and the fuel pressure regulator.

28. Connect the electrical connectors to the throttle position sensor.

29. Connect the throttle body coolant hoses.

30. Refill the cooling system and connect the negative battery cable.

31. Start the engine, bleed cooling system, and check for leaks.

4.1L Engine

1. Disconnect the negative battery cable.

✳✳ CAUTION

Never open, service or drain the radiator or cooling system when hot;

serious burns can occur from the steam and hot coolant. Coolant should be reused unless it is contaminated or is several years old.

2. Drain the cooling system.

3. Relieve the fuel system pressure.

4. Remove the intake air duct from the throttle body assembly.

5. Remove the intake manifold collector with the throttle body attached.

6. Disconnect the fuel injectors and lift up the fuel rail assembly with the injectors. Do not disconnect the fuel hose.

7. Remove the intake manifold mounting bolts and remove the intake manifold.

To install:

8. Clean the intake manifold mounting surface.

9. Use new gaskets and install the intake manifold. Tighten the bolts in sequence to 13–15 ft. lbs. (18–21 Nm).

10. Install the fuel tube assembly. Tighten the bolts in sequence first to 7–8 ft. lbs. (9–11 Nm), then in sequence to 15–20 ft. lbs. (21–26 Nm).

11. Clean the mounting surface and install the intake manifold collector using a new gasket. Tighten the bolts and nuts in sequence to 13–16 ft. lbs. (18–22 Nm).

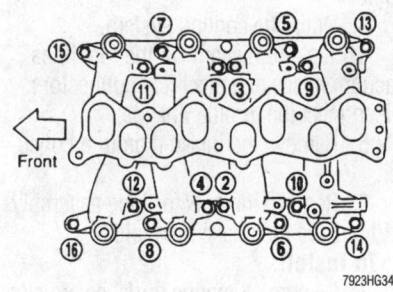

Tighten the intake manifold bolts in the proper sequence to prevent leakage

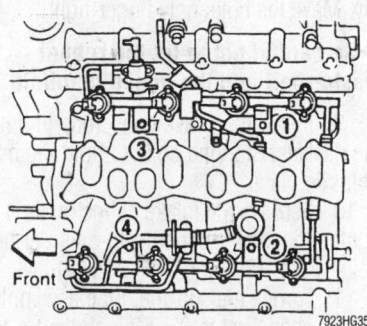

To prevent possible tube breakage, be sure to tighten the fuel tube mounting bolts according to the sequence shown

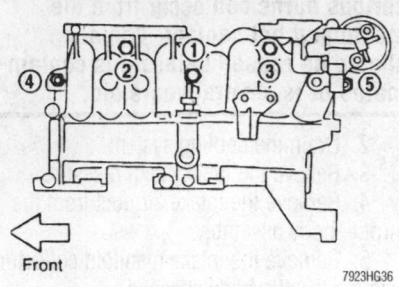

Tighten the intake manifold collector nuts and bolts in the proper sequence

12. Connect the air duct to the throttle body.
13. Connect the negative battery cable.
14. Refill and bleed the cooling system.

4.5L Engine

> ※※ **CAUTION**
>
> **The fuel injection system remains under pressure after the engine has been turned OFF. Properly relieve fuel pressure before disconnecting any fuel lines. Failure to do so may result in fire or personal injury.**

1. Properly relieve the fuel system pressure.
2. Disconnect the negative battery cable.
3. Drain the cooling system.
4. Tag and disconnect the fuel lines, vacuum hoses and electrical connectors. Disconnect the throttle linkage.
5. Remove the intake manifold collector.
6. Remove the injector tube assembly and remove the intake manifold.

To install:

7. Be sure all mating surfaces are clean prior to installation. Fit new gaskets and install the intake manifolds. Make the bolts only finger-tight at this time.
8. Install the injectors and tube assembly. Make the bolts only finger-tight.

➡ **Be careful not to let the rubber washer fall into the intake manifold.**

9. Install the intake collector without the gaskets. Start the bolts to hold the manifolds in place.
10. Remove the intake collector, then tighten the intake manifold-to-cylinder head bolts to 12–15 ft. lbs. (16–21 Nm).
11. Torque the sub-fuel tube assemblies in sequence, first to 37–52 inch lbs. (4.2–5.9 Nm), then to 74–96 inch lbs. (8.4–10.8 Nm).

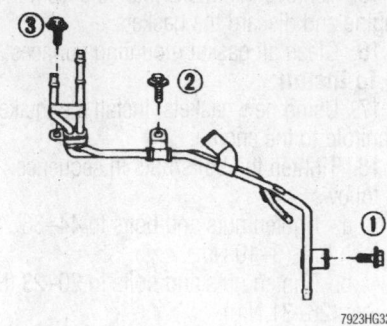

Sub-fuel tube bolt torque sequence—4.5L engine

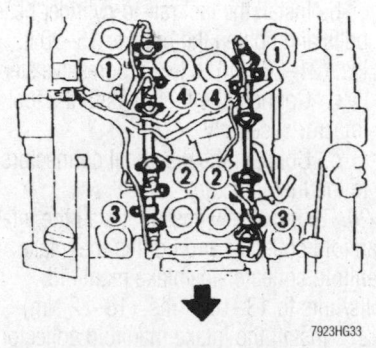

Fuel tube/injector assembly bolt torque sequence—4.5L engine

12. Torque the injector tube assemblies in sequence, first to 83–96 inch lbs. (9.3–10.8 Nm), then to 15–20 ft. lbs. (21–26 Nm).
13. Install the intake collector and tighten to 9–11 ft. lbs. (12–15 Nm).
14. Install the remaining components in the reverse order of their removal.

Exhaust Manifold

REMOVAL & INSTALLATION

2.0L Engine

1. Disconnect the negative battery cable. Raise and support the vehicle safely.
2. If equipped, remove the undercover and dust covers. Disconnect the exhaust pipe at the manifold flange.
3. If equipped, remove the Air Injection Valve (AIV), AIV tube, and the attaching bracket.
4. Disconnect the exhaust gas sensor electrical connection and remove the sensor.
5. Remove the exhaust manifold cover.
6. Remove the exhaust manifold nuts, starting at the outside and working towards the middle.
7. Remove the exhaust manifold and gasket.

To install:

8. Clean the gasket mating surface and install a new exhaust manifold gasket.
9. Install the exhaust manifold and tighten the manifold nuts in sequence, in two steps, to 27–35 ft. lbs. (37–48 Nm).
10. Install the exhaust manifold cover and exhaust gas sensor. Reconnect the sensor electrical connection.
11. Install the AIV, AIV tube, and the attaching bracket.
12. Install the exhaust pipe to the manifold flange and tighten the nuts to 30–35 ft. lbs. (41–48 Nm).
13. Lower the vehicle, start the engine, and check for leaks.

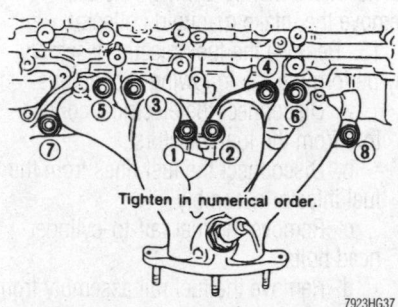

Be sure to tighten the exhaust manifold nuts in the proper sequence

3.0L (VG30DE) Engine

1. Disconnect the negative battery cable.
2. Raise and safely support the vehicle.

➡ **If necessary, soak the exhaust pipe retaining nuts with penetrating oil to loosen them.**

3. Disconnect the exhaust manifolds from the exhaust pipes.
4. Remove the protective covers from the manifolds.
5. Remove the exhaust manifold-to-engine mounting nuts. Remove the manifolds from the engine an discard the gaskets.

To install:

6. Clean all gasket mounting surfaces. Install new gaskets.
7. Install the exhaust manifold to the engine and tighten the mounting nuts in two progressive steps to 22–24 ft. lbs. (30–32 Nm).
8. Install the protective shields and tighten the mounting bolts in two progressive steps to 46–57 inch lbs. (5–7 Nm).
9. Install the exhaust manifolds to the exhaust pipes and tighten the mounting nuts to 32–37 ft. lbs. (43–50 Nm).

10. Connect the negative battery cable, start the engine, and check for exhaust leaks.

3.0L (VQ30DE) Engine

1. Disconnect the negative battery cable.

2. Raise and safely support the vehicle.

➡**If necessary, soak the exhaust pipe retaining nuts with penetrating oil to loosen them.**

3. Disconnect the exhaust manifolds from the exhaust pipes.

4. Remove the protective covers from the manifolds.

5. Remove the exhaust manifold-to-engine mounting nuts. Remove the manifolds from the engine an discard the gaskets.

To install:

6. Clean all gasket mounting surfaces. Install new gaskets.

7. Install the exhaust manifold to the engine and tighten the mounting nuts in two progressive steps to 22–24 ft. lbs. (30–32 Nm).

8. Install the protective shields and tighten the mounting bolts in two progressive steps to 46–57 inch lbs. (5.1–6.5 Nm).

9. Install the exhaust manifolds to the exhaust pipes and tighten the mounting nuts to 32–37 ft. lbs. (43–50 Nm).

10. Connect the negative battery cable, start the engine, and check for exhaust leaks.

4.1L and 4.5L Engines

1. Disconnect the negative battery cable. Raise and support the vehicle safely.

2. Remove the engine undercovers. Disconnect the exhaust pipe at the manifold flange.

3. If equipped, remove the heat shield from the exhaust manifold.

4. Disconnect the exhaust gas sensor electrical connection and if necessary, remove the sensor.

5. Remove the exhaust manifold nuts, starting at the ends and working towards the center.

6. Remove the exhaust manifold and gasket.

To install:

7. Clean the gasket mating surface and install a new exhaust manifold gasket.

8. Install the exhaust manifold and tighten the nuts in two steps and in sequence. Torque the nuts to 20–23 ft. lbs. (27–31 Nm).

9. If equipped, install the heat shield on the exhaust manifold.

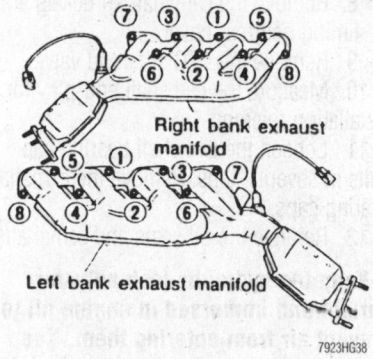

Be sure to tighten the exhaust manifold nuts in the correct sequence—4.5L engine

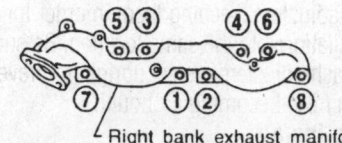

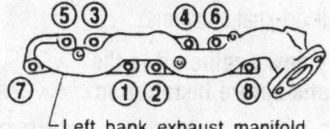

To avoid leaks, tighten the exhaust manifold nuts in the sequence shown—4.1L engine

10. If removed, install exhaust gas sensor and tighten to 30–37 ft. lbs. (40–50 Nm). Reconnect the sensor electrical connection.

11. Install the exhaust pipe to the manifold flange and tighten the nuts to 33–44 ft. lbs. (45–60 Nm).

12. Lower the vehicle and connect the negative battery cable.

13. Start the engine and check for leaks.

Camshaft and Valve Lifters

REMOVAL & INSTALLATION

2.0L Engine

✳✳ CAUTION

The fuel injection system remains under pressure, after the engine has been turned OFF. Properly relieve fuel pressure before disconnecting any fuel lines. Failure to do so may result in fire or personal injury.

1. Relieve the fuel system pressure.

2. Disconnect the negative battery cable. Remove the rocker cover and oil separator.

3. Rotate the crankshaft until the No. 1 piston is at TDC on the compression stroke

and the mating marks on the camshaft sprockets line up with the mating marks on the timing chain.

4. Remove the timing chain tensioner.

5. Remove the distributor.

6. Remove the timing chain guide.

7. Remove the camshaft sprockets. Use a wrench to hold the camshaft while loosening the sprocket bolt.

8. Loosen the camshaft bearing cap bolts in the opposite order of the tightening sequence.

9. Remove the camshaft from the cylinder head.

10. When removing the rocker arm, be careful not to drop the valve shims into the cylinder head. After removing the adjuster, set them upright or lay them down in a pan of clean engine oil. Do not lay them down on the bench or the oil will drain out and the adjuster will become air bound. Keep all of these parts in order so they can be installed in the same locations.

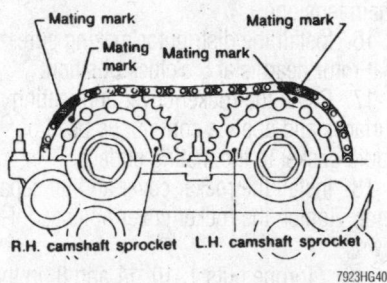

Be sure to align the marks on the sprockets with the marks on the chain—2.0L engine

To install:

11. Install the adjusters, shims and rockers into their original locations.

12. Clean the left-hand camshaft end bearing cap and coat the mating surface with liquid gasket. Install the camshafts, bearing caps, oil tubes and baffle plate. Ensure the left camshaft key is at 12 o'clock and the right camshaft key is at 10 o'clock.

13. The procedure for tightening bearing cap bolts must be followed exactly to prevent camshaft damage. Tighten bolts as follows:

a. Torque right camshaft bolts 9 and 10 (in that order) to 17 inch lbs. (2 Nm), then tighten bolts 1 through 8 (in that order) to the same specification.

b. Torque left camshaft bolts 11 and 12 (in that order) to 17 inch lbs. (2 Nm), then tighten bolts 1 through 10 (in that order) to the same specification.

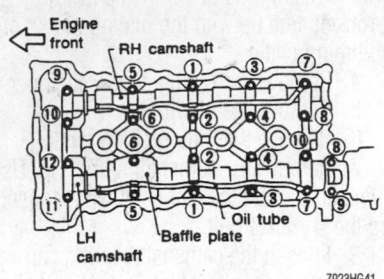

Camshaft bearing cap bolt torque sequence—2.0L engine

c. Torque all bolts in sequence to 52 inch lbs. (6 Nm).

d. Torque all bolts again in sequence to 6.5–8.5 ft. lbs. (9–12 Nm), then tighten bolts 8 and 9 on the left camshaft to 13–19 ft. lbs. (18–25 Nm).

14. Line up the mating marks on the timing chain and camshaft sprockets and install the sprockets. Torque sprocket bolts to 101–116 ft. lbs. (137–157 Nm).

15. Install the timing chain guide and chain tensioner.

16. Install the distributor making sure that rotor head is at 5 o'clock position.

17. Clean the rocker cover and mating surfaces and apply a continuous bead of liquid gasket to the mating surface.

18. Install the rocker cover and oil separator. Tighten the rocker cover bolts as follows:

a. Torque nuts 1, 10, 11 and 8, in that order to 36 inch lbs. (4 Nm).

b. Torque nuts 1 through 13 as indicated in the figure to 72–84 inch lbs. (8–10 Nm).

19. Connect the negative battery cable.

3.0L (VG30DE) Engine

❋❋ CAUTION

The fuel injection system remains under pressure after the engine has been turned OFF. Properly relieve fuel pressure before disconnecting any fuel lines. Failure to do so may result in fire or personal injury.

1. Relieve the fuel system pressure and disconnect the negative battery cable.

2. Drain the cooling system.

3. Remove the radiator.

4. Remove the intake manifold collector.

5. Remove the fuel injector assembly.

6. Remove the valve covers.

7. Remove the timing belt cover and the timing belt.

8. Remove the camshaft sprockets and the timing belt rear cover.

9. Remove the VTC solenoid valve.

10. Measure the camshaft end-play for installation reference.

11. Loosen the camshaft bearing cap bolts in several steps. Remove the camshaft bearing caps.

12. Remove the oil seals and camshafts.

➡ **Keep the hydraulic lash adjusters upright and immersed in engine oil to prevent air from entering them. Keep the lash adjusters in order and install them in their original positions.**

13. If necessary, remove the hydraulic lash adjusters, keeping them in order for installation into the same location. Be sure the lash adjusters are set upright to prevent them from becoming air bound.

To install:

14. Install the hydraulic lash adjusters into their original locations.

➡ **Apply new engine oil to the camshafts before installation.**

15. Be sure the crankshaft is set at TDC of No. 1 cylinder. Place the camshaft knock pins facing upwards and set the camshafts into place.

➡ **The left side exhaust camshaft has a spline for the camshaft position sensor.**

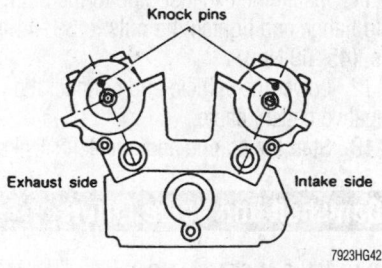

Positioning of the camshaft keys during installation—3.0L (VG30DE) engine

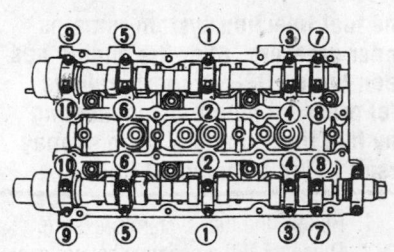

Camshaft bearing cap bolt tightening sequence—3.0L (VG30DE) engine

16. Apply liquid gasket to the front sealing surfaces of the front camshaft bearing caps. Install new camshaft oil seals.

17. Install the camshaft bearing caps and turn each bolt a little at a time to gradually draw the camshafts down against the valve springs. Torque the bolts in sequence to 90 inch lbs. (10 Nm).

18. Apply liquid gasket to the cylinder head and install the VTC solenoid valve.

19. Install the timing belt rear covers and camshaft sprockets.

a. Torque the exhaust camshaft sprocket bolts to 12 ft. lbs. (16 Nm).

b. Torque the large intake camshaft sprocket bolt to 95 ft. lbs. (128 Nm).

c. Install the spring, O-ring and cover plate on the intake camshaft sprocket (VTC).

20. Be sure all sprockets are correctly aligned and install the timing belt.

21. Install the valve covers with new gaskets.

22. Install the fuel injector assembly.

23. Install the timing belt covers.

24. Install the intake manifold collector and the radiator.

25. Install the remaining components and fill the cooling system.

26. Connect the negative battery cable, start the engine, and check for proper operation.

3.0L (VQ30DE) Engine

1. Relieve the fuel system pressure.

❋❋ CAUTION

The fuel injection system remains under pressure after the engine has been turned OFF. Properly relieve fuel pressure before disconnecting any fuel lines. Failure to do so may result in fire or personal injury.

2. Disconnect the negative battery cable.

3. Drain the engine oil and the cooling system. Be sure to drain the engine block and the radiator.

4. Remove the left side rocker cover ornament.

➡ **Before disconnecting any hoses or connectors, note the locations for reassembly.**

5. Remove the air duct to intake manifold hose, collector hose, blow-by hose, and vacuum hoses.

6. Remove the fuel hoses and disconnect the harness connections.

7. Disconnect the canister purge hoses.

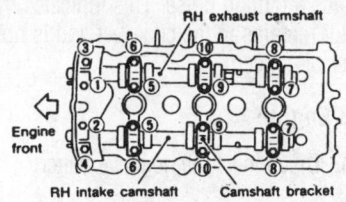

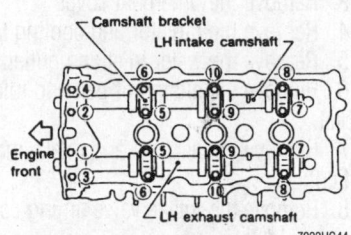

To avoid camshaft damage, loosen the bearing cap bolts in the sequence shown—3.0L (VQ30DE) engine

8. Remove the water hoses from the cylinder head and intake manifold.

9. Disconnect and remove all six ignition coils from the spark plugs.

10. Remove the spark plugs.

11. Remove the bolts that secure the EGR tube and remove the tube.

12. Remove the intake manifold collector supports and remove the collector.

13. Remove the bolts that secure the fuel tube and remove the fuel tube from the vehicle.

14. Remove the bolts that secure the intake manifold to the engine block and remove the manifold. Loosen the bolts in the reverse sequence of the tightening procedure.

15. Remove the LH and RH rocker covers from the cylinder head.

16. Remove the engine undercovers.

17. Remove the right front wheel and engine side covers.

18. Remove the drive belts and idler pulley.

19. Remove the power steering oil pump belt and remove the power steering oil pump assembly.

20. Remove the camshaft position sensor (PHASE) and crankshaft position sensors (REF)/(POS).

21. Set the No. 1 piston to TDC of compression stroke by rotating the crankshaft.

22. Remove the ring gear cover access plate.

23. Loosen the crankshaft pulley bolt while securing the ring gear so the crankshaft cannot rotate.

➡️**Use care not to damage the ring gear teeth.**

24. Using a suitable puller, remove the crankshaft pulley.

25. Remove the A/C compressor and bracket.

26. Remove the front exhaust pipe and its support.

27. Hang the engine at the right and left side engine slingers with a suitable hoist.

28. Support the transaxle with jack.

29. Remove the right side engine mounting, mounting bracket and nuts.

30. Remove the center crossmember assembly.

31. Remove the oil pan bolts and oil pans.

32. Remove the timing chain.

➡️**Remove the O-rings from the front of the engine block.**

33. Loosen the camshaft bearing caps in several steps. The bearing caps MUST be loosened in sequence.

➡️**Keep all bearing caps and camshafts in proper order for reinstallation.**

34. Remove the LH and RH camshaft tensioners from the cylinder head.

35. Remove the camshafts from the cylinder heads.

➡️**The valve adjusters have a replaceable shim on the top of the adjuster. Note the proper locations of each shim to adjuster and remove the shims from the adjusters.**

36. Using a magnet, remove the valve adjusting shim from the adjuster.

37. Remove the adjuster assembly from the bore. Be sure to note the locations from where each adjuster came.

38. Check the diameter of the valve adjuster and the valve adjuster guide bore.

39. The diameter of the adjuster should be 1.3764–1.3770 in. (34.960–34.975mm) and the diameter of the bore should be 1.3780–1.3788 in. (35.000–35.021mm).

40. Remove all traces of liquid gasket from the timing chain case and from the water pump covers.

41. Remove all traces of liquid gasket from the engine block.

42. Inspect the camshafts for excessive wear or damage and replace as necessary.

To install:

43. Lubricate the valve adjusters with clean engine oil and install the adjusters into the bore from which they were removed.

44. Lubricate the valve adjuster shims with clean engine oil and install the shims into the adjuster from which they were removed.

45. Turn the crankshaft clockwise until the No. 1 piston is set 240 degrees before TDC on compression stroke.

46. Install the camshaft tensioners on both sides of the cylinder heads. Tighten the tensioner mounting bolts to 75–96 inch lbs. (8.4–10.8 Nm).

➡️**The camshafts can be identified by the paint marks on the camshaft. The left cylinder head camshafts have a YELLOW paint mark and the right cylinder head camshafts have a WHITE paint mark. When installing the camshafts, position the camshaft keys at the 12 o'clock position in respect to the cylinder head angle.**

47. Install the exhaust and intake camshafts and install the bearing caps. Before installing the No. 1 bearing cap, apply liquid gasket to the corners of the cap.

48. Tighten the camshaft bearing caps as follows:

   a. Tighten bolts No. 7–10 to 17 inch lbs. (2 Nm).

   b. Tighten bolts No. 1–6 to 17 inch lbs. (2 Nm).

   c. Tighten bolts No. 1–10 to 52 inch lbs. (6 Nm).

   d. Tighten bolts No. 1–10 to 81–104 inch lbs. (9–11 Nm).

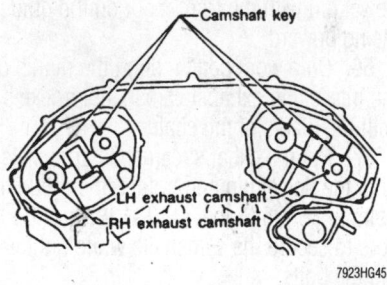

Positioning of the camshaft keys during installation—3.0L (VQ30DE) engine

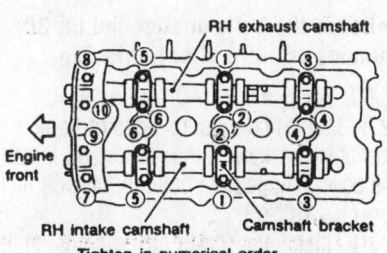

Be sure to tighten the camshaft bearing cap bolts in the correct sequence—3.0L (VQ30DE) engine, right cylinder head

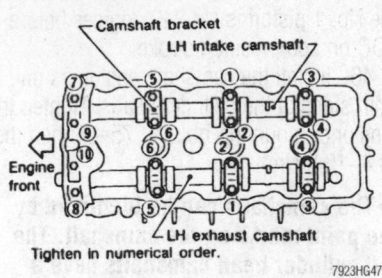

Tighten in numerical order.

7923HG47

Tighten the camshaft bearing cap bolts in the correct sequence—3.0L (VQ30DE) engine, left cylinder head

49. Install new O-rings to the front of the engine block.

50. Apply sealant to the hatched portion of the of the rear timing chain case.

51. Align the rear timing chain case with the dowel pins and install onto the cylinder heads and engine block.

52. Tighten the rear timing chain case mounting bolts in sequence to 105–121 inch lbs. (11.8–13.7 Nm).

53. Install the crankshaft sprocket with the mating mark facing out.

54. Rotate the crankshaft clockwise and position the crankshaft to TDC of compression stroke and align the dowels of the camshaft sprockets to the 12 o'clock position in respect to the cylinder head.

55. Install the lower chain guide on the dowel pin with the front mark on the guide facing upward.

56. On a work bench, align the marks on the intake and exhaust camshaft sprockets with the marks of the chain.

57. Put the exhaust camshaft sprockets onto the dowel pin and tighten the mounting bolts to 88–95 ft. lbs. (119–128 Nm). Be sure to secure the camshafts while tightening the bolts.

58. Align and install the timing chains and sprockets to the camshafts.

59. Install the timing cover evenly and gently. Be sure to align the dowel pin holes. Tighten the bolts in sequence

➡ **Leave the bolts unattended for 30 minutes or more after tightening.**

60. Apply a 0.091–0.130 in. (2.3–3.3mm) continuous bead of liquid gasket to the water pump cover and install the cover. Tighten the bolts to 84–108 inch lbs. (10–13 Nm).

61. Install the rocker covers. Tighten the mounting bolts in sequence.

62. Apply sealant to the front and rear seal of the oil pan. Install the bolts and tighten the bolts.

63. Install the center crossmember assembly.

64. Install the right side engine mounting bracket and mount assembly.

65. Remove the engine slinger assembly.

66. Install the front exhaust pipe and its support.

67. Install the A/C compressor and bracket.

68. Install the crankshaft pulley to the crankshaft and install the mounting bolt.

69. Tighten the mounting bolt to 14–22 ft. lbs. (20–29 Nm). Tighten the crankshaft bolt an additional 60–66 degrees clockwise. This is about the angle from one hexagon bolt head corner to another.

70. Install the ring gear cover plate.

71. Install the camshaft position sensor (PHASE) and crankshaft position sensors (REF)/(POS).

72. Install the power steering pump assembly, drive belts and the idler pulley.

73. Install the right front inner wheel cover and install the right front wheel.

74. Install the engine undercovers.

75. Using new gaskets, install the intake manifold. Tighten the nuts and bolts in sequence and in two stages as follows:

   a. Bolts and nuts—Tighten to 44–86 inch lbs. (5–10 Nm)

   b. Bolts and nuts—Tighten to 16–18 ft. lbs. (22–25 Nm)

76. Using new insulators, install the fuel tube assembly and tighten the bolts to 15–20 ft. lbs. (21–26 Nm). Tighten the bolts in several progressive steps.

77. Install the intake manifold collector gasket with the arrow facing forward.

78. Install the intake manifold collector assembly and support bracket. Tighten the mounting bolts to 16–18 ft. lbs. (22–25 Nm).

79. Using new gaskets, install the EGR tube and tighten the mounting bolts to 15–20 ft. lbs. (21–26 Nm) in two progressive steps.

80. Install the spark plugs, ignition coils and tighten the mounting bolts to 27–33 inch lbs. (2.9–3.8 Nm).

81. Install the water hoses to the cylinder head and intake manifold.

82. Connect the fuel hoses and wiring harness connections to the fuel rail.

83. Install and connect the air duct to intake manifold hose, collector hose, blow-by hose, and vacuum hoses.

84. Refill the engine oil and coolant with the proper type and amount of fluid.

85. Connect the negative battery cable.

86. Start the engine and run at 3000 RPM under no load to purge the air from the high pressure chamber. The engine may produce a rattling noise. This indicates that air still remains in the chamber and is not a matter of concern.

4.1L Engine

1. Disconnect the negative battery cable.

2. Relieve the fuel system pressure.

3. Remove the ornament cover.

4. Remove the radiator and cooling fan.

5. Remove the water inlet and outlet.

6. Remove the alternator belt and idler bracket.

7. Remove the air duct and intake manifold collector.

8. Remove the intake valve timing control solenoid.

9. Remove the left and right rocker arm covers.

10. Turn the crankshaft to position the No. 1 piston at TDC on compression.

11. Remove the camshaft position sensor.

12. Remove the upper front covers.

13. Paint alignment marks on the timing chain and camshaft sprockets.

14. Remove the upper chain tensioners.

15. Remove the camshaft sprockets.

16. Remove the camshaft bearing cap bolts in the proper sequence to prevent damage to the camshaft. Keep the caps in order so they can be installed in the correct locations.

17. Remove the camshafts, rocker arms and lash adjusters. The lash adjusters and rocker arms must be installed in their original positions. Keep them in order.

To install:

18. Install the lash adjusters and rocker arms.

19. Lubricate the camshafts with engine oil and place them on the cylinder head with the knock pins facing away from the crankshaft.

20. Install the bearing caps in their original positions and tighten the bolts using the following sup-steps:

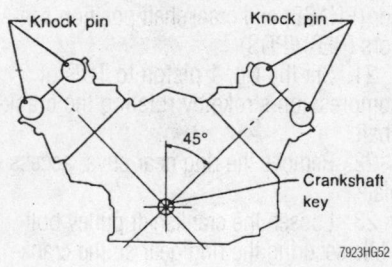

7923HG52

When installing the camshafts, position the knock pins as shown—4.1L engine

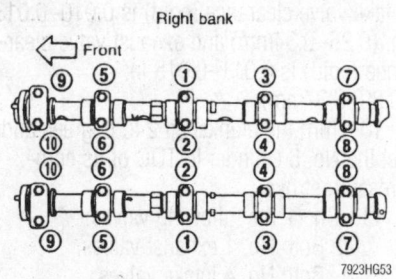

Camshaft bearing cap bolt numbered identification and tightening sequence—4.1L engine, right bank

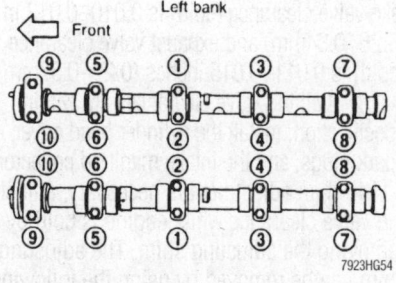

Camshaft bearing cap bolt numbered identification and tightening sequence—4.1L engine, left bank

    a. Tighten bolts 9 and 10, then 1 through 8 to 17 in lbs. (2 Nm).

    b. Tighten the bolts in order to 52 in lbs. (6 Nm).

21. Tighten the bolts in order to 8–11 ft. lbs. (11–14 Nm).

22. Install the camshaft sprockets. Tighten the intake sprocket bolt to 76–83 ft. lbs. (103–113 Nm) and the exhaust sprocket bolt to 12–15 ft. lbs. (16–21 Nm).

23. Install the chain tensioner. Tighten the bolts to 82–95 ft. lbs. (9–11 Nm).

24. Install the upper chain covers.

25. Install the rocker covers using new gaskets.

26. Install the valve timing control solenoid valve using a new O-ring. Tighten the solenoid to 18–25 ft. lbs. (25–34 Nm)

27. Install the remaining components in the reverse of removal.

4.5L Engine

1. Disconnect the negative battery cable.

2. Remove the engine and transmission assembly from the vehicle.

3. Remove the suspension member and engine mounts from the engine.

4. Remove the air compressor bracket.

5. Remove the cooling fan with coupling and the engine gusset.

6. Separate the engine from the transmission and mount the engine on a suitable workstand.

7. Remove the oil pan.

8. Remove the ornamental rocker cover and remove the ignition coils and spark plugs.

9. Bring the No. 1 piston to TDC on the compression stroke.

10. Use a suitable puller to remove the crankshaft pulley.

11. Remove the rocker cover.

12. Remove the crank angle sensor and the Valve Timing Control (VTC) solenoid.

13. Remove the chain tensioners and the upper front covers.

14. Remove the front timing chain cover.

➡**The timing chain will not be disengaged or dislocated from the crankshaft sprocket unless the front cover is removed. The cast portion of the front cover is located on the lower side of the crankshaft sprocket so the timing chain is not disengaged from the sprocket.**

15. Remove the VTC assembly and the camshaft sprocket.

16. Remove the oil pump chain and the timing chains.

➡**Do not attempt to disassemble the VTC assembly since they are difficult to reassemble accurately in the field. If it should be disassembled, the VTC assembly must be replaced with a new one.**

17. Remove the camshaft brackets and the camshafts. Mark the parts so they can be reinstalled in their original positions.

18. Remove the rocker arms. Be sure to identify each rocker arm so it can be reinstalled in it's original position.

To install:

19. Be sure all mating surfaces are clean before installation.

20. Install the rocker arms, camshafts and camshaft brackets on the right bank. Properly lubricate the rocker arms and camshafts prior to installation. Tighten the camshaft bracket bolts to 9–10 ft. lbs. (12–14 Nm) in the proper sequence.

21. Install the VTC assembly and the exhaust cam sprocket on the right bank.

22. Be sure the camshafts are still correctly positioned and the piston in the No. 1 cylinder is still at TDC.

23. Install the timing chain on the right bank, aligning the mating marks on the chain with those on the crankshaft and camshaft sprockets.

24. Install the chain tensioner on the right bank.

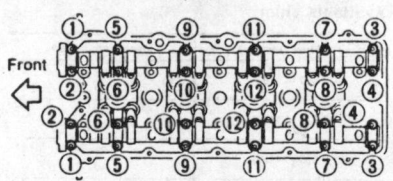

Tighten in numerical order.

Camshaft bearing cap bolt torque sequence—4.5L engine

25. Turn the crankshaft approximately 120 degrees clockwise from the point where the No. 1 piston is at TDC on the compression stroke. At this point, the valves on the left bank still remain closed.

26. Correctly position the camshafts and rocker arms for the left cylinder head. Properly lubricate the rocker arms and camshafts prior to installation. Tighten the camshaft bracket bolts to 9–10 ft. lbs. (12–14 Nm) in the proper sequence. Install the VTC assembly and the exhaust cam sprocket.

27. Install the timing chain on the left bank, aligning the mating marks on the chain with those on the crankshaft and camshaft sprockets.

28. Install the oil pump chain and sprockets.

29. Install the oil pump chain guides. Place a 0.04 in. (1.0mm) feeler gauge between the upper chain guide and chain before assembling the chain guides. The force applied to the chain is equivalent to the upper chain guide weight.

30. Apply suitable sealer and install the front covers.

31. Install the chain tensioner for the left bank.

32. Apply suitable sealer to the rubber plugs and install them on the cylinder head.

33. Install the crank angle sensor, VTC solenoid, rocker cover and crank pulley.

34. Installation of the remaining components is the reverse of the removal procedure.

Valve Lash

ADJUSTMENT

2.0L Engine

➡**A special gauge plate and collar will be needed to complete this procedure.**

1. Remove the camshafts.

2. Install the J38957–1 gauge plate to the cylinder head. Use the bolts supplied in the kit to secure the plate to the cam bearing journals.

Available shim

Thickness mm (in)	Identification mark
2.800 (0.1102)	28 00
2.825 (0.1112)	28 25
2.850 (0.1122)	28 50
2.875 (0.1132)	28 75
2.900 (0.1142)	29 00
2.925 (0.1152)	29 25
2.950 (0.1161)	29 50
2.975 (0.1171)	29 75
3.000 (0.1181)	30 00
3.025 (0.1191)	30 25
3.050 (0.1201)	30 50
3.075 (0.1211)	30 75
3.100 (0.1220)	31 00
3.125 (0.1230)	31 25
3.150 (0.1240)	31 50
3.175 (0.1250)	31 75
3.200 (0.1260)	32 00

7923HG56

Select the correct valve lash adjusting shim using the chart—2.0L engine

3. Install the collar J38957–2 on the dial indicator. Be sure the dished side of the collar is toward the gauge and tighten the set screw.

4. Place the gauge on the No. 1 intake valve (shim side). Be sure the shim has been removed. Place the tip of the dial gauge on the top of the valve stem and the collar on the gauge plate. Zero the dial gauge.

5. Move the dial gauge to the other intake valve (rocker guide side). Place the tip of the dial gauge on the rocker guide and the collar of the gauge plate. Record the measurement.

6. Select the correct size shim using the chart. Shims are available in 17 different sizes ranging from 0.1102 in. (2.800mm) to 0.1260 in. (3.200mm) in increments of 0.001 in. (0.025mm).

3.0L (VG30DE), 4.1L and 4.5L Engines

No valve lash adjustment is possible on this engine. If valve noise occurs, look for excessive wear on the camshaft, rocker arm and auto lash adjuster.

3.0L (VQ30DE) Engine

➥**Check and adjust the valve clearances while the engine is cold and not running.**

1. Remove the intake manifold collector.

2. Remove the left and right rocker covers.

3. Remove the spark plugs.

4. Set the No. 1 cylinder at TDC on its compression stroke. Align the pointer with the TDC mark on the crankshaft pulley. Check that the valve adjusters on the No. 1 cylinder are loose and valve adjusters on the No. 4 cylinder are tight. If not, turn the crankshaft one revolution (360 degrees) and align the pointer with the TDC mark on the crankshaft pulley.

5. Check the following valves:
   a. Both No. 1 intake valves.
   b. Both No. 2 exhaust valves.
   c. Both No. 3 exhaust valves.
   d. Both No. 6 intake valves.

6. Using a feeler gauge, measure the clearance between the valve adjuster and the camshaft. Record any valve clearance measurements which are out of specification. Intake valve clearance (cold) is 0.010–0.013 in. (0.26–0.34mm) and exhaust valve clearance (cold) is 0.011–0.015 in. 0.29–0.37mm).

7. Turn the crankshaft 240 degrees and set the No. 3 cylinder to TDC of its compression stroke.

8. Check the following valves:
   a. Both No. 2 intake valves.
   b. Both No. 3 intake valves.
   c. Both No. 4 exhaust valves.
   d. Both No. 5 exhaust valves.

9. Using a feeler gauge, measure the clearance between the valve adjuster and camshaft. Record any valve clearance mea-

surements which are out of specification. Intake valve clearance (cold) is 0.010–0.013 in. (0.26–0.34mm) and exhaust valve clearance (cold) is 0.011–0.015 in. (0.29–0.37mm).

10. Turn the crankshaft 240 degrees and set the No. 5 cylinder to TDC of its compression stroke.

11. Check the following valves:
   a. Both No. 1 exhaust valves.
   b. Both No. 4 intake valves.
   c. Both No. 5 intake valves.
   d. Both No. 6 exhaust valves.

12. Using a feeler gauge, measure the clearance between the valve adjuster and the camshaft. Record any valve clearance measurements which are out of specification. Intake valve clearance (cold) is 0.010–0.013 in. (0.26–0.34mm) and exhaust valve clearance (cold) is 0.011–0.015 inches (0.29–0.37mm).

13. If all the valve clearances are within specification, install the cylinder head cover, spark plugs, and the intake manifold collector.

14. If an adjustment is necessary, adjust the valve clearance while engine is cold by removing the adjusting shim. The adjusting shim can be removed by using the following procedures:
   a. Turn the crankshaft so the camshaft lobe of the valve to be adjusted is pointed straight up.
   b. Turn the adjuster so the notch is pointed towards the center of the cylinder head; this will facilitate the shim removal process.
   c. Using a depressor tool No. KV10115110 or equivalent, push down on the adjuster and insert a keeper tool on the edge of the adjuster to keep the adjuster in the depressed position.
   d. Remove the depressor tool and remove the shim with a magnet.

➥**Compressed air can be blown into the hole of the adjuster to separate the adjusting shim from the adjuster.**

RH cylinder head

③ ③

EXH

INT

⟵ Engine front

⑥ ⑥

INT

EXH

① ①

② ②

LH cylinder head

7923HG57

Valve lash checking sequence at TDC of cylinder No. 1—3.0L (VQ30DE) engine

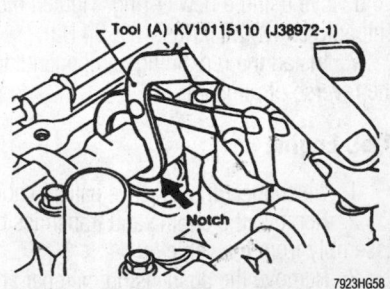

Tool (A) KV10115110 (J38972-1)

Notch

7923HG58

Install the depressor tool around the camshaft being careful not to damage the surfaces—3.0L (VQ30DE) engine

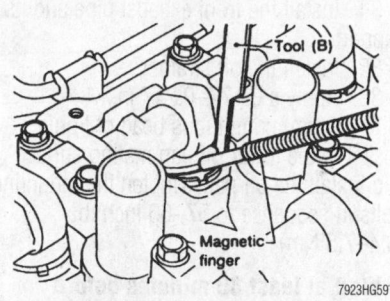

Use a magnet to remove the shim from the adjuster. Sometimes a shot of compressed air can help lift the shim up—3.0L (VQ30DE) engine

15. Determine the replacement adjusting shim size by using the following procedures and formula:

   a. Using a micrometer determine thickness of the removed shim.

   b. Calculate the thickness of a new adjusting shim so valve clearance is within the specified values.
- R= thickness of the removed shim
- N= thickness of the new shim
- M= measured valve clearance
- Calculate the Intake Shim as follows:

$N = R + M — 0.0118$ in. (0.30mm)
- Calculate the Exhaust Shim as follows:

$N = R + M — 0.0130$ in. (0.33mm)

16. Shims are available in 64 sizes from 0.0913–0.1161 in. (2.32–2.95mm) in steps of 0.004 in. (0.01mm). The thickness is stamped on the shim; this side is always installed facing down. Select new shims with thickness as close as possible to calculated valve and install it in the adjuster.

17. Install the new shim onto the adjuster.

18. Depress the adjuster and remove the keeper tool. Remove the depressor tool and recheck the valve clearance. Repeat this procedure for any other valves requiring adjustment.

19. When all valve adjustments are finished, install the cylinder head cover, spark plugs, and the intake manifold collector.

Oil Pan

REMOVAL & INSTALLATION

2.0L Engine

1. Raise and support the vehicle safely. Remove the engine undercover and drain the oil.

2. Remove the steel oil pan bolts in the reverse of the torque sequence. Remove the steel oil pan. Insert tool KV10111100 between steel oil pan and aluminum oil pan to break the seal.

3. Remove the oil baffle bolts and oil baffle.

4. Remove the front exhaust tube.

5. Support the transaxle with a suitable jack and raise the engine with an engine hoist.

6. Remove the center crossmember.

7. If equipped with an automatic transaxle, remove the transaxle shift control cable.

8. Remove the A/C compressor bracket gussets and the rear cover plate.

9. Remove the aluminum oil pan bolts in sequence.

10. Remove the two engine to transaxle bolts and install them into vacant bolt holes on the oil pan. Tighten the bolts to release the oil pan from the cylinder block. Use tool KV10111100 to break the remaining seal.

To install:

11. Remove the two bolts previously installed in the oil pan.

12. Clean the oil pan rail of all liquid gasket and apply a new bead of ⅛ inch thickness to the oil pan rail.

13. Install the aluminum oil pan and torque them in the numbered sequence. Torque bolts 1 through 16 to 12–14 ft. lbs. (16–19 Nm) and bolts 17 through 18 to 60–72 inch lbs. (6–8 Nm).

14. Install the two engine-to-transaxle bolts, rear cover plate, compressor bracket gussets, automatic transmission shift con-

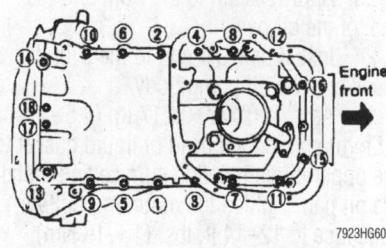

Tighten the aluminum oil pan mounting bolts in the sequence shown—2.0L engine

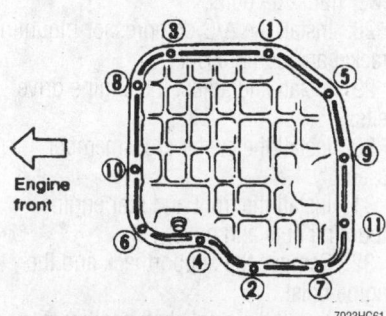

Be sure to tighten the steel oil pan mounting bolts in the proper order to prevent leakage—2.0L engine

trol cable (if equipped), center member, front exhaust tube and baffle plate.

15. Clean the oil pan rail of all liquid gasket and apply a new bead of ⅛ inch thickness to the oil pan rail.

16. Install the steel oil pan. Torque the bolts in numbered sequence to 56–66 inch lbs. (6.4–7.5 Nm). Wait 30 minutes before refilling crankcase with oil.

3.0L (VG30DE) Engine

1. Disconnect the negative battery cable. Raise and support the vehicle safely.

2. Remove the engine undercover and drain the engine oil and coolant.

3. Remove the radiator from the engine compartment.

4. Remove the air ducts.

5. Remove the upper and lower radiator shrouds. Disconnect the oil cooler lines on vehicles with automatic transmissions.

6. Remove the fan coupling.

7. Disconnect the power steering oil hoses.

8. Remove the power steering oil pump.

9. Remove the stabilizer bar.

10. Place a suitable support under the transmission.

11. Remove the engine mounting insulator lower attaching nuts from both sides of the engine. Hoist the engine so enough clearance is provided for oil pan removal.

12. Remove the oil pan bolts in the reverse order of installation. Tap the oil pan removal tool J-37228 or equivalent, between the cylinder block and oil pan. Tap the tool around oil pan to break the gasket seal.

13. Remove the oil strainer and lay it in the oil pan. Remove the oil pan.

To install:

14. Clean all gasket mating surfaces thoroughly.

15. Apply sealant to oil pump gasket and rear oil seal retainer gasket.

16. Apply a continuous bead of liquid gasket to the oil pan mating surface. Be sure the bead is ³⁄₁₆ inch wide.

17. Place oil pan under engine and install oil strainer.

18. Install oil pan and tighten bolts in the proper sequence to 60–72 inch lbs. (7–8 Nm).

➡**Wait at least 30 minutes before refilling the engine with oil**

19. Lower the engine and install the engine mount insulator bolts.

20. Install the stabilizer bar.

21. Install the power steering oil pump and hoses.

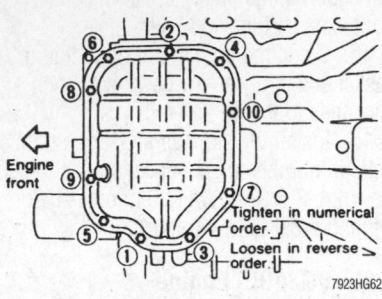

Tighten the steel oil pan mounting bolts in the order shown (loosen in reverse sequence)—3.0L (VG30DE) engine

22. Install the radiator, radiator hoses, shroud and transmission oil cooler lines.

23. Install the fan coupling.

24. Install the air ducts and engine undercover.

25. Refill the engine with coolant.

26. Fill the engine and power steering reservoir with the appropriate oils. Start the engine and allow it to reach normal operating temperature. Bleed the steering system. Check for leaks.

3.0L (VQ30DE) Engine

1. Disconnect the negative battery cable.

2. Drain the engine oil and remove the engine undercovers.

3. Remove the steel (lower) oil pan bolts in the reverse sequence of the torque sequence.

4. Insert a seal cutter between the steel and aluminum oil pan.

5. Tapping the cutter with a hammer, slide it around the entire edge of the oil pan. Be careful not to damage the aluminum mating surface of the upper oil pan.

6. Remove the steel oil pan and the oil strainer.

7. Remove the front exhaust pipe and its support.

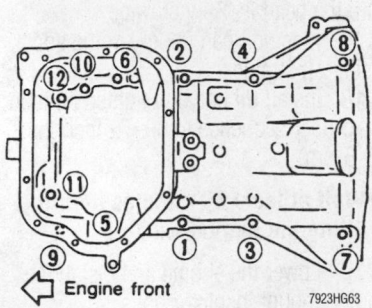

Aluminum oil pan torque sequence (loosen in reverse sequence)—3.0L (VQ30DE) engine

8. Hang the engine at the right and left side engine slingers with a suitable hoist.

9. Position a suitable jack under the transaxle.

10. Remove the crankshaft position sensors (REFERENCE and POSITION) from the oil pan.

11. Remove the front and rear engine mounting nuts and bolts.

12. Remove the center crossmember assembly.

13. Remove the engine drive belts.

14. Remove the air conditioner compressor and the compressor mounting bracket.

15. Remove the rear cover plate and the lower transaxle bolts.

16. Remove the aluminum (upper) oil pan bolts in the reverse sequence of the torque sequence.

17. Insert a seal cutter between the aluminum oil pan and the engine block.

18. Tapping the cutter with a hammer, slide it around the entire edge of the oil pan. Be careful not to damage the mating surfaces of the oil pan or engine block.

19. Remove the oil pan assembly.

20. Remove the bolts that secure the baffle plate and remove the baffle plate.

21. Remove the O-rings from the cylinder block and oil pump body.

To install:

22. Install the baffle plate to the oil pan and tighten the mounting bolts to 22–27 inch lbs. (2.5–3.1 Nm).

23. Apply sealant to the front and rear seal of the oil pan.

24. Install new O-rings to the cylinder block and the oil pump body.

25. Apply a 0.177–0.217 in. (4.5–5.5mm) continuous bead of liquid gasket to the upper oil pan mating surface and install the oil pan. Tighten the mounting bolts in sequence to 12–14 ft. lbs. (16–19 Nm).

26. Install the oil pan strainer and tighten the mounting bolts to 12–14 ft. lbs. (16–19 Nm).

27. Install the rear cover plate and the lower transaxle bolts.

28. Install the A/C compressor mounting bracket and compressor.

29. Install and adjust the engine drive belts.

30. Install the center crossmember assembly.

31. Install the front and rear engine mounting nuts and bolts.

32. Remove the support jack and the engine hoist.

33. Install the crankshaft position sensors (REFERENCE and POSITION) to the oil pan. Tighten the sensor mounting bolts to 75–96 in. lbs. (9–10 Nm).

34. Install the front exhaust pipe and its support.

35. Install the oil strainer.

36. Apply a 0.177–0.217 in. (4.5–5.5mm) continuous bead of liquid gasket to the lower oil pan mating surface and install the oil pan. Tighten the mounting bolts in sequence to 57–66 inch lbs. (6.4–7.5 Nm).

➡ **Wait at least 30 minutes before refilling the engine oil.**

37. Tighten the oil pan drain plug to 22–29 ft. lbs. (29–39 Nm).

38. Install the engine undercovers.

39. Refill the engine oil with the proper type and amount of fluid.

40. Start the engine and check for leaks.

4.1L Engine

1. Disconnect the negative battery cable.

2. Raise and safely support the vehicle.

3. Drain the engine oil.

4. Attach an engine support fixture to the engine so the right and left engine mounts can be removed.

5. Remove the following:
- Drive belts
- Cooling fan and coupling
- Power steering oil pump
- Front stabilizer bar brackets from the side members
- Right and left engine mounting bolts

6. Disconnect the steering shaft lower joint

7. Remove the bracket securing the power steering tube to the front suspension

8. Support the front suspension member with a jack, then remove the mounting bolts and lower it.

9. Remove the A/C compressor and bracket.

10. Remove the oil pan mounting bolts, then insert a tool into the notch on the oil

Use a suitable tool to break the seal between the oil pan and engine block—4.1L engine

pan and break the seal between the pan and engine block. Be careful not to damage the sealing surface.

11. Pull the oil pan out from the front while lowering the suspension as needed.

To install:

12. Clean all gasket mating surfaces thoroughly.

13. Apply a continuous bead of liquid gasket to the oil pan mating surface. Be sure the bead is ⅛ inch wide.

14. Install the oil pan and tighten bolts 1 through 21 in sequence to 12–14 ft. lbs. (16–19 Nm). Tighten bolts 22 and 23 to 56–65 in lbs. (6–7 Nm).

15. Wait at least 30 minutes for the sealant to cure before filling the engine with oil.

16. Install the A/C compressor and bracket.

17. Install the front suspension member. Tighten the nuts to 87–101 ft. lbs. (147–167 Nm).

18. Install the power steering tube on the suspension member.

19. Connect the lower steering shaft joint.

20. Install the stabilizer bar to the suspension member. Tighten the nuts to 35–46 ft. lbs. (47–62 Nm).

21. Install the engine mounting bolts. Tighten the nuts to 41–49 ft. lbs. (55–67 Nm).

22. Install the remaining components in the reverse of the removal.

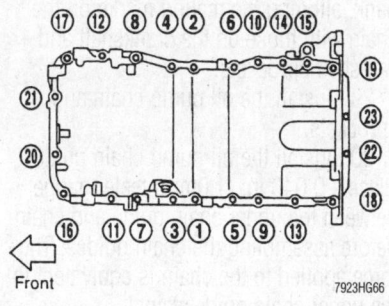

Tighten the oil pan bolts in the sequence shown—4.1L engine

4.5L Engine

1. Disconnect the negative battery cable. Raise and support the vehicle safely.

2. Remove the engine undercover and drain the engine oil.

3. Remove the fan coupling with the fan.

4. Remove the drive belts, alternator, air conditioning compressor and engine gusset.

5. Matchmark and remove the steering lower joint.

6. Place a suitable support under the transmission. Hoist the engine with engine sling (support fixture).

7. Remove the suspension member assembly.

8. Remove the oil pan bolts. Insert oil pan removal tool J-37228 or equivalent, between the cylinder block and oil pan. Tap the tool with a hammer around the oil pan to break the seal

9. Remove the oil pan.

To install:

10. Clean all gasket mating surfaces thoroughly.

11. Apply a continuous bead of liquid gasket to the oil pan mating surface. Be sure the bead is ⅛ inch wide.

12. Install oil pan and tighten the bolts in the proper sequence 60–72 inch lbs. (7–8 Nm). Wait at least 30 minutes for the sealant to cure before filling the engine with oil.

13. Install all remaining components in the reverse order of removal.

14. Fill the engine with oil. Start the engine and allow it to reach normal operating temperature. Check for leaks.

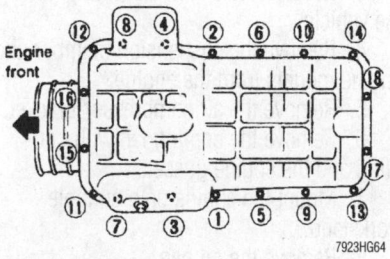

Oil pan bolt installation torque sequence—4.5L engine

Oil Pump

REMOVAL & INSTALLATION

2.0L Engine

⁑ CAUTION

The fuel injection system remains under pressure after the engine has been turned OFF. Properly relieve fuel pressure before disconnecting any fuel lines. Failure to do so may result in fire or personal injury.

1. Relieve the fuel system pressure and disconnect the negative battery cable.

2. Remove the drive belts.

3. Remove the cylinder head with the intake and exhaust manifolds attached.

4. Remove the oil pans.

5. Remove the oil strainer and baffle plate.

6. Remove the crankshaft pulley and the front cover assembly.

7. Remove the screws to remove the oil pump from the inside of the front cover.

To install:

8. Coat the oil pump gears with oil and fit the pump to the cover, using a new oil seal and O-ring.

9. Clean the mating surfaces of liquid gasket and apply a fresh bead of ⅛ inch sealer to the sealing surface of the front cover. Install the front cover assembly.

10. Install the crankshaft pulley.

11. Install the oil strainer, baffle plate, oil pans, cylinder head and drive belts.

12. Connect the negative battery cable.

3.0L (VG30DE) Engine

1. Raise and safely support the vehicle.

2. Drain the engine oil. Remove the oil level gauge (dipstick).

3. Remove the timing belt and crankshaft sprocket. Remove the oil pan.

4. Remove the oil pump mounting bolts and lift out the oil pump.

To install:

5. Always replace with a new oil seal and gasket. Apply oil to the inner and outer gears when installing.

6. Install the oil pump and tighten the long mounting bolt to 9–12 ft. lbs. (12–16 Nm) and the short bolts to 4.3–5.1 ft. lbs. (6–7 Nm).

7. Install the oil pan and oil level gauge.

8. Install the crankshaft sprocket and the timing belt.

9. Fill the engine with oil. Start the engine and check for leaks.

3.0L (VQ30DE) Engine

➡ **The oil pump bolts to the front of the engine block and is driven by the crankshaft. Removal of the timing cover and chains are necessary for oil pump service.**

1. Disconnect the negative battery cable and drain the engine oil.

2. Rotate the engine and position it to TDC compression stroke of cylinder No. 1.

3. Remove the drive belts.

4. Remove the camshaft position sensor (PHASE) and the crankshaft position sensor (REF/POS).

5. Remove the right front wheel and inner fender cover.

6. Remove the engine undercovers.

7. Remove the bolt that secures the crankshaft pulley and remove the pulley.

8. Remove the front exhaust pipe and its support.

9. Support the engine at the left and right side slingers with a suitable hoist.

10. Remove the engine right side mounting insulator and bracket nuts and bolts.

11. Remove the center crossmember assembly.

12. If equipped, remove the A/C compressor and the mounting bracket.

13. Remove the lower and upper oil pans.

14. Remove the oil strainer from the oil pump.

15. Remove the water pump cover and remove the front cover assembly.

16. Remove the lower timing chain assembly.

17. Remove the bolts that secure the oil pump to the engine block and remove the oil pump.

To install:

➡ **When installing the oil pump, be sure to apply engine oil to the gears.**

18. Install the oil pump to the engine block. Tighten the mounting bolts to 57 inch lbs. (6.5 Nm) and tighten the mounting screws to 33–44 inch lbs. (4–5 Nm).

19. Install the lower timing chain assembly.

20. Install the front timing cover and water pump covers.

21. Using a new gasket, install the oil strainer and tighten the mounting bolts to 12–14 ft. lbs. (16–19 Nm).

22. Install the upper and lower oil pans. Be sure to use new O-rings at the oil pump to upper oil pan mating surface.

23. If removed, install the A/C compressor mounting bracket and the install the compressor.

24. Install the center crossmember assembly.

25. Install the engine right side mounting insulator and bracket.

26. Remove the engine support hoist.

27. Install the front exhaust pipe and its support.

28. Install the crankshaft pulley.

29. Install the engine undercovers and install the right side inner fender cover.

30. Install the right front wheel.

31. Install the camshaft position sensor (PHASE) and the crankshaft position sensor (REF/POS).

32. Install and adjust the engine drive belts.

33. Refill the engine oil with the proper type and amount.

34. Start the engine, check the oil pressure, and check for oil leaks.

4.1L Engine

➡ **The oil pump is mounted in the cylinder block below the left bank and behind the left timing chain.**

1. Remove the timing chains.

2. Remove the oil pump assembly from the front of the engine.

To install:

3. Clean the oil pump mounting surface.

4. Install the oil pump using a new gasket. Tighten the short bolt to 56–66 ft. lbs. (6.4–7.5 Nm) and the long bolt to 12–14 ft. lbs. (16–19 Nm).

5. Install the timing chains.

4.5L Engine

1. The oil pump is mounted in the cylinder block below the left bank and behind the left timing chain. 2. Disconnect the negative battery cable.

3. Remove the engine assembly from the vehicle.

4. Remove the suspension member and engine mounts from the engine.

5. Remove the air compressor bracket.

6. Remove the cooling fan with coupling and the engine gusset.

7. Mount the engine on a suitable workstand.

8. Remove the oil pan.

9. Remove the ornamental rocker cover and remove the ignition coils and spark plugs.

10. Bring the No. 1 piston to TDC on the compression stroke.

11. Remove the rocker cover.

12. Use a suitable puller to remove the crankshaft pulley.

13. Remove the crank angle sensor and the Valve Timing Control (VTC) solenoid.

14. Remove the chain tensioners and the upper front covers.

15. Remove the front timing chain cover.

➡ **The timing chain will not be disengaged or dislocated from the crankshaft sprocket unless the front cover is removed. The cast portion of the front cover is located on the lower side of the crankshaft sprocket so the timing chain is not disengaged from the sprocket.**

16. Remove the VTC assembly and the camshaft sprocket.

17. Remove the oil pump chain and the timing chains.

18. Remove the mounting bolts and lift out the oil pump.

To install:

19. Thoroughly clean the mounting surfaces. Apply engine oil to the gears.

20. Install the oil pump with a new seal and gasket. Torque the long bolts to 12–15 ft. lbs. (16–20 Nm) and the short bolts to 3.3–4.3 ft. lbs. (4–6 Nm).

21. Be sure all mating surfaces are clean before installation.

22. Install the VTC assembly and the exhaust cam sprocket on the right bank.

23. Be sure the camshafts are still correctly positioned and the piston in the No. 1 cylinder is still at TDC.

24. Install the timing chain on the right bank, aligning the mating marks on the chain with those on the crankshaft and camshaft sprockets.

25. Install the chain tensioner on the right bank.

26. Turn the crankshaft approximately 120 degrees clockwise from the point where the No. 1 piston is at TDC on the compression stroke. At this point, the valves on the left bank still remain closed.

27. Correctly position the camshafts and rocker arms for the left cylinder head. Properly lubricate the rocker arms and camshafts prior to installation. Install the VTC assembly and the exhaust cam sprocket.

28. Install the timing chain on the left bank, aligning the mating marks on the chain with those on the crankshaft and camshaft sprockets.

29. Install the oil pump chain and sprockets.

30. Install the oil pump chain guides. Place a 0.040 in. (1.0mm) feeler gauge between the upper chain guide and chain before assembling the chain guides. The force applied to the chain is equivalent to the upper chain guide weight.

31. Apply suitable sealer and install the front covers.

32. Install the chain tensioner for the left bank.

33. Apply suitable sealer to the rubber plugs and install them on the cylinder head.

34. Install the crank angle sensor, VTC solenoid, rocker cover and crank pulley.

35. Install the drive belts.

36. Install the engine mounts and the suspension crossmember.

37. Install the engine assembly.

38. Connect all vacuum hoses, cables and electrical connections.

39. Refill the engine with coolant and engine oil.
40. Start the engine and check for leaks.

Rear Main Seal

All engines use a one-piece rear main oil seal.

1. Remove the transmission or transaxle.
2. Remove the drive plate from the crankshaft.
3. Carefully pry the seal out of the retainer without damaging the crankshaft or the seal retainer.

To install:

4. Lubricate the seal with engine oil.
5. Install the seal into the retainer using the appropriate seal driver.
6. Install the driveplate and transmission or transaxle.

Timing Chain, Sprockets, Front Cover and Seal

REMOVAL & INSTALLATION

2.0L Engine

✸✸ CAUTION

The fuel injection system remains under pressure after the engine has been turned OFF. Properly relieve fuel pressure before disconnecting any fuel lines. Failure to do so may result in fire or personal injury.

1. Relieve the fuel system pressure and disconnect the negative battery cable.
2. Raise and support the vehicle safely. Remove the engine under covers. Remove the right front wheel and engine side cover, then lower the vehicle.
3. Drain the cooling system and remove the radiator.
4. Remove the intake manifold air duct.
5. Remove the drive belts, water pump pulley, alternator and power steering pump.
6. Label and remove the vacuum hoses, fuel hoses and wiring harness connectors.
7. Remove the spark plugs.
8. Remove the cylinder head cover and oil separator.
9. Remove the intake manifold supports.
10. Remove the oil filter bracket and the power steering oil pump bracket.
11. Place the No. 1 piston at TDC on the compression stroke.
12. Remove the chain tensioner.
13. Remove the distributor. Do not turn the rotor while the distributor is removed.

14. Remove the timing chain guide.
15. Remove the camshaft sprockets.
16. Remove the camshafts, camshaft brackets, oil tubes and baffle plate.
17. Remove the starter.
18. Disconnect the heater hoses and the water hoses from the cylinder head.
19. Disconnect the knock sensor harness connector.
20. Remove the cylinder head outside bolts.
21. Use two or three steps and gradually remove the cylinder head bolts.
22. Remove the cylinder head with the intake and exhaust manifolds.
23. Raise and support the vehicle safely.
24. Remove the oil pans.
25. Remove the oil strainer and baffle plate.
26. Remove the crankshaft pulley.
27. Place a transmission jack under the main bearing beam and raise the engine slightly to take the weight off of the front engine mount.
28. Remove the front engine mount.
29. Remove the timing chain cover. Tap the seal out of the cover with a suitable seal driver.
30. Loosen and remove the timing chain sprocket bolts.
31. Remove the timing chain guides, then timing chain and sprockets.

To install:

32. Be sure all sealing surfaces are clean and prepared for assembly.
33. Install the crankshaft sprocket. Position the crankshaft so No. 1 piston is set at TDC (keyway at 12 o'clock, mating mark at 4 o'clock).
34. Fit the timing chain to crankshaft sprocket with the gold mating mark on the chain aligned with the mark on the sprocket. (The mating marks for the camshaft sprockets are silver.)
35. Install the timing chain guides and hang the chain off the left (front) guide. If necessary, secure the chain so it does not disengage from the crankshaft sprocket during assembly.
36. Install a new seal in the front cover and apply engine oil to the lip of the seal.
37. Apply a bead of liquid gasket to the front cover. Install the oil pump drive spacer and front cover. Torque the bolts evenly to 60 inch lbs. (6.7 Nm) and wipe away any excess liquid gasket.
38. Install the front engine mount.
39. Install the crankshaft pulley and temporarily tighten the bolt to hold the sprocket in place. The timing mark should align with the TDC mark.

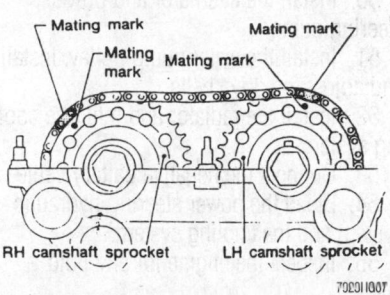

During disassembly, be sure to align the timing chain and camshaft sprocket mating marks—2.0L engine

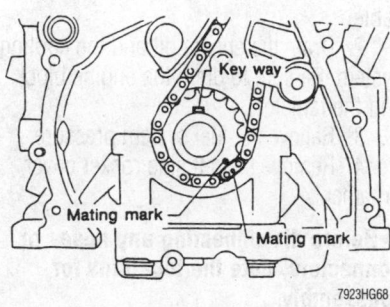

Crankshaft sprocket and timing chain alignment marks—2.0L engine

40. Install the oil strainer, baffle plate and oil pan.
41. Install the cylinder head, camshafts, oil tubes and baffles. Position the left camshaft key at 12 o'clock and the right camshaft key at 10 o'clock.
42. Install the camshaft sprockets by lining up the mating marks on the timing chain with the mating marks on the camshaft sprockets. Torque the camshaft bolts to 101–116 ft. lbs. (137–157 Nm). Torque the crankshaft pulley bolt to 105–112 ft. lbs. (142–152 Nm).
43. Install the upper timing chain guide and distributor. Ensure that the rotor is at the 5 o'clock position.
44. Before installing the chain tensioner, press the cam stopper down and the push in the sleeve until the hook can be engaged on the pin. When tensioner is bolted in position, the hook will release automatically. Ensure the arrow on the outside faces the front of the engine.
45. Install the oil filter bracket and the power steering pump bracket.
46. Install the intake manifold supports.
47. Install the oil separator and the cylinder head cover.
48. Install the spark plugs.
49. Connect the vacuum hoses, fuel hoses, and wiring harness connectors.

50. Install the alternator and power steering pump.

51. Install the water pump pulley. Install and adjust the drive belts.

52. Install the radiator and refill the cooling system.

53. Connect the negative battery cable.

54. Bleed the power steering hydraulic system and the cooling system.

55. Inspect the engine for any fluid leaks.

56. Replace the engine under covers, side cover and right front wheel.

3.0L (VQ30DE) Engine

1. Disconnect the negative battery cable.

2. Drain the engine oil and the cooling system. Be sure to drain the engine block and the radiator.

3. Relieve the fuel system pressure.

4. Remove the left side rocker cover ornament.

➡**Before disconnecting any hoses or connectors, note the locations for reassembly.**

5. Remove the air duct to intake manifold hose, collector hose, blow-by hose, and vacuum hoses.

❄❄ CAUTION

The fuel injection system remains under pressure after the engine has been turned OFF. Properly relieve fuel pressure before disconnecting any fuel lines. Failure to do so may result in fire or personal injury.

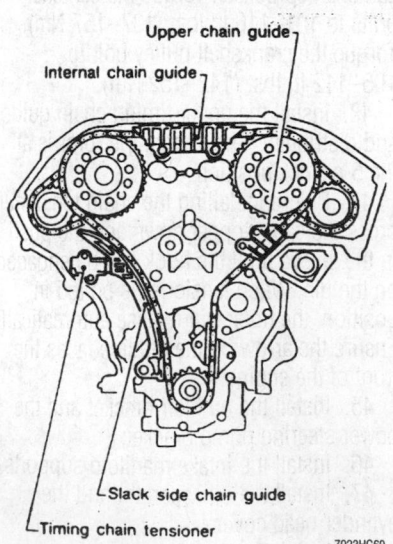

Timing chain tensioner and guide locations—3.0L (VQ30DE) engine

6. Remove the fuel hoses and disconnect the harness connections.

7. Disconnect the canister purge hoses.

8. Remove the water hoses from the cylinder head and intake manifold.

9. Disconnect and remove all six ignition coils from the spark plugs.

10. Remove the spark plugs.

11. Remove the bolts that secure the EGR tube and remove the tube.

12. Remove the intake manifold collector supports and remove the collector.

13. Remove the bolts that secure the fuel tube and remove the fuel tube from the vehicle.

14. Remove the bolts that secure the intake manifold to the engine block and remove the manifold. Loosen the bolts in the reverse sequence of the tightening procedure.

15. Remove the LH and RH rocker covers from the cylinder head.

16. Remove the engine undercovers.

17. Remove the right front wheel and the engine side covers.

18. Remove the drive belts and the idler pulley.

19. Remove the power steering oil pump belt and remove the power steering oil pump assembly.

20. Remove the camshaft position sensor (PHASE) and crankshaft position sensors (REF)/(POS).

21. Set the No. 1 piston to TDC of compression stroke by rotating the crankshaft.

22. Remove the ring gear cover access plate.

23. Loosen the crankshaft pulley bolt while securing the ring gear so the crankshaft cannot rotate.

➡**Use care not to damage the ring gear teeth.**

24. Using a suitable puller, remove the crankshaft pulley.

25. Remove the A/C compressor and bracket.

26. Remove the front exhaust pipe and its support.

27. Hang the engine at the right and left side engine slingers with a suitable hoist.

28. Support the transaxle with jack.

29. Remove the right side engine mounting, mounting bracket, and nuts.

30. Remove the center crossmember assembly.

31. Remove the steel (lower) oil pan bolts in the reverse sequence of the torque sequence.

32. Insert a seal cutter between the steel and aluminum oil pan.

33. Tapping the cutter with a hammer, slide it around the entire edge of the oil pan.

Be careful not to damage the aluminum mating surface of the upper oil pan.

34. Remove the steel oil pan and the oil strainer.

35. Remove the aluminum (upper) oil pan bolts in the reverse sequence of the torque sequence.

36. Remove the transaxle bolts that secure the oil pan.

37. Insert a seal cutter between the aluminum oil pan and the engine block.

38. Tapping the cutter with a hammer, slide it around the entire edge of the oil pan. Be careful not to damage the mating surfaces of the oil pan or engine block.

39. Remove the oil pan from the vehicle.

40. Remove the water pump cover and remove the bolts that secure the front timing chain case.

41. Using the seal cutter, remove the timing chain case cover.

42. Remove the internal timing chain guide and the upper chain guide.

43. Remove the timing chain tensioner and slack side chain guide.

44. Remove the left and right intake camshaft sprockets first. Be sure to hold the flats of the camshafts while removing the sprocket bolts.

45. Remove the lower timing chain assembly. Be sure to note the aligning marks of the chain before removal.

46. Insert a suitable stopper pin for the left and right camshaft tensioners.

47. Remove the left and right exhaust camshaft sprocket bolts. Be sure to hold the flats of the camshafts while removing the sprocket bolts.

48. Remove the upper timing chain assembly. Be sure to note the aligning marks of the chain before removal.

49. Remove the lower timing chain guide.

50. Remove the crankshaft sprocket.

51. Remove all traces of liquid gasket from the front timing chain case and from the water pump.

52. Inspect the timing chain for excessive wear or damage and replace as necessary.

To install:

53. Install the crankshaft sprocket with the mating mark facing out.

54. Position the crankshaft to TDC of compression stroke and align the dowels of the camshaft sprockets to the 12 o'clock position in respect to the cylinder head.

55. Install the lower timing chain guide. The front mark on the guide should face upwards.

56. On a work bench, align the marks on the intake and exhaust camshaft sprockets with the marks of the chain.

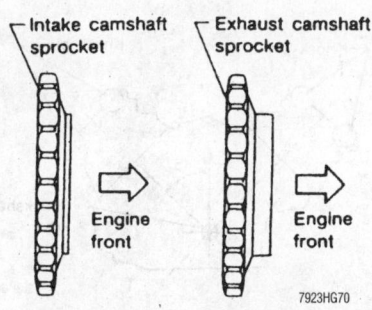

Identification of the intake and exhaust camshaft sprockets—3.0L (VQ30DE) engine

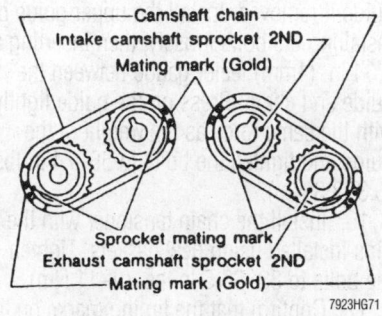

Upper timing chain alignment marks—3.0L (VQ30DE) engine

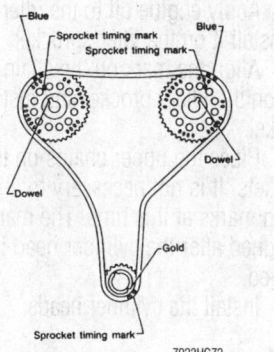

Lower timing chain alignment marks—3.0L (VQ30DE) engine

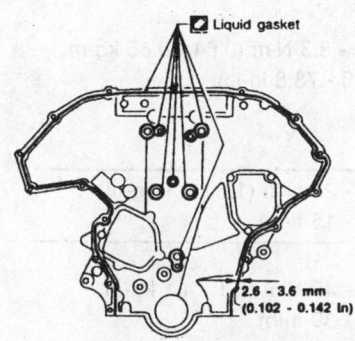

Application of liquid gasket to the front timing case—3.0L (VQ30DE) engine

57. Put the exhaust camshaft sprockets onto the dowel pin and tighten the mounting bolts to 88–95 ft. lbs. (119–128 Nm). Be sure to secure the camshafts while tightening the bolts.

58. Install the timing chains and sprockets to the intake camshafts. Be sure to align the timing chain and sprocket mating marks.

59. Remove the left and right camshaft tensioner stopper pins.

60. Align the mating mark on the crankshaft with the matchmark (gold link) on the lower timing chain.

61. Attach the lower timing chain to the water pump sprocket.

62. Working counterclockwise, install the lower timing chain camshaft sprockets. Be sure to align the sprocket marks with the blue links of the timing chain during installation.

63. Install the intake sprocket bolts and tighten to 88–95 ft. lbs. (119–128 Nm). Be sure to secure the camshafts while tightening the bolts.

64. Install the internal timing chain guide, upper timing chain guide, lower timing chain tensioner and slack side timing chain guide.

65. Tighten the tensioner mounting bolt to 75–96 inch lbs. (8.4–10.8 Nm) and tighten the guide bolts to 108–168 inch lbs. (13–19 Nm).

66. Apply a 0.102–0.142 in. (2.6–3.6mm) continuous bead of liquid gasket to all necessary areas as shown on the front timing cover.

67. Install the timing cover evenly and gently. Be sure to align the dowel pin holes.

68. Tighten the mounting bolts in sequence as follows:

    a. Bolts No. 1 and 2—Tighten to 19–23 ft. lbs. (26–31 Nm)

    b. Bolts No. 3 to 20—Tighten to 105–121 inch lbs. (11.8–13.7 Nm)

➡**Leave the bolts unattended for 30 minutes or more after tightening. This will allow the liquid gasket to cure sufficiently.**

69. Apply a 0.091–0.130 in. (2.3–3.3mm) continuous bead of liquid gasket to the water pump cover and install the cover. Tighten the bolts to 84–108 inch lbs. (10–13 Nm).

70. Apply a 0.12 in. (3mm) continuous bead of liquid gasket to the rocker covers and install the covers. Tighten the mounting bolts in sequence as follows:

    a. Bolts No. 1 to 10—Tighten to 9–26 inch lbs. (1–3 Nm)

    b. Bolts No. 1 to 10—Tighten to 52–69 inch lbs. (6–8 Nm)

71. Apply sealant to the front and rear seal of the oil pan.

72. Apply a 0.177–0.217 in. (4.5–5.5mm) continuous bead of liquid gasket to the upper oil pan mating surface and install the oil pan. Tighten the mounting bolts in sequence to 12–14 ft. lbs. (16–19 Nm).

73. Install the transaxle bolts that secure the oil pan.

74. Install the oil pan strainer and tighten the mounting bolts to 12–14 ft. lbs. (16–19 Nm).

75. Apply a 0.177–0.217 in. (4.5–5.5mm) continuous bead of liquid gasket to the lower oil pan mating surface and install the oil pan. Tighten the mounting bolts in sequence to 57–66 inch lbs. (6.4–7.5 Nm).

76. Tighten the oil pan drain plug to 22–29 ft. lbs. (29–39 Nm).

77. Install the center crossmember assembly.

78. Install the right side engine mounting bracket and mount assembly.

79. Remove the engine slinger assembly.

80. Install the front exhaust pipe and its support.

81. Install the A/C compressor and bracket.

82. Install the crankshaft pulley to the crankshaft and install the mounting bolt.

83. Tighten the mounting bolt to 14–22 ft. lbs. (20–29 Nm). Tighten the crankshaft bolt an additional 60–66 degrees clockwise. This is about the angle from one hexagon bolt head corner to another.

84. Install the ring gear cover plate.

85. Install the camshaft position sensor (PHASE) and crankshaft position sensors (REF)/(POS).

86. Install the power steering pump assembly.

87. Install the drive belts and the idler pulley.

88. Install the right front inner wheel cover and install the right front wheel.

89. Install the engine undercovers.

90. Using new gaskets, install the intake manifold. Tighten the nuts and bolts in sequence and in two stages as follows:

    a. Bolts and nuts—Tighten to 44–86 inch lbs. (5–10 Nm)

    b. Bolts and nuts—Tighten to 16–18 ft. lbs. (22–25 Nm)

91. Using new insulators, install the fuel tube assembly and tighten the bolts to 15–20 ft. lbs. (21–26 Nm). Tighten the bolts in several progressive steps.

92. Install the intake manifold collector gasket with the arrow facing forward.

93. Install the intake manifold collector assembly and tighten the mounting bolts to 16–18 ft. lbs. (22–25 Nm).

94. Install the intake manifold collector support brackets.

95. Using new gaskets, install the EGR tube and tighten the mounting bolts to 15–20 ft. lbs. (21–26 Nm) in two progressive steps.

96. Install the spark plugs.

97. Install the ignition coils and tighten the mounting bolts to 27–33 inch lbs. (2.9–3.8 Nm).

98. Install the rocker cover ornament on the left side.

99. Install the water hoses to the cylinder head and intake manifold.

100. Connect the canister purge hoses.

101. Connect the fuel hoses and wiring harness connections to the fuel rail.

102. Install and connect the air duct to intake manifold hose, collector hose, blow-by hose, and vacuum hoses.

103. Refill the engine oil and coolant with the proper type and amount of fluid.

104. Connect the negative battery cable.

105. Start the engine and run at 3000 RPM under no load to purge the air from the high pressure chamber. The engine may produce a rattling noise. This indicates that air still remains in the chamber and is not a matter of concern.

4.1L Engine

1. Remove the engine from the vehicle.

2. Remove the cylinder heads and the upper timing chains.

3. Remove the idler sprocket.

4. Remove the oil pan.

5. Remove the front cover bolts in the correct sequence, then remove the cover.

6. Compress the lower chain tensioners and install a pin through the hole to secure it, then remove the tensioner.

7. Remove the oil pump drive chain.

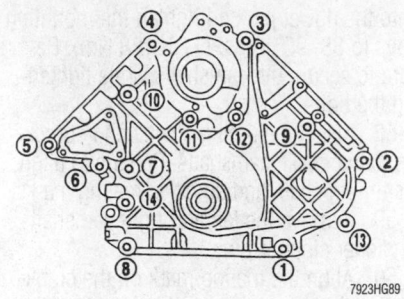

Remove the front cover bolts in the correct sequence—4.1L engine

8. Remove the slack and chain guides. Be sure to note the locations of the bolts so they can be installed in their original positions.

9. Remove the lower timing chains with the crankshaft sprockets.

To install:

10. Be sure the crankshaft key is pointing toward the center of the left bank. This should be a 45° angle from the center.

11. Install the lower right bank timing chain by aligning the mark on the chain with the mark on the sprocket and installing the sprocket with chain on the crankshaft. Be sure the thick side of the sprocket faces the cylinder block to provide clearance between the block and chain.

12. Install the slack and chain guides. Be sure to install the bolts in the correct locations. Tighten the bolts to 9–14 ft. lbs. (13–19 Nm).

13. Install the lower chain for the left bank in the same manner as the right one. Install the left slack and chain guide. Tighten the bolts to 9–14 ft. lbs. (13–19 Nm).

14. Install the oil pump drive chain and sprockets. Tighten the bolt on the driven gear to 22–30 ft. lbs. (30–40 Nm).

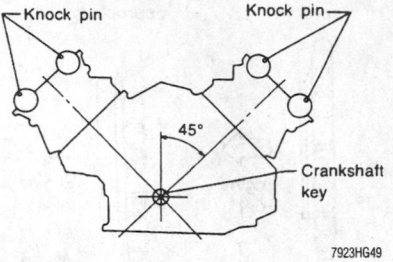

Before assembly, be sure to turn the crankshaft key towards the left cylinder head—4.1L engine

15. Install the lower oil pump drive chain guide if removed. Install the upper guide by installing the bolts loosely, then inserting a 0.04 in. (1mm) feeler gauge between the guide and chain. Press on the guide lightly with the same force as the weight of the guide and tighten the bolts to 56–74 in lbs. (6.3–8.3 Nm).

16. Install the chain tensioner with the pins installed using new gaskets. Tighten the bolts to 82–95.5 in lbs. (9–11 Nm).

17. Confirm that the timing marks on the crankshaft sprockets and chains are still aligned. Install the front cover. Be sure to install the bolts in the correct locations. Refer to the illustration as required.

18. Apply engine oil to the idler shaft and install it on the idler sprocket.

19. Align the mark on the chain with the mark on the idler sprocket and install the sprocket.

20. Place the upper chains on the idler sprockets. It is not necessary to align the mating marks at this time. The marks can be aligned after the cylinder head is installed.

21. Install the cylinder heads.

Tighten with oil filter bracket support

Position	Bolt dimensions	Tightening torque
①, ③, ⑤, ⑦	M6 x 45	
②	M6 x 47	
⑥, ⑩	M6 x 65	6.3 - 8.3 N·m (0.64 - 0.85 kg-m, 55.6 - 73.8 in-lb)
⑬	M6 x 67	
⑫	M6 x 84	
④, ⑧	M8 x 50	16 - 21 N·m (1.6 - 2.1 kg-m, 12 - 15 ft-lb)
⑨	M10 x 52	
⑭	M10 x 60	30 - 40 N·m (3.1 - 4.1 kg-m, 22 - 30 ft-lb)
⑪	M10 x 62	

Front cover bolt location and torque specifications—4.1L engine

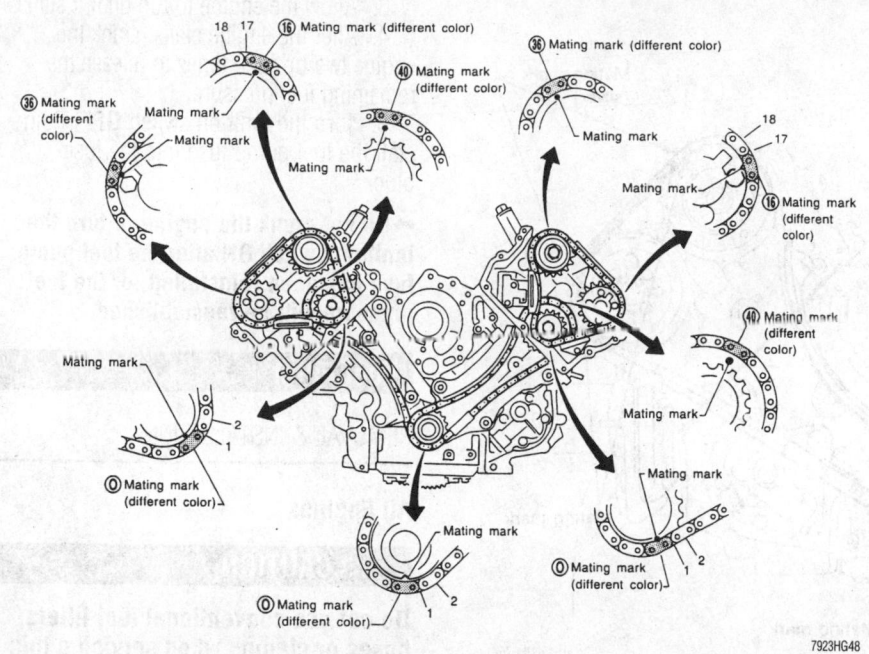

Timing chain and sprocket alignment marks—4.1L engine

22. Align the marks on the upper chains with the marks on the sprockets, then install the sprockets on the camshafts while keeping the marks aligned.

23. Install the idler shaft bolts. Tighten the bolts to 32–43 ft. lbs. (43–58 Nm).

24. Remove the lower chain tensioner pins.

25. Install the chain guide between the No. 1 camshaft bracket.

26. Align the upper timing chain mating marks with the marks on the sprockets and install the sprockets. Tighten the intake sprocket bolt to 76–83 ft. lbs. (103–113 Nm) and the exhaust sprocket bolt to 12–15 ft. lbs. (16–21 Nm).

27. Compress the upper chain tensioners, then install a pin through it to secure it in position.

28. Install the tensioners. Tighten the bolts to 82–95 ft. lbs. (9–11 Nm).

29. Lubricate the timing chains and related parts with clean engine oil and install the upper covers.

30. Install all remaining parts in the reverse of removal.

4.5L Engine

> ✳✳ **CAUTION**
>
> **The fuel injection system remains under pressure, after the engine has been turned OFF. Properly relieve fuel pressure before disconnecting any fuel lines. Failure to do so may result in fire or personal injury.**

1. Disconnect the negative battery cable.

2. Remove the engine and transmission assembly from the vehicle.

3. Remove the suspension member and engine mounts from the engine.

4. Remove the air compressor bracket.

5. Remove the cooling fan with coupling and the engine gusset.

6. Separate the engine from the transmission and mount the engine on a suitable workstand.

7. Remove the oil pan.

8. Remove the ornamental rocker cover and remove the ignition coils and spark plugs.

9. Bring the No. 1 piston to TDC on the compression stroke.

10. Use a suitable puller to remove the crankshaft pulley.

11. Remove the rocker covers.

12. Remove the crank angle sensor or camshaft position sensor on 1996 and later models.

13. Remove the left bank chain tensioner from the upper cover and remove both upper front covers.

14. Remove the lower front timing chain cover. Note how the dowels on the camshafts are aligned away from the crankshaft.

15. Loosen and remove the sprocket bolts.

> ✳✳ **WARNING**
>
> **Do not turn the crankshaft or the camshafts with the timing chains removed. The valves will contact the**

pistons and severe engine damage can occur.

16. Remove the right bank chain tensioner, then carefully pry the sprocket off.

17. Remove the oil pump chain and the timing chain. Keep the chains separated or tag them.

18. If necessary, carefully pry off the crankshaft sprockets.

To install:

19. Be sure all mating surfaces are clean before installation.

20. Install the any removed sprocket, making sure it is properly aligned.

21. Be sure the camshafts are still correctly positioned and the piston in the No. 1 cylinder is still at TDC. If removed, install the crankshaft chain sprockets with the collar towards the engine block. The mating mark for the right bank (inner) sprocket will be at the bottom when the crankshaft is at TDC.

22. Install the timing chain on the right bank, aligning the mating marks on the chain with those on the crankshaft and camshaft sprockets.

23. Install the timing chain on the left bank, aligning the mating marks on the chain with those on the crankshaft and camshaft sprocket.

24. Install the oil pump chain and sprocket.

25. Install the oil pump chain guides. Place a 0.040 in. (1.0mm) feeler gauge between the upper chain guide and chain before assembling the chain guides. The force applied to the chain is equivalent to the upper chain guide weight.

26. Apply a liquid gasket sealer and install the front covers. Be sure to seal around the bolt holes on the inside of the cover. Torque the bolts in sequence to 84 inch lbs. (9.5 Nm).

27. Install the chain tensioner for the left bank.

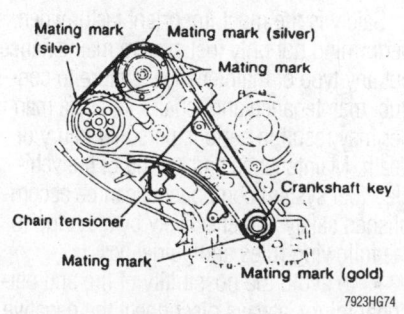

Timing chain mating mark alignment for right bank—4.5L engine

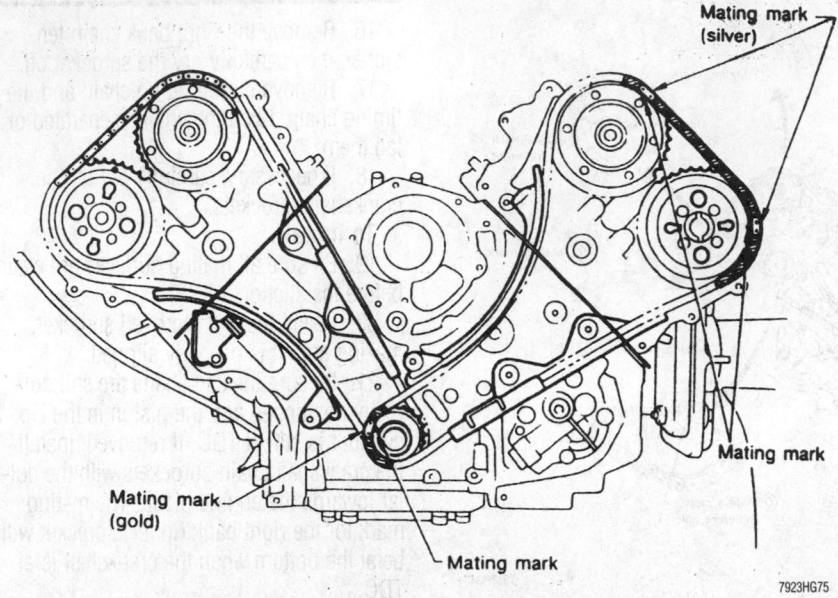

Timing chain mating mark alignment for left bank—4.5L engine

28. Apply suitable sealer to the rubber plugs and install them on the cylinder head.

29. Install the crank angle sensor or camshaft position sensor, VTC solenoid, rocker cover and crankshaft pulley. Torque the pulley bolt to 260–275 ft. lbs. (353–373 Nm).

30. Install the oil pan. Torque the bolts in the proper sequence to 4.6–6.1 ft. lbs. (6.3–8.3 Nm).

31. Install the transmission to the engine.

32. Install the engine mounts and the suspension member.

33. Install the engine and transmission assembly into the vehicle.

FUEL SYSTEM

Fuel System Service Precautions

Safety is the most important factor when performing not only fuel system maintenance but any type of maintenance. Failure to conduct maintenance and repairs in a safe manner may result in serious personal injury or death. Maintenance and testing of the vehicle's fuel system components can be accomplished safely and effectively by adhering to the following rules and guidelines.

• To avoid the possibility of fire and personal injury, always disconnect the negative battery cable unless the repair or test procedure requires that battery voltage be applied.

• Always relieve the fuel system pressure prior to disconnecting any fuel system component (injector, fuel rail, pressure regulator, etc.), fitting or fuel line connection. Exercise extreme caution whenever relieving fuel system pressure to avoid exposing skin, face and eyes to fuel spray. Please be advised that fuel under pressure may penetrate the skin or any part of the body that it contacts.

• Always place a shop towel or cloth around the fitting or connection prior to loosening to absorb any excess fuel due to spillage. Ensure that all fuel spillage (should it occur) is quickly removed from engine surfaces. Ensure that all fuel soaked cloths or towels are deposited into a suitable waste container.

• Always keep a dry chemical (Class B) fire extinguisher near the work area.

• Do not allow fuel spray or fuel vapors to come into contact with a spark or open flame.

• Always use a back-up wrench when loosening and tightening fuel line connection fittings. This will prevent unnecessary stress and torsion to fuel line piping. Always follow the proper torque specifications.

• Always replace worn fuel fitting O-rings with new. Do not substitute fuel hose or equivalent, where fuel pipe is installed.

Fuel System Pressure

RELIEVING

1. Remove the fuel pump fuse.
2. Start the engine.

3. Allow the engine to run until it stalls.

4. After the engine stalls, crank the engine two or three times to release the remaining fuel pressure.

5. Turn the ignition switch **OFF**. Reinstall the fuel pump fuse into the fuse block.

➡**Do not crank the engine or turn the ignition switch ON after the fuel pump fuse has been reinstalled, or the fuel pressure will be reestablished.**

Fuel Filter

REMOVAL & INSTALLATION

All Engines

❋❋ CAUTION

Do not use conventional fuel filters, hoses or clamps when servicing this fuel system. They are not compatible with the high pressures of the injection system and could fail, causing personal injury. Use only components specifically designed for fuel injection.

1. Relieve the fuel system pressure.
2. Disconnect the negative battery cable.
3. Disconnect the fuel hoses from the fuel filter, located at the right side of the engine compartment.
4. Remove the filter mounting screws and remove from the vehicle.

To install:
5. Inspect all hoses and clamps for damage of any type. Replace parts, as required.
6. The fuel filters are directional and should be installed with the arrow facing the direction of fuel flow.
7. Install a new filter in the bracket and install new hose clamps
8. Connect the negative battery cable.
9. Start vehicle and check for fuel leaks.

➡**On some vehicles, a code will be set and/or the check engine light will remain on after starting the vehicle. This is because a code was set for an open fuel pump circuit when the fuel pressure was released. If you did not disconnect the negative battery cable during this procedure, do it now so the code will be erased. The negative battery cable should be disconnected for at least 1 minute. Also, remember to reset the clock and radio stations when finished.**

Fuel Pump

REMOVAL & INSTALLATION

2.0L Engine

✳ CAUTION

The fuel injection system remains under pressure after the engine has been turned OFF. Properly relieve fuel pressure before disconnecting any fuel lines. Failure to do so may result in fire or personal injury.

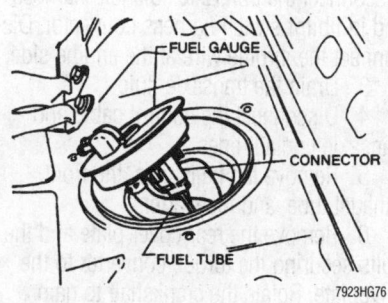

The fuel pump is located inside the tank—2.0L engine

1. Release the fuel system pressure.
2. Remove the inspection hole cover located beneath the rear seat.
3. Disconnect the connectors and fuel tubes.
4. Remove the fuel gauge locking ring using tool SST-X38879 or equivalent.
5. Lift out the fuel pump/gauge assembly and disconnect the tubes and connector.
6. Remove the fuel pump by sliding it out on an angle.
   To install:
7. Use a new O-ring on the locking ring.

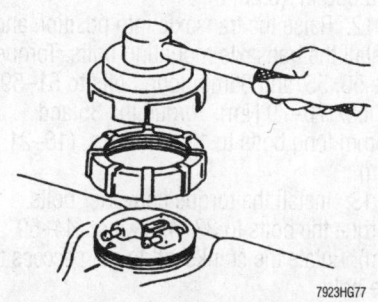

Locking ring removal and installation using special tool SST-X38879—2.0L engine

8. Install the fuel pump/gauge assembly and attach all fuel lines and connectors.
9. Using tool SST-X38879 or equivalent, tighten the locking ring to 22–26 ft. lbs. (30–35 Nm).
10. Install the inspection cover and test fuel system pressure at the injectors.

3.0L (VG30DE), 4.1L and 4.5L Engines

1. Relieve the fuel system pressure.
2. Disconnect the negative battery cable.
3. Remove the trunk front finish panel.
4. Disconnect the wiring harness connector and fuel tubes. Remove the fuel tank sender unit attaching bolts. Remove the fuel tank sender and discard the O-ring.
5. Remove the fuel pump from the sender unit.
   To install:
6. Install the new fuel pump on the sender unit assembly.
7. Using a new O-ring, install the sender unit in the fuel tank. Tighten the bolts to 1.4–1.9 ft. lbs. (2–2.5 Nm).
8. Connect the wiring harness connectors and fuel tubes.
9. Install the trunk room finish panel.
10. Connect the negative battery cable, start the engine and check for leaks.

3.0L (VQ30DE) Engine

1. Relieve the fuel system pressure

✳ CAUTION

The fuel injection system remains under pressure after the engine has been turned OFF. Properly relieve fuel pressure before disconnecting any fuel lines. Failure to do so may result in fire or personal injury.

2. Disconnect the negative battery cable.
3. Remove the access panel under the rear seat.

➡**If the vehicle has no fuel pump access cover, the fuel tank must be lowered or removed to gain access to the in-tank fuel pump.**

4. Disconnect the fuel gauge electrical connector and pump electrical connector.
5. Disconnect the fuel outlet and the return hoses. If necessary, remove the fuel tank.
6. Remove the fuel pump assembly-to-fuel tank bolts and lift the fuel pump assembly from the fuel tank. Discard the O-ring. Plug the fuel tank opening with a clean rag to prevent dirt from entering the system.

➡**When removing or installing the fuel pump assembly, be careful not to damage or deform it and always install a new O-ring.**

To install:
7. Remove the rag; using a new O-ring, install fuel pump assembly into the fuel tank. Torque the bolts to 1.4–1.9 ft. lbs. (2.0–2.5 Nm). If removed, install the fuel tank assembly
8. Connect the fuel lines and the electrical connectors. Always use new clamps when reconnecting fuel line hoses.
9. Install the fuel pump access cover.
10. Connect the negative battery cable.
11. Start the engine and check for fuel leaks.

➡**On some models, the Check Engine Light will stay ON after installation is completed. The memory code in the control unit must be erased. This code is stored for an open fuel pump circuit, this is caused when the fuel pressure is released. To erase the code, disconnect the battery cable for 10 seconds, then reconnect after installation of fuel pump.**

DRIVE TRAIN

Refer to the Unit Repair Section (URS) for information on servicing the driveshaft and halfshafts.

Transaxle Assembly

REMOVAL & INSTALLATION

Manual

G20

1. Disconnect the negative battery cable and remove the air duct.
2. Raise and safely support the vehicle.
3. Disconnect the clutch control cable and speedometer cable from the transaxle.
4. Disconnect the back-up light switch, neutral switch and ground harness connectors.
5. Remove the starter, shift control rod and support rod from the transaxle.
6. Drain the gear oil from the transaxle and remove the exhaust front tube.
7. Remove the halfshafts.
8. Support the engine with a suitable jack under the oil pan.
9. Remove the rear and left engine mount.

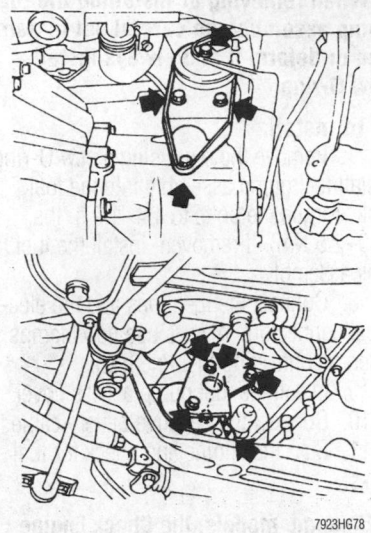

Transaxle mounting locations—G20 models

7923HG78

10. Raise the jack and remove the lower transaxle housing bolts. Lower jack and remove the upper housing bolts. Keep the bolts in order as they are different lengths and must be returned to the same position.

11. Lower the transaxle.

To install:

12. Raise the transaxle into place and install the attaching bolts. Torque the two shortest bolts to 22–30 ft. lbs. (30–40 Nm) and the remaining bolts to 51–59 ft. lbs. (70–79 Nm).

13. Install the rear and left engine mounts.

14. Install the driveshafts.

15. Install the shift control rods, support rod and starter on the transaxle.

16. Connect the back-up light switch, neutral switch and ground harness connectors.

17. Connect the clutch control cable and speedometer cable from the transaxle.

18. Refill the transaxle with oil and lower the vehicle to the floor.

19. Install the air duct and connect the negative battery cable. Road test the vehicle.

I30

1. Disconnect the negative and the positive battery cables.

2. Remove the battery, battery bracket and tray.

3. Remove the air cleaner assembly with the mass air flow sensor.

4. Raise and safely support the vehicle so there is clearance to remove the transaxle from underneath. Securely support the engine via the oil pan using a cushioning wooden block and a floor jack.

5. Remove the clutch operating cylinder; do not disconnect the hydraulic line from the cylinder.

6. Remove the clutch hose clamp.

7. Disconnect the speedometer pinion and the neutral position switch connectors and the ground harness connectors.

8. Remove the starter motor assembly from the transaxle.

9. Disconnect the back-up lamp switch and the neutral position switch.

10. Remove the crankshaft position sensor (POS) from the transaxle front side.

11. Remove the shifter control rod and the support rod bracket from the transaxle.

12. Drain the fluid from the transaxle.

13. Remove both driveshafts from the transaxle assembly. Securely support the transaxle with another jack.

14. Support the engine of the transaxle by placing a jack under the oil pan. Be sure to use a block of wood between the oil pan and jack.

15. Remove the bolts that secure the center crossmember.

16. Remove the LH engine mounts.

➡ **The transaxle bolts are of different lengths, be sure to note the location of the bolts for reassembly.**

17. Remove the transaxle bolts. Remove the transaxle from the vehicle by sliding the transaxle input shaft out of the clutch, lowering the rear of the transaxle, then lowering the transaxle from of the vehicle.

To install:

18. Install the transaxle assembly to the bell housing while aligning the output shaft of the transaxle with the clutch disc. Torque the transaxle bolts to specifications.

19. Install the LH engine mount and tighten the through-bolt to 32–41 ft. lbs. (43–55 Nm)

20. Install the center crossmember assembly and tighten the mounting bolts to 57–72 ft. lbs. (77–98 Nm).

21. Install both driveshafts to the transaxle assembly.

22. Install the shifter control rod and the support rod bracket to the transaxle.

23. Install the crankshaft position sensor (POS) to the transaxle front side.

24. Connect the back-up lamp switch and the neutral position switch.

25. Install the starter motor assembly to the transaxle.

26. Connect the speedometer pinion and the ground harness connectors.

27. Install the clutch hose clamp.

28. Install the clutch operating cylinder and tighten the bolts to 22–30 ft. lbs. (30–40 Nm).

29. Install the air cleaner assembly with the mass air flow sensor.

30. Install the battery tray, bracket and battery.

31. Connect the positive and the negative battery cables.

32. Refill the transaxle with proper amount and type of fluid.

33. Road test the vehicle for proper shift operation.

Automatic

G20

1. Disconnect the negative battery cable and the air duct.

2. Raise and support the vehicle safely. Disconnect the transaxle solenoid harness and inhibitor switch harness connector. Disconnect the throttle wire at the engine side.

3. Drain the transaxle fluid.

4. Disconnect the control cable and transaxle coolant lines.

5. Remove the halfshafts, the front exhaust tube, and the starter.

6. Remove the rear cover plate and the bolts securing the torque converter to the driveplate. Rotate the crankshaft to gain access to the bolts.

7. Support the engine with a suitable stand and use a suitable jack to support the transaxle.

➡ **Bolts are of different lengths, note the locations that the bolts are removed from.**

8. Remove the transaxle mounts.

9. Remove the transaxle mounting bolts and lower the transaxle.

To install:

10. Place a straightedge across the bell housing of the transaxle and measure the distance to the mounting bosses on the torque converter. The distance should be 0.626 in. (15.9mm). If not, the torque converter is not installed correctly.

11. Check the driveplate run-out with a dial indicator. Maximum allowable run-out is 0.008 in. (0.2mm).

12. Raise the transaxle into position and install the transaxle mounting bolts. Torque the 50, 55, and 65mm long bolts to 51–59 ft. lbs. (70–79 Nm). Torque the 35 and 45mm long bolts to 12–15 ft. lbs. (16–21 Nm).

13. Install the torque converter bolts. Torque the bolts to 33–43 ft. lbs. (44–59 Nm). Rotate the crankshaft to gain access to the bolts.

14. Install the transaxle mounts.

15. Install the halfshafts, the front exhaust tube, and the starter.

16. Connect the control cable and transaxle coolant lines.

17. Connect the transaxle solenoid harness and inhibitor switch harness connector. Connect the throttle wire at the engine side.

18. Fill the transaxle with fluid, connect the negative battery cable and road test the vehicle.

I30

➡ **The radio may contain a coded theft protection circuit. Always obtain the code number from the customer before disconnecting the battery.**

1. Disconnect the negative battery cable.
2. Raise the vehicle and support safely.
3. Remove the exhaust tube.
4. Drain the fluid from the transmission pan.
5. Remove the dipstick tube.
6. Remove the oil cooler lines.
7. Plug dipstick tube hole and oil cooler fittings after removing lines.
8. Remove the control linkage from the selector lever.
9. Disconnect the neutral safety switch and solenoid harness connectors.
10. Disconnect the speedometer cable.
11. Matchmark and remove the driveshaft. Insert plug into rear seal opening to prevent loss of fluid.
12. Support the transmission safely.
13. Remove the bolts securing the torque converter to the flexplate.
14. Remove the starter.
15. Remove the crankshaft position sensor.
16. Remove the gussets securing the transmission to the engine. Remove the bolts attaching the transmission to the engine.

➡ **The bolts securing the transmission to the engine are of different lengths. Note the length of the bolts as they are removed.**

17. Support the engine safely. Avoid jacking directly under the oil pan drain plug.
18. Remove the transmission from the vehicle.

To install:

19. Install the torque converter in the transmission. Be sure the torque converter is fully seated in the front pump assembly. The distance from the front edge of the transmission to the bolt hole of the torque converter should be 1.02 in. (26.0mm) or more.
20. Position the transmission to the engine and install a few bolts to hold the transmission in place. Do not fully tighten the bolts at this time.

21. Install the torque converter-to-flexplate bolts. Tighten to 33–43 ft. lbs. (44–59 Nm).
22. Secure the transmission to the engine. Torque the bolts as follows:
   • 58mm bolts to 29–36 ft. lbs. (39–49 Nm)
   • 47.5mm bolts to 29–36 ft. lbs. (39–49 Nm)
   • 25mm bolts to 22–29 ft. lbs. (29–39 Nm)
   • 20mm gusset bolts to 22–29 ft. lbs. (29–39 Nm)
23. Install the starter.
24. Install the crankshaft position sensor.
25. Align the matchmark and install the driveshaft.
26. Connect the speedometer cable.
27. Connect the neutral safety switch and solenoid harness connectors.
28. Install the control linkage to the selector lever.
29. Install the dipstick tube and oil cooler lines.
30. Connect the exhaust tube.
31. Lower the vehicle.
32. Fill the transmission with new fluid. Use the same amount of fluid that was drained before removal.
33. Connect negative battery cable and start the engine. Allow the engine to reach normal operating temperature and check the transmission fluid level. Add fluid as needed.

Transmission Assembly

REMOVAL & INSTALLATION

J30

1. Disconnect the negative battery cable.
2. Raise the vehicle and support safely.
3. Remove the exhaust tube.
4. Drain the fluid from the transmission pan.
5. Remove the dipstick tube.
6. Remove the oil cooler lines.
7. Plug dipstick tube hole and oil cooler fittings after removing lines.
8. Remove the control linkage from the selector lever.
9. Disconnect the neutral safety switch and solenoid harness connectors.
10. Disconnect the speedometer cable.
11. Matchmark and remove the driveshaft. Insert plug into rear seal opening to prevent loss of fluid.
12. Support the transmission safely.
13. Remove the bolts securing the torque converter to the flexplate.
14. Remove the starter.

15. Remove the gussets securing the transmission to the engine. Remove the bolts attaching the transmission to the engine.

➡ **The bolts securing the transmission to the engine are of different lengths. Note the length of the bolts as they are removed.**

16. Support the engine safely. Avoid jacking directly under the oil pan drain plug.
17. Remove the transmission from the vehicle.

To install:

18. Install the torque converter in the transmission. Be sure the torque converter is fully seated in the front pump assembly. The distance from the front edge of the transmission to the bolt hole of the torque converter should be 1.02 in. (26.0mm) or more.
19. Position the transmission to the engine and install a few bolts to hold the transmission in place. Do not fully tighten the bolts at this time.
20. Install the torque converter-to-flexplate bolts. Tighten to 33–43 ft. lbs. (44–59 Nm).
21. Secure the transmission to the engine. Torque the bolts as follows:
   • 58mm bolts to 29–36 ft. lbs. (39–49 Nm)
   • 47.5mm bolts to 29–36 ft. lbs. (39–49 Nm)
   • 25mm bolts to 22–29 ft. lbs. (29–39 Nm)
   • 20mm gusset bolts to 22–29 ft. lbs. (29–39 Nm)
22. Install the starter.
23. Install the crankshaft position sensor.
24. Align the matchmark and install the driveshaft.
25. Connect the speedometer cable.
26. Connect the neutral safety switch and solenoid harness connectors.
27. Install the control linkage to the selector lever.
28. Install the dipstick tube and oil cooler lines.
29. Connect the exhaust tube.
30. Lower the vehicle.
31. Fill the transmission with new fluid. Use the same amount of fluid that was drained before removal.
32. Connect negative battery cable and start the engine. Allow the engine to reach normal operating temperature and check the transmission fluid level. Add fluid as needed.

Q45

1. Disconnect the negative battery cable.
2. Raise the vehicle and support safely.

3. Remove the exhaust tubes.

4. Remove the fluid charging pipe.

5. Remove the oil cooler lines.

6. Plug fluid charging and oil cooler fittings after removing lines.

7. Remove the control linkage from the selector lever.

8. Disconnect the neutral safety switch and solenoid harness connectors.

9. Disconnect the speed sensor connection.

10. Matchmark and remove the driveshaft. Insert plug into rear seal opening to prevent loss of fluid.

11. Support the transmission safely.

12. Remove the bolts securing the torque converter to the flexplate.

13. Remove the gussets securing the transmission to the engine. Remove the bolts attaching the transmission to the engine.

➡ **The bolts securing the transmission to the engine are of different lengths. Note the length of the bolts as they are removed.**

14. Support the engine safely. Avoid jacking directly under the oil pan drain plug.

15. Remove the transmission from the vehicle.

To install:

16. Position the transmission in the vehicle and install the torque converter-to-flexplate bolts. Tighten to 33–43 ft. lbs. (44–59 Nm).

17. Secure the transmission to the engine. Torque the bolts as follows:

• 70mm bolts to 80–87 ft. lbs. (108–118 Nm)

• 30mm bolts to 51–58 ft. lbs. (69–78 Nm)

• 30mm (gusset) bolts to 51–58 ft. lbs. (69–78 Nm)

18. Install the torque converter-to-drive plate bolts and tighten in two steps to 33–43 ft. lbs. (44–59 Nm).

19. Align the matchmark and install the driveshaft.

20. Connect the speed sensor connection.

21. Connect the neutral safety switch and solenoid harness connectors.

22. Install the control linkage to the selector lever.

23. Install the fluid charging and oil cooler lines.

24. Connect the exhaust tubes.

25. Lower the vehicle.

26. Connect negative battery cable.

Clutch

ADJUSTMENT

G20

➡ **Clutch pedal height must be correct before the cable can be adjusted.**

1. Adjust pedal height by loosening the locknut on the pedal stopper or cruise control cancel switch. Pedal height should be 6.28–6.67 in. (159.5–169.5mm) from the top of the pedal to the floor, measured at a 90 degree angle to the top of the pedal.

2. To adjust the clutch cable:

a. Push the withdrawal lever until resistance is felt.

b. Tighten the adjusting the nut.

c. Turn the adjusting nut 2.5–3.5 turns back and then tighten the locknut. Free-play at the withdrawal lever should be 0.098–0.138 in. (2.5–3.5mm).

3. As a final check, measure pedal free travel at the center of the pedal pad. Pedal free travel should be 0.425–0.594 in. (10.8–15.1mm).

4. Make sure the starting switch on the pedal bracket still functions correctly. The starter should operate with the ignition switch when the pedal is depressed fully.

REMOVAL & INSTALLATION

G20

1. Disconnect the negative battery cable.

2. Raise and support the vehicle safely.

3. Remove the transaxle.

4. Insert tool KV30101000 or equivalent alignment tool into the clutch disc hub and loosen the pressure plate bolts in small increments using a star-type pattern.

5. Remove the pressure plate and clutch disc as an assembly.

6. Remove the release bearing by pulling the bearing retainers outward from the transaxle case.

7. Inspect the clutch disc for surface wear. Measure from the friction surface to the top of the rivets. Wear limit is 0.012 in. (0.3mm). Replace clutch disc as necessary.

8. Inspect the contact surface of the flywheel for burns or discoloration. Check flywheel run-out. Maximum run-out is 0.0059 in. (0.15mm).

9. Using tools ST20050100 and ST20050010 or equivalent, check pressure plate diaphragm springs. Measure from the pressure plate/flywheel mating surface to the top of the diaphragm spring. Height should be 1.201–1.280 in. (30.5–32.5mm). Replace pressure plate as necessary.

10. Inspect the release bearing for damage. Spin the bearing to see that it rolls freely.

To install:

11. Lightly lubricate the transaxle input shaft, input shaft collar, clutch lever assembly and the clutch release bearing with a lithium based grease.

➡ **Keep clutch disc and all clutch components clean during installation. Do not allow grease to contact the clutch disc.**

12. Insert tool KV30101000 or equivalent alignment tool into the clutch disc hub. Install the clutch disc and pressure plate on the tool and tighten the pressure plate bolts to 16–22 ft. lbs. (22–29 Nm) in 2–3 steps using a crisscross pattern. Remove the tool.

13. Install release bearing in the transaxle. Ensure that the bearing retainer clips are fully engaged.

14. Install the transaxle.

15. Adjust clutch pedal height and free-play.

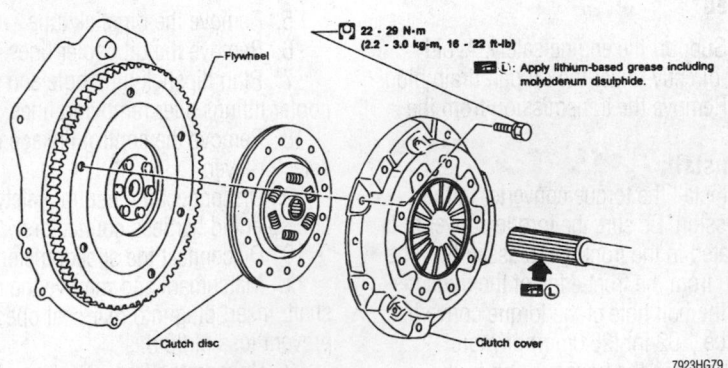

22 - 29 N·m (2.2 - 3.0 kg-m, 16 - 22 ft-lb)

Apply lithium-based grease including molybdenum disulphide.

Clutch disc and pressure plate—G20 models

7923HG79

16. Lower vehicle, connect negative battery cable and road test vehicle.

I30

1. Remove the battery and battery bracket.

2. Remove the air cleaner and the air flow meter.

3. Raise and safely support the vehicle so there is clearance to remove the transaxle from underneath. Securely support the engine via the oil pan using a cushioning wooden block and jack.

4. Drain the fluid from the transaxle.

5. Remove the transaxle from the engine and lower to the floor.

6. Insert a clutch aligning bar or similar tool all the way into the clutch disc hub. This must be done so as to support the weight of the clutch disc during removal. Mark the clutch assembly-to-flywheel relationship with paint or a center punch so the clutch assembly can be assembled in the same position from which it is removed.

7. Loosen the bolts in reverse order of tightening sequence, a turn at a time. Remove the bolts.

8. Remove the pressure plate and clutch disc.

9. Remove the release mechanism from the transaxle housing.

10. Inspect the pressure plate for wear, scoring, etc., and resurface or replace, as necessary.

11. Measure the thickness of the clutch plate lining to the rivet heads; if the it is worn to a minimum of 0.012 in. (0.3mm), replace the clutch plate.

12. Inspect the release bearing and replace as necessary.

13. Using a dial indicator, mount it to the engine and inspect the flywheel run-out; if the run-out exceeds 0.0059 in. (0.15mm), replace it.

To install:

14. Apply a small amount of grease to the transaxle input shaft splines. Install the disc on the splines and slide back and forth a few times. Remove the disc and remove excess grease on hub. Be sure no grease contacts the disc or pressure plate.

15. Apply lithium based molybdenum disulfide grease to the bearing sleeve inside groove, the contact point of the withdrawal lever and bearing sleeve, the contact surface of the lever ball pin and lever.

16. Install the release mechanism and release bearing.

17. Install the pressure plate and clutch disc, aligning it with a splined dummy shaft tool KV301010000 or equivalent.

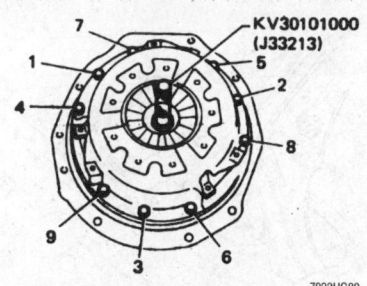

7923HG80

Tighten the pressure plate bolts according to the sequence shown—I30 models

18. Torque the pressure plate bolts in sequence to 25–33 ft. lbs. (34–44 Nm).

19. Remove the dummy shaft.

20. Install the transaxle in the correct position. Torque the transaxle-to-engine bolts.

21. Connect the rear and LH mounts.

22. Connect the speedometer cable.

23. Connect the electrical harness connector.

24. Install the clutch release cylinder.

25. Install the starter assembly.

26. Securely support the transaxle and install the driveshafts.

27. Refill the transaxle with the required amount of approved fluid.

28. Install the air flow meter and the air cleaner.

29. Install the battery and battery bracket.

30. Road test the vehicle for proper shift operation.

Hydraulic Clutch System

BLEEDING

The G20 model uses a cable actuated clutch system

I30

Bleeding is required to remove air trapped in the hydraulic system. The bleed screw is located on the clutch slave (release) cylinder.

1. Remove the bleed screw dust cap.

2. Attach a transparent vinyl tube to the bleed screw, immersing the free end in a clean container of clean brake fluid.

3. Fill the clutch master cylinder with the proper fluid.

4. Slowly depress the clutch pedal all the way several times and hold it down.

5. Have an assistant open the bleeder valve about ¾ turn to release the air. Then, close the bleeder valve while the pedal is still depressed.

6. Repeat the above procedure until no more air bubbles are seen in the fluid container.

7. Remove the bleed tube.

8. Replace the dust cap and refill the master cylinder.

9. Bleed the clutch damper, if equipped.

Halfshaft

REMOVAL & INSTALLATION

G20

1. Raise and support the vehicle safely. Remove the wheel bearing locknut.

2. Remove the brake caliper assembly and rotor. Using a piece of wire, position the caliper so it is not supported by the brake line.

3. Separate the tie-rod from the ball joint.

4. Separate the kingpin from the knuckle.

5. Remove the halfshaft from the wheel hub/knuckle by lightly tapping it with a wood drift. Take care not to damage the CV-boots.

6. Remove the halfshaft from the transaxle by prying outward with a suitable tool at the transaxle case.

7. On automatic transaxle models, remove the left halfshaft by tapping it out with a drift from the right side of the transaxle case. Take care not to damage the pinion mate shaft and side gear.

To install:

8. Drive a new oil seal into the transaxle. For the right side use tool KV38106800 or equivalent, along the inner circumference of the oil seal. For the left side use tool KV38106700

9. Insert the halfshaft into the transaxle. Ensure that the serration's are aligned. Remove the tool.

10. Push the halfshaft inward and install the circular clip in the groove of the side gear. After inserting the clip, pull outward on the flange of the slide joint to ensure the clip is properly meshed with the side gear. If it pulls out, the clip was not installed properly.

11. Install the halfshaft into the wheel hub/knuckle. Tighten the upper knuckle nut to 72–87 ft. lbs. (98–118 Nm) and wheel bearing locknut to 174–231 ft. lbs. (235–314 Nm).

12. Using a dial indicator, check wheel bearing axial end-play. Specification calls for 0.0020 in. (0.05mm) or less.

13. Install the rotor and brake caliper.

14. Install the wheel and lower the vehicle to the floor.

J30 and Q45

> ### ❋❋ CAUTION
>
> **The amount of force need to loosen the rear wheel bearing nut is high enough to cause the vehicle to fall off the jack. Loosen and tighten this nut with the vehicle on the ground.**

1. Remove the rear wheel cotter pin, adjusting cap and insulator. Loosen the wheel bearing nut with the brakes applied and the vehicle sitting on the ground.
2. Raise the vehicle and support safely.
3. Remove the rear wheel.
4. Remove the differential side flange bolts and nuts and separate shaft from the differential.
5. Remove the wheel bearing locknut and washer from halfshaft.
6. Remove the halfshaft by lightly tapping it with a copper hammer.
7. Remove the halfshaft assembly from the vehicle.

To install:

8. Insert halfshaft into wheel hub and install washer and wheel bearing locknut. Temporarily tighten the locknut.
9. Connect the halfshaft with the differential side flange. Install the nuts and bolts and tighten to 61–69 ft. lbs. (83–93 Nm).
10. Install the wheels and lower the vehicle to the ground.
11. Tighten the wheel bearing locknut with the brakes applied to 152–203 ft. lbs. (206–275 Nm). Install the insulator, adjusting cap and a new cotter pin.

I30

RIGHT HALFSHAFT

1. Raise and support the front of the vehicle safely and remove the wheels.
2. Remove the ABS wheel sensor and move it out of the way.
3. Remove the brake hose from the strut.
4. Remove wheel bearing locknut.
5. Matchmark and remove the bolts attaching the steering knuckle to the strut.

➡**Cover axle boots with waste cloth or equivalent so as not to damage them when removing halfshaft.**

6. Separate the halfshaft from the knuckle by slightly tapping it.
7. Pry the halfshaft from the transaxle with a flat bladed tool. Remove and discard the circlip on the end of the halfshaft.
8. Remove the seal from the transaxle.

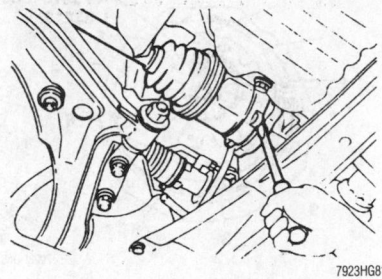

Separating the right halfshaft from the transaxle—I30 models

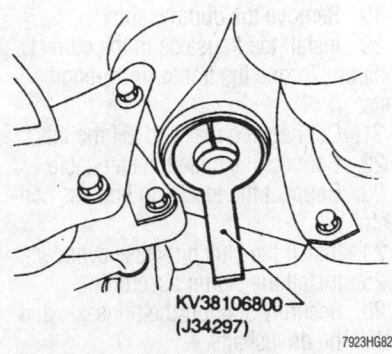

Right halfshaft alignment tool—I30 models

To install:

9. Install a new seal into the transaxle and install a halfshaft alignment tool KV38106800 or equivalent into the transaxle seal.
10. Install a new circlip to the halfshaft, then insert the halfshaft into the transaxle.
11. With the serration's aligned remove the alignment tool.
12. Push the halfshaft fully into the transaxle to seat the circlip. Try to pull the halfshaft from the transaxle by hand to verify that the circlip is properly seated.
13. Insert the halfshaft into the steering knuckle and install the hub locknut, do not tighten the hub nut at this time.
14. Connect the steering knuckle to the strut.
15. Install the strut mounting bolts and align the matchmarks. Torque the bolts to 103–117 ft. lbs. (140–159 Nm).
16. Install the brake hose to the strut.
17. Install the ABS wheel sensor and tighten the attaching bolt to 13–17 ft. lbs. (18–24 Nm).
18. Install the front wheels, lower the vehicle and tighten hub locknut to 174–231 ft. lbs. (235–314 Nm).
19. Check and/or adjust the wheel alignment as necessary.

LEFT HALFSHAFT

1. Raise and support the front of the vehicle safely and remove the wheels.
2. Remove the ABS wheel sensor and move it out of the way.
3. Remove the brake hose from the strut.
4. Remove wheel bearing locknut.
5. Matchmark and remove the bolts attaching the steering knuckle to the strut.

➡**Cover axle boots with waste cloth or equivalent so as not to damage them when removing halfshaft.**

6. Separate the halfshaft from the knuckle by slightly tapping it.
7. Loosen the bolts attaching the support bearing to the support bearing bracket.
8. If equipped with a manual transaxle, pry the halfshaft from the transaxle with a flat bladed tool.
9. If equipped with a automatic transaxle perform the following:
   a. Remove the right halfshaft from the vehicle.
   b. Insert a flat bladed tool into the transaxle where the right halfshaft was, place the end of the tool on the halfshaft, then drive the left shaft from the pinion side gear.
10. Remove the support bearing bolts and remove the halfshaft from the vehicle.
11. Remove and discard the circlip on the end of the halfshaft. Remove the seal from the transaxle.

To install:

12. Install a new seal into the transaxle and install a halfshaft alignment tool KV38106700 or equivalent into the transaxle seal.
13. Install a new circlip to the halfshaft, then insert the halfshaft into the transaxle.

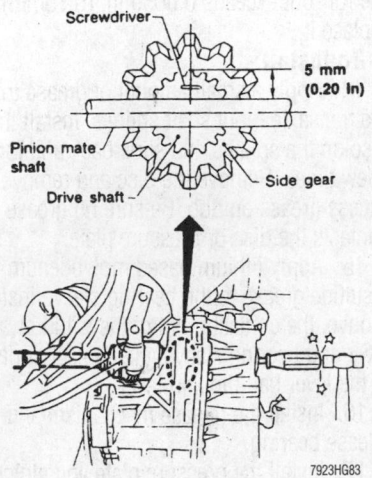

Separating the left halfshaft from an automatic transaxle—I30 models

14. With the serration's aligned remove the alignment tool.

15. Push the halfshaft fully into the transaxle to seat the circlip. Try to pull the halfshaft from the transaxle by hand to verify that the circlip is properly seated.

16. Install the support bearing bolts and tighten the bolts to 9–14 ft. lbs. (13–19 Nm).

17. Insert the halfshaft into the steering knuckle and install the hub locknut, do not tighten the hub nut at this time.

18. Connect the steering knuckle to the strut.

19. Install the strut mounting bolts and align the matchmarks. Torque the bolts to 103–117 ft. lbs. (140–159 Nm).

20. Install the brake hose to the strut.

21. Install the ABS wheel sensor and tighten the attaching bolt to 13–17 ft. lbs. (18–24 Nm).

22. Install the front wheels, lower the vehicle and tighten hub locknut to 174–231 ft. lbs. (235–314 Nm).

23. Check and/or adjust the wheel alignment as necessary.

STEERING AND SUSPENSION

Air Bag

✳✳ CAUTION

Some vehicles are equipped with an Air Bag system, also known as the Supplemental Inflatable Restraint (SIR) or Supplemental Restraint System (SRS). The system must be disabled before performing service on or around system components, steering column, instrument panel components, wiring and sensors. Failure to follow safety and disabling procedures could result in accidental Air Bag deployment, possible personal injury and unnecessary system repairs.

PRECAUTIONS

Several precautions must be observed when handling the inflator module to avoid accidental deployment and possible personal injury.

• Never carry the inflator module by the wires or connector on the underside of the module.

• When carrying a live inflator module, hold securely with both hands, and ensure that the bag and trim cover are pointed away.

• Place the inflator module on a bench or other surface with the bag and trim cover facing up.

• With the inflator module on the bench, never place anything on or close to the module which may be thrown in the event of an accidental deployment.

DISARMING

➥ All Air Bag electrical wiring harnesses and connectors are covered with YELLOW outer insulation. Do not use electrical test equipment on any circuit related to the Air Bag sensors. When installing Air Bag components, always install with the arrow marks facing the front of the vehicle.

1. Turn the ignition switch to the OFF position.

2. Disconnect both battery cables starting with the negative cable first and wait at least 10 minutes after the cables are disconnected. Be sure to insulate the battery terminal ends.

REARMING

1. Turn the ignition switch to the OFF position.

2. Connect both battery cables starting with the positive cable first.

➥ The Air Bag or Air Bag system is equipped with a self-diagnostic operation. After turning the ignition key to the ON or START position, the AIR BAG warning lamp will illuminate for 7 seconds. After 7 seconds, the AIR BAG lamp will extinguish if no malfunction is detected. If the AIR BAG lamp does not extinguish after 7 seconds, check the Air Bag self diagnostic system for a malfunction.

Strut

REMOVAL & INSTALLATION

Front

G20

1. Raise and support the vehicle safely. Remove the strut mounting bolt at the lower suspension member and the three nuts inside the engine compartment. Do not remove the piston rod locknut.

2. Remove the strut assembly and place in a suitable holding device.

3. Using a prybar to hold the upper spring mount, loosen but do not remove the piston rod locknut.

4. Compress the spring with a spring compressor so the strut mounting insulator can be turned by hand.

5. Remove the piston rod locknut. Remove the coil spring from strut assembly.

To install:

6. Inspect all components carefully for damage or wear. Replace as necessary.

7. Install the compressed coil spring on the strut assembly and tighten the locknut to 13–17 ft. lbs. (18–24 Nm).

8. Install the strut assembly in the vehicle. Ensure the bend in the lower shock bracket faces rearward on the left side and forward on the right side of the vehicle.

9. Install the upper spring seat with the cutout facing the inside of the vehicle.

10. Tighten the upper strut mounting bolts to 31–40 ft. lbs. (42–54 Nm) and the lower through-bolt to 82–93 ft. lbs. (112–126 Nm). Final tightening must take place with the suspension loaded (vehicle at normal ride height).

I30

1. Raise and safely support the vehicle.

2. Remove the wheel. Matchmark the position of the strut-to-steering knuckle location.

3. Disconnect the brake hose from the strut.

4. Remove the ABS wheel sensor and move it out of the way.

5. Matchmark and remove the bolts attaching the steering knuckle to the strut.

6. Open the hood and remove the strut attaching nuts while holding the strut.

✳✳ CAUTION

Do not remove the center locknut from the strut assembly until the strut is safely compressed.

7. Remove the strut from the vehicle.

8. Place the strut assembly in a vise with the special holding tool ST35652000 or in a spring compressor.

9. Loosen the piston rod locknut.

✳✳ CAUTION

Do not remove the piston rod locknut, the spring is under tension and can cause serious personal injury.

10. Compress the spring with the spring compressor, then remove the piston rod locknut.

➡**Before removing the strut from the coil spring, note the positioning of the strut in relationship to the coil spring for reassembly.**

11. Remove the strut mounting insulator bracket, strut mounting bearing, upper spring seat, and the upper spring rubber seat.

12. Remove the strut, leaving the coil spring compressed.

13. Remove the piston boot and rebound bumper from the strut.

To install:

14. Install the rebound bumper and the boot to the strut piston.

15. Install the strut into the coil spring, be sure the strut and spring are properly positioned.

16. Install the upper spring rubber seat, upper spring seat, strut mounting bearing, and the strut mounting insulator bracket. Be sure that the cutout on the upper spring seat is facing the outside of the vehicle.

17. Install the piston rod locknut, then remove the spring compressor.

18. Torque the piston rod locknut to 43–58 ft. lbs. (59–76 Nm).

19. Install the strut into the strut tower and install new attaching nuts. Torque the nuts to 29–40 ft. lbs. (39–54 Nm).

20. Install the bolts attaching the steering knuckle to the strut and align the matchmarks. Torque the bolts to 103–117 ft. lbs. (140–159 Nm).

21. Install the ABS wheel sensor and tighten the attaching bolt to 13–17 ft. lbs. (18–24 Nm).

22. Install the brake hose to the strut.

23. Install the front wheels and lower the vehicle.

24. Check and/or adjust the wheel alignment as necessary.

J30

1. Raise and support the vehicle safely.

2. Remove the front wheel.

3. Remove the brake caliper and hang it from the body with wire. Do not let the caliper hang by the brake hose.

4. Remove the nut and separate the stabilizer bar linkage from the strut.

5. Remove the lower ball joint nut and separate the ball joint from the strut.

6. Remove the nut and separate the tie rod end from the strut.

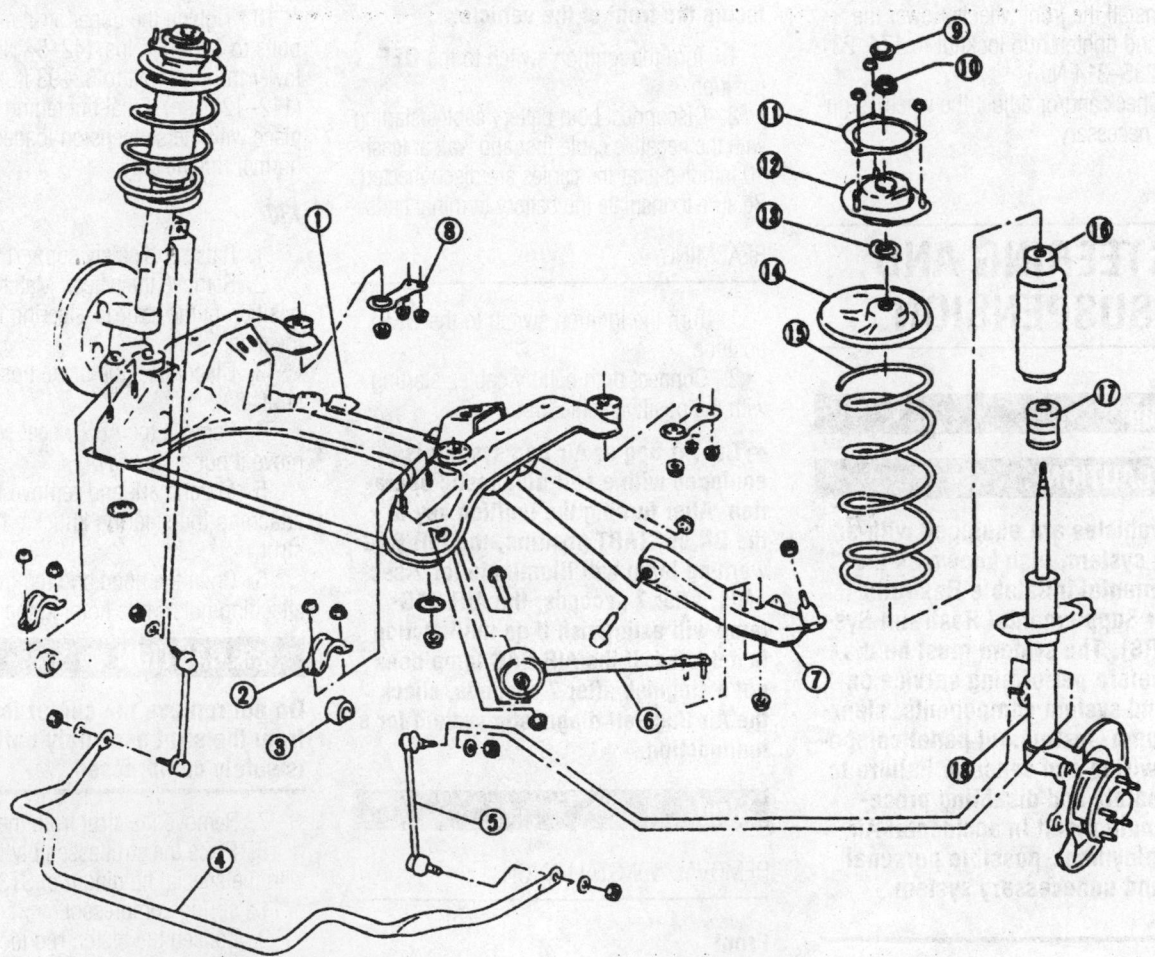

① Front suspension member	⑦ Transverse link	⑬ Upper plate
② Stabilizer bar clamp	⑧ Member stay	⑭ Spring upper seat
③ Bushing	⑨ Cap	⑮ Coil spring
④ Stabilizer bar	⑩ Lock nut	⑯ Dust cover
⑤ Stabilizer connecting rod	⑪ Gasket	⑰ Bound bumper
⑥ Tension rod	⑫ Strut mounting insulator	⑱ Strut assembly

Front strut and spring assembly—J30 models

7923HG84

✳✳ CAUTION

Do not remove the center nut from the strut piston rod unless the coil spring is compressed using the proper spring compressor.

7. Remove the nuts from the upper strut mount insulator and lower the strut assembly from the vehicle.

8. To disassemble the strut, secure the strut assembly in a vise using special tool ST-35652000, or equivalent, strut holding fixture.

9. Loosen, but do not remove the piston rod locknut.

10. Compress the spring with a spring compressor so the strut mounting insulator can be turned by hand.

11. Remove the piston rod locknut and coil spring.

12. To remove the internal shock absorber components, clean the top of the shock tube and remove the gland packing nut. Push the rod all the way down, then slowly pull it up again to withdraw the piston rod and guide.

To install:

13. Inspect the rubber parts for deterioration. If there is oil on the spring seat, the gland packing nut and O-ring should replaced. Some seepage on the rod or upper surface of the nut is normal.

14. Lubricate the sealing lip of the gland packing nut. Cover the piston rod with tape to prevent damage to the oil seal and install the nut.

15. Tighten the gland packing to 43–80 ft. lbs. (49–108 Nm) total torque. If a crows foot or other special tool is used to tighten the nut, its length must be calculated into the torque.

16. Set the spring into the lower spring seat. The flat portion of the spring goes on top.

17. Install the spring seat with its cutout facing the outer side of the vehicle. Install the upper plate and strut mount insulator and the piston rod nut. Torque the nut to 43–58 ft. lbs. (58–79 Nm).

18. Remove the spring compressor and fit the strut into the vehicle. Install the upper strut mount nuts and tighten to 30–35 ft. lbs. (41–48 Nm).

19. Fit the lower ball joint into place and tighten the nut to 71–88 ft. lbs. (96–120 Nm).

20. Connect the tie rod end and tighten the nut.

21. Connect the stabilizer bar linkage and tighten the nut to 54–67 ft. lbs. (74–90 Nm).

22. Install the brake caliper and wheel and check front wheel alignment.

Q45—STANDARD SUSPENSION

1. Remove the upper mounting insulator bolts.

2. Raise and safely support the vehicle.

3. Remove the lower strut mounting bolt and lift out the strut assembly.

4. Secure the strut in a suitable holding fixture.

5. Loosen the piston rod locknut. Do not remove the locknut.

6. Compress the spring with the proper tool so the strut assembly mounting insulator can be turned by hand.

7. Remove the piston rod locknut. Remove the spring assembly, dust cover and rubber seat. Remove the strut insert.

To install:

8. Inspect the rubber parts for deterioration. If the rubber is pulling away from the metal, the mounting insulator should be replaced.

9. Fit the spring into the lower seat, install the dust cover/bumper and upper seat and mounting insulator.

10. Install the piston rod locknut and tighten to 13–17 ft. lbs. (17–23 Nm).

11. Fit the strut into place and tighten the upper mounting nuts to 30–35 ft. lbs. (40–47 Nm).

12. Torque the lower mounting bolt to 80–94 ft. lbs. (108–128 Nm).

Q45—ACTIVE SUSPENSION

➡**The Nissan Consult or an equivalent scan tool that can issue commands to the control unit is required for bleeding the hydraulics in the Full Active Suspension system.**

1. Relieve the hydraulic pressure:

a. Raise all four wheels off the ground and wait at least 3 minutes for the system to stabilize.

b. Remove both front inner fenders and the rear pressure control unit cover.

c. Loosen the locknut and slowly open the bypass valve on each pressure control unit. Open the valves all the way and leave them open until the job is finished.

2. Remove the flange joint from the top of the actuator.

3. Install two 15mm bolts into the actuator in the flange joint mounting bolt holes.

4. Insert a bar between the bolts and loosen the joint adapter. Do not remove it yet.

5. Remove the upper mount insulator nuts.

6. Raise and safely support the vehicle.

7. Disconnect the hydraulic lines. Cap the lines to keep the system clean.

8. Remove the lower actuator mounting nut and remove the assembly.

9. Secure the actuator/spring assembly in a suitable holding fixture. Scribe alignment marks on the spring, upper mount insulator and actuator unit.

10. Compress the spring with the proper tool so the joint adapter can be turned by hand. Remove the joint adapter and lift off the mount insulator, spring, and any other components necessary.

To install:

11. If the actuator is being replaced, the rubber bumper should also be replaced. Fit the bumper, dust cover and rubber seat onto the actuator.

12. Fit the spring into the lower seat with the matchmarks aligned. Install the upper seat/mounting insulator with the marks aligned and start the joint adapter. The joint adapter will be tightened after installing the actuator assembly.

13. Fit the strut into place and tighten the upper mounting nuts to 30–41 ft. lbs. (40–55 Nm).

14. Torque the lower mounting bolt to 76–94 ft. lbs. (103–128 Nm).

15. Tighten the joint adapter to 63–72 ft. lbs. (85–98 Nm).

16. Install the flange adapter and tighten the bolts to 11–13 ft. lbs. (15–18 Nm).

17. Close the bypass valves on the pressure control units.

18. To bleed the system:

a. With all four wheels about 2 in. (50mm) off the ground, run the engine for about two minutes.

b. Connect the Consult scan tool and enter "WORK SUPPORT" mode. Select "4. AIR BLEEDING".

c. Check the fluid level in the reservoir. It should be slightly overfilled.

d. Touch "START" on the scan tool. The display will show a regular rise and fall in system pressure. When the pressure stabilizes, stop the engine.

e. Connect a clear tube to the air bleeder at the actuator and place the other end in a container.

➡**Do not allow the fluid to contact the body or the paint will be damaged.**

f. Open the bleeder and watch the fluid move through the tube. If there are still air bubbles in the fluid when the flow stops, check the fluid level, pressurize the system again and repeat the process.

Rear

G20

1. Raise and support the vehicle safely.

2. Remove the rear seat to gain access

to the top strut assembly bolts. Remove the three top mount bolts.

3. Remove the rear stabilizer connecting rod where it attaches the knuckle assembly.

4. Remove the strut through-bolts at the knuckle assembly and remove the shock absorber assembly.

5. Set the strut assembly in a vise using attachment ST-25652000 or equivalent.

✳✳ WARNING

Loosen the piston rod locknut but do not remove.

6. Compress the spring with a suitable tool so that the strut mounting insulator can be turned by hand.

7. Remove the piston rod locknut and spring with compressor attached.

To install:

8. Replace the bound rubber bumpers. Install the coil spring on the strut and tighten the piston rod locknut to 43–58 ft. lbs. (59–78 Nm). Gradually release the spring compressor. When the coil spring is located correctly, there should be two identification color codes on the lower side.

9. Tighten the strut assembly upper attaching bolts to 31–40 ft. lbs. (42–54 Nm); lower attaching bolts to 72–87 ft. lbs. (98–118 Nm) and the stabilizer bar connecting rod bolts to 30–35 ft. lbs. (41–47 Nm)

10. Check and adjust the wheel alignment if needed.

I30

1. Raise and safely support the vehicle.
2. Remove the rear wheels.
3. Support the rear torsion beam assembly with a jack.
4. Open the trunk and remove the two nuts attaching the strut to the vehicle.

✳✳ CAUTION

Do not remove the center locknut from the strut assembly until the strut is safely compressed.

5. Remove the bolt attaching the strut to the rear torsion beam assembly and remove the strut.

6. Place the strut assembly in a vise with the special holding tool HT71780000 or in a spring compressor.

7. Loosen the piston rod locknut.

✳✳ CAUTION

Do not remove the piston rod locknut, the spring is under tension and can cause serious personal injury.

8. Compress the spring with the spring compressor, then remove the piston rod locknut.

➡ **Before removing the strut from the coil spring, note the positioning of the strut in relationship to the coil spring for reassembly.**

9. Remove the bushing, strut mounting bracket, and the upper spring seat rubber.

10. Remove the strut, leaving the coil spring compressed.

11. Remove the bushing, bound bumper cover, and the bound bumper.

To install:

12. Install the bound bumper, bound bumper cover, and the bushing.

13. Install the strut into the coil spring, be sure the strut and spring are properly positioned.

14. Install the upper spring seat rubber, strut mounting bracket, and the bushing. Be sure that the mounting bracket is properly positioned.

15. Install the piston rod locknut, then remove the spring compressor.

16. Torque the piston rod locknut to 13–17 ft. lbs. (18–24 Nm).

17. Install the strut into the vehicle and

install new attaching nuts. Torque the nuts to 12–14 ft. lbs. (16–19 Nm).

18. Position the strut on the rear torsion beam and install the bolt. Torque the bolt attaching the strut to the torsion beam assembly to 72–87 ft. lbs. (98–118Nm).

19. Remove the support from the rear torsion beam.

20. Install the rear wheels and lower the vehicle.

21. Check the vehicle's alignment and adjust as necessary.

J30

1. Raise and support the vehicle safely.
2. Remove the rear parcel shelf to gain access to shock upper mounting nuts.

✳✳ CAUTION

Do not remove the piston rod locknut until the coil spring is compressed with a spring compressor.

3. Remove the strut upper mounting nuts.

4. Remove the lower strut mounting nut and bolt.

5. Remove the strut assembly.

6. Place the assembly into a suitable holding fixture.

① Gasket
② Strut mounting insulator
③ Upper spring seat
④ Dust cover
⑤ Coil spring
⑥ Bound bumper
⑦ Strut assembly
⑧ Connecting rod
⑨ Knuckle assembly
⑩ Baffle plate
⑪ Wheel hub bearing
⑫ Cotter pin
⑬ Cap
⑭ Parallel link
⑮ Mounting bracket
⑯ Bushing
⑰ Clamp
⑱ Stabilizer bar
⑲ Radius rod

Front

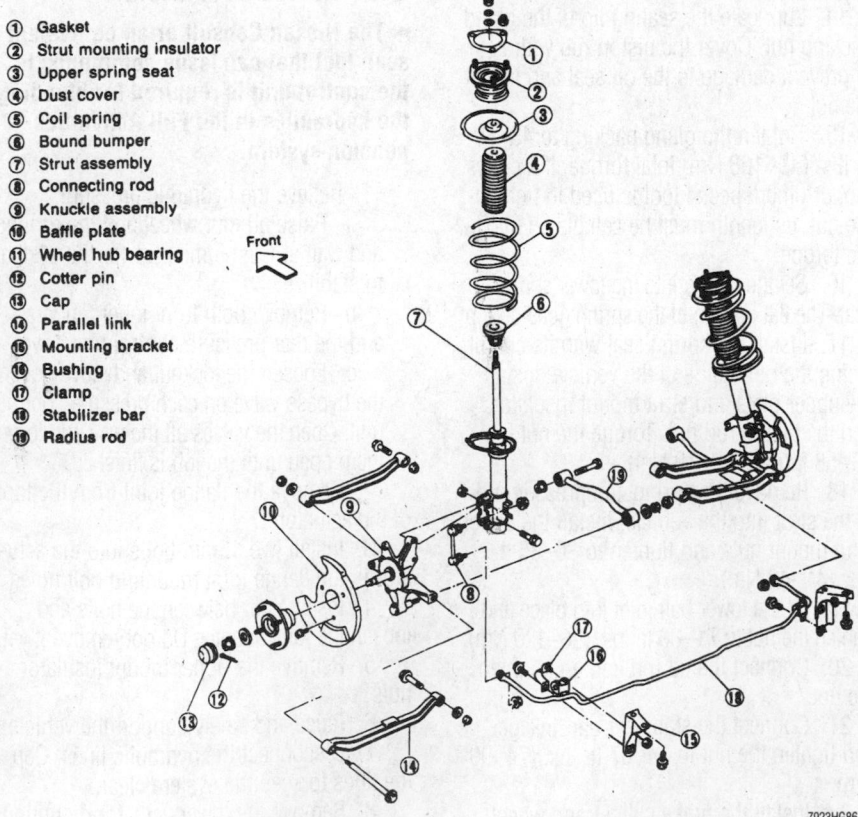

Exploded view of the rear suspension—J30 models

7923HG86

7. Compress the spring until the upper spring seat can be turned by hand.

8. Remove the locknut, spring seat components, spring, bushings and bumper.

To install:

9. Fit the bumper, spring, seat and other components onto the strut and install the locknut. Torque the nut to 13–17 ft. lbs. (18–24 Nm) and remove the spring compressor.

10. Install the strut assembly.

11. Torque the upper mounting nuts to 12–14 ft. lbs. (16–19 Nm) and the lower mounting bolt to 72–87 ft. lbs. (98–118 Nm).

12. Install the parcel shelf and lower the vehicle.

Q45—STANDARD SUSPENSION

✳✳ CAUTION

Do not remove piston rod locknut with the shock absorber on vehicle.

1. Remove the upper strut mounting nuts.

2. Raise and safely support the vehicle and remove the lower mounting bolt. Remove coil spring/strut absorber assembly.

3. Place the assembly into a suitable holding fixture and matchmark the spring, strut and upper seat. Loosen but do not remove the piston rod locknut.

4. Install a spring compressor and compress the spring until the upper spring seat can be turned by hand.

5. Remove the locknut, spring seat components, spring, bushings and bumper.

To install:

6. Fit the bumper, spring, upper seat and other components onto the strut with the matchmarks aligned. The top of the spring is flat.

7. Install the locknut and tighten it to 13–17 ft. lbs. (18–24 Nm) and remove the spring compressor.

8. Install strut assembly. Torque the upper shock mounting nuts to 12–14 ft. lbs. (16–19 Nm) and the lower shock mounting bolt to 57–72 ft. lbs. (77–98 Nm).

Q45—ACTIVE SUSPENSION

➡ **The Nissan Consult or an equivalent scan tool that can issue commands to the control unit is required for bleeding the hydraulics in the Full Active Suspension system.**

1. Relieve the hydraulic pressure:

a. Raise and safely support the vehicle with all four wheels off the ground and wait at least three minutes for the system to stabilize.

b. Remove both front inner fenders and the rear pressure control unit cover.

c. Loosen the locknut and slowly open the bypass valve on each pressure control unit. Do not open the bleeder valves.

d. Open the bypass valves all the way and leave them open until the job is finished.

2. Remove the upper mount insulator nuts.

3. Disconnect the hydraulic lines. Cap the lines to keep the system clean.

4. Remove the lower actuator mounting bolt and remove the actuator/spring assembly.

5. Secure the actuator/spring assembly in a suitable holding fixture. Scribe alignment marks on the spring, upper mount insulator and actuator unit.

6. Compress the spring with the proper tool. Remove the piston rod locknut lift off the mount insulator, hose joint adapter, spring, and any other components necessary.

To install:

7. Fit the bumper and dust cover onto the actuator.

8. Fit the spring into the lower seat with the matchmarks aligned. Install the upper seat, mounting insulator and other components with the marks aligned.

9. Install the locknut and tighten to 43–54 ft. lbs. (59–74 Nm).

10. Fit the assembly onto the vehicle and tighten the upper mounting nuts to 12–24 ft. lbs. (16–19 Nm).

11. Torque the lower mounting bolt to 58–72 ft. lbs. (78–98 Nm).

12. To bleed the system:

a. With all four wheels off the ground, run the engine for about two minutes.

b. Connect the Consult or equivalent scan tool and enter "WORK SUPPORT" mode. Select "4. AIR BLEEDING".

c. Check the fluid level in the reservoir and make it slightly overfilled.

d. Touch "START" on the scan tool. The display will show a regular rise and fall in system pressure that may last for several minutes. When the pressure stabilizes, stop the engine.

e. Connect a clear tube to the air bleeder at the actuator and place the other end in a container. Do not allow fluid to contact the body or the paint will be damaged.

f. Open the bleeder and watch the fluid move through the tube. If there are

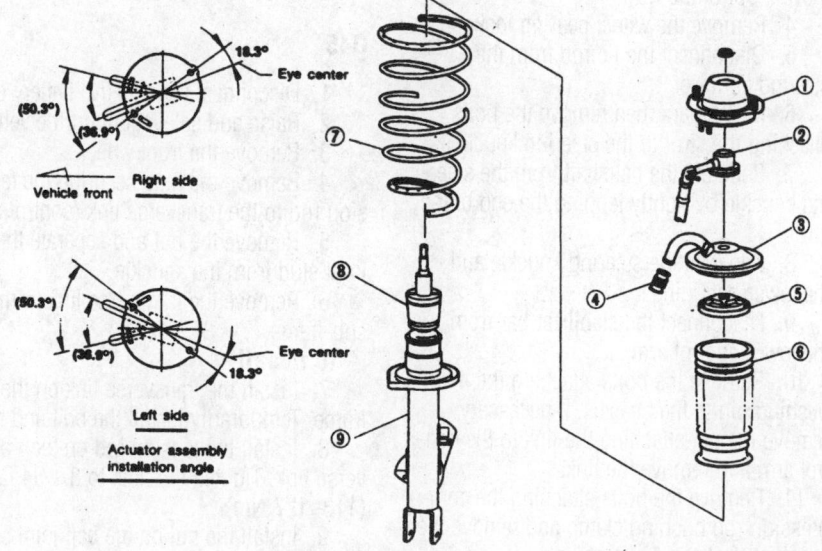

Figure shows rear left actuator.

① Mount insulator
② Rear joint hose
③ Spring upper seat
④ Air tube connector
⑤ Bound bumper cover
⑥ Rear actuator dust cover
⑦ Coil spring
⑧ Bound bumper
⑨ Rear actuator

7923HG85

Rear actuator removal with Active Suspension—Q45 models

still air bubbles in the fluid when the flow stops, close the bleeder and check the fluid level. Pressurize the system again and repeat the bleeding process.

Coil Spring

REMOVAL & INSTALLATION

Refer to the Strut removal and installation procedure for coil spring replacement.

Lower Ball Joint

REMOVAL AND INSTALLATION

The lower ball joint assembly is part of the lower control arm/transverse link. If replacement of the ball joint is required, the lower control arm needs to be replaced.

Lower Control Arms

REMOVAL AND INSTALLATION

G20

1. Raise and support the vehicle safely.
2. Remove the stabilizer bar.

➡**Take note of paint mark and clamp position when removing stabilizer bar for correct reinstallation.**

3. Support the steering knuckle with a suitable jack and remove the lower ball joint nut. Separate the ball joint from the knuckle.
4. Remove the bolts attaching the lower control arm to the chassis. Remove the lower control arm.
   To install:
5. If the lower ball joint is worn or damaged, the lower control arm must be replaced. The ball joint is not serviceable separately.
6. Reattach lower control arm to the chassis with the attaching bolts and nut.
7. Reinstall the ball joint stud in the knuckle and torque the nut to 52–64 ft. lbs. (71–86 Nm).
8. Install the stabilizer bar and wheel. Safely lower the vehicle.

➡**Final tightening must be done with the vehicle at normal ride height, tires on the ground and the chassis loaded.**

9. Torque front control arm bolts to 87–108 ft. lbs. (118–147 Nm) and rear gusset nut to 69–87 ft. lbs. (93–118 Nm).

J30

1. Raise and safely support the vehicle.
2. Remove the wheel and tire assembly.
3. Remove the nut and disconnect the transverse arm ball joint from the knuckle assembly, using the proper tool.
4. Remove the nuts and disconnect the tension rod from the transverse arm.
5. Remove the transverse arm from the front suspension member.
   To install:
6. Install the transverse arm to the suspension member. Do not tighten the bolt and nut until the vehicle is on the floor.

➡**Final tightening must be done with the vehicle at normal ride height with the weight of the vehicle on the tires.**

7. Connect the transverse arm ball joint to the knuckle. Torque the nut to 71–88 ft. lbs. (96–120 Nm).
8. Install the tension rod to the transverse arm. Torque the nuts to 72–87 ft. lbs. (98–118 Nm).
9. Install the wheel and tire assembly and lower the vehicle to the floor.
10. Torque the transverse arm mounting nut to 72–87 ft. lbs. (98–118 Nm).

I30

1. Raise and safely support the vehicle.
2. Remove the front wheels.
3. Remove the ABS wheel sensor and move it out of the way.
4. Remove the wheel bearing locknut.
5. Disconnect the tie rod from the steering knuckle.
6. Matchmark then remove the bolts attaching the strut to the steering knuckle.
7. Separate the halfshaft from the steering knuckle by lightly tapping the end of the shaft.
8. Separate the steering knuckle and the lower ball joint.
9. Disconnect the stabilizer bar from the lower control arm.
10. Remove the bolts attaching the link bushing pin to the chassis. If necessary, remove the nut attaching the link to the control arm and remove the link.
11. Remove the bolts attaching the compression rod bushing clamp and remove the lower control arm/traverse link.
    To install:
12. Install the lower control arm and the compression rod bushing clamp into the vehicle.
13. Install the link bushing pin, if removed from the control arm.
14. Tighten all bolts and nuts until they

are snug enough to support the weight of the vehicle but not fully tight, the bolts should be torqued to specification with the vehicle on the floor.
15. Install the steering knuckle to the lower control arm and connect the ball joint. Torque the ball joint nut to 46–56 ft. lbs. (62–76 Nm).

➡**Always use a new nut when installing the ball joint to the control arm.**

16. Connect the steering knuckle to the strut and to the halfshaft.
17. Install the strut mounting bolts and align the matchmarks. Torque the bolts to 103–117 ft. lbs. (140–159 Nm).
18. Install the tie rod ball joint and torque the nut to 46–54 ft. lbs. (63–73 Nm).
19. Install the wheel bearing locknut
20. Install the ABS wheel sensor and torque the attaching bolt to 13–17 ft. lbs. (18–24 Nm).
21. Install the front wheels, lower the vehicle and torque hub locknut to 174–231 ft. lbs. (235–314 Nm).
22. Torque the bolts attaching the compression rod bushing clamp and the link bushing pin, in the proper sequence to 87–108 ft. lbs. (118–147 Nm).
23. If the link bushing pin was removed from the control arm torque the attaching nut to 87–108 ft. lbs. (118–147 Nm).
24. Torque the sway bar attaching nut to 30–35 ft. lbs. (41–47 Nm).
25. Check the vehicle alignment.

Q45

1. Disconnect the negative battery cable.
2. Raise and safely support the vehicle.
3. Remove the front wheel.
4. Remove the nuts securing the tension rod to the transverse link (control arm).
5. Remove the nut and separate the ball joint stud from the knuckle.
6. Remove the transverse link from the sub-frame.
   To install:
7. Install the transverse link on the sub-frame. Temporarily install the bolt and nut.
8. Install the tension rod on the transverse link. Tighten the nuts to 87–94 ft. lbs. (118–127 Nm).
9. Install the nut on the ball joint stud. Tighten the nut to 71–88 ft. lbs. (96–120 Nm).
10. Install the front wheel and lower the vehicle to the floor.
11. Tighten the transverse link mounting bolt to 72–87 ft. lbs. (98–118 Nm).

Wheel Bearings

ADJUSTMENT

The front and rear wheel bearing assemblies on all models are pressed in and are not adjustable. If the bearing assembly does not turn smoothly or has more than 0.002 in. (0.05mm) of axial play, replace the bearing assembly.

REMOVAL & INSTALLATION

Front

G20

1. The axle nut torque is very high and should be loosened and tightened with the vehicle on the ground. Remove the cotter pin, adjusting cap and insulator and loosen the front axle nut.

2. Raise and safely support the vehicle.

3. Remove the brake caliper, carrier, and the rotor. Hang the caliper from the body with wire; do not let it hang by the brake hose.

4. Remove the cotter pin and nut and use a ball joint press to disconnect the tie rod end.

5. Remove the cap and the upper king-pin mounting nut and separate the kingpin from the third link.

6. Hold a block of wood against the axle stub and strike it with a hammer to release it from the hub. Withdraw the axle from the hub and fold the steering knuckle down on the ball joint.

7. Remove the cotter pin and nut and use a ball joint press to disconnect the ball joint. Remove the steering knuckle.

➡**Wheel bearings must be replaced any time the hub is removed.**

8. Pry the grease seals out of the steering knuckle.

9. Support the steering knuckle and press the hub out of the bearing.

10. Remove the snaprings and press the bearing out towards the inside of the knuckle.

To install:

11. Be sure all parts are clean and dry. The hub and steering knuckle should be inspected for cracks using dye or a magnetic crack detection process.

12. Install the inner snapring and carefully press the new bearing into the steering knuckle. Be sure the press tool contacts only the outer bearing race or the bearing will be damaged.

13. Install the outer snapring. Pack the new grease seals with clean grease and install them. If removed, install the splash guard.

14. Support the inner race on the press table and carefully press the hub into the bearing. Be sure the hub turns smoothly in both directions.

15. Fit the steering knuckle onto the lower ball joint and start the nut. Fit the axle shaft through the hub and start the nut.

16. Pack the king pin bearing housing with grease and fit the third link into place. Torque the kingpin nut to 72–87 ft. lbs. (98–118 Nm) and install the dust cap.

17. Torque the lower ball joint nut to 52–64 ft. lbs. (71–86 Nm). Install a new cotter pin.

18. Connect the tie rod end and tighten the nut to 22–29 ft. lbs. (29–39 Nm). Tighten as needed to install a new cotter pin but do not exceed 36 ft. lbs. (49 Nm).

19. Install the brake caliper, carrier, rotor, and the wheel and lower the vehicle to the ground.

20. Torque the front axle nut to 174–231 ft. lbs. (235–314 Nm). Install the insulator, adjusting cap and cotter pin.

I30

➡**Whenever the hub or bearing assembly is removed, the wheel bearing assembly must be replaced. Never reuse the old bearing assembly.**

1. Remove the knuckle assembly from

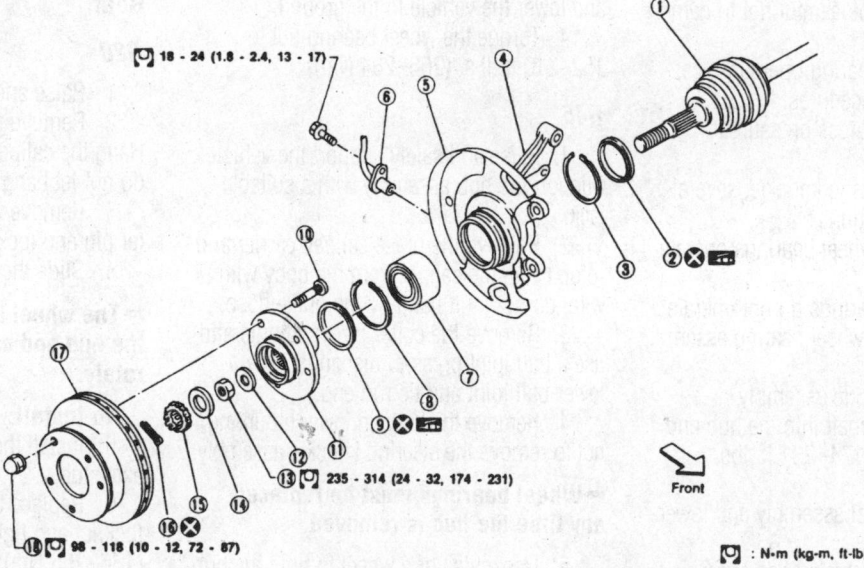

18 - 24 (1.8 - 2.4, 13 - 17)

235 - 314 (24 - 32, 174 - 231)

98 - 118 (10 - 12, 72 - 87)

Front

: N·m (kg-m, ft-lb)

① Drive shaft	⑦ Wheel bearing assembly	⑬ Wheel bearing lock nut
② Inner grease seal	⑧ Snap ring	⑭ Insulator
③ Snap ring	⑨ Outer grease seal	⑮ Adjusting cap
④ Knuckle	⑩ Hub bolt	⑯ Cotter pin
⑤ Baffle plate	⑪ Wheel hub	⑰ Disc rotor
⑥ ABS sensor	⑫ Plain washer	⑱ Wheel nut

7923HG87

Exploded view of the front knuckle assembly—I30 models

the vehicle by separating the ball joint and tie rod end, then removing the retaining hardware securing the knuckle to the strut.

2. Using a shop press and a suitable tool, press the hub with the inner race from the steering knuckle.

3. Using a shop press and a suitable tool, press the bearing inner race from the hub and remove the outer grease seal.

4. Use snapring pliers to remove the snaprings from the steering knuckle.

5. Inspect the hub, steering knuckle and snaprings for cracks and/or wear; if necessary, replace the damaged part(s).

To install:

6. Install the inner snapring in the steering knuckle groove.

7. Using a shop press and a suitable tool, press the new wheel bearing assembly into the steering knuckle, until it seats, using a maximum pressure of 3 tons.

8. Install the outer snapring.

9. Pack the new grease seal lips with multi-purpose grease.

10. Using a shop press and a suitable tool, press the new outer grease seal into the steering knuckle.

11. Using a shop press and a suitable tool, press the new inner grease seal into the steering knuckle.

12. Using a shop press and a suitable tool, press the hub into the steering knuckle, until it seats, using a maximum pressure of 5.5 tons; be careful not to damage the grease seal.

13. To check the bearing operation, perform the following procedures:

   a. Increase the press pressure to 3.5–5.0 tons.

   b. Spin the steering knuckle, several turns, in both directions.

   c. Be sure the wheel bearings operate smoothly.

14. If the wheel bearings do not operate smoothly, replace the wheel bearing assembly.

15. Install the knuckle assembly.

16. Install the halfshaft into the hub and tighten the locknut to 174–231 ft. lbs. (235–314 Nm).

17. Install the wheel assembly and lower the vehicle.

18. Road test the vehicle and verify proper operation.

J30

1. With the vehicle still on the ground, remove the hub cap and loosen the front wheel bearing nut.

2. Raise and safely support the vehicle. Remove the brake caliper, torque member and brake rotor. Support the caliper with wire, do not let it hang by the brake line.

3. Remove the knuckle assembly from the vehicle by separating the ball joint and tie rod end from the assembly and removing the hardware securing the spindle to the strut assembly.

4. Remove the nut and wheel hub from the spindle by forcing the hub out of the spindle assembly with a piece of wood and hammer.

5. Pry the grease seal out of the hub and remove the snapring.

➡**If the bearing is removed from the hub, it must be replaced.**

6. Using a suitable shop press, press the bearing out towards the inside of the hub.

To install:

7. Press a new wheel bearing assembly into the wheel hub. Be sure the press tool contacts only the outer bearing race or the bearing will be damaged.

8. Install the snapring.

9. Pack grease seal lip with multi-purpose grease.

10. Install the grease seal.

11. Install the knuckle assembly on to the strut. Connect the ball joint and tie rod end.

12. Install the hub and washer and start the nut.

13. Install the brake caliper and wheel and lower the vehicle to the ground.

14. Torque the wheel bearing nut to 152–210 ft. lbs. (206–284 Nm).

Q45

1. Raise and safely support the vehicle. Support the hub assembly with a suitable jack.

2. Remove the brake caliper, carrier and rotor. Hang the caliper from the body with wire, do not let it hang by the brake hose.

3. Remove the cotter pins and nuts and use a ball joint press to disconnect the lower ball joint and tie rod end.

4. Remove the kingpin lower mounting nut to remove the steering knuckle assembly.

➡**Wheel bearings must be replaced any time the hub is removed.**

5. Use a vise or a wheel to hold the hub and remove the hub cap and nut from the back of the hub. Remove the wheel speed sensor rotor.

6. Use a press or large drift pin to press the hub out of the steering knuckle.

7. Remove the snapring and press the bearings and grease seal out of the steering knuckle.

To install:

8. Be sure all parts are clean. Carefully press the new bearing into the steering knuckle. Be sure the press tool contacts only the outer bearing race or the bearing will be damaged.

9. Install a new grease seal and the snapring. If removed, install the splash guard.

10. Lightly lubricate the lips of the seal with clean grease. Be careful not to grease the bearing or hub mating surfaces.

11. Carefully press the hub into the bearing. Support the inner race on the press table or the bearing will be damaged. Do not exceed 3.9 tons (34.3 kN) pressure.

12. Install the speed sensor rotor and nut on the hub and tighten to 152–210 ft. lbs. (206–284 Nm). Stake the nut into place.

13. Lightly tap the cap into place and install the bolts.

14. Install the steering knuckle to the king pin. Torque the nut to 108–137 ft. lbs. (88–108 Nm).

15. Connect the lower ball joint and tighten the nut to 65–80 ft. lbs. (88–108 Nm). Install a new cotter pin.

16. Connect the tie rod end and tighten the nut to 22–29 ft. lbs. (29–39 Nm). Tighten as needed to install a new cotter pin but do not exceed 36 ft. lbs. (49Nm).

17. Install the rotor and the brake caliper.

18. Install the wheel and tire assembly.

Rear

G20

1. Raise and support the vehicle safely.

2. Remove the rear caliper and rotor. Hang the caliper from the body with wire, do not let hang by the brake hose.

3. Remove the rear wheel hub cap, cotter pin and locknut.

4. Slide the hub off the stub axle.

➡**The wheel bearing is integral with the hub and cannot be serviced separately.**

To install:

5. Install the new hub assembly onto the axle stub.

6. Replace the washer and wheel bearing locknut and tighten to 137–188 ft lbs. (186–255 Nm). Install a new cotter pin.

7. Reinstall the brake rotor, caliper and wheel. Lower the vehicle to the ground.

I30

➡**If the vehicle is equipped with ABS, the sensor must be removed to protect the sensor and its wiring.**

1. Raise and safely support the vehicle. Remove the rear wheel(s).

2. Remove the wheel speed sensor.

3. Perform the following procedures:

   a. Remove the brake caliper and hang it by a piece of wire.

   b. Remove the brake caliper support.

   c. Remove the disc brake pads.

   d. Remove the brake disc.

4. Remove the grease cap.

5. Remove the cotter pin, wheel bearing locknut, washer, and the wheel hub bearing assembly. A slide hammer may be needed to remove the hub bearing assembly.

➡ **The wheel hub bearing assembly is not repairable; it must be replaced when defective.**

To install:

6. Install the wheel hub bearing assembly, the washer and the wheel bearing locknut. Torque the wheel bearing locknut to 137–188 ft. lbs. (186–255 Nm).

7. Verify that the wheel bearings operate smoothly.

8. Install a new cotter pin into the spindle to hold the wheel bearing locknut.

9. Install a dial micrometer to the rear wheel hub bearing assembly and check the axial end-play; it should be less than 0.0020 in. (0.05mm).

10. Install the grease cap.

11. Install the ABS wheel sensor and its wiring.

12. Install the brake assembly and the wheels.

J30

1. Raise and safely support the vehicle.

2. Remove the cotter pin and adjusting cap and loosen the wheel bearing nut. Carefully tap the end of the axle shaft or use a puller to loosen the shaft from the hub.

3. Remove the brake caliper and rotor. Do not let the caliper hang by the brake hose, support it with wire.

4. Remove the parking brake assembly.

5. Remove the nuts and through-bolts to remove the axle housing from the suspension.

6. Secure the axle housing in a vise and use a pull hammer to remove the hub from the bearing. The bearing race may stay on the hub. Remove it with a standard bearing puller.

7. Remove the grease seal and snaprings and press the bearing out of the axle housing.

To install:

8. Install the inner snapring into the axle housing and press a new bearing into place. Be sure the press tool contacts only the outer bearing race or the bearing will be damaged. Install the outer snapring.

9. Lightly lubricate the lip of the new grease seals and press the seals into the axle housing.

10. Support the inner bearing race on the press table and carefully press the hub into the bearing.

11. Fit the axle housing onto the lower ball joint, tighten the nut to 58–69 ft. lbs. (78–93 Nm) and install a new cotter pin.

12. Fit the axle shaft into the hub and install the bolts through the suspension bushings. Tighten the bolts temporarily, they will be tightened with the vehicle resting on the wheels.

13. Install the brake components and apply the brake to hold the hub from turning.

14. Install the wheel bearing locknut and torque to 152–203 ft. lbs. (206–275 Nm). Install the insulator and adjusting cap and a new cotter pin.

15. Install the wheel and lower the vehicle to the ground. Torque the suspension bushing bolts to 57–72 ft. lbs. (77–98 Nm) and the shock absorber mounting bolt to 72–87 ft lbs. (98–113 Nm).

Q45

1. Raise and safely support the vehicle.

2. Remove the cotter pin and adjusting cap and loosen the wheel bearing nut. Carefully tap the end of the axle shaft or use a puller to loosen the shaft from the hub.

3. Remove the brake caliper and rotor. Do not let the caliper hang by the brake hose, support it with wire.

4. Remove the parking brake assembly.

5. Remove the nuts and through-bolts

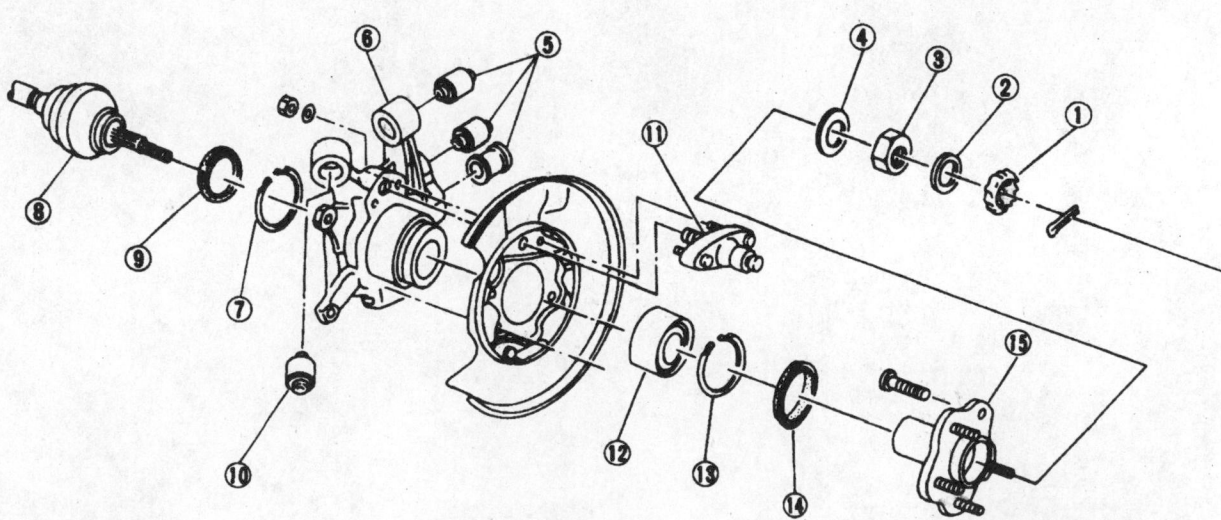

① Adjusting cap	⑥ Axle housing	⑪ Brake anchor pin
② Insulator	⑦ Snap ring	⑫ Wheel bearing
③ Wheel bearing lock nut	⑧ Drive shaft	⑬ Snap ring
④ Washer	⑨ Grease seal	⑭ Grease seal
⑤ Bushing	⑩ Bushing	⑮ Wheel hub

7923HG88

The rear wheel bearing and flange are supplied as an assembly—J30 models

to remove the axle housing from the suspension. If equipped with rear wheel steering, use a ball joint press to separate the tie rod end.

6. Remove the four bolts at the back and remove the bearing flange and hub from the bearing housing.

7. Press the hub out of the bearing flange and use a puller to remove the bearing from the hub. If it is not damaged, the hub can be used again but the bearing and flange are supplied as a single unit.

To install:

➡**The wheel bearing and flange are supplied as an assembly.**

8. Place the hub on a press table and press the new bearing and flange onto the hub. Be sure the press tool contacts only the inner bearing race and take care not to damage the seal.

9. Assemble the bearing flange onto the axle housing and tighten the bolts to 58–72 ft. lbs. (78–98 Nm).

10. Fit the axle housing onto the lower ball joint, tighten the nut to 58–69 ft. lbs. (78–93 Nm) and install a new cotter pin.

11. If equipped with rear wheel steering, tighten the tie rod end nut to 33–44 ft. lbs. (45–60 Nm) and install a new cotter pin.

12. Fit the axle shaft into the hub and

install the bolts through the suspension bushings. Tighten the bolts temporarily, they will be tightened with the vehicle resting on the wheels.

13. Install the brake components and apply the brake to hold the hub from turning.

14. Install the wheel bearing locknut and tighten to 152–203 ft. lbs. (206–275 Nm). Install the insulator and adjusting cap and a new cotter pin.

15. Install the wheel and lower the vehicle to the ground. Torque the suspension bushing bolts to 57–72 ft. lbs. (77–98 Nm).

KIA

Sephia

19

ENGINE REPAIR

➡**Disconnecting the negative battery cable on some vehicles may interfere with the functions of the on board computer systems and may require the computer to undergo a relearning process, once the negative battery cable is reconnected.**

Distributor

REMOVAL

1. Disconnect the negative battery cable.
2. Remove the air cleaner.
3. Remove the distributor cap and position aside, leaving the spark plug wires connected. Before removing the distributor, mark the position of the No. 1 spark plug wire tower on the distributor cap.
4. Detach the distributor electrical connectors.
5. Using a wrench on the crankshaft pulley, rotate the crankshaft to position the No. 1 piston at TDC on the compression stroke. The crankshaft pulley notch should align with the timing plate indicator and the distributor rotor should be pointing to the No. 1 spark plug tower position on the distributor cap.
6. Using chalk or paint, mark the position of the distributor housing on the cylinder head.
7. Remove the distributor hold-down bolts and remove the distributor.
8. Inspect the O-ring on the distributor housing and replace it if it is damaged or worn.

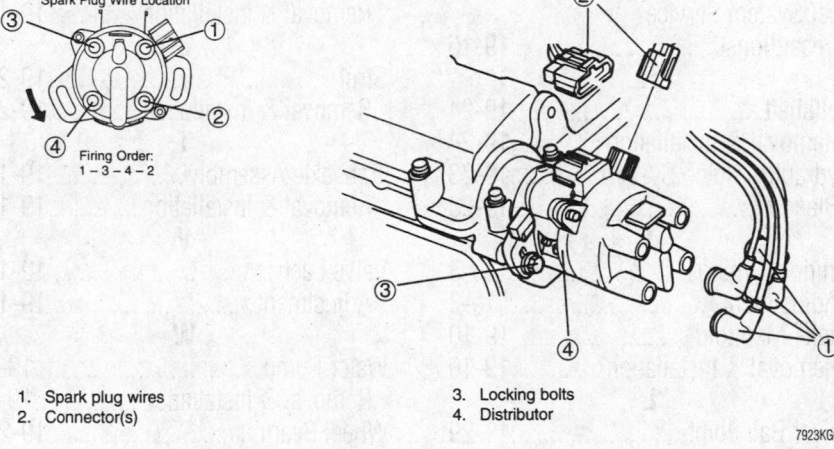

1. Spark plug wires
2. Connector(s)
3. Locking bolts
4. Distributor

7923KG01

Exploded view of the distributor component mounting

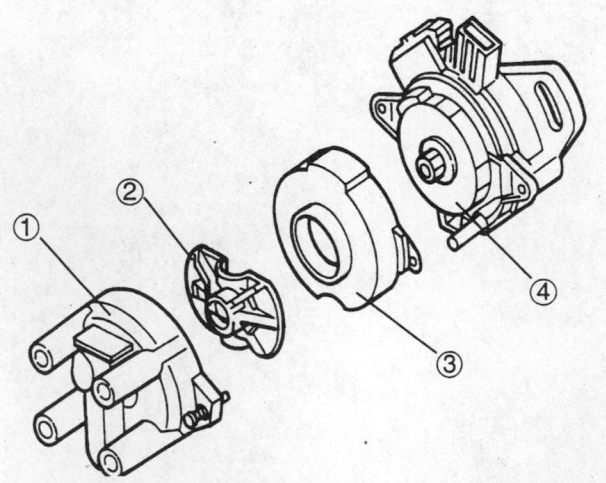

1. Distributor Cap
2. Rotor
3. Dust Cover
4. Camshaft position sensor

7923KG02

Exploded view of the distributor assembly

INSTALLATION

TIMING NOT DISTURBED

1. Using clean engine oil, lubricate the distributor O-ring.
2. Install the distributor. Be sure the distributor rotor aligns with the No. 1 spark plug tower position on the distributor cap and the distributor housing mark aligns with the cylinder head or cylinder block mark.

➡**There are existing marks on the distributor shaft and housing, which when aligned, indicate the No. 1 spark plug wire tower position.**

3. Install and loosely tighten the distributor hold-down bolts.
4. Attach the electrical connectors to their original locations. Install the distributor cap.
5. Install the air cleaner.
6. Connect the negative battery cable. Start the engine, and check or adjust the ignition timing.

TIMING DISTURBED

1. Using clean engine oil, lubricate the distributor O-ring.
2. Disconnect the spark plug wire from the No. 1 cylinder spark plug. Remove the spark plug from the No. 1 cylinder and press a thumb over the spark plug hole.
3. Using a wrench on the crankshaft pulley, rotate the crankshaft until pressure is felt at the spark plug hole, indicating the piston is approaching TDC on the compression stroke. Continue rotating the crankshaft until the crankshaft pulley mark aligns with the timing cover indicator.
4. Position the distributor rotor so it aligns with the No. 1 spark plug wire tower on the distributor cap.
5. Install the distributor. Align the mark that was made on the distributor housing with the mark that was made on the cylinder block. Loosely tighten the distributor hold-down bolts.
6. Attach the electrical connectors to their original locations. Install the distributor cap.
7. Install the spark plug in the No. 1 cylinder and connect the spark plug wire.
8. Install the air cleaner.
9. Connect the negative battery cable. Start the engine, and check or adjust the ignition timing.

Ignition Timing

ADJUSTMENT

1. Apply the parking brake. Place the shift lever in NEUTRAL on manual transaxles, and **P** on automatic transaxles.

2. Locate the timing marks on the crankshaft pulley and the timing indicator scale on the engine front cover. If the marks are hard to see, clean them with degreaser and a stiff brush.

3. Start the engine and bring to normal operating temperature. Be sure all accessories are OFF.

4. Connect timing light to the engine according to the manufacturer's instructions.

5. Locate the Data Link Connector (DLC).

6. Connect a jumper wire between the **ENGINE TEST** terminal and the **GND** terminal of the data link connector, as shown.

7. Verify that the timing mark (white) on the crankshaft pulley, and the T mark on the timing belt cover are aligned. Aim the timing light at the timing marks; the timing should be 9–11°.

8. If the timing marks are not aligned, loosen the distributor hold-down bolt and turn the distributor housing to adjust. When the marks align, tighten the distributor hold-

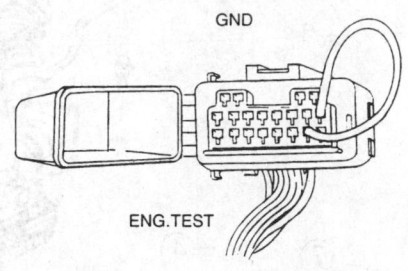

7923KG04

Jumper wire terminal locations

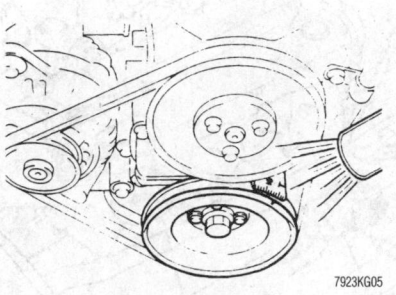

7923KG05

Clean the timing marks prior to starting the engine to get a more accurate view

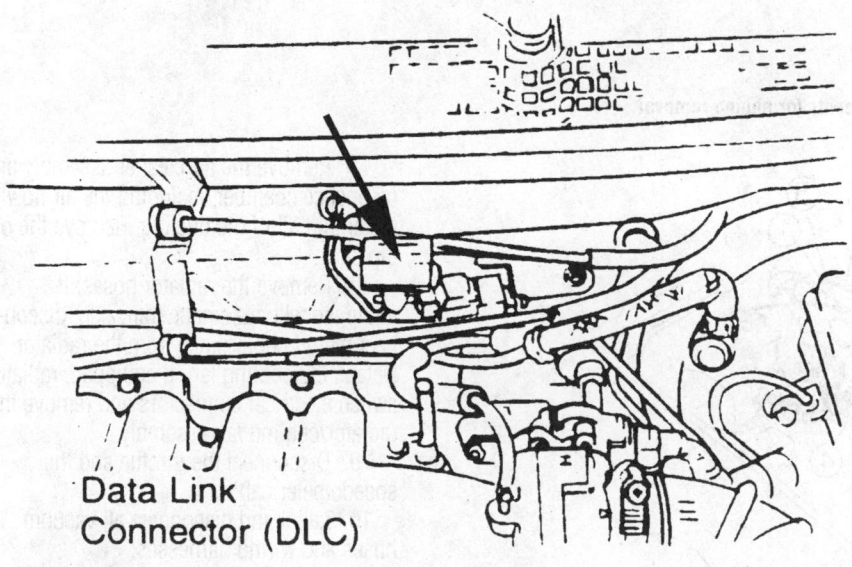

Data Link Connector (DLC)

7923KG03

Location of the data link connector in the engine compartment

down bolts to 14–19 ft. lbs. (19–25 Nm) and recheck the timing.

9. Remove the jumper wire.

10. Increase the engine speed and verify that the timing advances.

11. Remove all test equipment.

Engine Assembly

REMOVAL & INSTALLATION

➡The procedure for pulling the engine requires removing the transaxle along with it. As a result, when the half-shafts are pulled from the transaxle, a special plug/side gear holding tool is recommended.

✳✳ CAUTION

Observe all applicable safety precautions when working around fuel. Whenever servicing the fuel system, always work in a well ventilated area. Do not allow fuel spray or vapors to come in contact with a spark or open flame. Keep a dry chemical fire extinguisher near the work area. Always keep fuel in a container specifically designed for fuel storage; also, always properly seal fuel containers to avoid the possibility of fire or explosion.

1. Properly relieve the fuel system pressure. Raise and safely support the vehicle, as necessary.

2. Disconnect the battery cables and remove the battery and the battery tray.

3. Remove the hood.

4. Loosen the lug nuts on the front wheels.

5. Remove the front wheels.

6. Remove the splash shield(s) from under the vehicle and drain the engine and transaxle oil as well as the coolant.

✳✳ CAUTION

The EPA warns that prolonged contact with used engine oil may cause a number of skin disorders, including cancer! You should make every effort to minimize your exposure to used engine oil. Protective gloves should be worn when changing the oil. Wash your hands and any other exposed skin areas as soon as possible after exposure to used engine oil. Soap and water, or waterless hand cleaner should be used.

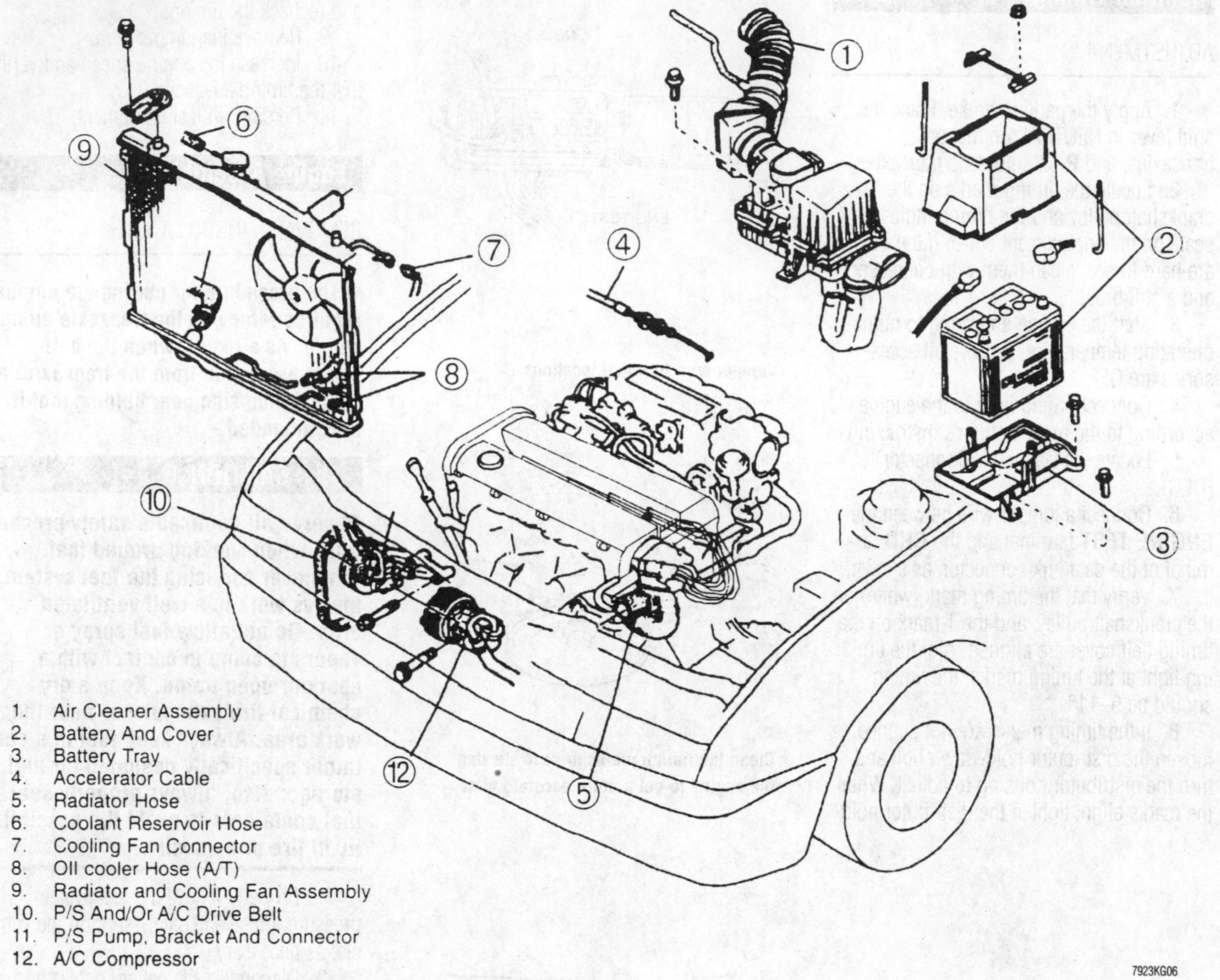

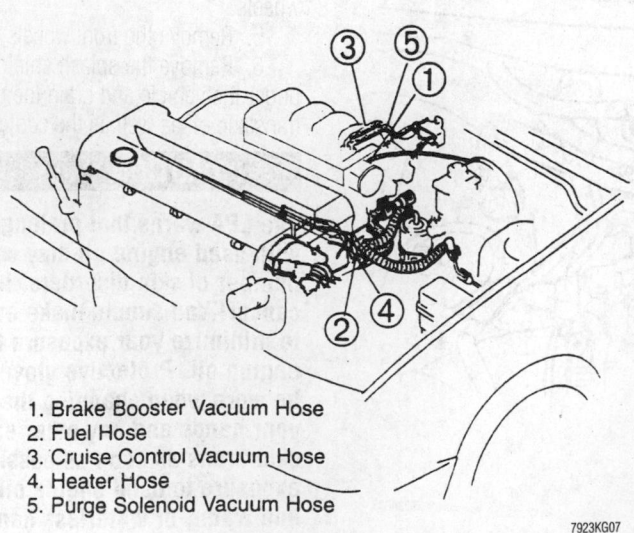

1. Air Cleaner Assembly
2. Battery And Cover
3. Battery Tray
4. Accelerator Cable
5. Radiator Hose
6. Coolant Reservoir Hose
7. Cooling Fan Connector
8. Oil cooler Hose (A/T)
9. Radiator and Cooling Fan Assembly
10. P/S And/Or A/C Drive Belt
11. P/S Pump, Bracket And Connector
12. A/C Compressor

7923KG06

Exploded view of typically removed external components for engine removal

1. Brake Booster Vacuum Hose
2. Fuel Hose
3. Cruise Control Vacuum Hose
4. Heater Hose
5. Purge Solenoid Vacuum Hose

7923KG07

Vacuum, fuel and water hose disconnect points for engine removal

7. Remove the air cleaner assembly and resonance chamber, including the air flow meter and all of the ducting. Remove the oil dip stick.

8. Remove the radiator hoses. If equipped with automatic transaxle, disconnect the oil cooler lines from the radiator. Detach the cooling fan, if equipped, radiator switch electrical connectors and remove the radiator/cooling fan assembly.

9. Disconnect the throttle and the speedometer cable.

10. Label and disconnect all vacuum hoses and wiring harnesses.

11. Disconnect the fuel supply and return hoses and the heater hoses.

12. Disconnect the exhaust pipe from the manifold.

13. Remove the accessory drive belt or belts.

14. Without disconnecting the hydraulic hoses, remove the power steering pump and hang it from the body with wire.

15. Without disconnecting the refrigerant lines, remove the air conditioning compressor and hang it from the body with wire.

16. If equipped with manual transaxle, disconnect the clutch cable and shift control rod. If equipped with hydraulic clutch, remove the slave cylinder from the transaxle without disconnecting the hydraulic line.

17. If equipped with automatic transaxle, disconnect the shift control cable.

18. Remove the nuts and disconnect the tie rod ends from the steering knuckles. Disconnect the stabilizer bar from the lower control arms.

19. Attach an engine lifting chain to the engine lifting eyes. Attach the chain to a suitable engine hoist and raise the hoist until there is tension on the chain.

20. Remove the engine mount nuts and the engine mount member bolts and nuts and remove the engine mount member.

➡**Be careful so the engine does not fall when removing the engine mount member.**

21. Remove the pinch bolts from the steering knuckle and pry the control arm down to slip the lower ball joint out of the knuckle.

22. If equipped, remove the bolts from the right side intermediate shaft support, using a suitable prybar, pry the intermediate shaft from the transaxle. Insert a suitable prybar between the inner CV-joint and transaxle case and carefully pry the inner CV-joints out of the transaxle. Suspend the halfshafts with wire.

23. Remove the dynamic damper from the right side engine mount, if equipped. Remove the engine/transaxle mount nuts/bolts and right engine, if equipped, left transaxle mounts. Carefully lift the engine/transaxle assembly from the vehicle.

24. Properly support the engine/transaxle assembly. Remove the intake manifold bracket, starter, torque converter nuts, stiffener, if equipped and No. 2 engine mount. Disconnect the throttle cable.

   a. Remove the set bolt and lock sensor switch.

   b. Remove the plug from the end of the motor and use a small flat bladed tool to turn the shift rod ½ turn clockwise.

   c. Remove the retaining bolts and the center differential lock motor.

25. Remove the transaxle mounting bolts and separate the transaxle from the engine.

To install:

26. Installation is the reverse of the removal procedure. Note the following important steps.

27. When possible, leave the engine mounting nuts/bolts loose (hand tight) until all mounts are aligned and bolted. This may help in aligning the engine and transmission assembly in the vehicle.

28. Install new circlips on the inner CV-joint stub shafts, if equipped, intermediate shaft. Grease the shaft splines before installing the halfshaft/intermediate shaft into the transaxle.

29. Always install new gaskets and/or O-rings. Use new self-locking nuts, especially on the exhaust.

✳✳ **WARNING**

Operating the engine without the proper amount and type of engine oil will result in severe engine damage.

30. Fill the engine and the transaxle with the proper types and quantities of oil. Fill the cooling system.

31. Connect the negative battery cable, start the engine and check for leaks. Check the ignition timing and the idle speed. Check all fluid levels.

Water Pump

REMOVAL & INSTALLATION

1. Disconnect the negative battery cable. Drain the cooling system.

2. Remove the timing belt covers, and remove the timing belt.

3. Disconnect the coolant inlet pipe and gasket.

4. Remove the timing belt idler pulleys still attached to the water pump.

5. Remove the water pump mounting bolts, and remove the water pump.

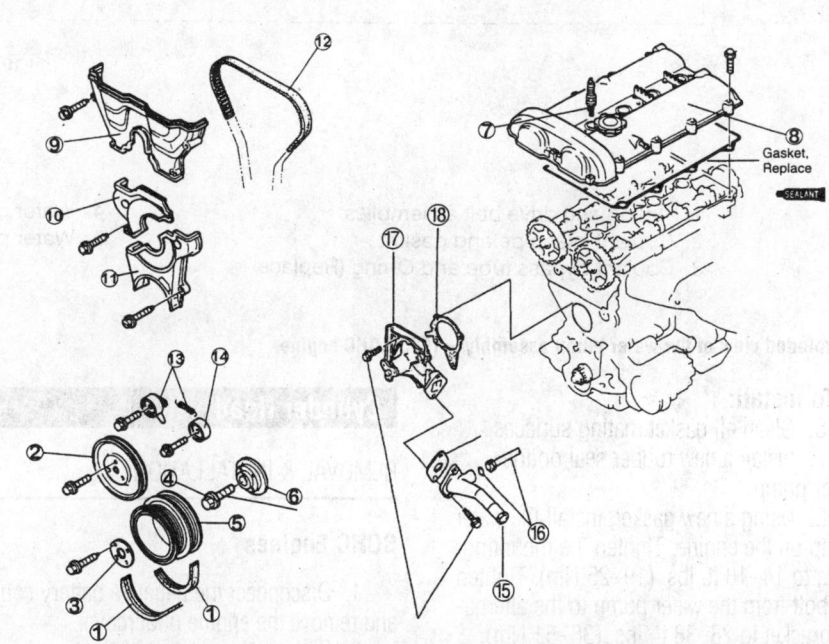

1. Drive Belts
2. Water Pump Pulley
3. Crankshaft Pulley Plate
4. Crankshaft Pulley Lock Bolt
5. Crankshaft Pulley
6. Crankshaft Pulley Boss
7. Spark Plug
8. Cylinder Head Cover
9. Upper Timing Belt Cover
10. Middle Timing Belt Cover
11. Lower Timing Belt Cover
12. Timing Belt
13. Tensioner Pulley, Tensioner Spring
14. Idler pulley
15. Coolant inlet pipe and gasket
16. Coolant bypass tube and O-ring (Replace)
17. Water pump assembly
18. Water Pump Gasket (Replace)

Exploded view of the water pump assembly—DOHC engine

7923KG08

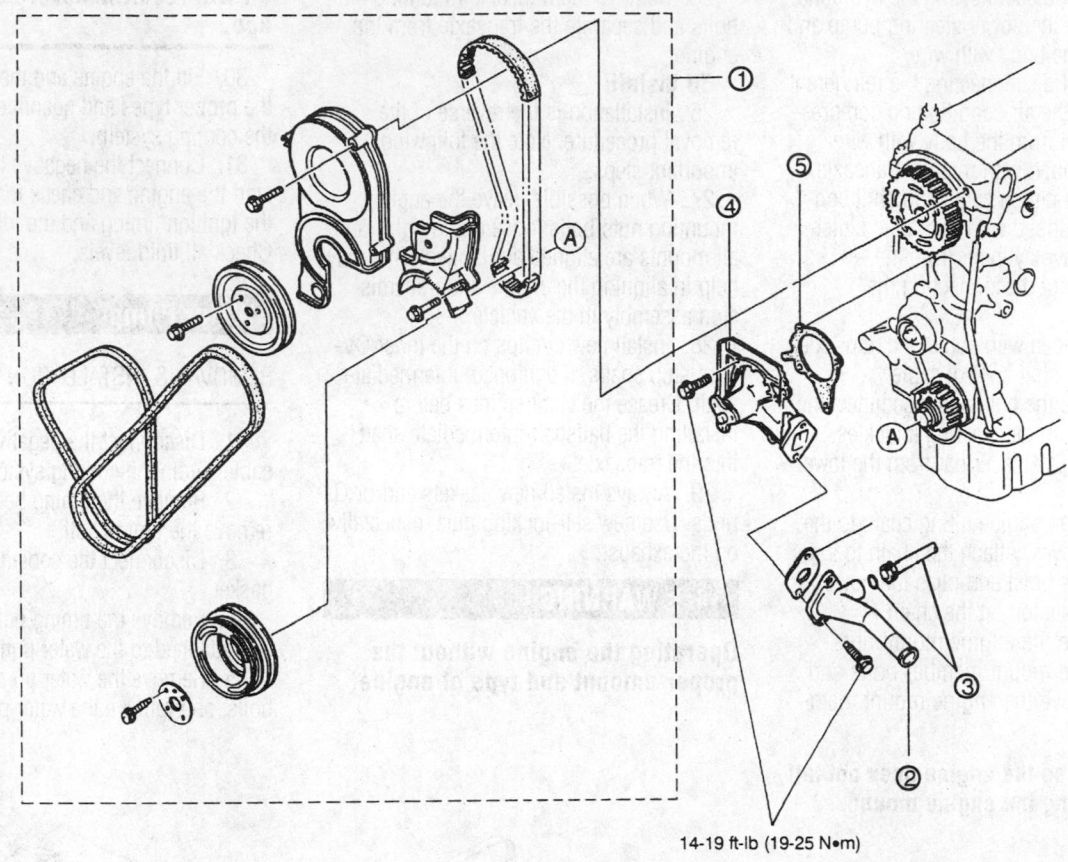

14-19 ft-lb (19-25 N•m)

1. Timing and drive belt assemblies
2. Coolant inlet pipe and gasket
3. Coolant bypass tube and O-ring (Replace)
4. Water pump assembly
5. Water pump gasket (Replace)

7923KG09

Exploded view of the water pump assembly—1995 SOHC engine

To install:

6. Clean all gasket mating surfaces.

7. Install a new rubber seal on the water pump.

8. Using a new gasket, install the water pump on the engine. Tighten the mounting bolts to 14–18 ft. lbs. (19–25 Nm). Tighten the bolt from the water pump to the alternator bracket to 28–38 ft. lbs. (38–51 Nm).

9. Install the timing belt idler pulleys that were removed.

10. Install the coolant inlet pipe, using a new gasket. Tighten the bolts to 14–18 ft. lbs. (19–25 Nm).

11. Install the timing belt and the timing belt covers.

12. Fill and bleed the cooling system. Connect the negative battery cable, start the engine and bring to normal operating temperature. Check for leaks.

Cylinder Head

REMOVAL & INSTALLATION

SOHC Engines

1. Disconnect the negative battery cable and remove the engine undercover.

2. Remove the air ducts from the air cleaner and throttle body.

3. Tag and disconnect the spark plug wires from the spark plugs. Remove the spark plugs and the distributor cap and wires assembly. Remove the distributor.

4. Drain the cooling system and disconnect the radiator and heater hoses.

5. Disconnect the exhaust pipe and remove the exhaust manifold.

6. Disconnect the accelerator cable.

7. Label and disconnect all necessary electrical connections and vacuum hoses. Disconnect the fuel lines.

8. Remove the intake manifold bracket and the intake manifold.

9. Remove the cylinder head cover bolts and the cylinder head cover.

10. Remove the timing belt cover(s). Rotate the crankshaft, in the normal direction of rotation, until the No. 1 cylinder piston is at TDC on the compression stroke. Be sure the timing marks on the crankshaft and camshaft sprocket(s) are properly aligned and mark the direction of rotation of the belt.

11. Loosen the timing belt tensioner and remove the belt. Do not rotate the crankshaft until the timing belt is reinstalled.

12. When everything is disconnected, loosen the cylinder head bolts in the reverse of the tightening sequence. Remove the bolts and lift the head off the engine.

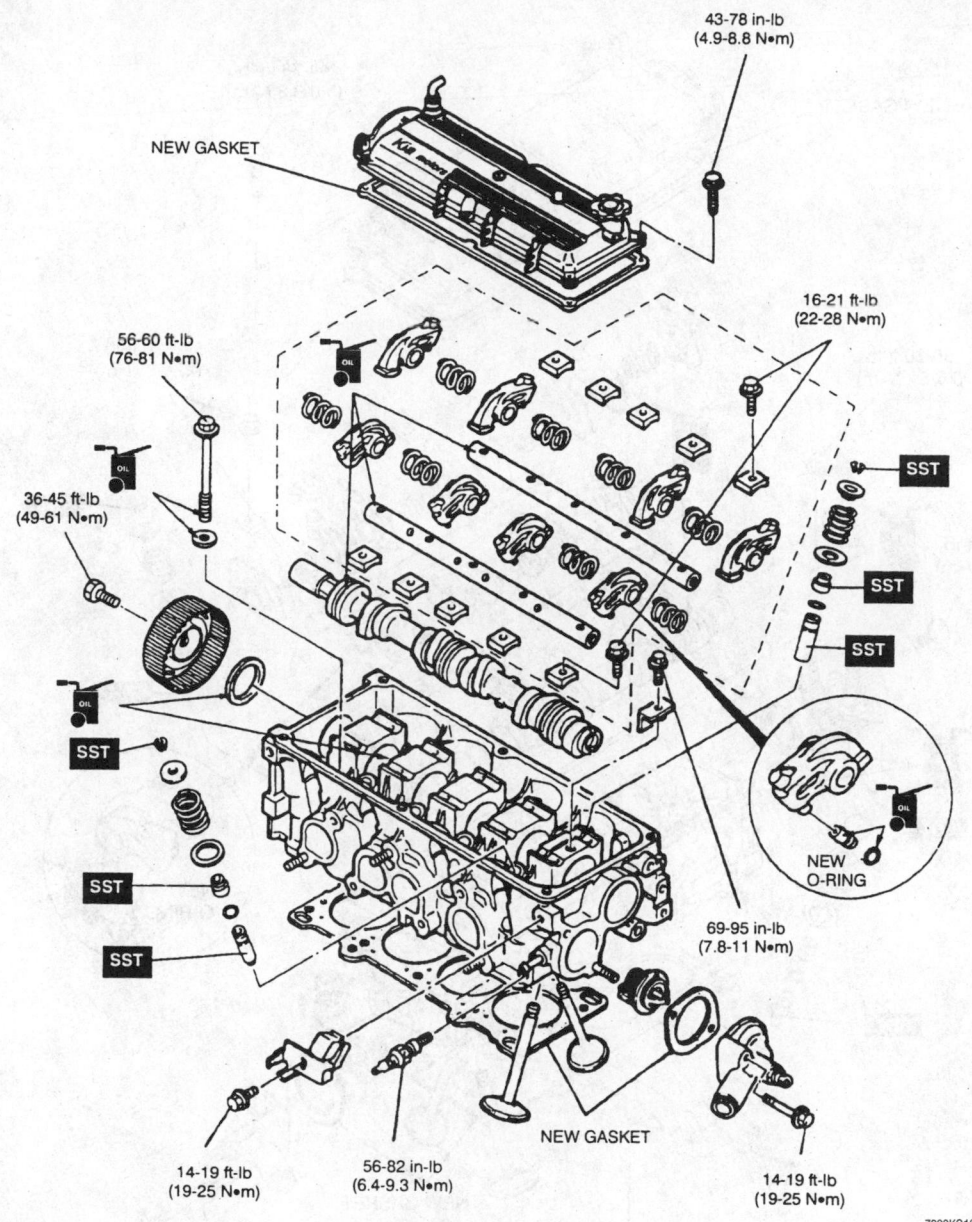

43-78 in-lb
(4.9-8.8 N•m)

NEW GASKET

56-60 ft-lb
(76-81 N•m)

16-21 ft-lb
(22-28 N•m)

36-45 ft-lb
(49-61 N•m)

SST

SST

SST

SST

SST

NEW O-RING

69-95 in-lb
(7.8-11 N•m)

14-19 ft-lb
(19-25 N•m)

56-82 in-lb
(6.4-9.3 N•m)

NEW GASKET

14-19 ft-lb
(19-25 N•m)

7923KG10

Exploded view of the cylinder head assembly—1995 SOHC 8-valve engine

To install:

13. Thoroughly, clean the cylinder head and the block contact surfaces. Examine the head gasket and check the cylinder head for cracks. Check the cylinder head for warpage using a feeler gauge and straightedge. The maximum allowable distortion is 0.004 in. (0.10mm).

14. Clean the cylinder head bolts and the threads in the block. Be sure the bolts turn freely in the block.

15. Install a new head gasket on the engine block. Be sure the camshaft sprocket timing marks are still aligned, as set during the removal procedure. Install the cylinder head.

16. Lubricate the bolt threads and seat surfaces with clean engine oil and install them. Tighten the bolts in 2–3 steps to 56–60 ft. lbs. (75–81 Nm) in the proper sequence.

17. Be sure the crankshaft and camshaft sprocket timing marks are aligned, install the timing belt and set the tension. Carefully rotate the crankshaft 2 turns to be sure the timing marks still line up.

18. Apply a thin bead of sealant to the cylinder head cover and install the new gasket. Install the cover and tighten the cover bolts to 78 inch lbs. (9 Nm).

19. Install the timing belt cover(s) and tighten the bolts to 95 inch lbs. (11 Nm).

20. Use new gaskets and install the manifolds. Tighten the intake manifold bolts/nuts to 19 ft. lbs. (25 Nm) and install the intake manifold bracket. Tighten the exhaust manifold nuts to 34 ft. lbs. (46 Nm).

21. Use a new gasket to connect the exhaust pipe and tighten the nuts to 34 ft. lbs. (46 Nm).

22. If removed, install the radiator and connect all cooling system hoses.

23. Install the distributor, spark plugs, distributor cap and wires.

24. Connect all vacuum and fuel system hoses and connect all wiring.

25. Connect the accelerator cable and install the air ducts and engine undercover.

26. Connect the negative battery cable. Fill and bleed the cooling system. Change the engine oil.

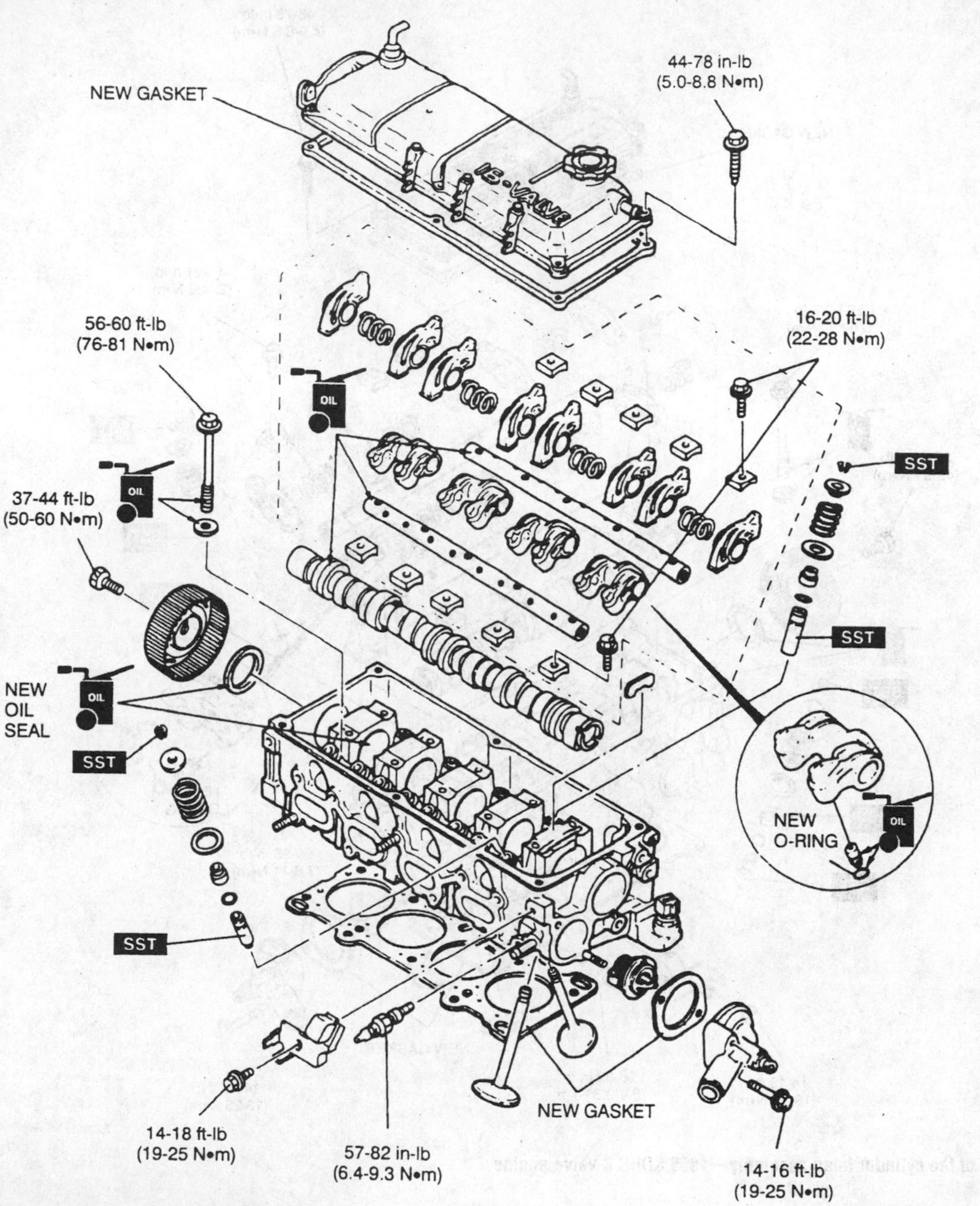

NEW GASKET

44-78 in-lb
(5.0-8.8 N•m)

56-60 ft-lb
(76-81 N•m)

16-20 ft-lb
(22-28 N•m)

37-44 ft-lb
(50-60 N•m)

OIL

SST

SST

NEW
OIL
SEAL

OIL

SST

SST

NEW
O-RING

OIL

SST

NEW GASKET

14-18 ft-lb
(19-25 N•m)

57-82 in-lb
(6.4-9.3 N•m)

14-16 ft-lb
(19-25 N•m)

7923KG11

Exploded view of the cylinder head assembly—1995 SOHC 16-valve engine

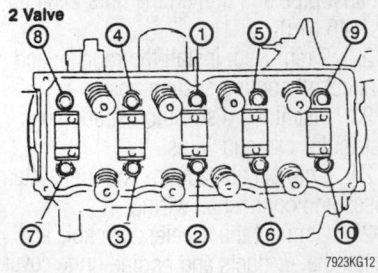

**Cylinder head bolt tightening sequence—
1995 SOHC 8-valve engine**

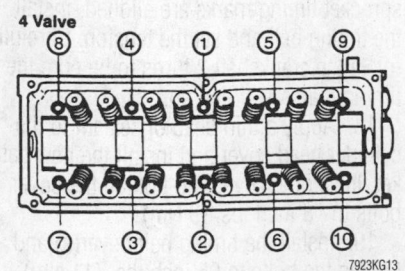

**Cylinder head bolt tightening sequence—
1995 SOHC 16-valve engine**

27. Start the engine and bring to normal operating temperature. Check for leaks. Check the ignition timing and idle speed.

DOHC Engines

1. Relieve the fuel system pressure and disconnect the negative battery cable. Drain the cooling system.

2. Raise and safely support the vehicle.

3. Remove the right front wheel and splash shield.

4. Loosen the water pump pulley attaching bolts.

B6 DOHC Shown
BP DOHC Similar

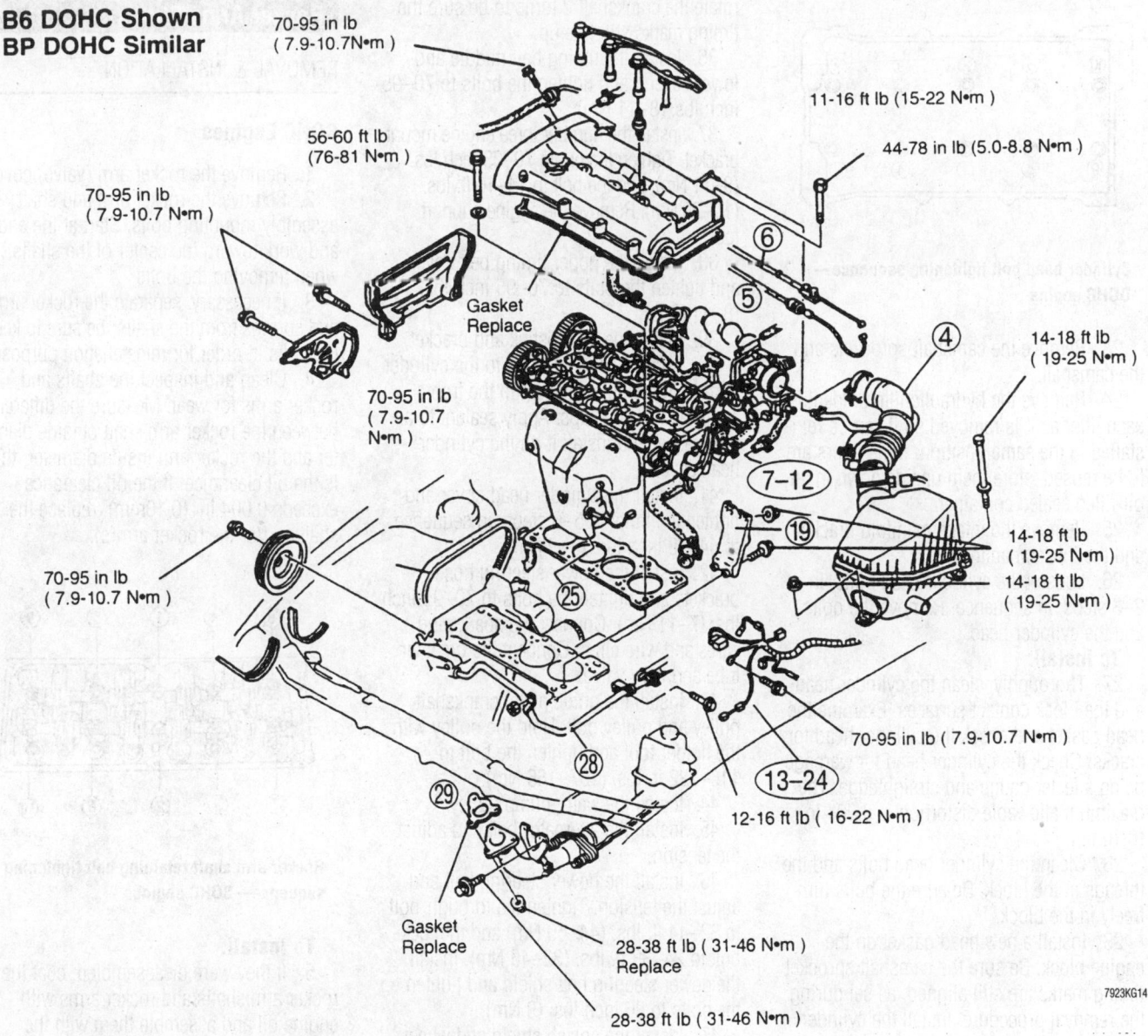

70-95 in lb
(7.9-10.7N•m)

11-16 ft lb (15-22 N•m)

56-60 ft lb
(76-81 N•m)

44-78 in lb (5.0-8.8 N•m)

70-95 in lb
(7.9-10.7 N•m)

Gasket
Replace

6

5

4

14-18 ft lb
(19-25 N•m)

70-95 in lb
(7.9-10.7
N•m)

7-12

19

14-18 ft lb
(19-25 N•m)

14-18 ft lb
(19-25 N•m)

70-95 in lb
(7.9-10.7 N•m)

25

70-95 in lb (7.9-10.7 N•m)

28

13-24

29

12-16 ft lb (16-22 N•m)

Gasket
Replace

28-38 ft lb (31-46 N•m)
Replace

28-38 ft lb (31-46 N•m)

7923KG14

Exploded view of the cylinder head assembly—DOHC engine

5. Remove the power steering belt shield. Loosen the power steering adjusting bolt, lockbolt and through-bolt and remove the power steering belt.

6. Loosen the alternator adjusting bolt and upper mounting bolt. Remove the alternator belt.

7. Remove the water pump pulley.

8. Using a holder tool, hold the crankshaft pulley and remove the pulley bolt. Use a suitable puller to remove the pulley, then remove the guide plate.

9. Remove the power steering hose brackets from the cylinder head cover. Label and disconnect the spark plug wires and wire clips.

10. Disconnect the breather tube and PCV valve from the cylinder head cover.

Remove the bolts, in 2 steps, in the reverse order of the tightening sequence. Remove the cylinder head cover.

11. Remove the oil dipstick and bracket.

12. Remove the timing belt upper cover.

13. Use a suitable engine support tool, and remove the number three engine mount bracket.

14. Remove the timing belt middle and lower covers.

15. Rotate the crankshaft, in the normal direction of rotation, until the No. 1 cylinder piston is at TDC on the compression stroke. Be sure the timing marks on the crankshaft and camshaft sprocket(s) are properly aligned and mark the direction of rotation of the belt.

16. Loosen the timing belt tensioner and

remove the belt. Do not rotate the crankshaft until the timing belt is reinstalled.

17. Remove air cleaner assembly and front pipe.

18. Remove the exhaust manifold.

19. Disconnect the accelerator and throttle cables.

20. Tag and disconnect the spark plug wires from the spark plugs. Remove the spark plugs and the distributor cap and wires assembly. Remove the distributor.

21. Disconnect the following hoses: Heater, brake vacuum, purge, fuel, water and upper radiator.

22. Label and detach the distributor/coil, engine coolant temperature sensor, cooling fan coolant temperature sensor, and temperature gauge sensor connector.

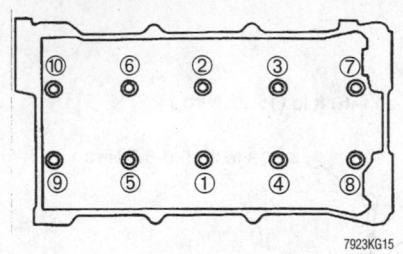

Cylinder head bolt tightening sequence—DOHC engine

23. Remove the camshaft sprockets and the camshaft.

24. Remove the hydraulic lifters. Identify each lifter as it is removed so it can be reinstalled in the same position. If the lifters are to be reused, store them upside down in an oil-filled sealed container.

25. Remove the intake manifold bracket and the intake manifold.

26. Loosen the cylinder head bolts, in 2–3 steps, in sequence. Remove the bolts and the cylinder head.

To install:

27. Thoroughly, clean the cylinder head and the block contact surfaces. Examine the head gasket and check the cylinder head for cracks. Check the cylinder head for warpage using a feeler gauge and straightedge. The maximum allowable distortion is 0.004 in. (0.10mm).

28. Clean the cylinder head bolts and the threads in the block. Be sure the bolts turn freely in the block.

29. Install a new head gasket on the engine block. Be sure the camshaft sprocket timing marks are still aligned, as set during the removal procedure. Install the cylinder head.

30. Lubricate the bolt threads and seat surfaces with clean engine oil and tighten the bolts in 2–3 steps to 56–60 ft. lbs. (75–81 Nm) in the proper sequence.

31. Use new gaskets and install the manifolds. Tighten the intake manifold bolts/nuts to 19 ft. lbs. (25 Nm) and install the intake manifold bracket. Tighten the exhaust manifold nuts to 34 ft. lbs. (46 Nm).

32. Use a new gasket to connect the exhaust pipe and tighten the nuts to 34 ft. lbs. (46 Nm).

33. Apply clean engine oil to the tappets and install them in their original positions.

34. Install the camshaft and sprockets.

35. Be sure the crankshaft and camshaft sprocket timing marks are aligned, install the timing belt and set the tension. Carefully

rotate the crankshaft 2 turns to be sure the timing marks still line up.

36. Install the timing belt middle and lower covers, and tighten the bolts to 70–95 inch lbs. (8–11 Nm).

37. Install the number three engine mount bracket. Tighten the nut to 70–95 inch lbs. (8–11 Nm), and the bolt to 14–16 ft. lbs. (19–22 Nm). Remove the engine support tool.

38. Install the upper timing belt cover and tighten the bolts to 70–95 inch lbs. (8–11 Nm).

39. Install the oil dipstick and bracket.

40. Apply silicone sealant to the cylinder surface in the area adjacent to the front camshaft bearing caps. Apply sealant to a new gasket and install it on the cylinder head cover.

41. Install the cylinder head cover and tighten the bolts in 5–6 steps, in sequence, to 44–78 inch lbs. (5–9 Nm).

42. Install the power steering hose brackets and tighten the bolts to 70–95 inch lbs. (7–11 Nm). Connect the spark plug wires and wire clips. Connect the breather tube and PCV valve.

43. Install the guide plate, crankshaft pulley and pulley bolt. Hold the pulley with the holder tool and tighten the bolt to 116–122 ft. lbs. (157–166 Nm).

44. Install the water pump pulley.

45. Install the alternator belt and adjust the tension.

46. Install the power steering belt and adjust the tension. Tighten the through-bolt to 32–44 ft. lbs. (44–60 Nm) and the lock-bolt to 24–33 ft. lbs. (32–46 Nm). Install the power steering belt shield and tighten the bolts to 86 inch lbs. (9 Nm).

47. Install the splash shield and wheel. Lower the vehicle.

48. Attach the electrical engine harness connectors.

49. Connect the heater, brake vacuum, purge, fuel, water and upper radiator hoses.

50. Install the distributor, spark plugs, distributor cap and wires.

51. Connect and adjust the accelerator and throttle cables.

52. Install the cleaner.

53. Connect the negative battery cable. Fill and bleed the cooling system. Change the engine oil.

54. Adjust the ignition timing.

55. Start the engine and bring to normal operating temperature. Check for leaks. Check the ignition timing and idle speed.

Rocker Arms/Shafts

REMOVAL & INSTALLATION

SOHC Engines

1. Remove the rocker arm (valve) cover.

2. Remove the rocker arm and shaft assembly mounting bolts. Start at the ends and work toward the center of the shafts, when removing the bolts.

3. If necessary, separate the rocker arms and springs from the shafts; be sure to keep the parts in order for reinstallation purposes.

4. Clean and inspect the shafts and rocker arms for wear. Measure the difference between the rocker arm shaft outside diameter and the rocker arm inside diameter; this is the oil clearance. If the oil clearance exceeds 0.004 in. (0.10mm), replace the shaft and/or the rocker arm(s).

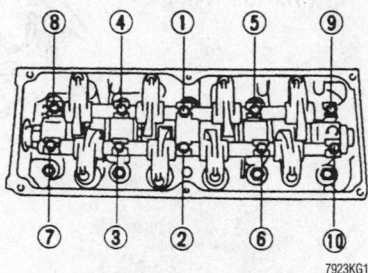

Rocker arm shaft retaining bolt tightening sequence—SOHC engines

To install:

5. If they were disassembled, coat the rocker arm shafts and rocker arms with engine oil and assemble them with the springs. When assembling and installing on the cylinder head, note the notches at the ends of the shafts; they are different on the intake and exhaust side and cannot be interchanged.

6. Install the rocker arm/shaft assemblies onto the cylinder head and tighten the rocker arm shaft-to-cylinder head bolts, in sequence, to 16–21 ft. lbs. (22–28 Nm), in several steps.

7. Install the rocker arm (valve) cover.

Intake Manifold

REMOVAL & INSTALLATION

1. Properly relieve the fuel system pressure. Disconnect the negative battery cable and drain the cooling system.

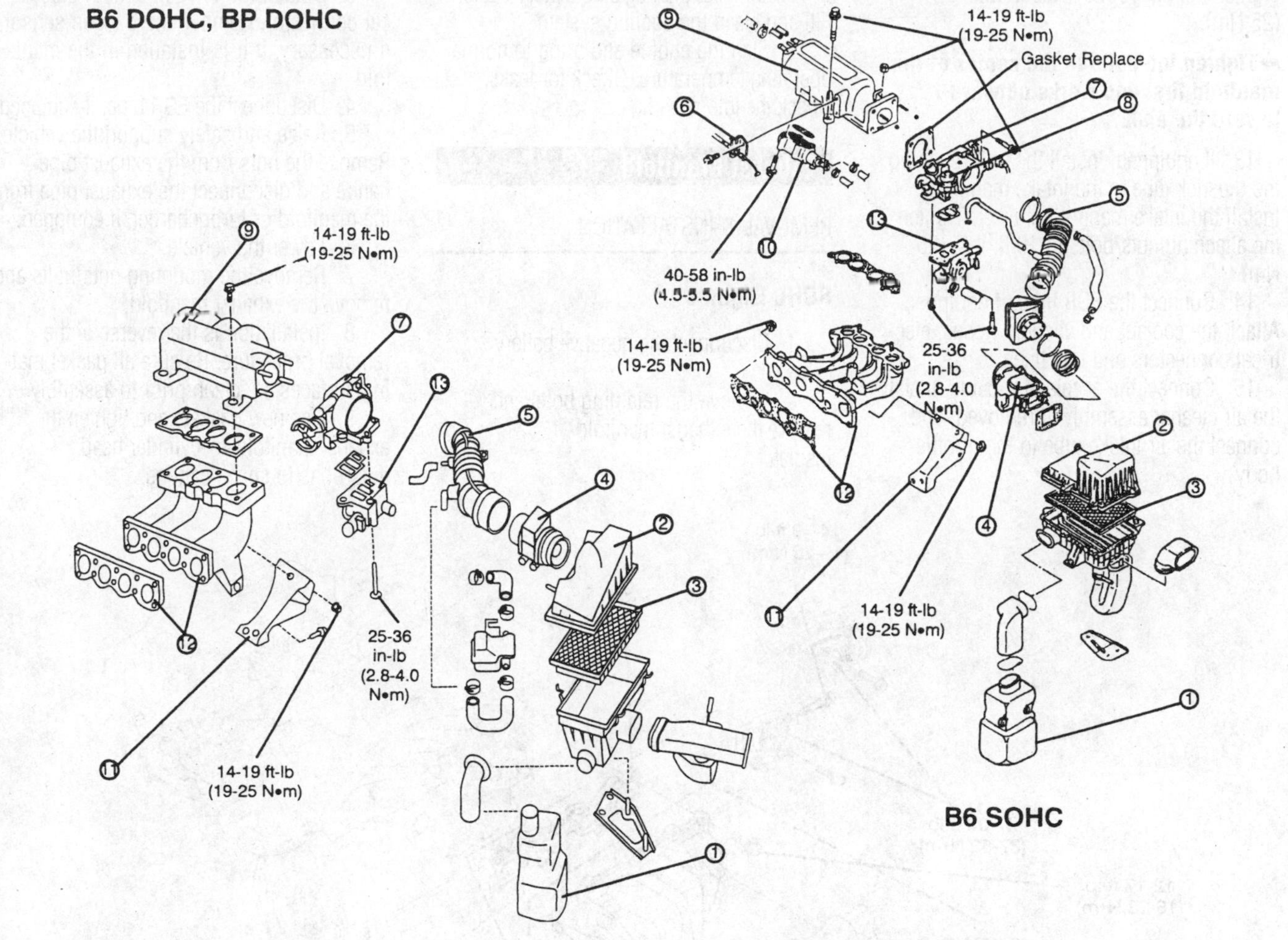

B6 DOHC, BP DOHC

14-19 ft-lb
(19-25 N•m)
Gasket Replace

40-58 in-lb
(4.5-5.5 N•m)

14-19 ft-lb
(19-25 N•m)

25-36
in-lb
(2.8-4.0
N•m)

14-19 ft-lb
(19-25 N•m)

14-19 ft-lb
(19-25 N•m)

25-36
in-lb
(2.8-4.0
N•m)

14-19 ft-lb
(19-25 N•m)

B6 SOHC

7923KG17

1. Resonance chamber
2. Upper air filter housing
3. Air filter (B6 SOHC)
4. Mass air flow (MAF) sensor (B6 DOHC, BP DOHC)/
   Volume air flow (VAF) sensor (B6 SOHC)
5. Intake air hose
6. Throttle cable
7. Throttle body
8. Dashpot (B6 SOHC)
9. Dynamic chamber
10. Air valve (B6 SOHC)
11. Intake manifold support bracket
12. Intake manifold and gasket (Replace)
13. Idle air control valve (B6 SOHC)/
    Bypass air control (BAC) valve (B6 DOHC, BP DOHC)

Exploded view of the intake manifold assembly

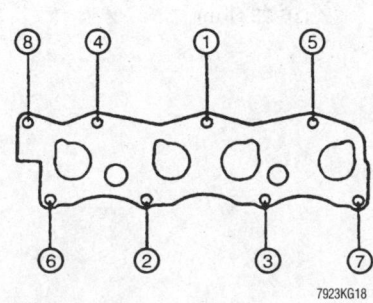

7923KG18

Intake manifold tightening sequence

2. Disconnect the air intake hose from the throttle body. Remove the hose and air cleaner assembly if necessary.

3. Disconnect the accelerator cable. Disconnect and plug the fuel lines.

4. Label and detach all necessary vacuum hoses and electrical connectors. Disconnect the coolant hoses.

5. Disconnect the EGR tube, if equipped.

6. If equipped, remove the air valve and remove the fuel rail attaching bolts. Remove the fuel rail and injectors as an assembly.

7. Remove the intake manifold support bracket. If necessary, remove the bolt retaining the dipstick tube bracket to the intake manifold.

8. Remove the intake manifold-to-cylinder bolts/nuts and remove the intake manifold assembly.

9. If necessary, remove the throttle body and separate the intake manifold upper and lower halves.

To install:

10. Clean all gasket mating surfaces.

11. If separated, connect the upper and lower intake manifolds using a new gasket. Tighten the nuts/bolts to 19 ft. lbs. (25 Nm). If removed, install the throttle body using a new gasket. Tighten the retaining nuts/bolts to 19 ft. lbs. (25 Nm).

12. Install the intake manifold assembly to the cylinder head using a new gasket.

Tighten the nuts/bolts to 19 ft. lbs. (25 Nm).

➡**Tighten the bolts in the center of the manifold first and works outward toward the ends.**

13. If equipped, install the bolt retaining the dipstick tube to the intake manifold. Install the intake manifold bracket. Tighten the attaching nuts/bolts to 19 ft. lbs. (25 Nm).

14. Connect the EGR tube, if equipped. Attach the coolant and vacuum hoses, electrical connectors and fuel lines.

15. Connect the accelerator cable. Install the air cleaner assembly, if removed, and connect the air intake tube to the throttle body.

16. Connect the negative battery cable. Fill and bleed the cooling system.

17. Start the engine and bring to normal operating temperature. Check for leaks. Check the idle speed.

Exhaust Manifold

REMOVAL & INSTALLATION

SOHC Engines

1. Disconnect the negative battery cable.

2. Remove the retaining bolts and remove the exhaust manifold insulator.

3. Detach the oxygen sensor electrical connector. Remove the oxygen sensor, if necessary, if it is installed in the manifold.

4. Disconnect the EGR pipe, if equipped.

5. Raise and safely support the vehicle. Remove the nuts from the exhaust pipe flange and disconnect the exhaust pipe from the manifold or turbocharger, if equipped.

6. Lower the vehicle.

7. Remove the mounting nuts/bolts and remove the exhaust manifold.

8. Installation is the reverse of the removal procedure. Be sure all gasket mating surfaces are clean prior to assembly.

9. Use new gaskets and tighten the exhaust manifold-to-cylinder head nuts/bolts to specifications.

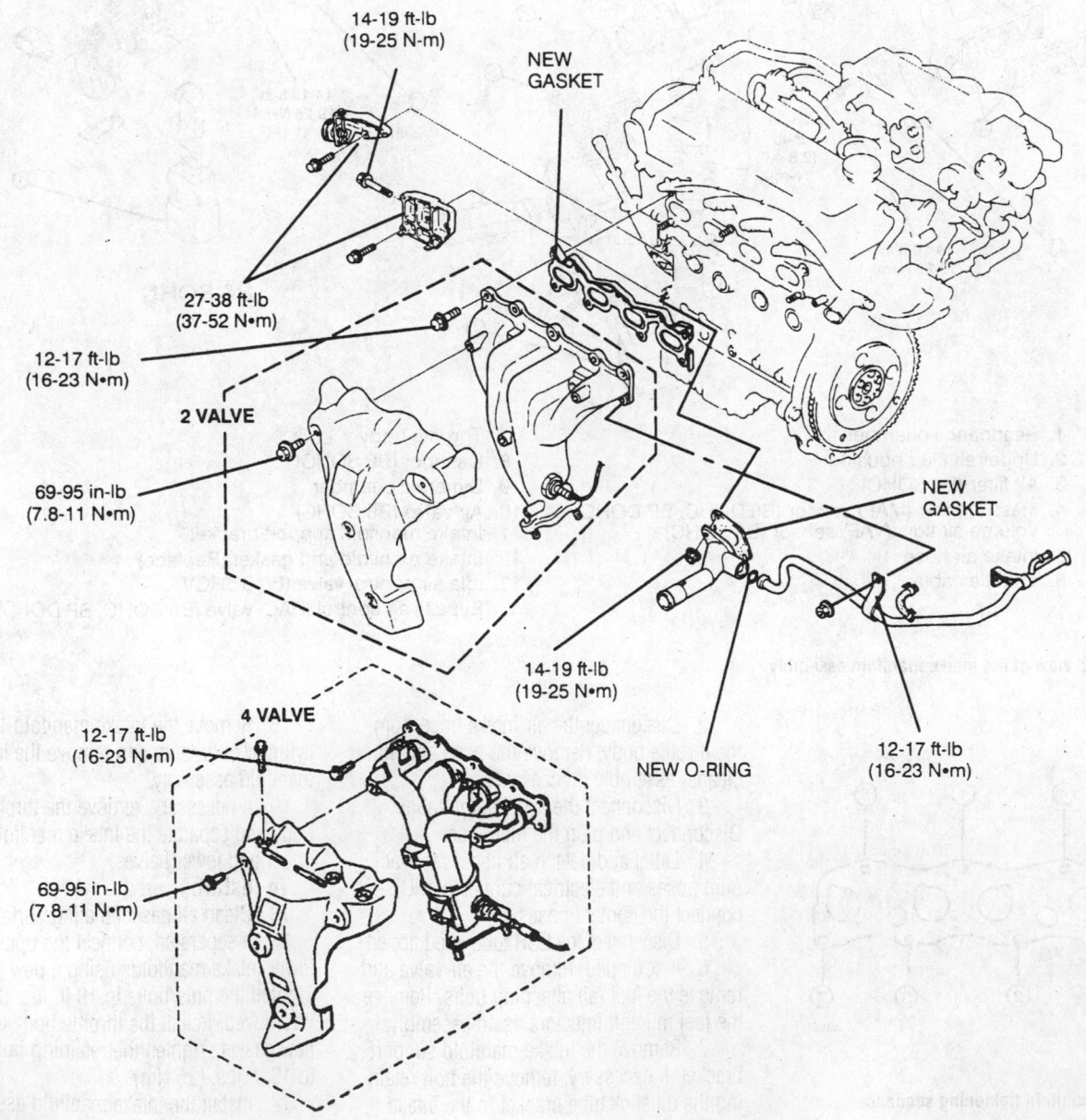

14-19 ft-lb (19-25 N·m)

NEW GASKET

27-38 ft-lb (37-52 N·m)

12-17 ft-lb (16-23 N·m)

2 VALVE

69-95 in-lb (7.8-11 N·m)

NEW GASKET

14-19 ft-lb (19-25 N·m)

NEW O-RING

12-17 ft-lb (16-23 N·m)

4 VALVE

12-17 ft-lb (16-23 N·m)

69-95 in-lb (7.8-11 N·m)

Exploded view of the exhaust manifold mounting—SOHC engines

7923KG19

10. Use a new gasket and tighten the exhaust pipe flange-to-exhaust manifold nuts to 34 ft. lbs. (46 Nm).

DOHC Engines

1. Disconnect the negative battery cable.
2. Remove the air cleaner, and disconnect the air hose.
3. Remove the water bypass pipe bolt.
4. Remove the exhaust manifold heat shield bolts and the heat shield.
5. Detach the oxygen sensor electrical connector.
6. Raise and safely support the vehicle.
7. Remove and discard the exhaust pipe-to-exhaust manifold nuts. Suspend the exhaust system with wire.
8. Disconnect the EGR pipe from the exhaust manifold and lower the vehicle.
9. Remove the nuts and bolts and remove the exhaust manifold. Discard the nuts.
   To install:
10. Clean all gasket mating surfaces.
11. Position a new exhaust manifold gasket over the studs and install the exhaust manifold. Tighten the mounting nuts and bolts to 29–34 ft. lbs. (39–47 Nm).
12. Raise and safely support the vehicle.
13. Connect the exhaust pipe to the manifold. Install new nuts and tighten to 38 ft. lbs. (52 Nm).
14. Attach the oxygen sensor connector.
15. Connect the EGR pipe to the back of the exhaust manifold and tighten to 34 ft. lbs. (47 Nm). Lower the vehicle.
16. Install the heat shield and tighten the bolts to 88 inch lbs. (10 Nm).
17. Install the water bypass pipe bolt, and tighten to 48–65 ft. lbs. (64–89 Nm).
18. Connect the air hose, and install the air cleaner.
19. Connect the negative battery cable.

Front Crankshaft Seal

REMOVAL & INSTALLATION

1. Disconnect the negative battery cable.
2. Raise and safely support the vehicle. On some models it may be necessary to remove the right front tire and splash shield.
3. Remove the crankshaft pulley.
4. Remove the timing belt covers and belt.
5. Remove the crankshaft pulley boss and sprocket. It may be necessary to use a

puller to remove the sprocket. Remove the key from the sprocket.
6. Protect the oil pump housing with a rag. Cut the oil seal lip with a knife. Using a small prybar, pry the oil seal from the engine block; be careful not to score the crankshaft or the seal seat.
   To install:
7. Lubricate the seal lip with clean engine oil and push the seal slightly in by hand.
8. Tap the seal in evenly using a seal installer. Install the seal until it is flush with the oil pump body.
9. Install the crankshaft sprocket. Install the sprocket key with the tapered side toward the oil pump body.
10. Install the remaining components in the reverse order of removal.

Camshaft

REMOVAL & INSTALLATION

The engines covered are not equipped with valve lifters, they utilize a Hydraulic Lash Adjuster (HLA) in the end of the rocker arm, on the SOHC engine, and in the bucket style on the DOHC engine.

SOHC Engines

➡**The camshaft is removed through the front of the cylinder head.**

1. Remove the cylinder head from the vehicle and position in a suitable holding fixture.

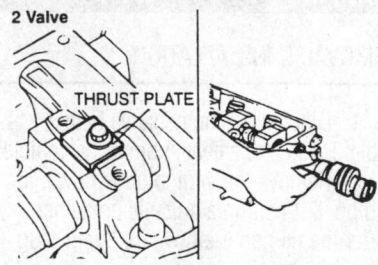

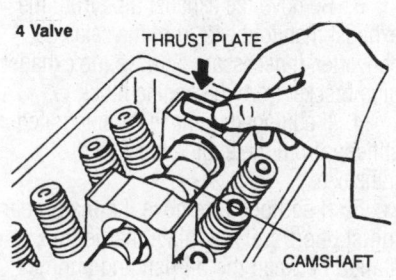

Remove the thrust plate, then slide the camshaft out of the cylinder head—SOHC engines

➡**Do not lay the cylinder head flat on the head gasket surface as the valves may be damaged.**

2. Hold the camshaft with a wrench on the hexagon cast into the front of the camshaft.
3. Remove the sprocket bolt and the sprocket.
4. Loosen the rocker arm shaft bolts in 2–3 steps, in the reverse of the torque sequence. Remove the rocker arm and shaft assemblies.
5. Pry out the camshaft seal using a small prybar, being careful not to damage the camshaft or seal bore.
6. Remove the thrust plate at the rear of the cylinder head.
7. Carefully slide the camshaft from the cylinder head, being careful not to damage the cylinder head bearing surfaces.
   To install:
8. Lubricate the camshaft lobes and journals and the cylinder head bearing surfaces with clean engine oil.
9. Carefully slide the camshaft into the cylinder head, being careful not to damage the bearing surfaces.
10. Install the camshaft thrust plate. On the 8-valve engine, tighten the thrust retaining bolt to 95 inch lbs. (11 Nm). On the 16-valve engine, the thrust plate is held in place by the rocker arm and shaft assembly.
11. Lubricate the lip of a new camshaft seal with clean engine oil and install in the cylinder head, using a seal installer.
12. Lubricate the rocker arms and valve stem tips with clean engine oil. Install the rocker arm and shaft assemblies and tighten the bolts, in 2–3 steps, in the proper sequence. The final torque should be 21 ft. lbs. (28 Nm).
13. Install the camshaft sprocket and retaining bolt. Hold the camshaft with the wrench on the hexagon and tighten the bolt to 45 ft. lbs. (61 Nm).
14. Install the cylinder head and the remaining components in the reverse order of removal.

DOHC Engines

1. Disconnect the negative battery cable.
2. Label and disconnect the spark plug wires and remove the spark plugs.
3. Disconnect the hoses from the cylinder head cover, if equipped.
4. Remove the cylinder head cover bolts and remove the cylinder head cover.
5. Remove the timing belt and the distributor.
6. Hold the camshaft with a wrench on

the hexagon cast into the camshaft. Remove the sprocket bolts and remove the sprockets.

7. Label the caps so they can be reinstalled in their original positions. Loosen the camshaft cap bolts in 2–3 steps in the reverse of the torque sequence, then remove the camshaft caps.

8. Remove the camshafts. Remove the camshaft oil seals from the camshafts.

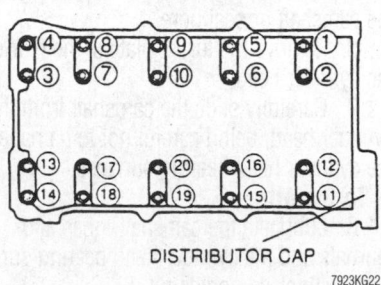

Camshaft bearing cap bolt loosening sequence—DOHC engine

To install:

9. Lubricate the camshaft journals and lobes with clean engine oil. Install the camshafts in the cylinder head.

10. Apply silicone sealant to the cylinder head on the front camshaft cap mating surfaces. Do not allow any sealant on the camshaft journals.

11. Install the camshaft caps in their original positions. Loosely install the cap bolts.

12. Tighten the camshaft cap bolts in 2–3 steps to 125 inch lbs. (14 Nm) in the proper sequence.

13. Apply clean engine oil to the lip of a new camshaft seal. Push the seal slightly in by hand. Tap the seal into position, using a

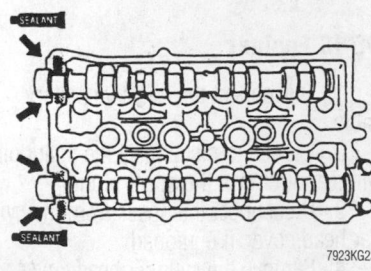

Apply silicone sealant to the cylinder head in the positions shown—DOHC engine

seal installer, until it is flush with the edge of the camshaft cap.

14. Turn the camshafts until the dowel pins face straight up. Install the camshaft sprockets and the sprocket bolts.

15. Hold the camshaft with the wrench on the cast hexagon and tighten the sprocket bolts to 44 ft. lbs. (60 Nm).

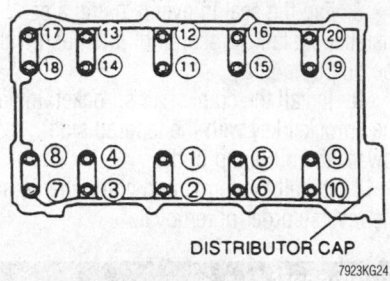

Camshaft bearing cap bolt tightening sequence—DOHC engine

16. Install the remaining components in the reverse order of removal.

Valve Lash

ADJUSTMENT

The valve lash on all engines is kept in adjustment hydraulically. No adjustment is necessary, or possible.

Oil Pan

REMOVAL & INSTALLATION

1. Disconnect the negative battery cable. Raise and safely support the vehicle.

2. Remove the engine undercover, if equipped. Position a suitable container under the oil pan. Remove the drain plug and drain the oil.

3. Remove the exhaust pipe from the exhaust manifold and from the catalytic converter. If necessary, remove the exhaust pipe bracket from the engine block.

4. If equipped, remove the integrated stiffener from the engine block and transaxle.

5. If equipped, remove the main bearing support/stiffener plate that is installed between the oil pan and engine block.

6. Remove the bolts and remove the oil

pan. It may be necessary to pry the pan away from the engine; be careful not to damage the gasket contact surfaces.

7. If necessary remove the oil strainer.

8. Remove the main bearing support/stiffener plate that is installed between the oil pan and engine block.

To install:

9. Clean all oil, dirt, old gasket material and sealer from the oil pan, support/stiffener plate, oil pan bolts and all gasket mating surfaces. If removed, clean the oil strainer.

10. If equipped with the main bearing support/stiffener plate, run a bead of silicone sealer around the perimeter of the plate, going inside the bolt holes. Install the plate and tighten the bolts.

➡**Be sure all old sealer is removed from the bolts prior to installation. Installing a bolt coated with old sealer could result in cracking of the bolt holes.**

11. If removed, install the oil strainer using a new gasket. Tighten the bolts.

12. If used, apply silicone sealer to new rubber end gaskets and press them into place on the engine.

13. Apply a bead of silicone to the perimeter of the oil pan, going around the inside of the bolt holes and install the pan to the engine. Install the oil pan bolts finger-tight.

14. Tighten the oil pan bolts.

➡**Be sure all old sealer is removed from the bolts prior to installation. Installing a bolt coated with old sealer could result in cracking of the bolt holes.**

15. If removed, install the integrated stiffener to the engine block and transaxle. Tighten the bolts to 38 ft. lbs. (52 Nm).

16. If removed, install the transverse member. Tighten the bolts to 93 ft. lbs. (126 Nm).

17. Install the front exhaust pipe bracket, if equipped. Install the front exhaust pipe, using new gaskets. Tighten the exhaust manifold flange nuts to 34 ft. lbs. (46 Nm).

18. Install the oil pan drain plug using a new gasket. Tighten the drain plug to 30 ft. lbs. (41 Nm).

19. Install the engine undercover and lower the vehicle.

20. Fill the engine with the proper type and quantity of oil.

21. Connect the negative battery cable. Start the engine and bring to normal operating temperature. Check for leaks.

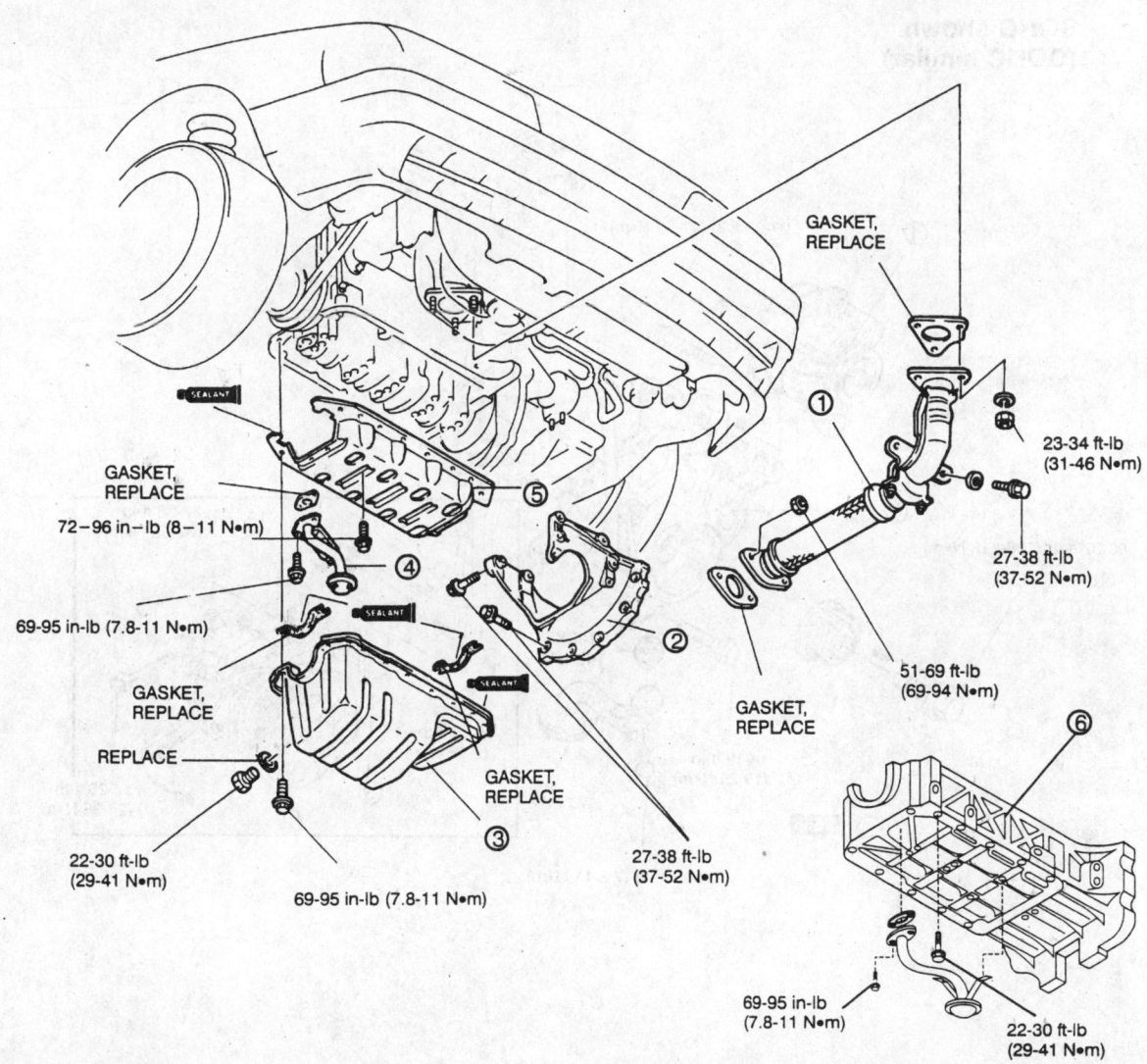

GASKET,
REPLACE

GASKET,
REPLACE

72–96 in–lb (8–11 N•m)

69-95 in-lb (7.8-11 N•m)

GASKET,
REPLACE

REPLACE

22-30 ft-lb
(29-41 N•m)

69-95 in-lb (7.8-11 N•m)

GASKET,
REPLACE

27-38 ft-lb
(37-52 N•m)

23-34 ft-lb
(31-46 N•m)

27-38 ft-lb
(37-52 N•m)

51-69 ft-lb
(69-94 N•m)

GASKET,
REPLACE

69-95 in-lb
(7.8-11 N•m)

22-30 ft-lb
(29-41 N•m)

1. Front exhaust pipe.
2. Integrated Stiffener (B6 SOHC)
3. Oil pan
4. Oil strainer
5. Main bearing support plate (B6 SOHC, BP DOHC)
6. Oil Pan Upper Block (B6 DOHC)

7923KG25

Exploded view of the oil pan and related components

Oil Pump

REMOVAL & INSTALLATION

1. Disconnect the negative battery cable.

2. Remove the crankshaft pulley, timing belt cover, belt and the crankshaft sprocket.

3. Remove the oil pan.

4. Remove the oil pick-up tube and discard the gasket.

5. Remove the oil pump attaching bolts and remove the oil pump.

6. Remove the front crankshaft seal from the oil pump if the pump is being replaced.

To install:

7. Clean the oil, dirt and old sealant from all contact surfaces. Replace the O-rings on the oil pump.

8. If the oil seal was removed from the oil pump, apply clean engine oil to the lip of the seal. Push the seal in lightly be hand. Press the seal, with a protrusion of 0.02–0.04 in. (0.5–1.0mm), into the oil pump with a suitable tool (49 B014 401 or equivalent).

9. Apply a bead of silicone to the oil pump at the cylinder block contact surface, going inside the bolt holes.

10. Install the oil pump and tighten the bolts to 14–18 ft. lbs. (19–25 Nm).

11. Install a new gasket and the oil pump pick-up tube. Tighten the mounting bolts to 70–95 inch lbs. (8–11 Nm).

12. Install the oil pan.

13. Install the crankshaft sprocket, timing belt and cover.

14. Install the crankshaft pulley.

15. Connect the negative battery cable.

16. Fill the engine with the proper type and quantity of oil. Run the engine and check for leaks.

**SOHC shown
(DOHC similar)**

27-38 ft-lb (37-52 N•m)

GASKET,
REPLACE

27-38 ft-lb
(37-52 N•m)

80-87 ft-lb (108-118 N•m)

14-19 ft-lb
(19-25 N•m)

GASKET,
REPLACE

14-19 ft-lb
(19-25 N•m)

17-26 ft-lb
(24-35 N•m)

SST

69-95 in-lb (7.8-11 N•m)

1. Generator
2. A/C compressor (if equipped)
3. A/C compressor bracket (if equipped)
4. Crankshaft pulley lock bolt
5. Timing belt pulley
6. Oil strainer
7. Oil pump

7923KG26

Exploded view of the oil pump

Rear Main Seal

REMOVAL & INSTALLATION

1. Disconnect the negative battery cable.
2. Raise and safely support the vehicle.
3. Remove the transaxle assembly.
4. If equipped with a manual transaxle, remove the clutch and flywheel assembly.
5. If equipped with an automatic transaxle, remove the flexplate-to-crankshaft bolts, the flexplate and shim plates.
6. Cut the oil seal lip with a knife. Install a rag to the housing and using a screwdriver, carefully pry the oil seal from

the oil seal housing. Clean the gasket mounting surfaces.

To install:

7. Clean the oil seal housing. Coat the oil seal and the housing with clean engine oil.
8. Press the oil seal into the housing and tap it evenly into place with a hammer and a large diameter piece of pipe. The seal must be flush with the edge of the rear cover.
9. Install flywheel assembly or the flexplate, as applicable, and tighten the mounting bolts to 71–76 ft. lbs. (97–102 Nm).
10. If applicable, install the clutch assembly.
11. Install the transaxle, lower the vehicle and connect the negative battery cable.

FUEL SYSTEM

Fuel System Service Precautions

Safety is the most important factor when performing not only fuel system maintenance but any type of maintenance. Failure to conduct maintenance and repairs in a safe manner may result in serious personal injury or death. Maintenance and testing of the vehicle's fuel system components can be accomplished safely and effectively by adhering to the following rules and guidelines.

• To avoid the possibility of fire and personal injury, always disconnect the negative battery cable unless the repair or test procedure requires that battery voltage be applied.

• Always relieve the fuel system pressure prior to disconnecting any fuel system component (injector, fuel rail, pressure regulator, etc.), fitting or fuel line connection. Exercise extreme caution whenever relieving fuel system pressure to avoid exposing skin, face and eyes to fuel spray. Please be advised that fuel under pressure may penetrate the skin or any part of the body that it contacts.

• Always place a shop towel or cloth around the fitting or connection prior to loosening to absorb any excess fuel due to spillage. Ensure that all fuel spillage (should it occur) is quickly removed from engine surfaces. Ensure that all fuel soaked cloths or towels are deposited into a suitable waste container.

• Always keep a dry chemical (Class B) fire extinguisher near the work area.

• Do not allow fuel spray or fuel vapors to come into contact with a spark or open flame.

• Always use a back-up wrench when loosening and tightening fuel line connection fittings. This will prevent unnecessary stress and torsion to fuel line piping.

• Always replace worn fuel fitting O-rings with new. Do not substitute fuel hose or equivalent, where fuel pipe is installed.

Fuel System Pressure

RELIEVING

❊❊ CAUTION

The fuel injection system remains under pressure after the engine has been turned OFF. Properly relieve fuel pressure before disconnecting any fuel lines. Failure to do so may result in fire or personal injury. Do not allow fuel spray or fuel vapors to come in contact with a spark or open flame. Keep a dry chemical fire extinguisher nearby. Never store fuel in an open container due to risk of fire or explosion.

1. Release the rear seat retainers (clips or catches) and remove rear seat cushion.
2. Remove the fuel pump cover.
3. Detach the fuel pump electrical connector.

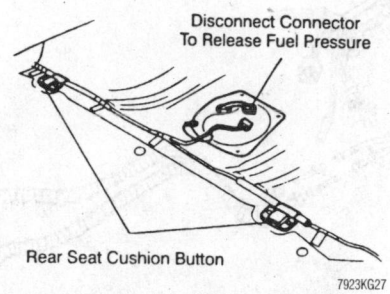

Disconnect Connector
To Release Fuel Pressure

Rear Seat Cushion Button

7923KG27

Be sure to leave the fuel pump connector detached while performing the service procedure

4. Start the engine, allowing it to idle until it runs out of fuel.
5. After the engine stalls, reattach the fuel pump connector and turn the ignition switch **OFF**.

Fuel Filter

REMOVAL & INSTALLATION

1. Relieve the fuel system pressure.
2. Disconnect the negative battery cable.
3. Remove the air intake hose.
4. Remove the two nuts an the fuel filter bracket, and remove the fuel tube clamps.
5. Disconnect the fuel lines from both ends of the fuel filter. Plug the lines to prevent leakage.
6. Remove the filter from the mounting bracket.
To install:
7. Position the filter in the mounting bracket.
8. Unplug the fuel lines and connect them to the filter.
9. Install the fuel tube clamps, and install and tighten the bracket nuts to 70–95 inch lbs. (8–11 Nm).
10. Install the air intake hose.
11. Connect the negative battery cable.
12. Run the engine and check for any fuel leaks.

Fuel Pump

REMOVAL & INSTALLATION

❊❊ CAUTION

Do not allow fuel spray or fuel vapors to come in contact with a spark or open flame. Keep a dry chemical fire extinguisher nearby. Never store fuel in an open container due to risk of fire or explosion.

1. Relieve the fuel system pressure.
2. Disconnect the negative battery cable.
3. Remove the rear seat cushion from the vehicle.
4. Remove any dirt that has accumulated around the fuel pump cover so it will not enter the tank during pump removal and installation.
5. Remove the fuel pump cover.
6. Detach the fuel gauge connector, hoses, and the gauge.
7. Detach the fuel pump electrical connector.
8. Remove the fuel pump from the bracket assembly. Remove and discard the seal ring.
To install:
9. Clean the fuel pump mounting flange, fuel tank mounting surface and seal ring groove.
10. Apply a light coating of grease on a new seal ring to hold it in place during assembly and install in the seal ring groove.
11. Install the fuel pump to the bracket assembly carefully to ensure the filter is not damaged. Be sure the seal ring remains in the groove.
12. Hold the pump assembly in place, and pull the fuel pump down so that it is tight against the bracket.
13. Attach the fuel pump electrical connector.
14. Install the fuel gauge, hoses, and gauge connector.
15. Install the fuel pump cover.
16. Install the rear seat cushion.
17. Connect the negative battery cable, start the engine and check for proper system operation and for fuel leaks.

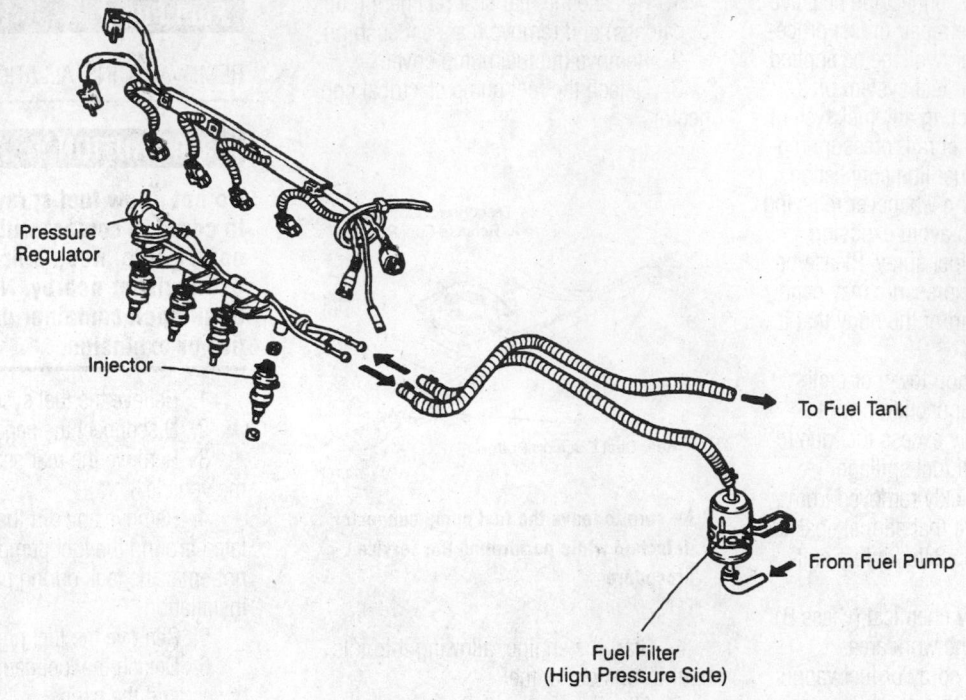

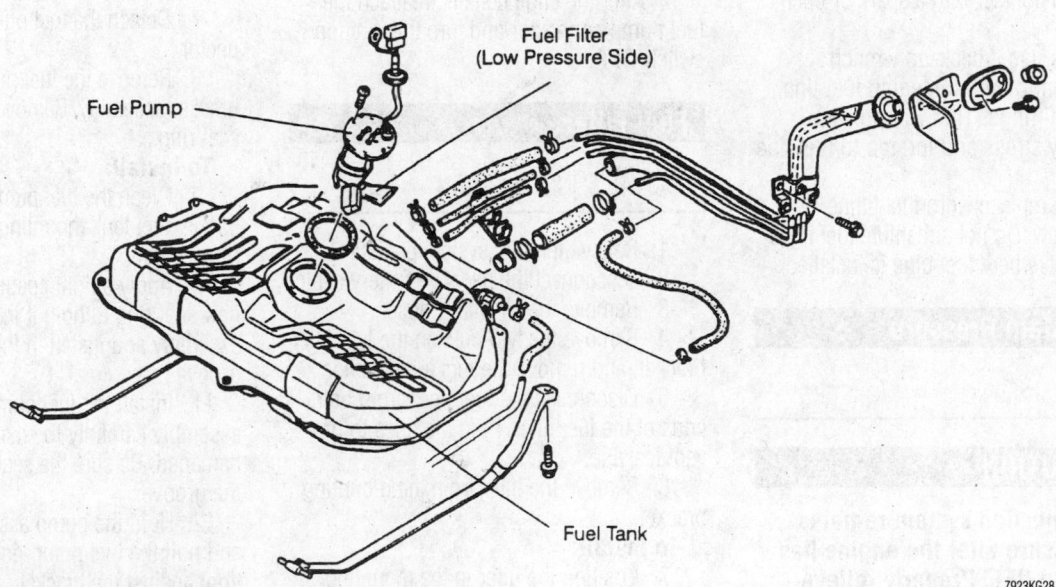

Pressure Regulator

Injector

To Fuel Tank

From Fuel Pump

Fuel Filter (High Pressure Side)

Fuel Filter (Low Pressure Side)

Fuel Pump

Fuel Tank

7923KG28

Exploded view of the fuel system

DRIVE TRAIN

Transaxle Assembly

REMOVAL & INSTALLATION

Manual

1. Drain the transaxle oil.
2. Remove the battery box and the battery.
3. Remove the air cleaner assembly.
4. Remove the battery carrier.
5. Disconnect the back-up light switch and remove the bracket.
6. Detach the neutral switch connector and the vehicle speedometer sensor connector.
7. Remove the harness bracket.
8. Remove the wheels.
9. Remove the splash shield.
10. Remove the transverse member.
11. Remove the extension bar and the change control rod.
12. Disconnect the tie-rod ends.
13. Remove the stabilizer control link.
14. Remove the halfshaft and the joint shaft.
15. Remove the intake manifold bracket.
16. Remove the starter.
17. Support the engine and remove the engine mounting member.
18. Remove the rear engine/transmission mount.
19. Remove the front engine/transmission mount.
20. Remove the clutch release cylinder.
21. Remove the side engine/transmission mount.
22. Support the transaxle on a jack and

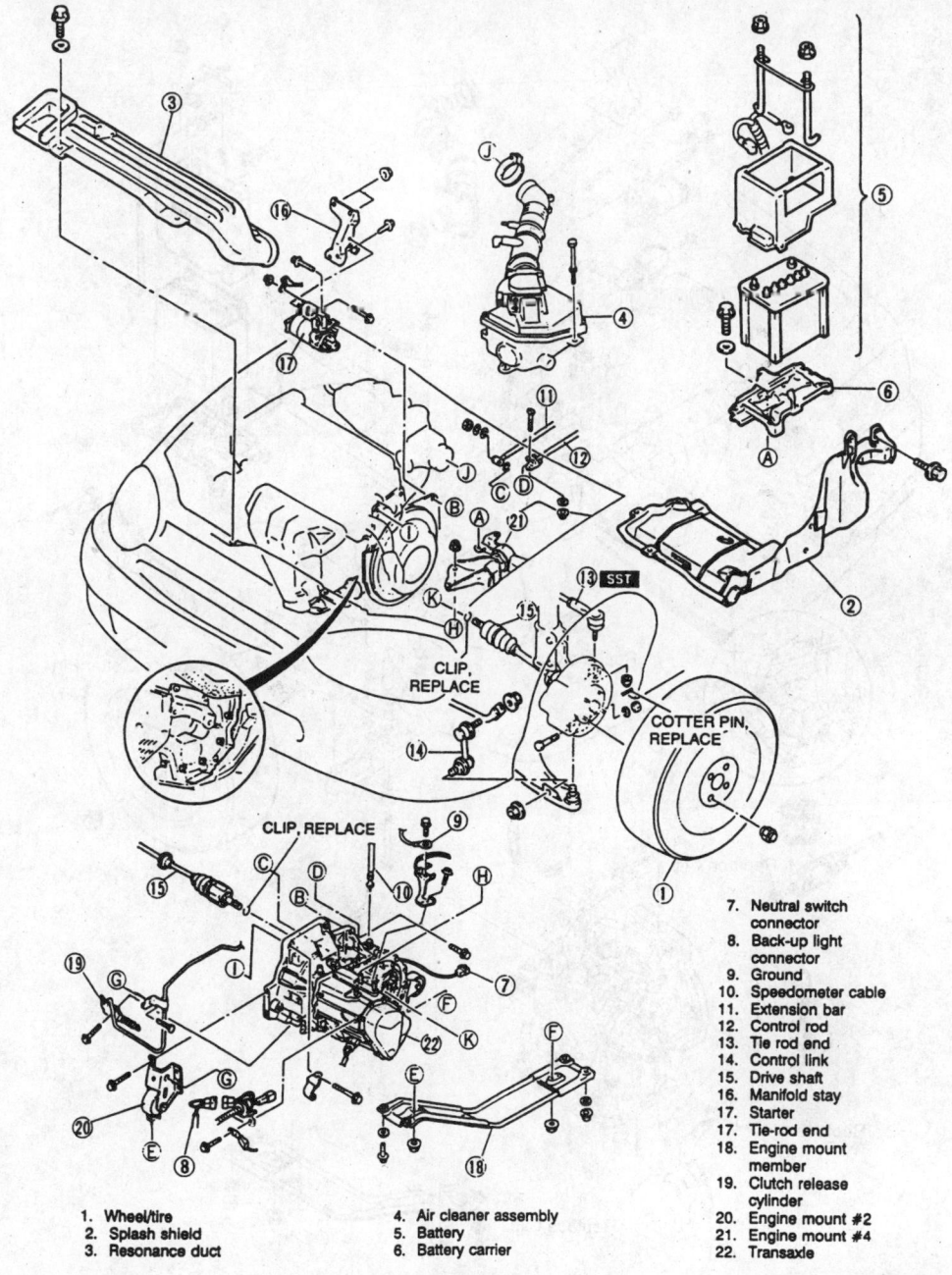

CLIP, REPLACE

COTTER PIN, REPLACE

CLIP, REPLACE

1. Wheel/tire
2. Splash shield
3. Resonance duct
4. Air cleaner assembly
5. Battery
6. Battery carrier
7. Neutral switch connector
8. Back-up light connector
9. Ground
10. Speedometer cable
11. Extension bar
12. Control rod
13. Tie rod end
14. Control link
15. Drive shaft
16. Manifold stay
17. Starter
17. Tie-rod end
18. Engine mount member
19. Clutch release cylinder
20. Engine mount #2
21. Engine mount #4
22. Transaxle

7923KG29

Exploded view of the manual transaxle assembly mounting and related components

remove the transmission mounting bolts.

23. Remove the transaxle.

To install:

24. Place the transaxle into position and install the mounting bolts. Tighten to 48–65 ft. lbs. (64–89 Nm).

25. Remove the support jack.

26. Install the side mount. Tighten the body side nuts and bolts to 32–44 ft. lbs. (44–60 Nm). Tighten the transmission side nuts to 50–68 ft. lbs. (67–93 Nm).

27. Install the clutch release cylinder.

28. Install the front mount, loosely tighten the mount nut and bolt.

29. Install the rear mount, align and set all bolts, then tighten to 50–68 ft. lbs. (67–93 Nm).

30. Install the engine mounting member. Tighten the four outer nuts and bolts to 50–65 ft. lbs. (67–89 Nm) and the two remaining nuts to 28–38 ft. lbs. (38–51 Nm).

31. Tighten the front mount nut and bolt to 50–68 ft. lbs. (67–93 Nm).

32. Install the starter.

33. Install the manifold bracket.

34. Install the joint shaft and the half-shaft.

35. Install the stabilizer control link.

36. Connect the tie-rod ends.

37. Connect the change control rod and the extension bar.

38. Install the transverse member.

39. Install the splash shield.

40. Install the wheels.

41. Install the harness bracket.

42. Install the vehicle speedometer sensor and the neutral switch connectors.

43. Install the back-up light switch connector bracket and the switch.

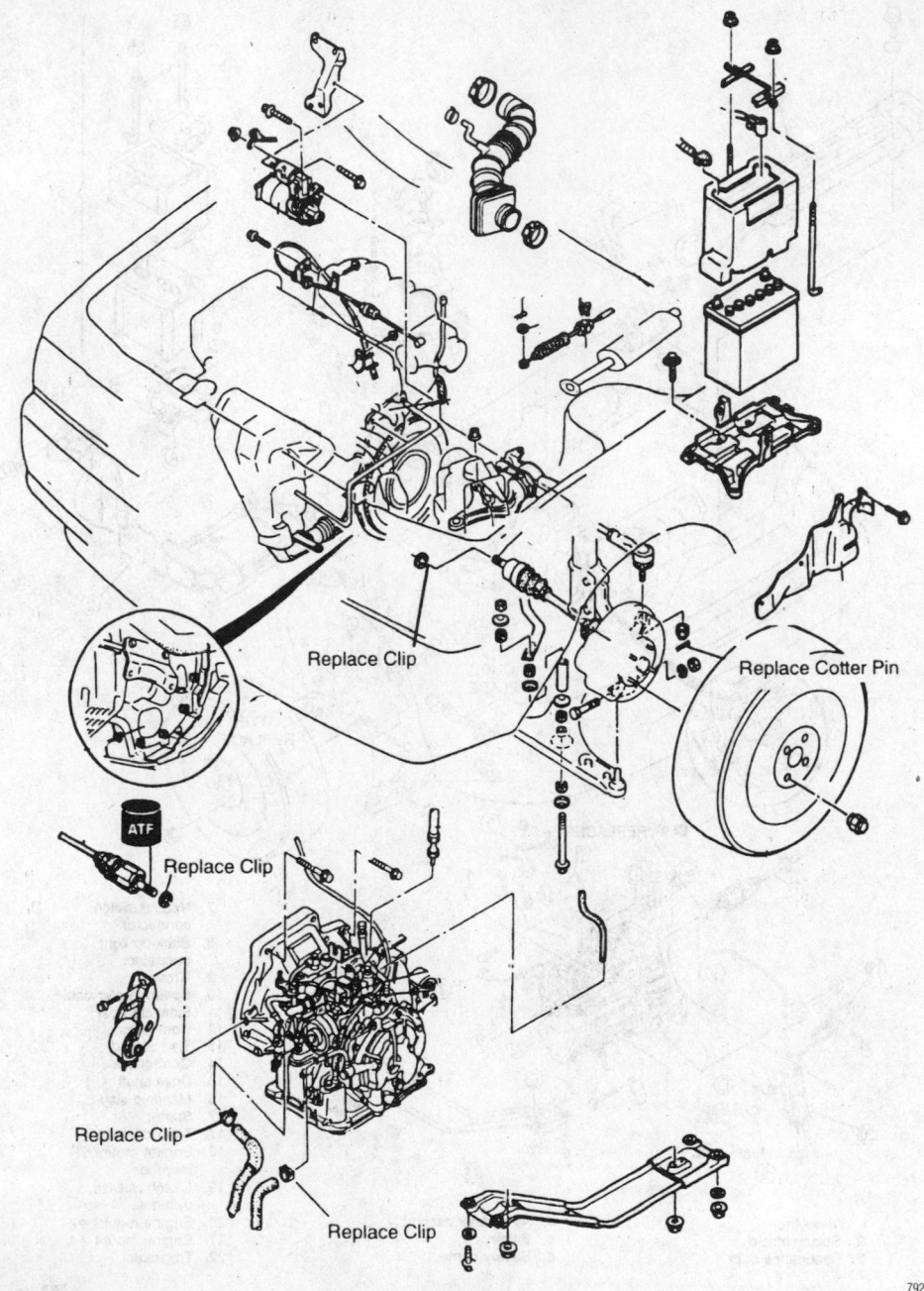

Replace Clip

Replace Cotter Pin

ATF

Replace Clip

Replace Clip

Replace Clip

7923KG30

Exploded view of the automatic transaxle assembly mounting and related components

44. Install the battery carrier.
45. Install the air cleaner assembly.
46. Install the battery and battery box.
47. Fill the transmission with the proper fluid.
48. Check for proper clutch operation.

Automatic

1. Raise the vehicle on a hoist.
2. Drain the transaxle fluid.
3. Remove the battery and battery cover.
4. Remove the air cleaner assembly.

5. Remove the battery carrier.
6. Detach the solenoid and the transaxle range switch connectors.
7. Disconnect the selector cable.
8. Detach the speedometer sensor connector.
9. Remove the harness bracket.
10. Disconnect the throttle cable.
11. Remove the front wheels.
12. Remove the splash shields.
13. Remove the transverse member, if equipped.
14. Disconnect the tie-rod ends.

15. Disconnect the stabilizer control links.
16. Disconnect the lower arm by removing the cinch bolt from the lower arm ball joints. Pry the lower arm out of the knuckle.
17. Support the engine and remove the engine mounting member.
18. Remove the left and right halfshaft and install Differential Side Gear holder K49A-4208-AT or equivalent to hold the side gears.
19. Remove the joint shaft.
20. Remove the manifold bracket.

21. Remove the starter.

22. Remove the front engine/transmission mount.

23. Remove the rear engine/transmission mount.

24. Disconnect the inner and outer oil hoses.

25. Remove the side engine/transmission mount.

26. Hold the drive plate and remove the converter nuts.

27. Support the transaxle on a jack and remove the mounting bolts.

28. Remove the transaxle.

To install:

29. Support the transaxle on a jack and lift it into place. Align the transaxle with the engine and install the mounting bolts. Tighten to 41–59 ft. lbs. (55–80 Nm).

30. Hold the driveplate and install the torque converter mount nuts. 26–36 ft. lbs. (35–49 Nm).

31. Install the side engine/transmission mount. Loosely tighten the nuts of the transaxle side. Tighten the nuts and bolts of the body side. Tighten to 32–44 ft. lbs. (44–60 Nm). Tighten the nuts of the transaxle side, tighten to 50–68 ft. lbs. (67–93 Nm).

32. Connect the inner and outer oil hoses.

33. Install the rear engine/transmission mount. Tighten the bolts to 50–68 ft. lbs. (67–93 Nm).

34. Install the front engine/transmission mount. Tighten the mount bracket to the transaxle to 28–38 ft. lbs. (38–51 Nm). Loosely tighten the nuts and bolts of the engine mount rubber, then tighten to 50–68 ft. lbs. (67–98 Nm).

35. Install the starter.

36. Install the manifold bracket.

37. Insert the joint shaft into the transaxle. Install the joint shaft to the cylinder block and tighten the bolts in sequence (counterclockwise). Tighten to 32–46 ft. lbs. (42–62 Nm).

38. Install the halfshafts, be sure that the shafts are properly installed and do not pull out.

39. Replace the engine mounting member. Install the mounting nuts/bolts and tighten the nuts/bolts at the far corners to 48–65 ft. lbs. (64–89 Nm), tighten the remaining two nuts to 28–38 ft. lbs. (38–51 Nm).

40. Connect the lower arm to the knuckle.

41. Install the stabilizer control link.

42. Connect the tie-rod ends.

43. Install the transverse member, if removed.

44. Install the splash shields.

45. Install the wheels.

46. Connect the throttle cable.

47. Install the harness bracket.

48. Attach the vehicle speedometer sensor connector.

49. Connect the selector cable.

50. Attach the transaxle range switch connector.

51. Attach the solenoid connector.

52. Install the battery carrier.

53. Install the air cleaner assembly.

54. Install the battery and battery cover.

55. Test drive the vehicle. Check for proper operation in all gear ranges.

Clutch

ADJUSTMENTS

Pedal Height

1. Measure the distance from the upper surface of the pedal pad to the carpet.

2. The distance should be 7.72–8.03 in. (196–204mm).

3. If the distance is not as specified, loosen the locknut on the stopper bolt or switch.

4. Turn the switch or bolt until the distance is correct, then tighten the locknut.

Free-Play

1. Depress the clutch pedal by hand until resistance is felt. The free-play should be 0.12–0.35 in. (3.0–9.0mm).

2. If the free-play is not correct, loosen the clutch master cylinder pushrod locknut and turn the pushrod to adjust.

REMOVAL & INSTALLATION

1. Disconnect the negative battery cable. Raise and safely support the vehicle.

2. Remove the transaxle.

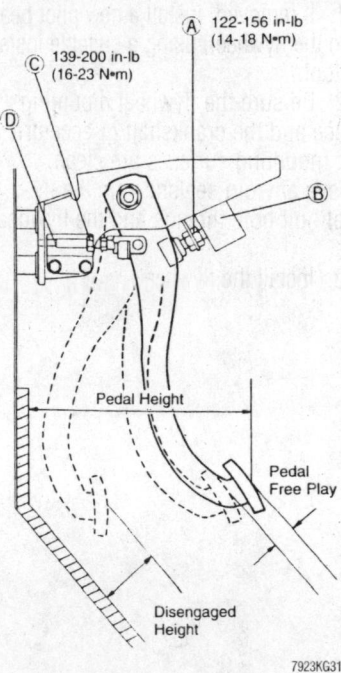

Clutch pedal measurement and adjustment points. (A) and (B) are for adjusting the pedal height, while (C) and (D) are for the free-play adjustment

3. Gradually loosen the clutch pressure plate bolts, in a crisscross pattern. Support the pressure plate and remove the bolts. Remove the pressure plate and clutch disc.

4. Inspect the pilot bearing. If it is worn or damaged and does not turn easily by hand, remove it using a puller/slide hammer.

5. Check the flywheel surface for scoring, cracks or burning and machine or replace, as necessary.

6. Install a flywheel holder to keep the flywheel from turning. Loosen the flywheel bolts evenly and gradually in a crisscross pattern. Remove the flywheel.

7. Inspect the clutch release bearing for wear. Replace it if it sticks or does not turn easily.

8. Inspect the release fork for wear or damage and replace as necessary.

To install:

9. Lubricate the release fork fingers and pivot with molybdenum grease and install in the release fork boot.

10. Install the clutch release bearing on the release fork.

11. If removed, install a new pilot bearing in the flywheel, using a suitable installation tool.

12. Be sure the flywheel mounting surface and the crankshaft or eccentric shaft mounting surfaces are clean. Remove any old sealant from the flywheel bolt hole threads and the flywheel bolts.

13. Install the flywheel.

14. Apply sealant to the flywheel bolt threads and install them hand tight. Install the flywheel holding tool. Tighten the bolts, in a crisscross pattern, to specification.

15. Apply a small amount of molybdenum grease to the clutch disc splines and install the clutch disc on the flywheel, spring side toward the transaxle. Install a suitable alignment tool in the pilot bearing to position the clutch disc.

16. Install the clutch pressure plate, aligning the dowel holes with the flywheel dowels. Install the pressure plate bolts and gradually tighten, in a crisscross pattern to 20 ft. lbs. (26 Nm). Remove the alignment tool.

17. Install the transaxle and lower the vehicle.

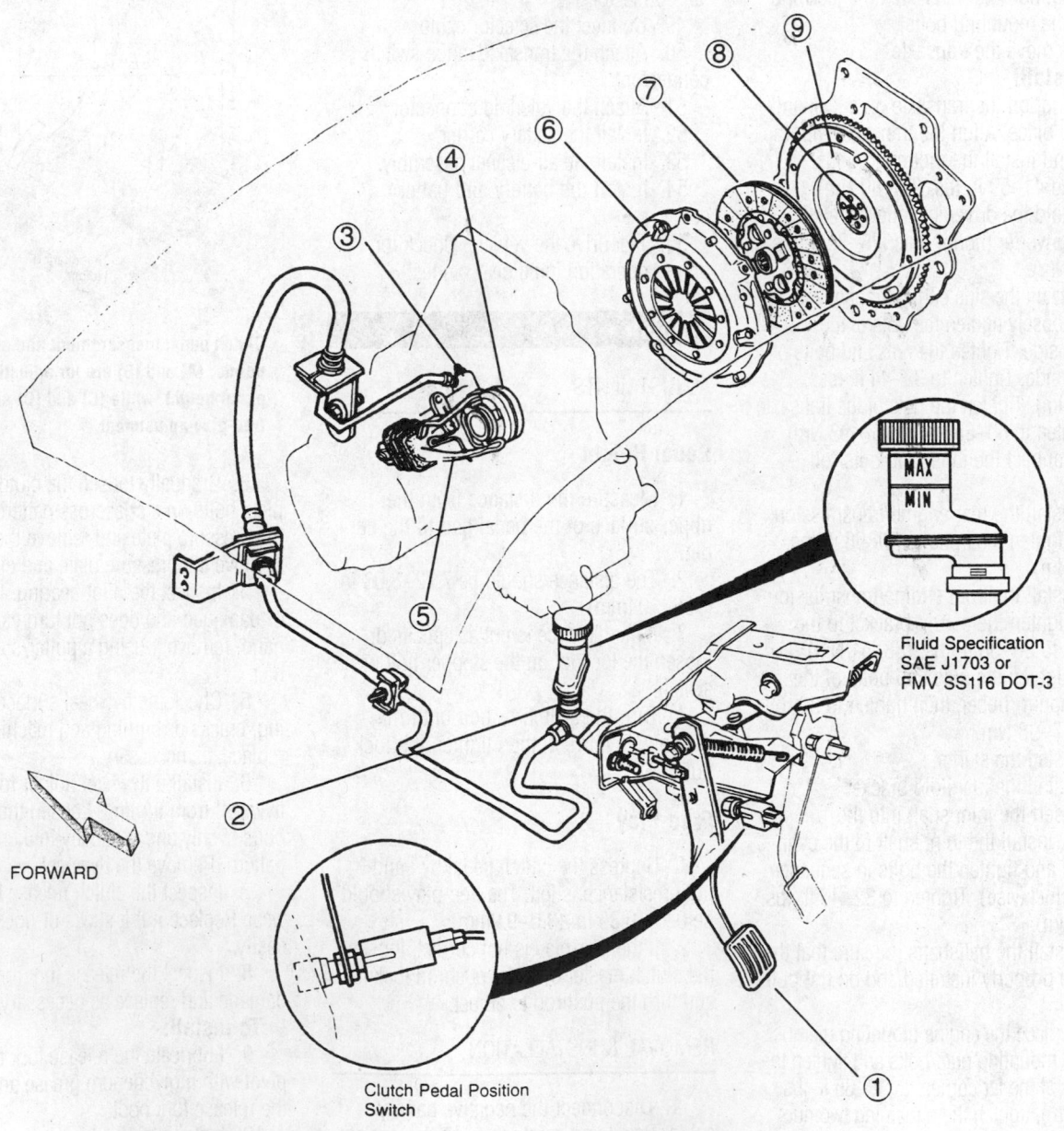

FORWARD

MAX
MIN

Fluid Specification
SAE J1703 or
FMV SS116 DOT-3

Clutch Pedal Position
Switch

1. Clutch pedal
2. Clutch master cylinder
3. Clutch release cylinder
4. Release bearing
5. Clutch release fork
6. Clutch cover
7. Clutch disc
8. Pilot bearing
9. Flywheel

Structural view of the hydraulic clutch system

7923KG32

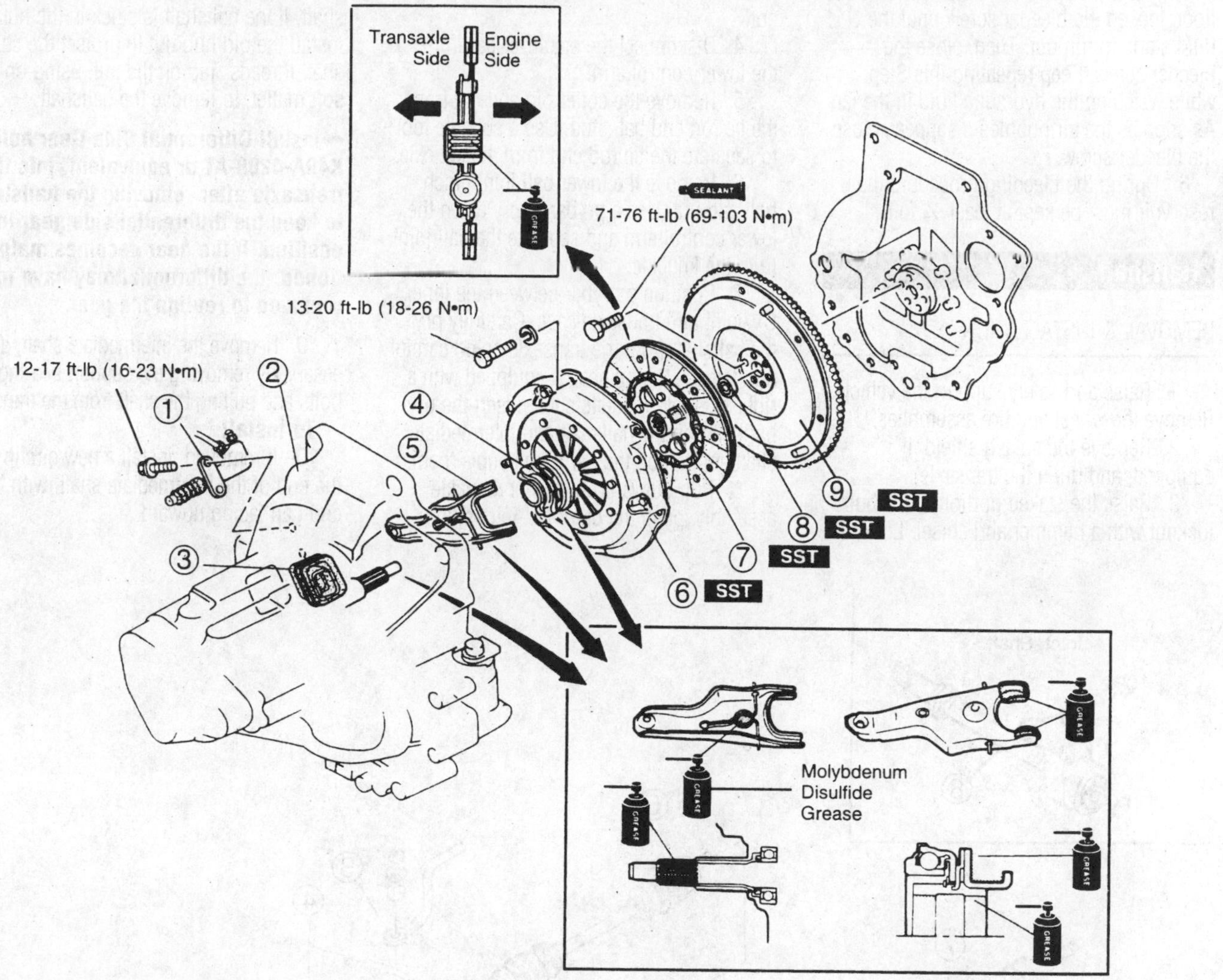

Transaxle Side | Engine Side

13-20 ft-lb (18-26 N•m)

12-17 ft-lb (16-23 N•m)

SEALANT

71-76 ft-lb (69-103 N•m)

① ② ③ ④ ⑤ ⑥ ⑦ ⑧ ⑨

SST

Molybdenum Disulfide Grease

1. Clutch release cylinder
2. Transaxle housing
3. Boot
4. Release bearing
5. Clutch release fork

6. Clutch cover
7. Clutch disc
8. Pilot bearing
9. Flywheel

7923KG33

Exploded view of the clutch assembly

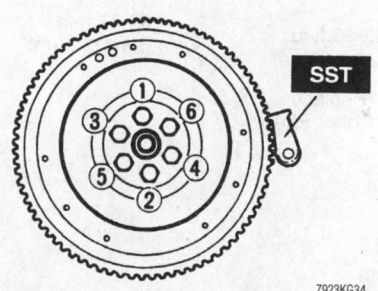

7923KG34

Flywheel tightening sequence

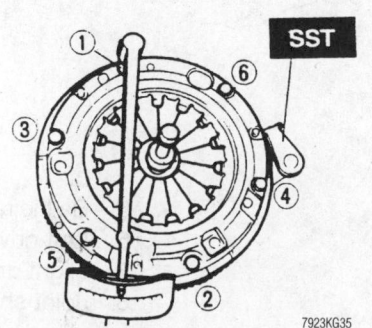

7923KG35

Pressure plate tightening sequence

Hydraulic Clutch System

BLEEDING

1. Remove the rubber cap from the bleeder screw on the release cylinder.
2. Place a bleeder tube over the end of the bleeder screw.
3. Submerge the other end of the tube in a jar half filled with hydraulic brake fluid.
4. Slowly pump the clutch pedal fully and allow it to return slowly, several times.

5. While pressing the clutch pedal to the floor, loosen the bleeder screw until the fluid starts to run out. Then, close the bleeder screw. Keep repeating this Step, while watching the hydraulic fluid in the jar. As soon as the air bubbles disappear, close the bleeder screw.

6. During the bleeding procedure the reservoir must be kept at least ¾ full.

Halfshaft

REMOVAL & INSTALLATION

1. Raise and safely support the vehicle. Remove the wheel and tire assemblies.

2. Remove the splash shield, if equipped, and drain the transaxle.

3. Raise the staked portion of the hub locknut with a hammer and chisel. Lock the hub by applying the brakes and remove the nut.

4. Disconnect the stabilizer bar from the lower control arm.

5. Remove the cotter pin and nut from the tie rod end ball stud. Use a suitable tool to separate the tie rod end from the knuckle.

6. Remove the lower ball joint pinch bolt and nut. Use a prybar to pry down the lower control arm and separate the ball joint from the knuckle.

7. Position a prybar between the inner CV-joint and transaxle case. Carefully pry the halfshaft from the transaxle being careful not damage the oil seal. If equipped with a right side intermediate shaft, insert the pry-bar between the halfshaft and intermediate shaft and tap on the bar to uncouple them.

8. Pull outward on the hub/knuckle assembly, push the outer CV-joint stub shaft through the hub, and remove the half-shaft. If the halfshaft is stuck in the hub, install the old hub nut to protect the stub shaft threads. Tap on the nut, using only a soft mallet, to remove the halfshaft.

→ **Install Differential Side Gear holder K49A-4208-AT or equivalent, into the transaxle after removing the halfshaft, to keep the differential side gear in position. If the gear becomes malpositioned, the differential may have to be removed to realign the gear.**

9. Remove the intermediate shaft, if necessary, by removing the support bearing bolts and pulling the shaft from the transaxle.

To install:

10. If removed, install a new circlip on the end of the intermediate shaft, with the end gap facing upward.

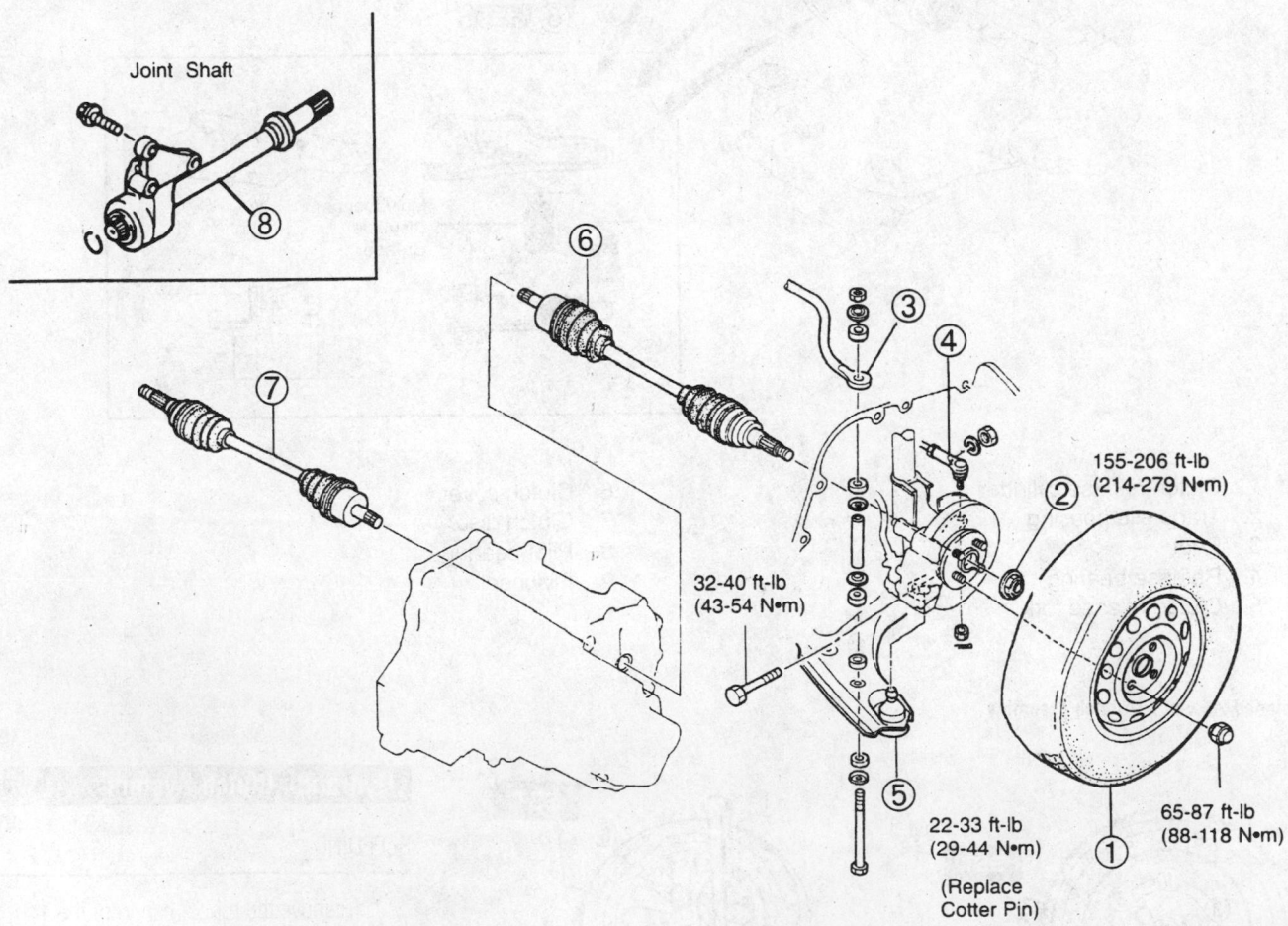

1. Wheel and tire
2. Locknut
3. Stabilizer bar
4. Tie-rod end
5. Ball joint
6. Left driveshaft
7. Right driveshaft
8. Joint shaft

Exploded view of the halfshafts and related components

7923KG36

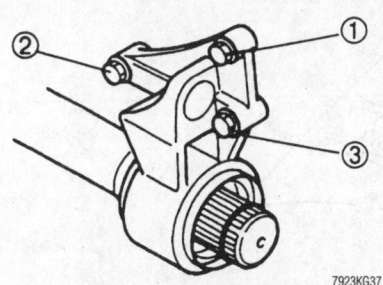

Support bearing bolt tightening sequence

7923KG37

11. Install the intermediate shaft in the transaxle, being careful not to damage the oil seals. Install the support bearing bolts and tighten, in sequence, to 45 ft. lbs. (61 Nm).

12. Install a new circlip on the end of the halfshaft, with the end gap facing upward. Insert the halfshaft into the transaxle, being careful not to damage the oil seal. If equipped, push the halfshaft into the intermediate shaft.

13. Insert the other end of the halfshaft through the hub. Loosely install a new locknut.

14. Install the lower ball joint into the knuckle. Install the pinch bolt and nut and tighten to 40 ft. lbs. (54 Nm).

15. Connect the tie rod end to the steering knuckle and tighten the nut to 42 ft. lbs. (57 Nm). Install a new cotter pin. Tighten the nut, if necessary, to align the ball stud hole with the nut castellation.

16. Connect the stabilizer bar to the lower control arm.

17. Install the splash shield and the wheel and tire assemblies. Lower the vehicle.

18. Lock the hub with the brakes. Tighten the new hub nut to 155–206 ft. lbs. (214–279 Nm). After tightening, stake the locknut using a hammer and dull bladed chisel.

19. Fill the transaxle with the proper type and quantity of fluid.

STEERING AND SUSPENSION

Air Bag

✳✳ CAUTION

Some vehicles are equipped with an air bag system, also known as the Supplemental Inflatable Restraint (SIR) or Supplemental Restraint System (SRS). The system must be disabled before performing service on or around system components, steering column, instrument panel components, wiring and sensors. Failure to follow safety and disabling procedures could result in accidental air bag deployment, possible personal injury and unnecessary system repairs.

PRECAUTIONS

Several precautions must be observed when handling the inflator module to avoid accidental deployment and possible personal injury.

• Never carry the inflator module by the wires or connector on the underside of the module.

• When carrying a live inflator module, hold securely with both hands, and ensure that the bag and trim cover are pointed away.

• Place the inflator module on a bench or other surface with the bag and trim cover facing up.

• With the inflator module on the bench, never place anything on or close to the module which may be thrown in the event of an accidental deployment.

• An air bag is an explosive device. Handle with extreme caution.

• Always disconnect the battery and the air bag connector before removing the steering wheel or beginning work on the air bag system.

• Air bag components must not be repaired or opened. Always use new parts, including the wiring harness.

• Always place a removed air bag unit with the horn pad facing up. Put it in a safe place where it will not be disturbed.

• The air bag unit must not be exposed to grease, fluids, or cleaning agents.

• The air bag unit must not be exposed to temperatures above 194° F (90° C) at any time. Even the heat of a soldering iron can damage or ignite the charge.

• Storage and transport of air bags is subject to rules governing explosive devices and should be done only in the original package.

• Failure to follow proper safety precautions may result in personal injury through accidental firing of the air bag, or through failure of the air bag in an accident.

DISARMING

1. Turn the ignition switch to the **LOCK** position.

2. Disconnect the negative battery cable.

3. Wait 10 minutes for the battery back-up power to discharge.

REARMING

1. Connect the negative battery cable.

2. Turn the ignition switch **ON**.

3. Verify that the air bag indicator illuminates for 4–8 seconds, then goes off.

Rack and Pinion Steering Gear

REMOVAL & INSTALLATION

Manual

1. Raise and safely support the vehicle. Disconnect the negative battery cable and remove the front wheels.

2. Remove the cotter pins from both steering tie rod ends and remove the nuts.

3. Use a tie rod press tool to press the tie rod out of the knuckle arm.

4. Remove the set plate from the firewall.

5. Remove the fixing bolt from the steering shaft to steering gear pinion shaft and separate the shaft from the steering gear.

6. Remove the steering gear mounting nuts and remove the steering gear to the right of the vehicle.

To install:

7. Install the steering gear to the vehicle and install the mounting nuts in the order shown. Tighten the nuts to 23–34 ft. lbs. (31–46 Nm).

8. Connect the steering shaft to the steering gear pinion shaft. Tighten the bolt/nut to 13–20 ft. lbs. (18–27Nm).

9. Install the set plate to the firewall.

10. Install the tie rod ends to the knuckle arm and tighten the nuts to 25–29 ft. lbs. (34–39 Nm). Install the cotter pins.

11. Install the wheels to the vehicle, lower the vehicle and connect the negative battery cable. Check the front end alignment.

Power

1. Raise and safely support the vehicle. Disconnect the negative battery cable and remove the front wheels.

2. Remove the cotter pins from both steering tie rod ends and remove the nuts.

3. Use a tie rod press tool to press the tie rod out of the knuckle arm.

4. Disconnect the pressure line and return pipe from the steering gear. Remove the set plate from the firewall.

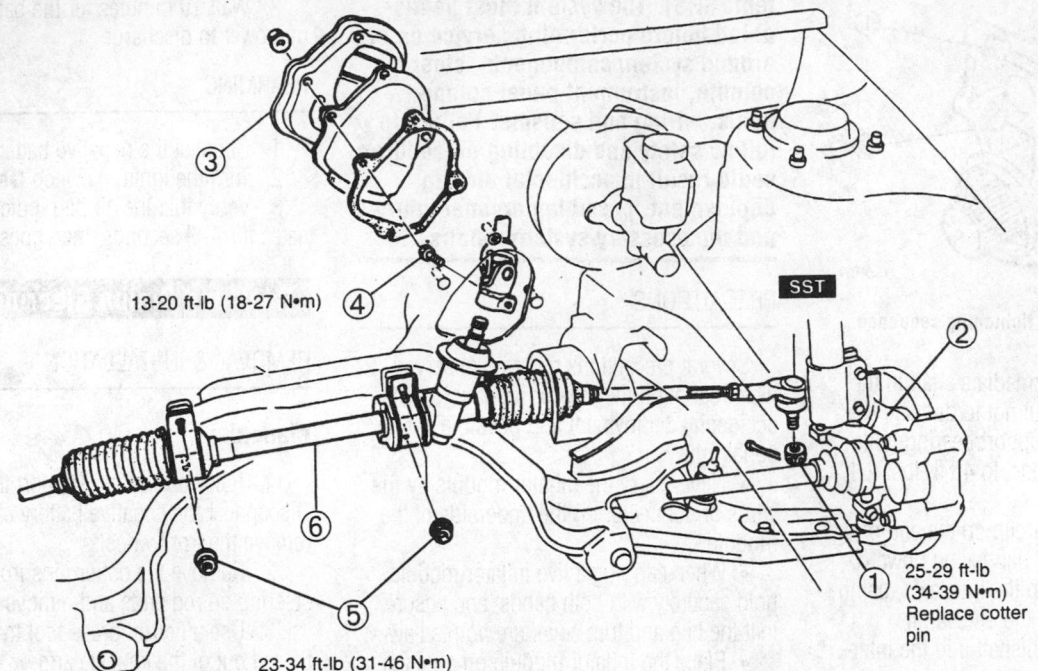

13-20 ft-lb (18-27 N•m)

SST

25-29 ft-lb
(34-39 N•m)
Replace cotter
pin

23-34 ft-lb (31-46 N•m)

1. Tie-rod end nut
2. Steering knuckle
3. Bulkhead sealing cover

4. Pinch bolts
5. Steering rack nuts
6. Steering rack

Exploded view of the manual steering gear assembly mounting

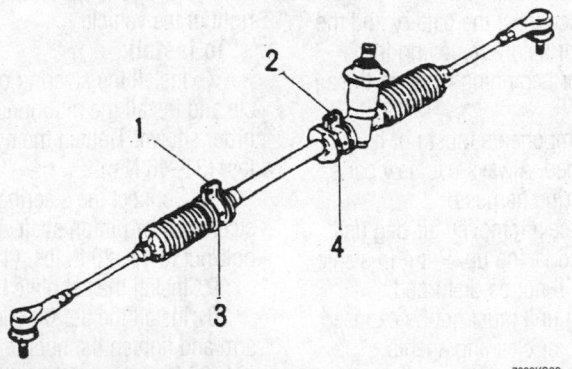

7923KG39

Rack mounting tightening sequence

5. Remove the fixing bolt from the steering shaft to steering gear pinion shaft and separate the shaft from the steering gear.

6. Disconnect the manual transmission shifter linkage if necessary.

7. Remove the steering gear mounting nuts and remove the steering gear to the right of the vehicle.

To install:

8. Install the steering gear to the vehicle and install the mounting nuts/bolts. Tighten the nuts to 23–34 ft. lbs. (31–46 Nm).

9. Connect the steering shaft to the steering gear pinion shaft. Tighten the bolt/nut to the specified torque.

10. Connect the manual transmission shift linkage if disconnected. Install the set plate to the firewall.

11. Connect the pressure line and return hose to the steering gear.

12. Install the tie rod ends to the knuckle arm and tighten the nuts to 22–33 ft. lbs. (29–44 Nm). Install the cotter pins.

13. Install the wheels to the vehicle, lower the vehicle and connect the negative battery cable. Check the power steering fluid and the front end alignment.

Strut

REMOVAL & INSTALLATION

Front

1. Raise and safely support the vehicle. Remove the wheel and tire assembly.

2. Support the lower control arm with a jack.

3. Remove the bolts or clips attaching the brake hose and/or ABS sensor harness to the strut.

4. Paint alignment marks on the upper strut mounting block and strut tower, and on the lower strut mount-to-steering knuckle so the strut can be reinstalled in the same position.

5. Remove the upper strut mounting block nuts and the strut-to-knuckle bolts and remove the strut assembly.

To install:

6. Install the strut into the strut tower, aligning the paint marks made during removal. Install the mounting nuts and tighten to 17–22 ft. lbs. (23–29 Nm).

7. Install the strut-to-knuckle bolts and tighten to 69–86 ft. lbs. (93–116 Nm).

8. Install the clips or bolts attaching the brake hose and/or ABS sensor harness.

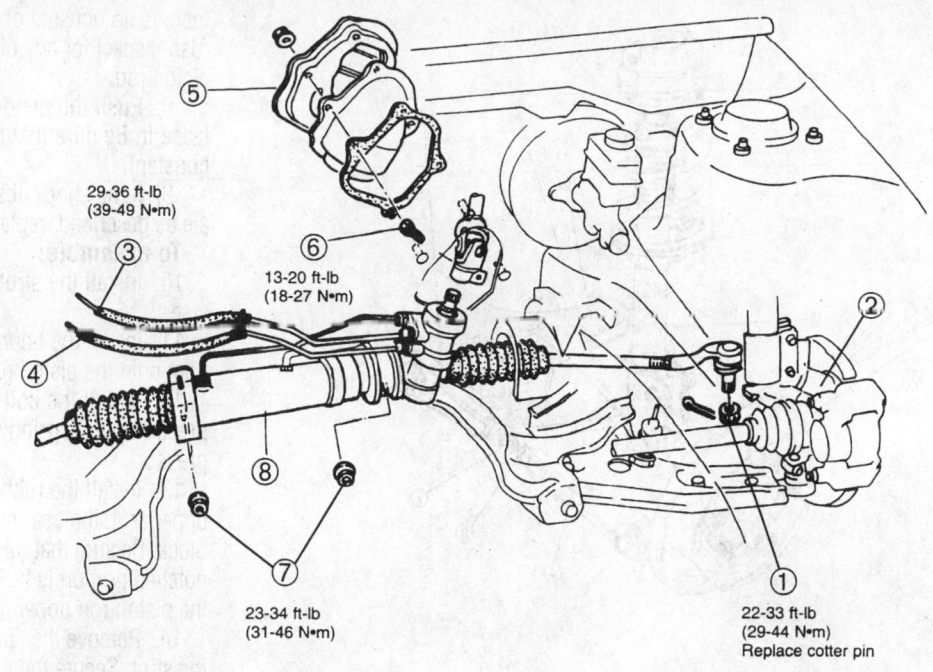

29-36 ft-lb
(39-49 N•m)

13-20 ft-lb
(18-27 N•m)

23-34 ft-lb
(31-46 N•m)

22-33 ft-lb
(29-44 N•m)
Replace cotter pin

1. Tie-rod end nut
2. Steering knuckle
3. High-pressure line
4. Return line
5. Sealing cover
6. Pinch bolt
7. Steering rack nuts
8. Steering rack and linkage

7923KG40

Exploded view of the power steering gear assembly mounting

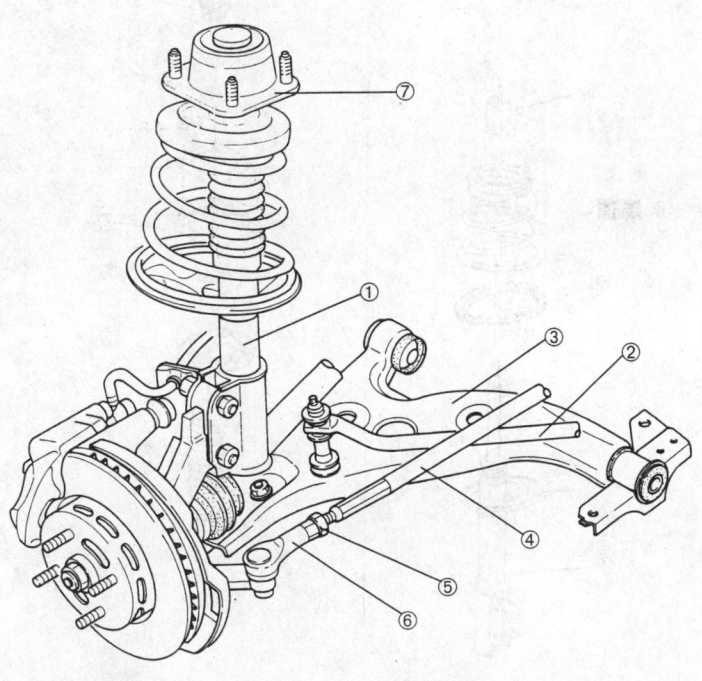

1. Front strut
2. Front stabilizer
3. Lower control arm
4. Tie-rod
5. Jam nut
6. Tie-rod end
7. Mounting block

7923KG41

Front suspension component identification

9. Install the wheel and tire assembly and lower the vehicle. Check the front end alignment.

Rear

1. As required, remove the side trim panels from the inside of the trunk or the rear seat and trim.

2. Loosen and remove the top mounting nuts from the strut mounting block assembly.

3. Raise and safely support the vehicle and remove the rear wheels. The suspension will drop when the weight lifts off the wheels.

4. Unclip the brake line or wiring retainers as required and unbolt the bottom strut mount. Remove the strut.

To install:

5. Install the strut into the strut tower. Install the mounting nuts and tighten to 17–22 ft. lbs. (23–29 Nm).

6. Install the strut-to-knuckle bolts and tighten to 69–86 ft. lbs. (93–116 Nm).

7. Install the clips or bolts attaching the brake hose and/or ABS sensor harness.

8. Install the wheel and tire assembly and lower the vehicle.

9. If removed, install the side trim panels from the inside of the trunk or the rear seat and trim.

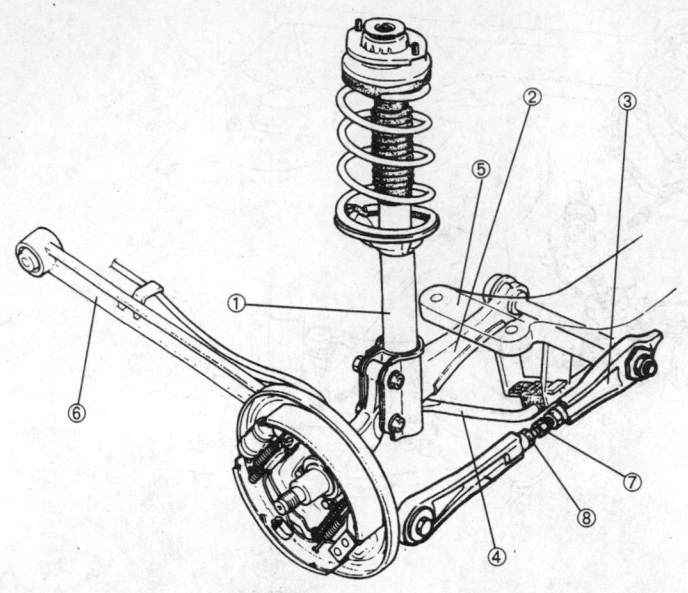

7923KG42

1. Rear strut
2. Front lateral link
3. Rear lateral link
   Rear stabilizer
5. Rear crossmember
6. Trailing link
7. Adjuster
8. Jam nuts

Rear suspension component identification

Coil Spring

REMOVAL & INSTALLATION

Front and Rear

1. Remove the strut from the vehicle.
2. Install the strut securely in a vise with either aluminum or copper plates to protect the strut.
3. Loosen the piston rod upper nut several turns but DO NOT REMOVE IT.
4. Install the lower end of the strut in the vise and install a coil spring compressor. Compress the coil spring and remove the upper nut.

✳✳ CAUTION

Failure to fully compress the spring and hold it securely can be extremely dangerous.

5. Slowly release the coil spring tension.
6. Remove the suspension support, dust seal, spring seat, spring insulators, coil spring and bumper.
7. While pushing on the piston rod, be sure that the pull stroke is even and that

there is no unusual noise or resistance. Also inspect for any oil leakage around the piston rod.

8. Push the piston rod in, then release it. Be sure that the return rate is constant.
9. If the shock absorber does not operate as described, replace it.

To assemble:

10. Install the strut assembly into a vise.
11. Install the bound stopper and dust boot onto the piston rod.
12. Install the coil spring and compress the coil spring with the spring compressor.
13. Install the rubber seat, the spring upper seat, the bearing and the mounting block. Be sure that the spring upper seat notched portion is facing inward and tighten the piston rod upper nut.
14. Remove the spring compressor from the strut. Secure the upper mounting block in the vise. Tighten the nut to 41–50 ft. lbs. (55–68 Nm) for the front strut and 47–59 ft. lbs. (64–80 Nm) for the rear strut.
15. Be sure that the spring is well seated in the upper seats.
16. Install the strut to the vehicle.

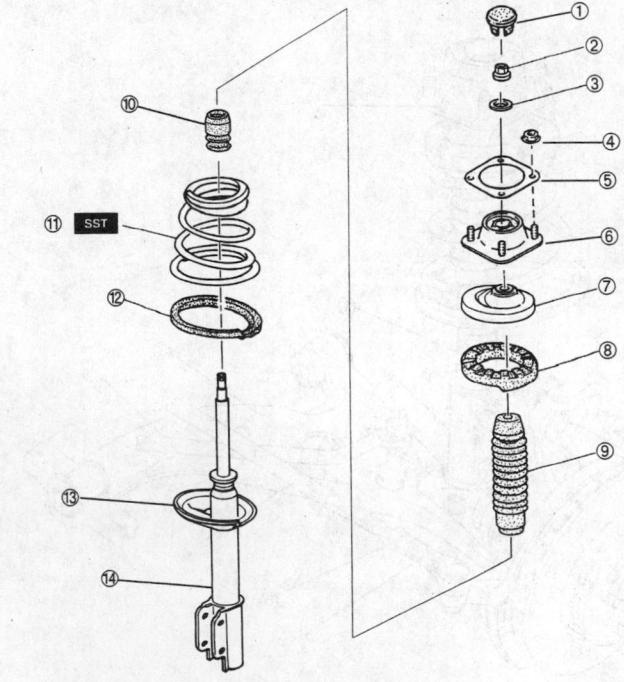

1. Dust cap
2. Piston retaining nut
3. Washer
4. Mounting nut
5. Gasket
6. Mounting block
7. Upper spring seat
8. Upper spring isolator
9. Dust boot
10. Rebound stopper
11. Coil spring
12. Lower spring isolator
13. Lower spring seat
14. Shock absorber

7932KG43

Exploded view of the front strut assembly

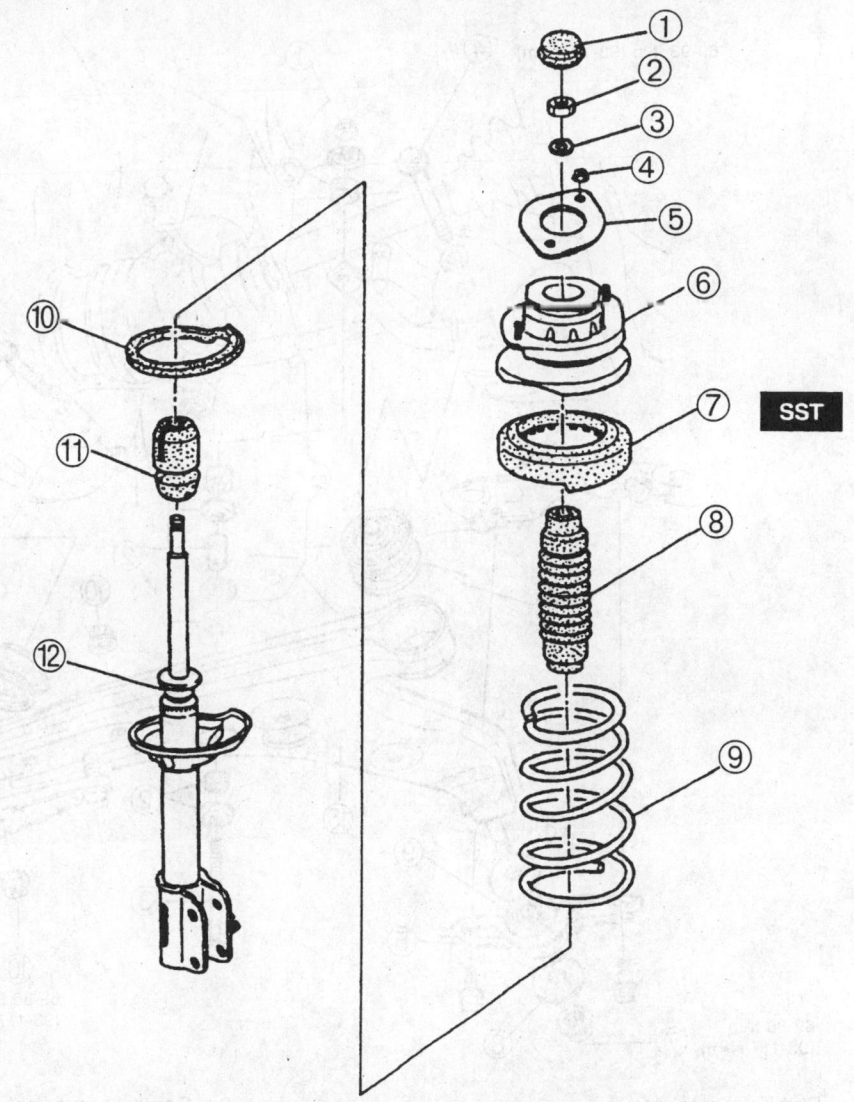

1. Dust cap
2. Piston retaining nut
3. Washer
4. Mounting nut
5. Gasket
6. Mounting block/upper spring seat

7. Upper spring isolator
8. Dust boot
9. Coil spring
10. Lower spring isolator
11. Rebound stopper
12. Strut

7923KG44

Exploded view of the rear strut

Lower Ball Joint

REMOVAL & INSTALLATION

1. Raise and safely support the vehicle. Remove the wheel and tire assembly.

2. Remove the ball joint stud pinch bolt and nut from the steering knuckle. Pry the lower control arm down from the knuckle, and separate the ball joint from the knuckle.

3. Remove the bolt and nut and remove the ball joint from the lower control arm.

4. Installation is the reverse of the removal procedure. Tighten the ball joint-to-lower control arm bolt and nut to 86 ft. lbs. (117 Nm). Tighten the ball joint pinch bolt and nut to 43 ft. lbs. (59 Nm). Check the front wheel alignment.

Wheel Bearings

ADJUSTMENT

The wheel bearings on these vehicles are not adjustable. To check if the bearing requires service, remove the wheel and tire assembly, brake caliper and disc brake rotor. Install a dial indicator with the indicator foot resting on the wheel hub. Try to move the hub in and out. If there is more than 0.002 in. (0.05mm) bearing play, check the wheel hub nut torque or replace the hub and bearing assembly.

REMOVAL & INSTALLATION

Front

1. Raise and safely support the vehicle. Remove the front wheels.

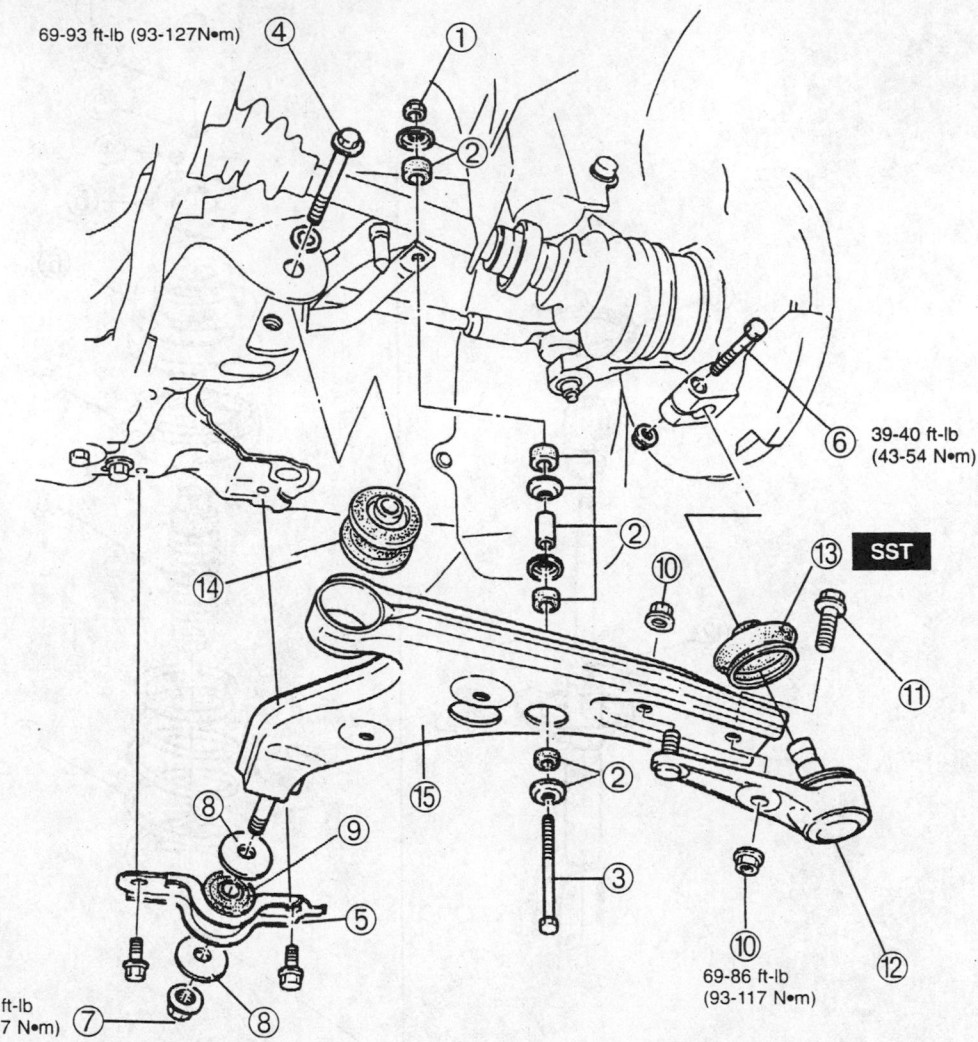

69-93 ft-lb (93-127N•m)

6 39-40 ft-lb (43-54 N•m)

SST

69-86 ft-lb (93-117 N•m) 7

69-86 ft-lb (93-117 N•m)

1. Stabilizer retaining nut
2. Stabilizer hardware - spacer, retainers, bushings
3. Stabilizer bolt
4. Pivot bolt
5. Mounting bolts
6. Pinch bolt,
7. Retaining nut
8. Washers
9. Control arm bushing - rear
10. Ball joint mounting nuts
11. Ball joint mounting bolt
12. Ball joint
13. Ball joint dust boot (Replace)
14. Control arm bushing - front
15. Control arm

7923KG45

Exploded view of the lower control arm with replaceable ball joint

2. Uncrimp the tab on the center locknut and remove the locknut. Discard the old locknut.

3. Remove the caliper assembly from the knuckle. Do not disconnect the brake lines. Support the caliper with a piece of wire. Do not allow the caliper to hang by the hose at any time. Remove the brake disc.

4. Remove the ABS speed sensor if so equipped.

5. Remove the tie rod end cotter pin and remove the tie rod end nut. Separate the tie rod end out of the knuckle assembly.

6. Remove the outer lower arm to ball joint mounting bolt and nut. Separate the lower arm from the knuckle assembly.

7. Using a plastic mallet, tap the knuckle assembly free of the halfshaft. Remove the knuckle assembly.

8. Clamp the knuckle in a vise with protected jaws.

9. Remove the inner oil seal from the knuckle.

10. Using the appropriate hub-puller remove the front wheel hub from the knuckle assembly.

11. Remove the bearing inner race from the front wheel hub.

➡ **If the bearing inner race still remains on the hub assembly, grind a section of the bearing inner race until about 0.02 in. (0.50mm) remains. Remove with a chisel.**

12. Remove the retaining ring from within the knuckle and using a wheel bearing removal tool, press the front wheel bearing from the knuckle.

13. Clean and inspect all parts but do not wash or clean the wheel bearing.

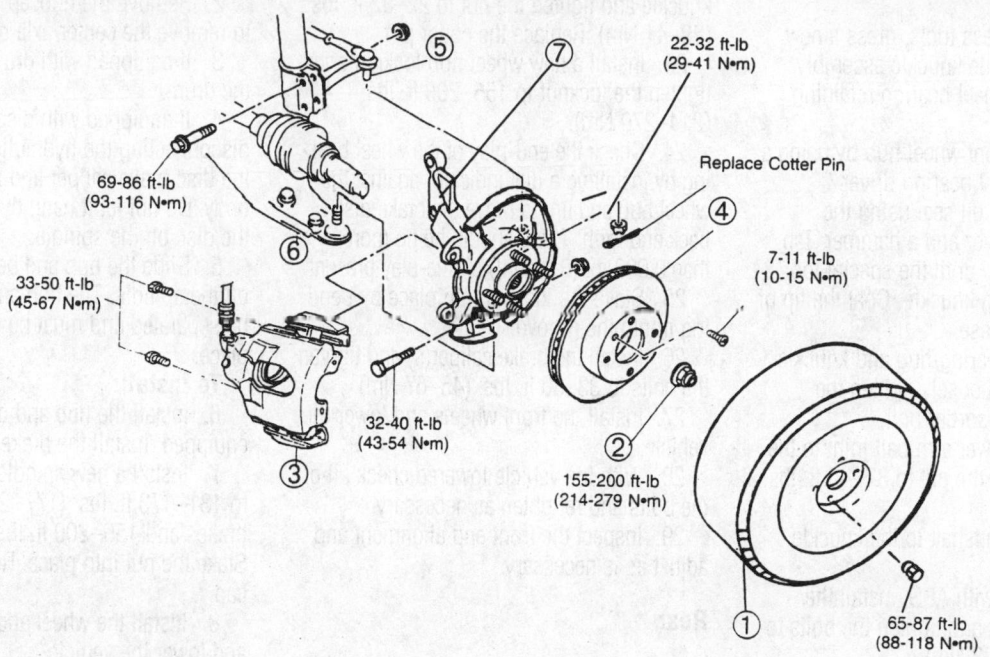

22-32 ft-lb
(29-41 N•m)

Replace Cotter Pin

69-86 ft-lb
(93-116 N•m)

7-11 ft-lb
(10-15 N•m)

33-50 ft-lb
(45-67 N•m)

32-40 ft-lb
(43-54 N•m)

155-200 ft-lb
(214-279 N•m)

65-87 ft-lb
(88-118 N•m)

1. Wheel and tire
2. Locknut (Replace)
3. Brake caliper assembly
4. Brake rotor

5. Tie-rod end
6. Ball joint
7. Steering knuckle/wheel hub

7923KG46

Exploded view of the front steering knuckle and related components

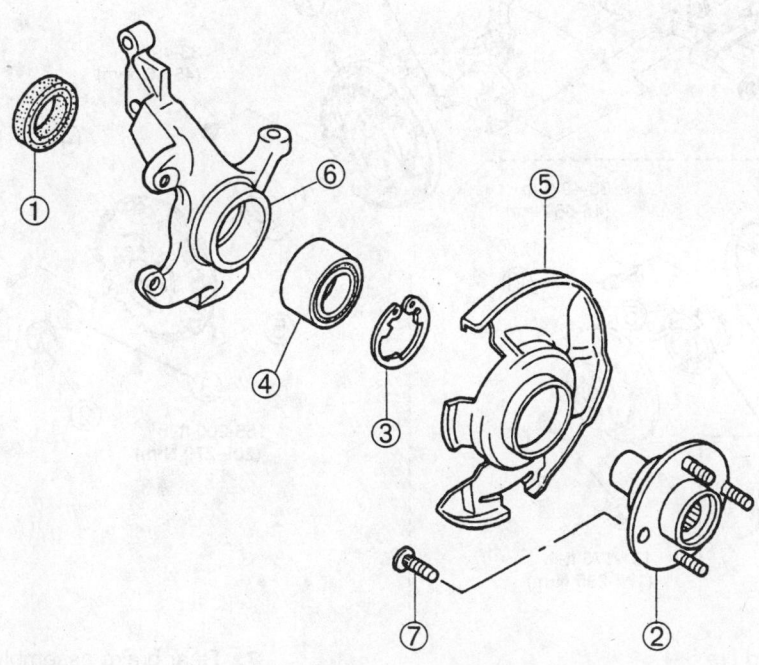

1. Oil seal (Replace)
2. Front wheel hub
3. Retaining ring (Replace)
4. Wheel bearing

5. Dust shield
6. Steering knuckle
7. Wheel stud

7923KG47

Exploded view of the front hub and bearing assembly

To install:

14. Using the press tools, press a new wheel bearing into the knuckle assembly.

15. Install the wheel bearing retaining ring.

16. Install the front wheel hub by using a press and the correct bearing driver.

17. Install a new oil seal using the appropriate seal driver and a hammer. Tap the oil seal in evenly until the special tool contacts the steering knuckle. Coat the lip of the oil seal with grease.

18. Install the bearing/hub and knuckle assembly in place. Loosely tighten the knuckle to shock absorber bolt.

19. Install the lower arm ball joint to the knuckle and tighten the nut to 32–40 ft. lbs. (43–54 Nm).

20. Install the halfshaft to the knuckle assembly.

21. If equipped with ABS, install the wheel speed sensor and tighten the bolts to 12–17 ft. lbs. (16–23 Nm).

22. Connect the tie rod end to the knuckle and tighten the nut to 22–32 ft. lbs. (29–41 Nm). Replace the cotter pin.

23. Install a new wheel hub locknut and tighten the locknut to 155–200 ft. lbs. (214–279 Nm).

24. Check the end-play of the wheel bearing by installing a dial indicator against the wheel hub and tire to move the brake disc back and forth. There should be no more than 0.002 in. (0.05mm) of free-play present.

25. Stake the locknut into place by bending it into the groove.

26. Install the brake caliper(s) and tighten the bolts to 33–50 ft. lbs. (45–67 Nm).

27. Install the front wheels and lower the vehicle.

28. With the vehicle lowered check all of the bolts and retighten as necessary.

29. Inspect the front end alignment and adjust as is necessary.

Rear

1. Raise and safely support the vehicle and remove the rear wheels.

2. Remove the hubcap. Hold the brake to remove the center axle nut.

3. If equipped with drum brakes, remove the drum.

4. If equipped with disc brakes, without disconnecting the hydraulic hose, remove the disc brake caliper and hang it from the body. Do not let it hang by the hose. Slide the disc off the spindle.

5. Slide the hub and bearing assembly off the spindle. The hub and bearing cannot be separated and must be replaced as one piece.

To install:

6. Install the hub and drum or rotor. If equipped, install the brake caliper.

7. Install a new spindle nut and tighten to 131–173 ft. lbs. (177–235 Nm) for disk brakes and 155–200 ft. lbs. (209–279 Nm). Stake the nut into place. Replace the hubcap.

8. Install the wheel and tire assembly and lower the vehicle.

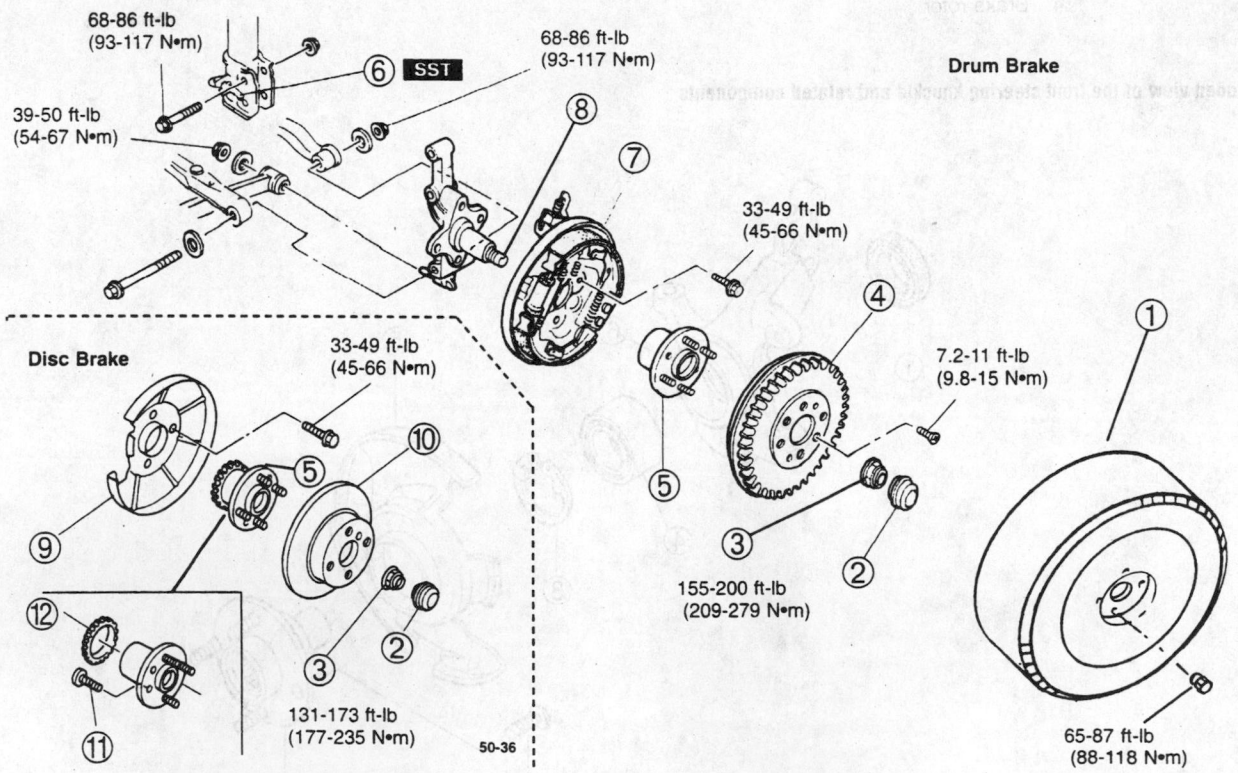

Disc Brake

68-86 ft-lb (93-117 N•m)

39-50 ft-lb (54-67 N•m)

68-86 ft-lb (93-117 N•m)

Drum Brake

33-49 ft-lb (45-66 N•m)

33-49 ft-lb (45-66 N•m)

7.2-11 ft-lb (9.8-15 N•m)

155-200 ft-lb (209-279 N•m)

131-173 ft-lb (177-235 N•m)

65-87 ft-lb (88-118 N•m)

50-36

1. Wheel and tire
2. Dust cap
3. Locknut (Replace)
4. Brake drum (or disc)
5. Hub with bearing assembly
6. Brake line
7. Rear brake assembly (drum or disc)
8. Spindle
9. Dust cover
10. Brake rotor
11. Hub bolt
12. ABS sensor rotor

Exploded view of the rear axle assembly

7923KG48

LEXUS

ES300 • GS300 • GS400 • LS400 • SC300 • SC400

20

ENGINE REPAIR

➡**Disconnecting the negative battery cable on some vehicles may interfere with the functions of the on board computer systems and may require the computer to undergo a relearning process, once the negative battery cable is reconnected.**

Distributor

REMOVAL

3.0L (1MZ-FE) Engine

The 1MZ-FE engine is not equipped with a distributor. The spark is controlled by the Engine Control Module (ECM) and spark is sent to the plugs via 6 separate ignition coils.

1995–97 3.0L (2JZ-GE) Engine

1. Disconnect the negative battery cable.
2. Set the No. 1 piston is at TDC of its compression stroke.
3. Tag and disconnect the spark plug wires at the distributor. Disconnect the distributor connector.
4. Loosen the hold-down bolt and remove the distributor.

1998–99 3.0L (2JZ-GE) Engine

The ignition system on this engine is a coil near plug design. The engine control module controls the coil primary circuit. No distributor is used.

1995–97 4.0L Engine

1. Disconnect the negative battery cable.

✵✵ CAUTION

Work must be started after 90 seconds from the time the ignition switch is turned to the LOCK position and the negative battery cable is disconnected.

2. Remove the No. 2 timing belt covers.
3. Disconnect the spark plug wires from the distributor caps.
4. Loosen the three bolts and remove the distributor cap. Remove both caps.
5. Remove the two rubber caps from each distributor cap.
6. Loosen the bolt and remove the distributor rotor. Remove both distributor rotors.

7. Remove the No. 2 distributor housing by removing the camshaft position sensor connector and removing the three bolts.
8. Remove the bolt, screw and camshaft position sensor.

1998–99 4.0L Engine

The ignition system on this engine is a coil on plug design. The engine control module controls the coil primary circuit. No distributor is used.

INSTALLATION

1995–97 3.0L (2JZ-GE) Engine

ENGINE UNDISTURBED

1. Lubricate a new O-ring with engine oil and install it.
2. Align the groove on the distributor housing with the protrusion on the drive gear. Insert the distributor so the center of the flange is aligned with that of the bolt hole on the cylinder head.
3. The rotor should be pointing towards the No. 1 tower on the distributor cap.
4. Lightly tighten the hold-down bolt.
5. Connect the distributor connector and the spark plug wires.
6. Connect the battery cable and check the ignition timing. When the timing is properly adjusted, tighten the hold-down bolt to 10 ft. lbs. (14 Nm).

ENGINE DISTURBED

1. Lubricate the new O-ring with engine oil, then install it.
2. Set the No. 1 cylinder to TDC of the compression stroke. Remove the oil filler cap. Rotate the crankshaft clockwise until the small end of the camshaft lobe can be seen through the hole. Turn the crankshaft counter-clockwise approximately 120°. Turn it as additional 10–40° clockwise until the

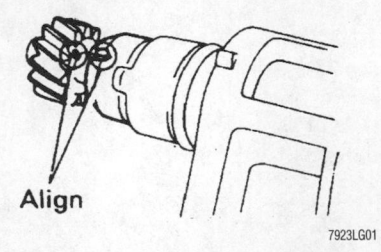

Aligning the marks on the distributor—3.0L (2JZ-GE) engine

timing marks on the crankshaft pulley and timing cover are aligned.
3. Align the groove on the distributor housing with the protrusion on the drive gear. Insert the distributor so the center of the flange is aligned with that of the bolt hole on the cylinder head.
4. Lightly tighten the hold-down bolt.
5. Connect the distributor connector and the spark plug wires.
6. Connect the battery cable and check the ignition timing. When the timing is properly adjusted, tighten the hold-down bolt to 10 ft. lbs. (14 Nm).

1995–97 4.0L Engine

ENGINE NOT DISTURBED

1. Install the camshaft position sensor with the bolt and screw. Install the both sensors and tighten the bolt to 34 inch lbs. (4 Nm).

➡**The No. 1 distributor housing is marked with an L and the No. 2 with an R.**

2. Install the left distributor housing with the three bolts. Tighten the bolts to 13 ft. lbs. (18 Nm). The two inner bolts are 3.15 in. and the outer bolt is 1.50 inches.
3. Install the right distributor housing with the three bolts. Tighten the bolts to 13 ft. lbs. (18 Nm). The two inner bolts are 3.78 in. and the outer bolt is 1.50 inches.
4. Align the protrusion of the distributor rotor with the groove of the camshaft timing pulley. Install both rotors and tighten the bolts to 34 inch lbs. (4 Nm).
5. Connect the camshaft position sensor connectors.
6. Install the distributor cap and tighten the three bolts to 34 inch lbs. (4 Nm). Install both caps. Install the two rubber caps to each distributor cap.
7. Connect the spark plug wires to the distributors.
8. Install the No. 2 timing belt covers.
9. Connect the negative battery cable. Start the engine and allow normal operating temperature to be reached.
10. Check the ignition timing.

ENGINE DISTURBED

1. Be sure that the No. 1 cylinder is in TDC of the compression stroke.
   a. Turn the crankshaft pulley and align its groove with the timing mark **0** of the timing chain cover.
   b. Check that the timing marks on the camshafts align with the timing marks.
   c. If not, turn the crankshaft 1 revolution (360 degrees) and align the crank-

shaft pulley groove with the timing mark **0** of the timing chain cover.

2. Install the camshaft position sensor with the bolt and screw. Install the both sensors and tighten the bolt to 34 inch lbs. (4 Nm).

➡**The No. 1 distributor housing is marked with an L and the No. 2 with an R.**

3. Install the left distributor housing with the three bolts. Tighten the bolts to 13 ft. lbs. (18 Nm). The two inner bolts are 3.15 in. and the outer bolt is 1.50 inches.

4. Install the right distributor housing with the three bolts. Tighten the bolts to 13 ft. lbs. (18 Nm). The two inner bolts are 3.78 in. and the outer bolt is 1.50 inches.

5. Align the protrusion of the distributor rotor with the groove of the camshaft timing pulley. Install both rotors and tighten the bolts to 34 inch lbs. (4 Nm).

6. Connect the camshaft position sensor connectors.

7. Install the distributor cap and tighten the three bolts to 34 inch lbs. (4 Nm). Install both caps. Install the two rubber caps to each distributor cap.

8. Connect the spark plug wires to the distributors.

9. Install the No. 2 timing belt covers.

10. Connect the negative battery cable. Start the engine and allow normal operating temperature to be reached.

11. Check the ignition timing.

Ignition Timing

ADJUSTMENT

3.0L (1MZ-FE) Engine

The ignition timing may be checked to verify the setting, but cannot be adjusted.

1. Allow the engine reach normal operating temperature.

2. Connect a Lexus Hand-Held Tester or OBD II scan tool to the DLC3.

3. Connect a timing light to the No. 1 high tension cord for No. 4 cylinder.

4. Race the engine speed at 2,500 rpm for at least 90 seconds and check that idle speed is 700 ±50 rpm.

5. Using SST 09843–18020, connect terminals TE1 and E1 of the DLC 1. With the timing light check the ignition timing. It should be: 8–12° BTDC at idle with the transmission in the neutral position. Remove the SST from the DLC1.

6. Further checking with the transmis-

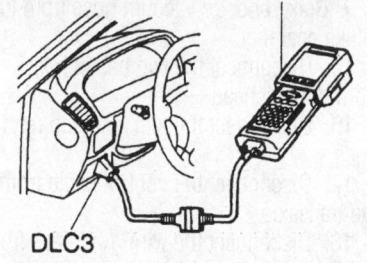

Connecting the scan tool—3.0L (1MZ-FE) engine

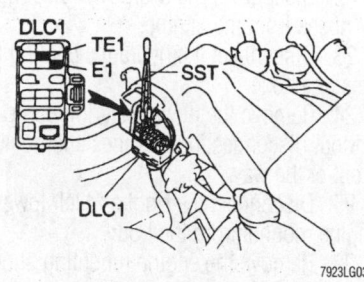

Using the SST tool to connect the terminals in the data link connector—3.0L (1MZ-FE) engine

sion in the neutral position and the engine at idle will be in the range of 7–24° BTDC.

7. Stop the engine and disconnect the timing light and the scan tool.

3.0L (2JZ-GE) Engine

➡**The ignition timing on the 1998–99 3.0L (2JZ-GE) is not adjustable, however it may be inspected using the following procedure.**

1. Allow the engine to reach normal operating temperature.

2. Connect a tachometer to terminal IG of the check connector.

➡**Never allow the tachometer test probe to touch ground as it could result in damage to the igniter and or ignition coil. Some tachometers are not compatible with this ignition system, always confirm the compatibility of the unit before use.**

3. With the transmission in **P** or **N**, raise the engine idle to 2500 rpm for 90 seconds, then release the throttle.

4. Check the idle speed.

5. Connect the proper jumper wire to terminals **TE₁** and **E₁** of the check connec-

tor in the engine compartment. On 1998–99 models, connect terminals **TC** and **E₁**.

6. Connect the timing light to spark plug wire for the No. 1 cylinder.

7. Start the engine and check the timing with the transmission in **N** position.

8. Check the ignition timing; it should be 10° BTDC on 1995–97 models and vary between 6–16° on 1998–99 models. On 1995–97 models, if the timing is not within specifications, loosen the hold-down bolt(s) and adjust the timing by turning the distributor.

9. Tighten the hold-down bolts and recheck the timing and idle speed, adjust as necessary.

10. Remove the jumper wires. Recheck the timing; it should fluctuate between 7–19°. The change in timing is normal and is controlled by the ECU when the jumper wire is removed.

11. Remove the remaining test equipment.

4.0L Engine

The ignition timing can be checked using this procedure. The timing, however, can not be adjusted.

1. Remove the upper spark plug wire cover.

2. Start the engine and allow it to reach normal operating temperature.

3. Connect a tachometer to terminal IG (–) of the check connector.

4. Set the tachometer to the 4 cylinder range.

➡**Never allow the tachometer test probe to touch ground as it could damage the igniter and or ignition coil. Some tachometers are not compatible with this ignition system; always confirm the compatibility of the unit before use.**

5. Check the idle speed, it should be 600–700 rpm.

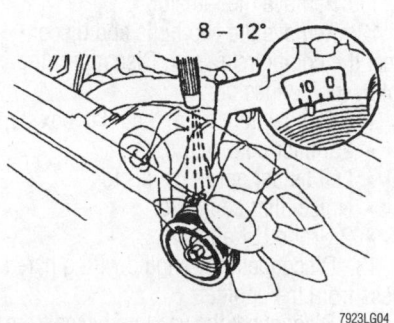

Checking the ignition timing—4.0L (1UZ-FE) engine

6. Connect a proper jumper wire to terminals **TE₁** and **E₁** of the check connector in the engine compartment.

7. Connect the timing light to spark plug wire for the No. 6 cylinder.

8. Check the ignition timing. Specification is 8–12° BTDC at idle in NEUTRAL.

➡**No adjustment is possible.**

9. If the timing is not within specification, check these items:

  a. The throttle valve must be fully closed.

  b. Continuity must exist between throttle position sensor terminals IDL1 (or IDL) and E2.

  c. The valve timing must be correct.

10. Remove the jumper from the data link connector. Disconnect and remove the test equipment.

Engine Assembly

REMOVAL & INSTALLATION

ES300

1. Release the fuel pressure.

2. With the ignition switch in the **LOCK** position, disconnect the negative battery terminal. If equipped with an air bag system, wait at least 90 seconds or longer before performing any other work.

3. Remove the battery and tray.

4. Matchmark the hood to the hinges and remove the hood.

5. Drain the engine coolant.

6. Drain the engine oil.

7. Disconnect the accelerator cable.

8. Disconnect the throttle cable.

9. Remove the air cleaner cover, volume air flow meter and air cleaner duct as an assembly.

10. If equipped, remove the cruise control actuator.

11. Remove the radiator.

12. Remove the two bolts and disconnect the engine relay box. Disconnect the following connectors:

  • 5 Connections from the relay box
  • 2 Igniter connectors
  • Left fender apron connector
  • Noise filter connector
  • 2 Ground straps

13. Disconnect the engine wiring harness from the engine.

14. Disconnect the vacuum hoses from the following connections:

  • Intake air control valve vacuum tank
  • Charcoal canister

  • Brake booster vacuum hose from the intake chamber

15. Disconnect the two heater hoses from the bulkhead.

16. Disconnect the fuel feed and return lines.

17. Disconnect the control cable from the transaxle.

18. Disconnect the wiring harness from the Engine Control Module (ECM) and route it through the bulkhead.

19. Remove the A/C compressor from the engine without disconnecting the lines and position it out of the way.

20. Remove the front exhaust pipe.

21. Remove the halfshafts.

22. Disconnect the two power steering air hoses from the engine.

23. Disconnect the hydraulic cooling fan pressure hose.

24. Remove the power steering pump without disconnecting the lines and position it out of the way.

25. Disconnect the right and left lower engine mounts from the body.

26. Remove the engine mounting shock absorber.

27. Remove the three front engine mounting bolts from the body.

28. Attach a lifting device to the engine.

29. Remove the coolant reservoir tank.

30. Remove the right engine mounting bracket.

31. Remove the engine moving control rod and right No. 2 engine mounting bracket.

32. Remove the engine and transaxle as an assembly.

✳✳ WARNING

Be careful not to hit the power steering gear housing or Park/Neutral switch.

33. Remove the engine mounting insulator below the oil filter.

34. Remove the right rear engine mounting insulator.

35. Remove the front exhaust pipe stay.

36. Disconnect the following connectors:

  • O/D solenoid
  • PNP switch speedometer
  • Starter terminal
  • Speed sensor

37. Disconnect the two wire clamps from the transaxle.

38. Remove the oil dipstick and guide.

39. Remove the starter.

40. Remove the flywheel housing cover.

41. Turn the crankshaft pulley to gain access to the eight torque converter bolts.

Secure the crankshaft and remove them as they become accessible.

42. Remove the two exhaust manifold stays and plate.

43. Remove the two bolts attaching the transaxle to the oil pan.

44. Remove the six transaxle mounting bolts.

45. Remove the transaxle.

To install:

46. Position the transaxle to the engine. Tighten the transaxle mounting bolts to 47 ft. lbs. (64 Nm).

47. Install the bolts that attach the transaxle to the oil pan bolts and tighten them to 34 ft. lbs. (46 Nm).

48. Install the exhaust manifold support. Tighten the bolts to 14 ft. lbs. (20 Nm).

49. Install the bolts that attach the flywheel to the torque converter. Coat the threads with a locking compound. Rotate the engine and tighten the bolts alternately to 30 ft. lbs. (41 Nm).

50. Install the starter to the engine.

51. Install the flywheel cover and tighten the bolts to 13 ft. lbs. (18 Nm).

52. Install the dipstick and tube with a new O-ring.

53. Connect the clamps and following connectors:

  • O/D solenoid
  • PNP switch speedometer
  • Starter terminal
  • Speed sensor

54. Install the exhaust pipe stay and tighten the bolts to 15 ft. lbs. (21 Nm).

55. Install the right rear insulator and tighten the bolts to 47 ft. lbs. (64 Nm).

56. Install the front engine mounting insulator and tighten the bolts to 47 ft. lbs. (64 Nm).

57. Lower the engine and transaxle into the engine compartment. Tilt the transaxle downward and clear the left mount.

58. Keep the engine level and align the right and left engine mounts.

59. Install the engine mounting bracket and moving control rod. Tighten the bolts to 47 ft. lbs. (64 Nm).

60. Install the right engine stay and tighten the bolts to 23 ft. lbs. (32 Nm).

61. Connect the ground straps.

62. Install the coolant reservoir.

63. Connect the front engine mounting insulator to the body and tighten the bolts to 59 ft. lbs. (81 Nm).

64. Install the engine mounting shock absorber and tighten the bolts to 35 ft. lbs. (48 Nm).

65. Connect the right engine mount and tighten the bolts to 48 ft. lbs. (66 Nm).

66. Connect the left engine mount and tighten the bolts to 47 ft. lbs. (64 Nm).

67. Remove the engine lifting device.

68. Install the power steering pump and the belt.

69. Connect the hydraulic cooling fan pressure hose and tighten the fitting to 33 ft. lbs. (44 Nm).

70. Connect the power steering air tube and hoses.

71. Install the halfshafts.

72. Install the front exhaust pipe with new gaskets.

73. Install the A/C compressor and tighten the bolts to 18 ft. lbs. (25 Nm).

74. Connect the harness to the ECM and assemble the instrument panel.

75. Connect the control cable to the transaxle.

76. Connect the fuel lines and tighten the fittings to 22 ft. lbs. (30 Nm).

77. Connect the heater hoses.

78. Connect the vacuum hoses to the following connections:
- Intake air control valve vacuum tank
- Charcoal canister
- Air intake chamber from the brake booster

79. Connect the engine wiring harness to the engine.

80. Connect the engine relay box. Install the two bolts and connect the following connectors:
- 5 Connections from the relay box
- 2 Ignitor connectors
- Left fender apron connector
- Noise filter connector
- 2 Ground straps

81. Install the radiator.

82. If equipped, install the cruise control actuator.

83. Install the air cleaner cover, volume airflow meter and air cleaner duct assembly.

84. Connect the throttle cable.

85. Connect the accelerator cable.

86. Fill the engine to the proper level with the recommended grade of oil.

87. Align the matchmarks and install the hood.

88. Fill the engine to the proper level with coolant.

89. Bleed the cooling system.

90. Install the battery and tray.

91. Check and/or adjust the ignition timing.

92. Start the engine and check for leaks.

93. Road test the vehicle.

94. Recheck the engine oil and coolant levels.

GS300

1. Release the fuel pressure.

✳✳ CAUTION

The Supplemental Inflatable Restraint (SIR) system must be disarmed before removing the engine assembly. Failure to do so may cause accidental deployment of the air bag, resulting in unnecessary SIR system repairs and/or personal injury.

2. Disconnect the negative battery cable. Wait at least 90 seconds before performing any other work.

3. Remove the hood insulator pad and remove the hood.

4. Remove the engine undercover, then drain the engine coolant and oil.

5. Drain the fuel from the tank.

6. Disconnect the accelerator cable, cruise control actuator cable and the A/T throttle control cable from the throttle body.

7. Remove the air cleaner assembly, volume air flow meter and the air intake hose. Remove the air cleaner duct.

8. Remove the drive belt.

9. Remove the radiator.

10. Disconnect the following wires and electrical connectors:
   a. Igniter
   b. Ignition coil
   c. Wiring harness from the wire clamp and coolant tank
   d. Alternator
   e. Ground strap from the left engine mount
   f. Starter

11. Disconnect the fuel lines from the intake and return lines.

12. Remove the power steering pump without disconnecting the lines and position it aside.

13. Remove the A/C compressor without disconnecting the A/C lines and position it aside.

14. Disconnect the brake booster vacuum hose.

15. Disconnect the EVAP hose.

16. Disconnect the heater hoses from the firewall.

17. Disconnect the heater valve and engine wire from the firewall.

18. Disconnect the electrical harness from the Engine Control Module (ECM) and route it through the firewall.

19. Disconnect and remove the sub heated oxygen sensor (if so equipped) from the front exhaust pipe.

20. Remove the front exhaust pipes and heat insulator.

21. Remove the rear center floor crossmember brace.

22. Disconnect the transmission control rod.

23. Remove the driveshaft.

24. Support the transmission with a jack. Use a piece of wood to prevent damage to the transmission oil pan.

25. Remove the rear transmission crossmember.

26. Attach a lifting device to the engine.

27. Remove the two hole plugs in the front crossmember. Remove the two nuts holding the engine insulators to the front crossmember.

28. Slowly and carefully, remove the engine and transmission from the engine compartment as an assembly. Take great care not to damage the A/C compressor or the cooling fan. Once removed, place the engine and transmission assembly onto a proper stand.

To install:

29. Lower the engine and transmission into the engine compartment.

30. Insert the stud bolts of the front engine mount into their bores in the front engine crossmember. Temporarily install the two nuts.

31. Remove the engine hoist.

32. Temporarily install rear engine support with the four nuts. Install the four support bolts and tighten them to 19 ft. lbs. (25 Nm). Tighten the nuts to 10 ft. lbs. (13 Nm).

33. Tighten the front engine crossmember to mount nuts to 54 ft. lbs. (74 Nm) and install the hole plugs.

34. Install the driveshaft.

35. Shift the transmission control shift rod into **N** by shifting the lever all the way back and returning it two notches. Connect the shift rod to the lever and tighten it to 9 ft. lbs. (13 Nm).

36. Install the rear center floor crossmember brace and tighten the bolts to 9 ft. lbs. (13 Nm).

37. Install the exhaust pipe heat insulator.

38. Install the front exhaust pipes.

39. Install the sub heated oxygen sensor, if equipped.

40. Connect the engine wiring harness to the ECM. Install the ECM and its cover. Reassemble the lower portion of the passenger side instrument panel, the vent, the carpet and the scuff panel.

41. Connect the heater water valve and engine wire to the cowl panel.

42. Connect the heater hoses.

43. Connect the EVAP hose.

44. Connect the brake booster hose.

45. Install the A/C compressor and tighten the Torx® bolt to 19 ft. lbs. (26 Nm). Tighten the nut and bolts to 38 ft. lbs. (52 Nm).

46. Install the power steering pump.

47. Connect the fuel lines with new gaskets and tighten the union bolts to 22 ft. lbs. (29 Nm).

48. Connect the following wires and electrical connectors:

    a. Igniter

    b. Ignition coil

    c. Wiring harness from the wire clamp and coolant tank

    d. Alternator

    e. Ground strap from the left engine mount

    f. Starter

49. Install the radiator.

50. Install the drive belt.

51. Install the air cleaner, volume air flow meter and intake air connector pipe as an assembly.

52. Install the air cleaner duct.

53. Connect the accelerator cable, cruise control cable and the A/T throttle control cable.

54. Fill the tank with fuel.

55. Fill the engine oil to the proper level.

56. Fill the engine coolant to the proper level.

57. Connect the negative battery cable.

58. Start the engine and check for leaks. Bleed the cooling system.

59. Check the automatic transmission fluid level.

60. Check and/or adjust the ignition timing.

61. Install the hood and the hood insulator pad.

62. Road test the vehicle.

63. Recheck the fluid levels.

LS400

1. Remove the battery clamp cover.

2. Relieve the fuel pressure from the fuel system.

3. Disconnect the battery cables, negative cable first.

4. Remove the battery from the engine compartment.

5. Remove the hood from the vehicle.

6. Raise and safely support the vehicle.

7. Remove the oil pan protector from the engine.

8. Drain the engine coolant from the engine.

9. Drain the engine oil from the engine.

10. Remove the air cleaner inlet.

11. Remove the air cleaner and intake air connector assembly.

12. Remove the drive belt, fan clutch and the fan pulley.

13. Disconnect the accelerator cable, cruise control actuator and automatic transmission throttle cable from the throttle body.

14. Remove the radiator assembly from the engine compartment.

15. Disconnect the following connectors, wires, straps, clamps and hoses:

  • Engine oil level sensor connector

  • Alternator connector and wire

  • Engine wire clamp from the bracket on the alternator

  • Two ignitor connectors

  • Engine wire clamp from the igniter bracket

  • Ground strap from the right-hand engine mounting bracket

  • Ground strap from under the left-hand fender apron

  • Engine wire clamp from the cowl panel

  • Radiator reservoir hose from the water bypass pipe

  • Brake booster vacuum hose from the air intake chamber

  • Heater hose from the heater water valve and water bypass pipe

  • Fuel inlet hose from the fuel inlet pipe

  • Fuel return hose to the return pipe

  • Power steering air hose from the air intake chamber

  • Two power steering air hoses from the clamp on the right-hand No. 3 timing belt cover

  • For California vehicles, remove the two EVAP hoses from the pipes (from charcoal canister)

  • For all vehicles except California, disconnect the EVAP hose from the pipe (from the charcoal canister)

16. Disconnect the engine wire from the cabin as follows:

    a. Remove the undercover from under the glove compartment.

    b. Remove the glove compartment.

    c. Disconnect the three ECM connectors.

    d. Disconnect the two cowl wire connectors from the connector on the bracket.

    e. Disconnect the wire clamp from the bracket.

    f. Disconnect the grommet from the cowl panel and pull the engine wire out.

17. Disconnect the power steering oil cooler pipe from the oil pan.

18. Disconnect the heated oxygen sensors from the front exhaust pipe.

19. Remove the front exhaust pipe.

20. Remove the two catalytic converters by removing the three nuts at each converter.

21. Remove the center exhaust pipe.

22. Remove the heat insulator from the rear side of the front exhaust pipe.

23. Remove the front center floor crossmember brace.

24. Remove the rear center floor crossmember brace.

25. Remove the driveshaft from the vehicle.

26. Remove the A/C compressor from the engine without disconnecting the A/C lines.

27. Disconnect the power steering pump from the engine by removing the nut and three bolts. Do not disconnect the power steering lines from the power steering pump. Set the power steering pump aside and support the pump.

28. Remove the heat insulators for the front side of the front exhaust pipe.

29. Remove the engine and transmission assembly as follows:

    a. Disconnect the heater water valve from the cowl panel by removing the two nuts.

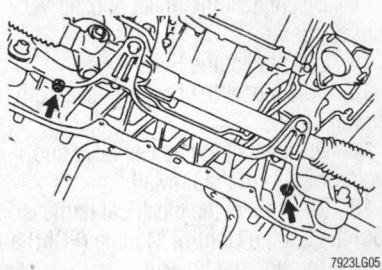

Engine mounting insulator fastener locations—LS400

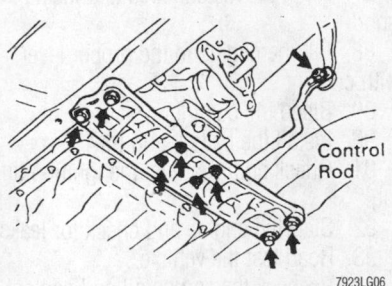

Engine mounting member bolt locations—LS400

b. Attach the engine chain hoist to the engine hangers.

c. Remove the engine mounting insulators from the engine suspension cross-member by removing the two nuts.

d. Disconnect the transmission control rod from the shift lever by removing the nut.

e. Remove the rear engine mounting member by removing the four nuts and four bolts.

f. Lift the engine and transmission assembly out of the vehicle slowly and carefully.

30. Disconnect the engine from the transmission as follows:

a. Disconnect the following:
- Vehicle speed sensor connector
- Park/Neutral switch connector
- Solenoid connector
- Direct clutch speed sensor connector
- Four engine wire clamps from the brackets

b. Remove the oil dipstick and guide from the transmission.

c. Remove the oil cooler pipes from the transmission and clamps.

d. Remove the flywheel housing undercover by removing the two bolts.

e. Turn the crankshaft pulley bolt to gain access to each torque converter bolt.

f. Hold the crankshaft pulley bolt with a wrench and remove the six torque converter bolts.

g. Remove the ten bolts holding the transmission to the engine.

h. Remove the transmission together with the torque converter clutch.

To install:

31. Install the transmission to the engine and install the 10 bolts. Tighten the bolts as follows:
- 14mm head bolt to 27 ft. lbs. (37 Nm)
- 17mm head bolt to 53 ft. lbs. (72 Nm)

32. Install the torque converter clutch bolts as follows:

a. Apply adhesive to two or three threads of the bolt end.

b. Hold the crankshaft pulley bolt with a wrench and install the six bolts evenly. Tighten the bolts to 30 ft. lbs. (41 Nm).

c. Install the flywheel housing undercover with the two bolts. Tighten the bolts to 14 ft. lbs. (19 Nm).

33. Install the oil cooler pipe for the transmission.

34. Install the dipstick guide and dipstick for the transmission.

35. Connect the engine wire to the transmission and connect the following:

- Vehicle speed sensor connector
- Park/Neutral switch connector
- Solenoid connector
- Direct clutch speed sensor connector
- Four wire clamps to the brackets

36. Install the engine and transmission assembly to the vehicle.

37. Install the rear engine mounting member to the vehicle and install the four bolts and four nuts. Tighten the bolts to 19 ft. lbs. (25 Nm) and the nuts to 10 ft. lbs. (14 Nm).

38. Connect the transmission control rod to the shift lever with the nut. Tighten the nut to 9 ft. lbs. (13 Nm).

39. Install the two nuts holding the engine mounting brackets to the front suspension crossmember. Tighten the two nuts to 52 ft. lbs. (70 Nm).

40. Install the heater water valve to the cowl panel with the two nuts.

41. Remove the engine hoist.

42. Install the heat insulators for the front side of the front exhaust pipe.

43. Install the power steering pump with the nut and three bolts. Tighten the nut to 32 ft. lbs. (43 Nm) and the bolts to 29 ft. lbs. (39 Nm).

44. Install the A/C compressor, tighten the bolts to 36 ft. lbs. (49 Nm) and the nut to 22 ft. lbs. (29 Nm).

45. Install the driveshaft to the vehicle.

46. Install the front center floor crossmember brace and tighten the bolts to 9 ft. lbs. (13 Nm).

47. Install the rear center floor crossmember brace and tighten the bolts to 9 ft. lbs. (13 Nm).

48. Install the heat insulator for the rear side of the front exhaust pipe.

49. Install the center exhaust pipe.

50. Install the two front catalytic converters with three new nuts each. Tighten the nuts to 46 ft. lbs. (62 Nm).

51. Install the front exhaust pipe.

52. Tighten the four bolts holding the pipe support bracket to the transmission. Tighten the bolts to 32 ft. lbs. (44 Nm).

53. Install the heated oxygen sensors and tighten the sensors to 33 ft. lbs. (44 Nm).

54. Install the power steering oil cooler pipe.

55. Connect the engine wire to the cabin as follows;

a. Push in the engine wire through the cowl panel. Install the grommet.

b. Connect the following:
- Three ECM connectors
- Two engine wire connectors to the connector on the bracket
- Engine wire clamp to bracket

c. Install the glove compartment and the dash undercover to the vehicle.

56. Attach the following connectors, wires, straps, clamps and hoses:
- Engine oil level sensor connector
- Alternator connector
- Alternator wire
- Engine wire clamp to the bracket on the alternator
- Two igniter connectors
- Engine wire clamp to the igniter bracket
- Ground strap to the right-hand engine mounting bracket
- Ground strap under the left-hand fender apron
- Engine wire clamp to the bracket on the cowl panel
- Radiator reservoir hose to the water bypass pipe
- Brake booster vacuum hose to the air intake chamber
- Heater hose to the heater water valve
- Heater hose to the water bypass pipe
- Fuel inlet hose to the fuel inlet pipe
- Fuel return hose to the return pipe
- Power steering air hose to the air intake chamber
- Two power steering air hoses to the clamp on the right-hand No. 3 timing belt cover
- For California vehicles, connect the two EVAP hoses to the pipe (from the charcoal canister)
- Except for California vehicles, connect the EVAP hose to the pipe (from the charcoal canister)

57. Install the radiator assembly.

58. Connect the accelerator cable, cruise control cable to the throttle body. If equipped with A/T, connect the throttle control cable to the throttle body.

59. Install the fan pulley, fan, fan clutch and the drive belt. Tighten the four nuts for the fan to 16 ft. lbs. (21 Nm).

60. Install the air cleaner and intake air connector assembly.

61. Install the air cleaner inlet.

62. Fill the engine coolant.

63. Install the battery to the vehicle.

64. Fill the engine with oil.

65. Connect the battery cables, positive cable first.

66. Install the battery cover.

67. Start the engine and check for leaks.

68. Install the engine undercover.

69. Install the oil pan protector.

70. Lower the vehicle.

71. Install the hood.

72. Recheck all the fluids and make all necessary engine adjustments.

SC300

1. Release the fuel system pressure.
2. Disconnect the negative battery cable. Wait at least 90 seconds before performing any other work.
3. Remove the battery and tray.
4. Remove the hood.
5. Remove the engine undercover, then drain the engine coolant and oil.
6. Drain the fuel from the fuel tank.

> ❊❊ **CAUTION**
>
> **Do not allow fuel spray or fuel vapors to come in contact with a spark or open flame. Keep a dry chemical fire extinguisher nearby. Never store fuel in an open container due to risk of fire or explosion.**

7. Disconnect the accelerator cable, the cruise control cable and the throttle control cable (A/T only) from the throttle body.
8. Remove the air cleaner assembly, resonator and the air intake hose.
9. Remove the drive belt, fan (with fluid coupling attached) and the water pump pulley.
10. Remove the radiator.
11. Disconnect the EVAP hoses(vacuum hose and air hose) from the charcoal canister and remove the charcoal canister.
12. Without disconnecting the hydraulic or refrigerant lines, remove the power steering pump and the A/C compressor and position them out of the way.
13. Tag and disconnect all wires, electrical leads and vacuum hoses from the block. Disconnect any wiring clips or brackets.
14. On models with manual transmission, remove the shift lever assembly.
15. Remove the undercover beneath the glove box. Remove the lower instrument panel, the trim panel and the glove box door.
16. Remove the right door sill trim

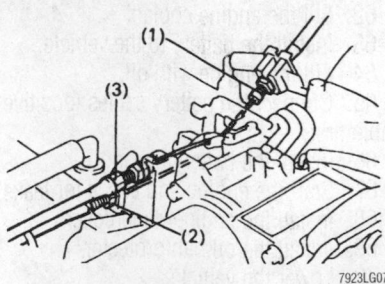

Accelerator cable (1), throttle control cable (2), cruise control cable (3) locations—SC300

(scuff plate). Lift the front edge of the carpet and remove the protective cover from the ECM.
17. Tag and disconnect the ECM connectors, the cowl wire connectors and the control unit connectors behind the glove box.
18. Remove the two nuts holding the harness to the firewall and carefully pull the engine harness into the engine compartment.

➡ **This is best done with an assistant guiding the harness inside the cabin.**

19. Remove the two clamp bolts and disconnect the power steering pipe from the engine block.
20. Remove the union bolt and two gaskets and disconnect the fuel inlet hose. Contain any dripping fuel; clean up spills immediately.
21. Disconnect the starter wiring and disconnect the starter wiring from the clip.
22. For manual transmissions, unbolt and remove the clutch master cylinder; move it aside without disconnecting the hydraulic lines.
23. Disconnect the front exhaust pipe, then remove the heat shield.
24. For automatic transmissions, disconnect the transmission control rod at the shift lever.
25. Remove the intermediate shaft. Place matchmarks on the flanges for proper reassembly. Some vehicles are not equipped with adjusting washers.
26. Attach an engine lift to the lift hooks.
27. Remove the two nuts holding the engine to the front suspension crossmember. Remove the four bolts and four nuts holding the engine to the rear crossmember; lift out the rear engine mount.
28. Slowly and carefully, lift the engine/transaxle assembly out of the engine compartment. Take great care to avoid damaging the A/C compressor or power steering solenoid.
29. Place the engine and transmission assembly onto a stand.
30. For vehicle with A/T, remove the oil dipstick guide and dipstick for the transmission.
31. For vehicle with A/T, remove the oil cooler tubes.
32. Disconnect the engine wire from the transmission.
33. Remove the starter from the engine by removing the bolts.
34. Separate the engine and transmission.
35. For vehicles with manual transmission, remove the clutch cover and disc.

To install:

36. For vehicles with manual transmission, install the clutch cover and disc.
37. Connect the engine and transmission.
38. Install the starter to the engine by installing the bolts..
39. Connect the engine wire to the transmission.
40. For vehicle with A/T, install the oil cooler tubes.
41. For vehicle with A/T, install the oil dipstick guide and dipstick to the transmission.
42. Carefully lower the engine into the engine compartment. With the engine level and all the mounts aligned with their brackets, install the rear mount. Tighten the four nuts to 10 ft. lbs. (13 Nm). Tighten the four bolts to 19 ft. lbs. (25 Nm).
43. Connect the intermediate shaft to the rear differential and tighten the bolts and nuts to 54 ft. lbs. (74 Nm). Install the center support bearing set bolts with the adjusting washers. Tighten the bolts to 36 ft. lbs. (49 Nm).
44. Connect the transmission control rod to the shift lever (automatic transmission only) by installing the nut.
45. Install the exhaust heat insulator by installing the four nuts.
46. Attach the No. 2 front exhaust pipe, tighten the nuts to 46 ft. lbs. (62 Nm). Install the pipe support bracket to the transmission with the two bolts. Tighten the bolts to 32 ft. lbs. (43 Nm). Install a new gasket and the No. 2 front exhaust pipe to the front exhaust pipe with the two bolts and nuts. Tighten the bolts to 32 ft. lbs. (43 Nm).
47. For manual transmissions, install the clutch release cylinder and tighten the bolts to 9 ft. lbs. (12 Nm).
48. Connect the starter wiring and secure the harness in the clips.
49. Connect the fuel inlet hose with two new gaskets and tighten the union bolt to 22 ft. lbs. (29 Nm).
50. Install the power steering pipe below the engine.
51. Carefully push the engine harness back into the cabin. Secure the bolts holding the harness.
52. Connect each connector to the proper ECM, controller, or relay. Make certain each connector is square and secure.
53. Install the ECM and cover; connect the wiring harnesses. Refit the carpet and install the scuff plate.
54. Install the lower instrument panel trim, the glove box door and the undercover.
55. Install the shift lever on models with

manual transmission. Install the upper and lower center console pieces.

56. Secure the engine wiring harness in the clips and retainers. Reconnect all wires, electrical leads and vacuum hoses. Make certain each wiring lead is firmly connected and is properly routed; double check the security of clamps and wiring retainers.

57. Install the A/C compressor and tighten the through-bolt to 19 ft. lbs. (26 Nm). Tighten the other bolt and nut to 38 ft. lbs. (52 Nm).

58. Install the power steering pump. Tighten the long bottom bolt to 43 ft. lbs. (58 Nm); tighten the others to 29 ft. lbs. (39 Nm). Connect the power steering air hoses.

59. Install the charcoal canister and connect the hoses.

60. Install the radiator, coolant hoses and the transmission lines.

61. Install the water pump pulley, the fan and the drive belt. Tighten the four pulley nuts to 12 ft. lbs. (16 Nm).

62. Install the air cleaner assembly, resonator and the air intake hose.

63. Connect the accelerator cable, throttle control cable (A/T only) and the cruise control cable to the throttle body.

64. Install battery tray and battery. Connect the battery cables.

65. Refill all fluids, including fuel.

66. Start the engine and check for leaks. Allow the engine to warm to normal operating temperature.

67. Check the automatic transmission fluid level.

68. Check the ignition timing.

69. Shut the engine off and install the engine undercovers.

70. Install the hood.

71. Road test the vehicle for proper operation. Recheck all fluid levels.

SC400

1. Disconnect the battery cables and remove the battery. Wait at least 90 seconds before proceeding with any other work.

❊❊ CAUTION

The fuel injection system remains under pressure after the engine has been turned off. Properly relieve fuel pressure before disconnecting any fuel lines. Failure to do so may result in fire or personal injury.

2. Relieve the fuel pressure from the fuel lines.

3. Remove the hood.

4. Drain the engine coolant from the cooling system.

5. Remove the V-bank cover, if equipped.

6. Raise and safely support the vehicle.

7. Remove the engine undercover and drain the engine oil. Lower the vehicle.

8. Loosen the drive belt tension by turning the belt tensioner counterclockwise. Remove the drive belt.

9. Disconnect the throttle body.

10. Remove the accelerator, transmission and cruise control cables from the throttle body.

11. Remove the air cleaner assembly.

12. Disconnect the vacuum hose (from the power steering air control valve) from the air intake chamber.

13. Remove the intake air connector.

14. Remove the coolant reservoir tank.

15. Remove the radiator.

16. Disconnect the igniter connectors, then remove the wire clamp from the body.

17. Disconnect the engine wires located next to the relay box. The relay box is located next to the left strut tower.

18. Disconnect the engine ground cable.

19. Disconnect the power steering solenoid valve connector.

20. Remove the alternator.

21. Disconnect the power steering tubes from the suspension crossmember.

22. Disconnect the power steering reservoir tank and bracket from the body by removing the three bolts.

23. Remove the power steering pump by removing the pump mounting bolts and nut. Do not disconnect the power steering lines and place the pump off to the side.

24. Disconnect the air conditioning compressor from the engine. Do not remove the compressor pressure lines.

25. Disconnect the following hoses and ground straps:
- Heater water hose from the water bypass hose.
- Heater water hose from the heater water valve.
- Brake booster hose from the union on the air intake chamber.
- Vacuum hose (from VSV for heater water valve) from the air intake chamber.
- Ground strap from the bracket on the body.
- Fuel inlet hose from fuel tube

26. Remove the charcoal canister from the engine.

27. Disconnect the engine wire from the cabin as follows:

   a. Remove the passenger side lower instrument panel undercover.

   b. Remove the four screws to the lower instrument panel finish panel and glove compartment door assembly. Remove the glove compartment and finish panel.

   c. Pull out the right scuff plate.

   d. Take out the front side of the floor carpet.

   e. Remove the two nuts and ECM protector.

   f. Remove the nut and disconnect the ECM from the floor panel.

   g. Disconnect the following connectors:
- Two connectors from the ECM
- Connector from the ABS and TRAC ECU
- Two connectors from the TRAC ECU
- Four connectors from connector cassette
- Connector from A/C control assembly

   h. Remove the bolt holding the engine wire clamp to the heater water valve bracket.

   i. Remove the two bolts holding the engine wire clamp to the body.

   j. Pull the engine wiring (through the cowl panel) from the vehicle cabin.

28. Disconnect the oxygen sensors from the front exhaust pipe.

29. Remove the front exhaust pipe.

30. Remove the front catalytic converter by removing the three nuts and gasket.

31. Remove the tailpipes.

32. Remove the center exhaust pipe by disconnecting the two hooks.

33. Remove the heat insulator by removing the four nuts.

34. Remove the center floor crossmember brace by removing the four bolts.

35. Remove the driveshaft from the vehicle using the proper tools (two of tool SST 09922–10010), loosen the adjusting nut on the driveshaft. Place matchmarks on the transmission flange and the flexible coupling.

36. Disconnect the transmission control rod from the shift lever by removing the nut.

37. Attach the engine chain hoist to the engine hangers.

38. Remove the two nuts holding the engine mounting insulators to the front suspension crossmember.

39. Remove the four bolts, four nuts and the rear engine mounting member. Disconnect the ground strap to the rear mounting member.

40. Lift the engine out of the vehicle slowly and carefully.

41. Separate the engine from the transmission as follows:

   a. Remove the oil dipstick guide and dipstick for transmission.

b. Remove the oil cooler pipes for the transmission.

c. Disconnect all the engine wiring.

d. Remove the engine bolts holding the transmission to the engine.

e. Disconnect the engine from the transmission.

To install:

42. Attach the transmission to the engine as follows:

a. Install the transmission to the engine and install the bolts. Tighten the bolts to 42 ft. lbs. (57 Nm).

b. Connect the engine wiring.

c. Install the oil cooler pipe for the transmission. Tighten the unions on the pipes to 25 ft. lbs. (34 Nm).

d. Install the engine oil dipstick guide and the dipstick for the transmission.

43. Install the engine and transmission to the vehicle.

44. Install the rear engine mounting member with the four bolts and four nuts. Tighten the bolts to 19 ft. lbs. (25 Nm) and the nuts to 10 ft. lbs. (13 Nm).

45. Install the two nuts holding the engine mounting brackets to the front suspension crossmember. Tighten the nuts to 43 ft. lbs. (59 Nm).

46. Remove the engine chain hoist.

47. Connect the transmission control rod to the shift lever by installing the nut.

48. Install the driveshaft.

49. Install the center floor crossmember brace by installing the four bolts. Tighten the bolts to 9 ft. lbs. (13 Nm).

50. Install the heat insulator for the front exhaust pipe by installing the four bolts.

51. Install the center exhaust pipe by installing the two hooks.

52. Install the tailpipe and tighten the two bolts to 14 ft. lbs. (19 Nm).

53. Install the front catalytic converter and tighten the nuts to 46 ft. lbs. (62 Nm).

54. Install the front exhaust. Tighten the four bolts and nuts holding the catalytic converter to the front exhaust pipe to 32 ft. lbs. (43 Nm). Tighten the two bolts and nuts holding the front exhaust pipe to the center exhaust pipe. Tighten the bolts to 32 ft. lbs. (43 Nm). Tighten the four bolts holding the pipe support bracket to the transmission. Tighten the bolt to 32 ft. lbs. (43 Nm).

55. Install the oxygen sensors to the front exhaust and tighten the sensors to 33 ft. lbs. (44 Nm).

56. Connect the engine wire to the cabin as follows:

a. Push in the engine wire through the cowl panel.

b. Install the engine wire retainer with the three bolts.

c. Connect the following connectors under the dash panel:

• Two connectors to the ECM

• Connector to the ABS and TRAC ECM

• Two connectors to the TRAC ECM

• Four connectors to the connector cassette

• Connector to the A/C control assembly

d. Install the ECM with the nut.

e. Install the ECM protector with the two nuts.

f. Install the floor carpet.

g. Install the scuff plate.

h. Connect the connectors.

i. Install the lower instrument panel finish panel and glove compartment door assembly with the four screws.

j. Install the instrument panel undercover with the two screws.

57. Install the charcoal canister.

58. Connect the following hoses and grounds:

• Heater water hose to the water bypass hose

• Heater water hose to the water valve

• Brake booster hose to the union on the air intake chamber

• Vacuum hose (from VSV for heater water valve) to the air intake chamber

• Ground strap to the bracket on the body

• Fuel inlet hose to the fuel tube. Tighten to 22 ft. lbs. (30 Nm)

• Fuel return hose to the return pipe

59. Install the air conditioning compressor with the nut and three bolts. Tighten the bolts to 36 ft. lbs. (49 Nm) and the nut to 22 ft. lbs. (29 Nm).

60. Install the power steering pump with the nut and three bolts. Tighten the bolts to 29 ft. lbs. (39 Nm) and the nut to 32 ft. lbs. (43 Nm).

61. Install the power steering reservoir tank and bracket with the three bolts.

62. Install the power steering tubes with the clamp and bolt.

63. Install the alternator, tighten the nut and bolt to 27 ft. lbs. (37 Nm).

64. Install the power steering solenoid valve connector.

65. Connect the engine wire connectors.

66. Connect the theft deterrent horn connector.

67. Install the ground cable to the body from the engine.

68. Connect the igniter connectors. Connect the yellow taped connector to the igniter on the rear side.

69. Install the radiator assembly. Install the reservoir tank and the inlet pipe to the

fan shroud and tighten the four bolts to 43 inch lbs. (5 Nm). Connect the two hydraulic lines for the fan motor and tighten the bolts to 47 ft. lbs. (64 Nm).

70. Connect the upper and lower radiator hoses to the radiator. Install the two oil cooler hoses for the transmission to the radiator.

71. Install the coolant tank.

72. Install the intake air connector.

73. Connect the vacuum hose (from the power steering air control valve) to the air intake chamber.

74. Install the air cleaner.

75. Connect the accelerator cable, transmission throttle control cable and the cruise control actuator cable to the engine.

76. Install the throttle cover and hose clamp with the cap nut and two bolts. Also install the EVAP hose to the hose clamp.

77. Install the drive belt to the engine.

78. Install the battery to the engine compartment and connect the electrical connectors.

79. Fill the engine coolant.

80. Install the V-bank cover if it was removed.

81. Fill the engine oil and check the transmission oil.

82. Start the engine, bleed the cooling system and check for leaks.

83. Install the engine undercover. Install the hood.

84. Perform a road test and recheck all fluids.

Water Pump

REMOVAL & INSTALLATION

3.0L (1MZ-FE) Engine

1. With the ignition switch in the **LOCK** position, disconnect the negative battery terminal. If equipped with an air bag system, wait at least 90 seconds or longer before performing any other work.

2. Drain the engine coolant.

3. Remove the timing belt.

4. Mark the left and right camshaft pulleys with a touch of paint. Using SST tools 09249–63010 and 09960–1000 or equivalents, remove the bolts to the right and left camshaft pulleys. Remove the pulleys from the engine. Be sure not to mix up the pulleys.

5. Remove the No. 2 idler pulley by removing the bolt.

6. Disconnect the three clamps and engine wire from the rear timing belt cover.

7. Remove the six bolts holding the

rear timing belt cover to the engine block.

8. Remove the four bolts and two nuts to the water pump.

9. Remove the water pump.

10. Remove all the old packing (sealant) and gasket material from the water pump and clean the mounting surfaces.

11. Scrape and clean all gasket material from the upper inner timing belt cover.

To install:

12. Check that the water pump turns smoothly. Also check the air hole for coolant leakage.

13. Using a new gasket, apply liquid sealer to the gasket, water pump and engine block.

14. Install the gasket and pump to the engine and install the four bolts and two nuts. Tighten the nuts and bolts to 53 inch lbs. (6 Nm).

15. Install the rear timing belt cover and tighten the six bolts to 74 inch lbs. (9 Nm).

16. Connect the engine wire with the three clamps to the rear timing belt cover.

17. Install the No. 2 idler pulley with the bolt. Tighten the bolt to 32 ft. lbs. (43 Nm). After torquing the bolt, be sure the idler pulley moves smoothly.

18. With the flange side **outward**, install the right-hand camshaft pulley to the engine. Be sure to align the knock pin hole on the camshaft pulley with the knock pin on the camshaft. Using the same tools as removal, tighten the camshaft bolt to 65 ft. lbs. (88 Nm).

19. With the flange side **inward**, install the left-hand camshaft pulley to the engine. Be sure to align the knock pin hole on the camshaft pulley with the knock pin on the camshaft. Using the same tools as removal, tighten the camshaft bolt to 94 ft. lbs. (125 Nm).

20. Install the timing belt to the engine.

21. Fill the engine coolant.

22. Connect the negative battery cable to the battery and start the engine.

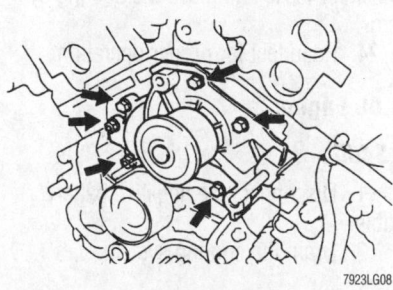

Water pump mounting bolts—ES300

23. Top off the engine coolant and check for leaks.

3.0L (2JZ-GE) Engine

GS300

1. Disconnect the negative battery cable. Wait at least 90 seconds before performing any other work.

2. Drain the cooling system.

3. Remove the lower engine cover.

4. Remove the nut, then remove the air cleaner duct.

5. Disconnect the high tension lead from the ignition coil.

6. Disconnect the high tension lead from the clamp on the VAF meter.

7. Disconnect the VAF meter electrical connector and disconnect the harness from the meter.

8. Disconnect the power steering air hose from the timing belt cover.

9. Disconnect the PCV hose from the cylinder head cover.

10. Loosen the hose clamp bolt securing the intake air connector pipe to the throttle body.

11. Remove the three bolts, air cleaner, VAF meter and intake air connector pipe assembly.

12. Remove the drive belt.

13. Place matchmarks on the cooling fan clutch and the cooling fan pulley. Remove the cooling fan by removing the four nuts.

14. Remove the water pump pulley.

15. Remove the radiator.

16. Remove the two nuts and disconnect the water inlet from the water pump.

17. Remove the thermostat.

18. Remove the distributor.

19. On all except California vehicles, remove the four nuts and exhaust manifold heat insulator.

20. Remove the two bolts, the No. 1 water bypass outlet and pipe. Discard the O-rings.

21. Remove the timing belt.

22. Remove the idler pulley.

23. Remove the mounting bolt and disconnect the engine wire bracket.

24. Remove the alternator mounting bolt and disconnect the alternator from the water pump.

25. Remove the two nuts and disconnect the No. 2 water bypass pipe from the water pump.

26. Remove the water pump mounting bolts. Note the position and type of each bolt for correct installation.

27. Lift out the water pump. If prying is necessary, use a protected blade and take great care not to damage the mating surfaces.

28. Remove all the old sealant and gasket material; clean the mounting surfaces.

To install:

29. Install a new O-ring to the cylinder block and a new gasket to the water pump.

30. Connect the water bypass pipe to the water pump. Do not install the nuts yet.

31. Install the water pump. Fit the bolts and hand-tighten them, then tighten the mounting bolts to 15 ft. lbs. (21 Nm).

➡ **Hand-tighten the A bolts prior to hand-tightening the B bolts.**

32. Install the nuts to the No. 2 water bypass pipe and tighten them to 15 ft. lbs. (21 Nm).

33. Install the alternator mounting bolt and nut and tighten them to 27 ft. lbs. (37 Nm).

34. Connect the engine wiring harness. Install and tighten the bolts.

35. Install the idler pulley.

36. Install the timing belt.

37. Install new O-rings to the No. 1 water bypass pipe and outlet. Install the bolts and tighten them to 78 inch lbs. (9 Nm).

38. Install the exhaust manifold heat insulator if it was removed and tighten the nuts to 13 ft. lbs. (18 Nm).

39. Install the distributor.

40. Install the thermostat and align the jiggle valve with the protrusion on the water inlet housing.

41. Install the water inlet housing; tighten the bolts to 78 inch lbs. (9 Nm).

42. Install the radiator.

43. Align the matchmarks and install the water pump pulley and fan assembly. Do not tighten the nuts at this time.

44. Install the drive belt and tighten the fan assembly nuts to 12 ft. lbs. (16 Nm).

45. Connect the intake air connector pipe to the throttle body.

46. Install the air cleaner, VAF meter and intake air connector pipe assembly with the three (3) bolts.

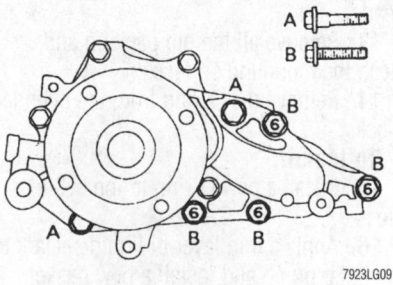

Water pump mounting bolt locations—GS300 and 1996–97 SC300

47. Install the hose clamp.
48. Connect the power steering air hose to the timing belt cover.
49. Connect the PCV hose to the cylinder head cover.
50. Connect the VAF meter harness and electrical connector.
51. Connect the high tension lead to the VAF meter and ignition coil.
52. Connect the air cleaner duct to the air cleaner.
53. Install the air cleaner duct.
54. Refill the cooling system.
55. Connect the negative battery cable.
56. Start the engine and check for leaks.
57. Check the fluid level on vehicles with A/T.
58. Inspect the engine ignition timing.
59. Install the lower engine cover.
60. Road test the vehicle.

1995 SC300

1. Disconnect the negative battery cable. Wait at least 90 seconds before performing any other work.
2. Drain the cooling system.
3. Remove the timing belt.
4. Remove the idler pulley.
5. Remove the thermostat.
6. Remove the two bolts, water bypass outlet and the No. 1 water bypass pipe. Discard the three O-rings.
7. Remove the mounting bolt and disconnect the engine wiring harness bracket above the alternator.
8. Remove the alternator mounting nut.
9. Remove the alternator mounting bolt and disconnect the alternator from the water pump.
10. Remove the two nuts and disconnect the No. 2 water bypass pipe from the water pump.
11. Remove the six water pump mounting bolts. The bolts are of different lengths and styles; note the correct position of each bolt during removal.
12. Lift out the water pump by carefully prying between the pump and the cylinder head.
13. Remove all the old packing and clean the mounting surfaces.
14. Remove the O-ring from the cylinder block.
   To install:
15. Install a new O-ring to the cylinder block.
16. Apply a thin layer of liquid sealant to the water pump and install a new gasket.

Be sure to use new O-rings when installing the water bypass pipe—SC300

17. Connect the No. 2 water bypass pipe to the water pump. Do not install the nuts yet.
18. Install the water pump. Install the bolts in the correct positions and tighten them finger-tight. Tighten the mounting bolts to 15 ft. lbs. (21 Nm).
19. Install the nuts to the No. 2 water bypass pipe and tighten them to 15 ft. lbs. (21 Nm).
20. Install the alternator mounting bolt and nut and tighten them to 27 ft. lbs. (37 Nm).
21. Connect the engine wiring harness. Install and tighten the bolts.
22. Install new O-rings to the No. 1 water bypass pipe and outlet. Install the bolts and tighten them to 78 inch lbs. (9 Nm).
23. Install the thermostat.
24. Install the idler pulley.
25. Install the timing belt.
26. Refill the cooling system.
27. Connect the negative battery cable.
28. Start the engine, bleed the cooling system and check for leaks.
29. Recheck the fluid levels and the ignition timing.

1996–99 SC300

1. Disconnect the negative battery cable. Wait at least 90 seconds before performing any work.
2. Drain the engine coolant and remove the radiator assembly.
3. Remove the air cleaner, MAF meter and the intake air connector pipe assembly.
4. Remove the timing belt.
5. Remove the idler pulley.
6. Remove the water inlet and the thermostat.
7. Remove the 2 bolts, the water bypass outlet and the No. 1 water bypass pipe. Remove the three O-rings from the water bypass outlet and the No. 1 water bypass pipe.
8. Remove the generator.
9. Remove the bolt and disconnect the engine wire bracket. Remove the bolt and disconnect the clamp bracket for the crankshaft position sensor connector. Remove the nuts and disconnect the No. 2 water bypass pipe from the water pump. Remove the six bolts and the water pump and gasket.
10. Remove the drain hose and the O-ring from the cylinder block.
   To install:
11. Install a new O-ring to the cylinder block.
12. Install the drain hose.
13. Install a new gasket to the water pump. Connect the water pump to the water bypass pipe. Do not install the nut yet. Install the water pump with the two bolts (A) and the four bolts (B).

➡**Hand-tighten the (A) bolts first. Tighten all six bolts to 15 ft. lbs. (21 Nm).**

14. Install the two nuts holding the No. 2 water bypass pipe to the water pump. Tighten the nuts to 15 ft. lbs. (21 Nm). Install the clamp bracket for the crankshaft position sensor connector.
15. Install the engine wire bracket.
16. Install the generator.
17. Install new O-rings to the No. 1 water bypass pipe. Install a new O-ring and the water bypass outlet with the two bolts and tighten them to 78 inch lbs. (9 Nm).
18. Install the thermostat and the water inlet.
19. Install the idler pulley.
20. Install the timing belt.
21. Install the air cleaner, the MAF meter and the intake air connector pipe assembly.
22. Install the radiator assembly.
23. Reconnect the negative battery cable. Refill the cooling system. Start the engine, check for leaks and bleed the cooling system.
24. Road test for proper operation.

4.0L Engine

LS400

1. Disconnect the negative battery cable.
2. Drain the cooling system.

3. Remove the timing belt and the No. 2 idler pulley.

4. Remove the right side ignition coil.

5. Remove the two water inlet housing to water pump bolts.

6. Disconnect the IAC valve bypass hose from the water inlet housing.

7. Remove the water inlet housing and discard the O-ring.

8. Remove the mounting bolts, studs and nut. Lift out the water pump by carefully prying between the pump and the cylinder head.

9. Remove all the old packing and clean all mounting surfaces. Remove the O-ring from the water bypass pipe.

To install:

10. Install new seal packing to the water pump groove and a new O-ring to the water bypass pipe.

11. Connect the water pump to the water bypass pipe end.

12. Install the water pump and tighten the mounting bolts to 13 ft. lbs. (18 Nm).

13. Apply sealant to the groove of the water inlet housing.

14. Install a new O-ring to the water inlet housing.

15. Push the water inlet housing end into the water pump hole.

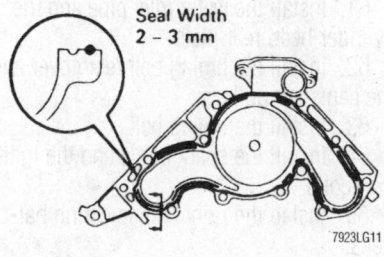

Apply seal packing (silicone sealant) to the water pump as shown—SC400

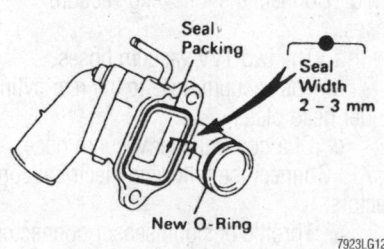

To avoid leakage, apply sealant to the water inlet housing as shown—SC400

16. Connect the IAC valve bypass hose to the water inlet housing.

17. Install the water inlet and housing assembly with the two bolts. Alternately tighten the bolts to 13 ft. lbs. (18 Nm).

18. Install the ignition coil.

19. Install the No. 2 idler pulley for the timing belt.

20. Install the timing belt.

21. Refill the cooling system.

22. Connect the negative battery cable.

23. Start the engine and check for leaks.

SC400

1. Disconnect the negative battery cable.

2. Drain the cooling system.

3. Remove the timing belt.

4. Remove the No. 2 idler pulley.

5. Remove the right side ignition coil by removing the ignition coil connector and two bolts.

6. Disconnect the bypass hose(s) from the water inlet housing.

7. Remove the two bolts holding the water inlet housing to water pump.

8. Remove the water inlet housing and discard the gasket.

9. Remove the mounting bolts, studs and the nut to the water pump. Remove the water pump by carefully prying between the pump and the cylinder head.

10. Remove all the old gasket and clean all mounting surfaces.

To install:

11. Install new seal packing to the water pump groove and a new O-ring to the water bypass pipe end.

12. Connect the water pump to the water bypass pipe end.

13. Install the water pump and tighten the mounting bolts and nut to 13 ft. lbs. (18 Nm).

14. Apply sealant to the groove of the water inlet housing.

15. Install a new O-ring to the water inlet housing.

16. Push the water inlet housing end into the water pump hole.

17. Install the water inlet and housing assembly with the two bolts. Alternately tighten the bolts to 13 ft. lbs. (18 Nm).

18. Connect the bypass hose(s) to the water inlet housing.

19. Replace the ignition coil by installing the two bolts and coil connector.

20. Install the No. 2 idler pulley.

21. Install the timing belt.

22. Refill the cooling system.

23. Connect the negative battery cable.

24. Start the engine, bleed the cooling system and check for leaks.

Cylinder Head

REMOVAL & INSTALLATION

3.0L (1MZ-FE) Engine

✳✳ CAUTION

The fuel injection system remains under pressure even after the engine has been turned OFF. The fuel system pressure must be relieved before disconnecting any fuel lines. Failure to do so may result in fire and/or personal injury.

1. Relieve the fuel pressure.

2. Disconnect the negative battery cable. If the vehicle is equipped with an air bag, wait at least 90 seconds from the time the ignition switch is turned to the **LOCK** position and the negative battery cable is disconnected before starting work.

3. Drain the cooling system.

4. Disconnect the accelerator cable and the throttle cable on vehicles equipped with an automatic transaxle.

5. Remove the air cleaner cover, air flow meter and the air duct.

6. Remove the cruise control actuator and bracket, if equipped.

7. Disconnect the two engine ground straps.

8. Remove the right engine mounting support.

9. Disconnect the radiator hoses.

10. Disconnect the two heater hoses.

11. Disconnect and plug the fuel feed and return lines from the fuel rail assembly.

12. Disconnect and plug the pressure hose from the hydraulic motor.

13. Remove the V-bank cover.

14. Disconnect the following vacuum hoses:

    a. Fuel pressure control VSV

    b. Fuel pressure regulator

    c. Cylinder head rear plate

    d. Intake air control valve VSV

    e. EGR vacuum modulator

    f. EGR valve

15. Disconnect the following connectors:

    a. Intake air control valve

    b. Fuel pressure regulator

    c. EGR VSV

16. Remove the two nuts and the emission control valve set.

17. Disconnect the following hoses:

    a. Brake booster vacuum hose

    b. PCV hose

    c. Intake air control valve vacuum hose

18. Remove the data link connector from the mounting bracket.

19. Remove the two ground straps from the intake chamber.

20. Remove the hydraulic motor pressure hose from the intake chamber.

21. Remove the right oxygen sensor connector from the P/S pressure tube.

22. Remove the two nuts and the P/S pressure tube from the intake chamber.

23. Disconnect the two P/S air hoses.

24. Remove the engine hanger and the intake chamber support.

25. Remove the EGR pipe and gaskets.

26. Disconnect the following connectors;
 a. Throttle position sensor connector
 b. IAC valve connector
 c. EGR gas temperature connector
 d. A/C idle up connector

27. Disconnect the following vacuum hoses:
 a. Two vacuum hoses from the TVV
 b. Vacuum hose from the cylinder head rear plate
 c. Vacuum hose from the charcoal canister

28. Disconnect the air assist hose and the two water bypass hoses.

29. Remove the air intake chamber.

30. Disconnect the left engine wiring harness and position it out of the way.

31. Remove the wiring harness from the rear of the engine.

32. Disconnect the right engine wiring harness and position it out of the way.

33. Remove the ignition coils and the spark plugs.

34. Remove the timing belt.

35. Remove the camshaft pulleys and the timing belt rear cover.

36. Remove the cylinder head rear plate.

37. Remove the water inlet pipe.

38. Remove the air assist hose and vacuum hose.

39. Remove the intake manifold and fuel rail assembly.

40. Remove the water outlet.

41. Remove the EGR pipe from the right exhaust manifold.

42. Remove the exhaust manifolds.

43. Remove the dipstick assembly and the P/S pump bracket.

44. Remove the valve covers and the camshaft position sensor.

45. Remove the camshafts.

46. Be sure the engine is at or near ambient temperature and remove the two (one on each head) 8mm recessed hex bolts. Loosen and remove the 8 head bolts evenly, in 3 passes, in the reverse order of the tightening sequence. Carefully lift the head from the engine; if it is necessary to pry the head loose, take great care not to damage the mating surfaces. Place the head on wood blocks in a clean work area.

➡ **If the cylinder head bolts are loosened out of sequence, warpage or cracking could result.**

47. Remove the cylinder head gasket. With a gasket scraper, carefully remove all the old gasket material from the cylinder head and engine block surfaces.

To install:
48. Place the new cylinder head gasket onto the cylinder block. Place the cylinder head onto the gasket.

49. Coat the threads of the 8 cylinder head bolts (12-sided) with clean engine oil and install the bolts into the cylinder head. Uniformly tighten the bolts in sequence in

three steps to an ultimate tighten of 40 ft. lbs. (54 Nm). If any of the bolts does not meet the torque, replace it.

50. Mark the forward edge of each bolt with paint, then retighten each bolt, in proper sequence, an additional 90 degrees. Check that each painted mark is now at a 90 degrees angle to the front. The paint mark should have been applied to the bolt in the 9 o'clock position and should now be in the 12 o'clock position.

51. Coat the threads of the two remaining 8mm bolts with engine oil and install them. Tighten to 13 ft. lbs. (18 Nm).

52. Install the camshafts and adjust the valves.

53. Apply sealant to the cylinder heads where the camshaft supports meet the cylinder heads.

54. Use new gaskets and install the cylinder head covers.

55. Install the dipstick and power steering pump bracket.

56. Install the exhaust manifolds. Tighten the nuts to 36 ft. lbs. (49 Nm).

57. Install the EGR pipe to the right exhaust manifold.

58. Install the water outlet.

59. Install the intake manifold and the fuel rail assembly. Tighten the intake manifold nuts and bolts to 11 ft. lbs. (15 Nm).

60. Install the air assist hose and the two water bypass hoses.

61. Install the water inlet pipe and the cylinder head rear plate.

62. Install the timing belt rear cover and the camshaft pulleys.

63. Install the timing belt.

64. Install the spark plugs and the ignition coils.

65. Install the right engine wiring harness.

66. Install the wiring harness to the rear of the engine.

67. Install the left engine wiring harness.

68. Install the air intake chamber.

69. Use new gaskets and install the EGR pipe.

70. Connect the following vacuum hoses:
 a. The two TVV vacuum hoses.
 b. The vacuum hose to the rear cylinder head plate.
 c. Charcoal canister vacuum hose.

71. Connect the following electrical connectors:
 a. Throttle position sensor connector.
 b. IAC valve connector.
 c. EGR gas temperature connector.
 d. A/C idle up connector.

72. Install the engine hanger and the intake chamber support.

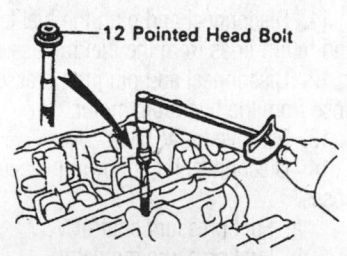
12 Pointed Head Bolt

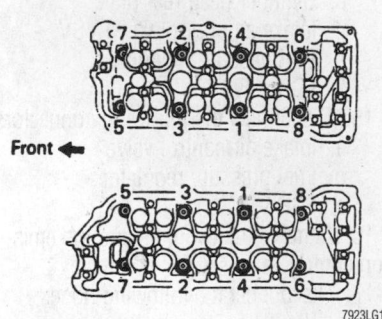
7923LG13

Cylinder head bolt tightening sequence—ES300

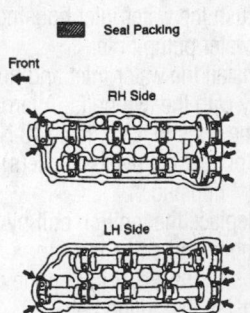
Seal Packing
Front
RH Side
LH Side
7923LG14

Apply sealant to the shaded areas on the cylinder head—3.0L (1MZ-FE) engine

73. Connect the two P/S air hoses.
74. Install the P/S pressure tube to the intake chamber.
75. Install the oxygen sensor connector to the pressure tube.
76. Install the two ground straps to the intake chamber.
77. Install the data link connector to the bracket.
78. Connect the following hoses:
　a. Power brake booster vacuum hose.
　b. PCV hose.
　c. IAC valve vacuum hose.
79. Install the emission control valve set and related vacuum hoses and connectors.
80. Install the V-bank cover.
81. Connect the pressure hose to the hydraulic motor.
82. Connect the fuel lines to the fuel rail assembly.
83. Connect the heater and radiator hoses.
84. Install the right engine mounting support.
85. Connect the two engine ground straps.
86. Install the cruise control actuator and bracket.
87. Install the air cleaner, air flow meter and air duct assembly.
88. Connect the accelerator cable and the throttle cable on vehicles equipped with an automatic transaxle.
89. Fill the cooling system to the proper level with coolant.
90. Connect the negative battery cable.
91. Start the engine and check for leaks. Bleed the air from the cooling system.
92. Adjust the ignition timing.
93. Road test the vehicle and check for unusual noise, shock, slippage, correct shift points and smooth operation.
94. Recheck the coolant and engine oil levels.

3.0L (2JZ-GE) Engine

1. Disconnect the negative battery cable. Wait at least 90 seconds before performing any other work.
2. Drain the engine coolant.
3. Relieve the fuel pressure from the fuel lines.

❋❋ CAUTION

The fuel injection system remains under pressure after the engine has been turned OFF. Properly relieve fuel pressure before disconnecting any fuel lines. Failure to do so may result in fire or personal injury.

4. Remove the undercovers.
5. Disconnect the accelerator, throttle control (A/T only) and cruise control cables from the throttle body.
6. Remove the air cleaner duct.
7. Remove the air cleaner, airflow meter and the intake air pipe.
8. Remove the drive belt, the fan and fluid coupling and the water pump pulley.
9. Remove the No. 2 front exhaust pipe.
10. If equipped with an exhaust manifold cover, remove the four nuts and remove the cover from the manifold.
11. Disconnect the 2 heated oxygen sensor connector(s). Remove the exhaust manifolds and gaskets by removing the eight bolts.
12. Remove the water bypass outlet and the No. 1 water bypass pipe.
13. Remove the power steering air hose from the No. 4 timing belt cover. Remove the power steering hose from the air intake chamber. Remove the two bolts and disconnect the vane pump from the pump bracket.

➡**Put aside the vane pump and suspend it. Remove the two bolts and the pump rear stay.**

14. Disconnect the fuel return hose from the fuel return pipe. Plug the hose end. Disconnect the fuel return hose from the oil dipstick guide.
15. Remove the bolt and bracket and disconnect the engine wire from the intake manifold stay.
16. Remove the throttle body and intake air connector assembly.
17. Remove the bolt, pull out the oil dipstick guide with the dipstick and remove the O-ring from the dipstick guide. If equipped with A/T, remove that transmission dipstick and guide.
18. Disconnect the connector from the No. 2 vacuum pipe. Disconnect the EGR gas temperature sensor connector from the wiring connector.
19. Remove the 2 nuts and disconnect the vacuum pipe from the air intake chamber and intake manifold. Remove the No. 2 vacuum pipe and VSV assembly.
20. Remove the nuts and disconnect the vacuum tank from the intake manifold. Disconnect the VSV connector and hoses, the vacuum hose(from the air intake chamber) from port B of the vacuum tank and the vacuum hose(from actuator) from the VSV. Remove the vacuum control valve set.
21. Remove the DLC1 bracket and VSV assembly.
22. Disconnect the vacuum hose from the brake booster union and the EVAP hose

from the No. 2 vacuum pipe. Remove the bolt holding the engine wire protector to the air intake chamber. Remove the five bolts, nut, air intake chamber and gasket.
23. Remove the No. 3 (top) timing belt cover by removing the oil filler cap and the 6 bolts using a 5mm hexagon wrench.
24. Using a 5mm hexagon wrench, remove the four bolts and the rear cylinder head cover.
25. Disconnect the spark plug cables and free the wires from the clips. Remove the spark plugs.
26. Remove the distributor with the spark plug wires attached.
27. Remove the spark plugs.
28. Remove the drive belt tensioner by removing the three bolts.
29. Set the engine to TDC/compression for cylinder No. 1. Turn the crankshaft pulley clockwise to align the pulley's groove with the timing mark **0** on the lower cover. Check that the timing marks of the camshaft pulleys are aligned with the marks on the rear belt cover. If the cam pulley marks do not align. rotate the crankshaft an additional full turn.
30. Alternately loosen the two bolts holding the timing belt tensioner; remove the tensioner and dust boot. Remove the timing belt from the camshaft pulleys. Support the belt so that it remains in contact with the crankshaft pulley.

➡**Protect the belt from fluids and grease. Take great care not to drop any objects into the lower timing belt cover.**

31. To disconnect the engine wire, remove and disconnect the following:
　a. The wire clamp from the bracket.
　b. The heated oxygen sensor connectors and the crankshaft position sensor connector.
　c. Remove the two bolts and disconnect the two ground straps from the intake manifold.
　d. Disconnect the ECT connector, the ECT sender gauge connector, the knock sensor connectors, the oil pressure switch connector, the oil level sensor connector, the AC compressor connector and the six injector connectors.
　e. Remove the three nuts and disconnect the engine wire protector from the intake manifold.
32. Disconnect the water bypass hose from the clamp on the oil filter bracket. Remove the two nuts, bolt and water outlet with the water bypass hose.
33. Remove the two bolts and the intake manifold stay.

34. Remove the fuel pressure pulsation damper.

35. Remove the clamp bolt from the intake manifold, remove the union bolt and gaskets and disconnect the fuel inlet pipe.

36. Remove the six bolts, two nuts, the intake manifold and the delivery pipe assembly and gasket.

37. Remove the cylinder head covers (valve covers).

38. Remove the camshaft timing pulleys.

39. Remove the rear (No. 4) timing belt cover.

40. Remove the camshafts.

41. Uniformly loosen the cylinder head bolts in several passes and in the reverse order of the tightening sequence. The cylinder head may crack or warp if the correct order is not followed.

42. Remove the head from the engine. If prying is necessary, use a protected blade; take great care not to damage the mating surfaces of the head and block.

43. Clean the head and block of all gasket material. Take great care not to gouge or scratch the mating surfaces.

To install:

44. Place a new gasket on the cylinder block. Be sure it is positioned correctly. Install the cylinder head in position.

45. Lightly coat the head bolt threads and plate washers with engine oil. Install plate washers and bolts to the head.

46. Uniformly tighten the head bolts in several passes in sequence to 25 ft. lbs. (34 Nm).

47. Mark the front (towards the front of the engine) of each bolt with a dot of paint. Following the correct order and tighten each bolt an additional 90 degrees. When complete, all the paint marks should face to the side of the engine.

48. Again following the correct order, tighten the bolts another 90 degrees of rotation. When complete, the paint marks

should face the rear of the engine, exactly 180 degrees away from the original starting point. Correct bolt torque is expressed as 25 ft. lbs. (34 Nm) + 90° + 90°.

➡**Correct bolt torque must be achieved in 3 steps; do not attempt to shorten the procedure by combining the two 90 degree steps.**

49. Coat the thrust portions of each camshaft with engine oil, then position them in the cylinder head with the cam lobes and the knock pins in the correct position.

50. Position the No. 3 and No. 7 bearing caps in place, coat the bolt threads with oil, then uniformly and alternately tighten them temporarily.

51. Coat new oil seals with multi-purpose grease, then slide them over the camshafts.

52. Clean the surfaces of the No. 1 bearing cap and cylinder head with cleaner. Apply seal packing to the No. 1 bearing cap.

53. Install the remaining bearing caps in their proper locations.

54. Coat the threads of each bolt with clean oil, then tighten them, in several passes, in the correct sequence, to 14 ft. lbs. (20 Nm).

55. Using SST tool 09316–60010 or equivalent, press the two oil seals in as far as it will go.

56. Rotate each camshaft until the forward straight (knock) pin is straight up. Loosen the exhaust Nos. 1, 2 and 6 bearing cap bolts until they can be turned by hand; retighten the bolts, in several passes, to 14 ft. lbs. (20 Nm). Loosen the intake Nos. 1, 2 and 5 bearing cap bolts and retighten the bolts, in several passes, to 14 ft. lbs. (20 Nm).

57. Turn each camshaft ⅓ of a revolution (120 degrees). Loosen the exhaust Nos. 4 and 7 bearing cap bolts; retighten the bolts, in several passes, to 14 ft. lbs. (20 Nm). Loosen the intake Nos. 4 and 6 bear-

ing cap bolts; retighten the bolts, in several passes, to 14 ft. lbs. (20 Nm).

58. Turn each camshaft an additional ⅓ of a revolution, loosen the exhaust bearing cap bolts Nos. 3 and 5, then retighten the bolts, in several passes, to 14 ft. lbs. (20 Nm). Loosen the intake bearing cap bolts Nos. 3 and 7, then retighten the bolts, in several passes, to 14 ft. lbs. (20 Nm).

59. Check and adjust the valve clearance.

60. Install the rear (No. 4) timing belt cover. Tighten the bolts to 78 inch lbs. (9 Nm).

61. Install the camshaft timing pulleys. Align the shaft pin with the pulley groove and slide the pulley on. Install the bolt temporarily. Hold the hex portion of the camshaft with a wrench and tighten the pulley bolt to 59 ft. lbs. (79 Nm).

62. Install the cylinder head covers.

63. Install the intake manifold and delivery pipe with a new gasket. Tighten the 6 bolts and 2 nuts to 20 ft. lbs. (27 Nm).

64. Install the fuel inlet pipe to the fuel rail. Tighten the union bolt to 30 ft. lbs. (42 Nm). Install the clamp bolt to the intake manifold.

65. Install the fuel pressure pulsation damper.

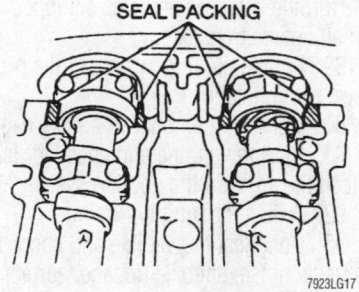

SEAL PACKING

7923LG17

Apply sealant to the areas indicated on the cylinder head before installing the cover— 3.0L (2JZ-GE) engine

66. Install the intake manifold stay and tighten the bolts to 29 ft. lbs. (39 Nm)

67. Install the water outlet and the bypass hose. Tighten the bolts to 15 ft. lbs. (21 Nm).

68. Install the engine wiring harness. Secure the wiring in all clamps and retainers. Connect each wiring lead to the proper sender, sensor or switch. Connect the injector leads.

69. Compress the timing belt tensioner in a vise and retain the pin with a 1.5mm hex wrench. Install the dust boot onto the tensioner.

70. Install the tensioner. Alternately tighten the bolts to 20 ft. lbs. (26 Nm).

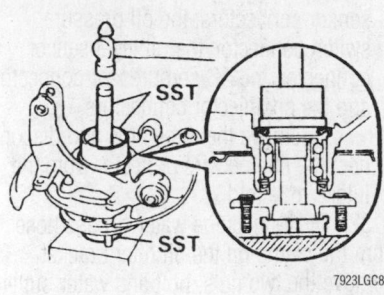

SST

SST

7923LGC8

Cylinder head bolt tightening sequence— 3.0L (2JZ-GE) engine

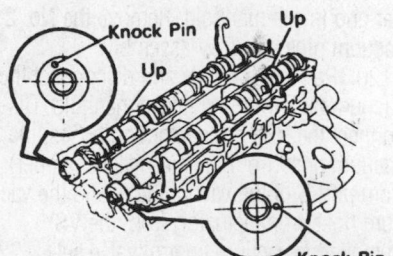

Knock Pin

Up

Up

Knock Pin

7923LG15

Position the knock pins as shown when installing the camshafts—3.0L (2JZ-GE) engine

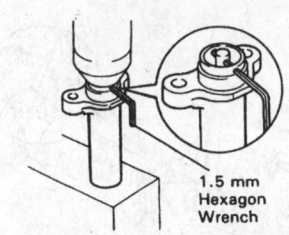

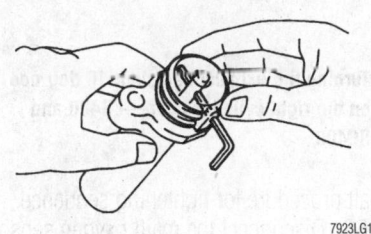

7923LG16

Compressing the timing belt tensioner— 3.0L (2JZ-GE) engine

Remove the hex wrench with a pair of pliers, allowing the tensioner to be applied to the timing belt.

71. Turn the crankshaft two full revolutions clockwise. Check that all timing marks align as before. If the marks (cam and crankshaft) do not align, remove the timing belt and reinstall it.

72. Install the accessory drive belt tensioner. Take great care not to drop the bolts inside the lower timing cover. Tighten the bolts to 15 ft. lbs. (21 Nm).

73. Double check that the engine is still set to TDC/compression for cylinder No. 1. Check the alignment of both the crank and camshaft timing marks. Install the timing belt.

74. Install the spark plugs.

75. Install the distributor, making sure all reference marks are aligned. Connect the wiring to the spark plugs.

76. Install the No. 3 timing belt cover.

77. Install the cylinder head rear cover.

78. Install the air intake chamber with a new gasket. Tighten the bolts to 20 ft. lbs. (27 Nm). Install the bolt to hold the engine wire protector to the air intake chamber. Connect the vacuum hose to the brake booster union and the EVAP hose to the No. 2 vacuum pipe.

79. Install the DLC connector and bracket and VSV connector.

80. Install the vacuum control set.

81. Install the No. 2 vacuum pipe assembly and connect the hoses. Tighten the nuts to 20 ft. lbs. (27 Nm).

82. Install the EGR gas temperature sensor. Tighten it to 14 ft. lbs. (20 Nm). Connect the vacuum hoses.

83. Install the dipstick tubes. Always use a new O-ring on each tube.

84. Install the intake chamber supports and tighten the bolts to 13 ft. lbs. (18 Nm). The supports are marked F and R for the front and rear positions.

85. Install the throttle body and intake air connector assembly.

86. Install the engine wire bracket.

87. Connect the fuel return hose.

88. Connect the vane pump to the pump bracket. Install the power steering air hose to the No. 4 timing belt cover and intake chamber.

89. Install the water bypass outlet and the bypass pipe. Always use new O-rings.

90. Install the exhaust manifolds with new gaskets. Tighten the bolts to 29 ft. lbs. (39 Nm). Connect the oxygen sensor leads.

91. Install the front exhaust pipe. Tighten the bolts to 46 ft. lbs. (62 Nm).

92. If equipped with an exhaust manifold cover, install the cover and four nuts.

93. Install the water pump pulley, the fan and coupling and the drive belt. Tighten the four nuts to 12 ft. lbs. (16 Nm).

94. Install the air cleaner, airflow meter and the intake air connector pipe.

95. Install the air cleaner duct.

96. Connect the control and accelerator cables to the throttle body.

97. Refill and bleed the engine coolant system.

98. Connect the negative battery cable. Start the engine and check for leaks.

99. Check the ignition timing.

100. Install the engine undercovers. Road test the vehicle. Recheck the engine coolant level.

4.0L Engine

LS400

1. Relieve the fuel system pressure.

2. Disconnect the negative battery cable. Wait at least 90 seconds before performing any other work.

3. Raise and safely support the vehicle.

4. Remove the oil pan protector.

5. Remove the engine undercover.

6. Drain the engine coolant from the radiator.

7. Remove the battery clamp cover.

8. Remove the air cleaner inlet.

9. Remove the V bank cover by removing the bolt and two cap nuts.

10. Remove the air cleaner and intake air connector assembly.

11. Remove the drive belt, fluid coupling and the fan pulley. The drive belt tension may be slackened by turning the tensioner counterclockwise. The pulley bolt for the drive belt tensioner has a left-handed thread.

12. Remove the radiator.

13. Remove the right-hand No. 3 timing belt cover.

14. Remove the left-hand No. 3 timing belt cover.

15. Remove the drive belt idler pulley by removing the pulley bolt and cover plate.

16. Remove the right-hand No. 2 timing belt cover.

17. Remove the left-hand No. 2 timing belt cover.

18. Remove the distributor housings.

19. Remove the No. 1 ignition coil.

20. Disconnect the A/C compressor from the engine.

21. Remove the fan bracket by removing the two bolts and two nuts.

22. Set the engine to TDC on cylinder No. 1. Turn the crankshaft pulley and align its groove with the timing mark **0** of the No. 1 timing cover. Check that the timing marks of the camshaft timing pulleys and timing belt rear plates are aligned. If not, turn the crankshaft 1 full revolution (360 degrees).

➡ **Since the thrust clearance of the camshaft is small, the camshaft must be kept level while it is being removed. If the camshaft is not kept level, the portion of the cylinder head receiving the shaft thrust may crack or be damaged, causing the camshaft to seize or break.**

23. Turn the crankshaft pulley approximately 50° clockwise and put the timing mark of the crankshaft pulley in line with the centers of the crankshaft pulley bolt and the idler pulley bolt.

✵✵ WARNING

If the timing belt is disengaged, having the crankshaft pulley at the wrong angle can cause the piston head and valve head to come into contact with each other when you remove the camshaft timing pulley. Always set the crankshaft pulley at the correct angle before removing the timing belt.

24. If the timing belt is to be reused, turn the crank pulley slowly; check that the 3 installation marks are present on the belt. If the marks are not present, make new installation marks before removing the belt. The marks should align with the timing marks on each camshaft pulley and the crank pulley.

25. Remove the timing belt tensioner. Alternately loosen the two bolts; remove the bolts, the tensioner and the dust protector.

26. Using the proper tool, 09278–54012

or equivalent, loosen the tension between the left side and the right side timing pulleys by slightly turning the left side camshaft clockwise.

27. Disconnect the timing belt from the camshaft timing pulleys. Using the proper tool, remove the bolt and the camshaft timing pulleys.

28. Disconnect the power steering pump from the engine. Do not disconnect the hoses or lines from the power steering pump. Support the power steering pump with a piece of wire. Do not allow the pump to hang.

29. Remove the front catalytic converter.

30. Remove the high tension spark plug wires, wire clamps and the wire cover assembly.

31. Remove the No. 2 ignition coil by removing the connector and the two bolts.

32. Remove the two bolts and the rear timing belt plate. Remove both plates.

33. Remove the intake chamber assembly.

34. Tag and disconnect the following:
• TPS connector
• With TRAC system, sub throttle position sensor connector
• With TRAC system, sub throttle actuator connector
• IAC valve connector
• EGR valve connector
• VSV connector for fuel pressure control
• VSV connector for EVAP system
• EGR gas temperature sensor connector

35. Disconnect the following hoses:
• Brake booster vacuum hose from the union on the air intake chamber
• PCV hose from the PCV valve on the left-hand cylinder head
• Water bypass hose (from the EGR valve) from the rear water bypass joint
• Water bypass hose (from the throttle body) from the rear water bypass joint
• Vacuum hose (from the VSV for fuel pressure control) from the fuel pressure regulator
• EVAP hose (from charcoal canister) from the VSV for EVAP

36. Disconnect the heater hose from the water bypass pipe.

❊❊ CAUTION

The fuel injection system remains under pressure after the engine has been turned OFF. Properly relieve fuel pressure before disconnecting any fuel lines. Failure to do so may result in fire or personal injury.

37. Disconnect the fuel inlet hose from the delivery pipe.

38. Disconnect the fuel return hose from the fuel return pipe.

39. Disconnect the engine wire from the delivery pipes and rear water bypass joint.

40. Disconnect the fuel hose from the fuel pressure regulator.

41. Remove the two bolts and fuel return pipe from the intake manifold.

42. Disconnect the eight injector connectors.

43. Remove the six bolts, four nuts, the intake manifold assembly and the two gaskets.

44. Remove the water inlet and inlet housing.

45. Remove the front water bypass joint.

46. Remove the rear water bypass joint and No. 1 EGR pipe assembly.

47. Remove the oil dipstick and guide for the automatic transmission.

48. Remove the oil dipstick and guide for the engine.

49. Remove the engine hangers.

50. Remove the right and left cylinder head covers by removing the eight bolts, seal washers and gaskets.

51. If necessary, remove the semi-circular plugs.

52. Remove the exhaust camshaft from the right side cylinder head. See the camshaft procedure for tightening sequence.

53. Remove the intake camshaft from the right side cylinder head. See the camshaft procedure for tightening sequence.

54. Remove the exhaust camshaft of the left side cylinder head. See the camshaft procedure for tightening sequence.

➡ **When removing the camshaft, be sure the torsional spring force of the subgear has been eliminated.**

55. Remove the intake camshaft from the left side cylinder head. See the cam-

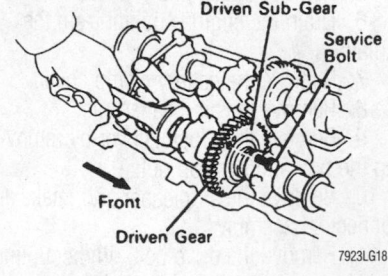

Securing the exhaust camshaft on the right cylinder head—LS400

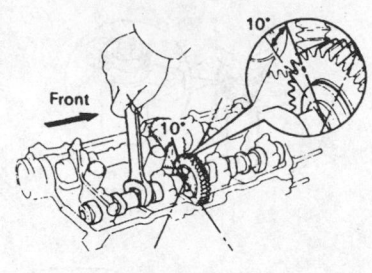

Turning the exhaust camshaft 10 degrees on the right cylinder head—LS400 and SC400

shaft procedure for tightening sequence.

56. Disconnect the main oxygen sensor connectors.

57. Remove the bolt and disconnect the ground cable from the right cylinder head.

58. Remove the bolt and disconnect the ground strap from the left cylinder head.

59. Remove the bolt and disconnect the engine wire protector from the left-hand cylinder head.

60. Remove the two bolts, seal washers, bearing cap and the camshaft housing plug from the right-hand cylinder head.

61. Remove the 10 cylinder head bolts and plate washers to each cylinder head. Loosen the bolts in the reverse order of the tightening sequence. Lift the heads from the dowels on the block with the exhaust manifolds attached. Place the heads on blocks of wood on the workbench.

➡ **Do not drop anything in the opening in the front of the right side cylinder head. The opening leads through the block and into the oil pan. If anything falls into the opening the oil pan will have to be removed in order to retrieve it.**

❊❊ WARNING

If necessary to pry the head loose, take great care not to damage the contact surfaces of the head or block.

62. Remove the two bolts, seal washers, bearing cap and camshaft housing plug from the right-hand cylinder head.

63. Remove the right exhaust manifold from the cylinder head by removing the heat insulator, eight nuts and the gasket.

64. Remove the left exhaust manifold from the cylinder head by removing the heat insulator, eight nuts and the gasket.

To install:

65. Install the right exhaust manifold. The new gasket must be installed with the white marks facing the manifold side.

Tighten the bolts to 33 ft. lbs. (44 Nm). Install the right oxygen sensor.

66. Install the left exhaust manifold. The new gasket must be installed with the white marks facing the manifold side. Tighten the bolts to 33 ft. lbs. (44 Nm). Install the left oxygen sensor.

67. Place the two new cylinder head gaskets in position on the engine block. Each gasket has a painted mark denoting the rear of the gasket. The gasket for the right bank has a white mark and the gasket for the left bank has a yellow mark. Double check the gasket position and placement.

68. Place the two cylinder heads in position on the block.

69. Apply a light coat of engine oil to the threads and under the head of each bolt. Temporarily install the plate washers and bolts.

70. Uniformly tighten the 10 bolts on 1 cylinder head in the order shown. Make several repetitive passes in the correct sequence, tightening each bolt slightly each time. Final torque for each bolt is 29 ft. lbs. (39 Nm). Repeat the tightening procedure for the other head.

71. Carefully mark the front of each bolt with a small dot of paint. Following the tightening sequence used earlier, retighten each bolt by turning it exactly 90 degrees. Check that each paint mark is now at a 90 degree angle to the front of the engine.

72. Connect the oxygen sensor connectors.

73. Install the engine wire to the right-hand cylinder head with the two bolts.

74. Install the ground cable to the right-hand cylinder head with the bolt.

75. Install the engine wire protector to the left-hand cylinder head with the bolt.

76. Install the ground cable to the left-hand cylinder head with the bolt.

77. Remove any old packing and apply new seal packing to the bearing caps.

78. Install the bearing cap on the right side cylinder head, marked **I1**, in position with the arrow mark facing the rear. Install the bearing cap on the left side cylinder head, marked **I6**, in position with the arrow mark facing the front.

79. Apply a light coat of oil on the threads of the cap bolts. Install the bearing cap bolts with new washers and alternately tighten each bolt to 12 ft. lbs. (16 Nm).

➡**Use silver colored bolts 1.50 in. (38mm) in length.**

80. Install new camshaft housing plugs on the cylinder heads. Be sure to face the cupped side forward.

81. Turn the crankshaft pulley clockwise or counterclockwise and put the timing mark of the crankshaft pulley in line with the centers of the crankshaft pulley bolt and the idler pulley bolt

➡**Since the thrust clearance of the camshaft is small, the camshaft must be kept level while it is being installed. If the camshaft is not kept level, the portion of the cylinder head receiving the shaft thrust may crack or be damaged, causing the camshaft to seize or break.**

82. Install the right side cylinder head intake camshaft. Tighten the bracket bolt in the reverse order of the loosening sequence.

83. Install the right side cylinder head exhaust camshaft. Tighten the bracket bolt in the reverse order of the loosening sequence.

84. Install the left side cylinder head intake camshaft. Tighten the bracket bolt in the reverse order of the loosening sequence.

85. Install the left side cylinder head exhaust camshaft. Tighten the bracket bolt in the reverse order of the loosening sequence.

86. Check and adjust the valve clearance.

87. Install the camshaft oil seals with the proper tool (SST 09223–46011). Be sure to apply MP grease to the new oil seal lip.

88. Install the semi-circular plugs.

89. Clean the cylinder head covers. Apply new sealant in the correct locations and install the gaskets.

90. Install the right cylinder head cover and bolts. Tighten the bolts to 52 inch lbs. (6 Nm).

91. Install the left cylinder head cover and bolts. Tighten the bolts to 52 inch lbs. (6 Nm).

92. Install the engine hanger with the two bolts. Install both engine hangers. Tighten the bolts to 27 ft. lbs. (37 Nm).

93. Install the oil dipstick guide for the engine.

94. Install the oil dipstick for the transmission.

95. Install the rear water bypass joint and No. 1 EGR pipe.

96. Install the front water bypass joints. Install two gaskets and alternately tighten the nuts to 13 ft. lbs. (18 Nm).

97. Install the water inlet and inlet housing, alternately tighten the bolts to 13 ft. lbs. (18 Nm).

98. Install the delivery pipe and intake manifold.

99. Install the return pipe with two new gaskets. Tighten the union bolt to 26 ft. lbs. (35 Nm).

100. Install the engine wire to the delivery pipes and rear water bypass joint.

101. Connect the fuel return hose to the fuel return pipe.

102. Connect the fuel inlet hose to the left-hand delivery pipe.

103. Connect the fuel hose to the fuel pressure regulator.

104. Install the air intake chamber assembly.

105. Connect the following hoses:
- Brake booster vacuum hose to the union on the air intake chamber
- PCV hose to the PCV valve on the left-hand cylinder head
- Water bypass hose (from EGR valve) to the rear water bypass joint
- Water bypass hose (from throttle body) to the rear water bypass joint
- Vacuum hose (from VSV for fuel pressure control) to the fuel pressure regulator
- EVAP hose (from charcoal canister) from the VSV for EVAP

106. Connect the following:
- TPS connector
- With TRAC system, sub TPS connector
- With TRAC system, sub throttle actuator connector
- IAC valve connector

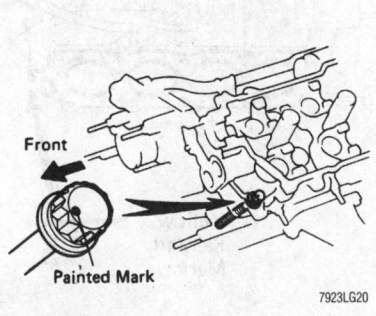

Apply a dot of paint at the front of each bolt—LS400

7923LG20

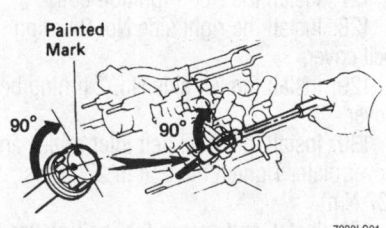

The paint mark must be 90 degrees from the starting point—LS400

7923LG21

- EGR valve connector
- EGR gas temperature sensor connector
- VSV connector for fuel pressure control
- VSV connector for EVAP

107. Install the accelerator bracket with the two bolts.

108. Connect the accelerator cable, A/T throttle control cable and the cruise control actuator cable.

109. Connect the spark plug wires and clamps to the right and left cylinder head cover.

110. Install the timing belt rear plates by installing the bolts. Tighten the bolts to 66 inch lbs. (8 Nm).

111. Install the No. 2 ignition coil.

112. Install a new gasket to the exhaust manifold and install the catalytic converters. Tighten the three nuts to each converter to 46 ft. lbs. (62 Nm).

113. Install the front exhaust pipe, tighten the bolts and nuts to 32 ft. lbs. (44 Nm). Tighten the four bolts holding the pipe support bracket to the transmission. Tighten the bolts to 32 ft. lbs. (44 Nm).

114. Install the power steering pump with the nut and three bolts. Tighten the nut to 32 ft. lbs. (43 Nm) and the bolts to 29 ft. lbs. (39 Nm).

115. Align the knock pin on the right side camshaft with the knock pin of the timing pulley. Slide on the timing pulley with the right side mark facing forward. Tighten the bolt to 80 ft. lbs. (108 Nm).

116. Align the knock pin on the left side camshaft with the knock pin of the timing pulley. Slide on the timing pulley with the left side mark facing forward. Tighten the bolt to 80 ft. lbs. (108 Nm).

117. Install the timing belt to the left side camshaft timing pulley:

a. Using the proper tool, slightly turn the left side timing pulley clockwise. Align the installation mark of the timing belt with the timing mark of the camshaft timing pulley and hang the timing belt on the left side camshaft pulley.

b. Align the timing marks of the left side camshaft pulley and the timing belt rear plate.

c. Check that the timing belt has tension between crankshaft timing pulley and the left side camshaft pulley.

118. Install the timing belt to the right side camshaft timing pulley by:

a. Using the proper tool, slightly turn the right side timing pulley clockwise. Align the installation mark of the timing belt with the timing mark of the camshaft timing pulley and hang the timing belt on the right side camshaft pulley.

b. Align the timing marks of the right side camshaft pulley and the timing belt rear plate.

c. Check that the timing belt has tension between crankshaft timing pulley and the right side camshaft pulley.

119. The timing belt tensioner must be set prior to installation. The tensioner can be set by:

a. Place a plate washer between the tensioner and a block. Using a press, press in the pushrod using 220–2205 lbs. of pressure.

b. Align the holes of the pushrod and housing, pass a 1.27mm Allen wrench through the holes to keep the setting position of the pushrod.

c. Release the press and install the dust boot to the tensioner.

120. Loosely install the tensioner. Evenly and alternately tighten the bolts to 20 ft. lbs. (26 Nm). Remove the tool from the tensioner.

121. Turn the crankshaft pulley two complete revolutions from TDC to TDC. Always turn the crankshaft clockwise. Check that all belt and pulley marks align with their reference marks. If any mark is out of perfect alignment, the timing belt must be removed and reinstalled.

122. Install the drive belt tensioner and tighten the bolt and nuts to 12 ft. lbs. (16 Nm).

123. Install both distributor housings and tighten the mounting bolts to 13 ft. lbs. (18 Nm). The distributors are marked L or R for correct installation.

124. Replace the distributor rotors and caps.

125. Install the fan bracket by installing the two bolts and two nuts. Tighten as follows:

- 12mm head to 12 ft. lbs. (16 Nm)
- 14mm head to 24 ft. lbs. (32 Nm)

126. Install the A/C compressor. Tighten the bolts to 36 ft. lbs. (49 Nm) and the nut to 22 ft. lbs. (29 Nm).

127. Install the No. 1 ignition coil.

128. Install the right side No. 2 timing belt cover.

129. Install the left side No. 2 timing belt cover.

130. Install the drive belt idler pulley and cover plate. Tighten the bolt to 27 ft. lbs. (37 Nm).

131. Install and secure the ignition wires. Make certain that all clips and retainers are securely engaged and that the wires are properly routed.

132. Install the right side No. 3 timing belt.

133. Install the left-hand No. 3 timing belt cover.

134. Install the radiator assembly.

135. Install the fan pulley, fan, fluid coupling and the drive belt.

136. Install the air cleaner and intake air connector assembly.

137. Install the V bank cover.

138. Fill the radiator with engine coolant.

139. Connect the negative battery cable to the battery.

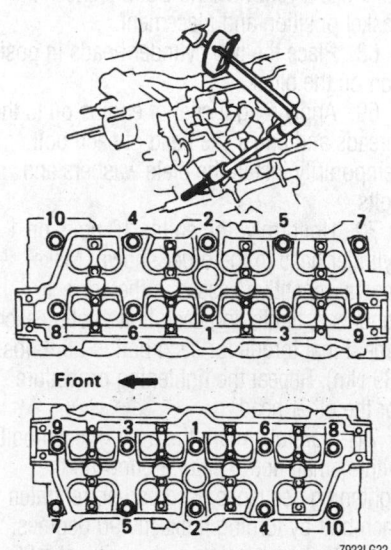

Tighten the cylinder head bolts using the correct sequence as shown—LS400 and SC400

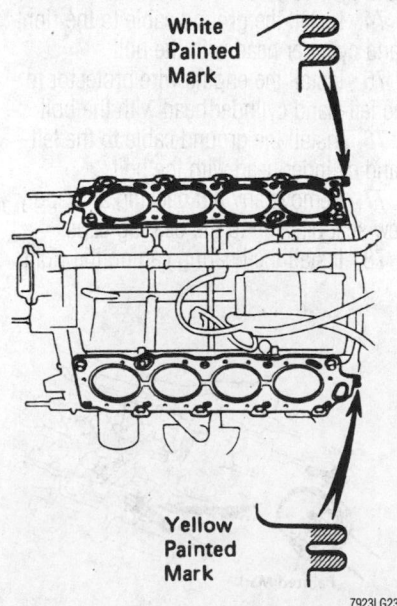

Position the paint marks as shown for correct head gasket installation—LS400 and SC400

140. Start the engine and check for leaks.

141. Bleed the cooling system and recheck the engine coolant level.

142. Make all the necessary engine adjustments.

143. Install the air cleaner inlet.

144. Install the battery clamp cover.

145. Install the engine undercover.

146. Install the oil pan protector.

147. Lower the vehicle.

SC400

1. Relieve the fuel system pressure.

2. If equipped, remove the V-bank cover.

3. Disconnect the negative battery cable. Wait at least 90 seconds before performing any other work. Disconnect the positive battery cable and remove the battery.

4. Remove the engine undercover.

5. Drain the cooling system.

6. Remove the accessory drive belt by turning the tensioner counterclockwise.

7. Remove the radiator.

8. Remove the engine coolant reservoir tank.

9. Disconnect the air conditioning compressor from the engine. Do not remove the compressor pressure lines.

10. Remove the intake air connector.

11. Remove the left ignition coil.

12. Remove the upper high tension cord cover by removing the two bolts.

13. Remove the right side engine wire cover by removing the bolt.

14. Remove the left side engine wire cover by removing the two bolts.

15. Remove the right side No. 3 timing cover by removing the two air control valve hoses and four bolts.

16. Remove the left No. 3 timing belt cover by removing the four bolts.

17. Disconnect and tag the vacuum hoses and remove the left side engine wire cover. Disconnect the spark plug wires.

18. Remove the pulley bolt, cover plate and idler pulley.

19. Remove the right-hand No. 2 timing belt cover.

20. Remove the left-hand No. 2 timing belt cover.

21. Remove the distributor caps and rotors. Mark both caps and rotors (left or right) for correct installation. Disconnect and remove both distributor housings.

22. Disconnect and remove the alternator from the engine.

23. Remove the drive belt tensioner.

24. Remove the spark plugs.

25. Set the engine to TDC on cylinder No. 1. Turn the crankshaft pulley and align its groove with the timing mark **0** of the No. 1 timing cover. Check that the timing marks of the camshaft timing pulleys and timing belt rear plates are aligned. If not, turn the crankshaft 1 full revolution (360 degrees).

26. If the timing belt is to be reused, turn the crank pulley slowly; check that the four installation marks are present on the belt. If the marks are not present, make new installation marks before removing the belt. The marks should align with the timing marks on each camshaft pulley, the crank pulley and a matchmark on the belt at the end of the fan bracket.

27. Remove the timing belt tensioner. Alternately loosen the two bolts; remove the bolts, the tensioner and the dust protector.

28. Using the proper tool, 09278–54012 or equivalent, loosen the tension between the left side and right side timing pulleys by slightly turning the left side camshaft clockwise.

29. Disconnect the timing belt from the camshaft timing pulleys. Using the proper tool, remove the bolt and the camshaft timing pulleys.

30. Remove the fan bracket by removing the two bolts and two nuts.

31. Remove the power steering pump.

32. Remove the front catalytic converter.

33. Disconnect the throttle body cover.

34. Remove the accelerator, transmission and cruise control cables from the throttle body.

35. Remove the right-hand ignition coil.

36. Remove the water inlet and inlet housing.

37. Remove the two bolts and rear timing belt plate. Remove both plates.

38. Remove the following vacuum hoses:

• Vacuum hose from the throttle body and VSV for the EVAP system.

• Vacuum hose from the fuel pressure regulator and VSV for the fuel pressure regulator.

• Vacuum hose from the VSV for fuel pressure and air intake chamber.

• Vacuum hose from the charcoal canister and VSV for the EVAP.

39. Remove the EGR valve and adapter.

40. Remove the idle air control valve from the engine.

41. Remove the heater water valve from the body.

42. Remove the throttle body assembly.

43. Remove the air intake chamber assembly.

44. Remove the transmission throttle cable bracket from the intake chamber by removing the four bolts and eight nuts.

45. Disconnect the DLC1 bracket from the air intake chamber.

46. Lift the air intake chamber assembly and disconnect the EGR gas temperature sensor connector.

47. Remove the air intake chamber assembly and four gaskets.

48. Disconnect the heater water hoses from the bypass pipes.

✳✳ CAUTION

The fuel injection system remains under pressure after the engine has been turned OFF. Properly relieve fuel pressure before disconnecting any fuel lines. Failure to do so may result in fire or personal injury.

49. Disconnect the fuel inlet hose from the left delivery pipe. Remove the pulsation damper.

50. Disconnect the fuel return hose from the fuel return pipe.

51. Disconnect the EGR pipe from the right-hand cylinder head by removing the bolt.

52. Disconnect the engine wire from the intake manifold by removing the two bolts.

53. Disconnect the engine wire from the delivery pipes, rear water bypass joint and the right-hand cylinder head.

54. Remove the fuel return pipe and rear fuel pipe.

55. Remove the delivery pipes and intake manifold assembly.

56. Remove the front water bypass joint.

57. Remove the oil dipstick guide and dipstick guide for the transmission.

58. Remove the bolt holding the bypass joint to the left-hand rear engine hanger. Remove the water bypass joint by removing the four bolts.

59. Remove the two bolts holding the EGR pipe to the right-hand exhaust pipe. Remove the EGR pipe and gasket.

60. Remove the engine hangers.

61. On the right cylinder head cover, release the clips and retainers holding the spark plug wires to the cylinder head cover. Remove the cylinder head cover.

62. On the left cylinder head cover:

a. Release the clips and retainers holding the spark plug wires.

b. Remove the eight bolts, seal washers, cylinder head cover and gaskets.

63. If necessary, remove the semi-circular plugs.

64. Remove the exhaust camshaft from the right side cylinder head.

65. Remove the intake camshaft from the right side cylinder head.

66. Remove the exhaust camshaft of the left side cylinder head.

➡**When removing the camshaft, be sure the torsional spring force of the subgear has been eliminated.**

67. Remove the intake camshaft from the left side cylinder head.

68. Remove the circular plugs by removing the two bolts, seal washers and the bearing caps. Remove both circular plugs. Arrange the bearing caps in order.

69. Disconnect the main oxygen sensor connectors.

70. Remove the 10 cylinder head bolts and plate washers to each cylinder head. Loosen the bolts in the reverse order of tightening. Lift the heads from the dowels on the block with the exhaust manifolds attached. Place the heads on blocks of wood on the workbench.

➡**Do not drop anything in the opening in the front of the right side cylinder head. The opening leads through the block and into the oil pan. If anything falls into the opening the oil pan will have to be removed in order to retrieve it.**

☆☆ WARNING

If necessary to pry the head loose, take great care not to damage the contact surfaces of the head or block.

71. Remove the main heated oxygen sensor from each exhaust manifold. Remove the manifolds and gaskets from the cylinder head.

To install:

72. Install the right exhaust manifold. The new gasket must be installed with the white marks facing the manifold side. Tighten the bolts to 33 ft. lbs. (44 Nm). Install the right oxygen sensor.

73. Install the left exhaust manifold. The new gasket must be installed with the white marks facing the manifold side. Tighten the bolts to 33 ft. lbs. (44 Nm). Install the left oxygen sensor.

74. Place the two new cylinder head gaskets in position on the engine block. Each gasket has a painted mark denoting the rear of the gasket. The gasket for the right bank has a white mark and the gasket for the left bank has a yellow mark. Double check the gasket position and placement.

75. Place the two cylinder heads in position on the block.

76. Apply a light coat of engine oil to the threads and under the head of each bolt. Temporarily install the 20 plate washers and bolts.

77. Uniformly tighten the 10 bolts on 1 cylinder head in the order shown. Make several repetitive passes in the correct sequence, tightening each bolt slightly each time. Final torque for each bolt is 29 ft. lbs. (39 Nm). Repeat the tightening procedure for the other head.

78. Carefully mark the front of each bolt with a small dot of paint. Following the tightening sequence used earlier, retighten each bolt by turning it exactly 90 degrees. Check that each paint mark is now at a 90 degree angle to the front of the engine.

79. Connect the oxygen sensor connectors.

80. Install new circular plugs on the heads with the cup side facing forward.

81. Remove any old packing and apply new seal packing to the bearing caps.

82. Install the bearing cap on the right side cylinder head, marked **I1**, in position with the arrow mark facing the rear. Install the bearing cap on the left side cylinder head, marked **I6**, in position with the arrow mark facing the front.

83. Apply a light coat of oil on the threads of the cap bolts. Install the bearing cap bolts with new washers and alternately tighten each bolt to 12 ft. lbs. (16 Nm).

➡**Use silver colored bolts 1.50 in. (38mm) in length.**

84. Install the right side cylinder head intake camshaft. See the camshaft procedure for tightening sequence.

85. Install the right side cylinder head exhaust camshaft. See the camshaft procedure for tightening sequence.

86. Install the left side cylinder head intake camshaft. See the camshaft procedure for tightening sequence.

87. Install the left side cylinder head exhaust camshaft. See the camshaft procedure for tightening sequence.

88. Check and adjust the valve clearance.

89. Install the camshaft oil seals with the proper tool (SST 09223–46011). Be sure to apply MP grease to the new oil seal lip.

90. Install the semi-circular plugs, remove any old packing material. Apply seal packing to the semi-circular plug grooves. Install the four semi-circular plugs to the cylinder heads.

91. Clean the cylinder head covers. Apply new sealant in the correct locations and install the gaskets.

92. Install the right cylinder head cover and bolts. Tighten the bolts to 52 inch lbs. (6 Nm).

93. Install the left cylinder head cover and bolts. Tighten the bolts to 52 inch lbs.

(6 Nm). Connect the wiring harness connectors and connect the fuel injector leads. Secure the engine wiring harness in the clamps on the delivery pipe.

94. Install the engine hanger with the two bolts. Install both engine hangers. Tighten the bolts to 27 ft. lbs. (37 Nm).

95. Install the engine wire to the cylinder heads as follows:

 a. Install the engine wire protector to the right-hand cylinder head with the two bolts.

 b. Install the ground strap to the right-hand cylinder head with the bolt.

 c. Install the engine wire protector to the left-hand cylinder head with the five bolts.

96. Install a new gasket and EGR pipe to the right-hand exhaust manifold with the two nuts.

97. Install the rear water bypass joint by installing the gasket and four nuts. Tighten the nuts to 13 ft. lbs. (18 Nm).

98. Install the oil dipstick guide for the engine.

99. Install the front water bypass joint by installing the gaskets and four nuts. Tighten the nuts to 13 ft. lbs. (18 Nm).

100. Install the engine wire protector to the front water bypass joint. Connect the ECT sensor connector and the ECT sender gauge connector.

101. Install the delivery pipe and intake manifold.

102. Install the return pipe with two new gaskets. Tighten the union bolt to 26 ft. lbs. (35 Nm).

103. Install the engine wire to the delivery pipes, rear water bypass joint and the right-hand cylinder head.

104. Install the engine wire to the intake manifold with the two bolts.

105. Temporarily install the EGR pipe to the right-hand cylinder head by installing the bolt. Do not tighten the bolt at this time.

106. Connect the fuel return hose to the fuel return pipe.

107. Connect the fuel inlet hose to the left-hand delivery pipe. Install the pulsation damper

108. Connect the water hose to the water bypass pipe and the water hose to the rear water bypass joint.

109. Install the air intake chamber assembly. Tighten the eight nuts and four bolts to the intake chamber to 13 ft. lbs. (18 Nm). Tighten the bolts holding the EGR pipe to the air intake chamber to 13 ft. lbs. (18 Nm). Tighten the bolt holding the EGR pipe to the right cylinder head to 13 ft. lbs. (18 Nm).

110. Install the accelerator bracket with the bolt and stud bolt.

111. Install the brake booster union by installing the gaskets and the union bolt. Tighten the bolt to 22 ft. lbs. (29 Nm).

112. Connect the following connectors and hoses:

   a. VSV connector for fuel pressure control.

   b. VSV connector for EVAP.

   c. Brake booster vacuum hose to the union on the air intake chamber.

   d. Vacuum hose (from the VSV for heater water valve) to the air intake chamber.

113. Install the throttle body.

114. Install the heater water valve.

115. Install the IAC valve with the two nuts. Tighten the nuts to 13 ft. lbs. (18 Nm).

116. Connect the water bypass hose (from the throttle body) to the IAC valve. Connect the IAC valve connector.

117. Install the EGR valve adapter. Tighten the bolts and nuts to 13 ft. lbs. (18 Nm).

118. Connect the PCV hose to the cylinder head.

119. Connect the EGR gas temperature sensor.

120. Install the EGR valve with the gasket and two nuts. Tighten the nuts to 13 ft. lbs. (18 Nm).

121. Connect water bypass hose to the IAC valve, then connect the water bypass hose to the water bypass pipe.

122. Connect the EGR valve.

123. Connect the following hoses:

   a. Vacuum hose to the throttle body and VSV for EVAP.

   b. Vacuum hose to the fuel pressure regulator and VSV for fuel pressure.

   c. Vacuum hose to the VSV for fuel pressure and air intake chamber.

   d. Vacuum hose to the charcoal canister and VSV for EVAP.

124. Install the rear timing belt plate with the two bolts. Install both sides. Tighten the bolts to 69 inch lbs. (8 Nm).

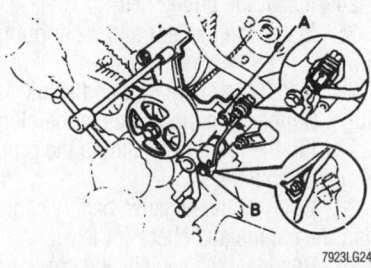

7923LG24

Be sure to install the hydraulic pump bolts in the proper locations—SC400

125. Install the water inlet and inlet housing. Tighten the bolts to 13 ft. lbs. (18 Nm).

126. Connect the water bypass hose (from the IAC valve) to the water inlet housing.

127. Connect the water hose (from the reservoir tank) to the water inlet housing.

128. Install the right ignition coil with the two bolts. Connect the ignition coil connector.

129. Connect the accelerator, transmission and cruise control cables to the throttle body.

130. Install the throttle body cover.

131. Install the power steering pump with the nut and three bolts. Tighten the bolts to 29 ft. lbs. (39 Nm) and the nut to 32 ft. lbs. (43 Nm).

132. Install the power steering reservoir tank and bracket with the three bolts.

133. Install the power steering tubes with the clamp and bolt.

134. Install the front catalytic converter.

135. Install the hydraulic pump. Tighten the 12mm fasteners to 12 ft. lbs. (16 Nm) and the other fasteners to 22 ft. lbs. (30 Nm).

136. Install the fan bracket with the two bolts and two nuts.

137. Align the knock pin on the right side camshaft with the knock pin of the timing pulley. Slide on the timing pulley with the right side mark facing forward. Tighten the bolt to 80 ft. lbs. (108 Nm).

138. Align the knock pin on the left side camshaft with the knock pin of the timing pulley. Slide on the timing pulley with the left side mark facing forward. Tighten the bolt to 80 ft. lbs. (108 Nm).

139. Turn the crankshaft pulley and align its groove with the **0** timing mark on the timing belt cover.

140. Turn each camshaft timing pulley and align the timing marks of the pulley with the timing belt rear plate.

141. Install the timing belt to the left side camshaft timing pulley.

142. Install the timing belt to the right side camshaft timing pulley.

143. The timing belt tensioner must be set prior to installation. The tensioner can be set by:

   a. Place a plate washer between the tensioner and a block. Using a press, press in the pushrod using 220–2205 lbs. of pressure.

   b. Align the holes of the pushrod and housing, pass a 1.27mm Allen wrench through the holes to keep the setting position of the pushrod.

   c. Release the press and install the dust boot to the tensioner.

144. Loosely install the tensioner. Tighten the bolts evenly and alternately to 20 ft. lbs. (26 Nm). Remove the tool from the tensioner.

145. Turn the crankshaft pulley two complete revolutions from TDC to TDC. Always turn the crankshaft clockwise. Check that all belt and pulley marks align with their reference marks. If any mark is out of perfect alignment, the timing belt must be removed and reinstalled.

146. Install the spark plugs and tighten to 13 ft. lbs. (18 Nm).

147. Install the drive belt tensioner and tighten the bolt to 12 ft. lbs. (16 Nm).

148. Install the alternator and engine wire bracket. Tighten the nut and bolt to 27 ft. lbs. (37 Nm). Connect the electrical connections at the alternator.

149. Install both distributor housings and tighten the mounting bolts to 13 ft. lbs. (18 Nm). The distributors are marked L or R for correct installation.

150. Install the distributor rotors and caps.

151. Install the right side No. 2 timing belt cover. Except California vehicles, install the camshaft position sensor connector to the ignition coil bracket.

152. Install the left side No. 2 timing belt cover.

153. Install the drive belt idler pulley and cover plate. Tighten the bolt to 27 ft. lbs. (37 Nm).

154. Install and secure the ignition wires. Make certain that all clips and retainers are securely engaged and that the wires are properly routed.

155. Install the right side No. 3 timing belt cover by install the gaskets, four bolts and the two air control valve hoses.

156. Install the left side No. 3 timing belt cover, connect the vacuum hose and the connectors.

157. Install the right side engine wire cover.

158. Install the left side engine wire cover and connect the vacuum hoses.

159. Install the upper high tension cord covers. Fit the front side claw groove of the upper cover to claw of the lower cover.

160. Install the left side ignition coil and connect the coil connector.

161. Install the intake air connector and connect the air hoses.

162. Install the A/C compressor to the engine.

163. Install the radiator.

164. Install the radiator reservoir tank.

165. Install the accessory drive belt.

166. Install the battery and connect the cables.

167. If removed, install the V-bank cover.
168. Fill the engine coolant and check the transmission fluid level.
169. Start the engine and check carefully for fluid leaks. Check the ignition timing.
170. Install the engine under-covers.
171. Road test the vehicle and verify proper operation.

Intake Manifold

✳✳ CAUTION

On models with an air bag, wait at least 90 seconds from the time that the ignition switch is turned to the LOCK position and the battery is disconnected before performing any further work.

REMOVAL & INSTALLATION

3.0L (1MZ-FE) Engine

1. Disconnect the negative battery cable. Drain the engine coolant.
2. Disconnect the throttle/accelerator cable from the throttle body.
3. Disconnect the air cleaner hose at the air intake chamber and remove it.
4. Remove the V-bank cover on the 3VZ-FE engine.
5. Tag and disconnect all lines and hoses, then remove both the ISC valve and the throttle body.
6. Remove the EGR valve and vacuum modulator. Remove the distributor.
7. On the 3VZ-FE engine, remove the emission control valve set, then disconnect the left side engine harness.
8. Remove the cylinder head rear plate.
9. Remove the intake chamber stays, any wires, then remove the air intake chamber.
10. Remove the fuel injection delivery pipe and the injectors.

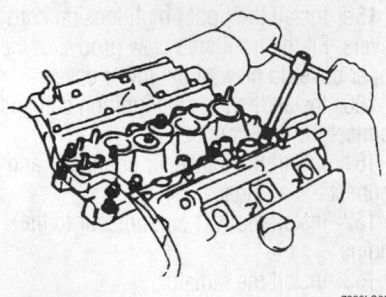

Intake manifold removal—3.0L (1MZ-FE) engine

7923LG25

11. Remove the water outlet and the bypass outlet.
12. Remove the 2 bolts and the No. 2 idler pulley bracket stay. Remove the 8 bolts and 4 nuts, then lift out the intake manifold.

To install:
13. Thoroughly clean the intake manifold and cylinder head surfaces. Using a machinist's straight edge and a feeler gauge, check the surface of the intake manifold for warpage. If the warpage is greater than 0.0039 in. (0.10mm), replace the intake manifold.
14. Place new gaskets onto the intake manifold and position the intake manifold between the cylinder heads. Tighten the nuts and bolts to 13 ft. lbs. (18 Nm). Tighten the No. 2 pulley bracket bolts to 13 ft. lbs. (18 Nm).
15. Install the water bypass outlet and tighten the bolts to 74 inch lbs. (8.3 Nm). Tighten the water outlet to 74 inch lbs. (8 Nm).
16. Install the injectors and delivery pipe.
17. Install the air intake chamber and tighten the 2 bolts and 2 nuts to 32 ft. lbs. (43 Nm); use an 8mm hex wrench. Install the chamber stays and tighten the mounting bolts to 29 ft. lbs. (39 Nm).
18. Install the remaining components. Tighten the emission control valve set to, 73 inch lbs. (8 Nm).
19. Unplug and connect all hoses.
20. If equipped with automatic transaxle, connect the accelerator cable and adjust it.
21. Fill the cooling system to the proper level and connect the negative battery cable.
22. Start the engine and inspect for leaks.

3.0L (2JZ-GE) Engine

1. Disconnect the negative battery cable.
2. Drain the cooling system.
3. Tag and disconnect the spark plug wires at the spark plugs.
4. Remove the spark plugs. Remove the distributor with the spark plug leads attached.
5. Remove the radiator, then remove the water pump pulley.
6. Place matchmarks on the timing belt and sprockets, support the belt, then slide it off the timing sprockets.
7. Remove the No. 2 front exhaust pipe. Disconnect the 2 O₂ sensor leads, remove the 4 nuts, then remove the manifold heat shield. Remove the exhaust manifolds.
8. Loosen the 2 bolts and remove the water bypass outlet and the No. 1 bypass pipe. Remove the 3 O-rings from the outlet and the pipe.

9. Loosen the 2 bolts and nut and remove the water outlet. Loosen the clamp and remove the No. 1 bypass hose.
10. Remove the vacuum control valve set and the No. 2 vacuum pipe.
11. Disconnect the fuel return hose from the oil dipstick guide, remove the mounting bolt and pull the guide and dipstick from the pan. Plug the hole.
12. Remove the air intake chamber. Remove the fuel delivery pipe, then pull out the injectors. Remove the No. 1 and 2 fuel pipes.
13. Disconnect the engine harness from the intake manifold.
14. Loosen the 2 bolts and remove the intake manifold stay. Loosen the 6 bolts and 2 nuts, then lift out the intake manifold.

To install:
15. Using a new gasket, install the intake manifold and tighten the bolts and nuts to 15 ft. lbs. (21 Nm).
16. Install the mounting stay and tighten the bolts to 29 ft. lbs. (39 Nm).
17. Connect the engine harness to the manifold.
18. Install the 2 fuel pipes and tighten the bolts to 78 inch lbs. (9 Nm). Install the delivery pipe and injectors. Tighten the pipe bolts to 15 ft. lbs. (21 Nm).
19. Install the air intake chamber and tighten it to 15 ft. lbs. (21 Nm). Install the 2 stays and tighten them to 13 ft. lbs. (18 Nm); The No. 1 stay is marked with an **F** and the No. 2 stay is marked with an **R**.
20. Use a new O-ring and install the oil dipstick and guide.
21. Install the VCV set and the vacuum pipe. Tighten the set mounting bolts to 15 ft. lbs. (21 Nm).
22. Install the water bypass outlet and the pipe, tighten the bolts to 78 inch lbs. (9 Nm).
23. Using a new gasket, install the exhaust manifolds. Tighten the bolts to 29 ft. lbs. (39 Nm). Install the heat shield and tighten it to 13 ft. lbs. (18 Nm). Install the No. 2 front pipe.
24. Install the timing belt.
25. Install the radiator and water pump pulley.
26. Install the distributor and spark plugs. Connect the plug wires to the plugs.
27. Fill the cooling system to the proper level with coolant.
28. Connect the negative battery cable. Start the engine and check for leaks.
29. Road test the vehicle and check for unusual noise, shock, slippage, correct shift points and smooth operation.
30. Recheck the coolant and engine oil levels.

4.0L Engine

1. Disconnect the negative battery cable. Drain the cooling system.

2. Remove the camshaft timing pulleys. Remove the cooling fan hydraulic pump on the SC400.

3. Disconnect the accelerator cable, the throttle control cable, if equipped with automatic transmission and the cruise control actuator cable.

4. Remove the high tension cord cover and the right side ignition coil.

5. Remove the water inlet housing mounting bolts and disconnect the water bypass hose from the ISC valve.

6. Remove the water inlet and inlet housing assemblies. Remove the O-ring from the water inlet housing.

7. Remove the EGR pipe.

8. Disconnect the following:

   a. VSV connector

   b. Vacuum pipe hose

   c. EGR water bypass pipe

   d. Fuel pressure VSV

9. Disconnect the EGR vacuum hoses and remove the EGR VSV.

10. Disconnect the following hoses:

   a. Water bypass pipe hose from the ISC valve.

   b. Water bypass joint hose.

   c. Vacuum pipe hoses.

11. Disconnect the EGR gas temperature sensor, California only. Remove the EGR valve adapter.

12. Disconnect the following:

   a. Fuel pressure regulator vacuum hose.

   b. Air intake chamber vacuum hose.

   c. Vacuum hose from the EVAP BVSV.

13. Remove the mounting bolts, hoses and the vacuum pipe.

14. Remove the ISC valve.

15. Remove the throttle body sensor connectors and the water bypass pipe from the rear water bypass joint.

16. Disconnect the PCV valve hose. Remove the throttle body and gasket.

17. Disconnect the accelerator cable bracket and the brake booster vacuum union and hose.

18. Disconnect the cold start injector connector and the cold start injector tube from the right side delivery pipe, if equipped.

19. Disconnect the check connector from the intake chamber and remove the mounting nuts and bolts.

20. Remove the air intake chamber and the cold start injector, if equipped, tube and wire assembly.

21. Disconnect the engine wire from the intake manifold and from the right side cylinder head. Disconnect the heater hoses.

22. Remove the delivery pipes and the fuel injectors. Remove the mounting bolts and nuts. Lift up the intake manifold.

To install:

23. Install the intake manifold, using new gaskets. Tighten the mounting nuts and bolts to 13 ft. lbs. (18 Nm).

➡**Align the port holes of the gasket and cylinder head. Be careful of the installation direction.**

24. Install the delivery pipes and fuel injectors. Install the fuel return pipe with new gaskets. Tighten the union bolt to 26 ft. lbs. (35 Nm).

25. Connect the fuel hoses and the injector connectors. Connect the engine wire to the delivery pipes.

26. Connect the connectors on the left side delivery pipe, the water temperature sensor connector, cold start injector time switch connector and the water temperature sender gauge connector.

27. Connect the heater hoses and engine wire bracket. Install the engine wire to the bracket.

28. Install the cold start injector, tube and wire assembly if equipped. Tighten the mounting bolts to 69 inch lbs. (8 Nm).

29. Install the air intake chamber with new gaskets and tighten the mounting bolts to 13 ft. lbs. (18 Nm).

30. Connect the cold start injector tube to the right side delivery pipe and tighten the union bolt to 11 ft. lbs. (15 Nm), if equipped.

31. Connect the cold start injector connector as necessary. Install the accelerator cable bracket.

32. Install the brake booster union and connect the vacuum hose. Tighten the union bolt to 22 ft. lbs. (29 Nm).

33. Connect the water bypass hose to the throttle body and the PCV hose to the cylinder head cover.

34. Install the throttle body, using a new gasket. Tighten the mounting bolts to 13 ft. lbs. (18 Nm).

35. Install the water bypass pipe and connect the sensor connectors. Install the ISC valve and tighten the mounting bolts to 13 ft. lbs. (18 Nm). Connect the water bypass hose.

36. Install the vacuum pipe and the following hoses:

   a. Fuel pressure regulator vacuum hose.

   b. Vacuum hose to the upper port of the EVAP BVSV.

   c. Air intake chamber vacuum hose.

   d. Throttle body vacuum hoses.

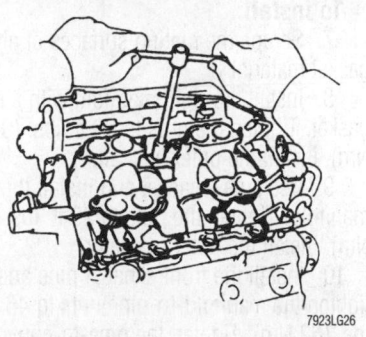

Intake manifold removal—4.0L (1UZ-FE) engine

37. Install the EGR valve adapter with a new gasket. Tighten the mounting bolts to 13 ft. lbs. (18 Nm).

38. Install the remaining components. Install the timing belt rear plates and tighten the bolts to 69 inch lbs. (8 Nm). Install the water inlet and inlet hosing and tighten the bolts to 13 ft. lbs. (18 Nm).

39. Fill the cooling system and connect the negative battery cable. Start the engine and check for leaks.

40. Recheck all the fluid levels. Road test the vehicle for proper operation.

Exhaust Manifold

✱✱ CAUTION

On models with an air bag, wait at least 90 seconds from the time that the ignition switch is turned to the LOCK position and the battery is disconnected before performing any further work.

REMOVAL & INSTALLATION

3.0L (1MZ-FE) Engine

1. Disconnect the negative battery cable.

2. Raise the vehicle, support it on safety stands, then remove the engine undercovers.

3. Remove the 2 front exhaust pipe stay bolts. Disconnect the front pipe from the center pipe and remove the gasket. Loosen the 3 nuts, then remove the front pipe.

4. Disconnect the O_2 sensor at the right side manifold. Remove the 3 mounting nuts and lift off the outside heat insulator.

5. Remove the 6 nuts and lift off the right side manifold and gasket.

6. Loosen the 2 nuts and bolt and lift off the left side heat insulator. Remove the 6 nuts and lift off the left side manifold and gaskets.

To install:

7. Scrape the mating surfaces of all old gasket material.

8. Install the right manifold with a new gasket. Tighten the nuts to 29 ft. lbs. (39 Nm). Install the outer insulator.

9. Use a new gasket and install the left manifold. Tighten the nuts to 29 ft. lbs. (39 Nm). Install the outer insulator.

10. Install the front exhaust pipe and tighten the manifold-to-pipe nuts to 46 ft. lbs. (62 Nm). Tighten the pipe-to-converter nuts to 32 ft. lbs. (43 Nm).

11. Connect the O_2 sensor, install the undercovers, then lower the vehicle. Connect the battery cable.

3.0L (2JZ-GE) Engine

1. Disconnect the negative battery cable.

2. Raise the vehicle, support it on safety stands, then remove the engine undercovers.

3. Remove the No. 2 front exhaust pipe bolts and disconnect it from the front exhaust pipe. Loosen the 4 nuts, then remove the front pipe.

4. Disconnect the two O_2 sensors at the manifold. Remove the 4 mounting nuts and lift off the outside heat insulator.

5. Remove the 4 nuts and disconnect the manifolds from the pipe. Loosen the mounting bolts and remove the two manifolds and the gasket.

To install:

6. Scrape the mating surfaces of all old gasket material.

7. Install the manifolds with a new gasket. Tighten the nuts to 29 ft. lbs. (39 Nm). Install the outer insulator and tighten the nuts to 13 ft. lbs. (18 Nm).

8. Use a new gasket and install the No. 2 front pipe. Tighten the nuts to 46 ft. lbs. (62 Nm).

9. Connect the front exhaust pipe and tighten the bolts and nuts to 32 ft. lbs. (43 Nm).

Removing the exhaust manifolds—SC300

7923LG27

10. Connect the O_2 sensors, install the undercovers, then lower the vehicle. Connect the battery cable.

4.0L Engine

1. Disconnect the negative battery cable. Drain the cooling system.

2. Remove the camshaft timing pulleys. Remove the cooling fan hydraulic pump on the SC400.

3. Disconnect the accelerator cable, the throttle control cable, if equipped with automatic transaxle and the cruise control actuator cable.

4. Remove the high tension cord cover and the right side ignition coil.

5. Remove the water inlet housing mounting bolts and disconnect the water bypass hose from the ISC valve.

6. Remove the water inlet and inlet housing assemblies. Remove the O-ring from the water inlet housing.

7. Remove the EGR pipe.

8. Disconnect the following:
   a. VSV connector
   b. Vacuum pipe hose
   c. EGR water bypass pipe
   d. Fuel pressure VSV

9. Disconnect the EGR vacuum hoses and remove the EGR VSV.

10. Disconnect the following hoses:
   a. Water bypass pipe hose from the ISC valve.
   b. Water bypass joint hose.
   c. Vacuum pipe hoses.

11. Disconnect the EGR gas temperature sensor, California only. Remove the EGR valve adapter.

12. Disconnect the following:
   a. Fuel pressure regulator vacuum hose.
   b. Air intake chamber vacuum hose.
   c. Vacuum hose from the EVAP BVSV.

13. Remove the mounting bolts, hoses and the vacuum pipe.

14. Remove the ISC valve.

15. Remove the throttle body sensor connectors and the water bypass pipe from the rear water bypass joint.

16. Disconnect the PCV valve hose. Remove the throttle body and gasket.

17. Disconnect the accelerator cable bracket and the brake booster vacuum union and hose.

18. Disconnect the cold start injector connector and the cold start injector tube from the right side delivery pipe, if equipped.

19. Disconnect the check connector from the intake chamber and remove the mounting nuts and bolts.

20. Remove the air intake chamber and the cold start injector, tube and wire assembly, if equipped.

21. Disconnect the engine wire from the intake manifold and from the right side cylinder head. Disconnect the heater hoses.

22. Remove the delivery pipes and the fuel injectors. Remove the mounting bolts and nuts. Lift up the intake manifold.

23. Remove the front and rear water bypass joint.

24. Raise and safely support the vehicle. Remove the front exhaust pipe and the main catalytic converters. Lower the vehicle.

25. Disconnect the right side oxygen sensor. Remove the mounting bolts and nuts and remove the right side exhaust manifold.

26. Remove the oil dipstick and guide. Disconnect the left side oxygen sensor.

27. Remove the mounting bolts and nuts and remove the left side exhaust manifold.

To install:

28. Install the right side exhaust manifold with a new gasket (the painted marks should face the manifold) and tighten the mounting bolts to 29 ft. lbs. (39 Nm). Connect the right side oxygen sensor connector.

29. Install the left side exhaust manifold with a new gasket (the painted marks should face the manifold) and tighten the mounting bolts to 29 ft. lbs. (39 Nm). Connect the left side oxygen sensor connector.

30. Install the oil dipstick and guide. Raise and safely support the vehicle.

31. Install the catalytic converters and front exhaust pipe. Lower the vehicle.

32. Install the front and rear water bypass joints. Tighten the mounting bolts to 13 ft. lbs. (18 Nm).

33. Install the intake manifold, using new gaskets. Tighten the mounting nuts and bolts to 13 ft. lbs. (18 Nm).

34. Install the delivery pipes and fuel injectors. Install the fuel return pipe with new gaskets. Tighten the union bolt to 26 ft. lbs. (35 Nm).

35. Connect the fuel hoses and the injector connectors. Connect the engine wire to the delivery pipes.

36. Connect the connectors on the left side delivery pipe, the water temperature sensor connector, cold start injector time switch connector and the water temperature sender gauge connector.

37. Connect the heater hoses and engine wire bracket. Install the engine wire to the bracket.

38. Install the cold start injector, tube and wire assembly. Tighten the mounting bolts to 69 inch lbs. (8 Nm), if equipped.

39. Install the air intake chamber with

new gaskets and tighten the mounting bolts to 13 ft. lbs. (18 Nm).

40. Connect the cold start injector tube to the right side delivery pipe and tighten the union bolt to 11 ft. lbs. (15 Nm), if equipped.

41. Connect the cold start injector connector, if necessary. Install the accelerator cable bracket.

42. Install the brake booster union and connect the vacuum hose. Tighten the union bolt to 22 ft. lbs. (29 Nm).

43. Connect the water bypass hose to the throttle body and the PCV hose to the cylinder head cover.

44. Install the throttle body, using a new gasket. Tighten the mounting bolts to 13 ft. lbs. (18 Nm).

45. Install the water bypass pipe and connect the sensor connectors. Install the ISC valve and tighten the mounting bolts to 13 ft. lbs. (18 Nm). Connect the water bypass hose.

46. Install the vacuum pipe and the assorted hoses. Install the remaining components.

47. Connect and adjust the accelerator cable, the automatic transmission throttle cable and the cruise control actuator cable. Install the cooling fan hydraulic pump on the SC400. Install the camshaft timing pulleys.

48. Fill the cooling system and connect the negative battery cable. Start the engine and check for leaks.

49. Recheck all the fluid levels. Road test for proper operation.

Camshaft and Valve Lifters

REMOVAL & INSTALLATION

The following procedures have the valve lash adjuster removal and installation incorporated in.

3.0L (1MZ-FE) Engine

1. Remove the timing belt and idler pulley.
2. Remove the camshaft timing pulleys.
3. Remove the cylinder head covers.

➡**The thrust clearance on both the intake and exhaust camshafts is very small, the camshafts must be kept level during removal. If the camshafts are removed without being kept level, the camshaft may be caught in the cylinder head causing the head to break or the camshaft to seize.**

4. To remove the exhaust and intake camshafts from the right side cylinder head:

a. Turn the camshaft with a wrench until the 2 pointed marks drive and driven gears are aligned. (The right camshaft gears have 2 marks apiece; the left side camshaft gears have one mark each.)

b. Secure the exhaust camshaft subgear to the main gear using a service bolt. A bolt 0.63–0.79 in. (16–20mm) long with a 6mm thread diameter and a 1mm pitch is recommended. When removing the exhaust camshaft be sure the subgear is not loaded; all the force must be eliminated.

c. Uniformly loosen and remove the exhaust camshaft bearing cap bolts in several passes and in the proper sequence. Remove the eight bearing cap bolts and remove the caps, keeping them in the correct order.

d. Remove the exhaust camshaft from the engine.

e. Uniformly loosen and remove the 10 bearing cap bolts in several passes, in the proper sequence. Remove the bearing caps, keeping them in order,

remove the oil seal, then lift out the intake camshaft.

5. To remove the exhaust and intake camshafts from the left side cylinder head:

a. Turn the camshaft with a wrench until the pointed marks on the drive and driven gears are aligned. (The right camshaft gears have 2 marks apiece; the left side camshaft gears have one mark each.)

b. Secure the exhaust camshaft subgear to the main gear using a service bolt. A bolt 0.63–0.79 in. (16–20mm) long with a 6mm thread diameter and a 1mm pitch is recommended. When removing the exhaust camshaft be sure the subgear is not loaded; all the force must be eliminated.

c. Uniformly loosen and remove the exhaust camshaft bearing cap bolts in several passes and in the proper sequence. Remove the eight bearing cap bolts and remove the caps. Keep the caps in the correct order.

d. Remove the exhaust camshaft from the engine.

e. Uniformly loosen and remove the 10 bearing cap bolts in several passes, in the proper sequence. Remove the

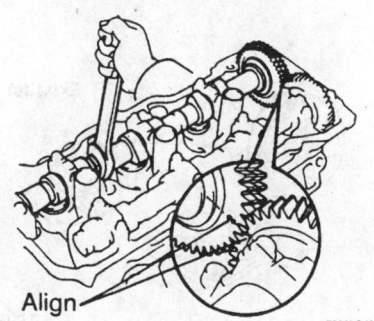

Aligning the right side camshaft timing marks—3.0L (1MZ-FE) engine

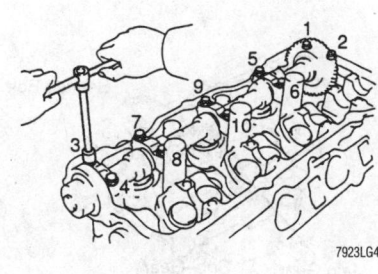

Right exhaust camshaft bearing loosening sequence—3.0L (1MZ-FE) engine

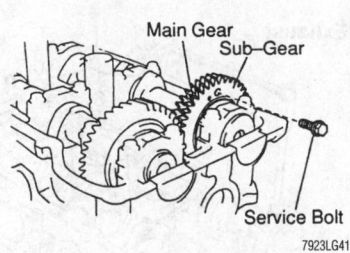

Securing the subgear and driven gear, right side—3.0L (1MZ-FE) engine

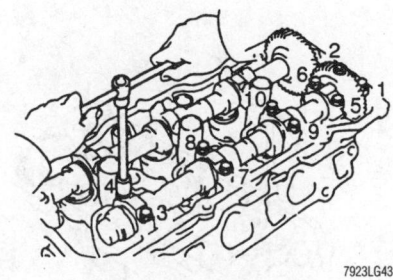

Right intake camshaft bearing loosening sequence—3.0L (1MZ-FE) engine

bearing caps, keeping them in order, remove the oil seal, then lift out the intake camshaft.

6. Remove the valve lash adjuster shims and hydraulic lash adjusters. Identify each lash adjuster and shim as it is removed so it can be reinstalled in the same position. If the lash adjusters are to be reused, store them upside down in a sealed container.

To install:

7. Install the valve lash adjusters into their original positions and install the shims. Check valve clearance and replace the shims as necessary.

8. When reinstalling, remember that the camshafts must be handled carefully and kept straight and level to avoid damage.

9. Before installing the camshafts in either cylinder head, apply multi-purpose grease to the thrust portions of each camshaft.

10. To install the right camshafts:

a. Position the intake camshaft on the head so that the alignment marks are at a 90 degree angle from vertical. The mark should be at the 3 o'clock position.

b. Apply sealant to the No. 1 bearing cap.

c. Apply a light coat of clean engine oil to the bolt threads and under the bolt

head. Install the bearing caps to their proper position. Tighten the bolts evenly and in several passes in the reverse order of loosening to 12 ft. lbs. (16 Nm) in the proper sequence.

d. Position the exhaust camshaft on the head so that the alignment marks are at a 90 degree angle from vertical. The mark should be at the 9 o'clock position and must align with the marks on the other gear.

e. Apply a light coat of clean engine oil to the bolt threads and under the bolt head. Install the bearing caps to their proper position. Tighten the bolts evenly

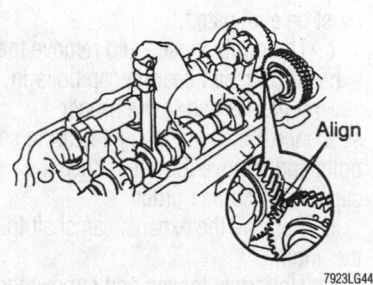

Aligning the left side camshaft timing marks—3.0L (1MZ-FE) engine

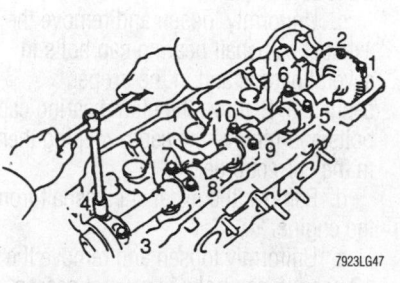

Left exhaust camshaft bearing cap bolt loosening sequence—3.0L (1MZ-FE) engine

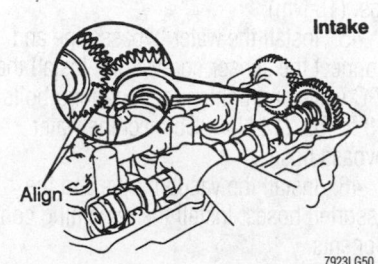

Intake camshaft installation position on the right cylinder head—3.0L (1MZ-FE) engine

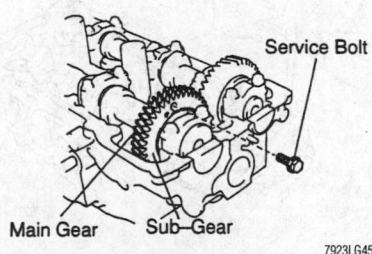

Securing the subgear and driven gear, left side—3.0L (1MZ-FE) engine

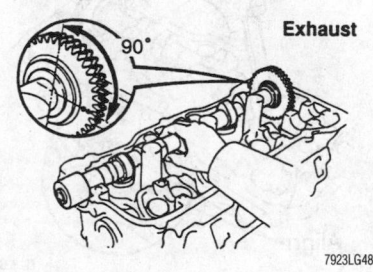

Exhaust camshaft installation position on the right cylinder head—3.0L (1MZ-FE) engine

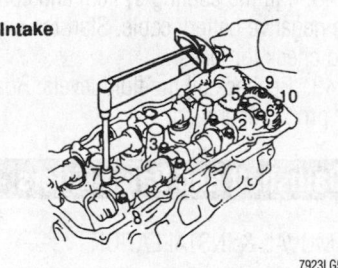

Intake camshaft bearing cap bolt tightening sequence on the right cylinder head—3.0L (1MZ-FE) engine

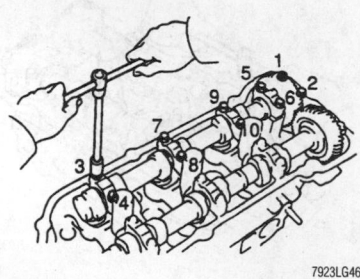

Left intake camshaft bearing cap bolt loosening sequence—3.0L (1MZ-FE) engine

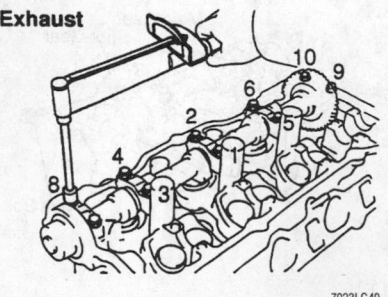

Exhaust camshaft bearing cap bolt tightening sequence on the right cylinder head—3.0L (1MZ-FE) engine

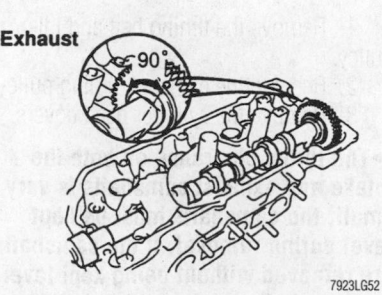

Exhaust camshaft installation position on the left cylinder head—3.0L (1MZ-FE) engine

and in several passes in the reverse order of loosening to 12 ft. lbs. (16 Nm) in the proper sequence.

f. Remove the service bolt.

11. To install the left camshafts:

a. Position the intake camshaft on the head so that the alignment mark is at a 90 degree angle from vertical. The mark should be at the 9 o'clock position.

b. Apply sealant to the No. 1 bearing cap.

c. Apply a light coat of clean engine oil to the bolt threads and under the bolt head. Install the bearing caps to their proper position. Tighten the bolts evenly

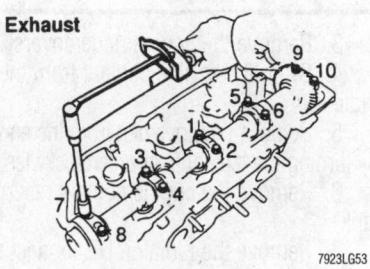

Exhaust camshaft bearing cap bolt tightening sequence on the right cylinder head—3.0L (1MZ-FE) engine

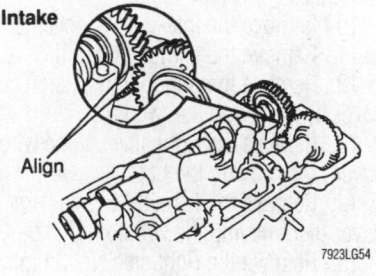

Intake camshaft installation position on the left cylinder head—3.0L (1MZ-FE) engine

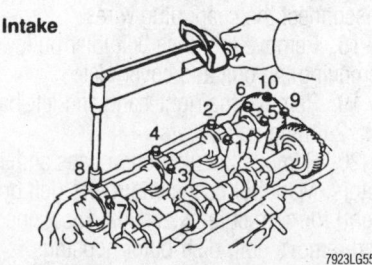

Intake camshaft bearing cap bolt tightening sequence on the right cylinder head—3.0L (1MZ-FE) engine

and in several passes to 12 ft. lbs. (16 Nm) in the proper sequence.

d. Position the exhaust camshaft on the head so that the alignment marks are at a 90 degree angle from vertical. The mark should be at the 3 o'clock position and must align with the marks on the other gear.

e. Apply a light coat of clean engine oil to the bolt threads and under the bolt head. Install the bearing caps to their proper position. Tighten the bolts evenly and in several passes to 12 ft. lbs. (16 Nm) in the proper sequence.

f. Remove the service bolt.

12. Apply multi-purpose grease to new camshaft oil seals. Install the seals.

13. Install the No. 3 (rear) timing belt cover.

14. Install the camshaft timing gears.

15. Install the idler pulley, timing belt and covers.

16. Check and adjust the valve clearance.

17. Install the cylinder head (valve) covers.

18. Start the engine. Check the ignition timing.

19. Test drive the vehicle.

20. Check all fluid levels.

3.0L (2JZ-GE) Engine

1. Disconnect the negative battery cable from the battery.

2. Remove the timing belt from the engine.

3. Remove the cylinder head covers.

4. While holding each camshaft with a wrench, loosen the camshaft sprocket bolt and remove the sprocket.

5. Remove the four bolts and lift out the No. 4 (inner) timing belt cover.

6. Uniformly loosen, then remove the four No. 1 camshaft bearing cap bolts. These are the bolts directly behind the sprockets. Remove the bearing caps.

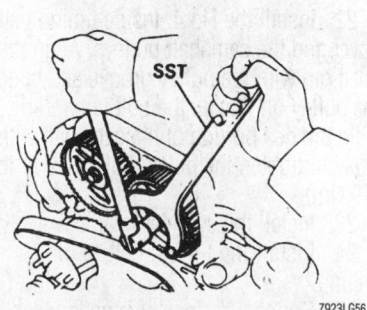

Removing the camshaft sprockets—3.0L (2JZ-GE) engine

7. Uniformly and in the correct sequence, loosen and remove the remaining bearing cap bolts. Note that there are separate sequences for the exhaust and intake camshafts. Lift off all 12 bearing caps.

8. Lift out the exhaust and intake camshafts.

9. Remove the valve lash adjuster shims and hydraulic lash adjusters. Identify each lash adjuster and shim as it is removed so it can be reinstalled in the same position. If the lash adjusters are to be reused, store them upside down in a sealed container.

To install:

10. Install the valve lash adjusters into their original positions and install the shims. Check valve clearance and replace the shims as necessary.

11. When reinstalling, remember that the camshafts must be handled carefully and kept straight and level to avoid damage.

12. Coat the thrust portions of each camshaft with engine oil, then position them in the cylinder head with the cam lobes and the knock pins in the correct position.

13. Position the No. 3 and No. 7 bearing caps in place, coat the bolt threads with oil, then tighten them temporarily.

14. Coat new oil seals with multi-purpose grease, then slide them over the camshafts.

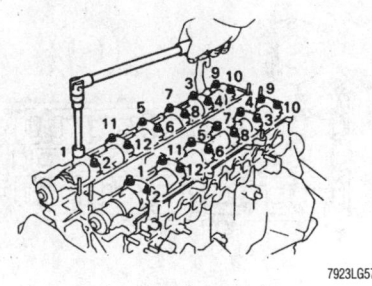

Camshaft bearing cap bolt removal sequence—3.0L (2JZ-GE) engine

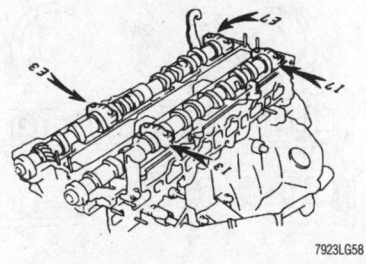

Installing No. 3 and 7 bearing caps—3.0L (2JZ-GE) engine

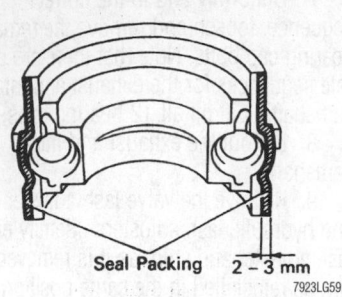

Applying sealant to the No. 1 bearing cap—3.0L (2JZ-GE) engine

Camshaft bearing cap bolt tightening sequence—3.0L (2JZ-GE) engine

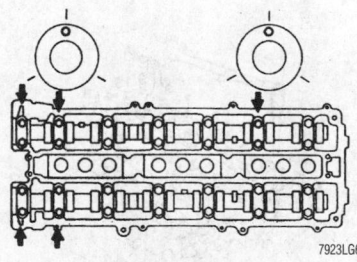

Retightening the camshafts (Step 1)—3.0L (2JZ-GE) engine

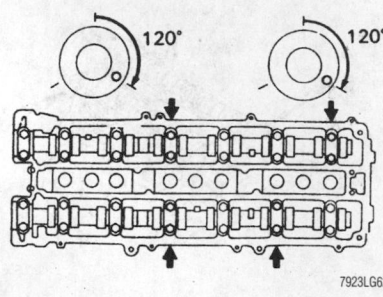

Retorquing the camshafts (Step 2)—3.0L (2JZ-GE) engine

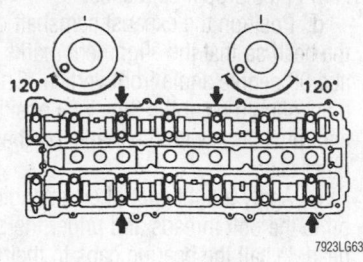

Retightening the camshafts (Step 3)—3.0L (2JZ-GE) engine

15. Clean the mating surfaces of the two No. 1 bearing caps, then apply some sealant. Install the bolts.

16. Install all remaining bearing caps, coat the threads of each bolt with clean oil, then tighten them, in several passes, in sequence, to 14 ft. lbs. (20 Nm). Note that there are separate sequences for the intake and exhaust sides.

17. Press the oil seal in as far as it will go.

18. Rotate each camshaft until the forward straight (knock) pin is straight up. Loosen exhaust Nos. 1, 2 and 6 bearing cap bolts until they can be turned by hand; retighten them to 14 ft. lbs. (20 Nm). Loosen intake Nos. 1 and 2 and re-tighten to 14 ft. lbs. (20 Nm).

19. Turn each camshaft ⅓ of a revolution (120 degrees). Loosen exhaust Nos. 4 and 7 bearing cap bolts; retighten them to 14 ft. lbs. (20 Nm). Loosen intake Nos. 4 and 6 bearing cap bolts; retighten them to 14 ft. lbs. (20 Nm).

20. Turn each camshaft an additional ⅓ of a revolution, loosen exhaust bearing cap bolts Nos. 3 and 5, then retighten them to 14 ft. lbs. (20 Nm). Loosen intake bearing cap bolts Nos. 3 and 7, then retighten them to 14 ft. lbs. (20 Nm).

21. Check and adjust the valve clearance.

22. Install the No 4. inside timing belt cover and the camshaft pulleys. Align the shaft pin with the pulley groove and slide the pulley on. Install the bolt temporarily. Hold the hex portion of the camshaft with a wrench; tighten the pulley bolt to 59 ft. lbs. (79 Nm).

23. Install the cylinder head covers.

24. Install the timing belt to the engine.

25. Connect the negative battery cable to the battery.

26. Check and/or adjust the ignition timing as necessary.

4.0L Engine

1995 SC400

1. Disconnect the negative battery cable. Wait at least 90 seconds before performing any other work. Disconnect the positive battery cable. Remove the battery.

2. Release the fuel pressure from the fuel lines.

✳✳ CAUTION

The fuel injection system remains under pressure after the engine has been turned OFF. Properly relieve fuel pressure before disconnecting any fuel lines. Failure to do so may result in fire or personal injury.

3. Remove the engine undercovers.

4. Drain the engine coolant from the radiator.

5. Remove the drive belt from the engine by turning the tensioner counterclockwise.

6. Remove the engine coolant reservoir tank.

7. Remove the radiator. Disconnect the two transmission oil cooler hoses from the radiator. Plug the hose ends.

8. Disconnect the suction and pressure hoses from the hydraulic pump.

9. Disconnect the air conditioning compressor from the engine. Do not remove the compressor pressure lines.

10. Remove the intake air connector.

11. Remove the left ignition coil.

12. Remove the upper high tension cord cover by removing the two bolts.

13. Remove the right side engine wire cover by removing the bolt.

14. Remove the left side engine wire cover by removing the two bolts.

15. Remove the right side No. 3 timing cover by removing the two air control valve hoses and four bolts.

16. Remove the left No. 3 timing belt cover by removing the four bolts.

17. Disconnect and tag the vacuum hoses and remove the engine wire covers. Disconnect the spark plug wires.

18. Remove the drive belt idler pulley by removing the bolt and cover plate.

19. Remove the right-hand and left-hand No. 2 timing belt covers.

20. Remove the distributor caps and rotors. Mark both caps and rotors (left or right) for correct reinstallation. Disconnect and remove both distributor housings.

21. Disconnect and remove the alternator from the engine.

22. Remove the drive belt tensioner by removing the bolt and two nuts.

23. Remove the spark plugs.

24. Set the engine to TDC on cylinder No. 1. Turn the crankshaft pulley and align its groove with the timing mark **0** of the No. 1 timing cover. Check that the timing marks of the camshaft timing pulleys and timing belt rear plates are aligned. If not, turn the crankshaft 1 full revolution (360 degrees).

25. If the timing belt is to be reused, turn the crank pulley slowly; check that the installation marks are present on the belt. If the marks are not present, make new installation marks before removing the belt. The marks should align with the timing marks on each camshaft pulley.

26. Remove the timing belt tensioner. Alternately loosen the two (2) bolts, then remove the bolts, the tensioner and the dust protector.

27. Using the proper tool, 09960–10010 or equivalent, loosen the tension spring between the left side and right side camshaft timing pulleys by slightly turning the left side camshaft clockwise.

28. Disconnect the timing belt from the camshaft timing pulleys.

29. Using the proper tool (SST 09960–10010 or equivalent), remove the bolt and the camshaft timing pulleys.

30. Remove the right-hand ignition coil.

31. Remove the timing belt rear plates by removing the two bolts for each plate.

32. Remove the throttle body cover.

33. Disconnect the accelerator cable, automatic transmission throttle control cable and the cruise control cable from the throttle body.

34. Disconnect the fuel return hose from the fuel return pipe.

35. Disconnect the PCV hose from the left cylinder head.

36. Remove the heater water valve from the firewall.

37. Remove the throttle body from the engine.

38. On the right cylinder head cover, release the clips and retainers holding the spark plug wires to the cylinder head cover. Remove the cylinder head cover by removing the eight bolts and eight washers.

39. On the left cylinder head cover:

    a. Release the clips and retainers holding the spark plug wires.

    b. Remove the bolt holding the fuel inlet hose to the delivery pipe.

    c. Remove the pulsation damper and two gaskets and disconnect the fuel inlet hose from the delivery pipe.

    d. Free the engine wiring harness from the clamps on the delivery pipe

and disconnect the four injector connectors.

    e. Remove the eight bolts, seal washers, cylinder head cover and gaskets.

40. Remove the semi-circular plugs, if necessary.

41. To remove the exhaust camshaft from the right side cylinder head :

    a. Position the service bolt hole of the drive subgear to the upright position. Secure the camshaft subgear to drive gear with a service bolt.

    b. Set the timing mark (single dot) on the camshaft drive gear at approximately 10 degrees. Turn the camshaft with a wrench on the hexagonal flats.

    c. Alternately loosen and remove the bearing cap bolts holding the intake camshaft side of the oil feed pipe to the cylinder head.

    d. Uniformly loosen (in several passes) and remove the bearing cap bolts, in sequence.

    e. Remove the oil feed pipe and the bearing caps. Remove the camshaft.

42. To remove the intake camshaft from the right side cylinder head:

    a. Set the timing mark (single dot) on the camshaft drive gear at approximately

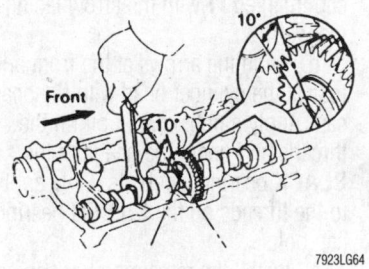

7923LG64

Turning the exhaust camshaft 10 degrees on the right cylinder head—1995 SC400

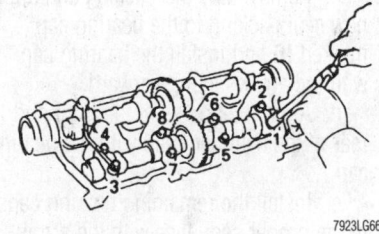

7923LG66

Bolt loosening sequence for the exhaust camshaft on the right cylinder head— LS400 and SC400

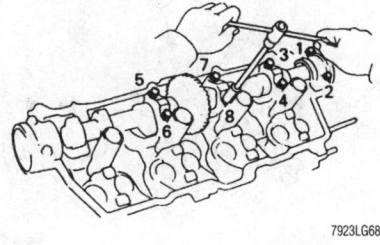

7923LG68

Loosen the bearing cap bolts for the intake camshaft on the right cylinder head in the sequence shown—LS400 and SC400

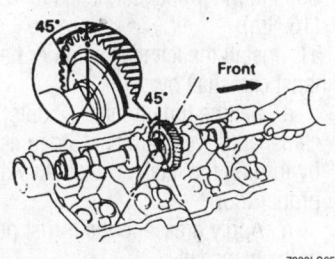

7923LG65

Turning the intake camshaft on the right cylinder head 45 degrees—1995 SC400

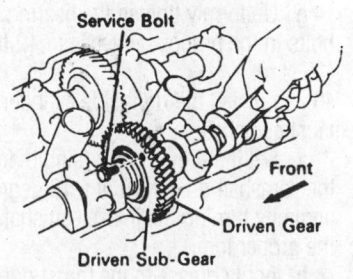

7923LG67

Securing the exhaust camshaft on the left cylinder head—1995 SC400

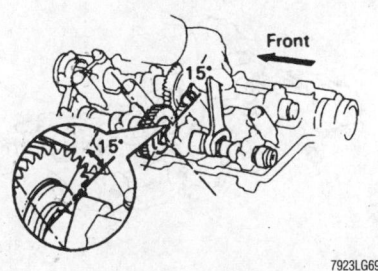

7923LG69

Turning the exhaust camshaft on the left cylinder head 15 degrees—SC400

45 degrees. Turn the camshaft with a wrench on the hexagonal flats.

b. Uniformly loosen (in several passes) and remove the bearing cap bolts in the proper sequence.

c. Remove the bearing caps, oil seal and the intake camshaft.

43. To remove the exhaust camshaft of the left side cylinder head:

a. Position the service bolt hole of the drive subgear to the upright position. Secure the camshaft subgear to drive gear with a service bolt.

➡**When removing the camshaft, be sure the torsional spring force of the subgear has been eliminated.**

b. Set the timing mark (2 dots) on the camshaft drive gear at approximately 15 degrees, by turning the camshaft with the proper tool.

c. Alternately loosen and remove the bearing cap bolts holding the intake camshaft side of the oil feed pipe to the cylinder head.

d. Uniformly loosen (in several passes) and remove the bearing cap bolts in the proper sequence.

e. Remove the oil feed pipe and the bearing caps. Remove the camshaft.

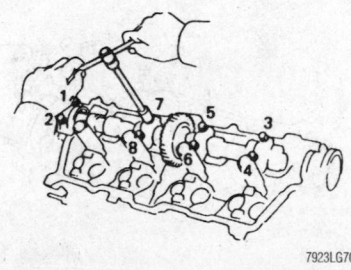

7923LG70

Bolt removal sequence for the intake camshaft on the left cylinder head—LS400 and SC400

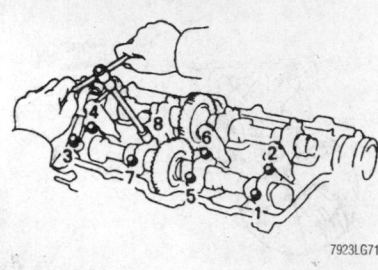

7923LG71

Bolt removal sequence for the exhaust camshaft on the left cylinder head—LS400 and SC400

44. To remove the intake camshaft from the left side cylinder head:

a. Set the timing mark (single dot) of the camshaft drive gear at approximately 60 degrees, by turning the camshaft with the proper tool.

b. Uniformly loosen (in several passes) and remove the bearing cap bolts, in sequence.

c. Remove the bearing caps, oil seal and the intake camshaft.

45. Remove the valve lash adjuster shims and hydraulic lash adjusters. Identify each lash adjuster and shim as it is removed so it can be reinstalled in the same position. If the lash adjusters are to be reused, store them upside down in a sealed container.

To install:

46. Install the valve lash adjusters into their original positions and install the shims. Check valve clearance and replace the shims as necessary.

47. When reinstalling, remember that the camshafts must be handled carefully and kept straight and level to avoid damage.

48. To install the right side cylinder head intake camshaft:

a. Apply grease to the thrust portion of the camshaft.

b. Place the intake camshaft at a 45 degree angle of the timing mark (single dot) on the cylinder head.

c. Remove any old packing and apply new seal packing to the bearing cap marked **I6** and install the bearing cap with the arrow facing rearward.

d. Align the arrows at the front and rear of the cylinder head with the bearing cap.

e. Install the remaining bearing caps in the proper sequence with the arrow mark facing rearward. Apply oil to the threads and under the heads of the **BLACK** colored bolts. Only apply oil to the threads of the **SILVER** colored bolts.

f. Install a new seal washer to the silver bearing cap bolts.

g. Uniformly tighten the bearing cap bolts in the proper sequence to 12 ft. lbs. (16 Nm).

49. To install the right side cylinder head exhaust camshaft:

a. Set the timing mark (single dot) on the camshaft drive gear at a 10 degree angle by turning the intake camshaft with the proper tool.

b. Apply grease to the thrust portion of the camshaft.

c. Align the timing marks (single dots) on the camshaft drive and driven gears.

d. Place the exhaust camshaft in the cylinder head. Install the rear bearing cap, marked **E1**, with the arrow mark facing rearward.

e. Align the arrow marks at the front and rear of the cylinder head with the mark on the bearing cap. Apply a light coat of oil on the threads of the bearing cap bolts.

f. Install the oil feed pipe and 10 bolts. Use bearing cap bolts 1.50 in. (38mm) and 2.05 in. (52mm) in length. Install the two 2.05 in. (52mm) bolts in the outside positions of the oil feed pipe. Install the eight 1.50 in. (38mm) bolts in the other positions.

g. Uniformly tighten the bearing cap bolts in the proper sequence to 12 ft. lbs. (16 Nm).

h. Bring the service bolt installed upward by turning the camshaft with the proper tool. Remove the service bolt.

50. To install the left side cylinder head intake camshaft:

a. Apply grease to the thrust portion of the camshaft.

b. Place the intake camshaft with the timing mark (single dot) at a 60 degree angle on the cylinder head.

c. Remove any old packing and apply new seal packing to the bearing cap marked **I1** and install the front bearing cap, marked **I1** with the arrow facing rearward.

d. Align the arrows at the front and rear of the cylinder head with the bearing cap. Apply a light coat of oil on the threads and under the heads of the **BLACK** bearing cap bolts. Only apply oil to the threads on the **SILVER** bearing cap bolts.

e. Install the remaining bearing caps in the proper sequence with the arrow mark facing rearward.

f. Install a new seal washer to the silver bearing cap bolts.

g. Uniformly tighten the bearing cap bolts in the proper sequence to 12 ft. lbs. (16 Nm).

51. Install the left side cylinder head exhaust camshaft by:

a. Set the timing mark (2 dots) on the camshaft drive gear at a 15 degree angle by turning the intake camshaft with the proper tool.

b. Apply grease to the thrust portion of the camshaft.

c. Align the timing marks (2 dots each) on the camshaft drive and driven gears.

d. Place the exhaust camshaft on the

cylinder head. Install the rear bearing cap, marked **E1** with the arrow mark facing forward.

e. Align the arrow marks at the front and rear of the cylinder head with the mark on the bearing cap. Apply a light coat of oil on the threads and under the heads of the **BLACK** bearing cap bolts. Only apply oil to the threads on the **SILVER** bearing cap bolts.

f. Install the oil feed pipe and 10 bolts. Use bearing cap bolts 1.50 in. (38mm) and 2.05 in. (52mm) in length. Install the two 2.05 in. (52mm) bolts in the outside positions of the oil feed pipe. Install the eight 1.50 in. (38mm) bolts in the other positions.

g. Uniformly tighten the bearing cap bolts in the proper sequence to 12 ft. lbs. (16 Nm).

h. Bring the service bolt installed upward by turning the camshaft with the proper tool. Remove the service bolt.

52. Install the camshaft oil seals with the proper tool (SST 09223–46011). Be sure to apply MP grease to the new oil seal lip.

53. Install the semi-circular plugs as follows:

a. Remove any old packing material.

b. Apply seal packing to the semi-circular plug grooves.

c. Install the four semi-circular plugs to the cylinder heads.

54. Clean the cylinder head covers. Apply new sealant in the correct locations and install the gaskets.

55. Install the left cylinder head cover, tighten the bolts to 52 inch lbs. (6 Nm).

56. Install the right cylinder head cover and bolts. Tighten the bolts to 52 inch lbs. (6 Nm).

57. On the right cylinder head cover, install the spark plug wires and connect the clips and retainers to the cylinder head cover.

58. Install the throttle body. Before installing the throttle body, place a new gasket in position on the air intake chamber. Connect the water bypass hose and the PCV hose to the throttle body. Tighten the bolts and nuts to 13 ft. lbs. (18 Nm).

a. Connect the water bypass pipe to the clamp on the engine wire cover.

b. Connect the water bypass hose to the IAC valve.

c. Connect the vacuum hose to the throttle body.

d. Connect the sensor wiring to the throttle body.

59. Install the heater water valve and connect the hoses and wiring.

60. Reconnect the PCV hose to the left cylinder head.

61. Connect the fuel return line to the pipe and connect the fuel inlet hose with the pulsation damper to the left delivery pipe. Tighten the bolts to 29 ft. lbs. (39 Nm).

62. Install the upper high tension cord cover.

63. Install the intake air connector. Connect the air connector from the throttle body and connect the air cleaner hose. Connect the hoses to the IAC valve and to the PS air control valve.

64. Connect the control cables to the throttle body.

65. Install the throttle body cover.

66. Install the two rear timing plates with the four bolts.

67. Install the right-hand ignition coil.

68. Align the knock pin on the right side camshaft with the knock pin of the timing pulley. Slide on the timing pulley with the right side mark facing forward. Tighten the bolt to 80 ft. lbs. (108 Nm).

69. Align the knock pin on the left side camshaft with the knock pin of the timing pulley. Slide on the timing pulley with the left side mark facing forward. Tighten the bolt to 80 ft. lbs. (108 Nm).

70. Turn the crankshaft pulley and align its groove with the **0** timing mark on the timing belt cover.

71. Turn each camshaft timing pulley and align the timing marks of the pulley with the timing belt rear plate.

72. Install the timing belt to the left side camshaft timing pulley:

a. Using the proper tool, slightly turn the left side timing pulley clockwise. Align the installation mark of the timing belt with the timing mark of the camshaft timing pulley and hang the timing belt on the left side camshaft pulley.

b. Align the timing marks of the left side camshaft pulley and the timing belt rear plate.

c. Check that the timing belt has tension between crankshaft timing pulley and the left side camshaft pulley.

73. Install the timing belt to the right side camshaft timing pulley:

a. Using the proper tool, slightly turn the right side timing pulley clockwise. Align the installation mark of the timing belt with the timing mark of the camshaft timing pulley and hang the timing belt on the right side camshaft pulley.

b. Align the timing marks of the right side camshaft pulley and the timing belt rear plate.

c. Check that the timing belt has ten-

sion between crankshaft timing pulley and the right side camshaft pulley.

74. The timing belt tensioner must be set prior to installation. The tensioner can be set by:

a. Place a plate washer between the tensioner and a block. Using a press, press in the pushrod using 220–2205 lbs. of pressure.

b. Align the holes of the pushrod and housing, pass a 1.27mm Allen wrench through the holes to keep the setting position of the pushrod.

c. Release the press and install the dust boot to the tensioner.

75. Loosely install the timing belt tensioner. Evenly and alternately tighten the bolts to 20 ft. lbs. (26 Nm). Remove the tool from the tensioner.

76. Turn the crankshaft pulley two complete revolutions from TDC to TDC. Always turn the crankshaft clockwise. Check that all belt and pulley marks align with their reference marks. If any mark is out of perfect alignment, the timing belt must be removed and reinstalled.

77. Install the remaining components. Install the accessory drive belt.

78. Install the battery and connect the cables.

79. Fill the engine coolant and check the transmission fluid level.

80. Start the engine. Check carefully for fluid leaks. Check the ignition timing.

81. Install the engine under-covers.

LS400 AND 1996–99 SC400

1. Disconnect the negative battery cable. Wait at least 90 seconds before performing any other work. Disconnect the positive battery cable and remove the battery.

2. Remove the air cleaner inlet.

3. Remove the V bank cover by removing the bolt and two cap nuts.

4. Relieve the fuel pressure from the fuel lines.

✳✳ CAUTION

The fuel injection system remains under pressure after the engine has been turned OFF. Properly relieve fuel pressure before disconnecting any fuel lines. Failure to do so may result in fire or personal injury.

5. Remove the air cleaner and intake air connector assembly.

6. Remove the drive belt, fluid coupling and the fan pulley. The drive belt tension may be slackened by turning the tensioner

counterclockwise. The pulley bolt for the drive belt tensioner has a left-handed thread.

7. Remove the radiator.

8. Remove the right-hand No. 3 timing belt cover.

9. Remove the left-hand No. 3 timing belt cover, for 1995 California vehicles and all 1996–99 vehicles, disconnect the EVAP hose clamp from the timing belt cover. For all other vehicles, disconnect the EVAP hose from the hose clamp on the timing belt cover. Remove the four bolts to the left-hand timing belt cover. Disconnect the cord grommet from the timing belt cover and remove the timing belt cover.

10. Remove the drive belt idler pulley by removing the pulley bolt and cover plate.

11. Remove the right-hand and left No. 2 timing belt covers.

12. Remove the distributor housings.

13. Remove the No. 1 ignition coil.

14. Disconnect the A/C compressor from the engine. Do not disconnect the A/C pressure lines.

15. Remove the fan bracket by removing the two bolts and two nuts.

16. Disconnect and remove the alternator from the engine.

17. Remove the drive belt tensioner.

18. Set the engine to TDC on cylinder No. 1. Turn the crankshaft pulley and align its groove with the timing mark **0** of the No. 1 timing cover. Check that the timing marks of the camshaft timing pulleys and timing belt rear plates are aligned. If not, turn the crankshaft 1 full revolution (360 degrees).

19. Turn the crankshaft pulley approximately 50° clockwise and put the timing mark of the crankshaft pulley in line with the centers of the crankshaft pulley bolt and the idler pulley bolt.

✵✵ WARNING

If the timing belt is disengaged, having the crankshaft pulley at the wrong angle can cause the piston head and valve head to come into contact with each other when you remove the camshaft timing pulley. Always set the crankshaft pulley at the correct angle before removing the timing belt.

20. If the timing belt is to be reused, turn the crank pulley slowly; check that the three installation marks are present on the belt. If the marks are not present, make new installation marks before removing the belt. The marks should align with the timing marks on each camshaft pulley and the crank pulley.

21. Remove the timing belt tensioner. Alternately loosen the two bolts; remove the bolts, the tensioner and the dust protector.

22. Using the proper tool, 09278–54012, or equivalent, loosen the tension between the left side and right side timing pulleys by slightly turning the left side camshaft clockwise.

23. Disconnect the timing belt from the camshaft timing pulleys.

24. Using the proper tool (SST 09960–10010 or equivalent), remove the bolt and the camshaft timing pulleys.

25. Remove the No. 2 ignition coil.

26. Remove the rear timing belt plates by removing the two bolts to each plate.

27. Remove the throttle body.

28. Remove the spark plug wires, wire clamps and the wire cover assembly from the right cylinder head.

29. Remove the right cylinder head cover by removing the eight bolts and eight washers.

30. Remove the transmission oil dipstick.

31. Disconnect the EVAP hose (from the charcoal canister) from the vacuum switching valve.

32. Disconnect the engine wire clamp from the wire bracket on the delivery pipe.

33. Disconnect the spark plug wires and clamps from the left-hand cylinder head cover.

34. Remove the left cylinder head cover by removing the eight bolts and eight seal washers.

35. Remove the semi-circular plugs, if necessary.

➡**Since the thrust clearance of the camshaft is small, the camshaft must be kept level while it is being removed. If the camshaft is not kept level, the portion of the cylinder head receiving the shaft thrust may crack or be damaged, causing the camshaft to seize or break.**

36. To remove the exhaust camshaft from the right side cylinder head:

a. Position the service bolt hole of the drive subgear to the upright position. Secure the camshaft subgear to drive gear with a service bolt.

b. Set the timing mark (single dot) on the camshaft drive gear at approximately 10 degrees. Turn the camshaft with a wrench on the hexagonal flats.

c. Alternately loosen and remove the bearing cap bolts holding the intake camshaft side of the oil feed pipe to the cylinder head.

d. Uniformly loosen (in several passes) and remove the bearing cap bolts, in sequence.

e. Remove the oil feed pipe and the bearing caps. Remove the camshaft.

37. To remove the intake camshaft from the right side cylinder head:

a. Set the timing mark (single dot) on the camshaft drive gear at approximately 45 degrees. Turn the camshaft with a wrench on the hexagonal flats.

b. Uniformly loosen (in several passes) and remove the bearing cap bolts in the proper sequence.

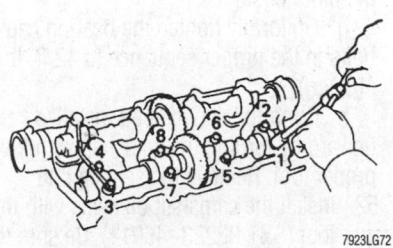

Right side exhaust camshaft bracket bolt removal sequence—LS400

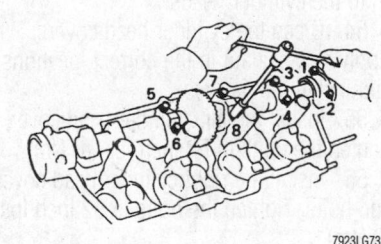

Right side intake camshaft bracket bolt removal sequence—LS400

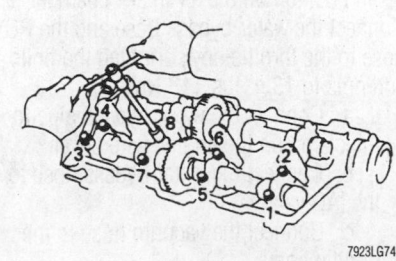

Left side exhaust camshaft bracket bolt removal sequence—LS400

c. Remove the bearing caps, oil seal and the intake camshaft.

38. To remove the exhaust camshaft of the left side cylinder head:

a. Position the service bolt hole of the drive subgear to the upright position. Secure the camshaft subgear to drive gear with a service bolt.

➡ **When removing the camshaft, be sure the torsional spring force of the subgear has been eliminated.**

b. Set the timing mark (2 dots) on the camshaft drive gear at approximately 15 degrees, by turning the camshaft with the proper tool.

c. Alternately loosen and remove the bearing cap bolts holding the intake camshaft side of the oil feed pipe to the cylinder head.

d. Uniformly loosen (in several passes) and remove the bearing cap bolts in the proper sequence.

e. Remove the oil feed pipe and the bearing caps. Remove the camshaft.

39. To remove the intake camshaft from the left side cylinder head:

a. Set the timing mark (single dot) of the camshaft drive gear at approximately 60 degrees, by turning the camshaft with the proper tool.

b. Uniformly loosen (in several passes) and remove the bearing cap bolts, in sequence.

c. Remove the bearing caps, oil seal and the intake camshaft.

40. Remove the valve lash adjuster shims and hydraulic lash adjusters. Identify each lash adjuster and shim as it is removed so it can be reinstalled in the same position. If the lash adjusters are to be reused, store them upside down in a sealed container.

To install:

41. Install the valve lash adjusters into their original positions and install the shims. Check valve clearance and replace the shims as necessary.

42. When reinstalling, remember that the camshafts must be handled carefully and kept straight and level to avoid damage.

43. Remove any old packing and apply new seal packing to the bearing caps.

44. Install the bearing cap on the right side cylinder head, marked **I1**, in position with the arrow mark facing the rear. Install the bearing cap on the left side cylinder head, marked **I6**, in position with the arrow mark facing the front.

45. Apply a light coat of oil on the threads of the cap bolts. Install the bearing

cap bolts with new washers and tighten to 12 ft. lbs. (16 Nm).

46. To install the right side cylinder head intake camshaft:

a. Apply grease to the thrust portion of the camshaft.

b. Place the intake camshaft at a 45 degree angle of the timing mark (single dot) on the cylinder head.

c. Remove any old packing and apply new seal packing to the bearing cap marked **I6** and install the front bearing cap, marked **I6** with the arrow facing rearward.

d. Align the arrows at the front and rear of the cylinder head with the bearing cap.

e. Install the remaining bearing caps in the proper sequence with the arrow mark facing rearward. Install the oil feed pipe and the mounting bolts.

f. Uniformly tighten the bearing cap bolts in the proper sequence to 12 ft. lbs. (16 Nm).

47. To install the right side cylinder head exhaust camshaft:

a. Set the timing mark (single dot) on the camshaft drive gear at a 10 degree angle by turning the intake camshaft with the proper tool.

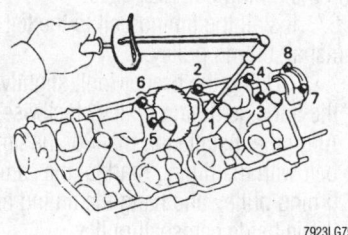

Right side intake camshaft bracket bolt tightening sequence—LS400

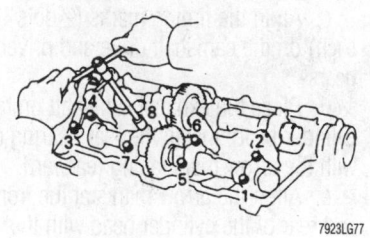

Left side exhaust camshaft bracket bolt removal sequence—1996–97 SC400

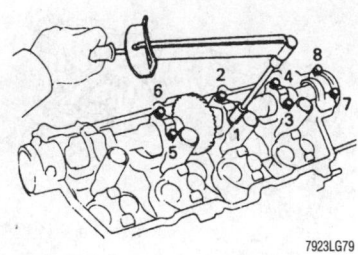

Right side intake camshaft bracket bolt tightening sequence—1996–97 SC400

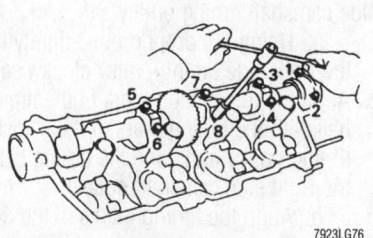

Right side intake camshaft bracket bolt removal sequence—1996–99 SC400

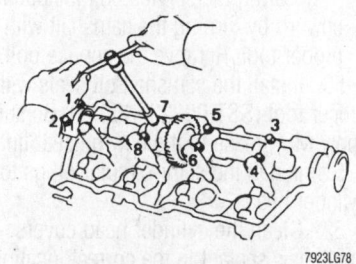

Left side intake camshaft bracket bolt removal sequence—1996–97 SC400

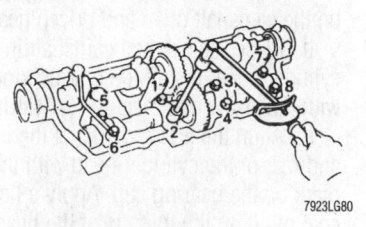

Right side exhaust camshaft bracket bolt tightening sequence—1996–97 SC400

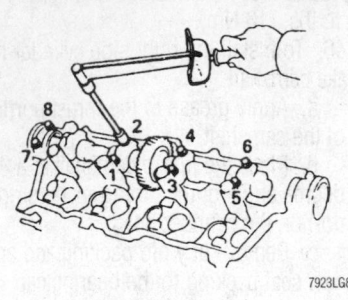

Left side intake camshaft bracket bolt tightening sequence—LS400 and SC400

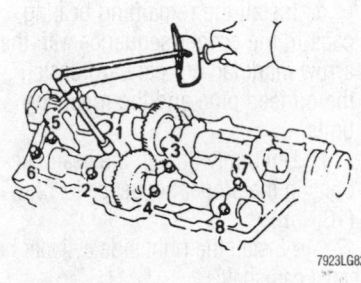

Left side exhaust camshaft bracket bolt tightening sequence—LS400 and SC400

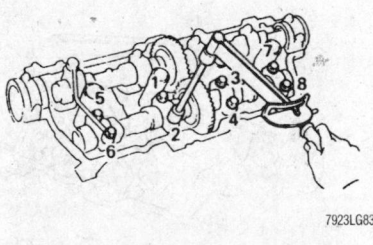

Right side exhaust camshaft bracket bolt tightening sequence—LS400

b. Apply grease to the thrust portion of the camshaft.

c. Align the timing marks (single dots) on the camshaft drive and driven gears.

d. Place the exhaust camshaft in the cylinder head. Install the rear bearing cap with the arrow mark facing rearward.

e. Align the arrow marks at the front and rear of the cylinder head with the mark on the bearing cap. Apply a light coat of oil on the threads of the bearing cap bolts.

f. Uniformly tighten the bearing cap bolts in the proper sequence to 12 ft. lbs. (16 Nm).

g. Bring the service bolt installed upward by turning the camshaft with the proper tool. Remove the service bolt.

48. To install the left side cylinder head intake camshaft:

a. Apply grease to the thrust portion of the camshaft.

b. Place the intake camshaft with the timing mark (single dot) at a 60 degree angle on the cylinder head.

c. Remove any old packing and apply new seal packing to the bearing cap marked **I6** and install the front bearing cap, marked **I1** with the arrow facing rearward.

d. Align the arrows at the front and rear of the cylinder head with the bearing cap. Apply a light coat of oil on the threads of the bearing cap bolts.

e. Install the remaining bearing caps in the proper sequence with the arrow mark facing rearward. Install the oil feed pipe and the mounting bolts.

f. Uniformly tighten the bearing cap bolts in the proper sequence to 12 ft. lbs. (16 Nm).

49. Install the left side cylinder head exhaust camshaft by:

a. Set the timing mark (2 dots) on the camshaft drive gear at a 15 degree angle by turning the intake camshaft with the proper tool.

b. Apply grease to the thrust portion of the camshaft.

c. Align the timing marks (2 dots each) on the camshaft drive and driven gears.

d. Place the exhaust camshaft on the cylinder head. Install the rear bearing cap with the arrow mark facing rearward.

e. Align the arrow marks at the front and rear of the cylinder head with the mark on the bearing cap. Apply a light coat of oil on the threads of the bearing cap bolts.

f. Uniformly tighten the bearing cap bolts in the proper sequence to 12 ft. lbs. (16 Nm).

g. Bring the service bolt installed upward by turning the camshaft with the proper tool. Remove the service bolt.

50. Install the camshaft oil seals with the proper tool (SST 09223–46011). Be sure to apply MP grease to the new oil seal lip.

51. Install the semi-circular plugs to the cylinder heads.

52. Clean the cylinder head covers. Apply new sealant in the correct locations and install the gaskets.

53. Install the left cylinder head cover and bolts. Tighten the bolts to 52 inch lbs. (6 Nm).

54. Connect the spark plug wires and clamps to the left cylinder head cover.

55. Connect the engine wire clamp to the wire bracket on the delivery pipe.

56. Connect the EVAP hose to the VSV.

57. Install the transmission oil dipstick.

58. Install the right cylinder head cover and bolts. Tighten the bolts to 52 inch lbs. (6 Nm).

59. Connect the spark plug wires and clamps to the right cylinder head cover.

60. Install the throttle body to the air intake chamber. Install the two bolts and two nuts and tighten to 13 ft. lbs. (18 Nm).

61. Install the timing belt rear plates by installing the bolts. Tighten the bolts to 66 inch lbs. (8 Nm).

62. Install the No. 2 ignition coil.

63. Align the knock pin on the right side camshaft with the knock pin of the timing pulley. Slide on the timing pulley with the right side mark facing forward. Tighten the bolt to 80 ft. lbs. (108 Nm).

64. Align the knock pin on the left side camshaft with the knock pin of the timing pulley. Slide on the timing pulley with the left side mark facing forward. Tighten the bolt to 80 ft. lbs. (108 Nm).

65. Turn the crankshaft pulley and align its groove with the **0** timing mark on the timing belt cover.

66. Turn each camshaft timing pulley and align the timing marks of the pulley with the timing belt rear plate.

67. Install the timing belt to the left side camshaft timing pulley:

a. Using the proper tool, slightly turn the left side timing pulley clockwise. Align the installation mark of the timing belt with the timing mark of the camshaft timing pulley and hang the timing belt on the left side camshaft pulley.

b. Align the timing marks of the left side camshaft pulley and the timing belt rear plate.

c. Check that the timing belt has tension between crankshaft timing pulley and the left side camshaft pulley.

68. Install the timing belt to the right side camshaft timing pulley:

a. Using the proper tool, slightly turn the right side timing pulley clockwise. Align the installation mark of the timing belt with the timing mark of the camshaft timing pulley and hang the timing belt on the right side camshaft pulley.

b. Align the timing marks of the right side camshaft pulley and the timing belt rear plate.

c. Check that the timing belt has tension between crankshaft timing pulley and the right side camshaft pulley.

69. The timing belt tensioner must be set prior to installation. The tensioner can be set by:

a. Place a plate washer between the tensioner and a block. Using a press, press in the pushrod using 220–2205 lbs. of pressure.

b. Align the holes of the pushrod and housing, pass a 1.27mm Allen wrench through the holes to keep the setting position of the pushrod.

c. Release the press and install the dust boot to the tensioner.

70. Loosely install the tensioner. Evenly and alternately tighten the bolts to 20 ft. lbs. (26 Nm). Remove the tool from the tensioner.

71. Turn the crankshaft pulley two complete revolutions from TDC to TDC. Always turn the crankshaft clockwise. Check that all belt and pulley marks align with their reference marks. If any mark is out of perfect alignment, the timing belt must be removed and reinstalled.

72. Install the remaining components. Install the V bank cover.

73. Fill the radiator with engine coolant.

74. Install the battery and battery tray. Connect the battery cables, positive cable first.

75. Start the engine and check for leaks.

76. Recheck the engine coolant level.

77. Install the air cleaner inlet.

78. Install the battery clamp cover.

79. Install the engine undercover.

80. Install the oil pan protector.

81. Lower the vehicle.

Valve Lash

ADJUSTMENT

✳✳ CAUTION

On models with an air bag, wait at least 90 seconds from the time that the ignition switch is turned to the LOCK position and the battery is disconnected before performing any further work.

3.0L (1MZ-FE) Engine

➡Adjust the valve clearance when the engine is cold.

1. Disconnect the negative battery cable.

2. Disconnect the accelerator/throttle cable from the throttle linkage.

3. Remove the air intake chamber.

4. On the 3VZ-FE engine, remove the V-bank cover with a 5mm Allen wrench.

5. Remove the cylinder head covers.

6. Turn the crankshaft pulley and align it's groove with the timing mark **0** of the No. 1 timing cover.

7. Check that the valve lash adjusters on the No. 1 intake are loose and the exhaust are tight. If not, turn the crankshaft on complete revolution (360 degrees).

8. Measure the clearance between the valve lash adjuster and the camshaft. Record the measurements on valves No. 1, 2, 3 and 6.

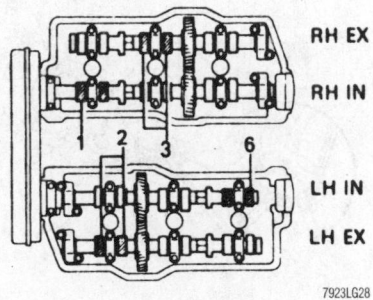

Adjust these valves FIRST—3.0L (1MZ-FE) ENGINE

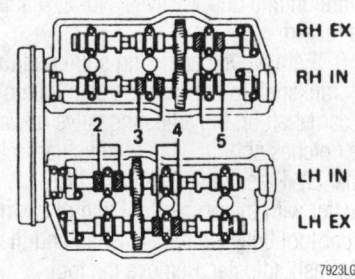

Adjust these valves SECOND—3.0L (1MZ-FE) ENGINE

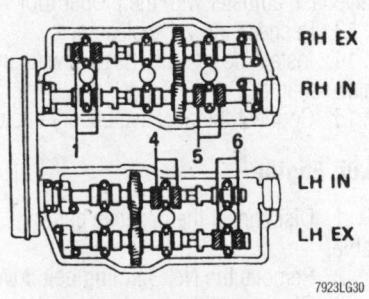

Adjust these valves THIRD—3.0L (1MZ-FE) ENGINE

a. The intake valve clearance cold is 0.005–0.009 in. (0.13–0.23mm).

b. The exhaust valve clearance cold is 0.011–0.015 in. (0.27–0.37mm).

9. Turn the crankshaft ⅔ of a revolution (240 degrees) and check the clearance on valves No. 2, 3, 4 and 5 and record.

10. Turn the crankshaft another ⅔ of a revolution and check valves; No. 1, 4, 5 and 6 and record.

11. Remove the adjusting shim and turn the crankshaft to position the cam lobe of the camshaft on the adjusting valve upward. Press down the valve lash adjuster with the proper tool and place the proper tool between the camshaft and the valve lash adjuster. Remove the tool.

12. Remove the adjusting shim with the proper tool.

13. Use the accompanying charts to determine the correct size replacement shim. Install the specified valve shim on the valve lash adjuster with the proper tool.

14. Recheck the valve clearance.

15. Install the cylinder head covers and intake chamber.

16. Connect the negative battery cable.

3.0L (2JZ-GE) Engine

➡Adjust the valve clearance when the engine is cold.

1. Disconnect the negative battery cable.

2. Disconnect the accelerator/throttle cable from the throttle linkage.

3. Remove the cylinder head covers.

4. Turn the crankshaft pulley and align it's groove with the timing mark **0** of the No. 1 timing cover.

5. Check that the timing marks on the camshaft sprockets are in alignment with the marks on the No. 4 timing cover. If not, turn the crankshaft 1 complete revolution (360 degrees).

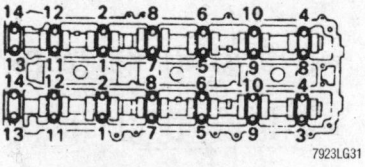

Camshaft bearing cap bolt tightening sequence—SC300

6. Uniformly tighten the camshaft bearing cap bolts in several passes, in the sequence, to 14 ft. lbs. (20 Nm).

7. Measure the clearance between the valve lash adjuster and the camshaft. Record the measurements on valves No. 1, 4 and 5.

    a. The intake valve clearance cold is 0.006–0.010 in. (0.15–0.25mm).

    b. The exhaust valve clearance cold is 0.010–0.014 in. (0.25–0.35mm).

8. Turn the crankshaft ⅔ of a revolution (240 degrees) and check the clearance on valves No. 3, 5 and 6 and record.

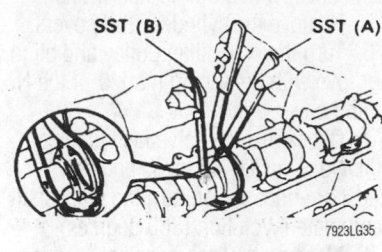

Press down the valve lash adjuster with a special tool—SC300 shown, others similar

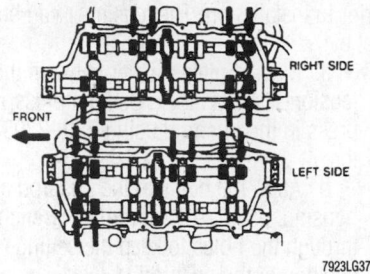

Adjust these valves FIRST—4.0L (1UZ-FE) engine

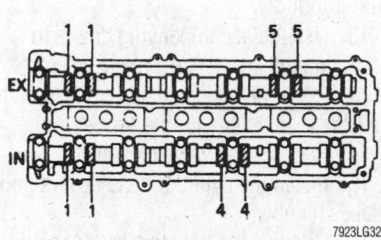

Adjust these valves FIRST—SC300

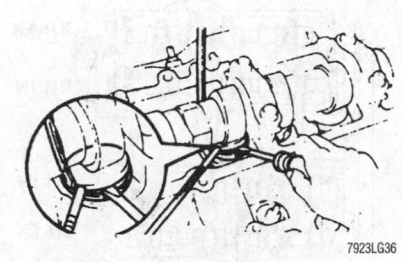

Removing the adjusting shim—SC300 shown, others similar

9. Turn the crankshaft another ⅔ of a revolution and check valves; No. 2, 4 and 6 and record.

10. Remove the adjusting shim and turn the crankshaft to position the cam lobe of the camshaft on the adjusting valve upward. The notches should be perpendicular to the camshaft. Press down the valve lash adjuster with the proper tool and place the proper tool between the camshaft and the valve lash adjuster. Remove the tool.

11. Remove the adjusting shim with the proper tool (a magnetic finger).

12. Use the accompanying charts to determine the correct size replacement shim. Install the specified valve shim on the valve lash adjuster with the proper tool.

13. Recheck the valve clearance.

14. Install the cylinder head covers and intake chamber.

15. Connect the negative battery cable.

4.0L Engine

1. Disconnect the negative battery cable.

2. Remove the No. 3 timing belt covers.

3. Disconnect the spark plug wires and remove the cylinder head covers.

4. Turn the crankshaft pulley and align it's groove with the timing mark **0** of the No.

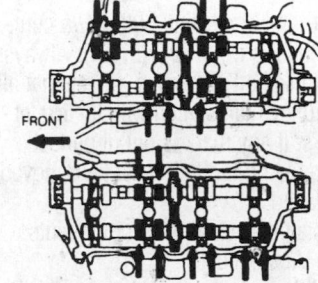

Adjust these valves SECOND—4.0L (1UZ-FE) engine

ADJUSTING SHIM SELECTION CHART

New shim thickness   mm (in.)

Shim No.	Thickness	Shim No.	Thickness
01	2.50 (0.0984)	38	2.95 (0.1161)
63	2.55 (0.1004)	43	3.00 (0.1181)
06	2.60 (0.1024)	48	3.05 (0.1201)
66	2.65 (0.1043)	51	3.10 (0.1220)
13	2.70 (0.1063)	77	3.15 (0.1240)
18	2.75 (0.1083)	56	3.20 (0.1260)
23	2.80 (0.1102)	80	3.25 (0.1280)
28	2.85 (0.1122)	61	3.30 (0.1299)
33	2.90 (0.1142)		

Valve shim selection chart—4.0L (1UZ-FE) engine

1 timing cover. Check that the timing marks of the camshaft timing pulleys and timing belt rear plates are aligned. If not, turn the crankshaft 1 revolution (360 degrees) and align the mark.

5. Measure the clearance between the valve lash adjuster and the camshaft on the valves in the first sequence and record.

    a. The intake valve clearance cold is 0.006–0.010 in. (0.15–0.25mm).

    b. The exhaust valve clearance cold is 0.010–0.014 in. (0.25–0.35mm).

6. Turn the crankshaft 1 full revolution (360 degrees) and align the mark.

7. Measure the clearance between the

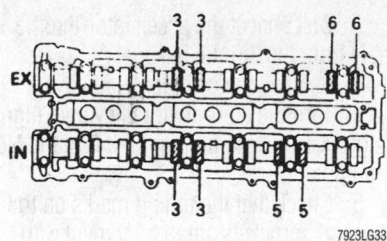

Adjust these valves SECOND—SC300

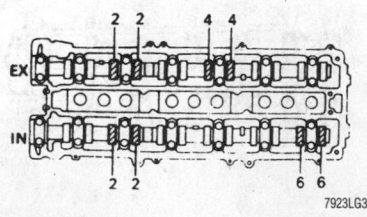

Adjust these valves THIRD—SC300

valve lash adjuster and the camshaft on the valves in the second sequence and record.

8. Remove the adjusting shim and turn the crankshaft to position the cam lobe of the camshaft on the adjusting valve upward. Position the hole in the shim toward the outside of the cylinder head. Press down the valve lash adjuster with the proper tool and place the proper tool between the camshaft and the valve lash adjuster. Remove the tool.

9. Remove the adjusting shim with the proper tool.

10. Use the accompanying charts to determine the correct replacement shim. Install the specified valve shim on the valve lash adjuster with the proper tool.

11. Recheck the valve clearance. Install the cylinder head covers.

12. Connect the spark plug wires and install the No. 3 timing belt covers.

13. Connect the negative battery cable.

Oil Pan

REMOVAL & INSTALLATION

3.0L (1MZ-FE) Engine

1. Disconnect the negative battery cable from the battery.

2. Raise and safely support the front of the vehicle.

3. Remove the right front wheel.

4. Remove the fender apron seal.

5. Remove the engine undercover.

6. Drain the engine oil from the engine.

7. Remove the front exhaust pipe.

8. Remove the front exhaust pipe bracket from the No. 1 oil pan.

9. Remove the flywheel housing undercover.

10. Remove the ten bolts and two nuts to the No. 2 oil pan.

11. Insert the blade of SST tool 09032–00100 or equivalent between the No. 1 and No. 2 oil pans. Tap the tool sideways to break the seal and remove the pan. Clean the surfaces of the oil pans.

12. Remove the oil strainer and gasket from the engine by removing the three nuts.

13. Remove the No. 1 oil pan as follows:

   a. Remove the two bolts to the flywheel housing undercover. Remove the flywheel undercover.

   b. Remove the 17 bolts and 2 nuts to the No. 1 oil pan. Make a note of the position of the each bolt. When replacing the bolts into the oil pan, place each bolt in the position from which it was removed.

c. Remove the oil pan by prying the portions between the cylinder block and the oil pan. Be careful not to damage the contact surfaces.

14. Remove the baffle plate from the No. 1 oil pan.

To install:

15. Clean all mating surfaces of the oil pans.

16. Install the baffle plate to the No. 1 oil pan and tighten to 69 inch lbs. (8 Nm)

17. Install the No. 1 oil pan as follows:

   a. Using a non residue solvent, clean both sealing surfaces to the oil pan.

   b. Apply liquid sealant to the oil pan and engine block.

   c. Install the oil pan with the 17 bolts and 2 nuts. Uniformly tighten the bolts and nuts in several passes.

   d. Tighten the bolts as follows:

- 10mm head bolt-69 inch lbs. (8 Nm)
- 12mm head bolt-14 ft. lbs. (20 Nm)
- 14mm head bolt-27 ft. lbs. (37 Nm)

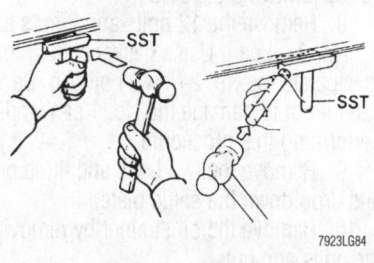

Use the special tool to break the seal and remove the oil pan—3.0L (1MZ-FE)

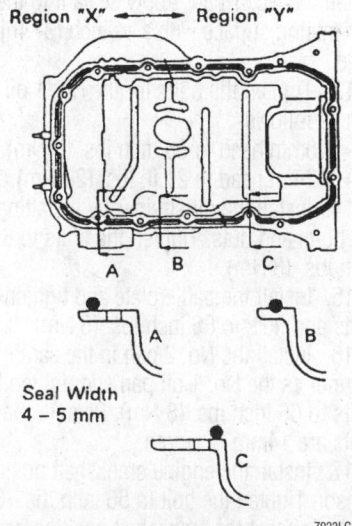

Apply sealant as shown to the No. 1 (upper) oil pan—3.0L (1MZ-FE) engine

e. Install the flywheel housing undercover with the two bolts. Tighten the bolts to 69 inch lbs. (8 Nm).

18. Install the oil strainer with the three nuts. Tighten the nuts to 69 inch lbs. (8 Nm).

19. Install the No. 2 oil pan as follows:

   a. Using a non residue solvent, clean both sealing surfaces to the oil pan.

   b. Apply liquid sealant to the oil pan and engine block.

   c. Install the No. 2 oil pan with the ten bolts and two nuts. Uniformly tighten the bolts and nuts in several passes. Tighten the bolts to 69 inch lbs. (8 Nm).

20. Install the flywheel housing undercover.

21. Install the front exhaust pipe bracket to the No. 1 oil pan. Tighten the bolts to 15 ft. lbs. (21 Nm).

22. Install the front exhaust pipe, Tighten the four nuts holding the exhaust manifolds to the front exhaust pipe. Tighten the four nuts to 46 ft. lbs. (62 Nm). Tighten the two bolts and two nuts holding the front exhaust pipe to the center exhaust pipe. Tighten the bolts and nuts to 41 ft. lbs. (56 Nm). Install the bracket with the two bolts and tighten to 14 ft. lbs. (19 Nm). Install the support stay with the two bolts and tighten to 22 ft. lbs. (30 Nm).

23. Install the engine undercover.

24. Install the right fender apron seal.

25. Install the right front wheel and lower the vehicle.

26. Fill the engine with oil.

27. Start the engine and check for leaks.

28. Recheck the engine oil.

3.0L (2JZ-GE) Engine

➡The No. 1 oil pan can not be removed with the engine in the vehicle. The engine/transmission assembly must be removed. The manufacturer does not provide any on vehicle information for the No. 2 oil pan removal and installation. If only the No. 2 oil pan is being serviced, the engine/transmission assembly can remain in the vehicle.

1. Remove the engine/transmission assembly, then separate the transmission from the engine.

2. With the engine on a stand, remove the timing belt, the idler pulley and the crankshaft timing pulley.

3. Remove the oil dipstick and guide.

4. Disconnect the oil sensor lead, remove the four attaching bolts and lift off the oil level sensor. Be careful not to drop this sensor.

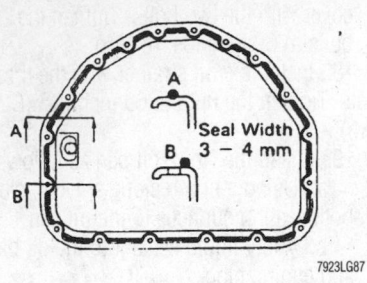

Lower oil pan sealant application—3.0L (2JZ-GE) engine

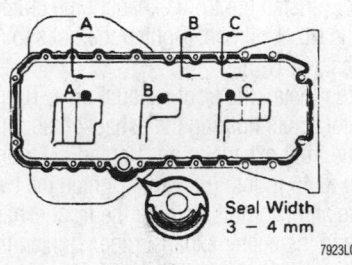

Upper oil pan sealant application—3.0L (2JZ-GE) engine

5. Remove the 14 bolts (16 bolts for GS300) and two nuts and pry off the lower (No. 2) oil pan. Be careful not to damage the No. 1 pan while performing this procedure.

6. Remove the bolt and two nuts and drop down the oil strainer and gasket.

7. Remove the five bolts and two nuts and drop down the baffle plate.

8. Remove the 22 bolts and the carefully pry off the upper (No. 1) oil pan. Remove the O-ring from the cylinder block.

To install:

9. Position a new O-ring in the block and scrape off any old sealant. Apply sealant to the pan mating surface with a ⅛ inch (3–4mm) bead. Install the upper pan and tighten the 12mm bolts to 15 ft. lbs. (21 Nm) and the 14mm bolts to 29 ft. lbs. (39 Nm).

10. Install the baffle plate and oil strainer. Tighten them both to 78 inch lbs. (9 Nm).

11. Install the lower pan in the same manner as the upper pan and tighten the bolts to 78 inch lbs. (9 Nm).

12. Using a new gasket, install the oil level sensor and tighten it to 48 inch lbs. (5 Nm).

13. Install the oil dipstick and guide, the

timing pulleys and belt and reconnect the transmission to the engine.

14. Install the engine and transmission.
15. Refill all fluids.
16. Start the engine and check for leaks.
17. Road test the vehicle.

4.0L Engine

LS400

1. Remove the engine/trans assembly. Separate the transmission from the engine.

2. With the engine on a stand, remove the timing belt, the idler pulleys and the crankshaft timing pulley.

3. Remove the oil dipstick and guide.

4. Disconnect the oil level sensor lead.

5. Remove the four bolts and lift off the oil level sensor. Be careful not to drop this sensor.

6. Remove the oil filter and the bracket assembly by removing the stud bolt and two nuts.

7. Disconnect the engine crankshaft position sensor connector. Remove the sensor by removing the bolt.

8. Remove the 12 bolts and 2 nuts to the No. 2 oil pan. Use a gasket cutting tool to separate the No. 2 (lower) oil pan. Be careful not to damage the No. 1 pan while performing this procedure.

9. Remove the two bolts and three nuts and drop down the baffle plate.

10. Remove the oil strainer by removing the bolts and nuts.

11. Remove the bolts, then carefully pry off the No. 1 oil pan. There are slots for inserting the prybar.

To install:

12. Scrape off any old sealant, then install the No. 1 pan. Apply sealant to the pan mating surface with a ⅛ inch (3–4mm) bead.

13. Tighten the bolts for the No. 1 oil pan as follows:
- 10mm head to 66 inch lbs. (8 Nm)
- 12mm head to 21 ft. lbs. (28 Nm)

14. Install the oil strainer by installing the bolts and nuts. Tighten the bolts to 66 inch lbs. (8 Nm).

15. Install the baffle plate and tighten the bolts and nuts to 66 inch lbs. (8 Nm).

16. Install the No. 2 pan in the same manner as the No. 1 oil pan and tighten the bolts to 66 inch lbs. (8 Nm). Be sure the bolts are 14mm in length.

17. Install the engine crankshaft position sensor. Tighten the bolt to 56 inch lbs. (6 Nm). Connect the crankshaft position sensor connector.

18. Place a new O-ring in position on

the oil filter bracket. Install the bracket and tighten the bolt and nuts to 13 ft. lbs. (18 Nm). Connect the wiring to the pressure switch.

19. Using a new gasket, install the oil level sensor and tighten the four bolts to 48 inch lbs. (5 Nm).

20. Install the dipstick and guide, the timing belt pulleys and the timing belt components.

21. Connect the transaxle to the engine.

22. Install the engine and transaxle, refill all fluids and road test the vehicle.

SC400

➡**The No. 1 oil pan cannot be removed with the engine in the vehicle. The engine and transmission must be removed as a unit, then separated. It may be possible to remove the No. 2 oil pan from the vehicle while the engine is still in the vehicle.**

1. Remove the engine/trans assembly. Separate the transmission from the engine.

2. Remove the oil dipstick and guide.

3. Remove the 12 bolts and 2 nuts. Use a gasket cutting tool to separate the No. 2 (lower) oil pan. Be careful not to damage the No. 1 pan while performing this procedure.

4. Remove the six bolts and two nuts; remove the baffle plate.

5. Remove the 16 bolts, then carefully pry off the No. 1 oil pan.

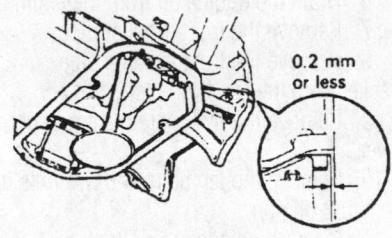

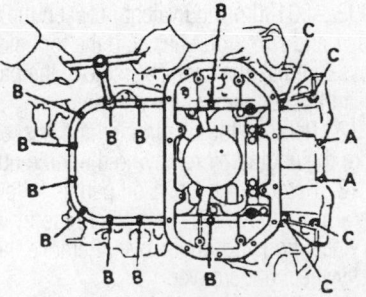

SC400 oil pan bolt locations—(A) 0.78 in (20mm), (B) 1.38 in. (35mm), (C) 0.78 in (20mm) with 12mm bolt heads

➡️**There are slots for inserting the pry-bar.**

To install:

6. Scrape off any old sealant, then install the No. 1 pan. Apply sealant to the pan mating surface with a ⅛ inch (3–4mm) bead.

7. Tighten the bolts as follows:
- 12mm bolts—69 inch lbs. (8mm)
- 14mm bolts—20 ft. lbs. (28 Nm)

8. Install the baffle plate and tighten the bolts and nuts to 69 inch lbs. (8 Nm).

9. Scrape off any old sealant, apply sealant to the pan mating surface with a ⅛ inch (3–4mm) diameter bead and install the No. 2 oil pan. Tighten the bolts to 69 inch lbs. (8 Nm). Be sure the bolts are 14mm in length.

10. Install the dipstick and guide.

11. Install the engine to the transaxle and install the assembly to the vehicle.

12. Refill all fluids, check for leaks and road test the vehicle.

Oil Pump

REMOVAL & INSTALLATION

3.0L (1MZ-FE) Engine

1. Disconnect the negative battery cable from the battery.

2. Raise and safely support the front of the vehicle.

3. Remove the right front wheel.

4. Remove the fender apron seal.

5. Remove the engine undercover.

6. Drain the engine oil from the engine.

7. Remove the front exhaust pipe.

8. Remove the front exhaust pipe bracket from the No. 1 oil pan.

9. Remove the alternator drive belt from the engine.

10. Disconnect the A/C compressor from the engine, without disconnecting the compressor lines.

11. Remove the power steering pump drive belt and adjusting strut.

12. Remove the timing belt from the engine.

13. Remove the timing belt pulleys.

14. Remove the rear timing belt cover from the engine by removing the wire clamps and six bolts.

15. Remove the A/C compressor housing bracket by removing the three bolts.

16. Remove the ten bolts and two nuts to the No. 2 oil pan.

17. Insert the blade of SST tool 09032–00100 or equivalent between the No. 1 and No. 2 oil pans. Remove the No. 2

oil pan from the engine. Clean the surfaces of the oil pans.

18. Remove the oil strainer and gasket from the engine by removing the three nuts.

19. Remove the No. 1 oil pan.

20. Remove the baffle plate from the No. 1 oil pan.

21. Remove the crankshaft position sensor by removing the connector and bolt.

22. Remove the oil pump as follows:

a. Remove the nine bolts. Make a note of the position of the each bolt. When replacing the bolts into the oil pump body, place each bolt in the position from which it was removed.

b. Remove the oil pump body by prying between the oil pump and main bearing cap.

c. Remove the O-ring from the cylinder block.

d. Remove the plug, gasket, spring and relief valve from the oil pump body.

e. Remove the nine screws, pump body cover, drive and driven rotors.

To install:

23. To install the oil pump:

a. Install the driven rotors, drive, pump body cover, then install the nine screws.

b. Install the oil pump relief valve, spring, gasket and the plug to the oil pump body.

c. Place a new O-ring on the cylinder block.

d. Using a non residue solvent, clean both sealing surfaces to the oil pump.

e. Apply liquid sealant to the oil pump as shown.

f. Install the oil pump to the engine block. Be sure to engage the spline teeth of the oil pump drive gear with the large teeth of the crankshaft.

g. Install the nine bolts to the oil pump and uniformly tighten the bolts in several passes. Tighten the bolts as follows:
- 10mm head-69 inch lbs. (8 Nm)
- 12mm head-14 ft. lbs. (20 Nm)

24. Install the crankshaft position sensor and install the bolt. Tighten the bolt to 69 inch lbs. (8 Nm).

25. Install the baffle plate to the No. oil pan and tighten to 69 inch lbs. (8 Nm).

26. Install the No. 1 oil pan as follows:

a. Using a non residue solvent, clean both sealing surfaces to the oil pan.

b. Apply liquid sealant to the oil pan and engine block.

c. Install the oil pan with the seventeen bolts and two nuts. Uniformly tighten the bolts and nuts in several passes.

d. Tighten the bolts as follows:

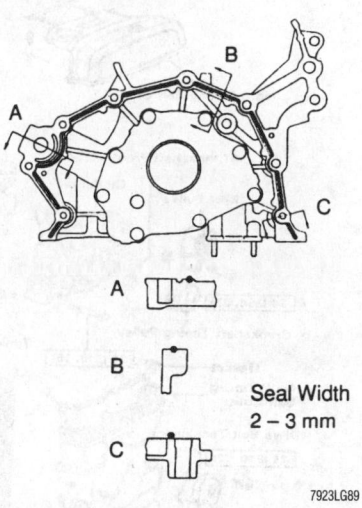

Seal Width
2 – 3 mm

7923LG89

Apply sealant to the mounting surface of the oil pump in the areas shown—3.0L (1MZ-FE) engine

- 10mm head bolt—69 inch lbs. (8 Nm)
- 12mm head bolt—14 ft. lbs. (20 Nm)
- 14mm head bolt—27 ft. lbs. (37 Nm)

e. Install the flywheel housing undercover with the two bolts. Tighten the bolts to 69 inch lbs. (8 Nm).

27. Install the oil strainer with the three nuts. Tighten the nuts to 69 inch lbs. (8 Nm).

28. Install the No. 2 oil pan as follows:

a. Using a non residue solvent, clean both sealing surfaces to the oil pan.

b. Apply liquid sealant to the oil pan and engine block.

c. Install the No. 2 oil pan with the ten bolts and two nuts. Uniformly tighten the bolts and nuts in several passes. Tighten the bolts to 69 inch lbs. (8 Nm).

29. Install the remaining components. Install the right front wheel and lower the vehicle.

30. Fill the engine with oil.

31. Connect the negative battery cable to the battery.

32. Start the engine and check for leaks.

33. Recheck the engine oil.

3.0L (2JZ-GE) Engine

1. Remove the engine and transmission.

2. Separate the transmission from the engine and mount the engine on a service stand.

3. Remove the timing belt.

4. Remove the idler pulley.

5. Remove the crankshaft timing pulley.

6. Remove the oil dipstick and tube.

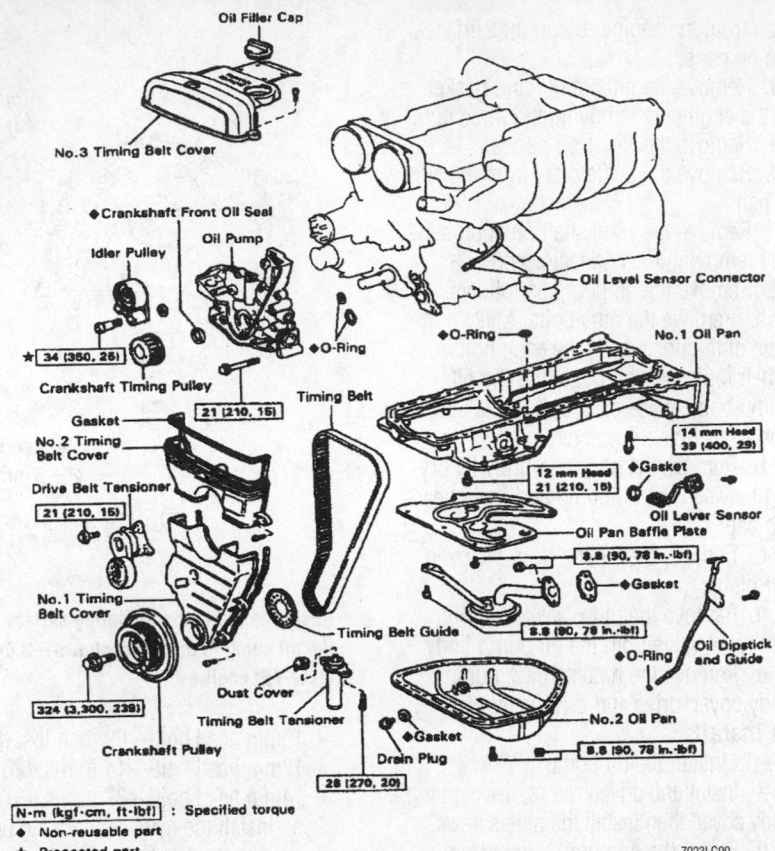

Exploded view of the oil pump removal—3.0L (2JZ-GE) engine

7. Remove the oil level sensor.

8. Remove the No. 2 (lower) oil pan.

9. Remove the oil strainer by removing the bolt and two nuts.

10. Remove the oil baffle plate by removing the six bolts.

11. Remove the No. 1 (upper) oil pan by removing the 22 bolts. Take note of bolt size and placement for correct reinstallation.

12. Remove the nine mounting bolts to the oil pump body. Carefully drive the pump off the cylinder block using a brass drift. Remove the two O-rings.

To install:

13. Position two new O-rings in the cylinder block. Scrape any old sealant from the mating surfaces. Draw a ⅛ inch (3–4mm) bead of sealant around the pump mating surface, taking great care around the oil passages. Install the pump and tighten the bolts to 15 ft. lbs. (21 Nm).

14. Place a new O-ring on the block. Remove all of the old sealant from the block and No. 1 oil pan. Apply a bead of sealant around the No. 1 oil pan. Avoid excessive application. Install the No. 1 oil pan and tighten the bolts with 12mm heads to 15 ft. lbs. (21 Nm). Tighten the bolts with 14mm heads to 29 ft. lbs. (39 Nm).

15. Install the oil baffle plate and tighten the nuts and bolts to 78 inch lbs. (9 Nm).

16. Install the oil strainer and tighten the nuts and bolts to 78 inch lbs. (9 Nm).

17. Remove all of the old sealant from the block and No. 2 oil pan. Apply a bead of sealant around the No. 2 oil pan. Avoid excessive application. Install the No. 2 oil pan and tighten the bolts to 78 inch lbs. (9 Nm).

18. Install the oil lever sensor with a new gasket and tighten the bolts to 48 inch lbs. (6 Nm).

19. Install the oil dipstick with a new O-ring.

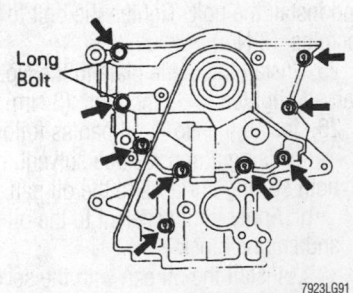

Oil pump mounting bolt installation locations—3.0L (2JZ-GE) engine

20. Install the remaining components. Fill all fluids.

21. Connect the negative battery cable.

22. Start the engine and check for leaks.

4.0L Engine

LS400

➡ **The oil pump cannot be removed with the engine in the vehicle. The engine and transmission must be removed as a unit, then separated.**

1. Remove the engine/transmission assembly. Separate the transmission from the engine.

2. With the engine on a stand, remove the timing belt, the idler pulleys and the crankshaft timing pulley.

3. Remove the oil dipstick and guide.

4. Disconnect the oil level sensor lead.

5. Remove the four bolts and lift off the oil level sensor. Be careful not to drop this sensor.

6. Remove the oil filter and filter bracket assembly by removing the stud bolt and two nuts.

7. Disconnect the engine crankshaft position sensor connector. Remove the sensor by removing the bolt.

8. Remove the 12 bolts and 2 nuts to the No. 2 oil pan. Use a gasket cutting tool to separate the No. 2 (lower) oil pan. Be careful not to damage the No. 1 pan while performing this procedure.

9. Remove the two bolts and three nuts and drop down the baffle plate.

10. Remove the oil strainer by removing the bolts and nuts.

11. Remove the bolts, then carefully pry off the No. 1 oil pan. There are slots for inserting the prybar.

12. Remove the eight bolts holding the oil pump to the engine.

➡ **Make certain to observe bolt position during removal. The bolts are different lengths and sizes. Record their position for proper reassembly.**

13. Carefully pry the oil pump away from the engine block. Use great care not to damage the contact faces of the block or pump. Remove the O-ring from the block.

To install:

14. Remove any old gasket material from the oil pump face and the engine block. Make certain the sealing groove is clean. Use a solvent which is free of residue to clean the surfaces.

➡ **Prior to installing the oil pump, lubricate the gears with clean engine oil.**

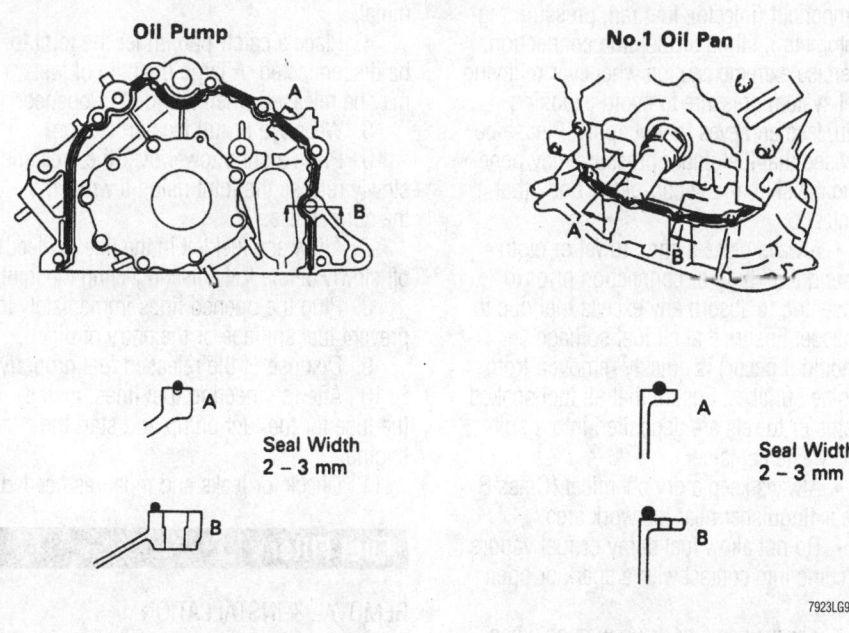

Oil Pump

Seal Width
2 – 3 mm

No.1 Oil Pan

Seal Width
2 – 3 mm

7923LG92

Apply sealant to the oil pump and the No. 1 oil pan, as shown, before installing the oil pump—LS400

15. Apply a bead of new sealant to the oil pump. The gasket bead should be 2–3mm wide (0.08–0.12 in). Avoid excessive application and be careful around oil passages.

➡ **Parts must be assembled within 5 minutes after the sealant is applied. If not, the sealant must be cleaned off and the parts re-coated.**

16. Place a new O-ring in position on the block. Mount the oil pump to the engine by engaging the spline teeth of the oil pump drive with the large teeth of the crankshaft. Slide the oil pump into position.

17. Install the eight bolts in their correct locations. Tighten the bolts with 12mm heads to 12 ft. lbs. (16 Nm) and the bolts with 14mm heads to 22 ft. lbs. (30 Nm).

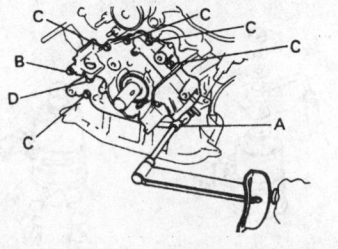

7923LG93

LS400 oil pump bolt locations, according to bolt lengths—(A) 1.97 in. (50mm), (B) 4.17 in. (106mm), (C) 1.18 in. (30mm) and (D) 1.57 in. (40mm)

18. Scrape off any old sealant, then install the No. 1 pan. Apply sealant to the pan mating surface with a ⅛ inch (3–4mm) bead.

19. Tighten the bolts for the No. 1 oil pan as follows:
• 10mm head to 66 inch lbs. (8 Nm)
• 12mm head to 21 ft. lbs. (28 Nm).

20. Install the oil strainer and tighten the bolts to 66 inch lbs. (8 Nm).

21. Install the baffle plate and tighten the bolts and nuts to 66 inch lbs. (8 Nm).

22. Install the remaining components. Connect the transaxle to the engine.

23. Install the engine and transaxle, refill all fluids and road test the vehicle.

SC400

➡ **The oil pump cannot be removed with the engine in the vehicle. The engine and transmission must be removed as a unit, then separated.**

1. Remove the engine/transmission assembly. Separate the transmission from the engine.

2. With the engine on a stand, remove the timing belt, the idler pulleys and the crankshaft timing pulley.

3. Remove the oil dipstick and guide.

4. Remove the main O_2 sensor bracket.

5. Disconnect the oil level sensor lead.

6. Remove the four bolts and lift off the oil level sensor. Be careful not to drop this sensor.

7. Remove the 12 bolts and 2 nuts. Use

a gasket cutting tool to separate the No. 2 (lower) oil pan. Be careful not to damage the No. 1 pan while performing this procedure.

8. Remove the six bolts and two nuts and drop down the baffle plate.

9. Remove the 16 bolts, then carefully pry off the No. 1 oil pan. There are slots for inserting the prybar.

10. Remove the oil strainer by removing the two bolts and two nuts.

11. Disconnect the oil pressure switch connector. Remove the bolts and stud, then remove the filter bracket and filter.

12. Disconnect the engine crankshaft position sensor connector. Remove the sensor by removing the bolt.

13. Remove the eight bolts holding the oil pump to the engine.

➡ **Make certain to observe bolt position during removal. The bolts are different lengths and sizes. Record their position for proper reassembly.**

14. Carefully pry the oil pump away from the engine block. Use great care not to damage the contact faces of the block or pump. Remove the O-ring from the block.

To install:

15. Remove any old gasket material from the oil pump face and the engine block. Make certain the sealing groove is clean. Use a solvent which is free of residue to clean the surfaces.

16. Apply a bead of new sealant to the oil pump. The gasket bead should be 2–3mm wide (0.08–0.12 in). Avoid excessive application and be careful around oil passages.

➡ **Parts must be assembled within 5 minutes after the sealant is applied. If not, the sealant must be cleaned off and the parts re-coated.**

17. Place a new O-ring in position on the block. Mount the oil pump to the engine by engaging the spline teeth of the oil pump

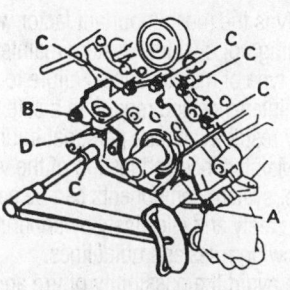

7923LG94

Oil pump bolt locations and length identification—(A) 1.97 in (50mm), (B) 4.17 in. (106mm), (C) 1.18 in (30mm), (D) 1.57 in. (40mm) with 14mm bolt heads

drive with the large teeth of the crankshaft. Slide the oil pump into position.

18. Install the eight bolts in their correct locations. Tighten the bolts with 12mm heads to 12 ft. lbs. (16 Nm) and the bolts with 14mm heads to 22 ft. lbs. (30 Nm).

19. Install the engine crankshaft position sensor. Tighten the bolt to 56 inch lbs. (6 Nm). Connect the crankshaft position sensor connector.

20. Place a new O-ring in position on the oil filter bracket. Install the bracket and tighten the bolts to 13 ft. lbs. (18 Nm). Connect the wiring to the pressure switch.

21. Install the oil strainer by installing the two bolts and two nuts. Tighten the bolts to 69 inch lbs. (8 Nm).

22. Scrape off any old sealant, then install the No. 1 pan. Apply sealant to the pan mating surface with a 1/8 inch (3–4mm) bead.

23. Tighten the bolts as follows:
- 12mm bolts—69 inch lbs. (8 Nm)
- 14mm bolts—20 ft. lbs. (28 Nm)

24. Install the baffle plate and tighten the bolts and nuts to 69 inch lbs. (8 Nm).

25. Install the No. 2 pan in the same manner as the No. 1 oil pan and tighten the bolts to 69 inch lbs. (8 Nm). Be sure the bolts are 14mm in length.

26. Using a new gasket, install the oil level sensor and tighten the four bolts to 48 inch lbs. (5 Nm).

27. Install the bracket to the oxygen sensor and install the bolt.

28. Install the dipstick and guide, the timing belt pulleys and the belt.

29. Connect the transaxle to the engine.

30. Install the engine and transaxle, refill all fluids and road test the vehicle.

FUEL SYSTEM

Fuel System Service Precautions

Safety is the most important factor when performing not only fuel system maintenance but any type of maintenance. Failure to conduct maintenance and repairs in a safe manner may result in serious personal injury or death. Maintenance and testing of the vehicle's fuel system components can be accomplished safely and effectively by adhering to the following rules and guidelines.

- To avoid the possibility of fire and personal injury, always disconnect the negative battery cable unless the repair or test procedure requires that battery voltage be applied.
- Always relieve the fuel system pressure prior to disconnecting any fuel system component (injector, fuel rail, pressure regulator, etc.), fitting or fuel line connection. Exercise extreme caution whenever relieving fuel system pressure to avoid exposing skin, face and eyes to fuel spray. Please be advised that fuel under pressure may penetrate the skin or any part of the body that it contacts.
- Always place a shop towel or cloth around the fitting or connection prior to loosening to absorb any excess fuel due to spillage. Ensure that all fuel spillage (should it occur) is quickly removed from engine surfaces. Ensure that all fuel soaked cloths or towels are deposited into a suitable waste container.
- Always keep a dry chemical (Class B) fire extinguisher near the work area.
- Do not allow fuel spray or fuel vapors to come into contact with a spark or open flame.
- Always use a back-up wrench when loosening and tightening fuel line connection fittings. This will prevent unnecessary stress and torsion to fuel line piping. Always follow the proper torque specifications.
- Always replace worn fuel fitting O-rings with new. Do not substitute fuel hose or equivalent, where fuel pipe is installed.

Fuel System Pressure

RELIEVING

❋ CAUTION

Failure to relieve fuel pressure before repairs or disassembly can cause serious personal injury and/or property damage. Fuel pressure is maintained within the fuel lines, even if the engine is OFF or has not been run in a period of time. This pressure must be safely relieved before any fuel-bearing line or component is loosened or removed. On vehicles equipped with inflatable restraints or air bag systems, wait at least 90 seconds after disconnecting the battery cable before performing any other work. The back-up power will keep the restraint system energized for a period of time after the battery is disconnected.

1. Remove the fuse for the electronic fuel pump.

2. Start the engine until the engine stalls.

3. Disconnect the negative battery terminal.

4. Place a catch-pan under the joint to be disconnected. A large quantity of fuel may be released when the joint is opened.

5. Wear eye or full face protection.

6. Place a shop towel over the area and slowly release the joint using a wrench of the correct size.

7. Allow any fuel left in the line to bleed off slowly before fully disconnecting the joint.

8. Plug the opened lines immediately to prevent fuel spillage or the entry of dirt.

9. Dispose of the released fuel properly.

10. After connecting fuel lines, install the fuse for the fuel pump and start the engine.

11. Check for leaks and repair as needed.

Fuel Filter

REMOVAL & INSTALLATION

The fuel filter on the ES300 is located under the hood, on the driver's side, by the fenderwell. The fuel filter SC300 is located under the vehicle, on the driver's side, in front of the rear axle.

The fuel filter for the LS400 and SC400 is located under the vehicle on the left side before the rear axle. The fuel filter on the GS300 is located under the vehicle, next to the left rear exhaust resonator.

1. Disconnect the negative battery cable. Wait at least 90 seconds before performing any other work.

2. Raise and safely support the vehicle if necessary.

3. On the GS300, remove the rear body protector.

4. Place a drain pan or plastic container under the fuel filter.

5. Slowly loosen the lower flare nut fitting until all the pressure is relieved and all the fuel is collected.

6. Loosen the union bolt on the upper

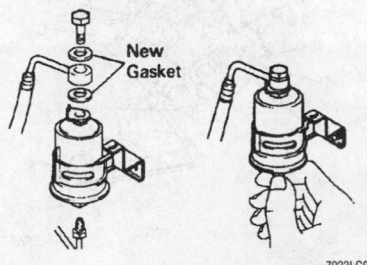

Exploded view of a typical fuel line connection at the filter

portion of the filter and remove the banjo fitting and two metal gaskets. Discard the gaskets.

7. Loosen the fuel filter bracket bolt, remove the fuel line with the flared nut from the filter and pull the filter from the mounting bracket.

To install:

8. Install a new fuel filter to the vehicle and tighten the bracket bolt.

9. Install the banjo fitting with a new metal gasket on each side and install the union bolt. Tighten the union bolt to 22 ft. lbs. (30 Nm).

10. Connect the flare nut to the lower connection. Tighten the flare nut to 22 ft. lbs. (30 Nm).

11. On the GS300, install the body protector.

12. Lower the vehicle if raised.

13. Remove the drain pan and/or rags and connect the negative battery cable.

14. Start the engine and visually inspect the upper and lower connections for leaks.

Fuel Pump

REMOVAL & INSTALLATION

ES300

> ⁂ **CAUTION**
>
> **The fuel injection system remains under pressure even after the engine has been turned OFF. The fuel system pressure must be relieved before disconnecting any fuel lines. Failure to do so may result in fire and/or personal injury.**

1. Relieve the fuel system pressure.

2. With the ignition switch in the **LOCK** position, disconnect the negative battery terminal. If equipped with an air bag system, wait at least 90 seconds or longer before performing any other work.

3. Remove the rear seat cushion.

4. Disconnect the fuel pump connector.

> ⁂ **CAUTION**
>
> **The Supplemental Inflatable Restraint (SIR) system must be disarmed before removing the fuel pump. Failure to do so may cause accidental deployment of the air bag, resulting in unnecessary SIR system repairs and/or personal injury.**

5. Remove the floor service hole cover.

➡ **Do not lift the fuel pump assembly up using the wiring harness.**

6. Remove the fuel filler cap. Disconnect the fuel outlet pipe and the return hose from the pump bracket.

7. Remove the 8 screws and lift out the pump/bracket assembly with gasket.

8. Remove the fuel pump lead wire.

9. Pull the lower end of the pump off of the bracket.

10. Disconnect the fuel hose from the pump and remove the pump.

11. Remove the rubber cushion from the pump.

To install:

12. Install the filter and rubber cushion on the new pump and install the pump on the bracket.

13. Connect the fuel hose and the wire connector on the pump.

14. Using a new gasket, install the pump and tighten the eight screws to 35 inch lbs. (4 Nm).

15. Connect the fuel pipe and return hose to the pump and tighten the bolts to 22 ft. lbs. (29 Nm).

16. Connect the wire, install the service cover and replace the rear seat.

17. Connect the negative battery cable.

18. Start the engine and check for leaks.

Except ES300

1. Disconnect the negative battery cable. Wait at least 90 seconds before performing any other work.

2. Remove the trunk floor mat.

3. Remove the trunk trim cover.

4. Disconnect the fuel pump electrical connector.

5. Remove the rear seat bottom and seat back.

6. Remove the partition cover.

7. Remove the mounting bolts and remove the fuel pump set plate.

8. Remove the three nuts and disconnect the fuel pump bracket from the tank.

9. Disconnect the fuel hose from the bracket. Remove the pump, bracket and set plate as an assembly.

To install:

10. Install a new gasket on the set plate. Connect the fuel hose to the pump and bracket,

11. Install the pump and bracket assembly with the three nuts; tighten the nuts to 48 inch lbs. (5 Nm). Install the set plate and tighten the bolts to 26 inch lbs. (3 Nm).

12. Install the panel partition.

13. Install the rear seat cushion and back.

14. Connect the fuel pump electrical connector.

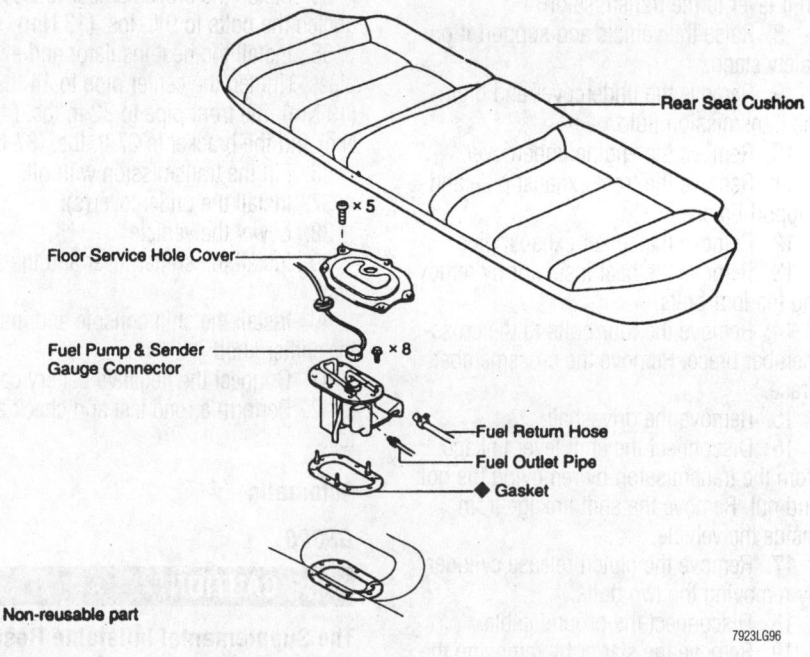

Rear Seat Cushion

Floor Service Hole Cover

Fuel Pump & Sender Gauge Connector

× 5

× 8

Fuel Return Hose

Fuel Outlet Pipe

◆ Gasket

◆ Non-reusable part

7923LG96

Exploded view of the fuel pump assembly—ES300

15. Install the trim panel.

16. Install the spare tire and the trunk floor mat.

17. Connect the negative battery cable.

18. Start the engine; check the fuel system for leaks

19. Road test the vehicle for proper operation.

DRIVE TRAIN

Transmission Assembly

REMOVAL & INSTALLATION

Manual

SC300

1. Disconnect the negative battery cable. Wait at least 90 seconds before performing any other work.

2. Pry out the rear side of the cup holder and remove the cup holder.

3. Unscrew the shift lever knob.

4. Pry up the upper rear console panel.

5. Remove the six mounting screws, then pry out the upper console panel.

6. Remove the four bolts and lift out the shift lever boots.

7. Remove the four bolts holding the shift lever to the transmission.

8. Raise the vehicle and support it on safety stands.

9. Remove the undercover and drain the transmission fluid.

10. Remove the engine undercover.

11. Remove the front exhaust pipe and support bracket.

12. Remove the center exhaust pipe.

13. Remove the heat insulator by removing the four bolts.

14. Remove the four bolts to the crossmember brace. Remove the crossmember brace.

15. Remove the driveshaft.

16. Disconnect the shift lever linkage from the transmission by removing the bolt and nut. Remove the shift linkage from inside the vehicle.

17. Remove the clutch release cylinder by removing the two bolts.

18. Disconnect the ground cable.

19. Remove the starter by removing the electrical wires and the two bolts.

20. Disconnect the back-up light switch connector and the speed sensor connector.

21. Remove the two lower transmission mounting bolts.

22. Raise the transmission slightly until its weight is off the rear support.

23. Remove the four nuts and bolts, then remove the rear mounting member.

24. Remove the five mounting bolts, lower the rear of the engine and remove the transmission.

To install:

25. Align the input spline with the clutch disc and install the transmission to the engine. Tighten the five mounting bolts to 53 ft. lbs. (72 Nm).

26. Install the rear engine mount and tighten the nuts to 10 ft. lbs. (13 Nm) and the bolts to 19 ft. lbs. (25 Nm).

27. Connect the speed sensor.

28. Connect the back-up light switch.

29. Install the two lower transmission mounting bolts and tighten it to 27 ft. lbs. (37 Nm).

30. Install the starter by installing the two bolts. Tighten the bolts to 29 ft. lbs. (39 Nm). Connect the electrical connectors.

31. Install the release cylinder and tighten the two bolts to 9 ft. lbs. (12 Nm). Connect and tighten the ground wire to 27 ft. lbs. (37 Nm).

32. Install the transmission shift lever inside the vehicle. From below the vehicle, connect the shift lever by installing the bolt and nut. Tighten the bolt and nut to 14 ft. lbs. (19 Nm).

33. Install the driveshaft.

34. Install the crossmember brace and tighten the bolts to 9 ft. lbs. (13 Nm).

35. Install the heat insulator and exhaust pipes. Tighten the center pipe to 14 ft. lbs. (19 Nm), the front pipe to 32 ft. lbs. (43 Nm) and the bracket to 27 ft. lbs. (37 Nm).

36. Fill the transmission with oil.

37. Install the undercover(s).

38. Lower the vehicle.

39. Install the shifter lever and the shifter boot.

40. Install the shift console and install the shifter knob.

41. Connect the negative battery cable.

42. Perform a road test and check all fluids.

Automatic

GS300

❊❊ CAUTION

The Supplemental Inflatable Restraint (SIR) system must be disarmed before removing the transmission assembly. Failure to do so may cause accidental deployment of the air bag, resulting in unnecessary SIR system repairs and/or personal injury.

1. Turn the ignition switch to the **LOCK** position and disconnect the negative battery cable. Wait at least 90 seconds or longer before doing any work on the vehicle.

2. Remove the transmission level gauge.

3. Remove the transmission dipstick and tube.

4. Disconnect the throttle cable from the throttle body.

5. Disconnect the oxygen sensor from the exhaust system.

6. Remove the left and right tail pipes.

7. Remove the front and center exhaust pipe.

8. Remove the exhaust heat insulator.

9. Remove the rear center floor crossmember brace.

10. Remove the shift control rod from the shift lever.

11. Remove the driveshaft.

12. Disconnect the following connectors;
    a. Overdrive and direct clutch speed sensor
    b. No. 1 vehicle speed sensor
    c. No. 2 vehicle speed sensor
    d. Solenoid wire
    e. Park/Neutral position switch

13. Disconnect the wiring from the starter.

14. Remove the oil cooler pipes as follows:
    a. Disconnect the two oil cooler union nuts.
    b. Disconnect the oil cooler hoses from the oil cooler pipes.
    c. Remove the front oil cooler pipe bracket.
    d. Remove the center and rear oil cooler pipe brackets.
    e. Remove the two oil cooler pipes.

15. Remove the torque converter inspection plate.

16. Turn the crankshaft to gain access to the torque converter bolts and remove them as they become accessible.

17. Support the transmission with a suitable jack.

18. Support the engine with a jack and a block of wood.

19. Remove the rear transmission mount.

20. Remove the wiring harness clamps.

21. Remove the starter.

22. Remove the nine transmission mounting bolts and transmission.

To install:

23. Install the transmission and tighten the bolts to 53 ft. lbs. (52 Nm)

24. Install the starter and tighten the bolts to 27 ft. lbs. (37 Nm).

25. Install the rear transmission mount and tighten the bolts to 19 ft. lbs. (25 Nm).

26. Install and tighten the torque converter bolts to 30 ft. lbs. (41 Nm) while rotating the crankshaft.

27. Install the converter inspection plate.

28. Connect the oil cooler lines as follows:

    a. Install the two oil cooler pipes.

    b. Install the center and rear oil cooler pipe brackets.

    c. Install the front oil cooler pipe bracket and tighten to 49 inch lbs. (5.5 Nm).

    d. Connect the oil cooler hoses to the oil cooler pipes.

    e. Install the two oil cooler union nuts and tighten to 32 ft. lbs. (44 Nm).

29. Connect the wiring to the starter.

30. Connect the transmission electrical connectors.

31. Install the driveshaft.

32. Install the remaining components. Connect the negative battery cable.

33. Install the transmission level gauge.

34. Fill the transmission to the proper level with Dexron II® or equivalent.

LS400

1. Disconnect the negative battery cable. Wait at least 90 seconds before performing any other work.

2. Remove the transmission dipstick and tube.

3. Disconnect the throttle cable.

4. Remove the driveshaft.

5. Remove the engine undercover.

6. Remove the shift control rod.

7. Remove the exhaust pipe support bracket by removing the two bolts.

8. Remove the catalytic converters by removing the six nuts.

9. Remove both side heat insulators.

10. Remove the oil cooler tube clamps and disconnect the tubes.

11. Remove the torque converter inspection plate by removing the two bolts.

12. Turn the crankshaft to gain access to the torque converter bolts and remove them as they become accessible.

13. Support the transmission with a suitable jack.

14. Remove the rear transmission mount by removing the four bolts and four nuts.

15. Tilt down the transmission and disconnect the following connectors:

    a. O/D direct clutch speed sensor connector.

    b. Vehicle speed sensor connector.

    c. Park/Neutral position switch connector.

    d. Solenoid connector.

16. Disconnect the three wiring harness clamp from the bracket on the transmission.

17. Remove the 10 transmission mounting bolts and the transmission.

To install:

18. Install the transmission and tighten the bolts as follows:

- 14mm bolts—27 ft. lbs. (37 Nm)
- 17mm bolts—53 ft. lbs. (72 Nm)

19. Install the three wiring harness clamp to the bracket on the transmission.

20. Connect the following connectors:

    a. Solenoid connector

    b. Park/Neutral position switch connector

    c. Vehicle speed sensor connector

    d. O/D direct clutch speed sensor connector

21. Install the rear transmission mount and tighten the bolts to 19 ft. lbs. (20 Nm) and the nuts to 10 ft. lbs. (13 Nm).

22. Install and tighten the torque converter bolts to 30 ft. lbs. (41 Nm).

23. Install the converter inspection plate.

24. Remove the support from the transmission.

25. Install the oil cooler pipes and tighten the union nuts to 32 ft. lbs. (44 Nm).

26. Install the side heat insulators.

27. Install the catalytic converters with new gaskets and new nuts. Tighten the nuts to 46 ft. lbs. (62 Nm).

28. Install the exhaust pipe support bracket with the two bolts and tighten the bolts to 32 ft. lbs. (44 Nm).

29. Install the shift control rod.

30. Install the engine undercover.

31. Install the driveshaft.

32. Connect and adjust the throttle control cable.

33. Install the transmission tube and dipstick.

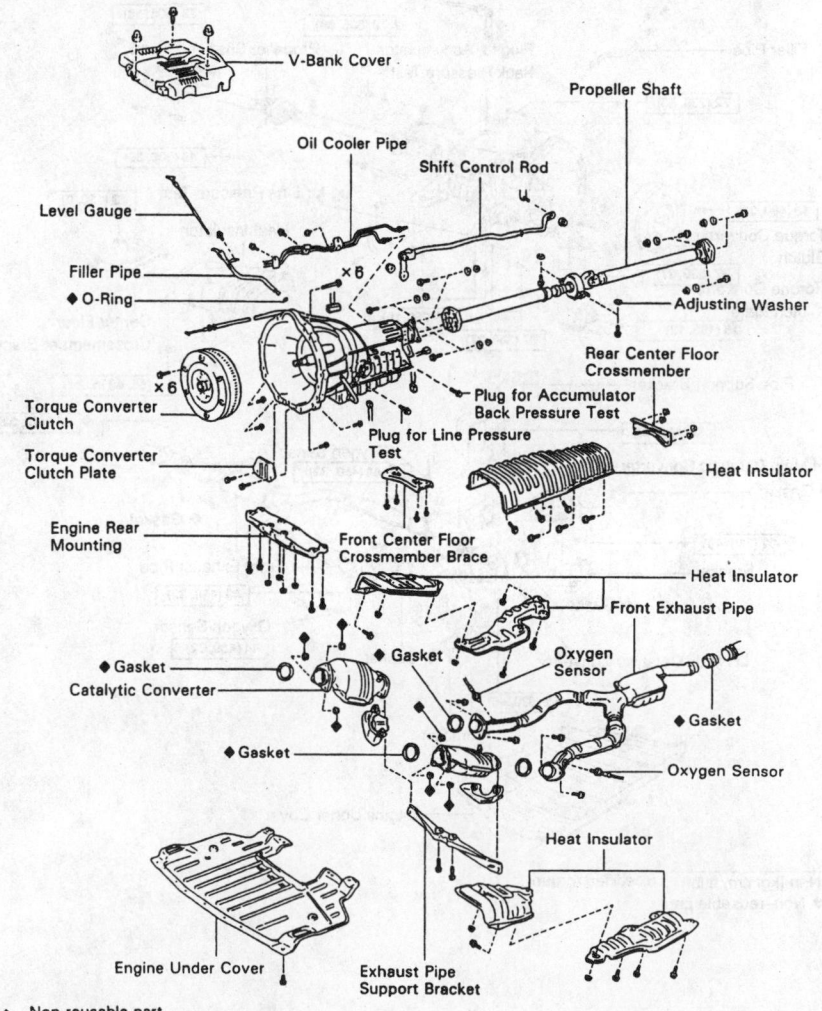

Exploded view of the transmission mounting—LS400 with the A340E transmission

7923LG97

34. Connect the negative battery cable.

35. Fill the transmission to the proper level with Dexron II® or equivalent.

SC300 AND SC400

1. Disconnect the negative battery cable. Wait at least 90 seconds before proceeding with any other work.

2. If equipped, remove the V-bank cover.

3. Remove the A/T oil level gauge if equipped.

4. Remove the transmission dipstick and tube.

5. Disconnect the throttle cable and clamps.

6. Remove the exhaust pipe and converters.

7. Remove the exhaust heat insulator.

8. Remove the rear center floor crossmember brace.

9. Remove the shift control rod.

10. Remove the driveshaft using two of tool SST 09922–10010 or equivalent, loosen the adjusting nut on the driveshaft. Place matchmarks on the transmission on the flanges prior to removal.

➡**The bolts inserted from the drive-shaft side should not be removed.**

11. Disconnect the electrical harness from the transmission.

12. Remove the oil cooler tube clamp and disconnect the tubes.

13. Remove the lower engine cover.

14. Remove the torque converter inspection plate.

15. Turn the crankshaft to gain access to the torque converter bolts and remove them as they become accessible.

16. Support the transmission with a suitable jack.

17. If necessary, remove the starter.

18. Remove the rear transmission mount by removing the bolts.

19. Remove the transmission mounting bolts and the transmission.

To install:

20. Before installing the transmission, use calipers and a straightedge to check the distance between the installed surface of the torque converter and the front edge of the transmission case. Correct distance is 0.673 in. (17.1mm). If this distance is not correct, check the torque converter installation.

21. Install the transmission and tighten the bolts as follows:

• SC300—14mm bolts 27 ft. lbs. (37 Nm)

• SC300—17mm bolts 53 ft. lbs. (52 Nm)

• SC400—14mm bolts 29 ft. lbs. (39 Nm)

• SC400—17mm bolts 42 ft. lbs. (57 Nm)

22. If removed, install the starter. Tighten the bolts to 27 ft. lbs. (37 Nm). Connect the electrical connectors to the starter.

23. Install the rear transmission mount and tighten the bolts to 19 ft. lbs. (20 Nm).

24. Install and tighten the torque converter bolts to 25 ft. lbs. (33 Nm) while rotating the crankshaft.

25. Install the converter inspection plate.

26. Install the lower engine cover.

27. Connect the oil cooler lines and tighten the lines to 25 ft. lbs. (34 Nm). Install the oil cooler pipe bracket and tighten the bolt.

28. Connect the transmission electrical connectors.

29. Connect the shift control rod and adjust the shift linkage. Tighten the nut to 12 ft. lbs. (16 Nm).

30. Install the rear center floor crossmember brace by installing the four bolts. Tighten the bolts to 9 ft. lbs. (13 Nm).

31. Install the heat insulator.

32. Install the transmission filler tube and dipstick.

33. Install the front exhaust pipe and converters with new gaskets.

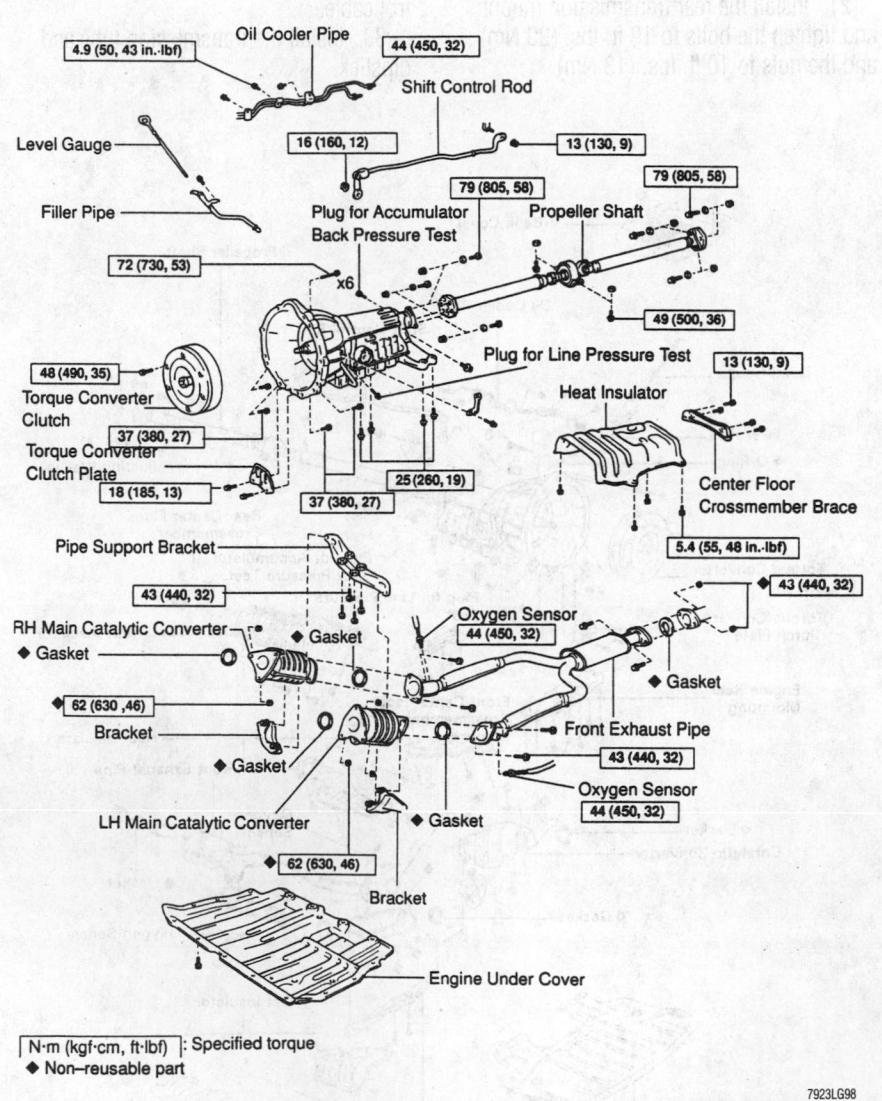

Oil Cooler Pipe
4.9 (50, 43 in.·lbf)
44 (450, 32)
Shift Control Rod
Level Gauge
16 (160, 12)
13 (130, 9)
Filler Pipe
79 (805, 58)
79 (805, 58)
Plug for Accumulator Back Pressure Test
Propeller Shaft
72 (730, 53)
x6
48 (490, 35)
Torque Converter Clutch
37 (380, 27)
Torque Converter Clutch Plate
18 (185, 13)
Plug for Line Pressure Test
13 (130, 9)
Heat Insulator
25 (260, 19)
37 (380, 27)
Center Floor Crossmember Brace
5.4 (55, 48 in.·lbf)
Pipe Support Bracket
43 (440, 32)
43 (440, 32)
RH Main Catalytic Converter
Oxygen Sensor
44 (450, 32)
◆ Gasket
◆ Gasket
62 (630, 46)
Bracket
◆ Gasket
Front Exhaust Pipe
43 (440, 32)
Oxygen Sensor
44 (450, 32)
LH Main Catalytic Converter
◆ Gasket
62 (630, 46)
Bracket
Engine Under Cover

N·m (kgf·cm, ft·lbf) : Specified torque
◆ Non–reusable part

7923LG98

Exploded view of the transmission mounting—SC300 and SC400 with the A650E transmission

34. Connect and adjust the throttle control cable.

35. Install the driveshaft. Align the matchmarks and connect the driveshaft to the transmission. Insert the bolts from the transmission side and tighten to 58 ft. lbs. (79 Nm). Align the matchmarks and install the driveshaft to the differential. Insert the bolts from the differential side and tighten to 58 ft. lbs. (79 Nm). Tighten the center bearing support bolts to 36 ft. lbs. (49 Nm). Using the same tools as removal, tighten the adjusting nut to 35 ft. lbs. (48 Nm).

36. Install the crossmember brace and tighten to 8 ft. lbs. (13 Nm).

37. Install the A/T oil level gauge.

38. Adjust the Park/Neutral switch.

39. Fill the transmission with ATF (Dexron II® or equivalent).

40. For 1996–97 vehicles, install the V-bank cover.

41. Connect the negative battery cable.

42. Start the engine and check the ATF fluid level. If necessary, add ATF (Dexron II® or equivalent) to the transmission to obtain the proper fluid level.

Transaxle Assembly

REMOVAL & INSTALLATION

Manual

ES300

❄❄❄ CAUTION

The Supplemental Inflatable Restraint (SIR) system must be disarmed before removing the transaxle assembly. Failure to do so may cause accidental deployment of the air bag, resulting in unnecessary SIR system repairs and/or personal injury.

1. Turn the ignition switch to the **LOCK** position and disconnect the negative battery cable. Wait at least 20 seconds or longer before doing any work on the vehicle.

2. Remove the air cleaner case and hose.

3. If equipped, remove the cruise control actuator and bracket.

4. Remove the clutch slave cylinder.

5. Remove the starter.

6. Remove the clutch accumulator and tube clamp.

7. Disconnect the back-up light switch electrical connector.

8. Disconnect the wire clamps from the transaxle.

9. Remove the clutch release cylinder bracket.

10. Disconnect the ground cables from the transaxle.

11. Disconnect the control cables from the transaxle.

12. Remove the three upper transaxle mounting bolts.

13. Disconnect the speed sensor electrical connector.

14. Raise and support the vehicle safely.

15. Remove the front wheels.

16. Remove the undercovers and side covers.

17. Drain the transaxle oil.

18. Remove the front exhaust pipe.

19. Remove the halfshafts.

20. Disconnect the sway bar bushing brackets.

21. Disconnect the steering gear from the front suspension crossmember and suspend it.

22. Remove the stiffener plate.

23. Remove the engine mounting absorber.

24. Remove the clutch inspection plate.

25. Remove the front engine mounting set bolts and nut.

26. Remove the rear engine mounting set nuts.

27. Remove the left engine mounting bolts and nuts.

28. Remove the power steering cooler pipe mounting bolts.

29. Remove the left and right fender liner set bolts.

30. Remove the front suspension member.

31. Support the engine with a suitable jack.

32. Remove the transaxle mounting bolts.

33. Lower the left side of the engine and remove the transaxle.

To install:

34. Install the transaxle and tighten the bolts to the following specifications:
- 10mm bolts—34 ft. lbs. (46 Nm)
- 12mm bolts—47 ft. lbs. (64 Nm)

35. Install the front suspension member and tighten the four main bolts to 134 ft. lbs. (181 Nm). Tighten the outer four bolts to 24 ft. lbs. (32 Nm) and nut to 27 ft. lbs. (36 Nm).

36. Install the left and right fender liner screws.

37. Install the power steering cooler pipe mounting bolts.

38. Install the left engine mount and tighten the bolts to 47 ft. lbs. (64 Nm) and nuts to 59 ft. lbs. (80 Nm). Install the hole plugs.

39. Install the front engine mount set bolts and tighten them to 59 ft. lbs. (80 Nm).

40. Install the engine shock absorber and tighten the bolts to 35 ft. lbs. (48 Nm).

41. Install the clutch inspection plate. Tighten the four outer bolts to 13 ft. lbs. (18 Nm) and two inner bolts to 27 ft. lbs. (37 Nm).

42. Connect the steering gear to the front suspension crossmember and tighten the bolts and nuts to 134 ft. lbs. (181 Nm).

43. Install the sway bar brackets and tighten the bolts to 14 ft. lbs. (19 Nm).

44. Install the halfshafts. Install the remaining components.

45. Fill the transaxle to the proper level with the recommended oil.

46. Install the front wheels.

47. Lower the vehicle.

48. Connect the negative battery cable.

49. Check the front wheel alignment.

Automatic

ES300

❄❄❄ CAUTION

The Supplemental Inflatable Restraint (SIR) system must be disarmed before removing the transaxle assembly. Failure to do so may cause accidental deployment of the air bag, resulting in unnecessary SIR system repairs and/or personal injury.

1. Turn the ignition switch to the **LOCK** position and disconnect the negative battery cable. Wait at least 90 seconds or longer before doing any work on the vehicle.

2. Remove the battery.

3. Remove the air cleaner assembly.

4. Disconnect the throttle cable from the throttle body.

5. Remove the cruise control actuator cover and disconnect the connector, if equipped.

6. Remove the ground wire.

7. Remove the starter.

8. Disconnect speed sensor connectors, direct clutch speed sensor and the Park/Neutral position switch connector on the transaxle.

9. Disconnect the solenoid connector on the transaxle.

10. Disconnect shift control cable.

11. Disconnect oil cooler hoses.

12. Remove the two front side transaxle mounting bolts.

13. Remove the two front engine mounting bolts.

14. Remove the oil cooler line mounting bolts from the front frame.

15. Remove the three upper transaxle to engine mounting bolts.

16. Install an engine support fixture. Tie steering gear housing to engine support fixture.

17. Raise and safely support the vehicle.

18. Drain the transaxle/differential fluid.

19. Remove the front wheels.

20. Remove the front exhaust pipe.

21. Remove the engine side covers and undercovers.

22. Disconnect both halfshafts.

23. Remove the front side engine mounting nut.

24. Remove the rear side engine mounting bolts (remove hole plugs).

25. Remove the four left side transaxle mounting bolts.

26. Remove the steering gear housing.

27. Remove the front frame assembly.

28. Properly support the transaxle assembly.

29. Remove the rear end plate mounting bolts.

30. Remove the torque converter cover.

31. Remove the torque converter retaining bolts.

32. Remove the remaining transaxle mounting bolts.

33. Carefully remove the transaxle assembly from the vehicle.

To install:

34. Install the transaxle aligning the two dowel pins on the block with the converter housing. Tighten the bolts as follows:
  • 10mm bolts—34 ft. lbs. (46 Nm)
  • 12mm bolts—47 ft. lbs. (64 Nm)

35. Coat the threads of the torque converter bolts with sealer. Install the bolts starting with the green bolt followed by the rest and tighten the bolts evenly to 20 ft. lbs. (27 Nm).

36. Install the rear end plate and tighten the bolts to 27 ft. lbs. (37 Nm).

37. Install the front frame assembly and tighten the fasteners as follows:
  • 12mm bolts—24 ft. lbs. (32 Nm)
  • 19mm bolts—134 ft. lbs. (181 Nm)
  • Nut—27 ft. lbs. (36 Nm)

38. Install the two fender liner set screws.

39. Connect the steering gear to the frame and tighten the bolts and nuts to 134 ft. lbs. (181 Nm).

40. Connect the sway bar brackets and toque the bolts to 14 ft. lbs. (19 Nm).

41. Install the left transaxle mounting bolts and tighten them to 38 ft. lbs. (52 Nm).

42. Install the rear side mounting bolts

and nuts and tighten them to 48 ft. lbs. (66 Nm). Install the plugs.

43. Install the front engine mounting nut and tighten it to 59 ft. lbs. (80 Nm).

44. Install the halfshafts.

45. Install the right and left engine side covers.

46. Install the lower engine cover.

47. Fill the transaxle/differential to the proper level with Dexron II® or equivalent.

48. Install the exhaust pipe to the engine with new gaskets and tighten the nuts to 46 ft. lbs. (62 Nm). Connect the exhaust pipe to the converter with a new gasket and tighten the nuts and bolts to 32 ft. lbs. (43 Nm).

49. Install the wheel.

50. Lower the vehicle.

51. Remove the engine support.

52. Install the four upper transaxle mounting bolts and tighten them to 47 ft. lbs. (64 Nm).

53. Install the oil cooler clamping bolts to the front frame.

54. Install the two front side engine mounting bolts and tighten them to 59 ft. lbs. (80 Nm).

55. Install the two front side transaxle mounting bolts and tighten them to 59 ft. lbs. (80 Nm).

56. Install the remaining components. Install the battery and connect the battery cables.

57. Check the transaxle/differential fluid level.

58. Check the front wheel alignment.

Clutch

REMOVAL & INSTALLATION

1. Disconnect the negative battery cable. Wait at least 90 seconds before proceeding with any other work.

2. Remove the transaxle/transmission assembly from the vehicle.

3. Place matchmarks on the flywheel and clutch cover.

4. Remove the clutch pressure plate retaining bolts. Loosen the bolts 1 turn at a time, in a crisscross pattern, until all the spring tension is released.

5. Remove the pressure plate and clutch disc. Do not drop the clutch disc or plate.

6. Remove the release bearing, fork and boot from the transmission.

To install:

7. Apply grease to the following;
  • Release fork and hub contact points
  • Release fork and pushrod contact point
  • Release fork pivot point
  • Clutch disc spline

8. Install a new release bearing on the fork and install the fork to the transaxle/transmission.

9. Install the clutch disc and pressure plate on the flywheel, aligning them with 09301–17010 or equivalent.

10. Tighten the pressure plate mounting bolts to 14 ft. lbs. (19 Nm) in X-type pattern.

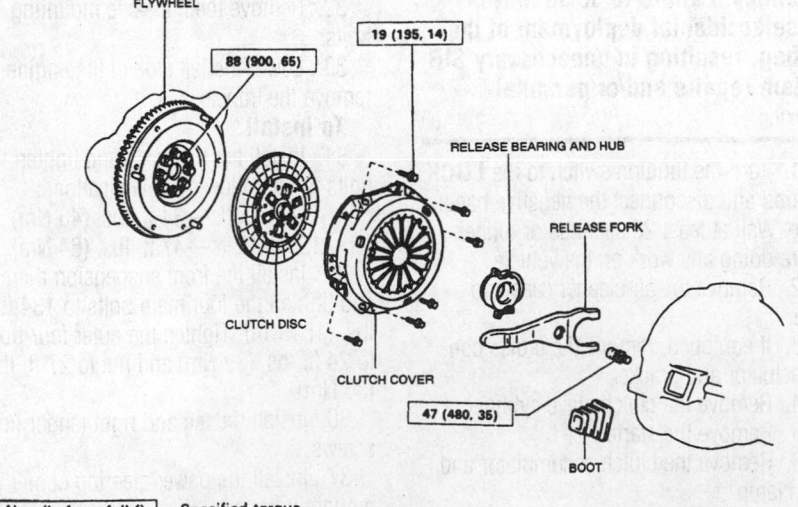

N·m (kgf·cm, ft·lbf) : Specified torque

Clutch disc and pressure plate assembly—ES300

7923LG99

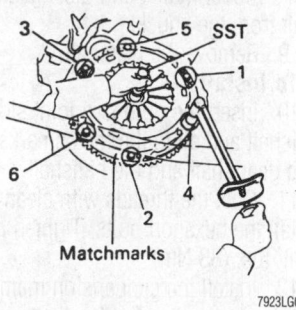

Tighten the pressure plate bolts in the sequence shown—ES300

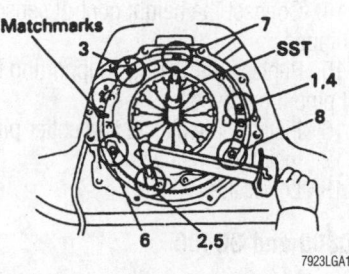

Pressure plate bolt tightening sequence—SC300

11. Install the transaxle/transmission.
12. Connect the negative battery cable.
13. Adjust the clutch pedal as necessary.
14. Road test the vehicle and verify proper clutch operation.

Hydraulic Clutch System

BLEEDING

➡**If any maintenance on the clutch system was performed or the system is suspected of containing air, bleed the system. Use care; brake fluid will remove the paint from any surface. If the brake fluid spills onto any painted surface, wash it off immediately with soap and water.**

1. Fill the clutch reservoir with brake fluid. Check the reservoir level frequently and add fluid as needed.
2. Connect one end of a vinyl tube to the bleeder plug on the slave cylinder and submerge the other end into a clear container half-filled with brake fluid.
3. Slowly pump the clutch pedal several times.
4. Have an assistant hold the clutch pedal down and loosen the bleeder plug

until fluid and/or air starts to run out of the bleeder plug. Close the bleeder plug while the pedal is held to the floor.

➡**Do not allow the pedal to rise back-up while the bleeder is still open. If this happens, it will allow air to re-enter the slave cylinder and cause the clutch system not to work properly.**

5. Repeat Steps 2 and 3 until all the air bubbles are removed from the system.
6. Tighten the bleeder plug when all the air is gone.
7. Refill the master cylinder to the proper level as required.
8. Check the system for leaks.

Halfshaft

REMOVAL & INSTALLATION

ES300

1. Disconnect the negative battery cable to the battery.
2. Raise and support the vehicle safely.
3. Remove the front wheel(s).
4. Remove the front fender apron seal.
5. Drain the transaxle.
6. Disconnect the tie rod end from the steering knuckle by removing the cotter pin and nut. Using tool SST 09628–62011 or equivalent, separate the tie rod from the steering knuckle.
7. Disconnect the stabilizer bar link from the lower control arm. Make note of the washers and cushions positions.
8. Disconnect the lower ball joint from the steering knuckle by removing the bolt and two nuts. Push down on the lower control arm and separate the steering knuckle from the ball joint.
9. Remove the cotter pin, lock cap and locknut holding the halfshaft to the steering knuckle.
10. Using a plastic hammer, disconnect the halfshaft from the steering knuckle.
11. Remove the left halfshaft from the transaxle as follows:
    a. Use a brass bar and hammer to tap the inner joint out of the transaxle.
    b. Remove the halfshaft.
    c. Once the halfshaft is removed from the vehicle, remove the snapring from the halfshaft.
12. Remove the right halfshaft from the transaxle as follows:
    a. Remove the bearing lockbolt. The lockbolt is located in the center of the halfshaft, near the dampener.
    b. Using snapring pliers, remove the

snapring and pull the halfshaft from the transaxle.

To install:

13. To install the right halfshaft to the transaxle:
    a. Coat the side gear shaft and differential case sliding surface with gear oil.
    b. Using snapring pliers, install the snapring to the halfshaft.
    c. Install the halfshaft and the bearing lockbolt. Tighten the lockbolt to 24 ft. lbs. (32 Nm).
14. To install the left halfshaft to the transaxle:
    a. Install a new snapring to the inner spline of the halfshaft.
    b. Coat the side gear shaft and differential case sliding surface with gear oil.
    c. Install the halfshaft to the transaxle with the snapring opening facing down. The halfshaft should click into place when installing.
    d. After installation of the halfshaft, check that the halfshaft cannot be removed by hand.
15. Connect the halfshaft to the steering knuckle, then install the locknut. Tighten the locknut to 217 ft. lbs. (294 Nm).
16. Install the lock cap and a new cotter pin to the halfshaft.
17. Connect the steering knuckle to the lower ball joint. Install the two nuts and bolt. Tighten the nuts and bolt to 94 ft. lbs. (127 Nm).
18. Connect the stabilizer bar link to the lower control arm. Tighten the nut to 29 ft. lbs. (39 Nm).
19. Connect the tie rod to the steering knuckle and tighten the nut to 36 ft. lbs. (49 Nm). Install a new cotter pin to the tie rod end.
20. Install the front fender apron seal.
21. Install the wheel(s) and lower the vehicle. Tighten the lug nuts to 76 ft. lbs. (103 Nm).
22. Refill the transaxle and check for leaks.
23. Connect the negative battery cable to the battery.

GS300

❊❊ CAUTION

The air bag system (SRS or SIR) must be disarmed before removing the halfshafts. Failure to do so may cause accidental deployment, property damage or personal injury.

1. Disconnect the negative battery cable from the battery.

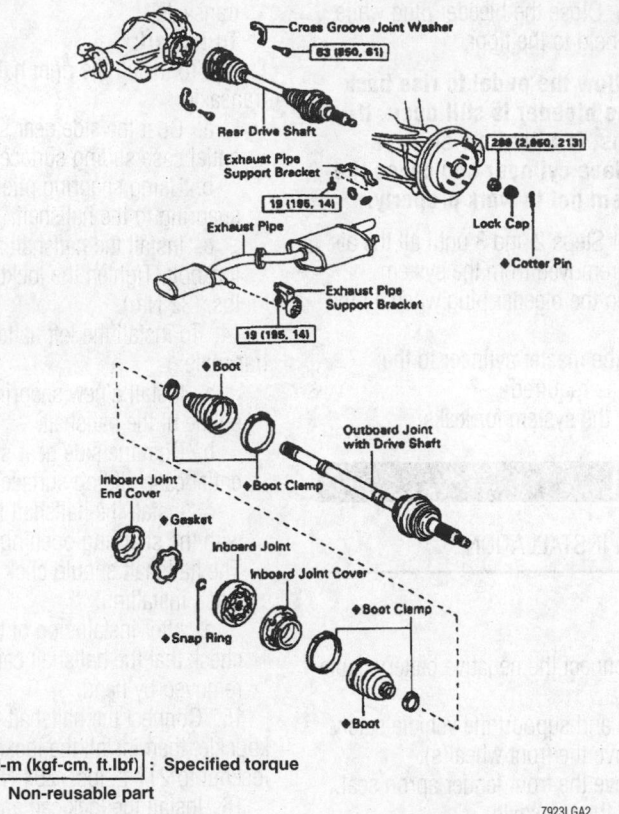

Cross Groove Joint Washer
83 (850, 61)

Rear Drive Shaft

Exhaust Pipe
Support Bracket
19 (185, 14)

288 (2,950, 213)

Lock Cap

Exhaust Pipe

Cotter Pin

Exhaust Pipe
Support Bracket
19 (185, 14)

Boot

Outboard Joint
with Drive Shaft

Inboard Joint
End Cover

Boot Clamp

Gasket

Inboard Joint

Inboard Joint Cover

Snap Ring

Boot Clamp

Boot

N-m (kgf-cm, ft.lbf) : Specified torque
♦ Non-reusable part

7923LGA2

Exploded view of the rear halfshaft and related components—GS300

2. Raise and safely support the vehicle.

3. Remove the rear tire and wheel assembly.

4. Remove the cotter pin, locknut cap and locknut while having someone pressing down on the brake pedal.

5. Secure the rear exhaust assembly with mechanics wire or equivalent.

6. Remove the two exhaust pipe support brackets.

7. Have someone press down on the brake pedal. Place matchmarks on the halfshaft and the side gear shaft. Remove the six hex bolts and two washers.

8. Hold the inboard joint side of the halfshaft so the outboard joint side does not bend too much. Tap the end of the halfshaft with a rubber mallet to loosen it from the axle hub and remove the halfshaft.

To install:

9. Insert the outboard joint side of the halfshaft through the axle hub. Align the matchmarks on the side gear shaft and the halfshaft.

10. Coat the threads with clean oil and install the hex bolts. Tighten the bolts to 61 ft. lbs. (83 Nm).

11. Install the exhaust pipe support brackets and tighten to 14 ft. lbs. (19 Nm).

12. Install the bearing locknut and have a helper apply the brakes. Tighten the locknut to 213 ft. lbs. (289 Nm).

13. Install the lock cap and a new cotter pin.

14. Replace the rear tire and wheel assembly.

15. Lower the vehicle and connect the negative battery cable.

LS400

1. Raise and safely support the rear of the vehicle.

2. Remove the rear wheel.

3. Remove the cotter pin, lock cap and the nut holding the halfshaft to the rear knuckle.

4. On some models it will be necessary to, remove the tail pipe O-rings and suspend the tail pipe, using a piece of wire.

5. Disconnect the height control sensor, if equipped.

6. Remove the suspension member brace by removing the two bolts.

7. Place matchmarks on the halfshaft and the side gear shaft. Remove the hexagon bolts and washers with the proper tool.

8. Hold the inboard joint side of the halfshaft so the outboard joint side does not bend too much. Tap the end of the halfshaft with a rubber mallet and disengage the halfshaft from the knuckle.

9. Remove the halfshaft.

To install:

10. Insert the outboard joint side of the halfshaft and align the matchmarks on the side gear shaft and the halfshaft.

11. Coat the threads with clean oil and install the hexagon bolts. Tighten bolts to 61 ft. lbs. (83 Nm).

12. Install the suspension member brace with the two bolts. Tighten the two bolts to 37 ft. lbs. (50 Nm).

13. Install the nut to hold the halfshaft to the rear knuckle. Tighten the nut to 253 ft. lbs. (344 Nm)on the 1993–94 models, 213 ft. lbs. (289 Nm) on the 1995–97 models.

14. Connect the height control sensor, if equipped.

15. Replace the O-rings supporting the tail pipe if removed.

16. Install the lock cap and cotter pin.

17. Install the rear wheel.

18. Lower the vehicle.

SC300 and SC400

1. Raise and safely support the vehicle.

2. Remove the rear tire and wheel assembly.

3. Remove the rear exhaust assembly.

4. Remove the cotter pin, locknut cap and the locknut holding the halfshaft to the rear axle carrier.

5. Remove the lower suspension arm brace by removing the four bolts.

6. Place matchmarks on the halfshaft and the differential side gear shaft. Remove the hexagon bolts and washers with the proper tool.

7. Hold the inboard joint side of the halfshaft so the outboard joint side does not bend too much. Tap the end of the halfshaft with a rubber mallet and disengage the halfshaft from the axle carrier.

8. Remove the halfshaft.

To install:

9. Insert the outboard joint side of the halfshaft and align the matchmarks on the side gear shaft and the halfshaft.

10. Coat the threads with clean oil and install the hexagon bolts. Tighten the bolts to 61 ft. lbs. (83 Nm).

11. Install the lower suspension arm brace and tighten the four bolts to 13 ft. lbs. (18 Nm).

12. Install the bearing locknut and tighten the locknut to 213 ft. lbs. (289 Nm).

13. Install the locknut cap and install a new cotter pin.

14. Install the rear exhaust assembly.
15. Replace the rear tire and wheel assembly.
16. Lower the vehicle.

STEERING AND SUSPENSION

Air Bag

✳✳ CAUTION

Some vehicles are equipped with an air bag system, also known as the Supplemental Inflatable Restraint (SIR) or Supplemental Restraint System (SRS). The system must be disabled before performing service on or around system components, steering column, instrument panel components, wiring and sensors. Failure to follow safety and disabling procedures could result in accidental air bag deployment, possible personal injury and unnecessary system repairs.

PRECAUTIONS

Several precautions must be observed when handling the inflator module to avoid accidental deployment and possible personal injury.

• Never carry the inflator module by the wires or connector on the underside of the module.

• When carrying a live inflator module, hold securely with both hands and ensure that the bag and trim cover are pointed away.

• Place the inflator module on a bench or other surface with the bag and trim cover facing up.

• With the inflator module on the bench, never place anything on or close to the module which may be thrown in the event of an accidental deployment.

DISARMING

To avoid personal injury when working on vehicles equipped with an air bag, the negative battery cable must be disconnected and at least 90 seconds must elapse before working on the system. Failure to do so may result in deployment of the air bag.

REARMING

To rearm the air bag system, simply reconnect the battery cable(s).

Rack and Pinion Steering Gear

REMOVAL & INSTALLATION

Power

ES300

1. Disconnect the negative battery cable and wait at least 90 seconds before working on the vehicle to disarm the air bag.
2. Secure the steering wheel in a straight forward position.
3. Raise and support the vehicle safely. Remove the front wheels.
4. Remove the left and right front fender apron seals by removing the two bolts.
5. Remove the cotter pin and nut holding the steering knuckle to the tie rod end. Using a tie rod puller, disconnect the tie rod end from the steering knuckle.
6. Place matchmarks on the intermediate shaft and the control valve shaft.
7. Loosen the upper bolt and remove the lower bolt holding the control valve shaft to the intermediate shaft. Disconnect the intermediate shaft from steering rack housing.
8. Remove the nut to the tube clamp. Remove the clamp from the vehicle.
9. Using SST No. 09631–22020 or equivalent, disconnect the return line and the pressure line from the control valve housing. Use a small plastic container to catch the fluid.
10. Remove the four stabilizer bar bolts and two nuts. Position the stabilizer bar out of the way. Do not remove the sway bar from the vehicle.
11. Remove the heated oxygen sensor (bank 1 sensor 1).
12. Remove the two steering gear mounting bolts and nuts. Remove the steering gear through the left side of the vehicle.
To install:
13. Position the steering gear on the vehicle and install the two mounting bolts and nuts. Tighten the nuts and bolts to 134 ft. lbs. (181 Nm).
14. Install the heated oxygen sensor. Tighten the sensor to 33 ft. lbs. (44 Nm).
15. Install the stabilizer bar bolts and nuts and tighten as follows:
   • Bolts: 14 ft. lbs. (19 Nm)
   • Nuts: 29 ft. lbs. (39 Nm)
16. Connect the pressure and return

lines and tighten the connectors to 18 ft. lbs. (25 Nm).
17. Connect the tube clamp and tighten the nut to 7 ft. lbs. (10 Nm).
18. Install the intermediate shaft to the steering rack and tighten the retaining bolts to 26 ft. lbs. (35 Nm).
19. Connect the tie rods to the steering knuckles with the castellated nuts. Tighten the nut to 36 ft. lbs. and install a new cotter pin. The prongs of the cotter pin should be firmly wrapped around the flats of the nut.
20. Install the front fender apron seals by installing the two bolts.
21. Install the front wheels and lower the vehicle.
22. Fill the power steering reservoir tank to the proper level with power steering fluid.
23. Connect the negative battery cable to the battery.
24. Release the steering wheel.
25. Bleed the system.
26. Check for leaks, adjust the toe-in and check the steering wheel center point.

GS300 AND GS400

✳✳ CAUTION

The air bag system (SRS or SIR) must be disarmed before removing the rack and pinion. Failure to do so may cause accidental deployment, property damage or personal injury.

1. Disconnect the negative battery cable from the battery.
2. Position the wheels in the straight ahead position and secure the steering wheel.
3. Raise and support the vehicle safely.
4. Remove the front wheels.
5. Matchmark the steering column universal joint to the control valve shaft.
6. Loosen the upper bolt and remove

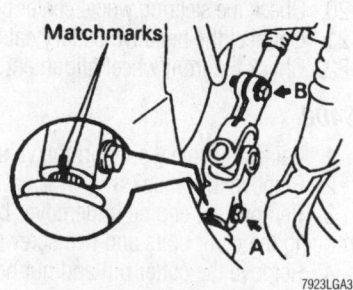

Matchmarking the intermediate shaft to the control valve shaft—GS300

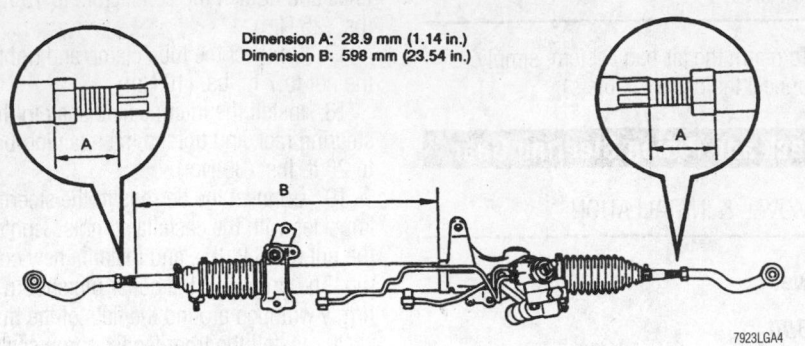

Dimension A: 28.9 mm (1.14 in.)
Dimension B: 598 mm (23.54 in.)

A   B   A

7923LGA4

Centering the rack and pinion—GS300

the lower bolt to the intermediate shaft universal joint.

7. Disconnect the intermediate shaft from the control valve shaft.

8. Disconnect the tie rod ends from the steering knuckle.

9. Disconnect the fluid lines from the rack and pinion and cap the lines.

10. Disconnect the two tube clamps by removing the bolt.

11. Remove the mounting bolts and nuts. Remove the rack and pinion.

To install:

12. Center the rack and pinion to the following dimensions.
- Dimension **A**—1.14 in. (28.9mm)
- Dimension **B**—23.54 in. (589mm)

13. Install the rack and tighten the bolts to 72 ft. lbs. (98 Nm).

14. Connect the two tube clamps and tighten the bolt to 12 ft. lbs. (17 Nm).

15. Align the matchmarks on the intermediate shaft and control valve shaft. Tighten the intermediate shaft bolts to 26 ft. lbs. (35 Nm).

16. Connect the fluid lines to the rack and pinion with new washers. Tighten the union bolts to 36 ft. lbs. (49 Nm).

17. Connect the ties rod ends.

18. Install the wheels.

19. Lower the vehicle.

20. Check the steering wheel center point.

21. Connect the negative battery cable.

22. Check the front wheel alignment.

LS400

1. Raise and safely support the vehicle.

2. Remove the wheel(s).

3. Remove the engine undercover by removing the eight bolts and five screws.

4. Remove the cotter pin and nut holding each tie rod to the steering knuckle.

5. Disconnect the tie rod end from the steering knuckle with a tie rod end puller.

6. Place matchmarks on the sliding yoke and control valve shaft.

7. Loosen the top bolt holding the sliding yoke to the intermediate shaft. Remove the bottom bolt holding the sliding yoke to the steering rack.

8. Disconnect the pressure feed and return lines to the rack and pinion.

9. Disconnect the power steering connector.

10. Remove the four mount bolts and nuts to the power steering rack.

11. Remove the two brackets and grommets.

12. Remove the power steering rack from the vehicle.

To install:

13. Install the power steering rack to the vehicle.

14. Install the two brackets and grommets to the power steering rack.

15. Install the four bolts and tighten the bolts to 56 ft. lbs. (76 Nm).

16. Connect the power steering solenoid connector.

17. Connect the pressure feed and return tubes. Tighten the union bolt to 36 ft. lbs. (49 Nm).

18. Align the matchmarks on the sliding yoke and control valve shaft.

19. Tighten the bolt holding the sliding yoke to the steering rack to 26 ft. lbs. (35 Nm).

20. Tighten the bolt holding the sliding yoke to the intermediate shaft to 26 ft. lbs. (35 Nm).

21. Connect the tie rod end to the steering knuckle. Tighten the nut to 48 ft. lbs. (65 Nm). Install a new cotter pin.

22. Install the engine undercover.

23. Bleed the power steering system.

24. Install the wheel(s) and check the front end alignment.

SC300 AND SC400

1. Disconnect the negative battery cable. Wait at least 90 seconds before performing any work.

2. Place the front wheels facing straight ahead.

3. Remove the steering wheel pad.

❊❊ WARNING

Keep the upper surface of the wheel pad pointed away from you at all times. Store the pad with the upper surface facing upward.

4. Disconnect the intermediate shaft.

5. Raise and safely support the vehicle.

6. Disconnect the right and left tie rod ends.

7. Remove the union bolt and gasket and remove the pressure tube.

8. Remove the union bolt and two gaskets; remove the return tube.

9. Disconnect the PPS solenoid connector.

10. On SC400 models, remove the bolt and disconnect the tube clamp.

11. Remove the two bolts and nuts and remove the bracket and grommet.

12. Remove the two bolts and nuts; remove the rack and pinion assembly.

To install:

13. Install the rack and pinion assembly with the two set bolts and nuts. Tighten the bolts to 56 ft. lbs. (76 Nm).

14. Install the bracket and grommet with the two bolts and nuts. Tighten the bolts to 56 ft. lbs. (76 Nm).

15. On SC400 models, reconnect the tube clamp with the bolt.

16. Reconnect the PPS solenoid.

17. Reconnect the return tube with the bolt and new gaskets. Tighten the union bolt to 36 ft. lbs. (49 Nm).

18. Connect the pressure tube with the union bolt and a new gasket. Tighten the union bolt to 36 ft. lbs. (49 Nm).

19. Connect the right and left tie rod ends.

20. Reconnect the intermediate shaft.

21. Position the front wheels facing straight ahead and safely lower the vehicle.

22. Align the matchmarks and install the steering wheel. Temporarily tighten the wheel set nut and connect the connector.

23. Reconnect the negative battery cable, refill the steering fluid and bleed the steering system.

24. Check the steering wheel center point and tighten the steering wheel set nut to 26 ft. lbs. (35 Nm).

25. Disconnect the negative battery cable. Wait at least 90 seconds before performing the next step.

26. Install the steering wheel pad.

27. Reinstall the negative battery cable and check the front wheel alignment.

28. Road test the vehicle for proper operation.

Strut and Coil Spring

REMOVAL & INSTALLATION

Front

ES300

1. Disconnect the negative battery cable from the battery.
2. Raise and safely support the vehicle.
3. Remove the tire and wheel assembly.
4. If equipped with ABS, disconnect the ABS speed sensor connector and brake line from the strut housing. Do not remove the brake line from the brake caliper.
5. Disconnect the strut assembly from the steering knuckle by removing the two nuts and bolts.
6. Remove the three upper mounting nuts from the strut tower and remove the strut assembly.

> **✳✳ CAUTION**
>
> **Do not remove the center nut to the strut at this time. The spring on the strut is under high pressure and can cause serious injury.**

7. Temporarily install the bolt and nuts to the lower bracket of the strut to support it and secure the strut in a vise.
8. Compress the coil spring using spring compressor 09727–30020 or equivalent.
9. Secure the spring seat using 09729–22031 or equivalent and remove the upper strut retaining nut.
10. Remove the suspension support, upper insulator, spring, bumper and the insulator.

To install:

11. Install the lower insulator.

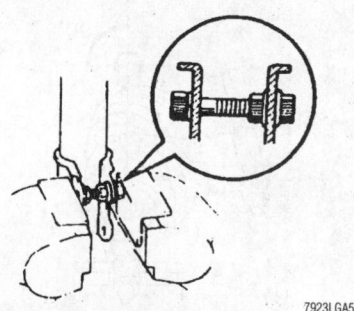

Temporarily install the support nuts and bolt to the strut—ES300

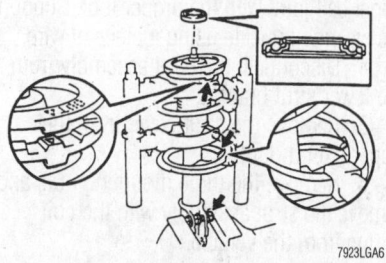

Align the out mark of the upper spring seat with the mark on the upper insulator— ES300 and LS400

12. Install the bumper to the piston rod.
13. Align the (compressed) coil spring end into the gap of the lower seat.
14. Install the upper insulator.
15. Install the upper support to the piston rod, aligning it with the groove in the strut rod.
16. Secure the spring seat and tighten the new upper strut retaining nut to 36 ft. lbs. (49 Nm).
17. Remove the spring compressor.
18. Remove the strut from the vise and disassemble the securing nuts and bolt.
19. Rotate the upper support so the lowest bolt on the support aligns with the projection part of the lower spring.
20. Install the strut and tighten the strut to body bolts to 59 ft. lbs. (80 Nm).
21. Connect the strut to the steering knuckle and tighten the bolts to 156 ft. lbs. (211 Nm).
22. Run the brake hose through the brake hose bracket and install the clip.
23. Connect the ABS speed sensor and tighten the mounting bolt to 48 inch lbs. (5 Nm).
24. Connect the brake line to the strut housing and tighten the bolt to 22 ft. lbs. (29 Nm).
25. Install the wheel.
26. Check the front alignment.
27. Connect the negative battery cable.

GS300 AND GS400

1. Disconnect the negative battery cable.
2. Raise and support the vehicle safely.
3. Remove the front wheel.
4. Loosen the three upper strut mounting nuts.
5. Loosen, but do not remove, the upper strut rod nut.

> **✳✳ CAUTION**
>
> **Do not remove the upper strut nut at this time.**

6. Remove the brake caliper, leaving the line attached and secure it out of the way.

➡**Never allow the brake caliper to hang freely from the brake hose.**

7. Disconnect the ABS speed sensor and harness.
8. Disconnect the upper suspension arm from the steering knuckle.
9. Disconnect the stabilizer bar from the link and remove the bracket.
10. Disconnect the strut from the lower suspension arm.
11. Remove the three upper strut mounting nuts and remove the strut.
12. Using a spring compressor, 09727–30020 or equivalent, compress the coil spring.
13. Remove the piston rod locknut.
14. Remove the suspension support, coil spring and bumper.

To install:

15. If disposing the strut, perform the following procedure:
    a. Fully extend the strut rod.
    b. Drill a hole near the bottom of the shock to remove the gas inside.

➡**The gas is harmless, but be careful of chips which may fly up when the gas is released.**

16. Match the bolt of the suspension support with the cut out portion of the insulator.
17. Install the spring bumper.
18. Install the compressed coil spring. Match the end of the coil into the recess of the strut spring seat.
19. Install the suspension support to the rod and temporarily install a new nut.
20. Turn the suspension support so one of the bolts on the support faces the same direction as shown in the illustration.

➡**Align the bolt so a line drawn between the rod and bolt would be at 90° to the direction of the lower bushing.**

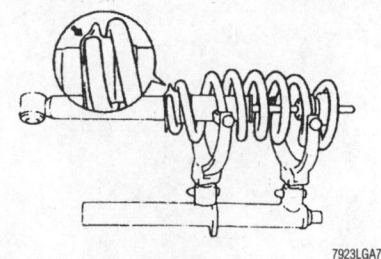

Matching the spring to the seat

21. Remove the spring compressor.

22. Install the strut and tighten the upper retaining nuts to 41 ft. lbs. (56 Nm).

23. Tighten the new upper strut rod nut to 20 ft. lbs. (27 Nm).

24. Connect the strut to the lower arm and temporarily tighten the nut and bolt.

25. Install the stabilizer bar bracket and tighten the bolts to 21 ft. lbs. (28 Nm).

26. Connect the stabilizer bar to the link and tighten the bolts to 29 ft. lbs. (39 Nm).

27. Connect the upper suspension arm to the steering knuckle. Tighten the nut to 64 ft. lbs. (87 Nm) and install a new cotter pin.

28. Install the ABS speed sensor and tighten the bolt to 69 inch lbs. (8 Nm).

29. Install the caliper.

30. Install the wheel.

31. Lower the vehicle.

32. Bounce the vehicle several times to stabilize the suspension.

33. Tighten the lower strut bolt and nut to 116 ft. lbs. (157 Nm).

34. Check the front wheel alignment.

LS400—WITHOUT AIR SUSPENSION

1. Raise and safely support the vehicle.
2. Remove the tire and wheel assembly.

3. Remove the steering knuckle from the upper ball joint with the proper tool. Support the steering knuckle using a piece of wire.

4. Disconnect the strut assembly from the lower strut bracket.

5. Remove the strut cover from the upper strut mount.

6. Remove the three mounting nuts and remove the strut assembly with the coil spring from the vehicle.

✳✳ CAUTION

Do not remove the center nut to the strut at this time. The spring on the strut is under high pressure and can cause serious injury.

7. Using compressor 09727–30020 or equivalent, compress the coil spring.

8. Remove the piston rod locknut.

9. Remove the suspension support, coil spring and the bumper.

To install:

10. If disposing the strut, perform the following procedure:

   a. Fully extend the strut rod.

   b. Drill a hole within the shaded area shown in the illustration to remove the gas inside.

➡The gas is harmless, but be careful of chips which may fly up when drilling.

   c. Properly dispose of the strut assembly.

11. Match the bolt of the suspension support with the cut out portion of the insulator.

12. Install the spring bumper.

13. Install the compressed coil spring. Match the end of the coil into the recess of the strut spring seat.

14. Install the suspension support to the rod and temporarily install a new nut.

15. Turn the suspension support so one of the bolts on the support faces the same direction as shown in the illustration.

➡Align the bolt so a line drawn between the rod and bolt would be at 90° to the direction of the lower bushing.

16. Tighten the strut rod nut to 20 ft. lbs. (27 Nm) and install the cap.

17. Remove the spring compressor.

18. Install the strut and tighten the upper retaining nuts to 43 ft. lbs. (58 Nm).

19. Connect the strut to the lower bracket and temporarily install the nut and bolt.

20. Connect the upper control arm to the steering knuckle. Tighten the nut to 48 ft. lbs. (65 Nm) and install a new cotter pin.

21. Install the wheel.

22. Lower the vehicle.

23. Bounce the vehicle several times to stabilize the suspension.

24. Tighten the lower strut bolt and nut to 116 ft. lbs. (157 Nm).

25. Check the front wheel alignment.

LS400—WITH AIR SUSPENSION

1. Move the height control switch to OFF.

2. Raise and support the vehicle safely.

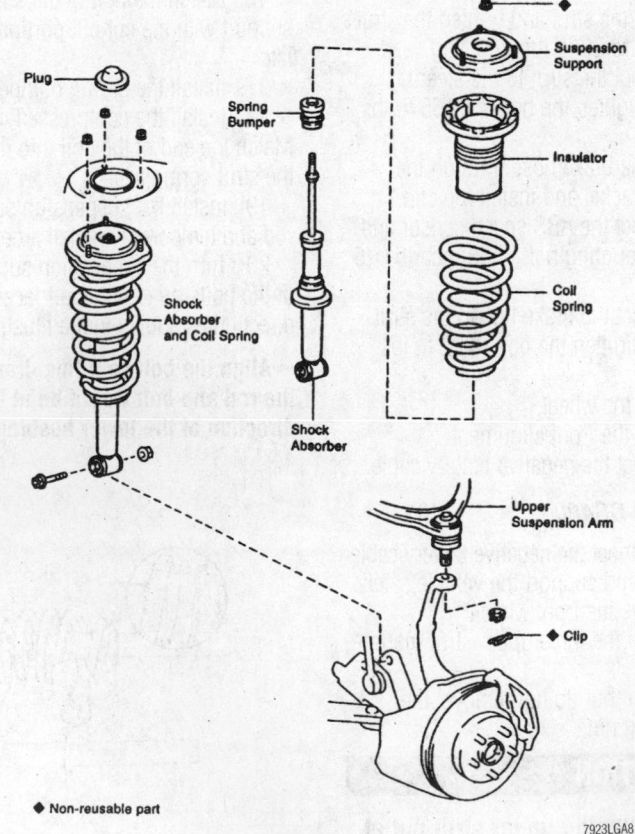

Plug
Spring Bumper
Shock Absorber and Coil Spring
Shock Absorber
Suspension Support
Insulator
Coil Spring
Upper Suspension Arm
◆ Clip
◆ Non-reusable part

7923LGA8

Exploded view of the strut and spring mounting—LS400, except with air suspension

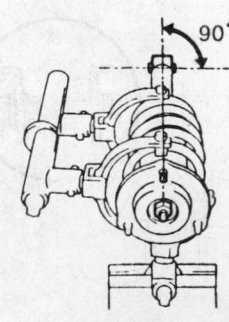

90°

7923LGA9

Be sure to align the suspension support to one of the upper mounting bolts as shown—LS400 without air suspension

3. Remove the wheel.

4. Bleed the air in the suspension.

5. Remove the height control sensor link from the lower strut bracket.

6. Remove the cotter pin and nut holding the upper control arm to the steering knuckle.

7. Disconnect the upper ball joint from the steering knuckle. Suspend the knuckle with wire to prevent excessive force on the brake line and ABS speed sensor.

8. Disconnect the pneumatic cylinder from the lower bracket by removing the through-bolt.

9. Disconnect the air tube from the strut.

10. Remove the three nuts holding the actuator cover to the strut tower. Remove the actuator cover.

⁂ CAUTION

Do not remove the center nut from the pneumatic cylinder.

11. Disconnect the actuator electrical connector.

12. Remove the two bolts to the suspension control actuator and position the actuator aside.

13. Remove the three upper mounting nuts and the strut from the vehicle.

To install:

14. If disposing the strut perform the following procedure:

a. Using a screwdriver, remove the air from inside the cylinder.

b. Fully extend the cylinder.

c. Drill a hole in the cylinder at a point above 1.57 in. (40mm) from the bottom of the strut assembly. This will release the gas charge in the strut. Do not puncture the pneumatic cylinder.

➡ **The gas coming out is harmless, but be careful of chips which may fly up while drilling.**

15. Install the strut and tighten the upper mounting nuts to 43 ft. lbs. (58 Nm).

16. Match the rods of the strut with the holes in the suspension control actuator.

17. Install the suspension control actuator and tighten the bolts. Tighten the two nuts to 13 ft. lbs. (17 Nm).

18. Install the suspension control actuator cover and tighten the nuts to 43 ft. lbs. (58 Nm).

19. Install two new O-rings to the air tube. Install the tube and tighten it to 13 ft. lbs. (17 Nm). Install the grommet.

20. Install the strut to the lower strut bracket and temporarily install the nut and bolt.

21. Connect the steering knuckle to the upper ball joint. Tighten the nut to 48 ft. lbs. (65 Nm) and install a new cotter pin.

22. Connect the height control sensor link and tighten a new nut to 48 inch lbs. (5 Nm).

23. Install the wheel.

24. Lower the vehicle.

25. Turn the height control switch ON.

26. Start the engine to fill the strut with air.

27. Bounce the vehicle several times to normalize the suspension.

28. Support the lower control arm with a jack.

29. Remove the front wheel.

30. Tighten the lower strut mounting nut and bolt to 76 ft. lbs. (106 Nm).

31. Install the wheel.

32. Lower the vehicle.

33. Check the front end alignment.

SC300 AND SC400

1. Raise and safely support the vehicle.

2. Remove the tire and wheel assembly.

3. Remove the brake caliper support bracket by removing the two bolts. Suspend it with a piece of wire.

4. Remove the fender apron, engine undercover and the front fender wheel opening molding.

5. If removing the left side strut, disconnect the windshield washer tank.

6. Remove the bolt and disconnect the ABS speed sensor at the steering knuckle. Remove the three bolts, then disconnect the wiring harness clamp in order to prevent the harness from being damaged when removing the through-bolt.

7. Remove the plug from the upper strut mount. Do not remove the center bolt.

⁂ CAUTION

Do not remove the center bolt to the strut at this time. The spring on the strut is under high pressure and can cause serious injury or vehicle damage.

8. Disconnect the upper control arm through-bolt from the subframe. Disconnect the upper control arm and turn the control arm completely around. It is not necessary to remove the upper ball joint.

9. Disconnect the strut at the lower control arm by removing the nut and bolt.

10. Remove the three upper mounting nuts and remove the strut assembly with the coil spring from the vehicle.

11. Using compressor 09727–30020 or equivalent, compress the coil spring.

12. Remove the piston rod locknut.

13. Remove the suspension support, coil spring and bumper.

To install:

14. If disposing the strut, perform the following procedure:

a. Fully extend the strut rod.

b. Drill a hole within the shaded area shown in the illustration to remove the gas inside and dispose the old strut.

➡ **The gas is harmless, but be careful of chips which may fly up when drilling.**

15. Match the bolt of the suspension support with the cut out portion of the insulator.

16. Install the spring bumper.

17. Install the compressed coil spring. Match the end of the coil into the recess of the strut spring seat.

18. Install the suspension support to the rod and temporarily install a new nut.

19. Turn the suspension support so one of the bolts on the support faces the same direction as shown in the illustration.

➡ **Align the bolt so a line drawn between the rod and bolt would be at 90° to the direction of the lower bushing.**

20. Remove the spring compressor.

21. Install the strut and tighten the three upper strut mount nuts to 26 ft. lbs. (35 Nm). Tighten the middle nut to 22 ft. lbs. (29 Nm) and install the plug.

22. Connect the lower end of the strut to the lower control arm. Do not tighten the bolt at this time.

23. Install the upper control arm and install the through-bolt and nut. Do not tighten the bolt at this time.

24. Connect the speed sensor, wiring harness and the washer tank.

25. Install the fender apron and the engine undercover.

26. Install the caliper support bracket and tighten the bolts to 87 ft. lbs. (118 Nm).

27. Install the tire and wheel assembly.

28. Lower the vehicle.

29. Bounce the vehicle a few times to stabilize the suspension, then tighten the strut to lower arm bolt to 106 ft. lbs. (143 Nm). Tighten the upper arm to 121 ft. lbs. (164 Nm).

30. Check the front end alignment.

Rear

ES300

1. Raise and safely support the vehicle.

2. Remove the tire and wheel assembly.

3. Disconnect the load sensing proportioning valve spring assembly from the lower arm.

4. Disconnect the ABS speed sensor harness and brake line from the strut assembly.

5. Disconnect the stabilizer bar link from the strut.

6. Loosen the two nuts attaching the strut to the axle carrier.

7. Support the axle carrier.

8. Remove the rear seat back and package tray trim.

9. Remove the upper mounting nuts.

10. Remove the two lower mounting bolts and remove the strut assembly.

11. Using spring compressor SST 09727–30020 or equivalent, compress the coil spring.

12. Temporarily install a bolt and two nuts on the bracket at the lower end of the strut and secure it in a vice.

13. Secure the upper support with SST 09729–22031 or equivalent and remove the strut rod retaining nut.

14. Remove the upper suspension support, upper insulator, coil spring, spring bumper and lower insulator.

To install:

15. If discarding the strut, perform the following:

   a. Fully extend the strut rod.

   b. Drill a hole in the side of the strut to release the gas.

❊❊ WARNING

The gas coming out is harmless, but be careful of chips which may fly up while drilling.

16. Install the lower insulator to the strut.

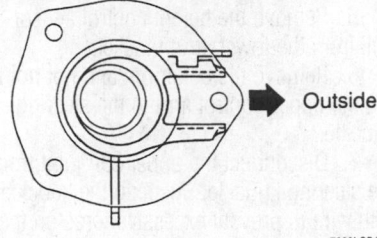

Outside

7923LGB1

Position the upper suspension support as shown when assembling the strut—ES300

17. Install the spring bumper to the strut piston rod.

18. Install the (compressed) coil spring.

19. Position the coil spring with the end butted against the gap in the lower seat.

20. Install the upper insulator and support matching the bolt of the support with the cut-off part of the insulator.

21. Install the upper suspension support.

22. Secure the upper suspension support and tighten a new strut piston rod nut to 36 ft. lbs. (49 Nm).

23. Remove the spring compressor.

24. Install the strut rod piston nut cap.

25. Install the strut and tighten the three nuts to 29 ft. lbs. (39 Nm).

26. Connect the strut to the axle carrier. Coat the nuts with engine oil and tighten the nuts and bolts to 188 ft. lbs. (255 Nm).

27. Connect the ABS harness to the strut and tighten the bolt to 48 inch lbs. (6 Nm).

28. Connect the brake line to the strut and tighten the retaining nut to 22 ft. lbs. (29 Nm).

29. Connect the spring to the lower arm and tighten the nut to 10 ft. lbs. (13 Nm).

30. Connect the LSPV to the lower arm. Tighten the nut to 9 ft. lbs. (12 Nm).

31. Install the rear wheel.

32. Lower the vehicle.

33. Install the rear seat and package tray.

GS300 AND GS400

1. Remove the front trunk compartment trim cover.

2. Raise and support the vehicle safely.

3. Remove the wheel(s).

4. Remove the brake caliper support bracket from the rear axle carrier by removing the two (2) bolts. Leave the brake line connected and position it out of the way.

➡**Never allow the brake caliper to hang freely from the brake hose.**

5. Remove the nut and disconnect the sway bar link from the lower control arm.

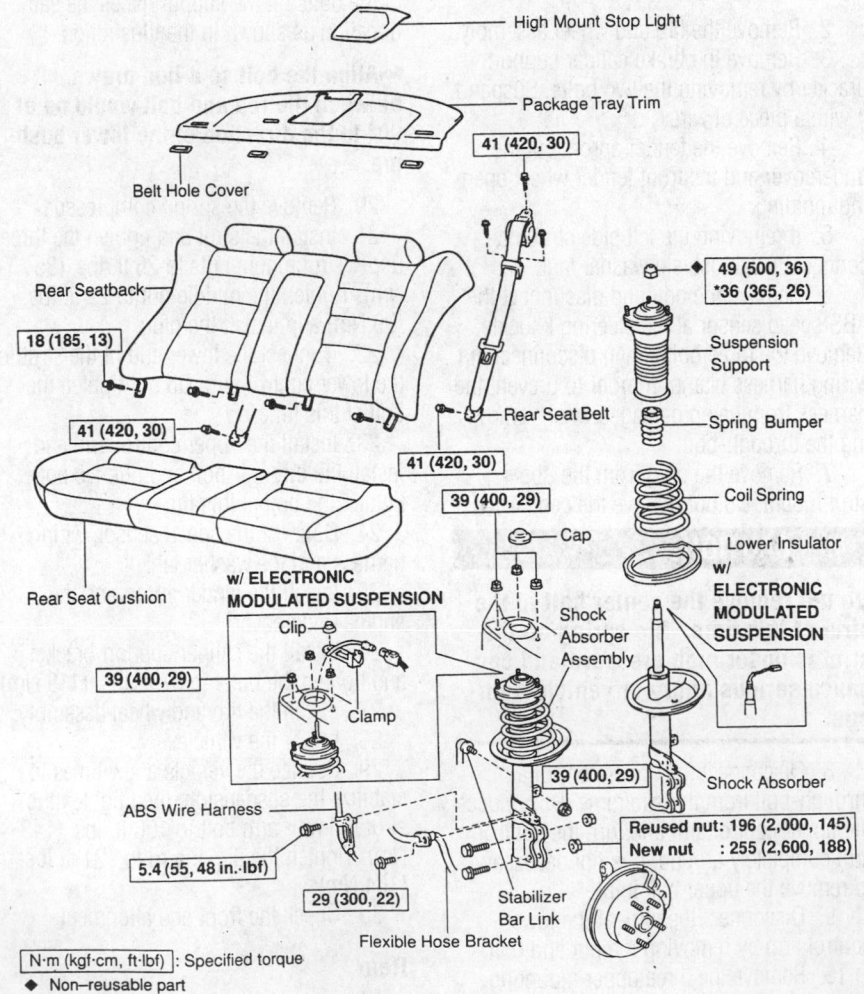

High Mount Stop Light

Package Tray Trim

41 (420, 30)

Belt Hole Cover

Rear Seatback

18 (185, 13)

41 (420, 30)

Rear Seat Belt

41 (420, 30)

39 (400, 29)

Cap

Rear Seat Cushion

w/ ELECTRONIC MODULATED SUSPENSION

Clip

39 (400, 29)

Clamp

ABS Wire Harness

5.4 (55, 48 in.·lbf)

29 (300, 22)

Stabilizer Bar Link

Flexible Hose Bracket

◆ 49 (500, 36)
* 36 (365, 26)

Suspension Support

Spring Bumper

Coil Spring

Lower Insulator
w/ ELECTRONIC MODULATED SUSPENSION

Shock Absorber Assembly

39 (400, 29)

Shock Absorber

Reused nut: 196 (2,000, 145)
New nut : 255 (2,600, 188)

N·m (kgf·cm, ft·lbf): Specified torque
◆ Non–reusable part
* For use with SST

7923LGA0

Exploded view of the rear strut and coil spring mounting—ES300

6. Remove the nut and bolt on the lower end of the strut.

7. Remove the three upper nuts and lift out the strut. Do not remove the center nut to the strut.

8. Using spring compressor SST 09727–30020 or equivalent, compress the coil spring.

9. Secure the upper support with SST 09729–22031 or equivalent and remove the strut rod retaining nut.

10. Remove the upper suspension support, upper insulator, coil spring, spring bumper and lower insulator.

To install:

11. If discarding the strut, perform the following:

   a. Fully extend the strut rod.

   b. Drill a hole in the side of the strut drain the gas inside

✷✷ CAUTION

The gas coming out is harmless, but be careful of chips which may fly up while drilling.

12. Install the lower insulator to the strut.

13. Install the spring bumper to the strut piston rod.

14. Install the (compressed) coil spring. Position the coil spring with the end butted against the gap in the lower seat.

15. Install the upper insulator and suspension support.

16. Secure the upper suspension support and tighten a new strut piston rod nut to 20 ft. lbs. (27 Nm).

17. Remove the spring compressor.

18. Install the strut to the vehicle and tighten the three upper mounting nuts to 14 ft. lbs. (20 Nm). Install the cap.

19. Install and tighten the lower strut bolt and nut to 101 ft. lbs. (137 Nm).

20. Connect the sway bar link to the lower control arm. Tighten the nut to 33 ft. lbs. (44 Nm).

21. Install the brake caliper to the rear axle carrier by installing the two (2) bolts. Tighten the bolts to 77 ft. lbs. (104 Nm).

22. Install the wheel(s).

23. Lower the vehicle.

24. Install the trunk compartment cover trim.

25. Check and adjust the vehicle alignment as necessary.

LS400—WITHOUT AIR SUSPENSION

1. Remove the rear seat cushion and seat back.

2. Remove the tray trim.

3. Raise and safely support the vehicle.

4. Remove the tire and wheel assembly.

5. Remove the rear halfshaft.

6. Disconnect the stabilizer bar link from the stabilizer bar.

7. Disconnect the ABS speed sensor and wiring harness.

8. Disconnect the brake caliper bracket from the axle carrier, leaving the brake line connected. Suspend the brake caliper aside with a piece of wire.

9. Remove the nut on the lower side of the strut. Do not remove the bolt.

10. Support the rear axle assembly with a lifting device.

11. Remove the strut cap by removing the three nuts.

12. Remove the three mounting nuts holding the strut assembly to the strut tower. Do not remove the center bolt.

✷✷ CAUTION

Do not remove the center nut to the strut at this time. The spring on the strut is under high pressure and can cause serious injury or vehicle damage.

13. Lower the rear axle assembly and remove the bolt on the lower side of the strut assembly.

14. Remove the strut assembly with the coil spring.

15. Using compressor 09727–30020 or equivalent, compress the coil spring.

16. Secure the strut housing in a vice.

17. Remove the strut rod retaining nut.

18. Remove the upper suspension support, upper insulator, coil spring, spring bumper and the lower insulator.

To install:

19. If discarding the strut, perform the following:

   a. Fully extend the strut rod.

   b. Drill a hole in the strut (about 1 in. above the strut lower mount) and drain the gas inside

➡**The gas coming out is harmless, but be careful of chips which may fly up while drilling.**

20. Install the lower insulator to the strut.

21. Install the spring bumper to the strut piston rod.

22. Install the (compressed) coil spring.

23. Position the coil spring with the end butted against the gap in the lower seat.

24. Install the upper insulator and support. Match the bolt of the support with the cut off part of the insulator.

25. Install the upper suspension support.

26. Temporarily install the upper strut rod retaining nut.

27. Rotate the suspension support so that the rod and one of the bolts on the suspension support are aligned with the lower bushing.

28. Remove the spring compressor.

29. Install the strut assembly to the vehicle and tighten the three nuts to 47 ft. lbs. (64 Nm).

30. Tighten the strut rod retaining nut to 20 ft. lbs. (27 Nm).

31. Install the strut assembly cap and install the three nuts.

32. Install the strut to the rear axle carrier. Install the bolt from the rear of the vehicle and temporarily tighten the nut.

33. Install the brake caliper and tighten the mounting bolts to 77 ft. lbs. (104 Nm).

34. Install the ABS speed sensor and wiring harness.

35. Connect the stabilizer link to the stabilizer bar and tighten the nut to 48 ft. lbs. (65 Nm).

36. Install the rear halfshaft.

37. Replace the tire and wheel assembly. Lower the vehicle.

38. Bounce the vehicle up and down to stabilize the suspension.

39. Raise and safely support the vehicle.

40. Remove the tire and wheel assembly.

41. Support the rear axle assembly with a lifting device. Tighten the lower strut bolt to 101 ft. lbs. (137 Nm).

42. Install the tire and wheel assembly.

43. Lower the vehicle.

44. Install the rear seat cushion and rear seat back.

45. Install the package tray trim.

46. Check the wheel alignment.

LS400—WITH AIR SUSPENSION

1. Remove the rear seat cushion and seat back.

2. Remove the package tray trim.

3. Remove the trunk trim panel. Move the height control switch, located in the trunk area, to the OFF position.

4. Raise and safely support the vehicle.

5. Remove the tire and wheel assembly.

6. Bleed the air system from the suspension.

7. Remove the rear halfshaft.

8. Disconnect the stabilizer links from the stabilizer bar.

9. Disconnect the ABS speed sensor and wiring harness.

10. Disconnect the brake caliper bracket from the rear axle carrier. Support the caliper using a piece of wire. Do not disconnect the brake line.

11. Place matchmarks on the height control sensor link and bracket. Disconnect

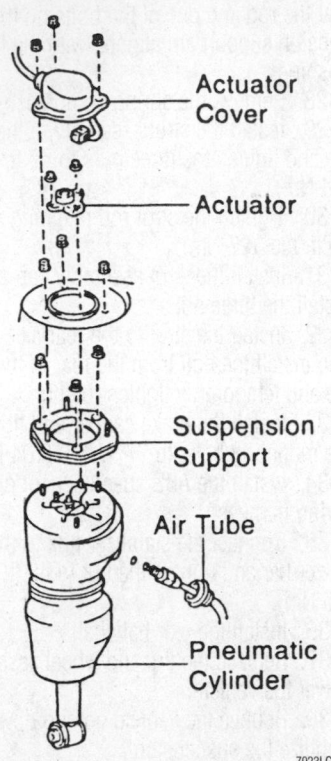

7923LGB2

Pneumatic cylinder (strut) component overview (air suspension)

the height control sensor link from the No. 1 lower control arm.

12. Remove the nut on the lower side of the shock absorber. Do not remove the bolt.

13. Support the rear axle assembly with a lifting device.

14. Remove the grommet and disconnect the air tube from the shock absorber.

15. Remove the actuator cover from the strut tower by removing the three nuts.

16. Disconnect the actuator electrical connector from the top of the strut.

17. Remove the actuator by removing the two nuts.

18. Remove the three upper mounting nuts holding the strut to the strut tower.

19. Lower the rear axle assembly.

20. Remove the bolt on the lower side of the shock absorber.

21. Remove the pneumatic cylinder strut assembly from the vehicle.

22. Remove the suspension support from the strut assembly by removing the three nuts.

To install:

23. If discarding the pneumatic cylinder, perform the following:

  a. Using a screwdriver, depressurize the air from inside the cylinder.

  b. Drill a hole in the shaded area shown in the illustration and remove the gas inside.

➡**The gas coming out is harmless, but be careful of chips which may fly up when drilling.**

24. Install the suspension support to the pneumatic cylinder (strut) and tighten the nuts to 27 ft. lbs. (36 Nm).

25. Install the strut assembly to the vehicle and tighten the upper mounting nuts to 47 ft. lbs. (64 Nm).

26. Match the holes in the pneumatic cylinder with the holes in the suspension control actuator.

27. Install the actuator and tighten the mounting nuts to 69 inch lbs. (8 Nm).

28. Install the actuator cover and tighten the three nuts to 18 ft. lbs. (25 Nm).

29. Install new O-rings and connect the air line to the shock absorber. Tighten the fitting to 13 ft. lbs. (18 Nm).

30. Install the strut to the rear axle carrier. Insert the bolt from the vehicle's rear and temporarily tighten the nut.

31. Align the matchmarks and connect the height control sensor link to the No. 1 lower control arm. Tighten the mounting nut to 48 inch lbs. (5 Nm).

32. Install the rear brake caliper to the rear axle carrier and tighten the mounting bolts to 77 ft. lbs. (104 Nm).

33. Install the ABS speed sensor and wiring harness.

34. Connect the stabilizer bar link and tighten the nut to 48 ft. lbs. (65 Nm).

35. Install the halfshaft.

36. Connect the actuator electrical connector to the top of the strut.

37. Stabilize the suspension by:

  a. Install the tire and lower the vehicle.

  b. Move the height control switch to the ON position. Start the engine an fill the pneumatic cylinder with air.

  c. Bounce the vehicle up and down several times to stabilize the suspension.

38. Turn the suspension height control to the OFF position.

39. Raise and safely support the vehicle.

40. Remove the tire and wheel assembly.

41. Support the rear axle carrier with a lifting device. Tighten the lower strut bolt to 101 ft. lbs. (137 Nm).

42. Replace the tire and wheel assembly.

43. Lower the vehicle.

44. Install the package tray trim.

45. Install the rear seat cushion and seat back.

46. Turn the suspension control switch to the ON position.

47. Check the wheel alignment.

SC300 AND SC400

1. Raise the rear of the vehicle and support it with safety stands.

2. Remove the wheel(s).

3. Remove the brake caliper support bracket by removing the two bolts. Leave the brake line connected and position it aside.

4. Remove the nut and bolt on the lower end of the strut.

5. Remove the cap nut on the upper end of the strut. Remove the three upper nuts and lift out the strut. Do not remove the center nut from the strut.

✳✳ CAUTION

Do not remove the center nut on the strut at this time. The spring on the strut is under high pressure and can cause serious injury.

6. Using compressor 09727–30020 or equivalent, compress the coil spring.

7. Secure the strut housing with two nuts and a bolt as shown in the illustration and secure it in a vice.

8. Secure the upper support with 09729–22031 or equivalent and remove the strut rod retaining nut.

9. Remove the upper suspension support, upper insulator, coil spring, spring bumper and the lower insulator.

To install:

10. If discarding the strut, perform the following:

  a. Fully extend the strut rod.

  b. Drill a hole in the strut in the shaded area shown in the illustration and drain the gas inside

➡**The gas coming out is harmless, but be careful of chips which may fly up while drilling.**

11. Install the lower insulator to the strut.

12. Install the spring bumper to the strut piston rod.

13. Install the (compressed) coil spring.

14. Position the coil spring with the end butted against the gap in the lower seat.

15. Install the upper insulator and support matching the bolt of the support with the cut off part of the insulator.

16. Install the upper suspension support.

17. Secure the upper suspension support and tighten the new strut piston rod nut to 20 ft. lbs. (27 Nm).

18. Remove the spring compressor.

19. Install the strut rod piston nut cap.

20. Install the strut and the three nuts.

Tighten the nuts to 19 ft. lbs. (25 Nm). Install the cap.

21. Install the lower bolt to hold the strut to the lower control arm. Do not tighten the bolt at this time.

22. Install the caliper support bracket and tighten the bolts to 77 ft. lbs. (104 Nm).

23. Install the wheel(s).

24. Lower the vehicle.

25. Bounce the vehicle several times to normalize the suspension.

26. Support the lower arm.

27. Tighten the lower strut mounting bolt to 106 ft. lbs. (143 Nm).

28. Check the alignment and adjust as necessary.

Upper Ball Joint

REMOVAL & INSTALLATION

The upper ball joint is an integral part of the upper arm and is not replaced separately. The upper ball joint replacement is accomplished by replacing the upper arm.

Upper Control Arm

REMOVAL & INSTALLATION

GS300

1. Disconnect the negative battery cable from the battery.

2. Raise and safely support the vehicle.

3. Remove the wheel.

4. Remove the strut and coil spring assembly as follows:

   a. Loosen the three upper strut mounting nuts.

   b. Loosen, but do not remove, the upper strut rod nut.

✳✳ CAUTION

DO NOT completely remove the upper strut nut at this time.

   c. Remove the brake caliper, leaving the line attached and secure it out of the way.

➡**Never allow the brake caliper to hang freely from the brake hose.**

   d. Disconnect the ABS speed sensor and harness.

   e. Remove the cotter pin and nut from the upper control arm.

   f. Using SST 09610–20010 or equivalent, disconnect the upper control arm from the steering knuckle.

   g. Disconnect the stabilizer bar from the link and remove the bracket.

   h. Remove the cotter pin and nut from the lower control arm.

   i. Disconnect the strut from the lower suspension arm.

   j. Remove the three upper strut mounting nuts and remove the strut.

5. Remove the mounting bolts holding the upper control arm to the frame.

6. Remove the upper control arm from the vehicle.

   To install:

7. Install the upper suspension arm and tighten the mounting bolts to 39 ft. lbs. (53 Nm).

8. Install the strut and spring assembly as follows:

   a. Install the strut and torque the upper retaining nuts to 41 ft. lbs. (56 Nm).

   b. Torque the new upper strut rod nut to 20 ft. lbs. (27 Nm).

   c. Connect the strut to the lower arm and temporarily tighten the nut and bolt.

   d. Install the stabilizer bar bracket and torque the bolts to 21 ft. lbs. (28 Nm).

   e. Connect the stabilizer bar to the link and torque the bolts to 29 ft. lbs. (39 Nm).

   f. Connect the upper suspension arm to the steering knuckle. Torque the nut to 64 ft. lbs. (87 Nm) and install a new cotter pin.

   g. Install the ABS speed sensor and torque the bolt to 69 inch lbs. (8 Nm).

   h. Install the caliper.

9. Install the front wheel.

10. Lower the vehicle.

11. Bounce the vehicle several times to stabilize the suspension.

12. Torque the lower strut bolt and nut to 116 ft. lbs. (157 Nm).

13. Check the front wheel alignment.

LS400

1. Raise and safely support the vehicle.

2. Remove the wheel.

3. Remove the strut or if equipped with air suspension, remove the pneumatic cylinder.

4. Disconnect the ABS speed sensor wire harness from the upper control arm by removing the bolt.

5. Remove the mounting bolts holding the upper control arm to the vehicle.

6. Remove the upper control arm.

   To install:

7. Install the upper control arm and torque the two mounting bolts to 83 ft. lbs. (113 Nm).

8. Connect the ABS speed sensor wire harness to the upper control arm with the attaching bolt.

9. Install the strut or if equipped with air suspension, install the pneumatic cylinder.

10. Install the wheel.

11. Lower the vehicle.

12. Check and adjust the wheel alignment as necessary.

SC300 and SC400

1. Raise the front of the vehicle and support it on safety stands.

2. Remove the wheel.

3. Remove the caliper support bracket by removing the two bolts. Leave the brake line connected and suspend it aside.

4. Remove the rotor.

5. Remove the front fender splash shield, fender liner and wheel opening molding.

6. If removing the left side arm, remove the washer tank.

7. Remove the bolt and disconnect the ABS speed sensor from the steering knuckle. Remove the three bolts and disconnect the wire harness clamp.

8. Remove the cotter pin and the nut from the upper ball joint; press the upper ball joint from the knuckle.

9. Remove the through-bolt, nut and the upper control arm.

   To install:

10. Install the upper control arm. Connect the upper control arm to the subframe and install the through-bolt. Do not torque the bolt at this time.

➡**The upper control arm mounting bolts are not torqued until the suspension has been assembled and vehicle is on the ground.**

11. Install the ball joint to the knuckle and tighten the nut to 76 ft. lbs. (103 Nm). Install a new cotter pin.

12. Connect the wire harness and ABS speed sensor. Tighten the speed sensor to knuckle bolt to 69 inch lbs. (8 Nm).

13. Install the washer tank, the fender liner, splash shield and molding.

14. Install the rotor.

15. Install the caliper support bracket and torque the bolts to 87 ft. lbs. (118 Nm).

16. Install the wheel.

17. Lower the vehicle.

18. Bounce the suspension several times to set the suspension.

19. Support the lower arm and tighten the upper control arm through-bolt and nut to 121 ft. lbs. (164 Nm).

20. Check the front wheel alignment and adjust as necessary.

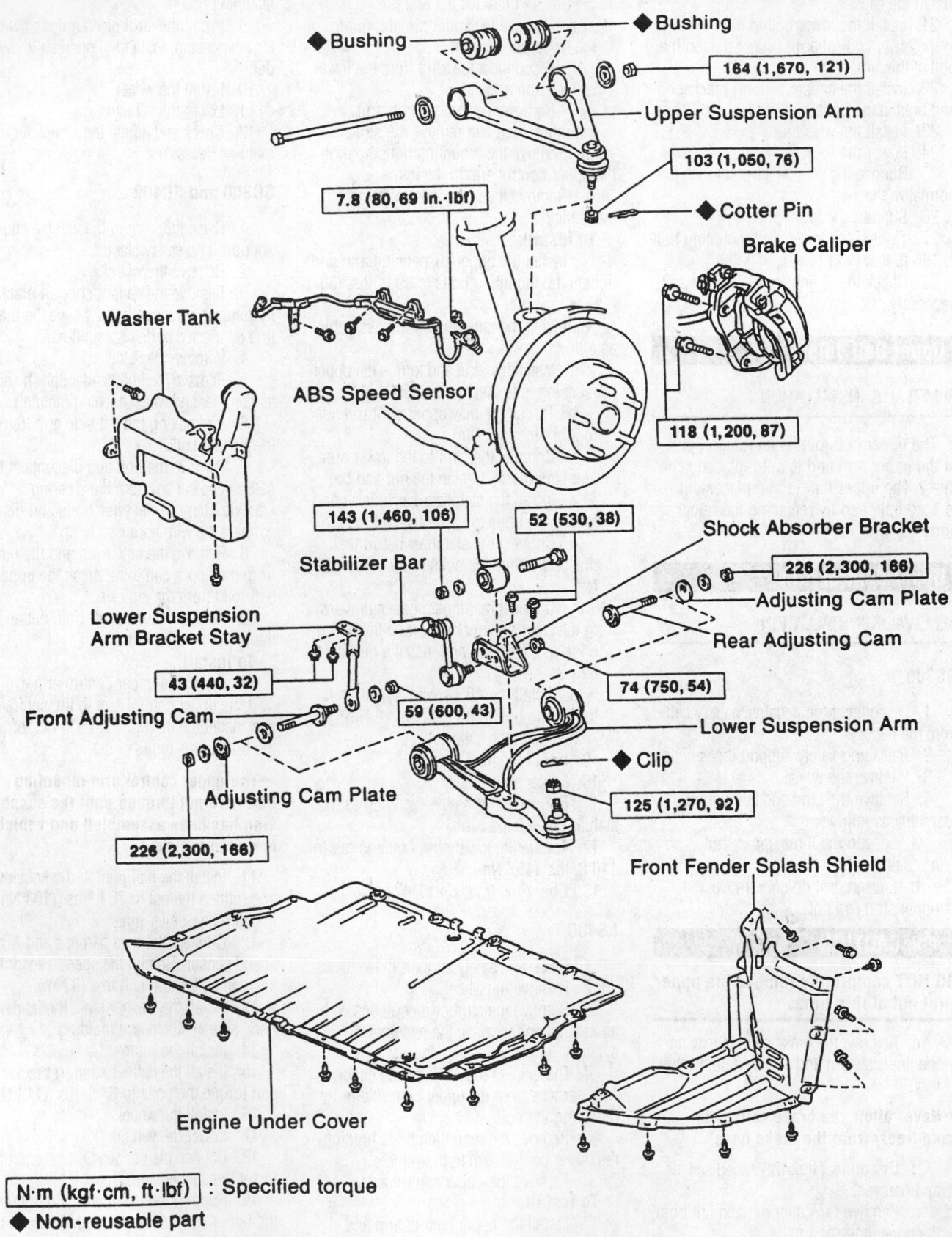

◆Bushing   ◆Bushing

164 (1,670, 121)

Upper Suspension Arm

103 (1,050, 76)

7.8 (80, 69 in.·lbf)

◆Cotter Pin

Brake Caliper

Washer Tank

ABS Speed Sensor

118 (1,200, 87)

143 (1,460, 106)   52 (530, 38)

Shock Absorber Bracket

226 (2,300, 166)

Stabilizer Bar

Adjusting Cam Plate

Rear Adjusting Cam

Lower Suspension
Arm Bracket Stay

43 (440, 32)

74 (750, 54)

Front Adjusting Cam

59 (600, 43)

Lower Suspension Arm

◆Clip

Adjusting Cam Plate

125 (1,270, 92)

226 (2,300, 166)

Front Fender Splash Shield

Engine Under Cover

N·m (kgf·cm, ft·lbf) : Specified torque

◆ Non-reusable part

7923LGC7

Exploded view of the front suspension control arms and related components—SC300 and SC400 models

Lower Ball Joint

REMOVAL & INSTALLATION

ES300

※※ CAUTION

The Supplemental Inflatable Restraint (SIR) system must be disarmed before removing the ball joint. Failure to do so may cause accidental deployment of the air bag, resulting in unnecessary SIR system repairs and/or personal injury.

1. Disconnect the negative battery cable.
2. Raise the front of the vehicle and support it safely.
3. Remove the front wheel(s).
4. Remove side fender apron seal.
5. Remove the steering knuckle with the axle hub, from the vehicle.
6. Pry the dust deflector from the knuckle.
7. Remove the cotter pin and the nut from the ball joint stud.
8. Using SST 09628–62011 or equivalent, remove the lower ball joint from the steering knuckle.

To install:
9. Install the lower ball joint onto the steering knuckle and tighten nut to 90 ft. lbs. (123 Nm). Install new cotter pin.
10. Align the hole in the dust deflector with the ABS speed sensor. Using the appropriate driver, install a new dust deflector.
11. Install the steering knuckle and hub onto the vehicle.
12. Install the fender apron seal.
13. Install the front wheel(s).
14. Connect the negative battery cable.

GS300 and GS400

※※ CAUTION

The Supplemental Inflatable Restraint (SIR) system must be disarmed before removing the ball joint. Failure to do so may cause accidental deployment of the air bag, resulting in unnecessary SIR system repairs and/or personal injury.

1. Disconnect the negative battery cable.
2. Raise and support the vehicle safely.
3. Remove the wheel(s).
4. Remove the caliper, leaving the brake line connected and suspend it out of the way.

➡ **Never allow the brake caliper to hang freely from the brake hose.**

5. Remove the rotor.
6. Remove the ABS speed sensor and harness.
7. Disconnect the tie rod end from the arm on the lower ball joint.
8. Remove the cotter pin and nut. Using SST 09610–20012 or equivalent, disconnect the upper control arm from the steering knuckle.
9. Remove the cotter pin and nut. Using SST 09628–62011 or equivalent, disconnect the steering knuckle from the lower control arm.
10. Remove the steering knuckle and ball joint assembly from the vehicle.
11. Remove the two ball joint mounting bolts, then remove the ball joint from the steering knuckle.

To install:
12. Install the ball joint and tighten the bolts to 83 ft. lbs. (113 Nm).
13. Connect the steering knuckle to the lower and upper suspension arms. Tighten the lower control arm nut to 95 ft. lbs. (127 Nm) and install a new cotter pin. Tighten the upper control arm to 64 ft. lbs. (87 Nm) and install a new cotter pin.
14. Connect the tie rod end to the ball joint arm. Tighten the nut to 64 ft. lbs. (87 Nm) and install a new cotter pin.
15. Install the rotor.
16. Install the caliper.
17. Install the ABS speed sensor and harness. Tighten the sensor retaining bolt to 69 inch lbs. (8 Nm).
18. Install the wheel(s).
19. Lower the vehicle and connect the negative battery cable.
20. Connect the negative battery cable.
21. Check the front wheel alignment.

LS400

1. If equipped with air suspension, move the height control switch (located in the trunk) to the **OFF** position.
2. Raise and safely support the vehicle.
3. Remove the tire and wheel assembly.
4. Remove the ABS speed sensor and wiring harness from the steering knuckle.
5. Disconnect the brake caliper support bracket by removing the two bolts. Leave the brake line connected. Support the caliper aside by using a piece of wire.
6. Loosen the two lower ball joint mounting bolts.

➡ **Do not remove the bolts.**

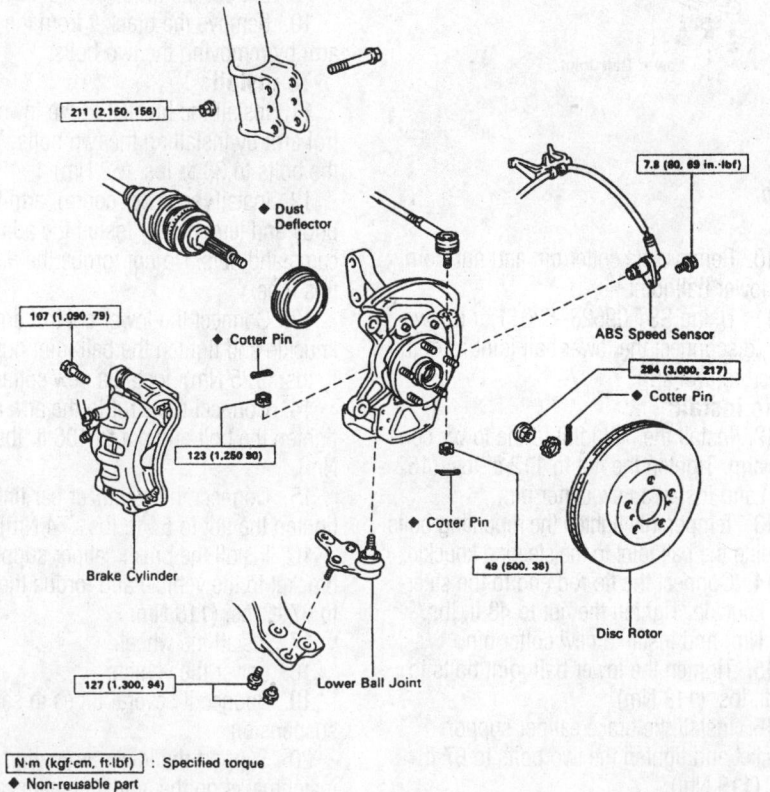

211 (2,150, 156)

Dust Deflector

107 (1,090, 79)

◆ Cotter Pin

7.8 (80, 69 in.·lbf)

ABS Speed Sensor

294 (3,000, 217)

◆ Cotter Pin

123 (1,250 90)

◆ Cotter Pin

Brake Cylinder

49 (500, 36)

Disc Rotor

127 (1,300, 94)

Lower Ball Joint

N·m (kgf·cm, ft·lbf) : Specified torque
◆ Non-reusable part

7923LGB3

Exploded view of the lower suspension—ES300

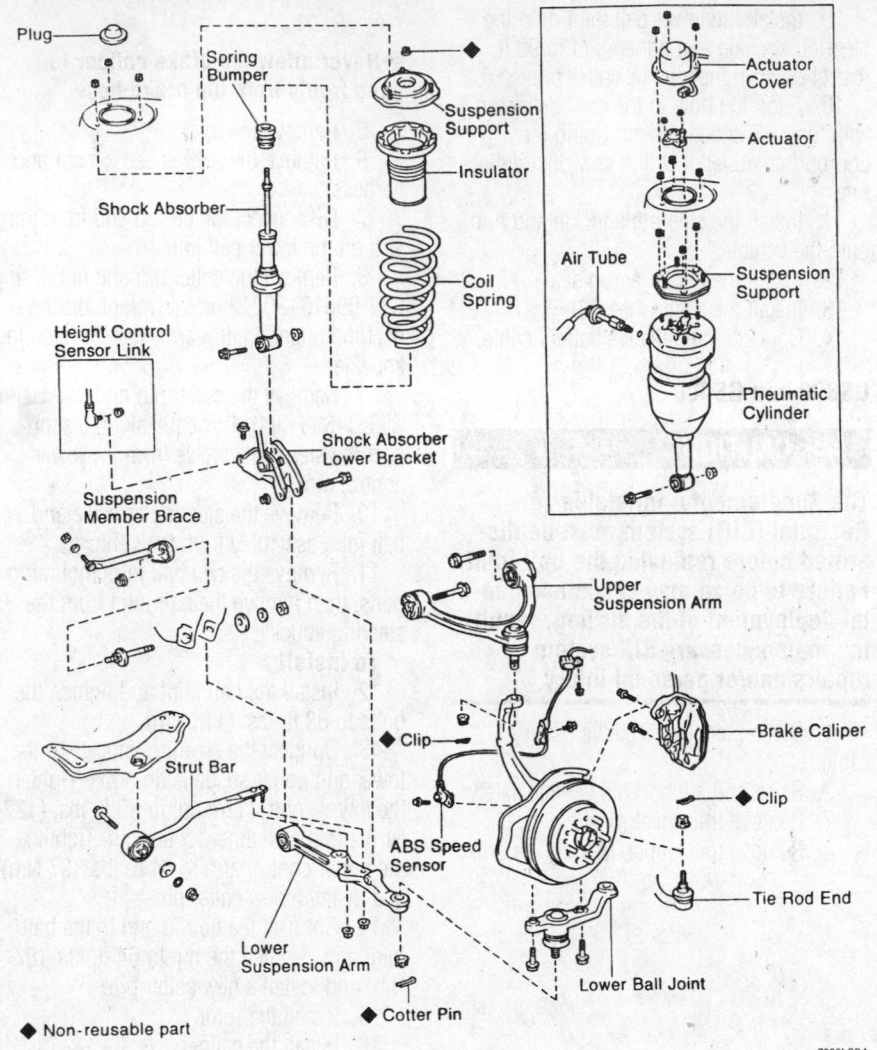

Exploded view of the lower ball joint mounting—LS400

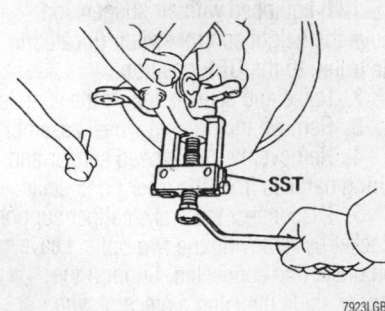

Disconnecting the ball joint from the lower suspension arm—LS400

7. Remove the clip and nut from the tie rod end.

8. Disconnect the tie rod end from the steering arm with the proper tool.

9. Remove the lower ball joint mounting bolts from the steering knuckle.

10. Remove the cotter pin and nut from the lower ball joint.

11. Using SST 09628–62011 or equivalent, disconnect the lower ball joint from the lower control arm.

To install:

12. Install the ball joint to the lower control arm. Tighten the nut to 112 ft. lbs. (152 Nm) and install a new cotter pin.

13. Temporarily tighten the mounting bolts holding the ball joint to the steering knuckle.

14. Connect the tie rod end to the steering knuckle. Tighten the nut to 48 ft. lbs. (65 Nm) and install a new cotter pin.

15. Tighten the lower ball joint bolts to 83 ft. lbs. (113 Nm).

16. Install the brake caliper support bracket and tighten the two bolts to 87 ft. lbs. (118 Nm).

17. Install the ABS speed sensor and wiring harness to the steering knuckle.

18. Install the wheel.

19. Lower the vehicle.

20. Turn the height control switch **ON**.

SC300 and SC400

The lower ball joint is not replaceable. If the lower ball joint is defective, replace the lower arm and ball joint as an assembly, as follows:

1. Raise the front of the vehicle and support it on safety stands.

2. Remove the wheel and the engine undercover.

3. Remove the caliper support bracket from the vehicle by removing the two bolts. Support the caliper and bracket with a wire. Do not let the assembly hang from the brake line.

4. Remove the nut and disconnect the stabilizer bar from the lower control arm.

5. Remove the cotter pin and nut from the lower ball joint. Press the lower ball joint out of the steering knuckle.

6. Disconnect the lower end of the strut by removing the nut and bolt.

7. Remove the nut, two bolts and the front lower arm bracket stay.

8. Matchmark the front and rear adjustment cams to the body and then remove the nuts and adjusting cams.

9. Lift out the lower control arm.

10. Remove the bracket from the control arm by removing the two bolts.

To install:

11. Install the bracket to the lower control arm by installing the two bolts. Torque the bolts to 38 ft. lbs. (52 Nm).

12. Install the lower control arm to the body and temporarily install the adjusting cams and nuts. Do not torque the nuts at this time.

13. Connect the lower control arm to the knuckle and tighten the ball joint nut to 92 ft. lbs. (125 Nm). Install a new cotter pin.

14. Connect the strut to the arm and tighten the bolt and nut to 106 ft. lbs. (143 Nm).

15. Connect the stabilizer bar link and tighten the nut to 54 ft. lbs. (74 Nm).

16. Install the brake caliper support bracket to the vehicle and torque the bolts to 87 ft. lbs. (118 Nm.

17. Install the wheel.

18. Lower the vehicle.

19. Bounce it several times to set the suspension.

20. Support the lower arm, align the matchmarks on the adjusting cams and tighten the nuts to 166 ft. lbs. (226 Nm).

21. Check the front end wheel alignment.

Wheel Bearings

ADJUSTMENT

Check the backlash in bearing shaft direction and the axle hub deviation. Maximum for backlash should be 0.0020 in. (0.05mm) and for axle hub deviation 0.020 in. (0.05mm).

➡**The front and rear wheel bearings are non-adjustable. If the wheel bearing is out of specifications, replace the wheel bearing.**

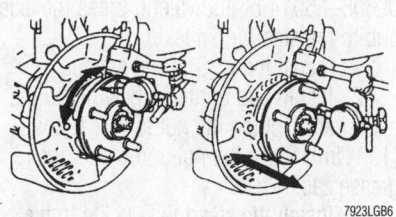

Checking wheel bearings for excessive play

REMOVAL & INSTALLATION

Front

ES300

1. Disconnect the negative battery cable.
2. Raise the vehicle and support safely.
3. Remove the front wheels.
4. Remove the fender apron seal.
5. Remove the cotter pin and lock cap from the end of the halfshaft.
6. While applying the front brakes, remove the halfshaft locknut.
7. Remove the brake caliper and use a wire to support it out of the way.

➡**Never allow the caliper to hang freely from the brake hose.**

8. Matchmark the rotor to the hub and remove the rotor.
9. If equipped with ABS brakes, remove the ABS speed sensor from the steering knuckle.
10. Loosen the nuts on the lower end of the strut.
11. Disconnect and separate the tie rod end from the steering knuckle.
12. Disconnect the lower control arm from the ball joint by removing the three bolts.

13. Remove the driveshaft from the axle hub. Secure the shaft out of the way using a wire. Be careful not to damage the shaft boot or ABS sensor rotor.
14. Remove the two nuts on the lower end of the strut and remove the steering knuckle.
15. Clamp the steering knuckle in a vise with soft jaws to protect the knuckle.
16. Carefully pry the dust deflector from the hub.
17. Remove the ball joint from the steering knuckle.
18. Using SST 09520–00031 or equivalent, remove the hub from the knuckle.
19. Using SST 09950–00020 or equivalent, remove the inner race from the hub.
20. Remove the four bolts to the dust cover, then remove dust cover.
21. Using snapring pliers, remove the snapring.
22. Take the inner race (removed from the hub) and install it on the outside of the bearing.
23. Using a bearing driver, drive the bearing from the steering knuckle.

To install:
24. Clean bearing seating surfaces with a clean, dry rag.
25. Using a press and SST 09608–

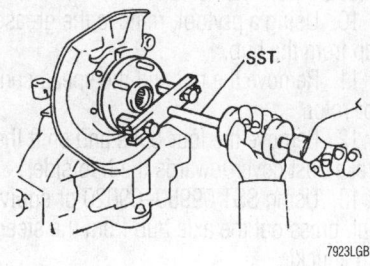

Removing the axle hub from the steering knuckle—ES300

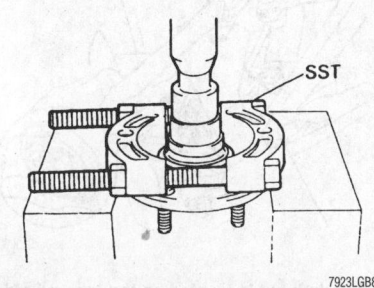

Using SST 09950–00020 or equivalent, remove the inner race from the hub— ES300

32010 or equivalent, install the bearing into the knuckle.
26. Install the snapring.
27. Install the dust cover. Tighten the four bolts to 74 inch lbs. (8.3 Nm).
28. Press the hub into the steering knuckle.
29. Install the lower ball joint to the steering knuckle. Tighten the nut to 90 ft. lbs. (123 Nm) and install a new cotter pin.
30. Align the hole in the dust deflector and the hole for the ABS speed sensor and install the dust deflector.
31. Position the knuckle to the lower strut and install the bolts.
32. Install the lower ball joint to the lower arm. Tighten the bolts to 94 ft. lbs. (127 Nm).
33. Connect the tie rod end to the steering knuckle. Tighten the nut to 36 ft. lbs. (49 Nm).
34. Install and tighten the nuts on the lower strut to 156 ft. lbs. (211 Nm).
35. Install the ABS speed sensor. Tighten the mounting bolt to 69 inch lbs. (8 Nm).
36. Align the matchmark and install the rotor on the hub. Install the brake caliper. Tighten the mounting bolts to 79 ft. lbs. (107 Nm).
37. Have a helper apply the brakes and install the axle locknut. Tighten the nut to 217 ft. lbs. (294 Nm). Install the lock cap and a new cotter pin.
38. Install front fender apron seal.
39. Install the wheel.
40. Turn the wheel by hand, verify that the wheel turns without noise and without binding.
41. Lower the vehicle.

GS300 AND GS400

1. Disconnect the negative battery cable.
2. Raise and support the vehicle safely.
3. Remove the front wheel.
4. Remove the caliper, leaving the brake line connected and suspend it out of the way.

➡**Never allow the brake caliper to hang freely from the brake hose.**

5. Remove the rotor.
6. Remove the ABS speed sensor and harness.
7. Disconnect the tie rod from the arm on the lower ball joint.
8. Remove the cotter pin and nut. Disconnect the upper suspension arm from the steering knuckle.
9. Remove the cotter pin and nut. Using SST 09628–62011 or equivalent, disconnect the steering knuckle from the lower control arm.

10. Remove the steering knuckle from the vehicle.

11. Remove the ball joint from the steering knuckle by removing the two bolts.

12. Pry out the front hub grease cap.

13. Clamp the hub in a soft jaw vise.

14. Using a hammer and chisel, loosen the staked part of the locknut.

15. Remove the locknut.

16. Remove the ABS speed sensor rotor.

➡ **Do not scratch the serration's of the sensor rotor.**

17. Remove the brake dust cover bolts and shift the cover toward the outside.

18. Using a puller SST 09950–40010 or equivalent, remove the hub from the steering knuckle.

19. Using the same puller, remove the inner bearing race from the hub shaft.

20. Remove the oil seal from the knuckle.

21. Remove the bearing snapring from the steering knuckle.

22. Using a press and a bearing driver, press the bearing from the steering knuckle.

To install:

23. Using a press and a bearing driver, press a new bearing into the steering knuckle.

➡ **If the inner race and balls come loose from the bearing outer race, be sure to install them on the same side as before.**

24. Install the snapring.

25. Install a new outside inner race and tap in the new seal using SST 09608–32010 or equivalent. Tap the seal until it is flush with the end surface of the steering knuckle.

26. Install the brake dust cover to the knuckle and tighten the bolts to 74 inch lbs. (8 Nm).

27. Press the hub into the steering knuckle.

28. Install the ABS speed sensor rotor.

29. Install the axle hub locknut. Tighten the nut to 147 ft. lbs. (199 Nm) and stake it.

30. Install the grease cap to the steering knuckle by tapping lightly around the circumference of the cap with a hammer.

31. Install the ball joint to the steering knuckle. Tighten the two bolts to 83 ft. lbs. (113 Nm).

32. Connect the steering knuckle to the upper and lower suspension arms. Tighten the upper nut to 64 ft. lbs. (87 Nm) and the lower nut to 95 ft. lbs. (127 Nm). Install a new cotter pin on the lower nut. Install the clip on the upper suspension arm nut.

33. Connect the tie rod end to the steering knuckle. Tighten the nut to 64 ft. lbs. (87 Nm) and install a new cotter pin.

34. Install the rotor, disc brake pads and the brake caliper.

35. Install the ABS speed sensor and harness. Tighten the sensor retaining bolt to 69 inch lbs. (8 Nm).

36. Install the wheel.

37. Lower the vehicle and connect the negative battery cable.

38. Check the front wheel alignment.

LS400

1. If equipped with air suspension, move the height control switch in the trunk area to the OFF position.

2. Raise and safely support the vehicle.

3. Remove the front tire and wheel assembly.

4. Disconnect the brake caliper bracket from the steering knuckle, leaving the brake line connected. Support the caliper with a piece of wire.

5. Remove the brake rotor.

6. Remove the ABS speed sensor from the steering knuckle.

7. Remove the steering knuckle from the lower ball joint by removing the two bolts.

8. Remove the steering knuckle from the upper ball joint.

9. Remove the steering knuckle with the axle hub from the vehicle.

10. Using a prytool, remove the grease cap from the hub.

11. Remove the nut and the speed sensor rotor.

12. Remove the four bolts and shift the brake dust cover towards the hub side.

13. Using SST 09950–00020 or equivalent, press out the axle hub from the steering knuckle.

14. Using SST 09950–00020 or equivalent, press out the outside inner race from the axle.

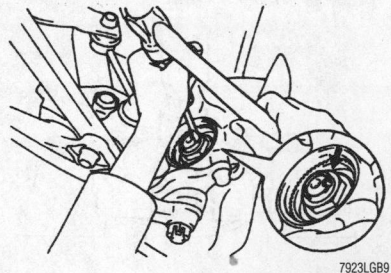

Axle hub nut is located on the inboard side of the knuckle assembly—LS400, SC300 and SC400

7923LGB9

15. Using SST 09308–00010 or equivalent, remove the oil seal from the steering knuckle.

16. Remove the snapring and using SST 09950–60010 or equivalent, remove the bearing from the steering knuckle.

To install:

17. Using the SST 09608–35014 or equivalent, install the bearing to the steering knuckle. Install the snapring.

18. Install the inner race (outside) and press in a new oil seal until it is flush with the end surface of the steering knuckle.

19. Install the brake dust cover to the steering knuckle and tighten the bolts to 74 inch lbs. (8.4 Nm).

20. Using SST 09608–32010 and 09608–35014 or equivalent, press the axle hub to the steering knuckle.

21. Install the ABS speed sensor.

22. Install and tighten a new nut to the axle shaft. Tighten the nut to 147 ft. lbs. (199 Nm). Stake the nut and install the grease cap.

23. Install the steering knuckle to the lower ball joint and tighten the bolts to 83 ft. lbs. (113 Nm).

24. Install the steering knuckle to the upper ball joint and tighten the nut to 48 ft. lbs. (65 Nm).

25. Install brake rotor.

26. Install the brake caliper and tighten the two bolts to 87 ft. lbs. (118 Nm).

27. Install the speed sensor to the steering knuckle.

28. Install the front tire and wheel assembly.

29. Lower the vehicle.

30. If equipped with air suspension, turn the height control switch to the ON position.

SC300 AND SC400

1. Raise and safely support the vehicle.

2. Remove the front tire and wheel assembly.

3. Remove the brake caliper support bracket, leaving the brake line connected and support it using a piece of wire.

4. Remove the rotor by removing the two screws.

5. Disconnect the ABS speed sensor.

6. Remove the cotter pin and nut and disconnect the tie rod from the steering knuckle.

7. Remove the cotter pin and nut and disconnect the steering knuckle from the upper control arm.

8. Remove the clip and nut and press the knuckle off the lower control arm.

9. Remove the steering knuckle from the vehicle.

10. Pry the hub bearing cap from the steering knuckle. Using a hammer and chisel, loosen the staked part of the hub nut and remove it.

11. Remove the ABS sensor rotor.

12. Remove the four bolts and shift the brake dust shield toward the hub (outside).

13. Using a two-arm puller, remove the axle hub from the knuckle.

14. With a puller, remove the inner bearing race from the axle hub. Pry out the oil seal.

15. Remove the bearing snapring, then position the inner race above the bearing on the inner side. Press the bearing out.

To install:

16. Press the bearing into the knuckle. If the inner race and balls come loose from the outer race, be sure to install them on the same side as before.

17. Install the snapring and inner race, then tap in a new oil seal until it is flush with the end surface of the knuckle.

18. Install the brake dust cover and tighten the bolts to 74 inch lbs. (8.3 Nm).

19. Press the hub into the knuckle and install the speed sensor.

20. Install a new locknut and tighten it to 147 ft. lbs. (199 Nm). Stake the nut with a chisel. Tap the bearing cap into place.

21. Connect the knuckle to the upper control arm and tighten the nut to 76 ft. lbs. (103 Nm). Install a new cotter pin.

22. Connect the knuckle to the lower control arm and tighten the nut to 92 ft. lbs. (125 Nm). Install a new clip.

23. Connect the tie rod end to the steering knuckle with the nut. Tighten the nut to 36 ft. lbs. (49 Nm). Install a new cotter pin.

24. Install the rotor by installing the two screws.

25. Install the caliper support bracket and tighten the bolt to 87 ft. lbs. (118 Nm).

26. Connect the speed sensor to the knuckle and tighten the bolt to 69 inch lbs. (8 Nm).

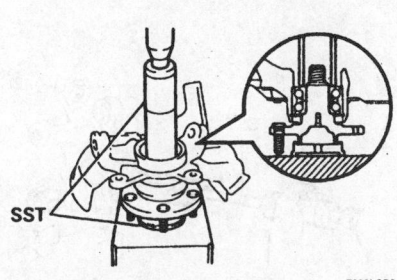

Pressing the hub into the knuckle—SC300 and SC400

27. Install the front wheel and tighten the lug nuts to 76 ft. lbs. (103 Nm).

28. Lower the vehicle.

29. Check the front end alignment and ABS speed sensor signal.

Rear

ES300

1. Raise and safely support the vehicle.

2. Remove the rear tire and wheel assembly.

3. If equipped with rear disc brakes, remove the caliper mounting bolts. Leave the brake line connected and suspend the assembly out of the way.

4. Remove the brake rotor or drum.

5. Remove the four bolts and pull off the rear axle hub. Remove the O-ring.

➡ **If it is necessary to replace the hub or bearing, replace the components as an assembly.**

To install:

6. Position the hub on the carrier and tighten the bolts to 59 ft. lbs. (80 Nm).

7. Install the rotor or drum.

8. If equipped with rear disc brakes, install the caliper and tighten the bolts to 34 ft. lbs. (64 Nm).

9. Install the wheel.

10. Lower the vehicle to the ground.

11. Road test the vehicle for proper operation.

GS300 AND GS400

1. Disconnect the negative battery cable.

2. Raise and safely support the vehicle.

3. Remove the rear tire and wheel assembly.

4. Disconnect the brake caliper support from the rear axle carrier and support it with a piece of wire.

5. Place matchmarks on the disc brake rotor and the axle hub. Remove the brake rotor.

6. Remove the speed sensor.

7. Remove the rear halfshaft.

8. Remove the parking brake shoes.

9. Remove the parking brake cable as follows:

    a. Remove the four bolts at the backing plate.

    b. Remove the shoe guide plate set bolt.

    c. Using a 14mm hexagon wrench, remove the hexagon bolt on the back of the backing plate.

    d. Slide the backing plate to the outside and disconnect the parking brake cable.

10. Remove the strut rod by removing the two bolts and two nuts.

11. Remove the No. 1 lower control arm as follows:

    a. Remove the parking brake cable bracket.

    b. Remove the exhaust support bracket.

    c. Place matchmarks on the adjusting cam and rear control crossmember.

    d. Remove the nut, adjusting cam and the washer to the No. 1 control arm.

    e. Disconnect the No. 1 lower control arm from the crossmember.

12. Disconnect the No. 2 lower control arm as follows:

    a. Loosen the nut holding the lower control arm to the axle carrier.

    b. Using SST 09610–20012 or equivalent, press the No. 2 lower control arm from the axle carrier.

    c. Remove the nut, then remove the No. 2 lower control arm from the axle carrier.

13. Remove the axle carrier as follows:

    a. Remove the nut holding the upper control arm to the axle carrier.

    b. Using SST 09628–62011 or equivalent, disconnect the upper control arm and remove the axle carrier.

    c. With the axle carrier out of the vehicle, remove the nut holding the No. 1 control arm to the axle carrier.

    d. Using SST 09628–10011 or equivalent, disconnect the No. 1 lower control arm from the axle carrier.

14. Remove the dust deflector.

15. Using a two-arm puller, remove the axle hub from the carrier. Remove the backing plate.

16. Pull out the inner race (outside), then remove the oil seal. Remove the snapring.

17. Install the inner race over the bearing and press the bearing out.

To install:

18. Install the bearing to the axle carrier.

➡ **If the inner races come loose from the bearing outer race, be sure to install them on the same side as before.**

19. Install the snapring. Install the inner race (outside) and a new oil seal.

20. Install the backing plate. Install the inner race (inside) and press in the axle hub with the proper tools.

21. Install the inner oil seal. Align the holes for the speed sensor in the dust deflector and axle carrier. Install the dust deflector.

22. Connect the No. 1 lower arm to the axle carrier and install a new nut. Tighten the nut to 43 ft. lbs. (59 Nm).

23. Install the upper control arm to the axle carrier. Tighten the new nut and bolt to 80 ft. lbs. (109 Nm).

24. Connect the No. 2 lower control arm to the axle carrier and tighten a new nut to 110 ft. lbs. (150 Nm).

25. Connect the No. 1 lower control arm to the rear crossmember. Tighten the nut to 136 ft. lbs. (184 Nm).

26. Connect the strut rod to the axle carrier. Tighten the nuts and bolts to 134 ft. lbs. (184 Nm).

27. Connect the parking brake cable and slide the backing plate to the inside. Install the hex bolt and tighten it to 132 ft. lbs. (180 Nm).

28. Install the shoe guide plate set bolt. Tighten the bolt to 13 ft. lbs. (18 Nm).

29. Install the four hub bolts and tighten them to 19 ft. lbs. (26 Nm).

30. Install the bolts at the speed sensor and tighten them to 69 inch lbs. (8 Nm).

31. Install the parking brake shoes.

32. Install the halfshafts. Apply the brakes and tighten the locknut to 213 ft. lbs. (289 Nm).

33. Install the brake rotor.

34. Connect the brake caliper support to the rear axle carrier. Tighten the bolts to 77 ft. lbs. (104 Nm).

35. Install the rear tire and wheel assembly.

36. Lower the vehicle and bounce it a few times to stabilize the suspension.

37. Connect the negative battery cable.

LS400

1. If equipped with air suspension, move the height control switch in the trunk area to the **OFF** position.

2. Disconnect the negative battery cable from the battery.

3. Raise and safely support the vehicle.

4. Remove the rear wheel(s).

5. If equipped with height control suspension, disconnect the height control sensor link from the lower control arm by removing the nut.

6. Remove the ABS speed sensor and wiring harness.

7. Disconnect the brake caliper bracket from the rear axle carrier by removing the two bolts. Support the caliper with a piece of wire.

8. Place matchmarks on the disc brake rotor and the axle hub. Remove the brake rotor.

9. Remove the parking brake shoes and cable.

10. Remove the halfshaft as follows:

a. Remove the cotter pin, lock cap and the nut holding the halfshaft to the rear axle.

b. Remove the suspension member brace by removing the two bolts.

c. Place matchmarks on the halfshaft and the side gear shaft. Remove the hexagon bolts and washers with the proper tool.

d. Hold the inboard joint side of the halfshaft so the outboard joint side does not bend too much. Tap the end of the halfshaft with a rubber mallet and disengage the axle hub.

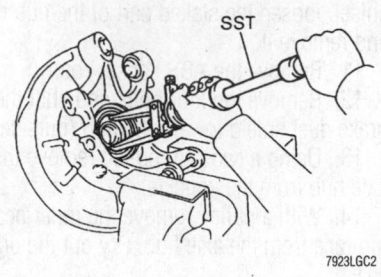

Removing the oil seal (inner)—LS400

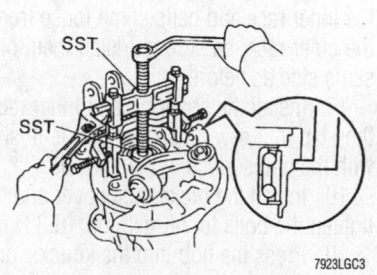

Removing the axle hub from the axle carrier—LS400

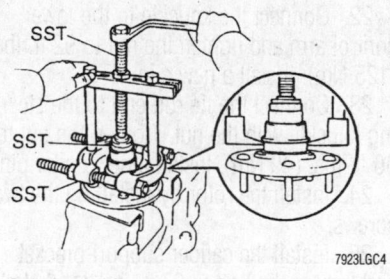

Removing the inner race (outside) from the axle hub—LS400

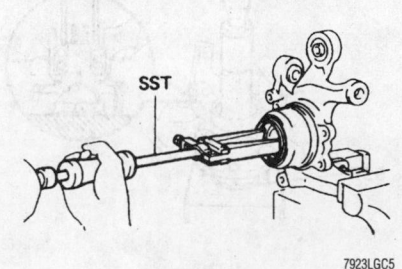

Removing the oil seal (outer)—LS400

◆Dust Deflector
◆Oil Seal
Axle Carrier
Backing Plate
◆Bearing
Snap Ring
◆Oil Seal
Axle Hub
◆Inner Race

◆ **Non-reusable part**

Exploded view of the axle carrier — GS300

e. Remove the halfshaft from the vehicle.

11. Remove the strut rod as follows:

a. Remove the nut and bolt and disconnect the strut rod from the rear axle carrier.

b. Remove the nut, bolt, and the strut rod from the body.

c. Remove the strut rod.

12. Remove the lower control arms as follows:

a. Place matchmarks on the adjusting cam and body for the No. 1 control arm. Remove the nut and adjusting cam.

b. Remove the nut on the axle carrier side of the No. 1 lower control arm.

c. Using a tie rod removal tool separate the control arm from the axle carrier. Remove the No. 1 lower control arm.

d. Disconnect the stabilizer bar link from the No. 2 lower control arm.

e. Place matchmarks on the adjusting cam and body. Remove the nut and adjusting cam from the No. 2 lower control arm.

f. Remove the nut and bolt holding the No. 2 lower control arm to the axle carrier.

g. Remove the No. 2 control arm from the vehicle.

13. Remove the nut and bolt on the lower side of the strut assembly.

14. Remove the two upper control arm set nuts and bolts.

15. Remove the axle carrier with the upper control arm.

16. Secure the axle carrier in a vise.

17. Remove the nut holding the upper control arm to the axle carrier and remove the control arm.

18. Using a suitable prytool, remove the dust deflector.

19. Using SST 09308–00010 or equivalent, remove the oil seal.

20. Remove the two bolts and nuts and shift the backing the plate towards the hub side (outside).

21. Using SST 09950–40010 and SST 09950–60010 or equivalent, press the axle hub out.

22. Remove the backing plate.

23. Using a press, remove the inner race (outside) from the axle hub.

24. Using SST 09308–00010 or equivalent, remove the oil seal (outer) from the axle.

25. Using snapring pliers, remove the snapring from inside the axle housing.

26. Using SST 09950–60010 and 09950–70010 or equivalent, remove the bearing from the axle housing.

To install:

27. Using SST 09527–17011 and 09608–32010 or equivalent, install a new bearing to the axle housing.

28. Using snapring pliers, install the snapring to the axle carrier.

29. Install a new outer oil seal. Coat the oil seal lip with multipurpose grease.

30. Install the backing plate to the axle housing. Do not install the bolts or nuts at this time.

31. Install the inner race (inside) to the axle housing.

32. Using a press, install the axle hub to the axle housing.

33. Place the backing plate in position install the bolts and nuts. Tighten the bolts and nuts to 43 ft. lbs. (59 Nm).

34. Install a new oil seal (inner) to the axle housing. Coat the oil seal lip with multipurpose grease.

35. Using a press, install a new dust deflector. Be sure to align the hose for the ABS speed sensor in the dust deflector and axle carrier.

36. Install the upper control arm to the axle carrier by installing the nut. Tighten the nut to 80 ft. lbs. (108 Nm).

37. Install the axle carrier and upper control arm to the vehicle as an assembly.

38. Install the two upper control arm set bolts and tighten the bolts to 121 ft. lbs. (164 Nm).

39. Install the bolt and nut holding the strut to the axle carrier. Tighten to 101 ft. lbs. (137 Nm).

40. Install the lower control arms as follows:

a. Install the bolt and nut connecting the No. 2 lower control arm to the axle carrier. Tighten the bolt to 60 ft. lbs. (81 Nm).

b. Install the nut and adjusting cam to hold the No. 2 lower control arm to the body. Align the adjusting cam marks and tighten the nut to 57 ft. lbs. (78 Nm).

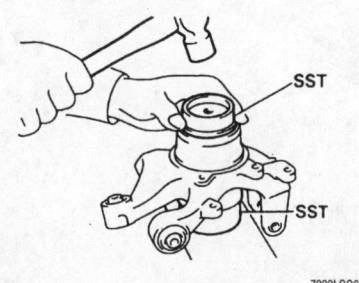

Installing the oil seal (outer)—LS400

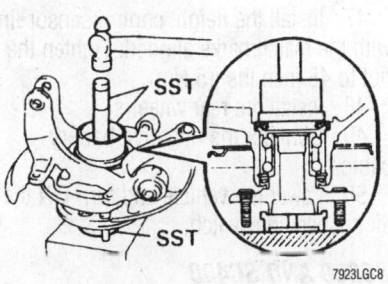

Installing the oil seal (inner)—LS400

c. Install the stabilizer bar link to the No. 2 lower control arm and tighten the nut to 48 ft. lbs. (65 Nm).

d. Install the No. 1 lower control arm to the axle carrier and body. Install the nut to hold the No. 1 lower control arm to the axle carrier. Tighten the nut to 43 ft. lbs. (59 Nm).

e. Install the nut and adjusting cam to hold the No. 1 lower control arm to the body. Align the matchmarks and tighten the nut to 57 ft. lbs. (78 Nm).

41. Install the strut rod as follows:

a. Install the strut rod to the axle carrier and body. Install the bolt and nut to hold the strut rod to the body. Tighten to 57 ft. lbs. (78 Nm).

b. With the strut connected to the axle carrier, install the bolt and nut to hold the strut rod to the axle carrier. Tighten to 136 ft. lbs. (184 Nm)

42. Install the parking brake shoes and cable.

43. Install the axle shaft to the vehicle as follows:

a. Insert the outboard joint side of the halfshaft and align the matchmarks on the side gear shaft and the halfshaft.

b. Coat the threads with clean oil and install the hexagon bolts. Tighten bolts to 61 ft. lbs. (83 Nm).

c. Install the suspension member brace with the two bolts. Tighten the two bolts to 37 ft. lbs. (50 Nm).

d. Install the nut to hold the halfshaft to the rear axle. Tighten the nut to 213 ft. lbs. (289 Nm).

e. Install the lock cap and cotter pin.

44. Install the brake disc to the axle hub with the matchmarks aligned. Install the two screws and tighten the screws to 48 inch lbs. (5 Nm).

45. Install the brake caliper to the vehicle and install the two bolts. Tighten the bolts to 77 ft. lbs. (104 Nm).

46. Install the ABS speed sensor and wiring harness.

47. Install the height control sensor link with the matchmarks aligned. Tighten the nut to 48 inch lbs. (5 Nm).

48. Install the rear wheel(s).

49. Connect the negative battery cable.

50. Lower the vehicle and turn **ON** the air suspension switch.

SC300 AND SC400

1. Raise and safely support the vehicle. Remove the rear tire and wheel assembly.

2. Disconnect the brake caliper support bracket from the rear axle carrier and support it with a piece of wire.

3. Place matchmarks on the disc brake rotor and the axle hub. Remove the brake rotor.

4. Remove the speed sensor.

5. Remove the rear halfshaft.

6. Remove the parking brake shoes.

7. Remove the two bolts at the parking brake cable. Remove the two hub bolts and the hex bolt. Slide the backing plate to the outside and disconnect the parking brake cable.

8. Disconnect the strut rod at the axle carrier.

9. Remove the nut, then press out the No. 1 lower suspension arm.

10. Remove the nut, then press out the No. 2 lower suspension arm.

11. Remove the nut, then press out the upper suspension arm. Remove the axle carrier.

12. Remove the dust deflector and pull out the oil seal.

13. Using a two arm puller, remove the axle hub from the carrier.

14. Remove the backing plate.

15. Press the inner race (outside) from the hub. Then, remove the oil seal and the snapring.

16. Place the inner race (outside) over the bearing and tap out the bearing and inner race (inside).

To install:

17. Install the bearing to the axle carrier.

➡**If the inner races come loose from the bearing outer race, be sure to install them on the same side as before.**

18. Install the snapring, the inner race (outside) and a new oil seal.

19. Install the backing plate. Install the inner race (inside) and press in the axle hub with the proper tools.

20. Install the inner oil seal. Align the holes for the speed sensor in the dust deflector and axle carrier. Install the dust deflector.

21. Install the upper arm to the axle carrier. Tighten the nut and bolt to 80 ft. lbs. (109 Nm).

22. Connect the No. 2 lower arm to the carrier and tighten a new nut to 110 ft. lbs. (150 Nm).

23. Connect the No. 1 lower arm to the carrier and tighten a new nut to 43 ft. lbs. (59 Nm).

24. Connect the strut rod to the carrier. Do not tighten the bolt at this time.

25. Connect the parking brake cable and slide the backing plate to the inside. Install the hex bolt and tighten it to 132 ft. lbs. (180 Nm). Install the two hub bolts and tighten them to 19 ft. lbs. (26 Nm).

26. Install the two bolts at the parking brake cable and tighten them to 69 inch lbs. (8 Nm). Install the parking brake shoes and the ABS sensor.

27. Install the halfshafts. Tighten the locknut to 213 ft. lbs. (289 Nm).

28. Install the brake rotor.

29. Connect the brake caliper to the rear axle carrier by installing the two bolts. Tighten the bolts to 77 ft. lbs. (104 Nm).

30. Replace the rear tire and wheel assembly. Lower the vehicle and bounce it a few times to stabilize the suspension. Raise the vehicle again, support the axle carrier and tighten the strut rod to 136 ft. lbs. (184 Nm).

MAZDA

323 • 626 • MX3 • MX6 • Miata • Millenia • Protege

ENGINE REPAIR

➡**Disconnecting the negative battery cable on some vehicles may interfere with the functions of the on board computer systems and may require the computer to undergo a relearning process, once the negative battery cable is reconnected.**

Distributor

REMOVAL

1. Disconnect the negative battery cable.

2. Remove the distributor cap and position it aside, leaving the ignition wires connected.

3. On SOHC engines, remove the air intake hose from it's position next to the distributor.

4. Disconnect the distributor electrical connector(s) from the side of the distributor.

5. Using a wrench on the crankshaft pulley, rotate the crankshaft to position the No. 1 piston on Top Dead Center (TDC) of the compression stroke; the crankshaft pul-

ley mark should align with the timing indicator and the distributor rotor should point towards the No 1. spark plug wire tower position of the cap.

6. Using chalk or paint, mark the position of the distributor housing on the cylinder head. Also mark the position of the distributor rotor in relation to the distributor housing.

7. Remove the distributor hold-down bolt(s).

8. On distributors attached to the end of the cylinder head (or inline with the camshaft), remove it by pulling it straight outward.

9. On distributors attached to the side of the cylinder head (or perpendicular with the camshaft), slowly pull it outward while watching the rotor. These distributors are gear driven and as you remove it, the gears will unmesh inside the engine, causing the rotor to rotate. when the rotor stops moving, stop pulling outward. Re-align the distributor body-to-cylinder head matchmark (do not push it back in to do this, simply rotate the body to align the marks). Place a third mark indicating the new rotor position-to-distributor body relation. When installing the distributor, align this mark and the body-to-head mark to properly position the distributor.

10. Inspect the O-ring on the distributor housing and replace it, if it is damaged or worn.

INSTALLATION

Engine Not Disturbed

1. Using engine oil, lubricate the O-ring.

2. Install the distributor, aligning the marks that were made in Step 6 of the removal procedure. Be sure to engage the drive gear or tangs with the camshaft gear or slot in the camshaft.

3. Tighten the distributor hold-down bolt(s).

4. Connect the electrical connector(s), if equipped, air intake hose. Install the distributor cap.

5. Connect the negative battery cable. Start the engine and check or adjust the ignition timing.

Engine Disturbed

1. Disconnect the spark plug wire from the No. 1 cylinder spark plug. Remove the spark plug from the No. 1 cylinder and press a thumb over the spark plug hole.

2. Using a wrench on the crankshaft pulley, rotate the crankshaft until pressure is

felt at the spark plug hole, indicating the piston is approaching TDC on the compression stroke. Continue rotating the crankshaft until the crankshaft pulley mark aligns with the timing cover indicator.

3. Place the distributor rotor in position so that it aligns with the No. 1 spark plug wire tower on the distributor cap.

4. Using engine oil, lubricate the O-ring.

5. Install the distributor. Be sure to engage the drive gear or tangs with the camshaft gear or slot. Align the mark that was made on the distributor housing with the mark that was made on the cylinder head. Tighten the distributor hold-down bolt(s).

6. Connect the electrical connector(s), if equipped, the air intake hose. Install the distributor cap.

7. Install the spark plug in the No. 1 cylinder and connect the spark plug wire.

8. Connect the negative battery cable. Start the engine and check or adjust the ignition timing.

Ignition Timing

ADJUSTMENT

➡**If the information given in the following procedures differs from that on the emission information label located in the engine compartment, follow the directions given on the label. The label often reflects production running changes made during the model year.**

Except 2.3L Engines

1. Apply the parking brake. If equipped with a manual transaxle, place the shifter in the neutral position. If equipped with an automatic transaxle, place the shift lever in **P**.

2. Locate the timing marks on the crankshaft pulley and timing belt lower cover. The engine may have to be cranked slightly to see the mark on the crankshaft pulley.

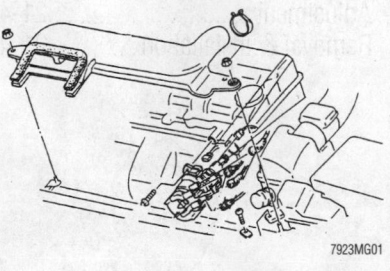

7923MG01

Exploded view of a typical side mounted distributor

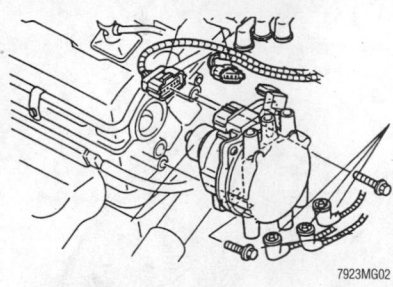

7923MG02

Exploded view of a typical end or inline mounted distributor

7923MG03

Jumper the connections shown on the data link—except 626 and MX6 with ATX

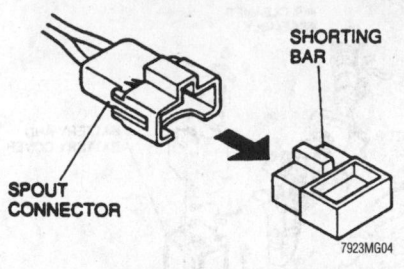

Remove the shorting bar from the spout connector—626 and MX6 with ATX

3. Start the engine and allow it to come to normal operating temperature. Be sure all accessories are **OFF**.

4. Check the idle speed and adjust, if necessary.

5. Turn the engine off.

6. On all engines except 1995–99 2.0L (FS) engines with an automatic transaxle, connect a jumper wire between the TEN terminal and the GND terminal at the underhood diagnosis connector.

7. On 1995–99 2.0L (FS) engines with an automatic transaxle, remove the shorting bar from the double wire SPOUT connector.

8. Connect an inductive timing light according to the manufacturers instructions.

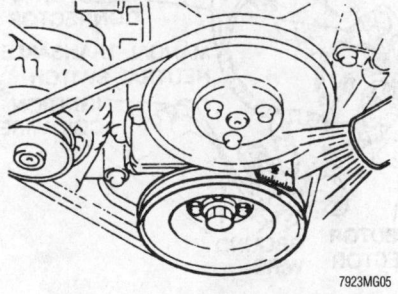

Connect an inductive timing light and aim it at the crankshaft pulley. Read the pulley mark against the scale

9. Start the engine and allow the idle to stabilize. Aim the timing light at the timing marks.

10. The mark on the crankshaft pulley should align with the specified BTDC degree mark on the timing cover scale, plus or minus 1 degree. If the marks are within alignment proceed with step 12. If the marks are not aligned, proceed to Step 11.

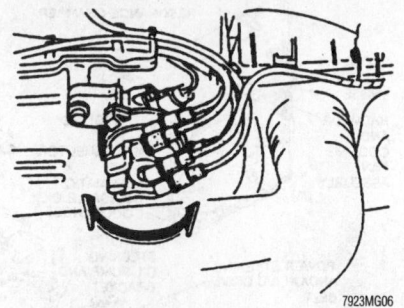

If adjustment is necessary, loosen the distributor lockbolts and rotate it until the mark is aligned

11. Loosen the distributor lockbolts just enough to turn the distributor. While aiming the timing light at the timing marks, turn the distributor until the marks are aligned. Tighten the distributor lockbolts to 14–19 ft. lbs. (19–25 Nm) and recheck the timing.

12. The ignition timing is now set. Disconnect the jumper wire from the underhood diagnosis connector or install the shorting bar from the double wire SPOUT connector.

13. Remove all test equipment.

2.3L Engines

The 2.3L engine utilizes individual ignition coils for each cylinder and the timing is controlled by the computer. Ignition timing adjustment is not possible or necessary.

Engine Assembly

REMOVAL & INSTALLATION

Except Miata

In the process of removing the engine, you will come across a number of steps which call for the removal of a separate component or system, such as "disconnect the exhaust system" or "remove the radiator." In most instances, a detailed removal procedure can be found elsewhere in this manual.

It is virtually impossible to list each individual wire and hose which must be disconnected, simply because so many different model and engine combinations have been manufactured. Careful observation and common sense are the best possible approaches to any repair procedure.

Removal and installation of the engine can be made easier if you follow these basic points:

• If you have to drain any of the fluids, use a suitable container.

• Always tag any wires or hoses, if possible, the components they came from before disconnecting them.

• Because there are so many bolts and fasteners involved, store and label the retainers from components separately in muffin pans, jars or coffee cans. This will prevent confusion during installation.

• After unbolting the transmission or transaxle, always be sure it is properly supported.

• If it is necessary to disconnect the air conditioning system, have this service performed by a qualified technician using a recovery/recycling station. If the system does not have to be disconnected, unbolt the compressor and set it aside.

• When unbolting the engine mounts, always be sure the engine is properly supported. When removing the engine, be sure that any lifting devices are properly attached to the engine. It is recommended that if your engine is supplied with lifting hooks, your lifting apparatus be attached to them.

• Lift the engine from its compartment slowly, checking that no hoses, wires or other components are still connected.

• After the engine is clear of the compartment, place it on an engine stand or workbench.

❊❊ CAUTION

When draining the coolant, keep in mind that cats and dogs are attracted by the ethylene glycol antifreeze, and are quite likely to drink any that is left in an uncovered container or in puddles on the ground. This will prove fatal in sufficient quantity. Always drain the coolant into a sealable container. Coolant should be reused unless it is contaminated or several years old.

➡**The procedure for pulling the engine requires removing the transaxle along with it. As a result, when the halfshafts are pulled from the transaxle, a special plug/side gear holding tool is recommended.**

❈❈ CAUTION

Observe all applicable safety precautions when working around fuel. Whenever servicing the fuel system, always work in a well ventilated area. Do not allow fuel spray or vapors to come in contact with a spark or open flame. Keep a dry chemical fire extinguisher near the work area. Always keep fuel in a container specifically designed for fuel storage; also, always properly seal fuel containers to avoid the possibility of fire or explosion.

1. Properly relieve the fuel system pressure. Raise and safely support the vehicle, as necessary.
2. Disconnect the battery cables and remove the battery and the battery tray.
3. Remove the hood.
4. Loosen the lug nuts on the front wheels.
5. Apply the parking brake, block the rear wheels, then raise and safely support the front of the vehicle securely on jackstands.
6. Remove the front wheels.
7. Remove the splash shield(s) from under the vehicle and drain the engine and transaxle oil as well as the coolant.

❈❈ CAUTION

The EPA warns that prolonged contact with used engine oil may cause a number of skin disorders, including cancer! You should make every effort to minimize your exposure to used engine oil. Protective gloves should be worn when changing the oil. Wash your hands and any other exposed skin areas as soon as possible after exposure to used engine oil. Soap and water, or waterless hand cleaner should be used.

8. Remove the air cleaner assembly and resonance chamber, including the air flow meter and all of the ducting. Remove the oil dip stick.
9. On the Millenia with the 2.3L engine, remove the charge air cooler, front grille, upper seal board (panels that the grille mounts to) and coolant overflow tank.

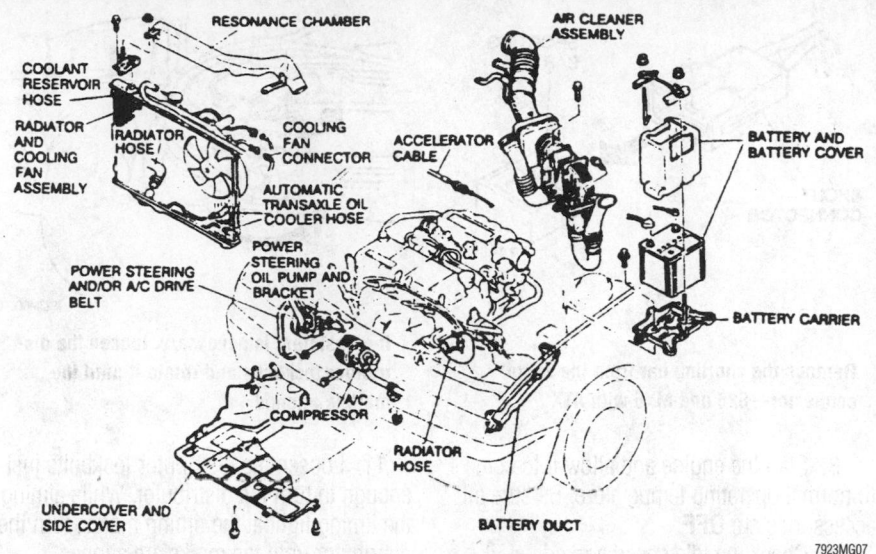

View of typically removed external components for engine removal—front wheel drive models

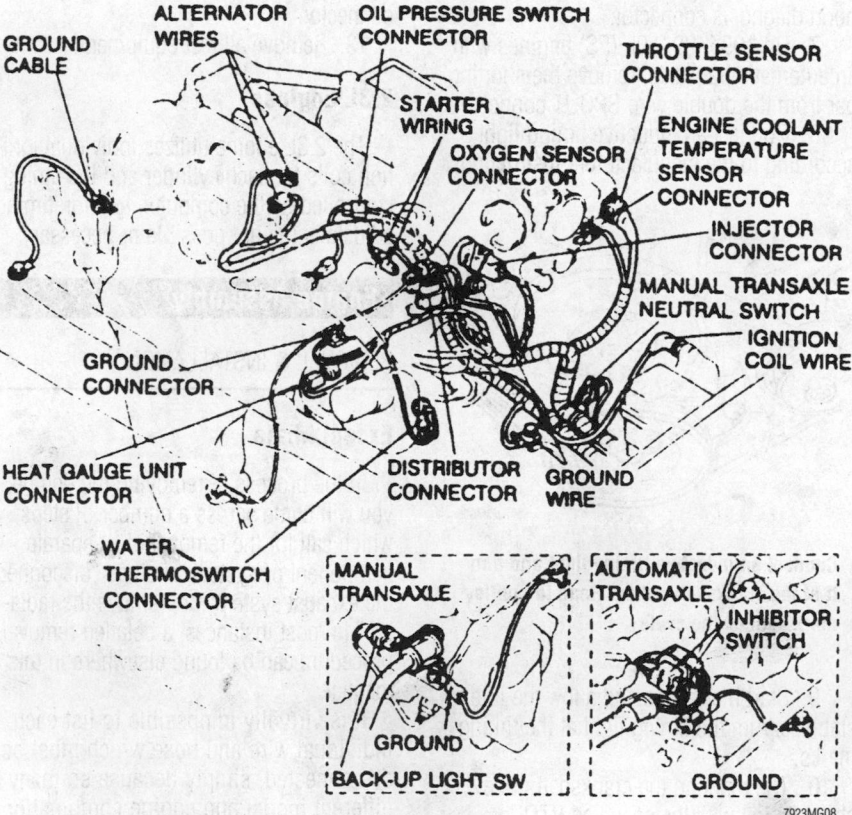

View of the common electrical harness plug disconnection points for engine removal—front wheel drive models

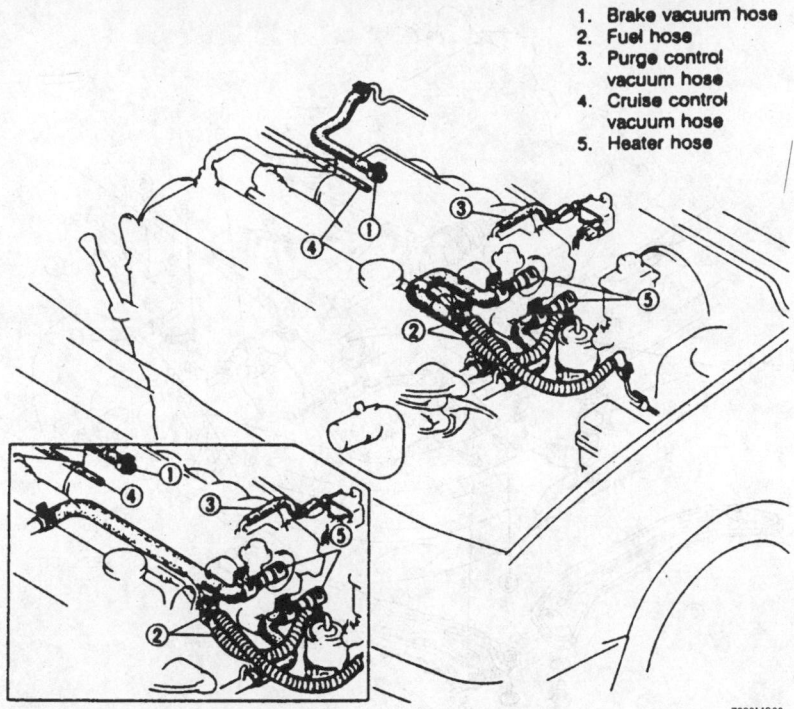

1. Brake vacuum hose
2. Fuel hose
3. Purge control vacuum hose
4. Cruise control vacuum hose
5. Heater hose

Typical vacuum, fuel and water hose disconnect points for engine removal—front wheel drive models

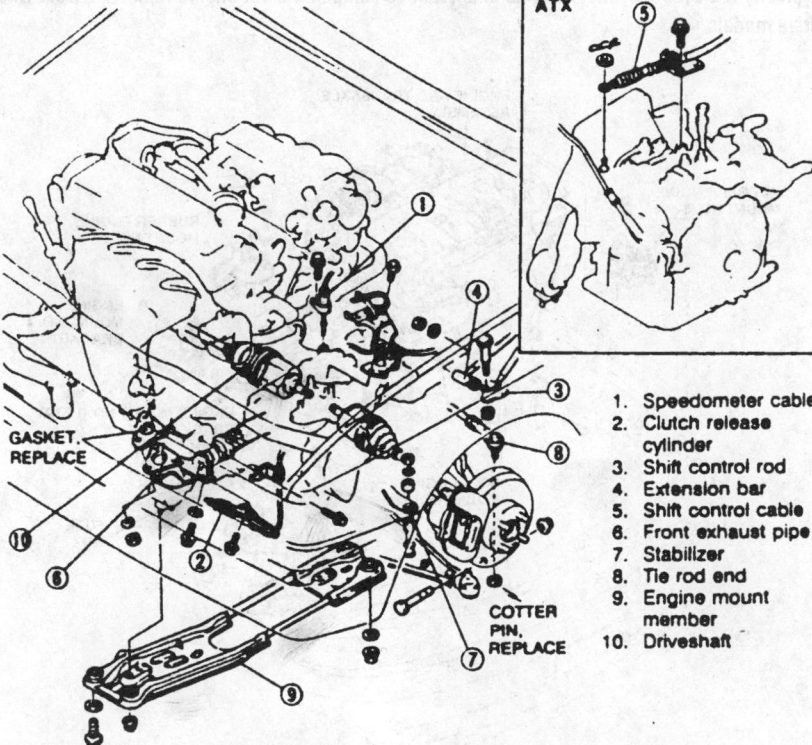

1. Speedometer cable
2. Clutch release cylinder
3. Shift control rod
4. Extension bar
5. Shift control cable
6. Front exhaust pipe
7. Stabilizer
8. Tie rod end
9. Engine mount member
10. Driveshaft

Typically removed 4-cylinder engine undervehicle components for engine removal—front wheel drive models

10. Remove the radiator hoses. If equipped with automatic transaxle, disconnect the oil cooler lines from the radiator. Disconnect the cooling fan, if equipped, radiator switch electrical connectors and remove the radiator/cooling fan assembly. On 4WD vehicles, remove the crossmember from the underside of the vehicle.

11. Disconnect the throttle and the speedometer cable.

12. Label and disconnect all vacuum hoses and wiring harnesses.

13. Disconnect the fuel supply and return hoses and the heater hoses.

14. Disconnect the exhaust pipe from the manifold. On 4WD vehicles, remove the exhaust manifold. If equipped, remove the water inlet pipe and gasket.

15. Remove the accessory drive belt or belts.

16. Without disconnecting the hydraulic hoses, remove the power steering pump and hang it from the body with wire.

17. Without disconnecting the refrigerant lines, remove the air conditioning compressor and hang it from the body with wire.

18. If equipped with manual transaxle, disconnect the clutch cable and shift control rod. If equipped with hydraulic clutch, remove the slave cylinder from the transaxle without disconnecting the hydraulic line.

19. If equipped with automatic transaxle, disconnect the shift control cable.

20. Remove the nuts and disconnect the tie rod ends from the steering knuckles. Disconnect the stabilizer bar from the lower control arms.

21. Attach an engine lifting chain to the engine lifting eyes. Attach the chain to a suitable engine hoist and raise the hoist until there is tension on the chain.

22. Remove the engine mount nuts and the engine mount member bolts and nuts and remove the engine mount member. On 4WD vehicles, remove the front transaxle mount.

➡**Be careful so the engine does not fall when removing the engine mount member.**

23. Remove the pinch bolts from the steering knuckle and pry the control arm down to slip the lower ball joint out of the knuckle.

24. If equipped, remove the bolts from the right side intermediate shaft support, using a suitable prybar, pry the intermediate shaft from the transaxle. Insert a suitable prybar between the inner CV-joint and transaxle case and carefully pry the inner CV-joints out of the transaxle. Suspend the halfshafts with wire.

25. If equipped with 4WD, mark the position of the driveshaft on the transaxle and rear axle flanges. Remove the drive-shaft, keeping all spacers, washers and bushings in order so they can be reinstalled in their original positions.

26. Remove the dynamic damper from the right side engine mount, if equipped. Remove the engine/transaxle mount nuts/bolts and right engine, if equipped, left transaxle mounts. Carefully lift the engine/transaxle assembly from the vehicle.

27. Properly support the engine/transaxle assembly. Remove the intake manifold bracket, starter, torque converter nuts, stiffener, if equipped and No. 2 engine mount. Disconnect the throttle cable.

28. If equipped with 4WD, remove the center differential lock motor as follows:

    a. Remove the set bolt and lock sensor switch.

    b. Remove the plug from the end of the motor and use a small flat bladed tool to turn the shift rod ½ turn clockwise.

    c. Remove the retaining bolts and the center differential lock motor.

29. Remove the transaxle mounting bolts and separate the transaxle from the engine.

To install:

30. Installation is the reverse of the removal procedure. Note the following important steps.

31. When possible, leave the engine mounting nuts/bolts loose (hand tight) until all mounts are aligned and bolted. This may help in aligning the engine and transmission assembly in the vehicle.

32. Install new circlips on the inner CV-joint stub shafts, if equipped, intermediate shaft. Grease the shaft splines before installing the halfshaft/intermediate shaft into the transaxle.

33. Always install new gaskets and/or O-rings. Use new self-locking nuts, especially on the exhaust.

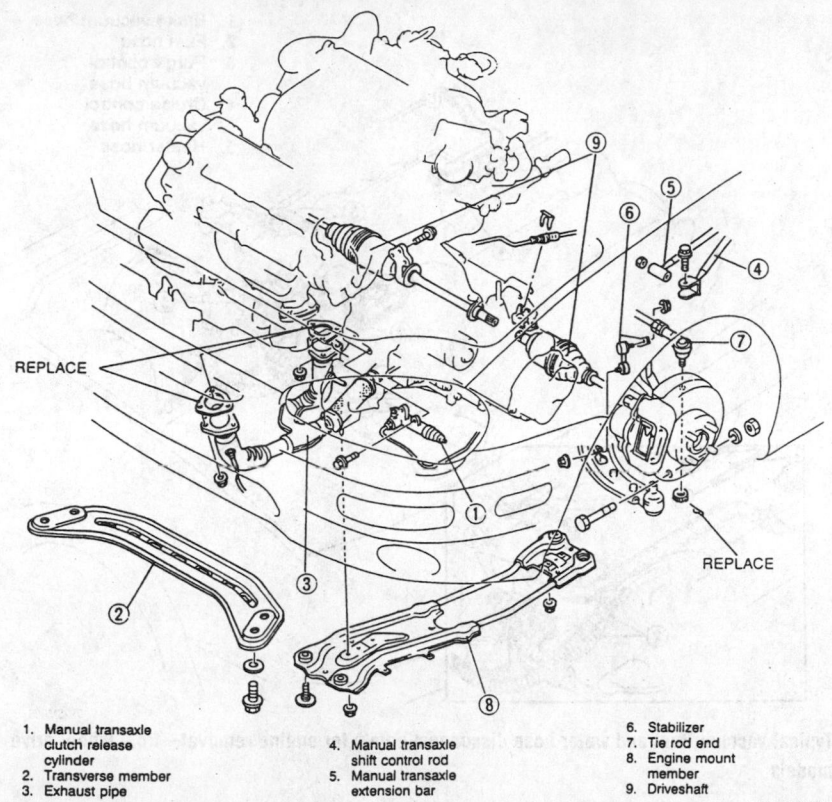

1. Manual transaxle clutch release cylinder	6. Stabilizer
2. Transverse member	7. Tie rod end
3. Exhaust pipe	8. Engine mount member
4. Manual transaxle shift control rod	9. Driveshaft
5. Manual transaxle extension bar	

7923MG11

Typically removed 6-cylinder engine undervehicle components for engine removal—front wheel drive models

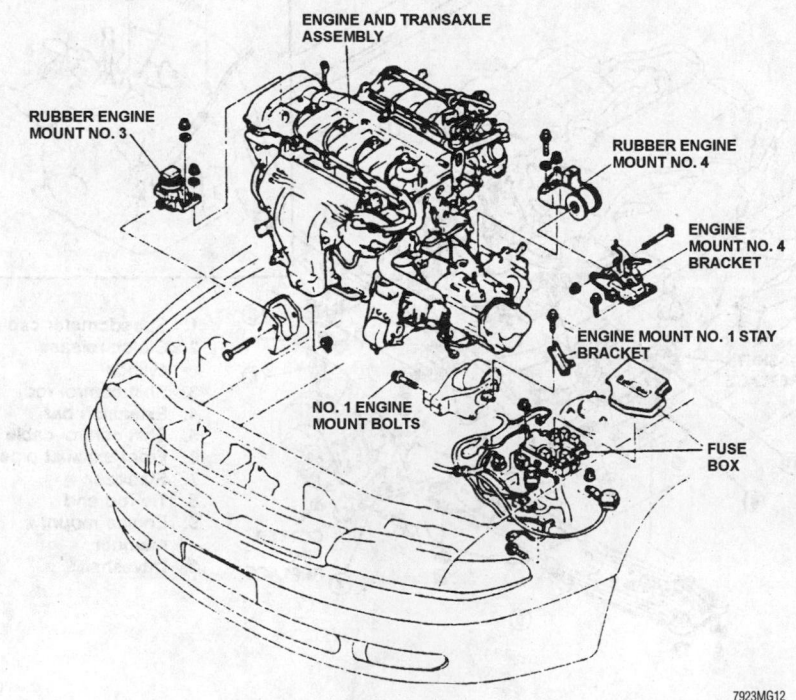

7923MG12

Once all components and mounts are unfastened, the engine is removed with the transaxle attached—front wheel drive

34. Fill the engine and the transaxle with the proper types and quantities of oil. Fill the cooling system.

35. Connect the negative battery cable, start the engine and check for leaks. Check the ignition timing and the idle speed. Check all fluid levels.

Miata

1. Properly relieve the fuel system pressure.

2. Raise the trunk lid and disconnect the negative battery cable. Remove the fresh air duct and the air cleaner/air flow meter assembly.

3. Remove the transmission.

4. Disconnect the accelerator cable from the throttle body.

5. Raise and safely support the vehicle and remove the undercover. Drain the engine oil and the coolant.

6. Disconnect the radiator hoses, including the coolant reservoir hose and the cooling fan electrical connector.

7. On automatic transmission vehicles only, remove the oil cooler hose.

8. Remove the radiator/cooling fan assembly.

9. Remove the accessory drive belts. Without disconnecting the hydraulic hoses, remove the power steering pump and secure it aside.

10. Without disconnecting the refrigerant, remove the air conditioner compressor and secure it aside.

11. Label and disconnect the following electrical connectors: Steering pressure sensor, throttle position sensor, idle air control valve, heated oxygen sensor, ignition coil, crankshaft position sensor, ground, fuel injector, alternator, oil pressure sensor and the starter.

12. Label and disconnect the following hoses: Brake vacuum, fuel, purge control vacuum, cruise control vacuum, water inlet and heater.

13. Disconnect the exhaust pipe from the exhaust manifold.

14. Install suitable lifting equipment onto the engine and be sure all hoses, wires and cables are disconnected.

15. Remove the engine mount nuts, and lift the engine from the vehicle.

To install:

16. Carefully and slowly install the engine assembly into the vehicle. Tilt the engine downward, and align the engine mounts with the crossmember mounting holes. Tighten the mount nuts to 42–57 ft. lbs. (57–78 Nm).

17. Use a new gasket and attach the exhaust pipe to the manifold. Torque the nuts to 34 ft. lbs. (46 Nm).

18. Install the following hoses: Brake vacuum, fuel, purge control vacuum, cruise control vacuum, water inlet and heater.

19. Install the following electrical connectors: Steering pressure sensor, throttle position sensor, idle air control valve, heated oxygen sensor, ignition coil, crankshaft position sensor, ground, fuel injector, alternator, oil pressure sensor and the starter.

20. Install the air conditioner compressor and power steering pump. Install and adjust the drive belts.

21. Install the radiator and fans, and connect all cooling system hoses.

22. Connect and adjust the accelerator cable as required.

23. Install the air cleaner and air flow meter assembly.

24. Install the transmission.

25. Check to be sure all wiring and hoses are properly connected. Fill and bleed the cooling system. Fill the engine and transmission with the proper type and quantity of oil.

26. Connect the negative battery cable, start the engine and bring to normal operating temperature. Check for leaks.

27. Check the ignition timing and idle speed. Check all fluid levels.

Water Pump

REMOVAL & INSTALLATION

1.5L, 1.6L and 1.8L (BP) Engines

1. Disconnect the negative battery cable. Drain the cooling system.

2. Remove the timing belt covers, and remove the timing belt.

3. Disconnect the coolant inlet pipe and gasket.

4. Remove the timing belt idler pulleys still attached to the water pump.

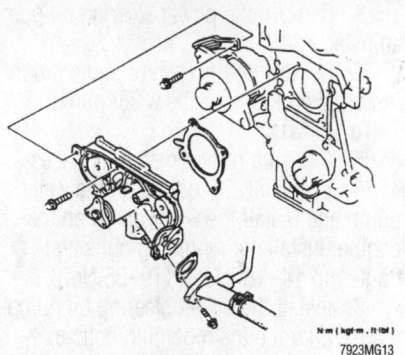

Exploded view of the water pump assembly—1.5L, 1.6L and 1.8L (BP) engines

5. Remove the water pump mounting bolts, and remove the water pump.

To install:

6. Clean all gasket mating surfaces.

7. Install a new rubber seal on the water pump.

8. Using a new gasket, install the water pump on the engine. Tighten the mounting bolts to 14–18 ft. lbs. (19–25 Nm). Tighten the bolt from the water pump to the alternator bracket to 28–38 ft. lbs. (38–51 Nm).

9. Install the timing belt idler pulleys that were removed.

10. Install the coolant inlet pipe, using a new gasket. Tighten the bolts to 14–18 ft. lbs. (19–25 Nm).

11. Install the timing belt and the timing belt covers.

12. Fill and bleed the cooling system. Connect the negative battery cable, start the engine and bring to normal operating temperature. Check for leaks.

2.0L Engines

1. Disconnect the negative battery cable. Drain the cooling system.

2. Remove the timing belt.

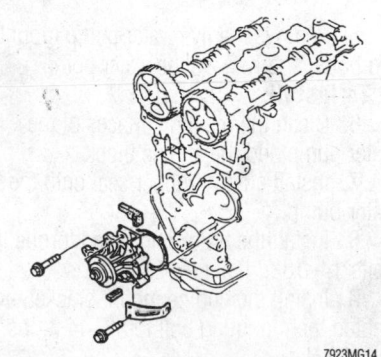

Exploded view of the water pump assembly—2.0L engine

3. Remove the power steering oil pump adjuster.

4. Remove the five water pump mounting bolts and remove the water pump.

To install:

5. Clean all gasket mating surfaces.

6. Install a NEW gasket on the water pump and install the water pump on the engine. Install the mounting bolts and tighten to 14–18 ft. lbs. (19–25 Nm).

7. Install the power steering oil pump adjuster, torque the mounting bolts to 12–16 ft. lbs. (16–22 Nm).

8. Install the timing belt.

9. Connect the negative battery cable. Fill and bleed the cooling system.

10. Start the engine and bring to normal operating temperature. Check for leaks.

1.8L (K8) and 2.5L Engines

1. Disconnect the negative battery cable. Drain the cooling system.

2. Remove the timing belt.

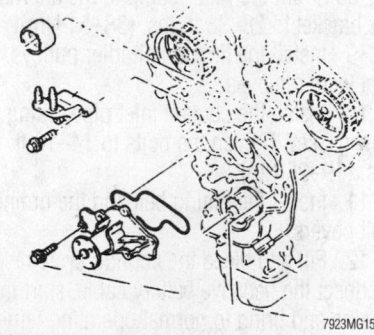

7923MG15

Exploded view of the water pump assembly—1.8L (K8) and 2.5L engines, 2.3L engine is similar

3. Remove the No.3 engine mount bracket.

4. Position a drain pan under the water pump.

5. Remove the five water pump mounting bolts, and remove the water pump.

To install:

6. Clean the mating surfaces of the water pump and the engine block.

7. Install a NEW rubber seal onto the water pump.

8. Install the water pump and torque the bolts 14–18 ft. lbs. (19–25 Nm).

9. Install the engine mount bracket, and tighten the mounting bolt to 32–44 ft. lbs. (44–60 Nm).

10. Install the timing belt.

11. Connect the negative battery cable. Fill and bleed the cooling system.

12. Start the engine and bring to normal operating temperature. Check for leaks.

2.3L Engine

1. Disconnect the negative battery cable. Drain the cooling system.

2. Remove the timing belt covers and the timing belt.

3. Use a pulley removal tool to hold the water pump pulley and remove the bolts. Remove the water pump pulley.

4. Position a drain pan under the water pump.

5. Remove the water pump mounting bolts, and remove the water pump.

To install:

6. Clean the mating surfaces of the water pump and the engine block.

7. Install a new O-ring onto the water pump.

8. Install the water pump and torque the bolts 18 ft. lbs. (25 Nm).

9. Install the water pump pulley with the bolts. Hold the pulley with the tool and tighten the bolts to 88 inch lbs. (10 Nm).

10. Install the timing belt and timing covers.

11. Connect the negative battery cable. Fill and bleed the cooling system.

12. Start the engine and bring to normal operating temperature. Check for leaks.

Cylinder Head

REMOVAL & INSTALLATION

➡**Before installing the cylinder head, thoroughly clean and inspect it, especially if the head needed to be removed to replace a blown gasket.**

1.5L and 1.8L (BP) Engines

1. Relieve the fuel system pressure and disconnect the negative battery cable. Drain the cooling system.

2. Raise and safely support the vehicle.

3. Remove the right front wheel and splash shield.

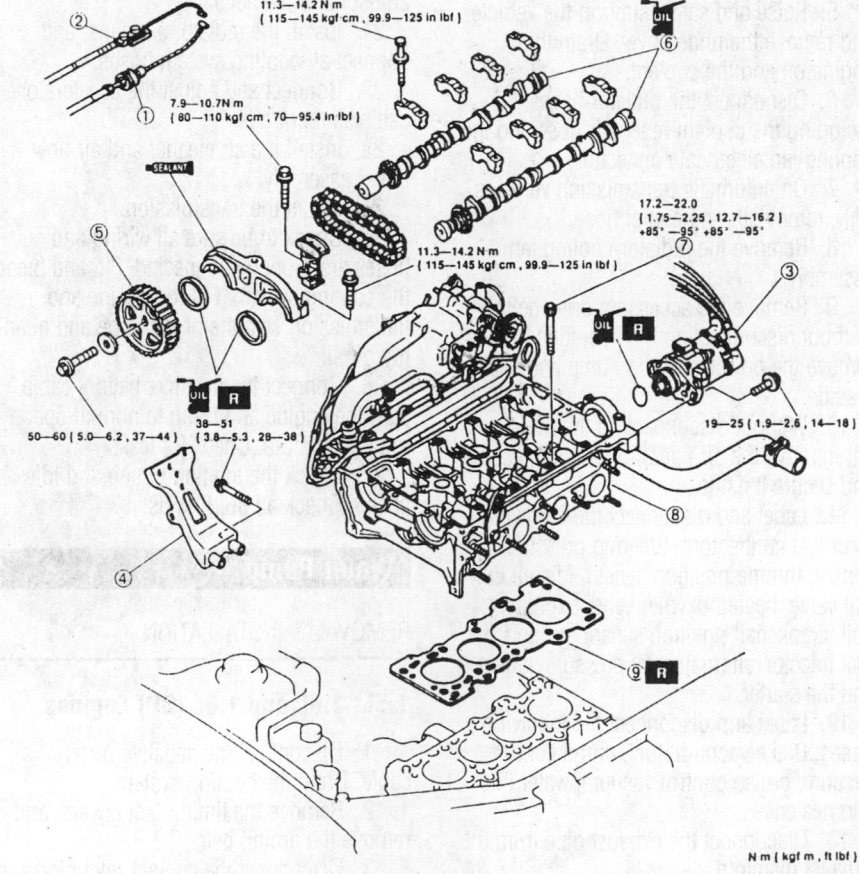

11.3—14.2 N·m (115—145 kgf·cm , 99.9—125 in lbf)

7.9—10.7 N·m (80—110 kgf·cm , 70—95.4 in lbf)

SEALANT

17.2—22.0 (1.75—2.25 , 12.7—16.2) +85° - 95° +85° - 95°

11.3—14.2 N·m (115—145 kgf·cm , 99.9—125 in lbf)

50—60 (5.0—6.2 , 37—44)

38—51 (3.8—5.3 , 28—38)

19—25 (1.9—2.6 , 14—18)

N·m (kgf·m , ft lbf)

1. Accelerator cable
2. Throttle cable (ATX)
3. Distributor
4. Intake manifold stay
5. Camshaft pulley
6. Camshaft
7. Cylinder head bolt
8. Cylinder head assembly
9. Cylinder head gasket

Exploded view of the cylinder head assembly—1.5L engine

7923MG19

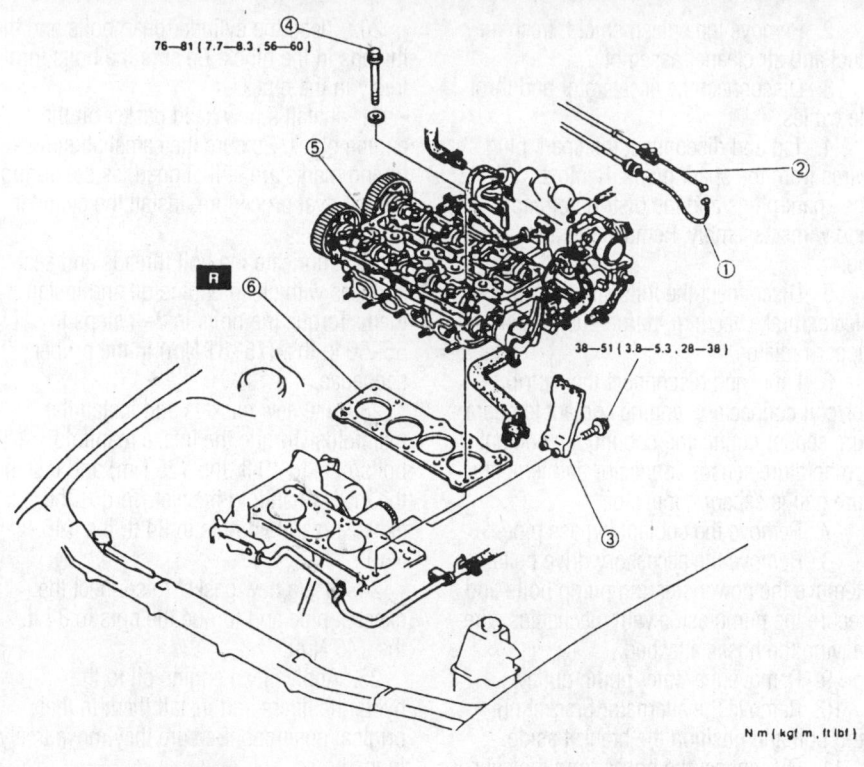

76—81 (7.7—8.3 , 56—60)

38—51 (3.8—5.3 , 28—38)

N m (kgf m , ft lbf)

1. Accelerator cable
2. Throttle cable (ATX)
3. Intake manifold stay
4. Cylinder head bolt
5. Cylinder head assembly
6. Cylinder head gasket

7923MG20

Exploded view of the cylinder head assembly for the 1.8L (BP) engine

4. Loosen the water pump pulley attaching bolts.

5. Remove the power steering belt shield. Loosen the power steering adjusting bolt, lockbolt and through-bolt and remove the power steering belt.

6. Loosen the alternator adjusting bolt and upper mounting bolt. Remove the alternator belt.

7. Remove the water pump pulley.

8. Using a holder tool, hold the crankshaft pulley and remove the pulley bolt. Use a suitable puller to remove the pulley, then remove the guide plate.

9. Remove the power steering hose brackets from the cylinder head cover. Label and disconnect the spark plug wires and wire clips.

10. Disconnect the breather tube and PCV valve from the cylinder head cover. Remove the bolts, in 2 steps, in the reverse order of the tightening sequence. Remove the cylinder head cover.

11. Remove the oil dipstick and bracket.

12. Remove the timing belt upper cover.

13. Use a suitable engine support tool, and remove the number three engine mount bracket.

14. Remove the timing belt middle and lower covers.

15. Rotate the crankshaft, in the normal direction of rotation, until the No. 1 cylinder piston is at TDC on the compression stroke. Be sure the timing marks on the crankshaft and camshaft sprocket(s) are properly aligned and mark the direction of rotation of the belt.

16. Loosen the timing belt tensioner and remove the belt. Do not rotate the crankshaft until the timing belt is reinstalled.

17. Remove air cleaner assembly and front pipe.

18. Remove the exhaust manifold.

19. Disconnect the accelerator and throttle cables.

20. Tag and disconnect the spark plug wires from the spark plugs. Remove the spark plugs and the distributor cap and wires assembly. Remove the distributor.

21. Disconnect the following hoses: Heater, brake vacuum, purge, fuel, water and upper radiator.

22. Label and disconnect the distributor/coil connectors, engine coolant temperature sensor connector, cooling fan coolant temperature sensor connector, and temperature gauge sensor connector.

23. Remove the camshaft sprockets and the camshaft.

24. On the 1.5L engine, remove the tappets. Identify each tappet as it is removed so it can be reinstalled in the same position.

25. On the 1.8L engine, remove the hydraulic lifters. Identify each lifter as it is removed so it can be reinstalled in the same position. If the lifters are to be reused, store them upside down in an oil-filled sealed container.

26. Remove the intake manifold bracket and the intake manifold.

27. Loosen the cylinder head bolts, in 2–3 steps, in sequence. Remove the bolts and the cylinder head.

To install:

28. Thoroughly, clean the cylinder head and the block contact surfaces. Examine the head gasket and check the cylinder head for cracks. Check the cylinder head for warpage using a feeler gauge and straightedge. The maximum allowable distortion is 0.004 in. (0.10mm).

29. Clean the cylinder head bolts and the threads in the block. Be sure the bolts turn freely in the block.

30. Install a new head gasket on the engine block. Be sure the camshaft sprocket timing marks are still aligned, as set during the removal procedure. Install the cylinder head.

31. Lubricate the bolt threads and seat surfaces with clean engine oil and install them as follows;

　a. On the 1.5L engines, torque the bolts in 2–3 steps to 13–16 ft. lbs. (17–22 Nm) in the proper sequence. Paint a reference mark on each bolt head and turn the bolts, in sequence, 90°, then an additional 90°.

　b. On the 1.8L (BPD) engines, torque the bolts in 2–3 steps to 56–60 ft. lbs. (75–81 Nm) in the proper sequence.

32. Use new gaskets and install the manifolds. Torque the intake manifold bolts/nuts to 19 ft. lbs. (25 Nm) and install the intake manifold bracket. Torque the exhaust manifold nuts to 34 ft. lbs. (46 Nm).

33. Use a new gasket to connect the exhaust pipe and torque the nuts to 34 ft. lbs. (46 Nm).

34. Apply clean engine oil to the tappets and install them in their original positions.

35. Install the camshaft and sprockets.

36. Be sure the crankshaft and camshaft sprocket timing marks are aligned, install the timing belt and set the tension. Carefully rotate the crankshaft 2 turns to be sure the timing marks still line up.

37. Install the timing belt middle and lower covers, and tighten the bolts to 70–95 inch lbs. (8–11 Nm).

38. Install the number three engine mount bracket. Tighten the nut to 70–95 inch lbs. (8–11 Nm), and the bolt to 14–16 ft. lbs. (19–22 Nm). Remove the engine support tool.

39. Install the upper timing belt cover and tighten the bolts to 70–95 inch lbs. (8–11 Nm).

40. Install the oil dipstick and bracket.

41. Apply silicone sealant to the cylinder surface in the area adjacent to the front camshaft bearing caps. Apply sealant to a new gasket and install it on the cylinder head cover.

42. Install the cylinder head cover and tighten the bolts in 5–6 steps, in sequence, to 61–95 inch lbs. (7–11 Nm) on the 1.5 (Z5D) engines and 44–78 inch lbs. (5–9 Nm) on the 1.8L (BPD) engines.

43. Install the power steering hose brackets and tighten the bolts to 70–95 inch lbs. (7–11 Nm). Connect the spark plug wires and wire clips. Connect the breather tube and PCV valve.

44. Install the guide plate, crankshaft pulley and pulley bolt. Hold the pulley with the holder tool and tighten the bolt to 116–122 ft. lbs. (157–166 Nm).

45. Install the water pump pulley.

46. Install the alternator belt and adjust the tension.

47. Install the power steering belt and adjust the tension. Tighten the through-bolt to 32–44 ft. lbs. (44–60 Nm) and the lockbolt to 24–33 ft. lbs. (32–46 Nm). Install the power steering belt shield and tighten the bolts to 86 inch lbs. (9 Nm).

48. Install the splash shield and wheel. Lower the vehicle.

49. Connect the electrical engine harness connectors.

50. Connect the heater, brake vacuum, purge, fuel, water and upper radiator hoses.

51. Install the distributor, spark plugs, distributor cap and wires.

52. Connect and adjust the accelerator and throttle cables.

53. Install the cleaner.

54. Connect the negative battery cable. Fill and bleed the cooling system. Change the engine oil.

55. Adjust the ignition timing.

56. Start the engine and bring to normal operating temperature. Check for leaks. Check the ignition timing and idle speed.

1.6L Engines

1. Relieve the fuel system pressure and disconnect the negative battery cable. Drain the cooling system.

2. Remove the splash shield, fresh air duct and air cleaner assembly.

3. Disconnect the accelerator and throttle cables.

4. Tag and disconnect the spark plug wires from the spark plugs. Remove the spark plugs and the distributor cap and wires assembly. Remove the distributor.

5. Disconnect the following hoses: Heater, brake vacuum, purge, fuel, water and upper radiator.

6. Label and disconnect the distributor/coil connectors, engine coolant temperature sensor connector, cooling fan coolant temperature sensor connector and temperature gauge sensor connector.

7. Remove the coolant bypass pipe.

8. Remove the accessory drive belts. Remove the power steering pump bolts and secure the pump aside with mechanics wire, leaving the hoses attached.

9. Remove the water pump pulley.

10. Remove the alternator bracket nut and bolt and position the bracket aside.

11. Disconnect the hoses from the cylinder head cover and loosen the cover bolts in 5–6 step sequences. Remove the cylinder head cover.

12. Remove the timing belt cover(s). Rotate the crankshaft, in the normal direction of rotation, until the No. 1 cylinder piston is at TDC on the compression stroke. Be sure the timing marks on the crankshaft and camshaft sprocket(s) are properly aligned and mark the direction of rotation of the belt.

13. Loosen the timing belt tensioner and remove the belt. Do not rotate the crankshaft until the timing belt is reinstalled.

14. Remove the camshaft sprockets and the camshaft.

15. Remove the hydraulic lifters. Identify each lifter as it is removed so it can be reinstalled in the same position. If the lifters are to be reused, store them upside down in an oil-filled sealed container.

16. Disconnect the exhaust pipe and remove the exhaust manifold.

17. Remove the intake manifold bracket and the intake manifold.

18. Loosen the cylinder head bolts, in 2–3 steps, in sequence. Remove the bolts and the cylinder head.

To install:

19. Thoroughly, clean the cylinder head and the block contact surfaces. Examine the head gasket and check the cylinder head for cracks. Check the cylinder head for warpage using a feeler gauge and straightedge. The maximum allowable distortion is 0.004 in. (0.10mm).

20. Clean the cylinder head bolts and the threads in the block. Be sure the bolts turn freely in the block.

21. Install a new head gasket on the engine block. Be sure the camshaft sprocket timing marks are still aligned, as set during the removal procedure. Install the cylinder head.

22. Lubricate the bolt threads and seat surfaces with clean engine oil and install them. Torque the bolts in 2–3 steps to 56–60 ft. lbs. (75–81 Nm) in the proper sequence.

23. Use new gaskets and install the manifolds. Torque the intake manifold bolts/nuts to 19 ft. lbs. (25 Nm) and install the intake manifold bracket. Torque the exhaust manifold nuts to 34 ft. lbs. (46 Nm).

24. Use a new gasket to connect the exhaust pipe and torque the nuts to 34 ft. lbs. (46 Nm).

25. Apply clean engine oil to the hydraulic lifters and install them in their original positions. Be sure they move freely in the bores.

26. Install the camshaft and sprockets.

27. Be sure the crankshaft and camshaft sprocket timing marks are aligned, install the timing belt and set the tension. Carefully rotate the crankshaft 2 turns to be sure the timing marks still line up.

28. Apply a thin bead of sealant to the cylinder head cover and install the new gasket. Install the cover and torque the cover bolts to 78 inch lbs. (9 Nm).

29. Install the timing belt cover(s) and tighten the bolts to 95 inch lbs. (11 Nm).

30. Install the alternator bracket. Tighten the bracket nut and bolt to 19 ft. lbs. (25 Nm).

31. Install the water pump pulley.

32. Install the alternator belt and adjust the tension.

33. Loosely install the power steering pump through and lockbolts. Connect the pump pressure switch connector.

34. Install the power steering pump belt and adjust the tension. Tighten the pump through-bolt to 45 ft. lbs. (61 Nm) and the lockbolt to 34 ft. lbs. (46 Nm).

35. Install the power steering pump belt shield and tighten the bolts to 86 inch lbs. (9 Nm). Install the power steering hose brackets to the cylinder head cover, and tighten the bolts to 88 inch lbs. (10 Nm).

36. Install the coolant bypass pipe.

37. Connect the electrical engine harness connectors.

38. Connect the heater, brake vacuum, purge, fuel, water and upper radiator hoses.

Cylinder head bolt tightening sequence for all 4-cylinder engines

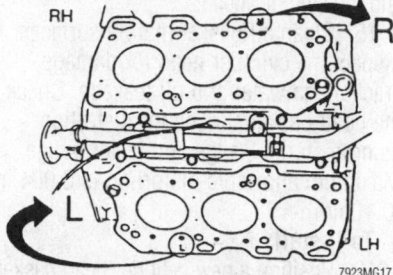

Cylinder head gasket positioning for 6-cylinder engines

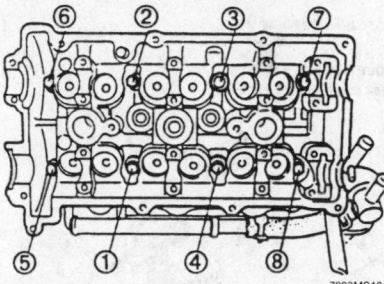

Cylinder head bolt tightening sequence for the 6-cylinder engines

39. Install the distributor, spark plugs, distributor cap and wires.

40. Connect and adjust the accelerator and throttle cables.

41. Install the air ducts, cleaner and splash shield.

42. Connect the negative battery cable. Fill and bleed the cooling system. Change the engine oil.

43. Adjust the ignition timing.

44. Start the engine and bring to normal operating temperature. Check for leaks. Check the ignition timing and idle speed.

1.8L (K8) and 2.5L Engines

1. Relieve the fuel system pressure and disconnect the negative battery cable. Drain the cooling system.

2. Remove the fresh air duct and the air cleaner assembly.

3. If additional clearance space is needed, remove the battery.

4. Disconnect the accelerator cable.

5. Disconnect the wiring harness from the cylinder heads.

6. Disconnect the fuel, heater and vacuum hoses.

7. Remove the intake manifold.

8. Remove the distributor.

9. Disconnect the ventilation pipe from the left cylinder head cover, remove the bolts and remove the cylinder head covers.

10. Remove the timing belt covers and the timing belt.

11. Remove the camshafts. Remove the 3 bolts and the seal plate from the front of the engine.

12. Remove the upper radiator hose. Raise and safely support the vehicle.

13. Disconnect the oxygen sensor connectors. Remove the exhaust pipe-to-manifold nuts and lower the exhaust pipes. Lower the vehicle.

14. Remove the hydraulic lifters. Identify each lifter as it is removed so it can be reinstalled in the same position. If the lifters are to be reused, store them upside down in an oil-filled, sealed container.

15. Loosen the cylinder head bolts, in 2–3 steps, in the reverse order of the torque sequence. Remove the bolts and remove the cylinder heads.

16. Clean all gasket mating surfaces. Inspect the cylinder head for damage, cracks, and water and oil leakage. Check the head gasket surface for distortion using a straight-edge and feeler gauge. Maximum allowable distortion is 0.004 in. (0.10mm).

To install:

17. Position NEW head gaskets on the cylinder block. The gaskets cannot be interchanged between sides and are marked R and L for right and left side.

18. Install the cylinder heads. Apply clean engine oil to the threads of new cylinder head bolts and install. Tighten the cylinder head bolts in 2–3 steps, in sequence, to 17–19 ft. lbs. (23–26 Nm).

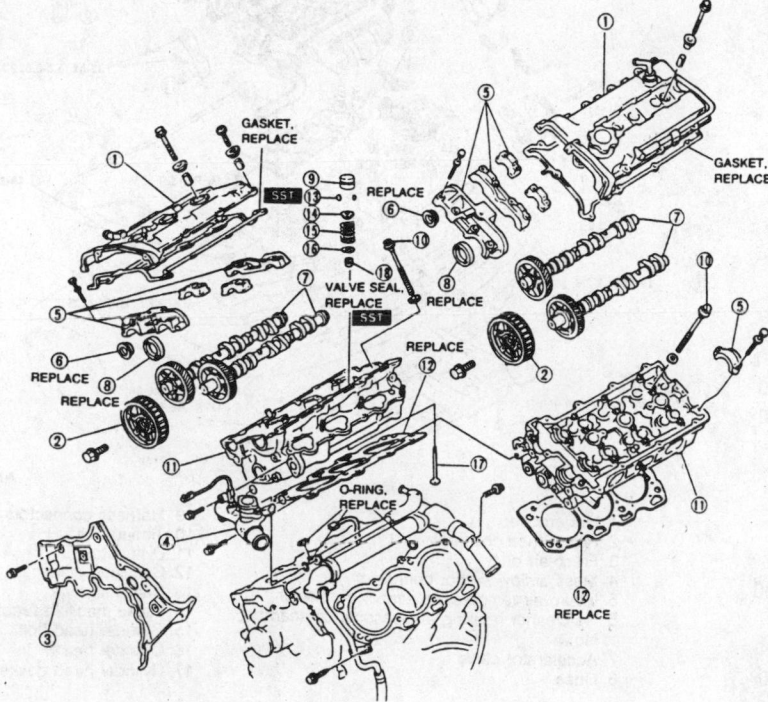

1. Cylinder head cover
2. Camshaft pulley
3. Seal plate
4. Water outlet
5. Camshaft cap
6. Blind cap
7. Camshaft
8. Camshaft oil seal
9. HLA
10. Cylinder head bolt
11. Cylinder head
12. Cylinder head gasket
13. Valve keeper
14. Valve spring seat, upper
15. Valve spring
16. Valve spring seat, lower
17. Valve
18. Valve seal

Exploded view of the cylinder head assemblies—1.8L (K8) and 2.5L engines

19. Paint a mark on the edge of each cylinder head bolt to use as a reference. Turn each bolt, in sequence, 90 degrees. Again, turn each bolt, in sequence, an additional 90 degrees.

20. Apply clean engine oil to the hydraulic lifters and install them in their original positions. Be sure they move freely in the bores.

21. Install the camshafts. Raise and safely support the vehicle.

22. Connect the exhaust pipes to the manifolds and tighten the nuts to 41 ft. lbs. (55 Nm). Connect the oxygen sensor connectors. Lower the vehicle.

23. Install the timing belt and timing belt covers.

24. Apply sealant to the cylinder head surface in the area of the front and rear camshaft caps. Install new gaskets and install the cylinder head covers. Tighten the bolts in 5–6 steps, in sequence, to 44–78 inch lbs. (5–9 Nm).

25. Install the intake manifold using new gaskets. Tighten the mounting bolts to 14–18 ft. lbs. (19–25 Nm).

26. Install the distributor:

a. Apply clean engine oil to a new O-ring, and position it on the distributor.

b. Apply clean engine oil to the drive blade. Install the distributor with the blade fit into the camshaft groove.

c. Hand-tighten the mounting bolts.

27. Connect the vacuum, heater and fuel hoses.

28. Connect the wiring harness to the cylinder heads.

29. Connect and adjust the accelerator cable.

30. If removed, install the battery.

31. Install the air cleaner assembly and the fresh air duct.

32. Connect the negative battery cable. Fill and bleed the cooling system. Adjust the ignition timing and idle speed. Run the engine and check for proper operation.

2.0L Engine

1. Relieve the fuel system pressure and disconnect the negative battery cable. Drain the cooling system.

2. Remove the splash shield, fresh air duct and air cleaner assembly.

3. Disconnect the accelerator cable.

4. Disconnect the following hoses: heater, brake vacuum, purge, fuel, water and upper radiator.

5. Remove the accessory drive belts. Remove the power steering pump bolts and secure the pump aside with mechanics wire, leaving the hoses attached.

6. Remove the alternator bracket nut and bolt and position the bracket aside. Detach the exhaust pipe from the manifold.

7. Label and disconnect the spark plug wires. Remove the power steering hose brackets from the cylinder head cover.

8. Label and detach the distributor/coil connectors, engine coolant temperature sensor connector, cooling fan coolant temperature sensor connector and temperature gauge sensor connector.

9. Disconnect the hoses from the cylinder head cover and loosen the cover bolts in 5–6 steps, in the reverse order of the tightening sequence. Remove the cylinder head cover.

10. Remove the timing belt cover and the timing belt.

11. Remove the coolant temperature sensor housing from the cylinder head. Remove the distributor.

12. Remove the camshaft sprockets and the camshaft.

13. Remove the hydraulic lifters. Identify each lifter as it is removed so it can be reinstalled in the same position. If the lifters are to be reused, store them upside down in an oil-filled sealed container.

14. Loosen the cylinder head bolts, in 2–3 steps, in sequence. Remove the bolts and the cylinder head.

15. Clean all gasket mating surfaces. Inspect the cylinder head for damage, cracks, and water and oil leakage. Check the head gasket surface for distortion using a straight-edge and feeler gauge. Maximum allowable distortion is 0.004 in. (0.10mm).

To install:

16. Position a new cylinder head gasket on the cylinder block, and install the cylinder head.

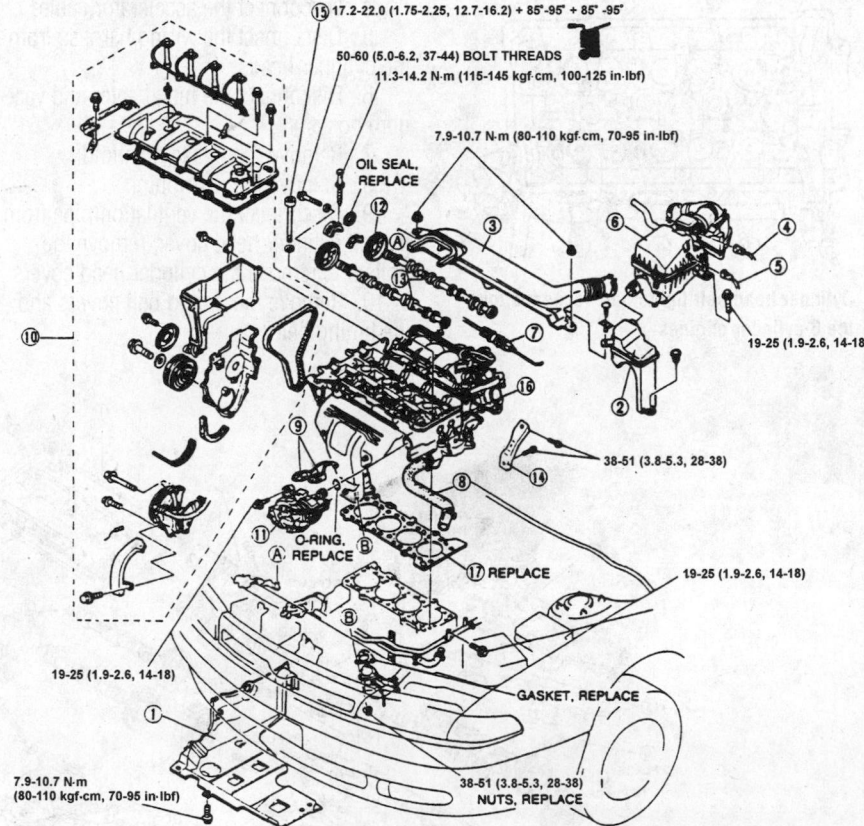

N·m (kgf·m, ft·lbf)

1. Undercover
2. Resonance chamber No.1
3. Fresh-air duct
4. Mass airflow sensor connector
5. Intake air temperature sensor connector
6. Air cleaner housing and resonance chamber No.2
7. Accelerator cable
8. Hose
9. Harness connectors
10. Timing belt
11. Distributor
12. Camshaft pulley
13. Camshaft
14. Intake manifold bracket
15. Cylinder head bolt
16. Cylinder head
17. Cylinder head gasket

Exploded view of the cylinder head assembly—2.0L engine

7923MG21

17. Apply clean engine oil to the bolt threads and seating faces. Install new cylinder head bolts and tighten in 2–3 steps, in sequence, to 13–16 ft. lbs. (17–22 Nm).

18. Paint a mark on the edge of each cylinder head bolt to use as a reference. Turn each bolt, in sequence, 90 degrees. Again, turn each bolt, in sequence, an additional 90 degrees.

19. Apply clean engine oil to the hydraulic lifters and install them in their original positions. Be sure they move freely in the bores.

20. Install the camshafts and sprockets. Install the distributor and connect the distributor/coil connectors.

21. Install the timing belt and cover.

22. Install a new cylinder head cover gasket on the cylinder head cover. Apply sealant to the cylinder head surface in the area adjacent to the front camshaft caps, then install the cover. Tighten the bolts in 5–6 steps, in sequence, to 61–95 inch lbs. (7–11 Nm).

23. Connect the hoses to the cylinder head cover. Connect the spark plug wires.

24. Install the exhaust manifold and the alternator bracket. Tighten the bracket nut and bolt to 19 ft. lbs. (25 Nm).

25. Install the alternator belt and adjust the tension.

26. Loosely install the power steering pump through and lockbolts. Connect the pump pressure switch connector.

27. Install the power steering pump belt and adjust the tension. Tighten the pump through-bolt to 45 ft. lbs. (61 Nm) and the lockbolt to 34 ft. lbs. (46 Nm).

28. Install the power steering pump belt shield and tighten the bolts to 86 inch lbs. (9 Nm). Install the power steering hose brackets to the cylinder head cover, and tighten the bolts to 88 inch lbs. (10 Nm).

29. Install the coolant temperature sensor housing with a new gasket. Tighten the bolts to 19 ft. lbs. (25 Nm). Connect the electrical connectors at the housing.

30. Connect and adjust the accelerator cable.

31. Connect the heater, brake vacuum, purge, fuel, water and upper radiator hoses.

32. Install the air cleaner assembly, fresh air duct and splash shield.

33. Connect the negative battery cable. Fill and bleed the cooling system. Run the engine and check for proper operation.

2.3L Engines

1. Relieve the fuel system pressure. Disconnect the negative battery cable.

2. Drain the engine coolant.

3. Raise and safely support the vehicle.

4. Disconnect the oxygen sensor connectors. Remove the exhaust pipe-to-manifold nuts and lower the exhaust pipes.

5. Remove the right-hand three-way catalytic converter. Lower the vehicle.

6. Remove the Lysholm compressor (supercharger).

7. Remove the intake manifold.

8. Remove the timing belt covers and timing belt.

9. Remove the spacer and O-ring from the front of the camshaft.

10. Remove the ignition coils.

11. Remove the cylinder head cover mounting bolts, in 5–6 steps, using the reverse of the tightening sequence. Remove the cylinder head cover.

12. Remove the camshaft sprockets.

13. Turn the camshafts so the knock pins are aligned with the marks on the camshaft caps. This will reduce the pressure on the adjustment shims.

14. Note the markings on the camshaft caps prior to removal, so they can be reinstalled in the same positions. The right-hand (rear) caps are marked with numbers and the left-hand (front) caps are marked with letters.

15. Loosen the front camshaft cap bolts in sequence, in 5–6 steps. Remove the front camshaft caps.

16. Remove the remaining camshaft cap bolts in the proper sequence. Remove the caps, being sure to remove the thrust caps last. Do not damage the cylinder head thrust bearing support.

17. Remove the camshafts and oil seals.

18. Remove the lifters and adjustment shims. Identify and mark each lifter as it is removed so it can be reinstalled in the same position.

19. Remove the lower radiator hose and water inlet pipe.

20. Remove the Lysholm compressor bracket.

21. Remove the alternator bracket bolt to gain additional clearance.

22. Remove the rubber insulator from the left-hand cylinder head.

23. Temporarily install the number three engine mount, which was removed with the timing belt, to support the engine. Remove the engine support device.

24. Loosen the cylinder head bolts, in 2–3 steps, in the reverse order of the torque sequence. Remove the bolts and remove the cylinder heads.

25. Remove the oil control plug O-rings.

26. Clean all gasket mating surfaces. Inspect the cylinder head for damage, cracks, and water and oil leakage. Check the head gasket surface for distortion using a straight-edge and feeler gauge. Maximum allowable distortion is 0.004 in. (0.10mm).

To install:

27. Apply clean engine oil to the O-rings, and install them onto the oil control plugs.

28. Position new head gaskets on the cylinder block. The gaskets cannot be interchanged between sides and are marked **R** and **L** for right and left side.

29. Install the cylinder heads. Apply clean engine oil to the threads of new cylinder head bolts and install. Tighten the cylinder head bolts in 2–3 steps, in sequence, to 17–19 ft. lbs. (23–26 Nm).

30. Paint a mark on the edge of each cylinder head bolt to use as a reference. Turn each bolt, in sequence, 90 degrees. Again, turn each bolt, in sequence, an additional 90 degrees.

31. Install the rubber insulator onto the left-hand cylinder head.

32. Fit the knock sensor harness into the drill hole on the cylinder block. Pass the harness under the rubber insulator.

33. Install an engine support device, and remove the number three engine mount.

34. Install the alternator bracket bolt. Tighten the mounting bolt to 12–16 ft. lbs. (16–22 Nm).

35. Install the Lysholm compressor bracket. Tighten the mounting bolts to 14–18 ft. lbs. (19–25 Nm).

36. Install the water inlet pipe. Tighten the mounting bolts to 14–18 ft. lbs. (19–25 Nm). Install the lower radiator hose.

37. Apply clean engine oil to the lifters, then install them in their original positions. Verify that they move smoothly in their bore.

38. Install new oil seals on the camshafts. Apply clean engine oil to the camshaft lobes, journals and supports.

39. Install the camshafts so the gear marks align.

40. Remove all oil and dirt from the mating surfaces between the front camshaft cap and the cylinder head.

41. Install the thrust caps. Tighten the thrust cap bolts, in 5–6 steps, until the caps are fully seated on the cylinder head.

42. Apply silicone sealant, at a thickness of 0.06–0.09 in. (1.5–2.5mm), to the cylinder head surface in the area forward of the camshaft gear cavity.

43. Install the remaining camshaft caps in their original positions. Tighten the caps, in sequence, in five equal steps, with the final step being 100–125 inch lbs. (11–14 Nm).

44. Apply clean engine oil to the lip of the new camshaft oil seal. Push the seal in lightly by hand. Tap the seal in evenly with a seal installer (49 F401 337A or equivalent) with a final protrusion of 0–0.02 in. (0–0.5mm). Tap in a new blind cap.

45. Install the camshaft sprockets. Tighten the mounting bolts to 91–103 ft. lbs. (123–140 Nm).

46. Measure and adjust valve clearances.

47. Remove any sealant and gasket material from the cylinder head cover contact surfaces.

48. Apply silicone sealant to the cylinder head in the area adjacent to the front and rear camshaft caps. Install a new gasket on the cylinder head.

49. Install the cylinder head cover. Tighten the bolts in 5–6 steps, in sequence, to 44–78 inch lbs. (5–9 Nm).

50. Using a new O-ring, install the distributor.

51. Install the ignition coils.

52. Install the spacer, using a new O-ring. Tighten the mounting bolt to 14–18 ft. lbs. (19–25 Nm).

53. Install the timing belt and timing belt cover.

54. Install the intake manifold.

55. Install the Lysholm compressor (supercharger).

56. Raise and safely support the vehicle.

57. Install the right-hand three-way catalytic converter.

58. Connect the exhaust pipes to the manifolds and tighten the nuts to 28–38 ft. lbs. (38–51 Nm). Connect the oxygen sensor connectors. Lower the vehicle.

59. Connect the negative battery cable.

60. Fill and bleed the coolant system.

61. Run the engine and check for leaks.

Rocker Arms/Shafts

REMOVAL & INSTALLATION

1.8L (BP) SOHC Engine

1. Remove the rocker arm (valve) cover.

2. Remove the rocker arm and shaft assembly mounting bolts. Start at the ends and work toward the center of the shafts, when removing the bolts.

3. If necessary, separate the rocker arms and springs from the shafts; be sure to keep the parts in order for reinstallation purposes.

4. Clean and inspect the shafts and rocker arms for wear. Measure the difference between the rocker arm shaft outside diameter and the rocker arm inside diameter; this is the oil clearance. If the oil clearance exceeds 0.004 in. (0.10mm), replace the shaft and/or the rocker arm(s).

To install:

5. If they were disassembled, coat the rocker arm shafts and rocker arms with engine oil and assemble them with the springs. When assembling and installing on the cylinder head, note the notches at the ends of the shafts; they are different on the intake and exhaust side and cannot be interchanged.

6. Install the rocker arm/shaft assemblies onto the cylinder head and torque the rocker arm shaft-to-cylinder head bolts, in sequence, to 16–21 ft. lbs. (22–28 Nm), in several steps.

7. Install the rocker arm (valve) cover.

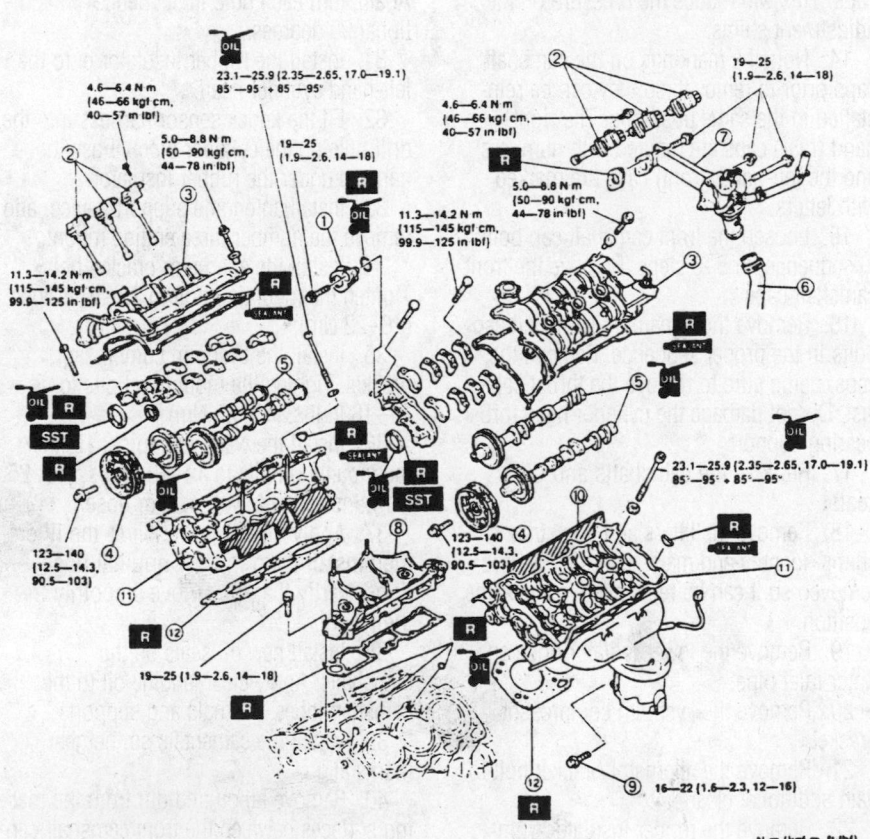

1. Spacer
2. Ignition coil
3. Cylinder head cover
4. Camshaft pulley
5. Camshaft
6. Lower radiator hose
7. Water inlet pipe
8. Lysholm compressor bracket
9. Generator bolt
10. Rubber insulator (LH)
11. Cylinder head
12. Cylinder head gasket

7923MG24

Exploded view of the cylinder head and related components—2.3L engine

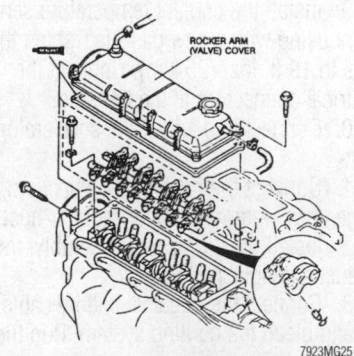

7923MG25

Exploded view of a typical rocker arm assembly

Supercharger

REMOVAL & INSTALLATION

2.3L Engine

> **⁂ CAUTION**
>
> **The fuel injection system remains under pressure after the engine has been turned OFF. Properly relieve fuel pressure before disconnecting any fuel lines. Failure to do so may result in fire or personal injury.**

1. Relieve the fuel system pressure, and disconnect the negative battery cable.
2. Drain the cooling system.
3. Remove the dynamic chamber cover.
4. Remove the charge air cooler air duct.
5. Label and disconnect the vacuum hoses and electrical connectors from the air cleaner housing. Remove the air cleaner assembly.
6. Remove the air and fresh air ducts.
7. Remove the mass air flow sensor and the air intake hose from the throttle body.
8. Remove the resonator.
9. Remove the right-hand charge air cooler.
10. Remove the left-hand charge air cooler.
11. Disconnect the accelerator cable.
12. Label and disconnect the necessary vacuum hoses from the rear of the intake manifold and EGR valve.
13. Remove the EGR valve.
14. Remove the air intake pipe assembly.
15. Remove the charge air cooler pipe.
16. Disconnect and plug the fuel supply line at the fuel rails and discard the copper crush washers. Disconnect the fuel and vacuum lines from the fuel pressure regulator.

> **⁂ CAUTION**
>
> **Do not allow fuel spray or fuel vapors to come in contact with a spark or open flame. Keep a dry chemical fire extinguisher nearby. Never store fuel in an open container due to risk of fire or explosion.**

17. Disconnect and plug the coolant hoses.
18. Remove the harness from the intake manifold.
19. Remove the intake manifold mounting nuts and bolts in 2–3 steps, then remove the intake manifold.
20. Label and disconnect the fuel hoses and electrical connectors from the throttle body. Remove the throttle body.
21. Remove the drive belt from the Lysholm compressor (supercharger).
22. Remove the mounting bolts from the Lysholm compressor, and remove the compressor from the vehicle.

To install:

23. Clean all gasket mating surfaces.
24. Position the rubber shield for the Lysholm compressor onto the compressor using double sided adhesive tape. Place the compressor onto the mounting studs and tighten the mounting nuts to 14–18 ft. lbs. (19–25 Nm).
25. Install and adjust the drive belt.

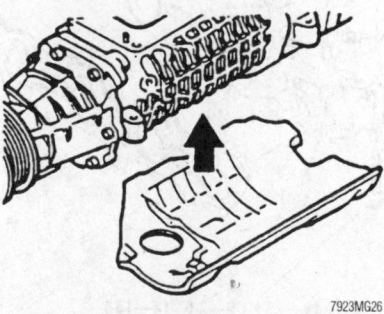

7923MG26

When installing the compressor, ensure that the rubber insulating pad is temporarily affixed to the compressor

26. Install the throttle body. Tighten the nuts and bolts to 14–18 ft. lbs. (19–25 Nm), and connect the fuel hoses and electrical connectors.
27. Position new gaskets and install the intake manifold. Tighten the nuts and bolts in 2–3 steps, from the center to the ends, to 14–18 ft. lbs. (19–25 Nm).
28. Install the harness onto the intake manifold.
29. Unplug and connect the coolant hoses.
30. Connect the fuel and vacuum lines to the fuel pressure regulator. Connect the fuel supply line to the fuel rail, using new copper crush washers.

31. Install the charge air cooler pipe.
32. Position the air intake pipe assembly using new gaskets. Hand-tighten the nuts and bolts in the order shown in the graphic until the air intake pipe contacts the intake manifold. Verify that the rubber gaskets are not twisted or distorted. Tighten the bolts marked **A** to 70–95 inch lbs. (8–11 Nm), and all others, in sequence, to 14–18 ft. lbs. (19–25 Nm).
33. Install the EGR valve using a new gasket.
34. Connect the vacuum hoses to the intake manifold and EGR valve.
35. Connect and adjust the accelerator cable.
36. Using new gaskets, position the left and right-hand charge air coolers. Hand-tighten the nuts and bolts in the order shown in the graphic until the air intake pipes and charge air coolers contact the intake manifold. Verify that the rubber gaskets are not twisted or distorted. Tighten the bolts marked **A** to 44–78 inch lbs. (5–9 Nm). Tighten the bolts marked **B** to 70–95 inch lbs. (8–11 Nm), and all others, in sequence, to 14–18 ft. lbs. (19–25 Nm).
37. Install the resonator. Tighten the nuts and bolts to 12–16 ft. lbs. (16–22 Nm)
38. Install the air intake hose onto the throttle body. Install the mass air flow sensor.
39. Install the fresh air and air ducts.
40. Install the air cleaner assembly and connect the vacuum hoses and electrical connectors to the air cleaner housing.
41. Install the charge air cooler air duct. Tighten the mounting bolts to 70–95 inch lbs. (8–11 Nm).
42. Install the dynamic chamber cover.
43. Connect the negative battery cable.
44. Fill and bleed the cooling system. Run the engine and check for leaks.

Intake Manifold

REMOVAL & INSTALLATION

1.5L and 1.8L (BP) Engines

1. Relieve the fuel system pressure, and disconnect the negative battery cable. Drain the cooling system.
2. Disconnect the mass air flow sensor electrical connector. Remove the air ducts, air cleaner assembly, mass air flow sensor and resonance chamber.

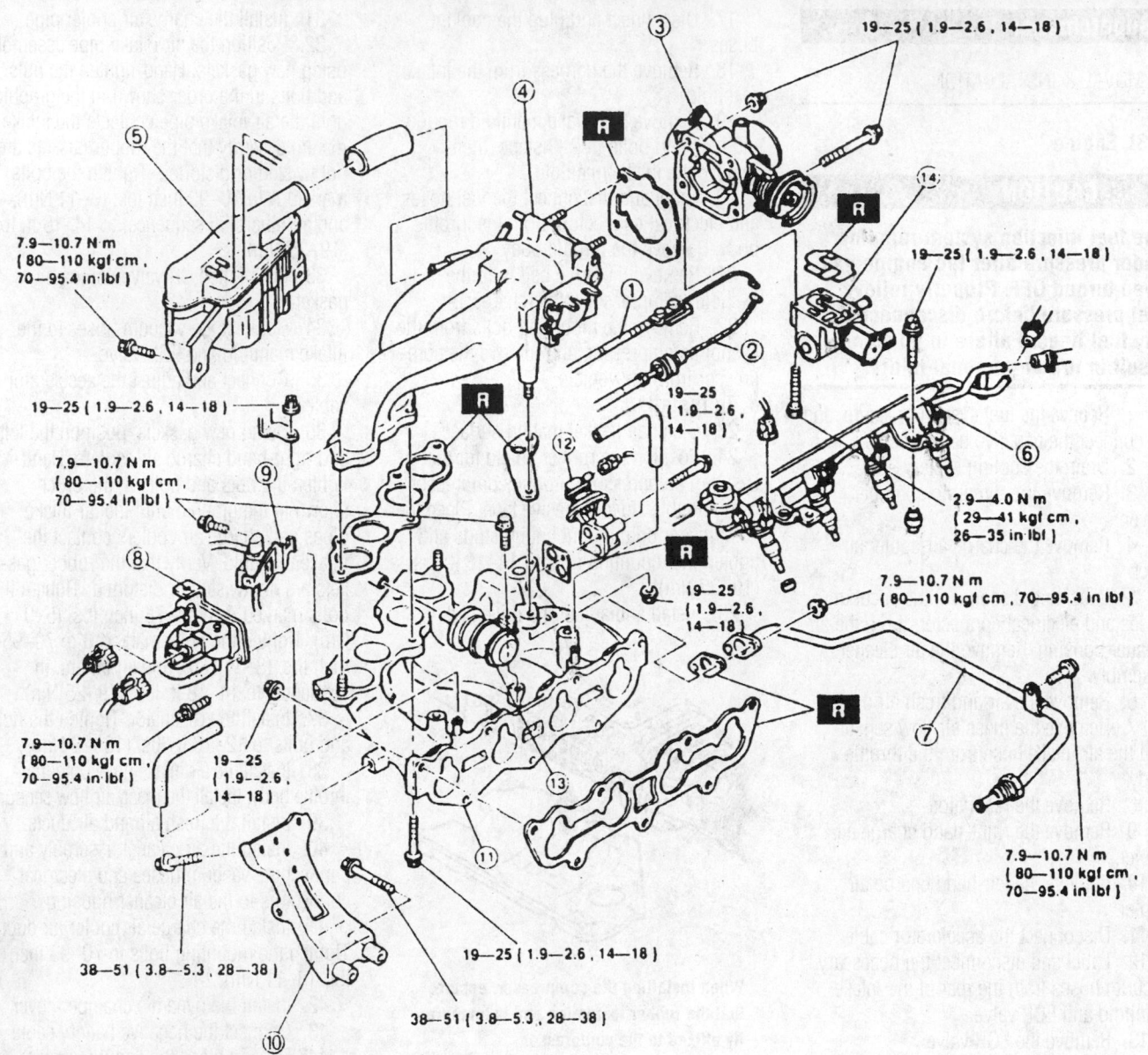

19—25 (1.9—2.6 , 14—18)

7.9—10.7 N·m
(80—110 kgf·cm ,
70—95.4 in·lbf)

19—25 (1.9—2.6 , 14—18)

19—25 (1.9—2.6 , 14—18)

7.9—10.7 N·m
(80—110 kgf·cm ,
70—95.4 in·lbf)

19—25
(1.9—2.6 ,
14—18)

2.9—4.0 N·m
(29—41 kgf·cm ,
26—35 in·lbf)

7.9—10.7 N·m
(80—110 kgf·cm , 70—95.4 in·lbf)

19—25
(1.9—2.6 ,
14—18)

7.9—10.7 N·m
(80—110 kgf·cm ,
70—95.4 in·lbf)

7.9—10.7 N·m
(80—110 kgf·cm ,
70—95.4 in·lbf)

19—25
(1.9—2.6 ,
14—18)

38—51 (3.8—5.3 , 28—38)

38—51 (3.8—5.3 , 28—38)

19—25 (1.9—2.6 , 14—18)

N·m (kgf·m , ft·lbf)

1. Throttle cable (ATX)
2. Accelerator cable
3. Throttle body
4. Dynamic chamber
5. Resonance chamber
6. Fuel distributor
7. EGR pipe

8. EGR solenoid valve stay
9. PRC solenoid valve stay
10. Intake manifold stay
11. Intake manifold
12. EGR valve
13. Air filter
14. BAC valve

7923MG27

Exploded view of the intake manifold and related components—1.5L engine

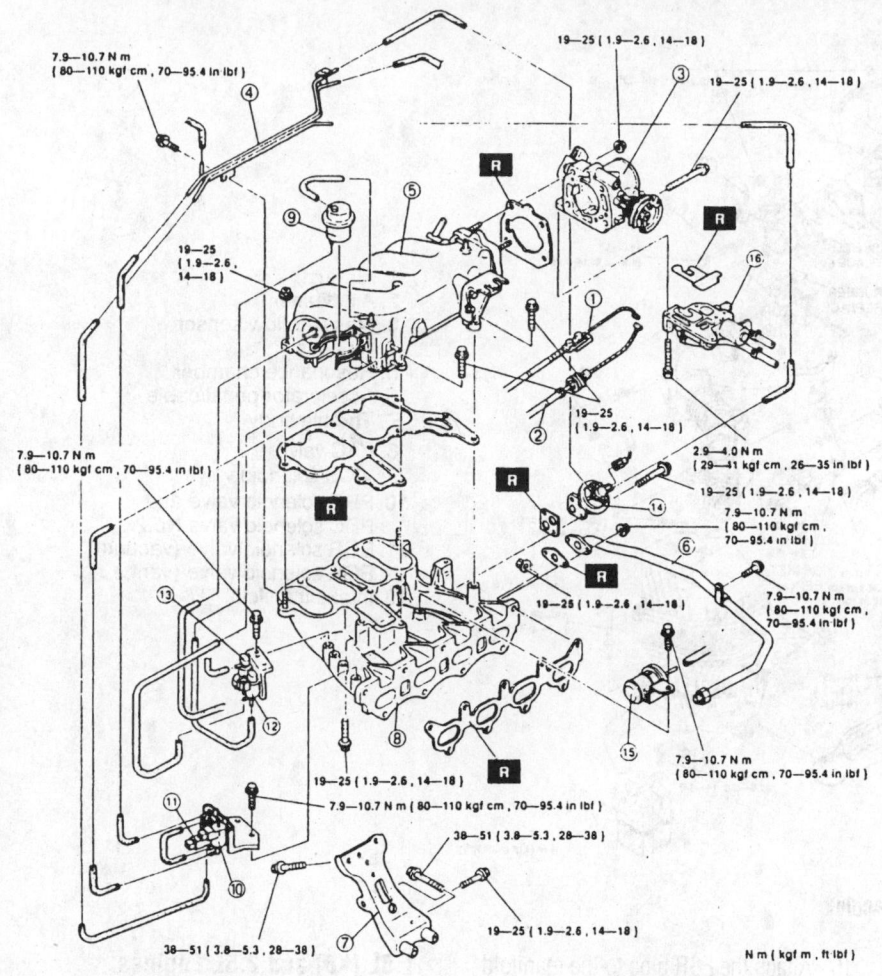

1. Throttle cable (ATX)
2. Accelerator cable
3. Throttle body
4. Vacuum pipe
5. Dynamic chamber
6. EGR pipe
7. Intake manifold stay
8. Intake manifold
9. Vacuum chamber
10. EGR solenoid valve (vacuum)
11. EGR solenoid valve (vent)
12. VICS solenoid valve
13. PRC solenoid valve
14. EGR valve
15. Air filter
16. BAC valve

7923MG28

Exploded view of the 1.8L (BP) engine intake manifold and related components

3. Disconnect the throttle and accelerator cables. Disconnect and plug the fuel lines.

4. Remove the throttle body as follows:
a. Disconnect the coolant hoses.
b. Label and disconnect the electrical connectors for the idle air control valve and the throttle position sensor.
c. Remove the mounting bolts/nuts from the throttle body, and remove the throttle body from the vehicle.

5. Label and disconnect the vacuum lines at the intake manifold.

6. Remove the dynamic chamber (upper intake manifold).

7. Disconnect and plug the fuel hoses from the fuel rail.

8. Label and disconnect the electrical connectors for the fuel injectors.

9. On some models you may need to remove the fuel rail with the injectors connected.

10. On the 1.5L engine, remove the EGR pipe from the intake manifold. Remove the EGR and pressure regulator control (PRC) solenoid valve brackets.

11. Remove the intake manifold support bracket.

12. Remove the bolts and nuts, and remove the intake manifold.

To install:

13. Clean all gasket mating surfaces.

14. Install the intake manifold, using a new gasket. Tighten the nuts and bolts to 14–18 ft. lbs. (19–25 Nm).

15. Attach the EGR pipe to the manifold and install the intake manifold support bracket. Tighten the support bracket bolts to 38 ft. lbs. (51 Nm).

16. If removed, install the EGR and pressure regulator control (PRC) solenoid valve brackets.

17. Install the fuel rail and injector assembly. Connect the electrical connectors to the injectors, and the fuel lines to the rail.

18. Install the upper intake manifold to the intake manifold using new gaskets. Tighten the nuts to 14–18 ft. lbs. (19–25 Nm).

19. Install the throttle body, using a new mounting gasket. Tighten the mounting bolts to 14–18 ft. lbs. (19–25 Nm).

20. Connect the electrical connectors for the idle air control valve and the throttle position sensor.

21. Connect the vacuum and coolant lines.

22. Connect and adjust the throttle and accelerator cables.

23. Connect the fuel lines.

24. Install the resonance chamber. Install the air cleaner assembly, mass air flow sensor and ducts. Connect the mass air flow sensor connector.

25. Connect the negative battery cable. Fill and bleed the cooling system. Run the engine and check for leaks.

1.6L Engine

1. Relieve the fuel system pressure and disconnect the negative battery cable. Drain the cooling system.

2. Disconnect the mass air flow sensor electrical connector. Remove the air ducts, air cleaner assembly and resonance chamber.

3. Remove the fuel line mounting bracket and disconnect the throttle cable. Disconnect and plug the fuel lines.

4. Remove the throttle body as follows:
a. Disconnect the coolant hoses.
b. Label and disconnect the electrical connectors for the idle air control valve and the throttle position sensor.

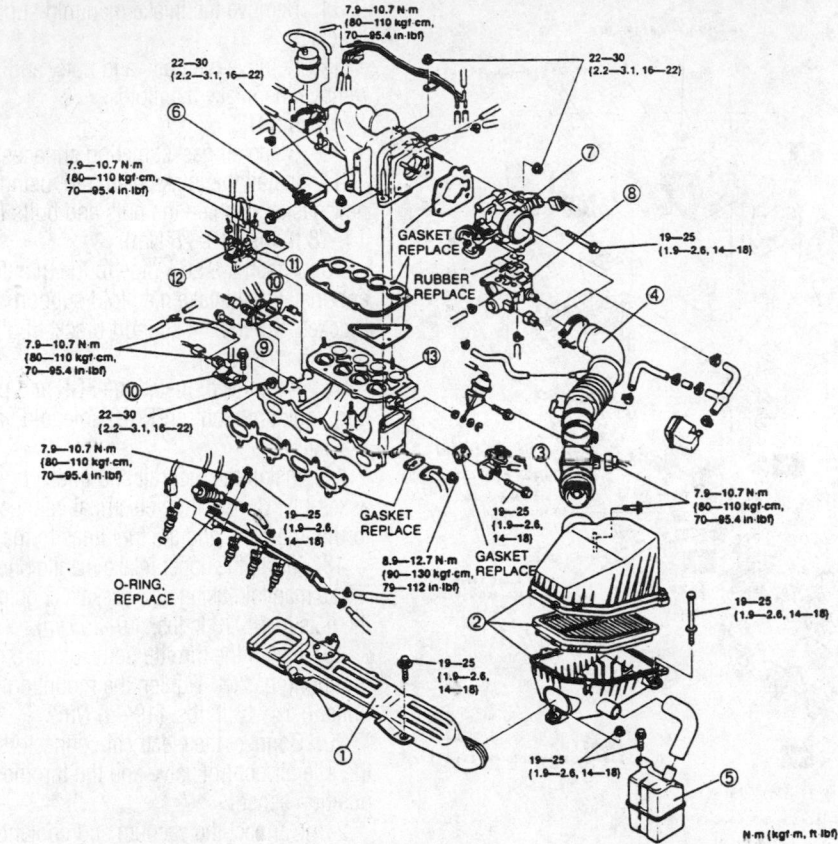

1. Air duct
2. Air cleaner
3. Mass air flow sensor
4. Air hose
5. Resonance chamber
6. Accelerator pedal/cable
7. Throttle body
8. BAC valve
9. VICS solenoid valve
10. PRC solenoid valve and PRC solenoid valve No.2
11. EGR solenoid valve (vacuum)
12. EGR solenoid valve (vent)
13. Intake manifold

Exploded view of the intake manifold for the 1.6L engine

c. Remove the mounting bolts/nuts from the throttle body, and remove the throttle body from the vehicle.

5. Label and disconnect the vacuum lines at the intake manifold.

6. Remove the bypass air control (BAC) valve, variable inertia charging system (VICS) solenoid valve, pressure regulator control (PRC) solenoid valves, and the EGR solenoid vent and vacuum valves from the intake manifold.

7. Disconnect and plug the fuel hoses from the fuel rail.

8. Label and disconnect the electrical connectors for the fuel injectors.

9. Remove the fuel rail with the injectors connected.

10. Remove the intake manifold support bracket and remove the EGR pipe from the intake manifold.

11. Lower the vehicle. Remove the bolts and nuts, and remove the intake manifold.

To install:

12. Clean all gasket mating surfaces.

13. Install the intake manifold, using a new gasket. Tighten the nuts and bolts to 16–22 ft. lbs. (22–30 Nm).

14. Raise and safely support the vehicle.

15. Attach the EGR pipe to the manifold and install the intake manifold support bracket. Tighten the support bracket bolts to 38 ft. lbs. (51 Nm).

16. Install the fuel rail and injector assembly. Connect the electrical connectors to the injectors, and the fuel lines to the rail.

17. Install the bypass air control (BAC) valve, variable inertia charging system (VICS) solenoid valve, pressure regulator control (PRC) solenoid valves, and the EGR solenoid vent and vacuum valves to the intake manifold.

18. Install the throttle body, using a new mounting gasket. Tighten the mounting bolts to 14–18 ft. lbs. (19–25 Nm).

19. Connect the electrical connectors, vacuum lines and coolant lines.

20. Connect the throttle cable and the fuel lines. Install the fuel line mounting bracket and tighten the bolt to 97 inch lbs. (11 Nm).

21. Install the resonance chamber. Install the air cleaner assembly and ducts. Connect the mass air flow sensor connector.

22. Connect the negative battery cable. Fill and bleed the cooling system. Run the engine and check for leaks.

1.8L (K8) and 2.5L Engines

1. Properly relieve the fuel system pressure.

2. Disconnect the negative battery cable and drain the cooling system.

3. On MX3, remove the upper strut bar.

4. Remove the air cleaner assembly and ducts.

5. Disconnect the accelerator cable. Label and disconnect the necessary electrical connectors and vacuum hoses.

6. Disconnect and plug the fuel lines. Disconnect the coolant hose from the air bypass valve.

7. Remove the intake manifold support bracket. Remove the intake manifold-to-cylinder head bolts and remove the intake manifold.

8. If necessary, remove the throttle body and air intake pipe from the manifold.

9. Check the intake manifold for cracks or other damage. Check the surface of the cylinder heads and intake manifold for warpage using a straightedge. Replace the intake manifold, as necessary.

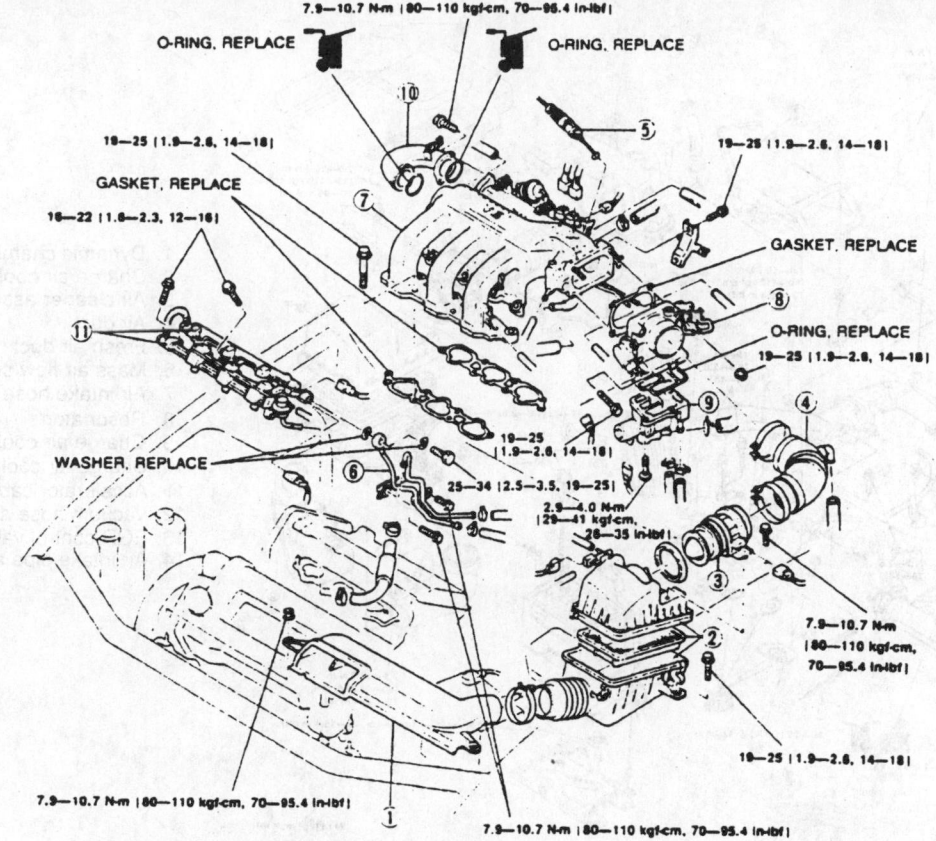

7.9—10.7 N·m (80—110 kgf-cm, 70—95.4 in-lbf)

O-RING, REPLACE

O-RING, REPLACE

19—25 (1.9—2.6, 14—18)

GASKET, REPLACE

16—22 (1.6—2.3, 12—16)

19—25 (1.9—2.6, 14—18)

GASKET, REPLACE

O-RING, REPLACE

19—25 (1.9—2.6, 14—18)

WASHER, REPLACE

19—25 (1.9—2.6, 14—18)

25—34 (2.5—3.5, 19—25)

2.9—4.0 N·m (29—41 kgf-cm, 26—35 in-lbf)

7.9—10.7 N·m (80—110 kgf-cm, 70—95.4 in-lbf)

19—25 (1.9—2.6, 14—18)

7.9—10.7 N·m (80—110 kgf-cm, 70—95.4 in-lbf)

7.9—10.7 N·m (80—110 kgf-cm, 70—95.4 in-lbf)

N·m (kgf-m, ft-lbf)

1. Fresh air duct
2. Air cleaner housing
3. Volume airflow sensor
4. Air intake hose
5. Accelerator cable
6. Fuel pipe
7. Intake manifold
8. Throttle body
9. BAC valve
10. Air intake pipe
11. Fuel distributor

7923MG30

Exploded view of the intake manifold assembly—1.8L and 2.5L engines

To install:

10. Clean all gasket mating surfaces.

11. If removed, install the throttle body using new gaskets. Tighten the nuts/bolts to 19 ft. lbs. (25 Nm).

12. If removed, apply clean engine oil to new O-rings and install the air intake pipe to the intake manifold. Tighten the bolts to 95 inch lbs. (10.8 Nm). On 1.8L engine, the bolts must be tightened in the proper sequence.

13. Position new gaskets and install the intake manifold to the cylinder head. Install the mounting bolts and tighten, in 2–3 steps, to 19 ft. lbs. (25 Nm), working from the center toward the ends of the manifold.

14. Install the intake manifold bracket and tighten the bolts to 19 ft. lbs. (25 Nm).

15. Connect the coolant hose to the air bypass valve. Connect the fuel lines.

16. Connect the vacuum hoses and electrical connectors. Connect the accelerator cable.

17. Install the air cleaner assembly and ducts. On MX3, install the upper strut bar.

18. Connect the negative battery cable. Fill and bleed the cooling system.

19. Start the engine and bring to normal operating temperature. Check for leaks. Check the idle speed.

2.3L Engine

1. Relieve the fuel system pressure, and disconnect the negative battery cable.

2. Drain the cooling system.

3. Remove the dynamic chamber cover.

4. Remove the charge air cooler air duct.

5. Label and disconnect the vacuum hoses and electrical connectors from the air cleaner housing. Remove the air cleaner assembly.

6. Remove the air and fresh air ducts.

7. Remove the mass air flow sensor and the air intake hose from the throttle body.

8. Remove the resonator.

9. Remove the right-hand charge air cooler.

10. Remove the left-hand charge air cooler.

11. Disconnect the accelerator cable.

12. Label and disconnect the necessary vacuum hoses from the rear of the intake manifold and EGR valve.

13. Remove the EGR valve.

14. Remove the air intake pipe assembly.

15. Remove the charge air cooler pipe.

16. Disconnect and plug the fuel supply line at the fuel rails and discard the copper crush washers. Disconnect the fuel and vacuum lines from the fuel pressure regulator.

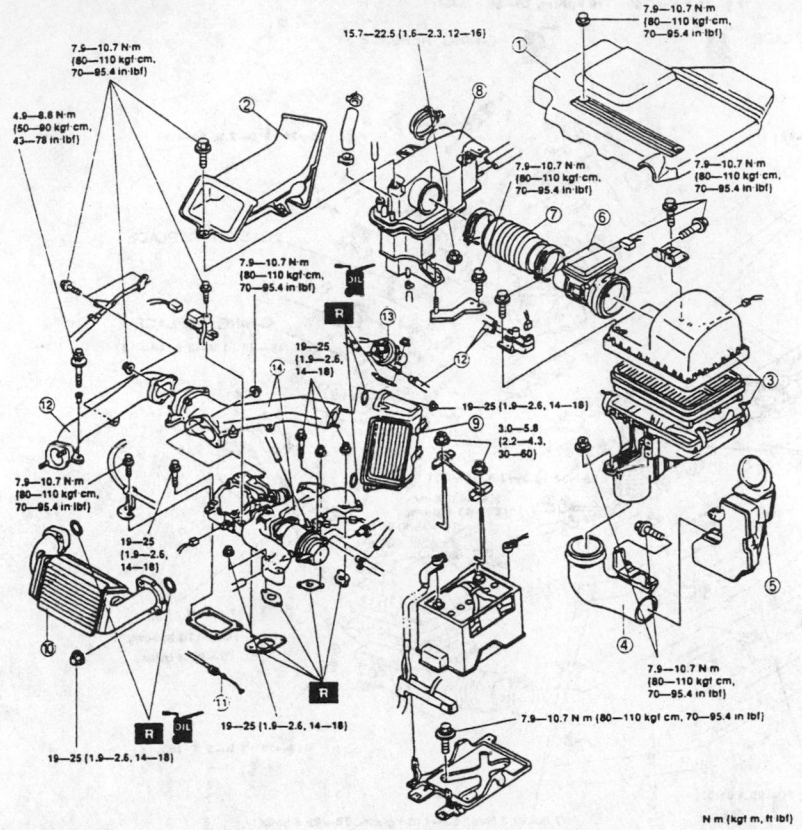

1. Dynamic chamber cover
2. Charge air cooler air duct
3. Air cleaner assembly
4. Air duct
5. Fresh air duct
6. Mass air flow sensor
7. Air intake hose
8. Resonator
9. Charge air cooler (RH)
10. Charge air cooler (LH)
11. Accelerator cable
12. Vacuum hose assembly
13. EGR control valve
14. Air intake pipe assembly

Exploded view of the intake manifold assembly (1 of 2)—2.3L engine

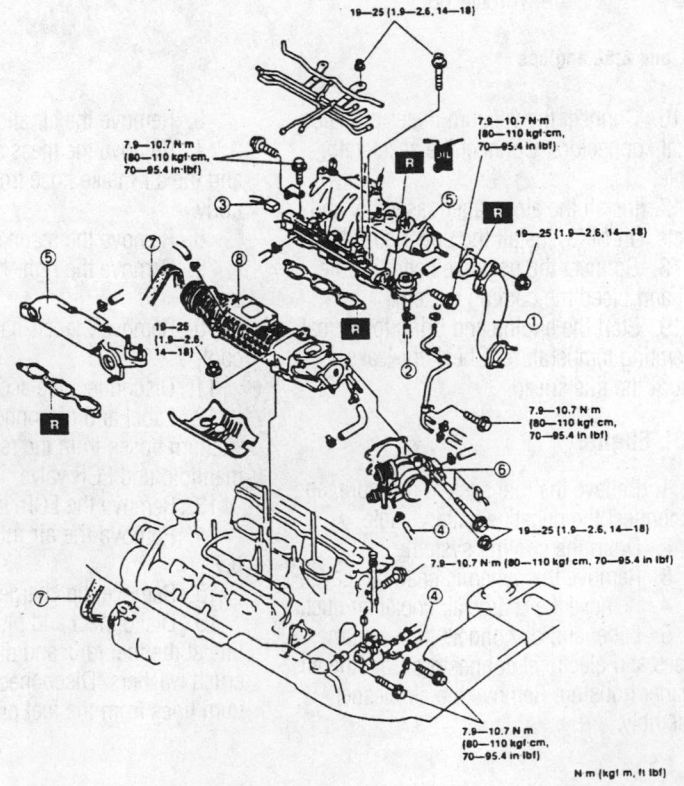

1. Charge air cooler pipe
2. Fuel hose
3. Fuel distributor connector
4. Coolant hose
5. Intake manifold assembly
6. Throttle body assembly
7. Drive belt
8. Lysholm compressor

Exploded view of the intake manifold assembly (2 of 2)—2.3L engine

17. Disconnect and plug the coolant hoses.

18. Remove the harness from the intake manifold.

19. Remove the intake manifold mounting nuts and bolts in 2–3 steps, then remove the intake manifold.

20. Label and disconnect the fuel hoses and electrical connectors from the throttle body. Remove the throttle body.

To install:

21. Clean all gasket mating surfaces.

22. Install the throttle body. Tighten the nuts and bolts to 14–18 ft. lbs. (19–25 Nm), and connect the fuel hoses and electrical connectors.

23. Position new gaskets and install the intake manifold. Tighten the nuts and bolts in 2–3 steps, from the center to the ends, to 14–18 ft. lbs. (19–25 Nm).

24. Install the harness onto the intake manifold.

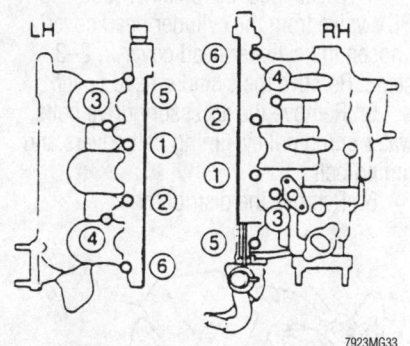

Be sure to tighten the intake manifold bolts in the sequence shown—2.3L engine

25. Unplug and connect the coolant hoses.

26. Connect the fuel and vacuum lines to the fuel pressure regulator. Connect the fuel supply line to the fuel rail, using new copper crush washers.

27. Install the charge air cooler pipe.

28. Position the air intake pipe assembly using new gaskets. Hand-tighten the nuts and bolts in the order shown in the graphic until the air intake pipe contacts the intake manifold. Verify that the rubber gaskets are not twisted or distorted. Tighten the bolts marked **A** to 70–95 inch lbs. (8–11 Nm), and all others, in sequence, to 14–18 ft. lbs. (19–25 Nm).

29. Install the EGR valve using a new gasket.

30. Connect the vacuum hoses to the intake manifold and EGR valve.

31. Connect and adjust the accelerator cable.

32. Using new gaskets, position the left and right-hand charge air coolers. Hand-tighten the nuts and bolts in the order shown in the graphic until the air intake pipes and charge air coolers contact the intake manifold. Verify that the rubber gaskets are not twisted or distorted. Tighten the bolts marked **A** to 44–78 inch lbs. (5–9 Nm). Tighten the bolts marked **B** to 70–95 inch lbs. (8–11 Nm), and all others, in sequence, to 14–18 ft. lbs. (19–25 Nm).

33. Install the resonator. Tighten the nuts and bolts to 12–16 ft. lbs. (16–22 Nm)

34. Install the air intake hose onto the throttle body. Install the mass air flow sensor.

35. Install the fresh air and air ducts.

36. Install the air cleaner assembly and connect the vacuum hoses and electrical connectors to the air cleaner housing.

37. Install the charge air cooler air duct. Tighten the mounting bolts to 70–95 inch lbs. (8–11 Nm).

38. Install the dynamic chamber cover.

39. Connect the negative battery cable.

40. Fill and bleed the cooling system. Run the engine and check for leaks.

Exhaust Manifold

REMOVAL & INSTALLATION

1.5L and 1.8L (BP) Engines

1. Disconnect the negative battery cable.

2. Remove the air cleaner, and disconnect the air hose.

3. Remove the water bypass pipe bolt.

4. Remove the exhaust manifold heat shield bolts and the heat shield.

5. Disconnect the oxygen sensor electrical connector.

6. Raise and safely support the vehicle.

7. Remove and discard the exhaust pipe-to-exhaust manifold nuts. Suspend the exhaust system with wire.

8. Disconnect the EGR pipe from the exhaust manifold and lower the vehicle.

9. Remove the nuts and bolts and remove the exhaust manifold. Discard the nuts.

To install:

10. Clean all gasket mating surfaces.

11. Position a new exhaust manifold gasket over the studs and install the exhaust manifold. Tighten the mounting nuts and bolts to 14–16 ft. lbs. (19–22 Nm) on the 1.5L, and for the 1.8L to 29–34 ft. lbs. (39–47 Nm).

12. Raise and safely support the vehicle.

13. Connect the exhaust pipe to the manifold. Install new nuts and tighten to 38

ft. lbs. (52 Nm). Connect the oxygen sensor connector.

14. Connect the EGR pipe to the back of the exhaust manifold and tighten to 34 ft. lbs. (47 Nm). Lower the vehicle.

15. Install the heat shield and tighten the bolts to 88 inch lbs. (10 Nm).

16. Install the water bypass pipe bolt, and tighten to 48–65 ft. lbs. (64–89 Nm).

17. Connect the air hose, and install the air cleaner.

18. Connect the negative battery cable.

1.8L (K8) and 2.5L Engines

1. Disconnect the negative battery cable. Raise and safely support the vehicle.

2. Disconnect the oxygen sensor connectors.

3. Remove the nuts from the front and rear exhaust pipes and lower the exhaust system. Both pipes must be disconnected, even if only one manifold is to be removed.

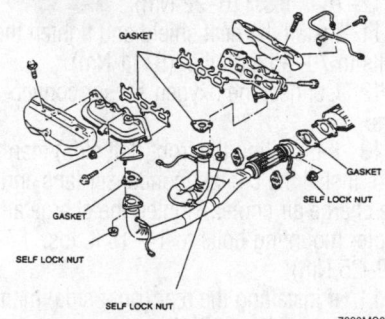

Exploded view of the exhaust manifolds— 1.8L (K8) and 2.5L engines, 2.3L engine is similar

4. If removing the rear (right side) manifold, disconnect the EGR pipe.

5. Remove the 3 heat shield bolts and remove the heat shield.

6. Remove the 2 nuts and 5 bolts and remove the exhaust manifold.

To install:

7. Clean all gasket mating surfaces.

8. Install the exhaust manifold, using a new gasket, and tighten the nuts to 15–20 ft. lbs. (20–28 Nm), and the bolts to 12–16 ft. lbs. (16–22 Nm).

9. Install the heat shield and tighten the bolts to 88 inch lbs. (10 Nm).

10. If installing the rear (right side) manifold, connect the EGR pipe.

11. Connect the exhaust pipes to the manifolds, using new gaskets and nuts, and tighten the nuts to 38 ft. lbs. (51 Nm).

12. Connect the oxygen sensor connectors and lower the vehicle.

13. Connect the negative battery cable.

2.3L Engine

1. Disconnect the negative battery cable.
2. Raise and safely support the vehicle.
3. Remove the nuts from the front and rear exhaust pipes and lower the exhaust system. Both pipes must be disconnected, even if only one manifold is to be removed.
4. If removing the rear (right side) manifold, disconnect the EGR pipe.
5. If removing the front (left side) manifold, remove the charge air cooler and the coolant/condenser fans.
6. Disconnect the front and rear oxygen sensor connectors.
7. Remove the three heat shield bolts and remove the heat shield.
8. Remove the two nuts and five bolts and remove the exhaust manifold.
   To install:
9. Clean all gasket mating surfaces.
10. Install the exhaust manifold, using a new gasket, and tighten the nuts and bolts to 12–16 ft. lbs. (16–22 Nm).
11. Install the heat shield and tighten the bolts to 70–95 inch lbs. (8–11 Nm).
12. Connect the oxygen sensor connectors.
13. If installing the front (left side) manifold, install the coolant/condenser fans and the charge air cooler. Tighten the charge air cooler mounting bolts to 14–18 ft. lbs. (19–25 Nm)
14. If installing the rear (right side) manifold, connect the EGR pipe.
15. Connect the exhaust pipes to the manifolds, using new gaskets and nuts, and tighten the nuts to 28–38 ft. lbs. (38–51 Nm).
16. Lower the vehicle.
17. Connect the negative battery cable.

1.6L and 2.0L Engines

1. Disconnect the negative battery cable.
2. Remove the retaining bolts and remove the exhaust manifold insulator.
3. Disconnect the oxygen sensor electrical connector. Remove the oxygen sensor, if necessary, if it is installed in the manifold.
4. Disconnect the EGR pipe, if equipped.
5. Raise and safely support the vehicle. Remove the nuts from the exhaust pipe flange and disconnect the exhaust pipe from the manifold or turbocharger, if equipped.
6. Lower the vehicle.
7. If equipped with turbocharger, proceed as follows:
   a. Drain the cooling system.
   b. Disconnect the air hose and coolant hoses from the turbocharger.
   c. Disconnect the oil feed and return lines.

8. Remove the mounting nuts/bolts and remove the exhaust manifold. On turbocharged vehicles, the manifold and turbocharger are removed as an assembly.
9. Installation is the reverse of removal procedure. Be sure all gasket mating surfaces are clean prior to assembly.
10. Use new gaskets and tighten the exhaust manifold-to-cylinder head nuts/bolts to 17 ft. lbs. (23 Nm) on 1.6L and 1.8L SOHC engines, 34 ft. lbs. (46 Nm) on 1.6L and 1.8L DOHC 4-cylinder engines. On 2.0L engine, tighten the nuts to 20 ft. lbs. (26 Nm) and the bolts to 16 ft. lbs. (22 Nm).
11. Use a new gasket and tighten the exhaust pipe flange-to-exhaust manifold nuts to 34 ft. lbs. (46 Nm).

Front Crankshaft Seal

REMOVAL & INSTALLATION

1. Disconnect the negative battery cable. Raise and safely support the vehicle.
2. Remove the timing belt. Remove the crankshaft damper retaining bolt, the damper and the timing belt sprocket. Remove the sprocket key from the crankshaft.
3. Using a small prybar, pry the oil seal from the engine block; be careful not to

score the crankshaft or the seal seat. Clean the seal bore.
   To install:
4. Using an oil seal installation tool or equivalent, lubricate the seal lip with clean engine oil and drive the new seal into the engine until it seats.
5. The rest of the installation is the reverse of the removal procedure. Tighten all bolts to the proper specification.

Camshaft

REMOVAL & INSTALLATION

1.5L Engine

1. Disconnect the negative battery cable.
2. Remove the power steering hose brackets from the cylinder head cover.
3. Label and disconnect the spark plug wires and spark plug wire clips.
4. Disconnect the breather tube and PCV valve from the cylinder head cover. Loosen the cylinder head cover in 2–3 steps. Remove the cylinder head cover.
5. Remove the accessory drive belts, water pump pulley, timing belt covers and timing belt.
6. Remove the distributor.

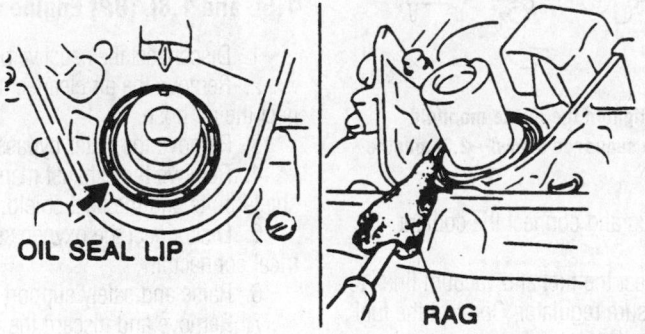

OIL SEAL LIP

RAG

7923MG35

Remove the front engine seal by cutting the seal lip, then, so as not to damage the crankshaft, carefully pry the seal out with a prybar

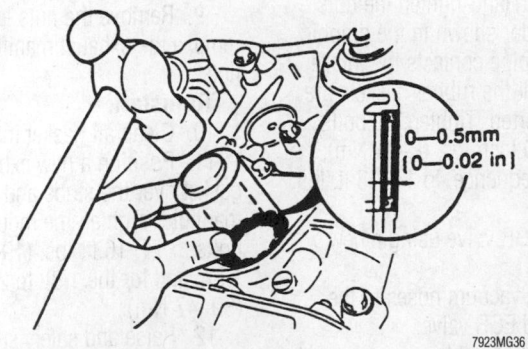

0—0.5mm
(0—0.02 in)

7923MG36

Install the seal using an appropriate driver, which fits over the crankshaft snout and presses on the outside edge of the seal

7. Hold the camshaft with a wrench on the cast hexagon, and loosen the camshaft sprocket mounting bolt. Remove the camshaft sprockets.

8. Remove the seal plate.

9. Rotate the camshafts clockwise so the cams don't press on the tappets.

10. Loosen the front camshaft cap bolts in 5–6 steps, starting on the two outside bolts and finishing on the two inside bolts. Remove the front camshaft bolts and caps.

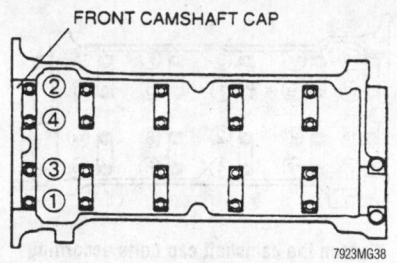

FRONT CAMSHAFT CAP

7923MG38

Front camshaft cap bolt loosening sequence—1.5L engine

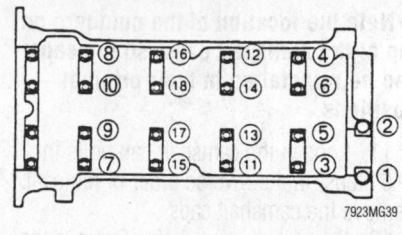

7923MG39

Camshaft cap bolt loosening sequence— 1.5L engine

11.3—14.2 N·m (115—145 kgf·cm , 99.9—125 in·lbf)

7.9—10.7 N·m (80—110 kgf·cm , 70—95.4 in·lbf)

OIL

OIL

R

38—51 (3.8—5.3 , 28—38)

3 SEALANT

11.3—14.2 N·m (115—145 kgf·cm , 99.9—125 in·lbf)

SEALANT

17.2—22.0 (1.75—2.25 , 12.7—16.2) + 85°—95° + 85°—95°

50—60 (5.0—6.2 , 37—44)

38—51 (3.8—5.3 , 28—38)

N·m (kgf·m , ft·lbf)

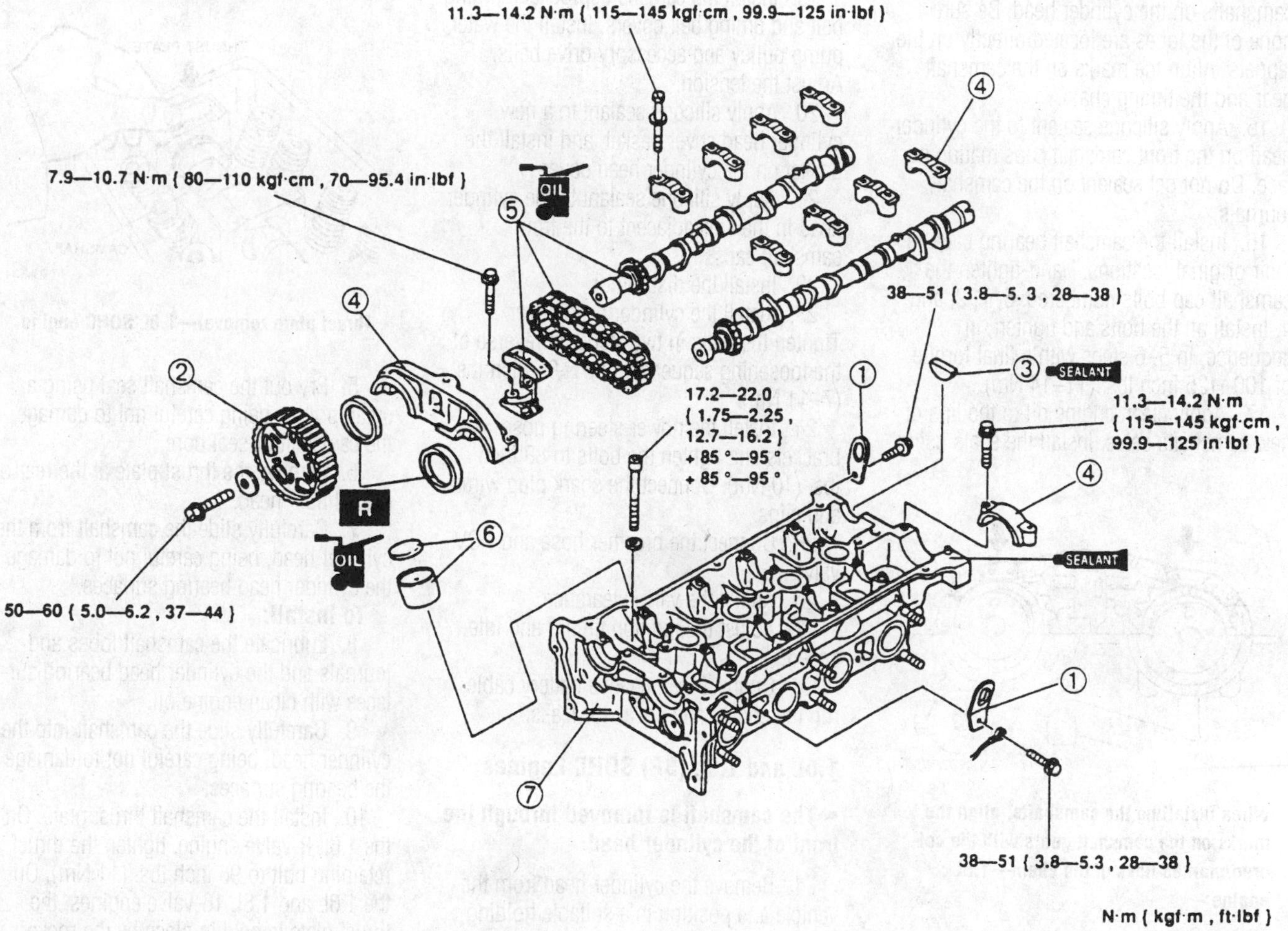

1. Engine hanger
2. Camshaft pulley
3. Seal cap
4. Camshaft cap
5. Camshaft, timing chain, and chain adjuster
6. Tappet and adjustment shim
7. Cylinder head

7923MG37

Exploded view of the camshaft assemblies—1.5L engines

➡**Note the location of the numbers on top of the camshaft caps, so the caps can be reinstalled in their original positions.**

11. Loosen the camshaft cap bolts in 5–6 steps, in the reverse order of removal. Remove the camshaft caps.

12. Remove the camshafts. Remove the chain and oil seals from the camshafts.

To install:

13. Insert the chain adjuster between the camshafts.

14. Lubricate the camshaft lobes and journals with clean engine oil and install the camshafts on the cylinder head. Be sure none of the lobes are located directly on the tappets. Align the marks on the camshaft gear and the timing chain.

15. Apply silicone sealant to the cylinder head on the front camshaft caps mating surface. Do not get sealant on the camshaft journals.

16. Install the camshaft bearing caps in their original locations. Hand-tighten the camshaft cap bolts numbered: 5, 7, 2, and 4. Install all the bolts and tighten, in sequence, in 5–6 steps with a final torque of 100–125 inch lbs. (11–14 Nm).

17. Apply clean engine oil to the lips of new camshafts seals. Install the seals using

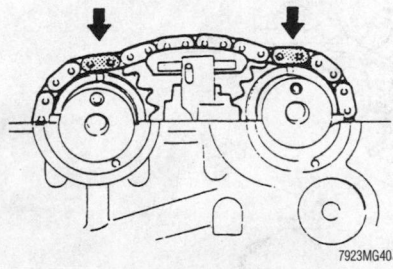

When installing the camshafts, align the marks on the camshaft gears with the colored/marked links of the chain—1.5L engine

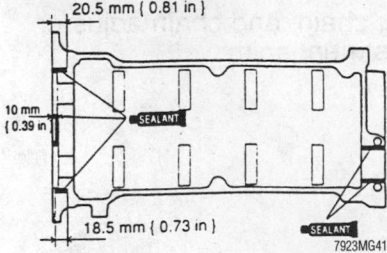

Apply sealant in the positions shown before installing the camshaft bearing caps—1.5L engine

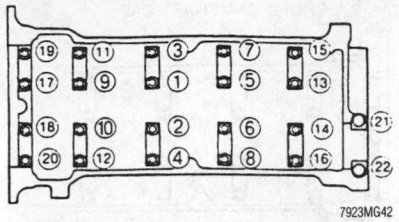

Tighten the camshaft cap bolts according to the sequence shown—1.5L engine

a suitable seal installer flush with the edge of the camshaft cap.

18. Install the seal plate.

19. Install the camshaft sprockets, timing belt and timing belt covers. Install the water pump pulley and accessory drive belts. Adjust the tension.

20. Apply silicone sealant to a new cylinder head cover gasket, and install the gasket on the cylinder head cover.

21. Apply silicone sealant to the cylinder head in the area adjacent to the front camshaft caps.

22. Install the distributor.

23. Install the cylinder head cover. Tighten the bolts in two steps, in reverse of the loosening sequence, to 61–95 inch lbs. (7–11 Nm).

24. Install the power steering hose brackets and tighten the bolts to 88 inch lbs. (10 Nm). Connect the spark plug wires and clips.

25. Connect the breather hose and PCV valve.

26. Adjust the valve clearance.

27. Adjust the ignition timing and idle speed.

28. Connect the negative battery cable, run the engine and check for leaks.

1.6L and 1.8L (BP) SOHC Engines

➡**The camshaft is removed through the front of the cylinder head.**

1. Remove the cylinder head from the vehicle and position in a suitable holding fixture.

➡**Do not lay the cylinder head flat on the head gasket surface as the valves may be damaged.**

2. Hold the camshaft with a wrench on the hexagon cast into the front of the camshaft.

3. Remove the sprocket bolt and the sprocket.

4. Loosen the rocker arm shaft bolts in 2–3 steps, in the reverse of the torque sequence. Remove the rocker arm and shaft assemblies.

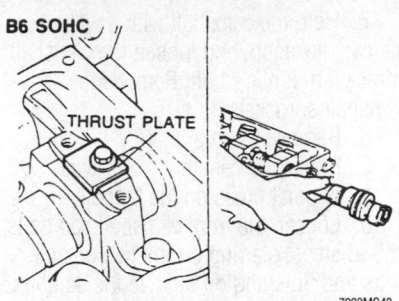

Remove the thrust plate, then slide the camshaft out of the cylinder head—1.6L SOHC engine

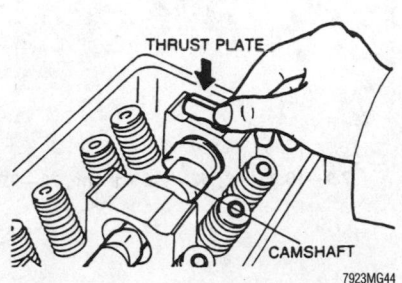

Thrust plate removal—1.8L SOHC engine

5. Pry out the camshaft seal using a small prybar, being careful not to damage the camshaft or seal bore.

6. Remove the thrust plate at the rear of the cylinder head.

7. Carefully slide the camshaft from the cylinder head, being careful not to damage the cylinder head bearing surfaces.

To install:

8. Lubricate the camshaft lobes and journals and the cylinder head bearing surfaces with clean engine oil.

9. Carefully slide the camshaft into the cylinder head, being careful not to damage the bearing surfaces.

10. Install the camshaft thrust plate. On the 1.6L 8-valve engine, tighten the thrust retaining bolt to 95 inch lbs. (11 Nm). On the 1.6L and 1.8L 16-valve engines, the thrust plate is held in place by the rocker arm and shaft assembly.

11. Lubricate the lip of a new camshaft seal with clean engine oil and install in the cylinder head, using a seal installer.

12. Lubricate the rocker arms and valve stem tips with clean engine oil. Install the rocker arm and shaft assemblies and tighten the bolts, in 2–3 steps, in the proper sequence. The final torque should be 21 ft. lbs. (28 Nm) on the 1.6L and 1.8L engines.

13. Install the camshaft sprocket and retaining bolt. Hold the camshaft with the wrench on the hexagon and tighten the bolt to 45 ft. lbs. (61 Nm).

14. Install the cylinder head and the remaining components in the reverse order of removal.

1.6L, 1.8L (BP) and 2.0L DOHC Engines

1. Disconnect the negative battery cable.

2. Label and disconnect the spark plug wires and remove the spark plugs.

3. Disconnect the hoses from the cylinder head cover, if equipped.

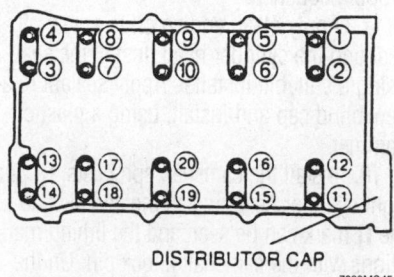

Camshaft cap bolt loosening sequence—1.6L, 1.8L and 2.0L DOHC engines

4. Remove the cylinder head cover bolts and remove the cylinder head cover. On 2.0L engine, loosen the bolts in 2–3 steps in the reverse of the torque sequence.

5. Remove the timing belt and the distributor.

6. Hold the camshaft with a wrench on the hexagon cast into the camshaft. Remove the sprocket bolts and remove the sprockets.

7. Label the caps so they can be reinstalled in their original positions. Loosen the camshaft cap bolts in 2–3 steps in the reverse of the torque sequence, then remove the camshaft caps.

8. Remove the camshafts. Remove the camshaft oil seals from the camshafts.

To install:

9. Lubricate the camshaft journals and lobes with clean engine oil. Install the camshafts in the cylinder head.

10. Apply silicone sealant to the cylinder head on the front camshaft cap mating surfaces. Do not allow any sealant on the camshaft journals.

11. Install the camshaft caps in their original positions. Loosely install the cap bolts.

12. Tighten the camshaft cap bolts in 2–3 steps to 125 inch lbs. (14 Nm) in the proper sequence.

13. Apply clean engine oil to the lip of a new camshaft seal. Push the seal slightly in by hand. Tap the seal into position, using a seal installer, until it is flush with the edge of the camshaft cap.

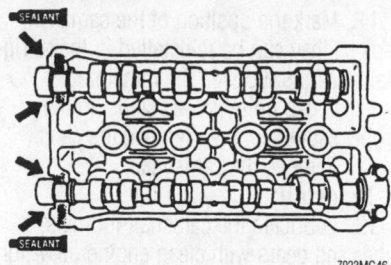

Apply silicone sealant to the cylinder head in the positions shown—1.6L, 1.8L and 2.0L DOHC engines

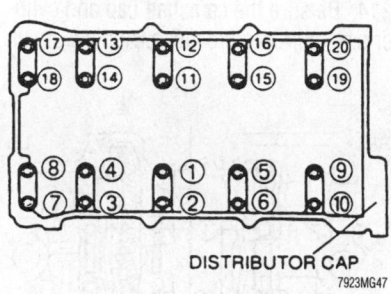

Camshaft cap bolt tightening sequence for the 1.6L, 1.8L and 2.0L DOHC engines

14. Turn the camshafts until the dowel pins face straight up. Install the camshaft sprockets and the sprocket bolts.

15. Hold the camshaft with the wrench on the cast hexagon and tighten the sprocket bolts to 44 ft. lbs. (60 Nm).

16. Install the remaining components in the reverse order of removal.

1.8L (K8) and 2.5L Engines

1. Properly relieve the fuel system pressure. Disconnect the negative battery cable and drain the cooling system.

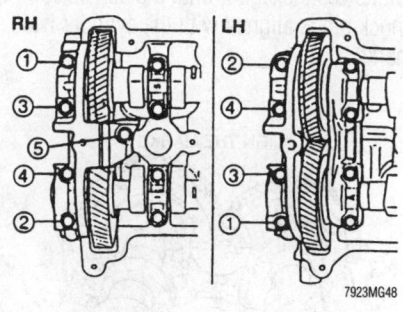

Front camshaft cap bolt loosening sequence—2.3L and 2.5L engines

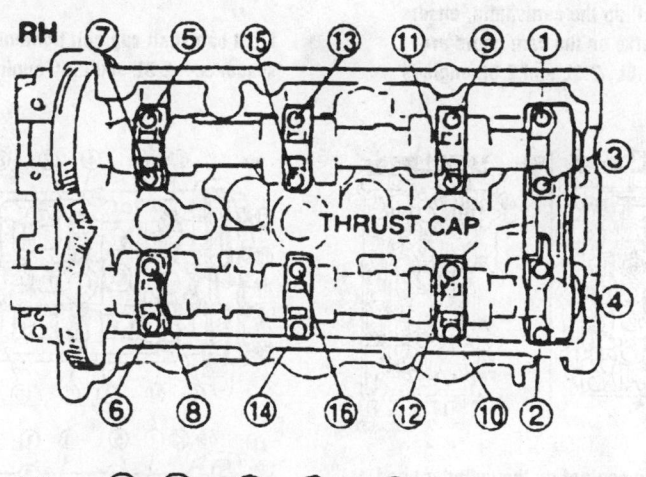

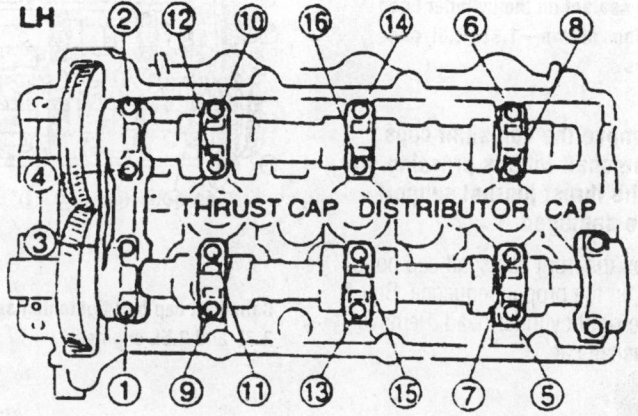

Camshaft cap bolt loosening sequence——2.3L and 2.5L engines

2. Remove the timing belt.

3. Disconnect the accelerator cable. On 1.8L engine, disconnect the throttle cable.

4. Label and disconnect the spark plug wires.

5. Label and disconnect the necessary wiring and hoses.

6. Remove the intake manifold and the cylinder head covers.

7. Remove the distributor.

8. Hold the camshaft with a wrench on the hexagon cast into the camshaft. Remove the sprocket bolt and remove the sprocket.

9. Turn the camshaft, using a wrench on the cast hexagon, until the camshaft knock pin is aligned with the cylinder head marks.

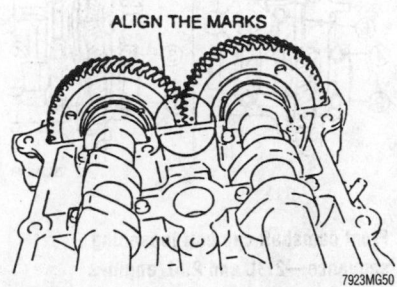

When installing the camshafts, ensure that the marks on the cam gears are aligned—1.8L, 2.3L and 2.5L engines

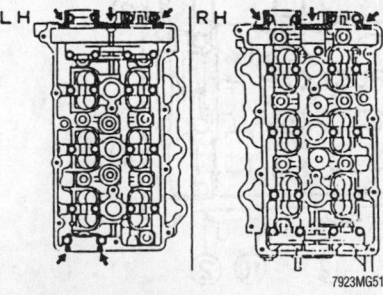

Put silicone sealant on the cylinder head at the positions shown—1.8L, 2.3L and 2.5L engines

➡ **Do not remove the camshaft caps when the camshaft lobe is pressing on a lifter, as the thrust journal support may become damaged.**

10. Loosen the front camshaft cap bolts in 5–6 steps, in the proper sequence. Bolt **A** is only on the right cylinder head. Remove the front camshaft cap.

11. Mark the position of the camshaft caps so they can be reinstalled in their original locations. Loosen the remaining camshaft cap bolts in 5–6 steps, in the proper sequence, then remove the caps.

12. Remove the camshafts.

To install:

13. Lubricate the camshaft journals, lobes and gears with clean engine oil. Align the intake and exhaust camshaft timing marks and install the camshafts.

➡ **The thrust plate positions for the right and left cylinder head camshafts are different.**

14. Be sure the camshaft cap and cylinder head surfaces are clean. Apply a small

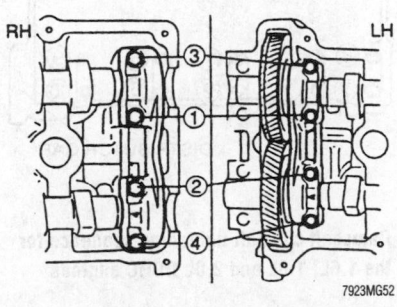

Front camshaft cap bolt tightening sequence—2.3L and 2.5L engines

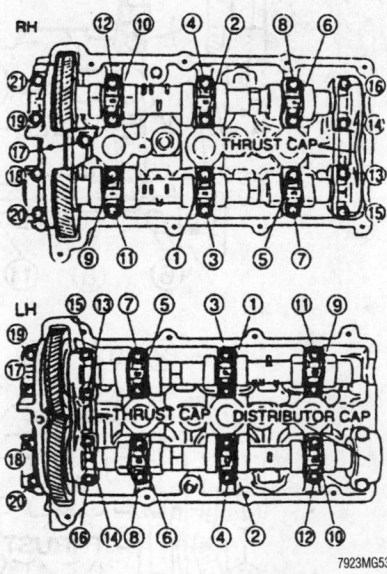

Camshaft cap bolt tightening sequence— 2.3L and 2.5L engines

amount of sealant to the mating surface of the front camshaft cap on both cylinder heads and the rear exhaust camshaft cap on the left cylinder head. Do not get any sealant on the camshaft rotating surfaces.

15. Install the front camshaft caps and thrust plate caps and tighten the bolts until the cap seats fully to the cylinder head. Install the remaining camshaft caps in their original locations and loosely tighten the bolts.

16. Tighten the camshaft cap bolts in 5–6 steps to 126 inch lbs. (14 Nm), in the proper sequence.

17. Apply clean engine oil to a new oil seal and the cylinder head. Install the seal, using a suitable installer. Apply sealant to a new blind cap and install, using a plastic hammer.

18. Install the camshaft sprockets. On the right cylinder head, install the sprocket so the **R** mark can be seen and the timing mark aligns with the camshaft knock pin. On the left cylinder head, install the sprocket so the **L** mark can be seen and the timing mark aligns with the camshaft knock pin.

19. Apply clean engine oil to the camshaft sprocket bolt threads and install. Hold the camshaft with a wrench on the cast hexagon and tighten the sprocket bolt to 103 ft. lbs. (140 Nm).

20. Coat a new gasket with sealant and install onto the cylinder head cover. Install the cover and tighten the bolts, in sequence, in 2–3 steps, to 78 inch lbs. (8.8 Nm). Install the ventilation pipe to the left cover.

21. Apply clean engine oil to a new O-ring and install on the distributor. Install the distributor with the blade fitting into the camshaft groove and loosely tighten the retaining bolt.

22. Install the intake manifold using a new gasket. Loosely install the bolts and nuts. Install the intake manifold stay and tighten the bolts to 19 ft. lbs. (25 Nm), then tighten the intake manifold bolts/nuts, in 2–3 steps, to 19 ft. lbs. (25 Nm).

23. Connect the wiring, hoses, and the fuel lines.

24. Connect the accelerator, if equipped, throttle valve cables.

25. Install the timing belt.

26. Connect the negative battery cable. Fill and bleed the cooling system.

27. Start the engine and bring to normal operating temperature. Check for leaks. Check the ignition timing and idle speed.

2.3L Engine

1. Relieve the fuel system pressure. Disconnect the negative battery cable.

2. Remove the timing belt covers and timing belt.

3. Remove the spacer and O-ring from the front of the camshaft.

4. Remove the ignition coils.

5. Remove the intake manifold.

6. Remove the bolts, in 5–6 steps, using the reverse of the tightening sequence. Remove the cylinder head cover.

7. Remove the camshaft sprockets.

8. Turn the camshafts so the knock pins are aligned with the marks on the camshaft caps. This will reduce the pressure on the adjustment shims.

9. Note the markings on the camshaft caps prior to removal, so they can be reinstalled in the same positions. The right-hand (rear) caps are marked with numbers and the left-hand (front) caps are marked with letters.

10. Loosen the front camshaft cap bolts in the reverse of the torque sequence, in 5–6 steps. Remove the front camshaft caps.

11. Remove the remaining camshaft cap bolts in the proper sequence. Remove the caps, being sure to remove the thrust caps last. Do not damage the cylinder head thrust bearing support.

12. Remove the camshafts and oil seals.

13. If necessary, remove the lifters and adjustment shims. Identify and mark each lifter as it is removed so it can be reinstalled in the same position.

To install:

14. Apply clean engine oil to the lifters, then install them in their original positions. Verify that they move smoothly in their bore.

15. Install new oil seals on the camshafts. Apply clean engine oil to the camshaft lobes, journals and supports.

16. Install the camshafts so the gear marks align.

17. Remove all oil and dirt from the mating surfaces between the front camshaft cap and the cylinder head.

18. Install the thrust caps. Tighten the thrust cap bolts, in 5–6 steps, until the caps are fully seated on the cylinder head.

19. Apply silicone sealant, at a thickness of 0.06–0.09 in. (1.5–2.5mm), to the cylinder head surface in the area forward of the camshaft gear cavity.

20. Install the remaining camshaft caps in their original positions. Tighten the caps, in sequence, in five equal steps, with the final step being 100–125 inch lbs. (11–14 Nm).

21. Apply clean engine oil to the lip of the new camshaft oil seal. Push the seal in lightly by hand. Tap the seal in evenly with a seal installer (49 F401 337A or equivalent) with a final protrusion of 0–0.02 in. (0–0.5mm). Tap in a new blind cap.

22. Install the camshaft sprockets. Tighten the mounting bolts to 91–103 ft. lbs. (123–140 Nm).

23. Measure and adjust valve clearances.

24. Remove any sealant and gasket material from the cylinder head cover contact surfaces.

25. Apply silicone sealant to the cylinder head in the area adjacent to the front and rear camshaft caps. Install a new gasket on the cylinder head.

26. Install the cylinder head cover. Tighten the bolts in 5–6 steps, in sequence, to 44–78 inch lbs. (5–9 Nm).

27. Using a new O-ring, install the distributor.

28. Install the ignition coils.

29. Install the intake manifold.

30. Install the spacer, using a new O-ring. Tighten the mounting bolt to 14–18 ft. lbs. (19–25 Nm).

31. Install the timing belt and timing belt cover.

32. Connect the negative battery cable. Run the engine and check for leaks.

Valve Lash

ADJUSTMENT

Except 1.5L, 2.3L, 1997–99 1.8L (BP) and 1998–99 2.0L and 2.5L Engines

The valve lash on all engines is kept in adjustment hydraulically. No adjustment is necessary, or possible.

1.5L, 2.3L, 1997–99 1.8L (BP) and 1998–99 2.0L and 2.5L Engines

These engines use solid cam followers with a removable adjustment shim. The valve lash clearance is measured with the original shim installed and checked against the specification. If adjustment is necessary, the original shim is removed, and a thicker or thinner shim is installed to obtain the proper clearance. Special tools are required in order to adjust the shim without removing the camshaft.

➡ **With the engine cold, standard valve clearance is 0.010-0.012 in. (0.25-0.31mm) on intake and exhaust sides.**

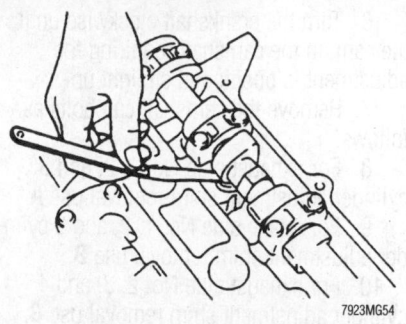

Ensure that the cam lobe faces away from the follower when checking the valve clearance

1. Remove the cylinder head cover.

2. Measure the valve clearance by turning the crankshaft clockwise until the No. 1 piston is at TDC.

3. Measure the valve clearance at **A**. If the clearance exceeds specifications, replace the adjustment shim.

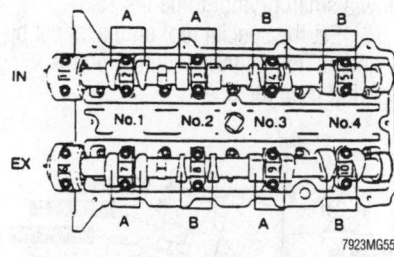

Mazda 4-cylinder engine valve clearance checking positions

4. Turn the crankshaft clockwise 360 degrees until the No. 4 piston is at TDC. Measure the valve clearance at **B**. If the clearance exceeds specifications, replace the adjustment shim.

5. Repeat this procedure for all the camshafts.

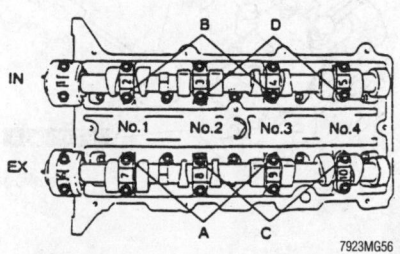

Mazda 4-cylinder engine cam bearing cap bolt removal positions

6. Turn the crankshaft clockwise until the cam on the camshaft requiring the adjustment is positioned straight up.

7. Remove the camshaft cap bolts as follows:

8. For exhaust side No. 1, 2, and 3 cylinder adjustment shim removal use **A**.

9. For intake side No. 1, 2, and 3 cylinder adjustment shim removal use **B**.

10. For exhaust side No. 2, 3, and 4 cylinder adjustment shim removal use **C**.

11. For intake side No. 2, 3, and 4 cylinder adjustment shim removal use **D**.

12. Install special tools 49-T012–002 and 003, using the camshaft cap bolt holes. Torque the bolts to 100-125 inch lbs. (11-14 Nm).

13. Align the mark on the 49-T012–002 (shaft) with the mark on the 49-T012–003 (clamp). Tighten special tool 49-T012–004 (bolt) to secure the shaft.

14. Position special tool 49-T012–001A toward the center of the cylinder head, and mount it on the shaft where the adjustment shim needs replacement.

15. Position the notch of the tappet to allow a small prytool to be inserted.

16. Set the special tool on the tappet by its notch. Tighten the mounting bolt **B** securing it on the shaft.

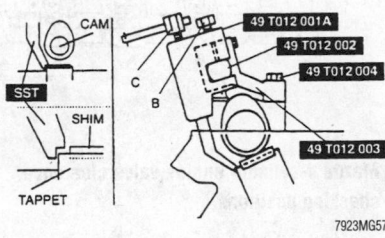

Mount the tappet depressor tool onto the shaft above the tappet which needs adjustment

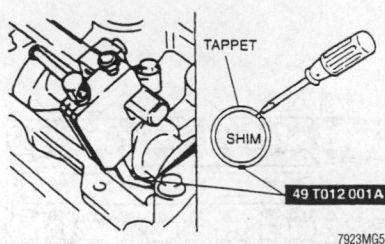

With the tappet depressed, use a small prytool to remove the adjustment shim

17. Tighten bolt **C**, and press down the tappet.

18. Using a small prytool, pry the adjustment shim upwards through the notch on the tappet. Remove the shim with a magnet.

19. Select and install the proper adjustment shim. Loosen bolt **C** to allow the tappet to move up, and loosen bolt **B** to remove special tool 49-T012–001A.

20. Remove special tools 49-T012–002, 003 and 004, and tighten the camshaft cap bolts to 100-125 inch lbs. (11-14 Nm).

21. Repeat the procedure for all necessary adjustment shims. Check the valve clearance.

2.3L and 1998–99 2.5L Engines

➡**With the engine cold, standard valve clearance is 0.011-0.012 in. (0.27-0.31mm) on intake and exhaust sides.**

1. Measure the valve clearance by turning the crankshaft clockwise until the No. 1 piston is at TDC.

2. Measure the valve clearance at **A**. Turn the crankshaft clockwise 240 degrees until the No. 3 piston is at TDC. Measure the valve clearance at **B**. Turn the crankshaft clockwise 240 degrees until the No. 5 piston is at TDC. Measure the valve clearance at **C**.

➡**If the valve clearance exceeds the standard, replace the adjustment shim.**

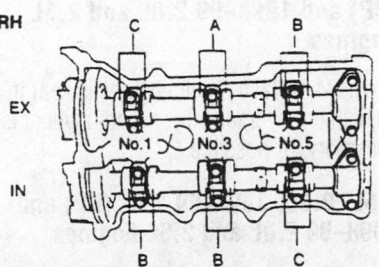

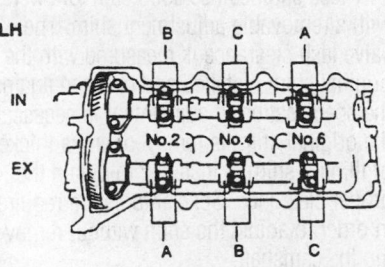

Mazda 6-cylinder engine valve clearance checking positions

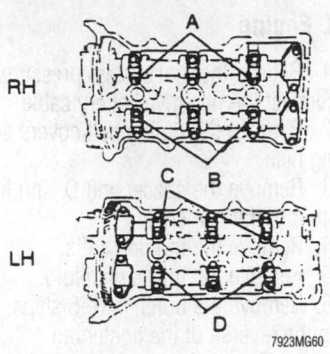

Mazda 6-cylinder engine camshaft cap bolt removal positions—refer to text

3. Turn the crankshaft clockwise until the cam, on the camshaft requiring the adjustment shim replacement, is positioned straight up.

4. Remove the camshaft cap bolts as follows:

a. For right-hand (RH) exhaust side shim removal use **1**.

b. For right-hand (RH) intake side shim removal use **2**.

c. For left-hand (LH) intake side shim removal use **3**.

d. For left-hand (LH) exhaust side shim removal use **4**.

5. Install special tools 49-T012–002 and 003, using the camshaft cap bolt holes.

6. Align the mark on the 49-T012–002 (shaft) with the mark on the 49-T012–003 (clamp).

7. Position special tool 49-T012–001 toward the center of the cylinder head, and mount it on the shaft where the adjustment shim needs replacement.

8. Position the notch of the tappet to allow a small prytool to be inserted.

9. Set the special tool on the tappet by its notch. Tighten the mounting bolt **B** securing it on the shaft.

10. Tighten bolt **C**, and press down the tappet.

11. Using a small prytool, pry the adjustment shim upwards through the notch on the tappet. Remove the shim with a magnet.

12. Select and install the proper adjustment shim. Loosen bolt **C** to allow the tappet to move up, and loosen bolt B to remove special tool 49-T012–001.

13. Remove special tools 49-T012–002, 003 and 004, and tighten the camshaft cap bolts to 100-125 inch lbs. (11-14 Nm).

14. Repeat the procedure for all necessary adjustment shims. Check the valve clearance.

Oil Pan

REMOVAL & INSTALLATION

1.5L and 1.8L (BP) Engines

1. Disconnect the negative battery cable. Raise and safely support the vehicle.

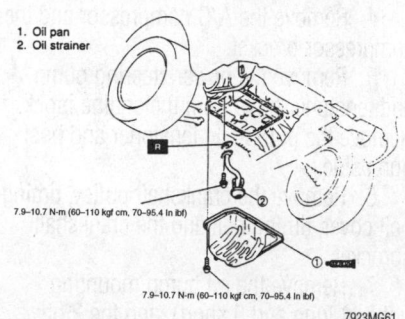

1. Oil pan
2. Oil strainer

7.9–10.7 N·m (60–110 kgf cm, 70–95.4 in lbf)

7.9–10.7 N·m (60–110 kgf cm, 70–95.4 in lbf)

7923MG61

Exploded view of the oil pan and related components—1.5L engine, 2.0L engine is similar

2. Remove the right-hand splash shield. Drain the engine oil into a suitable container.

3. Remove the transverse member.

4. Disconnect the oxygen sensor connector. Remove and discard the exhaust pipe-to-manifold nuts. Move the exhaust pipe aside and support it with a jack.

5. Remove the oil pan bolts and the oil pan.

To install:

6. Clean the oil pan. Clean all dirt, oil, gasket and old sealant from the oil pan and cylinder block contact surfaces.

7. Apply a continuous bead of silicone sealant on the gaskets and around the oil pan, going on the inside of the bolt holes.

8. Position the gaskets on the oil pan. Install the oil pan, and tighten the vertical bolts to 70–95 inch lbs. (8–11 Nm), and the horizontal bolts to 28–38 ft. lbs. (38–51 Nm).

9. Connect the exhaust pipe to the manifold with new nuts. Tighten the nuts to 28–38 ft. lbs. (38–51 Nm). Connect the oxygen sensor connector.

10. Install the transverse member, and tighten the mounting bolts to 69–97 ft. lbs. (94–131 Nm).

11. Install the right-hand splash shield and tighten the mounting bolts to 70–95 inch lbs. (8–11 Nm). Lower the vehicle.

12. Fill the engine with the proper type and quantity of engine oil. Connect the negative battery, run the engine and check for leaks.

1.8L (K8), 2.0L, 2.3L and 2.5L Engines

1. Disconnect the negative battery cable. Raise and safely support the vehicle.

2. Remove the passenger side splash shield. Drain the engine oil into a suitable container.

3. Disconnect the oxygen sensor. Remove the front exhaust pipe.

4. Remove the oil pan bolts and the oil pan.

To install:

5. Clean the oil pan. Clean all dirt, oil and old sealant from the oil pan and cylinder block contact surfaces.

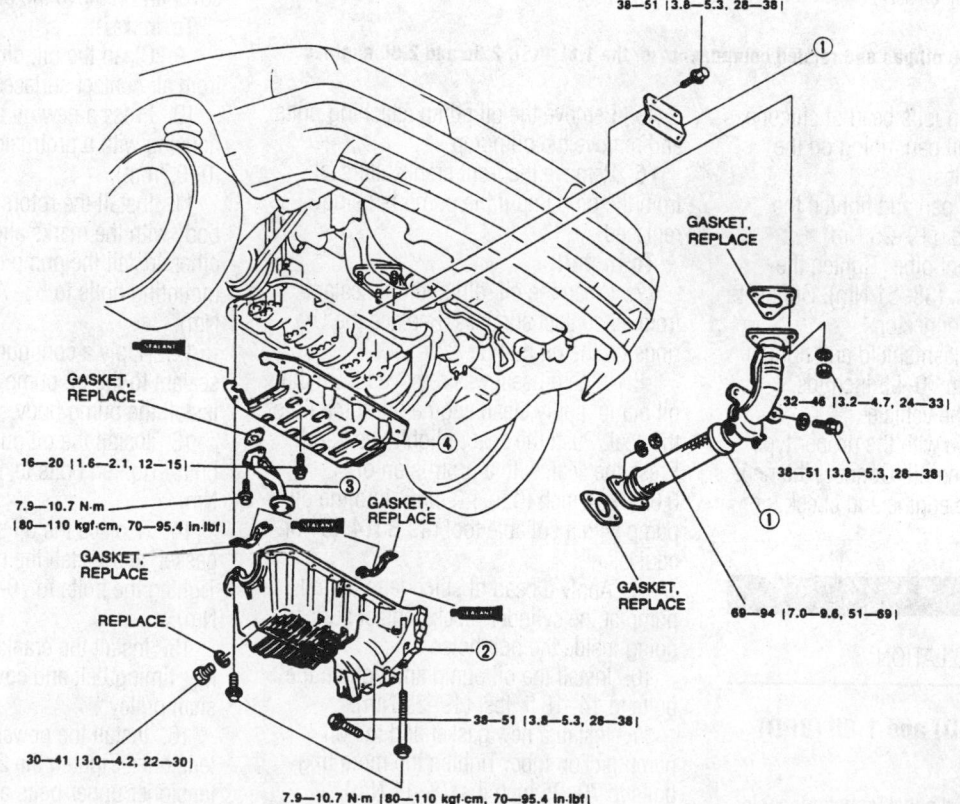

GASKET, REPLACE

GASKET, REPLACE

16–20 (1.6–2.1, 12–15)

7.9–10.7 N·m (80–110 kgf·cm, 70–95.4 in·lbf)

GASKET, REPLACE

GASKET, REPLACE

REPLACE

30–41 (3.0–4.2, 22–30)

7.9–10.7 N·m (80–110 kgf·cm, 70–95.4 in·lbf)

38–51 (3.8–5.3, 28–38)

38–51 (3.8–5.3, 28–38)

32–46 (3.2–4.7, 24–33)

38–51 (3.8–5.3, 28–38)

GASKET, REPLACE

69–94 (7.0–9.6, 51–69)

N·m (kgf·m, ft·lbf)

1. Front exhaust pipe and bracket
2. Oil pan
3. Oil strainer
4. Main bearing support plate (MBSP)

7923MG62

Exploded view of the oil pan and related components—1.8L (BP)

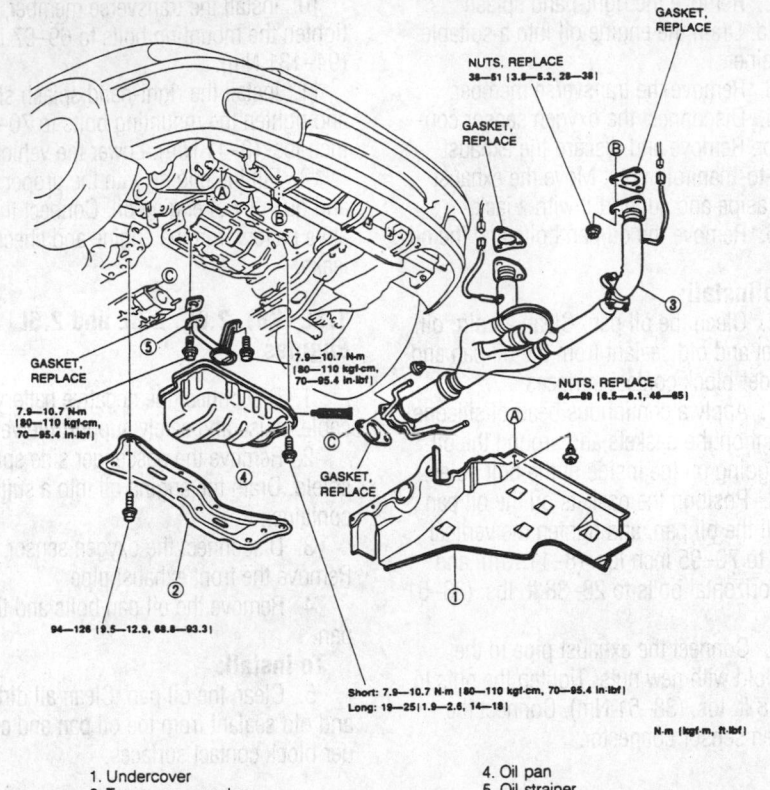

NUTS, REPLACE
38—51 (3.8—5.3, 28—38)

GASKET, REPLACE

GASKET, REPLACE

GASKET, REPLACE

7.9—10.7 N-m
(80—110 kgf-cm,
70—95.4 in-lbf)

GASKET, REPLACE

7.9—10.7 N-m
(80—110 kgf-cm,
70—95.4 in-lbf)

NUTS, REPLACE
64—89 (6.5—9.1, 48—65)

GASKET, REPLACE

94—126 (9.5—12.9, 68.8—93.3)

Short: 7.9—10.7 N-m (80—110 kgf-cm, 70—95.4 in-lbf)
Long: 19—25 (1.9—2.6, 14—18)

N-m (kgf-m, ft-lbf)

1. Undercover
2. Transverse member
3. Front exhaust pipe
4. Oil pan
5. Oil strainer

7923MG63

Exploded view of the oil pan and related components for the 1.8L (K8), 2.3L and 2.5L engines

6. Apply a continuous bead of silicone sealant around the oil pan, going on the inside of the bolt holes.

7. Install the oil pan and tighten the bolts to 14—18 ft. lbs. (19—25 Nm).

8. Install the front pipe. Tighten the nuts to 28—38 ft. lbs. (38—51 Nm). Connect the oxygen sensor connector.

9. Install the splash shield and tighten the mounting bolts to 70—95 inch lbs. (8—11 Nm). Lower the vehicle.

10. Fill the engine with the proper type and quantity of engine oil. Connect the negative battery, run the engine and check for leaks.

Oil Pump

REMOVAL & INSTALLATION

Protege 1.5L (Z5D) and 1.8L (BPD) Engines

1. Disconnect the negative battery cable.

2. Remove the crankshaft pulley, timing belt cover, belt and the crankshaft sprocket.

3. Remove the oil pan.

4. Remove the oil pickup tube and discard the gasket.

5. Remove the oil pump attaching bolts and remove the oil pump.

6. Remove the front crankshaft seal from the oil pump if the pump is being replaced.

To install:

7. Clean the oil, dirt and old sealant from all contact surfaces. Replace the O-rings on the oil pump.

8. If the oil seal was removed from the oil pump, apply clean engine oil to the lip of the seal. Push the seal in lightly be hand. Press the seal, with a protrusion of 0.02—0.04 inch (0.5—1.0 mm), into the oil pump with a suitable tool (49 B014 401 or equivalent).

9. Apply a bead of silicone to the oil pump at the cylinder block contact surface, going inside the bolt holes.

10. Install the oil pump and tighten the bolts to 14—18 ft. lbs. (19—25 Nm).

11. Install a new gasket and the oil pump pickup tube. Tighten the mounting bolts to 70—95 inch lbs. (8—11 Nm).

12. Install the oil pan.

13. Install the crankshaft sprocket, timing belt and cover.

14. Install the crankshaft pulley

15. Connect the negative battery cable.

16. Fill the engine with the proper type and quantity of oil. Run the engine and check for leaks.

MX3 1.8L (K8) and 626/MX6 2.5L (KL) Engines

1. Disconnect the negative battery cable.

2. Remove the oil pan.

3. Properly discharge the refrigerant from the A/C system.

4. Remove the A/C compressor and the compressor bracket.

5. Remove the power steering pump and tensioner bolts from the engine block. Remove the pump and tensioner and position aside.

6. Remove the crankshaft pulley, timing belt cover, timing belt and the crankshaft sprocket.

7. Remove the oil pump mounting bolts (4 long and 5 short), and the 2 oil strainer-to-pump bolts. Remove the oil pump body.

8. Press the oil seal from the housing. Remove the pump cover mounting bolts with an impact screwdriver. Remove the cover and remove the oil pump rotors.

To install:

9. Clean the oil, dirt and old sealant from all contact surfaces.

10. Press a new oil seal into the pump housing with a protrusion of 0—0.03 inch (0—0.7mm).

11. Install the rotors into the oil pump body with the marks aligned with each other. Install the pump cover and torque the mounting bolts to 53—78 in. lbs. (5.9—8.8 Nm).

12. Apply a continuous bead of silicone sealant to the oil pump mating surface, and install the pump body.

13. Install the oil pump body mounting bolts. Tighten bolts to 14—18 ft. lbs. (19—25 Nm).

14. Replace the oil strainer-to-pump gasket, and install the mounting bolts. Tighten the bolts to 70—95 inch lbs. (8—11 Nm).

15. Install the crankshaft sprocket and key, timing belt and cover. Install the crankshaft pulley.

16. Install the power steering pump and tensioner. Tighten the 2 power steering belt tensioner upper bolts and the power steering pump rear bracket bolt to 33 ft. lbs. (46 Nm). Tighten the tensioner lower bolt to 18 ft. lbs. (25 Nm).

17. Install the A/C compressor bracket and tighten the bolts to 38 ft. lbs. (51 Nm).

Install the A/C compressor and tighten the bolts to 38 ft. lbs. (51 Nm).

18. Install the oil pan. Fill the engine with the proper type and quantity of oil.

19. Connect the negative battery cable. Run the engine and check for leaks.

20. Evacuate and charge the A/C system.

MX3 1.6L (B6-ZE), Miata 1.8L (BPD) and 626/MX6 2.0L (FS) Engines

1. Disconnect the negative battery cable.

2. Remove the crankshaft pulley, timing belt cover, belt and the crankshaft sprocket.

3. Remove the A/C compressor and secure it aside, leaving the refrigerant lines attached. Remove the compressor mounting bracket.

4. Remove the oil pan.

5. Remove the oil pickup tube and discard the gasket.

6. Remove the oil pump body attaching bolts and remove the oil pump body.

7. Remove the front crankshaft seal from the oil pump if the pump is being replaced.

8. On the 2.0L (FS) engines, remove the pump cover mounting bolts with an impact screwdriver, remove the cover and the rotors.

To install:

9. Clean the oil, dirt and old sealant from all contact surfaces. Replace the O-rings on the oil pump.

10. If the oil seal was removed from the oil pump, apply clean engine oil to the lip of the seal. Push the seal in lightly be hand. Press the seal, with a protrusion of 0–0.02 inch (0–0.5 mm), into the oil pump.

11. Install the oil pump rotors into the pump body with the rotor marks aligned with each other.

12. On the 2.0L (FS) engines, install the pump cover, torque the bolts to 53–78 in. lbs. (5.9–8.8 Nm).

13. Apply a bead of silicone to the oil pump body-to-cylinder block contact surface, going inside the bolt holes.

14. Install the oil pump body and tighten the bolts to 14–18 ft. lbs. (19–25 Nm).

15. Install a new gasket and the oil pump pickup tube. Tighten the mounting bolts to 88 inch lbs. (10 Nm).

16. Install the oil pan.

17. Install the A/C compressor bracket and tighten the bolts to 38 ft. lbs. (52 Nm). Install the A/C compressor and tighten the bolts to 26 ft. lbs. (35 Nm).

18. Install the crankshaft sprocket, timing belt assembly and cover.

19. Install the crankshaft pulley

20. Connect the negative battery cable.

21. Fill the engine with the proper type and quantity of oil. Run the engine and check for leaks.

Millenia 2.3L (KJS) and 2.5L (KLD) engines

Due to space requirements, the engine assembly must be removed in order to replace the oil pump.

1. Disconnect the negative battery cable. Wait at least 90 seconds before performing any work. The backup power supply system for the SRS must deplete its stored energy.

2. Raise and safely support the vehicle.

3. Drain the engine oil and the engine coolant.

4. Remove and tag all electrical connections, hoses, and cables necessary to remove the engine assembly. Remove the front exhaust pipe.

5. Remove the engine assembly and secure to a suitable engine holding device.

6. Remove the oil pan, the oil strainer, and the oil pan baffle.

7. Remove the timing belt.

8. Using a suitable tool, remove the front timing belt pulley and key.

9. If equipped with a vacuum pump, remove the mounting bolts and nuts and remove the vacuum pump, O-ring, and gasket.

10. Remove the front seal, the mounting bolts, the oil pump, and the oil pump O-rings.

To install:

11. Make sure that the oil pump mating surfaces are clean. Install new O-rings coated with fresh engine oil to the oil pump cavity. Coat a new oil seal with fresh engine oil and install it to the oil pump. Apply silicone sealant to the oil pump contact surface and install the oil pump with the mounting bolts. Torque the mounting bolts to A: 16–22 ft. lbs. (22–30 Nm). All others: 14–18 ft. lbs. (19–25 Nm). Torque the bolts in sequence.

12. If equipped with a vacuum pump, reinstall the pump using a new gasket and O-ring.

13. Install the key and the timing belt pulley.

14. Reinstall the timing belt.

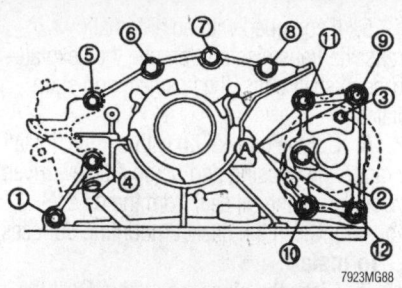

Tightening the oil pump mounting bolts in sequence—2.3L and 2.5L engines

15. Install the oil pan baffle, the oil pump strainer, and install the oil pan.

16. Remove the engine from the holding device and install the engine assembly into the vehicle.

17. Reinstall the front exhaust pipe. Reconnect all electrical connectors, cables, hoses, and install any components necessary to complete the engine installation.

18. Refill the engine with engine oil and coolant. Safely lower the vehicle.

19. Reconnect the negative battery cable. Start the engine, bleed the cooling system, make any necessary adjustments, check for leaks, and road test for proper operation.

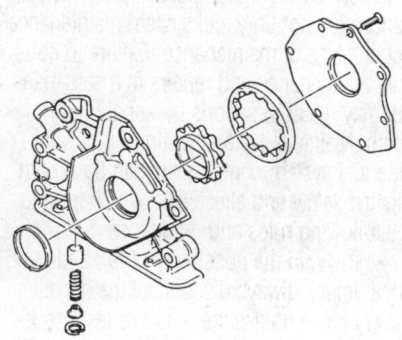

Exploded view of a typical oil pump assembly

Rear Main Seal

REMOVAL & INSTALLATION

1. Disconnect the negative battery cable.

2. Raise and safely support the vehicle.

3. Remove the transaxle/transmission assembly.

4. If equipped with a manual transaxle/transmission, remove the clutch and flywheel assembly.

5. If equipped with an automatic transaxle/transmission, remove the flexplate-to-crankshaft bolts, the flexplate and shim plates.

6. Cut the oil seal lip with a knife. Install a rag to the housing and using a screwdriver, carefully pry the oil seal from the oil seal housing. Clean the gasket mounting surfaces.

To install:

7. Clean the oil seal housing. Coat the oil seal and the housing with clean engine oil.

8. Press the oil seal into the housing and tap it evenly into place with a hammer and a large diameter piece of pipe. The seal must be flush with the edge of the rear cover.

9. Install the clutch and flywheel assembly or the flexplate, as applicable.

10. Install the transaxle/transmission, lower the vehicle and connect the negative battery cable.

FUEL SYSTEM

Fuel System Service Precautions

Safety is the most important factor when performing not only fuel system maintenance but any type of maintenance. Failure to conduct maintenance and repairs in a safe manner may result in serious personal injury or death. Maintenance and testing of the vehicle's fuel system components can be accomplished safely and effectively by adhering to the following rules and guidelines.

• To avoid the possibility of fire and personal injury, always disconnect the negative battery cable unless the repair or test procedure requires that battery voltage be applied.

• Always relieve the fuel system pressure prior to disconnecting any fuel system component (injector, fuel rail, pressure regulator, etc.), fitting or fuel line connection. Exercise extreme caution whenever relieving fuel system pressure to avoid exposing skin, face and eyes to fuel spray. Please be advised that fuel under pressure may penetrate the skin or any part of the body that it contacts.

• Always place a shop towel or cloth around the fitting or connection prior to loosening to absorb any excess fuel due to spillage. Ensure that all fuel spillage (should it occur) is quickly removed from engine surfaces. Ensure that all fuel soaked cloths or towels are deposited into a suitable waste container.

• Always keep a dry chemical (Class B) fire extinguisher near the work area.

• Do not allow fuel spray or fuel vapors to come into contact with a spark or open flame.

• Always use a back-up wrench when loosening and tightening fuel line connection fittings. This will prevent unnecessary stress and torsion to fuel line piping. Always follow the proper torque specifications.

• Always replace worn fuel fitting O-rings with new. Do not substitute fuel hose or equivalent, where fuel pipe is installed.

Fuel System Pressure

RELIEVING

✳✳ CAUTION

The fuel injection system remains under pressure after the engine has been turned OFF. Properly relieve fuel pressure before disconnecting any fuel lines. Failure to do so may result in fire or personal injury. Do not allow fuel spray or fuel vapors to come in contact with a spark or open flame. Keep a dry chemical fire extinguisher nearby. Never store fuel in an open container due to risk of fire or explosion.

323, Protege and MX3

1. Remove the rear seat cushion and locate the fuel pump connector.
2. Start the engine.
3. Detach the fuel pump connector.
4. After the engine stalls, turn the ignition switch **OFF** and reconnect the fuel pump connector.
5. Install the rear seat cushion.

MX6 and 626

1. Start the engine.
2. Remove the fuel pump relay from the relay box, located in the left side of the engine compartment.
3. After the engine stalls, turn the ignition switch **OFF** and reinstall the relay.

Millenia

2.3L ENGINES

1. Remove the cruise actuator mounting nuts, and move the cruise actuator to the side.
2. Start the engine.
3. Remove the fuel pump relay from the relay box, located in the right side of the engine compartment.
4. After the engine stalls, turn the ignition switch to **OFF** and reinstall the relay connector.
5. Position and install the cruise actuator.

2.5L ENGINES

To relieve the fuel system pressure on the 2.5L engine, refer to the MX6 and 626 procedure for the 2.0L and 2.5L engines.

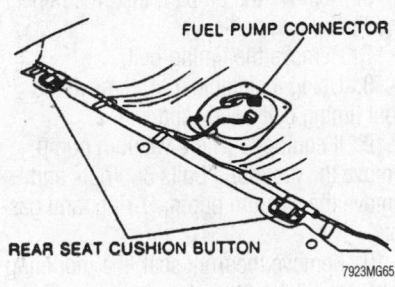

Fuel pump connector location—1995 MX3 1.6L and 1.8L (K8), and Protege 1.5L and 1.8L engines

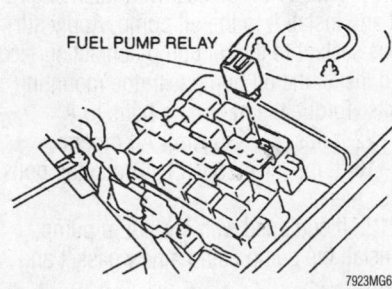

Fuel pump relay location—1995–99 626 and MX6 2.0L and 2.5L engines

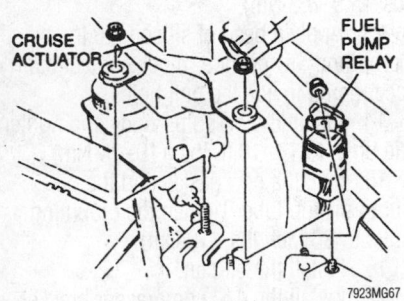

Fuel pump relay location—1995–97 Millenia 2.3L engines

Fuel Filter

On all models except the Millenia, the fuel filter is attached to a bracket located in the left rear of the engine compartment, next to or beneath the brake master cylinder fluid reservoir.

On the Millenia, the fuel filter is located beneath an access cover in the trunk. Access to the cover is achieved by removing the trunk mat to expose the cover.

The fuel filter should be serviced according to the Maintenance Intervals Chart at the end of this Section.

REMOVAL & INSTALLATION

✳✳ CAUTION

Do not allow fuel spray or fuel vapors to come in contact with a spark or open flame. Keep a dry chemical fire extinguisher nearby. Never store fuel in an open container due to risk of fire or explosion.

Except Millenia Models

1. Properly relieve the fuel system pressure.
2. Disconnect the negative battery cable.
3. If necessary, remove the air intake hose and/or filter housing.
4. If equipped, remove the fuel line clamps.
5. Disconnect the fuel lines from the filter and plug the ends to prevent leakage.
6. Loosen the bolt and nut and remove the fuel filter from its mounting bracket. Note the direction of the flow arrow on the filter so the replacement filter can be installed in the correct position.

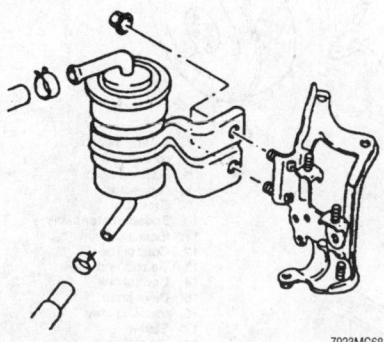

Exploded view of a common fuel filter and related components

To install:

7. Install the fuel filter in its mounting bracket, making sure the flow arrow is pointing in the proper direction. Tighten the bracket bolt and nut.
8. Unplug the fuel lines and connect them to the fuel filter.
9. If equipped, install the fuel line clamps.
10. If removed, install the air intake hose and/or filter housing.
11. Connect the negative battery cable.
12. Pressurize the fuel system and check all connections for leaks.

Millenia Models

1. Insure the ignition is **OFF**. Relieve the fuel system pressure.
2. Disconnect the negative battery cable.
3. Open the trunk, and remove the trunk mat.
4. Remove the service hole cover.
5. Disconnect the fuel lines from both ends of the fuel filter. Plug the lines to prevent leakage.
6. Remove the nut from the fuel filter bracket, and remove the filter and bracket from the vehicle.
7. Remove the filter from the mounting bracket.

To install:

8. Position the filter in the mounting bracket.
9. Install and tighten the bracket nut to 70–95 inch lbs. (8–11 Nm).
10. Unplug the fuel lines and connect them to the filter.
11. Install the service hole cover.
12. Replace the trunk mat, and close the trunk.
13. Connect the negative battery cable.
14. Run the engine and check for any fuel leaks.

Fuel Pump

REMOVAL & INSTALLATION

323 and Protege

1. Relieve the fuel pressure and disconnect the negative battery cable.
2. Depress the clips on each end of the rear seat cushion and remove the cushion.
3. Disconnect the electrical connector from the fuel pump/sending unit.

4. Remove the attaching screws from the fuel pump/sending unit access cover and remove the cover.
5. Disconnect the fuel supply and return hoses from the fuel pump/sending unit.
6. Remove the attaching screws and the fuel pump/sending unit from the fuel tank.
7. Disconnect the sending unit electrical connector, remove the sending unit attaching nuts and remove the sending unit from the fuel pump assembly.

To install:

8. Attach the sending unit to the fuel pump assembly and install the nuts. Connect the sending unit electrical connector.
9. Install the fuel pump/sending unit into the fuel tank with a new gasket and install the mounting screws.
10. Connect the fuel supply and return lines.
11. Install the access cover and the mounting screws.
12. Connect the sending unit electrical connector.
13. Position the rear seat cushion over the floor, making sure to align the retaining pins with the clips. Push down firmly until the 2 retaining pins are locked into the rear seat retaining clips.
14. Connect the negative battery cable, start the engine and check for proper system operation and for fuel leaks.

Miata

✳✳ CAUTION

Do not allow fuel spray or fuel vapors to come in contact with a spark or open flame. Keep a dry chemical fire extinguisher nearby. Never store fuel in an open container due to risk of fire or explosion.

1. Properly relieve the fuel pressure, and disconnect the negative battery cable.
2. Remove the rear package trim and remove the service hole cover.
3. Remove the fuel pump cover and disconnect the fuel pump connector. Disconnect the fuel hoses.
4. Remove the fuel pump and the fuel gauge sender unit as an assembly. Remove the fuel pump.
5. Installation is the reverse of removal. Use a NEW O-ring set when installing. After install the fuel pump to the bracket, pull the fuel pump down so that it is tight against the bracket.

MX3

❊❊ CAUTION

Do not allow fuel spray or vapors to come in contact with a spark or an open flame. Keep a dry chemical fire extinguisher nearby. Never store fuel in an open container due to risk of fire or explosion.

1. Relieve the fuel system pressure.
2. Disconnect the negative battery cable.
3. Remove the rear seat cushion from the vehicle.
4. Remove any dirt that has accumulated around the fuel pump cover so it will not enter the tank during pump removal and installation.
5. Remove the fuel pump cover.
6. Detach the fuel gauge connector, hoses, and the gauge.
7. Unplug the fuel pump electrical connector.
8. Remove the fuel pump from the bracket assembly. Remove and discard the seal ring.

To install:

9. Clean the fuel pump mounting flange, fuel tank mounting surface and seal ring groove.
10. Apply a light coating of grease on a new seal ring to hold it in place during assembly and install in the seal ring groove.
11. Install the fuel pump to the bracket assembly carefully to ensure the filter is not damaged. Be sure the seal ring remains in the groove.
12. Hold the pump assembly in place, and pull the fuel pump down so that it is tight against the bracket.
13. Attach the fuel pump electrical connector.
14. Install the fuel gauge, hoses, and gauge connector.
15. Install the fuel pump cover.
16. Install the rear seat cushion.
17. Connect the negative battery cable, start the engine and check for proper system operation and for fuel leaks.

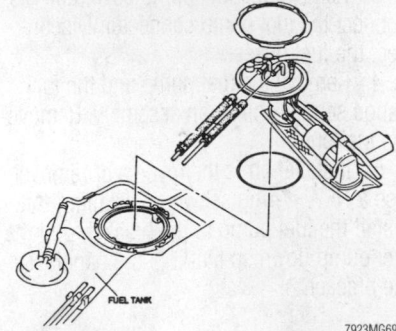

7923MG69

Exploded view of the 626 and MX6 fuel pump assembly

Millenia, 626 and MX6

1. Relieve the fuel system pressure and disconnect the negative battery cable.
2. Drain and remove the fuel tank.
3. Disconnect all fuel hoses from the fuel pump unit.
4. Turn the fuel pump ring counterclockwise and remove it.
5. Remove the fuel pump and gaskets from the fuel tank.

To install:

6. Install the fuel pump with a new gasket. Turn the fuel pump ring clockwise to tighten it until the flange hits the stopper.
7. Connect the fuel hoses to the fuel pump.
8. Install the fuel tank, add a minimum of 10 gallons and check for leaks.
9. Lower the vehicle and connect the negative battery cable.

DRIVE TRAIN

Refer to the Unit Repair Section (URS) for CV joint boot service information.

Transaxle Assembly

REMOVAL & INSTALLATION

Manual

EXCEPT MIATA

1. Disconnect the negative battery cable. Remove the air cleaner. Loosen the front wheel lug nuts.
2. Disconnect the speedometer cable or sensor wires from the transaxle.

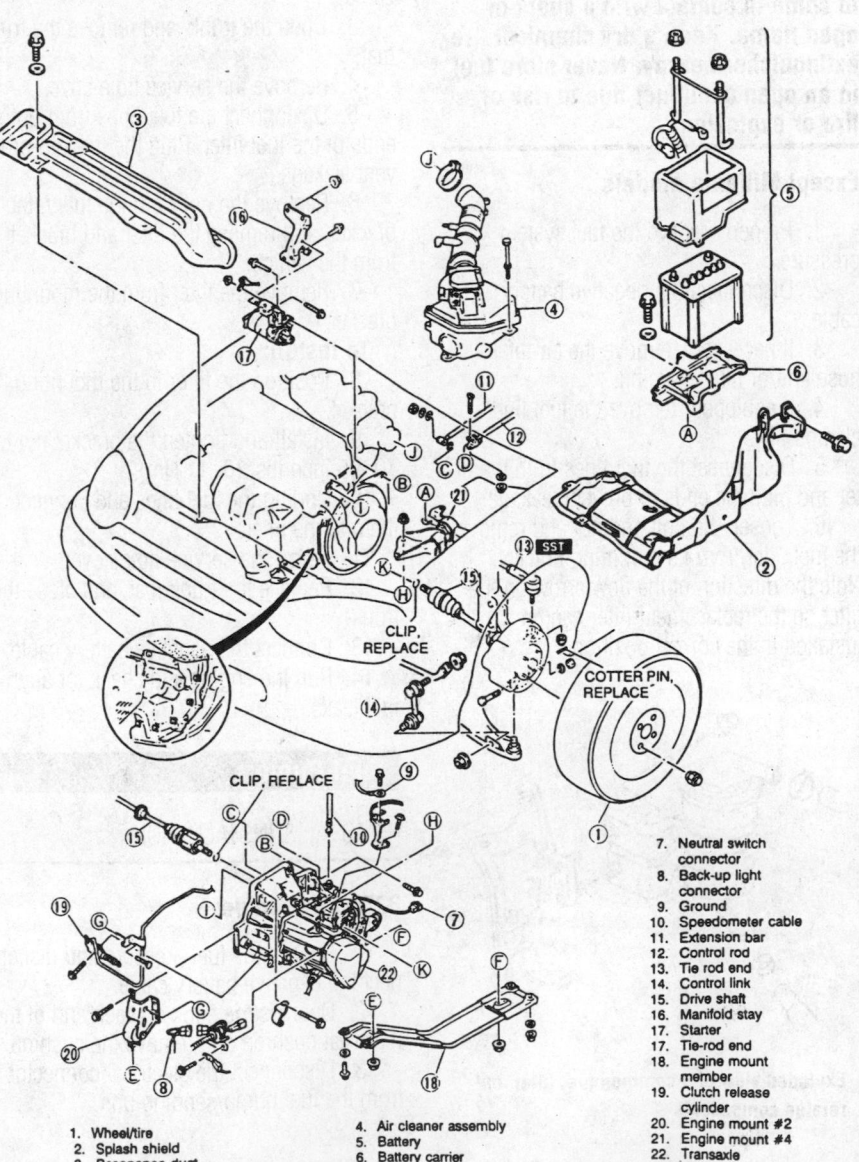

1. Wheel/tire
2. Splash shield
3. Resonance duct
4. Air cleaner assembly
5. Battery
6. Battery carrier
7. Neutral switch connector
8. Back-up light connector
9. Ground
10. Speedometer cable
11. Extension bar
12. Control rod
13. Tie rod end
14. Control link
15. Drive shaft
16. Manifold stay
17. Starter
17. Tie-rod end
18. Engine mount member
19. Clutch release cylinder
20. Engine mount #2
21. Engine mount #4
22. Transaxle

7923MG70

Exploded view of a typical transaxle assembly mounting and related components

3. On 4WD models, detach the neutral safety switch, back-up lamp switch, differential lock sensor switch and differential lock motor electrical connectors. Disconnect the transaxle shift and select control cables from the transaxle by removing the pins and cable retaining clips. Route the cables off to the side and out of the way.

4. Remove the clutch release (slave) cylinder from the transaxle.

5. Remove the water, secondary air, and EGR pipe brackets.

6. Remove the wiring harness clip. Disconnect the coupler for the neutral switch and back-up lamp switch. Detach the body ground connector.

7. Remove the two upper transaxle mounting bolts. Mount an engine support tool, 49-eR301–025A or equivalent, to the engine hanger.

8. Raise and support the vehicle safely. Drain the transaxle oil into a suitable container and remove the front wheels.

9. If necessary, remove the intake manifold support bracket.

10. Remove the engine under cover and side covers.

11. On 4WD models, remove the driveshaft and crossmember. Remove the oil filter and differential lock assembly (the differential lock assembly is fastened with three bolts).

12. Remove the halfshafts.

13. Insert differential side gear holder 49-B027–001 or its equivalent to hold the side gears in place and prevent misalignment.

14. Remove the transaxle crossmember. Separate the gear shift control rod from the transaxle. Remove the extension bar from the transaxle. Remove the wiring and the starter motor.

15. On 4WD models, remove the end plate bolts and connect a suitable hoist and lifting strap to the transaxle. Lift the transaxle and transfer carrier assembly out of the engine compartment.

16. On 2WD models, proceed as follows:

a. Remove the end plates. Lean the engine toward the transaxle side to lower the transaxle by loosening the engine support hook bolt. Support the transaxle with a suitable transaxle jack.

b. Remove the necessary engine brackets. Remove the remaining transaxle mounting bolt and engine brackets. Lower the jack and slide the transaxle out from under the vehicle.

To install:

17. Before installing the transaxle, lightly coat the splines of the primary shaft gear with molybdenum disulfide grease.

18. On 4WD models, connect a suitable hoist and lifting strap to the transaxle. Lower the transaxle and transfer carrier assembly into the engine compartment. Align the transaxle assembly to the engine.

19. On 2WD models, attach a thick rope to two places on the transaxle. Place a board on the jack and lower the transaxle onto the board. Using the jack, lift the transaxle into position and throw the end of the rope over the support fixture bar. Tension the rope to guide the transaxle onto its mounts while lifting the transaxle with the jack.

20. Once the transaxle is in place, have an assistant install and tighten all the transaxle-to-engine mounting bolts.

21. The remainder of the installation is the reverse of the removal procedure. Tighten all fasteners to specifications. Fill the transaxle with the proper amount and grade of fluid. Adjust any clutch and/or shifter linkages as necessary.

Automatic

EXCEPT MIATA

1. Raise and safely support the vehicle and remove the front wheels. Remove the battery and battery box and the air cleaner and ducting.

2. Remove the splash shield and drain the transaxle oil.

3. Disconnect the speedometer cable, throttle cable, shift cable and the wiring from the transaxle.

4. If necessary, properly relieve the fuel system pressure and disconnect the fuel lines.

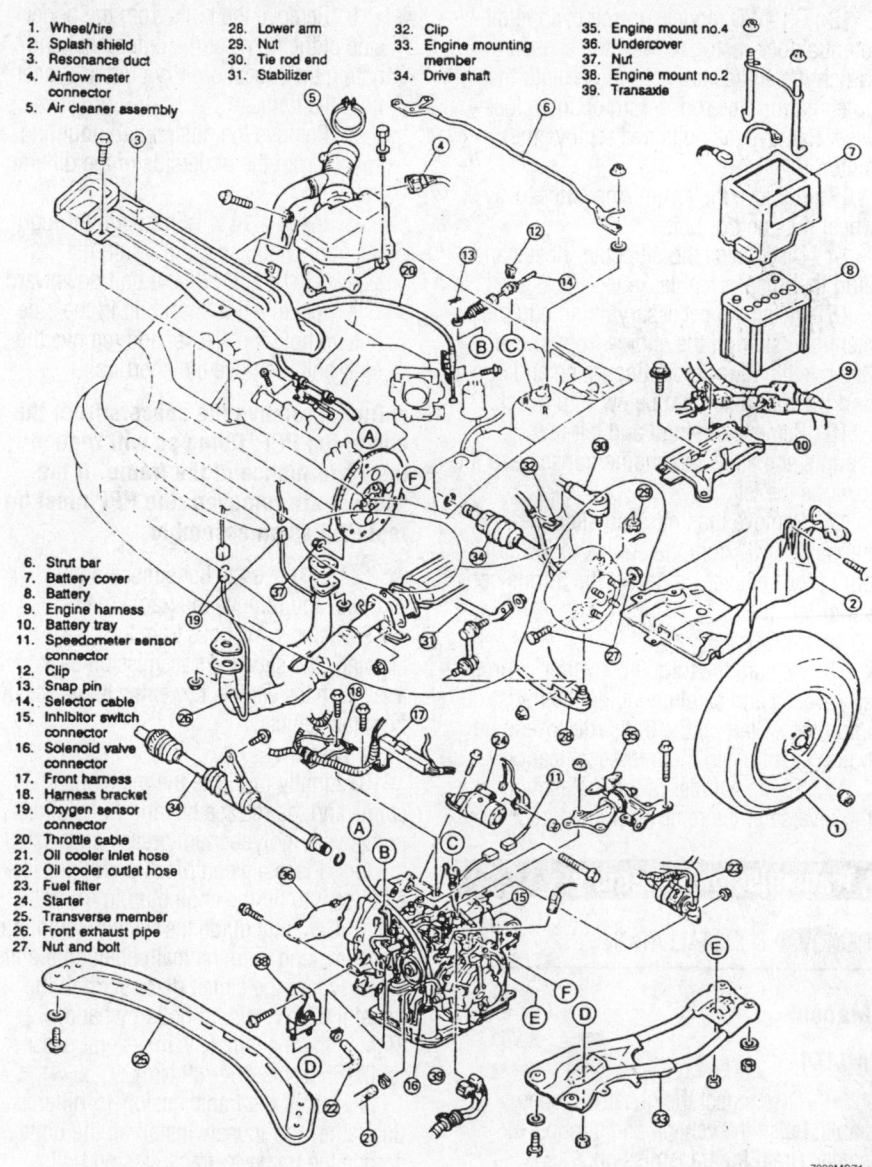

1. Wheel/tire
2. Splash shield
3. Resonance duct
4. Airflow meter connector
5. Air cleaner assembly
6. Strut bar
7. Battery cover
8. Battery
9. Engine harness
10. Battery tray
11. Speedometer sensor connector
12. Clip
13. Snap pin
14. Selector cable
15. Inhibitor switch connector
16. Solenoid valve connector
17. Front harness
18. Harness bracket
19. Oxygen sensor connector
20. Throttle cable
21. Oil cooler inlet hose
22. Oil cooler outlet hose
23. Fuel filter
24. Starter
25. Transverse member
26. Front exhaust pipe
27. Nut and bolt
28. Lower arm
29. Nut
30. Tie rod end
31. Stabilizer
32. Clip
33. Engine mounting member
34. Drive shaft
35. Engine mount no.4
36. Undercover
37. Nut
38. Engine mount no.2
39. Transaxle

Exploded view of a typical automatic transaxle assembly mounting and related components

7923MG71

5. If the fuel filter is mounted to the transaxle, unbolt its mounting bracket and position it aside.

6. Unbolt any exhaust crossover, coolant, vacuum or EGR pipe mounting brackets from the transaxle.

7. On the 2.0L engines, remove the intake manifold support bracket.

8. Disconnect the wiring and remove the starter.

9. On 4WD models, matchmark the flanges and remove the driveshaft.

10. Disconnect the exhaust pipe from the manifold and the catalytic converter and remove the pipe.

11. Disconnect the tie rod ends and lower ball joints and remove the halfshafts. Use special tool 49 G030 455 or equivalent to hold the differential side gears in place when the halfshafts are removed.

12. On 4WD models, to remove the differential lock motor, remove the sensor switch. Insert a small screwdriver into the hole and turn the rod ½ turn counterclockwise. Remove the bolts and remove the motor.

13. Remove the torque converter-to-flywheel nuts and/or bolts.

14. Disconnect the oil cooler hoses and plug them to prevent leakage.

15. Install the necessary lifting equipment and support the engine from above. Remove the lower mounting frame and support the transaxle from below with a jack.

16. Remove the front and left rear mounts and allow the engine/transaxle to tilt towards the left.

17. Remove the bolts and slide the transaxle away from the engine to lower it out of the vehicle. Do not let the torque converter fall out.

To install:

18. Be sure the torque converter is properly placed and carefully guide the transaxle into place. Start all the transaxle-to-engine bolts, then tighten them to specifications.

19. The remainder of the installation is the reverse of the removal procedure.

Transmission Assembly

REMOVAL & INSTALLATION

Manual

MIATA

1. Disconnect the negative battery cable. Raise the vehicle and support it safely. Drain the transmission.

2. Remove the shifter knob, and the center console. Remove the gearshift lever.

3. Remove the engine undercover and the performance rod.

4. Disconnect the exhaust pipe from the manifold. Remove the entire exhaust system as an assembly.

5. Matchmark the driveshaft flange at the rear, and remove the driveshaft.

6. Without disconnecting the hydraulic hose, remove the clutch release cylinder and set it aside. Disconnect the wiring and remove the starter.

7. Disconnect the speedometer cable, and remove the wiring from the frame member.

8. Support the transmission and differential with jacks. Remove the transmission-to-differential frame (also called the power plant frame or PPF) as follows:

   a. Remove the frame-to-transmission bracket.

   b. Remove the bolts from the underside of the frame at the differential end, noting their location. Pry out the spacer from the frame.

   c. Remove the differential mounting spacer from the underside of the differential.

   d. Insert a 14 **x** 1.5mm bolt through the frame hole, and turn it into the sleeve. Twist and pull the bolt downward.

   e. Install a 6 **x** 1mm bolt in the side hole to hold the sleeve, and remove the long bolt. Remove the short bolt.

➡**Do not remove the spacers from the end of the PPF. Doing so will reduce the performance of the frame. If the spacers are removed, the PPF must be replaced as an assembly.**

   f. Remove the transmission side bolts, and remove the frame member.

9. Remove the bolts from the clutch housing, and slide the transmission back away from the engine. Lower the transmission from the vehicle.

To install:

10. Lightly lubricate the main shaft spline and the release bearing fork contact points with molybdenum grease and install the fork. Place a wood block on a floor jack and use it to tilt the engine up in front.

11. Carefully guide the transmission into place, making sure the main shaft spline fits properly into the clutch disc. Start all the transmission retaining bolts by hand. Tighten them alternately, in several passes, to 48–65 ft. lbs. (64–89 Nm).

12. Install the transmission-to-differential frame, and loosely install all the bolts. Torque the frame-to-transmission bolts, then the frame-to-differential bolts to 76–91 ft. lbs. (104–124 Nm). Install the bracket,

and torque the bracket-to-transmission bolts to 27–40 ft. lbs. (37–54 Nm), the bracket-to-frame bolts to 77–91 ft. lbs. (104–124 Nm).

➡**After the frame installation, position a straightedge between the body frame members on each side of the vehicle. Measure the distance between the bottom of the frame to the straightedge; it should be 2.403–2.797 in. (61–71mm). If the distance is not as specified, reposition the frame member at the transmission.**

13. Connect the speedometer cable and wiring, and install the starter,.

14. Install the clutch release cylinder.

15. Install the driveshaft, and torque the nuts to 22 ft. lbs. (30 Nm).

16. Use a new gasket and install the exhaust system. Torque the nuts to 34 ft. lbs. (46 Nm).

17. Install the performance rod and engine undercover.

18. Lower the vehicle.

19. Fill the transmission with the proper type and quantity of oil.

20. Install the shift lever, rear console and the shift lever knob.

21. Connect the negative battery cable.

22. Verify proper operation of the transmission.

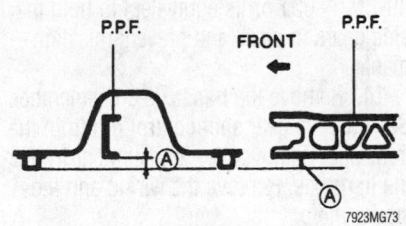

Measure distance A after installing the transmission-to-differential frame member—Miata with N4A-HL and NC4A-EL transmissions

Automatic

MIATA

1. Shift the selector lever to the **N** position.

2. Disconnect the negative battery cable. Raise and safely support the vehicle. Drain the transmission fluid.

3. Remove the engine undercover, and disconnect the shift rod.

4. Remove the performance rod.

5. Remove the complete exhaust system from the exhaust manifold.

6. Mark the position of the driveshaft on the rear axle flange, and remove the driveshaft.

7. Disconnect the speedometer cable, and disconnect the vacuum hose from the vacuum diaphragm.

8. Label and disconnect the electrical connectors from the: inhibitor switch, kick-down solenoid, overdrive cancel solenoid, oil pressure switch and lockup solenoid.

9. Remove the dipstick and dipstick tube. Disconnect the oil cooler lines.

10. Support the transmission and differential with jacks. Remove the transmission-to-differential frame as follows:

    a. Disconnect the wiring harness from the frame.

    b. Remove the bolts from the underside of the frame at the differential end, noting their location. Pry out the spacer from the frame.

    c. Remove the differential mounting spacer from the underside of the differential.

    d. Insert a 14 **x** 1.5mm bolt through the frame hole, and turn it into the sleeve. Twist and pull the bolt downward.

    e. Install a 6 **x** 1mm bolt in the side hole to hold the sleeve, and remove the long bolt. Remove the short bolt.

    f. Remove the transmission side bolts, and remove the frame member.

11. Remove the torque converter bolts, and the starter.

12. Remove the transmission mounting bolts and remove the transmission, being careful not to drop the torque converter.

To install:

13. Be sure the torque converter is fully installed in the transmission. The distance between 1 of the bolt hole lugs and a straightedge laid across the bell housing should be 0.89 in. (22.5mm).

14. Raise the transmission into position, and install the mounting bolts. Tighten to 48–66 ft. lbs. (64–89 Nm).

15. Install the starter. Install the torque converter bolts. Align the holes by turning the torque converter. Hold the flexplate with a small prybar, and tighten the bolts to 27–40 ft. lbs. (36–54 Nm).

16. Install the transmission-to-frame member as follows:

    a. Install the differential mounting spacer on the underside of the differential. Tighten the mounting bolts to 38 ft. lbs. (52 Nm).

    b. Position the jack under the transmission so the transmission is level.

    c. Position the frame and install the transmission side bolts. Snug the bolts.

---

d. Be sure the sleeve is installed in the block. Install the spacer and bolts with the reamer bolt in the front hole. Snug the bolts.

e. Tighten the transmission side bolts to 77–91 ft. lbs. (104–123 Nm), then tighten the differential bolts to the same specification.

f. Connect the wiring harness to the frame member and remove the jacks.

➡**After the frame installation, position a straightedge between the body frame members on each side of the vehicle. Measure the distance between the bottom of the frame to the straightedge; it should be 2.023–2.417 in.**

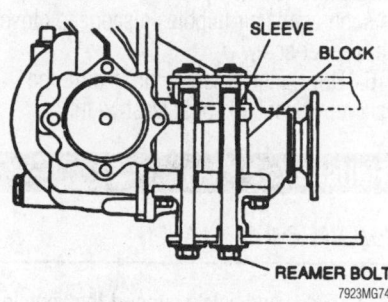

Sleeve and reamer bolt positioning— Miata with N4A-HL and NC4A-EL transmissions

---

(51.5–61.5mm). If the distance is not as specified, reposition the frame member at the transmission.

17. Connect the oil cooler lines, using new gaskets. Install the dipstick tube and dipstick.

18. Connect the electrical connectors and the vacuum hose. Connect the speedometer cable.

19. Install the driveshaft, aligning the marks made during removal. Tighten the bolts to 22 ft. lbs. (30 Nm).

20. Install the exhaust system. Connect the front pipe to the exhaust manifold, using a new gasket, and tighten the nuts to 34 ft. lbs. (46 Nm).

21. Install the performance rod.

22. Install the engine undercover and connect the shift rod.

23. Lower the vehicle and connect the negative battery cable. Fill the transmission with the proper type and quantity of fluid. Start the engine and check for leaks and proper operation.

Clutch

REMOVAL & INSTALLATION

1. Disconnect the negative battery cable. Raise and safely support the vehicle.

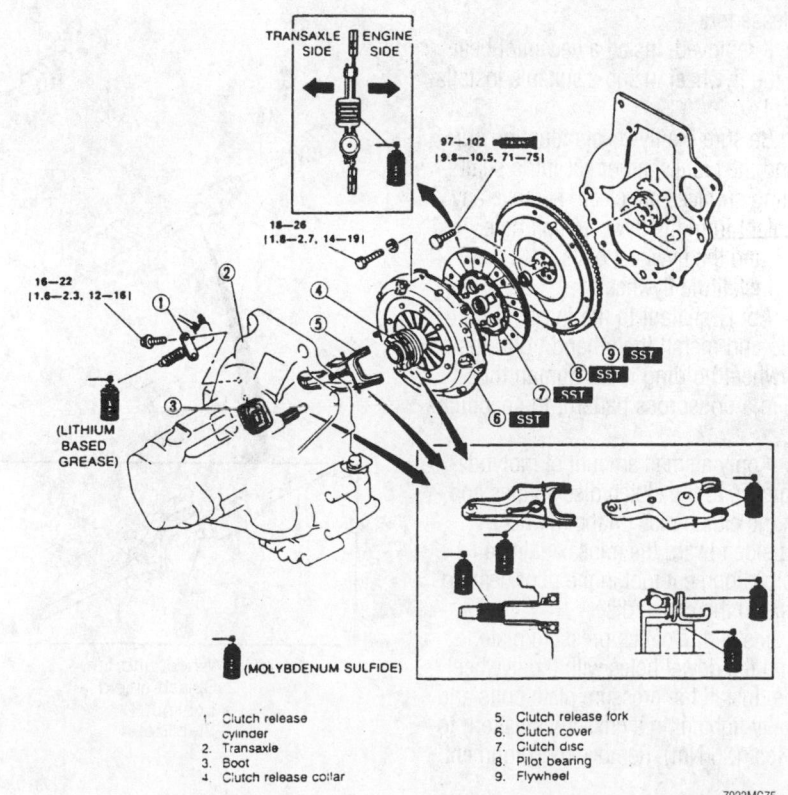

1. Clutch release cylinder
2. Transaxle
3. Boot
4. Clutch release collar
5. Clutch release fork
6. Clutch cover
7. Clutch disc
8. Pilot bearing
9. Flywheel

Exploded view of a typical clutch disc and pressure plate assembly with related components

2. Remove the transaxle.

3. Gradually loosen the clutch pressure plate bolts, in a crisscross pattern. Support the pressure plate and remove the bolts. Remove the pressure plate and clutch disc.

4. Inspect the pilot bearing. If it is worn or damaged and does not turn easily by hand, remove it using a puller/slide hammer.

5. Check the flywheel surface for scoring, cracks or burning and machine or replace, as necessary.

6. Install holder tool 49 E011 1A0 or equivalent, to keep the flywheel from turning. Loosen the flywheel bolts evenly and gradually in a crisscross pattern. Remove the flywheel.

7. Install holder tool 49 F011 101 or equivalent, to keep the flywheel from turning. Remove the locknut. Remove the flywheel, using a suitable puller and remove the key from the eccentric shaft.

8. Inspect the clutch release bearing for wear. Replace it if it sticks or does not turn easily.

9. Inspect the release fork for wear or damage and replace as necessary.

To install:

10. Lubricate the release fork fingers and pivot with molybdenum grease and install in the release fork boot.

11. Install the clutch release bearing on the release fork.

12. If removed, install a new pilot bearing in the flywheel, using a suitable installation tool.

13. Be sure the flywheel mounting surface and the crankshaft or eccentric shaft mounting surfaces are clean. Remove any old sealant from the flywheel bolt hole threads and the flywheel bolts.

14. Install the flywheel.

15. Apply sealant to the flywheel bolt threads and install them hand tight. Install the flywheel holding tool. Tighten the bolts, in a crisscross pattern, to specification.

16. Apply a small amount of molybdenum grease to the clutch disc splines and install the clutch disc on the flywheel, spring side toward the transaxle. Install a suitable alignment tool in the pilot bearing to position the clutch disc.

17. Install the clutch pressure plate, aligning the dowel holes with the flywheel dowels. Install the pressure plate bolts and gradually tighten, in a crisscross pattern to 20 ft. lbs. (26 Nm). Remove the alignment tool.

18. Install the transaxle and lower the vehicle.

Hydraulic Clutch System

BLEEDING

1. Remove the rubber cap from the bleeder screw on the release cylinder.

2. Place a bleeder tube over the end of the bleeder screw.

3. Submerge the other end of the tube in a jar half filled with hydraulic brake fluid.

4. Slowly pump the clutch pedal fully and allow it to return slowly, several times.

5. While pressing the clutch pedal to the floor, loosen the bleeder screw until the fluid starts to run out. Then, close the bleeder screw. Keep repeating this Step, while watching the hydraulic fluid in the jar. As soon as the air bubbles disappear, close the bleeder screw.

6. During the bleeding procedure the reservoir must be kept at least ¾ full.

Halfshafts

REMOVAL & INSTALLATION

1. Raise and safely support the vehicle. Remove the wheel and tire assemblies.

2. Remove the splash shield, if equipped, and drain the transaxle.

3. Raise the staked portion of the hub locknut with a hammer and chisel. Lock the hub by applying the brakes and remove the nut.

4. Disconnect the stabilizer bar from the lower control arm.

5. Remove the cotter pin and nut from the tie rod end ball stud. Use a suitable tool to separate the tie rod end from the knuckle.

6. On MX6/626 and Millenia, remove the transverse member.

7. Remove the lower ball joint pinch bolt and nut. Use a prybar to pry down the lower control arm and separate the ball joint from the knuckle.

8. If removing the left side shaft on MX3 and MX6/626 and Millenia with automatic transaxle, proceed as follows:

a. Suspend the engine using engine support tool 49 G017 5A0 or equivalent.

b. Remove the bolts and nuts and remove the engine mount member.

9. Position a prybar between the inner CV-joint and transaxle case. Carefully pry the halfshaft from the transaxle being careful not damage the oil seal. If equipped with a right side intermediate shaft, insert the prybar between the halfshaft and intermediate shaft and tap on the bar to uncouple them.

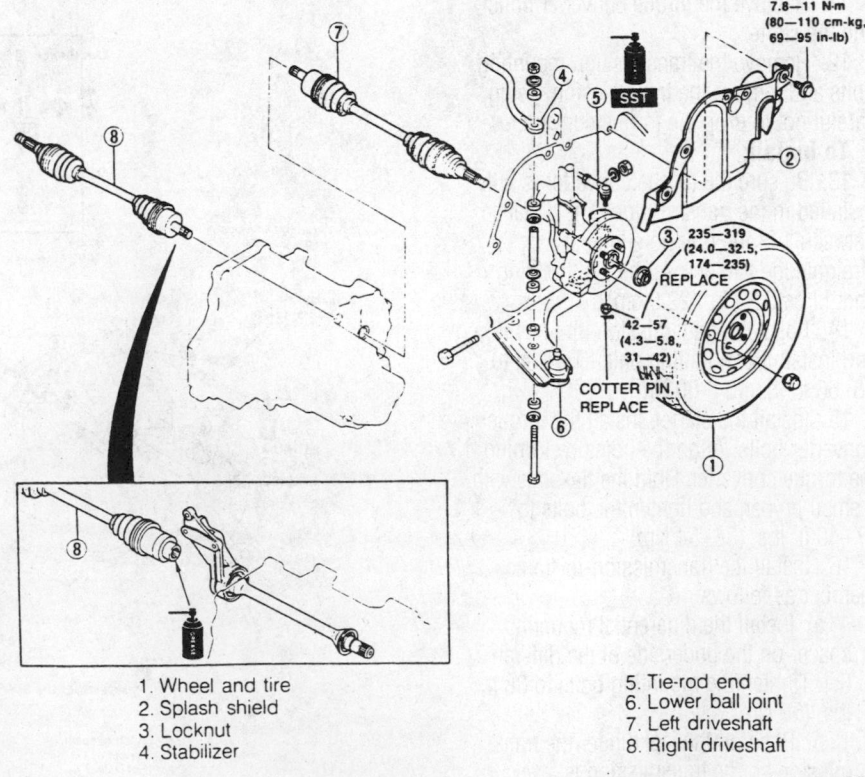

1. Wheel and tire
2. Splash shield
3. Locknut
4. Stabilizer
5. Tie-rod end
6. Lower ball joint
7. Left driveshaft
8. Right driveshaft

Exploded view of a typical halfshaft mounting

7923MG76

10. Pull outward on the hub/knuckle assembly, push the outer CV-joint stub shaft through the hub, and remove the half-shaft. If the halfshaft is stuck in the hub, install the old hub nut to protect the stub shaft threads. Tap on the nut, using only a soft mallet, to remove the halfshaft.

➡**Install plug tool 49 G030 455 or equivalent, into the transaxle after removing the halfshaft, to keep the differential side gear in position. If the gear becomes malpositioned, the differential may have to be removed to realign the gear.**

11. Remove the intermediate shaft, if necessary, by removing the support bearing bolts and pulling the shaft from the transaxle.

To install:

12. If removed, install a new circlip on the end of the intermediate shaft, with the end gap facing upward.

13. Install the intermediate shaft in the transaxle, being careful not to damage the oil seals. Install the support bearing bolts and tighten, in sequence, to 45 ft. lbs. (61 Nm).

14. Install a new circlip on the end of the halfshaft, with the end gap facing upward. Insert the halfshaft into the transaxle, being careful not to damage the oil seal. If equipped, push the halfshaft into the intermediate shaft.

15. Insert the other end of the halfshaft through the hub. Loosely install a new lock-nut.

16. If installing the left side shaft on MX3 and 1993–98 MX6/626 and Millenia with automatic transaxle, proceed as follows:

    a. Install the engine mount member. Tighten the mount member-to-body nuts and bolts to 66 ft. lbs. (89 Nm).

    b. On MX3, tighten the mount-to-mount member nuts to 38 ft. lbs. (52 Nm).

    c. On MX6/626 and Millenia, tighten the front mount-to-mount member nuts to 77 ft. lbs. (104 Nm) and the side mount bolts to 44 ft. lbs. (60 Nm).

    d. Remove the engine support tool.

17. Install the lower ball joint into the knuckle. Install the pinch bolt and nut and tighten to 40 ft. lbs. (54 Nm).

18. On MX6/626 and Millenia, install the transverse member and tighten the bolts to 96 ft. lbs. (132 Nm).

19. Connect the tie rod end to the steering knuckle and tighten the nut to 42 ft. lbs. (57 Nm) on all except 1993–98 MX6/626 and Millenia, where the torque is 32 ft. lbs.

(44 Nm). Install a new cotter pin. Tighten the nut, if necessary, to align the ball stud hole with the nut castellation.

20. Connect the stabilizer bar to the lower control arm.

21. Install the splash shield and the wheel and tire assemblies. Lower the vehicle.

22. Lock the hub with the brakes. Tighten the new hub nut to 174–235 ft. lbs. (235–318 Nm). After tightening, stake the locknut using a hammer and dull bladed chisel.

23. Fill the transaxle with the proper type and quantity of fluid.

STEERING AND SUSPENSION

Air Bag

✳✳ CAUTION

Some vehicles are equipped with an air bag system, also known as the Supplemental Inflatable Restraint (SIR) or Supplemental Restraint System (SRS). The system must be disabled before performing service on or around system components, steering column, instrument panel components, wiring and sensors. Failure to follow safety and disabling procedures could result in accidental air bag deployment, possible personal injury and unnecessary system repairs.

PRECAUTIONS

Several precautions must be observed when handling the inflator module to avoid accidental deployment and possible personal injury.

• Never carry the inflator module by the wires or connector on the underside of the module.

• When carrying a live inflator module, hold securely with both hands, and ensure that the bag and trim cover are pointed away.

• Place the inflator module on a bench or other surface with the bag and trim cover facing up.

• With the inflator module on the bench, never place anything on or close to the module which may be thrown in the event of an accidental deployment.

• An air bag is an explosive device. Handle with extreme caution.

• Always disconnect the battery and the air bag connector before removing the steering wheel or beginning work on the air bag system.

• Air bag components must not be repaired or opened. Always use new parts, including the wiring harness.

• Always place a removed air bag unit with the horn pad facing up. Put it in a safe place where it will not be disturbed.

• The air bag unit must not be exposed to grease, fluids, or cleaning agents.

• The air bag unit must not be exposed to temperatures above 194° F (90° C) at any time. Even the heat of a soldering iron can damage or ignite the charge.

• Storage and transport of air bags is subject to rules governing explosive devices and should be done only in the original package.

• Failure to follow proper safety precautions may result in personal injury through accidental firing of the air bag, or through failure of the air bag in an accident.

DISARMING

MX6 and 626

1. Disconnect the negative battery cable.

2. Locate and unplug the blue and orange connector. Protect this connector from accidental electrical contact, including static electricity.

3. The air bag is now disarmed and the battery can be connected to perform electrical testing on other systems.

4. When ready to reconnect the air bag; disconnect the battery, connect the air bag connector, install the cover or panel and be sure no one is in the vehicle when connecting the battery.

Protege

1. Turn the ignition switch to **LOCK**.
2. Disconnect the negative battery cable.

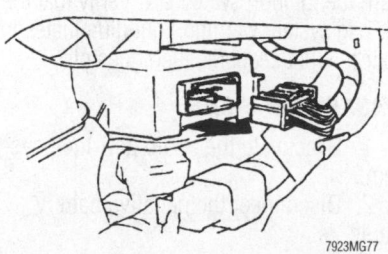

7923MG77

SAS unit and harness connector—Protege

3. Remove the interior right kick panel.

4. Disconnect the SAS unit connector.

Miata

DRIVER'S SIDE

❊❊ CAUTION

The air bag system must be disarmed before removing the steering wheel. Failure to do so may cause accidental deployment, property damage or personal injury.

1. Deactivate the audio anti-theft system.

2. Disconnect the negative battery cable.

3. Remove the lower panel under the steering wheel, then disconnect the orange and blue clock spring connectors.

4. The air bag is now disarmed and the battery can be connected to perform electrical testing on other systems.

5. If removing the unit, remove the mounting nuts, disconnect the support rope, disconnect the electrical connectors and remove the air bag.

6. When ready to reconnect the air bag; disconnect the battery, connect the air bag connectors, connect the support rope and mount the air bag. Torque the nuts in sequence to 35–52 inch lbs. (4.0–5.8 Nm).

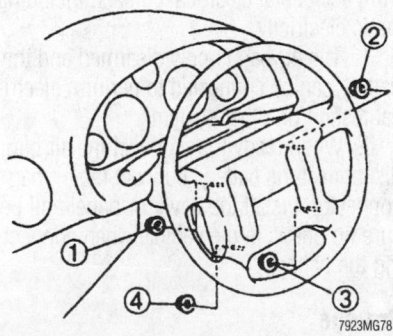

Driver's side air bag mounting nut torque sequence—1996 Miata

7. Connect the negative battery cable. Turn the ignition switch ON. Verify that the air bag system warning light illuminates for four to eight seconds, then goes off.

PASSENGER'S SIDE

1. Deactivate the audio anti-theft system.

2. Disconnect the negative battery cable.

3. Remove the glove compartment and the undercover.

4. Disconnect the orange and blue passenger side air bag module connectors.

5. The air bag is now disarmed and the battery can be connected to perform electrical testing on other systems.

6. If removing the unit, remove the mounting nuts, disconnect the electrical connectors and remove the air bag module.

7. When ready to reconnect the air bag; disconnect the battery, connect the air bag connectors and mount the air bag module. Torque the nuts in sequence to 13.1–20.2 ft. lbs. (17.7–27.4 Nm).

8. Connect the negative battery cable. Turn the ignition switch ON. Verify that the air bag system warning light illuminates for four to eight seconds, then goes off.

Millenia

1. Deactivate the audio anti-theft system.

2. Turn the ignition switch to LOCK.

3. Disconnect the negative battery cable and wait for more than one minute to allow the back-up power supply to deplete its stored power.

4. Remove the driver side undercover and lower dash panel. Disconnect the orange and blue clock spring connectors for the drivers side air bag.

5. Remove the glove compartment and disconnect the orange and blue passenger side air bag module connectors.

Rack and Pinion Steering Gear

REMOVAL & INSTALLATION

Manual

❊❊ CAUTION

Some models covered by this manual may be equipped with a Supplemental Restraint System (SRS), which uses an air bag. Whenever working near any of the SRS components, such as the impact sensors, the air bag module, steering column and instrument panel, disable the SRS, as described in Section 6.

1. Raise and safely support the vehicle. Disconnect the negative battery cable and remove the front wheels.

2. Remove the cotter pins from both steering tie rod ends and remove the nuts.

3. Use Mazda special tie rod press tool 49 0118 850C or equivalent and press the tie rod out of the knuckle arm.

4. Remove the set plate from the firewall.

5. Remove the fixing bolt from the steering shaft to steering gear pinion shaft and separate the shaft from the steering gear.

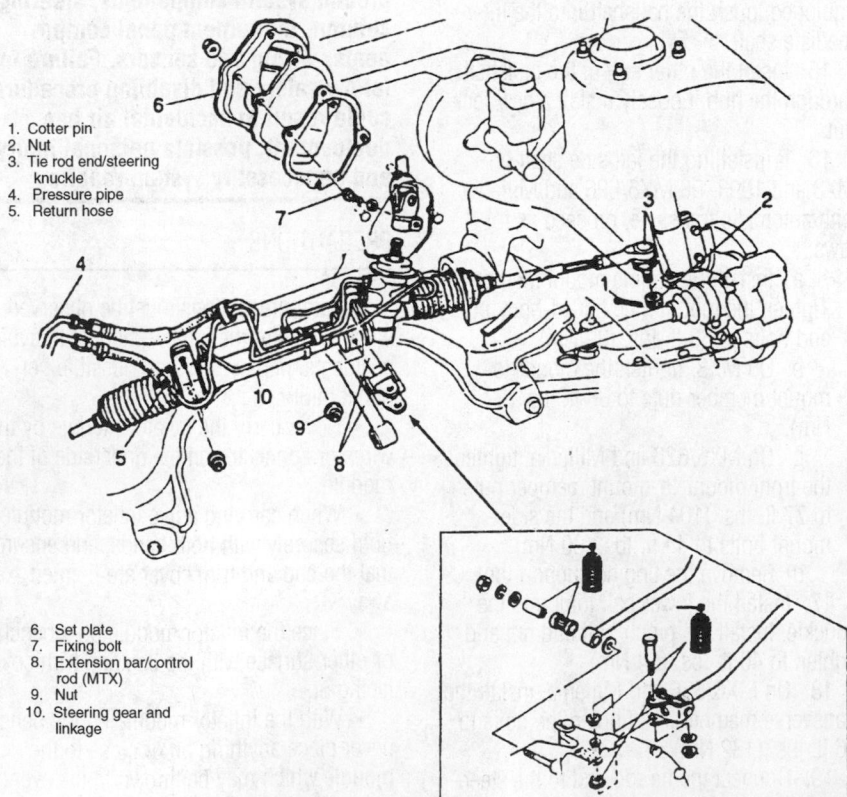

1. Cotter pin
2. Nut
3. Tie rod end/steering knuckle
4. Pressure pipe
5. Return hose
6. Set plate
7. Fixing bolt
8. Extension bar/control rod (MTX)
9. Nut
10. Steering gear and linkage

Exploded view of the manual steering gear assembly

6. Remove the steering gear mounting nuts and remove the steering gear to the right of the vehicle.

To install:

7. Install the steering gear to the vehicle and install the mounting nuts in the order shown. Tighten the nuts to 28–38 ft. lbs. (37–52Nm).

8. Connect the steering shaft to the steering gear pinion shaft. Tighten the bolt/nut to 13–20 ft. lbs. (18–27Nm).

9. Install the set plate to the firewall.

10. Install the tie rod ends to the knuckle arm and tighten the nuts to 31–42 ft. lbs. (42–57Nm). Install the cotter pins.

11. Install the wheels to the vehicle, lower the vehicle and connect the negative battery cable. Check the front end alignment.

Power

1. Raise and safely support the vehicle. Disconnect the negative battery cable and remove the front wheels.

2. Remove the cotter pins from both steering tie rod ends and remove the nuts.

3. Use Mazda special tie rod press tool 49 0118 850C or equivalent and press the tie rod out of the knuckle arm.

4. Disconnect the pressure line and return pipe from the steering gear. Remove the set plate from the firewall.

5. Remove the fixing bolt from the steering shaft to steering gear pinion shaft and separate the shaft from the steering gear.

6. Disconnect the shifter linkage if necessary.

7. Remove the steering gear mounting nuts and remove the steering gear to the right of the vehicle.

To install:

8. Install the steering gear to the vehicle and install the mounting nuts/bolts. Tighten the nuts to 28–38 ft. lbs. (37–52Nm).

9. Connect the steering shaft to the steering gear pinion shaft. Tighten the bolt/nut to the specified torque.

10. Connect the shift linkage if disconnected. Install the set plate to the firewall.

11. Connect the pressure line and return hose to the steering gear.

12. Install the tie rod ends to the knuckle arm and tighten the nuts to 31–42 ft. lbs. (42–57Nm). Install the cotter pins.

13. Install the wheels to the vehicle, lower the vehicle and connect the negative battery cable. Bleed the power steering system and check the front end alignment.

Strut

REMOVAL & INSTALLATION

Front

1. Raise and safely support the vehicle. Remove the wheel and tire assembly.

2. Support the lower control arm with a jack.

3. Remove the bolts or clips attaching the brake hose and/or ABS sensor harness to the strut.

4. On vehicles equipped with the Automatic Adjusting Suspension (AAS), unplug the electrical connector and remove the actuator from the top of the strut.

5. On 1990–92 626 and MX6, when removing the left side strut, remove the ignition coil bracket.

6. Paint alignment marks on the upper strut mounting block and strut tower, and on the lower strut mount-to-steering knuckle so the strut can be reinstalled in the same position.

7. Remove the upper strut mounting nuts and the strut-to-knuckle bolts, then remove the strut assembly.

To install:

8. Install the strut into the strut tower, aligning the paint marks made during removal. Install the mounting nuts and tighten to specifications.

9. Install the strut-to-knuckle bolts and tighten to specifications.

10. If equipped with AAS, install the actuator and engage the electrical connector.

11. Install the clips or bolts attaching the brake hose and/or ABS sensor harness.

12. Install the wheel and tire assembly and lower the vehicle. Check the front end alignment.

Rear

1. As required, remove the side trim panels from the inside of the trunk or the rear seat and trim.

2. If equipped with Automatic Adjusting Suspension (AAS) system, disconnect the wiring and remove the cap. Loosen and remove the top mounting nuts from the strut mounting.

3. Raise and safely support the vehicle and remove the rear wheels. The suspension will drop when the weight lifts off the wheels.

4. Unclip the brake line or wiring retainers as required and unbolt the bottom strut mount. Remove the strut.

5. Installation is the reverse of removal.

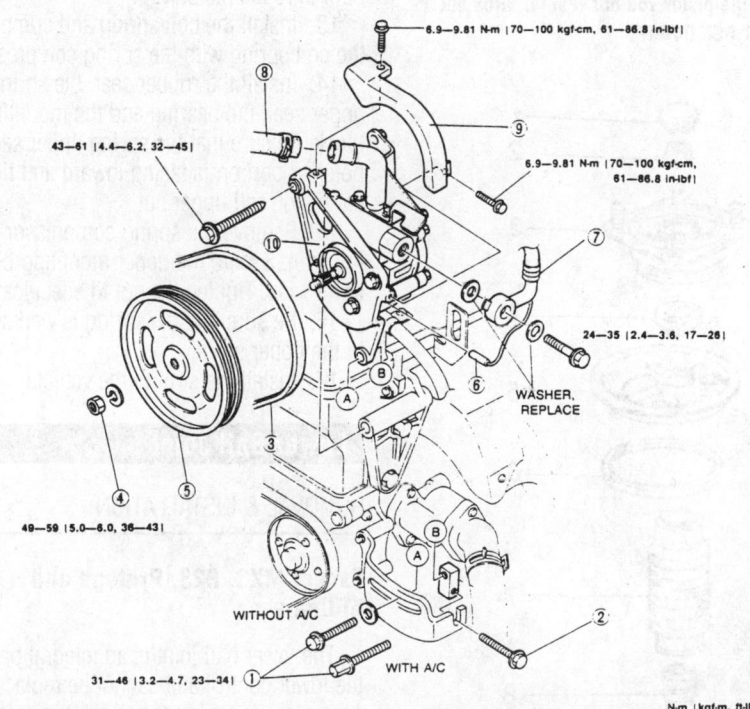

6.9—9.81 N·m |70—100 kgf-cm, 61—86.8 in-lbf|

43—61 |4.4—6.2, 32—45|

6.9—9.81 N·m |70—100 kgf-cm, 61—86.8 in-lbf|

24—35 |2.4—3.6, 17—26|

WASHER, REPLACE

49—59 |5.0—6.0, 36—43|

WITHOUT A/C

31—46 |3.2—4.7, 23—34| WITH A/C

N·m |kgf-m, ft-lbf|

1. Lock bolt
2. Adjusting bolt
3. Drive belt
4. Nut
5. Pulley
6. SPS connector
7. Pressure pipe
8. Return hose
9. Pulley cover
10. Power steering oil pump and bracket

7923MG80

Exploded view of the power steering gear assembly

Coil Spring

REMOVAL & INSTALLATION

1. Remove the strut from the vehicle.
2. If the vehicle is not equipped with AAS, remove the cap from the top of the strut.

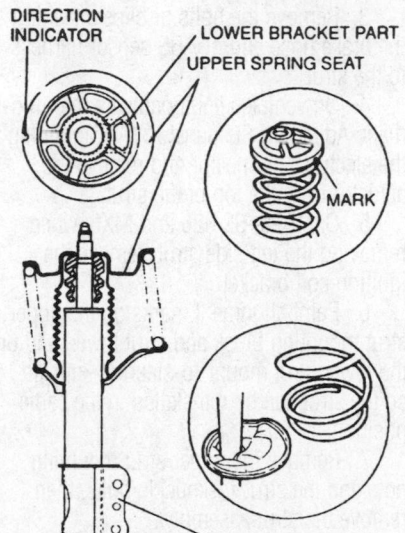

Be sure the end of the coil spring is in the step of the lower seat—626 and MX6

1. Cap
2. Piston rod nut
3. Mounting rubber
4. Thrust bearing
5. Upper spring seat
6. Upper rubber spring seat
7. Dust cover
8. Bound stopper

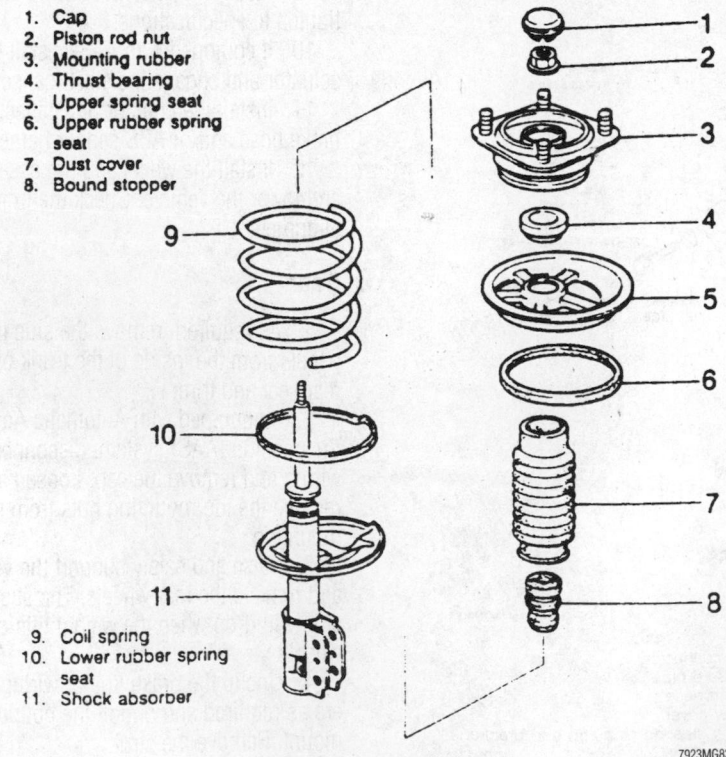

9. Coil spring
10. Lower rubber spring seat
11. Shock absorber

Exploded view of the front strut assembly—rear strut is similar

3. Install the strut securely in a vise with either aluminum or copper plates to protect the strut.
4. Loosen the piston rod upper nut one turn but DO NOT REMOVE IT.
5. Install the lower end of the strut in the vise and install a coil spring compressor. Compress the coil spring and remove the upper nut.

✳✳ CAUTION

Failure to fully compress the spring and hold it securely can be extremely dangerous.

6. Slowly release the coil spring tension.

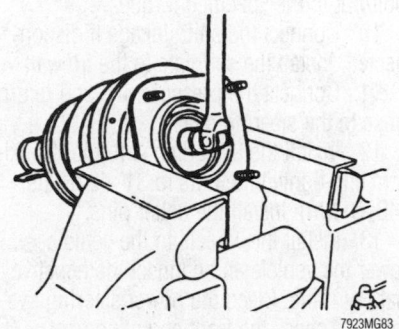

Secure the upper strut mount in a vise and loosen the piston rod nut several turns but DO NOT REMOVE IT

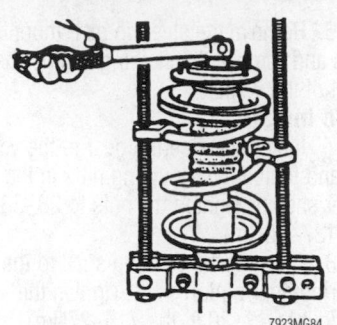

Use a coil spring compressor and relieve the spring tension from the upper mount, then remove the piston rod nut

7. Remove the suspension support, dust seal, spring seat, spring insulators, coil spring and bumper.
8. While pushing on the piston rod, be sure that the pull stroke is even and that there is no unusual noise or resistance. Also inspect for any oil leakage around the piston rod.
9. Push the piston rod in, then release it. Be sure that the return rate is constant.
10. If the shock absorber does not operate as described, replace it.

To assemble:

11. Install the strut assembly into a vise.
12. Install the bound stopper and dust boot onto the piston rod.
13. Install the coil spring and compress the coil spring with the spring compressor.
14. Install the rubber seat, the spring upper seat, the bearing and the mounting block. Be sure that the spring upper seat notched portion is facing inward and tighten the piston rod upper nut.
15. Remove the spring compressor from the strut. Secure the upper mounting block in the vise. Tighten the nut to specification.
16. Be sure that the spring is well seated in the upper seats.
17. Install the strut to the vehicle.

Lower Ball Joint

REMOVAL & INSTALLATION

Except MX3, 323, Protege and Millenia

The lower ball joint is an integral part of the lower control and cannot be replaced separately. If the lower ball joint is defective, the entire lower control arm must be replaced, as follows:

1. Raise and safely support the vehicle. Remove the wheel and tire assembly.
2. Remove the lower ball joint pinch bolt from the steering knuckle.

3. Disconnect the stabilizer bar link from the lower control arm.

4. Remove the lower control arm bolts and nuts and remove the lower control arm with the lower ball joint.

To install:

5. Install the lower control arm and loosely tighten the mounting nuts and bolts.

6. Connect the stabilizer link to the lower control arm and tighten the nut to 27–39 ft. lbs. (37–53 Nm).

7. Connect the lower ball joint to the steering knuckle. Install the pinch bolt and tighten to 26–41 ft. lbs. (35–56 Nm).

8. Install the wheel and tire assembly and lower the vehicle. With the vehicle at normal ride height, tighten the lower control arm mounting bolts. Tighten the front bushing through-bolt to 58–78 ft. lbs. (79–106 Nm) and the rear bushing strap bolts to 69–96 ft. lbs. (94–131 Nm).

9. Check the front wheel alignment.

MX3, 323, Protege, Miata and Millenia

1. Raise and safely support the vehicle. Remove the wheel and tire assembly.

2. Remove the ball joint stud pinch bolt and nut from the steering knuckle.

Pry the lower control arm down from the knuckle, and separate the ball joint from the knuckle.

3. Remove the bolt and nut and remove the ball joint from the lower control arm.

4. Installation is the reverse of the removal procedure. Tighten the ball joint-to-lower control arm bolt and nut to 86 ft. lbs. (117 Nm). Tighten the ball joint pinch bolt and nut to 43 ft. lbs. (59 Nm). Check the front wheel alignment.

Wheel Bearings

ADJUSTMENT

The front and rear wheel bearings are not adjustable. If the bearings become loose or make noise, they must be replaced.

REMOVAL & INSTALLATION

Front

1. Remove the steering knuckle from the vehicle.

2. Clamp the knuckle in a vise with protected jaws.

3. Remove the inner oil seal from the knuckle.

4. Use Mazda hub puller tools 49 G033 102, 49 G033 104 and 49 G033 105 or equivalent, and remove the front wheel hub from the knuckle assembly.

5. Remove the bearing inner race from the front wheel hub.

6. Remove the retaining ring from within the knuckle and using the hub puller tools, press the front wheel bearing from the knuckle.

7. Remove the brake dust shield.

8. Clean and inspect all parts but do not wash or clean the wheel bearing. The bearing must be replaced.

To install:

9. Using Mazda press tools 49 G033 107 and 49 H026 103 or equivalent, install a new dust shield cover assembly to the knuckle.

10. Using the press tools, press a new wheel bearing into the knuckle assembly.

11. Install the wheel bearing retaining ring, and install a new oil seal using installation tool 49 V001 795.

12. Install the front wheel hub by using the Mazda press tools or equivalent.

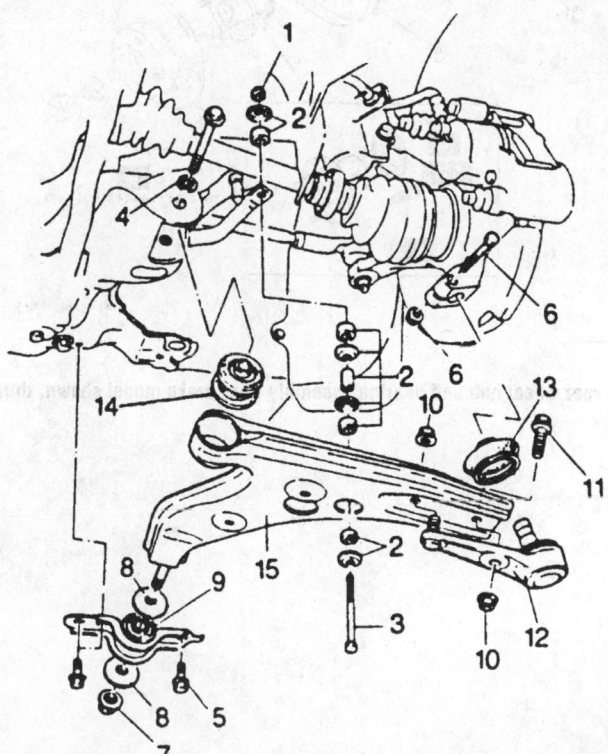

1. Stabilizer nut
2. Retainer, bushing and spacer
3. Stabilizer bolt
4. Bolt, washer
5. Bolt
6. Bolt, nut
7. Nut
8. Washer
9. Lower control arm bushing (rear)
10. Nut
11. Bolt
12. Lower arm ball joint
13. Ball joint dust boot
14. Lower arm bushing (front)
15. Lower arm

7923MG85

Exploded view of a common lower control arm with replaceable ball joint

Rear

1. Loosen the lug nuts on the rear wheels.
2. Block the front wheels, then raise and safely support the rear of the vehicle securely on jackstands.
3. Remove the rear wheels.
4. If equipped with drum brakes, remove the drum.
5. If equipped with disc brakes, remove the caliper and the rotor assembly from the hub.
6. Remove the hub dust cover.
7. Raise the staked portion of the hub retaining nut with a hammer and chisel. Remove and discard the nut.

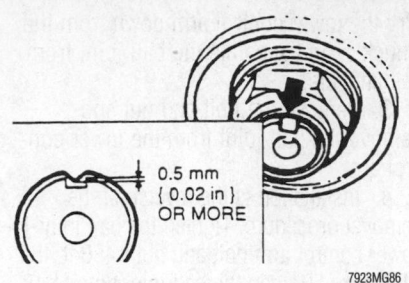

0.5 mm { 0.02 in } OR MORE

7923MG86

When installing the new hub nut, be sure to stake it into the notch on the spindle

8. Pull the hub and bearing assembly from the spindle.
9. The wheel bearings are not serviceable. If the bearings are bad, a new hub/bearing assembly must be installed.

To install:

10. Install the bearing assembly on the spindle using a new nut. Tighten the nut to 131–173 ft. lbs. (177–235 Nm).
11. Stake the nut into the groove in the spindle.
12. Install the dust cover.
13. Assembly the brakes.
14. Install the rear wheel and lower the vehicle to the floor.

1. Hub cap
2. Locknut
3. Brake caliper assembly
4. Disc plate
5. Wheel hub assembly
   Inspect for damage
   Inspect bearing for damage and rough rotation
6. ABS sensor rotor
7. Hub bolt
8. Wheel hub
9. Dust cover
   Inspect for damage and cracks
10. ABS wheel-speed sensor
11. Hub spindle
    Inspect for damage and cracks

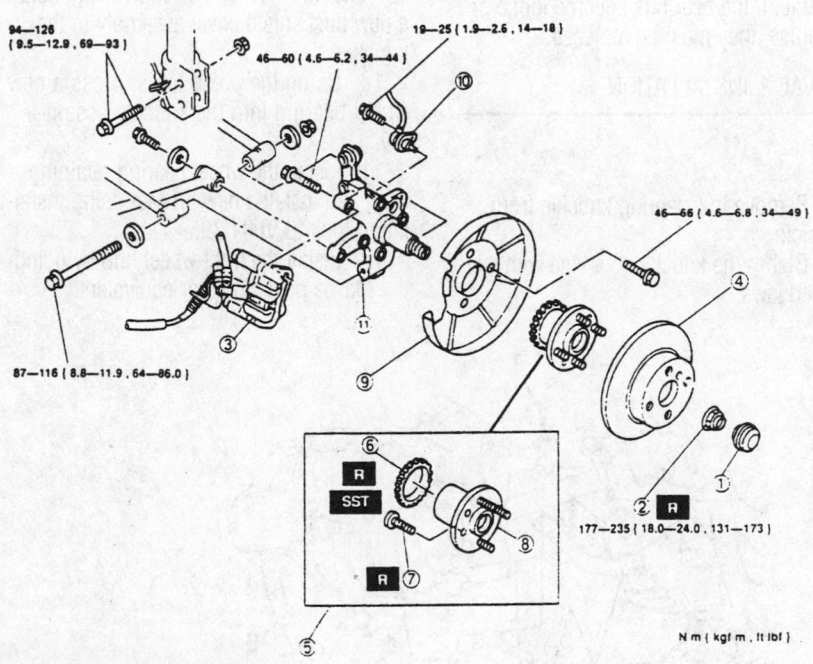

94—126 { 9.5—12.9 , 69—93 }

19—25 { 1.9—2.6 , 14—18 }

46—60 { 4.6—6.2 , 34—44 }

46—66 { 4.6—6.8 , 34—49 }

87—116 { 8.8—11.9 , 64—86.0 }

177—235 { 18.0—24.0 , 131—173 }

N m { kgf m , ft lbf }

7923MG87

Exploded view of the rear wheel hub and bearing assembly (disc brake model shown, drum is similar)

MERCEDES-BENZ

C • CLK • E • S • SL • SLK CLASSES

GASOLINE ENGINE REPAIR

➡ **Disconnecting the negative battery cable on some vehicles may interfere with the functions of the on board computer systems and may require the computer to undergo a relearning process, once the negative battery cable is reconnected.**

Ignition Timing

ADJUSTMENT

All engines except the 4.2L and 5.0L engines are equipped with a distributorless ignition system, no adjustment is necessary or possible. The 4.2L and 5.0L engines are equipped with an electronic ignition system. The ignition timing is controlled by the Electronic Ignition with Anti-Knock Retard (EZL/ARK) control unit. No ignition timing adjustment is possible.

Engine Assembly

REMOVAL & INSTALLATION

➡ **In all cases, Mercedes-Benz engines are removed as a unit with the transmissions.**

1. Remove the engine under cover.
2. Disconnect the negative battery cable.
3. Properly relieve the fuel system pressure.

✳✳ CAUTION

Never open, service or drain the radiator or cooling system when hot; serious burns can occur from the steam and hot coolant.

4. Drain and recycle the engine coolant.

✳✳ CAUTION

The EPA warns that prolonged contact with used engine oil may cause a number of skin disorders, including cancer! You should make every effort to minimize your exposure to used engine oil. Protective gloves should be worn when changing the oil. Wash your hands and any other exposed skin areas as soon as possible after exposure to used engine oil. Soap and water, or waterless hand cleaner should be used.

5. Drain and recycle the engine and transmission oil.
6. Remove the viscous coupling and fan. The viscous fan clutch assembly is equipped with left-hand threads.
7. Remove the radiator and fan shroud.
8. Disconnect all heater hoses and oil cooler lines. Plug all openings to keep out dirt.
9. Remove the hot film mass airflow sensor and the air intake assembly.
10. If equipped with a supercharger, remove the charge air pipes from the compressor.
11. Remove the accessory drive belts.
12. Guard the A/C condenser using a piece of sheet metal, ply-wood or plastic.
13. Label and disconnect the engine wiring harness.
14. Label and detach all ground straps and electrical connections.
15. Label and disconnect any vacuum hoses necessary to facilitate engine removal.
16. Label and detach the fuel supply and return lines.
17. Disconnect the accelerator linkage.
18. Remove the P/S fluid from the reservoir and disconnect the hoses from the P/S pump.
19. Unbolt the A/C compressor and position it aside, leaving the hoses attached.
20. Disconnect the transmission linkage.
21. If equipped with 4-MATIC, disconnect the front halfshafts from the front differential.
22. If equipped with a manual transmission, disconnect the clutch line at the slave cylinder.
23. Disconnect the exhaust system at the manifolds and remove.
24. Disconnect the driveshaft at the transmission and push it to the rear.
25. If equipped with a automatic transmission, ensure that all electrical and mechanical connections are detached from the transmission.
26. Support the engine and transmission assembly. Disconnect the engine and transmission mounts.
27. Using a chain hoist and cable, lift the engine and transmission upward and outward. An angle of about 45 degrees will allow the vehicle to be pushed backward while the engine is coming up.

To install:
28. Using a chain hoist and cable, position the engine and transmission into the vehicle. An angle of about 45 degrees will allow the vehicle to be pushed forward while the engine is going in.

29. Install the engine and transmission mounts and tighten the bolts as follows:
- Transmission mount bolts-to-crossmember to 18 ft. lbs. (25 Nm)
- Transmission crossmember-to-body bolts to 30 ft. lbs. (40 Nm)
- Transmission mount-to-transmission nut to 52 ft. lbs. (70 Nm)
- Engine mount-to-subframe mounting bolt to 30 ft. lbs. (40 Nm)
30. Connect the driveshaft.
31. If equipped with 4-MATIC, attach the front halfshafts to the front differential.
32. If equipped with a automatic transmission, ensure that all electrical and mechanical connections are attached to the transmission.
33. Connect the exhaust system.
34. If equipped with a manual transmission, connect the clutch hydraulic line to the slave cylinder.
35. Connect the transmission linkage.
36. Attach the A/C compressor.
37. Connect the P/S hoses and fill the reservoir.
38. Connect the accelerator linkage.
39. Connect the fuel supply and return lines.
40. Connect any vacuum hoses that were removed.
41. Attach all ground straps and electrical connections.
42. Connect the engine wiring harness.
43. Install the accessory drive belts.
44. If equipped with a supercharger, connect the charge air pipes to the compressor.
45. Install the hot film mass airflow sensor and the air intake assembly.
46. Remove the A/C condenser guard.
47. Install the radiator and connect the cooling hoses.
48. Remove the plugs from the oil cooler lines and connect to the appropriate fitting.
49. Install the viscous coupling and fan.

✳✳ WARNING

Operating the engine without the proper amount and type of engine oil will result in severe engine damage.

50. Fill the cooling system with coolant, and the engine and transmission with oil to the proper level.
51. Connect the negative battery cable.
52. Start the engine and check for leaks, then install the engine under cover.

Water Pump

REMOVAL & INSTALLATION

1. Disconnect the negative battery cable.

※※ **CAUTION**

Never open, service or drain the radiator or cooling system when hot; serious burns can occur from the steam and hot coolant.

2. Drain and recycle the engine coolant.

3. Remove the engine cooling fan and clutch, then the fan shroud.

4. Drain and recycle the engine coolant.

5. If equipped, remove the engine cover.

6. Remove the accessory drive belt.

7. If necessary, unbolt the power steering pump and position it aside leaving the hoses attached.

8. Disconnect the coolant hoses from the water pump.

9. If equipped, disconnect the coolant hoses from the oil-to-water heat exchanger.

10. Remove the belt pulley.

11. Remove the water pump mounting bolts, then the water pump.

12. Clean and dry the gasket mating surface for the water pump.

To install:

13. Install the water pump and gasket, and tighten the M6 bolts to 88 inch lbs. (10 Nm) and the M8 bolts to 177 inch lbs. (20 Nm).

14. Install the water pump belt pulley and tighten the mounting bolts to 88 inch lbs. (10 Nm).

15. Connect the coolant hoses to the water pump.

16. Install the power steering pump.

17. Install the accessory drive belt.

18. Install the engine cover.

19. Install the fan shroud and fan.

20. Fill the engine with coolant.

21. Connect the negative battery cable.

22. Start the vehicle and check for leaks.

1 Viscous fan
2 Poly V-belt
3 Fan shroud
4 Coolant hose
5 Coolant hose
6 Coolant hose at oil-water heat exchanger
7 Belt pulley of coolant pump
8 Coolant pump
9 Coolant pump gasket
10 Shock absorber
11 Bolts of shock absorber

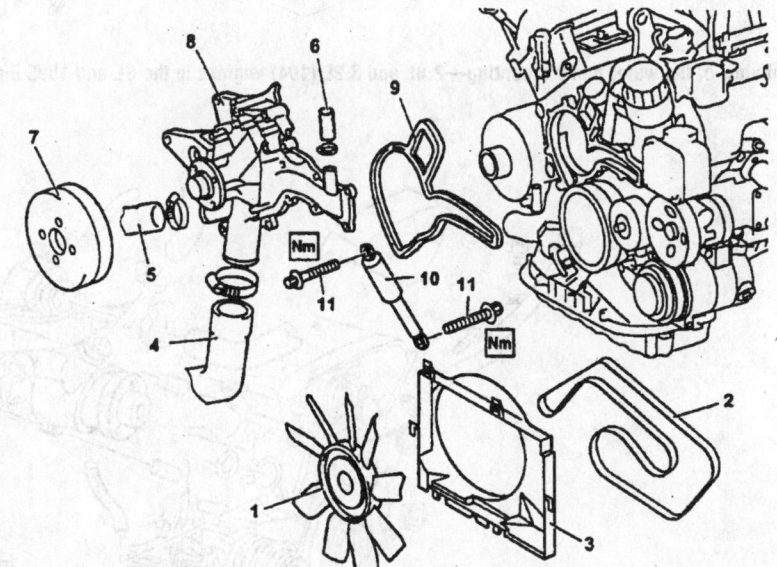

7923NG01

Exploded view of the water pump mounting—3.2L (112) and 4.3L (113) engines

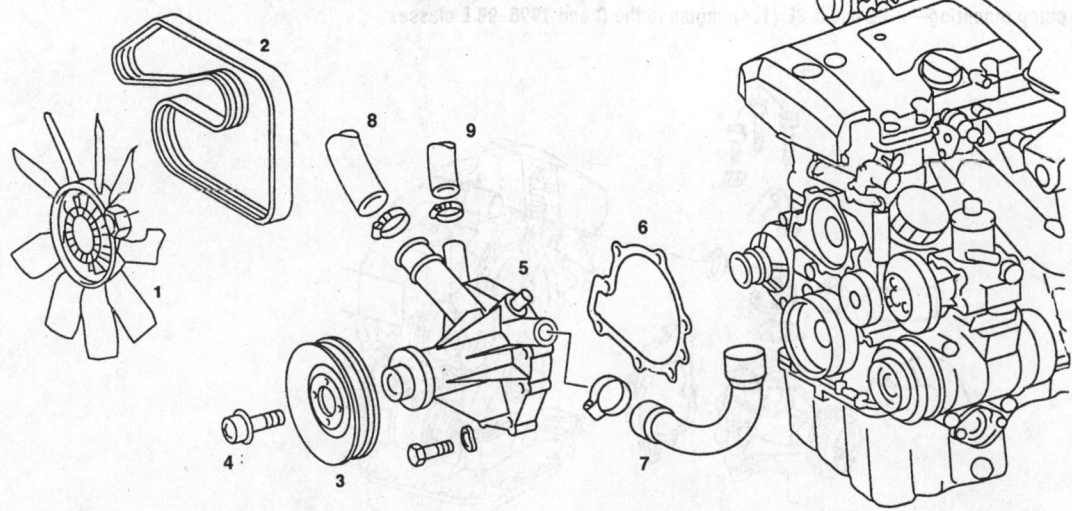

1. Viscous fan
2. Poly V-belt
3. Water pump pulley
4. Water pump pulley mounting
5. Water pump
6. Water pump gasket
7. Coolant hose
8. Heater hose
9. Heater hose

7923NG02

Exploded view of the water pump mounting—2.2L and 2.3L (111) engine

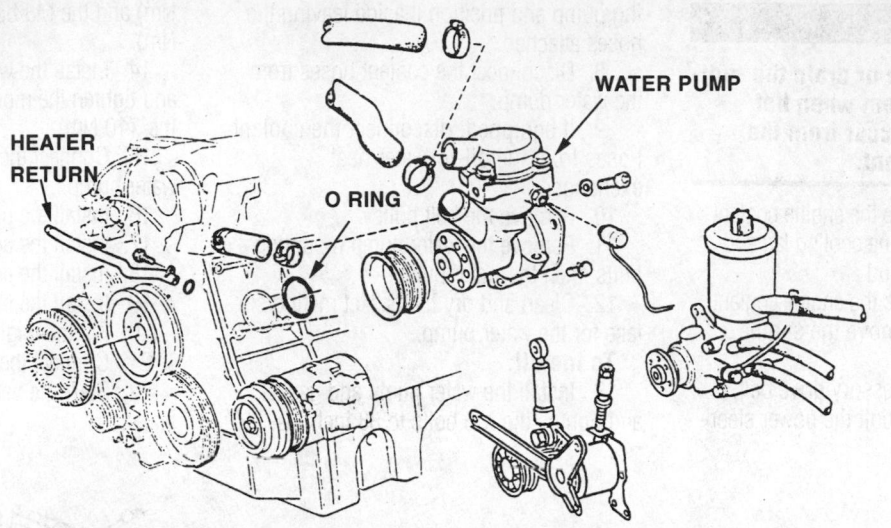

Exploded view of the water pump mounting—2.8L and 3.2L (104) engines in the SL and 1995 E classes

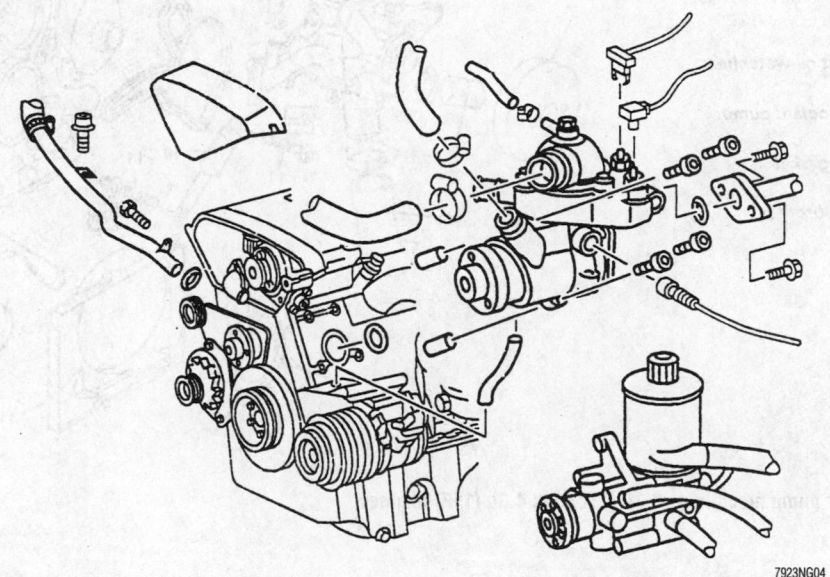

Exploded view of the water pump mounting—2.8L and 3.2L (104) engine in the C and 1996–99 E classes

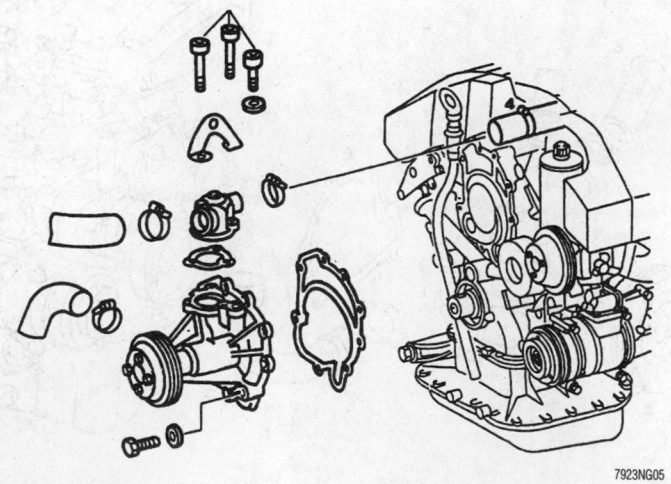

Exploded view of the water pump mounting—4.2L and 5.0L (119) engines

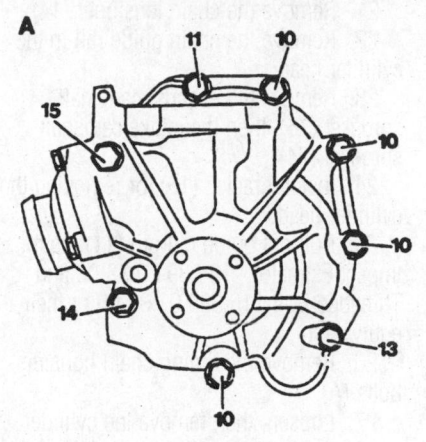

A   **Engine 119.96**

10	M8 × 60 bolt + washer
11	M8 × 65 bolt + washer
13	M8 × 85 bolt + washer
14	M8 × 90 bolt + washer
	(together with fan clutch carrier)
15	M8 × 135 bolt + washer

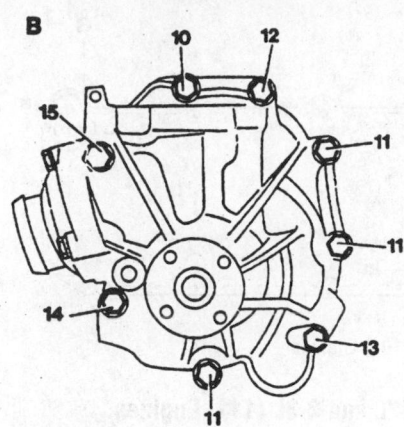

B   **Engine 119.97/98**

10	M8 × 60 bolt + washer
11	M8 × 65 bolt + washer
12	M8 × 75 bolt + washer
13	M8 × 85 bolt + washer
14	M8 × 90 bolt + washer
	(together with fan clutch carrier)
15	M8 × 135 bolt + washer

7923NG06

Water pump mounting bolt identification—4.2L and 5.0L (119) engines

Cylinder Head

REMOVAL & INSTALLATION

2.8L and 3.2L (104) Engine

1. Properly relieve the fuel system pressure.
2. Disconnect the negative battery cable.

✳✳ CAUTION

Never open, service or drain the radiator or cooling system when hot; serious burns can occur from the steam and hot coolant.

3. Drain and recycle the engine coolant.
4. Disconnect the coolant hoses from the cylinder head.
5. Position the No. 1 cylinder head at Top Dead Center (TDC).
6. Remove the upper timing cover.
7. Matchmark the camshaft sprocket to the timing chain.
8. Pull off pin (1) for the timing chain guide using Impact Extractor 116 589 20 33 00 and Threaded Bolt 116 589 01 34 00 or their equivalent.
9. Unbolt the guide sprocket (left-hand thread), then remove the bearing assembly.

10. Remove the timing chain from the camshaft sprockets and wire it aside.

➡️**Be sure the chain is securely wired so it will not slide into the engine.**

11. Label and detach any electrical connectors that would interfere with cylinder head removal.
12. Label and detach any vacuum hoses at the intake manifold.
13. Disconnect the exhaust system at the manifolds.
14. If equipped, disconnect the secondary air injection pipe at the exhaust manifold.
15. Unbolt the engine oil dipstick tube support bracket.
16. Detach the crankcase breather hose.
17. Open the fuel filler cap to release the vapor pressure in the tank, then close.
18. Disconnect the fuel lines at the manifold.
19. If equipped with an automatic transmission, unbolt the dipstick tube at the rear of the cylinder head.
20. Disconnect the throttle cable, if equipped with an automatic transmission, kickdown cable.
21. Disconnect the cruise control cable.

22. Loosen the head bolts in stages in the reverse order of the tightening sequence, then remove.

✳✳ CAUTION

Never, under any circumstances, use a prybar between the head and block to pry, as the head will be scarred badly and may be ruined.

23. Using an engine hoist, lift the cylinder head off the engine block.
24. Remove all gasket material from the sealing surfaces of the cylinder head and engine block. Be careful not to gouge or scratch the surface of the aluminum head. Be sure the cylinder head locating dowels are positioned in the engine block. Clean and dry the head bolt holes using compressed air.
25. Inspect length of the cylinder head bolt shaft, new bolt length is 6.30 in. (160mm) and the maximum permissible length is 6.44 in. (163.5mm). Replace bolts that measure grater than the maximum permissible length.

To install:

➡️**The head gasket is not watertight until the engine has reached operating temperature. Do not pressure test the cooling system until the engine has reached operating temperature.**

26. Rotate the camshafts so that the bottom edge of the holes in the camshaft flange are level with the top edge of the cylinder head.
27. Verify the TDC position of the No. 1 cylinder.
28. Clean the head bolt threads, then apply clean engine oil to the thread and head contact surfaces.
29. Install the new head gasket.
30. Install the cylinder head to the engine block and tighten the head bolts in sequence to 41 ft. lbs. (55 Nm) plus 90 degrees, then an additional 90 degrees.

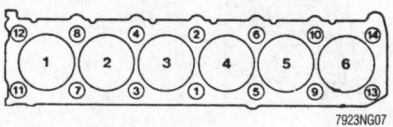

7923NG07

Cylinder head bolt tightening sequence—2.8L and 3.2L (104) Engines

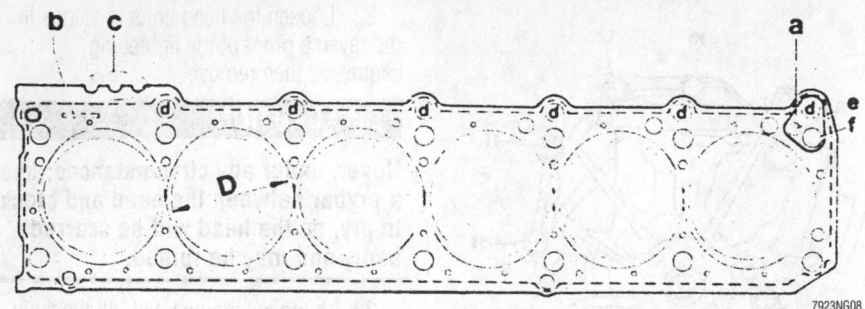

Cylinder head gasket identification—2.8L and 3.2L (104) Engines

31. Remove the wire and install the timing chain.

32. Install the upper timing chain guide.

33. Install the guide sprocket and bearing assembly, then tighten the bolt to 26 ft. lbs. (35 Nm).

34. Connect the throttle and cruise control cables.

35. If equipped with an automatic transmission, install the kickdown cable, and bolt the dip stick tube to the rear of the cylinder head.

36. Connect the fuel lines to the manifold.

37. Connect the crankcase breather hose.

38. Mount the engine oil dipstick tube support bracket.

39. If equipped, connect the secondary air injection pipe to the exhaust manifold.

40. Connect the exhaust system to the manifolds.

41. Attach any vacuum hoses and electrical connectors that were removed.

42. Install the pin into the timing chain guide.

43. Connect the coolant hoses that were removed.

44. Connect the negative battery cable.

45. Start the vehicle and check for leaks.

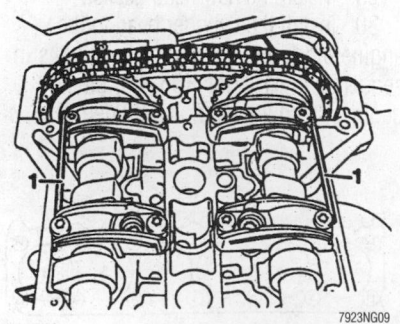

Verify the correct position of the camshafts for cylinder head installation using a 4mm pin (1)—2.8L and 3.2L (104) Engines

2.2L and 2.3L (111) Engines

❋❋ CAUTION

Never open, service or drain the radiator or cooling system when hot; serious burns can occur from the steam and hot coolant.

1. Drain and recycle the engine coolant from the block.

2. Disconnect the negative battery cable.

3. Unbolt the exhaust system from the manifolds.

4. Unbolt the exhaust system bracket at the transmission.

5. Remove the air intake tubes.

6. Remove the cover for the coolant temperature thermostat.

7. Unplug the electrical connections at the cylinder head.

8. Unbolt the intake manifold (19) and position it aside.

9. Disconnect the crankcase breather hose.

10. Unbolt bracket (2) from the manifold.

11. Disconnect the coolant hose (3) at the rear of the cylinder head.

12. Remove the valve cover.

13. Remove the thermostat housing.

14. Remove the cylinder head front cover.

15. Detach the oxygen sensor connector.

16. Unbolt the engine oil dipstick guide tube (7).

17. If equipped with an automatic transmission, unbolt the dipstick guide tube (8).

18. Position the crankshaft to 20° –30° after TDC for cylinder No. 1.

19. Lock the camshafts (75 and 78) with locking pins (01).

20. Matchmark the timing chain to the sprockets.

21. Remove the chain tensioner (14).

22. Remove the chain guide rail in the cylinder head.

23. Remove the exhaust camshaft sprocket (76), then the intake camshaft sprocket (74).

24. Install bracket (16) for removing the cylinder head.

25. Pull out guide rail pin (11), using Impact Extractor 116 589 20 33 00 and Threaded Bolt 116 589 01 34 00 or their equivalent.

26. Remove the timing chain housing bolts (A).

27. Loosen, then remove the cylinder head bolts in the reverse order of the tightening sequence.

❋❋ CAUTION

Never, under any circumstances, use a prybar between the head and block to pry, as the head will be scarred badly and may be ruined.

28. Using an engine hoist, lift the cylinder head off the engine block.

29. Remove all gasket material from the sealing surfaces of the cylinder head and engine block. Be careful not to gouge or scratch the surface of the aluminum head. Be sure the cylinder head locating dowels are positioned in the engine block. Clean and dry the head bolt holes using compressed air.

30. Inspect length of the cylinder head bolt shaft, new bolt length is 4.02 in. (102mm) and the maximum permissible length is 4.13 in. (105mm). Replace bolts that measure grater than the maximum permissible length.

To install:

➡**The head gasket is not watertight until the engine has reached operating temperature. Do not pressure test the cooling system until the engine has reached operating temperature.**

31. Verify the TDC position of the No. 1 cylinder.

32. Clean the head bolt threads, then apply clean engine oil to the thread and head contact surfaces.

33. Install the new head gasket.

34. Install the cylinder head to the engine block and tighten the head bolts in sequence to 41 ft. lbs. (55 Nm) plus 90 degrees, then an additional 90 degrees.

35. Install the reaming components in the reverse order of the installation.

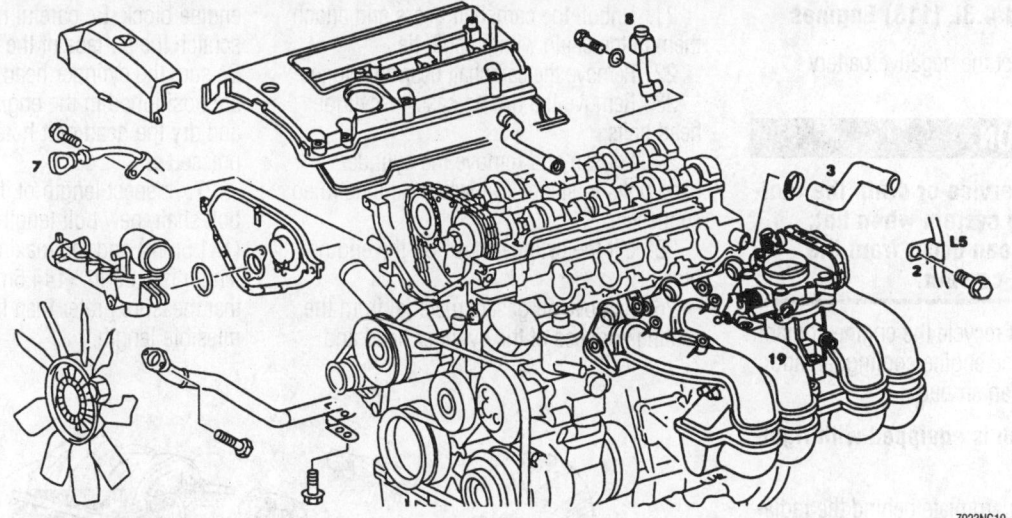

Exploded view of the cylinder head accessory components—2.2L and 2.3L (111) engines

7923NG10

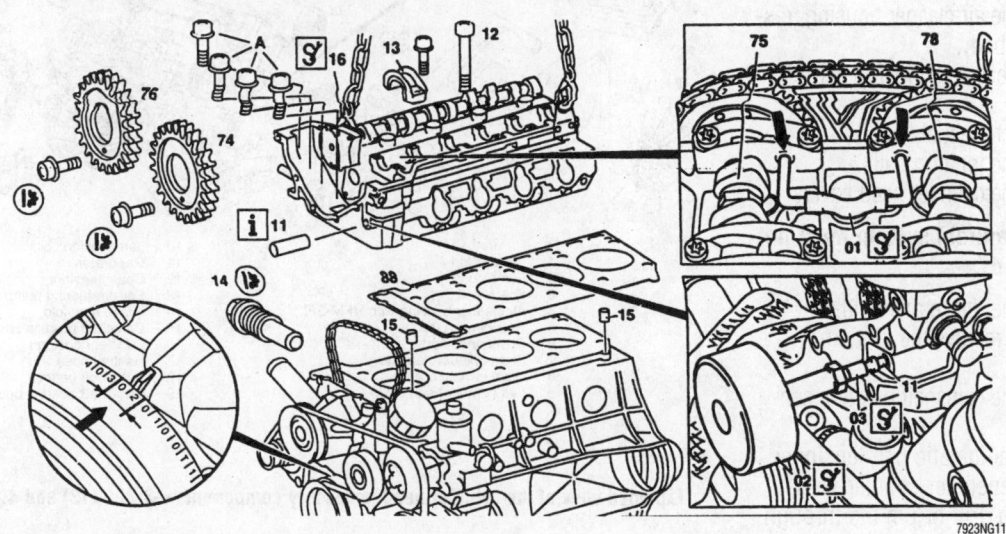

Exploded view of the cylinder head mounting—2.2L and 2.3L (111) engines

7923NG11

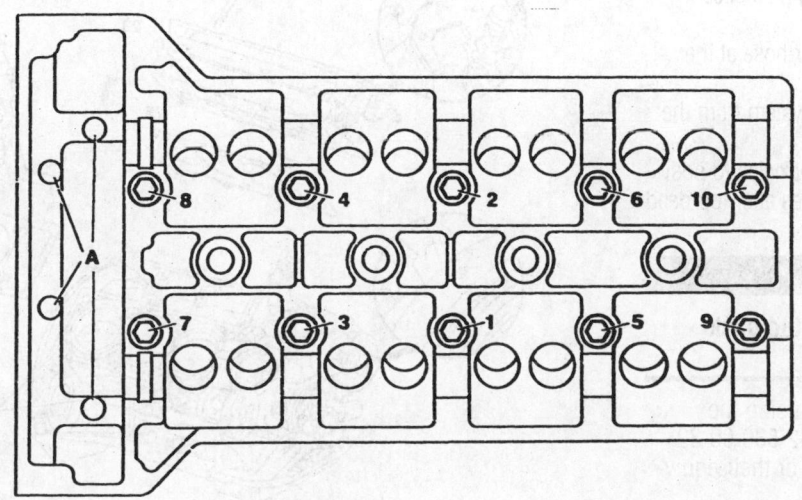

Cylinder head bolt removal sequence—2.2L and 2.3L (111) engines

7923NG12

3.2L (112) and 4.3L (113) Engines

1. Disconnect the negative battery cable.

2. Drain and recycle the engine coolant.
3. Remove the engine cooling fan and clutch, then the fan shroud.

➡ **The fan clutch is equipped with right-hand thread.**

4. Place a guard plate behind the radiator/condenser to protect it from damage during removal and installation.
5. Remove the engine cover.
6. Remove the air cleaner housing, resonance pipe and body.
7. Properly relieve the fuel system pressure.
8. Disconnect the fuel line.
9. Remove the ignition coils.
10. Remove the cylinder head covers.

➡ **The intake manifold system must not be disassembled.**

11. Remove the intake manifold.
12. Label and remove the vacuum switchover valve.
13. Remove the camshaft position sensor.
14. Lock the automatic belt tensioner by rotating the tensioner counterclockwise until a 5mm drift or pin fits through the tensioner, then remove the serpentine belt.
15. Remove the power steering pump and position it aside leaving the hoses attached.
16. Disconnect the heater hose at the firewall.
17. Detach the exhaust system from the exhaust manifolds.
18. Rotate the engine clockwise to position the crankshaft 40 degrees after top dead center.

19. Lock the camshafts using the Camshaft Locking tools 112 589 00 32 00 and 112 589 01 32 00, or their equivalent.
20. Remove the generator, then the timing chain tensioner.

21. Unbolt the camshaft gears and attach them to the chain with a cable tie.
22. Remove the camshaft bearing bridges.
23. Remove the timing case-to-cylinder head bolts.
24. Loosen and remove the cylinder head bolts in stages following the illustrated sequence.
25. Lift the cylinder head off the engine block.
26. Remove all gasket material from the sealing surfaces of the cylinder head and

engine block. Be careful not to gouge or scratch the surface of the aluminum head. Be sure the cylinder head locating dowels are positioned in the engine block. Clean and dry the head bolt holes using compressed air.

27. Inspect length of the cylinder head bolt shaft, new bolt length is 5.57 in. (141.5mm) and the maximum permissible length is 5.69 in. (144.5mm). Replace bolts that measure grater than the maximum permissible length.

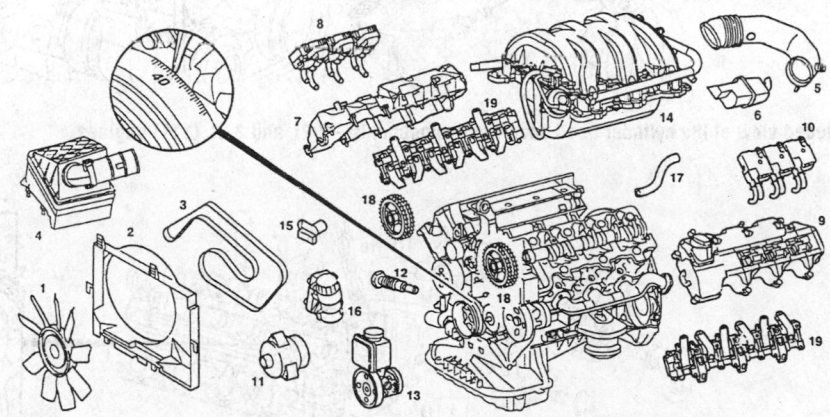

1	Viscous fan	10	Left ignition coils
2	Fan shroud	11	Generator
3	Poly V-belt	12	Chain tensioner
4	Air cleaner housing with HFM-SFI	13	Power steering pump with reservoir
5	Resonance pipe	14	Intake manifold
6	Resonance body	15	Camshaft position sensor
7	Right cylinder head cover	16	Oil filter housing
8	Right ignition coils	17	Heating hose
9	Left cylinder head cover	18	Camshaft gears
		19	Camshaft bearing bridges

7923NG45

Exploded view of the cylinder head accessory components—3.2L (112) and 4.3L (113) engines

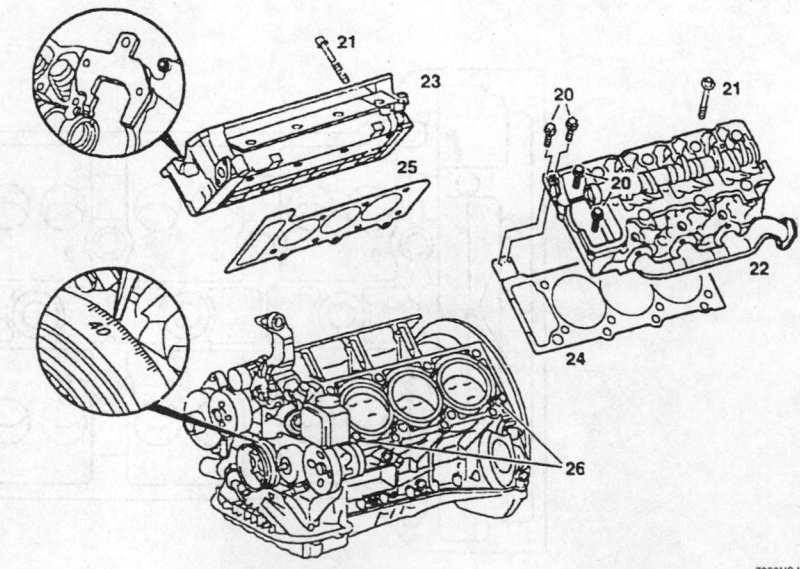

7923NG46

Exploded view of the cylinder head removal—3.2L (112) and 4.3L (113) engines

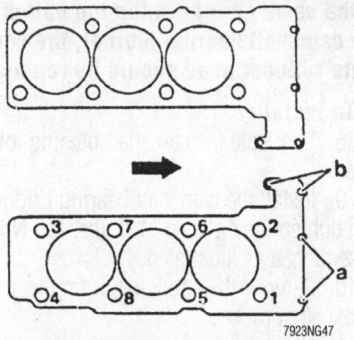

Cylinder head bolt removal sequence—
3.2L (112) engine

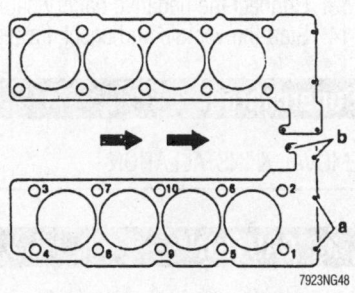

Cylinder head bolt removal sequence—
4.3L (113) engine

To install:

28. Clean the head bolt threads, then apply clean engine oil to the thread and head contact surfaces.

29. Install the cylinder head to the engine block and tighten the head bolts according to sequence as follows:
- Step 1. 15 ft. lbs. (20 Nm)
- Step 2. 37 ft. lbs. (50 Nm)
- Step 3. 60–70 degrees
- Step 4. additional 60–70 degrees

30. Install the timing case-to-cylinder head bolts and tighten to 15 ft. lbs. (20 Nm).

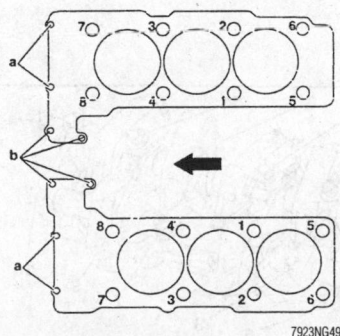

Cylinder head bolt tightening sequence—
3.2L (112) engine

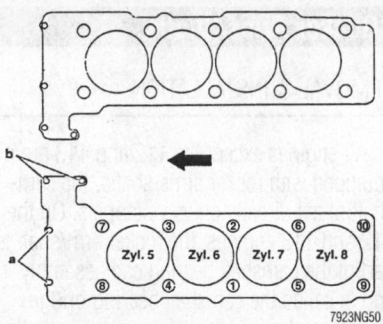

Cylinder head bolt tightening sequence—
4.3L (113) engine

31. Install the camshaft bearing bridges.

32. Cut the cable tie and install the camshaft gear and tighten the mounting bolt to 37 ft. lbs. (50 Nm) plus an additional 90 degrees. Check, if necessary, adjust the basic camshaft position.

33. Install the timing chain tensioner with a new gasket and tighten to 59 ft. lbs. (80 Nm).

34. Install the generator.

35. Remove the camshaft locking plates.

36. Connect the exhaust system to the manifolds and tighten the mounting nuts to 15 ft. lbs. (20 Nm).

37. Connect the heater hose to the cylinder head.

38. Install the power steering pump.

39. Install the serpentine belt and remove the locking pin.

40. Install the camshaft position sensor.

41. Install and connect the vacuum switchover valve.

42. Install the intake manifold.

43. Install the cylinder head covers and tighten the bolts to 88 inch lbs. (10 Nm).

44. Install the ignition coils and tighten the mounting bolts to 70 inch lbs. (8 Nm).

45. Connect the fuel pipe.

46. Install the air cleaner housing, resonance pipe and body.

47. Install the engine cover.

48. Remove the guard plate from the radiator/condenser.

49. Install the fan shroud, then the cooling fan.

➡The fan clutch is equipped with right-hand thread.

50. Fill the engine with coolant.
51. Connect the negative battery cable.
52. Start the vehicle and check for leaks.

4.2L and 5.0L (119) Engines

✵✵ CAUTION

Never open, service or drain the radiator or cooling system when hot; serious burns can occur from the steam and hot coolant.

1. Disconnect the negative battery cable and drain the engine coolant.

2. Remove the front covers.

3. Crank the No. 1 piston of the engine to 45 degrees BTDC. Look for the 4/5 on the timing indicator.

4. Mark all 4 camshaft timing gears and timing chain with colored dots at about 11 o'clock for the right outer and left inner camshaft sprocket and 1 o'clock for the right inner and left outer camshaft sprocket.

5. Remove the timing chain tensioner and top guide rails.

6. Unscrew exhaust camshaft gears and camshaft adjuster.

7. Remove the engine cover and intake manifold.

8. Label and disconnect all electrical wiring, hoses and cables from the manifolds and cylinder head(s).

9. If equipped with an automatic transmission, remove the dipstick guide tube.

10. Disconnect the exhaust system from the head.

11. Remove accessories as needed.

12. Remove the cylinder head cover very carefully.

➡Be sure the engine is cold before removing cylinder head bolts. On models 124 and 140, head bolt 10 and on model 210, head bolts 9 and 10 cannot be removed with cylinder head installed. Raise and secure the bolt when removing the cylinder head. On the "close-deck" crankcase, the cylinder head bolts have different lengths.

13. Loosen the cylinder head bolts in the reverse order of tightening. A special Torx® like socket is needed to remove the bolts.

14. Install a suitable lifting device onto the cylinder head and remove.

15. Remove all gasket material from the sealing surfaces of the cylinder head and engine block. Be careful not to gouge or scratch the surface of the aluminum head. Be sure the cylinder head locating dowels are positioned in the engine block. Clean and dry the head bolt holes using compressed air.

16. Inspect length of the cylinder head bolt shaft, new bolt length is 6.30 in. (160mm) and the maximum permissible length is 6.44 in. (163.5mm). Replace bolts that measure grater than the maximum permissible length.

To install:

➡️ **The head gasket is not watertight until the engine has reached operating temperature. Do not pressure test the cooling system until the engine has reached operating temperature.**

17. Clean the gasket mating surfaces and check for warpage.

18. Install the cylinder head gasket and head.

19. Clean the head bolt threads, then apply clean engine oil to the thread and head contact surfaces. Torque the bolts in sequence, in steps. The 1st step to 41 ft. lbs. (55 Nm), 2nd step to 90 degrees angle of rotation and 3rd step to 90 degrees angle of rotation. Torque the M8 bolts near the timing sprockets to 18 ft. lbs. (25 Nm).

➡️ **Bolts that are screwed into the front of the cylinder head must be coated with sealant when installed.**

20. Connect the exhaust pipe and torque to 20 ft. lbs. (25 Nm).

21. Install the remaining components.

22. Connect all wiring, hoses and cables.

23. Refill the engine with fluids, connect the battery cable, start the engine and check for leaks.

Rocker Arms/Shafts

REMOVAL & INSTALLATION

All engines except the 112 and 113 are not equipped with rocker arms/shafts, the camshaft(s) act directly on valve tappets. On the 112 and 113 engines, the rocker arm/shaft is part of the camshaft bearing cap assembly and is called the camshaft bearing bridge. This procedure is for removing and installing the camshaft bearing bridge.

3.2L (112) and 4.3L (113) Engines

1. Disconnect the negative battery cable.

2. Remove the cylinder head cover.

3. Rotate the engine clockwise to position the crankshaft 40 degrees after top dead center.

✳️ WARNING

Engine must not be rotated backwards.

4. Remove the generator.

5. Remove the timing chain tensioner.

6. Cable tie the timing chain to the camshaft sprocket.

7. Loosen the camshaft bearing bridge bolts in the reverse order of installation, starting at 16.

➡️ **The camshaft bearing bridge must not be disassembled. If damage exists**

at the valve gear or at the top half of the camshaft bearing journal, the complete cylinder head should be replaced.

To install:

8. Lubricate the camshaft bearing journals.

9. Install the camshaft bearing bridge and tighten the bolts to 11 ft. lbs. (15 Nm) in sequence as illustrated.

10. Remove the cable ties from the camshaft sprockets.

11. Install the timing chain tensioner with a new gasket and tighten to 59 ft. lbs. (80 Nm).

12. Install the cylinder head covers to 88 inch lbs. (10 Nm).

13. Connect the negative battery cable.

14. Start the vehicle and check for leaks.

Supercharger

REMOVAL & INSTALLATION

✳️ CAUTION

Never open, service or drain the radiator or cooling system when hot; serious burns can occur from the steam and hot coolant.

1. Drain and recycle the engine coolant.

2. Remove the intake air assembly and plug the openings of the supercharger.

3. Remove the sensor block cover at the front of the cylinder head.

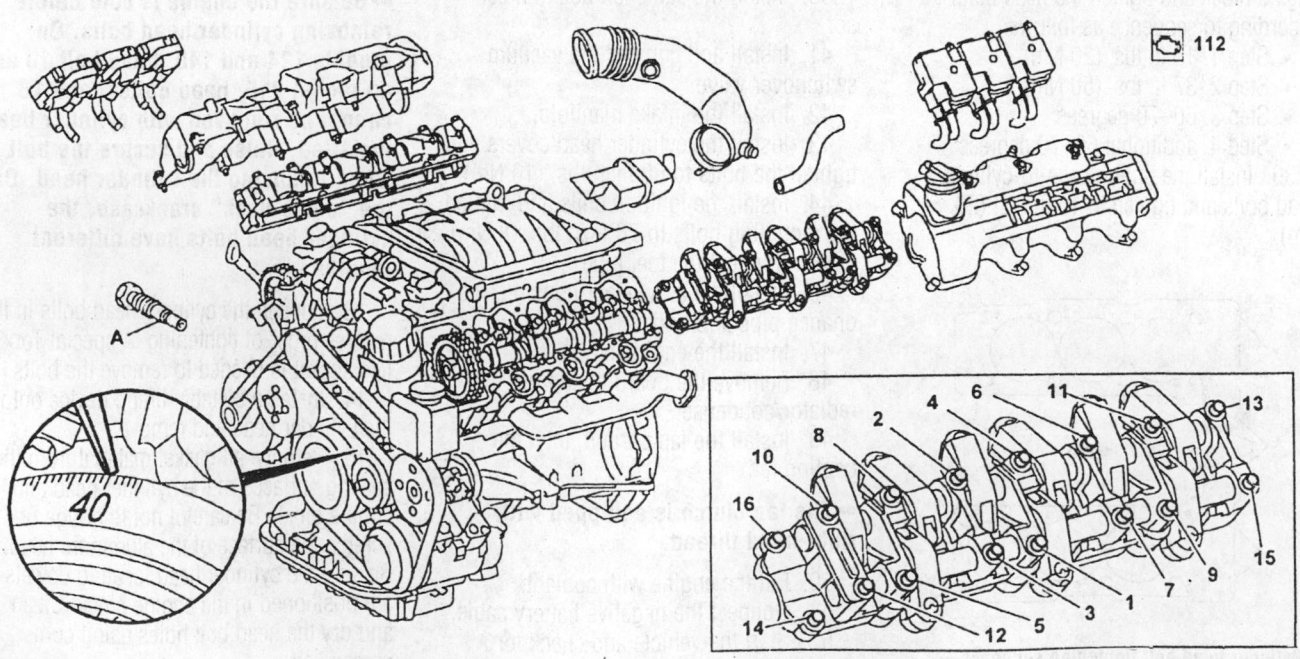

Exploded view of the camshaft bearing bridge mounting components—3.2L (112) and 4.3L (113) engines

7923NG79

4. Unplug the connector for the electro-magnetic clutch.

5. Remove the coolant hose from the radiator to water pump.

6. Remove the coolant return hose at the water pump.

7. Remove the fan and shroud.

8. Remove the drive belt for the super-charger.

9. Remove the pressure connections at the supercharger.

10. Unbolt and remove the supercharger.

To install:

➡**Install the bottom securing bolt in the supercharger before installing.**

11. Install the supercharger and tighten the mounting bolts to 16 ft. lbs. (21 Nm).

12. Install the pressure connections using new seals.

13. Install the drive belt.

14. Install the fan and shroud.

15. Connect any coolant hoses that were removed.

16. Attach the connector for the electro-magnetic clutch.

17. Install the sensor block cover at the front of the cylinder head.

18. Remove the plug from the inlet of the supercharger and install the intake air assembly.

19. Fill the cooling system.

Intake Manifold

REMOVAL & INSTALLATION

2.8L and 3.2L (104) Engines

1. Disconnect the negative battery cable.

2. Label and detach all vacuum, electrical and cable connectors from the intake manifold.

3. Properly relieve the fuel system pressure.

4. Unbolt, then remove the fuel rail with the injectors.

5. Remove the resonance intake manifold (throttle body) and clean the gasket surfaces.

6. Unbolt the intake manifold and remove.

7. Using clean shop rags, plug the openings in the cylinder head to prevent debris from entering.

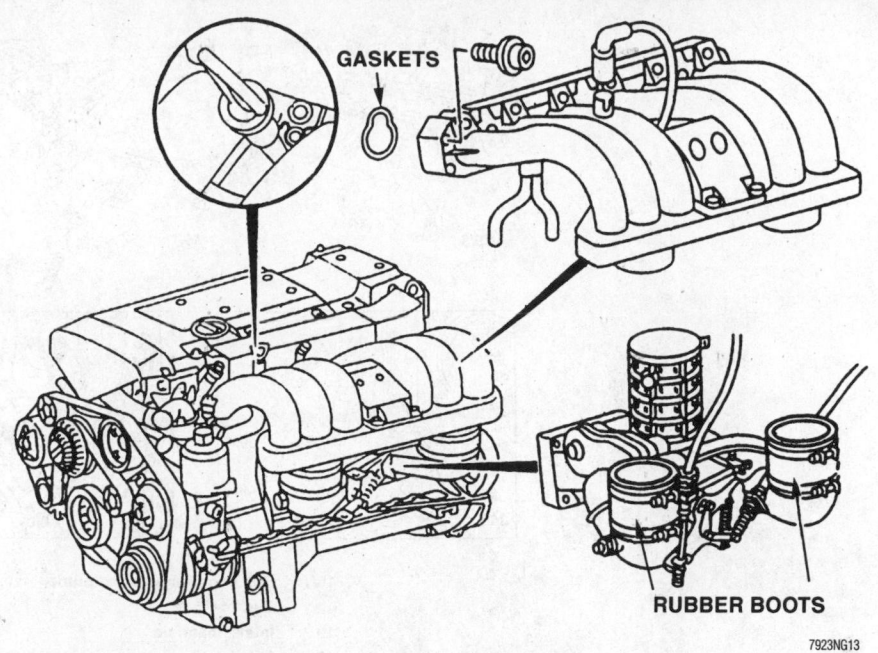

Exploded view of the intake manifold mounting and related components—2.8L and 3.2L (104) engines

To install:

8. Clean the gasket sealing surfaces for the intake manifold.

9. Remove the plugs from the manifold.

10. Install the manifold using new gaskets and tighten the mounting bolts to 18 ft. lbs. (25 Nm).

11. Install the resonance intake manifold using new gaskets.

12. Install the fuel rail with the injectors using new injector seals.

13. Reconnect all wiring, hoses and cables.

14. Check and adjust the throttle linkage.

15. Refill the engine fluids, connect the battery cable, start the engine and check for leaks.

2.2L and 2.3L (111) Engines

1. Properly relieve the fuel system pressure.

2. Disconnect the negative battery cable.

3. Remove the intake air cross pipe.

4. If necessary, unbolt the ignition coils.

5. If necessary, unbolt the power steering pump bracket mounting bolt.

6. Remove the fuel rail with the injectors.

7. Remove the intake manifold support bracket.

8. Label and disconnect all vacuum, electrical and cable connectors from the intake manifold.

9. Disconnect the throttle cable.

10. Loosen, then remove the manifold mounting nuts and bolts.

11. Clean the gasket mating surfaces and check for warpage.

To install:

12. Using a new gasket, install the intake manifold and tighten the mounting bolts/nuts to 15 ft. lbs. (20 Nm).

13. Check and adjust the throttle linkage.

14. Install the fuel rail with the injectors using new injector seals.

15. If removed, install the ignition coils and the power steering pump bracket mounting bolt.

16. Install the intake manifold support bracket.

17. Reconnect all wiring, hoses and cables.

18. Refill the engine fluids, connect the battery cable, start the engine and check for leaks.

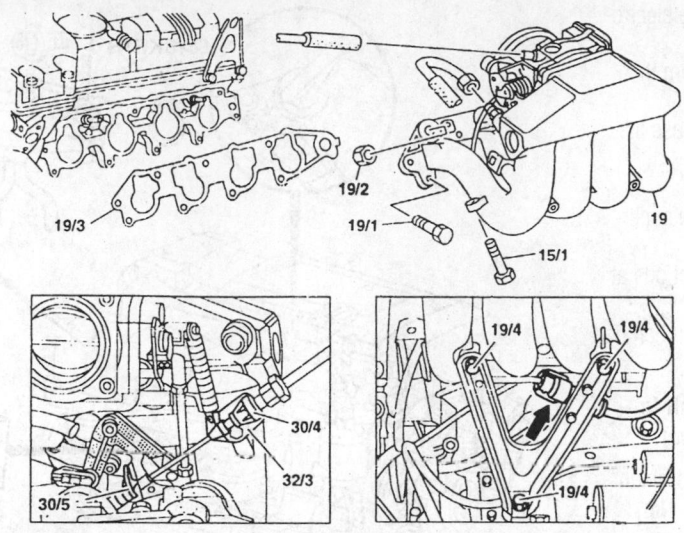

15/1	Bolt of power steering pump bucket
19	Intake manifold
19/1	Bolt
19/2	Nut
19/3	Gasket
19/4	Bolt of intake manifold support
30/4	Plastic clip
30/5	Guide piece
32/3	Bracket of throttle control lever
Arrow	Connector

7923NG14

Exploded view of the intake manifold mounting and related components—1995–96 2.2L and 2.3L (111) engines

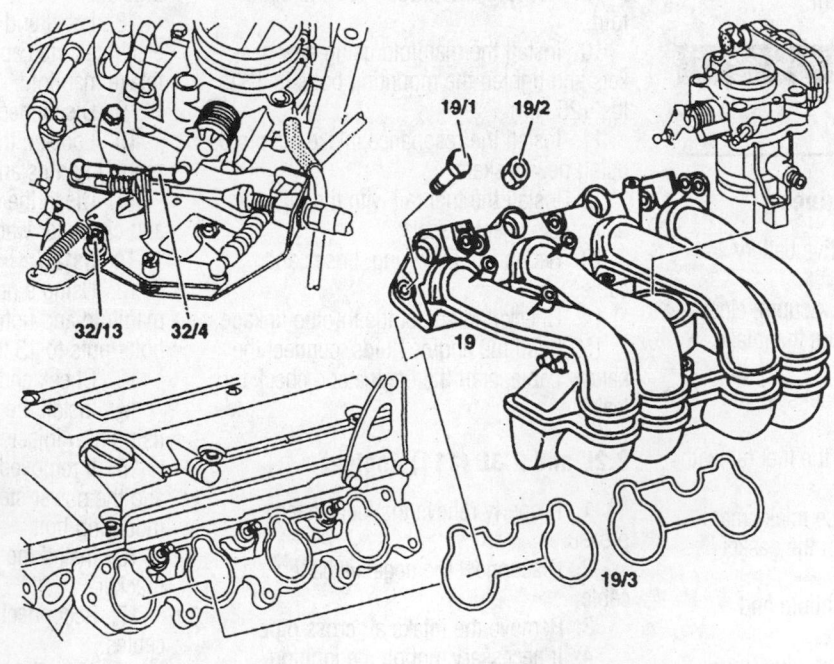

19	Intake manifold
19/1	Bolt
19/2	Nut
19/3	Moulded sealing ring
32/4	Connecting rod
32/13	Bolt

7923NG15

Exploded view of the intake manifold mounting and related components—1997–99 2.2L and 2.3L (111) engines

3.2L (112) and 4.3L (113) Engines

1. Disconnect the negative battery cable.

2. Remove the cylinder head cover.

3. Remove the hot film mass air flow sensor with the intake pipe.

4. Properly relieve the fuel system pressure.

5. Remove the fuel rail with the injectors.

6. Label and disconnect the vacuum lines from the intake manifold.

7. Label and detach any electrical connections to the intake manifold.

8. Disconnect the EGR valve.

9. Remove the combination valve, then the intake manifold mounting bolts.

10. Remove the intake manifold and gaskets.

11. Place clean shop rags into the intake passages to prevent dirt from entering. Clean the gasket mating surfaces.

To install:

12. Install the new gaskets and verify the secondary air injection passage opening in the gasket.

13. Remove the shop rags from the intake passages.

14. Install the intake manifold to the engine and tighten the mounting bolts to 15 ft. lbs. (20 Nm).

15. Install the combination valve and tighten the bolts to 15 ft. lbs. (20 Nm).

16. Connect the EGR valve.

17. Attach the electrical connections to the intake manifold.

18. Connect the vacuum lines to the manifold.

19. Install the fuel rail with the injectors.

20. Install the mass air flow sensor with the air intake pipe.

21. Install the cylinder head cover and tighten the bolts to 88 inch lbs. (10 Nm).

22. Connect the negative battery cable.

23. Start the vehicle and check for leaks.

4.2L and 5.0L (119) Engines

> **✳✳ CAUTION**
>
> **Never open, service or drain the radiator or cooling system when hot; serious burns can occur from the steam and hot coolant.**

1. Disconnect the negative battery cable. Partially drain the coolant.

2. Remove the air cleaner and engine cover.

3. Label and detach all vacuum lines and electrical connectors at the intake manifold.

4. Properly relieve the fuel system pressure.

5. Disconnect the fuel supply (17/3) and return (17/2) lines.

6. Remove the guide element (27).

7. Detach the throttle cable (30) at the control lever (9).

8. If equipped with an automatic transmission, detach the control pressure cable (98).

9. Detach the connector for the electronic accelerator actuator (M16/1x1) and the cable.

10. Disconnect the coolant hose (33) at the front of the manifold.

11. Disconnect the coolant hose at the rear of the manifold.

12. Remove the manifold mounting bolts, then the manifold.

13. Place clean shop rags into the intake passages to prevent dirt from entering. Clean the gasket mating surfaces.

14. Clean any remains of old gasket and sealant material.

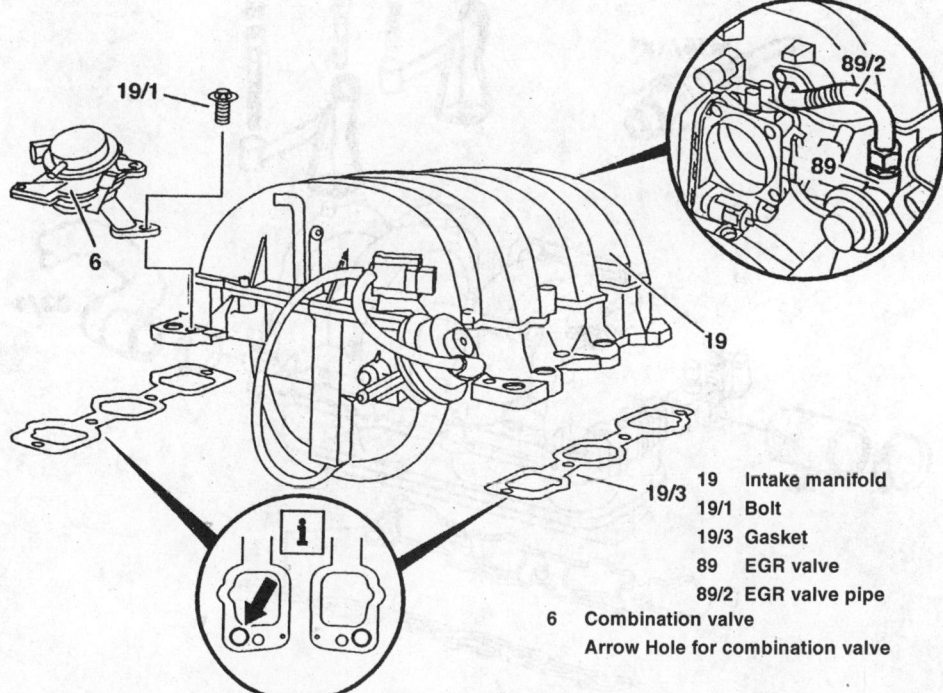

	19	Intake manifold
	19/1	Bolt
	19/3	Gasket
	89	EGR valve
	89/2	EGR valve pipe
6	Combination valve	
	Arrow Hole for combination valve	

7923NG16

Exploded view of the intake manifold mounting and related components—3.2L (112) and 4.3L (113) engines

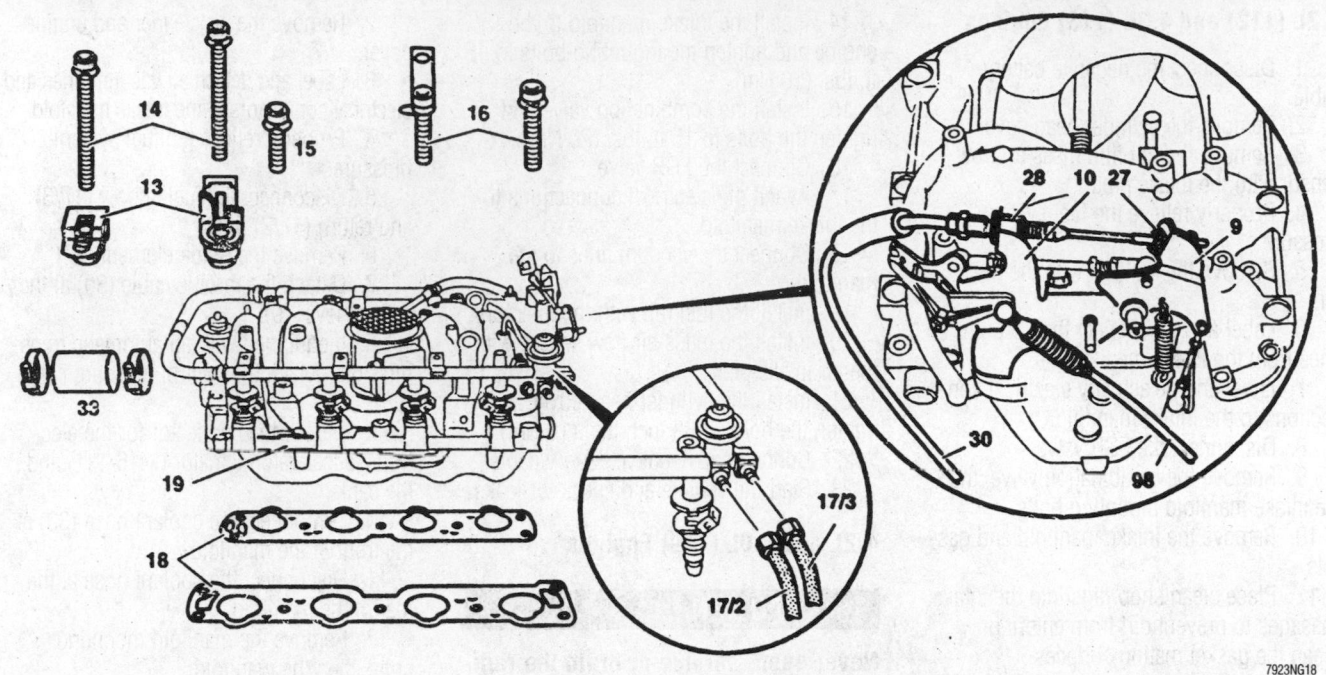

Exploded view of the intake manifold mounting and related components—1995 4.2L and 5.0L (119) engines

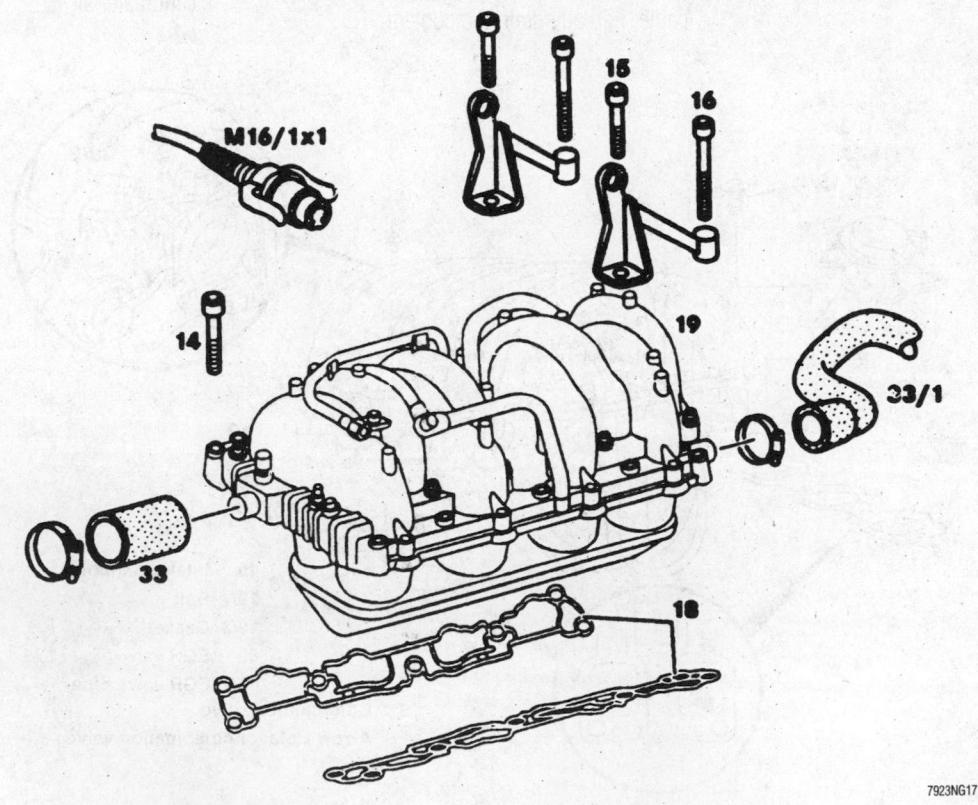

Exploded view of the intake manifold mounting and related components—1996–99 4.2L and 5.0L (119) engines

To install:

15. Remove the shop rags from the intake passages.

16. Using a new gasket, install the manifold and tighten the bolts in sequence to 18 ft. lbs. (25 Nm).

17. Connect the heater hoses to the manifold.

18. Connect the throttle and transmission cables.

19. Attach the fuel lines.

20. Attach all vacuum lines and electrical connectors to the intake manifold.

21. Install the air cleaner and engine cover.

22. Refill the engine fluids, connect the battery cable, start the engine and check for leaks.

Exhaust Manifold

REMOVAL & INSTALLATION

4-Cylinder and Inline 6-Cylinder Engines

1. Disconnect the negative battery cable.

2. Disconnect the exhaust pipe from the manifold.

3. Remove the exhaust support from the transmission, if so equipped.

4. Disconnect the air injection tube from the manifold, if so equipped.

5. Remove the exhaust manifold retaining bolts and manifold. Be careful with the manifold coupler tube between the 2 halves. Replace if damaged.

To install:

6. Install the manifold and torque the bolts in an even pattern. Use a new gasket.

7. Install the exhaust pipe with new self-locking nuts. Torque the pipe to 25 ft. lbs. (34 Nm).

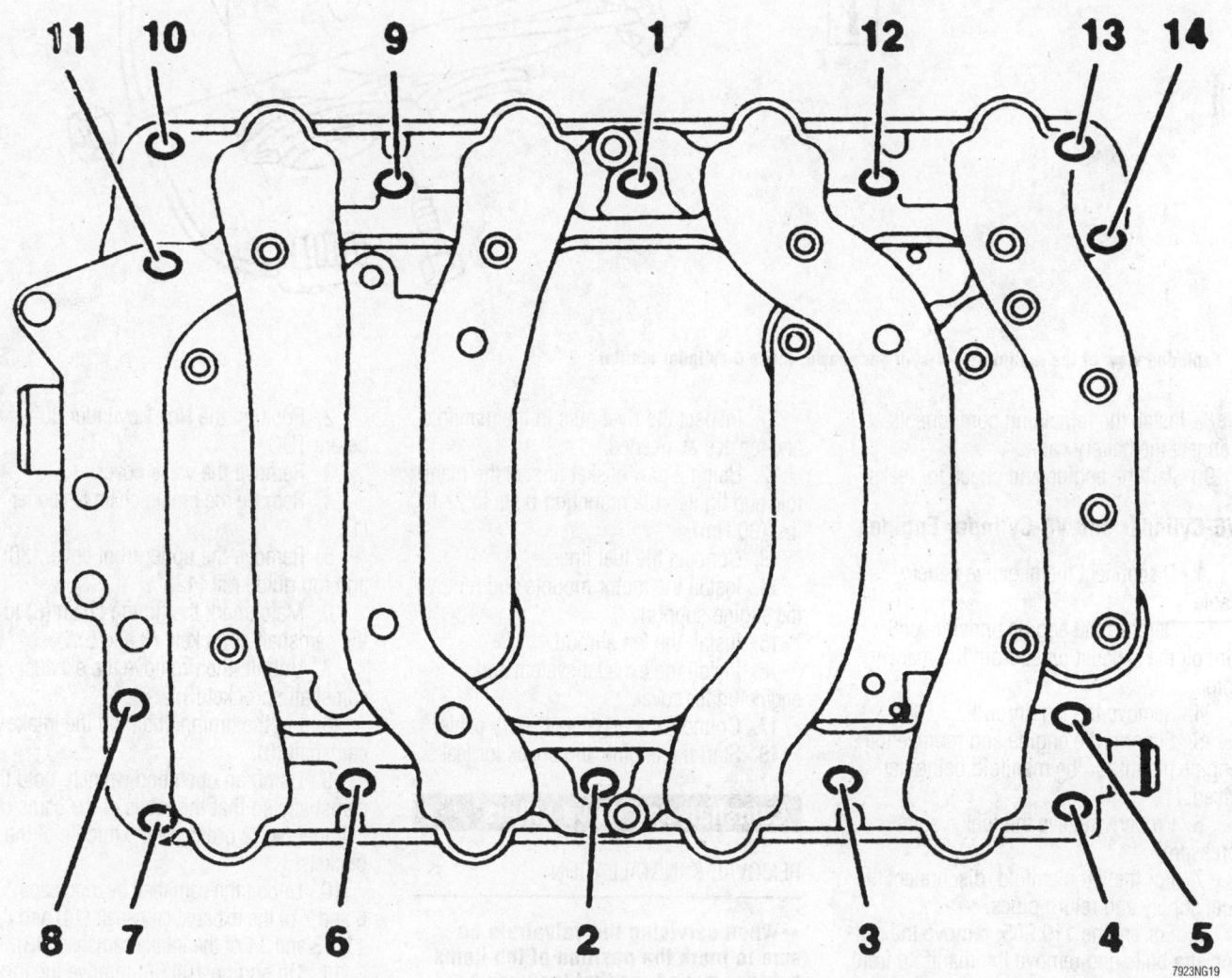

Intake manifold bolt tightening sequence—4.2L and 5.0L (119) engines

7923NG19

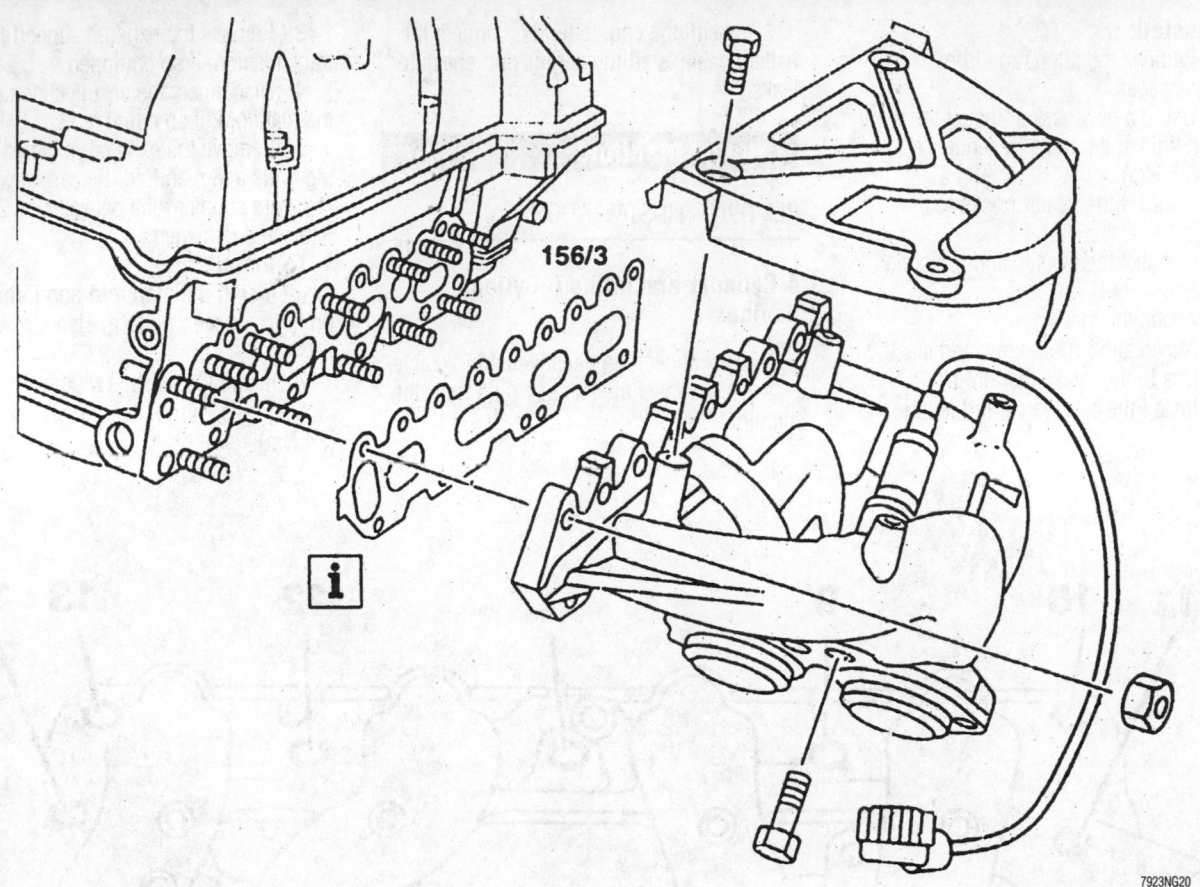

156/3

7923NG20

Exploded view of the mounting—4-cylinder shown, inline 6 cylinder similar

8. Install the remaining components and connect the battery cable.

9. Start the engine and check for leaks.

V6-Cylinder and V8-Cylinder Engines

1. Disconnect the negative battery cable.

2. Remove the engine undercover.3. Unbolt the exhaust pipes from the manifolds.

4. Remove the fan shroud.

5. Support the engine and remove the engine mount for the manifold being serviced.

6. Properly relieve the fuel system pressure.

7. For the left manifold, disconnect the fuel supply and return pipes.

8. For engine 119.985, remove the self-locking bolts and remove the manifold from below.

9. Remove the self-locking bolts, then lift out the manifold.

To install:

10. Clean the sealing surface of the manifold and cylinder head from old gasket material.

11. Inspect the rivet nuts in the manifold and replace as needed.

12. Using a new gasket, install the manifold and tighten the mounting bolts to 23 ft. lbs. (30 Nm).

13. Connect the fuel lines.

14. Install the motor mounts and remove the engine support.

15. Install the fan shroud.

16. Install the exhaust system and engine under cover.

17. Connect the negative battery cable.

18. Start the engine and check for leaks.

Camshaft and Valve Lifters

REMOVAL & INSTALLATION

➡**When servicing the valvetrain be sure to mark the position of the items being removed, so that they can be reinstalled into their original positions.**

2.8L and 3.2L (104) Engines

1. Disconnect the negative battery cable.

2. Position the No. 1 cylinder 30° before TDC.

3. Remove the valve cover.

4. Remove the timing chain tensioner (1).

5. Remove the upper front cover (30) and top guide rail (42).

6. Matchmark the timing chain (6) to the camshaft sprockets (3 and 5).

7. Unbolt, then remove the exhaust camshaft sprocket (3).

8. Lift the timing chain off the intake camshaft (5).

9. Using an open end wrench, hold the camshafts so that the lobes of the cams of cylinder No. 2 press on the middle of the buckets.

10. Unbolt the camshaft bearing caps 1, 4, 6 and 7 of the exhaust camshaft (14) and 8, 11, 13 and 14 of the intake camshaft (13).

11. On engine 104.98, remove the thrust washers for bearing caps 1, 8, 4 and 11.

12. Loosen the remaining bearing cap bolts one turn at a time until there is no counterpressure.

13. Remove the camshafts.

14. With the aid of a suction cup, remove the bucket tappet from its bore.

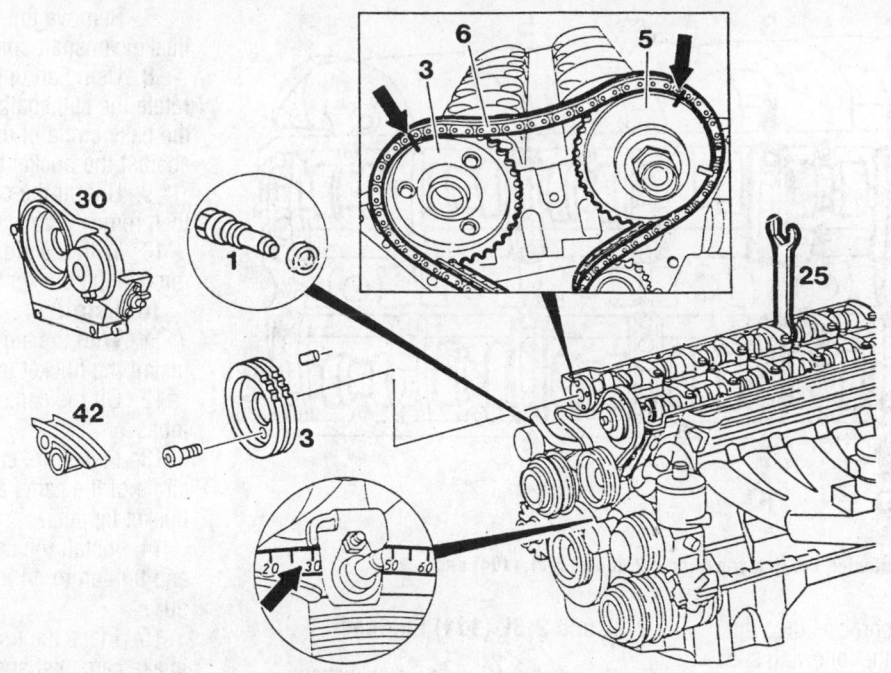

7923NG21

Position the crankshaft as shown to prevent piston-to-valve contact—2.8L and 3.2L (104) engines

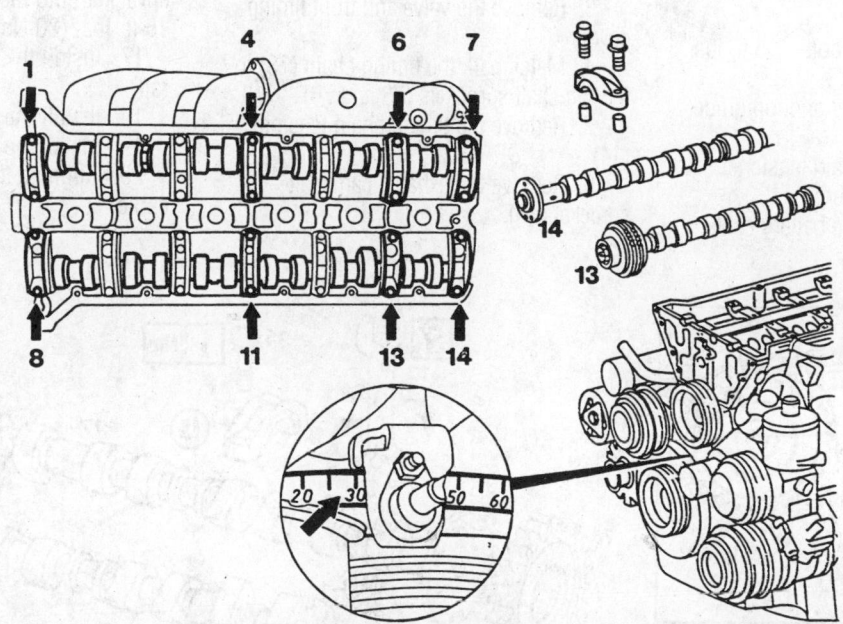

7923NG22

Be sure to remove the indicated bearing caps first—2.8L and 3.2L (104) engines

To install:

15. With the aid of a suction cup, install the bucket tappet into its bore.

16. Oil the camshaft journals and lobes.

17. Install the camshafts at 30° before TDC for cylinder No. 1. The lobes of the cams must be facing on the middle of the buckets for cylinder No. 2.

18. Install bearing caps 2, 3 and 5 for the exhaust camshaft, and 9, 10 and 12 for the intake camshaft, then tighten to 15 ft. lbs. (21 Nm).

19. Using an open end wrench, hold the camshafts so that the lobes of the cams of cylinder No. 2 press on the middle of the buckets.

20. On engine 104.98, install the thrust washers for bearing caps 1, 8, 4 and 11.

21. Install the remaining bearing caps and tighten them to 15.5 ft. lbs. (21 Nm).

22. Place the timing chain over the intake camshaft sprocket.

23. Install the exhaust camshaft sprocket on to the timing chain while aligning the

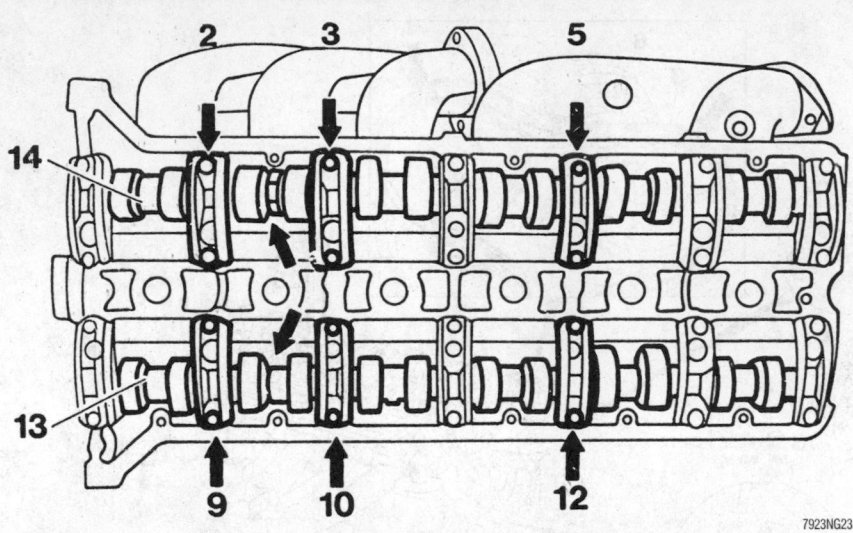

Be sure to install the indicated bearing caps first—2.8L and 3.2L (104) engines

7. Remove the timing chain from the intake camshaft sprocket (75).

8. Using an open end wrench (1), rotate the camshafts (77 and 78) so that the base circle of the cams are resting against the bucket tappets.

9. Unbolt the camshaft bearing caps, then remove the camshafts.

10. With the aid of a suction cup, remove the bucket tappet from its bore.

To install:

11. With the aid of a suction cup, install the bucket tappet into its bore.

12. Oil the camshaft journals and lobes.

13. Install the camshafts with the base circle of the cams are resting against the bucket tappets.

14. Install the camshaft bearing caps and tighten to 44 inch lbs. (5 Nm) plus 90°.

15. Place the timing chain over the intake camshaft sprocket.

16. Install the exhaust camshaft sprocket on to the timing chain while aligning the matchmarks. Install the sprocket onto the camshaft and tighten to 15 ft. lbs. (20 Nm) plus 90°.

17. Install the front cover and top guide rail.

18. Install the timing chain tensioner.

19. Install the valve cover.

20. Connect the negative battery cable.

matchmarks. Install the sprocket onto the camshaft and tighten to the following specification:

• M7 13.5mm T30 Torx® bolt—13 ft. lbs. (18 Nm)
• M7 13.5mm T40 Torx® bolt—16 ft. lbs. (22 Nm)
• M7 13mm T30 Torx® bolt—15 ft. lbs. (20 Nm) plus 60°

24. Install the front cover and top guide rail.

25. Install the timing chain tensioner.

26. Install the valve cover.

27. Connect the negative battery cable.

2.2L and 2.3L (111) Engines

1. Disconnect the negative battery cable.

2. Position the No. 1 cylinder 30° after TDC.

3. Remove the valve and front timing cover.

4. Matchmark the timing chain (73) to the camshaft sprockets (75 and 76).

5. Remove the timing chain tensioner (14).

6. Remove the exhaust camshaft sprocket (76).

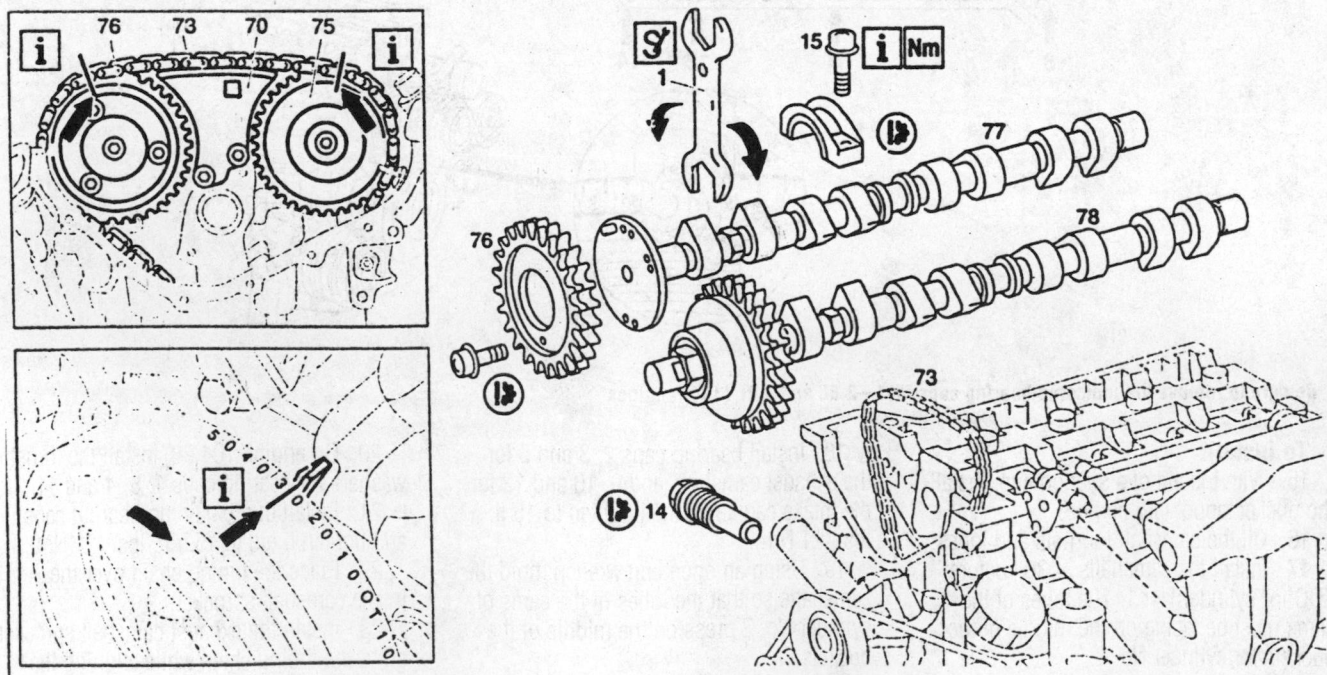

Exploded view of the camshaft mounting components and positioning—2.2L and 2.3L (111) engines

3.2L (112) and 4.3L (113) Engines

1. Disconnect the negative battery cable.
2. Remove the cylinder head cover.
3. Rotate the engine clockwise to position the crankshaft 40 degrees After Top Dead Center (ATDC).

❋❋ WARNING

Engine must not be rotated backwards.

4. Remove the generator.
5. Remove the timing chain tensioner (2).
6. Remove the camshaft Hall sensor (1).
7. Cable tie the timing chain to the camshaft sprocket (5).
8. Lock the camshafts using the Camshaft Locking (3) tools 112 589 00 32 00 and 112 589 01 32 00, or their equivalent.
9. Unbolt the camshaft gears.
10. Remove the camshaft bearing bridge (4).
11. Carefully remove the camshafts from the cylinder head.

To install:

➡ **Be sure to install the correct camshaft for the corresponding cylinder head.**

12. Apply clean engine oil to the camshaft contact surfaces, then install the camshaft.
13. Install the camshaft bearing bridge.

➡ **The camshafts can be rotated 40 degrees after top dead center of the No. 1 cylinder without the valves touching the pistons.**

14. Position the camshaft so that the groove points centered towards the contact surface of the cylinder head cover, then attach the camshaft fixing plate. Repeat this step for the other camshaft.

15. Install the camshaft sprockets and tighten the attaching bolt to 37 ft. lbs. (50 Nm) plus 90–100 °.
16. Remove the camshaft locking tools and the cable ties from the timing chain.
17. Install the camshaft position sensor and tighten the mounting bolt to 70 inch lbs. (8 Nm).
18. Install the timing chain tensioner with a new gasket and tighten to 59 ft. lbs. (80 Nm).
19. Install the cylinder head cover.
20. Connect the negative battery cable.

4.2L and 5.0L (119) Engines

1. Disconnect the negative battery cable.
2. Position the No. 1 cylinder 45° before TDC.
3. Remove the valve covers.
4. Remove the upper front covers (10 and 12).
5. Matchmark the position (relative) of the camshaft timing gears to the timing chain.
6. Remove the timing chain tensioner (13).
7. Remove the timing chain top slide rails (1, 1a, 2 and 2a).

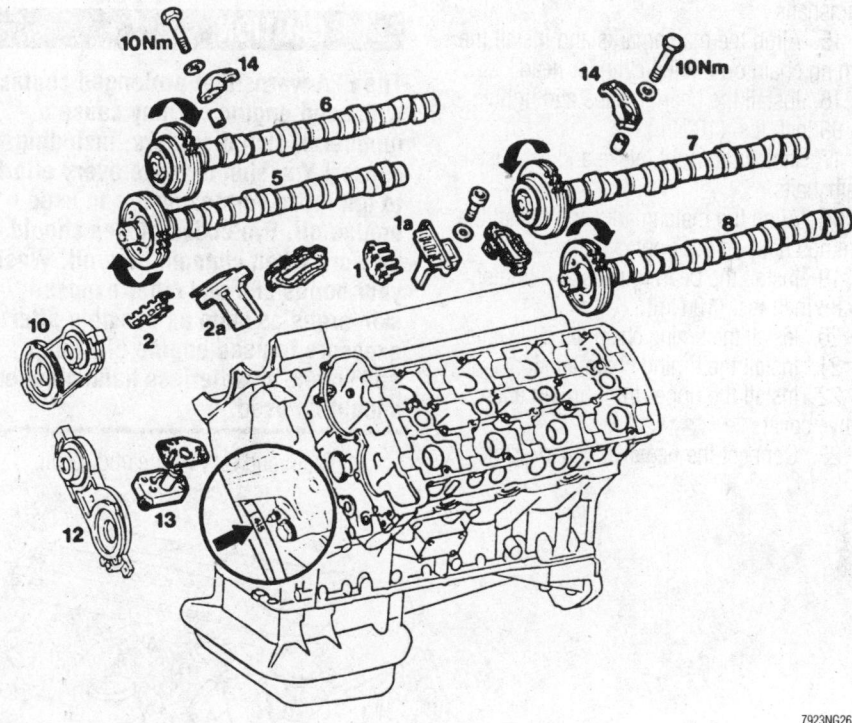

7923NG26

Exploded view of the camshaft mounting and related components—4.2L and 5.0L (119) engine

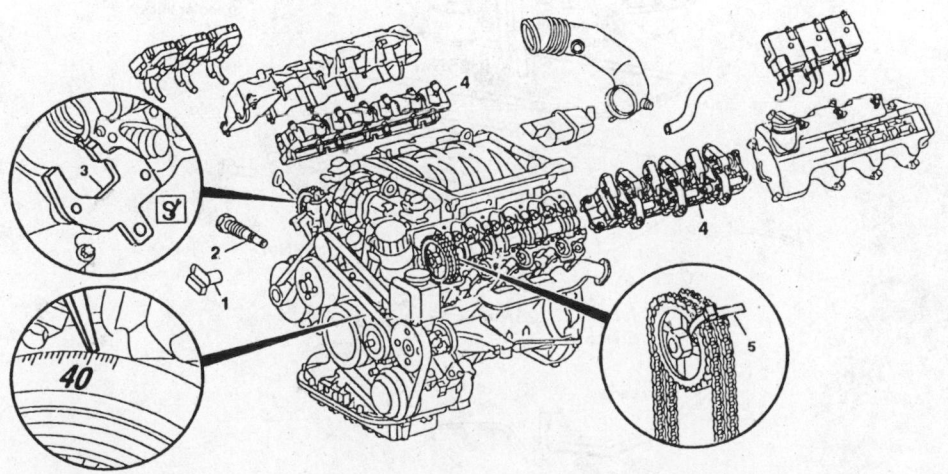

7923NG25

Exploded view of the camshaft mounting, showing related components—3.2L (112) and 4.3L (113) engines

8. Lift the timing chain off the camshaft sprockets, then rotate the cams so that the base circle of the cams are touching the bucket tappets.

9. Evenly loosen the camshaft bearing cap bolts.

10. Remove the camshaft bearing caps, then the camshafts with the sprockets.

11. With the aid of a suction cup, remove the bucket tappet from its bore.

To install:

12. With the aid of a suction cup, install the bucket tappet into its bore.

13. Oil the camshaft journals and lobes.

14. Install the left intake and exhaust camshafts.

15. Align the matchmarks and install the timing chain on the left cylinder head.

16. Install the bearing caps and tighten to 88 inch lbs. (10 Nm).

17. Install the right intake and exhaust camshafts.

18. Align the matchmarks and install the timing chain on the right cylinder head.

19. Install the bearing caps and tighten to 88 inch lbs. (10 Nm).

20. Install the timing chain top slide rails.

21. Install the timing chain tensioner.

22. Install the upper front timing and valve covers.

23. Connect the negative battery cable.

Valve Lash

ADJUSTMENT

All Mercedes-Benz engines use hydraulic valve lifters. There is no need or provision for valve clearance adjustments.

Oil Pan

REMOVAL & INSTALLATION

1. Disconnect the negative battery cable.

✳✳ CAUTION

The EPA warns that prolonged contact with used engine oil may cause a number of skin disorders, including cancer! You should make every effort to minimize your exposure to used engine oil. Protective gloves should be worn when changing the oil. Wash your hands and any other exposed skin areas as soon as possible after exposure to used engine oil. Soap and water, or waterless hand cleaner should be used.

2. Drain and recycle the engine oil.

3. Using a suitable engine support tool, support the weight of the engine.

4. Raise and safely support the vehicle.

5. Support the front subframe and remove the mounting bolts.

6. Lower the subframe enough to access the oil pan mounting bolts.

➡**It may be necessary to tap on the oil pan with a rubber mallet to dislodge it from the engine block.**

7. Remove the mounting bolts, then the pan.

8. Clean all remains of the old gasket and sealant from the oil pan and engine block.

To install:

9. Install a new pan gasket and install the oil pan.

10. Install the oil pan mounting bolts in a crisscross pattern.

11. Install the items that were removed.

✳✳ WARNING

Operating the engine without the proper amount and type of engine oil will result in severe engine damage.

12. Fill the crankcase with oil.

13. Connect the negative battery cable.

14. Start the engine and check for leaks.

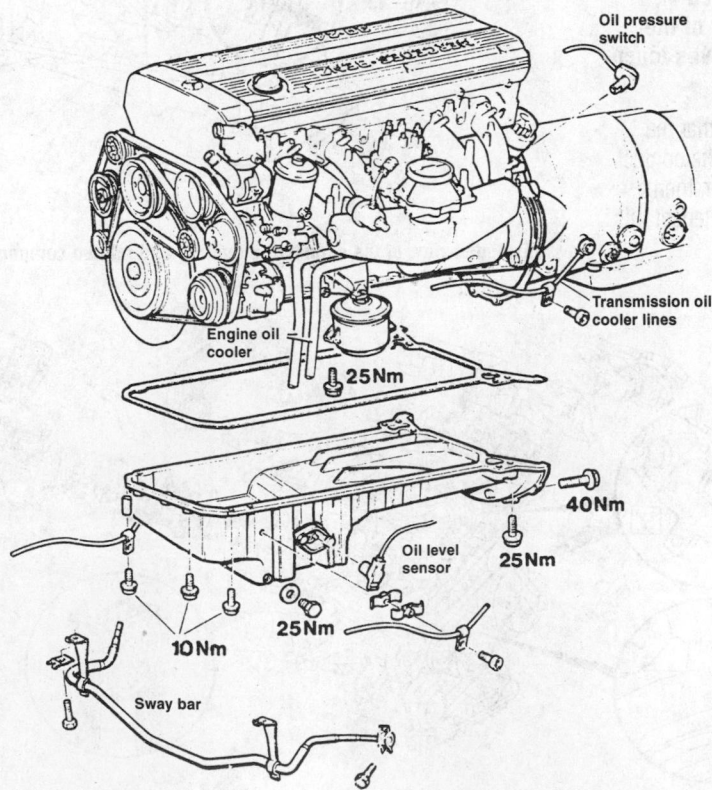

Exploded view of the oil pan mounting and related components—except 3.2L (112) and 4.3L (113) engines

7923NG27

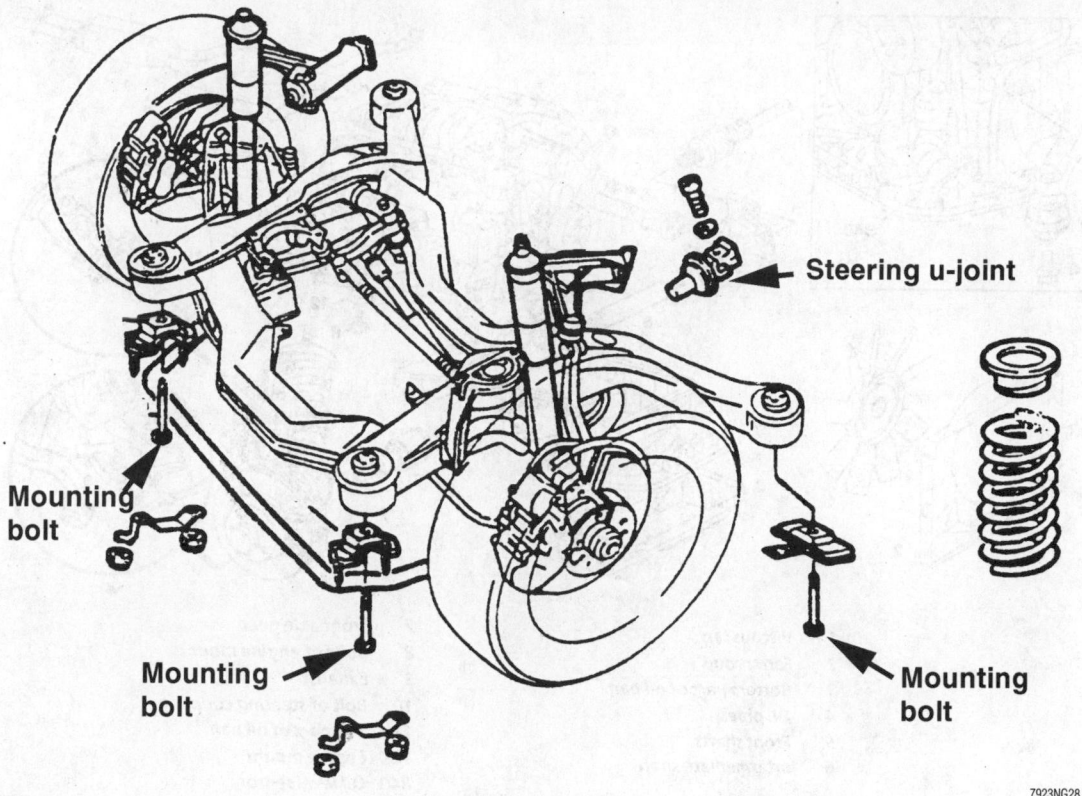

7923NG28

Lower the front subframe by removing the mounting bolts—except 3.2L (112) and 4.3L (113) engines

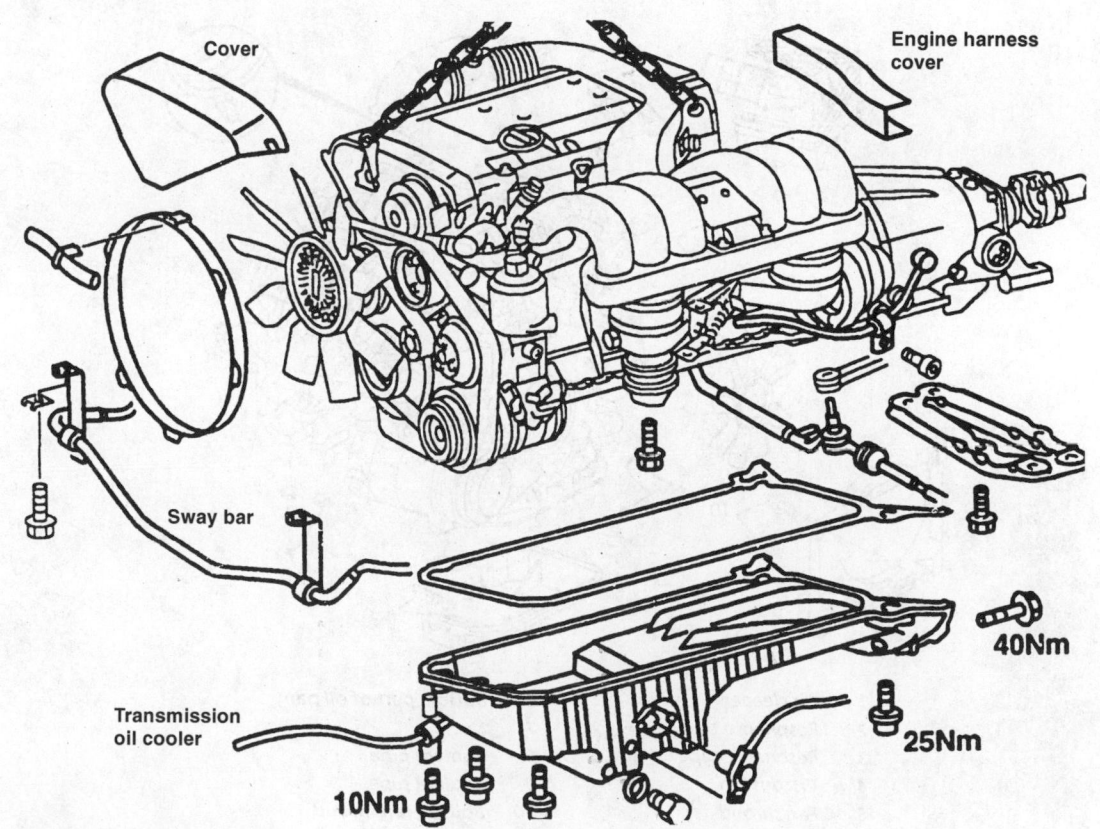

7923NG29

Exploded view of the oil pan mounting and related components—except 3.2L (112) and 4.3L (113) engines

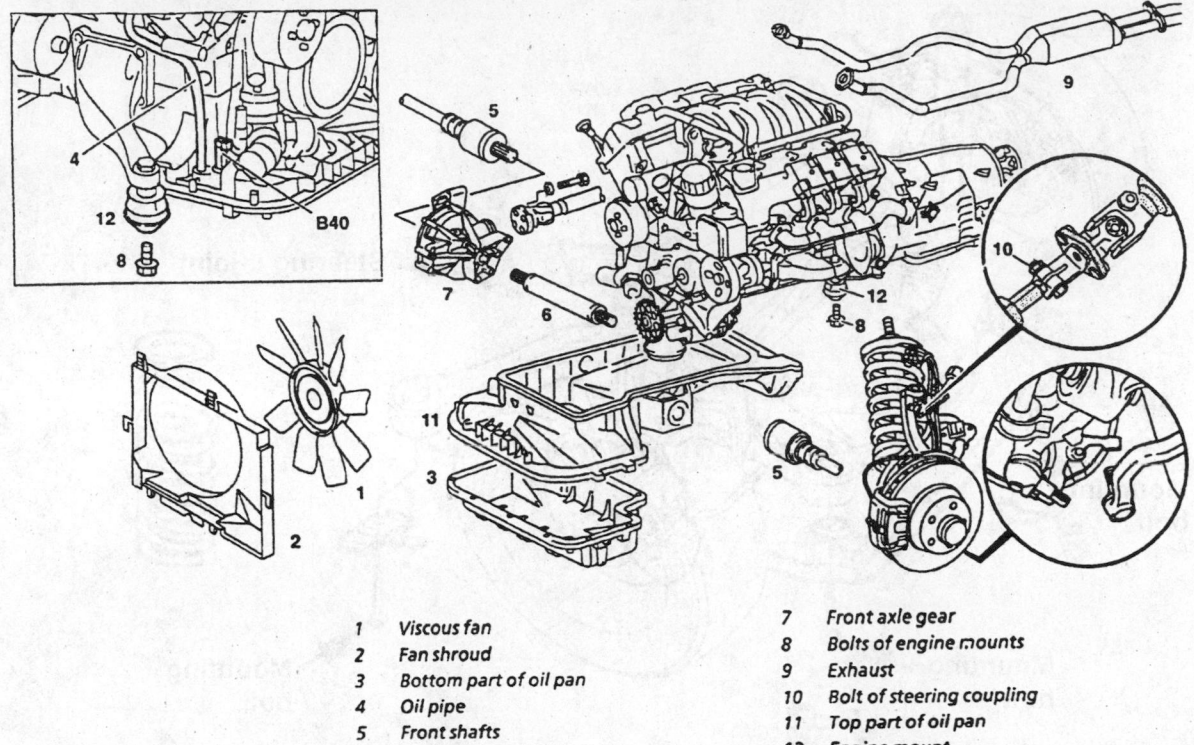

1	Viscous fan	7	Front axle gear
2	Fan shroud	8	Bolts of engine mounts
3	Bottom part of oil pan	9	Exhaust
4	Oil pipe	10	Bolt of steering coupling
5	Front shafts	11	Top part of oil pan
6	Intermediate shaft	12	Engine mount
		B40	Oil level sensor

7923NG30

Exploded view of the mounting of the upper oil pan and related components—3.2L (112) and 4.3L (113) engines

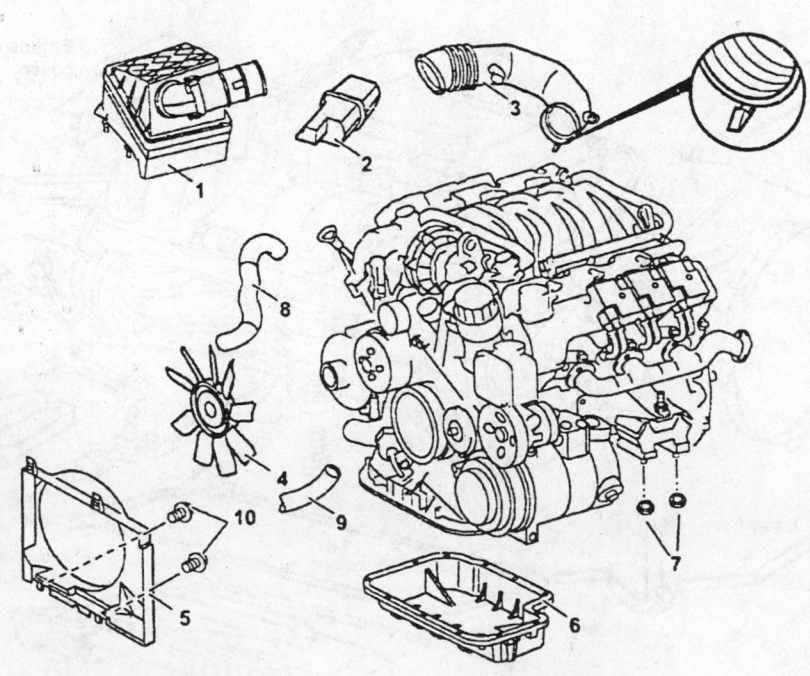

1	Air cleaner housing	6	Bottom part of oil pan
2	Resonance body	7	Nuts
3	Resonance pipe	8	Coolant pipe
4	Viscous fan	9	Coolant pipe
5	Fan shroud	10	Bolts of fan shroud

7923NG31

Exploded view of the mounting of the lower oil pan and related components—3.2L (112) and 4.3L (113) engines

Oil Pump

REMOVAL & INSTALLATION

1. Remove the oil pan.
2. Remove the oil pump drive gear, then remove the gear from the chain.
3. Remove the oil pump mounting bolts.

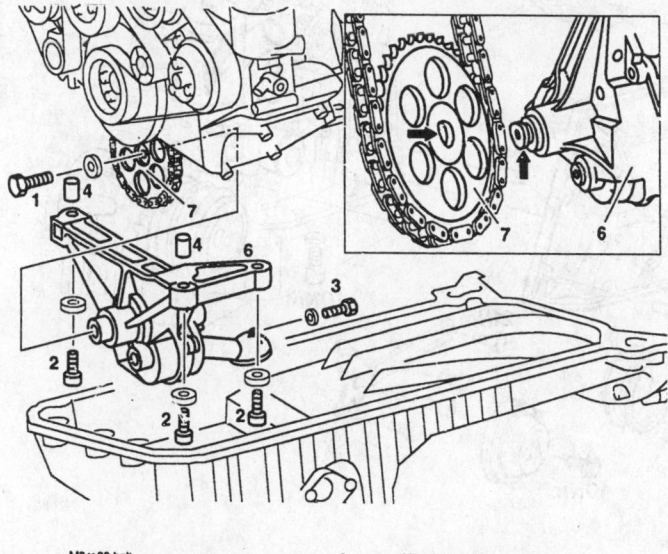

M8 × 20 bolt   3   M6 × 25 bolt + washer
M8 × 35 bolt + washer (hexagon socket)   4   Dowel sleeve

7923NG32

Exploded view of the oil pump mounting and related component—except 3.2L (112) and 4.3L (113) engines

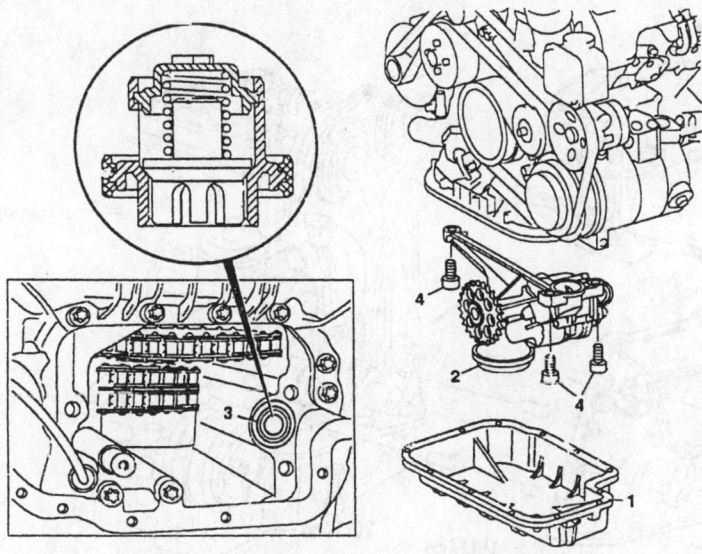

1. Oil pan
2. Oil pump
3. Oil return check valve
4. Oil pump mounting bolts

7923NG80

Exploded view of the oil pump mounting and related component—3.2L (112) and 4.3L (113) engines

To install:

➡️**Be sure to replace any O-rings or seals between the oil pump and engine block.**

4. Install the oil pump and tighten the mounting bolts as follows:
- M6 bolts to 88 inch lbs. (9 Nm)
- M8 bolts to 15 ft. lbs. (21 Nm)

5. Install the oil pump drive gear and tighten the mounting bolt to 21 ft. lbs. (28 Nm).
6. Install the oil pan.

Rear Main Seal

REMOVAL & INSTALLATION

➡️**All engines use a one piece radial seal.**

1. Disconnect the negative battery cable.
2. Remove the transmission and flywheel/flexplate.
3. Pry out the seal using a suitable pry-tool wrapped with a rag.

To install:

4. Coat the inner lip of the seal with engine oil, then, using a suitable seal driver such as 117589004300 or equivalent, install the seal into the retainer.

➡️**Be sure the seal is seated evenly, otherwise the seal will leak.**

5. Install the flywheel/flexplate.
6. Install the transmission.
7. Connect the negative battery cable.

Front Cover and Seal

REMOVAL & INSTALLATION

1. Disconnect the negative battery cable and drain the engine coolant.
2. Remove the engine cover, air cleaner and intake scoop.
3. Remove the viscous fan clutch, fan belt and mounting pulleys.
4. Disconnect all wiring and hoses from the front of the engine.
5. Remove the upper timing cover and mark the timing chain and sprockets.
6. Remove the timing chain tensioning device.
7. Remove the accessories as needed, referring to the illustration for the specific model.
8. Remove the TDC sensor, alternator and bracket.
9. Remove the crankshaft pulley.
10. Remove the lower timing covers.

To install:

11. Install the cover and torque the M6 bolts to 7 ft. lbs. (10 Nm). Torque the M8 bolts to 15 ft. lbs. (21 Nm).
12. Install the remaining components in reverse order.
13. Refill the engine with coolant, start the engine and check for leaks.

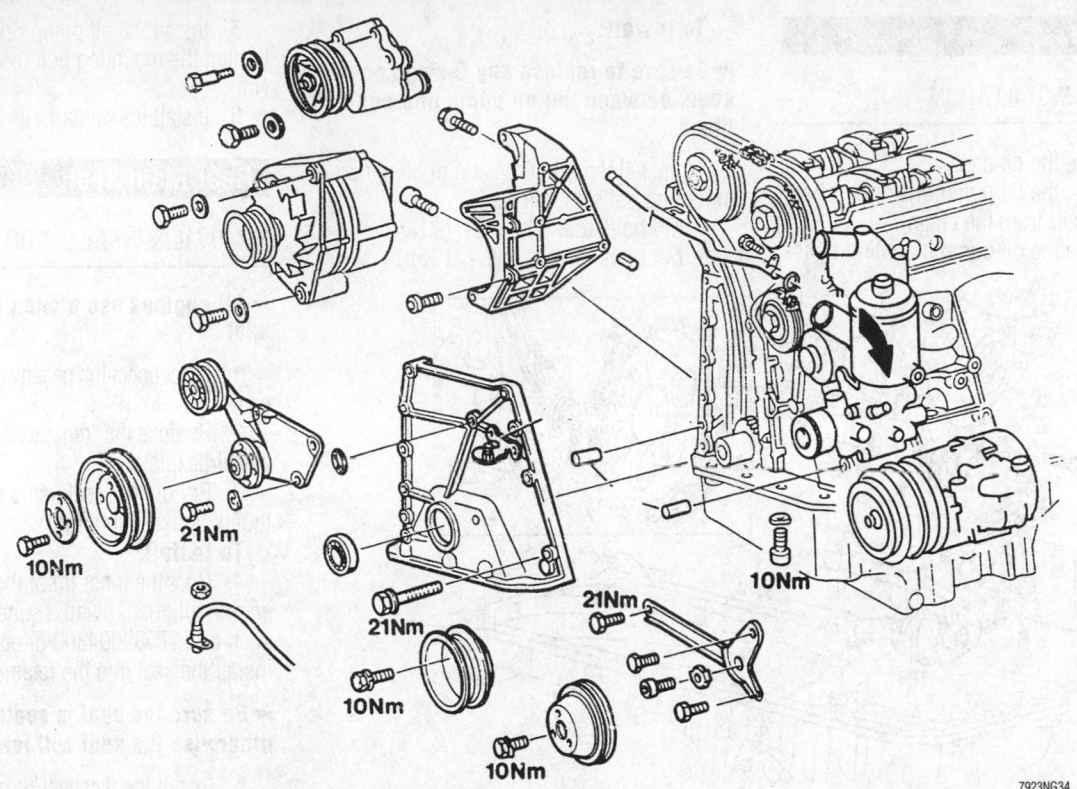

Exploded view of the lower timing cover mounting and related components—2.8L and 3.2L (104) engine in the 1995 E Class (124) and SL class (129)

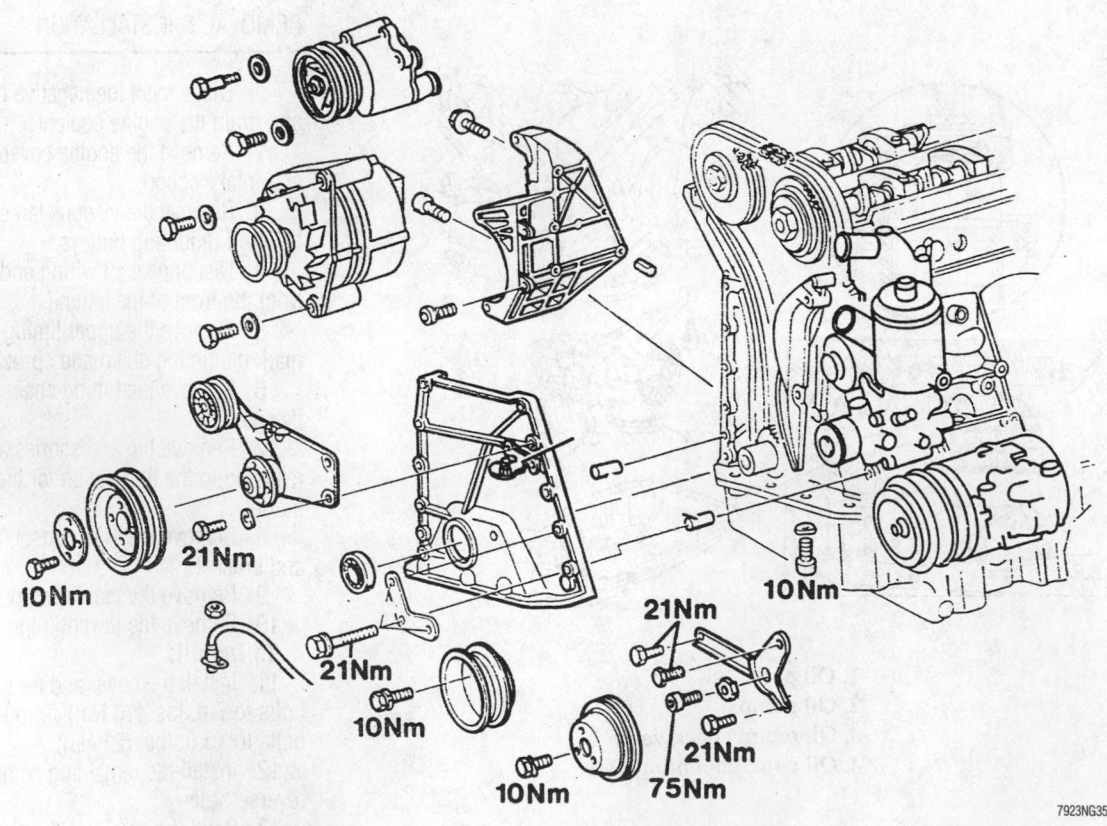

Exploded view of the lower timing cover mounting and related components—2.8L and 3.2L (104) engine in the S Class (140)

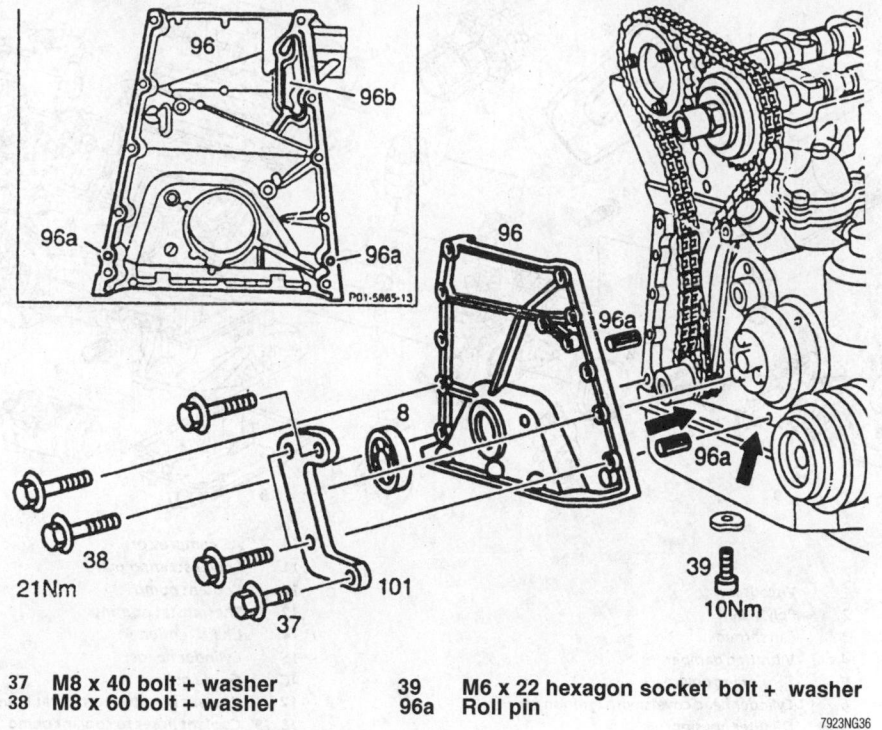

37	M8 x 40 bolt + washer	39	M6 x 22 hexagon socket bolt + washer
38	M8 x 60 bolt + washer	96a	Roll pin

Exploded view of the lower timing cover mounting and related components—2.8L and 3.2L (104) engine in the C Class (202) and 1996–99 E Class (210) models

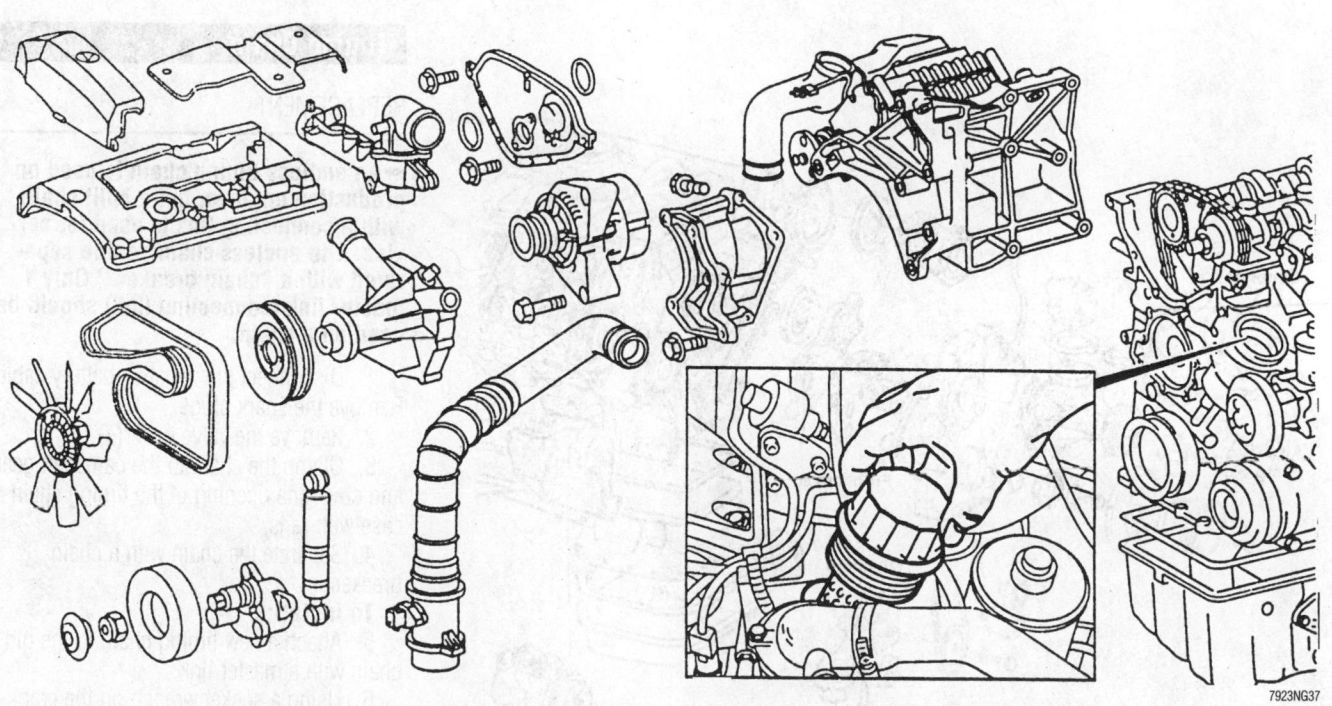

Exploded view of the lower timing cover mounting and related components—2.2L and 2.3L (111) engines in the 1995 E Class, CLK 230 (170), C Class (202) and 1996–99 E Class (210) models

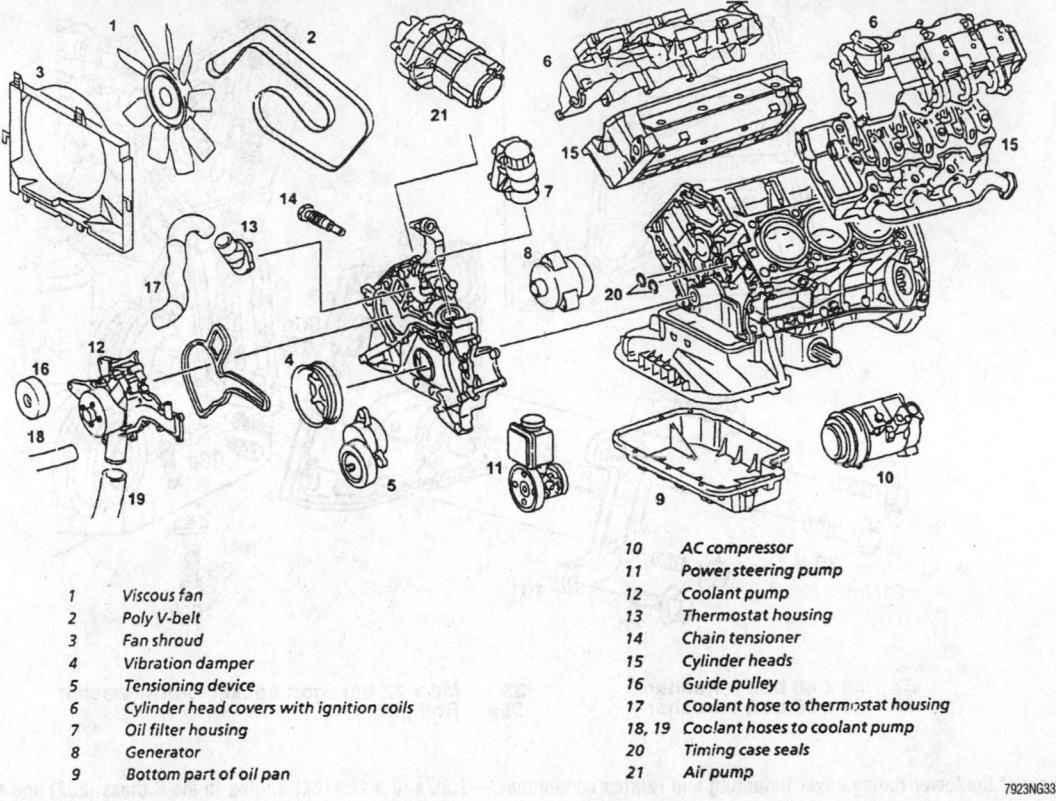

1	Viscous fan
2	Poly V-belt
3	Fan shroud
4	Vibration damper
5	Tensioning device
6	Cylinder head covers with ignition coils
7	Oil filter housing
8	Generator
9	Bottom part of oil pan
10	AC compressor
11	Power steering pump
12	Coolant pump
13	Thermostat housing
14	Chain tensioner
15	Cylinder heads
16	Guide pulley
17	Coolant hose to thermostat housing
18, 19	Coolant hoses to coolant pump
20	Timing case seals
21	Air pump

7923NG33

Exploded view of the lower timing cover mounting and related components—3.2L (112) and 4.3L (113) engines in the C Class (202) and 1996–99 E Class (210) models

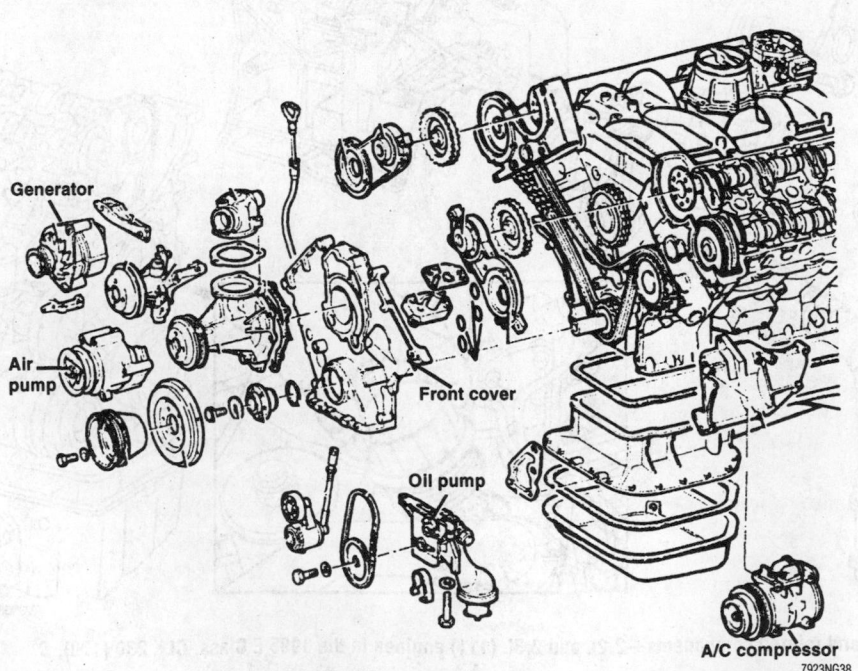

7923NG38

Exploded view of the lower timing cover mounting and related components—4.2L and 5.0L (119) engine

Timing Chain

REPLACEMENT

➡ **An endless timing chain is used on production engines, but a split chain with a connecting link is used for service. The endless chain can be separated with a "chain breaker." Only 1 master link (connecting link) should be used on a chain.**

1. Disconnect the negative battery cable. Remove the spark plugs.

2. Remove the valve cover(s).

3. Clamp the chain to the camshaft gear and cover the opening of the timing chain case with rags.

4. Separate the chain with a chain breaker.

To install:

5. Attach a new timing chain to the old chain with a master link.

6. Using a socket wrench on the crankshaft, slowly rotate the engine in the direction of normal rotation. Simultaneously, pull the old chain through until the master link is uppermost on the camshaft sprocket; be

sure to keep tension on the chain throughout this procedure.

7. Disconnect the old timing chain and connect the ends of the new chain with the master link. Insert the new connecting link from the rear so the lockwashers can be seen from the front.

8. Rotate the engine until the timing marks align. Check the valve timing. Once the new chain is assembled, rotate the engine, by hand, through a least 1 complete revolution to be sure everything is OK.

DIESEL ENGINE REPAIR

➡️**Disconnecting the negative battery cable may interfere with the functions of the on board computer systems and may require the computer to undergo a relearning process, once the negative battery cable is reconnected.**

Engine Assembly

REMOVAL & INSTALLATION

➡️**In all cases, Mercedes-Benz engines and transmissions are removed as a unit.**

1. Air conditioner lines should not be indiscriminately disconnected without taking proper precautions. It is best to swing the compressor aside while still connected to its hoses. Never do any welding around the compressor-heat may cause an explosion. Also, the refrigerant, while inert at normal room temperature, breaks down under high temperature into hydrogen fluoride and phosgene (among other products), which are highly poisonous.

❊❊ CAUTION

Never open, service or drain the radiator or cooling system when hot; serious burns can occur from the steam and hot coolant.

2. Remove the hood, drain the cooling system and disconnect the battery. While not strictly necessary, it is better to remove the battery completely to prevent breakage by the engine as it is lifted out.

3. Remove the fan shroud, radiator and disconnect all heater hoses and oil cooler lines.

4. Remove the air cleaner and all fuel, vacuum and oil hoses, e.g., power steering and power brakes.

5. Plug all openings to keep out dirt.

6. Remove the viscous coupling and fan.

7. Disconnect the accelerator linkage.

8. Disconnect all ground straps and electrical connections. It is a good idle to tag each wire for easy reassembly.

9. Detach the gearshift linkage and the exhaust pipes from the manifolds.

10. Loosen the steering relay arm and pull it aside, along with the center steering rod and hydraulic steering damper.

11. The hydraulic engine shock absorber should be removed.

12. Remove the hydraulic line from the clutch housing and the oil line connectors from the automatic transmission.

13. Unbolt the clutch slave cylinder from the bell housing after removing the return spring.

14. Remove the exhaust pipe bracket from the transmission. Support the bell housing or place a cable sling under the oil pan, to support the engine. On turbocharged models, disconnect the exhaust pipes at the turbocharger.

15. Mark the position of the rear engine support and unbolt the 2 outer bolts, then remove the top bolt at the transmission and pull the support out.

16. Disconnect the speedometer cable and the front driveshaft U-joint. Push the driveshaft back and wire it aside.

17. Unbolt the engine mounts on both sides.

18. Unbolt the power steering fluid reservoir and swing it aside;, then, using a chain hoist and cable, lift the engine and transmission upward and outward. An angle of about 45 degrees will allow the vehicle to be pushed backward while the engine is coming up.

To install:

19. With an assistant, install the engine and transmission into the vehicle using a chain hoist and cable. Lower the engine and transmission downward at an angle of about 45 degrees.

20. Connect the engine mounts on both sides and torque the bolts to 30 ft. lbs. (40 Nm).

21. Connect the speedometer cable and the front driveshaft U-joint.

22. Install the rear engine support and bolt the 2 outer bolts.

23. Install the exhaust pipe bracket to the transmission. On turbocharged models, connect the exhaust pipes at the turbocharger. Torque the pipe bolts to 25 ft. lbs. (34 Nm).

24. Install the clutch slave cylinder to the bell housing and install the return spring.

25. Install the hydraulic line to the clutch housing and the oil line connectors to the automatic transmission.

26. The hydraulic engine shock absorber should be installed.

27. Tighten the steering relay arm to 15 ft. lbs. (20 Nm). Connect the steering damper to the steering linkage.

28. Attach the gearshift linkage and the exhaust pipes to the manifolds. Torque the manifold bolts to 25 ft. lbs. (34 Nm).

29. Connect all ground straps and electrical connections.

30. Connect the accelerator linkage.

31. Install the viscous coupling and fan.

32. Install all fuel, vacuum and oil hoses, e.g., power steering, power brakes and air cleaner.

33. Install the radiator, fan shroud and connect all heater hoses and oil cooler lines.

34. Refill the engine coolant and install the hood.

35. Install the battery and connect.

36. Bleed the hydraulic clutch, power steering, power brakes and fuel system.

Water Pump

REMOVAL & INSTALLATION

➡️**After refilling the cooling system, the system may have to be bleed. Remove a cooling sensor or equivalent at the highest point in the engine's cooling system. Fill the radiator until coolant spills out of the hole. Apply thread sealing tape to the component and install. Finish filling the radiator to the proper level.**

1. Disconnect the negative battery cable. Remove the fan shroud.

2. Remove viscous fan clutch with the fan.

3. Remove fastening screws and pulley.

4. Remove hex nuts and magnet body.

➡️**The magnet carrier is glued to the water pump housing and should not be pulled off.**

5. Remove water pump housing.

6. Clean sealing surfaces.

To install:

7. Insert water pump with a new gasket and tighten combination screws to 7.5 ft. lbs. (10 Nm).

8. Mount water pump with a new gasket and tighten combination screws to 7.5 ft. lbs. (10 Nm).

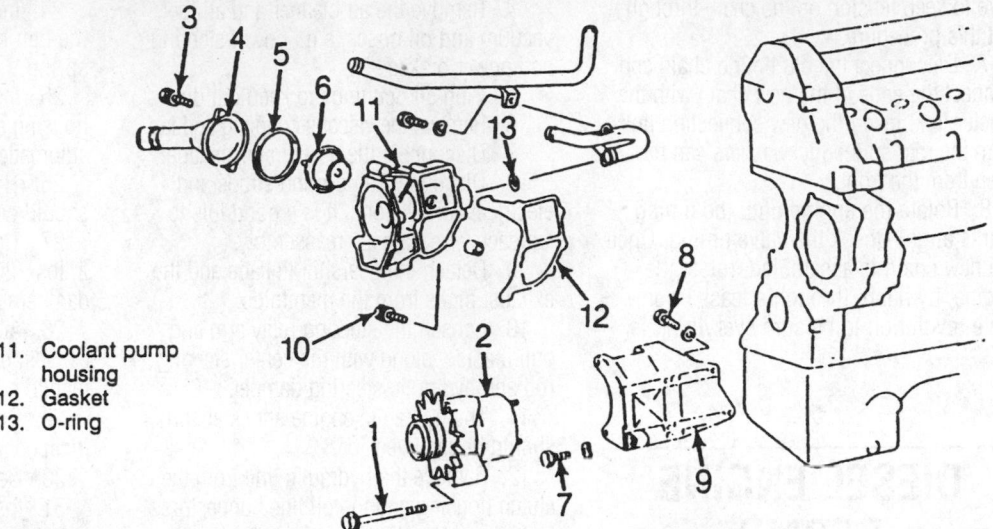

1. Collar screw—tightening torque 33 ft. lbs. (45 Nm)
2. Alternator
3. Screw—tightening torque 7 ft. lbs. (10 Nm)
4. Thermostat housing cap
5. Sealing ring
6. Thermostat
7. Screw—tightening torque 19 ft. lbs. (25 Nm)
8. Screw—tightening torque 19 ft. lbs. (25 Nm)
9. Carrier
10. Screw—tightening torque 7 ft. lbs. (10 Nm)
11. Coolant pump housing
12. Gasket
13. O-ring

Exploded view of the water pump housing assembly—Diesel engines

9. Mount magnet body and plug on cable.

10. Mount pulley and tighten fastening screws to 7.5 ft. lbs. (10 Nm).

11. Complete installation by reversing removal procedure.

Glow Plugs

REMOVAL & INSTALLATION

The glow plugs are located on the driver's side of the cylinder head.

1. Disconnect the negative battery cable.

2. Detach the electrical connectors from the glow plugs.

3. Remove the glow plugs from the cylinder head.

To install:

4. Apply anti-seize compound to the glow plug threads. Install the glow plugs and tighten to 15 ft. lbs. (20 Nm).

5. Attach the electrical connector and tighten the retaining nut to 35 inch lbs. (4 Nm).

6. Connect the negative battery cable.

Cylinder Head

REMOVAL & INSTALLATION

➡**Use care to ensure that the valve timing is not disturbed.**

1. Disconnect the negative battery cable. Drain the radiator and remove all hoses and wires. Tag all wires to ensure easy reassembly.

2. Remove the camshaft cover and associated throttle linkage.

3. Remove the camshaft sprocket nut.

4. Remove the rockers and their supports must be removed together.

5. Mark the chain, sprocket and cam for ease of assembly.

6. Using a suitable puller, remove the camshaft sprocket.

7. Remove the sprocket and chain and wire it aside.

➡**Be sure the chain is securely wired so it will not slide into the engine.**

8. Unbolt the manifolds and exhaust header pipe and push them aside.

9. Loosen the cylinder head hold-down bolts, in the reverse order of the torque sequence. It is good practice to loosen each bolt, a little at a time, working around the head, until all are free.

10. Reach into the engine compartment and gradually work the head loose from each end by rocking it. Never use prybar between the head and block to pry, as the head will be scarred badly and may be ruined.

※※ **WARNING**

All Diesel engines utilize cylinder head stretch bolts. These bolts undergo a permanent stretch each time they are tightened. When a maximum length is reached, they must be discarded and replaced with new bolts. Before tightening the head bolts on these engines refer to the illustration for the specific length of the bolts.

To install:

11. Clean the gasket mating surfaces. Install the cylinder head using an approved hoist.

12. Torque the cylinder head bolts in sequence as follows:

    a. First step: 11 ft. lbs. (15 Nm).

    b. Second step: 20 ft. lbs. (35 Nm).

    c. Third step: 90°

    d. Fourth step: wait 10 minutes

    e. Fifth step: 90°

Dimensions of cylinder head bolts and maximum permissible length (L)

New	Max. permissible length (L) in mm
M 10×80	83.6
M 10×102	105.6
M 10×115	118.6

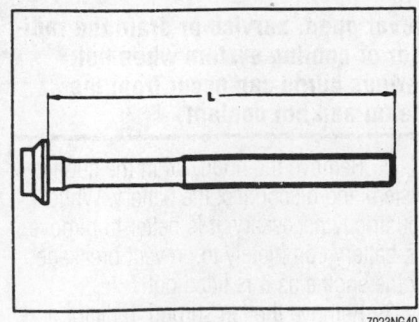

Be sure to replace bolts that are beyond maximum length—Diesel engines

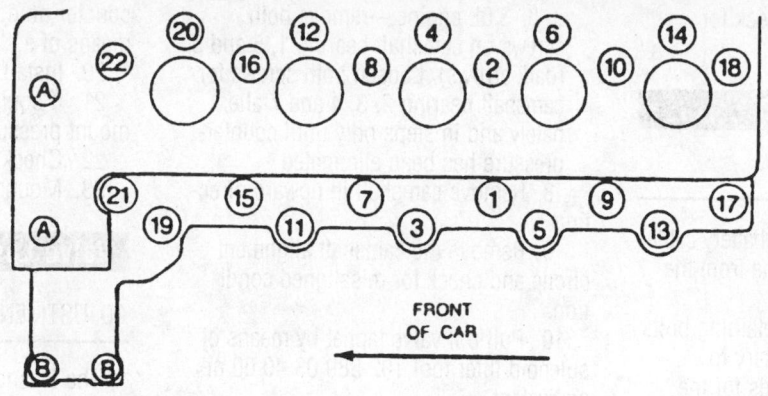

Cylinder head bolt tightening sequence—5-cylinder Diesel engines

7923NG41

Cylinder head bolt tightening sequence—6-cylinder Diesel engines

7923NG42

13. Install the manifolds and exhaust header pipe. Torque the pipe to 24 ft. lbs. (34 Nm).

14. Install the sprocket and chain.

15. Install the rockers and their supports. Torque to 15 ft. lbs. (20 Nm).

16. Install the camshaft sprocket nut.

17. Install the camshaft cover and associated throttle linkage.

18. Connect the negative battery cable. Refill the radiator and install all hoses and wires.

Rocker Arms/Shafts

REMOVAL & INSTALLATION

The Diesel engine does not use rocker arms. The camshaft acts directly on the hydraulic valve tappet.

Turbocharger

REMOVAL & INSTALLATION

1. Disconnect the negative battery cable. Remove the air filter.

2. Disconnect the electrical cable from the temperature switch.

3. Loosen the lower hose clamp on the air duct that connects the air filter with the compressor housing.

4. Remove the vacuum line and crankcase breather pipe.

5. Remove the air filter and air intake duct.

6. Disconnect the oil line at the turbocharger.

7. Remove the air filter mounting bracket.

8. Disconnect the turbocharger at the exhaust flange.

9. Disconnect and remove the pipe bracket on the automatic transmission.

10. Push the exhaust pipe rearward.

11. Remove the mounting bracket at the intermediate flange.

12. Unbolt and remove the turbocharger.

13. Remove the intermediate flange and oil return line at the turbocharger.

To install:

14. Before installing the turbocharger, install the oil return line and intermediate flange.

15. Install the flange gasket between the turbocharger and exhaust manifold with the reinforcing bead toward the exhaust manifold. Use only heat proof nuts and bolts.

16. Fill a new turbocharger with ¼ pint of engine oil through the engine oil supply bore before operating.

Intake Manifold

REMOVAL & INSTALLATION

1. Disconnect the negative battery cable.

2. Remove the engine ventilation tubes from the manifold.

3. Disconnect all electrical wiring, cables and hoses from the intake manifold.

4. Remove the manifold retaining bolts and manifold.

To install:

5. Clean the gasket mating surfaces and install a new gasket.

6. Install the manifold and bolts. Torque the bolts to 18 ft. lbs. (25 Nm), working from the middle outward.

7. Install the remaining components and battery cable.

8. Start the engine and check for leaks.

Exhaust Manifold

REMOVAL & INSTALLATION

1. Disconnect the negative battery cable.
2. Remove the exhaust pipe from the manifold.
3. Remove the manifold retaining bolts and manifold. It is not necessary to remove both exhaust manifolds for the 3.5L engine. Separate the gasket between cylinders 3 and 4.
   To install:
4. Clean the gasket mating surfaces and install a new gasket.
5. Install the manifold and bolts. Torque the bolts to 18 ft. lbs. (25 Nm), working from the middle outward.
6. Install the remaining components and battery cable.
7. Start the engine and check for leaks.

Camshaft and Valve Lifters

REMOVAL & INSTALLATION

1. Disconnect the negative battery cable. Remove cylinder head cover.
2. Set crankshaft to TDC of No. 1 cylinder.

➡**Do not rotate the engine on the fastening screw of the camshaft timing gear; do not rotate engine in reverse.**

3. Remove chain tensioner.
4. Mark the camshaft timing gear and timing chain in relation to each other.
5. Remove camshaft timing gear. To loosen screws, apply counter hold on camshaft by means of a mandrel.
6. If equipped with level control, remove pressure oil pump and place aside with lines connected.
7. To prevent damage to camshafts, be sure to apply the following sequence during assembly:
   a. 2.5L engine—remove both screws on camshaft bearing 1, 2 and 6 (dark arrows). Loosen both screws on the camshaft bearing 3, 4 and 5 alternately and in steps only until counterpressure has been eliminated.

b. 3.5L engine—remove both screws on camshaft bearing 1, 5 and 6 (dark arrows). Loosen both screws for camshaft bearing 2, 3, 4 and 7 alternately and in steps only until counterpressure has been eliminated.

8. Remove camshaft in upward direction.
9. Remove the camshaft alignment circlip and check for misaligned condition.
10. Pull out valve tappet by means of solenoid lifter tool 102 589 03 40 00 or equivalent.
11. Check valve tappet for condition, visual checkup, and renew, if required.
    To install:

➡**Install valve tappets only at the same spot where they were installed. If a new camshaft has been installed or if the cylinder head has been machined, check camshaft for easy operation.**

12. Insert circlip for axial locating into cylinder head.
13. Lubricate camshaft and place into cylinder head, without valve tappet.
14. Tighten camshaft bearing caps uniformly to 18.5 ft. lbs. (25 Nm). Pay attention to identification of bearing caps.
15. When checking for easy operation, the camshaft can be rotated by means of a hex socket screw M10 x 30, which is screwed in through camshaft timing gear instead of fastening screw. If the camshaft can be rotated with effort only, proceed as follows:
    a. Loosen camshaft bearing caps individually. Turn camshaft, if required.
    b. Repeat, until tight bearing point has been found.
    c. Check camshaft for runout.
16. Lubricate valve tappets and insert. Pay attention to sequence.
17. Lubricate camshaft and place into cylinder head so the TDC mark is vertical.
18. Install camshaft bearing caps opposite of the loosening sequence. Torque the bearing caps to 18 ft. lbs. (25 Nm).
19. Mount camshaft timing gear. Pay attention to color marks. Tighten fastening screw for camshaft timing gear to 48 ft. lbs. (65 Nm). For this purpose, apply

counter hold to camshaft timing gear by means of a steel pin or suitable tool.
20. Install chain tensioner.
21. If equipped with level control, mount pressure oil pump and driver.
22. Check engine for TDC marks.
23. Mount cylinder head cover.

Valve Lash

ADJUSTMENT

The 5- and 6-cylinder Diesel engines use hydraulic valve lifters. There is no need or provision for valve clearance adjustments.

Oil Pan

REMOVAL & INSTALLATION

1. Disconnect the negative battery cable.

✳ CAUTION

The EPA warns that prolonged contact with used engine oil may cause a number of skin disorders, including cancer! You should make every effort to minimize your exposure to used engine oil. Protective gloves should be worn when changing the oil. Wash your hands and any other exposed skin areas as soon as possible after exposure to used engine oil. Soap and water, or waterless hand cleaner should be used.

2. Drain and recycle the engine oil.
3. Using a suitable engine support tool, support the weight of the engine.
4. Raise and safely support the vehicle.
5. Support the front subframe and remove the mounting bolts.
6. Lower the subframe enough to access the oil pan mounting bolts.

➡**It may be necessary to tap on the oil pan with a rubber mallet to dislodge it from the engine block.**

7. Remove the mounting bolts, then the pan.
8. Clean all remains of the old gasket and sealant from the oil pan and engine block.

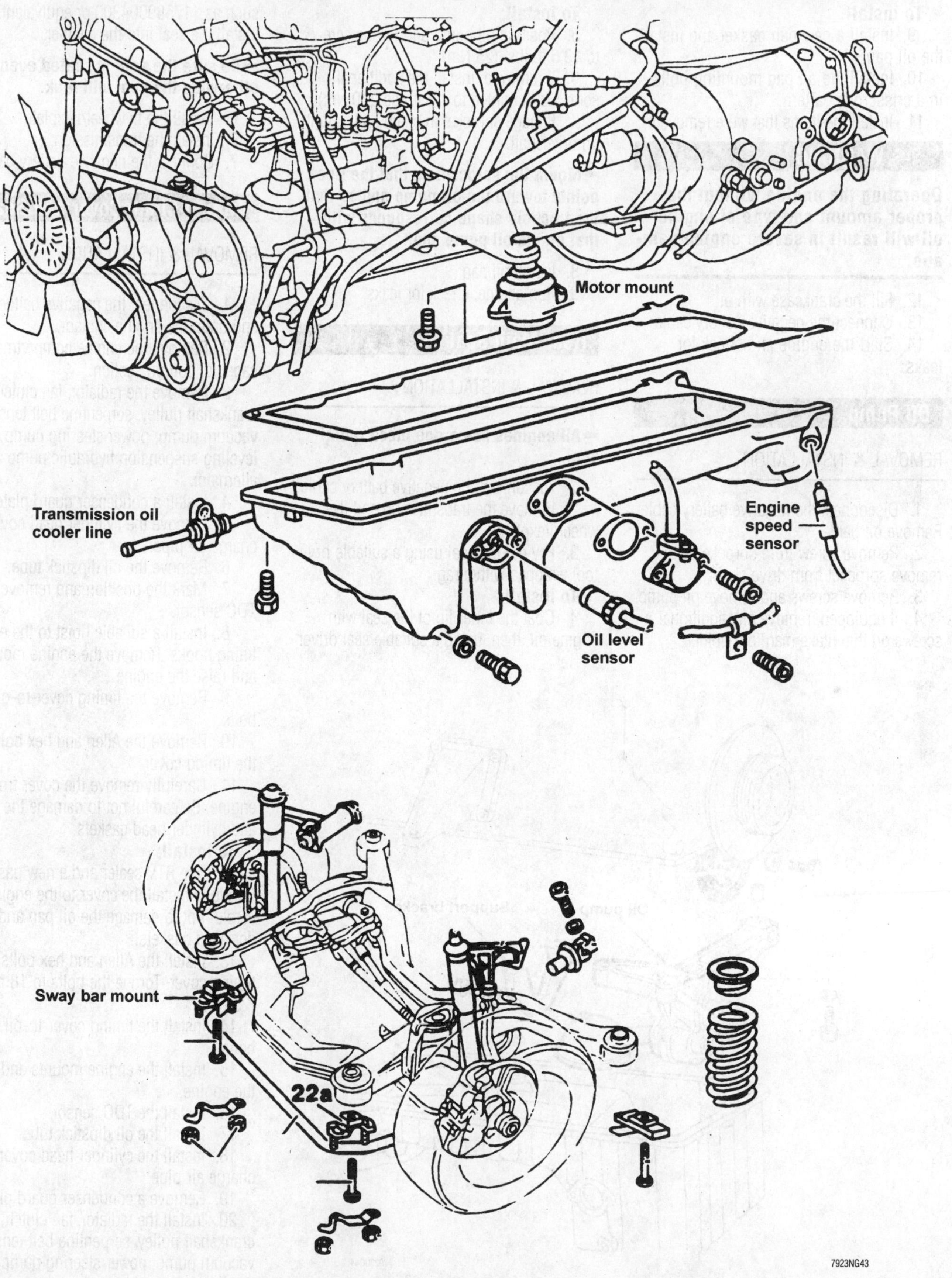

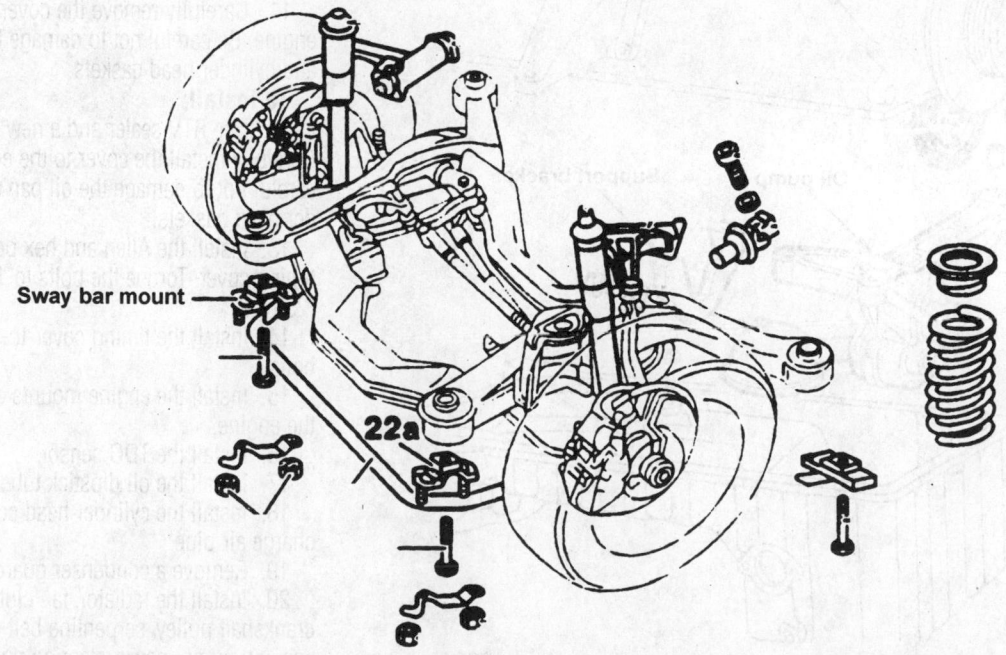

Motor mount

Transmission oil cooler line

Engine speed sensor

Oil level sensor

Sway bar mount

22a

7923NG43

Exploded view of the oil pan mounting and related removal and installation components—Diesel engines

To install:

9. Install a new pan gasket and install the oil pan.

10. Install the oil pan mounting bolts in a crisscross pattern.

11. Install the items that were removed.

✸✸ WARNING

Operating the engine without the proper amount and type of engine oil will result in severe engine damage.

12. Fill the crankcase with oil.

13. Connect the negative battery cable.

14. Start the engine and check for leaks.

Oil Pump

REMOVAL & INSTALLATION

1. Disconnect the negative battery cable. Remove oil pan.

2. Remove screw from sprocket and remove sprocket from drive shaft.

3. Remove screws and remove oil pump.

4. If equipped, remove any additional screws on the intake manifold holder.

To install:

5. Position oil pump and torque screw to 18.5 ft. lbs. (25 Nm).

6. If removed, install the additional screws and tighten to 7.5 ft. lbs. (10 Nm).

7. Engage sprocket in chain and mount on driveshaft.

➡**Mount the sprocket so that the rise points toward the oil pump and so that the trochoid shape corresponds with that on the oil pump shaft.**

8. Install oil pan.

9. Run engine, check for leaks.

Rear Main Seal

REMOVAL & INSTALLATION

➡**All engines use a one piece radial seal.**

1. Disconnect the negative battery cable.

2. Remove the transmission and flywheel/flexplate.

3. Pry out the seal using a suitable prytool wrapped with a rag.

To install:

4. Coat the inner lip of the seal with engine oil, then, using a suitable seal driver such as 117589004300 or equivalent, install the seal into the retainer.

➡**Be sure the seal is seated evenly, otherwise the seal will leak.**

5. Install the flywheel/flexplate.

6. Install the transmission.

7. Connect the negative battery cable.

Timing Chain Cover and Seal

REMOVAL & INSTALLATION

1. Disconnect the negative battery cable and drain the engine coolant.

2. Remove the engine compartment capsule, bottom section.

3. Remove the radiator, fan clutch, fan, crankshaft pulley, serpentine belt tensioner, vacuum pump, power steering pump, self-leveling suspension hydraulic pump and alternator.

4. Install a condenser guard plate.

5. Remove the cylinder head cover and charge air pipe.

6. Remove the oil dipstick tube.

7. Mark the position and remove the TDC sensor.

8. Install a suitable hoist to the engine lifting hooks. Remove the engine mounts and raise the engine.

9. Remove the timing cover-to-oil pan bolts.

10. Remove the Allen and hex bolts from the timing cover.

11. Carefully remove the cover from the engine. Be careful not to damage the oil pan and cylinder head gaskets.

To install:

12. Use RTV sealer and a new gasket. Carefully Install the cover to the engine. Be careful not to damage the oil pan and cylinder head gaskets.

13. Install the Alien and hex bolts to the timing cover. Torque the bolts to 18 ft. lbs. (25 Nm).

14. Install the timing cover-to-oil pan bolts

15. Install the engine mounts and lower the engine.

16. Install the TDC sensor.

17. Install the oil dipstick tube.

18. Install the cylinder head cover and charge air pipe.

19. Remove a condenser guard plate.

20. Install the radiator, fan clutch, fan, crankshaft pulley, serpentine belt tensioner, vacuum pump, power steering pump, self-leveling suspension hydraulic pump and alternator.

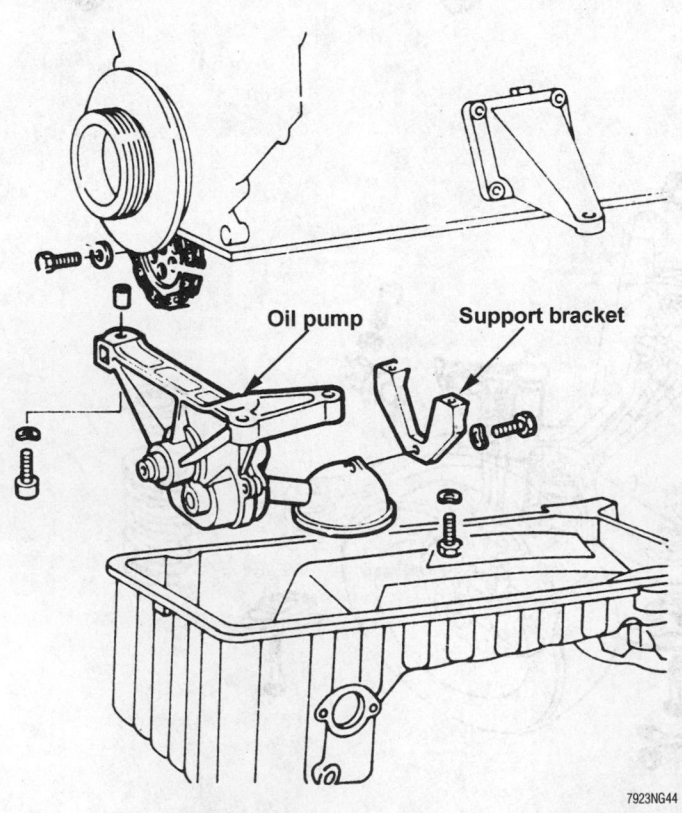

7923NG44

Exploded view of the oil pump mounting—Diesel engine

Oil pump   Support bracket

21. Install the engine compartment capsule, bottom section.

22. Connect the negative battery cable and refill the engine coolant.

Timing Chain

REMOVAL & INSTALLATION

1. Disconnect the negative battery cable. Remove cylinder head cover.

2. Remove injection nozzles.

3. Remove chain tensioner.

4. Remove fan and fan cover.

5. Connect new timing chain with connecting link to old timing chain.

6. Slowly, rotate crankshaft in rotating direction of engine, while simultaneously pulling up the old timing chain until the connecting link comes to rest against uppermost point of camshaft timing gear.

➡ **Timing chain should remain in mesh while rotating camshaft and crankshaft timing gears.**

7. Take off old timing chain and connect ends of new timing chain with connecting link. For this purpose, secure chain ends with wire on camshaft timing gear.

To install:

➡ **Use only a rivet-type connecting link. Do not use connecting link that use a retaining spring.**

8. Insert connecting link from the rear into timing chain.

9. Put separately enclosed outer flange of connecting link, with punched in IWIS identification, into pressing on tool. The outer flange is held magnetically.

10. Place pressing-on tool on connecting link and press on flange up to stop, while holding pressing-on tool on vertical level.

11. Rearrange plunger, of assembly tool, so the notch is pointing forward.

12. Hold assembly tool on handle and rivet chain bolts individually. Tightening torque of spindle approximately 22–26 ft. lbs. (30–35 Nm).

13. Check chain bolt rivet and rivet again, if required.

14. Install chain tensioner.

15. Rotate crankshaft and check adjusting mark at TDC position of engine.

➡ **If the adjusting mark is wrong, check timing of camshaft and begin of delivery of injection pump.**

16. Install cylinder head cover and tighten to 7.5 ft. lbs. (10 Nm).

17. Install fan and fan cover.

GASOLINE FUEL SYSTEM

Fuel System Service Precautions

Safety is the most important factor when performing not only fuel system maintenance but any type of maintenance. Failure to conduct maintenance and repairs in a safe manner may result in serious personal injury or death. Maintenance and testing of the vehicle's fuel system components can be accomplished safely and effectively by adhering to the following rules and guidelines.

• To avoid the possibility of fire and personal injury, always disconnect the negative battery cable unless the repair or test procedure requires that battery voltage be applied.

• Always relieve the fuel system pressure prior to disconnecting any fuel system component (injector, fuel rail, pressure regulator, etc.), fitting or fuel line connection. Exercise extreme caution whenever relieving fuel system pressure to avoid exposing skin, face and eyes to fuel spray. Please be advised that fuel under pressure may penetrate the skin or any part of the body that it contacts.

• Always place a shop towel or cloth around the fitting or connection prior to loosening to absorb any excess fuel due to spillage. Ensure that all fuel spillage (should it occur) is quickly removed from engine surfaces. Ensure that all fuel soaked cloths or towels are deposited into a suitable waste container.

• Always keep a dry chemical (Class B) fire extinguisher near the work area.

• Do not allow fuel spray or fuel vapors to come into contact with a spark or open flame.

• Always use a back-up wrench when loosening and tightening fuel line connection fittings. This will prevent unnecessary stress and torsion to fuel line piping.

• Always replace worn fuel fitting O-rings with new. Do not substitute fuel hose or equivalent, where fuel pipe is installed.

Fuel System Pressure

RELIEVING

1. Disconnect the negative battery cable.

2. Connect a fuel pressure gauge with a pressure release valve to the service port on the fuel supply rail.

3. Place the fuel release tube into a container and open the valve.

4. Remove the fuel pressure gauge from the service port on the fuel supply rail.

Fuel Filter

REMOVAL & INSTALLATION

➡ **Two types of filters are used, depending on the vehicle. Both are located between the rear axle and the fuel tank.**

1. Properly relieve the fuel system pressure.

2. Unscrew the cover box.

3. Remove the gas cap, fuel pump cover and pressure hoses.

4. Loosen the screws and remove the filter. Remove the connecting plug from the old filter and install it on a new filter using a new gasket.

To install:

5. Install a new filter in the direction of flow.

6. Replace the attaching screws.

7. Install the pressure hoses, pump cover and gas cap.

8. Install the fuel filter in the holder by positioning it in the center of the transparent holder. Be sure the plastic sleeve between the fuel filter and fuel pump is installed. Galvanic corrosion may occur in cases of direct contact between these components.

9. Replace the cover box and check for proper sealing.

Fuel Pump

REMOVAL & INSTALLATION

1. Disconnect the negative battery cable.

2. Properly relieve the fuel system pressure.

3. Raise and safely support the vehicle.

4. Remove and plug the intake, outlet and bypass lines from the pump.

5. Disconnect the electrical leads.

6. Unbolt and remove the fuel pump and vibration pads.

To install:

➡ **The V8 and later model engines utilize 2 fuel pumps connected in series.**

7. Install the fuel pump in the reverse order of removal.

8. Tighten the cap nuts and banjo bolts to 18–22 ft. lbs. (25–30 Nm).

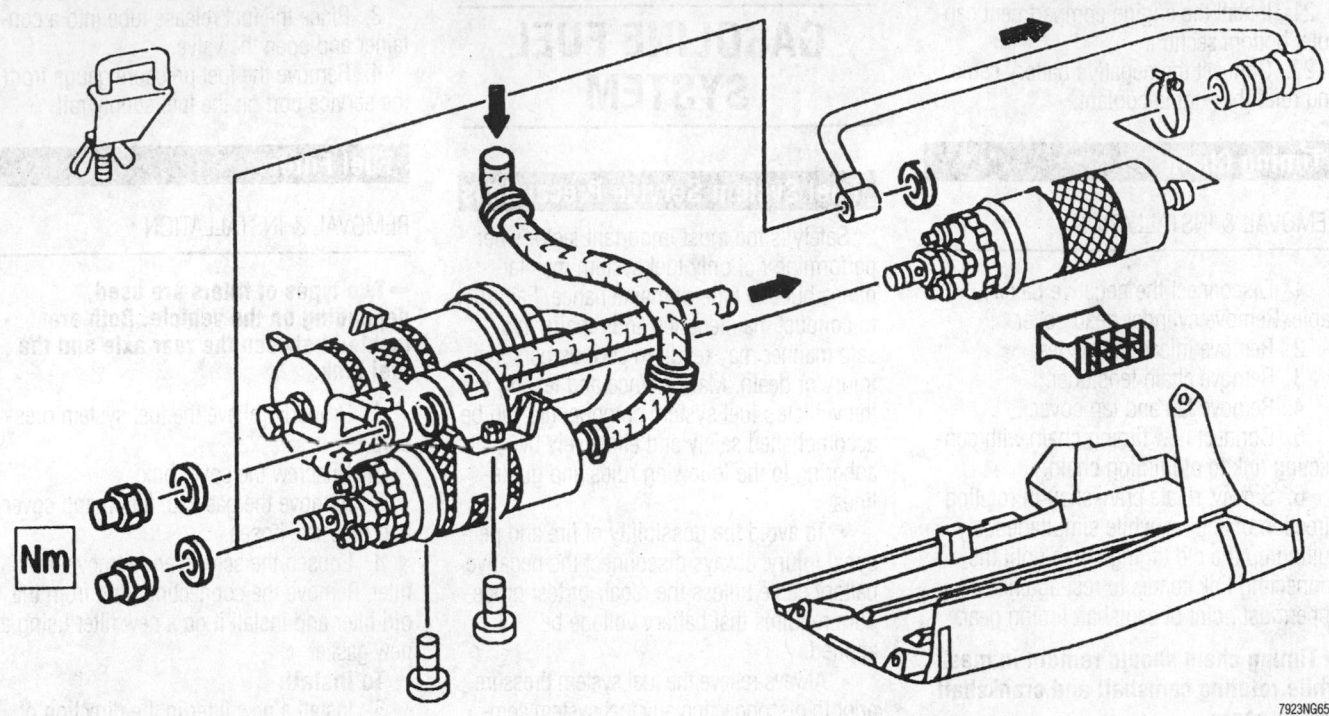

Typical fuel pumps and related components

➡Be sure the electrical leads are connected to the proper terminals. The negative wire (brown) is connected to the negative terminal (brown plastic plate) and the positive wire (black/red) is connect to the positive terminal (red plastic plate). If the terminals are reversed, the pump will operate in the reverse direction of normal rotation and will deliver no fuel.

DIESEL FUEL SYSTEM

Fuel System Service Precautions

Safety is the most important factor when performing not only fuel system maintenance but any type of maintenance. Failure to conduct maintenance and repairs in a safe manner may result in serious personal injury or death. Maintenance and testing of the vehicle's fuel system components can be accomplished safely and effectively by adhering to the following rules and guidelines.

• To avoid the possibility of fire and personal injury, always disconnect the negative battery cable unless the repair or test procedure requires that battery voltage be applied.

• Please be advised that fuel under pressure may penetrate the skin or any part of the body that it contacts.

• Always place a shop towel or cloth around the fitting or connection prior to loosening to absorb any excess fuel due to spillage. Ensure that all fuel spillage (should it occur) is quickly removed from engine surfaces. Ensure that all fuel soaked cloths or towels are deposited into a suitable waste container.

• Always keep a dry chemical (Class B) fire extinguisher near the work area.

• Do not allow fuel spray or fuel vapors to come into contact with a spark or open flame.

• Always use a back-up wrench when loosening and tightening fuel line connection fittings. This will prevent unnecessary stress and torsion to fuel line piping.

• Always replace worn fuel fitting O-rings with new. Do not substitute fuel hose or equivalent, where fuel pipe is installed.

Idle Speed

ADJUSTMENTS

These engines have electronically controlled idle speed, using a solenoid connected to the control unit. Idle speed and mixture adjustments are not recommended.

Fuel Filter/Water Separator

1. Loosen the center bolt while holding the filter cartridge.
2. Lubricate the gasket on the new filter cartridge.
3. Install the new cartridge, washer, O-ring and bolt.

➡Diesel engines in this section utilize a self-bleeding fuel pump, therefore the priming pump has been eliminated. There is no need to bleed the system.

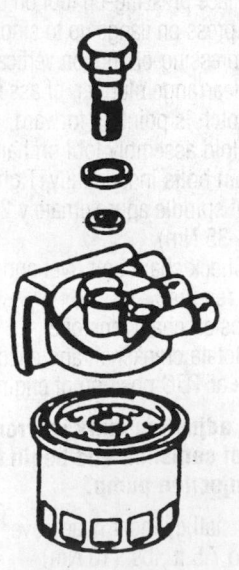

Fuel filter assembly—Diesel engines

Diesel Injection Pump

REMOVAL & INSTALLATION

1. Disconnect the negative battery cable.

➡ **The central fastening bolt uses left-hand threads, turn the bolt clockwise to loosen.**

2. Remove the cover and loosen the central fastening bolt while holding the crankshaft with a socket wrench.

3. Position the crankshaft pulley at 15 degrees after TDC. Fix the camshaft gear and injection pump in place with a cable strap, or equivalent.

4. Remove the timing chain tensioner.

5. Disconnect the fuel and injection lines from the fuel pump.

6. Detach the vacuum lines and two pin connector from the electronic idle speed control.

7. Disconnect the electrical cables from the control rod travel sensor and actuator.

8. Disconnect the pressure line at the ALDA unit.

9. Disconnect the throttle control linkage at the control lever on the pump.

10. Remove the vacuum control valve.

11. Remove the injection pump sprocket bolt, holding the crankshaft in place. The bolt is a left-hand thread.

12. Remove the sprocket bolt and the pump mounting bolts. Remove the injection pump while holding the timing device in place.

To install:

13. Be sure the crankshaft is still at 15° ATDC.

14. Remove the timing plug from the pump.

15. Using special tool 601 589 00 08 00 (serrated wrench) to turn the pump until the lug of the governor is visible through the hole.

16. Install the special tool 601 589 05 21 00 (locking bolt) in the timing plug hole to prevent the pump from turning while it is being installed.

17. Install the pump on the engine. Tighten the injection lines to 7–15 ft. lbs. (10–20 Nm), mounting bolts to 15–18 ft. lbs. (20–25 Nm) and the central bolt to 30–36 ft. lbs. (40–50 Nm).

18. Remove the locking bolt from the side of the injection pump. The pump will be damaged when the engine is started if this procedure is not followed.

19. Reconnect the timing chain and install remaining components.

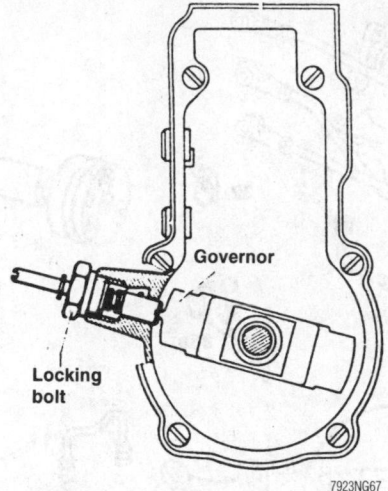

7923NG67

Install the locking bolt tool (023) to prevent the pump from turning during installation—Diesel engines

20. Crank the engine one revolution and recheck the TDC mark of the crankshaft and camshaft. Check the start of injector (injector pump timing) delivery with a digital tester.

Diesel Injection Timing

ADJUSTMENT

1. Remove the screw plug from the side of the injection pump governor housing. Be prepared to collect the lost oil.

2. Install a position sensor into the hole and connect the tester to the vehicle battery and position sensor.

3. Turn the crankshaft by hand until lamps A and B on the tester light simultaneously. Take the reading on the crankshaft pulley.

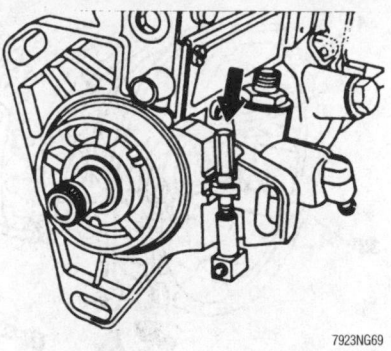

7923NG69

Turn the adjusting screw (arrow) to fine-tune the injection timing—Diesel engine

4. Loosen the injection pump and turn the adjusting screw until the timing pointer is at 15 degrees after TDC and both lamps are ON.

5. Install the oil and plug. Torque the plug to 22 ft. lbs. (30 Nm).

DRIVE TRAIN

Transmission Assembly

REMOVAL & INSTALLATION

Manual

1. Disconnect the negative battery cable.

2. Raise and safely support the vehicle.

3. Support the transmission with a jack and remove the support (19).

4. Remove the exhaust support bracket (10) and U-bolt (11).

5. Remove the heat shield (5).

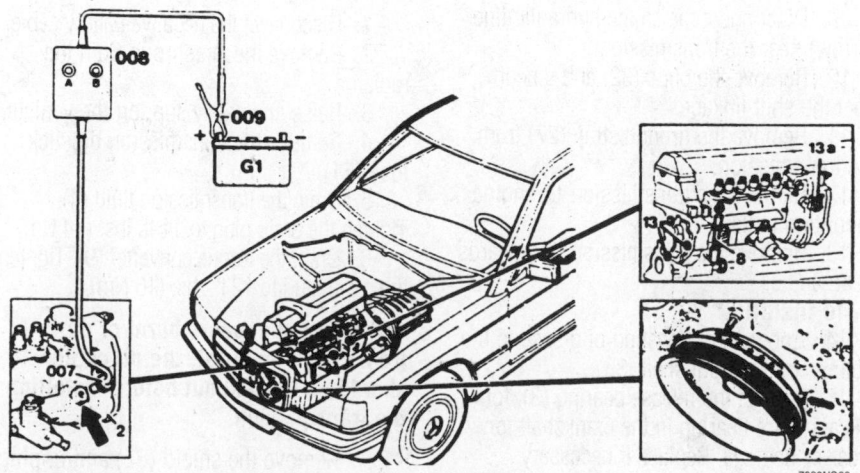

7923NG68

Remove the screw plug and install the injection timing position sensor

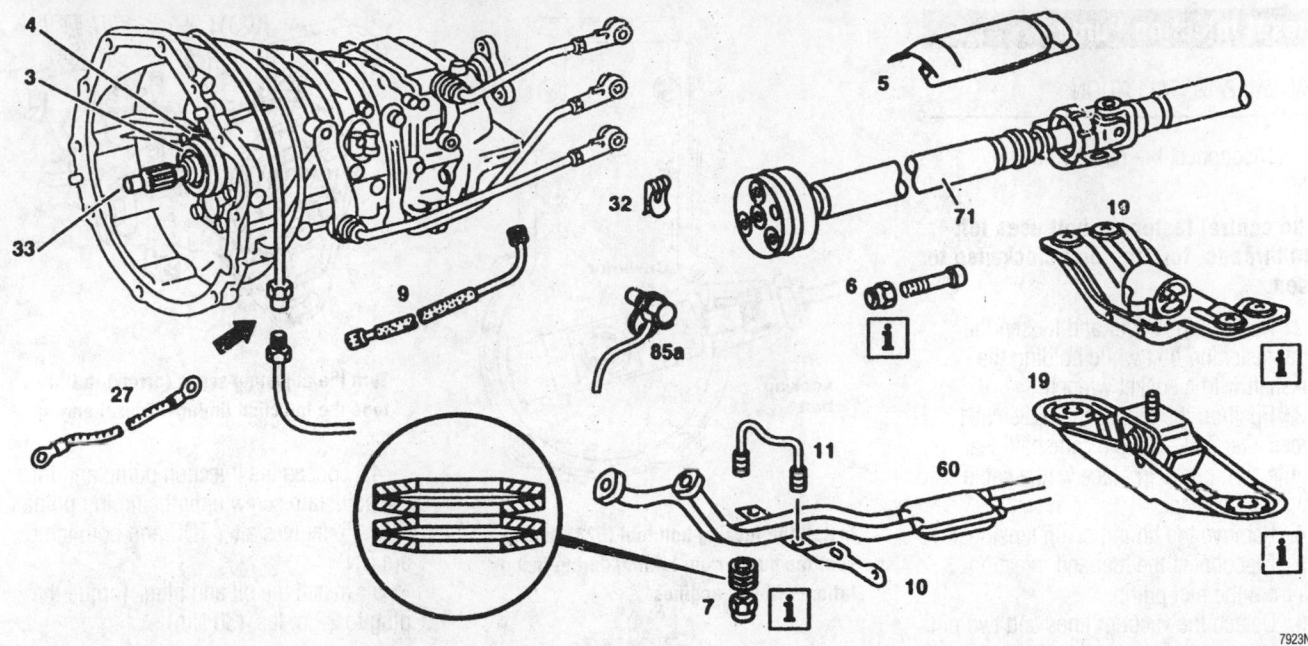

Exploded view of the components associated with manual transmission removal and installation

6. Remove the front drive shaft (71).

7. Detach the exhaust system (60) at the rear mount. Use a piece of wire to suspend it.

8. Remove the speedometer cable (9) or sensor (85a) from the transmission.

❊ CAUTION

Brake fluid contains polyglycol ethers and polyglycols. Avoid contact with the eyes and wash your hands thoroughly after handling brake fluid. If you do get brake fluid in your eyes, flush your eyes with clean, running water for 15 minutes. If eye irritation persists, or if you have taken brake fluid internally, IMMEDIATELY seek medical assistance.

9. Disconnect the clutch hydraulic line (arrow) near the transmission.

10. Remove the clips (32) and disconnect the shift linkage.

11. Remove the ground strap (27) from the transmission.

12. Remove the transmission-to-engine mounting bolts.

13. Remove the transmission downwards at an angle.

To install:

14. Apply a light coating of grease to the splines of the input shaft (33).

15. Inspect the release bearing (3), fork (4) and pilot bearing in the crankshaft for wear or damage. Replace if necessary.

16. Lift the transmission to the engine and install the mounting bolts.

17. Install the ground strap and connect the linkage.

18. Connect the clutch hydraulic line.

19. Install the speedometer cable or speed sensor.

20. Assemble the exhaust system using new gaskets if equipped.

21. Install the driveshaft.

22. Install the heat shield, exhaust support bracket and U-bolt.

23. Install the transmission support and remove the jack.

24. Check the transmission oil level and add if necessary.

25. Lower the vehicle to the floor and connect the negative battery cable.

26. Bleed the hydraulic clutch system.

Automatic

1. Disconnect the negative battery cable.

2. Remove the fan shroud from the radiator.

3. Raise and safely support the vehicle.

4. Remove the transmission dipstick tube (61).

5. Drain the transmission fluid (4). Tighten the drain plug to 11 ft. lbs. (14 Nm).

6. Drain the torque converter (9). Tighten the drain plug to 12 ft. lbs. (16 Nm).

➡ **If the fluid appears burnt or dark from clutch material, the oil cooler should be flushed out before installing the transmission.**

7. Remove the shield (62) and unplug the 13-pin connector at the transmission.

8. Disconnect the park interlock cable (80) from the transmission. Be sure the transmission selector lever is in PARK before removing the cable.

9. Remove the cover (81) and unbolt the torque converter from the flexplate. Turn the crankshaft or starter ring gear as needed to access the bolts.

10. Disconnect the oil cooler lines from the transmission.

11. Remove the clip and disconnect the shift rod (63).

12. Remove the exhaust mounting bracket (64).

13. Remove the exhaust system (94) as needed to provide clearance for transmission removal.

14. Support the transmission and remove the rear mount (65).

15. Remove the front half of the driveshaft (66).

16. Remove the ground strap (46).

❊ CAUTION

The torque converter may fall out of the transmission if tilted forward. Secure it to keep it from falling.

17. Remove the transmission-to-engine mounting bolts and remove the transmission.

To install:

18. Position the transmission on the engine and install the mounting bolts.

19. Install the ground strap.

20. Install the front half of the drive shaft.

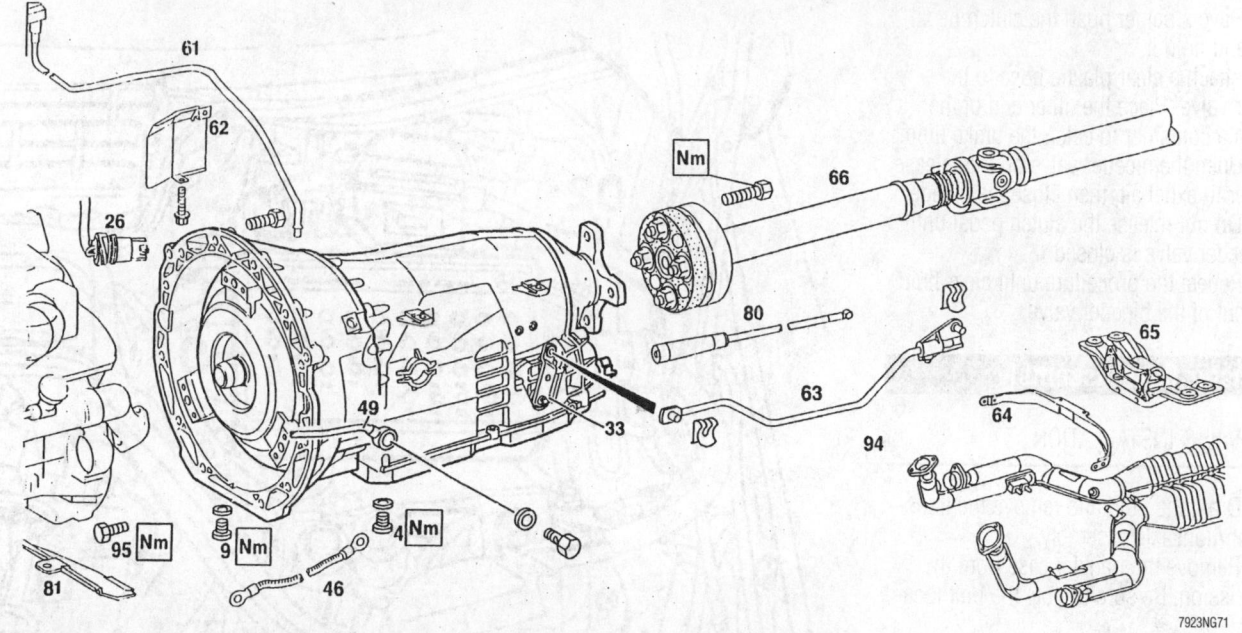

Exploded view of the components associated with automatic transmission removal and installation

7923NG71

21. Install the rear transmission mount and remove the jack.

22. Assemble the exhaust system if removed.

23. Install the exhaust system mounting bracket.

24. Connect the shift rod and the oil cooler lines.

25. Install the flexplate-to-torque converter bolts. Tighten the bolts to 31 ft. lbs. (42 Nm).

26. Install the park interlock cable.

27. Connect the 13-pin connector and install the shield.

28. Install the transmission dipstick tube.

29. Lower the vehicle to the floor.

30. Connect the negative battery cable.

Clutch

ADJUSTMENTS

All clutch release mechanisms are hydraulic, no periodic adjustment is necessary.

REMOVAL & INSTALLATION

1. Remove the transmission assembly.

2. Remove the pressure plate (6) mounting bolts (6). Be prepared to catch the pressure plate and clutch disc (5) after removing the last bolt.

3. Inspect the pilot bearing, release bearing (7), release fork (8) and flywheel. Replace as necessary.

To install:

4. Deglaze the flywheel with coarse abrasive cloth.

5. Position a new clutch disc on the flywheel. Place a line-up tool (012) through the disc into the pilot bearing to hold it in position.

6. Install a new pressure plate on the flywheel. Tighten the mounting bolts to 18 ft. lbs. (25 Nm).

7. Remove the line-up tool.

8. Lightly grease the friction surfaces (arrows) on the release fork.

9. Install the transmission assembly.

Hydraulic Clutch System

BLEEDING

1. Fill the reservoir with brake fluid. Do not allow brake fluid to contact painted surfaces.

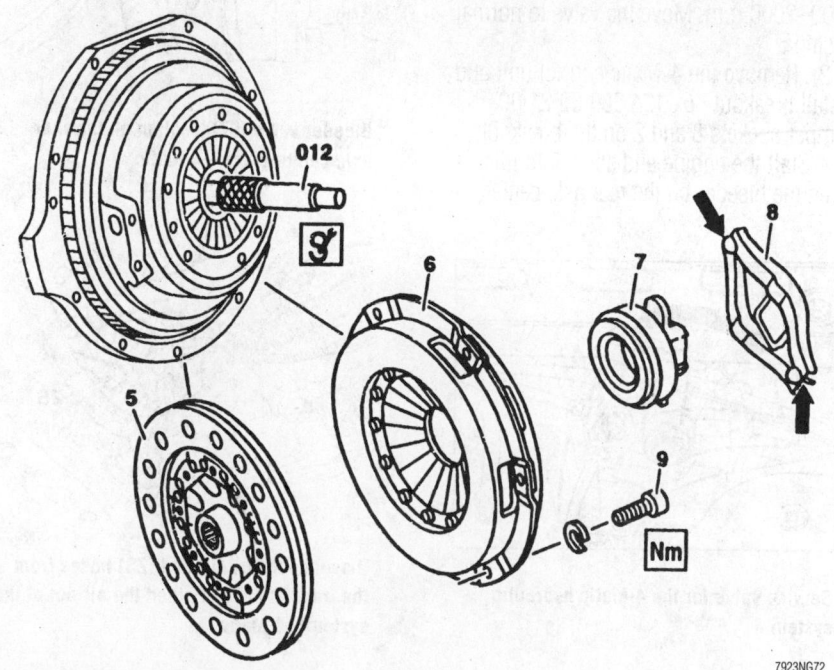

Exploded view of the clutch assembly

7923NG72

2. Have a helper push the clutch pedal down and hold it.

3. Attach a clear plastic hose to the bleeder valve. Place the other end of the hose in a container to catch the brake fluid.

4. Open the bleeder valve on the release cylinder to expel air, then close the bleeder valve. Do not release the clutch pedal until the bleeder valve is closed.

5. Repeat the procedure until clear fluid flows out of the bleeder valve.

Transfer Case Assembly

REMOVAL & INSTALLATION

1. Drain the fluid and remove the transmission/transaxle assembly.

2. Remove the transfer case from the transmission. Be sure to note the bolt locations.

To install:

3. Use a new gasket and install the transfer case on the transmission. Tighten the bolts to 20 ft. lbs. (27 Nm).

4. Install the transmission/transfer case assembly.

5. Perform the following to bleed the hydraulic system:

a. Fill the reservoir with fluid.

BLEEDING

1. On vehicles with service valve, move the valve to the test position and allow the engine to run for about 30 seconds at 1000–2000 rpm. Move the valve to normal position.

2. Remove the 4-Matic control unit and install breakout box 124 589 00 21 00. Jumper sockets 8 and 2 on the breakout box. Start the engine and allow it to idle. Open the bleeder on the rear axle center

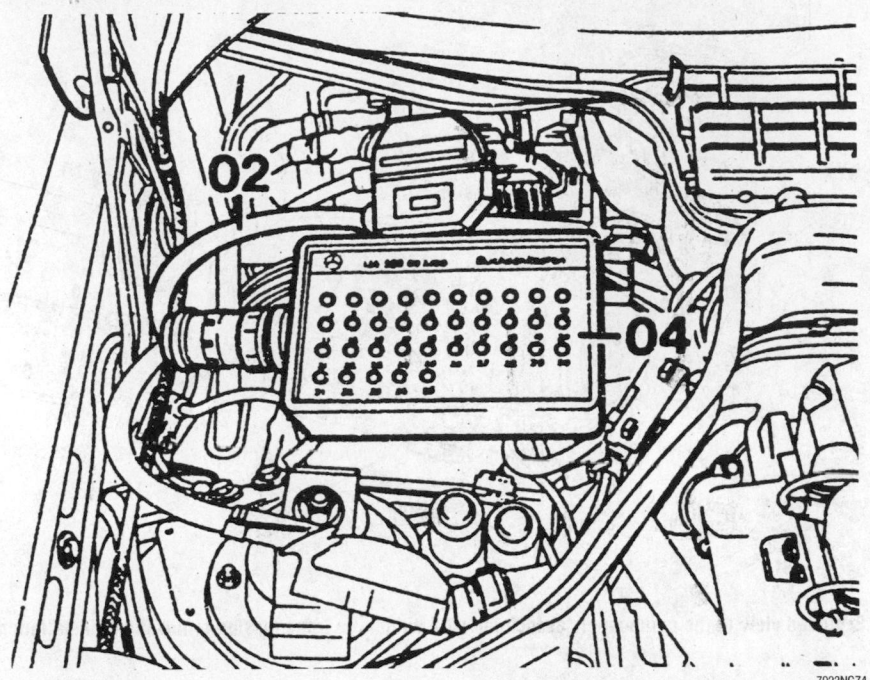

To bleed the transfer case, connect the breakout box (04) and adapter cable (02) in place of the 4-Matic control unit

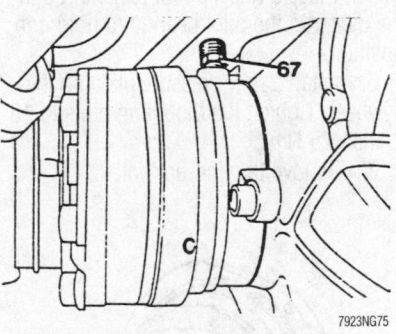

Bleeder valve (67) location on the rear axle center piece—4-Matic

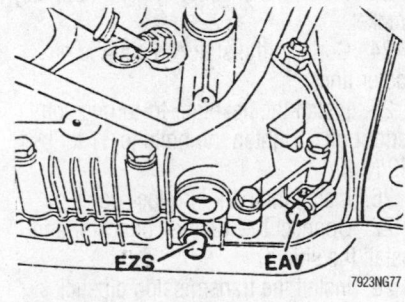

On transfer case No. 26 617, open the vent valves (EAV and EZS) to purge the air from the system—4-Matic

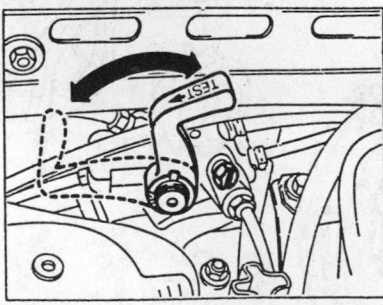

Service valve for the 4-Matic hydraulic system

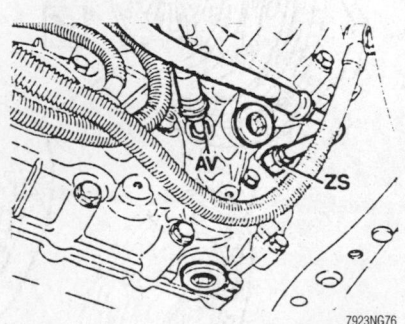

Disconnect the (AV) and (ZS) hoses from the transfer case to bleed the air out of the system—4-Matic

piece and allow the air to escape. Close the valve when the fluid is free of air. Turn the engine **OFF**.

3. Remove the hydraulic lines from the front axle drive train (AV) and center differential lock (ZS). Place the lines in a container and start the engine. Turn the engine **OFF** when the fluid is free of air (about 20 seconds). Connect the lines.

4. On transfer case No. 26 617, open the vent valves (EAV, EZS). Start the engine and allow any air to escape. Turn the engine **OFF** and close the valves.

5. Fill the reservoir with hydraulic oil to the full level.

Halfshaft

REMOVAL & INSTALLATION

Front (4-Matic)

1. Disconnect the negative battery cable.
2. Remove the front wheel.
3. Remove the 12-pointed collared nut (61) from the halfshaft.
4. Remove the engine undercover.
5. Remove the brake hose bracket (arrow) from the shock absorber.
6. Remove the stabilizer bar brackets (22b) from the lower control arms (4).
7. Remove the headlight leveling linkage (80).
8. Separate the lower ball joint (7) from the steering knuckle (5).
9. Remove the shock absorber (11) from the lower control arm (4).
10. Loosen the lower control arm mounting bolts at the axle carrier.
11. Remove the halfshaft from the differential.

To install:

12. Install the halfshaft into the differential.
13. Install the halfshaft through the hub.
14. Raise the lower control arm and install the lower ball joint.
15. Install the shock absorber on the lower control arm.
16. Install the headlight leveling linkage.
17. Attach the stabilizer bar brackets to the lower control arm.

18. Install the brake hose bracket on the shock absorber.
19. Install the engine undercover.
20. Lower the vehicle and install the 12-pointed nut. Tighten the nut to 162 ft. lbs. (220 Nm).
21. Install the nut lock.

Rear

➡ **The rubber covered joints are filled with special oil. If they are disassembled for any reason, they must be refilled with special oil.**

1. Raise and safely support the vehicle. Remove the wheel and center axle hold-down bolt (in hub).
2. Remove the brake caliper and suspend it from a hook.
3. Drain the differential oil and support the differential housing.
4. Unbolt the rubber mount from the chassis and the differential housing and remove the differential housing cover to expose the ring and pinion gears.
5. Press the shaft from the axle flange. If necessary, loosen the shock absorber.
6. Using a prybar, remove the axle lock ring inside the differential case.
7. Pull the axle from the housing by pulling the splined end from the side gears, with the spacer.

➡ **Axle shafts are stamped R and L for right and left units. Always use new lock rings.**

To install:

8. Install the axle shaft and seat the bearing. Torque the axle nut to 148–175 ft. lbs. (200–208>240 Nm).
9. Fill the rear axle with oil. New radial seal rings are used on all models. Lubricate the outside diameter of rubber covered radial sealing rings with hypoid gear lubricant prior to installation.

➡ **Check end-play of the lock ring in the groove. If necessary, install a thicker lock ring or spacer to eliminate all end-play, while still allowing the lock ring to rotate. Do not allow the joints in the axle shaft to hang free or the joint bearing may be damaged and leak.**

STEERING AND SUSPENSION

✳✳ CAUTION

Mercedes-Benz vehicles are equipped with an air bag system, also known as the Supplemental Inflatable Restraint (SIR) or Supplemental Restraint System (SRS). The system must be disabled before performing service on or around system components, steering column, instrument panel components, wiring and sensors. Failure to follow safety and disabling procedures could result in accidental air bag deployment, possible personal injury and unnecessary system repairs.

PRECAUTIONS

Several precautions must be observed when handling the inflator module to avoid accidental deployment and possible personal injury.

• Never carry the inflator module by the wires or connector on the underside of the module.

• When carrying a live inflator module, hold securely with both hands, and ensure that the bag and trim cover are pointed away.

• Place the inflator module on a bench or other surface with the bag and trim cover facing up.

• With the inflator module on the bench, never place anything on or close to the module which may be thrown in the event of an accidental deployment.

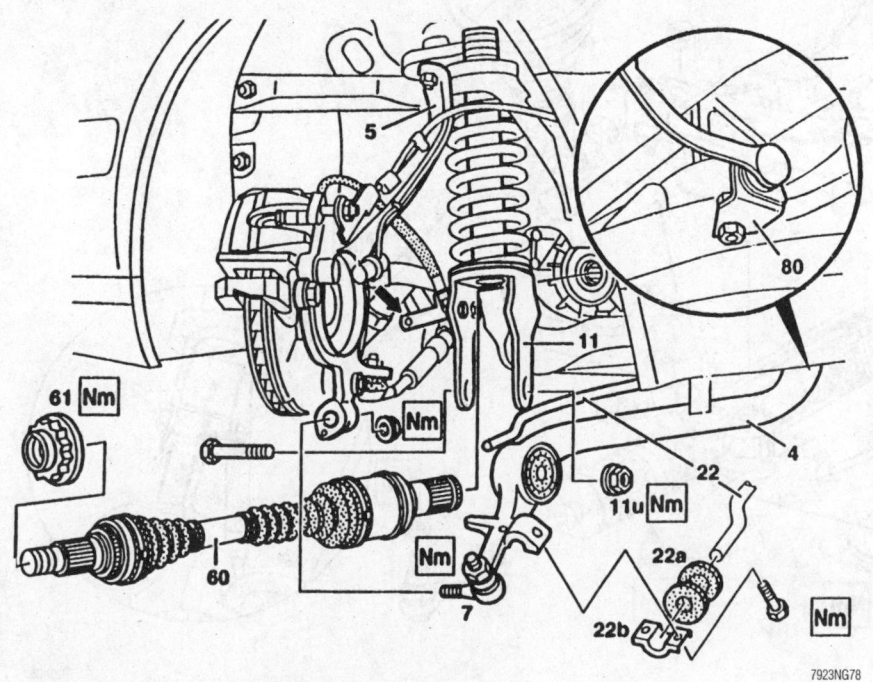

Exploded view of the front halfshaft mounted—4-Matic

7923NG78

DISARMING

To avoid personal injury when working on vehicles equipped with an air bag, the negative battery cable must be disconnected and insulated before working on the system. Failure to do so may result in accidental deployment of the air bag.

REARMING

To rearm the air bag system, reattach the battery cable(s).

Rack and Pinion Steering Gear

REMOVAL & INSTALLATION

1. Center the steering wheel and turn the key to the lock position. Allow the steering wheel to lock into position. Remove the key from the ignition switch.
2. Disconnect the negative battery cable.
3. Remove the cap from the power steering fluid reservoir and siphon out the fluid.
4. Raise and safely support the vehicle.
5. On models with level control, disconnect the feed line from the fluid reservoir.
6. Separate the tie rod ends from the steering knuckles.

7. Disconnect the pressure and return lines from the steering gear. If equipped, remove the heat shield to access the fluid lines.
8. Loosen the clip and remove the return line from the steering gear.
9. Remove the heat shield from the steering coupling.
10. Remove the pinch-bolt and carefully disconnect the steering coupling from the control valve. Don't use force to separate the joint. Use a flat bladed tool to expand the clamp if necessary.
11. If equipped with speed sensitive steering, detach the connector from the control valve.
12. Remove the bolts (23g) from the front of the steering gear retainer and remove the steering gear with the retainer.

To install:
13. Using new locking nuts, install the steering gear on the vehicle. Tighten the bolt to 36 ft. lbs. (50 Nm).
14. Attach the connector to the control valve if equipped with speed sensitive steering.
15. Install the coupling on the control valve. Tighten the bolt to 15 ft. lbs. (20 Nm).
16. Position the return line on the steering gear and replace the clip.
17. Use new O-rings and connect the fluid lines to the steering gear. Tighten the fittings to 30 ft. lbs. (40 Nm).

18. If equipped, replace the heat shield.
19. Use new nuts and install the tie rod ends to the steering knuckles. Tighten the nuts to 44 ft. lbs. (60 Nm).
20. If equipped with level control, connect the feel line to the reservoir.
21. Connect the negative battery cable.
22. Refill the reservoir with fresh fluid. Start the engine and turn the steering wheel from lock-to-lock several times. The system will bleed automatically.
23. Check for leaks and repair as needed.
24. Check the front wheel alignment and adjust if necessary.

Worm and Sector Steering Gear

REMOVAL & INSTALLATION

1. Center the steering wheel and turn the key to the lock position. Allow the steering wheel to lock into position. Remove the key from the ignition switch.
2. Disconnect the negative battery cable.
3. Remove the cap from the power steering fluid reservoir and siphon out the fluid.
4. Raise and safely support the vehicle.
5. Remove the trim panel if necessary to access the steering gear.
6. Separate the drag link and tie rod from the Pitman arm.

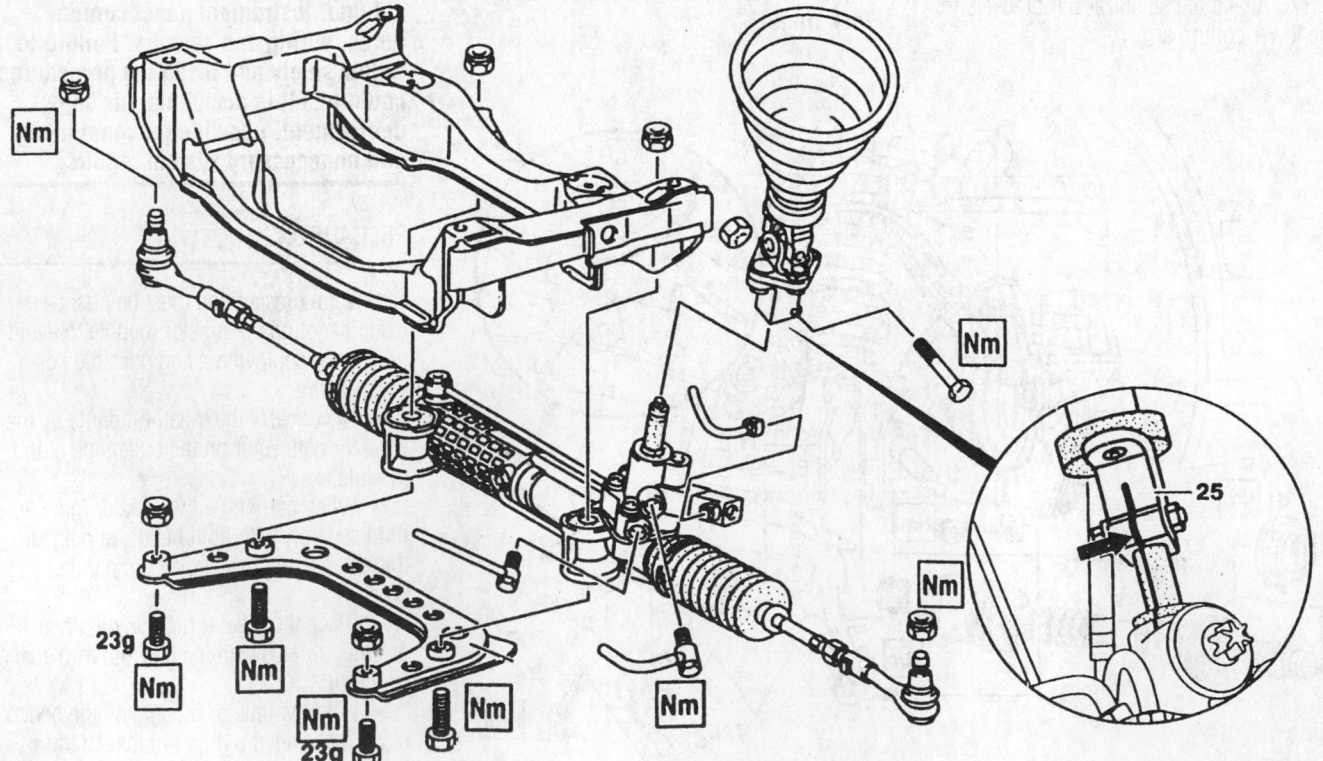

Exploded view of the rack and pinion steering gear mounting—E Class (210)

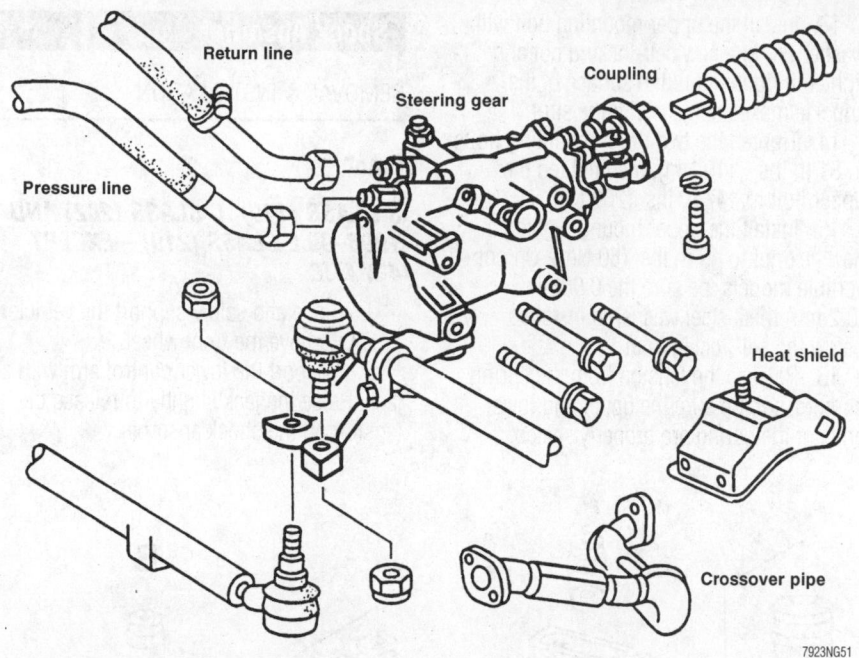

Exploded view of the worm and sector steering gear mounting

7. Remove the exhaust crossover pipe and the heat shield from the steering gear.

8. Disconnect the pressure and return lines from the steering gear. Cover the holes on the steering gear with plugs.

9. Remove the pinch bolts securing the steering shaft to the coupling on the steering gear.

10. Remove the three steering gear mounting bolts and the steering gear.

To install:

11. Center the steering gear by lining up the mark on the sector shaft with the mark on the housing.

✳✳ CAUTION

New stretch bolts must be used when installing the steering box and tightening the coupling.

12. Install the steering gear on the frame while inserting the steering shaft in the coupling. Tighten the new stretch bolts to 52–59 ft. lbs. (70–80 Nm).

13. Install a new bolt in the coupling. Tighten the bolt to 18 ft. lbs. (25 Nm).

14. Connect the pressure and return lines to the steering gear. Tighten the pressure line to 18 ft. lbs. (25 Nm) and the return line to 26–30 ft. lbs. (35–40 Nm).

15. Install the heat shield and the exhaust crossover pipe.

16. Install the drag link and tie rod on the Pitman arm with new self-locking nuts.

17. If removed, install the trim panel.

18. Lower the vehicle to the floor.

19. Connect the negative battery cable.

20. Refill the reservoir with fresh fluid. Start the engine and turn the steering wheel from lock-to-lock several times. The system will bleed automatically.

21. Check for leaks and repair if necessary.

Strut

REMOVAL & INSTALLATION

Front

SL CLASS (129) AND 1995 E CLASS (124)

1. Raise and safely support the vehicle.
2. Remove the front wheel.
3. Support the lower control arm with a jack.
4. Using a spring compressor, compress about eight coils of the spring to relieve the pressure on the lower control arm.

➡**Convertible models have a vibration damper mounted on top of the left strut mount. Installation of the nut is possible only with the use of socket 124 589 00 09 00 and torque wrench 001 589 67 21 00.**

5. Loosen the upper mounting nut while holding the piston rod with a hex wrench.
6. Remove the clip securing the ABS sensor wire to the strut.
7. Remove the lower mounting bolts that attach the strut to the knuckle.

✳✳ WARNING

Do not allow the knuckle assembly to hang by the brake hose, ABS wire or brake pad sensor wire.

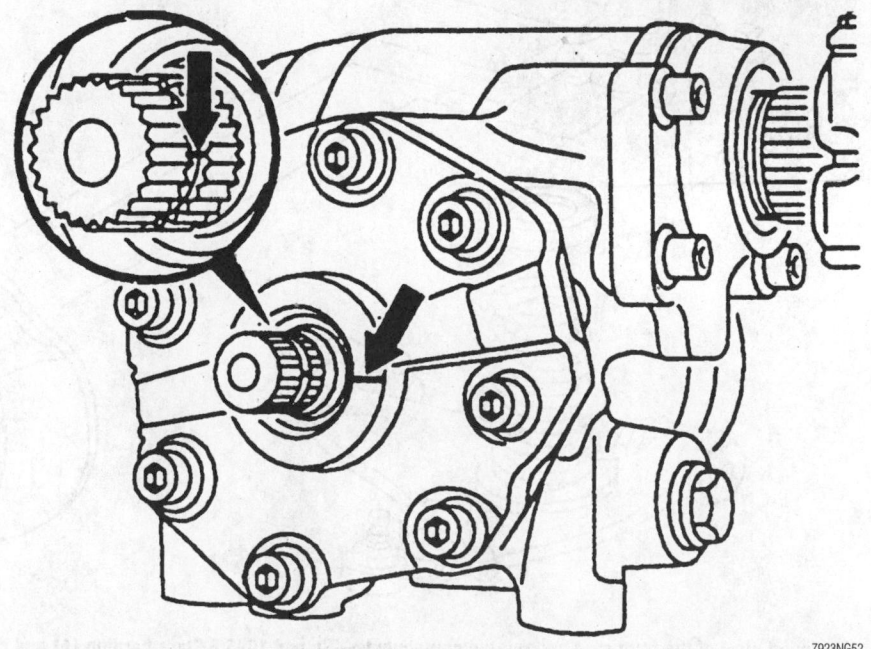

Align the mark on the shaft with the one on the housing to center the steering gear—worm and sector steering gear

8. Remove the upper nut and the strut from the vehicle. Be sure to support the knuckle assembly.

To install:

9. If installing a new strut, transfer all components from the old strut to the new one. Be sure to hook the boot on the two short collars of the rubber mount.

10. Clean the strut mounting surface of the knuckle.

11. Insert the strut through the rubber mount in the body.

12. Install the lower end of the strut on the knuckle. Tighten the bolt only until the heads make contact.

13. Install the upper mounting bolt with washer using a new self-locking nut and tighten it lightly until the surface of the knuckle makes contact with the strut.

14. Tighten the two lower mounting bolts to 81 ft. lbs. (110 Nm), then tighten the upper bolt to 147 ft. lbs. (200 Nm).

15. Install the upper mount. Tighten a new hex nut to 45 ft. lbs. (60 Nm). On convertible models, be sure the 0.008 in. (0.2mm) thick steel washer is installed below the self-locking nut.

16. Release the tension from the spring compressor. Be sure the upper and lower ends of the spring are properly seated.

Shock Absorber

REMOVAL & INSTALLATION

Front

S CLASS (140), C CLASS (202) AND 1996–99 E CLASS (210)—EXCEPT 4-MATIC

1. Raise and safely support the vehicle.
2. Remove the front wheel.
3. Support the lower control arm with a jack. Raise the jack slightly to release the tension on the shock absorber.

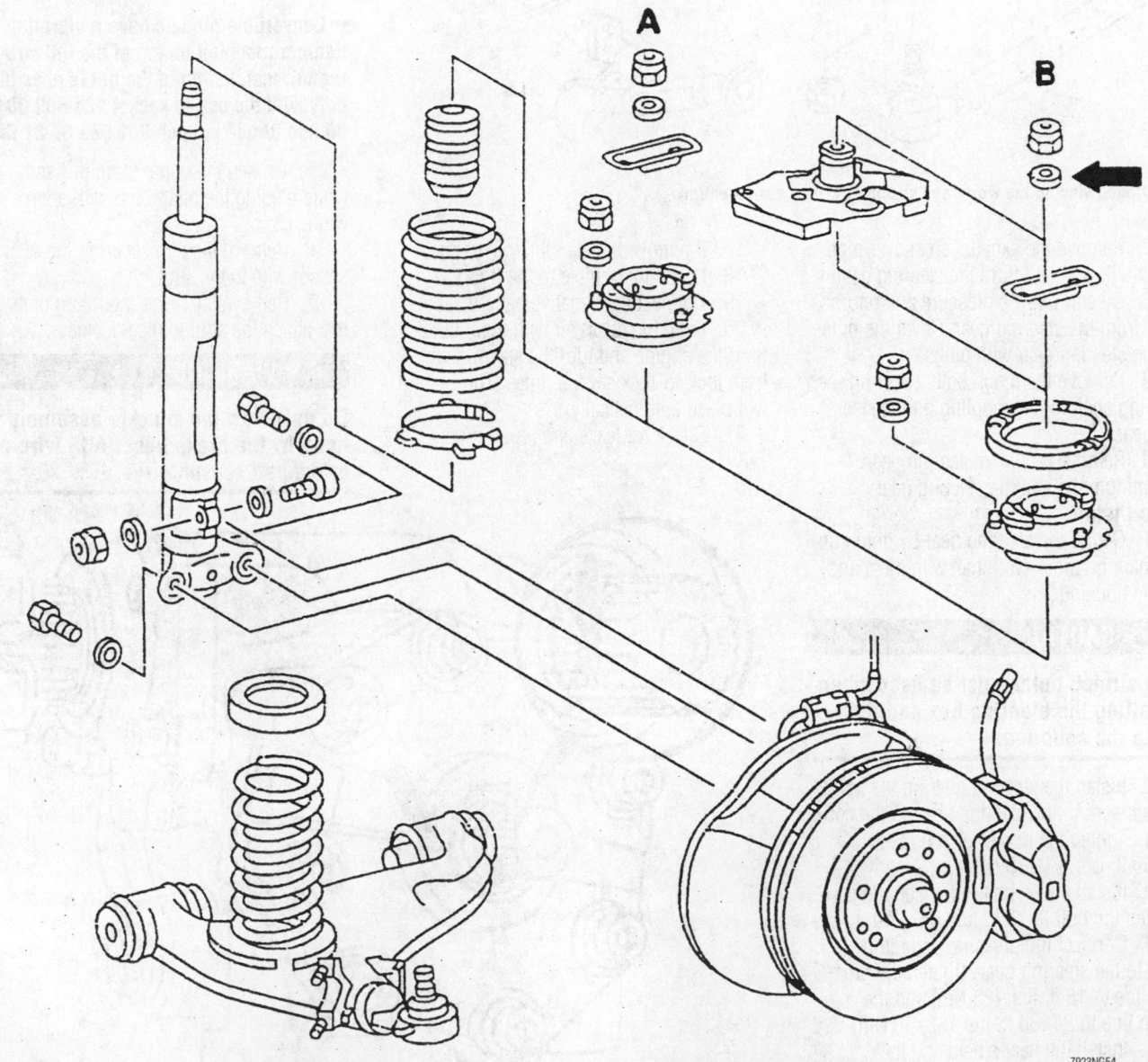

Exploded view of the front strut and related components—SL and 1995 E Class hardtop (A) and convertible (B) models

7923NG54

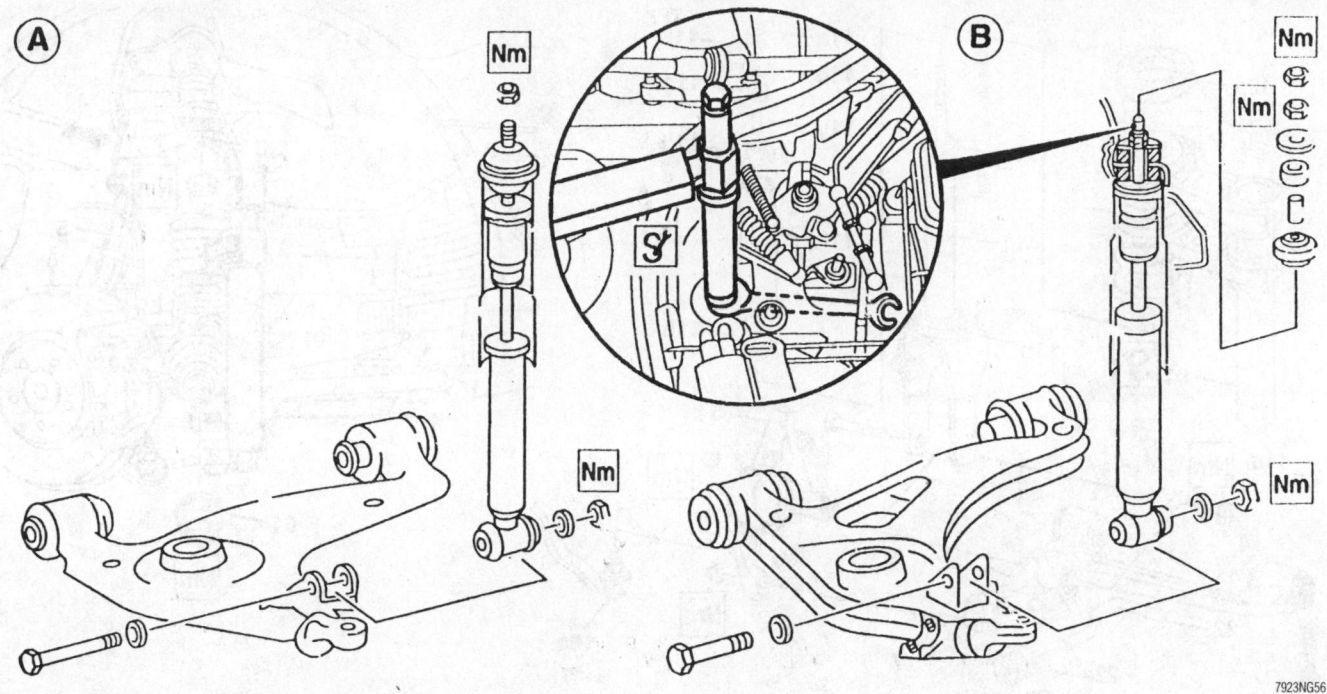

Front shock absorber mounting—S Class (A), C and E Class (B)

4. Remove the upper shock absorber mounting.

5. Remove the lower mounting bolt and remove the shock from the vehicle.

To install:

6. Position the shock absorber on the vehicle.

7. Install the lower mounting bolt. Tighten the bolt to 74 ft. lbs. (100 Nm).

8. Install the upper bushing, washer and nuts. Tighten the lower nut to 11–13 ft. lbs. (15–18 Nm) and the upper nut to 22 ft. lbs. (30 Nm).

9. Install the front wheel and lower the vehicle to the floor.

1996–99 E CLASS (210) 4-MATIC

1. Raise and safely support the vehicle.

2. Remove the front wheel.

3. Remove the 12 point nut from the axle shaft.

4. Remove the brake hose bracket from the shock absorber shackle.

5. Remove the speed sensor and brake lining sensor wires from the guide.

6. Separate the tie rod end from the steering knuckle.

7. Separate the supporting joint from the steering knuckle.

8. Pull the steering knuckle outward and move the front axle out of the axle shaft flange.

9. Separate the follower joint from the steering knuckle. Use wire to support the knuckle assembly.

10. Remove the engine undercover.

11. Press out the axle shaft.

12. Remove the stabilizer bar bracket from the lower control arm. Remove the bushing from the end of the bar.

13. Compress the coil spring and remove the upper shock absorber mounting.

14. Remove the lower shock absorber mounting and the shock absorber.

15. Remove the coil spring.

To install:

16. Place the coil spring into position on the shock absorber.

✳✳ WARNING

All suspension fasteners which pass through a rubber bushing should be tightened while the vehicle is at normal ride height. Premature bushing failure may occur if the fasteners are tightened while the suspension is hanging due to the twisting of the bushing when the suspension is returned to the normal position.

17. Install the shock absorber upper and lower mounting. On the upper mounting, tighten the bottom nut to 11–13 ft. lbs. (15–18 Nm) and the upper nut to 22 ft. lbs.

(30 Nm). Tighten the new self-locking nut on the lower mounting to 136 ft. lbs. (185 Nm).

18. Install the stabilizer bar on the lower control arm. Tighten the bracket bolts to 15 ft. lbs. (20 Nm).

19. Install the axle shaft and the engine undercover.

➡**Ensure that the axle shaft is in the proper position before connecting the follower joint.**

20. Install the follower joint to the steering knuckle. Tighten a new self-locking nut to 33 ft. lbs. (45 Nm).

21. Install the supporting joint to the steering knuckle. Tighten a new self-locking nut to 77 ft. lbs. (105 Nm).

22. Connect the tie rod to the steering knuckle. Tighten the new self-locking nut to 44 ft. lbs. (60 Nm).

23. Replace the speed and brake lining sensor wires in the guide.

24. Install the brake hose bracket on the shock absorber.

25. Install a new 12 point nut on the axle shaft. Tighten the nut to 162 ft. lbs. (220 Nm). Secure the nut with caulker (nut lock) so there is no gap between the groove and locking tab.

26. Install the front wheel and lower the vehicle to the floor.

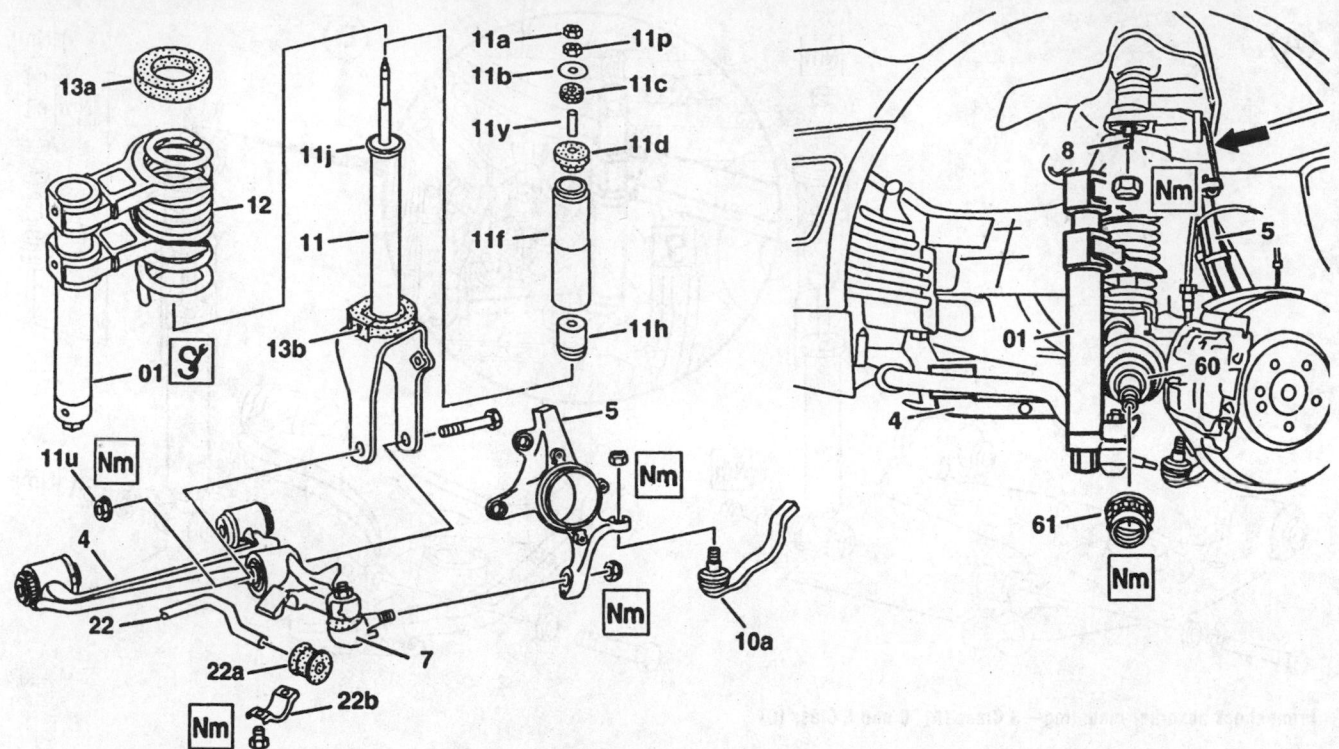

01. Spring compressor
4. Lower arm
5. Steering knuckle
7. Supporting joint
8. Follower joint
10a. Tie rod
11. Shock absorber
11a. Upper nut

11b. Washer
11c. Bushing
11d. Bushing
11f. Cover
11h. Bushing
11j. Cap
11p. Lower nut
11y. Sleeve

12. Coil spring
13a. Upper spring insulator
13b. Lower spring insulator
22. Stabilizer bar
22a. Stabilizer bar bushing
22b. Bracket
60. Axle shaft
61. 12 point nut

7923NG57

Exploded view of the front shock absorber/coil spring mounting—210 4-MATIC

Rear

C CLASS (202), E CLASS (124 & 210), S CLASS (140) AND SL CLASS (129)

1. Raise and safely support the vehicle.

2. Support the lower control arm with a jack.

✳✳ WARNING

Be sure that the piston rod does not turn when the removing the nut. The mount for the operating piston may

loosen and release the gas and oil from the shock absorber.

3. Remove the upper shock absorber mounting nut.

4. On convertible models, remove the switch for the rollover bar.

5. Remove the cover from the lower control arm.

6. Remove the lower shock absorber mounting bolt.

7. Press the shock absorber downward out of the body and remove the shock from the vehicle.

To install:

8. Install the lower rubber mount on the shock absorber. Compress the shock and position it in the vehicle. Install the lower bolt and tighten it to 40 ft. lbs. (55 Nm).

9. Install the lower control arm cover.

10. On convertible models, install the rollover bar switch.

11. Lower the vehicle and install the spacer, upper bushing, washer and new self-locking nuts. Tighten the bottom nut to 11–13 ft. lbs. (15–18 Nm) and the upper nut to 22 ft. lbs. (30 Nm).

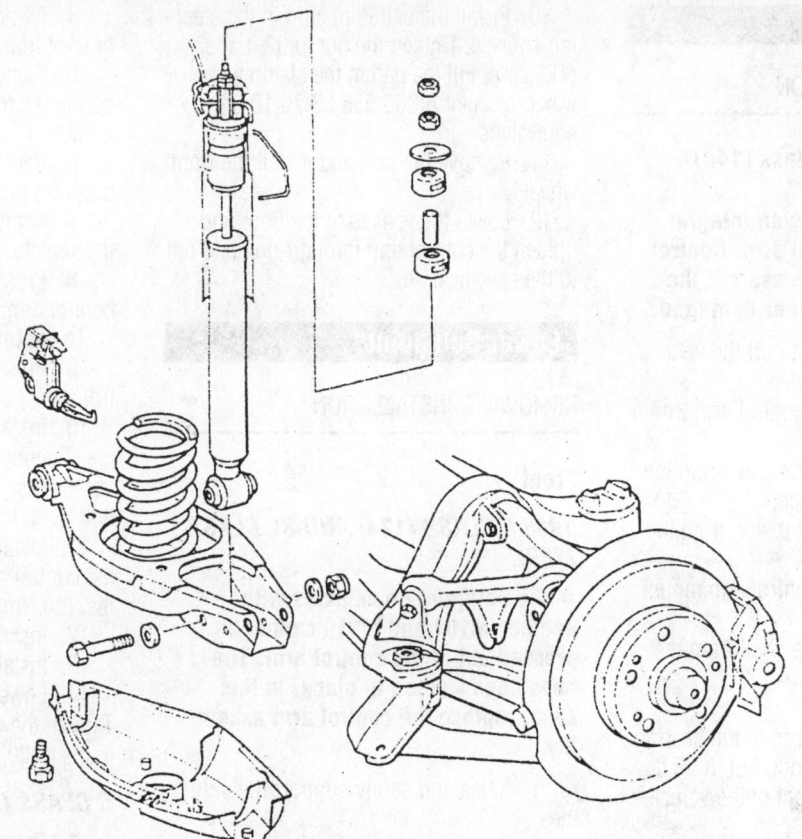

Exploded view of the rear shock absorber mounting—1995 E Class

Coil Spring

REMOVAL & INSTALLATION

Front

C CLASS (202), 1996–99 E CLASS (210), S CLASS (140) AND SL CLASS (129)—EXCEPT 4-MATIC

1. Raise and safely support the vehicle.
2. Remove the front wheel.
3. Install an internal spring compressor through the front coil spring.
4. Compress and remove the spring and upper seat towards the front of the vehicle.

To install:
5. Clean the spring mounting area on the lower control arm.
6. Compress the spring and position the spring on the control arm with the upper seat.
7. Carefully release the spring compressor until the spring is installed in the correct position.

8. Install the front wheel.
9. Lower the vehicle to the floor.
10. Check and adjust the headlamp position if necessary.

1996–99 E CLASS 210 4-MATIC
Refer to the shock absorber removal and installation procedure for coil spring service.

Rear

SL CLASS (129), S CLASS (140), C CLASS (202) AND E CLASS (124 & 210)

1. Raise and safely support the vehicle.
2. Remove the rear wheel.
3. On the SL Class, remove the rollover bar switch.
4. Remove the lower control arm cover.
5. Remove the cross brace.
6. Unclip the brake lining sensor wire from the lower control arm.
7. Jack up the lower control arm until the axle shaft is horizontal. Be careful not to raise the vehicle off the lift.

8. Compress the spring with an internal spring compressor. Do not use air tools to compress the spring.
9. Loosen the bolt securing the lower control arm to the axle carrier.
10. Remove the spring with the lower mount.

To install:
11. Clean the spring mounting surface.
12. Position the spring on the lower control arm and release the compressor.
13. Remove the jack from the lower control arm.
14. Attach the brake lining sensor wire to the control arm.
15. Install the cross brace.
16. Install the lower control arm cover.
17. On the SL Class, install the rollover bar switch.
18. Install the rear wheel and lower the vehicle to the floor.
19. Tighten the lower control arm-to-axle carrier nut to 52 ft. lbs. (70 Nm).
20. Check and adjust the headlamps if necessary.

7923NG55

Upper Ball Joint

REMOVAL & INSTALLATION

C Class (202) and S Class (140)

➡**The upper ball joint is an integral part of the upper control arm. Control arm replacement is necessary if the ball joint becomes worn or damaged.**

1. Raise and safely support the vehicle.
2. Remove the front wheel.
3. Support the lower control arm with a jack.
4. Remove the pinch bolt securing the ball joint stud to the knuckle.
5. Remove the ball joint stud from the knuckle.
6. Remove the four control arm mounting bolts.
7. Remove the bearing shell from the control arm.

To install:

8. Install the control arm in the bearing shell. Tighten the self-locking nut to 55 ft. lbs. (75 Nm) after the weight of the vehicle is on the suspension.
9. Install the new control arm assembly using four new microencapsulated bolts. Tighten the four bolts to 37 ft. lbs. (50 Nm).

10. Install the ball joint stud in the steering knuckle. Tighten the nut to 92 ft. lbs. (125 Nm). Fill the gap in the clamp with wax protectant A 000 986 33 70 10 or equivalent.
11. Remove the jack and install the front wheel.
12. Lower the vehicle to the floor and tighten the control arm through-bolt and nut to the specification.

Lower Ball Joint

REMOVAL & INSTALLATION

Front

1995 E CLASS (124) AND SL CLASS (129)

➡**The ball joint on vehicles with engines M104 and M119 cannot be pressed out of the control arm. The have been welded in place. In this case, replace the control arm assembly.**

1. Raise and safely support the vehicle.
2. Remove the front wheel.
3. Remove the coil spring using the recommended procedure.

4. Remove the stabilizer bar and bracket from the lower control arm.
5. Remove the bolts and nuts securing the lower control arm to the frame.
6. Remove the pinch bolt and nut securing the ball joint stud to the knuckle.
7. Remove the lower control arm from the vehicle.
8. Press the old ball joint out of the control arm.

To install:

9. Press a new ball joint into the control arm
10. Install the control arm in the vehicle. Temporarily tighten the control arm-to-frame nuts. Tighten the nut on the pinch bolt to 92 ft. lbs. (125 Nm).
11. Install the stabilizer bar bracket on the control arm. Tighten the bolts to 15 ft. lbs. (20 Nm).
12. Install the coil spring.
13. Install the front wheel.
14. Lower the vehicle to the floor. Tighten the control arm-to-frame nuts to 88 ft. lbs. (120 Nm).

C CLASS (202) AND S CLASS (140)

1. Raise and safely support the vehicle.
2. Remove the front wheel.

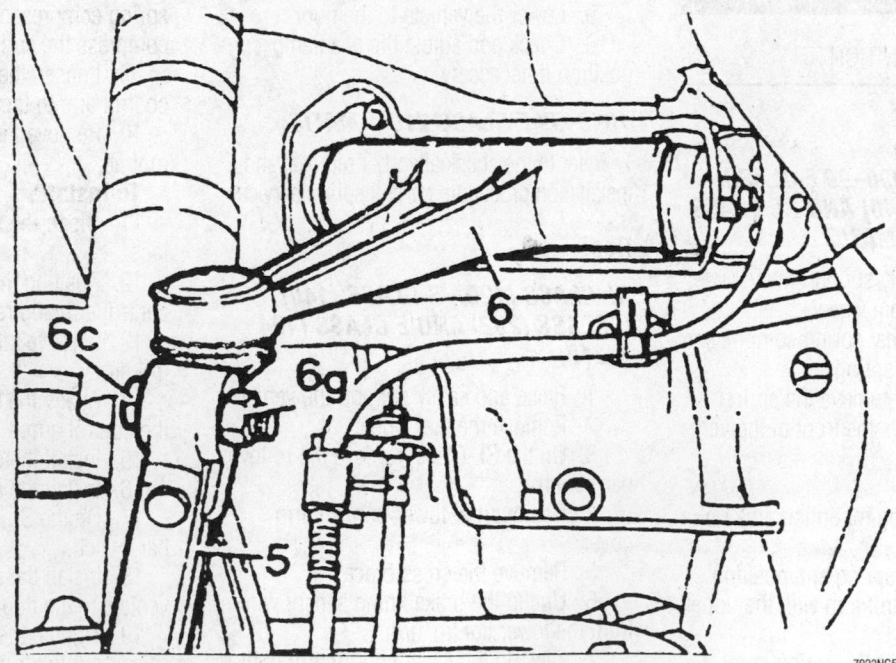

Steering knuckle (5), control arm (6), bolt (6c) and nut (6g)—S Class (140) and C Class (202)

7923NG58

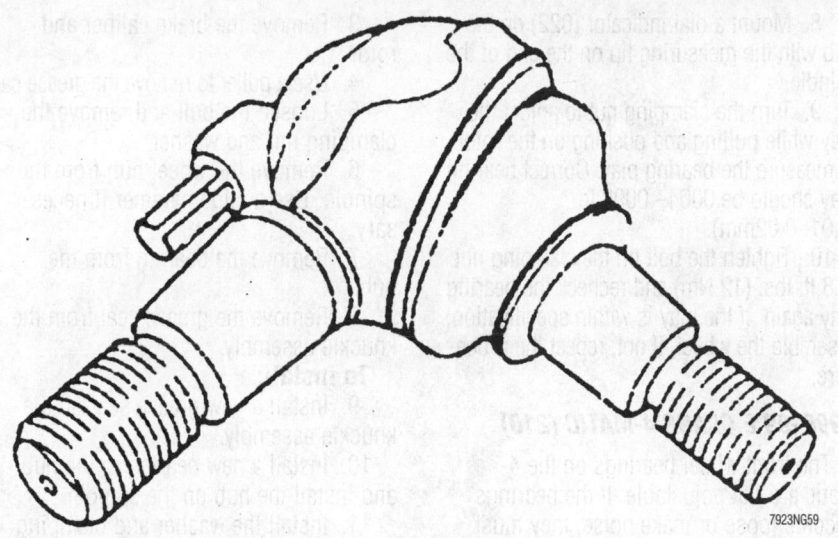

Common ball joint—C Class (202) and S Class (140)

3. Remove the disc brake rotor securing bolt and the backplate bolt.

4. Remove the nut securing the ball joint to the control arm.

5. Remove the nut securing the ball joint to the steering knuckle.

6. Separate the ball joint from the knuckle using a ball joint tool.

To install:

7. Install the ball joint in the knuckle and control arm. Tighten the control arm nut to 74 ft. lbs. (100 Nm) and the knuckle nut to 104 ft. lbs. (140 Nm).

8. Install the backplate bolt. Tighten the bolt to 6 ft. lbs. (8 Nm).

9. Install the brake rotor bolt. Tighten the bolt to 7 ft. lbs. (10 Nm).

10. Install the front wheel.

1996–99 E CLASS 4-MATIC (210)

1. Raise and safely support the vehicle.

2. Remove the front wheel.

3. Remove the caulker and the 12-point nut (61) from the axle shaft.

4. Remove the nut and press the ball joint out of the steering knuckle (5).

5. Pull the steering knuckle outward and remove the axle shaft (60) from the flange.

6. Support the knuckle assembly by placing a block of wood (arrow) between the coil spring and the knuckle.

7. Position the axle shaft out of the way using a piece of wire.

8. Remove the nut and press the ball joint (7) out of the lower control arm (4).

To install:

9. Install the ball joint in the lower control arm. Tighten a new self-locking nut to 77 ft. lbs. (105 Nm).

10. Assembly the axle shaft through the knuckle assembly.

11. Install the ball joint in the steering knuckle. Tighten a new self-locking nut to 77 ft. lbs. (105 Nm).

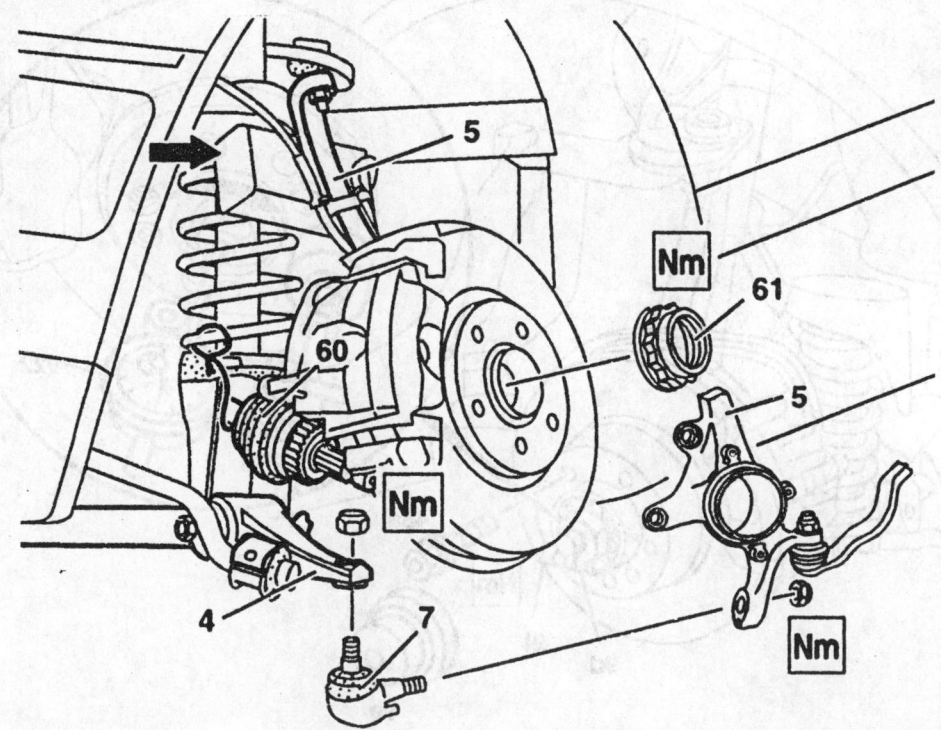

Lower ball joint mounting—1996–99 E class 4-Matic (210)

12. Install the 12-point nut on the axle. Tighten the nut to 162 ft. lbs. (220 Nm).

13. Install the front wheel.

14. Lower the vehicle to the floor.

Wheel Bearing

ADJUSTMENT

Front

EXCEPT 1996–99 E CLASS 4-MATIC (210)

1. Raise and safely support the vehicle.

2. Remove the front wheel.

3. Install one wheel bolt (48a) in the flange on the opposite side of the locking bolt.

4. Push the brake pads into the caliper until they don't touch the rotor.

5. Remove the grease cap (9e).

6. Loosen the bolt (9i) and tighten the clamping nut (9d) while turning the rotor until all bearing play is eliminated.

7. Loosen the clamping nut until play is detectable.

8. Mount a dial indicator (022) on the hub with the measuring tip on the end of the spindle.

9. Turn the clamping nut to adjust the play while pulling and pushing on the rotor to measure the bearing play. Correct bearing play should be .0004–.0008 in. (0.01–0.02mm).

10. Tighten the bolt on the clamping nut to 8 ft. lbs. (12 Nm) and recheck the bearing play again. If the play is within specification, assemble the wheel, if not, repeat the procedure.

1996–99 E CLASS 4-MATIC (210)

The front wheel bearings on the 4-Matic are not adjustable. If the bearings become loose or make noise, they must be replaced.

REMOVAL & INSTALLATION

Front

EXCEPT 1996–99 E CLASS 4-MATIC

1. Raise and safely support the vehicle.

2. Remove the front wheel.

3. Remove the brake caliper and rotor.

4. Use a puller to remove the grease cap.

5. Loosen the bolt and remove the clamping nut and washer.

6. Remove the wheel hub from the spindle. Use a slide hammer if necessary.

7. Remove the bearing from the hub.

8. Remove the grease seal from the knuckle assembly.

To install:

9. Install a new grease seal on the knuckle assembly.

10. Install a new bearing in the hub and install the hub on the spindle.

11. Install the washer and clamping nut. Adjust the bearing play and tighten the bolt on the clamp to 8 ft. lbs. (12 Nm).

12. Fill the grease cap to the flared edge with high temperature roller bearing grease and install the cap.

13. Install the rotor and caliper.

14. Install the front wheel and lower the vehicle to the floor.

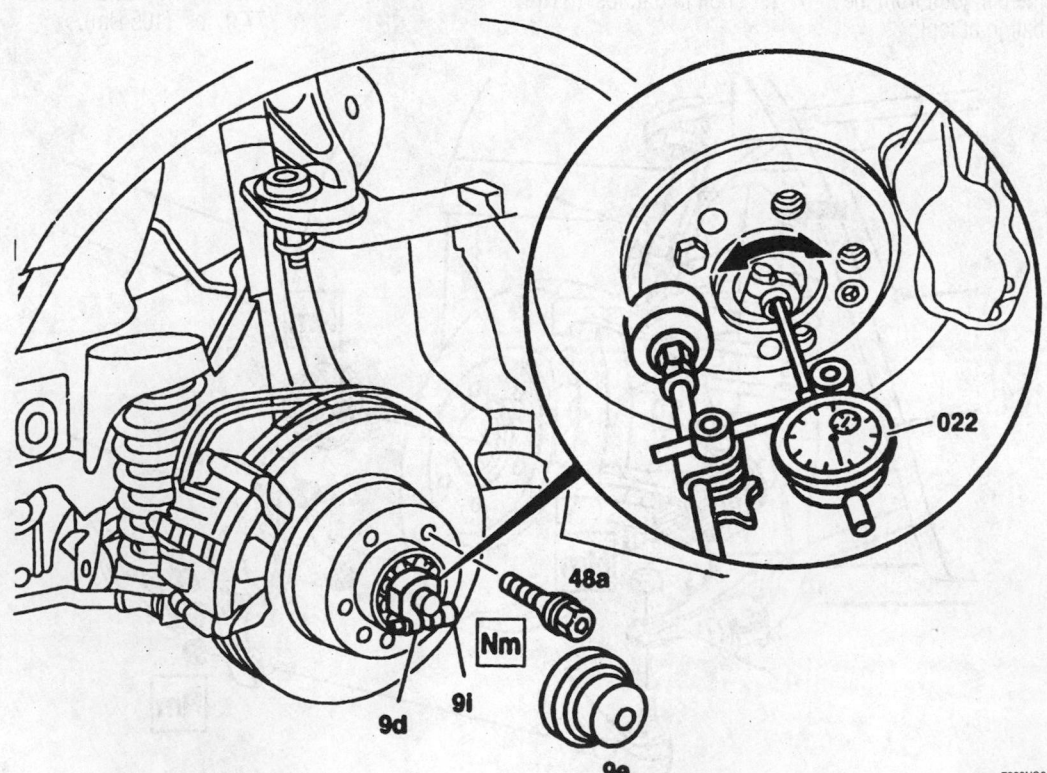

Mount a dial indicator on the rotor to measure the bearing play—except E Class 4-Matic

7923NG64

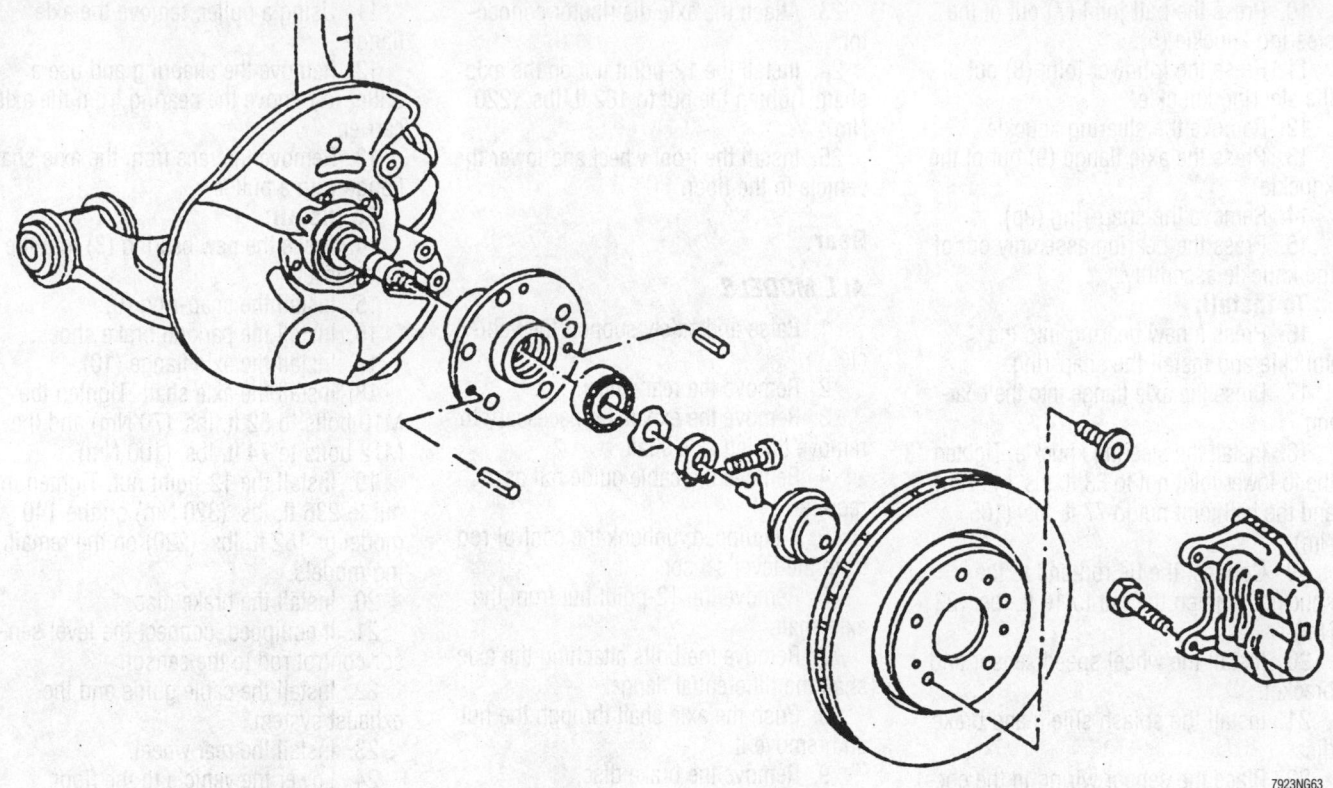

Exploded view of the front wheel bearing assembly—S class (140), SLK 230 (170), C Class (202), CLK 320 (208) and E Class (210)

1996–99 E CLASS 4-MATIC (210)

➡ **It is not necessary to remove the spring and shock absorber when removing the steering knuckle.**

1. Raise and safely support the vehicle.

2. Remove the front wheel.

3. Remove the 12-point nut (61) from the axle shaft.

4. Detach the front axle distributor connector.

5. Position the speed and brake lining sensor wires out of the way.

6. Remove the brake disc (34).

7. Remove the splash shield (35) from the knuckle.

8. Remove the speed sensor and bracket (5f).

9. Press the tie rod end (10a) out of the knuckle.

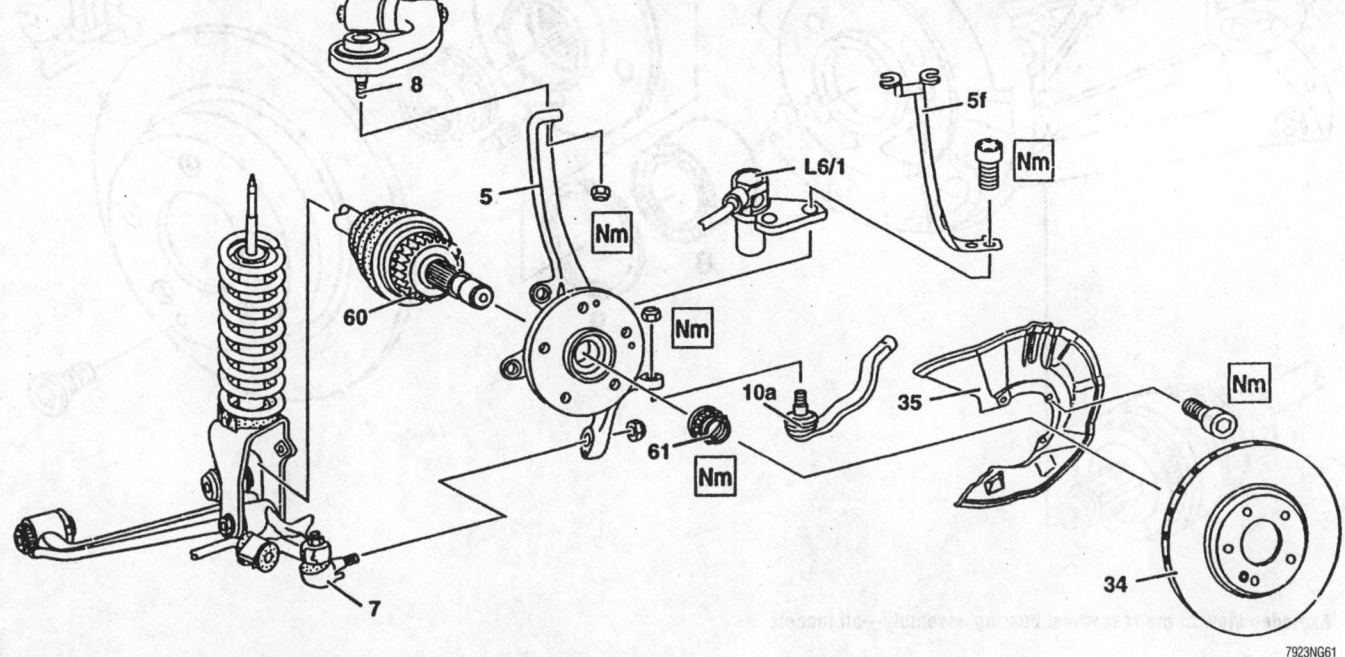

Steering knuckle and related components—E Class 4-matic (210)

10. Press the ball joint (7) out of the steering knuckle (5).

11. Press the follower joint (8) out of the steering knuckle.

12. Remove the steering knuckle.

13. Press the axle flange (9) out of the knuckle.

14. Remove the snap-ring (9p).

15. Press the bearing assembly out of the knuckle assembly.

To install:

16. Press a new bearing into the knuckle and install the snap-ring.

17. Press the axle flange into the bearing.

18. Install the steering knuckle. Tighten the follower joint nut to 33 ft. lbs. (45 Nm) and the ball joint nut to 77 ft. lbs. (105 Nm).

19. Connect the tie rod end to the knuckle. Tighten the nut to 16 ft. lbs. (22 Nm).

20. Install the wheel speed sensor and bracket.

21. Install the splash shield and brake disc.

22. Place the sensor wiring in the correct position.

23. Attach the axle distributor connector.

24. Install the 12-point nut on the axle shaft. Tighten the nut to 162 ft. lbs. (220 Nm).

25. Install the front wheel and lower the vehicle to the floor.

Rear

ALL MODELS

1. Raise and safely support the vehicle.

2. Remove the rear wheel.

3. Remove the exhaust if necessary to remove the left axle shaft.

4. Remove the cable guide rail as required.

5. If equipped, unhook the control rod from the level sensor.

6. Remove the 12-point nut from the axle shaft.

7. Remove the bolts attaching the axle shaft the differential flange.

8. Push the axle shaft through the hub and remove it.

9. Remove the brake disc.

10. Remove the parking brake shoes.

11. Using a puller, remove the axle flange.

12. Remove the snapring and use a puller to remove the bearing from the axle carrier.

13. Remove the race from the axle shaft flange with a puller.

To install:

14. Press the new bearing (8) into the axle carrier.

15. Install the snap-ring (9).

16. Install the parking brake shoes.

17. Install the axle flange (10).

18. Install the axle shaft. Tighten the M10 bolts to 52 ft. lbs. (70 Nm) and the M12 bolts to 74 ft. lbs. (100 Nm).

19. Install the 12-point nut. Tighten the nut to 236 ft. lbs. (320 Nm) on the 140 model or 162 ft. lbs. (220) on the remaining models.

20. Install the brake disc.

21. If equipped, connect the level sensor control rod to the sensor.

22. Install the cable guide and the exhaust system.

23. Install the rear wheel.

24. Lower the vehicle to the floor.

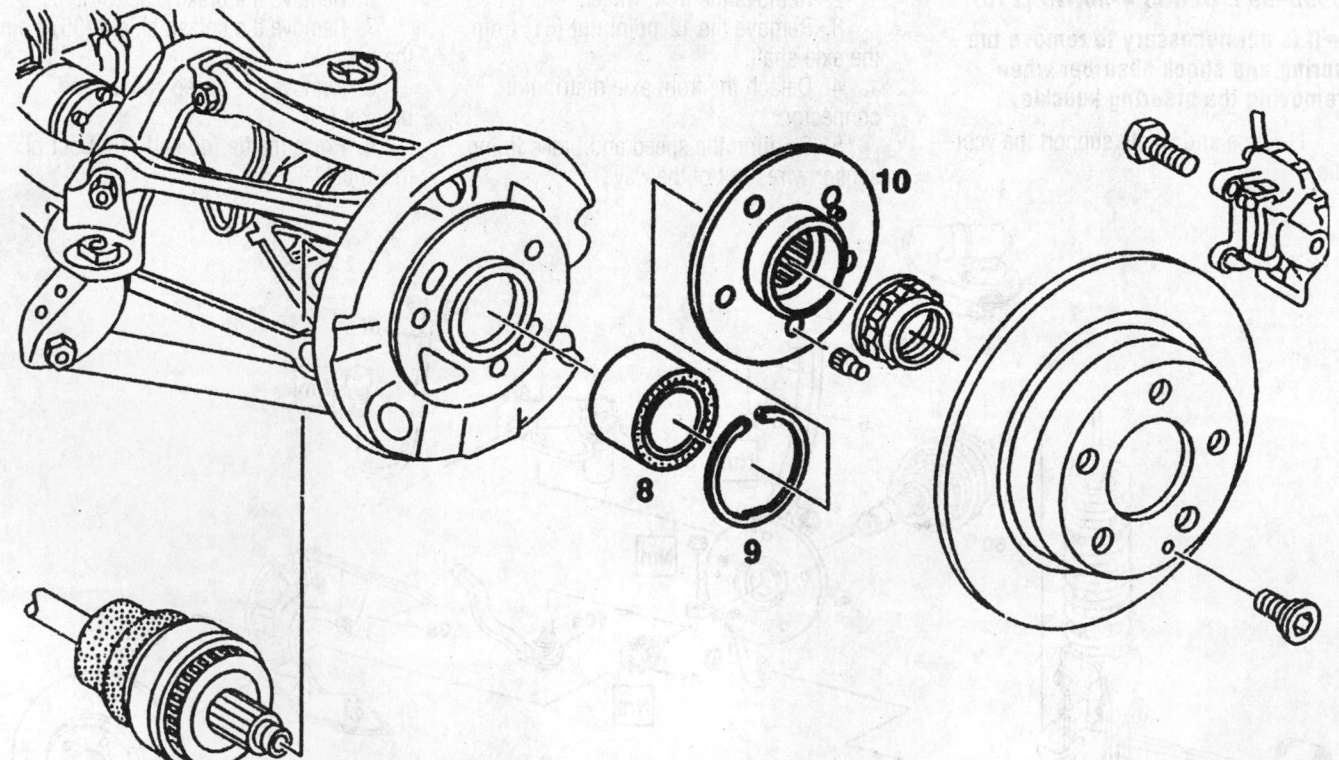

Exploded view of the rear wheel bearing assembly—all models

7923NG62

23

MITSUBISHI

3000GT • Diamante • Eclipse • Galant • Mirage

ENGINE REPAIR

→Disconnecting the negative battery cable on some vehicles may interfere with the functions of the on board computer systems and may require the computer to undergo a relearning process, once the negative battery cable is reconnected.

Distributor

REMOVAL

Before removing the distributor, position No. 1 cylinder at TDC on the compression stroke and align the timing marks.

1. Disconnect the negative battery cable. Remove the ignition wire cover, if equipped.
2. If necessary for access, tag and disconnect the spark plug wires from the distributor cap.
3. Detach the distributor harness connector.
4. Unscrew the distributor cap hold-down screws or release the clips and lift off the distributor cap with all ignition wires still connected. Remove the coil wire, if necessary.

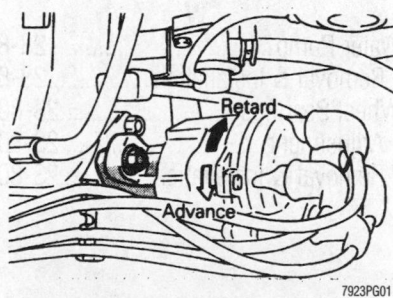

Adjusting the distributor—1.5L Mirage shown

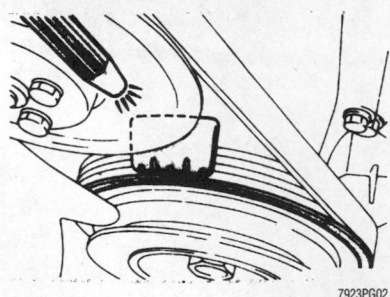

Checking the ignition timing—1.5L Mirage shown

5. Matchmark the rotor to the distributor housing and the distributor housing to the engine.

→Do not crank the engine during this procedure. If the engine is cranked, the matchmark must be disregarded.

6. Remove the hold-down nut.
7. Carefully remove the distributor from the engine.

→Some engines may be sensitive to the routing of the distributor sensor wires. If routed near the high-voltage coil wire or the spark plug wires, the electromagnetic field surrounding the high voltage wires could generate an occasional disruption of the ignition system operation.

INSTALLATION

Engine Not Disturbed

1. Install a new distributor housing O-ring and lubricate with clean oil.
2. Install the distributor in the engine so the rotor is aligned with the matchmark on the housing and the housing is aligned with the matchmark on the engine. Be sure the distributor is fully seated and the distributor shaft is fully engaged.
3. Install the hold-down nut.
4. Connect the distributor harness connectors.
5. Be sure the sealing O-ring is in place, install the distributor cap and tighten the screws or secure the clips.
6. Connect the negative battery cable.
7. Adjust the ignition timing and tighten the hold-down nut to 8 ft. lbs. (11 Nm).

Engine Disturbed

1. Install a new distributor housing O-ring and lubricate with clean oil.
2. Position the engine so the No. 1 piston is at TDC of its compression stroke and the mark on the vibration damper is aligned with **0** on the timing indicator.
3. Align the distributor housing and gear mating marks. Install the distributor in engine so the slot or groove of the distributor's installation flange aligns with the distributor installation stud in the engine block. Be sure the distributor is fully seated. Inspect alignment of the distributor rotor making sure the rotor is aligned with the position of the No. 1 ignition wire in the distributor cap.

→Be sure the rotor is pointing to where the No. 1 runner originates inside the cap, if equipped, and not where the No. 1 ignition wire plugs into the cap.

4. Install the hold-down nut.
5. Connect the distributor harness connectors.
6. Be sure the sealing O-ring is in place, install the distributor cap and tighten the screws or secure the clips.
7. Connect the negative battery cable.
8. Adjust the ignition timing and tighten the hold-down nut to 8 ft. lbs. (11 Nm).

Ignition Timing

ADJUSTMENT

3000GT and Diamante

The ignition timing is controlled by the engine control module and is not adjustable. However it can be inspected using a scan tool.

Eclipse

The ignition timing is controlled by the control module and is not adjustable. However it is possible to inspect the timing with a scan tool.

Galant

1. Set the parking brake, start and run the engine until normal operating temperature is obtained. Keep all lights and accessories off and the front wheels straight-ahead. Place the transaxle in **P** for automatic transaxle or neutral for manual transaxle.

→On Canadian vehicles the lights will remain on when the vehicle is running; this will not be a problem.

2. Locate the wire connector on the ignition coil connector. Insert a paper clip behind the TACH terminal connector to act as a tachometer adapter. Connect a tachometer to the paper clip. If not at specification, set the idle speed at the correct level.

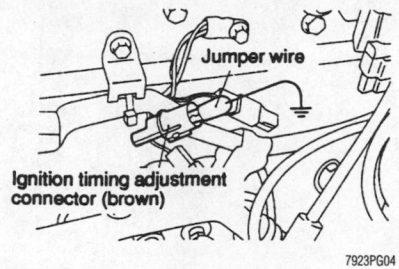

Timing connector identification—2.4L Galant

3. Turn the engine **OFF** and remove the waterproof cover from the ignition timing adjusting connector. This connector is a brown connector located near the center of the firewall. Connect a jumper wire from this terminal to a good ground.

4. Connect a conventional power timing light to the No. 1 cylinder spark plug wire. Start the engine and run at idle. Check the ignition timing by aiming the timing light at the timing scale located near the crankshaft pulley. Ignition timing should be 5° BTDC.

5. If ignition timing is not within specification, loosen the distributor hold-down nut just enough so the housing can be rotated.

6. Turn the housing in the proper direction until the specified timing is reached. Tighten the hold-down nut and recheck the timing. Turn the engine **OFF**.

7. Remove the jumper wire from the ignition timing adjusting terminal and install the waterproof cover.

8. Start the engine and check the actual timing without the terminal grounded. This reading should be approximately 5 degrees more than the basic timing. Actual timing may increase according to altitude. Also, actual timing may fluctuate because of slight variations accomplished by the ECU. As long as the basic timing is correct, the engine is timed correctly.

9. Turn the engine **OFF**. Disconnect the timing equipment and tachometer.

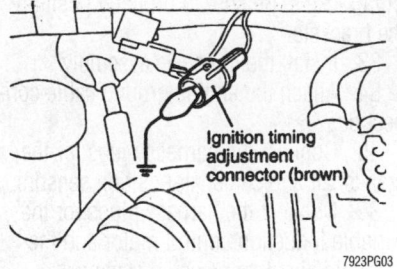

7923PG03

Timing connector identification—1.5L Mirage

Mirage

1. Attach the timing light according to the manufacturer's instructions.

2. Locate the timing tab line on the front of the engine and the notch on the crankshaft pulley. Mark them so they are easily recognizable with the timing light. Connect a tachometer to the engine.

3. **START** the engine and allow it to reach operating temperature.

4. Check that the engine idle is at specifications.

5. Turn the engine **OFF**.

6. Locate the ignition timing adjustment connector (brown) and connect a jumper wire to from the connector to a good ground connection.

➡**Grounding the ignition timing adjustment connector will set the engine to basic ignition timing.**

7. Point the timing light at the crankshaft pulley marks and inspect the ignition timing.

8. If the timing is not within specifications, loosen the distributor mounting nut and rotate the distributor slowly, in either direction, to align the timing marks.

9. Tighten the distributor mounting nut when the ignition timing is correct. Stop the engine and remove the timing light.

10. Adjust engine idle if needed.

Engine Assembly

REMOVAL & INSTALLATION

✳✳ CAUTION

The fuel injection system remains under pressure, after the engine has been turned OFF. Properly relieve fuel pressure before disconnecting any fuel lines. Failure to do so may result in fire or personal injury.

3000GT

1. Relieve fuel system pressure.
2. Disconnect the negative battery cable.

✳✳ CAUTION

Wait at least 90 seconds after the negative battery cable is disconnected to prevent possible deployment of the air bag.

3. Matchmark the hood and hinges and remove the hood assembly. Remove the air cleaner assembly and all adjoining air intake duct work.

4. Disconnect and remove the cruise control linkage and actuator assemblies.

5. Drain the engine coolant and remove the radiator assembly, coolant reservoir and intercooler.

6. Disconnect the heated oxygen sensor connection at the front exhaust pipe.

7. Unbolt and remove the front exhaust pipe assembly, discard gaskets.

8. Remove the transaxle assembly.

9. Disconnect the accelerator cable, breather hose and heater hose connections from the engine

10. Note locations and remove the vacuum hoses from the engine. Be sure to disconnect the brake booster vacuum supply.

11. Disconnect the fuel feed and return hoses.

12. Remove the solenoid valve assembly and disconnect ground cable.

13. Disconnect the purge hose and EGR temperature sensor, if equipped.

14. Remove the air conditioning and power steering drive belts.

15. Unbolt and remove the air conditioning compressor and the power steering pump assemblies.

➡**When removing the power steering pump and a/c compressor, it is not necessary to disconnect the hoses. Position the units aside and use rope or wire to secure.**

16. Disconnect the harness connections for the idle speed control, motor position sensor and throttle position sensor.

17. Disconnect the EGR temperature sensor (California).

18. For the turbocharger, disconnect the following:

    a. Connection for the booster vacuum hose.

    b. Connections for the oil cooler lines and discard the sealing rings.

    c. Connection for the oxygen sensor.

19. Disconnect the wiring at the oil pressure switch and oil pressure gauge unit.

20. Disconnect the fuel injection wiring harness plug.

21. Disconnect the wiring from the knock sensor and the crankshaft angle sensor.

22. Disconnect the coolant temperature switch, coolant temperature sensor and the coolant temperature gauge unit connections.

23. Disconnect the wiring to the ignition coil, condenser and the power transistor.

24. Disconnect the variable induction motor connection.

25. Open the cover of relay box and disconnect the alternator wiring.

26. Attach a hoist to the engine and support the engine weight. Remove the engine mount bracket.

27. Remove the front and rear roll stopper bracket mounting bolts.

28. Remove the engine assembly from the vehicle.

To install:

29. Install the engine and secure into position. Secure the engine mount bracket

to block and tighten bolts to 72–87 ft. lbs. (100–120 Nm). Install through-bolt and tighten bolt to 51 ft. lbs. (70 Nm).

30. Install the front and rear roll stopper through-bolt and tighten to 36–43 ft. lbs. (50–60 Nm).

31. Open the cover of relay box and connect alternator wiring.

32. Connect the variable induction motor connection.

33. Connect the fuel feed and return hoses. Using a new sealing ring, tighten pressure hose connection to 4 ft. lbs. (5 Nm).

34. Connect the wiring to the ignition coil, condenser and power transistor.

35. Connect the coolant temperature switch, coolant temperature sensor and the coolant temperature gauge unit connections.

36. Connect the wiring from the knock sensor and the crankshaft angle sensor.

37. Connect the fuel injection wiring harness plug.

38. Connect the wiring at the oil pressure switch and oil pressure gauge unit.

39. Connect the following for the turbocharger:

a. Connection for booster vacuum hose.

b. Connections for oil cooler lines using new sealing rings. Tighten fittings to 29–33 ft. lbs. (40–49 Nm).

c. Connection for oxygen sensor.

40. Connect EGR temperature sensor (California).

41. Connect harness connections for the idle speed control, motor position sensor and throttle position sensor.

42. Install the air conditioning compressor and the power steering pump assemblies.

43. Install the engine drive belts.

44. Connect the purge hose and the EGR temperature sensor, if equipped.

45. Install the solenoid valve assembly and connect ground cable to engine block.

46. Reconnect the vacuum hoses to the engine. Be sure to connect the brake booster vacuum supply.

47. Connect the accelerator cable, breather hose and heater hose connections to the engine.

48. Install the transaxle assembly.

49. Install the front exhaust pipe assembly, using new gaskets. Tighten manifold mounting bolts to 36 ft. lbs. (50 Nm).

50. Connect the heated oxygen sensor connection at the front exhaust pipe.

51. Replace the radiator assembly, coolant reservoir and intercooler. Refill the cooling system.

52. Install and connect the cruise control linkage and the actuator assemblies.

53. Install the hood assembly, air cleaner assembly and all adjoining air intake duct work.

54. Connect the negative battery cable and run engine.

55. Inspect all connections and check all fluid levels.

Diamante

1. Matchmark the hood and hinges and remove the hood assembly.

2. Relieve fuel system pressure.

3. Disconnect the negative, then the positive battery cable. Remove the battery from the vehicle.

4. Remove the air cleaner assembly and all adjoining air intake duct work.

5. Drain the engine coolant and remove the radiator assembly and coolant reservoir (and bracket).

6. Remove the engine undercover, if equipped.

7. Disconnect and remove the front exhaust pipe, main muffler and catalytic converter for 1993–94 Federal models and 1993 California models.

8. Disconnect and remove the front exhaust pipe, catalytic converter (front bank side) and catalytic converter (rear bank side) for 1995–96 Federal models and 1994–96 California models.

9. Remove the transaxle assembly.

10. Disconnect the accelerator cable from the throttle body.

11. Disconnect the brake booster vacuum hose from the booster.

12. Note the locations and disconnect the vacuum hoses.

13. Disconnect the high pressure fuel line and the fuel return line.

14. Note the locations and remove the vacuum hoses from the solenoid valves.

15. Remove the vacuum hoses from the purge canister.

16. Disconnect the heater hose connections from the engine.

17. If equipped, unplug the harness for the EGR temperature sensor connection.

18. Remove the engine drive belts.

19. Remove the power steering pump oil pressure switch connection from the pump.

20. Remove the power steering pump and wire aside. Do not disconnect the fluid hoses.

21. Remove the air conditioning compressor. Wire the compressor aside. Do not discharge or disconnect the A/C lines.

22. Disconnect the wiring to the alternator.

23. Note the locations and disconnect the harness plugs for the atmospheric pressure sensor, idle speed control motor, throttle position sensor connector, fuel injectors and knock sensor.

24. Note the locations and disconnect the harness plugs for the air conditioning engine coolant temperature switch, coolant temperature sensor and coolant temperature gauge.

25. Note the locations and disconnect the harness plugs for the ignition coil, condenser and ignition power transistor.

26. Note the locations and disconnect the harness plugs for the variable induction control motor and the manifold absolute pressure sensor.

27. Note the locations and disconnect the harness plugs for the crankshaft and camshaft position sensors.

28. Remove the radiator overflow tank and remove the mounting bracket.

29. Remove the ground cable connections.

30. Attach a hoist to the engine and take up the engine weight. Remove the engine mount bracket. Remove any torque control brackets (roll stoppers).

➡**Note that some engine mount components have arrows on them for proper assembly.**

31. Double check that all cables, hoses, harness connectors, etc., are disconnected from the engine. Lift the engine slowly and remove from the engine compartment.

To install:

32. Install the engine and secure all control brackets. Be sure to properly position the brackets.

33. Install the transaxle assembly.

34. Attach the engine ground cable connections.

35. Connect the harness plugs for the crankshaft and camshaft position sensors.

36. Connect the harness plugs for the variable induction control motor and the manifold absolute pressure sensor.

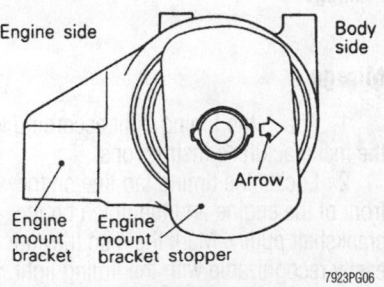

Alignment of the engine mount stopper bracket—Diamante shown

37. Connect the harness plugs for the ignition coil, condenser and ignition power transistor.

38. Connect the harness plugs for the air conditioning engine coolant temperature switch, coolant temperature sensor and coolant temperature gauge.

39. Connect the harness plugs for the atmospheric pressure sensor, idle speed control motor, throttle position sensor, fuel injectors and knock sensor.

40. Connect the wiring to the alternator.

41. Install the air conditioning compressor assembly.

42. Install the power steering pump assembly.

43. Connect the power steering pump oil pressure switch harness plug to the pump.

44. Install and adjust the engine drive belts.

45. If unplugged, connect the harness for the EGR temperature sensor.

46. Using new hose clamps, attach the heater hose connections to the engine.

47. Attach the vacuum hoses to the purge canister.

48. Connect the vacuum hoses to the solenoid valves.

49. Using new clamps or O-rings, connect the high pressure fuel line and the fuel return line.

50. Using the noted locations, connect the vacuum hoses.

51. Connect the brake booster vacuum hose to the booster.

52. Connect and adjust the accelerator cable to the throttle body.

53. Install the air cleaner assembly and all adjoining air intake duct work.

54. Install the radiator and coolant reservoir assembly.

55. Install the transaxle assembly.

56. Using new gaskets, connect the exhaust system to the engine.

57. Install the battery to the vehicle. Connect the positive, then the negative battery cables.

58. Install the engine undercover, if equipped.

59. Fill the engine with the proper amount of engine oil and coolant.

60. Start the engine, allow it to reach normal operating temperature. Check for leaks.

61. Check the ignition timing and adjust if necessary.

62. Align the matchmarks and install the hood.

63. Road test the vehicle and check all fluid levels and functions for proper operation.

Eclipse

2.0L (4G63) ENGINE

1. Relieve the fuel system pressure.

2. Disconnect the negative battery cable.

3. Matchmark the hood to the hinges and remove the hood.

4. Remove the intake air duct.

5. Drain the engine coolant.

6. Remove the radiator.

7. Remove the engine undercover.

8. Attach an engine lifting fixture to the engine and remove the transaxle assembly.

9. Disconnect the power steering pressure switch, oil pressure switch, oil pressure gauge sender and the alternator wiring connectors.

10. Remove the alternator.

11. Remove the power steering pump from the bracket and position the pump out of the way. It is not necessary to disconnect the fluid lines.

12. Remove the A/C compressor from the bracket and position it out of the way. Do not disconnect the hoses.

13. Disconnect the accelerator cable from the throttle body and mounting bracket.

14. Disconnect the following connectors;
- Idle air control (IAC) motor
- Knock sensor
- Heated oxygen sensor
- Engine coolant temperature gauge sender
- Engine coolant temperature sensor
- Ignition module (power transistor)
- Throttle position sensor
- Condenser
- Manifold differential pressure sensor
- Injectors
- Ignition coil
- Camshaft position sensor
- Crankshaft position sensor
- A/C compressor connector

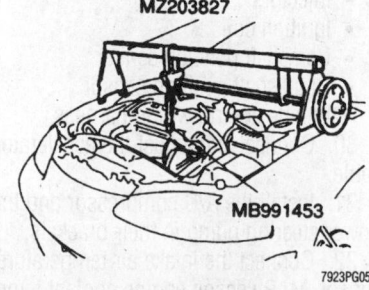

MZ203827

MB991453

7923PG05

Common method of supporting the engine using a fixture specifically designed for that purpose

- Engine control wiring harness

15. Disconnect the brake booster vacuum hose.

16. Disconnect the fuel lines from the fuel supply rail.

17. Disconnect the A and B water hose connections.

18. Label and disconnect the vacuum hoses.

19. Disconnect the front exhaust pipe from the turbocharger.

20. Support the engine under the oil pan with a floor jack and a piece of wood.

21. Remove the engine support fixture and replace it with a hoist. lift up the engine to take the weight off of the engine mount bracket.

22. Remove the engine mount bracket.

23. After checking that all cables, hoses and harness connectors are disconnected from the engine, raise the engine up slowly out of the engine compartment.

To install:

24. Slowly lower the engine assembly into the vehicle. Be sure that any wires, cables or hoses do not get damaged.

25. Position the floor jack under the oil pan with a piece of wood in between. Use the floor jack to adjust the height of the engine while installing the engine mount bracket.

26. Remove the chain hoist and install the engine support fixture.

27. Connect the front exhaust pipe to the turbocharger.

28. Connect the vacuum hoses.

29. Connect the A and B water hoses.

30. Install a new O-ring on the high pressure fuel line. Apply a small amount of clean engine oil to the O-ring and connect the fuel lines to the fuel supply rail.

31. Connect the brake booster vacuum hose.

32. Connect the following connectors;
- Idle air control (IAC) motor
- Knock sensor
- Heated oxygen sensor
- Engine coolant temperature gauge sender
- Engine coolant temperature sensor
- Ignition module (power transistor)
- Throttle position sensor
- Condenser
- Manifold differential pressure sensor
- Injectors
- Ignition coil
- Camshaft position sensor
- Crankshaft position sensor
- A/C compressor connector
- Engine control wiring harness

33. Install the A/C compressor and the power steering pump in their brackets.

34. Install the alternator.

35. Connect the oil pressure gauge sender, oil pressure switch and the power steering pressure switch connectors.

36. Install the transaxle assembly and remove the engine support fixture.

37. Install the engine undercover.

38. Install the radiator and connect the hoses.

39. Install the intake air duct.

40. Refill the engine with coolant.

41. Align the matchmarks and install the hood.

42. Connect the negative battery cable.

2.0L (420A) AND 2.4L (4G64) ENGINES

1. Relieve the fuel system pressure.

2. Disconnect the negative battery cable.

3. Matchmark the hood to the hinges and remove the hood.

4. Remove the intake air duct.

5. Drain the engine coolant.

6. Matchmark the hose clamps to the hoses and remove the radiator.

7. Remove the engine undercover.

8. Attach an engine lifting fixture to the engine and remove the transaxle assembly.

9. Disconnect the following connectors;
- A/C compressor
- Power steering pressure switch
- Heated oxygen sensor
- Engine coolant temperature gauge sender
- Engine coolant temperature sensor
- MAP sensor
- Intake air temperature sensor

10. Remove the power steering pump from the bracket and position the pump out of the way. It is not necessary to disconnect the fluid lines.

11. Remove the A/C compressor from the bracket and position it out of the way. Do not disconnect the hoses.

12. Disconnect the accelerator cable from the throttle body and mounting bracket.

13. Disconnect the following connectors:
- Idle air control (IAC) motor
- Knock sensor
- Ignition module (power transistor)
- EGR solenoid
- Oil pressure switch
- Throttle position sensor
- Condenser
- Manifold differential pressure sensor
- Injectors
- Ignition coil
- Camshaft position sensor
- Crankshaft position sensor
- Engine control wiring harness

14. Disconnect the heater hoses from the engine.

15. Disconnect the fuel lines from the fuel supply rail.

16. Disconnect the purge air hose and the brake booster vacuum hose.

17. Disconnect the front exhaust pipe from the manifold.

18. Place a floor jack against the oil pan with a piece of wood in between to protect the oil pan.

19. Raise the engine with the jack and remove the engine support fixture.

20. Install a chain hoist to the top of the engine.

21. Remove the engine mount bracket.

22. After checking that all cables, hoses and harness connectors are disconnected from the engine, raise the engine up slowly out of the engine compartment.

To install:

23. Slowly lower the engine assembly into the vehicle. Be sure that any wires, cables or hoses do not get damaged.

24. Position the floor jack under the oil pan with a piece of wood in between. Use the floor jack to adjust the height of the engine while installing the engine mount bracket.

25. Remove the chain hoist and install the engine support fixture.

26. Connect the front exhaust pipe to the manifold.

27. Connect the brake booster vacuum hose.

28. Install a new O-ring on the high pressure fuel line. Apply a small amount of clean engine oil to the O-ring and connect the fuel lines to the fuel supply rail.

29. Connect the following connectors;
- Idle air control (IAC) motor
- Knock sensor
- Ignition module (power transistor)
- EGR solenoid
- Oil pressure switch
- Throttle position sensor
- Condenser
- Manifold differential pressure sensor
- Injectors
- Ignition coil
- Camshaft position sensor
- Crankshaft position sensor
- Engine control wiring harness

30. Connect and adjust the accelerator cable.

31. Install the A/C compressor and the power steering pump in their brackets.

32. Connect the intake air temperature sensor, MAP sensor, engine coolant temperature sensor, engine coolant temperature gauge sender, heated oxygen sensor, power steering pressure switch and the A/C compressor harness connectors.

33. Install the radiator and connect the hoses. Position the hose clamps in the original position.

34. Install the transaxle and remove the engine support fixture.

35. Install the engine undercovers.

36. Install the intake air duct.

37. Connect the negative battery cable.

38. Align the matchmarks and install the hood.

39. Refill the engine with the proper amount of coolant.

Galant

The transaxle must be removed from the vehicle before removing the engine; they will not come out as a unit.

1. Disconnect the negative battery cable.

2. Drain the engine coolant.

3. Drain the engine oil and the transmission oil.

✳✳ CAUTION

Used motor oil may cause skin cancer if repeatedly left in contact with the skin for prolonged periods.

4. Safely relieve the pressure within the fuel injection system.

5. Matchmark the hood to the hinges and remove the hood.

6. Remove the transaxle assembly.

7. Remove the radiator, disconnecting the hoses at the engine.

8. Disconnect the accelerator cable and remove the bracket.

9. Disconnect the air intake and breather hoses.

10. Disconnect the heater hoses.

11. Disconnect the brake booster vacuum hose at the engine.

12. Label and disconnect the vacuum hoses at the throttle body.

13. Disconnect the high pressure fuel line and discard the O-ring; it is not reusable.

14. Remove the fuel return hose.

15. Disconnect the electrical connectors to the engine components. All wires and connectors should be labeled at the time of removal. The amount of time saved during reassembly makes the extra effort well worthwhile. Disconnect the following:
- power steering pressure switch
- alternator
- oil pressure switch
- A/C compressor
- each injector
- power transistor
- ignition coil

- throttle position sensor (TPS)
- idle air control motor (IAC)
- coolant temperature switch
- coolant temperature sensor
- EGR temperature sensor
- control wiring harness
- oxygen sensor
- crankshaft angle sensor
- camshaft position sensor
- refrigerant temperature switch
- condenser connection

16. Loosen the power steering drive belt and remove it. Remove the bolts holding the pump to its bracket and hang the pump out of the way. Do not disconnect the hoses and do not allow the pump to hang by the hoses.

17. Loosen the adjuster and remove the air conditioning drive belt. Remove the compressor from its mount and hang it from stiff wire out of the way. Note that the hoses are still attached; do not loosen them or discharge the system.

18. Remove the self-locking nuts and bolt at the exhaust system joint just below the manifold. Separate the exhaust pipes; discard the gasket and the two nuts.

19. Elevate the vehicle and support it safely. Install the engine hoist equipment and make certain the attaching points on the engine are secure. Draw tension on the hoist just enough to support the engine's weight but no more. Do not disturb the placement of the vehicle on the stands.

20. Remove the through-bolt from the rear (firewall side) roll stopper. Remove the through-bolt from the front engine roll stopper.

21. Remove the nuts and bolts holding the upper (left side) engine mount to the engine. Remove the through-bolt and remove the mount assembly. Also remove the support bracket below the mount.

22. Double check for any remaining cables, wires or hoses running to the engine. Elevate the hoist and remove the engine from the vehicle. Immediately place it on an engine stand or support it with wooden blocks. Do not allow it to rest on the oil pan or lie on its side. Never leave an engine hanging from a hoist.

To install:

After repairs, make certain the engine is fully reassembled before installation. All components removed with the engine out of the vehicle should be in place before reinstallation.

23. Install the engine into the vehicle and lower it until the bolt holes for the mounts and roll stoppers align with the brackets. Install the through-bolts and new self-locking nuts, tightening them just snug; they will be final tightened later.

24. Install the upper (left side) mount to the engine, tightening the nuts and bolts snug. Install the through-bolt and tighten it snug.

25. Slowly release tension on the hoist, allowing the weight of the engine to bear fully on the mounts. Once the hoist is slack, remove the lifting apparatus from the engine.

26. Connect the exhaust system to the manifold, using a new gasket and new locking nuts. Tighten the nuts and the small bolt to 33 ft. lbs. (44 Nm)

27. Final tighten the engine mount nuts and bolts. Correct torque values are:
- Upper mount to engine nuts 42 ft. lbs. (57 Nm)
- Upper mount to engine bolt 9 ft. lbs. (12 Nm)
- Upper mount through-bolt 72–87 ft. lbs. (98–118 Nm)
- Rear roll stopper through-bolt 32 ft. lbs. (44 Nm)
- Front roll stopper through-bolt 41 ft. lbs. (57 Nm)

28. Install the air conditioning compressor, tightening the mounting bolts to 18 ft. lbs. (25 Nm). Install the belt and adjust it.

29. Install the power steering pump, tightening the front bolts to 21 ft. lbs. (28 Nm) and the rear bolt to 16 ft. lbs. (22 Nm). Install and adjust the belt.

30. Connect the wiring and harness connectors to the engine. Make certain each terminal is clean and the connector is firmly seated to its mate. Do not route wires near hot surfaces or moving parts.

31. Install the fuel return hose and secure with the retaining clamp.

32. Using a new O-ring, connect the high pressure fuel line and tighten the bolts to 48 inch lbs. (6 Nm).

33. Connect the vacuum lines running to the throttle body.

34. Connect the heater hoses.

35. Install the accelerator cable bracket, tightening the bolts to 48 inch lbs. (6 Nm), and connect the accelerator cable.

36. Install the radiator and connect the hoses.

37. Install the transaxle.

38. Check the engine oil drain plug and secure it if necessary. Install the proper amount of engine oil.

39. Check the transaxle drain plug, tightening it if needed, and install the proper amount of transmission oil.

40. Check the radiator and engine drain cocks, closing them if necessary and refill the coolant system.

41. Double check all installation items, paying particular attention to loose hoses or hanging wires, loose nuts, poor routing of hoses and wires (too tight or rubbing) and tools left in the engine area.

42. Connect the negative battery cable. Start the engine and check for leaks.

43. Attend to all leaks immediately, remembering that fluids and metal surfaces may be hot. Adjust the drive belts to the correct tension. Adjust all cables (transmission, throttle, shift selector) and check the fluid levels. Check the operation of all gauges and dashboard lights.

44. With the help of an assistant, install the hood and align it for proper body fit and latching.

45. In a safe location at low speed, road test the vehicle for correct operation of steering brakes, transaxle, clutch and speedometer.

Mirage

1. Relieve fuel system pressure.

2. Disconnect the negative battery cable. Remove the undercover if equipped.

3. Matchmark the hood and hinges and remove the hood assembly. Remove the air cleaner assembly and all adjoining air intake duct work.

4. Drain the engine coolant and remove the radiator assembly, coolant reservoir and intercooler.

5. Remove the transaxle assembly.

6. Disconnect the ground cable, accelerator cable, breather hose and heater hose connections from the engine.

7. Note locations and remove vacuum hoses from engine. Be sure to disconnect brake booster vacuum supply.

8. Disconnect fuel feed and return hoses.

9. For 1.5L engines, disconnect the crankshaft and camshaft sensor wiring.

10. Disconnect oxygen sensor connections, coolant temperature gauge and coolant temperature sensor connections.

11. For 1.8L engines, disconnect the oil pressure switch connection.

12. On models with automatic transmissions, disconnect the thermo switch.

13. Disconnect harness connections for the idle speed control motor and throttle position sensor.

14. For 1.5L engines, disconnect harness connections for the intake air temperature sensor.

15. Disconnect EGR temperature sensor (California).

16. Note locations for reassembly and disconnect injector harness plugs.

17. Disconnect power transistor and the ignition coil connections.

18. Disconnect alternator and power steering switch wiring.

19. Remove the air conditioner drive belt and the air conditioning compressor. Leave the hoses attached. Do not discharge the system. Wire the compressor aside.

20. Remove the power steering pump and wire aside.

21. For 1.8L engines, remove the starter and alternator harness clamp.

22. Remove the exhaust manifold to head pipe nuts. Discard the gasket.

23. Attach a hoist to the engine and support the engine weight. Remove the engine mount bracket. Remove any torque control brackets (roll stoppers).

24. Remove the engine assembly from the vehicle.

To install:

25. Install the engine and secure in position. The front lower mount through-bolt nut should not be tightened until the full weight of the engine is on the mount. Tighten through-bolt to 72 ft. lbs. (100 Nm) and bracket mounting bolts to 42 ft. lbs. (58 Nm). Tighten bracket mounting nut to 38 ft. lbs. (53 Nm).

26. Using a new gasket, position exhaust pipe onto the manifold and tighten the flange nuts to 36 ft. lbs. (50 Nm).

27. Install power steering pump, alternator and air conditioner compressor. Install and adjust drive belts, tighten all mounting bolts.

28. Connect alternator and power steering wiring.

29. For 1.8L engines, secure the alternator and starter harness clamp.

30. Connect the ignition coil and power transistor connections.

31. Connect fuel injector harness connections.

32. On California models, connect EGR temperature sensor plug

33. For 1.5L engines, connect harness wiring for the intake air temperature sensor.

34. Connect wiring for idle speed control motor and throttle position sensor.

35. On automatic transmission models, connect the thermo switch.

36. For 1.8L engines, connect the oil pressure switch wiring.

37. Connect oxygen sensor, coolant temperature gauge and coolant temperature sensor.

38. For 1.5L engines, connect the crankshaft and camshaft sensor wiring.

39. Using new O-rings, connect fuel feed hose and tighten bolts to 44 inch lbs. (5 Nm).

40. Using a new hose clamp, connect the fuel return hose.

41. Connect noted vacuum hoses and connect brake booster vacuum supply.

42. Connect the breather hose, heater hoses, accelerator cable and ground cables. Inspect accelerator cable for proper adjustment.

43. Install the transaxle assembly.

44. Install radiator assembly and refill the cooling system, engine oil and transmission oil.

45. Install air cleaner and hood assembly.

46. Connect negative battery cable and run engine.

47. Inspect all connections and check all fluid levels.

Water Pump

REMOVAL & INSTALLATION

3000GT and Diamante

1. Disconnect the negative battery cable. Drain the cooling system.

2. Remove the engine undercover.

3. Disconnect the clamp bolt from the power steering hose.

4. Support the engine with the appropriate equipment and remove the engine mount bracket.

5. Remove the timing belt from the front of the engine.

6. Disconnect the coolant hoses from the pump, if equipped.

7. Remove the alternator brace.

➡️**The water pump bolts are different in size. Be sure to pay special attention to the bolts during the removal procedure.**

8. Remove the water pump, gasket and O-ring where the water inlet pipe joins the pump.

To install:

9. Thoroughly clean and dry both gasket surfaces of the water pump and block.

10. Install a new O-ring into the groove on the front end of the water inlet pipe. Do not apply oils or grease to the O-ring. Wet with water only.

➡️**Use care when aligning the water pump with the inlet water pipe.**

11. Using a new gasket, install the water pump assembly to the engine block. Torque the mounting bolts to 17 ft. lbs. (24 Nm).

12. Connect the hoses to the pump.

13. Reinstall the timing belt and related parts.

14. Install the engine drive belts and adjust.

15. Fill the system with coolant.

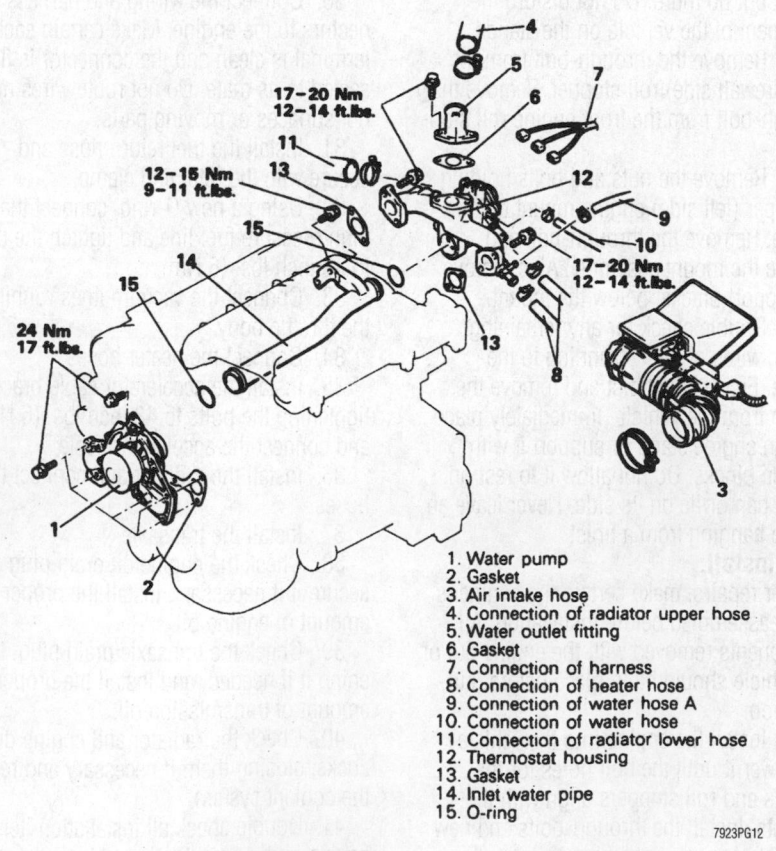

1. Water pump
2. Gasket
3. Air intake hose
4. Connection of radiator upper hose
5. Water outlet fitting
6. Gasket
7. Connection of harness
8. Connection of heater hose
9. Connection of water hose A
10. Connection of water hose
11. Connection of radiator lower hose
12. Thermostat housing
13. Gasket
14. Inlet water pipe
15. O-ring

Water pump and related components—DOHC Diamante shown, 3000GT similar

7923PG12

16. Connect the negative battery cable, run the vehicle until the thermostat opens and fill the radiator completely.

17. Once the vehicle has cooled, recheck the coolant level.

Eclipse

1. Disconnect the negative battery cable.
2. Drain the engine coolant.

3. Remove the timing belt.
4. If necessary, remove the alternator brace from the water pump.
5. If necessary, remove the timing belt rear cover.
6. Remove the water pump mounting bolts.
7. Remove the water pump, gasket and O-ring.

To install:

8. Install a new O-ring on the water inlet pipe. Coat the O-ring with water or coolant. Do not allow oil or other grease to contact the O-ring.

9. Use a new gasket and install the water pump to the engine block. Torque the mounting bolts to 8.7–11 ft. lbs. (12–15 Nm). Install the alternator brace on the water pump. Torque the brace pivot bolt to 17 ft. lbs. (24 Nm).

10. If removed, install the timing belt rear cover.

11. Install the timing belt.
12. Install the remaining components.
13. Refill the engine with coolant.
14. Connect the negative battery cable, start the engine and check for leaks.

Galant

1. Disconnect the negative battery cable.
2. Drain the cooling system.
3. Remove the engine undercover.
4. Disconnect the clamp bolt from the power steering hose.
5. Support the engine with the appropriate equipment and remove the engine mount bracket.
6. Remove the engine drive belts and the A/C tensioner bracket.
7. Remove the timing belt covers from the front of the engine.
8. Remove the camshaft and silent shaft timing belts.
9. Remove the alternator brace.
10. Remove the water pump, gasket and O-ring where the water inlet pipe(s) joins the pump.

To install:

11. Thoroughly clean and dry both gasket surfaces of the water pump and block.

12. Install a new O-ring into the groove on the front end of the water inlet pipe and wet with clean antifreeze only. Do not apply oils or grease to the O-ring.

13. Using a new gasket, install the water pump assembly. Tighten bolts with the head mark **4** to 10 ft. lbs. (14 Nm) and bolts with the head mark **7** to 18 ft. lbs. (24 Nm).

14. Reinstall the timing belt and related parts.

15. Install the engine drive belts and adjust.

16. Install the engine undercover.
17. Fill the system with coolant.
18. Connect the negative battery cable, run the vehicle until the thermostat opens and fill the radiator completely.

19. Once the vehicle has cooled, recheck the coolant level.

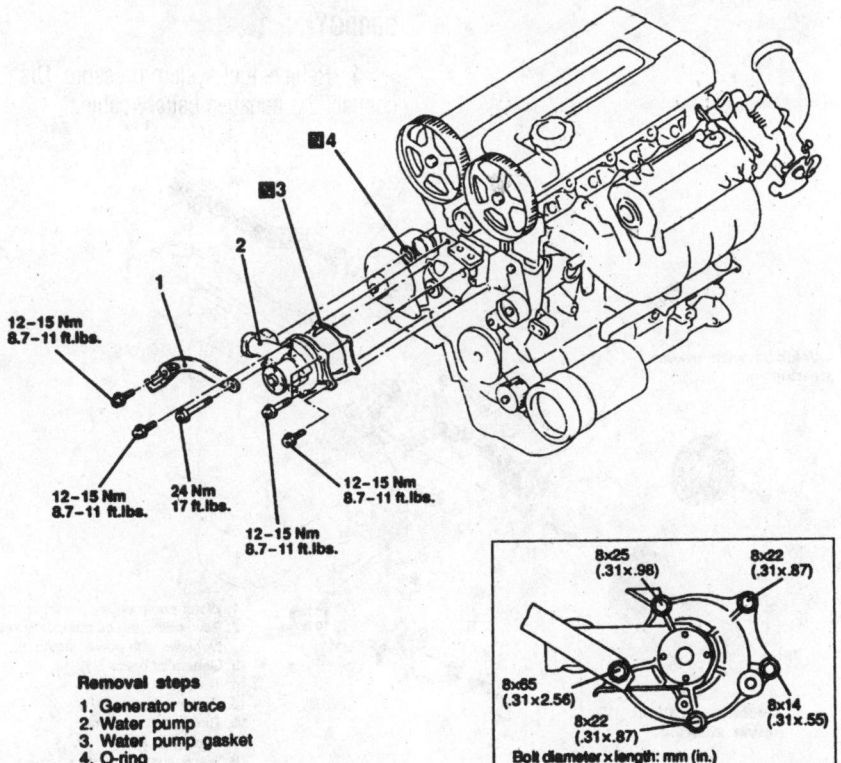

Removal steps
1. Generator brace
2. Water pump
3. Water pump gasket
4. O-ring

Water pump mounting—2.0L (4G63) engine

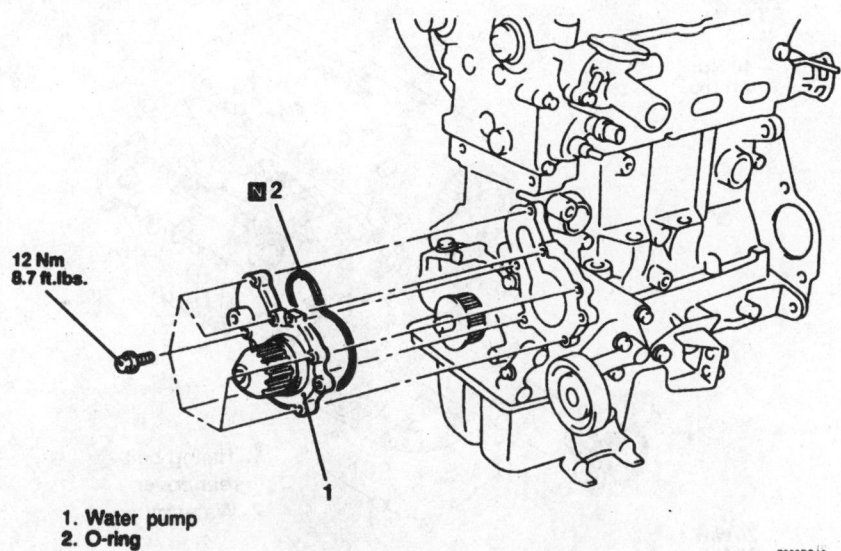

1. Water pump
2. O-ring

Water pump mounting—2.0L (420A) engines

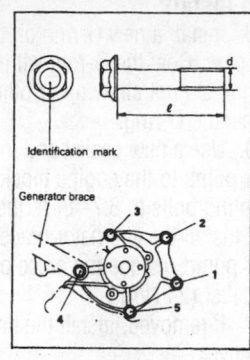

Water pump bolt identification—Galant

No.	Identification mark	Bolt diameter (d) x length (ℓ) mm (in.)	Torque Nm (ft.lbs.)
1	4	8 x 14 (.31 x .55)	
2	4	8 x 22 (.31 x .87)	12–15 (9–10)
3	4	8 x 30 (.31 x 1.18)	
4	7	8 x 65 (.31 x 2.56)	20–27 (15–19)
5	4	8 x 28 (.31 x 1.10)	12–15 (9–10)

7923PG11

Mirage

1. Disconnect the negative battery cable.
2. Rotate the engine and position the No. 1 piston to TDC of its compression stroke.
3. Drain the cooling system.
4. Remove the engine undercover.
5. Disconnect the clamp bolt from the power steering hose.
6. Remove the engine drive belts.
7. Support the engine with the appropriate equipment and remove the engine mount bracket.
8. Remove the timing belt from the front of the engine.
9. Remove the power steering pump bracket.
10. Remove the alternator brace.

➡ **The water pump mounting bolts are different in length, note their positioning for reassembly.**

11. Remove the water pump, gasket and O-ring where the water inlet pipe(s) joins the pump.

To install:

12. Thoroughly clean and dry both gasket surfaces of the water pump and block.
13. For 1.5L engines, install a new O-ring into the groove on the front end of the water inlet pipe. Do not apply oils or grease to the O-ring. Wet the O-ring with water only.
14. For 1.8L engines, apply a 0.09–0.12 in. (2.5–3.0mm) continuous bead of sealant to water pump and install the pump assembly. Install the water pump within 15 minutes of the application of the sealant. Wait 1 hour after installation of the water pump to refill the cooling system or starting the engine.
15. Install the gasket and pump assembly and tighten the bolts to specifications. Use care when aligning the water pump with the water inlet pipe.
16. Install the remaining components in the reverse order of removal.
17. Fill the system with coolant.

18. Connect the negative battery cable, run the vehicle until the thermostat opens and fill the radiator completely.
19. Once the vehicle has cooled, recheck the coolant level.

Cylinder Head

REMOVAL & INSTALLATION

3000GT

1. Relieve fuel system pressure. Disconnect the negative battery cable.

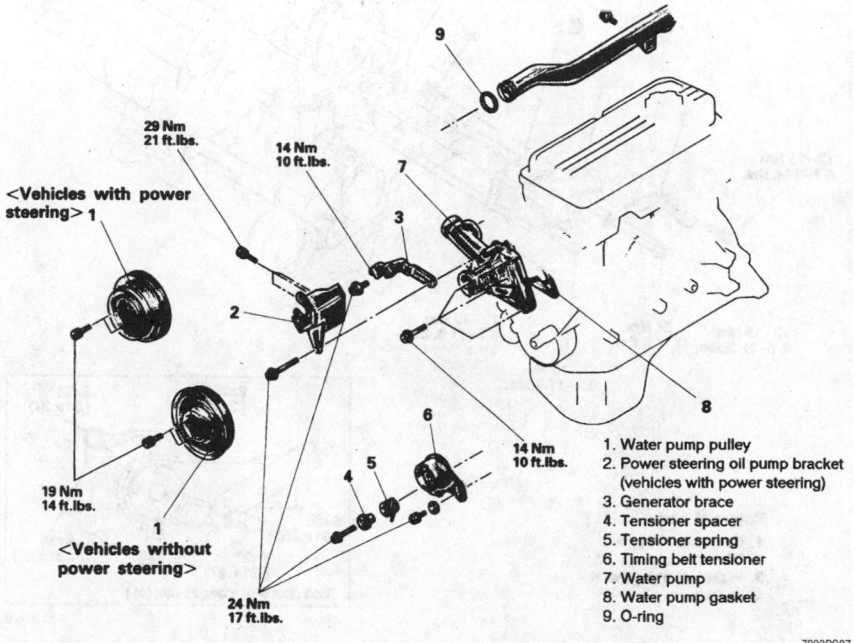

29 Nm
21 ft.lbs.

14 Nm
10 ft.lbs.

<Vehicles with power steering> 1

19 Nm
14 ft.lbs.

<Vehicles without power steering>

24 Nm
17 ft.lbs.

14 Nm
10 ft.lbs.

1. Water pump pulley
2. Power steering oil pump bracket (vehicles with power steering)
3. Generator brace
4. Tensioner spacer
5. Tensioner spring
6. Timing belt tensioner
7. Water pump
8. Water pump gasket
9. O-ring

7923PG07

Water pump and related components—Mirage with 1.5L (4G15) engine

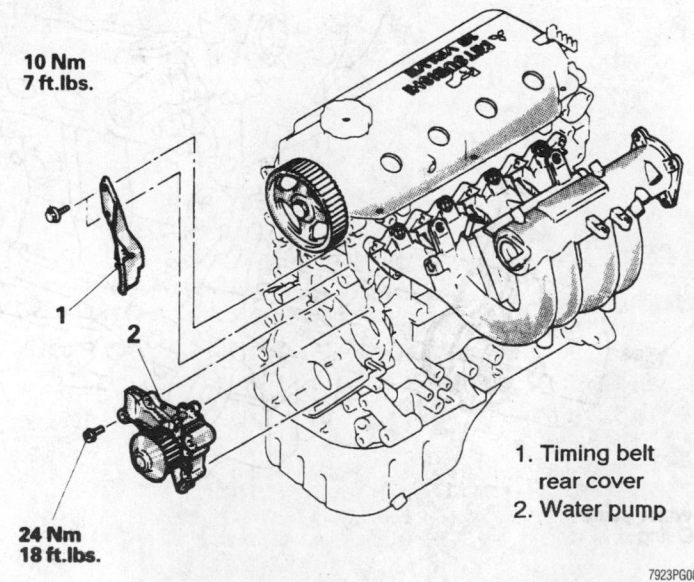

10 Nm
7 ft.lbs.

24 Nm
18 ft.lbs.

1. Timing belt rear cover
2. Water pump

7923PG08

Water pump and related components—Mirage with 1.8L (4G93) engines

Wait at least 90 seconds after the negative battery cable is disconnected to prevent possible deployment of the air bag.

2. Drain the cooling system.
3. Remove the air intake hoses.
4. Remove air intake plenum and intake manifold.
5. Remove the turbocharger, if equipped.
6. Remove the exhaust manifold.
7. Remove the timing belt.
8. Remove the triple pipe assembly across the top of the engine.
9. Remove the breather hose.
10. Remove the spark plug cable center cover and remove the spark plug cables.
11. When removing the valve cover, note that bolts for the front head are black and bolts for the rear head are green. Also, all bolts are 10mm long except the 1 closest to the sprockets on the rear head which is 20mm long.
12. To remove the intake camshaft sprocket, hold the camshaft with a wrench on the hexagon near the end of the camshaft and remove the bolt.
13. Remove the rear timing belt cover.
14. Remove the ignition coil.
15. Disconnect all water hoses from the thermostat housing and remove the housing.
16. Disconnect the water inlet from the front head and discard O-ring.
17. Loosen the cylinder head mounting bolts in three steps, starting from the outside and working inward. Lift off the cylinder head assembly and remove the head gasket.

To install:
18. Thoroughly clean and dry the mating surfaces of the head and block. Check the cylinder head for cracks, damage or engine coolant leakage. Remove scale, sealing compound and carbon. Clean oil passages thoroughly. Check the head for flatness. End to end, the head should be within 0.002 in. normally with 0.008 in. the maximum allowed out of true. The total thickness allowed to be removed from the head and block is 0.008 in. maximum.
19. Place a new head gasket on the cylinder block with the identification marks in the front top (upward) position. Do not use sealer on the gasket.
20. Carefully install the cylinder head on the block. Be sure the head bolt washers are installed with the chamfered edge upward. Using three even steps, torque the head

bolts in sequence, to 76–83 ft. lbs. (105–115 Nm) for non-turbocharged cold engine or 87–94 ft. lbs. (120–130 Nm) for turbocharged cold engine.
21. On turbocharged models, loosen all cylinder head bolts and retighten in sequence to 87–94 ft. lbs. (120–130 Nm).
22. Install new O-ring and connect the water inlet to the front head.
23. Replace the gaskets and install the thermostat housing and connect the hoses.
24. Install the ignition coil and center rear timing belt cover.
25. Install the intake camshaft sprocket. Use hex flange on camshaft to secure and torque the retaining bolt to 65 ft. lbs. (90 Nm).
26. Apply sealer to the lower edges of the half-round portions of the belt-side of the new gasket and install the valve cover. Be sure green bolts are installed on the rear head and black bolts are installed on the front head. Also, be sure the longest bolt is installed in its proper location closest to the sprockets on the rear head. Tighten the bolts in the proper sequence to 26 inch lbs. (3 Nm). Then, retighten bolts Nos. 1–6 to 35 inch lbs. (4 Nm).
27. Connect the spark plug cables and install the center cover.
28. Install the breather hose.
29. Install the triple pipe assembly across the top of the engine and tighten the retaining bolts to 7 ft. lbs. (10 Nm).
30. Install the timing belt and all related items.
31. Using all new gaskets, install the intake manifold, air intake plenum, turbocharger and exhaust manifold, following the proper torque sequences.
32. Install the air intake hoses.
33. Change the engine oil and oil filter.
34. Fill the system with coolant.
35. Connect the negative battery cable, run the vehicle until the thermostat opens, fill the radiator completely.
36. Adjust the accelerator cable. Check and adjust the idle speed and ignition timing.
37. Once the vehicle has cooled, recheck the coolant level.

Diamante

3.0L DOHC ENGINE

1. Relieve fuel system pressure. Disconnect the negative battery cable.
2. Drain the cooling system.
3. Remove the air intake hoses.
4. Remove air intake plenum and intake manifold.
5. Remove the exhaust manifold.
6. Remove the timing belt.

Once the timing belt is removed, DO NOT rotate the engine or camshafts. Internal engine damage will result.

7. Remove the breather hose.
8. Remove the spark plug cable center cover and remove the spark plug cables.
9. Remove the rocker covers.
10. Holding the flats of the camshaft, remove the intake camshaft sprocket holding bolts and remove the sprockets.
11. Remove the rear timing belt cover.
12. Remove the ignition coil assembly.
13. Disconnect all water hoses from the thermostat housing and remove the housing.
14. Disconnect the water inlet from the front head and discard O-ring.
15. Loosen the cylinder head mounting bolts in the reverse of the torque sequence and loosen the bolts in three steps. Lift off the cylinder head assembly and remove the head gasket.

To install:
16. Thoroughly clean and dry the mating surfaces of the head and block. Check the cylinder head for cracks, damage or engine coolant leakage. Remove scale, sealing compound and carbon. Clean oil passages thoroughly. Check the head for flatness. End to end, the head should be within 0.002 inches. The total thickness allowed to be removed from the head and block is 0.008 in. maximum.
17. Place a new head gasket on the cylinder block with the identification marks in the front top (upward) position. Do not use sealer on the gasket.
18. Carefully install the cylinder head on the block. Be sure the head bolt washers are installed with the chamfered edge upward. Using three even steps, torque the head bolts in sequence, to 76–83 ft. lbs. (105–115 Nm) for cold engine.

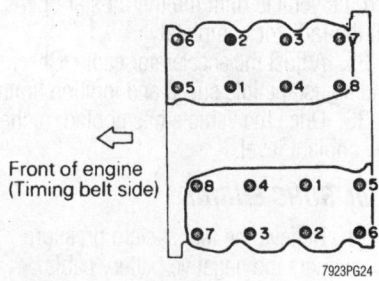

Front of engine (Timing belt side)

7923PG24

Cylinder head bolt torque sequence—Diamante and 3000GT

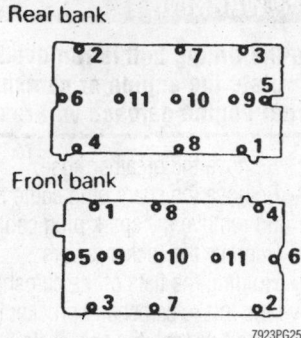

Rocker cover bolt torque sequence—Diamante 3.0L DOHC engine

19. Install new O-ring and connect the water inlet to the head. Tighten the mounting bolt to 9–11 ft. lbs. (12–15 Nm).

20. Replace the gaskets and install the thermostat housing. Tighten the mounting bolts to 12–14 ft. lbs. (17–20 Nm).

21. Using new hose clamps, connect the hoses to the thermostat housing.

22. Install the ignition coil and torque the mounting bolts to 7 ft. lbs. (10 Nm).

23. Install the rear timing belt cover and torque the mounting bolts to 17 ft. lbs. (24 Nm).

24. Install the intake camshaft sprocket. Use the hex flange on camshaft to secure and torque the retaining bolt to 65 ft. lbs. (90 Nm).

25. Apply sealer to the lower edges of the valve cover. Tighten the bolts in the proper sequence to 44–51 inch lbs. (5–6 Nm).

26. Connect the spark plug cables and install the center cover. Tighten the bolts that secure the center cover to 27 inch lbs. (3 Nm)

27. Install the breather hose.

28. Install the timing belt assembly.

29. Install the exhaust manifold assembly.

30. Using all new gaskets, install the intake manifold and air intake plenum.

31. Install the air intake hoses.

32. Change the engine oil and oil filter.

33. Fill the system with coolant.

34. Connect the negative battery cable, run the vehicle until the thermostat opens, fill the radiator completely.

35. Adjust the accelerator cable. Check and adjust the idle speed and ignition timing.

36. Once the vehicle has cooled, recheck the coolant level.

3.0L SOHC ENGINE

1. Relieve the fuel system pressure. Disconnect the negative battery cable.

2. Drain the cooling system.

3. Remove the air intake hose.

4. Remove the exhaust manifold.

5. Remove the air intake plenum and intake manifold.

6. Remove the timing belt.

7. Remove the camshaft sprocket and the rear timing belt cover.

8. Remove the power steering pump bracket. If removing the rear head, remove the alternator brace.

9. Disconnect the water inlet pipe.

10. Remove the purge pipe assembly.

11. Remove the valve cover.

12. Using the reverse sequence of the installation sequence, loosen the cylinder head mounting bolts in three steps. Lift off the cylinder head assembly and remove the head gasket.

To install:

13. Thoroughly clean and dry the mating surfaces of the head and block. Check the cylinder head for cracks, damage or engine coolant leakage. Remove scale, sealing compound and carbon. Clean oil passages thoroughly. Check the head for flatness. End to end, the head should be within 0.002 in. of true. The total thickness allowed to be removed from the head and block is 0.008 in. maximum.

14. Place a new head gasket on the cylinder block making sure the identification mark on the cylinder head gasket is in the front top (upward) location. Do not use sealer on the gasket. Be sure the gasket has the proper identification mark for the engine.

15. Carefully install the cylinder head on the block. Be sure the head bolt washers are installed with the chamfered edge upward. Using three even steps, torque the head bolts in sequence, to 76–83 ft. lbs. (105–115 Nm). This torque specification is for a cold engine.

16. Apply sealer to the lower edges of the half-round portions and install the valve cover. Tighten valve cover bolts to 7 ft. lbs. (9 Nm).

17. Install the purge pipe assembly.

18. Connect the water inlet pipe.

19. Install the power steering pump bracket and alternator brace.

20. Install the rear timing belt cover and cam sprocket. Torque the retaining bolt to 65 ft. lbs. (90 Nm).

21. Install the timing belt and all related items.

22. Using all new gaskets, install the intake manifold, air intake plenum and exhaust manifold, following the proper torque sequences.

23. Install the air intake hose.

24. Change the engine oil and oil filter.

25. Fill the system with coolant.

26. Connect the negative battery cable, run the vehicle until the thermostat opens, fill the radiator completely.

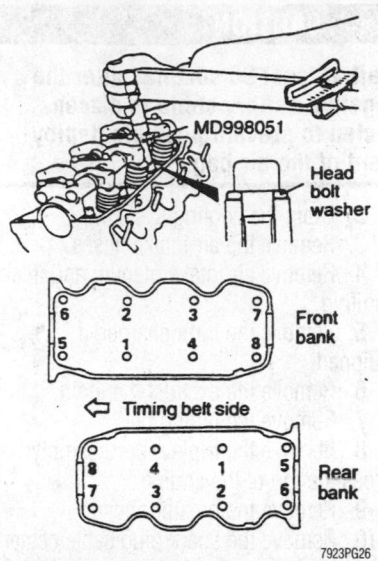

Tighten the cylinder head bolts according to the sequence shown—3.0L (SOHC and DOHC) engines

27. Check and adjust the idle speed and ignition timing.

28. Once the vehicle has cooled, recheck the coolant level.

3.5L ENGINE

1. Disconnect the negative battery cable.

2. Drain the engine coolant

3. Remove the timing belt.

4. Remove the intake and exhaust manifolds.

5. Remove the spark plug wires.

6. Remove the cylinder head covers.

7. Remove the timing belt rear center cover.

8. Loosen the cylinder head bolts gradually in three stages, in the opposite of the installation sequence.

9. Remove the cylinder head.

To install:

10. Clean and degrease the cylinder head and mounting surface on the engine block.

11. Install the cylinder head using a new gasket.

12. Tighten the bolts in sequence using three stages to 76–83 ft. lbs. (103–113 Nm).

13. Install the timing belt rear center cover.

14. Install the cylinder head covers using new gaskets. Tighten the bolts to 2–3 ft. lbs. (3–4 Nm).

15. Install the spark plug wires.

16. Install the intake and exhaust manifolds.

17. Install the timing belt.

18. Install any remaining components.

19. Refill the cooling system and change the engine oil and filter.

20. Connect the negative battery cable.

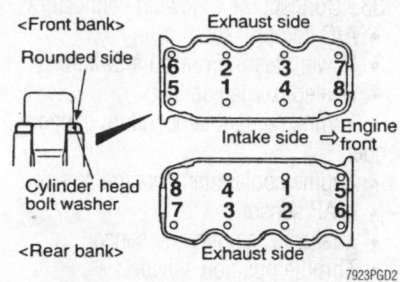

Cylinder head bolt tightening sequence—
3.5L engine

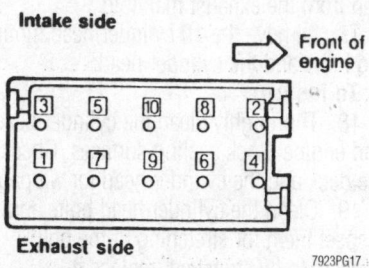

Cylinder head bolt removal sequence—
2.0L (4G63) engine

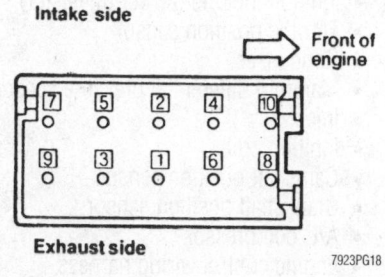

Cylinder head bolt torque sequence—2.0L
(4G63) engine

Eclipse

2.0L (4G63) ENGINE

1. Relieve the fuel system pressure.
2. Disconnect the negative battery cable.
3. Drain the engine coolant.
4. Drain the engine oil.
5. Disconnect the accelerator cable and remove the mounting bracket.
6. Disconnect the intake air duct (hose) from the throttle body.
7. Disconnect the following connectors;
- Idle air control (IAC) motor
- Knock sensor
- Heated oxygen sensor
- Engine coolant temperature gauge sender
- Engine coolant temperature sensor
- Ignition module (power transistor)
- Throttle position sensor
- Condenser
- Manifold differential pressure sensor
- Injectors
- Ignition coil
- Camshaft position sensor
- Crankshaft position sensor
- A/C compressor
- Engine control wiring harness
8. Remove the engine center cover.
9. Disconnect the spark plug wires.
10. Disconnect the brake booster vacuum hose.
11. Disconnect the fuel lines from the fuel supply rail.
12. Disconnect the bypass hose and the water hose connections.
13. Disconnect the vacuum hoses, breather hose and the PCV hose.
14. Remove the timing belt.
15. Remove the power steering pump.
16. Remove the cylinder head cover and the semi-circular packing.
17. Remove the heat protector.
18. Mark the position of the hose clamps on the hoses and disconnect the water hoses and the radiator hoses.

19. Remove the thermostat housing and the O-ring.
20. Remove the intake manifold stay.
21. Remove the turbocharger assembly from the exhaust manifold.
22. Gradually loosen the cylinder head bolts in two or three steps using the specified sequence and remove the bolts.
23. Remove the cylinder head and the gasket.

To install:
24. Thoroughly clean the deck surface of the engine block and the sealing surface of the cylinder head. Check the cylinder head for warpage.
25. Measure the length of the cylinder head bolts from below the head to the end, if the bolt measures more than 3.913 in. (99.4mm), replace the bolt.
26. Install a new gasket on the engine block with the identification mark facing upwards.
27. Carefully place the cylinder head on the engine. Apply clean engine oil to the bolts and install the bolts finger-tight.
28. Torque the bolts using the following procedure;
    a. Torque the bolts in sequence to 58 ft. lbs. (78 Nm).
    b. Loosen the bolts completely in the reverse order.
    c. Torque the bolts in sequence to 15 ft. lbs. (20 Nm).
    d. Make a paint mark on the head of the bolt and the cylinder head at the same spot. Tighten the bolt 90° or ¼ turn from the mark.
    e. Tighten the bolt an additional 90° or ¼ turn so that the mark on the head of the bolt is opposite the original mark on the cylinder head.
29. Use a new gasket and install the turbocharger to the exhaust manifold.
30. Install the intake manifold stay.

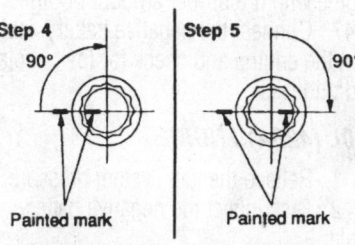

To ensure that the bolts are tightened exactly 90 degrees, mark the head bolt and cylinder head as shown—2.0L (4G63) engine

31. Install the thermostat housing and connect the hoses. Align the matchmarks and install the clamps.
32. Install the heat protector.
33. Apply sealant to the semi-circular packing and install it on the cylinder head.
34. Apply sealant at the front of the cylinder head where the camshaft oil seal retainer and the cylinder head come together and install the cylinder head cover using a new gasket.
35. Install the power steering pump.
36. Install the timing belt.
37. Connect the PCV, breather and vacuum hoses.
38. Connect the water hose and the bypass hose.
39. Use a new O-ring and connect the lines to the fuel supply rail. Apply a small amount of engine oil to the new O-ring.
40. Connect the brake booster vacuum hose.
41. Connect the spark plug wires and install the center cover.
42. Connect the following connectors;
- Idle air control (IAC) motor
- Knock sensor
- Heated oxygen sensor
- Engine coolant temperature gauge sender

- Engine coolant temperature sensor
- Ignition module (power transistor)
- Throttle position sensor
- Condenser
- Manifold differential pressure sensor
- Injectors
- Ignition coil
- Camshaft position sensor
- Crankshaft position sensor
- A/C compressor
- Engine control wiring harness

43. Connect the intake air hose to the throttle body.

44. Connect and adjust the accelerator cable.

45. Refill the engine with coolant.

46. Replace the oil filter and refill the engine with the proper amount of oil.

47. Connect the negative battery cable, start the engine and check for fuel, coolant and oil leaks.

2.0L (420A) ENGINE

1. Relieve the fuel system pressure.
2. Disconnect the negative battery cable.
3. Drain the engine coolant.
4. Drain the engine oil.
5. Remove the air cleaner and air intake duct.
6. Disconnect the following connectors;
- A/C compressor
- Power steering pressure switch
- Heated oxygen sensor
- Engine coolant temperature gauge sender
- Engine coolant temperature sensor
- MAP sensor
- Intake air temperature sensor
- Throttle position sensor
- Idle air control (IAC) motor
- Injector harness
- Ignition coil
- Camshaft position sensor
- EGR solenoid valve

7. Disconnect the accelerator cable from the throttle body.

8. Disconnect the heater hoses from the rear of the engine.

9. Disconnect the fuel lines from the fuel supply rail.

10. Disconnect the purge air hose and the brake booster vacuum hose connections.

11. Disconnect the overflow tube connection.

12. Mark the position of the clamp on the hose and disconnect the upper radiator hose and the water hose connections.

13. Remove the timing belt.

14. Remove the intake manifold stay.

15. Remove the intake and exhaust camshafts.

16. Disconnect the exhaust pipe connection from the exhaust manifold.

17. Remove the 10 cylinder head mounting bolts and the cylinder head.

To install:

18. Thoroughly clean the cylinder head and engine block sealing surfaces. Check the deck and the cylinder head for warpage.

19. Clean the cylinder head bolts and inspect them for stretching. If the bolt appears to by stretched, replace it.

20. Place a new head gasket on the engine block and carefully place the cylinder head on the engine.

21. Coat the threads of the bolts with clean engine oil and install the bolts finger-tight in the engine block. The short bolts go in the corners.

22. Torque the cylinder head bolts in the following sequence;
- Torque the center bolts 1 through 6 to 25 ft. lbs. (33 Nm), then torque the outer bolts 7 through 10 to 20 ft. lbs. (27 Nm)
- Torque the center bolts 1 through 6 to 50 ft. lbs. (67 Nm), then torque the outer bolts 7 through 10 to 20 ft. lbs. (27 Nm)
- Torque the center bolts 1 through 6 to 50 ft. lbs. (67 Nm), then torque the outer bolts 7 through 10 to 20 ft. lbs. (27 Nm)
- Turn all fasteners 1 through 10 ¼ turn (90°) more in sequence. Do not use a torque wrench for this step.

23. Use a new gasket and connect the front exhaust pipe to the exhaust manifold.

24. Install the camshafts.

25. Install the timing belts.

26. Install the intake manifold stay.

27. Connect the upper radiator hose. Install the clamp in the original position.

28. Connect the water hose to the water pipe.

29. Connect the overflow tube.

30. Connect the brake booster vacuum hose and purge air hose connection.

31. Use a new O-ring and connect the fuel lines to the fuel supply rail.

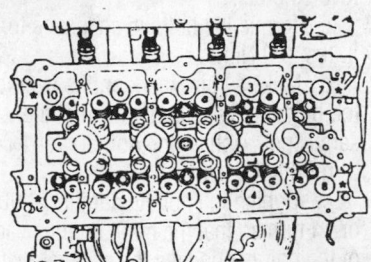

*Location of 110 mm (4.330 in.) short bolts.
7923PG20

Cylinder head bolt torque sequence—2.0L (420A) engine

32. Connect the heater hose.

33. Connect the following connectors;
- A/C compressor
- Power steering pressure switch
- Heated oxygen sensor
- Engine coolant temperature gauge sender
- Engine coolant temperature sensor
- MAP sensor
- Intake air temperature sensor
- Throttle position sensor
- Idle air control (IAC) motor
- Injector harness
- Ignition coil
- Camshaft position sensor
- EGR solenoid valve

34. Install and adjust the accelerator cable.

35. Install the air intake duct and the air cleaner assembly.

36. Refill the engine with oil and coolant. Replace the oil filter.

37. Turn the ignition to the **ON** position and check for fuel leaks. Then, start the engine and check for coolant leaks and proper operation.

2.4L (4G64) ENGINE

1. Relieve the fuel system pressure.
2. Disconnect the negative battery cable.
3. Remove the air cleaner with all air intake hoses.
4. Drain the cooling system.
5. Drain the engine oil.
6. Disconnect the accelerator cable. Remove cable mounting brackets and position the cable aside.
7. Remove the breather hose.
8. At the throttle body, disconnect the three small vacuum hoses, the coolant hoses, and the brake booster vacuum hose.
9. Disconnect and plug the high pressure fuel line.
10. Disconnect and plug the fuel return hose.
11. Disconnect the oxygen sensor, engine coolant temperature sensor, the engine coolant temperature gauge unit and the engine coolant temperature switch on vehicles with air conditioning.
12. Disconnect the ISC motor, throttle position sensor, distributor, fuel injectors, noise filter, EGR temperature sensor, ground cable and engine control wiring harness.
13. Remove the spark plug cables.
14. At the thermostat case assembly, remove the coolant hoses and unbolt the thermostat case from the engine.
15. Remove the timing belt:
   a. Remove the upper timing belt cover.

b. Align all timing marks.

c. Secure the timing belt to the camshaft sprocket.

d. Remove the camshaft sprocket.

16. Remove the valve cover and the half-round seal.

17. Disconnect the intake manifold stay bracket from the intake manifold.

18. Remove the exhaust pipe self-locking nuts and separate the exhaust pipe from the exhaust manifold. Discard the gasket.

19. Loosen the cylinder head mounting bolts in 3 steps, starting from the outside and working inward. Lift off the cylinder head assembly and remove the head gasket.

To install:

20. Thoroughly clean and dry the mating surfaces of the head and block. Check the cylinder head for cracks, damage or engine coolant leakage. Remove scale, sealing compound and carbon. Clean oil passages thoroughly. Check the head for flatness.

21. Place a new head gasket on the cylinder block with the identification marks at the front top (upward) position. Be sure the gasket has the proper identification mark for the engine. Do not use sealer on the gasket.

22. Inspect the cylinder head bolts shank length prior to installation. If the length exceeds 3.91 in. (99.4mm), the bolt must be replaced. Install the washer onto the bolt so the chamfer on the washer faces towards the head of the bolt.

23. Carefully install the cylinder head on the block and tighten the cylinder head bolts as follows:

a. Following the proper tightening sequence, tighten the cylinder head bolts to 58 ft. lbs. (78 Nm).

b. Loosen all bolts completely.

c. Torque bolts to 15 ft. lbs. (20 Nm).

d. Tighten bolts an additional ¼ turn.

e. Tighten bolts an additional ¼ turn.

24. Install the new exhaust pipe gasket and connect the exhaust pipe to the manifold. Tighten the self-locking bolts to 33 ft. lbs. (44 Nm).

25. Install the thermostat case and tighten the mounting bolts to 18 ft. lbs. (24 Nm).

26. Connect the coolant hoses to the thermostat case.

27. Apply sealer to the perimeter of the half-round seal and to the lower edges of the half-round portions of the belt-side of the new gasket. Install the valve cover.

28. Connect the intake manifold stay and tighten the mounting bolts to 22 ft. lbs. (30 Nm).

29. Install the timing belt and all related items.

30. Connect or install all previously disconnected hoses, cables and electrical connections.

31. Install the spark plug cable center cover.

32. Replace the O-rings and connect the fuel lines.

33. Install the air cleaner and intake hose. Connect the breather hose.

34. Replace the engine oil and oil filter.

35. Fill the system with coolant.

36. Connect the negative battery cable, run the vehicle until the thermostat opens, fill the radiator completely.

37. Check and adjust the idle speed and ignition timing.

38. Check for leaks and road test for proper operation.

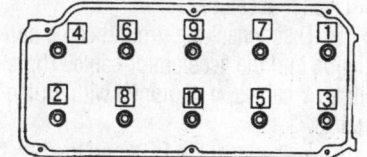

Cylinder head bolt removal sequence—2.4L (4G64) engine

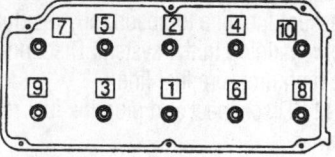

Cylinder head bolt installation sequence—2.4L (4G64) engine

Galant

1. Following proper procedure, relieve the fuel system pressure.

2. Remove the air cleaner with all air intake hoses.

3. Drain the cooling system.

4. Disconnect the accelerator cable. Remove cable mounting brackets and position the cable aside.

5. Remove the breather hose.

6. At the throttle body, disconnect the three small vacuum hoses, the coolant hoses, and the brake booster vacuum hose.

✳✳ CAUTION

Do not allow fuel spray or fuel vapors to come in contact with a spark or open flame. Keep a dry chemical fire extinguisher nearby. Never store fuel in an open container due to risk of fire or explosion.

7. Disconnect and plug the high pressure fuel line.

8. Disconnect and plug the fuel return hose.

9. Disconnect the oxygen sensor, engine coolant temperature sensor, the engine coolant temperature gauge unit and the engine coolant temperature switch on vehicles with air conditioning.

10. Disconnect the ISC motor, throttle position sensor, distributor, fuel injectors, noise filter, EGR temperature sensor, ground cable and engine control wiring harness.

11. Remove the spark plug cables.

12. At the thermostat case assembly, remove the coolant hoses and unbolt the thermostat case from the engine.

13. Remove the timing belt:

a. Remove the upper timing belt cover.

b. Align all timing marks.

c. Secure the timing belt to the camshaft sprocket.

d. Remove the camshaft sprocket.

14. Remove the valve cover and the half-round seal.

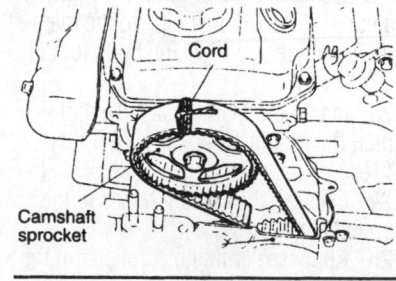

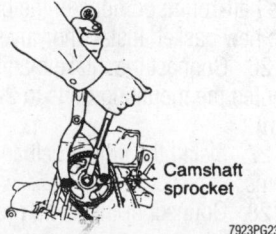

Secure the timing belt to the camshaft sprocket and remove the sprocket—Galant 2.4L (4G64) engine

15. Disconnect the intake manifold stay bracket from the intake manifold.

16. Remove the exhaust pipe self-locking nuts and separate the exhaust pipe from the exhaust manifold. Discard the gasket.

17. Loosen the cylinder head mounting bolts in 3 steps, starting from the outside and working inward. Lift off the cylinder head assembly and remove the head gasket.

To install:

18. Thoroughly clean and dry the mating surfaces of the head and block. Check the cylinder head for cracks, damage or engine coolant leakage. Remove scale, sealing compound and carbon. Clean oil passages thoroughly. Check the head for flatness.

19. Place a new head gasket on the cylinder block with the identification marks at the front top (upward) position. Be sure the gasket has the proper identification mark for the engine. Do not use sealer on the gasket. Replace the turbo gasket and ring, if equipped.

20. Inspect the cylinder head bolts shank length prior to installation. If the length exceeds 3.91 in. (99.4mm), the bolt must be replaced. Install the washer onto the bolt so the chamfer on the washer faces towards the head of the bolt.

21. Carefully install the cylinder head on the block and tighten the cylinder head bolts as follows:

    a. Following the proper tightening sequence, tighten the cylinder head bolts to 58 ft. lbs. (78 Nm).

    b. Loosen all bolts completely.

    c. Torque bolts to 15 ft. lbs. (20 Nm).

    d. Tighten bolts an additional ¼ turn.

    e. Tighten bolts an additional ¼ turn.

22. Install the new exhaust pipe gasket and connect the exhaust pipe to the manifold. Tighten the self-locking bolts to 33 ft. lbs. (44 Nm).

23. Install the thermostat case and tighten the mounting bolts to 18 ft. lbs. (24 Nm).

24. Connect the coolant hoses to the thermostat case.

25. Apply sealer to the perimeter of the half-round seal and to the lower edges of the half-round portions of the belt-side of the new gasket. Install the valve cover.

26. Connect the intake manifold stay and tighten the mounting bolts to 22 ft. lbs. (30 Nm).

27. Install the timing belt and all related items.

28. Connect or install all previously disconnected hoses, cables and electrical connections.

29. Install the spark plug cable center cover.

30. Replace the O-rings and connect the fuel lines.

31. Install the air cleaner and intake hose. Connect the breather hose.

32. Change the engine oil and oil filter.

33. Fill the system with coolant.

34. Connect the negative battery cable, run the vehicle until the thermostat opens, fill the radiator completely.

35. Check and adjust the idle speed and ignition timing.

36. Once the vehicle has cooled, recheck the coolant level.

Mirage

1.5L ENGINE

1. Relieve the fuel system pressure. Disconnect the negative battery cable.

2. Position the No. 1 piston to TDC of its compression stroke.

3. Drain the cooling system.

4. Remove the air intake hose and the air cleaner assembly.

5. Disconnect the ground cable connection and the accelerator cable. There will be 2 cables, if equipped with cruise control.

6. Disconnect the PCV and the breather hose connection.

7. Note the locations and disconnect the vacuum hoses from the intake and throttle body.

8. Disconnect the vacuum line for the brake booster.

9. Remove the upper radiator hose, throttle body hoses, bypass hose and heater hose connections.

10. Place a shop towel around the high pressure fuel line to absorb any residual fuel remaining in the system. Disconnect the high pressure fuel line.

11. Disconnect and plug the fuel return line.

12. Remove the spark plug cables.

13. Disconnect the electrical harness plugs from the crankshaft and camshaft position sensors.

14. Disconnect the electrical harness plugs from the oxygen sensor, engine coolant temperature gauge unit and the water temperature sensor.

15. Disconnect the electrical harness plugs from the idle speed control motor, throttle position sensor and air intake temperature sensor.

16. Disconnect electrical harness plugs from the ignition distributor, fuel injectors, EGR temperature sensor, power transistor and ground cable.

17. Disconnect the engine control wiring harness.

18. Remove the clamp that holds the power steering pressure hose to the engine mounting bracket.

19. Place a jack and wood block under the oil pan and carefully lift just enough to take the weight off the engine mounting bracket, remove the bracket.

20. Remove the valve cover and gasket.

21. Remove the timing belt front upper cover.

22. If not already done, rotate the crankshaft clockwise and align the timing marks.

23. While securing the sprocket, remove the sprocket bolt and remove the sprocket with the timing belt attached. Be sure to keep the sprocket and belt wired or tied together as one unit.

24. Remove the timing belt rear upper cover.

�֍ CAUTION

The crankshaft must always be rotated clockwise. Do not rotate engine by turning the camshaft.

25. Remove the exhaust pipe self-locking nuts and separate the exhaust pipe from the exhaust manifold. Discard the gasket.

26. Loosen the cylinder head mounting bolts in sequence using three steps. Lift off the cylinder head assembly and remove the head gasket.

To install:

27. Thoroughly clean and dry the mating surfaces of the head and block. Check the cylinder head for cracks, damage or engine coolant leakage. Remove scale, sealing compound and carbon. Clean oil passages thoroughly. Check the head for flatness. End to end, the head should be within 0.002 in. with 0.008 in. the maximum allowed out of true. The total thickness allowed to be removed from the head and block is 0.008 in. maximum.

28. Place a new head gasket on the cylinder block with the identification marks

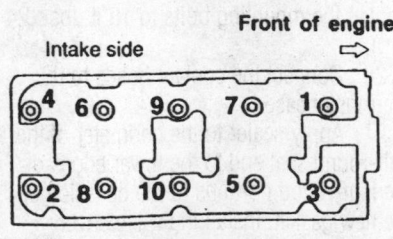

Front of engine ⇨

Intake side

Exhaust side

7923PG13

Cylinder head bolt loosening sequence—Mirage with 1.5L (4G15) engine

facing upward. Be sure the gasket has the proper identification mark for the engine. Do not use sealer on the gasket.

➡The cylinder head torque specifications are for a cold cylinder head.

29. Carefully install the cylinder head on the block. For 1995–96 models, tighten the cylinder head bolts in sequence to 53 ft. lbs. (73 Nm). For 1997–99 models, tighten the cylinder head bolts as follows:

    a. 36 ft. lbs. (49 Nm) in the correct sequence

    b. Loosen the bolts completely in the reverse order.

    c. Tighten the bolts in sequence to 14 ft. lbs. (20 Nm)

    d. Mark the front of each bolt with a dot or white paint

    e. Tighten each bolt in sequence 90°

    f. Tighten each bolt in sequence an additional 90°

    g. Be sure each dot of paint is now facing the rear of the engine.

30. Install a new exhaust pipe gasket and connect the exhaust pipe to the manifold.

31. Install the upper rear timing cover.

32. Align the timing marks and install the cam sprocket. Torque the retaining bolt to 51 ft. lbs. (70 Nm). Check the belt tension and adjust, if necessary. Install the outer timing cover.

33. Install a new valve cover gasket. Install the valve cover and torque the retaining bolts to 16 inch lbs. (1.8 Nm).

34. Install the engine mount bracket and remove the support jack.

35. Install the clamp that holds the power steering pressure hose to the engine mounting bracket.

36. Connect or install all previously disconnected hoses, cables and electrical connections. Adjust the throttle cable(s).

37. Replace the O-rings and connect the fuel lines.

38. Install the air cleaner assembly. Connect the breather hose.

39. Change the engine oil and oil filter.

40. Fill the system with coolant.

41. Connect the negative battery cable, run the vehicle until the thermostat opens, fill the radiator completely.

42. Check and adjust the idle speed and ignition timing.

43. Once the vehicle has cooled, recheck the coolant level.

1.8L ENGINE

1. Relieve fuel system pressure. Disconnect the negative battery cable.

2. Position the No. 1 piston to TDC of its compression stroke.

3. Remove the air cleaner assembly.

4. Drain the cooling system.

5. Disconnect the brake booster vacuum hose and PVC valve connection.

6. Note the locations and disconnect the vacuum hoses from the intake and throttle body.

7. Remove the upper radiator hose, overflow tube and the water hose from the thermostat to the throttle body.

8. Wrap the connection with a shop towel and disconnect the high pressure fuel line at the fuel rail. Discard the O-ring.

9. Disconnect the fuel return hose from the fuel pressure regulator.

10. Disconnect the accelerator cable connection from the throttle body.

11. Disconnect the electrical harnesses at the oil pressure switch, oxygen sensor, water temperature sensor connectors and distributor.

12. Disconnect the electrical harnesses at the idle air control motor and the EGR temperature sensor.

13. Disconnect the wiring from condenser, idle speed control, throttle position sensor and knock sensor.

14. Note harness plug connections for reassembly and disconnect fuel injectors.

15. Disconnect the spark plug cables from each spark plug.

16. Unbolt the control harness assembly and position aside.

17. Remove the thermostat housing, thermostat and the thermostat case with O-ring from the engine.

18. Remove the rocker cover.

19. Remove the timing belt upper cover.

20. If not already done, rotate the crankshaft in the clockwise direction to align the camshaft timing marks. Matchmark the camshaft sprocket and the timing belt. Tie the camshaft sprocket and the timing belt together so the sprocket will not move with respect to the timing belt.

21. While holding the camshaft sprocket in position using the appropriate wrench, remove the camshaft sprocket and with the belt attached. Wire the sprocket and belt aside making sure constant tension is maintained on the belt. Do not allow the belt to slacken or engine timing may be altered.

➡When removing the camshaft sprocket, do not allow the crankshaft to rotate. Confirm proper engine timing during installation.

22. Loosen the cylinder head bolts in two or three steps in the proper sequence.

23. Remove the cylinder head from the engine.

✲✲ CAUTION

When removing the cylinder head, take care not to bend or damage the plug guide. The plug guide can not be replaced.

24. Remove the cylinder head gasket from the block.

To install:

25. Thoroughly clean and dry the mating surfaces of the head and block. Check the cylinder head for cracks, damage or engine coolant leakage. Remove scale, sealing compound and carbon. Clean oil passages thoroughly. Check the head for flatness. end to end, the head should be within 0.002 in. with 0.008 in. the maximum allowed out of true. The total thickness allowed to be removed from the head and block is 0.008 in. maximum.

26. Place a new head gasket on the cylinder block with the identification marks facing upward. Be sure the gasket has the proper identification mark for the engine. Do not use sealer on the gasket.

27. Carefully install the cylinder head on the block.

28. Inspect the cylinder head bolt prior to installation. The length below the head of the bolts should not exceed 3.795 in. (96.4mm) If bolt shank length exceeds

---

Intake side ⟸ Front of engine

```
8    6    1    3    9

10   4    2    5    7
```

Exhaust side

7923PG14

Cylinder head bolt tightening sequence— Mirage with 1.5L (4G15) engine

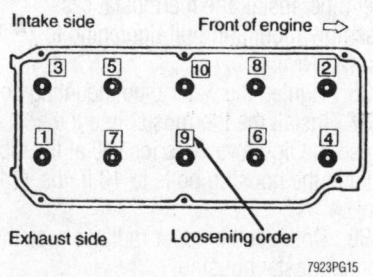

Cylinder head bolt loosening sequence— 1.8L engine

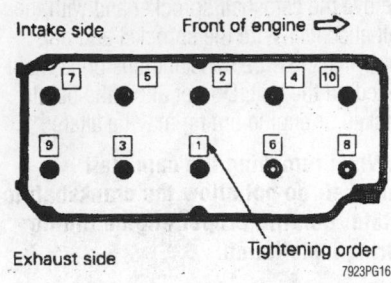

Intake side   Front of engine ⇨

Exhaust side   Tightening order

7923PG16

Cylinder head bolt torque sequence—1.8L engine

limit, bolt must be replaced. New bolts are always recommended.

29. Apply a small amount of engine oil to the thread section of the bolt and install so the chamfer of the washer faces upward.

30. Tighten the cylinder head bolts in the proper order as follows:

   a. In the proper tightening sequence, torque bolts to 54 ft. lbs. (75 Nm).

   b. In the reverse order of the tightening sequence, fully loosen bolts.

   c. In the proper tightening sequence, torque bolts to 14 ft. lbs. (20 Nm).

   d. In the proper tightening sequence, tighten bolts an additional ¼ turn (90 degrees).

   e. In the proper tightening sequence, tighten bolts an additional ¼ turn (90 degrees).

31. Install the camshaft sprocket and tighten bolt to 65 ft. lbs. (90 Nm), while holding the sprocket in place using the appropriate wrench. Confirm proper timing mark alignment.

32. Install the upper timing belt cover and rocker cover. Torque the rocker cover bolts to 29 inch lbs. (3.3 Nm).

33. Loosen the water pipe mounting bolt for ease of thermostat housing installation.

34. Apply a thin bead of sealant MD970389 or equivalent, to the water tube connection on the thermostat case.

35. Apply a small amount of water to the O-ring of the water inlet pipe and press the thermostat case assembly onto the water inlet pipe. Install the thermostat case assembly mounting bolt tightening to 16 ft. lbs. (22 Nm).

36. Tighten the water pipe mounting bolt.

37. Install the thermostat into the housing so the jiggle valve is located at the top. Tighten the housing bolts to 10 ft. lbs. (14 Nm).

38. Connect the upper radiator hose to the thermostat housing.

39. Connect or install all previously disconnected hoses, cables and electrical connections. Adjust the throttle cable(s).

40. Replace the O-ring for the high pressure hose and install a new clamp on the return hose and reconnect the fuel lines.

41. Install the air intake hose. Connect the breather hose and air cleaner case cover.

42. Reconnect the brake booster and the PCV vacuum hoses.

43. Change the engine oil and oil filter.

44. Fill the system with coolant.

45. Connect the negative battery cable, run the vehicle until the thermostat opens, fill the radiator completely.

46. Check and adjust the idle speed and ignition timing.

47. Check all systems for leaks. Allow the engine to cool and recheck the coolant level.

Rocker Arms/Shafts

REMOVAL & INSTALLATION

3000GT and Diamante

3.0L SOHC ENGINE

On this engine, the hydraulic lash adjusters are built into the rocker arms. If service is required, simply remove the lash adjuster from the bore in the rocker arm. It is recommended that all of the rocker arms and lash adjusters are replaced at the same time.

1. Disconnect the negative battery cable.

2. Remove the valve cover. Install lash adjuster retainer tools MD998443 or equivalent, to prevent the auto-lash adjuster from falling out of the rocker arm.

3. Rotate the engine clockwise and position number 1 cylinder at TDC compression stroke.

4. If necessary, remove the distributor adapter housing.

5. Remove the timing belt assembly.

6. Loosen rocker arm and shaft assembly evenly in several steps. Remove the rocker arm and shaft assembly as a complete unit.

7. Remove the rear camshaft bearing cap and slide rocker arms, springs and washers from shaft. Note location and positioning of all rocker shaft components.

8. Visually inspect the rocker arm roller and replace if damage or seizure is evident. Check the roller for smooth rotation. Replace if excess play or binding is present. Also, inspect valve contact surface for possible damage or seizure. It is recommended that all rocker arms and lash adjusters be replaced together.

To install:

9. Immerse the lash adjusters in clean diesel fuel. Using a small wire, move the plunger of the lash adjuster up and down 4 or 5 times while pushing down lightly on the check ball in order to bleed out the air. Install the lash adjusters in the rocker arms.

10. Using a light coat of engine oil, assembly the rocker arms to the shaft. Install the rear camshaft bearing cap.

11. Lubricate the camshaft and rocker shaft with heavy engine oil and position on the cylinder head.

12. Apply a drop of sealant to the rear edges of the end caps.

13. Install the assembly making sure the notches in the rocker shafts are facing up. Insert the bolts but do not tighten at this point.

14. Install the remaining cap bolts and tighten evenly and gradually to 14 ft. lbs. (20 Nm). Remove the lash adjuster retainers.

15. Install the distributor extension, if removed.

16. Install the valve cover with a new gasket and tighten to 84 inch lbs. (9 Nm).

17. Connect the negative battery cable.

13. Bearing cap
14. Rocker arm
15. Spring
16. Rocker arm
17. Spring
18. Bearing cap no. 3
19. Rocker arm
20. Spring
21. Rocker arm
22. Spring
23. Bearing cap no. 2
24. Rocker arm
25. Spring
26. Rocker arm
27. Spring
28. Rocker arm shaft
29. Rocker arm shaft
30. Bearing cap no. 1

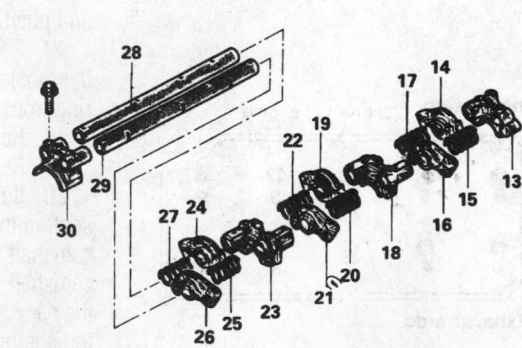

Rocker arm assembly—Diamante 3.0L SOHC engine

7923PG34

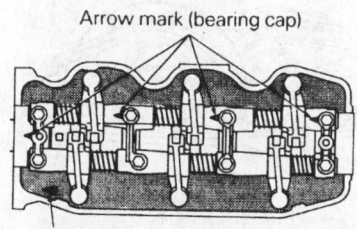

When installing the rocker arm/shaft assemblies, ensure that the arrow marks point in the same direction as the arrow stamped into the cylinder head—Diamante 3.0L SOHC engine

3.0L DOHC ENGINE

1. Relieve the fuel system pressure.

✷✷ CAUTION

The fuel injection system remains under pressure after the engine has been turned OFF. Properly relieve fuel pressure before disconnecting any fuel lines. Failure to do so may result in fire or personal injury.

2. Disconnect battery negative cable.
3. Remove the timing belt cover and timing belt.
4. Remove the center cover, breather and PCV hoses, and spark plug cables.
5. Remove the rocker cover, semi-circular packing, throttle body stay, both camshaft sprockets, and oil seals.
6. Remove the crank angle sensor and adapter from the rear of the camshaft.
7. Remove the intake and exhaust camshafts.
8. Remove rocker arms and lash adjusters from the head. It is recommended that all lash adjusters and rockers be replaced as a complete set.

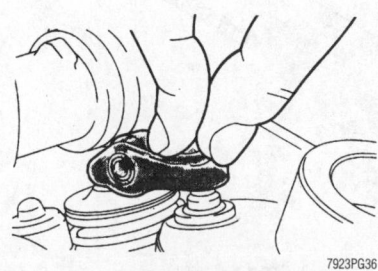

The rocker arm sits beneath the camshaft and is supported on one end by the valve stem and on the other end by the Hydraulic Lash Adjuster (HLA)—Diamante 3.0L DOHC engine

To install:

9. Immerse the lash adjusters in clean diesel fuel. Using a small wire, move the plunger of the lash adjuster up and down four or five times while pushing down lightly on the check ball in order to bleed out the air. Lubricate and install the lash adjusters in the cylinder head.
10. Lubricate the camshafts with heavy engine oil and position the camshafts on the cylinder head.
11. Be sure the dowel pin on both camshaft sprocket ends are positioned properly.
12. Install the bearing caps. Tighten the caps in sequence and in 2 or 3 steps. Caps 2, 3 and 4 have a front mark. Install with the mark aligned with the front mark on the cylinder head. Intake caps have **I** stamped on the cap and exhaust caps have **E**. Also, be sure the rocker arm is correctly mounted on the lash adjuster and the valve stem end. Torque the front and rear retaining cap bolts to 14 ft. lbs. (20 Nm) and tighten the center 3 retaining cap bolts to 8 ft. lbs. (11 Nm).

➡**If installing the camshaft to a cylinder head that is positioned on a workbench, the valves will protrude.**

13. Apply a coating of engine oil to the oil seals and install.

14. Install the timing belt, valve cover and all related parts.
15. Connect the negative battery cable and check for leaks.

3.5L ENGINE

1. Disconnect the negative battery cable.
2. Remove the timing belt.
3. Remove the rocker arm cover.
4. Install the lash adjuster clips on the rocker arms, then loosen the bearing cap bolts. Do not remove the bolts from the bearing caps.
5. Remove the rocker arms, shafts and bearing caps as an assembly.

To install:

6. Install the bearing caps/rocker arm assemblies. Tighten the bolts to 23 ft. lbs. (31 Nm).
7. Install the rocker arm cover using a new gasket.
8. Install the timing belt and remaining components.
9. Connect the negative battery cable.

Eclipse and Galant

1995 2.4L SOHC ENGINE

On this engine, the hydraulic lifters are built into the rocker arms. If lifter service is

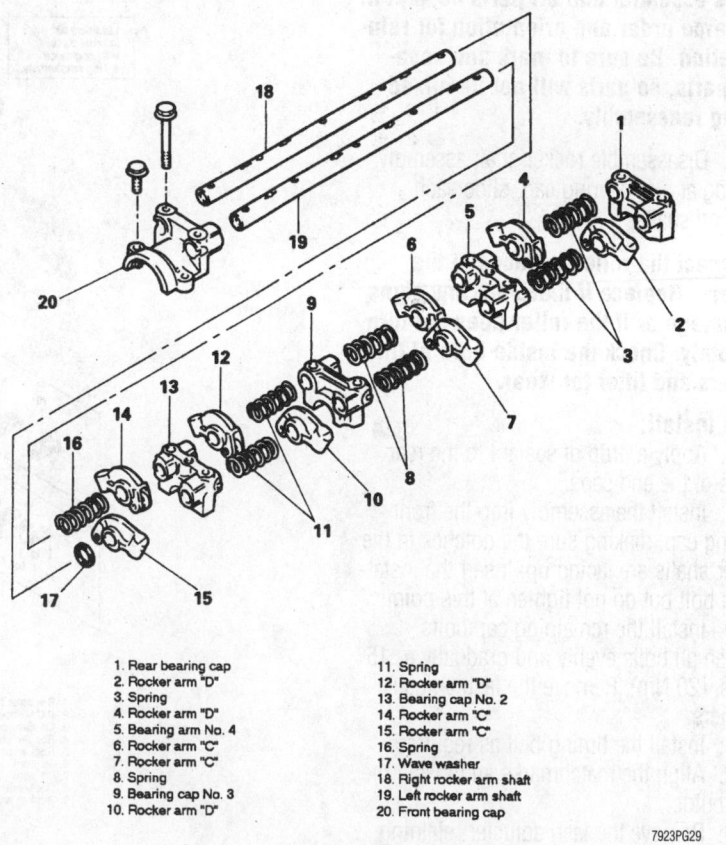

1. Rear bearing cap	11. Spring
2. Rocker arm "D"	12. Rocker arm "D"
3. Spring	13. Bearing cap No. 2
4. Rocker arm "D"	14. Rocker arm "C"
5. Bearing arm No. 4	15. Rocker arm "C"
6. Rocker arm "C"	16. Spring
7. Rocker arm "C"	17. Wave washer
8. Spring	18. Right rocker arm shaft
9. Bearing cap No. 3	19. Left rocker arm shaft
10. Rocker arm "D"	20. Front bearing cap

Exploded view of the rocker shaft assembly (8 valve engine)—2.4L (4G64) engines

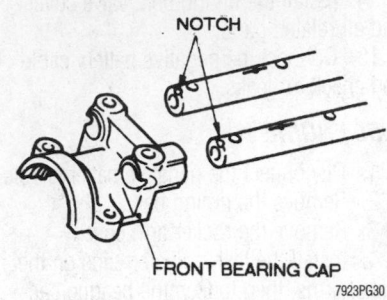

NOTCH

FRONT BEARING CAP

7923PG30

Aligning the front bearing cap to the rocker arm shaft (8 valve engine)—2.4L (4G64) engines

required, simply remove the lifter from the bore in the rocker arm. It is recommended that all of the rocker arms and lash adjusters be replaced at the same time.

1. Disconnect the negative battery cable.

2. Remove the valve cover.

3. Matchmark the distributor to the cylinder head and remove the distributor.

4. Remove camshaft timing belt.

5. Working in a crisscross pattern from the center outward, loosen the camshaft bearing caps in gradual steps.

6. Remove the rocker arms, shafts and bearing caps as an assembly.

➡ **It is essential that all parts be kept in the same order and orientation for reinstallation. Be sure to mark and separate parts, so parts will not be mixed during reassembly.**

7. Disassemble rocker shaft assembly. Starting at rear bearing cap, slide each piece off shafts.

➡ **Inspect the roller surfaces of the rockers. Replace if there are any signs of damage or if the roller does not turn smoothly. Check the inside bore of the rockers and lifter for wear.**

To install:

8. Apply a drop of sealant to the rear edges of the end caps.

9. Install the assembly into the front bearing cap, making sure the notches in the rocker shafts are facing up. Insert the installation bolt but do not tighten at this point.

10. Install the remaining cap bolts. Tighten all bolts evenly and gradually to 15 ft. lbs. (20 Nm). Remove the lash adjuster retainers.

11. Install the timing belt as required.

12. Align the matchmarks and install the distributor.

13. Remove the lash adjuster retaining tools.

14. Install the valve cover, with a new gasket and semi-circular packing in place.

15. Connect the negative battery cable.

16. Run the engine and check ignition timing.

1995 2.4L DOHC ENGINE

1. Disconnect the negative battery cable.

2. Remove the valve cover and discard the gasket.

3. Install lash adjuster retainer tools MD998443 or equivalent, to the rocker arm.

4. Remove the rocker shaft hold-down bolts gradually and evenly and remove the rocker shaft/arm assemblies.

5. If disassembly is required, keep all parts in the exact order of removal. Inspect the roller surfaces of the rockers. Replace if there are any signs of damage or if the roller does not turn smoothly. Check the inside bore of the rockers and the adjuster tip for wear.

To install:

6. Lubricate the rocker shaft with clean engine oil and install the rockers and springs in their proper places.

7. Install the rocker shaft assemblies on the engine. Tighten the bolts gradually and evenly to 21–25 ft. lbs. (29–35 Nm).

7923PG31

Rocker arm shaft installed position (16 valve engine)—2.4L (4G64) engines

➡ **When installing the rocker arm shaft, make certain the notch is properly located.**

8. Remove the lash adjuster retaining tools.

9. Install the valve cover with a new gasket.

10. Connect the negative battery cable.

1996–99 2.4L ENGINE

1. Disconnect the negative battery cable. Wait at least 90 seconds before performing any work.

2. Remove the battery.

3. Disconnect the accelerator cable, remove the cable clamp mounting screws and position the accelerator cable out of the way.

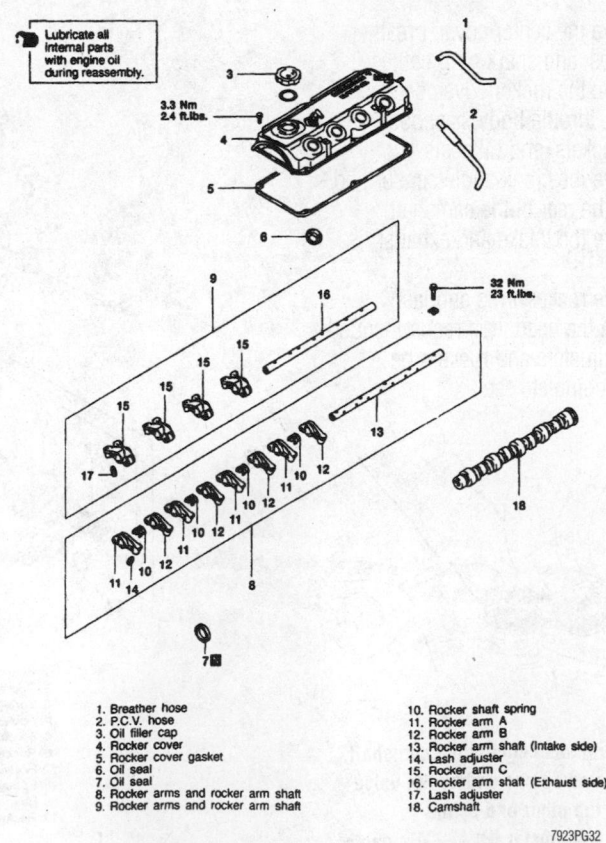

Lubricate all internal parts with engine oil during reassembly.

3.3 Nm
2.4 ft.lbs.

32 Nm
23 ft.lbs.

1. Breather hose
2. P.C.V. hose
3. Oil filler cap
4. Rocker cover
5. Rocker cover gasket
6. Oil seal
7. Oil seal
8. Rocker arms and rocker arm shaft
9. Rocker arms and rocker arm shaft
10. Rocker shaft spring
11. Rocker arm A
12. Rocker arm B
13. Rocker arm shaft (Intake side)
14. Lash adjuster
15. Rocker arm C
16. Rocker arm shaft (Exhaust side)
17. Lash adjuster
18. Camshaft

7923PG32

Rocker arm shafts and components—Eclipse and Galant 2.4L (4G64) engines

4. Remove the air intake hose.

5. Disconnect the breather hose and the PCV hose.

6. Disconnect the spark plug cables from the spark plugs.

7. Remove the rocker cover and gasket.

8. Install lash adjuster retainer tools MD998443 or equivalent, to the rocker arm.

9. Remove the rocker shaft hold-down bolts gradually and evenly and remove the rocker shaft/arm assemblies.

10. Disassemble the rockers and the rocker shaft springs from the rocker shafts. Remove the lash adjuster holding tool. Inspect the roller surfaces of the rockers. Replace if there are any signs of damage or if the roller does not turn smoothly. Check the inside bore of the rockers and the adjuster tip for wear. Check the rocker shafts for scoring. Inspect the lash adjusters for wear and smooth operation.

To install:

11. Immerse the lash adjusters in clean diesel fuel, and using a small wire, move the plunger up and down four or five times. while pushing down lightly on the check ball in order to bleed the air from the adjuster.

12. Install the lash adjusters to the rocker arms and attach the special holding tool. Be careful not to spill the diesel fuel from the adjuster.

13. Lubricate the rocker shaft with clean engine oil and install the rocker arms.

14. Temporarily tighten the rocker shaft assembly with the mounting bolts so that all rocker arms on the inlet valve side do not push on the valves.

15. Fit the rocker shaft springs from above and position them so that they are at right angles to the plug side. Install the rocker springs before installing the exhaust side rocker shaft and rocker arm assembly.

16. Install the exhaust side rocker shaft assembly in the engine. Tighten all the rocker shaft mounting bolts gradually and evenly to 23 ft. lbs. (32 Nm).

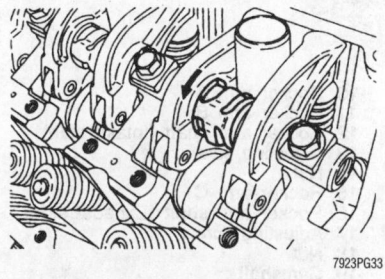

Installing the rocker shaft springs— Eclipse and Galant 2.4L (4G64) engines

➡**When installing the rocker arm shaft, make certain the notch is properly located.**

17. Remove the lash adjuster retaining tools.

18. Install the rocker cover with a new gasket and tighten the mounting bolts to 2.4 ft. lbs. (3.3 Nm).

19. Reinstall the spark plug cables to the spark plugs.

20. Reconnect the PCV and breather hoses.

21. Install the air intake hose.

22. Reattach the accelerator cable brackets with the screws and reconnect the accelerator cable.

23. Install the battery, reconnect the battery cables, start the engine and check for proper operation.

2.0L (4G63) ENGINE

1. Relieve the fuel system pressure following proper procedure. Disconnect negative battery cable.

2. Disconnect the accelerator cable, PCV hoses, breather hoses, spark plug cables and the remove the valve cover.

➡**Always rotate the crankshaft in a clockwise direction. Make a mark on the back of the timing belt indicating the direction of rotation so it may be reassembled in the same direction if it is to be reused.**

3. Rotate the crankshaft clockwise and align the timing marks so No. 1 piston will be at TDC of the compression stroke. At this time the timing marks on the camshaft sprocket and the upper surface of the cylinder head should coincide, and the dowel pin of the camshaft sprocket should be at the upper side.

4. Remove the timing belt upper and lower covers.

5. Remove the timing belt.

6. Remove the crank angle sensor.

➡**It is essential that all parts be kept in the same order and orientation for reinstallation. Mark and separate all parts. Do not mix parts. Valvetrain components that are to be reused must be installed in the same locations from which they were removed.**

7. Remove the camshafts, rocker arms and lash adjusters.

8. Visually inspect the rocker arm roller and replace if dent, damage or seizure is evident. Check the roller for smooth rotation. Replace if excess play or binding is present. Also, inspect valve contact surface

for possible damage or seizure. It is recommended that all rocker arms and lash adjusters be replaced together.

To install:

9. Install the lash adjusters and rocker arms into the cylinder head. Lubricate lightly with clean oil prior to installation.

10. Apply engine oil to the lobes and journals of each camshaft. Install the camshafts into the cylinder head taking care not to confuse the intake and the exhaust camshaft; the intake camshaft has a slit on its rear end for driving the crank angle sensor. Align shafts so dowel pins on camshaft sprocket end are located on the top.

11. Install the camshaft bearing caps and tighten the bolts in the proper sequence to specifications in three even steps.

12. Replace the camshaft oil seals and install the sprockets.

13. Locate the dowel pin on the sprocket end of the intake camshaft at the top position, if not already done.

14. Align the punch mark on the crank angle sensor housing with the notch on the sensor plate. Install the crank angle sensor into the cylinder head.

15. Install the timing belt, covers and related components.

16. Install the valve cover using new gasket. Reconnect all related components.

17. Reconnect the negative battery cable.

2.0L (420A) ENGINE

1. Relieve the fuel system pressure following proper procedure. Disconnect negative battery cable.

2. Disconnect the accelerator cable, PCV hoses, breather hoses, spark plug cables and the remove the cylinder head cover.

➡**Always rotate the crankshaft in a clockwise direction. Make a mark on the back of the timing belt indicating the direction of rotation so it may be reassembled in the same direction if it is to be reused.**

3. Rotate the crankshaft clockwise and align the timing marks so No. 1 piston will be at TDC of the compression stroke. At this time the timing marks on the camshaft sprocket and the upper surface of the cylinder head should coincide, and the dowel pin of the camshaft sprocket should be at the upper side.

4. Remove the timing belt upper and lower covers.

5. Remove the timing belt.

6. Remove the crank angle sensor.

➡It is essential that all parts be kept in the same order and orientation for reinstallation. Mark and separate all parts. Do not mix parts. Valvetrain components that are to be reused must be installed in the same locations from which they were removed.

7. Remove the camshafts, rocker arms (cam followers) and lash adjusters.

8. Visually inspect the rocker arm roller and replace if dent, damage or seizure is evident. Check the roller for smooth rotation. Replace if excess play or binding is present. Also, inspect valve contact surface for possible damage or seizure. It is recommended that all rocker arms and lash adjusters be replaced together.

To install:

9. Install the lash adjusters and rocker arms into the cylinder head. Lubricate lightly with clean oil prior to installation.

10. Apply engine oil to the lobes and journals of each camshaft. Install the camshafts into the cylinder head taking care not to confuse the intake and the exhaust camshaft; the intake camshaft has a slit on its rear end for driving the crank angle sensor. Align shafts so dowel pins on camshaft sprocket end are located on the top.

❊❊ WARNING

Piston should NOT be at TDC when installing the camshaft.

11. Install and tighten the four center camshaft bearing caps in the proper sequence to specifications in three even steps. Torque the bolts to 9 ft. lbs. (12 Nm).

12. Apply Loctite 51817® or equivalent to the front and rear bearing caps. Install the bearing caps and tighten the bolts to 20 ft. lbs. 28 Nm.

13. Replace the camshaft oil seals and install the sprockets.

14. Locate the dowel pin on the sprocket end of the intake camshaft at the top position, if not already done.

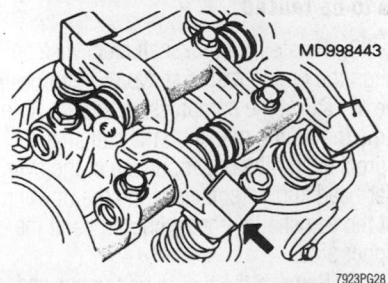

Install the auto lash adjuster holder to prevent them from falling out—2.0L (4G63) and 2.4L (4G64) engines

15. Align the punch mark on the crank angle sensor housing with the notch on the sensor plate. Install the crank angle sensor into the cylinder head.

16. Install the timing belt, covers and related components.

17. Install the valve cover using new gasket. Reconnect all related components.

18. Reconnect the negative battery cable.

Mirage

1. Disconnect the negative battery cable.

2. Rotate the engine and position the No. 1 piston to TDC of its compression stroke.

3. For 1.8L engines, label and disconnect the spark plug cables.

4. For 1.8L engines, disconnect the air flow sensor connector and remove the air cleaner case cover.

5. Disconnect the accelerator cable, breather hose and PCV hose connections.

6. Remove the rocker cover and discard the gasket.

7. Loosen both rocker arm shaft assemblies gradually and evenly and remove the rocket shafts from the vehicle. Do not disassembly rocker arms and rocker arm shaft assemblies.

8. If disassembly is required, keep all parts in the exact order of removal. Inspect

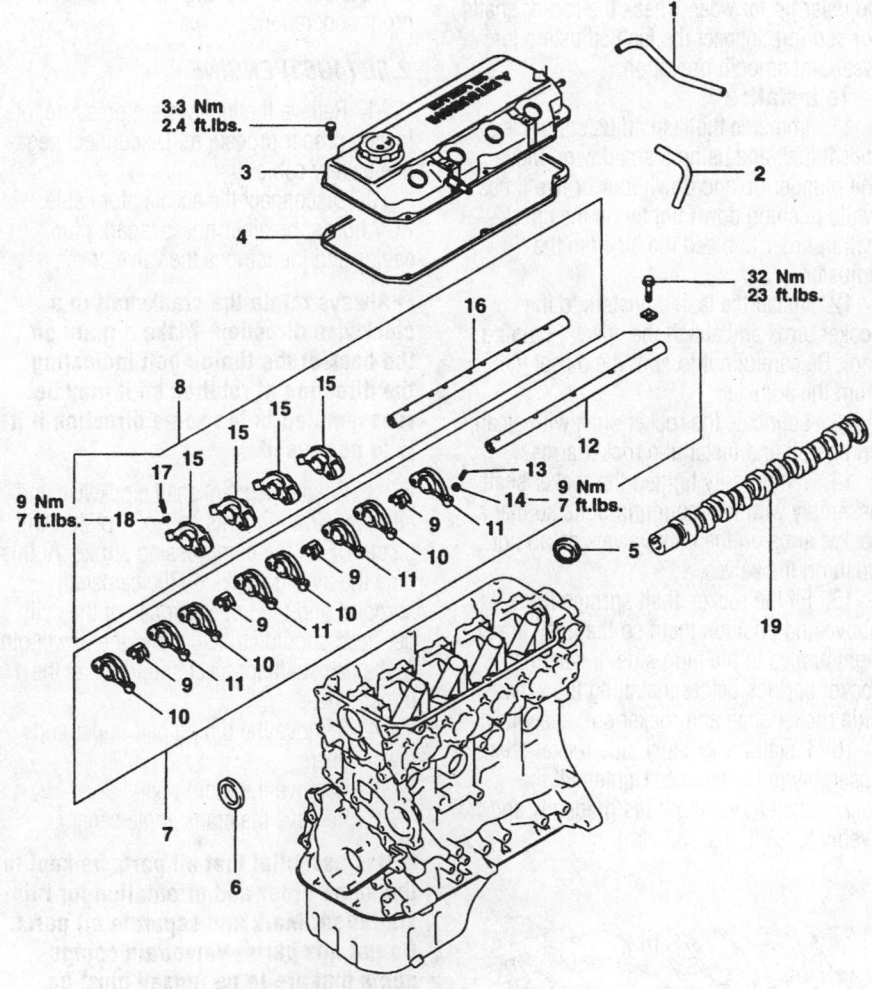

1. Breather hose
2. P.C.V. hose
3. Rocker cover
4. Rocker cover gasket
   Valve clearance pre-adjustment
5. Oil seal
6. Oil seal
7. Rocker arms and rocker arm shaft
8. Rocker arms and rocker arm shaft
9. Rocker shaft spring
10. Rocker arm A
11. Rocker arm B
12. Rocker arm shaft (Intake side)
13. Adjusting screw
14. Nut
15. Rocker arm C
16. Rocker arm shaft (Exhaust side)
17. Adjusting screw
18. Nut
19. Camshaft

Camshaft, rocker arm and shaft assemblies—Mirage 1.8L (4G93) engine

the roller surfaces of the rockers. Replace if there are any signs of damage or if the roller does not turn smoothly. Check the inside bore of the rockers and the adjuster tip for wear.

To install:

9. Lubricate the rocker shaft with clean engine oil and install the rockers and springs in their proper places.

10. Install the rocker arm and shaft assemblies. Tighten the rocker arm shaft retainer bolts to 23 ft. lbs. (32 Nm).

11. Check valve adjustment and install valve cover with a new gasket. Tighten the valve cover bolts to 16 inch lbs. (1.8 Nm) for the 1.5L engine or to 29 inch lbs. (3.3 Nm) for the 1.8L engine.

12. If detached, connect the spark plug cables.

13. Connect the accelerator cable, breather hose and PCV hose.

14. For the 1.8L engines, connect the air flow sensor connector and install the air cleaner case cover.

15. Connect the negative battery cable. Run the engine at idle until normal operating temperature is reached. Check idle speed and ignition timing and adjust as required.

Turbocharger

REMOVAL & INSTALLATION

3000GT

⁂ CAUTION

Work must be started after 90 seconds from the time the ignition switch is turned to the LOCK position and the negative battery cable is disconnected.

RIGHT SIDE (FRONT)

1. Disconnect the negative battery cable.
2. Remove the radiator.
3. Remove the right side transaxle bracket.
4. Remove the front exhaust pipe.
5. Carefully matchmark, diagram or photograph all air intake hoses and pipes along the front of the engine. It is imperative that all of these pieces are installed in the exact same positions when assembling. Remove the hoses and pipes and keep covered in a clean area.
6. Remove the alternator.
7. Remove the oil dipstick tube.
8. Remove the turbocharger heat protector.

9. Remove the water feed pipes.
10. Remove the oxygen sensor.
11. Remove the oil return line.
12. Remove the exhaust extension fitting and bracket.
13. Remove all air conditioning components preventing removal of the turbocharger.
14. Remove the oil feed tube.
15. Remove the turbocharger to exhaust manifold bolts and remove the turbocharger assembly.

To install:

16. Visually check the turbine wheel (hot side) and compressor wheel (cold side) for cracking or other damage. Check whether the turbine wheel and the compressor wheel can be easily turned by hand. Check for oil leakage. Check whether or not the wastage valve remains open. If any problem is found, replace the part.

17. Clean all mating surfaces. Pour clean engine oil through the oil pipe feed hole in the turbocharger.

18. Install a new gasket and ring a install the turbocharger to the manifold. Torque the bolts to 40–47 ft. lbs. (55–65 Nm).

19. Replace the eye-bolt rings and install the oil feed pipe.

20. Install the removed air conditioning components.

21. Install the exhaust extension fitting and bracket with a new gasket. Torque the nuts to 40–47 ft. lbs. (55–65 Nm).

22. Install the oil return line with new gaskets.

23. Install the oxygen sensor.

24. Replace the eye-bolt rings and install the water feed pipes.

25. Install the turbocharger heat protector.

26. Install the dipstick tube.

27. Install the alternator.

28. Install all air intake hoses and pipes along the front of the engine. Be sure all are in their proper positions.

29. Install a new gasket and connect the front exhaust pipe.

30. Install the right side transaxle bracket.

31. Install the radiator.

32. Fill the system with coolant.

33. Connect the negative battery cable and check for exhaust leaks.

LEFT SIDE (REAR)

1. Remove the battery.
2. Drain the coolant.
3. Remove the front exhaust pipe.
4. Disconnect the accelerator cable from the throttle body.
5. Remove the intake air hose, the air pipe across the top of the engine and its heat shield.

6. Remove the clutch booster vacuum hose and disconnect the accelerator cable from the pedal.
7. Remove the air intake hoses coming from the air cleaner box.
8. Remove the oxygen sensor and the turbocharger heat protector.
9. Remove the EGR pipe, if equipped.
10. Remove the oil feed pipe.
11. Remove the EGR valve, if equipped.
12. Remove the water feed pipes.
13. Remove the exhaust extension fitting and bracket.
14. Remove the inner heat protector.
15. Remove the oil return tube.
16. Remove the turbocharger to exhaust manifold nuts and remove the turbocharger assembly.

To install:

17. Visually check the turbine wheel (hot side) and compressor wheel (cold side) for cracking or other damage. Check whether the turbine wheel and the compressor wheel can be easily turned by hand. Check for oil leakage. Check whether or not the wastage valve remains open. If any problem is found, replace the part.

18. Clean all mating surfaces. Pour clean engine oil through the oil pipe feed hole in the turbocharger.

19. Install a new gasket and ring a install the turbocharger to the manifold. Torque the nuts to 40–47 ft. lbs. (55–65 Nm).

20. Install the oil return line with new gaskets.

21. Install the inner heat protector.

22. Install the exhaust extension fitting and bracket with a new gasket. Torque the nuts to 40–47 ft. lbs. (55–65 Nm).

23. Replace the eye-bolt rings and install the water feed pipes.

24. Install the EGR valve, if equipped.

25. Replace the eye-bolt rings and install the oil feed pipe.

26. Install the EGR pipe if equipped.

27. Install the turbocharger heat protector and oxygen sensor.

28. Install the air intake hoses coming from the air cleaner box. Be sure the triangular aligning marks are engaged.

29. Connect the accelerator cable to from the pedal and install the clutch booster vacuum hose.

30. Install the heat shield, the air pipe across the top of the engine and the air intake hose.

31. Connect the accelerator cable to the throttle body.

32. Install a new gasket and connect the front exhaust pipe.

33. Fill the system with coolant.

34. Install the battery.

35. Connect the negative battery cable and check for exhaust leaks.

Eclipse

> **✱✱✱ CAUTION**
>
> **The air bag system (SRS or SIR) must be disarmed before removing the turbocharger. Failure to do so may cause accidental deployment, property damage or personal injury.**

1. Disconnect the negative battery cable.
2. Drain the engine coolant.
3. Remove the condenser fan motor assembly if equipped with air conditioning.
4. Remove the heated oxygen sensor.
5. Remove the dipstick and tube assembly.
6. Remove the air cleaner and air intake hose assembly.
7. Disconnect the air intake hose from the turbocharger.
8. Disconnect the engine coolant hoses from the turbocharger.
9. Disconnect the oil supply pipe connection. Do not let dirt of foreign particles enter the oil pipe.
10. Remove the heat shields.
11. Remove the engine hanger.
12. Disconnect the front exhaust pipe from the turbocharger.
13. Remove the oil return pipe and gaskets.
14. Remove the flange bolts and nut that attach the turbo to the exhaust manifold. Take note of the positions of the coned disc springs and the washers.
15. Remove the turbocharger, gasket and ring.

To install:

16. Use a new gasket and install the turbo to the exhaust manifold. Be sure the coned disc spring and the washers are installed in their original positions. Torque the bolts and nut to 20–23 ft. lbs. (27–31 Nm). Further tighten the bolts and nuts 60° –70°.

17. Use a new gasket and connect the exhaust pipe to the turbo. Torque the mounting bolts to 40–47 ft. lbs. (54–64 Nm).
18. Using new gaskets, install the oil return pipe.
19. Install the engine hanger.
20. Install the heat shields.
21. Connect the oil supply pipe. Torque the flare nut fittings to 14 ft. lbs. (19 Nm).
22. Connect the engine coolant hoses to the turbo.
23. Connect the air hose.
24. Install the air cleaner and duct assembly.
25. Position a new O-ring on the dipstick tube and install the tube and dipstick.
26. Install the heated oxygen sensor.
27. Install the condenser fan assembly if removed.
28. Change the engine oil and filter.
29. Connect the negative battery cable and refill the engine with coolant.
30. Start the engine and let it idle. Do not race the engine until the oil reaches the turbo. Check for leaks.

Intake Manifold

REMOVAL & INSTALLATION

> **✱✱✱ CAUTION**
>
> **The fuel injection system remains under pressure after the engine has been turned OFF. Properly relieve fuel pressure before disconnecting any fuel lines. Failure to do so may result in fire or personal injury.**

3000GT

1. Relieve the fuel system pressure.
2. Disconnect battery negative cable and drain the cooling system.

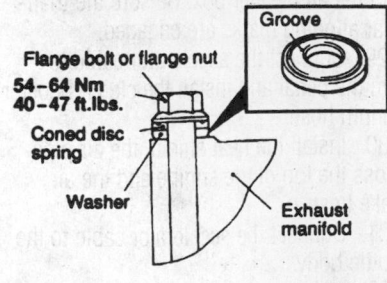

Flange bolt or flange nut
**54–64 Nm
40–47 ft.lbs.**
Coned disc spring
Washer
Exhaust manifold
Groove

7923PG37

Install the groove of the cone-shaped disc spring toward the flange bolt or nut— Eclipse

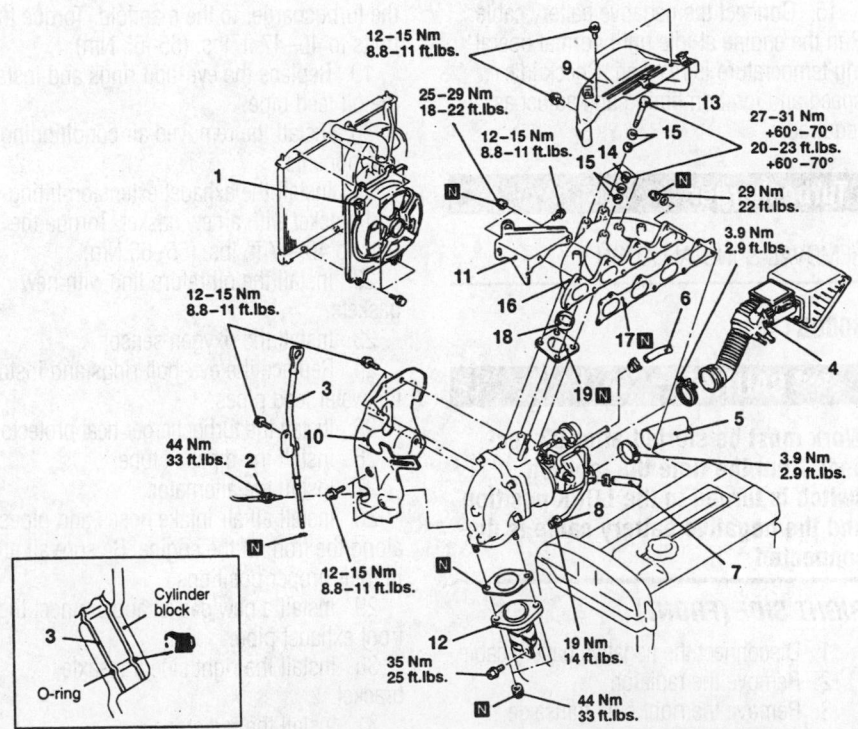

12–15 Nm
8.8–11 ft.lbs.

25–29 Nm
18–22 ft.lbs.

12–15 Nm
8.8–11 ft.lbs.

27–31 Nm
+60°–70°
20–23 ft.lbs.
+60°–70°

29 Nm
22 ft.lbs.

3.9 Nm
2.9 ft.lbs.

12–15 Nm
8.8–11 ft.lbs.

44 Nm
33 ft.lbs.

3.9 Nm
2.9 ft.lbs.

12–15 Nm
8.8–11 ft.lbs.

19 Nm
14 ft.lbs.

35 Nm
25 ft.lbs.

44 Nm
33 ft.lbs.

Cylinder block
O-ring

Removal steps
1. Condenser fan motor assembly <Vehicles with air conditioning>
2. Heated oxygen sensor <front>
3. Engine oil level gauge guide
4. Air cleaner and air intake hose assembly
5. Air hose (A) connection
6. Water hose connection
7. Water hose connection
8. Oil pipe (A) connection
9. Heat protector (A)
10. Heat protector (B)
11. Engine hanger
12. Front exhaust pipe connection
13. Flange bolts
14. Flange nut
15. Coned disc spring
16. Exhaust manifold
17. Exhaust manifold gasket
18. Ring
19. Gasket (A)

TSB Revision

7923PG47

Exploded view of the turbocharger mounting—Eclipse 2.0L engine

3. Remove the air intake hose(s).

4. Disconnect the accelerator control cables from the throttle body.

5. Matchmark and disconnect the vacuum hoses including the brake booster hose.

6. Disconnect the clutch booster vacuum hose connection, if equipped.

7. Disconnect all harness connectors.

8. Disconnect EGR components on California vehicles.

9. Remove the plenum retaining bracket.

10. Remove the plenum retaining nuts and bolts and remove the air intake plenum. Discard the gasket.

11. Disconnect the high pressure and return fuel hoses.

12. Matchmark and disconnect the vacuum hoses.

13. Disconnect the wiring harness connectors.

14. Remove the fuel rail with the injectors attached.

15. Remove the timing belt upper cover.

16. Remove the intake manifold mounting nuts; turbocharged engines have cone disc springs under some of the nuts which should be removed. Remove the intake manifold and discard the gaskets.

To install:

17. Check all items for cracks, clogging and warpage. Maximum warpage is 0.008 in. (0.2mm). Replace any questionable parts.

18. Thoroughly clean and dry the mating surfaces of the heads, intake manifold and air intake plenum.

19. Install new intake manifold gaskets to the heads with the adhesive side facing up.

20. Place the manifold on the heads and install the cone disc springs and/or lockwashers.

21. For Turbo engines, lubricate the studs lightly with oil, then install the nuts following this procedure:

   a. Tighten the nuts on the front bank to 4–6 ft. lbs. (3–5 Nm).

   b. Tighten the nuts on the rear bank to 14–17 ft. lb. (20–23 Nm).

   c. Tighten the nuts on the front bank to 14–17 ft. lbs. (20–23 Nm).

   d. Repeat Steps B and C.

22. On non-turbocharged engines only, tighten the nuts to a final torque of 14 ft. lbs. (18 Nm).

23. Install the timing belt upper cover.

24. Install the fuel rail assembly.

25. Connect the harness connector and vacuum hoses.

26. Replace the O-ring and connect the fuel hoses.

27. Install a new intake air plenum gasket and install the plenum. Tighten the retaining nuts and bolts evenly and gradually to 13 ft. lbs. (18 Nm).

28. Install the retaining bracket.

29. Connect EGR components on California vehicles.

30. Connect the harness connectors and vacuum hoses.

31. Connect and adjust the accelerator cables.

32. Install the air intake hose(s).

33. Fill the system with coolant.

34. Connect the negative battery cable, run the vehicle until the thermostat opens, fill the radiator completely.

35. Check and adjust the idle speed and ignition timing.

36. Once the vehicle has cooled, recheck the coolant level.

Diamante

1. Relieve the fuel system pressure.

2. Disconnect battery negative cable and drain the cooling system.

3. Remove the air intake hose(s).

4. Disconnect the accelerator control cables from the throttle body.

5. Tag and disconnect the vacuum hoses including the brake booster hose.

6. Tag and disconnect the wiring harness connectors.

7. Disconnect the high pressure and return fuel hoses.

8. Disconnect EGR pipe and remove the EGR valve and EGR temperature sensor from the intake plenum assembly.

9. If equipped, remove the manifold pressure sensor.

10. Remove the plenum retaining bracket.

11. Remove the plenum retaining nuts and bolts and remove the air intake plenum from the intake manifold. Discard the gasket.

12. Remove the upper timing belt covers.

13. Remove the water pump stay bracket.

➡**It is not necessary to remove the fuel injectors from the intake unless the manifold assembly is being replaced.**

14. Remove the fuel rail with the injectors attached.

15. Disconnect the coolant hoses from the intake manifold. Be sure to note the connections.

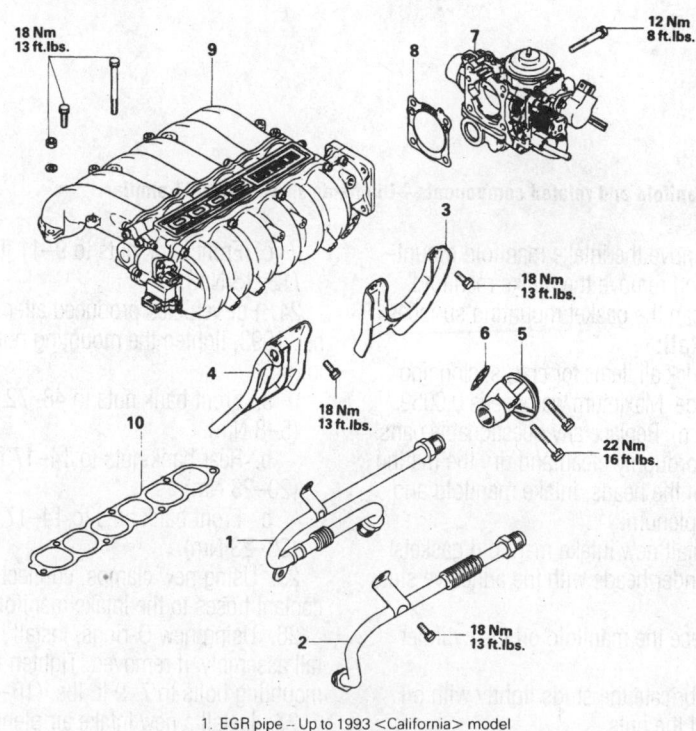

1. EGR pipe -- Up to 1993 <California> model
2. EGR pipe -- From 1994 <California> model
3. Intake manifold plenum stay, rear
4. Intake manifold plenum stay, front
5. EGR valve ⎫
6. EGR valve gasket ⎬ <For California>
7. Throttle body
8. Throttle body gasket
9. Intake manifold plenum
10. Intake manifold plenum gasket

Exploded view of air intake plenum assembly—Diamante shown, 3000GT similar

7923PG42

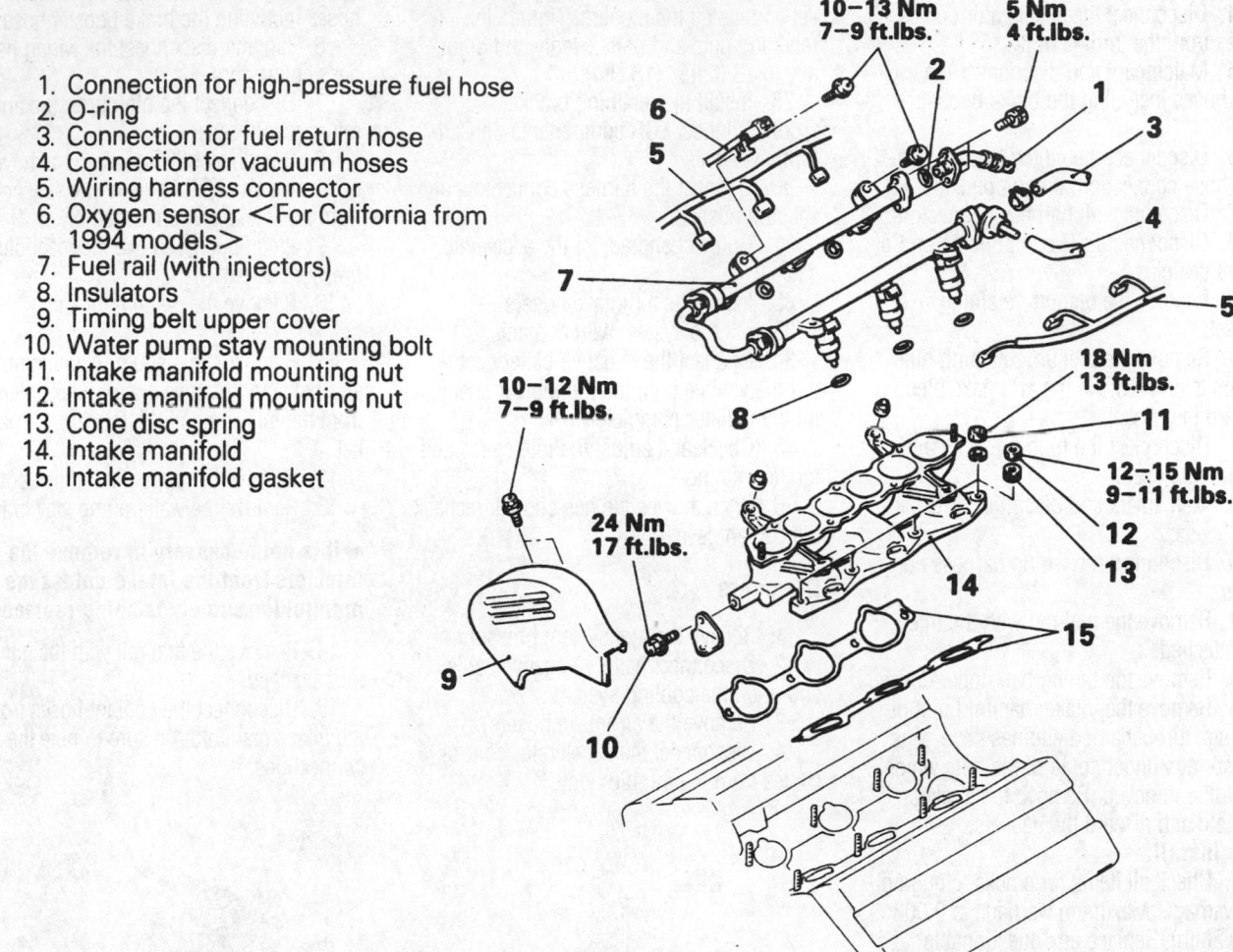

1. Connection for high-pressure fuel hose
2. O-ring
3. Connection for fuel return hose
4. Connection for vacuum hoses
5. Wiring harness connector
6. Oxygen sensor <For California from 1994 models>
7. Fuel rail (with injectors)
8. Insulators
9. Timing belt upper cover
10. Water pump stay mounting bolt
11. Intake manifold mounting nut
12. Intake manifold mounting nut
13. Cone disc spring
14. Intake manifold
15. Intake manifold gasket

Intake manifold and related components—Diamante shown, 3000GT similar

16. Remove the intake manifold mounting nuts and remove the intake manifold.

17. Clean the gasket mounting surfaces.

To install:

18. Check all items for cracks, clogging and warpage. Maximum warpage is 0.0059 in. (0.15mm). Replace any questionable parts.

19. Thoroughly clean and dry the mating surfaces of the heads, intake manifold and air intake plenum.

20. Install new intake manifold gaskets to the cylinder heads with the adhesive side facing up.

21. Place the manifold on the cylinder heads.

22. Lubricate the studs lightly with oil and install the nuts.

23. For vehicles produced up to and including November, 1993, tighten the mounting nuts as follows:

   a. Front bank nuts to 27–43 inch lbs. (3–5 Nm).

   b. Rear bank nuts to 9–11 ft. lbs. (12–15 Nm).

   c. Front bank nuts to 9–11 ft. lbs. (12–15 Nm).

24. For vehicles produced after November, 1993, tighten the mounting nuts as follows:

   a. Front bank nuts to 48–72 inch lbs. (5–8 Nm).

   b. Rear bank nuts to 14–17 ft. lbs. (20–23 Nm).

   c. Front bank nuts to 14–17 ft. lbs. (20–23 Nm).

25. Using new clamps, connect the coolant hoses to the intake manifold.

26. Using new O-rings, install the fuel rail assembly, if removed. Tighten the mounting bolts to 7–9 ft. lbs. (10–13 Nm).

27. Install a new intake air plenum gasket and install the plenum. Tighten the retaining nuts and bolts evenly and gradually to 13 ft. lbs. (18 Nm).

28. Install the retaining bracket and tighten the retaining bolts to 13 ft. lbs. (18 Nm).

29. If removed, install the manifold pressure sensor.

30. Using a new gasket, install the EGR valve and tighten the bolts to 16 ft. lbs. (22 Nm).

31. Install the EGR temperature sensor and tighten the fitting to 7–9 ft. lbs. (10–12 Nm).

32. Connect the EGR pipe and tighten the fittings to 43 ft. lbs. (60 Nm).

33. Replace the O-ring and connect the high pressure fuel hose. Tighten the retaining bolts to 48 inch lbs. (5 Nm).

34. Using a new hose clamp, connect the fuel return hose.

35. Install the water pump stay bracket.

36. Install the upper timing belt covers.

37. Connect the harness connector and vacuum hoses.

38. Connect and adjust the accelerator cables.

39. Install the air intake hose(s).

40. Fill the system with coolant.

41. Connect the negative battery cable, run the vehicle until the thermostat opens, fill the radiator completely.

42. Check and adjust the idle speed and ignition timing.

43. Once the vehicle has cooled, recheck the coolant level.

Eclipse

2.0L (4G63) ENGINE

1. Relieve the fuel system pressure.
2. Remove the battery.
3. Drain the engine coolant.
4. Disconnect the accelerator cable.
5. Remove the air intake hose.
6. Disconnect the ignition coil and the module wiring connectors.
7. Disconnect the Manifold Differential Pressure sensor.
8. Disconnect the condenser.
9. Disconnect the TPS and the IAC motor connectors.
10. Disconnect the knock sensor and the ECT sensor connectors.
11. Disconnect the camshaft position sensor and the crankshaft position sensor connectors.
12. Disconnect the A/C compressor connector.
13. Remove the engine control wiring harness retaining bracket and position the harness out of the way.
14. Label and disconnect the vacuum hoses.
15. Disconnect the spark plug wire from the ignition coil.
16. Disconnect the fuel lines from the fuel rail.
17. Disconnect the heater hoses.

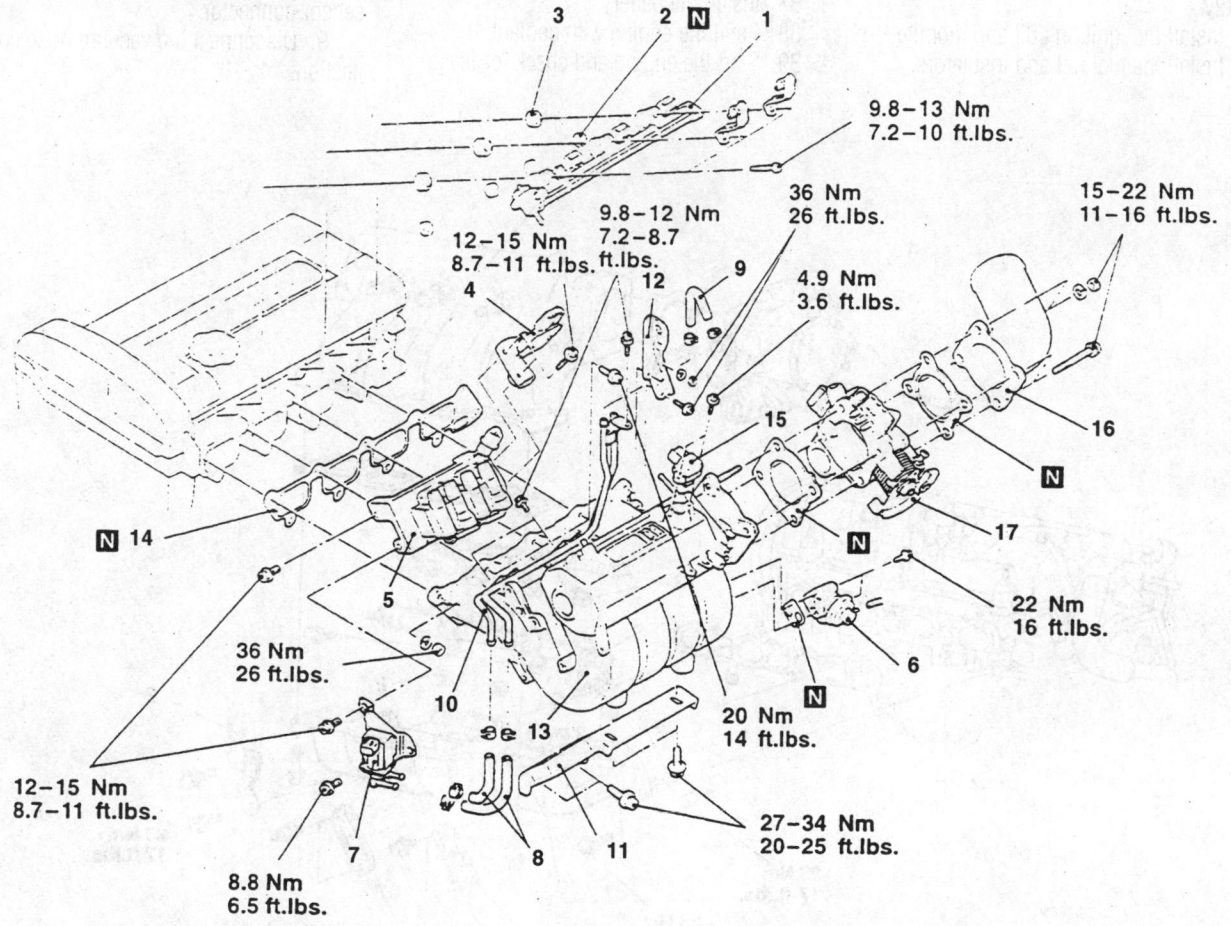

1 . Fuel rail, fuel injector and pressure regulator assembly
2 . Insulator
3 . Insulator
4 · Ignition power transistor
5 . Ignition coil
6 . EGR valve assembly
7 · Evaporative emission purge solenoid valve assembly
8 . Purge hoses
9 . Hose
10 . Vacuum pipe
11 . Intake manifold stay
12 . Engine hanger
13 . Intake manifold
14 . Intake manifold gasket
15 . Manifold differential pressure sensor
16 . Charge air cooler fitting
17 . Throttle body

Exploded view of the intake manifold and related components—2.0L (4G63) engine

7923PG40

18. Remove the fuel rail assembly and insulators.

19. Remove the ignition coil and module.

20. Remove the EGR valve assembly.

21. Remove the intake manifold stay and the engine hanger.

22. Remove the intake manifold and gasket.

To install:

23. Use a new gasket and install the intake manifold.

24. Install the intake manifold stay and the engine hanger.

25. Use a new gasket and install the EGR assembly.

26. Install the ignition coil and module.

27. Install the fuel rail and insulators.

28. Connect the heater hoses and fuel lines.

29. Connect the spark plug wires to the coil towers.

30. Connect the vacuum hoses.

31. Install the engine harness in the proper position.

32. Connect the harness to the proper sensors.

33. Connect the IAC motor.

34. Connect the ignition condenser.

35. Connect the ignition coil and the module connectors.

36. Install and adjust the accelerator cable.

37. Install the battery.

38. Refill the engine with coolant.

39. Start the engine and check for leaks.

2.0L (420A) ENGINE

1. Disconnect the negative battery cable.

2. Drain the engine coolant.

3. Remove the vacuum reservoir if equipped with cruise control.

4. Remove the air intake hose and breather hose.

5. Remove the accelerator cable from the bracket.

6. Disconnect the engine harness retaining clips.

7. Disconnect the MAP sensor connector.

8. Disconnect the charge temperature sensor connector.

9. Disconnect the vacuum hose connection.

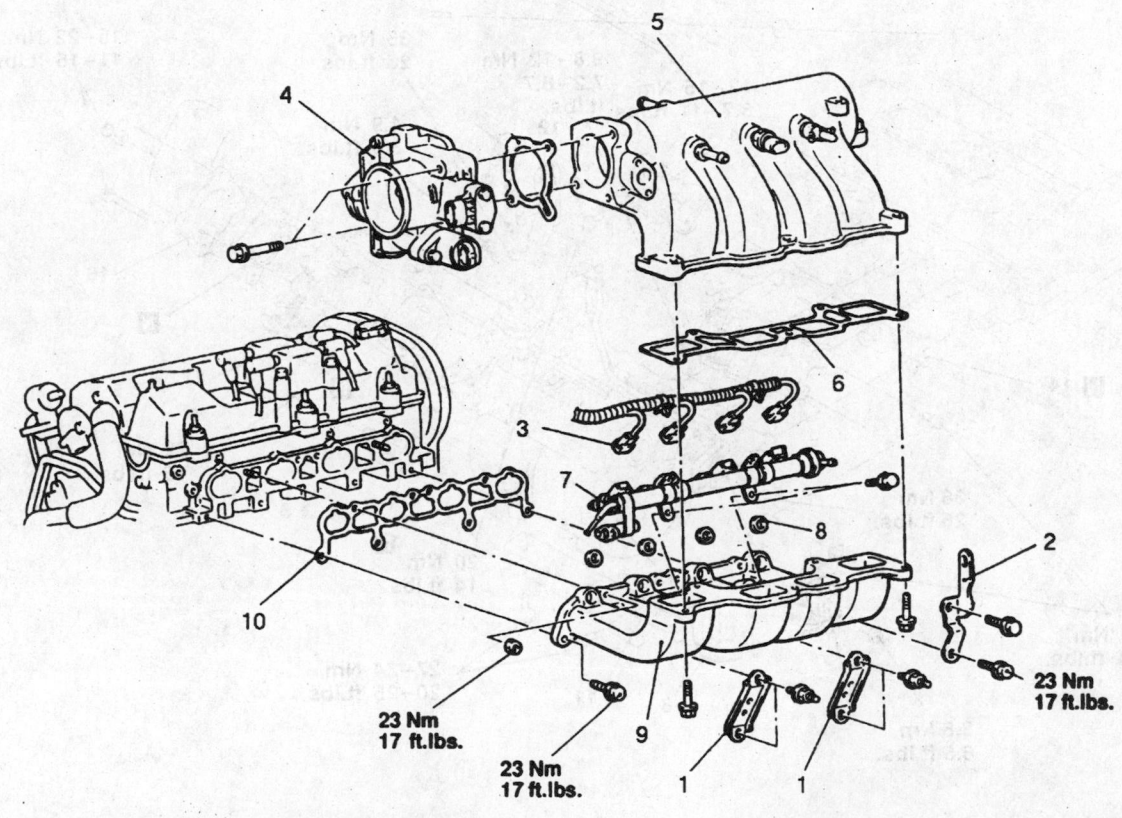

23 Nm
17 ft.lbs.

23 Nm
17 ft.lbs.

23 Nm
17 ft.lbs.

1. Intake manifold stay
2. Engine hanger
3. Injector connector
4. Throttle body
5. Intake manifold plenum
6. Intake manifold plenum gasket
7. Fuel rail, injector and pressure regulator assembly
8. O-ring
9. Intake manifold
10. Intake manifold gasket

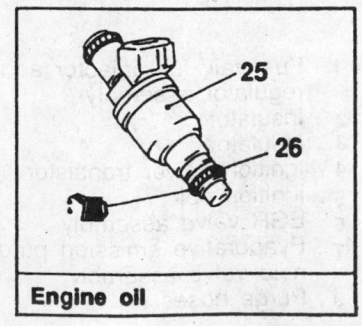

Engine oil

Intake manifold and related components—2.0L (420A) engine

7923PG39

10. Disconnect the TPS and the AIS motor connectors.

11. Position the engine control wiring harness out of the way.

12. Disconnect the alternator wiring harness connector.

13. Remove the PCV hose assembly.

14. Label and disconnect the vacuum hoses.

15. Disconnect the EGR pipe connection.

16. Disconnect the fuel lines from the fuel rail.

17. Remove the intake manifold stay and the engine hanger.

18. Remove the throttle body.

19. Remove the intake manifold plenum and gasket.

20. Disconnect the injector connectors.

21. Remove the fuel rail with the injectors.

22. Remove the intake manifold.

To install:

23. Use a new gasket and install the intake manifold.

24. Install the fuel rail assembly and connect the injectors.

25. Use a new gasket and install the intake plenum.

26. Use a new gasket and install the throttle body.

27. Install the intake manifold stay and the engine hanger.

28. Use a new O-ring and connect the fuel lines to the fuel rail.

29. Connect the EGR pipe.

30. Connect the vacuum hoses and the PCV hose assembly.

31. Connect the alternator wiring harness connector.

32. Reposition the engine control wiring harness and install the brackets and clips.

33. Connect the AIS motor and the TPS sensor connectors.

34. Connect the vacuum hose to the throttle body.

35. Connect the MAP and the charge temperature sensor connectors.

36. Install the accelerator cable in the bracket and connect it to the throttle body.

37. Connect the breather hose and the air intake hose.

38. Install the vacuum reservoir, if equipped.

39. Connect the negative battery cable.

40. Refill the engine with coolant.

41. Adjust the accelerator cable.

2.4L (4G64) ENGINE

1. Relieve the fuel system pressure.
2. Remove the battery.
3. Drain the engine coolant.
4. Disconnect the accelerator cable.

5. Remove the air intake hose.

6. Disconnect the ignition coil and the module wiring connectors.

7. Disconnect the Manifold Differential Pressure sensor.

8. Disconnect the condenser.

9. Disconnect the TPS and the IAC motor connectors.

10. Disconnect the heated oxygen sensor connector.

11. Disconnect the crankshaft position sensor connectors.

12. Disconnect the A/C compressor connector.

13. Remove the engine control wiring harness retaining bracket and position the harness out of the way.

14. Label and disconnect the vacuum hoses.

15. Disconnect the spark plug wire from the ignition coil.

16. Disconnect the fuel lines from the fuel rail.

17. Disconnect the heater hoses.

18. Disconnect the high-pressure fuel hose connection and remove the fuel rail assembly and insulators.

19. Remove the manifold differential pressure sensor.

20. Remove the ignition coil and module.

21. Remove the EGR valve assembly.

22. Remove the intake manifold stay and the engine hanger.

23. Remove the intake manifold and gasket.

24. Remove the throttle body assembly and gasket from the intake manifold.

To install:

25. Install the throttle body assembly with a new gasket to the intake manifold and tighten the mounting bolts to 11–16 ft. lbs. (15–22 Nm).

26. Use a new gasket and install the intake manifold. Torque the intake manifold bolts to 15 ft. lbs. (20 Nm).

27. Install the intake manifold stay and the engine hanger. Torque the mounting bolts to 19–24 ft. lbs. (26–33 Nm).

28. Use a new gasket and install the EGR assembly. EGR bolt torque is 16 ft. lbs. (22 Nm).

29. Install the ignition coil and module.

30. Install the fuel rail and insulators and reconnect the high-pressure fuel hose.

31. Connect the heater hoses and fuel lines.

32. Connect the spark plug wires to the coil towers.

33. Connect the vacuum hoses.

34. Install the engine harness in the proper position.

35. Connect the harness to the proper sensors.

36. Connect the IAC motor.

37. Connect the ignition condenser.

38. Connect the ignition coil and the module connectors.

39. Install and adjust the accelerator cable.

40. Install the battery.

41. Refill the engine with coolant.

42. Start the engine, check for leaks, and check for proper operation.

Galant

1. Relieve the fuel system pressure.

2. Disconnect battery negative cable and drain the cooling system.

3. Disconnect the accelerator cable, breather hose and air intake hose.

4. Disconnect the coolant hose from the throttle housing.

5. Disconnect the vacuum connection at the power brake booster and the PCV valve if still connected. Disconnect all remaining vacuum hoses and pipes as necessary.

✳✳ CAUTION

The fuel injection system remains under pressure after the engine has been turned OFF. Properly relieve fuel pressure before disconnecting any fuel lines. Failure to do so may result in fire or personal injury.

6. Disconnect the high pressure fuel line, fuel return hose and remove throttle control cable brackets.

7. Tag and disconnect the electrical connectors from the coolant temperature sensor, coolant temperature gauge, IAC valve, ignition coil, EGR temperature sensor, knock sensor, oxygen sensor, throttle position sensor, distributor, A/C temperature sensor, fuel injectors and ignition power transistor. Position the engine wiring harness aside.

8. Label and disconnect the spark plug wires, from the spark plugs.

9. Remove the intake manifold stay bracket.

10. Remove the intake manifold mounting bolts and remove the intake manifold assembly. Disassemble manifold on a work bench as required.

To install:

11. Clean all gasket material from the cylinder head intake mounting surface and intake manifold assembly. Check both surfaces for cracks or other damage. Check the intake manifold water passages and jet air passages for clogging. Clean if necessary.

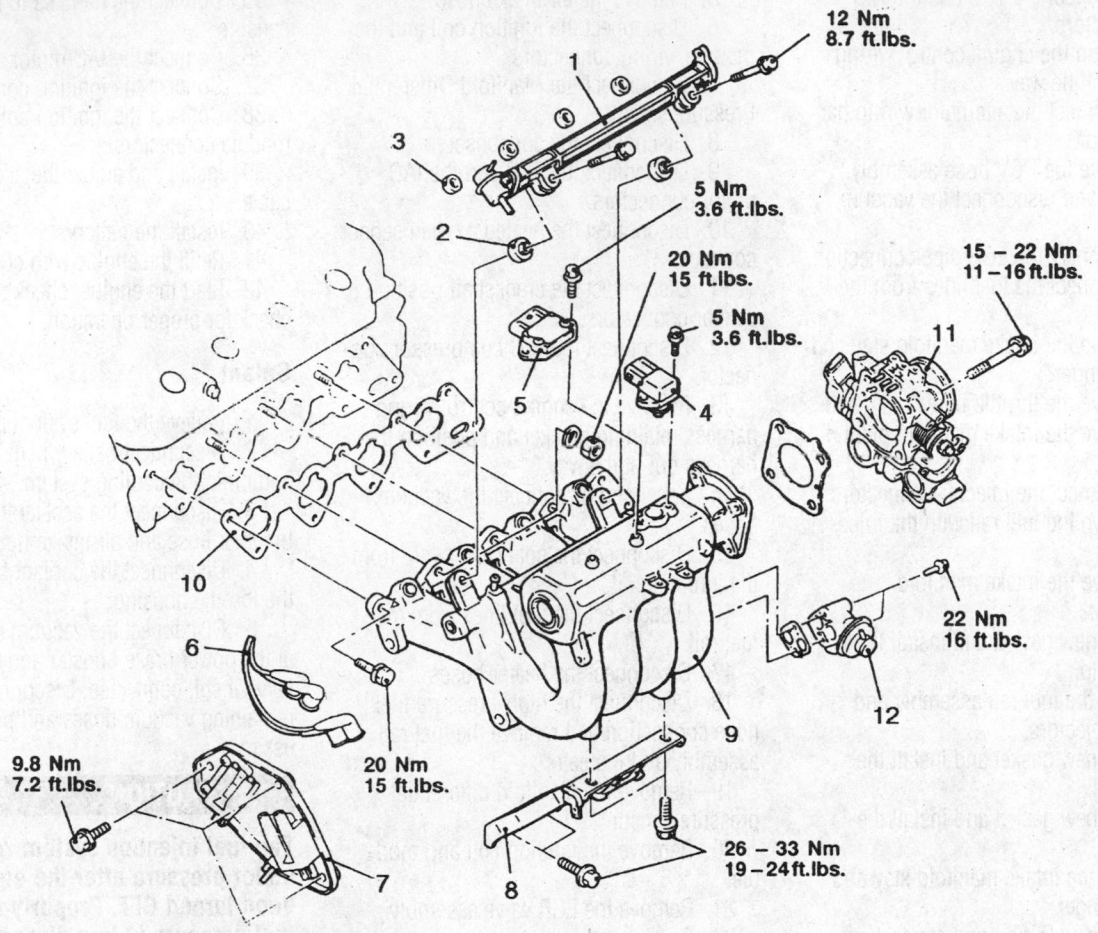

12 Nm
8.7 ft.lbs.

5 Nm
3.6 ft.lbs.

20 Nm
15 ft.lbs.

5 Nm
3.6 ft.lbs.

15 – 22 Nm
11 – 16 ft.lbs.

22 Nm
16 ft.lbs.

9.8 Nm
7.2 ft.lbs.

20 Nm
15 ft.lbs.

26 – 33 Nm
19 – 24 ft.lbs.

1. Fuel rail, fuel injector and pressure regulator assembly
2. Insulator
3. Insulator
4. Manifold differential pressure sensor
5. Ignition power transistor
6. Spark plug cable connection
7. Ignition coil
8. Intake manifold stay
9. Intake manifold
10. Intake manifold gasket
11. Throttle body
12. EGR valve assembly

7923PG41

Intake manifold and related components—Eclipse and Galant 2.4L (4G64) engines

12. Assemble the intake manifold assembly using all new gaskets.

13. Install a new intake manifold gasket to the head and install the manifold. Torque the manifold in a crisscross pattern, starting from the inside and working outwards to 15 ft. lbs. (20 Nm) for bolts, and 26 ft. lbs. (35 Nm) for nuts on the 1994 and 15 ft. lbs. (20Nm) on the 1995–96.

14. Install the fuel delivery pipe, injectors and pressure regulator to the engine. Torque the retaining bolts to 4 ft. lbs. (6 Nm).

15. Install the intake manifold brace bracket and tighten bolts to 21 ft. lbs. (29 Nm).

16. Connect or install all hoses, cables and electrical connectors that were removed or disconnected during the removal procedure.

17. Fill the system with coolant.

18. Connect the negative battery cable, run the vehicle until the thermostat opens, fill the radiator completely.

19. Adjust the accelerator cable. Check and adjust the idle speed and ignition timing.

20. Once the vehicle has cooled, recheck the coolant level.

Mirage

1.5L (4G15) ENGINE

1. Relieve the fuel system pressure.
2. Disconnect battery negative cable and drain the cooling system.
3. Disconnect the upper radiator hose, heater hose and water bypass hose.

4. Remove the thermostat housing from intake manifold.

5. Disconnect the accelerator cable, breather hose and air intake hose.

6. Remove all vacuum hoses and pipes as necessary, including the brake booster vacuum line.

7. Remove the throttle body assembly.

8. Disconnect the high pressure fuel line and the fuel return hose.

9. Tag and disconnect the electrical connectors from the oxygen sensor, coolant temperature sensor, intake air temperature, idle speed control connection, EGR temperature sensor, spark plug wires and distributor connectors.

10. Remove the fuel rail, fuel injectors, pressure regulator and insulators.

11. Remove the EGR valve from the intake manifold.

12. Remove the intake manifold support bracket and remove the engine mount support bracket.

13. Remove the intake manifold mounting bolts and remove the intake manifold assembly.

To install:

14. Clean all gasket material from the cylinder head intake mounting surface and intake manifold assembly. Check both surfaces for cracks or other damage. Check the intake manifold water passages and jet air passages for clogging. Clean if necessary.

15. Using a straight edge, measure the distortion of the intake manifold-to-cylinder head. Total distortion or warpage should be 0.006 in. (0.15mm or less).

16. Install a new intake manifold gasket to the head and install the manifold. Torque the manifold in a crisscross pattern, starting from the inside and working outwards to 13 ft. lbs. (18 Nm).

17. Install the intake manifold support bracket and tighten the mounting bolts to 16 ft. lbs. (22 Nm).

18. Install the engine mount support bracket and tighten the mounting bolts to 26 ft. lbs. (36 Nm).

19. Using a new gasket, install the EGR valve and tighten the mounting bolts to 15 ft. lbs. (21 Nm).

20. Using new insulators and O-rings, install the fuel delivery pipe, injectors and pressure regulator to the engine. Torque the retaining bolts to 7–9 ft. lbs. (10–13 Nm).

21. Connect the electrical connectors to the oxygen sensor, coolant temperature sensor, intake air temperature, idle speed control connection, EGR temperature sensor, spark plug wires and distributor connections.

22. Using a new O-ring for the feed pipe and a new clamp for the return pipe, install the fuel hoses.

23. Install the throttle body assembly.

24. Install the vacuum hoses and pipes as necessary, including the brake booster vacuum line.

25. Install and adjust the accelerator cable. Install the breather and air intake hose.

26. Using a new gasket, install the thermostat housing to the intake manifold and tighten the mounting bolts to 13 ft. lbs. (18 Nm).

27. Connect the upper radiator hose, heater hose and water bypass hose. Be sure to use new hose clamps.

28. Fill the system with coolant.

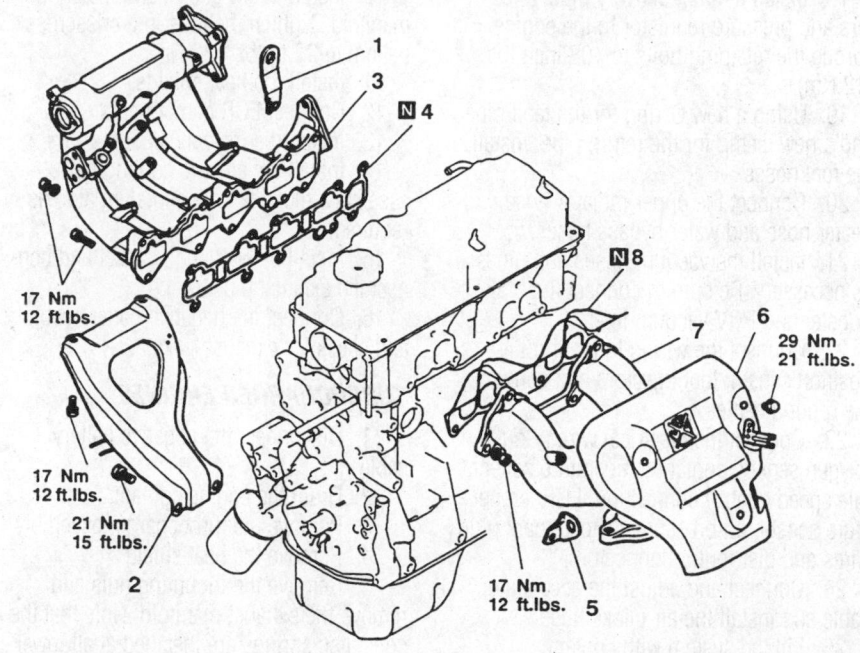

1. Engine hanger
2. Intake manifold stay
3. Intake manifold
4. Intake manifold gasket

5. Engine hanger
6. Exhaust manifold cover
7. Exhaust manifold
8. Exhaust manifold gasket

7923PG38

Exploded view of the intake and exhaust manifold mounting—Mirage 1.5L (4G15) engine

29. Connect the negative battery cable, run the vehicle until the thermostat opens, fill the radiator completely.

30. Check and adjust the idle speed and ignition timing.

31. Once the vehicle has cooled, recheck the coolant level.

1.8L (4G93) ENGINE

1. Relieve the fuel system pressure.

2. Disconnect battery negative cable and drain the cooling system.

3. Disconnect the accelerator cable and the air intake hose.

4. Tag and disconnect the electrical connectors from the oxygen sensor, coolant temperature sensor, idle speed control connection, EGR temperature sensor, oil pressure switch, spark plug wires and distributor connectors.

5. Disconnect the wiring from the throttle position sensor, fuel injectors and disconnect the ground cables.

6. Remove all vacuum hoses and pipes as necessary, including the brake booster and PCV vacuum lines.

7. Disconnect the upper radiator hose, heater hose and water bypass hose.

8. Disconnect the high pressure fuel line and the fuel return hose.

9. Remove the fuel rail, fuel injectors, pressure regulator and insulators.

10. Remove the intake manifold support bracket.

11. If the thermostat housing is preventing removal of the intake manifold, remove it.

12. Remove the intake manifold mounting bolts/nuts and remove the intake manifold assembly.

To install:

13. Clean all gasket material from the cylinder head intake mounting surface and intake manifold assembly. Check both surfaces for cracks or other damage. Check the intake manifold water passages and jet air passages for clogging. Clean if necessary.

14. Using a straight edge, measure the distortion of the intake manifold-to-cylinder head. Total distortion or warpage should be 0.006 in. (0.15mm or less).

15. Install a new intake manifold gasket to the head and install the manifold. Torque the manifold in a crisscross pattern, starting from the inside and working outwards to 14 ft. lbs. (20 Nm).

16. If removed, install the thermostat housing.

17. Install the intake manifold brace bracket.

18. Install the fuel delivery pipe, injectors and pressure regulator to the engine. Torque the retaining bolts to 108 inch lbs. (12 Nm).

19. Using a new O-ring for the feed pipe and a new clamp for the return pipe, install the fuel hoses.

20. Connect the upper radiator hose, heater hose and water bypass hoses.

21. Install the vacuum hoses and pipes as necessary. Be sure to connect the brake booster and PCV vacuum lines.

22. Connect the wiring to the throttle position sensor, fuel injectors and connect the ground cables.

23. Connect the electrical wiring to the oxygen sensor, coolant temperature sensor, idle speed control connection, EGR temperature sensor, oil pressure switch, spark plug wires and distributor connectors.

24. Connect and adjust the accelerator cable and install the air intake hose.

25. Fill the system with coolant.

26. Connect the negative battery cable, run the vehicle until the thermostat opens, fill the radiator completely.

27. Check and adjust the idle speed and ignition timing.

28. Once the vehicle has cooled, recheck the coolant level.

Exhaust Manifold

REMOVAL & INSTALLATION

3000GT

NON-TURBOCHARGED ENGINES

1. Disconnect battery negative cable.

2. Raise the vehicle and support safely.

3. Remove the exhaust pipe to exhaust manifold nuts and separate exhaust pipe. Discard gasket.

4. Lower vehicle.

5. Remove electric cooling fan assembly, if necessary. If removing the front manifold, remove the dipstick tube. If removing the front manifold from 3.0L DOHC engine, remove the alternator.

6. Disconnect necessary EGR components.

7. Disconnect the electrical connector and remove the oxygen sensor.

8. Remove the exhaust manifold mounting bolts, the inner heat shield and the exhaust manifold.

To install:

9. Clean all gasket material from the mating surfaces and check the manifold for damage.

10. Install a new gasket and install the manifold. Tighten the nuts in a crisscross pattern to 22 ft. lbs. (30 Nm).

11. Install the heat shields.

12. Connect EGR components.

13. Install the oxygen sensor.

14. Install the electric cooling fan assembly, dipstick tube and alternator, as required.

15. Install a new flange gasket and connect the exhaust pipe.

16. Connect the negative battery cable and check for exhaust leaks.

TURBOCHARGED ENGINES

1. Disconnect the negative battery cable.

2. Drain the engine coolant.

3. Remove the turbocharger assembly.

4. Remove the heat shield.

5. Remove the mounting nuts and remove the exhaust manifold. Note that the cone disc springs are installed at all lower mounting points.

To install:

6. Clean all gasket material from the mating surfaces and check the manifold for damage.

7. Install new gaskets and install the manifold. Be sure all cone disc springs are in their original locations with the grooved side facing the nut. Tighten the manifold nuts using the following procedure:

a. Tighten all but the outer two nuts to 22 ft. lbs. (30 Nm).

b. Tighten the outer two nuts to 34–38 ft. lbs. (47–53 Nm).

c. Loosen the outer two nuts, then tighten them to 22 ft. lbs. (30 Nm).

8. Install the heat shield.

9. Install the turbocharger assembly.

10. Fill the cooling system.

11. Connect the negative battery cable and check for exhaust leaks.

Diamante

✳✳ CAUTION

Do not attempt the work on the exhaust system until it has completely cooled.

1. Disconnect battery negative cable.

2. Raise the vehicle and support safely.

3. Remove the exhaust pipe to exhaust manifold nuts and remove the front exhaust pipe.

4. Lower the vehicle.

5. If removing the front manifold, remove condenser electric cooling fan assembly.

6. For the DOHC engine, if removing the front manifold, remove the alternator and mounting bracket from the vehicle.

7. For the DOHC engine, separate the A/C compressor from the mounting bracket. Leaving the hoses connected, position the compressor aside.

8. If removing the front manifold, remove the oil dipstick and tube from the engine.

9. For the DOHC engine, if removing the front manifold, remove the heat protector.

10. If removing the rear manifold, disconnect the EGR tube.

11. For the SOHC engine, if removing the rear manifold, remove the intake plenum stay and the roll stopper bracket.

12. Disconnect the electrical connector and remove the oxygen sensor.

13. Remove the exhaust manifold mounting bolts the manifold.

To install:

14. Clean all gasket material from the mating surfaces and check the manifold for damage.

15. Install a new gasket and install the manifold. Tighten the nuts in a crisscross pattern to 21 ft. lbs. (30 Nm) for the J- engine or to 14 ft. lbs. (19 Nm) for the H- engine.

16. Install the heat shields.

17. Connect the EGR tube and intake plenum stay and roll stopper bracket, if removed.

18. Install the oxygen sensor.

19. Install the electric cooling fan assembly, A/C compressor, dipstick tube and alternator, as required.

20. Install a new flange gasket and connect the exhaust pipe or converter assembly.

21. Install the drive belt(s) and adjust for proper tension.

22. Connect the negative battery cable and check for exhaust leaks.

Eclipse

2.0L (4G63) ENGINE

✳✳ CAUTION

The air bag system (SRS or SIR) must be disarmed before removing the exhaust manifold or turbocharger. Failure to do so may cause accidental deployment, property damage or personal injury.

1. Disconnect the negative battery cable.

2. Drain the engine coolant.

3. Remove the condenser fan motor assembly if equipped with air conditioning.

4. Remove the heated oxygen sensor.

5. Remove the dipstick and tube assembly.

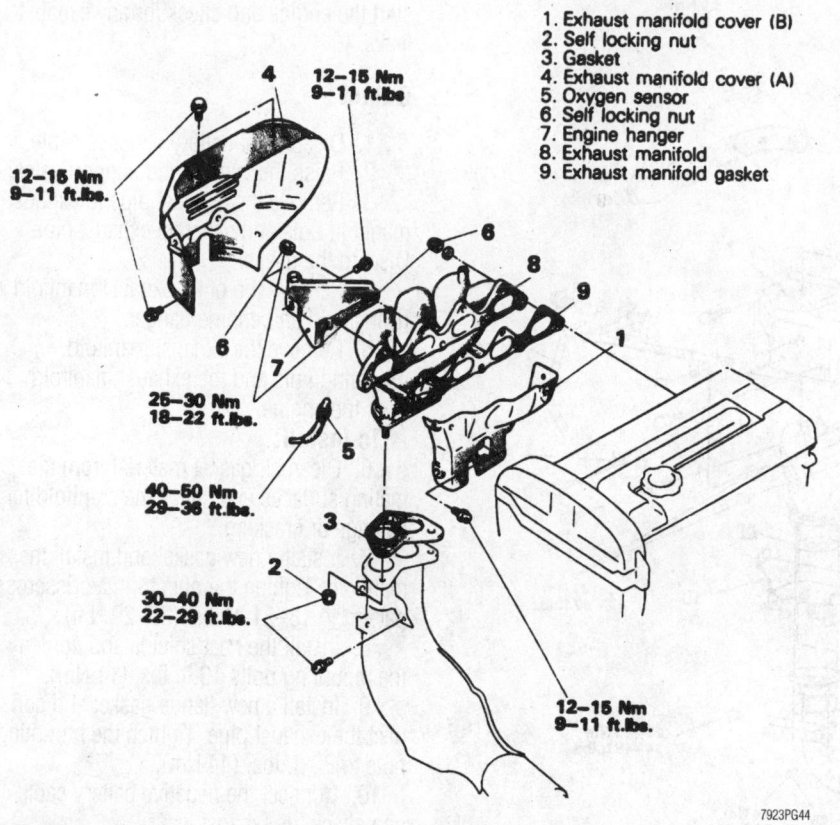

1. Exhaust manifold cover (B)
2. Self locking nut
3. Gasket
4. Exhaust manifold cover (A)
5. Oxygen sensor
6. Self locking nut
7. Engine hanger
8. Exhaust manifold
9. Exhaust manifold gasket

12–15 Nm
9–11 ft.lbs.

12–15 Nm
9–11 ft.lbs.

25–30 Nm
18–22 ft.lbs.

40–50 Nm
29–36 ft.lbs.

30–40 Nm
22–29 ft.lbs.

12–15 Nm
9–11 ft.lbs.

7923PG44

Exhaust manifold and related parts—Eclipse 2.0L (4G63) engine

6. Remove the air cleaner and air intake hose assembly.

7. Disconnect the air intake hose from the turbocharger.

8. Disconnect the engine coolant hoses from the turbocharger.

9. Disconnect the oil supply pipe connection. Do not let dirt of foreign particles enter the oil pipe.

10. Remove the heat shields.

11. Remove the engine hanger.

12. Disconnect the front exhaust pipe from the turbocharger.

13. Remove the oil return pipe and gaskets.

14. Remove the flange bolts, (washers if equipped), and nuts that attach the turbo to the exhaust manifold. Take note of the positions of the coned disc springs and the washers.

15. Remove the turbocharger, gasket and ring.

16. Remove the exhaust manifold and gasket.

To install:

17. Use a new gasket and install the exhaust manifold.

18. Use a new gasket and install the turbo to the exhaust manifold. Be sure the coned disc spring and the washers are installed in their original positions. Torque the bolts, (if equipped with washers, install them), and nuts to specification, loosen them and tighten them again.

19. Use a new gasket and connect the exhaust pipe to the turbo.

20. Using new gaskets, install the oil return pipe.

21. Install the engine hanger.

22. Install the heat shields.

23. Connect the oil supply pipe.

24. Connect the engine coolant hoses to the turbo.

25. Connect the air hose.

26. Install the air cleaner and duct assembly.

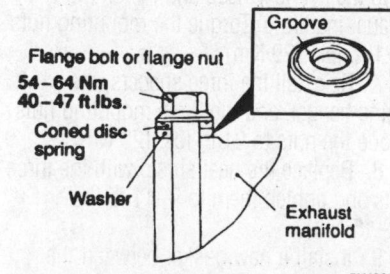

Flange bolt or flange nut
54–64 Nm
40–47 ft.lbs.
Coned disc spring
Washer
Groove
Exhaust manifold

7923PG37

Install the groove of the cone-shaped disc spring toward the flange bolt or nut—Eclipse 2.0L (4G63) engine

27. Position a new O-ring on the dipstick tube and install the tube and dipstick.

28. Install the heated oxygen sensor.

29. Install the condenser fan assembly if removed.

30. Change the engine oil and filter.

31. Connect the negative battery cable and refill the engine with coolant.

32. Start the engine and let it idle. Do not race the engine until the oil reaches the turbo. Check for leaks.

2.0L (420A) ENGINE

✸ CAUTION

The air bag system (SRS or SIR) must be disarmed before removing the exhaust manifold. Failure to do so may cause accidental deployment, property damage or personal injury.

1. Disconnect the negative battery cable.

2. Drain the engine coolant.

3. Remove the air intake hose.

4. Disconnect the upper radiator hose from the water outlet.

5. Disconnect the air hose connection.

6. Remove the engine control wiring harness from the rear of the engine.

7. Remove the water pipe assembly.

8. Remove the oil dip stick.

9. Remove the upper heat shield.

10. Remove the engine hanger.

11. Disconnect the pulsed secondary air injection (check valve) valve from the exhaust pipe. (manual transaxle only)

12. Disconnect the front exhaust pipe from the manifold.

13. Remove the lower heat shield.

14. Remove the exhaust manifold and gasket.

To install:

15. Use a new gasket and install the exhaust manifold. Torque the nuts and bolts to 17 ft. lbs. (23 Nm).

16. Install the lower heat shield.

17. Use a new gasket and connect the front exhaust pipe to the manifold.

18. On vehicles with manual transaxles, connect the pulsed secondary air injection valve to the exhaust pipe.

19. Install the engine hanger.

20. Install the upper heat shield.

21. Install the dip stick and the water pipe.

22. Attach the engine wiring harness to the rear of the engine.

23. Connect the air hose and the upper radiator hose.

24. Install the air intake hose.

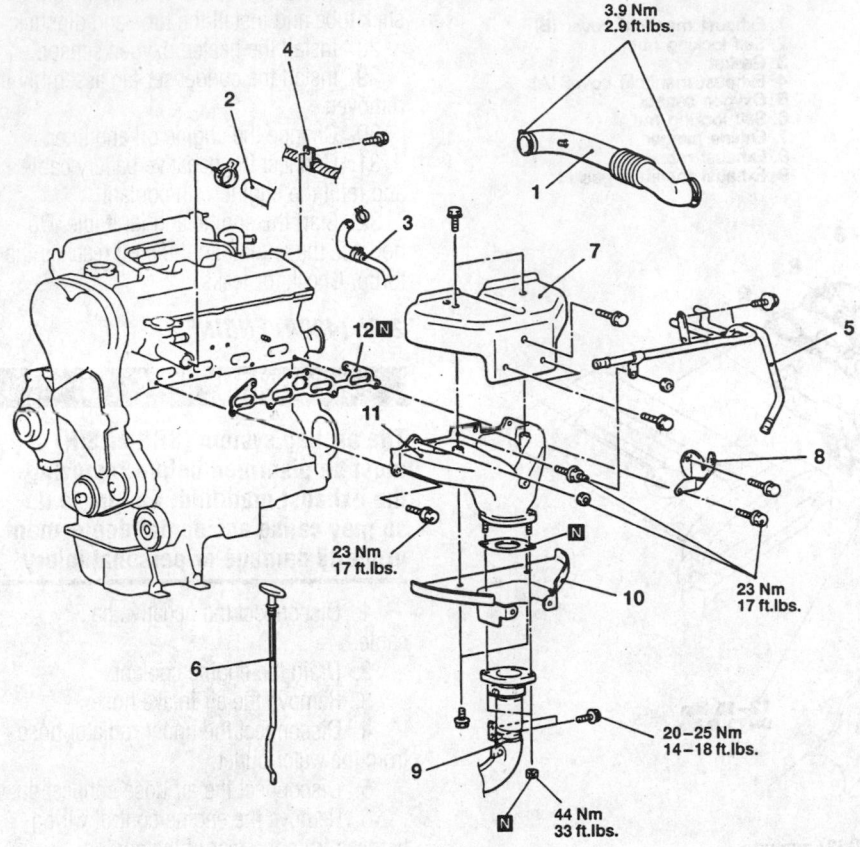

**3.9 Nm
2.9 ft.lbs.**

**23 Nm
17 ft.lbs.**

**23 Nm
17 ft.lbs.**

**20–25 Nm
14–18 ft.lbs.**

**44 Nm
33 ft.lbs.**

Removal steps

1. Air intake hose
2. Radiator upper hose connection
3. Air hose connection
4. Control wiring harness connection
5. Water pipe assembly
6. Engine oil level gauge
7. Heat protector
8. Engine hanger
9. Front exhaust pipe connection
10. Heat protector
11. Exhaust manifold
12. Exhaust manifold gasket

7923PG45

Exploded view of the exhaust manifold and related components—2.0L (420A) engine

25. Connect the negative battery cable.

26. Refill the engine with coolant, start the engine and check for leaks.

2.4L (4G64) ENGINE

> **⁎⁎ CAUTION**
>
> **The air bag system (SRS or SIR) must be disarmed before removing the exhaust manifold or turbocharger. Failure to do so may cause accidental deployment, property damage or personal injury.**

1. Disconnect the negative battery cable.

2. Remove the nuts and the gasket and disconnect the front exhaust pipe from the exhaust manifold.

3. Remove the three bolts and the heat shield.

4. Remove the three nuts, the engine hanger, and the three spacers from the exhaust manifold.

5. Remove the remaining mounting nuts, the exhaust manifold, and the exhaust manifold gasket.

To install:

6. Install a new exhaust manifold gasket to the cylinder head and install the exhaust manifold. Torque the mounting nuts to 21 ft. lbs. (29 Nm).

7. Reinstall the three spacers, the engine hanger, and the three mounting nuts. Torque the nuts to 21 ft. lbs. (29 Nm).

8. Replace the heat shield with the three bolts and tighten them to 9–11 ft. lbs. (12–15 Nm).

9. Install a new gasket between the exhaust manifold and the front exhaust pipe and reconnect the pipe with the three nuts. Torque the nuts to 32 ft. lbs. (34 Nm).

10. Reconnect the negative battery cable, start the engine and check for any exhaust leaks.

Galant

1. Disconnect battery negative cable.

2. Raise the vehicle and support safely.

3. Remove the exhaust pipe to exhaust manifold nuts and separate exhaust pipe. Discard the gasket.

4. Remove the outer exhaust manifold heat shield and engine hanger.

5. Remove the exhaust manifold mounting nuts and the exhaust manifold from the engine.

To install:

6. Clean all gasket material from the mating surfaces and check the manifold for damage or cracking.

7. Install a new gasket and install the manifold. Tighten the nuts to in a crisscross pattern to 18–21 ft. lbs. (25–29 Nm).

8. Install the heat shields and tighten the mounting bolts 10 ft. lbs. (14 Nm).

9. Install a new flange gasket and connect the exhaust pipe. Tighten the mounting nuts to 32 ft. lbs. (44 Nm).

10. Connect the negative battery cable and check for exhaust leaks.

Mirage

1. Disconnect battery negative cable.

2. Raise the vehicle and support safely.

3. Remove the exhaust pipe to exhaust manifold nuts and separate exhaust pipe. Discard gasket.

4. Lower vehicle.

5. Remove electric cooling fan assembly, if necessary.

6. If the oxygen sensor is located in the manifold, remove the sensor.

7. Disconnect necessary EGR components.

8. Remove outer exhaust manifold heat shield and engine hanger. Disconnect the electrical connector and remove the oxygen sensor.

9. Remove the exhaust manifold mounting bolts, the inner heat shield and the exhaust manifold.

To install:

10. Clean all gasket material from the mating surfaces and check the manifold for damage.

11. Using a new gasket and install the manifold. For 1.5L engines, tighten the nuts on a crisscross patter to 13 ft. lbs. (18 Nm). For 1.8L engines, tighten the inner nuts to in a crisscross pattern to 13 ft. lbs. (18 Nm) and tighten the two outer (larger) nuts to 22 ft. lbs. (30 Nm).

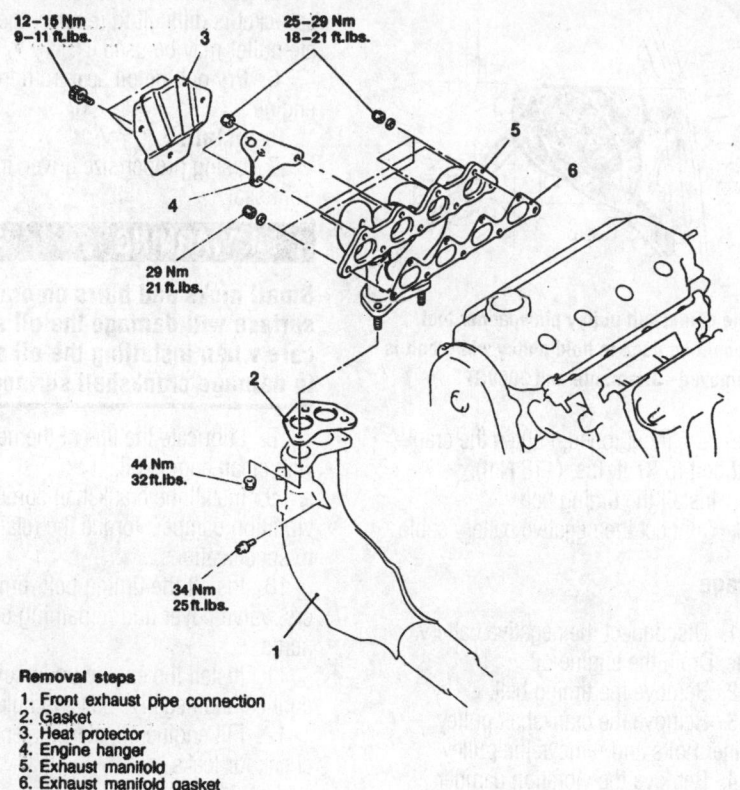

Removal steps
1. Front exhaust pipe connection
2. Gasket
3. Heat protector
4. Engine hanger
5. Exhaust manifold
6. Exhaust manifold gasket

7923PG46

Exploded view of the exhaust manifold mounting—Eclipse 2.4L (4G64) engine shown, Galant similar

12. Install the heat shields.

13. Connect EGR components.

14. If removed, install the oxygen sensor.

15. Install the electric cooling fan assembly as required.

16. Install a new flange gasket and connect the exhaust pipe.

17. Connect the negative battery cable and check for exhaust leaks.

Front Cover Seal

REMOVAL & INSTALLATION

3000GT

1. Disconnect the negative battery cable.

2. Remove the undercover.

3. Remove the cruise control pump and link assembly.

4. Remove the alternator assembly.

5. Raise and support the engine to take the weight off of the engine mount.

6. Remove the air hose and the air pipe.

7. Remove the power steering tensioner and drive belt.

8. Using a pin spanner, if available, hold the crankshaft pulley from turning and remove the crankshaft pulley bolt.

9. Remove the brake fluid level sensor and the upper timing belt cover.

10. Remove the engine mount bracket and the idler pulley for the alternator and A/C compressor drive belt.

11. Remove the engine support bracket and the lower timing belt cover.

12. Remove the timing belt and the auto tensioner.

13. Remove the crankshaft sprocket.

14. Pry out the crankshaft seal using a suitable tool.

To install:

15. Using a driver tool, install the new crankshaft seal.

16. Install the crankshaft sprocket, timing belt and the belt auto tensioner.

17. Install the lower timing belt cover and the engine support bracket.

18. Install the idler pulley for the alternator and A/C compressor and the engine mount bracket.

19. Install the upper timing belt cover and the brake fluid sensor.

20. Using the spanner tool, if available, install the crankshaft pulley.

21. Install the power steering drive belt and tensioner.

22. Install the air hose and the air pipe.

23. Install the alternator and the cruise control link and pump assembly.

24. Install the undercover and the negative battery cable.

Diamante

1. Disconnect the negative battery cable.

2. Position the engine to TDC of No. 1 cylinder compression stroke.

3. Remove the drive belts.

4. Using a pin spanner, if available, hold the crankshaft pulley and remove the retaining bolt. Remove the pulley.

5. Remove the timing belt covers and the timing belt.

6. If necessary, remove the crankshaft position sensor.

7. Remove the crankshaft sprocket, and if necessary, the sensing blade, spacer and Woodruff key®.

➡ **If sprocket is difficult to remove, an appropriate puller may be used.**

8. Pry the seal from the bore, using a suitable tool.

✳✳ WARNING

Use care not to nick or scratch the crankshaft, when removing the seal.

To install:

9. Using driver tool MD998717 or equivalent, install the new crankshaft seal. Lubricate the lips of the seal with clean engine oil.

10. Install the Woodruff key®, spacer, sensing blade (if necessary) and the crankshaft sprocket.

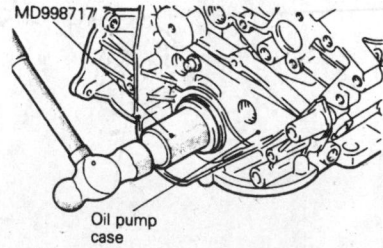

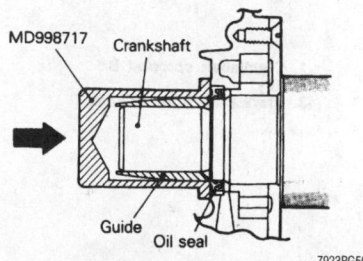

7923PG50

Crankshaft seal installation—Diamante and 3000GT

11. If removed, install the crankshaft position sensor and tighten the retaining bolts to 7 ft. lbs. (9 Nm).

12. Install the timing belt and the timing belt cover(s).

13. Install the crankshaft pulley and retaining bolt. Torque the retaining bolt to 130–137 ft. lbs. (180–190 Nm) for the DOHC engine or to 108–116 ft. lbs. (150–160 Nm) for the SOHC engine.

14. Install and adjust the drive belts.

15. Connect the negative battery cable and check for leaks.

Eclipse

2.0L AND 2.4L ENGINES

1. Disconnect the negative battery cable.
2. Remove the timing belt.
3. Remove the crankshaft sprocket using MB995027 or equivalent proper puller.
4. Carefully pry the oil seal out of the front case assembly using MB995020 or equivalent. Be careful not to damage the oil seal bore or the crankshaft sealing surface.

To install:

5. Apply clean engine oil to the oil seal lip. Using MB995022 or equivalent seal driver, install the oil seal.

6. Install the crankshaft sprocket using MB995035 and MB995026 or equivalent, if

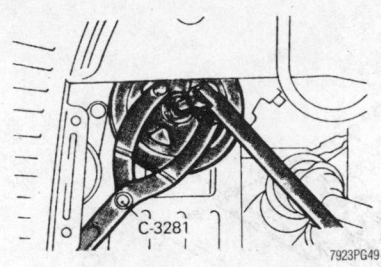

The crankshaft pulley pin spanner tool should be used to hold pulley while bolt is removed—Diamante and 3000GT

necessary. If equipped, tighten the crankshaft bolt to 87 ft. lbs. (118 Nm).

7. Install the timing belt.
8. Connect the negative battery cable.

Mirage

1. Disconnect the negative battery cable. Drain the engine oil.
2. Remove the timing belt.
3. Remove the crankshaft pulley retainer bolts and remove the pulley.
4. Remove the vibration damper retainer bolt and washer and remove damper. If difficult to remove, the appropriate puller may be used.

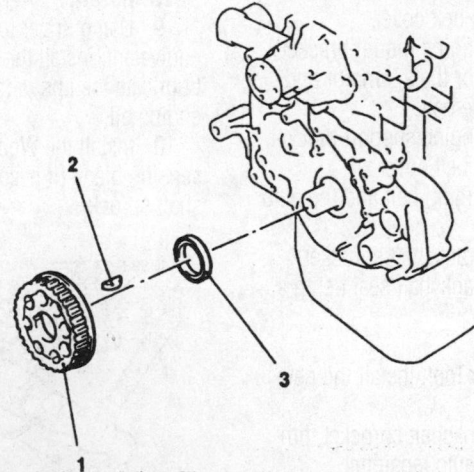

1. Crankshaft sprocket B
2. Key
3. Crankshaft front oil seal

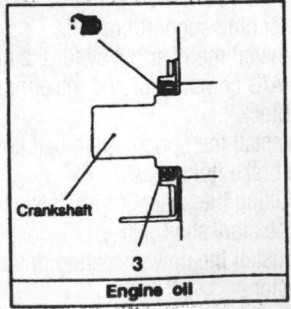

Front crankshaft oil seal—Eclipse 2.0L (4G63) engine shown

5. Remove the crankshaft sprocket. If sprocket is difficult to remove, the appropriate puller may be used.
6. Pry out the oil seal from front of engine.

To install:

7. Using proper size driver, install new front seal.

✱✱ WARNING

Small nicks and burrs on crankshaft surface will damage the oil seal. Use care when installing the oil seal not to damage crankshaft surface.

8. Lubricate the lips of the new seal with clean engine oil.
9. Install the crankshaft sprocket and vibration damper. Torque the retaining bolt to specifications.
10. Install the timing belt, timing covers, valve cover and remaining components.
11. Install the engine undercover and connect the negative battery cable.
12. Fill engine oil, start the engine and check for leaks.

Camshaft and Valve Lifters

REMOVAL & INSTALLATION

3000GT Engine

1. Relieve the fuel system pressure.

✱✱ CAUTION

The fuel injection system remains under pressure after the engine has been turned OFF. Properly relieve fuel pressure before disconnecting any fuel lines. Failure to do so may result in fire or personal injury.

2. Disconnect battery negative cable.

✱✱ CAUTION

Wait at least 90 seconds after the negative battery cable is disconnected to prevent possible deployment of the air bag.

3. Remove the timing belt cover and timing belt.
4. Remove the center cover, breather and PCV hoses, and spark plug cables.
5. Remove the rocker cover, semi-circular packing, throttle body stay, both camshaft sprockets, and oil seals.
6. Remove the crank angle sensor and adapter.

7. Remove the intake and exhaust camshafts.

To install:

8. Lubricate the camshafts with heavy engine oil and position the camshafts on the cylinder head.

➡**Do not confuse the intake camshaft with the exhaust camshaft. The intake camshaft has a J stamped on the hexagon and the exhaust camshaft has a K or N.**

9. Be sure the dowel pins on both camshaft sprocket ends are positioned properly.

10. Install the bearing caps. Tighten the caps gradually and in 2 or 3 steps. Caps 2, 3 and 4 have a front mark. Install with the mark aligned with the front mark on the cylinder head. Intake caps have **I** stamped on the cap and exhaust caps have **E**. Also, be sure the rocker arm is correctly mounted on the lash adjuster and the valve stem end. Torque the front and rear retaining cap bolts to 15 ft. lbs. (20 Nm) and tighten the center 3 retaining cap bolts to 8 ft. lbs. (11 Nm).

➡**If installing the camshaft to a cylinder head that is positioned on a workbench, the valves will protrude.**

11. Apply a coating of engine oil to the oil seals and install.

12. Install the timing belt, valve cover and all related parts.

13. Connect the negative battery cable and check for leaks.

Diamante

3.0L DOHC ENGINE

1. Relieve the fuel system pressure.
2. Disconnect negative battery cable.
3. Remove the intake manifold plenum.
4. Remove the timing belt cover and the timing belt.

✳✳ WARNING

DO NOT rotate the crankshaft or camshafts after the timing belt has been removed. If rotated, severe internal engine damage will result from the pistons hitting the valves.

5. Remove the center cover, breather, PCV hoses, and the spark plug cables.
6. Remove the rocker cover and the semi-circular packing.
7. Matchmark the positioning of the crankshaft position sensor at the rear of the camshaft and remove the sensor.
8. If equipped with a camshaft sensor, remove the sensor from the front of the engine.

9. Being sure to hold the flats of the camshaft, loosen the camshaft sprocket bolts.
10. Noting the positioning and location of the sprockets, remove the sprockets from the camshafts.

➡**Be sure to note the positioning of the knock pin at the end of the camshafts for reinstallation purposes.**

➡**Be sure to keep the valvetrain components labeled and in proper order for reassembly.**

11. Loosen the bearing cap bolts in 2–3 steps. Label and remove all camshaft bearing caps.

➡**If the bearing caps are difficult to remove, use a plastic hammer to gently tap the components.**

12. Mark the components and remove the intake and the exhaust camshafts.
13. Remove the rocker arms and the lash adjusters. Be sure to note the location of the valvetrain components for reinstallation purposes.
14. Check the camshaft journals for wear or damage. Check the cam lobes for damage. Also, check the cylinder head oil holes for clogging.

To install:

➡**Lubricate the valvetrain components with clean engine oil.**

15. Bleed and install the lash adjusters to the to the original bores in the cylinder head.
16. Install the rocker arms to the cylinder head.
17. Lubricate the camshafts with clean engine oil and position the camshafts on the cylinder head.

✳✳ WARNING

Be sure to properly position the knock pins of the camshaft to prevent valve to piston interference.

➡**Do not confuse the intake camshaft with the exhaust camshaft. The intake camshaft on the Diamante has a B or J stamped on the hexagon depending on the application. The exhaust camshaft on the Diamante has a D or K stamped on the hexagon depending on application.**

➡**Install the bearing caps according to the identification mark and cap number. Bearing caps No. 2, 3 and are marked as such. The caps also are marked I for intake or E for exhaust.**

18. Install the bearing caps. Tighten the caps in sequence and in 2 or 3 steps. Caps 2, 3 and 4 have a front mark. Install with the mark aligned with the front mark on the cylinder head. Torque the retaining bolts for caps No. 2, 3 and 4 to 8 ft. lbs. (11 Nm) and tighten the retaining bolts for the front and rear caps to 14 ft. lbs. (20 Nm).

19. Apply a coating of engine oil to the oil seals and install the oil seals to the front and rear of the camshafts.
20. Holding the flats of the camshaft, install and tighten the sprocket bolts to 65 ft. lbs. (90 Nm).
21. If removed, install the camshaft position sensor and tighten the mounting bolts to 78 inch lbs. (9 Nm).
22. Aligning the matchmark, install the crankshaft position sensor at the rear of the camshaft and tighten the mounting nut to 7 ft. lbs. (12 Nm).
23. Align the marks on the camshaft and crankshaft sprockets. Install the timing belt assembly.
24. Install the rocker cover and the semicircular packing.
25. Install the intake manifold plenum.
26. Install the spark plug cables, center cover, breather and PCV hoses.
27. Connect the negative battery cable and check for leaks.

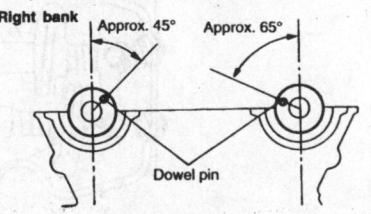
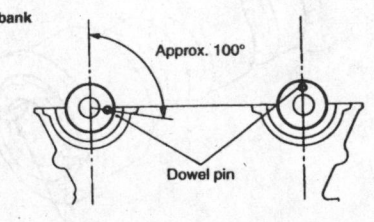

7923PG58

Proper positioning of the camshaft knock pins—Diamante 3.0L DOHC engine

3.0L SOHC ENGINE

1. Disconnect the negative battery cable.

2. Rotate and position the engine to TDC of compression stroke.

3. If removing the right side (front) camshaft, matchmark the distributor rotor and distributor housing to the engine block and remove the distributor.

4. Remove the intake manifold plenum stay bracket.

5. Remove the distributor housing adapter and discard the O-ring.

6. Remove the valve covers and the timing belt.

7. Using camshaft sprocket holding tool MB990767 and MD998719 or equivalent, hold the sprocket and loosen the bolt.

8. Remove the bolt and note the positioning of the of the knock pin at the end of the camshaft and remove the sprocket.

9. Install auto lash adjuster retainer tools MD998443 or equivalent, on the rocker arms.

➡**Be sure to note the position of the rocker arms, rocker shafts and bearing caps for reinstallation purposes.**

10. Remove the camshaft bearing caps but do not remove the bolts from the caps.

11. Remove the rocker arms, rocker shafts and bearing caps, as an assembly.

12. Remove the camshaft from the cylinder head.

13. Inspect the bearing journals on the camshaft, cylinder head, and bearing caps.

To install:

➡**The right bank camshaft is identified by a 4mm slit at the rear end of the camshaft.**

14. Lubricate the camshaft journals and camshaft with clean engine oil and install the camshaft in the cylinder head. Be sure to properly position the knock pin of the camshaft as noted during removal.

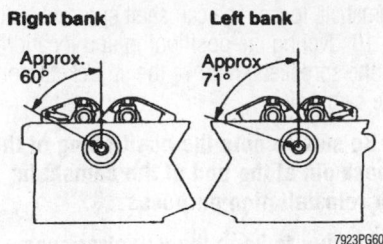

Right bank   **Left bank**

Approx. 60°   Approx. 71°

7923PG60

Proper positioning of the camshafts—Diamante 3.0L SOHC engine

15. Apply sealer at the ends of the bearing caps and install the rocker arms, rocker shafts and bearing caps as an assembly. Properly position the arrows on the bearing caps.

16. Torque the bearing cap bolts in the following sequence: No. 3, No. 2, No. 1 and No. 4 to 85 inch lbs. (10 Nm).

17. Repeat the sequence increasing the torque to 14 ft. lbs. (20 Nm).

18. Remove the auto lash adjuster retainer tools from the rocker arms.

19. Install the camshaft sprocket and bolt.

20. Using camshaft sprocket holding tool MB990767 and MD998719 or equivalent, hold the sprocket and tighten the bolt to 65 ft. lbs. (90 Nm).

21. Install the timing belt and valve covers.

22. Using a new O-ring, install the distributor extension housing.

23. Install the intake manifold plenum stay bracket.

24. Install the distributor assembly. Be sure to align the rotor and distributor housing matchmarks.

25. Connect the negative battery cable and check for leaks.

3.5L ENGINE

1. Disconnect the negative battery cable.
2. Remove the timing belt.
3. Remove the rocker arm cover.
4. Install the lash adjuster clips on the rocker arms, then loosen the bearing cap bolts. Do not remove the bolts from the bearing caps.
5. Remove the rocker arms, shafts and bearing caps as an assembly.
6. Remove the camshaft(s). Do not confuse the front and rear camshafts.

To install:

7. Lubricate the camshaft with engine oil and position it of the cylinder head.

8. Position the dowel pins as shown in the drawing.

9. Install the bearing caps/rocker arm assemblies. Tighten the bolts to 23 ft. lbs. (31 Nm).

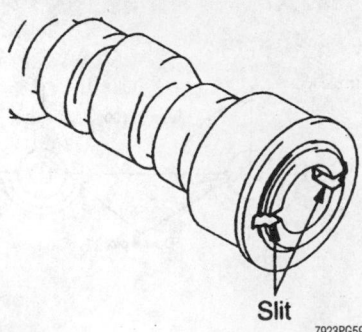

Slit

7923PG59

Right bank camshaft identification—Diamante 3.0L SOHC engine

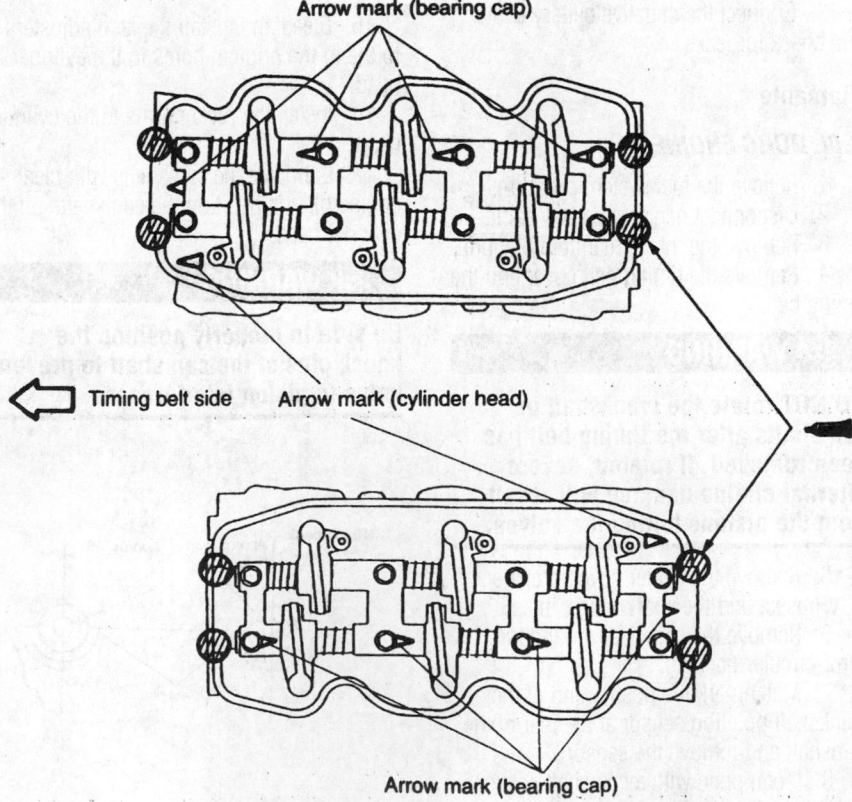

Arrow mark (bearing cap)

⬅ Timing belt side   Arrow mark (cylinder head)

Arrow mark (bearing cap)

7923PG61

Alignment of the rocker shafts and application of sealant—Diamante 3.0L SOHC engine

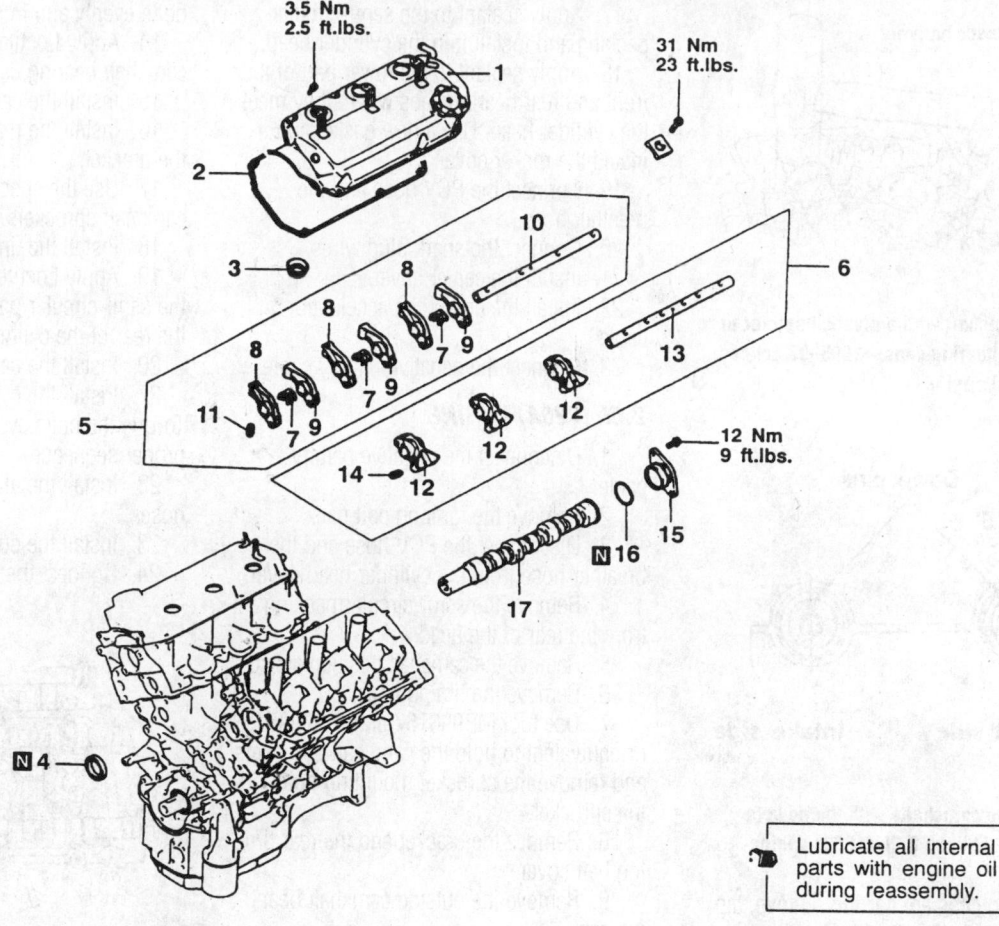

Removal steps

1. Rocker cover
2. Rocker cover gasket
3. Oil seal
4. Camshaft oil seal
5. Rocker arm, rocker arm shaft
6. Rocker arm, rocker arm shaft
7. Rocker shaft spring
8. Rocker arm A
9. Rocker arm B
10. Rocker arm shaft
11. Lash adjuster
12. Rocker arm C
13. Rocker arm shaft
14. Lash adjuster
15. Thrust case
16. O-ring
17. Camshaft

Lubricate all internal parts with engine oil during reassembly.

7923PGD3

Exploded view of the camshaft mounting—3.5L engine

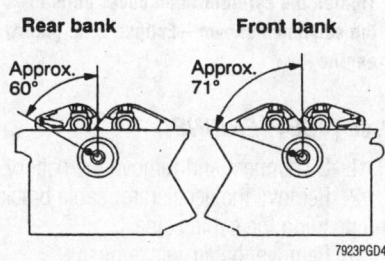

Camshaft dowel position during installation—3.5L engine

10. Install the rocker arm cover using a new gasket.

11. Install the timing belt and remaining components.

12. Connect the negative battery cable.

Eclipse

2.0L (4G63) ENGINE

1. Disconnect the negative battery cable.

2. Disconnect the accelerator cable from the throttle body and remove the cable bracket from the intake plenum.

3. Remove the engine center cover.

4. Disconnect the spark plug cables from the spark plugs. Label them if necessary.

5. Disconnect the breather hose and the PCV hose from the rocker cover.

6. Remove the rocker cover.

7. Position the No. 1 cylinder at TDC on compression.

8. Remove the timing belt.

9. Use a wrench on the hex shaped part of the camshaft to hold the cam and remove the camshaft sprockets.

10. Loosen the bearing cap bolts in two or three steps and remove the bearing caps.

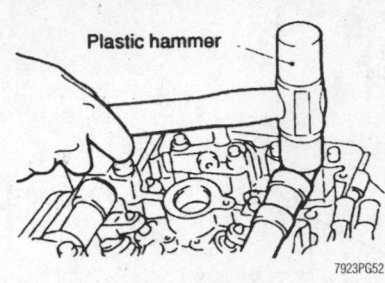

Tap the camshaft with a plastic hammer to loosen the bearing caps—1995–97 Eclipse 2.0L (4G63) engine

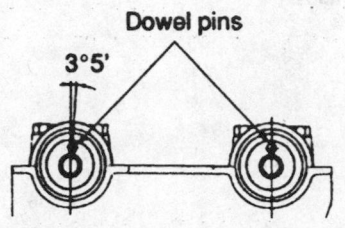

Position the camshafts with the dowels facing up—Eclipse 2.0L (4G63) engine

If the bearing caps are hard to remove, tap the rear of the camshaft with a plastic hammer.

11. Remove the camshaft(s) and the oil seals.

To install:

12. Apply engine oil or assembly lube to the camshafts and install them on the cylinder head.

❋❋ WARNING

If new camshaft(s) are being installed, remove the rocker arms and install the camshaft(s) and the bearing caps. Be sure the camshaft(s) can be turned by hand. After checking, remove the camshafts and install the rocker arms.

➡**Bearing caps and rocker arms must be installed in the same location that they were remove from.**

13. Install the bearing caps and tighten the bolts evenly in two or three steps to specifications.

14. Apply engine oil to the lip of the seal. Using MB998713, install the front oil seal.

15. Install the camshaft sprockets.

16. Install the timing belt.

17. Apply sealant to the semi-circular packing and install it in the cylinder head.

18. Apply sealant to the lower part of the front and rear bearing caps where they meet the cylinder head. Use a new gasket and install the rocker cover.

19. Connect the PCV hose and the breather hose.

20. Connect the spark plug wires.

21. Install the center cover.

22. Install the adjust the accelerator cable.

23. Connect the negative battery cable.

2.0L (420A) ENGINE

1. Disconnect the negative battery cable.

2. Remove the ignition coil pack.

3. Disconnect the PCV hose and the breather hose from the cylinder head cover.

4. Remove the semi-circular packing from the rear of the head.

5. Remove the camshaft position sensor.

6. Remove the timing belt.

7. Use tool MB990767 and MB998719 or equivalent to hold the camshaft sprockets and remove the sprocket mounting bolt and the sprocket.

8. Remove the bracket and the rear timing belt cover.

9. Remove the outside camshaft bearing cap.

10. Gradually loosen the camshaft bearing caps in sequence, one camshaft at a time and remove the bearing caps.

➡**Keep the bearing caps in order. They must be installed in the location that they were removed from.**

11. Mark the camshafts for later identification and remove the camshafts. The camshafts are not interchangeable.

To install:

12. Apply engine oil or assembly lube to the camshaft and install the camshafts.

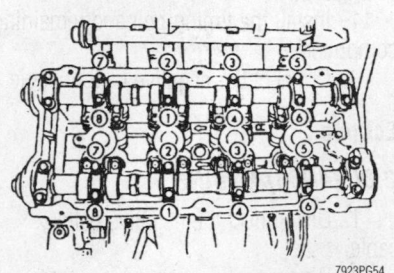

Camshaft bearing cap bolt removal sequence—Eclipse 2.0L (420A) engine

13. Install the bearing caps. Torque the bolts evenly and in sequence.

14. Apply Loctite 518® to the outside camshaft bearing caps and install them.

15. Install the camshaft oil seal.

16. Install the rear timing belt cover and the bracket.

17. Use the special tools and install the camshaft sprockets.

18. Install the timing belt.

19. Apply Loctite 5699® or equivalent to the semi-circular packing and install it in the rear of the cylinder head.

20. Install the camshaft position sensor.

21. Install the cylinder head cover. Torque the bolts evenly in three steps in the proper sequence.

22. Install the air, breather and PCV hoses.

23. Install the coil pack.

24. Connect the negative battery cable.

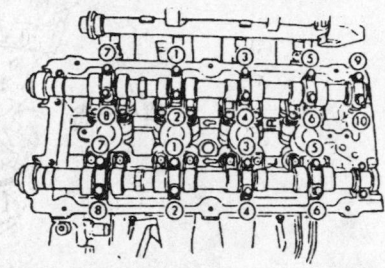

Camshaft bearing cap bolt installation sequence—Eclipse 2.0L (420A) engine

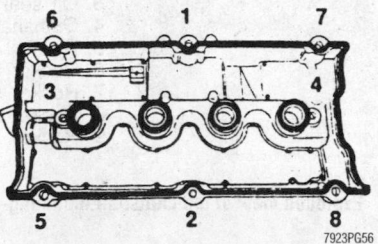

Tighten the cylinder head cover bolts in the sequence shown—Eclipse 2.0L (420A) engine

2.4L (4G64) ENGINE

1. Disconnect and remove the battery.

2. Remove the accelerator cable bracket and position the cable aside.

3. Remove the air intake hose.

4. Remove the breather hose and disconnect the PCV hose.

5. Label and disconnect the spark plug cables.

6. Remove the rocker cover.

7. Install lash adjuster retainer tools MD998443 or equivalent, to the rocker arm.

8. Remove the timing belt covers and the timing belt assembly.

9. While holding camshaft stationary, with an appropriate spanner wrench, remove the camshaft sprocket retainer bolt. Remove the sprocket from the shaft.

10. Remove the camshaft oil seal.

11. Remove both rocker arm shaft assemblies from the head. Do not disassembly rocker arms and rocker arm shaft assemblies.

12. Remove the camshaft from the cylinder head.

13. Inspect the bearing journals on the camshaft, cylinder head, and bearing caps.

To install:

14. Lubricate the camshaft journals and camshaft with clean engine oil and install the camshaft in the cylinder head.

15. Install the rocker arm and shaft assemblies. Tighten the rocker arm shaft retainer bolts to 21–25 ft. lbs. (29–35 Nm).

16. Apply a coating of engine oil to the oil seal. Using the proper size driver, press-fit the seal into the cylinder head.

17. Install camshaft sprocket and retainer bolt to 65 ft. lbs. (90 Nm).

18. Install the timing belt and belt covers.

19. Remove the lash adjuster retaining tools.

20. Install the rocker cover using new gasket material on mating surfaces.

21. Connect the spark plug cables.

22. Reinstall the air intake hose.

23. Install the breather hose and connect the PCV hose.

24. Install the battery.

25. Run the engine at idle until normal operating temperature is reached. Check idle speed and ignition timing; adjust as required.

Galant

2.4L (4G64) ENGINE

1. Relieve the fuel system pressure following proper procedure and disconnect negative battery cable.

2. Disconnect the accelerator cable, PCV hoses, breather hoses, spark plug cables and the remove the valve cover.

➡**Always rotate the crankshaft in a clockwise direction. Make a mark on the back of the timing belt indicating the direction of rotation so it may be reassembled in the same direction if it is to be reused.**

3. Rotate the crankshaft clockwise and align the timing marks so No. 1 piston will be at TDC of the compression stroke. At this time the timing marks on the camshaft sprocket and the upper surface of the cylinder head should coincide, and the dowel pin of the camshaft sprocket should be at the upper side.

4. Remove the timing belt upper and lower covers.

5. Remove the camshaft timing belt.

6. Use a wrench between No. 2 and No. 3 journals to hold the camshaft; remove the camshaft sprockets.

7. Loosen the bearing cap bolts in 2–3 steps. Label and remove all camshaft bearing caps.

➡**If the bearing caps are difficult to remove, use a plastic hammer to gently tap the rear part of the camshaft.**

8. Remove the intake and exhaust camshafts.

9. Remove the rocker arms and lash adjusters.

➡**It is essential that all parts be kept in the same order and orientation for reinstallation. In order to prevent confusion during installation, be sure to mark and separate all parts.**

To install:

10. Install the lash adjusters and rocker arms into the cylinder head. Lubricate lightly with clean oil prior to installation.

11. Lubricate the camshafts with heavy engine oil and position the camshafts on the cylinder head.

12. Check the camshaft journals and lobes for wear or damage. Also, check the cylinder head oil holes for clogging. Visually inspect the rocker arm roller and replace if dent, damage or seizure is evident. Check the roller for smooth rotation. Replace if excess play or binding is present. Also, inspect valve contact surface for possible damage or seizure. It is recommended

that all rocker arms and lash adjusters be replaced together.

➡**Do not confuse the intake camshaft with the exhaust camshaft. The intake camshaft has a split on the rear face for driving the crank angle sensor.**

13. Be sure the dowel pin on both camshaft sprocket ends are located on the top.

14. Install the bearing caps. Tighten the caps in sequence and in 2 or 3 steps. No. 2 and 5 caps are of the same shape. Check the markings on the caps to identify the cap number and intake/exhaust symbol. Only **L** (intake) or **R** (exhaust) is stamped on No. 1 bearing cap. Also, be sure the rocker arm is correctly mounted on the lash adjuster and the valve stem end. Torque the retaining bolts to 15 ft. lbs. (20 Nm).

15. Apply a coating of engine oil to the oil seal. Using the proper size driver, press-fit the seal into the cylinder head.

16. Install the camshaft sprockets. While holding the camshaft at its hexagon, between number 2 and 3 journals tighten sprocket bolts to 58–72 ft. lbs. (80–100 Nm).

17. Install the timing belt, covers and related components.

18. Install the valve cover, using new gasket, and reconnect all related components.

19. Reconnect the negative battery cable.

Mirage

1.5L (4G15) ENGINE

1. Disconnect the negative battery cable.

2. Rotate the engine and position the No. 1 piston to TDC of its compression stroke.

3. Disconnect the accelerator cable, breather hose and PCV hose connections.

4. Matchmark the positioning of the distributor housing and the positioning of the distributor rotor to the engine block and remove the distributor.

5. Remove the valve cover and discard the gasket.

6. Loosen both rocker arm assemblies gradually and evenly and remove the rocket shafts from the vehicle.

7. Remove the timing belt covers.

➡**DO NOT allow the camshaft or the crankshaft to rotate after the timing belt is removed.**

8. Remove the timing belt assembly.

9. Holding the camshaft sprocket from turning, loosen and remove the bolt that secures the sprocket.

Intake side   Exhaust side

Slits

7923PG57

Camshaft identification—2.4L (4G64) engine

10. Remove the camshaft sprocket from the camshaft. Note the positioning of the dowel pin at the end of the camshaft.

11. Remove the camshaft oil seal from the front of the cylinder head.

12. Remove the camshaft from the head.

13. Carefully check all parts for damage and wear.

To install:

14. Lubricate the camshaft with heavy engine oil and slide it into the head. Be sure to position the dowel pin at the 12 o'clock position.

15. Check the camshaft end-play between the thrust case and camshaft. The camshaft end-play should be 0.002–0.008 in. (0.05–0.20mm). If the end-play is not within specification, replace the camshaft thrust bearing.

16. Install a new camshaft oil seal. Be sure to lubricate the lips of the seal with clean engine oil.

17. Install the camshaft sprocket and install the mounting bolt. Tighten the bolt to 51 ft. lbs. (70 Nm) while holding the camshaft from turning.

18. Install the timing belt assembly.

19. Install the timing belt covers.

20. Install the rocker shaft assemblies. Torque the bolts gradually and evenly to 23 ft. lbs. (32 Nm).

21. Check valve adjustment and install the valve cover with a new gasket. Tighten the valve cover bolt to 16 inch lbs. (1.8 Nm).

22. Align the distributor marks and install the distributor.

23. Connect the accelerator cable, breather hose and PCV hose.

24. Connect the negative battery cable and check the ignition timing.

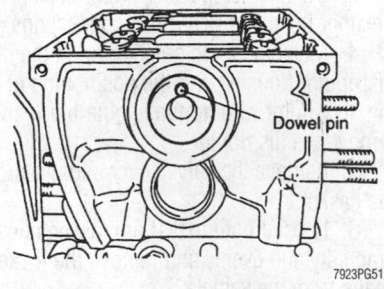

Positioning of the camshaft dowel pin—Mirage 1.5L (4G15) engine

1.8L (4G93) ENGINE

1. Disconnect the negative battery cable.

2. Rotate the engine and position the No. 1 piston to TDC of its compression stroke.

3. Label and disconnect the spark plug cables.

4. Matchmark the positioning of the distributor housing and the positioning of the distributor rotor to the engine block and remove the distributor.

5. Disconnect the air flow sensor connector and remove the air cleaner case cover.

6. Disconnect the accelerator cable, breather hose and PCV hose connections.

7. Remove the rocker cover and discard the gasket.

8. Loosen both rocker arm shaft assemblies gradually and evenly and remove the rocket shafts from the vehicle. Do not disassembly rocker arms and rocker arm shaft assemblies.

9. Remove the timing belt covers.

➡**DO NOT allow the camshaft or the crankshaft to rotate after the timing belt is removed.**

10. Remove the timing belt assembly.

11. Holding the camshaft sprocket from turning, loosen and remove the bolt that secures the sprocket.

12. Remove the camshaft sprocket from the camshaft. Note the positioning of the dowel pin at the end of the camshaft.

13. Remove the camshaft oil seal from the front of the cylinder head.

14. Remove the camshaft from the head.

15. Carefully check all parts for damage and wear.

To install:

16. Lubricate the camshaft journals and camshaft with clean engine oil and install the camshaft in the cylinder head. Be sure to position the dowel pin at the end of the camshaft as noted during the removal procedure.

17. Check the camshaft end-play between the thrust case and camshaft. The camshaft end-play should be 0.002–0.008 in. (0.05–0.20mm). If the end-play is not within specification, replace the camshaft thrust bearing.

18. Install a new camshaft oil seal. Be sure to lubricate the lips of the seal with clean engine oil.

19. Install camshaft sprocket and tighten the retainer bolt to 65 ft. lbs. (90 Nm). Be sure to secure the sprocket while tightening the bolt.

20. Install the timing belt assembly.

21. Install the timing belt covers.

22. Install the rocker arm and shaft assemblies. Tighten the rocker arm shaft retainer bolts to 23 ft. lbs. (32 Nm).

23. Check valve adjustment and install valve cover with a new gasket. Tighten the valve cover bolts to 29 inch lbs. (3.3 Nm).

24. Align the distributor marks and install the distributor.

25. Connect the spark plug cables.

26. Connect the accelerator cable, breather hose and PCV hose.

27. Connect the air flow sensor connector and install the air cleaner case cover.

28. Connect the negative battery cable. Run the engine at idle until normal operating temperature is reached. Check idle speed and ignition timing and adjust as required.

Valve Lash

ADJUSTMENT

Except 1995–97 Mirage

Valve clearance is not adjustable on these vehicles. If the valve makes noise, look for a leaky lash adjuster or excessive camshaft or rocker arm wear.

1995–97 Mirage

➡**Incorrect valve clearances will cause unsteady engine operation, excessive noise and reduced engine performance. Check the valve clearances and adjust as required while the engine is hot.**

1. Warm the engine to operating temperature, turn **OFF** and disconnect the negative battery cable.

2. Remove all spark plugs so engine can be easily turned by hand.

3. Remove the valve cover.

4. Turn the crankshaft clockwise to position the No. 1 cylinder at TDC of its compression stroke. The notch on the crankshaft pulley will be aligned with the **T** mark on the timing belt lower cover.

➡**When the No. 1 cylinder at TDC of its compression stroke, the No. 1 cylinder intake and exhaust valves will have valve lash.**

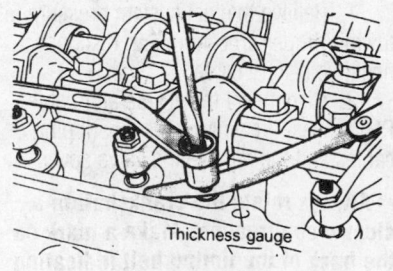

Proper method of adjusting valve clearance—Mirage

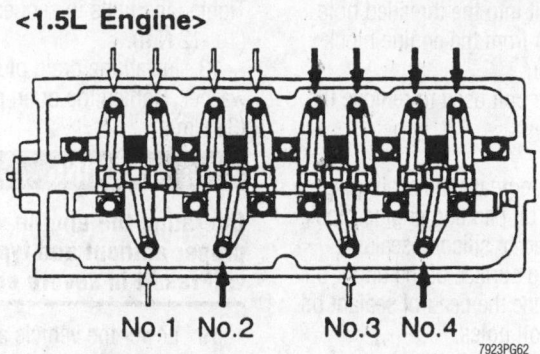

<1.5L Engine>

No.1 No.2   No.3 No.4

Adjust the valves with the white arrows when the No. 1 piston is at TDC on compression and the valves with the black arrows when the No. 4 piston is at TDC on compression—1.5L engine

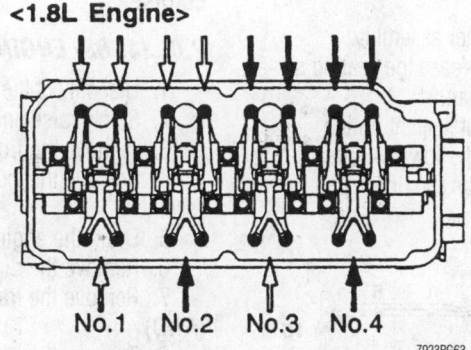

<1.8L Engine>

No.1   No.2   No.3   No.4

Adjust the valves with the white arrows when the No. 1 piston is at TDC on compression and the valves with the black arrows when the No. 4 piston is at TDC on compression—1.8L engine

5. Check the valve lash at cylinder No. 1 intake, cylinder No. 1 exhaust, cylinder No. 2 intake and cylinder No. 3 exhaust valves.

6. Correct valve lash for the 1.5L engine is as follows:
- Intake valve—0.008 in. (0.20mm)
- Exhaust valve—0.010 in. (0.25mm)

7. Correct valve lash for the 1.8L engine is as follows:
- Intake valve—0.008 in. (0.20mm)
- Exhaust valve—0.012 in. (0.30mm)

8. Rotate the crankshaft clockwise 1 complete turn and align the **T** mark with the notch on the crankshaft pulley. This will position the No. 4 cylinder at TDC of its compression stroke.

➡**When the No. 4 cylinder at TDC of its compression stroke, the No. 4 cylinder intake and exhaust valves will have valve lash.**

9. Check the valve lash at cylinder No. 2 exhaust, cylinder No. 3 intake and cylinder No. 4 intake and exhaust valves.

10. If the valve clearances are out of specification, loosen the rocker arm locknut and adjust the clearance using a feeler gauge while turning the adjusting screw. Be sure to hold the screw to prevent it from turning when

tightening the locknut. Tighten the locknut to 11 ft. lbs. (15 Nm) on the 1.5L engine or 7 ft. lbs. (9 Nm) on the 1.8L engine.

11. After adjusting the valves, install the valve cover with new gasket and spark plugs, and connect the negative battery cable.

Oil Pan

REMOVAL & INSTALLATION

3000GT

1. Disconnect the negative battery cable.
2. Raise the vehicle and support safely.
3. Remove the oil pan drain plug and drain the engine oil.
4. On vehicles equipped with AWD, remove the transfer assembly.
5. Disconnect and lower the exhaust pipe and on turbocharged engines, disconnect the return pipe for the turbocharger from the side of the oil pan.
6. Remove the oil pan mounting bolts.
7. Using the special tool, separate and remove the engine oil pan.
To install:
8. Thoroughly clean and dry the oil pan, cylinder block bolts and bolt holes.

9. Apply a thin bead of sealer around the surface of the oil pan.
10. Assemble the oil pan to the cylinder block within 15 minutes after applying the sealant.
11. Install the oil pan mounting bolts and tighten to 4–6 ft. lbs. (6–8 Nm).
12. Fill the engine with the proper amount of oil.
13. Connect the negative battery cable and check for leaks.
14. Safely lower the vehicle to the floor.

Diamante

3.0L ENGINES

1. Disconnect the negative battery cable.
2. Raise the vehicle and support safely.
3. Remove the oil pan drain plug and drain the engine oil.
4. Remove the left side crossmember. If equipped with 4WS, it will also be necessary to remove the right side crossmember.
5. Remove the starter motor.
6. Disconnect the roll stopper stay bracket, from the rear transaxle stay bracket. Remove the both transaxle stay brackets.
7. Remove the bell housing lower cover.
8. Remove the oil pan mounting bolts. Using special tool MD998727 or equivalent, separate and remove the engine oil pan.
To install:
9. Thoroughly clean and dry the oil pan, cylinder block bolts and bolt holes.
10. Apply a 0.16 in. (4mm) continuous bead of sealer around the surface of the oil pan.

➡**Assemble the oil pan to the cylinder block within 15 minutes after applying the sealant.**

11. Install the oil pan mounting bolts. Following proper sequence, tighten mounting bolts to 48 inch lbs. (6 Nm).
12. Install lower bell housing cover and the starter motor.

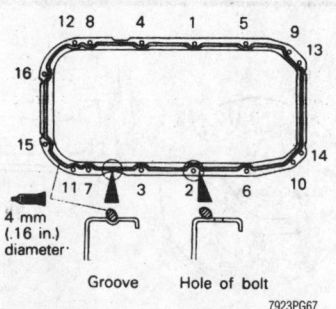

Oil pan bolt tightening sequence and application of sealant to the pan—Diamante 3.0L (J- and H-) engines

13. Install the transaxle stay brackets and connect the roll stopper bracket.

14. Install the crossmember(s) and tighten the mounting bolts to 43–51 ft. lbs. (60–70 Nm).

15. Fill the engine with the proper amount of oil.

16. Connect the negative battery cable and check for leaks.

3.5L ENGINE

1. Disconnect the negative battery cable.

2. Raise and safely support the vehicle.

❊❊ CAUTION

The EPA warns that prolonged contact with used engine oil may cause a number of skin disorders, including cancer! You should make every effort to minimize your exposure to used engine oil. Protective gloves should be worn when changing the oil. Wash your hands and any other exposed skin areas as soon as possible after exposure to used engine oil. Soap and water, or waterless hand cleaner should be used.

3. Drain the engine oil.

4. Remove the mounting bolts from the lower oil pan.

5. Place a block of wood against the side of the pan and tap the block with a hammer to break the seal and remove the lower pan.

6. Remove the starter.

7. Remove the dipstick tube.

8. Remove the two bolts along the rear (opposite of timing belt side) side of the upper oil pan. Remove all the remaining mounting bolts.

❊❊ WARNING

Do not pry or use seal breaker tool to remove the oil pan. Damage to the aluminum surface can result.

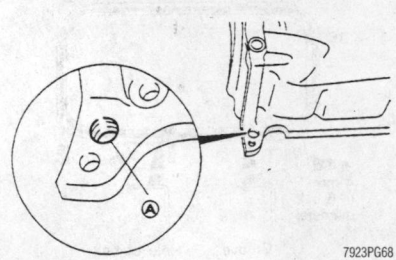

Install a bolt in the threaded hole to force the oil pan from the engine block—3.5L engine

9. Screw a bolt into the threaded hole to force the oil pan from the engine block and remove the pan.

10. Remove the bolt used to remove the pan.

To install:

11. Clean and degrease the sealing surfaces of the upper oil pan and engine block.

12. Apply a bead of silicone sealant along the mounting surface of the upper oil pan. Be sure to place the bead of sealant on the inside of the bolt holes.

13. Install the oil pan on the engine block. Tighten the bolts in sequence to 4 ft. lbs. (6 Nm).

14. Install the dipstick tube using a new O-ring.

15. Install the starter assembly.

16. Clean and degrease the sealing surface of the lower oil pan.

17. Place a bead of sealant on the mounting surface of the lower oil pan. Install the lower pan on the upper pan.

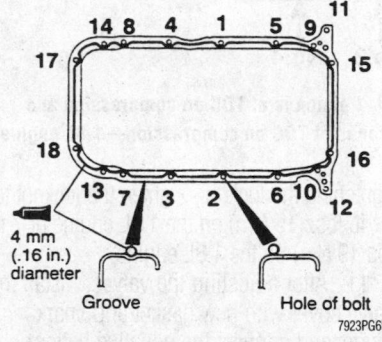

Apply sealant and tighten the bolts in the order shown—3.5L engine, upper oil pan

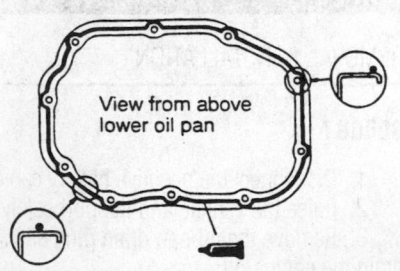

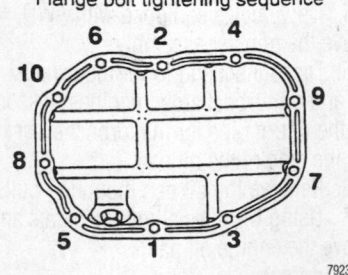

Apply sealant and tighten the bolts in the order shown—3.5L engine, lower oil pan

Tighten the bolts in sequence to 7–9 ft. lbs. (10–12 Nm).

18. Install the drain plug using a new washer. Tighten the drain plug to 29 ft. lbs. (39 Nm).

❊❊ WARNING

Operating the engine without the proper amount and type of engine oil will result in severe engine damage.

19. Lower the vehicle and fill the engine with the correct amount of new engine oil.

20. Connect the negative battery cable.

21. Start the engine and check for leaks.

Eclipse

2.0L (4G63) ENGINE

1. Disconnect the negative battery cable.

2. Safely raise and support the vehicle.

3. Remove the front exhaust pipe.

4. Remove the exhaust pipe and muffler assembly.

5. Drain the engine oil.

6. Remove the dipstick and tube.

7. Remove the transfer case assembly (AWD).

8. Remove the bell housing cover.

9. Disconnect the oil return pipe from the oil pan.

10. Remove the oil pan mounting bolts. Tap the oil pan seal breaker MB998727 or equivalent between the oil pan and the engine block to break the seal and remove the oil pan.

To install:

11. Clean the sealing surface on the oil pan and engine block. Apply a continuous bead of sealant MD970389 or equivalent to the oil pan.

12. Clean the oil pan mounting bolt holes in the oil seal case.

13. Install the oil pan to the engine block. Torque the mounting bolts to 5.1 ft. lbs. (6.9 Nm).

14. Use a new gasket and connect the oil return pipe to the oil pan.

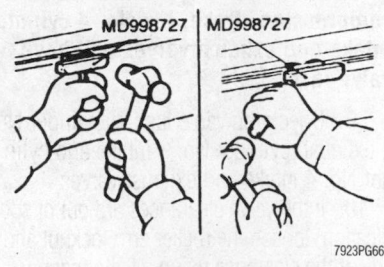

Using the special tool to remove the oil pan—Eclipse 2.0L (4G63) engine

15. Install the bell housing cover.
16. Install the transfer case assembly if equipped.
17. Install the dipstick and tube assembly.
18. Install the front exhaust pipe.
19. Install the exhaust pipe and muffler.
20. Install a new oil filter.
21. Safely lower the vehicle to the floor and add five quarts of oil to the crankcase.
22. Connect the negative battery cable.
23. Start the engine and check for leaks.

2.0L (420A) ENGINE

1. Disconnect the negative battery cable.
2. Raise and safely support the vehicle.
3. Drain the engine oil.
4. Remove the front exhaust pipe.
5. Remove the dipstick and tube assembly.
6. Remove the front plate.
7. Remove the oil pan mounting bolts.
8. Remove the oil pan and gasket.

To install:

9. Apply sealant at the point where the engine block meets the oil pump.
10. Use a new gasket and install the oil pan. Torque the mounting bolts to 8.9 ft. lbs. (12 Nm).
11. Install the front plate.
12. Install the front exhaust pipe.
13. Install the dipstick and tube assembly.
14. Safely lower the vehicle to the floor.
15. Refill the crankcase with oil to the proper level.
16. Connect the negative battery cable.
17. Start the engine and check for leaks.

2.4L (4G64) ENGINE

1. Remove the negative battery cable. Wait at least 90 seconds before performing any work.
2. Drain the engine oil.
3. Remove the engine dipstick and tube assembly.
4. Remove the front exhaust pipe.
5. Remove the bell housing inspection cover.
6. Remove the bolts attaching the oil pan to the cylinder block.
7. Using the special tool remove the oil pan assembly.

To install:

8. Be sure that the oil pan and cylinder block mating surfaces are free of any old sealing material. Apply the specified sealant or the equivalent.
9. Install the oil pan to the cylinder block and tighten the bolts to 5 ft. lbs. (7 Nm).
10. Reinstall the bell housing inspection cover. Torque the bolts to 7 ft. lbs. (9 Nm).
11. Install the front exhaust pipe.

12. Reinstall the engine dipstick and tube assembly using a new O-ring.
13. Refill the engine with oil. Reconnect the negative battery cable. Start the engine and check for leaks.

Galant

1. Disconnect the negative battery cable.
2. Raise the vehicle and support safely.
3. Remove the oil pan drain plug and drain the engine oil.
4. Remove the oil dipstick and tube assembly.
5. Disconnect the oxygen sensor connector. Remove the nuts and disconnect the front exhaust pipe from the exhaust manifold. Disconnect the front exhaust pipe from the catalytic converter and remove the front exhaust pipe from the vehicle.
6. Remove the four bolts securing the bell housing cover, to the engine and transmission.
7. Remove the oil pan retainer bolts. Using special tool MD998727 or equivalent, tap in between the engine block and the oil pan.

➡**Do not use a chisel, screwdriver or similar tool when removing the oil pan. Damage to engine components may occur.**

8. Inspect the oil pan for damage and cracks. Replace if faulty. While the pan is removed, inspect the oil screen for clogging, damage and cracks. Replace if faulty.

To install:

9. Using a wire brush or other tool, scrape clean all gasket surfaces of the cylinder block and the oil pan so that all loose material is removed. Clean sealing surfaces of all dirt and oil.
10. Apply sealant around the gasket surfaces of the oil pan in such a manner that all bolt holes are circled and there is a continuous bead of sealer around the entire perimeter of the oil pan.

➡**The continuous bead of sealer should be applied in a bead approximately 0.16 in. (4mm) in diameter.**

11. Install the oil pan onto the cylinder block within 15 minutes after applying sealant. Install the fasteners and tighten to 6 ft. lbs. (8 Nm).
12. Install the oil drain plug and tighten to 29 ft. lbs. (39 Nm), If not already done.
13. Install the bell housing cover, and tighten the mounting bolts to 7 ft. lbs. (9 Nm).
14. Reconnect the front exhaust pipe to the catalytic converter with a new gasket and tighten the bolts to 36 ft. lbs. (49 Nm).

Reinstall the front exhaust pipe with a new gasket to the exhaust manifold and tighten the nuts to 32 ft. lbs. (44 Nm).
15. Reconnect the oxygen sensor connector.
16. Lower the vehicle and fill the crankcase to the proper level with clean engine oil.
17. Connect the negative battery cable. Start the engine and check for leaks.

Mirage

1.5L (4G15) ENGINE

1. Disconnect the negative battery cable.
2. Raise the vehicle and support safely.
3. Remove the oil pan drain plug and drain the engine oil.
4. Remove the bell housing lower cover.
5. Remove the oil pan retainer bolts. Tap the oil pan with a rubber mallet to break seal.

➡**Do not use a chisel, screwdriver or similar tool when removing the oil pan. Damage to engine components may occur. If available, oil pan remover tool MD998727 or equivalent may be used break the seal.**

6. Inspect the oil pan for damage and cracks. Replace if faulty. While the pan is removed, inspect the oil screen for clogging, damage and cracks. Replace if faulty.

To install:

7. Using a wire brush or other tool, scrape clean all gasket surfaces of the cylinder block and the oil pan so that all loose material is removed. Clean sealing surfaces of all dirt and oil.
8. Apply sealant around the gasket surfaces of the oil pan in such a manner that all bolt holes are circled and there is a continuous bead of sealer around the entire perimeter of the oil pan.

➡**The continuous bead of sealer should be applied in a bead approximately 0.16 in. (4mm) in diameter.**

9. Install the oil pan onto the cylinder block within 15 minutes after applying sealant. Install the fasteners and tighten to 60 inch lbs. (7 Nm).
10. Install the bell housing cover.
11. Install the oil drain plug with a new seal and tighten to 29 ft. lbs. (40 Nm).
12. Lower the vehicle and fill the crankcase to the proper level with clean engine oil.
13. Connect the negative battery cable. Start the engine and check for leaks.

1.8L (4G93) ENGINE

1. Disconnect the negative battery cable.
2. Raise the vehicle and support safely.

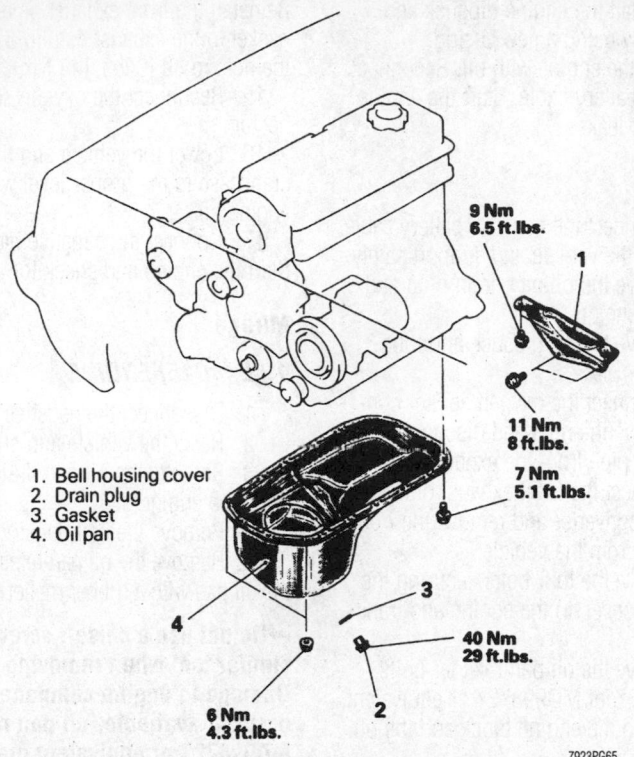

9 Nm
6.5 ft.lbs.

11 Nm
8 ft.lbs.

7 Nm
5.1 ft.lbs.

40 Nm
29 ft.lbs.

6 Nm
4.3 ft.lbs.

1. Bell housing cover
2. Drain plug
3. Gasket
4. Oil pan

7923PG65

Oil pan and related components—Mirage 1.5L (4G15) engine

3. Remove the oil pan drain plug and drain the engine oil.

4. Disconnect and lower the exhaust pipe from the engine manifold.

5. Remove the bell housing lower cover.

6. Remove the oil pan retainer bolts. Tap the oil pan with a rubber mallet to break the seal.

➡**Do not use a chisel, screwdriver or similar tool when removing the oil pan. Damage to engine components may occur. If available, oil pan remover tool MD998727 or equivalent may be used break the seal.**

7. Inspect the oil pan for damage and cracks. Replace if faulty. While the pan is removed, inspect the oil screen for clogging, damage and cracks. Replace if faulty.

To install:

8. Using a wire brush or other tool, scrape clean all gasket surfaces of the cylinder block and the oil pan so that all loose material is removed. Clean sealing surfaces of all dirt and oil.

9. Apply sealant around the gasket surfaces of the oil pan in such a manner that all bolt holes are circled and there is a continuous bead of sealer around the entire outside edge of the oil pan.

➡**The continuous bead of sealer should be applied in a bead approximately 0.16 in. (4mm) in diameter.**

10. Install the oil pan onto the cylinder block within 15 minutes after applying sealant. Install the fasteners and tighten to 60 inch lbs. (5 Nm).

11. Install the bell housing cover.

12. Connect the exhaust pipe from the engine manifold with new gasket in place. Tighten the exhaust pipe to manifold flange nuts to 33 ft. lbs. (45 Nm). Install and tighten the support bolt to 18 ft. lbs. (25 Nm).

13. Install the oil drain plug and tighten to 29 ft. lbs. (40 Nm).

14. Lower the vehicle and fill the crankcase to the proper level with clean engine oil.

15. Connect the negative battery cable. Start the engine and check for leaks.

Oil Pump

REMOVAL & INSTALLATION

3000GT and Diamante 3.0L Engines

3.0L ENGINES

➡**Whenever the oil pump is disassembled or the cover removed, the gear cavity must be filled with petroleum jelly to seal the pump and act as a prime. This allows the pump to draw oil as soon as the engine starts. Do not use grease.**

1. Disconnect the negative battery cable.

✳✳ CAUTION

Wait at least 90 seconds after the negative battery cable is disconnected to prevent possible deployment of the air bag.

2. Remove the front engine mount bracket and accessory drive belts.

3. Remove timing belt upper and lower covers.

4. Remove the timing belt and crankshaft sprocket.

5. Remove the oil pan.

6. Remove the oil screen and gasket.

7. Remove and tag the front cover mounting bolts. Note the lengths of the mounting bolts as they are removed for proper installation.

8. Remove the front case cover and oil pump assembly. Disassemble as required.

To install:

9. Thoroughly clean all gasket material from all mounting surfaces.

10. Apply engine oil to the entire surface of the gears or rotors.

11. Assemble the front case cover and oil pump assembly to the engine block using a new gasket.

12. Install the oil screen with new gasket.

13. Install the oil pan and timing belts.

14. Install the timing belt covers.

15. Install the drive belts and the front engine mount bracket.

16. Connect the negative battery cable, refill the crankcase and check for adequate oil pressure.

3.5L ENGINE

1. Disconnect the negative battery cable.

2. Remove the timing belt.

3. Raise and safely support the vehicle.

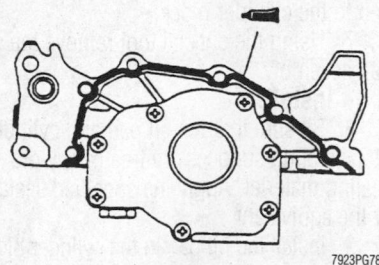

7923PG78

Apply sealant to the rear of the oil pump case—3.5L engine

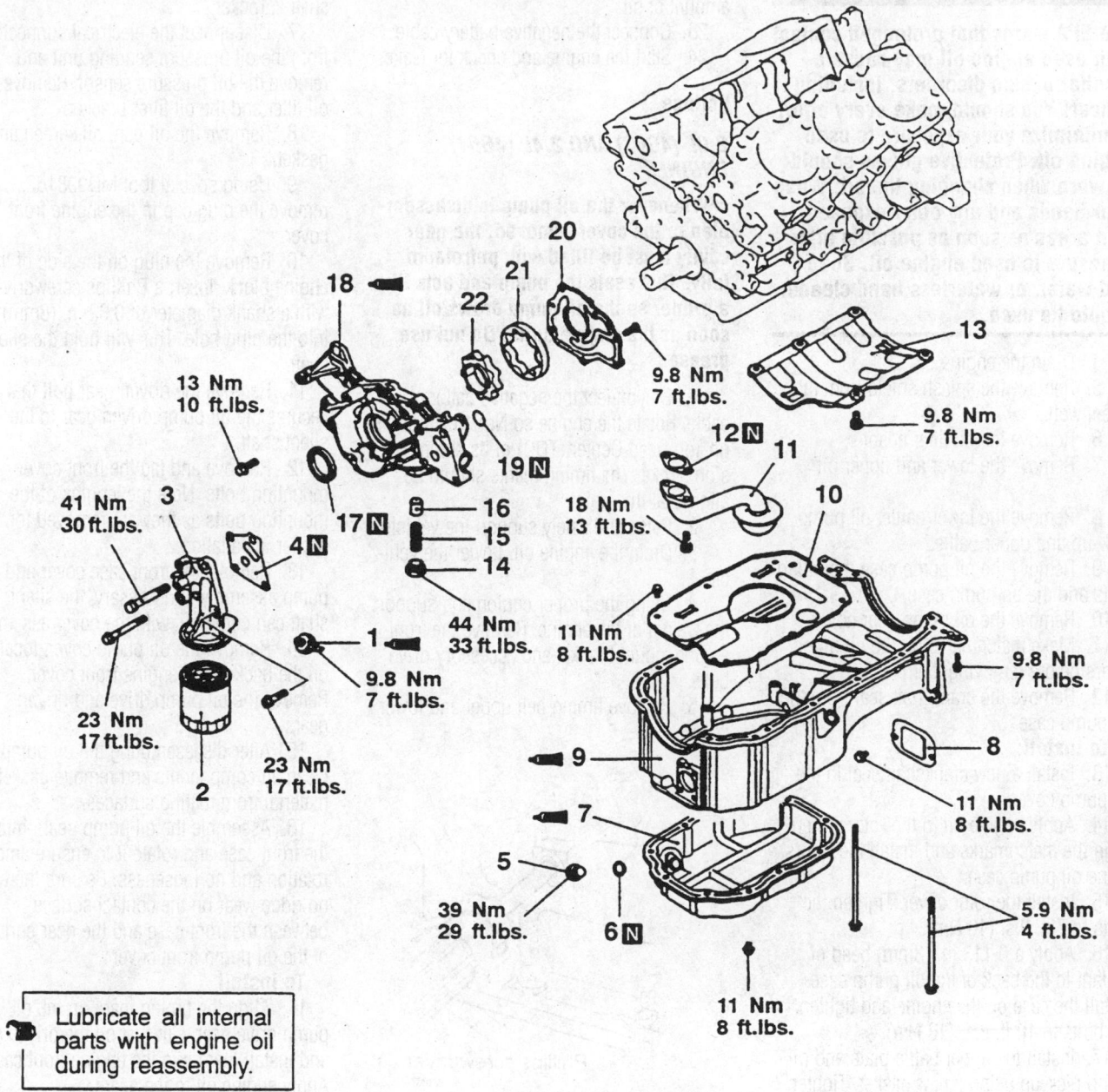

Lubricate all internal parts with engine oil during reassembly.

Removal steps

1. Oil pressure gauge unit
2. Oil filter
3. Oil filter bracket
4. Oil filter bracket gasket
5. Drain plug
6. Drain plug gasket
7. Oil pan, lower
8. Cover
9. Oil pan, upper
10. Baffle plate
11. Oil screen

12. Oil screen gasket
13. Baffle plate
14. Plug
15. Relief spring
16. Relief plunger
17. Crankshaft oil seal
18. Oil pump case
19. O-ring
20. Oil pump cover
21. Oil pump outer rotor
22. Oil pump inner rotor

Exploded view of the oil pump mounting—3.5L engine

7923PG77

4. Drain the engine oil.

5. Remove the splash shield from the wheel well.

6. Remove the oil filter adapter.

7. Remove the lower and upper oil pans.

8. Remove the lower baffle, oil pump pick-up and upper baffle.

9. Remove the oil pump case mounting bolts and the oil pump case.

10. Remove the oil pump gear cover.

11. Make matchmarks on the oil pump rotors before removing them.

12. Remove the crankshaft seal from the oil pump case.

To install:

13. Install a new crankshaft seal in the oil pump cover.

14. Apply engine oil to the rotors, then align the matchmarks and install the rotors in the oil pump case.

15. Install the rotor cover. Tighten the bolts to 7 ft. lbs. (10 Nm).

16. Apply a 0.113 in. (3mm) bead of sealant to the back of the oil pump case. Install the case on the engine and tighten the bolts to 10 ft. lbs. (13 Nm).

17. Install the upper baffle plate and oil pump pick-up using a new gasket. Tighten the baffle bolts to 7 ft. lbs. (10 Nm) and the pick-up bolts to 13 ft. lbs. (18 Nm).

18. Install the lower baffle in the upper oil pan. Tighten the bolts to 8 ft. lbs. (11 Nm).

19. Install the oil pans.

20. Install the oil filter adapter using a new gasket. Tighten the larger bolt to 30 ft. lbs. (41 Nm) and the smaller bolt to 17 ft. lbs. (23 Nm).

21. Install the timing belt and remaining components.

22. Fill the engine with the correct amount of oil.

23. Connect the negative battery cable.

24. Start the engine and check for leaks.

Eclipse

2.0L (4G63) AND 2.4L (4G64) ENGINES

➡ **Whenever the oil pump is disassembled or the cover removed, the gear cavity must be filled with petroleum jelly. This seals the pump and acts like a primer so the oil pump draws oil as soon as the engine turns. Do not use grease.**

1. Disconnect the negative battery cable. Rotate the engine so No. 1 cylinder is on Top Dead Center (TDC) of its compression stroke. The timing marks should be aligned at this point.

2. Raise and safely support the vehicle.

3. Drain the engine oil. Lower the vehicle.

4. Using the proper equipment, support the weight of the engine. Remove the front engine mount bracket and accessory drive belts.

5. Remove timing belt upper and lower covers.

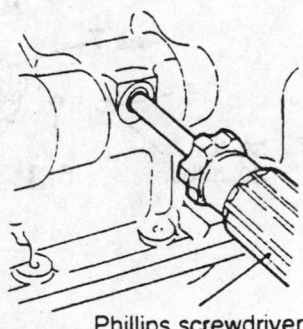

Phillips screwdriver

7923PG71

Holding the silent shaft for oil pump gear removal—Eclipse 2.0L (4G63) engine

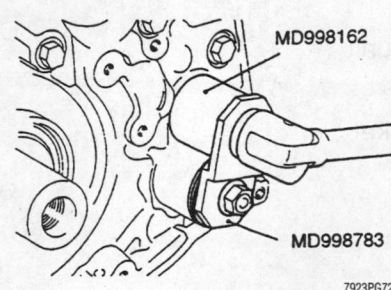

MD998162

MD998783

7923PG72

Use the special socket and holder to remove the balance shaft plug—Eclipse 2.0 (4G63) engine

6. Remove the timing belt and crankshaft sprocket.

7. Disconnect the electrical connector from the oil pressure sending unit and remove the oil pressure sensor. Remove the oil filter and the oil filter bracket.

8. Remove the oil pan, oil screen and gasket.

9. Using special tool MD998162, remove the plug cap in the engine front cover.

10. Remove the plug on the side of the engine block. Insert a Phillips screwdriver with a shank diameter of 0.32 in. (8mm) into the plug hole. This will hold the silent shaft.

11. Remove the driven gear bolt that secures the oil pump driven gear to the silent shaft.

12. Remove and tag the front cover mounting bolts. Note the lengths of the mounting bolts as they are removed for proper installation.

13. Remove the front case cover and oil pump assembly. If necessary, the silent shaft can come out with the cover assembly.

14. Remove the oil pump cover, located on the back of the engine front cover. Remove the oil pump drive and driven gears.

15. After disassembling the oil pump, clean all components and remove gasket material from mating surfaces.

16. Assemble the oil pump gears into the front case and rotate it to ensure smooth rotation and no looseness. Be sure there is no ridge wear on the contact surface between the front case and the gear surface of the oil pump front cover.

To install

17. Align the timing mark on the oil pump drive gear with that on the driven gear and install them into the engine front case. Apply engine oil to the gears.

18. Install the oil pump cover and tighten the retainer bolts to 13 ft. lbs. (18

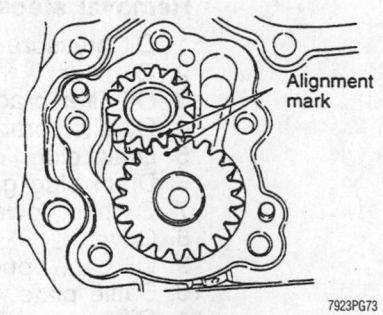

Alignment mark

7923PG73

Aligning oil pump timing marks—Eclipse 2.0L (4G63) engine

Nm) on Eclipse models and 17 ft. lbs. (24 Nm) on Galant models.

19. Using the appropriate driver, install a new crankshaft seal into the front case.

20. Position new front case gasket in place. Set seal guide tool MD998285 on the front end of the crankshaft to protect the seal from damage. Apply a thin coat of oil to the outer circumference of the seal pilot tool.

21. Install the front case assembly through a new front case gasket and temporarily tighten the flange bolts.

22. Mount the oil filter on the bracket with new oil filter bracket gasket in place. Install the bolts with washers and tighten to 14 ft. lbs. (19 Nm).

23. Insert a Phillips screwdriver into a hole in the left side of the engine block to lock the silent shaft in place.

24. Secure the oil pump drive gear onto the left silent shaft by installing and tightening the driven gear bolt to 27 ft. lbs. (37 Nm).

25. Install a new O-ring to the groove in the front case and install the plug cap. Using the special tool MD998162, tighten the cap to 17 ft. lbs. (24 Nm).

26. Install the oil screen in position with new gasket in place.

27. Clean both mating surfaces of the oil pan and the cylinder block. Apply sealant in the groove in the oil pan flange, keeping towards the inside of the bolt holes. The width of the sealant bead applied is to be about 0.016 in. (4mm) wide.

➡**After applying sealant to the oil pan, do not exceed 15 minutes before installing the oil pan.**

28. Install the oil pan to the engine and secure with the retainers. Tighten bolts to 5 ft. lbs. (7 Nm).

29. Install the oil pressure gauge unit and the oil pressure switch. Connect the electrical harness connector.

30. Install the oil cooler. Secure with oil cooler bolt tightened to 31 ft. lbs. (43 Nm).

31. Refill the crankcase. Install new oil filter.

32. Connect the negative battery cable and start the engine. Verify correct oil pressure. Inspect for leaks.

2.0L (420A) ENGINE

1. Disconnect the negative battery cable.

2. Raise the safely support the vehicle.

3. Drain the engine oil.

4. Remove the rear plate.

5. Remove the oil filter and adapter.

6. Remove the oil pan.

7. Remove the oil pick-up tube.

8. Remove the timing belt.

9. Using tool MB995027 or equivalent, remove the crankshaft sprocket.

✳✳✳ WARNING

Do not nick the crankshaft sealing surface or the seal bore.

10. Using tool MB995020 or equivalent, remove the crankshaft oil seal.

11. Remove the oil pump mounting bolts.

12. Remove the oil pump.

To install:

13. Apply a bead of the specified sealant to the sealing surface of the oil pump and install a new O-ring into the counterbore on the oil pump discharge passage.

14. Carefully install the oil pump on the crankshaft until seated to the engine block. Torque the bolts to 17 ft. lbs. (23 Nm).

15. Install a new crankshaft oil seal in the oil pump.

16. Install the crankshaft sprocket using the proper installation tools.

17. Install the timing belt and related components.

18. Install the oil pick-up tube.

19. Apply Loctite® 18718 or equivalent at the point where the oil pump meets the engine block.

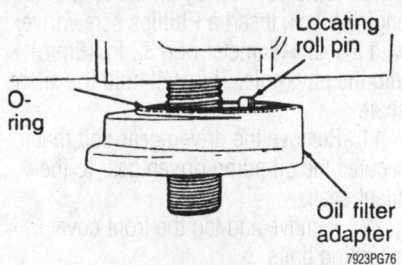

Oil filter adapter installation—Eclipse 2.0L (420A) engine

20. Install the oil pan using a new gasket. Torque the mounting bolts to 9 ft. lbs. (12 Nm).

21. Use a new O-ring and install the oil filter adapter to the engine. Made sure the roll pin aligns with the hole. Torque the assembly to 40 ft. lbs. (55 Nm).

22. Install a new oil filter.

23. Install the rear plate.

24. Safely lower the vehicle to the floor.

25. Refill the engine with the proper amount of oil.

26. Start the engine and check for leaks.

Galant

➡**Whenever the oil pump is disassembled or the cover removed, the gear cavity must be filled with petroleum jelly. This seals the pump and acts lime a prime so the oil pump draws oil as soon as the engine starts to turn. Do not use grease.**

1. Disconnect the negative battery cable. Rotate the engine so No. 1 cylinder is on Top Dead Center (TDC) of its compression stroke. The timing marks should be aligned at this point.

2. Raise and safely support the vehicle.

3. Drain the engine oil. Lower the vehicle.

4. Using the proper equipment, support the weight of the engine. Remove the front engine mount bracket and accessory drive belts.

5. Remove timing belt upper and lower covers.

6. Remove the timing belt and crankshaft sprocket.

7. Disconnect the electrical connector from the oil pressure sending unit and remove the oil pressure sensor. Remove the oil filter and the oil filter bracket.

8. Remove the oil pan, oil screen and gasket.

9. Using special tool MD998162, remove the plug cap in the engine front cover.

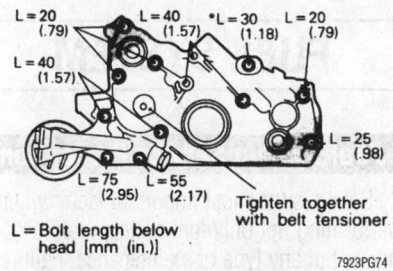

Front case bolt identification—Eclipse 2.0L (4G63) and 2.4L (4G64) engines

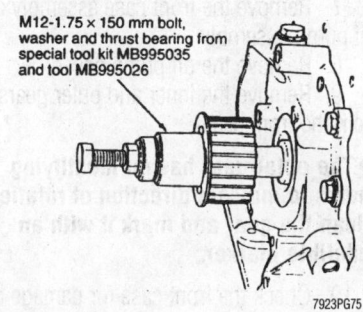

M12-1.75 × 150 mm bolt, washer and thrust bearing from special tool kit MB995035 and tool MB995026

Crankshaft sprocket installation—Eclipse 2.0L (420A) engine

10. Remove the plug on the side of the engine block. Insert a Phillips screwdriver with a shank diameter of 0.32 in. (8mm) into the plug hole. This will hold the silent shaft.

11. Remove the driven gear bolt that secures the oil pump driven gear to the silent shaft.

12. Remove and tag the front cover mounting bolts.

➡ **The mounting bolts are different lengths, make certain to identify their original location as they are removed, for proper installation.**

13. Remove the front case cover and oil pump assembly. If necessary, the silent shaft can come out with the cover assembly.

14. Remove the oil pump cover, located on the back of the engine front cover. Remove the oil pump drive and driven gears.

15. After disassembling the oil pump, clean all components and remove gasket material from mating surfaces.

16. Assemble the oil pump gears into the front case and rotate it to ensure smooth rotation and no looseness. Be sure there is no ridge wear on the contact surface between the front case and the gear surface of the oil pump front cover.

To install

17. Align the timing mark on the oil pump drive gear with that on the driven gear and install them into the engine front case. Apply engine oil to the gears.

18. Install the oil pump cover and tighten the retainer bolts to 17 ft. lbs. (24 Nm).

19. Using the appropriate driver, install a new crankshaft seal into the front case.

20. Position a new front case gasket in place. Set seal guide tool MD998285 on the front end of the crankshaft to protect the seal from damage. Apply a thin coat of oil to the outer circumference of the seal pilot tool.

21. Install the front case assembly through a new front case gasket and temporarily tighten the flange bolts.

22. Mount the oil filter on the bracket with new oil filter bracket gasket in place. Install the 3 bolts with washers and tighten to 16 ft. lbs. (22 Nm).

23. Insert a Phillips screwdriver into a hole in the left side of the engine block to lock the silent shaft in place.

24. Secure the oil pump drive gear onto the left silent shaft by installing and tightening the driven gear bolt to 29 ft. lbs. (40 Nm).

25. Install new O-ring to the groove in the front case and install the plug cap. Using the special tool MD998162, tighten the cap to 20 ft. lbs. (27 Nm).

26. Install the oil screen in position with new gasket in place.

27. Clean both mating surfaces of the oil pan and the cylinder block. Apply sealant in the groove in the oil pan flange, keeping towards the inside of the bolt holes. The width of the sealant bead applied is to be about 0.016 in. (4mm) wide.

➡ **After applying sealant to the oil pan, do not exceed 15 minutes before installing the oil pan.**

28. Install the oil pan to the engine and secure with the retainers. Tighten bolts to 6 ft. lbs. (8 Nm).

29. Install the oil pressure gauge unit and the oil pressure switch. Connect the electrical harness connector.

30. Refill crankcase with oil. Install new oil filter.

31. Install the timing belts and timing belt covers. Assemble the remaining components to the front of the engine.

32. Connect the negative battery cable and start the engine. Verify oil pressure. Inspect for leaks.

Mirage

➡ **Whenever the oil pump is disassembled or the cover removed, it is suggested the gear cavity is filled with petroleum jelly for priming purposes. Do not use grease.**

1. Disconnect the negative battery cable.

2. Remove the front engine mount bracket and accessory drive belts.

3. Remove timing belt upper and lower covers.

4. Remove the timing belt and crankshaft sprocket.

5. Remove the oil pan and remove the oil screen.

6. Remove and tag the front cover mounting bolts. Note the lengths of the mounting bolts as they are removed for proper installation.

7. Remove the front case assembly and oil pump assembly.

8. Remove the oil pump cover.

9. Remove the inner and outer gears from the front case.

➡ **The outer gear has no identifying marks to indicate direction of rotation. Clean the gear and mark it with an indelible marker.**

10. Check the front case for damage or cracks. Replace the front seal. Replace the oil screen O-ring. Clean all parts thoroughly with a safe solvent. Check the pump gears

for wear or damage, and ensure that the relief valve can slide freely in the case.

To install

11. Remove all gasket material from the mating surfaces and clean all parts.

12. Thoroughly coat both oil pump gears with clean engine oil and install them in the correct direction of rotation.

13. Install the pump cover and tighten the bolts to 84 inch lbs. (10 Nm).

14. Coat the relief valve and spring with clean engine oil, install them and tighten the plug to 33 ft. lbs. (45 Nm).

15. Install a new front crankshaft seal and coat the lips of the seal with clean engine oil.

16. Install the front case and oil pump assembly to the engine block using a new gasket. Use the noted locations of the mounting bolts for proper positioning and tighten the bolts to 10 ft. lbs. (14 Nm).

17. Install the oil screen with new gasket. Torque the screen bolts to 14 ft. lbs. (19 Nm).

18. Install the oil pan.

19. Install the sprocket, timing belt and pulley.

20. Connect the battery cable refill the engine oil, run the engine and check for leaks.

Rear Main Seal

1. Remove the transaxle using the recommended procedure.

2. Remove the flywheel/flexplate from the crankshaft.

3. Cut out a section of the oil seal lip.

4. Carefully pry the seal out of the oil seal case without damaging the sealing surface of the crankshaft.

To install:

5. Apply engine oil to the lip of the new seal and install the seal in the case using the proper size seal driver.

6. Install the flywheel/flexplate.

7. Install the transaxle using the recommended procedure.

FUEL SYSTEM

Fuel System Service Precautions

Safety is the most important factor when performing not only fuel system maintenance but any type of maintenance. Failure to conduct maintenance and repairs in a safe manner may result in serious personal injury or death. Maintenance and testing of the vehicle's fuel system components can be accomplished safely and effectively by adhering to the following rules and guide-lines.

• To avoid the possibility of fire and personal injury, always disconnect the negative battery cable unless the repair or test procedure requires that battery voltage be applied.

• Always relieve the fuel system pressure prior to disconnecting any fuel system component (injector, fuel rail, pressure regulator, etc.), fitting or fuel line connection. Exercise extreme caution whenever relieving fuel system pressure to avoid exposing skin, face and eyes to fuel spray. Please be advised that fuel under pressure may penetrate the skin or any part of the body that it contacts.

• Always place a shop towel or cloth around the fitting or connection prior to loosening to absorb any excess fuel due to spillage. Ensure that all fuel spillage (should it occur) is quickly removed from engine surfaces. Ensure that all fuel soaked cloths or towels are deposited into a suitable waste container.

• Always keep a dry chemical (Class B) fire extinguisher near the work area.

• Do not allow fuel spray or fuel vapors to come into contact with a spark or open flame.

• Always use a back-up wrench when loosening and tightening fuel line connection fittings. This will prevent unnecessary stress and torsion to fuel line piping. Always follow the proper torque specifications.

• Always replace worn fuel fitting O-rings with new. Do not substitute fuel hose or equivalent, where fuel pipe is installed.

Fuel System Pressure

RELIEVING

✳✳ CAUTION

The fuel injection system remains under pressure after the engine has been turned OFF. Properly relieve fuel pressure before disconnecting any fuel lines. Failure to do so may result in fire or personal injury.

1. Turn the ignition to the **OFF** position.
2. Loosen the fuel filler cap to release fuel tank pressure.
3. For the Mirage and Eclipse, remove the rear seat cushion, then remove the service cover and disconnect the fuel pump harness connector.
4. For the FWD Galant, Diamante and 3000GT, detach the fuel pump harness con-

nector located in the area of the fuel tank. It may be necessary to raise the vehicle to access the connector.

5. For the AWD Galant, remove the carpet from the trunk, locate the fuel tank wiring at the pump access cover, then detach the wiring.
6. Start the vehicle and allow it to run until it stalls from lack of fuel. Turn the key to the **OFF** position.
7. Disconnect the negative battery cable, then reconnect the fuel pump connector. Install the access cover, cushion or carpet as necessary.
8. Wrap shop towels around the fitting that is being disconnected to absorb residual fuel in the lines.
9. Place shop towels into proper safety container.

Fuel Filter

✳✳ CAUTION

The fuel injection system remains under pressure, after the engine has been turned OFF. Properly relieve fuel pressure before disconnecting any fuel lines. Failure to do so may result in fire or personal injury.

REMOVAL & INSTALLATION

✳✳ CAUTION

Do not use conventional fuel filters, hoses or clamps when servicing fuel injection systems. They are not compatible with the injection system and could fail, causing personal injury or damage to the vehicle. Use only hoses and clamps specifically designed for fuel injection.

3000GT and Diamante

1. Properly relieve the fuel pressure.
2. Disconnect the negative battery cable.

✳✳ CAUTION

Wait at least 90 seconds after the negative battery cable is disconnected to prevent possible deployment of the air bag.

3. The filter is located in the engine compartment, mounted on the inner fender panel.
4. Remove the air cleaner assembly and intake hoses. Remove the battery and battery tray with washer tank.

5. Separate the flare nut connection at the line.
6. Hold the fuel filter nut securely with a back-up or spanner wrench. Cover the hoses with shop towels and remove the eye bolts. Discard the gaskets.
7. Remove the mounting bolts and remove the fuel filter from the vehicle.

To install:
8. Install a new gaskets or O-rings whenever fuel connections have been disassembled.
9. Install the filter to its bracket only finger-tight. Movement of the filter will ease attachment of the fuel lines.
10. Install new gaskets and connect the high pressure hose and eye bolt, then the main pipe and eye bolt. While holding the fuel filter nut, tighten the eye bolts to 22 ft. lbs. (30 Nm). Tighten the flare nut to 25 ft. lbs. (35 Nm).
11. Tighten the mounting bolts fully.
12. Install the air cleaner assembly, battery and battery tray with washer tank, if removed.
13. Connect the negative battery cable, install the fuel filler cap, turn the key to the **ON** position to pressurize the fuel system and check for leaks. Release the fuel pressure and repair leaks as required.

Mirage, Galant and 1995 Eclipse

A replaceable fuel filter is located in the engine compartment.
1. Properly relieve the fuel system pressure.
2. Disconnect the negative battery cable.
3. If necessary, remove the air intake hose and the battery.

➡**Wrap shop towels around the fitting that is being disconnected to absorb residual fuel in the lines.**

4. Hold the fuel filter nut securely with a back-up or spanner wrench. Cover the hoses with shop towels and remove the eye bolt. Discard the gaskets.
5. Separate the flare nut connection at the filter. Discard the gaskets.
6. Remove the mounting bolts and the fuel filter from the vehicle.

To install:
7. If equipped with flare fitting, tighten the fitting by hand before installing the filter to the vehicle.
8. Install the filter to its bracket only finger-tight. Movement of the filter will ease attachment of the fuel lines.
9. Install new gaskets and connect the high pressure hose and eye bolt, then the main pipe. While holding the fuel filter nut,

tighten the eye bolts to 22 ft. lbs. (30 Nm). Tighten the flare nut to 27 ft. lbs. (37 Nm).

10. Tighten the filter mounting bolts fully.

11. Install the air intake hose and the battery.

12. Connect the negative battery cable, install the fuel filler cap, turn the key to the **ON** position to pressurize the fuel system and check for leaks.

13. Release the fuel pressure and repair leaks as required.

1996–99 Eclipse

1. Properly relieve the fuel system pressure.

2. Disconnect the negative battery cable. Wait at least 90 seconds before performing any work.

3. On turbo models and models equipped with the 2.4L engine, remove the battery and the air intake hose.

4. Raise and safely support the vehicle.

❊❊ WARNING

As there will be some pressure remaining in the fuel pipe line, cover it with a shop towel to prevent fuel from spraying out.

5. While holding the fuel filter with a suitable wrench, remove the eye bolt attaching the high pressure fuel line to the fuel filter. Discard the two washers.

6. Hold the fuel filter with a suitable wrench and remove the main fuel pipe from the fuel filter. On 2.0L non-turbo models, remove the fuel pipe from the connector.

7. On the 2.0L non-turbo engine models, remove the clamp and the hose from the fuel pressure regulator.

8. Remove the fuel filter mounting bracket bolts and remove the fuel filter.

9. Remove the bracket screw and remove the fuel filter from the mounting bracket.

10. On 2.0L non-turbo models, remove the following from the filter:

   a. The eye bolt and washer.

   b. The fuel connector and washer with the fuel pressure regulator.

To install:

11. On 2.0L non-turbo models, install the fuel connector with the fuel pressure regulator to the filter with two new washers and tighten the eye bolt to 22 ft. lbs. (36 Nm).

12. Install the fuel filter to the mounting bracket with the screw.

13. Reinstall the fuel filter to the vehicle with the bracket mounting bolts.

14. Reconnect the main fuel pipe to the fuel filter connector or the filter itself. Torque the flare nut to 27 ft. lbs. (36 Nm).

15. On the 2.0L non-turbo engine models reconnect the hose and clamp to the fuel pressure regulator.

16. Reconnect the high pressure fuel hose, using two new washers, to the fuel filter with the eye bolt. Torque the eye bolt to 22 ft. lbs. (29 Nm).

17. Safely lower the vehicle.

18. On 2.0L turbo and 2.4L engine models, reinstall the battery and the air intake hose.

19. Reconnect the negative battery cable, start the engine and check for fuel leaks.

Fuel Pump

REMOVAL & INSTALLATION

❊❊ CAUTION

The fuel injection system remains under pressure after the engine has been turned OFF. Properly relieve fuel pressure before disconnecting any fuel lines. Failure to do so may result in fire or personal injury.

➡**Cover all fuel hose connections with a shop towel, prior to disconnecting, to prevent splash of fuel that could be caused by residual pressure remaining in the fuel line.**

3000GT

1. Relieve fuel system pressure. Remove the fuel filler cap.

2. Disconnect the negative battery cable.

❊❊ CAUTION

Wait at least 90 seconds after the negative battery cable is disconnected to prevent possible deployment of the air bag.

3. The fuel pump is located in the fuel tank. Drain the fuel from the fuel tank.

4. Remove the fuel gauge cover located in the rear floor pan.

5. Remove the fuel pump and gauge electrical connector. Remove the overfill limiter (two-way valve).

6. Disconnect both sides of the high pressure fuel hose. When disconnecting the fuel pump side of the hose, hold the pump side nut with a wrench while turning the nut on the hose side. This will prevent any damage that will occur to the fittings and the hoses if two wrenches are not used.

7. Remove the fuel pump and gauge assembly from the tank.

To install:

8. Align the three projections on the packing with the holes on the fuel pump and the nipples on the pump facing the same direction as before removal.

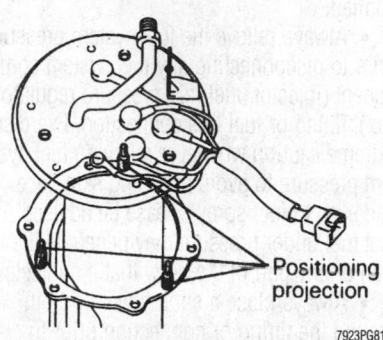

Positioning projection

7923PG81

Align the positioning projections—3000GT 3.0L engine

9. Temporarily tighten the flare nut on the high pressure hose by hand. Making sure the hose does not twist, tighten body side nut to 22 ft. lbs. (30 Nm) and the fuel pump side nut to 25 ft. lbs. (35 Nm).

10. Install the overfill limiter (2-way valve) with the long shouldered side of the valve facing the canister.

11. Connect the electrical connector to the pump assembly.

12. Reconnect the negative battery cable and check the entire system for leaks.

13. Install sealer to the rear floor pan and install the cover into place.

Diamante

1. Properly relieve the fuel system pressure.

2. Disconnect the negative battery cable.

3. Raise the vehicle and support it safely.

4. Remove the left rear wheel well liner, if equipped.

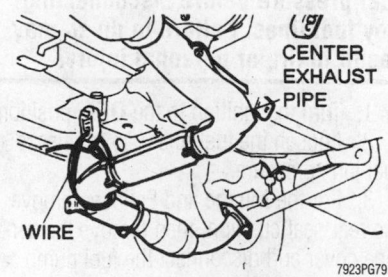

CENTER EXHAUST PIPE

WIRE

7923PG79

Proper method of supporting rear exhaust system—Diamante 3.0L engine

5. Disconnect the center exhaust system from the main muffler. Disconnect the rear exhaust hangers, lower the system and secure aside.

6. Remove the tank drain plug and drain the fuel into an approved container.

7. On models equipped with 4WS, the power cylinder must be lowered, in order to gain access to the fuel tank. Remove the mounting bolts and lower the rear steering gear.

8. Disconnect the fuel return hose, high pressure hose and all other hoses and connectors connected to the pump/sending unit.

❉❉ CAUTION

Cover all fuel hose connections with a shop towel, prior to disconnecting, to prevent splash of fuel that could be caused by residual pressure remaining in the fuel line.

9. Disconnect the filler and vent hoses. Place a support under the tank and remove the retaining nuts.

10. Lower the tank from the vehicle.

11. Remove the fuel pump retaining nuts and remove the assembly from the tank.

To install:

12. Install the pump assembly to the tank and tighten the retaining nuts to 2 ft. lbs. (3 Nm).

13. Install the fuel tank and connect the filler and vent hoses. Tighten the tank retaining nuts and bolts to 19 ft. lbs. (26 Nm).

14. Connect the return hose, high pressure hose and all other hoses and connectors connected to the pump/sending unit.

15. If equipped with 4WS, install the power cylinder unit to the crossmember and tighten the mounting bolts to 31 ft. lbs. (43 Nm).

16. Connect the exhaust pipe and secure in place with rear hangers.

17. Install the left rear wheel well liner, if removed.

18. Lower the vehicle and return fuel to the gas tank.

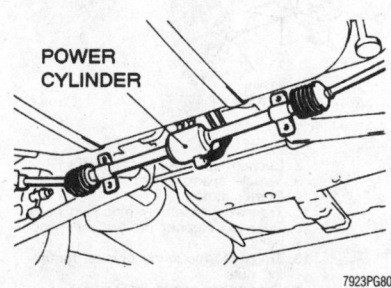

POWER CYLINDER

7923PG80

Power cylinder identification—Diamante 3.0L engine

19. Connect the negative battery cable and check the entire system for proper operation and leaks.

Eclipse

1. Relieve the fuel system pressure.

2. Disconnect the negative battery cable.

3. Remove the rear seat cushion by pulling the seat stopper near the floor and lifting the cushion up.

4. Remove the inspection cover on the right side of the vehicle.

5. Disconnect the harness connector and the fuel lines.

6. Remove the fuel pump assemble from the tank. Use MB991480 or equivalent to remove the locking ring on the AWD model.

To install:

7. Install the fuel pump in the tank.

8. Connect the hoses and the harness connector.

9. Install the inspection cover.

10. Install the rear seat.

11. Connect the negative battery cable.

Galant

1. Properly relieve the fuel system pressure. Disconnect the negative battery cable.

2. Remove the rear seat cushion, by pulling the seat stopper outward and lifting the lower cushion upward.

3. Remove the access cover.

4. Disconnect the fuel pump wiring.

5. Disconnect the return hose and the high pressure fuel hose.

6. Remove the pump mounting nuts and remove the pump assembly.

To install:

7. Install the fuel pump assembly to the tank and tighten the retaining nuts to 22 inch lbs. (2.5 Nm).

➡**Tilt the float to the left of the vehicle, when installing the pump assembly.**

8. Connect the high pressure hose, return hose and the fuel tank wiring.

9. Connect the negative battery cable.

10. Check the fuel pump for proper pressure and inspect the entire system for leaks.

11. Apply sealant to the access cover and install the cover.

12. Install the rear seat cushion.

Mirage

1. Properly relieve the fuel system pressure using proper procedures. Disconnect the negative battery cable.

2. Raise the vehicle and support safely.

3. Drain the fuel from the fuel tank into an approved container.

4. Disconnect the return hose, high pressure hose and vapor hoses from the fuel pump.

5. Disconnect the electrical connectors at the pump/sending unit.

6. Disconnect the filler and vent hoses.

7. Place a transmission jack under the center of the fuel tank and apply a slight upward pressure. Remove the fuel tank strap retaining nut.

8. Lower the tank slightly and disconnect any remaining electrical or hose connectors at the fuel tank.

9. Remove the fuel tank from the vehicle.

10. Remove the five nuts securing the access plate to the fuel tank and remove the pump assembly.

To install:

11. Install fuel pump into fuel tank, with new packing gasket, and tighten mounting nuts.

12. Install the fuel tank onto the transmission jack. Raise the tank in position under the vehicle. Leave enough clearance to attach the electrical and hose connections to the top of the fuel pump.

13. Attach all connections to the top of the tank.

14. Raise the tank completely and position the retainer straps around the fuel tank. Install new fuel tank self-locking nuts and tighten to 22 ft. lbs. (31 Nm).

15. Connect the return hose and high pressure hoses.

16. Install the vapor hose and the filler hose. Install the filler hose retainer screws to the fender, if removed.

17. Lower the vehicle and pour the drained fuel into the gas tank.

18. Connect the negative battery cable. Check the fuel pump for proper pressure and inspect the entire system for leaks.

DRIVE TRAIN

Transaxle Assembly

REMOVAL & INSTALLATION

Manual

3000GT

➡**If the vehicle is going to be rolled on its wheels while the halfshafts are out of the vehicle, obtain two outer CV-joints or proper equivalent tools and install to the hubs. If the vehicle is**

rolled without the proper torque applied to the front wheel bearings, the bearings will no longer be usable.

✳✳ CAUTION

Wait at least 90 seconds after the negative battery cable is disconnected to prevent possible deployment of the air bag.

1. Remove the battery and battery tray. Raise the vehicle and support safely. Drain the transaxle oil and the oil from the transfer case.

2. If equipped with AWD, disconnect the exhaust pipe. Remove the mounting bolts and lower the transfer case from the vehicle.

3. Remove the left side splash shield and engine undercover.

4. Remove the air cleaner assembly and all adjoining duct work.

5. Disconnect the shifter control cables and speedometer connector.

6. Remove the clutch release cylinder.

7. Disconnect the reverse light switch.

8. Support the weight of the transaxle and remove the transaxle mount through-bolt. Remove the access plug, remove the bolts for the bracket and remove the brackets.

9. Disconnect the transaxle ground cable.

10. Disconnect the tie rod end and ball joint from the steering knuckle.

11. Remove the right frame member.

12. Remove the starter motor.

13. Remove the halfshafts by inserting a prybar between the transaxle case and the driveshaft and prying the shaft from the transaxle. Do not pull on the driveshaft. Doing so damages the inboard joint. Use the prybar. Do not insert the prybar so far the oil seal in the case is damaged. On AWD, remove the right side shaft as just described. The left side shaft can be removed by tapping with a plastic hammer. Remove the shaft with the hub and knuckle as an assembly. Don't tap on the center bearing or it will be damaged. Tie the shafts aside. Note the circle clip on the end of the inboard shafts. These should not be reused.

14. Remove the transaxle brackets.

15. Remove the transaxle assembly. On turbocharged vehicles, take care to prevent damaging the lower radiator hose with the transaxle housing. Wind tape around the lower hose and put tape on the transaxle housing. Support the transaxle assembly using the proper jack, move the transaxle away from the engine and lower it.

To install:

16. Install the transaxle to the engine and install the mounting bolts.

17. When installing the halfshafts, use new circlips on the axle ends. Take care to get the inboard joint parts straight, not bent relative to the axle. Care must be taken to ensure that the oil seal lip of the transaxle is not damaged by the serrated part of the driveshaft.

18. Install the starter motor and cover.

19. Install the right side frame member.

20. Install the ball joint and tie rod to the steering knuckle.

21. Connect the transaxle ground cable.

22. Install the side mount brackets and install the access plug.

23. Connect the reverse light switch.

24. Install the clutch release cylinder.

25. Connect the shifter control cables and speedometer connector.

26. Install the transfer case and related items on AWD vehicles.

27. Install the air cleaner assembly and all adjoining duct work.

28. Install the left side splash shield.

29. Install the battery tray and battery.

30. Be sure the vehicle is level when refilling the transaxle. Use Hypoid gear oil or equivalent, GL-4 or higher.

31. Connect the negative battery cable and check the transaxle and transfer case for proper operation. Be sure the reverse lamps come ON when in reverse.

ECLIPSE

1. Remove the battery and the air intake hoses.

2. Remove the battery tray and support.

3. If equipped with cruise control, remove the auto-cruise actuator and bracket.

4. Drain the transaxle and transfer case fluid, if equipped, into a suitable container.

5. Remove the charcoal canister and bracket.

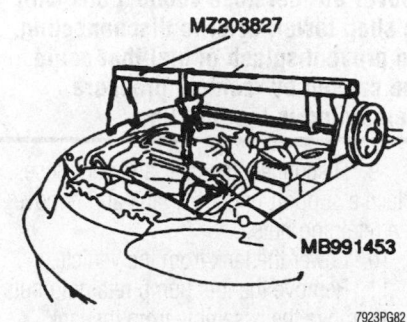

Proper method of supporting the engine assembly for transaxle removal

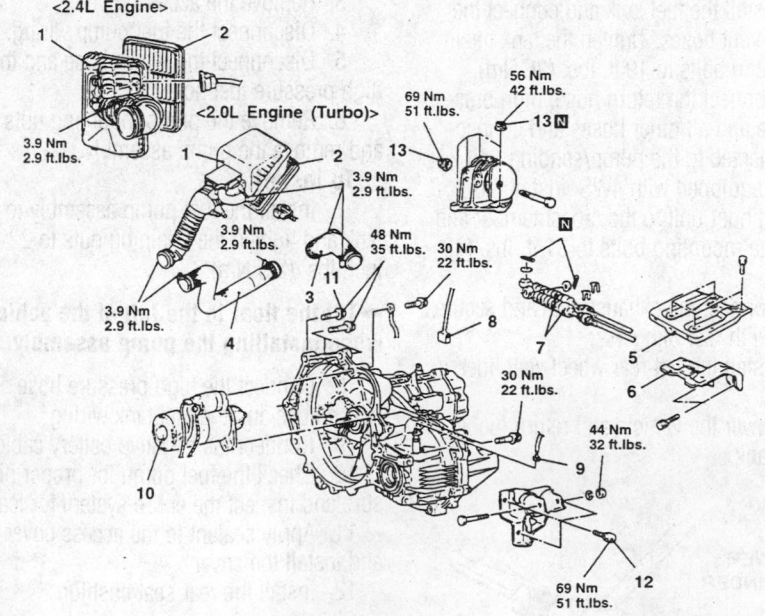

Removal steps
1. Air cleaner cover and air intake hose assembly
2. Air cleaner element
3. Air hose C <2.0L Engine (Turbo)>
4. Air hose A <2.0L Engine (Turbo)>
5. Battery tray
6. Battery tray stay
7. Shift cable and select cable connection
8. Backup light switch connector
9. Vehicle speed sensor connector
10. Starter motor
11. Transaxle assembly mounting bolts
12. Rear roll stopper bracket mounting bolts
13. Transaxle mounting bracket mounting nuts
• Supporting engine assembly

Exploded view of the manual transaxle mounting (1 of 2)—FWD Eclipse with 2.4L engine

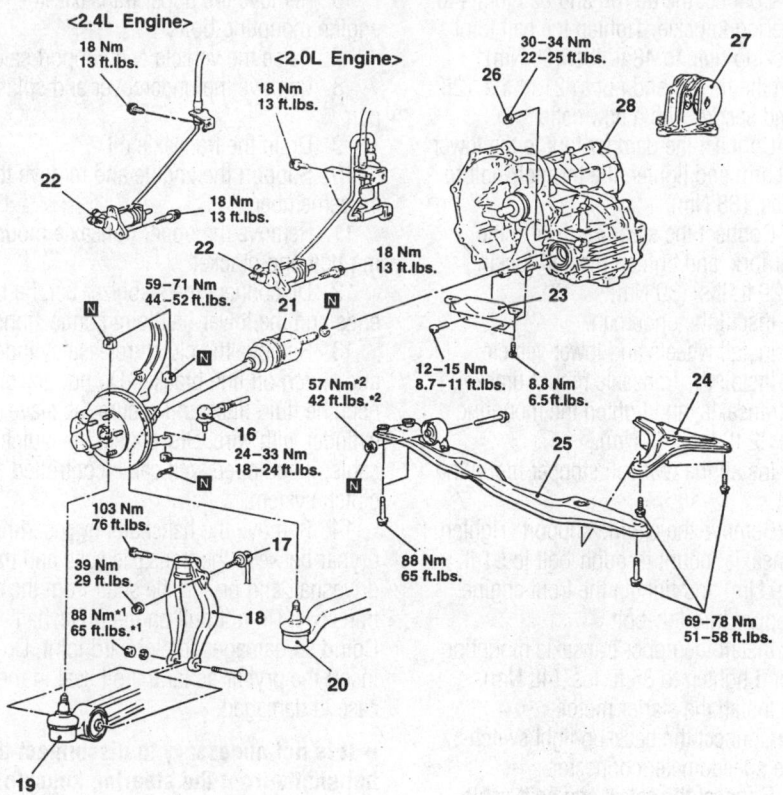

<2.4L Engine>
18 Nm
13 ft.lbs.

<2.0L Engine>
18 Nm
13 ft.lbs.

30–34 Nm
22–25 ft.lbs.

27

26

28

22

18 Nm
13 ft.lbs.

22

18 Nm
13 ft.lbs.

59–71 Nm
44–52 ft.lbs.

21

23

57 Nm*2
42 ft.lbs. *2

12–15 Nm
8.7–11 ft.lbs.

8.8 Nm
6.5 ft.lbs.

24

16

24–33 Nm
18–24 ft.lbs.

25

103 Nm
76 ft.lbs.

39 Nm
29 ft.lbs.

17

88 Nm*1
65 ft.lbs. *1

18

88 Nm
65 ft.lbs.

69–78 Nm
51–58 ft.lbs.

20

19

Lifiting up of the vehicle

16. Tie rod end ball joint and kunckle
connection
17. Stabilizer link connection
18. Damper fork
19. Lateral lower arm ball joint and
kunckle connection
20. Compression lower arm ball joint
and kunckle connection
21. Drive shaft connection
22. Clutch release cylinder connection
23. Bell housing cover
24. Stay (R.H.)
25. Center member assembly
26. Transaxle assembly mounting bolt

27. Transaxle mounting
28. Transaxle assembly

Caution
*1: **Indicates parts which should be temporarily
tightened, and then fully tightened with the
vehicle on the ground in the unladen condition.**
*2: **For tightening locations indicated by the symbol,
first tighten temporarily, and then make the final
tightening with the entire weight of the engine
applied to the vehicle body.**

7923PG87

Exploded view of the manual transaxle mounting (2 of 2)—FWD Eclipse with 2.4L engine

6. Disconnect the shift and select cables from the transaxle.

7. Disconnect the back-up light switch and the vehicle speed sensor connectors.

8. Remove the starter assembly.

9. Attach an engine support fixture to the engine and remove the transaxle mounting bolts.

10. Remove the rear roll stopper bracket mounting bolts.

11. Remove the transaxle mounting bracket mounting nuts.

12. Raise the vehicle and remove the engine undercovers.

13. If equipped with all wheel drive, remove the transfer case assembly.

❋❋ WARNING

Do not remove or install the axle shaft nut when the vehicle is on the

floor or damage to the bearings will occur.

14. Remove the axle shafts.

15. Remove the slave cylinder from the bell housing but do not disconnect the fluid line. Position it out of the way.

16. Remove the bell housing cover and the right-hand center member stay (support).

17. Remove the center member.

18. Place a transmission jack under the transaxle and remove the transaxle mounting bolt.

19. Remove the transaxle mounting and lower the transaxle.

To install:

20. Raise the transaxle into position and install the transaxle mounting. Torque the through-bolt to 50 ft. lbs. (69 Nm).

21. Install the transaxle assembly mounting bolt. Torque the bolt to 22–25 ft. lbs. (30–34 Nm).

22. Install the center member assembly and the right-hand stay.

23. Install the bell housing cover and the slave cylinder.

24. Install the axle shafts. Be sure to install the washer in the proper direction.

25. Install the engine undercovers and lower the vehicle.

26. Install the transfer case assembly if removed.

27. Install the transaxle mounting bracket mounting nuts.

28. Install the rear roll stopper bracket mounting bolts.

29. Install the transaxle assembly mounting bolts. Torque the mounting bolts to 35 ft. lbs. (48 Nm).

30. Remove the engine support fixture.

31. Install the starter assembly.

32. Connect the vehicle speed sensor and the back-up light connectors.

33. Install the cruise control actuator if removed.

34. Install the battery tray support and the tray.

35. Install the charcoal canister bracket and the canister.

36. Install the air duct and the air cleaner assembly.

37. Refill the transaxle and the transfer case if equipped with oil.

GALANT

1. Disconnect the negative battery cable and wait at least 90 seconds before performing any work.

2. Remove the air cleaner and intake hoses.

3. Drain the transaxle into a suitable waste container.

4. Remove the cotter pins and clips securing the select and shift cables and remove the cable ends from the transaxle.

5. If equipped with Active-ECS, disconnect the air compressor.

6. Disconnect the back-up light switch harness and position aside.

7. Disconnect the speedometer electrical connector, from the transaxle assembly.

8. Remove the starter motor and position aside.

9. Using special tool MZ203827 or equivalent, support the engine assembly.

10. Remove the rear roll stopper mounting bracket.

11. Remove the transaxle mount bracket.

12. Remove the upper transaxle mounting bolts.

13. Raise and safely support the vehicle.

14. Remove the front wheel assemblies.
15. Remove the right-hand undercover.
16. Remove the cotter pin and disconnect the tie rod end, from the steering knuckle.
17. Disconnect the stabilizer bar link, from the damper fork.
18. Disconnect the damper fork, from the lateral lower control arm.
19. Disconnect the later lower arm, and the compression arm, lower ball joints, from the steering knuckle.
20. Pry the halfshafts from the transaxle, and secure aside.
21. Remove the connection for the clutch release cylinder and without disconnecting the hydraulic line, secure aside.
22. Remove the cover from the transaxle bell housing.
23. Remove the engine front roll stopper through-bolt.
24. Remove the crossmember and the triangular right-hand stay.
25. Support the transaxle, using a transmission jack, and remove the transaxle lower coupling bolt.

➡ **The coupling bolt threads from the engine side, into the transaxle, and is located just above the halfshaft opening.**

26. Slide the transaxle rearward and carefully lower it from the vehicle.
To install:
27. Install the transaxle to the engine and install the mounting bolts and tighten to 35 ft. lbs. (48 Nm). Install the transaxle lower coupling bolt and tighten to 22–25 ft. lbs. (30–34 Nm).
28. Install the cover to the transaxle bell housing and tighten the mounting bolts to 7 ft. lbs. (9 Nm).
29. Install the crossmember and tighten the front mounting bolts to 65 ft. lbs. (88 Nm) and the rear bolt to 54 ft. lbs. (73 Nm). Install the front engine roll stopper through-bolt and lightly tighten. Once the full weight of the engine is on the mounts, tighten the bolt to 42 ft. lbs. (57 Nm).
30. Install the triangular stay bracket and tighten the mounting bolts to 65 ft. lbs. (88 Nm).
31. Connect the clutch release cylinder.
32. Install the halfshafts, using new circlips on the axle ends.

✳✳ WARNING

When installing the axle shaft, keep the inboard joint straight in relation to the axle, so not to damage the oil seal lip of the transaxle, with the serrated part of the halfshaft.

33. Connect the tie rod and ball joints to the steering knuckle. Tighten the ball joint self-locking nuts to 48 ft. lbs. (65 Nm). Tighten the tie rod end nut to 21 ft. lbs. (28 Nm) and secure with a new cotter pin.
34. Connect the damper fork to the lower control arm and tighten the through-bolt to 65 ft. lbs. (88 Nm).
35. Connect the stabilizer link to the damper fork, and tighten the self-locking nut to 29 ft. lbs. (39 Nm).
36. Install the underpan.
37. Install wheels and lower vehicle.
38. Install the transaxle mount bracket, to the transaxle, and tighten the mounting nuts to 32 ft. lbs. (43 Nm).
39. Install the rear roll stopper mounting bracket.
40. Remove the engine support. Tighten the transaxle mount through-bolt to 51 ft. lbs. (69 Nm) and tighten the front engine roll stopper through-bolt.
41. Install the upper transaxle mounting bolts and tighten to 35 ft. lbs. (48 Nm).
42. Install the starter motor.
43. Connect the back-up light switch and the speedometer connector.
44. Connect the select and shift cables and install new cotter pins.
45. Install the air cleaner and the air intake hose.
46. Connect the negative battery cable.
47. Be sure the vehicle is level, and refill the transaxle.
48. Check the transaxle for proper operation. Be sure the reverse lights come on when in reverse.

MIRAGE

➡ **If the vehicle is going to be rolled while the halfshafts are out of the vehicle, obtain 2 outer CV-joints or proper equivalent tools and install to the hubs. If the vehicle is rolled without the proper torque applied to the front wheel bearings, the bearings will no longer be usable. Also, the suspension components should not be tightened until the vehicles weight is resting on the ground.**

1. Disconnect the negative battery cable.
2. Remove the front wheels and the inner wheel panels.
3. Remove the air cleaner assembly and vacuum hoses.
4. Note the locations and disconnect the shifter cables.
5. Disconnect the back-up lamp switch connector, speedometer cable connection and remove the starter motor.

6. Remove the upper transaxle-to-engine mounting bolts.
7. Raise the vehicle and support safely.
8. Remove the undercover and splash pan.
9. Drain the transaxle oil.
10. Support the engine and remove the crossmember.
11. Remove the upper transaxle mounting bolt and bracket.
12. Disconnect the stabilizer bar, tie rod ends and the lower ball joint connections.
13. Remove the clutch release cylinder and clutch oil line bracket. Do not disconnect the fluid lines and secure the slave cylinder with wire. Disconnect the clutch cable, if equipped with cable controlled clutch system.
14. Remove the halfshafts by inserting a prybar between the transaxle case and the driveshaft and prying the shaft from the transaxle. Do not pull on the driveshaft. Doing so damages the inboard joint. Do not insert the prybar so far the oil seal in the case is damaged.

➡ **It is not necessary to disconnect the halfshafts from the steering knuckle. Remove the shaft with the hub and knuckle as an assembly. Tie the shafts aside. Note the circle clip on the end of the inboard shafts should not be reused.**

15. Remove the bell housing lower cover.
16. Remove the transaxle to engine bolts and lower the transaxle from the vehicle.
To install:

➡ **When installing the transaxle, be sure to align the splines of the transaxle with the clutch disc.**

17. Install the transaxle to the engine and install the mounting bolts. Torque the bolts to specifications.
18. Install the bell housing cover.

➡ **When installing the halfshafts, use new circlips on the axle ends. Care must be taken to ensure that the oil seal lip of the transaxle is not damaged by the serrated part of the driveshaft.**

19. Install and fully seat the halfshafts into the transaxle.
20. Install the slave cylinder.
21. Connect the ball joints, tie rod ends and the stabilizer bar connections.
22. Install the upper transaxle mounting bracket and bolt.
23. Install the crossmember.
24. Install the undercover.
25. Install the upper transaxle-to-engine mounting bolts.

26. Install the starter motor.
27. Connect the back-up light switch connector and speedometer cable.
28. Connect and adjust the shifter cables.
29. Install the air cleaner assembly.
30. Install the front wheels.
31. Be sure the vehicle is level when refilling the transaxle. Use Hypoid gear oil or equivalent, GL-4 or higher.
32. Connect the negative battery cable and check the transaxle for proper operation. Be sure the reverse lights operate when in reverse.

Automatic

3000GT

➡️If the vehicle is going to be rolled on its wheels while the halfshafts are out of the vehicle, obtain two outer CV-joints or proper equivalent tools and install to the hubs. If the vehicle is rolled without the proper torque applied to the front wheel bearings, the bearings will no longer be usable.

⁂ CAUTION

Wait at least 90 seconds after the negative battery cable is disconnected to prevent possible deployment of the air bag.

1. Disarm the air bag, if equipped. Remove the battery, battery tray and washer tank.
2. Remove the air cleaner assembly and adjoining duct work.
3. Disconnect the shifter control cable.
4. Disconnect and plug the oil cooler hoses.
5. Disconnect the inhibitor switch, kickdown servo switch, pulse generator, oil temperature sensor, shift control solenoid valve, and ground cable.
6. Disconnect the speedometer cable.
7. Raise the vehicle and support safely. Remove the undercovers.
8. Support the weight of the transaxle and remove the mount bracket. Remove the upper bell housing bolts.
9. Disconnect the tie rod end and ball joint from the steering knuckle.
10. Remove the right frame member.
11. Remove the starter.
12. Remove the halfshafts by inserting a prybar between the transaxle case and the driveshaft and prying the shaft from the transaxle. Do not pull on the driveshaft. Doing so damages the inboard joint. Use the prybar. Do not insert the prybar so far the oil seal in the case is damaged. Tie the halfshafts aside.

13. Remove the remaining mounting brackets.
14. Remove the bell housing cover plate.
15. Remove the special bolts holding the flexplate to the torque converter.
16. After removing the bolts, push the torque converter toward the transaxle so it doesn't stay on the engine side and allow oil to pour out the converter hub.
17. Remove the lower transaxle to engine bolts and remove the transaxle assembly.

To install:

18. After the torque converter has been mounted on the transaxle, install the transaxle assembly on the engine. Tighten the driveplate bolts to 34–38 ft. lbs. (46–53 Nm). Install the bell housing cover.
19. Install the mounting brackets.
20. Replace the circlips and install the halfshafts to the transaxle.
21. Install the starter and frame member.
22. Install the tie rods and ball joint to the steering arm.
23. Install the upper bell housing bolts.
24. Install the transaxle mounting bracket.
25. Install the undercovers.
26. Connect the speedometer cable.
27. Connect the inhibitor switch, kickdown servo switch, pulse generator, oil temperature sensor, shift control solenoid valve, and ground cable.
28. Connect the oil cooler hoses.
29. Connect the shifter control cable.
30. Install the air cleaner assembly and adjoining duct work.
31. Install the washer tank, battery tray and battery.
32. Refill with DEXRON II, Mopar ATF Plus type 7176 or equivalent, automatic transaxle fluid.
33. Start the engine and allow it to idle for two minutes. Apply parking brake and move selector through each gear position, ending in **N**. Recheck fluid level and add if necessary. Fluid level should be between the marks in the **HOT** range.

DIAMANTE

➡️If the vehicle is going to be rolled while the halfshafts are out of the vehicle, obtain 2 outer CV-joints or proper equivalent tools and install to the hubs. If the vehicle is rolled without the proper torque applied to the front wheel bearings, the bearings will no longer be usable.

1. Properly disarm the SRS system (air bag).
2. Raise and safely support the vehicle.
3. Remove the front wheels.

4. Remove the engine side and under-covers.
5. Drain the transaxle assembly.
6. If equipped, remove the front catalytic converter and exhaust pipe.
7. Remove the exhaust pipe, main muffler and catalytic converter.
8. Disconnect the tie rod end and ball joint from the steering knuckle.

➡️It will be necessary to unbolt the support bearing for the left side half-shaft.

9. Remove the halfshafts by inserting a prybar between the transaxle case and the driveshaft and prying the shaft from the transaxle. Do not pull on the driveshaft. Doing so damages the inboard joint. Do not insert the prybar so far that the oil seal in the case is damaged. Tie the halfshafts aside.
10. Remove the air cleaner assembly and adjoining duct work.
11. Detach the engine harness connection.
12. If the vehicle is equipped with active electronically controlled suspension (ACTIVE-ECS), remove the compressor assembly from the transaxle and suspend with wire. DO NOT disconnect the air hose from the compressor.
13. If equipped, remove the roll stopper stay bracket.
14. Disconnect the speedometer cable from the transaxle.
15. Remove the clip that secures the shifter and disconnect the shifter control cable from the transaxle.
16. Disconnect and plug the oil cooler hoses from the transaxle.
17. Disconnect the park/neutral switch electrical harness.
18. Disconnect the kickdown servo switch, pulse generator, oil temperature sensor electrical harness.
19. Disconnect the shift control solenoid valve harness.
20. Support the transaxle and remove the connection for the transaxle mounting bracket.

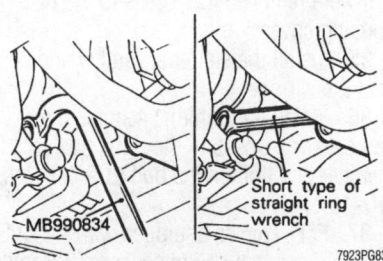

Location of 4-wheel steering oil pump mounting bolts—Diamante with a F4A33 automatic transaxle

21. Remove the three upper transaxle-to-engine mounting bolts.

22. For vehicles with 4WS remove the following

   a. Remove the heat shield for the 4WS oil pump.

   b. Without removing the oil hoses, remove the bolts that secure the oil pump and remove the pump.

   c. Secure the oil pump with a piece of wire.

23. For vehicles equipped with ACTIVE-ECS, disconnect the height sensor rod from the lower control arm.

24. Remove the bolt that secures the oxygen sensor harness to the right side crossmember.

25. Remove the starter assembly.

26. Remove the mounting brackets for access to the bell housing cover.

27. Remove the bell housing/oil pan covers assembly.

28. Remove the four bolts holding the flexplate to the torque converter. It will be necessary to rotate the engine by the front crankshaft bolt for access to all the torque converter bolts.

29. After removing the bolts, push the torque converter toward the transaxle so it does not stay on the engine side of the vehicle.

30. Remove the lower transaxle to engine bolts and remove the transaxle assembly.

To install:

➡**Be sure the torque converter is fully seated into the front of the transaxle before installing the transaxle.**

31. Install the transaxle assembly to the engine block and install the mounting bolts. Tighten the mounting bolts to specifications.

32. Install the four bolts that secure the torque convert to the driveplate. Tighten the driveplate bolts to 34–38 ft. lbs. (46–53 Nm).

33. Install the bell housing/oil pan covers.

34. Install the transaxle stay brackets that were removed for access to the bell housing cover.

35. Install the starter assembly and connect the wiring.

36. Install the bolt that secures the oxygen sensor harness to the right side crossmember and tighten the bolt to 7–9 ft. lbs. (10–12 Nm).

37. For vehicles equipped with ACTIVE-ECS, connect the height sensor rod from the lower control arm. Check the height sensor rod for a length (A) of 10.59–10.63 in. (269–270mm).

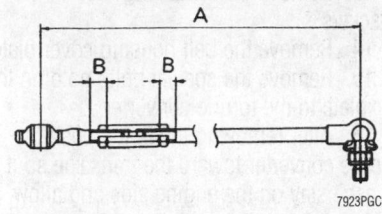

Height sensor rod adjustment—Diamante with a F4A33 automatic transaxle

➡**Be sure to keep the height sensor rod (B) equal on both sides of the adjuster and the ball joint at the tip of the rod should centered on the fulcrum.**

38. If removed, install the 4WS oil pump and tighten the mounting bolts to 17 ft. lbs. 24 Nm).

39. If removed, install the 4WS oil pump heat shield and tighten the mounting bolts to 17 ft. lbs. 24 Nm).

40. Install the three upper transaxle-to-engine mounting bolts. Tighten the mounting bolts to 54 ft. lbs. (75 Nm).

➡**One of the upper bolts has a grounding strap to secure under the bolt, DO NOT forget this strap.**

41. Install and connect the transaxle mounting bracket. Tighten the mounting nut and bolts to 51 ft. lbs. (70 Nm).

42. Connect the shift control solenoid valve harness.

43. Connect the kickdown servo switch, pulse generator and oil temperature sensor electrical harness.

44. Connect the park/neutral switch electrical harness.

45. Using new hose clamps, install the oil cooler hoses to the transaxle.

46. Install shifter control cable to the transaxle and secure the cable with clip.

47. Connect the speedometer cable to the transaxle.

48. If removed, install the roll stopper stay bracket and tighten the one through nut and bolt to 36–43 ft. lbs. (50–60 Nm). Tighten the two mounting bolts to 16 ft. lbs. (22 Nm).

49. If removed, install the ACTIVE-ECS compressor assembly. Tighten the mounting bolts to 48 inch lbs. (5 Nm) and connect the electrical harness.

50. Attach the engine harness connection.

51. Install the air cleaner assembly and adjoining duct work.

52. Using new circlips, install the half-shafts and seat halfshafts into the transaxle. Install the bolt that secure the left side support bearing and tighten the bolts to 33 ft. lbs. (45 Nm).

53. Connect the ball joint and tie rod end to the steering knuckle. Using new nuts,

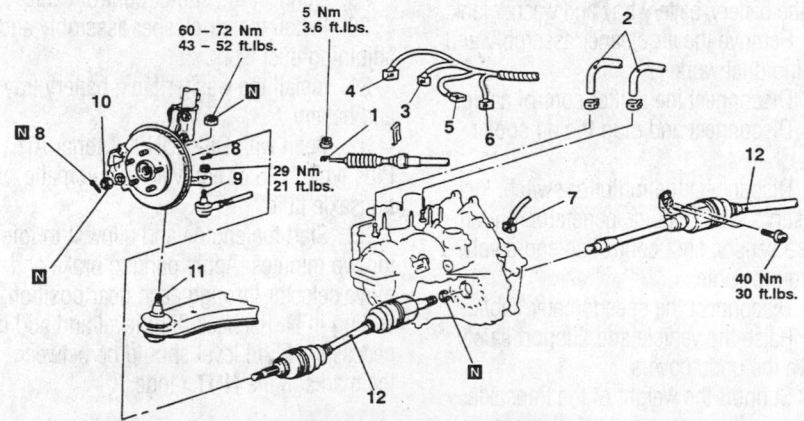

Removal steps
1. Transaxle control cable connection
2. Transaxle oil cooler hoses connection
3. PNP switch connector
4. A/T control solenoid valve connector
5. Input shaft speed sensor connector
6. Output shaft speed sensor connector
7. Vehicle speed sensor connector
8. Split pin
9. Connection of the tie rod end
10. Drive shaft nut
11. Connection for the lower arm ball joint
12. Drive shaft and inner shaft assembly (RH) and the drive shaft (LH)

Caution
Mounting locations marked by * should be provisionally tightened, and then fully tightened when the body is supporting the full weight of the engine.

Transaxle removal—Diamante with F4A51 transaxle 1 of 2

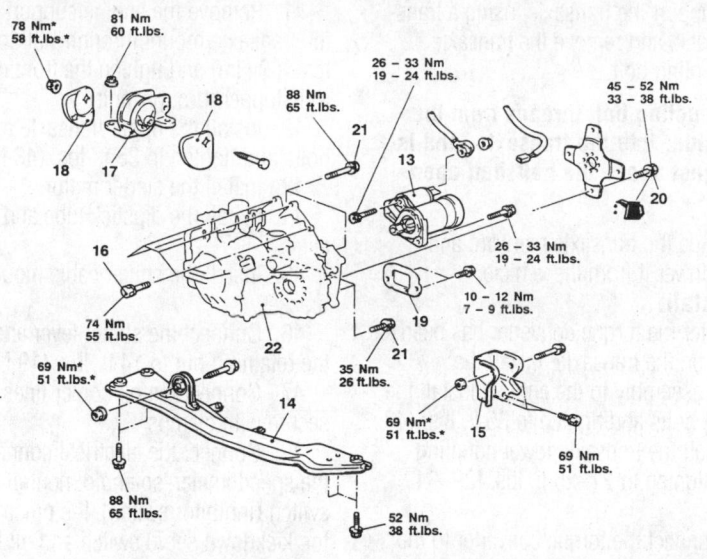

Lifting up of the vehicle
13. Starter motor
14. Center member assembly
15. Rear roll stopper bracket
16. Transaxle upper portion fixing bolt
17. Transaxle mounting bracket
18. Transaxle mount stopper
• Support the engine and transaxle assembly
19. Bell housing cover

20. Drive plate attaching bolt
21. Transaxle lower portion fixing bolt
22. Transaxle assembly

Caution
Mounting locations marked by * should be provisionally tightened, and then fully tightened when the body is supporting the full weight of the engine.

7923PG85

Transaxle removal—Diamante with F4A51 transaxle 2 of 2

tighten the ball joint castle nut to 43–52 ft. lbs. (60–72 Nm) and tighten the tie rod castle nut to 22 ft. lbs. (30 Nm). Install new cotter pins to both connections.

54. Using new gaskets, install the exhaust system.

55. If removed, install front catalytic converter and exhaust pipe. Be sure to use new gaskets.

56. Install the engine undercovers.

57. Connect the negative battery cable.

58. Refill with Dexron® II.

59. Start the engine and allow to idle for 2 minutes. Apply parking brake and move selector through each gear position, ending in **N**. Recheck fluid level and add if necessary. Fluid level should be between the marks in the **HOT** range.

60. Road test the vehicle.

ECLIPSE

1. Remove the battery and the air intake hoses.

2. Remove the battery tray and support.

3. If equipped with cruise control, remove the auto-cruise actuator and bracket.

4. Drain the transaxle and transfer case fluid, if equipped, into a suitable container.

5. Remove the charcoal canister and bracket.

6. Disconnect the shift and select cables from the transaxle.

7. Disconnect the back-up light switch and the vehicle speed sensor connectors.

8. Remove the dipstick and tube assembly.

9. Remove the starter assembly.

10. Disconnect the park/neutral switch, oil temperature sensor, kick down servo switch, solenoid valve, pulse generator and speedometer connections.

11. Attach an engine support fixture to the engine and remove the transaxle mounting bolts.

12. Remove the rear roll stopper bracket mounting bolts.

13. Remove the transaxle mounting bracket mounting nuts.

14. Raise the vehicle and remove the engine undercovers.

15. Remove the front exhaust pipe.

16. If equipped with all wheel drive, remove the transfer case assembly.

❊❊❊ WARNING

Do not remove or install the axle shaft nut when the vehicle is on the floor or damage to the bearings will occur.

17. Remove the axle shafts.

18. Remove the slave cylinder from the bell housing but do not disconnect the fluid line. Position it out of the way.

19. Remove the bell housing cover and the right-hand center member stay (support).

20. Remove the center member.

21. Remove the drive plate connecting bolts.

22. Place a transmission jack under the transaxle and remove the transaxle mounting bolt.

23. Remove the transaxle mounting and lower the transaxle.

To install:

24. Raise the transaxle into position and install the transaxle mounting bracket. Torque the through-bolt to 51 ft. lbs. (69 Nm).

25. Install the transaxle assembly mounting bolt. Torque the bolt to 22–25 ft. lbs. (29–34 Nm).

26. Install the drive plate connecting bolts. Torque the bolts to 33–38 ft. lbs. (45–52 Nm).

27. Install the center member assembly and the right-hand stay.

28. Install the bell housing cover and the slave cylinder.

29. Install the axle shafts. Be sure to install the washer in the proper direction.

30. Install the front exhaust pipe.

31. Install the engine undercovers and lower the vehicle.

32. Install the transfer case assembly if removed.

33. Install the transaxle mounting bracket mounting nuts.

34. Install the rear roll stopper bracket mounting bolts.

35. Install the transaxle assembly mounting bolts. Torque the bolts to 35 ft. lbs. (48 Nm).

36. Remove the engine support fixture.

37. Connect the park/neutral switch, oil temperature sensor, kick down servo switch, solenoid valve, pulse generator and speedometer connections.

38. Install the starter assembly.

39. Install the dipstick and tube assembly.

40. Connect the vehicle speed sensor and the back-up light connectors.

41. Install the cruise control actuator if removed.

42. Install the battery tray support and the tray.

43. Install the charcoal canister bracket and the canister.

44. Install the air duct and the air cleaner assembly.

45. Refill the transaxle and the transfer case if equipped with the proper fluid.

GALANT

1. Disconnect the negative battery cable.
2. Remove the air cleaner and intake hoses.
3. Drain the transaxle into a suitable waste container.
4. Remove the nut securing the shifter lever to the transaxle. Remove the cable retaining clip and remove the cable from the transaxle.
5. Remove the shifter cable mounting bracket.
6. Disconnect and tag the electrical connectors for the speedometer, solenoid, neutral safety switch (inhibitor switch), the pulse generator, kickdown servo switch, and the oil temperature sensor.
7. Disconnect and tag the oil cooler lines, at the transaxle.
8. Remove the bolt securing the fluid dipstick tube, to the transaxle. Remove the dipstick and tube from the transaxle.
9. Remove the starter motor and position it aside.
10. Using special tool MZ203827 or equivalent, support the engine assembly.
11. Remove the rear roll stopper mounting bracket.
12. Remove the transaxle mount bracket.
13. Remove the upper transaxle mounting bolts.
14. Raise and safely support the vehicle.
15. Remove the front wheel assemblies.
16. Remove the right-hand undercover.
17. Remove the cotter pin and disconnect the tie rod end, from the steering knuckle.
18. Disconnect the stabilizer bar link, from the damper fork.
19. Disconnect the damper fork, from the lateral lower control arm.
20. Disconnect the later lower arm, and the compression arm, lower ball joints, from the steering knuckle.
21. Pry the halfshafts from the transaxle, and secure aside.
22. Remove the cover from the transaxle bell housing.
23. Remove the engine front roll stopper through-bolt.
24. Remove the crossmember and the triangular right-hand stay.
25. Remove the bolts holding the flexplate to the torque converter with a box wrench. Rotate the engine to bring the bolts into a position appropriate for removal, one at a time. After removing the bolts, push the torque converter toward the transaxle. This will prevent the converter from remaining intact with the engine, possibly damaging the converter.

26. Support the transaxle, using a transmission jack, and remove the transaxle lower coupling bolt.

➡**The coupling bolt threads from the engine side, into the transaxle, and is located just above the halfshaft opening.**

27. Slide the transaxle rearward and carefully lower it from the vehicle.

To install:

28. After the torque converter has been mounted on the transaxle, install the transaxle assembly to the engine. Install the mounting bolts and tighten to 35 ft. lbs. (48 Nm). Install the transaxle lower coupling bolt and tighten to 21–25 ft. lbs. (29–34 Nm).
29. Connect the torque converter to the flexplate and tighten the bolts to 33–38 ft. lbs. (45–52 Nm).
30. Install the cover to the transaxle bell housing and tighten the mounting bolts to 7 ft. lbs. (9 Nm).
31. Install the crossmember and tighten the front mounting bolts to 65 ft. lbs. (88 Nm) and the rear bolt to 54 ft. lbs. (73 Nm). Install the front engine roll stopper through-bolt and lightly tighten. Once the full weight of the engine is on the mounts, tighten the bolt to 42 ft. lbs. (57 Nm).
32. Install the triangular stay bracket and tighten the mounting bolts to 65 ft. lbs. (88 Nm).
33. Install the halfshafts, using new circlips on the axle ends.

❈❈ WARNING

When installing the axle shaft, keep the inboard joint straight in relation to the axle, so as not to damage the oil seal lip of the transaxle, with the serrated part of the halfshaft.

34. Connect the tie rod and ball joints to the steering knuckle. Tighten the ball joint self-locking nuts to 48 ft. lbs. (65 Nm). Tighten the tie rod end nut to 21 ft. lbs. (28 Nm) and secure with a new cotter pin.
35. Connect the damper fork to the lower control arm and tighten the through-bolt to 65 ft. lbs. (88 Nm).
36. Connect the stabilizer link to the damper fork, and tighten the self-locking nut to 29 ft. lbs. (39 Nm).
37. Install the underpan.
38. Install wheels and lower vehicle.
39. Install the transaxle mount bracket, to the transaxle, and tighten the mounting nuts to 32 ft. lbs. (43 Nm).
40. Install the rear roll stopper mounting bracket.

41. Remove the engine support. Tighten the transaxle mount through-bolt to 51 ft. lbs. (69 Nm) and tighten the front engine roll stopper through-bolt.
42. Install the upper transaxle mounting bolts and tighten to 35 ft. lbs. (48 Nm).
43. Install the starter motor.
44. Install the dipstick tube and the dipstick.
45. Install the shifter cable mounting bracket.
46. Connect the shifter lever and tighten the retaining nut to 14 ft. lbs. (19 Nm).
47. Connect the oil cooler lines and secure with clamps.
48. Connect the electrical connectors for the speedometer, solenoid, neutral safety switch (inhibitor switch), the pulse generator, kickdown servo switch and oil temperature sensor.
49. Install the air cleaner and the air intake hose.
50. Connect the negative battery cable.
51. Be sure the vehicle is level, and refill the transaxle. Start the engine and allow to idle for 2 minutes. Apply parking brake and move selector through each gear position, ending in **N**. Recheck fluid level and add if necessary. Fluid level should be between the marks in the **HOT** range.
52. Check the transaxle for proper operation. Be sure the reverse lights come on when in reverse and the engine starts only in **P** or **N**.

MIRAGE

➡**If the vehicle is going to be rolled on its wheels while the halfshafts are out of the vehicle, obtain two outer CV-joints or proper equivalent tools and install to the hubs. If the vehicle is rolled without the proper torque applied to the front wheel bearings, the bearings will no longer be usable.**

1. Disconnect the negative battery cable.
2. Remove the battery and battery tray.
3. Remove the air hose and air cleaner assembly.
4. Raise the vehicle and support safely.
5. Remove the under guard pan.
6. Drain the transaxle oil.
7. Disconnect the control cable and cooler lines.
8. Disconnect the shift control solenoid valve connector.
9. Disconnect the inhibitor switch, kickdown servo switch, the pulse generator and oil temperature sensor, if equipped.
10. Disconnect the speedometer cable and remove the starter.

11. Remove the transaxle mounting bolts and bracket.

12. Disconnect the stabilizer bar from the lower control arm.

13. Disconnect the steering tie rod end and the ball joint from the steering arm.

14. Remove the halfshafts at the inboard side from the transaxle. Tie the joint assembly aside.

➡**It is not necessary to disconnect the halfshafts from the wheel hubs.**

15. Support the engine and remove the center member.

16. Remove the bell housing cover and remove the driveplate bolts.

17. Remove the transaxle assembly lower connecting bolt, located just over the halfshaft opening.

18. Properly support the transaxle assembly and lower it moving it to the right for clearance.

To install:

19. After the torque converter has been mounted on the transaxle, install the transaxle assembly on the engine. Install the mounting bolts and tighten to specifications.

20. Tighten the driveplate bolts to 33–38 ft. lbs. (46–53 Nm). Install the bell housing cover.

21. Install the center member and tighten the bolts to specifications.

22. Replace the circlips and install the halfshafts to the transaxle.

23. Install the tie rods, ball joints and stabilizer links to the steering arm.

24. Install the transaxle mounting bracket and bolts.

25. Install the starter.

26. Connect the speedometer cable.

27. Connect the inhibitor switch, kickdown servo switch, the pulse generator and oil temperature sensor, if disconnected.

28. Connect the shift control solenoid valve connector.

29. Connect the control cables and oil cooler lines.

30. Install the tension rod, if removed.

31. Install the air cleaner assembly.

32. Install the battery tray and battery. Connect the positive, then the negative terminal.

33. Refill with Dexron® II, Mopar ATF Plus type 7176 or equivalent, automatic transaxle fluid.

34. Start the engine and allow to idle for two minutes. Apply parking brake and move selector through each gear position, ending in **N**. Recheck fluid level and add if necessary. Fluid level should be between the marks in the **HOT** range.

Clutch

ADJUSTMENT

➡**The following adjustment is for cable actuated clutch systems. Hydraulic systems are self-adjusting and seldom require maintenance.**

Clutch pedal height

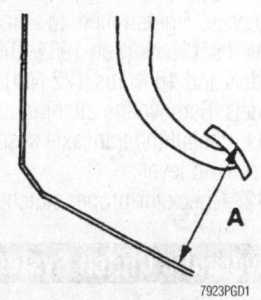

Clutch pedal height (A) measurement—Mirage

Mirage

1. Measure the clutch pedal height (measurement A). The specification is 6.38–6.50 in.(162–165mm).

➡**The clutch pedal height is not adjustable. If not within specifications, part replacement is required.**

2. Depress clutch pedal several times and check the pedal free-play (measurement B).

3. If measurement is not 0.67–0.87 in. (17–22mm), adjustment is required.

4. To adjust turn the outer cable adjusting nut, located at the firewall, until free-play is within range.

5. Depress clutch pedal several times and recheck measurement.

REMOVAL & INSTALLATION

3000GT

1. Disconnect the negative battery cable.

✳✳ CAUTION

Wait at least 90 seconds after the negative battery cable is disconnected to prevent possible deployment of the air bag.

2. Raise and safely support the vehicle.

3. Remove the transaxle assembly from the vehicle.

4. Remove the pressure plate attaching bolts. If the pressure plate is to be reused, loosen the bolts in succession, one or two

turns at a time to prevent warping the cover flange.

5. Remove the pressure plate release bearing assembly and the clutch disc. Do not use solvent to clean the bearing.

6. Inspect the condition of the clutch components and replace any worn parts.

To install:

7. Inspect the flywheel for heat damage or cracks. Resurface or replace the flywheel as required, using new bolts.

8. Using the proper alignment tool, install the clutch disc to the flywheel. Install the pressure plate assembly and tighten the pressure plate bolts evenly to 11–15 ft. lbs. (15–21 Nm). Remove the alignment tool.

9. Apply a very light coat of high temperature grease to the clutch fork at the ball pivot and where the fork contacts the bearing. Also a little bit of grease can be applied to end of the release cylinder's pushrod and to the pushrod hole on the fork. Apply a light coat of grease on the transaxle input shaft splines.

10. Install a new clutch release bearing. Pack its inner surface with high temperature grease.

11. Install the transaxle assembly.

12. Lower the vehicle and connect the negative battery cable.

13. Check the clutch for proper operation.

Eclipse, Galant and Mirage

1. Disconnect the negative battery cable. Raise and safely support the vehicle.

2. Remove the transaxle assembly from the vehicle.

3. Remove the pressure plate attaching bolts, pressure plate and clutch disc. If the pressure plate is to be reused, loosen the

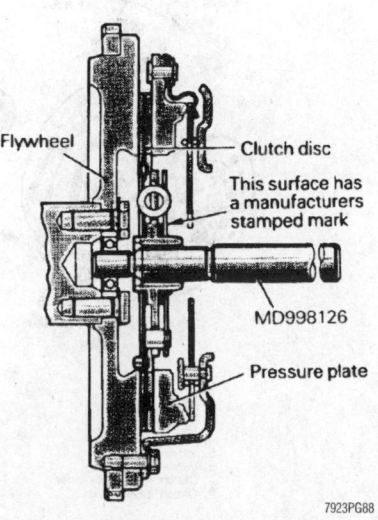

Use the alignment dowel to center the disc on the flywheel—Mirage

bolts in a diagonal pattern, 1 or 2 turns at a time. This will prevent warping the clutch cover assembly.

4. Remove the return clip and the pressure plate release bearing. Do not use solvent to clean the bearing.

5. Inspect the clutch release fork and fulcrum for damage or wear. If necessary, remove the release fork and unthread the fulcrum from the transaxle.

6. Carefully inspect the condition of the clutch components and replace any worn or damaged parts.

To install:

7. Inspect the flywheel for heat damage or cracks. Resurface or replace the flywheel as required. Install the flywheel using new bolts.

8. Install the fulcrum and tighten to 25 ft. lbs. (35 Nm). Install the release fork. Apply a coating of multi-purpose grease to the point of contact with the fulcrum and the point of contact with the release bearing. Apply a coating of multi-purpose grease to the end of the release cylinder's pushrod and the pushrod hole in the release fork.

➡ **When installing the clutch, apply grease to each part, but be careful not to apply excessive grease. Excessive grease will cause clutch slippage and shudder.**

9. Apply multi-purpose grease to the clutch release bearing. Pack the bearing inner surface and the groove with grease. Do not apply grease to the resin portion of the bearing. Place the bearing in position and install return clip.

10. Apply a coating of grease to the clutch disc splines, then use a brush to rub it in the grooves. Using a universal clutch disc alignment tool, position the clutch disc on the flywheel. Install the retainer bolts and tighten a little at a time, in a diagonal sequence. Tighten them to a final torque of 14 ft. lbs. (19 Nm) on 1994–97 Galant models and 16 ft. lbs. (22 Nm) on all other models. Remove the aligning tool.

11. Install the transaxle assembly and check fluid level.

12. Check for proper clutch operation.

Hydraulic Clutch System

BLEEDING

Galant, Eclipse, 3000GT and Mirage

✳✳ CAUTION

The clutch hydraulic system uses brake fluid. Use care, brake fluid is harmful to painted surfaces.

1. Fill the reservoir with clean brake fluid meeting DOT 3 specifications.

2. Press the clutch pedal to the floor, then open the bleeder screw on the slave cylinder.

3. Tighten the bleed screw and release the clutch pedal.

4. Repeat the procedure until the fluid is free of air bubbles.

➡ **It is suggested that a hose be attached to the bleeder with the other end immersed in a container at least half full of brake fluid during the bleeding operation. Do not allow the reservoir to run out of fluid during bleeding.**

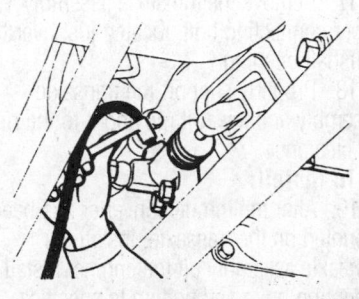

7923PG91

Bleeding a typical clutch hydraulic system

Transfer Case Assembly

REMOVAL & INSTALLATION

3000GT

1. Disconnect the negative battery cable.

✳✳ CAUTION

Wait at least 90 seconds after the negative battery cable is disconnected to prevent possible deployment of the air bag.

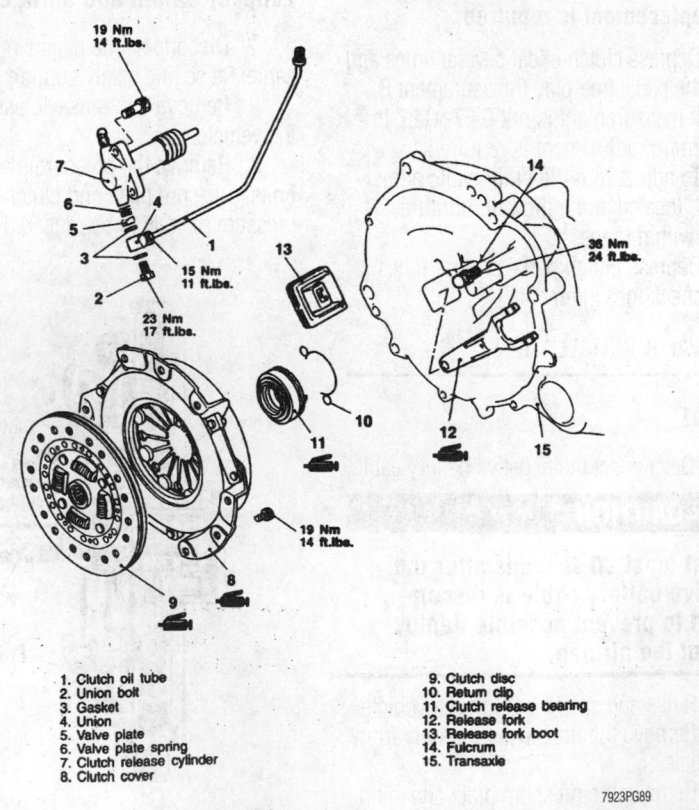

1. Clutch oil tube
2. Union bolt
3. Gasket
4. Union
5. Valve plate
6. Valve plate spring
7. Clutch release cylinder
8. Clutch cover

9. Clutch disc
10. Return clip
11. Clutch release bearing
12. Release fork
13. Release fork boot
14. Fulcrum
15. Transaxle

7923PG89

Exploded view of non-turbo clutch assembly—Eclipse with F5M31, F5M33, F5MC1 and W5M33 manual transaxles

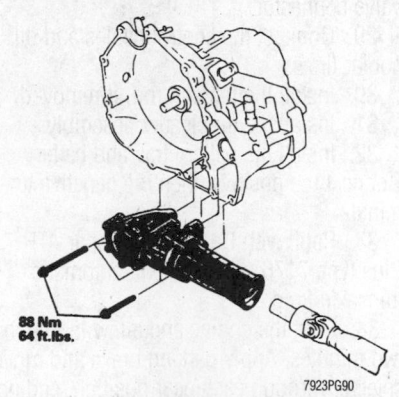

7923PG90

Transfer case assembly—3000GT with F5M33, W5MG1 and W6MG1 transfer cases

2. Raise the vehicle and support safely. Drain the transfer assembly oil.

3. Remove necessary front bumper components.

4. Disconnect the front exhaust pipe.

5. Unbolt the transfer case assembly and remove by sliding it off the rear driveshaft. Be careful not to damage the oil seal in the transfer case output housing. Do not let the rear driveshaft hang; suspend it from a frame piece. Cover the opening in the transaxle and transfer case to keep oil from dripping and to keep dirt out.

To install:

6. Lubricate the driveshaft sleeve yoke and oil seal lip on the transfer extension housing. Install the transfer case assembly to the transaxle. Use care when installing the rear driveshaft to the transfer case output shaft.

7. Tighten the transfer case to transaxle bolts to 18–22 ft. lbs. (25–29 Nm).

8. Install the exhaust pipe using a new gasket. Install removed bumper components.

9. Refill the transfer case and check the oil levels in the transaxle and transfer case.

10. Safely lower the vehicle and connect the negative battery cable.

Eclipse

1. Raise and safely support the vehicle.
2. Remove the engine undercovers.

3. Remove the front exhaust pipe.

4. Drain the transfer case fluid.

5. Remove the transfer case mounting bolts.

6. Support the driveshaft with wire or string and remove the transfer case from the transaxle.

To install:

7. Slide the driveshaft into the transfer case and install the transfer case to the transaxle. Torque the bolts to 40–44 ft. lbs. (54–59 Nm).

8. Install the front exhaust pipe.

9. Refill the transfer case with the proper gear oil.

10. Install the engine undercover.

11. Safely lower the vehicle to the floor.

Halfshaft

REMOVAL & INSTALLATION

3000GT

➡ **If the vehicle is going to be rolled on its wheels while the halfshafts are out of the vehicle, obtain two outer CV-joints or proper equivalent tools and install to the hubs. If the vehicle is rolled without the proper torque** applied to the front wheel bearings, the bearings will no longer be usable.

FRONT

1. Disconnect the negative battery cable.

2. With the vehicle on the floor and the brakes applied, remove the cotter pin, halfshaft nut and the washer.

3. Raise the vehicle and support safely. Remove the lower ball joint and the tie rod end from the steering knuckle.

4. On vehicles with an inner shaft, remove the center support bearing bracket bolts and washers.

5. On vehicles with an inner shaft, remove the halfshaft by setting up a puller on the outside wheel hub and pushing the halfshaft from the front hub. Then, tap the

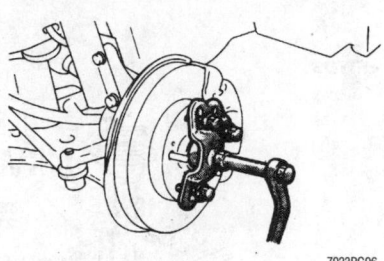

7923PG96

Front halfshaft removal—3000GT

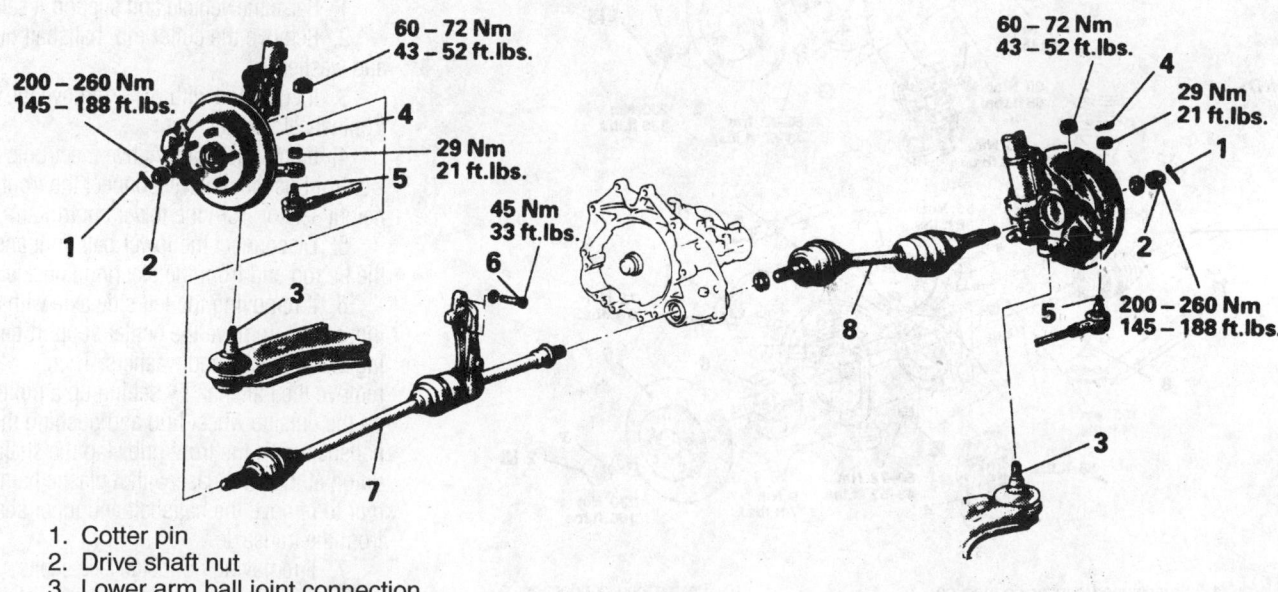

1. Cotter pin
2. Drive shaft nut
3. Lower arm ball joint connection
4. Cotter pin
5. Tie rod end connection
6. Center bearing bracket installation bolt
7. Drive and inner shaft assembly (L.H.)
8. Drive shaft (R.H.)
9. Circlip

Caution
In the case of AWD-vehicles with A.B.S., take care not to damage the rotor for A.B.S. installed to the B.J. outer race.

7923PG95

Front halfshafts and related components—3000GT

shaft union at the joint case with a plastic hammer to remove the halfshaft shaft and inner shaft from the transaxle.

6. On vehicles without an inner shaft, remove the halfshaft by setting up a puller on the outside wheel hub and pushing the halfshaft from the front hub. After pressing the outer shaft, insert a prybar between the transaxle case and the halfshaft and pry the shaft from the transaxle. Do not pull on the shaft. Doing so damages the inboard joint. Do not insert the prybar too far or the oil seal in the case may be damaged.

To install:

7. Inspect the halfshaft boot for damage or deterioration. Check the ball joints and splines for wear.

8. Replace the circlips on the ends of the halfshafts.

9. Insert the halfshaft into the transaxle.

Be sure it is fully seated.

10. Pull the strut assembly out and install the other end to the hub.

11. Install the center bearing bracket bolts and tighten to 33 ft. lbs. (45 Nm).

12. Install the washer so the chamfered edge faces outward. Install the nut and tighten temporarily.

13. Install the tie rod end and ball joint.

14. Install the wheel and lower the vehicle to the floor. Tighten the axle nut with the brakes applied. Tighten the nut to a maximum torque of 188 ft. lbs. (260 Nm). Install the cotter pin and bend to secure.

REAR

➡ **On vehicles with Limited Slip Differential, the right and left halfshafts are not the same. If both halfshafts are to be removed, be sure to mark one of the**

halfshafts (left or right) for proper installation.

1. Disconnect the negative battery cable. Raise the vehicle and support it safely.

2. Matchmark the halfshaft and the companion flange.

3. Remove the bolts that attach the rear halfshaft to the companion flange.

4. Use a prybar to pry the inner shaft out of the differential case. Don't insert the prybar too far or the seal could be damaged.

5. Remove the rear halfshaft from the vehicle.

To install:

6. Install a new circlip on the halfshaft and install it into the differential. Be sure it is fully seated.

7. Align the matchmarks and attach the halfshaft to the companion flange. Tighten the fasteners to 40–47 ft. lbs. (55–65 Nm).

8. Safely lower the vehicle and connect the negative battery cable.

Diamante, Mirage and Galant

➡ **If the vehicle is going to be rolled while the halfshafts are out of the vehicle, obtain 2 outer CV-joints or proper equivalent tools and install to the hubs. If the vehicle is rolled without the proper torque applied to the front wheel bearings, the bearings will no longer be usable.**

1. Raise the vehicle and support it safely.

2. Remove the cotter pin, halfshaft nut and washer.

3. If equipped with ABS, remove the front wheel speed sensor.

4. If equipped with Active Electronic Control Suspension, disconnect the front height sensor from the lower control arm.

5. Disconnect the lower ball joint and the tie rod end from the steering knuckle.

6. If removing the left side axle with an inner shaft, remove the center support bearing bracket bolts and washers. Then, remove the halfshaft by setting up a puller on the outside wheel hub and pushing the halfshaft from the front hub. Tap the shaft union at the joint case with a plastic hammer to remove the halfshaft and inner shaft from the transaxle.

7. If removing right side axle shafts without an inner shaft, remove the halfshaft by setting up a puller on the outside wheel hub and pushing the halfshaft from the front hub. After pressing the outer shaft, insert a prybar between the transaxle case and the halfshaft and pry the shaft from the transaxle.

➡ **Do not pull on the shaft; doing so damages the inboard joint. Do not**

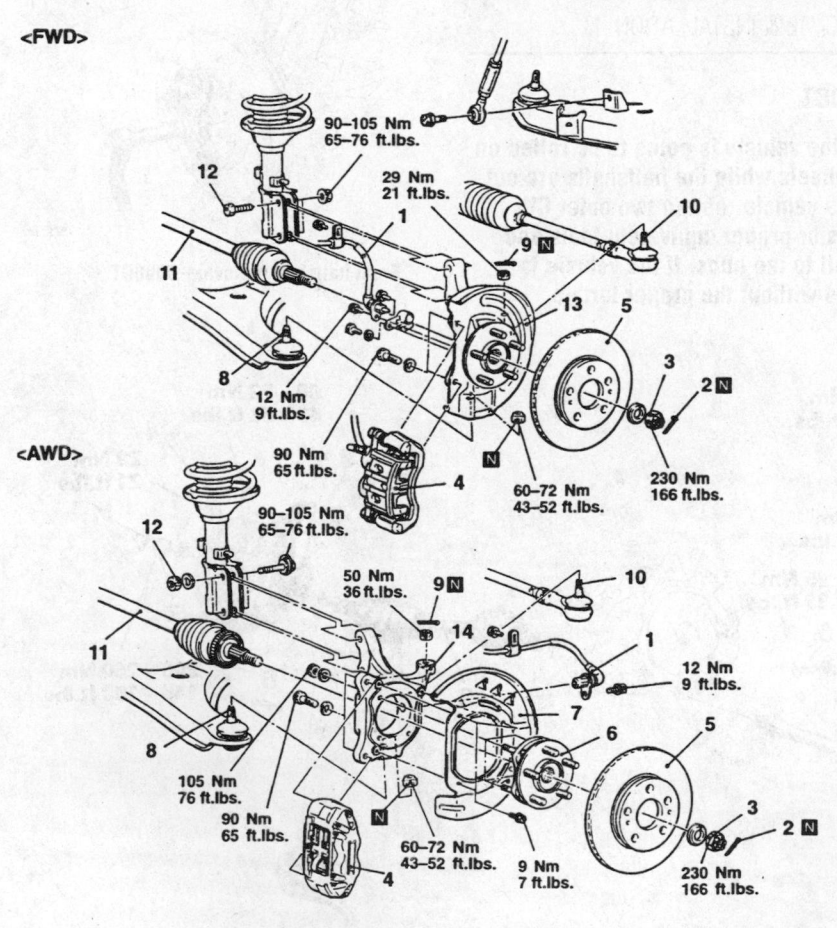

\<FWD\>

\<AWD\>

1. Front speed sensor connection \<Vehicles with ABS*\>
2. Cotter pin
3. Drive shaft nut
4. Caliper assembly
5. Brake disc
6. Front hub unit bearing
7. Dust shield
8. Lower arm ball joint connection
9. Cotter pin
10. Tie rod end connection
11. Drive shaft
12. Front strut mounting bolt or nut
13. Hub and knuckle
14. Hub

NOTE
*: Anti-lock braking system

Exploded view of the front suspension components—3000GT

7923PG97

DRIVE SHAFT

REMOVAL AND INSTALLATION

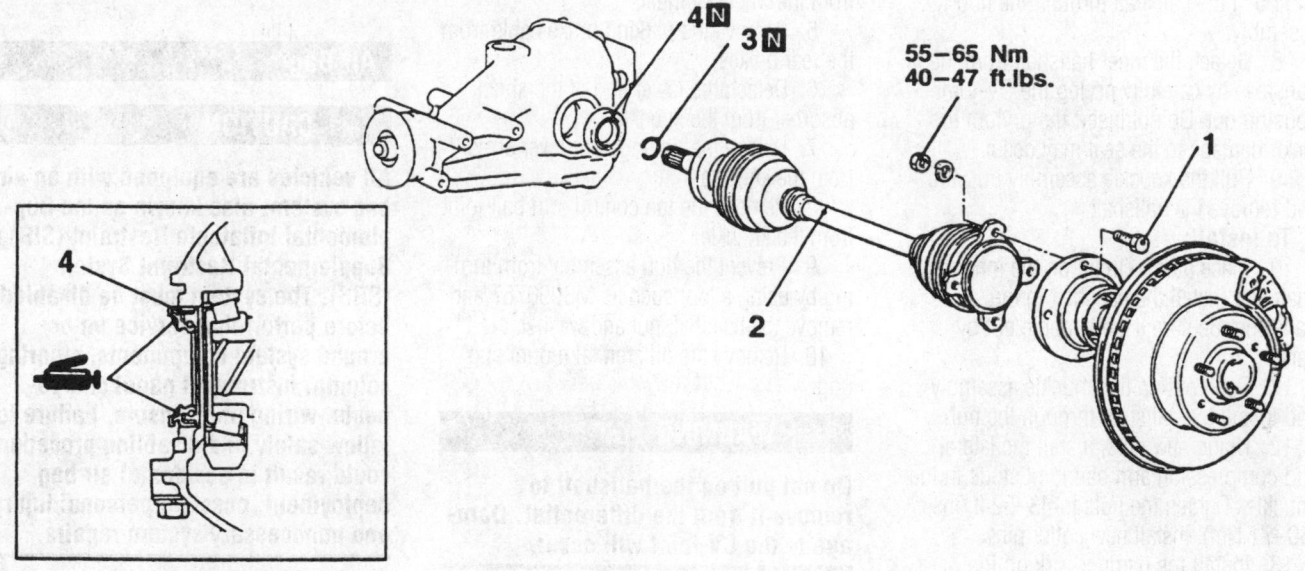

3. Circlip
4. Oil seal

7923PG98

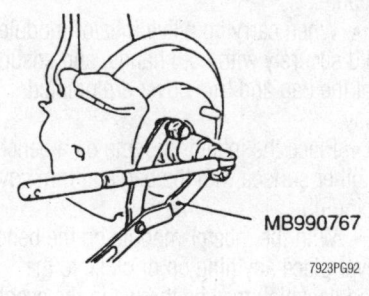

Removal steps
1. Bolt
2. Drive shaft

Rear halfshaft removal—AWD 3000GT

MB990767

7923PG92

Proper method for removing the halfshaft nut

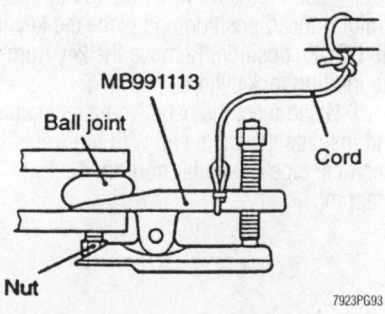

MB991113

Ball joint

Cord

Nut

7923PG93

Proper method for disconnecting ball joint studs and tie rod ends

insert the prybar too far or the oil seal in the case may be damaged.

To install:

8. Inspect the halfshaft boot for damage or deterioration. Check the ball joints and splines for wear.

9. Replace the circlips on the ends of the halfshafts.

10. Insert the halfshaft into the transaxle. Be sure it is fully seated.

11. Pull the strut assembly out and install the other end to the hub.

12. Install the center bearing bracket bolts and tighten to 33 ft. lbs. (45 Nm).

13. Install the washer so the chamfered edge faces outward. Install the nut and tighten temporarily.

14. Connect the ball joint to the steering knuckle. Torque the new retaining nut to 43–52 ft. lbs. (60–72 Nm) and secure with a new cotter pin.

15. Connect the tie rod end to the steering knuckle. Torque the retaining nut to 21 ft. lbs. (29 Nm) and secure with a new cotter pin.

16. If equipped with ABS, install the front wheel speed sensor.

17. If equipped with Active Electronic Control Suspension, connect the front height sensor to the lower control arm.

18. Install the wheel and lower the vehicle to the floor.

19. Tighten the axle nut with the brakes applied, to a maximum torque of 145–188 ft. lbs. (200–260 Nm) and secure with a new cotter pin.

20. Check the transaxle fluid and fill if necessary.

Eclipse

FRONT

1. Raise and safely support the vehicle.

2. Remove the front wheel.

✳✳ WARNING

Do not allow the weight of the vehicle to rest on the bearing assembly while removing the halfshaft outer nut. Damage to the wheel bearing may occur.

3. Prevent the hub assembly from turning by using a tool such as MB990767 and remove the halfshaft nut and washer.

4. Disconnect the tie rod end from the knuckle.

5. Remove the stabilizer link from the damper fork.

6. Disconnect the compression and lateral arm ball joint studs from the knuckle.

7. Mount a puller on the wheel studs and push the halfshaft through the hub assembly.

8. Detach the inner halfshaft from the transaxle by carefully prying the CV-joint housing out. Do not insert the prytool too far or damage to the seal may occur.

9. Pull the knuckle assembly outward and remove the halfshaft.

To install:

10. Place a new circlip on the inner halfshaft and install the halfshaft in the transaxle. Be sure it won't come out by hand.

11. Push out on the knuckle assembly and install the halfshaft through the hub.

12. Using new nuts, install the lateral and compression arm ball joint studs in the knuckle. Tighten the nuts to 43–52 ft. lbs. (59–71 Nm). Install new cotter pins.

13. Install the damper fork on the knuckle. Do nut tighten the nut at this time.

14. Attach the stabilizer link to the damper fork. Tighten the nut to 29 ft. lbs. (39 Nm).

15. Install the washer and nut on the halfshaft. Prevent the hub from turning and tighten the nut to 145–188 ft. lbs. (196–255 Nm).

16. Install the wheel and lower the vehicle to the floor. Tighten the damper fork nut to 65 ft. lbs. (88 Nm).

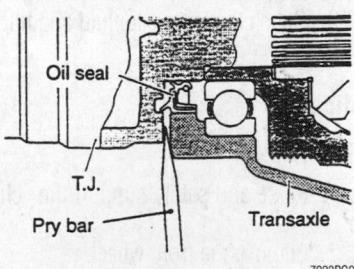

Oil seal

T.J.

Pry bar   Transaxle

7923PG94

Proper method for removing the inner halfshaft from the transaxle or differential

REAR

1. Raise and safely support the vehicle.
2. Remove the rear wheel.
3. Remove the rear wheel speed sensor, if equipped.

4. If equipped with disc brakes, remove the caliper and rotor. If equipped with drum brakes, remove the brake drum and shoes. Then, disconnect the brake hydraulic line from the wheel cylinder.

5. Remove the parking brake cable from the rear brakes.

6. Detach the lower end of the shock absorber from the knuckle.

7. Detach the trailing and lower arms from the knuckle.

8. Remove the toe control arm ball joint from the knuckle.

9. Prevent the hub assembly from turning by using a tool such as MB990767 and remove the halfshaft nut and washer.

10. Remove the differential mount support.

❋❋ WARNING

Do not pull on the halfshaft to remove it from the differential. Damage to the CV-joint will occur.

11. Push the lower part of the knuckle outward and pry the inner halfshaft out of the differential. Do not insert the prytool too far or damage to the seal may occur.

12. Push the outer end of the halfshaft through the hub/knuckle and remove it.

To install:

13. Install the outer end of the halfshaft through the hub/knuckle.

14. Place a new circlip on the inner halfshaft and install the halfshaft in the differential. Be sure it won't come out by hand.

15. Install the differential mount support.

16. Install the washer and a new nut on the end of the halfshaft. Tighten the nut to 145–188 ft. lbs. (196–255 Nm).

17. Install the toe control arm to the knuckle. Tighten the new nut to 20 ft. lbs. (28 Nm).

18. Connect the lower and trailing arms to the knuckle. Do not tighten the fasteners at this time.

19. Install the shock absorber. Tighten the bolt to 71 ft. lbs. (98 Nm).

20. Assemble the brake components.

21. Install the rear wheel speed sensor.

22. Install the rear wheel and lower the vehicle to the floor. Tighten the lower arm nut to 71 ft. lbs. (98 Nm) and the trailing arm nut to 85–99 ft. lbs. (118–137 Nm).

STEERING AND SUSPENSION

Air Bag

❋❋ CAUTION

All vehicles are equipped with an air bag system, also known as the Supplemental Inflatable Restraint (SIR) or Supplemental Restraint System (SRS). The system must be disabled before performing service on or around system components, steering column, instrument panel components, wiring and sensors. Failure to follow safety and disabling procedures could result in accidental air bag deployment, possible personal injury and unnecessary system repairs.

PRECAUTIONS

Several precautions must be observed when handling the inflator module to avoid accidental deployment and possible personal injury.

• Never carry the inflator module by the wires or connector on the underside of the module.

• When carrying a live inflator module, hold securely with both hands, and ensure that the bag and trim cover are pointed away.

• Place the inflator module on a bench or other surface with the bag and trim cover facing up.

• With the inflator module on the bench, never place anything on or close to the module which may be thrown in the event of an accidental deployment.

DISARMING

1. Position the front wheels in the straight-ahead position and place the key in the **LOCK** position. Remove the key from the ignition lock cylinder.

2. Disconnect the negative battery cable and insulate the cable end with high-quality electrical tape or similar non-conductive wrapping.

3. Wait at least one minute before working on the vehicle. The air bag system is designed to retain enough voltage to deploy the air bag for a short period of time after the battery has been disconnected.

REARMING

1. Connect the negative battery cable, turn the ignition switch to the **ON** position and check the SRS warning light for proper operation.

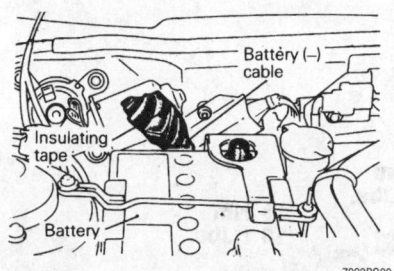

Insulate the negative battery cable to prevent accidental deployment of the air bag

Rack and Pinion Steering Gear

REMOVAL & INSTALLATION

Manual

MIRAGE

➡**If equipped with air bag, prior to removal of the steering rack, center the front wheels and remove the ignition key. Failure to do so may damage the SRS clockspring and render SRS system inoperative, risking serious driver injury. Be sure to properly disarm the air bag system.**

1. Disconnect the battery negative cable. Raise the vehicle and support safely and remove the wheels.
2. Disconnect the oxygen sensor and remove the front exhaust pipe.

➡**It may be easier on some vehicles, to disconnect all hangers and lower the complete exhaust system.**

3. Properly support the engine. Remove both roll stopper mounting bolts and the four center member installation bolts.
4. Remove the center member.

➡**Matchmark the pinion input shaft of the rack to the lower steering column joint for installation purposes.**

5. Remove the pinch bolt holding the lower steering column joint to the rack and pinion input shaft.
6. Remove the cotter pins and disconnect the tie rod ends from the steering knuckle.
7. Remove the rack and pinion steering assembly and its rubber mounts from the right side of the vehicle.

To install:

8. Align the matchmarks of the input shaft and install the rack to the vehicle.
9. Secure the rack using the retainer clamps and bolts. Torque the bolts to 51 ft. lbs. (70 Nm).
10. Torque the steering column pinch bolt to 13 ft. lbs. (18 Nm).
11. Install the center member.

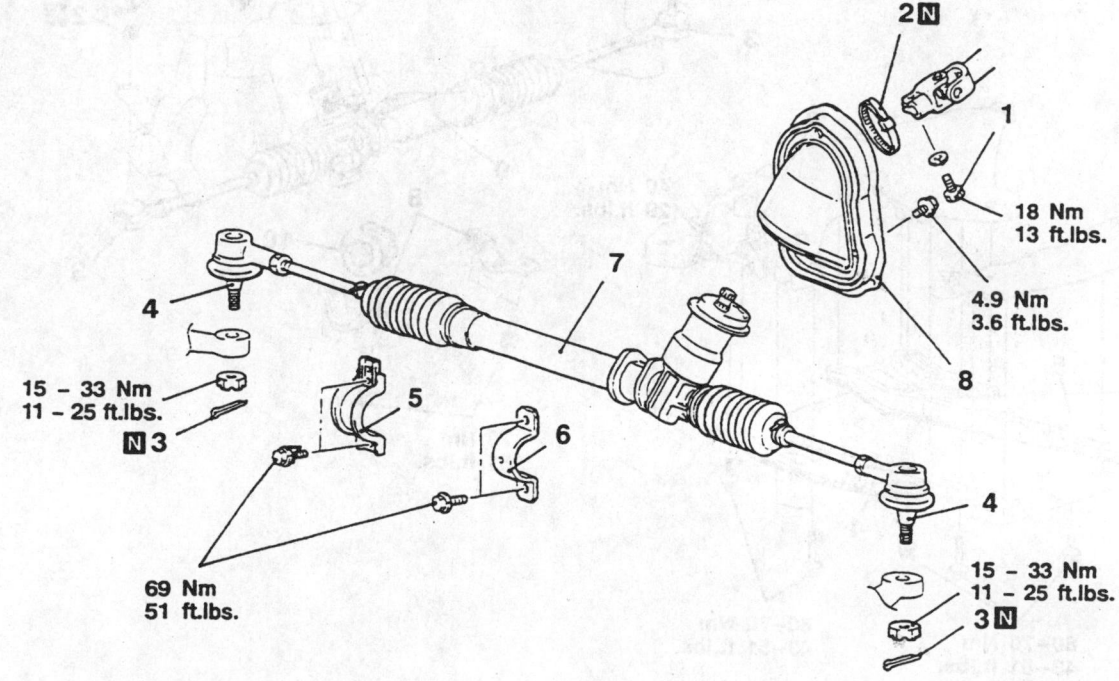

1. Steering shaft assembly and gear box connecting bolt
2. Band
3. Cotter pin
4. Tie-rod end and knuckle connection
5. Cylinder clamp
6. Gear housing clamp
7. Gear box assembly
8. Steering cover assembly

Exploded view of the manual steering gear mounting—Mirage

12. Install the front exhaust pipe.

13. Connect the tie rod ends to the steering knuckles and tighten the castle nuts to 25 ft. lbs. (34 Nm). Install new cotter pins.

14. Install the wheels and connect the negative battery cable.

15. Perform a front end alignment.

Power

3000GT AND DIAMANTE

➡Prior to removal of the steering gear box, center the front wheels and remove the ignition key. Failure to do so may damage the SRS clock spring and render SRS system inoperative, risking serious driver injury.

1. Disconnect the negative battery cable. Disarm the air bag.

❋❋ CAUTION

Work must be started after 90 seconds from the time the ignition switch is turned to the LOCK position and the negative battery cable is disconnected.

2. Disconnect the front exhaust pipe.

3. If equipped with AWD, remove the transfer case assembly.

4. Remove the bolt holding the lower steering column joint to the rack and pinion input shaft.

5. Remove the cotter pins and disconnect the tie rod ends.

6. Remove the left and right frame members.

7. Remove the stabilizer bar bracket.

8. If equipped with four-wheel steering, disconnect the lines going to the rear pump.

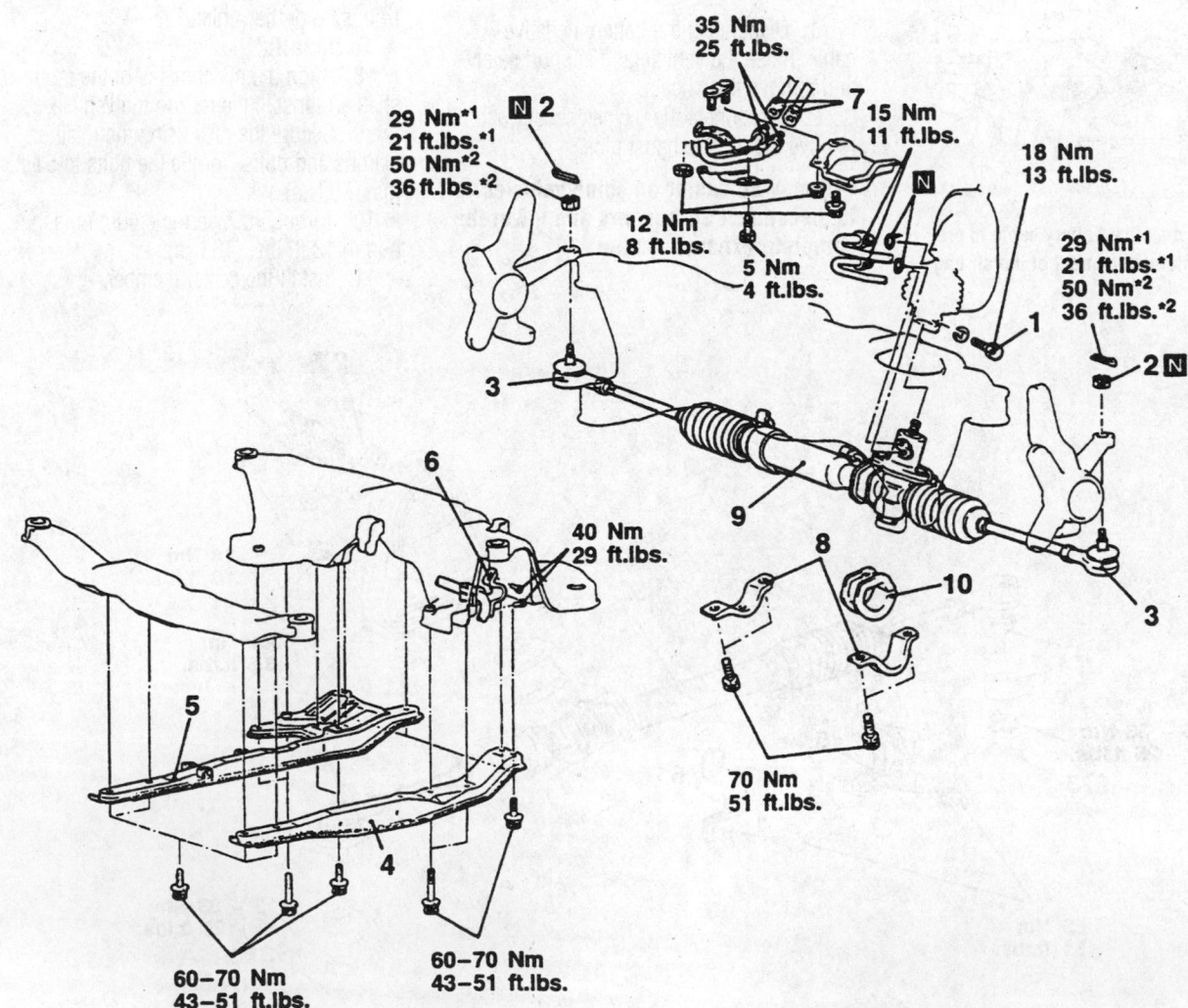

1. Joint assembly and gear box connecting bolt
2. Cotter pin
3. Tie-rod end and knuckle connecting nut
4. Left member
5. Right member
6. Stabilizer bar bracket
7. Connection of steering gear box with 4WS oil line
8. Clamp
9. Gear box assembly
10. Mounting rubber

NOTE
*1: FWD
*2: AWD

Exploded view of the power steering gear removal—3000GT

7923PGA5

9. Remove the rack and pinion steering assembly and its rubber mounts. Move the rack to the right to remove it from the crossmember. Use caution to avoid damaging the boots.

To install:

10. Install the rack and install the mounting bolts, tightening bolts to 51 ft. lbs. (70 Nm). When installing the rubber rack mounts, align the projection of the mounting rubber with the indentation in the crossmember. Install the pinch bolt.

11. Connect the lines going to the four-wheel steering rear pump and to the rack itself.

12. Install the frame members and tighten the bolts to 43–51 ft. lbs. (60–70 Nm).

13. Connect the tie rods and Install new cotter pins.

14. Install the transfer case and front exhaust pipe.

15. Refill the reservoir and bleed the system.

16. Perform a front end alignment.

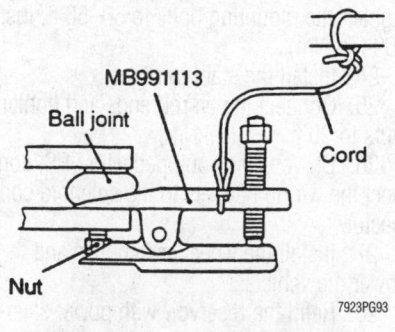

Proper tie-rod end removal method

ECLIPSE (NON-TURBO)

❋❋ CAUTION

The air bag system (SRS or SIR) must be disarmed before removing the rack and pinion. Failure to do so may cause accidental deployment, property damage or personal injury.

1. Center the front wheel and remove the ignition key from the switch.

2. Disconnect the negative battery cable.

3. Drain the power steering fluid.

4. Raise and safely support the vehicle.

5. Remove the stabilizer bar.

6. Remove the windshield washer reservoir.

7. Remove the pinch bolt from the joint assembly.

8. Disconnect the fluid lines from the steering rack.

9. Using the proper tools, disconnect the tie rod ends from the steering knuckles.

10. Remove the left and right stays (supports).

11. Support the engine and remove the center member.

12. Remove the clamp and the mounting bolts.

13. Disconnect the left lower compression arm from the body side of the vehicle and support it with wire or string.

14. Disconnect the steering rack from the joint assembly and remove the rack from the left side of the vehicle.

To install:

15. Position the steering rack in the vehicle and install the clamp and the mounting bolts. Be sure the rack is centered before connecting it to the joint assembly.

16. Install the left lower compression arm to the body.

17. Install the center member.

18. Install the left and right stays and remove the engine support fixture or jack.

19. Connect the tie rods to the steering knuckles.

20. Connect the fluid lines to the steering rack. Torque to specifications.

21. Install the pinch bolt in the joint assembly.

22. Install the stabilizer bar and the windshield washer reservoir.

23. Safely lower the vehicle.

24. Connect the negative battery cable.

25. Refill and bleed the power steering system.

26. Check wheel alignment.

ECLIPSE (TURBO)

1. Center the front wheel and remove the ignition key from the switch.

2. Disconnect the negative battery cable.

3. Drain the power steering fluid.

4. Raise and safely support the vehicle.

5. Remove the stabilizer bar.

6. Disconnect the fluid level sensor and remove the brake fluid reservoir and position it out of the way. Do not disconnect the brake hose.

7. Disconnect the electrical connector from the A/C compressor.

8. Remove the A/C compressor from the bracket and position it out of the way. Do not disconnect the hoses.

9. Remove the pinch bolt from the joint assembly.

10. Disconnect the fluid lines from the steering rack.

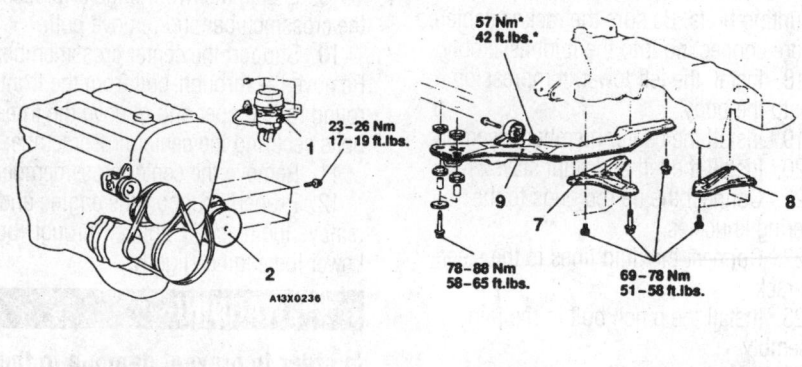

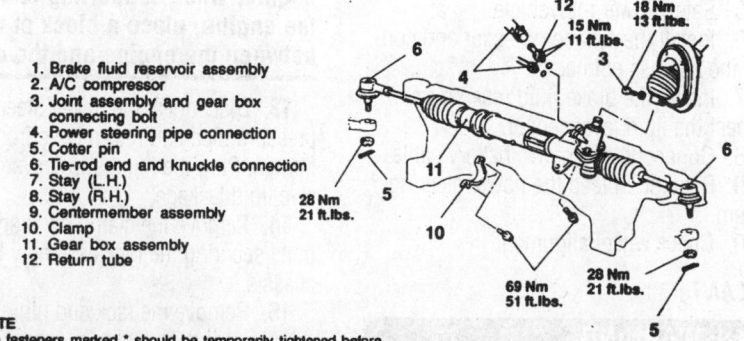

1. Brake fluid reservoir assembly
2. A/C compressor
3. Joint assembly and gear box connecting bolt
4. Power steering pipe connection
5. Cotter pin
6. Tie-rod end and knuckle connection
7. Stay (L.H.)
8. Stay (R.H.)
9. Centermember assembly
10. Clamp
11. Gear box assembly
12. Return tube

NOTE
The fasteners marked * should be temporarily tightened before they are finally tightened once the total weight of the engine has been placed on the vehicle body.

Power steering rack assembly and related components—Eclipse

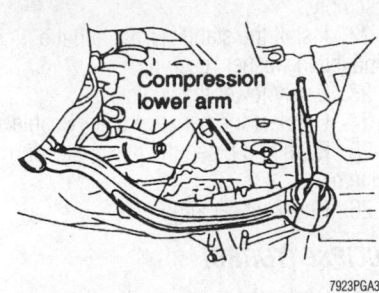

7923PGA3

Disconnect the lower compression arm from the body—Eclipse

11. Using the proper tools, disconnect the tie rod ends from the steering knuckles.

12. Remove the left and right stays (supports).

13. Support the engine and remove the center member assembly.

14. Remove the clamp and the mounting bolts.

15. Disconnect the left lower compression arm from the body side of the vehicle and support it with wire or string.

16. Disconnect the steering rack from the joint assembly and remove the rack from the left side of the vehicle.

To install:

17. Position the steering rack in the vehicle and install the clamp and the mounting bolts. Be sure the rack is centered before connecting it to the joint assembly.

18. Install the left lower compression arm to the body.

19. Install the center member assembly.

20. Install the left and right stays.

21. Connect the tie rod ends to the steering knuckles.

22. Connect the fluid lines to the steering rack.

23. Install the pinch bolt in the joint assembly.

24. Install the stabilizer bar.

25. Safely lower the vehicle.

26. Install the A/C compressor and connect the harness connector.

27. Install the brake fluid reservoir and connect the fluid level sensor.

28. Connect the negative battery cable.

29. Refill and bleed the power steering system.

30. Check wheel alignment.

GALANT

> ⁂ **WARNING**
>
> **Prior to removal of the steering gear box, center the front wheels and remove the ignition key. Failure to do so may damage the SRS clock spring and render SRS system inoperative, risking serious driver injury.**

1. Drain the power steering fluid as follows:

   a. Disconnect the power-steering return (low side) hose.

   b. Connect a suitable container to the hose.

   c. Properly disable the ignition system, by disconnecting the ignition coil wire and connecting it to a suitable ground.

   d. While cranking the engine, turn the wheels, several times, from side to side, until the fluid is removed.

2. Disarm the SRS system using the recommended procedure.

3. Raise and properly support the vehicle.

4. Remove both front wheel assemblies.

5. Remove the bolt holding lower steering column joint to the rack and pinion input shaft.

6. Remove the stabilizer bar.

7. Remove the cotter pins and using joint separator MB991113, disconnect the tie rod ends, from the steering knuckle.

8. On vehicles equipped with Electronic Control Power steering (EPS), disconnect the wiring harness, from the solenoid connector.

9. Locate the two triangular braces near the crossmember and remove both.

10. Support the center crossmember. Remove the through-bolt from the front round roll stopper and remove the three bolts securing the center crossmember.

11. Remove the center crossmember.

12. Properly support the engine and remove the rear roll stopper through-bolt. Lower the engine slightly.

> ⁂ **WARNING**
>
> **In order to prevent damage to the engine, when supporting and jacking the engine, place a block of wood between the engine and the oil pan.**

13. Disconnect the power steering fluid pressure pipe and return hose from the rack fittings. Plug the fittings to prevent excessive fluid leakage.

14. Remove the clamp bolts and the two bolts securing the rack assembly to the chassis.

15. Remove the rack and pinion steering assembly and its rubber mounts.

➡ **When removing the rack and pinion assembly, tilt the assembly to the vehicle side of the compression lower arm, and remove from the left side of the vehicle. Use caution to avoid damaging the boots.**

To install:

16. Align the rack assembly so the splines are inserted into the steering column shaft.

17. Install the rack and with the mounting bolts. Torque the mounting bolts to 51 ft. lbs. (69 Nm).

18. Install the pinch bolt and tighten the bolt to 13 ft. lbs. (18 Nm).

19. Connect the power steering fluid lines to the rack and tighten to high side fitting to 11 ft. lbs. (15 Nm). Secure the low side hose with the clamp.

20. Raise the engine into position. Install the rear roll stopper through-bolt and tighten to 32 ft. lbs. (43 Nm).

21. Raise the crossmember into position. Install the center member mounting bolts and tighten the front bolts to 58–65 ft. lbs. (78–88 Nm) and the rear bolt to 51–58 ft. lbs. (69–78 Nm).

22. Install the front roll stopper bolt and tighten the nut to 32 ft. lbs. (43 Nm).

23. Install the two triangular braces and tighten the mounting bolts to 50–56 ft. lbs. (69–78 Nm).

24. Install the stabilizer bar.

25. Connect the tie rod ends and tighten nuts to 20 ft. lbs. (27 Nm).

26. On vehicles equipped with EPS, connect the wiring harness to the solenoid connector.

27. Install the wheel assemblies and lower the vehicle.

28. Refill the reservoir with power steering fluid and bleed the system.

29. Perform a front end alignment.

MIRAGE

➡ **If equipped with air bag, prior to removal of the steering gear box, center the front wheels and remove the ignition key. Failure to do so may damage the SRS clockspring and render SRS system inoperative, risking serious driver injury.**

1. Drain power steering system:

   a. Disconnect the return hose at the reservoir and place into a suitable container.

   b. Disable the ignition system. While cranking the engine, turn the wheels several times, until system has been drained.

2. Disconnect the battery negative cable. Raise the vehicle and support safely.

3. Disconnect the oxygen sensor and remove the front exhaust pipe.

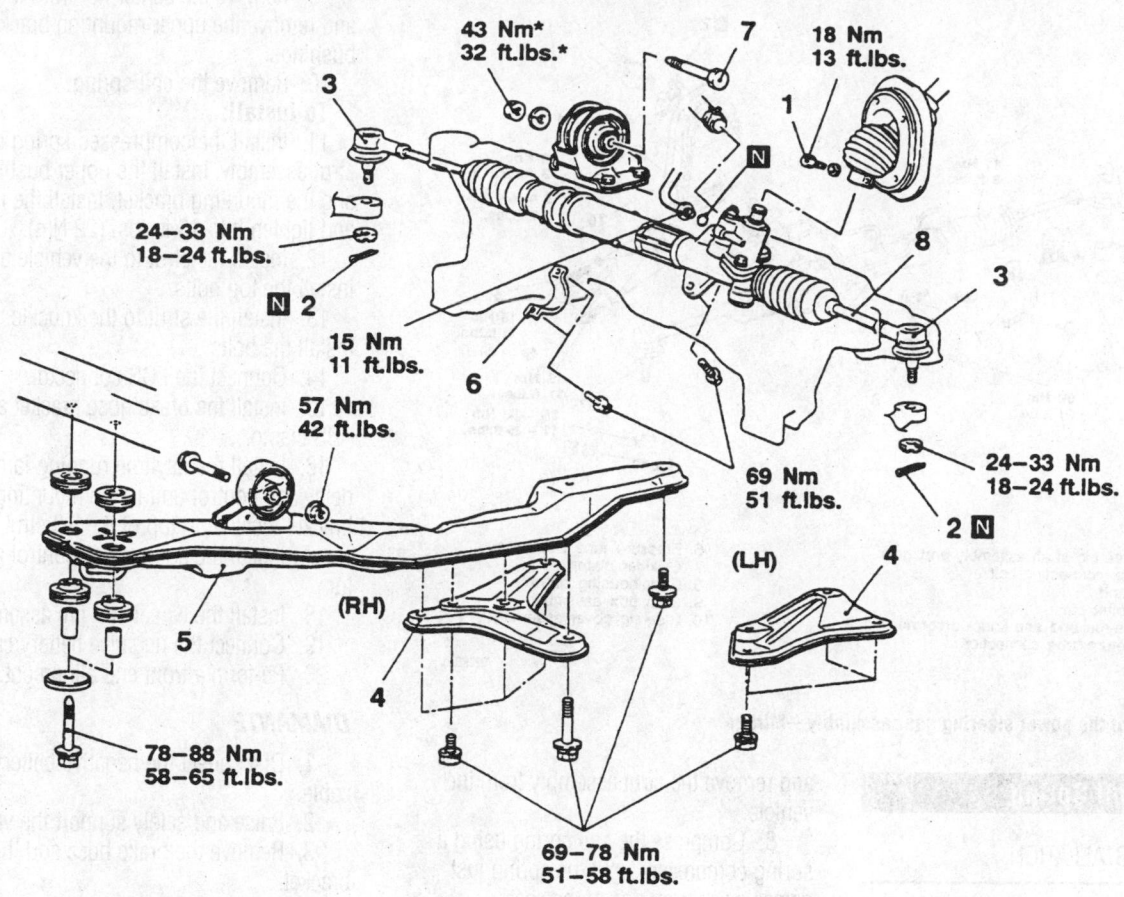

43 Nm*
32 ft.lbs.*

18 Nm
13 ft.lbs.

24–33 Nm
18–24 ft.lbs.

15 Nm
11 ft.lbs.

57 Nm
42 ft.lbs.

69 Nm
51 ft.lbs.

24–33 Nm
18–24 ft.lbs.

78–88 Nm
58–65 ft.lbs.

69–78 Nm
51–58 ft.lbs.

(RH)   (LH)

1. Joint assembly and gear box connecting bolt
2. Cotter pin
3. Connection for tie rod end and
   knuckle
4. Stay
5. Center member assembly
6. Clamp
7. Bolt

8. Gear box assembly

Caution
**The fasteners marked * should be temporarily
tightened before they are finally tightened once
the total weight of the engine has been placed
on the vehicle body.**

7923PGA4

Exploded view of the power steering gear removal procedure—Galant

➡**It may be easier on some vehicles,
to disconnect all hangers and lower the
complete exhaust system.**

4. Properly support the engine. Remove
both roll stopper mounting bolts and the
four center member installation bolts.
Remove the center member.

5. Remove the center member.

➡**Matchmark the pinion input shaft of
the rack to the lower steering column
joint for installation purposes.**

6. Remove the pinch bolt holding the
lower steering column joint to the rack and
pinion input shaft.

7. Remove the cotter pins and disconnect the tie rod ends from the steering
knuckle.

8. Disconnect the power steering fluid
pressure pipe and return hose from the rack
fittings.

9. Remove the rack and pinion steering
assembly and its rubber mounts from the
right side of the vehicle.

To install:

10. Align the matchmarks of the input
shaft and install the rack to the vehicle.

11. Secure the rack using the retainer
clamps and bolts. Torque the bolts to 51 ft.
lbs. (70 Nm).

12. Torque the steering column pinch
bolt to 13 ft. lbs. (18 Nm).

13. Using new O-rings, connect the
power steering fluid lines to the rack fittings.

14. Install the center member.

15. Install the front exhaust pipe.

16. Connect the tie rod ends to the steering knuckles and tighten the castle nuts to
25 ft. lbs. (34 Nm). Install new cotter pins.

17. Install the wheels and connect the
negative battery cable.

18. Refill the reservoir and bleed the
system.

19. Perform a front end alignment.

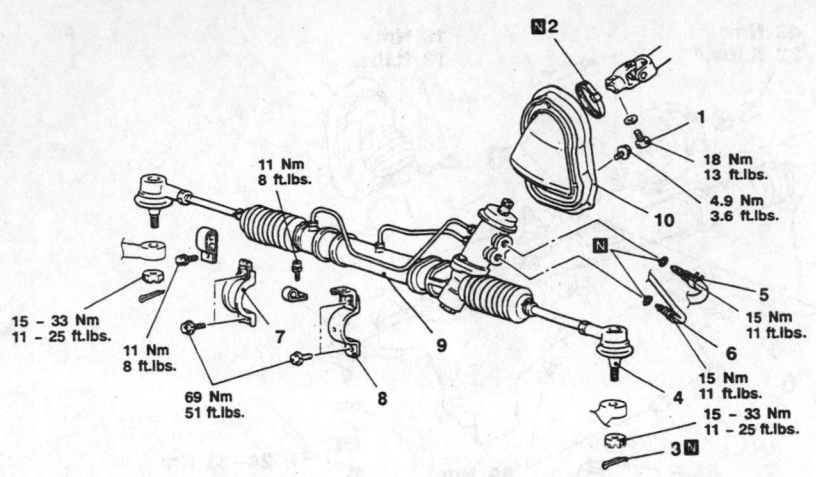

1. Steering shaft assembly and gear box connecting bolt
2. Band
3. Cotter pin
4. Tie-rod end and knuckle connection
5. Return tube connection
6. Pressure tube connection
7. Cylinder clamp
8. Gear housing clamp
9. Gear box assembly
10. Steering cover assembly

7923PGA1

Exploded view of the power steering gear assembly—Mirage

Strut and Coil Spring

REMOVAL & INSTALLATION

Front

3000GT

1. Disconnect the negative battery cable.

❊❊ CAUTION

Work must be started after 90 seconds from the time the ignition switch is turned to the LOCK position and the negative battery cable is disconnected.

2. Raise and safely support vehicle.
3. Remove the brake hose and tube bracket. Do not pry the brake hose and tube clamp away when removing it.
4. If equipped with ABS, disconnect the front speed sensor mounting clamp from the strut.
5. Support the lower arm and remove the strut to knuckle bolts. Use a piece of wire to suspend the knuckle to keep the weight off the brake hose.
6. Disconnect the ECS connector.
7. Before removing the top bolts, make matchmarks on the body and the strut insulator for proper reassembly. If this plate is installed improperly, the wheel alignment will be wrong. Remove the strut upper bolts

and remove the strut assembly from the vehicle.

8. Compress the coil spring using a spring compressor until the spring just comes away from one of the seats.

9. Remove the center nut from the strut and remove the upper mounting bracket and bushings.
10. Remove the coil spring.

To install:

11. Install the compressed spring on the strut assembly. Install the upper bushings and the mounting bracket. Install the nut and tighten it to 16 ft. lbs. (22 Nm).
12. Install the strut to the vehicle and install the top bolts.
13. Install the strut to the knuckle and install the bolts.
14. Connect the ECS connector.
15. Install the brake hose bracket and the ABS clamp.
16. Install the daytime running lamp delay and control unit to the mounting bracket located on top of the left strut tower.
17. Install the auto-cruise control actuator.
18. Install the wheel and tire assembly.
19. Connect the negative battery cable.
20. Perform a front end alignment.

DIAMANTE

1. Disconnect the negative battery cable.
2. Raise and safely support the vehicle.
3. Remove the brake hose and the tube bracket.

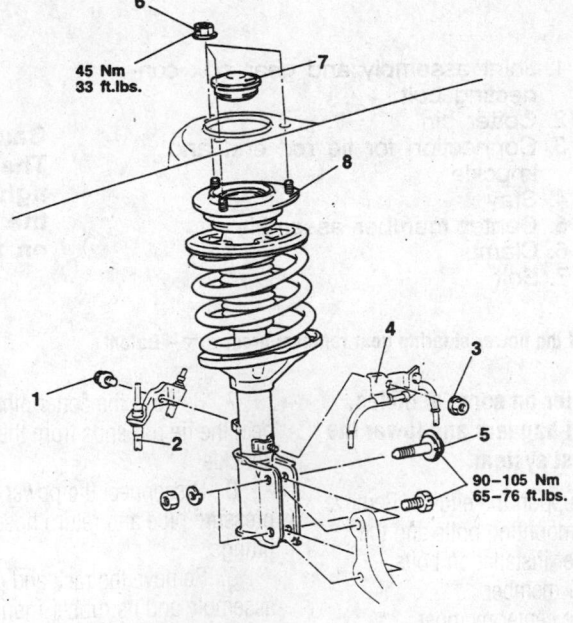

1. Brake hose tube clamp mounting bolt
2. Brake hose tube clamp
3. Front speed sensor clamp mounting nut <ABS>
4. Front speed sensor clamp <ABS>
5. Strut lower mounting bolt
6. Strut upper mounting bolt
7. Dust cover
8. Strut assembly

7923PGA7

Exploded view of the front strut assembly removal—3000GT

➡**Do not pry the brake hose and tube clamp away when removing it.**

4. If equipped with ABS, disconnect the front speed sensor mounting clamp from the strut.

5. Support the lower arm and remove the strut to knuckle bolts. Use a piece of wire to suspend the knuckle to keep the weight off the brake hose.

6. If equipped with Active-ECS, disconnect the air tubes and remove the O-rings from the actuator on top of the strut. Once the air line is removed, disconnect the ECS connector at the top of the strut assembly.

➡**Before removing the top bolts, make matchmarks on the body and the strut insulator for proper reassembly. If this plate is installed improperly, the wheel alignment will be wrong.**

7. Remove the strut upper nuts and remove the strut assembly from the vehicle.

8. Compress the coil spring using a spring compressor until the spring just comes away from one of the seats.

9. Remove the center nut from the strut and remove the upper mounting bracket and bushings.

10. Remove the coil spring.

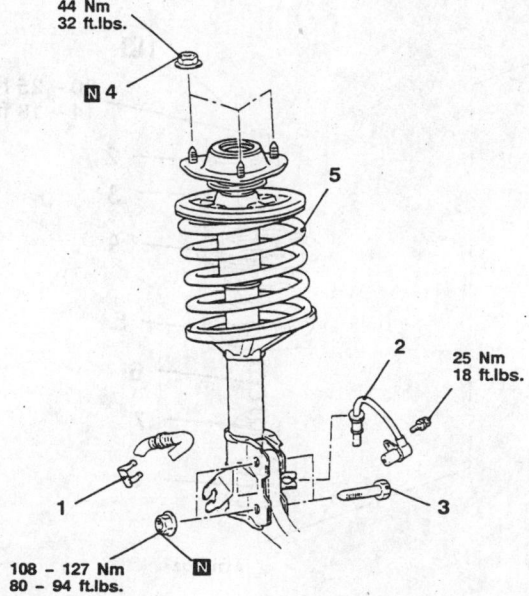

44 Nm
32 ft.lbs.

N 4

5

2

25 Nm
18 ft.lbs.

1

3

108 – 127 Nm
80 – 94 ft.lbs.   N

Removal steps
1. Brake hose clamp
2. Front speed sensor <Vehicles with ABS>
3. Bolts
4. Self-locking nut
5. Strut assembly

Caution
For vehicles with ABS, be careful when handling the pole piece at the tip of the speed sensor so as not to damage it by striking against other parts.

7923PGA6

Front strut assembly and related parts—Mirage

To install:

11. Install the compressed spring on the strut assembly. Install the upper bushings and the mounting bracket. Install the nut and tighten it to 43 ft. lbs. (59 Nm).

12. Install the strut to the vehicle and tighten the upper mounting nuts to 33 ft. lbs. (45 Nm).

13. Align the strut to the knuckle and connect with the mounting bolts. Torque the mounting bolts to 70–76 ft. lbs. (90–105 Nm).

14. If equipped with Active-ECS, lubricate the air tube and new O-ring. Connect the air line to the actuator and tighten the fitting to 7 ft. lbs. (9 Nm).

15. If equipped with Active-ECS, connect the actuator wiring.

16. Install the brake hose bracket and the ABS clamp, if equipped.

17. Install the wheel and tire assembly.

18. Perform a front end alignment.

MIRAGE

1. Disconnect the negative battery cable.
2. Raise and safely support vehicle.
3. Remove the brake hose and tube bracket retainer bolt and bracket from the front strut. Do not pry the brake hose and tube clamp away when removing.

4. If equipped with ABS, disconnect the front speed sensor mounting clamp from the strut.

5. Support the lower arm using floor jack or equivalent. Remove the lower strut to knuckle bolts. Once the mounting bolts have been removed, jack up the lower arm. Use a piece of wire to attach the brake hose, tube and driveshaft to the knuckle and to help keep the weight off. These components are not to be pulled.

6. Before removing the top bolts, make matchmarks on the body and the strut insulator for proper reassembly. If this plate is installed improperly, the wheel alignment will be wrong. Remove the strut upper mounting bolts. Remove the strut assembly from the vehicle.

7. Compress the coil spring using a spring compressor until the spring just comes away from one of the seats.

8. Remove the center nut from the strut and remove the upper mounting bracket and bushings.

9. Remove the coil spring.

To install:

10. Install the compressed spring on the strut assembly. Install the upper bushings and the mounting bracket. Install the nut and tighten it to 43 ft. lbs. (59 Nm).

11. Install the strut to the vehicle and install the top mounting bolts. Be sure the insulator is installed so the matchmarks made during disassembly are in alignment. Tighten the mounting bolts to 36 ft. lbs. (50 Nm) on Eclipse models and 29 ft. lbs. (40 Nm) on Mirage models.

12. Position the strut on the knuckle and install the mounting bolts. While holding the head of the lower mounting bolt, tighten the nuts to 80–101 ft. lbs. (110–140 Nm) on Eclipse models and 80–94 ft. lbs. (110–130 Nm) on Mirage models.

13. Install the brake hose bracket and the ABS clamp, if equipped.

14. Install the wheel and tire assembly. Perform a front end alignment.

Shock Absorber and Coil Spring

REMOVAL & INSTALLATION

Front

ECLIPSE

1. Raise and safely support the vehicle.
2. Remove the front wheel.
3. Remove the three upper shock absorber mounting nuts. Do not remove the larger nut in the center of the strut at this time.

4. Disconnect the stabilizer link from the damper fork.

5. Remove the damper fork mounting bolt.

6. Remove the shock absorber assembly from the vehicle.

7. Use a coil spring compressor and compress the coil spring. An air tool should not be used to tighten the spring compressor.

8. While holding the piston rod, remove the self-locking nut.

9. Remove the upper bracket assembly and spring pad.

10. Remove the collar, upper bushing, cup assembly, bump rubber and dust cover.

11. Remove the coil spring from the shock absorber.

To install:

12. Align the end of the coil spring with the stepped part of the spring seat and install the compressed coil spring on the shock.

13. Install the dust cover, bump rubber, cup assembly, upper bushing, collar, upper spring pad and bracket assembly on the strut.

14. Install the upper bushing and washer on the piston rod.

15. Install a new self-locking nut on the piston rod. Temporarily tighten the nut.

16. Carefully remove the spring compressor from the spring. Torque the self-locking nut to 16 ft. lbs. (25 Nm).

17. Position the shock absorber assembly in the damper fork and install the mounting bolt.

18. Pass the studs in the upper bracket assembly through the holes in the inner fender and install the three mounting nuts.

19. Connect the stabilizer link to the damper fork.

20. Install the wheel assembly.

21. Safely lower the vehicle to the floor.

22. Check and adjust the front wheel alignment if necessary.

GALANT

1. Disconnect the negative battery cable.

2. Raise and safely support vehicle.

3. Remove the appropriate wheel assembly.

4. Disconnect the sway bar link from the damper fork.

5. Remove the damper fork lower through-bolt and upper pinch bolt. Remove the damper fork assembly.

6. Remove the shock absorber upper nuts and remove the strut assembly from the vehicle.

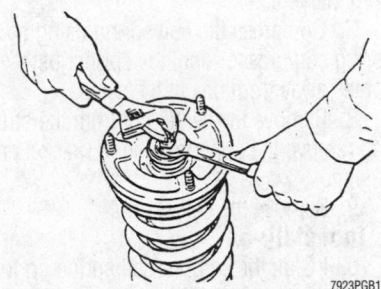

Removing the self-locking nut—Eclipse

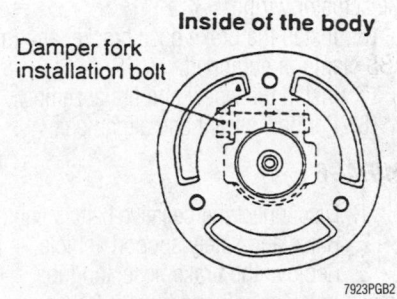

Inside of the body

Damper fork installation bolt

Upper bracket assembly alignment— Eclipse

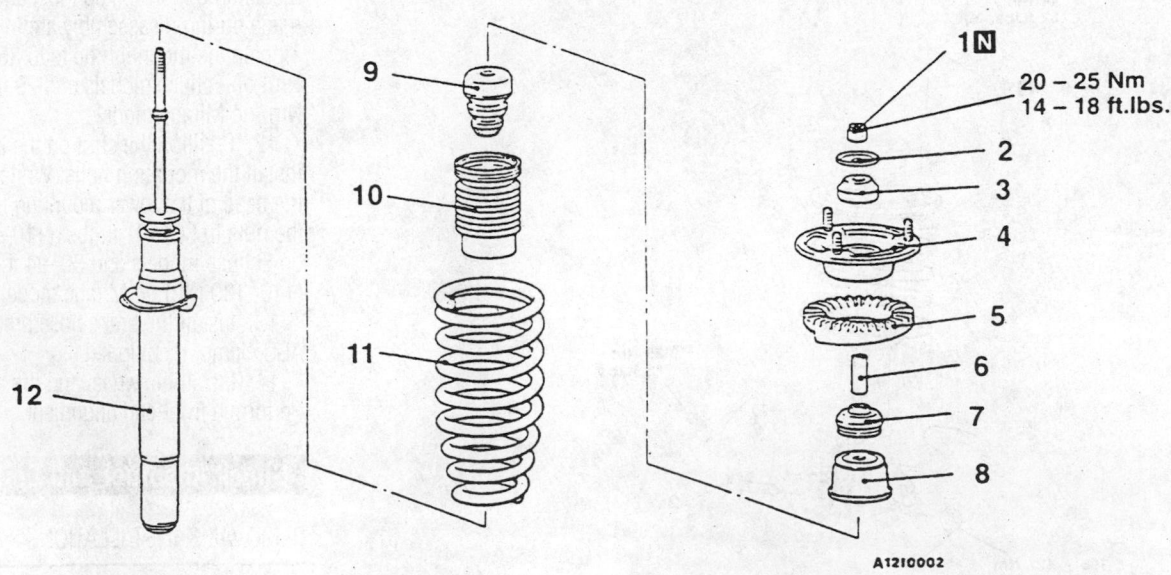

20 – 25 Nm
14 – 18 ft.lbs.

Disassembly steps

1. Self-locking nut
2. Washer
3. Upper bushing A
4. Upper bracket assembly
5. Upper spring pad
6. Collar
7. Upper bushing B
8. Cup assembly
9. Bump rubber
10. Dust cover
11. Coil spring
12. Shock absorber assembly

Exploded view of the coil spring removal procedure—Eclipse

7. Compress the coil spring with a special compression tool.

❊❊ CAUTION

Be sure that the coil spring compression tool is of an approved design. Great care should be utilized in the removal and installation of the coil spring. If the coil spring is not handled safely, personal injury and/ or property damage could result.

8. Remove the self-locking nut and washer. Remove the upper bushing, upper bracket assembly, the upper spring pad, and the collar.

9. Remove the other upper bushing, cup assembly, bump rubber, dust cover, and the coil spring. Carefully remove the coil spring compression tool.

To install:

10. Install the compressed coil spring to the shock absorber assembly. Be sure to align the edge of coil spring to the stepped part of the spring seat. Install the dust cover, bump rubber, cup assembly, upper bushing, collar, and upper spring pad.

11. Install the upper bracket assembly and position it so that the three bolts are in the correct position.

12. Install the upper bushing, washer, and locknut. Torque the locknut to 18 ft. lbs.

13. Install the shock absorber and tighten the upper mounting nuts to 32 ft. lbs. (44 Nm).

14. Align the shock to the damper fork and install the damper fork. Tighten the lower through-bolt/nut to 65 ft. lbs. (88 Nm) and the upper pinch bolt to 76 ft. lbs. (103 Nm).

15. Connect the sway bar link to the damper fork and tighten the link nut to 29 ft. lbs. (39 Nm).

16. Install the wheel and tire assembly.

17. Perform a front end alignment.

Rear

3000GT

1. Disconnect the negative battery cable and wait one minute for the air bag to disarm before working on the vehicle.

2. Raise and safely support the vehicle.

3. Remove the rear side trim in the luggage compartment and remove the ECS connector and cap.

4. Support the suspension and remove the upper shock absorber mounting bolts.

5. Remove the wheel and tire assembly and the lower shock mounting bolt.

6. Remove the shock absorber from the vehicle.

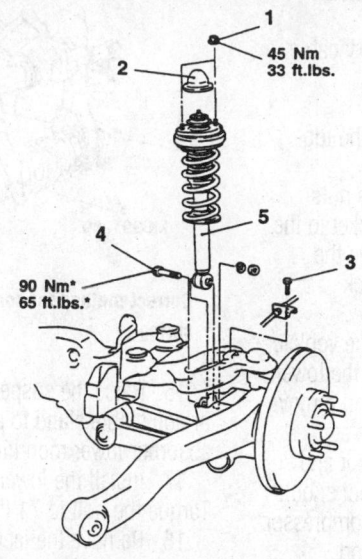

```
        1
        45 Nm
        33 ft.lbs.
  2

                    5
  4

90 Nm*
65 ft.lbs.*        3
```

1. Shock absorber upper mounting nut
2. Cap
3. Brake tube clamp bolt
4. Shock absorber lower mounting bolt
5. Shock absorber

Caution
*: **Indicates parts which should be temporarily tightened, and then fully tightened with the vehicle on the ground in an unladen condition.**

7923PGA8

Rear shock absorber removal—3000GT

To install:

7. Position the shock absorber in the trailing arm and temporarily install the lower mounting bolt. After the vehicle is on the ground at normal ride height, tighten the bolt to 65 ft. lbs. (90 Nm).

8. Guide the upper mounting studs through the body and tighten the upper mounting nuts to 33 ft. lbs. (45 Nm).

9. Install the cap and the ECS connector.

10. Install the wheel and tire assembly and connect the negative battery cable.

11. Lower the vehicle to the floor.

DIAMANTE

➡**For rear shock absorber replacement the vehicle chassis and axle weight must be supported separately, requiring the use of two separate lifting devices.**

1. Disconnect the negative battery cable.

2. Raise and properly support vehicle. Remove both rear wheels.

3. Working on one side at a time, lightly jack under the trailing arm for support.

4. Matchmark the positioning of the upper spring plate to the vehicle for reinstallation purposes.

5. If equipped with electronic control suspension (ECS) perform the following:

a. Loosen the nut that secures the air line to the to the top of the strut and discard the O-ring.

b. Remove the bolts that secure the actuator to the top of the strut and remove the component. Disconnect the wiring harness.

6. Remove the shock absorber lower mounting bolt and remove the two nuts that secure the shock upper plate to the vehicle.

7. Lower the support jack and remove the shock from the vehicle.

To install:

8. Position the upper spring plate and install the strut. Use the support jack to assist with installation.

9. Tighten the upper strut mounting nuts to 33 ft. lbs. (45 Nm).

10. Tighten the lower strut mounting bolt to 71 ft. lbs. (98 Nm).

11. If equipped with electronic control suspension (ECS) perform the following:

a. Using a new O-ring, tighten the nut that secures the air line to the to the top of the strut to 84 inch lbs. (9 Nm).

b. Install the actuator to the top of the shock absorber and secure with mounting bolts. Connect the wiring harness.

12. Remove the support jack, install wheels and lower vehicle.

13. Connect the negative battery cable.

ECLIPSE

1. Remove the service lid in the luggage compartment.

2. Remove the cap and flange nuts securing the upper mounting bracket to the body of the vehicle. Do not remove the larger nut in the center of the shock absorber.

3. Raise and safely support the vehicle.

4. Remove the bolt attaching the lower end of the shock to the knuckle and remove the shock absorber from the vehicle.

5. Use a coil spring compressor and compress the coil spring. An air tool should not be used to tighten the spring compressor.

6. While holding the piston rod, remove the self-locking nut.

7. Remove the upper bracket assembly and spring pad.

8. Remove the collar, upper bushing, cup assembly, bump rubber and dust cover.

9. Remove the coil spring from the shock absorber.

To install:

10. Align the end of the coil spring with the stepped part of the spring seat and install the compressed coil spring on the shock absorber.

11. Install the dust cover, bump rubber, cup assembly, upper bushing, collar, upper spring pad and bracket assembly on the shock absorber.

12. Install the upper bushing and washer on the piston rod.

13. Install a new self-locking nut on the piston rod. Temporarily tighten the nut.

14. Carefully remove the spring compressor from the spring. Torque the self-locking nut to 16 ft. lbs. (25 Nm).

15. Install the upper bracket of the shock to the vehicle. Torque the mounting nuts to 32 ft. lbs. (44 Nm).

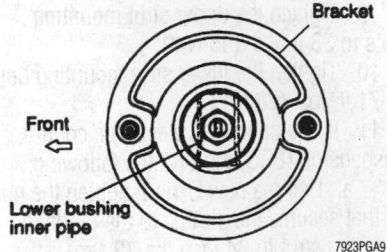

Correct upper bracket installed position— Eclipse

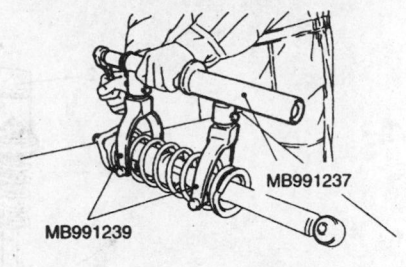

Correct method for compressing the coil spring

16. Raise the suspension up with a jack or adjustable stand to align the shock absorber lower mounting holes.

17. Install the lower mounting bolt. Torque the bolt to 71 ft. lbs.

18. Remove the jack or stand and safely lower the vehicle to the floor.

19. Install the cap and service lid.

GALANT

➡The shock absorber assembly is a load bearing component, therefore the vehicle chassis and axle weight must be supported separately, requiring the use of two separate lifting devices.

1. Raise and support the vehicle chassis.

2. Raise and support the lower control arm assembly slightly.

3. In order to gain access to the top mounting nuts, remove the rear seat as follows:

  a. While pulling the rear seat stopper outward, lift the lower cushion upward. Remove the lower cushion.

  b. Remove the seat back mounting bolts.

  c. Lift the seat back upward and remove the seat.

4. Remove the shock upper mounting nuts.

5. Remove the shock lower mounting bolt and remove the assembly from the vehicle.

6. Use a coil spring compressor and compress the coil spring. An air tool should not be used to tighten the spring compressor.

7. Remove the shock cap.

8. While holding the piston rod, remove the self-locking nut.

9. Remove the upper bracket assembly and spring pad.

10. Remove the collar, upper bushing, cup assembly, bump rubber and dust cover.

11. Remove the coil spring from the shock.

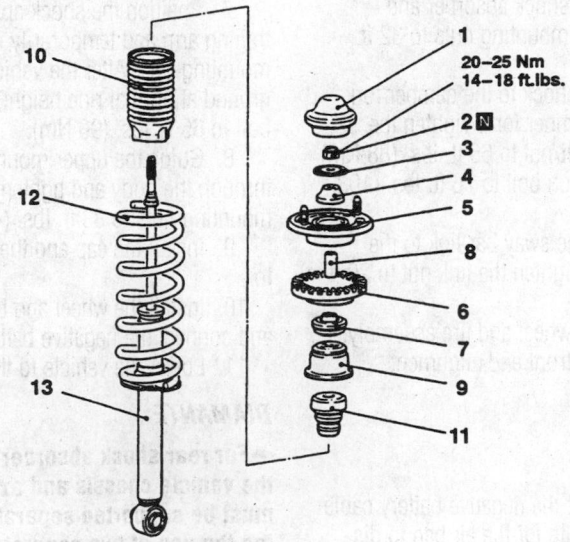

1. Cap
2. Self-locking nut
3. Washer
4. Upper bushing A
5. Bracket
6. Spring pad
7. Upper bushing B
8. Collar
9. Cup
10. Dust cover
11. Bump rubber
12. Coil spring
13. Shock absorber assembly

20–25 Nm
14–18 ft.lbs.

Exploded view of the rear shock absorber assembly—Galant and Mirage, 3000GT is similar

To install:

12. Align the end of the coil spring with the stepped part of the spring seat and install the compressed coil spring on the shock.

13. Install the dust cover, bump rubber, cup assembly, upper bushing, collar, upper spring pad and bracket assembly on the shock.

14. Install the upper bushing and washer on the piston rod.

15. Install a new self-locking nut on the piston rod. Temporarily tighten the nut.

16. Carefully remove the spring compressor from the spring. Torque the self-locking nut to 16 ft. lbs. (25 Nm).

17. Install the shock cap.

18. Position the shock assembly so that the lower mounting bolt can be installed and lightly tightened.

19. Use a jack to raise or lower the lower control arm, so that the top shock plate studs align through the body. Raise the jack to hold the shock assembly in position.

20. Install the top plate nuts on the studs and tighten the mounting nuts to 32 ft. lbs. (44 Nm).

21. With the vehicle on the ground, tighten the lower mounting bolt to 71 ft. lbs. (98 Nm).

22. Install the rear seat back and cushion.

MIRAGE

➡**The shock absorber assembly is a load bearing component, therefore the vehicle chassis and axle weight must be supported separately, requiring the use of two separate lifting devices.**

1. Remove the trunk interior trim to gain access to the top mounting nuts.

2. Remove the top cap and upper shock mounting nuts.

3. Raise and support vehicle chassis.

4. Raise and support the trailing arm assembly slightly.

➡**Matchmark the upper spring plate to the vehicle chassis for reassembly.**

5. Remove the shock lower mounting bolt and remove the assembly from the vehicle.

6. Compress the coil spring using the proper spring compressor.

✻ CAUTION

Do not use air tools to tighten the spring compressor.

7. Hold the piston rod with a wrench and remove the self-locking nut.

8. Remove the washer, upper bushing A, bracket, spring pad, upper bushing B, collar, cup, dust cover and bump rubber.

➡**Align the stepped part of the spring pad with the end of the spring.**

9. Remove the coil spring.

To install:

10. Install the coil spring on the shock.

11. Install the bump rubber, dust cover, cup, collar, upper bushing A, spring pad, bracket, upper bushing B and the washer.

12. Temporarily install a new self-locking nut, carefully release the spring from the compressor and tighten the self-locking nut to specifications.

13. Position the shock assembly so that lower mounting bolt can be installed and lightly tightened.

14. Use jack to raise or lower the axle assembly so that top shock plate studs aligns through body. Raise jack to hold the shock assembly in position. Be sure to properly position the upper spring plate.

15. Install top plate nuts on studs. Tighten the upper shock mounting nuts to 20 ft. lbs. (28 Nm) and the lower mounting bolt to 65 ft. lbs. (90 Nm).

16. Lower the vehicle. Install top cap and interior trim.

Upper Ball Joint

REMOVAL & INSTALLATION

The upper ball joints are an integral part of the upper control arm. If the ball joint becomes worn or damaged, the control arm must be replaced.

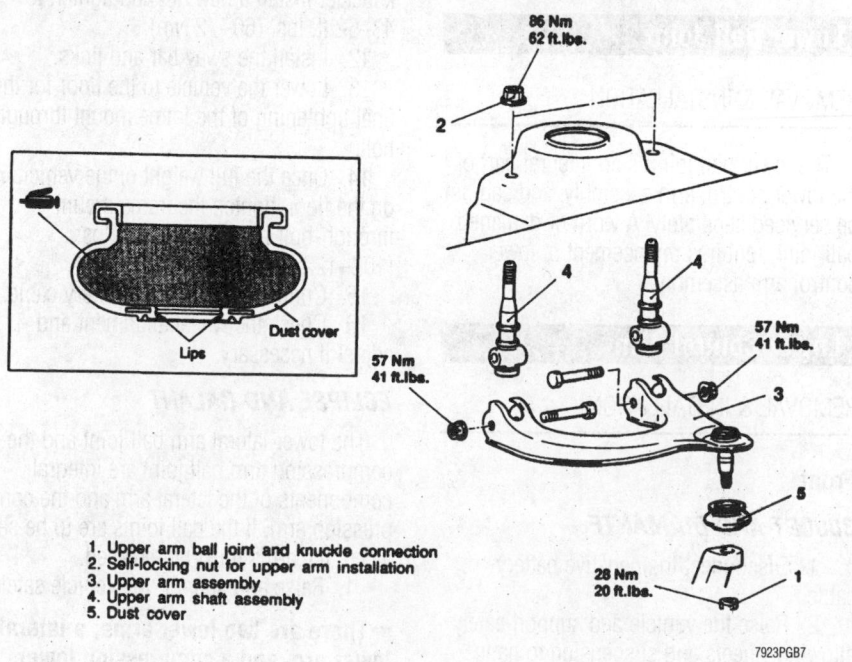

1. Upper arm ball joint and knuckle connection
2. Self-locking nut for upper arm installation
3. Upper arm assembly
4. Upper arm shaft assembly
5. Dust cover

Upper control arm assembly—Eclipse and Galant

Upper Control Arm

REMOVAL & INSTALLATION

3000GT, DIAMANTE AND MIRAGE

These vehicles use a strut type front suspension. No upper control arm is required.

ECLIPSE AND GALANT

1. Raise and safely support the vehicle.

2. Remove the front wheel(s).

3. Disconnect the upper arm ball joint from the steering knuckle.

4. Remove the upper arm shaft mounting nuts from the body.

5. Remove the upper arm.

6. Remove the through-bolts that attach the upper arm to the shafts.

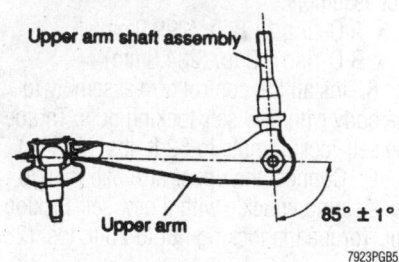

Correct angle of control arm and shafts— Eclipse and Galant

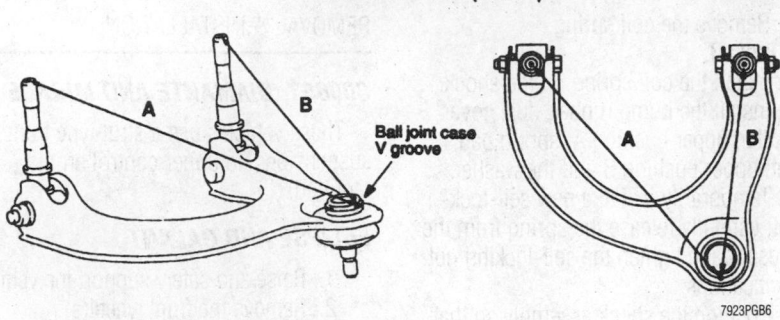

A : 299.9 mm (11.8 in.)
B : 234.0 mm (9.2 in.)

Ball joint case V groove

7923PGB6

Measure the dimensions A and B as shown—Eclipse and Galant

To install:

7. Assembly the upper arm to the shafts at the proper angle. Torque the through-bolts and nuts to 41 ft. lbs. (57 Nm).The proper angle is 84–86°. After the arm and the shafts are connected at the right angle, measure dimensions A and B to insure correct assembly.

- A O-ring 11.8 in. (299.9mm)
- B O-ring 9.2 in. (234.0mm)

8. Install the control arm assembly to the body with new self-locking nuts. Torque the self-locking nuts to 62 ft. lbs. (86 Nm).

9. Connect the upper arm ball joint to the steering knuckle with a new self-locking nut. Torque the locking nut to 20 ft. lbs. (28 Nm).

10. Install the front wheel(s).

11. Perform front wheel alignment and adjust if necessary.

12. Safely lower the vehicle to the floor.

Lower Ball Joint

REMOVAL & INSTALLATION

The lower ball joint is an integral part of the lower control arm assembly, and can not be serviced separately. A worn or damaged ball joint, requires replacement of lower control arm assembly.

Lower Control Arm

REMOVAL & INSTALLATION

Front

3000GT AND DIAMANTE

1. Disconnect the negative battery cable.

2. Raise the vehicle and support safely allowing wheels and suspension to hang freely.

3. Remove the sway bar links from the lower control arm.

4. Disconnect the ball joint stud from the steering knuckle.

5. Remove the inner mounting frame through-bolt and nut.

6. Remove the rear mount bolts. Remove the clamp if equipped.

7. Remove the rear rod bushing if servicing.

To install:

8. Assemble the control arm and bushing.

9. Install the control arm to the vehicle and install the through-bolt. Replace the nut and snug temporarily.

10. Install the rear mount clamp, bolts and replacement nuts. Torque the bolts to 72–87 ft. lbs. (100–120 Nm). Torque the nuts to 29 ft. lbs. (40 Nm).

11. Connect the ball joint stud to the knuckle. Install a new nut and tighten to 43–52 ft. lbs. (60–72 Nm).

12. Install the sway bar and links.

13. Lower the vehicle to the floor for the final tightening of the frame mount through-bolt.

14. Once the full weight of the vehicle is on the floor, tighten the frame mount through-bolt nuts to 75–90 ft. lbs. (102–122 Nm).

15. Connect the negative battery cable.

16. Check the wheel alignment and adjust if necessary.

ECLIPSE AND GALANT

The lower lateral arm ball joint and the compression arm ball joint are integral components of the lateral arm and the compression arm. If the ball joints are to be serviced, the arms must be replaced.

1. Raise and support the vehicle safely.

➡ **There are two lower arms, a lateral lower arm and a compression lower arm.**

2. Disconnect both ball joint studs from the steering knuckle.

3. To remove the lower lateral arm, remove the crossmember brackets.

4. Remove the inner lateral arm mounting bolts and nut.

5. Remove the arm from the vehicle.

6. Remove the two bolts holding the compression arm.

7. Remove the compression arm.

To install:

8. Assemble the control arms and bushings.

9. Install the lateral control arm to the vehicle and install the inner mounting bolts. Install a new nut and snug temporarily.

10. Install the compression arm to the vehicle with the two bolts.

11. Connect the ball joint studs to the knuckle. Install new nuts and tighten to 43–51 ft. lbs. (59–71 Nm).

12. Lower the vehicle to the floor for the final tightening.

13. Once the full weight of the vehicle is on the suspension, tighten the lateral arm rear bolt to 71–85 ft. lbs. (98–118 Nm) and the front bolt to the damper fork to 64 ft. lbs. (88 Nm).

14. Torque the bolts for the compression arm to 60 ft. lbs. (83 Nm).

15. Reinstall the crossmember brackets with their mounting bolts. Torque the mounting bolts to 51–58 ft. lbs. (69–78 Nm).

16. Inspect all suspension bolts, making sure they all have been fully tightened.

17. Perform an alignment on the vehicle.

MIRAGE

➡ **The suspension components should not be tightened until the vehicle's weight is resting on its wheels.**

1. Raise the vehicle and support safely.

2. Remove the wheel and tire assembly.

3. Remove sway bar links or mounting nuts and bolts from lower control arm. Remove the joint cups and bushings.

4. Disconnect the ball joint stud from the steering knuckle.

5. Remove the inner lower arm mounting bolt and nut.

6. Remove the rear mount bolts from the retaining clamp. Remove the rear retainer clamp if equipped.

7. Remove the arm from the vehicle.

To install:

8. Install the control arm to the vehicle and install the inner mounting bolt. Install new nut and tighten to 78 ft. lbs. (108 Nm).

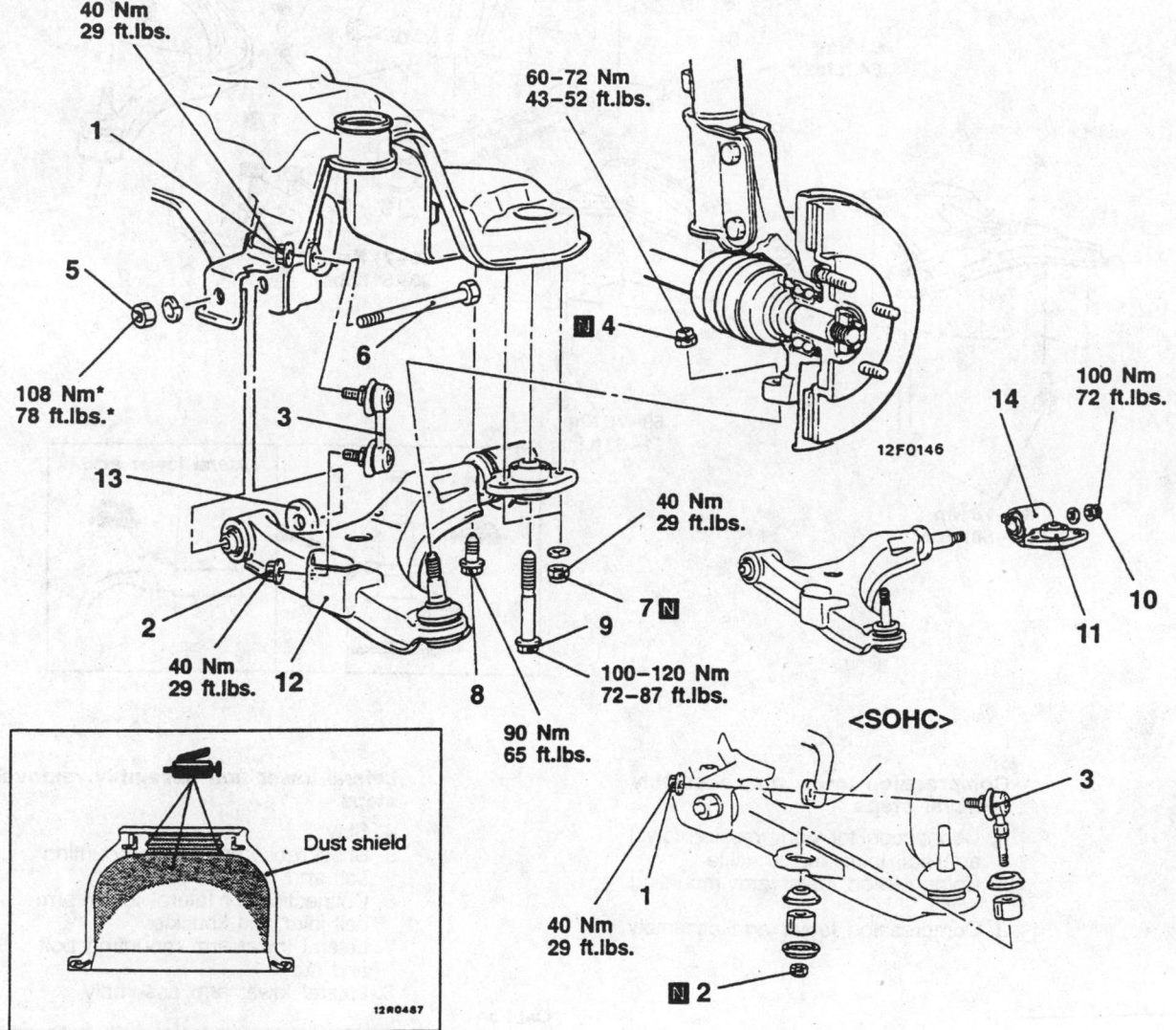

Removal steps

1. Stabilizer link mounting nut (stabilizer bar side)
2. Stabilizer link mounting nut (lower arm side)
3. Stabilizer link
4. Self-locking nut connecting lower arm ball joint to knuckle
5. Lower arm mounting nut
6. Lower arm mounting bolt
7. Clamp mounting self-locking nut
8. Clamp mounting bolt (small)
9. Clamp mounting bolt (large)
10. Lower arm clamp mounting self-locking nut
11. Lower arm mounting clamp
12. Lower arm
13. Stopper
14. Rod bushing

Caution
*: Indicates parts which should be temporarily tightened, and then fully tightened with the vehicle on the ground in an unladen condition.

Exploded view of the lower control arm removal procedure—3000GT shown, Diamante is similar

7923PGB9

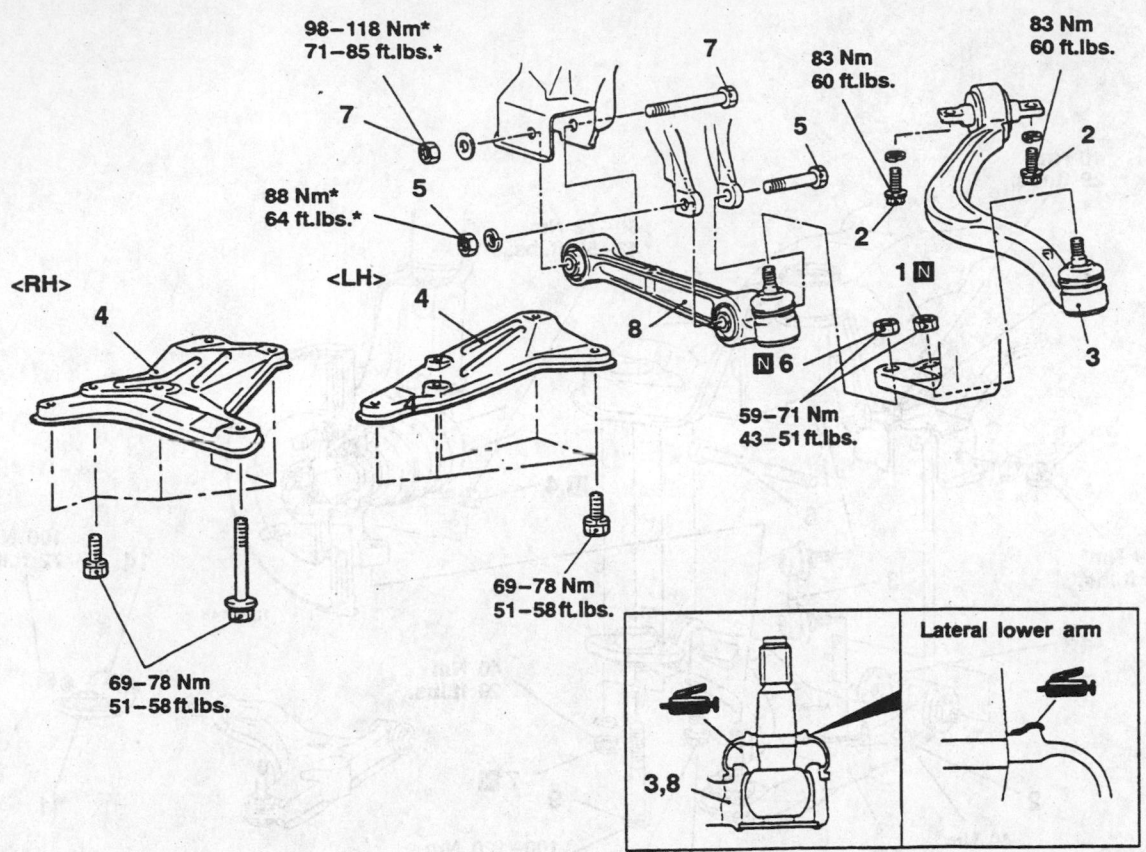

98–118 Nm*
71–85 ft.lbs.*

83 Nm
60 ft.lbs.

83 Nm
60 ft.lbs.

88 Nm*
64 ft.lbs.*

59–71 Nm
43–51 ft.lbs.

<RH>

<LH>

69–78 Nm
51–58 ft.lbs.

69–78 Nm
51–58 ft.lbs.

Lateral lower arm

3,8

7923PGB8

Compression lower arm assembly removal steps

1. Connection for compression lower arm ball joint and knuckle
2. Compression lower arm mounting bolt
3. Compression lower arm assembly

Lateral lower arm assembly removal steps

4. Stay
5. Shock absorber lower mounting bolt and nut
6. Connection for lateral lower arm ball joint and knuckle
7. Lateral lower arm mounting bolt and nut
8. Lateral lower arm assembly

Caution
*: Indicates parts which should be temporarily tightened, and then fully tightened with the vehicle on the ground in the unladen condition.

Exploded view of the lower control arms—Eclipse and Galant

9. Install the rear mount clamp and bolts. Torque the clamp mounting bolts to 65 ft. lbs. (90 Nm) on Mirage models and 34 ft. lbs. (47 Nm) on Eclipse models.

10. Connect the ball joint stud to the knuckle. Install a new nut and tighten to 43–52 ft. lbs. (60–72 Nm).

11. Install the sway bar and links.

12. Lower the vehicle to the floor for the final tightening of the inner frame mount bolt.

13. Inspect all suspension bolts, making sure they all have been fully tightened.

14. Install the wheel and tire assembly.

Wheel Bearings

ADJUSTMENT

The front and rear wheel bearings on these vehicles are not adjustable. If the bearings are noisy or become loose, they must be replaced.

REMOVAL & INSTALLATION

Front

3000GT

1. Disconnect the negative battery cable.

✳✳ CAUTION

Work must be started after 90 seconds from the time the ignition switch is turned to the LOCK position

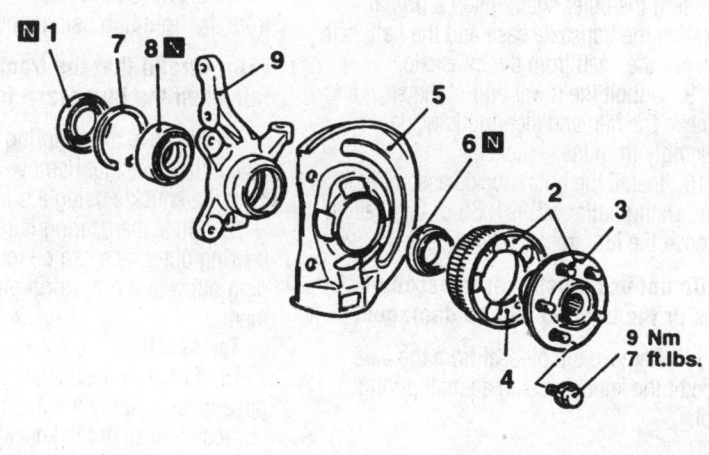

59 – 71 Nm
43 – 52 ft.lbs.

106 Nm*
78 ft.lbs.*

78 – 98 Nm
58 – 72 ft.lbs.

Removal steps
1. Lower arm ball joint connection
2. Self-locking nut
3. Stabilizer rubber
4. Stabilizer bar
5. Collar
6. Lower arm front bushing connection
7. Support bracket
8. Lower arm assembly

Caution
*: Indicates parts which should be temporarily tightened, and then fully tightened with the vehicle on the ground in the unladen condition.

7923PGB0

Lower control arm assembly and related components—Mirage

9 Nm
7 ft.lbs.

1. Oil seal (drive shaft side)
2. Hub and rotor

3. Hub
4. Rotor <Vehicles with ABS>

5. Dust shield
6. Oil seal (hub side)
7. Snap ring
8. Wheel bearing
9. Knuckle

7923PGC6

Exploded view of the front hub and bearing—3000GT

and the negative battery cable is disconnected.

2. Remove the cotter pin, halfshaft nut and washer.

3. Raise the vehicle and support safely. If equipped with ABS, remove the front wheel speed sensor. Remove the ball joint and tie rod end from the steering knuckle.

4. Remove the caliper and brake pads and suspend with a wire.

5. On vehicles with an inner shaft, remove the center support bearing bracket bolts and washers. Remove the halfshaft by setting up a puller on the outside wheel hub and pushing the halfshaft from the front hub. Then, tap the joint case with a plastic hammer to remove the halfshaft shaft and inner shaft from the transaxle.

6. On vehicles without an inner shaft, remove the halfshaft by setting up a puller on the outside wheel hub and pushing the halfshaft from the front hub. After pressing the outer shaft, insert a prybar between the transaxle case and the halfshaft and pry the shaft from the transaxle.

7. On vehicles with AWD, the front hub/bearing assembly can be serviced at this point as a unit. If the knuckle is being removed, proceed.

8. Unbolt the lower end of the strut and remove the hub and steering knuckle assembly.

9. Set up a puller with the knuckle/hub in a vise and pull the hub from the knuckle. Do not use a hammer to accomplish this or the bearing will be damaged.

10. Once the hub and outer bearing inner race are removed with a puller, the bearing outer races can be removed by tapping out with a brass drift pin and a hammer.

To install:

11. Assemble the hub/knuckle assembly with pressing tools, using new parts as required.

12. Install the knuckle assembly to the vehicle and install the strut bolts.

13. Apply a thin coat of grease to the outside of the outer races and install into the hub with a bearing driver.

14. Apply multi-purpose grease to the bearings, inside surface of the hub and the lip of the grease seal. Place the outside bearing into the knuckle and install the seal with a driver.

15. The hub is assembled to the knuckle with a puller. Draw the parts together firmly to seat the bearings. Use a small torque wrench to check the bearing turning torque. It should be 16 inch lbs. or less. Check that the bearings feel smooth when rotated.

16. Apply a thin coat of grease to the lip of the halfshaft side axle seal and drive into place until it contacts the inner bearing outer race.

17. Replace the circlips on the ends of the halfshafts.

18. Insert the halfshaft into the transaxle. Be sure it is fully seated.

19. Pull the strut assembly out and install the other end to the hub.

20. Install the center bearing bracket bolts and tighten to 33 ft. lbs. (45 Nm).

21. Install the washer so the chamfered edge faces outward. Install the nut and tighten temporarily.

22. Install the tie rod end and ball joint.

23. Install the wheel and lower the vehicle to the floor. Tighten the axle nut with the brakes applied to 166 ft. lbs. (230 Nm).

DIAMANTE AND MIRAGE

1. Disconnect the negative battery cable.

2. Remove the cotter pin from the driveshaft nut. With the brakes applied, loosen the halfshaft nut.

3. Raise the vehicle and support safely. Remove the halfshaft nut.

4. If equipped with ABS, remove the front wheel speed sensor.

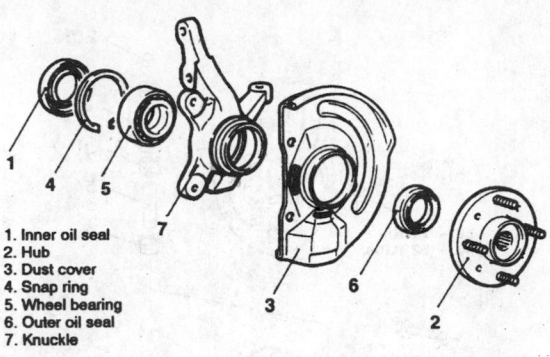

1. Inner oil seal
2. Hub
3. Dust cover
4. Snap ring
5. Wheel bearing
6. Outer oil seal
7. Knuckle

7923PGC1

Front wheel bearing assembly exploded view—Mirage and Diamante

5. If equipped with Active-ECS, disconnect the height sensor from the lower control arm.

6. Remove the caliper assembly and brake pads. Suspend the caliper with a wire.

7. Using tool MB991113 or equivalent, disconnect the ball joint and tie rod end from the steering knuckle.

➡**It is important to use proper methods of ball joint separation. Use of unproved techniques can result in damage to joint and possible failure.**

8. Remove the halfshaft by setting up a puller on the outside wheel hub and pushing the halfshaft from the front hub. After pressing the outer shaft, insert a prybar between the transaxle case and the halfshaft and pry the shaft from the transaxle.

9. Unbolt the lower end of the strut and remove the hub and steering knuckle assembly from the vehicle.

10. Install the hub/knuckle assembly in a vise. Using puller MB991056 or equivalent, remove the hub from the knuckle.

➡**Do not use a hammer to accomplish this or the bearing will be damaged.**

11. Remove the oil seal from the axle side of the knuckle using a small prying tool.

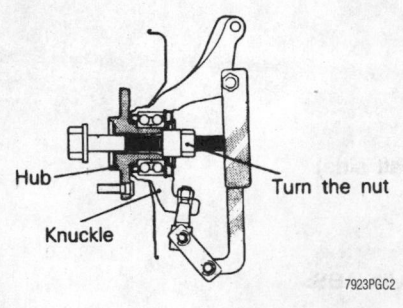

Hub

Knuckle

Turn the nut

7923PGC2

Use of press tool for hub removal—Mirage and Diamante

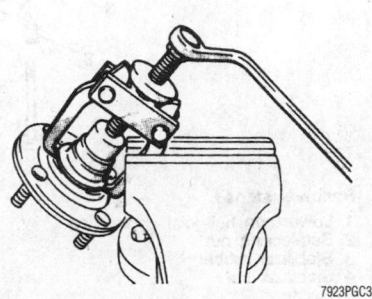

7923PGC3

Removing inner race from hub—Mirage and Diamante

12. Remove the wheel bearing inner race from the front hub using a puller.

➡**Be careful that the front hub does not fall when the inner race is removed.**

13. Remove the snapring from the axle side of the knuckle. Remove the bearing from the knuckle using a puller.

14. Once the bearing is removed, the bearing outer race can be removed by tapping out with a brass drift pin and a hammer.

To install:

15. Fill the wheel bearing with multipurpose grease. Apply a thin coating of multipurpose grease to the knuckle and bearing contact surfaces.

16. Press the wheel bearing into the knuckle using appropriate pressing tool. Once the bearing is installed, install the inner race using the proper driving tool.

17. Drive the oil seal into the knuckle by using the proper size driver. Drive seal into knuckle until it is flush with the knuckle end surface.

18. Using pressing tool MB990998 or equivalent, mount the front hub assembly into the knuckle. Tighten the nut of the pressing tool to 144–188 ft. lbs. (200–260 Nm). Rotate the hub to seat the bearing.

19. Install the hub and knuckle assembly onto the vehicle. Install the lower ball joint stud into the steering knuckle and install new nut. Tighten to 52 ft. lbs. (72 Nm).

20. Install the halfshaft into the transaxle extension housing and guide the outer end through the hub/knuckle assembly.

21. Install the two front strut lower mounting bolts and tighten to 80–94 ft. lbs. (110–130 Nm) on Mirage or 65–76 ft. lbs. (90–105 Nm) on Diamante models.

22. Install the connection for the tie rod end and tighten nut to 25 ft. lbs. (34 Nm) on all models except Diamante and 21 ft. lbs. (29 Nm) on Diamante models. Install new cotter pin and bend to locknut in position.

23. Install the brake disc and caliper assembly.

24. If equipped with Active-ECS, connect the height sensor and tighten the mounting bolt to 15 ft. lbs. (20 Nm).

25. Install the front speed sensor, if removed.

➡**When installing front speed sensor, be sure harness is routed in the original position and that it is not twisted.**

26. Install the washer and new locknut to the end of the halfshaft. Tighten the locknut snugly.

27. Install the tire and wheel assembly onto the vehicle. Lower the vehicle to the ground.

28. With the weight of the vehicle on the ground and the brakes applied, tighten the locknut to 144–188 ft. lbs. (200–260 Nm).

29. Install the cotter pin in the first matching holes and bend it securely.

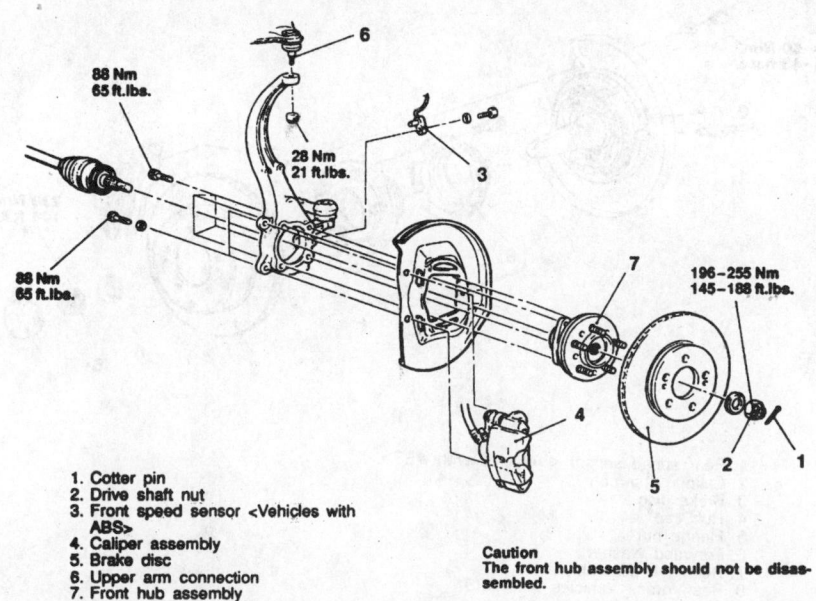

1. Cotter pin
2. Drive shaft nut
3. Front speed sensor <Vehicles with ABS>
4. Caliper assembly
5. Brake disc
6. Upper arm connection
7. Front hub assembly

88 Nm
65 ft.lbs.

28 Nm
21 ft.lbs.

88 Nm
65 ft.lbs.

196–255 Nm
145–188 ft.lbs.

Caution
The front hub assembly should not be disassembled.

7923PGC4

Front hub and related components—Eclipse

ECLIPSE

1. Raise and safely support the vehicle.
2. Remove the front wheel.
3. Use tool MB990767 or equivalent to hold the hub assembly while removing the axle nut.
4. On vehicles with ABS, remove the wheel speed sensor.
5. Remove the caliper and suspend it out of the way with wire or string.
6. Remove the brake rotor.
7. Disconnect the steering knuckle from the upper arm.
8. Pull the knuckle away from the vehicle to access the hub mounting bolts on the inboard side of the hub. Be careful not to damage the ball joint boot or the ABS rotor if equipped.
9. Remove the mounting bolts and the front hub assembly.

➡**Do not disassemble the hub assembly. If binding or damaged, it must be replaced as a unit.**

To install:

10. Install the hub to the knuckle. Torque the mounting bolts to 65 ft. lbs. (88 Nm).
11. Connect the knuckle to the upper arm.
12. Install the brake rotor and the caliper.
13. Install the wheel speed sensor if removed.
14. Install the axle nut washer in the proper direction. Install the axle nut and tighten to 145–188 ft. lbs. (196–255 Nm).
15. Install the wheel and lower the vehicle to the floor.

GALANT

1. Remove the cotter pin, halfshaft nut and washer.
2. Raise the vehicle and support safely.
3. Remove the appropriate wheel assembly.
4. If equipped with ABS, remove the vehicle speed sensor.
5. Remove the caliper and brake pads. Support the caliper out of the way using wire.
6. Remove the brake rotor from the hub assembly.
7. Disconnect the upper ball joint from the steering knuckle and pull the knuckle outward.

❊❊ WARNING

Use of improper methods of joint separation can result in damage to joint, leading to possible failure.

8. From the back of the knuckle, remove the four bolts securing the hub to the knuckle.
9. Remove the hub and bearing assembly from the knuckle.

➡**The hub assembly is not serviceable and should not be disassembled.**

To install

10. Install the hub to the steering knuckle and tighten the mounting bolts to 65 ft. lbs. (88 Nm).
11. Connect the upper ball joint to the steering knuckle and tighten the self-locking nut to 21 ft. lbs. (28 Nm).
12. Position the rotor on the hub. Install a couple of lug nuts and lightly tighten to hold rotor on hub.
13. Install the caliper holder and place brake pads in holder. Slide caliper over brake pads and install guide pins. Once caliper is secured, lug nuts can be removed.
14. If equipped with ABS, install the vehicle speed sensor.
15. Install the wheel assembly and lower the vehicle.
16. Install the wheel and lower the vehicle to the floor. Tighten the axle nut with the brakes applied. Tighten the nut to torque of 145–188 ft. lbs. (200–260 Nm).
17. Install a new cotter pin and bend to secure.

❊❊ WARNING

Pump the brake pedal until hard, before attempting to move the vehicle.

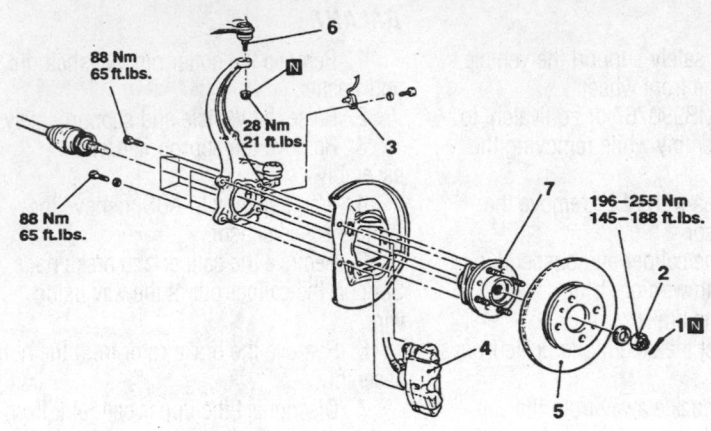

88 Nm
65 ft.lbs.

28 Nm
21 ft.lbs.

88 Nm
65 ft.lbs.

196–255 Nm
145–188 ft.lbs.

Removal steps
1. Cotter pin
2. Drive shaft nut
3. Front speed sensor <Vehicles with ABS>
4. Caliper assembly
5. Brake disc
6. Connection for upper arm
7. Front hub assembly

Caution
The front hub assembly should not be disassembled.

7923PGC5

Exploded view of the front hub removal—Galant

Rear

3000GT

➡**The hub assembly is not repairable, if defective replacement is the only option. If the hub is removed for any reason it must be replaced.**

1. Raise and support vehicle safely.
2. Remove the both of the rear wheels.
3. Remove the caliper and the brake disc. Support the caliper with wire to prevent stress to the brake hose.
4. If equipped with ABS, remove the bolt holding the speed sensor to the trailing arm and remove the sensor.

➡**The speed sensor has a pole piece projecting from it. This exposed tip must be protected from impact or scratches. Do not allow the pole piece to contact the toothed wheel during removal or installation.**

5. Remove the grease cap, self-locking nut and tongued washer.

➡**Do not use an air gun to remove the hub locknut.**

6. Remove the rear hub assembly from the spindle.
7. Remove the bolts that secure the ABS sensor ring to the hub and remove the ring from the hub.
   To install:
8. Secure the sensor ring to the hub assembly and tighten the mounting bolts.
9. Install the hub assembly, tongued washer and a new self-locking nut. Torque

the nut to 166 lbs. (230 Nm), align with the indentation in the spindle, and crimp.
10. Using a rope around the hub bolts and a spring balance, measure the resistance necessary to rotate the hub. If the resistance exceeds 7 lbs. (31 N), loosen and retighten the locknut. If the resistance still exceeds the specification, the hub must be replaced.
11. Using a dial indicator, measure the hub end-play. The end-play should be 0.002 in. (0.05mm) or less.
12. Install the brake rotor and caliper assembly.

13. Install the speed sensor to the knuckle.

➡**Route the speed sensor cable correctly. Improper installation may cause cable damage and system failure. Use the white stripe on the outer insulation to keep the sensor harness properly positioned.**

14. Use a brass or other non-magnetic feeler gauge to check the air gap between the tip of the pole piece and the toothed wheel. Correct gap is 0.008–0.028 in. (0.2–0.7mm). Tighten the sensor bracket nut with the sensor located so the gap is the same at several points on the toothed wheel. If the gap is incorrect, it is likely that the toothed wheel is worn or improperly installed.
15. Bleed the brake system and install the rear wheels.

ECLIPSE

The rear wheel bearing is not serviceable. If the wheel bearing must be replaced for any reason, the hub assembly must be replaced.

➡**The radio may contain a coded theft protection circuit. Always obtain the code number from the customer before disconnecting the battery.**

1. Disconnect the negative battery cable.
2. Raise and safely support the vehicle.
3. Remove the wheel and tire assembly.
4. Remove the rear wheel speed sensor if equipped with ABS.

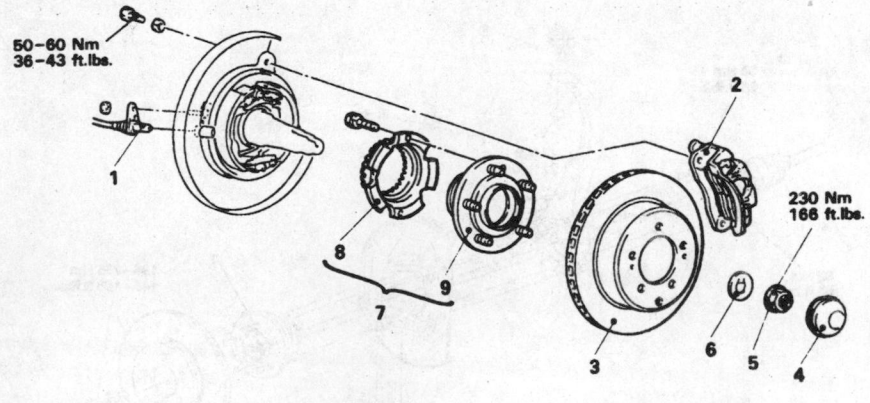

50–60 Nm
36–43 ft.lbs.

230 Nm
166 ft.lbs.

1. Rear speed sensor <Vehicles with ABS>
2. Caliper assembly
3. Brake disc
4. Hub cap
5. Flange nut
6. Tongued washer
7. Rear hub assembly
8. Rear rotor <Vehicles with ABS>
9. Rear hub unit bearing

7923PGC9

Exploded view of the rear hub/bearing assembly removal—Eclipse

5. Remove the caliper assembly and rotor, or drum. Suspend the caliper out of the way with wire.

6. On vehicles with rear disc brakes, remove the parking brake shoes.

7. On vehicles equipped with AWD, remove the axle shaft locking nut, and using a suitable tool, separate the hub from the axle shaft.

8. Remove the hub mounting bolts from behind the backing plate and remove the hub.

➡**The rotor for the ABS must be removed and installed using a press.**

9. Remove the through-bolt, lockwasher and nut, and disconnect the trailing arm.

10. Remove the bolt, washer, and locknut, and disconnect the lower control arm from the knuckle.

11. Remove the locknut and the toe control arm ball joint from the knuckle.

12. Remove the lower strut mounting bolt and disconnect the strut from the knuckle.

13. Remove the through-bolt, washer and locknut and disconnect the upper control arm from the knuckle. Remove the knuckle assembly.

To install:

14. Install the knuckle assembly to the upper control arm with the through-bolt, washer and locknut. Torque is 71 ft. lbs. (98 Nm).

15. Connect the lower strut mount to the knuckle and tighten the bolt to 71 ft. lbs. (98 Nm).

16. Install the toe control arm ball joint to the knuckle and tighten the mounting locknut to 20 ft. lbs. (28 Nm).

17. Reconnect the lower control arm to the knuckle with the through-bolt, washer, and locknut. Torque the bolt and nut to 71 ft. lbs.

18. Install the lower trailing arm to the knuckle with the bolt, washer, and locknut. Torque the nut and bolt to 85–99 ft. lbs. (118–137 Nm).

19. Press the rotor (ABS) to the hub.

20. On vehicles with AWD, engage the splines of the axle shaft with the hub assembly and tighten the axle shaft locking nut to 145–188 ft. lbs. (196–255 Nm).

21. Install the hub and tighten the mounting bolts to 54–65 ft. lbs. (74–88 Nm).

22. Install the parking brake shoes if equipped.

23. Install the rotor and caliper or drum.

24. Install the speed sensor if equipped.

25. Install the wheel and tire assembly.

26. Lower the vehicle to the floor.

27. Connect the negative battery cable.

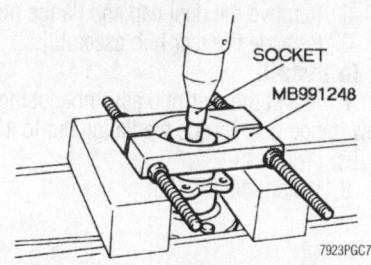

Use a press to remove the speed sensor rotor from the hub—Galant

GALANT AND DIAMANTE

1. Raise the vehicle and support safely.

2. Remove the appropriate wheel assembly.

3. If equipped with ABS, remove the vehicle speed sensor.

4. Remove the brake drum from the hub assembly.

5. From the back of the knuckle, remove the four bolts securing the hub to the knuckle.

6. Remove the hub and bearing assembly from the knuckle.

➡**The hub assembly is not serviceable and should not be disassembled.**

7. If replacing the hub, use special socket MB991248 and a press, to remove the wheel sensor rotor from the hub.

To install:

8. Press the wheel sensor rotor onto the hub.

9. Install the hub to the knuckle and tighten the mounting bolts to 54–65 ft. lbs. (74–88 Nm).

10. Install the brake drum on the hub.

11. If equipped with ABS, install the vehicle speed sensor.

12. Install the wheel assembly and lower the vehicle.

MIRAGE

➡**Never disassemble the rear hub bearing. The wheel bearing is serviced by replacement of the hub.**

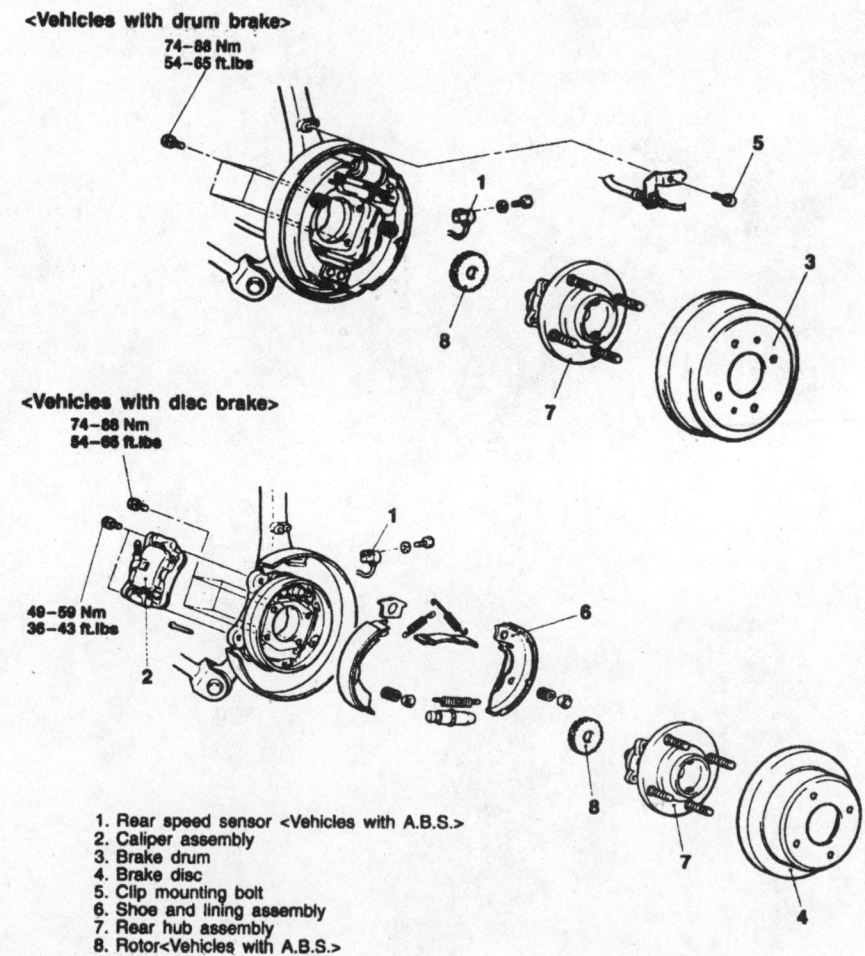

Exploded view of the rear hub/bearing assembly and related components—Galant

<Vehicles with drum brake>
74–88 Nm
54–65 ft.lbs

<Vehicles with disc brake>
74–88 Nm
54–65 ft.lbs

49–59 Nm
36–43 ft.lbs

1. Rear speed sensor <Vehicles with A.B.S.>
2. Caliper assembly
3. Brake drum
4. Brake disc
5. Clip mounting bolt
6. Shoe and lining assembly
7. Rear hub assembly
8. Rotor <Vehicles with A.B.S.>

1. If equipped with ABS, remove the wheel speed sensor.
2. Raise and safely support the vehicle.
3. Remove the rear wheel.
4. Remove the caliper and brake disc or brake drum.
5. Remove the dust cap and flange nut.
6. Remove the rear hub assembly.

To install:

7. Install the rear hub assembly using a new flange nut. Torque the flange nut to 130 ft. lbs. (180 Nm).
8. Install the dust cap.
9. Install the wheel speed sensor if removed. The air gap should be 0.012–0.035 in. (0.3–0.9mm).
10. Install the brake disc and caliper, or brake drum.
11. Install the rear wheel assembly and lower the vehicle to the floor.

ENGINE REPAIR

➡**Disconnecting the negative battery cable on some vehicles may interfere with the functions of the on board computer systems and may require the computer to undergo a relearning process, once the negative battery cable is reconnected.**

Distributor

REMOVAL

1.6L, 2.0L and 2.4L Engines

1. Disconnect the negative battery cable.
2. Set the engine to Top Dead Center (TDC) with the No.1 piston on compression stroke
3. Remove and label the distributor spark plug wires from the distributor cap.
4. Remove the distributor cap and scribe a mark on the engine block to show the rotor and distributor position prior to removal.
5. Detach and label the wiring connections to the distributor.
6. Remove the bolt(s) holding distributor to engine.
7. Pull the distributor upward to remove from cylinder block.

➡**Do not disturb the camshaft or crankshaft position after the distributor is removed from the engine. If any of these components are moved, TDC on cylinder No. 1 will have to be found again before reinstalling the distributor.**

INSTALLATION

1.6L, 2.0L and 2.4L Engines

ENGINE NOT DISTURBED

1. Install a new distributor housing O-ring.
2. Install the distributor in the engine so the rotor is aligned with the matchmark on the housing and the housing is aligned with the matchmark on the engine. Be sure the distributor is fully seated and the distributor gear is fully engaged.
3. Install and snug the hold-down bolt.

4. Connect the distributor pick-up lead wires.
5. Install the distributor cap and tighten the screws. Install the splash shield.
6. Install the spark plug wires.
7. Connect the negative battery cable.
8. After the ignition timing has been adjusted, tighten the hold-down bolt(s) to 80–104 inch lbs. (9–11 Nm) for GA16DE and VG30E engines or to 108–144 inch lbs. (13–16 Nm) for SR20DE and the Altima (KA24DE) engines. On the 240SX (KA24DE) engine, tighten to 34–39 inch lbs. (4–5 Nm).

ENGINE DISTURBED

1. Install a new distributor housing O-ring.
2. Position the engine so the No. 1 piston is at TDC of its compression stroke and the mark on the vibration damper is aligned with **0** on the timing indicator.
3. Install the distributor in the engine so the rotor is aligned with the position of the No. 1 ignition wire on the distributor cap. Be sure the distributor is fully seated and that the distributor shaft is fully engaged.
4. Install and snug the hold-down bolt.
5. Connect the distributor pick-up lead wires.
6. Install the distributor cap and tighten the screws. Install the splash shield, if equipped.
7. Install the spark plug wires.
8. Connect the negative battery cable.
9. After the ignition timing has been adjusted, tighten the hold-down bolt(s) to 80–104 inch lbs. (9–11 Nm) for GA16DE and VG30E engines or to 108–144 inch lbs. (13–16 Nm) for SR20DE and the Altima (KA24DE) engines. On the 240SX (KA24DE) engine, tighten to 34–39 inch lbs. (4–5 Nm).

Ignition Timing

ADJUSTMENT

1.6L, 2.0L and 2.4L Engines

Visually check the air cleaner, intake hoses, ducts, EGR valve operation and electrical connections prior to the adjustment of the ignition timing. Correct or repair any problem as required. Be sure to inspect the throttle valve and the throttle position sensor for proper operation.

1. Locate the timing marks on the crankshaft pulley and the front of the engine.
2. Clean the timing marks.
3. Using chalk or white paint, color the mark on the crankshaft pulley and the mark on the scale which will indicate the correct timing when aligned with the notch on the crankshaft pulley.
4. Attach a tachometer to the engine.
5. Attach a timing light to the engine, to No.1 cylinder's ignition wire.
6. Check to be sure all of the wires clear the fan; start the engine and allow it to reach normal operating temperatures.
7. Block the front wheels and set the parking brake. Shift the transmission into **NEUTRAL** for automatic and manual transaxles; do not stand in front of the vehicle when making adjustments.
8. Perform the following procedures:
   a. Race the engine at 2000 rpm for about two minutes under a no-load condition; be sure all of the accessories are turned off.

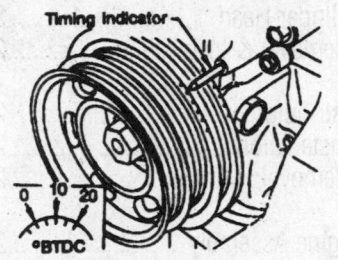

Point the timing light at the crankshaft pulley to see the timing marks—1.6L engine

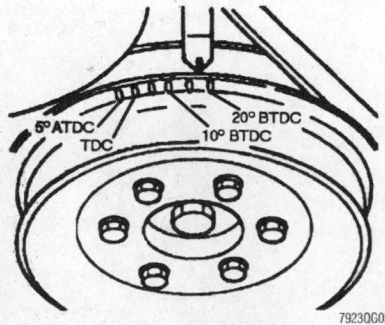

The timing marks are located on the crankshaft pulley—2.4L engine

b. Perform on board engine diagnostics and repair any fault code.

c. Race the engine 2–3 times under no-load, then run the engine it for one minute at idle.

d. Stop the engine and disconnect the throttle position sensor.

e. Race the engine at 2000 rpm for about two minutes under a no-load condition; be sure all of the accessories are turned **OFF**.

f. Run the engine at idle speed.

• 1.6L (GA16DE)—6–10 degrees BTDC

• 2.0L (SR20DE)—13–17 degrees BTDC

• 2.4L (KA24DE)—18–22 degrees BTDC

9. Aim the timing light at the timing marks. If the marks on the pulley and the engine are aligned when the light flashes, the timing is correct. Turn the engine **OFF** and remove the tachometer and the timing light. If the marks are not in alignment, proceed with the following steps.

10. Turn the engine **OFF**.

11. Loosen the bolts that secure the distributor just enough so it can be turned.

12. Start the engine. Keep the wires of the timing light clear of the cooling fan.

13. With the timing light aimed at the pulley and the marks on the engine, turn the distributor for the proper adjustment.

14. Race the engine 2–3 times under no-load, then run the engine it for one minute at idle.

15. Aim the timing light at the timing marks. If the marks on the pulley and the engine are aligned when the light flashes, the timing is correct.

16. Tighten the bolt that secures the distributor and recheck the timing.

17. Turn the engine **OFF** and remove the tachometer and the timing light.

18. Connect the throttle position sensor.

3.0L (VQ30DE) Engine

➡**The ignition timing is not adjustable. If not within specifications, further diagnostic inspection is required. The following procedure is for viewing the ignition timing setting.**

Visually check the air cleaner, intake hoses, ducts, EGR valve operation and electrical connections prior to the adjustment of the ignition timing. Correct or repair any problem as required. Be sure inspect the throttle valve and throttle position sensor for proper operation.

1. Locate the timing marks on the crankshaft pulley and the front of the engine.

2. Clean the timing marks.

➡**The ignition timing specification is 13–17° BTDC.**

3. Using chalk or white paint, color the mark on the crankshaft pulley and the mark on the scale which will indicate the correct timing when aligned with the notch on the crankshaft pulley.

4. Attach a tachometer to the engine.

5. Attach a timing light to the engine to number one cylinder ignition wire.

6. Turn all electrical equipment and accessories off.

7. Check to be sure all of the wires clear the fan, then, start the engine and allow it to reach normal operating temperatures.

8. Block the front wheels and set the parking brake. Shift the transmission into **NEUTRAL** for manual transmission and automatic transmissions. Do not stand in front of the vehicle when making adjustments.

9. Perform the following procedures:

a. Race the engine at 2000 rpm for about two minutes under a no-load condition; be sure all of the accessories are turned off.

b. Perform on board engine diagnostics and repair any fault code.

c. Race the engine at 2000 rpm for about two minutes under a no-load condition.

d. Turn the engine **OFF** and disconnect the throttle position sensor.

e. Start and race the engine 2–3 times under no-load, then run the engine at idle speed.

➡**The ignition timing specification is 13–17° BTDC.**

10. Aim the timing light at the timing marks. If the marks on the pulley and the engine are aligned when the light flashes, the timing is correct. Turn the engine **OFF** and remove the tachometer and the timing light. If the marks are not in alignment, proceed with the following steps.

11. Turn the engine **OFF**.

12. Check the camshaft position sensor (PHASE), crankshaft position sensor (REF) and crankshaft position sensor (POS). Replace if necessary.

13. If the ignition timing is still not correct, substitute a known good ECM.

➡**The ECM may be the cause of the problem but this is rarely the case.**

14. Turn the engine **OFF** and remove the tachometer and the timing light.

3.0L (VG30DE and VG30DETT) Engine

➡**The engine should be in good mechanical condition and all electrical connectors and vacuum hoses connected before making this adjustment.**

1. Start the engine and let it warm up to normal operating temperature.

2. Open the hood and run the engine under no load at about 2,000 rpm for about two minutes.

3. Perform Diagnostic Test Mode II and repair any causes of trouble codes as needed.

4. Run the engine under no load at 2,000 rpm for about two minutes. Rev the engine two or three times and let it idle for one minute.

5. Turn off the engine and detach the throttle position sensor connector. Remove the No. 1 ignition coil. Connect the coil to the spark plug using a spare piece of high-tension wire so you have a place to connect your timing light. Start the engine.

6. Run the engine under no load at 2,000 rpm for about two minutes. Rev the engine two or three times and let it idle.

7. Check the ignition timing and adjust if needed. Correct ignition timing is 8–12° BTDC on non-turbocharged

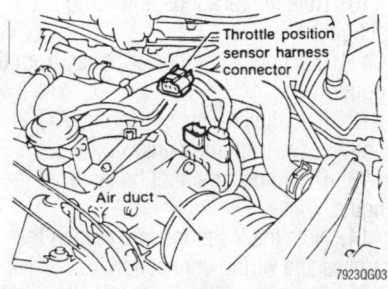

Throttle position sensor connector location—3.0L (VG30DE and VG30DETT) engines

vehicles; 13–17° BTDC on turbocharged vehicles. Adjustment is made by loosening the screws and turning the camshaft position sensor. Tighten the mounting screws and confirm ignition timing has not changed.

8. Turn the engine **OFF** and connect the throttle position sensor connector.

Engine Assembly

REMOVAL & INSTALLATION

Sentra and 200SX

➡The engine and transaxle are removed as one unit from the underside of the vehicle.

✳✳ CAUTION

The fuel injection system remains under pressure after the engine has been turned OFF. Properly relieve fuel pressure before disconnecting any fuel lines. Failure to do so may result in fire or personal injury.

1. Relieve the fuel system pressure.
2. Disconnect the negative and positive battery cables.
3. Remove the battery and battery tray from the vehicle.
4. Raise and safely support the vehicle.
5. Remove both front wheels.
6. Remove the engine undercovers and remove the engine side covers.
7. Drain the coolant from the radiator and the engine block.
8. Drain the engine oil.
9. Remove the air cleaner assembly and remove air duct.
10. Note the locations and remove the vacuum hoses.
11. Disconnect the heater hoses from the engine.
12. If equipped, disconnect the A/T cooler hoses from the transaxle.
13. Disconnect the fuel hoses from the engine.
14. Note the locations and detach the harness and wiring connections.
15. Disconnect the throttle cable and the cruise control cable.
16. If equipped with automatic transmission, disconnect the control cable.

17. Remove the cooling fans, radiator, and the recovery tank.
18. Remove the front driveshafts from the vehicle.
19. Remove the front exhaust pipe.
20. On the 1.6L engine, disconnect the control rod and support rod from the transaxle.
21. Remove the starter motor and intake manifold support brackets.
22. Remove the engine drive belts.
23. Remove the alternator and adjusting brackets.
24. Remove the power steering pump and A/C compressor. It is not necessary to disconnect the lines.
25. On the 1.6L engine, remove the cylinder head front mounting bracket.
26. Position a transmission jack under the transaxle and support the engine with engine slinger.
27. Remove the center crossmember.
28. On some models it may be necessary to remove the front stabilizer bar.
29. Remove the engine mounting bolts from both sides of the engine.
30. Slowly lower the jacking devices and remove the engine and transaxle from the vehicle.

To install:
31. Install the engine and transaxle assembly.
32. Install the mounting bolts to both sides of the engine.
33. For vehicles with manual transaxles, adjust the height of the mounting bracket (buffer rod). The distance between the two through-bolts should be 2.126–2.205 in. (54–56mm).
34. Install the center crossmember and remove the engine support jacks. Remove the engine slinger.

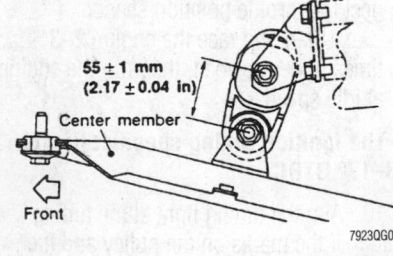

Height adjustment of the buffer rod—Sentra and 200SX with 1.6L engine

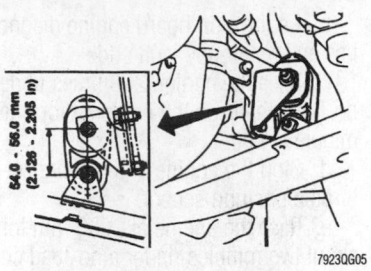

Be sure to adjust the height of the engine mount for M/T vehicles—2.0L engine

35. Install the remaining components in the reverse order of removal.
36. Tighten the control rod bolt to 10–13 ft. lbs. (14–18 Nm). Tighten the support rod bolt to 26–35 ft. lbs. (35–47 Nm).
37. Install the battery tray and install the battery.
38. Connect the positive, then the negative battery cables.
39. Start the engine and check for leaks. Make all the necessary adjustments.

Altima

✳✳ CAUTION

Release the fuel pressure in the system before disconnecting the fuel lines.

The fuel system will remain under pressure after the ignition has been turned OFF. Failure to do so may result in fire or personal injury.

➡The engine and transaxle must be removed as a single unit. The engine and transaxle are removed from under the vehicle.

1. Mark the location of the hinges on the hood and remove the hood from the vehicle.
2. Release fuel system pressure.
3. Disconnect the battery cables and remove the battery and battery tray.
4. Drain the coolant from the plug on the water pipe and drain the radiator.
5. If equipped with automatic transaxle, disconnect the cooler lines from the radiator.
6. Remove the upper and lower hoses from the radiator, then remove the radiator assembly.
7. Disconnect the heater hoses from the engine.

8. Disconnect the throttle cable and cruise control cable (if equipped).

9. Remove the air cleaner, air box, and the intake hose.

10. Disconnect the fuel feed and return hoses.

11. Detach and label all the necessary vacuum hoses and electrical connectors.

12. Disconnect the wiring from starter motor.

13. If equipped, disconnect the slave cylinder from the transaxle. It is not necessary to disconnect the hydraulic hose.

14. Remove the engine drive belts. Be sure to mark belts for reinstallation.

15. Remove the alternator, A/C compressor, and the power steering pump from the engine.

16. Remove the right and left driveshafts from the transaxle.

17. Disconnect the exhaust pipe from the exhaust manifold.

18. Disconnect the crankshaft position sensor from the engine block.

19. Support the engine with slinger and support the transaxle with proper jack.

20. Disconnect the left and right engine mounting through-bolts.

21. Remove the bolts that secure the crossmember to the vehicle and remove the crossmember.

22. Remove the front and rear engine mounts.

23. Lower the transaxle and engine assembly from the vehicle.

➡**The engine and transaxle assembly should be removed through the bottom of the vehicle. Do not attempt to remove the assembly from above.**

To install:

24. Raise the transaxle and engine assembly to the vehicle.

25. Install the front and rear engine mounts. Tighten the mounting bolts to specifications.

26. Install the crossmember and tighten the mounting bolts to 57–72 ft. lbs. (77–98 Nm).

27. Connect the left and right engine mounting through-bolts. Tighten the bolts to specifications.

28. Remove the engine and transaxle support jacks.

29. The balance of installation is the reverse of the removal procedure.

30. Refill the cooling system and the engine oil. Check all the fluid levels.

31. Start the engine, bleed the cooling system, and check for leaks. Make all the necessary adjustments.

32. Install the hood.

Maxima

It is recommended the engine and transaxle be removed as a single unit. If need be, the units may be separated after removal.

➡**The engine and transaxle assembly must be removed from the underside of the vehicle.**

1. Matchmark the hood hinge relationship and remove the hood.

2. Release the fuel system pressure. Disconnect the negative battery cable and raise and safely support the vehicle.

3. Drain the coolant from the cylinder block and the radiator. Drain the crankcase and the automatic transaxle, if equipped.

4. Remove the air cleaner, the air intake tube, the air flow meter and disconnect the throttle linkage.

5. Disconnect and/or remove the following:

- Drive belts
- Engine ground cable
- Electrical connector from the crank angle sensor
- Engine electrical harness connectors

✳ **CAUTION**

The fuel injection system remains under pressure after the engine has been turned OFF. Properly relieve the fuel pressure before disconnecting any fuel lines. Failure to do so may result in fire or personal injury.

- Fuel feed and fuel return hoses
- Upper and lower radiator hoses
- Heater inlet and outlet hoses
- Electrical connector from the crank angle sensor
- Engine vacuum hoses
- Carbon canister hoses
- Any interfering engine accessory: power steering pump, air conditioning compressor, and the alternator

6. Remove the carbon canister.

7. Remove the auxiliary fan, washer tank, and the radiator (with the fan assembly).

8. If equipped with a manual transaxle, remove the clutch release cylinder from the clutch housing.

9. On some models with a manual transaxle, disconnect the shift control rod and disconnect the shift support rod. On others with an automatic transaxle, disconnect the control cable from the transaxle.

10. Install engine slingers to the block and connect a suitable lifting device to the slingers. Do not tension the lifting device at this point.

11. Disconnect the exhaust pipe at both the manifold connections and remove the front exhaust pipe from the vehicle.

12. If equipped with a manual transaxle, drain the transaxle gear oil.

13. Support the engine and transaxle assembly with proper jack.

14. Disconnect the right and left side halfshafts from their side flanges and remove the bolt holding the radius link support.

15. Lower the shifter and selector rods and remove the bolts from the motor mount brackets. Remove the nuts holding the front and rear motor mounts to the frame.

16. On some models it will be necessary to remove the center crossmember assembly from the vehicle.

17. Lower the engine/transaxle assembly down and onto an engine stand.

To install:

18. Raise the engine/transaxle assembly into the vehicle. When raising the engine onto the mounts, be sure to keep it as level as possible.

19. After installing the motor mounts, adjust and install the buffer rods; the front should be 3.50–5.58 in. (89–91mm) and the rear should be 3.90–3.98 in. (99–101mm).

20. Check the clearance between the frame and clutch housing and be sure the engine mount bolts are seated in the groove of the mounting bracket.

21. Remove the transaxle and engine jack assembly.

22. The remaining components are installed in the reverse order from which they were removed.

23. Fill the transaxle, the engine, and the cooling system to the proper levels with the appropriate fluids.

24. Install the hood and connect the negative battery cable.

25. Make all the necessary engine adjustments. Charge the air conditioning system, if discharged. Road test the vehicle for proper operation.

240SX

➡**The engine assembly is removed from the top of the vehicle.**

1. Be sure it is on a flat and level surface and that wheels are tightly chocked.

2. Allow the exhaust system to cool completely before starting work to prevent burns and possible fire as fuel lines are disconnected.

✳✳ CAUTION

The fuel injection system remains under pressure after the engine has been turned OFF. Properly relieve the fuel pressure before disconnecting any fuel lines. Failure to do so may result in fire or personal injury.

3. Release fuel pressure from the fuel system before attempting to disconnect any fuel lines.

4. Mark the location of the hinges on the hood. Unbolt and remove the hood.

5. Disconnect the battery cables and remove the battery. Be sure to disconnect the negative cable first.

6. Remove the engine undercover.

7. Remove the transmission from the vehicle.

8. Drain the coolant from the radiator and the engine block.

9. Drain the engine oil from the vehicle.

10. Remove the radiator and radiator shroud after disconnecting the automatic transmission to radiator cooling tubes.

11. Remove the air cleaner.

12. Remove the engine drive belts.

13. Remove the fan and pulley.

14. Detach the electrical harness connectors at the water temperature sensor, oil pressure sending unit, and the starter motor. Disconnect the primary ignition wires.

15. Disconnect the fuel hoses.

✳✳ CAUTION

On all fuel injected models, the fuel pressure must be released before the fuel lines can be disconnected.

16. Detach the electrical connections at the alternator. Disconnect the heater hoses and throttle connections,

17. Disconnect the engine ground cable, thermal transmitter wire, wire to the fuel cut-off solenoid and the vacuum cut solenoid wire.

18. Remove the front exhaust pipe from the vehicle.

✳✳ CAUTION

On models with air conditioning, it is necessary to remove the compressor from the vehicle. DO NOT ATTEMPT TO UNFASTEN ANY OF THE AIR CONDITIONER HOSES BEFORE PROPERLY EVACUATING THE SYSTEM.

19. Remove the A/C compressor from the vehicle.

20. Remove the power steering pump from the engine.

21. Disconnect the power brake booster hose from the engine.

22. Attach a hoist to the lifting hooks on the engine (at either end of the cylinder head). Support the engine.

23. Remove the engine mounting nuts from both lower sides of the engine mounts.

➡**When lifting the engine out, guide it carefully to avoid hitting parts such as the master cylinder.**

24. Remove the engine from the vehicle.

To install:

25. With the engine assembly safely secured to the hoist, lower the assembly into the vehicle.

26. Tighten the engine mounting nuts to 51–58 ft. lbs. (69–78 Nm). It may be necessary to lower or raise the engine hoist to correctly position the engine assembly to line up with the mount holes. Remove the engine hoist.

27. Install the power steering pump to the engine. Install the power brake booster.

28. The balance of installation is the reverse of the removal procedure.

29. Refill the engine, transmission, and coolant levels with the proper type and amount of fluid.

30. Check all fluids to assure they are at the correct level. Start the engine and run at idle until normal operating temperature is reached. Check for fluid leaks and repair as required.

31. Road test the vehicle after you are sure there are no leaks. Check vehicle for proper operation. Recheck all fluid levels once road test is completed.

300ZX

WITH MANUAL TRANSMISSION

1. Release the fuel system pressure and disconnect the negative battery cable.

✳✳ CAUTION

The fuel injection system remains under pressure after the engine has been turned OFF. Properly relieve fuel pressure before disconnecting any fuel lines. Failure to do so may result in fire or personal injury.

2. Matchmark the hood hinge relationship and remove the hood.

3. Drain the coolant from the cylinder block and the radiator. Drain the crankcase and the transaxle.

4. Remove the air cleaner, the air intake tube, the air flow meter and disconnect the throttle linkage.

5. Note the locations and disconnect the vacuum hoses from the engine.

6. Disconnect the fuel feed and return hoses.

7. Note the wiring locations and detach the harness connections.

8. Raise and safely support the vehicle.

9. Remove the engine undercover(s).

10. Disconnect the front exhaust pipes from the engine.

11. Disconnect and remove the driveshaft from the rear of the transmission.

12. Remove the upper and lower radiator hoses.

13. Remove the radiator shroud and remove the radiator from the vehicle.

14. Remove the engine drive belts.

15. Remove the cooling fan and the cooling fan coupling.

16. Disconnect and remove the power steering pump, A/C compressor, and the alternator.

17. Disconnect the wiring to the starter motor and remove the starter assembly.

18. Remove the clutch slave cylinder from the transmission and position aside. It is not necessary to disconnect the fluid line.

19. Disconnect the A/C tube clamps from the A/C line.

20. Disconnect the steering column lower joint.

21. Remove the tension rod to lower control arm mounting nuts from both sides and disconnect the rod at the lower control arm.

22. Loosen but do not remove the lower control arm to crossmember bolts.

23. Position a suitable transmission jack under the crossmember and support the engine with slinger and hoist.

24. Remove the crossmember mounting bolts.

➡ **The crossmember will remain in the vehicle during engine removal.**

25. Remove the engine mounting bolts from both sides of the engine, then slowly lower the jack.

26. Remove the engine and transmission from the vehicle.

➡ **When lifting the engine from vehicle, be careful not to strike adjacent parts. Pay special attention to the accelerator wire casing, brake lines, and the master cylinder.**

To install:

27. Install the engine and transmission assembly.

28. Slowly raise the floor jack and install the engine mounting bolts.

29. Install the crossmember mounting bolts.

30. Remove the engine slinger and hoist.

➡ **All suspension bolts must be tightened with the full weight of the vehicle on the ground.**

31. Tighten the lower control arm to crossmember bolts.

32. Install the remaining components in the opposite order from which they were removed.

33. Tighten the traverse link and tension rod mounting bolts and nuts to 80–94 ft. lbs. (108–127 Nm).

34. Fill the engine oil, transmission, and the cooling system with the proper type and amount of fluid.

35. Install the hood and connect the negative battery cable.

36. Perform all necessary adjustments.

37. Start the engine and warm to full operating temperature, bleed cooling system, and check for leaks.

WITH AUTOMATIC TRANSMISSION

1. Release the fuel system pressure and disconnect the negative battery cable.

✳✳ CAUTION

The fuel injection system remains under pressure after the engine has been turned OFF. Properly relieve fuel pressure before disconnecting any fuel lines. Failure to do so may result in fire or personal injury.

---

2. Matchmark the hood hinge relationship and remove the hood.

3. Drain the coolant from the cylinder block and the radiator. Drain the crankcase and the automatic transaxle.

4. Remove the air cleaner, the air intake tube, the air flow meter and disconnect the throttle linkage.

5. Note the locations and disconnect the vacuum hoses from the engine.

6. Disconnect the fuel feed and return hoses.

7. Note the wiring locations and detach the harness connections.

8. Raise and safely support the vehicle.

9. Remove the engine undercover(s).

10. Disconnect the front exhaust pipes from the engine.

11. Disconnect and remove the driveshaft from the rear of the transmission.

12. Remove the upper and lower radiator hoses. Disconnect the heater core hoses.

13. Remove the radiator shroud and remove the radiator from the vehicle.

14. Remove the engine drive belts.

15. Disconnect and remove the power steering pump, A/C compressor, and the alternator.

16. Disconnect the wiring to the starter motor and remove the starter assembly.

17. Remove the transmission from the vehicle.

18. Install an engine slinger and hoist.

19. Support the weight of the engine with the hoist and disconnect the engine mounting bolts.

20. Lift the engine from the mounts and remove the engine assembly.

➡ **When lifting the engine from vehicle, be careful not to strike adjacent parts. Pay special attention to the accelerator wire casing, brake lines, and the master cylinder.**

To install:

21. Install the engine assembly to the engine compartment.

22. Install the engine mounting bolts.

23. Install the transmission assembly.

24. Install the starter motor and connect the wiring.

25. Install the remaining components in the reverse order of removal.

---

26. Refill the cooling system, transmission, and the crankcase with the proper type and amount of fluid.

27. Install the hood and connect the negative battery cable.

28. Perform all necessary adjustments.

29. Start the engine and warm it to full operating temperature, bleed cooling system, and check for leaks.

Water Pump

REMOVAL & INSTALLATION

1.6L Engine

1. Disconnect the negative battery cable.

✳✳ CAUTION

Never open, service or drain the radiator or cooling system when hot; serious burns can occur from the steam and hot coolant. Always drain coolant into a sealable container. Coolant should be reused unless it is contaminated or is several years old.

---

2. Drain the cooling system.

3. Remove the cylinder head front mounting bracket.

4. Loosen the water pump pulley bolts.

5. Remove the engine drive belts from the A/C compressor, power steering pump and alternator.

6. Remove the belt pulley from the water pump.

7. Detach electrical connectors and coolant hoses from the thermostat housing.

8. Unbolt and remove the water pump and thermostat housing from the engine.

➡ **Remove the thermostat housing with water pump assembly.**

9. Remove the bolts that secure the thermostat housing to the water pump.

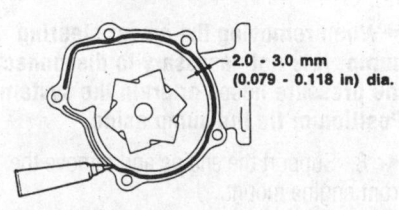

2.0 - 3.0 mm (0.079 - 0.118 in) dia.

79230G06

Apply RTV sealant to the water pump sealing surface as shown—1.6L engine

10. Remove all traces of gasket material from sealing surfaces.

To install:

11. Apply a continuous bead of liquid sealer to the sealing surface of the thermostat housing. The sealant should be 0.079–0.118 in. (2–3mm) diameter.

12. Install the thermostat housing to the water pump and tighten mounting bolts to 56–73 inch lbs. (7–8 Nm).

13. Apply a continuous bead of liquid sealer to the sealing surface of the water pump. The sealant should be 0.079–0.118 in. (2–3mm) diameter.

14. Install the water pump on the engine and tighten mounting bolts to 56–73 inch lbs. (7–8 Nm).

15. Install the pulley to the water pump and tighten the mounting bolts to 56–73 inch lbs. (7–8 Nm).

16. Connect electrical connectors and coolant hoses to the thermostat housing.

17. Install and adjust the alternator, power steering and A/C compressor drive belts.

18. Refill the cooling system and connect the negative battery cable.

19. Start the engine, bleed the cooling system, warm the engine to full operating temperature, and check for leaks.

20. If necessary, refill the cooling system when the engine has cooled.

2.0L Engine

1. Disconnect the negative battery cable.
2. Drain the radiator coolant.
3. Remove the cylinder block drain plug located at the left front of the engine and drain coolant.
4. Loosen the water pump pulley bolts.
5. Remove the power steering pump, alternator and A/C compressor drive belts (if equipped).
6. Remove the water pump pulley.
7. Note positioning of power steering pump adjusting bracket and remove the power steering pump adjusting bracket from the water pump. If necessary, remove the power steering pump for access to bracket.

➡**When removing the power steering pump, it is not necessary to disconnect the pressure hoses or drain the system. Position or tie the pump aside.**

8. Support the engine and remove the front engine mount.

9. Remove the mounting bolts from the water pump and remove the water pump.

10. Remove all traces of liquid gasket material from sealing surfaces.

To install:

11. Apply a continuous bead of liquid sealer to the mating surface of the water pump. Sealer should be 0.079–0.118 in. (2–3mm) wide.

12. Install the water pump assembly and tighten mounting bolts to 12–15 ft. lbs. (16–21 Nm).

➡**Be sure to properly position the adjusting bracket that was noted during removal.**

13. Install the front engine mount.
14. Install the power steering pump adjusting bracket and install the power steering pump if removed.
15. Install the water pump pulley and tighten mounting bolts to 55–73 inch lbs. (6–8 Nm).
16. Install and adjust the power steering pump, alternator and A/C compressor drive belts (if equipped).
17. Install the cylinder block drain plug located at the left front of the engine and tighten drain plug to 70–104 inch lbs. (8–12 Nm).
18. Refill the cooling system and connect the negative battery cable.
19. Start the engine, bleed the cooling system, and check for leaks.

2.4L Engine

ALTIMA

1. Disconnect the negative battery cable.
2. Drain the cooling system and water pipe, using the drain plugs.
3. Remove the upper radiator hose to provide working room and remove the drive belt(s) from the pulleys.
4. Remove the alternator and the A/C compressor.

➡**Do not disconnect the A/C compressor lines. Unbolt the compressor and lay it off to the side.**

5. Remove the water pump pulley.
6. Remove the mounting bolts and remove the water pump from the engine.

➡**The mounting bolts are different sizes and must be reinstalled in the correct location, therefore it is a good idea to arrange the bolts so that they can be easily identified during installation.**

To install:

7. Be sure all gasket surfaces are clean and properly apply a continuous bead of silicone sealer to the pump.

8. Install the pump to the engine and tighten the 6mm bolts to 57–66 inch lbs. (6–8 Nm) and the 8mm bolts to 12–14 ft. lbs. (16–19 Nm).

9. Install the water pump pulley and tighten the bolts to 57–66 inch lbs. (6–8 Nm).

10. Install the remaining components in the opposite order from which they were removed.

11. Fill and bleed the cooling system.

12. Start the engine and check for leaks.

240SX

1. Disconnect the negative battery cable.
2. Drain the cooling system and the engine block, using the radiator petcock and cylinder block drain plug.
3. Remove the upper radiator hose to provide working room and remove the drive belt(s) from the pulleys.
4. Remove the retaining screws and lift the fan shroud from the engine.
5. While holding the pulley, remove the nuts retaining the cooling fan and pulley to the water pump.
6. Remove the mounting bolts and remove the water pump from the engine.

To install:

7. Be sure all gasket surfaces are clean and properly apply liquid sealer to the pump.

8. Install the pump to the engine and tighten the bolts to 12–14 ft. lbs. (16–19 Nm).

9. Install the remaining components in the reverse order of removal.

10. Tighten the fan clutch, fan, and pulley mounting nuts to 66 inch lbs. (8 Nm).

11. Start the engine and check for leaks.

3.0L (VG30DE and VG30DETT) Engines

1. Disconnect negative battery cable.
2. Drain coolant from radiator and engine block.
3. Remove the undercover and the radiator.
4. Remove cooling fan assembly, water inlet, outlet and drive belts.
5. Remove crankshaft pulley and timing belt cover.
6. Remove water pump.

To install:

7. Remove all traces of gasket material.

8. Apply a continuous bead of liquid gasket to water pump mating surface.

9. Install water pump to engine block.

10. Tighten water pump bolts to 12–15 ft. lbs. (16–21 Nm).

11. Reinstall timing belt cover and tighten the cover bolts to 26–43 inch lbs. (3–5 Nm).

12. Replace crankshaft pulley and tighten the mounting bolt to 159–174 ft. lbs. (216–235 Nm).

13. Reinstall the drive belts, water inlet, water outlet, and cooling fan assembly.

14. Tighten cooling fan nuts to 51–86 inch lbs. (6–10 Nm).

15. Reinstall the radiator and undercover.

16. Refill and bleed the cooling system.

17. Reconnect the negative battery cable.

18. Start engine and check for proper operation.

3.0L (VQ30DE) Engine

1. Disconnect the negative battery cable.

2. Drain the coolant from the plugs on the radiator and both sides of the engine block.

3. Position a jack under the oil pan for support. Be sure to place a block of wood on the jack for protection to the engine parts.

4. Remove the right side engine mount and engine mounting bracket.

5. Remove the drive belts and the idler pulley bracket.

6. Remove the chain tensioner cover and the water pump cover.

7. Push the timing chain tensioner sleeve and apply a stopper pin so it does not return.

8. Remove the timing chain tensioner assembly.

9. Remove the three bolts that secure the water pump.

10. Rotate the crankshaft 20 degrees counterclockwise to provide timing chain slack.

11. Put M8 bolts to two M8 threaded holes of the water pump.

12. Tighten each bolt by turning alternately ½ turn until they reach the timing chain rear case. Be sure to turn each bolt ½ turn at a time to prevent damage.

13. Lift up the water pump and remove it.

14. When removing the water pump, do not allow the water pump gear to hit the timing chain.

15. Remove and discard the O-rings from the water pump.

16. Clean all traces of liquid gasket from the water pump and covers.

To install:

17. Using new O-rings, install the water pump to the engine block.

18. Tighten the three water pump mounting bolts evenly to 62–86 inch lbs. (7–10 Nm).

19. Rotate the crankshaft pulley to its original position by turning it 20 degrees clockwise.

20. Install the timing chain tensioner and tighten the mounting bolts to 75–96 inch lbs. (9–10 Nm).

21. Remove the stopper pin from the timing chain tensioner.

22. Apply a continuous 0.091–0.130 in. (2.3–3.3mm) bead of liquid sealant to the mating surfaces of the timing chain tensioner and water pump covers.

23. Install the timing chain tensioner and water pump covers to the engine block. Tighten the cover mounting bolts to 84–108 inch lbs. (10–13 Nm).

24. Install the drive belts and the idler pulley bracket.

25. Install the right side engine mounting bracket and the engine mount.

26. Remove the jack from under the engine and install the drain plugs to the cylinder block.

27. Connect the negative battery cable and refill the cooling system.

28. Start the engine, bleed the cooling system, and check for leaks.

Cylinder Head

REMOVAL & INSTALLATION

1.6L Engine

✳✳ CAUTION

The fuel injection system remains under pressure after the engine has been turned OFF. Properly relieve fuel pressure before disconnecting any fuel lines. Failure to do so may result in fire or personal injury.

1. Disconnect the negative battery cable, drain the cooling system, and relieve the fuel system pressure.

2. Remove all engine drive belts.

3. Remove the cylinder head cover and any related components.

4. Remove the distributor, plug wires and spark plugs.

5. Remove the spark plugs.

6. Remove the intake manifold and all related components.

7. Remove the idler pulley, camshaft sprockets and timing chains.

8. Remove the camshafts.

9. Loosen the cylinder head bolts in 2–3 steps in the reverse order of the tightening sequence to prevent warpage or cracking of the cylinder head assembly.

10. Carefully remove the cylinder head from the block, pulling the head up evenly from both ends. If the head seems stuck, do not pry it off. Tap lightly around the lower perimeter of the head with a rubber mallet to help break the seal. The cylinder head and the intake and exhaust manifolds are removed together. Remove the cylinder head gasket.

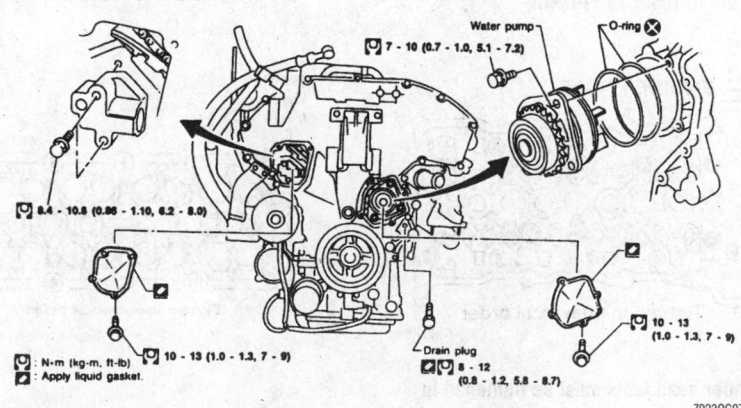

Water pump and timing cover assembly—3.0L (VQ30DE) engine

To install:

11. Thoroughly clean both the cylinder block and head mating surfaces. Avoid scratching either surface.

12. Coat the threads and the seating surface of the head bolts with clean engine oil. Install the cylinder head assembly (always replace the head gasket). Install head bolts (with washers) in their proper locations and tighten in sequence as follows:

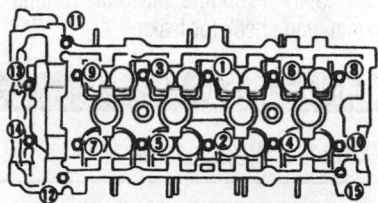

Tighten in numerical order.

79230G08

Be sure to tighten the cylinder head bolts according to the sequence shown—1.6L engine

a. Tighten bolts No. 1 through 10 in sequence to 22 ft. lbs. (29 Nm).

b. Tighten bolts No. 1 through 10 in sequence to 43 ft. lbs. (59 Nm).

c. Loosen bolts completely.

d. Tighten bolts No. 1 through 10 in sequence to 22 ft. lbs. (29 Nm).

e. Tighten bolts No. 1 through 10 to 50–55 degrees clockwise in sequence or if angle wrench is not available, tighten bolts to 40–47 ft. lbs. (54–64 Nm) in sequence.

f. Finally, tighten bolts No. 11 through 15 to 56–73 inch lbs. (6.3–8.3 Nm).

13. Install the camshafts.

14. Install the idler pulley, camshaft sprockets and timing chains.

15. Install the remaining components in the reverse order of removal. Refill and check all fluid levels.

16. Connect negative battery cable.

17. Start the engine, check for leaks, and adjust the ignition timing to specification.

18. Road test the vehicle for proper operation.

2.0L Engine

1. Release the fuel pressure. Disconnect the negative battery cable.

2. Drain the cooling system. Remove the radiator assembly.

3. Remove the cylinder head cover and oil separator.

4. On some models it may be necessary to disconnect the front exhaust pipe from exhaust manifold.

5. Remove the intake manifold.

6. Remove the distributor assembly.

7. Remove the timing chain, tensioner, chain guide and camshaft sprockets.

8. Remove the camshafts.

9. Remove the water hose from the cylinder block and water hose from the heater.

10. Remove the starter motor. Remove the water pipe bolt.

11. Remove the knock sensor harness connector and remove the EGR tube.

12. Remove the cylinder outside bolts. Remove the cylinder head bolts in two or three steps. Remove the cylinder head completely with manifolds attached.

To install:

13. Check all components for wear. Replace as necessary. Clean all mating surfaces and replace the cylinder head gasket.

➡ **If the length of any cylinder head bolt exceeds 6.228 in. (158.2mm), replace the bolt.**

14. Install cylinder head. Tighten cylinder head in the following sequence:

a. Tighten all bolts in sequence to 29 ft. lbs. (39 Nm).

b. Tighten all bolts in sequence to 58 ft. lbs. (78 Nm).

c. Loosen all bolts in sequence completely.

d. Tighten all bolts in sequence to 25–33 ft. lbs. (34–44 Nm).

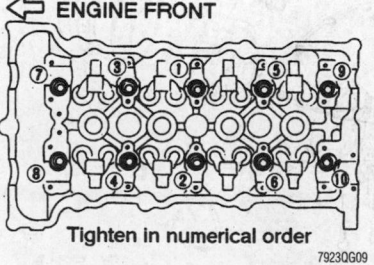

◁ **ENGINE FRONT**

Tighten in numerical order

79230G09

Cylinder head bolts must be tightened in sequence to prevent leakage—2.0L engines

e. Tighten all bolts to 90–100 degrees clockwise in sequence.

f. Tighten all bolts additional 90–100 degrees clockwise in sequence. Do not turn any bolt 180–200 degrees clockwise all at once.

15. Install the starter motor and connect the wiring.

16. Install the remaining components in the opposite order from which they were removed.

17. Install and connect the intake air hose.

18. Connect the negative battery cable.

19. Refill and check all fluid levels. Road test the vehicle for proper operation.

2.4L Engine

1. Disconnect the negative battery cable.

2. Drain coolant from the engine and radiator.

3. Relieve the fuel system pressure.

❊❊ CAUTION

The fuel injection system remains under pressure after the engine has been turned OFF. Properly relieve fuel pressure before disconnecting any fuel lines. Failure to do so may result in fire or personal injury.

4. Remove the intake manifold collector, exhaust manifold and all related components.

5. Remove the distributor assembly.

6. Using a block of wood, set a jack under the aluminum oil pan and remove the front engine mount.

7. Remove the cylinder head cover.

8. Remove the timing chain and camshaft sprockets.

9. Remove the camshafts.

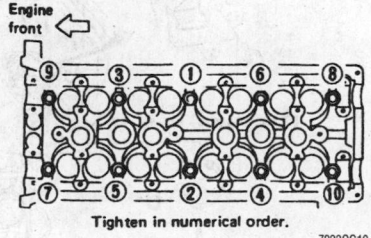

Engine front ◁

Tighten in numerical order.

79230G10

Cylinder head bolt tightening sequence—2.4L engines

➡ **The valvetrain components must be reassembled in their original positions.**

10. Loosen the cylinder head bolts in reverse order of tightening.

➡ **A warped or cracked cylinder head could result from loosening in incorrect order. The cylinder head bolts should be loosened in two or three steps.**

11. Remove the cylinder head and the intake manifold. Remove the cylinder head gasket. The lower timing chain will not be disengaged from crankshaft sprocket.

To install:

12. Clean the gasket surfaces.

13. Install new cylinder head gasket.

14. Install the cylinder head and temporarily tighten the cylinder head bolts. This is necessary to avoid damaging the cylinder head gasket. Be sure to install washers between the bolts and cylinder head.

15. Install the idler shaft assembly.

16. Install the upper timing chain and cover.

17. Tighten the cylinder head bolts in sequence as follows:

    a. Tighten all the bolts to 22 ft. lbs. (29 Nm).

    b. Tighten all the bolts to 59 ft. lbs. (79 Nm).

    c. Loosen all the bolts completely.

    d. Tighten all the bolts to 18–25 ft. lbs.(25–34 Nm).

    e. Turn all the bolts 86–91 degrees clockwise.

18. Install the camshafts.

19. Install the timing chains, chain tensioner, and camshaft sprockets.

20. The balance of installation is the reverse of the removal procedure.

21. Refill the engine coolant.

22. Connect the negative battery cable.

23. Start the engine and make the necessary adjustments. Check for proper operation and leaks.

3.0L (VG30DE and VG30DETT) Engines

✳✳ WARNING

After the timing belt has been removed, DO NOT rotate the crankshaft or camshafts separately. The valves and pistons will make contact thus causing severe engine damage.

✳✳ CAUTION

The fuel injection system remains under pressure after the engine has been turned OFF. Properly relieve fuel pressure before disconnecting any fuel lines. Failure to do so may result in fire or personal injury.

1. Relieve the fuel system pressure and disconnect the negative battery cable.

2. Remove the intake manifold collector and all related components.

➡ **After the intake manifold collector has been removed, cover the openings on the intake manifold to prevent foreign objects from entering the combustion chamber.**

3. Remove the fuel rail and injectors.

4. Remove the cylinder head covers.

5. Remove the accessory drive belts.

6. Remove the radiator and cooling fans.

7. Remove the timing belt, automatic tensioner and idler pulley assemblies.

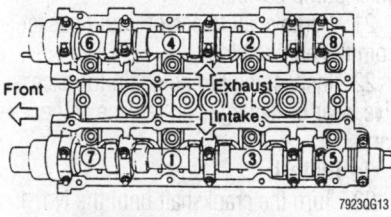

Right cylinder head bolt tightening sequence—3.0L (VG30DE and VG30DETT) engines

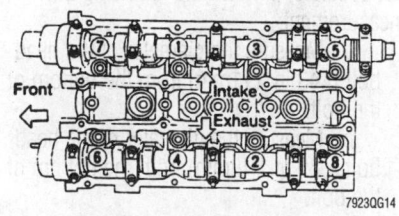

Left cylinder head bolt tightening sequence—3.0L (VG30DE and VG30DETT) engines

---

8. Remove the intake manifold assembly.

9. Disconnect the exhaust pipe from the exhaust manifold.

10. Loosen the cylinder head bolts (in reverse order of installation sequence) in 2–3 stages. Lift the cylinder head off the engine block with the exhaust manifolds attached. It may be necessary to tap the head lightly with a rubber mallet to loosen it.

To install:

11. Thoroughly clean the engine block and cylinder head surfaces.

12. Be sure the No. 1 cylinder is set at TDC on its compression stroke as follows:

    a. Align the crankshaft timing mark with the mark on the oil pump housing.

    b. Align camshaft sprocket timing mark with the mark on the rear timing belt cover.

13. Install the cylinder head with a new gasket. Apply clean engine oil to the threads and seats of the bolts and install the bolts with washers in the correct position. Be sure to position the head bolt washers with the flat side toward the cylinder head.

➡ **There is one special 6mm bolt per cylinder head. Follow the proper torque specifications for these bolts.**

14. Tighten the bolts in the proper sequence as follows:

    a. Tighten all bolts, in sequence, to 29 ft. lbs. (39 Nm).

    b. Tighten all bolts, in sequence, to 90 ft. lbs. (123 Nm).

    c. Loosen all bolts completely.

    d. Tighten all bolts, in sequence, to 25–33 ft. lbs. (34–44 Nm).

    e. Tighten all bolts, in sequence, to 90 ft. lbs. (123 Nm). If using an angle torque wrench, tighten them 70 degrees tighter rather than going to 90 ft. lbs. (123 Nm).

    f. Tighten the 6mm **X** bolts to 7–9 ft. lbs. (10–12 Nm). There is one of these bolts per head.

15. Install the remaining components in the opposite order from which they were removed.

16. Fill the cooling system to the proper level and connect the negative battery cable.

17. Run the engine and make all the necessary engine adjustments.

18. Road test the vehicle for proper operation.

3.0L (VQ30DE) Engine

1. Relieve the fuel system pressure.

☀☀ CAUTION

The fuel injection system remains under pressure after the engine has been turned OFF. Properly relieve fuel pressure before disconnecting any fuel lines. Failure to do so may result in fire or personal injury.

2. Disconnect the negative battery cable.
3. Drain the engine oil and the cooling system. Be sure to drain the engine block and the radiator.

➡**Before detaching any hoses or connectors, note the locations for reassembly.**

4. Remove the intake manifold collector.
5. Remove the fuel tube.
6. Remove the intake manifold.
7. Remove the cylinder head covers.
8. Remove the drive belts and idler pulley.
9. Remove the steel (lower) and aluminum (upper) oil pans.

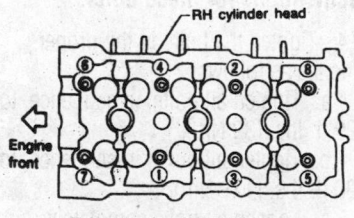

Right cylinder head bolt tightening sequence—3.0L (VQ30DE) engine

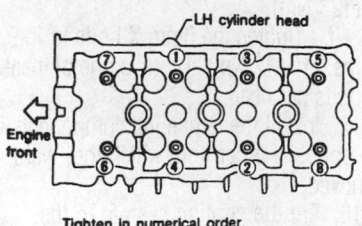

Left cylinder head bolt tightening sequence—3.0L (VQ30DE) engine

10. Remove the water pump cover.
11. Remove the timing chain case cover.
12. Remove the timing chains, camshaft sprockets and related components.
13. Remove the crankshaft sprocket.
14. Loosen the bolts that secure the rear timing chain case. The bolts must be loosened in the reverse order of installation sequence.
15. Using seal cutter tool, remove the rear timing case cover.

➡**Remove the O-rings from the front of the engine block.**

16. Remove the camshafts.
17. Remove the cylinder head bolts in the reverse order of the tightening sequence. The bolts should be loosened in 2–3 steps.

➡**A warped or cracked cylinder head could result from removing the bolts in incorrect order.**

18. Remove the cylinder heads from the vehicle.
19. Remove and discard the head gaskets.
20. Remove all traces of liquid gasket from the timing chain case and from the water pump covers.
21. Remove all traces of liquid gasket from the engine block.
22. Inspect the timing chain for excessive wear or damage and replace as necessary.

To install:
23. Turn the crankshaft until the No. 1 piston is set 240 degrees before TDC on compression stroke.
24. Using new head gaskets, install the cylinder heads.

➡**If possible, replacement of the head bolts is suggested.**

25. If replacement of the head bolts is not possible, perform the following bolt measurement:

   a. Measure the diameter of the head bolt 0.43 in. (11mm) from the bottom of the bolt.

   b. Measure the diameter of the head bolt 1.89 in. (48mm) from the bottom of the bolt.

   c. Whenever the size difference between the two measurements exceeds 0.0043 in. (0.11mm) the head bolts must be replaced.

26. Install the cylinder head bolts and tighten in sequence as follows:

   a. Tighten all bolts in sequence to 72 ft. lbs. (98 Nm).

   b. Completely loosen all bolts.

   c. Tighten all bolts in sequence to 25–33 ft. lbs. (24–44 Nm).

   d. Turn all bolts in sequence 90–95 degrees clockwise.

   e. Turn all bolts in sequence 90–95 degrees clockwise.

27. Install the camshafts and related components.
28. Install new O-rings to the front of the engine block.
29. Apply sealant to the hatched portion of the of the rear timing chain case.
30. Align the rear timing chain case with the dowel pins and install onto the cylinder heads and engine block.
31. Tighten the rear timing chain case mounting bolts in sequence to 105–121 inch lbs. (11.8–13.7 Nm).
32. The remaining components are installed in the reverse order from which they were removed.
33. Connect the negative battery cable.
34. Start the engine and run at 3000 RPM under no load to purge the air from the high pressure chamber. The engine may produce a rattling noise. This indicates that air still remains in the chamber and is not a matter of concern.
35. Verify that there are no leaks.

Rocker Arms

REMOVAL & INSTALLATION

Except 2.0L Engine

Nissan engines, with the exception of the 2.0L engine, do not utilize rocker arms. The valves are actuated directly by the camshafts.

2.0L Engine

1. Release the fuel pressure following the proper procedure.
2. Disconnect the negative battery cable.
3. Remove the rocker arm cover, gasket, and the oil separator.
4. Remove the intake manifold supports, oil filter bracket, and the power steering pump.

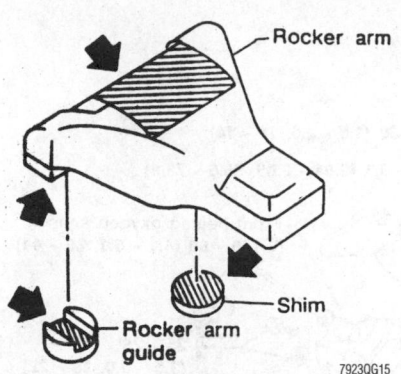

Rocker arm, guide and shim—2.0L engine

5. Set the No. 1 cylinder at TDC on the compression stroke.

6. Remove the timing chain tensioner from the side of the head.

7. Matchmark the position of the rotor and housing and remove the distributor.

8. Remove the timing chain guide. Remove the camshaft sprockets while holding the camshaft stationary with a large wrench. Secure the timing chain with wire so the timing is not lost. The front cover will have to be removed if the chain timing is lost.

➡**When removing the camshafts, loosen the journal caps in the opposite sequence of tightening. Camshaft damage may result if this step is not followed.**

9. Remove the camshafts, brackets, oil tubes, and the baffle plate. Label all components for proper installation.

➡**It is essential that all parts be kept in the same order and orientation for reinstallation. Be sure to mark and separate parts to keep them from getting mixed. This will aid assembly.**

10. Remove the rocker arms, shims, rocker arm guides, and the hydraulic lash adjusters. Label all components for proper installation.

➡**The valve lifters must be stored in the vertical position or submersed in clean oil to prevent air from entering the lifters.**

11. Inspect the surfaces of the rockers and replace if there are any signs of damage.

To install:

12. Lubricate the rocker arms, shims, rocker arm guides, and the hydraulic lash adjusters. Install them in their original locations.

13. Install the camshafts, brackets, oil tubes, and the baffle plate in the proper location.

14. Tighten the bolts in sequence as follows:

a. Tighten right camshaft bolts No. 9 and No. 10 to 17 inch lbs. (2 Nm).

b. Tighten right camshaft bolts No. 1-no. 8 to 17 inch lbs. (2 Nm).

c. Tighten left camshaft bolts No. 11 and No. 12 to 17 inch lbs. (2 Nm).

d. Tighten left camshaft bolts No. 1-no. 10 to 17 inch lbs. (2 Nm).

e. Tighten all camshaft bolts in numerical sequence to 52 inch lbs. (6 Nm).

f. Tighten all camshaft bolts in numerical sequence to 87–104 inch lbs. (10–11 Nm).

g. Tighten the rear two bolts of the LH camshaft to 13–19 ft. lbs. (18–25 Nm).

15. Install the camshaft sprockets while holding the camshaft stationary with a large wrench.

16. The remaining components are installed in the reverse order from which they were removed.

17. Connect the negative battery cable.

18. Check and adjust the ignition and valve timing. If there is air in the lifters, bleed the air by running the engine at 1000 rpm for 10 minutes.

Turbocharger

REMOVAL & INSTALLATION

Right Side

1. Remove the right side of the cowl panel and remove the battery.

2. Remove the air inlet hoses and pipes from the turbocharger.

3. Disconnect the lower pipe from the turbocharger.

4. Remove the automatic speed control bracket with wiper motor and solenoid valves.

5. Detach the oxygen sensor harness connector.

6. Disconnect the water tubes and oil inlet from the turbocharger. Cap the tubes to prevent contaminants from entering the system

7. Remove the front exhaust tube and the three-way catalyst assembly.

8. Remove the oil pressure switch.

9. Remove the oil filter.

10. Remove the turbocharger oil return tube. Cap the tube to prevent contaminants from entering the oil.

11. Disconnect the oil hose from the oil filter bracket.

12. Remove the rod pin from the wastegate valve actuator.

13. Remove the oil filter bracket.

14. Unbend the locking plates for turbocharger fastening nuts.

15. Remove the turbocharger fastening nuts and remove the turbocharger from the vehicle.

To install:

16. Using new gaskets, install the turbocharger unit. Tighten the turbocharger attaching nuts to 20–23 ft. lbs. (27–31 Nm).

17. Bend the locking tabs around the nuts that secure the turbocharger.

18. Reconnect the oil filter bracket. Tighten the bracket attaching bolts to 12–15 ft. lbs. (16–21 Nm).

19. Install the rod pin of wastegate valve actuator.

20. Using new gaskets, reconnect the oil hose to the turbocharger.

21. Using new gaskets, reconnect the oil return and water return tubes. Tighten the tube attaching bolts and nuts to 12–15 ft. lbs. (16–21 Nm).

22. Install a new oil filter.

23. Install the oil pressure switch.

24. Using a new gasket, install the front exhaust tube and three-way catalyst assembly. Tighten the catalyst-to-turbocharger bolts to 18–22 ft. lbs. (25–29 Nm).

25. Using new gaskets, install the water and oil inlet tubes. Tighten the nuts to 11–14 ft. lbs. (15–20 Nm).

26. The balance of installation is the reverse of the removal procedure.

27. Check the oil and coolant levels; add oil or coolant if necessary.

28. Start the engine and check the turbocharger for proper operation.

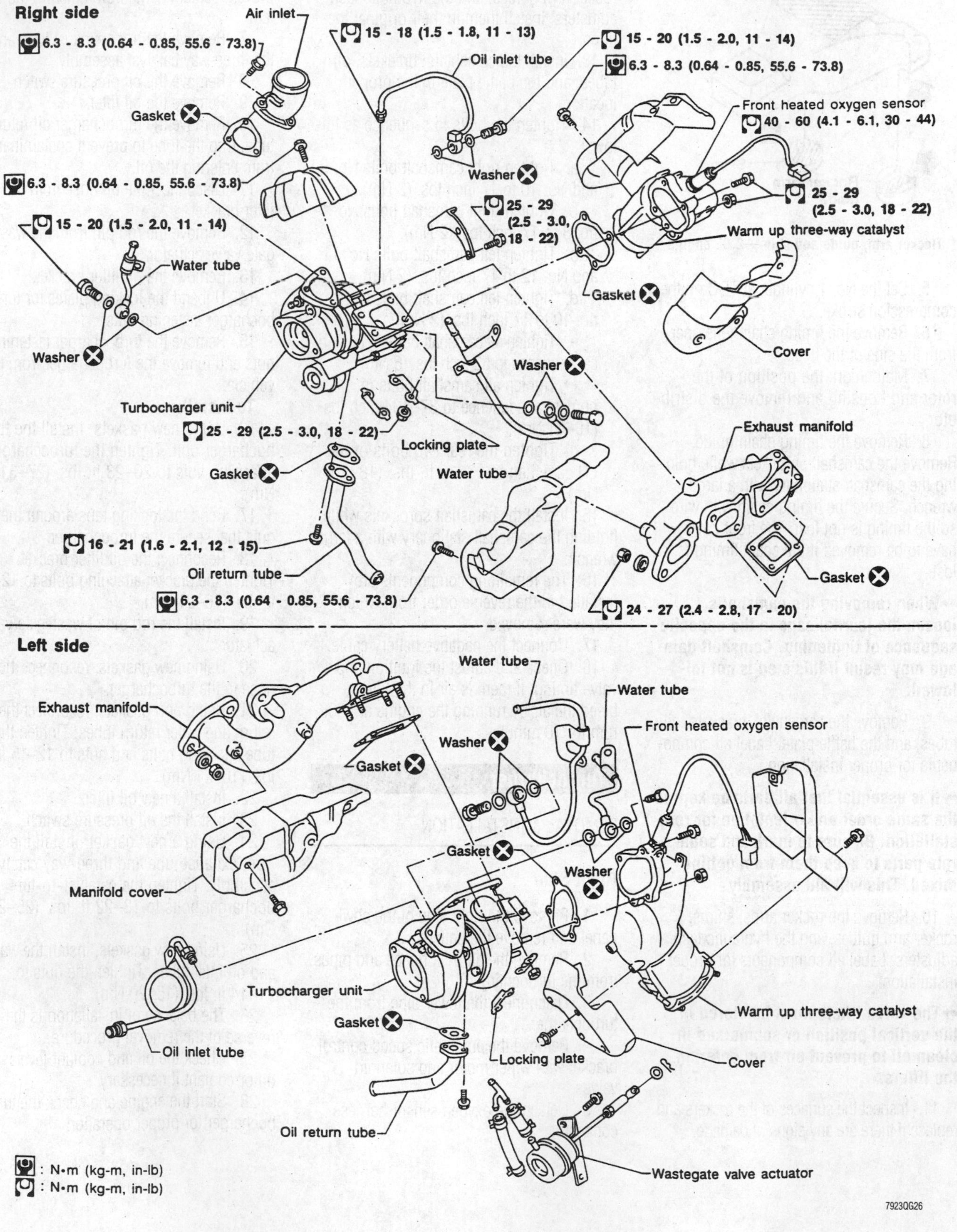

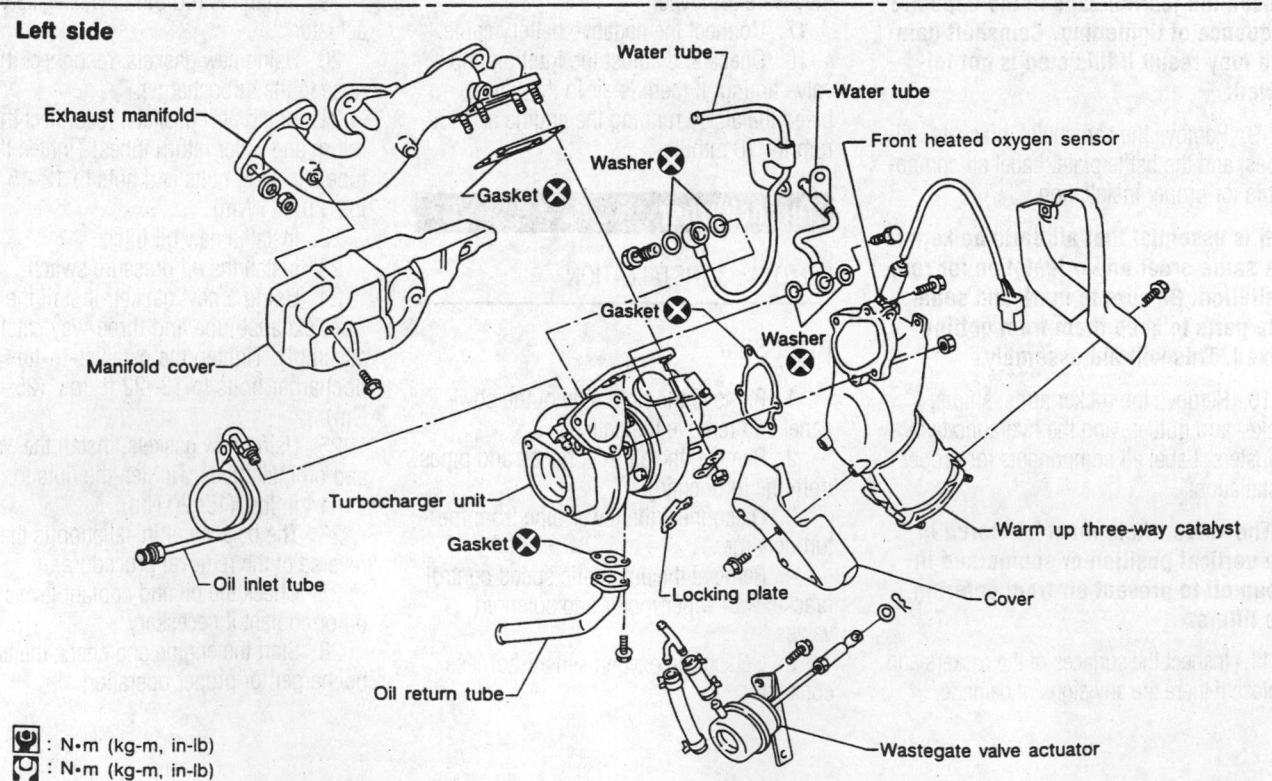

Right side

🔧 6.3 - 8.3 (0.64 - 0.85, 55.6 - 73.8)

Air inlet

🔧 15 - 18 (1.5 - 1.8, 11 - 13)

Oil inlet tube

🔧 15 - 20 (1.5 - 2.0, 11 - 14)

🔧 6.3 - 8.3 (0.64 - 0.85, 55.6 - 73.8)

Gasket ⊗

Front heated oxygen sensor

🔧 40 - 60 (4.1 - 6.1, 30 - 44)

🔧 6.3 - 8.3 (0.64 - 0.85, 55.6 - 73.8)

Washer ⊗

🔧 25 - 29 (2.5 - 3.0, 18 - 22)

🔧 15 - 20 (1.5 - 2.0, 11 - 14)

Water tube

🔧 25 - 29 (2.5 - 3.0, 18 - 22)

Warm up three-way catalyst

Washer ⊗

Washer ⊗

Gasket ⊗

Cover

Exhaust manifold

Turbocharger unit

🔧 25 - 29 (2.5 - 3.0, 18 - 22)

Locking plate

Water tube

Gasket ⊗

🔧 16 - 21 (1.6 - 2.1, 12 - 15)

Oil return tube

🔧 6.3 - 8.3 (0.64 - 0.85, 55.6 - 73.8)

Gasket ⊗

🔧 24 - 27 (2.4 - 2.8, 17 - 20)

Left side

Water tube

Exhaust manifold

Water tube

Washer ⊗

Front heated oxygen sensor

Gasket ⊗

Gasket ⊗

Washer ⊗

Manifold cover

Turbocharger unit

Warm up three-way catalyst

Gasket ⊗

Oil inlet tube

Locking plate

Cover

🔧 : N•m (kg-m, in-lb)
🔧 : N•m (kg-m, in-lb)

Oil return tube

Wastegate valve actuator

7923QG26

Exploded view of the turbocharger mounting—3.0L (VG30DETT) engine

Left Side

1. Remove the master cylinder and power brake booster.

2. Detach the oxygen sensor electrical connector.

3. Remove the air inlet hose and pipe.

4. Disconnect the lower pipe from the turbocharger.

5. Disconnect the water tubes and oil inlet tube from the turbocharger. Cap the tube to prevent contaminants from entering the system

6. Remove the front exhaust pipe and three way catalyst assembly.

7. Disconnect the lower steering joint.

8. Disconnect the turbocharger oil return tube. Cap the tube to prevent contaminants from entering the oil.

9. Disconnect the EGR tube and the turbocharger wastegate valve actuator bracket.

10. Remove the exhaust manifold heat shield.

11. Remove the left side exhaust manifold attaching nuts.

12. Remove the left side exhaust manifold and turbocharger as an assembly.

To install:

13. Using new gaskets, install the left side exhaust manifold and turbocharger assembly. Tighten the exhaust manifold bolts to 17–20 ft. lbs. (24–27 Nm).

14. Install the exhaust manifold heat shield.

15. Reconnect the EGR tube and the turbocharger wastegate valve actuator bracket.

16. Using new gaskets, reconnect the oil return tube. Tighten the tube attaching bolts to 12–15 ft. lbs. (16–21 Nm).

17. Reconnect the lower steering joint.

18. Using new gaskets, install the front exhaust pipe and three way catalyst assembly. Tighten the catalyst to turbocharger bolts to 18–22 ft. lbs. (25–29 Nm).

19. Using new gaskets, reconnect the water tubes and oil inlet tube to the turbocharger. Tighten the nuts to 11–14 ft. lbs. (15–20 Nm).

20. Install the remaining components in the reverse order of removal.

21. Start the engine and check the turbocharger for proper operation.

Intake Manifold

REMOVAL & INSTALLATION

1.6L Engine

1995 MODELS

❄ CAUTION

The fuel injection system remains under pressure after the engine has been turned OFF. Properly relieve fuel pressure before disconnecting any fuel lines. Failure to do so may result in fire or personal injury.

1. Relieve the fuel system pressure, disconnect the negative battery cable, and drain the cooling system.

2. Remove the air cleaner assembly.

3. Disconnect and tag the throttle linkage, electrical connections, fuel and vacuum lines from the throttle body or throttle chamber.

4. The throttle body/throttle chamber can be removed from the manifold at this point or can be removed as an assembly with the intake manifold.

5. Remove the bolts holding the upper portion of the intake to the lower portion. Remove the bolts in reverse order of the tightening sequence.

6. Remove the upper portion of the intake.

7. Loosen the intake manifold retaining bolts in the proper sequence and separate the manifold from the cylinder head. Remove the bolts in reverse order of the tightening sequence

8. Remove the intake manifold gasket and clean all the gasket contact surfaces thoroughly with a gasket scraper and suitable solvent. All traces of old gasket material must be removed to ensure proper sealing. Inspect the intake manifold for cracks. Using a metal straightedge, check the surface of the intake manifold for warpage.

To install:

9. Lay the new intake manifold gasket onto the cylinder head and position the lower intake manifold over the mounting studs and onto the gasket. Install the mounting nuts and tighten them to 12–15 ft. lbs. (16–21 Nm) in sequence.

10. Using a new gasket, install the upper portion of the intake manifold and tighten the bolts to 12–15 ft. lbs. (16–21 Nm) in sequence.

11. If removed, install the throttle body or throttle chamber and tighten the mounting bolts in a crisscross pattern. Tighten the bolts in two progressive steps to 15 ft. lbs. (21 Nm).

➡ **Be sure the gasket for the throttle body is positioned properly.**

12. Install the remaining components in the opposite order from which they were removed.

13. Fill the cooling system to the proper level and connect the negative battery cable.

14. Start the engine, bleed the cooling system, and check for leaks.

15. Road test the vehicle for proper operation.

1996–99 MODELS

❄ CAUTION

The fuel injection system remains under pressure after the engine has been turned OFF. Properly relieve fuel pressure before disconnecting any fuel lines. Failure to do so may result in fire or personal injury.

1. Relieve the fuel system pressure, disconnect the negative battery cable, and drain the cooling system.

2. Remove the air cleaner assembly.

3. Disconnect and tag the throttle linkage, electrical connections and vacuum lines from the throttle body.

4. Remove the intake manifold collector support brackets.

5. The throttle body can be removed from the manifold at this point or can be removed as an assembly with the intake manifold.

6. Remove the bolts holding the upper portion of the intake to the lower portion. Remove the bolts in reverse order of the tightening sequence.

7. Remove the upper portion of the intake.

8. Detach the fuel injector wiring harness connectors and the vacuum line from the fuel pressure regulator.

9. Disconnect the fuel hoses from the fuel rail assembly.

10. Remove the bolts that secure the fuel rail to the intake.

11. Remove the injectors with the fuel rail assembly.

12. Loosen the intake manifold retaining bolts in the proper sequence and separate the manifold from the cylinder head. Remove the bolts in reverse order of the tightening sequence

13. Remove the intake manifold gasket and clean all the gasket contact surfaces thoroughly with a gasket scraper and suitable solvent. All traces of old gasket material must be removed to ensure proper sealing. Inspect the intake manifold for cracks. Using a metal straightedge, check the surface of the intake manifold for warpage.

To install:

14. Lay the new intake manifold gasket onto the cylinder head and position the lower intake manifold over the mounting studs and onto the gasket. Install the mounting nuts and bolts; tighten them to 13–15 ft. lbs. (18–21 Nm) in sequence.

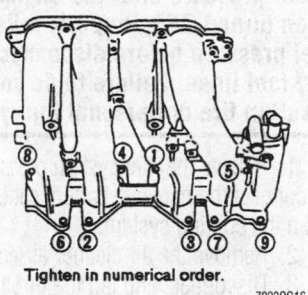

Tighten in numerical order.

79230G16

Tighten the lower intake manifold bolts in the order shown—1.6L engines

15. Install the injectors with the fuel rail assembly. Be sure to install the fuel rail insulators.

16. Install the bolts that secure the fuel rail to the intake. Tighten the bolts in two steps to 13–15 ft. lbs. (18–21 Nm).

17. Connect the fuel injector wiring harness connectors and the vacuum line from the fuel pressure regulator.

18. Using new hose clamps, connect the fuel hoses from the fuel rail assembly.

19. Using a new gasket, install the upper portion of the intake manifold and tighten the bolts to 13–15 ft. lbs. (18–21 Nm) in sequence.

20. If removed, install the throttle body or throttle chamber and tighten the mounting bolts in a crisscross pattern. Tighten the bolts in two progressive steps to 13–16 ft. lbs. (18–22 Nm).

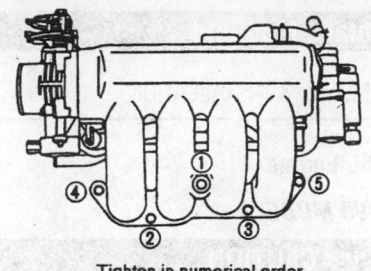

Tighten in numerical order.

79230G17

To prevent leaks or damage to the manifold tighten the upper intake manifold bolts in the order shown—1.6L engines

➡**Be sure to properly position the throttle body gasket with the cut out facing down.**

21. Install the intake manifold collector support brackets.

22. Connect the throttle linkage, electrical connections and vacuum lines.

23. Install the air cleaner.

24. Fill the cooling system to the proper level and connect the negative battery cable.

25. Start the engine, bleed the cooling system, and check for leaks.

26. Road test the vehicle for proper operation.

2.0L Engine

✳✳ CAUTION

The fuel injection system remains under pressure after the engine has been turned OFF. Properly relieve fuel pressure before disconnecting any fuel lines. Failure to do so may result in fire or personal injury.

1. Relieve the fuel system pressure, disconnect the negative battery cable, and drain the cooling system.

2. Remove the air cleaner assembly.

3. Disconnect the manifold support brackets.

4. Detach the throttle linkage, electrical connections, and vacuum lines from the throttle body. Be sure to note the locations of all connections.

5. Remove the EGR tube from the manifold.

6. Unbolt and remove the fuel rail assembly.

7. Remove the drive belts and water pump pulley.

8. Remove the alternator and power steering pump.

9. Remove the oil filter bracket and power steering bracket.

10. Loosen the intake manifold collector retaining bolts in the reverse of installation sequence and separate the collector from the manifold.

11. Loosen the intake manifold assembly retaining bolts in the reverse order of the tightening sequence and separate the manifold from the cylinder head.

12. Remove all gasket material and clean all the gasket contact surfaces thoroughly with a gasket scraper and a suitable solvent. All traces of old gasket material must be removed to ensure proper sealing. Inspect the intake manifold for cracks. Using a metal straightedge, check the surface of the intake manifold for warpage.

To install:

13. Using new gaskets, install the intake manifold assembly to the cylinder head and tighten the mounting bolts in sequence to 13–15 ft. lbs. (18–21 Nm).

14. Using new gaskets, install the intake manifold collector to the intake manifold assembly tighten the mounting bolts and nuts in sequence to 13–15 ft. lbs. (18–21 Nm).

15. Install the remaining components in the reverse order of removal.

16. Fill the cooling system to the proper level and connect the negative battery cable.

17. Bleed the cooling system and road test the vehicle for proper operation.

Tighten in numerical order.

79230G18

Tighten the lower intake manifold bolts in the sequence shown—200SX with 2.0L engines

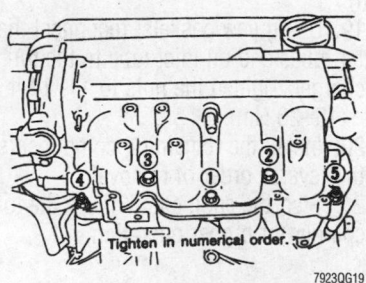

Tighten in numerical order.

79230G19

The upper intake manifold collector bolts must be tightened in the correct sequence—200SX with 2.0L engines

2.4L Engine

1. Relieve the fuel system pressure, disconnect the negative battery cable and drain the cooling system.

✳✳ CAUTION

The fuel injection system remains under pressure after the engine has been turned OFF. Properly relieve the pressure before disconnecting the fuel lines. Failure to do so may result in fire or personal injury.

2. Remove the air duct between the air flow meter and the throttle body.

3. Disconnect the throttle cable and the cruise control cable, if equipped.

4. Disconnect the fuel supply and return lines from the fuel injector assembly. Plug the lines to prevent leakage.

5. Detach and tag the electrical connectors and the vacuum hoses to the throttle body and intake manifold/collector assembly.

6. Remove the spark plug wires from the spark plugs.

7. Remove the throttle body assembly from the intake manifold.

8. Disconnect the EGR valve tube from the exhaust manifold.

9. Remove the intake manifold mounting brackets.

10. Remove the intake manifold collector to the intake manifold bolts/nuts in the reverse sequence of the tightening procedure and separate the intake manifold from the intake manifold collector.

11. Remove the bolts that secure the intake manifold to the cylinder head and remove the manifold. Be sure to loosen the bolts in the reverse sequence of the tightening procedure.

12. Using a putty knife or equivalent, clean the gasket mounting surfaces. Check the intake manifold/collector for cracks and warpage.

To install:

13. Using new gaskets, install the intake manifold to the cylinder head and tighten the mounting bolts in sequence to 12–14 ft. lbs. (16–19 Nm).

14. Using new gaskets, install the intake manifold collector to the intake manifold and tighten the mounting bolts/nuts in sequence to 12–14 ft. lbs. (16–19 Nm).

15. Install the intake manifold mounting brackets.

16. Connect the EGR valve tube to the exhaust manifold.

17. Using a new gasket, install the throttle body and tighten the mounting bolts in a

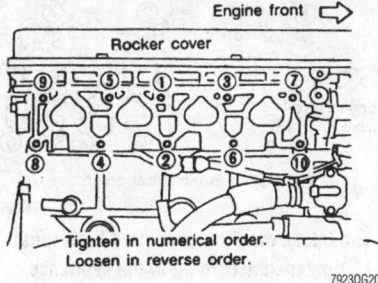

Be sure to tighten the intake manifold bolts in the order shown—2.4L engines

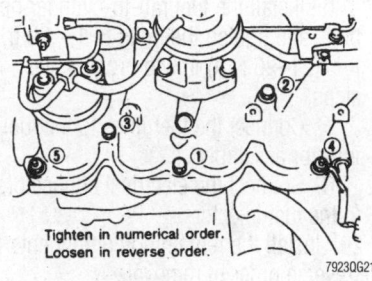

Tighten the intake manifold collector mounting bolts in the proper sequence—2.4L engines

crisscross pattern to 13–16 ft. lbs. (18–22 Nm). Be sure to tighten the bolts in two progressive steps.

18. The balance of the installation is the reverse of the removal procedure.

19. Fill the cooling system to the proper level and connect the negative battery cable.

20. Make all the necessary engine adjustments. Road test the vehicle for proper operation.

3.0L (VG30DE and VG30DETT) Engines

1. Disconnect the negative battery cable, drain the cooling system, and release the fuel system pressure.

2. Detach the electrical connectors from the throttle position sensor, exhaust gas temperature sensor, coolant temperature sensor, etc.

3. Label and disconnect the hoses from the throttle body, the EGR valve, the EGR control solenoid valve, the intake manifold collector, the power control solenoid valve and the power valve actuator (if equipped with a manual transaxle).

4. Disconnect the accelerator cable from the throttle body.

5. Remove the intake manifold collector-to-intake manifold bolts and remove the intake manifold collector.

6. Detach the electrical connectors from the ignition coils.

7. Detach the electrical connector from the crank angle sensor and the power transistor.

8. Detach the electrical connectors from the fuel injectors.

9. Disconnect the fuel injector assembly from the fuel lines.

10. Remove the fuel rail-to-cylinder head bolts.

11. Remove the fuel rail assembly from the engine.

12. Remove the intake manifold-to-engine bolts, in sequence, by reversing the tightening sequence.

13. Lift the intake manifold from the engine and discard the gasket.

To install:

14. Clean the gasket mounting surfaces.

15. Install the intake manifold and tighten the intake manifold-to-engine bolts and nuts in sequence.

16. Tighten the intake manifold bolts and nuts in two steps as follows:

   a. Tighten all bolt-and-nut combinations to 26–43 inch lbs. (3–5 Nm).

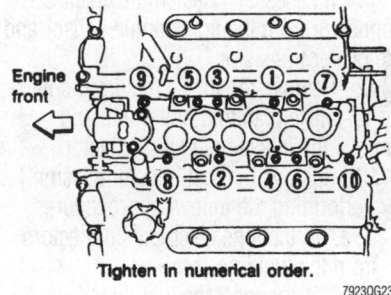

Intake manifold bolt tightening sequence—3.0L (VG30DE and VG30DETT) engines

   b. Tighten all bolts 12–14 ft. lbs. (16–20 Nm).

   c. Tighten all nuts to 17–20 ft. lbs. (24–27 Nm).

17. The balance of installation is the reverse of the removal procedure.

18. Using a new gasket, install the intake manifold collector and tighten the intake manifold collector-to-intake manifold bolts to 12–15 ft. lbs. (16–21 Nm).

19. Reconnect the negative battery cable.

20. Refill the cooling system.

21. Start the engine and check for leaks.

3.0L (VQ30DE) Engine

1. Disconnect the negative battery cable and drain the cooling system.
2. Release the fuel system pressure.

✳✳ CAUTION

The fuel injection system remains under pressure after the engine has been turned OFF. Properly relieve the fuel pressure before disconnecting any fuel lines. Failure to do so may result in fire or personal injury.

3. Remove the throttle body coolant hoses.
4. Label and detach the electrical connectors from the throttle position sensor.
5. Label and disconnect the hoses from the throttle body, the EGR valve, intake manifold collector, IAC valve, and the fuel pressure regulator.
6. Disconnect the canister purge hose and blow-by hose.
7. Disconnect the EGR guide tube.
8. Disconnect the accelerator cable from the throttle body.
9. Remove the intake manifold collector support brackets.
10. Detach the right side electrical connectors from the ignition coils.
11. If necessary, detach the electrical connector from the crank angle sensor and the power transistor.
12. Remove the intake manifold collector-to-intake manifold bolts/nuts and remove the intake manifold collector.
13. Remove the fuel injector assembly by performing the following procedures:
   a. Detach the electrical connectors from the fuel injectors.
   b. Disconnect the fuel lines from the fuel injector assembly.
   c. Remove the fuel rail-to-cylinder head bolts.
   d. Remove the fuel rail assembly from the engine.
14. Remove the intake manifold bolts/nuts in the reverse of the installation sequence.
15. Remove the intake manifold from the engine and discard the gaskets.
16. Clean all gasket mounting surfaces.
To install:
17. Using new gaskets, install the intake manifold to the engine.
18. Tighten the bolts/nuts in sequence as follows:
   a. Tighten nuts and bolts to 44–86 inch lbs. (5–10 Nm).
   b. Tighten nuts and bolts to 20–23 ft. lbs. (26–31 Nm).

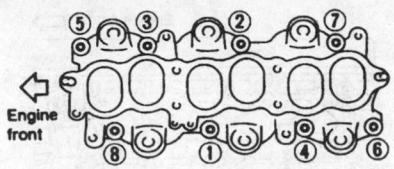

Tighten in numerical order.

79230G22

The intake manifold mounting bolts must be tightened according to the sequence shown—3.0L (VQ30DE) engine

19. Install the fuel injector assembly by performing the following procedures:
   a. Install the fuel rail assembly to the engine.
   b. Install the fuel rail-to-cylinder head bolts and tighten the bolts to 15–20 ft. lbs. (21–26 Nm) in two progressive steps.
   c. Connect the fuel lines to the fuel injector assembly.
   d. Connect the electrical connectors to the fuel injectors.
20. Install the remaining components in the reverse order of removal.
21. Refill the cooling system and connect the negative battery cable.
22. Start the engine, bleed cooling system, and check for leaks.

Exhaust Manifold

REMOVAL & INSTALLATION

1.6L and 2.0L Engine

1. Disconnect the negative battery cable. Raise and support the vehicle safely.
2. Remove the engine undercovers.
3. Remove the air cleaner or collector assembly.
4. Remove the heat shields from the manifold and front exhaust pipe.
5. Disconnect the front exhaust pipe from the exhaust manifold.
6. Remove or disconnect the temperature sensors, oxygen sensors, air induction pipes from the manifold.
7. Remove manifold support brackets.
8. Loosen and remove the exhaust manifold attaching nuts and remove the manifold from the block. Discard the exhaust manifold gaskets.
9. Clean the gasket surfaces and check the manifold for cracks and warpage.
To install:
10. Install the exhaust manifold with a new gasket. Tighten the manifold fasteners as follows from the center outward in several stages.

   a. GA16DE engines—Tighten mounting nuts with washers to 12–15 ft. lbs. (16–21 Nm).
   b. SR20DE engines—Tighten mounting nuts with washers to 27–35 ft. lbs. (37–48 Nm).
11. Install or connect the temperature sensors, oxygen sensors and air induction pipes.
12. Install the manifold support brackets.
13. Connect the exhaust pipe to the manifold using a new gasket. Tighten the manifold nuts to 21–25 ft. lbs. (28–33 Nm) for GA16DE engine models or 32–37 ft. lbs. (43–50 Nm) for SR20DE engine models.
14. Install the heat shields.
15. Install the air cleaner or collector assembly.
16. Install the engine undercovers.
17. Connect the negative battery cable.
18. Start the engine and check for exhaust leaks.

2.4L Engine

1. Disconnect the negative battery cable.
2. Raise and safely support the vehicle.

➡**Before loosening the exhaust pipe retaining nuts or bolts, soak the retaining nuts or bolts with penetrating oil.**

3. Disconnect the exhaust pipe from the exhaust manifold.

➡**On California models equipped with A/T and on all M/T equipped models, disconnect the exhaust pipe at the exhaust manifold collector.**

4. Detach the oxygen sensor electrical connector.
5. Remove the exhaust manifold cover.
6. Remove the EGR tube from the exhaust manifold.
7. Remove the exhaust manifold-to-engine bolts and nuts and discard the gaskets. Remove the retaining bolts and nuts in reverse of the tightening sequence.
8. Remove the exhaust manifold from the vehicle.
To install:
9. Clean all gasket mounting surfaces and install new gaskets.
10. Install the exhaust manifold to the engine and tighten the new nuts to 27–35 ft. lbs. (37–48 Nm). Tighten the bolts and nuts evenly in sequence until snug; now tighten the bolts and nuts in sequence to specification.
11. Install the EGR tube to the exhaust manifold, tighten the EGR tube nuts to 29–36 ft. lbs. (39–49 Nm).

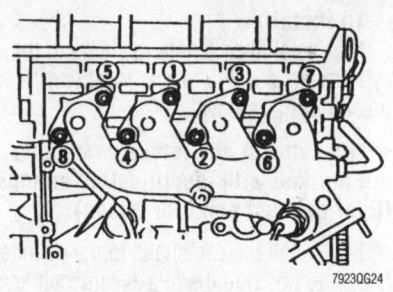

Exhaust manifold bolt tightening sequence—2.4L engine (except California models)

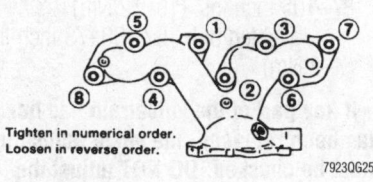

Tighten in numerical order.
Loosen in reverse order.

Exhaust manifold bolt tightening sequence—2.4L engine (California models)

12. Install the exhaust manifold cover.

13. Tighten the exhaust manifold cover bolts to 46–57 inch lbs. (5–7 Nm).

14. Connect the oxygen sensor electrical connector.

15. Using a new gasket, connect the exhaust pipe to the exhaust manifold and tighten the mounting nuts/bolts to 33–44 ft. lbs. (45–60 Nm).

16. Connect the negative battery cable.

17. Start the engine and check for exhaust leaks.

3.0L (VG30DE and VG30DETT) Engines

1. Disconnect the negative battery cable.

2. Raise and safely support the vehicle.

3. Disconnect the exhaust manifolds from the exhaust pipe.

4. Detach the oxygen sensor(s) electrical connections.

5. If equipped, remove the turbocharger assembly and three-way warm up catalyst from the exhaust manifold.

6. Remove the exhaust manifold cover(s).

7. If removing the left side exhaust manifold, remove the EGR tube.

➡Before loosening the exhaust manifold retaining nuts, soak the retaining nuts with penetrating oil.

8. Remove the retaining nuts from the exhaust manifold. Loosen all the retaining nuts evenly

9. Remove the manifold from the vehicle.

To install:

10. Clean all gasket mounting surfaces and install new gaskets.

11. Install the exhaust manifold to the engine and tighten the exhaust manifold-to-engine retaining nuts to 17–20 ft. lbs. (24–27 Nm) on non turbo engines, and tighten retaining nuts to 20–23 ft. lbs. (27–31 Nm) on turbo engines.

12. If equipped, install the turbocharger and three-way catalyst assembly to the exhaust manifold.

13. Tighten the turbocharger bolts to 32–40 ft. lbs. (43–54 Nm). Tighten the three-way catalyst bolts and nuts to 18–22 ft. lbs. (25–29 Nm).

14. Reconnect the oxygen sensor(s) electrical connectors.

15. If removed, reconnect the EGR tube to the EGR valve and the left side exhaust manifold. Tighten the EGR tube to EGR valve nut to 25–33 ft. lbs. (34–44 Nm).

16. Install the exhaust manifold covers.

17. Reconnect the exhaust pipe to manifold.

18. Lower the vehicle.

19. Connect the negative battery cable.

20. Start the engine and check for exhaust leaks.

3.0L (VQ30DE) Engine

1. Disconnect the negative battery cable.

2. Raise and safely support the vehicle.

➡If necessary, soak the exhaust pipe retaining nuts with penetrating oil to loosen them.

3. Disconnect the exhaust manifolds from the exhaust pipes.

4. Remove the protective covers from the manifolds.

5. Remove the exhaust manifold-to-engine mounting nuts. Remove the manifolds from the engine and discard the gaskets.

To install:

6. Clean all gasket mounting surfaces. Install new gaskets.

7. Install the exhaust manifold to the engine and tighten the mounting nuts in two progressive steps to 22–24 ft. lbs. (30–32 Nm).

8. Install the protective shields and tighten the mounting bolts in two progressive steps to 46–57 inch lbs. (5–7 Nm).

9. Install the exhaust manifolds to the exhaust pipes and tighten the mounting nuts to 32–37 ft. lbs. (43–50 Nm).

10. Connect the negative battery cable, start the engine, and check for exhaust leaks.

Front Crankshaft Seal

REMOVAL & INSTALLATION

➡The front crankshaft seal procedure is applicable to timing belt-equipped engines only. For the front seal on engines equipped with timing chains, refer to the timing chain, sprockets, front cover and seal procedure later in this section.

3.0L (VG30DE and VG30DETT) Engines

1. Disconnect the negative battery cable.

2. Position the engine to TDC of compression stroke.

3. Remove the engine undercover.

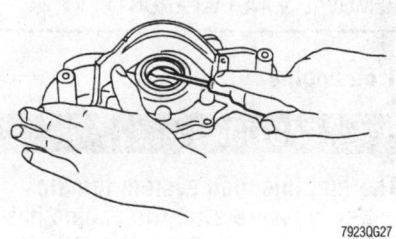

Carefully pry the crankshaft seal from the bore in the oil pump—3.0L (VG30DE and VG30DETT) engine

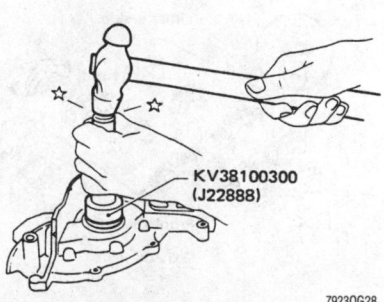

KV38100300
(J22888)

Use a suitable seal installer such as KV3800300 to install the seal in the oil pump—3.0L (VG30DE and VG30DETT) engine

4. Remove all accessory drive belts and remove the alternator from the vehicle.

5. Remove the crankshaft damper-to-crankshaft bolt and remove the damper.

6. Remove the timing belt covers and remove the timing belt.

7. Using a proper prying tool, remove the seal from the front oil pump housing.

✳✳ WARNING

Use care when prying the seal front the front cover so as not to damage the cover or crankshaft.

To install:

8. Using a proper seal driver, install the front crankshaft oil seal to the oil pump.

9. Install the timing belt and the timing belt covers.

10. Install the crankshaft damper and tighten the mounting bolt to 159–174 ft. lbs. (216–235 Nm).

11. Install the alternator assembly and install the engine drive belts.

12. Install the engine undercover.

13. Connect the negative battery cable and check all fluid levels.

14. Start engine, check ignition timing, and check for oil leaks.

Camshaft and Valve Lifters

REMOVAL & INSTALLATION

1.6L Engine

✳✳ CAUTION

The fuel injection system remains under pressure after the engine has been turned OFF. Properly relieve fuel pressure before disconnecting any fuel lines. Failure to do so may result in fire or personal injury.

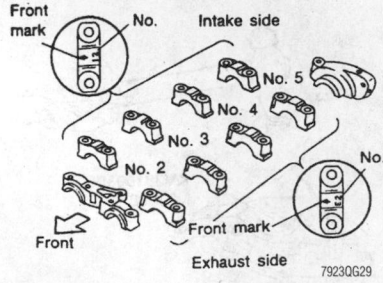

Be sure to install the camshaft bearing caps in their original positions—1.6L engine

1. Disconnect the negative battery cable, drain the cooling system, and relieve the fuel system pressure.

2. Remove all engine drive belts. Disconnect the exhaust pipe from the exhaust manifold.

3. Remove the power steering pulley and pump with the mounting bracket.

4. Remove the cylinder head cover.

5. Remove the distributor assembly.

6. Remove the timing chain tensioners and camshaft sprocket.

➡**Before the camshafts are removed from the cylinder head, note the positioning of the pins at the end of the camshafts for reassembly purposes.**

7. Remove the camshaft bearing caps in sequence and remove the camshafts from the cylinder head. Remove the idler sprocket bolt. These parts should be reassembled in their original position.

8. Remove the shims from the tops of the lifters. Be sure to note the position of each shim.

9. Remove the valve lifters from the bores in the cylinder head. Note the positioning of the lifters for reassembly.

10. Measure the diameter of the lifters. The diameter should be 1.1795–1.1801 in. (29.960–29.975mm).

11. Measure the diameter of the lifter bores. The diameter should be 1.1811–1.1819 in. (30.000–30.021mm).

12. Clearance between the lifter and bore should be 0.0010–0.0024 in. (0.025–0.061mm).

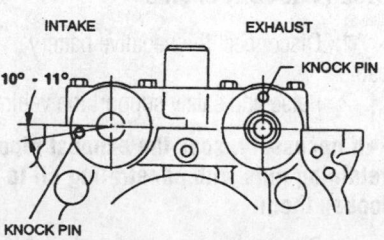

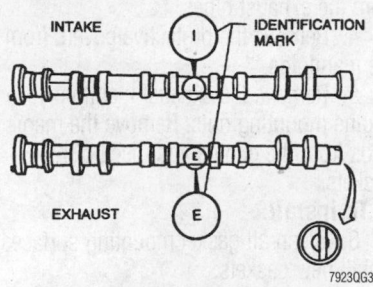

Positioning and identification of the camshafts—1.6L engine

To install:

13. Install the lifters and shims to the cylinder head in the proper locations as noted during removal.

➡**The exhaust and intake camshafts are marked with identification stamps. (E for exhaust and I for intake).**

14. Install the camshafts to the cylinder head and position the intake camshaft knock pin at the 9 o'clock position and the exhaust camshaft at the 12 o'clock position.

15. Install the camshaft bearing caps and tighten the mounting bolts as follows:

    a. Tighten bolts 11 through 15, then bolts 1 through 10 to 18 inch lbs. (2 Nm)

    b. Tighten bolts 1 through 15 to 53 inch lbs. (6 Nm)

    c. Tighten bolts 1 through 14 to 87–105 inch lbs. (10–12 Nm)

    d. Tighten bolt 15 to 56–73 inch lbs. (7–8 Nm)

➡**If any part of the valvetrain has been has been replaced, the valve adjustment must be checked. DO NOT adjust the valves or rotate the camshafts at this point. Internal engine damage will result.**

16. Install the camshaft sprockets with timing chains.

17. Install distributor assembly.

18. Check and adjust the valve clearance.

19. Install the cylinder head cover.

20. Install the remaining components. Refill and check all fluid levels.

21. Connect negative battery cable.

22. Start the engine, check for leaks, and adjust the ignition timing to specification.

23. Road test the vehicle for proper operation.

2.0L Engine

SENTRA

1. Disconnect the negative battery cable. Remove the rocker cover and oil separator.

2. Rotate the crankshaft until the No.1 piston is at TDC on the compression stroke. Then, rotate the crankshaft until the mating marks on the camshaft sprockets line up with the mating marks on the timing chain.

3. Remove the timing chain tensioner.

4. Remove the distributor.

5. Remove the timing chain guide.

6. Remove the camshaft sprockets. Use a wrench to hold the camshaft while loosening the sprocket bolt.

7. Loosen the camshaft bracket bolts in the opposite order of the tightening sequence.

8. Remove the camshaft.

To install:

9. Clean the left-hand camshaft end bracket and coat the mating surface with liquid gasket. Install the camshafts, camshaft brackets, oil tubes and baffle plate. Ensure the left camshaft key is at 12 o'clock and the right camshaft key is at 10 o'clock.

10. The procedure for tightening camshaft bolts must be followed exactly to prevent camshaft damage. Tighten bolts as follows:

   a. Tighten right camshaft bolts 9 and 10 (in that order) to 1.5 ft. lbs. (2 Nm), then tighten bolts 1–8 (in that order) to the same specification.

   b. Tighten left camshaft bolts 11 and 12 (in that order) to 1.5 ft. lbs. (2 Nm), then tighten bolts 1–10 (in that order) to the same specification.

   c. Tighten all bolts in sequence to 4.5 ft. lbs. (6 Nm).

   d. Tighten all bolts in sequence to 6.5–8.5 ft. lbs. (9–12 Nm) for type A, B and C bolts and 13–19 ft. lbs. (18–25 Nm) for type D bolts.

11. Line up the mating marks on the timing chain and camshaft sprockets and install the sprockets. Tighten sprocket bolts to 101–116 ft. lbs. (137–157 Nm).

12. The remaining components are installed in the reverse order from which they were removed.

13. Connect the negative battery cable. Refill all fluid levels. Road test the vehicle for proper operation.

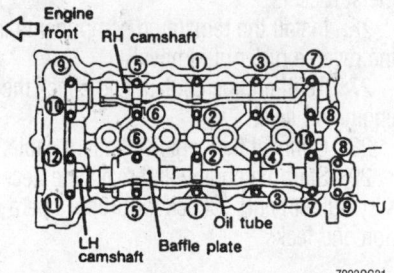

To prevent damage to the camshafts, tighten the bearing caps in the sequence shown—2.0L engines

200SX

❄❄ CAUTION

The fuel injection system remains under pressure after the engine has been turned OFF. Properly relieve fuel pressure before disconnecting any fuel lines. Failure to do so may result in fire or personal injury.

1. Release the fuel system pressure and disconnect the negative battery cable.

2. Raise and safely support the vehicle. Remove the engine undercovers.

3. Remove the right front wheel and the engine side cover.

4. Drain the cooling system and remove the radiator assembly.

5. Remove the air duct from the intake manifold.

6. Remove the drive belts and water pump pulley.

7. Remove the alternator and the power steering pump from the engine.

8. Note the locations for reassembly and remove the vacuum hoses, fuel hoses, wires, and the electrical connections.

9. Remove the distributor cap and ignition wires from the engine.

10. Remove all spark plugs.

11. Remove the rocker cover nuts in sequence.

12. Remove the rocker cover and oil separator.

13. Remove the intake manifold supports, oil filter bracket, and the power steering bracket.

14. Rotate the engine and set the No. 1 piston at TDC on the compression stroke. Rotate crankshaft until mating marks on camshaft sprockets are in the correct position.

15. Remove the timing chain tensioner.

16. Matchmark the position of the distributor rotor and housing to the engine block for reinstallation purposes. Unbolt and remove the distributor assembly.

17. Remove the timing chain guide.

➡**Wire the camshaft sprockets to the timing chain to maintain proper timing chain position.**

18. Holding the flats of the camshaft sprockets, remove the mounting bolts. Remove the sprockets from the camshafts.

➡**Note the positioning of the pins on the end of the camshafts for installation purposes.**

19. Remove the oil tubes, baffle plate, camshaft brackets and the camshafts. It is important that all parts are kept in order for correct installation.

➡**It is essential that the valvetrain components are kept in specific order for reassembly.**

20. Remove the rocker arms, shims, rocker arm guides and the hydraulic lash adjusters.

➡**When the lifters are removed, keep them straight up or soak them in clean engine oil to prevent air from entering them.**

21. Measure the diameter of the lifters. The diameter should be 0.6685–0.6690 in. (16.980–16.993mm).

22. Measure the diameter of the lifter bores. The diameter should be 0.6693–0.6701 in. (17.000–17.020mm).

23. Standard clearance between the lash adjuster and the guide hole should be 0.0003–0.0016 in. (0.007–0.040mm).

To install:

➡**The hydraulic lifters must be bled to remove the air from them. Air can not be bled from this type of lifter by running the engine.**

24. Bleed the lifters as follows:

   a. Submerse the lifter into a container of clean engine oil.

   b. While pushing the plunger, insert a thin rod into the check ball and lightly push the check ball.

   c. Air is completely bled when the plunger no longer moves.

25. Check all components for wear, replace as necessary and clean all mating surfaces.

➡**Apply clean engine oil to all components prior to installation. Always replace the rocker arm guide with a new one.**

26. At this point it is necessary to perform valve adjustment. Adjust the valves as follows:

➡**It will be necessary to determine the proper shim size when replacing the valve, cylinder head, shim, rocker arm guide, or the valve seat.**

27. Insert tool KV10115700 (J38957) with dial gauge into the lifter bore.

28. Before measuring, be sure the following parts are installed in the cylinder head.

   • Valve
   • Valve spring
   • Collet
   • Retainer
   • Rocker arm guide (except shim)

29. On the shim side, measure the difference between contact surfaces of the rocker arm guide and the valve stem end.

➡**When measuring, lightly pull dial indicator rod toward you to eliminate play in tool.**

30. Using this reading, select the proper shim size.

31. Shims are available in thickness' from 0.1102–0.1260 in. (2.800–3.200mm) in steps of 0.0010 in. (0.025mm).

32. Measure all the valves and select the proper shim sizes.

33. Remove the tool from cylinder head.

34. Install the valve lifters to the bores in the cylinder head.

35. Install the rocker arm guides, shims, and the rocker arms to the cylinder head.

36. Clean the left-hand camshaft end bracket and coat the mating surface with liquid gasket. Install the camshafts, camshaft brackets, oil tubes, and the baffle plate.

✳✳ WARNING

Ensure that the left camshaft key is at the 12 o'clock position and the right camshaft key is also at the 12 o'clock position.

37. The procedure for tightening camshaft bolts must be followed exactly to prevent camshaft damage. Tighten bolts as follows:

a. Tighten the right camshaft bolts No. 9 and 10 (in that order) to 18 inch lbs. (2 Nm), then tighten bolts 1–8 (in that order) to the same specification.

b. Tighten the left camshaft bolts No. 11 and 12 (in that order) to 18 inch lbs. (2 Nm), then tighten bolts 1–10 (in that order) to the same specification.

c. Tighten all bolts in sequence to 51 inch lbs. (6 Nm).

d. Tighten all bolts in sequence to 78–102 inch lbs. (9–12 Nm), then tighten the two bolts that secure the distributor housing cap to 13–19 ft. lbs. (18–25 Nm).

38. Line up the mating marks on the timing chain and camshaft sprockets and install the timing chain and sprockets. Tighten sprocket bolts to 101–116 ft. lbs. (137–157 Nm). Be sure to hold the flats of the camshaft when tightening the sprocket bolts.

39. The balance of installation is the reverse of the removal procedure.

40. Run the engine and reset the ignition timing.

41. Road test the vehicle for proper operation.

2.4L Engine

ALTIMA

1. Relieve the fuel system pressure.

✳✳ CAUTION

The fuel injection system remains under pressure after the engine has been turned OFF. Properly relieve fuel pressure before disconnecting any fuel lines. Failure to do so may result in fire or personal injury.

2. Disconnect the negative battery cable.

3. Drain coolant from the engine and radiator.

4. Remove the air intake ducts and the air cleaner assembly.

5. Remove the vacuum hoses, fuel hoses, wires, harness, and connectors that are necessary for removal of the rocker cover.

6. Remove the alternator and mounting bracket.

7. Remove the upper radiator hose and cooling fan.

8. Set the No. 1 piston at TDC on its compression stroke.

9. Disconnect the spark plug wires from the spark plugs.

10. Matchmark and note the positioning of the distributor rotor and housing to the engine block.

11. Remove the distributor assembly.

12. Remove the rocker cover by loosening the bolts in the reverse order of installation.

13. Wire the chain to the sprocket so the chain does not fall off during sprocket removal. Hold the flats of the camshaft with a wrench just behind the first camshaft bearing cap. Loosen the bolts and remove the sprockets.

14. Remove the camshaft sprockets.

➡**The stoppers on camshaft covers prevent the upper timing chain from disengaging from the idle sprocket. Also, after removal of the camshaft sprockets, note the positioning of the pins at the end of the camshafts for reinstallation purposes.**

15. Remove the camshaft bearing caps in reverse order of installation, then remove the camshafts. The camshaft brackets must be loosened in the correct sequence to prevent damage to the camshaft.

➡**All the valvetrain components must be reassembled in their original positions.**

16. Remove the valve lifter adjusting shims from the tops of the of the lifters. Be sure to note the location and positioning of each shim.

17. Remove the valve lifters from the bores in the cylinder heads. Be sure to note the location and positioning of each lifter.

18. Check the diameter of the valve lifter and the valve lifter bore and compare to the following specifications.

19. The valve lifter diameter should be 1.3370–1.3376 in. (33.960–33.975mm).

20. The lifter guide bore diameter should be 1.3386–1.3394 in. (33.960–33.975mm).

21. The valve lifter to lifter guide bore clearance should be 0.0010–0.0024 in. (0.025–0.061mm).

To install:

➡**When installing the valve components, apply a coat of clean engine oil to the component.**

22. Install the lifters into the lifter bores from which they were removed.

23. Install the valve shims to the lifters from which they came.

24. Install the camshafts in the same position as noted during removal and camshaft bearing caps; tighten cap bolts in the proper sequence as follows:

a. Tighten all bolts to 17 inch lbs. (2 Nm).

b. Tighten all bolts to 81–104 inch lbs. (9–12 Nm).

➡**When installing the timing chain and sprockets, align the marks on the sprockets with the colored links of the chain.**

25. Install the camshaft sprockets with the timing chain and tighten the sprocket bolts to 123–130 ft. lbs. (167–177 Nm). Install the chain guide between both camshaft sprockets. The alignment marks on the upper portion of the timing chain should now be aligned with the marks on the sprockets.

26. Install the remaining components in the reverse order of removal.

27. Refill engine coolant and check the engine oil level.

28. Connect the negative battery cable.

29. Start the engine and make the necessary adjustments. Check for proper operation and leaks.

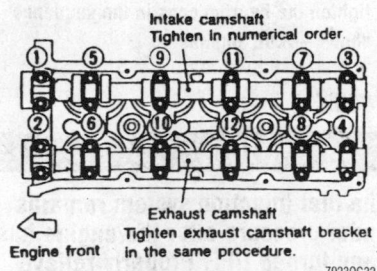

Tighten the camshaft bearing caps in sequence to prevent damage to the camshaft and cylinder head—2.4L engines

240SX

1. Disconnect negative battery cable.
2. Release fuel system pressure.

✳✳ CAUTION

The fuel injection system remains under pressure after the engine has been turned OFF. Properly relieve fuel pressure before disconnecting any fuel lines. Failure to do so may result in fire or personal injury.

3. Remove the air intake ducts and the air cleaner assembly.
4. Remove and tag vacuum hoses from valve cover. The fuel rail may need to be removed to access valve cover nuts.
5. Remove the cooling fan and the radiator fan shroud.
6. Remove spark plug wires.
7. Remove the valve cover and turn the crankshaft to align the timing marks at TDC on No. 1 cylinder.
8. Matchmark the position of the distributor rotor and the distributor. Remove the distributor assembly from the vehicle.
9. Remove the upper timing chain cover.
10. Remove the upper timing chain guide and tensioner.
11. Wire the chain to the sprocket so the chain does not fall off during sprocket removal. Hold the flats of the camshaft with a wrench just behind the first camshaft bearing cap. Loosen the bolts and remove the sprockets.

➡**The stoppers on camshaft covers prevent the upper timing chain from disengaging from the idle sprocket.**

➡**After removal of the camshaft sprockets, note the positioning of the pins at the end of the camshafts for reinstallation purposes.**

12. Remove the cam bearing caps in reverse order of the tightening sequence and remove the camshafts. The camshaft brackets must be loosened in the correct sequence to prevent damage to the camshaft.

➡**These parts should be reassembled in their original position. Bolts should be loosened in 2–3 steps (loosen all bolts in the reverse of the tightening order).**

13. Remove the valve lifter adjusting shims from the tops of the of the lifters.
14. Remove the valve lifters from the bores in the cylinder heads.
15. Check the diameter of the valve lifter and the valve lifter bore and compare to the following specifications.

16. The valve lifter diameter should be 1.3370–1.3376 in. (33.960–33.975mm).
17. The lifter guide bore diameter should be 1.3386–1.3394 in. (33.960–33.975mm).
18. The valve lifter to lifter guide bore clearance should be 0.0010–0.0024 in. (0.025–0.061mm).

To install:

➡**Apply a clean coat of new engine oil to the valvetrain components.**

19. Install the lifters into the lifter bores from which they were removed.
20. Install the valve shims to the lifters from which they came.
21. Install the camshafts in the same position as noted during removal and install the camshaft bearing caps.
22. Tighten bearing cap bolts in the proper sequence as follows:
   a. Tighten all bolts to 17 inch lbs. (2 Nm).
   b. Tighten all bolts to 81–104 inch lbs. (9–12 Nm).

➡**When installing the timing chain and sprockets, align the marks on the sprockets with the colored links of the chain.**

23. Install the camshaft sprockets and tighten the bolts to 123–130 ft. lbs. (167–177 Nm).
24. Install the timing chain, then install the remaining components in the opposite order from which they were removed.
25. Connect negative battery cable.
26. Start the engine and make the necessary adjustments. Check for proper operation and leaks.

3.0L (VG30DE and VG30DETT) Engines

✳✳ WARNING

After the timing belt has been removed, DO NOT rotate the crankshaft or camshafts. The valves and pistons will make contact thus causing severe engine damage.

1. Position the engine so that No. 1 piston is at TDC of the compression stroke.
2. Relieve the fuel system pressure and disconnect the negative battery cable.
3. Drain the cooling system. Be sure to drain the engine block.
4. Disconnect the accelerator cable from the throttle body linkage.
5. Disconnect the cruise control cable from the throttle body linkage.

6. Note the locations and disconnect the vacuum lines from the intake manifold collector.
7. Unbolt and remove the intake manifold collector from the intake manifold.

➡**After the intake manifold collector has been removed, cover the openings on the intake manifold to prevent foreign objects from entering the combustion chamber.**

8. Note the locations and disconnect the electrical wiring from the intake manifold and engine harness.
9. Remove the ignition coils and spark plugs from the engine.
10. Remove the mounting bolts from the fuel rail and remove the fuel injector pipe with injectors from the intake assembly.
11. Remove the valve covers.
12. Remove the radiator.
13. Remove the drive belts. Mark the drive belts for reinstallation purposes.
14. Remove the cooling fan and cooling fan coupling.
15. Mark the camshaft position sensor to the sensor mounting bracket and unbolt the sensor.
16. Remove the starter and lock the flywheel ring gear using a suitable locking device.
17. Remove the crankshaft pulley bolt.

➡**The engine is locked to prevent the crankshaft gear from turning during removal and installation. Remove the lock during engine timing portion of this procedure.**

18. Remove the crankshaft pulley using a suitable puller.
19. Remove the water inlet and outlet housings.
20. Remove the timing belt covers and gaskets.
21. Remove the crank angle sensor mounting bracket.
22. Install a suitable 6mm stopper bolt in the tensioner arm of the auto tensioner so the length of the pusher does not change.

➡**Mark the timing belt direction of rotation for reinstallation purposes.**

23. Remove the automatic tensioner and the timing belt.
24. Check the automatic tensioner for oil leaks in the pusher rod and diaphragm. If oil is evident, replace the automatic tensioner assembly.
25. Inspect the timing gear teeth for wear and if the sprockets show signs of wear they must be replaced.

26. Remove the idler pulley mounting nut, remove the idler pulley and remove the idler pulley stud.

27. Unbolt and remove the intake manifold assembly.

28. Disconnect the exhaust pipe from the exhaust manifold.

29. Loosen the cylinder head bolts (in reverse order of installation sequence) in 2–3 stages. Lift the cylinder head off the engine block with the exhaust manifolds attached. It may be necessary to tap the head lightly with a rubber mallet to loosen it.

30. Position the cylinder head on blocks of wood to protect the surface of the head and to protect the any valves that may extend below the surface of the cylinder head.

31. Remove the exhaust manifold from the cylinder head.

32. While holding the flats of the camshaft in position with a wrench, remove the mounting bolts from the camshaft sprockets.

➡️**Remove the front plate, O-ring and spring from the right (intake) camshaft to gain access to the sprocket bolt. The left camshaft sprocket is held in place by plate and four bolts.**

33. Note the positioning of the sprockets to the camshaft and remove the sprocket from the camshafts.

34. To remove the camshaft you will need to remove the rear timing belt cover which will be accessible after the camshaft sprockets are removed.

35. Mount a dial indicator and set the stylus of the indicator on the end of the camshaft. Zero the indicator and measure the camshaft end-play by moving the camshaft back and forth. End-play should be within 0.0012–0.0031 in. (0.03–0.08mm).

36. Remove the camshaft bearing caps. Loosen the bolts in the proper sequence (reverse the installation sequence) gradually in 2–3 stages.

37. Gently pry the camshaft oil seals from the cylinder head.

38. Remove the timing control solenoid valves.

39. Remove the camshafts from the cylinder head.

➡️**Be sure to note positioning of the exhaust and the intake camshafts before removing them.**

To install:

40. Install the camshafts so the knock pins are aligned properly. The exhaust side camshaft (left side) has a spline that accepts the crank angle sensor.

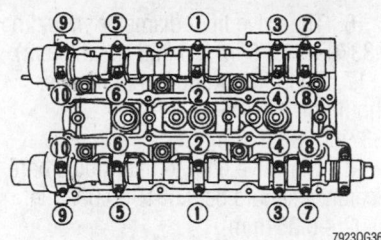

Camshaft bearing cap bolt tightening sequence—3.0L (VG30DE and VG30DETT) engines

41. Install the timing control solenoid valves. Tighten the bracket bolts to 12–18 ft. lbs. (16–25 Nm). Apply liquid gasket to the valve seating surface before installation.

42. Install the camshaft bearing caps. Tighten the bracket bolts in sequence to 7–9 ft. lbs. (9–12 Nm). Tighten the bolts gradually in 2–3 stages. When installing the front camshaft brackets, apply liquid gasket to the bracket seating surface.

43. Coat the lips of the new camshaft seals with clean engine oil and install the seals into the cylinder head.

44. Install the rear timing belt covers. Tighten the cover bolts to 5–6 ft. lbs. (6–8 Nm).

45. Install the camshaft sprockets.

46. Install the cylinder head with a new gasket.

47. Connect the exhaust tube to the exhaust manifold.

48. Install the remaining components in the reverse order of removal.

49. Fill the cooling system to the proper level and connect the negative battery cable.

50. Run the engine and make all the necessary engine adjustments.

51. Road test the vehicle for proper operation.

3.0L (VQ30DE) Engine

1. Relieve the fuel system pressure.

✳️✳️ CAUTION

The fuel injection system remains under pressure after the engine has been turned OFF. Properly relieve fuel pressure before disconnecting any fuel lines. Failure to do so may result in fire or personal injury.

2. Disconnect the negative battery cable.

3. Drain the engine oil and the cooling system. Be sure to drain the engine block and the radiator.

4. Remove the left side rocker cover ornament.

➡️**Before detaching any hoses or connectors, note the locations for reassembly.**

5. Remove the air duct to intake manifold hose, collector hose, blow-by hose, and vacuum hoses.

6. Remove the fuel hoses and detach the harness connections.

7. Disconnect the canister purge hoses.

8. Remove the water hoses from the cylinder head and intake manifold.

9. Disconnect and remove all six ignition coils from the spark plugs.

10. Remove the spark plugs.

11. Remove the bolts that secure the EGR tube and remove the tube.

12. Remove the intake manifold collector supports and remove the collector.

13. Remove the bolts that secure the fuel tube and remove the fuel tube from the vehicle.

14. Remove the bolts that secure the intake manifold to the engine block and remove the manifold. Loosen the bolts in the reverse sequence of the tightening procedure.

15. Remove the LH and RH rocker covers from the cylinder head.

16. Remove the engine undercovers.

17. Remove the right front wheel and engine side covers.

18. Remove the drive belts and idler pulley.

19. Remove the power steering oil pump belt and remove the power steering oil pump assembly.

20. Remove the camshaft position sensor (PHASE) and crankshaft position sensors (REF)/(POS).

21. Set the No. 1 piston to TDC of compression stroke by rotating the crankshaft.

22. Remove the ring gear cover access plate.

23. Loosen the crankshaft pulley bolt while securing the ring gear so the crankshaft cannot rotate.

➡️**Use care not to damage the ring gear teeth.**

24. Using a suitable puller, remove the crankshaft pulley.

25. Remove the A/C compressor and bracket.

26. Remove the front exhaust pipe and its support.

27. Hang the engine at the right and left side engine slingers with a suitable hoist..

28. Support the transaxle with jack.

29. Remove the right side engine mounting, mounting bracket and nuts.

30. Remove the center crossmember assembly.

31. Remove the steel (lower) oil pan bolts in the reverse of the installation sequence.

32. Insert a seal cutter between the steel and aluminum oil pan.

33. Tapping the cutter with a hammer, slide it around the entire edge of the oil pan. Be careful not to damage the aluminum mating surface of the upper oil pan.

34. Remove the steel oil pan and the oil strainer.

35. Remove the aluminum (upper) oil pan bolts in the reverse of the installation sequence.

36. Remove the transaxle bolts that secure the oil pan.

37. Insert a seal cutter between the aluminum oil pan and the engine block.

38. Tapping the cutter with a hammer, slide it around the entire edge of the oil pan. Be careful not to damage the mating surfaces of the oil pan or engine block.

39. Remove the oil pan from the vehicle.

40. Remove the water pump cover and remove the bolts that secure the front timing chain case cover.

41. Using the seal cutter, remove the timing chain case cover.

42. Remove the internal timing chain guide and the upper chain guide.

43. Remove the timing chain tensioner and slack side chain guide.

44. Remove the left and right intake camshaft sprockets first. Be sure to hold the flats of the camshafts while removing the sprocket bolts.

45. Remove the lower timing chain assembly. Be sure to note the aligning marks of the chain before removal.

46. Insert a suitable stopper pin for the left and right camshaft tensioners.

47. Remove the left and right exhaust camshaft sprocket bolts. Be sure to hold the flats of the camshafts while removing the sprocket bolts.

48. Remove the upper timing chain assembly. Be sure to note the aligning marks of the chain before removal.

49. Remove the lower timing chain guide.

50. Remove the crankshaft sprocket.

51. Loosen the bolts that secure the rear timing chain case. The bolts must be loosened in sequence.

52. Using seal cutter tool, remove the rear timing case cover.

➡**Remove the O-rings from the front of the engine block.**

53. Loosen the camshaft bearing caps in several steps. The bearing caps MUST be loosened in sequence.

➡**Keep all bearing caps and camshafts in proper order for reinstallation.**

54. Remove the LH and RH camshaft tensioners from the cylinder head.

55. Remove the camshafts from the cylinder heads.

➡**The valve lifters have a replaceable shim on the top of the lifter. Note the proper locations of each shim to lifter and remove the shims from the lifters.**

56. Using a magnet, remove the valve adjusting shim from the lifter.

57. Remove the lifter assembly from the bore. Be sure to note the locations from where each lifter came.

58. Check the diameter of the valve lifter and the valve lifter guide bore.

59. The diameter of the lifter should be 1.3764–1.3770 in. (34.960–34.975mm) and the diameter of the bore should be 1.3780–1.3788 in. (35.000–35.021mm).

60. Remove all traces of liquid gasket from the timing chain case and from the water pump covers.

61. Remove all traces of liquid gasket from the engine block.

62. Inspect the camshafts for excessive wear or damage and replace as necessary.

To install:

63. Lubricate the valve lifters with clean engine oil and install the lifters into the bore from which they were removed.

64. Lubricate the valve lifter shims with clean engine oil and install the shims into the lifter from which they were removed.

65. Turn the crankshaft clockwise until the No. 1 piston is set 240 degrees before TDC on compression stroke.

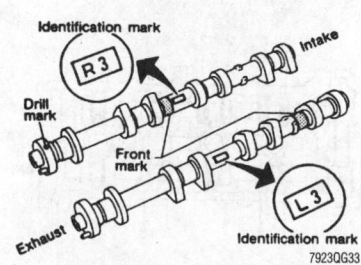

Camshaft identification marks—3.0L (VQ30DE) engines

66. Install the camshaft tensioners on both sides of the cylinder heads. Tighten the tensioner mounting bolts to 75–96 inch lbs. (8.4–10.8 Nm).

➡**The camshafts can be identified by the paint marks on the camshaft. The left cylinder head camshafts have a YELLOW paint mark and the right cylinder head camshafts have a WHITE paint mark.**

67. Install the exhaust and intake camshafts and install the bearing caps. Before installing the No. 1 bearing cap, apply liquid gasket to the corners of the cap.

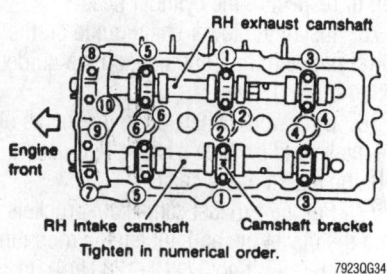

Right cylinder head camshaft bearing cap tightening sequence—3.0L (VQ30DE) engines

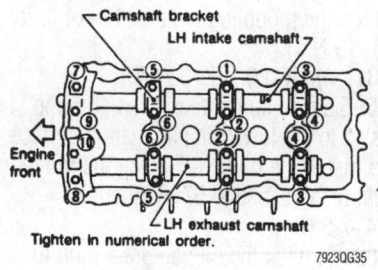

Left cylinder head camshaft bearing cap tightening sequence—3.0L (VQ30DE) engines

➡**When installing the camshafts, position the camshaft keys at the 12 o'clock position in respect to the cylinder head angle.**

68. Tighten the camshaft bearing caps as follows:

a. Tighten bolts No. 7–10 to 17 inch lbs. (2 Nm).

b. Tighten bolts No. 1–6 to 17 inch lbs. (2 Nm).

c. Tighten bolts No. 1–10 to 52 inch lbs. (6 Nm).

d. Tighten bolts No. 1–10 to 81–104 inch lbs. (9–11 Nm).

69. Install new O-rings to the front of the engine block.

70. Apply sealant to the hatched portion of the of the rear timing chain case.

71. Align the rear timing chain case with the dowel pins and install onto the cylinder heads and engine block.

72. Tighten the rear timing chain case mounting bolts in sequence to 105–121 inch lbs. (11.8–13.7 Nm).

73. Install the crankshaft sprocket with the mating mark facing out.

74. Rotate the crankshaft clockwise and position the crankshaft to TDC of compression stroke and align the dowels of the camshaft sprockets to the 12 o'clock position in respect to the cylinder head.

75. Install the lower chain guide on the dowel pin with the front mark on the guide facing upward.

76. On a work bench, align the marks on the intake and exhaust camshaft sprockets with the marks of the chain.

77. Put the exhaust camshaft sprockets onto the dowel pin and tighten the mounting bolts to 88–95 ft. lbs. (119–128 Nm). Be sure to secure the camshafts while tightening the bolts.

78. Install the timing chains, sprockets and related components.

79. Install the transaxle bolts that secure the oil pan.

80. Install the oil pan strainer and tighten the mounting bolts to 12–14 ft. lbs. (16–19 Nm).

81. Apply a 0.177–0.217 in. (4.5–5.5mm) continuous bead of liquid gasket to the lower oil pan mating surface and install the oil pan. Tighten the mounting bolts in sequence to 57–66 inch lbs. (6.4–7.5 Nm).

82. Tighten the oil pan drain plug to 22–29 ft. lbs. (29–39 Nm).

83. Install the center crossmember assembly.

84. Install the right side engine mounting bracket and mount assembly.

85. Remove the engine slinger assembly.

86. Install the front exhaust pipe and its support.

87. Install the A/C compressor and bracket.

88. Install the crankshaft pulley to the crankshaft and install the mounting bolt.

89. Tighten the mounting bolt to 14–22 ft. lbs. (20–29 Nm). Tighten the crankshaft bolt an additional 60–66 degrees clockwise. This is about the angle from one hexagon bolt head corner to another.

90. Install the remaining components in the opposite order from which they were removed.

91. Refill the engine oil and coolant with the proper type and amount of fluid.

92. Connect the negative battery cable.

93. Start the engine and run at 3000 RPM under no load to purge the air from the high pressure chamber. The engine may produce a rattling noise. This indicates that air still remains in the chamber and is not a matter of concern.

Valve Lash

ADJUSTMENT

1.6L and 2.4L Engines

CHECKING VALVE LASH

1. Run the engine until it reaches normal operating temperature and shut if off.

2. Remove the cylinder head cover and all the spark plugs.

3. Set the No. 1 cylinder at TDC on its compression stroke. Align the pointer with the TDC mark on the crankshaft pulley. Check that the valve lifters on the No. 1 cylinder are loose and valve lifters on the No. 4 cylinder are tight. If not, turn the crankshaft one revolution (360 degrees) and align the pointer with the TDC mark on the crankshaft pulley.

4. Check the following valves:
   a. Both No. 1 intake valves.
   b. Both No. 1 exhaust valves.
   c. Both No. 2 intake valves.
   d. Both No. 3 exhaust valves.

5. Using a feeler gauge, measure the clearance between the valve lifter and the camshaft. Record any valve clearance measurements which are out of specification.

6. Turn the crankshaft one revolution (360 degrees) and align the mark on the crankshaft pulley with the pointer. Check the following valves:
   a. Both No. 2 exhaust valves.
   b. Both No. 3 intake valves.
   c. Both No. 4 intake valves.
   d. Both No. 4 exhaust valves.

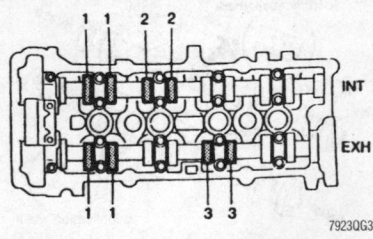

Measure the clearance of the valves indicated when the No. 1 piston is at TDC on compression—1.6L and 2.4L engines

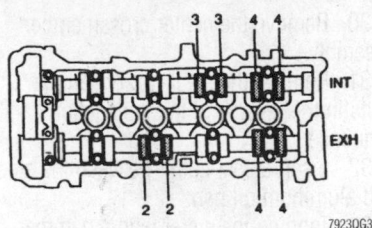

Measure the clearance of the valves indicated when the No. 4 piston is at TDC on compression—1.6L and 2.4L engines

7. Using a feeler gauge, measure the clearance between the valve lifter and the camshaft. Record any valve clearance measurements which are out of specification.

8. If all the valve clearances are within specification, install the cylinder head cover and the spark plugs.

ADJUSTING VALVE LASH

1. If an adjustment is necessary, adjust the valve clearance while engine is cold by removing the adjusting shim. The adjusting shim can be removed by using the following procedures:
   a. Turn the crankshaft so the camshaft lobe of the valve to be adjusted is pointed straight up.
   b. Turn the lifter so the notch is pointed towards the center of the cylinder head; this will facilitate the shim removal process.
   c. Using a depressor tool, push down on the lifter and insert a keeper tool on the edge of the lifter to keep the lifter in the depressed position.
   d. Remove the depressor tool and remove the shim with a magnet.

2. Determine the replacement adjusting shim size by using the following procedures and formula:
   a. Using a micrometer determine thickness of the removed shim.
   b. Calculate the thickness of a new adjusting shim so valve clearance is within the specified values.
   c. R = thickness of the removed shim.
   d. N = thickness of the new shim.
   e. M = measured valve clearance.
   • 1.6L engine—Intake shim determination formula: $N = R + (M - 0.0146$ in. or 0.37mm)
   • 1.6L engine—Exhaust shim determination formula: $N = R + (M - 0.0157$ in. or 0.40mm)
   • 2.4L engine—Intake shim determination formula: $N = R + (M - 0.0138$ in. or 0.35mm)
   • 2.4L engine—Exhaust shim determination formula: $N = R + (M - 0.0146$ in. or 0.37mm)

3. Shims are available different sizes from 0.0772–0.1055 in. (1.96–2.68mm) in increments of 0.0008 in. (0.02mm). The thickness is stamped on the shim; this side is always installed facing down. Select new shims with thickness as close as possible to calculated valve and install it in the lifter.

4. Install the new shim onto the lifter.

5. Depress the lifter and remove the keeper tool. Remove the depressor tool and recheck the valve clearance. Repeat this procedure for any other valves requiring adjustment.

6. Install the cylinder head cover and spark plugs when all valve adjustments are finished.

3.0L (VG30DE and VG30DETT) Engines

The engine is equipped with Hydraulic Lash Adjusters. The valve lash is not adjustable. If the valves make noise, check the cylinder head for mechanical damage or excessive wear of the camshaft or lash adjusters.

3.0L (VQ30DE) Engine

➡**Check and adjust the valve clearances while the engine is cold and not running.**

CHECKING VALVE LASH

1. Remove the intake manifold collector.

2. Remove the left and right rocker covers.

3. Remove the spark plugs.

4. Set the No. 1 cylinder at TDC on its compression stroke. Align the pointer with the TDC mark on the crankshaft pulley. Check that the valve lifters on the No. 1 cylinder are loose and valve lifters on the No. 4 cylinder are tight. If not, turn the crankshaft one revolution (360 degrees) and align the pointer with the TDC mark on the crankshaft pulley.

5. Check the following valves:
   a. Both No. 1 intake valves.
   b. Both No. 2 exhaust valves.
   c. Both No. 3 exhaust valves.
   d. Both No. 6 intake valves.

6. Using a feeler gauge, measure the clearance between the valve lifter and the camshaft. Record any valve clearance measurements which are out of specification. Intake valve clearance (cold) is 0.010–0.013 in. (0.26–0.34mm) and exhaust valve clearance (cold) is 0.011–0.015 in. (0.29–0.37mm).

7. Turn the crankshaft 240 degrees and set the No. 3 cylinder to TDC of its compression stroke.

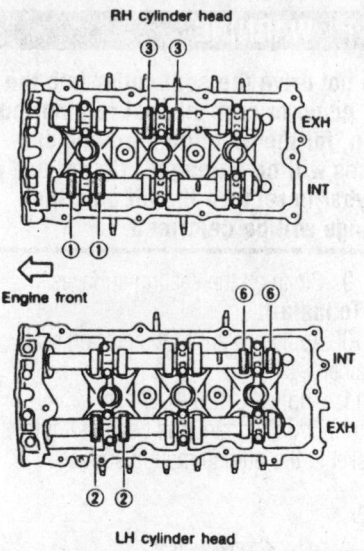

Measure the valves indicated while the No. 1 piston is at TDC on the compression stroke—3.0L (VQ30DE) engine

8. Check the following valves:
   a. Both No. 2 intake valves.
   b. Both No. 3 intake valves.
   c. Both No. 4 exhaust valves.
   d. Both No. 5 exhaust valves.

9. Using a feeler gauge, measure the clearance between the valve lifter and the camshaft. Record any valve clearance measurements which are out of specification. Intake valve clearance (cold) is 0.010–0.013 in. (0.26–0.34mm) and exhaust valve clearance (cold) is 0.011–0.015 in. (0.29–0.37mm).

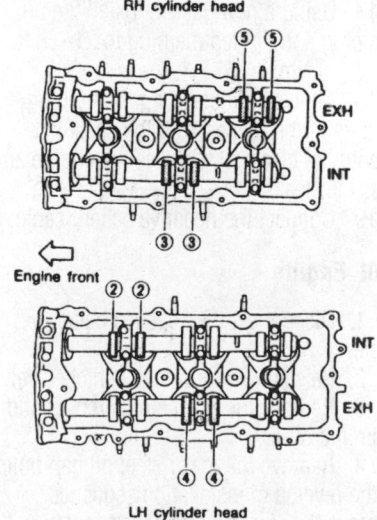

Measure the valves indicated while the No. 3 piston is at TDC on the compression stroke—3.0L (VQ30DE) engine

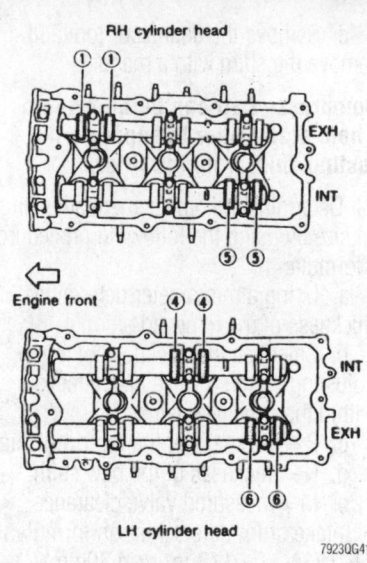

Measure the valves indicated while the No. 5 piston is at TDC on compression— 3.0L (VQ30DE) engine

10. Turn the crankshaft 240 degrees and set the No. 5 cylinder to TDC of its compression stroke.

11. Check the following valves:
   a. Both No. 1 exhaust valves.
   b. Both No. 4 intake valves.
   c. Both No. 5 intake valves.
   d. Both No. 6 exhaust valves.

12. Using a feeler gauge, measure the clearance between the valve lifter and the camshaft. Record any valve clearance measurements which are out of specification. Intake valve clearance (cold) is 0.010–0.013 in. (0.26–0.34mm) and exhaust valve clearance (cold) is 0.011–0.015 in. (0.29–0.37mm).

13. If all the valve clearances are within specification, install the cylinder head cover, spark plugs, and the intake manifold collector.

ADJUSTING VALVE LASH

1. If an adjustment is necessary, adjust the valve clearance while engine is cold by removing the adjusting shim. The adjusting shim can be removed by using the following procedures:

   a. Turn the crankshaft so the camshaft lobe of the valve to be adjusted is pointed straight up.

   b. Turn the lifter so the notch is pointed towards the center of the cylinder head; this will facilitate the shim removal process.

   c. Using a depressor tool No. KV10115110 or equivalent, push down on the lifter and insert a keeper tool on the edge of the lifter to keep the lifter in the depressed position.

d. Remove the depressor tool and remove the shim with a magnet.

➡**Compressed air can be blown into the hole of the lifter to separate the adjusting shim from the lifter.**

2. Determine the replacement adjusting shim size by using the following procedures and formula:

a. Using a micrometer determine thickness of the removed shim.

b. Calculate the thickness of a new adjusting shim so valve clearance is within the specified values.

c. R = thickness of the removed shim.

d. N = thickness of the new shim.

e. M = measured valve clearance.

• Intake shim determination formula:
N = R + (M—0.0118 in. or 0.30mm)

• Exhaust shim determination formula:
N = R + (M—0.0130 in. or 0.33mm)

3. Shims are available in 64 sizes from 0.0913–0.1161 in. (2.32–2.95mm) in steps of 0.004 in. (0.01mm). The thickness is stamped on the shim; this side is always installed facing down. Select new shims with thickness as close as possible to calculated valve and install it in the lifter.

4. Install the new shim onto the lifter.

5. Depress the lifter and remove the keeper tool. Remove the depressor tool and recheck the valve clearance. Repeat this procedure for any other valves requiring adjustment.

6. When all valve adjustments are finished, install the cylinder head cover, spark plugs, and the intake manifold collector.

Oil Pan

REMOVAL & INSTALLATION

1.6L Engine

1. Disconnect the negative battery cable.

2. Raise and safely support the vehicle.

3. Remove the undercovers.

4. Remove the oil pan plug and drain the oil into a container.

5. Remove the front exhaust tube.

6. Remove center crossmember assembly.

7. Remove the support brackets from the sides of the oil pan.

8. Remove the oil pan mounting bolts. Using the oil pan seal cutter tool KV10111100 or equivalent, separate the oil pan from the engine.

✻✻ WARNING

Do not drive the seal cutter into the oil pump or rear oil seal retainer portion, for the aluminum mating surfaces will be damaged. Do not use a prybar to remove the oil pan; the flange will be deformed.

9. Clean all the sealing surfaces.

To install:

10. Apply sealant to the rear oil seal retainer.

11. Apply a 0.128–0.177 in. (3.5–4.5mm) continuous bead of liquid gasket to the oil pan mating surface.

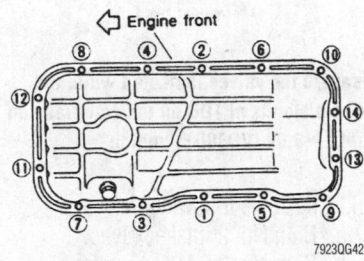

Tighten the oil pan bolts in the correct sequence to prevent oil leakage—1.6L engine

12. Install the oil pan and tighten the oil pan mounting bolts/nuts in sequence to 56–73 inch lbs. (6.3–8.3 Nm).

13. Install the oil pan support brackets.

14. Install the center crossmember.

15. Install the front exhaust tube.

16. Install the undercovers.

17. Using a new gasket, install the oil pan plug and tighten the plug to 21–28 ft. lbs. (7–8 Nm).

18. After 30 minutes of gasket curing time, refill the oil pan with the specified quantity of clean oil. Operate the engine and check for leaks.

19. Connect the negative battery cable.

2.0L Engine

1. Disconnect the negative battery cable.

2. Raise and support the vehicle safely.

3. Remove the engine undercover and drain the oil.

4. Remove the lower steel oil pan bolts in the reverse of installation sequence. Remove the steel oil pan. Insert tool KV10111100 or equivalent, between steel oil pan and aluminum oil pan. Tap the tool around the perimeter of the pan to cut the gasket material.

5. Remove the oil baffle bolts and oil baffle. Remove the front exhaust tube.

6. Set a suitable jack under the transaxle and raise the engine.

7. Remove the center crossmember from the vehicle.

8. If equipped with an automatic transaxle, remove the transaxle shift control cable.

9. Remove the compressor gussets and the rear cover plate.

10. Remove the aluminum oil pan bolts. Loosen aluminum oil pan bolts in reverse order of the tightening sequence.

11. Remove the two transaxle mounting bolts and refit the them into vacant holes at the bottom of the oil pan. Use tool KV10111100 or equivalent, to cut the gasket material.

12. Remove the two transaxle mounting bolts that were relocated and remove the pan from the vehicle.

To install:

13. Clean the oil pan rail of all liquid gasket and apply a new bead of ⁵⁄₃₂ in. (4.5mm) thickness to the aluminum oil pan rail.

14. Install the aluminum oil pan and tighten bolts No. 1–16 to 12–14 ft. lbs. (16–19 Nm) and bolts No. 17–18 to 5–6 ft.

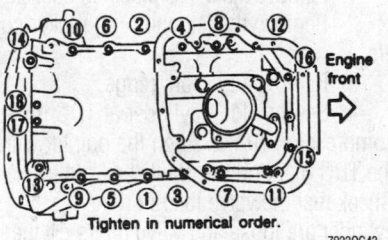

Aluminum oil pan bolt tightening sequence—2.0L engine

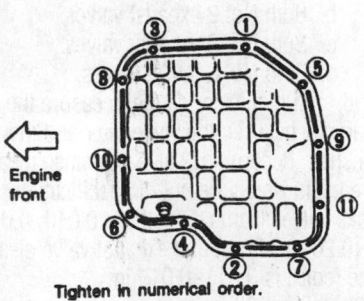

Steel oil pan bolt tightening sequence—2.0L engine

lbs. (6–8 Nm) in the opposite order of removal.

15. Install the two transaxle mounting bolts, rear cover plate, and the compressor gussets.

16. Install the automatic transmission shift control cable (if equipped).

17. Install the center crossmember member, front exhaust tube, and the baffle plate. Tighten the baffle plate mounting bolts to 56–66 inch lbs. (6.4–7.5 Nm).

18. Clean the steel oil pan rail of all liquid gasket and apply a new bead of ⁵⁄₃₂ in. (4.5mm) thickness to the steel oil pan rail.

19. Install the steel oil pan and install the bolts. Tighten the oil pan bolts in the proper sequence to 56–66 inch lbs. (6.4–7.5 Nm) and wait 30 minutes before refilling crankcase with oil.

20. After 30 minutes, refill the crankcase with oil and reconnect the negative battery cable.

21. Start the engine and check for leaks.

2.4L Engine

ALTIMA

1. Disconnect the negative battery cable.
2. Raise the front of the vehicle and support it safely.
3. Drain the oil from the oil pan.
4. Remove the engine undercover.
5. Remove the bolts securing the steel oil pan to the aluminum oil pan in reverse order of the tightening sequence.
6. Install a seal cutter between the steel oil pan and the aluminum oil pan
7. Tapping the cutter with a hammer, slide it around the entire edge of the oil pan. Take care not to damage the aluminum oil pan.
8. Remove the steel oil pan.
9. Remove the baffle plate and oil strainer.
10. Remove the front suspension member.

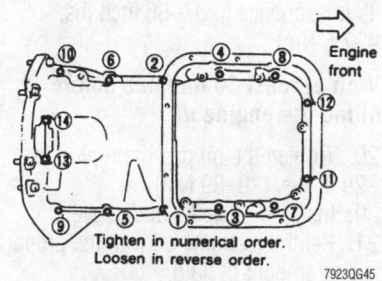

Oil pan bolt loosening and tightening sequence—2.4L engine

11. Remove the A/C compressor gussets.
12. Remove the rear cover plate.
13. Remove the aluminum oil pan retaining bolts in reverse order of the tightening sequence.
14. Insert a seal cutter between the oil pan and the cylinder block.
15. Tapping the cutter with a hammer, slide it around the entire edge of the oil pan. Take care not to damage the aluminum oil pan.
16. Lower the oil pan from the cylinder block and remove it from the engine.

To install:

17. Carefully scrape the old gasket material away from the pan and cylinder block mounting surfaces, then apply a continuous bead (3.5–4.5mm) of liquid gasket around the oil pan. Install the pan within five minutes or else this step will have to be repeated.

18. Install the aluminum oil pan and tighten the mounting bolts in sequence, to 13 ft. lbs. (17.5 Nm).

19. Install the baffle plate

20. Install the steel oil pan and tighten in sequence to 61 inch lbs. (7 Nm).

21. Wait 30 minutes before refilling the crankcase to allow for the sealant to cure properly.

22. Install the rear cover plate.
23. Install the front suspension member.
24. Install the A/C compressor gussets.
25. Install the front suspension member.
26. Install the engine undercovers.
27. Lower the vehicle.
28. Fill the crankcase to the proper level.
29. Connect the negative battery cable. Start the engine and check for leaks.

240SX

1. Disconnect the negative battery cable.
2. Raise the front of the vehicle and support it safely.
3. Position a hoist on the engine and support the engine.
4. Drain the oil pan.
5. Disconnect the tension rod bolts at the transverse link.
6. Separate the front stabilizer bar from the side member.
7. Remove the left and right engine mounting bolts.
8. Disconnect the lower steering joint.
9. Disconnect the power steering tube bracket at the left tension rod.
10. Remove the bolts and lower the front suspension member while supporting it with a jack. It is only necessary to lower the suspension member 2.36 in. (60mm).

11. Remove the oil pan retaining bolts.
12. Insert a seal cutter between the oil pan and the cylinder block.
13. Tapping the cutter with a hammer, slide it around the entire edge of the oil pan. Do not drive the seal cutter into the oil pump or rear seal retainer portion or the aluminum mating surface will be deformed.
14. Lower the oil pan from the cylinder block and remove it from the front side of the engine.

To install:

15. To install, carefully scrape the old gasket material away from the pan and cylinder block mounting surfaces, then apply a continuous bead 0.138–0.177 in. (3.5–4.5mm) of liquid gasket around the oil pan. Install the pan within five minutes or this step will have to be repeated.

16. Install the oil pan and tighten the mounting bolts. Start tightening the bolts from the center and work towards the ends.

17. Tighten the oil pan bolts to 12–14 ft. lbs. (16–19 Nm). Wait 30 minutes before refilling the crankcase to allow for the sealant to cure properly.

18. The remaining components are installed in the reverse order from which they were removed.

19. Fill the crankcase to the proper level.

20. Connect the negative battery cable. Start the engine and check for leaks.

21. Recheck the oil level.

3.0L (VG30DE and VG30DETT) Engine

1. Raise and safely support the vehicle.
2. Remove the oil pan plug and drain the oil into a container.
3. Remove the engine undercover.
4. Remove the engine oil filter and oil filter housing assembly.
5. Disconnect the A/C tube clamps.
6. Remove the tension rod securing bolts on both sides and loosen the traverse link bolts on both sides.
7. Disconnect the steering column lower joint.
8. Disconnect the engine mounts and raise engine using a suitable lifting device.
9. Remove the power steering gear bracket from the suspension crossmember.
10. Support the crossmember and remove the crossmember fixing bolts.
11. Remove the power steering tubes from crossmember.
12. Lower the crossmember.
13. Remove the oil pan bolts, in sequence.
14. Using the oil pan removal tool KV10111100 or equivalent, separate the oil pan from the engine.

Do not drive the seal cutter into the oil pump or rear oil seal retainer portion, for the aluminum mating surfaces will be damaged. Do not use a prybar or screwdriver, for the oil pan flange will be deformed.

15. Remove the oil pan.
16. Clean all the sealing surfaces.

To install:

17. Apply sealant to the oil pump gasket and the rear oil seal retainer.
18. Install new oil pan gasket. The use a continuous 0.138–0.177 in. (3.5 4.5mm) bead of liquid gasket. If using a liquid gasket, allow the gasket to cure for five minutes before assembly.
19. Install the oil pan and tighten the bolts in sequence as follows:
    a. Tighten the M6 bolts to 55–73 inch lbs. (6.3–8.3 Nm).
    b. Tighten the M8 bolts to 12–15 ft. lbs. (16–21 Nm).
20. Raise the center crossmember. Tighten the center crossmember fixing nuts/bolts to 29–36 ft. lbs. (39–49 Nm).
21. Reconnect the A/C lines and the power steering tubes to the crossmember.
22. Reconnect the power steering bracket to the suspension crossmember.
23. Lower the engine and connect the motor mounts.
24. Using a new washer, install the oil pan drain plug and tighten the plug to 22–29 ft. lbs. (29–39 Nm).
25. Install the tension rod bolts on both sides and tighten the mounting through-bolt to 80–100 ft. lbs. (108–135 Nm).
26. Connect the tension rod to the lower control arm and tighten mounting nuts as follows:
    a. Convertible models —69–83 ft. lbs. (93–113 Nm).
    b. Non-convertible models —80–94 ft. lbs. (108–127 Nm).
27. Connect the stabilizer links and tighten the mounting nuts to 41–47 ft. lbs. (56–64 Nm).
28. Reconnect the steering column lower joint and tighten the lower pinch bolt to 17–22 ft. lbs. (24–29 Nm).
29. Using a new gasket, install the oil filter and oil filter housing assembly. Tighten the oil filter housing bolts to 12–15 ft. lbs. (16–21 Nm).
30. Install the engine undercover.
31. Lower the vehicle.
32. Refill the oil pan with the specified quantity of clean oil.

➡️If a liquid gasket was used, wait at least 30 minutes for material to cure before refilling engine with oil.

33. Start the engine and check for leaks.

3.0L (VQ30DE) Engine

1. Disconnect the negative battery cable.
2. Drain the engine oil and remove the engine undercovers.
3. Remove the steel (lower) oil pan bolts in the reverse of the installation sequence.
4. Insert a seal cutter between the steel and aluminum oil pan.
5. Tapping the cutter with a hammer, slide it around the entire edge of the oil pan. Be careful not to damage the aluminum mating surface of the upper oil pan.
6. Remove the steel oil pan and the oil strainer.
7. Remove the front exhaust pipe and its support.
8. Hang the engine at the right and left side engine slingers with a suitable hoist.
9. Position a suitable jack under the transaxle.
10. Remove the crankshaft position sensors (REFERENCE and POSITION) from the oil pan.
11. Remove the front and rear engine mounting nuts and bolts.
12. Remove the center crossmember assembly.
13. Remove the engine drive belts.
14. Remove the air conditioner compressor and the compressor mounting bracket.
15. Remove the rear cover plate and the lower transaxle bolts.
16. Remove the aluminum (upper) oil pan bolts in the reverse of the installation sequence.
17. Insert a seal cutter between the aluminum oil pan and the engine block.
18. Tapping the cutter with a hammer, slide it around the entire edge of the oil pan. Be careful not to damage the mating surfaces of the oil pan or engine block.
19. Remove the oil pan assembly.
20. Remove the bolts that secure the baffle plate and remove the baffle plate.
21. Remove the O-rings from the cylinder block and oil pump body.

To install:

22. Install the baffle plate to the oil pan and tighten the mounting bolts to 22–27 inch lbs. (2.5–3.1 Nm).

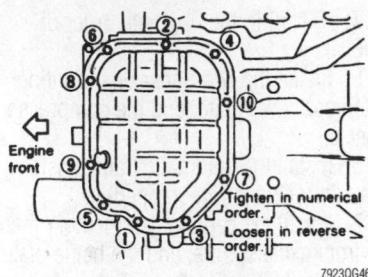

Bolt tightening sequence for the steel oil pan—3.0L (VQ30DE) engines

23. Apply sealant to the front and rear seal of the oil pan.
24. Install new O-rings to the cylinder block and the oil pump body.
25. Apply a 0.177–0.217 in. (4.5–5.5mm) continuous bead of liquid gasket to the upper oil pan mating surface and install the oil pan. Tighten the mounting bolts in sequence to 12–14 ft. lbs. (16–19 Nm).

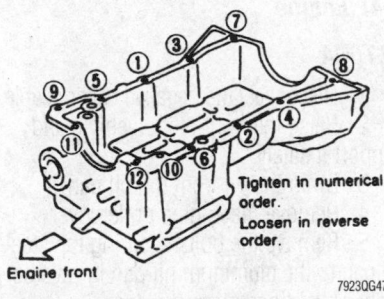

To prevent pan warpage, tighten the bolts in the sequence shown—3.0L (VQ30DE) engines

26. Install the oil pan strainer and tighten the mounting bolts to 12–14 ft. lbs. (16–19 Nm).
27. The balance of installation is the reverse of the removal procedure.
28. Apply a 0.177–0.217 in. (4.5–5.5mm) continuous bead of liquid gasket to the lower oil pan mating surface and install the oil pan. Tighten the mounting bolts in sequence to 57–66 inch lbs. (6.4–7.5 Nm).

➡️**Wait at least 30 minutes before refilling the engine oil.**

29. Tighten the oil pan drain plug to 22–29 ft. lbs. (29–39 Nm).
30. Install the engine undercovers.
31. Refill the engine oil with the proper type and amount of fluid.
32. Start the engine and check for leaks.

Rear Main Seal

REMOVAL & INSTALLATION

1.6L, 2.0L and 3.0L Engines

1. Remove the transaxle.
2. Remove the driveplate/ flywheel.
3. Carefully pry the seal from the retainer. Be sure not to scratch the sealing surface of the crankshaft or oil seal bore.

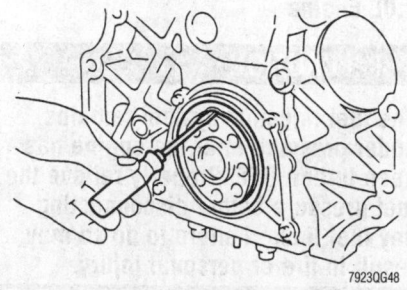

Carefully pry the rear main seal out of the retainer on the rear of the engine—1.6L and 2.0L engine

To install:

4. Apply clean engine oil to the new seal. Position the seal on the rear of the engine in the proper direction.
5. Using a suitable seal driver, tap the seal into position in the seal retainer.
6. Install the flywheel/flexplate.
7. Install the transaxle assembly.

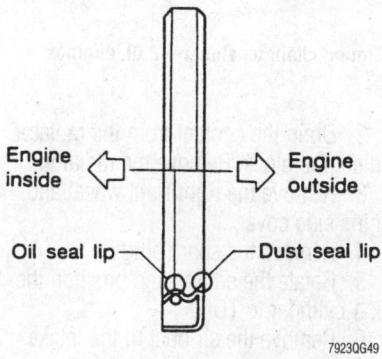

Be sure to install the seal in the correct orientation—1.6L and 2.0L engine

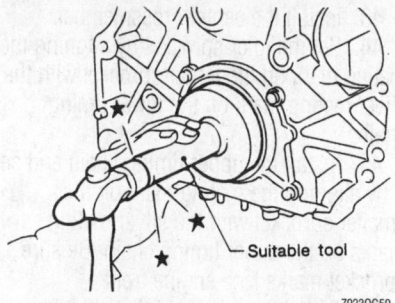

Carefully pry the rear main seal out of the retainer on the rear of the engine—1.6L and 2.0L engine

2.4L Engine

1. Remove the transaxle or transmission.
2. Remove the driveplate/ flywheel.
3. Remove the rear oil seal retainer with the oil seal.
4. Tap the oil seal out of the retainer with a hammer and drift.

To install:

5. Apply clean engine oil to the new seal.
6. Install the new seal in the retainer with a suitable seal driver.
7. Apply a continuos bead of RTV silicone sealant, 0.079–0.118 in. (2–3mm) wide to the seal retainer. Be sure to apply around the inner side of the bolt holes.
8. Using a suitable seal driver, tap the seal into position in the seal retainer.
9. Install the oil seal retainer. Tighten the bolts to 11 ft. lbs. (15 Nm).
10. Install the flywheel/flexplate.
11. Install the transmission or transaxle.

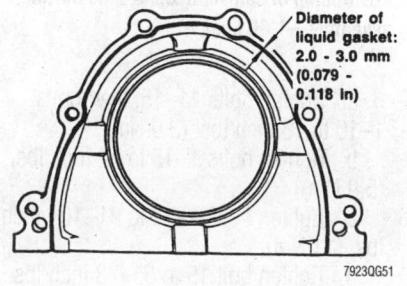

Apply sealant to the seal retainer as shown—2.4L engine

Timing Chain, Sprockets, Front Cover and Seal

REMOVAL & INSTALLATION

1.6L Engine

1. Disconnect the negative battery cable. Relieve the fuel pressure.
2. Drain the coolant from the radiator and cylinder block.
3. Remove the upper radiator hose from the engine.
4. Remove the engine drive belts.
5. Remove the power steering pulley and remove the pump with bracket.
6. Remove the air duct from the intake manifold collector.
7. Disconnect the vacuum hoses, wiring, and harness connectors.
8. Remove the right front wheel and remove the inner wheel covers.
9. Remove the engine undercovers.
10. Remove the front exhaust pipe.
11. Remove the cylinder head front mounting bracket.
12. Remove the cylinder head cover from the engine.
13. Remove the distributor.
14. Remove the water pump pulley from water pump.
15. Remove the complete thermostat housing assembly from the engine.
16. Remove the lower timing chain tensioner.
17. Remove the upper timing chain tensioner and slack side timing chain guide.
18. Loosen the idler sprocket bolt.
19. Remove the camshaft sprocket bolts and remove the sprockets from the camshafts. Be sure to mark the sprockets for proper reinstallation.
20. Remove the camshaft mounting caps by loosening the bolts in two or three steps. Remove the camshafts from the engine.
21. Remove the idler sprocket bolt.
22. Remove the cylinder head with the manifolds.
23. Remove the idler sprocket shaft from the rear side.
24. Remove the upper timing chain.
25. Support the engine assembly and remove the center crossmember.
26. Remove the oil pan and strainer assembly.
27. Remove the crankshaft pulley.

28. Remove the engine front mount.

29. Remove the engine front mount bracket.

30. Remove the bolts that secure the front timing cover and remove the cover from the engine. Once the timing chain cover is removed, drive out the old oil seal.

31. Remove the idler sprocket and remove the lower timing chain.

32. Remove the oil pump drive spacer and remove the crankshaft sprocket.

33. Remove the timing chain guide.

To install:

34. Drive a new oil seal into the front cover. Lubricate the oil seal lip with clean engine oil.

35. Confirm that No. 1 piston is set at TDC on compression stroke.

36. Install the crankshaft sprocket with the marks of the sprocket facing the front of the engine.

37. Install the oil pump drive spacer and install the chain guide.

38. Install the lower timing chain. Set the chain by aligning its mating mark with the one on the crankshaft sprocket. Be sure the sprocket's mating mark faces the front of the engine.

43. Install the center crossmember.

44. Set the idler sprocket by aligning the mating mark on the larger sprocket with the silver mating mark on the lower timing chain.

45. Install the upper timing chain and set it by aligning the mating mark on the smaller sprocket with the silver mating marks on the upper timing chain. Be sure sprocket marks face engine front.

46. Install the idler sprocket shaft from the rear side.

47. Install the cylinder head assembly.

48. Install the idler sprocket bolt. Be sure to lubricate the bolt with clean engine oil.

49. Install the exhaust and intake camshafts. The camshafts and marked **I** for intake and **E** for exhaust.

50. Position the intake camshaft knock pin at the 9 o'clock position and the exhaust camshaft knock pin at the 12 o'clock position.

51. Install the camshaft bearing caps and distributor bracket. Apply liquid sealant to the distributor bracket.

52. Tighten the mounting bolts in sequence as follows:

55. Install the upper timing chain tensioner. Before installation of the tensioner, install a suitable pin to hold the tensioner in the relaxed position. After installing the chain tensioner, remove the pin.

56. Install the lower timing chain tensioner. Be sure the notch of the gasket is positioned down.

57. Install the remaining components in the opposite order from which they were removed.

58. Connect the negative battery cable. Refill all fluid levels. Road test the vehicle for proper operation.

2.0L Engine

> ❄ **CAUTION**

The fuel injection system remains under pressure after the engine has been turned OFF. Properly relieve the fuel pressure before disconnecting any fuel lines. Failure to do so may result in fire or personal injury.

1. Relieve the fuel system pressure and remove the negative battery cable.

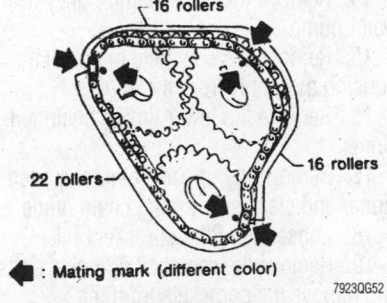

Be sure to align the camshaft sprockets with the timing chain—1.6L engine

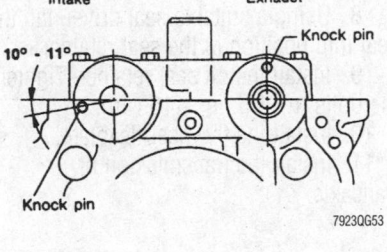

Positioning of camshaft knock pins during assembly—1.6L engine

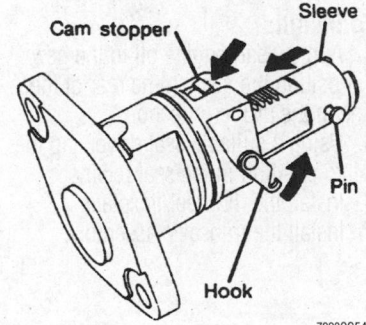

Timing chain tensioner—2.0L engines

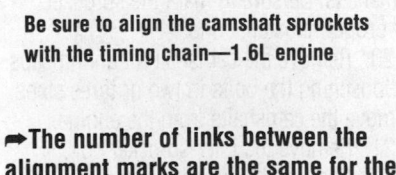

➡ **The number of links between the alignment marks are the same for the left and the right side.**

39. Install the crankshaft sprocket and the lower timing chain. Set the timing chain by aligning its mating mark with the one on the crankshaft sprocket. Be sure sprocket's mating mark faces engine front. The number of links between the alignment marks are the same for the left and right side.

40. Using liquid gasket, install the front cover assembly.

41. Install engine front mounting bracket and install the engine mount.

42. Install the oil strainer, oil pan assembly, and the crankshaft pulley.

a. Tighten bolts 11–15, then bolts 1–10 to 18 inch lbs. (2.0 Nm)

b. Tighten bolts 1–15 to 52 inch lbs. (5.9 Nm)

c. Tighten bolts 1–14 to 81–104 inch lbs. (11 Nm)

d. Tighten bolt 15 to 55–73 inch lbs. (11 Nm)

53. Install the camshaft sprockets with timing chain. Set the camshaft sprockets by aligning the mating marks of the timing chain with the marks on the camshaft sprockets.

54. Install the camshaft sprocket bolts and tighten them to 86 ft. lbs. (117 Nm). Be sure to lubricate the bolts with clean engine oil.

2. Drain the coolant from the radiator and engine block. Remove the radiator.

3. Remove the right front wheel and engine side cover.

4. Remove the spark plugs.

5. Rotate the engine and position the No. 1 cylinder to TDC.

6. Remove the air duct to the intake manifold.

7. Remove the drive belts and the water pump pulley.

8. Remove the alternator and the power steering pump from the engine.

9. Label and remove the vacuum hoses, fuel hoses, and the wiring harness connectors.

10. Remove the cylinder head cover.

11. Remove the intake manifold supports.

12. Unbolt and remove the oil filter bracket and the power steering pump bracket.

13. Remove the timing chain tensioner.

14. Remove the distributor.

15. Remove the timing chain guide.

16. Holding the flats of the camshaft sprockets, remove the bolts that secure the sprockets.

17. Remove the timing chain sprockets from the camshafts.

18. Remove the oil tubes, baffle plate, camshaft brackets and remove the camshafts from the cylinder head.

19. Remove the starter motor.

20. Remove the coolant hoses from the engine block.

21. Remove the knock sensor harness connector.

22. Remove the EGR tube.

23. Remove the cylinder head assembly.

24. Raise and support the vehicle safely.

25. Remove the oil pan, oil strainer, and the baffle plate.

26. Remove the crankshaft pulley using a suitable puller.

27. Remove the engine front mount.

28. Remove the front cover and oil pump drive spacer.

29. Remove the timing chain guides and timing chain. Check the timing chain for excessive wear at the roller links. Replace the chain if necessary.

To install:

30. Clean all gasket mating surfaces.

31. Install the crankshaft sprocket. Position the crankshaft so that No. 1 piston is set at TDC (keyway at 12 o'clock, mating mark at 4 o'clock) fit timing chain to crankshaft sprocket so the mating mark is in line with mating mark on crankshaft sprocket. The mating marks on timing chain for the camshaft sprockets should be silver. The mating mark on the timing chain for the crankshaft sprocket should be gold.

32. Install the timing chain to the crankshaft sprocket and install the timing chain guides. Tighten the timing chain guides to 9–14 ft. lbs. (13–19 Nm). Drape the timing chain over the left chain guide.

33. Install the oil pump drive spacer to the crankshaft.

34. Apply a continuous bead of liquid sealant to the front timing cover and install the cover. Tighten the front cover mounting bolts to 57–66 inch lbs. (6.4–7.5 Nm).

35. Install right front engine mount.

36. Install the crankshaft pulley and tighten the mounting bolt to 105–112 ft.

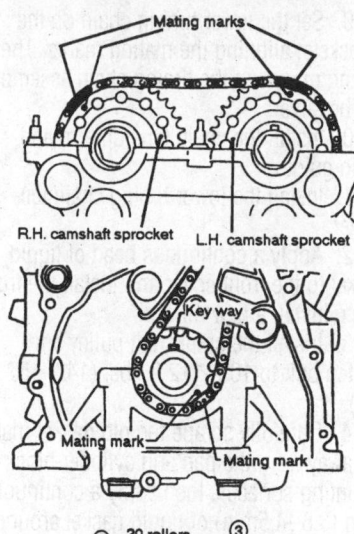

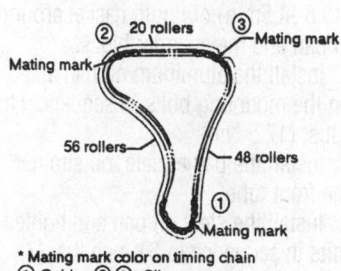

Timing chain sprocket alignment marks— 2.0L (SR20DE) engines

lbs. (142–152 Nm). Be sure the No. 1 piston is at TDC.

37. Install the oil strainer, baffle plate, and the oil pan assembly.

38. Install the cylinder head assembly. Be sure to apply a bead of sealant to the joint of the block and front timing cover.

39. Install the EGR tube.

40. Install the knock sensor harness connector.

41. Using new hose clamps, install the coolant hoses to the engine block.

42. Install the starter motor.

43. Install the camshafts, camshaft bearing caps, oil tubes, and the baffle plate.

➡**When installing the camshafts, be sure to position the LH and RH camshaft keys at 12 o'clock. Also be sure the camshaft brackets are facing in the correct direction.**

44. Install the camshaft sprockets by lining up the mating marks on the timing chain with the mating marks on the camshaft sprockets. Tighten the camshaft sprocket bolts to 101–116 ft. lbs. (137–157 Nm).

45. Install the timing chain guide and distributor.

46. Install the chain tensioner. Press the cam stopper down and the press-in sleeve

until the hook can be engaged on the pin. When tensioner is bolted in position the hook will release automatically.

➡**Ensure the arrow on the outside of the tensioner faces the front of the engine.**

47. Install the remaining components in the reverse order of removal.

48. Connect the negative battery cable, Refill fluid levels, start the engine, and bleed the cooling system. Check for leaks and road test the vehicle for proper operation.

2.4L Engine

ALTIMA

1. Disconnect the negative battery cable.

2. Drain the coolant from the engine and radiator.

3. Drain the engine oil.

4. Remove the engine undercover.

❊❊ CAUTION

The fuel injection system remains under pressure after the engine has been turned OFF. Properly relieve fuel pressure before disconnecting any fuel lines. Failure to do so may result in fire or personal injury.

5. Remove the vacuum hoses, fuel hoses, wires, harness, and connectors.

6. Remove the alternator and bracket, the upper radiator hose, the air duct, and the front exhaust tube.

7. Remove the intake manifold collector supports, intake manifold collector, and the exhaust manifold.

8. Set the No. 1 piston at TDC on its compression stroke.

9. Remove the distributor.

10. Using a block of wood, set a transmission jack under the aluminum oil pan and remove the front engine mounting.

11. Remove the rocker cover. Remove the rocker cover bolts in the proper sequence.

12. Remove the camshaft sprockets.

➡**The stoppers on camshaft covers prevent the upper timing chain from disengaging from the idle sprocket.**

13. Remove the cam bearing caps in sequence and remove the camshafts. The camshaft brackets must be loosened in reverse order of tightening to prevent damage to the camshaft.

➡**These parts must be reassembled in their original positions.**

14. Loosen the cylinder head bolts in the reverse order of installation.

➡**A warped or cracked cylinder head could result from loosening in incorrect order. The cylinder head bolts should be loosened in two or three steps.**

15. Remove the cam sprocket cover.

16. Remove the upper chain tensioner and upper chain guides.

17. Remove the upper timing chain.

18. Remove the idler sprocket bolt.

19. Remove the cylinder head and the intake manifold. Remove the cylinder head gasket. The lower timing chain will not be disengaged from crankshaft sprocket.

➡**The cast portion of the front cover is located on the lower side of the crankshaft sprocket, so the lower timing chain need not be disengaged from idler sprocket.**

20. Remove the steel oil pan bolts in the reverse sequence of the tightening procedure.

21. Install a seal cutter between the steel oil pan and the aluminum oil pan

22. Tapping the cutter with a hammer, slide it around the entire edge of the oil pan. Take care not to damage the aluminum oil pan.

23. Remove the steel oil pan.

24. Remove the baffle plate, oil strainer, and the front tube.

25. Support the transaxle with a jack and the engine with a engine hoist. Remove the front suspension member.

26. Remove the A/C compressor gussets.

27. Remove the rear cover plate.

28. Remove the aluminum oil pan retaining bolts in sequence.

29. Insert a seal cutter between the oil pan and the cylinder block.

30. Tapping the cutter with a hammer, slide it around the entire edge of the oil pan. Take care not to damage the aluminum oil pan.

31. Lower the oil pan from the cylinder block and remove it from the engine.

32. Remove the crankshaft pulley.

33. Remove the front timing chain cover.

34. Remove the oil pump drive spacer.

35. Remove the lower timing chain tensioner, tensioner arm, and lower timing chain guide.

36. Remove the lower timing chain and idler sprocket.

To install:

37. Install the crankshaft sprocket and oil pump drive spacer.

38. Install the idler sprocket and lower timing chain.

39. Set the lower timing chain on the sprockets, aligning the mating marks. The mating marks on the timing chain assembly will be silver.

40. Install the chain tension arm and chain guide.

41. Install the lower timing chain tensioner.

42. Apply a continuous bead of liquid gasket to the front cover and install the front cover. Install a new oil seal.

43. Install the crankshaft pulley and tighten bolt to 105–112 ft. lbs. (142–152 Nm).

44. Carefully scrape the old gasket material away from the pan and cylinder block mounting surfaces, then apply a continuous bead (3.5–4.5mm) of liquid gasket around the oil pan and the cylinder block.

45. Install the aluminum oil pan and tighten the mounting bolts in sequence, to 13 ft. lbs. (17.5 Nm).

46. Install the baffle plate, oil strainer, and the front tube.

47. Install the steel oil pan and tighten the bolts in sequence to 61 inch lbs. (7 Nm).

48. Reinstall the rear cover plate.

49. Install the A/C compressor gussets.

50. Reinstall the front suspension member.

51. Install the front engine mounting.

52. Remove the engine hoist and transaxle support.

53. Install new cylinder head gasket.

54. Reinstall the cylinder head and temporarily tighten the cylinder head bolts when installing the front cover. This is necessary to avoid damaging the cylinder head gasket. Be sure to install washers between the bolts and cylinder head.

55. Install the upper timing chain, chain tensioner and chain guide.

56. Set the upper timing chain on idler sprockets, aligning the mating marks.

57. Install cam sprocket cover. Apply a continuous bead of liquid gasket to front cover. Be careful not to damage the cylinder head gasket. Be careful that the upper timing chain does not slip or jump when installing cam sprocket cover.

58. Tighten cylinder head bolts.

59. Install the camshafts and camshaft bearing caps.

60. Install the camshaft sprockets, then tighten the sprocket bolts to 123–130 ft. lbs. (167–176 Nm). Install the chain guide between both camshaft sprockets. The alignment marks on the upper portion of the timing chain should now be aligned.

61. The balance of installation is the reverse of the removal procedure.

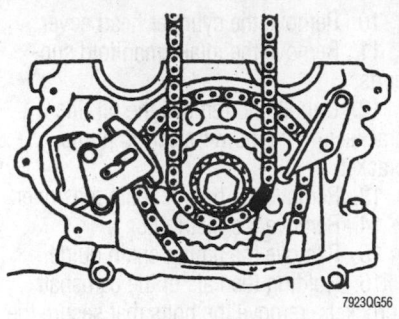

Be sure to align the mark on the idler sprocket with the mark on the upper chain—2.4L engines

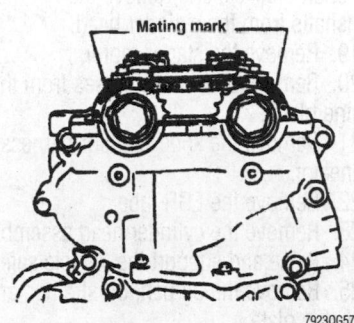

Align the marks on the camshaft sprockets with the upper portion of the upper timing chain and mating marks—2.4L engine

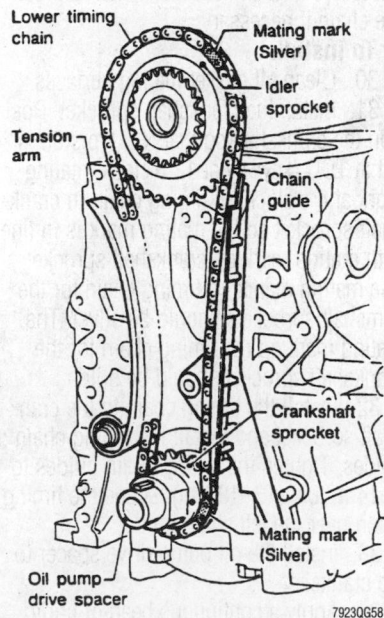

Lower timing chain alignment marks— 2.4L engine

62. Connect the negative battery cable.
63. Start the engine and make the necessary adjustments. Check for proper operation and leaks.

240SX

1. Disconnect the negative battery cable.
2. Relieve the fuel system pressure.
3. Drain the coolant from engine and radiator.
4. Drain the engine oil.
5. Remove the engine undercover.
6. Remove the air duct assembly.
7. Remove the fan shroud and the cooling fan.

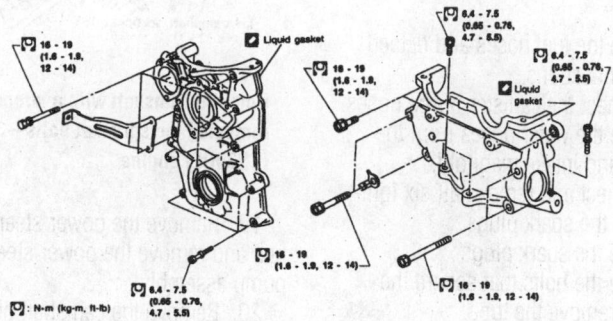

Exploded view of the timing chain front covers—2.4L engine

8. Remove the exhaust manifold cover.
9. Remove the front exhaust tube and if equipped, the A.I.V. pipe.

❊❊ CAUTION

The fuel injection system remains under pressure after the engine has been turned OFF. Properly relieve the fuel pressure before disconnecting any fuel lines. Failure to do so may result in fire or personal injury.

10. Disconnect and tag all of the vacuum hoses, fuel lines, and electrical connections.
11. Remove all of the spark plugs.
12. Remove the distributor cap with the spark plug wires attached.
13. Remove the injector tube assembly with the injectors.
14. Remove the rocker cover bolts, in reverse order of installation.
15. Set the No. 1 piston at TDC (top dead center) on its compression stroke.
16. Remove the distributor. Be sure to note the positioning of the distributor rotor before removing the distributor.
17. Remove the camshaft sprockets. Be sure to hold the flats of the camshafts when removing the sprocket bolts.

➡**The stoppers on the inside of the camshaft covers prevent the upper timing chain from disengaging from the idle sprocket.**

18. Remove the camshaft bearing caps, in reverse order of installation, then remove both of the camshafts.

➡**The camshaft bearing caps and camshafts should be kept in their original position for reassembly.**

19. Loosen the cylinder head bolts in the reverse order of installation.

➡**Head warpage or cracking could result from removing the head bolts in the incorrect order. The cylinder head bolts should be loosened in two or three steps.**

20. Remove the camshaft sprocket cover.
21. Remove the upper chain tensioner and upper chain guides.

➡**Compress the piston of the tensioner and insert a suitable pin into the pin hole.**

22. Remove the upper timing chain.
23. Remove the idler sprocket bolt.
24. Remove the cylinder head with the intake manifold and exhaust manifold assembly.
25. Remove the cylinder head gasket.
26. Remove the oil pan bolts.
27. Install a seal cutter between the oil pan and the engine block.
28. Tapping the cutter with a hammer, slide it around the entire edge of the oil pan. Take care not to damage the oil pan.
29. Remove the oil pan assembly.
30. Remove the baffle plate, oil strainer, and the front tube.
31. Remove the engine drive belts.

32. Remove the A/C compressor idler pulley.
33. Remove the crankshaft pulley.
34. Remove the front timing chain cover.
35. Remove the oil pump drive spacer.
36. Remove the lower timing chain tensioner, tensioner arm, and the lower timing chain guide.

➡**Compress the piston of the tensioner and insert a suitable pin into the pin hole.**

37. Remove the lower timing chain and crankshaft sprocket.

To install:

38. Check all components for wear. Replace as necessary. Clean all mating surfaces and replace the cylinder head gasket.
39. Install the crankshaft sprocket and the oil pump drive spacer.
40. Install the idler sprocket and the lower timing chain.

➡**Be sure that the mating marks of the crankshaft sprocket are facing front of engine.**

41. Set the lower timing chain on the sprockets, aligning the mating marks. The mating marks on the timing chain assembly will be silver.
42. Install the chain tension arm and the chain guide.
43. Install the lower timing chain tensioner.

➡**After installation of the tensioner, remove the pin to release the piston.**

44. Apply a continuous bead of liquid gasket to the front cover and install the front cover. Install a new oil seal.
45. Install the crankshaft pulley and tighten the bolt to 105–112 ft. lbs. (142–152 Nm).
46. Carefully scrape the old gasket material away from the pan and cylinder block mounting surfaces, then apply a continuous bead (3.5–4.5mm) of liquid gasket around the oil pan and the cylinder block.
47. Install the baffle plate, oil strainer, and the front tube.
48. Install the oil pan and tighten the mounting bolts, in sequence, to 57–66 inch lbs. (6.4–7.5 Nm).
49. Install the A/C compressor idler pulley.
50. Install a new cylinder head gasket and install the idler shaft.
51. Install the cylinder head and temporarily tighten the cylinder head bolts when installing the front cover. This is necessary to avoid damaging the cylinder head gasket. Be sure to install washers between the bolts and cylinder head.

❉❉ CAUTION

Do not fully tighten any of the cylinder head bolts at this time. tighten the bolts finger-tight only.

52. Install the upper timing chain, chain tensioner, and the chain guide. Be sure to align the mark on the timing chain with the idler.

➡ **After installation of the tensioner, remove the pin to release the piston.**

53. Apply a continuous bead of liquid sealant and install the camshaft sprocket cover. Tighten the mounting bolts to specifications.

54. Tighten the cylinder head bolts.

55. Install the camshafts and the camshaft bearing caps.

56. Install the camshaft sprockets and tighten the sprocket bolts to 123–130 ft. lbs. (167–176 Nm). Install the chain guide between both of the camshaft sprockets. The alignment marks on the upper portion of the timing chain should now be aligned.

57. The remaining components are installed in the reverse order from which they were removed.

58. Check and adjust the valve clearance.

59. Connect the negative battery cable.

60. Start the engine and check for leaks. Bleed the cooling system.

61. Install the engine undercover.

62. Check the ignition timing; adjust the timing as necessary.

3.0L (VQ30DE) Engine

1. Disconnect the negative battery cable.

2. Drain the engine oil and the cooling system. Be sure to drain the engine block and the radiator.

3. Relieve the fuel system pressure.

4. Remove the left side rocker cover ornament.

➡ **Before detaching any hoses or connectors, note the locations for reassembly.**

5. Remove the air duct to intake manifold hose, collector hose, blow-by hose, and vacuum hoses.

❉❉ CAUTION

The fuel injection system remains under pressure after the engine has been turned OFF. Properly relieve fuel pressure before disconnecting any fuel lines. Failure to do so may result in fire or personal injury.

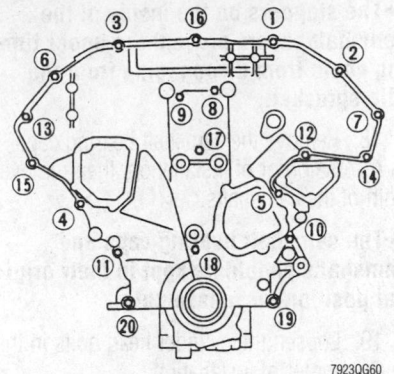

Remove the front timing chain case mounting bolts in the sequence shown—3.0L (VQ30DE) engines

6. Remove the fuel hoses and detach the harness connections.

7. Disconnect the canister purge hoses.

8. Remove the water hoses from the cylinder head and intake manifold.

9. Disconnect and remove all six ignition coils from the spark plugs.

10. Remove the spark plugs.

11. Remove the bolts that secure the EGR tube and remove the tube.

12. Remove the intake manifold collector supports and remove the collector.

13. Remove the bolts that secure the fuel tube and remove the fuel tube from the vehicle.

14. Remove the bolts that secure the intake manifold to the engine block and remove the manifold. Loosen the bolts in the reverse sequence of the tightening procedure.

15. Remove the LH and RH rocker covers from the cylinder head.

16. Remove the engine undercovers.

17. Remove the right front wheel and the engine side covers.

18. Remove the drive belts and the idler pulley.

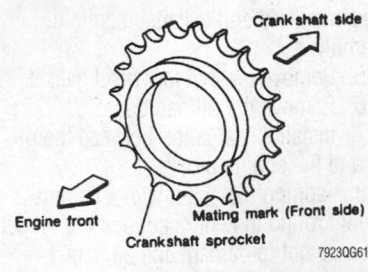

Crankshaft sprocket with mating marks—3.0L (VQ30DE) engine

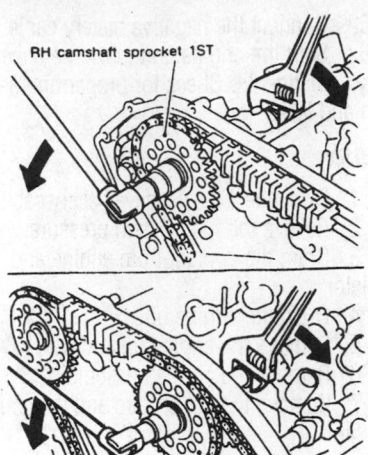

Hold the camshaft with a wrench while removing the sprocket bolts—3.0L (VQ30DE) engine

19. Remove the power steering oil pump belt and remove the power steering oil pump assembly.

20. Remove the camshaft position sensor (PHASE) and crankshaft position sensors (REF)/(POS).

21. Set the No. 1 piston to TDC of compression stroke by rotating the crankshaft.

22. Remove the ring gear cover access plate.

23. Loosen the crankshaft pulley bolt while securing the ring gear so the crankshaft cannot rotate.

➡ **Use care not to damage the ring gear teeth.**

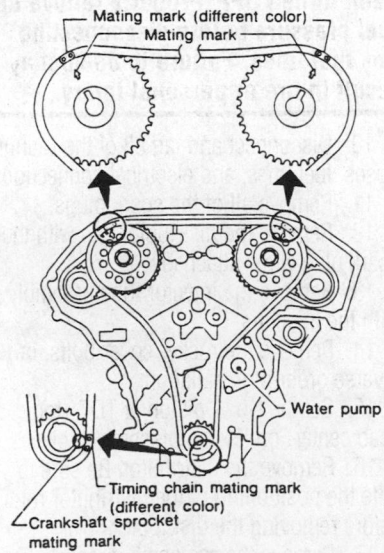

Timing chain alignment marks—3.0L (VQ30DE) engine

24. Using a suitable puller, remove the crankshaft pulley.

25. Remove the A/C compressor and bracket.

26. Remove the front exhaust pipe and its support.

27. Hang the engine at the right and left side engine slingers with a suitable hoist.

28. Support the transaxle with jack.

29. Remove the right side engine mounting, mounting bracket, and nuts.

30. Remove the center crossmember assembly.

31. Remove the steel (lower) oil pan bolts in the reverse of the installation sequence.

32. Insert a seal cutter between the steel and aluminum oil pan.

33. Tapping the cutter with a hammer, slide it around the entire edge of the oil pan. Be careful not to damage the aluminum mating surface of the upper oil pan.

34. Remove the steel oil pan and the oil strainer.

35. Remove the aluminum (upper) oil pan bolts in the reverse of the installation sequence.

36. Remove the transaxle bolts that secure the oil pan.

37. Insert a seal cutter between the aluminum oil pan and the engine block.

38. Tapping the cutter with a hammer, slide it around the entire edge of the oil pan. Be careful not to damage the mating surfaces of the oil pan or engine block.

39. Remove the oil pan from the vehicle.

40. Remove the water pump cover and remove the bolts that secure the front timing chain case.

41. Using the seal cutter, remove the timing chain case cover.

42. Remove the internal timing chain guide and the upper chain guide.

43. Remove the timing chain tensioner and slack side chain guide.

44. Remove the left and right intake camshaft sprockets first. Be sure to hold the flats of the camshafts while removing the sprocket bolts.

45. Remove the lower timing chain assembly. Be sure to note the aligning marks of the chain before removal.

46. Insert a suitable stopper pin for the left and right camshaft tensioners.

47. Remove the left and right exhaust camshaft sprocket bolts. Be sure to hold the flats of the camshafts while removing the sprocket bolts.

48. Remove the upper timing chain assembly. Be sure to note the aligning marks of the chain before removal.

49. Remove the lower timing chain guide.

50. Remove the crankshaft sprocket.

51. Remove all traces of liquid gasket from the front timing chain case and from the water pump.

52. Inspect the timing chain for excessive wear or damage and replace as necessary.

To install:

53. Install the crankshaft sprocket with the mating mark facing out.

54. Position the crankshaft to TDC of compression stroke and align the dowels of the camshaft sprockets to the 12 o'clock position in respect to the cylinder head.

55. Install the lower timing chain guide. The front mark on the guide should face upwards.

56. On a work bench, align the marks on the intake and exhaust camshaft sprockets with the marks of the chain.

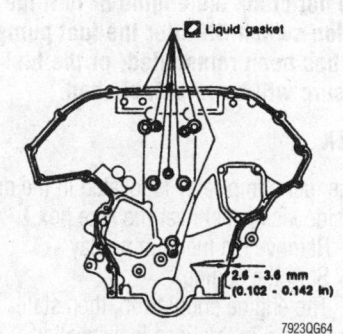

Application of liquid gasket to the front timing case—3.0L (VQ30DE) engines

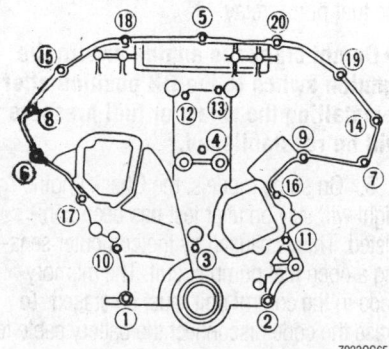

Tighten the front timing chain case bolts according to the sequence shown—3.0L (VQ30DE) engines

57. Put the exhaust camshaft sprockets onto the dowel pin and tighten the mounting bolts to 88–95 ft. lbs. (119–128 Nm). Be sure to secure the camshafts while tightening the bolts.

58. Install the timing chains and sprockets to the intake camshafts. Be sure to align the timing chain and sprocket mating marks.

59. Remove the left and right camshaft tensioner stopper pins.

60. Align the mating mark on the crankshaft with the matchmark (gold link) on the lower timing chain.

61. Attach the lower timing chain to the water pump sprocket.

62. Working counterclockwise, install the lower timing chain camshaft sprockets. Be sure to align the sprocket marks with the blue links of the timing chain during installation.

63. Install the intake sprocket bolts and tighten to 88–95 ft. lbs. (119–128 Nm). Be sure to secure the camshafts while tightening the bolts.

64. Install the internal timing chain guide, upper timing chain guide, lower timing chain tensioner and slack side timing chain guide.

65. Tighten the tensioner mounting bolt to 75–96 inch lbs. (8.4–10.8 Nm) and tighten the guide bolts to 108–168 inch lbs. (13–19 Nm).

66. Apply a 0.102–0.142 in. (2.6–3.6mm) continuous bead of liquid gasket to all necessary areas as shown on the front timing cover.

67. Install the timing cover evenly and gently. Be sure to align the dowel pin holes.

68. Tighten the mounting bolts in sequence as follows:

    a. Bolts No. 1 and 2—Tighten to 19–23 ft. lbs. (26–31 Nm)

    b. Bolts No. 3–20—Tighten to 105–121 inch lbs. (11.8–13.7 Nm)

➡**Leave the bolts unattended for 30 minutes or more after tightening. This will allow the liquid gasket to cure sufficiently.**

69. Apply a 0.091–0.130 in. (2.3–3.3mm) continuous bead of liquid gasket to the water pump cover and install the cover. Tighten the bolts to 84–108 inch lbs. (10–13 Nm).

70. The balance of installation is the reverse of the removal procedure.

71. Connect the negative battery cable.

72. Start the engine and run at 3000 RPM under no load to purge the air from the high pressure chamber. The engine may produce a rattling noise. This indicates that air still remains in the chamber and is not a matter of concern.

FUEL SYSTEM

Fuel System Service Precautions

Safety is the most important factor when performing not only fuel system maintenance but any type of maintenance. Failure to conduct maintenance and repairs in a safe manner may result in serious personal injury or death. Maintenance and testing of the vehicle's fuel system components can be accomplished safely and effectively by adhering to the following rules and guidelines.

• To avoid the possibility of fire and personal injury, always disconnect the negative battery cable unless the repair or test procedure requires that battery voltage be applied.

• Always relieve the fuel system pressure prior to disconnecting any fuel system component (injector, fuel rail, pressure regulator, etc.), fitting or fuel line connection. Exercise extreme caution whenever relieving fuel system pressure to avoid exposing skin, face and eyes to fuel spray. Please be advised that fuel under pressure may penetrate the skin or any part of the body that it contacts.

• Always place a shop towel or cloth around the fitting or connection prior to loosening to absorb any excess fuel due to spillage. Ensure that all fuel spillage (should it occur) is quickly removed from engine surfaces. Ensure that all fuel soaked cloths or towels are deposited into a suitable waste container.

• Always keep a dry chemical (Class B) fire extinguisher near the work area.

• Do not allow fuel spray or fuel vapors to come into contact with a spark or open flame.

• Always use a back-up wrench when loosening and tightening fuel line connection fittings. This will prevent unnecessary stress and torsion to fuel line piping.

Always follow the proper torque specifications.

• Always replace worn fuel fitting O-rings with new. Do not substitute fuel hose or equivalent, where fuel pipe is installed.

Fuel System Pressure

RELIEVING

Except 300ZX

The fuel pump fuse is located in the dash fuse box or in the engine compartment fuse box. Check the lid of the fuse box for exact location.

1. Remove the fuel pump fuse.
2. Start the engine.
3. Start the engine and run until the engine stalls.
4. After the engine stalls, try to restart the engine; if the engine will not start, the fuel pressure has been released.
5. Turn the ignition switch **OFF**. Reinstall the fuel pump fuse into the fuse block.

➡**Do not crank the engine or turn the ignition switch ON after the fuel pump fuse has been reinstalled, or the fuel pressure will be re-established.**

300ZX

The fuel pump relay is located in the drivers side kick panel near the fuse box.

1. Remove the fuel pump relay.
2. Start the engine.
3. The engine should run, then stall when the fuel in the lines is exhausted. When the engine stops, crank the starter a few times for about five seconds to be sure all pressure in the fuel lines is released.
4. Turn the ignition **OFF** and reinstall the fuel pump relay.

➡**Do not crank the engine or turn the ignition switch to the ON position after reinstalling the relay, or fuel pressure will be reestablished.**

5. On some models, the Check Engine Light will stay on after test has been completed. This is caused by the computer sensing a open fuel pump circuit. The memory code in the control unit must be erased. To erase the code disconnect the battery cable for 30 seconds, then reconnect the cable.

Fuel Filter

REMOVAL & INSTALLATION

All Models

⁂ **CAUTION**

Be sure to relieve the fuel system pressure on fuel injected engines before replacing the fuel filter.

1. Properly relieve fuel system pressure.
2. Disconnect the negative battery cable.
3. Loosen the fuel hose clamps and disconnect the hoses from the fuel filter.
4. Remove the bolt securing the filter to the bracket or just remove the filter from the bracket clips.
5. Remove the filter.

To install:

6. Install the new filter and secure the filter in the bracket.
7. If necessary, replace the fuel line hoses and hose clamps. Reconnect the fuel hoses and tighten the clamps.
8. Reconnect the negative battery cable.
9. Install the fuel pump fuse.
10. Start the engine and check for leaks.

Fuel Pump

REMOVAL & INSTALLATION

Sentra and 200SX

The fuel pump is located in the fuel tank on all vehicles. In-tank fuel pumps are accessible by lifting up the rear seat to gain access to the inspection cover.

1. Relieve the fuel system pressure.
2. Disconnect the negative battery cable.
3. Remove the rear seat from the vehicle.
4. Remove the inspection cover that is located under the rear seat.
5. Disconnect the inlet and outlet fuel lines from the fuel pump assembly.
6. Disconnect the fuel pump and gauge wiring connections.
7. Remove the six mounting bolts that secure the fuel pump assembly to the top of the fuel tank.

8. Raise up the fuel pump assembly and detach the fuel tubes and connector. Remove the fuel gauge assembly.

9. Remove the fuel pump with the fuel chamber.

10. Pull up the front of the fuel pump chamber and slide the chamber forward.

11. Remove the fuel pump from the chamber.

12. Discard the O-ring seal or gasket.

To install:

13. Install the fuel pump to the fuel pump chamber and slide chamber rearward.

14. Install the fuel pump with the fuel pump chamber.

15. Using a new O-ring, install the fuel gauge assembly and connect the fuel tubes and connector. Use new hoses and clamps.

16. Install the six mounting bolts to the top of the fuel gauge unit. Tighten the bolts to 28–37 inch lbs. (3–4 Nm).

17. Connect the fuel pump and gauge wiring connections.

18. Using new hoses and clamps, connect the inlet and outlet fuel lines to the fuel pump assembly.

19. Install the inspection cover and install the rear seat.

20. Connect the negative battery cable, start the engine, and check for leaks.

➡ **On some models, the Check Engine Light will stay on after the repair is complete. This is caused by an open fuel pump circuit when relieving the fuel pump pressure The memory code in the ECU must be erased. To erase the code, disconnect the battery cables for one minute, then reconnect.**

Altima

1. Relieve the pressure from the fuel system.

2. Disconnect the negative battery cable.

3. Remove the rear seat and remove the access cover.

4. Detach the fuel pump electrical connector.

5. Disconnect the fuel lines from the fuel pump assembly.

6. Remove the locking ring using special tool J38879 or equivalent.

7. Remove the fuel gauge assembly; detach the fuel tube and connector from the fuel gauge.

➡ **When the fuel sending unit needs to be removed, pull the tab upwards. The tab is located on the sending unit, opposite the end of the float. After the tab is pulled, the sending unit will lift straight out of the tank bracket.**

8. Remove the fuel pump by pinching the two locking tabs together. Lift the fuel pump assembly straight upward and out of fuel tank.

9. Remove the O-ring and discard. Place a clean rag in the hole to keep out dirt.

To install:

10. Remove the rag and install a new O-ring.

11. Install the fuel pump.

12. Connect the electrical connection and fuel tube to the fuel gauge sending unit.

13. Install the fuel sending unit into the tank.

➡ **Verify that the mark on the fuel tank and the components are aligned when installing the pump and fuel gauge sending unit.**

14. Install the locking ring and tighten ring to 22–26 ft. lbs. (30–35 Nm).

15. Connect the fuel lines and fuel pump electrical connector. Always install new clamps on the fuel lines.

16. Install the fuel pump access cover.

17. Install the rear seat.

18. Connect the negative battery cable.

19. Start the engine and check for leaks.

Maxima, 300ZX and 240SX

1. Relieve the fuel system pressure

2. Disconnect the negative battery cable.

3. Remove the rear seat or open the access panel in the trunk.

4. Detach the fuel gauge electrical connector and pump electrical connector.

5. Disconnect the fuel outlet and the return hoses. If necessary, remove the fuel tank.

6. On some 240SX models you need to remove the fuel pump assembly-to-fuel tank bolts and lift the fuel pump assembly from the fuel tank.

7. On other models you need to remove the locking ring with tool SST J38879-A or equivalent and raise the fuel pump from the tank. Disconnect the feed tube while raising the pump.

8. Discard the O-ring. Plug the fuel tank opening with a clean rag to prevent dirt from entering the system.

➡ **When removing or installing the fuel pump assembly, be careful not to damage or deform it and always install a new O-ring.**

To install:

9. Remove the rag; using a new O-ring, install fuel pump assembly into the fuel tank.

10. Install the fuel pump assembly-to-fuel tank bolts and tighten the bolts to 17–22 inch lbs. (2.0–2.5 Nm).

11. Install the locking ring assembly and tighten.

12. If removed, install the fuel tank assembly.

13. Connect the fuel lines and the electrical connectors. Always use new clamps when reconnecting fuel line hoses.

➡ **When installing the upper plate, be sure to align the mark with the center marks on the fuel tank.**

14. Install the fuel pump access cover.

15. Connect the negative battery cable.

16. Start the engine and check for fuel leaks.

➡ **On some models, the Check Engine Light will stay ON after installation is completed. The memory code in the control unit must be erased. This code is stored for an open fuel pump circuit, this is caused when the fuel pressure is released. To erase the code, disconnect the battery cable for 10 seconds, then reconnect after installation of fuel pump.**

DRIVE TRAIN

Transmission Assembly

REMOVAL & INSTALLATION

240SX

1. Disconnect the negative battery cable.

2. Raise and support the vehicle safely.

3. On automatic transmissions, drain the fluid and disconnect the cooler lines, then plug the openings.

➡ **It may be necessary disconnect to the left side cooler line after the transmission has been slightly lowered.**

4. On manual transmissions, place the shift lever in the **N** position and disconnect the shifter lever from the transmission. Remove the shifter lever with control housing from the transmission. On automatic transmissions, disconnect the shift linkage at the transmission.

5. If equipped, remove the crankshaft position sensor from the upper side of the transmission housing.

6. On manual transmissions, remove the clutch operating (slave cylinder) cylinder from the clutch housing.

7. Detach the electrical harness connections from the transmission and detach the rear heated oxygen sensor connection.

8. Matchmark, then unbolt the driveshaft at the rear and remove. If equipped with a center bearing, unbolt it from the crossmember. Plug the end of the transmission extension to prevent leakage.

9. Support the engine with a large wood block and a jack under the oil pan. Do not place the jack under the oil pan drain plug.

10. Unbolt the transmission from the crossmember. Support the transmission with a jack and remove the crossmember.

11. On automatic transmissions, mark the relationship between the torque converter and the drive plate. Remove the bolts that secure the torque converter to the drive plate. Gain access to the bolts by turning the front crankshaft bolt.

12. Lower the rear of the engine to allow clearance.

13. Disconnect the wiring to the starter motor and remove the starter from the transaxle.

14. Remove the exhaust tube mounting bracket from the transmission.

15. Unbolt the transmission assembly from the engine. Lower and move it to the rear.

➡ **Tagging the different length transmission bolts upon removal is necessary to ensure that they are installed in their original position.**

To install:

16. Clean the engine and transmission mating surfaces.

17. On manual transmissions, lightly lubricate the clutch disc splines and main drive gear splines. Also lubricate the control lever sliding surfaces with grease.

18. Properly support the transmission and install the transmission to the rear of the engine.

19. For manual transmissions, use the following torque specifications to bolt the transmission to the engine:

   a. Tighten bolts No. 1 and 2 to 29–36 ft. lbs. (39–49 Nm).

   b. Tighten bolts No. 3, 4 and 5 to 22–29 ft. lbs. (29–39 Nm).

20. For automatic transmissions tighten the transmission mounting bolts as follows:

   a. Tighten bolts No. 1 and No. 2 to 29–36 ft. lbs. (39–49 Nm)

   b. Tighten bolt No. 3 to 22–29 ft. lbs. (29–39 Nm)

   c. Tighten the gusset mounting bolts to 22–29 ft. lbs. (29–39 Nm).

21. Install the exhaust tube mounting bracket to the transmission.

22. Install the starter assembly.

23. Raise the rear of the engine to its original position.

24. Bolt the crossmember in place and remove the jack.

25. Install the remaining components in the reverse order of removal.

26. Connect the negative battery cable and check the transmission fluid level.

27. Lower the vehicle and road test the vehicle for proper operation.

300ZX

1. Disconnect the negative battery cable.

2. Raise and support the vehicle safely.

3. Remove the exhaust pipe section from the manifold and remove the support bracket from the transmission.

4. Matchmark the driveshaft and unbolt the shaft from the rear housing. Unbolt the driveshaft at the center bearing and remove the shaft from the rear of the transmission. Install a sealing plug in the end of the transmission extension housing to prevent leakage.

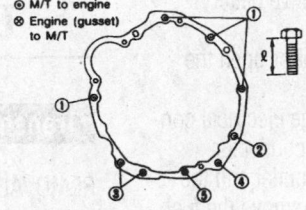

Bolt No.	Tightening torque N·m (kg-m, ft-lb)
①	39 - 49 (4.0 - 5.0, 29 - 36)
②	39 - 49 (4.0 - 5.0, 29 - 36)
③*	29 - 39 (3.0 - 4.0, 22 - 29)
④*	29 - 39 (3.0 - 4.0, 22 - 29)
⑤	29 - 39 (3.0 - 4.0, 22 - 29)
Gusset to engine	29 - 39 (3.0 - 4.0, 22 - 29)

*: With nut.

7923QG66

Transmission bolt tightening specifications—240SX with manual transmission

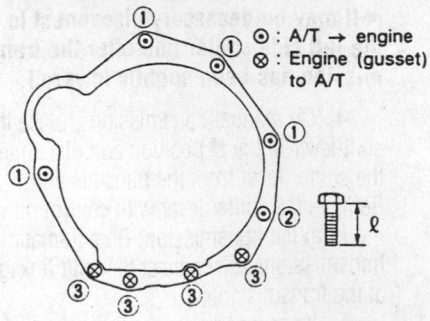

Bolt No.	Tightening torque N·m (kg-m, ft-lb)	Bolt length "ℓ" mm (in)
①	39 - 49 (4.0 - 5.0, 29 - 36)	40 (1.57)
②	39 - 49 (4.0 - 5.0, 29 - 36)	50 (1.97)
③	29 - 39 (3.0 - 4.0, 22 - 29)	25 (0.98)
Gusset to engine (4 bolts)	29 - 39 (3.0 - 4.0, 22 - 29)	20 (0.79)

7923QG67

Transmission bolt tightening specifications—240SX with automatic transmission

5. Disconnect the speedometer drive cable from the transmission.

6. On manual transmissions, disconnect the control rod from the lever. For automatics, disconnect the shift linkage.

7. On manual transmissions, without disconnecting the hydraulic line, remove the clutch operating cylinder from the clutch housing and position it aside.

8. Support the engine with a large wood block and a jack under the oil pan. Do not place the jack under the oil pan drain plug.

9. Unbolt the transmission from the crossmember. Support the transmission with a jack and remove the crossmember.

10. On automatics, remove the flexplate-to-torque converter bolts.

11. Lower the rear of the engine to allow clearance.

12. Unplug the back-up light, neutral and overdrive switch connectors.

13. Unbolt the transmission. Lower and remove it to the rear.

➡ **The transmission bolts are different lengths. Tagging the transmission-to-engine bolts upon removal will facilitate proper positioning during installation.**

To install:

14. Lubricate the output shaft of the transmission with high temperature grease.

15. Raise the transmission onto the engine and install the mounting bolts. Tighten bolts to correct torque specification.

➡ **The bolts that secure the transmission to the engine are of different length and require different torque. Be sure to properly position all bolts.**

16. On automatic transmissions, install the flexplate-to-torque converter bolts. Tighten the bolts to 33–43 ft. lbs. (45–58 Nm).

17. Connect the back-up light, and the neutral and overdrive switch connectors.

18. Install the crossmember.

19. Install the clutch slave cylinder and tighten the mounting bolts to 22–30 ft. lbs. (30–40 Nm).

20. On manual transmissions, install the control rod to the lever. On automatics, connect the shift linkage to the transmission.

21. Connect the speedometer drive cable.

22. Install the driveshaft. Tighten the flange bolts to 41–48 ft. lbs. (55–65 Nm) for non-turbo models or 47–54 ft. lbs. (64–74 Nm) for turbo models. Tighten the center bearing bracket mounting bolts to 43–58 ft. lbs. (59–78 Nm).

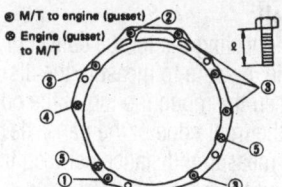

Bolt No.	Tightening torque N·m (kg-m, ft-lb)	ℓ mm (in)
①	39 - 49 (4.0 - 5.0, 29 - 36)	100 (3.94)
②	39 - 49 (4.0 - 5.0, 29 - 36)	65 (2.56)
③	39 - 49 (4.0 - 5.0, 29 - 36)	60 (2.36)
④	29 - 39 (3.0 - 4.0, 22 - 29)	55 (2.17)
⑤	29 - 39 (3.0 - 4.0, 22 - 29)	25 (0.98)

79230G68

Manual transmission mounting bolt identification and torque specifications—300ZX without turbocharger

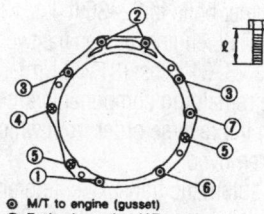

● Tighten all transmission bolts.

Bolt No.	Tightening torque N·m (kg-m, ft-lb)	ℓ mm (in)
①	39 - 49 (4.0 - 5.0, 29 - 36)	100 (3.94)
②	39 - 49 (4.0 - 5.0, 29 - 36)	55 (2.17)
③	39 - 49 (4.0 - 5.0, 29 - 36)	60 (2.36)
④	29 - 39 (3.0 - 4.0, 22 - 29)	55 (2.17)
⑤	29 - 39 (3.0 - 4.0, 22 - 29)	25 (0.98)
⑥	29 - 39 (3.0 - 4.0, 22 - 29)	60 (2.36)
⑦	39 - 49 (4.0 - 5.0, 29 - 36)	65 (2.56)

79230G69

Manual transmission mounting bolt identification and torque specifications—300ZX with turbocharger

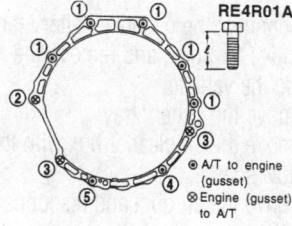

Bolt No.	Tightening torque N·m (kg-m, ft-lb)	Bolt length "ℓ" mm (in)
①	39 - 49 (4.0 - 5.0, 29 - 36)	47.5 (1.870)
②	39 - 49 (4.0 - 5.0, 29 - 36)	58 (2.28)
③	29 - 39 (3.0 - 4.0, 22 - 29)	25 (0.98)
④	29 - 39 (3.0 - 4.0, 22 - 29)	60 (2.36)
⑤	29 - 39 (3.0 - 4.0, 22 - 29)	65 (2.56)
Gusset to engine	29 - 39 (3.0 - 4.0, 22 - 29)	20 (0.79)

79230G70

Automatic transmission mounting bolt identification and torque specifications—300ZX with RE4R01A transmission

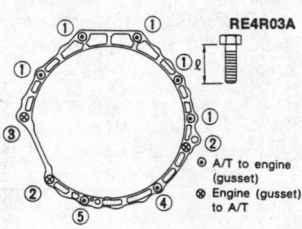

Bolt No.	Tightening torque N·m (kg-m, ft-lb)	Bolt length "ℓ" mm (in)
①	39 - 49 (4.0 - 5.0, 29 - 36)	65 (2.56)
②	29 - 39 (3.0 - 4.0, 22 - 29)	25 (0.98)
③	39 - 49 (4.0 - 5.0, 29 - 36)	58 (2.28)
④	29 - 39 (3.0 - 4.0, 22 - 29)	62 (2.44)
⑤	29 - 39 (3.0 - 4.0, 22 - 29)	100 (3.94)
Gusset to engine	29 - 39 (3.0 - 4.0, 22 - 29)	20 (0.79)

79230G71

Automatic transmission mounting bolt identification and torque specifications—300ZX with RE4R03A transmission

23. Connect the exhaust tube section to the manifolds and attach the support bracket to the transmission.

24. Remove the jack from underneath the engine.

25. Lower the vehicle and connect the negative battery cable.

26. On automatic transmissions, check the fluid level and add if necessary.

27. Road test the vehicle for proper operation.

Transaxle Assembly

REMOVAL & INSTALLATION

Sentra and 200SX

1. Disconnect the negative and positive battery cables.

2. Remove the battery and bracket from the vehicle.

3. On manual transaxles, remove the shifter linkage from the transaxle.

4. Remove the air duct between the throttle body and the air cleaner.

5. Tag and detach all electrical connectors from the transaxles.

6. Remove the crankshaft position sensor from the transaxle.

7. Drain the fluid from the transaxle.

8. Disconnect the control cable from the transaxle.

9. If equipped, disconnect the oil cooler lines from the transaxle.

10. Remove the halfshafts from the transaxle.

11. Remove the intake manifold support brackets.

12. Remove the starter motor from the transaxle.

13. Remove the upper bolts that secure the transaxle to the engine.

14. Using a block of wood, support the transaxle with a jack.

15. Remove the center crossmember.

16. If equipped, remove the front and rear gussets.

17. Remove the rear plate cover.

18. On automatic transaxles, remove the torque converter mounting bolts. It will be necessary to rotate the engine by hand to gain access to all bolts.

19. Remove the rear transaxle to engine bracket.

20. Remove the rear transaxle mount.

21. Remove the lower transaxle to engine mounting bolts.

22. Slide the transaxle away from the engine and lower the transaxle assembly.

To install:

When connecting the torque converter to the transaxle, be sure to measure the distance between the mounting lug of the converter and the front edge of the transaxle.

23. The measured distance between the converter and the front of the transaxle should be:

a. 0.831 in. (21.1mm) or more for GA16DE engine vehicles

b. 0.626 in. (15.9mm) or more for SR20DE engine vehicles

24. Raise the transaxle and install to engine drive plate.

25. Install the transaxle mounting bolts in the proper location as noted during removal. On the 1.6L engine, tighten the two bottom bolts to 12–15 ft. lbs. (16–21 Nm) and tighten all other bolts to 22–30 ft. lbs. (30–40 Nm).On the 2.0L engine, tighten the two bottom bolts to 23–31ft. lbs. (31–42 Nm).

26. The remaining components are installed in the reverse order from which they were removed.

27. On automatic transaxles, tighten the torque converter mounting bolts to 33–43 ft. lbs. (44–59 Nm).

28. Fill the transaxle to the proper level, start the engine, and recheck the fluid level.

29. Road test the vehicle and verify proper operation.

Altima

1. Disconnect the negative battery terminal, positive terminal, and remove the battery from the vehicle.

2. Remove the battery tray.

3. Remove the air cleaner box with the mass air flow sensor.

4. Remove the air duct and disconnect the shifter linkage.

5. Remove the front wheels.

6. Raise and safely support the vehicle so there is clearance to remove the transaxle from underneath. Securely support the engine with a jack. Place a wooden block between the jack and the oil pan. This will protect the oil pan from being damaged.

7. Drain the transaxle fluid.

8. Detach all wiring connectors from the transaxle.

9. Remove the crankshaft position sensor from the transaxle.

10. Remove the left-hand mounting bracket from the transaxle and body.

11. Disconnect the control cable from the side of the transaxle.

12. Remove both halfshafts from the transaxle assembly. Securely support the transaxle with another jack.

➡When removing the torque converter, turn the crankshaft for access to the bolts. Place alignment marks on the converter and drive plate, so the converter can be installed in its original position.

13. Remove the transaxle mounting bolts, pull the transaxle away from the engine and lower it from the vehicle.

14. On automatics, disconnect the oil cooler lines and remove the flexplate-to-torque converter bolts.

15. Remove the starter motor from the transaxle.

16. Remove the center crossmember from the vehicle.

➡The transaxle mounting bolts are different lengths. Tagging the bolts upon removal will facilitate proper tightening during installation.

To install:
➡When installing the torque converter to the transaxle, measure the depth of the converter to ensure proper installation.

Bolt No.	Tightening torque N·m (kg-m, ft-lb)	ℓ mm (in)
①	39 - 49 (4.0 - 5.0, 29 - 36)	45 (1.77)
②	30 - 36 (3.1 - 3.7, 22 - 27)	30 (1.18)
③	30 - 36 (3.1 - 3.7, 22 - 27)	40 (1.57)
④	74 - 83 (7.5 - 8.5, 54 - 61)	45 (1.77)
⑤	30 - 36 (3.1 - 3.7, 22 - 27)	80 (3.15)
⑥	30 - 36 (3.1 - 3.7, 22 - 27)	65 (2.56)

⊙ A/T to engine
⊗ Engine to A/T

79230G72

Be sure to install the bolts in the correct location and tighten them to specification—Altima with automatic transaxle

17. Using a straight edge across the mounting flange, measure the depth of the converter. The measurement is to the bolt mounting flange of the converter.

18. The depth measurement of the converter should be 0.75 in. (19mm) or more.

➡**The transaxle mounting bolts are different lengths and require special torque specifications. Use care when installing and tightening these bolts.**

19. Install the transaxle assembly into the vehicle. On manual transaxles, tighten the four lower mounting bolts to 22–30 ft. lbs. (30–40 Nm) and all remaining bolts to 29–36 ft. lbs. (39–49 Nm). Refer to the diagram for the automatic transaxle mounting bolt torque specifications.

20. On automatic transaxles, tighten the bolts holding the converter to the flexplate to 33–43 ft. lbs. (44–59 Nm).

21. Install the remaining components in the opposite order from which they were removed.

22. Connect the positive, then the negative battery terminals.

23. Refill the transaxle with the proper type and amount of fluid.

24. Road test the vehicle for proper operation and recheck the fluid level.

Maxima

1. Disconnect the negative and positive battery cables.

2. Remove the battery and the battery tray.

3. Remove the air cleaner and the resonator.

4. Raise and safely support the vehicle so there is clearance to remove the transaxle from underneath. Securely support the engine via the oil pan using a cushioning wooden block and a jack.

5. Detach the electrical connectors from the transaxle.

6. Remove the crankshaft position sensor (POS) from the transaxle.

7. Remove the LH mounting bracket from the transaxle and body.

8. Disconnect the control cable at the transaxle side.

9. Drain the fluid from the transaxle.

10. Remove both driveshafts from the transaxle assembly.

11. If equipped, disconnect and plug the oil cooler lines.

12. Remove the starter motor from the transaxle.

13. Support the transaxle with proper safety jack.

14. Remove the center crossmember assembly.

15. Remove the rear cover or access plate.

16. On automatic transaxles, remove the bolts that secure the flywheel to the torque converter. Rotate the flywheel to gain access to the three converter mounting bolts.

➡**When removing the torque converter, turn the crankshaft for access to the bolts. Place alignment marks on the converter and drive plate, so the converter can be installed in its original position.**

17. Remove the bolts that secure the transaxle to the engine block.

➡**The transaxle bolts are of different lengths, be sure to note the location of the bolts for reassembly.**

18. Lower the transaxle while supporting it with a jack. Be sure the torque converter remains with the transaxle.

To install:

19. On automatic transaxles, install the torque converter if removed. Be sure it is fully seated in the transaxle.

20. Install the transaxle assembly to the engine block while aligning the torque converter. Tighten the transaxle bolts to specifications.

21. On automatics, install the converter mounting bolts and tighten the bolts to 33–43 ft. lbs. (44–59 Nm).

22. If equipped, install the rear access cover plate to the transaxle.

23. Install the center crossmember assembly and tighten the mounting bolts to 57–72 ft. lbs. (77–98 Nm).

24. Install the LH engine mount and tighten the through-bolt to 32–41 ft. lbs. (43–55 Nm)

25. Install the remaining components in the reverse order of removal.

26. Lower the vehicle.

27. Refill the transaxle with the proper type and amount of fluid.

Clutch

REMOVAL & INSTALLATION

1. Remove the transmission/transaxle assembly.

2. Insert a clutch disc centering tool KV30101000 or equivalent, into the clutch disc hub for support.

3. Loosen the pressure plate bolts evenly in reverse order of the tightening sequence, a little at a time to prevent distortion.

4. Remove the clutch assembly.

5. Remove the throw-out bearing from the clutch lever.

To install:

6. Apply a light coating of chassis lube to the clutch disc splines, input shaft and pilot bearing. Use a disc centering tool to aid installation. Install the disc and pressure plate. Tighten the pressure plate bolts in a crisscross pattern and in several steps to 16–22 ft. lbs. (20–26 Nm).

7. Install the new throw-out bearing in the clutch release lever. Remove the clutch disc centering tool.

8. Install the transaxle into the vehicle. If the mating surfaces will not come together, do not force the units together. Remove the transaxle and recheck that the disc is centered.

➡**DO NOT draw the transaxle to the engine with the bolts. This may damage the clutch and/or transaxle. Also, be careful not to move the throw-out bearing when installing the transaxle.**

9. After the transaxle is installed, connect the clutch cable and check operation before complete reassembly.

10. Adjust the clutch pedal as necessary.

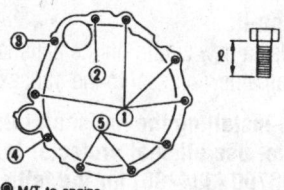

Bolt No.	Tightening torque N·m (kg-m, ft-lb)	"ℓ" mm (in)
①	70 - 79 (7.1 - 8.1, 51 - 59)	52 (2.05)
②	70 - 79 (7.1 - 8.1, 51 - 59)	65 (2.56)
③	70 - 79 (7.1 - 8.1, 51 - 59)	124 (4.88)
④	35.1 - 47.1 (3.58 - 4.80, 25.89 - 34.74)	40 (1.57)
⑤	35.1 - 47.1 (3.58 - 4.80, 25.89 - 34.74)	40 (1.57)

Ⓐ M/T to engine
Ⓑ Engine to M/T

③ with starter
④ with support rod bracket

7923QG73

Transaxle bolt torque specifications and locations—Maxima

Hydraulic Clutch System

BLEEDING

Bleeding is required to remove air trapped in the hydraulic system. The bleed screw is located on the clutch slave (operating) cylinder.

Some models are also equipped with a clutch damper mechanism. The clutch damper mechanism is bled in exactly the same manner as the operating cylinder. It should be bled along with the operating cylinder.

1. Remove the bleed screw dust cap.
2. Attach a transparent vinyl tube to the bleed screw, immersing the free end in a clean container of clean brake fluid.
3. Fill the master cylinder with the proper fluid.
4. Open the bleed screw about ¾ turn.
5. Depress the clutch pedal quickly. Hold it down. Have an assistant tighten the bleed screw. Allow the pedal to return slowly.
6. Repeat the above procedure until no more air bubbles are seen in the fluid container.
7. Remove the bleed tube.
8. Replace the dust cap and refill the master cylinder.
9. Bleed the clutch damper, if equipped.

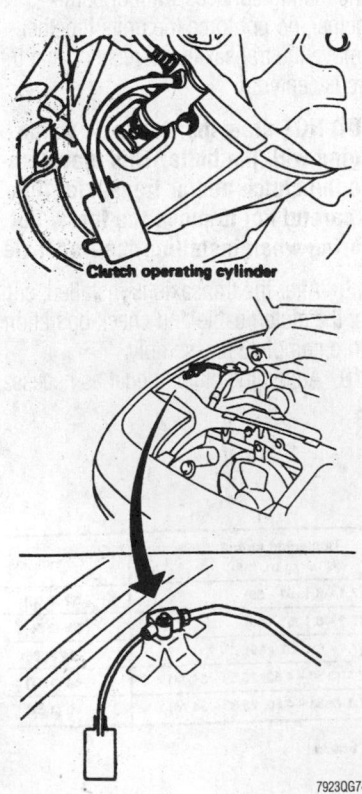

Clutch operating cylinder

Clutch system bleeding points—Altima, Maxima, 240SX and 300ZX

Halfshaft

REMOVAL & INSTALLATION

Sentra and 200SX

➡**The halfshafts will require a special tool for the spline alignment of the halfshaft end into the transaxle case. Do not perform this procedure without access to this tool. The Kent Moore tool Number is J-34296 and J-34297**

1. Raise the front of the vehicle and support it on jackstands, then remove the wheel and the tire assembly.
2. Using a bar to hold the wheel from turning, loosen and remove the hub nut.
3. Remove the clip and separate the brake hose from the strut.
4. Remove the caliper assembly and support it with a wire. Do not allow the caliper to hang from the brake hose.
5. Remove the bolts that secure the strut to the steering knuckle.

➡**Cover the halfshaft boots with shop towels to protect them during removal of the shaft.**

6. Separate the halfshaft from the knuckle by lightly tapping it with a hammer. If it is hard to remove, use a puller.
7. Remove the halfshaft from the transaxle as follows:
   a. Models without support bearing—Pry the halfshaft from the transaxle.
   b. Models with support bearing—Remove the support bearing bolts and pull the halfshaft from transaxle.

➡**When removing the halfshaft from the transaxle, do not pull on the halfshaft. The halfshaft will separate at the sliding joint (damaging the boot). Use a small prybar to remove it from the transaxle. Be sure to replace the oil seal in the transaxle.**

8. Remove the halfshaft from the vehicle.

To install:
9. Use a new circlip on the halfshaft and install a new oil seal to the transaxle.

➡**When installing the halfshaft into the transaxle, use oil seal protector tool KV38106700 (J34296) for the left side or tool KV38106800 (J34297) for the right side to protect the oil seal from damage; after installation of the shaft remove the tool.**

10. Install the halfshaft assembly into transaxle.

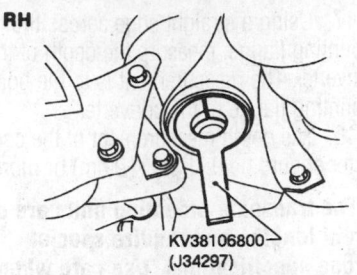

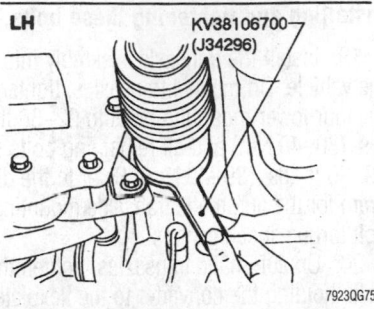

Halfshaft installation tools—Sentra and 200SX

➡**After installation of the halfshaft, try to pull the flange out by hand. If it pulls out, the circular clip is not locked into the transaxle.**

11. If removed, install the support bearing bracket bolts and tighten the mounting bolts to 19–26 ft. lbs. (25–35 Nm).
12. Lubricate the splines of the halfshaft and insert the shaft through the steering knuckle.
13. Align the steering knuckle with the lower strut mount and install the mounting bolts. Tighten the mounting bolts to 68–82 ft. lbs. (92–111 Nm).
14. Install the disc brake caliper and connect the brake hose to the strut with the clip.
15. Install the washer and hub nut to the halfshaft. Tighten the hub nut to 145–202 ft. lbs. (197–274 Nm).
16. Install the adjusting cap and a new cotter pin in drive axle.
17. Install the wheel and tire assembly and lower the vehicle.
18. Road test the vehicle for proper operation.

Altima

1. Raise and safely support the vehicle with the front wheels hanging freely.
2. Remove the front wheels from the vehicle.

➡**The brake caliper does not need to be disconnected from the knuckle.**

3. Pull out the cotter pin from the castellated nut on the wheel hub, then remove the wheel bearing locknut.

➡**Cover the CV-joint boots with a shop towel or waste cloth so not to damage them when removing the halfshaft.**

4. Remove the cotter pin and castle nut from the lower ball joint.

5. Strike the knuckle with a hammer and pull down on the transverse link to separate the lower ball joint from the knuckle.

6. Disconnect the tie rod end from the steering knuckle.

7. Separate the halfshaft from the steering knuckle by tapping it with a block of wood and a mallet.

8. Using a prybar, reach through the engine crossmember and carefully pry the right inner CV-joint from the transaxle.

9. If equipped with manual transaxle, carefully pry the left inner CV-joint from the transaxle.

10. If equipped with automatic transaxle, insert a long tool into the opening for the right halfshaft and strike the tool to with a hammer.

11. Remove the left halfshaft from the transaxle.

To install:

➡**Whenever the halfshafts are removed, the axle seals should be replaced.**

12. When installing the shafts into the transaxle, use a new oil seal, then install an alignment tool, KV38106700 for the left side or KV38106800 for the right side, along the inner circumference of the oil seal.

13. Insert the halfshaft into the transaxle, align the serrations, then remove the alignment tool.

14. Push the halfshaft, then press-fit the circular clip on the shaft into the clip groove on the side gear.

➡**After insertion, attempt to pull the flange out of the side joint to be sure the circular clip is properly seated in the side gear and will not come out.**

15. Insert the halfshaft into the steering knuckle.

16. Connect the lower ball joint and tie rod end in the correct position. Tighten the lower ball joint-to-control arm nuts to 52–64 ft. lbs. (71–86 Nm) and the tie rod end-to-steering knuckle nut to 22–29 ft. lbs. (29–39 Nm). Install new cotter pins to the castle nuts.

17. Install the axle nut and tighten the locknut to 174–231 ft. lbs. (235–314 Nm).

18. Install a new cotter pin on the wheel hub and install the wheel.

19. Install the front wheels to the vehicle.

20. Road test the vehicle for proper operation.

21. Check the transaxle fluid level and top off as necessary.

Maxima

1. Raise and support the front of the vehicle safely and remove the wheels.

2. Remove the ABS wheel sensor and move it out of the way.

3. Remove the brake hose from the strut.

4. Remove wheel bearing locknut.

5. Matchmark and remove the bolts attaching the steering knuckle to the strut.

➡**Cover axle boots with waste cloth or equivalent so as not to damage them when removing halfshaft.**

6. Separate the halfshaft from the knuckle by slightly tapping it.

7. Loosen the bolts attaching the support bearing to the support bearing bracket.

8. If equipped with a manual transaxle, pry the halfshaft from the transaxle with a flat bladed tool.

9. If equipped with a automatic transaxle perform the following:

   a. Remove the right halfshaft from the vehicle.

   b. Insert a flat bladed tool into the transaxle where the right halfshaft was, place the end of the tool on the halfshaft, then drive the left shaft from the pinion side gear.

10. Remove the support bearing bolts and remove the halfshaft from the vehicle.

11. Remove and discard the circlip on the end of the halfshaft. Remove the seal from the transaxle.

To install:

12. Install a new seal into the transaxle and install a halfshaft alignment tool KV38106700 into the transaxle seal.

13. Install a new circlip to the halfshaft, then insert the halfshaft into the transaxle.

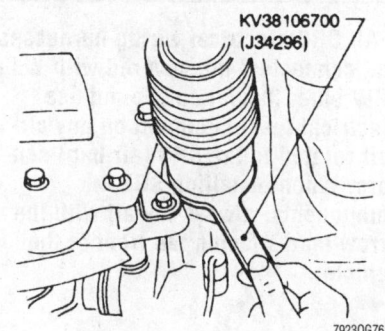

KV38106700
(J34296)

79230G76

Left halfshaft alignment tool—Maxima

14. With the serrations aligned remove the alignment tool.

15. Push the halfshaft fully into the transaxle to seat the circlip. Try to pull the halfshaft from the transaxle by hand to verify that the circlip is properly seated.

16. Install the support bearing bolts and tighten the bolts to 9–14 ft. lbs. (13–19 Nm).

17. Insert the halfshaft into the steering knuckle and install the hub locknut, do not tighten the hub nut at this time.

18. Connect the steering knuckle to the strut.

19. Install the strut mounting bolts and align the matchmarks. Tighten the bolts to 103–117 ft. lbs. (140–159 Nm).

20. Install the brake hose to the strut.

21. Install the ABS wheel sensor and tighten the attaching bolt to 13–17 ft. lbs. (18–24 Nm).

22. Install the front wheels, lower the vehicle and tighten hub locknut to 174–231 ft. lbs. (235–314 Nm).

23. Check and/or adjust the wheel alignment as necessary.

240SX

1. Raise and safely support the vehicle.

2. Remove the rear wheel(s).

3. Loosen the wheel bearing locknut and lightly tap on the axle shaft to loosen it from the steering knuckle. Remove the nut from the axle shaft.

4. Remove the brake caliper assembly.

➡**Support the brake caliper, do not allow the caliper to hang freely from the brake hose.**

5. Remove the ABS sensor from the steering knuckle.

6. Disconnect and separate the lower ball joint.

7. Pull the hub assembly outward.

8. Remove the shaft from hub assembly.

9. Unbolt and remove axle shaft from the differential.

To install:

10. Reconnect the shaft to the differential and tighten the mounting bolts to 25–33 ft. lbs. (34–44 Nm).

11. Insert the shaft into the hub assembly.

12. Reconnect the lower control arm and ball joint. Tighten the mounting nut to 52–64 ft. lbs. (71–86 Nm) and install a new cotter pin.

13. Replace the wheel bearing nut and tighten it to 152–202 ft. lbs. (206–274 Nm).

14. Install the ABS sensor to the steering knuckle and tighten the mounting bolt to 13–17 ft. lbs. (18–24 Nm).

15. Reinstall the brake caliper assembly. If the brake line was disconnected, bleed the brake system.

16. Install the wheel assembly, lower the vehicle, and perform a road test.

300ZX

➡When removing the driveshaft assembly, disconnect the ABS wheel sensor to prevent damage to the wheel sensor. When removing the driveshaft, cover the axle boots with a shop towel for protection.

1. Raise and safely support the vehicle.

2. Remove the rear wheel(s).

3. Unbolt the halfshaft from the flange at the rear housing.

4. Remove the cotter pin, adjusting cap, insulator, wheel bearing locknut and washer.

5. Remove the axle shaft by lightly tapping it with a copper hammer.

✳✳ WARNING

To avoid damaging the threads of the axle shaft, install the axle nut to protect the threads.

6. Remove the shaft from the hub assembly.

To install:

7. Insert the shaft into the hub assembly.

8. Install the washer and axle nut but do not tighten at this time.

9. Reconnect the shaft to the differential.

10. Tighten the mounting bolts and nuts with washers to:

    a. 51–58 ft. lbs. (69–78 Nm) for turbo models

    b. 47–58 ft. lbs. (64–78 Nm) for non-turbo models

11. Replace the wheel bearing nut and tighten it to 152–203 ft. lbs. (206–275 Nm).

12. Install the insulator, adjusting cap and install a new cotter.

13. Install the wheel assembly and lower the vehicle.

STEERING AND SUSPENSION

Air Bag

✳✳ CAUTION

Some vehicles are equipped with an air bag system, also known as the Supplemental Inflatable Restraint (SIR) or Supplemental Restraint System (SRS). The system must be disabled before performing service on or around system components, steering column, instrument panel components, wiring and sensors. Failure to follow safety and disabling procedures could result in accidental air bag deployment, possible personal injury and unnecessary system repairs.

PRECAUTIONS

Several precautions must be observed when handling the inflator module to avoid accidental deployment and possible personal injury.

• Never carry the inflator module by the wires or connector on the underside of the module.

• When carrying a live inflator module, hold securely with both hands, and ensure that the bag and trim cover are pointed away.

• Place the inflator module on a bench or other surface with the bag and trim cover facing up.

• With the inflator module on the bench, never place anything on or close to the module which may be thrown in the event of an accidental deployment.

DISARMING

➡All SRS electrical wiring harnesses and connectors are covered with YELLOW outer insulation. Do not use electrical test equipment on any circuit related to the SRS (air bag) sensors. When installing SRS components, always install with the arrow marks facing the front of the vehicle.

To disarm the SRS system turn the ignition switch to **OFF** position. Then, disconnect the both battery cables starting with the negative cable first and wait at least 10 minutes after the cables are disconnected. Be sure to insulate the battery terminal ends.

REARMING

To arm the SRS system turn the ignition switch to **OFF** position. Connect the both battery cables starting with the positive cable first.

➡The SRS or air bag system is equipped with a self-diagnostic operation. After turning the ignition key to the ON or START position, the AIR BAG warning lamp will illuminate for 7 seconds. After 7 seconds, the AIR BAG lamp will extinguish if no malfunction is detected. If the AIR BAG lamp does not extinguish after 7 seconds, check the SRS self diagnostic system for a malfunction.

Rack and Pinion Steering Gear

REMOVAL & INSTALLATION

Manual

SENTRA AND 200SX

1. Raise and support the vehicle on jackstands.

2. Remove the front wheels.

3. Remove the both tie rod ends from the steering knuckles.

4. Matchmark the steering column shaft to the lower joint and remove the pinch bolt from the joint.

5. Loosen and remove the steering gear mounting bolts.

6. Remove the mounting clamps from the steering gear and remove the steering gear by sliding it off the steering shaft.

7. Remove the steering gear from the vehicle.

To install:

8. Install the steering gear assembly to the vehicle. Be sure to align the matchmarks of the rack with the marks on the steering shaft.

9. Install the steering gear mounting clamps and tighten the bolts to 58 ft. lbs. (78 Nm).

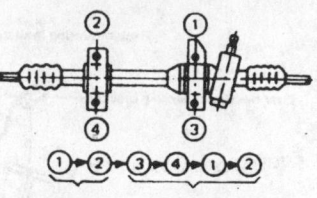

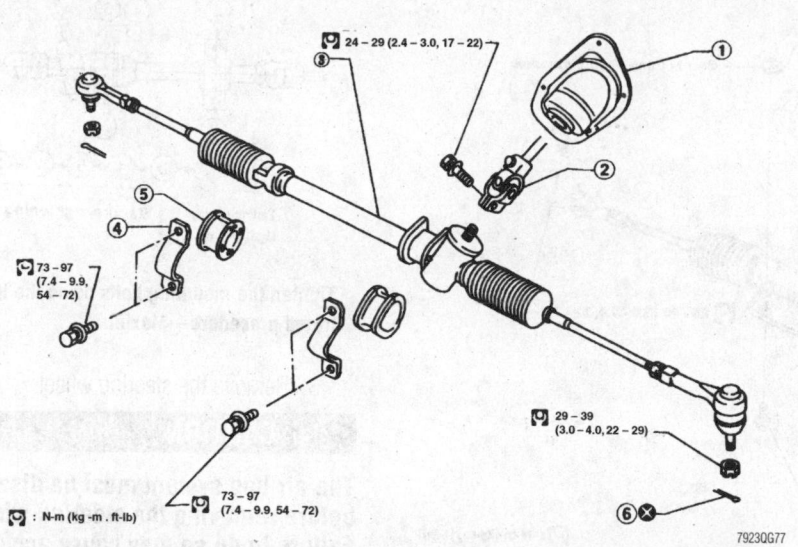

24 - 29 (2.4 - 3.0, 17 - 22)

73 - 97 (7.4 - 9.9, 54 - 72)

29 - 39 (3.0 - 4.0, 22 - 29)

73 - 97 (7.4 - 9.9, 54 - 72)

⬛ : N-m (kg -m . ft-lb)

79230G77

Exploded view of the manual rack and pinion steering gear mounting—Sentra and 200SX

Temporary tightening   Secure tightening

79230G78

Tighten the mounting bolts according to the sequence shown—Sentra and 200SX

10. Install the lower joint-to-steering column pinch bolt and tighten to 22 ft. lbs. (29 Nm).

11. Connect the tie rod end to the steering knuckle and install castle nut. Tighten the nut to 29 ft. lbs. (39 Nm) and install a new cotter pin.

➡If installing a new rack and pinion **assembly, transfer the lower steering joint to the new rack and pinion prior to installation. When installing the lower steering joint to the steering gear, be sure that the wheels are aligned with the vehicle (straight ahead position).**

12. To center the steering gear, turn it all the way to the lock position on one side. Now, count the number of turns it takes to get to the opposite side lock position. Turn the steering gear ½ the number of turns towards the original starting position. The steering rack should now be centered. When connecting the steering joint to the steering column shaft, be sure to align the matchmarks made during disassembly.

13. Install the front wheels, remove the jackstands, and lower the vehicle.

14. Check the vehicle's alignment.

Power

SENTRA AND 200SX

1. Raise and support the vehicle safely.

2. Disconnect the low pressure hose clamp and remove the low pressure hose at the steering gear. Be sure to use a pan to catch the fluid.

3. Disconnect the flare nut and the high pressure tube at the steering gear, then drain the fluid from the gear.

4. Remove the tie rod ends from the steering knuckle.

5. Place a floor jack under the transaxle and support it.

6. Remove the front exhaust pipe and remove the rear engine mount.

7. Position the front wheels so they are pointing straight ahead.

8. Matchmark the steering column lower joint to the steering gear.

➡**The steering gear splines have a flat spot or key way. Be sure to note this during removal.**

9. Remove the bolt that secures steering column lower joint.

10. Unbolt and remove the steering gear unit and the linkage.

To install:

11. Install the power steering gear assembly to the vehicle. Align the steering column to the steering gear.

➡**Be sure to align the flat spot or keyway during installation.**

12. Install the steering gear mounts and tighten the mounting bolts in sequence to 54–72 ft. lbs. (73–97 Nm).

13. Install the pinch bolt for the steering column-to-gear connection and tighten the bolt to 17–22 ft. lbs. (24–29 Nm).

14. Connect the tie rod ends to the steering knuckle and tighten the mounting nut to 22–29 ft. lbs. (29–39 Nm).

15. Tighten the tie rod mounting nut further so the groves in the nut align with first cotter pin hole. Install a new cotter pin.

16. Connect the power steering low pressure hose to the steering gear and tighten the fitting to 20–29 ft. lbs. (27–39 Nm).

17. Connect the power steering high pressure hose to the steering gear and tighten the fitting to 11–18 ft. lbs. (15–25 Nm).

18. Install the rear engine mount and remove the floor jack.

19. Using new gaskets, install the front exhaust pipe assembly.

20. Fill the power steering system and start the engine.

21. Bleed the power steering system and check the wheel alignment.

ALTIMA AND 240SX

❊❊ CAUTION

The air bag system must be disarmed before removing the rack and pinion. Failure to do so may cause accidental deployment, property damage or personal injury.

1. Disconnect the negative battery cable and disarm the air bag.

2. Raise and safely support the vehicle as necessary.

3. Remove the bolt securing the lower steering column shaft to the power steering gear assembly. Be sure to matchmark the shaft from the steering gear to the

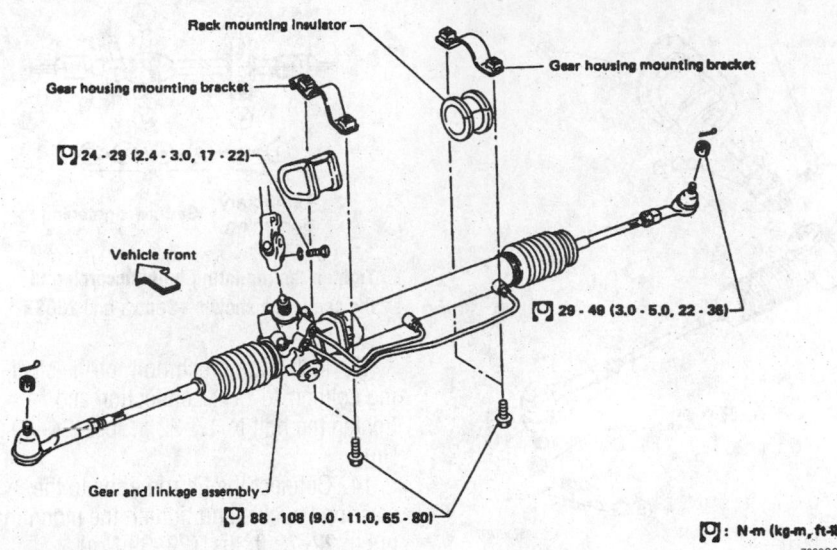

Exploded view of the power steering gear mounting—240SX

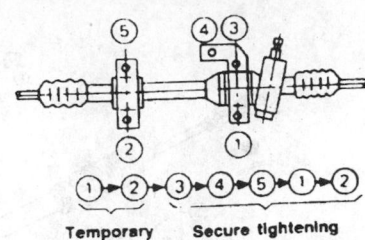

Tighten the mounting bolts using the illustrated procedure—Maxima

steering column joint for correct installation.

4. Disconnect the hoses from the power steering gear and plug the hoses to prevent leakage.

5. Remove the cotter pins and castle nuts from the tie rod ends.

6. Using a ball joint separating tool, remove tie rod ends from the steering knuckle.

7. Remove the front exhaust pipe mounting nuts and bolts. Remove the front exhaust pipe from the vehicle.

8. If necessary, disconnect the control cable or linkage from the transmission and position it out of the way.

9. Remove the power steering gear mounting bolts or nuts.

10. Remove the steering gear from the vehicle. Use care when separating the steering column joint.

11. Inspect the steering gear mount bushings and replace as necessary.

To install:

12. Align the steering column-to-steering gear matchmark and install the steering gear to the vehicle. Be sure to properly install the mounting bushings and hand-tighten the mounting nuts or bolts.

➡ **When installing the lower steering joint to the steering gear, be sure that the wheels are aligned straight and the steering joint slot is aligned.**

13. Tighten the steering gear mounts as follows:

   a. 240SX —65–80 ft. lbs. (88–108 Nm)

   b. 1995–96 Altima—54–72 ft. lbs. (73–97 Nm)

14. Install the pinch bolt securing the lower steering column shaft to the power steering gear assembly. Tighten the pinch bolt to 17–22 ft. lbs. (24–29 Nm)

15. Install the tie rod end to steering knuckle and tighten the castle nut to 22–29 ft. lbs. (29–39 Nm)

16. Tighten the castle nut further to align the slot in the castle nut with the cotter pin hole and install a new cotter pin.

17. If removed, connect the control cable or linkage to the transmission and position it out of the way.

18. Using new gaskets, install the front exhaust pipe assembly.

19. Connect the power steering hoses to the steering gear and fill the power steering reservoir.

20. Start the engine and refill the power steering reservoir.

21. Bleed the power steering system and perform a front end alignment.

22. Connect the negative battery cable.

23. If equipped, enable the air bag system.

MAXIMA

1. Disconnect both battery cables and wait at least 10 minutes after the battery cables are disconnected. This will disarm the air bag system so the steering wheel can be removed.

2. Point the front tires straight ahead and lock the steering in this position.

❊❊ **WARNING**

Do not turn the steering wheel or column with the lower joint removed from the steering column or the spiral cable may be damaged.

3. Remove the steering wheel.

❊❊ **CAUTION**

The air bag system must be disarmed before removing the steering wheel. Failure to do so may cause accidental deployment, property damage or personal injury.

➡ **The steering wheel must be removed before disconnecting the steering column lower joint to avoid damaging the SRS spiral cable.**

4. Raise and support the vehicle safely and remove the front wheels.

5. Disconnect the tie rod ends from the steering knuckles.

6. Remove the carbon canister from the vehicle.

7. Support the engine, then remove the bolts attaching the engine mounts to the engine mounting center member. Remove the engine mounting center member.

8. Remove the front stabilizer bar from the vehicle.

9. Remove the nuts attaching the hole cover to the bulkhead.

10. Move the hole cover aside and disconnect the lower joint from the rack and pinion. Matchmark the pinion shaft and the pinion housing to record the steering neutral position.

11. Disconnect the power steering fluid pipes from the rack and pinion.

12. Remove the bolts attaching the mounting brackets and remove the rack and pinion from the vehicle.

To install:

13. Position the rack and pinion in the vehicle and install the mounting brackets. Tighten the mounting nuts and bolts in the proper sequence to 54–72 ft. lbs. (73–97 Nm).

14. Install new O-rings to the power steering fluid pipes and connect them to the rack and pinion. Tighten the low pressure line 20–29 ft. lbs. (27–39 Nm). Tighten the high pressure line to 11–18 ft. lbs. (15–25 Nm).

15. Align the lower steering joint to the pinion shaft and install the joint onto the pinion shaft. Install the bolt and tighten the bolt to 17–22 ft. lbs. (24–29 Nm).

16. Properly position the hole cover and install the attaching nuts, tighten the nuts to 2.9–3.6 ft. lbs. (4–5 Nm).

17. Install the front stabilizer.

18. Install the engine mounting center member and tighten the attaching bolts to 57–72 ft. lbs. (77–98 Nm). Attach the engine mounts to the center member and tighten the bolts to 57–72 ft. lbs. (77–98 Nm). Remove the support from the engine.

19. Install the remaining components in the reverse order of removal.

20. Tighten the tie rod end nuts to 46–61 ft. lbs. (63–82 Nm), then install a new cotter pin.

21. Fill the power steering reservoir with fluid and bleed the air from the power steering system.

22. Check the vehicle front end alignment and adjust as necessary.

Strut and Coil Spring

REMOVAL & INSTALLATION

Front

SENTRA AND 200SX

1. Raise and support the vehicle on jackstands.

2. Remove the wheel.

3. Detach the brake tube from the strut. If equipped with ABS, disconnect the ABS wiring from the strut.

4. Support the transverse link with a jackstand.

5. Detach the steering knuckle from the strut.

➡ **Note the positioning of the strut alignment mark for reassembly purposes.**

6. Support the strut and remove the three upper attaching nuts. Remove the strut from the vehicle.

✲✲ CAUTION

Never loosen the center spring retaining nut until the coil spring is compressed, or serious injury or vehicle damage may occur.

7. Place the strut assembly in a vise with the special holding tool ST35652000 or in a spring compressor.

8. Loosen the piston rod locknut.

9. Compress the spring with the spring compressor, then remove the piston rod locknut.

➡ **Before removing the strut from the coil spring, note the positioning of the strut in relationship to the coil spring for reassembly.**

10. Remove the strut mounting insulator bracket, strut mounting bearing, upper spring seat, and the upper spring rubber seat.

11. Remove the strut, leaving the coil spring compressed.

12. Remove the piston boot and rebound bumper from the strut.

To install:

13. Install the rebound bumper and the boot to the strut piston.

14. Install the strut into the coil spring, be sure the strut and spring are properly positioned.

15. Install the upper spring rubber seat, upper spring seat, strut mounting bearing, and the strut mounting insulator bracket. Be sure that the cutout on the upper spring seat is facing the outside of the vehicle.

16. Install the piston rod locknut, then remove the spring compressor.

17. Tighten the piston rod locknut to 43–54 ft. lbs. (59–74 Nm).

➡ **When installing the strut, be sure to position the alignment mark toward the outside of the vehicle.**

18. Position the strut to the vehicle and install the three upper attaching nuts. Tighten the upper mounting nuts to 18–22 ft. lbs. (25–29 Nm).

19. Connect the steering knuckle to the strut and tighten mounting nuts of the mounting bolts to 68–82 ft. lbs. (92–111 Nm).

20. Connect the brake tube to the strut and connect the ABS wiring to the strut, if it was removed.

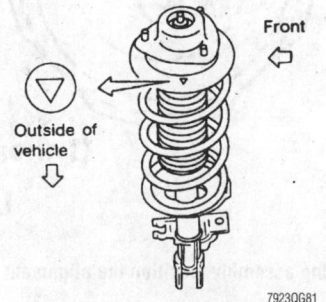

During assembly, be sure to point the alignment mark toward the outside of the vehicle—Sentra and 200SX

21. Bleed the brake system and install the wheel.

22. Perform a front end alignment.

ALTIMA

1. Raise and support the vehicle on jackstands.

2. Remove the wheel.

3. Detach the brake tube from the strut. If equipped with ABS, disconnect the ABS wiring from the strut.

4. Support the transverse link with a jackstand.

5. Detach the steering knuckle from the strut.

6. Support the strut and remove the three upper attaching nuts. Remove the strut from the vehicle.

✲✲ WARNING

Never loosen the center spring retaining nut until the coil spring is compressed, or serious injury or vehicle damage may occur.

To install:

➡ **When installing the strut, be sure to position the alignment mark toward the outside of the vehicle.**

7. Position the strut to the vehicle and install the three upper attaching nuts. Tighten the upper mounting nuts to 29–40 ft. lbs. (39–54 Nm).

8. Connect the steering knuckle to the strut and tighten mounting nuts to 87–108 ft. lbs. (118–147 Nm).

9. Connect the brake tube to the strut and connect the ABS wiring to the strut, if it was removed.

10. Bleed the brake system and install the wheel.

11. Lower the vehicle and perform a front end alignment.

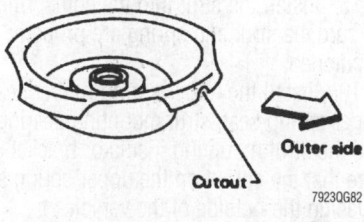

Position the alignment mark toward the outside of the vehicle—Altima

MAXIMA

1. Raise and safely support the vehicle.

2. Remove the wheel. Matchmark the position of the strut-to-steering knuckle location.

3. Disconnect the brake hose from the strut.

4. Remove the ABS wheel sensor and move it out of the way.

5. Matchmark and remove the bolts attaching the steering knuckle to the strut.

6. Open the hood and remove the strut attaching nuts while holding the strut.

✳✳ CAUTION

Do not remove the center locknut from the strut assembly until the strut is safely compressed.

7. Remove the strut from the vehicle.

8. Place the strut assembly in a vise with the special holding tool ST35652000 or in a spring compressor.

9. Loosen the piston rod locknut.

✳✳ CAUTION

Do not remove the piston rod locknut, the spring is under tension and can cause serious personal injury.

10. Compress the spring with the spring compressor, then remove the piston rod locknut.

➡Before removing the strut from the coil spring, note the positioning of the strut in relationship to the coil spring for reassembly.

11. Remove the strut mounting insulator bracket, strut mounting bearing, upper spring seat, and the upper spring rubber seat.

12. Remove the strut, leaving the coil spring compressed.

13. Remove the piston boot and rebound bumper from the strut.

To install:

14. Install the rebound bumper and the boot to the strut piston.

15. Install the strut into the coil spring, be sure the strut and spring are properly positioned.

16. Install the upper spring rubber seat, upper spring seat, strut mounting bearing, and the strut mounting insulator bracket. Be sure that the cutout on the upper spring seat is facing the outside of the vehicle.

17. Install the piston rod locknut, then remove the spring compressor.

18. Tighten the piston rod locknut to 43–58 ft. lbs. (59–76 Nm).

19. Install the strut into the strut tower and install new attaching nuts. Tighten the nuts to 29–40 ft. lbs. (39–54 Nm).

20. Install the bolts attaching the steering knuckle to the strut and align the matchmarks. Tighten the bolts to 103–117 ft. lbs. (140–159 Nm).

21. Install the ABS wheel sensor and tighten the attaching bolt to 13–17 ft. lbs. (18–24 Nm).

22. Install the brake hose to the strut.

23. Install the front wheels and lower the vehicle.

24. Check and/or adjust the wheel alignment as necessary.

240SX

1. Raise and support the vehicle on jackstands.

2. Remove the wheel.

3. Detach the brake tube from the strut. If equipped with ABS, disconnect the ABS wiring from the strut.

4. Support the transverse link with a jackstand.

5. Detach the steering knuckle from the strut by removing the two through-bolts.

6. Support the strut and remove the three upper attaching nuts. Remove the strut from the vehicle.

✳✳ WARNING

Never loosen the center spring retaining nut until the coil spring is compressed, or serious injury or vehicle damage may occur.

7. Compress the strut coil spring with a spring compressor.

✳✳ CAUTION

If coil spring is not properly compressed serious injury could result.

8. Remove the strut assembly center locknut.

➡Before removing the strut from the coil spring, note the positioning of the strut in relationship to the coil spring for reassembly.

9. Separate the strut from the coil spring. Keep the coil spring compressed.

To install:

10. Install the strut into the coil spring.

11. Install and tighten the center locknut to 43–58 ft. lbs. (59–78 Nm).

➡When installing the strut, be sure to position the alignment mark toward the inside of the vehicle.

12. Position the strut to the vehicle and install the three new upper attaching nuts.

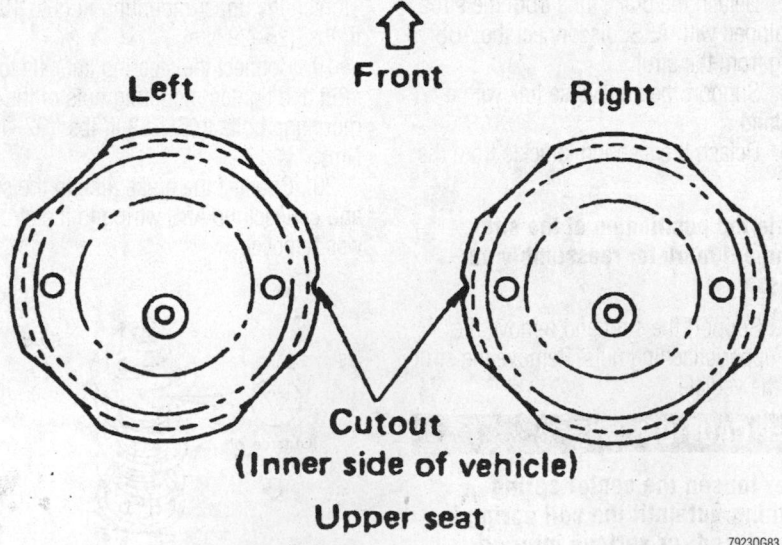

During assembly, position the alignment marks toward the inside of the vehicle—240SX

79230G83

Tighten the upper mounting nuts to 29–40 ft. lbs. (39–54 Nm).

13. Connect the steering knuckle to the strut and tighten mounting through-bolts to 90–112 ft. lbs. (123–152 Nm).

14. Connect the brake tube to the strut and if it was removed, connect the ABS wiring to the strut.

15. Bleed the brake system and install the wheel.

16. Perform a front end alignment.

300ZX

1. Raise and support the vehicle safely.
2. Remove the front wheels.
3. Lightly jack under the lower arm to unload the suspension.
4. Remove the nut connecting the lower portion of the strut to the third link.
5. From under the hood and remove the two nuts holding the top of the strut to the chassis.

❋❋ CAUTION

Do not remove the center nut from the strut piston. The center nut retains the coil spring, which is contained under considerable pressure. Improper servicing can lead to vehicle damage and serious injury.

6. If equipped with adjustable or Sonar suspension shocks, disconnect the electrical lead from the actuating unit.

7. Lower the jack slowly and cautiously until the strut assembly can be removed.

To install:

➡ **On vehicles with Sonar suspension, before installing the actuator ensure the output shaft of the actuating unit is aligned with the shock absorber control rod. If this is not done, the actuator will be damaged.**

8. Position the strut in the vehicle and using two new self-locking nuts, tighten the nuts holding the top of the strut to 25–33 ft. lbs. (34–44 Nm).

9. If equipped with adjustable or Sonar suspension shocks, connect the electrical lead to the actuating unit.

10. Connect the strut assembly to the knuckle arm and tighten the mounting nut to 72–87 ft. lbs. (98–118 Nm).

11. Install the front wheel and lower the vehicle.

12. Check front wheel alignment and adjust if necessary.

Rear

SENTRA AND 200SX

1. Raise the vehicle and support it safely.
2. Remove the rear wheel.
3. Remove the trim panel from the trunk to gain access to the upper mounting nuts of the strut.
4. Remove the protective cap from the upper portion of the strut.
5. Position a floor jack under the rear axle for support.

➡ **Note and mark the positioning of the upper strut plate to the vehicle body.**

❋❋ CAUTION

Never remove the center strut nut until the strut is removed from the vehicle and the spring is safely compressed.

6. Remove the lower strut mounting through-bolt.

7. Remove the two upper strut mounting nuts and remove the strut from the vehicle.

8. Place the strut assembly in a vise with the special holding tool ST35652000 or in a spring compressor.

9. Loosen the piston rod locknut.

10. Compress the spring with the spring compressor, then remove the piston rod locknut.

➡ **Before removing the strut from the coil spring, note the positioning of the strut in relationship to the coil spring for reassembly.**

11. Remove the strut mounting insulator bracket, strut mounting bearing, upper spring seat, and the upper spring rubber seat.

12. Remove the strut, leaving the coil spring compressed.

13. Remove the piston boot and rebound bumper from the strut.

To install:

14. Install the rebound bumper and the boot to the strut piston.

15. Install the strut into the coil spring, be sure the strut and spring are properly positioned.

16. Install the upper spring rubber seat, upper spring seat, strut mounting bearing, and the strut mounting insulator bracket. Be sure that the cutout on the upper spring seat is facing the outside of the vehicle.

17. Install the piston rod locknut and tighten the piston rod locknut to 13–17 ft. lbs. (18–24 Nm).

18. Remove the spring compressor from the coil spring.

19. Install the strut to the vehicle and tighten the two upper mounting nuts to 12–14 ft. lbs. (16–19 Nm).

20. Install the upper mount protective cap.

21. Install the through-bolt to the lower mount of the strut. Tighten the lower strut bolt to 72–87 ft. lbs. (98–118 Nm).

22. Install the trunk trim panel.

23. Install the rear wheel.

24. Lower the vehicle and perform an alignment.

ALTIMA

1. Raise and safely support the vehicle.

2. Remove the rear wheels from the vehicle.

3. Support the rear axle with a jack.

4. Remove the strut lower mounting through-bolts.

➡ **Be sure to note the position the strut upper plate to the vehicle for reinstallation purposes.**

5. Remove the two nuts from the top of the strut and remove the strut as an assembly.

❋❋ CAUTION

Do not remove the center locknut from the strut assembly until the strut is safely compressed.

6. Compress the strut coil spring with a spring compressor.

7. Remove the strut assembly center locknut.

➡ **Before removing the strut from the coil spring, note the positioning of the strut in relationship to the coil spring for reassembly.**

8. Remove the strut leaving the coil spring compressed.

➡ **Mark the coil spring position to the strut assembly for reinstallation purposes.**

9. To remove the spring from the strut assembly, perform the following steps:

a. Compress the coil spring with the proper compressor tool.

b. Remove the center retaining nut holding strut mounting insulator.

c. Slowly decompress the coil spring.

d. Remove the strut mounting insulator.

e. Remove coil spring.

To install:

10. Install the coil spring onto the strut assembly. Be sure to align the matchmarks made during the removal procedure.

11. Install the strut mounting insulator.

12. Compress the coil spring assembly.

➡**It will be necessary to use a new locknut for the center retaining nut of the coil spring.**

13. Install the center retaining nut and tighten the nut to 43–58 ft. lbs. (59–78 Nm). Be sure the spring is seated properly on the strut and in the mounting insulator.

14. Slowly remove the spring compressor tool.

15. Install the strut assembly into the vehicle.

16. Tighten the upper strut mounting nuts to 31–40 ft. lbs. (42–54 Nm).

17. Tighten the lower strut through-bolt to 87–108 ft. lbs. (118–147 Nm).

➡**Be sure to hold the through-bolt and tighten the nuts.**

18. Install the wheels, lower the vehicle and perform a front end alignment.

MAXIMA

1. Raise and safely support the vehicle.

2. Remove the rear wheels.

3. Support the rear torsion beam assembly with a jack.

4. Open the trunk and remove the two nuts attaching the strut to the vehicle.

※※ **CAUTION**

Do not remove the center locknut from the strut assembly until the strut is safely compressed.

5. Remove the bolt attaching the strut to the rear torsion beam assembly and remove the strut.

6. Place the strut assembly in a vise with the special holding tool HT71780000 or in a spring compressor.

7. Loosen the piston rod locknut.

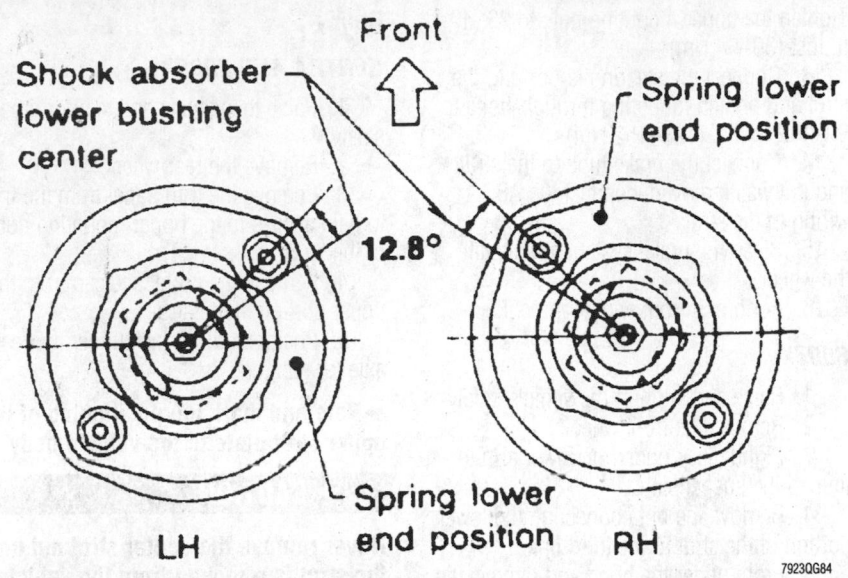

Positioning of the strut mounting brackets—Maxima

※※ **CAUTION**

Do not remove the piston rod locknut, the spring is under tension and can cause serious personal injury.

8. Compress the spring with the spring compressor, then remove the piston rod locknut.

➡**Before removing the strut from the coil spring, note the positioning of the strut in relationship to the coil spring for reassembly.**

9. Remove the bushing, strut mounting bracket, and the upper spring seat rubber.

10. Remove the strut, leaving the coil spring compressed.

11. Remove the bushing, bound bumper cover, and the bound bumper.

To install:

12. Install the bound bumper, bound bumper cover, and the bushing.

13. Install the strut into the coil spring, be sure the strut and spring are properly positioned.

14. Install the upper spring seat rubber, strut mounting bracket, and the bushing. Be sure that the mounting bracket is properly positioned.

15. Install the piston rod locknut, then remove the spring compressor.

16. Tighten the piston rod locknut to 13–17 ft. lbs. (18–24 Nm).

17. Install the strut into the vehicle and install new attaching nuts. Tighten the nuts to 12–14 ft. lbs. (16–19 Nm).

18. Position the strut on the rear torsion beam and install the bolt. Tighten the bolt attaching the strut to the torsion beam assembly to 72–87 ft. lbs. (98–118 Nm).

19. Remove the support from the rear torsion beam.

20. Install the rear wheels and lower the vehicle.

21. Check the vehicle's alignment and adjust as necessary.

240SX

1. Raise and safely support the vehicle.

2. Remove the rear wheels from the vehicle.

3. Support the rear control arm with a jack.

4. Remove the strut lower mounting bolt.

5. Mark the upper spring plate to the vehicle for reinstallation purposes.

6. Remove the two strut upper mounting nuts.

※※ **WARNING**

Do not remove the center piston locknut until the spring has been compressed.

7. Remove the strut from the vehicle.

8. Mark the positioning of the coil spring to the strut assembly.

9. Using a coil spring compressor, compress the spring.

✳✳ CAUTION

If coil spring is not properly compressed serious injury could result.

10. Remove the strut assembly center locknut.

➡ **Before removing the strut from the coil spring, note the positioning of the strut in relationship to the coil spring for reassembly.**

11. Remove the upper strut plate.

12. Remove the strut leaving the coil spring compressed.

13. Note the positioning of the spring and slowly release the spring compressor. Remove the coil spring assembly.

To install:

14. Properly position and compress the coil spring.

15. Install the strut to the coil spring.

16. Install the upper strut plate and tighten the center locknut to 13–17 ft. lbs. (18–24 Nm). Verify that the coil spring is seated properly on the strut.

17. Remove the coil spring compressor.

18. Install the strut assembly to the vehicle.

19. Using new locking nuts, install the top mounting nuts and tighten to 12–14 ft. lbs. (16–19 Nm).

➡ **Be sure that all final tightening is done with the full weight of the vehicle on the ground.**

20. Install the lower mounting bolt and tighten to 72–87 ft. lbs. (98–118 Nm).

21. Install the rear wheels.

22. Check the vehicle alignment and adjust it as necessary.

300ZX

1. Block the front wheels.

2. Raise and support the vehicle safely. Remove the rear wheel(s).

➡ **The vehicle should be far enough off the ground so the rear spring does not support any weight.**

3. Working inside the luggage compartment, turn and remove the caps above the strut mounts.

4. Detach the shock actuator sub-harness connector.

5. Remove the actuator fixing bolts.

6. Remove the strut mounting nuts.

✳✳ WARNING

Do not remove the center spring retaining nut until the coil spring is compressed or serious personal injury may result.

7. Remove the mounting bolt for the strut at the lower arm (transverse link), then lift out the strut.

8. Place the strut in a spring compressor and compress the spring.

9. Remove the center spring retaining nut. Decompress the spring and remove the spring seat, bushing, plate, and the spring.

10. Remove the strut from the spring compressor.

To install:

11. Install the strut into the spring compressor. Install the coil spring, plate, bushing, and upper spring seat.

12. Compress the coil spring and install the center spring retaining nut. Decompress the spring and tighten the nut to 13–17 ft. lbs. (18–24 Nm).

13. Remove the strut assembly from the coil spring compressor.

14. Install the strut upper end first and secure with the nuts snugged down, but not fully tightened. Attach the lower end of the strut to the transverse link and tighten the upper nuts to 12–14 ft. lbs. (16–19 Nm). Tighten the lower mounting bolt to 57–72 ft. lbs. (77–98 Nm).

15. Install the shock absorber actuator fixing bolts, tighten the bolts to 26–34 inch lbs. (3–4 Nm)

16. Reconnect the sub-harness connector.

17. Install the wheel assembly and lower the vehicle.

18. Install the strut mount caps in the trunk.

19. Inspect the vehicle alignment and adjust as necessary.

Shock Absorber and Coil Spring

REMOVAL & INSTALLATION

240SX

The 240SX uses a coil-over-shock absorber assembly that looks much like a

MacPherson strut. It is removed, installed and overhauled in much the same manner.

1. Remove the rear seat and package shelf to gain access to the upper shock attaching nuts.

2. Remove the upper attaching nuts.

3. Raise and safely support the vehicle.

4. Remove the rear wheel.

5. Support the lower control arm with a jack.

6. Remove the lower shock absorber mounting bolt.

7. Remove the shock absorber assembly from the vehicle.

✳✳ CAUTION

Coil springs are under extreme loads when compressed. Be sure to properly align the spring with the compressing tool to prevent personal injury from the spring releasing unexpectedly.

8. Compress the coil spring with the proper spring compressor tool.

9. Remove the center retaining nut holding the upper mounting insulator.

10. Slowly decompress the coil spring.

11. Remove the mounting insulator and the coil spring.

12. Remove coil spring.

To install:

13. Install the compressed coil spring onto the strut assembly. Be sure to the end of the spring is in the notch on the lower seat..

14. Install the upper mounting insulator.

➡ **It will be necessary to use a new locknut for the center retaining nut of the coil spring.**

15. Install the center retaining nut and tighten the nut to 13–17 ft. lbs. (18–24 Nm).

16. Slowly remove the spring compressor tool.

17. Position the shock absorber assembly in the vehicle and install the mounting fasteners. Tighten the lower mounting bolt to 72–87 ft. lbs. (98–118 Nm) and the upper nuts to 12–14 ft. lbs. (16–19 Nm).

18. Install the wheel.

19. Lower the vehicle.

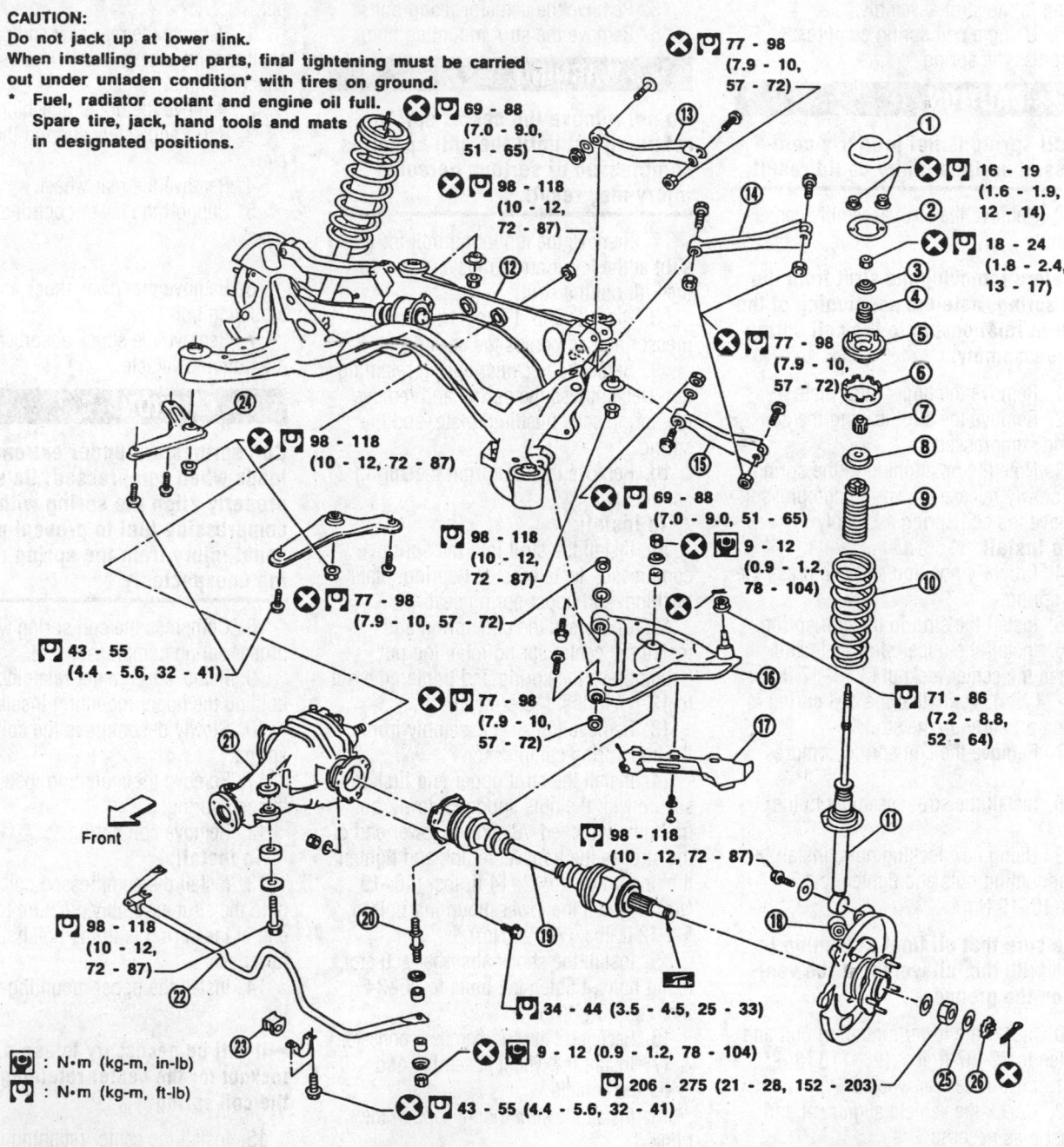

CAUTION:
Do not jack up at lower link.
When installing rubber parts, final tightening must be carried
out under unladen condition* with tires on ground.
* Fuel, radiator coolant and engine oil full.
  Spare tire, jack, hand tools and mats
  in designated positions.

77 - 98
(7.9 - 10,
57 - 72)

69 - 88
(7.0 - 9.0,
51 - 65)

98 - 118
(10 - 12,
72 - 87)

16 - 19
(1.6 - 1.9,
12 - 14)

18 - 24
(1.8 - 2.4,
13 - 17)

77 - 98
(7.9 - 10,
57 - 72)

98 - 118
(10 - 12, 72 - 87)

69 - 88
(7.0 - 9.0, 51 - 65)

9 - 12
(0.9 - 1.2,
78 - 104)

98 - 118
(10 - 12,
72 - 87)

77 - 98
(7.9 - 10, 57 - 72)

43 - 55
(4.4 - 5.6, 32 - 41)

77 - 98
(7.9 - 10,
57 - 72)

71 - 86
(7.2 - 8.8,
52 - 64)

Front

98 - 118
(10 - 12,
72 - 87)

98 - 118
(10 - 12, 72 - 87)

34 - 44 (3.5 - 4.5, 25 - 33)

9 - 12 (0.9 - 1.2, 78 - 104)

206 - 275 (21 - 28, 152 - 203)

43 - 55 (4.4 - 5.6, 32 - 41)

: N·m (kg-m, in-lb)

: N·m (kg-m, ft-lb)

① Cap	⑩ Coil spring	⑲ Drive shaft
② Gasket	⑪ Shock absorber	⑳ Connecting rod
③ Upper plate	⑫ Suspension member	㉑ Final drive
④ Bushing	⑬ Rear upper link	㉒ Stabilizer bar
⑤ Shock absorber mounting bracket	⑭ Front upper link	㉓ Bushing
⑥ Upper rubber seat	⑮ Lateral link	㉔ Member stay
⑦ Bushing	⑯ Lower arm	㉕ Insulator
⑧ Plate	⑰ Protector	㉖ Adjusting cap
⑨ Bumper rubber with dust cover	⑱ Axle housing	

7923QG88

Exploded view of the rear suspension—240SX

300ZX

1. Remove the strut from the vehicle.
2. Use the strut holding fixture and mount the strut in a vise.
3. **Loosen** but do not remove the piston rod locknut.
4. Use a coil spring compressor and compress the coil spring until the upper spring seat can be turned by hand.
5. Remove the piston rod locknut.
6. Remove the upper spring seat and the coil spring.

To install:

7. Install the compressed coil spring on the strut.
8. Install the upper spring seat on the coil spring.
9. Install a new self locking nut on the piston rod. Tighten the nut to 13–17 ft. lbs. (18–24 Nm).
10. Carefully remove the spring compressor from the spring.
11. Install the strut assembly in the vehicle.

Lower Ball Joint

REMOVAL & INSTALLATION

Sentra, Altima, 200SX and Maxima

The lower ball joint is not replaceable, if the ball joint is defective the lower control arm or transverse link must be replaced.

300ZX

➡**The lower ball joint is integral with the lower control arm (transverse link). They are removed and replaced as an assembly.**

1. Raise and support the vehicle safely.
2. Remove the front wheels.
3. Disconnect the ball joint stud from the steering knuckle.
4. Disconnect the tension rod from lower control arm.
5. Remove the mounting through-bolt connecting the lower control arm to the suspension crossmember.

➡**It may be necessary to turn the steering wheel to move the steering rack for removal of the through-bolt.**

To install:
➡**When installing the control arm, temporarily tighten the nuts and/or bolts securing the control arm to the suspension crossmember. Tighten them fully only after the vehicle's weight is resting on the wheels.**

6. Install the lower control arm and install the through-bolt with securing nut. At this time only snug the mounting nut.
7. Connect the ball joint stud to the knuckle and tighten the nut to 65–80 ft. lbs. (88–108 Nm). Install a new cotter pin.
8. Connect the tension rod to the control arm and tighten the nuts as follows:
   a. Convertible models —69–83 ft. lbs. (93–113 Nm).
   b. Non-convertible models —80–94 ft. lbs. (108–127 Nm).
9. Install the wheels and lower the vehicle to the floor.
10. Once the full weight of the vehicle is on the suspension, tighten the inner lower arm mounting bolt nut to 80–94 ft. lbs. (108–127 Nm).

Wheel Bearings

ADJUSTMENT

Front

SENTRA, 200SX, ALTIMA AND MAXIMA

➡**Whenever the hub or bearing assemblies are removed, the wheel bearing must be replaced. Never reuse the old bearing assembly.**

The wheel bearings are sealed and are not adjustable. If defective, replacement is the only option.

240SX

1. Raise and safely support the front wheels.
2. Remove the front wheels.
3. Remove the hub cap and wheel bearing nut cotter pin.
4. Tighten the wheel bearing locknut to 151–210 ft. lbs. (206–284 Nm).
5. Check that wheel bearings operate smoothly.
6. Check axial end-play. Axial end-play must be 0.0020 in. (0.05mm) or less. If axial end-play is not within specification or wheel bearing does not turn smoothly, replace the wheel bearing assembly.
7. Replace the wheel bearing locknut cotter pin and reinstall the hub cap.
8. Install the wheels and safely lower the vehicle.

300ZX

The front wheel bearing is not adjustable. Replace the bearing assembly if a growling noise is emitted during operation or if the bearing drags or turns roughly.

Rear

The rear wheel bearings on the 240SX and 300ZX are not adjustable. Replace the bearing assembly if a growling noise is emitted during operation or if the bearing drags or turns roughly.

SENTRA, 200SX, ALTIMA AND MAXIMA

If the wheel hub bearing assembly is removed, it must be replaced.

➡**The wheel hub bearing assembly is not repairable; it must be replaced when defective.**

1. Tighten the wheel bearing locknut to 138–188 ft. lbs. (187–255 Nm).
2. Verify that the wheel bearings operate smoothly.
3. Install a new cotter pin into the spindle to hold the wheel bearing locknut.
4. Install a dial indicator to the rear wheel hub bearing assembly and check the axial end-play; it should be less than 0.0020 in. (0.05mm).
5. Install the grease cap.
6. If the axial end-play exceeds specifications, the wheel bearing must be replaced.

REMOVAL & INSTALLATION

Front

SENTRA AND 200SX

➡**Whenever the hub or bearing assembly is removed, the wheel bearing assembly must be replaced. Never reuse the old bearing assembly.**

1. Raise the vehicle and support it safely. Remove the front wheel.
2. Remove the wheel bearing/axleshaft locknut while depressing the brake pedal.
3. Remove the brake caliper and support it with a piece of wire. It is not necessary to disconnect the brake line from the caliper.
4. Remove the ABS sensor from the steering knuckle.

➡**Do not depress the brake pedal or twist the brake line.**

5. Remove the tie rod end using a tie rod removing tool J25730A or equivalent.
6. Separate the halfshaft from the knuckle by slightly tapping with a soft hammer. Position the axle shaft nut on the threads of the shaft to protect them when lightly tapping.
7. Loosen the lower ball joint nut and separate using a ball joint separator J25730A or equivalent.

8. Remove the two strut-to-knuckle retaining bolts and separate.

9. Remove the steering knuckle from the vehicle.

10. Place the assembly in a vise. Drive the hub with the inner race from the knuckle with a suitable tool. Remove the inner and outer grease seals.

11. Remove the bearing inner race and outer grease seal from the hub.

12. Remove the snapring and press the bearing outer race to remove the bearing from the steering knuckle.

To install:

13. Press a new wheel bearing into the knuckle assembly not exceeding 3.3 tons (3,000 kg) pressure.

14. Install the snapring and pack the grease seal lips with chassis grease.

15. Install the inner and outer grease seals.

16. Press the wheel hub into the knuckle not exceeding 3.3 tons (3,000 kg) pressure.

17. Check bearing operation and by applying 3.9–5.5 tons of pressure to the hub assembly. Spin the hub several times in both directions.

18. Be sure the bearings rotate freely. If the bearings do not rotate freely, replace the bearings.

19. Install the knuckle and wheel hub assembly.

20. Install the lower ball joint and tighten the nut to 43–54 ft. lbs. (59–74 Nm). Install a new cotter pin.

21. Install the strut bolts and tighten them to 84–98 ft. lbs. (114–133 Nm) for 1991–94 vehicles or 68–82 ft. lbs. (92–118 Nm) for 1995–96 vehicles.

22. Connect the tie end and tighten the nut to 22–29 ft. lbs. (29–39 Nm). Install a new cotter pin.

23. Install the disc brake caliper.

24. Install and tighten the wheel bearing locknut to 145–203 ft. lbs. (196–275 Nm). Install a new cotter pin.

25. Install the front wheels and lower the vehicle.

26. Check the vehicle's alignment.

27. Road test the vehicle and verify proper operation.

MAXIMA AND ALTIMA

➡**Whenever the hub or bearing assembly is removed, the wheel bearing assembly must be replaced. Never reuse the old bearing assembly.**

1. Remove the knuckle assembly from the vehicle.

2. Using a shop press and a suitable tool, press the hub with the inner race from the steering knuckle.

3. Using a shop press and a suitable tool, press the bearing inner race from the hub and remove the outer grease seal.

4. Using a prybar, pry the inner grease seal from the steering knuckle.

5. Using snapring pliers to remove the inner and outer snaprings from the steering knuckle.

6. Using a shop press and a suitable tool, press the sealed bearing assembly from the steering knuckle.

7. Inspect the hub, steering knuckle and snaprings for cracks and/or wear; if necessary, replace the damaged part(s).

To install:

8. Install the inner snapring in the steering knuckle groove.

9. Using a shop press and a suitable tool, press the new wheel bearing assembly into the steering knuckle, until it seats, using a maximum pressure of three tons.

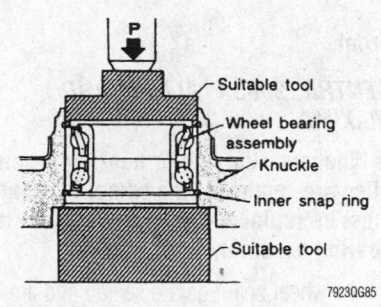

Typical method of installing the wheel bearing

10. Install the outer snapring.

11. Pack the new grease seal lips with multi-purpose grease.

12. Using a shop press and a suitable tool, press the new outer grease seal into the steering knuckle.

13. Using a shop press and a suitable tool, press the hub into the steering

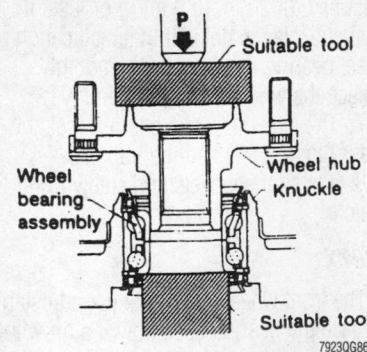

Use a press to install the hub into the knuckle assembly

knuckle, until it seats, using a maximum pressure of 5.5 tons; be careful not to damage the grease seal.

14. To check the bearing operation, perform the following procedures:

   a. Increase the press pressure to 3.5–5.0 tons.

   b. Spin the steering knuckle, several turns, in both directions.

   c. Be sure the wheel bearings operate smoothly.

15. If the wheel bearings do not operate smoothly, replace the wheel bearing assembly.

16. Install the knuckle assembly.

17. Install the halfshaft into the hub and tighten the locknut to 174–231 ft. lbs. (235–314 Nm).

18. Install the wheel assembly and lower the vehicle.

19. Road test the vehicle and verify proper operation.

240SX AND 300ZX

➡**If defective, the wheel bearing assembly can only be serviced by replacement of the hub.**

1. Raise and support the front of the vehicle safely and remove the wheels.

2. Remove the brake caliper assembly. Be sure not to twist the brake hose.

➡**When removing the brake caliper it is not necessary to disconnect the brake hose. Be sure to support the brake caliper after removal.**

3. If applicable, remove the ABS brake sensor from the steering knuckle.

4. Remove the brake rotor.

5. Remove the wheel bearing dust cap.

6. Remove the locking nut.

7. Remove the wheel bearing locknut and washer.

➡**The hub bearing is a sealed type and will be removed with the hub assembly.**

8. Pull hub assembly toward you to remove the hub assembly from the spindle.

9. Remove the tie rod joint and lower ball joint.

10. Disconnect the knuckle from the strut.

To install:

11. Connect the knuckle to the strut.

12. Install the tie rod joint and lower ball joint.

13. Lubricate the spindle and slide hub assembly onto spindle.

14. Install the washer and wheel bearing locknut.

15. Tighten the wheel bearing locknut to 152–210 ft. lbs. (206–284 Nm.).

16. Turn the hub assembly several times in both directions to seat the hub.

17. Measure the wheel bearing axial end-play.

    a. Axial end-play—0.0020 in. (0.05mm) or less

18. Clinch the locking nut with a hammer and chisel.

19. Install a new dust cap.

20. Install the ABS sensor and tighten the mounting bolt to 96–144 inch lbs. (11–16 Nm).

21. Install the brake rotor and caliper assembly.

22. Install the front wheels.

23. If the brake line was disconnected, bleed the brake system.

Rear

SENTRA, 200SX, ALTIMA AND MAXIMA

If the wheel hub bearing assembly is removed, it must be replaced.

➡**If the vehicle is equipped with ABS, the sensor must be removed to protect the sensor and its wiring.**

1. Raise and safely support the vehicle. Remove the rear wheel(s).

2. If equipped with disc brakes, perform the following procedures:

    a. Remove the brake caliper and hang it by a piece of wire.

    b. Remove the brake caliper support.

    c. Remove the disc brake pads.

    d. Remove the brake disc.

3. If equipped with drum brakes, perform the following procedures:

    a. Remove the brake drum.

    b. If necessary, remove the brake shoe assembly.

4. Remove the grease cap.

5. Remove the cotter pin, wheel bearing locknut, washer, and the wheel hub bearing assembly. A slide hammer may be needed to remove the hub bearing assembly.

➡**The wheel hub bearing assembly is not repairable; it must be replaced when defective.**

To install:

➡**If the vehicle is equipped with ABS, the sensor ring must be removed and installed on the new hub.**

6. Install the wheel hub bearing assembly, the washer and the wheel bearing locknut. Tighten the wheel bearing locknut to 138–188 ft. lbs. (187–255 Nm).

7. Verify that the wheel bearings operate smoothly.

8. Install a new cotter pin into the spindle to hold the wheel bearing locknut.

9. Install a dial micrometer to the rear wheel hub bearing assembly and check the axial end-play; it should be less than 0.0020 in. (0.05mm).

10. Install the grease cap.

11. If removed, install the ABS sensor and its wiring.

12. Install the brake assembly and the wheels.

240SX AND 300ZX

1. Raise and support the rear of the vehicle and remove the rear wheels. Remove the cotter pin, adjusting cap and insulator.

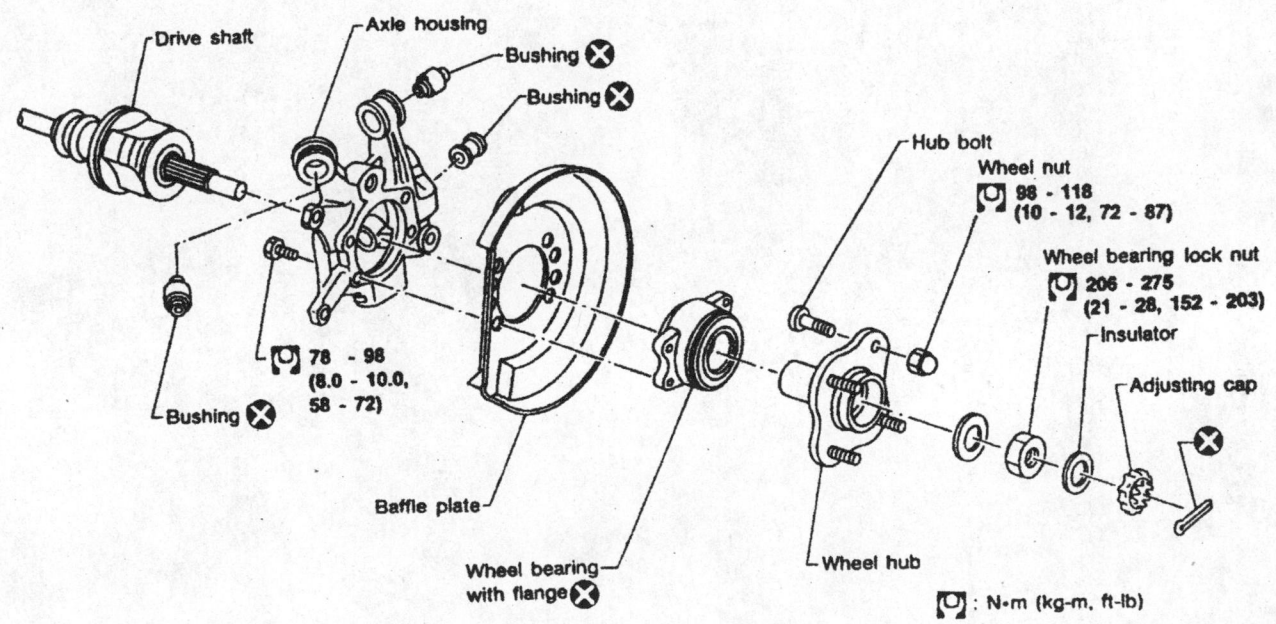

Exploded view of rear wheel bearing assembly—240SX

7923QG87

2. Apply the parking brake firmly to hold the rear halfshaft while removing the wheel bearing locknut. Remove the wheel bearing locknut.

3. Unbolt the caliper and move it aside. Do not disconnect the hose from the caliper. Do not allow the caliper to hang by the hose; support the caliper with a length of wire or rest it on a suspension member. Remove the brake disc.

4. Separate the halfshaft from the axle housing by lightly tapping it. Cover the driveshaft boots with a shop towel to prevent damage.

5. Remove the four bolts on the rear of the axle housing and that hold the wheel bearing, flange, and hub to the axle housing. Remove the wheel bearing, flange, and hub assembly.

6. Press the wheel bearing from the axle hub.

7. Mount the hub in a vise and remove the inner race, using bearing puller tool ST30031000, or equivalent. Discard the inner race and grease seals.

8. Clean all parts in a suitable solvent. Check the wheel hub and axle housing for cracks, preferably using the dye penetrant method. Check the wheel bearing seating surface for roughness, seizure or other damage that may interfere with proper bearing function. Check the rubber bushing for wear.

To install:

9. Place the hub on a block of wood and seat the inner race using a suitable drift. Be careful not to damage the grease seals during installation of the inner race.

10. Press the bearing into the hub using a suitable drift.

11. Mount the bearing/hub assembly to the axle housing and tighten the mounting bolts to 58–72 ft. lbs. (78–98 Nm).

12. Lubricate the halfshaft splines. Properly align the splines and insert the halfshaft into the wheel hub.

13. Install the wheel bearing locknut and tighten the nut to 152–203 ft. lbs. (206–275 Nm). Install the insulator and fit adjusting cap. Install a new cotter pin.

14. Check the axial end-play as follows before mounting the rear wheels. Mount a dial indicator so the stylus of the dial rests on the face of the hub and check the wheel bearing axial end-play by attempting to rock the wheel hub in and out. The end-play should be 0.0020 in. (0.05mm) or less.

15. Install the rotor and caliper assembly.

16. Mount the rear wheels and lower the vehicle.

PORSCHE

911 Carrera

ENGINE REPAIR

➡ **Disconnecting the negative battery cable on some vehicles may interfere with the functions of the on board computer systems and may require the computer to undergo a relearning process, once the negative battery cable is reconnected.**

Distributor

REMOVAL & INSTALLATION

911 Carrera

1. Disconnect the negative battery cable.
2. Position the No. 1 cylinder at Top Dead Center (TDC).
3. Remove the hot air pipe.
4. Remove the distributor cap, and detach any electrical connections at the distributor.
5. Loosen, then remove the distributor hold down nut and remove the distributor.

To install:
6. While installing the distributor, be sure to align the ignition rotors as shown.
7. Install the distributor and tighten the hold down nut to 10 ft. lbs. (15 Nm).
8. Attach the electrical connections to the distributor and install the distributor cap.
9. Install the hot air pipe.
10. Connect the negative battery cable.

Ignition Timing

ADJUSTMENT

The ignition timing is controlled by the engine control computer. No adjustment is necessary or possible.

Engine Assembly

REMOVAL & INSTALLATION

911 Carrera

The engine/transmission assembly is removed from the vehicle from below.
1. Properly relieve the fuel system pressure.
2. Disconnect the negative battery cable.
3. Raise and safely support the vehicle.

❋❋ WARNING

Cover the upper lock section protruding from the engine compartment cover with a piece of hose.

4. Remove the A/C compressor and position it aside leaving the hoses attached.
5. Remove the hot air blower and air cleaner assembly.
6. Disconnect the crankcase-to-oil tank vent hose.
7. Using a suction pump, drain the power steering fluid reservoir.
8. Detach the spark plug leads for cylinder No. 4 and 5. Place a drain pan

under the power steering pump and disconnect the hydraulic lines. Plug the openings in the pump and lines.
9. Label and detach the electrical connectors at the intake manifold.
10. Label and detach the following hoses and connectors from the left-side of the engine compartment, as illustrated.
11. Remove the engine compartment electrical system cover and detach multiple connectors.
12. Remove the right-hand electrical cover and detach the oxygen sensor electrical connector.

➡ **If the fuel tank is full, it may be necessary to pinch off the fuel lines to prevent excessive leakage.**

13. Disconnect the fuel supply and return lines.
14. Detach the throttle cable from the pedal return lever.
15. Remove the engine/transmission guard and rear undercover.
16. Remove the tail pipes, then the rear mufflers.
17. Disconnect the halfshafts from the transmission.
18. Disconnect the shifter at the transmission by removing the hex head screw and pulling the selector rod off the internal shift rod.
19. Remove the hot air hoses, pipes and air flaps.
20. Disconnect the front transverse strut.
21. Completely remove the sway-bar from the vehicle.
22. Matchmark the position of the toe adjustment.
23. Loosen the fastening nut (1), remove the toe eccentric (A), then slide the toe control arm off the crossmember.
24. Perform the previous step for the opposite side.
25. If equipped with an automatic transmission, disconnect the oil cooler lines from the transmission.

❋❋ CAUTION

The EPA warns that prolonged contact with used engine oil may cause a number of skin disorders, including cancer! You should make every effort to minimize your exposure to used engine oil. Protective gloves should be worn when changing the oil. Wash your hands and any other exposed skin areas as soon as possible after exposure to used engine oil. Soap and water, or waterless hand cleaner should be used.

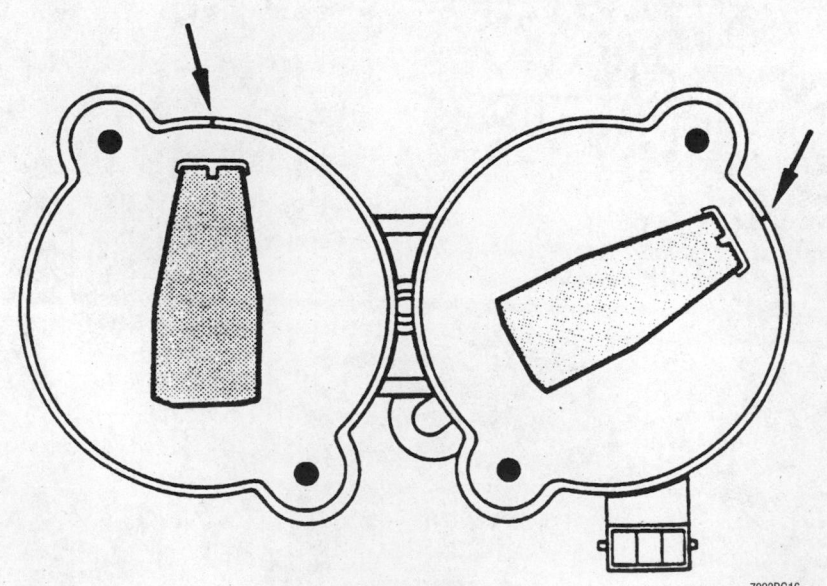

Ignition rotor alignment when the No. 1 cylinder is at TDC—911 Carrera

7923RG16

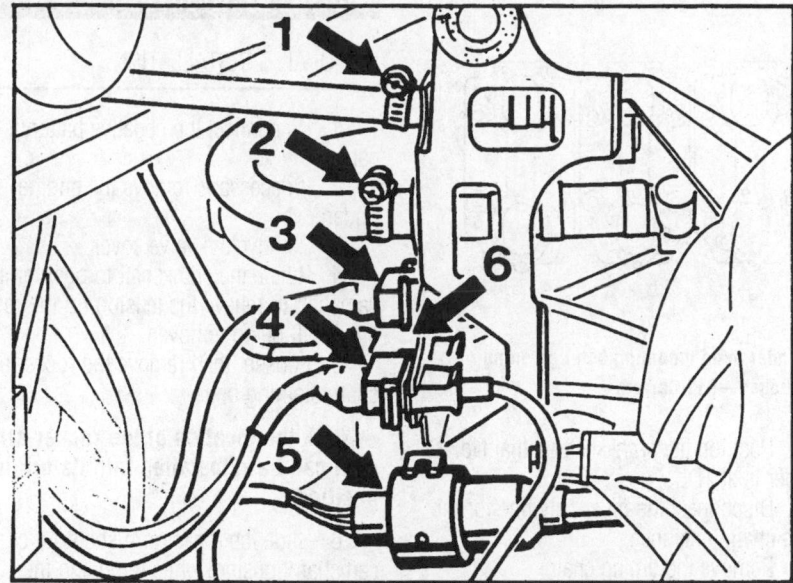

1 - to carbon canister

2 - to brake booster

3 - to temperature sensor II (cyl. no. 3)

4 - to knock sensor

5 - cruise control connector

6 - to reference mark sender

7923RG01

Left-side engine compartment connector and hose identification—911 Carrera

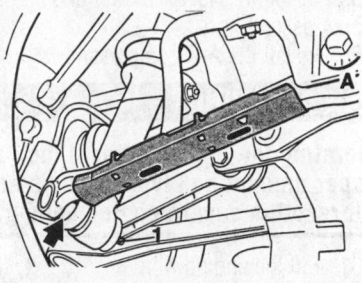

7923RG02

Be sure to matchmark the position of the toe eccentric before removing—911 Carrera

26. Drain the engine oil at the thermostat housing. If the engine is being serviced, drain the engine oil from the crankcase.

27. Disconnect the oil return line.

28. Disconnect the wiring from the starter.

29. Disconnect the ground strap between the starter and the body at the starter mounting.

30. Remove the throttle operating cable from its guide.

31. Using a suitable engine/transmission jack, support the engine.

32. Using a long reach socket, remove the transmission crossmember (6 bolts) and engine mount.

33. Slightly lower the engine.

34. If equipped with manual transmission, unbolt the clutch slave cylinder and position it aside leaving the lines connected.

35. Lower the engine/transmission further. Ensure that all vacuum and fluid line are disconnected, and that all wiring harness have been detached.

36. Slowly lower the engine out of the vehicle.

To install:

37. Raise the engine/transmission assembly into the vehicle.

38. If equipped with a manual transmission, install the clutch slave cylinder on to the transmission, then install the engine the remaining distance.

39. Install the transmission crossmember and tighten the mounting bolts to 34 ft. lbs. (46 Nm), then the engine mount and tighten to 63 ft. lbs. (85 Nm).

40. Install the throttle operating cable into its guide.

41. Install the ground strap between the starter and the body at the starter mounting.

42. Connect the oil return line.

43. Connect the wiring to the starter.

44. If equipped with an automatic transmission, connect the oil cooler lines to the transmission.

45. Install the toe control arm to the crossmember, and tighten the fastening nut and eccentric to 74 ft. lbs. (100 Nm).

46. Install the sway-bar into the vehicle, and tighten the sway-bar-to-crossmember mounting to 17 ft. lbs. (23 Nm) and the sway-bar-to-shock absorber mounting to 34 ft. lbs. (46 Nm).

47. Install the front transverse strut.

48. Install the hot air hoses, pipes and air flaps.

49. Connect the shifter linkage and tighten the hex head screw to 17 ft. lbs. (23 Nm).

50. Connect the halfshafts to the transmission and tighten the mounting bolts to 60 ft. lbs. (81 Nm).

51. Install the rear mufflers, then the tail pipes.

52. Be sure the drain plugs are installed and tighten to 37 ft. lbs. (50 Nm).

53. Connect the throttle cable to the pedal return lever.

54. Connect the fuel supply and return lines.

55. Attach all electrical components and connectors that were removed.

56. Connect all hoses that were removed.

57. Connect the power steering hydraulic lines and fill the reservoir.

58. Connect the crankcase-to-oil tank vent hose.

59. Install the hot air blower and air cleaner assembly.
60. Install the A/C compressor.

Operating the engine without the proper amount and type of engine oil will result in severe engine damage.

61. Fill the engine with oil.
62. Connect the negative battery cable.
63. Start and run the vehicle. Check for leak, then install the engine/transmission guard and rear undercover.

Cylinder Head

REMOVAL & INSTALLATION

911 Carrera

The cylinder heads can be removed in a group of 3, or independently if the camshaft housing is removed.

❊❊ **CAUTION**

If only the cylinder heads are removed during partial engine over-haul, the cylinder banks must be held in place to prevent damage to the O-rings at the cylinder base using Assembly Bracket 602 451 and 2 Supports 602 562 or their equivalents.

1. Remove the engine from the vehicle.
2. Remove the intake and exhaust manifolds.
3. Remove the rocker arm covers and gaskets.
4. Remove the rocker arms.

➡**Do not rotate the crankshaft or camshafts with the timing chains removed.**

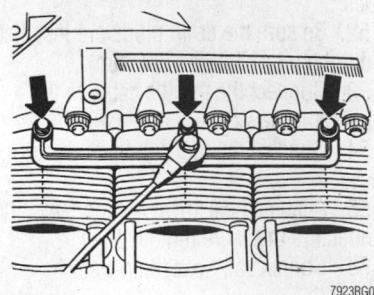

Knock sensor bridge mounting bolt locations—911 Carrera

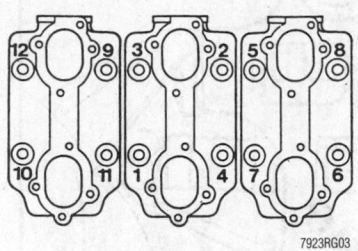

7923RG03

Cylinder head mounting bolt tightening sequence—911 Carrera

5. Position the crankshaft so that No. 1 cylinder is at TDC.
6. Disconnect the oil supply line for the timing chain tensioner.
7. Remove the timing chain tensioner(s) and the timing chain.
8. Remove the camshafts.
9. Remove the knock sensor bridge.
10. If a single cylinder is being serviced, remove the cylinder head retaining bolts for that cylinder.
11. If all the cylinders on one bank are being serviced, remove the timing chain housing-to-crankcase mounting nuts. Remove the cylinder head retaining bolts, then remove the cylinder heads with the timing chain housing.

To install:

➡**Use new gaskets/seals and mounting hardware when reassembling the engine.**

12. Apply a thin coat of Optimoly HT to the threads of the studs and nuts.
13. Install the cylinder head(s).
14. Install the knock sensor bridge retaining bolts.
15. Install the cylinder head retaining nuts and tighten in two steps following the illustrated sequence.
     a. Step 1—18 ft. lbs. (20 Nm)
     b. Step 2—90°
16. Tighten the knock sensor bridge retaining bolts to 7 ft. lbs. (9.7 Nm).
17. Install the camshafts.
18. Install the timing chain and tensioners.
19. Connect the oil supply line to the timing chain tensioner.
20. Install the rocker arms and covers.
21. Install the intake and exhaust manifolds.
22. Install the engine into the vehicle.

Rocker Arms/Shafts

REMOVAL & INSTALLATION

1. Disconnect the negative battery cable.
2. If necessary, remove the engine undercover.
3. Remove the valve cover.
4. Rotate the crankshaft to position the camshaft to relieve the tension on the rocker arm/shaft being removed.
5. Loosen, then remove the rocker arm shaft retaining bolts.

➡**Mark the location of the rocker arms so it can be reinstalled into its original position.**

6. Slide the rocker arm shaft out of the camshaft housing, while removing the rocker arm.

To install:

7. Oil the sliding surfaces of the rocker arm/shaft.
8. While holding the rocker arm in position, install the rocker arm shaft.
9. Install the rocker arm shaft retaining bolt and tighten to 10 ft. lbs. (13 Nm).
10. Install the rocker arm cover with a new gasket.
11. Connect the negative battery cable.
12. Start the vehicle and check for leaks.
13. If removed, install the engine undercovers.

Intake Manifold

REMOVAL & INSTALLATION

1. Properly relieve the fuel system pressure.
2. Disconnect the negative battery cable.
3. Disconnect the accelerator cable (1) at the pedal.
4. Remove the fresh air blower (2) and related hoses.
5. Remove the rear heater blower (3) and related hoses.
6. Remove the following vacuum hoses (4) from the intake manifold:
     • Active carbon canister (1)
     • Vacuum tank (2)
     • Recirculating-air flap for the heater and air conditioner (3)
     • Electronic switching valve for the intake manifold (4)

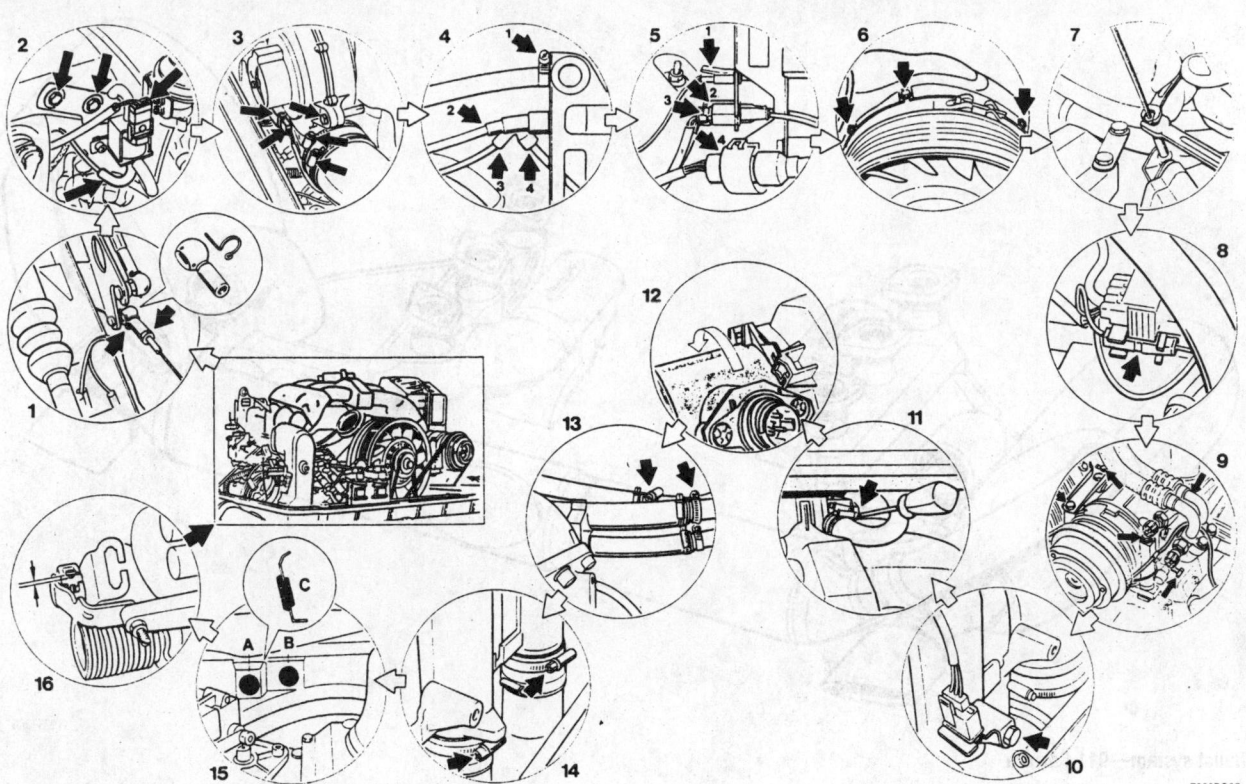

Intake manifold removal and installation procedure—911 Carrera

7. Separate (5) the reference mark sender (1), temperature sensor (2), knock sensor (3) and cruise control (4) electrical connectors.

8. Remove the intake cowl from the cooling fan (6).

9. Disconnect the cruise control actuator (7).

10. Detach the electrical connector for the take-up plate (8).

11. Remove the A/C compressor and position it aside (9).

12. Detach the plug at the knock sensor (10).

13. Disconnect and remove the mass air flow sensor and intake cowl (11 and 12).

14. Detach the oil hoses (13).

15. While removing the intake manifold, detach the accelerator cable from the front engine paneling (14).

To install:

16. Install the accelerator cable (15). Bore (A) is used for Tiptronic transmissions, and bore (B) is used for manual transmissions.

17. Connect the oil hoses.

18. Install the mass air flow sensor and intake cowl.

19. Connect the knock sensor.

20. Install A/C compressor.

21. Install the remaining components as they were removed.

22. Adjust the cruise control actuator (16).

23. Connect the negative battery cable.

Exhaust Manifold

REMOVAL & INSTALLATION

911 Carrera

The 911 Carrera does not have a exhaust manifold. It is equipped with a heat exchanger that is mounted to the cylinder heads.

1. Disconnect the negative battery cable.

2. Disconnect the heating hoses at the heat exchangers.

3. Raise and safely support the vehicle, then remove the engine/transmission undercovers.

4. Detach the hot air distributor pipe from the heat exchangers.

5. Disconnect and remove the oxygen sensors.

6. Remove the heat exchanger-to-exhaust system mounting bolts and remove the mufflers.

7. Remove the heat exchanger-to-cylinder head mounting bolts.

8. Remove heat exchangers from the vehicle.

To install:

9. Install the heat exchangers using new gaskets and mounting hardware, then tighten the mounting bolts to 21 ft. lbs. (28 Nm).

10. Install the mufflers and tighten the mounting bolts to 17 ft. lbs. (23 Nm).

11. Install and connect the oxygen sensors.

12. Connect the hot air distribution pipe to the heat exchangers.

13. Install the engine/transmission undercovers.

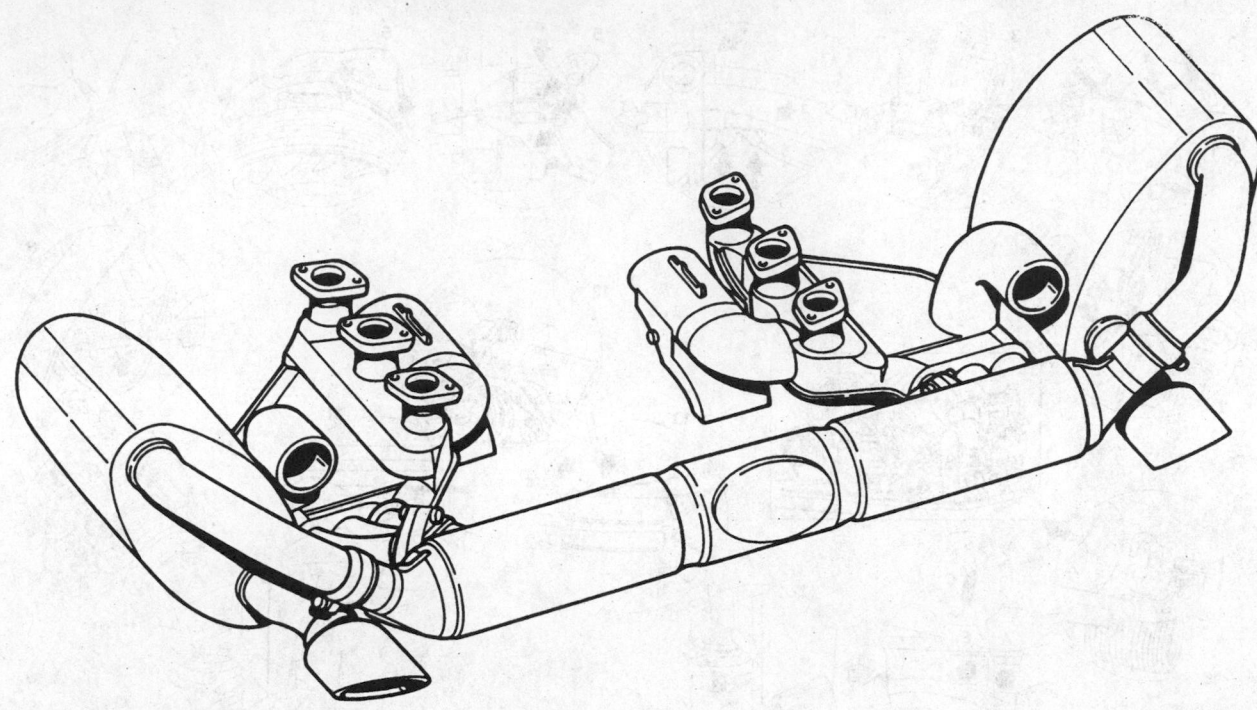

Exhaust system—911 Carrera

14. Lower the vehicle and connect the heating hoses to the heat exchanger.

15. Connect the negative battery cable.

Camshaft

REMOVAL & INSTALLATION

The 911 Carrera is not equipped with valve lifters. It is equipped with Hydraulic Lash Adjusters (HLA) that are fitted in the end of the rocker arm.

1. Remove the engine from the vehicle.

2. Remove the rocker arm covers and rocker arm assembly.

3. Remove the timing chain covers. Position the engine so that No. 1 cylinder is at TDC.

4. Remove the timing chain tensioners from the timing chain housing.

5. Lock the left camshaft using tool 9551 and the right camshaft using tool 9552, or their equivalents, then remove the camshaft sprocket center bolt.

6. Using the Straight Pin Remover tool P212 or equivalent, pull out the straight pin and remove the camshaft sprocket.

7. Remove the camshaft sprocket flange and end cover, then remove the camshaft from the camshaft housing.

To install:

8. Install the camshaft into the camshaft housing and tighten the end cover bolts to 10 ft. lbs. (15 Nm).

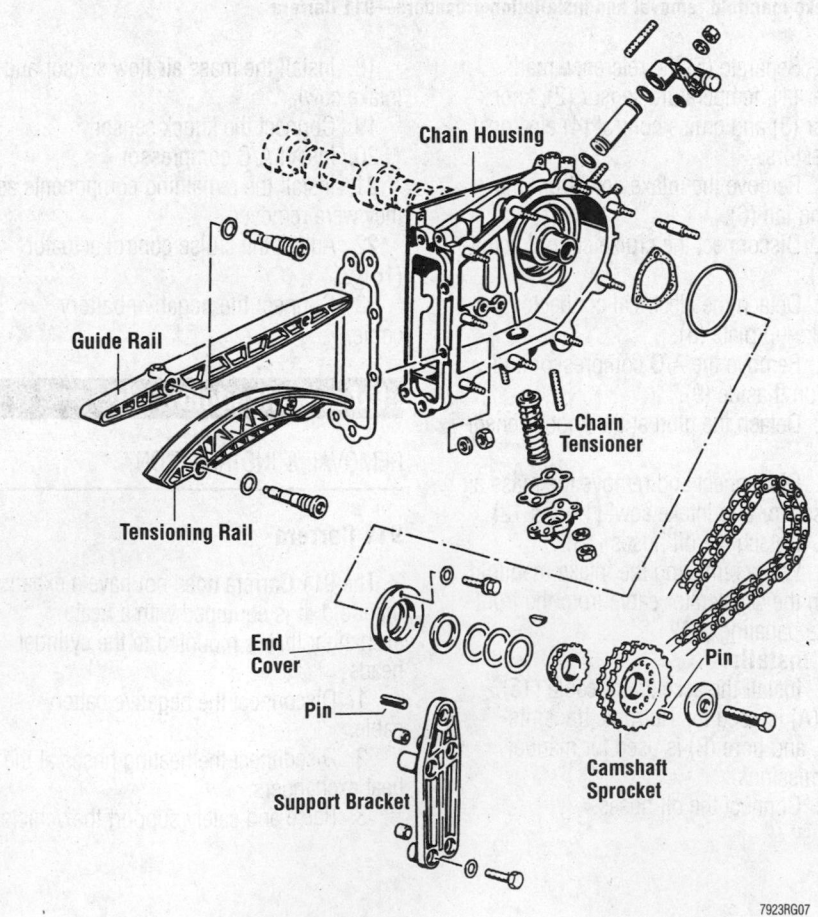

7923RG07

Exploded view of the camshaft drive and chain housing—911 Carrera

9. Install the camshaft sprockets and timing chain as outlined in the timing chain service procedure.

10. Install the timing chain tensioners and covers.

11. Install the rocker arm assemblies.

12. Install the engine into the vehicle.

Valve Lash

ADJUSTMENT

The 911 Carrera is equipped with HLA's. No adjustment is necessary or possible.

Timing Chain

REMOVAL & INSTALLATION

911 Carrera

1. Remove the engine from the vehicle.

2. Remove the muffler and rear cooling shroud.

3. Remove the timing chain covers.

4. Disconnect the timing chain tensioner oil supply lines.

5. Position No. 1 cylinder to TDC.

6. Remove the timing chain tensioners and guides.

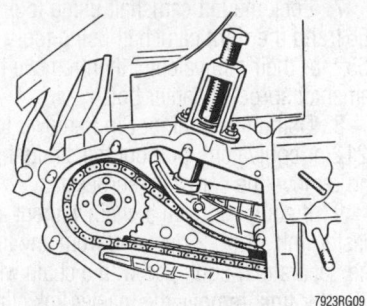

7923RG09

Using the Auxiliary Chain Tensioner 9401 to preload the timing chain guide—911 Carrera

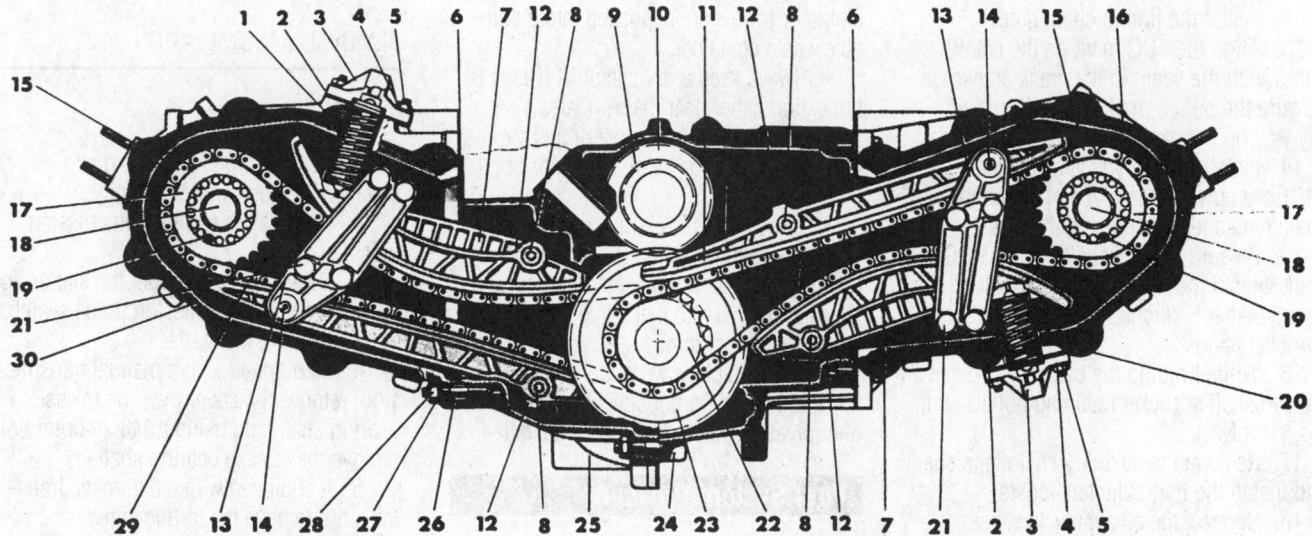

1 - Left-hand chain tensioner
2 - Chain tensioner gasket
3 - Chain tensioner cover
4 - Aluminum washer
5 - M 6 lock nut
6 - Chain housing gasket
7 - Tensioning rail
8 - Support stud
9 - Distributor drive gear
10 - Crankshaft sprocket z = 35
11 - Right-hand guide rail
12 - Spring-loaded thrust piece
13 - Bearing saddle
14 - Heavy type dowel pin
15 - Duplex roller chain
16 - Right-hand chain housing
17 - Chain sprocket z = 28
18 - Hexagon head bolt M 12 x 1.5
19 - Straight pin, 6 mm dia.
20 - Right-hand chain tensioner
21 - M 6 hexagon-head bolt
22 - Chain sprocket z = 24
23 - Intermediate shaft sprocket z = 60
24 - Right-hand crankcase section
25 - Left-hand crankcase section
26 - Fit sleeve
27 - Aluminum washer
28 - M 8 lock nut
29 - Left-hand guide rail
30 - Left-hand chain housing

7923RG08

Camshaft drive and chain housing component identification—911 Carrera

7. Lock the left camshaft using tool 9551 and the right camshaft using tool 9552, or their equivalents, then remove the camshaft sprocket center bolt.

8. Using the Straight Pin Remover tool P212 or equivalent, pull out the straight pin and remove the camshaft sprocket.

9. If equipped with a chain without a master link, grind 2 rivet end from any link and separate. If equipped with a chain with a master link, remove the master link clip and separate the chain.

10. Remove the timing chain.

To install:

11. Install the timing chain. If installing a chain with a master link, be sure the closed end of the link clip faces the direction of travel.

12. Install the timing chain guides.

13. Align the TDC mark on the crank pulley with the seam in the crank crankcase. Be sure the punch mark and/or Woodruff key way on the camshafts are straight up.

14. Install Auxiliary Chain Tensioner 9401 or equivalent in place of the timing chain tensioner and tension the chain.

15. Be sure the camshaft is at its TDC mark. Install the dowel pin in the hole in the sprocket that is aligned with the hole in the sprocket flange.

16. While holding the camshaft, tighten the camshaft sprocket retaining bolt to 88 ft. lbs. (120 Nm).

17. Remove the auxiliary chain tensioner and install the hydraulic tensioners.

18. Connect the oil supply lines to the timing chain tensioners and install the covers.

19. Install the cooling shroud and rear mufflers.

20. Install the engine into the vehicle.

FUEL SYSTEM

Fuel System Service Precautions

Safety is the most important factor when performing not only fuel system maintenance but any type of maintenance. Failure to conduct maintenance and repairs in a safe manner may result in serious personal injury or death. Maintenance and testing of the vehicle's fuel system components can be accomplished safely and effectively by adhering to the following rules and guidelines.

• To avoid the possibility of fire and personal injury, always disconnect the negative battery cable unless the repair or test procedure requires that battery voltage be applied.

• Always relieve the fuel system pressure prior to disconnecting any fuel system component (injector, fuel rail, pressure regulator, etc.), fitting or fuel line connection. Exercise extreme caution whenever relieving fuel system pressure to avoid exposing skin, face and eyes to fuel spray. Please be advised that fuel under pressure may penetrate the skin or any part of the body that it contacts.

• Always place a shop towel or cloth around the fitting or connection prior to loosening to absorb any excess fuel due to spillage. Ensure that all fuel spillage (should it occur) is quickly removed from engine surfaces. Ensure that all fuel soaked cloths or towels are deposited into a suitable waste container.

• Always keep a dry chemical (Class B) fire extinguisher near the work area.

• Do not allow fuel spray or fuel vapors to come into contact with a spark or open flame.

• Always use a back-up wrench when loosening and tightening fuel line connection fittings. This will prevent unnecessary stress and torsion to fuel line piping. Always follow the proper torque specifications.

• Always replace worn fuel fitting O-rings with new. Do not substitute fuel hose or equivalent, where fuel pipe is installed.

Fuel System Pressure

RELIEVING

1. Turn the ignition **OFF**.
2. Access the fuse panel.
3. Remove the fuel pump fuse. Start and run the engine until it stalls.
4. After stalling, crank the engine to ensure any residual fuel pressure is released.

Fuel Filter

REMOVAL & INSTALLATION

➡**The fuel filter is located behind the air filter housing.**

1. Properly relieve the fuel system pressure.
2. Disconnect the negative battery cable.
3. Remove the air cleaner assembly.
4. Place a rag under the filter to catch any remaining fuel. Remove the fuel lines attached the filter.
5. Loosen the fuel filter mounting clamp and remove the filter noting the direction of flow.

To install:

6. While observing the direction of flow, install the fuel filter and tighten the retaining clamp.

7. Connect the fuel lines to the filter.
8. Install the air cleaner assembly.
9. Connect the negative battery cable.
10. Start the vehicle and check for leaks.

DRIVE TRAIN

Transmission Assembly

REMOVAL & INSTALLATION

911 Carrera Model

The transmission is removed and installed as a unit with the engine.

1. Remove the engine/transmission unit.

2. Detach the wires from the starter and reverse light or transmission range switch.

3. Remove the starter.

4. If equipped with a manual transmission, remove the cover over the release bearing shaft, and using a M6 x 40mm bolt remove the release bearing shaft.

5. If equipped with a automatic transmission, remove the torque converter-to-flexplate mounting bolts, then install Torque Converter Holder 9325 or equivalent.

6. Remove the engine-to-transmission mounting bolts and remove the transmission.

To install:

7. For manual transmission, engage the release fork into the bearing, and using a suitable piece of tape, locate it in the installed position.

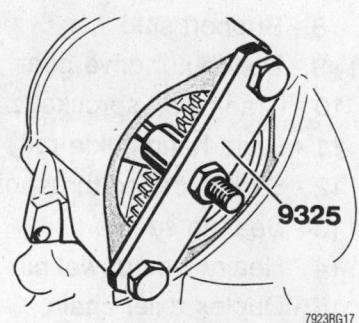

To prevent dropping the torque converter, install the Torque Converter Holder 9325 or equivalent as shown—911 Carrera

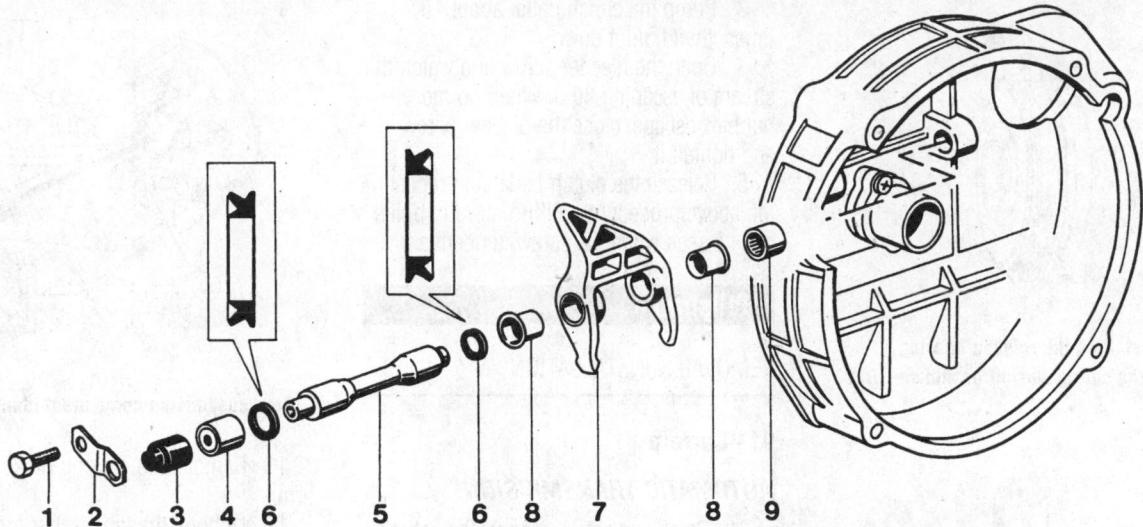

1. M6 x 16 bolt
2. Bracket
3. Bearing cover

4. Needle-roller bearing
5. Release lever shaft
6. Sealing ring

7. Release lever
8. Plastic bushing
9. Needle-roller bearing

7923RG19

Exploded view of the clutch control shaft—911 Carrera

8. Install the transmission to the engine and tighten the mounting bolts to 32 ft. lbs. (45 Nm).

9. For automatic transmission, remove the torque converter holder and install the converter-to-flexplate mounting bolts, then tighten the bolts to 34 ft. lbs. (46 Nm).

10. For manual transmission, install the release bearing shaft and tighten the retaining bolt to 17 ft. lbs. (23 Nm).

11. Install the starter, and attach the wires that were removed.

12. Install the engine/transmission unit.

Clutch

ADJUSTMENT

The 911 Carrera is equipped with a hydraulic clutch actuating system. The system is self-adjusting.

REMOVAL & INSTALLATION

1. Remove the transmission.

2. If reusing the thrust plate, match-mark the thrust plate-to-flywheel position.

3. Evenly remove the thrust plate-to-fly-wheel mounting bolts in a crisscross pattern.

4. Remove the thrust plate and clutch disk from the flywheel.

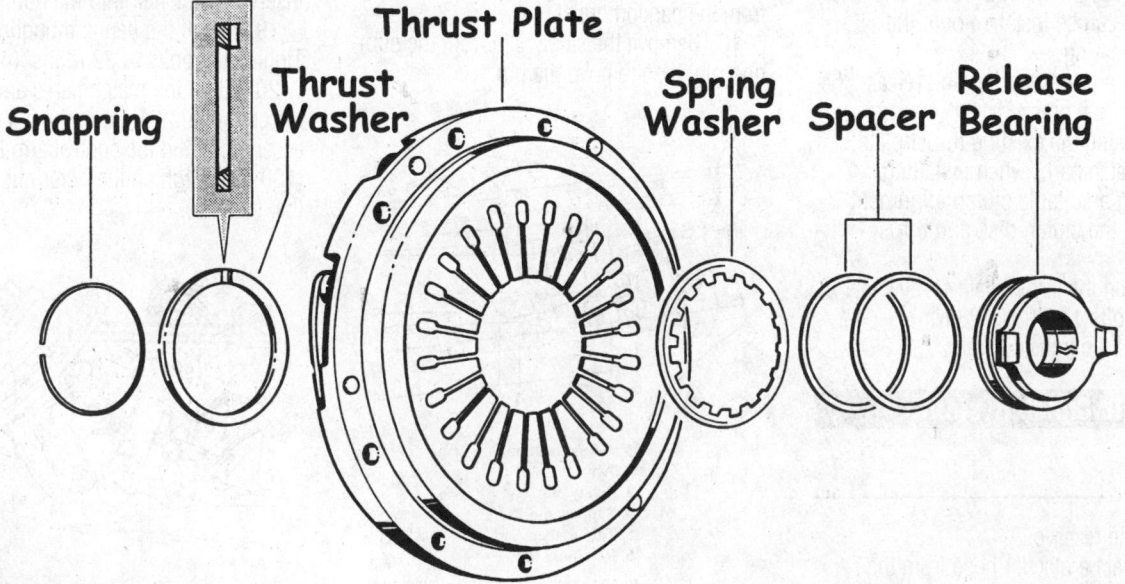

7923RG18

Exploded view of the thrust plate and release bearing assembly—911 Carrera

Before installing the release bearing, position the spring washer as shown—911 Carrera

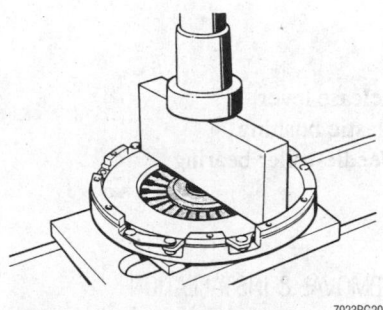

To install the release bearing snapring, press the thrust plate as shown—911 Carrera

To install:

➡**Be sure the spring washer is installed in the correct position.**

5. If installing a new thrust plate, assemble the release bearing as follows:

　a. Using a block of wood and a shop press, press on the thrust plate until the snapring can be installed onto the release bearing.

　b. Install the snapring and release the tension of the press.

6. If reusing the existing thrust plate, align the matchmarks when installing.

7. Using a suitable clutch alignment tool, install the clutch disk and thrust plate.

8. Tighten the thrust plate-to-flywheel mounting bolts 17 ft. lbs. (23 Nm).

9. Install the transmission.

Hydraulic Clutch System

BLEEDING

1. Fill the reservoir.

2. Connect a bleeder hose from the bleeder screw to a container filled with brake fluid so air cannot be drawn in during bleeding procedures.

3. Pump the clutch pedal about 10 times, then hold it down.

4. Open the bleeder screw and watch the stream of escaping fluid. When no more bubbles escape, close the bleeder screw and tighten it.

5. Release the clutch pedal and repeat the above procedure until no more bubbles can be seen when the screw is opened.

Halfshaft

REMOVAL & INSTALLATION

911 Carrera

AUTOMATIC TRANSMISSION

1. Raise and safely support the vehicle.

2. Remove the engine undercover and the rear wheel.

❄ WARNING

Do not move the vehicle after the halfshaft nut is loosened or damage to the bearing may occur.

3. Apply the brakes and remove the axle nut.

4. Remove the control arm cover.

5. Remove the ABS speed sensor and move the wiring out of the way.

6. Remove the brake caliper and position it out of the way with wire.

7. Remove the mount from the stabilizer bar. Do not remove the stabilizer bar.

8. Remove the cassette box and trim panel above the parking brake lever.

9. Remove the lock and adjusting nuts from the parking brake.

10. Remove the safety clip from the support pin, then remove the pin.

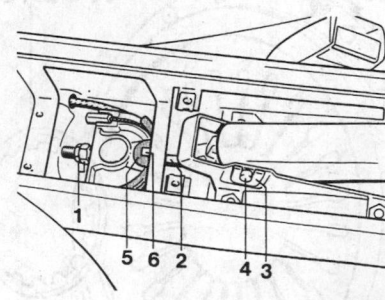

1. Adjustment nut
2. Arm
3. Safety clip
4. Support pin
5. Tab washer
6. Retaining lug

Parking brake lever component identification

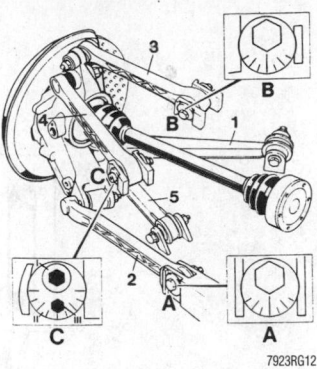

Rear suspension component identification

11. Remove the parking brake lever with arm.

12. Unhook the tab washer for the cable from the retaining lug at the top and bottom.

13. Remove the appropriate cable and pull out the cable from the rear.

14. Mark the position of the toe control arm eccentric (A), then remove the toe control arm. Lift the wheel carrier slightly with a jack if necessary for clearance.

15. Disconnect the lower control arm (1/5) from the sub-frame.

16. Remove the bolts securing the halfshaft to the transmission flange.

➡**When removing or installing the right halfshaft, disconnect the holder and move the oil line out of the way.**

17. Install a suitable lever on the wheel mounting studs and lift the carrier assembly while removing the halfshaft.

To install:

18. Grease the halfshaft splines and insert the halfshaft into the hub.

19. Install the flange mounting bolts. Tighten the bolts to 32 ft. lbs. (44 Nm).

20. Align the matchmarks and install the toe control arm. Tighten the nut on the eccentric of the toe control arm to 75 ft. lbs. (100 Nm). Tighten the outer nut to 63 ft. lbs. (85 Nm).

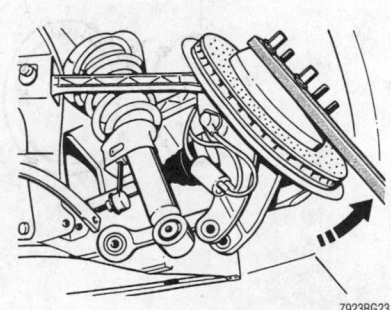

Lift the carrier assembly while removing the halfshaft

21. Install the lower control arm. Tighten the M14 nut to 147 ft. lbs. (200 Nm) and the M12 nut to 63 ft. lbs. (85 Nm).

22. Install the brake caliper and assembly the parking brake components.

23. Apply the brake and tighten the half-shaft nut to 340 ft. lbs. (460 Nm).

MANUAL TRANSMISSION—RIGHT HALFSHAFT

1. Raise and safely support the vehicle.
2. Remove the engine undercover and the rear wheel.

✳✳ WARNING

Do not move the vehicle after the halfshaft nut is loosened or damage to the bearing may occur.

3. Apply the brakes and remove the axle nut.
4. Remove the control arm cover.
5. Remove the ABS speed sensor and move the wiring out of the way.
6. Remove the brake caliper and position it out of the way with wire.
7. Remove the mount from the stabilizer bar. Do not remove the stabilizer bar.
8. Remove the cassette box and trim panel above the parking brake lever.
9. Remove the lock and adjusting nuts from the parking brake.
10. Remove the safety clip from the support pin, then remove the pin.
11. Remove the parking brake lever with arm.
12. Unhook the tab washer for the cable from the retaining lug at the top and bottom.
13. Remove the appropriate cable and pull out the cable from the rear.
14. Mark the position of the toe control arm eccentric (A), then remove the toe control arm. Lift the wheel carrier slightly with a jack if necessary for clearance.
15. Disconnect the lower control arm (1/5) from the sub-frame.
16. Remove the bolts securing the halfshaft to the transmission flange.

➡**When removing or installing the right halfshaft, disconnect the holder and move the oil line out of the way.**

17. Install a suitable lever on the wheel mounting studs and lift the carrier assembly while removing the halfshaft.

To install:
18. Grease the halfshaft splines and insert the halfshaft into the hub.
19. Install the flange mounting bolts. Tighten the bolts to 32 ft. lbs. (44 Nm).
20. Align the matchmarks and install the

toe control arm. Tighten the nut on the eccentric of the toe control arm to 75 ft. lbs. (100 Nm). Tighten the outer nut to 63 ft. lbs. (85 Nm).

21. Install the lower control arm. Tighten the M14 nut to 147 ft. lbs. (200 Nm) and the M12 nut to 63 ft. lbs. (85 Nm).

22. Install the brake caliper and assembly the parking brake components.

23. Apply the brake and tighten the half-shaft nut to 340 ft. lbs. (460 Nm).

MANUAL TRANSMISSION—LEFT HALFSHAFT

✳✳ WARNING

Do not move the vehicle after the halfshaft nut is loosened or damage to the bearing may occur.

1. Drive the vehicle onto a drive-on lift. Have a helper apply the brakes and remove the outer halfshaft nut. Do not roll the vehicle with this nut loosened or removed.
2. Remove the engine guard and the control arm cover.
3. Remove the No. 1 heating pipe.
4. Remove the bolts securing the halfshaft to the transmission flange.

➡**It may be necessary to use a puller to force the axle shaft through the hub assembly.**

5. Remove the halfshaft from the vehicle. If necessary for clearance, pull down on the vehicle to compress the suspension.

To install:
6. Install the halfshaft. Tighten the flange bolts to 32 ft. lbs. (44 Nm).
7. Install the No. 1 heating pipe.
8. Install the control arm cover and engine guard.
9. Tighten the axle shaft nut to 340 ft. lbs. (460 Nm) while the brakes are applied.

STEERING AND SUSPENSION

Air Bag

The Supplemental Restraint System (SRS/air bag) is a passive safety device designed to reduce the risk and severity of injuries to the front seat passengers of a vehicle involved in a frontal collision. The SRS is designed to be used in conjunction with the vehicle's seat belts. Airbags are less

effective in an accident if the passengers are not wearing seat belts. The SRS includes the driver's and passenger's air bag modules, a cable reel in the steering column, crash sensors, and a dedicated control module.

✳✳ CAUTION

The SRS is designed to operate on extremely low power levels. A back-up power circuit energizes the SRS if the vehicle's battery power is disconnected or interrupted in an accident. Due to the sensitive nature of the SRS, the air bag modules must be disabled if they, or any other part of the SRS, must be serviced or disconnected. Failing to disable the SRS before servicing its components may cause accidental air bag deployment and possible personal injury.

PRECAUTIONS

Several precautions must be observed when handling the inflator module to avoid accidental deployment and possible personal injury.

• Never carry the inflator module by the wires or connector on the underside of the module.

• When carrying a live inflator module, hold securely with both hands, and ensure that the bag and trim cover are pointed away.

• Place the inflator module on a bench or other surface with the bag and trim cover facing up.

• With the inflator module on the bench, never place anything on or close to the module which may be thrown in the event of an accidental deployment.

DISARMING

✳✳ CAUTION

The air bag must be disarmed before performing service around air bag components or air bag wiring. Failure to do so may cause accidental deployment of the air bag, resulting in unnecessary repairs and/or personal injury.

1. Position the vehicle with the front wheels in a straight ahead position and the key removed from the ignition switch.
2. Disconnect the negative battery cable.

3. Disconnect the positive battery cable.

4. Wait at least 1 minute for the air bag back-up power supply to drain before continuing.

5. Proceed with repair.

6. Once complete, connect the battery cables, negative cable last.

Power Rack and Pinion Steering Gear

REMOVAL & INSTALLATION

911 Carrera

➡**If the steering wheel is not locked in the straight ahead position, the steering wheel will have to be removed and the air bag contact unit (wind-up spring) centered as shown.**

1. Remove the floorboard of the pedal cluster to separate the universal joint (steering shaft) from the steering gear.

2. Raise and safely support the vehicle.

3. Remove the underside panel.

4. Separate the tie rod ends from the wheel carrier.

5. Undo the right and the left-hand tie rods from the steering gear, then remove the tie rods.

6. Disconnect the feed and return pipes from the steering gear. Drain any remaining fluid into a container. Plug the openings to prevent dirt from entering the system.

7. Unclip the power steering pipes from the bracket in the right-handed steering gear mount area. Then, undo the steering gear mounting bolts.

8. Retract the rack mounting clamps from the suspension crossmember. Retract the right side of the rack into the steering gear and lower the right side of the rack. Retract the left side of the rack into the steering gear. Remove the steering gear from the vehicle.

To install:

9. Install the steering gear into the vehicle and tighten the mounting bolts to 33 ft. lbs. (45 Nm).

10. Connect the feed and return lines to the steering gear.

11. Connect the tie rod ends.

12. Install the underside panel.

13. Connect the universal joint to the steering and install the remaining components as removed.

14. Check the front end alignment.

Strut

REMOVAL & INSTALLATION

911 Carrera

FRONT

1. Raise and safely support the vehicle.

2. Remove the front wheels and underpanel.

3. Detach the electrical connectors at the top of the strut.

4. Lock the brake pedal in the depressed position, disconnect the brake line at the strut mounting.

5. Remove the strut-to-wheel carrier bolts.

6. Remove the four upper strut mounting bolts, then remove the strut from the vehicle.

To install:

7. Install the strut into the vehicle.

➡**Be sure to position the red dot on the upper strut mounting facing toward the front of the vehicle.**

8. Install the upper strut mounting bolts and tighten to 24 ft. lbs. (33 Nm).

9. Install the lower strut mounting bolts and tighten the upper bolt to 88 ft. lbs. (120 Nm) and the lower mounting bolt to 148 ft. lbs. (200 Nm).

10. Attach the electrical connectors at the top of the strut.

11. Connect the brake line and bleed the brakes.

12. Install the underpanel and front wheels.

13. Lower the vehicle and check the front end alignment.

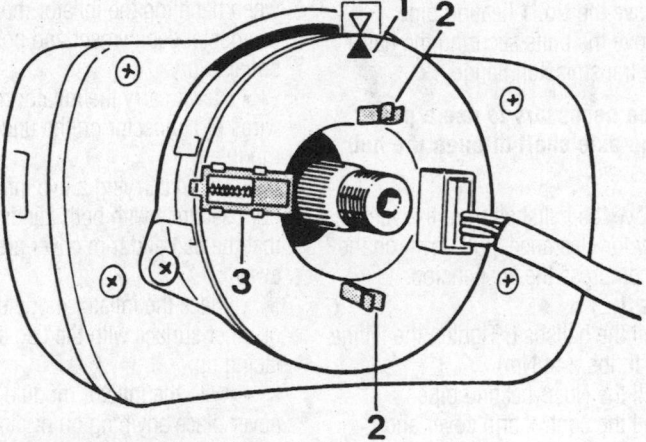

1 = Center position mark (arrows)

2 = Drivers engaging into the steering wheel

3 = Lock (rotation lock) becoming effective afer removal of the steering wheel.

7923RG10

Be sure to align the marks as shown for proper air bag contact unit alignment—911 Carrera

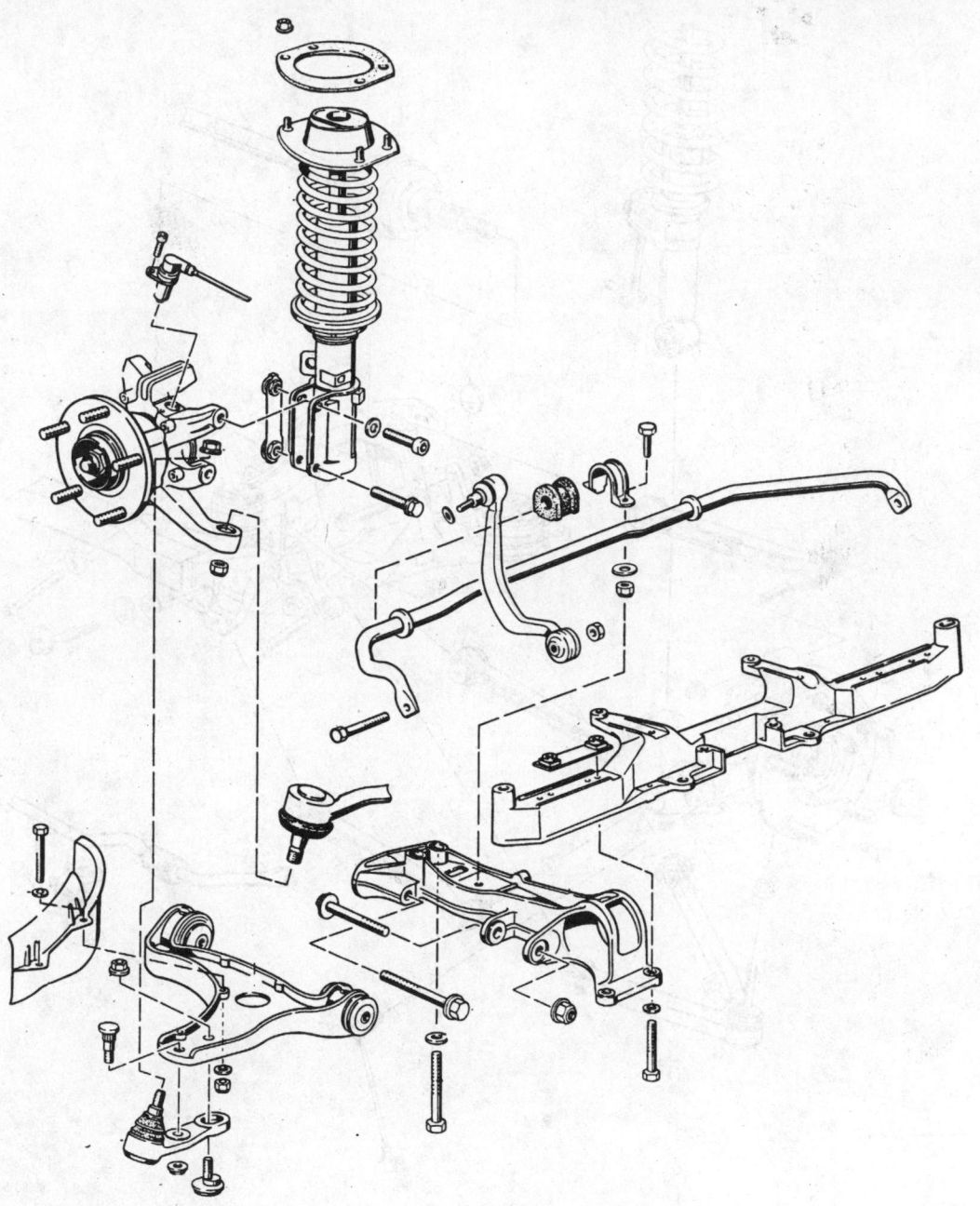

7923RG11

Exploded view of the front suspension—911 Carrera

REAR

1. Raise and safely support the vehicle.
2. Remove the rear wheels and the engine undercovers.
3. Disconnect the sway bar link from the sway bar.
4. Matchmark the position of the toe control arm (2) mounting bolt, then remove.
5. Loosen the toe control arm at the wheel carrier.
6. For the left strut, remove the heater blower.

7. For the right strut, remove the air filter housing.
8. Matchmark the position of the upper strut mounting for installation reference.
9. Remove the upper strut mounting bolts.
10. Remove the toe control arm, then the strut.

To install:

11. Install the strut into the vehicle, align the upper strut mounting and tighten the mounting bolts 24 ft. lbs. (33 Nm).

12. Install the air filter housing and heater blower.
13. Install the toe control arm through the strut mounting and tighten the retaining nut to 63 ft. lbs. (85 Nm).
14. Install the control arm to the cross-member, align the matchmarks and tighten the mounting bolt to 75 ft. lbs. (100 Nm).
15. Attach the sway bar link to the sway bar, and tighten the mounting nut to 34 ft. lbs. (46 Nm).

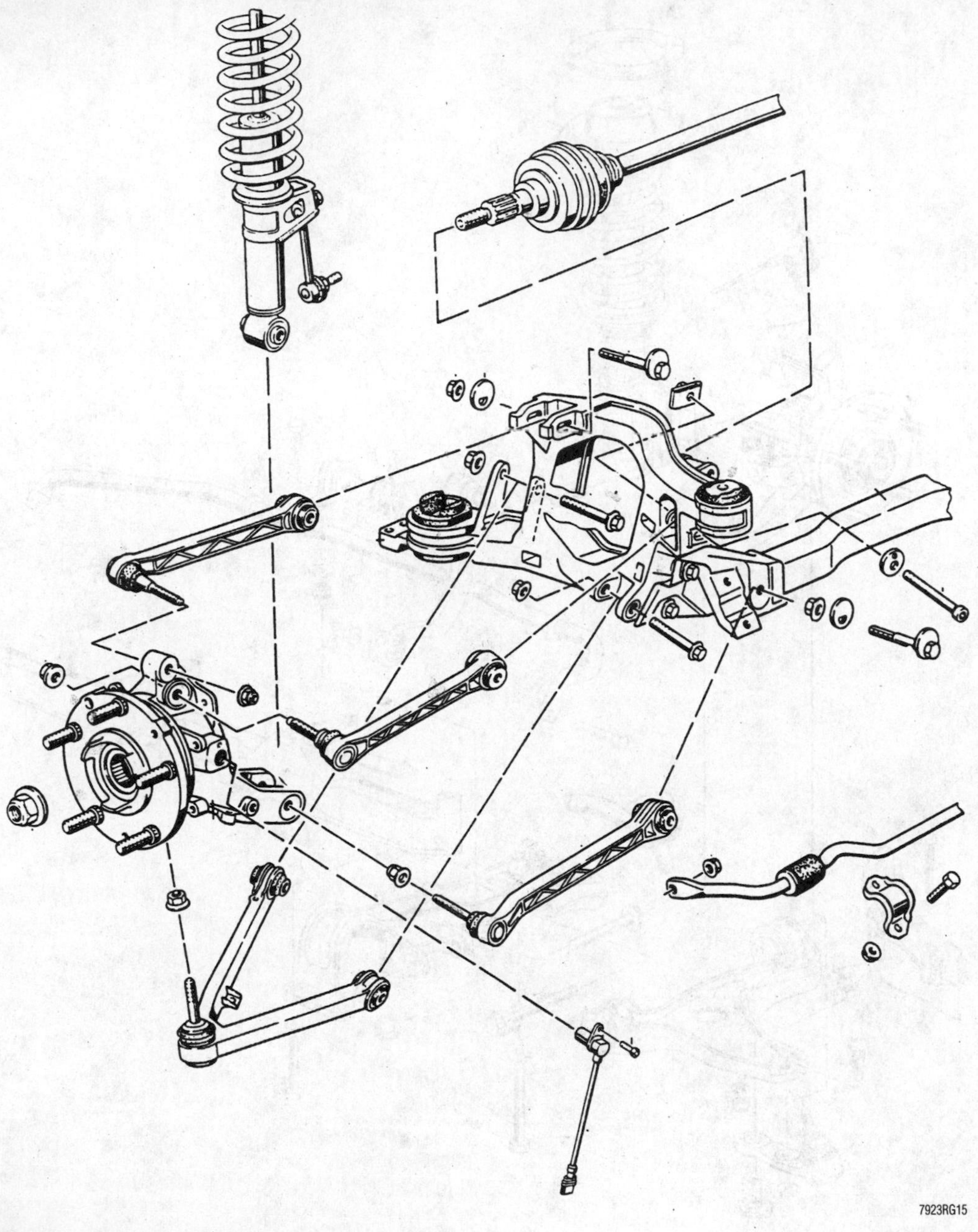

7923RG15

Exploded view of the rear suspension—911 Carrera

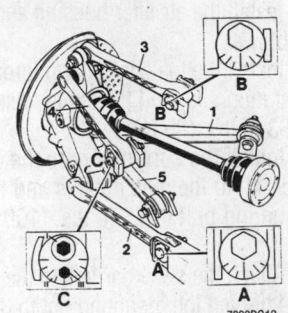

7923RG12

Be sure to matchmark the toe control arm adjusting bolt position for correct installation—911 Carrera

16. Install the engine undercover and rear wheels.

17. Lower the vehicle and check the rear suspension alignment.

Coil Spring

REMOVAL & INSTALLATION

911 Carrera Model

1. Remove the strut from the vehicle.

2. Install the strut into a suitable spring compressor.

3. Compress the spring and remove the strut mount retaining nut.

4. Remove the upper strut mount and remove the spring.

5. Slowly, release the tension of the coil spring.

To install:

6. Compress the spring enough to install the upper strut mount.

7. Install the mount and tighten the front mounting nut to 59 ft. lbs. (80 Nm) and the rear mounting nut to 43 ft. lbs. (58 Nm).

8. Remove the spring compressor.

9. Install the strut into the vehicle.

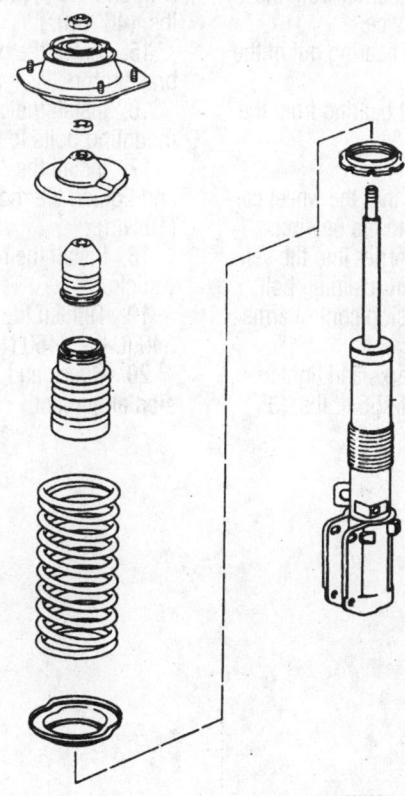

7923RG13

Exploded view of the front strut assembly—911 Carrera

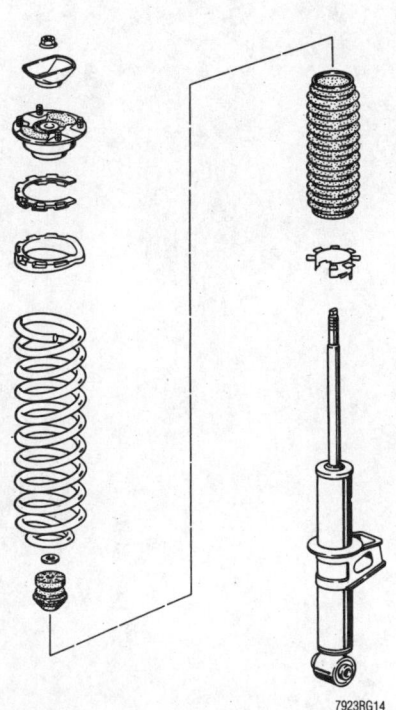

7923RG14

Exploded view of the rear strut assembly—911 Carrera

Lower Ball Joint

REMOVAL & INSTALLATION

911 Carrera

1. Raise and safely support the vehicle.
2. Remove the front wheels and the undercover.
3. If equipped, remove the air guide.
4. Separate the ball joint from the wheel carrier.
5. Remove the control arm-to-cross-member mounting bolt.
6. Remove the control arm from the vehicle.
7. Place the control arm in a vice and remove the ball joint off the arm.
   To install:
8. Install the ball joint to the control arm and tighten the M12 nut to 103 ft. lbs. (140 Nm) and the M10 nut to 59 ft. lbs. (80 Nm).
9. Install the control arm-to-cross-member mounting bolt and tighten, with suspension at normal height, to 63 ft. lbs. (85 Nm).
10. Install the control arm to the wheel carrier and tighten the mounting bolt to 55 ft. lbs. (75 Nm).

11. If removed, install the air guide.
12. Install the undercover and front wheels.
13. Check and adjust the front suspension alignment.

Wheel Bearings

ADJUSTMENT

911 Carrera

The wheel bearings are not adjustable. The wheel bearings must be replaced whenever the hub is removed.

REMOVAL & INSTALLATION

911 Carrera

FRONT

1. Raise and safely support the vehicle.
2. Remove the front wheels.
3. Remove the ABS wheel speed sensor.
4. Remove the caliper and position it side, then remove the brake rotor.
5. Remove the strut-to-wheel carrier mounting bolts and separate the lower ball joint.
6. Secure the wheel carrier in a vice and remove the hub mounting nut.
7. Press the hub assembly from the wheel carrier.
8. Press the wheel bearing out of the hub.
   To install:
9. Press the new wheel bearing into the hub.
10. Install the wheel hub onto the wheel carrier and tighten the mounting nut to 339 ft. lbs. (460 Nm).
11. Install the wheel carrier to the vehicle.
12. Install the lower ball joint and tighten the mounting nut to 55 ft. lbs. (75 Nm).
13. Install the lower strut mounting bolts and tighten the upper bolt to 88 ft. lbs. (120 Nm) and the lower mounting bolt to 148 ft. lbs. (200 Nm).
14. Install the brake rotor.
15. Install the caliper and tighten the mounting bolts to 63 ft. lbs. (85 Nm).
16. Install the ABS wheel speed sensor and tighten the mounting bolt to 7 ft. lbs. (10 Nm).
17. Install the front wheels and tighten the lug nuts to 96 ft. lbs. (130 Nm).
18. Lower the vehicle and check the front end alignment.

REAR

1. Loosen the halfshaft retaining nut while the vehicle is on the ground.

2. Raise and safely support the vehicle and remove the rear wheels.

3. Remove the ABS wheel speed sensor.

4. Remove the brake caliper and position it aside leaving the hose attached.

5. Remove the brake rotor and parking brake shoes.

6. Matchmark the positions of the suspension control arms.

7. Remove the control arms attached to the wheel carrier.

8. Remove the wheel carrier from the vehicle and mount it in a vice.

9. Press the hub and bearing out of the wheel carrier.

10. Separate the wheel bearing from the hub.

To install:

11. Install the bearing into the wheel carrier and press the hub onto the bearing.

12. Install the wheel carrier into the vehicle and install the halfshaft retaining bolt.

13. Attach the suspension control arms to the vehicle.

14. Align the matchmarks and tighten the M12 mounting bolts to 55 ft. lbs. (85 Nm) and the M10 mounting bolts to 34 ft. lbs. (46 Nm).

15. Install the parking brake shoes and brake rotors.

16. Install the caliper and tighten the mounting bolts to 63 ft. lbs. (85 Nm).

17. Install the ABS wheel speed sensor and tighten the mounting bolt to 7 ft. lbs. (10 Nm).

18. Install the rear wheels and lower the vehicle.

19. Tighten the halfshaft retaining nut to 340 ft. lbs. (460 Nm).

20. Check and adjust the rear suspension alignment.

SAAB

900 • 9000 • 9-3

26

ENGINE REPAIR

➡Disconnecting the negative battery cable on some vehicles may interfere with the functions of the on board computer systems and may require the computer to undergo a relearning process, once the negative battery cable is reconnected.

Ignition Timing

ADJUSTMENT

Saab vehicles are equipped with a Distributorless Ignition System (DIS). Ignition timing is controlled by the Electronic Control Module (ECM), no adjustment is necessary or possible.

Engine Assembly

REMOVAL & INSTALLATION

900 and 9–3 Models

➡The engine and transaxle are removed together as an assembly.

1. With all four wheels on the ground, loosen the axle hub nuts.
2. Drain the engine coolant and remove the battery.
3. Remove air cleaner assembly and attached hoses. Take off the resonator.
4. Disconnect throttle cable and move it to one side.
5. Disconnect cruise control wiring harness and cables from throttle body. Loosen retaining nuts on cruise control unit and remove.
6. Relieve the fuel system pressure.

⁂ CAUTION

The fuel injection system remains under pressure even after the engine is stopped. The fuel system pressure must be relieved before disconnecting any fuel lines. Failure to do so may result in fire and/or personal injury.

7. Disconnect the fuel lines at their connections at the front of the fuel injection manifold and on the fuel pressure regulator.
8. If equipped, disconnect the turbocharger pressure line from the turbocharger compressor and the intercooler/throttle housing.
9. Disconnect the vacuum hose from the secondary injection and disconnect the brake booster vacuum hose from the intake manifold.
10. If equipped, remove the boost pressure control unit.
11. Using a ⅜ in. ratchet extension relieve belt tensioner of pressure. Remove drive belt.
12. Disconnect the pressure and return pipe from the steering servo pump and have a plug handy to prevent oil escaping from the pipe.
13. Disconnect the cooling system hoses at the following connections:
    a. the heater control valve.
    b. the expansion tank.
    c. the bottom of the radiator.
    d. the thermostat housing.
14. Without disconnecting any hoses, remove the air conditioner compressor and it mounting bracket. Place them on the filter housing for the heater system. Secure the alternator so it will not drop or become damaged.
15. If not already done, raise and safely support the vehicle.
16. Disconnect the positive battery cable from the clips holding it to the body. Disconnect the ground cable from the transaxle.
17. Remove ignition cable and disconnect electrical connections from ignition coil.
18. Unplug the wiring connectors from the transaxle.
19. Detach the oxygen sensor wiring and the catalytic converter temperature sensor connector.
20. Under the vehicle, remove the taper pin from the gearshift rod joint. If equipped, disconnect the clutch cable and clutch pipe.
21. Inside the vehicle, pull back carpet under glove box to gain access to the central locking system and disconnect the control module. Feed the wires to the engine compartment through the grommet.
22. Remove hub nuts and remove both wheels. Remove front splash shield.
23. Remove both ball joint nuts and disconnect the ball joints from the struts. Be sure the halfshafts can slide out of the hubs. If necessary, use a wheel puller to push them out now.
24. Remove front pipe from exhaust manifold and take off catalytic converter.
25. Disconnect oil cooler lines and all lines connected to transaxle housing.
26. Position lifting table under vehicle so it is directly under front engine mount and gearbox. Take off subframe bolts and front engine mounts.
27. Lower lifting table slightly and separate the halfshafts from the hubs.
28. Lower lifting table fully and remove subframe.
29. Place engine on engine stand.
    To install:
30. Place engine/transaxle on lifting table.
31. Install subframe on rear engine mount and tighten to 35 ft. lbs. Fit halfshafts into hubs as engine is raised into engine compartment.

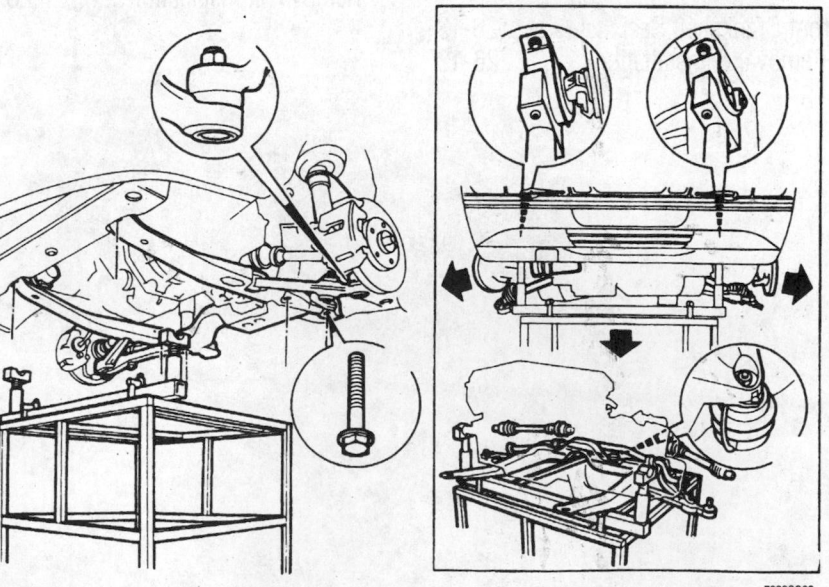

Lowering the entire powertrain assembly—900 and 9–3 models with 2.0L and 2.3L engines shown

7923SG02

32. Connect engine mount and subframe bolts and tighten in sequence:
- Front bolts to 85 ft. lbs.
- Middle subframe bolts to 141 ft. lbs.
- Rear subframe bolts to 90 ft. lbs.
- Rear engine mount to 54 ft. lbs.

33. Connect all oil cooler and hydraulic lines.

34. Reconnect all charge air cooler lines between the turbocharger and the cooler.

35. Install the front exhaust pipe to the exhaust manifold. Secure the catalytic converter to the rear exhaust section.

36. Fit engine shield to under side of vehicle.

37. Install hub nuts and wheels. Tighten wheel nuts to 89 ft. lbs. but do not tighten hub nuts yet.

38. Feed the wiring through the grommet and reconnect the central locking system control module.

39. As equipped, hook the clutch cable in place or bleed the clutch hydraulic cylinder.

40. Reconnect the taper pin to the gearshift selector.

41. Connect the oxygen and catalytic converter temperature sensor.

42. Connect all transaxle wiring.

43. Install A/C compressor, servo pump and the drive belt.

44. Reconnect power brake booster vacuum hose and pressure sensor connections.

45. Install cruise control including cables and electrical harness connections.

46. Install the battery and connect the cables.

47. With all four wheels on the ground, tighten the hub nuts to 215 ft. lbs. (290 Nm).

48. Fill coolant and oil to specified level and check.

49. Start engine and test drive vehicle.

9000

➡**The engine and transaxle assembly are removed together.**

1. Raise the vehicle and support it safely.

2. Drain the cooling system.

3. Disconnect the battery cables and remove the battery and tray.

4. Remove the connecting bolt for the expansion tank, disconnect the tank from the suction and remove the overflow hoses from the radiator.

5. Loosen the drive belt for the compressor by loosening the locknut, and loosening the adjusting nut under the locknut.

6. Disconnect the upper connection on the oil cooler, loosen the pipe clip on the radiator and slide the pipe down behind the radiator.

7. Unplug the connector to the electromagnetic clutch on the compressor and loosen the compressor mounting complete with the belt tensioner.

8. Place a protective cloth over the radiator member and rest the compressor on the radiator member. Secure the compressor to the radiator member.

9. If turbocharged, remove the turbocharger pressure pipe, situated between the turbocharger unit and the intercooler.

10. Detach the Lambda probe.

11. From the engine compartment, unbolt the flange joint between the exhaust pipe and the exhaust manifold. If turbocharged, disconnect the exhaust pipe coming from the turbocharger. Push the exhaust pipe to one side and unhook the rubber hangers from the exhaust system. Disconnect the bottom coolant hose from the water pump.

12. From under the vehicle, remove the bottom retaining bolt for the radiator fan.

13. Disconnect the speedometer drive from the gearbox.

14. If equipped with manual transmission, select the 4th gear and separate the rubber joint in the gear selector linkage.

15. If equipped with an automatic transmission, disconnect the gear selector cable from the selector lever. Do not separate at the ball joint.

16. Remove the clips on the rubber covers over the inboard universal joints and slide the covers off the drive axles.

17. Detach the electrical leads from the alternator and the starter motor. Unplug the connector for the oil pressure switch.

18. Remove the clips and remove the top radiator hose.

19. Disconnect the top radiator hose at the cylinder head.

20. Disconnect the high tension lead from the ignition coil at the distributor cap.

21. Remove the solenoid valve from the bracket on the radiator and unplug the electrical connections.

22. Remove the bolts from the top of the radiator fan. Disconnect the wiring loom and lift out the fan.

23. Pull the connector off the air mass meter. Detach the air mass meter from the air intake duct socket connector and the air cleaner. Leave the rubber socket connector attached to the turbocharger unit.

24. Remove the air intake duct by pulling it out of the aperture in the wing and twisting the ends inwards.

25. Remove the air cleaner top section first, then the remaining section.

26. Disconnect the relief valve hose from the turbocharger pressure pipe and remove the pipe, if equipped.

27. Detach the Hall Effect transducer, the earth lead from the gear box and the back-up lights.

28. Disconnect the end of the throttle cable and the throttle linkage.

29. Install a clamp to the hydraulic line to the slave cylinder and pinch the line tightly. With proper wrenches, open the line to the clutch slave cylinder.

30. Remove the front wheels.

31. From both sides of the vehicle, slacken the lower bolts retaining the steering swivel member to the strut assembly. Remove the 2 upper bolts.

32. Pivot the steering swivel member outwards to pull the inboard universal joint out of the halfshaft. Position dust covers over the exposed halfshaft cups.

33. Remove the engine mount bolt.

34. Remove the steering reservoir for the servo and position it within the engine compartment. Drain the fluid from the container.

35. Disconnect the large bore hose and the delivery hose from the steering servo pump and plug the open ends.

36. Disconnect the fuel return line from the pressure regulator.

37. Remove the nut from the rear engine mounting and back off the front mount bolts a few turns.

38. Attach the lifting sling (Saab 83–92–409) or equivalent to the rear lifting lug.

39. Lift the engine sufficiently to provide access for the removal of the components located between the engine and the firewall.

40. Disconnect and tag the vacuum hoses from the inlet manifold.

41. Remove the coolant hoses running between the heater core and the water pump pipe.

42. Separate the coupling between the fuel pipe and the fuel injection manifold. Do not allow the fuel to spill or collect.

43. Cut the clips securing the wiring looms to the oil pipe, water pipe. Disconnect and tag the vacuum hoses from the inlet manifold.

44. Unclip the wiring loom to the fuel injection manifold.

45. Detach the grounding connections and the electrical connectors from the wiring harness.

46. Unbolt the air cooled oil cooler and place it on top of the engine. The 2 lower bolts need only be loosened.

47. Disconnect the hood lifts and install hood extenders for extra clearance.

48. Carefully lift the engine from the

vehicle, taking care not to damage the radiator.

To install:

49. Install the engine in the vehicle and connect all the engine mounts.

50. Bolt the air cooled oil cooler in place. Tighten the 2 lower bolts.

51. Attach the grounds and the electrical connectors to the wiring harness.

52. Clip the wiring loom to the fuel injection manifold.

53. Secure the wiring looms to the oil pipe, water pipe, inlet manifold steady bar and the oil supply pipe.

54. Connect the coupling between the fuel pipe and the fuel injection manifold.

55. Connect the coolant hoses running between the heater core and the water pump pipe.

56. Connect the vacuum hoses to the correct connections on the inlet manifold.

57. Before removing the lifting sling, connect the components located between the engine and the firewall.

58. Attach the nut from the rear engine mounting and tighten the front mount bolts.

59. Connect the fuel return line to the pressure regulator.

60. Connect the large bore hose and the delivery hose to the steering servo pump.

61. Connect the steering reservoir for the servo. Fill the fluid container.

62. Connect the engine stay bolt.

63. Pivot the steering swivel member outwards to install the inboard universal joint into the halfshaft. Position dust covers over the exposed halfshaft cups and install clamps.

64. If equipped with manual transmission, select the 4th gear and connect the rubber joint to the gear selector linkage.

65. If equipped with an automatic transmission, connect the gear selector cable to the selector lever.

66. Connect the lower bolts retaining the steering swivel member to the strut assembly. Tighten the 2 upper bolts.

67. Install the front wheels.

68. With line wrenches, tighten the line to the clutch slave cylinder.

69. Connect the end of the throttle cable and the throttle linkage.

70. Connect the Hall Effect transducer, the negative battery cable to the gear box and the back-up light switch.

71. Connect the relief valve hose to the turbocharger pressure pipe, if equipped.

72. Install the air cleaner.

73. Install the air intake duct.

74. Push the connector on the air mass meter. Attach the air mass meter to the air intake duct socket connector and the air cleaner.

75. Connect the high tension lead to the ignition coil at the distributor cap.

76. Connect the top radiator hose at the cylinder head.

77. Attach the electrical leads to the alternator and the starter motor. Plug in the connector for the oil pressure switch.

78. Connect the speedometer drive to the gearbox.

79. From the engine compartment, connect the flange joint between the exhaust pipe and the exhaust manifold. If turbocharged, connect the exhaust pipe coming from the turbocharger. Hook the rubber hangers to the exhaust system.

80. Connect the bottom coolant hose to the water pump.

81. Connect the Lambda probe.

82. Plug the connector into the electromagnetic clutch on the compressor and tighten the compressor mounting complete with the belt tensioner.

83. Connect the upper connection on the oil cooler, tighten the pipe clip on the radiator.

84. Tighten the drive belt for the compressor.

85. Install the expansion tank, connect the tank to the suction and overflow hoses from the radiator.

86. Install the battery and tray. Connect the battery cables.

87. Remove the lift supports and lower the vehicle.

88. Fill the radiator with coolant, the engine with oil and the transmission with fluid. Start the engine and allow it to reach operating temperature. Check the ignition timing and all fluid levels. Test drive the vehicle.

Water Pump

REMOVAL & INSTALLATION

2.0L and 2.3L Engines

1. Disconnect the negative battery cable.

2. Loosen the expansion tank pressure cap.

3. Raise and safely support the vehicle.

4. Remove the center air deflector and properly drain the coolant.

5. Remove the right front wheel assembly and remove the front section of the inner fender panel.

6. Lower the vehicle.

7. Remove the expansion tank, then remove the coolant hoses and electrical connector. Pull hard on the drive belt and lock the tensioner using tool 83–94–488 or equivalent. Remove the belt from the water pump and the air conditioning compressor.

8. Protect the engine oil cooler and place a protective cover over the upper radiator crossmember.

9. Unplug the electrical connector to the air conditioning compressor and detach the compressor with lines attached and position it to the side. Remove the air conditioning compressor bracket.

10. Disconnect the coolant hoses from the water pump.

11. Remove the oxygen sensor wire from the clips.

12. Remove the coolant pipe from the turbocharger.

13. Raise and safely support the vehicle.

14. Remove the coolant pipe from the water pump and lower the vehicle.

15. Remove the 3 retaining screws securing the water pump to the timing cover.

16. Carefully pry the water pump loose. Start at the sleeve in the cylinder block and remove the water pump.

❊❊ WARNING

Use care not to damage the oxygen sensor.

17. Remove the water pump pulley and take the pump out of its housing.

To install:

18. Clean the sealing surfaces and install the water pump in the pump housing with a new gasket.

19. Install the water pump pulley and tighten the retaining bolts to 72 inch lbs. (8 Nm).

20. Lubricate the new O-rings with acid free petroleum jelly and fit the sleeve together with the water pump. Tighten the water pump bolts to 15 ft. lbs. (20 Nm).

21. Raise the vehicle and attach the coolant pipe to the water pump.

22. Lower the vehicle and attach the coolant pipe to the turbocharger.

23. Install the oxygen sensor wire in the clips.

24. Connect the hoses to the water pump.

25. Install the air conditioning compressor bracket. Install the air conditioning compressor and attach the electrical connector.

26. Remove the protective covers from the engine oil cooler and the upper radiator crossmember.

27. Install the belt, pull on it firmly and remove the locking pin from the tensioner. Check for proper belt alignment and tension.

28. Install the coolant hoses and electrical connector to the expansion tank, then install the expansion tank.

29. Check that the radiator drain plug is tightened and install the center air deflector.

30. Install the front section of the inner fender panel and install the wheel.

31. Properly fill the cooling system.

32. Connect the negative battery cable. Start the engine and check for proper cooling system operation.

2.5L Engine

1. Disconnect the negative battery cable.

2. Safely raise and support the vehicle.

3. Remove the lower center air deflector. Drain the engine coolant into a catch pan.

4. Lower the vehicle and remove the top engine covers, air filter assembly and mass air flow sensor.

5. Loosen the power steering pump and water pump pulley bolts. Remove the drive belt and tensioner. Remove the power steering pump, water pump pulley and timing cover.

6. Check the timing belt for damage and replace if necessary.

7. Remove the water pump.

To install:

8. Clean water pump mounting surface. Coat the water pump O-ring and sealing surface with acid-free petroleum jelly.

9. Install water pump with new O-ring and tighten the water pump bolts to 18 ft. lbs. (25 Nm)

10. Install the timing cover.

11. Install the water pump pulley. Tighten the water pump pulley bolts to 6 ft. lbs. (8 Nm).

12. Install power steering pump, drive belt tensioner and drive belt.

13. Install air filter assembly, mass air flow sensor. Install the upper engine covers and safely raise and support the vehicle.

14. Be sure the radiator drain plug is tightened and install the lower center air deflector.

15. Fill the radiator with clean coolant and check the cooling system for leaks

16. Bleed the cooling system.

17. Test drive the vehicle and check for any leaks.

3.0L Engine

1. Disconnect the negative battery cable.

2. Safely raise and support the vehicle.

3. Remove the lower center air deflector. Drain the engine coolant into a catch pan.

4. Lower the vehicle and remove the top engine covers.

5. Lift up the power steering reservoir.

6. Disconnect the connection on the torque arm engine mount and remove the torque arm.

7. Remove the power steering line clamp from the torque arm engine mount and remove the engine mount.

8. Remove the upper coolant hose.

9. Disconnect the hose from the coolant expansion tank and remove the upper alternator air intake.

10. Loosen the power steering pump and water pump pulley bolts. Remove the drive belt and tensioner. Remove the power steering pump, water pump pulley and timing cover.

11. Check the timing belt for damage and replace if necessary.

12. Remove the water pump.

To install:

13. Clean water pump mounting surface. Coat the water pump O-ring and sealing surface with acid-free petroleum jelly.

14. Install water pump with new O-ring and tighten the water pump bolts to 18 ft. lbs. (25 Nm)

15. Install the timing cover.

16. Install the water pump pulley. Tighten the water pump pulley bolts to 6 ft. lbs. (8 Nm).

17. Install power steering pump, drive belt tensioner and drive belt.

18. Install the upper alternator air intake.

19. Install the torque arm engine mount and attach the power steering line clamp.

20. Install the torque arm and bolt the connection to the torque arm engine mount.

21. Install the upper coolant hose.

22. Connect the upper hose to the coolant expansion tank and set the power steering reservoir back into position.

23. Install the upper engine covers and safely raise and support the vehicle.

24. Be sure the radiator drain plug is tightened and install the lower center air deflector.

25. Fill the radiator with clean coolant and check the cooling system for leaks

26. Bleed the cooling system.

27. Test drive the vehicle and check for any leaks.

Cylinder Head

REMOVAL & INSTALLATION

2.0L Engine

❊❊ CAUTION

Fuel injected systems remain under pressure, even after the engine has been turned OFF. The fuel system pressure must be relieved before disconnecting any fuel lines. Failure to do so may result in personal injury.

1. Remove the hood after scribing reference marks next to the mounting bolts to aid in reinstallation.

2. Remove the battery.

3. Drain the coolant from the radiator and cylinder block. Drain the oil from the crankcase.

4. Remove the exhaust manifold and turbocharger unit, if equipped.

5. Remove the tensioning pulley and drive belt for the air conditioner compressor.

6. Slacken the securing bolts for the steering pump bracket, remove the drive belt and push the pump aside.

7. Detach the wiring harness clips on the cylinder head.

8. Remove the 2 bolts in the timing cover, which are screwed into the cylinder head from underneath.

9. Remove the bolts in the right-hand engine mounting which are screwed into cylinder head, together with the spacer sleeves.

10. Disconnect the hose between the thermostat housing and the radiator at the thermostat housing.

11. Remove the fuel pressure regulator and disconnect the ground leads for the fuel injection system.

12. Remove the air conditioning compressor with lines still attached and carefully put it on the air intake for the heating system.

13. Remove the auxiliary air valve (not applicable to vehicles with a catalytic converter). Remove the bracket for the air conditioning compressor from the cylinder head.

14. Remove the intake manifold complete with injectors.

15. Disconnect the lead from the temperature sensor.

16. Remove the lid on the valve cover and the ignition cables together with the distributor cap.

17. Remove the valve cover. Disconnect the crankcase ventilation hose and remove the semicircular rubber plug halves from the cylinder head.

18. Remove the timing chain, tensioner and camshaft sprockets.

➡️**Do not rotate the camshaft or crankshaft after the sprockets have been removed, as rotation of one of the shafts can result in damage to the valves.**

19. Remove the cylinder head bolts and siphon off the oil from the cylinder head.

20. Install a guide pin (Saab special tool 83–92–128 or equivalent) in one of the bolt holes and lift off the cylinder head, making sure the pivoting guide for the timing chain is not damaged.

To install:

21. Align the **0** mark on the flywheel with the timing mark on the housing. Align the marks on the camshafts with their respective timing marks.

22. Install the cylinder head gasket, making sure it is held in position by the guide sleeves in the cylinder head flange.

23. Position the timing chain and pivoting guide for cylinder head installation.

24. With the guide pin still installed, carefully position the cylinder head. Use the guide pin as a pivot for the head, which must be turned slightly to enable it to pass the pivoting guide. Thereafter, alignment will be determined by the guide sleeves.

➡️**Remember to install the 2 M8 sized bolts in the underside of the cylinder head.**

25. Install the cylinder head bolts and tighten them in sequence and in 3 stages. Stage 1, tighten to 44 ft. lbs. (60 Nm) evenly. Stage 2, tighten to 59 ft. lbs. (80 Nm) evenly. Stage 3, another 90 degrees (¼ turn).

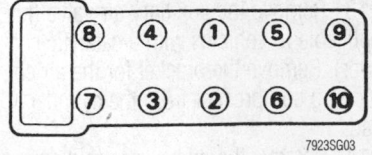

Cylinder head bolt torque sequence—2.0L and 2.3L engines

26. Unless the head is fitted with a new, special gasket which does not require retightening, the cylinder head bolts must be retightened. Run the engine for 30 minutes allow to cool, then retighten the head.

27. In the same sequence as before, slacken and retighten each bolt to 59 ft. lbs. (80 Nm).

28. Install the camshaft sprockets, timing chain and tensioner.

29. Install the semi-circular rubber plug halves in the cylinder head and install the valve cover tighten the retaining bolts to 11 ft. lbs. (14 Nm). Install the crankcase ventilation hose.

30. Install the spark plug wires and distributor cap in their original position and install the lid on the valve cover.

31. Connect the electrical lead to the temperature sensor.

32. Install the intake manifold complete with injectors.

33. Install the bracket for the air conditioning compressor and install the compressor.

34. Install the auxiliary air valve, if applicable.

35. Install the fuel pressure regulator and connect the ground leads for the fuel injection system.

36. Connect the upper radiator hose to the thermostat housing.

37. Install the bolts in the right-hand engine mounting. Screw them into the cylinder head, together with the spacer sleeves.

38. Connect the wiring harness clips on the cylinder head.

39. Install the power steering belt.

40. Install the tensioning pulley and drive belt for the air conditioner compressor.

41. Install the exhaust manifold and turbocharger unit, if equipped.

42. Fill the crankcase with new oil and properly fill the cooling system.

43. Install the battery.

44. Using the reference marks made earlier, align and install the hood.

45. Start the engine and check for proper operation. Check for leaks.

2.3L Engines

1. Disconnect the negative battery cable.

2. Raise and safely support the vehicle.

3. Remove the right front wheel assembly and the inner fender panel.

4. Drain the coolant. Remove the radiator expansion tank. Disconnect the power steering reservoir and set aside. Leave the hoses attached.

5. Loosen the compressor drive belt and remove the belt.

6. Disconnect the electrical leads from the air compressor.

7. Unbolt the compressor from its mounting bracket. Disconnect the top pipe connected to the air cooled oil cooler and push the pipe to one side. Carefully rest the compressor on the radiator crossmember. Unbolt the compressor mounting bracket and remove it.

8. Unbolt the front exhaust pipe flange and unhook the rubber hangers.

9. Remove the stay bar for the turbocharger unit and the oil return pipe (if vehicle is equipped with a turbocharger).

10. Disconnect the hose from the intercooler at the turbocharger unit. Disconnect the oil supply pipe from the turbocharger (if applicable).

11. Disconnect the hose between the air mass meter and the turbocharger unit (if applicable). Disconnect the coolant hose from the thermostat housing and the hose from the cylinder head.

12. Disconnect the oil supply hose or pipe so as not to obstruct the removal of the exhaust manifold. If necessary, remove the clip holding the pipe to the cylinder head and slave cylinder.

13. Unbolt and lift off the exhaust manifold complete with the turbocharger unit (if applicable), pushing the oil supply pipe aside at the same time.

14. Disconnect the lead to the temperature transducer.

15. Remove the engine stay bracket from its attachment point on the wing.

16. Remove the bolt securing the engine stay bracket to the cylinder head. Remove the intake manifold from the cylinder head.

17. Disconnect the breather hose for the crankcase ventilation from the camshaft cover.

18. Disconnect the vacuum hose and the Hall Effect transducer lead from the distributor and remove the distributor cap complete with the spark plug wires (if applicable).

19. Unscrew and remove the spark plug inspection plate and the clips for the spark plug wires.

20. Remove the camshaft cover.

21. Align the crankshaft with the **0** timing mark and check that the camshaft timing marks also coincide.

22. Using a wrench installed over the flats on the exhaust camshaft, hold the camshaft and remove the center bolt securing the camshaft sprocket. Repeat this step for the intake camshaft.

❋❋ WARNING

To remove or refit the timing chain, the camshafts and crankshaft must be lined up with their respective timing marks (No. 1 cylinder at top dead center). Never rotate the crankshaft or the camshaft once the timing chain has been detached. A fully opened valve could come in contact with a piston at top dead center.

23. Remove the timing chain tensioner. Remove the 2 cylinder head bolts adjacent to the timing cover, which are accessible from below.

24. Disconnect the starter motor lead from the clip on the thermostat housing.

25. Remove the Torx® type cylinder head bolts.

26. Install a guide pin in the drilled hole in the right top corner of the cylinder head. Be sure the timing chain is positioned such that the pivoting chain guide will not obstruct the cylinder head and carefully lift the cylinder head from the engine block.

To install:

27. Before installation, clean both the cylinder head and the engine block surfaces. Install a new gasket. Be sure the crankshaft is aligned in the **0** position and that the camshafts are align with their respective timing marks.

➡**When the pistons of the No. 1 and No. 4 cylinders are at TDC, the crankshaft 0 mark on the flywheel must be aligned with the mark on the clutch cover or the end plate, if the clutch cover has been removed. The marks on the camshafts must be aligned with those on the cam bearing caps. This indicates the exhaust valves for No. 1 and No. 4 cylinders are closed.**

28. Install a guide pin in the drilled hole in the top of the right corner of the cylinder head and lower the cylinder head carefully into position on the engine block. Locate the cylinder head on the guide sleeves.

29. Install the cylinder head bolts and tighten as follows:

    a. Tighten, in sequence, to 44 ft. lbs. (60 Nm).

    b. Tighten, in sequence, to 59 ft. lbs. (80 Nm).

    c. Tighten, in sequence, an additional 90 degrees (¼ turn).

    d. At the completion of cylinder head installation, run the engine until it reaches normal operating temperature,

then shut the engine **OFF** and allow to cool for 30 minutes.

    e. Slacken, then retighten the cylinder head bolts, in sequence, to 59 ft. lbs. (80 Nm).

    f. Tighten each bolt, in sequence, an additional 90 degrees (¼ turn).

30. To install the timing chain, run the chain up through the opening in the cylinder head. Install the chain and sprocket on the exhaust cam first. Be sure the chain is taut between the crankshaft and camshaft sprockets. Snug the bolt but do not tighten.

31. Install the chain and sprocket to the intake cam. Keep the chain taut between the cam sprockets while it is being installed. Snug the bolt but do not tighten. Be sure the chain is seated in the guide tensioner grooves.

32. Tension the chain tensioner by fully depressing the piston, then rotating it to the locked position.

33. Install the chain tensioner with the piston under tension. Be sure the copper gasket is in good condition and that the sealing surface is clean and free from burrs. Tighten the tensioner to 47 ft. lbs. (63 Nm).

34. Trigger the chain tensioner by pressing the pivoting chain guide against it, thereafter, press the pivoting guide against the chain to give the chain its basic tension. Check that the chain tensioner maintains tension on the chain when the pressure on the chain guide is released and that the basic setting stop for the tensioner holds the chain guide tight against the chain. A limited amount of play will be present until the hydraulic pressure takes over once the engine is running.

35. Check the setting by rotating the crankshaft 2 complete turns in its normal direction of rotation around to the timing mark. The basic setting of the cams should remain unaltered.

36. Lock the exhaust cam by using a wrench installed over the flats on the camshaft and tighten the sprocket bolt to 48 ft. lbs. (65 Nm). Repeat this step on the intake cam.

➡**The accuracy of the timing chain adjustment will depend on the condition of the chain.**

37. Install both halves of the split seal and the camshaft cover. Install the bolt at the distributor end and the middle bolt at the other end first. Tighten the bolts in sequence to 16 ft. lbs. (22 Nm).

38. Check that the timing marks for the distributor rotor are aligned, Install the dis-

tributor cap and connect the lead for the Hall Effect transducer. Connect all vacuum hoses.

39. Connect the spark plug wires. Secure the leads in the clips. Install the inspection plate and tighten the retaining screws.

40. Install the clip securing the starter motor lead to the thermostat housing.

41. Install a new gasket on the inlet manifold and install the manifold in place. Install the top securing bolts first, then install the lower bolts, using an extension bar.

42. Install the bolt for the engine stay bracket to the cylinder head and position the stay bracket in place. Install a new gasket onto the exhaust manifold and position the exhaust manifold to the cylinder head.

43. Install the oil supply pipe. Install the clip and the slave cylinder bolt. Install the oil return line and the steady bar for the turbocharger unit (if applicable).

44. Connect the hose between the turbocharger unit and the intercooler (if applicable). Connect the cooler hose to the thermostat housing and the hose to the cylinder head.

45. Install the air mass meter socket connector into the turbocharger unit and tighten the clip. Attach the hose between intercooler and the turbocharger unit (if applicable).

46. Install and tighten the nuts securing the front section of the exhaust pipe to the turbocharger compressor (if applicable). Bolt the air conditioning compressor mounting bracket onto the cylinder head and engine block.

47. Install the air conditioning compressor. Leave the coolant hose in the bracket when installing the compressor.

48. Connect the electrical leads and be sure the lead is clear of the compressor pulley. Install the steering servo reservoir. Install the coolant expansion tank and tighten the hose clip.

49. Connect the top pipe to the air cooled oil cooler and secure the cooler to the radiator. Install the overflow line between the expansion tank and the radiator.

50. Install the compressor belt, adjust the tension and tighten the belt tensioner bolt. Install the inner right wheel arch, and install the wheel.

51. Lower the vehicle and tighten the wheel. Connect the negative battery cable and properly fill the cooling system.

52. Start the engine.

53. Test the engine operation and check for leaks.

2.5L Engine

> ❋❋ **WARNING**
>
> To avoid damage to the valves, DO NOT rotate the camshafts. The crankshaft may only be turned between 0° and 60° BTDC when the camshafts are locked in position with the appropriate locking tool.

1. Disconnect the negative battery cable and drain the coolant from the radiator.

2. Safely raise and support vehicle and disconnect the rearward flange of the front exhaust pipe from the exhaust manifold.

3. Lower vehicle, remove engine covers, air filter assembly and mass air flow sensor.

4. Disconnect and label all hoses and electrical harness connections at the intake manifold. Remove the upper intake plenum and plug the center intake runner ports with paper.

5. Disconnect and plug the fuel lines.

6. Remove the center intake runners complete with fuel rail.

7. Remove lower section of the intake manifold and cover intake port openings.

8. Detach electrical connectors and upper radiator hose at the coolant bridge. Remove the coolant bridge.

9. Remove the spark plug wires.

10. Remove the oxygen sensor from it's holder and remove crankcase breather. Cover holes with paper.

11. Remove the ignition coil. Remove the bracket for the ignition coil, camshaft position sensor and oxygen sensor connectors and lay the bracket aside.

12. Disconnect the upper radiator hose from the engine. Remove the lifting eye.

13. Remove the right-front wheel and forward cover in the wheel housing.

14. Loosen the power steering pump, water pump pulley bolts and 6 outer crankshaft pulley bolts. Remove the drive belt and tensioner. Remove the power steering pump pulley, water pump pulley and timing cover.

➡ **When removing the crankshaft pulley, remove the 6 outer bolts only, DO NOT remove the center bolt.**

15. Remove the crankshaft pulley.

16. Set the engine at TDC No. 1 cylinder. The timing marks on both camshafts and crankshaft should be aligned. Install locking tools (83–94–926 or equivalent) between the camshaft sprockets to lock the camshafts in position.

17. Remove the timing belt, tensioner and camshaft sprockets.

> ❋❋ **WARNING**
>
> The wrench used to hold the camshaft must not have jaws that are too long. A wrench that is too long could damage the casting and lock the tappets.

18. Remove the bearing caps on the exhaust camshaft. Note the position markings on the bearing caps and loosen the bearing cap bolts in stages of ½ to 1 turn at a time. Remove the front camshaft seal and remove the camshaft.

➡ **The bearing caps located where valve tappets are compressed, should be removed last. Also, it is not necessary to remove the intake camshaft for cylinder head removal.**

19. Loosen the cylinder head bolts ¼ turn at a time in sequence, then ½ turn at a time. Remove cylinder head bolts and lift cylinder head off the engine.

> ❋❋ **WARNING**
>
> Be careful when setting down the cylinder head, since the intake camshaft is still installed, the valve stems could be accidentally bent.

20. With the cylinder head removed, the spark plugs, intake camshaft, tappets and exhaust manifold can be removed.

➡ **When removing the intake camshaft, use the same procedures used for the exhaust camshaft.**

To install:

21. Be sure the crankshaft is at 60° BTDC.

22. Be sure all gasket contact surfaces are clean. Note the positioning of the cylinder head gasket and be sure the words OBEN/TOP marked on the gasket are facing up.

23. Install the exhaust manifold and tappets, if removed from the cylinder head.

24. Install the cylinder head onto the engine using NEW cylinder head bolts.

25. Step tighten the cylinder head bolts in sequence as follows:
- Step 1: 18.5 ft. lbs. (25 Nm)
- Step 2: Tighten bolts 90°
- Step 3: Tighten bolts 90°
- Step 4: Tighten bolts 90°

26. Thoroughly lubricate the camshafts, install the camshafts and new front camshaft seals. Be sure the locating pins are properly positioned. Install the bearing caps in their proper location and position. Tighten the bearing cap bolts in sequence ½ to 1 turn at a time, to a torque of 6 ft. lbs. (8 Nm).

27. Install the timing cover and 2 retaining screws into cylinder head. Tighten retaining screws to 6 ft. lbs. (8 Nm).

28. Check to be sure that the camshaft locating pins are in the proper position.

➡ **On the rear or left bank cylinder head, the locating pins of both camshafts should be point towards the inboard bolts of the camshaft bearing caps. On the front or right bank cylinder head, the locating pin for the intake camshaft should be pointing downward in line with the inboard bolt of the bearing caps. The locating pin for the exhaust camshaft should be pointing upward in line with the edge of the camshaft sensor.**

29. Install the camshaft sprockets, the timing belt tensioner and timing belt.

30. Install crankshaft locking tool (83–94–926 or equivalent). Install tool 83–93–985 with a cut piece from an old timing belt, to measure belt tension.

31. Snug the center bolts of the adjusting rollers. Turn the lower adjusting roller counterclockwise, until a belt tension of 275–300 Nm is reached. Tighten adjusting roller center bolts to 30 ft. lbs. (40 Nm).

➡ **This adjustment of the timing belt is just preparation and should not be used as a final check.**

32. Adjust the tensioner pulley until the marks are aligned. Tighten the tensioner pulley to 15 ft. lbs. (20 Nm).

33. Remove camshaft locking tool on camshaft sprockets 1 & 2. Adjust the upper adjusting roller until sprocket No. 2 moves 1–2mm clockwise. Tighten the upper adjusting roller to 30 ft. lbs. (40 Nm) and remove the upper locking tool.

34. Rotate the engine two complete revolutions to just before 0° TDC and install the locking tool on the crankshaft. Carefully turn the crankshaft until the arm of the locking tool is against the water pump flange and tighten the locking tool. Set tool (83–94–926 or equivalent) into position on the front of the camshaft sprockets. Be sure that the timing marks on the camshaft sprockets are aligned with the marks on the tool and that the edge of the timing belt is flush with the edge of the camshaft sprockets.

➡ **Also check that the alignment marks on the tensioner pulley are still aligned.**

35. Install the crankcase ventilation housing and set the oxygen sensor connector back in it's holder. Be sure the valve

cover O-rings are still in position and clean. Lubricate the O-rings with soapy water, apply (81–52–381 or equivalent) sealer at the corners of the large end bearing caps and install the valve cover.

36. Install the spark plug wires.

37. Install the crankshaft pulley and tighten bolts to 15 ft. lbs. (20 Nm).

38. Install the timing cover. Install power steering pump pulley, water pump pulley, drive belt tensioner and drive belt.

39. Install the forward cover in the wheel housing and the right-front wheel. Tighten the wheel bolts to 89 ft. lbs. (120 Nm).

40. Install coolant bridge and tighten the bolts to 22 ft. lbs. (30 Nm). Install the radiator hose and electrical wiring harness connectors.

41. Install a new gasket, the lower intake manifold and tighten the bolts to 15 ft. lbs. (20 Nm).

42. Install the center intake runners complete with fuel rail and tighten the bolts to 15 ft. lbs. (20 Nm).

43. Connect the fuel lines.

44. Connect all hoses and electrical harness connections at the intake manifold, which should have been labeled at disassembly.

➡**Be sure the pressure regulator's vacuum hose is connected to front nipple on the throttle housing, in front of the butterfly. Install the upper intake plenum and tighten the bolts at the mating surface. Starting in the middle and working alternately outward to the ends. Tighten upper intake plenum bolts to 15 ft. lbs. (20 Nm).**

45. Install the mass air flow sensor, air filter assembly and engine covers.

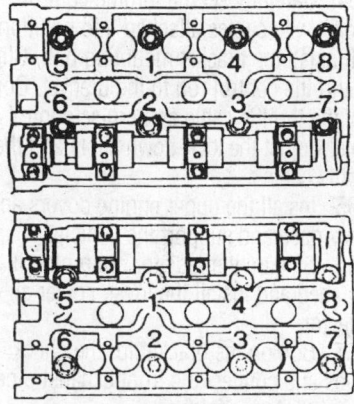

7923SG04

Cylinder head bolt torque sequence—2.5L and 3.0L engines

46. Safely raise and support vehicle and connect the rearward flange of the front exhaust pipe to the exhaust manifold.

47. Connect the negative battery cable.

48. Check all other oil and fluids for condition and proper level. Replace fluids and filters or top off as necessary.

49. Fill the radiator with clean coolant and check the cooling system for leaks

50. Bleed the cooling system.

51. Test drive the vehicle and check for any leaks.

3.0L Engine

✳✳ WARNING

To avoid damage to the valves, DO NOT rotate the camshafts. The crankshaft may only be turned between 0° and 60° BTDC when the camshafts are locked in position with the appropriate locking tool.

1. Disconnect the negative battery cable.

2. Safely raise and support the vehicle. Disconnect the front exhaust pipe from the exhaust manifolds. Disconnect the bracket for the check valves.

3. Remove the lower center air deflector. Drain the engine coolant into a catch pan.

4. Lower the vehicle and remove the top engine covers.

5. Disconnect the cruise control cable, throttle cable. Unclip the throttle control rod from the bracket. Remove the bracket mounting bolt and carefully set the bracket with cables attached to the side.

6. Loosen the clamps on the air intake pipes at the intake manifold and the mass air flow meter. Disconnect the pipes and raise them slightly. Disconnect the vacuum hoses and electrical connection. Lift the pipes with resonator attached carefully out.

7. Disconnect and label all electrical and vacuum connections on the intake manifold. Remove the intake plenum bolts.

8. Detach the IAC connector and fuel pressure regulator hose and remove the wiring harness from under the throttle body.

9. Detach the throttle position indicator, ignition coil and TCS connectors. Lift off the intake plenum and plug the intake runners with paper.

10. Detach the fuel injector and camshaft position connectors.

11. Disconnect and plug the fuel line connections.

12. Remove the center intake manifold, with fuel rails and set aside.

13. Mark the position of the lower intake manifold. Remove the lower intake manifold and plug the engine intake ports with paper.

14. Unbolt the coolant bridge and carefully bend it aside.

15. Disconnect the spark plug wires and the ignition coil. Bend the ignition coil aside and disconnect the ignition coil bracket.

16. Remove the lifting eye and the heat shield over the exhaust manifold.

17. Unbolt the resonator bracket and secondary air injection pipe from the exhaust manifold.

18. Lift up the power steering reservoir.

19. Disconnect the connection on the torque arm engine mount and remove the torque arm.

20. Remove the power steering line clamp from the torque arm engine mount and remove the engine mount.

21. Disconnect the hose from the coolant expansion tank and remove the upper alternator air intake.

22. Loosen the power steering pump, water pump pulley and 6 outer crankshaft pulley bolts. Remove the drive belt and tensioner. Remove the power steering pump, water pump pulley and timing cover.

➡**When removing the crankshaft pulley, remove the 6 outer bolts only, DO NOT remove the center bolt.**

23. Remove the crankshaft pulley.

24. Remove the timing belt, tensioner and camshaft sprockets.

25. Remove the valve cover. Be sure the O-rings stay in position and do not fall into the engine.

26. Remove the bearing caps on the exhaust camshaft. Note the position markings on the bearing caps and loosen the bearing cap bolts in stages of ½ to 1 turn at a time.

➡**The bearing caps located where valve tappets are compressed, should be removed last. Also, it is not necessary to remove the intake camshaft for cylinder head removal.**

27. Loosen the cylinder head bolts ¼ turn at a time in sequence, then ½ turn at a time. Remove cylinder head bolts and lift cylinder head off the engine.

✳✳ WARNING

Be careful when setting down the cylinder head, since the intake camshaft is still installed, the valve stems could be accidentally bent.

28. Once the cylinder head is removed from the engine assembly, the spark plugs,

camshafts, tappets and exhaust manifold can be removed.

To install:

29. Be sure the crankshaft is still positioned at 60° BTDC.

30. Be sure all gasket contact surfaces are clean. Note the positioning of the cylinder head gasket and be sure the words OBEN/TOP marked on the gasket are facing up.

31. Install the exhaust manifold and tappets, if removed from the cylinder head.

32. Install the cylinder head onto the engine using NEW cylinder head bolts.

Step tighten the cylinder head bolts in sequence as follows:
- Step 1: 18.5 ft. lbs. (25 Nm)
- Step 2: Tighten bolts 90°
- Step 3: Tighten bolts 90°
- Step 4: Tighten bolts 90°

➡**On the front or right bank cylinder head, the camshaft bearing caps are marked R1-r8 and the camshaft bearing seats in the head are numbered 1–8. On the rear or left bank cylinder head, the camshaft bearing caps are marked L1-L8, and the camshaft bearing seats in the head are numbered 1–8.**

33. Thoroughly lubricate the camshafts and install the camshafts with new front camshaft gaskets. Be sure the locating pins are properly positioned. Install the bearing caps in their proper location and position. Tighten the bearing cap bolts in sequence ½ to 1 turn at a time, to a torque of 6 ft. lbs. (8 Nm).

34. Install the timing cover and 2 retaining bolts.

35. Check to be sure that the camshaft locating pins are in the proper position. Check the locating pins, if they are hollow, replace them with solid pins.

➡**On the front or right bank cylinder head, the locating pin for the intake camshaft should be pointing downwards, in line with the inboard bolt of the bearing caps. The locating pin for the exhaust camshaft should be pointing upwards, in line with the edge of the camshaft sensor. On the rear or left bank cylinder head, the locating pins of both camshafts should point toward the inboard bolts of the camshaft bearing caps.**

36. Install the camshaft sprockets, tensioner and camshaft sprockets.

37. Install crankshaft locking tool KM-800–10 (or equivalent). Install tool 83–93–985 with a cut piece from an old timing belt, to measure belt tension.

38. Snug the center bolts of the adjusting rollers. Turn the lower adjusting roller counterclockwise, until a belt tension of 275–300 Nm is reached. Tighten adjusting roller center bolts to 30 ft. lbs. (40 Nm).

➡**This adjustment of the timing belt is just preparation and should not be used as a final check.**

39. Adjust the tensioner pulley until the marks are aligned. Tighten the tensioner pulley to 15 ft. lbs. (20 Nm).

40. Remove camshaft locking tool KM-800–1 (or equivalent) on camshaft sprockets 1 & 2. Adjust the upper adjusting roller until sprocket No. 2 moves 1–2mm clockwise. Tighten the upper adjusting roller to 30 ft. lbs. (40 Nm) and remove the upper locking tool.

41. Rotate the engine two complete revolutions to just before 0° TDC and install the locking tool on the crankshaft. Carefully turn the crankshaft until the arm of the locking tool is against the water pump flange and tighten the locking tool. Set tool KM-800–20 (or equivalent) into position. Be sure that the timing marks on the camshaft sprockets are aligned with the marks on the tool and that the edge of the timing belt is flush with the edge of the camshaft sprockets.

➡**Also check that the alignment marks on the tensioner pulley are still aligned.**

42. Install the crankcase ventilation housing and set the oxygen sensor connector back in it's holder. Be sure the valve cover O-rings are still in position and clean. Lubricate the O-rings with soapy water, apply 81–52–381 (or equivalent) sealer at the corners of the large end bearing caps and install the valve cover.

43. Install the timing cover.

44. Install the crankshaft pulley and tighten the bolts to 15 ft. lbs. (20 Nm).

45. Install the water pump pulley, power steering pump, drive belt tensioner and drive belt.

46. Install the upper alternator air intake.

47. Install the torque arm engine mount and attach the power steering line clamp.

48. Install the torque arm and bolt the connection to the torque arm engine mount.

49. Connect the upper hose to the coolant expansion tank and set the power steering reservoir back into position.

50. Install the lifting eye and secondary air injection pipe to the exhaust manifold. Install the exhaust manifold heat shield.

51. Install the resonator bracket.

52. Install the ignition coil bracket and ignition coil.

53. Install the spark plugs (if removed) and tighten to 18.5 ft. lbs. (25 Nm). Install the spark plug wires.

54. Carefully bend the coolant bridge back into position. Apply 74–96–284 (or equivalent) to the mounting bolts. Install the mounting bolts and tighten to 22 ft. lbs. (30 Nm).

55. Apply an acid-free petroleum jelly to the lower intake manifold and set the lower intake manifold onto the engine, noting the position marks made at disassembly. Apply thread locker 74–96–268 (or equivalent) to the lower intake manifold bolts. Install the lower intake manifold bolts and tighten to 15 ft. lbs. (20 Nm).

56. Install the center intake manifold with fuel rail. Apply thread locker 74–96–268 (or equivalent) to the lower intake manifold bolts. Install the center intake manifold bolts and tighten to 15 ft. lbs. (20 Nm).

57. Attach the fuel lines, crankshaft position sensor and fuel injector connectors.

58. Set the intake plenum into position and connect the TCS throttle body.

59. Tighten the intake plenum bolts to 15 ft. lbs. (20 Nm).

60. Attach the throttle position indicator, ignition coil and TCS connectors.

61. Route the wiring harness under the throttle body and connect the fuel pressure regulator hose and IAC connector.

62. Connect all labeled electrical and vacuum connections on the intake manifold.

63. Set the air intake pipes with resonator attached, carefully into position. Connect the vacuum hoses and electrical connection. Lower the air intake pipes and connect them to the mass air flow meter and intake manifold. Install and tighten the clamps on the air intake pipes at the intake manifold and the mass air flow meter.

64. Carefully set the throttle control bracket with cables attached into position and install the bracket mounting bolt. Clip the throttle control rod to the bracket. Connect the throttle cable and cruise control cable. Adjust the kick-down cable and the throttle cable.

65. Install the upper engine covers and safely raise and support the vehicle.

66. Be sure the radiator drain plug is tightened and install the lower center air deflector.

67. Bolt on the bracket for the check valves and connect the front exhaust pipe to the exhaust manifolds.

68. Connect the front exhaust pipe to the exhaust manifolds and lower the vehicle.

69. Connect the negative battery cable.

70. Check all other oil and fluids for

condition and proper level. Replace fluids and filters or top off as necessary.

71. Fill the radiator with clean coolant and check the cooling system for leaks

72. Bleed the cooling system.

73. Test drive the vehicle and check for any leaks.

Turbocharger

REMOVAL & INSTALLATION

2.0L Engine

1. Drain coolant from cooling system.

2. Raise and safely support the vehicle. Remove hose to the charge air cooler.

3. Remove the front exhaust pipe from the engine and be careful not to damage the oxygen sensor electrical wiring harness.

4. Break seal and remove locking ring on the wastegate.

5. Remove the oil return pipe from the turbocharger.

6. Disconnect the oil pipe from the oil filter housing.

7. Remove the hose to the boost control valve.

8. Disconnect the intake pipe and the bypass hose, take off the nuts to the wastegate diaphragm and remove the intake pipe together as one unit.

9. Remove the turbocharger coolant pipe with the oil pipe.

10. Remove the turbocharger.

To install:

11. Install the turbocharger with a new gasket and tighten the retaining nuts to 16 ft. lbs. (22 Nm).

12. Install turbocharger coolant and oil pipes.

13. Install the intake pipe for the turbocharger and mount the wastegate with the boost pressure control valve and pipes.

14. Fit the intake hose with the bypass hose.

15. Raise and safely support the vehicle.

16. Install the locking ring for the wastegate.

17. Install the front pipe to the turbocharger.

18. Install the hose connecting the charge air cooler with the turbocharger.

19. Fill the cooling system and check oil level.

20. Reseal the boost pressure control rod.

21. Test drive and check the boost pressure.

2.3L Engines

1. Disconnect the negative battery cable.

2. Raise and safely support the vehicle. Remove the center air deflector and properly drain the cooling system.

3. Disconnect the top pipe coupling on the air cooled oil cooler and disconnect the clips securing the pipe to the radiator.

4. Remove the solenoid valve from its mounting on the radiator and disconnect the electrical leads.

5. Disconnect the electrical leads at the radiator fan. Unbolt and remove the fan.

6. Remove the oxygen sensor cable clamps.

7. Disconnect the exhaust pipe from the turbocharger and bend it down slightly.

8. Unplug the electrical connectors for the air mass meter. Disconnect the toggle fasteners securing the air mass meter to the air cleaner cover and pull the rubber socket connector off the turbocharger unit.

9. Disconnect the turbocharger pressure pipe from the turbocharger compressor.

10. Remove the oil pipe to the turbocharger unit and remove the clip securing the oil pipe to the cylinder head. Disconnect the oil pipe banjo coupling from the block and remove the clip on the intake manifold.

11. Remove the coolant pipes from the turbocharger unit.

12. Remove the steady bar bracket between the engine block and the turbocharger compressor. Remove the securing bolts and loosen the oil return lines. Cap the opening to prevent washers or nuts from the exhaust manifold dropping inside during the removal.

13. Remove the turbocharger unit from the exhaust manifold.

To install:

14. Position the turbocharger unit to the exhaust manifold. Using new locknuts, tighten the retaining nuts to 16 ft. lbs. (22 Nm).

15. Fill the turbocharger interchamber passage with engine oil.

❋❋ WARNING

It is very important that there is oil in the turbocharger when the engine is started to avoid damage to the unit.

16. Install the clip holding the turbocharger oil supply pipe to the intake manifold. Connect and tighten the banjo

coupling to the engine block. Be sure the copper washers are in good condition. Secure the pipe to the turbocharger unit.

17. Install the return oil pipe and the steady bar bracket between the turbocharger unit and the crankcase. Connect the rubber hangers for the front exhaust hanger.

18. Lubricate the 3 studs on the turbocharger with Molykote 1000® and connect the exhaust pipe to the turbocharger compressor with new locknuts. Tighten the nuts to 19 ft. lbs. (25 Nm).

19. Install the coolant pipes to the turbocharger unit.

20. Install the turbocharger pressure pipe to the turbocharger compressor and assemble the air mass meter and rubber socket connector between the air cleaner body and the inlet side of the turbocharger compressor.

21. Assemble the fan and solenoid valve, securing the electrical leads into their clips. Connect the return hose to the solenoid valve. Install the air conditioning compressor.

22. Reconnect the oil pipe to the oil cooler and secure the pipe clip to the radiator.

23. Tighten the radiator drain plug and install the center air deflector.

24. Properly fill the cooling system and check the oil level and quality.

25. Connect the negative battery cable.

26. Start the engine and check for leaks and proper turbocharger system operation.

Intake Manifold

REMOVAL & INSTALLATION

2.0L and 2.3L Engines

❋❋ CAUTION

The fuel injection system remains under pressure, even after the engine has been turned OFF. The fuel system pressure must be relieved before disconnecting any fuel lines. Failure to do so may result in fire or personal injury.

1. Disconnect the negative battery cable. Drain the engine coolant.

2. If vehicle is equipped with a turbocharger, disconnect and remove the rubber elbow running between the throttle housing and the turbocharger.

3. Unplug the throttle position sensor.

Disconnect the hoses at the throttle housing, remove the three nuts and lift out the housing.

4. Unbolt the oil filler pipe bracket at the manifold and carefully position it out of the way.

5. Tag and disconnect all hoses and lines attached to the manifold.

6. Remove the AIC valve. Disconnect the fuel line from the pressure regulator.

7. Loosen the banjo fitting connecting the fuel line to the fuel rail. Cut the plastic tie and move the fuel line and pulsator out of the way. Be careful not to lose the seals.

8. Unplug each fuel injector electrical lead.

9. Disconnect the temperature sensor and the ground wires at the manifold.

10. Loosen the two screws on the cable clip underneath the manifold and move the harness assembly out of the way.

11. Detach the EGR pipe and all connectors.

12. Remove the eight mounting bolts and lift off the intake manifold.

To install:

13. Scrape off any excess gasket material, install a new gasket and install the manifold. Tighten the bolts in a crisscross pattern to 16 ft. lbs. (22 Nm).

14. Reposition the wire bundle and reconnect the EGR pipe.

15. Connect the ground wires and the temperature sensor.

16. Reconnect all injector leads.

17. Connect the fuel line to the pressure regulator. Connect the fuel line/pulsator to the fuel rail and secure it with a plastic tie.

18. Connect the oil filler pipe bracket to the manifold and install the AIC valve.

19. Install the throttle housing and all its attachments.

20. If vehicle is equipped with a turbocharger connect the rubber elbow between the turbocharger and the intake manifold.

21. Fill the cooling system.

22. Connect the negative battery cable.

23. Start the engine and check for leaks.

2.5L and 3.0L Engines

1. Remove the engine covers.

2. Disconnect the cruise control cable, detach the throttle cable and unclip the control rod and move it to aside.

3. Remove the hose clamps and intake tube from the mass air flow sensor assembly to the throttle plate housing. Detach the vacuum hoses and electrical connectors and remove complete assembly.

4. Disconnect all hoses and wiring from the upper intake manifold.

5. Remove the upper intake manifold bolts and remove the upper manifold.

6. Unplug the idle air control valve and the fuel pressure regulator hose. Remove harness from under the throttle plate housing.

7. Unplug the injector electrical harness and camshaft position sensor connector.

8. Disconnect and plug the fuel line.

9. Remove the bolts from the lower intake manifold and lift off engine.

To install:

10. Install the lower intake manifold with new gaskets and tighten retaining bolts to 15 ft. lbs. (20 Nm).

11. Connect the fuel lines, injector wiring and the camshaft position sensor.

12. Install the upper intake manifold with new gaskets and connect the throttle plate housing. Tighten the manifold bolts to 15 ft. lbs. (20 Nm).

13. Connect all wiring and hoses.

14. Install the idle air control valve and the hose to the fuel pressure regulator.

15. Install the mass air flow sensor assembly and hoses.

16. Connect the throttle control rod and associated cruise control cable.

17. Fit engine covers.

Exhaust Manifold

REMOVAL & INSTALLATION

2.0L Engine

1. Drain coolant from cooling system.

2. Raise and safely support the vehicle and remove the hose to the charge air cooler.

3. Remove the front exhaust pipe from the engine and be careful not to damage the oxygen sensor electrical wiring harness.

4. Break the seal and remove locking ring on the wastegate.

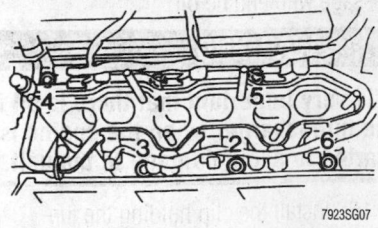

Intake manifold bolt torque sequence—2.5L and 3.0L engines

5. Remove the oil return pipe from the turbocharger.

6. Disconnect the oil pipe from the oil filter housing.

7. Remove the hose to the boost control valve.

8. Disconnect the intake pipe and the bypass hose, take off the nuts to the wastegate diaphragm and remove the intake pipe together as one unit.

9. Remove the turbocharger coolant pipe with the oil pipe.

10. Remove the turbocharger.

11. Using a ⅜ in. ratchet extension, relieve the pressure on the drive belt tensioner. Remove the drive belt.

12. Remove the power steering pump bracket and pump without disconnecting the hoses.

13. Remove the exhaust manifold studs, take off exhaust manifold and seal.

To install:

14. Clean mounting surface and install exhaust manifold seal and the manifold. Tighten the nuts to 18 ft. lbs. (25 Nm).

15. Fit power steering pump bracket with pump.

16. Install drive belt and release the pressure on the drive belt tensioner.

17. Install turbocharger tighten retaining nuts to 16 ft. lbs. (22 Nm).

18. Install turbocharger coolant and oil pipes.

19. Install the intake pipe for the turbocharger and mount the wastegate with the boost pressure control valve and pipes.

20. Fit the intake hose with the bypass hose.

21. Raise and safely supported the vehicle.

22. Install the locking ring for the wastegate.

23. Install the front pipe to the turbocharger.

24. Install the hose connecting the charge air cooler with the turbocharger.

25. Fill the coolant system with coolant and check oil level and quality.

26. Reseal the boost pressure control rod.

27. Test drive and check the boost pressure.

2.3L Engines

900

1. Unplug the oxygen sensor connector.

2. Safely raise and support the vehicle.

3. Disconnect the front pipe from exhaust manifold and lower the vehicle.

4. Using a ⅜ in. ratchet extension, relieve the drive belt tensioner and remove the belt.

5. Remove the power steering pump pulley and disconnect the pump bracket. Move pump to one side without disconnecting the hoses.

6. Remove the middle part of the exhaust manifold first, then remove the outer tubes.

To install:

7. Fit the outer tubes first, then mount middle part of exhaust manifold. Tighten the nuts to 18 ft. lbs. (25 Nm).

8. Install the power steering pump bracket and install pump and pulley.

9. Fit drive belt onto the pulleys and release the belt tensioner.

10. Raise the vehicle to connect front pipe to the exhaust manifold.

11. Connect the oxygen sensor.

9000

1. Disconnect the negative battery cable.

2. Raise and safely support the vehicle. Remove the center air deflector and properly drain the cooling system.

3. Disconnect the top pipe coupling on the air cooled oil cooler and disconnect the clips securing the pipe to the radiator.

4. Remove the solenoid valve from its mounting on the radiator and disconnect the electrical leads.

5. Disconnect the electrical leads at the radiator fan. Unbolt and remove the fan.

6. Remove the oxygen sensor cable clamps.

7. Disconnect the exhaust pipe from the turbocharger and bend it down slightly.

8. Unplug the electrical connectors for the air mass meter. Detach the toggle fasteners securing the air mass meter to the air cleaner cover and pull the rubber socket connector off of the turbocharger unit.

9. Disconnect the turbocharger pressure pipe from the turbocharger compressor.

10. Remove the oil pipe to the turbocharger unit and remove the clip securing the oil pipe to the cylinder head. Disconnect the oil pipe banjo coupling from the block and remove the clip on the intake manifold.

11. Remove the coolant pipes from the turbocharger unit.

12. Remove the steady bar bracket between the engine block and the turbocharger compressor. Remove the securing bolts and loosen the oil return lines. Cap the opening to prevent washers or nuts from the exhaust manifold dropping inside during the removal.

13. Remove the turbocharger unit from the exhaust manifold.

14. Release the tension on the air conditioning compressor belt by pulling hard upwards on the belt and installing 83–94–488 belt tensioner locking clamp tool or equivalent, on the tensioner and remove the belt.

15. Remove the air conditioning compressor mounting bolts. Insert a sheet of metal to protect the oil cooler and lift the air conditioning compressor towards the expansion tank.

16. Remove the nuts securing the exhaust manifold to the cylinder head.

17. Lift the exhaust manifold from the cylinder head complete with the gasket.

To install:

18. Clean the mating surfaces and install a new gasket over the studs for the exhaust manifold and install the manifold. Tighten the nuts to 19 ft. lbs. (25 Nm).

19. Install the air conditioning compressor and belt, remove the locking tool and check for proper belt tension.

20. Position the turbocharger unit to the exhaust manifold. Using new locknuts, tighten the retaining nuts to 16 ft. lbs. (22 Nm).

21. Fill the turbocharger interchamber passage with engine oil.

❋❋❋ WARNING

It is very important that there is oil in the turbocharger when the engine is started to avoid damage to the unit.

22. Install the clip holding the turbocharger oil supply pipe to the intake manifold. Connect and tighten the banjo coupling to the engine block. Be sure the copper washers are in good condition. Secure the pipe to the turbocharger unit.

23. Install the return oil pipe and the steady bar bracket between the turbocharger unit and the crankcase. Connect the rubber hangers for the front exhaust hanger.

24. Lubricate the 3 studs on the turbocharger with Molykote 1000® Connect the exhaust pipe to the turbocharger compressor with new locknuts and tighten the nuts to 19 ft. lbs. (25 Nm).

25. Install the coolant pipes to the turbocharger unit.

26. Install the turbocharger pressure pipe to the turbocharger compressor and assemble the air mass meter and rubber socket connector between the air cleaner body and the inlet side of the turbocharger compressor.

27. Assemble the fan and solenoid valve, securing the electrical leads into their clips.

Connect the return hose to the solenoid valve. Install the air conditioning compressor.

28. Reconnect the oil pipe to the oil cooler and secure the pipe clip to the radiator.

29. Tighten the radiator drain plug and install the center air deflector.

30. Properly fill the cooling system and check the oil level and quality.

31. Connect the negative battery cable.

32. Start the engine and check for leaks and proper turbocharger system operation.

2.5L and 3.0L Engines

1. Disconnect oxygen sensor cable.

2. Loosen top nuts on rear exhaust manifold.

3. Raise and safely support vehicle.

4. Disconnect catalytic converter and remove nuts on front pipe and remove pipe from exhaust manifold.

5. Remove lower nuts on rear exhaust manifold and remove.

6. Lower vehicle.

7. Remove dipstick holder.

8. Remove nuts on front exhaust manifold and remove.

To install:

9. Install front exhaust manifold and tighten nuts to 15 ft. lbs. (20 Nm).

10. Install the dipstick holder.

11. Raise and safely support vehicle.

12. Install rear exhaust manifold and tighten lower nuts to 15 ft. lbs. (20 Nm).

13. Install front pipe and tighten nuts to 30 ft. lbs. (40 Nm).

14. Install catalytic converter onto the front pipe.

15. Lower vehicle and tighten rear exhaust manifold top nuts to 15 ft. lbs. (20 Nm).

16. Connect oxygen sensor cable.

Front Crankshaft Seal

REMOVAL & INSTALLATION

2.5L and 3.0L Engines

1. Disconnect the negative battery cable.

2. Remove the air cleaner complete with hoses.

3. Using a 15mm wrench turn the serpentine drive belt tensioner towards the front of the vehicle, remove the belt from the water pump pulley and slowly release the belt tensioner.

4. Remove the belt tensioner and the power steering pump pulley.

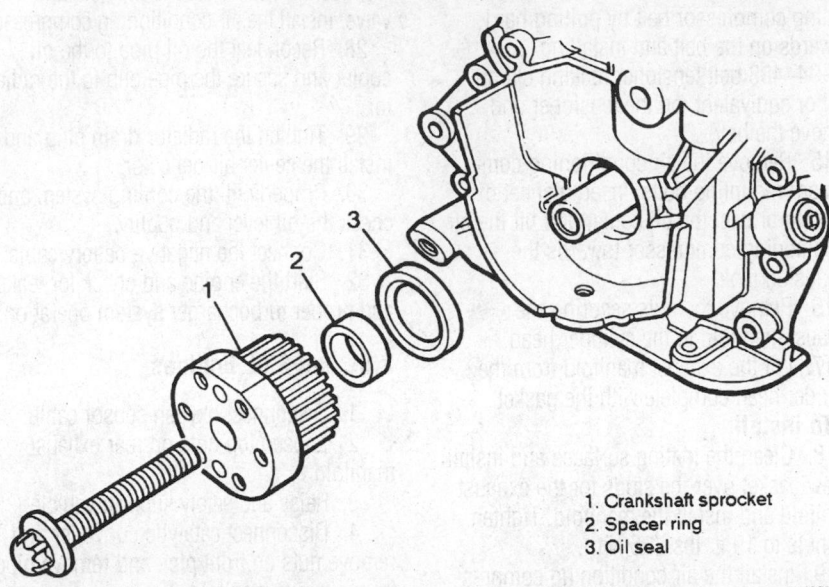

1. Crankshaft sprocket
2. Spacer ring
3. Oil seal

7923SG06

Remove the crankshaft sprocket for access to the front seal assembly—2.5L and 3.0L engines shown

5. Remove the water pump pulley. To do this, carefully pry the engine towards the left of the vehicle, using the engine bracket as a fulcrum.

6. Remove the timing cover.

7. Remove the right front wheel and the cover in the wheel well.

8. Remove the 6 bolts holding the crankshaft pulley and remove the pulley. Do not remove the center bolt.

9. Remove the tensioning roller and the two adjusting rollers and remove the timing belt.

10. Remove the crankshaft sprocket and spacer ring.

11. Using a small prybar, carefully remove the oil seal without marring the crankshaft stub end.

To install:

12. Lubricate and install a new oil seal using an appropriate seal installer.

13. Install the timing belt and belt tensioner.

14. Install the crankshaft pulley and tighten the retaining bolts to 15 ft. lbs. (20 Nm).

15. Install the timing belt cover tighten the bolts to 6 ft. lbs. (8 Nm).

16. Install the water pump pulley by prying the engine to the left, using the engine bracket as a fulcrum.

17. Install the power steering pump pulley and tighten the retaining bolts to 6 ft. lbs. (8 Nm).

18. Using a 15mm wrench turn the serpentine drive belt tensioner towards the

front of the vehicle, position the belt and slowly release the belt tensioner.

19. Install the air cleaner complete with hoses.

20. Connect the negative battery cable.

21. Start the engine and check the ignition timing.

Camshaft

REMOVAL & INSTALLATION

2.0L and 2.3L Engines

1. Disconnect the negative battery cable.

2. Remove the engine from the vehicle.

3. Remove the lid on the valve cover. Disconnect the spark plug wires and vacuum hose from the distributor and remove the distributor cap (if applicable).

4. Remove the valve cover and position the crankshaft for TDC. The **0** mark on the flywheel should align with the timing mark on the bell housing end plate. The camshafts should be lined up with their respective timing marks.

5. Remove the distributor (if applicable) and remove the oil pipes.

6. Remove the center bolts securing the camshaft sprockets. Use a proper holding tool to hold the camshafts from rotating. Always keep the camshafts in their correct basic setting. If the setting of the crankshaft or camshafts is altered at this stage the valves can be damaged.

7. Remove the timing chain tensioner. Remove the camshaft sprockets.

8. Mark the bearing cap positions and relation to the front of the engine. The caps must be installed in their original location.

9. Loosen the camshaft bearing cap bolts one turn at a time to avoid uneven valve spring pressure on the camshafts. When all bolts are loose, remove the bearing caps and lift out the camshafts.

To install:

10. Place the camshafts in their proper positions and install the bearing caps in their original location and position.

11. When installing the bolts, tighten them one turn at a time to draw the camshaft down evenly against the valve springs. Tighten the bearing cap bolts to 11 ft. lbs. (15 Nm).

➡**The black bolts have an oiling passage and must be installed on the spark plug side.**

12. Install the camshaft sprockets, fitting the sprocket for the exhaust cam first. Be sure the chain between the crankshaft sprocket and the camshaft sprocket is kept tight. Next install the intake cam sprocket. Keep the chain tight between the sprockets. Hand-tighten the center bolts securing the camshaft sprockets..

13. Install the chain tensioner with the piston under tension. Be sure the copper gasket is in good condition and that the sealing surface is clean and free from burrs. Tighten the tensioner to 47 ft. lbs. (63 Nm).

14. Trigger the chain tensioner by pressing the pivoting chain guide against it, thereafter, press the pivoting guide against the chain to give the chain its basic tension. Check that the chain tensioner maintains tension on the chain when the pressure on the chain guide is released and that the basic setting stop for the tensioner holds the chain guide tight against the chain. A limited amount of play will be present until the hydraulic pressure takes over once the engine is running.

15. Depress the pivoting guide to check that the tensioner is working. Rotate the crankshaft 2 complete turns clockwise, viewed from the transmission end. Be sure the crankshaft and camshaft timing marks still align properly.

16. Hold the camshafts in their proper position and tighten the cam sprocket bolts to 49 ft. lbs.

17. Install the oil pipes and the distributor (if applicable).

18. Install the semi-circular rubber plug halves in the cylinder head and install the valve cover tighten the retaining bolts to 11

ft. lbs. (14 Nm). Install the crankcase ventilation hose.

19. Install the spark plug wires and distributor cap in their original position (if applicable) and install the lid on the valve cover.

20. Install the engine assembly and connect the negative battery cable.

21. Start the engine and check for proper operation and leaks.

2.5L Engine

✳✳ WARNING

To avoid damage to the valves, DO NOT rotate the camshafts. The crankshaft may only be turned between 0° and 60° BTDC when the camshafts are locked in position with the appropriate locking tool.

1. Disconnect the battery and drain the coolant.

2. Remove engine covers, air filter assembly and mass air flow sensor.

3. Disconnect and label all hoses and electrical harness connections at the intake manifold. Remove the upper intake manifold plenum.

4. Remove spark plug wires.

5. Remove the right-front wheel and forward cover in the wheel housing.

6. Loosen the power steering pump, water pump pulley bolts and 6 outer crankshaft pulley bolts. Remove the drive belt and tensioner. Remove the power steering pump pulley, water pump pulley and timing cover.

➡**When removing the crankshaft pulley, remove the 6 outer bolts only, DO NOT remove the center bolt.**

7. Remove the crankshaft pulley.

8. Remove the valve covers. Be sure the O-rings stay in position and do not fall into the engine.

9. Remove the timing belt, tensioner and camshaft sprockets.

10. Rotate the crankshaft back to 60° BTDC, to prevent damage to the valves.

11. Remove the bearing caps on the camshafts. Note the position markings on the bearing caps and loosen the bearing cap bolts in stages of ½ to 1 turn at a time. Remove the front camshaft seal and remove the camshaft.

➡**The bearing caps located where valve tappets are compressed, should be removed last.**

To install:

12. Be sure the crankshaft is at 60° BTDC.

13. Thoroughly lubricate the camshafts, install the camshafts and new front camshaft seals. Be sure the locating pins are properly positioned. Install the bearing caps in their proper location and position. Tighten the bearing cap bolts in sequence ½ to 1 turn at a time, to a torque of 6 ft. lbs. (8 Nm).

14. Install the timing cover and retaining screws into cylinder head. Tighten retaining screws to 6 ft. lbs. (8 Nm).

15. Check to be sure that the camshaft locating pins are in the proper position.

➡**The locating pins of both camshafts 1 & 2 should be point towards the inboard bolts of the camshaft bearing caps. The locating pin for the intake camshaft 3 should be pointing downwards, in line with the inboard bolt of the bearing caps. The locating pin for the exhaust camshaft 4 should be pointing upwards, in line with the edge of the camshaft sensor.**

16. Install the camshaft sprockets. Use an open ended wrench on the hex flats of the camshaft to hold the camshaft in position when removing the timing sprockets. Check the proper position of the sprockets by locating pins and timing marks. Tighten the camshaft Torx® bolts to 37 ft. lbs. (50 Nm) plus 60°.

✳✳ WARNING

The wrench used to hold the camshaft must not have jaws that are too long. A wrench that is too long could damage the casting and lock the tappets.

17. Install the tensioner and timing belt.

18. Be sure the valve cover O-rings are still in position and clean. Lubricate the O-rings with soapy water, apply 81–52–381 (or equivalent) sealer at the corners of the large end bearing caps and install the valve covers.

19. Install the spark plug wires.

20. Install crankshaft pulley and tighten retaining bolts to 15 ft. lbs. (20 Nm).

21. Install the outer timing cover. Install power steering pump pulley, water pump pulley, drive belt tensioner and drive belt.

22. Install the forward cover in the wheel housing and the right front wheel. Tighten the wheel bolts to 89 ft. lbs. (120 Nm).

23. Install the upper intake plenum and tighten the bolts at the mating surface. Starting in the middle and working alternately outward to the ends. Tighten the upper intake plenum bolts to 15 ft. lbs. (20 Nm).

24. Connect all hoses and electrical harness connections at the intake manifold, which should have been labeled at disassembly.

➡**Be sure the pressure regulator's vacuum hose is connected to front nipple on the throttle housing, in front of the butterfly.**

25. Install the mass air flow sensor, air filter assembly and engine covers.

26. Connect the battery cable.

27. Fill coolant system and check all other fluids as necessary.

28. Start engine and check for leaks.

3.0L Engine

✳✳ WARNING

To avoid damage to the valves, DO NOT rotate the camshafts. The crankshaft may only be turned between 0° TDC and 60° BTDC when the camshafts are locked in position with the appropriate locking tool.

1. Disconnect the negative battery cable.

2. Safely raise and support the vehicle. Disconnect the front exhaust pipe from the exhaust manifolds. Disconnect the bracket for the check valves.

3. Remove the lower center air deflector. Drain the engine coolant into a catch pan.

4. Lower the vehicle and remove the top engine covers.

5. Disconnect the cruise control cable, throttle cable. Unclip the throttle control rod from the bracket. Remove the bracket mounting bolt and carefully set the bracket with cables attached to the side.

6. Loosen the clamps on the air intake pipes at the intake manifold and the mass air flow meter. Disconnect the pipes and raise them slightly. Disconnect the vacuum hoses and electrical connection. Lift the pipes with resonator attached carefully out.

7. Disconnect and label all electrical and vacuum connections on the intake manifold. Remove the intake plenum bolts.

8. Detach the IAC connector and fuel pressure regulator hose and remove the wiring harness from under the throttle body.

9. Detach the throttle position indicator, ignition coil and TCS connectors. Lift off the intake plenum and plug the intake runners with paper.

10. Detach the fuel injector and camshaft position connectors.

11. Detach and plug the fuel line connections.

12. Remove the center intake manifold, with fuel rails and set aside.

13. Disconnect the spark plug wires and the ignition coil. Bend the ignition coil aside and disconnect the ignition coil bracket.

14. Remove the lifting eye and the heat shield over the exhaust manifold.

15. Unbolt the resonator bracket and secondary air injection pipe from the exhaust manifold.

16. Lift up the power steering reservoir.

17. Disconnect the connection on the torque arm engine mount and remove the torque arm.

18. Remove the power steering line clamp from the torque arm engine mount and remove the engine mount.

19. Disconnect the hose from the coolant expansion tank and remove the upper alternator air intake.

20. Loosen the power steering pump, water pump pulley and 6 outer crankshaft pulley bolts. Remove the drive belt and tensioner. Remove the power steering pump, water pump pulley and timing cover.

➡**When removing the crankshaft pulley, remove the 6 outer bolts only, DO NOT remove the center bolt.**

21. Remove the crankshaft pulley.

22. Remove the timing belt, tensioner and camshaft sprockets.

23. Rotate the crankshaft back to 60° BTDC, to prevent damage to the valves.

24. Remove the valve cover. Be sure the O-rings stay in position and do not fall into the engine.

25. Remove the bearing caps on the camshafts. Note the position markings on the bearing caps and loosen the bearing cap bolts in stages of ½ to 1 turn at a time.

➡**The bearing caps located where valve tappets are compressed, should be removed last.**

26. Carefully lift the camshafts off of the cylinder head.

To install:

27. Be sure the crankshaft is still positioned at 60° BTDC.

28. Be sure all gasket contact surfaces are clean.

➡**The camshaft bearing caps are marked L1-l8 for the Front or Left bank and R1-r8 for the Rear or Right bank and the camshaft bearing seats in each head are numbered 1–8.**

29. Thoroughly lubricate the camshafts and install the camshafts with new front camshaft gaskets. Be sure the locating pins are properly positioned. Install the bearing caps in their proper location and position. Tighten the bearing cap bolts in sequence

½ to 1 turn at a time, to a torque of 6 ft. lbs. (8 Nm).

30. Install the timing covers and bolt to the head.

31. Check to be sure that the camshaft locating pins are in the proper position. Check the locating pins, if they are hollow, replace them with solid pins.

➡**The locating pins of both camshafts 1 & 2 should be point towards the inboard bolts of the camshaft bearing caps. The locating pin for the intake camshaft No, 3 should be pointing downwards, in line with the inboard bolt of the bearing caps. The locating pin for the exhaust No. 4 camshaft should be pointing upwards, in line with the edge of the camshaft sensor.**

32. Install the camshaft sprockets, tensioner and timing belt.

33. Install the crankcase ventilation housing and set the oxygen sensor connector back in it's holder. Be sure the valve cover O-rings are still in position and clean. Lubricate the O-rings with soapy water, apply (81–52–381 or equivalent) sealer at the corners of the large end bearing caps and install the valve cover.

34. Install the timing cover.

35. Install the crankshaft pulley and tighten the bolts to 15 ft. lbs. (20 Nm).

36. Install the water pump pulley, power steering pump, drive belt tensioner and drive belt.

37. Install the upper alternator air intake.

38. Install the torque arm engine mount and attach the power steering line clamp.

39. Install the torque arm and bolt the connection to the torque arm engine mount.

40. Connect the upper hose to the coolant expansion tank and set the power steering reservoir back into position.

41. Install the lifting eye and secondary air injection pipe to the exhaust manifold. Install the exhaust manifold heat shield.

42. Install the resonator bracket.

43. Install the ignition coil bracket and ignition coil.

44. Install the spark plugs (if removed) and tighten to 19 ft. lbs. (25 Nm). Install the spark plug wires.

45. Install the center intake manifold with fuel rail. Apply thread locker (74–96–268 or equivalent) to and install the center intake manifold bolts and tighten to 15 ft. lbs. (20 Nm).

46. Attach the fuel lines, crankshaft position sensor and fuel injector connectors.

47. Set the intake plenum into position and connect the TCS throttle body.

48. Tighten the intake plenum bolts to 15 ft. lbs. (20 Nm).

49. Attach the throttle position indicator, ignition coil and TCS connectors.

50. Route the wiring harness under the throttle body and attach the fuel pressure regulator hose and IAC connector.

51. Connect all labeled electrical and vacuum connections on the intake manifold.

52. Set the air intake pipes with resonator attached, carefully into position. Connect the vacuum hoses and electrical connection. Lower the air intake pipes and connect them to the mass air flow meter and intake manifold. Install and tighten the clamps on the air intake pipes at the intake manifold and the mass air flow meter.

53. Carefully set the throttle control bracket with cables attached into position and install the bracket mounting bolt. Clip the throttle control rod to the bracket. Connect the throttle cable and cruise control cable. Adjust the kick-down cable and the throttle cable.

54. Install the upper engine covers and safely raise and support the vehicle.

55. Be sure the radiator drain plug is tightened and install the lower center air deflector.

56. Bolt on the bracket for the check valves and connect the front exhaust pipe to the exhaust manifolds. Lower the vehicle.

57. Connect the negative battery cable.

58. Check all other oil and fluids for condition and proper level. Replace fluids and filters or top off as necessary.

59. Fill the radiator with clean coolant and check the cooling system for leaks

60. Bleed the cooling system.

61. Test drive the vehicle and check for any leaks.

Valve Lash

ADJUSTMENT

The hydraulic cam followers, which are used in all 1993–97 Saab engines, do not require adjusting. The cam followers keep the valve clearance within 0.7382–0.8189 in. (18.75–20.8mm). However, if the cam followers are making excessive noise or are diagnosed to be defective, perform the following procedure:

1. Disconnect the negative battery cable.

2. If a cam follower is noisy, it can be found by removing the valve cover and using a screw driver, gently pushing down on each cam follower until the defective follower(s) is found by exhibiting a spongy feeling.

3. Replace the defective cam follower(s); first removing the camshaft(s).

4. Reinstall the camshaft(s) and the valve cover.

5. Reconnect the negative battery cable.

Oil Pan

REMOVAL & INSTALLATION

900 and 9-3

On 900 Series vehicles the top of the transaxle is used as the engine oil pan (sump). To remove the engine oil pan, remove the engine and transaxle as an assembly. Then, separate the engine and transaxle once removed from the vehicle.

2.0L AND 2.3L ENGINES

1. Disconnect the negative battery cable.

2. Remove the dipstick and plug the hole with a shop towel.

3. Install engine lifting beam 83–94–850 or equivalent and slightly raise the engine.

4. Detach the oxygen sensor connector and unbolt the bracket.

5. Safely raise and support the vehicle, drain the engine oil and set a lifting table under the engine.

6. Remove the front wheels and front spoiler. Disconnect the front exhaust pipe at the exhaust manifold and intermediate pipe.

7. If equipped with turbocharger, unbolt the front exhaust pipe from the turbocharger. Disconnect the front pipe catalytic converter bracket and remove the front pipe.

8. Disconnect the lower ball joints from the steering knuckles. Unbolt the rear engine mount.

9. Raise the lifting table into position. Unbolt and remove the subframe.

10. Disconnect the oil level sensor and pull the sensor harness from it's clips.

11. Remove the flywheel inspection cover and oil pan.

12. Do not remove the guide sleeve from the block. Wipe off excess oil.

To install:

13. Thoroughly clean the flanges on the sump and block using a suitable solvent. Apply an even bead of Loctite® 518 sealant, or equivalent, along the oil pan flange.

14. Install the oil pan and loosely install the bolts.

15. Tighten the bolts, starting in the center on both sides, then crisscross out to the ends until snug,

16. Tighten the bolts in the same fashion to 16 ft. lbs. (22 Nm).

17. Install the flywheel inspection cover.

18. Set the oil level sensor harness back into it's clips and connect the oil level sensor.

19. Raise the subframe and install the subframe bolts. Tighten the subframe bolts as follows:

a. Front bolts to 85 ft. lbs. (115 Nm).

b. Center bolts to 140 ft. lbs. (190 Nm).

c. Rear bolts to 81 ft, lbs. (110 Nm) plus75° additional torque.

20. Connect the lower ball joints to the steering knuckles and tighten the nuts to 55 ft. lbs. (75 Nm).

21. Loosely connect the front exhaust pipe catalytic converter bracket. Apply Molykote 1000 ® to the studs of the exhaust manifold or turbocharger (if equipped) and install the front exhaust pipe.

22. Tighten the converter bracket and connect the front pipe to the mid pipe.

23. Install the front spoiler and front wheels. Tighten the wheel bolts to 89 ft. lbs. (120 Nm).

24. Be sure the oil drain plug is tight and lower the vehicle.

25. Install the dipstick, oxygen sensor connector bracket and attach the oxygen sensor.

26. Fill the engine with oil and remove the engine lifting beam.

27. Connect the negative battery cable. Warm up the engine and check for leaks.

2.5L ENGINE

1. Raise and safely support the vehicle.

2. Drain the engine oil.

3. Remove the engine cradle sub frame.

4. Unbolt and remove the oil pan.

5. Remove all gasket material and clean sealing surfaces.

To install:

6. Apply a sealing compound to the areas where the engine block mate to the oil pump and rear main bearing cap.

7. Install a new oil pan gasket. Install the oil pan to the engine block.

8. Using a bolt locking compound install oil pan retaining bolts and tighten to 11 ft. lbs. (15 Nm).

9. Install oil pan drain plug and tighten to 41 ft. lbs. (55 Nm).

10. Install the engine cradle sub frame and tighten bolts as follows:

a. Front bolts to 85 ft. lbs. (115 Nm).

b. Center bolts to 140 ft. lbs. (190 Nm).

c. Rear bolts to 81 ft. lbs. (110 Nm) plus 75° additional torque.

11. Lower the vehicle. Fill the engine with oil. Start the engine and check for leaks.

9000

2.3L ENGINES

1. Disconnect the negative battery cable.

2. Remove the oil dipstick and place a shop cloth into the end of the tube to prevent contamination.

3. Raise and safely support the vehicle.

4. Drain the oil into a suitable container.

5. Remove the right front wheel and inner wheel well.

6. Unbolt the front and rear engine mounts.

7. Remove the oxygen sensor and the front section of the exhaust pipe. Lower the vehicle.

8. Remove the tie rod between the wheel well and the subframe.

9. Install lifting beam tool 83–93–977 or equivalent, and slightly raise the engine.

10. Remove the bottom bolt holding the transmission to the oil pan.

11. Unplug the connector for the oil level sensor and remove the sensor.

12. Fold down the edge of the splash plate and remove the two rubber plugs in the back of the transmission case.

13. Remove the two bolts securing the oil pan to the block under the plugs.

14. Tap the guide sleeve into the block.

15. Remove the remaining oil pan bolts and remove the oil pan from the back first.

16. Remove the guide sleeve from the cylinder block.

To install:

17. Thoroughly clean the flanges on the sump and block using a suitable solvent. Apply an even bead of Permatex® Ultra Blue sealant, or equivalent, along the oil pan flange.

18. Install the rubber seal for the oil strainer in the groove on the oil pan.

19. Install the oil pan, front edge first, then the back.

➡**The longer bolt with the washer should be installed in the middle on the right-hand side.**

20. Install the bolts loosely, then, starting in the middle, tighten the bolts to 15 ft. lbs. (20 Nm).

21. Install the two rubber plugs in the back of the transmission and return the

edge of the splash plate to its original position.

22. Install the bolt securing the oil pan to the transmission case at the bottom and install the oil level sensor.

23. Align the engine over the mounts and lower it into position. Install the tie rod between the wheel well and the subframe. Install the dipstick.

24. Install the bolts in the front and rear engine mounts. Install the oxygen sensor and exhaust pipe.

25. Install the inner wheel well and wheel. Tighten the wheel bolts to 96 ft. lbs. (130 Nm).

26. Lower the vehicle, fill with oil and run the engine to normal operating temperature to check for leaks.

3.0L ENGINE

1. Raise and safely support the vehicle.

2. Drain the engine oil.

3. Unbolt front exhaust pipe from the exhaust manifolds, pull down slightly to allow for room to remove the oil pan.

4. Remove the flywheel inspection cover and remove the oil pan.

5. Remove all gasket material and clean sealing surfaces.

To install:

6. Apply a sealing compound to the areas where the engine block mate to the oil pump and rear main bearing cap.

7. Install a new oil pan gasket. Install the oil pan to the engine block.

8. Using a bolt locking compound install oil pan retaining bolts and tighten to 11 ft. lbs. (15 Nm).

9. Install oil pan drain plug and tighten to 41 ft. lbs. (55 Nm).

10. Install the flywheel inspection cover and tighten bolts to 6 ft. lbs. (8 Nm).

11. Lower the vehicle. Fill the engine with oil. Start the engine and check for leaks.

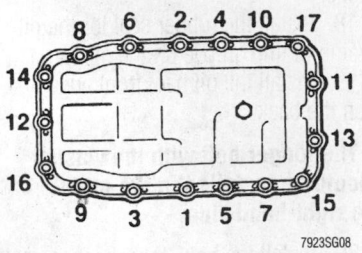

Oil pan bolt torque sequence—2.5L and 3.0L engines

Oil Pump

REMOVAL & INSTALLATION

900 and 9-3

2.0L AND 2.3L ENGINES

1. Disconnect the negative battery cable.

2. Drain the engine oil and the coolant.

3. Remove the bracket for the steering servo pump, complete with the pump and alternator.

4. Remove the belt tensioner and the water pump pipe.

5. Secure the flywheel with a flywheel locking segment, tool 83–92–987, and loosen the crankshaft pulley bolt.

6. Using a suitable puller remove the crankshaft pulley.

7. Remove the oil pipes and the water pump pulley.

8. Remove the oil pump.

To install:

9. Install the oil pump and tighten the retaining bolts to 6 ft. lbs. (8 Nm). Install the oil pipes and water pump pulley.

10. Install the crankshaft pulley and tighten the nut to 140 ft. lbs. (190 Nm).

11. Install the belt tensioner, water pump pipe, steering servo pump and alternator.

12. Fill the radiator with coolant and the crankcase with oil.

13. Start the engine and check for leaks.

2.5L ENGINE

1. Disconnect the negative battery cable.

2. Remove the air cleaner complete with hoses.

3. Using a 15mm wrench turn the serpentine drive belt tensioner towards the front of the vehicle, remove the belt from the water pump pulley and slowly release the belt tensioner.

4. Remove the belt tensioner and the power steering pump pulley.

5. Remove the water pump pulley. To do this you will have to carefully press the engine towards the left in the vehicle, using the engine bracket as a fulcrum.

6. Remove the timing cover.

7. Remove the right front wheel and the cover in the wheel well.

8. Remove the 6 bolts holding the crankshaft pulley and remove the pulley. Do not remove the center bolt.

9. Put the No. 1 cylinder in the top dead center position (on the compression stroke).

10. The timing marks on the crankshaft and camshafts should be in alignment with their respective marks on the engine. Insert the camshaft locking tool 83–94–926 and install, 83–94–868 the locking tool for the crankshaft.

11. If reusing the belt, mark the direction of its rotation. To help with refitting, the belt can be marked at the camshaft timing marks and also at the crankshaft timing mark.

12. Remove the tensioning roller and the two adjusting rollers and remove the timing belt.

13. Rotate the crankshaft back to 60° BTDC, to prevent damage to the valves.

14. Remove the bracket with the upper timing belt adjuster and tensioner rollers.

15. Remove inner timing cover bolts

16. Install an engine lifting beam.

17. Raise and safely support the vehicle. Remove the front spoiler and disconnect the front exhaust pipe.

18. Remove the flywheel inspection cover and install a flywheel stop.

19. Remove the crankshaft sprocket.

20. Remove the engine subframe.

21. Remove the alternator.

22. Remove the AC compressor and hang to aside.

23. Drain the engine oil.

24. Remove the oil pan and strainer.

25. Disconnect the oil pump pressure sensor wiring.

26. Remove the AC compressor bracket.

27. Remove the oil pump housing. Extract the oil pump impellers.

To install:

28. Install the oil pump impellers and fit the oil pump cover.

29. Install a new oil pump housing gasket with the housing.

30. Tighten the bolts to 53 inch lbs. (6 Nm).

31. Fit the AC compressor bracket.

32. Connect the oil pump pressure sensor wiring.

33. Install the oil strainer and fit the oil pan.

34. Tighten the oil pan bolts to 11 ft. lbs. (15 Nm).

35. Install the A/C compressor and alternator.

36. Install the subframe and bolts. Tighten the subframe bolts as follows:

    a. Front bolts to 85 ft. lbs. (115 Nm).

    b. Center bolts to 140 ft. lbs. (190 Nm).

    c. Rear bolts to 81 ft. lbs. (110 Nm) plus 75° additional torque.

37. Install crankshaft sprocket and tighten the center crankshaft bolt to 184 ft. lbs. (250 Nm) plus 45° additional torque.

38. Remove flywheel stop.
39. Install the front spoiler and connect the front exhaust pipe to the exhaust manifolds.
40. Lower the vehicle and remove the engine lifting beam.
41. Install the inner timing cover bolts.
42. Install the bracket with the upper timing belt adjuster and tensioner pulleys.
43. Install and locking tools 83–94–926 (or equivalent) between the camshaft sprockets to lock the camshafts of both heads in position.
44. Rotate the crankshaft forward to just before 0° TDC and install crankshaft locking tool 83–94–926 (or equivalent) on the crankshaft. Carefully rotate the engine until the arm of the tool is against the water pump flange. Be sure the crankshaft is at 0° TDC and all timing marks are aligned. Remove the locking tool.
45. If reusing the belt, install the timing belt according to its marked direction of rotation and timing marks. Adjust the tensioning roller loosely by hand to prevent the belt from slipping out of the cogs. Always adjust counterclockwise.
46. Measure the belt tension with tool 83–93–985.
47. Tighten the center bolts of the adjusting roller lightly. Adjust the adjusting rollers counterclockwise. Begin with the lower roller and adjust it to a belt tension of 275–300 Nm.

➡ **Adjustment of the belt tension is only a preparatory measure and must not be used as a check when the belt is finally adjusted.**

48. Continue to carry out the adjustment by means of the tensioning roller, mark against mark. Remove the locking tool for camshaft sprockets 1 and 2. Carry out the final adjustment with the upper center adjusting roller until camshaft sprocket No. 2 moves 0.04–0.08 in. (1–2mm) forward.
49. Remove the locking tool for camshaft sprockets 3 and 4 and also remove the crankshaft locking tool.
50. Tighten the tensioning roller to 15 ft. lbs. (20 Nm). Tighten the upper adjusting roller to 30 ft. lbs. (40 Nm) and tighten the lower adjusting roller to 15 ft. lbs. (20 Nm).
51. Turn the engine over two revolutions to the zero mark and refit the locking tool on the crankshaft. Check that the markings on the camshaft sprockets are in alignment with the markings on the timing cover. Check the positioning by installing the two camshaft locking tools, which should fit, and also by fitting tool 83–94–926 on

camshaft sprockets 1 and 2 and 3 and 4. Also check the tensioning roller to ensure that the marks are still in alignment.
52. Install the crankshaft pulley and tighten the retaining bolts to 15 ft. lbs. (20 Nm).
53. Install the timing belt cover tighten the bolts to 6 ft. lbs. (8 Nm).
54. Install the water pump pulley and tighten the retaining bolts to 6 ft. lbs. (8 Nm).
55. Install the power steering pump pulley and tighten the retaining bolts to 6 ft. lbs. (8 Nm).
56. Install the forward cover in the wheel housing and the right-front wheel. Tighten wheel bolts to 89 ft. lbs. (120 Nm).
57. Using a 15mm wrench turn the serpentine drive belt tensioner towards the front of the vehicle, position the belt and slowly release the belt tensioner.
58. Install the air cleaner complete with hoses.
59. Connect the negative battery cable.
60. Change the engine oil filter and fill the engine with oil.
61. Check all fluid levels and top off or change as necessary.
62. Start the engine and check the ignition timing.

9000

2.3L ENGINES

1. Remove air filter assembly.
2. Using a ⅜ in. ratchet extension release the pressure on the drive belt. Place a short drill bit in tensioner housing to hold in the non-tension position.
3. Raise the vehicle and support on safety stands remove left front wheel.
4. Remove crankshaft access cover. Remove crankshaft pulley.
5. Remove circlip to release oil pump. Lift off oil pump cover and extract oil pump gears.

To install:

6. Insert oil pump gears making sure that the marking on the oil pump ring gear faces outward.
7. Install the oil pump cover and fit a new circlip with the chamfer facing outward.
8. Install crankshaft pulley and tighten bolt to 130 ft. lbs. (175 Nm).
9. Install crankshaft cover and fit left front wheel.
10. Lower vehicle and release tensioner to fit drive belt.
11. Install air filter assembly.
12. Test drive vehicle and check drive belt is seated properly.

3.0L ENGINE

1. Disconnect the negative battery cable.
2. Safely raise and support the vehicle.
3. Remove the lower center air deflector. Drain the engine coolant into a catch pan.
4. Lower the vehicle and remove the top engine covers.
5. Lift up the power steering reservoir.
6. Disconnect the connection on the torque arm engine mount and remove the torque arm.
7. Remove the power steering line clamp from the torque arm engine mount and remove the engine mount.
8. Disconnect the hose from the coolant expansion tank and remove the upper alternator air intake.
9. Loosen the power steering pump, water pump pulley and 6 outer crankshaft pulley bolts. Remove the drive belt and tensioner. Remove the power steering pump, water pump pulley and timing cover.

➡ **When removing the crankshaft pulley, remove the 6 outer bolts only, DO NOT remove the center bolt.**

10. Remove the crankshaft pulley.
11. Remove the timing belt. Loosen the timing belt adjuster bolts.
12. Rotate the crankshaft back to 60° BTDC, to prevent damage to the valves.
13. Remove the bracket with the upper timing belt adjuster and tensioner pulleys.
14. Remove the water pump.

➡ **Keep a drain pan under the engine to catch any coolant draining out.**

15. Install camshaft locking tools (KM-800–1 and KM-800–2 or equivalents) between the camshaft sprockets, then remove the camshaft sprockets and inner timing covers.
16. Safely raise and support the vehicle. Disconnect the front exhaust pipe from the exhaust manifolds.
17. Remove the flywheel inspection cover and install holding tool (83–95–063 or equivalent) onto the flywheel.
18. Remove the crankshaft center bolt and crankshaft sprocket.
19. Remove the alternator.
20. Remove the A/C compressor and hang it aside.
21. Drain the engine oil. Remove the oil pan and strainer.
22. Disconnect the oil pressure switch.
23. Lower the vehicle and place a stand and wood block under the front edge of the engine block.

24. Remove the A/C compressor bracket.

25. Remove the oil pump housing.

26. Remove the oil pump housing cover and two oil pump impellers.

27. Inspect the oil pressure relief valve and oil pressure control valve.

To install:

28. Install the oil relief and control valves. Tighten the valves as follows:

   a. With a copper gasket, tighten to 21 ft. lbs. (28 Nm).

   b. With an aluminum gasket, tighten to 15 ft. lbs. (20 Nm).

29. Install the oil pump impellers into the housing. Apply a thread locking compound to the cover bolts and tighten the cover bolts to 53 inch lbs. (6 Nm).

➡**Be sure the alignment marks on the impellers are properly aligned.**

30. Position a new gasket and install the oil pump housing. Tighten the housing bolts to 53 inch lbs. (6 Nm).

31. Install the A/C compressor bracket.

32. Raise the vehicle and remove the stand and wood block from under the engine block.

33. Connect the oil pressure switch.

34. Apply a sealing compound to the areas where the engine block mate to the oil pump and rear main bearing cap.

35. Install the oil strainer. Apply a thread locking compound to the bolts and tighten the bolts to 6 ft. lbs. (8 Nm).

36. Install a new oil pan gasket. Install the oil pan to the engine block.

37. Using a bolt locking compound install oil pan retaining bolts and tighten to 11 ft. lbs. (15 Nm).

38. Install oil pan drain plug and tighten to 41 ft. lbs. (55 Nm).

39. Install the A/C compressor.

40. Install the alternator.

41. Install the crankshaft sprocket and tighten the crankshaft center bolt to 184 ft. lbs. (250 Nm) plus 45° additional bolt rotation.

42. Remove the crankshaft holding tool and install the flywheel inspection cover.

43. Connect the front exhaust pipe to the exhaust manifolds.

44. Be sure the radiator drain plug is installed properly. Install the lower center air deflector and lower the vehicle.

45. Install the inner timing covers and camshaft sprockets.

46. Install the bracket with the upper timing belt adjuster and tensioner pulleys.

47. Install the timing belt.

48. Clean water pump mounting surface. Coat the water pump O-ring and sealing surface with acid-free petroleum jelly.

49. Install water pump with new O-ring and tighten the water pump bolts to 18 ft. lbs. (25 Nm)

50. Install the crankshaft pulley and tighten the bolts to 15 ft. lbs. (20 Nm).

51. Install the timing cover.

52. Install the water pump pulley. Tighten the water pump pulley bolts to 6 ft. lbs. (8 Nm).

53. Install power steering pump.

54. Install the drive belt tensioner and tighten the bolts to 30 ft. lbs. (40 Nm).

55. Install the upper alternator air intake.

56. Install the torque arm engine mount and attach the power steering line clamp.

57. Install the torque arm and bolt the connection to the torque arm engine mount.

58. Connect the upper hose to the coolant expansion tank and set the power steering reservoir back into position.

59. Change the oil filter and fill the engine with oil.

60. Be sure the radiator drain plug is tightened and install the lower center air deflector.

61. Fill the radiator with clean coolant and check the cooling system for leaks

62. Check all other fluids and top off or change as necessary.

63. Bleed the cooling system.

64. Test drive the vehicle and check for any leaks.

Timing Chain, Sprockets, Front Cover and Seal

REMOVAL & INSTALLATION

2.0L Engine

1. Disconnect the negative battery cable.

2. Drain the engine oil and the coolant.

3. Remove the bracket for the steering servo pump, complete with the pump and alternator.

4. Remove the belt tensioner and the water pump pipe.

5. Secure the flywheel with a flywheel locking segment, tool 83–92–987, and loosen the crankshaft pulley bolt.

6. Using a suitable puller remove the crankshaft pulley.

7. Remove the oil pipes and the water pump pulley.

8. Remove the bolts and timing chain cover.

9. Using a suitable seal driver, tap out the oil seal.

10. The cam chain and crankshaft timing sprocket should now both be visible. From above, release the timing chain tensioner by pressing the pivoting guide firmly against it. Remove the chain tensioner.

11. Using a wrench to hold the cast hex bolt on the camshafts, remove the center bolts securing the camshaft sprockets. Throughout this procedure, keep the camshafts in their basic correct setting. If they are rotated out of position at any stage, especially without their sprockets and chain, the valves can be damaged.

12. Disconnect the timing chain from the sprockets and remove the chain, clearing it from the crankshaft sprockets.

To install:

13. Using a suitable seal installer, tap the new seal into the front cover until it is flush with the casting.

14. To install the timing chain, place the chain around the crankshaft sprocket. Run the chain up through the opening in the cylinder head. Install the chain and sprocket on the exhaust cam first. Be sure the chain is taut between the crankshaft and camshaft

1. Timing chain front cover
2. Oil pump assembly
3. Crankshaft pulley

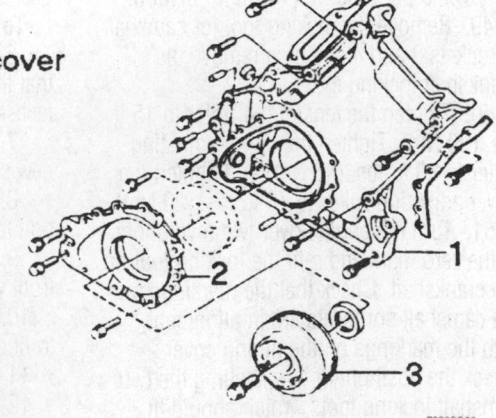

Exploded view of the timing chain front cover assembly—2.0L engine

7923SG09

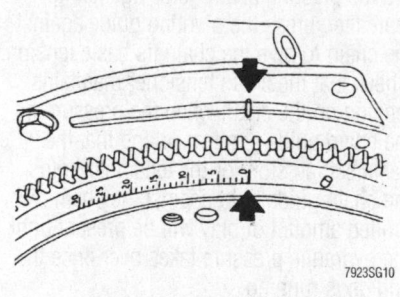

Be sure to align the crankshaft timing mark, as shown—2.0L engine

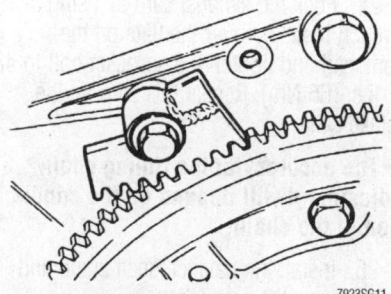

Secure the engine in position using flywheel locking tool 83–92–987—2.0L engine

Unthread the timing chain tensioner to remove it from the front cover—2.0L engine

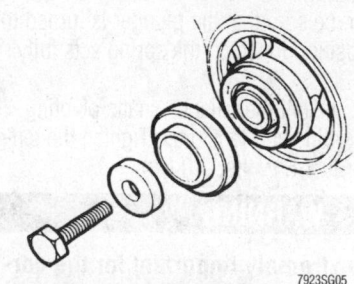

Installing the front crankshaft seal using installation tool 83–94–876—2.0L and 2.3L engines

sprockets. Install and snug the bolt but do not tighten.

15. Install the chain and sprocket to the intake cam. Keep the chain taut between the cam sprockets while it is being installed. Install and snug the bolt but do not tighten. Be sure the chain is seated in the guide tensioner grooves.

16. Tension the chain tensioner by fully depressing the piston, then rotating it to the locked position.

17. Install the chain tensioner with the piston under tension. Be sure the copper gasket is in good condition and that the sealing surface is clean and free from burrs. Tighten the tensioner to 47 ft. lbs. (63 Nm).

18. Trigger the chain tensioner by pressing the pivoting chain guide against it, thereafter, press the pivoting guide against the chain to give the chain its basic tension. Check that the chain tensioner maintains tension on the chain when the pressure on the chain guide is released and that the basic setting stop for the tensioner holds the chain guide tight against the chain. A limited amount of play will be present until the hydraulic pressure takes over once the engine is running.

19. Check the setting by rotating the crankshaft 2 complete turns in its normal direction of rotation around to the timing mark. The basic setting of the cams should remain unaltered.

20. Lock the exhaust cam by using a wrench on the cast hex bolt and tighten the sprocket bolt to 48 ft. lbs. (65 Nm). Repeat this on the intake cam.

21. Complete the procedure on the intake cam sprocket. When loosening or tightening the sprocket center bolts, hold the cam still using a wrench installed over the flats on the camshaft.

22. Install the timing cover and tighten the bolts to 15 ft. lbs. (20 Nm).

23. Install the oil pump and pipes and water pump pulley. Install the crankshaft pulley and tighten the nut to 140 ft. lbs. (190 Nm).

24. Install the belt tensioner, water pump pipe, steering servo pump and alternator.

25. Fill the radiator with coolant and the crankcase with oil.

26. Start the engine and check for leaks.

2.3L Engines

1. Disconnect the negative battery cable.

2. Raise and safely support the vehicle.

3. Lock the flywheel using tool 83–93–993, or equivalent.

4. Drain the coolant and the oil. Remove the right front wheel and wheel well.

5. Remove the serpentine belt and belt tensioner. Remove the tie bar between the wheel arch and the subframe.

6. Remove the steering servo pump, pump bracket and the alternator. Remove the top engine mounting bracket. Remove the torque arm.

7. Remove the top engine mounting bracket. Note the location of the bolts as the bolts are of different lengths.

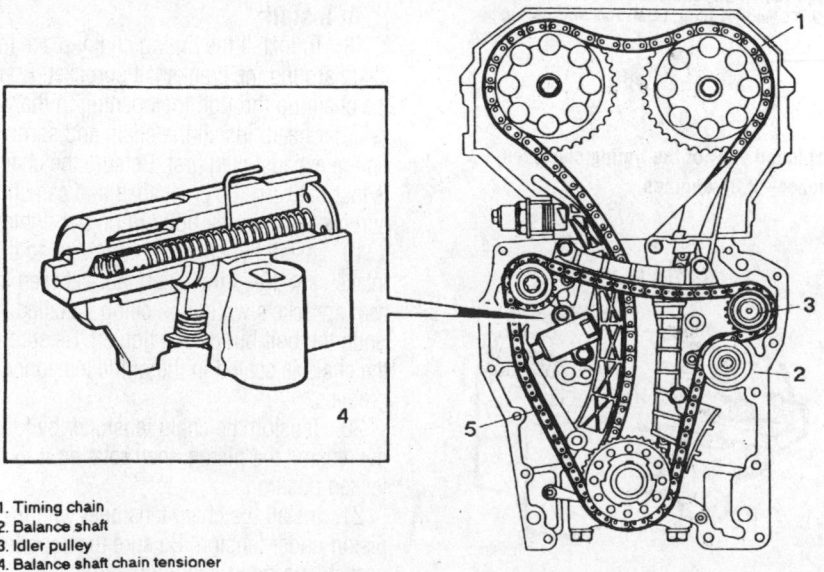

1. Timing chain
2. Balance shaft
3. Idler pulley
4. Balance shaft chain tensioner
5. Balance shaft chain

Timing and balance shaft chain assemblies' component identification—2.3L engines

8. Remove the top belt tensioner bracket, Without disconnecting any hoses, remove the air conditioner compressor and compressor bracket. Use a suitable rigid cover to prevent the oil cooler from being damaged.

9. Disconnect the coolant hoses and remove the water pump. Remove the crankshaft pulley and position the crankshaft sensor out of the way.

10. Move the coolant pipe aside and remove the oil pan. Remove the timing cover securing bolts. Note the locations of all bolts as they are of different lengths.

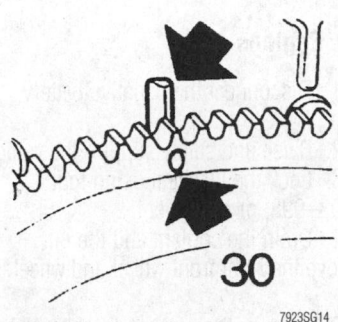

30

7923SG14

Be sure to align the flywheel timing mark, as shown—2.3L engines

3

1. Push down on the catch
2. Push in on tensioner arm
3. Tensioner plug, push rod and spring

7923SG15

Exploded view of the timing chain tensioner—2.3L engines

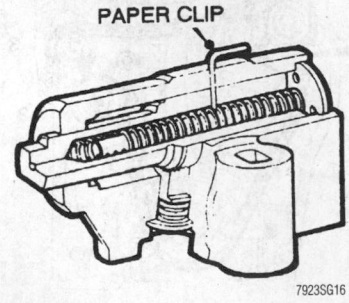

PAPER CLIP

7923SG16

Preparing the balance shaft's tensioner for installation—2.3L engine

Remove the bolts securing the timing cover to the cylinder head.

11. Carefully tap the cover off of the guide pin and remove the timing cover.

12. Remove the lid on the valve cover and remove the ignition wires. Remove the valve cover.

➡ **To remove or refit the timing chain, the camshafts and crankshaft must be lined up with their respective timing marks (No. 1 cylinder at top dead center).**

13. Remove the top chain guide and chain tensioner for the balance shaft chain.

14. Remove the idler sprocket and balance shaft chain.

❊❊ **WARNING**

Throughout this procedure, keep the camshafts in their basic correct setting. If they are rotated out of position at any stage, especially without their sprockets and chain, the valves can be damaged.

15. The camshaft chain and crankshaft timing sprocket should both be visible. From above, release the timing chain tensioner by pressing the pivoting guide firmly against it. Remove the chain tensioner.

16. Using a wrench installed over the flats on the exhaust camshaft, hold the camshaft and remove the center bolt securing the camshaft sprocket. Repeat this step for the intake camshaft.

17. Disconnect the timing chain from the sprockets and remove the chain, clearing it from the crankshaft sprockets.

To install:

18. To install the timing chain, place the chain around the crankshaft sprocket. Run the chain up through the opening in the cylinder head. Install the chain and sprocket on the exhaust cam first. Be sure the chain is taut between the crankshaft and camshaft sprockets. Snug the bolt but do not tighten.

19. Install the chain and sprocket to the intake cam. Keep the chain taut between the cam sprockets while it is being installed. Snug the bolt but do not tighten. Be sure the chain is seated in the guide tensioner grooves.

20. Tension the chain tensioner by fully depressing the piston, then rotating it to the locked position.

21. Install the chain tensioner with the piston under tension. Be sure the copper gasket is in good condition and that the sealing surface is clean and free from burrs.

22. Trigger the chain tensioner by press-

ing the pivoting chain guide against it, thereafter, press the pivoting guide against the chain to give the chain its basic tension. Check that the chain tensioner maintains tension on the chain when the pressure on the chain guide is released and that the basic setting stop for the tensioner holds the chain guide tight against the chain. A limited amount of play will be present until the hydraulic pressure takes over once the engine is running.

23. Check the setting by rotating the crankshaft 2 complete turns in its normal direction of rotation around to the timing mark. The basic setting of the cams should remain unaltered.

24. Lock the exhaust cam by using a wrench installed over the flats on the camshaft and tighten the sprocket bolt to 48 ft. lbs. (65 Nm). Repeat this step on the intake cam.

➡ **The accuracy of the timing chain adjustment will depend on the condition of the chain.**

25. Install the balance shaft chain and sprocket on the crankshaft.

26. Install the oil pump drive dog.

27. Install the balance shaft chain and idler wheel sprocket, ensuring that the aligning marks on the bearing housing and sprocket are in line. When installing, leave some slack in the chain in line with the tensioner, and keep the chain reasonably taut by means of the top chain guide.

➡ **There is an alternate way of installing the balance shaft chain. Install the top chain guide first, then adjust the run of the chain around the sprockets. Adjusting the chain is easier this way, although it will be more awkward to install the idler wheel sprocket.**

28. Cock the balance shaft chain tensioner and insert a paper clip through the hole in the cylinder to prevent the tensioner being triggered. Before installing the tensioner, be sure that the plunger is turned to the position in which the spring acts fully on it.

29. Install the balance shafts pivoting chain guide and tensioner. Tighten the tensioner to 7.5 ft. lbs. (10 Nm).

❊❊ **WARNING**

It is extremely important for the correct torque to be applied when installing the tensioner.

30. Install the top chain guide and trigger the tensioner by removing the paper clip.

31. Rotate the crankshaft a few times to ensure the balance shafts chain is installed correctly.

32. Ensure that the timing cover flange is absolutely clean.

33. Remove all traces of old sealant from the cover. Apply a 0.04 in. (1mm) bead of 45–3028972 sealant or equivalent to the flanges of the cover. Use sealant sparingly as excess sealant can get into the oil ways and do serious damage to the engine.

34. Install the timing cover taking care not to damage the head gasket. Install the bolts in their correct positions and tighten to 15 ft. lbs. (20 Nm).

35. Apply an even bead of Permatex® Ultra Blue sealant and fit the rubber seal for the oil strainer in the groove on the sump. Install the oil pan and tighten the bolts to 15 ft. lbs. (20 Nm).

36. Install the crankshaft pulley and tighten the retaining bolt to 140 ft. lbs. (190 Nm).

37. Install the coolant pipe and secure the crankshaft sensor.

38. Install the engine mountings, water pump and cooling hoses, air conditioner compressor, steering servo pump, pump bracket and the alternator. Install the top engine mounting bracket and torque arm.

39. Install the serpentine belt and belt tensioner. Install the tie bar between the wheel arch and the subframe.

40. Install the wheel well and wheel.

41. Fill the radiator with coolant and the engine with oil.

42. Connect the negative battery cable.

43. Start the engine and allow it to reach normal operating temperature and check for leaks.

FUEL SYSTEM

Fuel System Service Precautions

Safety is the most important factor when performing not only fuel system maintenance but any type of maintenance. Failure to conduct maintenance and repairs in a safe manner may result in serious personal injury or death. Maintenance and testing of the vehicle's fuel system components can be accomplished safely and effectively by adhering to the following rules and guidelines.

• To avoid the possibility of fire and personal injury, always disconnect the negative battery cable unless the repair or test proce-

dure requires that battery voltage be applied.

• Always relieve the fuel system pressure prior to disconnecting any fuel system component (injector, fuel rail, pressure regulator, etc.), fitting or fuel line connection. Exercise extreme caution whenever relieving fuel system pressure to avoid exposing skin, face and eyes to fuel spray. Please be advised that fuel under pressure may penetrate the skin or any part of the body that it contacts.

• Always place a shop towel or cloth around the fitting or connection prior to loosening to absorb any excess fuel due to spillage. Ensure that all fuel spillage (should it occur) is quickly removed from engine surfaces. Ensure that all fuel soaked cloths or towels are deposited into a suitable waste container.

• Always keep a dry chemical (Class B) fire extinguisher near the work area.

• Do not allow fuel spray or fuel vapors to come into contact with a spark or open flame.

• Always use a back-up wrench when loosening and tightening fuel line connection fittings. This will prevent unnecessary stress and torsion to fuel line piping. Always follow the proper torque specifications.

• Always replace worn fuel fitting O-rings with new. Do not substitute fuel hose or equivalent, where fuel pipe is installed.

Fuel System Pressure

RELIEVING

All Engines

✳✳ CAUTION

The fuel injection system remains under pressure, even after the engine has been turned OFF. The fuel system pressure must be relieved before disconnecting any fuel lines. Failure to do so may result in fire and/or personal injury.

1. Disconnect the negative battery cable.

2. Place clean shop rags under and around the fuel line banjo coupling to absorb any released fuel or fuel spray.

3. With the ignition switch in the **OFF** position, carefully relieve the fuel system pressure by firmly cracking open the banjo coupling at the inlet to the fuel injection manifold.

4. Be sure to always replace the banjo

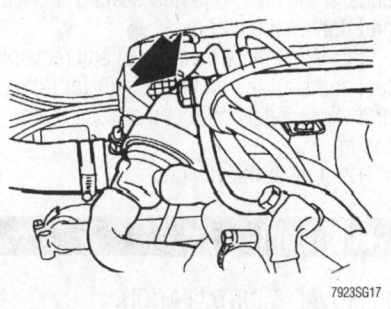

Location of fitting on fuel rail to relieve fuel pressure—900 and 9-3

coupling washers whenever you loosen or remove the banjo couplings.

Fuel Filter

REMOVAL & INSTALLATION

All Models

✳✳ CAUTION

The fuel injection system remains under pressure, even after the engine has been tuned OFF. The fuel system pressure must be relieved before disconnecting any fuel lines. Failure to do so may result in fire and/or personal injury.

1. Raise and safely support the vehicle.

2. Locate the fuel filter which is mounted under the vehicle, forward of the fuel tank.

3. Be sure that the ignition switch is in the **OFF** position.

4. Thoroughly clean the area around the banjo fittings before continuing.

5. Relieve the fuel pressure by placing a shop towel around one of the banjo fittings on the fuel filter and crack open the fitting until the pressure is relieved.

6. Once relieved, continue to remove the banjo fittings on both sides of the fuel filter, containing the fuel with the shop towel. Properly dispose of the towel once the job is complete.

7. Loosen the band-type clamp so the old filter can be removed.

To install:

8. Position the new fuel filter, making sure that the arrow is pointing in the correct direction of fuel flow.

9. Properly position the new filter and tighten the clamp.

10. Reinstall the banjo fitting to both

ends of the filter. Use new sealing washers and tighten the fittings.

11. Wipe the area clean of any remaining fuel and start the engine. Check for leakage and immediately correct if any leaks are found.

12. Lower the vehicle.

Fuel Pump

REMOVAL & INSTALLATION

900 and 9-3

> ❄❄ **CAUTION**
>
> **The fuel injection system remains under pressure, even after the engine has been tuned OFF. The fuel system pressure must be relieved before disconnecting any fuel lines. Failure to do so may result in fire and/or personal injury.**

1. Disconnect the negative battery cable. Tape or tie off the cable so it cannot contact the battery terminals.

2. Remove the carpet and floor panel in the trunk.

3. Remove the circular cover plate to gain access to the fuel pump.

4. Disconnect the electrical wiring at the fuel pump and move aside.

5. Relieve the fuel system pressure.

6. Remove the screw and protective plate over the fuel line fittings.

7. Remove the fuel lines at the fuel pump using angled needle-nose pliers or similar tool. Use a shop towel to absorb any fuel released.

8. Move the fuel lines and wiring harness to the side.

9. Place Saab tool 83–94–397 or equivalent over the fuel pump. Place the chain supplied with the tool through the two load securing brackets on the trunk floor

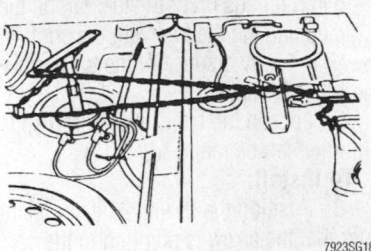

Positioning of tool and chains for tightening the retaining ring—900 and 9-3

and secure each end with a screwdriver or similar tool.

10. Loosen the large threaded ring that holds the pump in the tank using a wrench on the shaft of the tool.

11. Remove the chains and tool. Finish removal of the ring by hand.

12. Remove the rubber seal and remove the pump by tilting the top part of it forward and to the left. Allow the fuel to run out of the pump before removing it from the trunk.

13. Wipe up any fuel spilled.

To install:

14. Wipe clean the sealing area of the opening in the fuel tank.

15. Carefully place the pump into the tank to prevent damage to the fuel level float arm.

16. Once in the tank, line up the marks on the tank and pump.

17. Place a new seal with a thin coat of petroleum jelly into the large threaded ring and place on top of the fuel pump with the mark on the ring lined up with the mark on the fuel tank.

18. Press down hard and rotate the ring 90°.

19. Reinstall special tool 83–94–397 or equivalent, with the chain situated for tightening.

20. Tighten the ring until it will no longer turn.

21. Check that the pump did not rotate more than 30° from the alignment marks.

22. Connect the fuel lines using new O-rings. The return line is connected to the nipple closest to the front of the vehicle while the feed line is connected closest to the rear of the vehicle.

23. Connect the wiring and reinstall the protective plate over the fuel line fittings.

24. Reconnect the negative battery cable and check that the system is operating properly and that there are no fuel leaks.

25. Reinstall the access plate in the trunk floor. Pay careful attention that the plate seals properly so that no exhaust fumes can enter the cabin.

26. Reinstall the floor panel and carpet.

9000

> ❄❄ **CAUTION**
>
> **The fuel injection system remains under pressure, even after the engine has been tuned OFF. The fuel system pressure must be relieved before disconnecting any fuel lines. Failure to do so may result in fire and/or personal injury.**

1. Disconnect the negative battery cable. Tape or tie off the cable so that it cannot be reconnected during this procedure.

2. Relieve the fuel system pressure.

3. Gain access to the trunk floor. Remove the carpet in the trunk.

4. Remove the 2 screws securing the floor panel and remove the panel.

5. Remove the oblong cover plate to gain access to the fuel pump.

6. Release the clip and remove the electrical harness connector. Move the cover aside.

7. Place a shop towel around the fuel lines at the fuel pump and disconnect the lines. Place the fuel lines off to the side.

8. Soak up any fuel released with the shop towel.

9. Using Saab tool 83–94–462 or similar device, remove the large threaded ring that secures the pump in the tank.

10. Lift the fuel pump out of the tank by tilting the top to the right, being careful not to damage the arm for the fuel level sender.

11. Soak up any spilled fuel.

To install:

12. Clean the groove on the tank and place a new O-ring in the groove.

13. Place the fuel pump into the tank.

14. Line up the mark on the pump with the mark on the tank. This must be done for the fuel sender to operate properly.

15. Place the large threaded ring over the pump and tighten using tool 83–94–462 or equivalent, to 40 ft. lb. (55 Nm).

16. Check that the alignment of the marks are still in line. A tolerance of 5° is permissible.

17. Replace the O-rings on the inside of each fuel line, then reconnect to the fuel pump.

18. Reinstall the electrical connector and the clip.

19. Connect the negative battery cable.

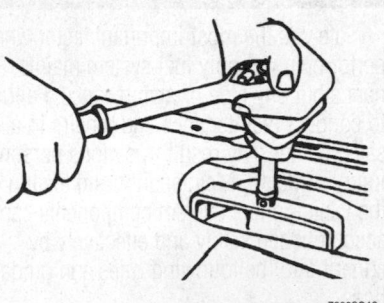

Removing threaded ring using Saab tool—9000

Check that the pump is operating properly and there are no leaks.

20. Reinstall the access plate in the trunk floor. Pay careful attention that the plate seals properly so that no exhaust fumes can enter the cabin.

21. Reinstall the floor panel, carpet and any accessories removed.

DRIVE TRAIN

Transaxle Assembly

REMOVAL & INSTALLATION

Manual

900 AND 9–3

➡**The engine and transaxle are removed together as an assembly.**

1. With all four wheels on the ground, loosen the axle hub nuts.
2. Drain the engine coolant and remove the battery.
3. Remove air cleaner assembly and attached hoses. Take off the resonator.
4. Disconnect throttle cable and move it to one side.
5. Disconnect cruise control wiring harness and cables from throttle body. Loosen retaining nuts on cruise control unit and remove.
6. Properly relieve the fuel system pressure.

✳✳ CAUTION

The fuel injection system remains under pressure even after the engine is stopped. The fuel system pressure must be relieved before disconnecting any fuel lines. Failure to do so may result in fire and/or personal injury.

7. Disconnect the fuel lines at their connections at the front of the fuel injection manifold and on the fuel pressure regulator.
8. If equipped, disconnect the turbocharger pressure line from the turbocharger compressor and the intercooler/throttle housing.
9. Disconnect the vacuum hose from the secondary injection and disconnect the brake booster vacuum hose from the intake manifold.
10. If equipped, remove the boost pressure control unit.

11. Using a ⅜ in. ratchet extension relieve belt tensioner of pressure. Remove drive belt.
12. Disconnect the pressure and return pipe from the steering servo pump and have a plug handy to prevent oil escaping from the pipe.
13. Disconnect the cooling system hoses at the following connections:
   a. the heater control valve.
   b. the expansion tank.
   c. the bottom of the radiator.
   d. the thermostat housing.
14. Without disconnecting any hoses, remove the air conditioner compressor and it mounting bracket. Place them on the filter housing for the heater system. Secure the alternator so it will not drop or become damaged.
15. If not already done, raise and safely support the vehicle.
16. Disconnect the positive battery cable from the clips holding it to the body. Disconnect the ground cable from the transaxle.
17. Remove ignition cable and disconnect electrical connections from ignition coil.
18. Unplug the wiring connectors from the transaxle.
19. Detach the oxygen sensor wiring and the catalytic converter temperature sensor connector.
20. Under the vehicle, remove the taper pin from the gearshift rod joint. If equipped, disconnect the clutch cable and clutch pipe.
21. Inside the vehicle, pull back carpet under glove box to gain access to the central locking system and disconnect the control module. Feed the wires to the engine compartment through the grommet.
22. Remove hub nuts and remove both wheels. Remove front splash shield.
23. Remove both ball joint nuts and disconnect the ball joints from the struts. Be sure the halfshafts can slide out of the hubs. If necessary, use a wheel puller to push them out now.
24. Separate front pipe from exhaust manifold and remove the catalytic converter.
25. Disconnect oil cooler lines and all lines connected to transaxle housing.
26. Position lifting table under vehicle so it is directly under front engine mount and gearbox. Take off subframe bolts and front engine mounts.
27. Lower lifting table slightly and separate the halfshafts from the hubs.
28. Lower lifting table fully and remove subframe.
29. Place engine on engine stand.

To install:

30. Place engine/transaxle on lifting table.
31. Install subframe on rear engine mount and tighten to 35 ft. lbs. Fit half-shafts into hubs as engine is raised into engine compartment.
32. Connect engine mount and subframe bolts and tighten in sequence:
   • Front bolts-85 ft. lbs.
   • Middle subframe bolts-141 ft. lbs.
   • Rear subframe bolts- to 90 ft. lbs.
   • Rear engine mount-54 ft. lbs.
33. Connect all oil cooler and hydraulic lines.
34. Reconnect all charge air cooler lines between the turbocharger and the cooler.
35. Install the front exhaust pipe to the exhaust manifold. Secure the catalytic converter to the rear exhaust section.
36. Fit engine shield to under side of vehicle.
37. Install hub nuts and wheels. Tighten wheel nuts to 89 ft. lbs. but do not tighten hub nuts yet.
38. Feed the wiring through the grommet and reconnect the central locking system control module.
39. As equipped, hook the clutch cable in place or bleed the clutch hydraulic cylinder.
40. Reconnect the taper pin to the gearshift selector.
41. Connect the oxygen and catalytic converter temperature sensor.
42. Connect all transaxle wiring.
43. Install A/C compressor, servo pump and the drive belt.
44. Reconnect power brake booster vacuum hose and pressure sensor connections.
45. Install cruise control including cables and electrical harness connections.
46. Install the battery and connect the cables.
47. With all four wheels on the ground, tighten the hub nuts to 215 ft. lbs. (290 Nm).
48. Fill coolant and oil to specified level and check.
49. Start engine and test drive vehicle.

9000 WITH GM65103 AND GM75701 TRANSAXLES

1. Disconnect the battery cables from the battery, negative cable first.
2. Remove the battery, washer fluid container and connectors, terminal blocks and battery tray. Release the stay for the hydraulic unit on ABS equipped vehicles.
3. Remove the 8 bolts for the bulkhead cover. Remove the rubber strip, lift the cover

and disconnect the washer hoses from the nozzle and remove the cover.

4. Separate the speedometer cable connector by first removing the washer hose, then the speedometer cable through the rubber grommet.

5. Detach the electrical connector from the air mass meter, and the intake air temperature sensor, then the hose on the delivery pipe from the bypass valve.

6. Remove the delivery pipe between the throttle housing and the intercooler. Also, remove the nuts retaining the starter motor. Remove the starter and place it on the steering gear.

7. Separate the gear selector rod universal joint and selector rod. Disconnect the slave cylinder pressure hose.

8. Remove the upper bolts for the stay at the wheel housing. Release the left-hand engine mounting. Attach a sling to the engine lifting beam. Raise and support the vehicle safely.

9. Remove the left wheel and wheel housing liner. Detach the reverse light connector from the transaxle.

10. Separate the suspension arm from the ball joint. Remove the sway bar. Remove the lower bolt for the stay at the wheel housing and the 3 bottom bolts holding the engine to the transaxle.

11. Remove the center and left skirts under the spoiler. Separate the subframe at the front and rear. Lower the subframe. Remove the universal joint and lower the vehicle.

12. Sling the transaxle from the work-shop hoist and remove the top nut and bolt. Remove the transaxle and separate the half-shafts. Carefully lower the transaxle to the floor.

To install:

13. Prior to installation, ensure that the halfshaft is in position and the aluminum tube is pressed into the seal.

14. Slide the transaxle into position, guiding the driver and input into place. Install the top bolt and nut into the engine/transaxle joint face. Release the transaxle from the hoist. Raise and support the vehicle safely.

15. Fit the 3 bottom bolts into the joint face and tighten to 40–74 ft. lbs. (54–100 Nm). Install the universal joint. Raise the subframe into position and secure to 31–42 ft. lbs. (43–57 Nm).

16. Install the sway bar and tighten to 30–40 ft. lbs. (40–52 Nm).

17. Secure the suspension arm to the ball joint and tighten to 14–20 ft. lbs. (20–27 Nm). Start the bolts to the bracket for the wheel housing stay. Do not tighten the wheel housing stay bolt at this time. Lower the vehicle.

18. Remove the lifting beam. Install the starter and the top mounting of the wheel housing. Now tighten all wheel housing bolts.

19. Tighten the left engine mount 37–67 ft. lbs. (49–91 Nm). Raise and support the vehicle safely.

20. Connect the negative battery cable and reverse light switch. Install the wheel housing liners, left wheel, under vehicle

skirts, selector rod universal joint, slave cylinder pressure pipe (remove the clamping tongs), speedometer cable, washer hose and the remainder of the components removed.

21. Bleed the slave cylinder, check the fluid level and road test the vehicle.

9000 WITH FM51001, FM54001, FM57001, FM57101 AND MX5 TRANSAXLES

1. Place vehicle in 4th gear.

2. Disconnect and remove battery.

3. Remove locking clips which hold the accelerator cable in the lead-though and turn cable to one side.

4. Disconnect and remove the front distribution box.

5. Remove the positive junction (without removing the cables).

6. Remove the positive cable's two clamps on battery shelf.

7. Remove the connect from the control module ABS.

8. Remove battery shelf.

9. Remove the bypass valve from the turbocharger pressure pipe.

10. Disconnect the temperature sensor from the turbocharger pressure pipe.

11. Remove the hose clips from the turbocharger pressure pipe.

12. Pull apart the connector on the cable for the speedometer sensor.

13. Remove the reverse light switch connector.

14. Disconnect the clutch slave cylinder hydraulic line.

15. Remove all the engine-to-transaxle bolts accessible from the top side, except the top bolt.

16. Place lifting beam on the edges of the engine compartment. Insert the hook into the engine's lifting eye and tighten the lifting beam wing nut slightly to support the engine and transaxle from above.

17. Safely raise and support the vehicle.

18. Remove left front wheel and the edging of the wheel housing.

19. Remove the front part of the inner fender and both left and middle spoiler units.

20. Remove the ground cable from the gear box.

21. Remove the two nuts on the gear selector universal joint and disconnect the shift linkage.

22. Remove the three bolts connecting the ball joint to the lower control arm.

23. Remove the anti-roll bar from the lower control arm.

24. Disconnect the front engine mounting nut from the bolt.

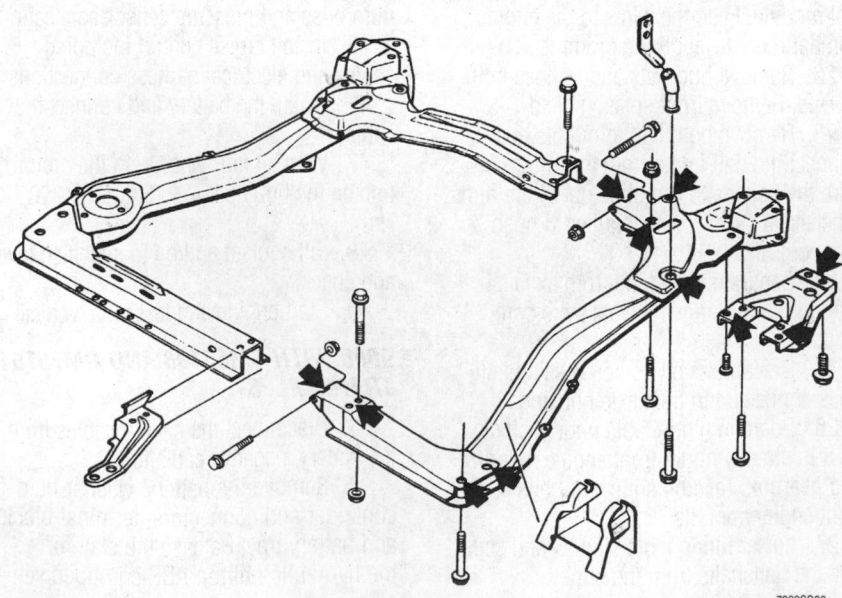

7923SG20

Subframe connection points—9000 models with either GM65103 or GM75701 transaxles

25. Remove the screw from the wheel housing stay and put the two washers in a safe place.

26. The transaxle side of the subframe must be folded down out of the way:

a. Remove the bolt in the front link of the subframe assembly and remove the two bolts that hold the front link.

b. Remove the bolt in the rear link of the subframe assembly.

c. Remove the two bolts that hold the rear link, one of which also holds the steering rack.

d. Remove the two bolts in the front corners of the subframe, then the four bolts in the rear corners.

e. Carefully fold down the subframe. Put the plate between the subframe and the chassis in a safe place.

27. Remove the clamp around the CV-joint boot and disconnect the CV-joint. With the ball joint disconnected, the strut will move with the halfshaft but be careful not to damage the boot.

28. Remove the protective plate for the transaxle.

29. Lower the vehicle.

30. Support the transaxle and remove the remaining engine-to-transaxle bolts. Carefully lower the transaxle out of the vehicle.

To install:

31. Fit two locating pins into the engine block.

32. Lift transaxle into position and onto the locating pins. Push the transaxle against the engine, making sure to align the clutch with the crankshaft, then start some of the bolts. Use the bolts to draw the transaxle up against the engine block.

33. Remove the locating pins and safely raise and support the vehicle.

34. Install the protection plate for the transaxle and install the remaining lower bolts. Tighten the bolts to 50 ft. lbs. (70 Nm).

35. Hang the subframe and secure each link with one bolt. Be sure the plate in the rear corner is positioned correctly.

36. Install the bolt in the front engine mount, making sure the washer is in the correct position.

37. Install all the remaining subframe bolts but do not tighten any of them yet.

38. When all bolts are installed, tighten all the bolts in the subframe to 41 ft. lbs. (55 Nm).

39. Install the bolt to the stay in the wheel housing, tighten to 37 ft. lbs. (50 Nm).

40. Install the bolts which hold the anti-roll bar bearing.

41. Install the nut which holds the anti-roll bar mount in the lower control arm.

42. Connect the halfshaft.

43. Install the three bolts of the lower ball joint to the control arm tighten to 25 ft. lbs. (34 Nm).

44. Install the CV-joint boot clamp.

45. Be sure the transaxle is in fourth gear, fit the two screws in the selector universal joint.

46. Fit the battery grounding cable to the gearbox.

47. Lower the vehicle and remove the lifting beam.

48. Install and tighten all the remaining engine-to transaxle bolts to 50 ft. lbs. (70 Nm).

49. Check the position of the washer in the engine mount and tighten nut to 50 ft. lbs. (70 Nm).

50. Connect the slave cylinder hydraulic line.

51. Attach the reverse light connector.

52. Reconnect the speedometer cable.

53. Install the turbocharger pressure pipe.

54. Fit the hose on the bypass valve.

55. Connect the wiring on the temperature sensor.

56. Install the battery shelf.

57. Connect the wiring on the ABS control module.

58. Fit the positive cable and the two clamps on the battery shelf.

59. Install the front distribution box.

60. Fit the accelerator cable and connect the locking clips in the lead-through and the clip on the cable.

61. Connect the battery cables.

62. Raise and safely support the vehicle.

63. Install the left and middle spoiler units.

64. Install the front part of the inner fender.

65. Install the edging of the wheel housing.

66. Install the left front wheel.

67. Bleed the clutch.

68. Check for any leakage. Check the oil level in the gearbox.

69. Test drive vehicle.

Automatic

900 AND 9–3

1. Disconnect battery and remove from vehicle.

2. Remove ground strap from transaxle housing.

3. Remove dipstick and sleeve, plug access hole in transaxle housing.

4. Remove vent hose from transaxle housing.

5. Disconnect the positive cable routing straps.

6. Remove the transaxle selector lever from housing.

7. Disconnect electrical harness on transaxle housing.

8. Disconnect the oxygen sensor and remove the securing straps.

9. Install the engine lifting beam. Insert the lifting bar in the eyelet's and connect to the engine holder.

10. Safely raise vehicle.

11. Remove front wheels.

12. Remove front exhaust pipes.

13. Remove front spoilers.

14. Disconnect the ball joints on both sides.

15. Using the proper jack remove the subframe assembly to gain access to transaxle housing.

16. Remove transaxle cooler lines from the transaxle.

17. Remove the left and right front drive-shafts and suspend shafts with securing straps.

18. Remove intermediate shaft bearing bracket and pull out shaft.

19. Remove actuator cover.

20. Release the torque converter's securing point on the actuator.

21. Remove transaxle mounting bracket.

22. Remove transaxle mounting bolts that secure the transaxle to the engine.

23. Remove the engine/transaxle surface mounting bolts.

24. Lower vehicle.

25. Remove the two outer bolts holding gear box to engine mounting surface.

26. Install lifting eye to transaxle and connect hoist to transaxle.

27. Remove last bolt and pull out transaxle assembly.

To install:

28. Using a hoist lift transaxle into position.

29. To help with installation use a guide pin to line up transaxle.

30. When transaxle is in position fit the three upper bolts into engine housing.

31. Tighten upper transaxle-to-engine mounting bolts to 55 ft. lbs. (75 Nm).

32. Install transaxle/engine lifting beam to hold unit in place.

33. Safely raise vehicle.

34. Fit transaxle/engine surface mounting bolts from under vehicle and tighten to 55 ft. lbs. (75 Nm).

35. Install transaxle mounting bracket to the transaxle housing and the frame assembly.

36. Install torque converter to the actua-

tor with retaining bolts, use locking fluid and tighten to 37 ft. lbs. (55 Nm).

37. Install actuator cover.

38. Install the intermediate halfshaft with the bracket bearing assembly.

39. Install the outer halfshafts.

40. Install transaxle cooler lines to transaxle assembly.

41. Using a proper jack install sub frame assembly. Fit sub frame bolts and tighten front bolts to 85 ft. lbs. (11 Nm), middle bolts to 141 ft. lbs. (190 Nm), and the rear bolts to 81 ft. lbs. (110 Nm) plus 75° bolt rotation. Tighten ball joints to 55 ft. lbs. (75 Nm).

42. Install front exhaust pipes.

43. Install front wheels.

44. Lower vehicle.

45. Fill transaxle with fluid.

46. Install selector lever cable and selector lever to the transaxle housing.

47. Connect the oxygen sensor electrical harness.

48. Install dipstick and sleeve to transaxle housing.

49. Fit the positive battery cable.

50. Install battery and connect ground cable.

51. Test drive vehicle.

9000

➡ **The engine and transaxle assembly are removed together.**

1. Raise the vehicle and support it safely.

2. Drain the cooling system.

3. Disconnect the battery cables and remove the battery and tray.

4. Remove the connecting bolt for the expansion tank, disconnect the tank from the suction and remove the overflow hoses from the radiator.

5. Loosen the drive belt for the compressor by loosening the locknut, and loosening the adjusting nut under the locknut.

6. Disconnect the upper connection on the oil cooler, loosen the pipe clip on the radiator and slide the pipe down behind the radiator.

7. Unplug the connector to the electromagnetic clutch on the compressor and loosen the compressor mounting complete with the belt tensioner.

8. Place a protective cloth over the radiator member and rest the compressor on the radiator member. Secure the compressor to the radiator member.

9. If turbocharged, remove the turbocharger pressure pipe, situated between the turbocharger unit and the intercooler.

10. Detach the Lambda probe connector leads.

11. From the engine compartment, unbolt the flange joint between the exhaust pipe and the exhaust manifold. If turbocharged, disconnect the exhaust pipe coming from the turbocharger. Push the exhaust pipe to one side and unhook the rubber hangers from the exhaust system. Disconnect the bottom coolant hose from the water pump.

12. From under the vehicle, remove the bottom retaining bolt for the radiator fan.

13. Disconnect the speedometer drive from the gearbox.

14. Disconnect the gear selector cable from the selector lever. Do not separate at the ball joint.

15. Remove the clips on the rubber gaiters over the inboard universal joints and slide the gaiters off the drive axles.

16. Detach the electrical leads from the alternator and the starter motor. Unplug the connector for the oil pressure switch.

17. Remove the clips and remove the top radiator hose.

18. Disconnect the top radiator hose at the cylinder head.

19. Disconnect the high tension lead from the ignition coil at the distributor cap.

20. Remove the solenoid valve from the bracket on the radiator and unplug the electrical connections.

21. Remove the bolts from the top of the radiator fan. Disconnect the wiring loom and lift out the fan.

22. Pull the connector off the air mass meter. Detach the air mass meter from the air intake duct socket connector and the air cleaner. Leave the rubber socket connector attached to the turbocharger unit.

23. Remove the air intake duct by pulling it out of the aperture in the wing and twisting the ends inwards.

24. Remove the air cleaner top section first, then the remaining section.

25. Disconnect the relief valve hose from the turbocharger pressure pipe and remove the pipe, if equipped.

26. Detach the Hall effect transducer, the earth lead from the gear box and the electrical connector for the back-up lights.

27. Disconnect the end of the throttle cable and disconnect the throttle linkage.

28. Install a clamp to the hydraulic line to the slave cylinder and pinch the line tightly. With proper wrenches, open the line to the clutch slave cylinder.

29. Remove the front wheels.

30. From both sides of the vehicle, slacken the lower bolts retaining the steering swivel member to the strut assembly. Remove the 2 upper bolts.

31. Pivot the steering swivel member outwards to pull the inboard universal joint out of the halfshaft. Position dust covers over the exposed halfshaft cups.

32. Remove the engine mount bolt.

33. Remove the steering reservoir for the servo and position it within the engine compartment. Drain the fluid from the container.

34. Disconnect the large bore hose and the delivery hose from the steering servo pump and plug the open ends.

35. Disconnect the fuel return lie from the pressure regulator.

36. Remove the nut from the rear engine mounting and back off the front mount bolts a few turns.

37. Attach the lifting sling (Saab 83–92–409) or equivalent to the rear lifting lug.

38. Lift the engine sufficiently to provide access for the removal of the components located between the engine and the firewall.

39. Disconnect and tag the vacuum hoses from the inlet manifold.

40. Remove the coolant hoses running between the heater core and the water pump pipe.

41. Separate the coupling between the fuel pipe and the fuel injection manifold. Do not allow the fuel to spill or collect.

42. Cut the clips securing the wiring looms to the oil pipe, water pipe. Disconnect and tag the vacuum hoses from the inlet manifold.

43. Unclip the wiring loom to the fuel injection manifold.

44. Detach the grounding connections.

45. Unbolt the air cooled oil cooler and place it on top of the engine. The 2 lower bolts need only be loosened.

46. Disconnect the hood lifts and install hood extenders for extra clearance.

47. Carefully lift the engine from the vehicle, taking care not to damage the radiator.

To install:

48. Install the engine in the vehicle and connect all the engine mounts.

49. Bolt the air cooled oil cooler in place. Tighten the 2 lower bolts.

50. Attach the grounding connections.

51. Clip the wiring loom to the fuel injection manifold.

52. Secure the wiring looms to the oil pipe, water pipe, inlet manifold steady bar and the oil supply pipe.

53. Connect the coupling between the fuel pipe and the fuel injection manifold.

54. Connect the coolant hoses running between the heater core and the water pump pipe.

55. Connect the vacuum hoses to the correct connections on the inlet manifold.

56. Before removing the lifting sling, connect the components located between the engine and the firewall.

57. Attach the nut from the rear engine mounting and tighten the front mount bolts.

58. Connect the fuel return line to the pressure regulator.

59. Connect the large bore hose and the delivery hose to the steering servo pump.

60. Connect the steering reservoir for the servo. Fill the fluid container.

61. Connect the engine stay bolt.

62. Pivot the steering swivel member outwards to install the inboard universal joint into the halfshaft. Position dust covers over the exposed halfshaft cups and install clamps.

63. Connect the gear selector cable to the selector lever.

64. Connect the lower bolts retaining the steering swivel member to the strut assembly. Tighten the 2 upper bolts.

65. Install the front wheels.

66. With line wrenches, tighten the line to the clutch slave cylinder.

67. Connect the end of the throttle cable and the throttle linkage.

68. Attach the Hall Effect transducer, the negative battery cable to the gear box and the electrical connector for the back-up lights.

69. Connect the relief valve hose to the turbocharger pressure pipe, if equipped.

70. Install the air cleaner.

71. Install the air intake duct.

72. Push the connector on the air mass meter. Attach the air mass meter to the air intake duct socket connector and the air cleaner.

73. Connect the high tension lead to the ignition coil at the distributor cap.

74. Connect the top radiator hose at the cylinder head.

75. Attach the electrical leads to the alternator and the starter motor. Plug in the connector for the oil pressure switch.

76. Connect the speedometer drive to the gearbox.

77. From the engine compartment, connect the flange joint between the exhaust pipe and the exhaust manifold. If turbocharged, connect the exhaust pipe coming from the turbocharger. Hook the rubber hangers to the exhaust system.

78. Connect the bottom coolant hose to the water pump.

79. Attach the Lambda probe connector leads.

80. Plug the connector into the electro-

magnetic clutch on the compressor and tighten the compressor mounting complete with the belt tensioner.

81. Connect the upper connection on the oil cooler, tighten the pipe clip on the radiator.

82. Tighten the drive belt for the compressor.

83. Install the expansion tank, connect the tank to the suction and overflow hoses from the radiator.

84. Install the battery and tray. Connect the battery cables.

85. Remove the lift supports and lower the vehicle.

86. Fill the radiator with coolant, the engine with oil and the transmission with fluid. Start the engine and allow it to reach operating temperature. Check the ignition timing and all fluid levels. Test drive the vehicle.

Clutch

ADJUSTMENT

The clutch cable utilized in the 900 Series vehicles is self-adjusting. Although it does not require periodic adjustments, when the clutch cable is first reinstalled in the vehicle a small procedure will insure that the self-adjuster is properly functioning.

1. Check the functioning of the clutch cable by moving the clutch lever forward in the car's direction of travel. The balancing spring should, then be compressed and the length of the clutch cable's cover reduced. When the clutch lever is released, the cover should regain its original length. Repeat three or four times.

2. Grip hold of the balancing spring and give it a small jerk to remove any free-play.

REMOVAL & INSTALLATION

900 and 9-3

1. Place vehicle on lift and engage in 4th gear.

2. Disconnect battery and remove from vehicle.

3. Remove ground strap from transaxle housing.

4. Disconnect the positive cable routing straps.

5. Detach the electrical harness connector for the rear light switch.

6. Disconnect the clutch cable from the clutch lever and release the cable's rubber damper from the fastener on the transaxle.

7. Detach the oxygen sensor connectors and remove the securing straps.

8. Install the engine holder. Insert the lifting bar in the eyelet's and connect to the engine holder.

9. Install transaxle locking pin to ensure the vehicle does not move out of gear.

10. Disconnect the shift rod and linkage from the transaxle.

11. Safely raise vehicle.

12. Remove front wheels.

13. Remove front exhaust pipes.

14. Remove front spoilers.

15. Disconnect the ball joints on both sides.

16. Using the proper jack remove the carrying frame assembly to gain access to transaxle housing.

17. Drain the transaxle.

18. Remove the left and right front half-shafts and suspend shafts with securing straps.

19. Remove intermediate shaft bearing bracket and pull out shaft.

20. Remove flywheel cover.

21. Remove transaxle mounting bracket.

22. Remove three transaxle bracket bolts on the transaxle

23. Remove the engine/transaxle surface mounting bolts.

24. Lower vehicle.

25. Remove the two outer bolts holding gear box to engine mounting surface.

26. Install lifting eye to transaxle and connect hoist to transaxle.

27. Remove last bolt and pull out transaxle assembly.

28. Place manual transaxle assembly on proper stand.

29. Lock flywheel in position and remove pressure plate retaining nuts.

30. Remove pressure plate and clutch disc.

To install:

31. Using a clutch disc alignment tool install clutch disc with pressure plate.

32. Tighten pressure plate retaining nuts to 16 ft. lbs. (22 Nm).

33. Using a hoist lift transaxle into position.

34. To help with installation use a guide pin to line up transaxle.

35. When transaxle is in position fit the three upper bolts into engine housing.

36. Tighten transaxle/engine surface mounting bolts to 65 ft. lbs. (90 Nm).

37. Safely raise vehicle.

38. Fit transaxle/engine surface mounting bolts from under vehicle and tighten to 65 ft. lbs. (90 Nm).

39. Install transaxle mounting bracket to the transaxle housing and the frame assembly.

40. Install flywheel cover.

41. Install intermediate driveshaft with bracket bearing assembly.

42. Install both halfshafts.

43. Using a proper jack install subframe assembly. Tighten ball joints to 55 ft. lbs. (75 Nm). Fit sub frame bolts and tighten front bolts to 85 ft. lbs. (11 Nm), tighten middle bolts to 141 ft. lbs. (190 Nm), and tighten rear bolts to 81 ft. lbs. (110 Nm) plus 75 degrees additional torque.

44. Install front exhaust pipes.

45. Install front wheels.

46. Lower vehicle.

47. Install shift rod and engage 4th gear, install shift rod mounting bolt to secure linkage.

48. Remove the engine holder and lifting rods.

49. Remove the transaxle locking pin.

50. Fill transaxle with fluid.

51. Attach the oxygen sensor electrical harness connectors.

52. Fit the positive battery cable.

53. Connect clutch cable, and the rear light switch harness.

54. Install battery and connect ground cable.

55. Test drive vehicle.

9000

1. Disconnect the battery cables from the battery, negative cable first.

2. Remove the battery, washer fluid container and connectors, terminal blocks, battery tray and release the stay for the hydraulic unit on ABS equipped vehicles.

3. Remove the 8 bolts for the bulkhead cover. If equipped, remove the rubber strip, lift the cover and disconnect the washer hoses from the nozzle and remove the cover.

4. If equipped, separate the speedometer cable connector by first removing the washer hose, then the speedometer cable through the rubber grommet.

5. Detach the electrical connector from the air mass meter, the intake air temperature sensor, then the hose on the delivery pipe from the bypass valve.

6. If equipped, remove the delivery pipe between the throttle housing and the intercooler. Also, remove the nuts retaining the starter motor. Remove the starter and place it on the steering gear.

7. Separate the selector rod universal joint and selector rod. Pinch the slave cylinder pressure hose with clamping tongs and disconnect the pressure line.

8. Remove the upper bolts for the stay at the wheel housing. Remove the left-hand engine mounting. Attach a sling to the engine lifting beam. Raise and support the vehicle safely.

9. Remove the left wheel and wheel housing liner. Detach the reverse light connector from the transaxle.

10. Separate the suspension arm from the ball joint. Remove the sway bar. Remove the lower bolt for the stay at the wheel housing and the 3 bottom bolts from the joint between the engine and the transaxle.

11. Disconnect the speedometer cable at the transaxle.

12. Remove the center and left skirts under the spoiler. Separate the subframe at the front and rear. Lower the subframe. Remove the universal joint and lower the vehicle.

13. Sling the transaxle from a suitable hoist and remove the top nut and bolt from the clutch housing face. Remove the transaxle by lowering to the floor.

14. Install a flywheel locking tool, if available, and remove the clutch assembly from the flywheel.

15. Inspect the flywheel for wear, scoring, cracking or other damage. Replace or machine, as necessary.

To install:

16. Use a centering arbor type tool or an appropriate input shaft to center the clutch disc to the flywheel.

17. Install the clutch pressure plate and tighten the bolts, alternately and evenly, in several steps, to 10–20 ft. lbs. (13–25 Nm). Remove the flywheel lock, if used.

18. Slide the transaxle assembly over the locating dowels, engaging the transaxle input shaft into the clutch plate splines.

19. Secure the transaxle to the engine with the necessary attaching bolts. Remove the lifting sling from the transaxle.

➡**Prior to installation, ensure that the halfshaft is in position and the aluminum tube is pressed into the seal.**

20. Fit the 3 bottom bolts into the joint face and tighten to 40–74 ft. lbs. (54–100 Nm). Install the universal joint. Raise the subframe into position and secure.

21. Install the sway bar, the suspension arm to the ball joint and the bracket for the wheel housing stay. Do not tighten the wheel housing stay bolt at this time.

22. Lower the vehicle.

23. Remove the lifting beam. Install the starter and the top mounting of the wheel housing. Now tighten all wheel housing bolts to 32–43 ft. lbs. (43–57 Nm).

24. Tighten the left engine mount to 37–67 ft. lbs. (41–91 Nm). Raise and support the vehicle safely.

25. Connect the negative battery cable.

26. Connect the reverse light switch.

27. Install the wheel housing liners, left wheel, under vehicle skirts and selector rod universal joint.

28. Connect the slave cylinder pressure pipe (remove the clamping tongs), speedometer cable, washer hose and the remainder of the components removed.

29. Bleed the slave cylinder, check the transaxle fluid level and road test the vehicle.

Hydraulic Clutch System

BLEEDING

1. Connect a hose to the slave cylinder bleeder valve. Place the other end of the hose in a suitable jar partially filled with brake fluid.

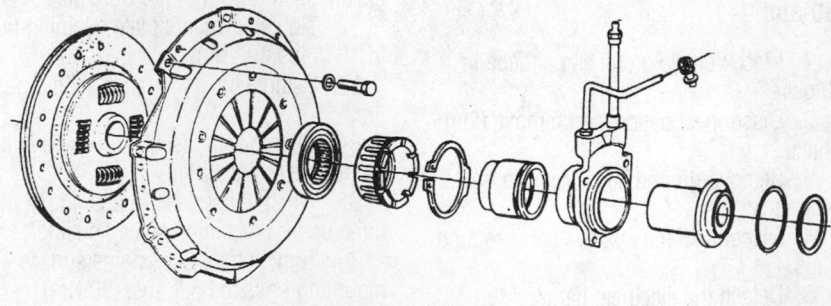

Exploded view of the clutch assembly—9000

7923SG21

2. Fill the master cylinder with brake fluid.

3. Open the bleeder valve on the slave cylinder ½ turn.

4. Place a cooling system pressure tester gauge over the opening of the master cylinder.

5. Pump the tester until all air is expelled from the hydraulic clutch system.

6. Close the slave cylinder bleeder valve.

7. Check that all air was removed from the system and the clutch is functioning properly. Adjust the fluid level, as required.

Halfshaft

REMOVAL & INSTALLATION

900 and 9-3

1. Disconnect the negative battery cable.

2. With the vehicle sitting on all four wheels, remove the hubcap and loosen the hub center nut. Do not remove the center nut at this time.

3. Carefully raise and support the vehicle safely. Remove the wheel.

4. Remove the hub center nut.

5. Remove the ball joint nut. Separate the ball joint from the knuckle using special tool 89–96–696 or equivalent.

6. Remove the sway bar nut. Remove the washer and rubber bushing.

7. Push down on the lower control arm.

8. Using a rubber mallet, tap the halfshaft out of the hub.

9. Move the strut to one side and withdraw the halfshaft. Be extremely careful not to stretch or break the ABS sensor cables or brake hoses. Place a drain pan under the transaxle to catch any fluid spillage.

10. Remove the inner halfshaft joint from the transaxle on the left-hand side. Remove the halfshaft joint from the intermediate

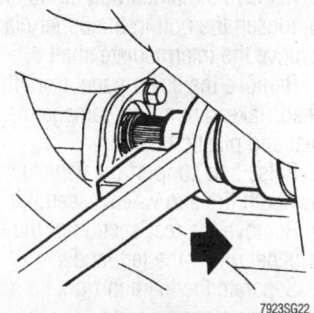

7923SG22

Removing the halfshaft joint from the intermediate shaft using special tool 89–96–654—900 and 9-3

shaft on the right-hand side. Use special tool 89–96–654 or equivalent to pry the joints out.

To install:

11. Place the halfshaft into the intermediate shaft on the right-hand side. Reposition the halfshaft into the transaxle on the left-hand side. If necessary, use a rubber mallet to secure the halfshaft into place.

12. Insert the halfshaft into the center of the hub. Install a new hub center nut but do not tighten yet.

13. Insert the ball joint into the bottom of the steering knuckle. Tighten the ball joint nut to 55 ft. lbs. (75 Nm).

14. Reconnect the sway bar to the lower control arm. Install the rubber bushing, a new washer and new retaining nut. Tighten the nut to 84 inch lbs. (10 Nm).

15. Install the wheels and lug nuts. Tighten the lug nuts to 80–90 ft. lbs. (108–122 Nm).

16. Lower the vehicle to the ground. With all four wheels on the ground, tighten the hub center nut to 214 ft. lbs. (290 Nm). Install the center hubcap.

17. Check the transaxle fluid level and top off if necessary.

18. Reconnect the negative battery cable.

9000

1. Disconnect the negative battery cable.

2. With all four wheels on the ground, loosen the center axle nut.

3. Raise and support the vehicle safely.

4. Remove the front wheel.

5. Remove the inner fender panel for working access.

6. Unbolt the strut from the steering swivel member and detach the flexible brake hose from the clip on the strut. If equipped with ABS, remove the ABS sensor and lead.

7. Loosen the clip on the rubber boot on the inboard CV-joint.

8. Separate the 2 halves of the joint. Install protective covers over the rubber boot and the drive axle.

9. Remove the hub center nut and withdraw the halfshaft from the steering swivel member.

To install:

10. Install the halfshaft and a new hub center nut. Do not tighten the nut yet.

11. Join the two halves of the joint and install a new rubber boot along with a new clamp.

12. Install the strut and tighten the strut-to-steering swivel bolts to 58–77 ft. lbs. (78–105 Nm).

13. If equipped with ABS, clean the end

of the sensor of any debris. Clean the sensor wheel of any debris. Install the sensor unit gently against the sensor wheel and tighten the retaining screw.

14. Install the brake hose to the clip on the strut and install the inner fender panel. Install the wheel and lug nuts.

15. Lower the vehicle and connect the negative battery cable.

16. With all four wheels on the ground, tighten the hub center nut to 207–221 ft. lbs. (280–300 Nm). Check the alignment and test drive the vehicle.

STEERING AND SUSPENSION

Air Bag

✳ CAUTION

Some vehicles are equipped with an air bag system, also known as the Supplemental Inflatable Restraint (SIR) or Supplemental Restraint System (SRS). The system must be disabled before performing service on or around system components, steering column, instrument panel components, wiring and sensors. Failure to follow safety and disabling procedures could result in accidental air bag deployment, possible personal injury and unnecessary system repairs.

PRECAUTIONS

Several precautions must be observed when handling the inflator module to avoid accidental deployment and possible personal injury.

• Never carry the inflator module by the wires or connector on the underside of the module.

• When carrying a live inflator module, hold securely with both hands, and ensure that the bag and trim cover are pointed away.

• Place the inflator module on a bench or other surface with the bag and trim cover facing up.

• With the inflator module on the bench, never place anything on or close to the module which may be thrown in the event of an accidental deployment.

DISARMING

All Models

✳✳ CAUTION

The air bag system must be disarmed before performing service around air bag system components or system wiring. Failure to do so may cause accidental deployment of the air bag, resulting in unnecessary air bag system repairs and/or personal injury.

Always disconnect the battery cables (negative cable first!) and wait 20 minutes prior to performing service around air bag system components or system wiring.

✳✳ CAUTION

Do not use any diagnostic instruments that are battery powered, such as buzzers, ohmmeters or diode testers, to diagnose faults in the steering wheel or electronic control unit. Using such devices may trigger the air bag. Also, ensure that the battery cables cannot accidentally come into contact with the battery terminals.

REARMING

To rearm the air bag system, reconnect the battery cables.

Rack and Pinion Steering Gear

REMOVAL & INSTALLATION

Manual

900 AND 9-3

1. Disconnect the negative battery cable.
2. Remove the left screen under the instrument panel and loosen the rubber bellows at the body lead through for the steering gear intermediate shaft, if required.
3. Raise and support the vehicle safely.
4. Remove the bolt holding the joint to the steering gear pinion or intermediate shaft.
5. Loosen the steering column tube from the body and separate the steering column joint from the pinion. Position the steering column so the wiring harness is not damaged.
6. Remove both tire and wheel assemblies. Remove the tie rod ends at the steer-

ing arms with the proper removal tool. Remove the 2 steering gear clamps.
7. Move the rack to the right as far as possible. Lift the steering gear to the right so the tie rod can be bent down in the opening of the engine compartment floor.
8. Pull the rack to the left and lift the steering gear down through the opening in the engine compartment floor.

To install:

9. Carefully lift the rack assembly in through the right side wheel arch. Install and tighten the mounting bolts.
10. Secure the pinion to intermediate shaft.
11. Screw the tie-rod ends back into the track rods and measure the distance between the tie-rod end and groove in the rack to ensure that both tie-rod ends are back in the same position as before removal.
12. Tighten the nut securing each tie-rod end to the steering swivel members. Tighten the clamp bolt to 26–30 ft. lbs. (35–42 Nm); the body to steering gear bolts to 44–60 ft. lbs. (60–80 Nm); the tie rod end nuts to 37–44 ft. lbs. (50–60 Nm).
13. Install both tire and wheel assemblies.
14. Lower the vehicle.
15. Refit the left screen and rubber bellows under the instrument panel.
16. Connect the negative battery cable.

Power

900 AND 9-3

1. Disconnect the hydraulic fluid return line and plug. Connect a length of hose to the fluid line and a suitable container to recover the fluid. Turn the steering wheel, stop to stop, until all the fluid is out of the system.
2. Disconnect the negative battery cable.
3. Remove the lower left dash finish panel. Remove the fuse box.
4. Remove the steering column to pinion shaft locknut.
5. Straighten the steering wheel and mark it with chalk to center.
6. Separate the column shaft and pinion shaft.
7. Disconnect the tracking rods from the center of the rack assembly.
8. Raise and safely support the vehicle.
9. Remove the front wheels.
10. Disconnect the tie rods from the strut/knuckle using tool 89–96–696 or equivalent.
11. Disconnect the rack assembly retaining clamps.

12. Disconnect the pressure and return lines using special wrench 87–91–287 or equivalent, from the rack assembly.
13. Maneuver the rack assembly out of the vehicle through the left wheel housing.

To install:

➡️**If a replacement rack assembly is being used, transfer all lines and bellows to the replacement unit as required.**

14. Maneuver the rack assembly in through the left wheel housing.
15. Connect the pressure and return lines using specialty wrench 87–91–287 from the rack assembly.
16. Connect the rack assembly retaining clamps.
17. Connect the tie rods to the strut/knuckle. Tighten to 44 ft. lbs. (60 Nm)
18. Connect the tracking rods to the center of the rack assembly. Tighten to 63–74 ft. lbs. (85–100 Nm).
19. Connect the column shaft and pinion shaft. Be sure the steering wheel was aligned with the chalk center mark.
20. Install the lower left dash finish panel. Install the fuse box.
21. Connect the negative battery cable.
22. Connect the hydraulic fluid return line. Fill the reservoir and turn the steering wheel, stop to stop, until all the air is out of the system.
23. Lower the vehicle.
24. Check the toe-in and steering wheel alignment.

9000

1. Disconnect the negative battery cable.
2. Remove the padding from under the instrument panel and the trim on the left side of the center tunnel, as required. Fold back the carpet where the steering column passes through the bulkhead. Remove the rubber boot from the intermediate shaft.
3. Remove the pinch bolt in the lower clamp, loosen the bolt in the upper clamp and remove the intermediate shaft.
4. Remove the cover panel from the bulkhead. Take care not to damage the gasket, seal and plastic bushing.
5. Raise and support the vehicle safely. Remove both tire and wheel assemblies.
6. Remove the rear section of the inner fender panel under the left fender.
7. Separate the left and right tie rod ends from the steering arms.
8. Drain the power steering fluid from the pump reservoir.
9. Disconnect the hoses from the pump

and reservoir. Plug the openings to prevent fluid from leaking out and dirt from entering.

10. Remove the retaining bolts from the rack and pinion assembly.

11. Remove the vertical brace between the engine subframe and the body.

12. Lift out the rack and pinion unit through the left fender inner panel opening. Do not damage the rubber boots or brake hose.

To install:

13. Maneuver the rack and pinion unit through the left fender inner panel opening. Do not damage the rubber boots or brake hose. Tighten the rack and pinion gear securing bolts to 46–56 ft. lbs. (60–80 Nm).

14. Install the vertical brace between the engine subframe and the body.

15. Connect the hoses from the pump and reservoir.

16. Connect the left and right tie rod ends to the steering arms.

17. Assemble the rear section of the inner fender panel under the left fender.

18. Install the wheel assemblies and lower the vehicle.

19. Install the cover panel to the bulkhead, do not damage the gasket, seal and plastic bushing.

20. Install the pinch bolt in the lower clamp. Tighten the steering column pinch bolt to 27–32 ft. lbs. (35–42 Nm).

21. Connect the negative battery cable.

22. Fill the reservoir and bleed the system by allowing the engine to run at idle and turning the steering wheel, stop to stop, two or more times.

Strut

REMOVAL & INSTALLATION

Front

900 AND 9-3

1. With all four wheels on the ground, loosen the front hub nut.

2. Raise the vehicle and support it safely.

3. Remove the front wheel.

4. Remove the hub nut and remove the wheel speed sensor.

5. Retract the caliper piston, unbolt and remove the caliper. Be sure the caliper is properly supported and not hanging from the brake hose.

6. Remove the rotor and backing plate.

7. Using tool 89–96–696 or equivalent, remove the tie rod end from the knuckle.

8. Disconnect the sway bar from the lower control arm.

9. Using tool 89–96–696 or equivalent, press the ball joint out of the steering knuckle.

10. Remove the three upper strut mounting bolts. Lift out the strut housing.

11. Place the strut housing in a vise.

12. Compress the spring with tool 89–18–809 or equivalent.

13. Remove the self-locking nut from the bearing plate and discard the nut.

14. Remove the coil spring, bellows and rubber snubber.

15. Remove the housing nut with spanner wrench 89–96–670 or equivalent.

16. Remove the strut cartridge from the housing.

To install:

17. Install the new strut cartridge into the housing and tighten the spanner nut to 159 ft. lbs. (215 Nm).

18. Position the coil spring in place. Be sure the end of the spring is up against the spring stop.

19. Position the upper spring seat with the notches properly aligned. Place the upper bearing assembly on the upper seat and secure it using a new self locking nut.

20. Remove the spring compressor and be sure the coil spring stays properly seated.

21. Position the strut assembly on the vehicle and tighten the upper mounting bolts 13 ft. lbs. (18 Nm).

22. Position the ball joint and install a new self locking nut and tighten it to 55 ft. lbs. (75 Nm).

23. Connect the tie rod end.

24. Tighten the sway bar at the lower control arm link to 8 ft. lbs. (10 Nm).

25. Install the backing plate.

26. Reassemble the brakes.

27. Install the wheel speed sensor.

28. Start the center hub nut.

29. Install the wheels and tighten the lugs to 77–95 ft. lbs. (105–130 Nm).

30. Lower the vehicle and with all four wheels on the ground, tighten the hub nut to 215 ft. lbs. (290 Nm).

31. Pump the brake pedal to position the brake caliper piston.

❈❈ CAUTION

Do not attempt to move the vehicle until a firm pedal is obtained.

32. Check front wheel alignment.

9000

1. With all four wheels on the ground, loosen the front hub nut.

2. Raise the vehicle and support it safely.

3. Remove the front wheel.

4. Remove the hub nut and remove the wheel speed sensor.

5. Retract the caliper piston, unbolt and remove the caliper. Be sure the caliper is properly supported and not hanging from the brake hose.

6. Remove the rotor and backing plate.

7. Using tool 89–96–696 or equivalent, press the tie rod end from the knuckle.

8. Disconnect the sway bar from the lower control arm.

9. Unbolt strut from lower steering swivel and pull strut from swivel assembly.

10. Remove the three upper strut mounting bolts and lower the strut out of the vehicle.

11. To remove the spring and shock absorber cartridge:

   a. Place the strut in a vise.

   b. Compress the spring with tool 89–18–809 or equivalent.

   c. Remove the self-locking nut from the bearing plate and discard the nut.

   d. Remove the coil spring, bellows and rubber snubber.

   e. Remove the upper flange nut with spanner wrench 89–96–670 or equivalent.

   f. Remove the strut cartridge from the housing.

To install:

12. To assemble the strut:

   a. Install the new strut cartridge into the housing and tighten the spanner nut to 159 ft. lbs. (215 Nm).

   b. Position the compressed coil spring in place. Be sure the end of the spring is up against the spring stop.

   c. Position the upper spring seat with the notches properly aligned. Place the upper bearing assembly on the upper seat and secure it using a new self locking nut.

   d. Remove the spring compressor and be sure the coil spring stays properly seated.

13. Position the strut assembly on the vehicle and tighten the upper mounting bolts 13 ft. lbs. (18 Nm).

14. Install mounting bolts to connect strut to steering swivel assembly.

15. Connect the tie rod end.

16. Tighten the sway bar at the lower control arm link to 96 inch lbs. (10 Nm).

17. Install the backing plate.

18. Reassemble the brakes.

19. Install the wheel sensor.

20. Start the center hub nut.

21. Install the wheel and tighten the lugs to 77–95 ft. lbs. (105–130 Nm).

22. Lower the vehicle. With all four wheels on the ground, tighten the hub nut to 215 ft. lbs. (290 Nm).

23. Pump the brake pedal to position the brake caliper piston.

✳✳ CAUTION

Do not attempt to move the vehicle until a firm pedal is obtained.

24. Check front wheel alignment.

Shock Absorber

REMOVAL & INSTALLATION

Rear

900 AND 9–3

1. Working inside the trunk, locate the upper mounting bolt. Cut out a flap in the carpeting to access the mounting bolt and bushings.

2. Remove the nut, washer and bushing from the mounting point.

3. Raise the rear of the vehicle and safely support it on jack stands.

4. Remove the rear tire and wheel assemblies.

5. Remove the lower shock mounting bolt and remove the shock.

To install:

6. Install the shock into the upper mounting. Be sure to install the lower part of the bushing on the shock.

7. Install the lower shock mounting bolt and tighten to 46 ft. lbs. (62 Nm).

8. Install the rear wheels and tighten the nuts to 77–96 ft. lbs. (105–130 Nm).

9. Lower the vehicle to the ground.

10. Install the upper bushing, washer and nut. Tighten to 15 ft. lbs. (20 Nm).

11. Reposition the carpeting.

9000

1. Disconnect the negative battery cable.

2. Raise and safely support the vehicle at the rear jacking point and remove the rear wheels.

3. Place jackstands under the rear end of the trailing arms where the they mount to the rear axle.

4. Position a jack at the rear jacking point. Raise the rear of the vehicle enough to relieve the load on the shock absorbers and anti-roll bar. Remove the lower shock absorber bolts.

5. From the trunk, pull back the carpet to locate the upper shock bolts. Remove the nut washer and bushing.

6. Remove the shock absorber from the vehicle.

To install:

7. Install the shock with the bushings in the proper orientation and tighten the upper shock bolts to 7–16 ft. lbs. (10–20 Nm).

8. Align the shock lower mounting with the anti-roll bar link. Tighten the lower shock bolt to 59–66 ft. lbs. (80–90Nm).

9. Install the rear wheels, remove the safety stands and lower the vehicle. Place the carpeting back into position.

Coil Spring

REMOVAL & INSTALLATION

Front

The front coil spring removal and installation procedure is covered in the strut removal and installation.

Rear

900 AND 9–3

1. Raise the vehicle and safely support it on jack stands. Do not place the stands under the rear axle assembly.

2. Remove the rear wheels.

3. Place a floor jack under the lower control arm and raise upward slightly.

➡**If the same spring is to be reinstalled, mark the rear of the spring to be sure it is reinstalled in the proper position.**

4. Remove the lower mounting bolt from the shock absorber.

5. Slowly lower the floor jack until the lower arm is relaxed.

6. Use a prybar to bring the lower arm down far enough to remove the spring.

To install:

7. Pry down the lower arm to install the spring.

8. Install the spring with the rubber cushions in place.

9. Raise the control arm with the floor jack, secure the lower shock mount and tighten to 46 ft. lbs. (62 Nm).

10. Install the wheels and tighten the nuts to 77–96 ft. lbs.105–130 Nm).

11. Lower the vehicle and road test.

9000

1. Raise and safely support the vehicle on jack stands.

2. Remove the rear wheels.

3. Disconnect the hand brake cable from the retaining bracket connected to the lower control arm.

4. Remove the ABS sensor wiring, if necessary, by remove the clip and releasing the cable.

5. Place a jack under the lower arm and disconnect the trailing-end bolt and lower shock absorber bolt.

6. Lower the jack slowly and remove the coil spring.

To install:

7. Position the coil spring in the vehicle.

8. Raise the lower arm until the bolt holes align. Install and tighten the bolts.

9. Install ABS wiring, if necessary, by connecting the clip and cable.

10. Install the wheels and lower the vehicle.

11. Road test the vehicle.

Lower Ball Joint

REMOVAL & INSTALLATION

900 and 9–3

➡**The ball joint cannot be removed from the control arm. To replace the ball joint, the control arm must be replaced.**

1. Raise and support the vehicle safely. Remove the wheel.

2. Remove the sway bar link bolt.

3. Loosen the ball joint nut. Use tool 89–96–696 or equivalent to press the joint out of the steering knuckle.

4. Remove the retaining nut at the support arm.

5. Remove the retaining bolt at the subframe.

To install:

6. Install the arm and install the bolt at the subframe.

7. Install the bolt at the support arm and tighten to 68 ft. lbs. (92 Nm).

8. Tighten the retaining bolt at the subframe to 85 ft. lbs. (115 Nm).

9. Connect the sway bar link and tighten to 7 ft. lbs. (10 Nm).

10. Connect the ball joint and tighten to 55 ft. lbs. (75 Nm)

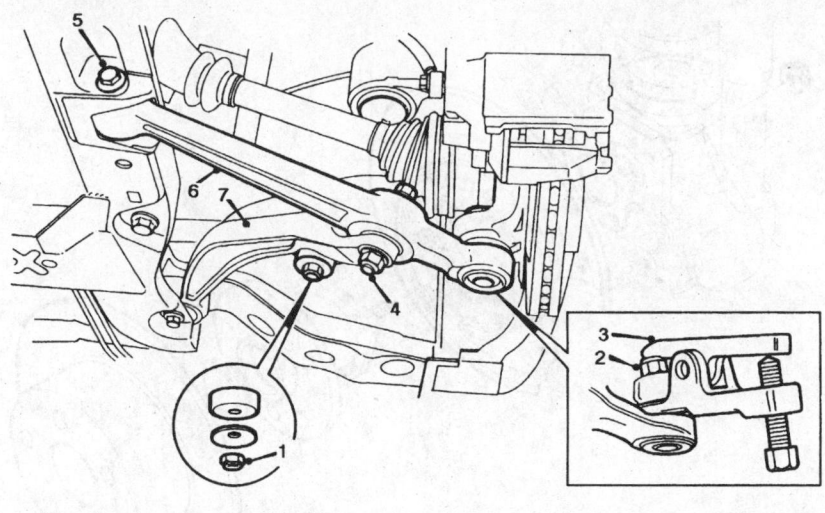

1. Sway bar nut
2. Ball joint nut
3. Ball joint press tool
4. Support arm connection
5. Subframe connection
6. Lower control arm
7. Support arm

7923SG23

Lower control arm connection points—900 and 9–3

11. Install the wheels and lower the vehicle to the ground.

12. Check front wheel alignment.

9000

1. Raise vehicle and support on safety stands.

2. Remove wheel.

3. Disconnect the three mounting bolts securing the ball joint to the lower control arm.

4. Remove the lockwasher from atop the steering swivel member.

5. Remove the bolt securing the ball joint to the steering swivel member.

6. Remove the ball joint.

To install:

7. Place the ball joint in the steering member socket. Install the bolt and tighten to 20 ft. lbs. (28 Nm).

8. Secure the locking ring atop the ball joint.

9. Fit the ball joint to the lower control arm and tighten the three mounting bolts to 22 ft. lbs. (30Nm).

10. Fit the wheel and tighten to 89 ft. lbs. (120 Nm).

11. Lower the vehicle and test drive.

Wheel Bearings

ADJUSTMENT

The wheel bearings found in all Saab vehicles are sealed units requiring no adjustment.

REMOVAL & INSTALLATION

Front

900 AND 9–3

1. With all four wheels on the ground, loosen the front hub nut.

2. Raise and safely support the vehicle.

3. Remove the front wheel.

4. Remove the hub nut and remove the wheel speed sensor.

5. Retract the brake caliper piston, unbolt and remove the caliper. Be sure the caliper is properly supported and not hanging from the brake hose.

6. Remove the rotor and backing plate.

7. Using tool 89–96–696 or equivalent, press the tie rod end from the knuckle.

8. Disconnect the sway bar from the lower control arm.

9. Using tool 89–96–696 or equivalent, press the ball joint out of the steering knuckle.

10. Remove the three upper strut mounting bolts. Lift out the strut housing.

11. Using the proper arbor press adapters, press the hub from the wheel bearing.

12. Remove the two circlips on each side of the wheel bearing.

13. Press out the wheel bearing using the proper size press adapters.

To install:

14. Install one circlip and press in wheel bearing until it contacts the circlip. Be sure

the press tool contacts only the outer bearing race.

15. Install other circlip.

16. Clean the bearing surface of the hub on a wire wheel. If the hub is pitted or damaged, it should be replaced.

17. Support the inner bearing race and press hub into wheel bearing. If the inner race is not properly supported, the bearing will fail quickly.

18. Position the strut assembly on the vehicle and tighten the upper mounting bolts 13 ft. lbs. (18 Nm).

19. Position the ball joint, install a new self locking nut and tighten it to 55 ft. lbs. (75 Nm).

20. Connect the tie rod end.

21. Tighten the sway bar at the lower control arm link to 96 inch lbs. (10 Nm).

22. Install the backing plate.

23. Reassemble the brakes.

24. Install the wheel sensor.

25. Start the center hub nut.

26. Install the wheel assemblies and tighten the lugs to 77–95 ft. lbs. (105–130 Nm).

27. Lower the vehicle. With all four wheels on the ground, tighten the hub nut to 215 ft. lbs. (290 Nm).

28. Pump the brake pedal to position the brake caliper piston.

✳✳ CAUTION

Do not attempt to move the vehicle until a firm pedal is obtained.

29. Check front wheel alignment.

9000

1. With all four wheels on the ground, loosen the hub center nut and the wheel bolts.

2. Raise the vehicle and support it safely.

3. Remove the tire and wheel assembly. Remove the hub center nut and thrust washer.

4. Remove the flexible brake hose from its support clip.

5. Unbolt the caliper and secure it to the suspension arm; do not let the caliper hang by the brake hose.

6. Unscrew the locating stud for the disc and remove it from the hub.

7. Push in on the halfshaft. If the CV-joint shaft does not push in, use puller 87–91–287 or equivalent to break it loose.

➡**Do not allow the puller to push the halfshaft in more the 2 in. or damage to the inboard joint may occur.**

8. Remove the 4 bolts securing the hub to the knuckle. All vehicles use socket head screws to retain the hub assembly. For easier removal, cut an Allen wrench to fit the head and turn it with an 8mm wrench.

9. Lift the hub and disc backing plate from the knuckle assembly.

10. Remove the two bolts connecting the knuckle to the strut.

To install:

11. Clean the bearing seat and lightly coat with molybdenum grease. Install the hub assembly. Draw the hub in by tightening the four retaining screws.

12. Tighten the hub retaining screws to 41–44 ft. lbs. (55–60 Nm).

13. Lift the knuckle into place and connect the two strut bolts with the nuts facing to the front of the vehicle. Tighten to 58–78 ft. lbs. (78–105 Nm).

14. Assemble the disc and brake assembly. Tighten the caliper bolts to 52–80 ft. lbs. (70–110 Nm). Connect the ABS speed sensor, if equipped.

15. Install the wheels and lower the vehicle.

16. With all four wheels on the ground, tighten the center hub nut to 207–221 ft. lbs. (280–300 Nm).

17. Pump the brakes to seat the brakes before road testing.

Rear

900 AND 9-3

1. Raise the vehicle and support it safely.

2. Remove the rear wheels.

3. Compress the caliper piston. Unbolt the caliper and secure it out of the way. Do not let the caliper hang by the brake hose.

4. Back off the adjuster on the parking brake shoes.

5. Remove the parking brake return spring and lever.

6. Remove the rotor retaining screws and the rotor.

7. Remove the 4 hub retaining nuts.

8. Disconnect the speed sensor.

9. Remove the hub and bearing as an assembly.

➡There is a spacer behind the brake backing plate.

To install:

10. Install the hub assembly, backing plate and spacer. Tighten the nuts to 37 ft. lbs. (50 Nm).

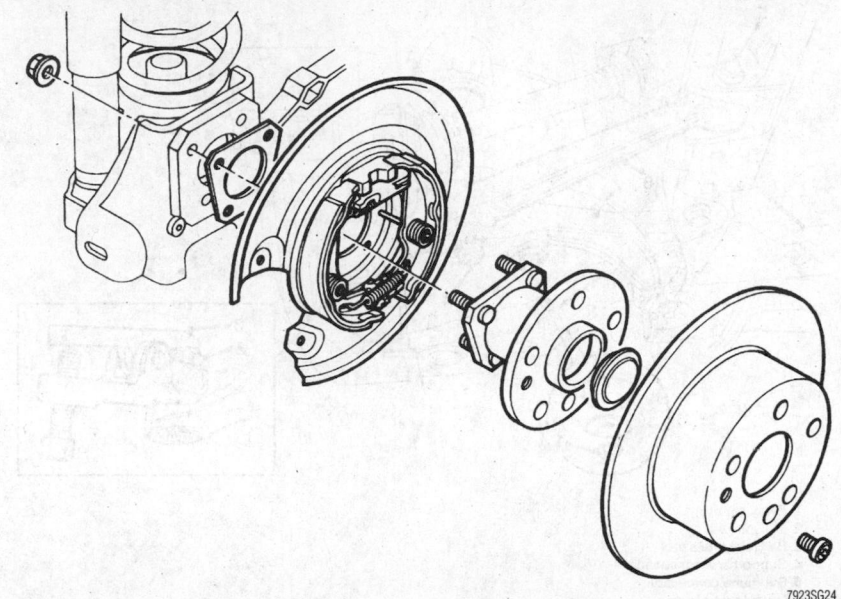

Exploded view of the rear hub and bearing components—900 and 9-3

11. Install the brake rotor, after applying low strength Loctite®, install the rotor retaining screw.

12. Install the parking brake lever and return spring.

13. Connect the speed sensor.

14. Install the Caliper and retaining bolts with Loctite®, applied.

15. Screw in the brake shoe adjusting screw until the rotor cannot turn. Back off the screw until the rotor can rotate freely.

16. Install the wheels and lower the vehicle. Tighten the wheel lugs to 75–95 ft. lbs. (105–130 Nm).

17. Pump the brakes to position the caliper piston.

✳✳ CAUTION

Do not attempt to move the vehicle until a firm pedal is obtained.

18. Road test the vehicle.

9000

1. With all four wheels on the ground, raise the stake with a cold chisel and loosen the rear hub nut.

2. Raise and safely support the rear of the vehicle. Remove the wheel.

3. Disconnect the hand brake cable from the caliper. Remove the adjuster screw plug and loosen the adjusting screw enough to allow the brake piston to slide back.

4. Unbolt and remove the brake caliper and backing plate. Remove the ABS sensor, if equipped. Do not allow the caliper to hang by the brake hose.

5. Remove the bolt and pull off the brake disc.

6. Pry off the hub nut dust cap. Remove the hub nut and thrust washer and pull off the hub assembly. Discard the used hub nut.

➡**Whenever the hub nut is removed, a new one must always be used because the locking device on the nut becomes ineffective.**

To install:

7. Check the spindle for damage and repair or replace as required. Install the hub assembly.

8. Install the thrust washer and nut but do not tighten the nut yet.

9. Install the brake disc. Install the caliper and tighten the bolts to 51–65 ft. lbs. (70–90 Nm).

10. Install the wheel and lower the vehicle.

11. With all four wheels on the ground, tighten the hub nut to 195–208 ft. lbs. (270–290 Nm). Stake the nut with a cold chisel and press the dust cap into place.

12. Pump the brake pedal several times before driving the vehicle.

SUBARU

Impreza • Impreza Outback • Impreza Outback Sport • Legacy • Legacy Outback • SVX

27

ENGINE REPAIR

→**Disconnecting the negative battery cable on some vehicles may interfere with the functions of the on board computer systems and may require the computer to undergo a relearning process, once the negative battery cable is reconnected.**

Ignition Timing

ADJUSTMENT

All 1995–99 Subaru models are equipped with distributorless ignition systems; the ignition timing is controlled by the Powertrain Control Module (PCM) and is not adjustable.

Engine Assembly

✳✳ CAUTION

Some models covered by this manual may be equipped with a Supplemental Restraint System (SRS), which uses an air bag. Whenever working near any of the SRS components, such as the impact sensors, the air bag module, steering column and instrument panel, properly disable the SRS.

REMOVAL & INSTALLATION

1.8L Engine

1. Open the hood and scribe alignment marks around the hood hinge. Remove the hinge retainer bolts and hood. Place the hood aside.
2. Remove the spare tire and tire bracket.

✳✳ CAUTION

Observe all applicable safety precautions when working around fuel. Whenever servicing the fuel system, always work in a well ventilated area. Do not allow fuel spray or vapors to come in contact with a spark or open flame. Keep a dry chemical fire extinguisher near the work area. Always keep fuel in a container specifically designed for fuel storage; also, always properly seal fuel containers to avoid the possibility of fire or explosion.

3. Properly relieve the fuel system pressure.
4. Disconnect the negative battery cable.
5. Remove the air cleaner assembly.
6. Tag and disconnect the fuel system hoses and the evaporative emissions system hoses.
7. Tag and disconnect the vacuum hoses for the cruise control, Master-Vac®, air intake shutter and the heater air intake door.
8. Disconnect the electrical wiring from the alternator, EGR, thermoswitch, cooling fan (if electric), A/C condenser and the ignition coil. Unfasten the main engine harness.
9. Tag and disconnect the spark plug wires, the engine ground strap and the fusible link assembly.
10. Disconnect the accelerator linkage. Remove the windshield washer reservoir and position it behind the right strut tower.
11. Remove the power steering pump as follows:
    a. Loosen the alternator pivot and mounting bolts, then remove the drive belt.
    b. Remove the pulley retainer bolts and pulley from the power steering pump.
    c. Remove the power steering pump-to-engine bolts and clamp.
    d. Remove the engine oil filler tube and brace.
    e. Remove the power steering pump and secure it to the bulkhead without disconnecting the pressure lines.
12. Loosen the air intake duct hose clamps and remove the duct. Seal the openings to keep dirt out of the intake passages.
13. Remove the air intake-to-flow meter line and cover the openings.
14. Remove the horizontal damper and clip.
15. Remove the center exhaust section by performing the following procedures:
    a. Disconnect the temperature sensor harness.
    b. Remove the rear cover.
    c. Remove the center exhaust section-to-transaxle bolt.
    d. Remove the hanger bolts, then carefully remove the exhaust pipe (clearance is tight) to avoid damage.
    e. Loosen the attaching bolts, then remove the torque converter/flywheel cover.
16. If equipped, disconnect the turbocharger oil supply and drain lines. Remove the turbo-to-exhaust bolts, the turbo assembly, the lower cover and the gasket.

17. Unfasten the electrical connector from the O2 sensor. Remove the torque converter/flywheel-to-drive plate bolts.
18. Using a hoist and suitable engine lift, connect the hoist to the crankshaft damper bracket and lifting eye. Support the engine, but do not lift. Remove the upper engine-to-transaxle bolts; leave the starter in place.

✳✳ CAUTION

Never open, service or drain the radiator or cooling system when hot; serious burns can occur from the steam and hot coolant.

19. Drain the engine coolant into a suitable container. Disconnect the upper/lower radiator hoses, oil cooler lines at the radiator, if equipped, ground wire and radiator.

✳✳ CAUTION

The EPA warns that prolonged contact with used engine oil may cause a number of skin disorders, including cancer! You should make every effort to minimize your exposure to used engine oil. Protective gloves should be worn when changing the oil. Wash your hands and any other exposed skin areas as soon as possible after exposure to used engine oil. Soap and water, or waterless hand cleaner should be used.

20. Disconnect the oil cooler lines from the engine. Drain the crankcase oil into a suitable container. Disconnect the heater hoses from the side of the engine.
21. Remove the front engine mount, then the lower engine-to-transaxle nuts.
22. Position a jack under the transaxle, then raise the engine/transaxle slightly. Pull the engine forward until the transaxle shaft clears the torque converter/clutch, then carefully raise the engine out of the engine compartment.

To install:

23. Carefully install the engine into the engine compartment. Push the engine rearward and engage it with the transaxle. Mark sure the clutch clears the transaxle shaft.
24. Install the front engine mount.
25. Connect the heater hoses to the engine and connect the oil cooler lines.
26. Install the radiator and connect the transaxle oil cooler lines, ground wire and the upper and lower radiator hoses.
27. Install the upper engine-to-transaxle bolts and remove the engine lifting fixture. Tighten the bolts to 34–40 ft. lbs. (44–52 Nm).

28. Install the torque converter-to-drive-plate bolts. Tighten the bolts one at a time to 17–20 ft. lbs. (22–26 Nm).

29. Attach the O_2 sensor electrical connector.

30. To install the center section of the exhaust pipe, perform the following procedures:

    a. Install the torque converter cover.

    b. Carefully install the exhaust pipe and install the hangar bolts.

    c. Install the center exhaust section-to-transaxle bolt.

    d. Install the rear cover.

    e. Attach the temperature sensor connector.

31. Install the remaining exhaust components, adhering to the following torque specifications;

• Exhaust system-to-transaxle bolt to 18–25 ft. lbs. (24–34 Nm)

• Exhaust system hanger bolts to 7–13 ft. lbs. (9–18 Nm)

• Rear exhaust pipe joint nuts to 7–13 ft. lbs. (9–18 Nm)

32. Install the horizontal damper and clip.

33. Install the air intake-to-flow meter line.

34. Install air intake the duct and tighten the air intake duct hose clamps. Install the upper cover.

35. To install the power steering pump, perform the following procedures:

    a. Install the power steering pump on the mounting bracket.

    b. Install the engine oil filler pipe brace.

    c. Install the power steering pump-to-engine bolts and clamp. tighten the bolts to 22–36 ft. lbs. (29–47 Nm).

    d. Install the pulley on the power steering pump. Tighten the retainer bolts to 31–46 ft. lbs. (40–60 Nm).

    e. Install the alternator belt and tension the belt. Tighten the alternator pivot and mounting bolts.

36. Install the windshield washer reservoir.

37. Install the crankshaft damper by tightening the nuts on the body side of the damper until the clearance is 0.08 in. (2mm). Tighten the locknuts to 6–9 ft. lbs. (8–12 Nm).

38. Connect the spark plug wires, the engine ground strap and the fusible link assembly.

39. Connect the main engine harness. Attach the electrical wiring connectors to the alternator, EGI, thermoswitch, electric fan, A/C condenser and the ignition coil.

40. Connect the vacuum hoses to the cruise control, Master-Vac®, air intake shutter and the heater air intake door.

41. Connect the fuel system hoses and the evaporative emissions system hoses.

42. Adjust the accelerator pedal so there is 0.4–1.2 in. (10–30mm) between the pin and stop. Adjust the cable for an end-play of 0–0.08 in. (0–2mm) on the actuator side.

43. Install the air cleaner assembly.

44. Connect the negative battery cable.

45. Install the spare tire and the spare tire bracket.

✳✳ WARNING

Operating the engine without the proper amount and type of engine oil will result in severe engine damage.

46. Refill the crankcase with oil, the radiator with coolant and check the transmission fluid. Start the engine and check for leaks.

2.2L and 2.5L Engines

✳✳ CAUTION

The fuel injection system remains under pressure after the engine has been turned OFF. Properly relieve fuel pressure before disconnecting any fuel lines. Failure to do so may result in fire or personal injury.

1. Relieve the fuel system pressure.

✳✳ CAUTION

Observe all applicable safety precautions when working around fuel. Whenever servicing the fuel system, always work in a well ventilated area. Do not allow fuel spray or vapors to come in contact with a spark or open flame. Keep a dry chemical fire extinguisher near the work area. Always keep fuel in a container specifically designed for fuel storage; also, always properly seal fuel containers to avoid the possibility of fire or explosion.

2. Disconnect the battery cables, negative first, then positive. Remove the battery from the vehicle.

✳✳ CAUTION

Never open, service or drain the radiator or cooling system when hot; serious burns can occur from the steam and hot coolant.

3. Drain the engine oil and coolant into suitable containers.

✳✳ CAUTION

The EPA warns that prolonged contact with used engine oil may cause a number of skin disorders, including cancer! You should make every effort to minimize your exposure to used engine oil. Protective gloves should be worn when changing the oil. Wash your hands and any other exposed skin areas as soon as possible after exposure to used engine oil. Soap and water, or waterless hand cleaner should be used.

4. Disconnect the radiator hoses and fan motor harness, then remove the radiator.

5. If equipped with A/C, discharge the system using an approved recovery/recycling machine. Disconnect and cap the lines from the compressor.

6. Remove the air intake duct.

7. Remove the air cleaner element and upper cover.

8. Remove the evaporator canister and bracket.

9. Unfasten the following electrical connectors:

• O_2 sensor

• Engine ground terminal

• Crank angle sensor connector

• Cam angle sensor connector

• Knock sensor connector

• Alternator connector and terminal

• A/C compressor connectors, if equipped

• Accelerator cable

• Cruise control cable, if equipped

• Clutch release spring, clutch cable and hill holder cable, if equipped with a manual transaxle

10. Disconnect the following hoses:

• Brake booster hose

• Heater inlet and outlet hoses

11. Remove the alternator drive belt.

12. Disconnect the wires from the spark plugs on the left side of the engine.

13. Remove the power steering pump line bracket.

14. Remove the power steering pump, leaving the lines connected and position it aside.

15. Raise and support the engine safely.

16. Remove the exhaust Y-pipe.

17. Remove the lower starter nuts.

18. Remove the lower engine-to-transaxle nuts.

19. Remove the front engine mount-to-crossmember nuts.

20. Lower the vehicle.

21. Remove the starter.

22. If equipped with an automatic transaxle, perform the following:

    a. Remove the torque converter service hole plug.

    b. Rotate the engine remove the torque converter-to-drive plate bolts as they become accessible.

23. Remove the pitching stopper.

24. Disconnect the fuel delivery, return and evaporation hoses.

25. Support the engine with a suitable lifting device attached to the engine lifting eyes.

26. Slightly raise the engine.

27. Raise the transaxle with a floor jack.

28. Slowly remove the engine from the vehicle.

To install:

29. Apply a small amount of grease to the splines of the mainshaft.

30. Position the engine in the engine compartment and align it with the transaxle.

31. Install the engine and tighten the upper bolts to 34–40 ft. lbs. (44–54 Nm).

32. Remove the lifting device and floor jack.

33. Install the pitching stopper and tighten the bolts to the following specifications:

- Body side—49 ft. lbs. (67 Nm)
- Bracket side—40 ft. lbs. (54 Nm)

34. If equipped with an automatic transaxle, perform the following:

    a. Install the torque converter-to-drive plate bolts while rotating the engine, and tighten to 20 ft. lbs. (26 Nm).

    b. Install the service hole cover.

35. Install the evaporator canister and bracket.

36. Install the power steering pump. Tighten the retainer bolts to 22–36 ft. lbs. (29–47 Nm).

37. Install and tension the drive belt.

38. Install the starter. Tighten the bolts to 34–40 ft. lbs. (44–52 Nm).

39. Raise and support the vehicle safely.

40. Install the lower engine-to-transaxle nuts and tighten them to 34–40 ft. lbs. (44–52 Nm).

41. Install the lower engine mounting nuts. Tighten them to 61 ft. lbs. (83 Nm) in the inner most elliptical hole in the front crossmember so the clearance is 0.16–0.24 in. (4–6mm).

42. Install the exhaust Y-pipe with new gaskets and nuts.

43. Connect the following hoses:

- Brake booster hose
- Heater inlet and outlet hoses

44. Attach the following connectors:

- Accelerator cable
- Cruise control cable, if equipped

- Clutch release spring, clutch cable, hill holder cable, if equipped with a manual transaxle

45. Fasten the following electrical connectors:

- Engine harness connectors
- O$_2$sensor
- Engine ground terminal
- Crank angle sensor connector
- Cam angle sensor connector
- Knock sensor connector
- Alternator connector and terminal
- A/C compressor connectors, if equipped

46. Install the air cleaner element and cover.

47. If equipped, connect the A/C lines with new O-rings and tighten the bolts to 23 ft. lbs. (31 Nm).

48. Install the radiator.

49. Install the engine cover.

50. Install the battery.

51. Fill the engine with the recommended oil.

52. Fill and bleed the cooling system.

53. Charge the A/C system using an approved recovery/recycling machine.

54. Adjust the clutch cable.

55. If equipped, check the automatic transaxle fluid level and add Dexron II® if necessary.

56. Start the engine and allow it to reach normal operating temperature. Check for leaks.

3.3L Engine

✲✲ CAUTION

The fuel injection system remains under pressure after the engine has been turned OFF. Properly relieve fuel pressure before disconnecting any fuel lines. Failure to do so may result in fire or personal injury.

1. Raise and safely support the vehicle.

2. Scribe alignment marks on the hood hinges, then remove the hinge retainer bolts and hood.

✲✲ CAUTION

Observe all applicable safety precautions when working around fuel. Whenever servicing the fuel system, always work in a well ventilated area. Do not allow fuel spray or vapors to come in contact with a spark or open flame. Keep a dry chemical fire extinguisher near the work area. Always keep fuel in a container specifically designed for fuel

storage; also, always properly seal fuel containers to avoid the possibility of fire or explosion.

3. Release the fuel system pressure.

4. Disconnect the negative battery cable.

✲✲ CAUTION

Never open, service or drain the radiator or cooling system when hot; serious burns can occur from the steam and hot coolant.

5. Remove the underbody cover and drain the engine coolant.

6. Remove the radiator and all coolant hoses.

7. If equipped with A/C, discharge the air conditioning system using an approved recovery/recycling machine.

8. Disconnect and plug the air conditioning lines.

9. Remove the air cleaner assembly.

10. Disconnect the accelerator cable.

11. Disconnect the cruise control cable.

12. Label and disconnect all wiring harnesses and cables.

13. Tag and remove the evaporation canister, vacuum hoses and bracket.

14. Disconnect the exhaust system from the exhaust manifold.

15. Place a drain pan beneath the power steering lines and disconnect.

16. Disconnect the automatic transaxle cooler lines from the radiator, if equipped.

17. Remove the nuts which secure the lower side of the engine to the transaxle.

18. Remove the starter.

19. Remove the nuts which attach the front cushion rubber to the subframe.

20. If equipped with an automatic transaxle, separate the torque converter from the driveplate by removing the converter retainer bolts.

21. Remove the pitching stopper and bracket.

22. Disconnect the fuel delivery hose, return hose and evaporation hoses.

23. Support the engine with a suitable lifting device and the transaxle with a suitable transmission jack.

24. Remove the bolts which secure the engine to the transaxle.

25. Slowly raise the engine and remove from the vehicle.

To install:

26. Install the engine to the transaxle and tighten the bolts which secure the right upper side of the engine to 37 ft. lbs. (50 Nm).

27. Remove the lifting device and transmission jack.

28. Install the pitching stopper and tighten to 37 ft. lbs. (50 Nm).

29. if equipped with an automatic transaxle, install the torque converter to drive plate bolts and tighten to 19 ft. lbs. (25 Nm).

30. Connect all hoses disconnected earlier.

31. Install the evaporation canister and bracket.

32. Install the radiator and coolant lines.

33. Install the nuts which hold the lower side of the engine to the transaxle. Tighten the bolts to 37 ft. lbs. (50 Nm).

34. Install the starter and wiring. Tighten the bolts to 37 ft. lbs. (50 Nm).

35. Install the nuts which hold the front cushion rubber to the subframe. Tighten to 50 ft. lbs. (84 Nm).

36. Connect the power steering and automatic transaxle cooler lines.

37. Install the exhaust system to the manifold.

38. Install the engine under cover.

39. Connect all electrical harnesses.

40. Connect the accelerator cable.

41. Connect the cruise control cable.

42. Connect the high pressure hoses to the A/C compressor. Tighten to 18 ft. lbs. (24 Nm).

43. Install the air intake system

44. Connect the negative battery cable.

45. Fill the cooling system with coolant.

46. Check the automatic transaxle oil level and add as required.

47. Check the power steering fluid level. Add as necessary and bleed all air from the system.

48. Check the engine oil level.

49. Charge the A/C system using an approved recovery/recycling machine.

50. Start the engine and allow it to reach normal operating temperature. Check for leaks.

Water Pump

REMOVAL & INSTALLATION

All Engines

> ⁙ **CAUTION**

Never open, service or drain the radiator or cooling system when hot; serious burns can occur from the steam and hot coolant.

1. Disconnect the negative battery cable.
2. If equipped, remove the engine undercover.

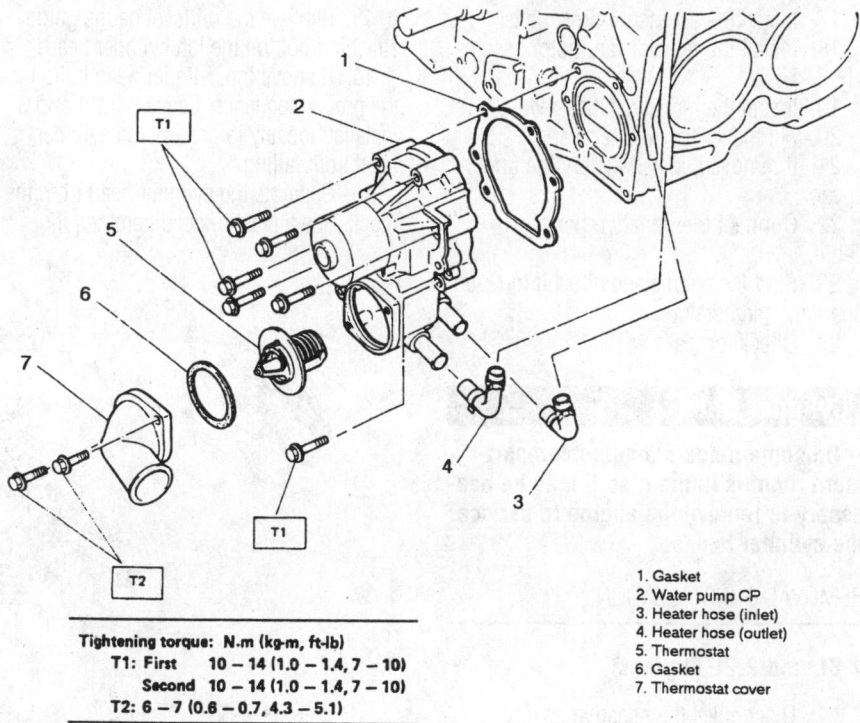

1. Gasket
2. Water pump CP
3. Heater hose (inlet)
4. Heater hose (outlet)
5. Thermostat
6. Gasket
7. Thermostat cover

7923TG01

Tightening torque: N.m (kg-m, ft-lb)		
T1: First	10 – 14 (1.0 – 1.4, 7 – 10)	
Second	10 – 14 (1.0 – 1.4, 7 – 10)	
T2: 6 – 7 (0.6 – 0.7, 4.3 – 5.1)		

Exploded view of the water pump assembly—all engines

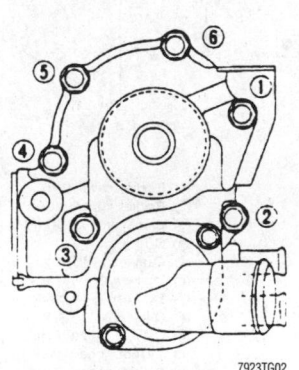

7923TG02

Water pump bolt tightening sequence— except 3.3L engine

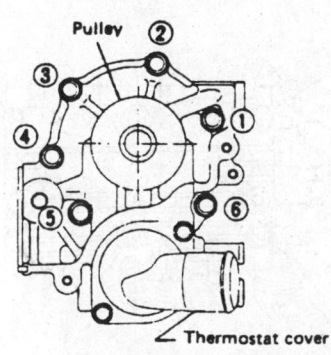

7923TG03

Water pump bolt tightening sequence— 3.3 engine

3. Drain the coolant into a suitable container.

4. Disconnect the radiator outlet hose.

5. Remove the radiator fan motor assembly.

6. Remove the accessory drive belts.

7. Remove the timing belt, tensioner and camshaft angle sensor.

8. Remove the left side camshaft pulley(s) and left side rear timing belt cover. Remove the tensioner bracket.

9. Disconnect the radiator hose and heater hose from the water pump.

10. Remove the water pump retainer bolts.

11. Remove the water pump.

To install:

12. Clean the gasket mating surfaces thoroughly. Always use new gaskets during installation.

13. Install the water pump and tighten the bolts, in sequence, to 7–10 ft. lbs. (10–14 Nm). After tightening the bolts once, retighten to the same specification again.

14. Inspect the radiator hoses for deterioration and replace as necessary. Connect the radiator hose and heater hose to the water pump.

15. Install the left side rear timing belt cover, left side camshaft pulley(s) and tensioner bracket.

16. Install the camshaft angle sensor, tensioner and timing belt.

17. Install the accessory drive belts.

18. Install the radiator ran motor assembly.

19. Install the radiator outlet hose.

20. Fill the system with coolant.

21. If removed, install the engine undercover.

22. Connect the negative battery cable.

23. Start the engine and allow it to reach operating temperature.

24. Check for leaks.

Cylinder Head

➡**On some models, engine compartment room is limited, so it may be necessary to remove the engine to service the cylinder heads.**

REMOVAL & INSTALLATION

1.8L and 2.2L Engines

1. Disconnect the negative battery cable.

2. Remove the drive belt.

3. Remove the power steering pump, alternator and bracket.

4. Remove the valve rocker cover.

5. Tag and disconnect the PCV hose and spark plug wires.

6. Remove the connector bracket attaching bolt.

7. Remove the crank angle and cam angle sensors.

8. Disconnect the oil pressure switch. Remove the knock sensor.

9. Disconnect the blow-by hose.

✳✳ CAUTION

Observe all applicable safety precautions when working around fuel. Whenever servicing the fuel system, always work in a well ventilated area. Do not allow fuel spray or vapors to come in contact with a spark or open flame. Keep a dry chemical fire extinguisher near the work area. Always keep fuel in a container specifically designed for fuel storage; also, always properly seal fuel containers to avoid the possibility of fire or explosion.

10. Relieve the fuel system pressure and disconnect the fuel pipes.

11. Remove the intake manifold and gasket. Remove the water pipe.

12. Remove the timing belt, camshaft sprocket and related components.

13. Remove the oil level gauge guide attaching bolt on the left cylinder head.

14. Remove the cylinder head bolts in the proper sequence. Leave bolts 1 and 3 installed loosely to prevent the cylinder head from falling.

15. Separate the cylinder head from the block, Use a plastic-faced hammer, if needed, to separate the head from the cylinder block.

16. Remove bolts 1 and 3. Remove the cylinder head and gasket.

17. Clean all gasket material from both mating surfaces.

To install:

18. Inspect the cylinder head for warpage. Warpage should not exceed 0.0020 in. (0.05mm).

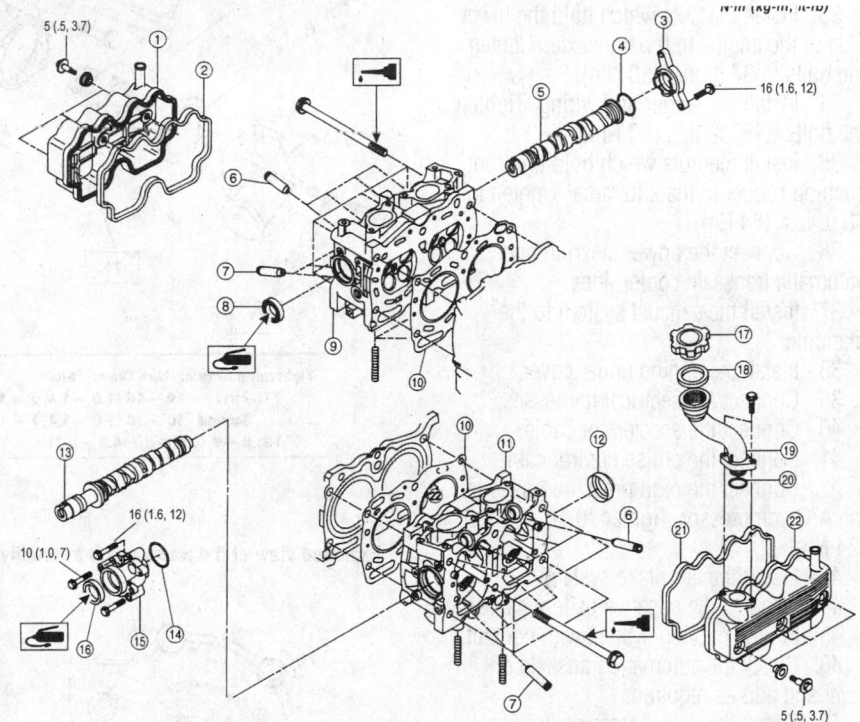

① Rocker cover (RH)	⑫ Plug
② Rocker cover gasket	⑬ Camshaft (LH)
③ Camshaft support (RH)	⑭ O-ring
④ O-ring	⑮ Camshaft support (LH)
⑤ Camshaft (RH)	⑯ Oil seal
⑥ Intake valve guide	⑰ Oil filler cap
⑦ Exhaust valve guide	⑱ Gasket
⑧ Oil seal	⑲ Oil filler pipe
⑨ Cylinder head (RH)	⑳ O-ring
⑩ Cylinder head gasket	㉑ Rocker gasket
⑪ Cylinder head (LH)	㉒ Rocker cover (LH)

7923TG04

Exploded view of the cylinder head assembly and related components—1.8L and 2.2L engines

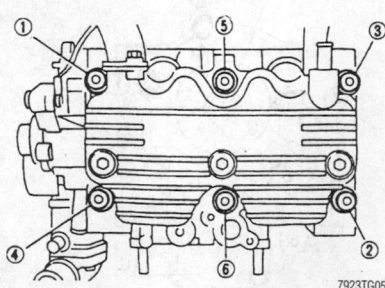

7923TG05

Cylinder head bolt loosening sequence—1.8L and 2.2L engines

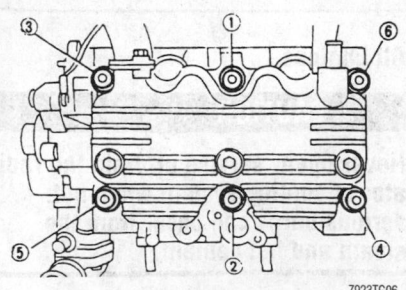

7923TG06

Cylinder head bolt tightening sequence—1.8L and 2.2L engines

19. Install the cylinder head on the block using a new gasket. Secure in place with the mounting bolts. Coat each bolts with clean engine oil, and hand-tighten.

20. Tighten the cylinder head bolts to the following specifications:

   a. Tighten all bolts in sequence to 22 ft. lbs. (29 Nm).

   b. Tighten all bolts in sequence to 51 ft. lbs. (69 Nm).

   c. Loosen all bolts by 180°, then loosen an additional 180°.

   d. Tighten bolts 1 and 2 to 25 ft. lbs. (34 Nm).

   e. Tighten bolts 3, 4, 5 and 6 to 11 ft. lbs. (15 Nm).

   f. Tighten all bolts in sequence 80–90°.

   g. Tighten all bolts in sequence an additional 80–90°.

➡ **Do not exceed 180° total tightening.**

21. Install the oil level gauge guide attaching bolt on the left cylinder head.

22. Install the timing belt, camshaft sprocket and related components.

23. Install the water pipe.

24. Install the intake manifold and tighten bolts to 21–25 ft. lbs. (28–34 Nm). Connect the fuel delivery pipes.

25. Connect the blow-by hose. Install the knock sensor.

26. Attach the oil pressure switch connector.

27. Install the crank and cam angle sensors.

28. Install the connector bracket attaching bolt.

29. Connect the spark plug wires. Connect the PCV hose.

30. Install the valve rocker cover and tighten bolts to 3.7 ft. lbs. (5 Nm).

31. Install the alternator, power steering pump and accessory drive belt.

32. Connect the negative battery cable. Start the engine and allow it to reach operating temperature. Check for leaks.

2.5L Engine

1. Disconnect the negative battery cable.
2. Remove the V-belt.
3. Remove the power steering pump, alternator and bracket.
4. Remove the valve rocker cover.
5. Remove the connector bracket attaching bolt.
6. Remove the crank angle and cam angle sensors.
7. Remove the coolant filler tank.

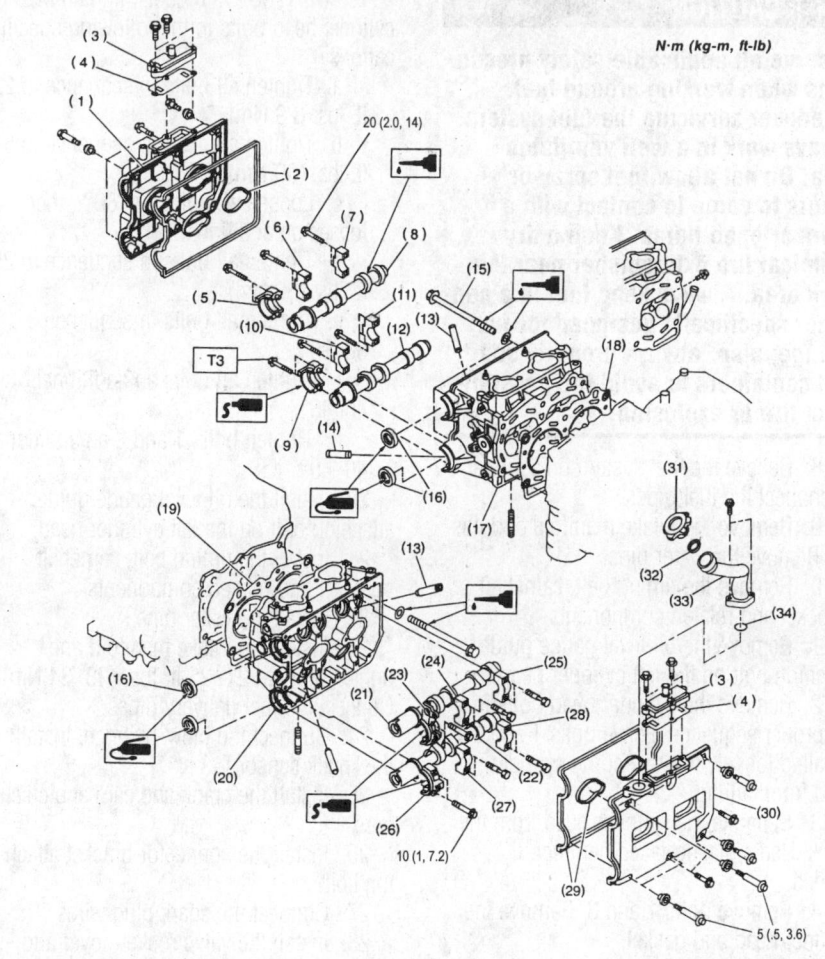

N·m (kg-m, ft-lb)

(1) Rocker cover (RH)	(15) Cylinder head bolt	(29) Rocker cover gasket (LH)
(2) Rocker cover gasket (RH)	(16) Oil seal	(30) Rocker cover (LH)
(3) Oil separator cover	(17) Cylinder head (RH)	(31) Oil filler cap
(4) Gasket	(18) Cylinder head gasket (RH)	(32) Gasket
(5) Intake camshaft cap (Front RH)	(19) Cylinder head gasket (LH)	(33) Oil filler duct
(6) Intake camshaft cap (Center RH)	(20) Cylinder head (LH)	(34) O-ring
(7) Intake camshaft cap (Rear RH)	(21) Intake camshaft (LH)	
(8) Intake camshaft (RH)	(22) Exhaust camshaft (LH)	
(9) Exhaust camshaft cap (Front RH)	(23) Intake camshaft cap (Front LH)	
(10) Exhaust camshaft cap (Center RH)	(24) Intake camshaft cap (Center LH)	
(11) Exhaust camshaft cap (Rear RH)	(25) Intake camshaft cap (Rear LH)	
(12) Exhaust camshaft (RH)	(26) Exhaust camshaft (Front LH)	
(13) Intake valve guide	(27) Exhaust camshaft cap (Center LH)	
(14) Exhaust valve guide	(28) Exhaust camshaft cap (Rear LH)	

7923TG07

Exploded view of the cylinder head and related components—2.5L engine

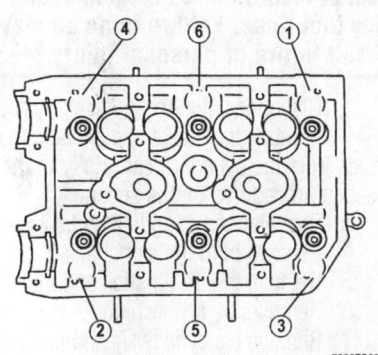

7923TG08

Cylinder head bolt loosening sequence—2.5L engine

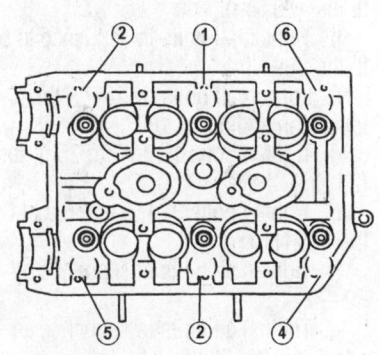

7923TG09

Cylinder head bolt tightening sequence—2.5L engine

8. Relieve the fuel system pressure and disconnect the fuel pipes.

9. Remove the intake manifold and gasket. Remove the water pipe.

10. Remove the timing belt, camshaft sprocket and related components.

11. Remove the oil level gauge guide attaching bolt on the left cylinder head.

12. Remove the cylinder head bolts in the proper sequence. Leave bolts 1 and 3 installed loosely to prevent the cylinder head from falling.

13. Separate the cylinder head from the block, Use a plastic-faced hammer, if needed.

14. Remove bolts 1 and 3. Remove the cylinder head and gasket.

15. Clean all gasket material from both mating surfaces.

To install:

16. Inspect the cylinder head for warpage. Warpage should not exceed 0.0020 in. (0.05mm).

17. Install the cylinder heads on the block using new gaskets. Secure in place with the mounting bolts. Coat each bolts with clean engine oil, and hand-tighten.

18. On 1996–97 models, tighten the cylinder head bolts to the following specifications:

   a. Tighten all bolts in sequence to 22 ft. lbs. (29 Nm).

   b. Tighten all bolts in sequence to 51 ft. lbs. (69 Nm).

   c. Loosen all bolts by 180°, then loosen an additional 180°.

   d. Tighten bolts 1 and 2 to 25 ft. lbs. (24 Nm).

   e. Tighten bolts 3, 4, 5 and 6 to 11 ft. lbs. (15 Nm).

   f. Tighten all bolts in sequence by 80–90°.

   g. Tighten all bolts in sequence an additional 80–90°.

➡**Do not exceed 180° total tightening.**

19. On 1998–99 models, tighten the cylinder head bolts to the following specifications:

   a. Tighten all bolts in sequence to 22 ft. lbs. (29 Nm).

   b. Tighten all bolts in sequence to 51 ft. lbs. (69 Nm).

   c. Loosen all bolts by 180°, then loosen an additional 180°.

   d. Tighten all bolts in sequence to 29 ft. lbs. (39 Nm).

   e. Tighten all bolts in sequence 80–90°.

   f. Tighten all bolts an additional 40–45°.

   g. Tighten bolts 1 and 2 an additional 40–45°.

20. Install the oil level gauge guide attaching bolt on the left cylinder head.

21. Install the timing belt, camshaft sprocket and related components.

22. Install the water pipe.

23. Install the intake manifold and tighten bolts to 21–25 ft. lbs. (28–34 Nm). Connect the fuel delivery pipes.

24. Connect the blow-by hose. Install the knock sensor.

25. Install the crank and cam angle sensors.

26. Install the connector bracket attaching bolt.

27. Connect the spark plug wires.

28. Install the valve rocker cover and tighten bolts to 4 ft. lbs. (9 Nm).

29. Install the alternator, power steering pump and accessory drive belt.

30. Connect the negative battery cable. Start the engine and allow it to reach operating temperature. Check for leaks.

3.3L Engine

1. Disconnect the negative battery cable.

2. Remove the intake and exhaust manifolds from the cylinder head. Refer to the needed procedures in this section.

3. Remove the timing belt and sprockets.

4. Remove both the camshafts.

5. Remove the oil dipstick and tube.

6. Remove the heater pipe.

7. Remove the cylinder head bolts in the proper sequence. DO NOT fully remove bolt number 5 or 8. One of these bolts should

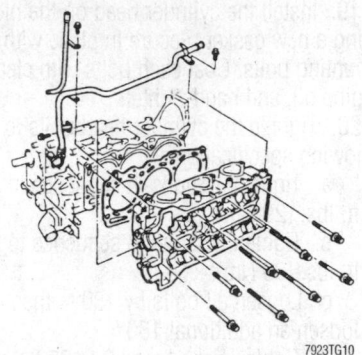

Exploded view of the cylinder head assembly—3.3L engine

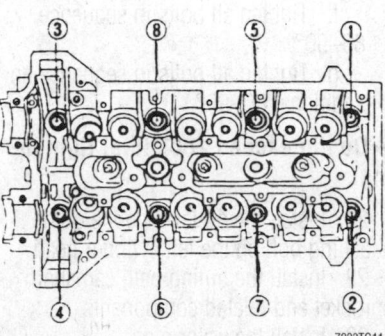

Cylinder head bolt loosening sequence—3.3L engine

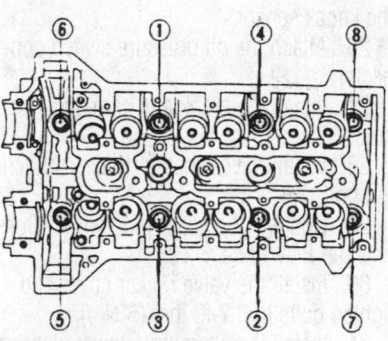

Cylinder head bolt tightening sequence—3.3L engine

remain three of four threads into the block to prevent the cylinder head from falling.

8. Free the cylinder head from the block, then remove the bolt left loosely in place. If the cylinder head will not come free easily, then tap the side of the head with a plastic-faced hammer.

9. Remove the cylinder head and gasket.

10. Clean all gasket material from both mating surfaces.

To install:

11. Install the cylinder head using a new gasket. Loosely install the cylinder head

mounting bolts, coating each with clean engine oil.

12. Tighten the mounting bolts as follows:

    a. Tighten the bolts in sequence to 22 ft. lbs. (29 Nm).

    b. Tighten the bolts in sequence to 51 ft. lbs. (69 Nm).

    c. Back off all the bolts 180°.

    d. Back off all bolts an additional 180°.

    e. Tighten the bolts in sequence to 20 ft. lbs. (27 Nm).

    f. Tighten bolts 1, 2, 3 and 4 an additional 90°.

    g. Tighten bolts 5, 6, 7, and 8 to 33 ft. lbs. (44 Nm).

    h. Tighten all bolts in sequence an additional 90°.

13. Install the camshafts.

14. Install the heater pipe.

15. Install the oil dipstick and tube.

16. Connect the intake and exhaust manifolds to the cylinder head.

17. Install the timing belt and sprockets.

18. Connect the negative battery cable.

Rocker Arms/Shafts

REMOVAL & INSTALLATION

1.8L and 2.2L Engines

1. Disconnect the PCV hose and remove the rocker cover.

2. Remove the valve rocker assembly by removing bolts 2 through 4 in numerical sequence.

3. Loosen bolt 1, but leave it engaged to retain the valve rocker assembly.

4. Remove bolts 5 through 8, taking care not to gouge the dowel pin.

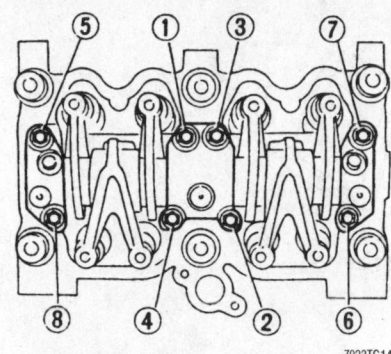

Rocker shaft bolt loosening/tightening sequence—1.8L and 2.2L engines

5. Remove the valve rocker assembly.

6. Place the valve rocker assembly with the air vent on the rocker arm facing upward into clean engine oil until ready to install. This is done to prevent damaging the hydraulic lash adjuster.

To install:

7. Install the valve rocker assembly on the cylinder head.

8. Temporarily tighten bolts 1 through 4 equally.

➡**Do not allow the valve rocker assembly to gouge the dowel pins.**

9. Tighten bolts 5 through 8 to 9 ft. lbs. (12 Nm).

10. Tighten bolts 1 through 4 to 9 ft. lbs. (12 Nm).

11. Install the rocker cover and connect the PCV hose.

2.5L and 3.3L Engines

These engines are not equipped with either rocker shafts or rocker arms. Instead, the camshaft drives the opening and closing of the individual valves directly.

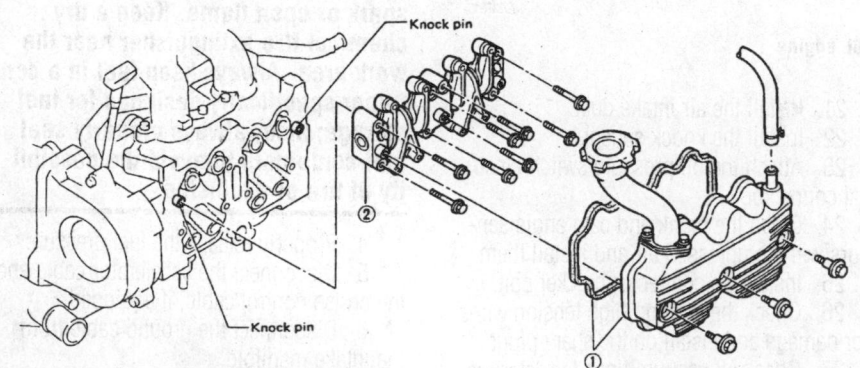

Exploded view of the cylinder head and rocker assembly—1.8L and 2.2L engines

Intake Manifold

REMOVAL & INSTALLATION

1.8L and 2.2L Engines

❊❊ CAUTION

Observe all applicable safety precautions when working around fuel. Whenever servicing the fuel system, always work in a well ventilated area. Do not allow fuel spray or vapors to come in contact with a spark or open flame. Keep a dry chemical fire extinguisher near the work area. Always keep fuel in a container specifically designed for fuel storage; also, always properly seal fuel containers to avoid the possibility of fire or explosion.

1. Release the fuel system pressure.

2. Disconnect the negative battery cable and remove the engine cover.

❊❊ CAUTION

Never open, service or drain the radiator or cooling system when hot; serious burns can occur from the steam and hot coolant.

3. Drain the cooling system into a suitable container.

4. Remove the accessory drive belt(s).

5. Remove power steering pump and or, alternator for added clearance.

6. Label and disconnect all electrical harnesses leading to the intake manifold.

7. Label and disconnect all vacuum hoses leading to the intake manifold. Disconnect the PCV and blow-by hoses.

8. Label and disconnect the ignition high tension wires at the spark plugs and lay them aside.

9. Remove the connector bracket attaching bolt.

10. Remove the crank angle sensor and cam angle sensor.

11. Detach the oil pressure switch connector.

12. Remove the knock sensor.

13. Disconnect the air intake duct.

14. Disconnect the fuel supply lines and accelerator linkage.

15. Remove the intake manifold bolts and remove the intake manifold and discard the gaskets.

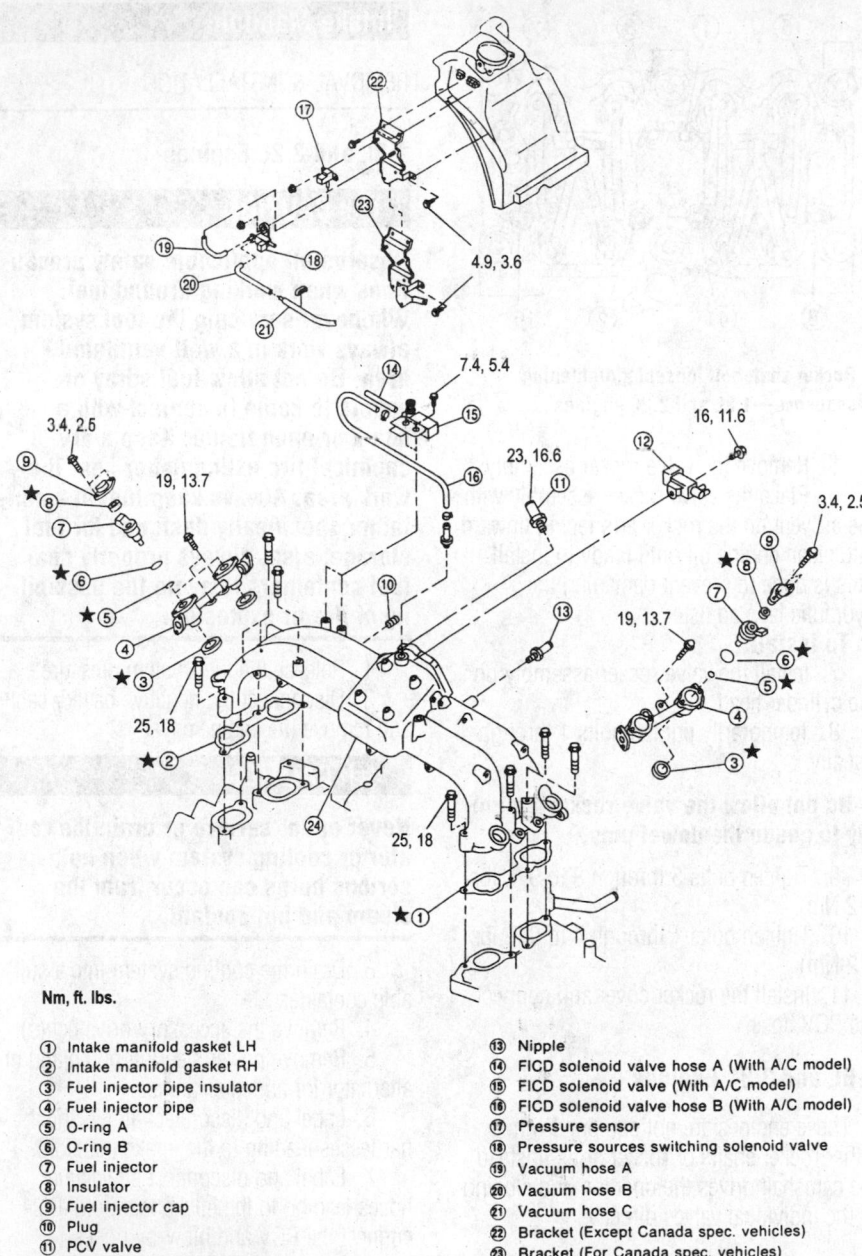

Nm, ft. lbs.

① Intake manifold gasket LH
② Intake manifold gasket RH
③ Fuel injector pipe insulator
④ Fuel injector pipe
⑤ O-ring A
⑥ O-ring B
⑦ Fuel injector
⑧ Insulator
⑨ Fuel injector cap
⑩ Plug
⑪ PCV valve
⑫ Purge control solenoid valve
⑬ Nipple
⑭ FICD solenoid valve hose A (With A/C model)
⑮ FICD solenoid valve (With A/C model)
⑯ FICD solenoid valve hose B (With A/C model)
⑰ Pressure sensor
⑱ Pressure sources switching solenoid valve
⑲ Vacuum hose A
⑳ Vacuum hose B
㉑ Vacuum hose C
㉒ Bracket (Except Canada spec. vehicles)
㉓ Bracket (For Canada spec. vehicles)
㉔ Intake manifold

7923TG15

Exploded view of the intake manifold assembly—1.8L engine

16. Remove the water pipe.
17. Clean all gasket material from both mating surfaces.
   To install:
18. Use a straightedge and a feeler gauge to inspect the intake manifold for flatness. Distortion should not exceed 0.020 in. (0.5mm).
19. Install the intake manifold and secure in place with retainer bolts. Tighten the short bolts to 21–25 ft. lbs. (28–34 Nm); the long bolts to 4–5 ft. lbs. (6–7 Nm).
20. Install the fuel lines and accelerator linkage.

21. Install the air intake duct.
22. Install the knock sensor.
23. Attach the oil pressure switch electrical connector.
24. Clean the crank and cam angle sensors with compressed air and install them.
25. Install the connector bracket bolt.
26. Check the ignition high tension wires for damage and install on the spark plugs.
27. Check all vacuum lines for deterioration and replace as necessary. Install the vacuum lines.
28. Check all electrical connectors for damage and replace as necessary. Install the electrical connectors.

29. Install the PCV valve and blow-by hose.
30. Install the power steering pump and alternator if removed.
31. Install the water pipes and fill the cooling system.
32. Install the engine cover.
33. Start the engine and allow it to reach operating temperature. Check for leaks and test drive the vehicle.

2.5L Engine

※※ **CAUTION**

The fuel injection system remains under pressure after the engine has been turned OFF. Properly relieve fuel pressure before disconnecting any fuel lines. Failure to do so may result in fire or personal injury.

1. Disconnect the negative battery cable.

※※ **CAUTION**

Never open, service or drain the radiator or cooling system when hot; serious burns can occur from the steam and hot coolant.

2. Drain the cooling system into a suitable container.
3. Remove the air intake duct, air cleaner upper cover and the air cleaner element.

※※ **CAUTION**

Observe all applicable safety precautions when working around fuel. Whenever servicing the fuel system, always work in a well ventilated area. Do not allow fuel spray or vapors to come in contact with a spark or open flame. Keep a dry chemical fire extinguisher near the work area. Always keep fuel in a container specifically designed for fuel storage; also, always properly seal fuel containers to avoid the possibility of fire or explosion.

4. Properly release the fuel pressure.
5. Disconnect the accelerator cable and the cruise control cable, if equipped.
6. Disconnect the ground cable from the intake manifold.
7. Disconnect the wiring harness from the throttle position sensor, fuel injectors, idle air control solenoid valve, purge control solenoid valve, and the exhaust gas recirculation solenoid valve.

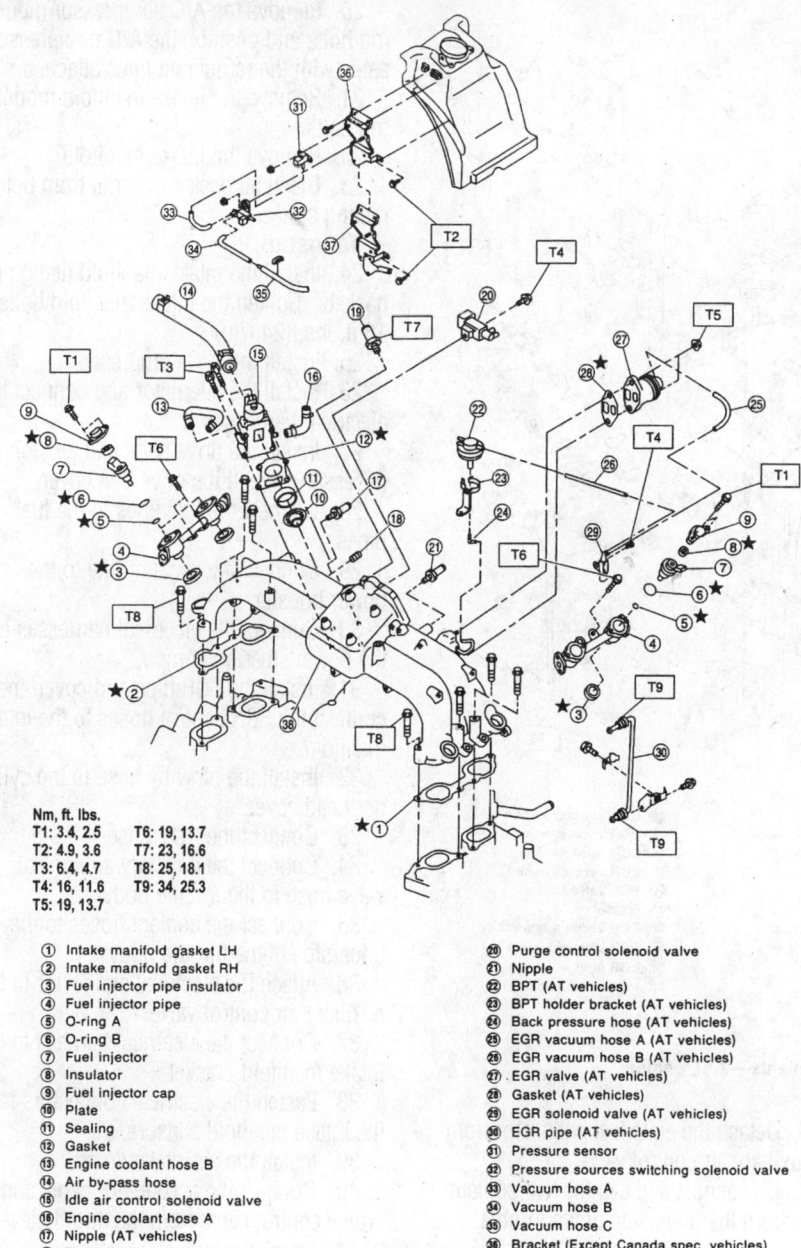

Nm, ft. lbs.
T1: 3.4, 2.5	T6: 19, 13.7
T2: 4.9, 3.6	T7: 23, 16.6
T3: 6.4, 4.7	T8: 25, 18.1
T4: 16, 11.6	T9: 34, 25.3
T5: 19, 13.7	

① Intake manifold gasket LH
② Intake manifold gasket RH
③ Fuel injector pipe insulator
④ Fuel injector pipe
⑤ O-ring A
⑥ O-ring B
⑦ Fuel injector
⑧ Insulator
⑨ Fuel injector cap
⑩ Plate
⑪ Sealing
⑫ Gasket
⑬ Engine coolant hose B
⑭ Air by-pass hose
⑮ Idle air control solenoid valve
⑯ Engine coolant hose A
⑰ Nipple (AT vehicles)
⑱ Plug
⑲ PCV valve

⑳ Purge control solenoid valve
㉑ Nipple
㉒ BPT (AT vehicles)
㉓ BPT holder bracket (AT vehicles)
㉔ Back pressure hose (AT vehicles)
㉕ EGR vacuum hose A (AT vehicles)
㉖ EGR vacuum hose B (AT vehicles)
㉗ EGR valve (AT vehicles)
㉘ Gasket (AT vehicles)
㉙ EGR solenoid valve (AT vehicles)
㉚ EGR pipe (AT vehicles)
㉛ Pressure sensor
㉜ Pressure sources switching solenoid valve
㉝ Vacuum hose A
㉞ Vacuum hose B
㉟ Vacuum hose C
㊱ Bracket (Except Canada spec. vehicles)
㊲ Bracket (For Canada spec. vehicles)
㊳ Intake manifold

7923TG16

Exploded view of the intake manifold assembly—2.2L engine

8. Disconnect the air bypass hose from the idle air control solenoid valve.

9. Remove the idle air control solenoid valve from the intake manifold.

10. Disconnect the engine coolant hoses from the throttle body.

11. Remove the throttle body from the intake manifold and discard the gasket.

12. Disconnect the fuel hoses from the fuel pipes.

13. Disconnect the EGR and the purge control solenoid valves.

14. Disconnect the wiring harness from the knock sensor, camshaft position sensor, crankshaft position sensor and the oil pressure switch.

15. Remove the intake manifold mounting bolts. Remove the manifold and discard the gaskets.

➡ **The intake manifold sits on pins that protrude from the cylinder heads. Be sure the pins remain in the cylinder heads.**

To install:

16. Using new gaskets, install the manifold to the engine. Tighten the mounting bolts to 19 ft. lbs. (26 Nm).

17. Connect the wiring to the knock sensor, camshaft position sensor, crankshaft position sensor and the oil pressure switch.

18. Connect the EGR and the purge control solenoid valves.

19. Connect the fuel hoses to the fuel pipes. Be sure to secure the hoses with new clamps.

20. Using new gaskets, install the throttle body to the intake manifold. Tighten the retaining bolts to 16 ft. lbs. (22 Nm).

21. Using new clamps, connect the engine coolant hoses to the throttle body.

22. Using a new gasket, install the idle air control solenoid valve to the intake manifold. Tighten the retaining bolts to 57 inch lbs. (6.4 Nm).

23. Connect the air bypass hose to the idle air control solenoid valve.

24. Connect the wiring harness to the throttle position sensor, fuel injectors, idle air control solenoid valve, purge control solenoid valve and the exhaust gas recirculation solenoid valve.

25. Connect the ground cable to the intake manifold.

26. Connect and adjust the accelerator cable and the cruise control cable.

27. Install the air cleaner assembly.

28. Connect the negative battery cable and refill the cooling system. Start the engine, and bleed the cooling system. Check for leaks.

3.3L Engine

✳ CAUTION

The fuel injection system remains under pressure after the engine has been turned OFF. Properly relieve fuel pressure before disconnecting any fuel lines. Failure to do so may result in fire or personal injury.

1. Relieve the fuel system pressure.

✳ CAUTION

Observe all applicable safety precautions when working around fuel. Whenever servicing the fuel system, always work in a well ventilated area. Do not allow fuel spray or vapors to come in contact with a spark or open flame. Keep a dry chemical fire extinguisher near the work area. Always keep fuel in a container specifically designed for fuel storage; also, always properly seal fuel containers to avoid the possibility of fire or explosion.

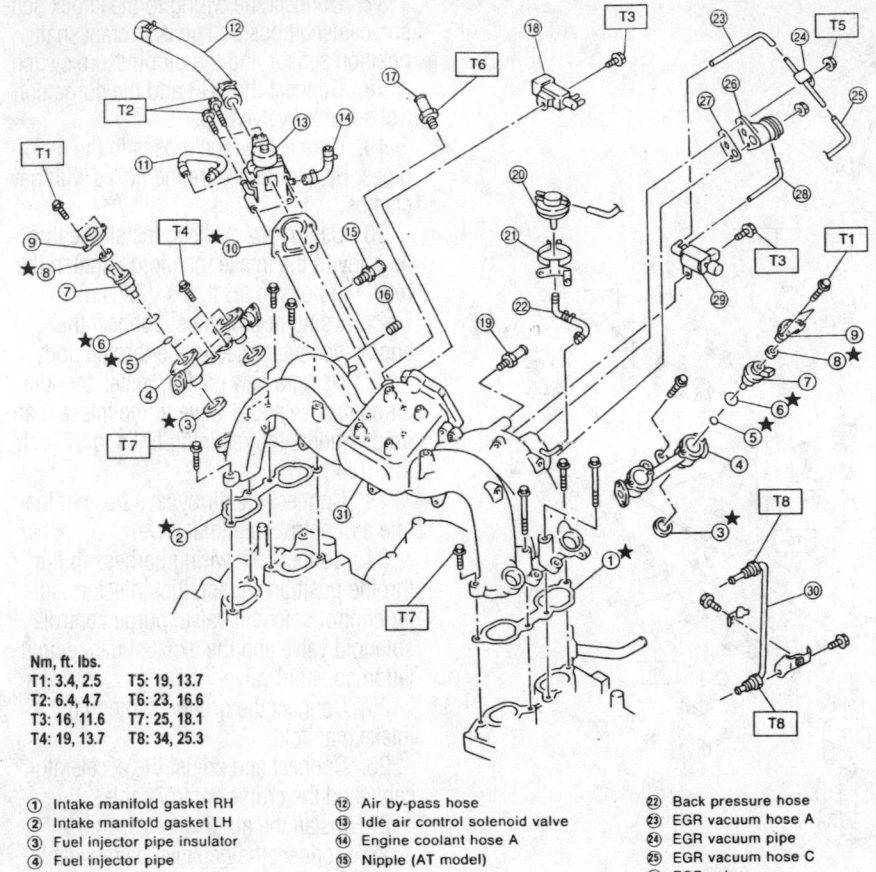

Nm, ft. lbs.	
T1: 3.4, 2.5	T5: 19, 13.7
T2: 6.4, 4.7	T6: 23, 16.6
T3: 16, 11.6	T7: 25, 18.1
T4: 19, 13.7	T8: 34, 25.3

① Intake manifold gasket RH
② Intake manifold gasket LH
③ Fuel injector pipe insulator
④ Fuel injector pipe
⑤ O-ring A
⑥ O-ring B
⑦ Fuel injector
⑧ Insulator
⑨ Fuel injector cap
⑩ Gasket
⑪ Engine coolant hose B

⑫ Air by-pass hose
⑬ Idle air control solenoid valve
⑭ Engine coolant hose A
⑮ Nipple (AT model)
⑯ Plug
⑰ PCV valve
⑱ Purge control solenoid valve
⑲ Nipple
⑳ BPT
㉑ BPT holder bracket

㉒ Back pressure hose
㉓ EGR vacuum hose A
㉔ EGR vacuum pipe
㉕ EGR vacuum hose C
㉖ EGR valve
㉗ Gasket
㉘ EGR vacuum hose B
㉙ EGR solenoid valve
㉚ EGR pipe
㉛ Intake manifold

7923TG17

Exploded view of the intake manifold and related components—2.5L engine

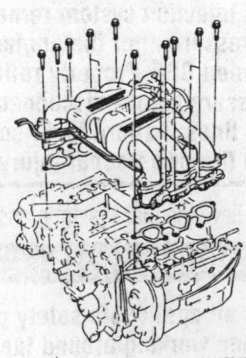

7923TG18

Exploded view of the intake manifold assembly—3.3L engine

2. Disconnect the negative battery cable.
3. Remove the collector cover.
4. Disconnect the accelerator cable and cruise control cable from the throttle lever, if equipped.
5. Remove the air intake ducts.
6. Disconnect the electrical harnesses from the intake manifold sensors and brackets.

7. Detach the electrical connector from the auxiliary air control valve.
8. Disconnect and cap the two coolant hoses from the underside of the throttle body.
9. Disconnect the auxiliary air control valve hose from the throttle body.
10. Disconnect the PCV hose.
11. Remove the blow-by hose from the cylinder head cover.
12. Disconnect the EGR control hoses from the intake manifold.
13. Remove the EGR pipe and cover.
14. Disconnect the power steering pump electrical harness
15. Disconnect the power brake booster vacuum hose.
16. Disconnect and cap the fuel lines from the fuel pipes.
17. Remove the drive belt cover and drive belts.
18. Unfasten the electrical connections from the alternator.
19. Remove the alternator.

20. Remove the A/C compressor mounting bolts and position the A/C compressor aside with the refrigerant lines attached.
21. Remove the intake manifold mounting bolts.
22. Remove the intake manifold.
23. Clean all gasket material from both mating surfaces.
To install:
24. Install the intake manifold using new gaskets. Tighten the intake manifold bolts to 18 ft. lbs. (24 Nm).
25. Install the A/C compressor.
26. Install the alternator and connect the electrical harnesses.
27. Install the drive belts and tension as necessary. Install the drive belt cover.
28. Connect the fuel lines to the fuel pipes.
29. Connect the vacuum line to the power booster.
30. Connect the electrical harnesses to the power steering pump.
31. Install the EGR pipe and cover and connect the EGR control hoses to the intake manifold.
32. Install the blow by hose to the cylinder head cover.
33. Connect the PCV hose.
34. Connect the auxiliary air control valve hose to the throttle body.
35. Connect the coolant hoses to the underside of the throttle body.
36. Attach the electrical connector to the auxiliary air control valve.
37. Connect the electrical harness to the intake manifold bracket.
38. Fasten the electrical connectors from the intake manifold sensors.
39. Install the air intake ducts.
40. Connect the accelerator cable and cruise control cable from the throttle lever.
41. Install the collector cover.
42. Connect the negative battery cable.
43. Pressurize the fuel system and check for leaks.

Exhaust Manifold

REMOVAL & INSTALLATION

Except 3.3L Engine

❊❊❊ CAUTION

The exhaust pipe may be hot; DO NOT perform any work until the system has completely cooled.

1. Disconnect the negative battery cable.
2. Detach the O₂ sensor electrical connector.

3. Remove the front under cover.

4. Remove the bolts securing the exhaust manifold covers and remove the covers.

5. Remove the front pipe-to-center pipe mounting nuts.

6. Remove the nuts that secure the exhaust pipe to the cylinder head and remove the exhaust pipe.

7. Discard the gaskets.

To install:

8. Clean all gasket surfaces completely.

9. Install the exhaust pipe to the cylinder head using new gaskets. Tighten the mounting nuts to 22–30 ft. lbs. (30–40 Nm).

10. Using new gaskets, connect the exhaust pipe to the center pipe. Tighten the mounting nuts to 9–17 ft. lbs. (13–23 Nm).

11. Install the exhaust manifold covers and cover mounting bolts.

12. Attach the O2sensor electrical connector.

13. Install the front under cover.

14. Lower the vehicle.

15. Connect the negative battery cable.

16. Start the engine and check for exhaust leaks.

3.3L Engine

1. Disconnect the negative battery cable.

2. Disconnect the O2sensor electrical harness.

3. Remove the front undercover, if equipped.

4. Remove the bolts securing the exhaust manifold covers and remove.

5. Remove the front pipe-to-exhaust manifold mounting nuts.

6. Disconnect the EGR pipe from the exhaust manifold, only if the right side manifold is being removed.

7. Remove the exhaust manifold mounting nuts, then remove the exhaust manifold.

8. Clean all gasket material from both mating surfaces.

To install:

9. Install the exhaust manifold using new gaskets. Install the mounting nuts and tighten the nuts to 25–33 ft. lbs. (32–43 Nm).

10. Connect the EGR pipe to the right manifold, if disconnected.

11. Connect the exhaust pipe to the exhaust manifold and tighten the mounting nuts to 22–29 ft. lbs. (29–38 Nm).

12. Install the exhaust manifold covers and cover mounting bolts. Tighten the bolts to 13–15 ft. lbs. (17–19 Nm).

13. Install the front under cover.

14. Connect the O2sensor electrical harness.

15. Lower the vehicle.

16. Connect the negative battery cable.

Front Crankshaft Seal

REMOVAL & INSTALLATION

1.8L Engine

1. Disconnect the negative battery cable.

2. Loosen the water pump pulley mounting bolts.

3. Loosen the alternator mounting bolts and remove the alternator belt.

4. Remove the water pump pulley bolts and remove the water pump pulley.

5. Disconnect the electrical harness from the oil pressure switch.

6. Remove the dipstick tube.

7. Remove the crankshaft pulley.

8. Remove the timing belt cover plate, if equipped.

9. Remove the timing belt cover bolts on the left, right, and center covers.

10. Remove the timing belt covers.

11. Remove the rubber cover seal. Clean all mating surfaces.

To install:

12. Install a new timing belt cover seal.

13. Install the timing belt covers on the engine using the mounting bolts. Tighten the bolts to 48 inch lbs. (5 Nm).

14. Install the timing belt cover plate, if removed.

15. Install the crankshaft pulley, and tighten the bolt to 69–76 ft. lbs. (93–103 Nm)

16. Install the dipstick tube.

17. Connect the electrical harness to the oil pressure switch.

18. Install the water pump pulley and loosely install the mounting bolts.

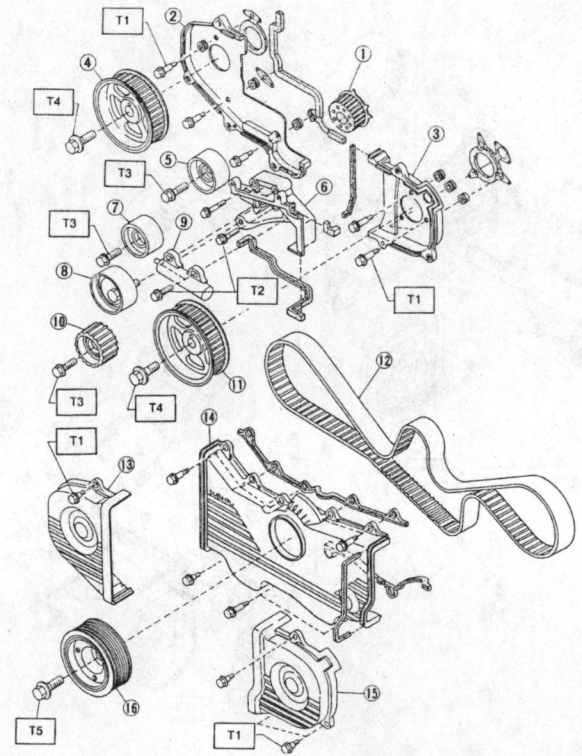

① Crankshaft sprocket
② Right-hand belt cover No. 2
③ Left-hand belt cover No. 2
④ Right-hand camshaft sprocket
⑤ Belt idler
⑥ Tensioner bracket
⑦ Belt idler
⑧ Belt tensioner
⑨ Tensioner adjuster
⑩ Belt idler No. 2
⑪ Left-hand camshaft sprocket
⑫ Timing belt

⑬ Right-hand belt cover
⑭ Front belt cover
⑮ Left-hand belt cover
⑯ Crankshaft pulley

Tightening torque: N·m (kg-m, ft-lb)
T1: 5 (0.5, 3.6)
T2: 23 — 26 (2.3 — 2.7, 17 — 20)
T3: 35 — 43 (3.6 — 4.4, 26 — 32)
T4: 74 — 83 (7.5 — 8.5, 54 — 61)
T5: 93 — 103 (9.5 — 10.5, 69 — 76)

7923TG19

Exploded view of the timing belt covers and components—1.8L engine

19. Install the drive belt and adjust the tension as necessary. Tighten the alternator mounting bolts.

20. Tighten the water pump pulley mounting bolts.

21. Connect the negative battery cable.

2.2L Engine

1. Disconnect the negative battery cable.

2. Remove the accessory drive belts.

3. Remove the power steering pump, alternator, air conditioner compressor brackets.

4. Secure the crankshaft pulley with tool No. 499977000 or equivalent.

5. Remove the crankshaft pulley bolt and pulley.

6. Remove the timing belt cover mounting bolts.

7. Remove the belt covers.

8. Remove the rubber cover seal. Clean all mating surfaces.

To install:

9. Install a new timing belt cover seal.

10. Install the belt covers and tighten the bolts to 36–48 inch lbs. (4–5 Nm).

11. Install the crankshaft pulley and tighten the bolt to 69–76 ft. lbs. (93–103 Nm).

12. Install the power steering pump, alternator, air conditioner compressor and associated brackets.

13. Install the accessory drive belts.

14. Connect the negative battery cable.

2.5L Engine

1. Disconnect the negative battery cable.

2. Detach the radiator electric fan motor wiring connectors.

3. Remove the coolant reservoir tank.

4. Remove the four bolts that secure the radiator shroud and remove the fan assembly.

5. Position the No. 1 piston to TDC of its compression stroke.

6. Remove the drive belt and the A/C compressor drive belt tensioner.

7. Remove the accessory drive belt cover.

8. Secure the crankshaft pulley with tool No. ST499977000 or equivalent.

9. Remove the crankshaft pulley bolt and pulley.

10. Remove the left timing belt cover mounting bolts and remove the left cover.

11. Remove the right timing belt cover mounting bolts and remove the right cover.

12. Remove the center timing belt cover mounting bolts and remove the center cover.

13. Remove the timing belt cover rubber seal and discard.

To install:

14. Install a new timing belt cover rubber seal.

15. Install the center, right, then the left timing belt covers and tighten the bolts to 44 inch lbs. (5 Nm).

16. Install the crankshaft pulley and tighten the bolt to 94 ft. lbs. (127 Nm).

17. Install the A/C compressor drive belt tensioner and install the drive belts.

18. Install the fan shroud and fan motor assembly.

19. Install the accessory drive belt cover.

20. Connect the negative battery cable.

3.3L Engine

1. Disconnect the negative battery cable.

2. Remove the drive belts.

3. Remove the nut securing the power steering pump pulley and remove the pulley.

4. Remove the alternator and belt cover bracket.

5. Remove the power steering pump bracket.

6. Remove the A/C belt tensioner bracket.

7. Remove the A/C compressor and set aside with the lines attached.

8. Remove the A/C compressor bracket.

9. Remove the crankshaft pulley bolt.

10. Remove the crankshaft pulley.

11. Remove the timing belt cover mounting bolts and remove the left, right and center covers.

12. Remove the timing belt cover seal.

To install:

13. Install a new timing belt cover seal.

14. Install the center, right and left covers. Tighten the cover mounting bolts to 48 inch lbs. (5 Nm).

15. Install the crankshaft pulley and tighten the mounting bolt to 115 ft. lbs. (157 Nm).

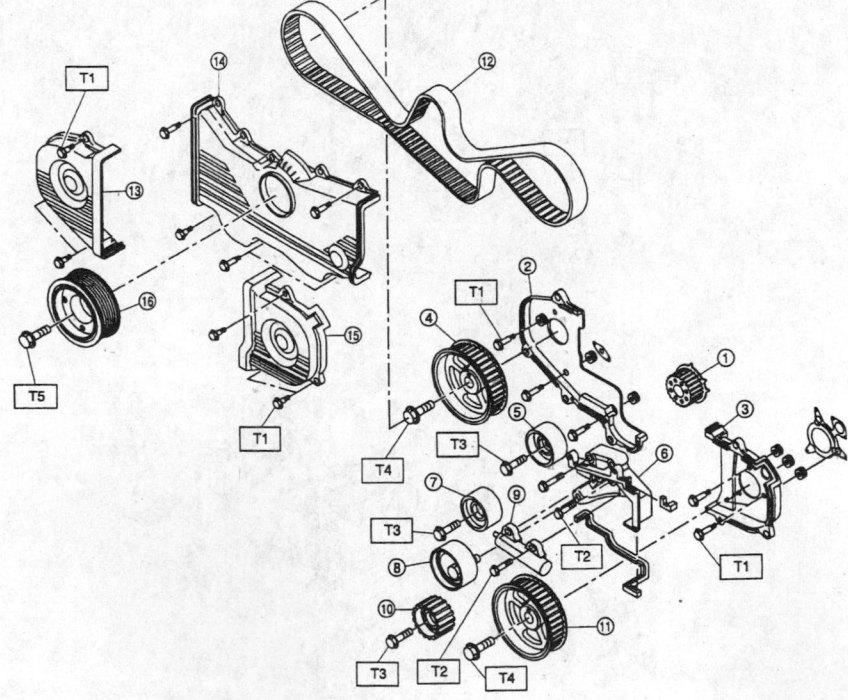

① Crankshaft sprocket
② Belt cover No. 2 (RH)
③ Belt cover No. 2 (LH)
④ Camshaft sprocket (RH)
⑤ Belt idler
⑥ Tensioner bracket
⑦ Belt idler
⑧ Belt tensioner
⑨ Tensioner adjuster
⑩ Belt idler No. 2
⑪ Camshaft sprocket (LH)
⑫ Timing belt

⑬ Belt cover (RH)
⑭ Front belt cover
⑮ Belt cover (LH)
⑯ Crankshaft pulley

Tightening torque: N·m (kg-m, ft-lb)
T1: 5 ± 1 $(0.5 \pm 0.1, 3.6 \pm 0.7)$
T2: 25 ± 2 $(2.5 \pm 0.2, 18.1 \pm 1.4)$
T3: 39 ± 4 $(4.0 \pm 0.4, 28.9 \pm 2.9)$
T4: 78 ± 5 $(8.0 \pm 0.5, 57.9 \pm 3.6)$
T5: 108^{+10}_{-5} $(11^{+1.0}_{-0.5}, 79.6^{+7.2}_{-3.6})$

7923TG20

Exploded view of the timing belt covers and components—2.2L engine

16. Install the A/C compressor bracket.
17. Install the A/C compressor.
18. Install the power steering pump bracket.
19. Install the alternator and belt cover bracket.
20. Install the power steering pump pulley and pulley mounting nut.

21. Install the drive belts.
22. Connect the negative battery cable.

Camshaft

On some models, it may be necessary to remove the engine from the vehicle to perform this service. The models covered are not equipped with valve lifters, they are equipped with Hydraulic Lash Adjuster's (HLA's) built into the end of the rocker arm. Servicing of the HLA's in covered under the rocker arm procedure.

REMOVAL & INSTALLATION

1.8L Engine

1. Disconnect the negative battery cable.
2. Tag and disconnect the spark plug wires from the distributor cap. Matchmark the distributor to the engine and remove the distributor.
3. Remove the timing belts covers and timing belts.

❊❊ CAUTION

Never open, service or drain the radiator or cooling system when hot; serious burns can occur from the steam and hot coolant.

4. Drain the cooling system.
5. Remove the water pipe.
6. Remove the oil fill pipe.
7. Disconnect the PCV hoses from the valve covers.
8. Remove the EGR pipe cover, clamps and EGR pipe, if equipped.
9. Remove the rocker arm covers.

➡**When removing the camshaft carrier the rocker arms may fall off the lash adjusters. Place a suitable container beneath the cylinder head to catch the rocker arms.**

10. Remove the camshaft case mounting bolts.
11. Remove the camshaft case and place on a clean surface.
12. Remove the camshaft retainer from the front of the camshaft carrier.
13. Slide the camshaft out from the carrier.
To install:
14. Slide the camshaft into the carrier and install the camshaft retainer and bolts. Tighten the retainer bolts to 5 ft. lbs. (7 Nm).
15. Apply grease to the spherical and sliding surface of each valve rocker, then secure the rockers to the valve lash adjusters and valves.

➡**Failure to apply grease to the rocker arms will result in the rocker arms falling off.**

16. Apply sealer to the groove of the camshaft carrier.
17. Install the carrier assembly using the mounting bolts. Tighten the bolts to 14 ft. lbs. (19 Nm).

Nm, ft. lbs.

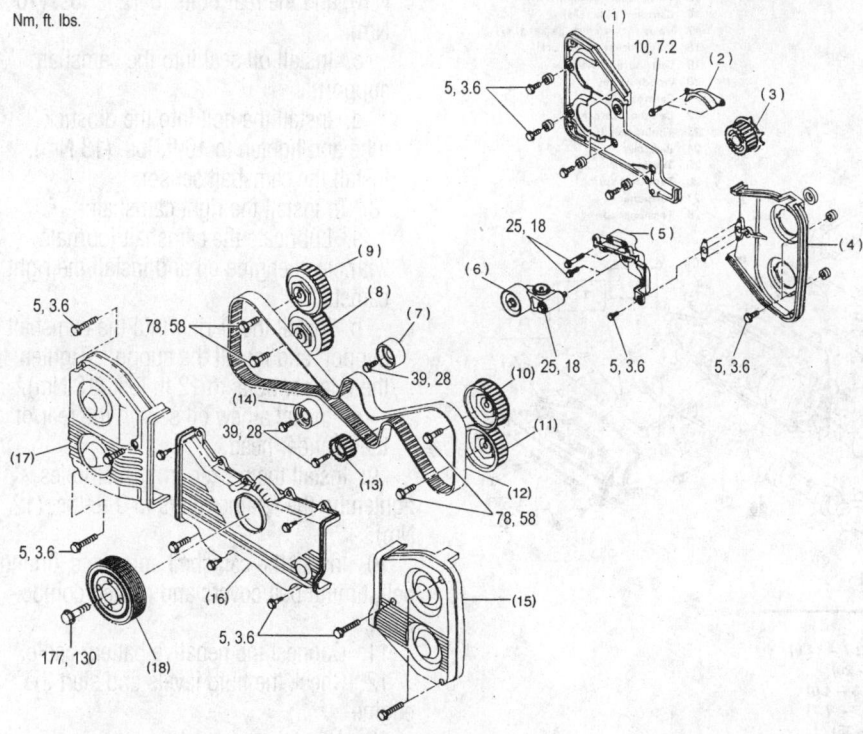

(1) Right-hand belt cover No. 2
(2) Timing belt guide (MT vehicles only)
(3) Crankshaft sprocket
(4) Left-hand belt cover No. 2
(5) Tensioner bracket
(6) Automatic belt tension adjuster ASSY
(7) Belt idler
(8) Right-hand exhaust camshaft sprocket
(9) Right-hand intake camshaft sprocket
(10) Left-hand intake camshaft sprocket
(11) Left-hand exhaust camshaft sprocket
(12) Timing belt
(13) Belt idler No. 2
(14) Belt idler
(15) Left-hand belt cover
(16) Front belt cover
(17) Right-hand belt cover
(18) Crankshaft pulley

7923TG21

Exploded view of the timing belt covers and components—2.5L engine

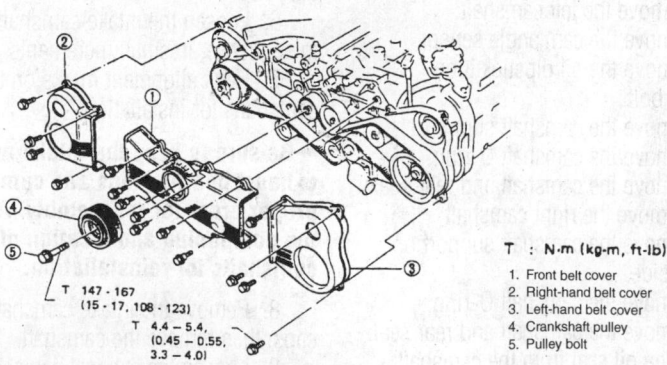

T : N·m (kg-m, ft-lb)

1. Front belt cover
2. Right-hand belt cover
3. Left-hand belt cover
4. Crankshaft pulley
5. Pulley bolt

7923TG22

Exploded view of the timing belt cover and components—3.3L engine

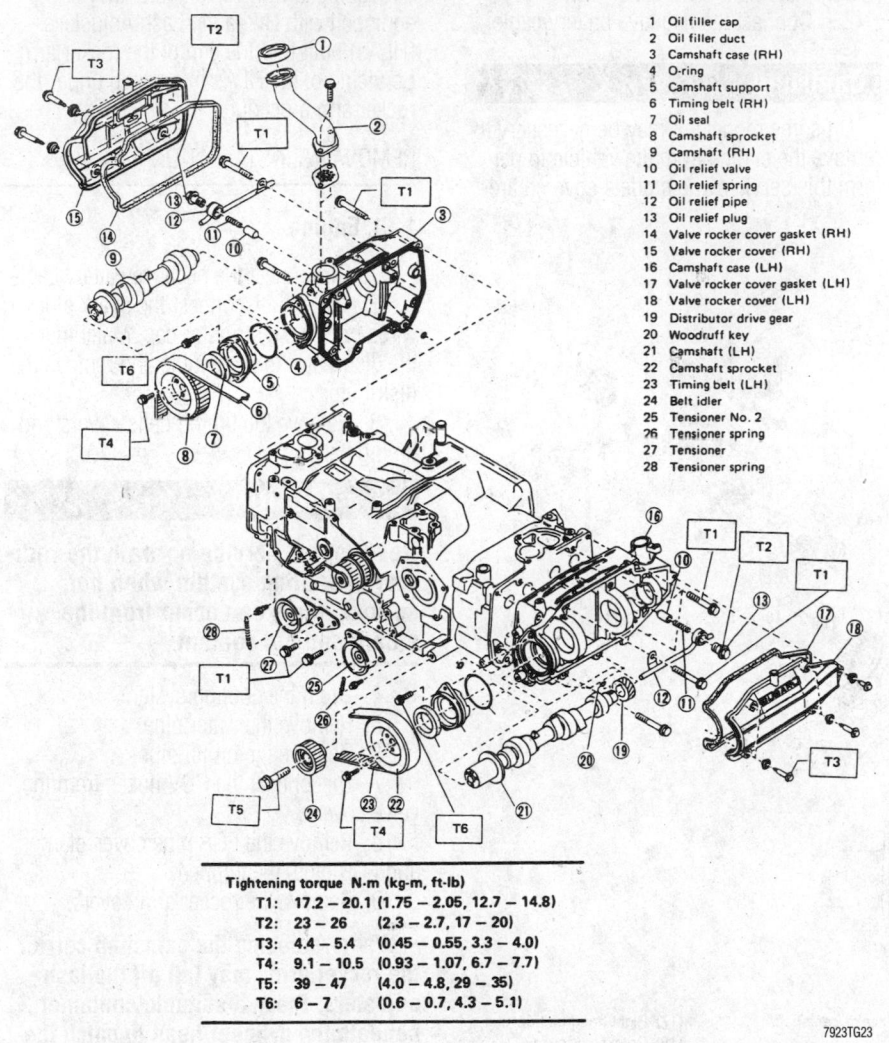

1	Oil filler cap
2	Oil filler duct
3	Camshaft case (RH)
4	O-ring
5	Camshaft support
6	Timing belt (RH)
7	Oil seal
8	Camshaft sprocket
9	Camshaft (RH)
10	Oil relief valve
11	Oil relief spring
12	Oil relief pipe
13	Oil relief plug
14	Valve rocker cover gasket (RH)
15	Valve rocker cover (RH)
16	Camshaft case (LH)
17	Valve rocker cover gasket (LH)
18	Valve rocker cover (LH)
19	Distributor drive gear
20	Woodruff key
21	Camshaft (LH)
22	Camshaft sprocket
23	Timing belt (LH)
24	Belt idler
25	Tensioner No. 2
26	Tensioner spring
27	Tensioner
28	Tensioner spring

Tightening torque N·m (kg-m, ft-lb)

	N·m	(kg-m, ft-lb)
T1:	17.2 – 20.1	(1.75 – 2.05, 12.7 – 14.8)
T2:	23 – 26	(2.3 – 2.7, 17 – 20)
T3:	4.4 – 5.4	(0.45 – 0.55, 3.3 – 4.0)
T4:	9.1 – 10.5	(0.93 – 1.07, 6.7 – 7.7)
T5:	39 – 47	(4.0 – 4.8, 29 – 35)
T6:	6 – 7	(0.6 – 0.7, 4.3 – 5.1)

7923TG23

Exploded view of the camshaft assembly—1.8L engine

18. Coat the camshaft with clean engine oil.

19. Install the rocker arm covers using new gaskets. Install the mounting bolts and tighten to 4 ft. lbs. (5 Nm).

20. Connect the PCV hose to the rocker arm cover.

21. Install the EGR pipe, clamps, and EGR pipe cover, if removed.

22. Install the oil filler pipe.

23. Install the water pipe.

24. Install the timing belts and timing belt covers.

25. Install the distributor and mounting bolts remembering to align the marks made during removal.

26. Connect the plug wires to the distributor.

27. Connect the negative battery cable.

28. Start the vehicle, adjust the ignition timing if needed and check for fluid leaks.

2.2L Engine

1. Disconnect the negative battery cable.

2. Remove the timing belt covers, timing belt and camshaft sprockets.

3. Remove the valve rocker covers.

4. Remove the rocker arm assemblies.

5. To remove the left camshaft:

a. Remove the cam angle sensor.

b. Remove the oil dipstick tube attaching bolt.

c. Remove the camshaft support.

d. Remove the camshaft O-ring.

e. Remove the camshaft and rear seal.

6. To remove the right camshaft:

a. Remove the camshaft support on the right side.

b. Remove the camshaft O-ring.

c. Remove the camshaft and rear seal. Remove the oil seal from the camshaft support.

To install:

7. To install the left camshaft:

a. Lubricate the camshaft journals with clean engine oil. Install the rear oil seal, then install the camshaft into the cylinder head.

b. Install the O-ring into the camshaft support and install the support. Tighten the front retainer bolts to 7 ft. lbs. (9 Nm), and the rear bolts to 12 ft. lbs. (16 Nm).

c. Install oil seal into the camshaft support.

d. Install the bolt into the dipstick tube and tighten to 10 ft. lbs. (13 Nm). Install the camshaft sensor.

8. To install the right camshaft:

a. Lubricate the camshaft journals with clean engine oil and install the right camshaft.

b. Install the O-ring into the camshaft support and install the support. Tighten the retainer bolts to 12 ft. lbs. (16 Nm).

c. Install a new oil seal in the rear of the cylinder head.

9. Install the rocker arm assemblies, tightening the retainer bolts to 9 ft. lbs. (12 Nm).

10. Install the camshaft sprockets, timing belt, timing belt covers and related components.

11. Connect the negative battery cable.

12. Check the fluid levels and start the engine.

13. Allow the engine to reach normal operating temperature and check for leaks.

2.5L Engine

1. Disconnect the negative battery cable.

2. Remove the timing belt covers, timing belt and camshaft sprockets.

3. Remove the camshaft position sensor.

4. Remove the ignition coils.

5. Remove the valve rocker covers and gaskets.

6. Loosen the intake camshaft cap bolts in sequence, in small increments.

7. Paint alignment marks on the camshafts for installation.

➡**Be sure to keep the intake and exhaust bearing caps and camshafts in proper order for reassembly. Also note the positioning and location of the camshafts for reinstallation.**

8. Remove the intake camshaft bearing caps, then remove the camshaft.

9. Loosen the exhaust camshaft cap bolts in sequence, in small increments.

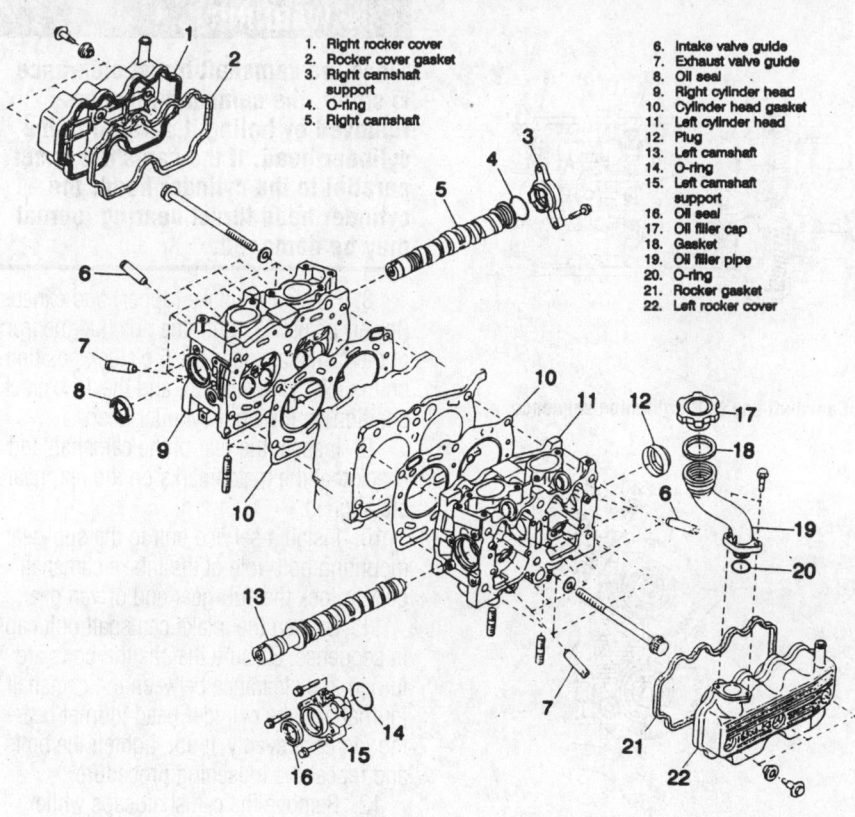

1. Right rocker cover
2. Rocker cover gasket
3. Right camshaft support
4. O-ring
5. Right camshaft

6. Intake valve guide
7. Exhaust valve guide
8. Oil seal
9. Right cylinder head
10. Cylinder head gasket
11. Left cylinder head
12. Plug
13. Left camshaft
14. O-ring
15. Left camshaft support
16. Oil seal
17. Oil filler cap
18. Gasket
19. Oil filler pipe
20. O-ring
21. Rocker gasket
22. Left rocker cover

Exploded view of the camshaft assembly—2.2L engine

7923TG24

10. Remove the exhaust camshaft bearing caps, then remove the camshaft.

To install:

➡ **Lubricate the camshaft bearings prior to camshaft installation.**

11. Install the camshafts so the base circle (non-lobe portion) of the camshafts are in contact with the lash adjusters. This will position the lobes of the camshafts away from the valves.

➡ **The left camshaft will need to be rotated for timing belt alignment.**

12. Apply liquid sealant to the front bearing cap mating surfaces, then install the bearing caps. Tighten the caps in sequence in two progressive steps to 14.5 ft. lbs. (20 Nm).

13. Install new oil seals to the camshafts using a suitable seal installation tool.

※※ WARNING

Only rotate camshafts the specified amount. If the camshafts are rotated beyond the specified amount, the valves will contact each other and cause severe internal damage.

14. For correct timing belt alignment, rotate the intake camshaft 80 degrees clock-

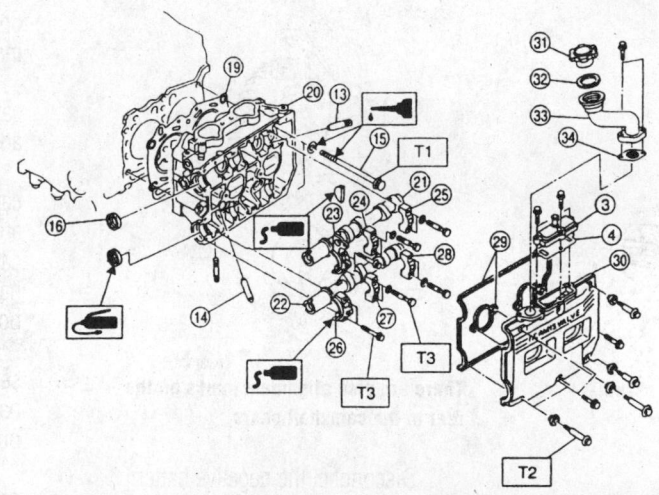

① Rocker cover (RH)
② Rocker cover gasket (RH)
⑨ Exhaust camshaft cap (Front RH)
⑩ Exhaust camshaft cap (Center RH)
⑫ Exhaust camshaft (RH)
⑮ Cylinder head bolt
⑯ Oil seal
⑲ Cylinder head gasket (LH)
⑳ Cylinder head (LH)
㉑ Intake camshaft (LH)
㉒ Exhaust camshaft (LH)
㉓ Intake camshaft cap (Front LH)
㉔ Intake camshaft cap (Center LH)

㉕ Intake camshaft cap (Rear LH)
㉖ Exhaust camshaft (Front LH)
㉗ Exhaust camshaft cap (Center LH)
㉘ Exhaust camshaft cap (Rear LH)
㉙ Rocker cover gasket (LH)
㉚ Rocker cover (LH)
㉛ Oil filler cap
㉜ Gasket
㉝ Oil filler duct
㉞ Gasket

Tightening torque: N·m (kg-m, ft-lb)
T1: Refer to [W4E1]☆2.
T2: 5 (0.5, 3.6)
T3: 10 (1.0, 7)

Exploded view of the camshaft and cylinder head assembly—2.5L engine

7923TG25

wise and the exhaust camshaft 45 degrees counterclockwise.

15. Using a new gasket, install the rocker covers. Be sure to apply liquid sealant to the front edges of the gasket at the camshaft opening.

16. Install the ignition coils.

17. Install the camshaft position sensor.

18. Install the camshaft sprockets and tighten the retaining bolts to 58 ft. lbs. (78 Nm). Be sure to secure the sprockets when tightening the bolts.

19. Check the timing sprockets for proper alignment and install the timing belt.

20. Connect the negative battery cable.

21. Check the fluid levels and start the engine.

22. Allow the engine to reach operating temperature and check for leaks.

3.3L Engine

❊ CAUTION

The fuel injection system remains under pressure after the engine has been turned OFF. Properly relieve fuel pressure before disconnecting any fuel lines. Failure to do so may result in fire or personal injury.

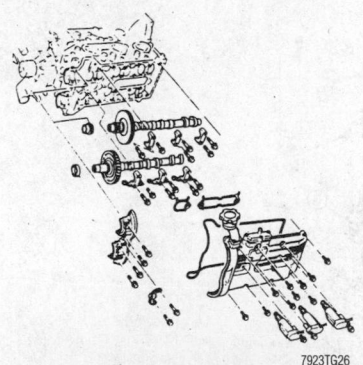

Exploded view of the camshaft assembly—3.3L engine

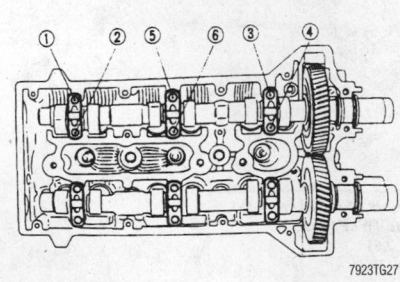

Camshaft cap bolt removal sequence

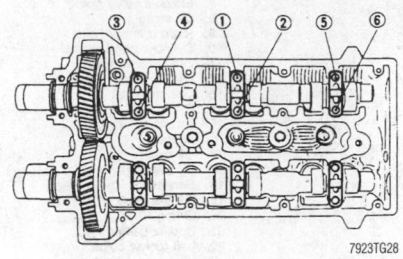

Camshaft cap bolt tightening sequence

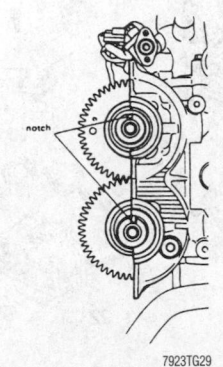

Align the notches on the front of the camshafts as shown

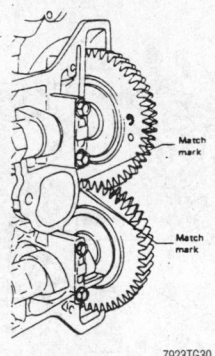

There are also alignment marks on the rear of the camshaft gears

1. Disconnect the negative battery cable.

2. Remove the timing belt followed by the camshaft sprockets.

3. Detach the cam angle sensor connector and remove the bracket.

4. Disconnect the ignition coil harness and remove the individual ignition coils.

5. Disconnect the blow-by hose and remove the cylinder head cover and gasket.

6. Remove the front camshaft cap.

7. Remove the camshaft oil seal and plug.

❊ WARNING

Since the camshaft thrust clearance is small, the camshaft must be removed by holing it parallel to the cylinder head. If the camshaft is not parallel to the cylinder head, the cylinder head thrust bearing journal may be damaged.

8. Rotate the intake (upper) and exhaust (lower) camshafts until the notch at the front of the camshafts face the 6 o'clock position on the left cylinder head, and the 12 o'clock position on the right cylinder head.

9. Inspect the rear of the camshaft and check that the matchmarks on the rear gears are aligned.

10. Install a service bolt to the sub-gear mounting bolt hole of the intake camshaft gear to lock the sub-gear and driven gear.

11. Loosen the intake camshaft bolt caps in sequence. Be sure that, as the bolts are turned, the clearance between the camshaft journal and the cylinder head journal bearing increase evenly. If not, tighten the bolts and repeat the loosening procedure.

12. Remove the camshaft caps while securing the intake camshaft with one hand. When the caps are removed, then lift out the intake camshaft.

13. Arrange the camshaft caps in the order they were removed. They must be installed to their original positions.

14. If you are removing the exhaust camshaft, rotate the camshaft clockwise for additional access.

15. Loosen the exhaust camshaft bolt caps in sequence. Be sure that, as the bolts are turned, the clearance between the camshaft journal and the cylinder head journal bearing increase evenly. If not, tighten the bolts and repeat the loosening procedure.

16. Remove the camshaft caps while securing the exhaust camshaft with one hand. When the caps are removed, then lift out the intake camshaft.

17. Arrange the camshaft caps in the order they were removed. They must be installed to their original positions.

18. Remove the hydraulic lash adjusters. Keep the lash adjusters in the order they were removed. They must be installed into their original positions.

To install:

19. Measure the thrust clearance of the camshaft with the hydraulic lash adjusters not installed. If thrust clearance exceeds 0.0051 in. (0.13mm) for the intake and 0.0047 in. (0.12mm) for the exhaust, replace the camshaft caps and the cylinder head as an assembly. If necessary replace the camshaft.

20. Measure the camshaft journal oil clearance with the lash adjusters not installed. If clearance exceeds 0.0039 in. (0.10mm), replace the camshaft.

21. Measure the camshaft gear backlash with the intake sub-gear not installed. If backlash exceeds 0.0118 in. (0.30mm), replace the camshafts as a set.

22. Lubricate and install the hydraulic lash adjusters.

23. Lubricate and install the camshafts with the notch on the front facing the 6 o'clock position for the left cylinder head camshafts and the 12 o'clock position for the right cylinder head camshafts. Ensure that the marks for both camshafts are facing the same position.

24. Install the camshaft caps and bolts and hand-tight.

25. Tighten the camshaft bolts on the caps equally, in small increments, in the correct sequence. Be sure that, as the bolts are turned, the clearance between the camshaft journal and the cylinder head journal bearing decreases evenly. If not, loosen the bolts and repeat the tightening procedure.

26. Tighten the camshaft cap bolts to a final torque of 7 ft. lbs. (10 Nm).

27. Ensure that the matchmarks on the rear side of the camshaft gears are aligned.

28. Remove the sub-gear securing bolt from the camshaft.

29. Install the front camshaft cover using new gaskets.

30. Lubricate and install new oil seals.

31. Install the camshaft plug.

32. Install the camshaft cover and connect the blow-by hose.

33. Connect the ignition coil harnesses and coils.

34. Connect the cam angle sensor and bracket.

35. Install the timing belt and camshaft sprockets.

36. Connect the negative battery cable.

Valve Lash

ADJUSTMENT

Except 1.8L Engine

The 2.2L, 2.5L and 3.3L engines are equipped with Hydraulic Lash Adjusters (HLA's) and do require periodic valve lash adjustment, nor is it possible.

1.8L Engine

The valve lash should be checked and adjusted every 15,000 miles (24,000 km). It

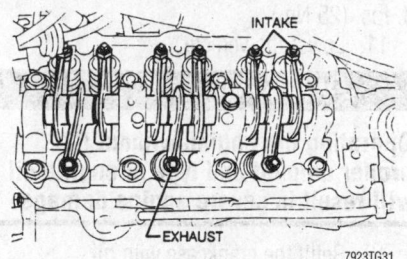

Valve arrangement—1.8L engine

is not necessary to adjust the valve lash on hydraulic lifter equipped vehicles.

➡ **Before adjusting the valve clearance, check the cylinder head torque.**

1. With the engine cold, rotate the engine so that the No. 1 piston is at Top Dead Center (TDC) of its compression stroke. The No. 1 piston is at top dead center when the distributor rotor is pointing to the No. 1 terminal (as though the distributor cap were in place) and the 0 mark on the flywheel or front pulley is opposite the pointer on the housing or front cover.

2. Check the clearance of both the intake and exhaust valves of the No. 1 cylinder by inserting a feeler gauge between each valve stem and rocker arm.

3. If the clearance is not within specifications, loosen the locknut with the proper size wrench and turn the adjusting stud either in or out until the valve clearance is correct.

➡ **Proper valve clearance is obtained when the feeler gauge slides between the valve stem and the rocker arm with a minimum amount of resistance.**

4. Tighten the locknut and recheck the valve stem-to-rocker clearance.

5. The rest of the valves are adjusted in the same way. Bring each piston to TDC of its compression stroke, then check and adjust the valves for that cylinder. The proper valve adjustment sequence is 1–3–2–4.

6. Rotate the crankshaft at least two revolutions, then recheck the valve clearance.

7. Tighten the rocker arm locknuts to 10–13 ft. lbs. (14–18 Nm).

8. Install the valve covers using new gaskets. Tighten the retaining nuts to 2–3 ft. lbs. (3–4 Nm).

Tightening torque N·m (kg-m, ft-lb)
T1: 31 – 37 (3.2 – 3.8, 23 – 27)
T2: 18 – 21 (1.8 – 2.1, 13 – 15)
T3: 22 – 27 (2.2 – 2.8, 16 – 20)
T4: 25 (2.5, 18)
T5: 4.4 – 5.4 (0.45 – 0.55, 3.3 – 4.0)

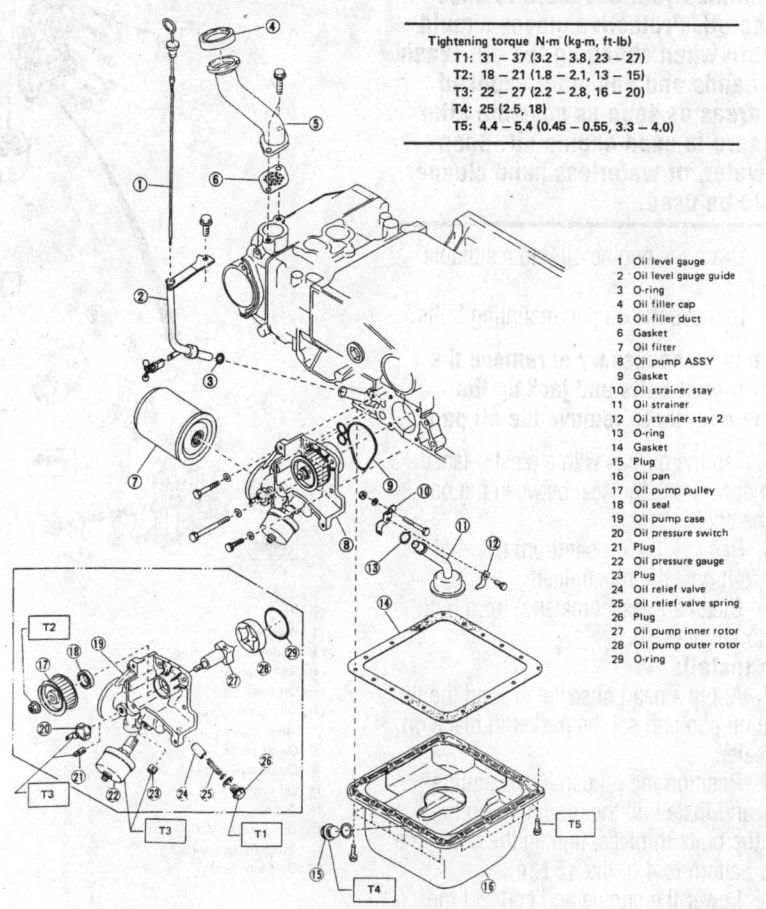

1 Oil level gauge
2 Oil level gauge guide
3 O-ring
4 Oil filler cap
5 Oil filler duct
6 Gasket
7 Oil filter
8 Oil pump ASSY
9 Gasket
10 Oil strainer stay
11 Oil strainer
12 Oil strainer stay 2
13 O-ring
14 Gasket
15 Plug
16 Oil pan
17 Oil pump pulley
18 Oil seal
19 Oil pump case
20 Oil pressure switch
21 Plug
22 Oil pressure gauge
23 Plug
24 Oil relief valve
25 Oil relief valve spring
26 Plug
27 Oil pump inner rotor
28 Oil pump outer rotor
29 O-ring

Exploded view of the oil pan and lubrication components—1.8L engine

Oil Pan

REMOVAL & INSTALLATION

✳ CAUTION

The EPA warns that prolonged contact with used engine oil may cause a number of skin disorders, including cancer! You should make every effort to minimize your exposure to used engine oil. Protective gloves should be worn when changing the oil. Wash your hands and any other exposed skin areas as soon as possible after exposure to used engine oil. Soap and water, or waterless hand cleaner should be used.

1.8L Engine

1. Raise and safely support the vehicle.

✳ CAUTION

The EPA warns that prolonged contact with used engine oil may cause a number of skin disorders, including cancer! You should make every effort to minimize your exposure to used engine oil. Protective gloves should be worn when changing the oil. Wash your hands and any other exposed skin areas as soon as possible after exposure to used engine oil. Soap and water, or waterless hand cleaner should be used.

2. Drain the engine oil into a suitable container.
3. Remove the oil pan mounting bolts.

➡ It may be necessary to remove the motor mount bolts and jack up the engine slightly to remove the oil pan.

4. Tap the oil pan with a plastic-faced hammer to break the seal between the pan and the engine block.
5. Remove the oil pan from the engine, guiding it past the oil strainer.
6. Clean all gasket material from both mating surfaces.

To install:

7. Apply a bead of sealer around the lip of the oil pan and set the gasket in place on the sealer.
8. Position the oil pan to the engine block and install all the mounting bolts. With the bolts in place, tighten the bolts in a cross pattern to 4 ft. lbs. (5 Nm).
9. Lower the engine and connect the engine mounts, if removed.

10. Tighten the oil pan drain plug to 18 ft. lbs. (25 Nm).
11. Lower the vehicle.

✳ WARNING

Operating the engine without the proper amount and type of engine oil will result in severe engine damage.

12. Refill the crankcase with oil.
13. Start the engine and check for leaks.

2.2L and 2.5L Engines

1. Raise and support the vehicle safely.

✳ CAUTION

The EPA warns that prolonged contact with used engine oil may cause a number of skin disorders, including cancer! You should make every effort

to minimize your exposure to used engine oil. Protective gloves should be worn when changing the oil. Wash your hands and any other exposed skin areas as soon as possible after exposure to used engine oil. Soap and water, or waterless hand cleaner should be used.

2. Drain the oil from the engine into a suitable container.
3. Install the drain plug with a new gasket and tighten it to 33–36 ft. lbs. (43–47 Nm).
4. Remove the air intake duct.
5. Detach the oxygen sensor electrical connector.
6. Remove the pitching stopper.
7. Remove the upper radiator brackets.
8. Remove the exhaust front Y-pipe.
9. Remove the nuts which secure the front engine mounts to the front crossmember.

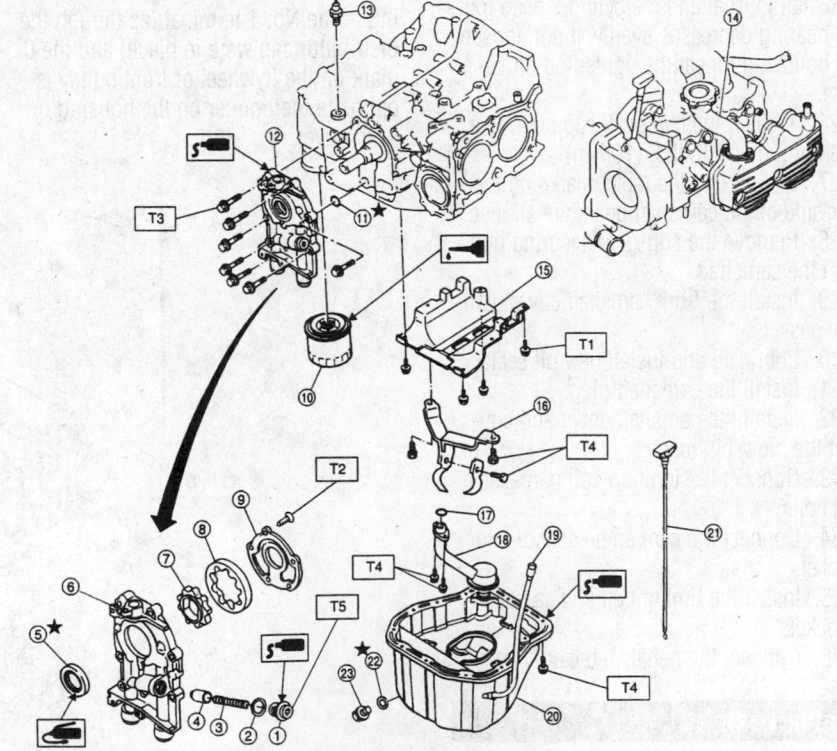

① Plug
② Washer
③ Relief valve spring
④ Relief valve
⑤ Oil seal
⑥ Oil pump case
⑦ Inner rotor
⑧ Outer rotor
⑨ Oil pump cover
⑩ Oil filter
⑪ O-ring
⑫ Oil pump ASSY
⑬ Oil pressure switch
⑭ Oil filler duct
⑮ Baffle plate
⑯ Oil strainer stay
⑰ O-ring
⑱ Oil strainer
⑲ Oil level gauge guide
⑳ Oil pan
㉑ Oil level gauge
㉒ Washer
㉓ Drain plug

Tightening torque: N·m (kg-m, ft-lb)
T1: 5 (0.5, 3.6)
T2: 5^{+1}_{-0} ($0.5^{+0.1}_{-0}$, $3.6^{+0.7}_{-0}$)
T3: 6.4 (0.65, 4.7)
T4: 9.8 (1.0, 7.0)
T5: 44.1 ± 3.4 (4.5 ± 0.35, 32.5 ± 2.5)

Exploded view of the oil pan and lubrication components—2.2L and 2.5L engines

7923TG33

10. Support the engine with a suitable lifting device.

11. Lift up the engine slightly.

12. Remove the bolts that secure the oil pan.

13. While supporting the oil pan, use a rubber mallet and tap the oil pan to free it from the engine. Be sure to support the oil pan.

14. Clean all gasket material from both mating surfaces.

To install:

15. Apply a continuous bead of sealer to a new oil pan gasket.

16. Install the oil pan assembly. Tighten the bolts to 3–4 ft. lbs. (4–5 Nm).

17. Lower the engine onto the front crossmember.

18. Install the front engine mount nuts and tighten to 61 ft. lbs. (83 Nm).

19. Remove the engine lifting device.

20. Install the front Y-pipe with new gaskets. Tighten the nuts that secure the pipe to the engine to 23 ft. lbs. (30 Nm).

21. Attach the oxygen sensor electrical connector.

22. Install the pitching stopper and tighten the bolts as follows:
- Front bolt—40 ft. lbs. (54 Nm)
- Rear bolt—49 ft. lbs. (67 Nm)

23. Install the upper radiator brackets.

24. Install the air intake duct.

25. Install the wheels.

※※ WARNING

Operating the engine without the proper amount and type of engine oil will result in severe engine damage.

26. Fill the engine to the proper level with the recommended oil and run the engine. Check for leaks.

3.3L Engine

1. Disconnect the negative battery cable.

2. Detach the left side O_2 sensor connector.

3. Remove the bolts securing the dipstick tube to the cylinder head.

4. Raise and safely support the vehicle.

※※ CAUTION

The EPA warns that prolonged contact with used engine oil may cause a number of skin disorders, including cancer! You should make every effort to minimize your exposure to used engine oil. Protective gloves should be worn when changing the oil. Wash your hands and any other exposed

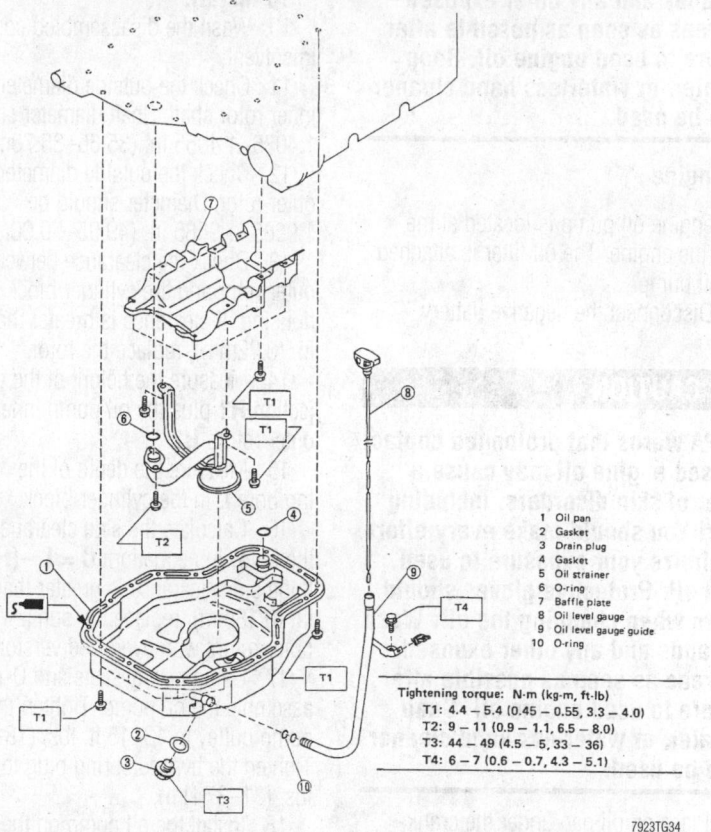

Exploded view of the oil pan and lubrication components—3.3L engine

1 Oil pan
2 Gasket
3 Drain plug
4 Gasket
5 Oil strainer
6 O-ring
7 Baffle plate
8 Oil level gauge
9 Oil level gauge guide
10 O-ring

Tightening torque: N·m (kg-m, ft-lb)
T1: 4.4 — 5.4 (0.45 — 0.55, 3.3 — 4.0)
T2: 9 — 11 (0.9 — 1.1, 6.5 — 8.0)
T3: 44 — 49 (4.5 — 5, 33 — 36)
T4: 6 — 7 (0.6 — 0.7, 4.3 — 5.1)

7923TG34

skin areas as soon as possible after exposure to used engine oil. Soap and water, or waterless hand cleaner should be used.

5. Drain the engine oil into a suitable container.

6. Remove the engine under covers.

7. Remove the left side exhaust manifold cover, front pipe and left exhaust manifold.

8. Remove the rack and pinion mounting bolts.

9. Disconnect the dipstick tube from the oil pan.

10. Remove the oil pan mounting bolts.

11. Remove the oil pan and gasket.

12. Clean all gasket material from both mating surfaces.

To install:

13. Clean all gasket surfaces completely.

14. Install the oil pan using a new gasket. Tighten the mounting bolts to 4 ft. lbs. (5 Nm).

15. Connect the dipstick tube to the oil pan.

16. Install the rack and pinion mounting bolts.

17. Install the left side exhaust manifold, front exhaust pipe and left side manifold cover.

18. Install the engine under covers.

19. Lower the vehicle.

20. Install the bolts securing the dipstick tube to the cylinder head.

21. Connect the left side O_2 sensor.

※※ WARNING

Operating the engine without the proper amount and type of engine oil will result in severe engine damage.

22. Refill the engine with oil.

23. Connect the negative battery cable.

Oil Pump

REMOVAL & INSTALLATION

※※ CAUTION

The EPA warns that prolonged contact with used engine oil may cause a number of skin disorders, including cancer! You should make every effort to minimize your exposure to used engine oil. Protective gloves should be worn when changing the oil. Wash

your hands and any other exposed skin areas as soon as possible after exposure to used engine oil. Soap and water, or waterless hand cleaner should be used.

1.8L Engine

The engine oil pump is located at the front of the engine. The oil filter is attached to the oil pump.

1. Disconnect the negative battery cable.

❊❊ CAUTION

The EPA warns that prolonged contact with used engine oil may cause a number of skin disorders, including cancer! You should make every effort to minimize your exposure to used engine oil. Protective gloves should be worn when changing the oil. Wash your hands and any other exposed skin areas as soon as possible after exposure to used engine oil. Soap and water, or waterless hand cleaner should be used.

2. Place an oil pan under the crankcase, remove the drain plug and drain the oil from the crankcase.
3. Remove the left and right front timing belt covers.
4. Loosen the tensioner mounting bolts on the No. 1 cylinder.
5. Turn the tensioner to fully loosen the belt, then tighten the mounting bolts.
6. Using a piece of chalk, mark the rotating direction of the timing belt, then remove the belt from the vehicle.
7. Remove the oil pump-to-engine bolts and the oil pump along with the oil filter from the cylinder block.
8. Remove the oil pump's outer rotor from the cylinder block.
9. Disassemble the oil pump.

To install:

10. Wash the disassembled components in solvent.
11. Check the outside diameter of the inner rotor shaft. Shaft diameter should be 1.4035–1.4055 in. (35.65–35.70mm).
12. Check the outside diameter of the outer rotor. Diameter should be 1.9665–1.9685 in. (49.95–50.00mm).
13. Check the clearance between the outer rotor and the cylinder block rotor housing. If clearance is greater than 0.0087 in. (0.22mm), replace the rotor.
14. Measure the height of the case projection **H1** plus the oil pump inner and outer rotors **H2**.
15. Measure the depth of the rotor housing bore **L** in the cylinder block.
16. Calculate the side clearance **C** using the following equation: $C = L - (H1 + H2)$. If the side clearance is greater than 0.00071 in. (0.18mm), replace the pump inner and outer rotors with oversized versions.
17. Using new gaskets and O-rings, assemble the oil pump. Tighten the oil pump pulley to 13–15 ft. lbs. (18–21 Nm). Tighten the bypass spring plug to 23–27 ft. lbs. (31–37 Nm).
18. Install the oil pump on the cylinder block.
19. Install the timing belt and related components.

❊❊ WARNING

Operating the engine without the proper amount and type of engine oil will result in severe engine damage.

20. Fill the engine with oil. Connect the negative battery cable.
21. Start the engine and allow it to reach operating temperature. Check for adequate oil pressure. Check for leaks.
22. Adjust the ignition timing as required.

2.2L Engine

1. Disconnect the negative battery cable.

❊❊ CAUTION

The EPA warns that prolonged contact with used engine oil may cause a number of skin disorders, including cancer! You should make every effort to minimize your exposure to used engine oil. Protective gloves should be worn when changing the oil. Wash your hands and any other exposed skin areas as soon as possible after exposure to used engine oil. Soap and water, or waterless hand cleaner should be used.

2. Drain the engine oil into a suitable container.
3. Drain the coolant into a suitable separate container.
4. Remove the water pump.
5. Remove the oil pump mounting bolts.
6. Carefully pry the oil pump from the engine block.

❊❊ WARNING

Use extreme care not to damage the engine block or the oil pump during removal of the pump.

7. Remove the front crankshaft seal.

To install:

8. Measure the tip clearance of the rotors. If clearance is greater than 0.0071 in. (0.18mm), replace the rotors.
9. Measure the clearance between the outer rotor and the cylinder block rotor housing. If clearance exceeds 0.0079 in. (0.20mm), replace the rotor.
10. Measure the side clearance between the oil pump inner rotor and the pump cover. If clearance exceeds 0.0047 in. (0.12mm), replace the rotor or pump body.

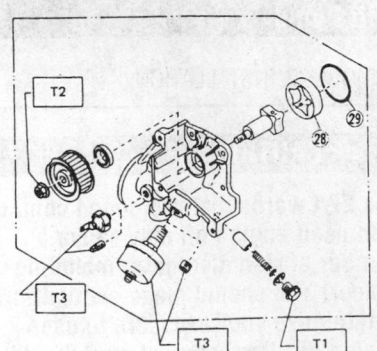

Tightening torque N·m (kg-m, ft-lb)
T1: 31 – 37 (3.2 – 3.8, 23 – 27)
T2: 18 – 21 (1.8 – 2.1, 13 – 15)
T3: 22 – 27 (2.2 – 2.8, 16 – 20)
T4: 39 – 44 (4.0 – 4.5, 29 – 33)
T5: 4.4 – 5.4 (0.45 – 0.55, 3.3 – 4.0)

7923TG35

Exploded view of the oil pump and components—1.8L engine

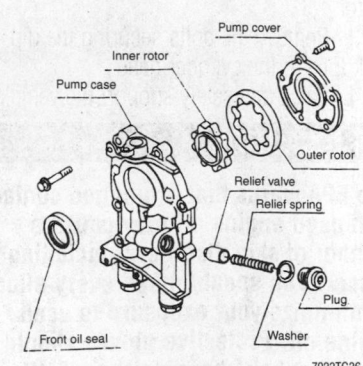

7923TG36

Exploded view of the oil pump and components—2.2L and 2.5L engines

11. Install a new front oil seal on the pump cover using a driver.

12. Assemble the oil pump.

13. Install a new front crankshaft seal using installer 499587100 or equivalent.

14. Apply sealant and a new O-ring to the oil pump.

15. Install the oil pump and tighten the bolts to 5 ft. lbs. (7 Nm).

16. Install the water pump.

※※ WARNING

Operating the engine without the proper amount and type of engine oil will result in severe engine damage.

17. Fill the crankcase to the proper level with the recommended oil.

18. Fill and bleed the cooling system.

19. Connect the negative battery cable.

20. Start the engine and check for leaks.

2.5L Engine

1. Disconnect the negative battery cable.

※※ CAUTION

Never open, service or drain the radiator or cooling system when hot; serious burns can occur from the steam and hot coolant.

2. Drain the cooling system into a suitable container.

3. Raise and support the vehicle safely.

※※ CAUTION

The EPA warns that prolonged contact with used engine oil may cause a number of skin disorders, including cancer! You should make every effort to minimize your exposure to used engine oil. Protective gloves should be worn when changing the oil. Wash your hands and any other exposed skin areas as soon as possible after exposure to used engine oil. Soap and water, or waterless hand cleaner should be used.

4. Drain the engine oil into a separate container.

5. Remove the belt covers, timing belt and related parts.

6. Remove the belt tensioner bracket.

7. Remove the engine coolant pipe.

8. Remove the water pump assembly.

9. Remove the oil pump mounting bolts and pry the pump from the engine block.

➡**Use extreme care not to damage the engine block or the oil pump during removal of the pump.**

To install:

10. Install a new front seal to the oil pump.

11. Apply a continuous bead sealant to the mating surfaces of the oil pump.

12. Install a new O-ring to the oil pump.

13. Install the oil pump and tighten the bolts to 56 inch lbs. (6.4 Nm).

14. Install the water pump.

15. Install the engine coolant pipe.

16. Install the belt tensioner bracket, timing belt and the covers.

※※ WARNING

Operating the engine without the proper amount and type of engine oil will result in severe engine damage.

17. Fill the engine to the proper level with oil and coolant.

18. Connect the negative battery cable.

19. Fill and bleed the cooling system.

3.3L Engine

1. Disconnect the negative battery cable.

2. Raise and safely support the vehicle.

※※ CAUTION

The EPA warns that prolonged contact with used engine oil may cause a number of skin disorders, including cancer! You should make every effort to minimize your exposure to used engine oil. Protective gloves should be worn when changing the oil. Wash your hands and any other exposed skin areas as soon as possible after exposure to used engine oil. Soap and water, or waterless hand cleaner should be used.

3. Drain the engine oil into a suitable container.

4. Remove the engine under covers.

5. Remove the bolts securing the power steering oil cooler pipe. DO NOT remove the pipe.

6. Remove the cooling fan assemblies.

7. Remove the drive belt cover and drive belt.

8. Remove the crank angle sensors.

9. Remove the crankshaft pulley.

10. Remove the timing belt, tensioners and tensioner brackets.

11. Remove the oil pump mounting bolts.

12. Separate the oil pump from the block by inserting a suitable prying tool between the pump and the block.

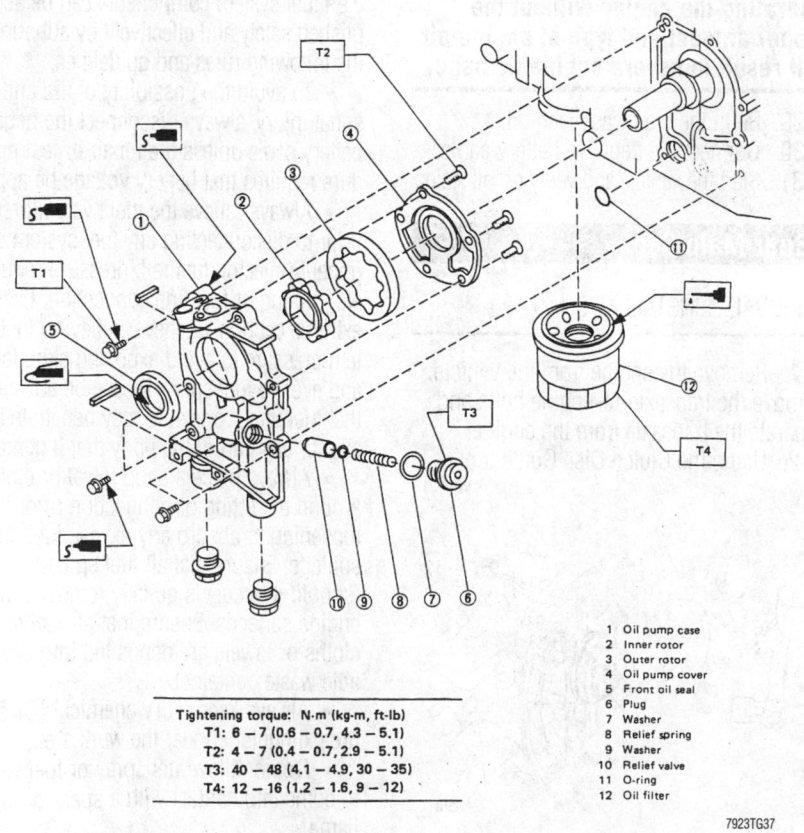

Tightening torque: N·m (kg·m, ft-lb)
T1: 6 – 7 (0.6 – 0.7, 4.3 – 5.1)
T2: 4 – 7 (0.4 – 0.7, 2.9 – 5.1)
T3: 40 – 48 (4.1 – 4.9, 30 – 35)
T4: 12 – 16 (1.2 – 1.6, 9 – 12)

1	Oil pump case
2	Inner rotor
3	Outer rotor
4	Oil pump cover
5	Front oil seal
6	Plug
7	Washer
8	Relief spring
9	Washer
10	Relief valve
11	O-ring
12	Oil filter

7923TG37

Oil pump and components—3.3L engine

13. Remove the oil pump and gasket.
14. Remove the front seal from the oil pump.

To install:

15. Measure the tip clearance of the rotors. If clearance is greater than 0.0071 in. (0.18mm), replace the rotors.

16. Measure the clearance between the outer rotor and the cylinder block rotor housing. If clearance exceeds 0.0079 in. (0.20mm), replace the rotor.

17. Measure the side clearance between the oil pump inner rotor and the pump cover. If clearance exceeds 0.0047 in. (0.12mm), replace the rotor or pump body.

18. Install a new oil seal in the oil pump.
19. Clean all gasket surfaces completely.
20. Install the oil pump using a new gasket. Tighten the mounting bolts to 5 ft. lbs. (7 Nm).
21. Install the timing belt components.
22. Install the crankshaft pulley.
23. Install the crank angle sensors.
24. Install the drive belts and drive belt cover.
25. Install the cooling fan assembly.
26. Install the bolts securing the power steering oil cooler pipe.
27. Install the engine under covers.
28. Lower the vehicle.

❈❈ WARNING

Operating the engine without the proper amount and type of engine oil will result in severe engine damage.

29. Refill the crankcase with oil.
30. Connect the negative battery cable.
31. Start the vehicle and verify no oil leaks.

Rear Main Seal

REMOVAL & INSTALLATION

1. Remove the engine from the vehicle. Remove the transaxle-to-engine bolts and separate the transaxle from the engine.
2. Using the Clutch Disc Guide tool

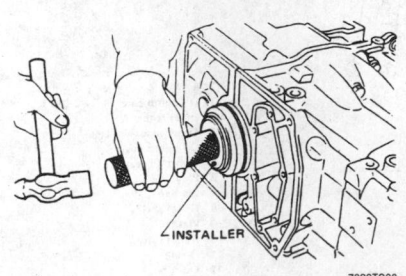

Installing the rear main seal

499747000 or equivalent, remove the clutch assembly/flywheel (MT). If equipped with an AT, remove the torque converter flexplate from the crankshaft.

3. Remove the flywheel housing from the engine. Using a small prybar, pry the oil seal from the housing.

To install:

4. Install the new oil seal and press it into the flywheel housing using the appropriate driver.

5. Install the flywheel housing using new gaskets and sealant where necessary. Tighten the bolts to specification.

6. Install the flywheel and tighten the bolts to specification.

7. Join the engine and transaxle. Install the assembly in the vehicle.

FUEL SYSTEM

Fuel System Service Precautions

Safety is the most important factor when performing not only fuel system maintenance but any type of maintenance. Failure to conduct maintenance and repairs in a safe manner may result in serious personal injury or death. Maintenance and testing of the vehicle's fuel system components can be accomplished safely and effectively by adhering to the following rules and guidelines.

• To avoid the possibility of fire and personal injury, always disconnect the negative battery cable unless the repair or test procedure requires that battery voltage be applied.

• Always relieve the fuel system pressure prior to disconnecting any fuel system component (injector, fuel rail, pressure regulator, etc.), fitting or fuel line connection. Exercise extreme caution whenever relieving fuel system pressure to avoid exposing skin, face and eyes to fuel spray. Please be advised that fuel under pressure may penetrate the skin or any part of the body that it contacts.

• Always place a shop towel or cloth around the fitting or connection prior to loosening to absorb any excess fuel due to spillage. Ensure that all fuel spillage (should it occur) is quickly removed from engine surfaces. Ensure that all fuel soaked cloths or towels are deposited into a suitable waste container.

• Always keep a dry chemical (Class B) fire extinguisher near the work area.

• Do not allow fuel spray or fuel vapors to come into contact with a spark or open flame.

• Always use a back-up wrench when loosening and tightening fuel line connection fittings. This will prevent unnecessary stress and torsion to fuel line piping.

• Always replace worn fuel fitting O-rings with new. Do not substitute fuel hose or equivalent, where fuel pipe is installed.

Fuel System Pressure

RELIEVING

➠**This procedure must be performed prior to servicing any component of the fuel injection system.**

1. Disconnect the fuel pump harness at the fuel pump.
2. Crank the engine for 5 seconds or more to relieve the fuel pressure. If the engine starts during this time, allow it to run until it stalls.
3. After performing the required service, connect the fuel pump harness.

Fuel Filter

The fuel filter should be replaced and all fuel system hoses and connection should be inspected at the 30,000 mile (48,300 km) intervals. When the vehicle is operated in extremely cold or hot conditions, contamination of the filter may occur and the filter should be replaced more often.

REMOVAL & INSTALLATION

❈❈ CAUTION

Observe all applicable safety precautions when working around fuel. Whenever servicing the fuel system, always work in a well ventilated area. Do not allow fuel spray or vapors to come in contact with a spark or open flame. Keep a dry chemical fire extinguisher near the work area. Always keep fuel in a container specifically

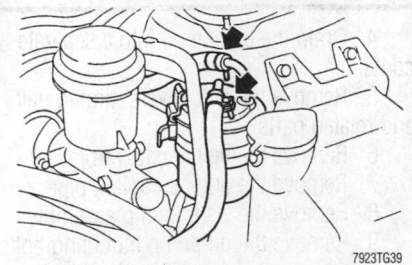

Be sure to replace any fuel lines that are leaking or showing signs of deterioration

designed for fuel storage; also, always properly seal fuel containers to avoid the possibility of fire or explosion.

1. Locate the fuel filter in the engine compartment on the left inside fender.
2. Properly relieve the fuel system pressure.
3. Disconnect the negative battery cable.
4. Loosen the hose clamp screws and slide the hoses off the filter.
5. Remove the filter from the bracket.

To install:
6. Inspect the hoses for wear or cracks, and replace if needed.
7. Install the new filter into the bracket and tighten the hose clamp screws.
8. Lower the vehicle and connect the negative battery cable.
9. Start the engine and check for leaks.

Fuel Pump

REMOVAL & INSTALLATION

1.8L Engine

❋❋ CAUTION

Observe all applicable safety precautions when working around fuel. Whenever servicing the fuel system, always work in a well ventilated area. Do not allow fuel spray or vapors to come in contact with a spark or open flame. Keep a dry chemical fire extinguisher near the work area. Always keep fuel in a container specifically designed for fuel storage; also, always properly seal fuel containers to avoid the possibility of fire or explosion.

1. Release the fuel system pressure.
2. Disconnect the negative battery cable.
3. Keep the fuel pump harness disconnected after releasing the fuel system pressure.

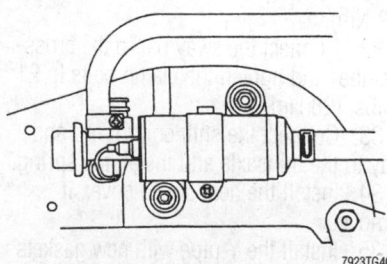

Fuel pump assembly mounting location—1.8L engine

4. Raise and safely support the vehicle.
5. Clamp the middle portion of the hose connecting the pipe to the fuel pump to prevent fuel from flowing out of the tank.
6. Loosen the hose clamp and disconnect the hose.
7. Remove the three pump bracket mounting bolts and remove the pump together with the pump damper.

To install:
8. If the pump and damper have been removed from the bracket, reinstall and tighten the bolts securely.
9. Install the hose and tighten the clamp screw to 9–13 inch lbs. (1.0–1.5 Nm).
10. Install the pump bracket in position to the vehicle body and secure it with the bolts.

➡**Take care to position the rubber cushion properly.**

11. Attach the pump harness connector.
12. Connect the negative battery cable and test the fuel pump for proper operation.
13. Check for leaks.

2.2L and 2.5L Engines

❋❋ CAUTION

The fuel injection system remains under pressure after the engine has been turned OFF. Properly relieve fuel pressure before disconnecting any fuel lines. Failure to do so may result in fire or personal injury.

1. Relieve the fuel system pressure.

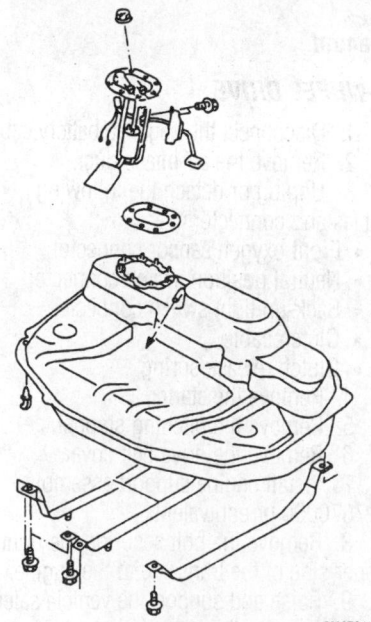

Exploded view of the fuel pump assembly—2.2L and 2.5L engines

❋❋ CAUTION

Observe all applicable safety precautions when working around fuel. Whenever servicing the fuel system, always work in a well ventilated area. Do not allow fuel spray or vapors to come in contact with a spark or open flame. Keep a dry chemical fire extinguisher near the work area. Always keep fuel in a container specifically designed for fuel storage; also, always properly seal fuel containers to avoid the possibility of fire or explosion.

2. Disconnect the negative battery cable.
3. Remove the rear seat bottom to reach the fuel pump access cover.
4. On Legacy models, fold the seat back, then roll the floor mat back.
5. Remove the fuel pump cover mounting bolts, then remove the fuel pump cover.
6. Disconnect the electrical harness from the pump assembly.
7. Tag and disconnect the fuel lines from the fuel pump.
8. Remove the eight fuel pump mounting nuts.
9. Remove the fuel pump assembly from the tank.

To install:
10. Using a new gasket, install the fuel pump assembly into the fuel tank and install the fuel pump mounting nuts. Tighten the nuts to 20–33 inch lbs. (2–4 Nm) on Impreza models, 24–48 inch lbs. (3–6 Nm) for Legacy models.
11. Connect the electrical harness to the fuel pump assembly.
12. Connect the fuel lines to the pump assembly and tighten the clamps and fittings.
13. Install the fuel pump service cover and cover mounting bolts.
14. Install the rear seat bottom.
15. Connect the negative battery cable.
16. Start the engine and check for leaks.

3.3L Engine

❋❋ CAUTION

The fuel injection system remains under pressure after the engine has been turned OFF. Properly relieve fuel pressure before disconnecting any fuel lines. Failure to do so may result in fire or personal injury.

1. Relieve the fuel system pressure.

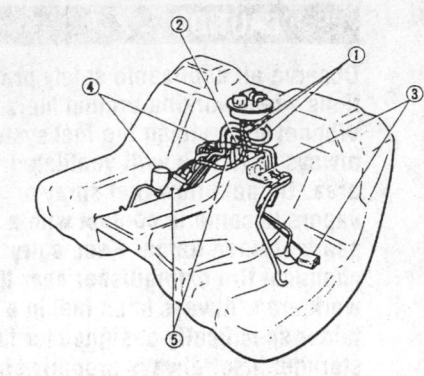

(1) Disconnect hoses and harness connector, and remove fuel tank cap.
(2) Remove bracket cover for installing each assembly bracket onto tank inner.
(3) Take out fuel meter unit LH.
(4) Take out fuel meter unit RH.
(5) Take out fuel pump ASSY.

7923TG42

Exploded view of the fuel pump assembly—3.3L engine

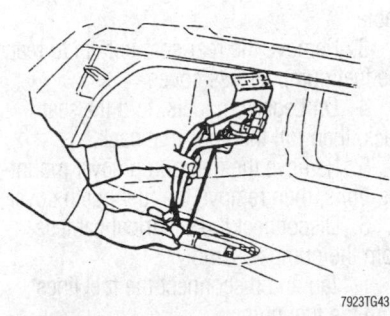

7923TG43

Carefully remove the fuel pump assembly from the tank

❉❉ CAUTION

Observe all applicable safety precautions when working around fuel. Whenever servicing the fuel system, always work in a well ventilated area. Do not allow fuel spray or vapors to come in contact with a spark or open flame. Keep a dry chemical fire extinguisher near the work area. Always keep fuel in a container specifically designed for fuel storage; also, always properly seal fuel containers to avoid the possibility of fire or explosion.

2. Disconnect the negative battery cable.
3. Disconnect and tag the four hoses from the fuel sender assembly.
4. Remove the fuel sender retaining cap using 42911PA000 or equivalent.
5. Take out the left side fuel meter assembly.
6. Take out the right side fuel meter assembly.
7. Remove the fuel pump assembly.
To install:
8. Install the fuel pump assembly.
9. Install the right side fuel meter assembly.

10. Install the left side fuel meter assembly.
11. Install the bracket covers.
12. Install the fuel sender retaining cap using 42911PA000, or an equivalent replacer.
13. Connect the four hoses to the sender assembly.
14. Connect the electrical harness to the sender assembly.
15. Install the fuel pump access cover.
16. Connect the negative battery cable.

DRIVE TRAIN

Transaxle Assembly

REMOVAL & INSTALLATION

Manual

2-WHEEL DRIVE

1. Disconnect the negative battery cable.
2. Remove the air intake duct.
3. Unplug or detach the following cables and connectors:
   • Front oxygen sensor connector
   • Neutral position switch connector
   • Back-up light switch connector
   • Clutch cable
   • Clutch release spring
4. Remove the starter.
5. Remove the pitching stopper.
6. Remove the drive belt cover.
7. Install engine support assembly 927670000 or equivalent.
8. Remove the bolt securing the right upper side of the transaxle to the engine.
9. Raise and support the vehicle safely.
10. Remove the front Y-pipe.
11. Detach the rear oxygen sensor connector.

12. Remove the hanger bracket from the right side of the transaxle.
13. Remove the spring and disconnect the shifter stay and rod from the transaxle.
14. Remove the bolts securing the sway bar clamps to the crossmember.
15. Disconnect the ball joints from the steering knuckle.
16. Separate the halfshafts from the transaxle.
17. Remove the 2 nuts securing the lower side of the transaxle to the engine.
18. Support the transaxle with a jack.
19. Remove the rear transaxle crossmember.
20. Remove the transaxle from the vehicle.
To install:
21. Secure the transaxle in place onto the transaxle jack.
22. Raise the transaxle assembly in place to the engine block and install it. Take care when mating the input shaft to the clutch assembly.
23. Install and tighten the crossmember to the following specifications:
Impreza:
   • T1—20 ft. lbs. (26 Nm)
   • T2—35 ft. lbs. (47 Nm)
   • T3—61 ft. lbs. (83 Nm)
Legacy:
   • T1—40–62 ft. lbs. (54–84 Nm)
   • T2—87–115 ft. lbs. (117–157 Nm)
24. Remove the transmission jack.
25. Install the nuts securing the lower portion of the engine to the transaxle and tighten them to 40 ft. lbs. (54 Nm).
26. Install the bolt securing the right upper side of the transaxle to the engine and tighten it to 40 ft. lbs. (54 Nm).
27. Remove the engine support.
28. Install the drive belt cover.
29. Install the pitching stopper and tighten the bolts to the following specifications:
   • T1—40 ft. lbs. (54 Nm)
   • T2—49 ft. lbs. (67 Nm)
30. Insert the halfshafts into the transaxle and install new roll pins.
31. Connect the ball joint to the steering knuckle and tighten the bolt to 22 ft. lbs. (29 Nm).
32. Connect the sway bar to the crossmember and tighten the clamp bolts to 21 ft. lbs. (28 Nm).
33. Connect the shift control rod and stay to the transaxle and install the spring.
34. Install the heat shield cover, if removed.
35. Install the Y-pipe with new gaskets and nuts.
36. Install the hanger bracket on the right side of the transaxle, if removed.

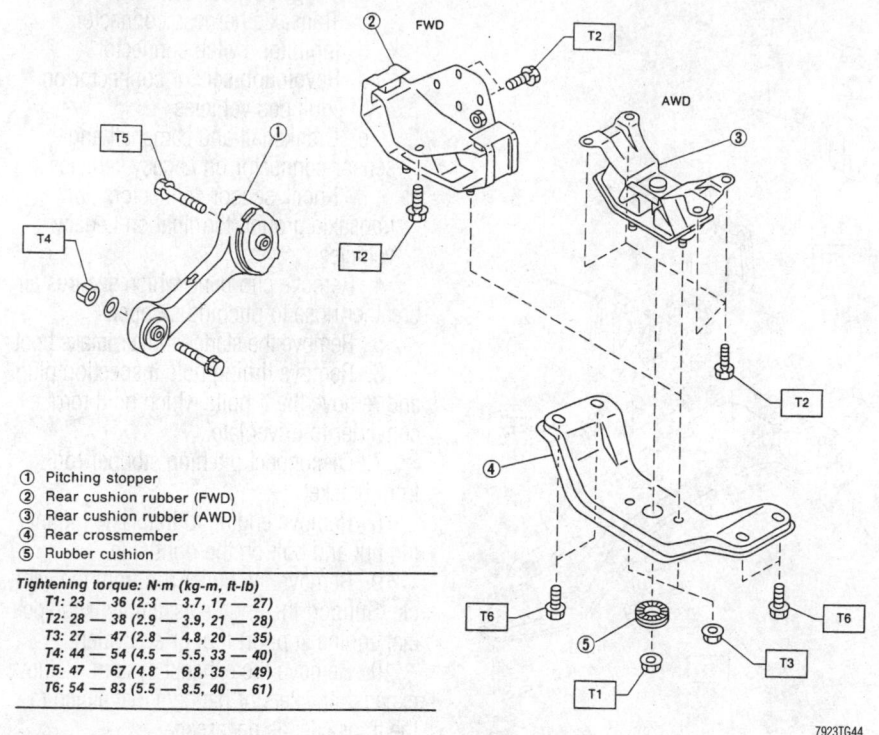

① Pitching stopper
② Rear cushion rubber (FWD)
③ Rear cushion rubber (AWD)
④ Rear crossmember
⑤ Rubber cushion

Tightening torque: N·m (kg-m, ft-lb)
T1: 23 — 36 (2.3 — 3.7, 17 — 27)
T2: 28 — 38 (2.9 — 3.9, 21 — 28)
T3: 27 — 47 (2.8 — 4.8, 20 — 35)
T4: 44 — 54 (4.5 — 5.5, 33 — 40)
T5: 47 — 67 (4.8 — 6.8, 35 — 49)
T6: 54 — 83 (5.5 — 8.5, 40 — 61)

Exploded view of the transaxle mounting—Impreza models

37. Attach the rear oxygen sensor connector.
38. Install the transaxle connectors bracket.
39. Install the drive belt cover.
40. Install the pitching stopper.
41. Install the starter.
42. Engage the following cables and connectors:
- Front oxygen sensor connector
- Neutral position switch connector
- Back-up light switch connector
- Clutch cable
- Clutch release spring
43. Install the air intake duct and attach the airflow sensor connector.
44. Connect the negative battery cable.

4-WHEEL DRIVE

1. Disconnect the negative battery cable.
2. Remove the air intake duct.
3. Unplug or detach the following cables and connectors:
- Front oxygen sensor connector
- Neutral position switch connector
- Back-up light switch connector
- Clutch cable
- Clutch release spring
4. Remove the starter.
5. Remove the pitching stopper.
6. Remove the drive belt cover.
7. Install engine support assembly 927670000 or equivalent.

8. Remove the bolt securing the right upper side of the transaxle to the engine.
9. Raise and support the vehicle safely.
10. Remove the front Y-pipe.
11. Detach the rear oxygen sensor connector.
12. Remove the center exhaust pipe and the heat shield cover.
13. Remove the hanger bracket from the right side of the transaxle.
14. Remove the driveshaft.
15. Remove the spring and disconnect the shifter stay and rod from the transaxle.
16. Remove the bolts securing the sway bar clamps to the crossmember.
17. Disconnect the ball joints from the steering knuckle.
18. Separate the halfshafts from the transaxle.
19. Remove the 2 nuts securing the lower side of the transaxle to the engine.
20. Support the transaxle with a jack.
21. Remove the rear transaxle crossmember.
22. Remove the transaxle from the vehicle.
To install:
23. Install the transaxle assembly and secure to the engine block.
24. Install and tighten the crossmember to the following specifications:
Legacy:
- T1—40–62 ft. lbs. (54–84 Nm)
- T2—87–115 ft. lbs. (117–157 Nm)

Impreza:
- T1—20 ft. lbs. (26 Nm)
- T2—35 ft. lbs. (47 Nm)
- T3—61 ft. lbs. (83 Nm)
25. Remove the transmission jack.
26. Install the nuts securing the lower portion of the engine to the transaxle and tighten them to 40 ft. lbs. (54 Nm).
27. Install the bolt securing the right upper side of the transaxle to the engine and tighten it to 40 ft. lbs. (54 Nm).
28. Remove the engine support.
29. Install the drive belt cover.
30. Install the pitching stopper and tighten the bolts to the following specifications:
- T1—40 ft. lbs. (54 Nm)
- T2—49 ft. lbs. (67 Nm)
31. Insert the halfshafts into the transaxle and install new roll pins.
32. Connect the ball joint to the steering knuckle and tighten the bolt to 22 ft. lbs. (29 Nm).
33. Connect the sway bar to the crossmember and tighten the clamp bolts to 21 ft. lbs. (28 Nm).
34. Connect the shift control rod and stay to the transaxle and install the spring.
35. Install the driveshaft.
36. Install the heat shield cover, if removed.
37. Install the Y-pipe with new gaskets and nuts.
38. Install the hanger bracket on the right side of the transaxle, if removed.
39. Attach the rear oxygen sensor connector.
40. Install the transaxle connectors bracket.
41. Install the drive belt cover.
42. Install the pitching stopper.
43. Install the starter.
44. Attach the following cables and connectors:
- Front oxygen sensor connector
- Neutral position switch connector
- Back-up light switch connector
- Clutch cable
- Clutch release spring
45. Install the air intake duct and attach the airflow sensor connector.
46. Connect the negative battery cable.

Automatic

1. Disconnect the negative battery cable.
2. Remove speedometer cable or electronic wiring connector from speed sensor.
3. Unplug the following electrical harness connections on the automatic transaxle:

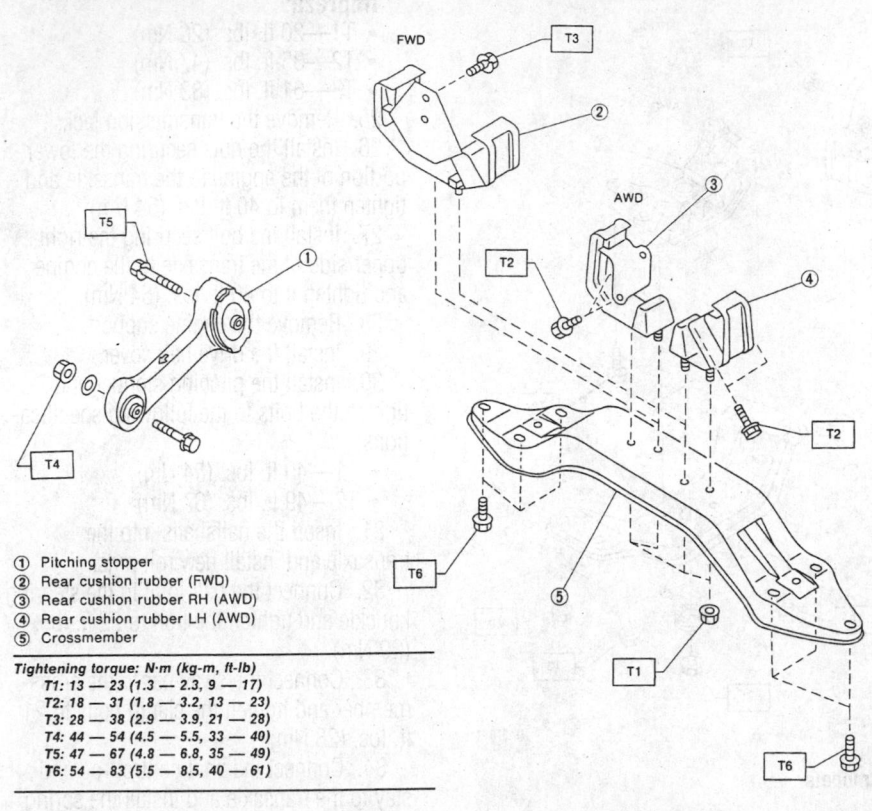

① Pitching stopper
② Rear cushion rubber (FWD)
③ Rear cushion rubber RH (AWD)
④ Rear cushion rubber LH (AWD)
⑤ Crossmember

Tightening torque: N·m (kg-m, ft-lb)
T1: 13 — 23 (1.3 — 2.3, 9 — 17)
T2: 18 — 31 (1.8 — 3.2, 13 — 23)
T3: 28 — 38 (2.9 — 3.9, 21 — 28)
T4: 44 — 54 (4.5 — 5.5, 33 — 40)
T5: 47 — 67 (4.8 — 6.8, 35 — 49)
T6: 54 — 83 (5.5 — 8.5, 40 — 61)

7923TG45

Exploded view of the engine and transaxle mounts—Impreza and Legacy models

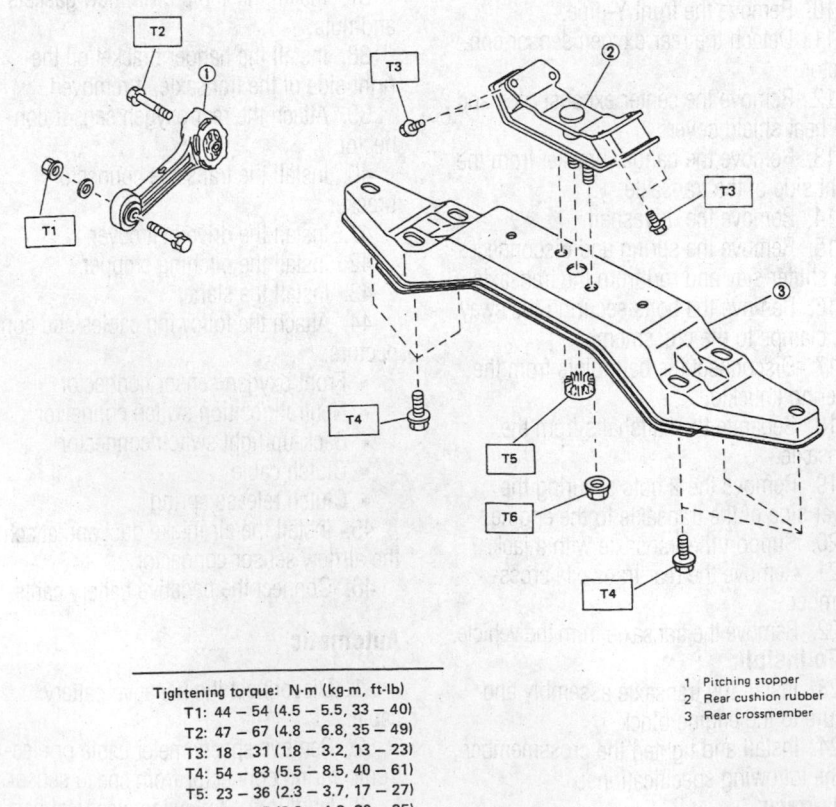

Tightening torque: N·m (kg-m, ft-lb)
T1: 44 — 54 (4.5 — 5.5, 33 — 40)
T2: 47 — 67 (4.8 — 6.8, 35 — 49)
T3: 18 — 31 (1.8 — 3.2, 13 — 23)
T4: 54 — 83 (5.5 — 8.5, 40 — 61)
T5: 23 — 36 (2.3 — 3.7, 17 — 27)
T6: 27 — 47 (2.8 — 4.8, 20 — 35)

1 Pitching stopper
2 Rear cushion rubber
3 Rear crossmember

7923TG46

Exploded view of the engine and transaxle mount assembly—SVX model

a. Oxygen sensor connector
b. Transaxle harness connector
c. Inhibitor switch connector
d. Revolution sensor connector on 4WD equipped vehicles
e. Crankshaft and camshaft angle sensor connector on Legacy vehicles
f. Knock sensor connectors and transaxle ground terminal on Legacy vehicles

4. Remove clip band which secures air breather hose to pitching stopper.

5. Remove the starter and air intake boot.

6. Remove timing hole inspection plug and remove the 4 bolts which hold torque converter to driveplate.

7. Disconnect pitching stopper rod from bracket.

8. Remove engine to transaxle mounting nut and bolt on the right side.

9. Remove the buffer rod from the vehicle. Support the engine assembly with special engine support tool or equivalent.

10. Remove the exhaust system. Remove exhaust brackets or hangers that attach to the transaxle, as necessary.

11. Matchmark and remove the driveshaft on 4WD vehicles. Plug the opening at the rear of extension housing to prevent oil from flowing out.

12. Disconnect the gear shift cable from the transaxle select lever.

13. Remove stabilizer from transverse link.

14. Remove parking brake cable bracket from transverse link and bolt holding transverse link to crossmember on each side. Lower the transverse link.

15. Remove spring pin and separate halfshaft from transaxle on each side.

➡Use a suitable tool to remove spring pin. Discard old spring pin and always install a new pin.

16. Disconnect the halfshaft from transaxle on each side. Be sure to remove axle shaft from transaxle by pushing the rear of tire outward.

17. Remove engine to transaxle mounting nuts.

18. Disconnect oil cooler hoses.

19. Place transaxle jack or equivalent, under transaxle. Always support transaxle case with a transaxle jack.

➡Do not place jack under oil pan otherwise oil pan may be damaged.

20. Remove rear cushion rubber mounting nuts and rear crossmember.

21. Move torque converter and transaxle as a unit away from the engine. Remove the transaxle.

To install:

22. Install transaxle to engine and temporarily tighten engine to transaxle mounting nuts.

23. Install rear crossmember to rear cushion rubber mounts. Align rear cushion guide with rear crossmember guide hole and tighten nuts.

24. Install rear crossmember to chassis; be careful not to damage threads. Tighten rear crossmember bolts to 39–49 ft. lbs. (53–66 Nm).

25. Tighten engine to transaxle retaining nuts to 34–40 ft. lbs. (46–54 Nm). Remove transaxle jack from the vehicle.

26. Remove the engine support tool and install buffer rod.

27. Install axle shaft to transaxle and install spring pin into place.

➡ **Always use new spring pin. Be sure to align the axle shaft and shaft from the transaxle at chamfered holes and install shaft splines correctly.**

28. Install transverse link temporarily to front crossmember by using bolt and self locking nut. Do not complete final torque at this point.

29. Install stabilizer temporarily to transverse link. Install parking brake cable bracket to transverse link.

30. Lower vehicle to floor. Tighten transverse link to front crossmember mounting bolts and transverse link to stabilizer mounting bolts with the tires placed on the ground when the vehicle is not loaded. Tighten the transverse link to front crossmember (self-locking nuts) to 43–51 ft. lbs. (58–69 Nm) and the transverse link to stabilizer to 14–22 ft. lbs. (19–30 Nm).

31. Raise and safely support the vehicle. Reconnect the gear shift cable to the select lever. Be sure the lever operates smoothly all across the operating range.

32. Install driveshaft on 4WD vehicles. Tighten driveshaft-to-rear differential retaining bolts to 17–24 ft. lbs. (23–33 Nm) and center bearing location retaining bolts to 25–33 ft. lbs. (34–45 Nm).

33. Connect oil cooler hoses.

34. Tighten engine to transaxle bolts to 34–40 ft. lbs. (46–54 Nm).

35. Install starter.

36. Install pitching stopper. Be sure to tighten the bolt for the body side first, then the 1 for engine or transaxle side. Tightening tighten for chassis side is 27–49 ft. lbs. (37–66 Nm) and for engine or transaxle side is 33–40 ft. lbs. (45–54 Nm).

37. Install and tighten torque converter-to-driveplate mounting bolts to 17–20 ft. lbs. (23–27 Nm).

38. Install timing hole inspection plug, air intake boot and air breather hose to pitching stopper.

39. Engage the following electrical harness connections on the automatic transaxle:

    a. Oxygen sensor connector

    b. Transaxle harness connector

    c. Inhibitor switch connector

    d. Revolution sensor connector on 4WD equipped vehicles

    e. Crankshaft and camshaft angle sensor connector on Legacy

    f. Knock sensor connectors and transaxle ground terminal on Legacy

40. Reconnect the speedometer cable. Manually tighten cable nut all the way, then turn it approximately 30 degrees more with a tool.

41. Install exhaust system and exhaust brackets or hangers that attach to the transaxle, as necessary.

42. Connect the battery ground cable. Refill and check transaxle oil level.

43. Road test vehicle for proper operation across all operating ranges.

Clutch

ADJUSTMENT

Some models are equipped with a mechanical clutch system, which is adjustable. Other models are equipped with an hydraulic system, which is not adjustable.

Cable Adjustment

The clutch cable can be adjusted at the cable bracket where the cable is attached to the side of the transaxle housing.

1. Remove the circlip and clamp.

2. Slide the cable end in the direction desired, then replace the circlip and clamp into the nearest gutters on the cable end.

➡ **The cable should not be stretched out straight nor should it have right angle kinks in it. Any straightening should be gradual.**

3. Check the clutch for proper operation.

Pedal Height

Adjust the pedal with the return stop bolt, so that its pad is on the same level as the brake pedal pad.

Check to be sure that the stroke of the pedal is 5.04–5.43 in. (128–138mm). Check the clutch release fork stroke. It should be 0.67 in. (17mm).

Free-Play Adjustment

1. Remove the clutch fork return spring and loosen the locknut on the fork adjusting nut.

2. Turn the adjusting nut (wing nut) until a release fork free-play of 0.14–0.18 in. (3.5–4.5mm) is obtained.

3. Tighten the locknut.

4. Check the pedal free-play. It should be 0.12–0.16 in. (3.0–4.0mm).

5. Adjust the pedal free-play, as necessary, with the pedal adjusting bolt.

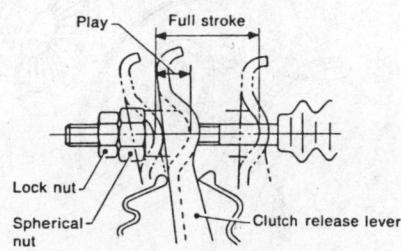

7923TG47

Be sure to tighten the locknut after making the necessary adjustments—mechanical clutch

REMOVAL & INSTALLATION

❊❊ CAUTION

The clutch driven disc may contain asbestos, which has been determined to be a cancer causing agent. Never clean clutch surfaces with compressed air! Avoid inhaling any dust from any clutch surface! When cleaning clutch surfaces, use a commercially available brake fluid.

1. Disconnect the negative battery cable. Remove the transaxle.

2. Gradually unscrew the six bolts (6mm) which hold the pressure plate assembly on the flywheel. Loosen the bolts only one turn at a time, working around the pressure plate. Do not unscrew all the bolts on one side at one time.

3. When all of the bolts have been removed, remove the clutch plate and disc.

❊❊ WARNING

Do not get oil or grease on the clutch facing.

4. Remove the two retaining springs and remove the throwout bearing and the release fork.

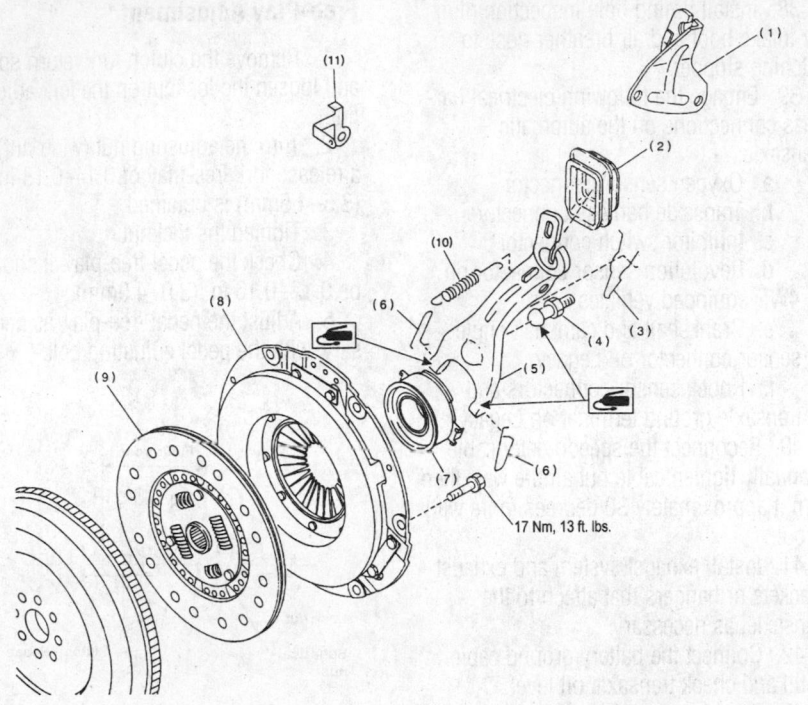

(1) Clutch cable bracket	(6) Clip	(10) Return spring (Models without
(2) Clutch release lever sealing	(7) Clutch release bearing	hill holder only)
(3) Retainer spring	(8) Clutch cover	(11) Clutch return spring bracket
(4) Pivot	(9) Clutch disc	
(5) Pivot		

7923TG48

Exploded view of the clutch system components—mechanical clutch

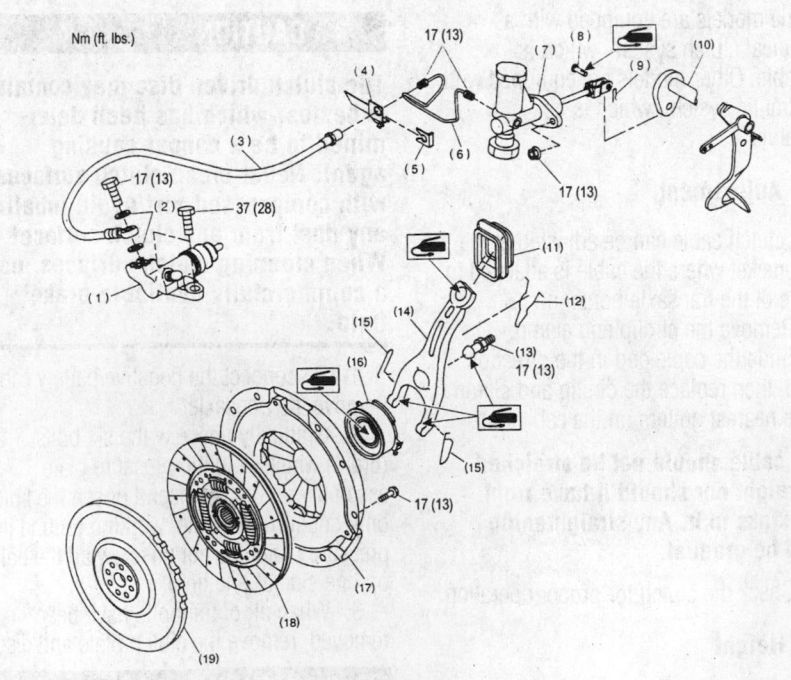

Nm (ft. lbs.)

(1) Operating cylinder	(8) Clevis pin	(14) Release lever
(2) Washer	(9) Snap pin	(15) Clip
(3) Clutch hose	(10) Lever	(16) Release bearing
(4) Bracket	(11) Clutch release lever sealing	(17) Clutch cover
(5) Clamp	(12) Retainer spring	(18) Clutch disc
(6) Pipe	(13) Pivot	(19) Flywheel
(7) Master cylinder ASSY		

7923TG49

Exploded view of the clutch system components—hydraulic clutch

➡Do not disassemble either the clutch cover or disc. Inspect the parts for wear or damage and replace any parts as necessary. Replace the clutch disc if there is any oil or grease on the facing. Do not wash or attempt to lubricate the throwout bearing. If it requires replacement, the bearing may be removed and a new one installed in the holder by means of a press.

To install:

5. Fit the release fork boot on the front of the transaxle housing. Install the release fork.

6. Insert the throwout bearing assembly and secure it with the two springs. Coat the inside diameter of the bearing holder and the fork-to-holder contact points with grease.

7. Insert a pilot shaft through the clutch cover and disc, then insert the end of the pilot into the needle bearing.

8. Tighten the pressure plate bolts gradually, one turn at a time, until the proper torque is reached. Tighten to 13 ft. lbs. (17 Nm).

❊❊ WARNING

When installing the clutch pressure plate assembly, be sure that the O marks on the flywheel and the clutch pressure plate assembly are at least 120° apart. These marks indicate the direction of residual unbalance. Also, be sure that the clutch disc is installed properly, noting the FRONT and REAR markings.

9. After installation of the transaxle in the car, perform the adjustments outlined above.

Hydraulic Clutch System

BLEEDING

➡To properly bleed the system, it must be bled at the slave cylinder and at the damper. Each of these has an air bleeder on it.

1. Connect a vinyl tube to the air bleeder on the damper and put the other end in a jar with clean clutch fluid.

2. With the help of an assistant depressing the clutch pedal, slowly open the bleeder valve. Close the bleeder valve and release the pedal. Repeat this process until no air bubbles appear in the jar.

3. Move the tube to the bleeder on the slave cylinder and repeat the process.

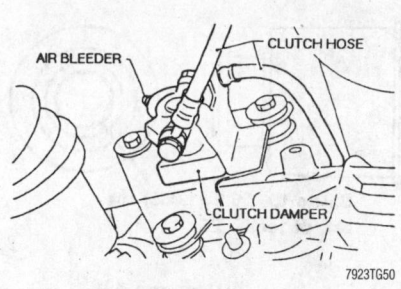

Bleeding the hydraulic clutch at the clutch damper

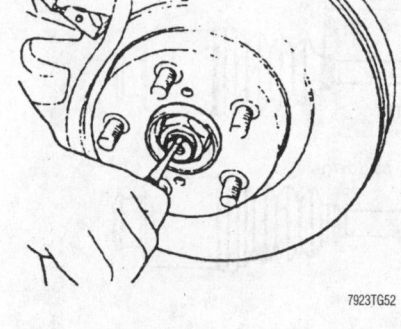

Unstaking the axle nut

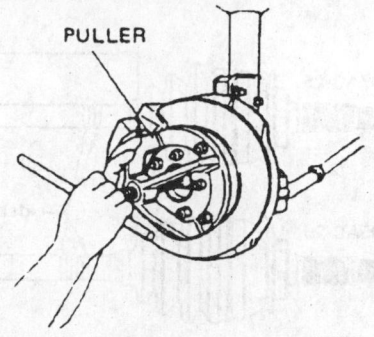

Using a special puller tool, press the axle shaft from the spindle housing

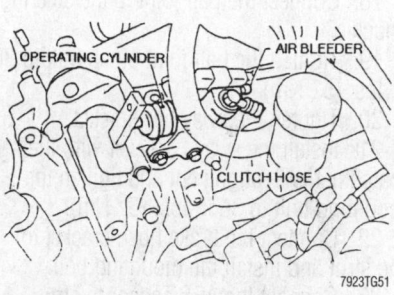

Bleeding the hydraulic clutch at the slave cylinder

Check the operation of the clutch after the bleed procedure is complete.

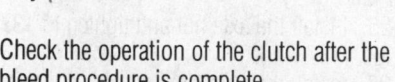
Transfer Case Assembly

The transfer case must be removed as an assembly with the transaxle.

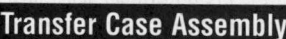
Halfshafts

REMOVAL & INSTALLATION

Except SVX

1. Disconnect the negative battery cable.
2. Raise and support the vehicle safely.
3. Remove the wheel.
4. Unstake and remove the axle nut.
5. Remove the transverse link arm from the front crossmember.
6. Remove the halfshaft-to-transaxle roll pin and discard it.
7. Remove the sway bar bracket.
8. Disconnect the halfshaft from the transaxle.
9. Using puller 92707000 or equivalent, remove the halfshaft from the hub.

To install:

10. Insert the halfshaft into the hub.

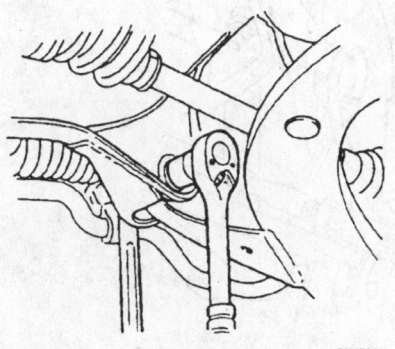

Remove the transverse link arm from the crossmember

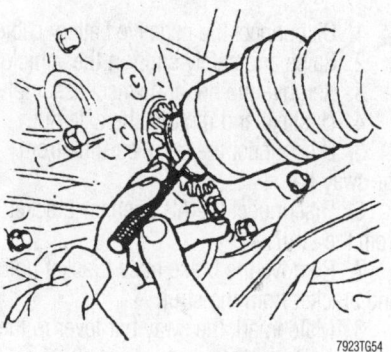

Drive out the halfshaft-to-transaxle roll pin

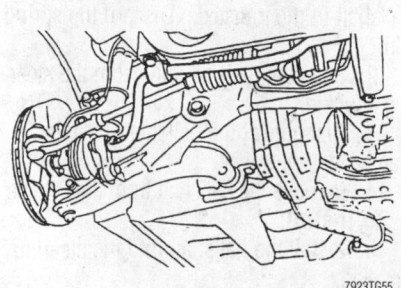

Remove the sway bar bracket

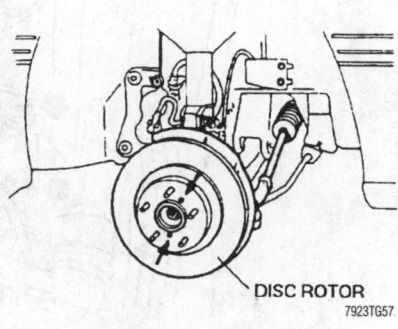

Use two 8mm bolts (arrows) to loosen the rotor from the spindle housing

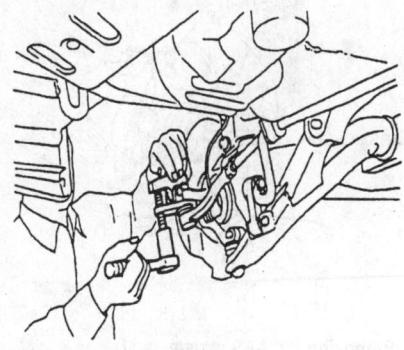

Using a special tool to separate the tie rod end from the steering knuckle

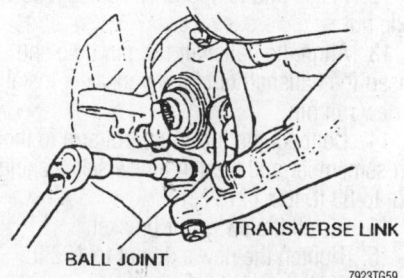

Remove the transverse link arm from the spindle housing

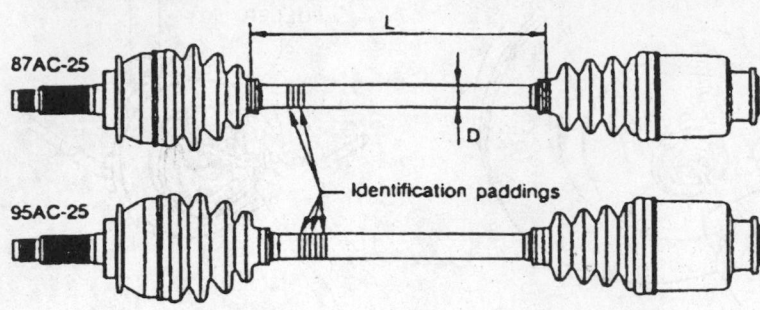

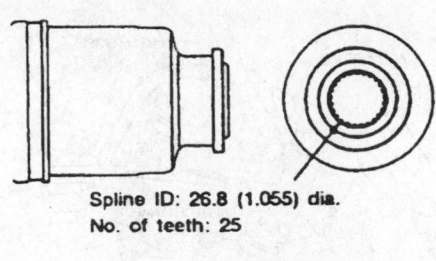

Spline ID: 26.8 (1.055) dia.
No. of teeth: 25

Unit: mm (in)

7923TG60

Be sure to identify the correct halfshaft

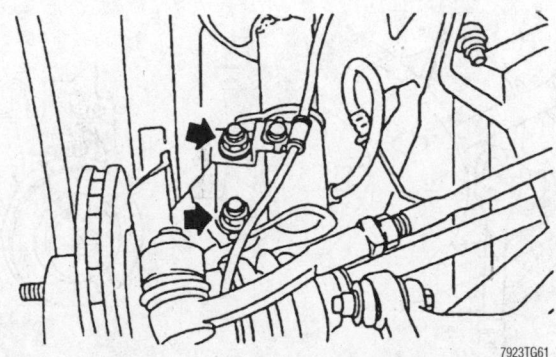

7923TG61

Before loosening the strut-to-housing bolts (arrows), matchmark the camber adjustment bolt and strut

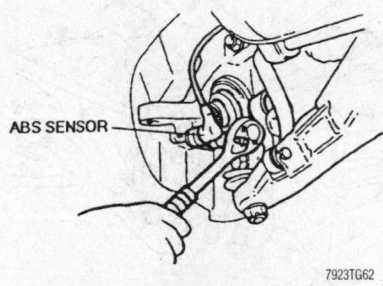

ABS SENSOR

7923TG62

Removing the ABS sensor

11. Using installer 922431000 and adapter 927390000 or equivalent, pull the halfshaft through the hub.

12. Install and temporarily tighten a new axle nut.

13. Align the halfshaft roll pin hole and insert the halfshaft onto the transaxle. Install a new roll pin.

14. Connect the lower control arm to the crossmember and tighten a new self-locking nut to 83 ft. lbs. (113 Nm).

15. Install the sway bar bracket.

16. Tighten the new axle nut to 152 ft. lbs. (206 Nm) and stake the nut.

17. Install the wheel.

18. Lower the vehicle.

19. Connect the negative battery cable.

SVX

1. Disconnect the negative battery cable.

2. Raise and safely support the vehicle.

3. Remove the tire and wheel assembly.

4. Unstake and remove the axle nut.

5. Disconnect the sway bar link from the sway bar.

6. Disconnect the ABS sensor bracket from the strut.

7. Remove the brake hose bracket bolt and bracket from the strut.

8. Matchmark the sway bar lever to the sway bar. Remove the sway bar lever.

9. Loosen the sway bar bracket.

10. Remove the lower ball joint pinch bolt.

11. Drive out the spring pin securing the halfshaft to the transaxle. Discard the spring pin.

12. Separate the ball joint from the steering knuckle.

13. Disconnect the halfshaft from the transaxle.

14. Remove the halfshaft from the vehicle.

To install:

15. Install the halfshaft into the steering knuckle.

16. Connect the inboard end on the halfshaft to the transaxle. Be sure the spring pin holes line up.

17. Install a new spring pin.

18. Connect the ball joint to the steering knuckle.

19. Tighten the ball joint pinch bolt to 38 ft. lbs. (52 Nm).

20. Tighten the sway bar bracket.

21. Install the sway bar lever with the matchmarks in alignment and tighten the mounting bolt to 38 ft. lbs. (52 Nm).

22. Connect the brake hose bracket to the strut and install the mounting bolt.

23. Connect the ABS sensor to strut bracket.

24. Connect the sway bar to the sway bar link.

25. Install the axle nut and tighten to 137 ft. lbs. (187 Nm).

26. Install the dust cap.

27. Install the tire and wheel assembly.

28. Lower the vehicle.

29. Connect the negative battery cable.

30. Check the front end alignment and adjust as necessary.

STEERING AND SUSPENSION

Air Bag

❊❊ CAUTION

Some vehicles are equipped with an air bag system, also known as the Supplemental Inflatable Restraint (SIR) or Supplemental Restraint System (SRS). The system must be disabled before performing service on or around system components, steering column, instrument panel components, wiring and sensors. Failure to follow safety and disabling procedures could result in accidental air

bag deployment, possible personal injury and unnecessary system repairs.

PRECAUTIONS

Several precautions must be observed when handling the inflator module to avoid accidental deployment and possible personal injury.

• Never carry the inflator module by the wires or connector on the underside of the module.

• When carrying a live inflator module, hold securely with both hands, and ensure that the bag and trim cover are pointed away.

• Place the inflator module on a bench or other surface with the bag and trim cover facing up.

• With the inflator module on the bench, never place anything on or close to the module which may be thrown in the event of an accidental deployment.

DISARMING

1. Disconnect the negative battery cable.
2. Disconnect the positive battery cable.
3. Wait more than 20 seconds before starting work.

Power Rack and Pinion Steering Gear

REMOVAL & INSTALLATION

Legacy and Legacy Outback Models

1. Disconnect the negative battery cable.
2. Raise and support the vehicle safely.
3. Remove the front tire and wheel assemblies.
4. Detach the electrical connector from the oxygen sensor. Remove the front exhaust pipe assembly.
5. Remove the tie rod end cotter pin and loosen the castle nut. Using a ball joint puller, separate the tie rod ends from the steering knuckle arm.
6. Remove the jack up plate and the front stabilizer bar.
7. From the power steering rack, remove the center pressure pipe, connect a vinyl hose to the pipe and joint, then turn the steering wheel to discharge the fluid into a container.

➡When discharging the power steering fluid (line A and B), turn the steering wheel fully, left and right. Be sure to

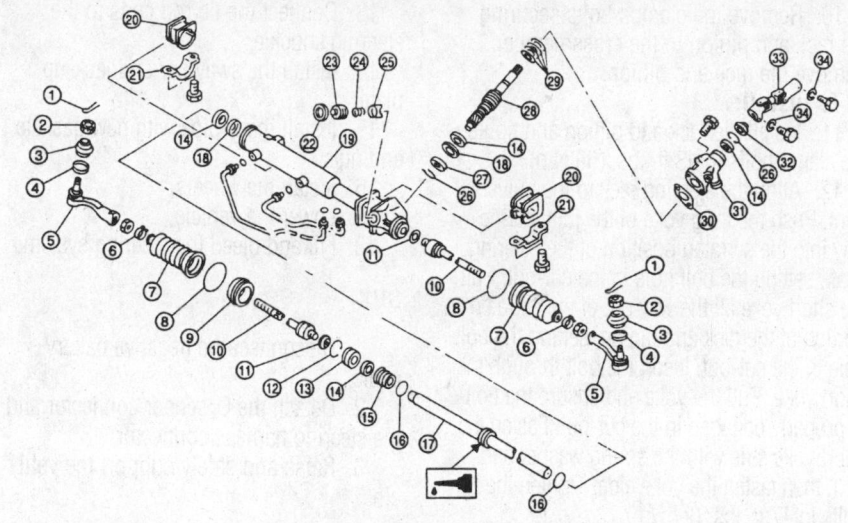

① Cotter pin	⑬ Rack stopper	㉔ Spring
② Castle nut	⑭ Oil seal	㉕ Sleeve
③ Dust cover	⑮ Rack bushing	㉖ C-ring
④ Clip	⑯ O-ring	㉗ Ball bearing
⑤ Tie-rod end	⑰ Rack	㉘ Valve
⑥ Clip	⑱ Back-up washer	㉙ Seal ring
⑦ Boot	⑲ Rack housing	㉚ Packing
⑧ Clip	⑳ Adapter	㉛ Valve housing
⑨ Spacer	㉑ Clamp	㉜ Dust seal
⑩ Tie-rod	㉒ Lock nut	㉝ Universal joint
⑪ Lock washer	㉓ Adjusting screw	㉞ Spring washer
⑫ Circlip		

7923TG63

Exploded view of the steering rack assembly—Legacy models

disconnect the other pipe and drain the fluid in the same manner.

8. From the control valve of the gearbox assembly, remove the power steering **C** and **D** pressure pipes. Remove pipe **D** first and pipe **C** second.

9. If not disconnected when draining the fluid from the control valve of the gearbox assembly, remove the power steering **A** and **B** pressure pipes. Remove pipe **A** first and pipe **B** second.

10. Remove the power steering gearbox to crossmember assembly bolts. Remove the gearbox assembly from the vehicle.

To install:

11. Install the power steering rack and tighten the rack to crossmember bolts to 35–52 ft. lbs. (47–70 Nm). When installing the universal joint assembly, be sure to align the matchmarks.

12. Tighten the power steering pressure pipes 7–12 ft. lbs. (10–16 Nm), the universal joint assembly to power steering gearbox bolts 16–19 ft. lbs. (22–24 Nm) and the universal joint assembly to steering shaft bolts 16–19 ft. lbs. (22–24 Nm).

13. Tighten the tie rod end to steering knuckle nut 18–22 ft. lbs. (25–29 Nm).

After tightening this nut, turn it up to 60 degrees further to align the cotter pin hole. Install a new cotter pin.

14. Install the tires and tighten the wheel lug nuts to specification.

15. Refill and bleed the power steering system.

16. Check and adjust the toe-in and the steering angle.

Impreza, Impreza Outback and Impreza Outback Sport Models

1. Disconnect the negative battery cable.
2. Raise and support the vehicle safely.
3. Remove the front wheels.
4. Remove the front Y-pipe.
5. Remove the tie rod end cotter pin and nut. Using a puller, disconnect the tie rod ends from the steering knuckle.
6. Remove the jack-up plate and front sway bar.
7. Disconnect the fluid lines from the rack and pinion.
8. Matchmark the universal joint to the serration in the steering rack for installation reference.
9. Remove the universal joint bolts and lift the joint upward disconnecting it from the rack and pinion shaft.

10. Remove the clamps bolts securing the rack and pinion to the crossmember. Remove the rack and pinion.

To install:

11. Install the rack and pinion and tighten the clamp bolts to 43 ft. lbs. (59 Nm).

12. Align the steering rack to the universal joint. Push the long yoke of the joint all the way into the serrated position of the steering shaft, setting the bolt hole in the cut-out. Pull the short yoke all the way out of the serrated portion of the rack and pinion, setting the bolt hole in the cut-out. Insert the bolt through the short yoke. Pull the yoke and ensure the bolt is properly engaged in the cut-out. Fasten the short yoke side with the spring washer and bolt, then fasten the yoke side. Tighten the bolts to 17 ft. lbs. (24 Nm).

13. Connect the tie rod ends to the steering knuckle.

14. Install the sway bar and jack-up plate.

15. Install the Y-pipe with new gaskets and nuts.

16. Install the wheels.

17. Lower the vehicle.

18. Fill and bleed the steering system.

SVX

1. Disconnect the negative battery cable.

2. Detach the O2 sensor connector and the steering harness connector.

3. Raise and safely support the vehicle.

4. Remove the front under cover.

5. Disconnect the O2 sensor harness from the clip.

6. Remove the collector cover and rear catalytic converter cover.

7. Disconnect the front exhaust pipe.

8. Remove the cotter pin and castle nut from the outer tie rod ends and separate the tie rods from the steering knuckles using a suitable puller.

9. Remove the spring pin securing the halfshaft to the transaxle.

10. Disconnect the halfshaft from the transaxle.

11. Disconnect the power steering lines one at a time. After disconnecting each allow the line to hang into a drain pan while you rotate the wheel lock to lock. This will

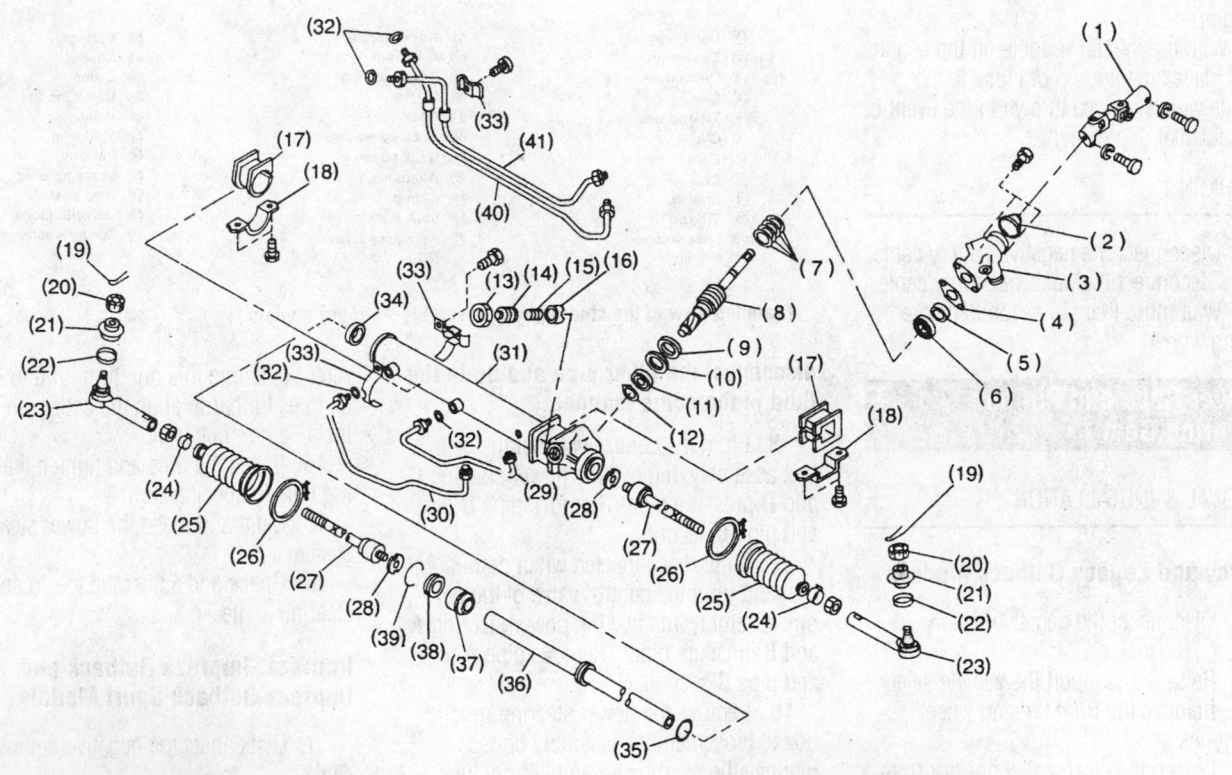

(1)	Universal joint	(15)	Spring	(29)	Pipe B
(2)	Dust cover	(16)	Sleeve	(30)	Pipe A
(3)	Valve housing	(17)	Adapter	(31)	Steering body
(4)	Gasket	(18)	Clamp	(32)	O-ring
(5)	Oil seal	(19)	Cotter pin	(33)	Clamp
(6)	Special bearing	(20)	Castle nut	(34)	Oil seal
(7)	Seal ring	(21)	Dust cover	(35)	Piston ring
(8)	Pinion and valve ASSY	(22)	Clip	(36)	Rack
(9)	Oil seal	(23)	Tie-rod end	(37)	Rack bushing
(10)	Back-up washer	(24)	Clip	(38)	Rack stopper
(11)	Ball bearing	(25)	Boot	(39)	Circlip
(12)	Snap ring	(26)	Band	(40)	Pipe E
(13)	Lock nut	(27)	Tie-rod	(41)	Pipe F
(14)	Adjusting screw	(28)	Lock washer		

Exploded view of the steering rack assembly—Impreza models

7923TG64

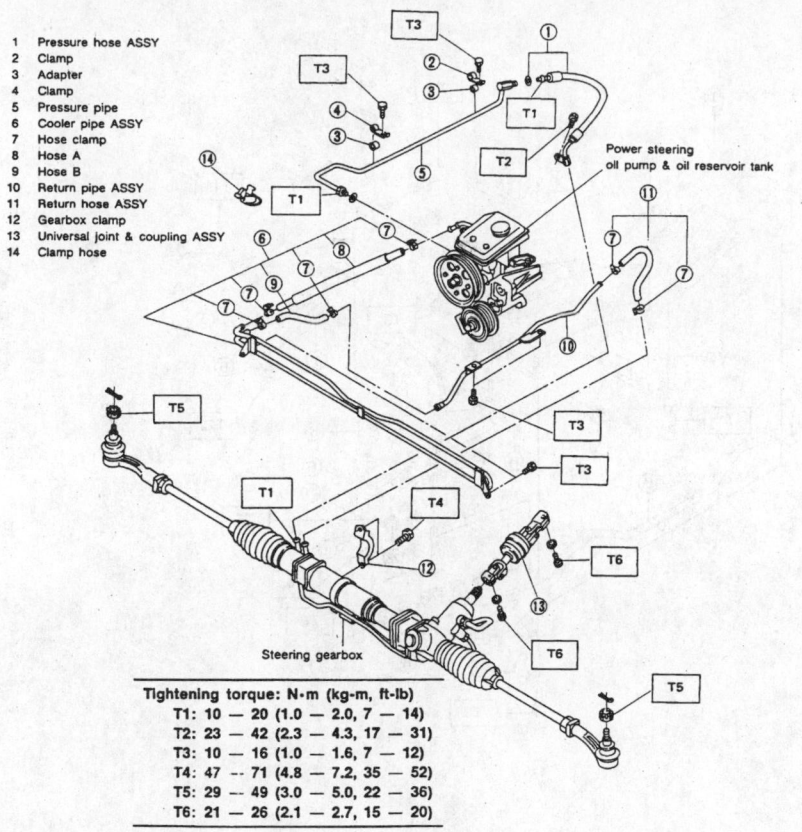

Tightening torque: N·m (kg-m, ft-lb)
T1: 10 — 20 (1.0 — 2.0, 7 — 14)
T2: 23 — 42 (2.3 — 4.3, 17 — 31)
T3: 10 — 16 (1.0 — 1.6, 7 — 12)
T4: 47 — 71 (4.8 — 7.2, 35 — 52)
T5: 29 — 49 (3.0 — 5.0, 22 — 36)
T6: 21 — 26 (2.1 — 2.7, 15 — 20)

1. Pressure hose ASSY
2. Clamp
3. Adapter
4. Clamp
5. Pressure pipe
6. Cooler pipe ASSY
7. Hose clamp
8. Hose A
9. Hose B
10. Return pipe ASSY
11. Return hose ASSY
12. Gearbox clamp
13. Universal joint & coupling ASSY
14. Clamp hose

7923TG65

Exploded view of the power steering assembly—SVX model

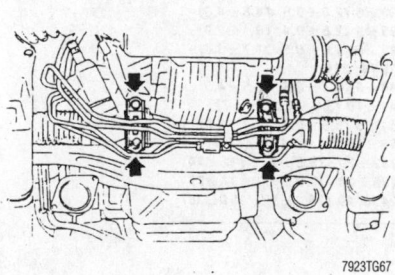

7923TG67

Removing the steering rack retainer bolts

purge the power steering fluid from the rack and pinion unit.

12. Disconnect and remove the performance rod.

13. Remove the lower ball joint pinch bolt on the right side. Separate the ball joint from the steering knuckle.

14. Matchmark the universal joint to the rack and pinion unit and the steering shaft.

15. Remove the upper and lower pinch bolts on the universal joint and push the joint up to remove it from the rack and pinion unit. Once it clears the rack pull it down and off the steering shaft.

16. Remove the bolts securing the rack and pinion unit to the vehicle underbody.

17. Remove the rack by performing the following:

 a. Turn the rack so the control valve faces the rear.

 b. Move the rack to the right so the left tie rod can be removed from the rack assembly.

 c. Remove the rack from the vehicle.

To install:

18. Install the rack and pinion unit into the vehicle.

19. Install the mounting bolts and tighten to 44 ft. lbs. (59 Nm).

20. Install the universal with the matchmarks in line onto the steering shaft first, then down over the rack pinion shaft. Install the pinch bolts and tighten them to 17 ft. lbs. (24 Nm)

21. Connect the ball joint to the steering knuckle and tighten the pinch bolt to 38 ft. lbs. (52 Nm).

22. Install the performance rod.

23. Connect the power steering lines to the rack and pinion unit.

24. Connect the halfshaft to the axle and install a new spring pin.

25. Connect the tie rod ends to the steering knuckles and tighten the castle nuts to 29 ft. lbs. (39 Nm).

26. Install the front exhaust pipe.

27. Install the exhaust covers removed.

28. Connect the O_2sensor harness to the clip.

29. Install the front under cover.

30. Install the tire and wheel assembly.

31. Lower the vehicle.

32. Attach the steering harness and O_2sensor connector.

33. Connect the negative battery cable.

34. Check the front end alignment and adjust as necessary.

Strut

REMOVAL & INSTALLATION

Front

EXCEPT SVX—WITH STANDARD STRUTS

➡**Do not remove the large nut on top of the strut assembly unless the coil spring is properly compressed with a suitable spring compressor.**

1. Disconnect the negative battery cable.

2. Raise and support the vehicle safely.

3. Remove the front wheel assembly.

4. Disconnect the ABS sensor, if equipped.

5. Remove the caliper, leaving the line connected and suspend it out of the way with a piece of wire or string.

6. Remove the clip attaching the brake line to the strut housing.

7. Matchmark the camber adjustment bolt to the strut housing as reference for installation.

8. If equipped with ABS, remove the bolt securing the sensor harness.

9. Remove the two bolts and nuts securing the strut to the steering knuckle. Notice that the shaft of the top bolt is not round. This bolt is used for camber adjustment, and most always be installed in the top hole.

10. Remove the three nuts securing the strut to the body in the engine compartment.

11. Remove the strut and coil spring assembly from the vehicle.

To install:

12. Install the strut assembly into the vehicle.

13. Install the upper strut retainer nuts, and tighten the nuts to 15 ft. lbs. (20 Nm).

14. If equipped, install the ABS sensor harness, and tighten the bolt to 14 ft. lbs. (20 Nm).

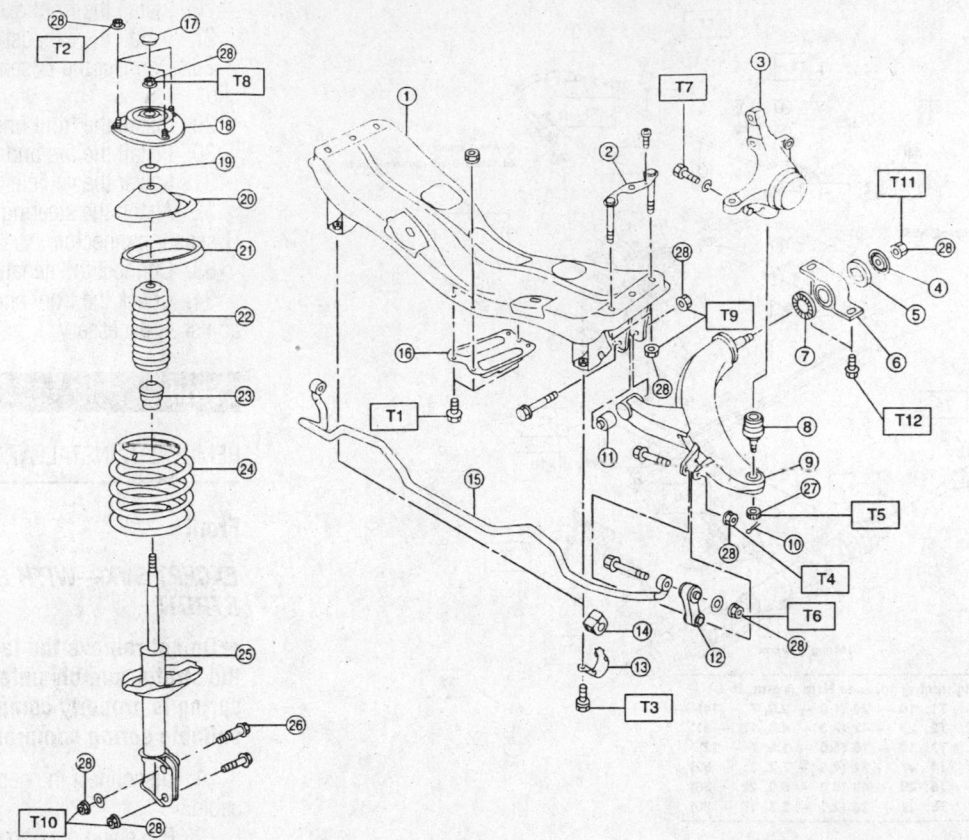

① Crossmember	⑮ Stabilizer
② Bolt ASSY	⑯ Jack-up plate
③ Housing	⑰ Dust seal
④ Washer	⑱ Strut mount
⑤ Stop rubber (Rear)	⑲ Spacer
⑥ Rear bushing	⑳ Upper spring seat
⑦ Stop rubber (Front)	㉑ Rubber seat
⑧ Ball joint	㉒ Dust cover
⑨ Transverse link	㉓ Helper
⑩ Cotter pin	㉔ Coil spring
⑪ Front bushing	㉕ Damper strut
⑫ Stabilizer link	㉖ Adjusting bolt
⑬ Clamp	㉗ Castle nut
⑭ Bushing	㉘ Self-locking nut

Tightening torque: N·m (kg-m, ft-lb)
T1: 18 ± 5 (1.8 ± 0.5, 13.0 ± 3.6)
T2: 20 ± 6 (2.0 ± 0.6, 14.5 ± 4.3)
T3: 25 ± 4 (2.5 ± 0.4, 18.1 ± 2.9)
T4: 29 ± 5 (3.0 ± 0.5, 21.7 ± 3.6)
T5: 39 (4, 29)
T6: 44 ± 6 (4.5 ± 0.6, 32.5 ± 4.3)
T7: 49 ± 10 (5.0 ± 1.0, 36 ± 7)
T8: 54 ± 5 (5.5 ± 0.5, 39.8 ± 3.6)
T9: 98 ± 15 (10.0 ± 1.5, 72 ± 11)
T10:152 ± 20 (15.5 ± 2.0, 112 ± 14)
T11:186 ± 10 (19.0 ± 1.0, 137 ± 7)
T12:245 ± 49 (25.0 ± 5.0, 181 ± 36)

7923TG68

Exploded view of the front suspension assembly—Legacy and Impreza models

15. Install the lower strut nuts and bolts. Be sure the alignment adjustment bolt is installed in the top mounting hole. Tighten the nuts, while securing the bolts to 112 ft. lbs. 152 Nm).

16. Install the caliper.

17. Attach the brake line to the strut and install the clip.

18. Install the front wheel.

19. Lower the vehicle to the floor.

20. Connect the negative battery cable.

21. Have the front wheel alignment checked by a qualified professional.

EXCEPT SVX—WITH PNEUMATIC STRUTS

1. Disconnect the negative battery cable.
2. Raise and support the vehicle safely.

3. Remove the front wheel assembly.

4. From inside the engine compartment, disconnect the air line and height sensor harness from the strut assembly.

5. Disconnect the ABS sensor, if equipped.

6. Remove the caliper, leaving the line connected and suspend it out of the way with a piece of wire or string.

7. Remove the clip attaching the brake line to the strut housing.

8. Matchmark the camber adjustment bolt to the strut housing as reference for installation.

9. If equipped with ABS, remove the bolt securing the sensor harness.

10. Remove the two bolts and nuts securing the strut to the steering knuckle.

Notice that the shaft of the top bolt is not round. This bolt is used for camber adjustment, and most always be installed in the top hole.

11. Remove the three nuts securing the strut to the body in the engine compartment.

12. Remove the strut and coil spring assembly from the vehicle.

To install:

13. Install the strut assembly into the vehicle.

14. Install the upper strut retainer nuts, and tighten the nuts to 15 ft. lbs. (20 Nm).

15. If equipped, install the ABS sensor harness, and tighten the bolt to 14 ft. lbs. (20 Nm).

16. Install the lower strut nuts and bolts. Be sure the alignment adjustment bolt is

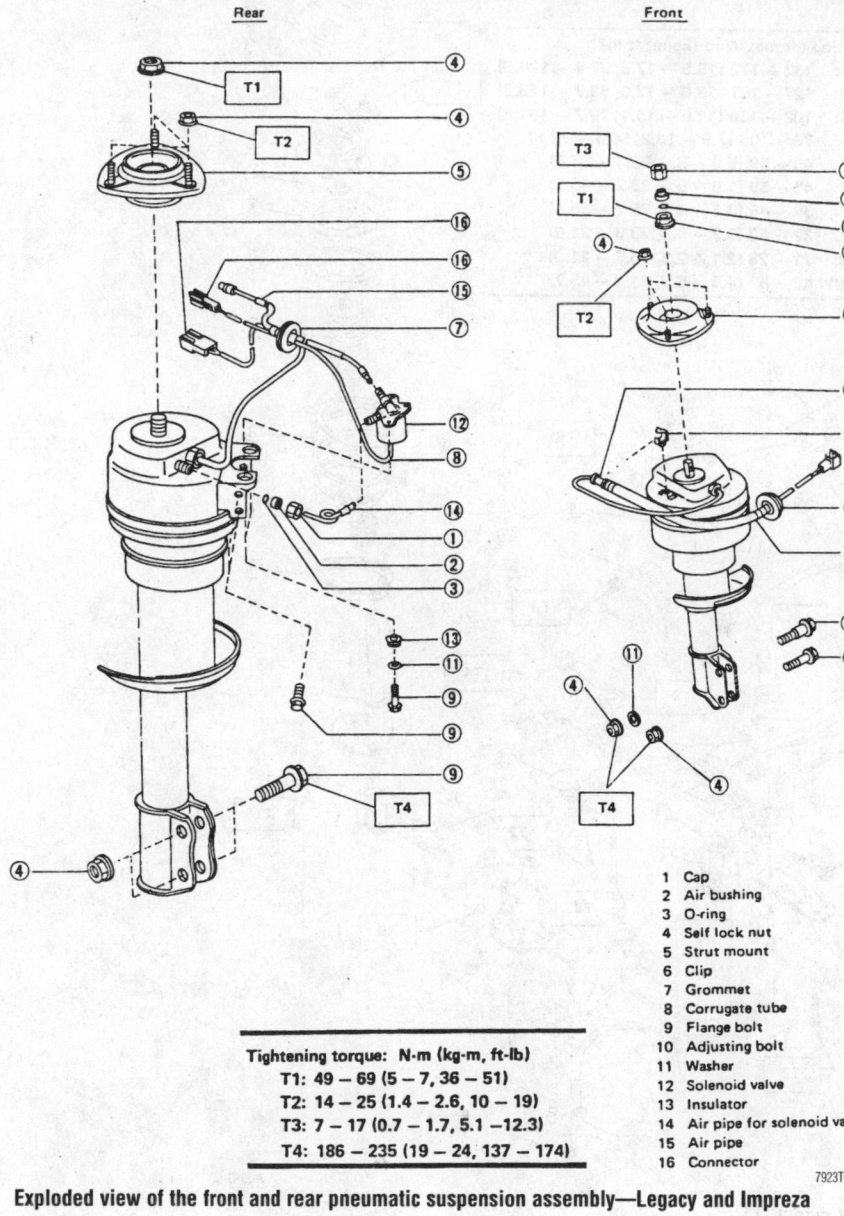

Rear

Front

Tightening torque: N·m (kg-m, ft-lb)
T1: 49 – 69 (5 – 7, 36 – 51)
T2: 14 – 25 (1.4 – 2.6, 10 – 19)
T3: 7 – 17 (0.7 – 1.7, 5.1 –12.3)
T4: 186 – 235 (19 – 24, 137 – 174)

1	Cap
2	Air bushing
3	O-ring
4	Self lock nut
5	Strut mount
6	Clip
7	Grommet
8	Corrugate tube
9	Flange bolt
10	Adjusting bolt
11	Washer
12	Solenoid valve
13	Insulator
14	Air pipe for solenoid valve
15	Air pipe
16	Connector

7923TG69

Exploded view of the front and rear pneumatic suspension assembly—Legacy and Impreza models

installed in the top mounting hole. Tighten the nuts, while securing the bolts to 112 ft. lbs. 152 Nm).

17. Install the caliper.

18. Attach the brake line to the strut and install the clip.

19. Attach the height sensor harness and air line.

20. Install the front wheel.

21. Lower the vehicle to the floor.

22. Connect the negative battery cable.

23. Start the vehicle and allow enough time for the strut to pressurize before driving.

24. Have the front wheel alignment checked by a qualified professional.

SVX

➡ **Do not remove the large nut on top of the strut assembly unless the coil spring is properly compressed with a suitable spring compressor.**

1. Disconnect the negative battery cable.

2. Raise and safely support the vehicle.

3. Remove the tire and wheel assembly.

4. Disconnect the sway bar from the strut assembly.

5. Remove the bolt and bracket securing the ABS sensor wire to the strut assembly.

6. Remove the bolt and bracket securing the brake hose to the strut.

7. Scribe matchmarks on the camber adjusting bolt and the steering knuckle for installation purposes.

8. Remove the nuts from the lower strut mounting bolts. Remove the lower mounting bolt. The upper bolt **MUST** remain in place.

9. Support the lower control arm under the ball joint with a suitable jack.

10. Remove the strut mount cap in the engine compartment.

11. Remove the three strut plate mounting nuts.

12. Lower the jack about an in. and remove the strut-to-steering knuckle upper mounting bolt.

13. Remove the strut from the vehicle.

14. If the strut is to be replaced, remove the sway bar link.

To install:

15. Install the sway bar link on the strut and tighten the mounting nut to 28 ft. lbs. (37 Nm).

16. Install the strut assembly into the vehicle and install the upper mounting nuts and tighten to 30 ft. lbs. (41 Nm).

17. Install the strut mount cap.

18. Connect the strut to the steering knuckle and install the lower mounting through-bolts and loosely install the nuts.

19. Remove the jack.

20. Rotate the camber adjusting bolt so the matchmarks are in alignment. Tighten the mounting nuts to 112 ft. lbs. (152 Nm).

21. Install the brake hose bracket and bolt on the strut.

22. Install the ABS sensor bracket and bolt on the strut.

23. Connect the sway bar to the sway bar link and tighten the mounting nut to 28 ft. lbs. (37 Nm).

24. Install the tire and wheel assembly.

25. Lower the vehicle.

26. Connect the negative battery cable.

27. Check the front end alignment and adjust as necessary.

Rear

EXCEPT SVX—WITH STANDARD STRUTS

❈❈ CAUTION

Do not remove the large nut on top of the strut assembly unless the coil spring is properly retained with a spring compressor.

1. On the Sedan, remove the rear seat assembly.

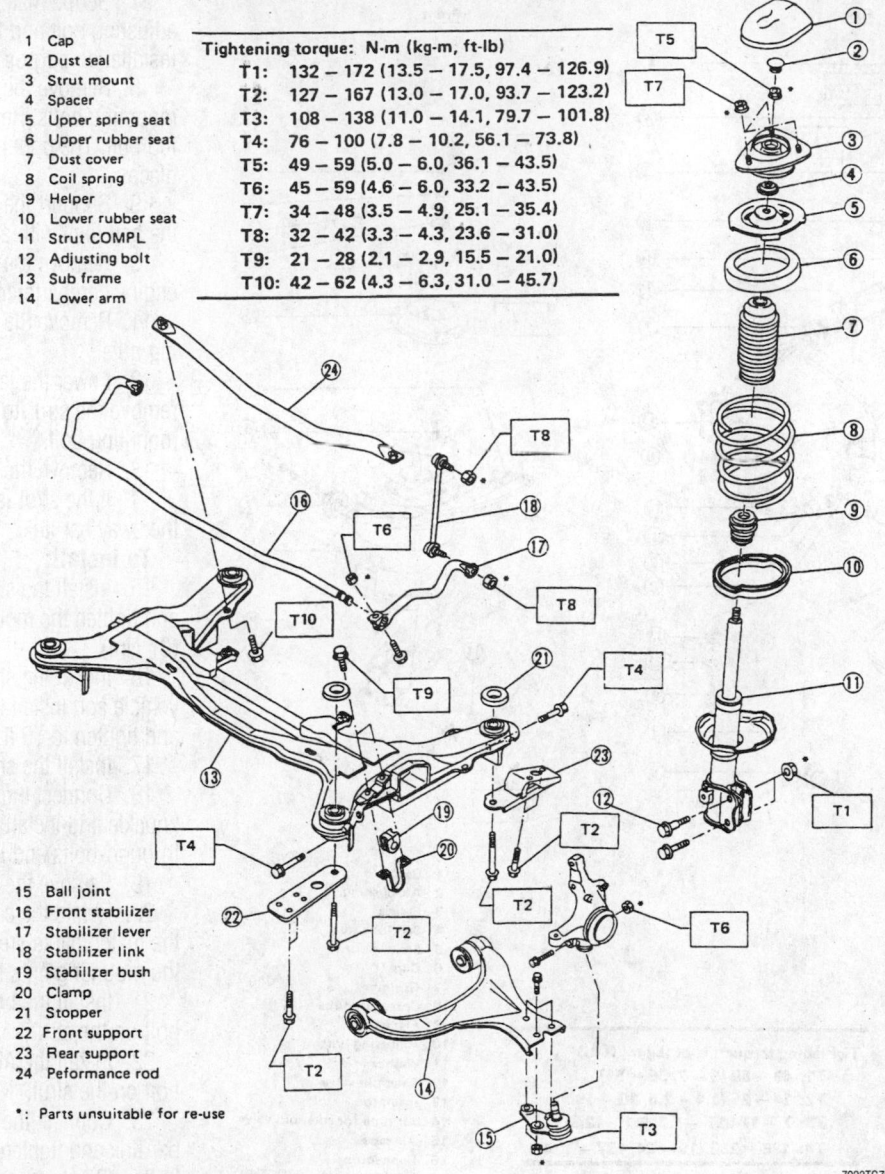

1 Cap
2 Dust seal
3 Strut mount
4 Spacer
5 Upper spring seat
6 Upper rubber seat
7 Dust cover
8 Coil spring
9 Helper
10 Lower rubber seat
11 Strut COMPL
12 Adjusting bolt
13 Sub frame
14 Lower arm

Tightening torque: N·m (kg-m, ft-lb)
T1: 132 – 172 (13.5 – 17.5, 97.4 – 126.9)
T2: 127 – 167 (13.0 – 17.0, 93.7 – 123.2)
T3: 108 – 138 (11.0 – 14.1, 79.7 – 101.8)
T4: 76 – 100 (7.8 – 10.2, 56.1 – 73.8)
T5: 49 – 59 (5.0 – 6.0, 36.1 – 43.5)
T6: 45 – 59 (4.6 – 6.0, 33.2 – 43.5)
T7: 34 – 48 (3.5 – 4.9, 25.1 – 35.4)
T8: 32 – 42 (3.3 – 4.3, 23.6 – 31.0)
T9: 21 – 28 (2.1 – 2.9, 15.5 – 21.0)
T10: 42 – 62 (4.3 – 6.3, 31.0 – 45.7)

15 Ball joint
16 Front stabilizer
17 Stabilizer lever
18 Stabilizer link
19 Stabilzer bush
20 Clamp
21 Stopper
22 Front support
23 Rear support
24 Peformance rod

*: Parts unsuitable for re-use

Exploded view of the front suspension assembly—SVX model

2. On the Wagon, remove the rear speaker grille and service hole cap.

3. Remove the strut mount cap.

4. Raise and safely support the vehicle.

5. Remove the wheel and tire assembly.

6. Remove the brake hose clip.

7. Remove the union bolt from the brake caliper. Move the brake hose out of the way.

8. Remove the lower nuts and bolts securing the strut to the rear wheel housing.

9. From inside the vehicle, loosen and remove the retainer nuts securing the strut bearing cap to the strut tower.

10. Lower and remove the strut from the vehicle.

To install:

11. Install the strut on to the vehicle, making sure to position the strut properly in the upper strut tower mounts. Refer to the illustration if needed. Install the retainer nuts, and tighten to 11 ft. lbs. (15 Nm).

12. Connect the strut to the rear wheel knuckle assembly, using the retainer nuts and bolts, and tighten the bolts to 145 ft. lbs. (196 Nm).

13. Install the brake union bolt, and tighten to 13 ft. lbs. (18 Nm).

14. Insert the brake hose clip.

15. Bleed the brakes.

16. Install the wheel.

17. Lower the vehicle.

18. Install the strut mount cap.

19. On Sedan, install the rear seat.

20. On Wagon, install the speaker grille.

EXCEPT SVX—WITH PNEUMATIC STRUTS

1. Disconnect the negative battery cable.

2. On the Sedan, remove the rear seat assembly.

3. On the Wagon, remove the rear speaker grille and service hole cap.

4. Remove the strut mount cap.

5. Disconnect the air line from the top of the strut assembly.

6. Disconnect the height sensor and solenoid valve wiring harnesses from the strut assembly.

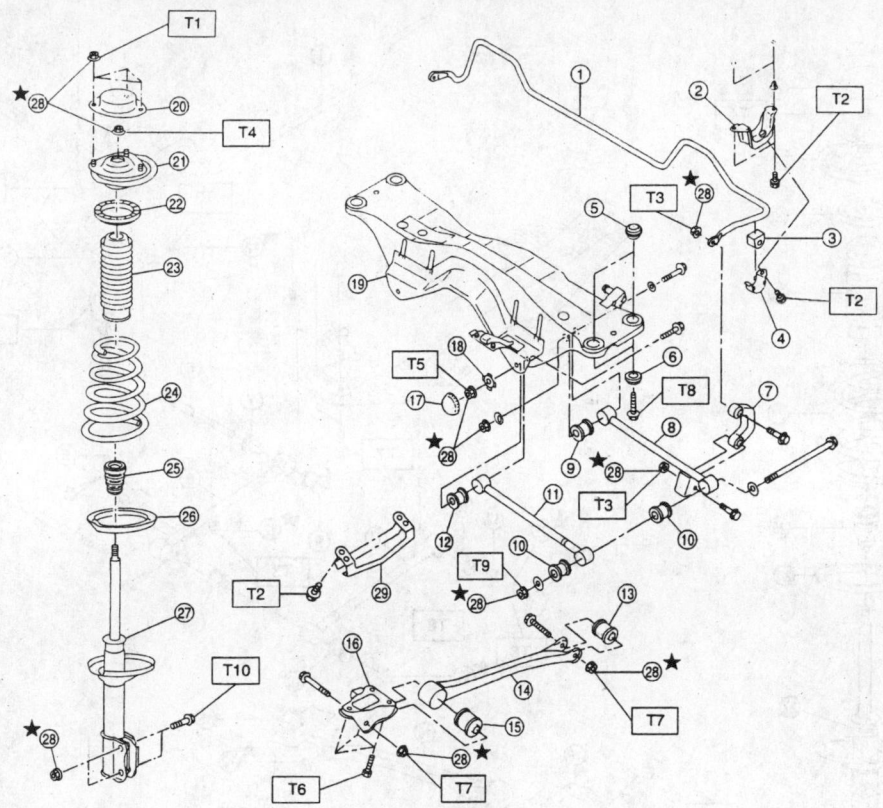

① Stabilizer	⑯ Trailing link bracket
② Stabilizer bracket	⑰ Cap (Protection)
③ Stabilizer bushing	⑱ Washer
④ Clamp	⑲ Crossmember
⑤ Floating bushing	⑳ Strut mount cap
⑥ Stopper	㉑ Strut mount
⑦ Stabilizer link	㉒ Rubber seat upper
⑧ Rear lateral link	㉓ Dust cover
⑨ Bushing (C)	㉔ Coil spring
⑩ Bushing (A)	㉕ Helper
⑪ Front lateral link	㉖ Rubber seat lower
⑫ Bushing (B)	㉗ Damper strut
⑬ Trailing link rear bushing	㉘ Self-locking nut
⑭ Trailing link	㉙ Crossmember reinforcement
⑮ Trailing link front bushing	lower (Sedan model only)

Tightening torque: N·m (kg-m, ft-lb)
T1: 20 ± 6 (2.0 ± 0.6, 14.5 ± 4.3)
T2: 25 ± 7 (2.5 ± 0.7, 18.1 ± 5.1)
T3: 44 ± 6 (4.5 ± 0.6, 32.5 ± 4.3)
T4: 59 ± 10 (6.0 ± 1.0, 43 ± 7)
T5: 98 ± 15 (10.0 ± 1.5, 72 ± 11)
T6: 98 ± 20 (10.0 ± 2.0, 72 ± 14)
T7: 113 ± 15 (11.5 ± 1.5, 83 ± 11)
T8: 127 ± 20 (13.0 ± 2.0, 94 ± 14)
T9: 137 ± 20 (14.0 ± 2.0, 101 ± 14)
$T10: 196^{+39}_{-10} (20.0^{+4.0}_{-1.0}, 145^{+29}_{-7})$

7923TG71

Exploded view of the rear standard strut assembly—AWD Legacy and Impreza models

7. Raise and safely support the vehicle.

8. Remove the wheel and tire assembly.

9. Remove the brake hose clip.

10. Remove the union bolt from the brake caliper. Move the brake hose out of the way.

11. Remove the lower nuts and bolts securing the strut to the rear wheel housing.

12. From inside the vehicle, loosen and remove the retainer nuts securing the strut bearing cap to the strut tower.

13. Lower and remove the strut from the vehicle.

To install:

14. Install the strut on to the vehicle, making sure to position the strut properly in the upper strut tower mounts. Refer to the illustration if needed. Install the retainer nuts, and tighten to 11 ft. lbs. (15 Nm).

15. Connect the strut to the rear wheel knuckle assembly, using the retainer nuts and bolts, and tighten the bolts to 145 ft. lbs. (196 Nm).

16. Install the brake union bolt, and tighten to 13 ft. lbs. (18 Nm).

17. Insert the brake hose clip.

18. Bleed the brakes.

19. Install the wheel.

20. Lower the vehicle.

21. Attach the height sensor, and solenoid valve wiring harnesses to the strut.

22. Attach the air line to the top of the strut.

23. Install the strut mount cap.

24. On Sedan, install the rear seat.

25. On Wagon. install the speaker grille.

26. Connect the negative battery cable.

27. Start the vehicle, and allow enough time for the shock to pressurize before driving the vehicle.

SVX

1. Raise and safely support the vehicle.

2. Remove the tire and wheel assembly.

3. Remove the rear quarter interior trim panel.

4. Remove the bolt securing the brake hose bracket to the strut and position the bracket out of the way.

5. Remove the knuckle-to-strut mounting nuts and remove the lower bolt. Leave the upper bolt in place.

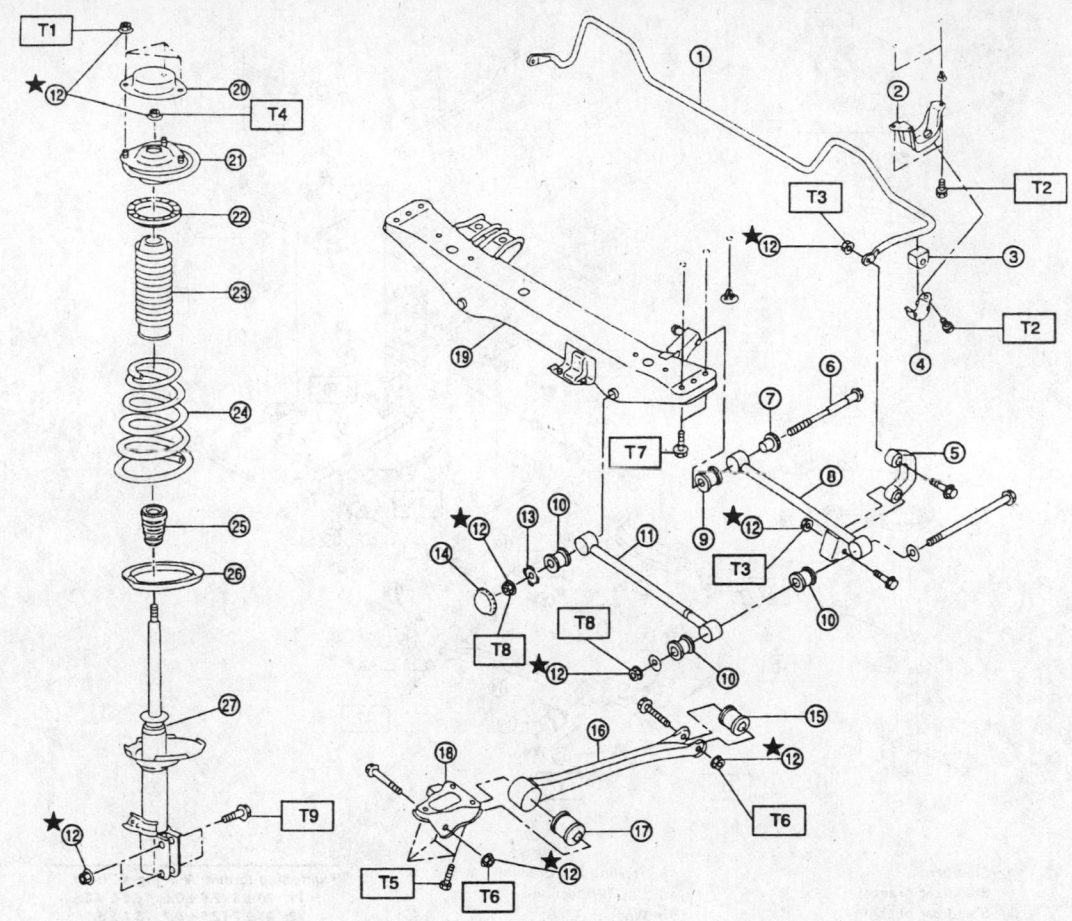

① Stabilizer
② Stabilizer bracket
③ Stabilizer bushing
④ Clamp
⑤ Stabilizer link
⑥ Adjusting bolt
⑦ Adjusting wheel
⑧ Rear lateral link
⑨ Bushing (D)
⑩ Bushing (A)
⑪ Front lateral link
⑫ Self-locking nut
⑬ Washer
⑭ Cap (Protection)

⑮ Trailing link rear bushing
⑯ Trailing link
⑰ Trailing link front bushing
⑱ Trailing link bracket
⑲ Crossmember
⑳ Strut mount cap
㉑ Strut mount
㉒ Rubber seat upper
㉓ Dust cover
㉔ Coil spring
㉕ Helper
㉖ Rubber seat lower
㉗ Damper strut

Tightening torque: N·m (kg-m, ft-lb)
T1: 20 ± 6 (2.0 ± 0.6, 14.5 ± 4.3)
T2: 25 ± 7 (2.5 ± 0.7, 18.1 ± 5.1)
T3: 44 ± 6 (4.5 ± 0.6, 32.5 ± 4.3)
T4: 59 ± 10 (6.0 ± 1.0, 43 ± 7)
T5: 98 ± 20 (10.0 ± 2.0, 72 ± 14)
T6: 113 ± 15 (11.5 ± 1.5, 83 ± 11)
T7: 127 ± 20 (13.0 ± 2.0, 94 ± 14)
T8: 137 ± 20 (14.0 ± 2.0, 101 ± 14)
T9: $196\,^{-39}_{-10}$ $(20.0\,^{+4.0}_{-1.0},\ 145\,^{-29}_{-7})$

7923TG72

Exploded view of the rear standard suspension assembly—FWD Legacy and Impreza models

6. Support the rear knuckle assembly with a suitable jack.

7. Remove the three upper strut plate mounting nuts.

8. Lower the jack about and in. and remove the upper mounting bolt from the knuckle assembly.

9. Remove the strut from the vehicle.
To install:

10. Install the strut in the vehicle and install the three upper mounting nuts. Tighten the nuts to 13 ft. lbs. (18 Nm).

11. Raise the jack to line up the knuckle with the strut bracket, and install the mounting bolts and nuts. Tighten the mounting nuts to 112 ft. lbs. (152 Nm).

12. Remove the jack.

13. Connect the brake hose bracket to the strut and install the mounting bolt.

14. Install the rear quarter interior trim.

15. Install the tire and wheel assembly.

16. Lower the vehicle.

17. Check the alignment and adjust as necessary.

Coil Spring

REMOVAL & INSTALLATION

Front

➡ **Do not remove the large nut on top of the strut assembly unless the coil spring is properly compressed with a suitable spring compressor.**

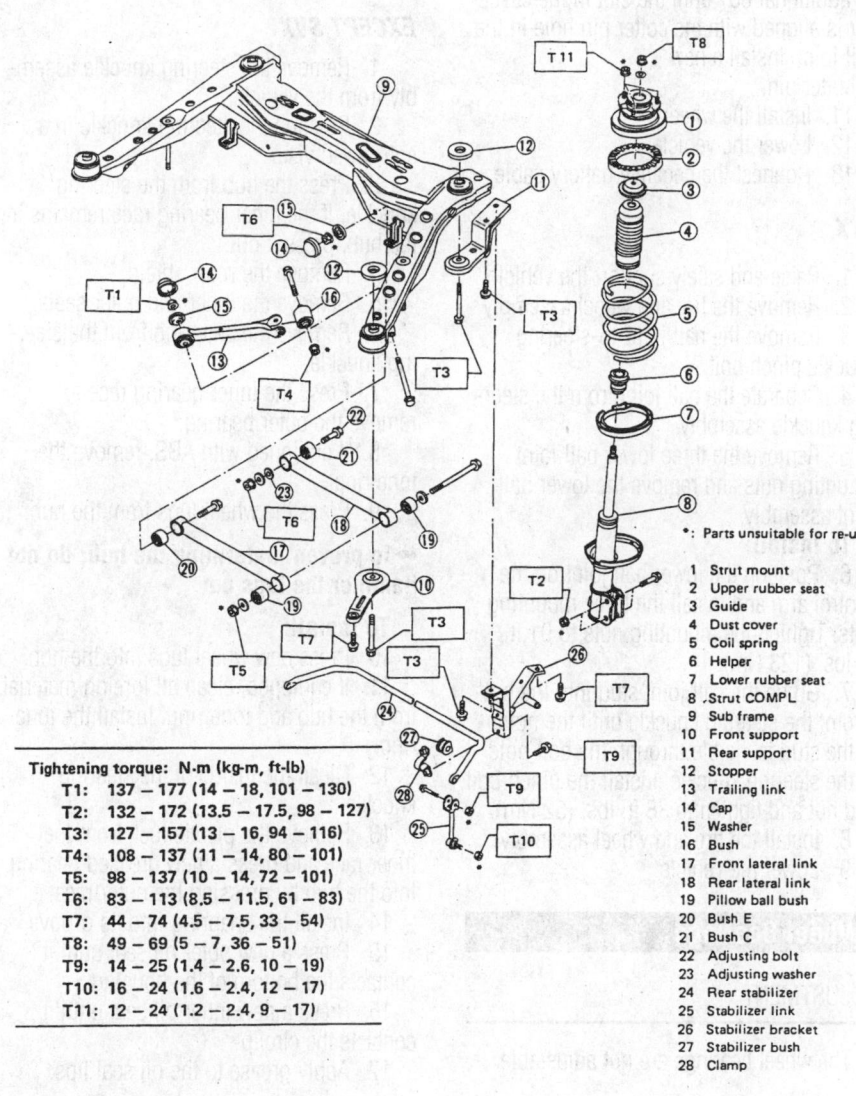

Tightening torque: N-m (kg-m, ft-lb)
T1: 137 — 177 (14 — 18, 101 — 130)
T2: 132 — 172 (13.5 — 17.5, 98 — 127)
T3: 127 — 157 (13 — 16, 94 — 116)
T4: 108 — 137 (11 — 14, 80 — 101)
T5: 98 — 137 (10 — 14, 72 — 101)
T6: 83 — 113 (8.5 — 11.5, 61 — 83)
T7: 44 — 74 (4.5 — 7.5, 33 — 54)
T8: 49 — 69 (5 — 7, 36 — 51)
T9: 14 — 25 (1.4 — 2.6, 10 — 19)
T10: 16 — 24 (1.6 — 2.4, 12 — 17)
T11: 12 — 24 (1.2 — 2.4, 9 — 17)

*: Parts unsuitable for re-use

1   Strut mount
2   Upper rubber seat
3   Guide
4   Dust cover
5   Coil spring
6   Helper
7   Lower rubber seat
8   Strut COMPL
9   Sub frame
10 Front support
11 Rear support
12 Stopper
13 Trailing link
14 Cap
15 Washer
16 Bush
17 Front lateral link
18 Rear lateral link
19 Pillow ball bush
20 Bush E
21 Bush C
22 Adjusting bolt
23 Adjusting washer
24 Rear stabilizer
25 Stabilizer link
26 Stabilizer bracket
27 Stabilizer bush
28 Clamp

7923TG74

Exploded view of the rear strut and suspension assembly—SVX

1. Remove the strut assembly from the vehicle.

⁂ WARNING

Remove the strut from the vehicle and install a spring compressor before removing the strut center nut.

2. Place the strut assembly in a vise with a holding tool and install a spring compressor.

3. Compress the spring slightly.

4. Loosen but do not remove the bearing cap locknut.

5. Compress the spring with the spring compressor, then remove the locknut.

6. Remove the strut bearing cap, mounting insulator bracket and upper spring seat.

7. Remove the coil assembly, leaving the spring compressed.

8. Remove the strut boot and rebound bumper from the strut. Inspect and replace if worn.

9. Remove the strut retainer nut using a suitable wrench. Remove the strut insert from the assembly.

To install:

10. Install the strut into the chamber, and install the retainer nut. Tighten the nut snugly.

11. Install the rebound bumper and the boot to the strut piston rod.

12. Install the coil spring on the strut assembly. Be sure the spring is properly positioned on the lower bracket.

13. Install the upper spring seat, mounting insulator and bearing cap. Be sure the upper spring seat is facing the proper direction.

14. Install the locknut, and tighten the locknut to 36–43 ft. lbs. (47–56 Nm).

15. Loosen and remove the spring compressor from the coil spring.

16. Install the strut to the vehicle.

Rear

EXCEPT SVX

⁂ CAUTION

Do not remove the large nut on top of the strut assembly unless the coil spring is properly retained with a spring compressor.

1. Remove the strut assembly from the vehicle and secure in a soft jawed vise.

2. Compress the coil spring with a spring compressor until the upper spring seat can be turned by hand.

3. Remove the self-locking nut on the top of the strut assembly, then remove the upper spring seat.

4. Remove the coil spring and compressor. If the spring is being replaced, slowly release the spring from the compressor and compress the new coil spring.

To install:

5. Place the proper end of the coil spring on the lower spring seat on the strut.

6. Install the insulator, upper spring seat and strut mount on the strut piston. Install a new self-locking nut. Tighten the nut to 43 ft. lbs. (59 Nm).

7. Slowly release the spring compressor.

8. Install the strut on to the vehicle.

SVX

⁂ CAUTION

Do not remove the large nut on top of the strut assembly unless the coil spring is properly retained with a spring compressor.

1. Remove the strut assembly from the vehicle and secure in a soft jawed vise.

2. Compress the coil spring with a spring compressor until the upper spring seat can be turned by hand.

3. Remove the self locking nut on the top of the strut assembly, then remove the upper spring seat.

4. Remove the coil spring and compressor. If the spring is being replaced, slowly release the spring from the compressor and compress the new coil spring.

To install:

5. Place the proper end of the coil spring on the lower spring seat on the strut.

6. Install the insulator, upper spring seat and strut mount on the strut piston. Install a new self-locking nut. Tighten the nut to 43 ft. lbs. (59 Nm).

7. Slowly release the spring compressor.

8. Install the strut into the vehicle.

Lower Ball Joint

REMOVAL & INSTALLATION

Except SVX

1. Disconnect the negative battery cable.

2. Raise and support the vehicle safely.

3. Remove the front wheel and tire assembly.

4. Remove the ball joint castle nut cotter pin. Discard the cotter pin.

5. Loosen and remove the castle nut.

6. Using a suitable puller or prytool, disconnect the ball joint from the lower control arm assembly.

7. Remove the bolt securing the ball joint to the steering knuckle. Use a suitable wedge to expand the steering knuckle connection point, and remove the ball joint.

To install:

8. Install the ball joint to the steering knuckle.

9. Install the bolt, and tighten the retaining bolt to 36 ft. lbs. (49 Nm).

10. Connect the ball joint to the lower control arm, and tighten the castle nut to 29 ft. lbs. (39 Nm). Then, tighten the castle nut an additional 60° until the slot in the castle nut is aligned with the cotter pin hole in the ball joint. Install a new cotter pin.

11. Install the wheel.

12. Lower the vehicle.

13. Connect the negative battery cable.

SVX

1. Raise and safely support the vehicle.

2. Remove the tire and wheel assembly.

3. Remove the ball joint-to-steering knuckle pinch bolt.

4. Separate the ball joint from the steering knuckle assembly.

5. Remove the three lower ball joint mounting nuts and remove the lower ball joint assembly.

To install:

6. Position the lower ball joint on the control arm and install the three mounting nuts. Tighten the mounting nuts to 91 ft. lbs. (123 Nm).

7. Guide the ball joint stud into the bottom of the steering knuckle until the notch in the stud is visible through the bolt hole in the steering knuckle. Install the pinch bolt and nut and tighten to 38 ft. lbs. (52 Nm).

8. Install the tire and wheel assembly.

9. Lower the vehicle.

Wheel Bearings

ADJUSTMENT

The wheel bearings are not adjustable.

REMOVAL & INSTALLATION

Front

EXCEPT SVX

1. Remove the steering knuckle assembly from the vehicle.

2. Position the steering knuckle in a soft-jawed vise.

3. Press the hub from the steering knuckle. If the inner bearing race remains in the hub, press it out.

4. Remove the rotor shield.

5. Remove the inner and outer seals.

6. Remove the snapring from the steering knuckle.

7. Press the Inner bearing race to remove the outer bearing.

8. If equipped with ABS, remove the tone ring.

9. Press the wheel lugs from the hub.

➡**To prevent deforming the hub, do not hammer the lugs out.**

To install:

10. Press new wheel lugs into the hub.

11. If equipped, clean all foreign material from the hub and tone ring. Install the tone ring.

12. Clean the inside of the steering knuckle.

13. Remove the plastic lock from the inner race and press a new greased bearing into the hub by pressing the outer race.

14. Install the snapring into its groove.

15. Press a new outer oil seal until it contacts the bottom of the housing.

16. Press a new inner oil seal until it contacts the circlip.

17. Apply grease to the oil seal lips.

*: Parts unsuitable for re-use

1 Ball joint
2 Lower arm
3 Housing

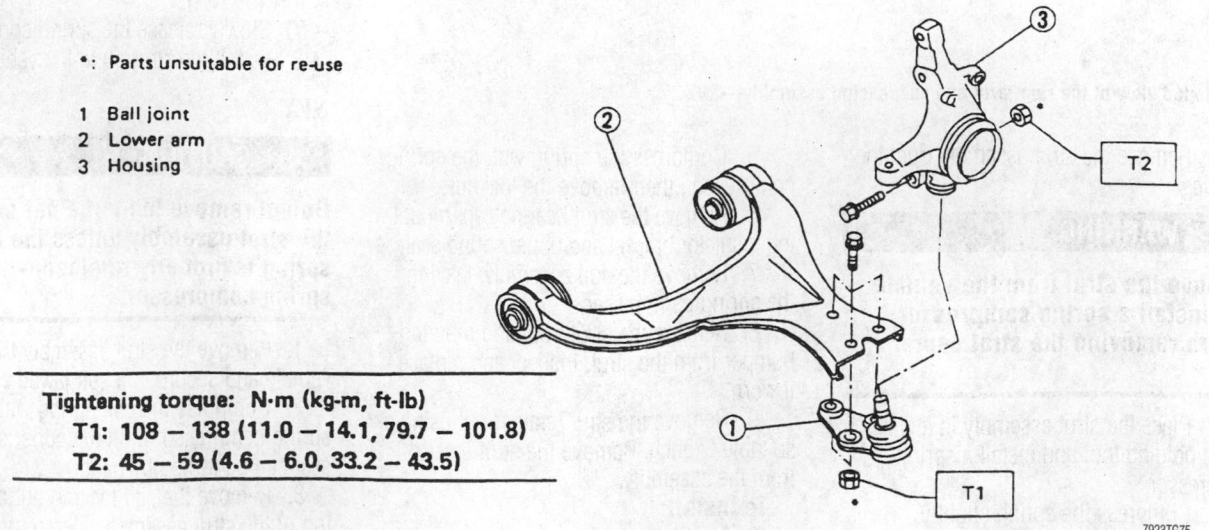

Tightening torque: N·m (kg-m, ft-lb)
T1: 108 — 138 (11.0 — 14.1, 79.7 — 101.8)
T2: 45 — 59 (4.6 — 6.0, 33.2 — 43.5)

Exploded view of the ball joint, steering knuckle and control arm assembly—SVX model

7923TG75

18. Install the rotor shield and tighten the bolts to 10 ft. lbs. (14 Nm).

19. Attach the hub to the steering knuckle.

20. Press a new bearing into the hub by driving the inner race.

21. Install the steering knuckle on the vehicle.

SVX

1. Remove the steering knuckle assembly from the vehicle.

2. Using a hub stand, 28099PA080 or equivalent, support the steering knuckle assembly.

3. Drive the hub out of the steering knuckle using tool 28099PA040, or equivalent hub remover.

4. Remove the backing plate from the steering knuckle.

5. Remove the inner and outer wheel seals using a suitable prying tool.

6. Remove the snapring from the rear of the steering knuckle.

7. Using a bearing installer, 28099PA000 or equivalent remove the outer bearing.

To install:

8. Using bearing installer 28099PA000 or equivalent install the outer bearing.

9. Install the snapring in the groove.

10. Install the inner and outer oil seals in the steering knuckle.

11. Install the backing plate.

12. Install the hub into the bearing assembly using a 28099PA020 bearing installer or equivalent.

13. Install the steering knuckle assembly onto the vehicle.

Rear

EXCEPT SVX—WITH ALL WHEEL DRIVE

1. Disconnect the negative battery cable.

2. Loosen the parking brake adjustment.

3. Raise and support the vehicle safely.

4. Remove the wheel assembly.

5. Unstake and remove the axle nut.

6. Remove the caliper, leaving the line connected, and suspend it aside, then remove the rotor.

7. Disconnect the parking brake cable.

8. Remove the sway bar clamp.

9. Remove the bolt securing the lateral link to the housing.

10. Remove the bolts securing the trailing link to the housing.

11. Remove the halfshaft.

12. Remove the bolts securing the strut to the housing.

13. If equipped with ABS, remove the speed sensor from the backing plate.

14. Remove the housing assembly.

15. Using Hub Stand 92708000 and Puller 927420000 or equivalent, remove the hub from the rear housing.

16. Remove the backing plate from the housing.

17. Remove the outer, inner and sub oil seals.

18. Remove the snapring.

19. Remove the bearing by pressing the inner race.

To install:

20. Clean the housing thoroughly.

➡**Do not remove the plastic lock from the inner race when installing the bearing.**

21. Install a new bearing into the housing by pressing the outer race.

22. Pack the bearing with grease.

23. Install the snapring and ensure it fits properly.

24. Using Installer 927460000 or equivalent, press in a new outer seal until it comes in contact with the snapring.

25. Using Installer 927450000 or equivalent, press in a new inner seal until it contacts the bottom.

26. Install a new sub oil seal.

27. Apply grease to the oil seal lip.

28. Install the backing plate and tighten the bolts to 43 ft. lbs. (58 Nm).

29. Using installer 927450000 or equivalent, press in the hub into the housing.

30. Connect the housing to the strut and tighten the bolts to 119 ft. lbs. (162 Nm).

31. If equipped with ABS, install the speed sensor.

32. Install the halfshaft.

33. Connect the trailing link to the housing and tighten the bolt and new nut to 94 ft. lbs. (127 Nm).

34. Connect the lateral link to the housing and tighten the bolt and new nut to 116 ft. lbs. (157 Nm).

35. Install the sway bar clamp.

36. Connect the parking brake cable.

37. Install the rear brake assembly.

38. Install a new axle nut and tighten it to 152 ft. lbs. (206 Nm). Stake the nut.

39. Install the wheel.

40. Lower the vehicle.

41. Adjust the parking brake cable.

42. Connect the negative battery cable.

EXCEPT SVX—WITH FRONT WHEEL DRIVE

1. Disconnect t he negative battery cable.

2. Loosen the parking brake adjustment.

3. Raise and support the vehicle safely.

4. Remove the wheel assembly.

5. Unstake and remove the axle nut.

6. If equipped, remove the caliper, leaving the line connected and suspend it aside. Remove the rotor. Or if equipped with rear drum, remove the drum brake assembly.

7. Disconnect the parking brake cable.

8. Remove the bolt securing the lateral link to the spindle.

9. Remove the bolts securing the trailing link to the spindle.

10. Remove the bolts securing the strut to the spindle.

11. Remove the spindle assembly.

12. Remove the hub and bearing assembly from the spindle.

To install:

13. Pack the oil seal located and the rear of the hub with grease.

14. Install the hub on the spindle and temporarily tighten a new nut and washer to hold the hub in place.

15. Connect the spindle to the strut and tighten the bolts to 119 ft. lbs. (162 Nm).

16. Connect the trailing link to the spindle and tighten the bolt and new nut to 94 ft. lbs. (127 Nm).

17. Connect the lateral link to the spindle and tighten the bolt and new nut to 116 ft. lbs. (157 Nm).

18. Connect the parking brake cable.

19. Install the rear brake assembly.

20. Install a new axle nut and tighten it to 152 ft. lbs. (206 Nm). Stake the nut.

21. Install the wheel.

22. Lower the vehicle.

23. Adjust the parking brake cable.

24. Connect the negative battery cable.

SVX

1. Set the parking brake.

2. Raise and safely support the vehicle.

3. Remove the rear wheel assembly.

4. Remove the dust cap and remove the axle nut.

5. Return the parking brake lever and remove the console box lid.

6. Loosen the parking brake adjuster nut.

7. Remove the stabilizer link.

8. Remove the ABS sensor and clamp.

9. Disconnect the parking brake cable clamp.

10. Disconnect the brake hose from the strut.

11. Loosen the caliper assembly securing bolts.

12. Disconnect the trailing link from the knuckle.

13. Remove the nut securing the lateral link to the knuckle.

14. Remove the two nuts securing the strut to the knuckle.

15. Remove the caliper assembly and fasten to the strut using wire.

16. Remove the rotor.

17. Remove the parking brake shoes and disconnect the cable from the shoe.

18. Remove the parking brake cable clamp and remove the cable from the back plate.

19. Separate the knuckle from the half-shaft.

20. Remove the mounting bolts and remove the knuckle.

21. Remove the bearings.

To install:

22. Install the bearings.

23. Install the knuckle and the mounting bolts.

24. Install the halfshaft into knuckle. Tighten the axle nut to 123–152 ft. lbs. (167–206 Nm).

25. Replace the parking brake cable to the back plate and install the cable clamp.

26. Connect the cable to the brake shoe and install the shoes.

27. Install the brake rotor and the caliper assembly.

28. Install the axle nut and the dust cap.

29. Install the nuts securing the strut to the knuckle. Tighten the nuts to 98–127 ft. lbs. (132–172 Nm).

30. Install the nut securing the lateral link to the knuckle.

31. Tighten the caliper assembly mounting bolts.

32. Connect the brake hose to the strut and connect the parking brake cable clamp.

33. Install the ABS sensor and clamp.

34. Install the stabilizer link.

35. Tighten the parking brake cable adjusting nut.

36. Install the console box lid.

37. Install the rear wheel.

38. Lower the vehicle.

SUZUKI

Esteem • Swift

28

ENGINE REPAIR

➡ **Disconnecting the negative battery cable on some vehicles may interfere with the functions of the on board computer systems and may require the computer to undergo a relearning process, once the negative battery cable is reconnected.**

Distributor

REMOVAL & INSTALLATION

1995–97 Models

1. Disconnect the negative battery cable.
2. Disconnect the distributor electrical coupler.
3. Remove the distributor cap.
4. Mark the position of the distributor rotor in relation to the distributor body, and the distributor body in relation to the cylinder head.
5. Remove the hold-down bolts and the distributor from the cylinder head. Do not rotate the engine after the distributor has been removed.

To install:

➡ **Before installing the distributor, check to be sure the O-ring is in good condition. If using a new O-ring, apply a light coat of engine oil to it.**

6. Install the distributor into the cylinder head. The tabs of the distributor coupling are offset. Therefore, if the tabs can not be fitted into the slots, turn the distributor shaft 180 degrees and try again.
7. Align the reference marks made during removal.
8. Connect the electrical coupler to the distributor.
9. Install the distributor cap.
10. Connect the negative battery cable.
11. Check and adjust the ignition timing as necessary.

1998–99 Models

The 1998–99 models utilize a distributorless ignition system. With this system, the ECM determines proper ignition timing and time for the primary ignition coil circuit to turn ON and OFF.

Ignition Timing

ADJUSTMENT

1995–97 Models

1. Start the engine and allow it to reach normal operating temperature. Prior to any adjustment, be sure all electrical accessories including the A/C are **OFF**.
2. After the engine is reaches normal operating temperature, check and be sure the idle speed is:
   a. a minimum of 800 rpm for automatic transmission equipped models.
   b. a minimum of 750 rpm for manual transmission equipped models.
3. Be sure manual transmission equipped models are in **NEUTRAL** and automatic transmission equipped modes are in **PARK**. Also be sure the parking brake is fully applied.
4. Remove the cap from monitor coupler and connect terminals **D** and **E** with a jumper wire.
5. Connect a timing light according to manufactures instruction, using No. 1 cylinder spark plug wire as an ignition pick-up.

➡ **When terminals D and E are connected, observe if ignition timing is varying. If ignition timing is varying, this indicates ungrounded D terminal which prevents accurate inspection and adjustment. Be sure to ground the D terminal securely.**

6. On 1.6L models, open the air cleaner upper case cover and position the upper case and hose out of the way to observe the timing marks on the crankshaft pulley.
7. With the engine running, direct the timing light to the crankshaft pulley. If the timing mark on the timing tab are aligned with the timing notch on the crankshaft pulley, the ignition is properly timed.
8. Initial ignition timing should be 5 degrees BTDC at 800 rpm for automatic models, 750 rpm for manual models.
9. To adjustment the timing, loosen the distributor flange bolt and turn the distributor housing to advance or retard the timing.
10. After adjusting, tighten the flange bolt and recheck the timing. Tighten the flange bolt to 9–13 ft lbs. (12–18 Nm).
11. After checking or adjusting, disconnect the service wire from monitor coupler.
12. With the engine idling and test terminals ungrounded, check that the ignition timing is 12° BTDC. A constant variation of timing is an indication that the computer controlled timing is working correctly.

1998–99 Models

The correct ignition timing is controlled by the ECM which receives signals from various sensors mounted on the engine. No ignition timing adjustment is possible.

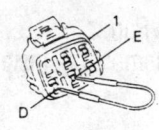

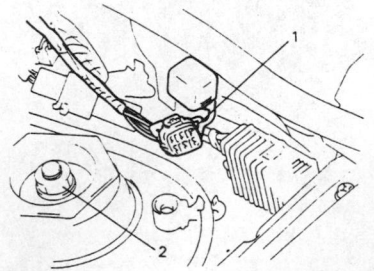

1. Monitor coupler
2. Front strut of left side
D: Ground terminal
E: Test switch terminal

7923UG01

Monitor coupler location and terminal identification—1.3L engine

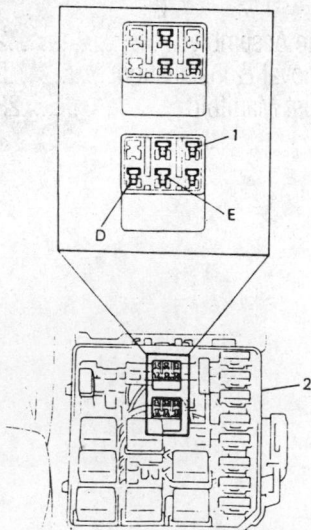

D: Ground terminal   1. Diagnosis connector
E: Test switch terminal   2. Reray box

7923UG02

Monitor coupler location and terminal identification—1.6L engine

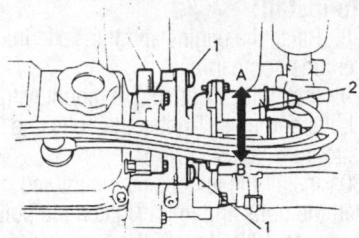

A: To be advanced
B: To be retarded
1. Distributor flange bolt
2. Distributor

7923UG03

Slightly turn the distributor housing to change the ignition timing—all engines

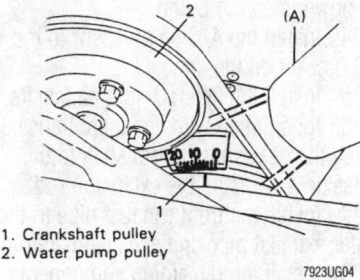

1. Crankshaft pulley
2. Water pump pulley

7923UG04

Timing marks on crankshaft pulley and timing belt cover—1.0L and 1.3L engines shown, 1.6L engine is similar

Engine Assembly

REMOVAL & INSTALLATION

1.0L and 1.3L Engine

✳✳ CAUTION

The fuel injection system remains under pressure after the engine has been turned OFF. Properly relieve fuel pressure before disconnecting any fuel lines. Failure to do so may result in fire or personal injury.

1. Properly relieve the fuel system pressure.
2. Disconnect the negative battery cable.
3. Remove the battery and tray.
4. Disconnect the windshield washer hose from the hood. Using a grease pencil or marker, mark the hood hinge to hood outline. With the aid of an assistant, remove the hood.
5. Drain the cooling system.

6. Remove the air cleaner assembly with the MAF sensor outlet hose.
7. Remove the radiator and cooling fan.
8. Disconnect the following electrical wires and release the wiring harness from the clamps:
    a. Ignition coil wire from the distributor cap.
    b. Distributor electrical wires.
    c. EGR solenoid vacuum valve.
    d. Radiator fan temperature switch.
    e. Engine coolant temperature gauge sensor.
    f. ECT sensor.
    g. IAC actuator.
    h. Ground wires from the intake manifold.
    i. TP sensor.
    j. Fuel injector.
    k. Oxygen sensor.
    l. Oil pressure gauge sensor.
    m. Alternator.
    n. Starter.
    o. Back-up light switch.
    p. Negative battery cable from the transaxle.
    q. Vehicle speed sensor.
    r. Noise filter ground wire.
    s. EGRT sensor, if equipped.
    t. EVAP SP valve.
9. Disconnect the following vacuum hoses:
    a. Brake booster hose from the intake manifold.
    b. Canister purge hose.
    c. Air conditioning SV valve hose.
10. Disconnect the fuel return hose and the fuel feed hose from the throttle body.
11. Disconnect the heater inlet and outlet hoses.
12. Disconnect the following cables:
    a. Accelerator cable from the throttle body.
    b. Clutch cable from the transaxle, if equipped.
    c. Speedometer cable from the transaxle, if equipped.
    d. Shift switch, if equipped with an automatic transaxle.
    e. Vehicle speed sensor, if equipped.
13. Remove the EVAP canister from the vehicle.
14. Safely raise and support the vehicle.
15. Remove the fender apron extensions.
16. Disconnect the exhaust pipe from the exhaust manifold.
17. Disconnect the control shaft and extension rod from the transaxle.
18. Drain the engine and transaxle oil.
19. Remove the left and right halfshafts.

➡**For engine and transaxle removal, it is not necessary to remove the halfshafts from the steering knuckles.**

20. If equipped with air conditioning, remove the air conditioning compressor from its mounting bracket with the hoses still attached.

➡**Suspend the compressor where no damage will occur during engine removal and installation.**

21. If equipped with power steering, disconnect the power steering hoses from the power steering pump.

➡**Plug the power steering hose, pipe and pump ports to minimize fluid loss.**

22. If equipped with a automatic transaxle, remove the rear torque rod bracket from the transaxle.
23. If equipped with a manual transaxle, remove the rear mount from the body.
24. If equipped with a automatic transaxle, remove the rear mounting nut.
25. Lower the vehicle.
26. Install an engine lifting device.
27. Remove the rear mount from the body.
28. Remove the left side engine mounting bracket bolts and bracket.
29. Remove the right side engine mount from its bracket.
30. Before lifting the engine and assembly check to be sure that all the hoses, electric wires and cables are disconnected.
31. Remove the engine with the transaxle from the vehicle.

To install:
32. Lower the engine and transaxle into the engine compartment but do not remove the lifting device.
33. Install the rear mount to the body.
34. Install the left side engine mounting bracket and bolts.
35. Install the right side engine mount to its bracket.
36. If equipped with an automatic transaxle, install the rear mounting nut.
37. Tighten the engine mounting nuts and bolts to specification.
38. Remove the lifting device.
39. Safely raise and support the vehicle.
40. If equipped with air conditioning, install the air conditioning compressor to its mounting bracket. Tighten the mounting bolts to 13–20 ft. lbs. (18–28 Nm).
41. If equipped with power steering, connect the power steering hose and pipe to the power steering pump.
42. Install the left and right halfshafts.
43. Connect the control shaft and the

extension rod to the transaxle. Tighten the control shaft nuts and bolts to 11–14 ft. lbs. (15–20 Nm) and tighten the extension rod nut to 19–29 ft. lbs. (25–40 Nm).

44. Connect the exhaust pipe to the exhaust manifold. Tighten the bolts to 29–36 ft. lbs. (40–50 Nm).

45. Fill the transaxle with gear oil.

46. Install the remaining components.

47. Adjust the clutch pedal free-play.

48. Adjust the accelerator cable free-play.

49. Install the air cleaner assembly.

50. Fill the engine with engine oil and the cooling system with coolant.

51. Install the hood and connect the windshield washer hose.

52. Install the battery and tray, and connect the negative battery cable

53. Fill the power steering reservoir and bleed the power steering system.

54. Run the engine and verify that there are no fuel, coolant, transmission or exhaust leaks.

1.6L Engine

✳✳ CAUTION

The fuel system pressure must be relieved before disconnecting any fuel lines. Failure to do so may result in personal injury.

1. Disconnect the battery cables from the battery, negative cable first.

2. Mark the position of the hood on the hinges for installation reference, then remove the hood with the aid of an assistant.

3. Drain the cooling system.

4. Remove the radiator and cooling fan.

5. Remove the air cleaner outlet hose.

6. Remove the air cleaner case by removing the fastening bolts.

7. Disconnect the following cables:

    a. Accelerator cable from the throttle body

    b. If equipped with M/T, disconnect the clutch cable from the transaxle

    c. If equipped with A/T, disconnect the gear select cable from the transaxle.

8. Disconnect the following vacuum hoses:

    a. Brake booster hose from the intake manifold

    b. Canister purge hose from the EVAP canister purge valve

    c. MAP sensor hose from the intake manifold

9. Disconnect the following electrical connectors:

    a. Distributor coil wire

    b. Camshaft position sensor

    c. Engine oil pressure switch

    d. EGR solenoid vacuum valve

    e. EVAP canister purge valve

    f. Engine coolant temperature sensor

    g. Fuel injectors

    h. Power steering pressure switch

    i. Heated oxygen sensor

    j. Back-up light switch (M/T)

    k. Shift switch (A/T)

    l. Forward clutch revolution sensor (A/T)

    m. A/T vehicle speed sensor

    n. Alternator

    o. Starter

    p. Battery negative cable from the transaxle

    q. Vehicle speed sensor

    r. Throttle position sensor

    s. Idle air control valve

    t. Manifold absolute sensor

10. Remove the engine wires from the engine.

11. Disconnect the fuel feed hose from the feed pipe and remove the return hose from the fuel pressure regulator.

12. Disconnect the heater inlet and outlet hoses.

13. Raise and safely support the front of the vehicle.

14. Remove the right and left engine undercovers.

15. Remove the front exhaust pipe from the exhaust manifold and center exhaust pipe.

16. If equipped with a manual transaxle, remove the gear shift control shaft from the transaxle and remove the extension rod.

17. Drain the engine and transaxle oil.

18. Remove the left and right halfshafts.

19. If equipped with A/C, remove the A/C compressor from the compressor bracket with the hoses still attached. Position the A/C out of the way from the engine.

20. If equipped with power steering, drain the power steering pump of fluid.

21. If equipped with power steering, disconnect the power steering hose from the power steering pump.

22. Install a lifting device to the engine.

23. Remove the center member from the vehicle by removing the seven nuts and four bolts.

24. Remove the left engine mount.

25. Remove the right engine mount and bracket.

26. Check to be sure all cooling hoses, vacuum hoses and electrical wires are disconnected from the engine.

27. Lower the engine with the transaxle from the vehicle.

To install:

28. Raise the engine and transaxle into the engine compartment.

29. Install the right engine mount with the bolts and nuts. Tighten the bolts and nuts to 40 ft. lbs. (55 Nm).

30. Install the left engine mount and install the bolts and nuts. Tighten the bolts and nuts to 40 ft. lbs. (55 Nm).

31. Install the center member using the seven nuts and four bolts. Tighten the bolts and nuts as follows:

• Center member to the radiator support to 33 ft. lbs. (45 Nm).

• Center member to crossmember to 33 ft. lbs. (45 Nm).

• Engine mounts to center member nuts to 40 ft. lbs. (55 Nm).

32. Remove the lifting device.

33. Connect the power steering hose to the power steering pump.

34. Install the A/C compressor to the A/C bracket on the engine.

35. Install the left and right halfshafts.

36. If equipped with a manual transaxle, install the gear shift control shaft to the transaxle and install the extension rod.

37. Install the front exhaust pipe to the center exhaust pipe and exhaust manifold.

38. Install the remaining components.

39. Fill the cooling system, engine, transaxle and power steering pump.

40. Adjust all cables and check all connections.

41. Connect the negative battery cable to the battery.

42. Start the vehicle and check for leaks.

Water Pump

REMOVAL & INSTALLATION

1.0L and 1.3L Engines

1. Disconnect the negative battery cable.

2. Drain the cooling system into a suitable container and tighten the drain plug.

3. Remove the air cleaner assembly and the MAF sensor and outlet hose.

4. Remove the air cleaner bracket.

5. Raise and safely support the vehicle.

6. Remove the right side fender apron clips by pushing the center pin.

➡**Do not push the center pin too far in, or it will fall off into the fender.**

7. If equipped, remove the power steering and air conditioning belt.

8. Loosen the water pump pulley bolts.

9. Remove the alternator drive belt.

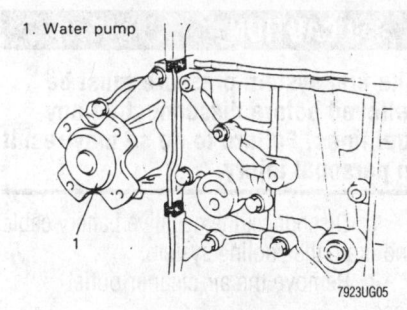

1. Water pump

7923UG05

Water pump location—1.3L engine

10. Remove the water pump pulley.
11. To remove the crankshaft pulley perform the following:

    a. If equipped with a manual transaxle, insert a suitable flat bladed tool into the hole in the bell housing next to the exhaust pipe. This will lock the crankshaft in place.

    b. If equipped with a automatic transaxle, hold a suitable flat bladed tool in line with the oil pan and insert into the teeth of the drive plate. This will lock the crankshaft in place.

    c. Loosen the crankshaft pulley bolts.

    d. Remove the crankshaft timing belt pulley bolt with special tool 09919–16020 or a 17mm socket.

    e. Remove the pulley from the crankshaft.

    f. Install the crankshaft bolt.

    g. Remove the flat bladed tool that was used to lock the crankshaft in place.

➡**To remove the crankshaft pulley with the engine assembly mounted on the body, it is necessary to remove the crankshaft timing belt pulley bolt. If the engine assembly is dismounted, the bolt does not need to be removed.**

12. Remove the resonator and the timing belt outside cover.
13. Loosen the right engine mounting bolt.
14. Remove the timing belt.

❊❊ **CAUTION**

After the timing belt is removed never turn the camshafts or the crankshaft. Interference may occur between the pistons and the valves causing component damage.

15. Remove the timing belt inside cover.
16. Remove the water pump belt adjusting arm.

17. Carefully remove the rubber seal between the water and oil pumps, and remove the seal between the water pump and the cylinder head.
18. Remove the water pump bolts and remove the water pump.

To install:

19. Clean the water pump mounting surface of old gasket material.
20. Install a new water pump gasket to the cylinder block.
21. Install the water pump to the cylinder block and tighten the bolts to 7–9 ft. lbs. (10–13 Nm).
22. Install the rubber seal between the water pump and the oil pump. Install the seal between the water pump and the cylinder head.
23. Install the water pump belt adjusting arm.
24. Install the timing belt inside cover.
25. With the crankshaft locked in position, remove the crankshaft bolt and install the crankshaft pulley. Tighten the crankshaft pulley bolts to 10–13 ft. lbs. (14–18 Nm). Using special tool 09919–16020 or a 17mm socket, tighten the crankshaft timing belt pulley bolt to 76–83 ft. lbs. (105–115 Nm).
26. Install the timing belt.
27. Install the water pump pulley and drive belt. Tighten the water pump pulley bolts to 7–8 ft. lbs. (9–12 Nm).
28. Install the remaining components.
29. Fill the cooling system.
30. Connect the negative battery cable.
31. Start the engine and top off the coolant as necessary.
32. Check the cooling system for leaks.
33. Check the ignition timing.

1.6L Engine

1. Disconnect the negative battery cable.
2. Drain the cooling system into a resealable container and tighten the drain plug.
3. Remove the timing belt.
4. Remove the alternator adjusting shim.
5. Remove the oil dipstick guide and dipstick.
6. Remove the water pump bolts, gasket, the water pump and rubber seal.

To install:

7. Clean the water pump mounting surface of old gasket material.
8. Install a new water pump gasket to the cylinder block.
9. Install the water pump to the cylinder

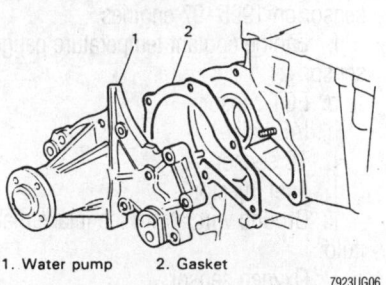

1. Water pump   2. Gasket

7923UG06

Exploded view of the water pump mounting—1.6L engine

block and tighten the bolts to 7–9 ft. lbs. (10–13 Nm).

10. Install the rubber seal between the water pump and the oil pump. Install the seal between the water pump and the cylinder head.
11. Install the timing belt.
12. Install the alternator adjusting arm.
13. Using a new O-ring, install the oil dipstick guide and dipstick.
14. Lower the vehicle.
15. Fill the cooling system with engine coolant.
16. Connect the negative battery cable.
17. Start the engine and top off the coolant as necessary.
18. Check the ignition timing.

Cylinder Head

REMOVAL & INSTALLATION

1.0L and 1.3L Engines

❊❊ **CAUTION**

The fuel injection system remains under pressure after the engine has been turned OFF. Properly relieve fuel pressure before disconnecting any fuel lines. Failure to do so may result in fire or personal injury.

1. Disconnect the negative battery cable and drain the cooling system.
2. Remove the air cleaner assembly.
3. Disconnect the following electrical wires and release the wiring harness from the clamps:

    a. Ignition coil wire from the distributor cap.

    b. Distributor electrical wires.

    c. EGR solenoid vacuum valve.

    d. Radiator fan thermo switch.

e. Engine coolant temperature gauge sensor on 1995–97 engines.

f. Engine coolant temperature gauge sensor.

g. ECT sensor.

h. IAC valve.

i. TP sensor.

j. Fuel injector.

k. Ground wires from the intake manifold.

l. Oxygen sensor.

4. Disconnect the radiator hose from the thermostat housing.

5. Disconnect the heater hose from the intake manifold.

6. Disconnect the throttle body coolant outlet hose from the throttle body.

7. Disconnect the following vacuum hoses:

a. MAP sensor hose from the intake manifold.

b. Canister hose from its pipe.

c. Canister purge hose from the intake manifold.

d. Brake booster from the intake manifold.

8. Disconnect the fuel return hose and the fuel feed hose from the throttle body.

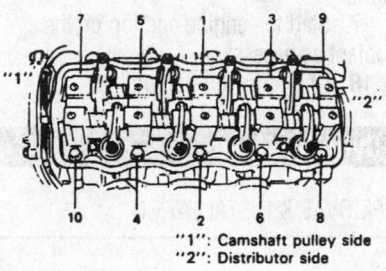

"1": Camshaft pulley side
"2": Distributor side

7923UG07

Cylinder head bolt tightening sequence—1.3L engine

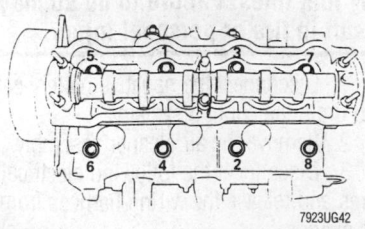

7923UG42

Cylinder head bolt tightening sequence—1.0L engine

9. Disconnect the throttle cable from the throttle body.

10. Remove the water pump and crankshaft pulleys. Remove the timing belt.

11. Remove the rubber seal between the cylinder head and the water pump.

12. Disconnect the exhaust pipe from the exhaust manifold.

13. Remove the spark plug wire clamps from the cylinder head cover and disconnect the PCV hose.

14. Remove the cylinder head cover.

15. Loosen all of the valve adjusting screws and allow the valves to close.

16. Remove the cylinder head bolts in the reverse order of the tightening sequence.

17. Remove the cylinder head with the distributor, intake manifold and exhaust manifold.

18. Remove the distributor, intake manifold and exhaust manifold from the cylinder head.

19. Clean the cylinder block mating surface of any old gasket material and clean any engine coolant from the cylinders.

To install:

20. Install the intake and exhaust manifolds.

21. Install the distributor to the cylinder head.

22. Install a new cylinder head gasket with the top mark facing up and toward the crankshaft pulley.

23. Install the cylinder head to the engine block. Coat the cylinder head mounting bolts threads with clean engine oil and install them. Tighten the bolts evenly in 3 equal steps and in sequence to 51–54 ft. lbs. (70–75 Nm).

24. Install the rubber seal between the cylinder head and the water pump.

25. Install the timing belt.

26. Connect the exhaust pipe to the exhaust manifold, tighten the attaching bolts to 26–36 ft. lbs. (35–50 Nm).

27. Install the crankshaft and water pump pulleys.

28. Connect the throttle cable to the throttle body.

29. Connect the fuel return hose and the fuel feed hose to the throttle body.

30. Install the remaining components.

31. Refill the cooling system with coolant.

32. Connect the negative battery cable.

33. Adjust the ignition timing.

34. With the engine running, be sure that there are no fuel, coolant or exhaust leaks.

1.6L Engine

✳✳ CAUTION

The fuel system pressure must be relieved before disconnecting any fuel lines. Failure to do so may result in personal injury.

1. Disconnect the negative battery cable and drain the cooling system.

2. Remove the air cleaner outlet hose.

3. Remove the intake manifold rear stiffener bolt, alternator adjustment arm reinforcement bolt and right mounting bracket stiffener from the intake manifold.

4. Disconnect the heated oxygen sensor coupler and release its clamps.

5. Remove the exhaust from the manifold.

6. Disconnect the electrical connectors from the following components:

a. Distributor

b. Engine coolant temperature sensor and gauge

c. Engine ground wire from intake manifold

d. EGR solenoid vacuum valve

e. Fuel injectors

f. Throttle position sensor

g. Idle air control valve

h. Heated oxygen sensor

i. Evaporative emissions solenoid purge valve

7. Label and disconnect the vacuum hoses from the following:

a. EVAP canister purge hose

b. Brake booster supply hose

8. Disconnect the fuel feed and return hoses from each pipe.

9. Remove the cylinder head cover.

10. Fully loosen all the valve lash adjusting screws.

11. Disconnect the following engine cooling water hose:

a. Radiator inlet hose

b. Heater inlet hose

c. IAC valve outlet

12. Remove the timing belt.

13. Loosen the cylinder head bolts in reverse order of tightening. Once each bolt is loose, remove the bolts from the cylinder head.

14. Check to be sure all components are removed or disconnected before removing the cylinder head.

15. Remove the cylinder head with the intake manifold, exhaust manifold and distributor as an assembly.

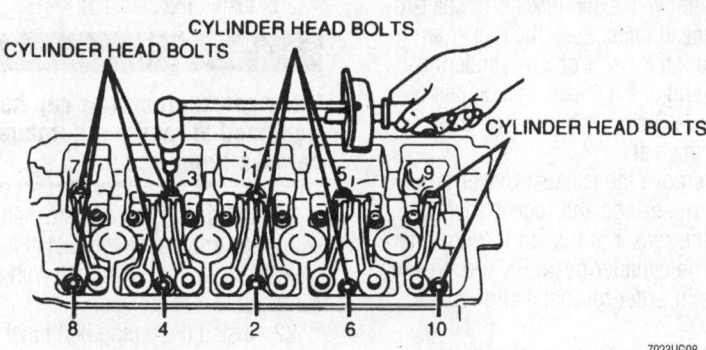

Tighten the cylinder head bolts in the correct sequence as shown—1.6L engine

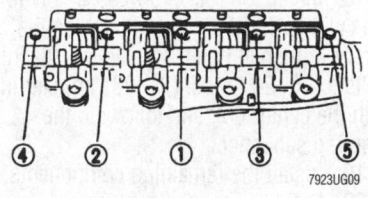

Tighten the rocker arm shaft bolts in the order shown—1.3L engine

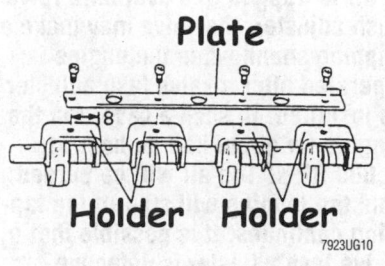

Special holder tool installed on the rocker arm and shaft assembly—1.3L engine

To install:

16. Install a new cylinder head gasket and the cylinder head with the distributor case onto the cylinder block. Tighten the cylinder head bolts, in sequence, in 3 Steps:
   a. Step 1: 26 ft. lbs. (35 Nm)
   b. Step 2: 41 ft. lbs. (55 Nm)
   c. Step 3: 49 ft. lbs. (68 Nm)
17. Install the timing belt.
18. Connect the engine cooling water hoses.
19. Adjust the valve lash.
20. Install the cylinder head cover.
21. Connect the fuel feed and return hoses to each pipe.
22. Connect the vacuum hoses.
23. Connect the remaining electrical components.
24. Install the remaining components.
25. Fill the engine coolant and check all fluids.
26. Connect the negative battery cable to the battery.
27. Start the engine and check for leaks.

Rocker Arms/Shafts

REMOVAL & INSTALLATION

Only the 1.3L engine uses rocker arms/shafts, on the other engines the camshaft directly actuates the valves.

1.3L Engine

1. Disconnect the negative battery cable.
2. Drain the cooling system into a re-sealable container.
3. Remove the air cleaner assembly.
4. Remove the spark plug wire clamps from the cylinder head.
5. Disconnect the PCV hose.
6. Remove the cylinder head cover from the cylinder head.
7. Remove the distributor assembly.

8. Remove the rocker arm shaft screws. Be careful not to drop them into the engine.
9. Remove the rocker arm shafts, arms and springs. Keep all of the parts in order so they can be reinstalled in their original locations.
10. Inspect the rocker arms and shafts for wear and/or damage and replace parts as necessary.
11. Remove the hydraulic valve lash adjuster from the rocker arm if necessary.

➡Do not remove the valve lash adjusters unless they need air bleeding or replacement. Be careful not to scratch the valve lash adjuster. Never disassemble the valve lash adjusters. Immerse the removed valve lash adjusters in clean engine oil until reinstallation. If a valve lash adjuster is left in the air, place it with its body facing down. Do not place the valve lash adjusters on their side or the body facing up.

   a. If the tip of the valve lash adjuster is badly worn, replace the adjuster. Check the O-ring for breakage or deterioration, replace as necessary.
   b. Using a valve lash adjuster air bleeding tool, bleed the air from the adjusters in kerosene.
   c. After filling the valve lash adjuster with fresh kerosene, compress the plunger and body with your finger (about 5 kg. or 11 lbs.) for a moment and inspect that its stroke is 0–0.02 in. (0–0.5mm). If its stroke is more than specified, bleed the air again and recheck the stroke. If the stroke is not within specification, replace the lash adjuster.

To install:

12. Install the hydraulic valve lash adjusters, if the valve lash adjusters were removed from the rocker arms.
13. Apply engine oil to the rocker arms and the rocker arm shafts.

14. Install the rocker arms, springs, and the rocker shafts.

➡The two rocker arm shafts are different, they are distinguishable by the stepped ends of the shafts. Looking at the screw holes in the intake shaft, one end of the shaft will have the two sides that are stepped, and this end is installed toward the camshaft pulley. Looking at the screw holes in the exhaust rocker shaft, one end will have a step on one side only, and this end is installed toward the distributor.

15. Install the intake rocker arms, springs, and washers to the intake side rocker shaft. Install the holder, special tool part 09916–56030 to hold the rocker arms, springs, and washers in place on the rocker shaft, then install the plate to the holders.
16. Install the rocker arm assembly to the cylinder head and evenly tighten the rocker shaft bolts. Tighten the attaching bolts to 16–20 ft. lbs. (22–28 Nm) starting with the center bolt and following the required sequence.
17. Install the exhaust rocker arms, springs, and washers to the exhaust side rocker shaft. Install the holder, special tool part 09916–56030 to hold the rocker arms,

springs, and washers in place on the rocker shaft, then install the plate to the holders.

18. Install the rocker arm assembly to the cylinder head and evenly tighten the rocker arm shaft bolts. Tighten the attaching bolts to 16–20 ft. lbs. (22–28 Nm) starting with the center bolt and following the required sequence.

19. Install the remaining components.
20. Refill the cooling system.
21. Connect the negative battery cable.
22. Adjust the ignition timing.
23. Top off the engine coolant as necessary.

✶✶ WARNING

If air is trapped in a hydraulic valve lash adjuster, the valve may make a tapping sound when the engine is operated after a valve lash adjuster is installed. In such a case, run the engine for about half an hour at 2,000 RPM. The air will be purged and the tapping will stop. If the tapping continues, it is possible that a valve lash adjuster is defective.

1.6L Engine

1. Disconnect the negative battery cable.
2. Remove the timing belt.

✶✶ WARNING

After the timing belt is removed, never turn the camshaft and crankshaft independently more than 90° in either direction. If turned, interference may occur among the piston and valves causing possible damage to the effected parts.

3. Using camshaft sprocket holding tool (09917–68220) or equivalent, hold the sprocket stationary and remove the camshaft sprocket bolt.
4. Remove the cylinder head cover.
5. Remove the distributor cap and distributor assembly from the engine.
6. Loosen all of the valve adjusting screw locknuts until the rocker arms move freely.
7. Remove the camshaft.
8. Remove the rocker arm shaft plug from the cylinder head.
9. Remove the timing belt inner cover-to-cylinder head bolts and the cover.
10. Remove all intake rocker arms and clips from the rocker shaft. Keep all parts in order so they can be reinstalled in their original locations.

11. Remove the six rocker arm shaft-to-cylinder head bolts. Push the rocker arm shaft through the rear of the cylinder head until the end of the rocker shaft appears. Remove the O-ring from the rear of the rocker arm shaft.
12. Remove the exhaust rocker arms, rocker arm springs and rocker shaft by pulling the rocker arm shaft through the front of the cylinder head. Be sure to keep the parts in order for installation purposes.
13. Clean and inspect all parts for wear and/or damage; replace parts as necessary.

To install:
14. Lubricate the rocker arms and shafts with clean engine oil before installation.
15. Push the rocker shaft into the front of the cylinder head; install the exhaust rocker arms and springs as the rocker arm shaft is being installed into the cylinder head.
16. Push the rocker arm shaft through the rear of the cylinder head. Install a new O-ring onto the rocker shaft.
17. Rotate the rocker arm shaft so the flat machined surface is horizontal and facing downward, parallel with the cylinder head mating surface and slide the shaft back into the cylinder head.
18. Install the 6 rocker arm shaft bolts and tighten the rocker arm-to-cylinder head bolts to 89 inch lbs. (10 Nm). Fill the rocker arm shaft bolt holes with clean engine oil.
19. Install the intake rocker arms and clips onto the rocker arm shaft.

➡**The camshaft carrier caps are embossed with numbers and arrows to ensure correct assembly. The No. 1 camshaft carrier cap must be installed at the front of the cylinder head with the remaining carrier caps following in numerical order. The directional arrows must always point toward the front of the cylinder head.**

20. Install the camshaft.

✶✶ WARNING

If the camshaft carrier cap bolts are tightened at random, damage to the camshaft may occur.

21. Lubricate the new camshaft seal lip with clean engine oil and install it into the cylinder head until it is flush with the camshaft carrier surface.
22. Install the timing belt inner cover and tighten the cover-to-cylinder head bolts to 89 inch lbs. (10 Nm). Install the rocker arm shaft plug into the cylinder head and tighten to 24 ft. lbs. (33 Nm).

➡**During camshaft timing belt sprocket installation, align the camshaft dowel pin with the slot in the camshaft timing belt gear designated as "E".**

23. Install the camshaft sprocket. Using holding tool (09917–68220) or equivalent, to hold the sprocket in place, tighten the camshaft sprocket bolt to 44 ft. lbs. (60 Nm).

➡**When installing the timing belt, the directional arrows on the timing belt must be matched with the rotation of the crankshaft; if not, excessive wear and timing belt failure may occur.**

24. Install the timing belt.
25. Apply RTV silicone rubber sealant to the surface of the distributor case that mates with the rear of the rocker arm shaft. Install the distributor case and tighten the 3 case-to-cylinder head bolts to 89 inch lbs. (10 Nm).

➡**With the timing marks aligned on the sprockets and the timing belt installed, the number four piston is at TDC of the compression stroke.**

26. Install the distributor into the distributor case. Be sure the rotor is aligned with

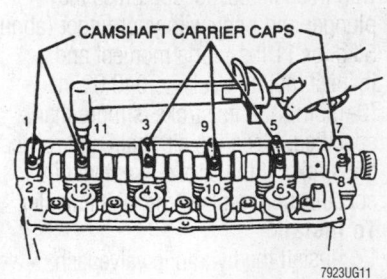

Camshaft carrier cap bolt torque sequence—1.6L engine

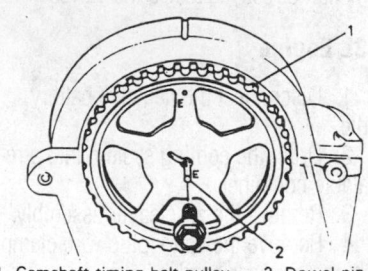

1. Camshaft timing belt pulley   2. Dowel pin

Align the dowel pin with the ``E'' slot in the sprocket—1.6L engine

the No. 4 tower on the distributor cap. Install the distributor cap.

27. Adjust the valve lash.

28. Install the cylinder head cover onto the cylinder head, in the reverse order of removal. Clean all sealing surfaces and use a new gasket and O-rings. Tighten the cylinder head cover bolts to 89 inch lbs. (10 Nm).

29. Connect the negative battery cable.

30. Start the engine; allow it to reach normal operating temperature and check for leaks.

31. Check and adjust the ignition timing as necessary.

Intake Manifold

REMOVAL & INSTALLATION

⁂ CAUTION

The fuel system pressure must be relieved before disconnecting any fuel lines. Failure to do so may result in personal injury.

1. Properly relieve the fuel system pressure.

2. Disconnect the negative battery cable.

3. Drain the coolant from the vehicle.

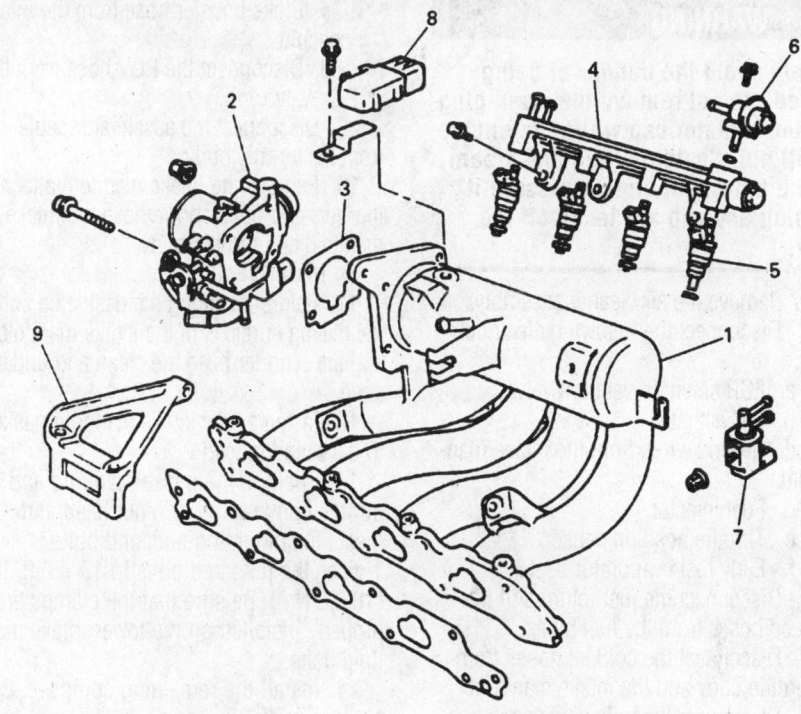

1. Intake manifold
2. Throttle body
3. Gasket
4. Fuel delivery pipe
5. Fuel injector
6. Fuel pressure regulator
7. EVAP canister purge valve
8. MAP sensor
9. Intake manifold upper stiffener

7923UG14

Intake manifold and related components—1998–99 1.3L engine

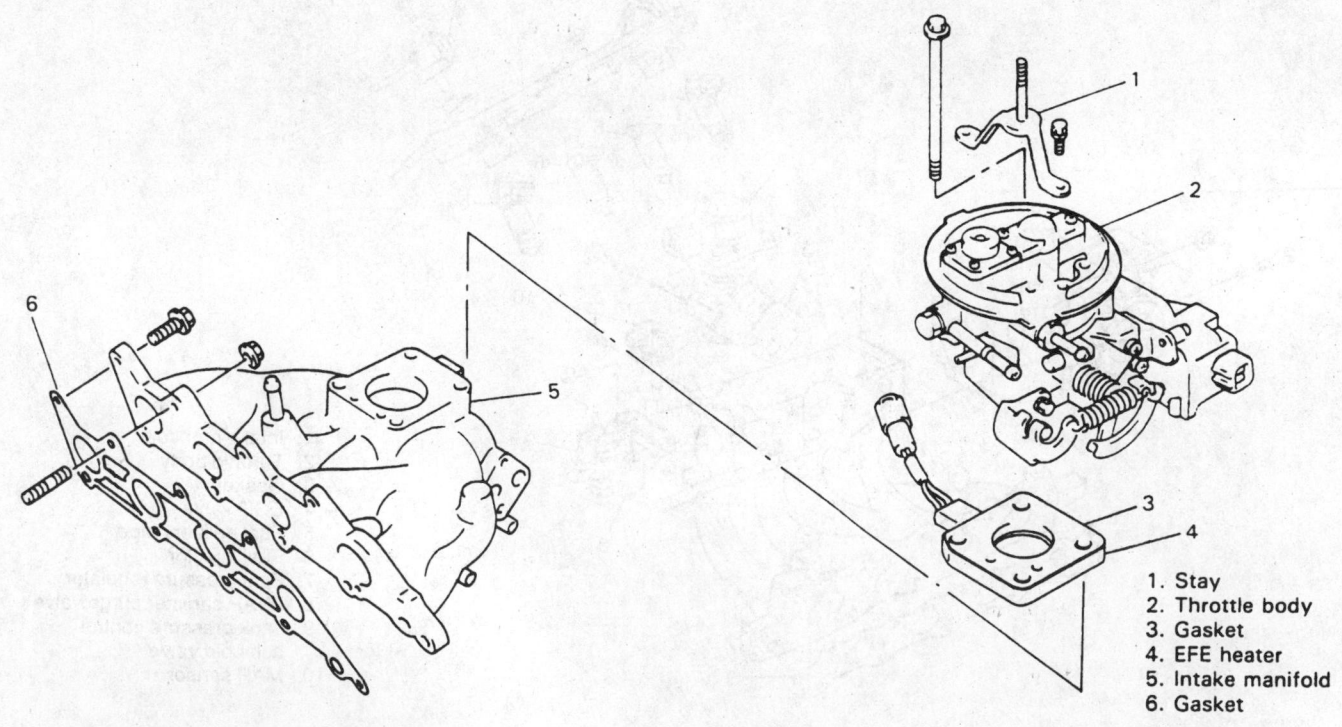

1. Stay
2. Throttle body
3. Gasket
4. EFE heater
5. Intake manifold
6. Gasket

7923UG13

Intake manifold and related components—1995–97 1.3L engine

4. Remove the air cleaner assembly.

5. Disconnect the following electrical wires:

   a. EGR solenoid vacuum valve.

   b. ISC actuator.

   c. Ground wires from the Intake manifold.

   d. Fuel injector.

   e. Throttle position sensor.

   f. Early fuel evaporator heater.

6. Disconnect the fuel return and the fuel feed hoses from the fuel pipes.

7. Disconnect the coolant hoses from the throttle body and the intake manifold.

8. Disconnect the following vacuum hoses:

   a. Canister purge hose from the intake manifold.

   b. MAP sensor hose from the intake manifold.

   c. Brake booster hose from the intake manifold.

   d. Disconnect the PCV hose from the PCV valve.

9. Disconnect the accelerator cable from the throttle body.

10. Remove the intake manifold attaching nuts and bolts and remove the intake manifold and throttle body.

To install:

11. Before installing the gasket be sure the mating surfaces of the intake manifold and the cylinder head are clean and undamaged.

12. Install a new intake manifold gasket to the cylinder head.

13. Position the intake manifold and throttle body on the cylinder head and install the mounting nuts and bolts. Tighten the nuts and bolts to 13–20 ft. lbs. (18–28 Nm). Be sure that the clamps are properly installed on the lower intake manifold bolts.

14. Install the remaining components.

15. Refill the cooling system.

16. Connect the negative battery cable.

17. Start the engine and check for fuel and cooling system leaks.

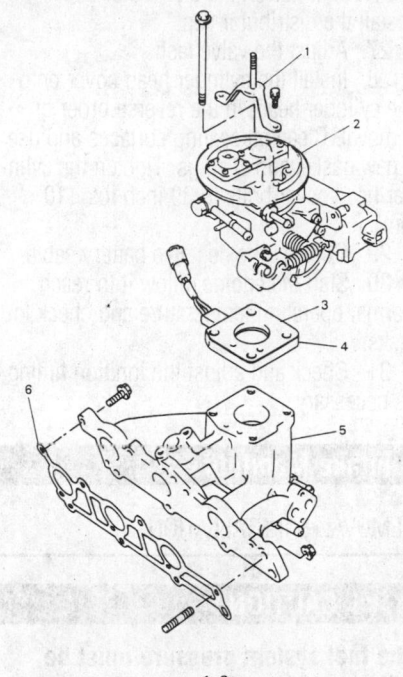

1. Stay
2. Throttle body
3. Gasket
4. EFE heater
5. Intake manifold
6. Gasket

7923UG43

Intake manifold and related components—1.0L engine

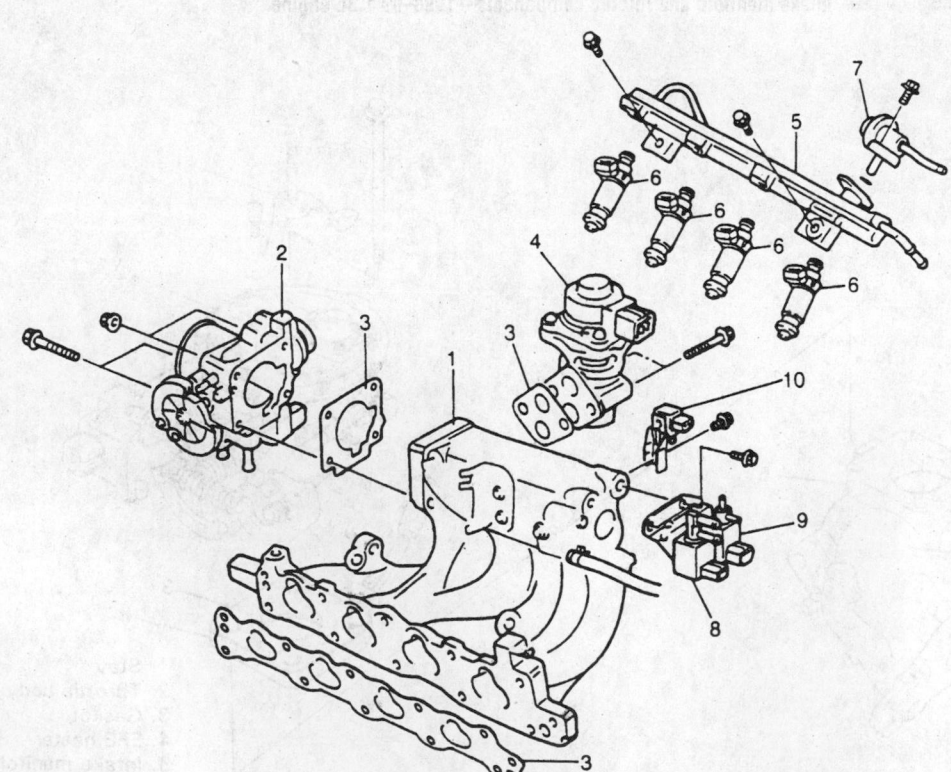

1. Intake manifold
2. Throttle body
3. Gasket
4. EGR valve
5. Fuel delivery pipe
6. Fuel injector
7. Fuel pressure regulator
8. EVAP canister purge valve
9. Tank pressure control solenoid valve
10. MAP sensor

7923UG15

Intake manifold and related components—1.6L engine

Exhaust Manifold

REMOVAL & INSTALLATION

1.0L and 1.3L Engine

✳✳ CAUTION

To avoid the danger of being burned, do not service the exhaust system while it is hot. Service should be performed only after the system cools down.

1. Disconnect the negative battery cable.
2. Disconnect the oxygen sensor electrical coupler and release the wire from its clamps.
3. Remove the exhaust manifold cover,
4. Remove the exhaust manifold stiffener bolt.
5. Remove the two bolts attaching the exhaust pipe to the exhaust manifold.
6. Remove the exhaust manifold mounting nuts and bolts.
7. Remove the exhaust manifold and the gasket.
   To install:
8. Before installing any components check the exhaust manifold and the engine

for deterioration or damage and replace as necessary.

9. Install the manifold gasket and the exhaust manifold to the engine. Tighten the nuts and bolts to 13–20 ft. lbs. (18–28 Nm).
10. Install the exhaust pipe gasket to the exhaust pipe and position the exhaust pipe.
11. Install the two bolts that attach the exhaust pipe to the exhaust manifold and tighten the bolts to 25–36 ft. lbs. (35–50Nm).
12. Install the exhaust manifold stiffener and tighten the bolt to 29–43 ft. lbs. (40–60 Nm).
13. Install the remaining components.
14. Connect the negative battery cable.
15. Run the engine and check for exhaust leaks.

1.6L Engine

1. Disconnect the negative cable.
2. Raise and safely support the vehicle.
3. Disconnect the two exhaust pipe bolts connecting the exhaust pipe to the exhaust manifold. Lower the vehicle after the exhaust manifold is disconnected from the exhaust pipe.
4. Disconnect the oxygen sensor lead wire at the coupler.
5. Disconnect the exhaust manifold

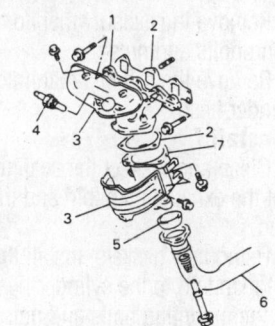

For vehicle with WU-TWC

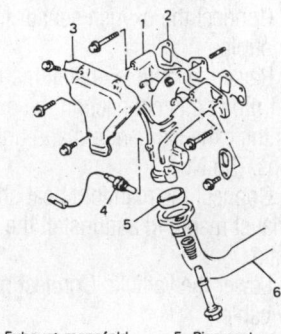

For vehicle without WU-TWC

1. Exhaust manofold
2. Gasket
3. Cover
4. Heated oxygen sensor
5. Pipe seal
6. Exhaust pipe
7. WU-TWC (if equipped)

7923UG44

Exhaust manifold and related components—1.0L engine

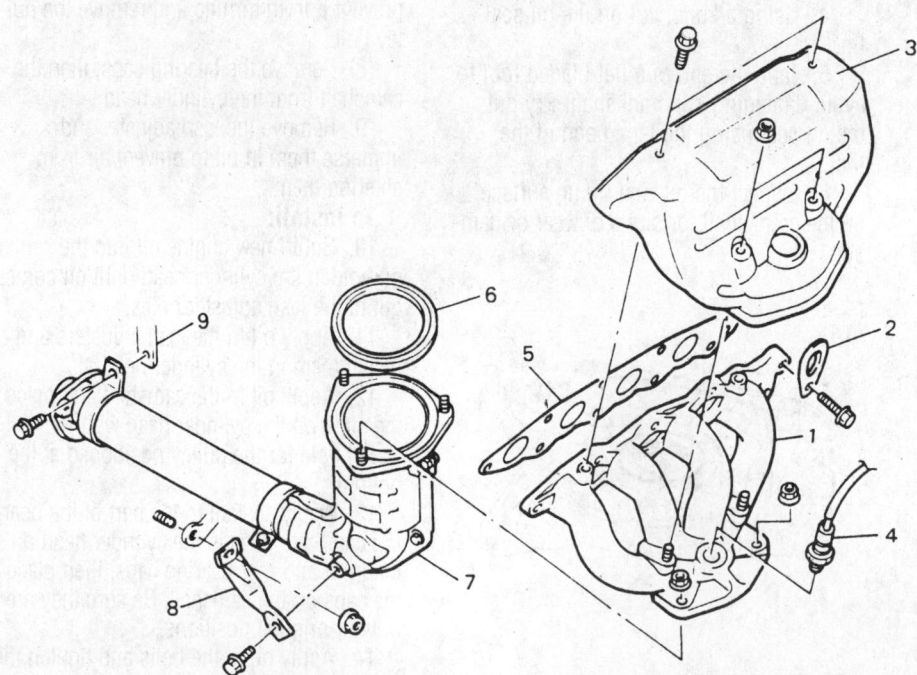

1. Exhaust manifold
2. Engine hook
3. Exhaust manifold cover
4. Heated oxygen sensor
5. Exhaust manifold gasket
6. Exhaust No.1 pipe gasket
7. Exhaust No.1 pipe
8. Exhaust pipe front stiffener
9. Exhaust pipe rear stiffener

7923UG16

Exhaust manifold and related components—1.6L engine

heat shield from the manifold by removing the nut and bolt.

6. Remove the exhaust manifold mounting bolts and nuts.

7. Remove the exhaust manifold from the cylinder head.

To install:

8. Clean and inspect the sealing surfaces of the exhaust manifold and the cylinder head.

9. Using new gaskets, install the exhaust manifold to the cylinder head and tighten the mounting bolts and nuts to 17 ft. lbs. (23 Nm).

10. Connect the oxygen sensor lead wire at the coupler.

11. Raise the vehicle. Install the three exhaust pipe bolts connecting the exhaust pipe to the exhaust manifold and tighten to 36 ft. lbs. (50 Nm).

12. Connect the manifold heat shield to the exhaust manifold and install the nuts and bolts.

13. Lower the vehicle. Connect negative battery cable.

14. Check for exhaust leaks when finished.

Front Crankshaft Seal

REMOVAL & INSTALLATION

1.0L and 1.3L Engines

1. Disconnect the negative battery cable.
2. Remove the timing belt.
3. Remove the crankshaft sprocket bolt using a suitable gear stopper to hold the flywheel. Remove the sprocket bolt, sprocket and key.
4. Use a suitable tool to remove the seal from the oil pump housing.

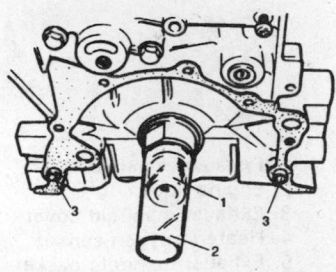

1 Crankshaft
2 Oil seal guide (Vinyl resin) (special tool 09926-18210)
3 Oil pump pin

7923UG17

Oil seal guide tool—1.3L engine

➡ **Be careful not to damage the crankshaft or the oil pump sealing surfaces when removing or installing the seal.**

5. Clean and inspect the surfaces of the crankshaft and the oil pump assembly.

To install:

6. Lubricate the new seal with clean engine oil.

7. Install the new seal over the crankshaft and into the oil pump, making sure the oil seal lip is not turned up. Use oil seal guide tool 09926-18210 or equivalent.

8. Install the crankshaft sprocket and timing belt.

9. Connect the negative battery cable.

1.6L Engine

➡ **The front oil seal can be removed from the engine without removing the oil pump.**

1. Disconnect the negative battery cable from the battery.
2. Remove the timing belt.
3. Remove the crankshaft timing belt sprocket.

✳✳ WARNING

When removing the front seal, be extremely careful not to damage the crankshaft.

4. Using a knife, cut off the oil seal lip.

5. Tape the end of a flat bladed tool to avoid damaging the crankshaft. Pry out the oil seal using the taped end of the tool.

6. Inspect the oil seal riding surface on the crankshaft for signs of wear or damage.

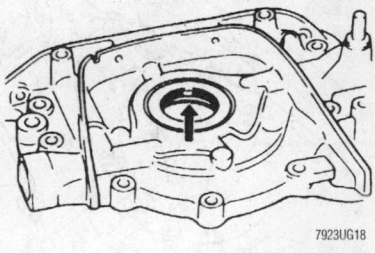

7923UG18

Front crankshaft oil seal location—1.6L engine

To install:

7. Wipe the seal bore with a clean rag.

8. Apply multipurpose grease to the lip of a new oil seal.

9. Drive the oil seal into place using a seal installer tool. Be sure the seal surface is flush with the oil pump case edge. Work from the front of the cover. Be extremely careful not to damage the seal.

10. Install the crankshaft sprocket.

11. Install the timing belt.

12. Connect the negative battery cable to the battery.

13. Start the engine and check for leaks.

Camshaft and Valve Lifters

REMOVAL & INSTALLATION

1.0L Engine

1. Properly relieve the fuel system pressure.

2. Disconnect the negative battery cable.

3. Remove the cylinder head cover.

4. If equipped, remove the distributor from the cylinder head.

5. Remove the timing belt.

6. Turn the crankshaft so the keyway is 60° to the left of the 12 O'clock position as shown.

7. Insert a rod through the camshaft to prevent it from turning and remove the pulley bolt.

8. Remove the bearing caps, then the camshaft from the cylinder head.

9. Remove the lash adjuster and immerse them in oil to prevent air from entering them.

To install:

10. Squirt new engine oil into the center oil hole in the cylinder head until oil comes out of the lash adjuster holes.

11. Apply oil to the lash adjusters and install them in the cylinder head.

12. Apply oil to the camshaft and place the shaft on the cylinder head with the dowel hole for the pulley positioned at the bottom.

13. Apply sealant to the part of the bearing cap that contacts the cylinder head on the front and rear bearing caps, then place the caps on the camshaft. Be sure they are in their original positions.

14. Apply oil to the bolts and tighten the caps gradually in several passes to 7–8.5 ft. lbs. (9–12 Nm).

1. Cylinder head cover
2. Gasket
3. Cylinder head
4. Camshaft housing No.1
5. Camshaft housing No.2
6. Camshaft housing No.3
7. Camshaft
8. Oil seal
9. Valve lash adjuster

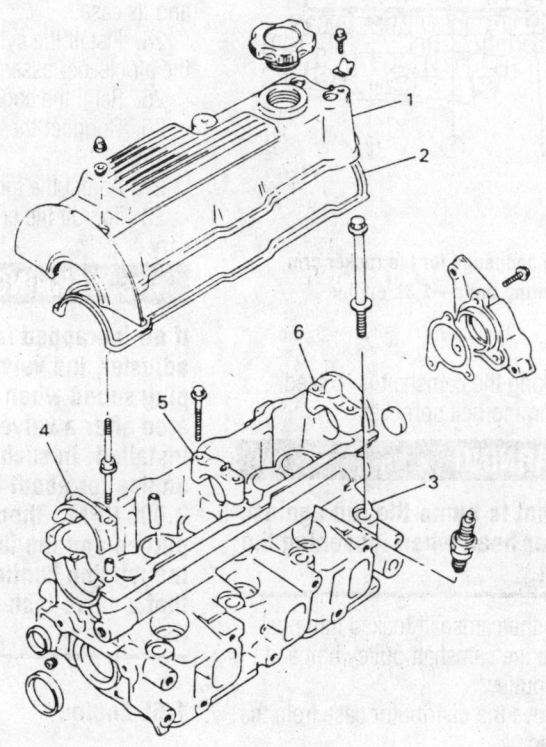

7923UG45

Camshaft and related components—1.0L engine

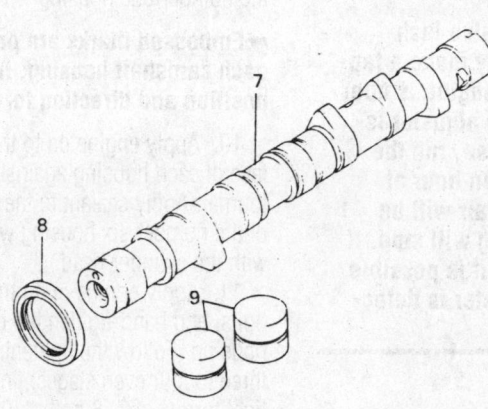

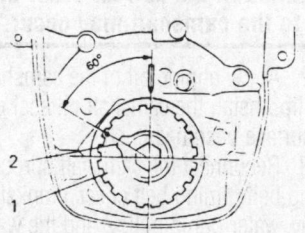

1. Crankshaft timing belt pulley   2. Key

7923UG46

To avoid valve-to-piston contact, turn the crankshaft as shown after removing the timing belt—1.0L engine

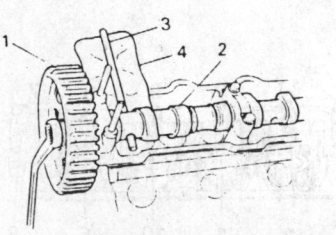

1. Camshaft timing   3. Rod
   belt pulley   4. Shop cloth
2. Camshaft

7923UG47

Insert a rod through the camshaft when removing or installing the pulley bolt—1.0L engine

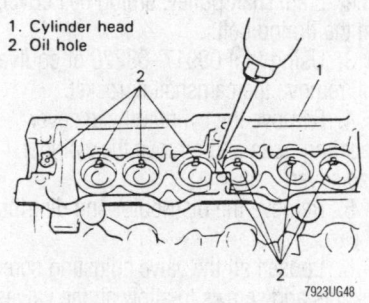

1. Cylinder head
2. Oil hole

7923UG48

Squirt engine oil into the camshaft journal holes before installing the lash adjusters—1.0L engine

15. Apply oil to the seal and install the seal until it is flush with the housing.

16. Install the camshaft pulley. Tighten the bolt to 41–46 ft. lbs. (56–64 Nm).

17. Install the cylinder head cover, timing belt and remaining components.

1.3L Engine

1. Disconnect the negative battery cable.

2. Drain the cooling system into a re-sealable container.

3. Remove the air cleaner assembly.

4. Remove the spark plug wire clamps from the cylinder head.

5. Disconnect the PCV hose.

6. Remove the cylinder head cover from the cylinder head.

7. Remove the distributor assembly.

8. Remove the rocker arm shaft bolts. Be careful not to drop them into the engine.

9. Remove the intake and the exhaust rocker arm shafts.

10. Remove the crankshaft pulley and timing belt.

11. The camshaft must be locked in position to remove the camshaft pulley. Lock the camshaft by inserting a proper sized rod into a hole in the camshaft. The

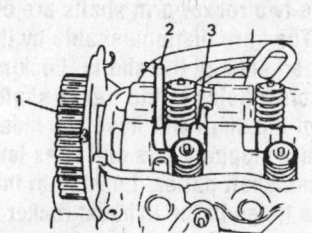

1. Camshaft timing belt pulley   3. Camshaft
2. Proper size rod

7923UG19

Locking the camshaft—1.3L engine

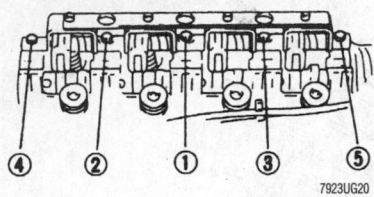

Tightening sequence for the rocker arm shaft mounting bolts—1.3L engine

hole for locking the camshaft is located toward the timing belt sprocket.

❋❋ WARNING

Use care not to bump the rod against the cylinder head when loosening the pulley bolt.

12. With the camshaft locked into position, remove the camshaft pulley bolt and remove the pulley.

13. Remove the distributor case from the cylinder head.

14. Remove the camshaft from the cylinder head.

To install:

15. Apply engine oil to the camshaft lobes and journals and to the camshaft oil seal on the cylinder head.

16. Install the camshaft into the cylinder head from the transmission case side.

17. Install the distributor case to the cylinder head.

18. Lock the camshaft in place.

19. Install the camshaft timing belt pulley by fitting the pulley pin on the camshaft into the slot on the camshaft pulley. Install the attaching bolt and tighten the bolt to 41–46 ft. lbs. (56–64 Nm)

20. Install the timing belt.

21. Apply engine oil to the rocker arms and the rocker arm shafts.

➥ **The two rocker arm shafts are different. They are distinguishable by the stepped ends of the shafts. Looking at the screw holes in the intake shaft, one end of the shaft will have two sides that are stepped; this end goes toward the camshaft pulley. Looking at the screw holes in the exhaust rocker shaft, one end will have a step on one side only; this end goes toward the distributor.**

22. Install the intake and the exhaust rocker arms and shafts, Tighten the bolts to

16–20 ft. lbs. (22–28 Nm) in the proper sequence.

23. Install the distributor to the camshaft and its case.

24. Install the cylinder head cover and the air cleaner assembly.

25. Refill the cooling system.

26. Connect the negative battery cable.

27. Adjust the ignition timing.

28. Top off the engine coolant as necessary.

❋❋ WARNING

If air is trapped in a valve lash adjuster, the valve may make a tapping sound when the engine is operated after a valve lash adjuster is installed. In such a case, run the engine for about half an hour at 2,000 RPM,, then the air will be purged and the tapping will stop. If the tapping continues it is possible that a valve lash adjuster is defective.

1.6L Engine

1. Disconnect the negative battery cable.

2. Remove the water pump belt and pulley, crankshaft pulley, timing belt cover, and the timing belt.

3. Using tool 09917–68220 or equivalent, remove the camshaft sprocket.

4. Remove the cylinder head cover mounting bolts and remove the cylinder head cover.

5. Remove the distributor and distributor case.

6. Loosen all the valve adjusting screw locknuts and screws to allow all the valves to close.

7. Remove the camshaft housing bolts, housings, and the camshaft.

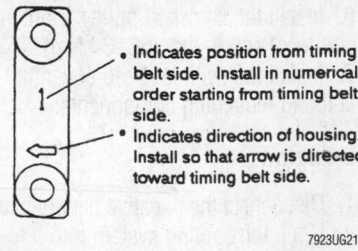

Camshaft bearing cap identification—1.6L engine

❋❋ CAUTION

The camshaft housing bolts must be removed in the reverse order of installation or damage to the camshaft may occur.

To install:

8. Lubricate the lobes and journals of the camshaft with clean engine oil.

9. Install the camshaft on the cylinder head. Install the camshaft housing to the camshaft and cylinder head, starting with the number one housing.

➥ **Embossed marks are provided on each camshaft housing, indicating position and direction for installation.**

10. Apply engine oil to the sliding surface of each housing against the camshaft journal. Apply sealant to the mating surface of the number six housing which will mate with the cylinder head.

11. Apply engine oil to the housing bolts, and hand-tighten the bolts into the housing. Follow the tightening sequence in three to four even stages, finishing with a final torque of 7–8 ft. lbs. (9–12 Nm).

❋❋ CAUTION

The camshaft housing bolts must be tightened in the correct order or damage to the camshaft may occur.

12. Apply engine oil to the camshaft oil seal lip. Install the camshaft oil seal until the surface becomes flush.

13. Reconnect the camshaft sprocket, timing belt, timing belt cover, crankshaft pulley, water pump pulley, and the water pump belt. Be sure the pin on the camshaft fits into the slot at the **E** mark on the camshaft sprocket. Tighten the sprocket bolt to 41–46 ft. lbs. (56–64 Nm).

14. Prior to installation, apply sealant to the area of the distributor housing that cov-

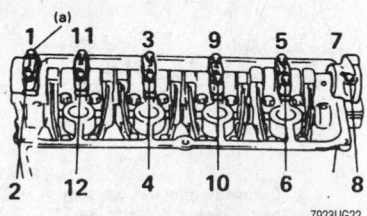

Camshaft housing bolt tightening sequence—1.6L engine

ers the rear of the rocker arm shaft on the cylinder head. Install the distributor and distributor housing to the cylinder head. Be sure the distributor is facing the correct firing position.

15. Adjust the valve lash.

16. Using a new gasket, install the cylinder head cover.

17. Connect the negative battery cable to the battery.

18. Start the engine and check for any water or oil leaks when finished.

19. Check and/or adjust the ignition timing as necessary.

Valve Lash

ADJUSTMENT

1.0L and 1.3L Engines

Hydraulic Valve Lash (HVL) adjusters are used to adjust the valve clearance to 0 (zero) lash automatically at all times. Adjustment is not required.

1.6L Engine

1. Disconnect the negative battery cable.

2. Remove the cylinder head cover.

3. Safely raise and support the vehicle.

4. Remove the right front wheel and fender apron.

5. Turn the crankshaft pulley clockwise until the **V** mark on the pulley is aligned with the 0 calibration on the timing belt cover.

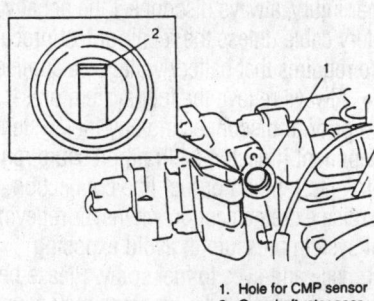

1. Hole for CMP sensor
2. Camshaft rotor gear

Valve numbered locations and camshaft rotor gear mark—1998-99 1.6L engines

7923UG23

6. On 1995–97 models, remove the distributor cap and be sure the ignition rotor is aligned with cylinder number one's ignition wire. If the ignition rotor is not aligned, turn the crankshaft one rotation clockwise and realign the timing marks. On 1998–99 models, remove the camshaft position sensor and look in the hole for the notch on the camshaft rotor gear, if the notch is not visible in the hole, rotate the crankshaft one complete revolution until the notch is visible.

7. The valve lash is measured between the rocker arm adjusting screw and the valve stem. Use a thickness gauge to measure the gap.

8. Check the valve lash for the following valves:

   a. Intake valve of cylinder number one (ID 1).

   b. Intake valve of cylinder number two (ID 2).

   c. Exhaust valve of cylinder number one (ID 5).

   d. Exhaust valve of cylinder number three (ID 7).

9. If the valve lash is out of specification, adjust the specification after loosening the locknut and turning the adjusting screw. Hold the screw stationary while tightening the locknut. Recheck the specification after tightening the locknut.

10. Rotate the crankshaft one rotation clockwise and realign the timing marks.

11. Check the valve lash for the following valves:

   a. Intake valve of cylinder number three (ID 3).

   b. Intake valve of cylinder number four (ID 4).

   c. Exhaust valve of cylinder number two (ID 6).

   d. Exhaust valve of cylinder number four (ID 8).

12. Adjust the valves that are out of specification and recheck after tightening the locknut.

13. Install the remaining components.

14. Connect the negative battery cable.

Oil Pan

REMOVAL & INSTALLATION

All Models

1. Disconnect the negative battery cable.

2. Safely raise and support the vehicle.

3. On 1.6L models, remove the engine undercovers.

4. Drain the engine oil into a suitable container.

5. On 1.3L models, remove the lower plate, from the clutch housing if equipped with a manual transaxle, or from the torque converter housing if equipped with a automatic transaxle.

6. On 1.6L models, perform the following sub-steps;

   a. Remove the front exhaust pipe from the vehicle.

   b. Remove the transaxle stiffener plate from the engine and transaxle.

   c. Support the transmission and engine.

   d. Remove the vehicle center member by removing the seven nuts and four bolts from the center member.

7. Remove the crankshaft position sensor from the oil pan, if equipped.

8. Remove the oil pan retainer bolts. Remove the oil pan from the cylinder block.

9. Remove the oil pump pick-up.

To install:

10. Clean the mating surfaces of the oil pan and the engine block.

11. Install the oil pump strainer. Tighten the strainer bolt first at the bracket. Tighten the bolts to 7–8 ft. lbs. (9–12 Nm).

12. Apply silicon sealant to the oil pan mating surface in one continuous bead.

13. Fit the oil pan to the engine block and install the bolts. Start tightening the bolts at the center and move outward. Tighten the bolts to 7–8 ft. lbs. (9–12 Nm).

14. Install the crankshaft position sensor to the oil pan.

15. Install the drain plug and drain plug gasket to the oil pan.

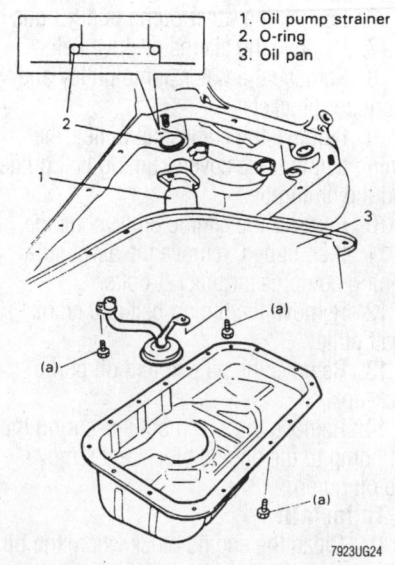

1. Oil pump strainer
2. O-ring
3. Oil pan

Oil pan mounting—1.3L and 1.6L engines

7923UG24

16. On 1.3L models, install the lower plate, to the clutch housing if equipped with a manual transaxle, or to the torque converter housing if equipped with an automatic transaxle.

17. On 1.6L models, do the following;

a. Install the center member to the vehicle. Tighten the bolts and nuts at the engine mounts to 40 ft. lbs. (55 Nm) and all other nuts to 33 ft. lbs. (45 Nm).

b. Install the transaxle stiffener plate and tighten the bolts to 37 ft. lbs. (50 Nm).

c. Install the exhaust pipe and tighten the nuts and bolts to 37 ft. lbs. (50 Nm).

d. Install the engine undercovers.

18. Lower the vehicle.

19. Refill the engine with oil.

20. Connect the negative battery cable.

21. Start the engine and check for leaks.

Oil Pump

REMOVAL & INSTALLATION

All Models

1. Disconnect the negative battery cable.

2. Safely raise and support the vehicle.

3. Drain the engine oil.

4. Remove the right side fender apron clips by pushing the center pin.

✸✸ WARNING

Do not push the center pin too far in, or it will fall off into the fender.

5. If equipped, remove the power steering and air conditioning belt.

6. Loosen the water pump pulley bolts.

7. Remove the alternator drive belt.

8. Remove the water pump pulley and alternator bracket.

9. Remove the crankshaft pulley, the timing belt outside covers, timing belt guide and the timing belt.

10. Remove the engine oil level gauge.

11. If equipped, remove the air conditioning compressor bracket bolts.

12. Remove the timing belt and crankshaft pulley.

13. Remove the oil pan and oil pump pick-up.

14. Remove the seven bolts securing the oil pump to the engine block and remove the oil pump.

To install:

15. Clean the engine block where the oil pump mounts, then install the oil pump gasket and the two oil pump alignment pins.

16. To prevent damage to the oil seal when installing the oil pump, fit special tool 09926–18210 to the crankshaft and apply a thin coating of engine oil to the special tool.

17. Install the oil pump to the crankshaft and the engine block. Install a long bolt to the lowest bolt hole on the intake manifold side of the engine. Install two long bolts to the two lowest bolt holes on the exhaust manifold side of the engine. Install the four short bolts to the other four bolt holes in the oil pump. Tighten all of the bolts to 7–8 ft. lbs. (9–12 Nm). Check that the oil seal lip is not turned up, then remove the special tool.

18. If the of the oil pump gasket bulges where the oil pan attaches cut the excess off with a sharp knife.

19. Install the oil pan and the oil pump pick-up.

20. Install the rubber seal between the oil pump and the water pump.

21. Install the crankshaft key, timing belt pulley and the crankshaft pulley pin.

22. With the crankshaft locked, install and tighten the pulley bolt to 80 ft. lbs. (110 Nm).

23. Install the timing belt guide so that the concave side faces the oil pump, then install the crankshaft key and the timing belt pulley.

24. Install the remaining components.

25. Connect the negative battery cable.

26. Start the engine and check the engine oil pressure.

27. Check that no leaks are present.

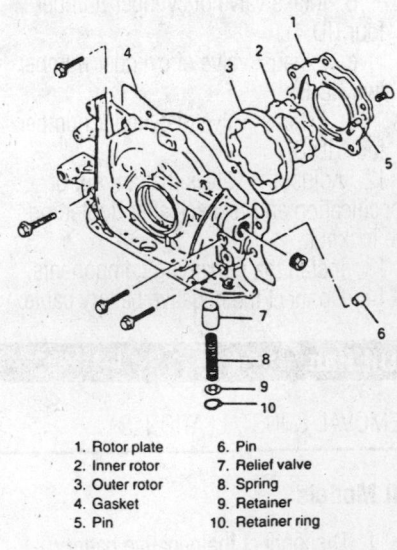

1. Rotor plate
2. Inner rotor
3. Outer rotor
4. Gasket
5. Pin
6. Pin
7. Relief valve
8. Spring
9. Retainer
10. Retainer ring

7923UG25

Exploded view of the oil pump—1.3L and 1.6L engine

Rear Main Seal

REMOVAL & INSTALLATION

All Engines

1. Remove the transaxle assembly.

2. Remove the flexplate/flywheel from the crankshaft.

3. Carefully pry the oil seal out of the retainer without scratching the sealing surface of the crankshaft.

To install:

4. Apply engine oil the lip of the new seal.

5. Install the seal in the retainer using a suitable seal driver.

6. Install the flexplate/flywheel.

7. Install the transaxle assembly.

FUEL SYSTEM

Fuel System Service Precautions

Safety is the most important factor when performing not only fuel system maintenance but any type of maintenance. Failure to conduct maintenance and repairs in a safe manner may result in serious personal injury or death. Maintenance and testing of the vehicle's fuel system components can be accomplished safely and effectively by adhering to the following rules and guidelines.

• To avoid the possibility of fire and personal injury, always disconnect the negative battery cable unless the repair or test procedure requires that battery voltage be applied.

• Always relieve the fuel system pressure prior to disconnecting any fuel system component (injector, fuel rail, pressure regulator, etc.), fitting or fuel line connection. Exercise extreme caution whenever relieving fuel system pressure to avoid exposing skin, face and eyes to fuel spray. Please be advised that fuel under pressure may penetrate the skin or any part of the body that it contacts.

• Always place a shop towel or cloth around the fitting or connection prior to loosening to absorb any excess fuel due to spillage. Ensure that all fuel spillage (should it occur) is quickly removed from the engine surfaces. Ensure that all fuel soaked cloths or towels are deposited into a suitable waste container.

• Always keep a dry chemical (Class B) fire extinguisher near the work area.

- Do not allow fuel spray or fuel vapors to come into contact with a spark or open flame.
- Always use a back-up wrench when loosening and tightening fuel line connection fittings. This will prevent unnecessary stress and torsion to fuel line piping. Always follow the proper torque specifications.
- Always replace worn fuel fitting O-rings with new. Do not substitute fuel hose or equivalent, where fuel pipe is installed.

Fuel System Pressure

RELIEVING

❋❋ CAUTION

Care should be used when working around the fuel system. DO NOT smoke or expose the fuel system to any open flames. Keep a fire extinguisher handy.

1. Disconnect the negative battery cable from the battery.
2. Place the vehicle in **PARK** for A/T or **NEUTRAL** for M/T.
3. Remove the fuel filler cap from the filler neck to release the fuel vapor pressure in the fuel tank.
4. Remove the main fuse box cover and engine coolant reservoir tank from its bracket.
5. Detach the main fuse box from the body and disconnect the electrical connector from the fuel pump relay.
6. Connect the negative battery cable.
7. Start the vehicle and allow the engine to run until it stalls.
8. Crank the engine for three more times to eliminate any remaining pressure in the fuel lines.
9. Disconnect the negative battery cable.
10. Connect the electrical connector to the fuel pump relay.
11. Install the main fuse box to the body.
12. Install the engine coolant reservoir tank to its bracket.
13. After servicing the fuel system, connect the negative battery cable.
14. Start the engine and check for leaks in the system.

Fuel Filter

REMOVAL & INSTALLATION

All Models

❋❋ CAUTION

The fuel system pressure must be relieved before disconnecting any fuel lines. Failure to do so may result in personal injury.

1. Properly relieve the fuel system pressure.
2. Disconnect the negative battery cable.
3. Raise and safely support the vehicle.
4. Place a container under the fuel filter.
5. Disconnect the fuel inlet hose from the fuel filter.

❋❋ CAUTION

A small amount of fuel may be released after the fuel hose is disconnected. Cover the hose and pipe with a shop towel.

6. Disconnect the outlet hose from the fuel feed pipe.
7. Remove the two fuel filter mounting bracket bolts and remove the fuel filter from the frame with the outlet hose attached.
8. Remove the fuel filter from the bracket by removing the mounting bolt.
9. Disconnect the outlet hose from the fuel filter.

To install:
10. Install the fuel filter on the bracket and install the mounting bolt.
11. Install the remaining components.
12. Connect the inlet and outlet hoses to the filter.
13. Connect the negative battery cable.
14. With the ignition **ON** and the engine **OFF** check for leaks.

Fuel Pump

REMOVAL & INSTALLATION

All Models

1. Relieve the pressure from the fuel system.
2. Disconnect the negative battery cable.

3. Remove the rear seat cushion by performing the following:
   a. Remove the spare tire.
   b. Remove the seat back by removing the two center mounting nuts and the four mounting screws.
   c. Remove the fitting screws from the rear of the seat cushion.
   d. Lift the front of the seat cushion and remove the cushion.
4. Disconnect the fuel level gauge and the fuel pump lead wire couplers and detach the wire tape.
5. Raise and safely support the vehicle on jackstands.
6. Disconnect the fuel filler hose from the fuel tank and disconnect the breather hose from the filler neck.
7. Drain the fuel from the tank by pumping the fuel out through the fuel tank filler.

❋❋ CAUTION

Use a gasoline safe hand operated pump device to drain the fuel tank.

8. Disconnect the fuel hoses from the fuel pipes, located near the fuel filter.

❋❋ CAUTION

A small amount of fuel may be released after the fuel hose is disconnected. Cover the hose and pipe to be disconnected with a shop cloth.

9. Install a support (for example, a transmission jack) under the fuel tank.
10. Remove the fuel tank mounting hardware and remove the tank from the vehicle.
11. Disconnect the fuel lines from the fuel pump and sender assembly.
12. Remove the twelve screws that secure the fuel pump and fuel gauge assembly to the tank and remove the pump and sender assembly.
13. Disconnect the fuel pump electrical connectors.
14. Remove the fuel strainer.
15. Remove the fuel pump.
To install:
16. Install the fuel pump to the fuel gauge assembly.
17. Install the fuel strainer on the fuel pump.

➡**Always install a new fuel pump strainer when replacing the fuel pump.**

18. Connect the electrical connectors to the fuel pump.

19. Install the remaining components.

20. Connect the negative battery cable.

21. Turn the ignition switch to the **ON** position, but leave the engine **OFF** and check for fuel leaks.

DRIVE TRAIN

Transaxle Assembly

REMOVAL & INSTALLATION

Manual

1. Disconnect the negative battery cable, then remove the battery and tray.

2. Remove the clutch cable adjusting nut, joint pin from the cable and cable from the bracket.

3. Disconnect and tag all the wiring harness clamps and connectors involved with the transaxle removal.

4. Remove the speedometer cable boot, case clip and cable from the case.

5. On Swift models, remove the radiator outlet pipe from the transmission side cover.

6. Remove the transaxle retaining bolts.

7. Remove the starter.

8. Raise and support the vehicle safely. Drain the transaxle oil.

9. Remove the fender apron extension on the left side.

10. Remove the bolts connecting the exhaust pipe to the exhaust manifold and disconnect the joint.

11. On Esteem models, with the engine supported, remove the vehicle center mounting member by removing the seven nuts and four bolts.

12. Remove the gearshift control shaft nut and bolt, then disconnect the control shaft from the gear shift control shaft.

13. Remove extension rod nut and washers.

14. Remove the clutch housing lower plate.

15. Remove the sway bar.

16. Remove the left and right front wheels.

17. Disconnect the left and right ball joints.

18. Remove the left and right halfshafts.

19. Remove the transaxle stiffener.

20. Remove the transmission to engine bolt and nut.

21. Remove the engine rear mounting bracket bolts.

22. Install a engine support.

23. Support the transaxle with a suitable jack.

24. Remove the left engine mounting bracket and stiffener.

25. Lower the transaxle with the engine attached. Pull the transaxle straight out toward the left side.

26. Lower and remove the transaxle.

To install:

27. Install the transaxle from the left side of the vehicle. Use care when inserting the pilot shaft into the clutch assembly. If the spline on the input shaft does not align with the clutch assembly spline, turn the crankshaft slightly to aid in spline alignment.

28. Raise the transaxle and engine.

29. Install the left engine mounting bracket and stiffener. Tighten the bolts to 29–43 ft. lbs. (40–60 Nm).

30. Install the rear engine mounting bracket bolts and tighten them to 29–43 ft. lbs. (40–60 Nm).

➡️**Before installing the bolts into the rear mounting bracket, apply sealant to the bolt threads.**

31. Install the transmission-to-engine bolt and nut. Tighten the nut and bolt to 29–43 ft. lbs. (40–60 Nm).

32. On Esteem models, install the center member to the vehicle and install the seven nuts and four bolts. Tighten the bolts and nuts as follows:

    a. Center member to the radiator support to 33 ft. lbs. (45 Nm).

    b. Center member to crossmember to 33 ft. lbs. (45 Nm).

    c. Engine mounts to center member nuts to 40 ft. lbs. (55 Nm).

33. Install the transaxle stiffener.

34. Lower the transaxle supporting jack.

35. Install the left and right halfshafts.

36. Install the ball joints.

37. Connect the sway bar.

38. Install the clutch housing lower plate.

39. Install the extension rod nut and washers. Tighten the rod nut to 18–28 ft. lbs. (25–40 Nm).

40. Install the control shaft to gear shift and install the gear shift control shaft bolt and nut. Tighten the gear shift control shaft bolt and nut to 11–14 ft. lbs. (15–20 Nm).

41. Connect the exhaust pipe to the manifold and install the bolts. Tighten the bolts to 29–36 ft. lbs. (40–50 Nm).

42. Install the left fender apron extension.

43. Refill the transaxle with the recommended lubricant.

44. Lower the vehicle.

45. Remove the engine support fixture.

46. Install the starter.

47. Install the transaxle retaining bolts. Tighten the retaining bolts to 29–43 ft. lbs. (40–60 Nm).

48. Install the remaining components.

49. Install the negative battery cable and the ground strap to the transaxle.

Automatic

1. Disconnect the negative battery cable from the battery and transaxle.

2. Disconnect the speedometer cable.

3. Disconnect the electrical connector for the solenoids, vehicle speed sensor, shift lever switch, forward clutch cylinder revolution sensor and vehicle speed sensor (for A/T).

4. Remove the wiring harness from the clamps on the transaxle.

5. Disconnect the select cable from the transaxle.

6. Drain the cooling system.

7. Remove the cooling system pipe from the transaxle.

8. Remove the top transaxle-to-engine bolts.

9. Remove the starter.

10. Remove the exhaust manifold cover. Disconnect the front exhaust pipe from the exhaust manifold.

11. Support the engine.

12. Lift the vehicle and safely support.

13. Drain the transaxle fluid from the transaxle.

14. Remove the engine undercovers, if equipped.

15. Place an oil pan under the transaxle and disconnect the cooler hoses.

16. On Esteem models, with the engine supported, remove the vehicle center mounting member by removing the seven nuts and four bolts.

17. Remove the front exhaust pipe from the vehicle.

18. Remove the transaxle stiffener plate by removing the bolts.

19. Remove the transaxle housing lower plate.

20. Remove the torque converter bolts. To lock the drive plate, engage a flat bladed tool in the flywheel.

21. Disconnect the sway bar from the control arms.

22. Remove the left and right front wheels.

23. Disconnect the left and right ball joints from the steering knuckles.

24. Remove the left and right halfshafts.

25. Remove the engine rear mount and bracket.

26. After removing the rear mount, remove the engine-to-transaxle bolt and nut located behind the rear bracket. Remove all bolts holding the engine to the transaxle.

27. Support the transaxle with a transaxle jack.

28. Remove the bolts from the engine left-hand mount.

29. Remove the transaxle with the torque converter from the engine compartment.

➡When removing the transaxle from the engine, move it parallel with the crankshaft and use care so not to apply excessive force to the drive plate and torque converter.

❋❋ CAUTION

Be sure to keep the transaxle with the torque converter horizontal or facing up throughout the work. Should it be tilted with converter down, the converter may fall off and cause personal injury.

To install:

30. Install the transaxle to the engine assembly and install the attaching nuts and bolts.

31. Install the left-hand mounting bolts. Tighten the bolts to 40 ft. lbs. (55 Nm).

32. Install the engine-to-transaxle bolt and nut before installing the rear transaxle mount. Tighten the nut to 65 ft. lbs. (90 Nm).

33. Install all bolts for the transaxle. Tighten the bolts to 65 ft. lbs. (90 Nm).

34. Install the left halfshafts.

35. Install the ball joints.

36. Connect the sway bar to the control arms.

37. Install the torque converter bolts and tighten the bolts to 14 ft. lbs. (19 Nm).

38. Install the transaxle housing lower plate.

39. Install the stiffener plate with the four bolts. Tighten the bolts to 40 ft. lbs. (55 Nm).

40. Install the exhaust pipe to the center pipe and tighten the bolts to 37 ft. lbs. (50 Nm).

41. On Esteem models, install the center member to the vehicle and install the seven nuts and four bolts. Tighten the bolts and nuts as follows:

a. Center member to the radiator support to 33 ft. lbs. (45 Nm).

b. Center member to crossmember to 33 ft. lbs. (45 Nm).

c. Engine mounts to center member nuts to 40 ft. lbs. (55 Nm).

42. Connect the oil hoses for the transaxle.

43. Install the engine undercovers.

44. Lower the vehicle.

45. Remove the engine support.

46. Connect the exhaust pipe to the exhaust manifold and install the nuts.

47. Install the exhaust manifold cover.

48. Install the starter motor.

49. Install the remaining components.

50. Fill the cooling system and the transaxle. Check all fluids.

51. Connect the negative battery cable to the transaxle and battery.

Clutch

ADJUSTMENT

1. Depress the clutch pedal lightly until tension on the clutch cable can be felt.

2. Measure the clutch pedal free-play; it should be 0.6–0.8 in. (15–20mm).

3. Adjust the clutch pedal free-play by tightening or loosening the clutch cable adjustment nut.

4. Measure the clutch lever free-play; it should be 0.0–0.08 in. (0–2mm). If the clutch release lever free-play exceeds specification, inspect the release shaft return spring for cracks or weakness.

➡Be sure the marks on the clutch release lever and release shaft are aligned. If they are not, remove the lever from the shaft, align the marks and repeat the free-play adjustment procedure.

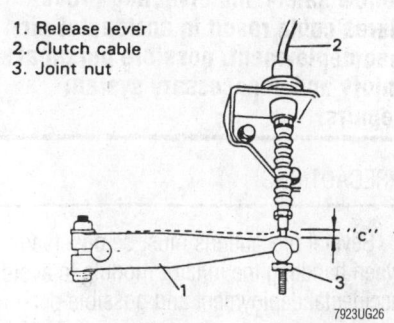

1. Release lever
2. Clutch cable
3. Joint nut

7923UG26

Clutch cable free-play adjustment

REMOVAL & INSTALLATION

1. Remove the transaxle.

2. Hold the flywheel stationary with tool 09924–17810 or equivalent.

3. Matchmark the pressure plate and flywheel for installation reference.

4. Loosen the pressure plate attaching bolts, one turn at a time (evenly) until the spring pressure is released.

5. Remove the clutch disc and pressure plate.

To install:

6. Clean the flywheel mating surfaces of all oil, grease and metal deposits. Inspect flywheel for cracks, heat checking or other defects and replace or resurface as necessary.

7. Check the wear on the facings of the clutch disc by measuring the depth of each rivet head depression. Replace clutch disc when rivet heads are 0.02 in. (0.5mm) below the surface of clutch surface.

8. Check the diaphragm spring and pressure plate for wear or damage. If the spring or plate is excessively worn, replace the pressure plate assembly.

9. Check the pilot bearing for smooth operation. If the bearing does not spin freely, replace it.

10. Position the clutch disc and pressure plate with the matchmarks aligned and install a clutch alignment tool 09923–36330 or equivalent.

11. Install the pressure plate bolts. Tighten the mounting bolts evenly and in a crisscross pattern to 13–20 ft. lbs. (18–28 Nm). Remove the alignment tool and the flywheel holding tool.

12. Lightly lubricate the transaxle input shaft splines, pilot bearing surface of the input shaft, and the release bearing with grease.

13. Install the transaxle.

14. Adjust the clutch cable.

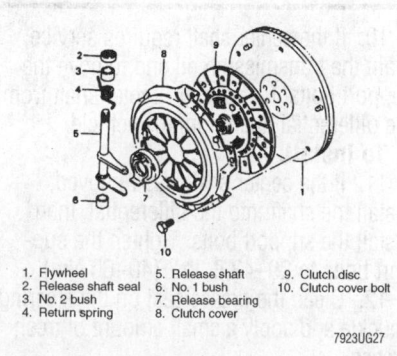

1. Flywheel
2. Release shaft seal
3. No. 2 bush
4. Return spring
5. Release shaft
6. No. 1 bush
7. Release bearing
8. Clutch cover
9. Clutch disc
10. Clutch cover bolt

7923UG27

Clutch component identification

Halfshaft

REMOVAL & INSTALLATION

1. Disconnect the negative battery cable.
2. Undo the caulking on the halfshaft nut, then remove the nut and washer.
3. Safely raise and support the vehicle.
4. Drain the oil from the transmission.
5. Use two large prybars, release the snapring fitting on the halfshaft inner joint from the differential.
6. Remove the sway bar attaching nut, washer and bushing from the suspension arm.
7. Remove the ball joint stud bolt and nut.
8. Pull the inboard joint from the differential, then disconnect the outer joint from the steering knuckle.
9. Remove the halfshaft from the vehicle.

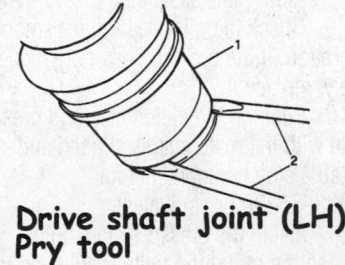

1 Drive shaft joint (LH)
2 Pry tool

7923UG28

Disconnecting the left inboard joint

✳✳ WARNING

To prevent breakage of the boots, be careful not to bring the boots in contact with other components when removing the shaft assembly.

10. If the center shaft requires service, drain the transmission oil and remove the support bolts. Remove the center shaft from the differential, then from the vehicle.

To install:
11. If the center shaft was removed, install the shaft into the differential, then install the support bolts. Tighten the support bolts to 29–43 ft. lbs. (40–60 Nm).
12. Clean the grease seal on the steering knuckle and apply a small amount of fresh grease.
13. Install the wheel side joint to the steering knuckle, then the differential side joint to the differential. Seat the differential joint by hand, making sure that the snapring is seated. Install the halfshaft nut to the outer joint loosely, to hold it in position.

✳✳ WARNING

Do not hit the joints with a hammer to seat them use your hands only or component damage may occur.

14. Connect the suspension arm to the steering knuckle and install the ball joint stud.
15. Position the sway bar on the suspension arm and install the sway bar bushing, washer and nut. Tighten the nut to 17–20 ft. lbs. (23–28 Nm).
16. If equipped with a manual transaxle fill the transaxle with the specified gear oil.
17. Lower the vehicle.
18. Tighten the halfshaft nut to 109–145 ft. lbs. (150–200 Nm).
19. Connect the negative battery cable.
20. If equipped with a automatic transaxle, fill the transaxle with the specified transmission oil.

STEERING AND SUSPENSION

Air Bag

✳✳ CAUTION

Vehicles are equipped with an air bag system, also known as the Supplemental Inflatable Restraint (SIR) or Supplemental Restraint System (SRS). The system must be disabled before performing service on or around system components, steering column, instrument panel components, wiring and sensors. Failure to follow safety and disabling procedures could result in accidental air bag deployment, possible personal injury and unnecessary system repairs.

PRECAUTIONS

Several precautions must be observed when handling the inflator module to avoid accidental deployment and possible personal injury.

• Never carry the inflator module by the wires or connector on the underside of the module.
• When carrying a live inflator module, hold securely with both hands, and ensure that the bag and trim cover are pointed away.
• Place the inflator module on a bench or other surface with the bag and trim cover facing up.
• With the inflator module on the bench, never place anything on or close to the module which may be thrown in the event of an accidental deployment.

DISARMING

✳✳ WARNING

When performing service on or around the air bag system components or wiring, disable the air bag system. Failure to follow the procedures could result in possible deployment, personal injury or unneeded system repairs.

1. Disconnect the negative battery cable.
2. Turn the steering wheel so the wheels are pointing straight ahead.
3. Turn the ignition switch to the **LOCK** position and remove the key.
4. Remove the **AIR BAG-IG** fuse from the air bag fuse box located near the junction/fuse box.

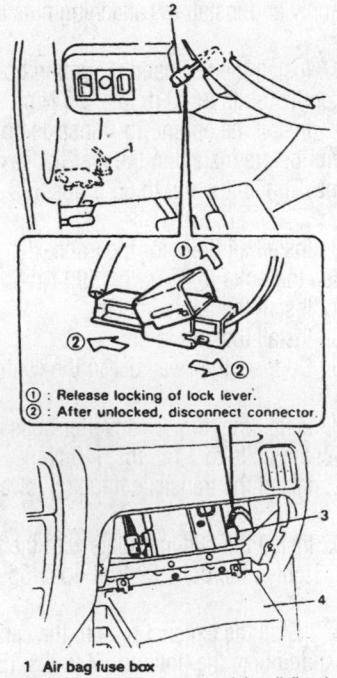

① : Release locking of lock lever.
② : After unlocked, disconnect connector.

1 Air bag fuse box
2 Yellow connector of driver air bag (inflator) module
3 Yellow connectors of passenger air bag (inflator) module
4 Glove box

7923UG29

Air bag connector locations

5. Remove the left side steering wheel side cap and disconnect the yellow connector for the driver side air bag (inflator) module.

6. Pull out the glove box while pushing in on the stoppers from the left and the right sides. Disconnect the yellow connector for the passenger air bag (inflator) module.

REARMING

✷✷ WARNING

When performing service on or around the air bag system components or wiring, disable the air bag system. Failure to follow the procedures could result in possible deployment, personal injury or unneeded system repairs.

1. Connect the negative battery cable.
2. Turn the ignition switch to the **LOCK** position and remove the key.
3. Connect the yellow connector for the passenger side air bag (inflator) module and the yellow connector for the driver air bag (inflator) module. Be sure to lock each connector with the lock lever.
4. Install the glove box assembly.
5. Install the left side steering wheel side cover.
6. Install the **AIR BAG-IG** fuse to the air bag fuse box.
7. Turn the ignition **ON** and verify that the **AIR BAG** warning lamp flashes seven times, then turns off. If the system does not operate as described, diagnosis and repairs to the air bag system are necessary.

Rack and Pinion Steering Gear

REMOVAL & INSTALLATION

Manual

✷✷ WARNING

Be sure to set the front wheels straight ahead and remove the ignition key from the cylinder before starting repairs. If equipped with an air bag the contact coil of the air bag system may get damaged if the key is not removed and the wheels are not straight ahead.

1. Disconnect the negative battery cable.
2. Slide the driver's seat back as far as possible.
3. Pull back the front part of the floor mat on the driver's side and remove the

steering shaft joint cover.
4. Loosen the steering shaft upper joint bolt, but do not remove.
5. Remove the steering shaft lower joint bolt and disconnect the lower joint from the pinion.
6. Raise and support the vehicle safely.
7. Remove the front wheels.
8. Disconnect the tie rod ends from the steering knuckles.
9. Remove the steering gear mounting bolts and the brackets, then remove steering gear case from the vehicle.
To install:
10. Install the steering gear, brackets and mounting bolts. Tighten bolts to 14–21 ft. lbs. (20–30 Nm).
11. Connect the tie rod ends to the steering knuckles.
12. Be sure the steering wheel is straight and the front wheels are pointing straight ahead.
13. Connect the steering shaft to the steering gear. Install the lower steering shaft-to-steering gear clinch bolt and tighten both steering joint bolts (upper and lower) to 14–21 ft. lbs. (20–30 Nm).
14. Install the remaining components.
15. Connect the negative battery cable.
16. Check and adjust the front wheel alignment.

Power

✷✷ WARNING

Be sure to set the front wheels straight ahead and remove the ignition key from the cylinder before starting repairs. If equipped with an air bag the contact coil of the air bag system may get damaged if the key is not removed and the wheels are not straight ahead.

1. Disconnect the negative battery cable.
2. Remove the steering column joint covers.
3. Loosen the steering shaft upper joint bolt, but do not remove.
4. Remove the steering shaft lower joint bolt and disconnect the lower joint from the pinion.
5. Raise and support the vehicle safely.
6. Remove the front wheels.
7. Disconnect the tie rod ends from the steering knuckles.
8. Remove the front exhaust pipe.
9. If equipped with a manual transaxle, disconnect the gear shift control shaft and extension rod from the transaxle.
10. Remove the rear engine mount

together with the bracket from the engine and suspension member.
11. Remove the mounting member from the suspension frame by removing the two bolts.
12. Disconnect the high and low pressure lines from the rack and pinion.

➡**When the lines are disconnected plug the lines or place a oil pan under the vehicle.**

13. Remove the cylinder lines from the rack and pinion.
14. Remove the rack and pinion mounting bolts and brackets.
15. Remove rack and pinion case from the vehicle.
To install:
16. Install the rack and pinion, brackets and mounting bolts. Tighten the bolts to 40 ft. lbs. (55 Nm).
17. Install the cylinder lines to the rack and pinion and tighten their fittings to 14–21 ft. lbs. (20–30 Nm).
18. Connect the high and low pressure lines to the rack and pinion. Tighten the fittings to 22–28 ft. lbs. (30–40 Nm).
19. If equipped with a manual transaxle, connect the gear shift control shaft and extension rod to the transaxle. Tighten the extension rod nut to 18–28 ft. lbs. (25–40 Nm) and tighten the control shaft nut and bolt to 11–14 ft. lbs. (15–20 Nm).
20. Connect the tie rod ends to the steering knuckles.
21. Be sure the steering wheel is straight and the front wheels are pointing straight ahead.
22. Connect the steering shaft to the rack and pinion. Install the lower steering shaft-to-rack and pinion clinch bolt and tighten both steering joint bolts (upper and lower) to 14–21 ft. lbs. (20–30 Nm).
23. Install the front wheels.
24. Connect the negative battery cable.
25. Bleed the power steering system.
26. Lower the vehicle.
27. Check and adjust the front wheel alignment.

Strut

REMOVAL & INSTALLATION

Front

1. Raise and support the vehicle safely.
2. Remove the wheels.
3. If equipped with ABS, disconnect the ABS wheel speed sensor.

4. Remove the brake hose clip, then the hose from the strut.

5. Support the lower control arm with a floor jack.

6. Remove the strut bracket bolts.

7. Remove the upper strut support nuts from the engine compartment, hold the strut by hand so it will not fall.

> ❋❋ **WARNING**
>
> **Do not loosen the center nut at this time or serious injury or vehicle damage may result.**

8. Remove the strut assembly from the vehicle.

9. Install a pair of coil spring compressors to the coil spring on the strut assembly. Turn the spring compressors alternately until the spring tension is released from the strut assembly. If the spring can be turned slightly, then it has been collapsed enough.

> ❋❋ **CAUTION**
>
> **This procedure requires the use of a spring compressor; it cannot be performed without one. If you do not access to this special tool, do not attempt to disassemble the strut. The coil spring is retained under considerable pressure. It can exert enough force to cause serious injury. Exercise extreme caution.**

10. Keeping the spring collapsed remove the strut center nut and remove the other components from the top of the strut assembly.

11. Remove the spring from the strut.

To install:

12. If installing a new spring, compress the spring with a pair of spring compressors. Be sure that the spring compresses to 9 in. (230mm) for installation.

13. Position the coil spring on the strut making sure that the end of the spring is mated to the stepped part of the lower seat.

14. Install the bump stop to the strut rod.

15. Install the strut cover, spring seat, upper spring seat, and the bearing spacer. Align the strut bracket to the mark on the upper spring seat.

16. Clean the bearing lower washer and install it to the strut rod.

17. Clean the strut bearing and apply fresh grease to the bearing. Install the bearing to the lower washer.

18. Clean the bearing upper washer and install it.

19. Install these components in the following order: the bearing upper seal, bearing seat, strut support, inner spacer, washer and strut nut. Tighten the nut to 29–43 ft. lbs. (40–60 Nm). Apply a water proof coating (paint or lacquer) to the nut and strut rod threads.

20. Loosen the spring compressors alternately, checking that the stepped part of the spring seat and spring are properly positioned.

21. Install the strut assembly onto the vehicle. Install the upper support nuts loosely. Tighten the upper strut support nuts to 16–23 ft. lbs. (22–33 Nm).

22. Install the strut bracket nuts and bolts and tighten to 51–65 ft. lbs. (70–90 Nm).

23. If equipped with ABS, connect the ABS wheel speed sensor.

24. Install the brake hose clip.

25. Install the wheels.

26. Lower the vehicle.

27. Check the front end alignment.

Rear

ESTEEM

1. Raise and safely support the vehicle. Remove the wheel and tire assembly.

2. Place a jack under the lower control arm to support the suspension.

3. Disconnect the brake line from the brake hose at the strut.

4. Remove the E ring securing the brake hose to the strut.

5. Remove the strut upper support nuts and push the strut down.

6. Remove the two bolts and nuts holding the strut to the rear knuckle.

7. Remove the strut from the knuckle, then remove the strut from the vehicle.

8. Install a pair of coil spring compressors to the coil spring on the strut assembly. Turn the spring compressors alternately until the spring tension is released from the strut assembly. If the spring can be turned slightly, then it has been collapsed enough.

> ❋❋ **CAUTION**
>
> **This procedure requires the use of a spring compressor; it cannot be performed without one. If you do not access to this special tool, do not attempt to disassemble the strut. The coil spring is retained under considerable pressure. Id can exert enough force to cause serious injury. Exercise extreme caution.**

9. Keeping the spring collapsed remove the strut center nut and remove the other components from the top of the strut assembly.

10. Remove the spring from the strut.

To install:

11. If installing a new spring, compress the spring with a pair of spring compressors, be sure that the spring compresses to 11 in. (290mm) for installation.

12. Position the coil spring on the strut making sure that the end of the spring is mated to the stepped part of the lower seat.

13. Install the remaining components.

14. Install the strut support with the center nut. Tighten the nut to 40 ft. lbs. (55 Nm).

15. Loosen the spring compressors alternately, checking that the stepped part of

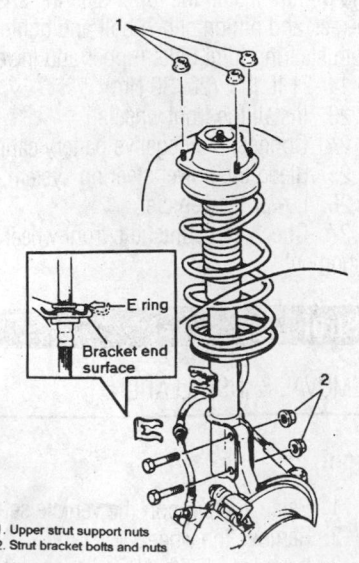

1. Upper strut support nuts
2. Strut bracket bolts and nuts

7923UG30

Strut assembly mounting

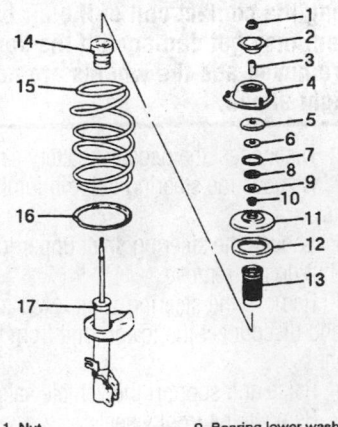

1. Nut
2. Stopper
3. Inner spacer
4. Support comp.
5. Bearing seat
6. Bearing upper washer
7. Bearing seal
8. Bearing
9. Bearing lower washer
10. Bearing spacer
11. Coil spring upper seat
12. Coil spring seat
13. Strut cover
14. Bump stopper
15. Coil spring
16. Coil spring lower seat
17. Strut

7923UG31

Exploded view of the strut assembly

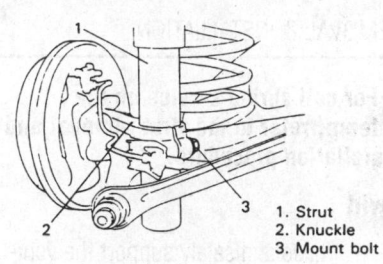

WARNING:
Strut with "GAS FILLED" stamp is filled with gas and oil. When handling it, make sure to observe the precautions (WARNING) of p. 3E-6.

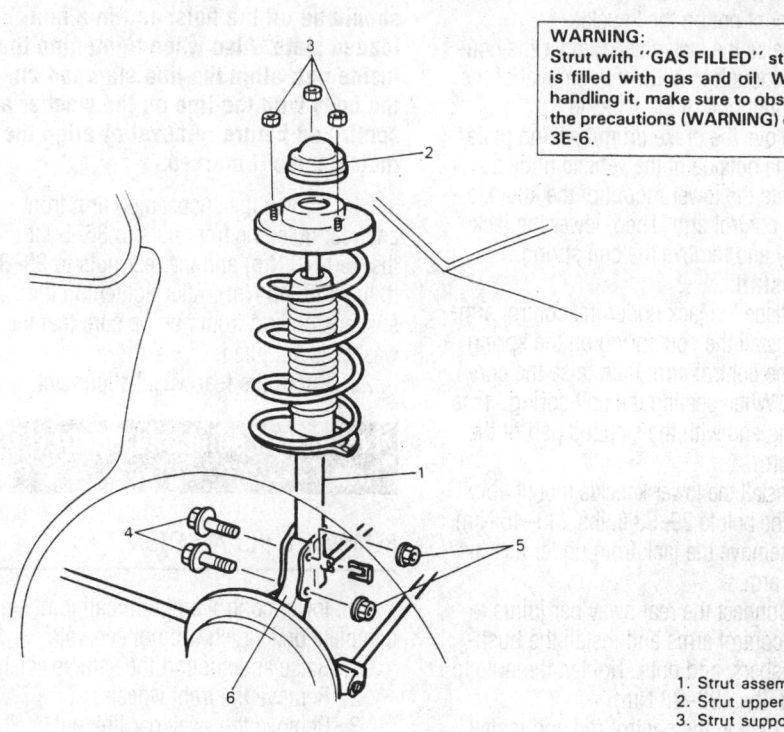

1. Strut assembly
2. Strut upper cap
3. Strut support nut
4. Strut bracket bolt
5. Strut bracket nut
6. Rear knuckle

7923UG32

Rear strut assembly mounting—Esteem

the spring seat and spring are properly positioned.

16. Install the strut in the vehicle.

17. Install the two bolts and nuts to hold the strut to the rear knuckle. Tighten the nuts 65 ft. lbs. (90 Nm).

18. Fully extend the strut and position the upper part of the strut into the vehicles body. If the upper part of the strut does not reach the vehicle body, raise the jack under the control arm a little.

19. Install the upper support nuts and tighten them to 20–27 ft. lbs. (28–38 Nm).

20. Connect the brake hose to the strut and install the E clip.

21. Connect the brake hose to the brake line.

22. Remove the jack from under the control arm.

23. Install the wheels.

24. Fill the master cylinder with brake fluid and bleed the brake system.

25. Lower the vehicle.

SWIFT

1. Raise and safely support the vehicle. Remove the wheels.

2. Place a jack under the lower control arm to support the suspension.

❈❈ CAUTION

The coil spring is under extreme pressure. Be sure the lower control arm is firmly supported with a hydraulic jack before continuing with procedure. If this caution is not observed, serious bodily injury may result.

3. Remove the strut support nuts and push the strut down.

4. Remove the strut lower mounting bolt.

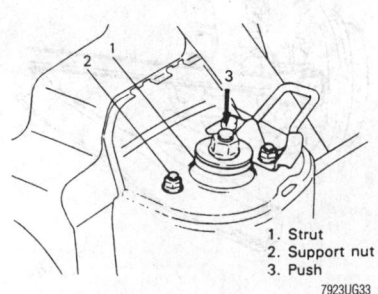

1. Strut
2. Support nut
3. Push

7923UG33

Rear strut upper mounting nuts—Swift

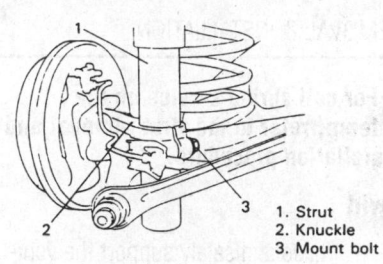

1. Strut
2. Knuckle
3. Mount bolt

7923UG34

Rear strut lower mounting—Swift

5. Remove the strut from the knuckle. Compress the strut as short as possible for removal. If the strut is hard to remove, open the slit of the knuckle by inserting a wedge.

➡**Do not open the knuckle slit wider than necessary. Do not lower the jack more than necessary during the strut removal to prevent the coil spring from coming off, or the brake flexible hose from stretching.**

To install:

6. Install the strut in the vehicle. Position the bottom of the alignment projection inside the knuckle opening.

7. Install strut lower mounting bolt and tighten it to 36–50 ft. lbs. (50–70 Nm).

8. Fully extend the strut and position the upper part of the strut into the vehicles body. If the upper part of the strut does not reach the vehicle body raise the jack under the control arm a little.

9. Install the upper support nuts and tighten them to 20–27 ft. lbs. (28–38 Nm).

10. Remove the jack from under the control arm.

11. Install the wheels.

12. Lower the vehicle.

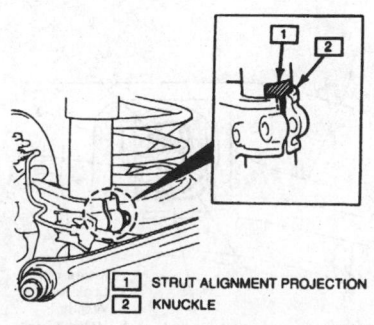

1. STRUT ALIGNMENT PROJECTION
2. KNUCKLE

7923UG35

Shock-to-steering knuckle alignment—Swift

Coil Spring

REMOVAL & INSTALLATION

➡ **For coil spring service on the Esteem, refer to the strut removal and installation procedure.**

Swift

1. Raise and safely support the vehicle.

2. Remove the rear wheels.

➡ **To facilitate the toe-in adjustment after reinstallation, confirm which one of the lines stamped on the washer is in the closest alignment with the stamped line on the control rod. If not marked, add matchmarks.**

3. Remove the control rod inside bolt (body center side).

4. Remove the outside (wheel side) of the control rod from the rear knuckle stud and disconnect the control rod from the knuckle.

5. Remove the nuts, washers, and bushings connecting the rear sway bar to the rear lower control arms.

6. Loosen the rear mount nut on the control arm, but do not remove the bolt.

7. Loosen the front nut of the control arm.

8. If equipped with ABS, disconnect the wheel speed sensor.

✹✹ CAUTION

The coil spring is under extreme pressure. Be sure control arm is firmly supported with a hydraulic jack before continuing with procedure. If this precaution is not observed, serious bodily injury may result.

9. Loosen the lower mount nut on the knuckle. Place a jack under the control arm

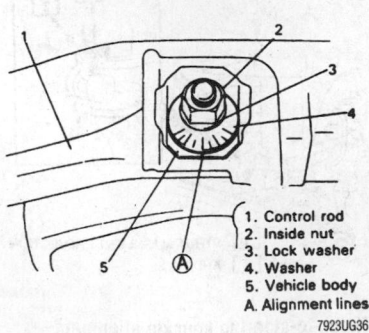

1. Control rod
2. Inside nut
3. Lock washer
4. Washer
5. Vehicle body
A. Alignment lines

7923UG36

Control rod inside mount—Swift

to prevent it from lowering and remove the lower mount nut on the knuckle.

10. Raise the jack placed under the control arm enough to allow the removal of the lower mount bolt of the knuckle.

11. Move the brake drum/backing plate toward the outside of the vehicle body so as to separate the lower mount of the knuckle from the control arm. Then, lower the jack gradually and remove the coil spring.

To install:

12. Place the jack under the control arm.

13. Install the coil spring on the spring seat of the control arm, then raise the control arm. When seating the coil spring, mate the spring end with the stepped part of control arm.

14. Install the lower knuckle mount bolt. Tighten the bolt to 29–33 ft. lbs. (40–45 Nm).

15. Remove the jack from under the suspension arm.

16. Connect the rear sway bar joints to the rear control arms and install the bushings, washers, and nuts. Tighten the nuts to 16–20 ft. lbs. (22–28 Nm).

17. Position the control rod and install the inside control rod bolt and the outside control rod nut, but do not tighten them at this time.

18. If equipped with ABS, install the wheel speed sensor.

19. Install the wheels and lower the vehicle.

20. Tighten the control rod inside and outside nuts to 51–65 ft. lbs. (70–90 Nm).

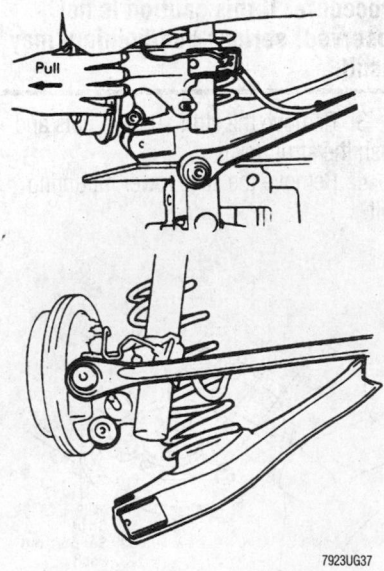

7923UG37

Disconnecting the knuckle and removing the coil spring

➡**When tightening the nuts, the vehicle should be off the hoist and in a non-loaded state. Also when tightening the inside nut, align the line stamped on the body with the line on the washer as confirmed before removal or align the matchmarks if marked.**

21. Tighten the suspension arm front and rear nuts: the front nuts to 36–50 ft. lbs. (50–70 Nm) and the rear nuts to 29–33 ft. lbs. (40–45 Nm). After tightening the suspension arm front nut, be sure that the washer is not tilted.

22. Check the rear wheel alignment.

Lower Ball Joint and Control Arm

REMOVAL & INSTALLATION

The lower control arm and ball joint are a complete unit which will not separate.

1. Raise and support the vehicle safely.

2. Remove the front wheels.

3. Remove the sway bar link nut, washer, and cushion.

4. Remove the ball joint stud bolt and nut.

5. Remove the lower control arm front bushing bolt.

6. Remove the two bolts holding the lower control arm bracket to the vehicle.

7. Separate the lower control arm from the steering knuckle and remove the lower control arm from the vehicle.

To install:

8. Install the ball joint to the knuckle and secure it with the nut and bolt.

9. Install the lower control arm rear bracket and bolts.

10. Install the lower control arm front mounting bolt.

11. Connect the sway bar link to the control arm and install the cushion, washer and nut.

12. Tighten control arm rear mounting bracket bolts to 27 ft. lbs. (37 Nm); front mounting bolt to 65 ft. lbs. (90 Nm); ball joint nut and bolt to 44 ft. lbs. (60 Nm) and the sway bar link nut to 18 ft. lbs. (26 Nm).

13. Install wheels.

14. Lower the vehicle.

15. Check the front wheel alignment.

Wheel Bearings

ADJUSTMENT

The front and rear wheel bearings are a cartridge type design and cannot be adjusted.

REMOVAL & INSTALLATION

Front

➡️**Always replace bearing races as a complete set.**

1. Raise and support the vehicle.
2. Remove the front wheel.
3. Remove the brake caliper, carrier and disc from the steering knuckle.
4. If equipped with ABS, remove the speed sensor.
5. Remove the tie rod from the steering knuckle.
6. Remove the hub from the steering knuckle, then remove the steering knuckle.
7. Remove the wheel bearing outside race from the hub using a suitable bearing puller.
8. Remove the outside oil seal, snapring, outside bearing, inside oil seal and the inside bearing in that order.

➡️**Once the bearing outer race is removed, the bearing set (outer race, bearings and inner races) should be replaced.**

9. Remove the bearing outer race from the knuckle by pressing the race out of the knuckle with the aid of tool 09913–75520.

To install

➡️**When installing the oil seals, be careful not to deform or tilt. Damage to the rubber part of the seal may occur.**

10. Press the new bearing outer race into the knuckle using a press and the following tools: bearing installer handle 09924–74510, bearing and oil seal installer 09944–68210 and a bearing installer support 0994–78210
11. Apply lithium grease to the bearing races, bearings and oil seal lips.
12. Install the outside bearing to the steering knuckle. Install the snapring to

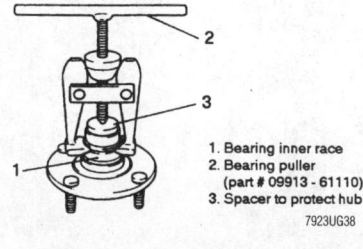

1. Bearing inner race
2. Bearing puller (part # 09913 - 61110)
3. Spacer to protect hub

7923UG38

Use a two or three jaw puller to remove the bearing race from the hub

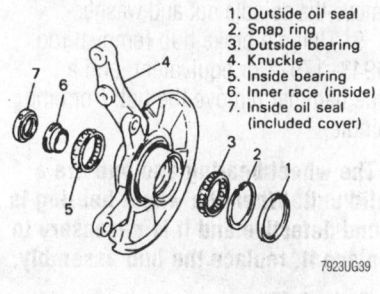

1. Outside oil seal
2. Snap ring
3. Outside bearing
4. Knuckle
5. Inside bearing
6. Inner race (inside)
7. Inside oil seal (included cover)

7923UG39

Steering knuckle bearing components—Swift

hold the outside bearing in place, then install the outside oil seal.

13. Install the inside bearing to the steering knuckle. Install the inside race and oil seal to the steering knuckle. When installing the inside oil seal, drive the oil seal in until it contacts the steering knuckle.

❋❋ CAUTION

If equipped with ABS use caution when installing the oil seal, because the seal has a hole that must align with the speed sensor position.

14. Install the outside race to the wheel hub using a bearing installer.
15. Using a press and the proper tools, press the hub into the steering knuckle. After installation, be sure the hub is installed straight and turns freely.
16. Install the steering knuckle in the vehicle.
17. Connect the tie rod end.
18. Install the brake caliper, carrier and disc to the steering knuckle.
19. Install the front wheel.
20. Lower the vehicle.

Rear

WITH WHEEL HUBS

1. Set the parking brake.
2. Raise and safely support the vehicle.
3. Remove the rear wheels.
4. Remove the rear brake drums.
5. Use a brass drift and knock the wheel bearings from the drum assembly.

To install:

6. Position the inner wheel bearing on the drum with the sealed side facing out. Using a rear wheel bearing installer 09913–76010 or equivalent, install the rear wheel bearing. Install the wheel bearing spacer into the drum.
7. Install the outer wheel bearing with

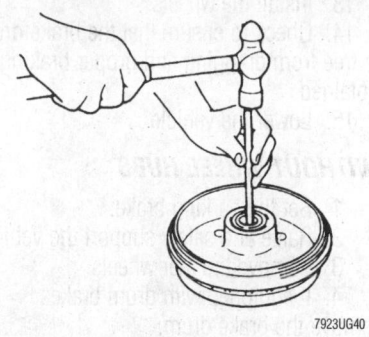

7923UG40

Use a hammer and drift to remove the wheel bearings from the hub—Swift

the sealed side facing out, using a wheel bearing installer.

8. Fill the space in the brake drum in between the wheel bearings to about 40% capacity with wheel bearing grease.
9. Install the brake drum.
10. Install the spindle washer and a new spindle nut. Tighten the spindle nut to 58–86 ft. lbs. (80–120 Nm).
11. Caulk the spindle nut and the spindle dust cap.

➡️**When installing the spindle cap, hammer lightly several times on the collar of the cap until the collar comes closely into contact with the brake drum. If the fitting part of the cap is deformed or damaged or if it fits loose, replace the cap with a new one.**

12. Depress the brake pedal with about 66 lbs. (30 kg) of force three to five times to obtain proper drum to shoe clearance.

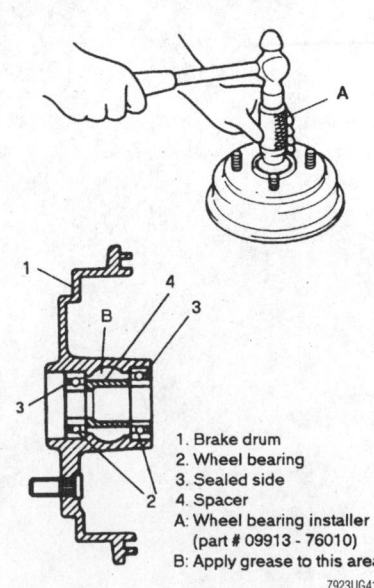

1. Brake drum
2. Wheel bearing
3. Sealed side
4. Spacer
A: Wheel bearing installer (part # 09913 - 76010)
B: Apply grease to this area

7923UG41

Wheel bearing installation—Swift

13. Install the wheels.

14. Check to ensure that the brake drum is free from dragging and proper braking is obtained.

15. Lower the vehicle.

WITHOUT WHEEL HUBS

1. Set the parking brake.

2. Raise and safely support the vehicle.

3. Remove the rear wheels.

4. If equipped with drum brakes, remove the brake drum.

5. If equipped with rear disc brakes, remove the caliper, carrier and disc.

6. Release the parking brake.

7. Remove the spindle cap without deforming it.

8. Uncalk the spindle nut, then remove the spindle nut and washer.

9. Using a brake hub removal tool 09943–17911 or equivalent, and a slide hammer remove the hub from the spindle.

➡ **The wheel bearing and hub are a solid unit. When the wheel bearing is found defective and it is necessary to replace it, replace the hub assembly.**

To install:

10. Install the wheel hub, washer and a new spindle nut. Tighten the spindle nut to 108–144 ft. lbs. (150–200 Nm).

11. Caulk the spindle nut and install the spindle cap.

12. If equipped with rear drum brakes, install the brake drums.

13. If equipped with rear disc brakes, install the brake caliper carrier and disc.

14. Depress the brake pedal with about 66 lbs. (30 kg) of force three to five times to obtain proper drum/rotor to shoe/pad clearance.

15. Install the wheel and tighten the lug nuts to 36–58 ft. lbs. (50–80 Nm).

16. Check to ensure that the brakes are free from dragging and that proper braking is obtained.

17. Lower the vehicle.

TOYOTA

Avalon • Camry • Celica • Corolla • Paseo • Supra • Tercel

29

ENGINE REPAIR

➡️**Disconnecting the negative battery cable on some vehicles may interfere with the functions of the on board computer systems and may require the computer to undergo a relearning process, once the negative battery cable is reconnected.**

Distributor

REMOVAL

1. Disconnect the negative battery cable. On vehicles equipped with an air bag, wait at least 90 seconds before proceeding.
2. If equipped, disconnect the electrical connector from the air flow meter. Disconnect the air cleaner hose from the throttle body and remove the air cleaner cover, air flow meter and air duct as one unit.
3. If equipped, remove the air intake to provide clearance.
4. If equipped, remove the intercooler to provide clearance. Disconnect the air temperature sensor, cruise control actuator cable, and the air cleaner hose, if more clearance is necessary.
5. Disconnect the spark plug wires from the distributor cap. Be careful to pull the spark plug wires from the boot, not the wire. Using a suitable tool, lift up the lock claw and disconnect the holder from the distributor cap.
6. Disconnect the distributor connector.
7. Remove the distributor mounting bolt and pull the distributor out.

➡️**The marks on the distributor drive gear and distributor housing should be aligned. If the marks are not aligned, mark the distributor housing and rotor position.**

8. Remove the O-ring from the distributor housing.

INSTALLATION

Engine Disturbed

1. On the 2JZ-GE engine, remove the No. 3 timing belt cover.
2. Turn the crankshaft clockwise, and position the slit of the intake camshaft as required to bring No. 1 cylinder to TDC on the compression stroke.
3. Apply a light coat of engine oil to a new O-ring and install it to the distributor housing.

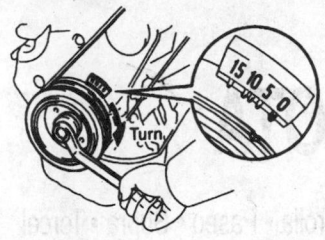

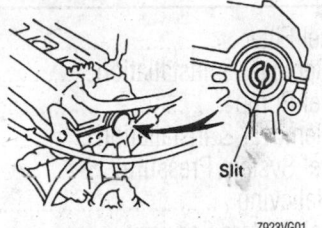

Common method of properly positioning the distributor drive mechanism to TDC

4. Align the cut out portion of the coupling with the groove of the housing. Insert the distributor, aligning the bolt hole of the flange with that of the bolt hole on the cylinder head. Tighten the mounting bolts.
5. If removed, reinstall the intercooler or the air intake and associated parts.
6. Reinstall the spark plug wires to the distributor cap, making sure the firing order is correct.
7. On the 2JZ-GE engine, reinstall the No. 3 timing belt cover.
8. Install the air flow meter, air hose and the air cleaner cover if removed.
9. Connect the distributor and air flow meter connectors.
10. Install the air flow meter, air hose and the air cleaner cover.
11. Connect the negative battery cable.
12. Start the engine and check the ignition timing.

Engine Not Disturbed

1. Apply a light coat of engine oil to a new O-ring and install it to the distributor housing.
2. Align the cut out portion of the coupling with the groove of the housing. Insert the distributor, aligning the bolt hole of the flange with that of the bolt hole on the cylinder head. Tighten the mounting bolts.
3. Be sure the mark on the engine aligns with the mark on the distributor made during removal. Also be sure the rotor is in the same position as removal.
4. If removed, reinstall the intercooler or the air intake and associated parts.
5. Install the distributor cap.
6. Reinstall the spark plug wires to the distributor cap, making sure the firing order is correct.

7. Install the air flow meter, air hose and the air cleaner cover if removed.
8. Connect the distributor and air flow meter connectors.
9. Connect the negative battery cable.
10. Start the engine and check the ignition timing.

Ignition Timing

ADJUSTMENT

➡️**The timing on engines equipped with Distributorless Ignition Systems (DIS) is not adjustable.**

1. Start the engine and run it at idle until the engine reaches normal operating temperature. Remove the cap from the check connector (DLC1) which is usually located in the engine compartment.
2. Connect the tachometer and timing light to the engine. Connect the tachometer tester probe to terminal IG of the DLC1.

➡️**Never allow the tachometer test probe to touch ground as it could result in damage to the igniter and or ignition coil.**

➡️**Not all tachometers are compatible with this system. Be sure to confirm compatibility of your unit before use.**

3. Using a jumper connector, connect terminals TE1 and E1 of the DLC1.
4. Check the idle speed.

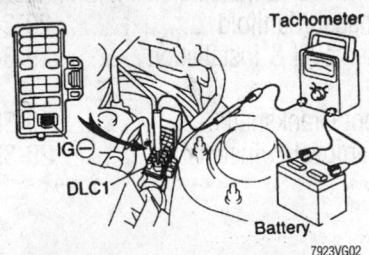

Attach the tachometer test probe to the IG terminal of the data link connector

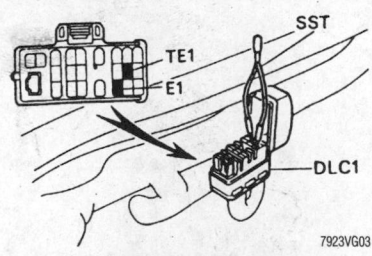

A jumper wire is used to connect terminals TE1 and E1 of the DLC

5. Aim the timing light at the timing indicator and check the ignition timing. Timing should be 10° BTDC at idle.

6. If adjustment is necessary, loosen the distributor hold-down bolt and adjust by turning. Tighten the hold-down bolt and recheck the timing.

7. Remove the jumper connector. Check that the ignition timing advances.

8. Disconnect the tachometer and timing light from the engine.

Engine Assembly

REMOVAL & INSTALLATION

1.5L (5E-FE) Engine

1. Disconnect the negative battery cable from the battery. On vehicles equipped with an air bag, wait at least 90 seconds before proceeding.

2. Remove the hood from the vehicle.

3. Remove the undercovers from the vehicle.

4. Drain the cooling system and remove the radiator.

5. If equipped with automatic transaxle, disconnect and plug the transaxle fluid lines and disconnect the accelerator cable.

6. Disconnect the throttle cable.

7. Drain the transaxle fluid and the engine oil.

8. Remove the air cleaner assembly and bracket.

9. Remove the charcoal canister.

✴✴ CAUTION

To avoid personal injury, properly release the fuel pressure on any fuel injected model before disconnecting any fuel lines.

10. Disconnect the fuel return and inlet hoses.

11. Disconnect the speedometer cable.

12. Disconnect the idle up air hoses from the power steering air control valve.

13. Disconnect and label the oxygen sensor wire, the oil pressure switch wire, the coolant fan switch wire, the water temperature gauge wire, the back-up light switch and neutral safety switch wires.

14. Disconnect and tag any remaining wiring harnesses connected to the engine. Remove the wiring harness from the engine.

15. Disconnect and tag the PCV hoses and any other vacuum hoses that prevent the removal of the engine.

16. Disconnect and tag the starter wires.

17. If equipped with cruise control, remove the actuator assembly.

18. Disconnect and tag the heater hoses.

19. If equipped with manual transaxle, remove the clutch release cylinder.

20. Remove the clips and washers, then disconnect the transaxle control cables.

21. If equipped with power steering, remove the power steering pump and position it aside. Do not disconnect the power steering hoses.

22. If equipped with air conditioning, remove the air conditioning compressor and position it aside. Leave the refrigerant lines connected.

23. Disconnect the front exhaust pipe.

24. Remove the halfshafts.

25. Support the engine/transaxle assembly properly.

26. Connect a suitable lifting device to the engine lifting hooks.

27. If equipped with manual transaxle, remove the rear mounting through-bolt and the rear mounting assembly. If equipped with automatic transaxle, remove the front mounting through-bolt and front mounting assembly.

28. Remove the right and left side mounting bolts and brackets.

29. Carefully lift the engine/transaxle assembly out of the vehicle.

30. Disconnect the starter from the engine/transaxle by removing the starter stay bracket, then the two bolts to the starter.

31. For automatic transaxles, disconnect the torque converter clutch mounting bolts.

32. Separate the engine from the transaxle by removing the bolts.

To install:

33. Attach the transaxle to the engine and install the bolts. On automatic transaxles, install the torque converter clutch and mounting bolts. Install the gray bolt first, then install the other five bolts. Torque the bolts to 20 ft. lbs. (27 Nm).

34. Install the starter with the two bolts and tighten the bolts to 29 ft. lbs. (39 Nm).

35. Attach an engine sling to the engine hangers and lower the engine and transaxle assembly in the vehicle.

36. Tilt the transaxle downward, lower the engine and clear the left side mounting.

37. Keep the engine level and align the right side and left side mountings with the body bracket.

38. Attach the right side mounting insulator to the mounting bracket and the body and temporarily install the through-bolt, two bolts and nut.

39. Install the left side mounting bracket to the transaxle and mounting insulator with

the five bolts and tighten the bolts. Torque the bracket to transaxle bolts with the head marked (NT) to 47 ft. lbs. (64 Nm), and the bracket to insulator bolts with the bolt head marked (7T) to 35 ft. lbs. (48 Nm).

40. Connect the ground strap. Torque the ground strap to 35 ft. lbs. (49 Nm).

41. Install the rear mounting bracket to the transaxle with the three bolts (M/T) or two bolts (A/T) and tighten to 35 ft. lbs. (48 Nm).

42. Install and tighten the rear insulator through-bolt to 47 ft. lbs. (64 Nm).

43. Install the halfshafts.

44. Remove the engine sling from the engine.

45. Install the two bolts, nut and through-bolt of the right side mounting insulator. Torque the two bolts and nut to 47 ft. lbs. (64 Nm) and the through-bolt to 54 ft. lbs. (73 Nm).

46. Connect the front exhaust pipe.

47. If equipped with air conditioning, install the air conditioning compressor.

48. If equipped with power steering, install the power steering pump to the engine.

49. Install the transaxle control cables with the washers and clips.

50. If equipped with manual transaxle, install the clutch release cylinder. Torque the bolts to 9 ft. lbs. (12 Nm).

51. Connect the idle up air hoses to the power steering air control valve.

52. Connect the speedometer cable to the transaxle.

53. Connect the heater hoses.

54. If equipped with cruise control, connect the actuator assembly.

55. Connect the starter wires and nut.

56. Connect vacuum hoses to the engine.

57. Connect the oxygen sensor wire, oil pressure switch, coolant fan switch wire, water temp. gauge wire, back-up light switch, and the neutral safety switch wires.

58. Connect any other wiring to the engine.

59. Connect the fuel line hoses.

60. Connect the charcoal canister.

61. Install the air cleaner assembly and bracket.

62. Install the radiator. Connect the coolant hoses and on vehicles equipped with A/T, connect the cooling lines.

63. Refill all fluids to specifications.

64. Install the engine undercovers.

65. Install the hood.

66. Connect the negative battery cable to the battery.

67. Start the engine and check for leaks.

1.6L (4A-FE) and 1.8L (7A-FE and 1ZZ-FE) Engines

COROLLA

1. Relieve the fuel system pressure.
2. Disconnect the negative battery cable. On vehicles equipped with an air bag, wait at least 90 seconds before proceeding.
3. Remove the battery.
4. Remove the hood.
5. Remove the engine undercover, then drain the engine coolant and oil.
6. Drain the transaxle assembly.
7. Disconnect the accelerator cable from the accelerator bracket.
8. If equipped with A/T, disconnect the throttle cable from the accelerator cable.
9. Remove the radiator assembly with the cooling fan.
10. Remove the air cleaner assembly.
11. Remove the two bolts and the coolant reservoir tank stay.
12. Disconnect the electrical connector, the hose, the mounting bolt, and remove the washer tank.
13. On models with cruise control, remove the actuator cover, unplug the connector, remove the three bolts,, then disconnect the actuator with the bracket.
14. Disconnect or remove the following components:
    a. The MAP sensor vacuum hose from the gas filter on the intake manifold
    b. The brake booster vacuum hose from the intake manifold
    c. With A/C: the A/C vacuum hose from the actuator
    d. With P/S: the air hose from the air pipe
    e. With A/C: the A/C actuator connector
15. Disconnect the following wires and connectors from the RH fender apron:
    a. The ground strap connector
    b. The MAP sensor connector
    c. With A/C: the A/C pressure switch
    d. The engine wiring harness from the fender apron
16. Disconnect the DLC 1 connector and ground strap from the LH fender apron.
17. Remove the two bolts and remove the engine relay box.
18. Disconnect the four connectors from the engine relay box.
19. Disconnect the hose from the charcoal canister and remove the canister from the bracket.
20. Disconnect the two heater hoses from water inlet housing.

21. Disconnect the fuel inlet and the fuel return hoses.
22. On M/T model, remove the three bolts and remove the clutch release cylinder without disconnecting the pipe.
23. Disconnect the transaxle control cable(s) from the transaxle.
24. Disconnect and remove the following components to disconnect the engine wiring harness:
    a. The LH and RH front door scuff plate
    b. The lower finish panel
    c. The lower panel with the glove compartment
    d. The radio and center cluster finish panel
    e. The rear console box
    f. M/T: the shift lever knob an on A/T: the shifting hole bezel
    g. The lower center finish panel
    h. The floor carpet bracket
    i. The three ECM connectors and cowl wire connector
25. Pull out the wiring harness from the cowl.
26. If equipped w/ A/C, disconnect the A/C compressor connector. Remove the drive belt and remove the four mounting bolts and the A/C compressor. Hint: securely hang the compressor out of the way.
27. Disconnect the front exhaust pipe from the exhaust manifold.
28. Remove the halfshafts from the transaxle.
29. If equipped with P/S, remove the drive belt and remove the two mounting bolts and the P/S pump. Hint: securely hang the pump out of the way.
30. Remove the engine mounting center member.
31. Remove the through-bolt and nut holding the mounting insulator to the mounting bracket.
32. Remove the engine and transaxle assembly from the vehicle.
33. Lift the engine and transaxle out of the vehicle slowly and carefully.
34. Remove the two bolts and the front engine mounting bracket.

35. Remove the three bolts (except the A131L A/T) or the two bolts on the A131L A/T; remove the mounting bracket.
36. Remove the starter and separate the transaxle from the engine.

To install:

37. Assemble the engine to the transaxle.
38. Install the starter.
39. Install the rear engine mounting bracket with two bolts on the A131L A/T or the three bolts (except the A131L A/T). The bolts are tightened to 57 ft. lbs. (77 Nm).
40. Install the front engine mounting bracket to the transaxle and tighten the two bolts to 57 ft. lbs. (77 Nm).
41. Install the engine and transaxle assembly into the vehicle.
42. Install the engine mounting center member.
43. Tighten the through-bolt and nut holding the front engine mounting insulator to the mounting bracket. Tighten the bolt to 64 ft. lbs. (87 Nm).
44. Install the halfshafts.
45. Install the front exhaust pipe.
46. If equipped with P/S, install the pump with the two bolts and tighten to 29 ft. lbs. (39 Nm). Install the drive belt.
47. If equipped with A/C, install the compressor with the four mounting bolts and tighten the bolts to 18 ft. lbs. (25 Nm). Install the drive belt and reconnect the connector.
48. To install and connect the engine wiring harness, perform the following:
    a. Push the wire through the cowl
    b. Connect the three ECM connectors
    c. Attach the cowl wire connector
    d. Install the floor carpet bracket
    e. Install the center lower finish panel
    f. With A/T, install the shifting hole bezel, with M/T, install the shift lever knob
    g. Install the rear console box
    h. Install the center cluster finish panel and the radio
    i. Install the lower panel with the glove compartment door
    j. Install the RH and LH door scuff plate
    k. Install the lower finish panel
49. If equipped with M/T, install the clutch release cylinder with the tube and the three bolts.
50. Connect the transaxle control cable(s) to the transaxle.
51. Connect the fuel return hose and the fuel inlet hose. Torque the bolt to 22 ft. lbs. (29 Nm).

52. Connect the two heater hoses to the water inlet housing.

53. Connect the hose to the charcoal canister and install the canister to the bracket.

54. Connect the following wires and connectors on the LH fender apron:

    a. The four connectors to the engine relay box

    b. Install the engine relay box with the two bolts

    c. The DLC1 connector

    d. The connector from the fender apron

    e. The ground strap from the fender apron

55. Connect the following wires and connectors on the RH fender apron:

    a. The ground strap connector

    b. The MAP sensor connector

    c. With A/C, the A/C pressure switch

    d. The engine wire from the fender apron

56. Connect the following hoses and connectors:

    a. With A/C, the A/C actuator connector

    b. With P/S, the air hoses to the air pipe

    c. The vacuum hose from the MAP sensor to the gas filter to the intake chamber

    d. The brake booster vacuum hose to the air intake chamber

    e. With A/C, the A/C vacuum hose from the actuator

57. If equipped with cruise control, install the actuator and bracket with the three bolts. Connect the connector, connect the actuator cable to the actuator, and install the cover.

58. Install the connector and the vinyl hose. Install the washer tank with the bolt.

59. Install the coolant reservoir tank stay with the bolts.

60. Install the air cleaner assembly.

61. Install the radiator with the cooling fan.

62. If equipped with A/T, connect the throttle cable and adjust.

63. Install the accelerator cable and adjust it.

64. Fill the radiator with engine coolant.

65. Fill the engine with oil.

66. Fill the transaxle assembly with oil.

67. Connect the negative battery cable, start the engine, and check for leaks.

68. Perform engine adjustments, install the engine undercovers, and install the hood.

69. Road test the vehicle for proper operation.

70. Recheck all fluid levels.

CELICA

1. Release the fuel system pressure.

2. Disconnect the negative battery cable. On vehicles equipped with an air bag, wait at least 90 seconds before proceeding.

3. Remove the battery.

4. Remove the hood.

5. Remove the engine undercover, then drain the engine coolant and oil.

6. Drain the oil from the transaxle.

7. Disconnect the accelerator cable from the throttle body, cable bracket, and the clamps.

8. Remove the air cleaner assembly.

9. Disconnect the cruise control actuator cable from the clamps.

10. Remove the radiator assembly.

✳✳ CAUTION

The fuel injection system remains under pressure after the engine has been turned OFF. Properly relieve fuel pressure before disconnecting any fuel lines. Failure to do so may result in fire or personal injury.

11. Disconnect or remove the following components:

- The MAP sensor vacuum hose from the gas filter on the intake manifold
- The P/S air hose from the intake manifold
- The P/S hose from the air the air pipe
- The brake booster vacuum hose from the intake manifold
- The A/C idle-up valve
- The A/C idle-up valve hose from the intake manifold
- The A/C idle-up valve hose from the air pipe
- The DLC1 from the bracket
- The engine wiring harness from the bracket
- The ground cable from the body and the ground strap from the body
- The two heater hoses from the water outlet
- The heater hose from the water bypass pipe
- The fuel inlet hose from the fuel filter and the fuel return hose from the return pipe
- The EVAP hose from the charcoal canister

12. Disconnect the two connectors, remove the two relay box covers, and disconnect the engine wiring harness from the engine compartment relay box.

13. To remove the engine wiring harness from the vehicle cabin, remove or disconnect the following:

- The scuff plate
- The cowl side trim
- The finish panel from the lower instrument panel
- Remove the front side of the floor carpet
- The wiring harness from the clamp of the ECM bracket
- The three ECM connectors
- The circuit opening relay connector
- The three connectors from the connectors on the bracket
- The A/C amplifier connector
- The MAP sensor connector
- The MAP sensor wire from the clamp on the bracket
- The wire clamp from the bracket
- The two nuts holding the engine wiring harness to the cowl

14. Remove the front exhaust pipe.

15. Remove the halfshafts.

16. Remove the alternator drive belt.

17. Remove the A/C drive belt, disconnect the A/C compressor connector, remove the four bolts, and the A/C compressor. Do not disconnect the A/C lines and position the compressor safely out of the way.

18. Remove the drive belt and remove the four bolts that secure the P/S pump. Without disconnecting the lines, securely hang the pump out of the way.

19. Remove the two bolts and disconnect the A/C relay box from the body.

20. On M/T equipped vehicles, remove the bolt and disconnect the bracket from the transaxle. Remove the two mounting bolts and the clutch release cylinder from the transaxle.

21. Disconnect the transaxle control cable(s) from the transaxle.

22. On A/T equipped vehicles, remove the two bolts and disconnect the transaxle control cable from the engine mounting center member.

23. Remove the exhaust pipe support bracket.

24. To remove the engine mounting center member, remove the following components:

- The two dust covers from the rear side of the member
- The A/C pipe from the bracket
- The bolt and nut holding the front engine mounting bracket to the mounting insulator
- The bolt holding the rear engine mounting bracket to the insulator
- The bolt and two nuts holding the rear engine mounting insulator to the front suspension member
- The two bolts and the rear engine mounting bracket, and the center member with the rear mounting insulator.

25. Attach an engine chain hoist or suitable equivalent to the engine hangers. Remove the two bolts and nut and disconnect the LH engine mounting bracket from the mounting insulator.

26. Remove the through-bolt and the LH mounting insulator.

27. Disconnect the ground strap connector.

28. Remove the bolt, two nuts, and disconnect the RH engine mounting bracket from the mounting insulator.

29. Lift the engine and transaxle assembly from the vehicle slowly and carefully making sure that it is clear of all the wiring, cables, and hoses.

30. Separate the transaxle from the engine assembly.

To install:

31. Install the transaxle to the engine assembly.

32. Attach an engine chain hoist to the engine hangers and slowly lower the engine into the engine compartment. Tilt the transaxle downward, lower the engine and clear the LH body mounting.

33. Keeping the engine level, align the RH and the LH mounts with the body mounts and attach the RH engine mounting bracket to the mounting insulator. Temporarily install the three nuts.

34. Temporarily install the LH engine mounting insulator to the body with the through-bolt. Attach the LH engine mounting bracket to the mounting insulator and install the two bolts and nut. Torque the bolts and nut to 47 ft. lbs. (64 Nm).

35. Tighten the LH engine mounting through-bolt to the body. Torque the bolt to 54 ft. lbs. (73 Nm). Tighten the three nuts holding the RH mounting bracket to the insulator. Torque the 12mm nut to 21 ft. lbs. (28 Nm) and the 14mm nut to 38 ft. lbs. (52 Nm).

36. Connect the engine ground strap connector and remove the engine hoist.

37. To install the engine mounting center member, perform the following:
- Attach the center member together with the rear engine mounting insulator to the front suspension member
- Temporarily install the two bolts and nut holding the center member to the body
- Install the rear engine mounting bracket with the two bolts. Torque the bolts to 58 ft. lbs. (78 Nm
- Temporarily install the bolt and two nuts holding the rear engine mounting insulator to the front suspension member
- Temporarily install the bolt holding the rear engine mounting bracket to the insulator

- Temporarily install the bolt and nut holding the front engine mounting bracket to the insulator
- Torque the two bolts holding the center member to the body to 26 ft. lbs. (35 Nm)
- Torque the bolt and two nuts holding the rear mounting insulator to the front suspension member to 59 ft. lbs. (80 Nm)
- Torque the bolt holding the rear engine mounting bracket to the insulator to 65 ft. lbs. (88 Nm)
- Torque the bolt and nut holding the front engine mounting bracket to the insulator to 65 ft. lbs. (88 Nm
- Install the A/C pipe to the bracket and install the two dust covers to the center member

38. Install the exhaust pipe support bracket with the two bolts and the nut. Torque the bolts to 14 ft. lbs. (19 Nm).

39. Connect the transaxle control cable(s) to the transaxle.

40. On A/T vehicles, install the transaxle control cable to the engine mounting center member with the two clamps and two bolts.

41. On M/T vehicles, install the clutch release cylinder. Tighten the bolts to 9 ft. lbs. (12 Nm), then attach the bracket with the bolt.

42. Connect the A/C relay box to the body.

43. Install the P/S pump with the four bolts. Torque the 12mm bolts to 14 ft. lbs. (19 Nm) and tighten the 14mm bolts to 29 ft. lbs. (39 Nm). Install the drive belt and tighten the adjusting bolt to 29 ft. lbs. (39 Nm).

44. Install the A/C compressor with the four bolts and tighten them to 18 ft. lbs. (25 Nm). Install the A/C drive belt with the adjusting bolt and tighten the idler pulley locknut to 29 ft. lbs. (39 Nm). Connect the connector.

45. Install the alternator drive belt.

46. Install the halfshafts.

47. Install the front exhaust pipe.

48. To install the engine wiring harness to the vehicle cabin, perform the following:
- Push the harness through the cowl panel, install the retainer to the cowl with the two nuts and install the wire clamp to the bracket
- Connect the harness to the clamp on the ECM
- Connect the three ECM connectors and the circuit opening relay connector
- Connect the three connectors to the connectors on the bracket
- Connect the A/C amplifier connector
- Install the floor carpet, the lower instrument panel finish panel, the cowl side trim panel, and the scuff plate

49. Connect the engine wiring harness with the two connectors to the engine compartment relay box and install the relay box covers.

50. Install and/or connect the following:
- The MAP sensor connector
- The MAP sensor wire to the clamp on the bracket
- The MAP sensor vacuum hose to the gas filter on the intake manifold
- The brake booster vacuum hose to the intake manifold
- The A/C idle-up valve connector
- The A/C idle-up valve hose to the intake manifold
- The A/C idle-up valve hose to the air pipe
- The DLC1 to the bracket
- The engine harness protector to the bracket
- The ground cable and the ground strap to the body
- The heater hose to the water outlet and the heater hose to the water bypass pipe
- Connect the fuel inlet hose to the fuel filter
- Connect the fuel inlet hose with two new gaskets and the union bolt-tighten the union bolt to 22 ft. lbs. (30 Nm).
- Connect the fuel return hose to the return pipe and connect the EVAP hose to the charcoal canister
- Connect the P/S air hoses to the intake manifold and the air pipe.

51. Install the radiator assembly.

52. On the model equipped with cruise control, install the actuator cable to the clamps.

53. Install the accelerator cable to the throttle body, cable bracket, and the clamps.

54. Install the air cleaner assembly.

55. Install the battery tray and battery.

56. Install the hood.

57. Refill the transaxle assembly.

58. Fill the engine with oil and coolant. Connect the negative battery cable, start the engine, bleed the cooling system, and check for any leaks.

59. Install the engine undercover.

60. Road test the vehicle for any abnormal noise and verify proper operation.

2.2L (5S-FE) Engine

CAMRY

1. Disconnect the negative battery cable. On vehicles equipped with an air bag, wait at least 90 seconds before proceeding.

2. Remove the battery and the battery tray.

3. Remove the hood.

4. Remove the engine undercover, then drain the engine coolant and oil.

5. Disconnect the accelerator cable from the throttle body. On models with A/T, disconnect the throttle cable.

6. Remove the air cleaner assembly, resonator, and the air intake hose.

7. On models with cruise control, remove the actuator cover, unplug the connector, remove the three bolts,, then disconnect the actuator with the bracket.

8. Disconnect the ground strap at the battery carrier.

9. Remove the radiator, then disconnect the coolant reservoir hose.

10. Remove the washer tank and disconnect the electrical lead and hose.

11. Tag and disconnect the following:

a. The five connectors to the engine relay box

b. The igniter connector

c. The noise filter connector

d. The connector at the LH fender apron

e. The two ground straps from the LH and RH fender aprons

f. The data link connector (DLC1)

g. Disconnect the MAP sensor connector

12. Inside the vehicle, remove the dash panel undercover, the glove compartment door, the glove compartment, disconnect the cowl harness connectors and the two ECM connectors.

✳✳ CAUTION

The fuel injection system remains under pressure after the engine has been turned OFF. Properly relieve fuel pressure before disconnecting any fuel lines. Failure to do so may result in fire or personal injury.

13. Disconnect the heater hoses, the fuel return hose, and the fuel inlet hose.

14. On models with M/T, remove the starter and the clutch release cylinder. Don't disconnect the hydraulic line, simply hang the cylinder out of the way.

15. Disconnect the transaxle control cables at the transaxle.

16. Tag and disconnect all remaining vacuum hoses and connectors.

17. Remove the two nuts and pull out the engine wire from the cowl panel.

18. Without disconnecting the refrigerant lines, remove the A/C compressor and carefully position it out of the way.

19. Loosen the two bolts and disconnect the front exhaust pipe bracket. Use a deep 14mm socket and remove the three nuts attaching the front pipe to the manifold. Dis-

connect the front pipe from the exhaust manifold.

20. Remove the halfshafts.

21. Without disconnecting the hydraulic lines, remove the power steering pump and carefully position it aside.

22. Remove the three bolts (M/T) or four bolts (A/T), then disconnect the left engine mounting insulator. Remove the access plugs, remove the three nuts, then remove the right rear engine mounting insulator. Remove the three bolts and disconnect the front right engine mounting insulator.

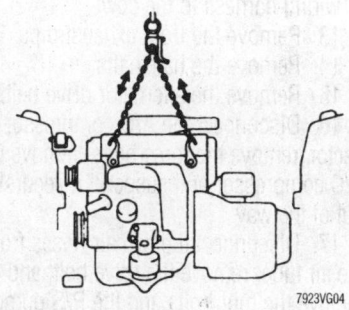

Use a hoist to remove the engine assembly—Camry 5S-FE engine

23. Attach an engine lifting device to the lift hooks. Remove the three bolts and disconnect the control rod. Slowly and carefully, lift the engine/transaxle assembly out of the engine compartment.

24. If equipped with A/T, remove the starter. Separate the engine assembly from the transaxle.

To install:

25. Connect the engine assembly to the transaxle. On vehicles equipped with A/T, install the starter.

26. Carefully lower the engine and transaxle assembly into the engine compartment. With the engine level and all the mounts aligned with their brackets, install the engine control rod. Tighten the three bolts, in the sequence to 47 ft. lbs. (64 Nm).

27. Connect the front right engine mount and tighten the bolts to 59 ft. lbs. (80 Nm). Connect the rear mount and tighten the nuts to 48 ft. lbs. (66 Nm). Don't forget the access plugs.

28. Connect the left mount and tighten the bolts (3 or 4) to 47 ft. lbs. (64 Nm).

29. Install the power steering pump and tighten the bolts to 31 ft. lbs. (43 Nm). Install the drive belt and connect the two air hoses to the air pipe.

30. Install the halfshafts.

31. Connect the front pipe to the manifold and tighten the new nuts to 46 ft. lbs. (62 Nm).

32. Install the A/C compressor and tighten the bolts to 20 ft. lbs. (27 Nm).

33. Feed the engine harness through the cowl and reattach the clamp to the cowl. Make the following connections:

a. The two ECM connectors

b. The two cowl wire connectors

c. Install the glove compartment and door

d. Install the lower instrument panel and the undercover

34. Connect the vacuum hoses and the transaxle control cables.

35. On M/T vehicles, install the release cylinder and the starter.

36. Connect the fuel inlet hose and tighten it to 22 ft. lbs. (29 Nm). Connect the return hose and the two heater hoses.

37. Connect the following:

a. Attach the five connectors to the relay box

b. The connectors from the LH fender apron

c. Install the engine relay box

d. The igniter connector

e. On California models, the ignition coil connector

f. The noise filter connector

g. The two ground straps from the LH and RH fender apron

h. The data link connector (DLC1)

i. The MAP sensor connector.

38. Install the washer tank and connect the electrical lead and hose.

39. Install the coolant reservoir hose and the radiator.

40. If equipped with cruise control, install the actuator and bracket with the three bolts. Connect the actuator connector and install the cover.

41. Connect the ground strap to the battery carrier.

42. Install the air cleaner assembly.

43. On California models, connect the air hose to the air cleaner assembly and connect the air intake temperature sensor connector.

44. On vehicles equipped with A/T, connect and adjust the throttle cable.

45. Connect and adjust the accelerator cable.

46. Install the battery tray and battery.

47. Install the hood.

48. Fill the engine with oil and coolant. Connect the negative battery cable, start the engine, bleed the cooling system, and check for any leaks.

49. Install the engine undercover.

50. Road test the vehicle for any abnormal noise and verify proper operation.

CELICA

1. Release the fuel system pressure.
2. Disconnect the negative battery cable. On vehicles equipped with an air bag, wait at least 90 seconds before proceeding.
3. Remove the battery.
4. Remove the hood.
5. Remove the engine undercover, then drain the engine coolant and oil.
6. Disconnect the accelerator cable from the throttle body, the cable bracket and clamps.
7. Remove the air cleaner assembly.
8. On models with cruise control, unplug the connector, remove the three bolts, then disconnect the actuator from the bracket.
9. Remove the radiator assembly.
10. Disconnect the following connectors, wiring harness, cables, and hoses:
   a. The MAP sensor
   b. The MAP sensor wire from the clamp
   c. The MAP sensor vacuum hose from the gas filter on the intake manifold
   d. The brake booster vacuum hose from the intake manifold
   e. The DLC1 from the bracket
   f. The igniter connector
   g. On California models, the ignition coil connector
   h. On California models, the ignition coil high tension wire, the noise filter, the wire clamp from bracket, the ignition coil and igniter assembly, and disconnect the wire from the bracket
   i. The ground cable from the body
   j. The ground strap from the body
   k. The heater hose from the water outlet and bypass pipe

✳✳ CAUTION

The fuel injection system remains under pressure after the engine has been turned OFF. Properly relieve fuel pressure before disconnecting any fuel lines. Failure to do so may result in fire or personal injury.

   l. The fuel inlet hose from the fuel filter and the fuel return hose from the return pipe
   m. The EVAP hose from the charcoal canister
11. Disconnect the two connectors, remove the two relay box covers, and disconnect the engine wiring harness from the engine compartment relay box.
12. To remove the engine wiring harness from the vehicle cabin, remove or disconnect the following:

   a. The scuff plate
   b. The cowl side trim
   c. The finish panel from the lower instrument panel
   d. Remove the front side of the floor carpet
   e. The wiring harness from the clamp of the ECM bracket
   f. The three ECM connectors
   g. The circuit opening relay connector
   h. The three connectors from the connectors on the bracket
   i. The A/C amplifier connector
   j. The wire clamp from the bracket
   k. The two nuts holding the engine wiring harness to the cowl
13. Remove the front exhaust pipe.
14. Remove the halfshafts.
15. Remove the alternator drive belt.
16. Disconnect the A/C compressor connector, remove the three bolts, remove the A/C compressor and suspend it securely out of the way.
17. Disconnect the two air hoses from the air tube, remove the drive belt, and remove the four bolts and the P/S pump. Securely hang the pump out of the way.
18. On M/T model, remove the starter.
19. On M/T, remove the clutch release cylinder and associated components.
20. Disconnect the transaxle control cable(s) from the transaxle.
21. On A/T model, remove the two bolts and disconnect the transaxle control cable from the engine mounting center member.
22. Remove the exhaust pipe support bracket.
23. To remove the engine mounting center member, remove the following components:

   a. The two dust covers from the rear side of the member
   b. The A/C pipe from the bracket
   c. The bolt and nut holding the front engine mounting bracket to the mounting insulator
   d. The bolt holding the rear engine mounting bracket to the insulator
   e. The bolt and two nuts holding the rear engine mounting insulator to the front suspension member
   f. The two bolts (A/T) or three bolts (M/T) and rear engine mounting bracket. Remove the center member with the rear mounting insulator.
24. Attach an engine chain hoist or suitable equivalent to the engine hangers. Remove the two nuts and bolt (M/T) or the two bolts (A/T), and disconnect the LH engine mount bracket from the mounting insulator.
25. Remove the through-bolt and the LH mounting insulator.

26. Disconnect the ground strap connector.
27. Remove the bolt, two nuts, and disconnect the RH engine mount bracket from the mounting insulator.
28. Remove the halfshaft bearing bracket.
29. Lift the engine and transaxle assembly from the vehicle slowly and carefully making sure that it is clear of all the wiring, cables and hoses. Separate the transaxle from the engine assembly.

To install:

30. Install the transaxle to the engine assembly and reattach the halfshaft bearing bracket with the three bolts. Torque the bolts to 47 ft. lbs. (64 Nm).
31. Attach an engine chain hoist to the engine hangers, and slowly lower the engine into the engine compartment. Tilt the transaxle downward, lower the engine and clear the LH body mounting.
32. Keeping the engine level, align the RH and the LH mounts with the body mounts, and attach the RH engine mount bracket to the mounting insulator, and temporarily install the bolt and two nuts.
33. Temporarily install the LH engine mounting insulator to the body with the through-bolt. Attach the LH engine mount bracket to the mounting insulator and install the two nuts and bolt (M/T) or the two bolts (A/T). Torque the bolts and nuts to 47 ft. lbs. (64 Nm).
34. Tighten the LH engine mounting through-bolt to the body. The torque is 54 ft. lbs. (73 Nm). Tighten the bolt and two nuts holding the RH mounting bracket to the insulator. Torque the bolt to 27 ft. lbs. (37 Nm) and the nut to 38 ft. lbs. (52 Nm).
35. Connect the engine ground strap connector and remove the engine hoist.
36. To install the engine mounting center member, perform the following:

   a. Attach the center member together with the rear engine mounting insulator to the front suspension member
   b. Temporarily install the two bolts holding the center member to the body
   c. Install the rear engine mounting bracket with the two bolts (A/T) or the three bolts (M/T). Tighten the bolt to 58 ft. lbs. (79 Nm).
   d. Temporarily install the bolt and two nuts holding the rear engine mounting insulator to the front suspension member
   e. Temporarily install the bolt holding the rear engine mounting bracket to the insulator
   f. Temporarily install the bolt and nut holding the front engine mounting bracket to the insulator

g. Tighten the two bolts holding the center member to the body to 26 ft. lbs. (35 Nm).

h. Tighten the bolt and two nuts holding the rear mounting insulator to the front suspension member to 59 ft. lbs. (80 Nm).

i. Tighten the bolt holding the rear engine mounting bracket to the insulator to 65 ft. lbs. (88 Nm).

j. Tighten the bolt and nut holding the front engine mounting bracket to the insulator to 65 ft. lbs. (88 Nm).

k. Install the A/C pipe to the bracket and install the two dust covers to the center member

37. Install the exhaust pipe support bracket with the two bolts and the nut. Tighten the bolts to 14 ft. lbs. (19 Nm).

38. Connect the transaxle control cable(s) to the transaxle.

39. On the A/T vehicles, install the transaxle control cable to the engine mounting center member with the two clamps and two bolts.

40. On M/T model only, install and/or connect the following:

a. The clutch release cylinder-tighten the two bolts to 9 ft. lbs. (12 Nm).

b. The bracket with the bolt and the tube to the bracket with the clamp

c. The tube with the clamp and bolt

d. The back-up switch connector

41. On M/T model only, install the starter, connect the starter wire and cable with the nut, and connect the starter connector.

42. Temporarily install the P/S pump with the four bolts. Tighten the three bolts (except the pivot bolt). Tighten the bolts to 32 ft. lbs. (44 Nm). Install the drive belt and tighten the adjusting bolt to 29 ft. lbs. (39 Nm).and the pivot bolt to 32 ft. lbs. (44 Nm). Connect the two hoses to the air tube.

43. Install the A/C compressor with the three bolts and tighten them to 18 ft. lbs. (24 Nm) and connect the connector.

44. Install the alternator drive belt.

45. Install the halfshafts.

46. Install the front exhaust pipe.

47. To install the engine wiring harness to the vehicle cabin, perform the following:

a. Push the harness through the cowl panel, install the retainer to the cowl with the two nuts and install the wire clamp to the bracket

b. Connect the harness to the clamp on the ECM

c. Connect the three ECM connectors and the circuit opening relay connector

d. Connect the three connectors to the connectors on the bracket

e. Connect the A/C amplifier connector

f. Install the floor carpet, the lower instrument panel finish panel, the cowl side trim panel, and the scuff plate

48. Connect the engine wiring harness with the two connectors to the engine compartment relay box and install the relay box covers.

49. Install and/or connect the following:

a. The MAP sensor connector

b. The MAP sensor wire to the clamp on the bracket

c. The MAP sensor vacuum hose to the gas filter on the intake manifold

d. The brake booster vacuum hose to the intake manifold

e. The DLC1 to the bracket

f. The engine harness protector to the bracket

g. On California models, the engine harness clamp to the bracket and the ignition coil and igniter assembly with the three bolts, and install the harness to the bracket

h. The igniter connector

i. On California model, the ignition coil connector, high tension wire to the coil, and the noise filter

j. The ground cable and the ground strap to the body

k. The heater hose to the water outlet and the heater hose to the water bypass pipe

l. Connect the fuel inlet hose to the fuel filter

m. Connect the fuel inlet hose with two new gaskets and the union bolt-tighten the union bolt to 22 ft. lbs. (30 Nm).

n. Connect the fuel return hose to the return pipe and connect the EVAP hose to the charcoal canister

50. Install the radiator assembly.

51. On models equipped with cruise control, install the actuator with the three bolts and connect the connector.

52. Install the accelerator cable to the throttle body, the cable bracket, and the clamps.

53. Install the air cleaner assembly.

54. Install the battery tray and battery.

55. Install the hood.

56. Fill the engine with oil and coolant. Connect the negative battery cable, start the engine, bleed the cooling system, and check for any leaks.

57. Install the engine undercover.

58. Road test the vehicle for any abnormal noise and verify proper operation.

3.0L (1MZ-FE) Engine

CAMRY AND AVALON

1. Release the fuel system pressure.

2. Turn the ignition switch **OFF**. Disconnect the battery cables, negative cable first. On vehicles equipped with an air bag, wait at least 90 seconds before proceeding.

3. Matchmark the hood hinges and remove the hood. Remove the battery and battery tray.

4. Drain the engine oil and cooling system.

5. Disconnect the accelerator and throttle cables. Remove the cruise control actuator, if equipped.

6. Remove the air cleaner assembly, mass air flow meter and air cleaner hose.

7. Remove the radiator.

8. Remove the two bolts and disconnect the engine relay box. Disconnect the five connectors to the engine relay box.

9. Disconnect the following connectors:

a. Two igniter connectors

b. Noise filter connector

c. Connector from the left-hand fender apron

d. Disconnect the two ground straps and any other electrical connections keeping them from being removed.

10. Disconnect all vacuum hoses from the engine.

✳✳ CAUTION

The fuel injection system remains under pressure after the engine has been turned OFF. Properly relieve fuel pressure before disconnecting any fuel lines. Failure to do so may result in fire or personal injury.

11. Disconnect the fuel inlet and return hoses.

12. Disconnect the heater hoses.

13. Disconnect the transaxle control cable from the transaxle.

14. Remove the instrument panel undercover, the lower instrument panel and glove box assembly.

15. Disconnect the three ECM connectors, the five cowl wire connectors, and the cooling fan ECM connector. Push the engine wire through the cowl panel.

16. Remove the front exhaust pipe.

17. Remove the halfshafts from the vehicle.

18. Disconnect the power steering pressure tube.

19. Remove the power steering pump.

20. Remove the air conditioning compressor without disconnecting the hoses.

21. Remove the left-hand engine mounting insulator by removing the four bolts.

22. Remove the right-hand engine mounting insulator by removing the two hole plugs, then removing the four nuts.

23. Remove the four bolts to the engine mounting shock absorber, then remove the absorber.

24. Remove the front right engine mounting insulator by removing the three bolts.

25. Attach a hoist chain to the engine hangers.

26. Disconnect the coolant reservoir hose and remove the reservoir tank.

27. Remove right-side engine mounting stay bracket. Remove the engine control rod and bracket assembly.

➡ **Make certain all wires, connectors and hoses are cleared from the engine.**

28. Using an engine hoist, carefully lift the engine/transaxle assembly from the vehicle.

To install:

29. Carefully lower the engine position. Keep the engine level while aligning the engine mounts.

30. Install the engine control rod and bracket. Tighten to 47 ft. lbs. (64 Nm).

31. Install the right engine mount stay bracket. Tighten to 23 ft. lbs. (31 Nm).

32. Connect the engine ground straps. Install the coolant reservoir tank.

33. Install the front engine insulator. Tighten to 48 ft. lbs. (66 Nm).

34. Install the engine mounting shock absorber. Tighten to 35 ft. lbs. (48 Nm).

35. Install the left and right engine mounts. Tighten to 48 ft. lbs. (66 Nm).

36. Install the power steering pump and air conditioning compressor.

37. Connect the power steering pressure tube.

38. Install the halfshafts and front exhaust pipe.

39. Push the engine wires through the cowl panel and connect all wires and connectors.

40. Connect the transaxle control cable to the transaxle.

41. Connect the fuel hoses and heater hoses.

42. Connect all vacuum hoses, wiring and connectors.

43. Install the radiator.

44. Install the cruise control actuator, if equipped. Connect the throttle cable and accelerator cable.

45. Install the Mass Air Flow meter, the air cleaner assembly, and air cleaner hose.

46. Fill the cooling system with the proper coolant/water mixture. Fill the engine with engine oil.

47. Install the battery tray and battery. Connect the battery cables; negative cable last.

48. Align the marks and install the hood.

49. Start the engine and check for leaks.

50. Perform a road test.

51. Recheck all fluid levels.

3.0L (2JZ-GE) and (2JZ-GTE) Engines

✳✳ CAUTION

The Air Bag System must be disarmed before performing this procedure. Failure to do so may cause accidental deployment of the air bag, resulting in unnecessary system repairs and/or personal injury.

1. Turn the ignition switch **OFF**. Disconnect the negative battery cable from the battery. Do not start any work for at least 90 seconds to prevent accidental deployment of the air bag.

2. Remove the hood.

3. Remove the radiator assembly from the vehicle.

4. Relieve the fuel pressure from the fuel lines.

✳✳ CAUTION

The fuel injection system remains under pressure even after the engine has been turned OFF. The fuel system pressure must be relieved before disconnecting any fuel lines. Failure to do so may result in fire and/or personal injury.

5. Drain the engine oil.

6. Disconnect the accelerator cable and cruise control actuator.

7. Remove the air cleaner assembly, volume air flow meter, and the air intake hose.

8. Remove the drive belt by turning the tensioner clockwise.

9. Remove the fan, fan clutch, and the water pump pulley by removing the four nuts.

10. Remove the charcoal canister.

11. Disconnect the heater water hoses.

12. Disconnect the brake booster vacuum hose.

13. Disconnect the EVAP hose.

14. Disconnect the following the connectors and wires:
- Noise filter connector
- Ignition coil connector
- Engine wire from the wire clamp
- Rubber cap, nut and wire from the alternator

- Engine room main wire
- Igniter connector
- Theft deterrent horn connector
- Engine wire from the two wire clamps
- Wire clamp and power steering solenoid valve connector
- Ground strap from the cylinder block by removing the bolt
- Rubber cap, nut, and the wire from the starter

15. Disconnect the fuel inlet hose from the engine by removing the union bolt and two gaskets. Suspend the hose union upward.

16. Disconnect the fuel return hose from the fuel return guide.

17. Disconnect the fuel return hose from the fuel return hose. Plug the hose end.

18. Remove the bolt and bracket and disconnect the engine wire from the intake manifold stay.

19. Disconnect the power steering pump.

20. Disconnect the power steering pressure tube from the engine by removing the two bolts.

21. Disconnect the A/C compressor without disconnecting the hoses.

22. Disconnect the engine wire from the cowl panel.

23. Disconnect the engine wire from the cabin as follows:

    a. Remove the scuff plate from the right door.

    b. Take out the front side of the floor carpet.

    c. Remove the two nuts and ECM protector.

    d. Remove the nut and disconnect the ECM from the floor panel.

    e. Disconnect the two connectors from the ECM.

    f. Disconnect the connector from the instrument panel wire.

    g. Disconnect the connector from the connector cassette.

    h. Pull out the engine wire from the cabin.

24. For vehicles equipped with an manual transmission, remove the upper console panel, shift lever boots, and holding bolts.

25. If equipped with manual transmission, disconnect the clutch release cylinder and the ground strap from the transmission.

26. Remove the No. 2 front exhaust pipe.

27. Remove the exhaust pipe heat insulator.

28. Remove the driveshaft.

29. For vehicles with automatic transmissions, disconnect the control rod from the shift lever by removing the nut.

30. Support the transmission with a jack.

31. Remove the rear support member by removing the eight bolts.

32. Attach the engine hoist chain to the engine and raise the engine slightly.

33. Remove the two nuts holding the engine front mounting insulators to the front suspension crossmember.

34. Lift the engine out of the vehicle. While lifting the engine, be sure the engine is clear of all wiring, hoses and cables.

35. Remove the oil dipstick guide from the transmission.

36. Disconnect the engine wire from the transmission.

37. Disconnect the starter connector, two bolts, engine wire bracket, and the starter.

38. For vehicles with automatic transmissions, remove the oil cooler tubes from the transmission.

39. For vehicle with automatic transmissions, remove the torque converter clutch mounting bolts.

40. Remove the six bolts from the transmission and remove the transmission from the engine.

To install:

41. Assemble the engine and transmission by installing the six bolts. Tighten the bolts as follows:
- 14mm-29 ft. lbs. (39 Nm)
- 17mm-43 ft. lbs. (72 Nm)

42. If equipped with an automatic transmission, install the torque converter clutch mounting bolts by first installing the gray bolt;, then install the other five bolts. Tighten the bolts to 25 ft. lbs. (33 Nm).

43. If equipped with automatic transmission, install the oil cooler tubes to the transmission. Tighten the union nuts to 25 ft. lbs. (33 Nm).

44. Install the starter.

45. Connect the engine wire to the transmission.

46. If equipped with automatic transmission, install the oil dipstick guide and dipstick form the transmission.

47. Install the engine and transmission as an assembly to the vehicle. Keep slight tension on the engine until the mounting bolts and nuts are installed.

48. Install the two nuts holding the engine front mounting insulators to the front suspension crossmember. Tighten the nuts to 43 ft. lbs. (59 Nm).

49. Install the four bolts holding the support member to the body. Tighten the bolts to 19 ft. lbs. (25 Nm).

50. Install the four nuts holding the support member to the engine rear mounting insulator. Tighten the nuts to 10 ft. lbs. (13 Nm).

51. Remove the engine hoist from the engine.

52. Install the driveshaft.

53. For vehicles equipped with automatic transmissions, connect the transmission control rod as follows:

a. Shift the shift lever to the **N** position.

b. Fully turn the control shaft lever back and return two notches. The control shaft is now in the neutral position.

c. Connect the control rod to the shift lever with the nut. Tighten the nut to 9 ft. lbs. (13 Nm).

54. Install the exhaust pipe heat insulator.

55. Install the No. 2 front exhaust pipe.

56. For vehicle with manual transmission, install the clutch release cylinder and ground strap. Tighten the clutch release cylinder bolts to 9 ft. lbs. (13 Nm) and the ground strap bolt to 27 ft. lbs. (37 Nm).

57. For vehicles with manual transmission, install the upper console panel, shift lever boots, and holding bolts.

58. Connect the engine wire to the cabin as follows:

a. Push in the engine wire through the cowl panel.

b. Connect the connector to the connector cassette.

c. Connect the connector to the instrument panel wire connector.

d. Connect the two connectors to the ECM.

e. Insert the ECM bracket into the stay on the floor panel.

f. Install the ECM with the nut.

g. Install the ECM protector with the two nuts.

h. Install the floor carpet.

i. Install the scuff plate.

59. Connect the engine wire to the cowl panel.

60. Install the A/C compressor to the engine.

61. Install the power steering tube with the two clamp bolts.

62. Install the power steering pump as follows:

a. Install the pump bracket with the two bolts. Tighten the lower bolt to 43 ft. lbs. (58 Nm) and the upper bolt to 29 ft. lbs. (39 Nm).

b. Install the pump rear stay with the two bolts. Tighten the bolts to 29 ft. lbs. (39 Nm).

c. Install the pump housing to the pump bracket.

d. Connect the air hose to the No. 4 timing belt cover.

e. Connect the air hose to the air intake chamber and secure the hose to the No. 4 timing belt cover.

f. Install the front pump bracket with the two bolts. Tighten the bolts to 43 ft. lbs. (58 Nm).

g. Install the plate washer and bolt to the power steering fluid pump.

63. Install the engine wire bracket.

64. Connect the fuel return hose to the fuel return pipe.

65. Install the fuel return hose to the clamp of the oil dipstick guide.

66. Install the fuel inlet hose with the two new gaskets and the union bolt. Tighten the bolt to 22 ft. lbs. (29 Nm).

67. Connect the wires and connectors.

68. Connect the EVAP hose.

69. Connect the brake booster vacuum hose.

70. Connect the heater hoses.

71. Install the charcoal canister.

72. Install the water pump pulley, fan, and the fan clutch.

73. Install the drive belt to the engine.

74. Install the air cleaner, VAF meter, and the intake air connector pipe assembly.

75. Connect the control cables to the throttle body.

76. Fill the engine with oil.

77. Install the radiator assembly.

78. Connect the negative battery cable to the battery.

79. Start the engine, bleed the cooling system, and check for leaks.

80. Install the hood.

81. Road test the vehicle and check all fluids.

Water Pump

REMOVAL & INSTALLATION

1.5L (5E-FE) Engine

1. Disconnect the negative battery cable. On vehicles equipped with an air bag, wait at least 90 seconds before proceeding.

2. Drain the engine coolant.

3. Remove the alternator.

4. For engines with distributorless ignition, remove the intake manifold stay bracket by disconnecting the wire clamps and removing the two nuts.

5. For engine with distributor ignition, remove the intake manifold stay bracket by removing the two nuts and two bolts.

6. Remove the water inlet pipe as follows:

a. Disconnect the water inlet hose.

b. Disconnect the heater hose.

c. Disconnect the bypass hose.

d. Remove the bolt, water inlet pipe, and O-ring.

7. Remove the oil dipstick guide.

8. Remove the alternator adjusting bar.

9. Remove the water pump attaching bolt and nuts. Remove the water pump assembly.

To install:

10. Scrape any remaining gasket material off the pump mating surface. Apply a 2–3mm (0.08–0.12 in.) bead of sealant to the groove in the pump.

11. Replace the O-ring on the water inlet pipe and lubricate the O-ring with a little soap and water. Install the pump assembly. Tighten the bolts to 13 ft. lbs. (17 Nm).

12. Replace the O-ring on the oil dipstick guide and install the assembly. Install the alternator adjusting bar and dipstick guide clamp bolt.

13. Connect the water inlet pipe to the cylinder block with a bolt. Tighten the bolt to 65 inch lbs. (7.5 Nm).

14. Connect the water bypass, heater inlet and water inlet hoses.

15. For distributorless ignition engines, install the intake manifold bracket by installing the two bolts and the wire clamp. Tighten the bolts to 15 ft. lbs. (20 Nm).

16. For distributor ignition, install the intake manifold bracket by installing the two bolts. Tighten the bolts to 15 ft. lbs. (20 Nm).

17. Install the alternator and belt.

18. Refill the engine with coolant.

19. Connect the negative battery cable and start the engine.

20. Check for coolant leaks.

1.6L (4A-FE) and 1.8L (7A-FE) Engines

1. Disconnect the negative battery cable. On vehicles equipped with an air bag, wait at least 90 seconds before proceeding.

2. Drain the engine coolant into a suitable container.

3. Remove the RH engine mounting insulator.

4. Remove No. 2 and No. 3 timing belt covers.

5. If equipped with power steering, safely raise and support the engine. Remove the hole cover and remove the two mounting bolts from the front engine mount insulator. Remove the nut and the through-bolt and remove the insulator.

6. If equipped with power steering, remove the electric cooling fan.

7. Remove the bolt and two nuts and remove the engine wire.

8. On the 7A-FE engine, disconnect crankshaft position sensor connector from the dipstick guide.

9. Remove the mounting bolt and pull out the dipstick guide and the dipstick.

10. Disconnect the water temperature sender gauge connector.

11. Remove the two nuts and the No. 2 water inlet from the water inlet hose.

12. Remove the three water pump bolts, the water pump, and the O-ring from the block.

To install:

13. Install a new O-ring on the block and install the water pump with the three bolts. Tighten the bolts to 10 ft. lbs. (14 Nm).

14. Connect the inlet hose to the water pump and install the water inlet No. 2 to the cylinder head with the two nuts. Tighten the nuts to 11 ft. lbs. (15 Nm).

15. Connect the water temperature sender gauge connector.

16. After applying a small amount of oil to the O-ring, install a new O-ring on the oil dipstick guide. Install the guide mounting bolt and tighten it to 82 inch lbs. (9 Nm).

17. On the 7A-FE engine, connect the crankshaft position sensor connector.

18. Connect the engine wire with the two nuts and the bolt.

19. If equipped with power steering, install the electric cooling fan.

20. If equipped with power steering, install the front mounting insulator through-bolt and nut. Tighten the nut to 64 ft. lbs. (87 Nm).

21. Install the engine insulator mounting bolts and tighten the two bolts to 47 ft. lbs. (64 Nm). Install the hole cover and safely lower the engine.

22. Install the No. 2 and No. 3 timing belt covers.

23. Install the RH engine mounting insulator.

24. Refill the cooling system with coolant and connect the negative battery cable. Start the engine and bleed the cooling system. Check for cooling system leaks and proper system operation.

1.8L (1ZZ-FE) Engine

1. Remove the right-hand engine under cover.

2. Drain the engine coolant.

3. Turn the tensioner bolt clockwise to loosen the belt tension, then remove the belt. Slowly release the tensioner.

4. Remove the water pump mounting bolts, then remove the pump.

To install:

5. Place a new o-ring on the timing belt cover and install the water pump. Tighten the bolts marked A (short) to 80 inch lbs. (9

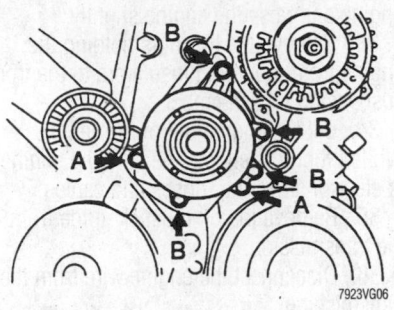

Water pump bolt identification—1.8L (1ZZ-FE) engine

Nm) and the bolts marked B (long) to 8 ft. lbs. (11 Nm).

6. Install the drive belt.

7. Install the right engine under cover.

8. Refill the engine with coolant.

9. Start the engine and check for leaks.

10. Allow the engine to cool and recheck the coolant level.

2.2L (5S-FE) Engines

1. Disconnect the negative battery cable. On vehicles equipped with an air bag, wait at least 90 seconds before proceeding.

2. Raise and safely support the vehicle.

3. Remove the right engine undercover.

4. Drain the engine coolant into a suitable container. Disconnect the lower radiator hose from the water outlet.

5. Remove the timing belt, timing belt tension spring, and the No. 2 idler pulley.

6. Remove the alternator, drive belt and the adjusting bar if necessary.

7. Remove the two nuts holding the water pump to the water bypass pipe and remove the three bolts in sequence.

8. Disconnect the water pump cover from the water bypass pipe and remove the water pump cover assembly.

9. Remove the gasket and two O-rings from the water pump and the bypass pipe.

10. Remove the water pump from the water pump cover by removing the three bolts in sequence.

To install:

11. Cleaned the gasket mating surfaces.

12. Install a new gasket and assemble the water pump to the water pump cover. Tighten the bolts to 78 inch lbs. (9 Nm) in proper sequence.

13. Install a new O-ring and gasket to the water pump cover and install a new O-ring on the water bypass pipe. Connect the water pump cover to the water bypass pipe, but do not install the nuts yet.

14. Install the water pump and tighten the three bolts in sequence. Tighten the

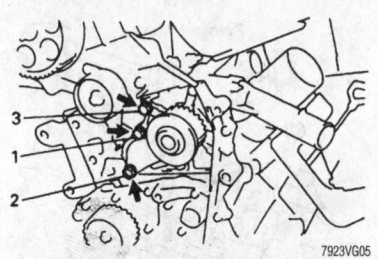

Install the three water pump bolts in this sequence—2.2L (5S-FE) engine

bolts to 78 inch lbs. (9 Nm). Install the two nuts holding the water pump cover to the water bypass pipe and tighten them to 82 inch lbs. (9 Nm).

15. Install the alternator drive belt adjusting bar with the bolt and tighten the bolt to 13 ft. lbs. (18 Nm).

16. Install the No. 2 idler pulley and the timing belt tension spring.

17. Connect the lower radiator hose.

18. Install the timing belt.

19. Install the right undercover and safely lower the vehicle.

20. Fill the cooling system with coolant and connect the negative battery cable. Start the engine and bleed the cooling system. Check the cooling system for leaks and proper operation.

3.0L (1MZ-FE) Engine

1. Disconnect the negative battery cable from the battery. On vehicles equipped with an air bag, wait at least 90 seconds before proceeding.

2. Drain the engine coolant.

3. Remove the timing belt.

4. Mark the left and right camshaft pulleys with a touch of paint. Using SST tools 09249–63010 and 09960–10000 or equivalents, remove the bolts to the right and left camshaft pulleys. Remove the pulleys from the engine. Be sure not to mix up the pulleys.

5. Remove the No. 2 idler pulley by removing the bolt.

6. Disconnect the three clamps and engine wire from the rear timing belt cover.

7. Remove the six bolts holding the rear timing belt cover to the engine block.

8. Remove the four bolts and two nuts to the water pump.

9. Remove the water pump and the gasket from the engine.

To install:

10. Check that the water pump turns smoothly. Also check the air hole for coolant leakage.

11. Using a new gasket, apply liquid sealer to the gasket, water pump and engine block.

12. Install the gasket and pump to the engine and install the four bolts and two nuts. Tighten the nuts and bolts to 53 inch lbs. (6 Nm).

13. Install the rear timing belt cover and tighten the six bolts to 74 inch lbs. (9 Nm).

14. Connect the engine wire with the three clamps to the rear timing belt cover.

15. Install the No. 2 idler pulley with the bolt. Tighten the bolt to 32 ft. lbs. (43 Nm). After tightening the bolt, be sure the idler pulley moves smoothly.

16. With the flange side **outward** , install the right-hand camshaft pulley to the engine. Be sure to align the knock pin hole on the camshaft pulley with the knock pin on the camshaft. Using the same tools as removal, tighten the camshaft bolt to 65 ft. lbs. (88 Nm).

17. With the flange side **inward** , install the left-hand camshaft pulley to the engine. Be sure to align the knock pin hole on the camshaft pulley with the knock pin on the camshaft. Using the same tools as removal, tighten the camshaft bolt to 94 ft. lbs. (125 Nm).

18. Install the timing belt to the engine.

19. Fill the engine coolant.

20. Connect the negative battery cable to the battery and start the engine.

21. Top off the engine coolant and check for leaks.

3.0L (2JZ-GE and 2JZ-GTE) Engines

1. Disconnect the negative battery cable from the battery. On vehicles equipped with an air bag, wait at least 90 seconds before proceeding.

2. Remove the air cleaner and MAF meter assembly.

3. Remove the radiator assembly from the vehicle.

4. If equipped with manual transmission, remove the drive belt tensioner damper by removing the two nuts.

5. Loosen the four nuts holding the fan clutch to the water pump.

6. Loosen the drive belt tension by turning the drive belt tensioner clockwise. Remove the drive belt from the engine.

7. Remove the four nuts, the fan, fan clutch, and the water pump pulley.

8. Remove the water inlet, lower radiator hose assembly, and the thermostat.

9. Remove the timing belt.

10. Remove the alternator from the engine.

11. On turbo models disconnect the turbo water hoses from the water outlet.

12. Except for the California vehicles, remove the exhaust manifold heat insulator.

13. Remove the water outlet and No. 1 water bypass pipe.

14. Disconnect the No. 2 water bypass from the water pump by disconnecting the two nuts.

15. Disconnect the No. 3 turbo water hose from the water pump.

16. Remove the six bolts securing the water pump and remove the water pump from the engine. Be sure to replace the each bolt to its original position.

17. Clean the surface of the engine and remove the O-ring from the cylinder block.

To install:

18. Install the O-ring to the cylinder block.

19. Apply a thin layer of liquid sealant to the engine and water pump. Install a new gasket to the water pump.

20. Connect the water pump to the water bypass pipe. Do not install the nut at this time.

21. Install the water pump with the six bolts. Be sure to replace the bolts to their original positions. Tighten the bolts to 15 ft. lbs. (21 Nm).

22. Install the two nuts holding the No. 2 water bypass pipe to the water pump. Tighten the nuts to 15 ft. lbs. (21 Nm).

23. Connect the No. 3 turbo water hose to the water pump.

24. Install the water bypass outlet and No. 1 water bypass pipe.

25. Connect the turbo water hoses to the water outlet.

26. Install the alternator to the engine.

27. Except for California vehicle, install the exhaust manifold heat insulator.

28. Install the engine wire bracket with the bolt.

29. Install the timing belt.

30. Install the thermostat, water inlet, and the lower radiator hose assembly.

31. Install the water pump pulley, fan, fluid clutch assembly, and the drive belt. Tighten the fan nuts to 12 ft. lbs. (16 Nm).

32. If equipped with manual transmission, install the drive belt tensioner damper.

33. Install the radiator assembly to the vehicle.

34. Install the air cleaner and MAF meter assembly.

35. Install the No. 1 air hose.

36. Connect the negative battery cable to the battery.

37. Fill and bleed the cooling system.

38. Start the engine and check for leaks.

Cylinder Head

REMOVAL & INSTALLATION

1.5L (5E-FE) Engine

1. Disconnect the negative battery cable. On vehicles equipped with an air bag, wait at least 90 seconds before proceeding.

2. Relieve the fuel pressure.

3. Remove the right engine undercover.

4. Drain the cooling system.

5. Disconnect the front exhaust pipe from the exhaust manifold by removing the two bolts and two compression springs.

6. Disconnect the accelerator and on vehicles equipped with automatic transmissions, disconnect the throttle cable.

7. Remove the air cleaner and air intake collector assembly.

8. Disconnect the fuel inlet and return hoses from the delivery pipe. Plug the hoses to prevent fuel leakage.

9. If equipped with power steering, remove the power steering pump and bracket. Do not disconnect the power steering hoses. Set and safely support the power steering pump aside.

10. Remove the ignition coils, spark plug wires, and spark plugs. Be sure to mark the position of the spark plug wires.

11. Tag and disconnect all electrical wire and vacuum hoses that interfere with removal of the cylinder head.

12. Remove the EGR pipe, EGR valve, and vacuum modulator.

13. Disconnect the radiator, water inlet and heater hoses.

14. Remove the water inlet and outlet housing.

15. Remove the throttle body assembly.

16. Remove the exhaust manifold.

17. Remove the fuel rail and injector assembly.

18. Remove the intake manifold.

19. Remove the valve cover.

20. Remove the No. 2 timing belt cover by removing the four bolts.

21. Remove the alternator belt, then the No. 3 timing belt cover from the No. 1 timing belt cover.

22. Turn the crankshaft pulley and align its groove with the timing mark 0 on the No. 1 timing belt cover.

23. Check that the hole of the camshaft timing pulley on the side with the 5E-FE mark is aligned with the timing mark on the No. 1 bearing cap. If the marks do not line up, turn the crankshaft pulley one (1) revolution and check marks.

24. Place matchmarks on the timing belt, loosen the No. 1 idler pulley and carefully remove the belt.

25. Remove the No. 2 idler pulley.

➡Keep tension on the belt so it does not shift position and do not allow the crankshaft to rotate. This includes allowing the vehicle to roll while in gear. Do not allow anything to drop inside the belt cover, including dirt. The belt can be damaged. Do not let the belt come into contact with water, oil or grease.

26. Remove the camshaft timing pulley.

27. Remove the camshafts following the proper sequences and procedures.

28. Loosen the cylinder head bolts in several passes and in the reverse order of the installation sequence.

29. Remove the cylinder head from the engine. There are two different bolt lengths. Make note of there positions and replace the cylinder bolts in their original position.

※※ WARNING

Failure to loosen the bolts as described can result in cylinder head warpage or cracking.

To install:

30. Clean the gasket mating surfaces using care not to damage the aluminum components, replace the gasket, then lower the cylinder head onto the engine. Be sure the dowel pins are aligned and no hoses or wires are between the head and cylinder block.

➡The head bolts stretch and must be replaced once removed.

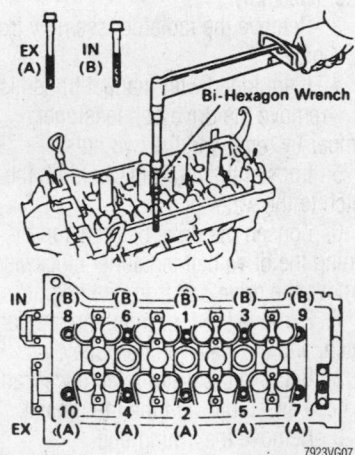

Cylinder head bolt tightening sequence and bolt identification—1.5L (5E-FE) engine

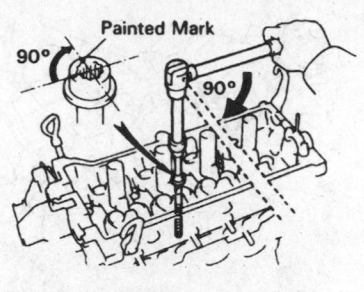

Turn each bolt an additional 90° in the correct sequence—1.5L (5E-FE) engine

31. Lightly oil and place the two different size head bolts in their correct positions and tighten them in several passes in the proper sequence, evenly until arriving at a tighten of 33 ft. lbs. (44 Nm).

32. Mark each bolt with a reference mark and tighten each bolt in sequence an additional 90 degrees.

33. Install the camshafts following the proper sequences and procedures.

34. Install the camshaft timing pulley in its original position and tighten the bolt to 37 ft. lbs. (50 Nm).

35. Install the No. 2 idler pulley, tightening the bolt to 20 ft. lbs. (27 Nm).

36. Install the timing belt on the matchmarks and tension it properly.

37. Install the No. 3 timing belt cover, the alternator belt and the No. 2 timing belt cover with its gasket and four bolts.

38. Install the valve cover with the proper gasket and sealer and tighten the nuts to 61 inch lbs. (7 Nm).

39. Install the intake manifold and tighten the nuts and bolts evenly to 14 ft. lbs. (19 Nm).

40. Install the intake manifold stay with the bolt and nut. Tighten the nut and bolt to 15 ft. lbs. (20 Nm).

41. Install the MAP sensor vacuum line and the brake booster.

42. Install the fuel injector rail assembly, using new grommets and O-rings. Lightly lubricate the O-rings with gasoline and check that the injectors can be rotated smoothly once pressed in.

43. Using a new gasket, install the exhaust manifold to the engine. Evenly tighten the nuts in several passes, then tighten each nut to 35 ft. lbs. (48 Nm).

44. Place the exhaust manifold stay on the engine and exhaust manifold. Tighten the bolt, then the two (2) nuts to 29 ft. lbs. (40 Nm).

45. Install the exhaust manifold heat insulator with the three bolts and tighten to 69 inch lbs. (8 Nm).

46. Install the water inlet and outlet housings.

47. Connect the radiator, heating and water inlet hoses.

48. Install the EGR valve, and vacuum modulator. Tighten the EGR bolts to 13 ft. lbs. (18 Nm).

49. Install the EGR pipe and tighten the union nut to 29 ft. lbs. (40 Nm) and the two nuts to 22 ft. lbs. (30 Nm).

50. Install the throttle body.

51. Reconnect all remaining electrical and vacuum hose fittings.

52. Install the ignition coils and spark plugs.

53. Install the power steering pump bracket. Tighten the three bolts to 32 ft. lbs. (43 Nm). Install the power steering belt and adjust the belt.

54. Connect the fuel return hose and the union bolt for the fuel inlet hose. Tighten the union bolt to 22 ft. lbs. (29 Nm).

55. Install the air cleaner assembly with the air intake connector.

56. Install and adjust the accelerator cable.

57. If equipped with A/T, install and adjust the throttle cable.

58. Using a new gasket, install the exhaust pipe using two springs and two bolts. Tighten the bolts to 46 ft. lbs. (62 Nm).

59. Install the right-hand engine undercover.

60. Connect the battery cable, refill all fluids, and start the engine. Check the ignition timing and check for engine leaks.

1.6L (4A-FE) and 1.8L (7A-FE) Engines

COROLLA

1. Release the fuel system pressure.

2. Disconnect the negative battery cable. On vehicles equipped with an air bag, wait at least 90 seconds before proceeding.

3. Drain the engine coolant into a suitable container.

4. Remove the air cleaner and cap assembly.

5. Disconnect the accelerator cable bracket from the throttle body.

6. Remove the alternator.

7. Disconnect the spark plug wires and remove the distributor assembly.

8. Disconnect the oxygen sensor connector and remove the front exhaust pipe.

9. Remove the exhaust manifold.

10. Disconnect the radiator inlet hose and remove the water outlet.

11. Disconnect the electrical connectors and hoses from the water inlet and the water inlet housing and remove the housing.

12. Disconnect the ground strap connector.

13. Disconnect all of the hoses from the air chamber.

14. If equipped with and EGR, remove the EGR VSV and remove the intake manifold stay.

15. Remove the air pipe.

16. Remove the throttle body assembly and remove the air intake chamber.

❄❄ CAUTION

The fuel injection system remains under pressure after the engine has been turned OFF. Properly relieve fuel pressure before disconnecting any fuel lines. Failure to do so may result in fire or personal injury.

17. Remove the fuel delivery pipe and the fuel injectors.

18. Disconnect the engine wiring harness, if equipped with A/C, disconnect the A/C compressor connector. Disconnect the crankshaft position sensor connector.

19. Remove the intake manifold.

20. Safely support the engine assembly and raise the engine to remove the RH engine mounting insulator.

21. Remove the cylinder head cover and the spark plugs.

22. Remove the No. 3 and the No. 2 timing belt covers.

23. Set No. 1 cylinder to TDC/compression.

24. Place matchmarks on the camshaft timing pulley and timing belt. Remove the plug from the No. 1 timing belt cover, loosen the idler pulley and push the pulley as far left as possible. Remove the timing belt from the camshaft timing pulley.

25. Remove the camshaft timing pulley.

26. Remove the alternator bracket.

27. Remove the oil dipstick guide and the dipstick.

28. Remove the No. 2 water inlet.

29. Remove the intake and exhaust camshafts following the proper sequences and procedures.

30. Using an SST 09205–16010 tool or its equivalent, uniformly loosen the bolts in the reverse order of the installation sequence. Remove the 10 cylinder head bolts in several passes and in sequence.

➡**Cylinder head warpage or cracking could result from removing the bolts in the incorrect order.**

31. Lift the cylinder head assembly from the cylinder block dowels and remove the cylinder head.

To install:

32. Clean the gasket mating surfaces using care not to damage the aluminum components, replace the gasket, then lower the cylinder head onto the engine. Be sure the dowel pins are aligned and no hoses or wires are between the head and cylinder block.

33. The cylinder head bolts are tightened in three progressive steps. Apply a light coat of engine oil to the cylinder head bolts. Uniformly tighten the 10 cylinder head bolts in several passes and in sequence.

➡**The cylinder head bolts are in lengths of 3.54 in. (90mm) and 4.25 in. (108mm). The 3.54 in. (90mm) bolts (A) are to be installed in the intake side of the cylinder head. The 4.25 in. (108mm) bolts (B) are to be installed in the exhaust manifold side of the cylinder head.**

34. Mark the front of the cylinder head bolt with paint. tighten the cylinder head bolts by 90° in sequence. Tighten an additional 90° and be sure that the paint mark is now positioned toward the rear.

35. Install the intake and exhaust camshafts following the proper sequences and procedures.

36. Check and adjust valve clearance.

37. Install the No. 2 water inlet and connect the water inlet hose.

38. Install the oil dipstick guide, and the dipstick.

39. Install the alternator bracket.

40. Install the camshaft timing pulley.

41. Install the timing belt to the camshaft timing pulley aligning the matchmarks and properly tensioning the belt with the tensioning pulley. Loosen the pulley bolt ½ of a turn and turn the crankshaft clockwise two turns. Be sure that each pulley align with the timing marks.

42. Install the No. 2 and the No. 3 timing belt covers.

43. Install the spark plugs.

44. Install the semi-circular plug with sealant to the cylinder head and install the cylinder head cover with a new gasket to the cylinder head.

45. Install the RH engine mounting insulator and safely lower the engine.

46. Install the intake manifold and ground strap.

47. Connect the engine wiring harness to the cylinder head cover, if equipped with A/C, fasten the compressor connector. Also fasten the oil switch connector and the crankshaft position sensor connector.

48. Install the injectors and the delivery pipe.

49. Install the air intake chamber.
50. Install the throttle body assembly.
51. Install the air pipe.
52. Install the intake manifold stay and if equipped with an EGR, install the EGR VSV.
53. Connect all of the hoses to the air intake chamber.
54. Connect the ground strap connector.
55. Install the water inlet and the water inlet housing and connect all hoses and connectors.
56. Install the water outlet.
57. Install the exhaust manifold.
58. Install the front exhaust pipe and connect the oxygen sensor connectors.
59. Install the distributor and connect the spark plug wires.
60. Install the alternator.
61. Connect the accelerator cable bracket to the throttle body.
62. Install the air cleaner and cap assembly.
63. Connect the negative battery cable, fill the engine with coolant, start the engine, warm up, and check for leaks.
64. Install the RH engine undercover, check ignition timing, and road test for proper operation.

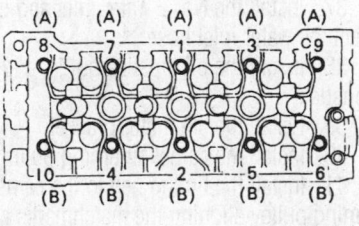

Cylinder head bolt positioning and tightening sequence—1.6L (4A-FE) and 1.8L (7A-FE) engines

CELICA

1. Release the fuel system pressure.
2. Disconnect the negative battery cable. On vehicles equipped with an air bag, wait at least 90 seconds before proceeding.
3. Drain the engine coolant into a suitable container.
4. Remove the air cleaner and cap assembly.
5. Disconnect the accelerator cable bracket from the throttle body. Disconnect the throttle cable if equipped with A/T.
6. Remove the alternator.
7. If equipped with A/C, remove the A/C drive belt and idler pulley.
8. Remove the P/S pump adjusting bracket and drive belt.

9. Disconnect the spark plug wires and remove the distributor assembly.
10. Remove the front TWC.
11. Remove the exhaust manifold.
12. Disconnect the radiator inlet hose and remove the water outlet.
13. Disconnect the electrical connectors and hoses from the water inlet and the water inlet housing and remove the housing.
14. Disconnect all of the hoses from the air chamber.
15. If equipped with an EGR, remove the EGR VSV and remove the intake manifold stay.
16. Remove the air pipe.
17. Remove the throttle body assembly and remove the air intake chamber.

✷✷ CAUTION

The fuel injection system remains under pressure after the engine has been turned OFF. Properly relieve fuel pressure before disconnecting any fuel lines. Failure to do so may result in fire or personal injury.

18. Remove the fuel delivery pipe and the fuel injectors.
19. Disconnect the engine wiring harness, if equipped with A/C, disconnect the A/C compressor connector. Disconnect the crankshaft position sensor connector.
20. Remove the intake manifold.
21. Remove the mounting bolt attaching the oil dipstick guide and the engine wiring harness bracket.
22. Remove the No. 2 water inlet.
23. Remove the cylinder head cover and the spark plugs.
24. Remove the No. 3 and the No. 2 timing belt covers.
25. Set No. 1 cylinder to TDC/compression.
26. Place matchmarks on the camshaft timing pulley and timing belt. Remove the plug from the No. 1 timing belt cover, loosen the idler pulley and push the pulley as far left as possible. Remove the timing belt from the camshaft timing pulley.
27. Remove the camshaft timing pulley.
28. Remove the alternator bracket.
29. Remove the camshafts following the proper sequences and procedures.
30. Using an SST 09205–16010 tool or its equivalent, uniformly loosen the bolts in the reverse order of the installation sequence. Remove the 10 cylinder head bolts in several passes and in sequence.

➡**Cylinder head warpage or cracking could result from removing the bolts in the incorrect order.**

31. Lift the cylinder head assembly from the cylinder block dowels and remove the cylinder head.

To install:
32. Clean the gasket mating surfaces using care not to damage the aluminum components, replace the gasket, then lower the cylinder head onto the engine. Be sure the dowel pins are aligned and no hoses or wires are between the head and cylinder block.
33. The cylinder head bolts are tightened in three progressive steps. Apply a light coat of engine oil to the cylinder head bolts. Uniformly tighten the 10 cylinder head bolts in several passes and in sequence to 22 ft. lbs. (29 Nm).

➡**The cylinder head bolts are in lengths of 3.54 in. (90mm) and 4.25 in. (108mm). The 3.54 in. (90mm) bolts (A) are to be installed in the intake side of the cylinder head. The 4.25 in. (108mm) bolts (B) are to be installed in the exhaust manifold side of the cylinder head.**

34. Mark the front of the cylinder head bolt with paint. Tighten the cylinder head bolts by 90° in sequence. Tighten an additional 90° and be sure that the paint mark is now positioned toward the rear.
35. Install the camshafts following the proper sequences and procedures.
36. Check and adjust valve clearance.
37. Install the No. 2 water inlet and connect the water inlet hose.
38. Install the oil dipstick guide, and the dipstick.
39. Install the alternator bracket.
40. Install the camshaft timing pulley.
41. Install the timing belt to the camshaft timing pulley aligning the matchmarks and properly tensioning the belt with the tensioning pulley. Loosen the pulley bolt ½ of a turn and turn the crankshaft clockwise two turns. Be sure that each pulley align with the timing marks.
42. Install the No. 2 and the No. 3 timing belt covers.
43. Install the spark plugs.
44. Install the semi-circular plug with sealant to the cylinder head and install the cylinder head cover with a new gasket to the cylinder head.
45. Install the intake manifold and the ground strap. Tighten the seven bolts and two nuts to 14 ft. lbs. (19 Nm).
46. Install the intake manifold stay and connect the engine wiring harness.
47. Install the injectors and the delivery pipe.
48. Install the air intake chamber.

49. Install the throttle body assembly.
50. Install the air pipe.
51. Install the intake manifold stay and if equipped with an EGR, install the EGR VSV.
52. Connect all of the hoses to the air intake chamber.
53. Install the water inlet and the water inlet housing and reconnect all hoses and connectors.
54. Install the water outlet.
55. Install the exhaust manifold.
56. Install the front TWC.
57. Install the front exhaust pipe.
58. Install the distributor and reconnect the spark plug wires.
59. Temporarily install the water pump pulley.
60. Install the P/S pump adjusting bracket and drive belt.
61. If equipped with A/C, install The A/C idler pulley and drive belt.
62. Install the alternator and drive belt.
63. Tighten the water pump pulley bolts.
64. Reconnect the accelerator cable bracket to the throttle body. Reconnect the throttle cable if equipped with A/T. If equipped with cruise control, install the cruise control actuator cable.
65. Install the air cleaner and cap assembly.
66. Reconnect the negative battery cable, fill the engine with coolant, start the engine, warm up, and check for leaks.
67. Install the RH engine undercover, check ignition timing, and road test for proper operation.

1.8L (1ZZ-FE) Engine

1. Disconnect the negative battery cable.
2. Drain the engine coolant.
3. Remove the drive belt and alternator.
4. Remove the air intake duct.
5. Disconnect the accelerator cable.
6. Disconnect the exhaust pipe from the manifold.
7. Remove the exhaust manifold support bracket.
8. Remove the exhaust manifold.
9. Remove the spark plug wires, then remove the ignition coils. Tag the wires so they can be returned to their original positions.
10. Remove the spark plugs.
11. Remove the PCV hoses.
12. Remove the throttle body assembly.
13. Remove the two bolts securing the wiring harness protector to the brackets on the intake manifold.
14. Unplug the wiring connectors and remove the ground wires from the cylinder head.

15. Remove the intake manifold.
16. Remove the engine hanger, camshaft position sensor and the ECT sensor.
17. Remove the PCV valve and grommet.
18. Remove the oil filler cap.
19. Remove the timing belt and camshaft sprockets.
20. Remove the camshafts.
21. Disconnect the hoses from the water hose union.
22. Remove the bolt attaching the water bypass pipe to the cylinder head.
23. Gradually remove the cylinder head bolts in sequence using a bi-hexagon wrench. To prevent damage to the cylinder head, loosen each bolt about ¼ of a turn during each pass until the bolts are loose.
24. Remove the cylinder head from the engine and place it on wooden blocks on the workbench.

To install:
25. Clean and degrease the surface of the engine block.
26. Place a new gasket on the engine block with the Lod No. stamp facing up.
27. Carefully place the cylinder head on the engine. Be careful not to damage the gasket.
28. Apply a light coat of oil to threads and under the heads of the cylinder head bolts. Replace any bolt that appears deformed.

29. Tighten the cylinder head bolts in sequence to 36 ft. lbs. (49 Nm).
30. Mark the front of each bolt with a dot of white paint.
31. Tighten each bolt in sequence an additional 90°. Be sure that each paint mark is now facing 90° from the front of the engine.
32. Install the water bypass pipe to the cylinder head. Tighten the bolt to 80 inch lbs. (9 Nm). Connect the hoses to the pipe.
33. Install the camshafts.
34. Check and adjust the valve clearance.
35. Install the camshaft timing sprockets.
36. Install the oil filler cap.
37. Install the grommet and PCV valve.
38. Install the ECT and camshaft position sensors.
39. Install the engine hanger. Tighten the bolt to 28 ft. lbs. (38 Nm).
40. Install the intake manifold.
41. Reconnect the harness wiring to the cylinder head and install the harness protector with the two bolts.

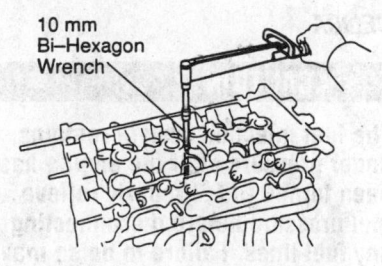

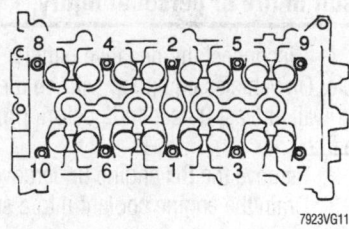

Be sure to tighten the cylinder head bolts in the sequence shown—1.8L (1ZZ-FE) engine

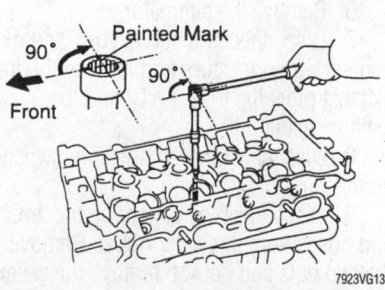

Tighten each bolt an additional 90° after making a paint mark on the front—1.8L (1ZZ-FE) engine

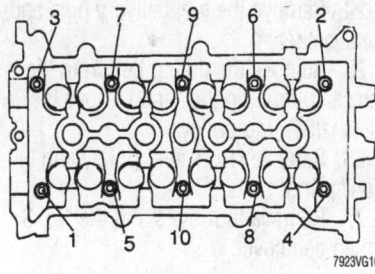

Be sure to remove the cylinder head bolts in the sequence shown—1.8L (1ZZ-FE) engine

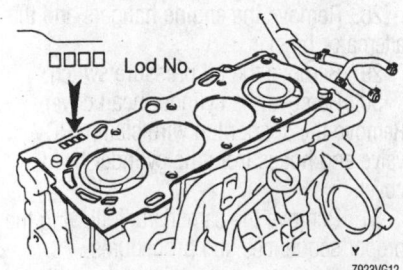

Position the head gasket correctly on the cylinder head—1.8L (1ZZ-FE) engine

42. Install the fuel injectors, throttle body and the PCV hoses.

43. Install the spark plugs and the ignition coils. Tighten the two nuts and two bolts to 80 inch lbs. (9 Nm).

44. Install the exhaust manifold.

45. Install the exhaust manifold and support bracket. Tighten the bolts alternately to 37 ft. lbs. (49 Nm).

46. Connect the front exhaust pipe to the manifold. Tighten the bolts to 46 ft. lbs. (62 Nm).

47. Using a new gasket and two new nuts, install the oxygen sensor. Tighten the nuts to 14 ft. lbs. (20 Nm).

48. Connect the accelerator cable and air intake duct.

49. Install the alternator and drive belt.

50. Refill the cooling system with coolant.

51. Connect the negative battery cable.

52. Start the engine and check for leaks.

53. Recheck the coolant and oil levels after the engine has cooled.

2.2L (5S-FE) Engine

CELICA

> ✳✳ **CAUTION**
>
> **The fuel injection system remains under pressure after the engine has been turned OFF. Properly relieve fuel pressure before disconnecting any fuel lines. Failure to do so may result in fire or personal injury.**

1. Disconnect the negative battery cable. On vehicles equipped with an air bag, wait at least 90 seconds before proceeding.

2. Remove the RH engine undercover.

3. Drain the engine coolant into a suitable container.

4. Remove the air cleaner and cap assembly.

5. Remove the spark plug wires and the distributor assembly.

6. Remove the alternator.

7. Disconnect the sub oxygen and oxygen sensor connectors and remove the front exhaust pipe, the front TWC, and the exhaust manifold.

8. Disconnect the oil pressure switch connector.

9. Disconnect the sensor connectors and hoses from the water outlet. Remove the two nuts and gasket; remove the water outlet.

10. Disconnect the hoses; remove the heat protector and the water bypass pipe.

11. Remove the throttle body assembly.

12. Disconnect the vacuum hoses from the intake manifold and remove the A/C idle-up valve.

13. Remove the EGR valve, vacuum modulator, vacuum hoses and gasket.

14. Remove the intake manifold stay and disconnect the A/T throttle control cable.

15. Remove the air hoses from the air tube and remove the air tube assembly.

16. Except California vehicles, disconnect the sensing hoses and remove the vacuum pipe.

17. Disconnect the knock sensor connector, remove the bolt and the ground cable from the intake manifold.

18. On California vehicles, remove the VSV for fuel pressure control and EGR. On all vehicles (except California), remove the VSV for EGR.

19. Disconnect the PCV hose from the intake manifold and disconnect the A/T throttle control cable and bracket from the intake manifold.

20. Disconnect the engine wiring harness from the starter bracket. Disconnect the VSS sensor connector and the wiring harness protector from the LH side of the manifold.

21. Disconnect the fuel inlet hose from the delivery pipe and disconnect the fuel return hose from the return pipe.

22. Remove the six bolts, two nuts, and the intake manifold and gasket.

23. On California vehicles, remove the air hose for the air assist system.

24. Remove the fuel delivery pipe and the injectors.

25. Remove the timing belt from the camshaft timing pulley and remove the camshaft timing pulley.

26. Remove No. 1 idler pulley and tension spring.

27. Remove four bolts and the No. 3 timing belt cover.

➡ **Support the timing belt, so that the meshing of the crankshaft timing pulley and the timing belt does not shift. Be careful not to drop anything inside the timing belt cover.**

28. Remove the engine hangers and the alternator bracket.

29. Remove the oil pressure switch.

30. Remove the cylinder head cover. Remove the spark plug wire clamp, PCV valve and hoses from the cylinder head cover.

31. Remove the camshafts following the proper sequences and procedures.

32. Uniformly loosen and remove the cylinder head bolts in several passes and in the reverse order of the installation sequence. Lift the cylinder head from the cylinder block disengaging the cylinder head from the block dowel pins.

To install:

33. Clean the gasket mating surfaces using care not to damage the aluminum components, replace the gasket, then lower the cylinder head onto the engine. Be sure the dowel pins are aligned and no hoses or wires are between the head and cylinder block.

34. The cylinder head bolts are tightened in two progressive steps. Apply a light coat of engine oil to the cylinder head bolts. Uniformly tighten the 10 cylinder head bolts in several passes and in sequence to 36 ft. lbs. Mark the front of the cylinder head bolt with paint. Tighten the cylinder head bolts by 90° in sequence. Tighten an additional 90° and be sure that the paint mark is now positioned toward the rear.

35. Install the camshafts following the proper sequences and procedures.

36. Check and adjust valve clearance.

37. Install sealant to the two new semicircular seals and install the seals to the cylinder head.

38. Install the PCV valve and hoses and the spark plug wire clamp to the cylinder head cover.

39. Install the cylinder head cover with a new gasket, the four grommets and nuts and uniformly tighten the nuts in several passes. Tighten the nuts to 17 ft. lbs. (23 Nm).

40. Install the oil pressure switch.

41. Install the alternator bracket and the engine hangers.

42. Install the No. 3 timing belt cover and temporarily install the No. 1 idler pulley and tension spring.

43. Install the camshaft timing pulley and install the timing belt to the pulley. Correctly tension the timing belt and be sure that the belt timing is correct.

44. Install the fuel injectors and the delivery pipe.

45. On California cars, install the air hose for the air assist system.

46. Install the intake manifold and insert the engine wiring harness between the head and the intake manifold, install a new gaskets the six bolts and two nuts, and tighten the intake manifold to 25 ft. lbs. (34 Nm)

47. Connect the fuel inlet and return hoses to the delivery and return pipes.

48. Install the engine wire protector and harness to the intake manifold and connect the four fuel injector connectors. Connect the VSS connector.

49. Install the cable bracket to the intake manifold and install the accelerator, A/T control cables and bracket.

50. Connect the PCV hose to the intake manifold.

51. On California vehicles, install the VSV assembly for fuel pressure control and EGR.

52. Except for California vehicles, install the VSV for EGR.

53. Connect the knock sensor connector and install the ground cable with the bolt.

54. Except for California vehicles, install the sensing hoses to the vacuum pipe and mount the pipe with the bolt.

55. Install the air tube and the bracket for the EGR. Connect the hoses to the air pipe.

56. Install the A/T throttle control cable to the clamp on the rear side of the intake manifold. Install the manifold stay and tighten the bolt to 15 ft. lbs. (21 Nm) and tighten the nut to 32 ft. lbs. (44 Nm).

57. Install the EGR valve and the vacuum modulator using a new gasket, connect the two vacuum hoses to the VSV for the EGR, and connect the EGR gas temperature sensor connector. Connect the EVAP hose to the charcoal canister.

58. Install the vacuum sensor hose to the gas filter and the brake booster vacuum hose to the intake manifold.

59. Install the A/C idle up valve.

60. Install the throttle body.

61. Connect the water bypass pipe to the water pump cover; install the water bypass pipe and connect the hoses to the water bypass pipe.

62. Install the water outlet with a new gasket and the two nuts and install the water hoses, vacuum hoses, and connect the sensor connectors.

63. Connect the oil pressure switch connector.

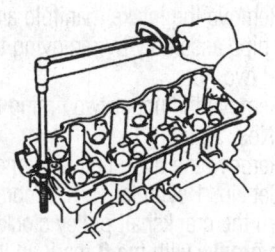

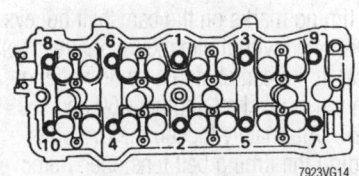

Tighten the cylinder head bolts in the sequence shown—2.2L (5S-FE) engine

64. Assemble the TWC to the exhaust manifold and install the main oxygen sensor to the exhaust manifold. Install the sub oxygen sensor to the TWC.

65. Install the exhaust manifold to the cylinder head and connect the oxygen sensors connectors.

66. Install the front exhaust pipe and install the alternator with the drive belt.

67. Install the distributor assembly and connect the spark plug wires.

68. Install the air cleaner and cap assembly.

69. Connect the negative battery cable, fill the engine with coolant, start the engine, warm up, and check for leaks. Bleed the cooling system and top off coolant as necessary.

70. Install the RH engine undercover, check ignition timing, and road test the vehicle for proper operation.

CAMRY

1. Release the fuel system pressure.

2. Disconnect the negative battery cable. On vehicles equipped with an air bag, wait at least 90 seconds before proceeding.

3. Drain the engine coolant into a suitable container.

4. Disconnect the A/T throttle control cable and accelerator cable.

5. Remove the air cleaner and cap assembly.

6. Remove the alternator.

7. Remove the spark plug wires and the distributor assembly.

8. Disconnect the sub oxygen and oxygen sensor connectors and remove the front exhaust pipe, the front TWC, and the exhaust manifold.

9. Disconnect the oil pressure switch connector.

10. Disconnect the sensor connectors and hoses from the water outlet.

11. Remove the two nuts, water outlet and gasket.

12. Disconnect the hoses and remove the water bypass pipe.

13. Remove the throttle body assembly.

14. Remove the EGR valve, vacuum modulator, vacuum hoses assembly, and the gasket.

15. Disconnect the PCV hose from the intake manifold.

16. Remove the air hoses from the air tube and remove the air tube assembly.

17. Remove the air tube as follows:

   a. Remove the two power steering hoses from the air tube.

   b. Remove the power steering air hose from the intake manifold.

   c. Remove the air hose from the fuel pressure regulator.

d. Remove the three bolts and air tube. Disconnect the ground cable and clamp bracket for the engine wire.

18. Remove the bolt and disconnect the ground cable from the intake manifold.

19. Disconnect the knock sensor connector.

20. Remove the bolt and disconnect the engine wire protector from the left-hand side of the intake manifold.

21. Disconnect the four injector connectors.

22. Disconnect the engine wire protector from the two brackets on the front side of the intake manifold.

23. Disconnect the engine wire protector from the two mounting bolts of the No. 2 timing belt cover.

24. Remove the four bolts, wire bracket, No. 1 air intake chamber and intake manifold stays.

25. Remove the six bolts and two nuts and remove the intake manifold and gasket.

❋❋ CAUTION

The fuel injection system remains under pressure after the engine has been turned OFF. Properly relieve fuel pressure before disconnecting any fuel lines. Failure to do so may result in fire or personal injury.

26. Remove the fuel delivery pipe and the injectors.

27. Remove the timing belt from the camshaft timing pulley. Remove the camshaft timing pulley.

➡**Support the timing belt, so that the meshing of the crankshaft timing pulley and the timing belt does not shift. Be careful not to drop anything inside the timing belt cover.**

28. Remove No. 1 idler pulley and tension spring.

29. Remove four bolts and the No. 3 timing belt cover.

30. Remove the engine hangers and the alternator bracket.

31. Remove the oil pressure switch.

32. Remove the cylinder head cover.

33. Remove the camshafts following the proper sequences and procedures.

34. Uniformly loosen and remove the cylinder head bolts in several passes and in the reverse order of the installation sequence. Lift the cylinder head from the cylinder block disengaging the cylinder head from the block dowels.

To install:

35. Clean the gasket mating surfaces, but be careful not to damage the aluminum

components, replace the gasket, then lower the cylinder head onto the engine. Be sure the dowel pins are aligned and no hoses or wires are between the head and cylinder block.

36. The cylinder head bolts are tightened in two progressive steps. Apply a light coat of engine oil to the cylinder head bolts. Uniformly tighten the 10 cylinder head bolts in several passes and in sequence to 36 ft. lbs. (49 Nm). Mark the front of the cylinder head bolt with paint. Tighten the cylinder head bolts an additional 90° in the proper sequence. The paint mark should now be 90° from the front.

37. Install the camshafts following the proper sequences and procedures.

38. Check and adjust valve clearance.

39. Install sealant to the two new semicircular seal and install the seals to the cylinder head.

40. Install the cylinder head cover with a new gasket, the four grommets and nuts and uniformly tighten the nuts in several passes. Tighten the nuts to 17 ft. lbs. (23 Nm).

41. Install the oil pressure switch.

42. Install the alternator bracket and the engine hangers.

43. Install the No. 3 timing belt cover and temporarily install the No. 1 idler pulley and tension spring.

44. Install the camshaft timing pulley and install the timing belt to the pulley. Correctly tension the timing belt and be sure that the belt timing is correct.

45. Install the fuel injectors and the delivery pipe.

46. Install the intake manifold.

47. Install the engine wire protector and harness to the intake manifold and connect the four fuel injector connectors.

48. Install the air tube.

49. Connect the PCV hose to the intake manifold.

50. Install the EGR valve and the vacuum modulator using a new gasket, connect the two vacuum hoses to the VSV for the EGR, and connect the EGR gas temperature sensor connector.

51. Install the throttle body.

52. Connect the water bypass pipe to the water pump cover, install the water bypass pipe and reconnect the hoses to the water bypass pipe.

53. Install the water outlet with a new gasket and the two nuts. Tighten the nuts to 11 ft. lbs. (15 Nm), Install the water hoses, vacuum hoses, and reconnect the sensor connectors.

54. Connect the oil pressure switch connector.

55. Install the exhaust manifold to the cylinder head. Uniformly tighten the nuts in several passes to 36 ft. lbs. (49 Nm).

56. Install the manifold stay with the bolt and nut. Tighten the bolt and nut to 31 ft. lbs. (42 Nm).

57. Install the No. 1 manifold stay with the bolt and nut. Tighten the bolt and nut to 31 ft. lbs. (42 Nm).

58. Install the manifold upper heat insulator with the four bolts.

59. Install the main oxygen sensor to the exhaust manifold. Install the sub oxygen sensor to the TWC.

60. Install the front exhaust pipe to the catalytic converter and tighten the nuts to 46 ft. lbs. (62 Nm).

61. Install the alternator and drive belt.

62. Install the distributor assembly and reconnect the spark plug wires.

63. Install the air cleaner and cap assembly.

64. For vehicles with A/T, connect and adjust the throttle cable and accelerator cable.

65. Fill the engine with coolant.

66. Check all fluids.

67. Connect the negative battery cable to the battery.

68. Start the engine and check for leaks.

3.0L (2JZ-GTE) Engine

❊❊ CAUTION

The fuel injection system remains under pressure after the engine has been turned OFF. Properly relieve fuel pressure before disconnecting any fuel lines. Failure to do so may result in fire or personal injury.

1. Relieve the fuel pressure in the fuel line before disconnecting any fuel lines.

2. Disconnect the negative battery cable from the battery. On vehicles equipped with an air bag, wait at least 90 seconds before proceeding.

3. Drain the engine coolant from the engine and radiator.

4. Remove the turbocharger from the engine.

5. Remove the exhaust manifold by removing the 12 nuts and two gaskets.

6. If equipped with manual transmission, remove the drive belt tensioner damper by removing the two nuts.

7. Remove the drive belt by turning the tensioner clockwise.

8. Remove the water outlet and the No. 1 water bypass pipe.

9. Disconnect the power steering pump without disconnecting the hoses as follows:

a. Disconnect the air hose from the throttle body.

b. Disconnect the air hose from the air intake chamber.

c. Remove the power steering pump housing from the pump bracket by removing the two bolts.

d. Suspend the pump housing without disconnecting the hoses.

10. Disconnect the fuel return hose from the fuel return pipe. Plug the hose end.

11. Remove the air intake chamber assembly.

12. Disconnect the six injector connectors.

13. Disconnect the two camshaft position sensor connectors.

14. Disconnect the three engine wire clamps from the injector holders.

15. Disconnect the VSV connector for the EVAP.

16. Remove the two ground straps from the intake manifold by removing the two bolts.

17. Disconnect the engine wire protector from the intake manifold by removing the nut.

18. Remove the starter.

19. Remove the pressure tank and VSV assembly.

❊❊ CAUTION

The fuel injection system remains under pressure, even after the engine has been turned OFF. The fuel system pressure must be relieved before disconnecting any fuel lines. Failure to do so may result in fire and/or explosion.

20. Remove the fuel pressure pulsation damper.

21. Remove the fuel inlet pipe by disconnecting the union bolt and clamp bolt.

22. Remove the intake manifold and delivery pipe assembly by removing the four bolts and two nuts.

23. Remove the upper two timing belt covers (Nos. 2 and 3).

24. Remove the drive belt tensioner.

25. Set No. 1 cylinder to TDC/compression. Turn the crankshaft pulley clockwise to align the groove with the **0** mark on the lower (No. 1) timing belt cover. Check that the timing marks on the camshaft pulleys are aligned with the marks on the rear belt cover. If the camshaft marks do not align, turn the crankshaft another 360 degrees.

26. Alternately loosen the two bolts holding the timing belt tensioner. Remove the bolts and remove the tensioner.

27. Remove the timing belt from the camshaft pulleys. If the belt is to be reused,

place matchmarks on the belt and gears before removing the belt. Mark the belt with an arrow to show direction of rotation.

28. Remove the ignition coils.

29. Remove the spark plugs.

30. Remove the cylinder head covers.

31. While holding each camshaft with a wrench, remove the camshaft bolts and remove the gears.

32. Remove the four bolts and lift out the No. 4 (inner) timing belt cover.

33. Remove the camshafts following the proper sequences and procedures.

34. Remove the cylinder head from the engine block as follows:

a. Using a 10mm bi-hexagon wrench, uniformly loosen and remove the 14 cylinder head bolts. Loosen the bolts in several passes and in the reverse order of the installation sequence.

b. Remove the 14 plate washers.

c. Lift the cylinder head from the dowels on the cylinder block.

d. Place the head on wooden blocks on a bench.

35. Remove the engine hangers and the ground strap.

36. Remove the camshaft position sensors.

37. Remove the EGR cooler.

38. Remove the valve lifters and shims. Be sure to make a note to the positions of the lifters and shims. When installing the lifters and shims, install them in the same position as removal.

39. Using SST 09202–70010 or equivalent, compress the valve spring and remove the two keepers.

40. Remove the spring retainer, valve spring, valve and spring seat.

41. Using needle nose pliers, remove the oil seal.

To install:

42. Install new valve oil seals and assemble the cylinder head.

43. Install the engine hangers and the ground strap. Tighten the mounting bolts to 29 ft. lbs. (39 Nm).

44. Install the camshaft position sensors and the EGR cooler. Tighten the cooler and sensor mounting bolts to 78 inch lbs. (9 Nm).

45. Install the cylinder head as follows:

a. Clean the cylinder head and block.

b. Place a new cylinder head gasket in position on the cylinder block.

c. Place the cylinder head in position on the cylinder head gasket.

d. Lightly coat the head bolt threads and plate washers with engine oil. Install plate washers and bolts to the head.

e. Uniformly tighten the head bolts in several passes in the correct order.

Tighten the bolts to 25 ft. lbs. (34 Nm).

f. Mark the front (towards the front of the engine) of each bolt with a dot of paint. Following the correct order, tighten each bolt an additional 90 degrees. When complete, all the paint marks should face to the side of the engine.

g. Again following the correct order, tighten the bolts another 90 degrees of rotation. When complete, the paint marks should face the rear of the engine, exactly 180 degrees away from the original starting point.

➡**Correct bolt torque must be achieved in three steps; do not attempt to shorten the procedure by combining the two 90 degree steps.**

46. Install the cylinder head covers.

47. Install the spark plugs.

48. Install the ignition coils.

49. Install the timing belt.

50. Install the intake manifold and delivery pipe assembly by installing a new gasket, engine wire and the four bolts and two nuts. Tighten the nuts and bolts to 20 ft. lbs. (27 Nm).

51. Install the fuel inlet pipe by installing a new gasket and the union bolt. Tighten the union bolt to 30 ft. lbs. (42 Nm). Install the fuel inlet pipe clamp bolt to the intake manifold.

52. Install the fuel pressure pulsation damper.

53. Install the pressure tank and VSV assembly.

54. Install the starter.

55. Connect the engine wire as follows:

a. Install the engine wire protector to the intake manifold with the nut.

b. Install the two ground straps to the intake manifold with the bolts.

c. Connect the following connectors and clamps:

• VSV connector for EVAP

• Six injector connectors

• Two camshaft position sensor connectors

• Three engine wire clamps to injector holders

56. Install the air intake chamber assembly.

57. Connect the fuel return hose

58. Install the power steering pump. Tighten the bolts to 43 ft. lbs. (58 Nm).

59. Install the water outlet and No. 1 water bypass pipe as follows:

a. Install the two O-rings to the No. 1 water bypass pipe.

b. Apply soapy water to the O-rings.

c. Install the No. 1 water bypass pipe to the water pump.

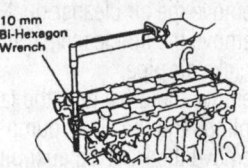

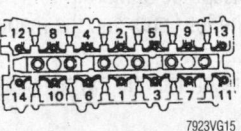

7923VG15

To prevent leaks and damage to the cylinder head, tighten the bolts in the sequence shown—3.0L (2JZ-GE and 2JZ-GTE) engines

d. Install a new gasket and the water outlet with the two bolts. Tighten the bolts to 15 ft. lbs. (21 Nm).

e. Connect the ECT sensor and sender gauge connectors.

f. Connect the upper radiator hose to the water outlet.

60. Install the drive belt.

61. If equipped with manual transmission, install the drive belt tensioner damper. Tighten the two nuts to 14 ft. lbs. (20 Nm).

62. Place two new gaskets to the cylinder head.

63. Install the exhaust manifold with 12 new nuts and tighten the nuts in several passes. Tighten the bolts to 29 ft. lbs. (39 Nm).

64. Install the turbocharger.

65. Fill the engine with coolant.

66. Connect the negative battery cable to the battery.

67. Start the engine and check for leaks.

3.0L (2JZ-GE) Engine

1. Relieve the fuel pressure in the fuel lines.

✳✳ CAUTION

The fuel injection system remains under pressure after the engine has been turned OFF. Properly relieve fuel pressure before disconnecting any fuel lines. Failure to do so may result in fire or personal injury.

2. Disconnect the negative battery cable. Wait at least 90 seconds before performing any other work.

3. Drain the engine coolant.

4. Remove the undercovers from the vehicle.

5. Disconnect the accelerator, throttle control (A/T only) and cruise control cables from the throttle body.

6. Remove the air cleaner duct.

7. Remove the air cleaner, airflow meter, and the intake air pipe.

8. Remove the drive belt, the fan and fluid coupling, and the water pump pulley.

9. Remove the front exhaust pipe.

10. Except for California vehicles, remove the four nuts and the manifold heat insulator.

11. Disconnect the oxygen sensor connector(s).

12. Remove the exhaust manifolds and gaskets by removing the eight bolts.

13. Disconnect the power steering pump without disconnecting the hoses as follows:

a. Disconnect the air hose from the No. 4 timing belt cover.

b. Disconnect the air hose from the intake chamber.

c. Disconnect the power steering pump housing from the pump bracket by removing the two bolts.

d. Remove the two bolts and the pump rear stay.

14. Disconnect the brake booster vacuum hose.

15. Disconnect the EVAP hose.

16. Remove the throttle body and intake air connector assembly.

17. Remove the air intake chamber stays by removing the nut and bolt to each stay.

18. Remove the No. 2 vacuum pipe and VSV assembly.

19. Remove the No. 3 timing belt cover by removing the oil filler cap and six bolts.

20. Remove the cylinder head rear cover by removing the four bolts.

21. Disconnect the spark plug wires from the cylinder head covers.

22. Remove the distributor and wires from the engine.

23. Remove the spark plugs.

24. Remove the timing belt.

25. Remove the water bypass outlet and No. 1 bypass pipe.

26. Disconnect the fuel return hose from the fuel return pipe.

27. Disconnect the fuel return hose from the oil dipstick guide.

28. Remove the engine wire bracket from the intake manifold by removing the bolt.

29. Remove the oil dipstick and guide for the engine.

30. Remove the starter.

31. Remove the air intake chamber as follows:

a. Except for California vehicles, disconnect the vacuum sensing hose from the fuel pressure regulator.

b. Remove the bolt holding the engine wire protector to the air intake chamber.

c. Remove the five bolts, nut, air intake chamber, and the gasket.

32. Remove the vacuum control valve set by disconnecting the VSV connector, then removing the two nuts.

33. Disconnect the engine wire from the intake manifold as follows:

a. Remove the bolt and disconnect the engine wire bracket from the water pump.

b. Remove the two bolts and disconnect the two ground straps from the intake manifold.

c. Remove the two bolts and disconnect the two wire clamps from the intake manifold.

d. Disconnect the following connectors:
• Six injector connectors
• ECT sensor connector
• ECT sender gauge connector

e. Disconnect the engine wire protector from the intake manifold by removing the three nuts.

34. Remove the water outlet and No. 1 bypass hose assembly by removing the bolt and two nuts.

35. Remove the intake manifold stay by removing the two bolts.

36. Remove the fuel pressure pulsation damper.

37. Remove the fuel inlet pipe.

38. Remove the two nuts and six bolts to the intake manifold.

39. Remove the intake manifold and gaskets.

40. Remove the cylinder head covers (valve covers).

41. Remove the camshaft timing pulleys. Hold the hexagon portion of the camshaft with a wrench and remove the pulley mounting bolt and camshaft pulley.

42. Remove the No. 4 timing belt cover.

43. Remove the camshafts following proper sequences and procedures.

44. Uniformly loosen the cylinder head bolts in several passes and in the reverse order of the installation sequence.

☀ WARNING

The cylinder head may crack or warp if the correct order is not followed.

45. Remove the head from the engine. If prying is necessary, use a protected blade; take great care not to damage the mating surfaces of the head and block.

46. Clean the head and block of all gasket material. Take great care not to gouge or scratch the mating surfaces.

To install:

47. Place a new gasket on the cylinder block. Be sure it is positioned correctly. Install the cylinder head in position.

48. Lightly coat the head bolt threads and plate washers with engine oil. Install plate washers and bolts to the head.

49. Uniformly tighten the head bolts in several passes in the correct order. Tighten the bolts to 25 ft. lbs. (34 Nm).

50. Mark the front (towards the front of the engine) of each bolt with a dot of paint. Following the correct order and tighten each bolt an additional 90 degrees. When complete, all the paint marks should face to the side of the engine.

51. Again following the correct order, tighten the bolts another 90 degrees of rotation. When complete, the paint marks should face the rear of the engine, exactly 180 degrees away from the original starting point.

➡**Correct bolt torque must be achieved in three steps; do not attempt to shorten the procedure by combining the two 90 degree steps.**

52. Coat the camshaft with engine oil, then position them in the cylinder head with the cam lobes and the knock pins in the correct position.

53. Position the No. 3 and No. 7 bearing caps in place, coat the bolt threads with oil, then uniformly and alternately tighten them temporarily.

54. Coat the new oil seals with multipurpose grease, then slide them over the camshafts.

55. Clean the surfaces of the No. 1 bearing cap and cylinder head with cleaner. Apply seal packing to the No. 1 bearing cap.

56. Install the camshafts following the proper sequences and procedures.

57. Using SST tool 09316–60010 or equivalent, press the two oil seals in as far as it will go.

58. Rotate each camshaft until the forward straight (knock) pin is straight up. Loosen exhaust No. 1, 2, and 6 bearing cap bolts until they can be turned by hand; tighten, in several passes, to 14 ft. lbs. (20 Nm). Loosen intake No. 1, 2, and 5 and tighten, in several passes, to 14 ft. lbs. (20 Nm).

59. Turn each camshaft ⅓ of a revolution (120 degrees). Loosen exhaust Nos. 4 and 7 bearing cap bolts; tighten, in several passes, to 14 ft. lbs. (20 Nm). Loosen intake No. 4 and 6 bearing cap bolts; tighten, in several passes, to 14 ft. lbs. (20 Nm).

60. Turn each camshaft an additional ⅓ of a revolution, loosen exhaust bearing cap bolts Nos. 3 and 5, then tighten them, in several passes, to 14 ft. lbs. (20 Nm). Loosen intake bearing cap bolts No. 3 and 7, then tighten them, in several passes to 14 ft. lbs. (20 Nm).

61. Check and adjust the valve clearance.

62. Install the No. 4 timing belt cover. Tighten the bolts to 78 inch lbs. (9 Nm).

63. Install the camshaft timing pulleys. Align the shaft pin with the pulley groove and slide the pulley on. Install the bolt temporarily. Hold the hex portion of the camshaft with a wrench and tighten the pulley bolt to 59 ft. lbs. (79 Nm).

64. Install the cylinder head covers.

65. Install the intake manifold with a new gasket. Tighten the bolts to 20 ft. lbs. (27 Nm).

66. Install the injectors and the delivery pipe. Tighten the bolts holding the pipe to the manifold to 20 ft. lbs. (27 Nm).

67. Install the fuel inlet pipe with two new gaskets and tighten the union bolt. Tighten the union bolt to 30 ft. lbs. (42 Nm). Install the clamp bolt to the intake manifold.

68. Install the fuel pressure pulsation damper.

69. Install the intake manifold stay and tighten the bolts to 29 ft. lbs. (39 Nm).

70. Install the water outlet and No. 1 bypass hose assembly.

71. Connect the engine wire as follows:

 a. Install the engine wire protector to the intake manifold with the three nuts.

 b. Connect the six injector connectors.

 c. connect the ECT sensor connector.

 d. Connect the ECT sender gauge connector.

 e. Install the two wire clamps to the intake manifold with the bolts.

 f. Install the two ground straps to the intake manifold with the bolts.

 g. Install the engine wire bracket to the water pump with the bolt.

72. Install the vacuum control valve set. Tighten the bolts to 15 ft. lbs. (21 Nm). Connect the VSV connector.

73. Install the air intake chamber as follows:

 a. Install a new gasket and the intake chamber with the nut and five bolts. Tighten the bolts and nut to 20 ft. lbs. (27 Nm).

 b. Install the bolt holding the engine wire protector to the air intake chamber.

 c. Except for California vehicles, connect the vacuum sensing hose to the fuel pressure regulator.

74. Install the starter.

75. Install the oil and transmission dipstick tubes. Always use a new O-ring on each tube.

76. Install the engine wire bracket.

77. Connect the fuel return hose.

78. Install the water bypass outlet and No. 1 water bypass pipe.

79. Install the timing belt as follows:

 a. Turn the crankshaft pulley and align its groove with the timing mark, **0** on the No. 1 timing belt cover.

 b. Align the timing marks on the camshaft timing gears and the No. 4 timing belt cover.

 c. Install the timing belt.

 d. Double check that all the timing marks for the crankshaft pulley and the camshaft gears are aligned as they were during disassembly.

 e. Set the timing belt tensioner:

• Use a press to slowly push in the pushrod on the tensioner. This will require between 220–2200 pounds of pressure.

• Align the holes of the pushrod and housing. Place a 1.5mm hex wrench through the holes to keep the pushrod retracted.

• Release the press and install the dust boot onto the tensioner.

 f. Install the tensioner; alternately tighten the bolts to 20 ft. lbs. (26 Nm).

 g. Remove the hex wrench from the tensioner with a pair of pliers.

 h. Turn the crankshaft pulley two full turns clockwise. Check that each pulley's timing marks align correctly after the two turns. If any mark does not align, remove the timing belt and reinstall it.

80. Install the drive belt tensioner by installing the three bolts. Tighten the bolts to 15 ft. lbs. (21 Nm).

81. Install the No. 2 timing belt cover.

82. Install the spark plugs.

83. Install the distributor and spark plug wires to the engine.

84. Connect the spark plug wires to the cylinder head covers.

85. Install the No. 3 timing belt cover.

86. Install the cylinder head rear cover.

87. Install the No. 2 vacuum pipe and VSV assembly.

88. Install the air intake chamber stays with the nut and bolt for each stay. Tighten the bolt and nut to 13 ft. lbs. (18 Nm).

89. Install the throttle body and intake air connector assembly.

90. Connect the EVAP hose.

91. Connect the brake booster vacuum hose.

92. Install the power steering pump.

93. Install the exhaust manifolds.

94. Install the No. 2 front exhaust pipe.

95. Install the water pump pulley, fan, fluid coupling assembly, and the drive belt. Tighten the four bolts for the pulley to 12 ft. lbs. (16 Nm).

96. Install the air cleaner, VAF meter, and the intake air connector pipe assembly.

97. Install the air cleaner duct.

98. Connect the cruise control cable, throttle control, and accelerator cables to the throttle body.

99. Fill the engine with coolant.

100. Connect the negative battery cable to the battery.

101. Start the engine and check for leaks.

102. Check the ignition timing.

103. Install the engine undercover.

104. Perform a road test.

3.0L (1MZ-FE) Engine

✸✸ CAUTION

The fuel injection system remains under pressure even after the engine has been turned OFF. The fuel system pressure must be relieved before disconnecting any fuel lines. Failure to do so may result in fire and/or personal injury.

1. Relieve the fuel pressure.

2. Disconnect the negative battery cable. On vehicles equipped with an air bag, wait at least 90 seconds before proceeding.

3. Drain the cooling system.

4. Disconnect the accelerator cable and the throttle cable on vehicles equipped with an automatic transaxle.

5. Remove the air cleaner cover, air flow meter, and the air duct.

6. Remove the cruise control actuator and bracket, if equipped.

7. Disconnect the two engine ground straps.

8. Remove the right engine mounting support.

9. Disconnect the radiator hoses.

10. Disconnect the two heater hoses.

11. Disconnect and plug the fuel feed and return lines from the fuel rail assembly.

12. Disconnect and plug the pressure hose from the hydraulic motor.

13. Remove the V-bank cover.

14. Disconnect the following vacuum hoses:

 a. Fuel pressure control VSV

 b. Fuel pressure regulator

 c. Cylinder head rear plate

 d. Intake air control valve VSV

 e. EGR vacuum modulator

 f. EGR valve

15. Disconnect the following connectors:

 a. Intake air control valve

 b. Fuel pressure regulator

 c. EGR VSV

16. Remove the two nuts and the emission control valve set.

17. Disconnect the following hoses;

a. Brake booster vacuum hose

b. PCV hose

c. Intake air control valve vacuum hose

18. Remove the data link connector from the mounting bracket.

19. Remove the two ground straps from the intake chamber.

20. Remove the hydraulic motor pressure hose from the intake chamber.

21. Remove the right oxygen sensor connector from the P/S pressure tube.

22. Remove the two nuts and the P/S pressure tube from the intake chamber.

23. Disconnect the two P/S air hoses.

24. Remove the engine hanger and the intake chamber support.

25. Remove the EGR pipe and gaskets.

26. Disconnect the following connectors;

a. Throttle position sensor connector

b. IAC valve connector

c. EGR gas temperature connector

d. A/C idle up connector

27. Disconnect the following vacuum hoses:

a. Two vacuum hoses from the TVV

b. Vacuum hose from the cylinder head rear plate

c. Vacuum hose from the charcoal canister

28. Disconnect the air assist hose and the two water bypass hoses.

29. Remove the air intake chamber.

30. Disconnect the left engine wiring harness and position it out of the way.

31. Remove the wiring harness from the rear of the engine.

32. Disconnect the right engine wiring harness and position it out of the way.

33. Remove the ignition coils and the spark plugs.

34. Remove the timing belt.

35. Remove the camshaft pulleys and the timing belt rear cover.

36. Remove the cylinder head rear plate.

37. Remove the water inlet pipe.

38. Remove the air assist hose and vacuum hose.

39. Remove the intake manifold and fuel rail assembly.

40. Remove the water outlet.

41. Remove the EGR pipe from the right exhaust manifold.

42. Remove the exhaust manifolds.

43. Remove the dipstick assembly and the P/S pump bracket.

44. Remove the valve covers and the camshaft position sensor.

45. Remove the camshafts following the proper sequences and procedures.

46. Be sure the engine is at or near ambient temperature and remove the two

(one on each head) 8mm recessed hex bolts. Loosen and remove the eight head bolts evenly, in three passes, in the reverse order of the installation sequence. Carefully lift the head from the engine; if it is necessary to pry the head loose, take great care not to damage the mating surfaces. Place the head on wood blocks in a clean work area.

➡️**If the cylinder head bolts are loosened out of sequence, warpage or cracking could result.**

47. Remove the cylinder head gasket. With a gasket scraper, carefully remove all the old gasket material from the cylinder head and engine block surfaces.

To install:

48. Place the new cylinder head gasket onto the cylinder block. Place the cylinder head onto the gasket.

49. Coat the threads of the eight cylinder head bolts (12-sided) with clean engine oil and install the bolts into the cylinder head. Uniformly tighten the bolts in sequence in three steps to an ultimate torque of 40 ft. lbs. (54 Nm), using the proper sequence.

50. Mark the forward edge of each bolt with paint, then tighten each bolt, in proper sequence, an additional 90 degrees. Check that each painted mark is now at a 90 degrees angle to the front. The paint mark should have been applied to the bolt in the 9 o'clock position and should now be in the 12 o'clock position.

51. Coat the threads of the two remaining 8mm bolts with engine oil and install them. Tighten to 13 ft. lbs. (18 Nm).

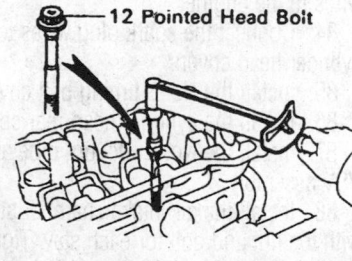

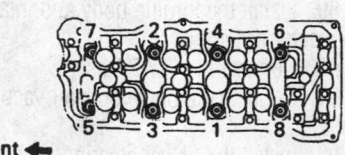

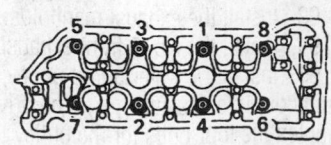

Front ⬅️

7923VG16

Cylinder head tightening sequence—3.0L (1MZ-FE) engine

52. Install the camshafts following the proper sequences and procedures.

53. Check and adjust the valves.

54. Apply sealant to the cylinder heads where the camshaft supports meet the cylinder heads.

55. Use new gaskets and install the cylinder head covers.

56. Install the dipstick and power steering pump bracket.

57. Install the exhaust manifolds. Tighten the nuts to 36 ft. lbs. (49 Nm).

58. Install the EGR pipe to the right exhaust manifold.

59. Install the water outlet.

60. Install the intake manifold and the fuel rail assembly. Tighten the intake manifold nuts and bolts to 11 ft. lbs. (15 Nm).

61. Install the air assist hose and the two water bypass hoses.

62. Install the water inlet pipe and the cylinder head rear plate.

63. Install the timing belt rear cover and the camshaft pulleys.

64. Install the timing belt.

65. Install the spark plugs and the ignition coils.

66. Install the right engine wiring harness.

67. Install the wiring harness to the rear of the engine.

68. Install the left engine wiring harness.

69. Install the air intake chamber.

70. Use new gaskets and install the EGR pipe.

71. Connect the following vacuum hoses:

a. The two TVV vacuum hoses.

b. The vacuum hose to the rear cylinder head plate.

c. Charcoal canister vacuum hose.

72. Connect the following electrical connectors:

a. Throttle position sensor connector.

b. IAC valve connector.

c. EGR gas temperature connector.

d. A/C idle up connector.

73. Install the engine hanger and the intake chamber support.

74. Connect the two P/S air hoses.

75. Install the P/S pressure tube to the intake chamber.

76. Install the oxygen sensor connector to the pressure tube.

77. Install the two ground straps to the intake chamber.

78. Install the data link connector to the bracket.

79. Connect the following hoses:

a. Power brake booster vacuum hose.

b. PCV hose.

c. IAC valve vacuum hose.

80. Install the emission control valve set and related vacuum hoses and connectors.

81. Install the V-bank cover.

82. Connect the pressure hose to the hydraulic motor.

83. Connect the fuel lines to the fuel rail assembly.

84. Connect the heater and radiator hoses.

85. Install the right engine mounting support.

86. Connect the two engine ground straps.

87. Install the cruise control actuator and bracket.

88. Install the air cleaner, air flow meter, and air duct assembly.

89. Connect the accelerator cable and the throttle cable on vehicles equipped with an automatic transaxle.

90. Fill the cooling system to the proper level with coolant.

91. Connect the negative battery cable.

92. Start the engine and check for leaks. Bleed the air from the cooling system.

93. Adjust the ignition timing.

94. Road test the vehicle and check for unusual noise, shock, slippage, correct shift points and smooth operation.

95. Recheck the coolant and engine oil levels.

Turbocharger

REMOVAL & INSTALLATION

3.0L (2JZ-GTE) Engine

1. Disconnect the negative battery cable from the battery.

2. Drain the engine coolant from the engine.

3. Disconnect the cruise control actuator cable from the throttle body.

4. Remove the No. 1 air hose.

5. Remove the air cleaner and MAF meter assembly as follows:

   a. Remove the three bolts to the air assembly.

   b. loosen the hose clamp and disconnect the air hose from the intake air connector.

   c. Disconnect the MAF meter wire from the clamp on the air cleaner case.

   d. Disconnect the MAF meter connector and remove the air cleaner and MAF meter assembly.

6. Disconnect the theft deterrent horn from the body.

7. Raise and safely support the vehicle.

8. Remove the front lower arm bracket stay by removing the two bolts, nut, and plate washer.

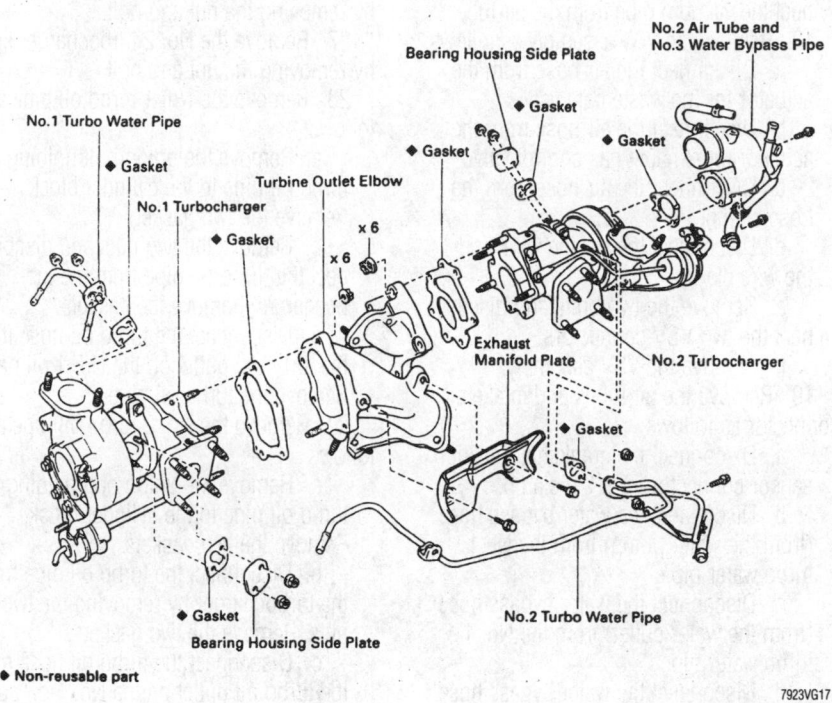

Exploded view of the turbocharger component assembly—Supra with 3.0L (2JZ-GTE) engine

9. Remove the front upper crossmember extension by removing the two bolts and two nuts.

10. Remove the No. 2 front exhaust pipe as follows:

   a. Remove the front exhaust pipe to the No. 2 front exhaust pipe by removing the two bolts and nuts.

   b. Remove the pipe support bracket by removing the two bolts.

   c. Disconnect the front exhaust pipe from the No. 2 exhaust pipe and remove the gasket.

   d. Remove the three nuts, then remove the No. 2 front exhaust pipe and gasket.

11. Remove the heat insulator for the No. 2 front exhaust pipe by removing the two bolts and two nuts.

12. If equipped with automatic transmission, disconnect the A/T oil cooler tubes from the engine.

13. Disconnect the engine wire protector from the body by removing the two bolts.

14. Disconnect the heater hose from the No. 3 water bypass pipe.

15. Disconnect the EVAP hose from the No. 1 vacuum pipe.

16. Disconnect the IAC valve pipe from the No. 2 air tube as follows:

   a. Disconnect the engine wire from the clamp.

   b. Disconnect the air hose (from the No. 1 vacuum pipe) from the IAC valve pipe.

   c. Disconnect the air hose from the No. 2 air tube.

   d. Disconnect the IAC valve pipe from the clamp.

17. Disconnect the No. 1 vacuum pipe from the air tubes as follows:

   a. Disconnect the VSV connector for the intake air control valve.

   b. Disconnect the VSV connector fro the exhaust bypass valve.

   c. Disconnect the engine wire from the three clamps.

   d. Disconnect the following hoses:

   • Air hose from the No. 4 air tube•   Air hose from the No. 1 air tube

   • Air hose (from the VSV for the waste gate valve) from the vacuum pipe

   • Air hose (from the VSV for the exhaust has control valve) from the vacuum pipe

   • Vacuum hose (from the air bypass valve) from the No. 1 air tube

   • Two air hoses (from the VSV for exhaust bypass valve) from the vacuum pipe

   • Air hose (from the No. 2 air tube) from the vacuum pipe

- Two air hoses (from the pressure tank) from the vacuum pipe

e. Remove the three bolts and disconnect the vacuum pipe from the air tubes.

18. Remove the VSV assembly as follows:

a. Disconnect the air hose from the actuator for the waste gate valve.

b. Disconnect the air hose from the actuator for exhaust gas control valve.

c. Disconnect the air hose from the hose clamp.

d. Disconnect the engine wire from the wire clamp.

e. Remove the two bolts and disconnect the two VSV connectors.

f. Remove the VSV assembly.

19. Remove the air tubes and intake air connector as follows:

a. Disconnect the crankshaft position sensor connector from the clamp.

b. Disconnect the water bypass hose (from the water pump) from the No.1 turbo water pipe.

c. Disconnect the water bypass hose (from the water outlet) from the No. 1 turbo water pipe.

d. Disconnect the water bypass hose (from the water outlet) from the No. 2 turbo water pipe.

e. Remove the bolt and disconnect the No. 2 turbo water pipe from the No. 4 air tube.

f. Disconnect the No. 1 air tube from the No. 1 turbocharger by removing the two bolts.

g. Remove the two bolts holding the No. 4 air tube to the No. 1 turbocharger.

h. Disconnect the air hose from the No. 4 air tube.

i. Disconnect the air hose from the intake air connector.

j. Remove the No. 4 air tube and air bypass valve assembly.

k. Remove the intake air control valve and gasket by removing the two nuts.

l. Disconnect the air hose from the No. 2 air tube.

m. Disconnect the PCV hose from the No. 2 cylinder head cover.

n. Remove the intake air connector and No. 1 air tube assembly.

20. Remove the air inlet duct and cable bracket by removing the bolt and two nuts.

21. Remove the heat insulator for the turbocharger by removing the four bolts.

22. Remove the exhaust bypass pipe and gasket by removing the four nuts.

23. Remove the exhaust gas control valve stay by removing the nut and bolt.

24. Remove the main heated oxygen sensor by disconnecting the electrical connector and two nuts.

25. Remove the exhaust gas control valve by removing the three nuts.

26. Remove the No. 1 turbocharger stay by removing the nut and bolt.

27. Remove the No. 2 turbocharger stay by removing the nut and bolt.

28. Remove the No. 1 turbo oil pipe as follows:

a. Remove the union bolt holding the turbo oil pipe to the cylinder block. Remove the two gaskets.

b. Remove the two nuts and disconnect the turbo oil pipe from the turbocharger. Remove the gaskets.

c. Disconnect the turbo oil hose from the turbo oil outlet on the No. 1 oil pan. Remove the turbo oil pipe.

29. Remove the No. 2 turbo oil pipe as follows:

a. Remove the union bolt holding the turbo oil pipe to the cylinder block. Remove the two gaskets.

b. Disconnect the turbo oil pipe from the turbocharger by removing the two nuts. Remove the two gaskets.

c. Disconnect the turbo oil hose from the turbo oil outlet on the No. 1 oil pan and remove the turbo oil pipe.

30. Remove the turbochargers and turbine outlet elbow assembly as follows:

a. Disconnect the heater hose (from the No. 3 water bypass pipe) from the No. 2 water bypass pipe.

b. Disconnect the water bypass hose (from the No. 2 turbo water pipe) from the No. 2 water bypass pipe.

c. Remove the eight nuts holding the turbochargers to the exhaust manifold.

d. Remove the two turbochargers and turbine outlet elbow assembly.

e. Remove the two gaskets.

31. Remove the No. 1 vacuum pipe from the No. 2 turbocharger by disconnecting the two air hoses from the actuator.

32. Remove the No. 2 air tube and No. 3 water bypass pipe assembly from the No. 2 turbocharger by removing the two bolts and air tube.

33. Remove the exhaust manifold plate from the turbine outlet elbow.

34. Remove the No. 2 turbo water pipe from the No. 2 turbocharger by removing the two nuts.

35. Remove the bearing housing side plate from the No. 1 turbocharger by removing the two nuts.

36. Remove the No. 1 turbo water pipe from the No. 1 turbocharger by removing the two nuts.

37. Remove the bearing housing side plate from the No. 2 turbocharger by removing the two nuts.

38. Remove the No. 1 turbocharger from the turbine outlet elbow by removing the six nuts.

39. Remove the No. 2 turbocharger from the turbine outlet elbow by removing the six nuts.

To install:

40. Install the No. 2 turbocharger to the turbine outlet elbow by installing a new gaskets and six new nuts. Tighten the nuts to 18 ft. lbs. (25 Nm).

41. Install the No. 1 turbocharger to the turbine outlet elbow by installing a new gasket and six new nuts. Tighten the nuts to 18 ft. lbs. (25 Nm).

42. Install the bearing housing side plate to the No. 2 turbocharger by installing a new gasket and two nuts. Tighten the nuts to 78 inch lbs. (9 Nm).

43. Install the No. 1 turbo water pipe to the No. 1 turbocharger by installing a new gasket and two nuts. Tighten the nuts to 78 inch lbs. (9 Nm).

44. Install the bearing housing side plate to the No. 1 turbocharger by installing a new gasket and two nuts. Tighten the nut to 78 inch lbs. (9 Nm).

45. Install the No. 2 turbo water pipe to the No. 2 turbocharger by installing a new gasket and two nuts. Tighten the nuts to 78 inch lbs. (9 Nm).

46. Install the exhaust manifold plate to the turbine outlet elbow and install the two bolts.

47. Install the No. 2 air tube and No. 3 water bypass pipe assembly to the No. 2 turbocharger and install the gasket and two bolts. Tighten the bolts to 15 ft. lbs. (21 Nm).

48. Install the No. 1 vacuum pipe to the No. 2 turbocharger by installing the two hoses.

49. Install the turbochargers and turbine outlet elbow assembly as follows:

a. Install two new gaskets.

b. Install the turbochargers and turbine outlet elbow assembly to the exhaust manifold.

c. Install eight new nuts and uniformly tighten the nuts in several passes. Tighten the nuts to 40 ft. lbs. (54 Nm).

d. Connect the water bypass hose (from the No. 2 turbo water pipe) to the No. 2 water bypass pipe.

e. Connect the heater hose (from the No. 3 water bypass pipe) to the No. 2 water bypass pipe.

50. Install the No. 2 turbo oil pipe as follows:

a. Install the turbo oil pipe and connect the turbo oil hose to the turbo oil outlet on the No. 1 oil pan.

b. Using a new gasket, install the turbo oil pipe to the turbocharger by installing the two nuts. Tighten the nuts to 15 ft. lbs. (21 Nm). Be sure to align the oil holes of the gasket and turbocharger housing.

c. Using two new gaskets, connect the union bolt to hold the turbo oil pipe to the cylinder block. Tighten the union bolt to 29 ft. lbs. (39 Nm).

51. Install the No. 1 turbo oil pipe as follows:

a. Install the turbo oil pipe and connect the turbo oil hose to the turbo oil outlet on the No. 1 oil pan.

b. Using a new gasket, install the turbo oil pipe to the turbocharger by installing the two nuts. Tighten the nuts to 15 ft. lbs. (21 Nm). Be sure to align the oil holes of the gasket and turbocharger housing.

c. Using two new gaskets, connect the union bolt to hold the turbo oil pipe to the cylinder block. Tighten the bolts to 29 ft. lbs. (39 Nm).

52. Install the No. 2 turbocharger stay by installing the nut and bolt. Tighten the nut and bolt to 32 ft. lbs. (43 Nm).

53. Install the No. 1 turbocharger stay by installing the nut and bolt. Tighten the nut and bolt to 32 ft. lbs. (43 Nm).

54. Install the exhaust gas control valve by installing two new gaskets and the three nuts. Tighten the three nuts to 51 ft. lbs. (69 Nm).

55. Install the main heated oxygen sensor by connecting the electrical connector and installing the two nuts. Tighten the nuts to 14 ft. lbs. (20 Nm).

56. Install the exhaust gas control valve stay by installing the bolt and nut. Tighten the bolt and nut to 32 ft. lbs. (43 Nm).

57. Install the exhaust bypass pipe by installing two new gaskets and four new nuts. Tighten the nuts to 18 ft. lbs. (25 Nm).

58. Install the heat insulator for the turbocharger by installing the four bolts.

59. Install the air inlet duct by installing the cable bracket, bolt, and two nuts.

60. Connect the air tubes and intake air connector as follows:

a. Install the intake air connector and No. 1 air tube assembly.

b. Connect the PCV hose to the No. 2 cylinder head cover.

c. Connect the air hose to the No. 2 air tube.

d. Install the intake air control valve and gasket by installing the two nuts. Tighten the nut to 15 ft. lbs. (21 Nm).

e. Install the No. 4 air tube and air bypass valve assembly.

f. Connect the air hose to the intake air connector.

g. Connect the air hose to the No. 4 air tube.

h. Install the two bolts holding the No. 4 air tube to the No. 1 turbocharger. Tighten the bolt to 15 ft. lbs. (21 Nm).

i. Connect the No. 1 air tube to the No. 1 turbocharger and install the two bolts. Tighten the bolts to 15 ft. lbs. (21 Nm).

j. Install the No. 2 turbo water pipe to the No. 4 air tube by installing the bolt.

k. Connect the water bypass hose (from the water outlet) to the No. 2 turbo water pipe.

l. Connect the water hose (from the water outlet) to the No. 1 turbo water pipe.

m. Connect the water bypass hose (from the water pump) to the No. 1 turbo water pipe.

n. Connect the crankshaft position sensor connector to the clamp.

61. Install the VSV assembly as follows:

a. Connect the VSV assembly and connect the two VSV connectors.

b. Install the two bolts.

c. Connect the engine wire to the wire clamp.

d. Connect the air hose to the hose clamp.

e. Connect the air hose to the actuator for the exhaust gas control valve.

f. Connect the air hose to the actuator for the waste gate valve.

62. Connect the No. 1 vacuum pipe to the air tubes as follows:

a. Install the vacuum pipe to the air tubes and install the three bolts.

b. Connect the two air hoses (from the vacuum pressure tank) to the vacuum pipe.

c. Connect the air hose to the VSV for the intake air control valve.

d. Connect the air hose (from the No. 2 air tube) from the vacuum pipe.

e. Connect the two air hoses (from the VSV for exhaust bypass valve) to the vacuum pipe.

f. Connect the vacuum hose (from the air bypass valve) to the No. 1 air tube.

g. Connect the air hose (from the VSV for exhaust gas control valve) to the vacuum pipe.

h. Connect the air hose (from the VSV for waste gate valve) to the vacuum pipe.

i. Connect the air hose to the No. 1 air tube.

j. Connect the air hose to the No. 4 air tube.

k. Connect the engine wire to the three clamps.

l. Connect the VSV connector for the exhaust bypass valve.

m. Connect the VSV connector for the intake air control valve.

63. Connect the IAC valve pipe to the No. 2 air tube as follows:

a. Connect the IAC valve pipe to the clamp.

b. Connect the air hose to the No. 2 air tube.

c. Connect the air hose (from the No. 1 vacuum pipe) to the IAC valve pipe.

d. Connect the engine wire to the clamp.

64. Connect the EVAP hose to the No. 1 vacuum pipe.

65. Connect the heater hose to the No. 3 water bypass pipe.

66. Connect the engine wire protector to the body by installing the two bolts.

67. Connect the A/T oil cooler tubes to the engine.

68. Install the heat insulator for the No. 2 front exhaust pipe by installing the two bolts and two nuts.

69. Install the No. 2 front exhaust pipe as follows:

a. Using a new gasket, install the No. 2 front exhaust pipe and install the three nuts. Tighten the nuts to 46 ft. lbs. (62 Nm).

b. Connect the front exhaust pipe to the No. 2 exhaust pipe.

c. Install the pipe support bracket and install the two bolts. Tighten the bolts to 32 ft. lbs. (43 Nm).

d. Install the two bolts and nuts to hold the front exhaust pipe to the No. 2 front exhaust pipe. Tighten the bolts and nuts to 43 ft. lbs. (58 Nm).

70. Install the upper front crossmember extension by installing the two bolts and two nuts. Tighten the bolts to 22 ft. lbs. (29 Nm) and the nuts to 25 ft. lbs. (33 Nm).

71. Install the front lower arm bracket stay by installing the plate washer, nut, and two bolts. Tighten the bolts to 33 ft. lbs. (44 Nm) and the nut to 43 ft. lbs. (59 Nm).

72. Connect the theft deterrent horn to the body.

73. Connect the air cleaner and MAF meter assembly as follows:

a. Install the MAF meter assembly and air cleaner. Connect the MAF meter connector.

b. Connect the MAF meter wire to the clamp on the air cleaner case.

c. Connect the air hose to the intake air connector and tighten the clamp.

d. Install the three bolts.

74. Install the air cleaner duct.
75. Install the No. 1 air hose.
76. Install the cruise control actuator cable to the throttle body.
77. Install the engine undercover.
78. Fill the engine coolant.
79. Check all fluids.
80. Connect the negative battery cable to the battery.

Intake Manifold

REMOVAL & INSTALLATION

1.5L (5E-FE) Engine

1. Disconnect the negative battery cable. On vehicles equipped with an air bag, wait at least 90 seconds before proceeding.
2. Drain the cooling system.
3. Disconnect the accelerator cable and if equipped with an automatic transmission, disconnect the throttle cable.
4. Remove the PCV hose.
5. Remove the air cleaner to throttle body hose.
6. Tag and disconnect all electrical wires and vacuum hoses that interfere with removal of the intake manifold.
7. Remove the EGR pipe from the intake manifold and EGR valve.
8. Remove the throttle body assembly.
9. Disconnect the air intake chamber stay by removing the bolt and nut.
10. For vehicles with distributor ignition, disconnect the air pipe by removing the two bolts.
11. Disconnect the engine wire clamps from the intake manifold stay.
12. Remove the bolt(s) and nut(s) to the intake manifold stay. Remove the intake manifold stay from the engine.
13. Remove the intake manifold by removing two bolts and three nuts. Remove the intake manifold gasket.
   To install:
14. Be sure the intake manifold surface is clean.
15. Using a new gasket, install the intake manifold and tighten the nuts and bolts evenly to 14 ft. lbs. (19 Nm).
16. Install the intake manifold stay and tighten the bolt(s) and nut(s) to 15 ft. lbs. (20 Nm).
17. For vehicles with distributor ignition, install the vacuum hoses, then install the air pipe. Tighten the bolts to 48 inch lbs. (6 Nm).
18. Install the air intake chamber stay by installing the bolt and nut.

19. Install the throttle body, using a new gasket and tightening the nuts and bolts evenly to 9 ft. lbs. (13 Nm). Install all components to the throttle body.
20. Connect the EGR pipe to the intake manifold and EGR valve.
21. Connect all electrical wires and vacuum hoses removed from the intake manifold.
22. Install the air cleaner hose to the throttle body.
23. Refill all fluids.
24. Connect the negative battery cable, start the engine and check the ignition timing. Check for engine coolant leaks.

1.6L (4A-FE) and 1.8L (7A-FE) Engines

1. Disconnect the negative battery cable. On vehicles equipped with an air bag, wait at least 90 seconds before proceeding.

✳✳ CAUTION

The fuel injection system remains under pressure after the engine has been turned OFF. Properly relieve fuel pressure before disconnecting any fuel lines. Failure to do so may result in fire or personal injury.

2. Drain the engine coolant into a suitable container.
3. Disconnect the throttle body from the air intake chamber.
4. Disconnect the ground strap connector.
5. Tag and disconnect the hoses from the intake chamber.
6. On vehicles with A/C remove the hose from the idle-up valve.
7. On vehicles with P/S remove the air hose from the air pipe.
8. Using a 6mm hexagon wrench, remove the three bolts, two nuts, and the air intake chamber cover and gasket.
9. If equipped with EGR, remove the EGR VSV.
10. Remove the intake manifold stay.
11. Remove the air pipe.
12. Disconnect the fuel injector connectors.
13. Remove the union bolt and two gaskets and disconnect the fuel inlet hose from the delivery pipe. Place a shop towel under the connection to absorb the fuel.
14. Disconnect the fuel return hose from the fuel pressure regulator and the air hose from the IAC valve to the air pipe.
15. Remove the two bolts and the delivery pipe together with the four injectors.
16. Remove the seven bolts, four nuts, ground strap, intake manifold and the two gaskets.

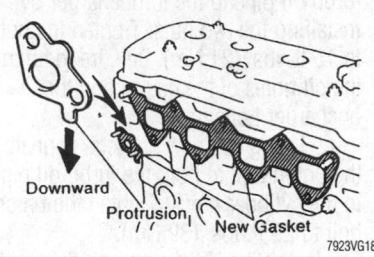

Installing the EGR and intake manifold gaskets—1.6L (4A-FE) and 1.8L (7A-FE) engines

To install:
17. Place a new intake manifold gasket to the cylinder head. Place a new EGR gasket to the cylinder head, with the protrusion facing down.
18. Install the intake manifold with seven bolts, four nuts, and the ground strap. Uniformly tighten the bolts and nuts in several passes. Tighten the **A** nuts to 9 ft. lbs. (13 Nm). Tighten all other nuts and bolts to 14 ft. lbs. (19 Nm).
19. Install the injectors and the delivery pipe.
20. If equipped with EGR, install the EGR VSV.
21. Install the air intake chamber cover with a new gasket.
22. Install the throttle body.
23. Install the air pipe and fuel inlet hose with the two bolts and nut. Install new gaskets on the fuel inlet hose and secure the clamp of the inlet hose to the intake manifold.
24. Connect the fuel return hose to the pressure regulator and the air hose from the IAC valve to the air pipe.
25. Install the intake manifold stay. Tighten the 12mm head bolt to 14 ft. lbs. (19 Nm) and the 14mm head bolt to 29 ft. lbs. (39 Nm).
26. Connect the hoses to the intake chamber.
27. If equipped, install the air hoses the idle-up valve and the air pipe.
28. Connect the ground strap connector and connect the fuel injector connectors.
29. Refill the cooling system with coolant and connect the negative battery cable. Start the engine, check for leaks, and road test the vehicle for proper operation.

1.8L (1ZZ-FE) Engine

1. Disconnect the negative battery cable.
2. Drain the engine coolant.
3. Remove the drive belt and alternator.
4. Remove the air intake duct.

5. Disconnect the accelerator cable.

6. Disconnect the exhaust pipe from the manifold.

7. Remove the exhaust manifold support bracket.

8. Remove the spark plug wires, then remove the ignition coils. Tag the wires so they can be returned to their original positions.

9. Remove the spark plugs.

10. Remove the PCV hoses.

11. Remove the throttle body assembly.

12. Remove the two bolts securing the wiring harness protector to the brackets on the intake manifold.

13. Unplug the wiring connectors and remove the ground wires from the cylinder head.

14. Remove the two bolts and the intake manifold support bracket.

15. Remove the two mounting bolts, two mounting nuts and brackets, then remove the intake manifold and gasket.

To install:

16. Position a new intake manifold gasket on the cylinder head.

17. Install the nuts, bolts and brackets. Tighten the fasteners evenly, in several passes to 14 ft. lbs. (18.5 Nm).

18. Reconnect the harness wiring to the cylinder head and install the harness protector with the two bolts.

19. Install the fuel injectors, throttle body and the PCV hoses.

20. Install the spark plugs and the ignition coils. Tighten the two nuts and two bolts to 80 inch lbs. (9 Nm).

21. Install the exhaust manifold and support bracket. Tighten the bolts alternately to 37 ft. lbs. (49 Nm).

22. Connect the front exhaust pipe to the manifold. Tighten the bolts to 46 ft. lbs. (62 Nm).

23. Using a new gasket and two new nuts, install the oxygen sensor. Tighten the nuts to 14 ft. lbs. (20 Nm).

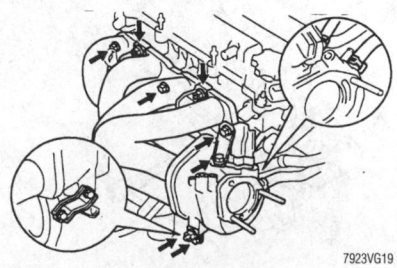

Intake manifold mounting fastener locations—1.8L (1ZZ-FE) engine

24. Connect the accelerator cable and air intake duct.

25. Install the alternator and drive belt.

26. Refill the cooling system with coolant.

27. Connect the negative battery cable.

28. Start the engine and check for leaks.

29. Recheck the coolant and oil levels after the engine has cooled.

2.2L (5S-FE) Engine

CELICA

1. Relieve the fuel system pressure.

2. Disconnect the negative battery cable. On vehicles equipped with an air bag, wait at least 90 seconds before proceeding.

3. Drain the engine coolant into a suitable container.

4. Disconnect and/or remove the following components:

   a. The IAT sensor connector.

   b. The high tension spark plug wire from air cleaner hose.

   c. The accelerator cable from the clamp and the cruise control actuator cable from the clamps.

   d. For California vehicles, disconnect the air hose for the idle up from the air cleaner hose

   e. The four clamps and the and the air cleaner cap from the air cleaner case.

   f. The air cleaner hose from the throttle body.

   g. The air cleaner cap and hose assembly.

5. Remove the throttle body.

6. Disconnect the vacuum sensor hose from the gas filter on the intake manifold.

7. Remove the brake booster vacuum line from the intake manifold.

8. If equipped with A/C, remove the A/C idle-up valve.

9. Disconnect the EGR temperature sensor connector from the bracket on the intake manifold, and disconnect the sensor connector from the wiring connector. Disconnect the EVAP hose from the charcoal canister and disconnect the hose clamp from the bracket on the air tube.

10. Disconnect the two vacuum hoses from the VSV for the EGR. Disconnect the vacuum modulator from the clamp on the intake manifold.

11. Loosen the union nut of the EGR pipe, and remove the two nuts, the EGR valve, vacuum modulator, vacuum hoses assembly and gasket.

12. Remove the bolt, nut, and the intake manifold stay.

13. Disconnect the A/T throttle control cable from the clamp on the rear side of the intake manifold.

14. Remove the following hoses:

• Two P/S air hose(s) from the air tube and the intake manifold

• Air hose from the fuel pressure regulator

15. Remove the two bolts, hose bracket (for EGR), and the air tube.

16. Remove the bolt and the vacuum pipe.

17. Disconnect the knock sensor connector. Remove the bolt and disconnect the ground cables from the intake manifold.

18. Remove the VSV assembly for fuel pressure control and EGR.

19. Disconnect the PCV hose, accelerator, A/T throttle control cables and bracket from the intake manifold.

20. Disconnect the engine wiring harness protector from the bracket on the starter, disconnect the Vehicle Speed Sensor (VSS) connector, remove the bolt and disconnect the engine wiring harness protector from the LH side of the intake manifold.

21. Disconnect the four fuel injector connectors and disconnect the engine wiring harness protector from the two brackets on the front side of the intake manifold.

22. Disconnect the engine wiring harness protector from the two mounting bolts of the No. 2 timing belt cover in sequence.

> **✳✳ CAUTION**
>
> **The fuel injection system remains under pressure after the engine has been turned OFF. Properly relieve fuel pressure before disconnecting any fuel lines. Failure to do so may result in fire or personal injury.**

23. Remove the union bolt and two gaskets and disconnect the fuel inlet hose from the delivery pipe. Disconnect the fuel return hose from the return pipe.

24. Remove the six bolts and two nuts and disconnect the engine wiring harness between the intake manifold and the cylinder head. Remove the intake manifold and gasket.

To install:

25. Insert the intake manifold between the cylinder head and the firewall. Insert the engine wiring harness between the intake manifold and the cylinder head.

26. Install a new gasket and the intake manifold with the six bolts and two nuts. Uniformly tighten the bolts and nuts in several passes. Tighten them to 14 ft. lbs. (19 Nm).

27. Connect the fuel inlet pipe to the fuel delivery hose using two new gaskets and the union bolt. Tighten the union bolt to 25

ft. lbs. (34 Nm). Connect the fuel return hose to the return hose.

28. Install the engine wiring harness protector to the two mounting bolts of the No. 2 timing belt cover in reverse of removal sequence. Install the harness protector to the two brackets on the front side of the intake manifold.

29. Connect the fuel injector connectors. The Nos. 1 and 3 injector connectors are brown; the No. 2 and 4 connectors are gray.

30. Install the engine wiring harness protector to the LH side of the intake manifold with the bolt and connect the VSS sensor connector.

31. Reinstall the engine wiring harness protector to the starter bracket.

32. Install the cable bracket to the intake manifold with the two bolts, and install the accelerator and A/T throttle control cables to the four clamps.

33. Connect the PCV hose to the intake manifold.

34. Install the VSV for fuel pressure control and EGR.

35. Connect the knock sensor connector and install the ground cable to the intake manifold with the bolt.

36. Install the air tube and hose bracket (for EGR) with the two bolts and connect these hoses:
  • The P/S air hose(s) to the air tube and the intake manifold
  • The vacuum sensing hose to the fuel pressure regulator

37. Install the A/T throttle control cable to the clamp on the rear side of the intake manifold.

38. Install the intake manifold stay bolt with the bolt and nut. Tighten the bolt to 15 ft. lbs. (21 Nm) and the nut to 32 ft. lbs. (44 Nm).

39. Install a new gasket, the EGR valve with the union bolt and two nuts. Tighten the union bolt to 43 ft. lbs. (59 Nm) and the nuts to 9 ft. lbs. (13 Nm). Install the vacuum modulator to the clamp on the intake manifold.

40. Connect the vacuum hoses to the VSV for the EGR and install the hose clamp to the bracket on the air tube.

41. If equipped with A/C, install the A/C idle-up valve.

42. Connect the EGR gas temperature sensor connector and install the sensor connector to the bracket on the intake manifold. Connect the EVAP hose to the charcoal canister.

43. Install the vacuum sensor hose to the gas filter on the intake manifold and install the brake booster vacuum hose.

44. Install the throttle body assembly.

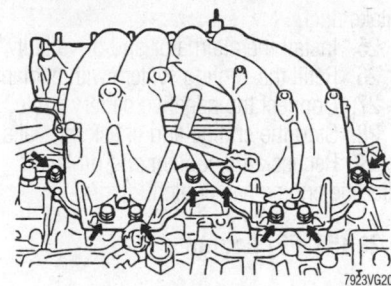

Intake manifold mounting fastener locations—2.2L (5S-FE) engine

45. Install the air filter, the air cleaner cap and hose assembly, the air cleaner hose to the throttle body, and secure the air cleaner cap with the four clamps.

46. Install the accelerator and cruise control cables to the clamps. Connect the high tension spark plug wire to the air cleaner hose.

47. Connect the IAT sensor connector.

48. Refill the cooling system and connect the negative battery cable. Start the engine, check for leaks, and road test the vehicle for proper operation.

CAMRY

1. Disconnect the negative battery cable. On vehicles equipped with an air bag, wait at least 90 seconds before proceeding.

> ※※ **CAUTION**

The fuel injection system remains under pressure after the engine has been turned OFF. Properly relieve fuel pressure before disconnecting any fuel lines. Failure to do so may result in fire or personal injury.

2. Drain the engine coolant into a suitable container.

3. Disconnect the accelerator cable from the throttle body. If equipped with A/T, disconnect the throttle cable from the throttle body.

4. Disconnect the intake air temperature sensor connector

5. On California models, disconnect the air cleaner hose.

6. Loosen the air cleaner hose clamp bolt, disconnect the four air cleaner cap clips, disconnect the air hose from the throttle body, and remove the air cleaner cap together with the resonator and the air cleaner hose.

7. Tag and remove the electrical connections and hoses from the throttle body.

8. Remove the throttle body. Type A throttle bodies are secured with four bolts

and Type B throttle bodies are secured with two bolts and two nuts.

9. Remove the vacuum hose bracket and the engine wiring harness from the intake manifold.

10. Remove the EGR valve.

11. Remove the four bolts, the wire bracket, the No. 1 air intake chamber, and the manifold stays. Remove the six bolts, two nuts, the intake manifold, and the gasket.

To install:

12. Install the intake manifold to the cylinder head with a new gasket. Tighten the six bolts and the two nuts in several passes to 14 ft. lbs. (19 Nm). Install the two wire clamps to the wire brackets on the intake manifold.

13. Install the vacuum hose bracket and the engine wiring harness.

14. Install the No. 1 air intake chamber and manifold stays with the four bolts. Tighten the 14mm bolts to 31 ft. lbs. (42 Nm) and tighten the 12mm bolts to 16 ft. lbs. (22 Nm).

15. Install the EGR valve.

16. Install the throttle body with a new gasket on the intake chamber. Connect the hoses and electrical connections to the throttle body.

➡ **The protrusion on the gasket should be facing down and the water hose connections on the throttle body should also face down.**

17. On type A throttle body, tighten the four bolts to 14 ft. lbs. (19 Nm). Bolt A is 45mm in length and bolt B is 55mm

18. On type B throttle body, tighten the two bolts and the two nuts to 14 ft. lbs. (19 Nm).

19. Make the following connections to the throttle body:
  • The PCV hose
  • The two vacuum hoses from the EGR modulator

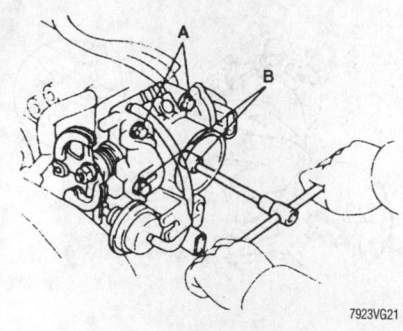

Mounting bolt length identification for the throttle body—2.2L (5S-FE) engine

• The vacuum hose from the TVV (for EVAP)

• The IAC valve connector

• The throttle position sensor connector

20. Connect the air cleaner hose to the throttle body and install the air cleaner cap together with the resonator and the air cleaner hose. Connect the intake air temperature sensor connector.

21. On California models, connect the air hose to the air cleaner hose.

22. If equipped with A/T, connect and adjust the throttle cable. Connect the accelerator cable.

23. Fill the cooling system with coolant and connect the negative battery cable. Start the engine, check for leaks, and road test the vehicle for proper operation.

3.0L (1MZ-FE) Engine

✳✳ CAUTION

The fuel injection system remains under pressure even after the engine has been turned OFF. The fuel system pressure must be relieved before disconnecting any fuel lines. Failure to do so may result in fire and/or personal injury.

1. Relieve the fuel pressure from the fuel lines.

2. With the ignition switch in the **LOCK** position, disconnect the negative battery terminal. If equipped with an air bag system, wait at least 90 seconds or longer before performing any other work.

3. Remove the air cleaner hose from the engine compartment.

4. Remove the V-bank cover from the engine.

5. Remove the air cleaner chamber assembly as follows:

　a. Drain the engine coolant from the vehicle.

　b. Disconnect the following connectors and cables:

• Accelerator cable

• A/T throttle cable

• TPS connector

• IAC valve connector

• EGR gas temperature sensor connector

• A/C idle up valve connector

• VSV connector for the ACIS

• VSV connector for the fuel pressure control

• Disconnect the VSV for the EVAP

• VSV connector for the EGR

• DLC1 from the bracket on the intake air control valve

　c. Remove the power steering pressure tube from the No. 1 engine hanger by removing the two bolts.

　d. Disconnect the following hoses, clamps, and cables:

• Brake booster vacuum hose from the intake air control valve for the ACIS

• PCV hose from the PCV valve on the right-hand cylinder head

• Ground strap and cable from the air intake air control valve from the ACIS

• Ground cable from the air intake chamber

• Vacuum hose clamp from fuel pipe

• Two bypass hoses from the throttle body

• Two power steering air hoses to the air intake chamber.

• Air assist hose from the throttle body

• Remove the EVAP hose from the pipe on emission control valve set

• Two vacuum hoses from the pipes on the cylinder head rear plate

• Vacuum sensing hose from the fuel pressure regulator

• Engine wire clamp from emission control valve set

　e. Remove the two bolts and the No. 1 engine hanger.

　f. Remove the two bolts and the air intake chamber stay.

　g. Remove the No. 2 EGR pipe and two gaskets by removing the four nuts.

　h. Disconnect the hose from the VSV from the EVAP.

　i. Using an 8mm hexagon wrench, remove the two bolts. Remove the two nuts, the air intake chamber assembly and gasket.

6. Disconnect the fuel injector connectors.

7. Remove the air assist hoses and pipe.

8. Disconnect the fuel return hose from the No. 1 fuel pipe.

9. Disconnect the fuel inlet hose from the fuel filter. Catch any fuel leaking from the filter in a shop rag. Dispose of the rag properly.

10. Remove the delivery pipes and injectors from the engine as follows:

　a. Loosen the two union bolts holding the No. 2 fuel pipe to the delivery pipes.

　b. Disconnect the fuel return hose from the fuel pressure regulator.

　c. Remove the union bolt and two gaskets for the right-hand delivery pipe.

　d. Remove the two bolts to the left-hand delivery pipe, then remove the left-hand delivery pipe, three injectors, and the No. 2 fuel pipe as an assembly.

　e. Remove the union bolt and two gaskets from the left-hand delivery pipe.

Disconnect the No. 2 fuel pipe from the left-hand delivery pipe.

　f. Remove the right-hand delivery pipe by removing the three bolts. Remove the delivery pipe, injectors, and the fuel inlet hose as an assembly.

　g. Remove the four spacers from the intake manifold.

　h. Pull out the six injectors from the delivery pipes.

　i. Remove the two O-rings and two grommets from each injector.

11. Remove the heater hoses from the intake manifold.

12. Remove the nine bolts, two nuts, and two plate washers. Remove the intake manifold.

To install:

13. Thoroughly clean the intake manifold and cylinder head surfaces.

14. Using new gaskets, install the intake manifold with the two plate washers, nine bolts, and two nuts. Tighten the nuts and bolts to 11 ft. lbs. (15 Nm).

15. Install the heater hoses to the intake manifold.

16. Install the injectors as follows:

　a. Install two new grommets to each injector.

　b. Apply a light coat of spindle oil or gasoline to the two O-rings and install them to each injector.

　c. Install the injector into the delivery pipe by turning the injector back and forth. Install all the injectors into the delivery pipes. Be sure to position the injector electrical connector outward.

　d. Place the four spacers in position on the intake manifold.

　e. Place the right-hand delivery pipe and the No. 1 fuel pipe together with the three injectors in position on the intake manifold.

　f. Temporarily install the two bolts holding the right-hand delivery pipe to the intake manifold.

　g. Temporarily install the bolt holding the No. 1 fuel pipe to the intake manifold.

　h. Place the left-hand delivery pipe and the No. 2 fuel pipe together with the three injectors in position.

　i. Connect the fuel return hose to the fuel pressure regulator.

　j. Temporarily install the two bolts holding the left-hand delivery pipe to the intake manifold.

　k. Temporarily install the No. 2 fuel pipe to the left and right-hand delivery pipes with the union bolts and two new gaskets.

　l. Check that the injectors rotate smoothly. If the injectors do not rotate smoothly, the probable cause is incorrect installation of the O-rings. Replace the O-rings.

m. Tighten the four bolts holding the delivery pipes to the intake manifold. Tighten the bolts to 7 ft. lbs. (10 Nm).

n. Tighten the bolt holding the No. 1 fuel pipe to the intake manifold. Tighten the bolts to 14 ft. lbs. (20 Nm).

o. Tighten the two union bolts holding the No. 2 fuel pipe to the delivery pipes. Tighten the bolts to 24 ft. lbs. (33 Nm).

17. Connect the fuel inlet hose to the fuel filter by installing the union bolt. Use two new gaskets when installing the union bolt.

18. Connect the fuel return hose to the No. 1 fuel pipe. When routing the fuel return hose, pass the hose under the heater hoses.

19. Connect the air assist hoses to the intake manifold, then install the air assist pipe to the bracket on the No. 1 fuel pipe.

20. Connect the injector connectors.

21. Install the air intake chamber assembly as follows:

a. Using an 8mm hexagon wrench, install the air intake chamber with a new gasket. Install the two bolts and two nuts. Uniformly tighten the bolts and nuts in several passes, then tighten the bolts and nuts to 32 ft. lbs. (43 Nm).

b. Connect the hose to the VSV for the EVAP system.

c. Install the two new gaskets and No. 2 EGR pipe with the four nuts. Tighten the nuts to 9 ft. lbs. (12 Nm).

d. Install the No. 1 engine hanger with the two bolts. Tighten the bolts to 19 ft. lbs. (39 Nm).

e. Install the air intake chamber stay with the two bolts. Tighten the bolts to 14 ft. lbs. (20 Nm).

f. Connect the hoses, clamp, and cables as follows:

• Brake booster vacuum hose to the intake air control valve for the ACIS.

• PCV hose to the PCV valve on the right-hand cylinder head.

• Ground strap and cable to the intake air control valve for the ACIS.

• Connect the ground cable and strap with the nut. Tighten the nut to 10 ft. lbs. (15 Nm)

• Ground cable to air intake chamber.

• Vacuum hose clamp to fuel pipe.

• Two water bypass hoses to the throttle body.

• Air assist hose to the throttle body.

• Two power steering air hoses to the air intake chamber.

• Connect the EVAP hose to the pipe on the emission control valve set

• Two vacuum hoses to the pipes on the cylinder head rear plate.

• Vacuum sensing hose to the fuel pressure regulator.

• Engine wire clamp to the emission control valve set.

22. Install the power steering pressure tube with the two nuts.

23. Connect the following connectors and cables:

• TPS sensor connector
• IAC valve connector
• EGR gas temperature sensor connector
• A/C idle up valve connector
• VSV connector for the ACIS
• VSV connector for the fuel pressure control
• For California vehicles, install the VSV connector for the EVAP
• VSV connector for the EGR
• DLC1 to the bracket on the intake air control valve
• Accelerator cable
• A/T throttle cable

24. Install the V-bank cover.

25. Install the air cleaner hose to the engine.

26. Fill the engine coolant.

27. Connect the negative battery cable to the battery and start the engine.

28. Check the engine coolant level and check for leaks.

3.0L (2JZ-GE) Engine

1. Disconnect the negative battery cable. Wait at least 90 seconds before proceeding with any other work.

2. Drain the engine coolant from the engine.

3. Remove the VSV connector.

4. Remove the vacuum sensing hose from the fuel pressure control and the air intake chamber.

5. For vehicles with automatic transmissions, remove the transmission dipstick and guide.

6. Remove the EGR pipe.

7. Disconnect the EGR gas temperature sensor connector from the No. 2 vacuum pipe and wiring connector.

8. Disconnect the No. 2 vacuum pipe from the air intake chamber and from the intake manifold by removing the two nuts.

9. Remove the air intake chamber as follows:

a. Remove the two nuts holding the throttle body bracket to the cylinder head.

b. Remove the intake air connector from the air intake chamber by removing the four bolts and two nuts.

c. Remove the bolt and disconnect the engine wire protector from the air intake chamber.

d. Disconnect the:

• Three vacuum hoses from the No. 1 vacuum pipe.

• Vacuum hose from the air intake chamber

• Power steering air hose from the air intake chamber

• Brake booster vacuum hose from the air intake chamber by removing the union bolt and two gaskets

• Vacuum hose (from the actuator for ACIS) from the No. 1 vacuum pipe

• Except for California vehicles, disconnect the vacuum sensing hose (from the fuel pressure regulator) from the air intake chamber

e. Loosen the two nuts holding the air intake chamber stays to the cylinder head

f. Remove the two bolts and disconnect the two air intake chamber stays from the air intake chamber.

g. Remove the air intake chamber by removing the nut and five bolts. Remove the gaskets.

10. Relieve the fuel pressure in the fuel line before disconnecting any fuel lines.

✳✳ CAUTION

The fuel injection system remains under pressure, even after the engine has been turned OFF. The fuel system pressure must be relieved before disconnecting any fuel lines. Failure to do so may result in fire and/or explosion.

11. Disconnect the fuel return hose.

12. Remove the oil dipstick and guide for the engine.

13. Disconnect the engine wire from the intake manifold as follows:

a. Remove the bolt and disconnect the engine wire bracket from the water pump.

b. Remove the two bolts and disconnect the two ground straps from the intake manifold.

c. Remove the two bolts and disconnect the two wire clamps from the intake manifold.

d. Disconnect the following connectors:

• Six injector connectors
• ECT sensor and the ECT sender gauge connectors

e. Disconnect the engine wire protector from the intake manifold by removing the three nuts.

14. Remove the intake manifold stay by removing the two bolts.

15. Remove the fuel inlet pipe as follows:

a. Remove the clamp bolt from the intake manifold.

b. Remove the union bolt and disconnect the fuel inlet pipe.

16. Remove the two nuts and six bolts to the intake manifold.

17. Remove the intake manifold and gaskets.

To install:

18. Inspect the contact surfaces of the head and manifold. Remove any traces of gasket material. Install a new gasket, then install the intake manifold. Tighten the bolts and nuts to 20 ft. lbs. (27 Nm).

19. Install the fuel inlet pipe as follows:

a. Install the inlet pipe with new gaskets. Tighten the union bolt to 30 ft. lbs. (42 Nm).

b. Install the clamp bolt to the intake manifold.

20. Install the intake manifold stay and tighten the bolts to 29 ft. lbs. (39 Nm).

21. Connect the engine wire as follows:

a. Install the engine wire protector to the intake manifold with the three nuts.

b. Connect the following connectors:

• Six injector connectors

• ECT sender gauge and the ECT sender gauge connectors

c. Install the two wire clamps to the intake manifold with the bolts.

d. Install the two ground straps to the intake manifold with the bolts.

e. Install the engine wire bracket to the water pump with the bolt.

22. Install the intake air chamber as follows:

a. Install a new gasket to the air intake chamber facing the protrusion on the gasket rearward.

b. Install the intake air chamber and install the nut and five bolts.

c. Install the two air intake chamber stays to the air intake chamber and install the two bolts. Tighten the bolts to 13 ft. lbs. (18 Nm).

d. Tighten the two nuts holding the air intake chamber to the cylinder head. Tighten the nuts to 13 ft. lbs. (18 Nm).

e. Connect the following hoses:

• Except California vehicles, connect the vacuum sensing hose (from the fuel pressure regulator) to the air intake chamber.

• Vacuum hose (from the actuator for ACIS) to the No. 1 vacuum pipe.

• Brake booster vacuum hose to the air intake chamber. Tighten the union bolt to 22 ft. lbs. (29 Nm).

• Power steering air hose to the air intake chamber

• Vacuum hose (from the No. 2 vacuum pipe) to the air intake chamber

• Three vacuum hoses (from the No. 2 vacuum pipe) to the No. 1 vacuum pipe

f. Install the engine wire protector to the air intake chamber and install the bolt.

g. Install the four bolts and two nuts holding the intake air connector to the air intake chamber.

h. Install the throttle body bracket to the cylinder head and install the two nuts. Tighten the nuts to 15 ft. lbs. (21 Nm).

23. Connect the No. 2 vacuum pipe to the air intake chamber and intake manifold. Tighten the two nuts to 20 ft. lbs. (27 Nm).

24. Connect the EGR gas temperature sensor connector.

25. Install the EGR pipe. Tighten the union nut to 47 ft. lbs. (64 Nm) and the two bolts to 20 ft. lbs. (27 Nm).

26. Install the oil dipstick guide and dipstick.

27. Connect the fuel return hose.

28. If equipped with automatic transmission, install the transmission oil dipstick guide and dipstick.

29. Install the VSV for the fuel pressure control.

30. Connect the negative battery cable to the battery.

31. Fill the cooling system, start the engine, and check for leaks. Bleed the cooling system.

32. Road test the vehicle for proper operation.

3.0L (2JZ-GTE) Engine

❋❋ CAUTION

The fuel injection system remains under pressure after the engine has been turned OFF. Properly relieve fuel pressure before disconnecting any fuel lines. Failure to do so may result in fire or personal injury.

1. Relieve fuel system pressure.

2. Disconnect the negative battery cable to the battery. On vehicles equipped with an air bag, wait at least 90 seconds before proceeding.

3. Remove the engine undercover.

4. Remove the throttle body as follows.

a. Drain the engine coolant from the radiator and engine.

b. Disconnect the following hose, cable and connectors from the throttle body:

• Air hose

• Accelerator and cruise control actuator cables

• Throttle position sensor connector

• Sub throttle position sensor connector

• Sub throttle actuator connector

c. Disconnect the throttle body from the air intake chamber by removing the two bolts and two nuts.

d. Remove the gasket from the throttle body.

e. Disconnect the following hoses from the throttle body and remove the throttle body from the engine:

• EVAP hose

• Water bypass hose (from the No. 4 water bypass pipe)

• Power steering air hose

5. For vehicles with automatic transmissions, remove the transmission dipstick and guide.

6. Remove the oil dipstick and guide for the engine.

7. Remove the air intake chamber stay by removing the nut and bolts.

8. Disconnect the control cable bracket from the air intake chamber by removing the two bolts.

9. Disconnect the following connectors and hose:

• IAC valve connector

• Turbo pressure sensor connector

• VSV connector for the fuel pressure control

• VSV connector for the EGR valve

10. Remove the bolt and disconnect the engine wire protector from the vehicle body.

11. Disconnect the following hoses:

• Disconnect the IAC valve pipe from the clamp on the cylinder head cover and disconnect the air hose form the IAC valve

• Disconnect the air hose (from the air intake chamber) from the vacuum pipe on the IAC valve pipe

• Air hose for the EGR from the valve pipe

• PCV hose from the PCV valve

• Vacuum sensing hose from the fuel pressure regulator

• Water bypass hose (from the IAC valve) from the No. 4 water bypass pipe

• EVAP hose (from the air intake chamber) from the vacuum pipe on the manifold stay

• EVAP hose (from the vacuum pipe on the No. 4 water bypass pipe) from the No. 2 vacuum pipe

• EVAP hose (from the charcoal canister) from the No. 2 vacuum pipe

• Power steering air hose from the air intake chamber

• Brake booster vacuum hose from the union on the air intake chamber

12. Disconnect the EGR gas temperature sensor connector from the No. 2 vacuum pipe and wiring connector.

13. Remove the EGR pipe.

14. Remove the No. 4 water bypass pipe by removing the two bolts.

15. Remove the intake manifold stay by removing the two bolts.

16. Remove the air intake chamber assembly as follows:

   a. Disconnect the ground cable from the intake manifold by removing the bolt.

   b. Remove the two bolts holding the engine wire protector to the intake manifold.

   c. Disconnect the two clamps for the engine wire protector from the brackets.

   d. Remove the five bolts, two nuts, and the engine wire bracket.

   e. Disconnect the air intake air chamber assembly from the intake manifold.

   f. Disconnect the water bypass hose from the IAC valve.

   g. Remove the gasket.

17. Disconnect the six injector connectors.

18. Disconnect the two camshaft position sensor connectors.

19. Disconnect the three engine wire clamps from the injector holders.

20. Disconnect the fuel inlet pipe from the delivery pipe.

21. Disconnect the fuel return pipe from the fuel pressure regulator.

22. Remove the intake manifold and delivery pipe assembly by removing the four bolts and two nuts.

To install:

23. Install a new gasket, the intake manifold and delivery pipe assembly with the four bolts and two nuts. Tighten the six bolts and two nuts to 20 ft. lbs. (27 Nm).

24. Connect the fuel return pipe to the fuel pressure regulator and tighten the union bolts to 20 ft. lbs. (27 Nm).

25. Connect the fuel inlet pipe to the delivery pipe and tighten the union bolt to 30 ft. lbs. (41 Nm).

26. Connect the three engine wire clamps to the injector holders.

27. Install the two camshaft position sensor connectors.

28. Install the six injector connectors. The No. 1, 3, and 5 injector connectors are dark gray; the No. 2, 4, and 6 injector connectors are gray.

29. Install the air intake chamber assembly as follows:

   a. Install the gasket.

   b. Connect the water bypass hose to the IAC valve.

   c. Connect the air intake chamber assembly to the intake manifold and install the five bolts and two nuts. Tighten the nuts and bolts in several passes to 20 ft. lbs. (27 Nm).

   d. Connect the two clamps of the engine wire protector to the brackets.

   e. Install the two bolts to hold the engine wire protector to the intake manifold.

   f. Connect the ground cable to the intake manifold by installing the bolt.

30. Install the intake manifold stay and install the two bolts. Tighten the bolts to 29 ft. lbs. (39 Nm).

31. Install the No. 4 water bypass pipe with the two bolts.

32. Install the EGR pipe as follows:

   a. Install the EGR pipe and gasket with the two bolts. Tighten the bolts to 20 ft. lbs. (27 Nm).

   b. Install the union bolt holding the EGR pipe to the EGR valve and tighten the union bolt to 47 ft. lbs. (64 Nm).

33. Connect the EGR gas temperature sensor connector.

34. Install the following hoses:

• Brake booster vacuum hose to the union on the air intake chamber

• Power steering air hose to the air intake chamber

• EVAP hose (from the charcoal canister) to the No. 2 vacuum pipe

• EVAP hose (from the vacuum pipe on the No. 4 water bypass pipe) to the No. 2 vacuum pipe

• EVAP hose (from the air intake chamber) to the vacuum pipe on the manifold stay

• Water bypass hose (from the IAC valve) to the No. 4 water bypass pipe

• Vacuum sensing hose to the fuel pressure regulator

• PCV hose to the PCV valve

• Air hose for the EGR to the valve pipe

• Air hose (from the air intake chamber) to the vacuum pipe on the IAC valve pipe

• Connect the IAC valve pipe to the clamp on the cylinder headcover

• Connect the air hose to the IAC valve

35. Connect the engine wire protector to the vehicle body with the bolt.

36. Connect the following connectors:

• VSV connector for the EGR valve and the fuel pressure control

• Turbo pressure sensor and the IAC valve connectors

37. Connect the control cable bracket to the intake chamber by installing the two bolts. Tighten the bolts to 14 ft. lbs. (19 Nm).

38. Install the intake chamber stay by installing the bolt and nut. Tighten the nut and bolt to 14 ft. lbs. (19 Nm).

39. Install the oil dipstick guide and dipstick to the engine. Use a new O-ring for the dipstick guide.

40. Install the transmission oil dipstick guide and dipstick to the transmission. Use a new O-ring for the dipstick guide.

41. Install the throttle body and install the following hoses:

• Power steering air hose

• Water bypass hose (from the cylinder head)

• Water bypass hose (from the No. 4 water bypass pipe)

• EVAP hose

42. Install the gasket and throttle body to the air intake chamber and install the two bolts and two nuts. Tighten the nuts and bolts to 15 ft. lbs. (21 Nm).

43. Install the following hose, cables, and connectors to the throttle body:

• Sub throttle actuator connector

• Sub throttle position sensor connector

• Throttle position sensor connector

• Cruise control actuator and the accelerator cables

• Air hose

44. Fill engine coolant and install the engine undercover.

45. Connect the negative battery cable to the battery.

Exhaust Manifold

REMOVAL & INSTALLATION

1.5L (5E-FE) Engine

PASEO AND TERCEL

1. Disconnect the negative battery cable from the battery. On vehicles equipped with an air bag, wait at least 90 seconds before proceeding.

2. Tag and disconnect all electrical wires and vacuum hoses that interfere with removal of the exhaust manifold.

3. Remove the three bolts to the exhaust heat insulator, then remove the insulator from the engine.

4. Remove the exhaust pipe stay by removing the bolt and two nuts.

5. Disconnect the exhaust pipe from the manifold by removing the two bolts and two compression springs.

6. Remove the six nuts to the exhaust manifold, then remove the exhaust manifold.

To install:

7. Clean the gasket surfaces and install the exhaust manifold, tightening the six bolts to 35 ft. lbs. (47 Nm).

8. Install the exhaust stay to the engine

and exhaust manifold. Tighten the bolt and two nuts to 29 ft. lbs. (40 Nm).

9. Reconnect the exhaust manifold heat insulator with the three bolts. Tighten the bolts to 69 inch lbs. (8 Nm).

10. Connect the exhaust pipe to the exhaust manifold with the two compression springs and two bolts. Tighten the bolts to 46 ft. lbs. (62 Nm).

11. Connect all electrical wires and vacuum hoses that were disconnected for removal of the exhaust manifold.

12. Connect the battery cable, start the engine and check for exhaust leaks.

1.6L (4A-FE) and 1.8L (7A-FE) Engine

1. Disconnect the negative battery cable. On vehicles equipped with an air bag, wait at least 90 seconds before proceeding.

2. Raise and safely support the vehicle.

3. Working from under the vehicle, remove the bolts holding the front exhaust pipe to the mounting bracket.

4. Using a 14mm deep socket wrench, remove the nuts and the gasket and disconnect the front exhaust pipe from the manifold.

5. Remove the main oxygen sensor connector.

6. Remove the bolts and the upper heat insulator.

7. Remove the nuts, the exhaust manifold, and the gasket.

8. Remove the bolts and lower heat insulator from the exhaust manifold.

To install:

9. Install the lower heat insulator to the exhaust manifold with the bolts.

10. Install a new gasket and the exhaust manifold with the nuts. Uniformly tighten the nuts in several passes. Tighten the nuts to 25 ft. lbs. (34 Nm).

11. Install the upper heat insulator with the bolts.

12. Install the front exhaust pipe with a new gasket to the exhaust manifold. Install the nuts using a 14mm deep socket wrench. Tighten the nuts to 46 ft. lbs. (62 Nm).

13. Secure the front exhaust pipe to the exhaust pipe bracket with the bolts.

14. Connect the main oxygen sensor connector.

15. Lower the vehicle safely and reconnect the negative battery cable.

16. Start the engine and be sure that there are no exhaust leaks.

1.8L (1ZZ-FE) Engine

1. Disconnect the negative battery cable.
2. Drain the engine coolant.
3. Remove the drive belt and alternator.
4. Remove the air intake duct.

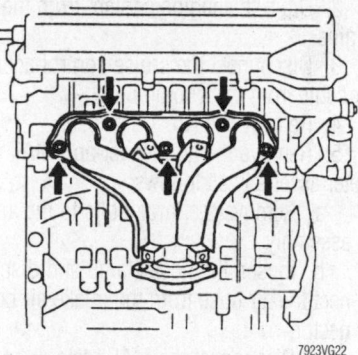

7923VG22

Exhaust manifold mounting nut locations—1.8L (1ZZ-FE) engine

5. Disconnect the accelerator cable.

6. Disconnect the exhaust pipe from the manifold.

7. Remove the exhaust manifold support bracket.

8. Remove the heat insulator from the dash panel.

9. Remove the upper heat insulator.

10. Remove the five mounting nuts and the exhaust manifold.

11. If necessary, remove the lower heat insulator from the exhaust manifold.

To install:

12. If removed, install the lower heat insulator on the exhaust manifold. Tighten the three bolts to 9 ft. lbs. (12 Nm).

13. Install the exhaust manifold using a new gasket. Tighten the nuts in several passes to 27 ft. lbs. (37 Nm).

14. Install the upper heat insulator. Tighten the six bolts to 9 ft. lbs. (12 Nm).

15. Install the heat insulator on the dash panel.

16. Install the exhaust manifold and support bracket. Tighten the bolts alternately to 37 ft. lbs. (49 Nm).

17. Connect the front exhaust pipe to the manifold. Tighten the bolts to 46 ft. lbs. (62 Nm).

18. Using a new gasket and two new nuts, install the oxygen sensor. Tighten the nuts to 14 ft. lbs. (20 Nm).

19. Connect the accelerator cable and air intake duct.

20. Install the alternator and drive belt.

21. Refill the cooling system with coolant.

22. Connect the negative battery cable.

23. Start the engine and check for leaks.

24. Recheck the coolant and oil levels after the engine has cooled.

2.2L (5S-FE) Engine

1. Disconnect the negative battery cable. On vehicles equipped with an air bag, wait at least 90 seconds before proceeding.

2. Raise and safely support the vehicle.

3. Working from under the vehicle, remove the bolts holding the front exhaust pipe to the mounting bracket.

4. Using a 14mm deep socket wrench, remove the nuts and the gasket and disconnect the front exhaust pipe from the manifold.

5. Disconnect the main oxygen sensor connector and the sub oxygen sensor connector.

6. Remove the bolt and nut and remove the LH exhaust manifold stay.

7. Remove the bolts and the upper manifold heat insulator.

8. Remove the bolts, nuts, and the RH exhaust manifold stay.

9. Remove the nuts, the exhaust manifold, and the three-way catalytic converter assembly.

10. Remove the nuts, oxygen sensor and gasket from the exhaust manifold.

11. Remove the sub oxygen sensor from the three-way catalytic converter.

12. Remove the bolts and the lower heat insulator from the exhaust manifold.

13. Remove the bolts and the three-way catalytic converter heat insulators.

14. Remove the bolts, nuts, TWC, gasket, retainer and cushion from the exhaust manifold.

To install:

15. Place the cushion, retainer, and a new gasket on the TWC and reinstall it to the exhaust manifold with the bolts and the nuts. Tighten the bolts and nuts to 22 ft. lbs. (30 Nm).

16. Install the lower manifold heat insulator with the bolts and install the TWC heat insulators with the bolts.

17. Install the main oxygen sensor to the exhaust manifold with a new gasket and new nuts. Tighten the nuts to 14 ft. lbs. (19 Nm).

18. Install the sub oxygen sensor to the front TWC and tighten to 33 ft. lbs. (45 Nm).

19. Install a new gasket, the exhaust manifold and front TWC assembly to the engine with the nuts. Uniformly tighten the nuts in several passes. Tighten the nuts to 36 ft. lbs. (49 Nm).

20. Install the RH exhaust manifold stay with the bolts and new nuts. Tighten the bolts and nuts to 31 ft. lbs. (42 Nm).

21. Install the upper heat insulator with the bolts.

22. Install the LH exhaust manifold stay with the bolt and nut. Tighten the bolt to 29 ft. lbs. (39 Nm) and tighten the nut to 31 ft. lbs. (42 Nm).

23. Connect the main and the sub oxygen sensors connectors.

24. Install the front exhaust pipe with a new gasket to the TWC. Install the nuts using a 14mm deep socket wrench. Tighten the nuts to 46 ft. lbs. (62 Nm).

25. Secure the front exhaust pipe to the exhaust pipe bracket with the bolts. Tighten the bolts to 14 ft. lbs. (19 Nm).

26. Lower the vehicle safely and reconnect the negative battery cable.

27. Start the engine and be sure that there are no exhaust leaks.

3.0L (1MZ-FE) Engine

1. Disconnect the negative battery cable from the battery.

2. Raise and safely support the vehicle.

3. Remove the engine undercovers from the vehicle.

4. From below the engine, disconnect the front exhaust pipe from the exhaust manifold by removing the two nuts.

5. If necessary, lower the vehicle to access the top of the engine.

6. Remove the EGR pipe from the exhaust manifold by removing the four nuts.

7. Disconnect the heated oxygen sensor connector to the right exhaust manifold.

8. Remove the exhaust manifold stay by removing the bolt and nut.

9. Remove the six nuts to the exhaust manifold and remove the exhaust manifold from the engine.

To install:

10. Using a new gasket, install the exhaust manifold to the engine and install the six nuts. Uniformly tighten, then tighten the bolts to 36 ft. lbs. (49 Nm).

11. Install the exhaust manifold stay and install the bolt and nut. Tighten the bolt and nut to 15 ft. lbs. (20 Nm).

12. Connect the heated oxygen sensor connector to the right exhaust manifold.

13. Using new gaskets, install the EGR pipe to the exhaust manifold and the engine. Tighten the four nuts to 9 ft. lbs. (12 Nm).

14. Raise the vehicle and safely support.

15. Connect the front exhaust pipe to the exhaust manifold. Use a new gasket and tighten the two nuts to 46 ft. lbs. (62 Nm).

16. Install the engine undercovers to the vehicle.

17. Lower the vehicle.

18. Connect the negative battery cable to the battery.

3.0L (2JZ-GTE) Engine

1. Disconnect the negative battery cable from the battery. On vehicles equipped with an air bag, wait at least 90 seconds before proceeding.

2. Drain the engine coolant from the engine.

3. Disconnect the cruise control actuator cable from the throttle body.

4. Remove the No. 1 air hose.

5. Remove the air cleaner and MAF meter assembly as follows:

a. Remove the three bolts to the air assembly.

b. loosen the hose clamp and disconnect the air hose from the intake air connector.

c. Disconnect the MAF meter wire from the clamp on the air cleaner case.

d. Disconnect the MAF meter connector and remove the air cleaner and MAF meter assembly.

6. Disconnect the theft deterrent horn from the body.

7. Raise and safely support the vehicle.

8. Remove the engine undercover.

9. Remove the front lower arm bracket stay by removing the two bolts, nut, and the plate washer.

10. Remove the front upper crossmember extension by removing the two bolts and two nuts.

11. Remove the No. 2 front exhaust pipe.

12. Remove the heat insulator for the No. 2 front exhaust pipe by removing the two bolts and two nuts.

13. If equipped with automatic transmission, disconnect the A/T oil cooler tubes from the engine.

14. Disconnect the engine wire protector from the body by removing the two bolts.

15. Disconnect the heater hose from the No. 3 water bypass pipe.

16. Disconnect the EVAP hose from the No. 1 vacuum pipe.

17. Disconnect the IAC valve pipe from the No. 2 air tube as follows:

a. Disconnect the engine wire from the clamp.

b. Disconnect the air hose (from the No. 1 vacuum pipe) from the IAC valve pipe.

c. Disconnect the air hose from the No. 2 air tube.

d. Disconnect the IAC valve pipe from the clamp.

18. Disconnect the No. 1 vacuum pipe from the air tubes as follows:

a. Disconnect the VSV connector for the intake air control valve.

b. Disconnect the VSV connector for the exhaust bypass valve.

c. Disconnect the engine wire from the three clamps.

d. Disconnect the following hoses:
- Air hose from the No. 4 air tube
- Air hose from the No. 1 air tube

- Air hose (from the VSV for the waste gate valve) from the vacuum pipe
- Air hose (from the VSV for the exhaust gas control valve) from the vacuum pipe
- Vacuum hose (from the air bypass valve) from the No. 1 air tube
- Two air hoses (from the VSV for exhaust bypass valve) from the vacuum pipe
- Air hose (from the No. 2 air tube) from the vacuum pipe
- Two air hoses (from the pressure tank) from the vacuum pipe

e. Remove the three bolts and disconnect the vacuum pipe from the air tubes.

19. Remove the VSV assembly as follows:

a. Disconnect the air hose from the actuator for the waste gate valve.

b. Disconnect the air hose from the actuator for exhaust gas control valve.

c. Disconnect the air hose from the hose clamp.

d. Disconnect the engine wire from the wire clamp.

e. Remove the two bolts and disconnect the two VSV connectors.

f. Remove the VSV assembly.

20. Remove the air tubes and intake air connector as follows:

a. Disconnect the crankshaft position sensor connector from the clamp.

b. Disconnect the water bypass hose (from the water pump) from the No.1 turbo water pipe.

c. Disconnect the water bypass hose (from the water outlet) from the No. 1 turbo water pipe.

d. Disconnect the water bypass hose (from the water outlet) from the No. 2 turbo water pipe.

e. Remove the bolt and disconnect the No. 2 turbo water pipe from the No. 4 air tube.

f. Disconnect the No. 1 air tube from the No. 1 turbocharger by removing the two bolts.

g. Remove the two bolts holding the No. 4 air tube to the No. 1 turbocharger.

h. Disconnect the air hose from the No. 4 air tube.

i. Disconnect the air hose from the intake air connector.

j. Remove the No. 4 air tube and air bypass valve assembly.

k. Remove the intake air control valve and gasket by removing the two nuts.

l. Disconnect the air hose from the No. 2 air tube.

m. Disconnect the PCV hose from the No. 2 cylinder head cover.

n. Remove the intake air connector and No. 1 air tube assembly.

21. Remove the air inlet duct and cable bracket by removing the bolt and two nuts.

22. Remove the heat insulator for the turbocharger by removing the four bolts.

23. Remove the exhaust bypass pipe and gasket by removing the four nuts.

24. Remove the exhaust gas control valve stay by removing the nut and bolt.

25. Remove the main heated oxygen sensor by disconnecting the electrical connector and two nuts.

26. Remove the exhaust gas control valve by removing the three nuts.

27. Remove the No. 1 turbocharger stay by removing the nut and bolt.

28. Remove the No. 2 turbocharger stay by removing the nut and bolt.

29. Remove the No. 1 turbo oil pipe as follows:

    a. Remove the union bolt holding the turbo oil pipe to the cylinder block. Remove the two gaskets.

    b. Remove the two nuts and disconnect the turbo oil pipe from the turbocharger. Remove the gaskets.

    c. Disconnect the turbo oil hose from the turbo oil outlet on the No. 1 oil pan. Remove the turbo oil pipe.

30. Remove the No. 2 turbo oil pipe as follows:

    a. Remove the union bolt holding the turbo oil pipe to the cylinder block. Remove the two gaskets.

    b. Disconnect the turbo oil pipe from the turbocharger by removing the two nuts. Remove the two gaskets.

    c. Disconnect the turbo oil hose from the turbo oil outlet on the No. 1 oil pan and remove the turbo oil pipe.

31. Remove the turbochargers and turbine outlet elbow assembly as follows:

    a. Disconnect the heater hose (from the No. 3 water bypass pipe) from the No. 2 water bypass pipe.

    b. Disconnect the water bypass hose (from the No. 2 turbo water pipe) from the No. 2 water bypass pipe.

    c. Remove the eight nuts holding the turbochargers to the exhaust manifold.

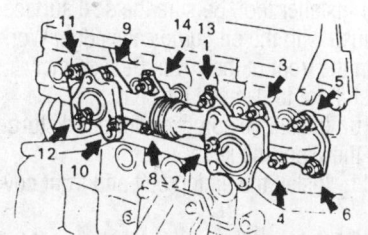

7923VG23

Exhaust manifold bolt installation sequence—3.0L (2JZ-GTE) engine

    d. Remove the two turbochargers and turbine outlet elbow assembly.

    e. Remove the two gaskets.

32. Remove the exhaust manifold by removing the 12 nuts and two gaskets.

To install:

33. Install the exhaust manifold with two new gaskets and install the nuts. Tighten the nuts in several passes and in sequence to 29 ft. lbs. (39 Nm).

34. Install the turbochargers and turbine outlet elbow assembly as follows:

    a. Install two new gaskets.

    b. Install the turbochargers and turbine outlet elbow assembly to the exhaust manifold.

    c. Install eight new nuts and uniformly tighten the nuts in several passes. Tighten the nuts to 40 ft. lbs. (54 Nm).

    d. Connect the water bypass hose (from the No. 2 turbo water pipe) to the No. 2 water bypass pipe.

    e. Connect the heater hose (from the No. 3 water bypass pipe) to the No. 2 water bypass pipe.

35. Install the No. 2 turbo oil pipe as follows:

    a. Install the turbo oil pipe and connect the turbo oil hose to the turbo oil outlet on the No. 1 oil pan.

    b. Using a new gasket, install the turbo oil pipe to the turbocharger by installing the two nuts. Tighten the nuts to 15 ft. lbs. (21 Nm). Be sure to align the oil holes of the gasket and turbocharger housing.

    c. Using two new gaskets, connect the union bolt to hold the turbo oil pipe to the cylinder block. Tighten the union bolt to 29 ft. lbs. (39 Nm).

36. Install the No. 1 turbo oil pipe as follows:

    a. Install the turbo oil pipe and connect the turbo oil hose to the turbo oil outlet on the No. 1 oil pan.

    b. Using a new gasket, install the turbo oil pipe to the turbocharger by installing the two nuts. Tighten the nuts to 15 ft. lbs. (21 Nm). Be sure to align the oil holes of the gasket and turbocharger housing.

    c. Using two new gaskets, connect the union bolt to hold the turbo oil pipe to the cylinder block. Tighten the bolts to 29 ft. lbs. (39 Nm).

37. Install the No. 2 turbocharger stay by installing the nut and bolt. Tighten the nut and bolt to 32 ft. lbs. (43 Nm).

38. Install the No. 1 turbocharger stay by installing the nut and bolt. Tighten the nut and bolt to 32 ft. lbs. (43 Nm).

39. Install the exhaust gas control valve by installing two new gaskets and the three nuts. Tighten the three nuts to 51 ft. lbs. (69 Nm).

40. Install the main heated oxygen sensor by connecting the electrical connector and installing the two nuts. Tighten the nuts to 14 ft. lbs. (20 Nm).

41. Install the exhaust gas control valve stay by installing the bolt and nut. Tighten the bolt and nut to 32 ft. lbs. (43 Nm).

42. Install the exhaust bypass pipe by installing two new gaskets and four new nuts. Tighten the nuts to 18 ft. lbs. (25 Nm).

43. Install the heat insulator for the turbocharger by installing the four bolts.

44. Install the air inlet duct by installing the cable bracket, bolt, and two nuts.

45. Connect the air tubes and intake air connector as follows:

    a. Install the intake air connector and No. 1 air tube assembly.

    b. Connect the PCV hose to the No. 2 cylinder head cover.

    c. Connect the air hose to the No. 2 air tube.

    d. Install the intake air control valve and gasket by installing the two nuts. Tighten the nut to 15 ft. lbs. (21 Nm).

    e. Install the No. 4 air tube and air bypass valve assembly.

    f. Connect the air hose to the intake air connector.

    g. Connect the air hose to the No. 4 air tube.

    h. Install the two bolts holding the No. 4 air tube to the No. 1 turbocharger. Tighten the bolt to 15 ft. lbs. (21 Nm).

    i. Connect the No. 1 air tube to the No. 1 turbocharger and install the two bolts. Tighten the bolts to 15 ft. lbs. (21 Nm).

    j. Install the No. 2 turbo water pipe to the No. 4 air tube by installing the bolt.

    k. Connect the water bypass hose (from the water outlet) to the No. 2 turbo water pipe.

    l. Connect the water hose (from the water outlet) to the No. 1 turbo water pipe.

    m. Connect the water bypass hose (from the water pump) to the No. 1 turbo water pipe.

    n. Connect the crankshaft position sensor connector to the clamp.

46. Install the VSV assembly as follows:

    a. Connect the VSV assembly and connect the two VSV connectors.

    b. Install the two bolts.

    c. Connect the engine wire to the wire clamp.

    d. Connect the air hose to the hose clamp.

e. Connect the air hose to the actuator for the exhaust gas control valve.

f. Connect the air hose to the actuator for the waste gate valve.

47. Connect the No. 1 vacuum pipe to the air tubes as follows:

a. Install the vacuum pipe to the air tubes and install the three bolts.

b. Connect the two air hoses (from the vacuum pressure tank) to the vacuum pipe.

c. Connect the air hose to the VSV for the intake air control valve.

d. Connect the air hose (from the No. 2 air tube) from the vacuum pipe.

e. Connect the two air hoses (from the VSV for exhaust bypass valve) to the vacuum pipe.

f. Connect the vacuum hose (from the air bypass valve) to the No. 1 air tube.

g. Connect the air hose (from the VSV for exhaust gas control valve) to the vacuum pipe.

h. Connect the air hose (from the VSV for waste gate valve) to the vacuum pipe.

i. Connect the air hose to the No. 1 air tube.

j. Connect the air hose to the No. 4 air tube.

k. Connect the engine wire to the three clamps.

l. Connect the VSV connector for the exhaust bypass valve.

m. Connect the VSV connector for the intake air control valve.

48. Connect the IAC valve pipe to the No. 2 air tube as follows:

a. Connect the IAC valve pipe to the clamp.

b. Connect the air hose to the No. 2 air tube.

c. Connect the air hose (from the No. 1 vacuum pipe) to the IAC valve pipe.

d. Connect the engine wire to the clamp.

49. Connect the EVAP hose to the No. 1 vacuum pipe.

50. Connect the heater hose to the No. 3 water bypass pipe.

51. Connect the engine wire protector to the body by installing the two bolts.

52. Connect the A/T oil cooler tubes to the engine.

53. Install the heat insulator for the No. 2 front exhaust pipe by installing the two bolts and two nuts.

54. Using a new gasket, install the No. 2 front exhaust pipe and install the three nuts. Tighten the nuts to 46 ft. lbs. (62 Nm).

55. Connect the front exhaust pipe to the No. 2 exhaust pipe and install the pipe support bracket and install the two bolts. Tighten the bolts to 32 ft. lbs. (43 Nm).

56. Install the two bolts and nuts to hold the front exhaust pipe to the No. 2 front exhaust pipe. Tighten the bolts and nuts to 43 ft. lbs. (58 Nm).

57. Install the upper front crossmember extension by installing the two bolts and two nuts. Tighten the bolts to 22 ft. lbs. (29 Nm) and the nuts to 25 ft. lbs. (33 Nm).

58. Install the front lower arm bracket stay by installing the plate washer, nut, and two bolts. Tighten the bolts to 33 ft. lbs. (44 Nm) and the nut to 43 ft. lbs. (59 Nm).

59. Safely lower the vehicle.

60. Connect the theft deterrent horn to the body.

61. Connect the air cleaner and MAF meter assembly as follows:

a. Install the MAF meter assembly and air cleaner. Connect the MAF meter connector.

b. Connect the MAF meter wire to the clamp on the air cleaner case.

c. Connect the air hose to the intake air connector and tighten the clamp.

d. Install the three bolts.

62. Install the air cleaner duct.

63. Install the No. 1 air hose.

64. Install the cruise control actuator cable to the throttle body.

65. Install the engine undercover.

66. Fill the engine coolant.

67. Check all fluids.

68. Connect the negative battery cable to the battery.

69. Start the engine, bleed the cooling system, and check for exhaust leaks.

3.0L (2JZ-GE) Engine

1. Disconnect the negative battery cable. Wait at least 90 seconds before performing any other work.

2. Raise the vehicle and support it safely.

3. Disconnect the two O_2 sensors at the exhaust manifolds.

4. If equipped, remove the four mounting nuts and lift off the outside heat insulator.

5. Remove the four nuts to disconnect the manifold from the front exhaust pipe. Loosen the mounting nuts to the exhaust manifolds and remove the two manifolds and the gasket.

To install:

6. Scrape the mating surfaces of all old gasket material.

7. Install the manifolds with a new gasket. Tighten the nuts to 29 ft. lbs. (39 Nm).

8. Use a new gasket and install the front exhaust pipe to the exhaust manifolds. Tighten the nuts to 46 ft. lbs. (62 Nm).

9. If equipped, install the outer insulator and tighten the nuts to 13 ft. lbs. (18 Nm).

10. Connect the O_2 sensors.

11. Install the undercovers.

12. Lower the vehicle.

13. Connect the battery cable.

14. Start the engine and check for exhaust leaks.

Front Crankshaft Seal

REMOVAL & INSTALLATION

1.5L (5E-FE) Engine

PASEO AND TERCEL

➡The front oil seal can be removed from the engine without removing the oil pump.

1. Disconnect the negative battery cable from the battery. On vehicles equipped with an air bag, wait at least 90 seconds before proceeding.

2. Remove the front covers and the timing belt.

3. Remove the crankshaft timing belt sprocket.

✳ WARNING

When removing the front seal, be extremely careful not to damage the crankshaft.

4. Using a knife, cut off the oil seal lip.

5. Tape the end of a flat bladed tool to avoid damaging crankshaft. Pry out the oil seal using the taped end of the tool.

6. Inspect the oil seal riding surface on the crankshaft for signs of wear or damage.

To install:

7. Wipe the seal bore with a clean rag.

8. Apply multipurpose grease the lip of a new oil seal.

9. Drive the oil seal into place using tool SST 09309–37010 or an equivalent seal installer tool. Be sure the seal surface is flush with the oil pump case edge. Work from the front of the cover. Be extremely careful not to damage the seal.

10. Install the sprocket without disturbing the Woodruff key.

11. Install the timing belt and front covers.

12. Connect the negative battery cable to the battery.

13. Start the engine and check for leaks.

1.6L (4A-FE) and 1.8L (7A-FE and 1ZZ-FE) Engines

1. Disconnect the negative battery cable. On vehicles equipped with an air bag, wait at least 90 seconds before proceeding.
2. Remove the timing belt and crankshaft sprocket.
3. Using a cutting tool, cut off the oil seal lip.

❋❋ WARNING

When removing the oil seal, be careful not to damage the crankshaft or the oil pump housing.

4. Carefully pry out the oil seal from the oil pump housing.

To install:

5. Apply clean engine oil to the new oil seal.
6. Coat the inside of the new oil seal with multi-purpose grease.
7. Install the new oil seal to the oil pump using a suitable seal driver.
8. Install the timing belt and crankshaft sprocket.
9. Connect the negative battery cable. Check the engine oil level.
10. Run the engine and check for leaks.

2.2L (5S-FE) Engines

1. Disconnect the negative battery cable. On vehicles equipped with an air bag, wait at least 90 seconds before proceeding.

➥**The front oil seal can be removed from the engine without removing the oil pump.**

2. Disconnect the negative battery cable from the battery. On vehicles equipped with an air bag, wait at least 90 seconds before proceeding.
3. Remove the timing belt covers and the timing belt from the engine.
4. Using SST tool 09950–50010 or equivalent (crankshaft gear puller), remove the front crankshaft gear from the crankshaft. Be sure not to damage any part of the crankshaft.
5. Using a knife, cut off the oil seal lip.
6. Using a suitable tool, pry out the oil seal. Wrap the edge of the tool with a rag or tape to prevent damaging the crankshaft. Be careful not to damage the crankshaft.

To install:

7. Using a new seal, apply a thin layer of liquid sealer to the outside of the seal.
8. Apply multi purpose grease to the new oil seal lip.
9. Using SST tool 09223–00010 or equivalent (oil seal installer) and a hammer,

tap in the oil seal until its surface is flush with the oil pump body edge.
10. Install the timing belt and the timing belt covers.
11. Install all other components, then connect the negative battery cable to the battery.
12. Start the engine and check for leaks.

3.0L (1MZ-FE) Engine

AVALON AND CAMRY

1. With the ignition switch in the **LOCK** position, disconnect the negative battery terminal. If equipped with an air bag system, wait at least 30 seconds or longer before performing any other work.
2. Remove the timing belt.
3. Remove the crankshaft timing gear.
4. Cut out the lip portion of the oil seal.
5. Tape the end of a suitable prybar to protect the crankshaft and carefully remove the oil seal.

❋❋ WARNING

Be careful not to damage the crankshaft sealing surface.

To install:

6. Apply multi-purpose grease to the lip of a new oil seal. Also apply a light coating of liquid sealant to the outside of the oil seal.
7. Lightly tap the oil seal with a hammer and installer 09309–37010 or equivalent, until its surface is flush with the oil pump case edge.
8. Install the crankshaft timing gear.
9. Install the timing belt.
10. Reconnect the negative battery cable.

3.0L (2JZ-GE and 2JZ-GTE) Engines

1. Remove the timing belt.
2. Remove the crankshaft timing pulley.
3. Cut the oil seal lip.
4. Tape the end of a small prying tool and remove the oil seal.

❋❋ WARNING

Be careful not to damage the crankshaft.

To install:

5. Apply multi-purpose grease to a new oil seal lip.
6. Lightly tap the oil seal with a hammer and installer 09316–60010, or equivalent, until its surface is flush with the oil pump case edge.
7. Install the crankshaft timing pulley.
8. Install the timing belt.
9. Start the engine and check for leaks.

Camshaft

REMOVAL & INSTALLATION

1.5L (5E-FE) Engine

1. Disconnect the negative battery cable from the battery. On vehicles equipped with an air bag, wait at least 90 seconds before proceeding.
2. Remove the valve cover.
3. Remove the timing belt assembly.
4. Remove the camshaft timing sprocket.

➥**Due to the relatively small amount of camshaft thrust clearance, the camshaft must be kept level during removal. If the camshaft is not level on removal, the portion of the head receiving the thrust may crack or be damaged.**

5. Set the intake camshaft so the service bolt holes of the intake camshaft gears are directly up.
6. Remove each front bearing cap of the intake and exhaust camshafts.
7. Secure the intake camshaft sub-gear to the main gear with a 6mm diameter bolt, 16–20mm long and with a pitch of 1.0mm Be sure that the torsional spring force of the subgear has been eliminated by the above operation.
8. In correct sequence, loosen and remove the eight bolts of the four bearing caps of the exhaust camshaft and remove the exhaust camshaft. If the camshaft is not being lifted out straight, reinstall the middle bearing cap and loosen it evenly to keep the camshaft straight.
9. In the reverse order of the installation sequence, loosen and remove the eight bolts of the four bearing caps of the intake camshaft and remove the intake camshaft. If the camshaft is not being lifted out straight, reinstall the middle bearing cap and loosen it evenly to keep the camshaft straight.
10. Using a small screwdriver and a magnetic finger, remove the adjusting shim.

To install:

11. Install the valve shim to the engine.
12. Apply engine oil to the surface of the intake camshaft.
13. Place the intake camshaft on the cylinder head so the service bolt points directly up.
14. Install the four rearward bearing caps in their original order and temporarily tighten them evenly. Do not tighten the bolts at this time.

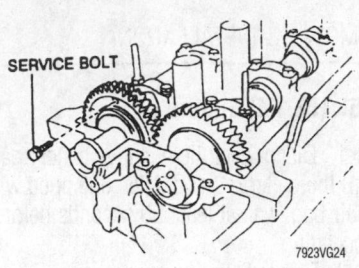

Securing the intake camshaft subgear to the main gear—1.5L (5E-FE) engine

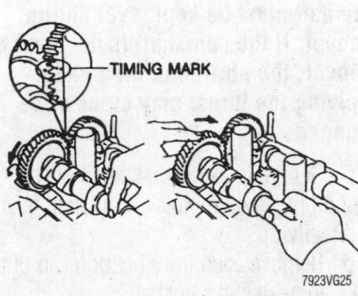

Exhaust and intake camshaft gear timing marks—1.5L (5E-FE) engine

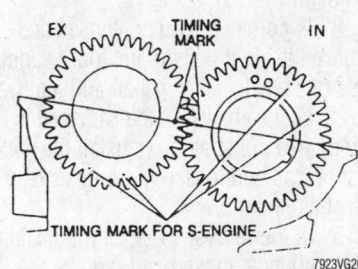

Correct exhaust and intake camshaft gear timing mark alignment—1.5L (5E-FE) engine

Tighten the intake camshaft bearing cap bolts in sequence—1.5L (5E-FE) engine

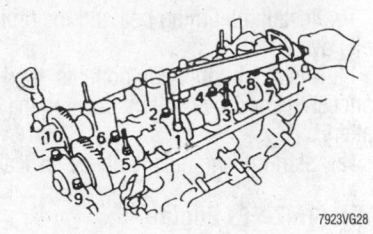

Tighten the exhaust bearing cap bolts according to the sequence shown—1.5L (5E-FE) engine

15. Apply engine oil to the portion of the exhaust camshaft.

16. Engage the exhaust camshaft gear to the intake camshaft gear by matching the proper timing marks on each gear.

17. Roll down the exhaust camshaft onto the bearing journals while engaging the gears with each other.

➡ **There are other marks present for the "S" engine. Do not use these marks.**

18. Install the four rearward intake bearing caps in their original order and tighten evenly. Do not tighten the bolts at this time.

19. Remove the service bolt.

20. Clean the mating surfaces of the No. 2 bearing cap and apply sealer. Install and temporarily tighten the bolts. Install the camshaft housing plug.

21. Now tighten the intake camshaft cap bolts to 9 ft. lbs. (13 Nm), in the proper sequence.

22. Apply grease to a new camshaft oil seal lip and install it as far as the deepest part of the cylinder head.

23. Install the No. 1 bearing cap and temporarily tighten the bolts.

24. Now tighten the exhaust camshaft bearing cap bolts to 9 ft. lbs. (13 Nm) evenly and in the proper sequence.

25. Turn the camshaft one revolution and check that the timing marks of the camshaft gears are aligned.

26. Check and adjust the valve clearance. Install the valve cover.

27. Install the camshaft timing sprocket and tighten the bolt to 37 ft. lbs. (50 Nm).

28. Install the timing belt. start the engine and check for leaks.

29. Connect the negative battery cable to the battery.

1.6L (4A-FE) and 1.8L (7A-FE) Engines

1. Disconnect the negative battery cable. On vehicles equipped with an air bag, wait at least 90 seconds before proceeding.

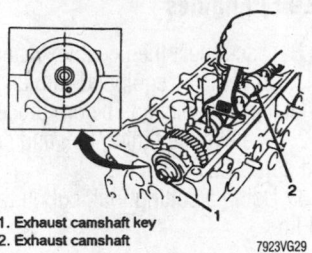

1. Exhaust camshaft key
2. Exhaust camshaft

Positioning the exhaust camshaft for removal—1.6L (4A-FE) and 1.8L (7A-FE) engines

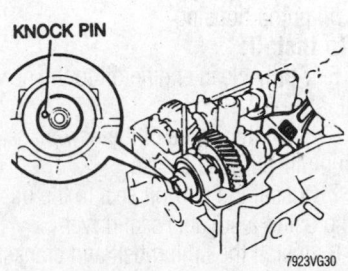

Setting the knockpin to remove the intake camshaft—1.6L (4A-FE) engine

2. Disconnect the spark plug wires from the spark plugs. Be sure to make note of the proper firing order for easier installation.

3. Remove the valve cover.

4. Remove the timing belt covers.

5. Remove the timing belt and idler pulley.

6. Set the exhaust camshaft so that the knock pin is slightly above the cylinder head. This angle allows the No. 1 and No. 3 cylinder cam lobes of the intake camshaft to push their valve lifters evenly.

7. Remove the two bolts and the front bearing cap of the intake camshaft.

8. Secure the intake camshaft end gear to the sub-gear with a service bolt. The service bolt should match the following specifications:

- Thread diameter: 6.0mm
- Thread pitch: 1.0mm
- Bolt length: 16mm

9. Uniformly loosen each intake camshaft bearing cap bolt in several passes in the reverse order of the installation sequence.

❋❋ WARNING

The camshaft must be held level while it is being removed. If the camshaft is not kept level, the por-

tion of the cylinder head receiving the thrust may crack or become damaged. In turn, this could cause the camshaft to bind or break. Before removing the intake camshaft, be sure the rotational force has been removed from the sub-gear; that is, the gear should be in a neutral or "unloaded" state.

10. Remove the four bearing caps and remove the intake camshaft.

➡️If the camshaft cannot be removed straight and level, install and tighten the No. 3 bearing cap. Alternately loosen the bolts on the bearing cap a little at a time while pulling upwards on the camshaft gear. DO NOT attempt to pry or force the cam loose with tools.

11. With the intake camshaft removed, turn the exhaust camshaft approximately 105°, so that the guide pin in the end is just past the 5 o'clock position. This angle allows the No. 1 and the No. 3 cylinder cam lobes of the exhaust camshaft to push their valve lifters evenly.

12. Loosen the camshaft bearing cap bolts a little at a time and in the reverse order of the installation sequence.

13. Remove the bearing caps and remove the exhaust camshaft. After removal, label each cap.

➡️If the camshaft cannot be removed straight and level, install and tighten the No. 3 bearing cap. Alternately loosen the bolts on the bearing cap a little at a time while pulling upwards on the camshaft gear. DO NOT attempt to pry or force the cam loose with tools.

14. Remove the valve lifter shims and hydraulic lifters. Identify each lifter and shim as it is removed so it can be reinstalled in the same position. If the lifters are to be reused, store them upside down in a sealed container.

To install:

15. Install the valve lifters into their original positions and install the shims. Check valve clearance and replace the shims as necessary.

16. When reinstalling, remember that the camshafts must be handled carefully and kept straight and level to avoid damage.

17. Apply multi-purpose grease to the portion of the camshaft.

18. Place the exhaust camshaft on the cylinder head so that the cam lobes press evenly on the lifters for cylinders Nos. 1 and

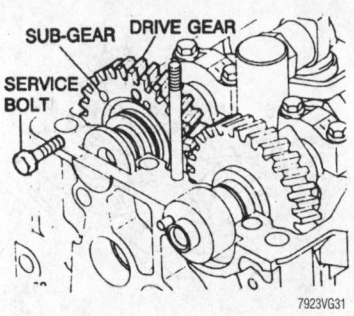

Securing the intake camshaft subgear to the drive gear—1.6L (4A-FE) engine

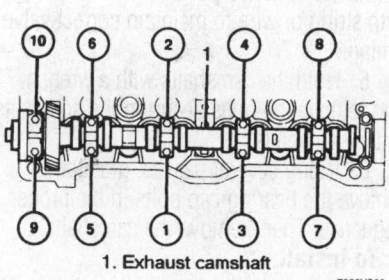

1. Exhaust camshaft

Exhaust camshaft bearing cap tightening sequence—1.6L (4A-FE) and 1.8L (7A-FE) engines

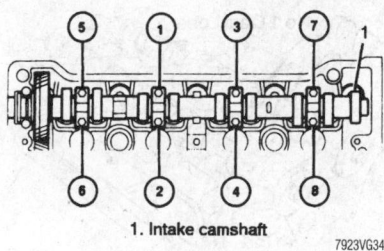

1. Intake camshaft

Intake camshaft bearing cap tightening sequence—1.6L (4A-FE) and 1.8L (7A-FE) engines

3. This will place the guide pin on the camshaft slightly counter clockwise from the vertical axis (about 5 o'clock).

19. Apply a light coat of clean engine oil to the camshaft bearing cap bolts. Install the five bearing caps in position according to the number cast into the cap. The arrow should point towards the pulley end (front) of the motor.

20. Tighten the bearing cap bolts uniformly and in several passes in the proper sequence to 9 ft. lbs. (13 Nm).

21. Apply multi-purpose grease to a new exhaust camshaft oil seal.

22. Install the exhaust camshaft oil seal using a seal driver. Be very careful not to install the seal on a slant or allow it to tilt during installation.

23. Set the exhaust camshaft so that the guide pin is slightly above the cylinder head.

24. Apply multi-purpose grease to the portion of the intake camshaft.

25. Hold the intake camshaft next to the exhaust camshaft and engage the gears by matching the installation marks on each gear.

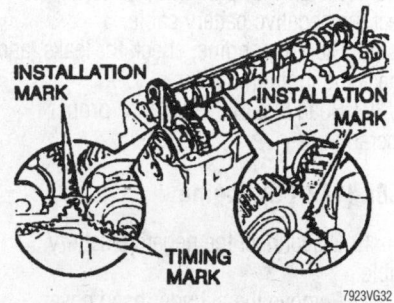

Engaging the intake camshaft gear to the exhaust camshaft gear, matching the installation marks—1.6L (4A-FE) engine

➡️DO NOT use the TDC timing marks for the timing belt.

26. Keeping the gears engaged, roll the intake camshaft down and into its bearing journals. This angle allows the No. 1 and the No. 3 cylinder cam lobes of the intake camshaft to push their valve lifters evenly.

27. Apply a light coat of clean engine oil to the camshaft bearing cap bolts and install the four bearing caps. Observe the numbers on each cap and make certain the arrows point to the pulley end (front) of the motor.

28. Uniformly tighten each of the eight bearing cap bolts in several passes in the proper sequence. Tighten each bolt to 9 ft. lbs. (13 Nm).

29. Remove any retaining pins or bolts in the intake camshaft gears.

30. Apply a light coat of clean engine oil to the camshaft bearing cap bolts and install the No. 1 bearing cap for the intake camshaft. Tighten the bearing cap bolts to 9 ft. lbs. (13 Nm).

➡️If the No. 1 bearing cap does not fit properly, push the camshaft gear backwards by prying apart the cylinder head and camshaft gear with a suitable tool.

31. Turn the exhaust camshaft clockwise, and set it with the guide pin facing upward. Check that the timing marks of the camshaft

gears are aligned. The camshaft assembly installation marks should now be in the 12 o'clock position.

32. Secure the exhaust camshaft and install the timing belt pulley. Tighten the bolt to 43 ft. lbs. (59 Nm).

33. Check and adjust valve clearance.

34. Be sure that both the crankshaft and camshaft positions are set correctly, insuring that they are both set to TDC/compression for No. 1 cylinder.

35. Install the timing belt.

36. Install the timing belt covers and the valve cover.

37. Install the spark plug wires and connect the negative battery cable.

38. Start the engine, check for leaks, and check the ignition timing.

39. Road test the vehicle for proper operation.

1.8L (1ZZ-FE) Engine

1. Disconnect the negative battery cable.

2. Remove the cylinder head cover.

3. Turn the crankshaft so that the No. 1 piston is at TDC on the compression stroke. Check to see that the point marks on the camshaft sprockets are facing each other, if not, rotate the crankshaft one full revolution.

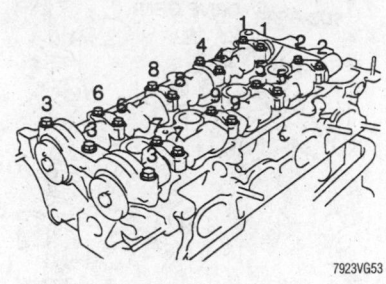

Camshaft bearing cap bolt removal sequence—1.8 (1ZZ-FE) engine

4. Tie the timing chain to each sprocket with string or wire to maintain correct valve timing.

5. Hold the camshafts with a wrench and remove the bolts securing the sprockets to the camshafts.

6. Using several passes, gradually remove the bearing cap bolts in the proper sequence. Then, remove the camshafts,

To install:

7. Lubricate the camshafts with clean engine oil and place them on the cylinder head. Be sure to position the lobes for the No. 1 cylinder as shown in the illustration.

8. Install the bearing caps in their original positions. Apply clean engine oil to the threads and under the heads of the bearing cap bolts. After tightening the bolts on the No. 1 bearing cap to 17 ft. lbs. (23 Nm), tighten the remaining bolts in sequence using several passes to 10 ft. lbs. (13 Nm).

9. Check the valve clearance and make adjustments as needed.

10. Install the camshaft sprockets and the chain.

11. Install the cylinder head cover.

2.2L (5S-FE) Engine

1. Disconnect the negative battery cable. On vehicles equipped with an air bag, wait at least 90 seconds before proceeding.

2. Disconnect the spark plug wires from the spark plugs. Make a note of the proper firing order for installation.

3. Remove the timing belt, gears, and the covers.

4. Remove or disconnect any wire connectors, clamps, cables, or components necessary in order to remove the cylinder head cover.

5. Remove the four nuts, grommets, head cover, and the gasket.

6. Set the No. 1 cylinder to TDC. Turn the crankshaft pulley and align its groove

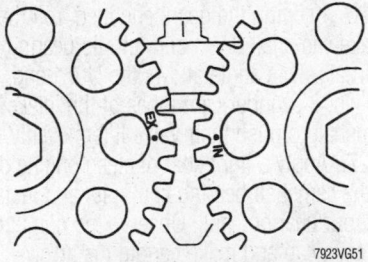

The sprocket marks will align when the No. 1 piston is at TDC on the compression stroke—1.8 (1ZZ-FE) engine

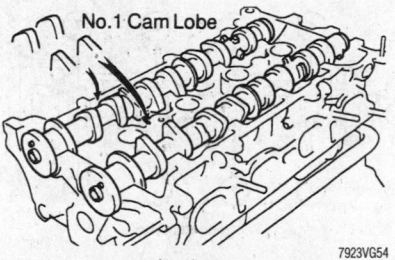

When installing the camshafts, position the lobes for the No. 1 cylinder as shown—1.8 (1ZZ-FE) engine

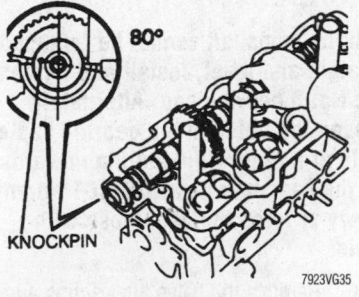

Intake camshaft removal and installation positioning—2.2L (5S-FE) engine

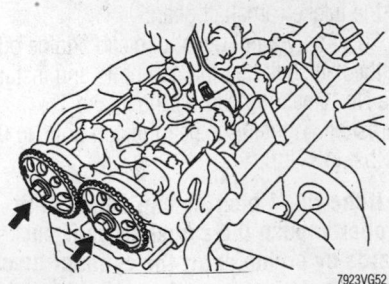

Hold the camshaft with a wrench while removing the sprocket bolt—1.8 (1ZZ-FE) engine

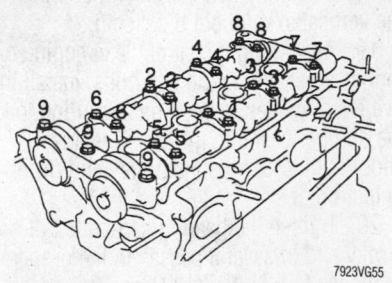

Camshaft bearing cap bolt tightening sequence—1.8 (1ZZ-FE) engine

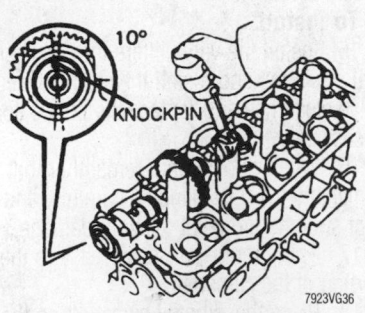

Exhaust camshaft removal and installation positioning—2.2L (5S-FE) engine

with the timing mark 0 of the No. 1 timing belt cover. Check that the valve lifters on the No. 1 cylinder are loose and valve lifters on the No. 4 cylinder are tight. If not, rotate the crankshaft 360°.

➡Since the thrust clearance on both the intake and exhaust camshafts is small, the camshafts must be kept level during removal. If the camshafts are removed without being kept level, the camshaft may damage the bearing surface, causing the camshaft to seize during engine operation.

7. To remove the exhaust camshaft proceed as follows:

a. Set the knock pin of the intake camshaft at 10–45° BTDC of camshaft angle on the cylinder head. This angle will help to lift the exhaust camshaft level and evenly by pushing the No. 2 and No. 4 cylinder camshaft lobes of the exhaust camshaft toward their valve lifters.

b. Secure the exhaust camshaft sub-gear to the main gear using a service bolt. The manufacturer recommends a bolt 0.63–0.79 in. (16–20mm) long with a thread diameter of 6mm and a 1mm thread pitch. When removing the exhaust camshaft be sure that the torsional spring force of the sub-gear has been eliminated.

c. Remove the No. 1 and No. 2 rear bearing cap bolts and remove the cap. Uniformly loosen and remove the bearing cap bolts on the No. 1, No. 2, and No. 4 bearing caps in several passes and in the reverse order of the installation sequence. Do not remove bearing cap bolts to No. 3 bearing cap at this time. Remove the No. 1, 2, and 4 bearing caps.

d. Alternately loosen and remove the bearing cap bolts on the No. 3 bearing cap. As these bolts are loosened check to see that the camshaft is being lifted out straight and level.

➡If the camshaft is not lifted out straight and level, tighten the No. 3 bearing cap bolts. Reverse the order of Steps 7c through 7a and reset the intake camshaft knock pin to 10–45° BTDC, then repeat Steps 7a through 7c. Do not attempt to pry the camshaft from its mounting.

e. Remove the No. 3 bearing cap and exhaust camshaft from the engine.

8. To remove the intake camshaft, proceed as follows:

a. Set the knock pin of the intake camshaft at 80–115° BTDC of the camshaft angle on the cylinder head.

This angle will help to lift the intake camshaft level and evenly by pushing No. 1 and No. 3 cylinder camshaft lobes of the intake camshaft toward their valve lifters.

b. Remove the two front bearing cap bolts and remove the front bearing cap and oil seal. If the cap will not come apart easily, leave it in place without the bolts.

c. Uniformly loosen and remove the bearing cap bolts to No. 1, No. 3, and the No. 4 bearing caps in several phases and in the reverse order of the installation sequence. Do not remove bearing cap bolts to the No. 2 bearing cap at this time. Remove No. 1, 3, and 4 bearing caps.

d. Alternately loosen and remove bearing cap bolts to the No. 2 bearing cap. As these bolts are loosened and after breaking the adhesion on the front bearing cap, check to see that the camshaft is being lifted out straight and level.

➡If the camshaft is not lifting out straight and level tighten the No. 2 bearing cap bolts. Reverse Steps 8b through 8d, then start over from Step 8b. Do not attempt to pry the camshaft from its mounting.

e. Remove the No. 2 bearing cap with the intake camshaft from the engine.

9. Remove the valve lifter shims and hydraulic lifters. Identify each lifter and shim as it is removed so it can be reinstalled in the same position. If the lifters are to be reused, store them upside down in a sealed container.

To install:

10. Install the valve lifters into their original positions and install the shims.

11. Before installing the intake camshaft, apply multi-purpose grease to the camshaft.

12. To install the intake camshaft, proceed as follows:

a. Position the camshaft at 80–115° BTDC of camshaft angle on the cylinder head.

b. Apply sealant to the front bearing cap.

c. Coat the bearing cap bolts with clean engine oil.

d. Tighten the camshaft bearing caps evenly in sequence and in several passes to 14 ft. lbs. (19 Nm).

e. Apply MP grease to a new oil seal lip, and by using a suitable tool, tap a new oil seal into place.

13. To install the exhaust camshaft, proceed as follows:

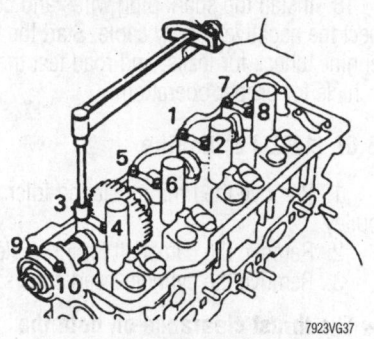

Intake camshaft bearing cap bolt tightening sequence—2.2L (5S-FE) engine

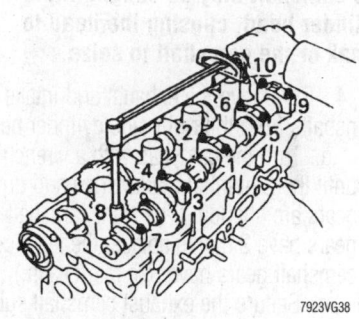

Exhaust camshaft bearing cap bolt tightening sequence—2.2L (5S-FE) engine

a. Set the knock pin of the camshaft at 10–45° BTDC of camshaft angle on the cylinder head.

b. Apply multipurpose grease to the camshaft.

c. Position the exhaust camshaft gear with the intake camshaft gear so that the timing marks are in alignment with one another. Be sure to use the proper alignment marks on the gears. Do not use the assembly reference marks.

d. Turn the intake camshaft clockwise or counterclockwise little by little until the exhaust camshaft sits in the bearing journals evenly without rocking the camshaft on the bearing journals.

e. Coat the bearing cap bolts with clean engine oil.

f. Tighten the camshaft bearing caps evenly in sequence and in several passes to 14 ft. lbs. (19 Nm). Remove the service bolt from the assembly.

14. Check and adjust valve clearance.

15. Install the cylinder head cover with the grommets and the four nuts.

16. Install the timing belt and related components.

17. Connect the electrical connectors, cables, brackets, and components attached to the cylinder head cover.

18. Install the spark plug wires and connect the negative battery cable. Start the engine, check for leaks, and road test the vehicle for proper operation.

3.0L (1MZ-FE) Engine

1. Remove the timing belt and idler pulley.
2. Remove the camshaft timing pulleys.
3. Remove the cylinder head covers.

➡**The thrust clearance on both the intake and exhaust camshafts is very small; the camshafts must be kept level during removal. If the camshafts are removed without being kept level, the camshaft may be caught in the cylinder head, causing the head to break or the camshaft to seize.**

4. To remove the exhaust and intake camshafts from the right side cylinder head:

a. Turn the camshaft with a wrench until the 2 pointed marks drive and driven gears are aligned. (The right camshaft gears have 2 marks apiece; the left side camshaft gears have one mark each.)

b. Secure the exhaust camshaft sub-gear to the main gear using a service bolt. A bolt 0.63–0.79 in. (16–20mm) long with a 6mm thread diameter and a 1mm pitch is recommended. When removing the exhaust camshaft be sure the sub-gear is not loaded; all the force must be eliminated.

c. Uniformly loosen and remove the exhaust camshaft bearing cap bolts in several passes and in the proper sequence. Remove the eight bearing cap bolts and remove the caps, keeping them in the correct order.

d. Remove the exhaust camshaft from the engine.

e. Uniformly loosen and remove the 10 bearing cap bolts in several passes, in the proper sequence. Remove the bearing caps, keeping them in order, remove the oil seal, then lift out the intake camshaft.

5. To remove the exhaust and intake camshafts from the left side cylinder head:

a. Turn the camshaft with a wrench until the pointed marks on the drive and driven gears are aligned. (The right camshaft gears have 2 marks apiece; the left side camshaft gears have one mark each.)

b. Secure the exhaust camshaft sub-gear to the main gear using a service bolt. A bolt 0.63–0.79 in. (16–20mm) long with a 6mm thread diameter and a 1mm pitch is recommended. When removing the exhaust camshaft be sure

the sub-gear is not loaded; all the force must be eliminated.

c. Uniformly loosen and remove the exhaust camshaft bearing cap bolts in several passes and in the proper sequence. Remove the eight bearing cap bolts and remove the caps. Keep the caps in the correct order.

d. Remove the exhaust camshaft from the engine.

e. Uniformly loosen and remove the 10 bearing cap bolts in several passes, in the reverse order of the installation sequence. Remove the bearing caps, keeping them in order, remove the oil seal, then lift out the intake camshaft.

6. Remove the valve lifter shims and hydraulic lifters. Identify each lifter and shim as it is removed so it can be reinstalled in the same position. If the lifters are to be reused, store them upside down in a sealed container.

To install:

7. Install the valve lifters into their original positions and install the shims. Check valve clearance and replace the shims as necessary.

8. When reinstalling, remember that the camshafts must be handled carefully and kept straight and level to avoid damage.

9. Before installing the camshafts in either cylinder head, apply multi-purpose grease to each camshaft.

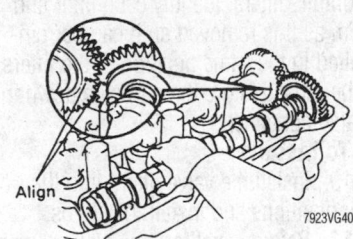

Aligning the camshaft gear timing marks for the right camshafts—3.0L (1MZ-FE) engine

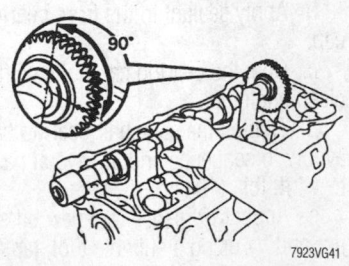

Camshaft installation for the right exhaust camshaft—3.0L (1MZ-FE) engine

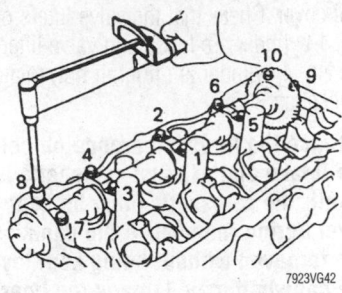

Bearing cap bolt tightening sequence for the right exhaust camshaft—3.0L (1MZ-FE) engine

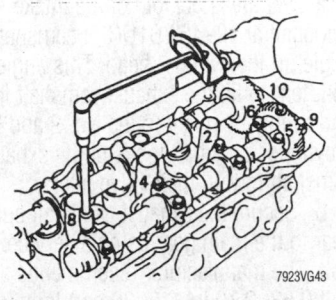

Bearing cap bolt tightening sequence for the right intake camshaft—3.0L (1MZ-FE) engine

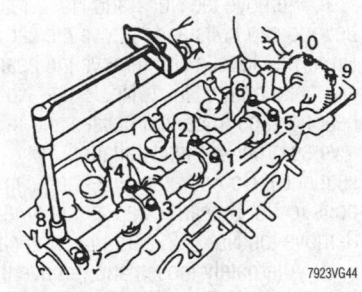

Bearing cap bolt tightening sequence for the left exhaust camshaft—3.0L (1MZ-FE) engine

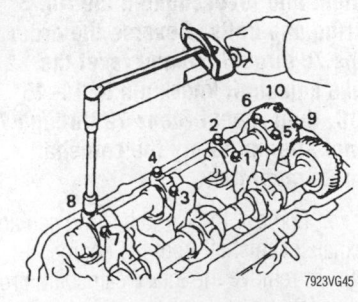

Bearing cap bolt tightening sequence for the left intake camshaft—3.0L (1MZ-FE) engine

10. To install the right camshafts:

a. Position the intake camshaft on the head so that the alignment marks are at a 90 degree angle from vertical. The mark should be at the "3 o'clock" position.

b. Apply sealant to the No. 1 bearing cap.

c. Apply a light coat of clean engine oil to the bolt threads and under the bolt head. Install the bearing caps to their proper position. Tighten the bolts evenly and in several passes to 12 ft. lbs. (16 Nm) in the proper sequence.

d. Position the exhaust camshaft on the head so that the alignment marks are at a 90 degree angle from vertical. The mark should be at the " o'clock" position and must align with the marks on the other gear.

e. Apply a light coat of clean engine oil to the bolt threads and under the bolt head. Install the bearing caps to their proper position. Tighten the bolts evenly and in several passes to 12 ft. lbs. (16 Nm) in the proper sequence.

f. Remove the service bolt.

11. To install the left camshafts:

a. Position the intake camshaft on the head so that the alignment mark is at a 90 degree angle from vertical. The mark should be at the "9 o'clock" position.

b. Apply sealant to the No. 1 bearing cap.

c. Apply a light coat of clean engine oil to the bolt threads and under the bolt head. Install the bearing caps to their proper position. Tighten the bolts evenly and in several passes to 12 ft. lbs. (16 Nm) in the proper sequence.

d. Position the exhaust camshaft on the head so that the alignment marks are at a 90 degree angle from vertical. The mark should be at the "3 o'clock" position and must align with the marks on the other gear.

e. Apply a light coat of clean engine oil to the bolt threads and under the bolt head. Install the bearing caps to their proper position. Tighten the bolts evenly and in several passes to 12 ft. lbs. (16 Nm) in the proper sequence.

f. Remove the service bolt.

12. Apply multi-purpose grease to new camshaft oil seals. Install the seals.

13. Install the No. 3 (rear) timing belt cover.

14. Install the camshaft timing gears.

15. Install the idler pulley, timing belt and covers.

16. Check and adjust the valve clearance.

17. Install the cylinder head (valve) covers.

18. Start the engine. Check the ignition timing.

19. Test drive the vehicle.

20. Check all fluid levels.

3.0L (2JZ-GTE and 2JZ-GE) Engines

1. Disconnect the negative battery cable from the battery. On vehicles equipped with an air bag, wait at least 90 seconds before proceeding.

2. Remove the timing belt from the engine.

3. Remove the cylinder head covers.

4. While holding each camshaft with a wrench, loosen the camshaft sprocket bolt and remove the sprocket.

5. Remove the four bolts and lift out the No. 4 (inner) timing belt cover.

6. Uniformly loosen, then remove the four No. 1 camshaft bearing cap bolts. These are the bolts directly behind the sprockets. Remove the bearing caps.

7. Uniformly, and in the reverse order of the installation sequence, loosen and remove the remaining bearing cap bolts. Note that there are separate sequences for the exhaust and intake camshafts. Lift off all 12 bearing caps.

8. Lift out the exhaust and intake camshafts.

9. Remove the valve lifter shims and hydraulic lifters. Identify each lifter and shim as it is removed so it can be reinstalled in the same position. If the lifters are to be reused, store them upside down in a sealed container.

To install:

10. Install the valve lifters into their original positions and install the shims. Check valve clearance and replace the shims as necessary.

11. When reinstalling, remember that the camshafts must be handled carefully and kept straight and level to avoid damage.

12. Coat each camshaft with engine oil, then position them in the cylinder head with the cam lobes and the knock pins in the correct position.

13. Position the No. 3 and No. 7 bearing caps in place, coat the bolt threads with oil, then tighten them temporarily.

14. Coat new oil seals with multi-purpose grease, then slide them over the camshafts.

15. Clean the mating surfaces of the two No. 1 bearing caps, then apply some sealant. Install the bolts.

16. Install all remaining bearing caps, coat the threads of each bolt with clean oil, then tighten them, in several passes, in the correct sequence, to 14 ft. lbs. (20 Nm). Note that there are separate sequences for the intake and exhaust sides.

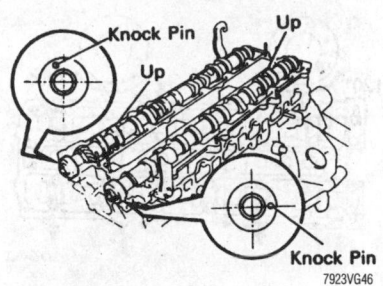

During installation, position the knock pins as shown—3.0L (2JZ-GE and 2JZ-GTE) engines

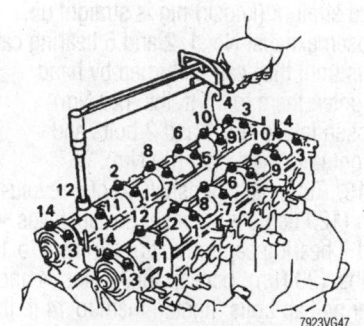

Camshaft bearing cap bolt tightening sequence—3.0L (2JZ-GE and 2JZ-GTE) engines

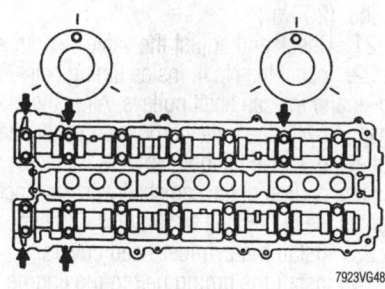

Retightening the camshafts (Step 1)—3.0L (2JZ-GE and 2JZ-GTE) engines

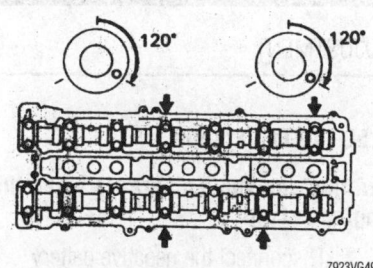

Retightening the camshafts (Step 2)—3.0L (2JZ-GE and 2JZ-GTE) engines

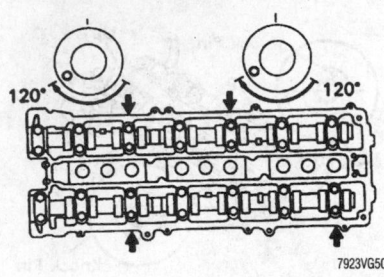

7923VG50

Retightening the camshafts (Step 3)—3.0L (2JZ-GE and 2JZ-GTE) engines

17. Press the oil seal in as far as it will go.

18. Rotate each camshaft until the forward straight (knock) pin is straight up. Loosen exhaust No. 1, 2 and 6 bearing cap bolts until they can be turned by hand; retighten them to 14 ft. lbs. (20 Nm). Loosen intake No. 1 and 2 bolts and retighten to 14 ft. lbs. (20 Nm).

19. Turn each camshaft ⅓ of a revolution (120 degrees). Loosen exhaust Nos. 4 and 7 bearing cap bolts; tighten them to 14 ft. lbs. (20 Nm). Loosen intake Nos. 4 and 6 bearing cap bolts; tighten them to 14 ft. lbs. (20 Nm).

20. Turn each camshaft an additional ⅓ of a revolution, loosen exhaust bearing cap bolts No. 3 and 5, then tighten them to 14 ft. lbs. (20 Nm). Loosen intake bearing cap bolts No. 3 and 7, then tighten them to 14 ft. lbs. (20 Nm).

21. Check and adjust the valve clearance.

22. Install the No 4. inside timing belt cover and the camshaft pulleys. Align the shaft pin with the pulley groove and slide the pulley on. Install the bolt temporarily. Hold the hex portion of the camshaft with a wrench; tighten the pulley bolt to 59 ft. lbs. (79 Nm).

23. Install the cylinder head covers.

24. Install the timing belt to the engine.

25. Connect the negative battery cable to the battery.

26. Check and/or adjust the ignition timing as necessary.

Valve Lash

ADJUSTMENT

1.5L (5E-FE) Engine

➡**Adjust the valve clearance when the engine is cold.**

1. Disconnect the negative battery cable. On vehicles equipped with an air bag, wait at least 90 seconds before proceeding.

2. Remove the cylinder head covers.

3. Turn the crankshaft pulley and align its groove with the timing mark **0** of the No. 1 timing cover.

4. Check that the timing marks on the camshaft sprockets are in alignment with the marks on the No. 4 timing cover. If not, turn the crankshaft one complete revolution (360° degrees).

5. Measure the clearance between the valve lifter and the camshaft. Record the measurements on the intake valves No. 1 and 2. Measure the exhaust valves at No. 1 and 3.

   a. The intake valve clearance cold is 0.006–0.010 in. (0.15–0.25mm).

Adjusting Shim Selection Chart

Shim No.	New shim thickness mm (in.) Thickness	Shim No.	Thickness
02	2.500 (0.0984)	20	2.950 (0.1161)
04	2.550 (0.1004)	22	3.000 (0.1181)
06	2.600 (0.1024)	24	3.050 (0.1201)
08	2.650 (0.1043)	26	3.100 (0.1220)
10	2.700 (0.1063)	28	3.150 (0.1240)
12	2.750 (0.1083)	30	3.200 (0.1260)
14	2.800 (0.1102)	32	3.250 (0.1280)
16	2.850 (0.1122)	34	3.300 (0.1299)
18	2.900 (0.1142)		

7923VG56

Adjusting shim chart (intake and exhaust)—1.5L (5E-FE) engine

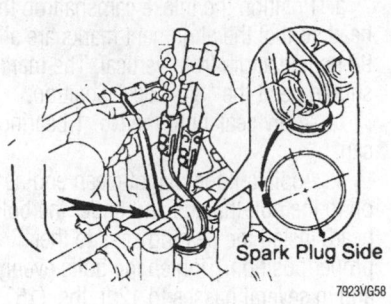

Spark Plug Side

7923VG58

Common method of removing valve shims

   b. The exhaust valve clearance cold is 0.012–0.016 in. (0.31 0.41mm).

6. Turn the crankshaft pulley one revolution (360°) and align the groove with the timing mark **0** of the No.1 timing belt cover.

7. Measure the clearance between the valve lifter and the camshaft. Record the measurements on the intake valves No. 3 and 4. Measure the exhaust valves at No. 2 and 4.

   a. The intake valve clearance cold is 0.006–0.010 in. (0.15–0.25mm).

   b. The exhaust valve clearance cold is 0.012–0.016 in. (0.31–0.41mm).

8. To adjust the valve clearance:

Adjusting Shim Selection Chart

Shim No.	New shim thickness mm (in.) Thickness	Shim No.	Thickness
1	2.500 (0.0984)	10	2.950 (0.1161)
2	2.550 (0.1004)	11	3.000 (0.1181)
3	2.600 (0.1024)	12	3.050 (0.1201)
4	2.650 (0.1043)	13	3.100 (0.1220)
5	2.700 (0.1063)	14	3.150 (0.1240)
6	2.750 (0.1083)	15	3.200 (0.1260)
7	2.800 (0.1102)	16	3.250 (0.1280)
8	2.850 (0.1122)	17	3.300 (0.1299)
9	2.900 (0.1142)		

HINT: New shims have the thickness in millimeters imprinted on the face.

7923VG57

Adjusting shim chart (intake and exhaust)—1.6L (4A-FE), 1.8L (7A-FE), 2.2L (5S-FE), 3.0L (1MZ-FE), 3.0L (2JZ-GE and 2JZ-GTE) engines

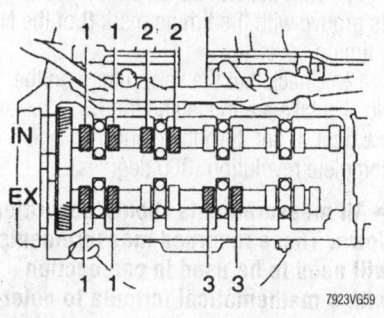

Intake valves (1 and 2) and exhaust valves (1 and 3)—5E-FE engine

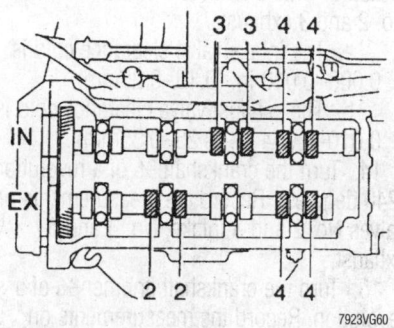

Intake valves (3 and 4) and exhaust valves (2 and 4)—5E-FE engine

a. Remove the adjusting shim and turn the crankshaft to position the cam lobe of the camshaft on the valve to be adjusted upward.

b. Turn the valve lifter so that the notch is perpendicular to the camshaft and facing the spark plug side.

c. Using SST 09248–55040 (valve lifter press) or equivalent, hold the camshaft in place.

d. Using SST 09248–55040 (valve lifter press) or equivalent, press down the valve lifter and place SST 09248–05420 (valve lifter stopper) or equivalent between the camshaft and valve lifter.

e. Remove the SST 09248–44040 tool.

f. Using a small screwdriver and a magnetic finger, remove the adjusting shim.

9. Determine the replacement adjusting shim size by either using the chart or the following formula:

- Intake: $N = T + A - 0.008$ in. (0.20mm)
- Exhaust: $N = T + A - 0.014$ in. (0.36mm)
- T = Thickness of removed shim
- A = Measured valve clearance
- N = Thickness of new shim

10. Install a new shim.
11. Recheck the valve clearance.
12. Install the cylinder head covers.
13. Connect the negative battery cable.

1.6L (4A-FE) and 1.8L (7A-FE and 1ZZ-FE) Engine

➡**Adjust the valve clearance when the engine is cold.**

1. Disconnect the negative battery cable.

2. Remove the cylinder head covers.

3. Turn the crankshaft pulley and align its groove with the timing mark **0** of the No. 1 timing cover.

4. Check that the timing marks on the camshaft sprockets are in alignment with the marks on the No. 4 timing cover. If not, turn the crankshaft one complete revolution (360° degrees).

5. Measure the clearance between the valve lifter and the camshaft. Record the measurements on the intake valves No. 1 and 2. Measure the exhaust valves at No. 1 and 3.

a. The intake valve clearance cold is 0.006–0.010 in. (0.15–0.25mm).

b. The exhaust valve clearance cold is 0.010–0.014 in. (0.25–0.35mm).

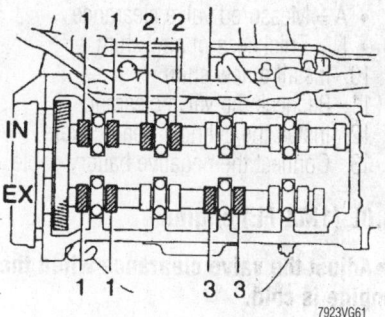

Intake valves (1 and 2) and exhaust valves (1 and 3)—1.6L (4A-FE) and 1.8L (7A-FE and 1ZZ-FE) engines

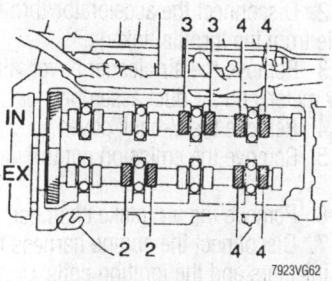

Intake valves (3 and 4) and exhaust valves (2 and 4)—1.6L (4A-FE) and 1.8L (7A-FE and 1ZZ-FE) engines

6. Turn the crankshaft pulley one revolution (360°) and align the groove with the timing mark **0** of the No.1 timing belt cover.

7. Measure the clearance between the valve lifter and the camshaft. Record the measurements on the intake valves No. 3 and 4. Measure the exhaust valves at No. 2 and 4.

a. The intake valve clearance cold is 0.006–0.010 in. (0.15–0.25mm).

b. The exhaust valve clearance cold is 0.010–0.014 in. (0.25–0.35mm).

8. To adjust the intake valve clearance:

a. Remove the intake camshaft.

b. Using a small screwdriver and a magnetic finger, remove the adjusting shim.

c. Determine the replacement adjusting shim size by either using the chart or the following formula:

- Intake: $N = T + A - 0.008$ in. (0.20mm)
- T = Thickness of removed shim
- A = Measured valve clearance
- N = Thickness of new shim
d. Install a new shim.
e. Install intake camshaft.
f. Recheck the valve clearance.

9. To adjust the exhaust valve clearance:

a. Turn the crankshaft to position the cam lobe of the camshaft on the valve to be adjusted, upward.

b. Turn the valve lifter so that the notch is perpendicular to the camshaft and facing the spark plug side.

c. Using SST 09248–55040 (valve lifter press) or equivalent, hold the camshaft in place.

d. Using SST 09248–55040 (valve lifter press) or equivalent, press down the valve lifter and place SST 09248–05420 (valve lifter stopper) or equivalent between the camshaft and valve lifter.

e. Remove the SST 09248–44040 tool.

f. Using a small screwdriver and a magnetic finger, remove the adjusting shim.

10. Determine the replacement adjusting shim size by either using the chart or the following formula:

- Exhaust: $N = T + A - 0.014$ in. (0.36mm)
- T = Thickness of removed shim
- A = Measured valve clearance
- N = Thickness of new shim

11. Install a new shim.
12. Recheck the valve clearance.
13. Install the cylinder head covers.
14. Connect the negative battery cable.

2.2L (5S-FE) Engine

➡**Adjust the valve clearance when the engine is cold.**

1. Disconnect the negative battery cable. On vehicles equipped with an air bag, wait at least 90 seconds before proceeding.

2. Remove the cylinder head covers.

3. Turn the crankshaft pulley and align its groove with the timing mark **0** of the No. 1 timing cover.

4. Check that the timing marks on the camshaft sprockets are in alignment with the marks on the No. 4 timing cover. If not, turn the crankshaft one complete revolution (360° degrees).

5. Measure the clearance between the valve lifter and the camshaft. Record the measurements on the intake valves No. 1 and 2. Measure the exhaust valves at No. 1 and 3.

   a. The intake valve clearance cold is 0.007–0.011 in. (0.19–0.29mm).

   b. The exhaust valve clearance cold is 0.011–0.015 in. (0.28–0.38mm).

6. Turn the crankshaft pulley one revolution (360°) and align the groove with the timing mark **0** of the No.1 timing belt cover.

7. Measure the clearance between the valve lifter and the camshaft. Record the mea-surements on the intake valves No. 3 and 4. Measure the exhaust valves at 2 and 4.

   a. The intake valve clearance cold is 0.007–0.011 in. (0.19–0.29mm).

   b. The exhaust valve clearance cold is 0.011–0.015 in. (0.29–0.38mm).

8. To adjust the valve clearance:

   a. Turn the crankshaft to position the cam lobe of the camshaft on the valve to be adjusted, upward.

   b. Turn the valve lifter so that the notch is perpendicular to the camshaft and facing the spark plug side.

   c. Using SST 09248–55040 (valve lifter press) or equivalent, hold the camshaft in place.

   d. Using SST 09248–55040 (valve lifter press) or equivalent, press down the valve lifter and place SST 09248–05420 (valve lifter stopper) or equivalent between the camshaft and valve lifter.

   e. Remove the SST 09248–44040 tool.

   f. Using a small screwdriver and a magnetic finger, remove the adjusting shim.

9. Determine the replacement adjusting shim size by either using the chart or the following formula:

- Intake: $N = T + A - 0.009$ in. (24mm)
- Exhaust: $N = T + A - 0.013$ in. (0.33mm)
  - T = Thickness of removed shim
  - A = Measured valve clearance
  - N = Thickness of new shim

10. Install a new shim.

11. Recheck the valve clearance.

12. Install the cylinder head covers.

13. Connect the negative battery cable.

3.0L (1MZ-FE) engine

➡**Adjust the valve clearance when the engine is cold.**

1. With the ignition switch in the **LOCK** position, disconnect the negative battery terminal. If equipped with an air bag sys-tem, wait at least 90 seconds or longer before performing any other work.

2. Disconnect the accelerator/throttle cable from the throttle linkage.

3. Remove the air cleaner cover, air flow meter, and air duct assembly.

4. Remove the V-bank cover.

5. Remove the emission control valve set.

6. Remove the air intake chamber.

7. Disconnect the engine harness from the injectors and the ignition coils.

8. Remove the ignition coils and keep them in order for reassembly.

9. Remove the spark plugs.

10. Remove the cylinder head covers.

11. Turn the crankshaft pulley and align its groove with the timing mark **0** of the No. 1 timing cover.

12. Check that the valve lifters on the No. 1 intake are loose and the No. 1 exhaust are tight. If not, turn the crankshaft one complete revolution (360 degrees).

➡**All measurements should be written down. These recorded measurements will need to be used in conjunction with a mathematical formula to deter-mine the thickness of the replacement shims.**

13. Measure the clearance between the valve lifters and the camshaft. Record the measurements on valves No. 1 and 6 intake; No. 2 and 3 exhaust.

   a. The intake valve clearance cold is 0.006–0.010 in. (0.15–0.25mm).

   b. The exhaust valve clearance cold is 0.010–0.014 in. (0.25–0.35mm).

14. Turn the crankshaft ⅔ of a revolution (240 degrees). Record the measurements on valves No. 2 and 3 intake; No. 4 and 5 exhaust.

15. Turn the crankshaft another ⅔ of a revolution. Record the measurements on valves No. 4 and 5 intake; No. 1 and 6 exhaust.

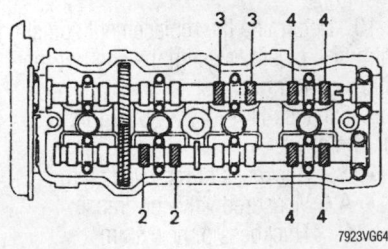

Intake valves (1 and 2) and exhaust valves (1 and 3)—2.2L (5S-FE) engine

7923VG63

Intake valves (3 and 4) and exhaust valves (2 and 4)—2.2L (5S-FE) engine

7923VG64

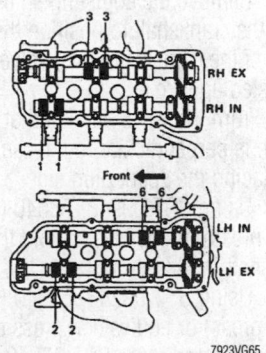

7923VG65

Adjust these valves during the 1st step—3.0L (1MZ-FE) engine

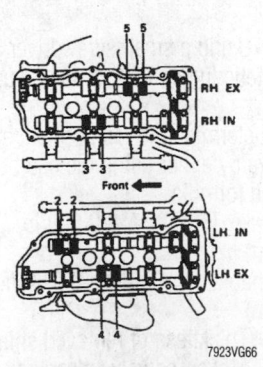

7923VG66

Adjust these valves during the 2nd step—3.0L (1MZ-FE) engine

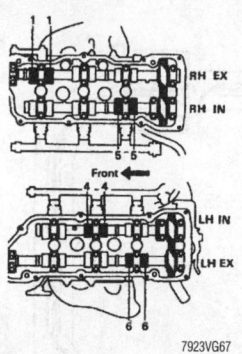

Adjust these valves during the 3rd step— 3.0L (1MZ-FE) engine

16. Remove the adjusting shim by turning the crankshaft to position the cam lobe of the camshaft in the up position on the valve to be adjusted. Using a small thin flat bladed tool, turn the valve lifter so that the notches are perpendicular to the camshaft. Press down the valve lifter with SST 09248–55010 part A or equivalent. Place SST 09248–55010 part B between the camshaft and the valve lifter; remove part A.

17. Remove the adjusting shim with a magnet and a small screwdriver.

18. Determine the replacement adjusting shim size by either using the charts or the following formulas:
- Intake: $N = T + (A—0.008$ in./0.020mm)
- Exhaust: $N = T + (A—0.012$ in./0.30mm)
- T = Thickness of removed shim
- A = Meassured valve clearance
- N = Thickness of new shim

19. Select a new shim with a thickness as close as possible to the calculated value. Install the new replacement shim.

➡**Shims are available in 17 sizes in increments of 0.0020 in. (0.050mm), from 0.0984 in. (2.500mm) to 0.1299 in. (3.300mm).**

20. Recheck the valve clearance.
21. Install the cylinder head covers
22. Install the spark plugs and the ignition coils.
23. Connect the engine wiring harness to the injectors and the coils.
24. Install the intake chamber.
25. Install the emission control valve set.
26. Install the V-bank cover.
27. Install the air flow meter, air duct, and air cleaner cover.
28. Connect the negative battery cable.

3.0L (2JZ-GE and 2JZ-GTE) Engines

➡**Adjust the valve clearance when the engine is cold.**

1. Disconnect the negative battery cable. On vehicles equipped with an air bag, wait at least 90 seconds before proceeding.
2. Remove the cylinder head covers.
3. Turn the crankshaft pulley and align its groove with the timing mark **0** of the No. 1 timing cover.
4. Check that the timing marks on the camshaft sprockets are in alignment with the marks on the No. 4 timing cover. If not, turn the crankshaft one complete revolution (360° degrees).
5. Measure the clearance between the valve lifter and the camshaft. Record the measurements on the intake valves No. 1, 2 and 4. Measure the exhaust valves at No. 1, 3 and 5.
   a. The intake valve clearance cold is 0.006–0.010 in. (0.15–0.25mm).
   b. The exhaust valve clearance cold is 0.010–0.014 in. (0.25–0.35mm).
6. Turn the crankshaft pulley one revolution (360°) and align the groove with the timing mark **0** of the No.1 timing belt cover.
7. Measure the clearance between the valve lifter and the camshaft. Record the measurements on the intake valves No. 3, 5 and 6. Measure the exhaust valves at No. 2, 4, and 6.
   a. The intake valve clearance cold is 0.006–0.010 in. (0.15–0.25mm).
   b. The exhaust valve clearance cold is 0.010–0.014 in. (0.25–0.35mm).

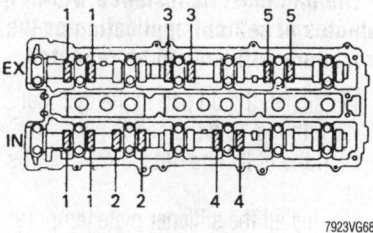

Valve clearance inspection (before turning crankshaft 360 degrees)—2JZ-GE and 2JZ-GTE engines

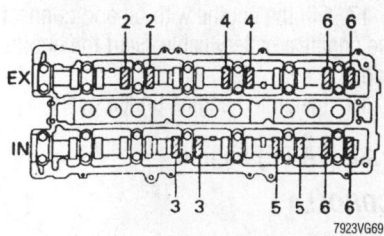

Valve clearance inspection (after turning crankshaft 360 degrees)—2JZ-GE and 2JZ-GTE engines

8. To adjust the valve clearance:
   a. Remove the adjusting shim and turn the crankshaft to position the cam lobe of the camshaft on the adjusting valve upward.
   b. Turn the valve lifter so that the notches are perpendicular to the camshaft.
   c. Using SST 09248–55040 (valve lifter press) or equivalent, hold the camshaft in place.
   d. Using SST 09248–55040 (valve lifter press) or equivalent, press down the valve lifter and place SST 09248–05420 (valve lifter stopper) or equivalent between the camshaft and valve lifter.
   e. Remove the SST 09248–44040 tool.
   f. Using a small screwdriver and a magnetic finger, remove the adjusting shim.
9. Determine the replacement adjusting shim size by either using the chart or the following formula:
- Intake: $N = T + (A—0.008$ in./0.20mm)
- Exhaust: $N = T + (A—0.012$ in./0.30mm)
- T = Thickness of removed shim
- A = Meassured valve clearance
- N = Thickness of new shim
10. Recheck the valve clearance.
11. Install the cylinder head covers.
12. Connect the negative battery cable.

Oil Pan

REMOVAL & INSTALLATION

1.5L (5E-FE) Engine

1. Disconnect the negative battery cable. On vehicles equipped with an air bag, wait at least 90 seconds before proceeding.
2. Raise the vehicle and support it safely, then drain the oil.
3. Remove the hood.
4. Remove the oil dipstick.
5. Remove the timing belt.
6. Suspend the engine with a hoist.

➡**Do not raise the engine more than necessary, since wiring and other components can be damaged.**

7. Remove the crankshaft timing sprocket and oil pump sprocket.
8. Remove the air conditioning compressor and mounting bracket by removing the bolts.
9. For vehicles with distributor ignition, disconnect the texhaust pipe stay.
10. Disconnect the oxygen sensor.

11. Disconnect the front exhaust pipe by removing the two bolts and two compression springs.

12. Remove the ten oil pan bolts, then remove the oil pan.

To install:

13. Clean all gasket mating surfaces.

14. Using a new gasket and sealer, install the oil pan. Tighten the bolts to 97–10 ft. lbs. (13 Nm).

15. Connect the front exhaust pipe using a new gasket. Tighten the two bolts to 46 ft. lbs. (62 Nm).

16. Connect the oxygen sensor connector.

17. Install the exhaust pipe stay, if removed. the bolts to 14 ft. lbs. (19 Nm).

18. Install the A/C compressor mounting bracket and tighten the bolts to 20 ft. lbs. (27 Nm).

19. Connect the A/C compressor to the bracket using the four bolts. Tighten the bolts to 18 ft. lbs. (25 Nm).

20. Install the crankshaft timing and oil pump sprockets.

21. Install the timing belt.

22. Install the oil dipstick.

23. Fill the engine with oil.

24. Connect the negative battery cable to the battery.

25. Install the hood.

26. Start the engine check for leaks.

1.6L (4A-FE) Engine

COROLLA

1. Disconnect the negative battery cable. On vehicles equipped with an air bag, wait at least 90 seconds before proceeding.

2. Raise and safely support the vehicle.

3. Remove the undercovers from the vehicle.

4. Drain the engine oil.

5. Remove the front exhaust pipe.

6. On Celica, remove the following items:
- Lower suspension crossmember
- Center engine mount
- Oil level gauge

7. Remove the five set bolts and remove the stiffener plate.

8. Remove the 19 bolts and the two nuts holding the oil pan.

9. Insert the blade of the SST 09032–00100 tool between the oil pan and the cylinder block; cut off the applied sealer and remove the oil pan.

➡**Do not use the tool for the oil pump body side and rear oil seal retainer.**

To install:

10. Remove any old sealant from the oil pan flange and thoroughly clean both sealing surfaces.

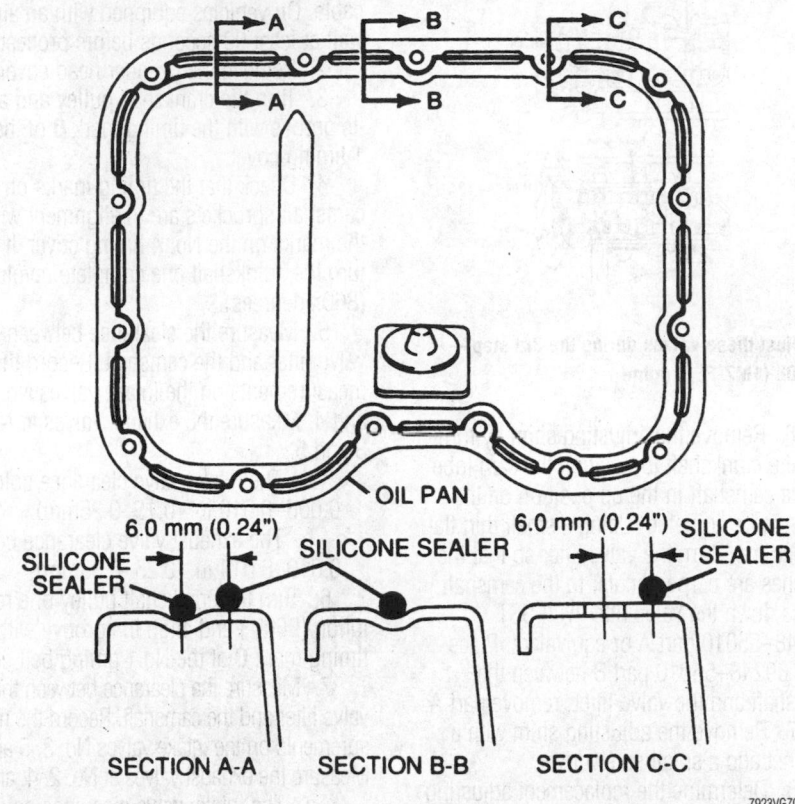

Oil pan sealer application pattern—1.6L (4A-FE) and 1.8L (7A-FE) engines

11. Apply a 3–5mm bead of sealant to the No. 1 oil pan flange.

➡**The pan must be installed within five minutes of sealant application or the procedure will have to be repeated.**

12. Install the oil pan to the cylinder block with the 19 bolts and the two nuts. Tighten the bolts and nuts to 43 inch lbs. (5 Nm).

13. Install the stiffener plate temporarily with No. 1 bolt and tighten the remaining bolts to 17 ft. lbs. (23 Nm).

14. On Celica, install the lower suspension crossmember, center engine mount, and oil level gauge.

15. Install the front exhaust pipe.

16. Safely lower the vehicle.

17. Fill the engine with oil and connect the negative battery cable. Start the engine and check for leaks.

18. Install the engine undercovers.

1.8L (7A-FE) Engine

COROLLA

1. Disconnect the negative battery cable. On vehicles equipped with an air bag, wait at least 90 seconds before proceeding.

2. Remove the hood.

3. Raise and safely support the vehicle.

4. Remove the undercovers from the vehicle.

5. Drain the engine oil.

6. Remove the front exhaust pipe.

7. Remove the 13 bolts and 2 nuts and remove the No. 2 oil pan.

8. Insert the blade of the SST 09032–00100 tool between the oil pan and the cylinder block, and cut off the applied sealer and remove the oil pan.

➡**Do not use the tool for the oil pump body side and rear oil seal retainer.**

9. Remove the three bolts holding the No. 1 oil pan to the transaxle, remove the six bolts, and using a 5mm hexagon wrench, remove the 14 bolts and the No. 1 oil pan.

To install:

10. Remove any old sealant from the oil pan flange and thoroughly clean both sealing surfaces.

11. Apply a 3–5mm bead of sealant to the No. 1 oil pan flange. The pan must be installed within five minutes of sealant application or the procedure will have to be repeated.

12. Using a 5mm hexagon wrench, install the No. 1 oil pan with 14 new bolts. Tighten the bolt A to 12 ft. lbs. (16 Nm).

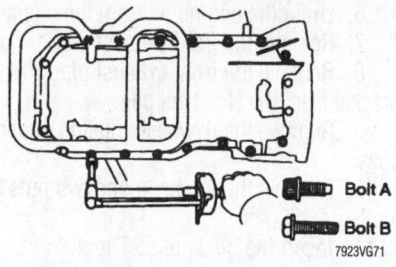

Mounting bolts A and B for the oil pans—1.8L (7A-FE) engine

Install the six bolts and tighten bolt B to 69 inch lbs. (8 Nm). Reinstall the three bolts holding the No. 1 oil pan to the transaxle and tighten the bolts to 17 ft. lbs. (23 Nm).

13. Apply a 3–5mm bead of sealant to the No. 2 oil pan flange. The pan must be installed within five minutes of sealant application or the procedure will have to be repeated.

14. Install the No. 2 oil pan with the 13 bolts and the 2 nuts. Tighten the bolts and nuts to 43 inch lbs. (5 Nm).

15. Install the front exhaust pipe.

16. Safely lower the vehicle.

17. Reinstall the hood.

18. Fill the engine with oil, reconnect the negative battery cable, start the engine and check for leaks.

19. Reinstall the engine undercovers.

CELICA

1. Disconnect the negative battery cable. On vehicles equipped with an air bag, wait at least 90 seconds before proceeding.

2. Remove the hood.

3. Raise and safely support the vehicle.

4. Remove the RH front wheel.

5. Remove the undercovers from the vehicle.

6. Drain the engine oil.

7. Remove the front exhaust pipe.

8. If equipped with A/T, disconnect the transaxle control cable from the engine mounting center member.

9. If equipped with A/C, disconnect the A/C pressure pipe from the engine center mounting member.

10. Remove the front exhaust pipe support bracket.

11. Raise and safely support the engine assembly.

12. Remove the rear engine mounting insulator and the engine mounting center member.

13. Install an engine hanger and suspend the engine with an engine sling device or equivalent.

14. Remove the oil dipstick guide and the dipstick.

15. Remove the 13 bolts and two (2) nuts and remove the No. 2 oil pan.

16. Insert the blade of the SST 09032–00100 tool, or equivalent, between the oil pan and the cylinder block; cut off the applied sealer and remove the oil pan.

➡ **Do not use the tool for the oil pump body side and rear oil seal retainer.**

17. Remove the two bolts and two nuts securing the baffle plate. Remove the baffle plate from the engine.

18. Remove the three nuts, the oil strainer, and the strainer gasket.

19. Remove the three bolts holding the No. 1 oil pan to the transaxle and remove the six **B** bolts. Remove the 14 bolts and the No. 1 oil pan.

To install:

20. Remove any old sealant from the oil pan flange and thoroughly clean both sealing surfaces.

21. Apply a 3–5mm bead of sealant to the No. 1 oil pan flange.

➡ **The pan must be installed within five minutes of sealant application or the procedure will have to be repeated.**

22. Install the No. 1 oil pan with 14 new bolts. Tighten the **A** bolts to 12 ft. lbs. (16 Nm). Install the six **B** bolts and tighten the bolts to 69 inch lbs. (8 Nm). Install the three bolts holding the No. 1 oil pan to the transaxle and tighten the bolts to 17 ft. lbs. (23 Nm).

23. Install the oil strainer and gasket with three new nuts. Tighten the nuts to 82 inch lbs. (9 Nm).

24. Install the oil pan baffle plate. Tighten the two bolts and two nuts to 69 inch lbs. (8 Nm).

25. Apply a 3–5mm bead of sealant to the No. 2 oil pan flange.

➡ **The pan must be installed within five minutes of sealant application or the procedure will have to be repeated.**

26. Install the No. 2 oil pan with the 13 bolts and the two (2) nuts. Tighten the bolts and nuts to 43 inch lbs. (5 Nm).

27. Remove the engine sling device, safely lower the engine, remove the hanger, and install the rear engine mounting insulator and engine mounting center member.

28. Install the remaining components in the reverse order they were removed.

29. Fill the engine with oil and connect the negative battery cable. Start the engine and check for leaks.

30. Install the engine undercovers.

1.8L (1ZZ-FE) Engine

1. Disconnect the negative battery cable. On vehicles equipped with an air bag, wait at least 90 seconds before proceeding.

2. Raise and safely support the vehicle.

3. Remove the undercovers from the vehicle.

4. Drain the engine oil.

5. Remove the front exhaust pipe.

6. Remove the oil pan mounting bolts and nuts.

7. Insert the blade of the SST 09032–00100 tool between the oil pan and the cylinder block, and cut off the applied sealer and remove the oil pan.

To install:

8. Remove any old sealant from the oil pan flange and thoroughly clean the sealing surface.

9. Apply a 3–5mm bead of sealant to the oil pan flange. The pan must be installed within five minutes of sealant application or the procedure will have to be repeated.

10. Tighten the 14 bolts and 2 nuts in several passes to 80 inch lbs. (9 Nm).

11. Install the front exhaust pipe.

12. Safely lower the vehicle.

13. Fill the engine with oil, reconnect the negative battery cable, start the engine and check for leaks.

14. Reinstall the engine undercovers.

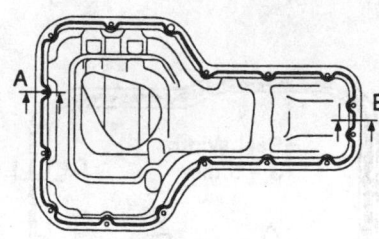

Seal Width
4 – 5 mm

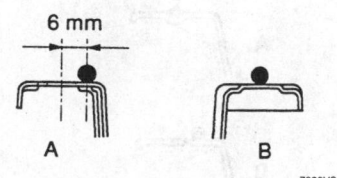

Apply sealant to the oil pan as shown—1.8L (1ZZ-FE) engine

2.2L (5S-FE) Engine

1. Disconnect the negative battery cable. On vehicles equipped with an air bag, wait at least 90 seconds before proceeding.

2. Raise and safely support the vehicle.

3. Drain the engine oil and remove the engine undercovers.

4. Remove the front exhaust pipe.

5. Safely support the engine assembly and remove the engine mounting center member.

6. On Celica, remove the TWC as follows:

   a. Disconnect the sub oxygen sensor connector.

   b. Remove the RH exhaust manifold stay.

   c. Remove the three bolts, two nuts, the TWC with gasket, retainer, and the cushion.

7. Remove the rear end stiffener plate.

8. Remove the oil dipstick and remove the 17 bolts and two nuts attaching the oil pan to the engine.

9. Insert the blade of the SST 09032–00100 tool, or equivalent, between the oil pan and the cylinder block; cut off the applied sealer and remove the oil pan.

➡**Do not use the tool for the oil pump body side and rear oil seal retainer.**

To install:

10. Remove any old sealant from the oil pan flange and thoroughly clean both sealing surfaces.

11. Apply a 3–5mm bead of sealant to the oil pan flange.

➡**The pan must be installed within five minutes of sealant application or the procedure will have to be repeated.**

12. Install the oil pan with the 17 bolts and two nuts. Uniformly tighten the bolts and nuts in several passes. Tighten the bolts and nuts to 48 inch lbs. (5.4 Nm) and install the oil dipstick.

13. Install the rear end stiffener plate.

14. On Celica, install the TWC.

15. Install the engine mounting center member and safely lower the engine.

16. Install the front exhaust pipe.

17. Fill the engine with oil and connect the negative battery cable. Start the engine and check for leaks.

18. Recheck the engine oil level and reinstall the engine undercovers.

3.0L (1MZ-FE) Engine

1. Disconnect the negative battery cable from the battery. On vehicles equipped with an air bag, wait at least 90 seconds before proceeding.

2. Raise and safely support the front of the vehicle.

3. Remove the right front wheel.

4. Remove the fender apron seal.

5. Remove the engine undercover.

6. Drain the engine oil from the engine.

7. Remove the front exhaust pipe.

8. Remove the front exhaust pipe bracket from the No. 1 oil pan.

9. Remove the flywheel housing undercover.

10. Remove the ten bolts and two nuts to the No. 2 oil pan.

11. Insert the blade of SST tool 09032–00100 or equivalent between the No. 1 and No. 2 oil pans. Clean the surfaces of the oil pans.

12. Remove the oil strainer and gasket from the engine by removing the three nuts.

13. Remove the No.1 oil pan as follows:

   a. Remove the two bolts to the flywheel housing undercover. Remove the flywheel undercover.

   b. Remove the 17 bolts and 2 nuts to the No. 1 oil pan. Make a note of the position of the each bolt. When replacing the bolts into the oil pan, place each bolt in the position from which it was removed.

   c. Remove the oil pan by prying the portions between the cylinder block and the oil pan. Be careful not to damage the contact surfaces.

14. Remove the baffle plate from the No. 1 oil pan.

To install:

15. Clean all mating surfaces of the oil pans.

16. Install the baffle plate to the No. 1 oil pan and tighten to 69 inch lbs. (8 Nm).

17. Install the No. 1 oil pan as follows:

   a. Using a non residue solvent, clean both sealing surfaces to the oil pan.

   b. Apply liquid sealant to the oil pan and engine block.

   c. Install the oil pan with the 17 bolts and 2 nuts. Uniformly tighten the bolts and nuts in several passes.

   d. Tighten the bolts as follows:

   • 10mm head bolt-69 inch lbs. (8 Nm)
   • 12mm head bolt-14 ft. lbs. (19.5 Nm)
   • 14mm head bolt-27 ft. lbs. (37.2 Nm)

   e. Install the flywheel housing undercover with the two bolts. Tighten the bolts to 69 inch lbs. (7.8 Nm).

18. Install the oil strainer with the three nuts. Tighten the nuts to 69 inch lbs. (7.8 Nm).

19. Install the No. 2 oil pan as follows:

   a. Using a non residue solvent, clean both sealing surfaces to the oil pan.

   b. Apply liquid sealant to the oil pan and engine block.

   c. Install the No. 2 oil pan with the ten bolts and two nuts. Uniformly tighten the bolts and nuts in several passes. Tighten the bolts to 69 inch lbs. (7.8 Nm).

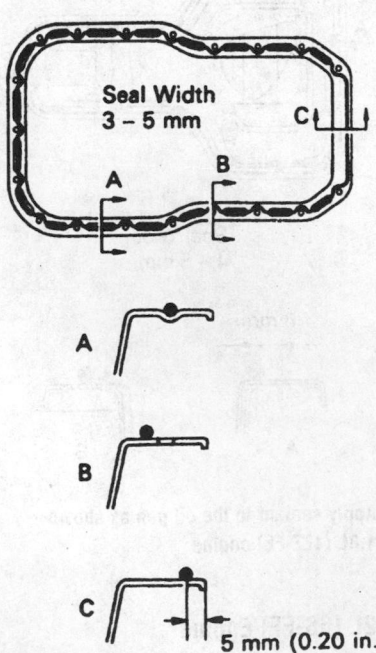

Oil pan sealing diagram—2.2L (5S-FE) engine

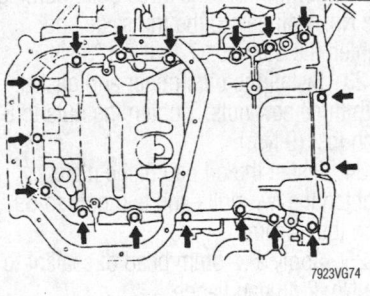

No. 1 oil pan mounting bolt locations— 3.0L (1MZ-FE) engine

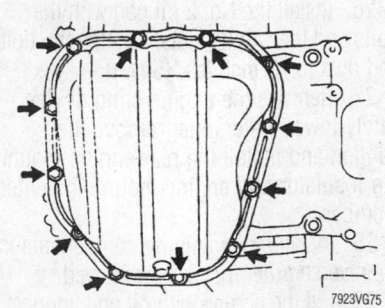

No. 2 oil pan mounting bolt locations— 3.0L (1MZ-FE) engine

20. Install the flywheel housing under-cover.

21. Install the front exhaust pipe bracket to the No. 1 oil pan. Tighten the bolts to 15 ft. lbs. (21 Nm).

22. Install the front exhaust pipe as follows:

a. Temporarily install the three new gaskets and the front exhaust pipe with the two bolts and six nuts.

b. Tighten the four nuts holding the exhaust manifolds to the front exhaust pipe. Tighten the four nuts to 46 ft. lbs. (62 Nm).

c. Tighten the two bolts and two nuts holding the front exhaust pipe to the center exhaust pipe. Tighten the bolts and nuts to 41 ft. lbs. (56 Nm).

d. Install the bracket with the two bolts and tighten to 14 ft. lbs. (19 Nm).

e. Install the support stay with the two bolts and tighten to 22 ft. lbs. (29 Nm).

23. Install the engine undercover.

24. Install the right fender apron seal.

25. Install the right front wheel and lower the vehicle.

26. Fill the engine with oil.

27. Start the engine and check for leaks.

3.0L (2JZ-GE and 2JZ-GTE) Engines

➡The No. 1 oil pan can not be removed with the engine in the vehicle. The engine/transmission assembly must be removed. If only the No. 2 oil pan is being serviced, the engine/transmission assembly can remain in the vehicle.

1. Remove the engine/transmission assembly, then separate the transmission from the engine.

2. With the engine on a stand, remove the timing belt, the idler pulley and the crankshaft timing pulley.

3. Remove the oil dipstick and guide.

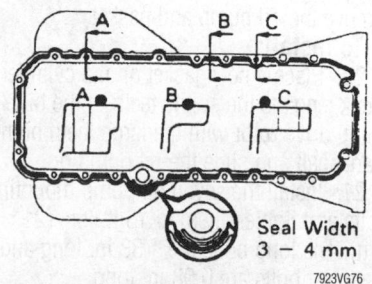

Upper oil pan sealant application—3.0L (2JZ-GE and 2JZ-GTE) engines

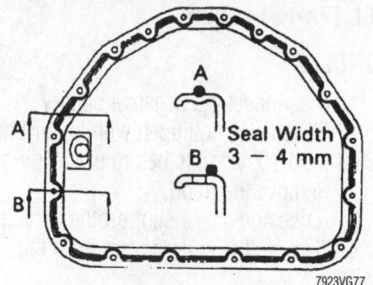

Lower oil pan sealant application—3.0L (2JZ-GE and 2JZ-GTE) engines

4. Disconnect the oil sensor lead, remove the four attaching bolts and lift off the oil level sensor. Be careful not to drop this sensor.

5. Remove the 14 bolts and two nuts and pry off the lower (No. 2) oil pan. Be careful not to damage the No. 1 pan while performing this procedure.

6. Remove the bolt and two nuts and drop down the oil strainer and gasket.

7. Remove the five bolts and two nuts and drop down the baffle plate.

8. On turbocharged engines, disconnect the turbo oil outlet pipe as follows:

a. Disconnect the two turbo oil outlet hoses.

b. Remove the two nuts, oil outlet pipe and gasket.

9. Remove the twenty two bolts and the carefully pry off the upper (No. 1) oil pan. Remove the O-ring from the cylinder block.

To install:

10. Position a new O-ring in the block and scrape off any old sealant. Apply sealant to the pan mating surface with a ⅛ inch (3–4mm) bead. Install the upper pan and tighten the 12mm bolts to 15 ft. lbs. (21 Nm) and the 14mm bolts to 29 ft. lbs. (39 Nm).

11. On turbocharged engines, install the turbo oil outlet pipe as follows:

a. Install the oil outlet pipe and gasket with the two nuts. Tighten the nuts to 20 ft. lbs. (27 Nm).

b. Install the two turbo oil outlet hoses.

12. Install the baffle plate and oil strainer. Tighten them both to 78 inch lbs. (9 Nm).

13. Install the lower pan in the same manner as the upper pan and tighten the bolts to 78 inch lbs. (9 Nm).

14. Using a new gasket, install the oil level sensor and tighten it to 48 inch lbs. (5.4 Nm).

15. Install the oil dipstick and guide, the timing pulleys and belt, and reconnect the transmission to the engine.

16. Install the engine and transmission.

17. Refill all fluids.

18. Start the engine and check for leaks.

19. Road test the vehicle.

Oil Pump

REMOVAL & INSTALLATION

1.5L (5E-FE) Engine

1. Disconnect the negative battery terminal. On vehicles equipped with an air bag, wait at least 90 seconds before proceeding.

2. Remove the hood.

3. Raise the vehicle and support it safely, then drain the oil.

4. Remove the oil dipstick.

5. Remove the timing belt.

6. Suspend the engine with a hoist.

✷✷ WARNING

Do not raise the engine more than necessary, since wiring and other components can be damaged.

7. Remove the necessary items for access, then remove the oil pan.

8. Remove the three bolts and oil strainer with the O-ring.

9. Remove the pressure regulator valve.

10. Remove the oil pump by removing the nine bolts and tension spring bracket. Use a soft faced hammer to remove the oil pump.

11. Using a vise, remove the nut and oil pump pulley.

To install:

12. Clean all gasket mating surfaces.

13. Install the oil pump pulley by first placing the driven rotors into the pump body with the marks facing the front.

14. Align the pulley and oil pump driveshaft.

15. Using a vise, install the oil pump pulley and nut. Tighten the nut to 27 ft. lbs. (37 Nm).

16. Install the pressure regulator valve and tighten to 22 ft. lbs. (30 Nm).

17. Using a new O-ring, install the oil strainer with the three bolts. Tighten the bolts to 7 ft. lbs. (10 Nm).

18. Using a new gasket and sealer, install the oil pan.

19. Install all remaining components in the reverse order they were removed.

20. Fill the engine with oil.

21. Connect the negative battery cable to the battery.

22. Install the hood.

23. Start the engine check for leaks.

1.6L (4A-FE) Engine

COROLLA

1. Disconnect the negative battery cable. On vehicles equipped with an air bag, wait at least 90 seconds before proceeding.

2. Remove the hood.

3. Raise and safely support the vehicle.

4. Remove the undercovers from the vehicle.

5. Drain the engine oil.

6. Remove the front exhaust pipe.

7. Remove the timing belt.

8. Remove the bolt and remove the idler pulley and tension spring.

9. Using a suitable tool, remove the crankshaft timing pulley.

10. Remove the oil dipstick guide, dipstick and oil pan.

11. Remove the two bolts and nuts, the oil strainer and gasket.

12. Remove the seven mounting bolts, the oil pump, and the gasket.

To install:

13. Place a new gasket on the cylinder block, engage the splined teeth of the oil pump drive rotor with the large teeth of the crankshaft, and slide the oil pump in place.

14. Install the seven mounting bolts and tighten them to 16 ft. lbs. The long bolts are 1.38 in. and the short bolts are 0.98 in.

15. Install a new gasket and the oil strainer with the 2 bolts and nuts. Tighten to 82 inch lbs.

16. Install the oil pan.

17. Install the idler pulley and tension spring.

18. Install the timing belt.

19. Install the front exhaust pipe.

20. Safely lower the vehicle.

21. Install the oil dipstick guide and the dipstick.

22. Install the crankshaft timing pulley. Align the timing pulley set key with the groove of the pulley. Be sure the flange side of the pulley faces inward.

23. Install the hood.

24. Fill the engine with oil, reconnect the negative battery cable, start the engine and check for leaks.

25. Install the engine undercovers.

1.8L (7A-FE) Engine

COROLLA

1. Disconnect the negative battery cable. On vehicles equipped with an air bag, wait at least 90 seconds before proceeding.

2. Remove the hood.

3. Raise and safely support the vehicle.

4. Remove the undercovers from the vehicle.

5. Drain the engine oil.

6. Remove the timing belt.

7. Remove the bolt and remove the idler pulley and tension spring.

8. Using a suitable tool, remove the crankshaft timing pulley.

9. Remove the oil dipstick guide, the crankshaft position sensor, and the dipstick.

10. Remove the front exhaust pipe.

11. Remove the No. 2 oil pan, baffle plate, oil strainer and No. 1 oil pan.

➡**Be careful not to damage the contact surfaces of the cylinder block and the No. 1 oil pan.**

12. Remove the bolt and the crankshaft position sensor.

13. Remove the seven bolts and the oil pump.

To install:

14. To install the oil pump, place a new gasket to the cylinder block. Engage the splined teeth of the oil pump drive rotor with the large teeth of the crankshaft, and slide the oil pump on.

15. Install the seven oil pump mounting bolts and tighten them to 16 ft. lbs. The long bolts are 1.38 in. and the short bolts are 0.98 in.

16. Install the crankshaft position sensor.

17. Install the No. 1 oil pan, strainer, baffle plate and No. 2 oil pan.

18. Install a new O-ring to the dipstick guide, push the dipstick guide and the dipstick in together, and attach the crankshaft position sensor to the dipstick guide mounting bolt. Tighten the bolt to 82 inch lbs. (9 Nm).

19. Install the idler pulley and the tensioner spring with the mounting bolt, do not tighten the bolt, but push the pulley as far left as possible.

20. Align the timing pulley set key with the groove of the pulley and slide the pulley on with the flange facing in.

21. Install the idler pulley and the tensioner spring with the mounting bolt, do not tighten the bolt, but push the pulley as far left as possible.

22. Install the timing belt.

23. Install the front exhaust pipe.

24. Safely lower the vehicle.

25. Reinstall the hood.

26. Fill the engine with oil, reconnect the negative battery cable, start the engine and check for leaks.

27. Reinstall the engine undercovers.

CELICA

1. Disconnect the negative battery cable. On vehicles equipped with an air bag, wait at least 90 seconds before proceeding.

2. Remove the hood.

3. Raise and safely support the vehicle.

4. Remove the RH front wheel.

5. Remove the undercovers from the vehicle.

6. Drain the engine oil.

7. Remove the front exhaust pipe.

8. If equipped with A/T, disconnect the transaxle control cable from the engine mounting center member.

9. If equipped w/ A/C, disconnect the A/C pressure pipe from the engine mounting center member.

10. Remove the front exhaust pipe support bracket.

11. Raise and safely support the engine assembly.

12. Remove the rear engine mounting insulator and the engine mounting center member.

13. Install an engine hanger and suspend the engine with an engine sling device or equivalent.

14. Remove the timing belt.

15. Remove the bolt and remove the idler pulley and tension spring.

16. Using a suitable tool, remove the crankshaft timing pulley.

17. Remove the oil dipstick guide and the dipstick.

18. Remove the No. 2 oil pan, baffle plate, oil strainer, and the No. 1 oil pan.

19. Remove the oil pump seven attaching bolts and by carefully tapping the oil pump body with a plastic tipped hammer, remove the oil pump and gasket.

To install:

20. Place a new gasket on the cylinder block, engage the spline teeth of the oil pump drive rotor with the large teeth of the crankshaft and slide the oil pump on.

21. Install the seven oil pump mounting bolts and tighten them to 16 ft. lbs. (21 Nm). The long bolts are 1.38 in. long and the short bolts are 0.98 in. long.

22. Install the No. 1 oil pan, oil strainer, baffle plate and the No. 2 oil pan.

23. Install the timing belt.

24. Remove the engine sling device, safely lower the engine, remove the hanger, and install the rear engine mounting insulator and engine mounting center member.

25. Install the front exhaust pipe support bracket.

26. If equipped with A/C, connect the A/C pressure pipe to engine mounting center member clamp.

27. If equipped with A/T, connect the transaxle control cable to the engine mounting center member.

28. Install the front exhaust pipe.

29. Install the RH front wheel and safely lower the vehicle.

30. Install the hood.

31. Fill the engine with oil, connect the negative battery cable, start the engine, and check for leaks.

32. Recheck the engine oil level and install the engine undercovers.

1.8L (1ZZ-FE) Engine

COROLLA

1. Disconnect the negative battery cable.

2. Drain the engine oil.

3. Remove the timing chain and crankshaft sprocket.

4. Remove the five mounting bolts, oil pump and gasket.

To install:

5. Clean the mounting surface.

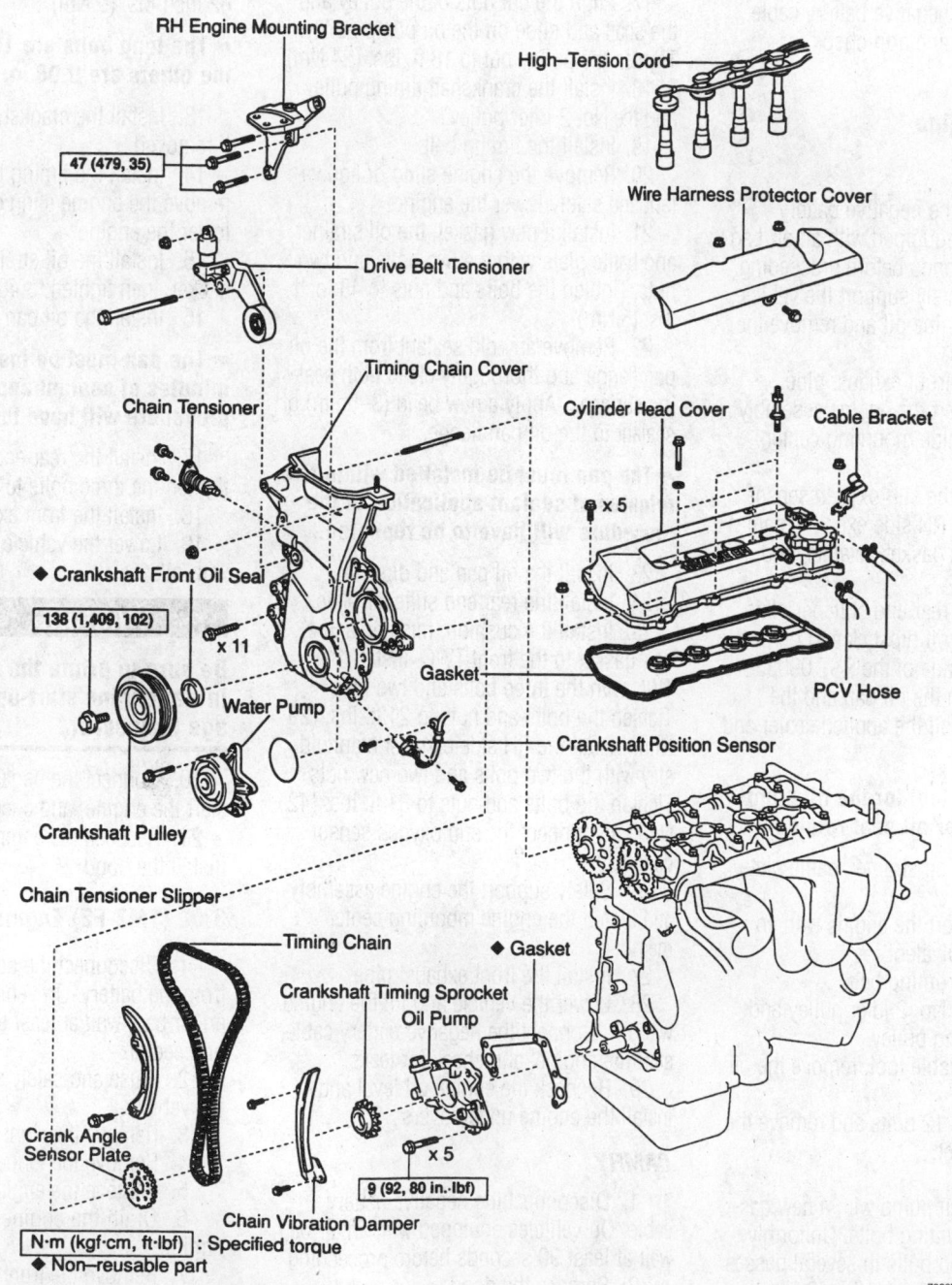

Exploded view of the oil pump mounting—1.8L (1ZZ-FE) engine

RH Engine Mounting Bracket

47 (479, 35)

High–Tension Cord

Wire Harness Protector Cover

Drive Belt Tensioner

Timing Chain Cover

Chain Tensioner

Cylinder Head Cover

Cable Bracket

◆ Crankshaft Front Oil Seal

138 (1,409, 102)

x 11

Water Pump

Gasket

Crankshaft Position Sensor

PCV Hose

x 5

Crankshaft Pulley

Chain Tensioner Slipper

Timing Chain

◆ Gasket

Crankshaft Timing Sprocket

Oil Pump

Crank Angle Sensor Plate

9 (92, 80 in.·lbf)

x 5

Chain Vibration Damper

N·m (kgf·cm, ft·lbf) : Specified torque
◆ Non–reusable part

7923VGB0

6. Install the oil pump using a new gasket. Tighten the bolts to 80 in. lbs. (9 Nm).

7. Install the crankshaft sprocket and timing chain.

✳✳ WARNING

Operating the engine without the proper amount and type of engine oil will result in severe engine damage.

8. Fill the engine with the correct amount of oil.

9. Connect the negative battery cable.

10. Start the engine and check for leaks.

2.2L (5S-FE) Engine

CELICA

1. Disconnect the negative battery cable. On vehicles equipped with an air bag, wait at least 90 seconds before proceeding.

2. Raise and safely support the vehicle.

3. Drain the engine oil and remove the engine undercovers.

4. Remove the front exhaust pipe.

5. Safely support the engine assembly and remove the engine mounting center member.

6. Disconnect the sub oxygen sensor wiring. Remove the RH side exhaust manifold stay, TWC with gasket, retainer, and cushion.

7. Remove the rear end stiffener plate.

8. Remove the oil dipstick and oil pan.

9. Insert the blade of the SST 09032–00100 tool between the oil pan and the cylinder block; cut off the applied sealer and remove the oil pan.

➡ **Do not use the tool for the oil pump body side and rear oil seal retainer.**

10. Remove the oil strainer, baffle plate, and the gasket.

11. Safely support the engine with an engine sling or equivalent.

12. Remove the timing belt.

13. Remove the No. 2 idler pulley and the crankshaft timing pulley.

14. Using a suitable tool, remove the oil pump pulley.

15. Remove the 12 bolts and remove the oil pump and gasket.

To install:

16. Install the oil pump with a new gasket and the 12 mounting bolts. Uniformly tighten the oil pump bolts in several passes. Tighten the bolts to 78 inch lbs. (9 Nm). Bolt A is 0.98 in. long and bolt B is 1.38 in. long.

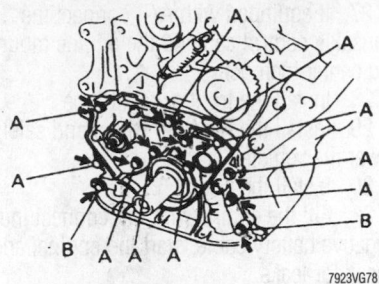

Oil pump mounting bolt identification—2.2L (5S-FE) engine

17. Align the cut outs of the pulley and the shaft and slide on the oil pump pulley. Tighten the pulley nut to 18 ft. lbs. (24 Nm).

18. Install the crankshaft timing pulley and the No. 2 idler pulley.

19. Install the timing belt.

20. Remove the engine sling or equivalent and safely lower the engine.

21. Install a new gasket, the oil strainer and baffle plate with the two bolts and two nuts. Tighten the bolts and nuts to 48 inch lbs. (5 Nm).

22. Remove any old sealant from the oil pan flange and thoroughly clean both sealing surfaces. Apply a new bead (3–5mm) of sealant to the oil pan flange.

➡ **The pan must be installed within five minutes of sealant application or the procedure will have to be repeated.**

23. Install the oil pan and dipstick.

24. Install the rear end stiffener plate.

25. Install the cushion, retainer, and a new gasket to the front TWC. Install the TWC with the three bolts and two nuts. Tighten the bolts and nuts to 21 ft. lbs. (29 Nm). Install the RH side exhaust manifold stay with the two bolts and two new nuts. Tighten the bolts and nuts to 31 ft. lbs. (42 Nm) and connect the sub oxygen sensor connector.

26. Safely support the engine assembly and install the engine mounting center member.

27. Install the front exhaust pipe.

28. Lower the vehicle and fill the engine with oil. Connect the negative battery cable, start the engine, and check for leaks.

29. Recheck the engine oil level and install the engine undercovers.

CAMRY

1. Disconnect the negative battery cable. On vehicles equipped with an air bag, wait at least 90 seconds before proceeding.

2. Remove the hood.

3. Raise and safely support the vehicle.

4. Drain the engine oil.

5. Remove the front exhaust pipe.

6. Remove the rear end stiffener plate.

7. Remove the oil dipstick and oil pan.

8. Remove the oil strainer and gasket.

9. Carefully suspend the engine with a sling device or equivalent and remove the timing belt and pulleys.

10. If equipped, remove the crankshaft position sensor.

11. Remove the 12 mounting bolts, the oil pump, and the gasket.

To install:

12. Install a new gasket and the oil pump with the 12 bolts. Tighten the bolts to 82 inch lbs. (9 Nm).

➡ **The long bolts are 1.38 in. and all the others are 0.98 in.**

13. Install the crankshaft position sensor, if removed.

14. Install the timing belt and pulleys, remove the engine sling device and safely lower the engine.

15. Install the oil strainer with a new gasket, then tighten to 48 inch lbs. (5 Nm).

16. Install the oil pan and dipstick.

➡ **The pan must be installed within five minutes of sealant application or the procedure will have to be repeated.**

17. Install the rear end stiffener plate and tighten the three bolts to 27 ft. lbs. (37 Nm).

18. Install the front exhaust pipe.

19. Lower the vehicle and fill the engine with oil.

✳✳ WARNING

Be sure to prime the oil pump prior to initial engine start-up or engine damage may occur.

20. Connect the negative battery cable, start the engine, and check for leaks.

21. Recheck the engine oil level and install the hood.

3.0L (1MZ-FE) Engine

1. Disconnect the negative battery cable from the battery. On vehicles equipped with an air bag, wait at least 90 seconds before proceeding.

2. Raise and safely support the front of the vehicle.

3. Remove the right front wheel.

4. Remove the fender apron seal.

5. Remove the engine undercover.

6. Drain the engine oil from the engine.

7. Remove the front exhaust pipe.

8. Remove the front exhaust pipe bracket from the No. 1 oil pan.

9. Remove the alternator drive belt from the engine.

10. Disconnect the A/C compressor from the engine.

11. Remove the power steering pump drive belt and adjusting strut.

12. Remove the timing belt and belt pulleys from the engine.

13. Remove the rear timing belt cover from the engine by removing the wire clamps and six bolts.

14. Remove the A/C compressor housing bracket by removing the three bolts.

15. Remove the No. 2 oil pan, oil strainer, No.1 oil pan and baffle plate.

16. Remove the crankshaft position sensor by removing the connector and bolt.

17. Remove the oil pump as follows:

a. Remove the nine bolts. Make a note of the position of the each bolt. When replacing the bolts into the oil pump body, place each bolt in the position from which it was removed.

b. Remove the oil pump body by prying between the oil pump and main bearing cap.

c. Remove the O-ring from the cylinder block.

d. Remove the plug, gasket, spring, and relief valve from the oil pump body.

e. Remove the nine screws, pump body cover, drive, and driven rotors.

To install:

18. To install the oil pump:

a. Install the driven rotors, drive, pump body cover,, then install the nine screws.

b. Install the oil pump relief valve, spring, gasket, and the plug to the oil pump body.

c. Place a new O-ring on the cylinder block.

d. Using a non residue solvent, clean both sealing surfaces to the oil pump.

e. Apply liquid sealant to the oil pump and engine block.

f. Install the oil pump to the engine block. Be sure to engage the spline teeth of the oil pump drive gear with the large teeth of the crankshaft.

g. Install the nine bolts to the oil pump and uniformly tighten the bolts in several passes. Tighten the bolts as follows:
- 10mm head-69 inch lbs. (8 Nm)
- 12mm head-14 ft. lbs. (20 Nm)

19. Install the crankshaft position sensor and install the bolt. Tighten the bolt to 69 inch lbs. (8 Nm).

20. Install the baffle plate to the No. oil pan and tighten to 69 inch lbs. (8 Nm).

21. Install the No. 1 oil pan, oil strainer and No. 2 oil pan.

22. Install the A/C compressor housing bracket and three bolts. Tighten the three bolts to 18 ft. lbs. (25 Nm).

23. Install the rear timing belt cover with the six bolts and three wire clamps. Tighten the bolts to 74 inch lbs. (9 Nm).

24. Install the timing belt pulleys.

25. Install the timing belt.

26. Install the adjusting strut and power steering drive belt as follows:

a. Temporarily install the adjusting strut with the bolt and nut.

b. Install the drive belt, then install the pivot and adjusting bolts. Tighten the bolt and nut to 32 ft. lbs. (43 Nm).

27. Install the A/C compressor.

28. Install the alternator drive belt.

29. Install the front exhaust pipe bracket to the No. 1 oil pan. Tighten the bolts to 15 ft. lbs. (21 Nm).

30. Install the front exhaust pipe.

31. Install the engine undercover.

32. Install the right fender apron seal.

33. Install the right front wheel and lower the vehicle.

34. Fill the engine with oil.

35. Connect the negative battery cable to the battery.

36. Start the engine and check for leaks.

37. Recheck the engine oil.

3.0L (2JZ-GE and 2JZ-GTE) Engines

1. Disconnect the negative battery cable. On vehicles equipped with an air bag, wait at least 90 seconds before proceeding.

2. Remove the engine and transmission.

3. Separate the transmission from the engine and mount the engine on a service stand.

4. Remove the timing belt.

5. Remove the idler pulley.

6. Remove the crankshaft timing pulley.

7. Remove the oil dipstick and tube, oil level sensor, No. 2 (lower) oil pan, oil strainer and oil baffle plate.

8. If the engine is turbocharged, remove the turbo oil outlet pipe by disconnecting the two turbo oil hoses and two nuts.

9. Remove the No.1 (upper) oil pan.

10. Remove the nine mounting bolts to the oil pump body. Carefully drive the pump off the cylinder block using a brass drift. Remove the two O-rings.

To install:

11. Position two new O-rings in the cylinder block. Scrape any old sealant from the mating surfaces. Draw a ⅛ inch (3–4mm) bead of sealant around the pump mating surface, taking great care around the

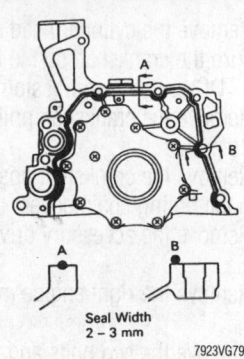

Seal Width 2 – 3 mm

7923VG79

Oil pump sealant application—3.0L (2JZ-GE and 2JZ-GTE) engines

oil passages. Install the pump and tighten the bolts to 15 ft. lbs. (21 Nm).

12. Place a new O-ring on the block. Remove all of the old sealant from the block and No. 1 oil pan. Apply a bead of sealant around the No. 1 oil pan. Avoid excessive application. Install the No.1 oil pan and tighten the bolts with 12mm heads to 15 ft. lbs. (21 Nm). Tighten the bolts with 14mm heads to 29 ft. lbs. (39 Nm).

13. If removed, install the turbo oil outlet pipe by installing the two bolts and two hoses. Tighten the bolts to 20 ft. lbs. (27 Nm).

14. Install the oil baffle plate, oil strainer, No. 2 oil pan, oil lever sensor and oil dipstick with a new O-ring.

15. Install the crankshaft and idler pulley.

16. Install the timing belt.

17. Connect the engine and transmission.

18. Install the engine and transmission.

19. Fill all fluids.

20. Connect the negative battery cable.

21. Start the engine and check for leaks.

Timing Chain, Sprockets, Front Cover and Seal

REMOVAL & INSTALLATION

Corolla

1. Disconnect the negative battery cable.

2. Drain the engine coolant.

3. Remove the right front wheel and the right engine cover.

4. Remove the accessory drive belt and generator.

5. Remove the power steering pump from the engine and position it to the side without disconnecting the hoses.

6. Raise the engine slightly using a jack with a block of wood on it. Remove the right engine mount.

7. Remove the cylinder head cover.

8. Turn the crankshaft so the No. 1 piston is at TDC on the compression stroke.

9. Remove the crankshaft pulley using a suitable puller.

10. Remove the crankshaft position sensor from the timing chain cover.

11. Remove the accessory drive belt tensioner.

12. Remove the right engine mounting bracket.

13. Remove the two bolts and the chain tensioner.

14. Remove the water pump.

15. Remove the timing chain cover.

16. Remove the crankshaft angle sensor plate.

17. Remove the timing chain tensioner slipper.

18. Remove the timing chain and crankshaft timing sprocket.

To install:

19. Install the crankshaft sprocket with timing chain. Be sure to align the No. 1 mark link with the mark on the sprocket.

20. Install the timing chain on the camshaft sprockets. Align the Nos. 53 and 67 mark links with the marks on the camshaft sprockets.

21. Install the chain tensioner slipper. Tighten the bolt to 14 ft. lbs. (18.5 Nm).

22. Install the crankshaft angle sensor plate on the crankshaft.

23. Install a new seal in the front cover and lubricate the seal lip with engine oil.

24. Clean the mounting surfaces of the engine and timing chain cover. Apply silicone sealant to the cover as illustrated and install the timing chain cover and water pump. Tighten the bolts marked "C" to 80 inch lbs. (9 Nm) and tighten the remaining bolts to 14 ft. lbs. (18.5 Nm). Be sure to install the bolts in their original locations. The bolts lengths are as follows:

• A—1.77 in (45mm)
• B—1.38 in (35mm)
• C—1.18 in (30mm)
• D—0.98 in (25mm)

25. Apply sealant to the threads of the bolts, then install the right engine mounting bracket. Do not apply sealant near the tip of the bolt. Tighten the bolts to 35 ft. lbs. (47 Nm).

26. Inspect the drive belt tensioner. Replace it if there are any cracks or damage.

27. Install the drive belt tensioner. Tighten the bolt to 51 ft. lbs. (69 Nm) and the nut to 21 ft. lbs. (29 Nm).

28. Install the crankshaft position sensor. Tighten to 80 inch lbs. (9 Nm).

29. Install the crankshaft pulley. Tighten the bolt to 102 ft. lbs. (138 Nm).

30. Release the ratchet pawl and compress the chain tensioner. Place the hook on the pin to keep the tensioner compressed.

31. Install the tensioner using a new o-ring. Tighten the bolts to 80 in lbs. (9 Nm).

32. Turn the crankshaft counterclockwise and remove the hook from the pin. Turn the crankshaft clockwise and be sure the slipper is pushed by the plunger.

33. check the valve timing by turning the crankshaft clockwise until the mark of the pulley is aligned with the mark on the timing chain cover. The marks on the camshaft sprockets should be facing each other as shown.

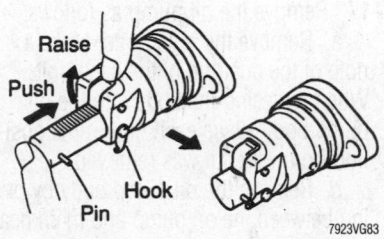

Compress the timing chain tensioner and place the hook on the pin—1.8L (1ZZ-FE) engine

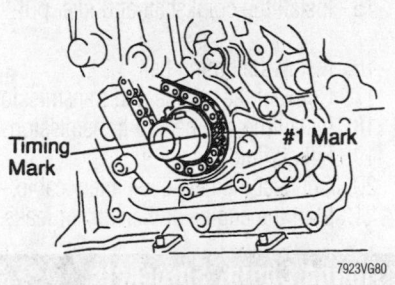

Align the No. 1 mark link with the mark on the crankshaft sprocket—1.8L (1ZZ-FE) engine

Align mark links Nos. 53 and 67 with the marks on the camshaft sprockets as shown—1.8L (1ZZ-FE) engine

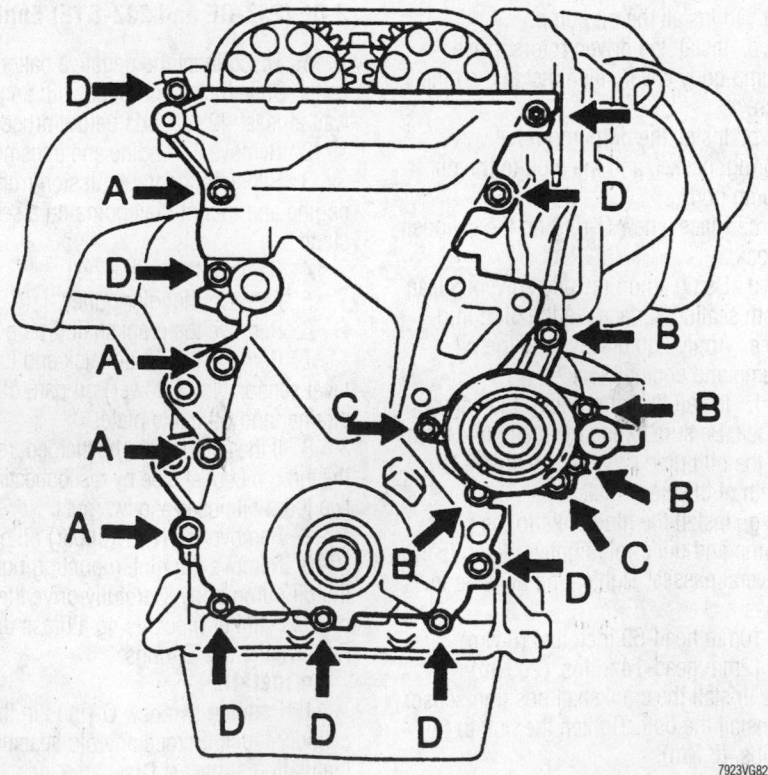

Timing chain cover bolt identification—1.8L (1ZZ-FE) engine

34. Apply silicone sealant to the two areas where the timing chain cover meets the cylinder head.

35. Install the cylinder head cover. Tighten the bolts with washers in the sequence shown to 80 in lbs. (9 Nm) and without washers to 8 ft. lbs. (11 Nm).

36. Install the right engine mount. Tighten the bolts marked A to 47 ft. lbs. (64 Nm), marked B to 19 ft. lbs. (26 Nm) and tighten the nut to 38 ft. lbs. (52 Nm).

37. Install the power steering pump, generator and drive belt.

38. Install the right engine under cover and the front wheel.

39. Connect the negative battery cable.

40. Install the washer tank and refill the engine with coolant.

41. Start the engine and check for leaks.

FUEL SYSTEM

Fuel System Service Precautions

Safety is the most important factor when performing not only fuel system maintenance but any type of maintenance. Failure to conduct maintenance and repairs in a safe manner may result in serious personal injury or death. Maintenance and testing of the vehicle's fuel system components can be accomplished safely and effectively by adhering to the following rules and guidelines.

• To avoid the possibility of fire and personal injury, always disconnect the negative battery cable unless the repair or test procedure requires that battery voltage be applied.

• Always relieve the fuel system pressure prior to disconnecting any fuel system component (injector, fuel rail, pressure regulator, etc.), fitting or fuel line connection. Exercise extreme caution whenever relieving fuel system pressure to avoid exposing skin, face and eyes to fuel spray. Please be advised that fuel under pressure may penetrate the skin or any part of the body that it contacts.

• Always place a shop towel or cloth around the fitting or connection prior to loosening to absorb any excess fuel due to spillage. Ensure that all fuel spillage (should it occur) is quickly removed from engine surfaces. Ensure that all fuel soaked cloths or towels are deposited into a suitable waste container.

• Always keep a dry chemical (Class B) fire extinguisher near the work area.

• Do not allow fuel spray or fuel vapors to come into contact with a spark or open flame.

• Always use a back-up wrench when loosening and tightening fuel line connection fittings. This will prevent unnecessary stress and torsion to fuel line piping. Always follow the proper torque specifications.

• Always replace worn fuel fitting O-rings with new. Do not substitute fuel hose or equivalent, where fuel pipe is installed.

Fuel System Pressure

RELIEVING

❊ CAUTION

Failure to relieve fuel pressure before repairs or disassembly can cause serious personal injury and/or property damage. Fuel pressure is maintained within the fuel lines, even if the engine is OFF or has not been run in a period of time. This pressure must be safely relieved before any fuel-bearing line or component is loosened or removed. On vehicles equipped with inflatable restraints or air bag systems, wait at least 90 seconds after disconnecting the battery cable before performing any other work. The back-up power will keep the restraint system energized for a period of time after the battery is disconnected.

1. Remove the fuse for the fuel pump.

2. Start the engine until the engine stalls.

3. Disconnect the negative battery terminal.

4. Place a catch-pan under the joint to be disconnected. A large quantity of fuel may be released when the joint is opened.

5. Wear eye or full face protection.

6. Place a shop towel over the area and slowly release the joint using a wrench of the correct size.

7. Allow the any fuel left in the line to bleed off slowly before fully disconnecting the joint.

8. Plug the opened lines immediately to prevent fuel spillage or the entry of dirt.

9. Dispose of the released fuel properly.

10. After connecting fuel lines, install the fuse for the fuel pump and start the engine.

11. Check for leaks and repair as needed.

Fuel Filter

REMOVAL & INSTALLATION

All Models

1. Disconnect the negative battery cable. On vehicles equipped with an air bag, wait at least 90 seconds before proceeding.

2. Unbolt the retaining screws and remove the protective shield for the fuel filter, if equipped.

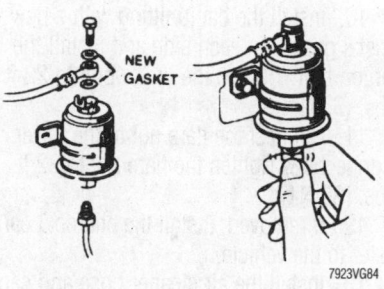

Always use new gaskets when replacing a fuel filter—Tercel shown

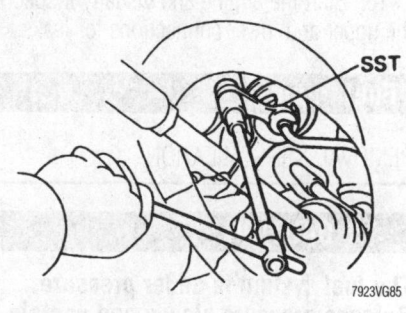

A line wrench with an extension may be needed to loosen the inlet line at the filter—1995–99 Corolla

❊ CAUTION

The fuel injection system remains under pressure after the engine has been turned OFF. Properly relieve fuel pressure before disconnecting any fuel lines. Failure to do so may result in fire or personal injury.

3. Place a drain pan or plastic container under the fuel filter.

4. If necessary, remove the air cleaner hose and cap.

5. If necessary, remove the charcoal canister.

6. Slowly loosen the lower flare nut fitting until all the pressure is relieved and all the fuel is collected.

7. Loosen the union bolt on the upper portion of the filter and remove the banjo fitting and two metal gaskets. Discard the gaskets.

8. Loosen the fuel filter bracket bolt, remove the fuel line with the flared nut from the filter, and pull the filter from the mounting bracket.

To install:

9. Install a new fuel filter and tighten the bracket bolt.

10. Install the banjo fitting with a new metal gasket on each side and install the union bolt. Tighten the union bolt to 22 ft. lbs. (30 Nm).

11. Connect the flare nut to the lower connection. Tighten the flare nut to 22 ft. lbs. (30 Nm).

12. If removed, install the charcoal canister to the vehicle.

13. Install the air cleaner hose and cap.

14. If removed, install the protective shield.

15. Remove the drain pan and/or rags and connect the negative battery cable.

16. Start the engine and visually inspect the upper and lower connections for leaks.

Fuel Pump

REMOVAL & INSTALLATION

❋❋ CAUTION

The fuel system is under pressure. Release pressure slowly and contain spillage. Observe no smoking/no open flame precautions. Have a Class B-C (dry powder) fire extinguisher within arm's reach at all times.

Celica, Corolla, Paseo and Tercel

1. Relieve the fuel system pressure.

2. Disconnect the negative battery cable. On vehicles equipped with an air bag, wait at least 90 seconds before proceeding.

3. Remove the access plate-to-fuel tank bolts, then pull out the plate/fuel pump assembly.

4. On all other engines, remove the rear seat cushion and remove the four screws and the floor service hole cover.

a. Disconnect the fuel pump sender and fuel pump connector.

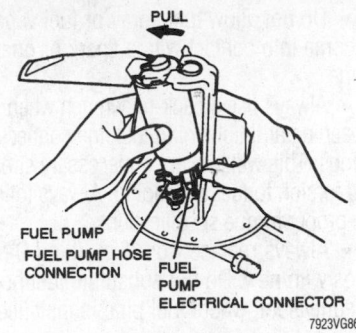

Pull the pump off the sender unit; the filter is still attached to the pump

b. Disconnect the outlet pipe from the fuel pump bracket and disconnect the return hose from the pump bracket.

c. Remove the eight bolts and pull the fuel pump bracket assembly from the fuel tank.

5. Separate the fuel pump from the fuel pump bracket as follows:

a. Pull off the lower side of the fuel pump from the pump bracket.

b. Disconnect the fuel pump connector.

c. Disconnect the fuel hose from the fuel pump and remove the rubber cushion from the pump.

d. Remove the fuel filter from the pump by removing the small clip.

To install:

6. Install a new cushion to the fuel pump. Install a new fuel filter and a new clip to the fuel pump. Reconnect the fuel hose to the fuel pump, reconnect the fuel pump connector and reinstall the fuel pump to the bracket.

7. Install the fuel pump bracket assembly to the fuel tank using a new gasket and the eight bolts. Tighten the bolts to 30 inch lbs. (3 Nm).

8. Connect the fuel return hose and the fuel outlet pipe to the fuel pump bracket.

9. Connect the fuel pump and fuel pump sender connector.

10. On the 3S-GTE engines, install the fuel tank.

11. Connect the negative battery cable, start the engine, and check for leaks and proper operation.

12. Install the floor service hole cover with the four screws and reinstall the rear seat cushion.

Avalon and Camry

1. Relieve the fuel system pressure.

2. Disconnect the negative battery cable from the battery. On vehicles equipped with an air bag, wait at least 90 seconds before proceeding.

3. Position a suitable waste container under the fuel tank and drain the fuel from the tank. Next, remove the fuel tank.

4. On some models it will be necessary to remove the rear seat cushion.

5. Remove the floor service hole cover by removing the retaining screws.

6. Disconnect the electrical connector at the fuel pump assembly.

7. Disconnect the fuel outlet pipe from the fuel pump bracket.

8. Disconnect the return hose from the fuel pump bracket.

9. Disconnect the fuel pump bracket assembly from the fuel tank by removing the bolts.

10. Pull out the pump bracket assembly.

11. Remove the fuel pump from the fuel bracket.

To install:

12. Connect the fuel pump to the fuel bracket.

13. Install the pump bracket assembly.

14. Connect the fuel pump to the fuel tank by installing the bolts. Tighten the bolts to 35 inch lbs. (4 Nm).

15. Connect the return hose to the fuel pump bracket.

16. Connect the outlet pipe to the fuel pump bracket. Tighten to 21 ft. lbs. (28 Nm).

17. Install the service hole cover to the fuel tank and install the screws.

18. Connect the fuel pump connector.

19. Install the rear seat cushion.

20. Install the fuel tank.

21. Fill the fuel tank and check for leaks.

22. Connect the negative battery cable to the battery.

23. Check the operation of the fuel pump and check for leaks.

Supra

1. Relieve the fuel pressure in the fuel line before disconnecting any fuel lines.

❋❋ CAUTION

The fuel injection system remains under pressure, even after the engine has been turned OFF. The fuel system pressure must be relieved before disconnecting any fuel lines. Failure to do so may result in fire and/or explosion.

2. Disconnect the negative battery cable from the battery. Wait at least 90 seconds before proceeding with any other work.

3. Open the hatchback and remove the following components:

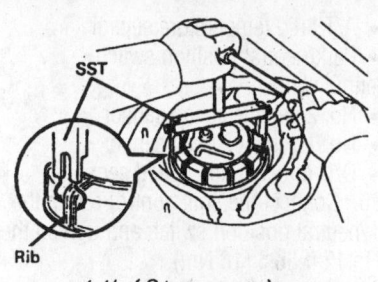

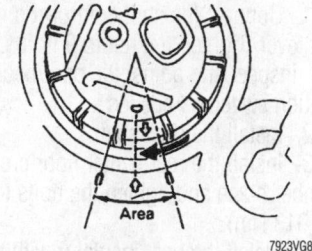

7923VG87

Tighten the fuel pump retainer until the arrow mark on the retainer is within the lines on the fuel tank—Supra

    a. Floor carpet
    b. Spare wheel cover
    c. Spare wheel
    d. Service hole cover by removing the six nuts

4. Disconnect the fuel pump electrical connector from the fuel pump.

5. Disconnect the outlet hose to the fuel pump by removing the union bolt and two gaskets.

6. Disconnect the fuel return hose from the fuel pump.

7. Disconnect the fuel breather clamp.

8. Using SST 09808–14010 or equivalent, loosen the retainer to the fuel pump.

9. Disconnect the fuel return hose from the return port of the fuel pump bracket.

10. Remove the retainer, fuel pump, sender gauge assembly, and the gasket as a unit.

To install:

11. Install a new gasket to the fuel pump.

12. Insert the fuel pump and sender gauge assembly into the fuel tank.

13. Align the arrow marks of the fuel pump bracket and the fuel tank.

14. Using the same tool as removal, install and tighten the fuel pump retainer until the arrow mark on the retainer is within the lines on the fuel tank.

15. Check that the arrow marks of the fuel pump bracket and fuel tank are aligned.

16. Install the retainer clamp to the fuel pump.

17. Connect the fuel pump electrical connector to the fuel pump.

18. Connect the outlet hose with two new gaskets and the union bolt. Tighten the union bolt to 22 ft. lbs. (29 Nm).

19. Connect the fuel return hose to the fuel pump.

20. Connect the fuel breather hose to the fuel pump.

21. Connect the negative battery cable to the vehicle. Start the engine and check for fuel leaks.

22. Install the service hole cover with the six nuts.

23. Install the following components:
    a. Spare wheel
    b. Spare wheel cover
    c. Floor carpet

DRIVE TRAIN

Transmission Assembly

REMOVAL & INSTALLATION

Supra

MANUAL

1. Disconnect the negative battery cable. On vehicles equipped with an air bag, wait at least 90 seconds before proceeding.

2. Remove the fan shroud set bolts.

3. Remove the shift lever knob. Using a flat bladed tool, pry out the upper console panel. Remove the four mounting bolts. Remove the shift and select lever boot No. 1 and No. 2. Remove the four mounting bolts.

4. Raise and support the vehicle.

5. Drain the transmission fluid.

6. Remove the oxygen sensor, exhaust front pipe and pipe support bracket.

7. Remove the exhaust center pipe. Remove the heat insulator.

8. Remove the center floor crossmember brace.

9. Matchmark and remove the driveshafts. Cap the end of the transmission to prevent leakage.

10. Remove the transmission lever bolt and nut, remove the shift lever.

11. Remove the two bolts and the clutch release cylinder. Remove the bolt, ground cable and flexible hose bracket.

12. Disconnect the starter connector.

13. Disconnect the back-up light switch and vehicle speed sensor connectors.

14. If equipped with a V160 transmission, remove the clutch cover set bolts and the service hole cover. Place matchmarks on the flywheel and clutch cover. Remove the six bolts.

15. Jack up the transmission slightly.

16. Remove the rear engine mounting member.

17. Lower the engine rear side and remove the two starter bolts and remove the starter.

18. Remove the remaining transmission mounting bolts. Lower the engine rear side and remove the transmission from the engine.

19. If equipped with a V160 transmission, remove the four bolts and the shift lever retainer from the transmission.

20. Remove the rear engine mounting from the transmission.

To install:

21. Install the rear engine mount to the transmission, tighten the bolts to 18 ft. lbs. (25 Nm).

22. If equipped with a V160 transmission, install the shift lever retainer. Tighten the bolts to 14 ft. lbs. (19 Nm). Tighten the through-bolt and nut to 18 ft. lbs. (25 Nm).

23. Raise the engine front side, align the input spline with the clutch disc and install the transmission to the engine. Tighten the mounting bolts to 53 ft. lbs. (72 Nm).

24. Install the starter, tighten the bolts to 29 ft. lbs. (39 Nm).

25. Install the rear engine mounting member, tighten the nuts to 10 ft. lbs. (13 Nm) and the bolts to 19 ft. lbs. (25 Nm).

26. If equipped with a V160 transmission, align matchmarks, then install the clutch cover set bolts, tighten to 14 ft. lbs. (19 Nm). Install the service hole cover and tighten the bolts to 9 ft. lbs. (12 Nm).

27. Connect the vehicle speed sensor and the back-up light switch connectors.

28. Connect the starter wire connector.

29. Install the clutch release cylinder, tighten the bolts to 9 ft. lbs. (12 Nm). Install and tighten the bolt with the clamp and ground cable to 53 ft. lbs. (72 Nm).

30. Install the transmission shift lever, tighten the bolts to 14 ft. lbs. (19 Nm).

31. Apply grease to the flexible coupling centering bushings. Install the driveshafts.

32. Inspect the driveshaft joint angle.

33. Install the center floor crossmember brace, tighten the bolts to 9 ft. lbs. (13 Nm).

34. Install the heat insulator, tighten the bolts to 48 inch lbs. (5 Nm).

35. Using new gaskets, install the center pipe. Tighten the bolts to 14 ft. lbs. (19 Nm).

36. Using new gaskets, install the front pipe. Tighten the bolts to 43 ft. lbs. (58 Nm).

37. Install the pipe support bracket and tighten the bolts to 27 ft. lbs. (37 Nm).

38. Install the front pipe set bolts and nuts, tighten to 32 ft. lbs. (43 Nm).

39. Install the oxygen sensor and cover using a new gasket. Tighten to 13 ft. lbs. (18 Nm).

40. Fill the transmission with proper gear oil. Lower the vehicle.

41. Install the shift lever mount bolts, tighten to 69 inch lbs. (8 Nm). Install the shift and select lever boot No. 1 and No. 2. Install the mounting bolts.

42. Fit the upper console panel to the console box with the four clips.

43. Install the shift lever knob.

44. Install the fan shroud set bolts.

45. Install the negative battery cable.

AUTOMATIC

1. Disconnect the negative battery cable from the battery. On vehicles equipped with an air bag, wait at least 90 seconds before proceeding.

2. Remove the transmission oil level gauge and filler pipe.

3. Raise and safely support the vehicle.

4. Remove the engine undercover.

5. Disconnect the oxygen sensor from the exhaust by removing the two nuts.

6. Remove the exhaust pipe.

7. Remove the heat insulator by removing the four bolts (normal roof) or six bolts (sport roof).

8. Remove the rear center floor cross-member brace.

9. Remove the driveshaft.

10. Disconnect the control rod from the shift lever by removing the nut.

11. Remove the shift control rod from the park/neutral position switch by removing the nut.

12. Disconnect the following:
- No. 1 vehicle speed sensor
- No. 2 vehicle speed sensor
- Solenoid wire
- Park/neutral position switch
- A/T fluid temperature sensor
- If equipped, O/D direct clutch speed sensor

13. Remove the starter electrical connector to the starter, then remove the nut and cable from the starter.

14. Remove the transmission oil cooler pipes as follows:
   a. Loosen the two oil cooler union nuts.
   b. Remove the center and rear oil cooler pipe brackets in front and in back of the lower control arm.
   c. Remove the front oil cooler pipe bracket next to the crankshaft pulley.
   d. Disconnect the two oil cooler pipes.

15. On turbocharged models, remove the intercooler pipe by loosening the two clamps and removing the two bolts.

16. Remove the torque converter clutch mounting bolts.

17. Support the transmission with a jack.

18. Remove the rear mounting by removing the four outer bolts.

19. Lower the transmission and remove the four wiring harness clamps from the retainer.

20. Remove the starter and transmission set bolts.

21. Remove the transmission from the engine and vehicle.

To install:

22. Install the transmission and starter and install the nine bolts. Tighten the bolts as follows:
- 14mm head bolts to 27 ft. lbs. (37 Nm.)
- 17mm head bolts to 53 ft. lbs. (72 Nm).

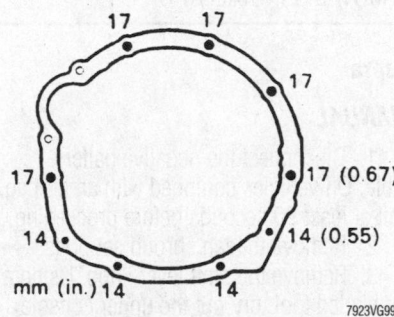

Transmission bolt identification—Supra

23. Install the starter cable and nut, then install the electrical connector to the starter.

24. Install the four wiring harness clamps to the retainer.

25. Install the rear mounting bolts and tighten the bolts to 19 ft. lbs. (25 Nm).

26. Install the torque converter clutch mounting bolts.

27. On turbocharged models, install the intercooler pipe by installing the two bolts and tightening the two clamps.

28. Connect the transmission oil cooler pipes as follows.
   a. Install and connect the two oil cooler pipes. Connect the two oil cooler union nuts and tighten to 25 ft. lbs. (34 Nm).
   b. Install the oil cooler pipe bracket next to the crankshaft pulley. Tighten the bolt to 7 ft. lbs. (10 Nm).
   c. Connect the center and rear oil cooler pipe brackets and tighten the bolts to 7 ft. lbs. (34 Nm).

29. Connect the following:

- A/T fluid temperature sensor
- Park/neutral position switch•
Solenoid wire
- No. 2 vehicle speed sensor
- No. 1 vehicle speed sensor
- O/D direct clutch speed sensor

30. Connect the shift control rod to the park/neutral position switch and tighten the nut to 12 ft. lbs. (16 Nm).

31. Connect the shift control rod to the shift lever. Tighten the nut to 9 ft. lbs. (13 Nm). Inspect and adjust the park/neutral position switch as needed.

32. Install the driveshaft.

33. Install the rear center floor cross-member bracc and tighten the bolts to 9 ft. lbs. 913 Nm).

34. Install the heat insulator with the bolts and tighten to 48 inch lbs. (5.4 Nm).

35. Install the exhaust pipe. Tighten the bolts as follows:
- Bracket to transmission housing to 27 ft. lbs. (37 Nm)
- No. 2 exhaust pipe to center exhaust pipe to 43 ft. lbs. (58 Nm)

36. Connect the oxygen sensor with the two nuts.

37. Install the engine undercover.

38. Install the transmission filler pipe and level gauge.

39. Connect the negative battery cable to the battery.

40. Check all fluids.

Transaxle Assembly

REMOVAL & INSTALLATION

Manual

TERCEL

1. Disconnect the negative battery cable, then the positive. On vehicles equipped with an air bag, wait at least 90 seconds before proceeding.

2. Remove the battery.

3. Remove the air cleaner case assembly with the air hose.

4. Remove the clutch release cylinder and tube clamp.

5. Remove the two bolts and ground cable from the transaxle side and engine left mounting bracket side.

6. Disconnect the back-up light switch electrical connector.

7. Remove the clips and washers that connect the shifter control cables to the transaxle.

8. Remove the retainers that attach the shifter control cables to the transaxle and position the cables out of the way.

9. Remove the upper transaxle attaching bolts.

10. Safely raise and support the vehicle.

11. Remove the front wheels.

12. Remove the engine undercovers, if equipped.

13. Drain the oil from the transaxle.

14. Remove the left and right halfshafts.

15. Remove the front exhaust pipe, if necessary.

16. Disconnect the speedometer cable from the transaxle.

17. Remove the starter electrical connectors and mounting bolts. Remove the starter from the engine.

18. Remove the rear engine mounting insulator and bracket.

19. Using a block of wood and a floor jack, place it under the engine's oil pan and support it.

20. Using a floor jack, position it under the transaxle and support its weight.

21. Remove the left engine mount.

22. Remove the transaxle attachment bolts from the engine.

23. Lower the left side of the engine and remove the transaxle from the vehicle.

To install:

24. Align the input shaft spline with the clutch disc and install the transaxle to the engine. Install the transaxle attaching bolts and tighten the bolts to specification.

25. Install the left engine mounting bracket with the two bolts. Tighten the bolts to 36 ft. lbs. (48 Nm).

26. Install the rear mounting insulator with the six bolts. Tighten the bolts as follows:

• A—Body side bolt through brackets to 58 ft. lbs. (78 Nm)

• B—Through-bolt to 47 ft. lbs. (64 Nm)

• C—Body side bolt though insulator to 67 ft. lbs. (90 Nm)

27. Remove the jacks from under the transaxle and engine.

28. Install the starter and electrical connectors. Tighten the starter bolts to 29 ft. lbs. (39 Nm).

29. Connect the speedometer to the transaxle.

30. Install the front exhaust pipe, if removed.

31. Install the right and left halfshafts.

32. Fill the transaxle with 75W-90 or 80W-90 gear oil.

33. Install the undercovers, if removed.

34. Install the wheels.

35. Lower the vehicle.

36. Position the shift control cables and install the retainers, the washers and the clips.

37. Connect the back-up light switch electrical connector.

38. Install the clutch release cylinder and tube clamp.

39. Install the air cleaner assembly with the air hose.

40. Connect the two ground cable to the transaxle.

41. Install the battery.

42. Road test the vehicle.

PASEO

1. Disconnect the negative battery cable. On vehicles equipped with an air bag, wait at least 90 seconds before proceeding.

2. Remove the air cleaner case assembly with the air hose.

3. Remove the clutch release cylinder and tube clamp.

4. Disconnect the back-up light switch electrical connector.

5. Remove the transaxle control cable end clips and washers, then remove the retainer clips from the cables. Remove the transaxle shift control cables from the transaxle.

6. Remove the clutch release cylinder bracket and the ground cable.

7. Remove the bolts from the upper transaxle mount.

8. Raise the vehicle and support it safely.

9. Remove the engine undercovers and drain the transaxle fluid.

10. Disconnect the speedometer cable from the transaxle.

11. Disconnect both halfshafts from the transaxle.

12. Remove the engine rear mounting bracket.

13. Remove the starter assembly.

14. Support the engine and transaxle assembly using the proper equipment.

15. Disconnect the left engine mount.

16. Remove the bolts mounting the transaxle to the engine.

17. Carefully remove the transaxle assembly from the vehicle. Lower the engine left side to aid in the transaxle removal.

To install:

18. Align the input shaft spline with the clutch disc, and install the transaxle to the engine. Tighten the bolts to the following specifications:

• Bolt A: 47 ft. lbs. (64 Nm)

• Bolt B: 34 ft. lbs. (46 Nm)

• Bolt C: 65 inch lbs. (7 Nm)

• Bolt D: 17 ft. lbs. (24 Nm)

19. Install the rear engine mounting brackets. Tighten the bolts as follows:

• Bolt A: 67 ft. lbs. (90 Nm)

• Bolt B: 58 ft. lbs. (78 Nm)

• Bolt C: 47 ft. lbs. (64 Nm)

20. Install the left engine mount and tighten the bolts to 35 ft. lbs. (48 Nm).

21. Install the starter and electrical connectors. Tighten the bolts to 29 ft. lbs. (39 Nm).

22. Connect the speedometer cable to the transaxle.

23. Install the intake manifold stay. Tighten the bolts to 14 ft. lbs. (20 Nm).

24. Install the sway bar and tighten the bracket bolts to 14 ft. lbs. (20 Nm).

25. Install the front exhaust pipe.

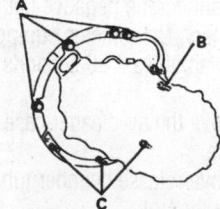

Torque A: 64 Nm (650 kgf.cm, 47 ft.lbf)
Torque B: 46 Nm (470 kgf.cm, 34 ft.lbf)
Torque C: 7 Nm (75 kgf.cm, 65 ft.lbf)

7923VG88

Tighten the transaxle mounting bolts to specification—Tercel

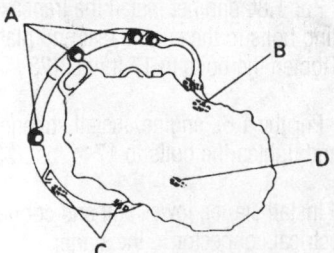

Torque A: 64 N·m (650 kgf·cm, 47 ft·lbf)
Torque B: 46 N·m (470 kgf·cm, 34 ft·lbf)
Torque C: 7 N·m (75 kgf·cm, 65 in.·lbf)
Torque D: 24 N·m (240 kgf·cm, 17 ft·lbf)

7923VG89

Torque specifications for mounting the transaxle to the engine—Paseo

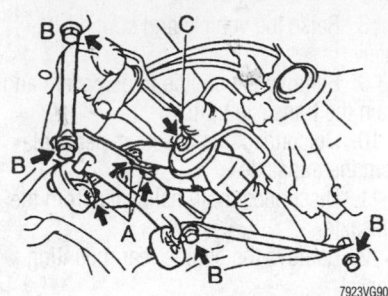

Rear engine mounting bolt identification— Paseo

7923VG90

26. Connect the halfshafts to the transaxle.

27. Fill the transaxle with gear oil.

28. Install the undercovers.

29. Lower the vehicle.

30. Connect the two ground cables with the two bolts.

31. Connect the transaxle shift control cables and install the retainers to the cables. Connect the cables to the shift linkage with the washers and clips.

32. Connect the back-up light switch electrical connector.

33. Install the clutch release cylinder and tube clamp.

34. Install the air cleaner assembly with the air hose.

35. Connect the negative battery cable.

36. Adjust the clutch and perform a road test of the vehicle.

COROLLA

1. Disconnect the negative battery cable. On vehicles equipped with an air bag, wait at least 90 seconds before proceeding.

2. Remove the air cleaner case assembly with hose.

3. If necessary, remove the coolant reservoir tank.

4. Remove release cylinder tube bracket by removing the bolt.

5. Remove the clutch release cylinder by removing the two bolts.

6. Disconnect the back-up light switch connector.

7. Remove the ground cable.

8. Disconnect shift cables from the transaxle.

9. Disconnect the vehicle speed sensor connector or the speedometer cable.

10. Disconnect the engine wire clamps.

11. Remove the starter set bolt from the transaxle upper side.

12. Remove the two transaxle upper mounting bolts.

13. Remove the engine left mounting stay by removing the two bolts.

14. Remove the engine left mounting set bolt from the rear side.

15. Install engine support fixture. Raise and safely support the vehicle.

16. Remove the front wheels. Remove the undercovers from the vehicle.

17. Drain the transaxle oil.

18. Disconnect the lower ball joint from the lower arm by removing the bolt and two nuts.

19. Remove the halfshafts.

20. Remove the front exhaust pipe.

21. Remove the hole cover.

22. Remove the engine front mounting set bolts.

23. Disconnect engine rear mounting by removing the three set nuts.

24. Place a jack under the engine center support member.

25. Remove the engine center support member by removing the eight bolts.

26. Remove the starter by disconnecting the electrical leads and removing the lower bolt.

27. For the 1.6L engine, remove stiffener plate.

28. Raise the transaxle and engine slightly with a jack.

29. For the 1.8L engine, remove the transaxle mounting bolts from the engine rear end plate side.

30. Remove the engine left mounting set bolts from the front side.

31. Remove the transaxle mounting bolts from the engine front side. Remove the transaxle mounting bolts from the engine rear side. Lower the engine left side and remove the transaxle from the engine.

To install:

32. Align the input shaft with the clutch disc and install the transaxle to the engine and tighten the engine to transaxle bolts to 47 ft. lbs. (64 Nm) for 12mm bolts and 34 ft. lbs. (46 Nm) for 10mm bolts.

33. Raise the transaxle and engine slightly and install the left engine mounting. Install the left engine to transaxle mount bolts. Tighten the bolts to 41 ft. lbs. (56 Nm).

34. For 1.8L engine, install the transaxle mounting bolts to the engine rear end plate side. Tighten the bolts to 17 ft. lbs. (23 Nm).

35. For the 1.6L engine, install stiffener plate and tighten the bolts to 17 ft. lbs. (23 Nm).

36. Install starter, lower bolt and connect the electrical connector to the starter. Tighten the bolt to 29 ft. lbs. (39 Nm).

37. Install engine center support member and tighten the support member to radiator support bolts to 45 ft. lbs. (61 Nm).

Tighten the support member to frame bolts to 152 ft. lbs. (206 Nm).

38. Connect the engine rear mounting and tighten bolts to 35 ft. lbs. (48 Nm).

39. Connect the engine front mounting and tighten bolts to 47 ft. lbs. (64 Nm). Install hole covers.

40. Install front exhaust pipe.

41. Install the halfshafts.

42. Connect the lower ball joint to lower arm. Tighten the bolt and nuts to 105 ft. lbs. (142 Nm).

43. Fill the transaxle with the correct gear oil.

44. Install undercovers.

45. Remove the engine support fixture.

46. Install front wheels and lower the vehicle.

47. Install the engine left mounting set bolt to the rear side. Tighten the bolt to 41 ft. lbs. (56 Nm).

48. Install engine left mounting stay. Tighten the bolt to 15 ft. lbs. (21 Nm).

49. Install the two transaxle upper side mounting bolts and tighten to 29 ft. lbs. (39 Nm).

50. Install the starter set bolt to the transaxle upper side. Tighten the bolt to 29 ft. lbs. (39 Nm).

51. Connect the engine wire clamps.

52. Connect the vehicle speed sensor connector or the speedometer cable.

53. Connect the transaxle shift cables and install ground cable.

54. Connect the back-up light switch connector.

55. Install release cylinder and release cylinder tube bracket. Tighten the bolts to 9 ft. lbs. (12 Nm).

56. If removed, install the coolant reservoir tank.

57. Install the air cleaner case assembly.

58. Connect the negative battery cable and check the front wheel alignment.

59. Road test the vehicle and check for abnormal noise and smooth shifting.

CELICA

1. Disconnect the negative battery cable from the battery. On vehicles equipped with an air bag, wait at least 90 seconds before proceeding.

2. Remove the air cleaner case assembly with hose.

3. Remove release cylinder tube bracket by removing the bolt.

4. Remove the clutch release cylinder by removing the two bolts.

5. Disconnect the back-up light switch connector.

6. Remove the ground cable on the transaxle by removing the bolt.

7. Disconnect shift cables from the transaxle.

8. Disconnect the vehicle speed sensor connector or the speedometer cable.

9. Disconnect the engine wire clamps.

10. Remove the starter set bolt from the transaxle upper side.

11. Install a engine support fixture. Raise and safely support the vehicle.

12. Remove the front wheels.

13. Remove the undercovers from the vehicle.

14. Drain the transaxle oil.

15. Remove the halfshafts.

16. Remove the front exhaust pipe and support bracket.

17. Remove the starter.

18. Place a jack under the engine center support member.

19. Remove the engine center support member.

20. Disconnect engine rear mounting by removing the three set nuts.

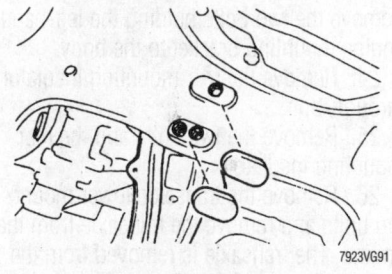

Rear mounting insulator set bolt locations—Celica

21. Remove the engine front mounting bracket and insulator by removing the through-bolt and two set bolts.

22. Raise the transaxle and engine slightly with a jack.

23. Remove the engine left mounting bracket by removing the three set bolts.

24. Remove the transaxle mounting bolts from the engine rear end plate side.

25. Remove the transaxle case protector by removing the two bolts.

26. Lower the engine left side and remove the three upper transaxle bolts.

27. Remove the transaxle from the engine.

To install:

28. Connect the transaxle to the engine and raise the engine right side. Align the input shaft with the clutch disc and install the transaxle to the engine. Tighten the three upper transaxle bolts to 47 ft. lbs. (64 Nm).

29. Install the transaxle case protector and tighten the two bolts to 9 ft. lbs. (13 Nm).

30. Install and tighten the four transaxle lower bolts as follows:

- Bolt A: 17 ft. lbs. (23 Nm)
- Bolt B: 34 ft. lbs. (46 Nm).

31. Raise the transaxle and engine slightly. Install the left engine mounting bracket to the engine left mounting insulator. Install the three set bolts and tighten to 47 ft. lbs. (64 Nm).

32. Install the engine front mounting bracket and insulator and tighten the two bracket bolts to 57 ft. lbs. (77 Nm). Tighten the through-bolt to 64 ft. lbs. (87 Nm).

33. Install the engine rear mounting bracket and insulator and tighten the bracket bolts to 57 ft. lbs. (77 Nm). Tighten the through-bolt to 64 ft. lbs. (87 Nm).

34. Install engine center support member.

35. Install the starter.

36. Install front exhaust pipe and support bracket.

37. Install the halfshafts.

38. Fill the transaxle with the correct gear oil.

39. Install undercovers.

40. Remove the engine support fixture.

41. Install front wheels and lower the vehicle.

42. Install the starter set bolt to the transaxle upper side. Tighten the bolt to 29 ft. lbs. (39 Nm).

43. Connect the engine wire clamps.

44. Connect the vehicle speed sensor connector or the speedometer cable.

45. Connect the transaxle shift cables and install ground cable.

46. Connect the back-up light switch connector.

47. Install the release cylinder.

48. Install air cleaner case assembly.

49. Connect the negative battery cable and check the front wheel alignment.

50. Road test the vehicle and check for abnormal noise and smooth shifting.

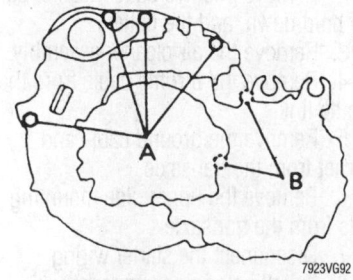

Upper transaxle mounting bolt locations—Celica

CAMRY

1. Disconnect the negative battery cable. On vehicles equipped with an air bag, wait at least 90 seconds before proceeding.

2. Remove the air cleaner assembly.

3. If equipped with cruise control, remove the cruise control actuator.

4. Remove the clutch release cylinder and tube clamp.

5. Remove the starter.

6. Disconnect the back-up light switch connector and ground strap.

7. Disconnect the wires clamp.

8. Remove the clips and washers that attach the transaxle control cables to the control levers. Remove the retaining clips and disconnect the transaxle control cables.

9. Disconnect the speed sensor connector.

10. Install a engine support fixture.

11. Tie the steering gear housing to the engine support fixture with a cord.

12. Raise the vehicle and support it safely.

13. Remove the engine undercovers and the front wheels.

14. Drain the fluid from the transaxle.

15. Remove the left and right halfshafts.

16. Disconnect the steering gear housing from the front suspension member as follows:

   a. Remove the four bolts.

   b. Remove the stabilizer bar bushing bracket.

   c. Remove the two set bolts and nuts.

   d. Disconnect the steering gear box from the suspension member and suspend it securely.

17. Remove the exhaust pipe.

18. Remove the stiffener plate.

19. Disconnect the engine front mounting from the suspension member by removing the two bolts.

20. Disconnect the engine rear mounting from the front suspension member by removing the two grommets and the three nuts.

21. With a transaxle jack and block of wood, raise the transaxle and engine slightly and disconnect the left engine mounting.

22. Disconnect the steering cooler pipe from the suspension member.

23. Remove the two fender liner set screws.

24. Disconnect the front suspension member as follows:

   a. Remove the two bolts and four nuts located on the outside of the brackets.

   b. Remove the four larger bolts holding the suspension member to the vehicle body.

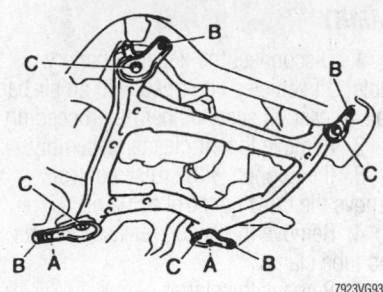

Front suspension member and fastener locations—Camry

c. Remove the two front lower braces, rear braces, and the front suspension member.

25. Remove the six transaxle mounting bolts from the engine.

26. Lower the left side of the engine and remove the transaxle.

27. Clean the mating surfaces of grease and dirt in preparation for reinstallation.

To install:

28. Move the transaxle into position so that the input shaft spline is aligned with the clutch disc.

29. Install the transaxle into the engine and secure with the lower mounting bolts. Tighten the 10mm mounting bolts to 47 ft. lbs. and 12mm bolts to 34 ft. lbs.

30. Install the front suspension member to the vehicle and install the two front lower braces and rear lower braces. Install the four large bolts that hold the suspension member to the vehicle. Tighten the bolts to 134 ft. lbs. (181 Nm).

31. Install the two outside bolts and four outside nuts. Tighten the bolts to 24 ft. lbs. (32 Nm).

32. Install the two fender liner set screws.

33. Connect the steering cooler pipe to the suspension member.

34. Raise the transaxle and engine slightly with a jack and wooden block.

35. Install the engine left mounting as follows:

a. Install the engine left mounting, then install the three bolts. Tighten the bolts to 38 ft. lbs. (52 Nm).

b. Install the two nuts and two grommets. Tighten the nuts to 59 ft. lbs. (80 Nm).

36. Install the engine rear mounting to the front suspension member by installing the three nuts and two grommets. Tighten the nuts to 59 ft. lbs. (80 Nm).

37. Install the engine front mounting to the suspension member and install the nut and two bolts. Tighten the bolt to 59 ft. lbs. (80 Nm).

38. Install the stiffener plate and tighten the bolts to 27 ft. lbs. (37 Nm).

39. Install the exhaust pipe.

40. Install the steering gear housing to the front suspension member as follows:

a. Lower the steering gear housing onto the suspension member.

b. Install the two set bolts and nuts. Tighten the bolts and nuts to 134 ft. lbs. (181 Nm).

c. Install the stabilizer bar bushing bracket.

d. Install the four bolts and tighten to (14 ft. lbs. (19 Nm).

41. Install the right and left halfshafts.

42. Install the engine undercovers.

43. Install the wheels and lower the vehicle.

44. Fill the transaxle with transaxle fluid.

45. Untie the steering gear housing from the engine support fixture.

46. Remove the engine support fixture from the vehicle.

47. Connect the vehicle speed sensor.

48. Connect the control cables by installing the washers and clips.

49. Connect the clamp that retains the wires to the transaxle.

50. Connect the back-up light switch connector and ground cables.

51. Install the starter to the vehicle and install the two bolts. Tighten the bolts to 29 ft. lbs. (39 Nm).

52. Install the pipe clamp and clutch release cylinder to the transaxle. Tighten the bolts to 9 ft. lbs. (13 Nm).

53. Install the cruise control actuator.

54. Install the air cleaner case assembly.

55. Connect the negative battery cable to the battery.

Automatic

TERCEL

1. Disconnect the negative battery cable. On vehicles equipped with an air bag, wait at least 90 seconds before proceeding.

2. Remove positive battery cable, battery hold down, and the battery.

3. Remove the air cleaner assembly.

4. Remove the throttle cable from the throttle link.

5. Remove the ground cable and bracket from the transaxle.

6. Remove the upper side mounting bolts from the transaxle.

7. Disconnect the starter wiring.

8. Install an engine support device.

9. Safely raise and support the vehicle.

10. Remove the undercovers from the vehicle.

11. Drain the fluid from the transaxle assembly and the differential.

12. Disconnect the right and left half-shafts from the transaxle.

13. Disconnect the speedometer cable from the transaxle.

14. Remove the clip holding the shift control cable to the body.

15. Disconnect the shift control cable from the control lever by removing the nut.

16. Disconnect the park/neutral position switch electrical connector.

17. Disconnect the O/D solenoid connector, if equipped.

18. Disconnect the oil cooler hose from the transaxle.

19. Remove the starter assembly.

20. Remove the front exhaust pipe.

21. Remove the transaxle converter cover and turn the crankshaft to gain access to each converter bolt.

22. Hold the crankshaft pulley nut with a socket and breaker bar and remove the six converter bolts.

23. Support the engine and transaxle with two jacks or a chain block and jack. Remove the two bolts holding the left-hand engine mounting bracket to the body.

24. Remove the rear mounting insulator through-bolt.

25. Remove the five bolts and the rear mounting insulator.

26. Remove the transaxle lower mounting bolts and remove the transaxle from the engine. The transaxle is removed from the bottom of the vehicle.

27. Remove the torque converter from the transaxle.

To install:

28. Install the torque converter to the transaxle. If the torque converter has been drained and flushed, refill with ATF.

➡**Fluid type: ATF Dexron® II**

29. Using a suitable jack, install the transaxle assembly to the engine aligning the two engine block knock pins with the converter housing. Install two bolts, but do not tighten at this time.

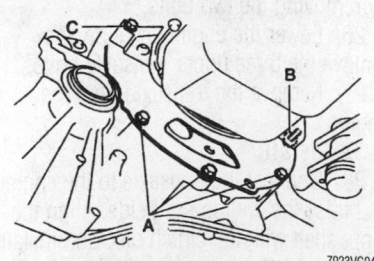

Tighten the lower transaxle mounting bolts to the proper specification—Tercel

30. Install the remaining transaxle mounting bolts and tighten the bolts to specifications.

31. Install the LH engine mounting bracket and tighten the two bolts to 35 ft. lbs. (48 Nm).

32. Install the rear engine mounting bracket and tighten the five bolts as follows:
- Outside bolts through brackets to 58 ft. lbs. (78 Nm)
- Inside bolts though rear mounting to 69 ft. lbs. (92 Nm)

33. Install the rear mounting insulator through-bolt and tighten the bolt to 47 ft. lbs. (64 Nm).

34. Install the six torque converter mounting bolts. Tighten the bolts to 20 ft. lbs. (27 Nm).

35. Remove the jacks supporting the transaxle and install the torque converter cover.

36. Install the front exhaust pipe.

37. Install the starter assembly.

38. Connect the oil cooler hose to the transaxle.

39. If equipped, connect the O/D solenoid connector.

40. Connect the park/neutral switch electrical connector.

41. Install the shift control cable to the lever and secure with the nut.

42. Connect the shift control cable to the body by installing the clip.

43. Connect the speedometer cable to the transaxle.

44. Install the left and right halfshafts.

45. Install the engine undercovers.

46. Safely lower the vehicle.

47. Remove the engine support.

48. Install the ground cable and bracket to the transaxle.

49. Install the starter upper bolt. Tighten the bolt to 29 ft. lbs. (39 Nm).

50. Install the transaxle upper side mounting bolts and tighten the bolts to 47 ft. lbs. (64 Nm).

51. Install and adjust the throttle cable.

52. Install the air cleaner assembly.

53. Install the battery with hold-down. Connect the battery cables.

54. Refill the transmission and differential.

➡ **Use ATF Dexron® II**

55. Road test the vehicle for proper operation and check for leaks.

PASEO

1. Disconnect the negative battery cable from the battery. On vehicles equipped with an air bag, wait at least 90 seconds before proceeding.

2. Remove the transaxle oil level gauge electrical connector.

3. Remove the air duct (leading from the front of the engine compartment to the air cleaner) by removing the two (2) bolts, nuts, and screw.

4. Remove the air cleaner upper cover, air cleaner, and lower case assembly.

5. Remove the throttle cable from the engine.

6. Remove all electrical connections on top of the transaxle and remove the wiring harness.

7. Remove the upper side bolt from the starter and the two (2) upper side mounting bolts.

8. Attach an engine chain hoist to the engine hangers.

9. Raise and safely support the vehicle.

10. Remove the engine undercovers from the vehicle.

11. Drain the fluid from the transaxle and differential by removing the drain plugs.

12. Remove the halfshaft.

13. Disconnect the tie rod end from the steering knuckle.

14. Disconnect the lower ball joint from the lower control arm.

15. Using a rubber mallet, disconnect the halfshaft from the steering knuckle.

16. Using a brass bar and hammer, drive out the halfshaft from the transaxle.

17. Remove the snaprings from the inboard joints of the halfshafts.

18. Pull out the clip from the shift cable.

19. Disconnect the solenoid connector from the transaxle.

20. Remove the locknut and disconnect the shift cable from the control lever.

21. Disconnect the connector from the park/neutral position switch.

22. Using a line wrench, loosen the nuts connecting the oil cooler tubes to the transaxle. Remove the oil cooler lines from the transaxle.

23. Remove the two (2) bolts and (2) clamps to the oil cooler tube, then remove the oil cooler.

24. If equipped, remove the ground wire from the transaxle.

25. Remove the two (2) bolts from the left engine mounting bracket.

26. Remove the starter by removing the two (2) nuts, electrical lines and the lower starter bolt.

27. Remove the speedometer cable from the transaxle.

28. Disconnect the No. 2 vehicle speed sensor.

29. Raise the transaxle and hold the transaxle with the a transaxle jack.

30. Remove the through-bolt from the rear mounting bracket.

31. Remove the hole plug and the three (3) cover plate bolts.

32. Turn the crankshaft to gain access to each torque converter bolt.

33. Hold the crankshaft pulley nut with a wrench and remove the six (6) converter bolts.

34. Remove the transaxle housing mounting bolts.

35. Remove the transaxle from the engine. Lower the rear of the engine slightly to help with the removal of the transaxle.

36. Remove the torque converter clutch from the transaxle.

To install:

37. Install the torque converter clutch in the transaxle.

➡ **If the torque converter clutch has been drained and washed, refill with new ATF fluid.**

38. Using a transaxle jack, install the transaxle to the engine. Be sure to align the two (2) knock pins of the engine block with the transaxle housing holes. Leave the jack supporting the transaxle until mounts are installed.

39. Install the transaxle housing mounting bolts and tighten the bolts as shown:
- Bolt A to 34 ft. lbs. (46 Nm).
- Bolt B to 66 inch lbs. (8.0 Nm)
- Bolt C to 29 ft. lbs. (39 Nm)
- Bolt D 18 ft. lbs. (25 Nm)
- Bolt E to 47 ft. lbs. (64 Nm)

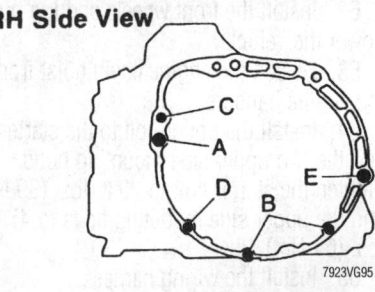

RH Side View

Transaxle mounting bolt locations—Paseo

40. Install all the torque converter clutch bolts to the transaxle. Tighten the bolts to 13 ft. lbs. (18 Nm).

41. Install the hole plug and three cover plate bolts.

42. Install the through-bolt to the rear mount bracket.

43. Release the transaxle jack.

44. Connect the vehicle speed sensor.

45. Install the speedometer cable to the transaxle.

46. Install the starter, then tighten the bolt to 29 ft. lbs. (39 Nm).

47. Install the two bolts to the left engine mount side bracket. Tighten the two bolts to 32 ft. lbs. (43 Nm).

48. If removed, install the transaxle ground wire with the bolt. Tighten the bolt to 32 ft. lbs. (43 Nm).

49. Install the oil cooler tubes to the transaxle.

50. Connect the electrical connector to the park/neutral position switch.

51. Install the shift cable to the control lever and install the locknut.

52. Install the clip keeping the shift cable to the vehicle body.

53. Connect the solenoid connector.

54. Install new snaprings to the inboard joints of the halfshafts.

55. Install the halfshafts to the transaxle.

56. Connect the halfshaft to the axle hub.

57. Connect the lower ball joint to the lower control arm by install the bolt and two nuts. Tighten the bolt and nuts to 59 ft. lbs. (80 Nm).

58. Connect the tie rod end to the steering knuckle. Install the nut and tighten to 36 ft. lbs. (49 Nm). Install a new cotter pin. If the cotter pin does not line up with a hole, tighten the nut by the smallest amount possible and install the pin.

59. Install the halfshaft locknut and tighten to 159 ft. lbs. (216 Nm).

60. Install the lock cap and a new cotter pin.

61. Install the engine undercovers.

62. Install the front wheels and lug nuts. Lower the vehicle.

63. Detach the engine chain hoist from the engine hangers.

64. Install the upper bolt to the starter and the two upper side mounting bolts. Tighten the starter bolt to 29 ft. lbs. (39 Nm) and the upper side mounting bolts to 47 ft. lbs. (64 Nm).

65. Install the wiring harness.

66. Connect all electrical connectors on the top of the transaxle.

67. Install and adjust the throttle cable.

68. Install the air cleaner lower case, air cleaner element, and upper cover.

69. Install the air duct by installing the two bolts, nuts, and screw.

70. Install the oil level gauge electrical connector.

71. Install the differential drain plug and the transaxle drain plug.

72. Fill the transaxle.

73. Connect the negative battery cable and start the engine.

74. Check for leaks, then check fluid level.

COROLLA

1. Disconnect the negative battery cable. On vehicles equipped with an air bag, wait at least 90 seconds before proceeding.

2. Disconnect the negative battery cable from the transaxle.

3. Remove the transaxle level gauge.

4. If equipped with A245E transaxle, remove the reservoir tank and air cleaner assembly.

5. Remove the throttle cable from the bracket.

6. Remove the engine left mounting upper side bolts.

7. Remove the engine left mounting stay.

8. Remove the ground cable from the transaxle.

9. Disconnect wiring harness clamp and throttle cable clamp.

10. If equipped with A245E transaxle, remove the starter upper side mounting bolt and the two transaxle mounting bolts from transaxle side.

11. Raise and safely support the vehicle.

12. Remove the undercovers from the vehicle.

13. Remove the left and right halfshafts.

14. Support the transaxle with a jack.

15. Remove the front exhaust pipe.

16. Install an engine support fixture.

17. Remove the suspension member by removing the grommet, 14 bolts, and three nuts.

18. Remove the starter.

19. Disconnect the vehicle speed sensor connector.

20. Disconnect the solenoid connector and park/neutral position switch connector. Remove the wiring harness clamps.

21. Remove the nut from the manual shift lever, then disconnect the control cable from the bracket by removing the clip.

22. Loosen the two clips and disconnect the two oil cooler hoses.

23. Remove the transaxle filler tube.

24. If equipped with A131L transaxle, remove the stiffener plate.

25. If equipped with A245E transaxle, remove the converter cover.

26. Turn the crankshaft to gain access to the torque converter bolts.

27. Remove the six torque converter bolts.

28. Remove the five (A245E) or four (A131L) transaxle mounting bolts.

29. Disconnect the transaxle from the engine and lower the transaxle to the ground.

To install:

30. Raise the transaxle into position.

31. On vehicles equipped with the A245E transaxle, install the five lower transaxle bolts. Tighten the bolts as follows:

- Bolt A: 17 ft. lbs. (23 Nm)
- Bolt B: 18 ft. lbs. (25 Nm)
- Bolt C: 34 ft. lbs. (46 Nm)
- Bolt D: 47 ft. lbs. (64 Nm)

32. On vehicles equipped with the A131L transaxle, install the four lower transaxle bolts. Tighten the bolts to 47 ft. lbs. (64 Nm).

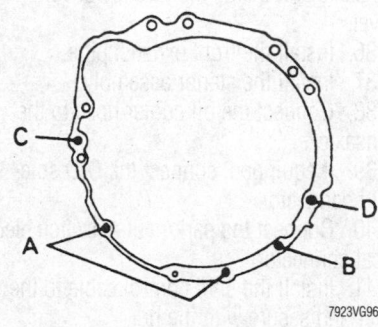

Transaxle mounting bolt torque specifications—Corolla A245E transaxle

33. Install the torque converter bolts to the transaxle and install the six bolts. Tighten the bolt to 18 ft. lbs. (25 Nm).

34. On vehicles equipped with the A131L transaxle, install the stiffener plate and tighten the mounting bolts to 13 ft. lbs. (18 Nm).

35. On vehicles equipped with the A245E transaxle, install the torque converter cover.

36. Install the transaxle filler pipe.

37. Connect the two oil cooler hoses and replace the clips to their original positions.

38. Connect the control cable for the transaxle to the bracket and install the clip. Connect the control cable to the manual shaft lever by installing the nut.

39. Connect the solenoid connector and park/neutral position switch connector. Connect the wiring to the clamps.

40. Connect the vehicle speed sensor wiring.

41. Install the starter by installing the lower bolt and the electrical connector. Tighten the lower bolt to 29 ft. lbs. (39 Nm).

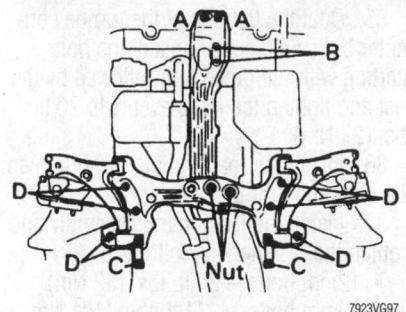

Suspension member fastener identification—Corolla

42. Install the suspension member with the 14 bolts and three nuts. Tighten the bolts and nuts as follows:
- Bolt A: 45 ft. lbs. (61 Nm)
- Bolt B: 47 ft. lbs. (64 Nm)
- Bolt C: 152 ft. lbs. (206 Nm)
- Bolt D: 152 ft. lbs. (206 Nm)
- Nuts: 42 ft. lbs. (57 Nm)

43. Once the suspension member is in place, remove the engine support fixture.
44. Install the front exhaust pipe.
45. Remove the jack from the transaxle.
46. Install the left and right halfshafts.
47. Install the engine undercovers.
48. Lower the vehicle.
49. Install the two transaxle mounting bolts to the transaxle side.
50. Install the starter upper side mounting bolt. Tighten the bolt to 29 ft. lbs. (39 Nm).
51. Connect the wiring harness clamp and throttle cable clamp.
52. Install the ground cable and tighten the bolt to 13 ft. lbs. (18 Nm).
53. Install the engine left mounting stay and tighten the bolts to 15 ft. lbs. (21 Nm).
54. Install the engine left mounting upper side bolts and tighten the bolts to 38 ft. lbs. (52 Nm).
55. Install the throttle cable.
56. Install the air cleaner and the reservoir tank, if removed.
57. Install the transaxle level gauge.
58. Fill the transaxle with fluid and adjust the throttle cable, shift cable and park/neutral position switch.
59. Connect the negative battery cable to the battery.

CELICA

1. Disconnect the negative and positive battery cables from the battery. On vehicles equipped with an air bag, wait at least 90 seconds before proceeding.
2. Disconnect the throttle cable from the engine.

3. Remove the cruise control actuator by disconnecting the connector and removing the three bolts. The cruise control actuator and bracket should be removed as an assembly.
4. Remove the air cleaner assembly and battery.
5. Disconnect the vehicle speed sensor and the transaxle ground strap.
6. Remove the engine left mounting upper side bolt.
7. Disconnect the wiring and remove the starter assembly by removing the two bolts.
8. Disconnect the park/neutral position switch connector.
9. Disconnect the two solenoid connectors.
10. Remove the three upper transaxle retaining bolts.
11. Disconnect the transaxle oil cooler hoses by loosening the two clips.
12. Install a engine support fixture.
13. Raise and support the vehicle safely.
14. Remove the front wheel.
15. Remove the engine undercover.
16. Drain the transaxle fluid.
17. Disconnect both halfshafts.
18. Support the transaxle with a jack.
19. Disconnect the shift control cable from the control shaft lever and body bracket.
20. Remove the engine rear mounting through-bolt.
21. Remove the front exhaust pipe.
22. Remove the suspension and center members as follows:
    a. Hold up the suspension and center members with a jack.
    b. Disconnect the air conditioner pipe bracket by removing the bolt.
    c. Remove the two shift cable mounting bolts and disconnect the cable from the suspension member.
    d. Remove the two power steering gear assembly set bolts and nuts.
    e. Remove the three grommets from the center crossmember.
    f. Remove the 13 bolts and 2 nuts holding the suspension and center crossmembers.
    g. Lower the jack and remove the crossmembers from the vehicle.
23. Remove the No. 1 manifold stay by removing the nut and bolt.
24. Remove the stiffener plate by removing the nut and six bolts.
25. Turn the crankshaft to gain access to the torque converter bolts. Remove all six bolts from the torque converter.

26. Remove the three bolts from the transaxle and lower the transaxle from the engine.

To install:
27. Raise and connect the transaxle to the engine.
28. Install the three transaxle bolts to the engine and tighten as follows:
- 10mm bolt to 34 ft. lbs. (46 Nm)
- 12mm bolt to 47 ft. lbs. (64 Nm)
29. Apply silicone to the transaxle torque converter bolts.
30. Turn the transaxle to install the six torque converter bolts. Tighten the bolts to 18 ft. lbs. (25 Nm).
31. Install the stiffener plate and install the six bolts. Alternately tighten the six bolts and tighten the bolts as follows:
- 12mm bolts to 15 ft. lbs. (21 Nm)
- 14mm bolts to 32 ft. lbs. (43 Nm)
32. Install the No. 1 manifold stay with the nut and bolt. Tighten the bolt to 15 ft. lbs. (21 Nm) and the nut to 32 ft. lbs. (43 Nm).
33. Install the suspension member and center member as follows:
    a. Raise the suspension member into position and install the two bolts to hold the suspension to the body. Tighten the bolts to 94 ft. lbs. (127 Nm).
    b. Install the three bolts to hold the rear of the lower control arms to the subframe and body. Tighten the bolt that goes through the lower control arm to 123 ft. lbs. (167 Nm) and the other two bolts to 130 ft. lbs. (175 Nm). Install the bolts to both sides.
    c. Install the center member.
    d. Install the engine rear mount and bracket by installing the bolt and two nuts. Tighten the fasteners as follows:
- Two nuts to 59 ft. lbs. (80 Nm)
- Bolt to 65 ft. lbs. (88 Nm)
    e. Install the engine front mount by installing the two bolts. Tighten the bolts to 59 ft. lbs. (80 Nm).
    f. Install the two front bolts to connect the center mount to the radiator support. Tighten the bolts to 26 ft. lbs. (35 Nm).
    g. Install the grommets to the center member.
    h. Install the two power steering gear assembly set bolts and nuts. Tighten the nuts and bolts to 94 ft. lbs. (127 Nm).
    i. Install the two shift cable mounting bolts.
    j. Connect the air conditioner pipe bracket.
34. Install the front exhaust pipe.
35. Install the engine rear mounting bolt. Tighten the bolt to 64 ft. lbs. (88 Nm).

36. Connect the shift control cable to the control shaft lever and body bracket. Install the clips.

37. Remove the jack holding the transaxle up.

38. Install the left and right halfshafts.

39. Install the engine undercovers.

40. Install the front wheel and lower the vehicle.

41. Remove the engine support fixture.

42. Connect the two oil cooler hoses with the two clips.

43. Install the three upper transaxle mounting bolts. Tighten the bolts to 47 ft. lbs. (64 Nm).

44. Connect the two solenoid connectors.

45. Connect the park/neutral position switch connector.

46. Connect the vehicle speed sensor connector and the ground strap to the transaxle.

47. Install the starter with the two bolts. Tighten the bolts to 29 ft. lbs. (39 Nm). Connect the starter electrical connectors.

48. Install the left mounting upper side bolt and tighten the bolt to 47 ft. lbs. (64 Nm).

49. Install the air cleaner assembly.

50. Install the cruise control actuator.

51. Install the battery.

52. Install the adjust the throttle cable.

53. Adjust the shift control cable.

54. Adjust the park/neutral position switch.

55. Fill the transaxle with fluid.

56. Check all fluid levels.

57. Check front wheel alignment and ABS speed sensor signal.

58. Connect the positive and negative battery cable.

CAMRY AND AVALON

1. Turn the ignition switch to the LOCK position and disconnect the negative battery cable. Wait at least 90 seconds or longer before doing any work on the vehicle.

2. Remove the battery.

3. Remove the air cleaner assembly.

4. Disconnect the throttle cable from the throttle body.

5. Remove the cruise control actuator cover and disconnect the connector, if equipped.

6. Remove the ground wire.

7. Remove the starter.

8. Disconnect speed sensor connectors, direct clutch speed sensor, and the park/neutral position switch connector on the transaxle.

9. Disconnect the solenoid connector on the transaxle.

10. Disconnect shift control cable.

11. Disconnect oil cooler hoses.

12. Remove the two front side transaxle mounting bolts.

13. Remove the two front engine mounting bolts.

14. Remove the oil cooler line mounting bolts from the front frame.

15. Remove the three upper transaxle to engine mounting bolts.

16. Install a engine support fixture. Tie steering gear housing to engine support fixture.

17. Raise and safely support the vehicle.

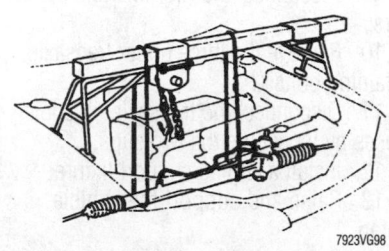

7923VG98

Tie the steering rack to the engine support fixture components, as shown—Avalon and Camry

18. Drain the transaxle/differential fluid.

19. Remove the front wheels.

20. Remove the front exhaust pipe.

21. Remove the engine side covers and undercovers.

22. Disconnect both halfshafts.

23. Remove the front side engine mounting nut.

24. Remove the rear side engine mounting bolts (remove hole plugs).

25. Remove the four left side transaxle mounting bolts.

26. Remove the steering gear housing.

27. Remove the front frame assembly.

28. Properly support the transaxle assembly.

29. Remove the rear end plate mounting bolts.

30. Remove the torque converter cover.

31. Remove the torque converter retaining bolts.

32. Remove the remaining transaxle mounting bolts.

33. Carefully remove the transaxle assembly from the vehicle.

To install:

34. Install the transaxle aligning the two dowel pins on the block with the converter housing. Tighten the bolts as follows:

- 10mm bolts—34 ft. lbs. (46 Nm)
- 12mm bolts—47 ft. lbs. (64 Nm)

35. Coat the threads of the torque converter bolts with sealer. Install the bolts starting with the green bolt followed by the rest and tighten the bolts evenly to 20 ft. lbs. (27 Nm).

36. Install the rear end plate and tighten the bolts to 27 ft. lbs. (37 Nm).

37. Install the front frame assembly and tighten the fasteners as follows:

- 12mm bolts—24 ft. lbs. (32 Nm)
- 19mm bolts—134 ft. lbs. (181 Nm)
- Nut—27 ft. lbs. (36 Nm)

38. Install the two fender liner set screws.

39. Connect the steering gear to the frame and tighten the bolts and nuts to 134 ft. lbs. (181 Nm).

40. Connect the sway bar brackets and toque the bolts to 14 ft. lbs. (19 Nm).

41. Install the left transaxle mounting bolts and tighten them to 38 ft. lbs. (52 Nm).

42. Install the rear side mounting bolts and nuts and tighten them to 48 ft. lbs. (66 Nm). Install the plugs.

43. Install the front engine mounting nut and tighten it to 59 ft. lbs. (80 Nm).

44. Install the halfshafts.

45. Install the right and left engine side covers.

46. Install the lower engine cover.

47. Fill the transaxle/differential to the proper level with Dexron II® or equivalent.

48. Install the exhaust pipe and tighten the nuts to 46 ft. lbs. (62 Nm). Connect the exhaust pipe to the converter and tighten the nuts and bolts to 32 ft. lbs. (43 Nm). Always use new gaskets.

49. Install the wheel.

50. Lower the vehicle.

51. Remove the engine support.

52. Install the four upper transaxle mounting bolts and tighten them to 47 ft. lbs. (64 Nm).

53. Install the oil cooler clamping bolts to the front frame.

54. Install the two front side engine mounting bolts and tighten them to 59 ft. lbs. (80 Nm).

55. Install the two front side transaxle mounting bolts and tighten them to 59 ft. lbs. (80 Nm).

56. Connect the oil cooler hoses.

57. Connect and adjust the shift control cable.

58. Connect the solenoid electrical connector.

59. Connect the park/neutral switch electrical connector.

60. Connect the speed sensor and the direct clutch speed sensor connectors.

61. Install the starter.

62. Connect the ground strap.
63. If equipped, install the cruise control actuator and cover.
64. Connect the throttle cable to the engine and tighten the nuts to 11 ft. lbs. (15 Nm).
65. Install the air cleaner.
66. Install the battery and connect the battery cables.
67. Check the transaxle/differential fluid level.
68. Check the front wheel alignment.

Clutch

ADJUSTMENTS

Hydraulic clutch actuating systems used in Toyota vehicles do not require adjustment.

REMOVAL & INSTALLATION

Paseo, Tercel, Supra and Camry

1. Disconnect the negative battery cable from the battery. On vehicles equipped with an air bag, wait at least 90 seconds before proceeding.
2. Raise and safely support the vehicle.
3. Remove the transaxle assembly from the vehicle.
4. Matchmark the clutch cover to the flywheel.
5. Loosen each set bolt one turn at a time until spring tension is released.
6. Once the tension on the springs are released, remove the clutch pressure plate retaining bolts.
7. Remove the clutch cover.
8. Remove the clutch disc.
9. Remove the retaining clip and withdraw the release bearing from the transaxle.

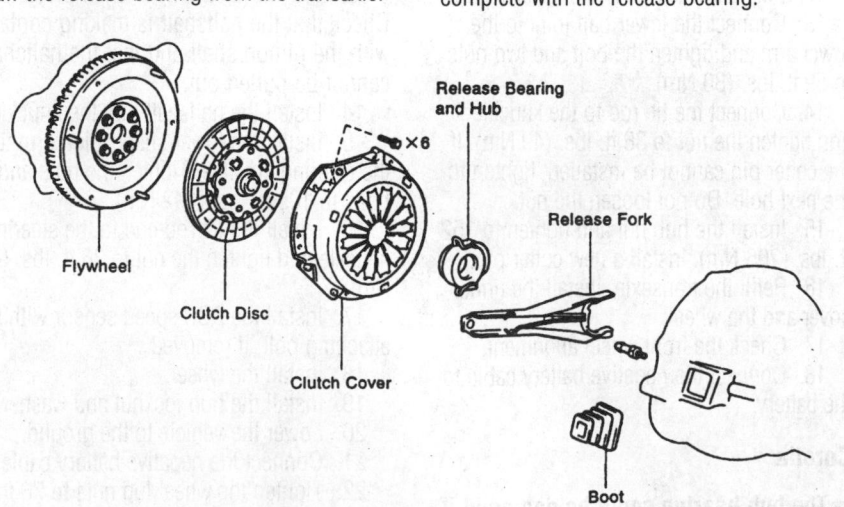

Clutch component assembly—Camry shown, others similar

10. Remove the release fork and boot assembly.

To install:
11. Using a suitable clutch disc alignment tool, install the clutch disc onto the flywheel.
12. Position the clutch cover onto the flywheel and align the matchmarks.
13. Install the clutch cover retaining bolts. Tighten the bolts in a crisscross pattern to 14 ft. lbs. (19 Nm).
14. Lubricate the release fork pivot contact points and the release bearing, bearing hub and input shaft spline surfaces with a suitable molybdenum disulfide lithium based or multi-purpose grease.
15. Install the boot, release fork, hub and bearing assemblies.
16. Install the transaxle to the vehicle.
17. Lower the vehicle and check the clutch is working properly.
18. Connect the negative battery cable to the battery.

Corolla and Celica

➡ **Do not allow grease or oil to get on any part of the disc, pressure plate, or flywheel surfaces.**

1. Disconnect the negative battery cable. On vehicles equipped with an air bag, wait at least 90 seconds before proceeding.
2. Raise and safely support the vehicle.
3. Remove the transaxle.
4. Make matchmarks on the clutch cover (pressure plate) and flywheel so that the pressure plate can be returned to its original position during installation.
5. Unfasten the release fork bearing clips. Withdraw the release bearing hub, complete with the release bearing.

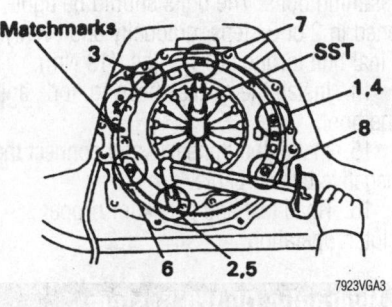

Tighten the pressure plate bolts according to the sequence shown—Camry

Be sure to tighten the pressure plate bolts in the correct order—Supra

6. Remove the release fork and support.
7. Slowly unfasten the bolts which attach the pressure plate. Loosen each bolt one turn at a time until the spring tension is released.

❄❄ CAUTION

If the bolts are released improperly the clutch assembly could fly apart, causing possible injury.

8. Separate the pressure plate from the clutch cover/spring assembly.
9. Inspect the disc, pressure plate and flywheel for damage and wear using a caliper to measure depth and width and a dial indicator to measure runout.
   a. The minimum clutch disc rivet head depth is 0.012 in. (0.3mm).
   b. The maximum clutch disc runout is 0.031 in. (0.8mm).
   c. The maximum pressure plate spring depth is 0.024 in. (0.6mm).
   d. The maximum pressure plate spring width is 0.197 in. (5.0mm).
   e. The maximum flywheel runout is 0.004 in. (0.1mm).
10. Replace or machine parts as necessary.

To install:

11. When reassembling, apply a thin coating of multipurpose grease to the release bearing hub and release fork contact points. Also, pack the groove inside the clutch hub with multipurpose grease and lubricate the pivot points of the release fork.

12. Align the matchmarks on the clutch cover and flywheel which were made during disassembly. Install the clutch disc and pressure plate assembly and tighten the retaining bolts just finger-tight.

13. Center the clutch disc by using a clutch pilot tool or an old input shaft. Insert the pilot into the end of the input shaft front bearing, wiggle it gently to align the clutch disc and pressure plate and tighten the retaining bolts. The bolts should be tightened in 2 or 3 steps, gradually and evenly. Final bolt torque is 14 ft. lbs. (19 Nm).

14. Install the release bearing, fork, and the boot.

15. Install the transaxle and connect the negative battery cable.

16. Road test and check for proper clutch operation.

Hydraulic Clutch System

BLEEDING

➡ **If any maintenance on the clutch system was performed or the system is suspected of containing air, bleed the system. Use care; brake fluid will remove the paint from any surface. If the brake fluid spills onto any painted surface, wash it off immediately with soap and water.**

1. Fill the clutch reservoir with brake fluid. Check the reservoir level frequently and add fluid as needed.

2. Connect one end of a vinyl tube to the bleeder plug on the slave cylinder and submerge the other end into a clear container half-filled with brake fluid.

3. Slowly pump the clutch pedal several times.

4. Have an assistant hold the clutch pedal down and loosen the bleeder plug until fluid and/or air starts to run out of the bleeder plug. Close the bleeder plug while the pedal is held to the floor.

➡ **Do not allow the pedal to rise back-up while the bleeder is still open. If this happens, it will allow air to re-enter the slave cylinder and cause the clutch system not to work properly.**

5. Repeat Steps 2 and 3 until all the air bubbles are removed from the system.

6. Tighten the bleeder plug when all the air is gone.

7. Refill the master cylinder to the proper level as required.

8. Check the system for leaks.

Halfshaft

REMOVAL & INSTALLATION

Paseo and Tercel

1. Disconnect the negative battery cable to the battery. On vehicles equipped with an air bag, wait at least 90 seconds before proceeding.

2. Raise and support the vehicle safely.

3. Remove the left engine undercover and drain the transaxle.

4. If equipped with ABS brakes, disconnect the ABS speed sensor by removing bolt.

5. Remove the cotter pin and lock cap from the halfshaft.

6. Apply the brake and remove the hub nut.

7. Disconnect the tie rod from the steering knuckle.

8. Disconnect the lower ball joint from the lower control arm by removing bolt and two nuts.

9. Using a plastic hammer, disconnect the halfshaft from the axle hub.

10. Using a brass bar and hammer, tap the inner joint out of the transaxle and remove the halfshaft.

To install:

11. Using a new snapring, push the halfshaft into the transaxle until it clicks into position. Pull on the inner joint to be sure it is fully installed.

12. Push the outer joint into the axle hub.

13. Connect the lower ball joint to the lower arm and tighten the bolt and two nuts to 59 ft. lbs. (80 Nm).

14. Connect the tie rod to the knuckle and tighten the nut to 36 ft. lbs. (49 Nm). If the cotter pin cannot be installed, tighten to the next hole. Do not loosen the nut.

15. Install the hub nut and tighten to 152 ft. lbs. (206 Nm). Install a new cotter pin.

16. Refill the transaxle. Install the undercover and the wheel.

17. Check the front wheel alignment.

18. Connect the negative battery cable to the battery.

Corolla

➡ **The hub bearing could be damaged if subjected to the full weight of the vehicle, such as if the vehicle is moved**

without the halfshafts. If it is absolutely necessary to place the full vehicle weight on the hub bearing, first support the bearing with SST No. 09608–16041, or equivalent.

1. Disconnect the negative battery cable. On vehicles equipped with an air bag, wait at least 90 seconds before proceeding.

2. Remove the wheel cover.

3. Remove the cotter pin, locknut cap, and bearing locknut.

4. Loosen the wheel nuts.

5. Raise and safely support the vehicle. Remove the engine undercovers and drain gear oil or transaxle fluid.

6. Remove the wheel.

7. If equipped with ABS, remove the bolt and the speed sensor.

8. Separate the tie rod ball joint from the steering knuckle.

9. Disconnect the lower ball joint from the lower suspension arm.

10. Using a plastic hammer or equivalent, drive the halfshaft from the knuckle.

➡ **The halfshaft can be separated from the knuckle using a brass or plastic hammer; some others may require the use of a puller. Be careful not to damage the inner oil seal, the ABS sensor rotor, or the halfshaft.**

11. Using a suitable prying tool, pry the halfshaft from the transaxle. Remove the halfshaft from the vehicle.

To install:

12. Coat gear oil to the inboard joint tulip and position the snapring opening side facing downward.

13. Install the halfshaft into the transaxle. After installing the halfshaft to the transaxle, check that there is 0.08–0.12 in. (2–3mm) of axial play. Check that the halfshaft is making contact with the pinion shaft and that the halfshaft cannot be pulled out.

14. Install the halfshaft into the knuckle.

15. Install the lower suspension arm to the steering knuckle. Tighten the nuts and bolts to 105 ft. lbs. (142 Nm).

16. Install the tie rod end to the steering knuckle and tighten the nut to 36 ft. lbs. (49 Nm).

17. Install the ABS speed sensor with the attaching bolt, if removed.

18. Install the wheel.

19. Install the hub locknut and washer.

20. Lower the vehicle to the ground.

21. Connect the negative battery cable.

22. Tighten the wheel lug nuts to 76 ft. lbs. (103 Nm). Tighten the hub locknut to 159 ft. lbs. (216 Nm).

23. Install the locknut cap and a NEW cotter pin. Fill transaxle with gear oil or transaxle fluid if necessary.

24. Install engine cover. Check front wheel alignment and check ABS speed sensor signal.

Celica

➡The hub bearing could be damaged if subjected to the full weight of the vehicle, such as if the vehicle is moved without the halfshafts. If it is absolutely necessary to place the full vehicle weight on the hub bearing, first support the bearing with SST No. 09608–16041, or equivalent.

1. Disconnect the negative battery cable. On vehicles equipped with an air bag, wait at least 90 seconds before proceeding.

2. Remove the wheel cover.

3. Remove the cotter pin, locknut cap, and the bearing locknut.

4. Loosen the wheel lug nuts.

5. Raise and safely support the vehicle. Remove the engine undercovers and drain the gear oil or transaxle fluid.

6. Remove the wheel.

7. Separate the tie rod ball joint from the steering knuckle.

8. Disconnect the stabilizer bar link from the lower suspension arm.

9. Disconnect the lower ball joint from the lower suspension arm.

10. Using a plastic hammer or equivalent, tap the halfshaft from the knuckle.

➡Be careful not to damage the inner oil seal or the ABS sensor rotor on the halfshaft.

11. To remove the left side halfshaft, use a suitable prying tool and separate the halfshaft from the transaxle. Remove the halfshaft from the vehicle.

12. To remove the right side halfshaft perform the following steps:

   a. Remove the two bolts of the center bearing bracket

   b. Pull the halfshaft out together with the center bearing case and the center halfshaft.

   c. Remove the center shaft with the RH halfshaft from the transaxle through the bearing bracket.

➡Do not damage the oil seal lip.

To install:

13. Coat gear oil to the inboard joint tulip and position the snapring opening side facing downward.

14. To install the left side halfshaft, simply insert the halfshaft into the transaxle.

15. To install the right side halfshaft, insert the halfshaft with the bearing case and center shaft into the transaxle. Attach the center bearing case to and tighten the two bolts to 47 ft. lbs. (64 Nm).

16. After installing either halfshaft, check that there is 0.08–0.12 in. (2–3mm) of axial play. Check that the halfshaft is making contact with the pinion shaft and that the halfshaft cannot be pulled out.

17. Install the halfshaft into the knuckle.

18. Connect the lower suspension arm to the lower ball joint. Tighten the ball joint bolt and nuts to 94 ft. lbs. (127 Nm).

19. Connect the tie rod end to the steering knuckle and tighten the nut to 36 ft. lbs. (49 Nm).

20. Install the stabilizer bar link to the lower suspension arm. Tighten the nuts to 33 ft. lbs. (44 Nm).

21. Install the wheel and tighten the wheel lugs to 76 ft. lbs.

22. Install the locknut and washer. Tighten the locknut to 159 ft. lbs. (216 Nm).

23. Lower the vehicle to the ground.

24. Connect the negative battery cable.

25. Install the locknut cap and a new cotter pin. Fill transaxle with gear oil or transaxle fluid if necessary.

26. Install engine cover, check front wheel alignment, and check ABS speed sensor signal.

Camry and Avalon

1. Disconnect the negative battery cable to the battery. On vehicles equipped with an air bag, wait at least 90 seconds before proceeding.

2. Raise and support the vehicle safely.

3. Remove the front wheel(s).

4. Remove the front fender apron seal.

5. Drain the transaxle.

6. Disconnect the tie rod end from the steering knuckle by removing the cotter pin and nut. Using tool SST 09628–62011 or equivalent, separate the tie rod from the steering knuckle.

7. Disconnect the stabilizer bar link from the lower control arm. Make note of the washers and cushions positions.

8. Disconnect the lower ball joint from the steering knuckle by removing the bolt and two nuts. Push down on the lower control arm and separate the steering knuckle from the ball joint.

9. Remove the cotter pin, lock cap and locknut holding the halfshaft to the steering knuckle.

10. Using a plastic hammer, disconnect the halfshaft from the steering knuckle.

11. Remove the left halfshaft from the transaxle as follows:

   a. Use a brass bar and hammer to tap the inner joint out of the transaxle.

   b. Remove the halfshaft.

   c. Once the halfshaft is removed from the vehicle, remove the snapring from the halfshaft.

12. Remove the right halfshaft from the transaxle as follows:

   a. Remove the bearing lockbolt. The lockbolt is located in the center of the halfshaft, near the dampener.

   b. Using snapring pliers, remove the snapring and pull the halfshaft from the transaxle.

To install:

13. To install the right halfshaft to the transaxle:

   a. Coat the side gear shaft and differential case sliding surface with gear oil.

   b. Using snapring pliers, install the snapring to the halfshaft.

   c. Install the halfshaft and the bearing lockbolt. Tighten the lockbolt to 24 ft. lbs. (32 Nm).

14. To install the left halfshaft to the transaxle:

   a. Install a new snapring to the inner spline of the halfshaft.

   b. Coat the side gear shaft and differential case sliding surface with gear oil.

   c. Install the halfshaft to the transaxle with the snapring opening facing down. The halfshaft should click into place when installing.

   d. After installation of the halfshaft, check that the halfshaft cannot be removed by hand.

15. Connect the halfshaft to the steering knuckle, then install the locknut. Tighten the locknut to 217 ft. lbs. (294 Nm).

16. Install the lock cap and a new cotter pin to the halfshaft.

17. Connect the steering knuckle to the lower ball joint. Install the two nuts and bolt. Tighten the nuts and bolt to 94 ft. lbs. (127 Nm).

18. Connect the stabilizer bar link to the lower control arm. Tighten the nut to 29 ft. lbs. (39 Nm).

19. Connect the tie rod to the steering knuckle and tighten the nut to 36 ft. lbs. (49 Nm). Install a new cotter pin to the tie rod end.

20. Install the front fender apron seal.

21. Install the wheel(s) and lower the vehicle. Tighten the lug nuts to 76 ft. lbs. (103 Nm).

22. Refill the transaxle and check for leaks.

23. Connect the negative battery cable to the battery.

Supra

1. Disconnect the negative battery cable. On vehicles equipped with an air bag, wait at least 90 seconds before proceeding.

2. Raise and safely support the vehicle.

3. Remove the rear tire and wheel assembly.

4. Remove the rear exhaust assembly.

5. Remove the cotter pin, locknut cap, and the locknut holding the halfshaft to the rear axle carrier.

6. Remove the lower suspension arm brace by removing the four bolts.

7. Place matchmarks on the halfshaft and the differential side gear shaft. Remove the hexagon bolts and washers with the proper tool.

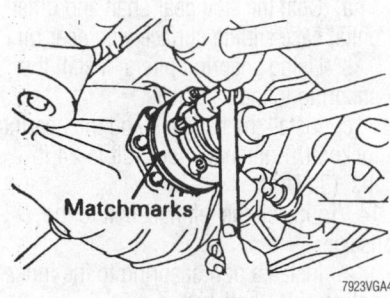

Whenever removing the halfshaft, be sure to matchmark it to the differential side gear—Supra

8. Hold the inboard joint side of the halfshaft so the outboard joint side does not bend too much. Tap the end of the halfshaft with a rubber mallet and disengage the half-shaft from the axle carrier.

9. Remove the halfshaft.

To install:

10. Insert the outboard joint side of the halfshaft and align the matchmarks on the side gear shaft and the halfshaft.

11. Coat the threads with clean oil and install the hexagon bolts. Tighten the bolts to 61 ft. lbs. (83 Nm).

12. Install the lower suspension arm brace and tighten the four bolts to 13 ft. lbs. (18 Nm).

13. Install the bearing locknut and tighten the locknut to 213 ft. lbs. (289 Nm).

14. Install the locknut cap and install a new cotter pin.

15. Install the rear exhaust assembly.

16. Replace the rear tire and wheel assembly.

17. Lower the vehicle.

18. Connect the negative battery cable.

STEERING AND SUSPENSION

Air Bag

✳✳ CAUTION

Some vehicles are equipped with an air bag system, also known as the Supplemental Inflatable Restraint (SIR) or Supplemental Restraint System (SRS). The system must be disabled before performing service on or around system components, steering column, instrument panel components, wiring and sensors. Failure to follow safety and disabling procedures could result in accidental air bag deployment, possible personal injury and unnecessary system repairs.

PRECAUTIONS

Several precautions must be observed when handling the inflator module to avoid accidental deployment and possible personal injury.

• Never carry the inflator module by the wires or connector on the underside of the module.

• When carrying a live inflator module, hold securely with both hands, and ensure that the bag and trim cover are pointed away.

• Place the inflator module on a bench or other surface with the bag and trim cover facing up.

• With the inflator module on the bench, never place anything on or close to the module which may be thrown in the event of an accidental deployment.

DISARMING

To avoid personal injury when working on vehicles equipped with an air bag, the negative battery cable must be disconnected and at least 90 seconds must elapse before working on the system. Failure to do so may result in deployment of the air bag.

REARMING

After vehicle service is completed, reattach the battery cables (positive cable first!) to rearm the air bag system.

Rack and Pinion Steering Gear

REMOVAL & INSTALLATION

Manual

1. Position the front wheels straight ahead.

2. Disconnect the negative battery cable. On vehicles equipped with an air bag, wait at least 90 seconds before proceeding.

3. If equipped with an air bag, disable the system and secure the steering wheel.

4. Remove sliding yoke on rack and pinion assembly.

5. Raise and safely support the vehicle.

6. Remove the front wheels.

7. Disconnect the tie rod ends.

8. On Corolla, remove the engine rear mount insulator by removing the bolt and three nuts. Remove the engine rear mount bracket by removing the three bolts.

9. Remove rack and pinion assembly bracket bolts and rack and pinion assembly from the vehicle.

To install:

10. Install the rack and pinion assembly into the vehicle with the attaching bolts. Tighten the bolts to 43 ft. lbs. (58 Nm).

11. On Corolla, install the engine rear mount bracket and install the three bolts. Tighten the bolts to 57 ft. lbs. (77 Nm). Install the engine rear mount insulator by installing the bolt and three nuts. Tighten the bolt to 64 ft. lbs. (87 Nm) and the nuts to 35 ft. lbs. (48 Nm).

12. Connect the tie rod ends to the knuckle arms.

13. Connect sliding yoke onto rack and pinion assembly. On Tercel, tighten the bolts to 19 ft. lbs. (26 Nm). On Corolla, tighten the lower bolt to 26 ft. lbs. (35 Nm) and the upper bolt to 20 ft. lbs. (27 Nm).

14. Install the front wheels.

15. Safely lower the vehicle.

16. Connect the negative battery cable.

17. Check the front end alignment. The toe adjustment may have to be reset.

Power

TERCEL AND PASEO

1. Position the front wheels straight ahead.

2. Disconnect the negative battery cable. On vehicles equipped with an air bag, wait at least 90 seconds before proceeding.

3. If equipped with an air bag, disable the system and secure the steering wheel.

4. Disconnect the tie rod ends from the knuckle arm using a tie rod separator or equivalent.

5. Remove the column hole cover. Matchmark the sliding yoke and control valve shaft for installation.

6. Loosen the upper bolt and disconnect the lower bolt to the control valve shaft. Slide the shaft upward and disconnect the control valve shaft from the steering rack.

7. Disconnect the oxygen sensor and the exhaust pipe.

8. If necessary for additional access, remove the stabilizer bar.

9. Remove the engine rear mount insulator bolts.

10. Disconnect the two vacuum hoses.

11. If equipped with a manual transaxle, disconnect the transmission control cables.

12. Disconnect the power steering hoses and drain the fluid into a container.

13. Disconnect the tube clamp.

14. Remove the two brackets and grommets.

15. Remove the housing-to-frame retaining bolts and remove the assembly. Slide the housing out the left-hand side of the vehicle.

To install:

16. Line up the steering splines, then install the unit.

17. Install the two grommets and brackets. Tighten the two bolts and nuts to 43 ft. lbs. (58 Nm).

18. Connect the pressure feed and return tubes and tighten to 26 ft. lbs. (36 Nm).

19. Connect the tube clamp and tighten the bolt to 9 ft. lbs. (13 Nm).

20. Connect the two vacuum hoses.

21. Install the rear brackets and tighten the bolts to 35 ft. lbs. (48 Nm).

22. Install the engine rear mount insulator. Tighten the through-bolt to 47 ft. lbs. (64 Nm) the support bracket bolts to 58 ft. lbs. (78 Nm).

23. If applicable, install the stabilizer bar to the vehicle.

24. Connect the sliding yoke to the control valve shaft. Tighten the bolts to 21 ft. lbs. (28 Nm).

25. Install the column hole cover with the three nuts and tighten the nuts to 43 inch lbs. (5 Nm).

26. Connect the right and left tie rod ends.

27. Install the wheels to the vehicle and lower the vehicle.

28. Connect the negative battery cable to the battery.

29. Fill the power steering reservoir to specification and bleed the system.

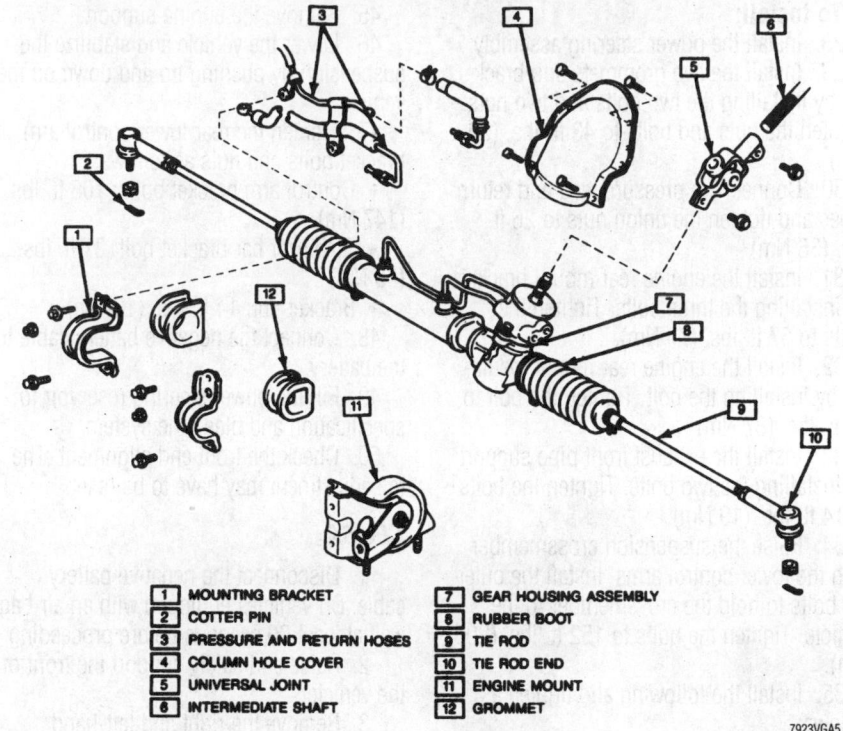

1	MOUNTING BRACKET	7	GEAR HOUSING ASSEMBLY
2	COTTER PIN	8	RUBBER BOOT
3	PRESSURE AND RETURN HOSES	9	TIE ROD
4	COLUMN HOLE COVER	10	TIE ROD END
5	UNIVERSAL JOINT	11	ENGINE MOUNT
6	INTERMEDIATE SHAFT	12	GROMMET

7923VGA5

Exploded view of a typical power rack and pinion steering gear unit

30. Check the front end alignment. The toe adjustment may have to be reset.

COROLLA

1. Position the front wheels straight ahead.

2. Disconnect the negative battery cable.

3. If equipped with an air bag, disable the system and secure the steering wheel.

4. Place a drain pan under the steering rack.

5. Remove the steering column hole cover by removing the five bolts.

6. Loosen the upper pinch bolt to the sliding yoke. Remove the lower pinch bolt at the pinion shaft.

7. Loosen the wheel lug nuts.

8. Raise and safely support the vehicle.

9. Remove both front wheels.

10. Remove the left and right engine undercovers.

11. Disconnect the left and right tie rod ends.

12. Install an engine support and tension it to support the engine without raising it.

❋❋ CAUTION

The engine hoist is now in place and under tension. Use care when repositioning the vehicle and make necessary adjustments to the engine support.

13. Disconnect the lower control arms from the ball joints.

14. If equipped with a stabilizer bar, disconnect the stabilizer bar links from both lower control arms.

15. Remove the nut and three bolts and remove the right rear control arm bushing retaining bracket. Do this for both lower control arms.

16. Remove the stabilizer bar from the vehicle.

17. Remove the grommet in the crossmember.

18. Remove the bolt and four nuts holding in the middle of the crossmember.

19. Support the suspension crossmember with a jack.

20. Remove the six bolts from the outer side of the suspension crossmember.

21. Remove the suspension crossmember with the lower suspension arms.

22. Remove the exhaust front pipe support by removing the two bolts.

23. Remove the engine rear mount insulator by removing the bolt.

24. Remove the engine rear mount bracket by removing the three bolts.

25. Disconnect the pressure feed and return tubes.

26. Remove the two brackets and grommets to the power steering rack by removing the two bolts and nuts.

27. Slide the power steering gear assembly to the right side of the vehicle.

To install:

28. Install the power steering assembly.

29. Install the two grommets and brackets by installing the two bolts and two nuts. Tighten the nuts and bolts to 43 ft. lbs. (59 Nm).

30. Connect the pressure feed and return tubes and tighten the union nuts to 26 ft. lbs. (36 Nm).

31. Install the engine rear mount bracket by installing the three bolts. Tighten the bolts to 57 ft. lbs. (77 Nm).

32. Install the engine rear mount insulator by installing the bolt. Tighten the bolt to 64 ft. lbs. (87 Nm).

33. Install the exhaust front pipe support by installing the two bolts. Tighten the bolts to 14 ft. lbs. (19 Nm).

34. Raise the suspension crossmember with the lower control arms. Install the outer six bolts to hold the crossmember to the vehicle. Tighten the bolts to 152 ft. lbs. (206 Nm).

35. Install the following and tighten as follows:
- Center crossmember-to-radiator support bolts: 45 ft. lbs. (61 Nm)
- Lower A frame-to-center bolts: 161 ft. lbs. (218 Nm)
- Lower A frame-to-outer bolts: 109 ft. lbs. (147 Nm)
- Front, center and rear mount bolts: 45 ft. lbs. (61 Nm)

36. Install the grommet to the crossmember.

37. Install the stabilizer bar to the vehicle.

38. Install the lower control arm bushing retaining bracket and install the nut and three bolts. Do not tighten the bolts or nut at this time.

39. Connect the lower control arm to the lower ball joint by installing the bolt and two nuts. Tighten the bolt and nuts to 105 ft. lbs. (142 Nm). Connect both lower control arms to the ball joints.

40. Connect the stabilizer bar links to the lower control arms and tighten the nuts to 33 ft. lbs. (44 Nm).

41. Connect the sliding yoke to the pinion shaft. Install the lower bolt and tighten the bolt to 26 ft. lbs. (35 Nm). Tighten the upper bolt to the sliding yoke to 20 ft. lbs. (27 Nm).

42. Install the steering column hole cover by installing the five bolts. Tighten the bolts to 43 inch lbs. (5 Nm).

43. Install the left and right-hand tie rod ends and tighten the nuts to 36 ft. lbs. (49 Nm).

44. Install the wheels and lower the vehicle to the ground.

45. Remove the engine support.

46. Lower the vehicle and stabilize the suspension by pushing up and down on the vehicle.

47. Tighten the rear lower control arm bracket bolts and nuts as follows:
- Control arm bracket bolts: 108 ft. lbs. (147 Nm)
- Stabilizer bar bracket bolt: 37 ft. lbs. (50 Nm)
- Bracket nut: 14 ft. lbs. (19 Nm)

48. Connect the negative battery cable to the battery.

49. Fill the power steering reservoir to specification and bleed the system.

50. Check the front end alignment. The toe adjustment may have to be reset.

CELICA

1. Disconnect the negative battery cable. On vehicles equipped with an air bag, wait at least 90 seconds before proceeding.

2. Raise and safely support the front of the vehicle.

3. Remove the right and left-hand engine undercovers.

4. Disconnect the left and right-hand tie rod ends by removing the cotter pins and nuts. Separate the tie rod using a puller.

5. Remove the oxygen sensor.

6. Remove the front exhaust pipe and brackets.

7. Place matchmarks on the steering column intermediate shaft and the steering gear control valve shaft.

8. Loosen the upper bolt and remove the lower bolt to the intermediate shaft.

9. Disconnect the intermediate shaft from the control valve shaft.

10. Using a line wrench, disconnect the pressure feed and return tubes from the steering rack. Be sure to place a pan under the tubes to catch any fluid.

11. Disconnect the tube clamp bracket by removing the two bolts.

12. Support the engine and transaxle with a support fixture.

13. Disconnect the lower control arms from the lower ball joints by removing the bolt and two nuts on each side.

14. Remove the through-bolts to the front and rear mounting insulators.

15. Remove the three bolts holding the rear of the lower control arm to the sub-frame and body. Remove the bolts on both sides of the sub-frame.

16. Support the front sub-frame with a jack.

17. Remove the two bolts holding the sub-frame to the body. Lower the front sub-frame with the lower suspension arms and steering gear.

18. Remove the power steering gear assembly by removing the set bolts and nuts from the sub-frame.

To install:

19. Install the power steering gear assembly to the sub-frame by installing the two bolts and two nuts. Tighten to 94 ft. lbs. (127 Nm).

20. Raise the sub-frame into position and install the two bolts to hold the sub-frame to the body. Tighten the bolts to 94 ft. lbs. (127 Nm).

21. Install the three bolts to hold the rear of the lower control arms to the sub-frame and body. Tighten the bolt that goes through the lower control arm to 123 ft. lbs. (167 Nm) and the other two bolts to 130 ft. lbs. (175 Nm). Install the bolts to both sides.

22. Connect the two through-bolts to the front and rear mounting insulators and tighten to 64 ft. lbs. (88 Nm).

23. Connect the lower control arms to the lower ball joints by installing the bolt and two nuts on each side. Tighten the nuts and bolts to 94 ft. lbs. (127 Nm).

24. Remove the support from under the engine and transaxle.

25. Connect the pressure feed and return tubes to the steering rack. Tighten the tubes to 26 ft. lbs. (36 Nm).

26. Connect the tube clamp bracket and tighten the two bolts to 9 ft. lbs. (13 Nm).

27. For the intermediate shaft on the steering column, align the matchmarks on the intermediate shaft and control valve shaft.

28. Install the lower bolt and tighten the upper and lower bolts to 26 ft. lbs. (35 Nm).

29. Install the front exhaust pipe.

30. Install the oxygen sensor.

31. Connect the tie rod ends to the steering knuckles and install the nuts.

32. Install the right and left-hand engine undercovers by installing the bolts.

33. Lower the vehicle.

34. Connect the negative battery cable to the battery.

35. Fill the power steering reservoir to specification and bleed the system.

36. Check the front end alignment. The toe adjustment may have to be reset.

CAMRY AND AVALON

1. Position the front wheels straight ahead.

2. Disconnect the negative battery cable. On vehicles equipped with an air bag, wait at least 90 seconds before proceeding.

3. If equipped with an air bag, disable the system and secure the steering wheel.

4. Raise and support the vehicle safely. Remove the front wheels.

5. Remove the left and right front fender apron seals by removing the two bolts.

6. Remove the cotter pin and nut holding the steering knuckle to the tie rod end. Using a tie rod puller, disconnect the tie rod end from the steering knuckle.

7. Place matchmarks on the intermediate shaft and the control valve shaft.

8. Loosen the upper bolt and remove the lower bolt holding the control valve shaft to the intermediate shaft. Disconnect the intermediate shaft from steering rack housing.

9. Remove the nut to the tube clamp. Remove the clamp from the vehicle.

10. Disconnect the return line and the pressure line from the control valve housing. Use a small plastic container to catch the fluid.

11. Remove the four stabilizer bar bolts and two nuts. Position the stabilizer bar out of the way. Do not remove the bar from the vehicle.

12. If necessary, remove the rear engine mounting and bracket for additional clearance.

13. On the V6 engine, remove the oxygen sensor.

14. Remove the two steering gear mounting bolts and nuts. Remove the steering gear.

To install:

15. Position the steering gear on the vehicle and install the two mounting bolts and nuts. Tighten the nuts and bolts to 134 ft. lbs. (181 Nm).

16. On the V6, install the oxygen sensor.

17. If applicable, install the rear engine mounting bracket and tighten the retaining bolts to 38 ft. lbs. (52 Nm).

18. Install the stabilizer bar bolts and nuts.

19. Connect the tube clamp and tighten the nut to 7 ft. lbs. (10 Nm).

20. Install the intermediate shaft to the steering rack and tighten the retaining bolts to 26 ft. lbs. (35 Nm).

21. Connect the tie rods to the steering knuckles with the castellated nuts.

22. Install the front fender apron seals by installing the two bolts.

23. Install the front wheels and lower the vehicle.

24. Connect the negative battery cable to the battery.

25. Fill the power steering reservoir to specification and bleed the system.

26. Check the front end alignment. The toe adjustment may have to be reset.

SUPRA

1. Position the front wheels straight ahead.

2. Disconnect the negative battery cable. On vehicles equipped with an air

bag, wait at least 90 seconds before proceeding.

3. If equipped with an air bag, disable the system and secure the steering wheel.

4. Raise and safely support the vehicle. Remove the front wheels.

5. Remove the engine undercovers and front suspension member protection.

6. Place matchmarks on the universal joint and the control valve shaft.

7. Loosen the bolt on the upper side of the intermediate shaft.

8. Remove the bolt on the lower side of the intermediate shaft and disconnect the universal joint from the steering rack.

9. Disconnect and plug the hydraulic lines to the rack assembly by removing the union bolts and gaskets.

10. Disconnect the tie rod ends from the steering knuckles.

11. Disconnect the solenoid wiring from the rack and pinion unit.

12. Remove the bolts and brackets holding the steering rack to the frame.

13. Remove the rack assembly.

To install:

14. Install the rack. Tighten mounting bracket bolts to 55 ft. lbs. (75 Nm).

15. Connect the solenoid wiring.

16. Connect the tie rod ends to the steering knuckles.

17. Connect the fluid lines to the rack and pinion with new washers. Tighten the union bolts to 36 ft. lbs. (49 Nm).

18. Align the matchmarks on the universal joint and the control valve shaft. Tighten the upper and lower bolts to 26 ft. lbs. (35 Nm).

19. Install the front suspension member protection and the engine undercovers.

20. Install the front wheels to the vehicle and lower the vehicle to the ground.

21. Connect the negative battery cable to the battery.

22. Fill the power steering reservoir to specification and bleed the system.

23. Check the front end alignment. The toe adjustment may have to be reset.

Strut and Coil Spring

REMOVAL & INSTALLATION

Front

EXCEPT SUPRA

1. Disconnect the negative battery cable. On vehicles equipped with an air bag, wait at least 90 seconds before proceeding.

2. Loosen the lug nuts.

3. Raise and support the vehicle safely.

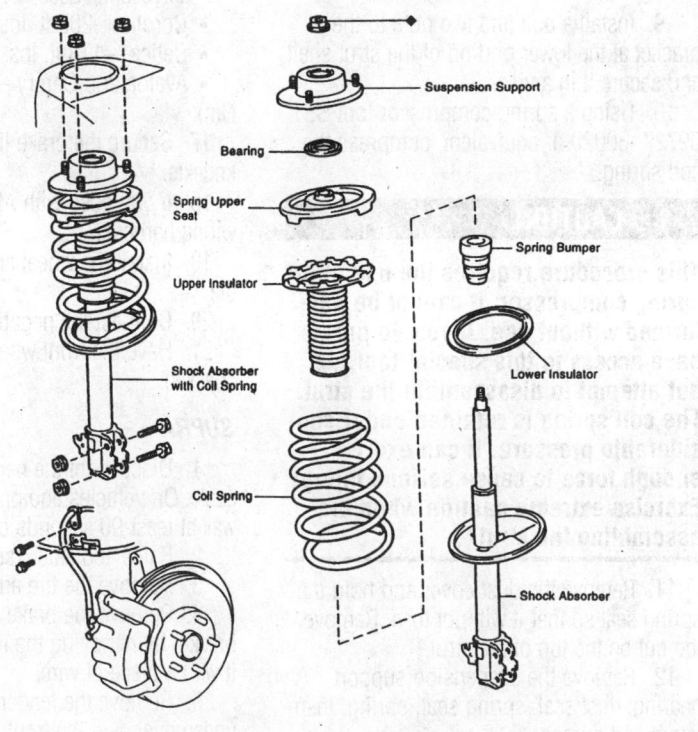

◆ Non-reusable part

7923VGA6

Common coil spring and strut component assembly—except Supra

※ **WARNING**

Do not support the weight of the vehicle on the suspension arm; the arm will deform under its weight.

4. Unfasten the lug nuts and remove the wheel.

5. Remove the bolt and disconnect the brake hose from the strut.

6. If equipped with ABS brakes, disconnect the wiring harness from the strut.

7. Remove the bolts and disconnect the strut from the steering knuckle.

8. Remove the strut from the body.

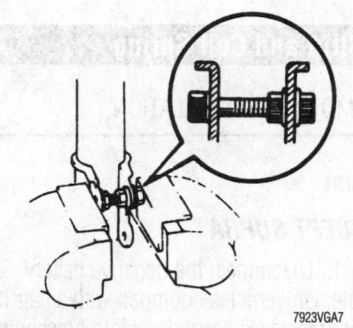

7923VGA7

Proper method of supporting the strut in a vise

9. Install a bolt and two nuts to the bracket at the lower portion of the strut shell and secure it in a vise.

10. Using a spring compressor tool SST 09727–30020 or equivalent, compress the coil spring.

※ **CAUTION**

This procedure requires the use of a spring compressor; it cannot be performed without one. If you do not have access to this special tool, do not attempt to disassemble the strut. The coil spring is retained under considerable pressure. It can exert enough force to cause serious injury! Exercise extreme caution when disassembling the strut.

11. Remove the dust cover and hold the spring seat so that it will not turn. Remove the nut on the top of the strut.

12. Remove the suspension support, bearing, dust seal, spring seat, spring, insulators and bumper.

To install:

13. To assemble the strut:

a. Install the spring bumper to piston.

b. Using a spring compressor, compress the spring.

c. Install the coil spring to the strut. Fit the lower end of the coil spring into the gap of the lower seat.

d. Install the spring seat with the insulator.

e. Install the dust seal on the spring seat.

f. Install the suspension support and tighten the new suspension nut to 35 ft. lbs. (47 Nm). After the nut has been tighten, release the compressor tool tension.

g. Pack multipurpose grease into the suspension support. Install the dust cover.

➡ **Do not use an impact wrench to tighten the nut. Also, check that the bearing fits into the recess in the suspension support.**

14. Install the nuts holding the strut to the strut tower. Tighten the nuts to 29 ft. lbs. (39 Nm), except on Avalon, Camry and Celica. On Avalon, Camry and Celica, tighten the nuts to 59 ft. lbs. (80 Nm).

15. Connect the steering knuckle to the strut lower bracket.

16. Insert the two bolts from the rear side and tighten the strut-to-steering knuckle arm bolts. Tighten as follows:

• Tercel and Paseo—166 ft. lbs. (226 Nm)
• Corolla—203 ft. lbs. (275 Nm)
• Celica—113 ft. lbs. (153 Nm)
• Avalon and Camry—156 ft. lbs. (211 Nm)

17. Secure the brake line to the steering knuckle.

18. If equipped with ABS, secure the wiring harness.

19. Install the wheel and lower the vehicle.

20. Connect the negative battery cable.

21. Have the front wheel alignment checked.

SUPRA

1. Disconnect the negative battery cable. On vehicles equipped with an air bag, wait at least 90 seconds before proceeding.

2. Raise and safely support the vehicle.

3. Remove the tire and wheel assembly.

4. Remove the brake caliper support bracket by removing the two bolts. Suspend it with a piece of wire.

5. Remove the fender apron, engine undercover, and the front fender wheel opening molding.

6. If removing the left side strut, disconnect the windshield washer tank.

7. Remove the bolt and disconnect the ABS speed sensor at the steering knuckle. Remove the three bolts, then disconnect the wiring harness clamp in order to prevent the harness from being damaged when removing the through-bolt.

8. Remove the plug from the upper strut mount. Do not remove the center bolt.

※ **CAUTION**

Do not remove the center bolt to the strut at this time. The spring on the strut is under high pressure and can cause serious injury or vehicle damage.

9. Disconnect the upper control arm through-bolt from the sub-frame. Disconnect the upper control arm and turn the control arm completely around. It is not necessary to remove the upper ball joint.

10. Disconnect the strut at the lower control arm by removing the nut and bolt.

11. Remove the three upper mounting nuts and remove the strut assembly with the coil spring from the vehicle.

12. Using compressor 09727–30020 or equivalent, compress the coil spring.

13. Remove the piston rod locknut.

14. Remove the suspension support, coil spring and bumper.

To install:

15. Match the bolt of the suspension support with the cut out portion of the insulator.

16. Install the spring bumper.

17. Install the compressed coil spring. Match the end of the coil into the recess of the strut spring seat.

18. Install the suspension support to the rod and temporarily install a new nut.

19. Turn the suspension support so one of the bolts on the support faces the same direction as shown in the illustration.

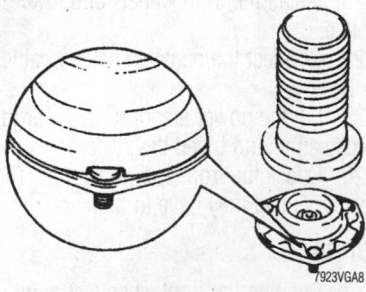

7923VGA8

Aligning the insulator to the support— Supra

→**Align the bolt so a line drawn between the rod and bolt would be at 90° to the direction of the lower bushing.**

20. Remove the spring compressor.

21. Install the strut and tighten the three upper strut mount nuts to 26 ft. lbs. (35 Nm). Tighten the middle nut to 22 ft. lbs. (29 Nm) and install the plug.

22. Connect the lower end of the strut to the lower control arm. Do not tighten the bolt at this time.

23. Install the upper control arm and install the through-bolt and nut. Do not tighten the bolt at this time.

24. Connect the speed sensor, wiring harness, and the washer tank.

25. Install the fender apron and the engine undercover.

26. Install the caliper support bracket and tighten the bolts to 87 ft. lbs. (118 Nm).

27. Install the tire and wheel assembly.

28. Lower the vehicle.

29. Bounce the vehicle a few times to stabilize the suspension, then tighten the strut-to-lower arm bolt to 106 ft. lbs. (143 Nm). Tighten the upper arm to 121 ft. lbs. (164 Nm).

30. Connect the negative battery cable.

31. Have the front end alignment checked.

Rear

1. Disconnect the negative battery cable from the battery. On vehicles equipped with an air bag, wait at least 90 seconds before proceeding.

2. Remove the rear seat cushion and any trim necessary to access the strut towers.

3. Loosen the lug nuts to the wheel on the side being serviced.

4. Raise and safely support the vehicle. Place jackstands under the vehicle frame.

5. Support the axle beam with a jack.

6. Remove the wheel on the side being serviced.

7. On Supra, remove the brake caliper support bracket by removing the two bolts. Leave the brake line connected and position it aside.

8. If equipped with ABS, disconnect the sensor wire from the strut.

9. If equipped, disconnect the stabilizer bar from the strut.

10. On Camry, disconnect the Load Sensing Proportioning Valve (LSPV) from the lower control arm by removing the bolt.

11. Loosen the fasteners securing the strut to the axle carrier. Do not remove the bolts at this time.

12. Support the axle carrier with a jack.

13. Disconnect the strut from the strut tower by the nuts.

⁜⁜⁜ CAUTION

Do not loosen the center nut on the top of the strut piston.

14. Remove the strut from the vehicle.

15. To disassemble, proceed as follows:

⁜⁜⁜ CAUTION

This procedure requires the use of a spring compressor; it cannot be performed without one! If you do not have access to this special tool, do not attempt to disassemble the strut. The coil springs are retained under considerable pressure. They can exert enough force to cause serious injury! Exercise extreme caution when disassembling the strut.

a. Place the strut assembly in a pipe vise or strut vise.

→**Do not attempt to clamp the strut assembly in a flat jaw vise as this will result in damage to the strut tube.**

b. Attach a spring compressor and compress the spring until the upper suspension support is free of any spring tension. Do not over-compress the spring.

c. Hold the upper support, then remove the nut on the end of the shock piston rod.

d. Remove the support, coil spring, insulator, and bumper.

16. Inspect the strut as follows:

a. Check the shock absorber by moving the piston shaft through its full range of travel. It should move smoothly and evenly throughout its entire travel without any trace of binding or notching.

b. Use a small straightedge to check the piston shaft for any bending or deformation.

c. Inspect the spring for any sign of deterioration or cracking. The waterproof coating on the coils should be intact to prevent rusting.

To install:

→**Never reuse a self-locking nut. Always replace self-locking nuts and cotter pins as applicable.**

17. Assemble the strut as follows:

a. Loosely assemble all components onto the strut assembly. Be sure the spring end aligns with the hollow in the lower seat.

b. Align the upper suspension support with the piston rod and install the support.

c. Align the suspension support with the strut lower bracket. This assures the spring will be properly seated top and bottom.

d. Compress the spring to expose the strut piston rod threads.

e. Install a new strut piston nut and tighten to the following: Tercel and Paseo to 25 ft. lbs. (34 Nm), Corolla, Celica, Camry and Avalon to 36 ft. lbs. (49 Nm) and Supra to 20 ft. lbs. (27 Nm).

f. Remove the spring compressor. Be sure the paint mark on the upper support faces the outside of the strut.

18. Place the strut on the vehicle and install the nuts to hold the strut to the strut tower. Tighten the nuts to 29 ft. lbs. (39 Nm), except on Supra models. On Supra models, tighten to 19 ft. lbs. (26 Nm).

19. Connect the strut to the axle carrier and install the bolt and nut. Do not tighten at this time.

20. Connect the stabilizer link to the strut.

21. On Camry, connect the load sensing proportioning valve to the control arm. Tighten to 94 ft. lbs. (130 Nm).

22. On Supra, install the brake caliper.

23. Install the wheel and remove the jackstands. Bounce the vehicle up and down to stabilize the suspension.

24. With the vehicle weight on the suspension, tighten the bolt holding the strut to the axle carrier as follows: Tercel and Paseo to 50 ft. lbs. (68 Nm), Corolla to 105 ft. lbs. (142 Nm), Celica, Camry and Avalon to 188 ft. lbs. (255 Nm) and Supra to 106 ft. lbs. (143 Nm).

25. Install the rear seat cushion and any applicable trim.

26. Connect the negative battery cable to the battery.

Upper Ball Joint

REMOVAL & INSTALLATION

The upper ball joint (used only on the Supra) is an integral part of the upper arm and is not replaced separately. The upper ball joint replacement is accomplished by replacing the upper arm, as follows:

Upper Control Arm

REMOVAL AND INSTALLATION

This procedure is applicable only for the Supra.

1. Disconnect the negative battery cable. On vehicles equipped with an air bag, wait at least 90 seconds before proceeding.

2. Raise the front of the vehicle and support it on safety stands.

3. Remove the wheel.

4. Remove the caliper support bracket by removing the two bolts. Leave the brake line connected and suspend it aside.

5. Remove the rotor.

6. Remove the front fender splash shield, fender liner, and wheel opening molding.

7. If removing the left side arm, remove the washer tank.

8. Remove the bolt and disconnect the ABS speed sensor from the steering knuckle. Remove the three bolts and disconnect the wire harness clamp.

9. Remove the cotter pin and the nut from the upper ball joint; press the upper ball joint from the knuckle.

10. Remove the through-bolt, nut, and the upper control arm.

To install:

11. Install the upper control arm. Connect the upper control arm to the sub-frame and install the through-bolt. Do not tighten the bolt at this time.

➡**The upper control arm mounting bolts are not tightened until the suspension has been assembled and vehicle is on the ground.**

12. Install the ball joint to the knuckle and tighten the nut to 76 ft. lbs. (103 Nm). Install a new cotter pin.

13. Connect the wire harness and ABS speed sensor.

14. Install the washer tank, the fender liner, splash shield, and molding.

15. Install the rotor.

16. Install the caliper support bracket and tighten the bolts to 87 ft. lbs. (118 Nm).

17. Install the wheel.

18. Lower the vehicle.

19. Bounce the suspension several times to set the suspension.

20. Support the lower arm and tighten the upper control arm through-bolt and nut to 121 ft. lbs. (164 Nm).

21. Connect the negative battery cable.

22. Have the front wheel alignment checked.

Lower Ball Joint

REMOVAL & INSTALLATION

All Models

➡**On the Supra, the lower ball joint is not replaceable. If the lower ball joint is defective, replace the lower arm and ball joint as an assembly.**

1. Disconnect the negative battery cable. On vehicles equipped with an air bag, wait at least 90 seconds before proceeding.

2. Raise and support the vehicle safely. Remove the front wheels.

3. Remove the cotter pin from the bearing locknut cap, then remove the cap.

4. Depress the brake pedal and loosen the axle nut.

5. Remove the brake caliper attaching hardware, position the caliper aside with the hydraulic line still attached and suspend it with a wire.

6. Remove the ABS speed sensor, if equipped.

7. Remove the brake rotor.

8. Loosen the two nuts holding the strut to the steering knuckle assembly. Do not remove at this time.

9. Remove the cotter pin and nut from the tie rod end. Using a tie rod end removal tool, separate the tie rod end from the steering knuckle.

10. Remove the ball joint bolt and two nuts to disconnect the steering knuckle from the lower arm.

11. Remove the two nuts and bolts holding the strut to the steering knuckle and separate the knuckle from the strut assembly.

12. Remove the axle nut and grasp the hub and knuckle assembly. With a plastic hammer tap the axle shaft to remove knuckle and hub.

➡**Cover the halfshaft boot with a shop rag to protect it from any damage.**

13. Clamp the steering knuckle in a vise and remove the dust deflector. Remove the nut holding the steering knuckle to the ball joint. Press the ball joint out of the steering knuckle.

To install:

14. Reattach the ball joint to the steering knuckle. Tighten the ball joint-to-steering knuckle nut to 72 ft. lbs. (97 Nm) on Tercel and Paseo, 87 ft. lbs. (118 Nm) on Celica and Corolla or 90 ft. lbs. (123 Nm) on Avalon and Camry. Install a new cotter pin. Drive the deflector shield onto the knuckle.

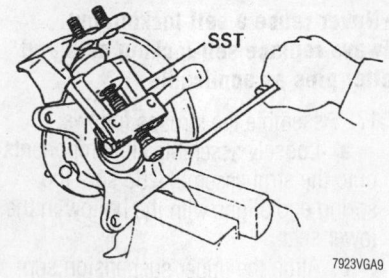

Removing the ball joint from the knuckle

15. Install the knuckle and hub assembly to the axle and temporarily tighten the axle nut.

16. Connect the knuckle assembly to the lower strut bracket. Temporarily insert the mounting bolts from the rear and install the nuts.

17. Connect the lower ball joint to the lower arm and tighten the fasteners to 59 ft. lbs. (79 Nm) on Tercel and Paseo or 94 ft. lbs. (127 Nm) on Celica, Corolla, Avalon and Camry.

18. Connect the tie rod end to the knuckle.

19. Tighten the bolts on the lower side of the strut assembly.

20. If equipped, install the ABS speed sensor.

21. Install the brake disc and the caliper.

22. Tighten the axle nut.

23. Connect the negative battery cable.

24. Have the front end alignment checked.

Wheel Bearings

ADJUSTMENT

Front

All models use a non-adjustable wheel bearing. To determine the condition of the wheel bearing, check the backlash in bearing shaft direction and the axle hub deviation. Maximum for backlash should be as follows:

• Corolla, Supra, Avalon and Camry— 0.0020 in. (0.05mm)

• 1995–97 Celica—0.0031 in. (0.08mm)

Maximum axle hub deviation is:

• Corolla—0.0028 in. (0.07mm)

• 1994–97 Celica—0.0028 in. (0.07mm)

• Supra, Avalon and Camry—0.0020 in. (0.05mm)

If the wheel bearing is out of specifications, replace the wheel bearing.

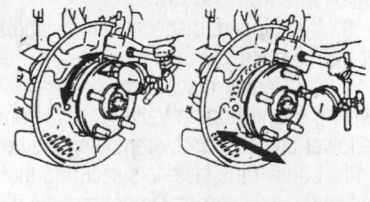

Checking the wheel bearings for deviation and free-play

Rear

PASEO AND TERCEL

1. Disconnect the negative battery cable. On vehicles equipped with an air bag, wait at least 90 seconds before proceeding.
2. Raise and support the vehicle safely.
3. Remove the rear wheels.
4. Remove the locknut cap and cotter pin. Remove the locknut.
5. Install the bearing locknut and tighten it to 22 ft. lbs. (29 Nm) while spinning the drum.
6. Spin the brake drum several times to snug down the bearing, then loosen the bearing locknut until it can be turned by hand.

➡ **There must be absolutely no brake drag at this time.**

7. Retighten the bearing locknut until there is a bearing preload of 0.9–2.2 lbs. (3.2–9.8 N) while turning the wheel. Measure with a spring scale hooked to one of the studs.
8. Install the locknut lock, a new cotter pin, and the cap. If the cotter pin hole does not align properly, align the holes by tightening the nut to the next hole. Do not loosen the nut.
9. Install the rear wheel and lower the vehicle.
10. Connect the negative battery cable.

EXCEPT PASEO AND TERCEL

Check the backlash in bearing shaft direction and the axle hub deviation. Maximum for backlash should be 0.0020 in. (0.05mm). Maximum axle hub deviation is 0.0028 in. (0.07mm), except on Supra, which is 0.020 in. (0.05mm).

➡ **The wheel bearing is non-adjustable. If the wheel bearing is out of specifications, replace the wheel bearing.**

REMOVAL & INSTALLATION

Front

EXCEPT SUPRA

1. Disconnect the negative battery cable from the battery. On vehicles equipped with an air bag, wait at least 90 seconds before proceeding.
2. Raise and support the vehicle safely. Remove the front wheels.
3. Remove the cotter pin from the axle nut cap, then remove the cap.
4. Depress the brake pedal and loosen the axle nut.

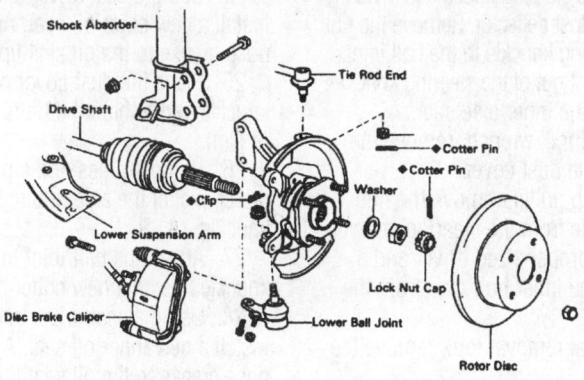

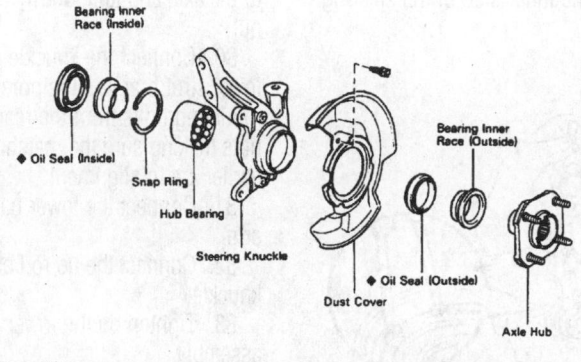

Steering knuckle and hub assembly—except Supra

5. Remove the brake caliper attaching hardware, position the caliper aside with the hydraulic line still attached and suspend it with a wire.
6. Remove the ABS speed sensor, if equipped.
7. Remove the brake rotor.
8. Loosen the nuts on the lower side of the strut assembly. Do not remove at this time.
9. Remove the cotter pin and nut from the tie rod end. Using a tie rod end removal tool, separate the tie rod end from the steering knuckle.
10. Place matchmarks on the strut assembly lower mounting bracket and the camber adjustment cam.
11. Remove the ball joint bolt and two nuts and disconnect the steering knuckle from the lower control arm.
12. Remove the two nuts and bolts on the lower side of the strut mount and separate the knuckle from the strut assembly.
13. Remove the axle nut and grasp the hub and knuckle assembly. With a plastic hammer tap the axle shaft to remove knuckle and hub.

➡ **Cover the halfshaft boot with a shop rag to protect it from any damage.**

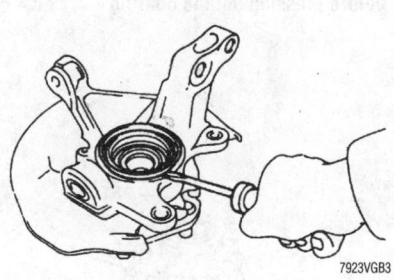

Removing the inner axle seal from the hub assembly

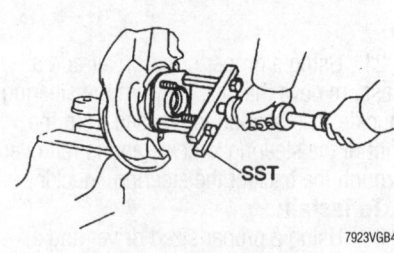

Removing the axle hub from the knuckle

14. Clamp the steering knuckle in a vise and remove the dust deflector. Remove the nut holding the steering knuckle to the ball joint. Press the ball joint out of the steering knuckle.

15. Remove the inner axle seal.

16. Using a Torx® wrench, remove the bolts securing the dust cover.

17. Using hub puller, remove the hub and backing plate from the steering knuckle.

18. Using a proper sized driver and a press, remove the inner hub race from the axle hub.

19. Using seal removal tool, remove the outer axle seal.

20. Using snapring pliers, remove the snapring from the inner side of the steering knuckle.

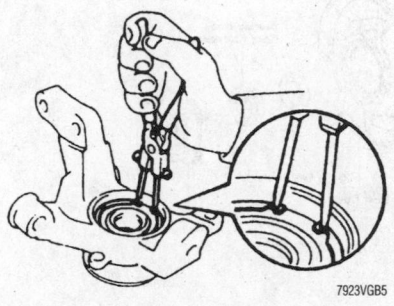

Removing the snapring from the knuckle before pressing out the bearing

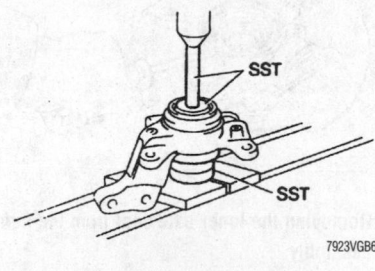

Removing the bearing from the steering knuckle using a press

21. Using a proper sized driver and a press, remove the bearing from the steering knuckle. The bearing is pressed from the front of the steering knuckle and is removed through the back of the steering knuckle.

To install:

22. Using a proper sized driver and a press, install a new bearing to the steering knuckle.

23. Install the snapring to the steering knuckle using snapring pliers.

24. Using a seal driver and a hammer, install a new outer oil seal. Apply multipurpose grease to the oil seal lip.

25. Place the dust cover on the steering knuckle and tighten the bolts to 78 inch lbs. (9 Nm).

26. Using a press and a proper sized driver, install the axle hub to the steering knuckle.

27. Attach the ball joint to the steering knuckle. Install a new cotter pin.

28. Using a seal driver and a hammer, install a new inner oil seal. Apply multipurpose grease to the oil seal lip.

29. Install the knuckle and hub assembly to the axle and temporarily tighten the axle nut.

30. Connect the knuckle assembly to the lower strut bracket. Temporarily insert the mounting bolts from the rear and install the nuts making sure the matchmarks made earlier are in alignment.

31. Connect the lower ball joint to lower arm.

32. Connect the tie rod end to the knuckle.

33. Tighten on the lower side of the strut assembly.

34. If equipped, install the ABS speed sensor.

35. Install the brake disc and the caliper.

36. Tighten the axle nut while someone depresses the brake pedal. Install the adjusting nut cap and insert a new cotter pin.

37. Install the wheels to the vehicle. Verify that the wheel turns freely.

38. Lower the vehicle. Connect the negative battery cable to the battery.

39. Have the wheel alignment checked.

SUPRA

1. Disconnect the negative battery cable. On vehicles equipped with an air bag, wait at least 90 seconds before proceeding.

2. Raise and safely support the vehicle.

3. Remove the front tire and wheel assembly.

4. Remove the brake caliper support bracket, leaving the brake line connected and support it using a piece of wire.

5. Remove the rotor by removing the two screws.

6. Disconnect the ABS speed sensor.

7. Remove the cotter pin and nut and disconnect the tie rod from the steering knuckle.

8. Remove the cotter pin and nut and disconnect the steering knuckle from the upper control arm.

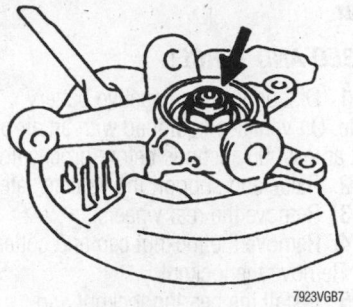

Axle hub locknut location—Supra

9. Remove the clip and nut and press the knuckle off the lower control arm.

10. Remove the steering knuckle from the vehicle.

11. Pry the hub bearing cap from the steering knuckle. Using a hammer and chisel, loosen the staked part of the hub nut and remove it.

12. Remove the ABS sensor rotor.

13. Remove the four bolts and shift the brake dust shield toward the hub (outside).

14. Using a two armed puller, remove the axle hub from the knuckle.

15. With a puller, remove the inner bearing race from the axle hub. Pry out the oil seal.

16. Remove the bearing snapring, then position the inner race above the bearing on the inner side. Press the bearing out.

To install:

17. Press the bearing into the knuckle. If the inner race and balls come loose from the outer race, be sure to install them on the same side as before.

18. Install the snapring and inner race, then tap in a new oil seal until it is flush with the end surface of the knuckle.

19. Install the brake dust cover and tighten the bolts to 74 inch lbs. (9 Nm).

20. Press the hub into the knuckle and install the speed sensor.

21. Install a new locknut and tighten it to 147 ft. lbs. (199 Nm). Stake the nut with a chisel. Tap the bearing cap into place.

22. Connect the knuckle to the upper control arm and tighten the nut to 76 ft. lbs. (103 Nm). Install a new cotter pin.

23. Connect the knuckle to the lower control arm and tighten the nut to 92 ft. lbs. (125 Nm). Install a new clip.

24. Connect the tie rod end to the steering knuckle with the nut.

25. Install the rotor by installing the two screws.

26. Install the caliper support bracket.

27. Connect the speed sensor to the knuckle.

28. Install the front wheel and lower the vehicle.

29. Connect the negative battery cable.

30. Have the front end alignment checked.

Rear

PASEO AND TERCEL

1. Disconnect the negative battery cable. On vehicles equipped with an air bag, wait at least 90 seconds before proceeding.

2. Raise and support the vehicle safely.

3. Remove the rear wheels.

4. Remove the locknut cap and cotter pin. Remove the locknut.

5. Carefully pull off the brake drum along with the outer wheel bearing and thrust washer. Do not drop the bearing.

6. Pry the inner bearing oil seal out of the brake drum assembly, then remove the inner bearing.

7. Drive out the bearing races, as required.

To install:

8. Press the new outer bearing races into the brake drum and add a liberal amount of bearing grease to the inside of the hub and the bearing cap.

9. Clean and pack the bearings with grease.

10. Position the inner bearing into the brake drum, then drive in a new oil seal to the original position. Lightly coat the seal with grease.

11. Position the brake drum onto the axle shaft. Install the outer bearing and position the thrust washer. Install the bearing locknut and tighten it to 22 ft. lbs. (29 Nm) while spinning the drum.

12. Spin the brake drum several times to snug down the bearing, then loosen the bearing locknut until it can be turned by hand.

➡**There must be absolutely no brake drag at this time.**

13. Retighten the bearing locknut until there is a bearing preload of 0.9–2.2 lbs. (3.2–9.8 N) while turning the wheel. Measure with a spring scale hooked to one of the studs.

14. Install the locknut lock, a new cotter pin, and the cap. If the cotter pin hole does not align properly, align the holes by tightening the nut to the next hole. Do not loosen the nut.

15. Install the rear wheel and lower the vehicle.

16. Connect the negative battery cable.

CELICA, COROLLA, CAMRY AND AVALON

1. Disconnect the negative battery cable. On vehicles equipped with an air bag, wait at least 90 seconds before proceeding.

2. Raise and safely support the vehicle.

3. Remove the wheel.

4. Remove the brake drum or rotor.

5. If equipped with ABS brakes, disconnect and remove the ABS wheel speed sensor.

6. Remove the four bolts securing the hub to the knuckle and remove the hub.

7. Remove the O-ring from the backing plate.

To install:

8. Install a new O-ring onto the backing plate. Coat the O-ring with multipurpose grease.

9. Install the hub to the knuckle with the mounting bolts and tighten to 59 ft. lbs. (80 Nm).

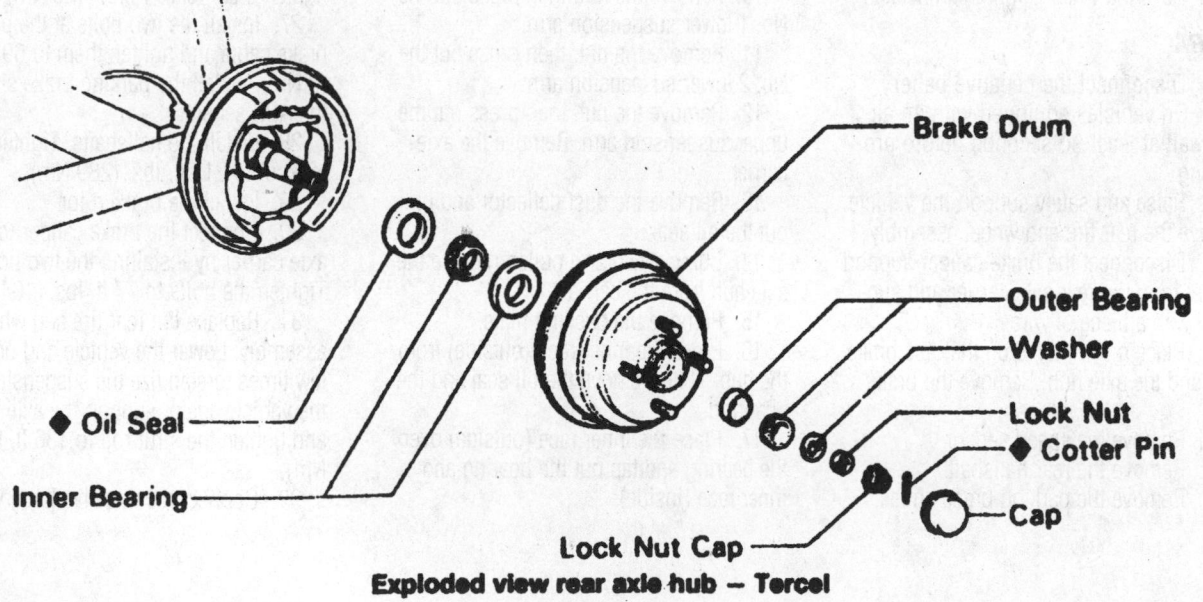

Exploded view rear axle hub — Tercel

7923VGB8

Exploded view of the rear wheel bearing assembly—Tercel shown, Paseo similar

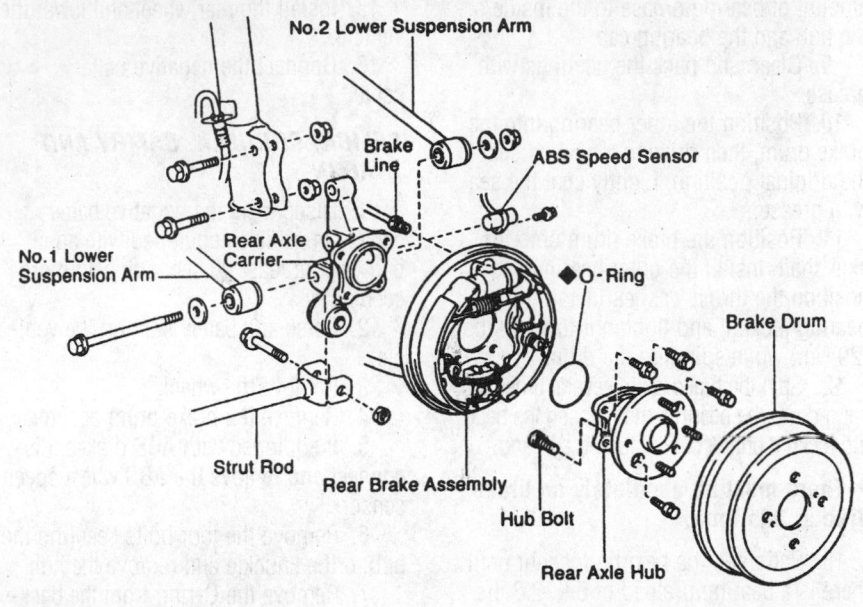

No.2 Lower Suspension Arm

Brake Line

ABS Speed Sensor

No.1 Lower Suspension Arm

Rear Axle Carrier

◆O-Ring

Brake Drum

Strut Rod

Rear Brake Assembly

Hub Bolt

Rear Axle Hub

◆ Non-reusable part

7923VGB9

Exploded view of the hub and wheel bearing assembly—Corolla shown, Celica similar

10. If equipped with ABS brakes, install the ABS wheel speed sensor.

11. Install the brake drum or rotor.

12. Install the wheel and lower the vehicle.

13. Connect the negative battery cable.

14. Have the wheel alignment checked.

SUPRA

1. Disconnect the negative battery cable. On vehicles equipped with an air bag, wait at least 90 seconds before proceeding.

2. Raise and safely support the vehicle. Remove the rear tire and wheel assembly.

3. Disconnect the brake caliper support bracket from the rear axle carrier and support it with a piece of wire.

4. Place matchmarks on the disc brake rotor and the axle hub. Remove the brake rotor.

5. Remove the speed sensor.

6. Remove the rear halfshaft.

7. Remove the parking brake shoes.

8. Remove the two bolts at the parking brake cable. Remove the two hub bolts and the hex bolt. Slide the backing plate to the outside and disconnect the parking brake cable.

9. Disconnect the strut rod at the axle carrier.

10. Remove the nut, then press out the No. 1 lower suspension arm.

11. Remove the nut, then press out the No. 2 lower suspension arm.

12. Remove the nut, then press out the upper suspension arm. Remove the axle carrier.

13. Remove the dust deflector and pull out the oil seal.

14. Using a two arm puller, remove the axle hub from the carrier.

15. Remove the backing plate.

16. Press the inner race (outside) from the hub. Then, remove the oil seal and the snapring.

17. Place the inner race (outside) over the bearing and tap out the bearing and inner race (inside).

To install:

18. Install the bearing to the axle carrier.

➡ **If the inner races come loose from the bearing outer race, be sure to install them on the same side as before.**

19. Install the snapring, the inner race (outside), and a new oil seal.

20. Install the backing plate. Install the inner race (inside) and press in the axle hub with the proper tools.

21. Install the inner oil seal. Align the holes for the speed sensor in the dust deflector and axle carrier. Install the dust deflector.

22. Install the upper arm to the axle carrier. Tighten the nut and bolt to 80 ft. lbs. (109 Nm).

23. Connect the No. 2 lower arm to the carrier and tighten a new nut to 110 ft. lbs. (150 Nm).

24. Connect the No. 1 lower arm to the carrier and tighten a new nut to 43 ft. lbs. (59 Nm).

25. Connect the strut rod to the carrier. Do not tighten the bolt at this time.

26. Connect the parking brake cable and slide the backing plate to the inside. Install the hex bolt and tighten it to 132 ft. lbs. (180 Nm). Install the two hub bolts and tighten them to 19 ft. lbs. (26 Nm).

27. Install the two bolts at the parking brake cable and tighten them to 69 inch lbs. (8 Nm). Install the parking brake shoes and the ABS sensor.

28. Install the halfshafts. Tighten the locknut to 213 ft. lbs. (289 Nm).

29. Install the brake rotor.

30. Connect the brake caliper to the rear axle carrier by installing the two bolts. Tighten the bolts to 77 ft. lbs. (104 Nm).

31. Replace the rear tire and wheel assembly. Lower the vehicle and bounce it a few times to stabilize the suspension. Raise the vehicle again, support the axle carrier and tighten the strut rod to 136 ft. lbs. (184 Nm).

32. Connect the negative battery cable.

VOLKSWAGEN

Cabrio • Golf • GTI • Jetta • Passat

30

GASOLINE ENGINE REPAIR

➡ **Disconnecting the negative battery cable on some vehicles may interfere with the functions of the on board computer systems and may require the computer to undergo a relearning process, once the negative battery cable is reconnected.**

Distributor

REMOVAL

1. Disconnect the coil high tension wire and the connector plug at the distributor. Disconnect vacuum lines, if equipped.
2. Unsnap the cap retainer clips, and remove the cap and static shield as a unit.
3. At the front crankshaft pulley bolt, turn the engine to Top Dead Center (TDC) on No. 1 piston. Make a chalk or paint mark where the rotor points to the rim of the distributor; some vehicles already have a notch there. Also matchmark the distributor to the engine block or head.
4. Remove the bolt and distributor clamp and lift the distributor straight out.

INSTALLATION

Timing Not Disturbed

1. On some vehicles, the distributor engages its drive with an offset slot and is easy to reinstall in the reverse order of removal, even if the crankshaft or camshaft has been turned. Gently rotate the rotor while pushing the distributor into place. Install the hold-down bolt and adjust the ignition timing.
2. On engines with the drive gear on the distributor, be sure the engine is still at TDC and insert the distributor with the matchmarks aligned.
3. Install the hold-down clamp and bolt, connector plug, cap and static shield, and high tension wires.
4. Check and adjust the ignition timing.

Timing Disturbed

1. Rotate the crankshaft to TDC of No. 1 piston.
2. With a suitable tool, turn the oil pump drive so it is parallel with the crankshaft.
3. Install the rotor onto the distributor and align it with the No. 1 mark on the rim of the body.

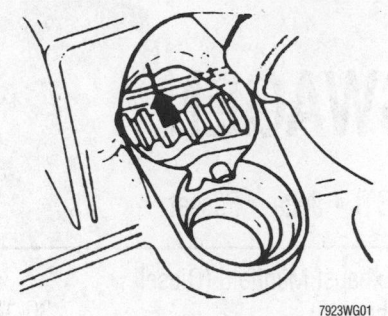

Timing mark locations on the bellhousing

4. Install the distributor, making sure the rotor still aligns with the mark when the distributor is all the way in.
5. With the distributor installed, install the hold-down clamp and bolt, connector plug, cap and static shield, and high tension wires.
6. Check and adjust the ignition timing.

Ignition Timing

ADJUSTMENT

➡ **The ignition timing is controlled by the engine control module and is not adjustable. However the timing can be monitored on a scan tool connected to the data link connector in the vehicle. No specification has been given by the manufacturer.**

Engine Assembly

REMOVAL & INSTALLATION

1. The engine and transaxle are lifted from the vehicle as an assembly.
Disconnect the battery cables and remove the battery.
2. Open the fuel filler cap to relieve tank pressure, then loosen the fuel filter fitting to relieve system pressure. Be sure to take the appropriate fire safety precautions.
3. Remove the air filter and disconnect the accelerator cable from the injection pump.
4. Remove the radiator cap. Turn the heater temperature control all the way towards warm and remove the thermostat housing to drain the coolant.
5. Remove the upper radiator hose and disconnect the wiring from the radiator fan motor and switches. Remove the mounting nuts or bolts and lift out the radiator and fan shroud as an assembly.

6. Begin disconnecting electrical connections and vacuum lines, carefully labeling each one. Don't forget ground connections that are screwed to the body.
7. If equipped with power steering, remove pump and secure it to the body. Do not disconnect the hydraulic lines. If equipped with air conditioning, remove the compressor and secure it aside without disconnecting the lines.
8. Disconnect the fuel inlet and outlet lines from the injection pump and plug the holes to keep the pump clean. Note the outlet fitting has a special orifice.
9. On turbocharged engines, disconnect the exhaust pipe and the oil lines from the turbocharger and cap the oil line fittings on the turbocharger. Unbolt the turbocharger and lift it out of the engine.
10. If equipped with an automatic transaxle, place the selector lever in **P** and disconnect the selector cable at the transaxle.
11. On manual transaxle shift linkage, remove the 2 rods with the plastic socket ends and unbolt the remaining linkage from the case as required. Disconnect the clutch cable, lift it out of the case and set it aside.
12. Disconnect the wiring from the starter, the back-up light switch and the ground cable from the transaxle. Remove the speedometer cable from the transaxle and plug the hole in the case.
13. Attach an engine sling tool VW-2024A or equivalent, to the engine and attach the sling to a suitable lifting device.
14. Remove the nuts or spring clamps holding the exhaust pipe to the manifold or turbocharger.

✵✵ CAUTION

On some models, special tools are required for removing and installing the exhaust pipe-to-manifold spring clamps; VW3140/1 and /2 or equivalent. This is a set of different sized wedges for spreading the spring clamps in steps. The installed spring clamp has considerable tension and could cause damage or injury if not properly removed. Clamps with wedges installed are also under high tension and should be handled carefully.

15. Unbolt the halfshafts from the flanges and hang them from the body with wire.
16. Be sure everything is disconnected and unbolt the mounts. Remove the starter first and the front mount with it.

17. With all mounts unbolted, slightly lower the engine/transaxle assembly and tilt it towards the transaxle side. Then, carefully lift the assembly out of the vehicle.

To install:

18. Carefully install the engine/transaxle assembly and be sure all mounts are securely bolted to the engine/transaxle. Start all nuts and bolts that secure the mounts to the body but don't tighten them yet.

19. With all mounts installed and the engine safely in the vehicle, allow some slack in the lifting equipment. With the vehicle safely supported, shake the engine/transaxle as a unit to settle it in the mounts. Torque all mounting bolts, starting at the rear and working forward. Torque to 33 ft. lbs. (41 Nm) for 10mm bolts or 54 ft. lbs. (73 Nm) for 12mm bolts.

20. Install the starter and torque the bolts to 33 ft. lbs. (45 Nm).

21. Connect the halfshafts to the flanges and torque the bolts to 33 ft. lbs. (45 Nm).

22. Install the exhaust pipe and use new self-locking nuts to secure the flange. Torque the nuts to 30 ft. lbs. (40 Nm). If equipped with spring clamps, the clamps can be used again.

23. Connect the shift linkage and the clutch cable, if equipped. Make any necessary adjustments.

24. Install the fuel injector lines and torque to 18 ft. lbs. (25 Nm). Be careful not to over torque the line nuts. If a line is damaged or clogged, replace all lines as a set.

25. Connect the inlet and outlet lines to the injector pump. Note the special outlet fitting has the word "OUT" printed on the top. Use new gaskets.

26. Install the air conditioning compressor and/or power steering pump, if equipped. Install and adjust the drive belts.

27. Connect the wiring and vacuum hoses.

28. Install the radiator, fan and heater hoses. Use a new O-ring on the thermostat and torque the thermostat housing bolts to 7 ft. lbs. (10 Nm).

29. Fill and bleed the cooling system. Check the adjustment of the accelerator cable.

Water Pump

REMOVAL & INSTALLATION

Except 2.8L Engine

1. To drain the cooling system, remove the thermostat housing from under the water pump housing.

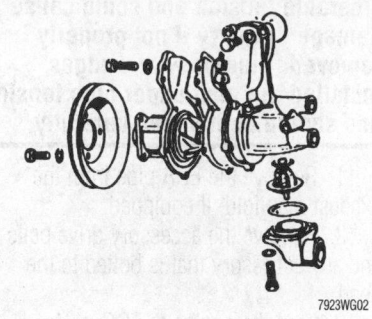

Water pump and thermostat housing— except 2.8L engine

2. Raise and safely support the vehicle. Loosen but don't remove the bolts holding the pulley to the water pump.

3. Remove the timing belt cover.

4. Loosen the alternator and/or steering pump as required to remove the water pump drive belt.

5. Remove the water pump pulley. On some vehicles, the crankshaft pulley must also be removed by removing the bolts holding the pulley to the timing belt sprocket.

6. All the bolts are now accessible and the water pump can be removed from its housing.

To install:

7. Be sure to clean the housing before installing the new gasket. Install the pump into the housing and torque the pump-to-housing bolts to 7 ft. lbs. (10 Nm).

8. Install the water pump drive pulley and torque the bolts to 15 ft. lbs. (20 Nm). If the crankshaft drive pulley was removed, install it and torque the bolts to 15 ft. lbs. (20 Nm).

9. Adjust drive belt tension and install the thermostat and housing. Torque the bolts to 7 ft. lbs. (10 Nm).

2.8L Engine

1. Obtain the security code for the radio.

2. Disconnect the negative battery cable.

3. Disconnect the front exhaust pipe from the catalytic converter.

❊❊ CAUTION

Never open, service or drain the radiator or cooling system when hot; serious burns can occur from the steam and hot coolant.

4. Drain the engine coolant.

5. Remove the accessory drive belt.

6. Remove the air intake duct.

7. Disconnect the ignition wires from the coils and unclip them from the retainers.

8. Remove the ignition wire guide above coil assembly.

9. Disconnect the vacuum hose from the fuel pressure regulator.

10. Remove the Intake Air Temperature (IAT) sensor from the upper intake manifold.

11. Without disconnecting the hoses, remove and place the coolant expansion tank to the side.

12. Install an engine support fixture to the lifting eyes on the left and right sides of the cylinder head. Lift the engine slightly to remove the weight from the mounts.

13. Remove the right and left rear engine/transaxle mount center bolts.

14. Remove the front engine mounting center bolts.

15. Carefully raise the engine to gain access to the water pump pulley mounting bolts.

16. Remove the water pump pulley using wrench VAG 1590 or equivalent. Modify the wrench as shown in necessary to fit the bolt.

17. Remove the mounting bolts and the water pump.

To install:

18. Using a new O-ring, install the water pump. Tighten the mounting bolts to 15 ft. lbs. (20 Nm).

19. Install the water pump pulley. Tighten the bolt to 18 ft. lbs. (25 Nm).

20. Lower the engine and install the engine/transaxle mount bolts. Tighten the mounting bolts to 44 ft. lbs. (60 Nm). Tighten the front and right rear mounts first, then the left rear mount.

21. Install the expansion tank. Tighten the bolts to 7 ft. lbs. (10 Nm).

22. Install the IAT sensor in the upper intake manifold.

23. Connect the vacuum hose to the fuel pressure regulator.

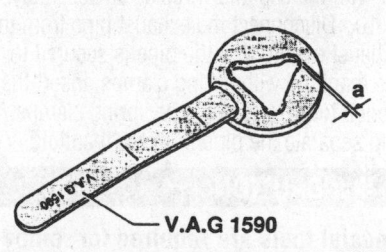

Modify wrench VAG 1590 or equivalent to fit the water pump pulley bolt as needed

24. Install the ignition wires and the wire guide.

25. Install the air duct and accessory drive belt.

26. Refill the engine with coolant.

27. Start the engine and check for leaks.

28. Recheck the coolant level after the engine has cooled and add if necessary.

Cylinder Head

REMOVAL & INSTALLATION

Except 2.8L Engine

1. Disconnect the negative battery cable.

➡**On some of the 16V models, removing the battery may make the job easier.**

2. Open the radiator cap and remove the thermostat housing to drain the cooling system.

3. Disconnect the throttle cable. Label and disconnect all wiring and vacuum lines from the intake manifold. On the 16V engine, remove the upper half of the intake manifold.

4. On vehicles with CIS-E fuel injection, remove the injectors and the cold start valve without disconnecting the fuel lines and cap them. Secure all the lines aside.

5. On vehicles with Digifant fuel injection, the injectors and fuel rail assembly may be left on the head. Disconnect the fuel supply and return lines and the wiring connector for the injectors.

6. Disconnect the radiator and heater hoses.

7. Disconnect and label wiring for oil pressure and temperature sensors.

8. On vehicles with CIS-E fuel injection, remove the auxiliary air regulator from the intake manifold, if equipped.

9. Remove the distributor cap and wires. On 16V engines, remove the distributor with the cap and wires as an assembly.

10. Disconnect the exhaust pipe from the exhaust manifold. If the pipe is secured to the manifold with spring clamps, insert the wedge tools to remove the spring clamps and separate the pipe from the manifold.

✳✳ CAUTION

Special tools are required for removing and installing the clamps; VW3140/1 and /2 or equivalent. This is a set of different sized wedges for spreading the spring clamps in steps. The installed spring clamp has con-

siderable tension and could cause **damage or injury if not properly removed. Clamps with wedges installed are also under high tension and should be handled carefully.**

11. Remove the EGR pipe from the exhaust manifold, if equipped.

12. Remove the accessory drive belts and any accessory that is bolted to the head.

13. Turn the engine to TDC of No. 1 cylinder, if possible, and remove the cylinder head cover, timing belt cover and belt.

14. Loosen the cylinder head bolts in the reverse of the tightening sequence.

15. Remove the bolts and lift the head straight off.

To install:

16. Before reinstalling the head, check the flatness of the head and block in both width and length, then diagonally from each corner.

17. Install the new cylinder head gasket with the word TOP or OBEN facing upward; do not use any sealing compound.

18. Carefully fit the head in place and install the bolts in positions 8 and 10 in the torque sequence. These holes are smaller and will properly locate the gasket and cylinder head.

19. Install the remaining bolts. Torque the bolts in sequence in 3 steps: 29 ft. lbs. (39 Nm), 44 ft. lbs. (60 Nm) and an additional ½ turn. Two ¼ turns are allowed.

20. Install the camshaft drive belt and adjust the tension.

21. Connect the exhaust pipe to the manifold. Use new gaskets and self-locking nuts and torque to 18 ft. lbs. (25 Nm). On vehicles that use spring clamps, install the clamps and carefully remove the wedge tools.

22. Connect the EGR pipe, if equipped.

23. On 16V engines, install the distributor. Install the distributor cap and wires.

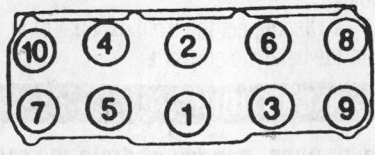

Tighten the cylinder head bolts in the sequence shown—except 2.8L engine

24. On vehicles with CIS-E fuel injection, install the auxiliary air regulator to the intake manifold, if equipped.

25. Connect wiring to the oil pressure and temperature sensors.

26. Install the ignition system components.

27. Connect the radiator and heater hoses.

28. Connect the throttle cable and all wiring and vacuum lines.

29. On vehicles with Digifant fuel injection, connect the fuel supply and return lines. Connect the wiring connector for the injectors.

30. Install the thermostat with a new O-ring. Torque the housing bolts to 7 ft. lbs. (10 Nm). Refill the cooling system.

31. On vehicles with CIS-E fuel injection, install the injectors and the cold start valve.

32. Install the accessory drive belts and adjust the tension.

33. On the 16V engine, install the upper half of the intake manifold. Torque the manifold retaining bolts to 18 ft. lbs. (25 Nm).

34. Connect all wiring and vacuum lines disconnected from the intake manifold. Connect the throttle cable.

35. Install the battery, if removed. Connect the negative and positive battery cables.

36. Refill and bleed the cooling system.

37. When everything has been properly installed and connected, be sure to change the oil and filter before starting the engine.

2.8L Engine

This procedure requires special tool 3268 or equivalent. This is a setting tool that holds the camshafts in the correct position for installing the timing chains. Before removing the cylinder head, be sure new bolts are available. The cylinder head bolts are made to stretch and cannot be used again.

1. Disconnect the battery cables and remove the battery.

2. Disconnect the wiring and vacuum lines as required to remove the air cleaner, air mass sensor and duct.

3. Open the radiator cap and remove the drain plug from the coolant pipe below the intake manifold to drain the cooling system.

4. Remove the engine trim cover. Remove the distributor cap, ignition wires and wire guide as an assembly.

5. Disconnect the throttle cable. Label and disconnect the wiring and vacuum lines from the intake manifold and remove the upper manifold.

6. The injectors and fuel rail assembly may be left on the manifold. Disconnect the fuel supply and return lines and the wiring connector for the injectors.

7. Disconnect the radiator and heater hoses.

8. Thread a long 8 **x** 10mm bolt into the accessory drive belt tensioner to release the tension. Move the tensioner only as required to remove the belt.

9. Remove the alternator and the belt tensioner.

10. Remove the heatshield and the bolts to disconnect the 2 piece exhaust manifold from the engine. Note the position of the gaskets.

11. Remove the distributor and the timing chain tensioner bolt from the timing chain cover.

12. Remove the cylinder head cover, upper timing chain cover and the retaining plate.

13. If possible, rotate the crankshaft to TDC of No. 1 piston. Clean the oil off the chain and sprockets and mark the direction of rotation for assembly.

14. Hold the camshafts at the flats with a 24mm wrench and remove the bolts to remove the sprockets and chain. Note the position of the distributor drive on the short camshaft.

➡**Do not use the setting tool to hold the camshafts when removing or installing the sprocket bolts. The camshafts and the tool will be damaged.**

15. Carefully check to be sure all necessary wires, hoses and brackets and components have been removed.

16. Loosen the cylinder head bolts in the reverse of the torque sequence. Remove and discard the bolts.

17. Remove the cylinder head.

To install:

18. Carefully clean the old gasket material from the head and the block. Before reinstalling the head, check the flatness of the head and block in both width and length, then diagonally from each corner. Maximum allowable distortion is 0.004 in. (0.1mm).

19. If the new head gasket already has sealant in the small holes at the timing chain end, remove the sealant. Apply a silicone sealer to the timing chain end and install the gasket onto the block with the word TOP or OBEN facing up.

20. Fit the cylinder head over the locating dowels and set the head onto the engine. Install new bolts and hand-tighten them. Do not attempt to re-use the old bolts.

21. Torque the bolts in sequence as described in the following sub-steps:
   a. 30 ft. lbs. (40 Nm)
   b. 44 ft. lbs. (60 Nm)
   c. Tighten an additional ¼ turn.
   d. Tighten an additional ¼ turn.

22. Be sure the crankshaft is at TDC on No. 1 piston. Install the setting tool to lock the camshafts in place, then install the timing chain and sprockets. Be sure they are positioned to rotate in the original direction.

23. Hold the camshaft with a 24mm wrench and install the sprocket bolt. Be sure the distributor drive is correctly positioned and torque the bolts to 74 ft. lbs. (100 Nm).

24. Install the tensioner shoe and temporarily install the upper timing chain cover. Install the tensioner bolt and remove the setting tool. Rotate the crankshaft 4 full turns and stop at TDC of No. 1 piston. The setting tool should fit into the camshafts.

25. Remove the tensioner bolt and upper timing chain cover again. Apply new sealant as required, install the cover and torque the bolts to 82 inch lbs. (10 Nm). Install the tensioner bolt and torque to 15 ft. lbs. (20 Nm).

26. Install the cylinder head cover.

27. Use new gaskets and install the intake and exhaust manifolds. Torque the nuts and bolts to 18 ft. lbs. (25 Nm).

28. Install the alternator belt and adjust tension.

29. Install the accessory drive belt and adjust tension.

30. Connect the radiator and heater hoses.

31. Install the injectors, fuel rail assembly and manifold. Connect the fuel supply and return lines. Connect the wiring connector for the injectors.

32. Install the upper manifold. Torque the manifold bolts to 18 ft. lbs. (25 Nm).

33. Connect the wiring and vacuum lines disconnected from the intake manifold. Connect the throttle cable.

34. Install the distributor cap, ignition wires and wire guide as an assembly. Install the engine trim cover.

35. Disconnect the battery cables and remove the battery.

36. Refill and bleed the cooling system.

37. When everything has been properly installed and connected, be sure to change the oil and filter before starting the engine.

Turbocharger

REMOVAL & INSTALLATION

1998–99 1.8L Engine

1. Disconnect the negative battery cable.

2. Remove the engine undercover, and unbolt the A/C compressor.

3. Unbolt the turbocharger support bracket.

4. Disconnect the oil return line at the turbocharger.

5. Remove the air hoses from the turbocharger.

6. Disconnect the oil feed line at the turbocharger.

7. Disconnect the hose for the boost pressure regulation valve vacuum diaphragm.

8. Unbolt the bracket for the coolant supply line at the boost pressure regulation valve vacuum diaphragm.

9. Using Clamp 3094 or equivalent, pinch off the coolant supply hose.

10. Remove the intake air duct between the cowl and the air cleaner housing.

11. Remove the air cleaner housing cover.

12. Label and detach the following lines and electrical connectors:
   • Wastegate bypass regulator valve
   • Evaporative Emission (EVAP) canister purge regulator valve
   • Power outage stage

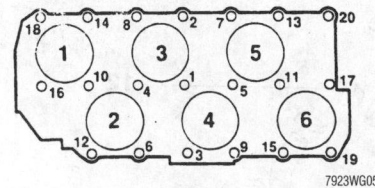

7923WG05

Be sure to tighten the cylinder head bolts in the sequence shown—2.8L engine

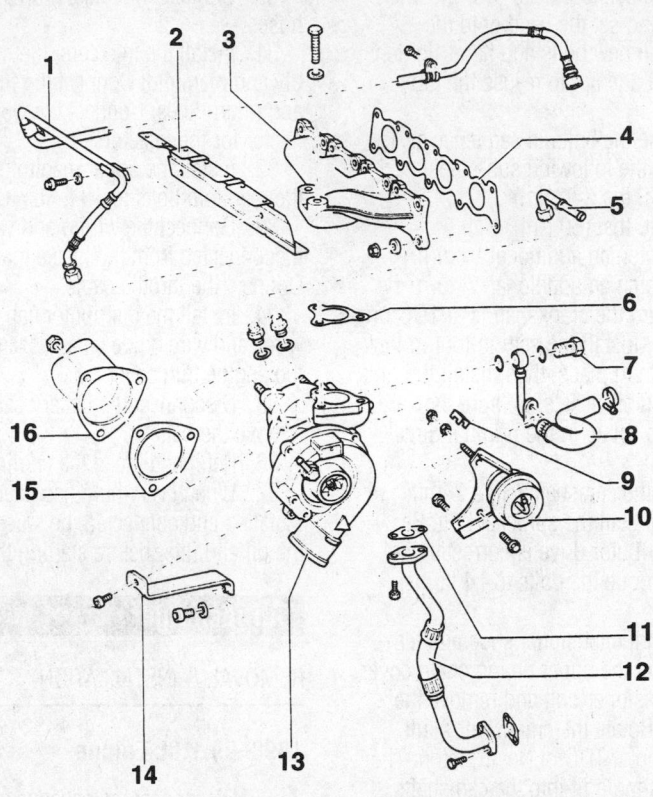

1. Oil supply line
2. Heat shield
3. Exhaust manifold
4. Exhaust manifold gasket
5. Coolant return line
6. Exhaust manifold-to-turbo gasket
7. Banjo bolt
8. Coolant supply hose
9. Fuse
10. Vacuum diaphragm for the wastegate
11. Gasket
12. Oil return line
13. Turbocharger
14. Support bracket
15. Gasket
16. Three Way Catalytic Converter (TWC)

7923WG07

Exploded view of the turbocharger and related components—1.8L engine

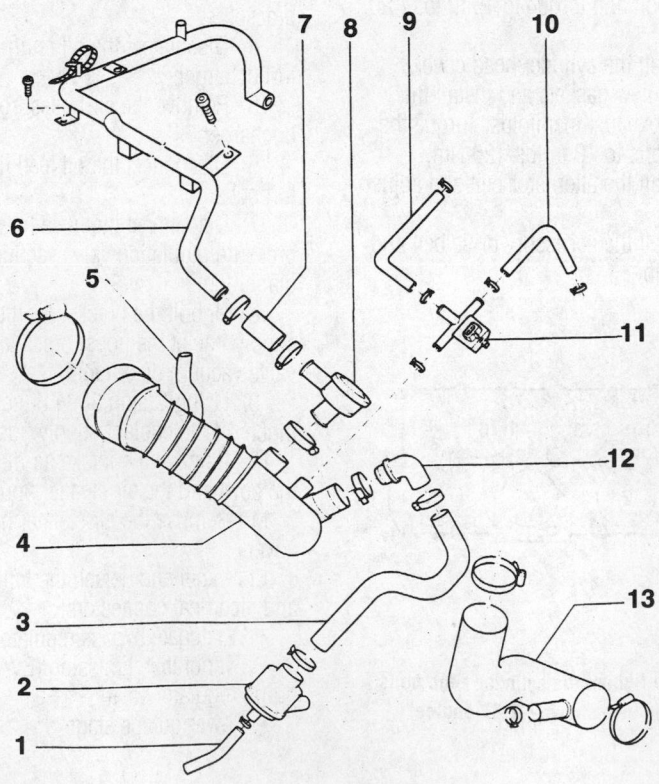

1. Vacuum hose
2. Boost pressure recirculation valve
3. Hose
4. Intake air duct
5. EVAP hose
6. Crankcase ventilation hose
7. Crankcase ventilation hose
8. PCV valve
9. Hose
10. Wastegate vacuum hose
11. Wastegate bypass regulator valve
12. Elbow
13. Hose to the turbocharger

7923WG08

Exploded view of the hoses related to the turbocharger—1.8L engine

- Mass Air Flow (MAF) sensor

13. Remove the air cleaner housing and the engine cover.

14. Disconnect the crankcase breather hose at the valve cover.

15. Disconnect the oil supply line at the turbocharger.

16. Remove the heat shield, and sleeve from the coolant return hose.

17. Using Clamp 3094 or equivalent, pinch off the coolant return hose, then remove.

➡**The exhaust flexpipe may be damaged if bent more than 10°.**

18. Disconnect the Three Way Catalytic Converter (TWC) from the turbo.

19. Unbolt the turbo from the exhaust manifold.

20. Move the turbocharger aside to disconnect the coolant supply banjo fitting.

21. Remove the turbocharger.

To install:

22. Install the turbocharger.

23. Connect the coolant supply banjo fitting and tighten to 18 ft. lbs. (25 Nm).

24. Using new gaskets, install the turbocharger to the exhaust manifold , coat the bolts with Hot Bolt Paste G 052 112 A3 (or equivalent), then tighten the mounting bolts to 26 ft. lbs. (35 Nm). Tighten the turbo support bracket mounting bolts to 33 ft. lbs. (45 Nm).

25. Attach the Three Way Catalytic Converter (TWC) to the turbo.

26. Connect the coolant supply hose.

27. Install the sleeve and heat shield to the return hose.

28. Connect the oil return hose.

29. Add oil to the turbo through the oil feed line.

30. Connect the oil supply line to the turbo and tighten to 18 ft. lbs. (25 Nm).

31. Connect the crankcase breather, and install the engine cover and air cleaner housing.

32. Attach the following lines and electrical connectors:

- Mass Air Flow (MAF) sensor
- Power outage stage
- Wastegate bypass regulator valve

33. Connect the hoses and brackets for the boost pressure regulation valve vacuum diaphragm.

34. Connect the air hoses to the air cleaner assembly and the turbo.

35. Install the A/C compressor and engine undercovers.

36. Refill the coolant system and check the oil level.

37. Connect the negative battery cable.

38. Start the vehicle and check for leaks, then let the engine idle for approx. 1 minute without increasing the engine speed. This ensures adequate oil supply to the turbo.

Intake Manifold

REMOVAL & INSTALLATION

1. Disconnect the negative battery cable.

Remove the air duct from the throttle valve body and disconnect the accelerator cable.

2. On Digifant fuel injection systems, remove the idle stabilizer valve, fuel pump pressure switch and the fuel injector wiring harness. Disconnect the fuel supply and return lines.

3. On CIS-E fuel injection systems, remove the auxiliary air regulator. Remove the fuel injectors and the cold start valve from the cylinder head without disconnecting the fuel lines.

4. Label and disconnect the vacuum hoses as required.

5. Label and disconnect any remaining wiring as required.

6. If equipped, disconnect the EGR pipe.

7. On 16V and VR6 engines, remove the bolts to remove the upper intake manifold.

8. Remove the bolts and remove the manifold from the cylinder head.

To install:

9. On 16V and VR6 engines, install the lower intake manifold to the cylinder head with a new gasket. Torque the bolts to 18 ft. lbs. (25 Nm).

10. On the VR6 engine, if the fuel injectors were removed, examine the injector O-rings and replace as required. Install the injectors and rail.

11. Use new gaskets and fit the manifold or the upper manifold in place. Torque the bolts to 18 ft. lbs. (25 Nm).

12. Connect fuel system hoses or install the injectors now to protect the system.

13. Connect all vacuum hoses and wiring.

14. If equipped, connect the EGR pipe.

15. Connect and adjust the throttle cable as required.

16. Install the remaining components

and run the engine to check idle speed and ignition timing.

Exhaust Manifold

REMOVAL & INSTALLATION

✳✳ CAUTION

On some models, special tools are required for removing and installing the exhaust pipe-to-manifold spring clamps. Special tools VW3140/1 and /2, or equivalent, are a set of different sized wedges for spreading the spring clamps in steps. The installed spring clamp has considerable tension and could cause damage or injury if not properly removed. Clamps with wedges installed are under high spring pressure and should be handled carefully.

1. Disconnect the oxygen sensor wiring and remove any heatshields that may be in the way.

2. Remove the emissions sample tap, if equipped, disconnect the EGR pipe from the exhaust manifold.

3. On models with manifold studs, remove the self-locking nuts and lower the exhaust pipe.

4. Expand the spring clamp by pushing the exhaust pipe to one side and insert the starter wedge into the clamp all the way up to the shoulder.

5. Push the pipe to the other side and install another wedge in the opposite clamp. Continue to work the pipe side to side while pushing the wedges into the clamps until the clamps are spread far enough to lift off easily.

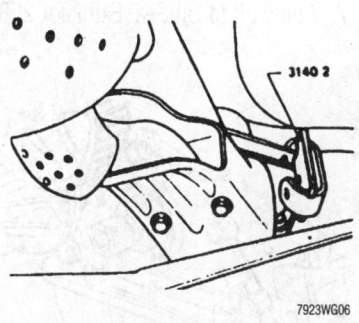

Exhaust pipe clamp removal tool

⁕⁕ CAUTION

The removed spring clamps with wedges in them are under spring tension, if miss-handled, could fly apart with enough force to cause serious injury. Store the removed clamps in a safe area where they won't be disturbed.

6. Remove the self-locking nuts and remove the manifold. On the VR6 engine, the exhaust manifold is two sections. Note the position of the gaskets.

To install:

7. Installation is the reverse of removal. Use new gaskets and self-locking nuts and torque to 18 ft. lbs. (25 Nm).

8. If the exhaust pipe is bolted to the manifold, install a new gasket and use new self-locking nuts. Torque the nuts to 30 ft. lbs. (40 Nm).

9. If equipped with spring clamps, hold the pipe in position with a new gasket and install the clamps. Carefully remove the wedge tools.

Front Crankshaft Seal

REMOVAL & INSTALLATION

1. Disconnect the negative battery cable.
2. Remove the accessory drive belts.
3. Remove the timing belt.
4. Hold the crankshaft sprocket with tool 3099 and remove the center bolt and sprocket.
5. Unscrew the inner part of Oil Seal Extractor 2085 or equivalent out of the outer part about two turns.
6. Install the socket head bolt into the crankshaft to guide the tool.
7. Apply oil to Oil Seal Extractor 2085

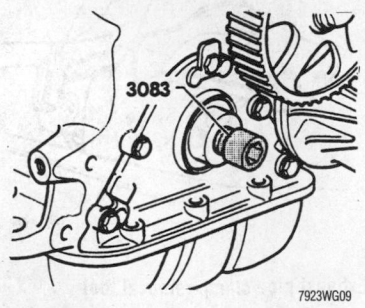

Install the socket head bolt (3083) in the crankshaft to guide the Seal Extractor—1.8L and 2.0L engines

or equivalent. Apply firm pressure and screw the tool into the oil seal as far as it will go.

8. Loosen the knurled screw and turn the inner part against the crankshaft until the seal is removed.

9. Remove the socket head bolt.

To install:

10. Install the guide sleeve from tool 3083 on the end of the crankshaft.

11. Apply oil to the lip of the seal and slide the seal over the guide sleeve.

12. Press the seal in using the thrust sleeve from tool 3083 and a socket head bolt.

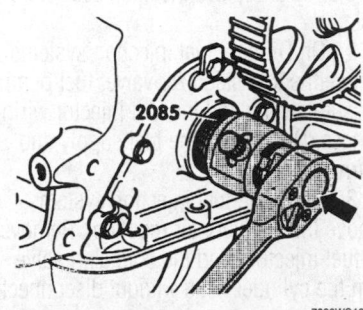

Screw the Seal Extractor (2085) into the oil seal while applying pressure—1.8L and 2.0L engines

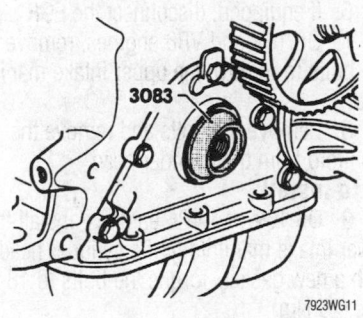

Install the guide sleeve (3083) on the crankshaft—1.8L and 2.0L engines

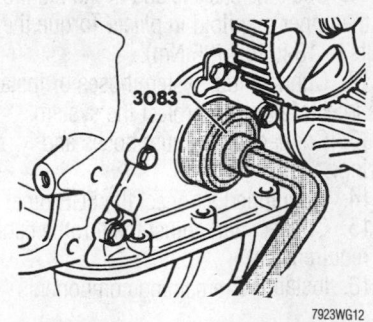

Tighten the socket head bolt to press the seal into place—1.8L and 2.0L engines

13. Remove the tools and install the crankshaft sprocket. Tighten the bolt to 66 ft. lbs. (90 Nm) plus ¼ turn.

14. Install the timing belt.

15. Install the accessory drive belt and remaining components.

16. Connect the negative battery cable.

Camshaft and Valve Lifters

REMOVAL & INSTALLATION

1.8L (SOHC) and 2.0L Engines

1. Disconnect the negative battery cable. Remove the timing belt cover(s), the timing belt, camshaft sprocket and cylinder head cover.

2. Number the bearing caps from front to back. If the cap does not already have one, scribe an arrow pointing towards the front of the engine. The caps are offset and must be installed correctly. Factory numbers on the caps are not always on the same side.

3. Remove the front and rear bearing caps. Loosen the remaining bearing cap nuts diagonally, in several steps, starting from the outside caps near the ends of the head and working toward the center.

4. Remove the bearing caps and the camshaft.

5. If required, remove the lifters from the valves. Keep them in order so they can be installed in their original positions. Place them in a bath of oil or place them upside down to prevent air from entering them.

To install:

6. Install a new oil seal and end plug in the cylinder head. Lubricate the camshaft bearing journals and lobes and set the camshaft in place.

7. Install the bearing caps in the correct position with the arrow pointing towards the front of the engine. Tighten the cap nuts diagonally and in several steps until they are torqued to 15 ft. lbs. (20 Nm). Do not over-torque.

8. Install the drive sprocket and torque the bolt to 58 ft. lbs. (80 Nm).

9. Align the timing marks, install the timing belt and adjust the tension.

10. On engines with hydraulic lifters, wait at least ½ hour after installing the camshaft before starting the engine to allow the lifters to leak down. Observe the following values:

- Camshaft shaft end-play—0.006 in. (0.15mm)
- Bearing cap bolts—15 ft. lbs. (20 Nm)
- Camshaft sprocket bolt—58 ft. lbs. (80 Nm)

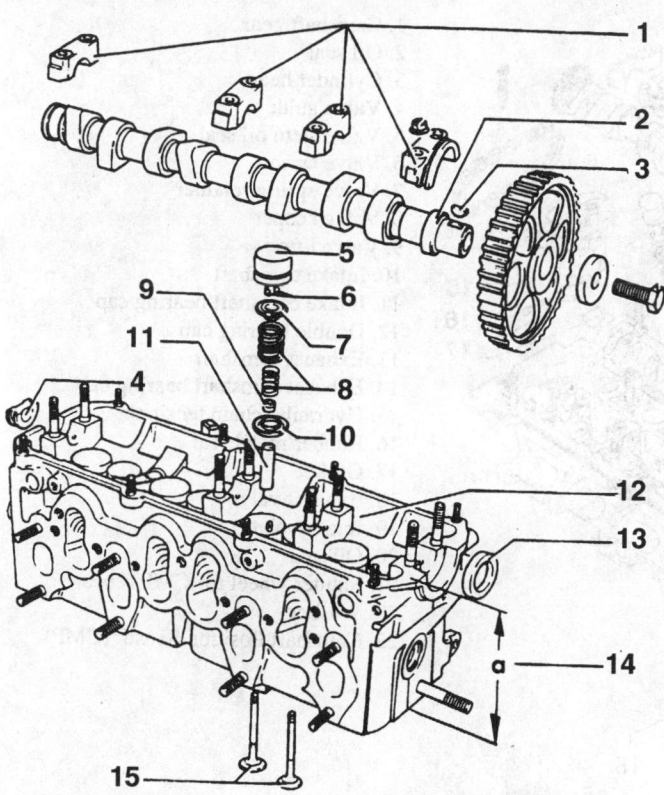

1. Bearing cap
2. Camshaft
3. Woodruff key
4. End cap
5. Valve lifter
6. Valve keeper
7. Upper spring seat
8. Valve spring
9. Valve stem oil seal
10. Lower spring seat
11. Valve guide
12. Cylinder head
13. Oil Seal
14. Cylinder head machining dimension
15. Valves

7923WG13

Exploded view of the camshaft and related components—1.8L (SOHC) and 2.0L engines

1.8L (DOHC) Engine

1. Place the lock carrier into the service position as follows:

a. Remove the front bumper.

b. Tag and remove any wiring or connector that would inhibit locking the carrier.

c. Remove the three quick-release screws on the front noise insulation panel.

d. Unbolt the air guide between the lock carrier and the air filter.

e. If installed, remove the retaining clamps for the wiring harness at the left side of the radiator frame.

f. Remove the No. 2 bolts and install Support tool 3369 or equivalent.

g. Remove the remaining bolts and pull the lock carrier out to the stop.

h. To secure the lock carrier, install the appropriate M6 bolts into the rear of the lock carrier and fender.

2. Turn the ignition switch to the **OFF** position, then disconnect the negative battery cable.

3. Remove the accessory drive belts.

4. Remove the engine covers.

5. Remove timing belt upper cover.

6. Turn the crankshaft, in the direction of rotation (clockwise), until the No. 1 cylinder is at Top Dead Center (TDC).

7. Using Torx® wrench T45, loosen the timing belt tensioner.

8. Push down on the tensioner, and remove the belt from the camshaft gear.

9. Remove the Torx® bolt and swing the tensioner assembly bracket forward.

10. Remove the valve cover.

11. Using Retainer tool 3036 or equivalent, loosen the cam gear retaining bolt.

12. Remove the camshaft gear.

13. Remove the housing for Camshaft Position (CMP) sensor and shutter wheel.

14. Secure the hydraulic chain tensioner with Bracket-Tensioner 3366 or equivalent.

15. Verify that the camshafts are at Top Dead Center (TDC) for the No. 1 cylinder. Both camshaft markings must align with arrows on the bearing caps.

16. Clean the drive chain and the cam chain gears opposite both arrows on the bearing caps. Matchmark the installed position using paint.

➡The distance between the two arrows/paint marks is equivalent to 16 drive chain rollers, and the notch on the exhaust camshaft is slightly offset inward toward the drive chain roller.

17. Remove bearing caps 3 and 5 from the intake and exhaust camshafts.

18. Remove the double bearing cap.

19. Remove both bearing caps from the chain gears on the intake and exhaust camshafts.

20. Remove the hydraulic chain tensioner retaining bolts.

21. In an alternating and diagonal sequence, loosen the bearing caps 2 and 4 of the intake and exhaust manifold, then remove.

22. Remove the camshafts with the hydraulic chain tensioner.

To install:

✳✳ CAUTION

After installing the lifters or the camshaft(s), the engine must NOT be started for at least 30 minutes. Otherwise the valves could strike the pistons. Rotate the engine by hand, at least two revolutions, to ensure that the valves do not strike the pistons.

23. Replace the rubber/metal chain tensioner gasket and apply sealant to the hatched area, as shown.

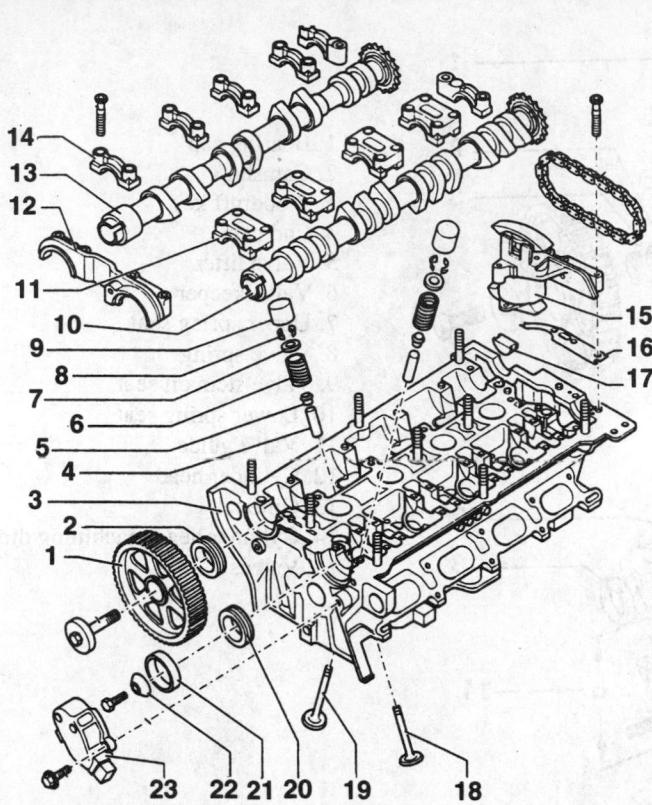

1. Camshaft gear
2. Oil seal
3. Cylinder head
4. Valve guide
5. Valve stem oil seal
6. Valve spring
7. Valve spring retainer
8. Valve keeper
9. Valve lifter
10. Intake camshaft
11. Intake camshaft bearing cap
12. Double bearing cap
13. Exhaust camshaft
14. Exhaust camshaft bearing cap
15. Hydraulic chain tensioner
16. Rubber/metal seal
17. Gasket
18. Exhaust valve
19. Intake valve
20. Oil seal
21. Shutter wheel for CMP sensor
22. Washer
23. Camshaft Position Sensor (CMP)

7923WG14

Exploded view of the camshaft and related components—1.8L (DOHC) engine

24. Install the drive chain on the camshaft as follows:

   a. If installing the old chain, align the paint marks with the camshaft marks.

   b. If installing a new chain, the distance between the notches A and B on the camshafts must equal the distance between 16 drive chain rollers.

25. Slide the hydraulic chain tensioner between the drive chain.

26. Install the camshafts with the chain tensioner into the cylinder head.

27. Oil the camshaft contact surfaces.

➡ **When installing the bearing caps, verify the markings on the caps are readable from the intake side of the cylinder head.**

28. Tighten the bearing caps 2 and 4 of the intake and exhaust camshafts in an alternating diagonal sequence to 7 ft. lbs. (10 Nm).

29. Install both bearing caps on the chain sprockets of the intake and exhaust camshafts and tighten to 7 ft. lbs. (10 Nm).

30. Verify the correct positions of the camshafts.

31. Remove the bracket-tensioner 3366.

32. Lightly coat the cylinder head mating surface of the double bearing cap with sealant, then install.

33. Install the remaining bearing caps and tighten to 7 ft. lbs. (10 Nm).

34. Install the camshaft gear and tighten the retaining bolt to 48 ft. lbs. (65 Nm).

35. Install the CMP shutter wheel and housing cover.

36. Install the valve cover.

37. Align the camshaft gear and the vibration damper with the TDC markings.

38. Install the timing belt.

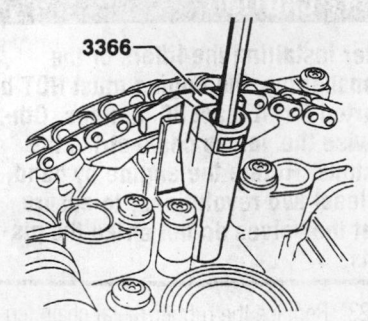

3366

7923WG15

Overtightening will damage the chain tensioner (3366)—1.8L (DOHC) engine

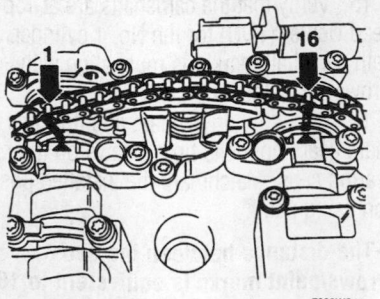

7923WG16

To ensure proper installation, matchmark the chain-to-camshaft position—1.8L (DOHC) engine

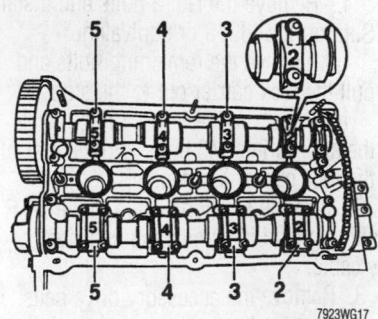

7923WG17

Camshaft bearing cap identification—1.8L (DOHC) engine

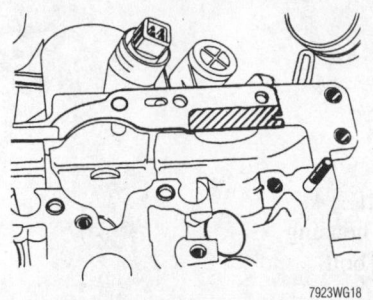

To ensure a proper seal, be sure to apply sealant to the hatched area—1.8L (DOHC) engine

39. Install the accessory drive belts, then the engine cover.

40. Reinstall the lock carrier.

41. Connect the negative battery cable.

42. Fully close all power windows to stop, operate all window switches for at least one second in the close direction to activate the one-touch opening/closing function

43. Check the oil level before starting the engine

44. Set the clock to the correct time.

➡**Diagnostic Trouble Codes (DTCs) are stored when harness connectors are detached.**

45. Read the DTCs and clear the fault codes.

46. Adjust the headlights.

2.8L Engine

This procedure requires special tool 3268 or equivalent. This is a setting tool that holds the camshafts in the correct position for installing the timing chains.

1. Remove the distributor cap, wires and wire guide as an assembly.

2. Remove the upper intake manifold.

3. Remove the cylinder head cover.

4. Remove the timing chain tensioner bolt and the upper timing chain cover.

5. Rotate the crankshaft to TDC of No. 1 piston.

6. Mark the direction of travel on the upper camshaft drive chain. Remove the tensioner shoe and the chain.

7. Hold the camshafts at the flats with a 24mm wrench and remove the bolts to remove the sprockets. Note the position of the distributor drive on the short camshaft.

➡**Do not use the setting tool to hold the camshafts when removing or installing the sprocket bolts. The camshafts or the tool will be damaged.**

8. On the long camshaft, remove the end bearing caps. Loosen the center cap nuts in a diagonal pattern 2 turns at a time until the valve springs are relieved. Remove the camshaft.

9. On the short camshaft, remove the center bearing cap and loosen the nuts on the end caps in a diagonal pattern 2 turns at a time. When the valve springs are relieved, remove the camshaft.

To install:

10. Lubricate the long camshaft and the cylinder head bearing surfaces and set the camshaft in place. Install bearing caps 3 and 5 and tighten the bolts 2 turns at a time in a diagonal pattern to draw the camshaft down against the valve springs.

11. Install the other bearing caps and torque all the nuts to 15 ft. lbs. (20 Nm).

12. Repeat the process with the short camshaft, using bearing caps 2 and 6 to draw the camshaft down against the springs.

13. Hold the camshaft with a 24mm wrench and install the sprockets. Be sure the distributor drive is correctly positioned and torque the bolts to 74 ft. lbs. (100 Nm).

14. Be sure the crankshaft is at TDC on No. 1 piston. Install the setting tool and install the timing chain.

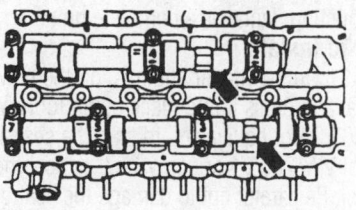

Hold the camshafts on the wrench flats when removing or installing the sprocket bolts

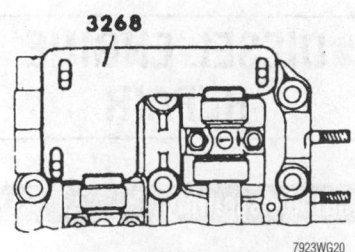

The camshaft setting tool holds the camshafts in place when installing the chain—do not use this tool to loosen or tighten the sprocket bolts

15. Install the tensioner shoe and temporarily install the upper timing chain cover. Install the tensioner bolt and remove the setting tool. Rotate the crankshaft 4 full turns and stop at TDC of No. 1 piston. The setting tool should fit into the camshafts.

16. Remove the tensioner bolt and upper timing chain cover again. Clean the old sealant off the cylinder head gasket and apply new sealant.

17. Install the upper timing chain cover and torque the bolts to 82 inch lbs. (10 Nm). Install the tensioner bolt and torque to 15 ft. lbs. (20 Nm).

18. Install the cylinder head cover, upper intake manifold and ignition system components.

Valve Lash

ADJUSTMENT

All engines are equipped with hydraulic valve lash adjusters. No periodic valve lash adjustment is necessary.

Oil Pan

REMOVAL & INSTALLATION

The oil pan can be removed with the engine in the vehicle.

1. Raise and safely support the vehicle and drain the oil.

2. Loosen and remove the bolts retaining the oil pan.

3. Lower the pan from the engine.

To install:

4. Be sure the gasket surface is flat and install the pan with a new gasket.

5. Torque the retaining bolts in a crisscross pattern to 15 ft. lbs. (20 Nm) on the 4-cylinder engine or 11 ft. lbs. (15 Nm) on the 6-cylinder engines.

✳✳ WARNING

Operating the engine without the proper amount and type of engine oil will result in severe engine damage.

6. Refill the engine with oil. Start the engine and check for leaks.

Oil Pump

REMOVAL & INSTALLATION

1. Raise and safely support the vehicle and remove the oil pan.

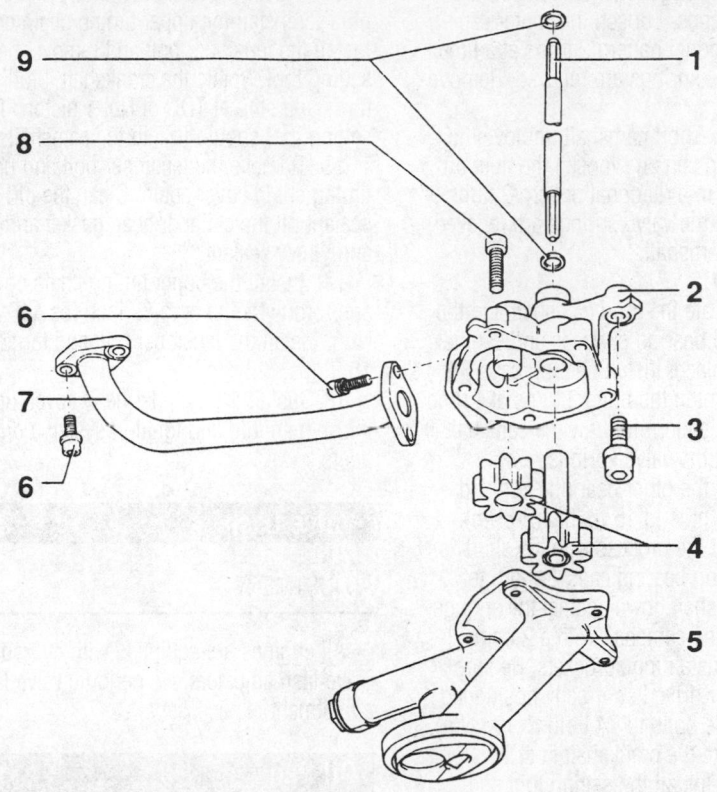

1. Drive shaft
2. Oil pump housing
3. Mounting bolt
4. Gears
5. Oil pump cover with pressure relief valve
6. Pressure pipe mounting bolt
7. Pressure pipe
8. Oil pump housing bolt
9. O-ring

7923WG21

Exploded view of the oil pump assembly—2.8L engine

2. Remove the mounting bolts and lower the pump from the engine.

3. Remove the bottom cover and disassemble the pump. The pressure relief valve is in the bottom cover.

4. Clean and inspect all parts for wear and replace as needed.

5. After reassembling the pump, prime it with oil and install in the reverse order of removal.

6. Observe the following values:
- Oil pump bottom cover bolts—7 ft. lbs. (10 Nm)
- Oil pump suction foot bolts—7 ft. lbs. (10 Nm)
- Oil pump retaining bolts—18 ft. lbs. (25 Nm)

Rear Main Seal

REMOVAL & INSTALLATION

The rear main oil seal is located in a housing on the rear of the cylinder block.

To replace the seal on all vehicles it is necessary to remove the transaxle and flywheel.

1. Remove the transaxle and flywheel.

2. On 6-cylinder engines, pry the old seal out of the support ring. On four cylinder engines, remove the oil seal with the mounting flange as a complete unit.

To install:

3. On 6-cylinder engines, oil the new seal and press it into place using tool VW-2003/2A or equivalent, to start the seal and tool VW-2003/1 or equivalent, to seat the seal. Be careful not to damage the seal or score the crankshaft. On four cylinder engines, install a new mounting flange with seal using a new gasket. Tighten the mounting flange bolts to 7 ft. lbs. (10 Nm).

4. Install the flywheel and transaxle.

DIESEL ENGINE REPAIR

Engine Assembly

REMOVAL & INSTALLATION

1. The engine and transaxle are lifted from the vehicle as an assembly. Disconnect the battery cables and remove the battery.

2. Open the fuel filler cap to relieve tank pressure, then loosen the fuel filter fitting to relieve system pressure. Be sure to take the appropriate fire safety precautions.

3. Remove the air filter and disconnect the accelerator cable from the injection pump.

4. Remove the radiator cap. Turn the heater temperature control all the way towards warm and remove the thermostat housing to drain the coolant.

5. Remove the upper radiator hose and disconnect the wiring from the radiator fan motor and switches. Remove the mounting nuts or bolts and lift out the radiator and fan shroud as an assembly.

6. Begin disconnecting electrical connections and vacuum lines, carefully labeling each one. Don't forget ground connections that are screwed to the body.

7. If equipped with power steering, remove pump and secure it to the body. Do not disconnect the hydraulic lines. If equipped with air conditioning, remove the compressor and secure it aside without disconnecting the lines.

8. Disconnect the fuel inlet and outlet lines from the injection pump and plug the holes to keep the pump clean. Note the outlet fitting has a special orifice.

9. On turbocharged engines, discon-

nect the exhaust pipe and the oil lines from the turbocharger and cap the oil line fittings on the turbocharger. Unbolt the turbocharger and lift it out of the engine.

10. If equipped with an automatic transaxle, place the selector lever in **P** and disconnect the selector cable at the transaxle.

11. On manual transaxle shift linkage, remove the 2 rods with the plastic socket ends and unbolt the remaining linkage from the case as required. Disconnect the clutch cable, lift it out of the case and set it aside.

12. Disconnect the wiring from the starter, the back-up light switch and the ground cable from the transaxle. Remove the speedometer cable from the transaxle and plug the hole in the case.

13. Attach an engine sling tool VW-2024A or equivalent, to the engine and attach the sling to a suitable lifting device.

14. Remove the nuts or spring clamps holding the exhaust pipe to the manifold or turbocharger.

❊❊ CAUTION

On some models, special tools are required for removing and installing the exhaust pipe-to-manifold spring clamps; VW3140/1 and /2 or equivalent. This is a set of different sized wedges for spreading the spring clamps in steps. The installed spring clamp has considerable tension and could cause damage or injury if not properly removed. Clamps with wedges installed are also under high tension and should be handled carefully.

15. Unbolt the halfshafts from the flanges and hang them from the body with wire.

16. Be sure everything is disconnected and unbolt the mounts. Remove the starter first and the front mount with it.

17. With all mounts unbolted, slightly lower the engine/transaxle assembly and tilt it towards the transaxle side. Then, carefully lift the assembly out of the vehicle.

To install:

18. Carefully install the engine/transaxle assembly and be sure all mounts are securely bolted to the engine/transaxle. Start all nuts and bolts that secure the mounts to the body but don't tighten them yet.

19. With all mounts installed and the engine safely in the vehicle, allow some slack in the lifting equipment. With the vehicle safely supported, shake the

engine/transaxle as a unit to settle it in the mounts. Torque all mounting bolts, starting at the rear and working forward. Torque to 33 ft. lbs. (41 Nm) for 10mm bolts or 54 ft. lbs. (73 Nm) for 12mm bolts.

20. Install the starter and torque the bolts to 33 ft. lbs. (45 Nm).

21. Connect the halfshafts to the flanges and torque the bolts to 33 ft. lbs. (45 Nm).

22. Install the exhaust pipe and use new self-locking nuts to secure the flange. Torque the nuts to 30 ft. lbs. (40 Nm). If equipped with spring clamps, the clamps can be used again.

23. Connect the shift linkage and the clutch cable, if equipped. Make any necessary adjustments.

24. Install the fuel injector lines and torque to 18 ft. lbs. (25 Nm). Be careful not to over torque the line nuts. If a line is damaged or clogged, replace all lines as a set.

25. Connect the inlet and outlet lines to the injector pump. Note the special outlet fitting has the word "OUT" printed on the top. Use new gaskets.

26. Install the air conditioning compressor and/or power steering pump, if equipped. Install and adjust the drive belts.

27. Connect the wiring and vacuum hoses.

28. Install the radiator, fan and heater hoses. Use a new O-ring on the thermostat and torque the thermostat housing bolts to 7 ft. lbs. (10 Nm).

29. Fill and bleed the cooling system. Check the adjustment of the accelerator cable.

Water Pump

REMOVAL & INSTALLATION

On some diesel engines, the belt tension is adjusted with shims between the outer and inner halves of the pulley. On others, the alternator swivels to adjust belt tension.

1. To drain the cooling system, remove the thermostat housing from under the water pump housing.

2. Raise and safely support the vehicle.

3. Working under the vehicle, loosen but don't remove the bolts holding the pulley to the water pump.

4. On vehicles with a movable alternator, loosen the alternator and remove the drive belt.

5. Remove the water pump pulley and remove the pump.

To install:

6. Installation is the reverse of removal. Be sure to clean the pump housing before

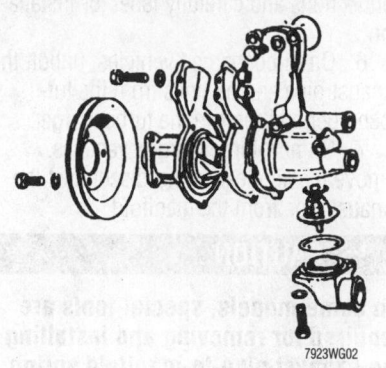

Water pump assembly—1.9L Diesel

installing the new gasket. Torque the following:

- Water pump-to-housing—7 ft. lbs. (10 Nm)
- Water pump drive pulley—15 ft. lbs. (20 Nm)
- Thermostat housing—7 ft. lbs. (10 Nm)
- Alternator mounting bolts—18 ft. lbs. (25 Nm)

Diesel Glow Plugs

REMOVAL & INSTALLATION

1. Remove the bus-bar connecting the glow plugs and determine which plugs need replacement.

2. Remove the defective plugs.

3. When installing new plugs, tighten them to 22 ft. lbs. (30 Nm).

➡**Diesel glow plugs have an air gap much like a spark plug to prevent overheating of the plug. Over-tightening the glow plug will close the gap and cause the plug to burn out.**

Cylinder Head

REMOVAL & INSTALLATION

➡**The cylinder head bolts on all diesel vehicles are stretch bolts and must be replaced when removed.**

1. Disconnect the battery ground cable.

2. Remove the thermostat and drain the cooling system.

3. Remove the fuel lines from the injectors and the pump as an assembly. Put the lines where they will stay clean; protect the injector and pump fittings with caps.

4. Disconnect the radiator and heater hoses.

5. Disconnect all vacuum and electrical

connections and carefully label for installation.

6. On turbocharged vehicles, unbolt the exhaust pipe and oil lines from the turbocharger and remove the turbocharger.

7. On non-turbocharged vehicles, remove the air cleaner and disconnect the exhaust pipe from the manifold.

✳✳ CAUTION

On some models, special tools are required for removing and installing the exhaust pipe-to-manifold spring clamps; VW3140/1 and /2 or equivalent. This is a set of different sized wedges for spreading the spring clamps in steps. The installed spring clamp has considerable tension and could cause damage or injury if not properly removed. Clamps with wedges installed are also under high tension and should be handled carefully.

8. Remove the cylinder head cover and camshaft drive belt cover.

9. Turn the engine to TDC of No. 1 cylinder, if possible, and remove the camshaft drive belt.

10. Remove the head bolts in the reverse order of installation sequence and lift the head out of the vehicle. The torque sequence is the same as for gasoline engines.

To install:

11. On these engines, the pistons actually project above the deck of the block. If the crankshaft and pistons are not to be removed, examine the old head gasket to see how many notches are on the edge near the oil return hole, between No. 2 and 3 cylinders. Replace the gasket with the same thickness.

12. If the pistons were removed or if the old gasket is not available, the piston height (pop up) must be measured to select the proper head gasket. Use a dial indicator or caliper to obtain the measurement.

Pop-up on engines with solid lifters:
- 0.026–0.031 in. (0.67–0.80mm)—1 notch
- 0.032–0.035 in. (0.81–0.90mm)—2 notches
- 0.036–0.040 in. (0.91–1.02mm)—3 notches

Pop-up on engines with hydraulic lifters:
- 0.026–0.034 in. (0.66–0.86mm)—1 notch
- 0.034–0.035 in. (0.87–0.90mm)—2 notches
- 0.036–0.040 in. (0.91–1.02mm)—3 notches

Measure piston pop-up to determine required head gasket thickness—1.9L Diesel engine

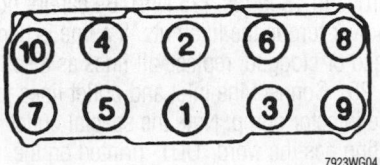

Tighten the cylinder head bolts in the correct order as shown—1.9L Diesel engine

13. Install the new cylinder head gasket with the word TOP or OBEN facing upward. Do not use any sealing compound.

14. Turn the crankshaft to TDC of No. 1 cylinder, then back about ¼ turn to bring all pistons about even.

15. Carefully lower the head on and install new head bolts into No. 8 and 10 first. These holes are smaller and will properly locate the gasket and cylinder head.

16. Install the remaining bolts and torque in the proper sequence in 3 steps: 29 ft. lbs. (40 Nm), 44 ft. lbs. (60 Nm), then a full ½ turn more. Two quarter turns are allowed.

17. Installation of the remaining parts is the reverse of removal, be sure to change the oil and filter. Install the camshaft drive belt and set injection pump timing.

18. Install the fuel injector lines and torque to 18 ft. lbs. (25 Nm). Be careful not to over torque the line nuts. If a line is damaged or clogged, replace all lines as a set.

19. After the engine has be run about 1000 miles, the cylinder head bolts must be re-torqued. Remove the cylinder head cover and turn each head bolt, in sequence, an additional ¼ turn in 1 movement. This can be done on a cold or warm engine.

Turbocharger

REMOVAL & INSTALLATION

1. Disconnect the negative battery cable.
2. Remove the exhaust pipe from the turbocharger outlet.
3. Clean the oil supply fitting on the top of the turbocharger and remove the supply line and bracket.
4. Remove the inlet air hose.
5. Under the vehicle, remove the oil return line and the turbocharger mounting bracket.
6. Still underneath, remove the turbo-to-manifold bolts. Lift the turbocharger out from the top.

To install:

7. Installation is the reverse of removal. Before installing the oil supply line, fill the connection on the turbocharger with engine oil. Torque the following:
- Turbocharger-to-exhaust manifold—33 ft. lbs. (45 Nm)
- Mounting bracket nuts—18 ft. lbs. (25 Nm)
- Turbocharger outlet nuts—18 ft. lbs. (25 Nm)
- Oil return line—22 ft. lbs. (30 Nm)

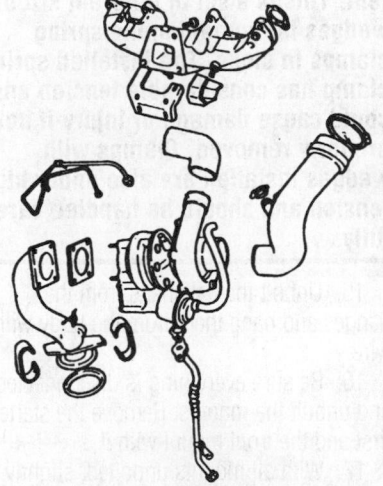

Diesel engine turbocharger and exhaust manifold

Intake Manifold

REMOVAL & INSTALLATION

1. Disconnect the hose and wiring from the blow-off valve.
2. Disconnect the air inlet hose.
3. Remove the bolts to remove the intake manifold.

4. Installation is the reverse of removal. Use a new gasket and torque the bolts to 18 ft. lbs. (25 Nm).

Exhaust Manifold

REMOVAL & INSTALLATION

→**On some models, special tools are required for removing and installing the exhaust pipe-to-manifold spring clamps; VW3140/1 and /2 or equivalent. This is a set of different sized wedges for spreading the spring clamps in steps. The installed spring clamp has considerable tension and could cause damage or injury if not properly removed. Clamps with wedges installed are also under high tension and should be handled carefully.**

1. Disconnect the negative battery cable and remove any heatshields that may be in the way.

2. On turbocharged engines, unbolt the exhaust pipe from the turbocharger outlet.

3. On non-turbocharged engines, expand the spring clamp by pushing the exhaust pipe to one side and insert the starter wedge into the clamp all the way up to the shoulder.

4. Push the pipe to the other side and install another wedge in the opposite clamp. Continue to work the pipe side to side while pushing the wedges into the clamps until the clamps are spread far enough to lift off easily.

❊ CAUTION

The removed spring clamps with wedges in them are under spring pressure, if miss handled, could fly apart with enough force to cause serious injury. Store the removed clamps in a safe area where they won't be disturbed.

5. On turbocharged engines, remove the turbocharger oil lines and the turbocharger.

6. Remove the manifold locking nuts and lift the manifold off the head.

7. Installation is the reverse of removal. Use new gaskets and locking nuts and torque to 18 ft. lbs. (25 Nm).

Camshaft and Valve Lifters

REMOVAL & INSTALLATION

1. Disconnect the negative battery cable. Remove the timing belt cover(s), the timing belt, cylinder head cover and the camshaft sprocket.

2. Number the bearing caps from front to back. If the cap does not already have one, scribe an arrow pointing towards the front of the engine. The caps are offset and must be installed correctly. Factory numbers on the caps are not always on the same side.

3. Remove the front and rear bearing caps. Loosen the remaining bearing cap nuts a little at a time to avoid bending the camshaft. Start from the outside caps near the ends of the head and work toward the center.

4. Remove the bearing caps and the camshaft.

To install:

5. Install a new oil seal and end plug in the cylinder head. Lubricate the camshaft bearing journals and lobes and set the camshaft in place.

6. Install the bearing caps in the correct position with the arrow pointing towards the front of the engine. Tighten the cap nuts diagonally and in several steps until they are torqued to 15 ft. lbs. (20 Nm). Do not over torque. Camshaft shaft end-play should be about 0.006 in. (0.15mm).

7. Install the drive sprocket and timing belt. Wait at least ½ hour after installing the camshaft before starting the engine to allow the lifters to leak down.

Valve Lash

ADJUSTMENT

All vehicles have hydraulic valve lifters and require no adjustment. On these vehicles there will be a sticker under the hood indicating hydraulic lifters.

Oil Pan

REMOVAL & INSTALLATION

The oil pan can be removed with the engine in the vehicle.

1. Raise and safely support the vehicle and drain the oil.

2. Loosen and remove the bolts retaining the oil pan.

3. Lower the pan from the engine.

To install:

4. Be sure the gasket surface is flat and install the pan with a new gasket.

5. Torque the retaining bolts in a criss-cross pattern to 14 ft. lbs. (20 Nm). Do not over-torque.

6. Refill the engine with oil. Start the engine and check for leaks.

Oil Pump

REMOVAL & INSTALLATION

1. Raise and safely support the vehicle and remove the oil pan.

2. Remove the mounting bolts and lower the pump from the engine.

3. Remove the bottom cover and disassemble the pump. The pressure relief valve is in the bottom cover.

4. Clean and inspect all parts for wear and replace as needed.

5. After reassembling the pump, prime it with oil and install in the reverse order of removal.

6. Observe the following values:
- Oil pump bottom cover bolts—7 ft. lbs. (10 Nm)
- Oil pump suction foot bolts—7 ft. lbs. (10 Nm)
- Oil pump retaining bolts—18 ft. lbs. (25 Nm)

Rear Main Seal

REMOVAL & INSTALLATION

The rear main oil seal is located in a housing on the rear of the cylinder block. To replace the seal on all vehicles it is necessary to remove the transaxle and flywheel.

1. Remove the transaxle and flywheel.

2. Remove the oil seal with the mounting flange as a complete unit.

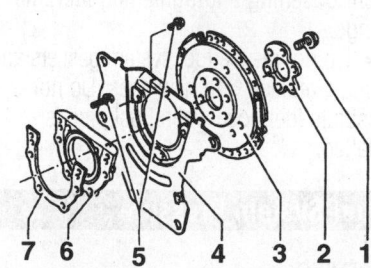

1. Bolt
2. Washer
3. Drive plate
4. Intermediate plate
5. Intermediate plate mounting bolt
6. Mounting flange with seal
7. Gasket

7923WG22

Rear main oil seal and related components—1.9L Diesel engine

To install:

3. Install a new mounting flange with seal using a new gasket. Tighten the mounting flange bolts to 7 ft. lbs. (10 Nm).

4. Install the flywheel and transaxle.

GASOLINE FUEL SYSTEM

Fuel System Service Precautions

Whenever working on or around gasoline or the fuel delivery system, heed the following precautions:

• Do not allow fuel spray or fuel vapors to come into contact with a heating element or open flame. Do not smoke while working on the fuel system.

• Always disconnect the negative battery cable unless the repair or test procedure requires that battery voltage be applied.

• Always relieve the fuel system pressure prior to disconnecting any fitting or fuel line connection.

• To control fuel spray when relieving system pressure, place a shop towel around the fitting prior to loosening to catch the spray. Ensure that all fuel spillage is quickly wiped up and that all fuel soaked rags are deposited into a proper fire safety container.

• Always keep a dry chemical (Class B) fire extinguisher near the work area.

• Always use a back-up wrench when loosening and tightening fuel line fittings.

• Do not re-use fuel system gaskets and O-rings, replace with new ones. Do not substitute fuel hose where fuel pipe is installed.

Fuel System Pressure

RELIEVING

On CIS systems, fuel pressure can be vented at the cold start injector line, either at the fuel distributor end or the injector end. Lay a rag over the fitting and use a socket or line wrench to crack the fitting.

On Digifant systems, pressure can be vented at the fuel pump switch in front of the throttle body. Lay a rag over the switch and loosen the clamp.

Fuel Filter

REMOVAL & INSTALLATION

On the Digifant system, the fuel filter is a lifetime unit and only needs to be changed in the event of contamination. It is mounted under the vehicle, near the rear axle. The pump, accumulator, filter and reservoir are all part of a single assembly, but the filter can be removed separately.

On the CIS system, the fuel filter is mounted under the hood, sometimes on the fuel distributor. To make the job easier, open the clips holding the air filter housing and lift the whole assembly.

1. Disconnect the negative battery cable.

2. On vehicles with the Digifant system, raise and safely support the vehicle.

3. Relieve the fuel system pressure.

4. Remove the fuel lines, the mounting bracket nut and the filter.

5. Installation is the reverse of removal. Be sure to use the new sealing rings and torque the fuel lines to the filter to 14 ft. lbs. (20 Nm).

Fuel Pump

REMOVAL & INSTALLATION

1. The main fuel pump is located under the vehicle in front of the rear axle or in front of the tank on the right side. Disconnect the negative battery cable.

2. Raise and safely support the vehicle.

3. Disconnect the electrical connector.

4. Relieve the fuel system pressure.

5. Remove the mounting bolts and the fuel pump.

6. Installation is the reverse of removal. Be sure to use new sealing rings and/or gaskets.

DIESEL FUEL SYSTEM

Fuel System Service Precautions

Whenever working on or around Diesel fuel or the fuel delivery system, heed the following precautions:

• Do not allow fuel spray or fuel vapors to come into contact with a heating element or open flame. Do not smoke while working on the fuel system.

• Always disconnect the negative battery cable unless the repair or test procedure requires that battery voltage be applied.

• Always relieve the fuel system pressure prior to disconnecting any fitting or fuel line connection.

• To control fuel spray when relieving system pressure, place a shop towel around the fitting prior to loosening to catch the spray. Ensure that all fuel spillage is quickly wiped up and that all fuel soaked rags are deposited into a proper fire safety container.

• Always keep a dry chemical (Class B) fire extinguisher near the work area.

• Always use a back-up wrench when loosening and tightening fuel line fittings.

• Do not re-use fuel system gaskets and O-rings, replace with new ones. Do not substitute fuel hose where fuel pipe is installed.

Idle Speed

ADJUSTMENT

Diesel engines have both an idle speed and a maximum speed adjustment. The maximum speed adjustment is a high idle speed that prevents the engine from over-

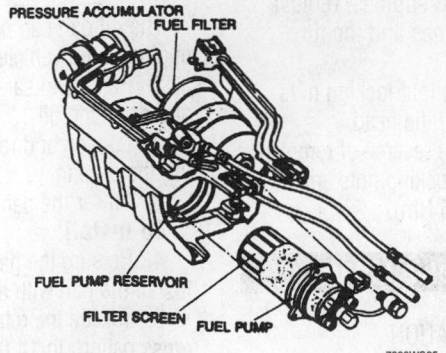

Fuel pump and reservoir assembly—Digifant system

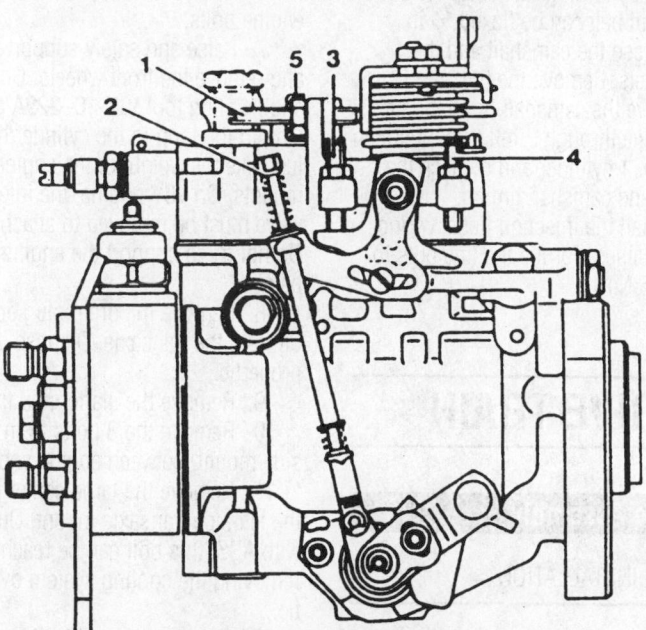

1. Previous idle adjustment screw
2. Linkage with cap nut for idle adjustment
3. Stop screw for minimum idle speed
4. Stop screw for idle speed boost
5. Tamper-proof cap

7923WG27

Low idle speed adjustment is made at the linkage cap—Diesel engine

revving when the control lever is in the full speed position but there is no load on the engine. No increase in power is available through this adjustment. The control lever idle stop screw is no longer used for idle speed adjustment. The idle speed boost linkage includes an adjustment for basic idle speed.

1. If the vehicle has no tachometer, connect a suitable diesel engine tachometer as per the manufacturer's instructions.
2. Run the engine to normal operating temperature.
3. Be sure the manual cold start/idle speed boost knob is pushed in all the way.
4. Turn the linkage cap nut to adjust idle speed to 820–880 rpm, at a point where there is the least vibration.
5. Advance the control lever to full speed. The high idle speed is 5300–5400 rpm. Adjust as needed and secure the locknut with sealer.

Fuel Filter/Water Separator

DRAINING WATER

Although diesel fuel and water do not readily mix, fuel does tend to entrap mois-

ture from the air each time it is moved from one container to another. Eventually every diesel fuel system collects enough water to become a potential hazard. Fortunately, when it's allowed to settle out, the water will always drop to the bottom of the tank or filter housing. Some diesel fuel filters are equipped with a water drain; a bolt or petcock at the bottom of the housing. All ECO Diesel models sold in North America are equipped with a water separator, located in front of the fuel tank under the right side of the vehicle. It's purpose is to allow water to settle from the fuel right at the tank and to alert the driver when draining is required. When the water level in the separator reaches a certain point, a sensor turns on the glow plug indicator light on the dashboard, causing it to blink continuously.

At The Water Separator

1. Raise and safely support the vehicle. Remove the fuel filler cap.
2. At the separator, connect a hose from the separator drain to a catch pan.
3. Open the drain valve (3 turns) and drain the separator until a steady stream of fuel flows from the separator, then close the valve. Don't forget to install the filler cap.

At The Filter

1. If the filter is equipped with a water drain at the bottom, place a pan under the drain to catch the water and fuel.
2. If equipped, loosen the vent bolt on the filter base. If there is no vent, loosen the return line at the pump, the line not connected to the filter.
3. Loosen the bolt or valve. When fuel flows in a clean stream, close the drain and tighten the vent or return line.

REMOVAL & INSTALLATION

✳✳ WARNING

Do not allow diesel fuel to contact the coolant hoses. If this happens, wipe it off and wash the hoses with soap and water immediately.

1. Remove the retaining clip (5).
2. Remove the control valve from the filter with the fuel lines attached.
3. Disconnect the hoses from connections (1) and (2).
4. Remove the filter assembly.

To install:
5. Use a new O-ring and install the control valve on the filter.
6. Install the retaining clip (5).
7. Connect the hoses to connections (1) and (2) and secure them with clamps.
8. Start the engine and check for leaks.

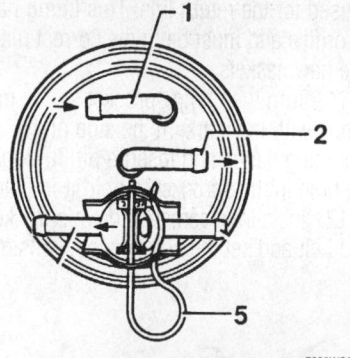

7923WG26

Fuel filter assembly—Diesel engine

Diesel Injection Pump

REMOVAL & INSTALLATION

➡**Special tools are required for injection pump installation. Do not remove the pump without these tools on hand.**

1. Disconnect the negative battery cable and remove the air cleaner, cylinder head cover and timing belt cover.

2. Turn the engine to TDC of No. 1 cylinder and insert a setting bar into the slot on the rear of the camshaft, VW tool 2065A or equivalent, to hold the camshaft in place. Remove the timing belt. Be careful to not turn the engine while the belt is removed.

3. Loosen the pump drive sprocket nut but don't remove it yet. Install a puller on the sprocket and apply moderate tension.

4. Rap the puller bolt with light hammer taps until the sprocket jumps off the tapered shaft, then remove the puller and sprocket. Be careful not to lose the Woodruff key.

5. Hold the pump fittings with a wrench and using a line wrench, remove the injection lines from the pump. Cap the pump fittings to keep dirt out. It may be easier to remove the lines from the injectors also and set them aside as an assembly. Cap the injector fittings to keep dirt out.

6. Disconnect the control cables, fuel solenoid wire and fuel supply and return lines.

7. Remove the pump mounting bolts and lift the pump from the vehicle.

To install:

8. When reinstalling, align the marks on the top of the mounting flange and the pump and torque the mounting bolts to 18 ft. lbs. (25 Nm).

9. Install the Woodruff key and sprocket and torque the nut to 33 ft. lbs. (45 Nm).

10. When reinstalling the supply and return lines, be sure the fitting marked OUT is used for the return line. This fitting has an orifice and must be in the correct place. Use new gaskets.

11. Turn the pump sprocket so the mark aligns with the mark on the side of the mounting flange and insert a pin through the hole in the sprocket to hold it in place.

12. Install the camshaft drive sprocket and belt and set the belt tension. Tension

the drive belt by turning the tensioner pulley clockwise until belt can be flexed ½ in. (13mm) between the camshaft and the pump sprockets. Remove the pin.

13. Remove the camshaft holding bar. Turn the engine through 2 full turns, return to TDC of No. 1 cylinder and recheck the belt tension and camshaft timing.

14. Reinstall the injection lines, wiring and control cables. Torque the line nuts to 18 ft. lbs. (25 Nm).

DRIVE TRAIN

Transaxle Assembly

REMOVAL & INSTALLATION

Manual

1995–97 PASSAT

➡**If equipped with electronically theft-protected radio, obtain the security code before disconnecting the battery.**

1. Disconnect the negative battery cable.

2. Disconnect the back-up light switch connector and the speedometer cable from the transaxle, plug the speedometer cable hole.

3. Remove the clutch slave cylinder without disconnecting the hydraulic line. Hang the cylinder from the body with wire.

4. On the cable shift linkage, remove the back-up light switch bracket. Disconnect the cable from the relay lever but remove the gearshift lever with the cable still attached. Remove the cable support and set the cables aside.

5. If necessary, remove the intake hose from the air flow sensor.

6. Remove the upper transaxle-to-engine bolts.

7. Raise and safely support the vehicle and remove the front wheels. Connect the engine sling tool VW-10–222A or equivalent, to the loop in the cylinder head and just take the weight of the engine off the mounts. On 16V engine, the idle stabilizer valve must be removed to attach the tool. Do not try to support the engine from below.

8. Remove the drain plug and drain the oil from the transaxle. Dispose of the oil properly.

9. Remove the starter and front mount.

10. Remove the 3 bolts from the right side mount, between engine and firewall.

11. Remove the large center bolt from the left side transaxle mount. On vehicles with ABS, this bolt can be reached by removing the cooling system overflow bottle.

12. Remove the radiator fan shroud and fan as an assembly.

13. Remove the long transaxle support bracket which connects the front and rear mounts on the left side.

14. Remove the heatshield for the right side inner CV-joint.

15. Disconnect the halfshafts from the output flanges and hang them from the body.

16. Remove the left rear transaxle mount. It may be necessary to push the engine/transaxle rearward to get the lower bolt out.

17. Lower the transaxle slightly.

18. Remove the bell housing cover and position a jack under the transaxle.

19. Remove the last transaxle-to-engine bolts and gently pry the transaxle away from the engine. Lower it carefully from the vehicle.

To install:

20. Press the clutch release lever towards the transaxle housing and secure it with a pin or 8mm bolt.

21. Coat the input shaft lightly with molybdenum grease and carefully fit the transaxle in place. If necessary, put the transaxle in any gear and turn an output flange to align the input shaft spline with the clutch spline.

22. Install the engine-to-transaxle bolts and torque to 59 ft. lbs. (80 Nm).

23. When installing the mounts to the transaxle, torque the left and rear bracket-to-transaxle bolts to 18 ft. lbs. (25 Nm). Torque the remaining mount-to-transaxle bolts to 44 ft. lbs. (60 Nm). Don't forget the balance weight. Install but do not torque the bolts that go into the rubber mounts.

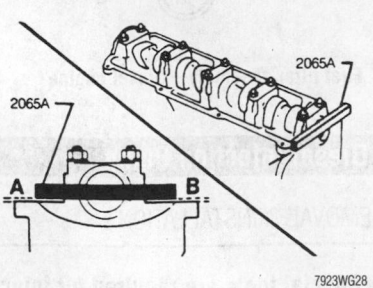

Install the bar to hold the camshaft in position during injection pump service

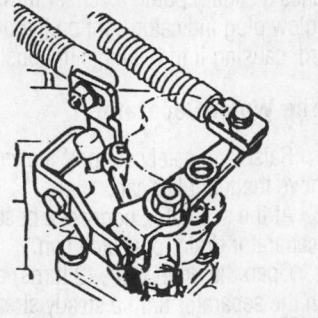

Cable shift linkage—Passat, the relay lever is on the left

24. Install the starter and front mount.

25. With all mounts installed and the transaxle safely in the vehicle, allow some slack in the lifting equipment. With the vehicle safely supported, shake the engine/transaxle as a unit to settle it in the mounts. Torque all mounting bolts, starting at the rear and working forward. Torque the bolts that go into the rubber transaxle mounts to 44 ft. lbs. (60 Nm).

26. Install the halfshafts and torque the bolts to 33 ft. lbs. (45 Nm). Install the heatshield.

27. Remove the pin or bolt from the release lever and install the clutch slave cylinder. Torque the bolts to 18 ft. lbs. (25 Nm).

28. Lubricate the shift linkage lightly with molybdenum grease and install it. Torque the bolts to 18 ft. lbs. (25 Nm). Adjust the linkage as required.

29. Install the radiator fan assembly and connect the wiring.

30. Complete the installation and refill the transaxle with oil.

1998–99 PASSAT

1. Obtain the security code for the radio, then disconnect the negative battery cable.

2. Remove the engine undercover.

3. Disconnect the front exhaust pipes from the engine. Loosen the U-bolt and push to the rear.

4. Remove the starter.

5. Disconnect the shift rod from the transaxle.

6. Detach the speed sensor and left back-up light connectors from the transaxle.

7. Support the transaxle with a jack.

8. Remove the right and left transaxle mounts.

9. Remove the right and left halfshaft from the transaxle.

10. Remove the halfshaft shield.

11. Remove the transaxle-to-engine mounting bolts.

12. Pry the transaxle from the engine and lower it about 6 in. (13cm) to access the slave cylinder.

13. Remove the slave cylinder with bracket without disconnecting the fluid line.

14. Lower and remove the transaxle assembly.

To install:

15. Clean the input shaft and apply a thin film of No. 000 100 high-performance grease or equivalent to the splines.

16. Lubricate the plunger contact surface on the release lever with Dow Corning® CU-7439 plus copper paste or equivalent.

17. Raise the transaxle into position and install the slave cylinder. Tighten the mounting bolts to 18 ft. lbs. (25 Nm).

18. Install the transaxle-to-engine bolts. Tighten the M8 bolts to 18 ft. lbs. (25 Nm), M10 bolts to 33 ft. lbs. (45 Nm) and the M12 bolts to 48 ft. lbs. (65 Nm).

19. Install the transaxle mounts. Tighten the mounting bolts to 30 ft. lbs. (40 Nm).

20. Install the halfshafts. Tighten the M8 bolts to 33 ft. lbs. (45 Nm) and the M10 bolts to 59 ft. lbs. (80 Nm).

21. Install the halfshaft shield.

22. Install the shift rod. Tighten the bolts to 15 ft. lbs. (20 Nm).

23. Install the starter.

24. Assembly the exhaust system.

25. Replace the engine undercover.

26. Connect the negative battery cable.

EXCEPT PASSAT

➡ **If equipped with electronically theft-protected radio, obtain the security code before disconnecting the battery.**

1. Disconnect the negative battery cable.

2. Disconnect the back-up light switch connector and the speedometer cable from the transaxle; plug the speedometer cable hole.

3. Remove the upper engine-to-transaxle bolts.

4. Remove the 3 right side engine mount bolts, between engine and firewall.

5. To disconnect the shift linkage, pry open the ball joint ends and remove the shift and relay shaft rods.

6. Remove the center bolt from the left transaxle mount.

7. Raise and safely support the vehicle and remove the front wheels. Connect the engine sling tool VW-10–222A or equivalent, to the loop in the cylinder head and just take the weight of the engine off the mounts. On 16V engine, the idle stabilizer valve must be removed to attach the tool. Do not try to support the engine from below.

8. Remove the drain plug and drain the oil from the transaxle. Dispose of the oil properly.

9. Remove the left inner fender liner.

10. Disconnect the halfshafts from the inner drive flanges and hang them from the body.

11. Remove the clutch cover plate and the small plate behind the right halfshaft flange.

12. Remove the starter and front engine mount.

13. Disconnect the clutch cable and remove it from the transaxle housing.

14. Remove the remaining transaxle mount bolts and mounts.

15. Place a jack under the transaxle and remove the last bolts holding it to the engine. Carefully pry the transaxle away from the engine and lower it from the vehicle.

To install:

16. Coat the input shaft lightly with molybdenum grease and carefully fit the transaxle in place. If necessary, put the transaxle in any gear and turn an output flange to align the input shaft spline with the clutch spline.

17. Install the engine-to-transaxle bolts and torque to 55 ft. lbs. (75 Nm).

18. When installing the mounts to the transaxle, torque the rear bracket-to-engine bolts and the transaxle support bolts to 18 ft. lbs. (25 Nm). Torque the left bracket-to-transaxle bolts to 25 ft. lbs. (35 Nm) and the remaining mounting bolts to 44 ft. lbs. (60 Nm). Install but do not torque the bolts that go into the rubber mounts.

19. Install the starter and front mount.

20. With all mounts installed and the transaxle safely in the vehicle, allow some slack in the lifting equipment. With the vehicle safely supported, shake the engine/transaxle as a unit to settle it in the mounts. Torque all mounting bolts, starting at the rear and working forward. Torque the bolts that go into the rubber mounts to 44 ft. lbs. (60 Nm).

21. Install the halfshafts and torque the bolts to 33 ft. lbs. (45 Nm). Install the clutch cover plates.

22. Connect the shift linkage and clutch cable and adjust as required.

23. Install the inner fender and complete the remaining installation. Refill the transaxle with oil.

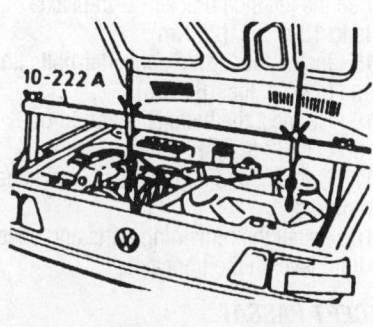

Supporting the engine to remove the transaxle

Automatic

PASSAT

1. If equipped with electronically theft-protected radio, obtain the security code before disconnecting the battery.

2. Remove the battery and disconnect the wiring from the transaxle.

3. Remove the upper engine-to-transaxle bolts.

4. Raise and safely support the vehicle and remove the front wheels. Connect the engine sling tool VW-10-222A or equivalent, to the cylinder head and just take the weight of the engine off the mounts. On Passat, the idle stabilizer valve must be removed to attach the tool. Do not try to support the engine from below.

5. Put the shifter in **P** and disconnect the shift cable.

6. Clamp and remove the hoses at the transaxle cooler.

7. Remove the starter and the engine's left and right mounts.

8. Remove the skid plate and disconnect the halfshafts from the drive flanges. Hang them from the body with wire.

9. Remove the torque converter plate and turn the engine as needed to remove the torque converter-to-flywheel bolts.

10. Remove the remaining transaxle mounts and lower the hoist slightly.

11. Support the transaxle with a jack and remove the remaining engine-to-transaxle bolts. Be careful to secure the torque converter so it does not fall out of the transaxle.

12. Carefully lower the transaxle out of the vehicle.

To install:

13. Fit the transaxle into the vehicle and be sure the guide pins fit properly between the engine and transaxle. Install the bolts and torque the 12mm bolts to 59 ft. lbs. (80 Nm), the 10mm bolts to 44 ft. lbs. (60 Nm).

14. Install the transaxle mounts and torque the bolts to 44 ft. lbs. (60 Nm). Torque the left side bracket-to-transaxle bolts to 18 ft. lbs. (25 Nm).

15. Install the torque converter bolts and torque to 44 ft. lbs. (60 Nm).

16. Connect the halfshafts and torque the bolts to 33 ft. lbs. (45 Nm).

17. Connect and adjust the shift linkage as required.

18. Install the remaining parts and check the fluid level in the transaxle.

EXCEPT PASSAT

1. If equipped with electronically theft-protected radio, obtain the security code before disconnecting the battery.

2. Disconnect the battery and the speedometer drive and plug the hole in the transaxle.

3. On Golf and Jetta, with the vehicle on the ground, remove the front axle nuts.

➡**When loosening or tightening an axle nut, be sure the vehicle is on the ground. Axle nut torque is high enough that attempting to loosen it may cause the vehicle to fall.**

4. Raise and safely support the vehicle and remove the front wheels. Connect the engine sling tool VW-10-222A or equivalent, to the cylinder head and just take the weight of the engine off the mounts. On 16V engine, the idle stabilizer valve must be removed to attach the tool. Do not try to support the engine from below.

5. Remove the driver's side rear transaxle mount and support bracket.

6. On Golf and Jetta, remove the front mount bolts from the transaxle and from the body and remove the mount as a complete assembly.

7. Remove the selector and accelerator cables from the transaxle lever but leave them attached to the bracket. Remove the bracket assembly to save the adjustment.

8. Unbolt the halfshafts from the drive flanges. On Golf and Jetta, the shafts must be removed, which may require separating the ball joints from the wheel bearing housing to gain the necessary clearance. Remove the ball joint clamping bolt.

9. Remove the heatshield and brackets and remove the starter. On Cabrio, the front mount comes off with the starter.

10. Turn the engine as needed to remove the torque converter-to-flywheel bolts.

11. Remove the remaining transaxle mounts, on Golf and Jetta, the subframe bolts and allow the subframe to hang free.

12. Support the transaxle with a jack and remove the remaining engine-to-transaxle bolts. Be careful to secure the torque converter so it does not fall out of the transaxle.

13. Carefully lower the transaxle from the vehicle.

To install:

14. When reinstalling, be sure the torque converter is fully seated on the pump shaft splines. The converter should be recessed into the bell housing and turn by hand. Keep checking that it still turns while drawing the engine and transaxle together with the bolts.

15. Install the engine-to-transaxle bolts and torque to 55 ft. lbs. (75 Nm).

16. Install all mount and subframe bolts before tightening any on them. Tighten the

bolts starting at the rear and work forward. Torque the smaller bolts to 25 ft. lbs. (34 Nm) and the larger bolts to 58 ft. lbs. (80 Nm). Remove the lifting equipment when all mounts are installed.

17. Install the torque converter-to-flywheel bolts and torque them to 26 ft. lbs. (35 Nm).

18. Install the starter and torque the bolts to 14 ft. lbs. (20 Nm). Install the heat-shields.

19. If the halfshafts were removed, be sure the splines are clean and apply a thread locking compound to the splines before sliding it into the hub. Connect the halfshafts to the drive flanges and torque the bolts to 37 ft. lbs. (50 Nm). Install new axle nuts but do not fully torque them until the vehicle is on the ground.

20. If removed, fit the ball joints to the control arm and torque the clamping bolt to 37 ft. lbs. (50 Nm).

21. Connect and adjust the shift linkage as required.

22. When assembly is complete and the vehicle is on its wheels, torque the axle nuts to 195 ft. lbs. (265 Nm).

Clutch

ADJUSTMENT

On cable operated clutches with an adjustable cable, special tool US5043 or equivalent, is available to make it easier to determine proper adjustment. The tool is a simple go or no-go gauge, but proper adjustment can be accomplished without it.

1. Depress the clutch pedal several times.

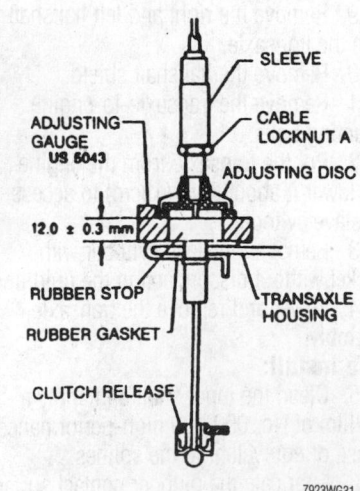

Clutch cable adjustment using gauge

2. Pull the cable adjusting sleeve up at the transaxle until resistance is felt and insert the gauge or measure the clearance.

3. Loosen the locknut and turn the adjusting sleeve until there is no free-play at the gauge. Without the gauge, this distance should be 0.472 in. (12mm).

4. Tighten the locknut and operate the pedal several times. Recheck the adjustment.

REMOVAL & INSTALLATION

Passat

1. Raise and safely support the vehicle and remove the transaxle.

2. Matchmark the flywheel and pressure plate if the pressure plate is going to be reused.

3. Gradually loosen the pressure plate bolts 1–2 turns at a time in a crisscross pattern to prevent distortion.

4. Remove the pressure plate and disc.

5. Check the clutch disc for uneven or excessive lining wear. Examine the pressure plate for cracking, scorching or scoring. Replace any questionable components.

To install:

6. Install the clutch disc and pressure plate with the springs on the disc towards the plate. Use an alignment tool to keep the clutch disc centered.

7. Gradually tighten the pressure plate-to-flywheel bolts in a crisscross pattern. Tighten the bolts to 18 ft. lbs. (24 Nm).

8. Install the clutch release bearing.

9. Install the transaxle.

Except Passat

1. Raise and safely support the vehicle and remove the transaxle.

2. Attach a toothed flywheel holder tool VW-558 or equivalent, to the flywheel and gradually loosen the flywheel-to-pressure plate bolts a few turns at a time. Use a crisscross pattern to prevent distortion.

3. Remove the flywheel and the clutch disc.

4. Use a small prybar to remove the release plate retaining ring. Remove the release plate.

To install:

5. Use new bolts to attach the pressure plate to the crankshaft. Use a thread locking compound and torque the bolts in a diagonal pattern to 72 ft. lbs. (100 Nm).

6. Lightly lubricate the clutch disc splines, release plate contact surface and pushrod socket with multi-purpose grease. Install the release plate, retaining ring and clutch disc.

7. Install a centering tool VW-547 or equivalent, to align the clutch disc.

8. Install the flywheel, tightening the bolts 1–2 turns at a time in a crisscross pattern to prevent distortion. Torque the bolts to 14 ft. lbs. (20 Nm).

9. Remove the alignment tool, reinstall the transaxle and adjust the clutch cable.

Hydraulic Clutch System

BLEEDING

1. The clutch and brakes share the same reservoir. Clean all dirt and grease from the cap to be sure no foreign substances enter the system.

2. Remove the cap and diaphragm and fill the reservoir to the top with the approved DOT 3 or 4 brake fluid. Fully loosen the bleed screw which is in the slave cylinder body next to the inlet connection.

3. At this point bubbles of air will appear at the bleed screw outlet. When the slave cylinder is full and a steady stream of fluid comes out of the slave cylinder bleeder, tighten the bleed screw.

4. Refill the reservoir and cap it. Exert a light load of about 20 lbs. to the slave cylinder piston by pushing the release lever

towards the cylinder and loosen the bleed screw. Maintain a constant light load; fluid and any air that is left will be expelled through the bleed port. Tighten the bleed screw when a steady flow of fluid with no air is being expelled.

5. Fill the reservoir fluid level back to normal capacity, if necessary repeat Step 4.

6. Exert a light load to the release lever but do not open the bleeder screw as the piston in the slave cylinder will move slowly down the bore. Repeat this operation 2–3 times; the fluid movement will force any air left in the system into the reservoir. The hydraulic system should now be fully bled.

7. Check the operation of the clutch hydraulic system and repeat this procedure, if necessary. Check the pushrod travel at the slave cylinder to insure the minimum travel is 0.57 in. (15mm).

Halfshaft

REMOVAL & INSTALLATION

➡**When loosening or tightening axle nuts, be sure the vehicle is on the ground. Axle nut torque is high enough that attempting to loosen it may cause the vehicle to fall off the jackstands.**

1. With the vehicle on the ground, remove the front axle nut.

2. Raise and safely support vehicle and remove the front wheels.

3. Remove the socket head bolts retaining the halfshaft to the transaxle flange.

4. Separate the strut from the control arm:

  a. On Fox, matchmark the ball joint to the control arm and remove the nuts to disconnect the ball joint from the control arm.

  b. On Passat, remove the bolts securing the ball joint to the control arm.

  c. On all other vehicles, remove the ball joint clamping bolt and push the control arm down, away from the ball joint.

5. Remove the transaxle side of the halfshaft from the drive flange and secure it out of the way. Do not let it hang unsupported.

6. Push the halfshaft out of the hub. A wheel puller may be required.

To install:

7. Fit the halfshaft to the drive flange and install the bolts. It is not necessary to torque them yet.

8. Apply a thread locking compound to the outer ¼ in. of the spline. Slip the spline through the hub and loosely install a new axle nut.

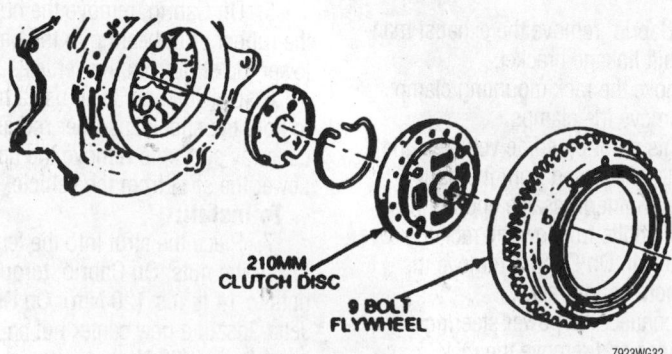

210MM CLUTCH DISC

9 BOLT FLYWHEEL

7923WG32

Exploded view of the clutch components—Cabrio, Golf, GTI and Jetta

9. Assemble the front suspension, being careful to align the matchmarks.

a. On Passat, torque the ball joint bolts to 26 ft. lbs. (35 Nm).

b. On all other models, torque the ball joint clamping bolt to 37 ft. lbs. (50 Nm).

10. Install the wheel and hold it to keep the axle from turning. Torque the inner axle bolts to 33 ft. lbs. (45 Nm).

11. With the vehicle on the ground, torque the axle nut:

- Cabrio—175 ft. lbs. (240 Nm)
- Golf, GTI, Jetta and Passat—195 ft. lbs. (265 Nm)

12. Check and adjust the front wheel alignment.

STEERING AND SUSPENSION

Air Bag

✳✳ CAUTION

Some vehicles are equipped with an air bag system, also known as the Supplemental Inflatable Restraint (SIR) or Supplemental Restraint System (SRS). The system must be disabled before performing service on or around system components, steering column, instrument panel components, wiring and sensors. Failure to follow safety and disabling procedures could result in accidental air bag deployment, possible personal injury and unnecessary system repairs.

PRECAUTIONS

Several precautions must be observed when handling the inflator module to avoid accidental deployment and possible personal injury.

- Never carry the inflator module by the wires or connector on the underside of the module.
- When carrying a live inflator module, hold securely with both hands, and ensure that the bag and trim cover are pointed away.
- Place the inflator module on a bench or other surface with the bag and trim cover facing up.
- With the inflator module on the bench, never place anything on or close to the module which may be thrown in the event of an accidental deployment.

DISARMING

To avoid personal injury when working on vehicles equipped with an air bag, the negative battery cable must be disconnected before working on the system. Failure to do so may result in deployment of the air bag.

Rack and Pinion Steering Gear

REMOVAL & INSTALLATION

Manual

1. Raise and safely support the vehicle.
2. Remove both front wheels and disengage both tie rod ends.
3. At the steering column, remove the boot clamp, push the boot towards the body and remove the clamp bolt from the universal joint.
4. Remove the rack mounting nuts and remove the rack from its mounts.
5. At this point on some vehicles, the rack cannot be removed from the body. Support the engine/transaxle and remove the subframe bolts or the rear transaxle mount and bracket to allow the rack to move towards the rear.
6. Installation is the reverse of removal. Torque the subframe bolts to 96 ft. lbs. (130 Nm).

Power

1. Raise and safely support the vehicle.
2. Remove both front wheels and disengage both tie rod ends.
3. Remove the low pressure (suction) hose from the pump and drain the system into a catch pan. Properly discard the fluid.
4. At the steering column, remove the boot clamp, push the boot towards the body and remove the clamp bolt from the universal joint.
5. On Cabrio, remove the exhaust manifold and shift linkage bracket.
6. Remove the rack mounting clamp nuts and remove the clamps.
7. At this point on some vehicles, the rack cannot be removed from the body. Support the engine/transaxle and remove the subframe bolts to allow the rack to move towards the rear. On Cabrio, remove the transaxle mount and bracket.
8. Disconnect the power steering hydraulic lines and remove the rack.

To install:

9. Be sure the mounting bushings are

in good condition. Fit the rack assembly into place and torque the clamp nuts to 22 ft. lbs. (32 Nm).

10. Install any subframe bolts that were removed.

11. Connect the hydraulic lines and install the steering column universal joint bolt.

12. Fill the system with new fluid and run the engine to check for leaks and bleed the system.

Strut

REMOVAL & INSTALLATION

Front

The upper strut-to-steering knuckle bolt may have an eccentric washer for adjusting wheel camber. Use a wire brush to clean the area and use a cold chisel to mark a fine line on the washer and the strut together. This matchmark may be enough to preserve the front wheel camber adjustment. It will at least be accurate enough to allow driving the vehicle to a shop for a proper front wheel alignment. If there is no eccentric washer, a new bolt and eccentric washer can be substituted. The parts are available through the dealer.

A special tool is required to remove the upper strut nut on Golf and Jetta. If necessary, it can be made by cutting away part of a 22mm socket.

1. Raise and safely support the vehicle and remove the front wheels.
2. Detach the brake line from the strut and remove the caliper. DO NOT let the caliper hang by the hydraulic line, hang it from the body with wire.
3. Clean and matchmark the position of the strut-to-steering knuckle bolt.
4. Remove the bolts and push the steering knuckle down away from the strut. Support the knuckle so it is not hanging on the outer CV-joint.
5. On Cabrio, remove the nuts holding the rubber strut bearing to the body and lower the strut from the vehicle.
6. On Golf and Jetta, use a hex wrench to hold the shock absorber rod and use the cut away socket to remove the upper nut. Lower the strut from the vehicle.

To install:

7. Place the strut into the fender and install the nuts. On Cabrio, torque the 3 nuts to 14 ft. lbs. (20 Nm). On Golf and Jetta, install a new center nut and torque it to 44 ft. lbs. (60 Nm).
8. Fit the knuckle into the strut and

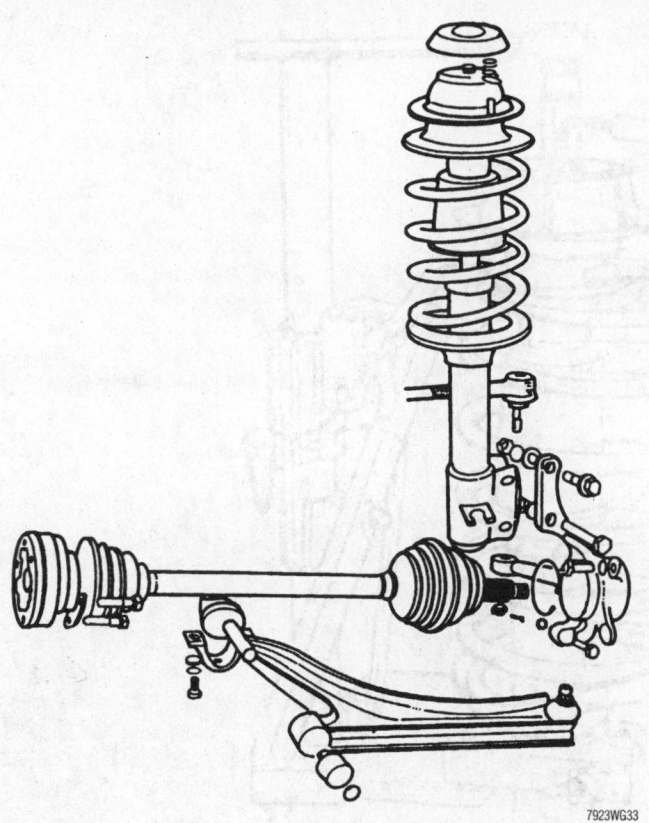

The MacPherson strut is mounted between the steering knuckle and body

Cut away socket for removing the upper strut nut

install the bolts. Be sure the matchmarks are aligned and install the nuts.

   a. On Golf and Jetta, the strut-to-knuckle bolts are 2 different wrench sizes. Torque the 19mm bolts to 70 ft. lbs. (95 Nm) and the 18mm bolts to 59 ft. lbs. (80 Nm).

   b. On Cabrio, torque the strut-to-knuckle bolts to 70 ft. lbs. (95 Nm).

9. Install the brake caliper and torque the bolts to 44 ft. lbs. (60 Nm).

10. Install the wheel and align the front wheels.

Rear

➡**Do not remove both suspension struts at the same time or the axle beam will be hanging on the brake lines.**

1. Working inside the vehicle, remove the cap from the top shock mount and remove the top nut, washer and rubber bushings.

2. Remove the second nut.

3. Slowly lift the vehicle and safely support it on jack stands. Do not place the stands under the axle beam.

4. Unbolt the strut from the axle and carefully remove the strut and spring from the vehicle. It may be necessary to press the axle down slightly.

To install:

5. Install the shock on the axle assembly. Do not tighten the nut until the vehicle is on the floor at normal riding height.

6. Install the upper end of the strut to the body. Tighten the lower nut to 11 ft. lbs. (15 Nm) and the upper nut to 18 ft. lbs. (25 Nm).

7. Install the wheel and lower the vehicle to the floor.

8. Tighten the lower strut mounting nut to 52 ft. lbs. (70 Nm).

Coil Spring

REMOVAL & INSTALLATION

1. Remove the strut from the vehicle.

2. Clamp the Spring Compressor VAG 1752/2 or equivalent in a vise.

3. Install the strut into the spring compressor.

4. Pry off the mounting bolt cap.

5. Compress the coil spring and remove the self-locking nut from the piston rod.

6. Matchmark the position of the spring retainer and spring mount.

7. Remove the spring seat and related components noting the order of removal.

8. Remove the strut from the spring compressor.

9. Release the tension on the coil spring, and remove the spring out of the compressor.

To install:

10. Install the new spring into the compressor.

11. Compress the spring and insert the strut through the spring.

12. Install the spring seat and related components in the reverse order as they were removed and aligning the matchmarks.

13. Install a new self-locking nut.

14. Reinstall the mounting bolt cap.

15. Release the spring compressor and install the strut into the vehicle.

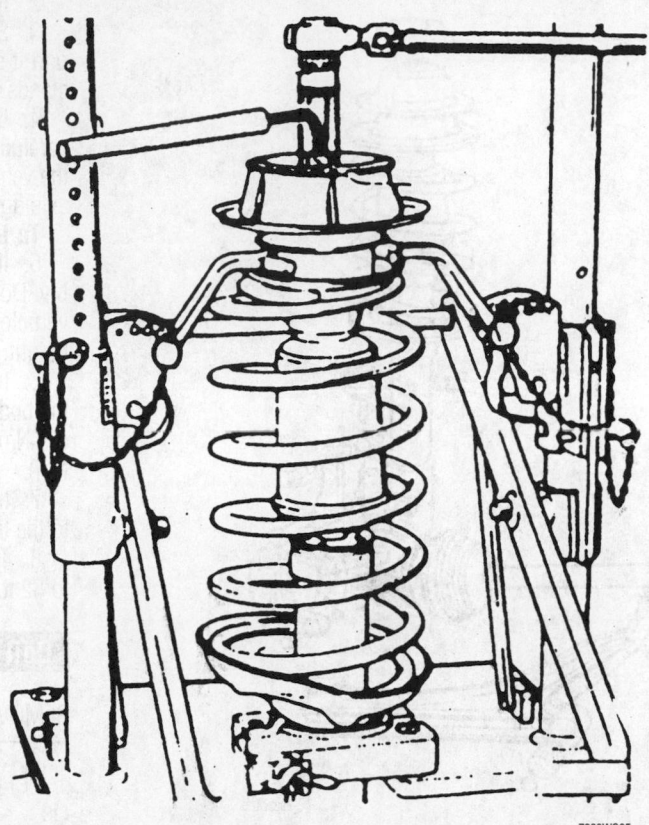

7923WG35

Compress the coil spring before removing the upper strut rod nut

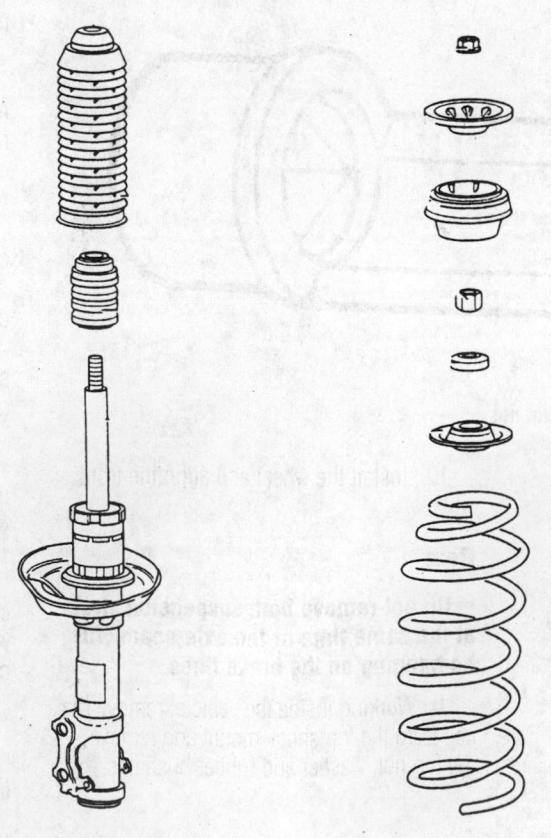

7923WG36

Exploded view of the front strut—1995–97 Passat

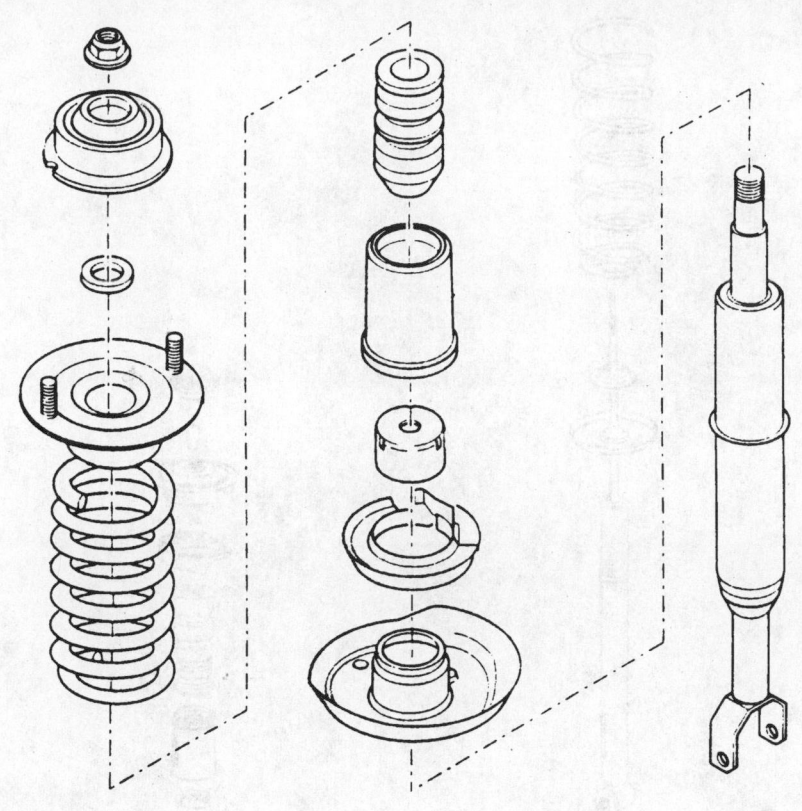

7923WG37

Exploded view of the front strut—1998–99 Passat

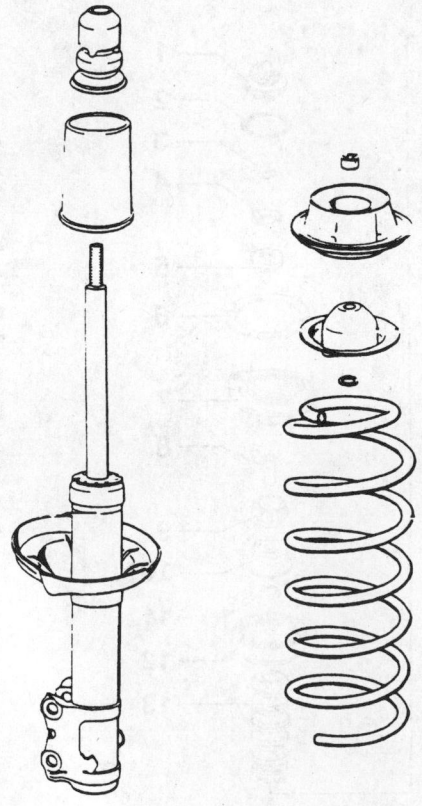

7923WG38

Exploded view of the front strut—except Passat

7923WG39

Exploded view of the rear strut—except Passat

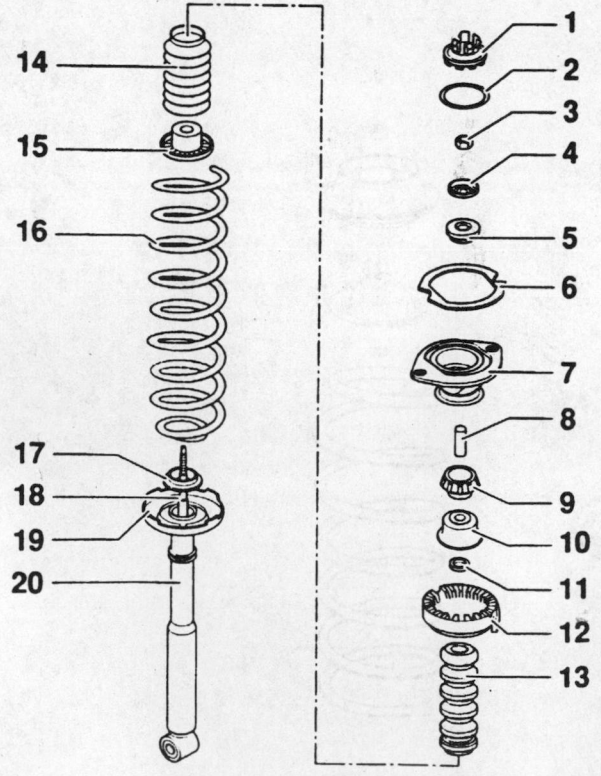

1. Cap
2. O-ring
3. Self-locking nut
4. Dished washer
5. Upper rubber mounting
6. Foam gasket
7. Strut bearing
8. Spacer tube
9. Lower rubber mounting
10. Metal cap
11. Washer
12. Spring seat
13. Stop buffer
14. Protective tube
15. Plastic cap
16. Coil spring
17. Packing
18. Circlip
19. Lower spring plate

Exploded view of the rear strut—1995–97 Passat

7923WG40

Upper Ball Joint

REMOVAL & INSTALLATION

1998–99 Passat

The 1998–99 Passat front suspension is equipped with two separate upper ball joints that are not replaceable, the upper link (front or rear) must be replaced as follows:

1. Raise and safely support the vehicle, and remove the front wheels.
2. Remove clip No. 1 as shown. The clip does not have to be replaced.
3. Remove the pinchbolt and pull both control arms upward and out.
4. Cover the steering gear boot.
5. Remove the guide link ball joint and press off the joint.
6. Detach the ABS wheel speed sensor wire front the bracket on the brake caliper.
7. Support the suspension from excessive rebound travel.
8. Remove the lower strut mounting bolt.
9. Swing the wheel bearing housing aside.

10. Raise the hood and remove the rubber grommets from the plenum chamber.
11. Remove the upper strut-to-body mounting nuts.
12. Remove the strut together with the mounting bracket.
13. Clamp the strut in a vise with the protective jaw covers.
14. Remove the upper link bolts and detach both of the links.
15. Remove the bracket-to-strut mounting nuts, then separate.

To install:

16. Position the brackets and links as shown, and tighten the bracket-to-strut mounting nuts to 15 ft. lbs. (20 Nm).
17. Align the links as shown, then tighten to 37 ft. lbs. (50 Nm) plus ¼ turn (90°).
18. Install the strut with mounting bracket into the vehicle and tighten the upper strut-to-body mounting nuts to 48 ft. lbs. (75 Nm).
19. Install the lower strut mounting bolt and tighten to 66 ft. lbs. (90 Nm).
20. Install the nut on the ball joint and tighten to 74 ft. lbs. (100 Nm).
21. Install the upper links to the wheel

bearing housing and tighten the pinchbolt to 30 ft. lbs. (40 Nm).
22. Attach the ABS wiring to the brake caliper bracket.
23. Install the wheel, lower the vehicle, and check the front wheel alignment.

Lower Ball Joint

REMOVAL & INSTALLATION

1. Raise and safely support the vehicle, allowing the front wheels to hang. Remove the front wheels.
2. Remove the ball joint clamping bolt.
3. Pry the lower control arm down to remove the ball joint from the steering knuckle.
4. Remove the ball joint-to-lower control arm retaining nuts and bolts or drill out the rivets with a ¼ in. (6mm) drill. Remove the ball joint.

To install:

5. Install the new ball joint in the reverse order of removal. If no parts were installed other than the ball joint, no camber adjustment is necessary. Tighten the two control

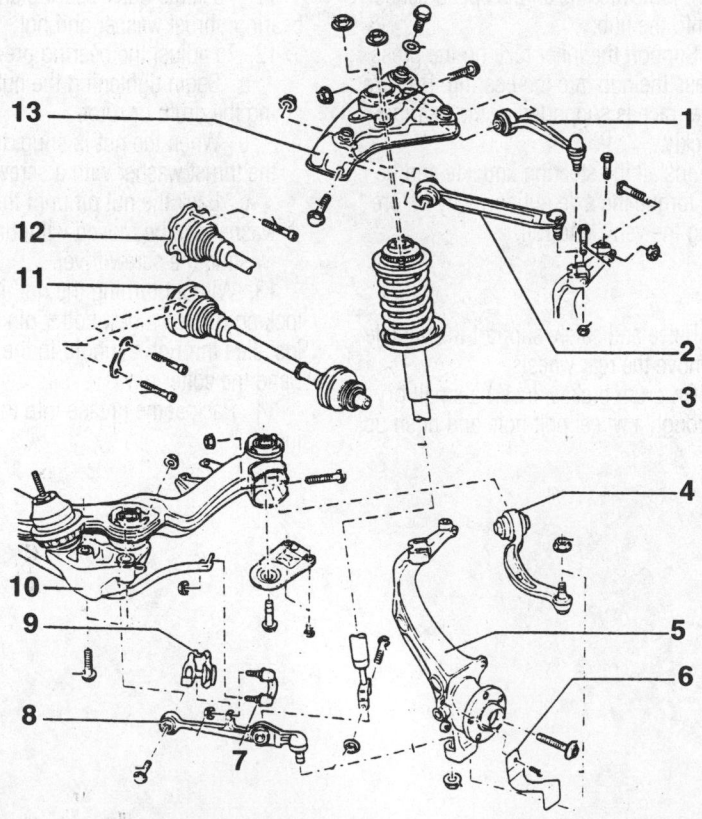

1. Upper link, rear
2. Upper link, front
3. Suspension strut
4. Guide link
5. Wheel bearing housing
6. Splash shield
7. Connecting link
8. Lower track control link
9. Clamp
10. Subframe
11. Halfshaft w/CV joint
12. Halfshaft w/triple-rotor joint
13. Mounting bracket

Exploded view of the front suspension—1998–99 Passat

7923WG41

arm-to-ball joint bolts to 18 ft. lbs. (25 Nm) and the ball joint clamping bolt to 37 ft. lbs. (50 Nm).

Wheel Bearings

ADJUSTMENT

Front

The front wheel bearings are sealed, no adjustment is necessary or possible.

Rear

1. Raise and support the vehicle safely.
2. Remove the grease cap.
3. Remove the cotter pin and the locking nut.
4. While turning the wheel, so the wheel bearing does not jam, tighten the adjusting nut firmly.
5. Back the nut off slightly. The nut is properly adjusted when it is possible to pry the thrust washer side to side with some drag by using finger pressure on the tool.
6. Install the locking nut and a new cotter pin. When installing the cap, be sure it is securely in place.

REMOVAL & INSTALLATION

Front

➡**The hub and bearing are pressed into the knuckle and the bearing cannot be reused once the hub has been removed.**

1. With the vehicle on the ground, remove the front axle nut. Raise and safely support the vehicle and remove the steering knuckle.
2. .Clamp the upper knuckle–to–strut bolt boss in a vice.
3. Install the special press tool onto the hub as shown and press the hub out of the bearing.
4. If the inner bearing race stayed on the hub, clamp the hub in a vise and use a bearing puller to remove it.
5. On the knuckle, remove the splash shield and internal snaprings from the bearing housing.
6. After removing the snapring, the same press tool can be used to push the bearing out of the knuckle.
7. Clean the bearing housing and hub with a wire brush and inspect all parts. Replace parts that have been distorted or discolored from heat. If the hub is not absolutely prefect where it contacts the inner bearing race, the new bearing will fail quickly.

To install:

8. The new bearing is pressed in from the hub side using a regular arbor press. Install the snapring and support the steering knuckle on the press.
9. Using the old bearing as a press tool, press the new bearing into the housing up against the snapring. Be sure the press tool contacts only the outer race of the bearing.
10. Install the outer snapring and splash shield. If removed, install the speed sensor rotor onto the hub.
11. Support the inner race on the press and press the hub into the bearing. Be sure the inner race is supported or the bearing fail quickly.
12. .Install the steering knuckle and be sure to torque the axle nut correctly before allowing the vehicle to roll.

Rear

1. Raise and safely support the vehicle and remove the rear wheels.
2. On drum brakes, insert a small pry-tool through a wheel bolt hole and push up on the adjusting wedge to slacken the rear brake adjustment.
3. On disc brakes, remove the caliper.
4. Remove the grease cap, cotter pin, locking ring, axle nut and thrust washer. Carefully remove the bearing and put all these parts where they will stay clean.
5. Remove the brake drum or rotor and pry out the inner seal to remove the inner bearing.
6. Clean all the grease off the bearings using solvent. If the bearings appear worn or damaged, they must be replaced.
7. To remove the bearing races, support the drum or rotor and carefully drive the race out with a long drift pin. They can also be removed on a press.

To install:

8. Carefully press the new race into the drum or rotor. The old race can be used as a press tool but be sure it does not become stuck in the hub.
9. Pack the inner bearing with clean wheel bearing grease and fit it into the inner race. Press a new axle seal into place by hand.
10. Lightly coat the stub axle with grease and install the drum or rotor. Be careful not to damage the axle seal.
11. Pack the outer bearing and install the bearing, thrust washer and nut.
12. To adjust the bearing pre-load:
    a. Begin tightening the nut while turning the drum or rotor.
    b. When the nut is snug, try to move the thrust washer with a screwdriver.
    c. Back the nut off until the thrust washer can be moved without prying or twisting the screwdriver.
13. Without turning the nut, install the locking ring so a new cotter pin can be installed through the hold in the stub axle. Bend the cotter pin.
14. Pack some grease into the cap and install it.

ENGINE REPAIR

➡**Disconnecting the negative battery cable on some vehicles may interfere with the functions of the on-board computer systems and may require the computer to undergo a relearning process, once the negative battery cable is reconnected.**

Distributor

REMOVAL & INSTALLATION

4-Cylinder Engines

1. Disconnect the negative battery cable.
2. Remove the protective cover and distributor cap.
3. Remove the rotor, dust cover and O-ring.
4. Remove the camshaft center bolt and rotor holder. Pull the distributor housing forward until the base rests against the rotor holder. Tap the distributor housing lightly to release the rotor holder from the shaft.

To install:

5. Install the distributor and tighten the rotor holder bolt to 52–66 ft. lbs. (70–90 Nm).
6. Install the distributor cap and protective cover. Connect the negative battery cable.

5-Cylinder Engines

1. Disconnect the negative battery cable.
2. Disconnect the electrical leads to the distributor.
3. Remove the distributor cap. Mark the position of the rotor in relation to the cylinder head.
4. Remove the distributor retaining bolts and pull the distributor out.

To install:

5. Clean the cylinder head where the distributor installs.
6. Install the distributor with the scribe marks aligned and secure in place with the retaining bolts.
7. Install the distributor cap and connect the electrical leads.
8. Connect the negative battery cable.

Ignition Timing

ADJUSTMENT

The B230 engine is equipped with a Bosch LH 2.4 control system or Bosch LH Lambda control system.

The B6304 engine is equipped with a Motronic 4.4 control system. And the B5234 and B5254 engines are equipped with a Motronic 4.3 control system. Each control system adjusts the ignition timing continuously. Manual adjustment of the ignition timing is not possible.

Engine Assembly

REMOVAL & INSTALLATION

4-Cylinder Engines

1. Disconnect the battery cables, negative lead first. Remove the battery.
2. If equipped with a manual transmission, remove the 4 retaining clips and lift up the shifter boot. Remove the snapring from the shifter.
3. Disconnect the windshield washer hose and engine compartment light wire. Scribe marks around the hood mount brackets on the underside of the hood for later alignment. Remove the hood.

❊❊ CAUTION

Never open, service or drain the radiator or cooling system when hot; serious burns can occur from the steam and hot coolant.

4. Remove the overflow tank cap. Drain the cooling system.
5. Remove the upper and lower radiator hoses. Disconnect the overflow hoses at the radiator. Disconnect the PCV hose at the cylinder head.
6. If equipped with an automatic transmission, disconnect the oil cooler lines at the radiator.
7. Remove the radiator and fan shroud.
8. Remove the air cleaner.
9. If equipped, disconnect the hoses at the air pump. Remove the air pump and drive belt.
10. Remove the vacuum pump and hoses. Disconnect the power brake booster vacuum hose.
11. Remove the power steering pump, drive belt and bracket. Position aside without disconnecting the hydraulic lines.
12. If equipped with A/C, remove the crankshaft pulley and compressor drive belt. Then, install the pulley again for reference. Disconnect the air conditioning wiring and remove the compressor from the bracket. Position the compressor aside without disconnecting the hoses. Remove the bracket.

13. Disconnect the vacuum hoses from the engine. Disconnect the carbon canister hoses.
14. Disconnect the distributor wire connector, high tension lead, starter cables and the clutch cable clamp.
15. Disconnect the wiring harness at the voltage regulator. Disconnect the throttle cable at the pulley and the wire for the A/C at the manifold solenoid.
16. Remove the gas cap. Disconnect the fuel lines at the filter and return pipe.
17. At the firewall, disconnect the electrical connectors for the ballast resistor and relays. Disconnect the heater hoses.
18. Disconnect the micro-switch connectors at the intake manifold and all remaining harness connectors to the engine.

❊❊ CAUTION

The EPA warns that prolonged contact with used engine oil may cause a number of skin disorders, including cancer! You should make every effort to minimize your exposure to used engine oil. Protective gloves should be worn when changing the oil. Wash your hands and any other exposed skin areas as soon as possible after exposure to used engine oil. Soap and water, or waterless hand cleaner should be used.

19. Drain the crankcase.
20. Remove the exhaust manifold flange retaining nuts. Loosen the exhaust pipe clamp bolts and remove the bracket for the front exhaust pipe mount.
21. From underneath, remove the front motor mount bolts.
22. If equipped with an automatic transmission, place the gear selector lever in **PARK** and disconnect the gear shift control rod from the transmission.
23. On manual transmission vehicles, disconnect the clutch controls, then drive out the pivot pin and remove the shifter from the control rod.
24. Disconnect the speedometer cable and the driveshaft from the transmission.
25. On overdrive equipped vehicles, disconnect the control wire from the shifter.
26. Raise and support the vehicle safely. Use a floor jack and a wooden block and support the weight of the engine beneath the transmission.
27. Remove the bolts for the rear transmission mount. Remove the transmission support crossmember.
28. Lift out the engine using the proper lifting equipment.

To install:

29. Install the engine assembly in the vehicle and tighten all engine mounting bolts. Install the transmission crossmember and remove the floor jack.

30. Install the driveshaft, speedometer cable, clutch controls (manual transmission) and gear selector mechanism.

31. Install the exhaust system.

32. Install the A/C compressor and related accessory drive units. Install all accessory drive belts. Install the vacuum pump.

33. Install the radiator and shroud. Install all vacuum, coolant and fuel lines and hoses. Connect all electrical connectors previously disconnected.

34. Install the hood, windshield wipers and battery with cables.

⁂ WARNING

Operating the engine without the proper amount and type of engine oil will result in severe engine damage.

35. After installing the remaining components, fill the engine with oil, the radiator with coolant and the transmission with fluid.

36. Adjust the reverse lock clamp and the gear selector. Adjust the throttle valve/pulley, automatic transmission kickdown cable and link rod.

37. Start the engine and allow it to reach normal operating temperature. Check the ignition timing. Check for leaks.

5-Cylinder Engines

1. Disconnect the negative battery cable.
2. Remove the battery and tray.
3. Raise and safely support vehicle.
4. On vehicles with automatic transmissions remove the air baffle from below the engine.

⁂ CAUTION

Never open, service or drain the radiator or cooling system when hot; serious burns can occur from the steam and hot coolant.

5. Remove the radiator expansion cap. Drain the coolant into a suitable container.
6. Remove the front wheels and disconnect both track rods from the axle. Remove both ball joints from the control arm.
7. Remove the ABS/brake hose bracket bolt. Remove both halfshafts.
8. Remove the right side engine mount retainer bolts.
9. Remove the torque rod bolt in the gearbox

➡Install transmission plugs in axle shaft holes to prevent fluid leakage.

10. Remove the front exhaust pipe lower nut and bolt from the bracket. Remove the two carriage bolts and skid plate. Disconnect the speedometer connection and remove the front and rear lower engine mount bolts. Lower the vehicle.
11. Remove the fresh air intake to the air cleaner, coil wires, throttle pulley cover and throttle cable from the pulley.
12. Tag and disconnect the throttle body inlet hose, idle air control valve, crankcase ventilation, preheat hoses and mass air flow sensor connector.
13. Disconnect the torque rod from the bracket and firewall.
14. Disconnect the ground strap from the firewall.
15. Unfasten the heated oxygen sensors and clips.
16. Remove the brake booster hose from the engine.
17. Remove the radiator and fan.
18. Remove the upper air charge pipe and fresh air intake from the radiator, then disconnect the vacuum hoses to the turbocharger and EGR regulator. Protect the radiator with a piece plywood.
19. Remove the radiator, expansion tank and coolant hoses
20. Remove the clutch slave cylinder retaining ring, if equipped. Be sure that the piston does not slip out. Remove the gear cable selector, after marking the position.
21. On automatic transmissions, mark the position, then remove the gear selector cable.
22. Remove the accessory drive belt
23. Remove the A/C compressor without disconnecting the lines and set it aside
24. Remove the starter.
25. Remove the fuel distribution manifold cover, injector covers, upper and lower fuel line clips and engine ground strap.
26. Install holders 999–5533 or equivalent on the injectors. Disconnect the fuel pressure regulator vacuum hose. Lift the fuel distribution manifold off and lay it aside.

➡Be sure that the injectors and needles are not damaged.

27. Disconnect and remove the wiring harness from the engine.
28. Lift up the air pump and lay it to one side. Install engine lifting yoke 999–2810 and arm 999–5428, or equivalents, and connect to hoist. Remove the front engine mount when the engine/transmission is secured.
29. Lift the engine out of the vehicle. On vehicles with automatic transmissions remove the turbo oil cooler lines and valve from the right side of the oil sump.

To install:

30. Install the engine using the lifting yoke and arm. Adjust the yoke to properly balance the engine as it goes in.
31. Tighten the front engine mount nut.
32. Install the A/C compressor, air pump and shield.
33. Install the wiring harness and bracket.
34. Lubricate the O-rings on the fuel injectors and install. Tighten the fuel distribution manifold.
35. Install the fuel distribution manifold cover and upper and lower clips.
36. Install the accessory drive belt.
37. Install the radiator, expansion tank, cooling fan and hoses. If equipped, connect the upper and lower oil cooler hoses.
38. Install the starter motor.
39. If equipped with a manual transmission, install the clutch slave cylinder and clips.
40. On vehicles with manual transmissions, install the reverse light switch wiring, rubberized gear selector cable mount, and the gear selector cables to its mount on the transmission. They are color coded for easy installation.
41. On vehicles with automatic transmissions, connect the electrical connectors and install the gear selector cables.

➡If the vehicle is equipped with cruise control, the vacuum hoses and motor wiring must be connected before the battery tray is installed.

42. Install the speedometer connector.
43. Install the front and rear engine mount bolts and tighten to 37 ft. lbs. (50 Nm).
44. Install a new torque rod bolt and nut in the transmission and tighten to 26 ft. lbs. (35 Nm) plus 90 degrees
45. Install the halfshafts, followed by the ball joints.
46. Install the remaining engine components making sure all wire connections are made, and all vacuum lines are attached.
47. Lower the vehicle and tighten the three exhaust pipe nuts. Tighten the bolts to 18 ft. lbs. (25 Nm)

⁂ WARNING

Operating the engine without the proper amount and type of engine oil will result in severe engine damage.

48. Fill the radiator and crankcase with new anti-freeze and engine oil.

49. Start the engine and run it until the thermostat opens.

50. Check the engine for leaks and add fluids as necessary.

6-Cylinder Engines

1. Disconnect the negative battery cables, negative first, then remove the battery.

2. Disconnect the ground lead connection to the body at the top of side member.

3. Remove the drive belt.

4. Remove the cooling fan.

5. Release the upper bolts and unfasten the connector at the relay in front of the battery. Disconnect the ground lead at the right-hand ground terminal.

✳✳ CAUTION

Never open, service or drain the radiator or cooling system when hot; serious burns can occur from the steam and hot coolant.

6. Drain the cooling system.

7. Remove the upper and lower radiator hoses from the engine. Remove the radiator overflow hose.

8. Remove the transmission cooler lines from the radiator.

9. Remove the top nut on both left and right side engine mounts.

10. Disconnect and remove the large and small crankcase ventilation hoses and the idle air hose. Disconnect the idle air valve wiring.

11. Disconnect and remove the two EVAP valve hoses at the intake manifold. Disconnect the air mass meter connector, air preheater hose and throttle pulley cover.

12. Remove the servo pump mounting bolts.

13. Disconnect and remove the fuel return line at the regulator and fuel line at the bulkhead. Remove the throttle cable, cruise control vacuum hose and fuel line snap catches.

14. Remove the engine wiring harness cover and disconnect the harness. Disconnect the relay connector. Remove the harness duct retaining nuts.

15. Disconnect the heater hoses at the bulkhead, ECC hoses at the intake manifold and brake servo vacuum hose. Disconnect the timing pick up and camshaft sensor connectors.

16. Support the engine at the rear using engine removal tool assembly 5033, 5006, 5115, 5428 and 5429, or equivalent that will support the engine from above.

17. Remove the splash guard and air baffle under the engine.

18. Remove the radiator.

✳✳ CAUTION

The EPA warns that prolonged contact with used engine oil may cause a number of skin disorders, including cancer! You should make every effort to minimize your exposure to used engine oil. Protective gloves should be worn when changing the oil. Wash your hands and any other exposed skin areas as soon as possible after exposure to used engine oil. Soap and water, or waterless hand cleaner should be used.

19. Drain the engine oil.

20. Disconnect the hose at the oil thermostat in the cylinder block.

21. Disconnect the A/C compressor wiring. Remove the compressor from the mount and set it aside without disconnecting the hoses.

22. Remove the exhaust pipe flanges at the manifold. Remove the lower section of the air preheater pipe and remove the exhaust pipe shield.

23. Remove the oil pipe connections at the gearbox. Plug the openings.

24. Remove the clips between the gear selector lever and control rod/reaction arm. Withdraw the rods from their mounting.

25. Disconnect and remove the oxygen sensor wiring.

➡ **Before separating the driveshaft, mark the coupling halves for reassembly.**

26. Disconnect the driveshaft and transmission support member.

27. Install engine lifting tool (2810 or equivalent) and adjust the lifting yoke to ensure the engine is balanced. Position the wiring harnesses so as to avoid damage when lifting.

28. Remove the engine and transmission assembly from the vehicle.

To install:

29. Install the engine into the vehicle, guiding the engine mounting into position.

30. Install the mounting nuts. Tighten to 37 ft. lbs. (50 Nm).

31. Position a jack to support the transmission and disconnect the engine from the hoist.

32. Support the rear of the engine using support rails and lifting beam assembly. Remove the jack from beneath the transmission.

33. Using the transmission lifting tool (5972 or equivalent), raise the transmission. Tighten the bolted joints between the support member and side members. Tighten the transmission bump stop nut to 37 ft. lbs. (50 Nm).

34. Attach the control rod and reaction arm to the gear selector lever mounting. Install the locking clip.

35. Connect the oxygen sensor lead.

36. Install the driveshaft. Tighten the front and rear couplings, noting the marks made during removal.

37. Connect the preheater pipe to the exhaust pipe. Tighten the sump bolts.

38. Install the A/C compressor.

39. Reconnect the hoses to the oil cooler. Tighten to 22 ft. lbs. (30 Nm).

40. Remove the lifting tools from rear of engine.

41. Install the heater hoses.

42. Install the timing pick up and camshaft position sensor connectors.

43. Connect the engine connector to the wiring harness at the left side wheel housing. Connect the relay and install the wiring duct retaining nuts.

44. Connect the fuel hoses, vacuum hoses and electrical connectors.

45. Install the throttle cable and throttle pulley cover.

46. Install the servo pump. Install the accessory drive belts.

47. Install the radiator and fan with the coolant hoses and transmission oil pipes.

48. Install the battery. Connect the lead at the right side wheel housing and battery positive leads.

49. Install the remaining components.

50. Fill the radiator and crankcase with new anti-freeze and engine oil.

51. Connect the negative battery cable. Start the engine and check for leaks.

Water Pump

REMOVAL & INSTALLATION

4-Cylinder Engines

1. Disconnect the negative battery cable.

2. Set the heater control to MAX heat.

✳✳ CAUTION

Never open, service or drain the radiator or cooling system when hot; serious burns can occur from the steam and hot coolant.

3. Remove the expansion tank cap. Open the draincock on the right-hand side of the engine block and on the radiator and drain the coolant into a suitable container.

4. Close the draincocks when the coolant is completely drained.

5. Remove the radiator shroud and fan

6. Remove the lower radiator hose at the water pump. If required, remove the retaining bolt for the coolant pipe beneath the exhaust manifold and pull the pipe rearward.

7. Remove the drive belts and water pump pulleys.

8. Remove the water pump bolts, washers and nuts. Remove the water pump assembly.

To install:

9. Clean the gasket contact surfaces thoroughly and use a new gasket and O-rings.

10. Install the water pump and tighten the bolts to 11–15 ft. lbs. (14–19 Nm). Install the coolant pipe and lower radiator hose. Install the accessory drive belts and water pump pulley.

11. Install the fan and shroud. Connect the negative battery cable.

12. Fill the coolant system with coolant. Start the engine and allow it to reach normal operating temperature. Check for leaks. Add coolant as necessary.

5-Cylinder Engines

1. Disconnect the negative battery cable.

2. Raise and safely support vehicle. Remove the splash guard from below the engine.

✳✳ CAUTION

Never open, service or drain the radiator or cooling system when hot; serious burns can occur from the steam and hot coolant.

3. Drain the cooling system.

4. Remove the following:
- Spark plug cover
- Fuel line clips
- Expansion tank
- Front timing cover
- Accessory belts

5. Remove the timing belt.

6. Remove the water pump and clean the cylinder block where the two mate.

To install:

7. Install the new water pump and gasket, and tighten the bolts to 15 ft. lbs. (20 Nm).

8. Install the timing belt.

9. Install the following:
- The two fuel line clips
- Front timing cover
- Accessory belts
- Spark plug cover
- Vibration damper guard

- Wheel well panel
- Wheel

10. Connect the negative battery cable.

11. Fill the cooling system. Run the engine to normal operating temperature. Top off as necessary and check for leaks.

6-Cylinder Engines

1. Disconnect the negative battery cable.

✳✳ CAUTION

Never open, service or drain the radiator or cooling system when hot; serious burns can occur from the steam and hot coolant.

2. Drain the cooling system by opening the draincock on the right side of the cylinder block.

3. Remove the timing belt.

4. Remove the water pump retaining bolts (7) and remove the water pump.

To install:

5. Before installing the water pump, clean the mating surfaces.

6. Install the water pump, using a new gasket. Tighten and tighten the mounting bolts 15 ft. lbs. (20 Nm).

7. Install the timing belt.

8. Fill the cooling system. Connect the negative battery cable.

9. Start the engine and check for leaks.

Cylinder Head

REMOVAL & INSTALLATION

4-Cylinder Engines

1. Disconnect the battery.

✳✳ CAUTION

Never open, service or drain the radiator or cooling system when hot; serious burns can occur from the steam and hot coolant.

When installing the cylinder head with the camshaft set at TDC, be sure the water pump O-ring sits correctly in the groove—4-cylinder engine

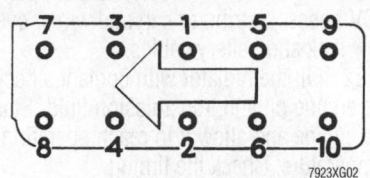

Cylinder head bolt torque sequence—4-cylinder engine

2. Remove the overflow tank cap and drain the coolant. Disconnect the upper radiator hose.

3. Remove the distributor cap and wires.

4. Remove the PCV hoses.

5. Remove the EGR valve and vacuum pump.

6. Remove the air pump, if equipped, and air injection manifold. Disconnect and remove all hoses to the turbocharger, if equipped. Plug all open hoses and holes immediately.

7. Remove the exhaust manifold and header pipe bracket.

8. Remove the intake manifold.

9. Remove the fuel injectors.

10. Remove the valve cover.

11. Remove the fan and shroud.

12. Remove the timing belt

13. Loosen the cylinder head bolts by reversing the torque sequence. Remove the cylinder head.

To install:

14. Check the position of the crankshaft. No. 1 piston should be at TDC. Check the position of the camshaft for cylinder No. 1. Both lobes should be in such a position that if the head were installed, the valves would be closed.

15. Install the cylinder head gasket and the cylinder head. Ensure that the O-ring for the water pump is in place. Apply a light coat of oil to the head bolts and install.

16. Tighten the head bolts in three steps using the proper sequence.

a. Step 1—Tighten all bolts to 14 ft. lbs. (20 Nm)

b. Step 2—Tighten all bolts to 43 ft. lbs. (60 Nm)

c. Step 3—Angle tighten all bolts an additional 90 degrees

17. Install the timing belt.

18. Install the shroud and fan. Install the drive belts and pulleys.

19. Install the intake manifold, fuel injection system, throttle cable and valve covers.

20. Install the exhaust manifold and header pipe. Install the air pump assembly. If equipped with a turbocharger, install the turbocharger and related parts.

21. Install the EGR valve, vacuum pump, PCV hoses, distributor cap and wires, over-flow tank and battery cables.

22. Fill the radiator with coolant, check the engine oil and transmission fluid. Start the engine and allow it to reach operating temperature. Check the timing.

5-Cylinder Engines

1. Disconnect the negative battery cable.

❊❊❊ CAUTION

Never open, service or drain the radiator or cooling system when hot; serious burns can occur from the steam and hot coolant.

2. Raise and safely support vehicle. Remove the splash guard below the engine. Drain the coolant into a suitable container. Disconnect the exhaust pipe from the manifold.

3. Remove the exhaust manifold. Remove the timing belt. Disconnect the fuel distribution manifold and lift it and the injectors off to one side. Use 999–5533 holders or equivalent to separate them. Disconnect the two ground straps from the engine.

➡**Be sure that the injectors and needles are not damaged.**

4. Remove the engine cooling fan.
5. Remove the intake manifold.
6. Remove the upper radiator hose from thermostat housing.
7. Remove the camshaft sprockets. Mark them intake or exhaust.
8. Remove the inner timing cover bolt.
9. Remove the air cleaner and hoses.
10. Remove the camshaft position sensor and damper.
11. Remove the distributor cap, wiring and rotor.
12. Remove the extension arm and brackets.
13. Working inwards from each end, loosen the bolts on the upper half of the cylinder head.
14. Gently tap the upper half with a soft mallet on the edges and front of the camshafts. Remove the bolts and upper half of the cylinder head.
15. Mark the camshafts and remove.
16. Remove the coolant pipe bolts. Remove the cylinder head bolts working inwards.
17. Remove the lower portion of the cylinder head and head gasket.
18. Clean all mating surfaces thoroughly.

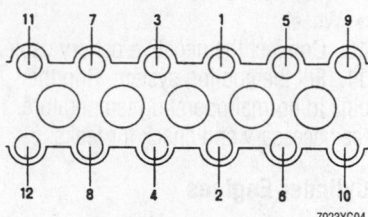

Lower half cylinder head torque order—5-cylinder engines

❊❊❊ WARNING

Do not use a metal scraper. Use a soft putty knife and gasket solvent cleaner with an exhaust fan. The surfaces must be totally clean to assure a tight seal.

To install:

19. Align the crankshaft timing marks. Install crankshaft locking tool 999–5451 or equivalent and turn the crankshaft counter-clockwise until it stops.

20. Install a new cylinder head gasket and the lower cylinder head. Apply a small amount of oil to the bolts.

21. Tighten the lower cylinder head in three stages, starting on the inside and working outward as follows:
    a. 15 ft. lbs. (20 Nm)
    b. 44 ft. lbs. (60 Nm)
    c. angle tighten an additional 130° using an angle gauge

22. Install the coolant pipe using a new gasket. Replace the O-rings in the spark plug wells. Remove No. 1 and No. 5 spark plugs.

23. Using a roller, apply liquid gasket 161–059–9 or equivalent to the upper cylinder head.

➡**Be sure that no liquid gasket gets into the oil passages. Only a thin coating is required.**

24. Install the camshafts and lock them in place using tools 999–5453 (front) and 999–5452 (rear) or equivalents.

25. Install the upper cylinder head. Pull the head down using press tools 999–5453 or 5454 (2) or equivalents. Tighten the upper half working from the inside outward. Tighten to 13 ft. lbs. (17 Nm). Remove tools 999–5453 and 999–5454 or equivalents.

26. Install the camshaft seals using an appropriate seal driver.

27. Mount the upper timing cover. Install the camshaft sprockets and line up the camshaft timing marks.

28. Install two camshaft sprocket bolts furthest from the timing mark and tighten

until they are just touching the sprocket. Remove the upper timing cover. Be sure that the remaining camshaft sprocket bolt hole is centered.

29. Install the tensioner pulley lever and tighten to 18 ft. lbs. (25 Nm). Install the idler pulley and tighten to 18 ft. lbs. (25 Nm).

30. Compress the tensioner fully with tool 999–5456 or equivalent and install a lock pin 2mm in diameter in the piston. If the tensioner leaks, has no resistance or will not compress, replace it.

31. Install the timing belt. Install the third camshaft sprocket bolt and tighten the bolts to 15 ft. lbs. (20 Nm). Remove the tensioner lock pin.

32. Remove the crankshaft locking tool from the flywheel end of the block and install the plug in the hole. Install the starter motor. Remove the camshaft locking tool 999–5452 or its equivalent.

33. Turn the crankshaft two complete revolutions and check that the timing marks are lined up.

34. Install the rear camshaft seal using driver 999–5450 or equivalent.

35. Install the upper timing cover.

36. Install the remaining engine components

37. Connect the negative battery cable. Change the engine oil. Fill the cooling system.

38. Start the engine and run it until the thermostat opens, top it off as necessary.

39. Check the engine for leaks.

6-Cylinder Engines

1. Disconnect the negative battery cable.

❊❊❊ CAUTION

Never open, service or drain the radiator or cooling system when hot; serious burns can occur from the steam and hot coolant.

2. Drain the cooling system.
3. Remove the front exhaust pipe, heat shield and exhaust manifold(s).

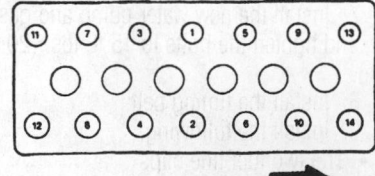

Tightening sequence for cylinder head bolts—6-cylinder engine

4. Remove the coolant pipe bolts.

5. Remove the timing belt.

6. Remove the transmission mounting plate bolt.

7. Remove the air mass meter and intake hose.

8. Remove the throttle pulley cover, throttle cable and cable bracket.

9. Disconnect the throttle switch lead and vacuum hoses at throttle housing and cruise control servo.

10. Remove the intake manifold outer section.

11. Mark the positions and remove the ignition coils.

12. Mark the camshaft pulleys (intake and exhaust sides) and remove the pulleys, using holding tool 5199 or equivalent.

13. Remove the camshaft sensor, ground terminals and temperature sensor connector. Remove the coolant hose at rear.

14. Carefully tap the top half of the cylinder head upwards, using a soft mallet.

15. Tap the joint lugs and camshaft front ends. Remove the camshafts.

16. Remove the cylinder head bolts, starting at the outside and working inwards. Lift the cylinder head from the engine. Remove the gasket.

17. Clean and inspect the cylinder head and block mating surface.

To install:

18. Align the crankshaft timing mark by removing the starter motor and installing the crankshaft locking tool 5451 or equivalent. Turn the crankshaft until it is stopped by the tool.

19. Fit a new cylinder head gasket and install the bottom half of the cylinder head. Oil the cylinder head bolts; install and tighten in sequence as follows:

   a. Stage 1—15 ft. lbs. (20 Nm)

   b. Stage 2—44 ft. lbs. (60 Nm)

   c. Stage 3—angle tighten 130 degrees

20. Install new O-rings in the spark plug wells and oil the camshaft bearing seats.

21. Apply sealing compound (Part No. 1161059–9 or equivalent) to the upper section of the cylinder head.

➡**Do not allow any compound to penetrate the coolant or oil passages.**

22. Install the camshaft.

23. Place the upper section of the cylinder head into position. Install the press tools (5454 or equivalent) and tighten against the lower section. Install the bolts, working from the inside outwards. Tighten to 13 ft. lbs. (17 Nm). Remove the tools.

24. Grease the camshaft front seal and tap the seal into place.

25. Place the upper timing cover into position. Install the camshaft pulleys while aligning the timing marks.

26. Temporarily install and tighten the pulley mounting bolts.

27. Remove the timing cover and install the mounting plate bolt.

28. Install the timing belt.

29. Loosen the camshaft pulley bolts and withdraw the tensioner locking pin.

Insert the remaining camshaft pulley bolt. Hold the pulley using the counterhold tool 5199 or equivalent and tighten all bolts alternately to 15 ft. lbs. (20 Nm).

30. Remove the crankshaft locking tool. Install the protective plug and install the starter motor.

31. Install the upper timing cover.

32. Check that the timing marks on the crankshaft and camshaft pulleys are correctly aligned.

33. Install the camshaft sensor, ground terminals and temperature sensor connector. Install the coolant hose at rear.

34. Install the remaining components.

35. Change the engine oil. Fill the cooling system.

36. Connect the negative battery lead. Start the engine and check for leaks.

37. Recheck the cooling system level.

Rocker Arms/Shafts

REMOVAL & INSTALLATION

These vehicles are not equipped with rocker arms/shafts. The camshaft is positioned so that the valves are directly actuated by the camshaft.

Turbocharger

REMOVAL & INSTALLATION

4-Cylinder Engines

1. Disconnect the negative battery cable.

2. Disconnect the expansion tank from the retainer. Remove the expansion tank retainer.

3. Remove preheater hose to the air cleaner. Remove the pipe and rubber bellows between the air/fuel control unit and the turbocharger unit. Pull out the crankcase ventilation hose from the pipe.

4. Remove the pipe and pipe connector between the turbocharger unit and the intake manifold.

➡**Cover the turbocharger intake and outlet ports to keep dirt out of the system.**

5. Disconnect the exhaust pipe and place aside.

6. Disconnect the spark plug wires at the plugs.

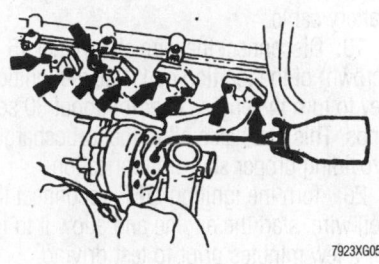

7923XG05

Fill the turbocharger inlet with oil, and tighten the bolts shown—4-cylinder engine

7. Remove the upper heat shield. Remove the brace between the turbocharger unit and the manifold.

8. Remove the lower heat shield by removing the retaining screw under the manifold.

9. Remove the oil pipe clamp, retaining screws on the turbo unit and the pipe connection screw in the cylinder block under the manifold. Do not allow any dirt to enter the oil passages.

10. Remove the manifold retaining nuts and washers. Leave one nut in place to keep the manifold in position.

11. Remove the oil delivery pipe. Cover the opening on the turbo unit.

12. Disconnect the air/fuel control unit by loosening the clamps. Move the unit with the lower section of the air cleaner up to the right side wheel housing. Place a cover over the wheel housing as protection.

13. Remove the remaining nut and washer on the manifold. Lift the assembly forward and up. Remove the manifold gaskets. Disconnect the return oil pipe O-ring from the cylinder block.

14. Disconnect the turbocharger unit from the manifold.

To install:

15. Be sure to use a new gasket for the exhaust manifold and a new O-ring to the return oil pipe. Keep everything clean during assembly and use extreme care to keep dirt out of the various turbo inlet and outlet pipes and hoses.

16. Install the turbocharger on the exhaust manifold and tighten the bolts as follows:

   a. Step 1—0.7 ft. lbs. (0.9 Nm)

   b. Step 2—30 ft. lbs. (40 Nm)

   c. Step 3—Tighten all bolts an additional 120 degrees (⅓ turn).

17. Install the exhaust manifold and turbocharger assembly on the engine. Connect

all oil pipes from and to the turbocharger using new O-rings.

18. Install the air/fuel control unit and air cleaner. Install the heat shields, spark plug wires, exhaust pipes, preheater assembly and expansion tank. Connect the negative battery cable.

19. Disconnect the wire at terminal 15 (brown) of the ignition coil. Use the ignition key to turn the engine over for about 30 seconds. This circulates oil to the turbocharger, providing proper start-up lubrication.

20. Turn the ignition **OFF**, reconnect the coil wire, start the engine and allow it to idle for a few minutes prior to test driving.

5-Cylinder Engines

1. Disconnect the negative battery cable.

2. Remove the heat shield from over the exhaust manifold.

3. Remove the upper air charge pipe and rubber hose from the turbo and move it to one side. Remove the fresh air intake hose and inner heat shield. Disconnect the upper turbo coolant return pipe and clamp off the hose, move it to the side. Disconnect the oil inlet pipe nipple.

4. Raise and safely support the vehicle. Remove or disconnect the following from under side:

- clamp between the pipes
- oil return pipe
- exhaust pipe bracket and bolt
- exhaust pipe to turbo nut

- exhaust manifold to turbo nuts
5. From the top side remove the exhaust pipe to turbo nuts. Disconnect the coolant inlet pipe to the turbo. Remove the turbo/exhaust manifold nuts.

6. Disconnect the following hoses from the turbo:

- red boost pressure
- white bypass valve
- yellow pressure regulator

7. Remove the turbo and the old pin bolts from the exhaust manifold.

To install:

8. Install new pin bolts with thread locking compound and tighten to 15 ft. lbs. (20 Nm). Install the turbo and connect the red, white, and yellow hoses to it.

9. Install the upper exhaust manifold nuts and tighten them lightly.

10. Working from under the vehicle, install the lower exhaust manifold nuts and tighten them to 18 ft. (25 Nm). On the top side, tighten the upper exhaust manifold nuts to 18 ft. lbs. (25 Nm). Tighten the exhaust manifold/turbo nuts to 22 ft. lbs. (30 Nm) and check that they are mated properly.

11. Under the vehicle install the oil pipe, grease the O-ring. Install the exhaust pipe bracket bolt.

12. Lower the vehicle and install or connect the following:

- oil inlet pipe
- inlet and outlet coolant pipes (be sure the clamps are removed)

- fresh air intake hose
- inner heat shield
- upper air charge pipe
- outer heat shield

➡**Replace the copper coolant pipe and upper oil pipe washers.**

13. Raise the vehicle and remove the clamp from coolant return hose.

14. Connect the negative battery cable.

15. Run the engine to check the boost pressure. Check oil and coolant levels.

➡**It may be necessary to reset a fault code after replacing the turbocharger.**

Intake Manifold

REMOVAL & INSTALLATION

4-Cylinder Engines

1. Disconnect the negative battery cable. Remove the air cleaner and all necessary hoses.

2. Remove the PCV valve.

3. Disconnect the wiring and the fuel hose from the cold start injector. If necessary, remove the cold start injector.

4. Disconnect the wiring and the hoses at the auxiliary valve. If necessary, remove the auxiliary valve.

5. Remove the intake manifold brace.

6. Label and disconnect the vacuum hoses at the intake manifold.

7. Loosen the clamp for the rubber connecting pipe on the air-fuel control unit.

8. Remove the manifold bolts and manifold.

To install:

9. Clean the gasket mating surfaces thoroughly. Install the intake manifold, using new gaskets, and tighten the bolts to 15 ft. lbs. (20 Nm).

10. Install the intake manifold brace and the air-fuel control unit connecting pipe.

11. Install and connect the auxiliary valve, cold start injector and PCV valve.

12. Connect all vacuum hoses and electrical connectors.

13. Connect the negative battery cable, start the engine and bring it to normal operating temperature.

5-Cylinder Engines

1. Disconnect the negative battery cable.

2. Remove the injector cover and throttle cable. Unfasten the connectors and clips from the injectors. Remove the two clips holding the fuel line. Remove the distribution manifold mounting bolts.

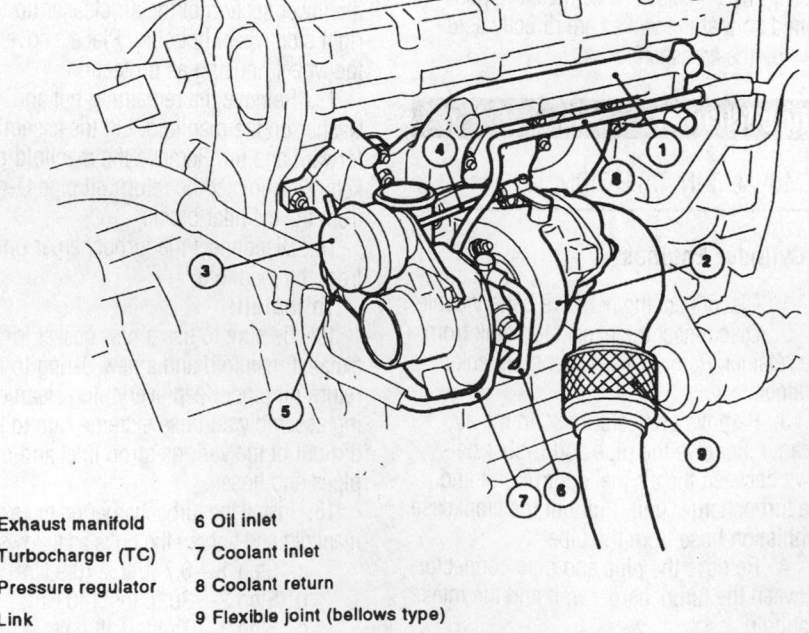

1 Exhaust manifold	6 Oil inlet
2 Turbocharger (TC)	7 Coolant inlet
3 Pressure regulator	8 Coolant return
4 Link	9 Flexible joint (bellows type)
5 Bypass valve	

7923XG06

Turbocharger and exhaust manifold component identification—5-cylinder engines

3. Install the five 999 5533 or equivalent injector holders and carefully pull the injectors off with the distribution manifold.

4. Disconnect the hose to the purge valve. Carefully lay the distribution manifold and injectors on the engine.

⁂ WARNING

Be sure that the injectors and needles are not damaged.

5. Disconnect the throttle linkage from the pulley.

6. Disconnect the intake air hose to the throttle body.

7. Remove the multi-nipple.

8. Remove the throttle pulley with bracket and idle speed control valve

9. Remove the EGR hose clamp on turbo models.

10. Remove the pressure line to turbo instrumentation/EGR valve control.

11. Disconnect the vacuum hose.

12. Disconnect the brake booster hose.

13. Loosen the dipstick bracket and intake manifold lower bracket bolt. Loosen the lower intake manifold bolts several turns.

➡The lower intake manifold bolts are not through-bolts.

14. Remove the upper intake manifold bolts. Remove the intake manifold by lifting it up 0.75 in. (20mm).

15. Be sure the mating surfaces of the cylinder head and intake manifold is clean.

To install:

16. Install a new intake gasket.

17. Install the intake manifold and upper bolts. Tighten all bolts from inside to outside to 15 ft. lbs. (20 Nm).

18. Install the EGR valve (if equipped) with a new gasket.

19. Install the throttle body with a new gasket.

20. Install the throttle pulley and bracket.

21. Install the multi-nipple and connect the hoses.

22. Install the fuel distribution manifold.

23. Install the wiring and injector cover.

24. Install the remaining components.

25. Connect the negative battery cable.

26. Test run engine and check for leaks.

6-Cylinder Engines

1. Disconnect the negative battery lead.

2. Disconnect the harness at the air mass meter.

3. Disconnect the idle air valve wiring and air hose. Remove the flame trap holder and remove the intake hose.

4. Remove the throttle pulley cover.

5. Disconnect and remove the throttle switch wiring, throttle cable and bracket, cruise control vacuum servo and vacuum hoses at throttle the housing.

6. Remove the injector cover plate and distribution manifold retaining bolts (3).

7. Disconnect the pressure regulator vacuum hose and fuel line bracket.

8. Carefully lift out the injector and distribution manifold assembly.

9. Remove the air preheater hose. Remove the left and right side power stage connectors on the bottom of the manifold. Remove the manifold bottom mounting.

10. Disconnect the brake servo hose and vacuum hoses under the manifold.

11. Cut away the clamps securing the rubber sleeves between the manifold sections, and lift out the outer manifold section.

12. Remove the upper bolts and loosen the lower bolts. Remove the inner section of the manifold.

To install:

13. Install the inner section of the manifold, using a new gasket. Install the rubber sleeves on the inner section and lubricate the free ends with petroleum jelly. Install the mounting bolts and tighten to 15 ft. lbs. (20 Nm).

14. Route the wiring between the second and third branches of the outer manifold section. Place the manifold against the lower section and connect the crankcase ventilation hoses.

15. Insert the manifold branches in the rubber sleeves. Secure with new Oetiker clamps.

16. Tighten the manifold lower mounting. Reconnect the vacuum hoses, brake servo hose, power stage connectors and air preheater hose.

17. Inspect the injector O-rings. Lubricate with petroleum jelly.

18. Reconnect the fuel pressure regulator vacuum hose.

19. Press the fuel distribution manifold into position. Tighten the manifold.

20. Reconnect the injector harnesses and EGR vacuum hoses. Install the injector cover.

21. Install the throttle cable, throttle pulley cover and vacuum hoses (cruise control and throttle housing).

22. Install the cable bracket at the throttle pulley. Reconnect the PCV, idling valve wiring, air hose, air mass meter and throttle housing connector.

23. Connect the negative battery lead. Start the engine and check operation.

Exhaust Manifold

REMOVAL & INSTALLATION

4-Cylinder Engines

1. Disconnect the negative battery cable. Remove the air cleaner and all necessary hoses.

2. Remove the EGR valve pipe from the manifold.

3. Remove the exhaust pipe from the exhaust manifold.

4. Remove the manifold bolts and manifold.

To install:

5. Position and install the manifold using a new gasket. Tighten the manifold bolts to 10–20 ft. lbs. (14–27 Nm).

6. Install the remaining components.

7. Connect the negative battery cable.

5-Cylinder Engines

1. Disconnect the negative battery cable.

2. Raise and safely support the vehicle. Disconnect the exhaust pipe from the manifold by removing the nuts on the flanged joint. Remove the carriage bolts from the manifold.

3. Remove the two heat shields from the exhaust manifold. Remove the exhaust manifold bolts. Turn the manifold 90° to the right and lift it out from the top.

⁂ WARNING

When removing or installing the exhaust manifold, be careful not to damage the air conditioning pressure switch, if so equipped.

To install:

4. Check the gasket surface of the cylinder head, clean if necessary. Install the exhaust manifold using new gaskets. Line up the exhaust manifold with the pipe using the carriage bolts.

5. Install the exhaust manifold bolts using a locking compound on the threads. Tighten the bolts to 18.5 ft. lbs. (25 Nm).

6. Install the heat shields. Tighten the carriage bolts using thread sealing compound. Tighten the nuts to no more than 86 inch lbs. (10 Nm). Remember to install the springs and washers with the nuts.

7. Connect the negative battery cable.

8. Run the engine and check for leaks.

6-Cylinder Engines

1. Disconnect the negative battery lead.

2. Remove the exhaust pipe mounting nuts at the manifold joints.

3. Remove the heat shield retaining bolts and heat shield.

4. Remove the exhaust manifold mounting nuts. Remove the exhaust manifold and gasket.

To install:

5. Before installation, clean the manifold and cylinder head mating surfaces.

6. Fit a new gasket and place the exhaust manifold into position. Install the mount lifting lug on the studs between the 3rd and 4th exhaust branches. Tighten the studs to 15 ft. lbs. (20 Nm) and mounting nuts to 18 ft. lbs. (25 Nm).

7. Install the heat shield to the rear manifold. Tighten to 11 ft. lbs. (15 Nm).

8. Install the front exhaust pipe to manifold. Using thread locking compound, tighten to 44 ft. lbs. (60 Nm).

➡**Loosen the joint at the catalytic converter and re-tighten to 18 ft. lbs. (25 Nm). This is necessary to prevent stresses in the system.**

9. Connect the negative battery lead. Start the engine and check for leaks.

Front Crankshaft Seal

REMOVAL & INSTALLATION

4-Cylinder Engines

1. Disconnect the negative battery cable.

2. Remove the cooling fan and shroud.

3. Remove the drive belts and water pump pulley.

4. Remove upper timing belt cover.

5. Set the crankshaft to TDC and remove the timing belt.

➡**Do not turn the crankshaft or camshaft. Pistons may strike valves.**

6. Carefully pry loose the seal to be replaced. Do not damage the contact face.

To install:

7. Clean the contact faces. Lubricate the seal and seat,, then press the seal into position.

8. Install the timing belt. Be sure the timing belt is correctly positioned.

9. Install the timing belt cover and vibration damper. Tighten the crankshaft center bolt to 45 ft. lbs. (60 Nm) plus an additional 60 degrees.

10. Install the drive belts and water pump pulley.

11. Install the cooling fan and shroud.

12. Connect the negative battery cable. Start the engine. Check for leaks and proper operation.

5-Cylinder Engines

1. Disconnect the negative battery cable.

2. Remove the fuel line clips.

3. Lift the coolant expansion tank and place it on top of the engine.

4. Remove the drive belts.

5. Remove the front timing cover.

6. Raise and safely support the vehicle.

7. Remove the right front wheel and loosen the inner fender liner. Remove the vibration damper guard and turn crankshaft pulley until all timing marks align.

8. Remove the timing belt.

✻✻ WARNING

Do not turn the crankshaft or camshafts once the timing belt has been removed.

9. Install a universal puller so the claws pull against the bolts and not the sprocket. Pull the sprocket off.

✻✻ WARNING

Be sure that the puller does not damage the sprocket teeth.

10. Remove the front seal using a groove cut chisel. Clean the mating surface.

To install:

11. Install the crankshaft timing belt sprocket using the nut and a spacer.

12. Install the timing belt.

13. Turn the crankshaft two complete revolutions and be sure the timing marks on the crankshaft and camshaft pulleys align properly.

14. Install the two fuel line clips.

15. Install the remaining components.

16. Install the wheel.

17. Connect the negative battery cable.

18. Test run the engine.

6-Cylinder Engines

1. Disconnect the negative battery cable.

2. Remove the timing belt.

3. Remove the crankshaft pulley, using a suitable puller.

4. Carefully pry out the old seal.

To install:

5. Before installing the new seal, thoroughly clean the crankshaft face.

6. Lubricate the new seal and tap the seal into place, using tool 5455 or equivalent.

7. Install the timing belt.

8. Connect the negative battery cable.

Camshaft

REMOVAL & INSTALLATION

4-Cylinder Engines

1. Disconnect the negative battery cable.

2. Remove the drive belts.

3. Remove the timing belt.

4. Remove the valve cover.

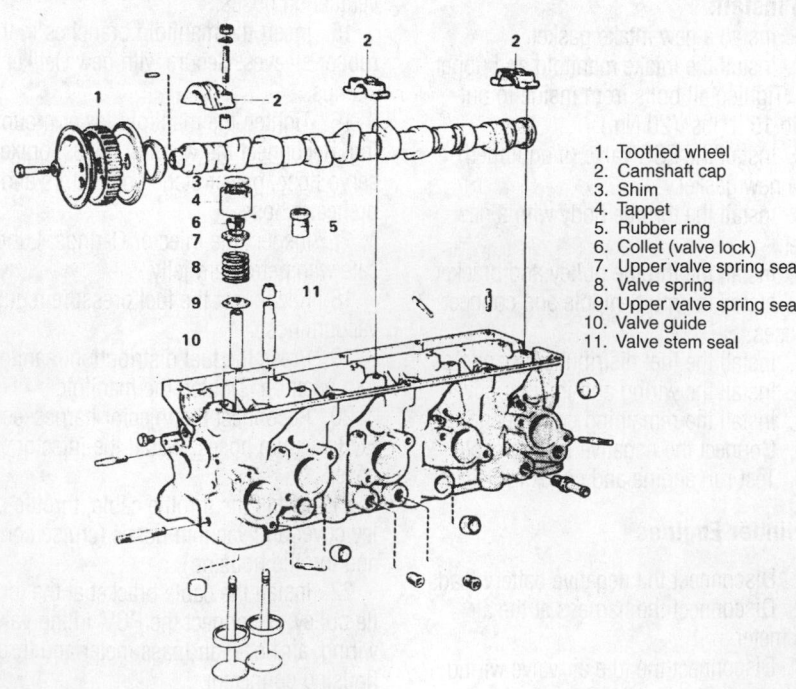

1. Toothed wheel
2. Camshaft cap
3. Shim
4. Tappet
5. Rubber ring
6. Collet (valve lock)
7. Upper valve spring seat
8. Valve spring
9. Upper valve spring seat
10. Valve guide
11. Valve stem seal

Exploded view of the cylinder head—4-cylinder engine

7923XG07

5. Remove the camshaft center bearing cap. Install camshaft press tool 5021 or equivalent over the center bearing journal to hold the camshaft in place while removing the other bearing caps.

6. Remove the 4 remaining bearing caps.

7. Remove the seal from the forward edge of the camshaft.

8. Release camshaft press tool and lift out the camshaft.

❊❊❊ WARNING

Do not rotate the crankshaft while the camshaft is removed from the cylinder head.

To install:

9. Apply sealant to the outer sealing surfaces of the front and rear caps. Lubricate and install the camshaft into position. The guide pin for the timing gear should face up.

10. Install the rear bearing cap. Slide the camshaft back and forth to check the camshaft end-play. End-play should be 0.004–0.016 in. (0.1–0.4mm).

11. Install the camshaft press tool. Install the camshaft seal. Lubricate and install the remaining caps. Tighten bolts to 14 ft. lbs. (20 Nm).

12. Lubricate the front seal and install, using tool 5025 or equivalent.

13. Install the camshaft gear and spacer washer. Install the timing belt.

14. Install the remaining components.

15. Connect the negative battery cable.

5-Cylinder Engines

1. Disconnect the negative battery cable.

2. Remove the drive belt.

3. Remove the timing belt.

4. Remove the ignition coils cover.

➡**Do not turn the crankshaft while the belt is removed.**

5. Remove the camshaft position sensor and shutter at the right rear of camshaft assembly. Remove the switch holder and shield at the left rear of assembly.

6. Remove the ignition coils. Mark their locations.

7. Remove the camshaft pulleys, using the holding tool 5199 or equivalent. Mark the pulleys intake and exhaust so they can be returned to their original side.

8. Remove the top half of the cylinder head. Tap the joint lugs and camshaft front ends lightly.

9. Remove the camshafts.

10. Thoroughly clean the mating sur-

faces between the upper and lower halves of the cylinder head.

❊❊❊ WARNING

Do not use a metal scraper. Use a soft putty knife and gasket solvent cleaner with an exhaust fan. The surfaces must be totally clean to assure a tight seal.

To install:

11. Lubricate the camshafts and bearing seats. Place the camshafts into position.

12. Install the holding tool 5453 or equivalent to the front end and the locking tool 5452 or equivalent to the rear end of the cylinder head upper section.

13. Remove No. 1 and No. 5 spark plugs

14. Using a roller, apply liquid gasket 161 059–9 or equivalent to the upper half of the cylinder head.

➡**Be sure that no liquid gasket gets into the oil passages. Only a thin coating is required.**

15. Install the upper cylinder head section and tighten against the lower section, using the press tools 5454 or equivalent.

16. Install and tighten the retaining bolts to 13 ft. lbs. (17 Nm), starting from the inside and working outwards. Remove the tools.

17. Lubricate the camshaft front seals and tap into place.

18. Mount the upper timing cover. Install the camshaft sprockets and line up the camshaft timing marks.

19. Install two camshaft sprocket bolts furthest from the timing mark and tighten until they are just touching the sprocket. Remove the upper timing cover. Be sure that the remaining camshaft sprocket bolt hole is centered.

20. Turn all the idler pulleys listening for bearing noise. Check to see that the contact surfaces are clean and smooth. Remove the tensioner pulley lever and idler pulley, lubricate the contact surfaces and bearing with grease. If the tensioner pulley lever or idler is seized, replace it.

21. Install the tensioner pulley lever and tighten to 18 ft. lbs. (25 Nm). Install the idler pulley and tighten to 18 ft. lbs. (25 Nm).

22. Compress the tensioner fully with tool 999 5456 or equivalent.

23. Install the timing belt.

24. Install the rear camshaft seal using drift 999 5450 or equivalent and press it carefully into position flush with the inner chamfer edge.

25. Install the remaining components

26. Connect the negative battery cable.

27. Start the engine and run it until the thermostat opens.

28. Check the engine for leaks.

6-Cylinder Engines

1. Disconnect the negative battery cable.

2. Remove the drive belts.

3. Remove the timing belt.

➡**Do not turn the crankshaft while the belt is removed.**

4. Remove the camshaft pulleys, using the holding tool 5199 or equivalent.

5. Remove the top half of the cylinder head. Tap the joint lugs and camshaft front ends lightly.

6. Remove the camshafts.

To install:

7. Lubricate the camshafts and bearing seats. Place the camshafts into position. Install the holding tool 5453 or equivalent to the front end and the locking tool 5452 or equivalent to the rear end of the cylinder head upper section.

8. Install the upper cylinder head section and tighten against the lower section, using the press tools 5454 or equivalent.

9. Install and tighten the retaining bolts to 13 ft. lbs. (17 Nm), starting from the inside and working outwards. Remove the tools.

10. Lubricate the camshaft front seals and tap into place.

11. Install the camshaft pulleys. Tighten the bolts alternately to 15 ft. lbs. (20 Nm).

12. Install the timing belt.

13. Install the tensioner and tighten the bolts to 18 ft. lbs. (25 Nm). Check that the timing marks on the crankshaft and camshaft pulleys are correctly aligned.

14. Install the remaining components.

15. Install the drive belts.

16. Connect the negative battery cable.

Oil Pan

REMOVAL & INSTALLATION

4-Cylinder and 6-Cylinder Engines

1. Disconnect the negative battery cable. Raise and support the vehicle safely.

❊❊❊ CAUTION

The EPA warns that prolonged contact with used engine oil may cause a number of skin disorders, including cancer! You should make every effort to minimize your exposure to used engine oil. Protective gloves should

be worn when changing the oil. Wash your hands and any other exposed skin areas as soon as possible after exposure to used engine oil. Soap and water, or waterless hand cleaner should be used.

2. Drain the engine oil.
3. Remove the splash guard, if equipped.
4. On 2.3L engines, perform the following steps;
   a. Remove the engine mount retaining nuts.
   b. Remove the lower bolt and loosen the top bolt on the steering column yoke.
   c. Slide the yoke assembly up on the steering shaft.
5. Raise and safely support the front of the engine.
6. Remove the retaining bolts for the front axle crossmember.
7. Remove the crossmember.
8. Remove the left engine mount.
9. Remove the pan support bracket.
10. Remove the pan bolts and remove the pan.
To install:
11. Clean the gasket mating surfaces thoroughly. Install the oil pan and using new gaskets, tighten the bolts to 8 ft. lbs. (11 Nm).
12. Lower the engine and install all engine mounts. Install the front crossmember and install the bolts.
13. On 2.3L engines, install the yoke assembly on the steering shaft and tighten the bolts to 18 ft. lbs. (24 Nm).

❈❈ WARNING

Operating the engine without the proper amount and type of engine oil will result in severe engine damage.

14. Install the splash guard, if equipped. Lower the vehicle and connect the negative battery cable. Fill the engine with oil.
15. Start the engine and allow it to reach operating temperature. Check for leaks.

5-Cylinder Engines

➡ **This procedure is performed with engine removed from the vehicle.**

1. Remove the oil filter.
2. Remove the oil pan bolts.
3. Carefully tap the oil pan to break the seal and remove the oil pan.
4. Remove the oil passage O-rings.
To install:
5. Thoroughly clean the mating surfaces of the cylinder block and oil pan.
6. Install new oil passage O-rings.

7. Apply a thin layer of gasket sealant to the engine block.
8. Install the oil pan and pan bolts.
9. Tighten the pan bolts to 12 ft. lbs. (17 Nm).
10. Install the oil filter.

Oil Pump

REMOVAL & INSTALLATION

4-Cylinder Engines

1. Remove the oil pan.
2. Remove the 2 oil pump retaining bolts.
3. Remove the oil pump and pull the delivery tube from the block.
To install:
4. Fit new sealing rings at either end of the delivery tube.
5. Install the pump with the delivery tube attached. Align the pipe to the block so that the seal does not become damaged. Tighten the two oil pump retaining bolts.
6. Attach the clamp for the oil trap drain hose to the oil pump bolts. Be sure the hose is securely clamped behind the oil pump shoulder. Do not shorten the hose.
7. Install the oil pan.
8. Fill the engine with oil. Start the vehicle and check the oil level.

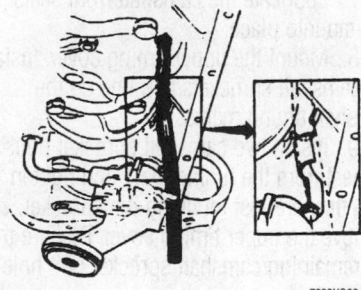

Be sure the hose from the oil trap is securely clamped behind the oil pump shoulder—4-cylinder engine

5-Cylinder Engines

The oil pump is on the front of the crankshaft.

1. Remove spark plug cover.
2. Remove the drive belts.
3. Remove the front timing cover and timing belt.
4. Raise and safely support the vehicle.

❈❈ WARNING

Do not turn the crankshaft or camshafts once the timing belt has been removed.

5. Remove the crankshaft damper, using tool 999 5433 or equivalent to counterhold it from moving.
6. Remove the crankshaft sprocket.

❈❈ WARNING

Be sure the puller does not damage the sprocket teeth.

7. Remove the old front seal using a groove cut chisel. Clean the mating surface where the seal lies.
8. Remove the four bolts retaining the oil pump. There are tabs on the oil pump housing located at the 6 o'clock and 11 o'clock positions.
9. Carefully pry out the oil pump using a groove cut chisel. Clean the surfaces where the pump mates to the engine.
To install:
10. Install the new oil pump using tool 999–5455 or equivalent using the bolts to guide it in. Use the crankshaft nut to press it in. Tighten the bolts alternately to 7 ft. lbs. (10 Nm). Install the crankshaft timing belt sprocket using the nut and a spacer.
11. Install the timing belt and cover.
12. Install the drive belts.

❈❈ WARNING

Operating the engine without the proper amount and type of engine oil will result in severe engine damage.

13. Fill the engine with clean engine oil.
14. Start the engine and check for leaks.

6-Cylinder Engines

1. Disconnect the negative battery cable.

❈❈ CAUTION

Never open, service or drain the radiator or cooling system when hot; serious burns can occur from the steam and hot coolant.

2. Drain the cooling system.
3. Remove the drive belts, front timing belt cover, cooling fan and splash guard. Remove the radiator.
4. Remove the timing belt.
5. Remove the crankshaft pulley, using a suitable puller.
6. Remove the oil pump mounting bolts and remove the oil pump.
To install:
7. Before installing the oil pump, thoroughly clean the mating surfaces.
8. Transfer the snow shield.
9. Place a new gasket into position,, then install the oil pump using tool 5455 or

equivalent. Use the mounting bolts as a guide. Pull in the pump using the crankshaft center nut.

10. Apply thread locking compound to the pump mounting bolts and install the bolts. Tighten alternately to 7 ft. lbs. (10 Nm).

11. Install the crankshaft pulley, using the center bolt and spacer.

12. Install the timing belt.

13. Install the tensioner. Align the timing marks and install the ignition coil cover.

14. Install the radiator.

15. Install the remaining components.

16. Connect the negative battery cable.

✵✵ WARNING

Operating the engine without the proper amount and type of engine oil will result in severe engine damage.

17. Check the engine oil level.

Rear Main Seal

REMOVAL & INSTALLATION

4-Cylinder Engines

1. Disconnect the negative battery terminal.

2. Remove the transmission.

3. Remove the clutch and pressure plate, if equipped.

4. Remove the pilot bearing snapring and remove the bearing.

5. Remove the flywheel or driveplate, as equipped.

➡**Be careful not to press in the activator pins for the timing device.**

6. Remove the rear oil pan brace.

7. Remove the 2 center bolts from the pan that bolt into the seal housing.

8. Loosen 2 bolts on either side of the 2 in the seal housing.

9. Remove the 6 seal housing bolts and remove the seal housing.

➡**Be careful not to damage the oil pan gasket when removing the seal housing.**

10. Remove the seal using special tool 2817 or a suitable replacement.

To install:

11. Use a new gasket on the seal housing and coat the seal with oil prior to installation. Install the seal.

12. Install the seal housing.

13. Install the rear oil pan brace and flywheel. Tighten the flywheel bolts to 47–54 ft. lbs. (64–73 Nm). When installing the fly-

wheel turn the crankshaft to bring the No. 1 piston to TDC. The lower flywheel pin should be installed approximately 15 degrees from the horizontal and opposite the starter.

14. Install the pilot bearing.

15. Install the clutch assembly and transmission, as required.

16. Connect the negative battery cable.

17. Fill the transmission with fluid.

18. Start the engine and allow it to reach operating temperature.

19. Check for leaks.

5-Cylinder engines

1. Disconnect the negative battery cable.

2. Raise and safely support the vehicle.

3. Remove the transmission.

4. Remove the flywheel if equipped with manual transaxle, or the flexplate if equipped with automatic transmission.

5. Using a seal puller or other suitable tool, remove the old seal. Take care not to damage the block surface during removal or new seal could leak.

To install:

6. Thoroughly clean sealing surface on the block.

7. Using special tools 999–5430 and 999–1801 or equivalent, install the new seal into the engine block.

8. Install the flywheel/flexplate, using thread locking compound on the bolts.

9. Tighten all the bolts in two stages:
   a. Tighten to 33 ft lbs. (45 Nm).
   b. Angle tighten 50°.

10. Install the transmission.

11. Lower the vehicle.

12. Connect the negative battery cable.

6-Cylinder Engines

1. Disconnect the negative battery cable.

2. Remove the transmission from the vehicle.

3. Remove the flexplate.

4. Carefully pry out the seal, taking care not to damage the sealing faces on the shaft and in seat.

To install:

5. Before installing the seal, thoroughly clean the seat and inspect for signs of wear.

6. Lubricate the mating surface between the seat and seal. Oil the seal lips and press the new seal into place, using a suitable seal installer tool 5430 and 1801 or equivalent.

7. Install the flexplate. Use new bolts and thread locking compound. Tighten the bolts in 2 stages: first to 33 ft. lbs. (45 Nm);, then tighten an additional 50 degree turn.

8. Install the transmission.

9. Connect the negative battery cable.

FUEL SYSTEM

Fuel System Service Precautions

Safety is the most important factor when performing not only fuel system maintenance but any type of maintenance. Failure to conduct maintenance and repairs in a safe manner may result in serious personal injury or death. Maintenance and testing of the vehicle's fuel system components can be accomplished safely and effectively by adhering to the following rules and guidelines.

• To avoid the possibility of fire and personal injury, always disconnect the negative battery cable unless the repair or test procedure requires that battery voltage be applied.

• Always relieve the fuel system pressure prior to disconnecting any fuel system component (injector, fuel rail, pressure regulator, etc.), fitting or fuel line connection. Exercise extreme caution whenever relieving fuel system pressure to avoid exposing skin, face and eyes to fuel spray. Please be advised that fuel under pressure may penetrate the skin or any part of the body that it contacts.

• Always place a shop towel or cloth around the fitting or connection prior to loosening to absorb any excess fuel due to spillage. Ensure that all fuel spillage (should it occur) is quickly removed from engine surfaces. Ensure that all fuel soaked cloths or towels are deposited into a suitable waste container.

• Always keep a dry chemical (Class B) fire extinguisher near the work area.

• Do not allow fuel spray or fuel vapors to come into contact with a spark or open flame.

• Always use a back-up wrench when loosening and tightening fuel line connection fittings. This will prevent unnecessary stress and torsion to fuel line piping. Always follow the proper tighten specifications.

• Always replace worn fuel fitting O-rings with new. Do not substitute fuel hose or equivalent, where fuel pipe is installed.

Fuel System Pressure

RELIEVING

1. Connect adapter 999–5484 or equivalent to fuel drainage unit 981–2270, 2273 and 2282 or suitable equivalent.

2. Remove protective cap from the valve on the fuel rail.

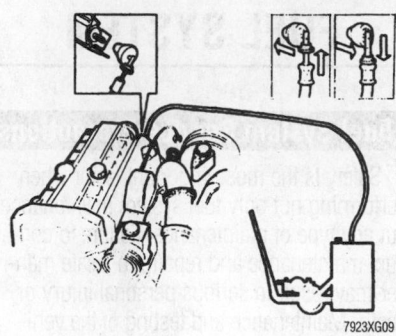

Connecting the adapter and drainage unit to the fuel rail—6-cylinder engine

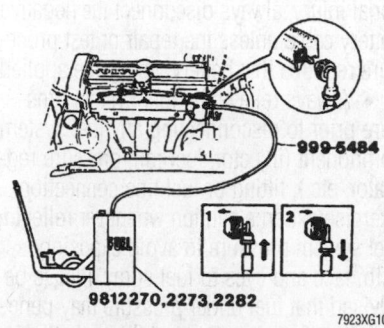

Connecting the adapter and fuel drainage unit—5-cylinder engines

3. Connect the adapter in the locked or closed position to the valve on the fuel rail.

4. Start the fuel drainage unit.

5. Unlock or open the adapter valve.

6. Raise and safely support the vehicle.

7. Remove the fuel filter valve cap.

8. Connect vent hose 999–5480 to the upstream valve of the fuel filter.

9. Drain the system for approximately 2 minutes.

10. When the system is drained, disconnect vent hose and install the valve cap.

11. Lower the vehicle and disconnect the adapter from the fuel rail. Install the valve cap.

12. Install the protective cap for the fuel rail and throttle pulley cover.

Fuel Filter

REMOVAL & INSTALLATION

4-Cylinder and 6-Cylinder Engines

> ✷✷ **CAUTION**
>
> **The following procedure will produce fuel vapors and slight fuel spillage. Be sure that there is proper ventilation and take appropriate fire safety precautions.**

The fuel filter is either on the left side of the firewall or next to the fuel pump near the left side of the fuel tank.

1. Disconnect the negative battery cable.

2. Relieve the fuel system pressure.

3. Remove the fuel filler cap.

4. Place a suitable container in position.

5. Hold a rag around the connections and loosen the fuel filter connections with flare nut wrenches.

6. Remove the clamp retaining the fuel filter to the bracket.

To install:

7. Transfer the bracket to the new filter.

8. Note the direction on the fuel filter and install the filter to the bracket.

9. Connect the fuel lines to the fuel filter. Check to ensure the copper seals are correctly installed.

10. Install the fuel filler cap.

11. Reconnect the negative battery cable.

5-Cylinder Engines

1. Disconnect the negative battery cable.

2. Relieve the fuel system pressure.

3. Raise and safely support vehicle.

4. Remove the quick disconnect couplers from the fuel filter. Remove the bolt for the filter bracket and remove the filter.

To install:

5. Install the new filter in bracket and attach with bolt.

6. Push the quick-connectors onto both ends of the fuel filter.

7. Lower the vehicle. Connect the negative battery cable. Turn the ignition switch **ON** to pressurize the fuel system. Check for leaks.

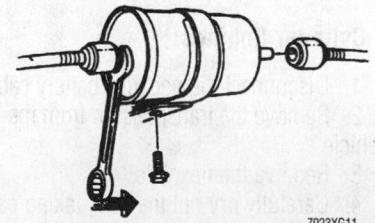

Use a 17mm open end wrench to push the quick disconnect coupler sleeves back—5-cylinder engines

Fuel Pump

REMOVAL & INSTALLATION

4-Cylinder and 6-Cylinder Engines

1. Disconnect the negative battery cable.

2. Raise and support the vehicle safely.

3. Relieve the fuel system pressure.

4. Remove the fuel tank.

5. Loosen the lock ring at the top of the fuel tank and remove the sending unit with the transfer pump attached. Note the direction of the float in the tank.

6. Remove the transfer pump from the sending unit.

To install:

7. Install the transfer pump on the sending unit. Install the sending unit in the fuel tank and tighten the lock-ring. Do not overtighten the lock ring, as the plastic threads on some fuel tanks are easily stripped.

8. Install the fuel tank in the vehicle.

9. Lower the vehicle.

10. Connect the negative battery cable, start the engine and check for leaks.

5-Cylinder Engines

1. Disconnect the negative battery cable.

2. Relieve the fuel system pressure.

3. Tilt the rear seat forward and remove or fold back the trunk compartment carpet over the right-hand wheel well panel.

4. Disconnect the fuel pump electrical connections.

➡**Take note of the color markings on the hoses; colored tape should identify hose locations on the pump.**

5. Disconnect the quick-connect couplers for the fuel delivery and return hoses.

6. Remove the pump unit by unscrewing the retaining nut using tool 999–5485 or equivalent. Lift the pump out carefully and remove the rubber seal. When lifting the pump out, do not grab the connections with pliers of any other sharp tools that might cause damage and result in fuel leakage.

To install:

> ✷✷ **WARNING**
>
> **Install the retaining nut while the pump is removed, otherwise the tank**

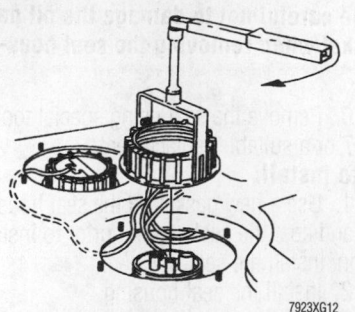

Be sure to use a new seal when installing the fuel pump retaining nut—5-cylinder engines

connection may swell and the nut will be difficult to install

7. Install a new dry seal, making sure that it is seated properly. Lubricate the top and outer side of the seal with petroleum jelly.

8. Install the pump with the heater connection facing towards the right side of the vehicle.

9. Install the fuel pump retaining nut and tighten it to 30 ft. lbs. (40 Nm) using tool 999–5484 or equivalent.

10. Apply a small amount of petroleum jelly to the delivery and return hose ends and install them on the pump. The delivery line is marked with yellow tape, which should be matched to the yellow marked pump outlet. Be sure that the quick-connectors are properly seated on the pump.

11. Connect the electrical connections, making sure that they are in the correct position. Install the panels and carpets.

12. Connect the negative battery cable.

13. Run the engine and check its function.

DRIVE TRAIN

Transmission Assembly

REMOVAL & INSTALLATION

Manual

940, S90 AND V90 MODELS

1. Disconnect the battery. At the firewall, disconnect the back-up light connector.

2. Raise the front of the vehicle and install jackstands. Under the vehicle, loosen the setscrew and drive out the pin for the shifter rod. Disconnect the shift lever from the rod.

3. Drain the gear oil from the transmission's drain plug.

4. Inside the vehicle, pull up the shift boot. Remove the fork for the reverse gear detent. Remove the snapring and lift up the shifter. If overdrive-equipped, disconnect the engaging switch wire.

5. Disconnect the clutch controls and return spring at the fork.

6. Disconnect the exhaust pipe bracket from the flywheel cover. Remove the oil pan splash guard.

7. Using a floor jack and a block of wood, support the engine beneath the oil pan. Remove the transmission support crossmember.

8. Disconnect the driveshaft. Disconnect the speedometer cable.

9. Remove the starter.

❈❈ CAUTION

Support the transmission weight with a second jack or hoist before removing. Do not allow the transmission to hang partially removed on the input shaft.

10. Support the transmission using another floor jack. Remove the flywheel housing-to-engine bolts and remove the transmission by pulling it straight back.

To install:

11. Prior to installation, inspect the condition of the clutch and throwout bearing. Replace the bearing if it is scored or has been noisy in operation.

12. Carefully fit the transmission into place and start all the mounting bolts. After all bolts are started, tighten to 30 ft. lbs. (39 Nm).

13. Reinstall the transmission crossmember. When secure, remove the jack from beneath the engine. Replace the splash guard and attach the exhaust bracket to the flywheel housing.

14. Install the starter.

15. Connect the driveshaft, the speedometer cable and if necessary, the overdrive wiring.

16. Fill the transmission with the correct type and amount of gear oil.

17. Reconnect the clutch controls and return spring to the fork.

18. Inside the vehicle, connect the shifter and the reverse gear detent fork. Connect the wiring for the overdrive switch and install the shift boot and cover.

19. Under the vehicle, connect the shifter rod to the shift lever. Don't forget to tighten the setscrew.

20. Connect the reverse light wiring and attach the negative battery cable.

Automatic

940, 960, S90 AND V90 MODELS

❈❈ CAUTION

If the vehicle has been driven within the last 3–5 hours, the transmission oil may still be hot. Use extreme care when draining the oil or handling components.

1. Disconnect the battery ground cable.

2. Place the gear selector in the **P** position. Disconnect the kickdown cable at the throttle pulley on the engine.

3. Disconnect the oil filler tube at the oil pan and drain the transmission oil.

4. Disconnect the control rod at the transmission lever and disconnect the reaction rod at the transmission housing.

5. On AW71 transmissions, disconnect the wire at the solenoid slightly to the rear of the transmission output flange.

6. Matchmark the transmission-to-driveshaft flange and unbolt the driveshaft.

7. Place a jack or transmission dolly under the transmission and support the unit. Remove the transmission crossmember assembly.

8. Disconnect the exhaust pipe at the joint and remove the exhaust pipe bracket from the exhaust pipe. Remove the rear engine mount with the exhaust pipe bracket.

9. Remove the starter motor.

10. Remove the cover plate at the torque converter housing.

11. Disconnect the oil cooler lines at the transmission.

12. Remove the upper bolts at the torque converter cover. Remove the oil filler tube.

➡ **It is helpful to have another person steadying and guiding the transmission during the removal process.**

13. Remove the lower bell housing bolts.

14. Remove the bolts retaining the torque converter to the drive plate. Pry the torque converter back from the drive plate with a small prybar.

15. Slowly lower the transmission while pulling it back to clear the input shaft.

❈❈ WARNING

Do not tilt the transmission forward or the torque converter may fall out.

To install:

16. Install the two lower bolts in the casing as soon as the transmission is in place. On 2.3L engines, adjust the panel between the starter motor and torque converter casing and install the bolts for the starter.

17. Mount the oil filler tube at the oil pan but do not tighten the nut.

18. Install the tube bracket and the two upper bolts in the converter casing. Tighten the nut for the oil tube to 65 ft. lbs. (85 Nm).

19. Install the bolts for the torque converter. Hand-tighten the bolts first,, then tighten in a crisscross pattern to 32 ft. lbs. (42 Nm).

20. Install the rear engine mount with the exhaust pipe bracket and reconnect the exhaust system.

21. Install the transmission crossmember. When it is securely bolted in place, the supporting jack may be removed.

22. Reinstall the driveshaft.

23. Making sure that both the transmission linkage and the shift selector in the vehicle are in the **PARK** position. Attach the actuator rod and the reaction rod. Adjust the shift linkage as necessary.

24. On AW71 models, install and connect the wiring to the solenoid valve.

25. Connect the kickdown cable at the throttle pulley. Adjust the cable if necessary.

26. Fill the transmission with oil. Connect the negative battery cable.

27. Apply the parking brake. Start the engine and allow to idle. Move the selector lever through all gear positions.

28. Place the selector lever in **P**. Wait 2 minutes and check the fluid level. Fill, as required.

Transaxle Assembly

REMOVAL & INSTALLATION

Manual

850, C70, S70 AND V70 MODELS

1. With all four wheels on the ground, loosen the front axle shaft locknuts.

2. Place the transaxle in **NEUTRAL** and set the parking brake.

3. Disconnect and remove the battery, air cleaner air intake ducts. Remove the battery tray. On turbocharged models, disconnect the timing valve from air cleaner and the turbocharger air duct clamp and hose.

4. Disconnect the gear selector cables from the brackets and lever. Remove the selector link plate by tapping out the lock pin. Disconnect the reverse light switch connector.

5. On turbocharged models remove the control pulley cover. Disconnect the turbocharger inlet pipe and tie it back out of the way. Disconnect the upper coolant hose to the engine oil cooler.

6. Disconnect the clutch slave cylinder and remove the clip. Remove the ground strap from the transaxle.

7. Loosen the rear engine mount and splash guard nut. Remove the bolts connecting the engine, transaxle and starter. Disconnect the transaxle ground strap.

8. Disconnect the ground strap from the firewall. Remove the torque arm bolt.

9. Secure the engine from above with an engine support that rests on the inner edges of the engine compartment.

10. Lift the engine up slightly to take weight off the engine mounts.

11. Raise and safely support the vehicle and remove the wheels. Disconnect the ABS sensor from the left side axle shaft, but do not unfasten the connector.

12. Drain the gear oil from the transaxle.

13. Disconnect all brackets for the front brake lines and ABS wiring for both sides of the vehicle.

14. Remove the plastic inner fender liners on both sides. Remove and discard the axle shaft locknuts.

15. Separate the ball joint from control arm, being careful not to damage the boots. Disconnect the sway bar links on both sides.

16. Remove the mounting screws holding the cable to the front of the subframe and disconnect the cable from the subframe. Disconnect the carbon filter hoses.

17. Disconnect the exhaust pipe clamp behind the catalytic converter.

18. Remove the left and right halfshafts.

➡**Be careful not to damage the transaxle seal.**

19. Install seal plugs in the transaxle.

20. Loosen the two right side subframe-to-body bolts approximately ½ in. (15mm). Remove the subframe-to-body bolts on the left side.

➡**Be sure the steering gear bolts come out of the subframe and the control arm is free of the axle shaft boot on the right side.**

21. Remove the jack and let the frame hang down from the right side bolts.

22. Tie the left side of the steering gear to the left side frame rail for support. Remove the steering gear engine mount bolt and nut at the top of the mount and remove.

➡**Be sure the steering gear is properly secured so the lower steering shaft does not slide out of the steering column.**

23. Disconnect the oxygen sensor wiring clamps from the cover, as well as the connector and wiring to the vehicle speed sensor. Remove the cover at the back of the engine and the mount from the transaxle.

24. Lower the engine and transaxle with the lifting hook.

✳✳ WARNING

If the engine is lowered too far, the exhaust pipe will be crushed against the steering rack. Be careful not to pinch any wiring or hoses and be sure that the engine dipstick tube is free of the fan.

25. Remove the seven remaining transaxle-to-engine bolts. Pull the gearbox away from the engine. Lower the jack and move the transaxle away.

To install:

26. Secure the throwout bearing fork to the transaxle. Be sure the mating surfaces on the transaxle and engine are clean and that the dowel pins are in place on the engine.

➡**Do not grease the primary shaft or throwout bearing sleeve. Be sure there are no breaks in the clutch plate.**

27. Lift the transaxle into place and mate to the engine. Install the seven bolts securing the engine and transaxle and tighten them a little at a time to draw the transaxle into place. Tighten the bolts to 37 ft. lbs. (50 Nm) and remove the transaxle jack.

28. Lift the engine and transaxle up until the distance between the engine support beam and spark plug cover is 0.20 in. (5mm).

29. Install the rear transaxle mount and bolts. Tighten the rear two bolts to 37 ft. lbs. (50 Nm),, then remove the front bolt. Install the cover.

30. Install the engine mount by fitting its guide pin into the cover. Install a new nut and hand-tighten. Install the steering rack engine mount bolt but do not tighten. Remove the support for the steering gear.

31. Reconnect the oxygen sensor wiring and clamps on the cover. Install the vehicle speed sensor connector and wiring and connect the transaxle ground strap.

32. Install the subframe using new 4 x M14 bolts and apply grease to the threads. Starting on the left side, lift the frame with a jack. Mount the support brackets on both sides. Tighten the frame bolts to 78 ft. lbs. (105 Nm), then tighten an additional 120 degrees. Tighten the bracket bolts to 37 ft. lbs. (50 Nm). Remove the jack and repeat the procedure for the right side.

33. Remove the engine support tool and lifting eyelet from engine. Tighten the engine mount nut to 37 ft. lbs. (50 Nm).

34. Install five new nuts on the steering rack and tighten them to 37 ft. lbs. (50 Nm). Install the front engine mount nut,, then tighten the front and rear bolts to 37 ft. lbs. (50 Nm).

35. Install the torque rod mount on the transaxle using new bolts. On earlier vehicles equipped with M18 bolts, tighten the bolts to 13 ft. lbs. (18 Nm), then an additional 90 degrees. On later models with M10 bolts, tighten to 26 ft. lbs. (35 Nm), then an additional 40 degrees.

36. Install the oil line bracket bolts and tighten to 19 ft. lbs. (25 Nm). Tighten the exhaust pipe clamp while rocking the pipe back and forth to seat it properly.

37. Install the right and left halfshafts.

➡**Be sure the transaxle axle seal and axle boot are not damaged.**

38. Connect the control arms to the ball joints using new nuts.

39. Connect the brake line and ABS cable bracket on both sides. Install the ABS sensor on the axle shaft and clean if needed. Tighten the sensor to 7.4 ft. lbs. (10 Nm).

40. Attach the cable pipe and carbon filter container to the subframe.

41. Install the front splash guard. Install the wheels.

42. Install the starter and tighten the bolts to 30 ft. lbs. (40 Nm). Connect the cable conduit and oxygen sensor connectors. Install the dipstick tube with a new O-ring and tighten the bolt to 19 ft. lbs. (25 Nm).

43. Connect the slave cylinder and clips.

44. Connect the reverse light connector, shift lever plate and pin. Install the cables and lubricate the levers, cables, washers and clips with grease.

45. Connect ground strip to the firewall. Install a new bolt and nut for the extension arm and torque rod.

46. Connect the oil cooler hose to the cooler, if equipped. Install the throttle body and cover over control pulley. Connect the intake manifold to the turbocharger.

47. Install the coolant expansion tank, battery tray, air cleaner and connectors. Connect the control valve to air cleaner on turbocharged models. Install the battery and attach leads.

48. Tighten the axle shaft nut to 89 ft. lbs. (120 Nm),, then tighten an additional 60 degrees. Lock the axle shaft nut by notching its flange into the axle shaft groove.

49. Fill the transaxle with the specified amount of oil. Reinstall the plug. Check the function of the clutch before driving.

Automatic

850, C70, S70 AND V70 MODELS

1. Pull the steering wheel adjustment lever out and adjust the wheel in and up as far as it will go. Then, lock it in position.

2. Put the transaxle gear selector in **N** and set the parking brake.

3. Disconnect and remove the battery and air cleaner assembly.

4. Remove the battery tray.

5. On turbo models, disconnect the control valve from air cleaner and the air charge manifold clamp and hose. Also remove the turbocharger air cleaner intake.

6. Disconnect the transaxle cable and connector from the transaxle. Be careful not to damage the rubber seal.

7. Remove the wiring harness and ground from the control system cover.

8. On early models, disconnect the transaxle vent hose.

9. On later models, disconnect the wiring and oxygen sensor from the transaxle brackets.

10. Disconnect the transaxle cooling lines from the quick disconnects and drain the transmission fluid.

11. Remove the dipstick and tube.

12. On vehicles with an EGR valve, disconnect the hoses to the valve.

13. On turbo models, remove the cover over the control pulley and disconnect the intake to the throttle body and pull it to one side so the throttle body is free. Seal all oil connections to prevent dirt from entering.

14. Remove the bolts connecting the engine and transaxle and starter.

15. Disconnect the transaxle ground strap.

16. Lift off the radiator overflow tank and let it hang.

17. Remove the torque rod extension arm bolt and swing it out of the way.

➡**It will be necessary to support the engine from above and still be able raise and lower the vehicle.**

18. Install lifting yoke 999–5534 or equivalent in place of the torque rod extension arm to lift from.

19. Install support tool 999–5033 on the inside fender rail, lifting beam support 999–5006 or equivalent placing the beam directly over the eye let for lifting yoke.

20. Install the lifting hook 999–5460 or equivalent and pull it up slightly until the load is taken off of the engine mounts. Measure the distance between the beam and spark plug cover and make note of it.

21. Raise and safely support the vehicle.

22. Remove the front wheels.

23. Disconnect the ABS sensor from the left side axle shaft but do not disconnect.

24. Disconnect all brackets for the brake lines and ABS wiring on both sides.

25. Remove the plastic inner fender wells on both sides.

26. Remove the transfer case (V70 AWD models only).

27. Remove the left and right side half-shafts.

28. Install a seal plug in the transaxle.

29. Remove all the splash guards.

30. Separate the ball joints from the control arms, being careful not to damage the boots.

31. Disconnect the sway bar links on both sides.

32. Remove the subframe cable mounting screws and disconnect them from the subframe.

33. Disconnect the carbon filter and hoses from the subframe. Cut the wire tie and hang the holder on the body.

34. Disconnect the exhaust pipe clamp behind the catalytic converter.

35. Remove the oil line bracket screws and torque rod holder mounting screws.

36. Back the engine mounting/steering gear bolt off one turn.

37. Remove the five steering gear mounting nuts in the subframe.

38. Position a jack under the left-hand side of the subframe so that it is barely touching.

39. Remove the subframe bracket bolts on the body. Back the 15mm bolts between the frame and body on the right-hand side several turns. Remove the bolts from the left.

➡**Be sure the steering gear bolts come out of the subframe.**

40. Remove the jack and let the frame hang down from the right side bolts.

41. Secure the end of the right side driveshaft on the oil lines.

42. Remove the steering gear engine mount bolt and nut at the top of the engine mount and remove the mount.

➡**Be sure the steering gear is properly secured so the lower steering shaft does not slide out of the steering column.**

43. Disconnect the oxygen sensor wiring clamps from the cover and the connector and wiring to the vehicle speed sensor.

44. Remove the cover at the back of the engine and the mount from the transaxle.

45. Lower the engine and transaxle with the lifting hook until the distance between the beam and spark plug cover is 12.6 in. (320mm).

✳✳ WARNING

If the engine is lowered too far, the exhaust pipe will be crushed against the steering rack. Be careful not to pinch any wiring or hoses and be sure that the engine dipstick tube is free of the fan.

46. Install transaxle fixture 5463 on the transaxle jack or equivalent, using the torque rod mounting bolts to hold it in place. At the same time, fit tool 5463–1 support plate or equivalent and raise the jack so that it is making light contact.

47. Remove the six torque converter bolts using a TX50 Torx® socket.

48. Remove the lower plastic nut and fold out the inner fender well on the right-hand side.

49. Then, turn the crankshaft with a socket and ratchet.

50. Remove the seven bolts between the engine and transaxle.

51. Remove the torque converter bolts from the flywheel.

52. Remove the transaxle making sure the torque converter comes out with it and does not slip off the shaft. Use the hole in the torque converter cover to press the torque converter in to keep it from sliding off.

✳✳ WARNING

Do not pry against carrier plate rim, as damage may result.

To install:

53. Flush the oil lines with clean transmission fluid.

54. Install the line and hose on the transaxle using new O-rings on the quick-connectors.

55. Install the hose on the upper transaxle cooler line catch pan under the return line.

56. Inspect all components before installing.

57. Apply a small amount of grease to the torque converter guide pin and install, making sure the converter is all the way into the transaxle. The distance between the cover and converter bolt flange should be 0.55 in. (14mm).

58. Install the transaxle securing in place with the seven bolts between the engine and transaxle. Tighten them to 37 ft. lbs. (50 Nm).

59. Install new torque converter bolts and torque to 22 ft. lbs. (30 Nm) using a TX50 Torx® socket.

➡**Remove the socket from the crankshaft.**

60. Install the rear transaxle mount and three bolts. Torque the rear two bolts to 37 ft. lbs. (50 Nm), then remove the front bolt.

61. Install the cover against the mount and torque the bolt to 37 ft. lbs. (50 Nm).

62. Install the engine mount guide pin into the cover. Install a new nut and hand-tighten.

63. Install the steering rack engine mount bolt, but do not tighten.

64. Reconnect the oxygen sensor.

65. Install the vehicle speed sensor connector and connect the transaxle ground strap.

66. Install the subframe using new bolts. Apply grease to the threads.

67. Starting on the left, lift the frame with a transaxle jack. Mount the support brackets on both sides.

68. Tighten the frame bolts to 78 ft. lbs. (105 Nm) plus an additional 120 degrees.

69. Tighten the bracket bolts to 37 ft. lbs. (50 Nm).

70. Remove the jack and repeat the procedure for the right side.

71. Install five new nuts on the steering rack and tighten them to 37 ft. lbs. (50 Nm).

72. Install the front engine mount nut, then tighten the front and rear bolts to 37 ft. lbs. (50 Nm).

73. Install the torque rod mount on the transaxle using new bolts, tighten the M18 bolts (early models) to 13 ft. lbs. (18 Nm) plus an additional 90 degrees or M10 bolts (later models) to 26 ft. lbs. (35 Nm) plus an additional 40 degrees.

74. Install the transfer case (if equipped).

75. Install the right and left halfshafts.

➡**Be sure the transaxle axle seal and axle boot are not damaged.**

76. Connect the control arms to the ball joints using new nuts.

77. Install the sway bar link using new nuts and tighten to 37 ft. lbs. (50 Nm).

78. Attach the cable pipe and carbon filter container to the subframe. Tie the hoses with a strip clamp in subframe.

79. Install engine splash guard on early models.

80. Install the front splash guard by pressing in the guides and installing the screws.

81. Install the five transaxle bolts on top side and tighten to 37 ft. lbs. (50 Nm).

82. Install the starter.

83. Connect the cable conduit and oxygen sensor connectors.

84. Install the dipstick tube with a new O-ring and tighten to 19 ft. lbs. (25 Nm).

85. Connect the wiring harness, ground lead, transaxle connectors.

86. Connect the transaxle vent and EGR hoses.

87. Attach the transaxle cable and adjust.

88. Connect the ground strip to the firewall.

89. Install a new bolt and nut for the extension arm and torque rod.

90. Tighten early model M8 bolts to 13 ft. lbs. (18 Nm) plus an additional 120 degrees. Torque later model M10 bolts to 26 ft. lbs. (35 Nm) plus an additional 90 degrees.

91. Install the remaining components.

92. Fill the transaxle with fluid.

93. Check the fluid level after the engine has reached normal operating temperature to assure that it is correct.

Clutch

ADJUSTMENT

The Volvo vehicles covered are equipped with a hydraulic clutch system that is self-adjusting.

REMOVAL & INSTALLATION

940, S90 and V90 Models

1. Remove the transmission.

2. Scribe alignment marks on the clutch and flywheel. Slowly loosen the bolts holding the clutch to the flywheel in a diagonal pattern. Remove the bolts and lift off the clutch and pressure plate.

3. Inspect the pressure plate for heat damage, cracks, scoring or any other damage.

4. Place a ruler diagonally over the pressure plate friction surface and measure the distance between the straightedge of the ruler and the inner diameter of the pressure plate. This measurement must not be greater than 0.008 in. (0.2mm). In addition, there must be no clearance between the straightedge and the outer diameter of the pressure plate. This check should be made at several points.

To install:

5. Clean the pressure plate and flywheel with solvent to remove any traces of oil and wipe them clean with a cloth.

6. Position the clutch assembly with the longest side of the hub facing away from the engine. Fit it to the flywheel and align the bolt holes. Insert centering tool 5111 or equivalent or an input shaft from an old transmission of the same type, through the clutch assembly and flywheel. This centers the assembly and pilot bearing.

7. Install the clutch retaining bolts and tighten them in a diagonal pattern, a few turns at a time. After all the bolts are tightened, remove the centering tool.

8. Install the transmission.

850, C70, S70 and V70 Models

1. Remove the transaxle.

2. Lock the flywheel in position. Remove the six bolts retaining the pressure plate and disc, loosen them in rotation. Remove the pressure plate and disc.

3. Remove the throwout bearing from the sleeve and fork. Remove the fork and dust cap from the transaxle.

4. Clean and check the throwout bearing. It should rotate freely and quietly. Clean and check the fork for cracks and wear and that the dust cap is intact.

5. Check the pressure plate carefully for signs of overheating, cracks, scoring or other damage to the friction surface. Be sure that the diaphragm spring is not split or damaged. If any part of the pressure plate is damaged, it must be replaced.

6. Check the pressure plate for warpage by laying a straightedge across the contact surface and checking the distance with a feeler gauge. The maximum width is 0.008 in. (0.2mm).

➡**Warpage is permitted in one direction only.**

7. Check the flywheel for cracks and heavy scoring. If it is damaged it must be replaced. Check the clutch disc for oil or dirt and clean if necessary.

To install:

8. Install the throwout bearing fork, and lubricate the ball joint with grease. Install the throwout bearing and dust cap. Secure the fork to the transaxle, so that it cannot be moved during installation.

9. Install the clutch disc and pressure plate using centering drift 999 5487 or equivalent clutch alignment tool. Install the clutch bolts, tightening them in rotation so that the clutch slides over the locating pins and lies evenly against the flywheel. Then, tighten the bolts to 18 ft. lbs. (25 Nm). Remove the centering drift.

➡**Do not apply any grease to the splines on the transaxle input shaft.**

10. Install the transaxle and check the clutch operation.

Hydraulic Clutch System

BLEEDING

✷✷ CAUTION

Use only DOT 4 brake fluid. Never reuse brake fluid. Always keep brake fluid well sealed in its original container.

1. Turn the ignition switch **OFF** and if equipped with ABS, remove the key to prevent accidental pump activation.

2. Raise and safely support the vehicle.

3. Clean the fluid reservoir filler cap. Remove the cap and fill the reservoir completely. Replace the reservoir cap.

4. Depress the clutch pedal a few times to purge the air bubbles in the master cylinder.

➡**Repeat this step if multiple bleeding steps are needed. Check the reservoir**

fluid level during bleeding, add as necessary.

5. Connect a hose from a drain bottle to the nipple on the slave cylinder.

6. While the clutch pedal is depressed to the floor, open the bleed nipple.

7. Hold the pedal to the floor to allow brake fluid and air bubble to exist through the hose. Close the nipple.

8. Repeat this procedure until no air bubbles are visible in the escaping fluid.

9. Be sure the bleed nipple is tight.

10. Pump the clutch pedal a few times to build pressure in the system.

11. Check the fluid reservoir. The fluid level should not be above the MAX level.

12. Road test the vehicle.

Transfer Case Assembly

REMOVAL & INSTALLATION

AWD 850, C70, S70 and V70 Models

1. Raise and safely support the vehicle.

2. Remove the right front wheel.

3. Remove the passenger side halfshaft from the transfer case.

4. Remove the transfer case vibration damper with support bracket.

5. Mark the flange, and remove the rear driveshaft from the transfer case.

6. Remove the transfer case-to-transaxle retaining bolts.

7. Remove the transfer case.

To install:

8. Install the transfer case on the transaxle. Be sure that the coupling sleeve is between the transaxle and the transfer case.

9. Tighten the retaining bolts to 37 ft lbs. (50 Nm).

10. Install the rear driveshaft.

11. Install the vibration damper and bracket.

12. Install the passenger side halfshaft.

13. Install the right front wheel. Tighten the lug nuts to 81 ft lbs. (110 Nm).

14. Lower the vehicle.

Halfshaft

REMOVAL & INSTALLATION

940, 960, S90 and V90 Models

1. With the vehicle on all four wheels, loosen the large axle nut in the center of the wheel hub.

2. Raise and safely support the rear of the vehicle.

3. Remove the wheel and remove the axle nut.

4. Remove the eight bolts holding the upper and lower sections of the final drive subframe section.

5. Remove the bolts holding the halfshaft to the final drive unit (differential) and remove the shaft from the wheel bearing housing.

6. When the shaft is removed, inspect the rubber boots for any sign of splitting or cracking. A light coat of silicone or vinyl protectorant applied to a CV-boot will extend its life.

To install:

7. When reinstalling, fit the axle into the hub first,, then position and secure the inboard end. Always use new, lightly oiled bolts and tighten them to 70 ft. lbs. (91 Nm). Do not tighten the axle nut yet.

8. Reinstall the lower section of the subframe section. Before tightening the eight mounting bolts, Install two long 12mm bolts or drift pins in the centering holes and align the panel. This is essential to insure correct wheel alignment when finished.

9. Tighten the eight mounting bolts to 52 ft. lbs. (68 Nm). plus an additional 30 degrees of rotation.

10. Use a new, lightly oiled axle nut and install it on the threaded end of the shaft. Tighten it until it is snug but do not attempt to apply final tightening.

11. Install the wheels. Lower the vehicle to the ground.

12. Apply the hand brake and tighten the axle nut to 103 ft. lbs. (134 Nm) plus an additional 60 degrees of rotation.

850, C70, S70 and V70 Models

1. With the vehicle sitting on all four wheels, loosen the axle shaft nut.

2. Raise and safely support the vehicle. Remove the wheels. Disconnect the ABS sensor from the halfshaft, but do not disconnect the harness.

3. Disconnect all brackets for brake lines and ABS wiring on both sides and let them hang.

4. Remove the axle nut. Push the end of the halfshaft from hub using a soft drift and a mallet.

5. Disconnect the sway bar from the link. Remove all splash guards.

6. Separate the ball joint from control arm, being careful not to damage the boots.

7. For the right side halfshaft, remove the bearing cap and pull the shaft out of the transmission while holding the strut out of the way. Install a seal plug in the transmission.

➡**Be careful not to damage the transmission seal.**

8. For left side, remove the halfshaft by carefully prying between the transmission and the halfshaft. Hold the strut assembly out of the way. Install a seal plug in the transmission.

To install:

9. Install the right halfshaft and tighten the bearing cap to 19 ft. lbs. (25 Nm). Install the splash guard.

➡**Be sure that the transmission axle seal and axle boot are not damaged.**

10. Clean the ABS wheel if necessary. Apply metal adhesive to the halfshaft splines. Carefully press shaft in so that the lock ring engages with the differential gear. Check it by carefully pulling on the shaft joint housing. Install the axle nut and hand-tighten.

11. Connect the ball joints using new nuts. Install the sway bar link using new nuts.

12. Connect the brake line and ABS cable bracket on both sides. Install the ABS sensor on the halfshaft and clean it with a soft brush.

13. Install the wheels.

14. With all four wheels on the ground, tighten the axle nut to 89 ft. lbs. (120 Nm) plus an additional 60 degrees. Lock the nut by staking its flange into the driveshaft groove.

STEERING AND SUSPENSION

Air Bag

❋❋ CAUTION

Some vehicles are equipped with an Air Bag system, also known as the Supplemental Inflatable Restraint (SIR) or Supplemental Restraint System (SRS). The system must be disabled before performing service on or around system components, steering column, instrument panel components, wiring and sensors. Failure to follow safety and disabling procedures could result in accidental Air Bag deployment, possible personal injury and unnecessary system repairs.

PRECAUTIONS

Several precautions must be observed when handling the inflator module to avoid accidental deployment and possible personal injury.

• Never carry the inflator module by the wires or connector on the underside of the module.

• When carrying a live inflator module, hold securely with both hands, and ensure that the bag and trim cover are pointed away.

• Place the inflator module on a bench or other surface with the bag and trim cover facing up.

• With the inflator module on the bench, never place anything on or close to the module which may be thrown in the event of an accidental deployment.

• Before beginning work which could affect the SRS system, turn the ignition switch **OFF**, disconnect the negative battery cable and tape the cable end to avoid accidentally recharging the Air Bag power supply.

• When working around the instrument panel or steering column, take special care to ensure that the SRS wires are not pinched, chafed or penetrated by bolts/screws etc. This is most likely to happen when installing the sound insulation, knee bolsters, ignition lock, steering column cover or additional telephone or stereo equipment.

• Never service the steering shaft or steering gear without first locking the contact reel and removing the steering wheel.

• If it is necessary to connect the battery with the Air Bag removed, install the special tool 998 8695 or an equivalent resistor. This tool has the same resistance as the Air Bag assembly and will prevent setting a fault code in the SRS control unit.

• Do not connect an ohmmeter across the Air Bag connector terminals. The current in the meter is enough to fire the Air Bag explosive charge.

DISARMING

❋❋ WARNING

Before beginning work on vehicles equipped with a Supplemental Restraint System (SRS), the system must be properly disarmed and the Air Bag must be removed and properly stored. Unintended Air Bag deployment can cause serious or fatal injury.

1. Disconnect the negative battery cable and tape the cable end. Allow sixty (60) seconds for the reserve power to dissipate.

2. Turn the ignition key to unlock the steering.

3. Turn the steering wheel slightly in order to reach the two Torx® bolts in back of the steering wheel. Unplug the connector and remove the Air Bag.

➡**Do not turn the ignition switch ON while the Air Bag assembly is removed. This will set a fault code in the SRS control unit.**

4. Store the Air Bag module where it will not be disturbed. Set it face up and do not place anything on top that may become a projectile in case of accidental deployment.

❋❋ CAUTION

When connecting the battery, be sure that no one is in the vehicle in case of an SRS malfunction causing accidental Air Bag deployment.

Rack and Pinion Steering Gear

REMOVAL & INSTALLATION

Power

850, C70, S70 AND V70 MODELS

❋❋ CAUTION

On vehicles with Air Bags, the front wheels must be pointing straight ahead with the steering wheel locked. If this is not done, the contact reel of the Air Bag system will reach its end position and deploy the Air Bag.

➡**The front subframe must be lowered. The bolts cannot be used again once loosened: new subframe bolts are required.**

1. Disconnect the negative battery cable. Disarm the Air Bag system, if equipped.

2. Install support rail 5033, bracket 5006 and lifting hook or equivalents over the engine.

3. Lift the engine up slightly so that there is no pressure on the engine mounts.

4. Raise and safely support the vehicle. Remove the front wheels. Disconnect the tie rod ends.

5. Remove the splash guard from below the engine. Disconnect the power steering fluid lines brackets and clamps from the front and rear of the subframe. Remove the five nuts holding the steering rack to the subframe.

6. Position a jack below the rear part of the subframe and remove the following:

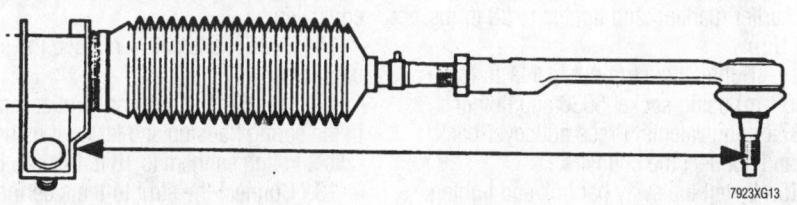

Measuring tie rod distance from the rack housing—850, C70, S70 and V70 models

• the four bolts holding the subframe to the body on both sides

• the two bolts and washers holding the bracket to the subframe

• loosen the front subframe bolts so the frame lowers 0.59–0.79 in. (15–20mm)

7. Lower the subframe using the jack, and place a spacer between the frame and the body at the rear edge so the frame will not pop up.

8. Position a catch pan under the steering rack and disconnect the power steering lines from the rack.

9. Remove the steering column joint bolt and press it up from the steering rack. Remove the bolt holding the rack to the engine mount.

10. Remove rack from the right side.

To install:

11. Transfer the thermal protection plate and center attachment mount, but do not tighten the mounting bolts. Install the protective plugs in the line connection holes.

12. Install the tie rod ends.

13. Install the steering rack from the right side and let it rest on the rear engine mount.

14. Raise the rack up on the right side so that it is straight in relation to the frame and tighten the engine mount bolt to 37 ft. lbs. (50 Nm).

15. Connect the fluid lines and brackets loosely using new O-rings on the lines.

16. Align the fluid lines in the bracket and tighten them in the steering rack.

17. Fit the steering rack onto the steering shaft joint and tighten the bolt to 15 ft. lbs. (20 Nm). Install the bolt lock clip.

18. Lift the rear of the subframe up using a jack and line up the steering rack mount bolts in the frame.

19. Install new subframe bolts loosely. Move the jack to the front and replace the bolts with new ones, but do not tighten. Tighten the bolts on the left side of the subframe to 77 ft. lbs. (105 Nm) plus an addi-

tional 120 degrees. Tighten the right side bolts the same way. Finally, tighten the bracket bolts on both sides to 37 ft. lbs. (50 Nm).

20. Install new nuts on the steering rack and tighten them to 37 ft. lbs. (50 Nm). Tighten the steering rack center bolt to 59 ft. lbs. (80 Nm). Install the front and rear steering fluid line brackets and tighten them.

21. Install the engine splash guard below the engine.

22. Install the wheels.

23. Fill the power steering fluid reservoir with fluid.

24. Check the fluid level once again. Lower the vehicle and check the toe-in.

940, 960, S90 AND V90 MODELS

1. Disconnect the negative battery cable.

2. Raise and support the vehicle safely.

3. Remove the splash guard and the small jacking panel on the front crossmember.

4. Disconnect the lower steering shaft from the steering gear.

5. At the lower universal joint, remove the snaprings and loosen the upper clamp bolt.

6. Remove the lower clamp bolt and slide the joint up on the shaft.

7. Use a ball joint separator and disconnect the tie rods at the outer ends.

8. Disconnect the fluid lines from the steering gear. Catch the spilled fluid in a pan and install plugs in the lines.

9. Remove the sway bar mounting brackets from the side members and move them out of the way.

10. Remove the steering gear retaining bolts and lower the assembly out of the vehicle.

To install:

11. When reinstalling, position the rack in position and install the retaining bolts. Tighten them to 32 ft. lbs. (44 Nm).

12. Install the sway bar mounting brackets.

13. Use new copper washers and connect the fluid lines to the assembly.

14. Connect the tie rods and tighten their nuts to 44 ft. lbs. (60 Nm).

15. Slide the lower universal joint down the shaft and into position.

16. Tighten the lower clamp bolt first, then the upper. Both bolts are tightened to 15 ft. lbs.(20 Nm).

17. Install the snaprings.

18. Reinstall the jacking plate and the splash guard.

19. Fill the reservoir with ATF.

20. Start the engine and smoothly turn the steering wheel from lock to lock 3 or 4 times.

21. Bleed the system and recheck the fluid level in the reservoir.

22. Lower the vehicle.

Strut

REMOVAL & INSTALLATION

Front

940, 960, S90 AND V90 MODELS

✳✳ CAUTION

A coil spring compressor is required to remove the spring. Improper removal procedures may cause serious injury.

1. Raise and safely support vehicle. Remove the wheel.

2. Disconnect the tie rod end.

3. Place a floor jack under the control arm. Disconnect the sway bar from the link. Unbolt the brake lines from the bracket and detach them from the clips.

4. Remove the cover over the strut nut. Disconnect the coil wire and place it out of the way.

5. Hold the strut shaft with tool 5037 or equivalent and loosen the nut a few turns with tool 5036 or equivalent.

6. Mark the position of the upper mount in the housing,, then remove the nuts and washers. Carefully lower the jack and pull the strut and spring out of the housing.

✳✳ WARNING

Be careful not to damage the fender when removing the strut assembly. Use retaining hook 5045 or equivalent attached to the anti-sway bar to prevent it from falling.

7. To remove the spring:

a. Attach spring compressor tool 5040 or equivalent to the spring. The two parts of the tool should be opposite each other and have three coils between the claws.

b. Compress each side alternately until the strut is loose inside the spring.

c. Hold the strut shaft with tool 5037 or equivalent and remove the nut with tool 5036 or equivalent and lift off the upper mount, spring retainer, spring and rubber bumper.

➡**On vehicles equipped with gas pressure struts, the bumper has been replaced by a rubber bellow and disc.**

8. To remove the strut, unscrew the retaining nut and pull the shock insert out of the casing, using tool 5039 or equivalent for standard struts, or tool 5173 or equivalent for gas struts.

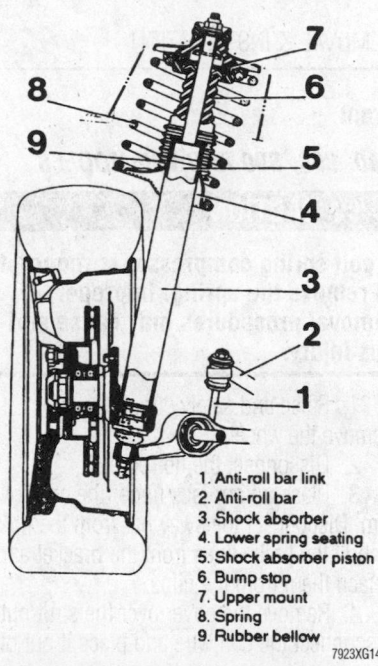

1. Anti-roll bar link
2. Anti-roll bar
3. Shock absorber
4. Lower spring seating
5. Shock absorber piston
6. Bump stop
7. Upper mount
8. Spring
9. Rubber bellow

7923XG14

Spring strut assembly—940, 960, S90 and V90 models

To install:
9. Insert the strut insert into the housing and tighten the retaining nut to 119 ft. lbs. (160 Nm). Install the bumper or bellows and disc on the strut, making sure that the top of the bumper is lower than the top of the strut shaft.

10. Install the spring so the compressor tool bolt holes face upwards. Install the upper mount, washer and nut but do not tighten fully.

11. Remove the compressor loosening the bolts alternately and be sure that the ends of the spring fit correctly into the upper and lower plates.

12. Guide the strut assembly into the body.

13. Install the upper mount according to the earlier marking and tighten to 30 ft. lbs. (40 Nm).

14. Tighten the strut nut to 111 ft. lbs. (150 Nm) using socket 5036 and holder 5037 or equivalents. Press nut cover back on and connect the coil wire.

15. Install the sway bar link and tighten it until the distance between the washers is 1.65 in. (42mm). Install the brake line bracket and clips. Be sure that the brake lines are sitting correctly in the wheel well.

16. Install the tie rod end. Install the wheel. Lower the vehicle and test.

850, C70, S70 AND V70 MODELS

1. Raise and safely support vehicle. Remove the wheel.

2. Disconnect the sway bar link from the strut. Remove the ABS sensor lead from the strut and brake bracket, but do not disconnect.

3. Install support tool 5466 or equivalent under the control arm.

⁂ WARNING

If this tool is not installed, the axle joint may be damaged from excessive downward pressure.

4. Remove the upper nuts attaching the strut attachment to the body.

5. Remove the two nuts and bolts holding the strut to the steering knuckle. Remove the spring and strut assembly.

6. Mount the spring and strut assembly in a vice and secure it. Install spring compressing tool 5407 or equivalent and alternately compress the spring.

7. Remove the bolt and washer from the strut attachment using socket 5467 and counterhold 5468 or equivalents.

8. Remove the strut nut using socket 5469 and counterhold 5468 or equivalents. Remove the spring seat, rubber stopper, boot, and check them for damage. Remove the compressed spring from the strut.

To install:
9. Compress the spring to about 12 in. (300mm) in length.

10. Install the rubber stopper.
11. Install the washer.
12. Install the spring (compressed).
13. Install the spring seat.

➡**Be sure that the spring ends are properly seated.**

14. Install the strut nut and tighten it to 52 ft. lbs. (70 Nm) using socket 5469 and counterhold 5468 or equivalents.

15. Install the spring attachment, washer, and nut and tighten it to 52 ft. lbs. (70 Nm) using socket 5467 and counterhold 5468 or equivalents.

16. Slowly, alternately remove the spring compressor.

17. Install the spring and strut assembly in the spring housing and fasten it using new nuts and tighten them to 18 ft. lbs. (25 Nm).

18. Connect the strut to the steering knuckle using new bolts and nuts.

19. Tighten them to 48 ft. lbs. (65 Nm) and angle tighten 90°.

20. Connect the sway bar to the strut using new nuts.

21. Install the ABS sensor lead to the strut and brake pipe bracket. Remove the support tool.

22. Install the wheel.

Shock Absorber

REMOVAL & INSTALLATION

Rear

940, 960 WAGON AND V90 MODELS

1. Raise and safely support the vehicle on jackstands.

2. Place a jack under the lower arm and lift the rear suspension to unload the shock.

3. Remove the lower nut and bolt securing the shock to the rear axle.

4. Lower the vehicle enough to remove the shock.

5. Remove the upper shock absorber through-bolt. Remove the old shock absorber.

To install:
6. Install the shock absorber in the upper mount and hand-tighten the through-bolt.

7. Raise the rear end and attach the shock using the lower mount bolt. Hand-tighten the nut.

8. Tighten the lower bolt to 63 ft. lbs. (85 Nm).

9. Remove the jack.
10. Lower the vehicle.
11. Tighten the upper bolt to 63 ft. lbs. (85 Nm).

960 SEDAN, S90 AND AWD V70 MODELS

1. Raise and safely support the rear of the vehicle with jackstands. Place the jackstands so they do not interfere with the support arm of the vehicle.

2. Remove the wheel.

3. Use a floor jack and raise up the support arm.

4. Disconnect the shock absorber from the support arm.

5. Disconnect the shock absorber from the body. Remove shock absorber.

To install:

6. Install the shock absorber and connect it to the body using a new bolt. Tighten the bolt to 59 ft. lbs. (80 Nm).

7. With the support arm raised, connect the shock absorber to the support arm and install a new bolt. Tighten the bolt to 59 ft. lbs. (80 Nm).

8. Carefully lower the lift from the support arm.

9. Install the wheels.

10. Lower the vehicle.

850, C70, S70 AND FWD V70 MODELS

1. On early four door Series, remove the plastic side panel,, then fold the back seat forward and fold back the trunk carpet. Remove the support panel under the edge of the carpet and detach the side panels at the front and fold them over. Remove the back seat catch and panel mounting clip.

2. On later four door Series, remove the support panel over the shock mount. Make small cuts in the panel if necessary to fold it up.

3. On five door Series, remove the front floor panel bolts and pull the panel back to free it from the front mount.

4. Remove the panel.

5. Raise and safely support the vehicle so the rear wheels are off the ground.

6. Remove the two upper shock mount bolts.

7. Using a floor jack, press the trailing arm up to unload the shock absorber.

8. Disconnect the shock absorber from the lower mount,, then pull it off of the trailing arm.

9. Lower the vehicle and lift the shock assembly out from the top.

10. Check the shock upper mount bushing for damage and replace if necessary.

To install:

11. Install the upper mount on the new shock as follows:

• Tighten standard shock absorbers to 30 ft. lbs. (40 Nm)

• Tighten gas shock absorber M12 nuts to 30 ft. lbs. (40 Nm)

• Tighten gas shock absorber M10 nuts to 15 ft. lbs. (20 Nm) plus an additional 90 degrees

12. Install the shock absorber in the vehicle and turn the upper mount bolts a few turns.

13. Raise the vehicle and position the jack under the trailing arm and lift up.

14. Connect the shock to the lower mount, making sure that the shock is seated correctly in the upper mount. Tighten the nut to 59 ft. lbs. (80 Nm).

15. Lower the vehicle and tighten the upper shock mount bolts to 18 ft. lbs. (25 Nm).

16. Install the front edge of the panels using the clips.

17. Install the back seat catches and bolts with thread locking compound, tightening to 15 ft. lbs. (20 Nm).

18. Replace the cover plate and trunk carpet.

19. On later four door Series, fold down the cover over the shock mount and install the carpeting.

20. On five door Series, line up the front edge of the floor panel and install the bolt at the rear edge. Line up the panel with the rear floor panel and tighten the bolts.

Coil Spring

REMOVAL & INSTALLATION

Front

The front coil spring removal and installation procedure is included in the strut removal and installation procedure.

Rear

The 960 sedan and S90 is equipped with a multi-link rear suspension that does not utilize a coil spring.

940, 960 WAGON AND V90

❊❊ CAUTION

A coil spring compressor is required to remove the spring. Improper removal procedures may cause serious injury.

1. Raise and safely support the vehicle on the frame and remove the rear wheels.

2. Place a hydraulic jack beneath the rear axle housing and raise the housing sufficiently to compress the spring.

3. Install the spring compressor and tighten. Be sure there are at least 3 coils of spring between the attachment points of the compressor.

4. Loosen the nuts for the upper and lower spring attachments.

5. Disconnect the shock absorber at the upper attachment. Lower the jack enough to remove the spring.

To install:

6. Be sure the coil spring is compressed.

7. Position the retaining bolt and inner washer for the upper attachment inside the spring. While holding the outer washer and rubber spacer to the upper body attachment, install the spring and inner washer to the upper attachment sandwiching the rubber spacer. Tighten the retaining bolt to 35 ft. lbs. (48 Nm).

8. Raise the jack and secure the bottom of the spring to the lower attachment with the washer and retaining bolt tightened to 63 ft. lbs. (85 Nm). Slowly remove the spring compressor.

9. Connect the shock absorber to its upper attachment. Install the wheels.

10. Lower the vehicle.

850, C70, S70, AND FWD V70 MODELS

1. Raise and safely support vehicle.

2. Remove the wheels.

3. Use a jack to press the trailing arm up to unload the shock absorber.

4. Remove the shock lower mount bolt and pull the shock off of its mount.

5. Lower the jack and remove the spring mounting nut.

6. Remove the spring from the vehicle.

To install:

7. Transfer the rubber spring spacer and lower mount if installing a new spring.

8. Install the spring in the trailing arm recess and center the mount washer guide pin in the hole.

9. Install a new nut and tighten it to 30 ft. lbs. (40 Nm).

10. Position the jack under the trailing and lift it up.

11. Install the shock on the lower mount, making sure that the spring is correctly seated in the upper mount.

12. Tighten the shock nut to 59 ft. lbs. (80 Nm).

13. Install the wheels.

14. Lower the vehicle.

AWD V70 MODELS

1. Raise and safely support the vehicle.

2. Press lower control arm using jack or suitable tool to unload the pressure.

3. Remove the rear shocks.

4. Remove the sway bar.

5. Remove the control arm inner and outer mounting nuts. Using a suitable tool, knock the mounting bolts out.

6. Remove the front support arm bolts.

7. Remove the jacking point.

8. Lower the control arm to release the support arm and bracket.

9. Install the jacking point to protect the brake lines.

10. Lower the control arm until spring is fully relieved of pressure, and remove the spring.

To install:

11. Replace the spring rubber inserts if worn.

12. Install the spring in the control arm.

13. Raise the control arm up slightly.

14. Remove the jacking point.

15. Raise the control arm all the way and install the support arm bracket, tighten the bolts to 48 ft. lbs. (65 Nm).

16. Install the jacking point, tighten the bolt to 78 ft. lbs. (105 Nm).

17. Install the outer and inner control arm mounting bolts, tighten the nuts to 59 ft. lbs. (80 Nm).

18. Install the sway bar.

19. Install the shocks.

20. Remove the jack from under the control arm.

21. Lower the vehicle.

Lower Ball Joint

REMOVAL & INSTALLATION

940, 960, S90 and V90 Models

1. Raise the vehicle and support it safely.

2. Mark the position of the wheel stud nearest to the valve. Wheel is marked to facilitate installation and to avoid the need for rebalancing.

3. Remove the wheel and tire assembly.

4. Remove the bolt which holds the anti-roll bar link to the control arm.

5. Remove the cotter pin, nut and washer for the ball joint stud.

6. Pull the control arm from the ball joint using a suitable puller (5259 or equivalent).

7. Remove the bolts holding the ball joint to the spring strut.

8. Press the control arm downwards and remove the ball joint.

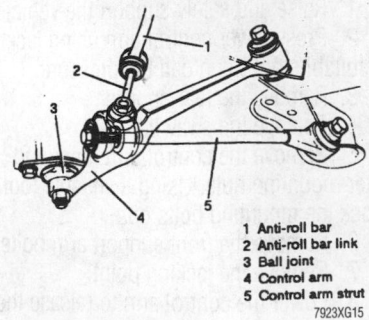

1 Anti-roll bar
2 Anti-roll bar link
3 Ball joint
4 Control arm
5 Control arm strut

7923XG15

Front suspension assembly—940, 960, S90 and V90 models

To install:

9. Install the new ball joint.

10. Use new bolts and apply sealing fluid to the threads. Check that the bolt heads sit flat on the ball joint. Tighten the bolts to 22 ft. lbs. (30 Nm) PLUS angle-tighten 90 degrees.

11. Install the control arm to ball joint.

12. Install the washer and nut.

13. Tighten ball joint stud (nut) to 44 ft. lbs. 60 Nm.

14. Install the cotter pin.

15. Install the anti-roll bar.

16. Install the wheel on the hub assembly, while aligning the marking made earlier. Alternately tighten the nuts to specifications.

17. Lower the vehicle.

850, C70, S70 and V70 Models

1. Raise and safely support the vehicle.

2. Remove the wheel.

3. Remove the three nuts holding the ball joint to the lower control arm.

4. Remove the clamping bolt and nut from the steering knuckle where the ball joint is mounted.

5. Spread the ball joint apart and remove it from the hub housing.

To install:

6. Clean the control arm and steering knuckle where the new ball joint is fitted.

7. Install the new ball joint and clamping bolt and nut. Tighten the bolt to 37 ft. lbs. (50 Nm).

8. Connect the ball joint to the lower control arm and fasten it with new nuts. Apply rust proofing compound to the nuts. Starting from inside, working outward, tighten the nuts to 13 ft. lbs. (18 Nm), then angle tighten 120°.

9. Install the wheel.

10. Lower the vehicle.

Wheel Bearing

ADJUSTMENT

Front

940, 960, S90 AND V90 MODELS

The front wheel bearings are not adjustable on the rear drive vehicles. If the lateral runout on the hub with the disc removed exceeds 0.0012 in. (0.030mm), the hub must be replaced.

850, C70, S70 AND V70 MODELS

The front wheel bearings are not adjustable on the front drive vehicles. If the lateral runout on the hub with the disc removed exceeds 0.0007 in. (0.020mm), the hub must be replaced.

Rear

The rear wheel bearings are sealed, pressed-in units, and no adjustment is possible.

REMOVAL & INSTALLATION

Front

940, 960, S90 AND V90 MODELS

1. Raise and support the vehicle safely.

2. Remove the brake caliper. Hang the caliper out of the way with a piece of stiff wire. Do not let the caliper hang by the brake hose.

3. Pry off the grease cap. Remove the cotter pin and castle nut.

4. Remove the hub and brake disc assembly. Use a bearing puller 2722 or equivalent to remove the inner bearing if it stays on the spindle.

➡ **If the vehicle is equipped with separate brake disc and hub, the guide pin and brake disc must be removed from the hub prior to bearing replacement.**

5. Use a brass drift and carefully tap out the grease seal and inner bearing race.

6. Remove the outer bearing race, using a suitable handle and drift 2725 or equivalent.

To install:

7. Press in a new inner bearing race, using a suitable handle and drift 5005 or equivalent.

8. Press in a new outer bearing race, using a suitable handle and drift 2724 or equivalent.

9. Pack the wheel bearing between the cage and inner race with as much grease as possible. Also smear grease on the outer side of the bearing and bearing races inside the hub. Fill the space in the hub with grease to a diameter of the smallest ball races.

10. On hubs with integrated brake disc, position the inner bearing seal in the hub and press the seal in so the edge lies in the same plane as the hub.

11. On hub with separate hub and brake disc:

a. Press the sealing ring onto the spindle, making sure that the seal ring is square. The sealing ring lip should face outwards.

b. Install the inner bearing in the hub. Press in the sealing washer.

12. Install the hub, outer race and castle nut.

➡**On vehicles with separate hub and brake disc, install the brake disc and guide pin.**

13. To adjust the bearing pre-load:
   a. Spin the hub and simultaneously tighten the center nut to 42 ft. lbs. (57 Nm).
   b. Loosen the nut 1/2 turn,, then tighten the nut by hand, approximately 1 ft. lb. (1.5 Nm).
   c. Install the cotter pin. If the pin hole in the spindle does not align with the pin hole in the nut, unscrew the nut slightly to the nearest pin hole.
   d. Install the protective cap.
14. Install the brake caliper.
15. Install the wheel. Lower the vehicle.

850, C70, S70 AND V70 MODELS

1. Raise and safely support vehicle. Remove the wheel.
2. Disconnect the ABS sensor from the axle shaft but do not disconnect the sensor connector. Hang it out of the way.
3. Remove the caliper, carrier and rotor. Hang the caliper safely out of the way.
4. Remove the halfshaft.
5. Separate the ball joint from the control arm. Disconnect the sway bar link.
6. Remove the four Torx® E14 bolts retaining the hub. Remove the hub.

To install:
7. Clean the axle shaft and hub mating surfaces. Clean the ABS sensor with a soft brush.
8. Install the new hub and tighten the bolts alternately to 33 ft. lbs. (45 Nm) plus an additional 60 degrees.
9. Insert the axle shaft into the hub and fit the splines. Tighten the new axle shaft nut by hand.
10. Connect the ball joint to the control arm using new nuts.
11. Connect the sway bar link.
12. Install the brake rotor, carrier and caliper.
13. Tighten the axle nut to 89 ft. lbs. (120 Nm) plus an additional 60 degrees, using tool 5461 or equivalent to counterhold. Lock the axle shaft nut using a chisel to tap the flange into the groove.
14. Clean the ABS sensor and its seat with a soft brush. Tighten the sensor to 7 ft. lbs. (10 Nm).
15. Install the wheels. Lower the vehicle.

Rear

940, 960 WAGON AND V90 MODELS

1. With the vehicle sitting on all four wheels, loosen the rear axle nut.

2. Raise and support the vehicle safely. Do not allow the lifting arms to interfere with the support arms.
3. Remove the wheels. Remove the brake caliper and use a piece of wire to hang the caliper out of the way.
4. Remove the brake disc and parking brake shoes.
5. Disconnect and remove the parking brake cable from the wheel bearing housing.
6. Remove the retaining bolt for the support arm at the housing. Tap the support arm loose.
7. Remove the nut and bolt holding the lower link arm to the housing.
8. Remove the retaining bolt for the track rod at the bearing housing and use a small claw-type puller to remove the track rod.
9. Remove the axle nut.
10. Remove the retaining nut for the upper link at the bearing housing. The wheel bearing housing can now be removed as a unit.

➡**There are shims between the bearing housing and the upper link arm. Collect them when the housing is removed.**

11. Mount the housing assembly in a vise. Place counterhold tool 5340 or equivalent between the hub and bearing housing. Press out the hub with a proper sized drift.
12. Remove the circlip retaining the bearing in the wheel bearing housing and press the bearing out. Press against the inner race.
13. Use bearing puller 2722 or equivalent to pull the inner race off the hub.
To install:
14. Press the new bearing into the housing. Be sure the press tool contacts only the outer bearing race or the bearing will be damaged. Install the circlip.
15. Support the inner race and press the hub into the bearing. If the inner race is not properly supported, the bearing will be damaged.
16. Install the wheel bearing housing onto the housing and install the axle nut hand-tight.
17. Install the shims between the upper link and the wheel bearing housing, then install the retaining nut at the upper link.
18. Pull the wheel bearing housing outwards at the top and tighten the upper link arm nut to 85 ft. lbs. (110 Nm). This is essential to insure correct wheel alignment when completed.
19. Tilt the bearing housing outwards at the bottom as necessary to refit the lower link arm and its retaining bolt. When in place, pull the bottom of the bearing hous-

ing in towards the center of the vehicle and tighten the link arm to 36 ft. lbs. (47 Nm) plus an additional 90 degrees of rotation.
20. Install the support arm and its bolt.
21. Install the track rod and tighten to 63 ft. lbs. (82 Nm).
22. Reinstall the parking brake cable at the bearing housing.
23. Reinstall the parking brake shoes, the brake disc as marked and the brake caliper.
24. Install the wheels. Lower the vehicle.
25. With all four wheels on the ground, tighten the axle nut to 103 ft. lbs. (134 Nm) plus an additional 60 degrees of rotation.

960 SEDAN, S90 AND AWD V70 MODELS

1. With the vehicle sitting on all four wheels, loosen the rear axle nut.
2. Raise and safely support vehicle. Position a lift or jackstands so they do not interfere with the suspension arms. Remove the wheel.
3. Use tool 999 5577 or equivalent to compress the suspension slightly against the spring.
4. Remove the damper bolt and pull the damper from the support arm. Remove the anti-sway bolts in both of the support arms.
5. Raise the suspension up into normal position. Remove the axle shaft nut. Remove the brake caliper and support it safely out of the way. Mark the position of the brake disc and remove the disc.
6. Remove the parking brake shoes and disconnect the adjuster from the parking brake cable. Disconnect the parking brake cable from the wheel bearing housing and remove the nut for the upper link bushing. Tap the bushing bolt free of the wheel bearing housing.
7. Remove the bolt to the track rod and separate the rod from the wheel bearing housing. Remove the support arm bolt and tap if free of the bushing. Remove the link nut and tap the bolt out using a brass drift. Remove the wheel bearing housing.

➡**The bearing must be replaced any time the hub is pressed out.**

8. Position the wheel bearing housing in a press so the hub can be pressed out. Using an appropriate sized drift, press the hub off.
9. Remove the circlip holding the bearing in the housing and press the bearing out.
10. Position the drift in the inner bearing race. Using the puller 999 2722 and counter hold 999 5310, pull the inner race out of the hub.

To install:

11. Properly support the bearing housing and press the new bearing in. Be sure the press tool contacts only the outer bearing race or the bearing will be damaged. Install the circlip.

12. Support the inner bearing race on the press table and press the hub into the bearing. Be sure the inner bearing race is supported or the bearing will be damaged.

13. Fit the wheel bearing housing to the upper link and driveshaft. Then, install all the nuts and bolts before tightening any of them. Tighten the nut for the upper link to 85 ft. lbs. (115 Nm) and all others to 63 ft. lbs. (80 Nm).

14. Connect the parking brake cable to the wheel bearing housing and fasten with clip.

15. Install the adjuster for the cable with arrow on the upper side pointing up.

16. Install the parking brake shoes, retainers, and spring. Install the brake disc.

17. Install a new axle nut but do not tighten it yet.

18. Install the brake caliper using new bolts.

19. Install new anti-sway bolts on both sides, but do not tighten them fully.

20. Clean the face of the brake disc and the back side of the wheel where the two mate. Lubricate the guide pin with rust proofing compound.

21. Install the wheels.

22. Lower the vehicle. Tighten the anti-sway bar bolts on both sides to 63 ft. lbs. (80 Nm).

23. With all four wheels on the ground, tighten the axle nut to 130 ft. lbs. (140 Nm) plus an additional 60°.

850, C70, S70 AND FWD V70 MODELS

➡**The bearing and hub are replaced as a single component. The bearing is not available separately.**

1. Raise and safely support vehicle. Remove the wheels.

2. Remove the caliper and the brake line from the mounting clip. When the left brake caliper is removed, the three-way brake line connector mounting bolt must also be removed. Remove the caliper and

hang it from the spring to prevent brake hose damage.

3. Back the parking brake adjustment off so the disc can be removed. Remove the disc. Remove the cap, hub nut and hub.

To install:

4. Clean the sub axle thoroughly. Install the hub using a new nut and tighten it to 89 ft. lbs. (120 Nm) plus an additional 30 degrees. Be sure that there is no play in the bearings after installation.

5. Install the dust cap using an appropriate tool. Clean the face of the hub and back side of the brake disc where the two mate.

6. Install the disc and guide pin. Tighten the pin to 7 ft. lbs. (9 Nm).

7. Adjust the parking brake shoes until the disc cannot be turned,, then back it off four to six notches.

8. Install the brake caliper.

9. Install the brake line and mounting clips and the three-way connector on left-hand side, if applicable.

10. Install the wheels. Lower the vehicle.

Total Car Care, continued

Sentra/Pulsar/NX 1982-96
PART NO. 8263/52700
Stanza/200SX/240SX 1982-92
PART NO. 8262/52750
240SX/Altima 1993-98
PART NO. 52752
Datsun/Nissan Z and ZX 1970-88
PART NO. 8846/52800
RENAULT
Coupes/Sedans/Wagons 1975-85
PART NO. 58300
SATURN
Coupes/Sedans/Wagons 1991-98
PART NO. 8419/62300
SUBARU
ff-1/1300/1400/1600/1800/Brat 1970-84
PART NO. 8790/64300
Coupes/Sedans/Wagons 1985-96
PART NO. 8259/64302
SUZUKI
Samurai/Sidekick/Tracker 1986-98
PART NO. 66500

TOYOTA
Camry 1983-96
PART NO. 8265/68200
Celica/Supra 1971-85
PART NO. 68250
Celica 1986-93
PART NO. 8413/68252
Corolla 1970-87
PART NO. 8586/68300
Corolla 1988-97
PART NO. 8414/68302
Cressida/Corona/Crown/MkII 1970-82
PART NO. 68350
Cressida/Van 1983-90
PART NO. 68352
Toyota Trucks 1970-88
PART NO. 8578/68600
Pick-Ups/Land Cruiser/4Runner 1989-96
PART NO. 8163/68602

Previa 1991-98
PART NO. 68640
Tercel 1984-94
PART NO. 8595/68700
VOLKSWAGEN
Air-Cooled 1949-69
PART NO. 70200
Air-Cooled 1970-81
PART NO. 70202
Front Wheel Drive 1974-89
PART NO. 8663/70400
Golf/Jetta/Cabriolet 1990-93
PART NO. 8429/70402
VOLVO
Coupes/Sedans/Wagons 1970-89
PART NO. 8786/72300
Coupes/Sedans/Wagons 1990-98
PART NO. 8428/72302

Total Service Series

Auto Detailing
PART NO. 8394
Auto Body Repair
PART NO. 7898
Automatic Transmission Repair
1980-84
PART NO. 7890
Automatic Transmissions/
Transaxles Diagnosis and Repair
PART NO. 8944
Brake System Diagnosis and Repair
PART NO. 8945
Chevrolet Engine Overhaul Manual
PART NO. 8794
Easy Car Care
PART NO. 8042
Engine Code Manual
PART NO. 8851
Ford Engine Overhaul Manual
PART NO. 8793
Fuel Injection and Feedback
Carburetors 1977-85
PART NO. 7488
Fuel Injection Diagnosis
and Repair
PART NO. 8946
Motorcycle Repair
PART NO. 9099
Small Engine Repair
(Up to 20 Hp)
PART NO. 8325

Collector's Hard-Cover Manuals

Auto Repair Manual 1993-97
PART NO. 7919
Auto Repair Manual 1988-92
PART NO. 7906
Auto Repair Manual 1980-87
PART NO. 7670
Auto Repair Manual 1972-79
PART NO. 6914
Auto Repair Manual 1964-71
PART NO. 5974
Auto Repair Manual 1954-63
PART NO. 5652

Auto Repair Manual 1940-53
PART NO. 5631
Import Car Repair Manual 1993-97
PART NO. 7920
Import Car Repair Manual 1988-92
PART NO. 7907
Import Car Repair Manual 1980-87
PART NO. 7672
Truck and Van Repair Manual 1993-97
PART NO. 7921
Truck and Van Repair Manual 1991-95
PART NO. 7911

Truck and Van Repair Manual 1986-90
PART NO. 7902
Truck and Van Repair Manual 1979-86
PART NO. 7655
Truck and Van Repair Manual 1971-78
PART NO. 7012
Truck Repair Manual 1961-71
PART NO. 6198
Motorcycle and ATV Repair Manual
1945-85
PART NO. 7635

System-Specific Manuals

Guide to Air Conditioning Repair and
Service 1982-85
PART NO. 7580
Guide to Automatic Transmission
Repair 1984-89
PART NO. 8054
Guide to Automatic Transmission
Repair 1984-89
Domestic cars and trucks
PART NO. 8053

Guide to Automatic Transmission
Repair 1980-84
Domestic cars and trucks
PART NO. 7891
Guide to Automatic Transmission
Repair 1974-80
Import cars and trucks
PART NO. 7645
Guide to Brakes, Steering, and
Suspension 1980-87
PART NO. 7819

Guide to Fuel Injection and Electronic
Engine Controls 1984-88
Guide to Electronic Engine Controls
1978-85
PART NO. 7535
Guide to Engine Repair and Rebuilding
PART NO. 7643
Guide to Vacuum Diagrams 1980-86
Domestic cars and trucks
PART NO. 7821

Multi-Vehicle Spanish Repair Manuals

Auto Repair Manual 1992-96
PART NO. 8947
Import Repair Manual 1992-96
PART NO. 8948
Truck and Van Repair Manual
1992-96
PART NO. 8949
Auto Repair Manual 1987-91
PART NO. 8138
Auto Repair Manual
1980-87
PART NO. 7795
Auto Repair Manual
1976-83
PART NO. 7476